WATER TREATMENT PLANT OPERATION

Volume II

Sixth Edition

A Field Study Training Program

prepared by

Office of Water Programs
College of Engineering and Computer Science
California State University, Sacramento

Kenneth D. Kerri, Chief Project Consultant

2015

In recognition of the need to preserve natural resources, this manual is printed using recycled paper. The text paper is composed of 30% post-consumer waste and the cover is composed of 10% post-consumer waste. The Office of Water Programs will strive to increase its commitment to sustainable printing practices.

Funding for the production of this operator training manual was provided by University Enterprises, Inc., California State University, Sacramento. Mention of trade names or commercial products does not constitute endorsement or recommendation for use by the project director; the Office of Water Programs; or California State University, Sacramento.

First edition published 1983. Sixth Edition 2015.

Printed in the United States of America

21 20 19 18 17 16 15 1 2 3 4 5

ISBN
978-1-59371-068-2

www.owp.csus.edu

OFFICE OF WATER PROGRAMS

The Office of Water Programs is a nonprofit organization operating under University Enterprises, Inc., California State University, Sacramento, to provide distance learning courses for persons interested in the operation and maintenance of drinking water and wastewater facilities. These training programs were developed by people who explain, through the use of our manuals, how they operate and maintain their facilities. The university, fully accredited by the Western Association of Schools and Colleges, administers and monitors these training programs, under the direction of Dr. Ramzi J. Mahmood.

Our training group develops and implements programs and publishes manuals for operators of water treatment plants, water distribution systems, wastewater collection systems, and municipal and industrial wastewater treatment and reclamation facilities. We also offer programs and materials for pretreatment facility inspectors, environmental compliance inspectors, and utility managers. All training is offered as distance learning, using correspondence, video, or computer-based formats with opportunities for continuing education and contact hours for operators, supervisors, managers, and administrators.

Materials and opportunities available from our office include manuals in print, CD, DVD, or video formats, and enrollments for courses providing Continuing Education Unit (CEU) contact hours. Here is a sample:

- Water Treatment Plant Operation, 2 volumes (print, course enrollment)
- Water Distribution System Operation and Maintenance (print, CD, course enrollment, five online courses)
- Small Water System Operation and Maintenance (print, video, CD, course enrollment, five online courses)
- Water Systems Operation and Maintenance Video Training Series (print, video, DVD, course enrollment)
- Utility Management (print, course enrollment)
- Manage for Success (print, course enrollment)
- and more

These and other materials may be ordered from:

Office of Water Programs
California State University, Sacramento
6000 J Street
Sacramento, CA 95819-6025
(916) 278-6142 – phone
(916) 278-5959 – FAX

or

visit us on the web at www.owp.csus.edu

ADDITIONAL VOLUMES OF INTEREST

Water Treatment Plant Operation, Volume I

The Water Treatment Plant Operator
Water Sources and Treatment
Reservoir Management and Intake Structures
Coagulation and Flocculation
Sedimentation
Filtration
Disinfection
Corrosion Control
Taste and Odor Control
Plant Operation
Laboratory Procedures

Water Distribution System Operation and Maintenance

The Water Distribution System Operator
Storage Facilities
Distribution System Facilities
Water Quality Considerations in Distribution Systems
Distribution System Operation and Maintenance
Disinfection
Safety
Distribution System Administration

Small Water System Operation and Maintenance

The Small Water System Operator
Water Sources and Treatment
Wells
Small Water Treatment Plants
Disinfection
Safety
Laboratory Procedures
Setting Water Rates and System Security for Small Water Utilities

Manage for Success

Supervising
Communicating
Human Relations
Planning and Organizing
Training and Teaching Skills
Problem-Solving Skills (Looking for Opportunities)
Decision Making
Technical Issues and Regulatory Compliance
Financial Management
Computers in Managing a Utility
Emergency Planning
Health and Safety Programs
Community Relations
Personal and Professional Skills

PREFACE

This volume of is a continuation of Volume I, which emphasized the knowledge and skills needed by operators of conventional surface water treatment plants. Volume II focuses on information needed by these operators and includes details on specialized water treatment processes for iron and manganese control, fluoridation, softening, arsenic removal, trihalomethanes, membrane filtration, demineralization, and the handling and disposal of process wastes. This manual covers topics of importance to operators of all types of water treatment plants, including maintenance, instrumentation, safety, advanced laboratory procedures, water quality regulations, administration, and water treatment plant math calculations.

The material in this manual will be easier to understand if you understand the meaning of words used by water treatment plant operators. To assist you with this, words specific to a chapter are defined in the "Words" section at the beginning of each chapter as well as in footnotes throughout. The Appendix includes a "Water Words" section that contains definitions of all the words used in both volumes of this series. You may wish to concentrate your studies on those chapters that apply to your type of water treatment plant. Upon successful completion of both volumes, you will have gained a broad and comprehensive knowledge of the water treatment field.

The Office of Water Programs is indebted to the many operators and others who contributed to this manual. Every effort was made to acknowledge material from the many excellent references in the water treatment field. We appreciate the contributions of consultants, reviewers, and advisors to previous editions, all of whose efforts are represented in the material presented in this new edition.

KENNETH D. KERRI

OBJECTIVES OF THIS MANUAL

Proper installation, inspection, operation, maintenance, repair, and management of water treatment plants have a significant impact on the operation and maintenance costs and effectiveness of the plants. The objective of this manual is to provide water treatment plant operators with the knowledge and skills required to operate and maintain water treatment plants effectively, thus eliminating or reducing the following problems:

1. Health hazards created by the production or output of unsafe water from the plant
2. System failures that result from the lack of proper installation, inspection, preventive maintenance, surveillance, and repair programs designed to protect the public's investment in the plant
3. Taste and odor complaints from consumers
4. Turbid or colored waters that are unacceptable to consumers
5. Corrosion damages to pipes, equipment, tanks, and structures at the water treatment plant and in the distribution system
6. Complaints from the public or local officials due to the unreliability or failure of the water treatment plant to perform as designed
7. Fire damage caused by insufficient water at a time of need

SCOPE OF THIS MANUAL

Operators with the responsibility for the operation of water treatment plants will find this manual very helpful. Topics include controlling iron and manganese, fluoridating and softening water, controlling trihalomethanes, removing arsenic, treating suspended and dissolved solids in water, and handling process wastes. Other topics include maintaining plant equipment, troubleshooting instrumentation, developing safety programs, laboratory procedures for analyzing water samples, water quality regulations, and plant administration.

Material in this manual covers situations encountered by water treatment plant operators in most areas. The goal is to provide you with an understanding of basic operational and maintenance concepts for water treatment plants and with the ability to analyze and solve problems when they occur. Operation and maintenance programs for water treatment plants vary with the age of the plant, the extent and effectiveness of existing programs, and local conditions. You can adapt the information and procedures to your particular situation.

Technology is advancing rapidly in the field of water treatment plant operation. To keep pace with scientific and industry advances, the material in this program is periodically revised and updated. This means that you, the water treatment plant operator, must be aware of new advances and recognize the need for continuous professional training beyond what is offered in this field training program. Such opportunities exist in your daily work experience, from your co-workers, and from attending meetings, workshops, conferences, and classes.

USES OF THIS MANUAL

This manual was developed to serve the needs of operators in several different situations. The format used was developed to serve as a home-study or self-paced instruction course for operators in remote areas or persons unable to attend formal classes either due to shift work, personal reasons, or the unavailability of suitable classes. This home-study training program uses the concepts of self-paced instruction where you are your own instructor and work at your own speed. In order to certify that a person has successfully completed this program, objective tests and special answer sheets for each chapter are provided when a person enrolls in this course.

Also, this manual can serve effectively as a textbook in the classroom. Many colleges and universities have used this manual as a text in formal classes (often taught by operators). In areas where colleges are not available or are unable to offer classes in the operation of water distribution systems, operators and utility agencies can join together to offer their own courses using this manual.

Cities or utility agencies can use this manual in several types of on-the-job training programs. In one type of program, a manual is purchased for each operator. A senior operator or a group of operators are designated as instructors. These operators help answer questions when the persons in the training program have questions or need assistance.

This manual was prepared to help operators run their water treatment plants. Please feel free to use it in the manner that best fits your training needs and the needs of other operators. We will be happy to work with you to assist you in developing your training program. Please feel free to contact:

Project Director
Office of Water Programs
California State University, Sacramento
6000 J Street
Sacramento, CA 95819-6025
(916) 278-6142 – phone
(916) 278-5959 – FAX
wateroffice@csus.edu – e-mail

INSTRUCTIONS TO PARTICIPANTS IN HOME-STUDY COURSE

The procedures for reading the lessons and answering the questions in this manual are contained in this section.

To progress steadily through this training program, establish a regular study schedule. Some of the chapters are longer and more difficult than others, so many of them are divided into two or more lessons. The time required to complete a lesson will depend on your background and experience. The important thing is that you understand the material in the lesson before starting the next one.

Each lesson is arranged for you to read a short section, write your answers to the questions at the end of the section, and check your answers against the suggested answers. You can then decide if you understand the material sufficiently to continue or whether you need to read the section again. You may find that this procedure is slower than reading a typical textbook, but you will probably remember much more when you have finished.

Discussion and review questions follow each lesson in the chapters. These questions are provided to help you review the important points covered in the lesson. Write your answers to the discussion and review questions to help yourself retain the material better.

In the appendix at the end of this manual, you will find comprehensive review questions and suggested answers. These questions and answers are provided for you to review how well you remember the material. You will probably need to review the entire manual before attempting to answer these questions. Some of the questions are essay-type questions, which are used by some states for higher-level certification examinations. After you have answered all the questions, compare your answers with those provided and determine the areas in which you might need additional review before your next certification or civil service examination. Please do not send your answers to the Office of Water Programs at California State University, Sacramento.

You are your own teacher in this training program. You could merely look up the suggested answers to the questions at the end of the chapters and in the comprehensive review section, but doing so will not help you understand the material. Consequently, you would not be able to apply the material to the performance of your job or recall it during an examination for certification or a civil service position. You will get out of this program what you put into it, so we encourage you to make the most of the material presented.

WATER TREATMENT PLANT OPERATION

Volume II

Sixth Edition

CHAPTER 12

IRON AND MANGANESE CONTROL

by

Jack Rossum

With a Special Section by

Gerald Davidson

TABLE OF CONTENTS

Chapter 12. IRON AND MANGANESE CONTROL

LEARNING OBJECTIVES

Chapter 12. IRON AND MANGANESE CONTROL

Following completion of Chapter 12, you should be able to:

1. Identify and describe the various processes used to control iron and manganese.
2. Collect samples for analysis of iron and manganese.
3. Safely operate and maintain the following iron and manganese control processes:
   a. Phosphate treatment
   b. Ion exchange
   c. Oxidation by aeration
   d. Oxidation with chlorine
   e. Oxidation with permanganate
   f. Greensand
   g. Proprietary processes
4. Troubleshoot red water problems.

WORDS

Chapter 12. IRON AND MANGANESE CONTROL

ACIDIFICATION (uh-SID-uh-fuh-KAY-shun) ACIDIFICATION

The addition of an acid (usually nitric or sulfuric) to a sample to lower the pH below 2.0. The purpose of acidification is to fix a sample so it will not change until it is analyzed.

AQUIFER (ACK-wi-fer) AQUIFER

A natural, underground layer of porous, water-bearing materials (sand, gravel) usually capable of yielding a large amount or supply of water.

BACKFLOW BACKFLOW

A reverse flow condition, created by a difference in water pressures, that causes water to flow back into the distribution pipes of a potable water supply from any source or sources other than an intended source. Also see BACKSIPHONAGE.

BACKSIPHONAGE BACKSIPHONAGE

A form of backflow caused by a negative or below atmospheric pressure within a water system. Also see BACKFLOW.

BENCH-SCALE ANALYSIS (TEST) BENCH-SCALE ANALYSIS (TEST)

A method of studying different ways or chemical doses for treating water or wastewater and solids on a small scale in a laboratory. Also see JAR TEST.

BREAKPOINT CHLORINATION BREAKPOINT CHLORINATION

Addition of chlorine to water or wastewater until the chlorine demand has been satisfied. At this point, further additions of chlorine will result in a free chlorine residual that is directly proportional to the amount of chlorine added beyond the breakpoint.

CHELATION (key-LAY-shun) CHELATION

A chemical complexing (forming or joining together) of metallic cations (such as copper) with certain organic compounds, such as EDTA (ethylene diamine tetracetic acid). Chelation is used to prevent the precipitation of metals (copper). Also see SEQUESTRATION.

COLLOIDS (KALL-loids) COLLOIDS

Very small, finely divided solids (particles that do not dissolve) that remain dispersed in a liquid for a long time due to their small size and electrical charge. When most of the particles in water have a negative electrical charge, they tend to repel each other. This repulsion prevents the particles from clumping together, becoming heavier, and settling out.

DIVALENT (dye-VAY-lent) DIVALENT

Having a valence of two, such as the ferrous ion, Fe^{2+}. Also called bivalent.

GREENSAND GREENSAND

A mineral (glauconite) material that looks like ordinary filter sand except that it is green in color. Greensand is a natural ion exchange material that is capable of softening water. Greensand that has been treated with potassium permanganate ($KMnO_4$) is called manganese greensand; this product is used to remove iron, manganese, and hydrogen sulfide from groundwaters.

INSOLUBLE (in-SAWL-yoo-bull) INSOLUBLE

Something that cannot be dissolved.

ION EXCHANGE ION EXCHANGE

A water or wastewater treatment process involving the reversible interchange (switching) of ions between the water being treated and the solid resin contained within an ion exchange unit. Undesirable ions are exchanged with acceptable ions on the resin or recoverable ions in the water being treated are exchanged with other acceptable ions on the resin.

ION EXCHANGE RESINS ION EXCHANGE RESINS

Insoluble polymers, used in water or wastewater treatment, that are capable of exchanging (switching or giving) acceptable cations or anions to the water being treated for less desirable ions or for ions to be recovered.

RESINS RESINS

See ION EXCHANGE RESINS.

SEQUESTRATION (SEE-kwes-TRAY-shun) SEQUESTRATION

A chemical complexing (forming or joining together) of metallic cations (such as iron) with certain inorganic compounds, such as phosphate. Sequestration prevents the precipitation of the metals (iron). Also see CHELATION.

ZEOLITE ZEOLITE

A type of ion exchange material used to soften water. Natural zeolites are siliceous compounds (made of silica) that remove calcium and magnesium from hard water and replace them with sodium. Synthetic or organic zeolites are ion exchange materials that remove calcium or magnesium and replace them with either sodium or hydrogen. Manganese zeolites are used to remove iron and manganese from water.

CHAPTER 12. IRON AND MANGANESE CONTROL

12.0 NEED TO CONTROL IRON AND MANGANESE

Like the cities of Minneapolis and St. Paul, iron and manganese are referred to as a pair. They are, in fact, two distinct elements and are often found in water separately. Neither of them has any direct adverse health effects. Indeed, both are essential to the growth of many plants and animals, including humans; however, the iron and manganese found in drinking water have no nutrient value for humans. Even if they were available in beneficial amounts, the presence of iron and manganese in drinking water would still be objectionable.

Clothes laundered in water containing iron and manganese above certain levels come out stained. When bleach is added to remove the stains, they are only intensified and become fixed so that no amount of further washing with iron-free water will remove the stains. They can be removed by treatment with oxalic acid, but this is rather hard on fabrics, or by the use of commercial rust removers. Excessive amounts of iron and manganese are also objectionable because they stain plumbing fixtures, bathtubs, and sinks.

Perhaps the most troublesome consequence of iron and manganese in the water is that they promote the growth of a group of microorganisms known as iron bacteria. These organisms obtain energy for their growth from the chemical reaction that spontaneously occurs between iron and manganese and dissolved oxygen. These bacteria form thick slimes on the walls of the distribution system mains. Such slimes are rust colored from iron and black from manganese. Variations in flow cause these slimes to come loose and create dirty water (a big source of consumer complaints). Furthermore, these slimes will cause foul tastes and odors in the water.

The growth of iron bacteria is controlled by chlorination. However, when water containing iron is chlorinated, the iron is converted into rust particles, and manganese is converted into a jet-black compound called manganese dioxide. These materials form a loosely adherent coating on the pipe walls. Pieces of this coating will break loose from the pipe walls when there are changes or reversals of flow in the distribution system.

Iron and manganese in water can be easily detected by observing the color of the inside walls of filters and the filter media. If the raw water is prechlorinated, there will be black stains on the walls below the water level and a black coating over the top portion of the sand filter bed. This black color will usually indicate a high level of manganese in the raw water while a brownish-black stain indicates the presence of both iron and manganese.

The generally acceptable limit for iron in drinking water is 0.3 mg/L and that for manganese is 0.05 mg/L. However, if the water contains more than 0.02 mg/L of manganese, the operator should initiate an effective flushing program to avoid complaints. By regularly flushing the water mains, the buildup of black manganese dioxide can be prevented.

QUESTIONS

Please write your answers to the following questions and compare them with those on page 24.

12.0A What problems are caused by iron and manganese in drinking water?

12.0B How can the growth of iron bacteria be controlled?

12.0C What are the generally acceptable limits for iron and manganese in drinking water?

12.1 MEASUREMENT OF IRON AND MANGANESE

12.10 Occurrence of Iron and Manganese

Because both iron and manganese react with dissolved oxygen to form *INSOLUBLE COMPOUNDS*,[1] they are not found in high concentrations in waters containing dissolved oxygen except as *COLLOIDAL SUSPENSIONS*[2] of the oxides. Accordingly, surface waters are generally free from both iron and manganese. One exception to this rule is that manganese up to one mg/L or higher may be found in shallow reservoirs and may come and go several times a year.

Iron and manganese are most frequently found in water systems supplied by wells and springs. Horizontal wells under rivers are notoriously prone to produce water containing iron. Bacteria will reduce iron oxides in soil to the soluble, *DIVALENT*[3] form of iron (Fe^{2+}), which will produce groundwater with a high iron content.

Iron bacteria can make use of the ferrous ion (Fe^{2+}). These bacteria will oxidize the iron and use the energy for reducing

1. *Insoluble* (in-SAWL-yoo-bull). Something that cannot be dissolved.
2. *Colloids* (KALL-loids). Very small, finely divided solids (particles that do not dissolve) that remain dispersed in a liquid for a long time due to their small size and electrical charge. When most of the particles in water have a negative electrical charge, they tend to repel each other. This repulsion prevents the particles from clumping together, becoming heavier, and settling out.
3. *Divalent* (dye-VAY-lent). Having a valence of two, such as the ferrous ion, Fe^{2+}. Also called bivalent.

carbon dioxide to organic forms (slimes). The manganous ion (Mn^{2+}) is used in a similar fashion by certain bacteria. Very small concentrations of iron and manganese in water can cause problems because bacteria obtain the nutrients (iron and manganese) from water in order to grow, even when the concentrations are very low.

Iron bacteria are found nearly everywhere. They are frequently found in iron water pipes and everywhere else that a combination of dissolved oxygen and dissolved iron is usually or frequently present. Only one cell of iron bacteria is needed to start an infestation of iron bacteria in a well or a distribution system. Unfortunately, it is almost impossible to drill a well and maintain sterile conditions to prevent the introduction of iron bacteria.

12.11 Collection of Iron and Manganese Samples

The best way to determine if there is an iron and manganese problem in a water supply is to look at the plumbing fixtures in a couple of houses. If the fixtures are stained, then there is a problem. Determination of the concentrations of iron and manganese in water is useful when evaluating well waters for use and treated waters for effectiveness of treatment processes.

The results of tests for iron and manganese are wrong more often than they are right. This is because samples for these substances are difficult to collect. Both iron and manganese form loosely adherent (not firmly attached) scales on pipe walls, including the sample lines. When the sample tap is opened, particles of scale may be dislodged and enter the sample bottle. If many of these particles enter the sample bottle, the error can become very large. Furthermore, unless the sample is acidified (enough nitric acid added to drop the pH to less than 2), both iron and manganese tend to form an adherent scale on the walls of the sample bottle in the few days that sometimes elapse before the analysis is started. When the sample is poured from the bottle for testing, most of the iron and manganese will then remain inside the sample bottle.

To avoid this situation, samples should be taken from a plastic sample line located as close to the well or other source as possible. Open the sampling tap slowly so that the flow rate is suitable for filling the sample bottle. Allow the sample water to flow for at least one minute for each 10 feet (3 m) of sample line before the sample is collected.

Samples for iron and manganese should be tested within 48 hours unless they have been acidified. If the sample contains any clay or if any particles of rust are picked up from a steel pipe or fitting, an acidified sample will dissolve the iron in these substances and the results will be too high. If clay or rust particles are observed in a sample, do not acidify and ask the laboratory to analyze the sample immediately. Furthermore, many laboratories fail to be sure that iron and manganese are in the divalent form (Fe^{2+} or Mn^{2+}) by adding enough nitric acid prior to the tests to lower the pH to less than 2, so laboratory errors may be even greater than sampling errors.

12.12 Analysis for Iron and Manganese

The preferred method of testing for iron and manganese is atomic absorption, but for the small plant the equipment is too expensive. With careful attention to laboratory procedures, colorimetric methods (comparing colors of unknowns with known standards) can provide sufficient accuracy in most instances. Colorimetric methods use either a spectrophotometer, a filter photometer, or the less satisfactory set of matched Nessler tubes with standards. Good results have been obtained by the use of properly calibrated colorimeters (Figure 12.1). For detailed procedures on how to use a spectrophotometer to measure iron and manganese, see Chapter 21, "Advanced Laboratory Procedures."

QUESTIONS

Please write your answers to the following questions and compare them with those on page 24.

12.1A How do iron and manganese form insoluble compounds?

12.1B Why must iron and manganese samples be acidified when they are collected?

12.1C Where should a sample for iron and manganese testing be collected?

12.2 REMEDIAL ACTION

Several methods are available to control iron and manganese in water. This section discusses how to operate the most common treatment processes.

12.20 Alternate Source

The construction of a plant to remove iron and manganese will cost as much as or more than a new well so it pays to investigate the possibility of obtaining an alternate supply of water that is free from iron and manganese. This investigation should include samples from nearby private wells, discussions with well drillers who have been active in the locality, and discussions with engineers in the state agency responsible for the regulation of well drilling.

If the water produced by the well contains dissolved oxygen along with iron and manganese, this is an indication that water is being drawn from more than one *AQUIFER*.[4] One or more of the aquifers must be producing water containing dissolved oxygen but is free of iron and manganese since oxygen reacts with both elements to form insoluble compounds. Furthermore, it is highly probable that the iron- or manganese-bearing water is from deeper aquifers. It may be possible to cure the problem simply by sealing off these deeper aquifers.

4. *Aquifer* (ACK-wi-fer). A natural, underground layer of porous, water-bearing materials (sand, gravel) usually capable of yielding a large amount or supply of water.

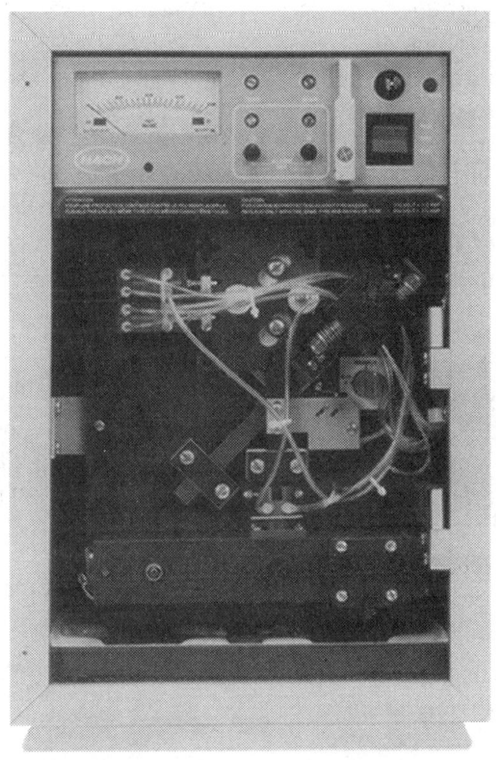

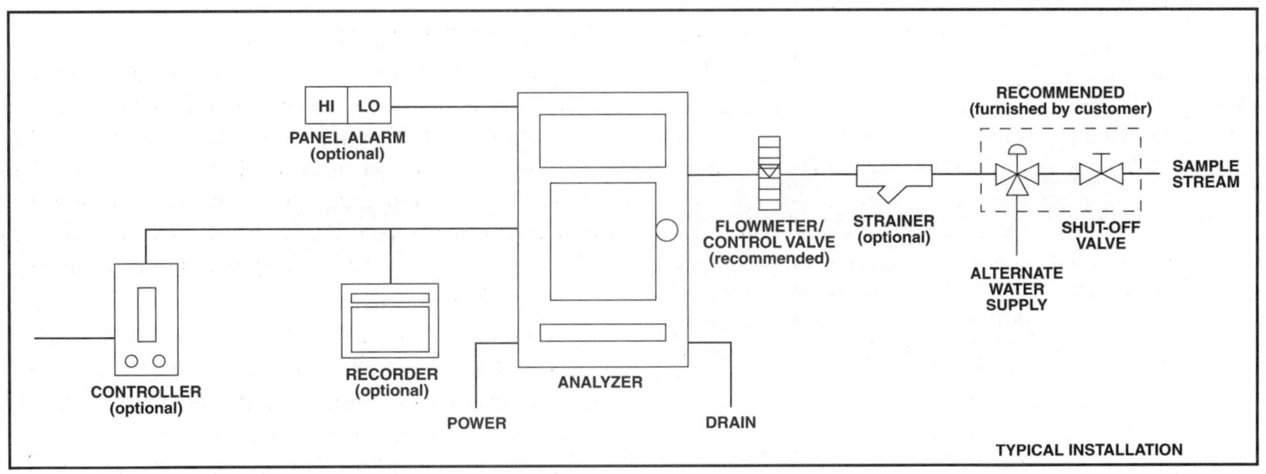

Fig. 12.1 Typical continuous on-line pump colorimeter analyzer for iron, manganese, or permanganate

(Permission of HACH)

12.21 Phosphate Treatment

If the water contains manganese up to 0.3 mg/L and less than 0.1 mg/L of iron, an inexpensive and reasonably effective control can be achieved by feeding the water with one of the three polyphosphates mentioned below. Chlorine usually must be fed along with the polyphosphate to prevent the growth of iron bacteria. The effect of the polyphosphate is to delay the precipitation of oxidized manganese for a few days so that the scale that builds up on the pipe walls is greatly reduced.

The chlorine dose for phosphate treatment should be sufficient to produce a free chlorine residual of approximately 0.25 mg/L after a five-minute contact time (a higher chlorine dose may be required with some water to maintain a free chlorine residual of at least 0.2 mg/L throughout the distribution system).

Any of the three polyphosphates (pyrophosphate, tripolyphosphate, and metaphosphate) can be used, but sodium metaphosphate is effective in lower concentrations than the others. The proper phosphate dose is determined by laboratory *BENCH-SCALE TESTS* [5] in the following manner.

1. Treat a series of samples with a standard chlorine solution to determine the chlorine dose required to produce the desired chlorine residual.
2. Prepare a standard polyphosphate solution by dissolving 1.0 gram of polyphosphate in a liter of distilled water.
3. Treat another series of samples with varying amounts of polyphosphate solution. One mL of the standard polyphosphate solution (0.1% solution) in a liter sample is equivalent to 8.34 pounds of polyphosphate per million gallons (see Examples 2 and 3 on page 17). Stir to ensure that the polyphosphate has been well mixed, and continue stirring while adding the previously determined chlorine dose so as to minimize the creation of high concentrations of chlorine.
4. Observe the samples daily against a white background, noting the amount of discoloration. The proper polyphosphate dose is the lowest dose that delays noticeable discoloration for a period of four days.

Samples for the above bench-scale test should be as fresh as possible and should be kept away from direct sunlight to avoid heating.

Polyphosphate treatment to control iron and manganese is usually most effective when the polyphosphate is added upstream from the chlorine, but satisfactory results may be obtained by feeding them together. The chlorine should never be fed ahead of the polyphosphate because the chlorine will oxidize the iron and manganese (cause insoluble precipitates to form too soon).

If you are able to install a one-half inch (12 mm) polyethylene hose in the well so that it discharges a few inches below the suction screen, you can construct a very satisfactory semiautomatic feed system. Use a gas chlorinator whose water supply is obtained from the well discharge downstream from the check valve (Figure 12.2). In this way, the chlorinator operates only when the pump is running. The chlorine solution is fed down the polyethylene tube. Polyphosphate is fed down the same tube by means of a plastic tee. The phosphate is fed by means of an electrically operated solution feeder connected so as to run when the well pump runs.

The chlorine solution flowing through the polyethylene tube is extremely corrosive. If the tube does not discharge into flowing water, the corrosive effect of the solution on a metal surface can be disastrous. Wells have been destroyed by corrosion from this cause. The following simple test should be made at least once every three months.

Example 1

1. Calculate the time required for water to flow from the pump suction to the pump discharge.

   a. Record the following information:

      (1) Length of pump column from suction to discharge in feet, 324 ft.

      (2) Diameter of pump column in inches, 8 in.

      (3) Discharge rate from pump in gallons per minute, 423 GPM.

   b. Calculate the volume of the pump column in cubic inches (in³).

$$\text{Volume} = (0.785)(\text{Diameter, in})^2(\text{Length, ft})\left(12\ \frac{\text{in}}{\text{ft}}\right)$$

$$= (0.785)(8\ \text{in})^2(324\ \cancel{\text{ft}})\left(12\ \frac{\text{in}}{\cancel{\text{ft}}}\right)$$

$$= 195{,}333\ \text{in}^3$$

   c. Convert the pump column volume from cubic inches (in³) to gallons.

$$\text{Volume} = \frac{\text{Volume, in}^3}{231\ \text{in}^3/\text{gal}}$$

$$= \frac{195{,}333\ \text{in}^3}{231\ \text{in}^3/\text{gal}}$$

$$= 846\ \text{gallons}$$

   d. Determine the time (in minutes) required for the water to flow from the pump suction to the pump discharge.

$$\text{Time} = \frac{\text{Volume}}{\text{Pump Discharge}}$$

$$= \frac{846\ \text{gallons}}{423\ \text{gal/min}}$$

$$= 2.0\ \text{minutes}$$

5. *Bench-Scale Analysis (Test).* A method of studying different ways or chemical doses for treating water or wastewater and solids on a small scale in a laboratory. Also see JAR TEST.

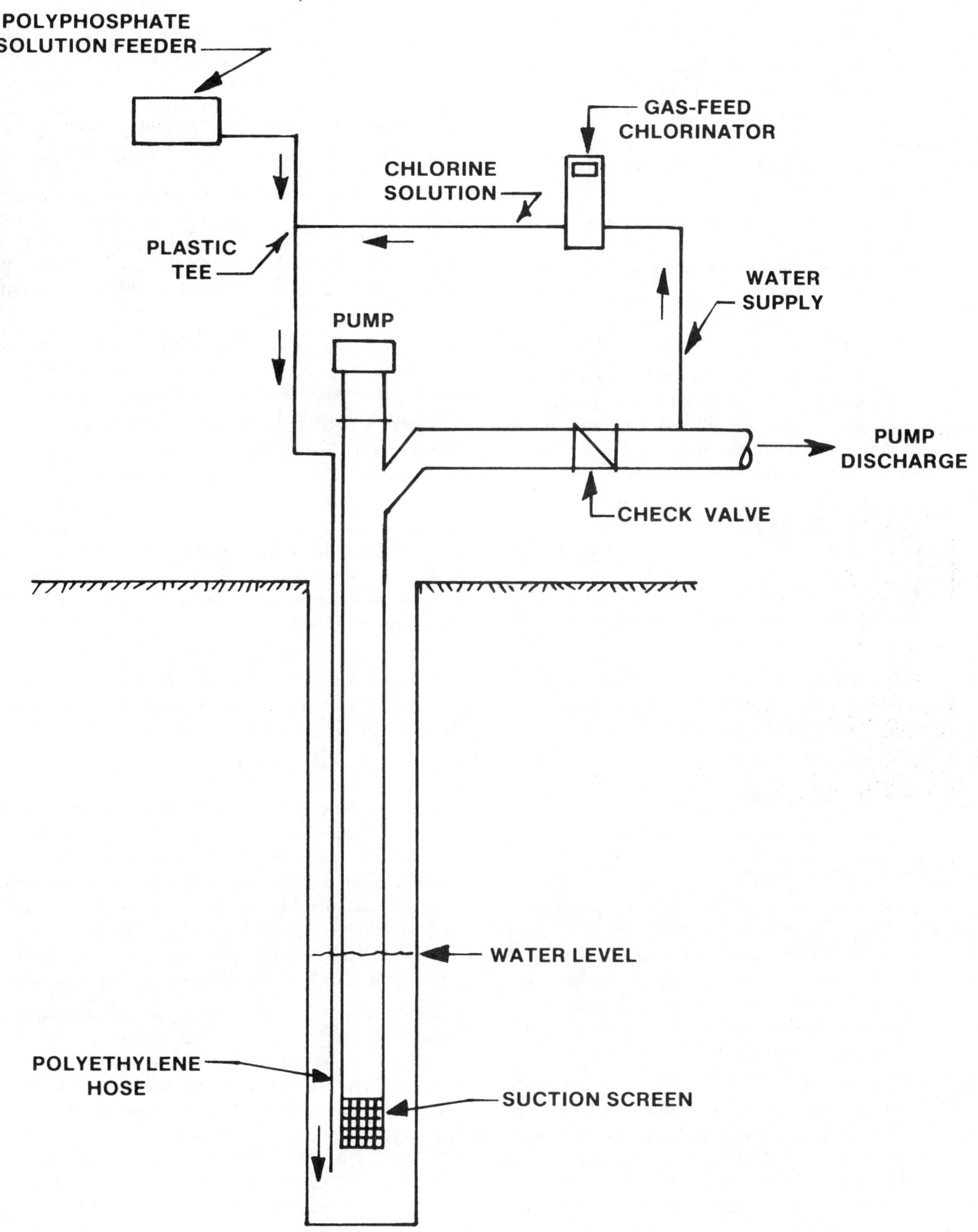

Fig. 12.2 Polyphosphate and chlorine system

2. Turn off the water supply to the chlorinator. Since some of the chlorine in the feed line will drain into the well, it may take several minutes for the chlorine residual to disappear. Check to be sure there is no chlorine residual in the water.
3. Turn the chlorinator back on, noting the exact time. Take samples for chlorine residual every 15 seconds. If the chlorine residual has reached its proper value within 30 seconds of the calculated time (two minutes in this example), you can be sure that the polyethylene tube is properly positioned below the suction screen.

Solutions of polyphosphate containing more than one-half pound per gallon (60 g/L) may be very viscous (thick like molasses), depending upon which of the polyphosphates is used. Do not use a solution much over 48 hours old because the polyphosphates react slowly with water to form orthophosphates, which are much less effective in preventing manganese deposits.

There are some reports in the literature of the successful use of tetrasodium pyrophosphate on iron-bearing waters, but many attempts to control iron have failed. The available information is not clear as to whether the process works only under special conditions or whether the reports of success are in error.

QUESTIONS

Please write your answers to the following questions and compare them with those on page 24.

12.2A How could you find out if nearby wells produce water containing iron and manganese?

12.2B If a well produces water containing dissolved oxygen as well as iron and manganese, how would you attempt to solve the iron and manganese problem?

12.2C What are bench-scale tests?

12.2D Why should polyphosphate solutions over 48 hours old not be used?

12.22 Removal by Ion Exchange[6]

Iron and manganese ion exchange units are similar to a downflow pressure filter. The water to be treated enters the unit through an inlet distributor located in the top. The water is forced (usually by a pump) down through the ion exchange resin into an underdrain structure. From the underdrain structure, the treated water flows out of the unit to the next treatment process, to storage, or into the distribution system.

The actual location of the *ION EXCHANGE RESINS*[7] with respect to other water treatment processes will depend on the raw water quality and the design engineer. If the water to be treated contains no oxygen, both iron and manganese may be removed by ion exchange using the same resins that are used for water softening. If the water being treated contains any dissolved oxygen, the resin becomes fouled with iron rust or insoluble manganese dioxide. The resin can be cleaned but this is expensive and the capacity is reduced. Well water may contain no oxygen in normal operating conditions except immediately after the well is first turned on. If this is the case, provisions should be made to run the well to waste until the oxygen is no longer present.

In one eastern city, an ion exchange plant had operated for seven years, reducing iron from 52 mg/L to 0.1 mg/L and manganese from 1.3 mg/L to zero. When a pump was repaired, a gasket on the suction side of the pump was improperly installed, allowing air to enter the raw water. Within three months, the resin was fouled by iron oxide.

The main advantage of ion exchange for iron and manganese removal is that the plant requires little attention. The disadvantages are the danger of fouling the ion exchange resin with oxides and high initial cost.

To operate an ion exchange unit, try to operate as close as possible to design flows. Monitor the treated water for iron and manganese on a daily basis. When iron and manganese start to appear in the treated water, the unit must be regenerated. Regenerate with a brine solution that is treated with 0.01 pound of sodium bisulfite per gallon (1.2 g/L) of brine to remove oxygen present. After regeneration is complete, dispose of the brine in an approved manner.

See Chapter 14, "Softening," for procedures on how to calculate the frequency for regenerating the unit. Details are given in Chapter 17, "Handling and Disposal of Process Wastes," on how to properly dispose of brine solutions.

12.23 Oxidation by Aeration

Iron can be oxidized by aerating the water to form insoluble ferric hydroxide. As shown in Figure 12.3, this reaction is accelerated by an increase in pH. The rates indicated in Figure 12.3 were determined at 25°C (77°F) under laboratory conditions. If the water contains any organic substances, the rates will be significantly lower; reduced temperatures will also lower the rates. The oxidation of manganese by aeration is so slow that this process is not used on waters with high manganese concentrations.

Since pH is increased by the removal of carbon dioxide, it is important that the aeration (which removes carbon dioxide) be as efficient as possible. Lime is sometimes added to the water to increase the pH as well as remove carbon dioxide. The higher the pH, the shorter the time required for aeration, as shown in Figure 12.3.

6. *Ion Exchange.* A water or wastewater treatment process involving the reversible interchange (switching) of ions between the water being treated and the solid resin contained within an ion exchange unit. Undesirable ions are exchanged with acceptable ions on the resin or recoverable ions in the water being treated are exchanged with other acceptable ions on the resin.
7. *Ion Exchange Resins.* Insoluble polymers, used in water or wastewater treatment, that are capable of exchanging (switching or giving) acceptable cations or anions to the water being treated for less desirable ions or for ions to be recovered.

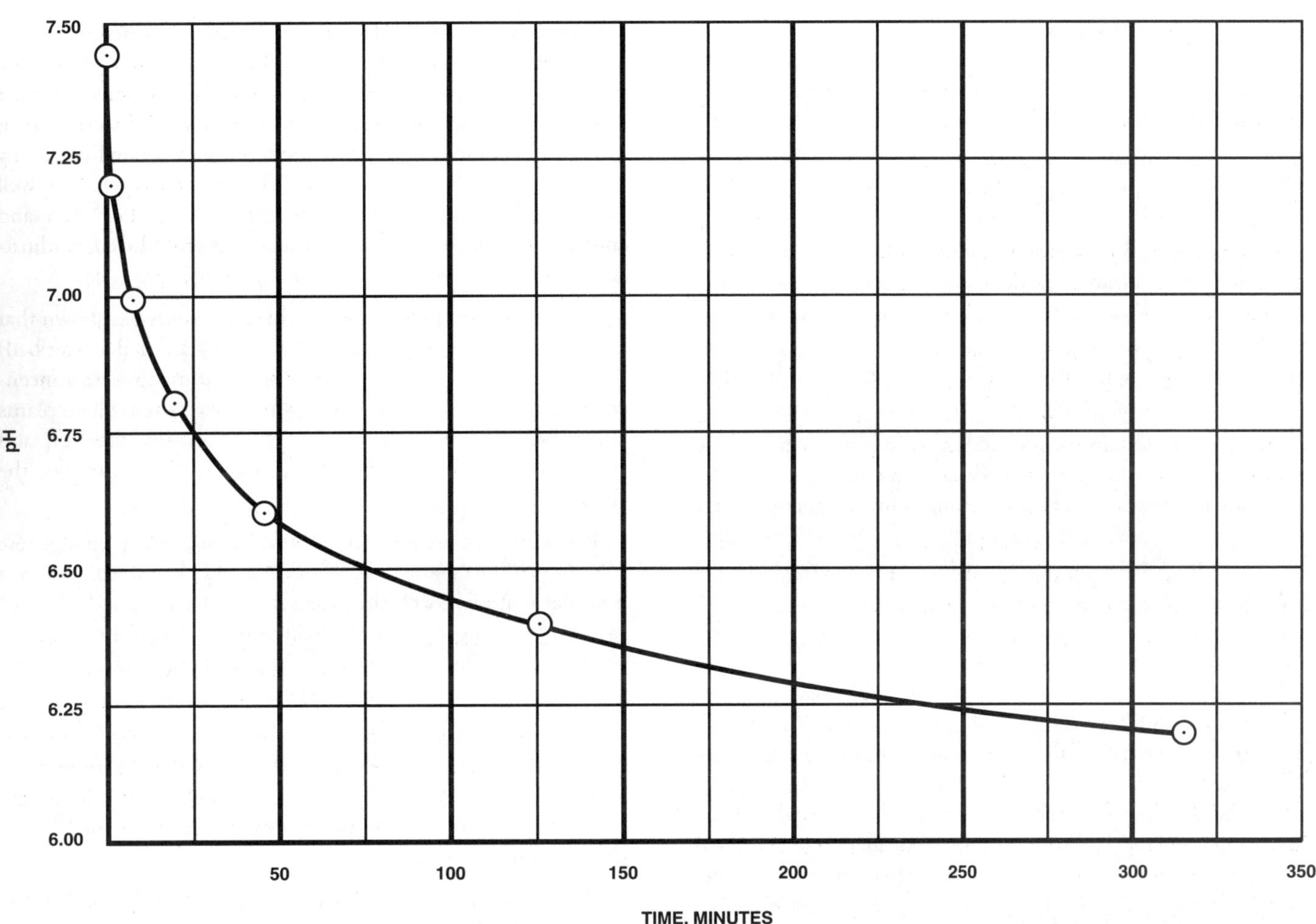

Fig. 12.3 pH versus time to oxidize 99 percent iron

Operation of the aeration process to remove iron and manganese requires careful control of the flow through the process. If the flow becomes too great, not enough time will be available for the reactions to occur. Flows are controlled by the use of variable speed pumps or the selection of the proper number or combinations of pumps. Carefully monitor the iron and manganese content of the treated water. If iron is detected, the flows may have to be reduced.

There are several methods of providing aeration. Either the water being treated is dispersed (scattered) into the air or air is bubbled through the water. Aeration may be achieved by the use of compressed air that passes through diffusers in the water. These diffusers produce many small bubbles that allow the transfer of oxygen in the air to dissolved oxygen in the water.

Other aeration techniques include forced draft, multiple trays, cascades, and sprays. These methods may cause slime growths to develop on surfaces or coatings on media. Slime growths and coatings on media should be controlled to prevent the development of tastes and odors in the product water and the sloughing off of the slimes. Chlorination may be used to control slime growths and coatings. Regularly inspect aeration equipment for the development of anything unusual.

A reaction basin (sometimes called a collection or detention basin) follows the aeration process. The purpose of the reaction basin is to allow time for the oxidation reactions to take place. The aeration process should produce sufficient dissolved oxygen for the iron to be oxidized to insoluble ferric hydroxide. A minimum recommended detention time is 20 minutes with desirable detention times ranging from 30 to 60 minutes (see Example 4, page 17). As shown in Figure 12.3, the pH of the water strongly influences the time for the reaction to take place. Sometimes chlorine is added before the reaction basin.

The reaction basin may be a cylindrically (circular) shaped basin similar to a clarifier. Often the basin is baffled to prevent short-circuiting and the deposition of solids. Since there are no provisions for sludge removal, the basin must be drained and cleaned regularly. If the basins are not cleaned, slugs of deposits or sludge or mosquito and fly larvae (young of any insect) could reach the filters in the next process and cause them to plug.

Operators of reaction basins must always be on the alert for potential sources of contamination. These basins should have covers and protective lids to keep out rain, stormwater runoff, rodents, and insects. All vents must be properly screened. The outlet to the drain must not be connected directly to a sewer or stormwater drain. There must be an air gap or some other protective device to prevent contamination from *BACKFLOW*.[8]

After the ferric hydroxide is formed, it is removed by sedimentation or by filtration alone. If only filtration is used, water from the reaction basin is usually pumped to pressure filters for filtration. The water may also be pumped or flow by gravity to rapid sand filters. For details on how to operate and maintain filters, see Volume I, Chapter 6, "Filtration."

The main advantage of this method is that no chemicals are required; however, lime may be added to increase the pH. The major disadvantage is that small changes in raw surface water quality may affect the pH and soluble organics levels and slow the oxidation rates to a point where the capacity of the plant is reduced.

12.24 Oxidation with Chlorine

Chlorine will oxidize manganese to the insoluble manganese dioxide and will oxidize iron to insoluble ferric hydroxide, which can then be removed by filtration. The higher the chlorine residual, the faster this reaction occurs. Some very compact plants have been constructed by treating the water to a free chlorine residual of 5 to 10 mg/L, filtering, and dechlorinating to a residual suitable for domestic use. Do not use high doses of chlorine if the water contains a high level of organic color because excessive concentrations of total trihalomethanes (TTHMs) could develop. The water is dechlorinated by the use of reducing agents such as sulfur dioxide (SO_2), sodium bisulfite ($NaHSO_3$), and sodium sulfite (Na_2SO_3). Bisulfite is commonly used because it is cheaper and more stable than sulfite. When dechlorinating with reducing agents, be very careful not to overdose because inadequate disinfection could result (no chlorine residual remains) and if the dissolved oxygen level in the water is depleted, fish kills could occur in home aquariums. Frequently, a reaction basin (as described in Section 12.23, "Oxidation by Aeration") is installed between the chlorination processes to allow time for the reactions to occur.

12.25 Oxidation with Permanganate

Potassium permanganate oxidizes iron and manganese to insoluble oxides, and can be used to remove these elements in the same way chlorine is used. The dose of potassium permanganate must be exact. Bench-scale tests are required to determine the proper dosage. Too small a dose will not oxidize all the manganese in the water and too large a dose will allow permanganate to enter the system and may produce a pink color in the water. Actual observations of the water being treated will tell you if any adjustments of the chemical feeder are necessary. Most well waters have relatively consistent concentrations of iron and manganese. Therefore, once you set the chemical feeder, adjustments of the dosage usually are not necessary.

Experience from many water treatment plants has shown that a regular filter bed (a rapid sand filter or a dual media filter bed) can remove manganese as long as iron and manganese concentrations are both less than one milligram per liter. These plants use either chlorine or permanganate to oxidize the iron and manganese before the water being treated flows through the filter bed.

Potassium permanganate is often used with manganese *ZEOLITE*[9] or manganese *GREENSAND*.[10] Greensand is a granular material. After the greensand has been treated with potassium permanganate, it can oxidize both iron and manganese to their insoluble oxides. The greensand also acts as a filter and must be backwashed to remove the insoluble oxides.

Manganese greensand filters can be operated in three modes: continuous regeneration (CR), intermittent regeneration (IR), and catalytic regeneration. The method used will depend on the concentrations of iron and manganese in the water and the pH of the water.

The manganese greensand continuous regeneration process can be used for waters containing iron concentrations as high as 15 mg/L or more, but with such high concentrations, frequent backwashing will be necessary. Generally, waters having iron concentrations in the range of 0.5 to 3.0 mg/L can be treated with more acceptable run lengths of 18 to 36 hours before backwashing.

In the CR process, chlorine and potassium permanganate are typically added to the raw water ahead of the manganese greensand bed. Chlorine is added first to oxidize most of the iron and any sulfide. A slight excess of potassium permanganate is then added to oxidize the remaining iron and soluble manganese. This reaction produces insoluble oxides. When the raw water passes through the manganese greensand bed, two things occur: (1) the insoluble iron oxide particles are filtered out, and (2) any remaining permanganate is reduced to manganese oxides by the greensand. These manganese oxides attach to the grains of greensand, thereby continuously regenerating the manganese

8. *Backflow.* A reverse flow condition, created by a difference in water pressures, that causes water to flow back into the distribution pipes of a potable water supply from any source or sources other than an intended source. Also see BACKSIPHONAGE.
9. *Zeolite.* A type of ion exchange material used to soften water. Natural zeolites are siliceous compounds (made of silica) that remove calcium and magnesium from hard water and replace them with sodium. Synthetic or organic zeolites are ion exchange materials that remove calcium or magnesium and replace them with either sodium or hydrogen. Manganese zeolites are used to remove iron and manganese from water.
10. *Greensand.* A mineral (glauconite) material that looks like ordinary filter sand except that it is green in color. Greensand is a natural ion exchange material that is capable of softening water. Greensand that has been treated with potassium permanganate ($KMnO_4$) is called manganese greensand; this product is used to remove iron, manganese, and hydrogen sulfide from groundwaters.

greensand. As the run progresses, the filter bed becomes clogged with insoluble oxides and the differential pressure increases. When head loss reaches a predetermined point or a certain number of gallons of water have been treated, the filter must be backwashed to remove the filtered particulates.

The intermittent regeneration (IR) process is suitable for raw water containing only manganese or mostly manganese with small amounts of iron. The raw water flows through a manganese greensand bed where oxidation of manganese occurs directly on the grains of greensand. Some iron will also be oxidized directly on the manganese greensand but if the iron concentrations are high, iron oxides will quickly coat or foul the media. To prevent fouling of the media, the iron is sometimes converted to its insoluble form before the water enters the greensand bed by adding chlorine ahead of the filter or aerating the water before it enters the greensand bed. After treating a certain number of gallons of water or when head loss reaches a predetermined point, the capacity of the greensand to oxidize manganese and iron is used up and the media must be backwashed and regenerated. Regeneration consists of the downflow passage of a dilute potassium permanganate solution through the bed using 1.5 ounces (43 milligrams) of $KMnO_4$ per cubic foot of media, followed by thorough rinsing of the media.

When the well water contains low concentrations of iron and manganese (less than 1.0 mg/L) and the pH is greater than 7.0, the catalytic regeneration mode of operation may be a suitable method for removing iron and manganese. In this process, sufficient chlorine is applied ahead of the filter to maintain a chlorine residual of 0.5 to 1.0 mg/L through a bed containing a specially refined grade of greensand. The chlorine keeps the greensand continuously regenerated. Manganese in the raw water can usually be oxidized directly by the manganese oxide coating on the greensand, provided that the pH is neutral and adequate detention time is provided. The advantages of catalytic oxidation are that the filter run lengths are longer than for other process modes, higher filtration rates can be used, the backwash waste suspended solids are low because the manganese is oxidized on the manganese greensand grains, and the chemical operating costs are minimal because only chlorine is required. For more information on the manganese greensand process, see Section 12.3, "Operation of an Iron and Manganese Removal Plant."

12.26 Operation of Filters

When iron and manganese are oxidized to insoluble forms by aeration, chlorination, or permanganate, the oxidation processes are usually followed by filters to remove the insoluble material. In addition to the procedures for operating and maintaining filters that were outlined in Volume I, Chapter 6, "Filtration," the procedures discussed in this section apply to filters used to remove iron and manganese.

Iron tests should be made monthly on the water entering a filter to be sure the iron is in the ferric (Fe^{3+}) state. Collect a sample of the water and pass the water through a filter paper. Run an iron test on the water that has passed through the filter. If the iron is still in the soluble ferrous (Fe^{2+}) state, there will be iron in the water. If aeration is being used to oxidize the iron from the soluble ferrous to the insoluble ferric state and iron is still present in the soluble state in the water entering the filter, try adding chlorine or potassium permanganate. If chlorine or potassium permanganate is being used and soluble iron is in the water entering the filter, try increasing the chemical dose. If potassium permanganate is being used, the sand may be replaced by greensand to improve the efficiency of the process.

If oxidation is being accomplished by either aeration or chlorination, a free chlorine residual must be maintained in the effluent of the filter to prevent the insoluble ferric iron from returning to the soluble ferrous form and passing through the filter.

Most iron removal treatment plants are designed so that the filters are backwashed according to head loss. If iron breakthrough is a problem, filters should be backwashed when breakthrough occurs or just before breakthrough is expected. Accurate records can reveal when breakthrough occurs and when breakthrough can be expected.

12.27 Proprietary Processes
by Bill Hoyer

Several patented processes are available for iron and manganese control. The best way to learn about the effectiveness and maintenance requirements of these processes is to contact someone who has one. Once you are operating one of these processes, the manufacturer is a good source of help when troubleshooting. Remember that various sources of raw water are different and what works for one operator may not work at your water treatment plant.

Electromedia iron and manganese removal systems are generally used on groundwater supplies at individual well sites because of their compactness and simplicity of treatment. The system uses reaction vessels where chemical reactions take place and an adsorptive media that requires no regeneration by special chemicals. Chlorine is used as the oxidizing chemical because of its cost and efficiency. (Any suitable oxidizing chemical can be used.) Almost 30 percent less chlorine, pound for pound, is required to perform the same amount of oxidation as potassium permanganate.

After oxidation with chlorine, a small dose of sulfur dioxide (0.25 to 0.50 mg/L) is introduced prior to the second reaction vessel. This dosage is factory set according to the general mineral analysis of the raw water. Dosage should not be altered. The sulfur dioxide is used to accelerate the oxidation of any sulfur compounds in the water to form compounds having no objectionable taste or odor.

The water is then sent to a filter operating at a preset rate of up to 15 gallons per minute per square foot (10 liters per second per square meter or 10 millimeters per second). In the filter vessel, the iron and manganese are adsorbed on the surface of the media until backwashing. The media can withstand a very high backwash rate (20 gallons per minute per square foot, 13.6 L/sec/m^2 or 13.6 mm/sec) and requires only a four-minute backwash to obtain thorough cleaning.

The filter effluent can be sampled by a continuously monitoring analyzer that drives a 30-day strip-chart recorder. The recorder may have a color-coded indicating strip to direct the operator in the proper chemical dosage. If the recorder trace falls out of the green, the operator increases the chlorine dosage. The dosage is adjusted by turning one knob and can be read immediately. The effect of the change can be seen on the chart trace in five minutes and will reach a steady state within 10 minutes. Thus, the operator can quickly determine the proper dosage. With the chemical dosage set properly, a free chlorine residual exists in the filter effluent providing the required disinfection in the distribution system. Variations in water quality are quickly reflected in the chart tracing. Since no permanganate is used, there are no black water or pink water complaints from accidental underdosage or overdosage of chemicals.

The process uses *BREAKPOINT CHLORINATION* [11] and the very effective adsorptive qualities of the media. Each system is provided with an automatic control panel that permits adjustment of any of the filter cycles simply by rotating a timer knob. Status of the system is displayed on the front panel with pilot lights for easy viewing. The automatic control panel operates the manually set chemical feed system using gaseous chlorine and gaseous sulfur dioxide. Backwash is accomplished automatically by using a process signal and filtration timer with a differential pressure override.

Maintenance on the system is quite limited. Most systems are built with an automatic standby for the chemicals that will switch from an empty container to a full container. Table 12.1 lists the recommended maintenance.

TABLE 12.1 RECOMMENDED MAINTENANCE FOR THE ELECTROMEDIA PROCESS

	Daily	*Weekly*	*Monthly*
1. Inspection of chart paper for proper chemical dosage	X		
2. Free chlorine residual test	X		
3. Total chlorine residual test	X		
4. Addition of buffer solution in the analyzer		X	
5. Colorimetric analysis of the influent and effluent for iron concentration		X	
6. Laboratory tests for analysis of influent and effluent for iron and manganese concentration			X
7. Changing of chart paper			X
8. Routine maintenance checks associated with valves, pipes, and pumps			X

Each application for iron and manganese removal is based on the general mineral analysis of the raw water. The required chemical treatment is provided for iron, manganese, and sulfide treatment. Additional equipment may be provided where corrosivity or chelating compounds are found to be present. Aeration may precede the process where methane extraction or carbon dioxide removal is required. Plant operators are directed to the operation and maintenance instructions provided with the equipment for additional details.

12.28 Monitoring of Treated Water

When controlling iron and manganese by aeration or with chemicals, the product water must be monitored closely. If laboratory facilities are available, the treated water can be analyzed for iron and manganese to be sure treatment is adequate. A quick way to monitor treated water is to collect a sample and add a dose of chlorine. If a brown- or rust-colored floc develops, then the treatment is inadequate. You will either have to increase the chemical doses or reduce the flows. If a pink color appears in the product water when using permanganate, then the dose is too high and must be reduced until the pink color is no longer visible.

12.29 Summary

Small iron and manganese water treatment plants can be very difficult to operate. If your plant is not operating as desired, talk to other operators in your area and see if they have any suggestions. If you have problems, you will have to try different chemical doses and procedures. Keep accurate records so you can evaluate the effectiveness of your efforts.

A lot of iron complaints in drinking water are caused by old steel or cast-iron water mains. A possible solution to this problem is to inject polyphosphates directly into the distribution mains. See Section 12.5, "Troubleshooting Red Water Problems," for additional ideas on how to solve problems.

11. *Breakpoint Chlorination.* Addition of chlorine to water or wastewater until the chlorine demand has been satisfied. At this point, further additions of chlorine will result in a free chlorine residual that is directly proportional to the amount of chlorine added beyond the breakpoint.

FORMULAS

A standard polyphosphate solution is prepared by mixing and dissolving a known amount of polyphosphate in a container and adding distilled water to the one-liter mark. Use the following formulas to determine the settings on polyphosphate chemical feeders:

1. Prepare a series of samples and test with polyphosphate.
2. Select the optimum dosage in mg/L.
3. Calculate the chemical feeder setting in pounds of polyphosphate per day.

$$\text{Stock Solution, mg/mL} = \frac{(\text{Polyphosphate, grams})(1{,}000\ \text{mg/gram})}{(\text{Solution, liter})(1{,}000\ \text{mL/L})}$$

$$\text{Dose, mg/L} = \frac{(\text{Stock Solution, mg/mL})(\text{Volume Added, mL})}{\text{Sample Volume, L}}$$

$$\text{Dose, lb/MG} = \frac{(\text{Dose, mg/L})(3.785\ \text{L/gal})(1{,}000{,}000/\text{Million})}{(1{,}000\ \text{mg/g})(454\ \text{g/lb})}$$

$$\text{Chemical Feeder Setting, lb/day} = (\text{Flow, MGD})(\text{Dose, mg/L})(8.34\ \text{lb/gal})$$

Example 2

A standard polyphosphate solution is prepared by mixing and dissolving 1.0 gram of polyphosphate in a container and adding distilled water to the one-liter mark. Determine the concentration of the stock solution in milligrams per milliliter. If 5 milliliters of the stock solution are added to a one-liter sample, what is the polyphosphate dose in milligrams per liter and pounds per million gallons?

Known		Unknown
Polyphosphate, g	= 1.0 g	1. Stock Solution, mg/L
Solution, L	= 1.0 L	2. Dose, mg/L
Stock Solution, mL	= 5 mL	3. Dose, lb/MG
Sample, L	= 1 L	

1. Calculate the concentration of the stock solution in milligrams per milliliter (mg/mL).

$$\text{Stock Solution} = \frac{(\text{Polyphosphate, g})(1{,}000\ \text{mg/g})}{(\text{Solution, L})(1{,}000\ \text{mL/L})}$$

$$= \frac{(1.0\ \text{g})(1{,}000\ \text{mg/g})}{(1.0\ \text{L})(1{,}000\ \text{mL/L})}$$

$$= 1.0\ \text{mg/mL}$$

2. Determine the polyphosphate dose in the sample in milligrams per liter (mg/L).

$$\text{Dose} = \frac{(\text{Stock Solution, mg/mL})(\text{Volume Added, mL})}{\text{Sample Volume, L}}$$

$$= \frac{(1.0\ \text{mg/mL})(5\ \text{mL})}{1\ \text{L}}$$

$$= 5.0\ \text{mg/L}$$

3. Determine the polyphosphate dose in the sample in pounds of polyphosphate per million gallons (MG).

$$\text{Dose} = \frac{(\text{Dose, mg/L})(3.785\ \text{L/gal})(1{,}000{,}000/\text{Million})}{(1{,}000\ \text{mg/g})(454\ \text{g/lb})}$$

$$= \frac{(5.0\ \text{mg/L})(3.785\ \text{L/gal})(1{,}000{,}000/\text{Million})}{(1{,}000\ \text{mg/g})(454\ \text{g/lb})}$$

$$= 42\ \text{lb/MG}$$

Example 3

Determine the chemical feeder setting in pounds of polyphosphate per day if 0.4 MGD is treated with a dose of 5 mg/L.

Known		Unknown
Flow, MGD	= 0.4 MGD	Chemical Feeder, lb/day
Dose, mg/L	= 5 mg/L	

Determine the chemical feeder setting in pounds per day (lb/day).

$$\text{Chemical Feeder Setting} = (\text{Flow, MGD})(\text{Dose, mg/L})(8.34\ \text{lb/gal})$$

$$= (0.4\ \text{MGD})(5\ \text{mg/L})(8.34\ \text{lb/gal})$$

$$= 17\ \text{lb/day}$$

FORMULAS

Use the following formulas to calculate the average detention time for a reaction basin:

1. Determine the dimensions of the basin.
2. Measure and record the flow of water being treated.

$$\text{Basin Volume, ft}^3 = (0.785)(\text{Diameter, ft})^2(\text{Depth, ft})$$

$$\text{Basin Volume, gal} = (\text{Basin Volume, ft}^3)(7.48\ \text{gal/ft}^3)$$

$$\text{Detention Time, min} = \frac{(\text{Basin Vol, gal})(24\ \text{hr/day})(60\ \text{min/hr})}{\text{Flow, gal/day}}$$

Example 4

A reaction basin 12 feet in diameter and 5 feet deep treats a flow of 200,000 gallons per day. What is the average detention time in minutes?

Known		Unknown
Diameter	= 12 ft	Detention Time, min
Depth	= 5 ft	
Flow	= 200,000 GPD	

1. Calculate the basin volume in cubic feet (ft^3).

$$\text{Basin Vol, ft}^3 = (0.785)(\text{Diameter, ft})^2(\text{Depth, ft})$$

$$= (0.785)(12\ \text{ft})^2(5\ \text{ft})$$

$$= 565\ \text{ft}^3$$

2. Convert the basin volume from cubic feet (ft^3) to gallons.

$$\text{Basin Vol, gal} = (\text{Basin Vol, ft}^3)(7.48\ \text{gal/ft}^3)$$
$$= (565\ \text{ft}^3)(7.48\ \text{gal/ft}^3)$$
$$= 4{,}226\ \text{gal}$$

3. Determine the average detention time in minutes for the reaction basin.

$$\text{Detention Time} = \frac{(\text{Basin Vol, gal})(24\ \text{hr/day})(60\ \text{min/hr})}{\text{Flow, gal/day}}$$
$$= \frac{(4{,}226\ \text{gal})(24\ \text{hr/day})(60\ \text{min/hr})}{200{,}000\ \text{gal/day}}$$
$$= 30\ \text{minutes}$$

QUESTIONS

Please write your answers to the following questions and compare them with those on page 24.

12.2E What happens if water being treated for iron and manganese by ion exchange contains any dissolved oxygen?

12.2F How does the pH of the water influence the rate of oxidation of iron to insoluble ferric hydroxide?

12.2G What is the purpose of a reaction basin following the aeration process?

12.2H After chlorine has been added to oxidize iron and manganese, how is the water dechlorinated?

12.2I What are the three modes of operation for a manganese greensand process called?

12.3 OPERATION OF AN IRON AND MANGANESE REMOVAL PLANT

by Gerald Davidson

The very low recommended limits for iron (0.3 mg/L) and manganese (0.05 mg/L) in water make these contaminants difficult to treat and sometimes the processes are expensive. Because of this, operators should know how the processes work and what to check for when something goes wrong and the limits are exceeded. This section describes the operation of an iron and manganese removal plant using the continuously regenerated (CR) manganese greensand process.

12.30 Description of Equipment and Process

The filter is the most important piece of treatment equipment in an iron and manganese removal plant. Figure 12.4 illustrates a typical filter consisting of layers of graded gravel, manganese greensand, and anthracite coal. Some of the differences between greensand filters and conventional filters include the following: (1) the greensand is finer than filter sand; (2) the filtration rate is slower (2 to 5 GPM/ft^2 (1.3 to 3.4 liters per sec/m^2 or 1.3 to 3.4 mm/sec)); and (3) the backwash rate is lower (12 GPM/ft^2 at 55°F versus 15 GPM/ft^2 for sand/anthracite filters).

Figure 12.5 shows a typical flow diagram for a continuous regeneration (CR) manganese greensand plant. Treatment with

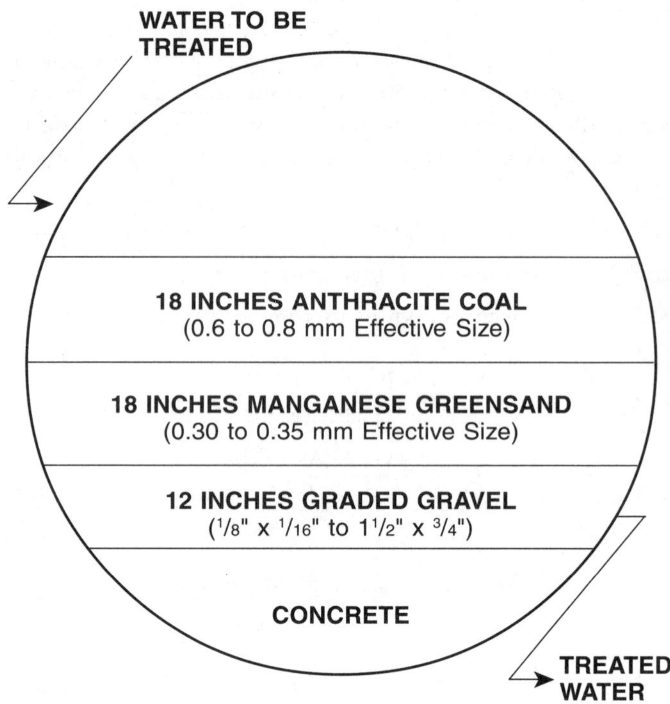

Fig. 12.4 Multimedia manganese greensand filter (horizontal)

manganese greensand can remove 95 percent of both iron and manganese. However, if the iron concentration is above 10 mg/L or manganese is above 5 mg/L, the efficiency of the greensand (or any other type of filter) drops very quickly as insoluble oxide particles accumulate on the grains of greensand and within the media bed. Pretreatment may be required to oxidize and remove a major portion of the iron and any sulfide present.

At the plant shown in Figure 12.5, raw water being pumped from a 50-foot (15 m) deep well contains 3.0 mg/L iron and 0.75 mg/L manganese (see Chapter 21 for test procedures) and it has a pH of 6.2. The pH of the water has a significant effect on iron removal. The oxidation potential of chlorine and potassium permanganate decreases as the pH increases, although the rate of reaction increases significantly with the increase in pH. Because the pH of the water in this example is 6.2, which is the minimum recommended pH for use of the CR process, the pH is adjusted to within the range of 6.5 to 6.8 by adding sodium carbonate or sodium hydroxide or by passing the water through a calcite neutralizing filter.

Next, the water is injected with chlorine, flash mixed, and flocculated for a period of ten minutes in order to oxidize most of the iron. The chlorine dosage can be calculated using the following formula:

$$Cl_2\ \text{Required, mg/L} = 1 \times \text{Fe Conc, mg/L}$$

Therefore,

$$Cl_2\ \text{Required, mg/L} = 1 \times 3.0\ \text{mg/L}$$
$$= 3\ \text{mg/L}$$

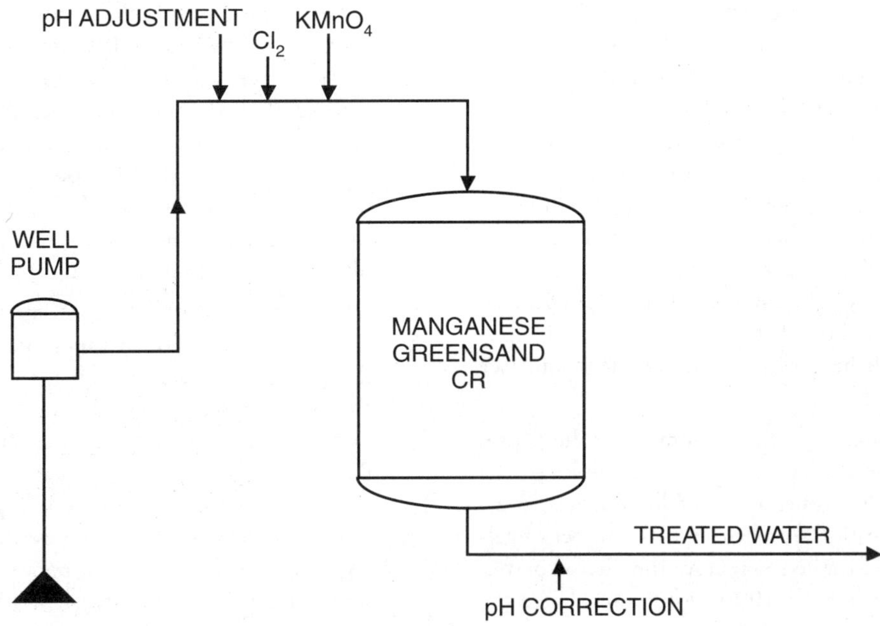

Fig. 12.5 Typical flow diagram of a continuous regeneration (CR) greensand process
(Permission of Hungerford & Terry, Inc.)

After chlorination, the water is injected with potassium permanganate ($KMnO_4$) to complete the oxidation of any remaining iron and soluble manganese. Sodium hydroxide (30 mg/L)[12] may also be added if needed for corrosion control (pH correction, Figure 12.5). The $KMnO_4$ dosage can be calculated using the following formula:

$$KMnO_4 \text{ Required, mg/L} = (0.2 \times \text{Fe Conc, mg/L}) + (2 \times \text{Mn Conc, mg/L})$$

Therefore,

$$KMnO_4 \text{ Required, mg/L} = (0.2 \times 3.0 \text{ mg/L}) + (2 \times 0.75 \text{ mg/L})$$

$$= 2.1 \text{ mg/L}$$

NOTE: The calculated 2.1 mg/L potassium permanganate dose is the minimum dose. This dose assumes there are no other oxidizable compounds in the raw water. However, typical oxidizable compounds usually found include organic color, bacteria, and even hydrogen sulfide (H_2S). Therefore, the actual dose may be higher. A bench-scale test should be performed to determine the required dose.

To prepare a stock solution of potassium permanganate, mix 2 to 4 ounces (56 to 113 mg) of potassium permanganate chemical per gallon of water (15 to 30 mg/L) in a solution tank. Mix thoroughly to make sure the chemical dissolves completely. $KMnO_4$ is soluble up to 8.3 ounces (235 mg) per gallon (62 mg/L) of water at 68°F so you should have little trouble dissolving the chemical.

Problems will develop in the iron and manganese removal process using manganese greensand if you have too short a detention time for the chemicals to react. That is, it takes a little time for the chemicals to start working. If the plant you are operating does not allow sufficient detention time for the chemical reactions to take place, you should perform jar tests to see if a flash mix will improve performance. In some plants, the injection of the potassium permanganate solution in the volute of the pump will produce complete mixing of potassium permanganate.

The potassium permanganate feed system in our example plant consists of a 50-gallon (190 L) polyethylene solution vat, two ¼ HP mixers, liquid level switches, and a metering pump. The mixers do not have to run continuously because of the solubility of potassium permanganate in water.

The influent flow to the manganese greensand/anthracite filter is approximately 1,500 to 2,000 GPM. If the chemical

12. *NOTE:* Addition of 30 mg/L of sodium hydroxide (NaOH) will increase the sodium content of the water by 17 mg/L. If you are trying to keep the sodium level below 20 mg/L, then the sodium in the raw water must be below 3 mg/L.

dosage rate for $KMnO_4$ has been correctly calculated, the filter influent will have a slight pink/orange color due to the presence of a slight excess of permanganate. As the water is treated in the filter, this color will disappear because the permanganate will be reduced to manganese oxide by the manganese greensand. The manganese oxides will then precipitate on the grains, thus continuously regenerating the manganese greensand. It is important not to underfeed permanganate for an extended period of time. Without adequate permanganate, you will eventually exhaust the oxidative capacity of the media. If this happens, iron and manganese may pass through the filter into the distribution system and you will have to recharge the greensand (see Section 12.31).

As the filter run progresses and the amount of insoluble precipitates builds up in the media, the pressure drop across the filter will gradually increase. When the head loss reaches 10 psi (0.07 pascals), or after treating a predetermined number of gallons, the bed must be backwashed to remove the filtered particulates. Backwashing usually takes 10 to 15 minutes. Since the media is continuously regenerated during operation, there is no need to regenerate the greensand after backwashing. The used backwash water can be collected and reused after allowing the precipitates to settle out. The settled precipitates may be disposed of in a sanitary sewer or sent to a drying bed.

In addition to backwashing, some plants use an air washing system to more thoroughly clean the manganese greensand grains on a weekly basis. Air wash is particularly useful for plants treating water that has a relatively high concentration of manganese. Manganese hydroxide oxidation products tend to be sticky or gummy and may cause the grains of greensand to stick together in clumps, especially in long filter runs. Backwashing alone may not always separate these grains and, over a period of time, mudballs may form. Agitating the greensand with an air wash (or, better still, an air-water wash) helps to keep the filter media in a free, loose, fluid condition. This will result in cleaner media and longer runs by minimizing cementation of the media bed.

Daily tests the operator should perform include iron, manganese, pH, and chlorine residual. The iron and manganese test tells the operator if the treatment plant is working and meeting state and federal water quality requirements. The pH test is also very important because of the relationship between pH and the corrosivity of water and because effective iron and manganese removal by the CR manganese greensand process requires a pH of 6.2 to 6.8.

12.31 Regeneration of Manganese Greensand

If the potassium permanganate charge is lost in the filter bed due to underdosing of potassium permanganate, the operator must regenerate the greensand. There are two ways to regenerate the bed:

1. Shut down the process and pour a saturated solution of potassium permanganate (about 5 percent) into the filters; let the saturated solution sit for approximately 24 hours. After 24 hours, backwash the filters at a normal rate to flush out the excess potassium permanganate.
2. Recharge the greensand by increasing the potassium permanganate dosage until pink water flows out of the greensand media. Then, decrease the potassium permanganate until you have a slight pink color before filtration. There should be no pink water after filtration when the water is being pumped into the distribution system. If there is still pink water after filtration, keep decreasing the potassium permanganate dose until no pink water is present in the water after filtration. The pink color is the best indication that the greensand is regenerated or recharged.

   One problem with this method is that you might reduce the permanganate level too far and pass water with iron and manganese. This could cause red-colored water in the distribution system and stain clothes or bathroom fixtures. Because of the importance of the potassium permanganate in the greensand process, it is highly recommended that some type of fail-safe system be installed to prevent filtering water in the event the potassium permanganate solution vat reaches a low level. Without permanganate, the greensand could lose its charge and iron and manganese will enter the distribution system. When a low level is reached, the plant should be automatically shut down. Typical fail-safe systems include low-level alarms in the vat or an automatic switchover system to another vat when the level drops too low in the vat in use.

Some operators find that method No. 1 is more effective. Since all treatment plants are unique in some respect, one method or the other or some modification may work best for your plant. Therefore, procedures and methods should be developed through actual experience. These methods should be adopted only if they provide the desired results without eliminating any concepts of design or of good operating practice.

12.32 Troubleshooting

See Table 12.2, "Troubleshooting Manganese Greensand Systems," for a listing of troubles, causes, and remedies.

QUESTIONS

Please write your answers to the following questions and compare them with those on page 24.

12.3A How does a water's pH affect the manganese greensand process?

12.3B How can an operator quickly tell whether enough potassium permanganate is being fed in a continuous regeneration manganese greensand process?

12.3C Why should a greensand plant be shut down when the permanganate solution level in the vat gets low?

12.4 MAINTENANCE OF A CHEMICAL FEEDER

In small water treatment plants that remove iron and manganese, a hypochlorite solution may be used to provide chlorine instead of using chlorine gas. Commercial sodium hypochlorite solutions (such as Clorox) contain an excess of caustic (sodium hydroxide, NaOH). When the solution is diluted with water

TABLE 12.2 TROUBLESHOOTING MANGANESE GREENSAND SYSTEMS

Trouble	Cause	Remedy
SYSTEM TYPE: CR/IR[a]		
1. Filter effluent clear, iron low, manganese higher than raw water.	Manganese being leached from greensand grains; bed is insufficiently regenerated.	Increase frequency of regeneration. Regenerate bed with sufficient $KMnO_4$ (1½ oz/ft^3) so heavy purple color comes through bed. Make sure proper influent chemicals (Cl_2 or $KMnO_4$) are being continuously fed.
2. Filter effluent turbid with yellow to brownish color. Iron and manganese high.	Too much alkali being fed ahead of filter.	Reduce alkali feed. Maintain correct pH prior to filter at 6.2–6.5. Post pH correct if higher pH required in system.
	Polyphosphate being fed ahead of filter.	Discontinue polyphosphate feed.
	Channeling through filters.	Check bed surface for mounds, pockets, and channeling. Backwash and air-scrub if possible.
	Iron organically bound: reactions with oxidizing agent produce a nonfilterable colloid.	Feed alum or other coagulant prior to filter. Amount determined in field.
3. Excessive pressure drop across bed immediately after backwashing.	Accumulation of greensand fines at surface of bed.	Remove fines by scraping after backwashing. In severe cases, bed replacement may be required.
		Check maximum pressure differential recommended and do not exceed 10 psi ΔP as additional fines may be formed.
	Backwash rate too low.	Increase backwash rate to 10–12 GPM/ft^2.
	Filter bed cemented, as evidenced by mounding around periphery of vessel.	Break up cemented areas with air-water wash combination. Bed replacement may be necessary.
	Well throwing fine sand, silt, and colloidal clay.	Check well supply, especially immediately after pump start-up. Allow well to pump overboard at start of pumping cycle.
4. On multiple unit installations, water quality good on some units, bad on others.	Unequal distribution of prefeed chemicals.	Inject chemical at a point where thorough mixing of chemicals with raw water occurs before diversion to the various filters.
SYSTEM TYPE: CR[a]		
5. Iron breakthrough before recommended maximum ΔP of 10 psi is reached.	Some iron waters filter in depth and do not build up head loss.	Backwash should be initiated by total number of gallons treated rather than by head loss. Use ΔP as a backup to prevent exceeding ΔP of 10 psi.
6. Faint pink color in filter effluent.	Permanganate feed rate is too high.	Operate filter for 1–2 hours with $KMnO_4$ feed off. Then reset feeder at slightly lower setting.
SYSTEM TYPE: IR[a]		
7. Low capacity.	Manganese oxide coating stripped from greensand grains due to insufficient regeneration. May be especially troublesome with high sulfide water.	Increase frequency of regeneration. Prefeed Cl_2 with sulfide waters. Replace bed if required.
	Manganese greensand heavily iron-fouled.	Use CR method with dual-media anthracite/manganese greensand bed to prevent iron fouling.
	Excessive grain growth due to high manganese oxide buildup.	Increase frequency of regeneration. Bed replacement may eventually be required.

a CR—Continuous Regeneration
IR—Intermittent Regeneration

(Reprinted with permission of Hungerford & Terry, Inc., Clayton, NJ) containing carbonate alkalinity,[13] the resulting solution becomes supersaturated with calcium carbonate. This calcium carbonate tends to form a coating on the poppet valves in the solution feeder. The coated valves do not seal properly and the feeder fails to pump the hypochlorite solution properly.

Calcium carbonate scale can be removed from poppet valves by using the following procedure:

1. Fill a one-quart (946 mL) Mason jar half full of tap water.
2. Add one fluid ounce (44 mL) of 30- to 37-percent hydrochloric acid (HCl) to the Mason jar.
3. Finish filling the jar to the top with tap water.
4. Place the suction hose of the hypochlorinator in the jar and pump the entire contents of the jar through the system.
5. Return the suction hose to the solution tank and resume normal operation.

The hydrochloric acid (HCl), also called muriatic acid, can be obtained from stores selling swimming pool supplies.

One way to avoid the formation of calcium carbonate coatings is to obtain the dilution water for the hypochlorite from an ordinary home water softener.

For additional information on the operation and maintenance of various types of chemical feeders, see Chapter 13, "Fluoridation."

QUESTIONS

Please write your answers to the following questions and compare them with those on page 25.

12.4A What problems may develop in a chemical feeder pumping sodium hypochlorite?

12.4B How can the problems caused by calcium carbonate scale on a hypochlorinator's poppet valves be solved?

12.5 TROUBLESHOOTING RED WATER PROBLEMS

The first step when troubleshooting red or dirty water complaints is to be sure the iron and manganese treatment processes are working properly. If the iron and manganese are being removed by the treatment processes, investigate the distribution system for sources of the problem.

Red water or dirty water problems may be caused by iron or manganese in the water, corrosive water, or iron bacteria in the distribution system. When an unstable or corrosive water (see Volume I, Chapter 8, "Corrosion Control") is pumped into the distribution system, the water attacks cast-iron pipes and metal service lines, picks up iron, and causes red water complaints. All water treatment plants should run a Marble Test (see Chapter 21, "Advanced Laboratory Procedures"). If the test indicates that the water is corrosive, the addition of caustic (sodium hydroxide, NaOH) to the water to increase the pH will help. When the water becomes stable (according to the Marble Test), some of the red water complaints could be eliminated.

The growth of iron bacteria inside water mains causes one of the most troublesome and most difficult to eliminate red water problems. These bacteria are not harmful. They live and accumulate in the iron in the water flowing through the distribution system. As the bacterial growths increase, slimes will build up in the mains and eventually slough off into the water. When these slimes come out a consumer's water tap, you can expect complaints of red water and slimes.

Slime growths can be controlled by maintaining a free chlorine residual throughout the distribution system. Sometimes the residual is very difficult to maintain. If bacterial growths have been in the distribution system for a long time and are flourishing, it is very difficult to maintain a free chlorine residual at the extremes or in dead ends of the system. Also, if the water has a natural high chlorine demand, your chlorination equipment may not be capable of feeding enough chlorine to maintain a free chlorine residual. Remember that frequently when consumers complain about a strong chlorine taste or swimming pool-tasting water, the solution is to add more chlorine in order to get past the breakpoint.

One way to rid a distribution system of iron bacteria is to develop a flushing program.[14] Flushing should start at the location where the water enters the distribution system, such as an elevated tank. Flush the water mains by working toward the extremes or most distant points of the distribution system. Usually, only one portion of the distribution system is flushed, followed by another portion until the entire system has been flushed.

A common practice is to open a hydrant at the extreme end of the system at the start of the flushing job to be sure the water being flushed will carry the sediment and insoluble precipitates in the desired direction and out of the system. Flushing is often done late at night when water demands are low so facilities will not be overworked and consumers will not be inconvenienced.

Valves will have to be opened and closed in the proper sequences to be sure the desired mains are being flushed and that no one will be without water. Hydrants that are opened to allow flushing must be of sufficient size to produce flushing velocities (2.5 up to 5.0 ft/sec preferred or 0.75 to 1.5 m/sec) in the mains. Also, the mains providing the flushing flows must have sufficient capacity to deliver the desired flows.

When flushing a system, be sure the pressure in the distribution system does not drop below 20 psi (1.4 kg/cm^2 or 138 kPa). If a four-inch (100 mm) water main is flushed using a six-inch (150 mm) hydrant, the water pressure in the main downstream from the hydrant could become dangerously low. When this happens, the distribution system could be subject to contamination by *BACKSIPHONAGE*.[15] Never allow a backsiphon condition to develop in a distribution system.

---

13. See Chapter 14, "Softening," for a discussion of carbonate alkalinity.
14. See *Water Distribution System Operation and Maintenance* in this series of manuals, Chapter 5, Section 5.706, "Pipe Flushing."
15. *Backsiphonage.* A form of backflow caused by a negative or below atmospheric pressure within a water system. Also see BACKFLOW.

In summary, to minimize red water or dirty water problems and complaints, you must provide adequate treatment to control iron and manganese. This is necessary to ensure that the water pumped into the distribution system contains little or no iron and manganese. The water must be stable (noncorrosive) so that iron will not be picked up in the distribution system. Corrosion control treatment processes can produce a stable water. If bacterial growths are a problem, a free chlorine residual must be maintained in all water throughout the distribution system. If red or dirty water problems exist in a distribution system, a thorough flushing program can be very helpful.

12.6 ARITHMETIC ASSIGNMENT

Turn to the Arithmetic Appendix at the back of this manual. Read and work the problems in Section A.30, "Iron and Manganese Control." Check the arithmetic in this section using an electronic calculator. You should be able to get the same answers. Section A.50 contains similar problems using metric units.

12.7 ADDITIONAL READING

Manual of Water Utilities Operations (also called *Texas Manual*), Chapter 11 (depends on edition), "Special Water Treatment (Iron and Manganese Removal)." This publication is available from Texas Water Utilities Association at www.twua.org or phone (888) 367-8982.

12.8 ACKNOWLEDGMENT

Portions of this chapter were revised by Raymond Ellis, Hungerford & Terry, Inc.

QUESTIONS

Please write your answers to the following questions and compare them with those on page 25.

12.5A List the possible causes of red or dirty water complaints.

12.5B How can slime growths be controlled in water distribution systems?

Please answer the discussion and review questions next.

DISCUSSION AND REVIEW QUESTIONS

Chapter 12. IRON AND MANGANESE CONTROL

Please write your answers to the following discussion and review questions to determine how well you understand the material in the chapter.

1. Why should iron and manganese be controlled in drinking water?
2. Why are accurate results of tests for iron and manganese difficult to obtain?
3. Why is chlorine usually fed with polyphosphates when controlling iron and manganese?
4. How do polyphosphates control manganese?
5. How is the proper polyphosphate dose determined?
6. What happens when an ion exchange resin becomes fouled with iron rust or insoluble manganese dioxide?
7. When should iron and manganese ion exchange units be regenerated?
8. How would you determine whether or not to adjust the flows to an oxidation by aeration process to remove iron?
9. Why should reaction basins be drained and cleaned?
10. What are the advantages and disadvantages of the oxidation by aeration process to remove iron?
11. What happens if the dose of potassium permanganate to remove iron and manganese is not exact?
12. How does an operator decide when to backwash a manganese greensand filter?
13. What must you do if the potassium permanganate charge is lost in the filter bed?
14. How would you attempt to prevent red or dirty water complaints?

SUGGESTED ANSWERS

Chapter 12. IRON AND MANGANESE CONTROL

Answers to questions on page 7.

12.0A When clothes are washed in water containing iron and manganese, they will come out stained. Excessive iron and manganese in water can also stain plumbing fixtures. Iron bacteria will cause thick slimes to form on the walls of water mains. These slimes are rust colored from iron and black from manganese. Variations in flow cause these slimes to slough, which results in dirty water. Furthermore, these slimes will cause foul tastes and odors in the water.

12.0B The growth of iron bacteria is easily controlled by chlorination. However, when water containing iron is chlorinated, the iron is converted into rust particles, and manganese is converted into a jet-black compound called manganese dioxide.

12.0C The generally acceptable limit for iron in drinking water is 0.3 mg/L and that for manganese is 0.05 mg/L.

Answers to questions on page 8.

12.1A Iron and manganese react with dissolved oxygen to form insoluble compounds.

12.1B Iron and manganese samples are acidified when they are collected to prevent the formation of iron and manganese scales on the walls of the sample bottles.

12.1C Samples for iron and manganese testing should be collected as close to the well or source of water as possible.

Answers to questions on page 12.

12.2A To determine if nearby wells contain iron and manganese, samples could be collected and analyzed from nearby private wells. Also, discussions with well drillers who have been active in the locality and with engineers in the state agency responsible for the regulation of well drilling will be helpful.

12.2B If a well produces water containing dissolved oxygen as well as iron and manganese, the iron and manganese are probably coming from the deeper aquifers. Try to seal off the deeper aquifers.

12.2C Bench-scale tests are a method of studying different ways or chemical doses for treating water or wastewater and solids on a small scale in a laboratory.

12.2D If polyphosphate solutions are much over 48 hours old, they will react slowly with water to form orthophosphates, which are much less effective in preventing manganese deposits.

Answers to questions on page 18.

12.2E If water being treated for iron and manganese by ion exchange contains any dissolved oxygen, the resin becomes fouled with iron rust or insoluble manganese dioxide.

12.2F The higher the pH, the faster the rate of oxidation of iron to insoluble ferric hydroxide.

12.2G The purpose of the reaction basin is to allow time for the oxidation reactions to take place. The aeration process should produce sufficient dissolved oxygen for the iron to be oxidized to insoluble ferric hydroxide.

12.2H Water can be dechlorinated by the use of reducing agents such as sulfur dioxide (SO_2), sodium bisulfite ($NaHSO_3$), and sodium sulfite (Na_2SO_3). Bisulfite is commonly used because it is cheaper and more stable than sulfite.

12.2I The three process modes for manganese greensand are continuous regeneration, intermittent regeneration, and catalytic regeneration.

Answers to questions on page 20.

12.3A The oxidation potential of chlorine and potassium permanganate used in the manganese greensand process decreases as the pH increases. To operate efficiently, the CR manganese greensand process requires a pH range of 6.2 to 6.8.

12.3B When adequate potassium permanganate is being fed in a manganese greensand process, the filter influent will have a slight pink/orange color.

12.3C A greensand plant should be shut down when the permanganate solution level in the vat gets low because of the importance of the permanganate in the process. Without permanganate, the greensand could lose its charge and iron and manganese will enter the distribution system.

Answers to questions on page 22.

12.4A When a chemical feeder pumps sodium hypochlorite, calcium carbonate coatings may develop on the poppet valves if the dilution water contains carbonate alkalinity. Coated valves do not seal properly and the feeder fails to pump the hypochlorite solution properly.

12.4B The problems caused by calcium carbonate scale on a hypochlorinator's poppet valves can be solved in two ways:

1. A hydrochloric acid solution can be pumped through the system.
2. The dilution water for the hypochlorite can be obtained from an ordinary home water softener.

Answers to questions on page 23.

12.5A Red or dirty water complaints may be caused by iron or manganese in the water, corrosive water, or iron bacteria in the distribution system.

12.5B Slime growths in distribution systems can be controlled by maintaining a free chlorine residual throughout the distribution system and by a distribution system flushing program.

CHAPTER 13

FLUORIDATION

by

Harry Tracy

TABLE OF CONTENTS

Chapter 13. FLUORIDATION

LEARNING OBJECTIVES

Chapter 13. FLUORIDATION

Following completion of Chapter 13, you should be able to:

1. Explain the reason for fluoridating drinking water.
2. Describe how fluoridation programs are implemented.
3. List the compounds used to furnish fluoride ion.
4. Review designs and specifications of fluoridation equipment.
5. Inspect fluoridation equipment.
6. Start up a chemical feeder.
7. Operate and maintain a chemical feeder.
8. Calculate fluoride dosages.
9. Prepare fluoride solutions.
10. Develop and keep accurate fluoride log sheets.
11. Prevent overfeeding of fluoride.
12. Shut down chemical feed systems.
13. Safely handle fluoride compounds.

WORDS

Chapter 13. FLUORIDATION

BATCH PROCESS BATCH PROCESS

A treatment process in which a tank or reactor is filled, the water is treated or a chemical solution is prepared, and the tank is emptied. The tank may then be filled and the process repeated.

DAY TANK DAY TANK

A tank used to store a chemical solution of known concentration for feed to a chemical feeder. A day tank usually stores sufficient chemical solution to properly treat the water being treated for at least one day. Also called an age tank.

ENDEMIC (en-DEM-ick) ENDEMIC

Something peculiar to a particular people or locality, such as a disease that is always present in the population.

FLUORIDATION (floor-uh-DAY-shun) FLUORIDATION

The addition of a chemical to increase the concentration of fluoride ions in drinking water to a predetermined optimum limit to reduce the incidence (number) of dental caries (tooth decay) in children. Defluoridation is the removal of excess fluoride in drinking water to prevent the mottling (brown stains) of teeth.

GRAVIMETRIC FEEDER GRAVIMETRIC FEEDER

A dry chemical feeder that delivers a measured weight of chemical during a specific time period.

POSITIVE DISPLACEMENT PUMP POSITIVE DISPLACEMENT PUMP

A type of piston, diaphragm, gear, or screw pump that delivers a constant volume with each stroke. Positive displacement pumps are used as chemical solution feeders.

SATURATOR (SAT-yoo-ray-tor) SATURATOR

A device that produces a fluoride solution for the fluoridation process. The device is usually a cylindrical container with granular sodium fluoride on the bottom. Water flows either upward or downward through the sodium fluoride to produce the fluoride solution.

VOLUMETRIC FEEDER VOLUMETRIC FEEDER

A dry chemical feeder that delivers a measured volume of chemical during a specific time period.

CHAPTER 13. FLUORIDATION

13.0 IMPORTANCE OF FLUORIDATION

During the period 1902 to 1931, Dr. Frederick S. McKay, a dentist practicing in Colorado Springs, noted what seemed an *ENDEMIC*[1] brown stain on the teeth of his patients. McKay devoted much of his time researching the case of mottled (brown, chalky deposits) tooth enamel but it was not until 1931 that the cause was found to be excessive fluoride in the water supplies (2 to 13 mg/L). During this period, McKay had also noted that the mottled teeth seemed to have fewer dental caries (decay or cavities).

The next logical step was to add fluoride to waters that were deficient in fluoride and to discover if children drinking water treated with fluoride had fewer cavities. In 1945, controlled fluoridation was started in the cities of Grand Rapids, Michigan, and Newburgh, New York, with control cities of Muskegon, Michigan, and Kingston, New York.

Finally, in 1955, the results were in and they showed a 60 percent reduction of caries in the children who drank fluoridated water compared to the children in the control cities.

The progress of fluoridation did not go smoothly. Antifluoridationists became increasingly vocal and were able to stop fluoridation in many cities through action in the political arena. Although dentists practicing in areas with naturally high fluoride waters noted that their patients had remarkably few cavities, there were still those disfiguring brown stains. Mottling of the teeth occurs when the fluoride level exceeds about 1.5 mg/L. Fluoride concentrations in excess of 1,000 mg/L have been found in waters from volcanic regions. Waters with fluoride concentrations more than 1.4 to 2.4 mg/L should be treated to reduce the level to approximately 1.0 mg/L. Recommended control limits for fluoride depend on the annual average of maximum daily air temperatures (Table 13.1). Typically, the higher the average maximum daily air temperature, the more water people drink. The more water people drink, the more fluoride they consume in their water. Therefore, the recommended fluoride control limits are lower in warmer climates, where people drink more water, and higher in cooler climates. The maximum contaminant level (MCL) is 4.0 mg/L.

QUESTIONS

Please write your answers to the following questions and compare them with those on page 66.

13.0A What happens if a person drinks water with an excessive concentration of fluoride?

13.0B What happens if children drink a recommended dose of fluoride?

13.1 FLUORIDATION PROGRAMS

Generally speaking, fluoridation programs start with citizens' inquiries about fluoridation of their water supplies and encouragement of the local dental society. These requests for the addition of fluoride to prevent dental caries are passed along to appropriate governing agencies. The governing body will usually rely upon a vote of the people, or it may be forced into a vote by

TABLE 13.1 PRIMARY DRINKING WATER REGULATIONS FOR FLUORIDE AND RECOMMENDED LEVELS

Annual Average Maximum Daily Air Temperatures[a]		Recommended Control Limits of Fluoride Levels, mg/L			Maximum Contaminant Level, mg/L
°F	°C	Lower	Optimum	Upper	
53.7 & Below	12.0 & Below	0.9	1.2	1.7	4.0
53.8 to 58.3	12.1 to 14.6	0.8	1.1	1.5	4.0
58.4 to 63.8	14.7 to 17.6	0.8	1.0	1.3	4.0
63.9 to 70.6	17.7 to 21.4	0.7	0.9	1.2	4.0
70.7 to 79.2	21.5 to 26.2	0.7	0.8	1.0	4.0
79.3 to 90.5	26.3 to 32.5	0.6	0.7	0.8	4.0

a Contact your local Weather Service Office to determine the "Annual Average Maximum Daily Air Temperature" for your service area.

1. *Endemic* (en-DEM-ick). Something peculiar to a particular people or locality, such as a disease that is always present in the population.

threat of a referendum. If the decision is made to fluoridate, the water department or water company will almost always make the final decisions as to types of fluoride chemicals and feeding equipment to be used.

13.2 COMPOUNDS USED TO FURNISH FLUORIDE ION

The most commonly used compounds to fluoridate water systems are sodium fluoride, sodium fluorosilicate, and hydrofluosilicic (HI-dro-FLOO-o-suh-LIS-ick) acid. There are also a few systems using such compounds as hydrofluoric acid and ammonium silicofluoride. All of these chemicals are refined from minerals found in nature and they yield fluoride ions identical to those found in natural waters. Hydrofluosilicic acid (also called fluorosilicic acid) is the compound most commonly used in several states, including California, Florida, and Illinois.

The American Water Works Association (AWWA) has developed standards for the purity and composition of the fluoride compounds most commonly used. In order for you to be confident of the fluoride compound you are using, insist that your supplier furnish only compounds meeting the appropriate AWWA specifications.

The plant should keep copies of the appropriate AWWA standard for reference. Standards can be purchased from AWWA at www.awwa.org or phone (800) 926-7337.

Standard	Chemical	Order No.
B701-11	Sodium Fluoride	42701
B702-11	Sodium Fluorosilicate	42702
B703-11	Fluorosilicic Acid	42703

Also, NSF International's[2] "Drinking Water Additives Standards Set," NSF Standard 60, "Drinking Water Treatment Chemicals - Health Effects," most current edition, lists companies by chemical name that supply chemicals approved for use in drinking waters. Fluoridation drinking water treatment chemicals approved by NSF include:

1. Fluosilicic acid
2. Hydrofluosilicic acid
3. Sodium fluoride

Only the fluoride ion in these compounds is of any importance in the fluoridation of water; therefore, pound for pound, each compound will provide a different final fluoride level. If you switch from one type of fluoride compound to another, you will have to make separate calculations for each type. Table 13.2 summarizes the important properties of fluoride compounds.

When selecting a fluoridation chemical, several important factors must be considered. If powdered or crystal forms are used, solubility of the chemical in water will determine how readily it dissolves in water and how well it remains in solution. Operator safety and ease of handling must be given serious consideration. Storage and feeding requirements as well as costs must also be studied when selecting any chemical. Remember, all fluoridation chemicals are poisonous at high levels.

TABLE 13.2 FLUORIDE COMPOUNDS

	Sodium Silicofluoride Na_2SiF_6	**Sodium Fluoride NaF**	**Hydrofluosilicic Acid H_2SiF_6**
1. Form	Powder	Powder or Crystal	Liquid
2. Molecular Weight	188.1	42.0	144.1
3. Commercial Purity, %	98–99	95–98	22–30 by Weight
4. Fluoride Ion, % (Purity, %)	60.7 (100%) 59.8 (98.5%)	45.3 (100%) 43.4 (96%)	79.2 (100%) 23.8 (30%)
5. Density	55–72 lb/ft^3	65–90 lb/ft^3	10.5 (30%) lb/gal
6. Solubility in Water, % (gram/100 mL water at 77°F or 25°C)	0.76	4.05	100[a]
7. pH of Saturated Solution	3.5	7.6	1.2 (1% Solution)

a Infinite because we are dealing with a liquid.

Hydrofluosilicic acid is usually the easiest fluoridation chemical to feed. However, hydrofluosilicic acid produces poisonous fumes that must be vented and is very irritating to your skin. Sodium fluoride is easier to feed than the other powdered fluoridation chemicals because it is more soluble in water.

Operators can receive instructions from the manufacturer on how to make up the chemical solutions and how much solution to meter per million gallons. See Section 13.61, "Preparation of Fluoride Solution," for calculations and procedure details.

Before fluoridation, the water should be checked for its natural fluoride level. If there is natural fluoride in the water, it is only necessary to add enough to bring the total to the desired level recommended by the local health authorities.

QUESTIONS

Please write your answers to the following questions and compare them with those on page 66.

13.1A Who makes the final decisions as to types of fluoride chemicals and feeding equipment to be used?

13.2A List the three compounds most commonly used to fluoridate water.

2. NSF International, www.nsf.org, phone 800-NSF-MARK (800-673-6275).

13.3 FLUORIDATION SYSTEMS

Drinking waters may come to contain fluoride ions by three different types of situations. First, the raw water source may have adequate or excessive fluoride ions naturally present. Second, sometimes two water sources are blended together to produce an acceptable level of fluoride ions. This can occur when one source has a higher than acceptable level of fluoride ions and the other is below the desired level. This chapter is mainly concerned with the third situation in which fluoride ions must be added to the water to achieve an acceptable level of fluoride ions.

13.30 Chemical Feeders

Fluoride ions are added to water by either chemical solution feeders or dry feeders. Solution feeders are *POSITIVE DISPLACEMENT*[3] diaphragm pumps (Figure 13.1), peristaltic pumps (Figure 13.2), or electronic pumps (Figure 13.3), that deliver a fixed amount of liquid fluoridation chemical with each stroke or pulse. The dry feeders are either *VOLUMETRIC*[4] or *GRAVIMETRIC*[5] types of chemical feeders. Volumetric feeders (Figures 13.4 and 13.5) are usually simpler, less expensive, less accurate, and feed smaller amounts of chemicals than gravimetric feeders. Gravimetric feeders (Figure 13.6) are usually more accurate than volumetric feeders; however, they are more expensive and require more space for installation. The amount fed is measured on the basis of the weight of chemical to be fed to the system. Fluoride chemical feeders must be very accurate.

Whatever the type of feeding system or chemical used, the design should be planned by an engineer experienced in developing feeding systems (Figure 13.7). The design must incorporate means to prevent both overfeeding and backsiphonage along with a means to monitor the amount of chemical used. It is also desirable to incorporate some means of feeding fluoride that is adjusted (paced) according to the plant flow rate. Also, a means to continuously measure the finished water's fluoride ion content with an adjustable high-fluoride alarm is desirable. Fluoride doses must never be metered against a negative or suction head.

13.31 Saturators[6]

The saturator (Figure 13.8) is a special application of a solution feeder. A small pump delivers a saturated solution of sodium fluoride into the water system. The principle of a saturator is that a saturated solution will result if water is allowed to trickle through a bed containing sodium fluoride.

Although saturated solutions of sodium fluoride can be manually prepared, generally the easiest and best way is an automatic feed device. Only crystal-grade (20 to 60 mesh) sodium fluoride should be fed with a saturator. Sodium fluoride has a nearly constant solubility at normal temperature ranges and thus produces a fluoride solution of uniform strength. Sodium silicofluoride is not recommended to feed through the saturator because of its very low solubility in water (see Table 13.2). Maintain a depth of 6 to 10 inches (150 to 250 mm) in the sodium fluoride bed of the saturator. Saturators should be stirred every day to prevent fluoride solids from building up on the bottom. The use of powdered sodium fluoride is not recommended because this chemical will clog the saturator.

Many dry chemical feed systems include a mixer, dissolving tank, and *DAY TANK*[7] (Figure 13.9). The mixer mixes a known amount of chemicals with a measured amount of water in a dissolving tank or solution chamber. Because the chemicals are mixed with a specified amount of water rather than being mixed in flowing water, this is a"batch mixed" process. A "batch process" involves filling a tank or reactor and treating the water or preparing a chemical solution before the tank is emptied and ready to repeat the process. The chemical feed dissolving tank allows the chemicals to become dissolved in water. The resulting solution is then either continuously applied to the water being treated or is stored in a day tank or storage tank. The day tank usually stores enough chemical solution to properly treat the plant flow for at least one day. The chemical solution is fed from the day tank to the water being treated by a chemical feeder (feed pump) whose feed rate continuously adjusts to the flow being treated (flow-paced).

When working with fluoridation systems using a sodium fluoride solution, the hardness of the water is very important. Hard water can produce problems in systems using saturators and dissolving tanks through the formation of low-solubility deposits of calcium and magnesium fluoride compounds. If the dilution water has a hardness of less than 10 mg/L hardness as $CaCO_3$, there will be no problem. If the hardness range is above 10 mg/L and below 75 mg/L, scaling will continue to occur and the cost of ongoing cleaning and maintenance must be considered. Above a hardness of 75 mg/L, a softened water must be used for the water to prepare a fluoride solution in order to prevent severe scaling in the equipment.

3. *Positive Displacement Pump.* A type of piston, diaphragm, gear, or screw pump that delivers a constant volume with each stroke. Positive displacement pumps are used as chemical solution feeders.
4. *Volumetric Feeder.* A dry chemical feeder that delivers a measured volume of chemical during a specific time period.
5. *Gravimetric Feeder.* A dry chemical feeder that delivers a measured weight of chemical during a specific time period.
6. *Saturator* (SAT-yoo-ray-tor). A device that produces a fluoride solution for the fluoridation process. The device is usually a cylindrical container with granular sodium fluoride on the bottom. Water flows either upward or downward through the sodium fluoride to produce the fluoride solution.
7. *Day Tank.* A tank used to store a chemical solution of known concentration for feed to a chemical feeder. A day tank usually stores sufficient chemical solution to properly treat the water being treated for at least one day. Also called an age tank.

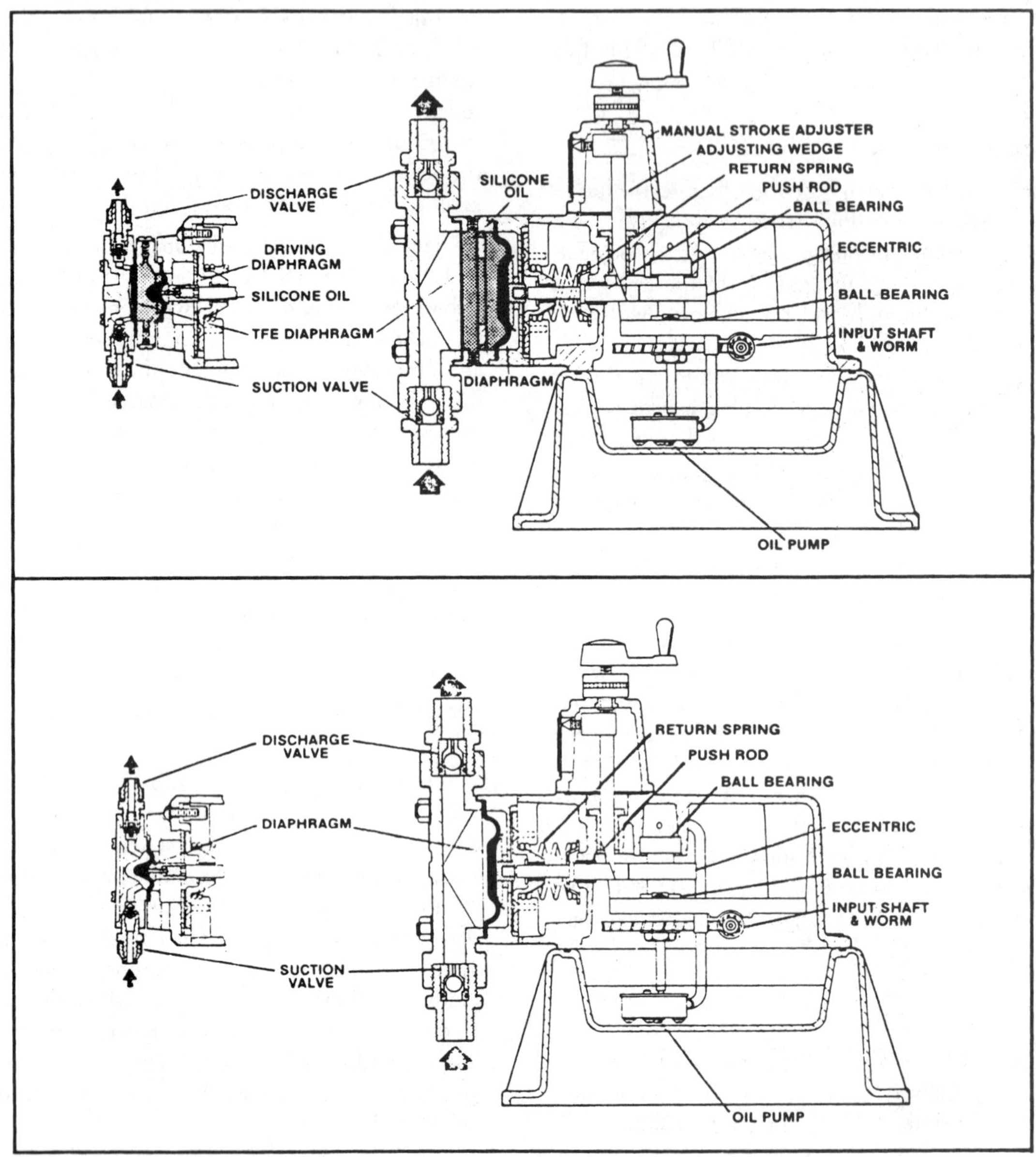

Fig. 13.1 Positive displacement diaphragm pumps

(Permission of Wallace & Tiernan, Inc.)

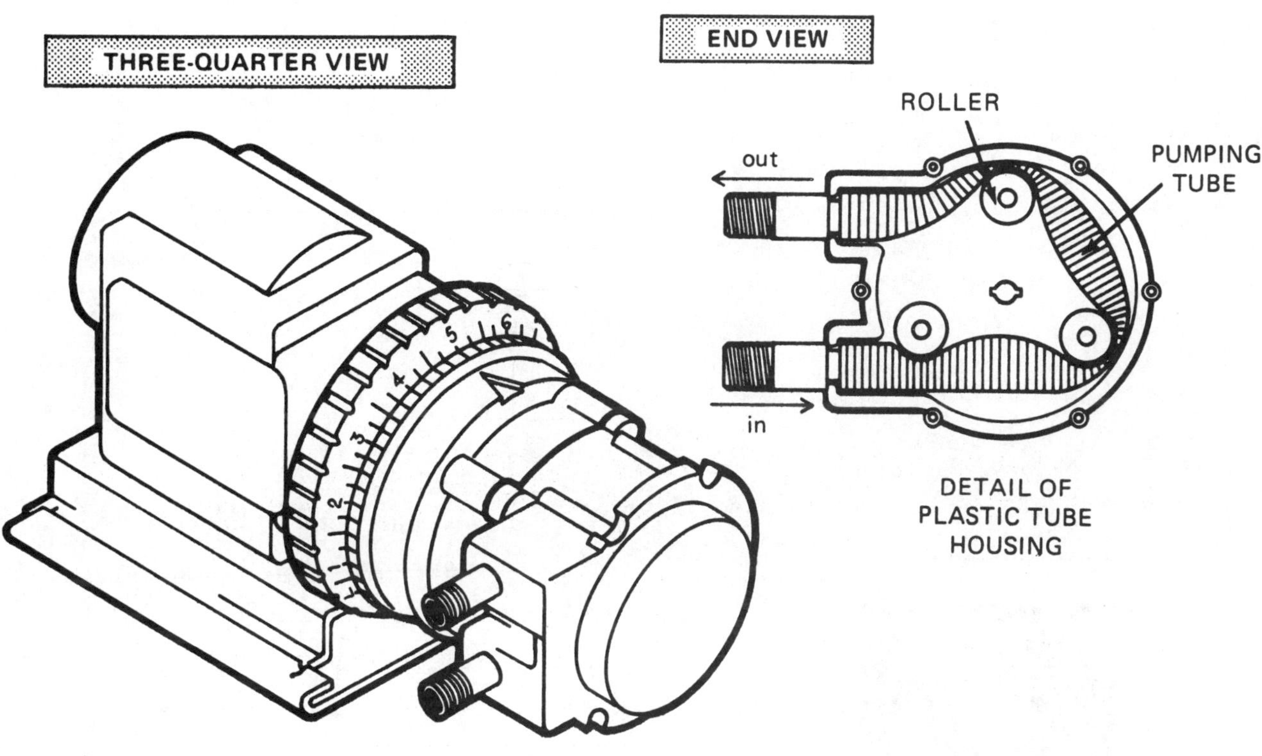

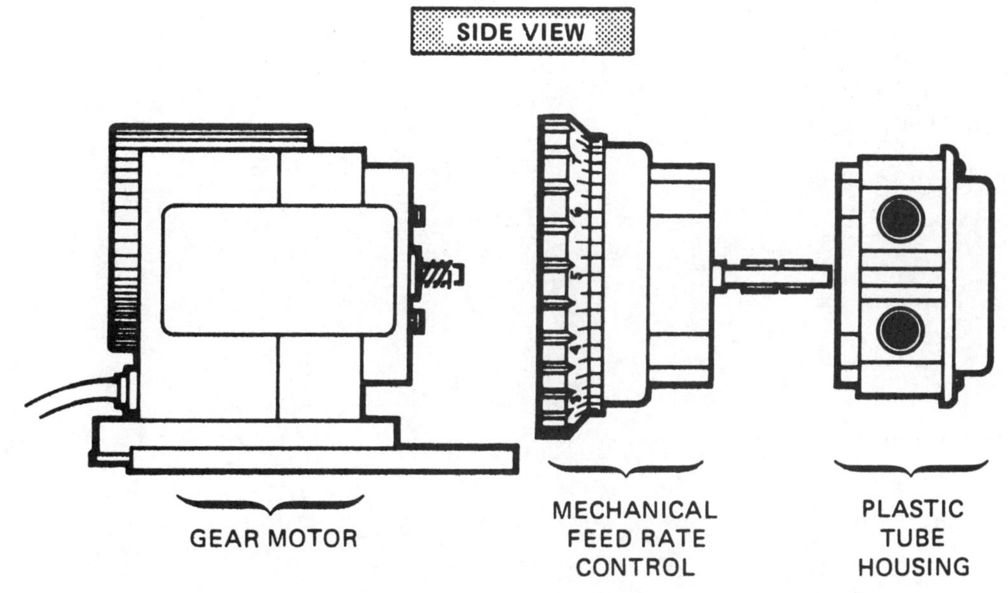

Fig. 13.2 Peristaltic feeder

(Reproduced from *Water Fluoridation, A Training Course Manual for Engineers and Technicians,* by permission of the Dental Disease Prevention Activity, US Public Health Service)

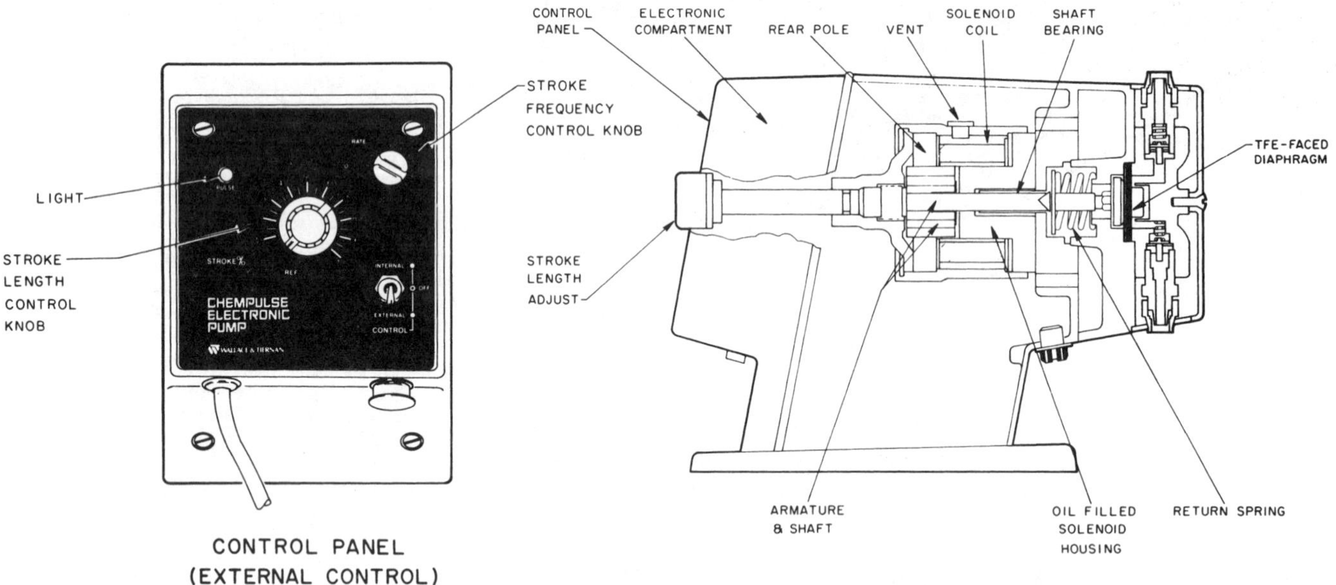

Fig. 13.3 Electronic feeder
(Permission of Wallace & Tiernan, Inc.)

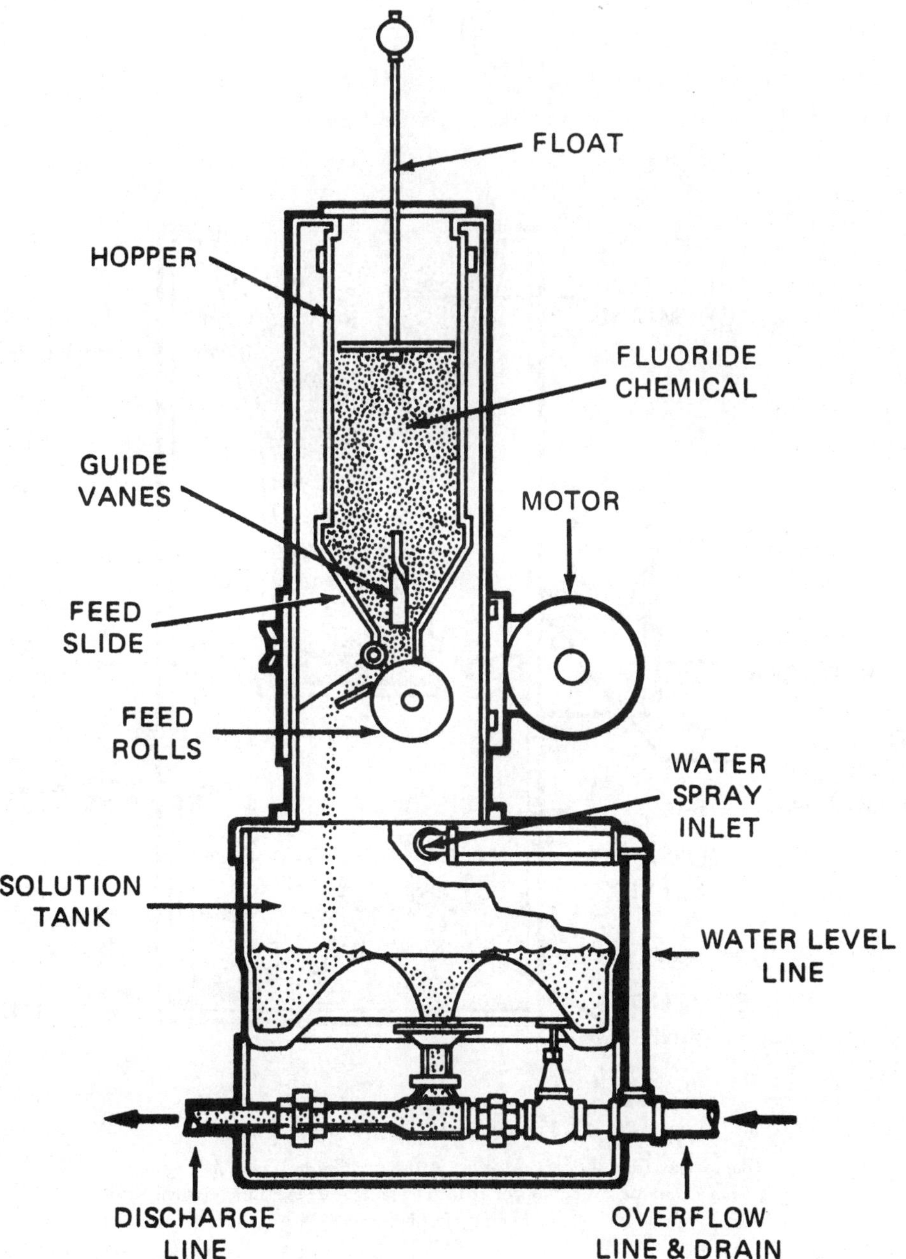

Fig. 13.4 Volumetric feeder, roll-type

(Reproduced from *Water Fluoridation, A Training Course Manual for Engineers and Technicians,* by permission of the Dental Disease Prevention Activity, US Public Health Service)

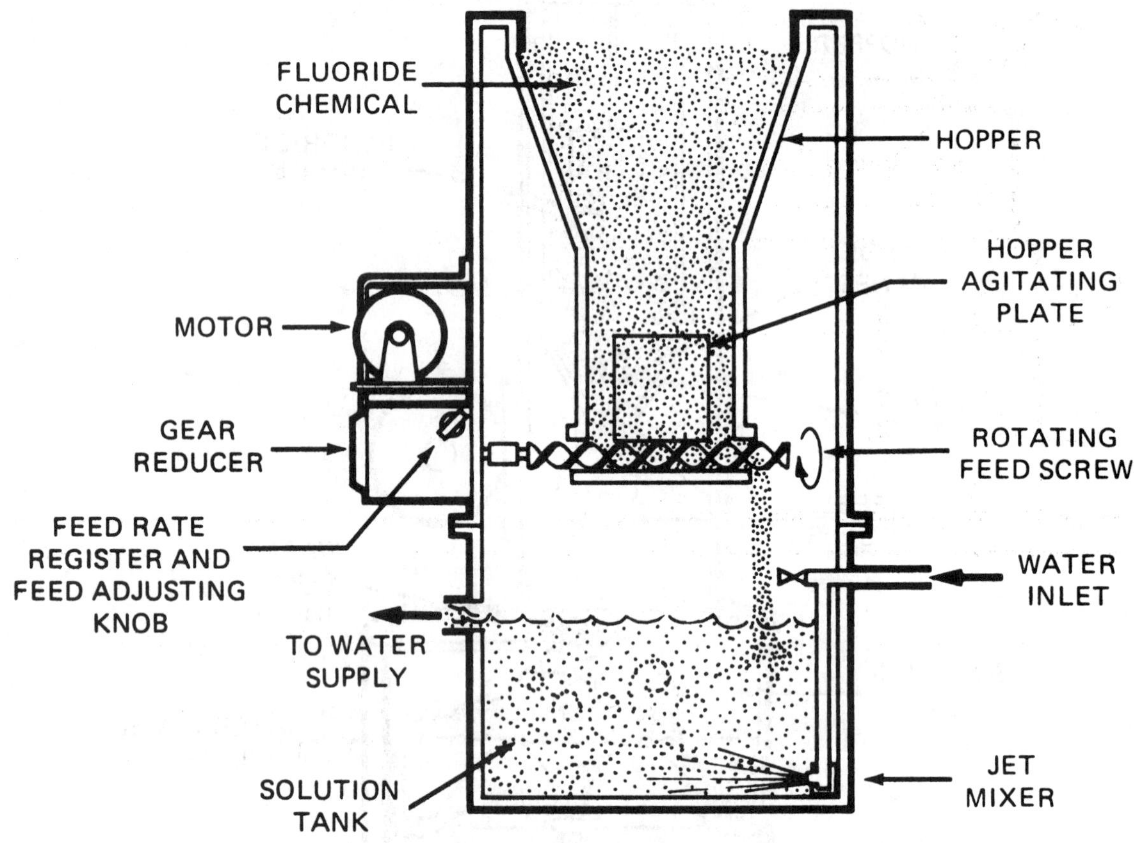

Fig. 13.5 Volumetric feeder, screw-type

(Reproduced from *Water Fluoridation, A Training Course Manual for Engineers and Technicians,* by permission of the Dental Disease Prevention Activity, US Public Health Service)

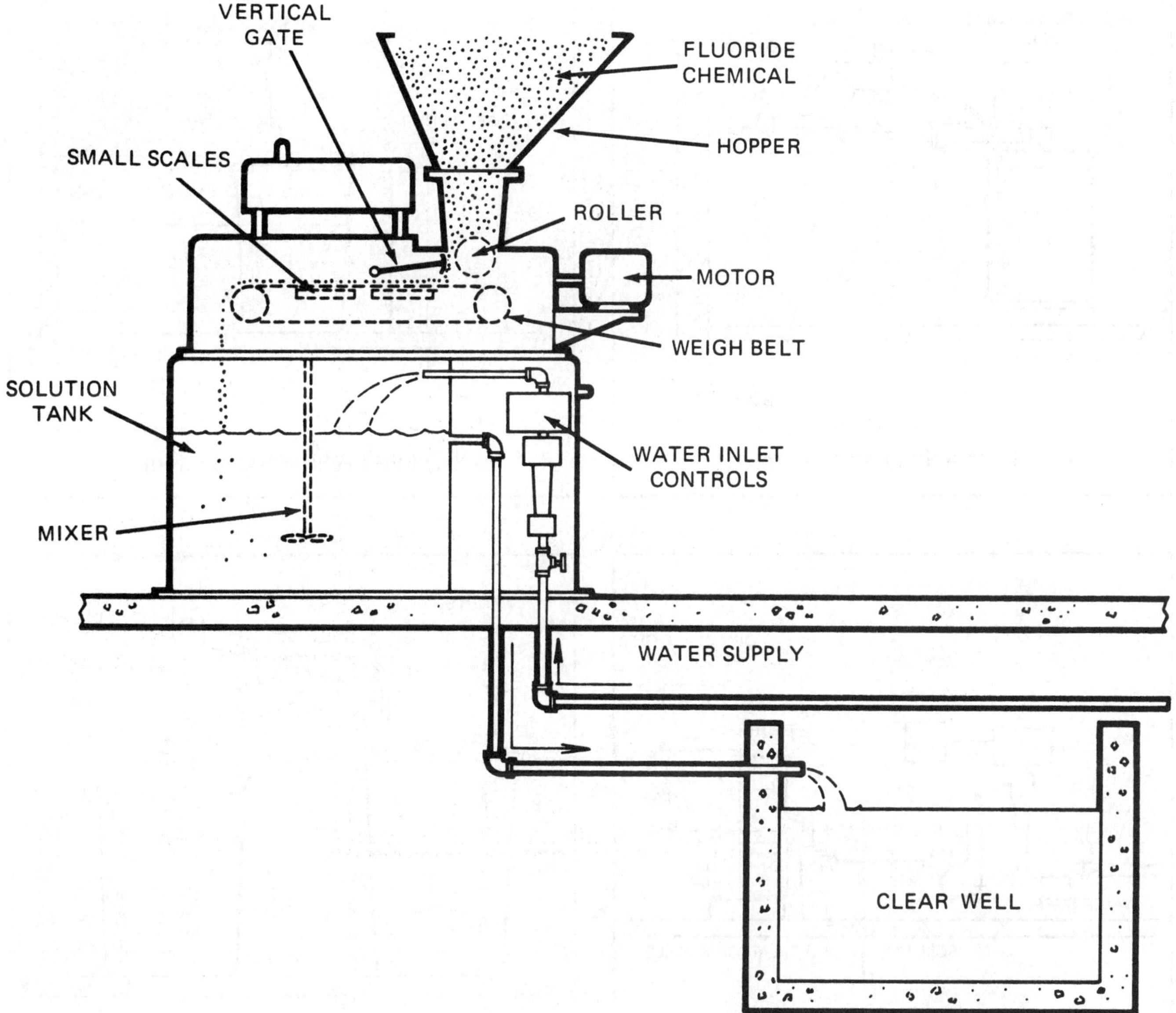

Fig. 13.6 Gravimetric feeder, belt-type

(Reproduced from *Water Fluoridation, A Training Course Manual for Engineers and Technicians,* by permission of the Dental Disease Prevention Activity, US Public Health Service)

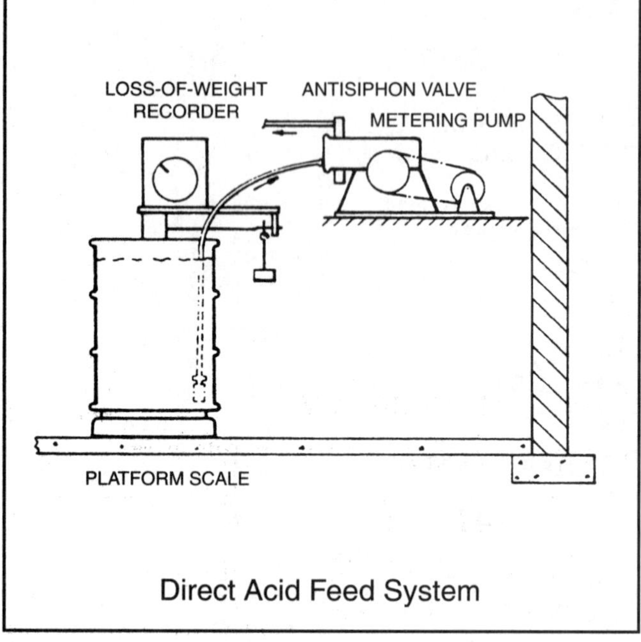

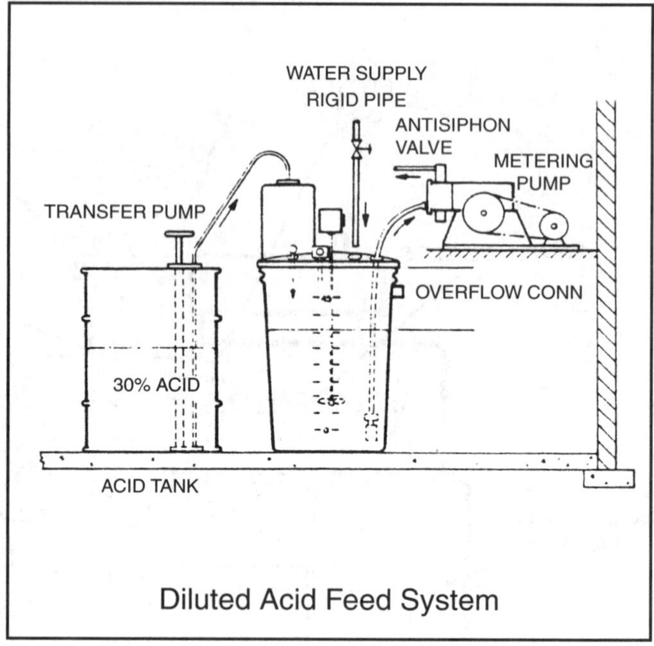

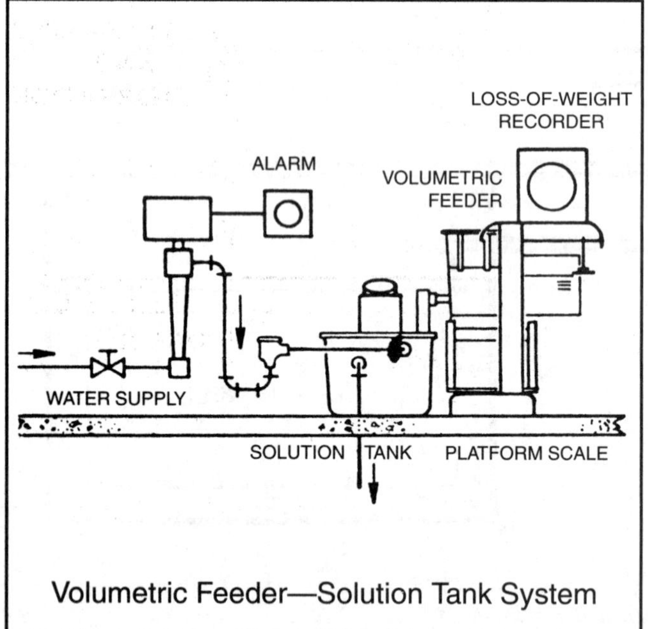

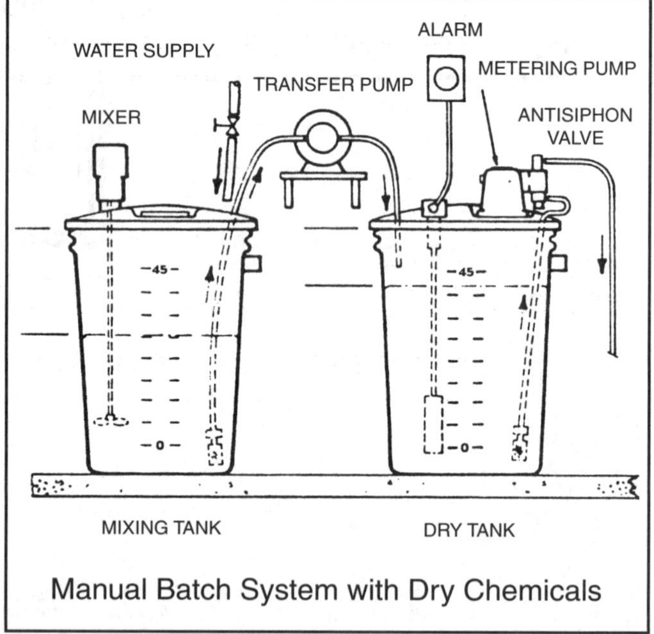

Fig. 13.7 Typical fluoridation systems
(Permission of Wallace & Tiernan, Inc.)

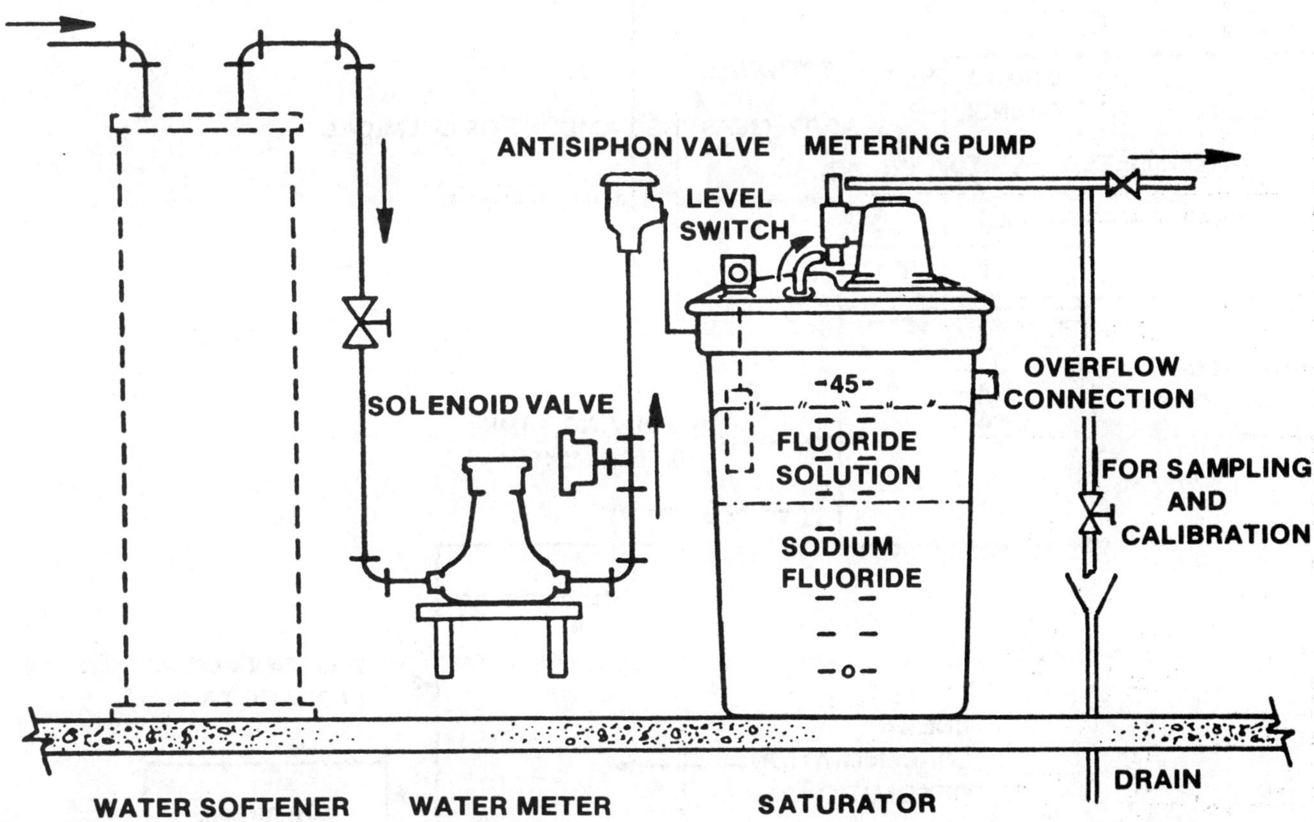

Fig. 13.8 Fluoride saturator

(Permission of Wallace & Tiernan, Inc.)

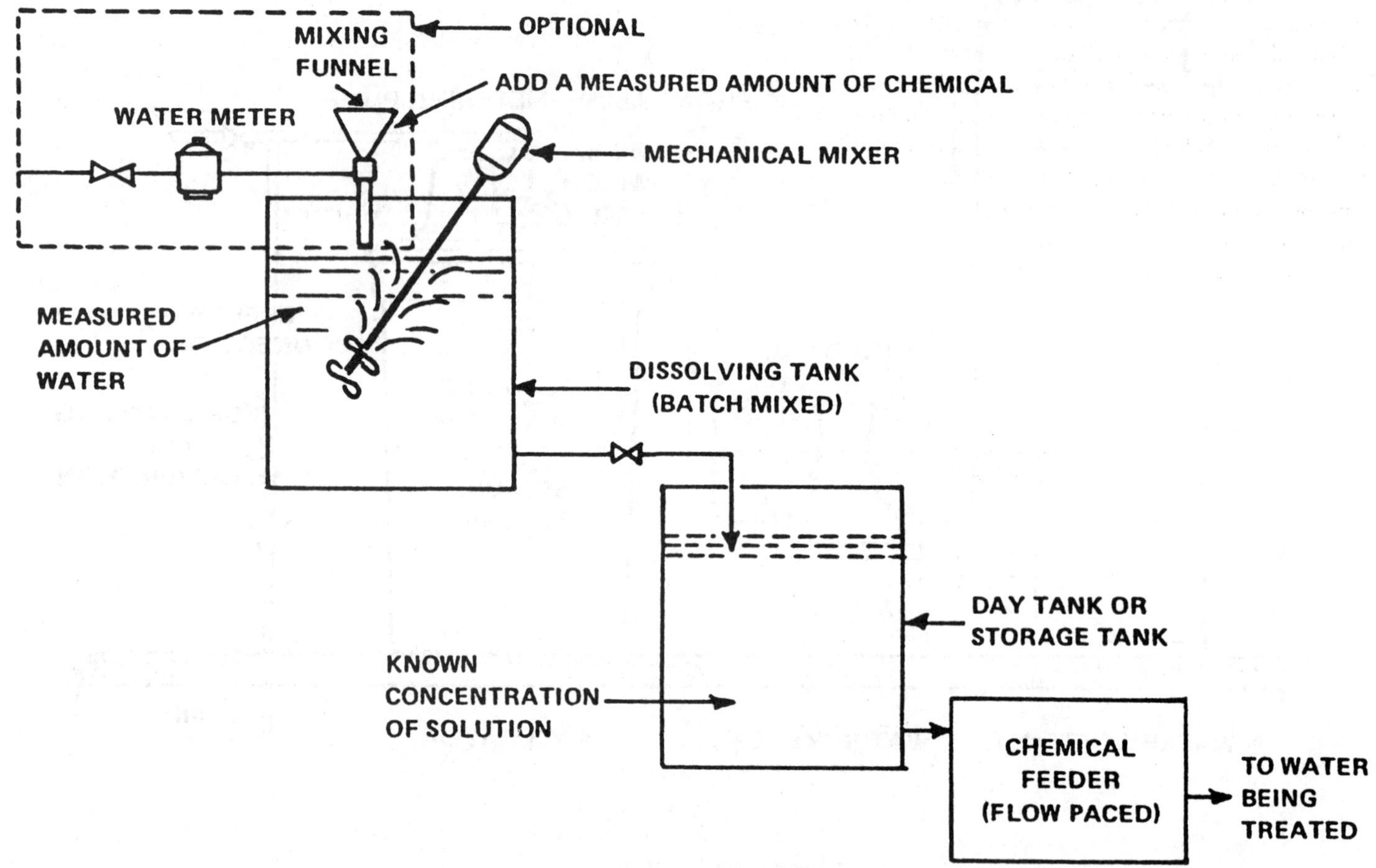

Fig. 13.9 Dry chemical dissolver, day tank, and feeder

Water can be softened before use with fluoridation equipment by the use of zeolite ion exchangers. As an alternative to the zeolite water softener, polyphosphates (at 7 to 15 mg/L) may be used to prevent plugging by scale in the feed system. The polyphosphates are added to the day tank to prevent scale deposits. If neither a zeolite softener nor polyphosphates are used, plugging may occur at any point in the feed system, including valves, saturator bed, and injection point. Remove these hardness deposits by flushing the system with vinegar or a five-percent solution of hydrochloric acid (muriatic acid). The saturator beds also may require the removal of water hardness deposits.

13.32 Downflow Saturators

There are two kinds of saturators, the upflow saturator and the downflow saturator. In the downflow saturator (Figure 13.10), the solid sodium fluoride is held in a plastic drum or barrel and is isolated from the prepared solution by a plastic cone or a pipe manifold. A filtration barrier is provided by layers of sand and gravel to prevent particles of undissolved sodium fluoride from infiltrating the solution area under the cone or within the pipe manifold. The feeder pump draws the solution from within the cone or manifold at the bottom of the plastic drum. Downflow saturators require clean gravel and sand. In some systems, the gravel and sand must be cleaned every day or two.

13.33 Upflow Saturators

In an upflow saturator (Figure 13.11), undissolved sodium fluoride forms its own bed below which water is forced upward under pressure. No barrier is used since the water comes up through the bed of sodium fluoride and the specific gravity of the solid material keeps it from rising into the area of the clear solution above. A spider-type water distributor located at the bottom of the tank contains hundreds of very small slits. Water, forced under pressure through these slits, flows upward through the sodium fluoride bed at a controlled rate to ensure the desired 4 percent solution. The feeder pump intake line floats on top of the solution in order to avoid withdrawal of undissolved sodium fluoride. The water pressure requirements are 20 psi (138 kPa) minimum to 125 psi (862 kPa) maximum and the flow is regulated at 4 GPM (0.25 L/sec). Since introduction of water to the bottom of the saturator constitutes a definite cross-connection, a mechanical siphon-breaker must be incorporated into the water line; or, better still, the saturator can be factory modified to include an air gap and a water feed pump.

Figures 13.8 and 13.11 show two configurations of upflow-type saturators that feed and prepare constant-strength fluoride solution from granular sodium fluoride. The upflow type is the preferred type over the downflow saturator, as it is much

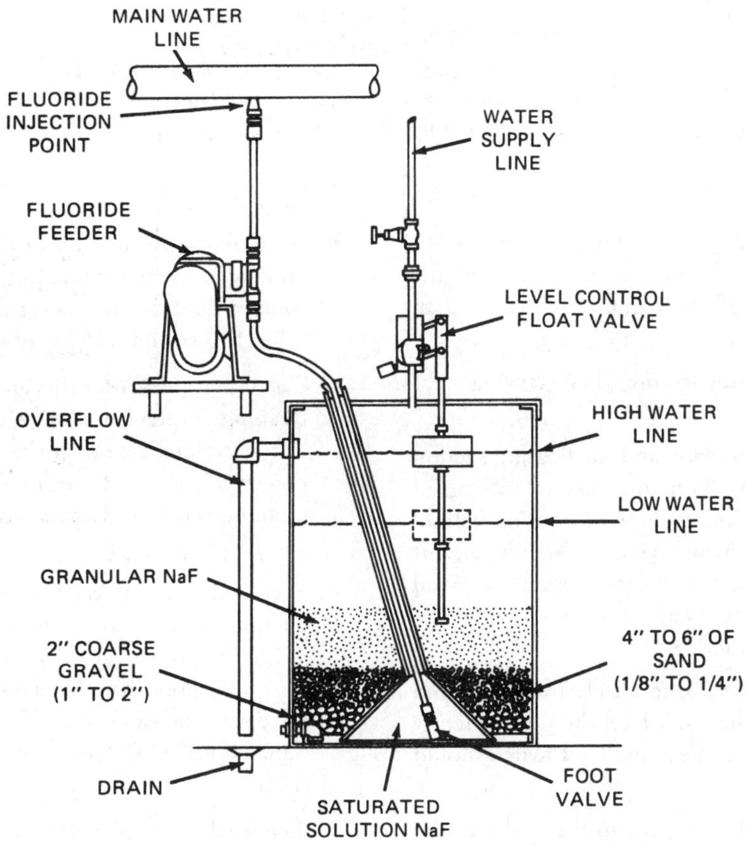

Fig. 13.10 Downflow saturator

(Reproduced from *Water Fluoridation, A Training Course Manual for Engineers and Technicians,* by permission of the Dental Disease Prevention Activity, US Public Health Service)

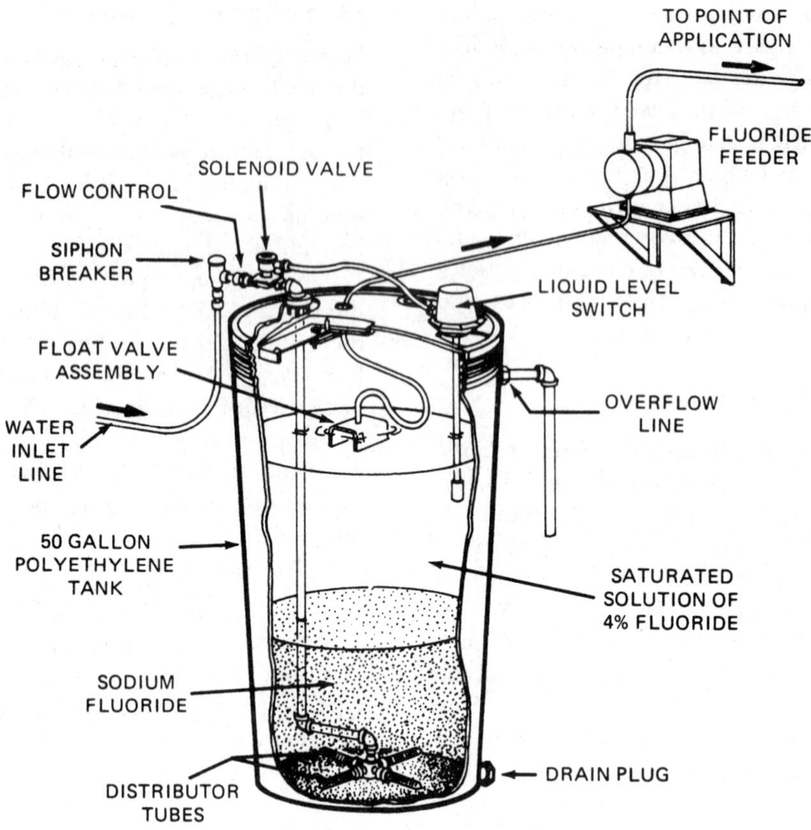

Fig. 13.11 Upflow saturator

(Reproduced from *Water Fluoridation, A Training Course Manual for Engineers and Technicians,* by permission of the Dental Disease Prevention Activity, US Public Health Service)

easier to clean and maintain. Under normal conditions, it should need cleaning only once a year. For these reasons, the construction and use of an upflow saturator is discussed in some detail.

To prepare an upflow saturator for use, the following steps should be taken.

1. With the distributor tubes in place, and the floating suction device removed, add 200 to 300 pounds (91 to 136 kg) of sodium fluoride directly to the tank. Any type of sodium fluoride can be used, from coarse crystal to fine crystal, but fine crystal will dissolve better than coarse material. If crystal is not available, powder can be used, but it is not as desirable as a crystal form of sodium fluoride.
2. Connect the solenoid water valve to an electric outlet and turn on the water supply. The water level should be slightly below the overflow; if it is not, the liquid level switch should be adjusted.
3. Replace the intake float and connect it to the feeder intake line. The saturator is now ready to use.
4. By looking through the translucent wall of the saturator tank, you should be able to see the level of undissolved sodium fluoride. Whenever the level is low enough, add another 100 pounds (45 kg) of fluoride.
5. The water distributor slits are supposed to be essentially self-cleaning. The accumulation of insolubles and precipitates is not as serious a problem as it is in a downflow saturator; however, a periodic cleaning is still required. Frequency of cleaning is dictated by the severity of use and the rate of accumulation of debris.
6. Because of the thicker bed of sodium fluoride attainable in an upflow saturator, higher withdrawal rates are possible. With 300 pounds (136 kg) of sodium fluoride in the saturator tank, more than 15 gallons per hour (58 L/hr) of saturated solution can be fed. This rate is sufficient to treat about 5,000 GPM (135 L/sec) of water to a fluoride level of 1.0 mg/L.
7. The fixed water inlet rate of 4 GPM (0.25 L/sec) should register satisfactorily on a ⅝-inch (16 mm) meter.

From a financial point of view, many water agencies will want to design their fluoridation plants for unattended operation. In such cases, the system design will include provisions for automatic shutdown and an alarm system. For the sake of the operator who has to respond at all hours, alarm lights should be wired to indicate reasons for plant shutdown. For example, alarms could be set to indicate low water flow in the main pipeline, high fluoride flow, high or low fluoride levels, low injection water pressure, power outage, and a running time meter to indicate downtime. These warning systems are particularly helpful in large systems.

13.34 Large Hydrofluosilicic Acid Systems

A more complicated system for fluoridation is a closed-loop control feeding system using hydrofluosilicic acid (Figure 13.12). This system finds use in large installations where the hydrofluosilicic acid can be delivered by tanker truck of around 4,000 gallons (15,140 L), although, of course, smaller amounts can be purchased. The installation depicted in Figure 13.12 can treat up to 285 MGD (1,079 mL/day). The advantages in this system are the elimination of dust hazards and minimal labor requirements.

The storage tanks are made of filament-wound fiberglass. The interior lining consists of a surfacing fabric with a veil-type covering serving as a final barrier. Steel tanks lined with at least 3⁄32 inch (2.34 mm) polyvinyl chloride sheet or neoprene sheet secured to the metal surface with adhesive can also be used. Hydrofluosilicic acid storage tanks constructed of plastic should be housed in enclosures to protect the tanks from vandalism. Leaks could be dangerous to passersby and will kill surrounding vegetation. The tanks must be vented to the outside as the fumes from the acid are highly corrosive.

Inspection of internal conditions of the hydrofluosilicic tanks should be made on two-year intervals as some lining deterioration can be expected over a period of time. Should small leaks occur in the PVC piping, repairs should be made at once; they will only become worse and any acid dripping on concrete surfaces will dissolve the surface fairly quickly.

The use of a closed-loop control system in an unattended plant using a fluoride analyzer as one of the controls is not recommended. The problem is that if the analyzer goes haywire and incorrectly indicates low fluoride levels, the system will try to correct itself and increase the addition of fluoride chemical. The net result will be to actually over-fluoridate the water supply.

QUESTIONS

Please write your answers to the following questions and compare them with those on page 66.

13.3A List the three different types of situations whereby drinking waters may contain fluoride ions.

13.3B List three important design features of fluoridation systems.

13.3C What is a saturator?

13.3D What problems can be created by hard water in systems using saturators and dissolving tanks?

13.4 FINAL CHECKUP OF EQUIPMENT

13.40 Avoid Overfeeding

The operator must be certain that there will be no overfeed of fluoride ions. A gross overfeed can cause illness and bad public relations. Of all the chemicals used in the water treatment plant, fluoride concentrations probably require the closest attention to maximum dosages.

13.41 Review of Designs and Specifications

When reviewing fluoride feeding system designs and specifications, the operator should check the items listed below.

1. Review the results of predesign tests to determine the fluoride rate for both the present and future. The fluoridator should be sized to handle the full range of chemical doses or provisions should be made for future expansion.
2. Determine if sampling points are provided to measure chemical feeder output.
3. Examine plans for valving to allow flushing the system with water before removing from service.
4. Be sure corrosion-resistant drip pans and drains are provided to prevent chemical leaks from reaching the floor, for example, drips from pump packing.
5. Check that all piping, valves, and fittings are made of corrosion-resistant materials such as PVC or Stainless Steel Type 316.
6. Determine the amount of maintenance required. The system should require a minimum of maintenance. Equipment should be standard, with replacement parts readily available.
7. Consider the effect of changing head conditions (both feeder suction and discharge head conditions) on the chemical feeder output. Changing head conditions will not affect the output if the proper chemical feeder has been specified and installed.
8. Determine whether locations for monitoring readouts and dosage controls are convenient to the operation center and easy to read and record.
9. Check that any switches that throw the equipment from automatic into a hand or manual mode are equipped with a color-coded red warning light to indicate that the equipment is on "hand" or "manual." This can easily be accomplished by a double-throw, double-pole toggle switch. Lights with different colors can be used to indicate normal or automatic operation as well as ON or OFF in order to avoid confusion.

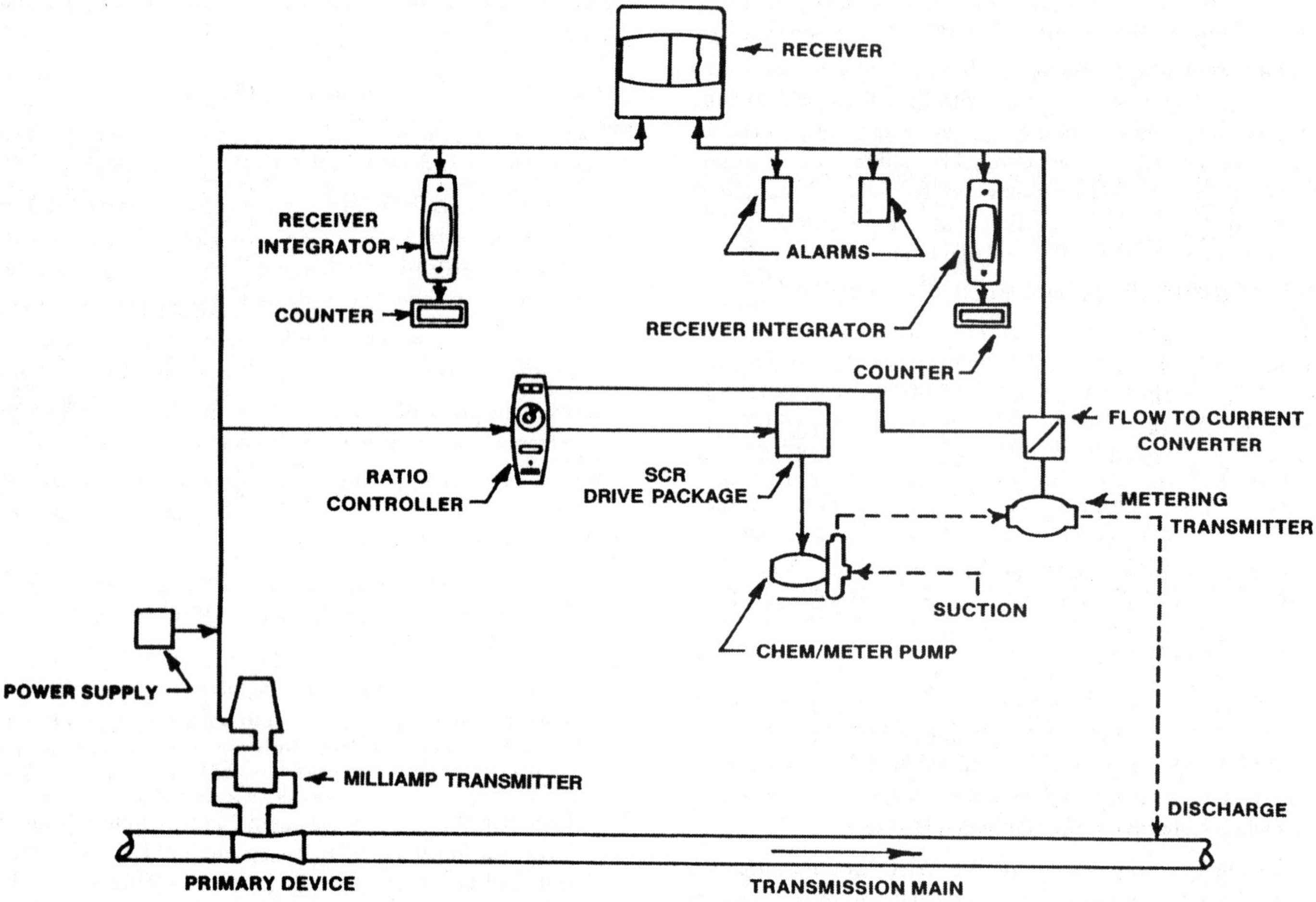

Fig. 13.12 Large automatic hydrofluosilicic acid system

10. Check that the location where fluoride is added to the water is where there will be the least possible removal of fluoride by other chemicals added to the water (after filtration and before the clear well).
11. Be sure the chemical hoppers are located where there is plenty of room so they can be conveniently and safely filled with the fluoride chemical.
12. Determine that dust exhaust systems are installed where substantial amounts of dry chemicals are handled.
13. Ensure that scales are provided. In any fluoridation system, except the sodium fluoride saturator, scales are necessary for weighing the quantity of chemical (including solution) fed per day.
14. Ensure that alarms are provided. Alarms are important to signal and prevent both the loss of feed and overfeeding.
15. Ensure that any potable water line connected to a chemical feeder unit is provided with a vacuum breaker or an air gap to prevent a cross-connection.

13.5 CHEMICAL FEEDER STARTUP

After the chemical feed system has been purchased and installed, carefully check the system out before startup. Even if the contractor who installed the system is responsible for ensuring that the equipment operates as designed, the operation by plant personnel, the functioning of the equipment, and the results from the process are the responsibility of the chief operator. Therefore, before startup, check the items listed below:

1. Inspect the electrical system for proper voltage, for properly sized overload protection, for proper operation of control lights on the control panel, for proper safety lockout switches and operation, and for proper equipment rotation.
2. Confirm that the manufacturer's lubrication and startup procedures are being followed. Equipment may be damaged in minutes if it is run without lubrication.
3. Examine all fittings, inspection plates, and drains to ensure that they will not leak when placed into service.
4. Determine the proper positions for all valves. A positive displacement pump will damage itself or rupture lines in seconds if allowed to run against a closed valve or system.
5. Be sure that the type of fluoride to be fed is available and in the hopper or feeder. A progressive cavity pump will be damaged in minutes if it is allowed to run dry.
6. Inspect all equipment for binding or rubbing.
7. Confirm that safety guards are in place.
8. Examine the operation of all auxiliary equipment, including the dust collectors, fans, cooling water, mixing water, and safety equipment.
9. Check the operation of alarms and safety shutoffs. If it is possible, operate these devices by manually tripping each one. Examples of these devices are alarms and shutoffs for high water, low water, high temperature, high pressure, and high chemical levels.
10. Be sure that safety equipment such as eye wash stations, drench showers, dust masks, face shields, gloves, and vent fans are in place and functional.
11. Record all important nameplate data and place it in the plant files for future reference.

QUESTIONS

Please write your answers to the following questions and compare them with those on page 66.

13.4A Why must overfeeding of fluoridation chemicals be prevented?

13.4B What should be the capacity or size of the fluoridator?

13.5A What items should be considered when inspecting the fluoridation electrical system?

13.5B List the safety equipment that should be available near a fluoridation system.

13.6 CHEMICAL FEEDER OPERATION

13.60 Fine-Tuning

Once the chemical feed equipment is in operation and the major bugs are worked out, the feeder will need to be fine-tuned. To aid in fine-tuning and build confidence in the entire chemical feed system, the operator must maintain accurate records (see Section 13.62, "Fluoridation Log Sheets").

A comment or remarks section should be used to note abnormal conditions, such as a feeder plugged for a short time, related equipment that malfunctions, and other problems. Daily logs should be summarized into a form that operators can use as a future reference.

13.61 Preparation of Fluoride Solution

To learn how to make up a fluoride solution, let us assume a hypothetical case using the following data:

1. Flow to be treated is 10 MGD.
2. Hydrofluosilicic acid 20 percent is the chemical to be used.
3. The unfluoridated water contains 0.05 mg/L (ppm) fluoride ion (F^-).
4. The desired fluoride concentration in the treated water is 1.0 mg/L.

What should the feed rate be?

A treatment chart, such as those on pages 50 and 51, can be used to determine the feed rate. For example, using "Treatment Chart I, Hydrofluosilicic Acid," on page 50, locate the 10 MGD flow on the left-hand scale of the graph and follow that line to the right until it intersects the 20 percent diagonal line. Project this point down vertically to the intersection of the bottom line indicating gallons per day (or gallons per hour) required to produce a

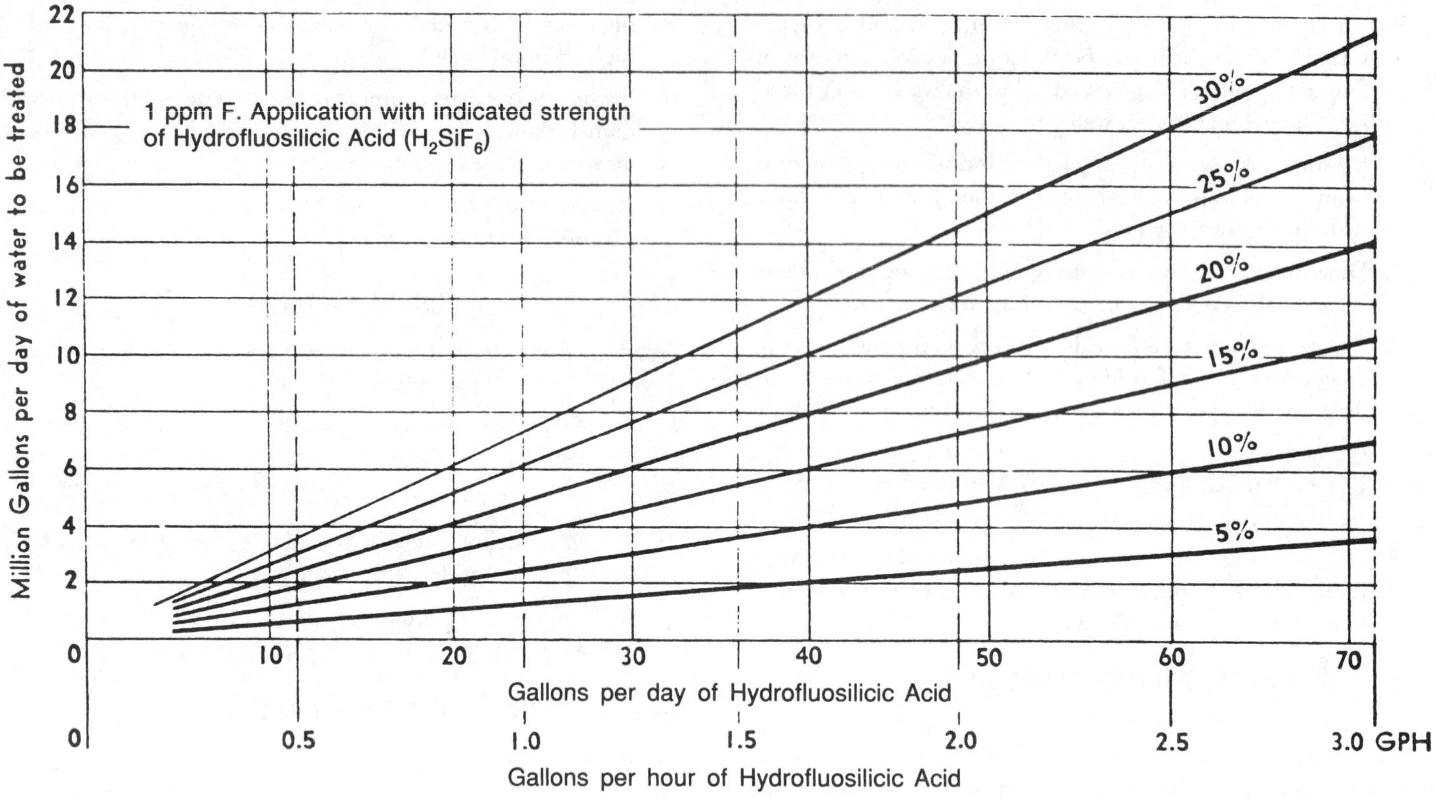

1 ppm F. Application with indicated strength of Hydrofluosilicic Acid (H_2SiF_6)
Million Gallons per day of water to be treated
0
2
4
6
8
10
12
14
16
18
20
22
30%
25%
20%
15%
10%
5%
10
20
30
40
50
60
70
Gallons per day of Hydrofluosilicic Acid
0.5
1.0
1.5
2.0
2.5
3.0 GPH
Gallons per hour of Hydrofluosilicic Acid

TREATMENT CHART I
Hydrofluosilicic Acid

1 ppm F. Application with indicated strength of Hydrofluosilicic Acid (H_2SiF_6)
Million Gallons per day of water to be treated
0
10
20
30
40
50
60
30%
25%
20%
15%
10%
5%
25
50
75
100
125
150
175
200
Gallons per day of Hydrofluosilicic Acid
1
2
3
4
5
6
7
8 GPH
Gallons per hour of Hydrofluosilicic Acid

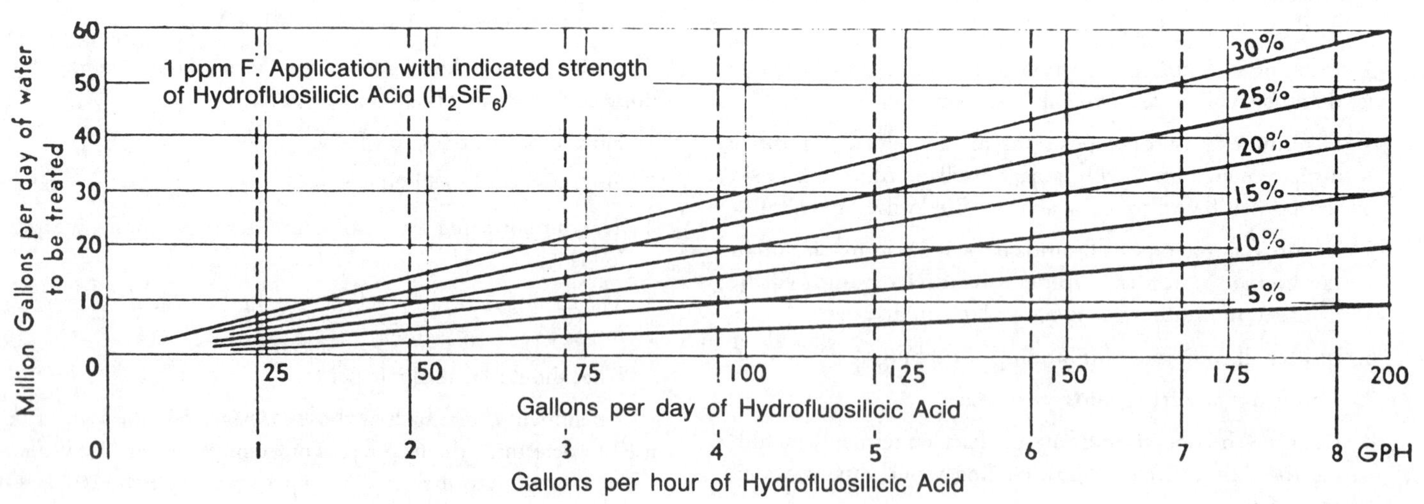

TREATMENT CHART II
Hydrofluosilicic Acid

(Permission of Wallace & Tiernan, Inc.)

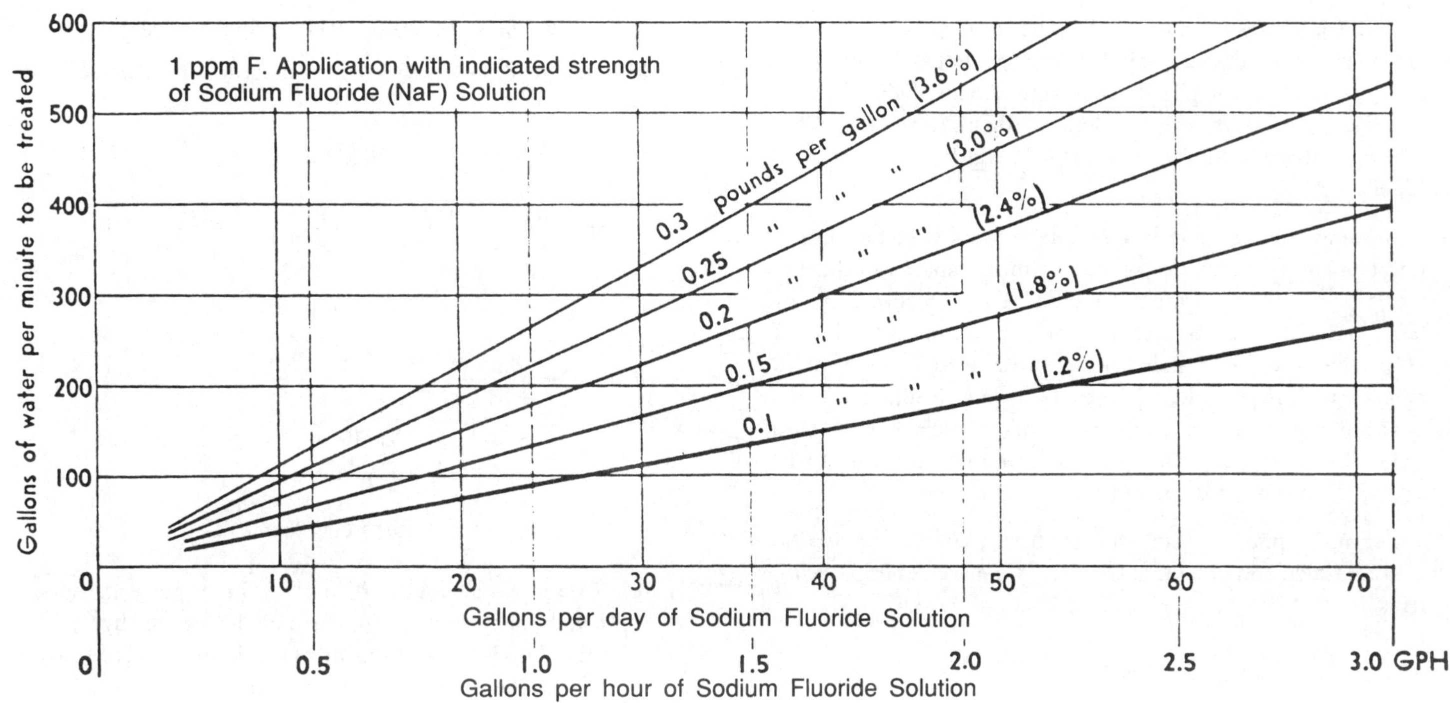

TREATMENT CHART III

Sodium Fluoride

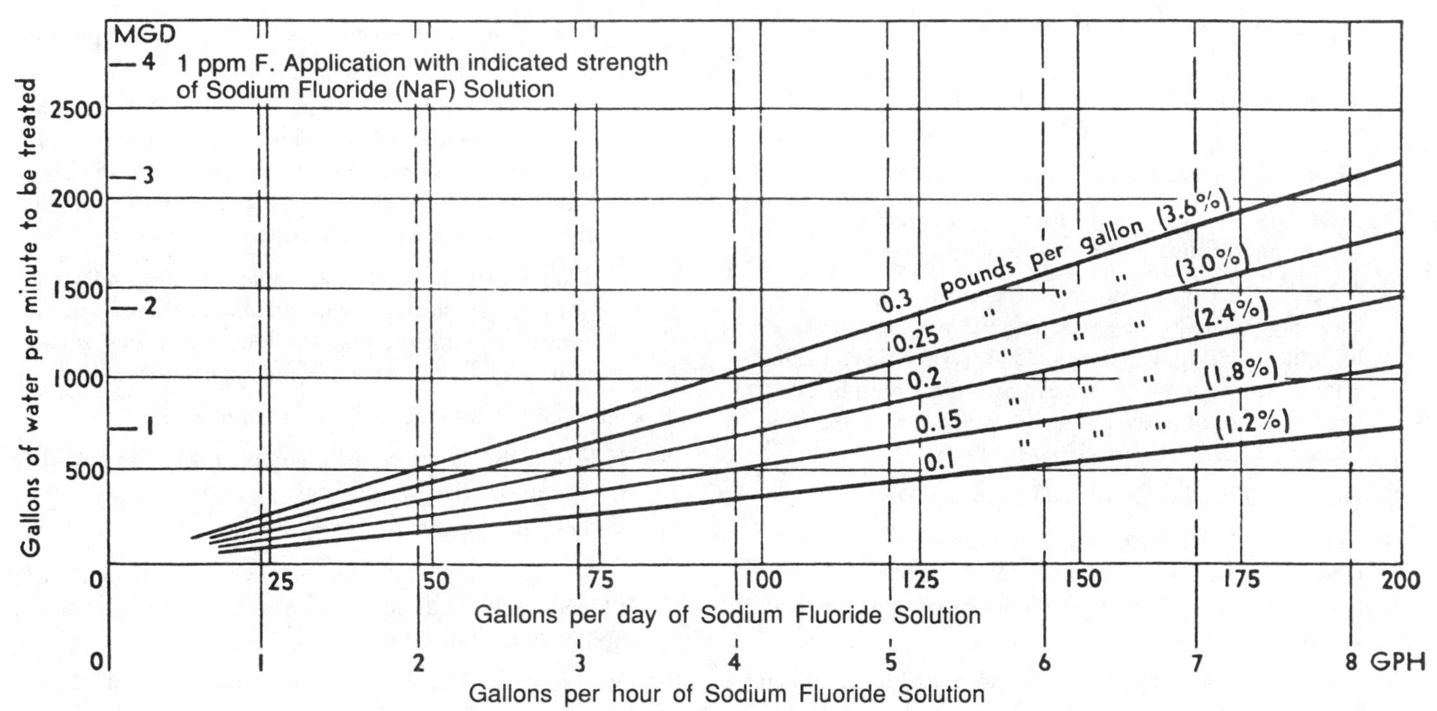

TREATMENT CHART IV

Sodium Fluoride

(Permission of Wallace & Tiernan, Inc.)

1 mg/L (ppm) dose of fluoride (F). The answer is 50 gallons per day or a little less than 2.1 GPH (gallons per hour). Multiply the 50 by (1.00 – 0.05) to give the needed treatment of 47.5 GPD or 2 GPH. The 1.00 is the desired dose of 1.00 mg/L and the 0.05 is the actual fluoride concentration of 0.05 mg/L in the untreated water.

In some cases, it might be desirable to use a weaker acid solution to avoid feed rates below the minimum capacity of the metering pump. Dilution then is in order. The concentration may be reduced by volumetric proportions, for example 1 gallon of 20 percent acid plus 1 gallon of water results in 2 gallons of 10 percent acid. If possible, try to avoid having to dilute acid because of potential errors and problems, especially with hard water. Peristaltic and electronic feeder pumps (Figures 13.12 and 13.3) may be used when the feed rates are low.

For more practice using the treatment charts, see Section 13.12, "Calculating Fluoride Dosages," for eleven example problems.

13.62 Fluoridation Log Sheets

You will probably want to design your own log sheets so they will be consistent with the installation features at your plant. Sample log sheets are shown on Figures 13.13, 13.14, and 13.15 (see pages 53, 55, and 56).

13.620 Hydrofluosilicic Acid

Figure 13.13 shows a typical log sheet from a hydrofluosilicic acid station. An explanation of the various columns is given below.

1. "Date" refers to calendar date when readings were logged or the date a shipment of fluoride was received.
2. "Time" refers to time event happened.
3. "Tank" that is supplying the feeder is circled. "Gals." refers to the gauge reading of the amount of acid in the tank.
4. "% & Sp. Gr." Each delivery is accompanied by a vendor's laboratory analysis. The specific gravity is not measured until the tank is ready to be placed in service. When mixing acids of varying strengths, the end percentage must be calculated and entered in the proper column. See Example 10 in Section 13.12, "Calculating Fluoride Dosages."
5. For each tank follow the directions given in Step 4.
6. "Tank Loss Gals." refers to the amount of feed during the reporting period. In the sample, 2,930 – 2,600 = 330 gallons. The feeding equipment should be equipped with an acid totalizer readout.
7. The "ratio" column indicates the feed setting computed using the acid strength, specific gravity, and required dosage. The following steps illustrate how to calculate the feed setting for a specific piece of equipment.
   (a) (Sp. Gr.)(lb/gal water)(% H_2SiF_6)(% F^- (in H_2SiF_6)) = lb F^-/gal.
   (b) Substituting figures in the above formula.
   (c) (1.226)(8.34)(0.229)(0.791) = 1.85 lb F^-/gal.
   (d) Dosage: 8.34 ÷ 1.85 = 4.51 gallons acid/MG water.

Explanation:

1 liter weighs 1 million milligrams.

$$\text{Dosage, } \frac{mg}{L} = \frac{mg}{1\text{ M mg}} = \frac{lb}{1\text{ M lb}} = \frac{gal}{1\text{ M Gal}}$$

or

$$1\text{ mg/L} = \frac{8.34\text{ lb}}{1\text{ M Gal}}$$

$$= \frac{8.34\text{ lb/M Gal Water}}{1.85\text{ lb/gal Acid}}$$

$$= 4.51\text{ gal Acid/M Gal Water}$$

   (e) To compensate for the .05 mg/L F^- in the raw water supply, the above figure of 4.51 should be reduced by 5 percent. This is the relationship of the desired level of 1 mg/L F^- to the raw water level of .05 mg/L F^-.
   (f) 4.51 – (.05 × 4.51) = 4.51 – .23 = 4.28 gal/MG.
   (g) Ratio setting therefore is 4.28 ÷ 4.80 or 0.89.
   (h) The flow capacity of the pipeline water meter at 100 percent is 300 MGD.
   (i) The flow capacity of the acid feed pump is 1,440 gallons of H_2SiF_6/day.

   The ratio of the above two 100 percent capacities is 1,440 ÷ 300 or 4.8 gal/MG.

   Note the difference of the setting of 0.88 and the calculated figure of 0.89. This adjustment is made in order that the fluoride dosage will agree with the laboratory results. In all instances, the laboratory results should determine the feed settings.

   The small difference in calculated setting and actual setting can also result from accumulated errors in the control equipment, that is, flow transmitter, square root ($\sqrt{}$) extractor, and ratio controller.
8. The H_2SiF_6 column records the totalizer values.
9. "H_2SiF_6 Gals." is the actual amount of acid fed into the system and is derived as follows:

   885,005.50 – 884,676.08 = 329.42 gallons

   This figure should be fairly close to the reading obtained at Step 6. If it is not, look for errors in readings, leaks, or equipment malfunctioning.
10. "Water Meter Totalizer" is the cumulative total of the amount of water being treated measured by a Venturi or some other type of primary water meter.
11. "Water M/Gals." is the actual amount of water passing through the water meter for the time period involved and again is derived by simple subtraction:

    268.00 – 191.01 = 76.99 MG.

BYPASS TUNNEL FLUORIDE STATION

Week Ending *December 11, 2013*

DATE	TIME	ACID STORAGE TANK 1 GALS.	TANK 1 % & SP. GR.	TANK 2 GALS.	TANK 2 % & SP. GR.	TANK 3 GALS.	TANK 3 % & SP. GR.	TANK 4 GALS.	TANK 4 % & SP. GR.	TANK LOSS GALS.	RATIO	H_2SIF_6 TOTALIZER	H_2SIF_6 GALS.	WATER METER TOTALIZER	WATER M/GALS.	ACID GAL/MG	F RESID. PPM	DOWN TIME	OBS. BY
END OF PREVIOUS WEEK		2930	1.226 22.9	4370	23.0	120	23.9	2315	23.1		0.88	884676.08		191.01			0.98		
12/5	11:00A	2600								330	0.88	885005.50	329.42	268.00	76.99	4.28	0.97		[illegible]

FLOW BIF METER	F & P METER	REMARKS:

Fig. 13.13 Log sheet for bypass tunnel fluoride station

12. "Acid Gal/MG" is the rate of treatment for hydrofluosilicic acid and is obtained by dividing the figures from Step 9 by the figure from Step 11.

    329.42 ÷ 76.99 = 4.28

13. "F Resid. PPM" is actual fluoride content in the treated water as read by continuous flow fluoride ion analyzer.

14. "Down Time." The equipment should be equipped with a running time meter reading in seconds that begins to operate any time the plant shuts down. This will give reasons for low feed as indicated by the readings in Step 12. The operator should know why this deviation occurred.

15. "Obs. By." This will be the operator's initials.

13.621 Sodium Silicofluoride

1. Figure 13.14 is a typical log sheet for a gravimetric feeder feeding powdered sodium silicofluoride.

2. "Date" and "Time" are entered in the first two columns.

3. "Totalizer Reading (lbs.)." This reading indicates a cumulative reading of the amount of silicofluoride that has been fed by the machine.

4. "Weight Loss per 24 hrs. (lbs.)" is the amount of silicofluoride actually fed during the time frame and is determined from the readings observed in Step 3 as follows:

    44,165.5 – 43,276.9 = 888.6.

5. "Mach. Feed Setting" is the feed rate being used. This rate may vary from machine to machine depending upon gear ratios and other devices used to control the rate of chemical feed. If the laboratory tests indicate low or high fluoride ion level, the adjustment is made with this setting on a percentage basis. If the laboratory fluoride ion level is 10 percent low, then this setting should be raised 10 percent.

6. "Chem. Added to Bin (lbs.)" is the amount of chemical taken from storage and dumped into the feeder hopper.

7. "Chemical Left in Storage (lbs.)" is the amount of fluoride in the bulk storage. This is useful in programming supply orders and in checking the accuracy of the feeder over a period of time. To check accuracy, compare this amount with the amount indicated in Step 4 over a 6-month or 1-year period.

8. "Pump Operating" is useful if several pumps are available to inject the dissolved sodium silicofluoride into the water main.

9. "Water Meter Reading (m.g.)" is the reading from the main line water meter.

10. "Water Treated (m.g.)" is the amount of water actually treated and is derived from Step 9 data as follows:

    39,829.73 – 39,762.41 = 67.32.

11. "Dosage (lbs. per m.g.)" represents the actual dosage of fluoride in the water. This figure should be constant, barring downtime, changes in machine feed setting, Step 5, or equipment malfunctioning. The value is derived from the weight loss (Step 4) divided by the water treated (Step 10) as follows:

    888.6 ÷ 67.32 = 13.2.

12. "Plant Down Time (Hrs.)" is the period of time fluoride was not being fed and the plant was shut down because of power failure or automatic shutdown. The installation should be equipped with a resettable running time meter reading hours and tenths.

13. Figure 13.15 is a typical small plant log sheet used in plants using sodium silicofluoride. This log sheet is provided through the courtesy of the City of Palo Alto, California.

13.63 Equipment Check Procedures

Figure 13.16 is a flowchart indicating the procedures an operator could follow when starting water fluoridation equipment. The flowchart procedures offer potential ways of correcting problems that may develop and help operators achieve a system with a good operating efficiency.

QUESTIONS

Please write your answers to the following questions and compare them with those on page 67.

13.6A What should be the feed rate in gallons per day for treating 6 MGD with hydrofluosilicic acid 20 percent if the desired fluoride ion concentration is 1.2 mg/L? Assume the raw water does not contain any fluoride ion.

13.6B What could be the causes of differences between the recorded volume of acid used from a storage tank and the volume of acid actually fed into the system as measured by a flowmeter?

13.7 PREVENTION OF OVERFEEDING

1. Operators must ensure that no overfeeding occurs because no additional benefits result from overfeeding and there is a waste of chemicals and money. Excessive overfeeding could be harmful to consumers. Overfeeding can be prevented by proper operation and continuous monitoring of the product water.

2. If the size of the installation warrants, a continuous fluoride ion analyzer should be installed in the treated water line located downstream a sufficient distance so that adequate mixing is ensured.

3. In a large plant involving shift operation, grab samples can be analyzed for the fluoride level during each shift; otherwise, once-a-day checks will suffice.

FLUORIDE STATION REPORT

Station *Sunset Supply Line* Week Ending *December 11, 2013*

Date	Time	Totalizer Reading (lbs.)	Weight Loss per 24 hrs. (lbs.)	Mach. Feed Setting	Chem. Added to Bin (lbs.)	Chemical Left in Storage (lbs.)	Pump Operating	Water Meter Reading (m.g.)	Water Treated (m.g.)	Dosage (lbs. per m.g.)	Plant Down Time (Hrs.)	Feeder Time Lapse (sec.)	Observ
END OF PREVIOUS WEEK		43276.9		52.8		42000	1A	39762.41		13.2			
12/5	10:20A	44165.5	888.6	52.8	400	41600	1A	39829.73	67.32	13.2		12735	[illegible]

Fig. 13.14 Log sheet for fluoride station report

WEEKLY WATER PRODUCTION AND TREATMENT LOG

Week Ending Wed.________________, 20___.   Station:____________________

	Sacks on hand	FLUORIDE						
		Feeder Lbs.				Feed	Calc	Timer
		Start	Add	Final	Used	Setting	mg/l	Setting
Total								
Wed.								
Tues.								
Mon.								
Fri.								
Thurs.								
Previous								

WATER (CCF)			SAND (ml)
Meter Record			
		Consumption	

Pump A ☐ No. motor starts _____ + _____ + _____ = _____
No. running hours _______ - _______ = _____
Lead Pump ↕
No. motor starts _____ + _____ + _____ = _____
Pump B ☐ No. running hours _______ - _______ = _____

OTHER METER READINGS			
Meter Function			
Present Reading			
Previous Reading			
Difference			

	Cylinders on hand	CHLORINE						
		Cylinder, Lbs.						OTO
				Net				
		Gross	Tare	#1	#2	Used	Feed Rate	resid. mg/l
Previous								
Thurs.								
Fri.								
Mon.								
Tues.								
Wed.								
Total								

MISCELLANEOUS MEASUREMENTS			
Well static level, ft.			
Well pumping level, ft.			
Head against well, pump, psi			
Amp draw reading	L	C	R

REMARKS

Fig. 13.15 Weekly water production and treatment log

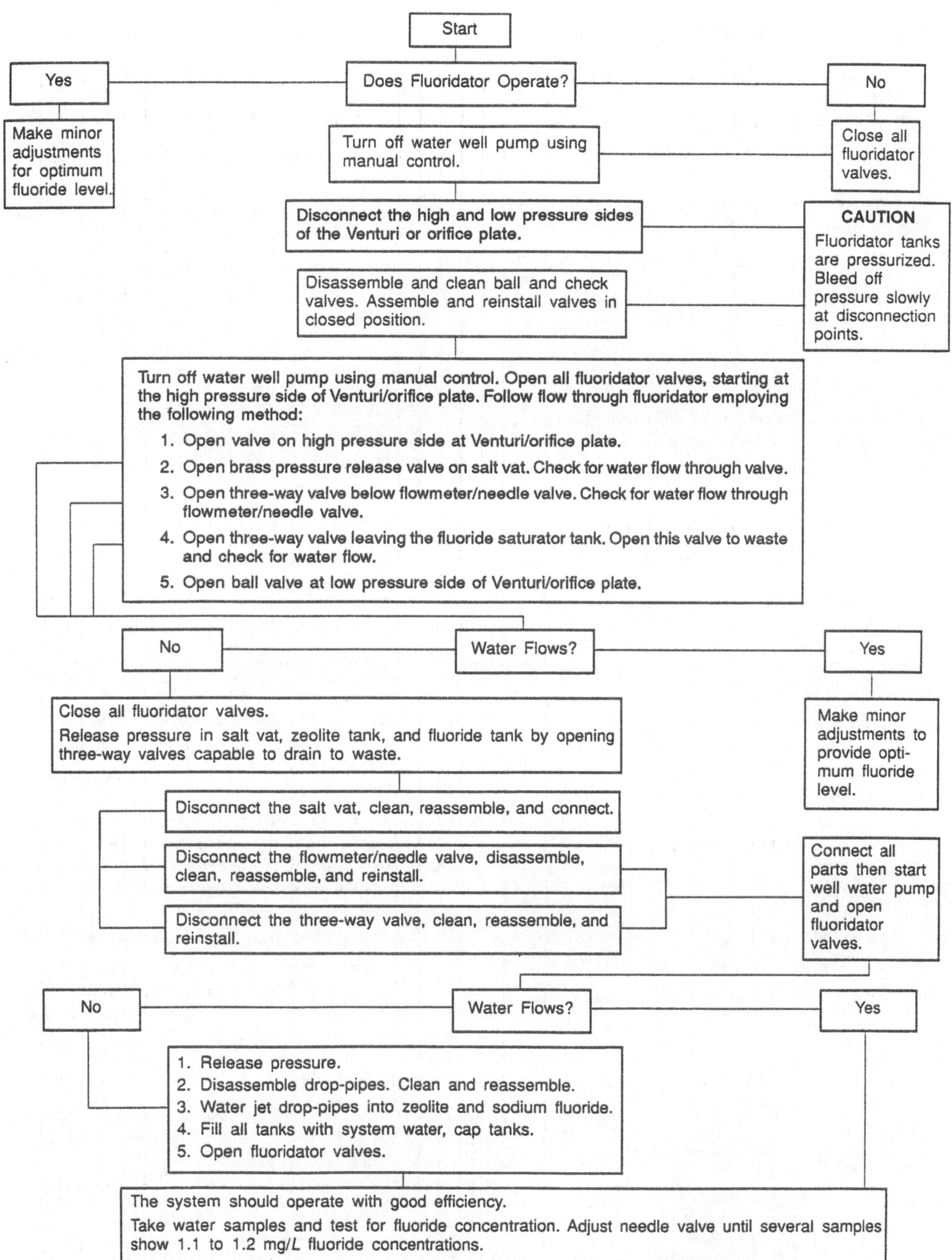

Fig. 13.16 Water fluoridation equipment check procedures

(Reproduced with permission from *Operator's Drumbeat,* Volume 1, Number 6, March–April 1989. Albuquerque Area Indian Health Service, J. R. Olguin, Publisher)

4. If the plant uses one of the solid fluoride compounds and the operator questions whether there is total solubility, the fluoride feeder can be shut down and the lack of fluoride traced out in the distribution system. There should be a sudden drop to zero fluoride or to the background level if total solubility is not being achieved (the undissolved solid fluoride compound will settle out).
5. All liquid systems should be checked for positive protection against backsiphonage from fluoride storage tanks.
6. Shut down the plant if there is any significant overfeeding. Start flushing the affected mains and notify the local and state health departments. The water department and the health departments will then decide if public notification should be undertaken.

13.8 UNDERFEEDING

In contrast to the chlorination process where continuous operation must be ensured, fluoridation does not have to be continuous. Shutdowns for cleaning, adjustments, or due to safety controls can be tolerated for short time periods. This does not mean that sloppy operation and maintenance is desirable. Every attempt should be made to maintain constant feeding. For example, the installation of standby electrical generating equipment just to maintain fluoridation equipment in operation would not be warranted. If the standby generator had to be purchased for other reasons, then the emergency circuit may also include the fluoride feeding equipment. Underfeeding should not be allowed because this results in a very significant reduction of the benefits of fluoridation.

Daily inspection of the fluoridation equipment, fluoride tests on the treated water, and calculation of the dosage from water treatment and chemical use data can greatly minimize the possibility of both overfeeding and underfeeding.

QUESTIONS

Please write your answers to the following questions and compare them with those on page 67.

13.7A Why should overfeeding be prevented?

13.7B What should be done if significant overfeeding occurs?

13.8A Why might a fluoridation operation be shut down?

13.9 SHUTTING DOWN CHEMICAL SYSTEMS

If the fluoridation equipment is going to be shut down for an extended length of time, it should be cleaned out to prevent corrosion or the solidifying of the chemical. Lines and equipment could be damaged when restarted if chemicals left in them solidify. Operators could be seriously injured if they open a chemical line that has not been properly flushed out.

The following items should be included in your checklist for shutting down the chemical system:

1. Flush out the chemical supply with water.
2. Run dry chemicals completely out of the equipment and clean equipment by using a vacuum cleaner.
3. Flush out all the solution lines with water until the lines are clean.
4. Shut off the electrical power.
5. Shut off the water supply and protect both the water supply and the chemical system from freezing.
6. Drain and clean the mix and feed tanks.
7. Padlock (lock out) the main electric switch box to the fluoride equipment.

13.10 MAINTENANCE

Maintenance should follow the same routine as with any similar chemical feeder, including regular cleanup and painting of the equipment and appurtenant metal piping and conduits. In order to give the plant a fresher look and hold down on painting, consider using all plastic piping even though it is needed only for the water supply. Conduit and fittings should also be plastic for the same reason. Vacuum any gears and other similar parts to remove fluoride dust.

Since fluoride solutions are extremely corrosive, be constantly on the lookout for drips or leaks and any other evidence of corrosion. Repair these conditions as quickly as possible. Also look for the buildup of insoluble deposits in feed lines and equipment. Schedule the removal of insoluble deposits on a regular basis to prevent buildups from creating any problems.

All containers of fluoride chemicals must be disposed of in an acceptable manner. Thoroughly rinse all containers with water to remove all traces of chemicals before allowing containers to leave your plant. You may burn the containers if a nuisance will not be created. Remember that fluoride fumes can kill vegetation and are harmful to people.

You do not need to be too concerned about checking the feed rate by catching a given amount of fluoride over a time period. The log will show long-period discrepancies and the daily laboratory tests will indicate any drifting from the desired fluoride concentration in the treated water.

Either you or the laboratory personnel must analyze the fluoridated water daily. Check the results for any deviations from the norm and take corrective action. Handheld colorimeters are available for measuring fluoride in water. See Chapter 21, "Advanced Laboratory Procedures," for details on how to analyze samples for the fluoride ion.

An important part of your maintenance program is the prevention of any sanitary defects that could adversely affect the safety or quality of your treated drinking water. Sanitary defects that could develop in fluoridation systems include the following:

1. Lack of or inadequate start-stop controls
2. Inadequate feed rate control equipment
3. No analyzer to measure fluoride ion levels in treated water
4. Lack of or inadequate backflow safeguards
5. Fluoridation chemical not meeting AWWA specifications
6. Inadequate free chlorine residual in treated water

13.11 SAFETY IN HANDLING FLUORIDE COMPOUNDS[8]

From the operator's viewpoint, fluoride chemicals have one thing in common with all other chemicals found in treatment plants: Fluoride chemicals can seriously injure or kill the careless or untrained operator. Safety should be of special concern to you because it is your own health that is at stake.

13.110 Avoid Overexposure

One of the major causes of overexposure is the inhalation of fluoride dust. This usually occurs while a dry feeder or saturator is being loaded. To protect yourself from the dust of dry fluoride compounds, be sure the dust collector system works properly. Even with the use of dust collector systems, dust will circulate in the air. Always use approved respirators equipped with cartridges for organic dusts and vapors, protective coveralls, and gloves when emptying sacks or cleaning up equipment and plant surfaces.

When loading a saturator, dust will be minimized if crystalline sodium fluoride is fed instead of powdered sodium fluoride. When loading a dry feeder, you should wear a mask, apron, and rubber gloves to minimize exposure.

When the protection gear is removed, the remaining small traces of chemical should also be removed from your body. Some large water plants have dust collection systems that use a partial vacuum to draw dust from your body and vent it to the outside air after filtering.

Care should be taken when emptying bags of chemicals into a feeder hopper. The bags should be opened carefully at the top and the contents poured gently to minimize dust. Care should also be taken during storage of the bags. Bags should be stored in a dry place, preferably off the floor. If bags are stacked too high there is the possibility of them falling and breaking open.

If a saturator is used, you should be cautious about allowing the solution to come in contact with skin and clothing. If this does happen, the affected area should be washed immediately with water. This also applies to other fluoride solutions (such as the dissolving water used in a dry feeder).

If a fluoride acid is being fed, extra precaution must be taken. Fluoride acid is probably the most corrosive chemical found in a water plant. The pH of fluoride acid is approximately 1.2 and it will eat through glass faster than chlorine. Special care should be taken to keep fumes to a minimum. If the acid does come in contact with your skin, you may not be able to wash it off fast enough to prevent a burn. If this happens, standard first aid should be administered as soon as possible.

A good pair of safety goggles should be worn at all times when working around fluoridation equipment where there is any possibility of splashing fluoride solutions. Be especially cautious around the fluoride acids as the concentrated acid can dissolve the whites of one's eyes in addition to the usual burns associated with acids. Another "must" is a safety shower. This must be located within easy access to both the unloading operation and points of liquid usage.

Another safety precaution that should be followed is the labeling of all feeders and solution tanks. Proper labeling will help prevent placement of chemicals in the wrong feeder. If possible, fluoride chemical should be tinted blue to differentiate it from other water treatment chemicals.

13.111 Symptoms of Fluoride Poisoning

In the event that someone is poisoned, it is vitally important to recognize the early symptoms.

Some of the obvious signs of poisoning are vomiting, stomach cramps, and diarrhea. Usually, the person will become very weak, have trouble speaking, be very thirsty, and have poor color vision. In cases of extreme poisoning, there are strong, jerky muscle contractions in the arms and legs leading to convulsions. If poisoning is not treated immediately, the person may die. Fatal doses range from 4 to 5 g, or about a tablespoon. This equals about 2,000 times the amount of fluoride swallowed by a person from a water supply.

If a person is poisoned by inhaling fluoride, the first symptoms will be a sharp, biting pain in the nose followed by a runny nose or nosebleed. It is doubtful that a person could inhale enough fluoride to produce the same effects as encountered from drinking a large amount of fluoride. However, the sudden presence of bad stomach cramps and pains in the nose and eyes should not be ignored.

---

8. Portions of the material in this section were adapted from "Safety Procedures Necessary During Fluoridation Process," by Ed Hansen. Reproduced from *Opflow,* Volume 9, No. 7, (July 1983) by permission. Copyright 1983, The American Water Works Association.

The victim should see a doctor immediately, and the water treatment practices should be checked to determine the source of the fluoride poisoning. It is probably a good idea to check out treatment practices occasionally.

The importance of quick treatment for fluoride poisoning cannot be overemphasized. In such cases, a doctor should be called immediately, and if the poisoning is severe, an ambulance should be called.

13.112 Basic First Aid

Once it is established that fluoride is the cause of the poisoning, first aid should be started while waiting for medical help. The following are recommended first-aid procedures:

1. Move the person away from any contact with fluoride and keep warm.
2. Inhalation: Remove exposed person to an uncontaminated area immediately. If breathing has stopped, start artificial respiration at once. Oxygen should be provided for an exposed person having difficulty breathing (but only by an authorized person) until exposed person is able to breathe easily unassisted. Exposed person should be examined by a physician.
3. Eye Contact: Flush eyes for at least 15 minutes with large amounts of water. Eyelids should be held apart during the flushing to ensure contact of water with all accessible tissue of the eyes and lids. Medical attention should be given as soon as possible.
4. Skin Contact: Exposed person should be removed to an uncontaminated area and subjected immediately to a drenching shower of water for a minimum of 15 to 20 minutes. Remove all contaminated clothing while under shower. Medical attention should be given as soon as possible for all burns, regardless of how minor they seem.
5. Ingestion: If conscious, give the exposed person large quantities of water immediately to dilute the acid. Do not induce vomiting. If hydrofluosilicic acid is swallowed, vomiting may cause the acid to go the wrong way into the lungs and cause more serious problems. Milk may be given for its soothing effect. If sodium fluoride is swallowed, vomiting may be induced. A physician should be contacted immediately.
6. Take the person to the hospital as soon as possible.

If common sense and good safety practices are used, the hazard to the water plant operator should be as small as the hazard to the water consumer.

> Fluoridation chemicals are poisonous. Protect yourself from these toxic chemicals.

13.113 Protecting Yourself and Your Family

Avoid swallowing fluoridation chemicals. Do not eat, drink, or smoke in or around chemical storage or feed areas. Do not inhale chemical dusts or vapors. Wear a respirator. Be sure exhaust fans and dust collectors are operating properly. Prevent hydrofluosilicic acid from coming in contact with your skin or eyes because hydrofluosilicic acid is very corrosive. If any hydrofluosilicic acid touches you, flood the contact area with plenty of water. If you are acutely poisoned by a fluoride chemical, you may be thirsty, vomit, and have stomach cramps, diarrhea, difficulty in speaking, and disturbed color vision. If any of these symptoms occur, consult a physician immediately.

When leaving the fluoride plant, wash your hands and change coveralls so that fluoride dust is not carried home.

13.114 Training

Special safety training must be given to all operators who will handle fluoride compounds. Training must include how to safely receive compounds from supplier, store until needed, prepare solutions, load feeders, and dose water being treated.

QUESTIONS

Please write your answers to the following questions and compare them with those on page 67.

13.9A Why should fluoridation equipment be cleaned out if the equipment is going to be shut down for an extended length of time?

13.10A How can fluoride dust be removed from gears?

13.10B During maintenance of fluoridation equipment, how concerned should you be about checking the feed rate by catching a given amount of fluoride over a time period?

13.11A What are the symptoms of acute fluoride poisoning?

13.12 CALCULATING FLUORIDE DOSAGES

FORMULAS

1. Treatment charts, such as the ones located on pages 50 and 51, can be used to determine feed rates. The feed rate is usually based on a dose of 1 mg/L; therefore actual feed rates must be adjusted.

$$\text{Actual Feed Rate, GPD} = \frac{(\text{Chart Feed Rate, GPD})(\text{Actual Dose, mg/L})}{1 \text{ mg/L}}$$

2. Feed rates may be calculated on the basis of pounds per day or gallons per day. Consideration must be given to the pounds of fluoride ion per pound of commercial chemical.

$$\text{Feed Rate, lb/day} = \frac{(\text{Flow, MGD})(\text{Dose, mg/L})(8.34 \text{ lb/gal})(100\%)}{\text{Solution, \% F}}$$

or

$$\text{Feed Rate, lb/day} = \frac{\text{Feed Rate, lb F/day}}{\text{lb F/lb Commercial Chemical}}$$

or

$$\text{Feed Rate, gal/day} = \frac{\text{Feed Rate, lb/day}}{\text{Chemical Solution, lb/gal}}$$

3. If the water being treated contains some fluoride ion, but is insufficient, then a feed dose must be calculated.

$$\text{Feed Dose, mg/L} = \text{Desired Dose, mg/L} - \text{Actual Concentration, mg/L}$$

4. Commercial chemicals usually are not 100 percent pure. Also, the chemical only contains a portion of the ion of concern (fluoride ion in this chapter).

$$\text{Portion F} = \frac{(\text{Commercial Purity, \%})(\text{Fluoride Ion, \%})}{(100\%)(100\%)}$$

The portion F is the pounds of F per pound of commercial chemical. For example, 0.6 pound F per 1 pound of commercial sodium silicofluoride.

5. To calculate the fluoride dosage, or any chemical dosage, you need to know the pounds of chemical and volume of water in million gallons.

$$\text{Dosage, mg/L} = \frac{\text{Chemical, lb}}{(\text{Water, M Gal})(8.34\ \text{lb/gal})}$$

$$= \frac{\text{lb Chemical}}{\text{Million lb Water}}$$

If we substitute milligrams for pounds, we get

$$= \frac{\text{mg Chemical}}{\text{Million mg Water}}$$

One million milligrams of water occupy a volume of 1 liter.

$$= \frac{\text{mg Chemical}}{\text{Liter of Water}}$$

$$= \text{mg/L}$$

6. To determine the amount of feed solution in either gallons or gallons per day to treat a water, you need to know the amount of water to be treated in gallons or gallons per day, the feed dose in milligrams per liter, and the feed solution in milligrams per liter.

$$\text{Feed Solution, gal} = \frac{(\text{Flow, gal})(\text{Feed Dose, mg/L})}{\text{Feed Solution, mg/L}}$$

NOTE: If the "Feed Solution" is in gallons per day instead of gallons, then the "Flow" must be in gallons per day also instead of gallons.

7. When mixing the same two acids or chemicals, but of different strengths, the volumes or flows of the chemicals and their strengths must be known.

$$\text{Mixture Strength, \%} = \frac{(\text{Vol 1, gal})(\text{Strength 1, \%}) + (\text{Vol 2, gal})(\text{Strength 2, \%})}{\text{Volume 1, gal} + \text{Volume 2, gal}}$$

NOTE: The "Volumes" may be in gallons or treated as flows in GPD or MGD. The "Strengths" may be in percentages or concentrations such as mg/L.

8. When using chemicals for fluoridation, we need to know the percentage fluoride ion purity. This information will allow us to convert the pounds of chemical dosage to pounds of fluoride ion available.

$$\text{Fluoride Ion Purity, \%} = \frac{(\text{Molecular Weight of Fluoride})(100\%)}{\text{Molecular Weight of Chemical}}$$

Example 1

A flow of 4 MGD is to be treated with a 20 percent solution of hydrofluosilicic acid (H_2SiF_6). The water to be treated contains no fluoride and the desired fluoride concentration is 1.8 mg/L. What should be the feed rate of hydrofluosilicic acid? Use the treatment charts.

Known		Unknown
Flow, MGD	= 4 MGD	1. Feed Rate, gal/day
Acid Solution, %	= 20%	2. Feed Rate, gal/hr
Desired F, mg/L	= 1.8 mg/L	

1. Use "Treatment Chart I" on page 50 because we are treating a relatively small flow (4 MGD).
2. Start on the left side at the 4 MGD value and move horizontally to the right to the 20 percent diagonal line.
3. At this point, drop vertically downward to the bottom lines and read the feed rates for 1 mg/L (ppm).
   a. Feed Rate, gallons per day = 19 gal/day
   b. Feed Rate, gallons per hour = 0.8 gal/hr
4. Calculate the feed rate to produce the desired fluoride concentration of 1.8 mg/L.

$$\text{Feed Rate, GPD} = \frac{(\text{Feed Rate, GPD})(\text{Desired F, mg/L})}{1\ \text{mg/L}}$$

$$= \frac{(19\ \text{GPD})(1.8\ \text{mg/L})}{1\ \text{mg/L}}$$

$$= 34.2\ \text{gal/day}$$

$$\text{Feed Rate, gal/hr} = \frac{(\text{Feed Rate, gal/hr})(\text{Desired F, mg/L})}{1\ \text{mg/L}}$$

$$= \frac{(0.8\ \text{gal/hr})(1.8\ \text{mg/L})}{1\ \text{mg/L}}$$

$$= 1.44\ \text{gal/hr}$$

Example 2

A flow of 4 MGD is to be treated with a 20 percent solution of hydrofluosilicic acid (H_2SiF_6) that contains a fluoride purity of 79.2 percent. The water to be treated contains no fluoride and the desired fluoride concentration is 1.8 mg/L. Assume the hydrofluosilicic acid weighs 9.8 pounds per gallon. What should be the feed rate of hydrofluosilicic acid? Calculate the feed rate.

Known		Unknown
Flow, MGD	= 4 MGD	1. Feed Rate, gal/day
Acid Solution, %	= 20%	2. Feed Rate, gal/hr
Acid, lb/gal	= 9.8 lb/gal	
Purity, %	= 79.2%	
Desired F, mg/L	= 1.8 mg/L	

1. Calculate the hydrofluosilicic acid feed rate in pounds per day.

$$\text{Feed Rate, lb/day} = \frac{(\text{Flow, MGD})(\text{Desired F, mg/L})(8.34\ \text{lb/gal})(100\%)(100\%)}{(\text{Acid Solution, }\%)(\text{Purity, }\%)}$$

$$= \frac{(4\ \text{MGD})(1.8\ \text{mg/L})(8.34\ \text{lb/gal})(100\%)(100\%)}{(20\%)(79.2\%)}$$

$$= 379\ \text{lb Acid/day}$$

2. Determine the feed rate of the acid in gallons per day.

$$\text{Feed Rate, gal/day} = \frac{\text{Feed Rate, lb/day}}{\text{Acid, lb/gal}}$$

$$= \frac{379\ \text{lb Acid/day}}{9.8\ \text{lb Acid/gal}}$$

$$= 39\ \text{gal Acid/day}$$

NOTE: We obtained a feed rate of 34 gallons of acid per day from "Treatment Chart I." The differences result from the problems of drawing and reading the chart accurately.

3. Calculate the feed rate in gallons of acid per hour.

$$\text{Feed Rate, gal/hr} = \frac{\text{Feed Rate, gal/day}}{24\ \text{hr/day}}$$

$$= \frac{39\ \text{gal Acid/day}}{24\ \text{hr/day}}$$

$$= 1.6\ \text{gal Acid/hr}$$

Example 3

A flow of 200 GPM is to be treated with a 2.4 percent (0.2 lb/gallon) solution of sodium fluoride (NaF). The water to be treated contains 0.7 mg/L of fluoride ion and the desired fluoride ion concentration is 1.6 mg/L. What should be the feed rate of sodium fluoride? Use the treatment charts.

Known		Unknown
Flow, GPM	= 200 GPM	1. Feed Rate, gal/day
NaF Solution, %	= 2.4%	2. Feed Rate, gal/hr
Desired F, mg/L	= 1.6 mg/L	
Actual F, mg/L	= 0.7 mg/L	

1. Use "Treatment Chart III" on page 51 because we are treating 200 GPM.
2. Start on the left side at the 200 GPM value and move horizontally to the right to the 2.4 percent diagonal line.
3. At this point, drop vertically downward to the bottom lines and read the feed rates for 1 mg/L (ppm).
   a. Feed Rate, gallons per day = 26.5 gal/day
   b. Feed Rate, gallons per hour = 1.1 gal/hr
4. Calculate the feed rate to produce the desired fluoride concentration of 1.6 mg/L.

$$\text{Feed Dose, mg/L} = \text{Desired F, mg/L} - \text{Actual F, mg/L}$$

$$= 1.6\ \text{mg/L} - 0.7\ \text{mg/L}$$

$$= 0.9\ \text{mg/L}$$

$$\text{Feed Rate, GPD} = \frac{(\text{Feed Rate, GPD})(\text{Feed Dose, mg/L})}{1\ \text{mg/L}}$$

$$= \frac{(26.5\ \text{gal/day})(0.9\ \text{mg/L})}{1\ \text{mg/L}}$$

$$= 23.8\ \text{GPD}$$

$$\text{Feed Rate, gal/hr} = \frac{(\text{Feed Rate, gal/hr})(\text{Feed Dose, mg/L})}{1\ \text{mg/L}}$$

$$= \frac{(1.1\ \text{gal/hr})(0.9\ \text{mg/L})}{1\ \text{mg/L}}$$

$$= 0.99\ \text{gal/hr or 1 gal/hr}$$

Example 4

A flow of 200 GPM is to be treated with a 2.4 percent (0.2 pound per gallon) solution of sodium fluoride (NaF). The water to be treated contains 0.7 mg/L of fluoride ion and the desired fluoride ion concentration is 1.6 mg/L. What should be the feed rate of sodium fluoride? Calculate the feed rate. Assume the sodium fluoride has a fluoride ion purity of 45.3 percent.

Known		Unknown
Flow, GPM	= 200 GPM	1. Feed Rate, gal/day
NaF Solution, %	= 2.4%	2. Feed Rate, gal/hr
NaF Solution, lb/gal	= 0.2 lb/gal	
Desired F, mg/L	= 1.6 mg/L	
Actual F, mg/L	= 0.7 mg/L	
Purity, %	= 45.3%	

1. Convert flow from gallons per minute to million gallons per day.

$$\text{Flow, MGD} = \frac{(\text{Flow, gal/min})(60\ \text{min/hr})(24\ \text{hr/day})}{1{,}000{,}000/\text{Million}}$$

$$= \frac{(200\ \text{gal/min})(60\ \text{min/hr})(24\ \text{hr/day})}{1{,}000{,}000/\text{Million}}$$

$$= 0.288\ \text{MGD}$$

2. Determine the fluoride feed dose in milligrams per liter.

$$\text{Feed Dose, mg/L} = \text{Desired F, mg/L} - \text{Actual F, mg/L}$$
$$= 1.6 \text{ mg/L} - 0.7 \text{ mg/L}$$
$$= 0.9 \text{ mg/L}$$

3. Calculate the feed rate in pounds of fluoride ion per day.

$$\text{Feed Rate, lb F/day} = (\text{Flow, MGD})(\text{Feed Dose, mg/L})(8.34 \text{ lb/gal})$$
$$= (0.288 \text{ MGD})(0.9 \text{ mg/L})(8.34 \text{ lb/gal})$$
$$= 2.16 \text{ lb F/day}$$

4. Convert the feed rate from pounds of fluoride per day to gallons of sodium fluoride solution per day.

$$\text{Feed Rate, gal/day} = \frac{(\text{Feed Rate, lb F/day})(100\%)}{(\text{NaF Solution, lb F/gallon})(\text{Purity, \%})}$$
$$= \frac{(2.16 \text{ lb F/day})(100\%)}{(0.2 \text{ lb F/gal})(45.3\%)}$$
$$= 23.8 \text{ gal/day}$$

NOTE: We obtained a feed rate of 23.8 gal/day in both Examples 3 and 4. Sometimes you may get slightly different results using these two methods of calculating the feed rate. The differences could result from inaccurately preparing and reading the treatment chart as well as the assumed purity of fluoride ion in the sodium fluoride.

5. Convert the feed rate from gallons per day to gallons per hour.

$$\text{Feed Rate, gal/hr} = \frac{\text{Feed Rate, gal/day}}{24 \text{ hr/day}}$$
$$= \frac{23.8 \text{ gal/day}}{24 \text{ hr/day}}$$
$$= 1.0 \text{ gal/hr}$$

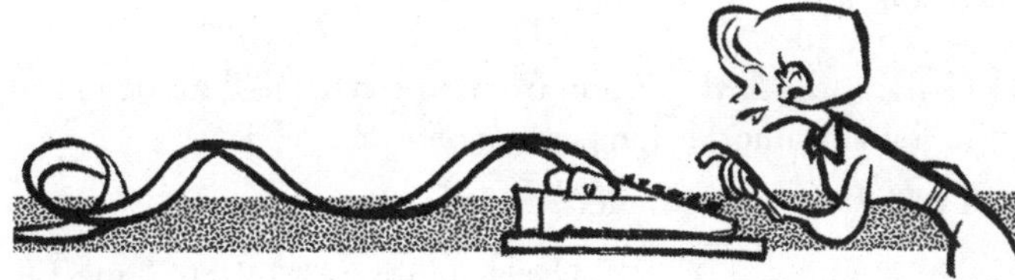

Example 5

A flow of 1 MGD is treated with sodium silicofluoride (Na_2SiF_6) to provide a fluoride ion dose of 1.4 mg/L. What is the feed rate in pounds per day? Commercial sodium silicofluoride has a purity of 98.5 percent and the fluoride ion purity of sodium silicofluoride is 60.7 percent.

Known		Unknown
Flow, MGD	= 1 MGD	Feed Rate, lb/day
Dose, mg/L	= 1.4 mg/L	
Na_2SiF_6 Purity, %	= 98.5%	
Fluoride Ion Purity, %	= 60.7%	

1. Calculate the portion of fluoride ion in the commercial sodium silicofluoride.

$$\text{Portion F} = \frac{(Na_2SiF_6 \text{ Purity, \%})(\text{Fluoride Ion Purity, \%})}{(100\%)(100\%)}$$
$$= \frac{(98.5\%)(60.7\%)}{(100\%)(100\%)}$$
$$= 0.598 \text{ lb F/lb Commercial } Na_2SiF_6$$

This says that there is 0.598 pound of fluoride ion in a pound of commercial sodium silicofluoride.

2. Calculate the pounds of fluoride required per day.

$$\text{Fluoride, lb/day} = (\text{Flow, MGD})(\text{Dose, mg/L})(8.34 \text{ lb/gal})$$
$$= (1 \text{ MGD})(1.4 \text{ mg/L})(8.34 \text{ lb/gal})$$
$$= 11.7 \text{ lb F/day}$$

3. Determine the chemical feed rate for the commercial sodium silicofluoride in pounds per day.

$$\text{Feed Rate, lb/day} = \frac{\text{Fluoride, lb/day}}{\text{Fluoride, lb/lb Commercial } Na_2SiF_6}$$
$$= \frac{11.7 \text{ lb F/day}}{0.598 \text{ lb F/lb Commercial } Na_2SiF_6}$$
$$= 19.6 \text{ lb/day Commercial } Na_2SiF_6$$

Example 6

A flow of 1.4 MGD is treated with sodium silicofluoride. The raw water contains 0.4 mg/L of fluoride ion and the desired fluoride ion concentration is 1.6 mg/L. What should be the chemical feed rate in pounds per day? Assume each pound of commercial sodium silicofluoride (Na_2SiF_6) contains 0.6 pound of fluoride ion.

Known		Unknown
Flow, MGD	= 1.4 MGD	Feed Rate, lb/day
Raw Water F, mg/L	= 0.4 mg/L	
Desired F, mg/L	= 1.6 mg/L	
Chemical, lb F/lb	= 0.6 lb F/lb	

1. Determine the fluoride feed dose in milligrams per liter.

$$\text{Feed Dose, mg/L} = \text{Desired F, mg/L} - \text{Raw Water F, mg/L}$$
$$= 1.6 \text{ mg/L} - 0.4 \text{ mg/L}$$
$$= 1.2 \text{ mg/L}$$

2. Calculate the fluoride feed rate in pounds per day.

$$\text{Feed Rate, lb F/day} = (\text{Flow, MGD})(\text{Feed Dose, mg/L})(8.34 \text{ lb/gal})$$
$$= (1.4 \text{ MGD})(1.2 \text{ mg/L})(8.34 \text{ lb/gal})$$
$$= 14.0 \text{ lb F/day}$$

3. Determine the chemical feed rate in pounds of commercial sodium silicofluoride per day.

$$\text{Feed Rate, lb/day} = \frac{\text{Feed Rate, lb F/day}}{\text{lb F/lb Commercial } Na_2SiF_6}$$

$$= \frac{14.0 \text{ lb F/day}}{0.6 \text{ lb F/lb Commercial } Na_2SiF_6}$$

$$= 23.3 \text{ lb/day Commercial } Na_2SiF_6$$

Example 7

The totalizer for a water treatment plant indicated that a total of 100,000 gallons of water had been treated with 3 pounds of 98 percent pure sodium fluoride (NaF). The fluoride ion purity for sodium fluoride is 45.3 percent. What was the added fluoride ion dosage in milligrams per liter?

Known		Unknown
Water Treated, MG	= 0.1 M Gal	Fluoride Dosage, mg/L
NaF, lb	= 3 lb	
NaF Purity, %	= 98%	
F Ion Purity, %	= 45.3%	

1. Calculate the portion of fluoride ion in the commercial sodium fluoride.

$$\text{Portion F} = \frac{(\text{NaF Purity, \%})(\text{Fluoride Ion Purity, \%})}{(100\%)(100\%)}$$

$$= \frac{(98\%)(45.3\%)}{(100\%)(100\%)}$$

$$= 0.444$$

$$\text{or} = 0.444 \text{ lb F/lb Commercial NaF}$$

2. Calculate the pounds of fluoride used.

$$\text{Fluoride, lb} = (\text{Commercial NaF, lb})(0.444 \text{ lb F/lb Comm NaF})$$

$$= (3 \text{ lb Comm NaF})(0.444 \text{ lb F/lb Comm NaF})$$

$$= 1.33 \text{ lb F}$$

3. Calculate the fluoride dosage in milligrams per liter.

$$\text{Fluoride Dosage, mg/L} = \frac{\text{Fluoride, lb F}}{(\text{Water Treated, M Gal})(8.34 \text{ lb/gal})}$$

$$= \frac{1.33 \text{ lb F}}{(0.1 \text{ M Gal})(8.34 \text{ lb/gal})}$$

$$= \frac{1.33 \text{ lb F}}{0.834 \text{ Million lb Water}}$$

$$= \frac{1.6 \text{ lb F}}{1 \text{ M lb Water}}$$

$$= 1.6 \text{ mg/L}$$

Example 8

Determine the percentage of fluoride ion in the feed solution from a saturator. The saturator contains 95 percent pure sodium fluoride, the maximum water solubility for sodium fluoride is four percent, and sodium fluoride is 45.3 percent fluoride ion.

Known		Unknown
Commercial NaF Purity, %	= 95%	Solution, % F
NaF Solubility, %	= 4%	
F Ion Purity, %	= 45.3%	

Calculate the percentage of fluoride ion in the feed solution.

$$\text{Solution, \% F} = \frac{(\text{NaF Solubility, \%})(\text{F Ion Purity, \%})}{100\%}$$

$$= \frac{(4\%)(45.3\%)}{100\%}$$

$$= 1.8\%$$

NOTE: In a saturator, the commercial NaF purity of 95 percent does not enter into the calculations because the four percent solubility is all NaF.

Example 9

The feed solution from a saturator containing 1.8 percent fluoride ion is used to treat a total flow of 400,000 gallons of water. The raw water has a fluoride ion content of 0.5 mg/L and the desired fluoride in the finished water is 1.8 mg/L. How many gallons of feed solution are needed?

Known		Unknown
Flow Vol, gal	= 400,000 gal	Feed Solution, gallons
Raw Water F, mg/L	= 0.5 mg/L	
Desired F, mg/L	= 1.8 mg/L	
Feed Solution, % F	= 1.8% F	

1. Convert the feed solution from a percentage fluoride ion to milligrams fluoride ion per liter of water.

   1.0% F = 10,000 mg F/L

$$\text{Feed Solution, mg/L} = \frac{(\text{Feed Solution, \%})(10{,}000 \text{ mg/L})}{1\%}$$

$$= \frac{(1.8\%)(10{,}000 \text{ mg/L})}{1\%}$$

$$= 18{,}000 \text{ mg/L}$$

2. Determine the fluoride feed dose in milligrams per liter.

$$\text{Feed Dose, mg/L} = \text{Desired F, mg/L} - \text{Raw Water F, mg/L}$$

$$= 1.8 \text{ mg/L} - 0.5 \text{ mg/L}$$

$$= 1.3 \text{ mg/L}$$

3. Calculate the gallons of feed solution needed.

$$\text{Feed Solution, gal} = \frac{(\text{Flow Vol, gal})(\text{Feed Dose, mg/L})}{\text{Feed Solution, mg/L}}$$

$$= \frac{(400{,}000 \text{ gal})(1.3 \text{ mg/L})}{18{,}000 \text{ mg/L}}$$

$$= 28.9 \text{ gallons}$$

Example 10

A hydrofluosilicic acid (H_2SiF_6) tank contains 300 gallons of acid with a strength of 18 percent. A commercial vendor delivers 2,000 gallons of acid with a strength of 20 percent to the tank. What is the resulting strength of the mixture as a percentage?

Known		Unknown
Tank Contents, gal	= 300 gal	Mixture Strength, %
Tank Strength, %	= 18%	
Vendor, gal	= 2,000 gal	
Vendor Strength, %	= 20%	

Calculate the strength of the mixture as a percentage.

$$\text{Mixture Strength, \%} = \frac{(\text{Tank, gal})(\text{Tank, \%}) + (\text{Vendor, gal})(\text{Vendor, \%})}{(\text{Tank, gal} + \text{Vendor, gal})}$$

$$= \frac{(300 \text{ gal})(18\%) + (2{,}000 \text{ gal})(20\%)}{(300 \text{ gal} + 2{,}000 \text{ gal})}$$

$$= \frac{5{,}400 + 40{,}000}{2{,}300}$$

$$= \frac{45{,}400}{2{,}300}$$

$$= 19.7\%$$

Example 11

Sodium silicofluoride (Na_2SiF_6) is used as the chemical to fluoridate a water supply. What is the fluoride ion purity as a percentage?

Known	Unknown
Atomic Weights	Fluoride Ion Purity, %
Na = 22.99	
Si = 28.09	
F = 19.00	

1. Determine the molecular weight of the fluoridation chemical, sodium silicofluoride, Na_2SiF_6.

Symbol	No. Atoms	×	Atomic Wt.[a]	=	Molecular Wt.
Na_2	2	×	22.99	=	45.98
Si	1	×	28.09	=	28.09
F_6	6	×	19.00	=	114.00
Molecular Wt of Chemical				=	188.07

a Atomic weight values can be obtained from a chemistry book.

2. Calculate the fluoride ion purity as a percentage.

$$\text{Fluoride Ion Purity, \%} = \frac{(\text{Molecular Weight of Fluoride})(100\%)}{\text{Molecular Weight of Chemical}}$$

$$= \frac{(114.00)(100\%)}{188.07}$$

$$= 60.62\%$$

This means that there is 0.6062 pound of fluoride ion in every pound of sodium silicofluoride.

13.13 ARITHMETIC ASSIGNMENT

Turn to the Arithmetic Appendix at the back of this manual. Read and work the problems in Section A.31, "Fluoridation." Check the arithmetic in this section using an electronic calculator. You should be able to get the same answers. Section A.51 contains similar problems using metric units.

13.14 ADDITIONAL READING

1. *Texas Manual*, chapter on "Special Water Treatment (Fluoridation)."
2. *Water Fluoridation Principles and Practices* (M4). Obtain from American Water Works Association (AWWA) at www.awwa.org or phone (800) 926-7337. Order No. 30004. ISBN 978-1-58321-311-7.
3. *Water Fluoridation: A Manual for Engineers and Technicians.* Obtain from the Centers for Disease Control and Prevention (CDC) at stacks.cdc.gov/view/cdc/13103.

13.15 ACKNOWLEDGMENTS

The author wishes to acknowledge the assistance graciously given by Robert A. Hewitt, Assistant Water Quality Engineer, San Francisco Water Department, San Francisco, California, and Tom Reeves, Centers for Disease Control and Prevention, Center for Prevention Services, Dental Disease Prevention Activity, Atlanta, Georgia.

Please answer the discussion and review questions next.

DISCUSSION AND REVIEW QUESTIONS

Chapter 13. FLUORIDATION

Please write your answers to the following discussion and review questions to determine how well you understand the material in the chapter.

1. Why are drinking waters fluoridated?
2. What factors would you consider when selecting a fluoridation chemical?
3. How can water be softened before use with fluoridation equipment?
4. What items should be considered when reviewing plans and specifications for the location of fluoride chemical hoppers?
5. Why is it important for operators to ensure that no overfeeding of fluoridation chemicals occurs? How can overfeeding be prevented?
6. What should be done if significant overfeeding occurs?
7. What should be done when fluoridation equipment is going to be shut down for an extended length of time?
8. How would you dispose of fluoride chemical containers?
9. How would you protect yourself from the dust of dry fluoride compounds?

SUGGESTED ANSWERS

Chapter 13. FLUORIDATION

Answers to questions on page 33.

13.0A If a person drinks water with an excessive amount of fluoride, the teeth become mottled (brown, chalky deposits).

13.0B Children who drink a recommended dose of fluoride have fewer dental caries (decay or cavities).

Answers to questions on page 34.

13.1A The water department or water company makes the final decisions as to types of fluoride chemicals and feeding equipment to be used.

13.2A The three compounds most commonly used to fluoridate water are sodium fluoride, sodium fluorosilicate, and hydrofluosilicic acid.

Answers to questions on page 47.

13.3A Drinking waters may come to contain fluoride ions by three different types of situations:

1. Raw water source may have adequate or excessive fluoride ions naturally present.
2. Two water sources may be blended together (one higher and one lower than acceptable level) to produce an acceptable level.
3. Fluoride ions must be added to the water to achieve an acceptable level.

13.3B Fluoridation systems must incorporate means to prevent both overfeeding and backsiphonage along with means to monitor the amount of chemical used.

13.3C A saturator is a device that produces a fluoride solution for the fluoridation process. The device is usually a cylindrical container with granular sodium fluoride on the bottom. Water flows either upward or downward through the sodium fluoride to produce the fluoride solution.

13.3D Hard water can produce problems in systems using saturators and dissolving tanks through the formation of low-solubility deposits of calcium and magnesium fluoride compounds.

Answers to questions on page 49.

13.4A Overfeeding of fluoridation chemicals must be prevented to avoid illness and bad public relations.

13.4B The fluoridator should be sized to handle the full range of both present and future doses or provisions should be made for future expansion.

13.5A When inspecting the fluoridation electrical system, inspect the system for proper voltage, properly sized overload protection, proper operation of control lights on the control panel, proper safety lockout switches and operation, and proper equipment rotation.

13.5B Safety equipment that should be available near a fluoridation system include eye wash stations, drench showers, dust masks, face shields, gloves, and vent fans.

Answers to questions on page 54.

13.6A What should be the feed rate in gallons per day for treating 6 MGD with hydrofluosilicic acid 20 percent if the desired fluoride ion concentration is 1.2 mg/L? Assume the raw water does not contain any fluoride ion.

Known		Unknown
Flow, MGD	= 6 MGD	Feed Rate, gal/day
Conc Fluoride, mg/L	= 1.2 mg/L	
Hydrofluosilicic Acid, %	= 20%	

1. Use "Treatment Chart I, Hydrofluosilicic Acid." Start at the left side with 6 MGD and move horizontally to the right to the intersection of the 20 percent diagonal line.
2. Drop down vertically to the chemical feed rate of 30 gallons per day, for 1 mg/L fluoride application.
3. Adjust the flow rate for a dose of 1.2 mg/L.

$$\text{Flow Rate, gal/day} = \frac{\left(\text{Flow Rate from Chart, gal/day}\right)\left(\text{Conc Fluoride, mg/L}\right)}{1 \text{ mg/L}}$$

$$= \frac{(30 \text{ gal/day})(1.2 \text{ mg/L})}{1 \text{ mg/L}}$$

$$= 36 \text{ gal/day}$$

13.6B Differences between the volume of acid used from a storage tank and the volume actually fed into the system could be caused by errors in readings, leaks, or equipment malfunctioning.

Answers to questions on page 58.

13.7A Overfeeding should be prevented because no additional benefits result from overfeeding and there is a waste of chemicals and money. Excessive overfeeding could be harmful to consumers.

13.7B If significant overfeeding occurs, the plant should be shut down. The affected mains should be flushed and the local and state health departments notified.

13.8A A fluoridation operation could be shut down for cleaning, adjustments, or due to safety controls.

Answers to questions on page 60.

13.9A If fluoridation equipment is going to be shut down for an extended length of time, it should be cleaned out to prevent corrosion or the solidifying of the chemical. Lines and equipment could be damaged when restarted if chemicals left in them solidify.

13.10A Fluoride dust can be removed from gears by the use of a vacuum cleaner.

13.10B During maintenance of fluoridation equipment, you do not need to be too concerned about checking the feed rate by catching a given amount of fluoride over a time period. The log will show long-period discrepancies and the daily laboratory tests will indicate any drifting from the desired fluoride concentration in the treated water. Analyze the fluoridated water daily. Check the results for any deviations from the norm and take corrective action. Handheld colorimeters are available for measuring fluoride in water.

13.11A If you are acutely poisoned by a fluoride chemical, you may be thirsty, vomit, and have stomach cramps, diarrhea, difficulty in speaking, and disturbed color vision.

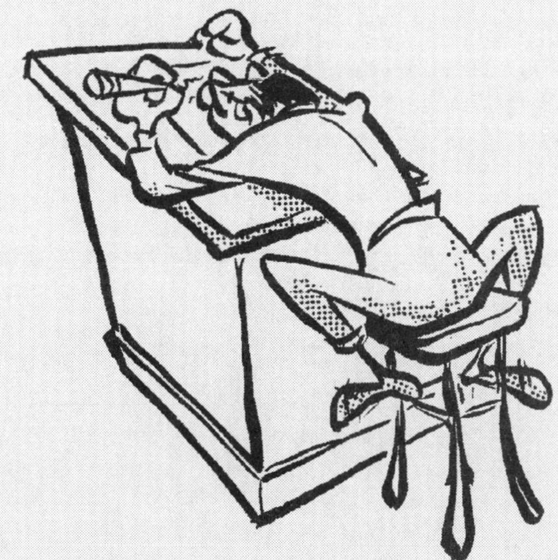

CHAPTER 14

SOFTENING

by

Don Gibson

and

Marty Reynolds

TABLE OF CONTENTS

Chapter 14. SOFTENING

ION EXCHANGE SOFTENING by Marty Reynolds

LESSON 2

LEARNING OBJECTIVES

Chapter 14. SOFTENING

Following completion of Chapter 14, you should be able to:

1. Explain what makes water hard and the advantages of softening.
2. Describe the processes used to soften water.
3. Prepare chemical doses to soften water with considerations given to coagulants and stability.
4. Safely handle softening chemicals.
5. Dispose of process sludges and brines.
6. Keep neat and accurate softening records.
7. Perform jar tests and apply results.
8. Operate and maintain chemical precipitation and ion exchange softening processes.
9. Start up and shut down water softening units.
10. Blend softened waters with unsoftened waters (split treatment) for delivery to consumers.

WORDS

Chapter 14. SOFTENING

ALKALINITY (AL-kuh-LIN-it-tee) ALKALINITY

The capacity of water or wastewater to neutralize acids. This capacity is caused by the water's content of carbonate, bicarbonate, hydroxide, and occasionally borate, silicate, and phosphate. Alkalinity is expressed in milligrams per liter of equivalent calcium carbonate. Alkalinity is not the same as pH because water does not have to be strongly basic (high pH) to have a high alkalinity. Alkalinity is a measure of how much acid must be added to a liquid to lower the pH to 4.5.

ANION (AN-EYE-en) ANION

A negatively charged ion in an electrolyte solution, attracted to the anode under the influence of a difference in electrical potential. Chloride ion (Cl^-) is an anion.

CALCIUM CARBONATE ($CaCO_3$) EQUILIBRIUM CALCIUM CARBONATE ($CaCO_3$) EQUILIBRIUM

A water is considered stable when it is just saturated with calcium carbonate. In this condition, the water will neither dissolve nor deposit calcium carbonate. Thus, in this water the calcium carbonate is in equilibrium with the hydrogen ion concentration.

CALCIUM CARBONATE ($CaCO_3$) EQUIVALENT CALCIUM CARBONATE ($CaCO_3$) EQUIVALENT

An expression of the concentration of specified constituents in water in terms of their equivalent value to calcium carbonate. For example, the hardness in water that is caused by calcium, magnesium, and other ions is usually described as calcium carbonate equivalent. Alkalinity test results are usually reported as mg/L $CaCO_3$ equivalents. To convert chloride to $CaCO_3$ equivalents, multiply the concentration of chloride ions in mg/L by 1.41, and for sulfate, multiply by 1.04.

CATION (KAT-EYE-en) CATION

A positively charged ion in an electrolyte solution, attracted to the cathode under the influence of a difference in electrical potential. Sodium ion (Na^+) is a cation.

DIVALENT (dye-VAY-lent) DIVALENT

Having a valence of two, such as the ferrous ion, Fe^{2+}. Also called bivalent.

EQUIVALENT WEIGHT EQUIVALENT WEIGHT

That weight that will react with, displace, or is equivalent to 1 gram atom of hydrogen.

HARDNESS, WATER HARDNESS, WATER

A characteristic of water caused mainly by the salts of calcium and magnesium, such as bicarbonate, carbonate, sulfate, chloride, and nitrate. Excessive hardness in water is undesirable because it causes the formation of soap curds, increased use of soap, deposition of scale in boilers, damage in some industrial processes, and sometimes causes objectionable tastes in drinking water.

HARD WATER HARD WATER

Water having a high concentration of calcium and magnesium ions. A water may be considered hard if it has a hardness greater than the typical hardness of water from the region. Some textbooks define hard water as a water with a hardness of more than 100 mg/L as calcium carbonate.

HYDRATED LIME — HYDRATED LIME

Limestone that has been burned and treated with water under controlled conditions until the calcium oxide portion has been converted to calcium hydroxide ($Ca(OH)_2$). Hydrated lime is quicklime combined with water. $CaO + H_2O \rightarrow Ca(OH)_2$. Also called slaked lime. Also see QUICKLIME.

INSOLUBLE (in-SAWL-yoo-bull) — INSOLUBLE

Something that cannot be dissolved.

ION — ION

An electrically charged atom, radical (such as SO_4^{2-}), or molecule formed by the loss or gain of one or more electrons.

ION EXCHANGE — ION EXCHANGE

A water or wastewater treatment process involving the reversible interchange (switching) of ions between the water being treated and the solid resin contained within an ion exchange unit. Undesirable ions are exchanged with acceptable ions on the resin or recoverable ions in the water being treated are exchanged with other acceptable ions on the resin.

ION EXCHANGE RESINS — ION EXCHANGE RESINS

Insoluble polymers, used in water or wastewater treatment, that are capable of exchanging (switching or giving) acceptable cations or anions to the water being treated for less desirable ions or for ions to be recovered.

LANGELIER INDEX (LI) — LANGELIER INDEX (LI)

An index reflecting the equilibrium pH of a water with respect to calcium and alkalinity. This index is used in stabilizing water to control both corrosion and the deposition of scale.

$$\text{Langelier Index} = pH - pH_s$$

where pH = actual pH of the water

pH_s = pH at which water having the same alkalinity and calcium content is just saturated with calcium carbonate

METHYL ORANGE ALKALINITY — METHYL ORANGE ALKALINITY

A measure of the total alkalinity in a water sample. The alkalinity is measured by the amount of standard sulfuric acid required to lower the pH of the water to a pH level of 4.5, as indicated by the change in color of methyl orange from orange to pink. Methyl orange alkalinity is expressed as milligrams per liter equivalent calcium carbonate.

NPDES PERMIT — NPDES PERMIT

National Pollutant Discharge Elimination System permit is the regulatory agency document issued by either a federal or state agency that is designed to control all discharges of potential pollutants from point sources and stormwater runoff into US waterways. NPDES permits regulate discharges into US waterways from all point sources of pollution, including industries, municipal wastewater treatment plants, sanitary landfills, large animal feedlots, and return irrigation flows.

pH (pronounce as separate letters) — pH

pH is an expression of the intensity of the basic or acidic condition of a liquid. Mathematically, pH is the logarithm (base 10) of the reciprocal of the hydrogen ion activity.

$$pH = \text{Log}\,\frac{1}{\{H^+\}}$$

If $\{H^+\} = 10^{-6.5}$, then pH = 6.5. The pH may range from 0 to 14, where 0 is most acidic, 14 most basic, and 7 neutral.

PHENOLPHTHALEIN (FEE-nol-THAY-leen) ALKALINITY — PHENOLPHTHALEIN ALKALINITY

The alkalinity in a water sample measured by the amount of standard acid required to lower the pH to a level of 8.3, as indicated by the change in color of phenolphthalein from pink to clear. Phenolphthalein alkalinity is expressed as milligrams per liter of equivalent calcium carbonate.

PRECIPITATE (pre-SIP-uh-TATE) PRECIPITATE

(1) An insoluble, finely divided substance that is a product of a chemical reaction within a liquid.

(2) The separation from solution of an insoluble substance.

QUICKLIME QUICKLIME

A material that is mostly calcium oxide (CaO) or calcium oxide in natural association with a lesser amount of magnesium oxide. Quicklime is capable of combining with water, that is, becoming slaked. Also see HYDRATED LIME.

RECARBONATION (re-kar-bun-NAY-shun) RECARBONATION

A process in which carbon dioxide is bubbled into the water being treated to lower the pH. The pH may also be lowered by the addition of acid. Recarbonation is the final stage in the lime–soda ash softening process. This process converts carbonate ions to bicarbonate ions and stabilizes the solution against the precipitation of carbonate compounds.

RESINS RESINS

See ION EXCHANGE RESINS.

SATURATION SATURATION

The condition of a liquid (water) when it has taken into solution the maximum possible quantity of a given substance at a given temperature and pressure.

SLAKE SLAKE

To mix with water so that a true chemical combination (hydration) takes place, such as in the slaking of lime.

SLAKED LIME SLAKED LIME

See HYDRATED LIME.

SUPERSATURATED SUPERSATURATED

An unstable condition of a solution (water) in which the solution contains a substance at a concentration greater than the saturation concentration for the substance.

TITRATE (TIE-trate) TITRATE

To titrate a sample, a chemical solution of known strength is added drop by drop until a certain color change, precipitate, or pH change in the sample is observed (end point). Titration is the process of adding the chemical reagent in small increments (0.1–1.0 milliliter) until completion of the reaction, as signaled by the end point.

CHAPTER 14. SOFTENING

Lime–Soda Ash Softening by Don Gibson

(Lesson 1 of 2 Lessons)

14.0 WHAT MAKES WATER HARD?[1]

Water hardness is a measure of the soap- or detergent-consuming power of water. Technically, hardness is caused by *DIVALENT*[2] metallic cations that are capable of reacting with soap (detergent) to form precipitates and with certain anions present in water to form scale.

Cations Causing Hardness	Most Common Anions
Calcium, Ca^{2+}	Bicarbonate, HCO_3^-
Magnesium, Mg^{2+}	Sulfate, SO_4^{2-}
Strontium, Sr^{2+}	Chloride, Cl^-
Iron, Fe^{2+}	Nitrate, NO_3^-
Manganese, Mn^{2+}	Silicate, SiO_3^{2-}

Calcium and magnesium are usually the only cations that are present in significant concentrations. Therefore, hardness is generally considered to be an expression of the total concentration of the calcium and magnesium ions that are present in the water. However, if any of the other cations listed are present in significant amounts, they should be included in the hardness determination.

Table 14.1 describes various levels of hardness. Different textbooks will use similar classifications. Hardness levels in source waters, local conditions, and local usage will influence consumers' attitudes toward the hardness of their water.

To help you understand this chapter on water softening, some of the terms used are defined below.

HARD WATER is a water having a high concentration of calcium and magnesium ions. A water may be considered hard if it has a hardness greater than the typical hardness of water from the region. Some textbooks define hard water as a water with a hardness of more than 100 mg/L as calcium carbonate.

HARDNESS is a characteristic of water caused mainly by the salts of calcium and magnesium, such as bicarbonate, carbonate, sulfate, chloride, and nitrate. Excessive hardness in water is undesirable because it causes the formation of soap curds, increased use of soap, deposition of scale in boilers, damage in some industrial processes, and sometimes causes objectionable tastes in drinking water.

TABLE 14.1 DESCRIPTION OF VARIOUS LEVELS OF HARDNESS[a]

Description	Hardness in Terms of mg/L as Calcium Carbonate
1. Extremely soft to soft	0–45
2. Soft to moderately hard	46–90
3. Moderately hard to hard	91–130
4. Hard to very hard	131–170
5. Very hard to excessively hard	171–250
6. Too hard for ordinary domestic use	Over 250

a From L. A. Lipe and M. D. Curry, "Ion Exchange Water Softening," a discussion for water treatment plant operators, 1974–75 seminar series sponsored by Illinois Environmental Protection Agency.

CALCIUM HARDNESS is caused by calcium ions (Ca^{2+}).

MAGNESIUM HARDNESS is caused by magnesium ions (Mg^{2+}).

TOTAL HARDNESS is the sum of the hardness caused by both calcium and magnesium ions.

CARBONATE HARDNESS is caused by the alkalinity present in water up to the total hardness. This value is usually less than the total hardness.

NONCARBONATE HARDNESS is that portion of the total hardness in excess of the alkalinity.

ALKALINITY (AL-kuh-LIN-it-tee) is the capacity of water or wastewater to neutralize acids. This capacity is caused by the water's content of carbonate, bicarbonate, hydroxide, and occasionally borate, silicate, and phosphate. Alkalinity is expressed in milligrams per liter of equivalent calcium carbonate. Alkalinity is not the same as pH because water does not have to be strongly basic (high pH) to have a high alkalinity. Alkalinity is a measure of how much acid must be added to a liquid to lower the pH to 4.5.

1. Portions of the material covered in the first three sections of this chapter were provided by Don Gibson, Marty Reynolds, Susumu Kawamura, Terry Engelhardt, Jack Rossum, and Mike Curry.
2. *Divalent* (dye-VAY-lent). Having a valence of two, such as the ferrous ion, Fe^{2+}. Also called bivalent.

CALCIUM CARBONATE ($CaCO_3$) EQUIVALENT is an expression of the concentration of specified constituents in water in terms of their equivalent value to calcium carbonate. For example, the hardness in water that is caused by calcium, magnesium, and other ions is usually described as calcium carbonate equivalent. Alkalinity test results are usually reported as mg/L $CaCO_3$ equivalents. To convert chloride to $CaCO_3$ equivalents, multiply the concentration of chloride ions in mg/L by 1.41, and for sulfate, multiply by 1.04.

14.1 WHY SOFTEN WATER?

The dissolved minerals (calcium and magnesium ions) in water cause difficulties in doing the laundry and in dishwashing in the household. These ions also cause a coating to form inside the hot water heater similar to that in a tea kettle after repeated use.

Hardness, in addition to inhibiting the cleaning action of soaps, will tend to shorten the life of fabrics that are washed in hard water. The scum or curds may become lodged in the fibers of the fabric and cause them to lose their softness and elasticity.

In industry, hardness can cause even greater problems. Many processes are affected by the hardness content of the water used. Industrial plants using boilers for processing steam or heat must remove the hardness from their makeup water, even beyond what a water treatment plant would do. The reason for this is that the minerals will plate out on the boiler tubes and form a scale. This scale forms an insulation barrier that prevents proper heat transfer, thus causing excessive energy requirements to fire the boilers. The problems associated with process water softening are too numerous to go into; however, everything from food processing to intricate manufacturing processes is affected by the hardness of water.

In addition to the removal of hardness from water, some other benefits of softening include:

1. Removal of iron and manganese
2. Control of corrosion when proper stabilization of water is achieved
3. Disinfection due to high pH values when using lime (especially the excess lime softening process)
4. Sometimes a reduction in tastes and odors
5. Reduction of some total solids content by the lime treatment process
6. Removal of radioactivity

Possible limitations of softening might include:

1. Free chlorine residual is predominantly hypochlorite at pH levels above 7.5 and is a less powerful disinfectant.
2. Costs and benefits must be carefully weighed to justify softening.
3. Ultimate disposal of process wastes.
4. At the pH levels associated with softening chemical precipitation, the trihalomethane fraction in the treated water may increase (depends on several other factors).
5. Production of an aggressive water that would tend to corrode metal ions from the distribution system piping. Hard waters usually do not corrode pipe. However, excessively hard water can cause scaling on the inside of pipes and thereby restrict flow.

In many cases, the decision to soften the water is left up to each community as softening is done mostly as a customer service. Hard water does not have an adverse effect on health, but can create several unwanted side effects:

1. Over a period of time, the detergent-consuming power of hard water can be very costly.
2. Scale problems on fixtures will be more noticeable.
3. The lifetime of several types of fabrics will be reduced with repeated washing in hard water. Also, a residue can be left in clothing, creating a dirty appearance.

Once the decision is made to soften, a method must be selected. The two most common methods used to soften water are chemical precipitation (lime–soda ash) and *ION EXCHANGE*.[3] Ion exchange softening can best be applied to waters high in noncarbonate hardness and where the total hardness does not exceed 350 mg/L. This method of softening can produce a water of zero hardness, as opposed to lime softening where zero hardness cannot be reached.

Ion exchange softening will also remove noncarbonate hardness without the addition of soda ash, which is required with lime softening. Ion exchange is a nonselective method of softening. This means it will remove total hardness (the sum of carbonate and noncarbonate hardness) making it a very desirable means of water softening.

Limitations of the ion exchange softening process include an increase in the sodium content of the softened water if the ion exchanger is regenerated with sodium chloride. The sodium level should not exceed 20 mg/L in treated water because of the potentially harmful effect on persons susceptible to hypertension. Also, the ultimate disposal of spent brine and rinse waters from softeners can be a major problem for many installations.

QUESTIONS

Please write your answers to the following questions and compare them with those on page 115.

14.0A What causes hardness in water?

14.0B Why is excessive hardness undesirable in a domestic water supply?

14.1A What are some of the limitations of the ion exchange softening process?

3. *Ion Exchange.* A water or wastewater treatment process involving the reversible interchange (switching) of ions between the water being treated and the solid resin contained within an ion exchange unit. Undesirable ions are exchanged with acceptable ions on the resin or recoverable ions in the water being treated are exchanged with other acceptable ions on the resin.

14.2 CHEMISTRY OF SOFTENING

To understand how water hardness is described and how hardness is removed from water by softening processes, operators need to have an idea of the chemical reactions that take place in water. In this section, hardness, pH, and alkalinity reactions in water will be discussed.

14.20 Hardness

Hardness is due to the presence of divalent metallic cations in water, but the 22nd edition of *Standard Methods*[4] identifies only calcium and magnesium as hardness constituents. Hardness is a factor commonly measured by *TITRATION*[5] as described in Volume I (see Chapter 11, Section 11.3, "5. Hardness"). Individual divalent cations may be measured in the laboratory using an atomic absorption (AA) spectrophotometer for very accurate work.

Hardness is usually reported as *CALCIUM CARBONATE ($CaCO_3$) EQUIVALENT.*[6] This procedure allows us to combine or add up the hardness caused by both calcium and magnesium and report the results as total hardness.

$$\text{Calcium Hardness, mg/L as } CaCO_3 = (\text{Calcium, mg/L})\left(\frac{\text{Equivalent Weight of } CaCO_3}{\text{Equivalent Weight of Calcium}}\right)$$

$$= (\text{Ca, mg/L})\left(\frac{50}{20}\right)$$

$$= 2.50(\text{Ca, mg/L})$$

This equation indicates that if the calcium concentration in milligrams per liter is multiplied by 2.50, the result is the calcium hardness in milligrams per liter as calcium carbonate. The *EQUIVALENT WEIGHT*[7] of most elements or chemical radicals (SO_4^{2-} is a radical) can be obtained by dividing the molecular weight by the valence.

$$\text{Equivalent Weight of Calcium} = \frac{\text{Atomic Weight}}{\text{Valence}}$$

$$= \frac{40}{2}$$

$$= 20$$

To express the magnesium hardness of water as calcium carbonate equivalent, use the following formula.

$$\text{Magnesium Hardness, mg/L as } CaCO_3 = (\text{Magnesium, mg/L})\left(\frac{\text{Equivalent Weight of } CaCO_3}{\text{Equivalent Weight of Magnesium}}\right)$$

$$= (\text{Mg, mg/L})\left(\frac{50}{12.15}\right)$$

$$= 4.12(\text{Mg, mg/L})$$

The total hardness of water is the sum of the calcium and magnesium hardness as $CaCO_3$.

$$\text{Total Hardness, mg/L as } CaCO_3 = \text{Calcium Hardness, mg/L as } CaCO_3 + \text{Magnesium Hardness, mg/L as } CaCO_3$$

Example 1

Determine the total hardness as $CaCO_3$ for a sample of water with a calcium content of 30 mg/L and a magnesium content of 20 mg/L.

Known		Unknown
Calcium, mg/L	= 30 mg/L	Total Hardness, mg/L as $CaCO_3$
Magnesium, mg/L	= 20 mg/L	

Calculate the total hardness as milligrams per liter of calcium carbonate equivalent.

$$\text{Total Hardness, mg/L as } CaCO_3 = \text{Calcium Hardness, mg/L as } CaCO_3 + \text{Magnesium Hardness, mg/L as } CaCO_3$$

$$= 2.50(\text{Ca, mg/L}) + 4.12(\text{Mg, mg/L})$$

$$= 2.50(30 \text{ mg/L}) + 4.12(20 \text{ mg/L})$$

$$= 75 \text{ mg/L} + 82.4 \text{ mg/L}$$

$$= 157.4 \text{ mg/L as } CaCO_3$$

Total hardness is also described as the sum of the carbonate hardness (temporary hardness) and noncarbonate hardness (permanent hardness).

$$\text{Total Hardness, mg/L as } CaCO_3 = \text{Carbonate Hardness, mg/L as } CaCO_3 + \text{Noncarbonate Hardness, mg/L as } CaCO_3$$

---

4. *Standard Methods for the Examination of Water and Wastewater,* 22nd Edition. Obtain from American Water Works Association (AWWA) at www.awwa.org or phone (800) 926-7337. Order No. 10085. ISBN 978-0-87553-013-0.
5. *Titrate* (TIE-trate). To titrate a sample, a chemical solution of known strength is added drop by drop until a certain color change, precipitate, or pH change in the sample is observed (end point). Titration is the process of adding the chemical reagent in small increments (0.1–1.0 milliliter) until completion of the reaction, as signaled by the end point.
6. *Calcium Carbonate* ($CaCO_3$) *Equivalent.* An expression of the concentration of specified constituents in water in terms of their equivalent value to calcium carbonate. For example, the hardness in water that is caused by calcium, magnesium, and other ions is usually described as calcium carbonate equivalent. Alkalinity test results are usually reported as mg/L $CaCO_3$ equivalents. To convert chloride to $CaCO_3$ equivalents, multiply the concentration of chloride ions in mg/L by 1.41, and for sulfate, multiply by 1.04.
7. *Equivalent Weight.* That weight that will react with, displace, or is equivalent to 1 gram atom of hydrogen. The equivalent weight of an element (such as Ca^{2+}) is equal to the atomic weight divided by the valence.

$$\text{Equivalent Weight of } CaCO_3 = \frac{\text{Molecular Weight}}{\text{Valence}}$$

$$= \frac{100}{2}$$

$$= 50$$

The amount of carbonate and noncarbonate hardness depends on the alkalinity of the water. This relationship can be described as follows.

1. When the alkalinity (expressed as calcium carbonate equivalent) is greater than the total hardness, all the hardness is in the carbonate form.

   $$\text{Carbonate Hardness, mg/L as } CaCO_3 = \text{Total Hardness, mg/L as } CaCO_3$$

2. When the total hardness is greater than the alkalinity, the alkalinity is carbonate hardness, and noncarbonate hardness is the difference between total hardness and alkalinity.

   $$\text{Carbonate Hardness, mg/L as } CaCO_3 = \text{Alkalinity, mg/L as } CaCO_3$$

   $$\text{Noncarbonate Hardness, mg/L as } CaCO_3 = \text{Total Hardness, mg/L as } CaCO_3 - \text{Alkalinity, mg/L as } CaCO_3$$

14.21 pH

pH is an expression of the intensity of the basic or acidic condition of a liquid. Mathematically, pH is the logarithm (base 10) of the reciprocal of the hydrogen ion activity.

$$pH = \text{Log} \frac{1}{\{H^+\}}$$

If $\{H^+\} = 10^{-6.5}$, then pH = 6.5. The pH may range from 0 to 14, where 0 is most acidic, 14 most basic, and 7 neutral. Natural waters usually have a pH between 6.5 and 8.5.

Table 14.2 shows the relationship between pH and hydrogen and hydroxide ions.

TABLE 14.2 RELATIONSHIP BETWEEN pH AND HYDROGEN AND HYDROXIDE IONS

pH	Hydrogen Ion (H^+), Moles/Liter	Hydroxide Ion (OH^-), Moles/Liter
0	1.0	0.000 000 000 000 01
1	0.1	0.000 000 000 000 1
2	0.01	0.000 000 000 001
3	0.001	0.000 000 000 01
4	0.000 1	0.000 000 000 1
5	0.000 01	0.000 000 001
6	0.000 001	0.000 000 01
7	0.000 000 1	0.000 000 1
8	0.000 000 01	0.000 001
9	0.000 000 001	0.000 01
10	0.000 000 000 1	0.000 1
11	0.000 000 000 01	0.001
12	0.000 000 000 001	0.01
13	0.000 000 000 000 1	0.1
14	0.000 000 000 000 01	1.0

When treating waters, the pH is very important. The pH of water may be increased or decreased by the addition of certain chemicals used to treat water (Table 14.3). In many instances, the effect on pH of adding one chemical is neutralized by the addition of another chemical. When softening water by chemical precipitation processes (lime–soda softening for example), the pH must be raised to 11 for the desired chemical reactions to occur. The levels of carbon dioxide, bicarbonate ion, and carbonate ion in waters are very sensitive to pH.

TABLE 14.3 INFLUENCE OF WATER TREATMENT CHEMICALS ON pH

Lowers pH	Increases pH
Aluminum Sulfate (Alum), $Al_2(SO_4)_3 \cdot 18H_2O$	Calcium Hypochlorite, $Ca(OCl)_2$
Carbon Dioxide, CO_2	Caustic Soda, NaOH
Chlorine, Cl_2	Hydrated Lime, $Ca(OH)_2$
Ferric Chloride, $FeCl_3$	Soda Ash, Na_2CO_3
Hydrofluosilicic Acid, H_2SiF_6	Sodium Aluminate, $NaAlO_2$
Sulfuric Acid, H_2SO_4	Sodium Hypochlorite, NaOCl

The stability of treated water is determined by measuring the pH and calculating the "Langelier Index" (see Volume I, Chapter 8, Section 8.322, "Calcium Carbonate Saturation"). This index reflects the equilibrium pH of a water with respect to calcium and alkalinity.

$$\text{Langelier Index (LI)} = pH - pH_s$$

where pH = actual pH of the water

pH_s = pH at which water having the same alkalinity and calcium content is just saturated with calcium carbonate

A negative Langelier Index indicates that the water is corrosive and a positive index indicates that the water is scale forming. After water has been softened, the treated water distributed to consumers must be stable (neither corrosive nor scale forming).

14.22 Alkalinity

Alkalinity is the capacity of water or wastewater to neutralize acids. This capacity is caused by the water's content of carbonate, bicarbonate, hydroxide, and occasionally borate, silicate, and phosphate. Alkalinity is expressed in milligrams per liter of equivalent calcium carbonate. Alkalinity is not the same as pH because water does not have to be strongly basic (high pH) to have a high alkalinity. Alkalinity is a measure of how much acid must be added to a liquid to lower the pH to 4.5.

Alkalinity is measured in the laboratory by the addition of color indicator solutions and the alkalinity is then determined by the amount of acid required to reach a titration end point (specific color change) (see Volume I, Chapter 11, "Laboratory

Procedures," Section 11.3, "1. Alkalinity"). The P (phenolphthalein) end point is at pH 8.3. When the pH is below 8.3, there is no P alkalinity present. When the pH is above 8.3, P alkalinity is present. No carbon dioxide is present when the pH is above 8.3, so there is no carbon dioxide in the water when P alkalinity is present. Also, hydroxide and carbonate alkalinity are not present when pH is below 8.3.

The relationship between the various alkalinity constituents (bicarbonate (HCO_3^-), carbonate (CO_3^{2-}), and hydroxide (OH^-)) can be based on the P (phenolphthalein) and T (total or methyl orange) alkalinity as shown in Table 14.4 and Figure 14.1.

TABLE 14.4 ALKALINITY CONSTITUENTS

	Alkalinity, mg/L as $CaCO_3$		
Titration Result	**Bicarbonate**	**Carbonate**	**Hydroxide**
P = 0	T	0	0
P is less than ½T	T–2P	2P	0
P = ½T	0	2P	0
P is greater than ½T	0	2T–2P	2P–T
P = T	0	0	T

where P = phenolphthalein alkalinity
T = total alkalinity

When the pH is less than 8.3, all alkalinity is in the bicarbonate form and is commonly referred to as natural alkalinity. When the pH is above 8.3, the alkalinity may consist of bicarbonate, carbonate, and hydroxide. As the pH increases, the alkalinity progressively shifts to carbonate and hydroxide forms.

Total alkalinity is the sum of the bicarbonate, carbonate, and hydroxide. Each of these values can be determined by measuring the P and T alkalinity in the laboratory and referring to Table 14.4. Alkalinity is expressed in milligrams per liter as calcium carbonate equivalents. Alkalinity is influenced by chemicals used to treat water as shown in Table 14.5.

TABLE 14.5 INFLUENCE OF WATER TREATMENT CHEMICALS ON ALKALINITY

Lowers Alkalinity	Increases Alkalinity
Aluminum Sulfate (Alum), $Al_2(SO_4)_3 \cdot 18H_2O$	Calcium Hypochlorite, $Ca(OCl)_2$
Carbon Dioxide, CO_2	Caustic Soda, NaOH
Chlorine Gas, Cl_2	Hydrated Lime, $Ca(OH)_2$
Ferric Chloride, $FeCl_3$	Soda Ash, Na_2CO_3
Ferric Sulfate, $Fe_2(SO_4)_3$	Sodium Aluminate, $NaAlO_2$
Sulfuric Acid, H_2SO_4	

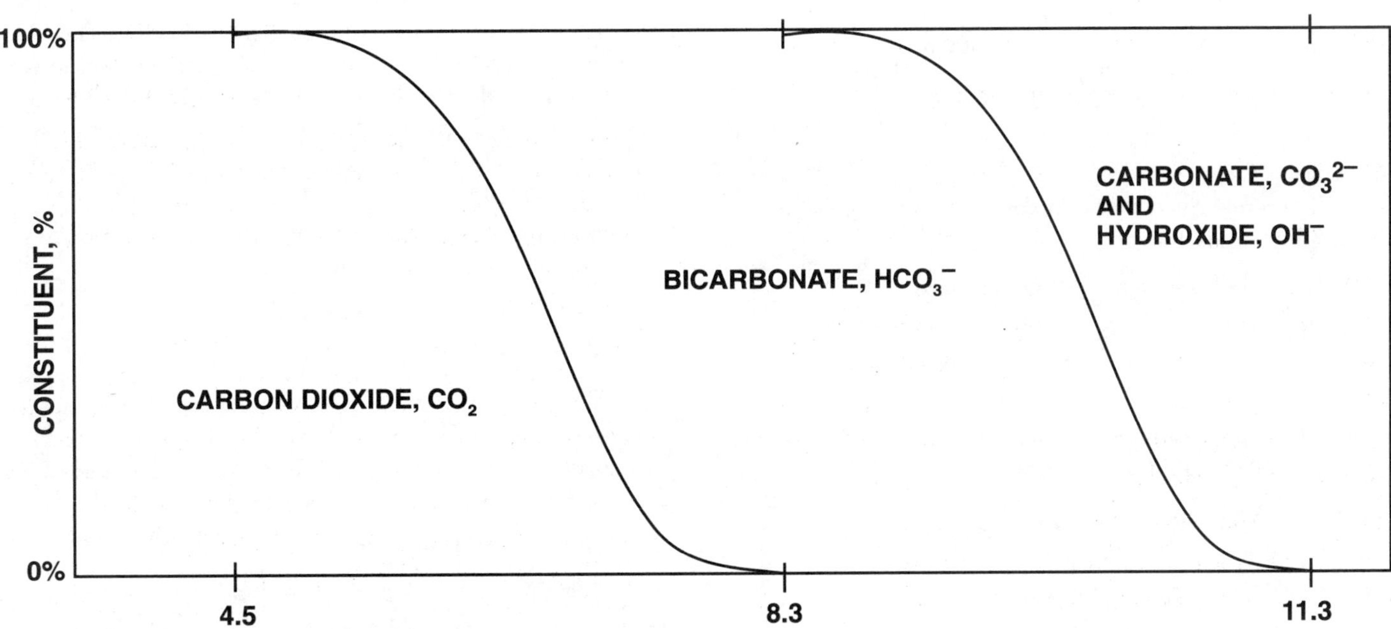

Fig. 14.1 Relationship between pH and alkalinity constituents (CO_2, HCO_3^-, CO_3^{2-}, and OH^-)

Example 2

Results from alkalinity titrations on a raw water sample were as follows:

Known

Sample Size, mL	=	100 mL
mL Titrant Used to pH 8.3, A	=	0 mL
Total mL of Titrant Used, B	=	8.2 mL
Acid Normality, N	=	$0.02\ N\ H_2SO_4$

Unknown

1. Total Alkalinity, mg/L as $CaCO_3$
2. Bicarbonate Alkalinity, mg/L as $CaCO_3$
3. Carbonate Alkalinity, mg/L as $CaCO_3$
4. Hydroxide Alkalinity, mg/L as $CaCO_3$

See Volume I, Chapter 11, "Laboratory Procedures," Section 11.3, "1. Alkalinity" for details and formulas.

1. Calculate the phenolphthalein alkalinity in mg/L as $CaCO_3$.

$$\text{Phenolphthalein Alkalinity, mg/L as } CaCO_3 = \frac{A \times N \times 50{,}000}{\text{mL of Sample}}$$

$$= \frac{(0\text{ mL}) \times (0.02\ N) \times (50{,}000)}{100\text{ mL}}$$

$$= 0\text{ mg/L as } CaCO_3$$

2. Calculate the total alkalinity in mg/L as $CaCO_3$.

$$\text{Total Alkalinity, mg/L as } CaCO_3 = \frac{B \times N \times 50{,}000}{\text{mL of Sample}}$$

$$= \frac{(8.2\text{ mL}) \times (0.02\ N) \times (50{,}000)}{100\text{ mL}}$$

$$= 82\text{ mg/L as } CaCO_3$$

3. Refer to Table 14.4 for alkalinity constituents. The first row indicates that since P = 0, the total alkalinity is equal to the bicarbonate alkalinity.

$$\text{Bicarbonate Alkalinity, mg/L as } CaCO_3 = \text{Total Alkalinity, mg/L as } CaCO_3$$

$$= 82\text{ mg/L as } CaCO_3$$

The first row also indicates that since P = 0, the carbonate and hydroxide alkalinities are also zero.

$$\text{Carbonate Alkalinity, mg/L as } CaCO_3 = 0\text{ mg/L as } CaCO_3$$

$$\text{Hydroxide Alkalinity, mg/L as } CaCO_3 = 0\text{ mg/L as } CaCO_3$$

QUESTIONS

Please write your answers to the following questions and compare them with those on pages 115 and 116.

14.2A What laboratory procedures are used to measure hardness?

14.2B Determine the total hardness as $CaCO_3$ for a sample of water with a calcium content of 25 mg/L and a magnesium content of 14 mg/L.

14.2C Which water treatment chemicals lower the pH when added to water?

14.2D Results from alkalinity titrations on a sample of water were as follows: sample size, 100 mL; mL titrant used to pH 8.3, 1.2 mL; total mL of titrant used, 5.6 mL; and the acid normality was $0.02\ N\ H_2SO_4$. Calculate the total, bicarbonate, carbonate, and hydroxide alkalinity as $CaCO_3$.

14.3 HOW WATER IS SOFTENED

14.30 Basic Methods of Softening

The two basic methods of softening a municipal water supply are chemical precipitation and ion exchange. Ion exchange will be discussed in the second portion of this chapter in Sections 14.10 through 14.21, pages 99 through 114. We will begin here with the chemical precipitation methods, mainly lime–soda ash softening and variations of this process.

Hardness is not completely removed by the chemical precipitation methods used in water treatment plants. That is, hardness is not reduced to zero. Water having a hardness of 150 mg/L as $CaCO_3$ or more is usually treated to reduce the hardness to 80 to 90 mg/L when softening is chosen as a water treatment option.

The minimum hardness that can be achieved by the lime–soda ash process is around 30 to 40 mg/L as $CaCO_3$. The effluent from an ion exchange softener could contain almost zero hardness. Regardless of the method used to soften water, consumers usually receive a blended water with a hardness of around 80 to 90 mg/L as $CaCO_3$ when softening is used in water treatment plants.

Lime–soda softening may produce benefits in addition to the softening of water. These advantages include:

1. Removal of iron and manganese
2. Reduction of solids
3. Removal and inactivation of bacteria and viruses due to high pH
4. Control of corrosion and scale formation with proper stabilization of treated water
5. Removal of excess fluoride

Limitations of the lime–soda softening process include:

1. Inability to remove all hardness

2. A high degree of operator control must be exercised for maximum efficiency in cost, hardness removal, and water stability
3. Color removal may be complicated by the softening process due to high pH levels
4. Large quantities of sludge are created that must be handled and disposed of in an acceptable manner

QUESTIONS

Please write your answers to the following questions and compare them with those on page 116.

14.3A What is the minimum hardness that can be achieved by the lime–soda ash process?

14.3B List some of the benefits that could result from the lime–soda softening process in addition to softening the water.

14.31 Chemical Reactions

In the chemical precipitation process, the hardness-causing ions are converted from soluble to insoluble forms. Calcium and magnesium become less soluble as the pH increases. Therefore, calcium and magnesium can be removed from water as insoluble precipitates at high pH levels.

Addition of lime to water increases the hydroxide concentrations, thus increasing the pH. Addition of lime to water also converts alkalinity from the bicarbonate form to the carbonate form, which causes the calcium to be precipitated as calcium carbonate ($CaCO_3$). As additional lime is added to the water, the phenolphthalein (P) alkalinity increases to a level where hydroxide becomes present (excess causticity) allowing magnesium to precipitate as magnesium hydroxide.

Following the chemical softening process, the pH is high and the water is *SUPERSATURATED*[8] with excess caustic alkalinity in either the hydroxide or carbonate form. Carbon dioxide can be used to decrease the causticity and scale-forming tendencies of the water prior to filtration.

The chemical reactions that take place in water during the chemical precipitation process are described in the remainder of this section. The procedures for softening water depend on whether the hardness to be removed is carbonate or noncarbonate hardness. Carbonate hardness (also called "temporary hardness") can be removed by the use of lime only. Removal of noncarbonate hardness (also called "permanent hardness") requires both lime and soda.

14.310 Lime

The lime used in the chemical precipitation softening process may be from either *HYDRATED LIME*[9] ($Ca(OH)_2$, calcium hydroxide, or "slaked" lime) or calcium oxide (CaO, *QUICKLIME*[10] or "unslaked" lime). The hydrated lime may be used directly. The calcium oxide or quicklime must first be *SLAKED*.[11] This involves adding the calcium oxide (CaO) pellets to water and heating to cause "slaking" (the formation of calcium hydroxide ($Ca(OH)_2$)) before use. Small facilities commonly use hydrated lime ($Ca(OH)_2$). Large facilities may find it more economical to use quicklime (CaO) and slake it on site.

14.311 Removal of Carbon Dioxide

The application of lime for the removal of carbonate hardness also removes carbon dioxide. Carbon dioxide does not contribute to hardness and therefore does not need to be removed. However, carbon dioxide will consume a portion of the lime to be used and therefore must be considered. Equation (1) describes the reaction of carbon dioxide with lime.

(1) Carbon Dioxide + Lime → Calcium Carbonate ↓ + Water

$$CO_2 + Ca(OH)_2 \rightarrow CaCO_3 \downarrow + H_2O$$

14.312 Removal of Carbonate Hardness

The equations below describe the removal of carbonate hardness.

(2) Calcium Bicarbonate + Lime → Calcium Carbonate ↓ + Water

$$Ca(HCO_3)_2 + Ca(OH)_2 \rightarrow 2CaCO_3 \downarrow + 2H_2O$$

(3) Magnesium Bicarbonate + Lime → Calcium Carbonate ↓ + Magnesium Carbonate + Water

$$Mg(HCO_3)_2 + Ca(OH)_2 \rightarrow CaCO_3 \downarrow + MgCO_3 + H_2O$$

(4) Magnesium Carbonate + Lime → Calcium Carbonate ↓ + Magnesium Hydroxide ↓

$$MgCO_3 + Ca(OH)_2 \rightarrow CaCO_3 \downarrow + Mg(OH)_2 \downarrow$$

When lime is added to water, any carbon dioxide present is converted to calcium carbonate if enough lime is added (Equation (1)). With the addition of more lime, the calcium bicarbonate will be precipitated as calcium carbonate. To remove both the calcium and magnesium bicarbonate, an excess of lime must be used.

8. *Supersaturated.* An unstable condition of a solution (water) in which the solution contains a substance at a concentration greater than the saturation concentration for the substance.
9. *Hydrated Lime.* Limestone that has been burned and treated with water under controlled conditions until the calcium oxide portion has been converted to calcium hydroxide ($Ca(OH)_2$). Hydrated lime is quicklime combined with water. $CaO + H_2O \rightarrow Ca(OH)_2$. Also called slaked lime. Also see QUICKLIME.
10. *Quicklime.* A material that is mostly calcium oxide (CaO) or calcium oxide in natural association with a lesser amount of magnesium oxide. Quicklime is capable of combining with water, that is, becoming slaked. Also see HYDRATED LIME.
11. *Slake.* To mix with water so that a true chemical combination (hydration) takes place, such as in the slaking of lime.

14.313 Removal of Noncarbonate Hardness

Magnesium noncarbonate hardness requires the addition of both lime and soda ash (sodium carbonate, Na_2CO_3).

(5) Magnesium Sulfate + Lime → Magnesium Hydroxide↓ + Calcium Sulfate

$$MgSO_4 + Ca(OH)_2 \rightarrow Mg(OH)_2\downarrow + CaSO_4$$

(6) Calcium Sulfate + Soda Ash → Calcium Carbonate↓ + Sodium Sulfate

$$CaSO_4 + Na_2CO_3 \rightarrow CaCO_3\downarrow + Na_2SO_4$$

Equation (6) is also one of the equations for the removal of calcium noncarbonate hardness. Similar equations can be written for the removal of noncarbonate hardness caused by calcium and magnesium chloride.

14.314 Stability

The main chemical reaction products from the lime–soda softening process are $CaCO_3\downarrow$ and $Mg(OH)_2\downarrow$. The water thus treated has been chemically changed and is no longer stable because of pH and alkalinity changes. Lime–soda softened water is usually supersaturated with calcium carbonate ($CaCO_3$). The degree of instability and excess calcium carbonate depends on the degree to which the water is softened. Calcium carbonate hardness is removed at a lower pH than magnesium carbonate hardness. If maximum carbonate hardness removal is practiced (thus requiring a high pH to remove the magnesium carbonate hardness), the water will be supersaturated with calcium carbonate and magnesium hydroxide. Under these conditions, deposition of precipitates will occur in filters and pipelines.

Excess lime addition to remove magnesium carbonate hardness results in supersaturated conditions and a residual of lime, which will produce a pH of about 10.9. The excess lime is called caustic alkalinity since it raises the pH. If the pH is then lowered, better precipitation of calcium carbonate and magnesium hydroxide will occur. Alkalinity will be lowered also. This is usually accomplished by pumping carbon dioxide (CO_2) gas into the water. This addition of carbon dioxide to the treated water is called *RECARBONATION.*[12]

Recarbonation may be carried out in two steps. The first addition of carbon dioxide would follow excess lime addition to lower the pH to about 10.4 and encourage the precipitation of calcium carbonate and magnesium hydroxide. A second addition of carbon dioxide after treatment removes noncarbonate hardness. This would again lower the pH to about 9.8 and would encourage precipitation. By carrying out recarbonation prior to filtration, the buildup of excess lime and calcium carbonate and magnesium hydroxide precipitates in the filters will be prevented or minimized. The recarbonation reaction for excess lime removal is shown below.

(7) Calcium Hydroxide + Carbon Dioxide → Calcium Carbonate↓ + Water

$$Ca(OH)_2 + CO_2 \rightarrow CaCO_3\downarrow + H_2O$$

Care must be exercised when using recarbonation. Feeding excess carbon dioxide may result in no lowering of the hardness by causing calcium carbonate precipitates to go back into solution and cause carbonate hardness.

(8) Calcium Carbonate + Carbon Dioxide + Water → Calcium Bicarbonate

$$CaCO_3 + CO_2 + H_2O \rightarrow Ca(HCO_3)_2$$

14.315 Caustic Soda Softening

An alternative method in the lime–soda softening process is the use of sodium hydroxide (NaOH, often called caustic soda) in place of soda ash. The chemical reactions of sodium hydroxide with carbonate and noncarbonate hardness are listed below.

(9) Carbon Dioxide + Sodium Hydroxide → Sodium Carbonate + Water

$$CO_2 + 2NaOH \rightarrow Na_2CO_3 + H_2O$$

(10) Calcium Bicarbonate + Sodium Hydroxide → Calcium Carbonate↓ + Sodium Carbonate + Water

$$Ca(HCO_3)_2 + 2NaOH \rightarrow CaCO_3\downarrow + Na_2CO_3 + 2H_2O$$

(11) Magnesium Bicarbonate + Sodium Hydroxide → Magnesium Hydroxide↓ + Sodium Carbonate + Water

$$Mg(HCO_3)_2 + 4NaOH \rightarrow Mg(OH)_2\downarrow + 2Na_2CO_3 + 2H_2O$$

(12) Magnesium Sulfate + Sodium Hydroxide → Magnesium Hydroxide↓ + Sodium Sulfate

$$MgSO_4 + 2NaOH \rightarrow Mg(OH)_2\downarrow + Na_2SO_4$$

These chemical reactions show that in removing carbon dioxide and carbonate hardness, sodium carbonate (Na_2CO_3, soda ash) is formed, which will react to remove the noncarbonate hardness. Not only will sodium hydroxide substitute for soda ash, but it may replace all or part of the lime ($Ca(OH)_2$) requirement for removal of the carbonate hardness. The use of caustic soda (usually as a 50 percent solution) may have several advantages, including stability in storage, less sludge formation, and ease of handling and storage.

Safe handling procedures for caustic soda must be used at all times. A 50 percent caustic solution is very dangerous. Caustic soda is a strong base and will attack fabrics and leather and cause severe burns to the skin. Rubber gloves, respirator, safety goggles, and a rubber apron must be worn when handling caustic soda. A safety shower and an emergency eye wash station must be readily available at all times.

The decision to use caustic soda rather than soda ash depends on the quality of the source water and the delivered costs of the various chemicals.

QUESTIONS

Please write your answers to the following questions and compare them with those on pages 116 and 117.

14.3C What causes the pH to increase during the lime–soda softening process?

12. *Recarbonation* (re-kar-bun-NAY-shun). A process in which carbon dioxide is bubbled into the water being treated to lower the pH. The pH may also be lowered by the addition of acid. Recarbonation is the final stage in the lime–soda ash softening process. This process converts carbonate ions to bicarbonate ions and stabilizes the solution against the precipitation of carbonate compounds.

14.3D How can the scale-forming tendencies be decreased in water after the chemical softening process?

14.3E What is recarbonation?

14.3F Under what conditions might caustic soda softening be used?

14.316 Calculation of Chemical Dosages

There are several different approaches to calculating chemical doses for the lime–soda softening process. This section illustrates one step-by-step procedure. To use this procedure, you need to obtain a chemical analysis of the water you are softening. From this analysis, obtain the known values for your water similar to the "Knowns" listed in Example 3. Then calculate the dosages of chemicals for your water by following the steps in the example.

To help you understand where some of the numbers come from in the formulas, we have listed the molecular weights of the major chemical components involved in the chemical precipitation softening process.

Quicklime, CaO	= 56
Hydrated Lime, $Ca(OH)_2$	= 74
Magnesium, Mg^{2+}	= 24.3
Carbon Dioxide, CO_2	= 44
Magnesium Hydroxide, $Mg(OH)_2$	= 58.3
Soda Ash, Na_2CO_3	= 106
Alkalinity, as $CaCO_3$	= 100
Hardness, as $CaCO_3$	= 100

FORMULAS

1. The lime dosage for softening can be estimated by using the following formula:

$$\text{Quicklime (CaO) Feed, mg/L} = \frac{(A + B + C + D)1.15}{\text{Purity of Lime, as a decimal}}$$

Where A = CO_2 in source water (mg/L as CO_2)(56/44)

B = Bicarbonate alkalinity removed in softening (mg/L as $CaCO_3$)(56/100)

C = Hydroxide alkalinity in softener effluent (mg/L as $CaCO_3$)(56/100)

D = Magnesium removed in softening (mg/L as Mg^{2+})(56/24.3)

1.15 = Excess lime dosage (using a 15 percent excess)

NOTE: If hydrated lime ($Ca(OH)_2$) is used instead of quicklime, substitute 74 for 56 in A, B, C, and D.

2. The soda ash dosage to remove noncarbonate hardness can be estimated by using the formula below.

$$\text{Soda Ash } (Na_2CO_3) \text{ Feed, mg/L} = (\text{Noncarbonate Hardness, mg/L as } CaCO_3)(106/100)$$

3. The dosage of carbon dioxide required for recarbonation can be estimated using the formula below.

$$\text{Total } CO_2 \text{ Feed, mg/L} = (Ca(OH)_2 \text{ excess, mg/L})(44/74) + (Mg^{2+} \text{ residual, mg/L})(44/24.3)$$

Example 3

Calculate the hydrated lime ($Ca(OH)_2$) with 90 percent purity, soda ash, and carbon dioxide dose requirements in milligrams per liter for the water shown below.

Known

Constituents	Source Water	Softened Water After Recarbonation and Filtration
CO_2, mg/L	= 6 mg/L	= 0 mg/L
Total Alkalinity, mg/L	= 170 mg/L as $CaCO_3$	= 30 mg/L as $CaCO_3$
Total Hardness, mg/L	= 280 mg/L as $CaCO_3$	= 70 mg/L as $CaCO_3$
Mg^{2+}, mg/L	= 21 mg/L	= 3 mg/L
pH	= 7.5	= 8.8
Lime Purity, %	= 90%	

Unknown

1. Hydrated Lime, mg/L
2. Soda Ash, mg/L
3. Carbon Dioxide, mg/L

1. Calculate the hydrated lime ($Ca(OH)_2$) required in milligrams per liter.

$$A = (CO_2\text{, mg/L})(74/44) = (6 \text{ mg/L})(74/44) = 10 \text{ mg/L}$$

$$B = (\text{Alkalinity, mg/L})(74/100) = (170 \text{ mg/L} - 30 \text{ mg/L})(74/100) = 104 \text{ mg/L}$$

$C = 0$ Hydroxide Alkalinity = 0

$$D = (Mg^{2+}\text{, mg/L})(74/24.3) = (21 \text{ mg/L} - 3 \text{ mg/L})(74/24.3) = 55 \text{ mg/L}$$

$$\text{Hydroxide Lime } (Ca(OH)_2) \text{ Feed, mg/L} = \frac{(A + B + C + D)1.15}{\text{Purity of Lime, as a decimal}}$$

$$= \frac{(10 \text{ mg/L} + 104 \text{ mg/L} + 0 + 55 \text{ mg/L})1.15}{0.90}$$

$$= \frac{(169 \text{ mg/L})(1.15)}{0.90}$$

$$= 216 \text{ mg/L}$$

2. Calculate the soda ash required in milligrams per liter.

$$\text{Total Hardness Removed, mg/L as } CaCO_3 = \text{Total Hardness, mg/L as } CaCO_3 - \text{Total Hardness Remaining, mg/L as } CaCO_3$$

$$= 280 \text{ mg/L} - 70 \text{ mg/L}$$

$$= 210 \text{ mg/L as } CaCO_3$$

$$\text{Noncarbonate Hardness, mg/L as } CaCO_3 = \text{Total Hardness Removed, mg/L as } CaCO_3 - \left(\text{Carbonate Hardness, mg/L as } CaCO_3 - \text{Carbonate Hardness Remaining, mg/L as } CaCO_3\right)$$

$$= 210 \text{ mg/L} - (170 \text{ mg/L} - 30 \text{ mg/L})$$

$$= 70 \text{ mg/L as } CaCO_3$$

$$\text{Soda Ash } (Na_2CO_3) \text{ Feed, mg/L} = \left(\text{Noncarbonate Hardness, mg/L as } CaCO_3\right)(106/100)$$

$$= (70 \text{ mg/L})(106/100)$$

$$= 74 \text{ mg/L}$$

3. Calculate the dosage of carbon dioxide required for recarbonation.

$$\text{Excess Lime, mg/L} = (A + B + C + D)(0.15)$$

$$= (10 \text{ mg/L} + 104 \text{ mg/L} + 0 + 55 \text{ mg/L})(0.15)$$

$$= (169 \text{ mg/L})(0.15)$$

$$= 25 \text{ mg/L}$$

$$\text{Total } CO_2 \text{ Feed, mg/L} = (Ca(OH)_2 \text{ excess, mg/L})(44/74) + (Mg^{2+} \text{ residual, mg/L})(44/24.3)$$

$$= (25 \text{ mg/L})(44/74) + (3 \text{ mg/L})(44/24.3)$$

$$= 15 \text{ mg/L} + 5 \text{ mg/L}$$

$$= 20 \text{ mg/L}$$

QUESTION

Please write your answer to the following question and compare it with that on page 117.

14.3G Calculate the hydrated lime ($Ca(OH)_2$) with 90 percent purity, soda ash, and carbon dioxide dose requirements in milligrams per liter for the water shown below.

Constituents	Source Water	Softened Water After Recarbonation and Filtration
CO_2, mg/L	= 5 mg/L	= 0 mg/L
Total Alkalinity, mg/L	= 150 mg/L as $CaCO_3$	= 20 mg/L as $CaCO_3$
Total Hardness, mg/L	= 240 mg/L as $CaCO_3$	= 50 mg/L as $CaCO_3$
Mg^{2+}, mg/L	= 16 mg/L	= 2 mg/L
pH	= 7.4	= 8.8
Lime Purity, %	= 90%	

14.32 Lime Softening

Water having hardness caused by calcium and magnesium bicarbonate (carbonate hardness) can usually be softened to an acceptable level using only lime. The lime reacts with the bicarbonate to form calcium carbonate, which will precipitate and settle out (convert from soluble to insoluble form) at a pH above 10, and magnesium carbonate, which will remain in solution. The magnesium carbonate reacts with additional lime at a pH above 11 to form magnesium hydroxide, which will precipitate.

In practice, if enough hardness can be removed by reacting lime with the calcium bicarbonate, softening can be accomplished at less expense. This procedure is sometimes called partial lime softening (no magnesium removal), which removes hardness caused by calcium ions. This may be referred to as calcium hardness. On the other hand, if some of the magnesium is to be removed, additional lime will be required.

Figure 14.2 is a flow diagram of a typical straight lime softening treatment plant. Settling should be provided after the addition of carbon dioxide (recarbonation) to ease the load on the filters. Recarbonation is used to lower the pH of the water. When properly recarbonated, the water is still supersaturated with calcium carbonate ($CaCO_3$). If the pH is much above 9, the water will usually cause scale to form. By recarbonation, the pH can be lowered to a range between 8.8 and 8.4 and the Langelier Index will still be positive; therefore there will be little or no corrosion. A polyphosphate is sometimes added to the water to prevent excessively heavy calcium carbonate scale deposits from forming. A polyphosphate may not be necessary if recarbonation is properly controlled. Addition of acid will not accomplish the same things as recarbonation and the addition of a polyphosphate; however, adding acid will lower the pH.

14.33 Split Lime Treatment

The amount of calcium and magnesium in source waters may vary. When the water contains a high level of magnesium, a method known as split lime treatment may be used (Figure 14.3). Split lime treatment can be used in lime treatment only or lime–soda ash treatment. In split lime treatment, a portion of the water (say 80 percent) is treated with an excess amount of lime to remove the magnesium at a pH of over 11. Then, source water (the other 20 percent) is added in the next basin to neutralize (lower the pH) the excess-lime-treated portion. The percentages will vary depending upon the water hardness, treatment layout, and desired results.

Split lime treatment softening can eliminate the need for recarbonation as well as offer a significant savings in lime feed. Since the fraction of the water that is treated has a high lime dose, magnesium is almost completely removed from this portion. When this water is mixed with the unsoftened water, the carbon dioxide and bicarbonate in the unsoftened fraction of the water tend to recarbonate in the final blend or mix of the treated water (effluent).

If the water shown in Figure 14.4 were treated by conventional treatment (not split treatment), it would require a lime dose of 400 mg/L as $CaCO_3$ (which is 25 percent higher) and a

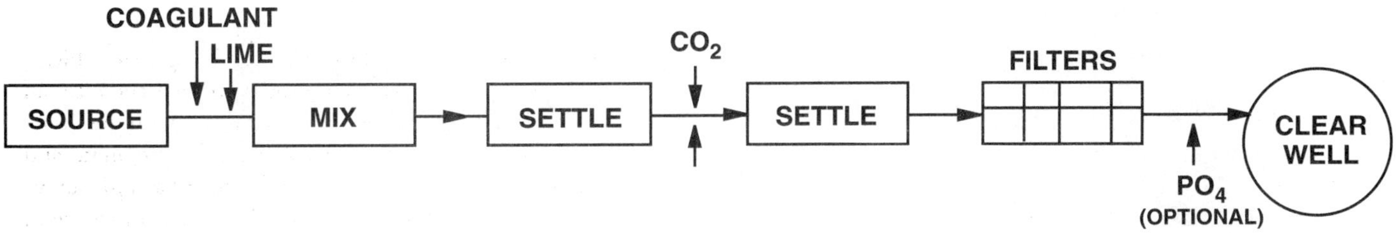

COAGULANT
LIME
CO_2
FILTERS
SOURCE
MIX
SETTLE
SETTLE
CLEAR WELL
PO_4
(OPTIONAL)

Fig. 14.2 Straight lime treatment

LIME
CO_2
FILTERS
SOURCE
MIX
SETTLE
MIX
SETTLE
CLEAR WELL
BYPASS
PO_4
PO_4
(OPTIONAL)
COAGULANT

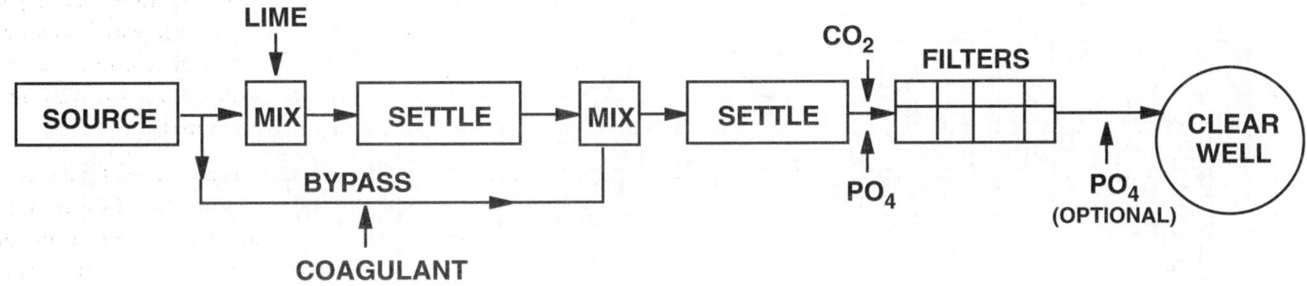

Fig. 14.3 Split lime treatment

carbon dioxide dose of 145 mg/L as $CaCO_3$ to produce a water having a hardness of 61 mg/L as $CaCO_3$ and a pH of 8.63.

While split treatment may be used in the lime–soda process, it is often advantageous to use a lime–ion exchange process (see Section 14.10). The salt used to remove noncarbonate hardness in the ion exchange process is much less expensive than the soda ash required in the lime–soda ash process.

The curves shown in Figure 14.4 assume that carbonate equilibrium has been achieved. In practice, it is not possible to attain equilibrium, but if the reactions take place in solids contact units, the results are very close to carbonate equilibrium.

The proper fraction of water to bypass is rather critically dependent on the lime dose and chemical composition of the unsoftened water. The proper fraction may be calculated, but the calculations are very complex. An experienced water chemist can perform the calculations.

QUESTIONS

Please write your answers to the following questions and compare them with those on page 117.

14.3H What compounds are formed when calcium and magnesium are precipitated out of water in the lime softening process?

14.3I What hardness is removed by partial lime softening (no magnesium removal)?

14.3J What is split lime treatment?

14.34 Lime–Soda Ash Softening

Let us look now at hardness requiring lime–soda ash treatment for removal (Figure 14.5).

When water cannot be softened to the desired level with lime only, it no doubt contains noncarbonate hardness. Noncarbonate hardness requires the addition of a compound that increases carbonate concentration, usually soda ash (sodium carbonate).

A water could contain only calcium hardness, yet require both lime and soda ash treatment. This would occur if the hardness were only calcium bicarbonate, sulfate, or chloride. In other words, all of the hardness is calcium carbonate and calcium noncarbonate hardness. This would not require split treatment (Figure 14.6).

14.35 Caustic Soda Softening

An alternative to the lime–soda ash process is the use of caustic soda (sodium hydroxide, NaOH) instead of soda ash.

The reactions of caustic soda with the carbonate and noncarbonate hardness are given in Section 14.315. Recall that caustic soda reacts with the carbonate hardness to form soda ash (sodium carbonate), which will react with calcium sulfate to form calcium carbonate ($CaCO_3$), as shown previously.

The advantages of using liquid caustic soda include ease of handling and feeding, lack of deterioration in storage, and less calcium carbonate sludge to handle and dispose of. Caustic soda is capable of removing both carbonate and noncarbonate hardness. Therefore, caustic soda may be used instead of soda ash,

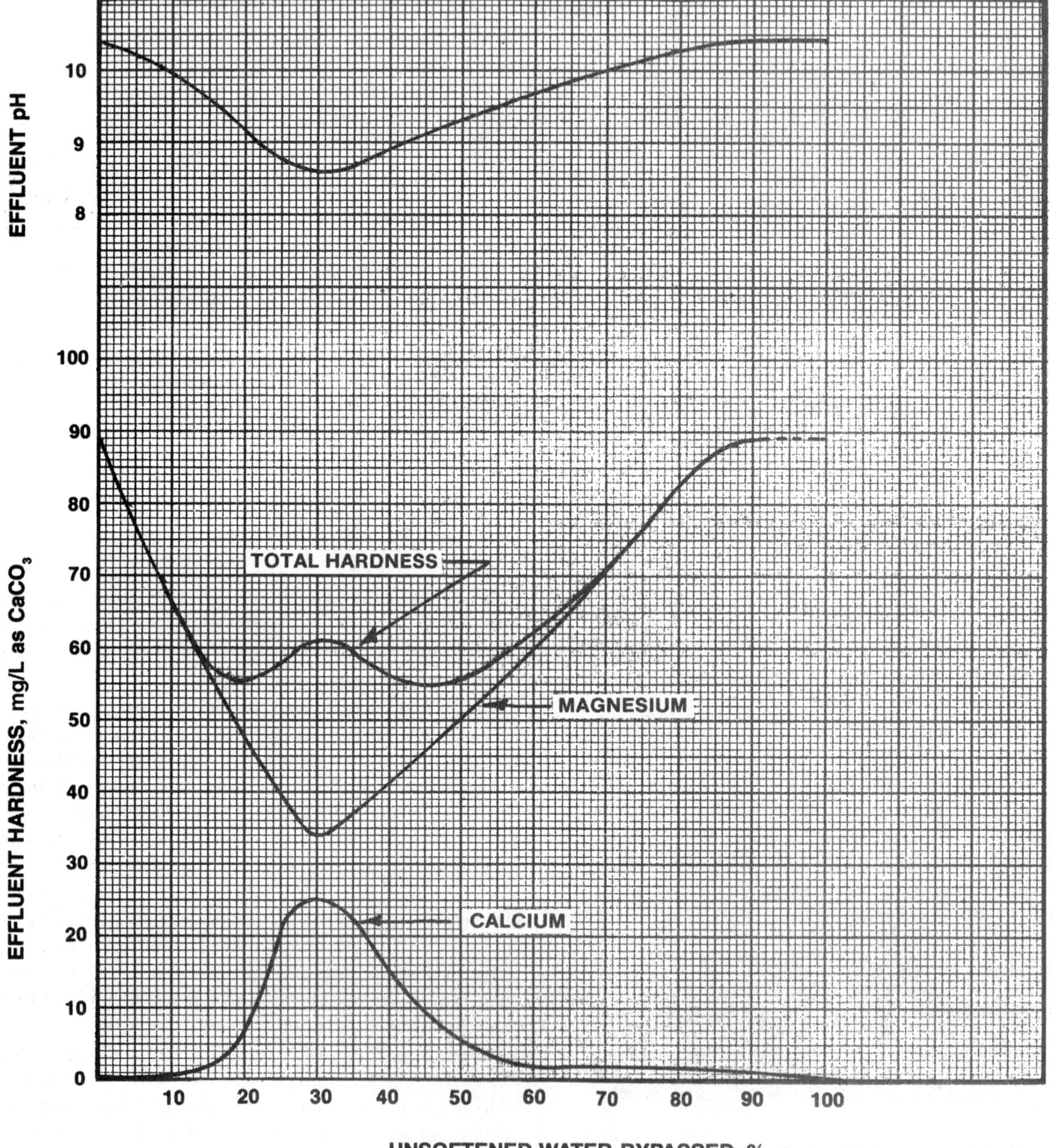

Fig. 14.4 Split treatment softening

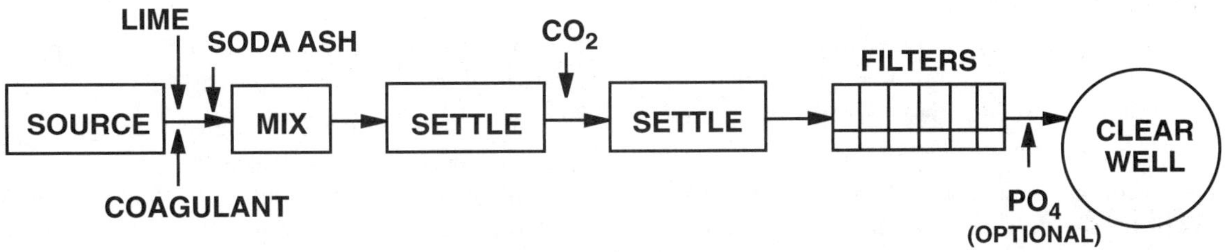

Fig. 14.5 Lime–soda ash treatment

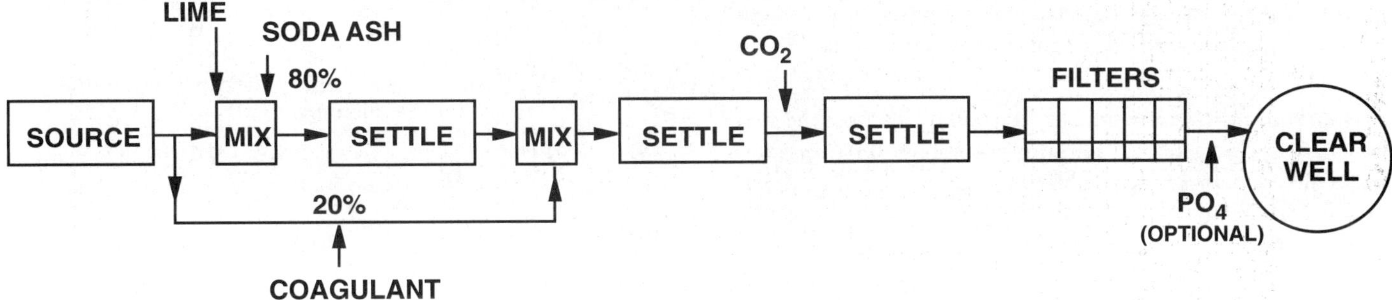

Fig. 14.6 Lime–soda ash split treatment

but also in place of part or all of the lime requirement. The use of caustic soda depends on a comparison of the costs of caustic soda, lime, and soda ash, and the characteristics of the source water.

Two points to observe are that sodium does not contribute to hardness, thus all the reactions having sodium compounds as an end product are nonhardness-producing compounds. However, sodium levels in drinking water should be less than 20 mg/L. The second point is that the precipitated compounds ($CaCO_3\downarrow$ and $Mg(OH)_2\downarrow$) are the desired end products whether lime or lime–soda ash or caustic soda treatment is used.

QUESTIONS

Please write your answers to the following questions and compare them with those on page 118.

14.3K Under what conditions would lime–soda ash softening be used?

14.3L What chemical is used to remove noncarbonate hardness in the chemical precipitation softening process?

14.36 Handling, Application, and Storage of Lime

Where the daily requirements for lime are small, lime is usually delivered to the water treatment plant in bags. At larger treatment plants, either quicklime (CaO) or hydrated lime ($Ca(OH)_2$) is delivered in bulk quantities. Truckloads of lime are commonly transferred to weathertight bins or silos by mechanical or pneumatic conveying systems.

Storage areas for bagged lime must be covered to prevent rain from wetting the bags. Bagged quicklime (CaO or calcium oxide) should never be stored close to combustible materials because considerable heat will be generated if the lime accidentally gets wet. Quicklime may be stored as long as 6 months, but in general should not be stored over 3 months. Hydrated lime should not be stored for more than 3 months before use.

Lime may be applied by dry feeding techniques using volumetric or gravimetric feeders. Lime is too insoluble to make solution feeding by pump feeders practical because of the accumulation of carbonate precipitation. See Chapter 13, "Fluoridation," for additional details and pictures of the various types of chemical feeders.

Operator safety must be considered before attempting to work with lime. A properly designed lime feeding system can minimize or eliminate lime dust problems. If lime dust is a problem, operators must wear protective clothing to avoid burns from contact with lime. Protective clothing includes long-sleeved shirt with sleeves and collar buttoned, trousers with legs down over tops of shoes or boots, head protection, and gloves.

Clothing should not fit too tightly around your neck, wrists, or ankles. A protective cream should be applied to exposed parts of the body, especially your neck, face, and wrists. You should wear a lightweight filter mask and tight-fitting safety glasses with side shields to protect yourself from the lime dust.

If lime comes in contact with your skin or eyes, immediately flush the affected areas with water and consult a physician if necessary. Do not rub your eyes if they are irritated with lime dust because rubbing will make the irritation worse. Keep any lime burns covered with a bandage during healing to prevent infection.

After handling lime, you should take a shower. If your clothes are covered with dust, or splattered with a lime slurry, take them off and have them washed. If possible, wear clean clothes on every shift.

For additional information regarding lime, contact the National Lime Association at lime.org or phone (703) 243-5463 and request a copy of their publication, *Lime: Handling, Application, and Storage in Treatment Processes*, 7th Edition, 1995. Lime, as well as other water treatment chemicals, should comply with the Standards of the American Water Works Association.

QUESTIONS

Please write your answers to the following questions and compare them with those on page 118.

14.3M How is lime delivered to plants where the daily requirements are small?

14.3N Why should quicklime be kept dry?

14.3O What types of chemical feeders are used to apply dry lime?

14.4 INTERACTIONS WITH COAGULANTS

Coagulation is discussed in detail in Volume I, Chapter 4, "Coagulation and Flocculation." However, the interactions of lime and soda ash with metallic coagulants such as alum, iron salts (ferric chloride, ferric sulfate, and ferrous sulfate), sodium aluminate, and many polymers are important.

Alum and iron salts are acidic and react with the alkalinity in water to cause a demand the same way that free carbon dioxide will. Therefore, this acidic condition must be met before softening can occur. In other words, extra lime will be required as the alum or iron feed rate goes up and therefore less lime will be required as the alum or iron feed rate is reduced.

Cationic polymers are not very pH sensitive and are often used as coagulant aids in softening plants rather than alum or iron salts. On the other hand, when sodium aluminate (a basic rather than an acidic compound) is the coagulant, the lime required to achieve a specific hardness reduction will be less, and will vary the opposite of alum or iron salts. The proportion of lime required in either instance is directly related to the coagulant dosage as well as the hardness removal desired. Approximate relationships can be calculated; however, experimentation is in order since plant equipment and source water variations are primary factors in the efficiencies of each waterworks. Jar tests are discussed later in Section 14.9.

If you are treating highly colored waters, these waters must be coagulated for color removal at low pH values. Alum is a good coagulant under these conditions. Ozone, permanganate, and chlorine may be tried along with alum to oxidize color. The high pH values required during softening tend to set the color, which then becomes very difficult to remove.

QUESTIONS

Please write your answers to the following questions and compare them with those on page 118.

14.4A What happens to the lime dose when the alum dose is increased for coagulation?

14.4B How can color be removed from water?

14.5 STABILITY

In nature, most waters are more or less stable. That is, they are in chemical balance. If a water is not stable, either scale will form on the filter sand, distribution mains, and household plumbing; or the water will be corrosive. When lime is added, the chemical balance is changed. The calcium carbonate ($CaCO_3$) formed in lime treatment is scale forming unless the exact chemical balance is achieved, which is seldom the case.

Under most conditions, a slight excess of lime is fed to cause a caustic condition to ensure complete reactions and achieve the desired results. In order to prevent scale formation on the filter sand, distribution mains, and household plumbing, the excess caustic and unprecipitated carbonate ions (pin floc) must be converted to soluble forms. Recarbonation is the most common way to do this. Again, as with all chemical treatment, recarbonation must be controlled to achieve the desired results.

Recarbonation lowers the pH to about 8.8 and thus converts some of the carbonate (CO_3^{2-}) back to the original bicarbonate (HCO_3^-) that existed in the source water. Recarbonation can be accomplished, to a degree, by using the source water in the split treatment mode discussed earlier. Usually, this is not adequate so further recarbonation is required. One reason for using source water as a neutralizing agent is that the recarbonation process is much less costly than if a high caustic water (high pH) is neutralized by chemical addition.

Use of carbon dioxide gas is the most common method of recarbonation. The reactions are:

1. $Ca(OH)_2 + CO_2 + H_2O \rightarrow Ca(HCO_3)_2$
2. $CaCO_3 + CO_2 + H_2O \rightarrow Ca(HCO_3)_2$

These reactions may be looked at as lime softening in reverse and will increase the hardness slightly. In these reactions, you are producing bicarbonate ions, which were removed in softening as carbonate hardness. This process tends to move the water back toward its original state, thus rendering it more stable.

The use of acids such as sulfuric or hydrochloric instead of recarbonation with carbon dioxide (CO_2) does not produce the same results. When carbon dioxide is added to a water containing calcium ions (Ca^{2+}) and hydroxide ions (OH^-), a calcium carbonate ($CaCO_3$) precipitate will form and the water will be saturated (or supersaturated) with calcium carbonate. If a strong acid is added to neutralize the softened water, which is highly basic, these reactions will not take place.

The marble test is the simplest method of measuring stability in the laboratory. Run the marble test[13] as outlined below:

1. Collect a sample of tap water that has been softened and stopper the sample bottle (avoid splashing into the flask).
2. To an identical sample, add one gram of powdered calcium carbonate. Mix and let stand for an hour or so.
3. Filter both samples (so they are both exposed to the same conditions).
4. Run pH and alkalinity tests on both samples.
5. The goal is to have the sample of softened tap water as nearly matched to the softened sample treated with calcium carbonate as possible. The plant treatment must be controlled to permit this condition to exist. If the pH and alkalinity in the softened sample are higher than in the softened sample treated with calcium carbonate, you are probably overtreating your supply and have scale-forming water. But, if the pH and alkalinity in your untreated softened sample are lower than in the treated one (calcium carbonate added), you are undertreating your supply. If they are similar, then stability is near.

Another way to check your water is to suspend a couple of nails on strings in your filter. Observe the nails occasionally to see if they are rusting or scaling up. To further protect the distribution system as well as prevent scale formation in the filter bed, 0.7 to 1.0 mg/L polyphosphate could be fed far enough ahead of the filters to allow mixing before the water reaches the filters. Addition of polyphosphate can prevent the formation of scale on filter media and in distribution system mains, but polyphosphate does not prevent corrosion. The Langelier Index (see Volume I, Chapter 8, Section 8.322, "Calcium Carbonate Saturation") is another approach to determining the corrosivity of water.

Caution should be exercised when using polyphosphate compounds. If they are converted to the orthophosphate form, they will lose their effectiveness. With the addition of phosphorus to water, there could be an increase in bacterial growths in the distribution system. Also, some wastewater treatment plants have phosphorus discharge limitations and polyphosphates added to drinking water can cause wastewater treatment plants to violate their discharge requirements.

QUESTIONS

Please write your answers to the following questions and compare them with those on page 118.

14.5A What problems are sometimes created when a slight excess of lime is fed during softening to cause a caustic condition to ensure complete reactions?

14.5B How can excess caustic and unprecipitated carbonate ions (pin floc) be removed from softened water?

14.5C What test is used to determine if a water is stable?

14.5D How can nails be used to determine if a water is stable?

14.6 SAFETY

When quicklime reacts with water in the slaking process (Figure 14.7), it gets hot enough to cause serious burns. Also, being caustic in nature, it can harm your eyes and skin. Always wear goggles or a face shield when working with lime that has been or is in the process of slaking. Flush with water if exposed to lime. Seek medical attention if it gets in your eyes. As for hands or face burns, immediately wash the affected areas and consult a physician if the burns appear serious.

Feeding equipment has moving parts. All moving machinery is a potential safety hazard. A paste-type slaker is particularly dangerous. This type of slaker will eat you alive. Never put your hand in or near the slaker paddles while the slaker is running. Use wooden paddles as cleaning tools on any slaker in operation. A metal tool will damage the slaker and could even injure the operator if dropped by accident. However, a wooden paddle will likely be broken up with no damage to the equipment or the operator.

Types of equipment vary greatly. Usually, the operator has little or no input in this area. Engineers usually design a plant and specify the type of chemical feed equipment.

Equipment suppliers are usually quite cooperative in advising any operator in the use and care of their equipment in your treatment plant. Detailed startup, shutdown, and maintenance procedures are available in the equipment manuals supplied by the manufacturer of the equipment.

Another important safety precaution is to avoid using the same conveyor or bin for alternately handling both quicklime and one of the coagulants containing water such as alum, ferric sulfate, or copperas. This water may be withdrawn by the quicklime and could generate enough heat to cause a fire. Explosions have been reported to have been caused by lime–alum mixtures in enclosed bins. Therefore, always clean facilities before switching from one chemical to another.

13. For additional information on the marble test, see Chapter 21, "Advanced Laboratory Procedures," Test Procedure 9, "Marble Test (Calcium Carbonate Saturation Test)."

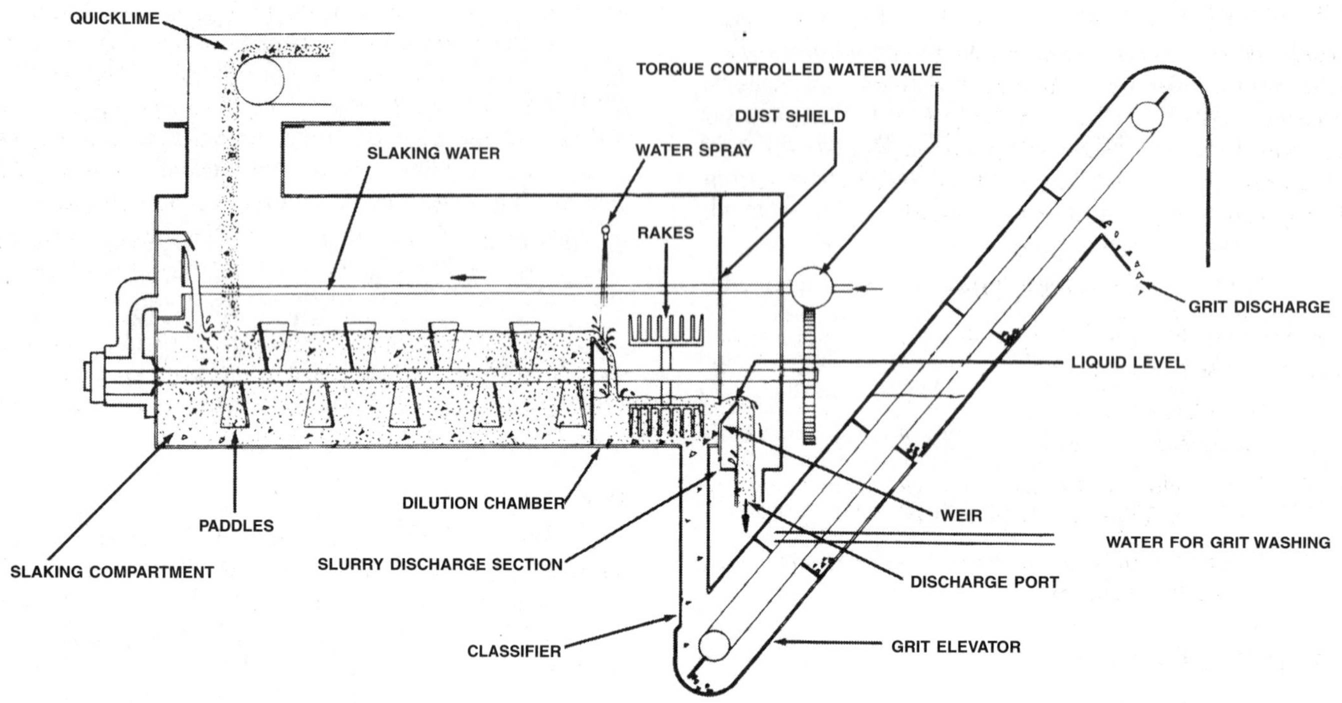

Fig. 14.7 Lime-slaking system
(Permission of Wallace & Tiernan, Inc.)

QUESTIONS

Please write your answers to the following questions and compare them with those on page 118.

14.6A Why should wooden paddles be used as cleaning tools on any slaker in operation?

14.6B Where would you look for information on how to safely maintain equipment?

14.7 SLUDGE RECIRCULATION AND DISPOSAL

Considerable sludge may be produced by the lime and lime–soda softening processes. When calcium and magnesium hardness are converted from soluble forms to insoluble precipitates (calcium carbonate and magnesium hydroxide), these precipitates form sludge. This sludge is removed from the bottom of settling basins and may be recirculated or must be disposed of by an acceptable procedure.

In some instances, sludge is recirculated back into the primary mix area of conventional plants to help seed the process. The advantages are that recirculation speeds up the precipitation process, and some reduction of chemical requirements may result. One disadvantage is that an increase in magnesium could result. Only trial and error will really determine if sludge recirculation will serve a useful purpose in your plant.

Sludge disposal is a problem everywhere. Perhaps the most common method is landfill disposal. This is accomplished by dewatering the sludge (drying beds or mechanical means) and then hauling the sludge to landfill sites developed solely for sludge disposal or sanitary landfills. Generally, a sludge with a Ca:Mg ratio of less than 2:1 will be difficult to dewater, whereas a sludge with a Ca:Mg ratio of greater than 5:1 will dewater relatively easily. The less water in the sludge, the less volume to transport to the disposal site and the less space required in the landfill.

To a lesser degree, sanitary sewer disposal is sometimes used. This only moves the sludge to another location for someone else to deal with. Some work has also been done with land application as a substitute for agricultural lime to increase the pH of highly acidic soils. The lime sludge is applied at a rate that will produce the optimum soil pH for the crops to be planted. For additional information on sludge disposal, see Chapter 17, "Handling and Disposal of Process Wastes."

14.8 RECORDS

Records should be kept on the amounts of chemicals ordered and the amounts fed. Laboratory results should be recorded in a permanent laboratory book. See Chapter 18, "Maintenance," for details on how to keep equipment maintenance schedules and records. Again, every plant is different, but accurate records will help a good operator be a better operator.

QUESTIONS

Please write your answers to the following questions and compare them with those on page 118.

14.7A What is a disadvantage of recirculating sludge back to the primary mix area?

14.7B How could you determine if sludge recirculation will serve a useful purpose in your plant?

14.8A What types of records should be kept regarding treatment plant chemicals?

14.9 JAR TESTS

14.90 Typical Procedures

Approximate amounts of chemicals required can be calculated (see Section 14.316, "Calculation of Chemical Dosages"); however, the best method of determining the proper dosages is by the use of the jar test. See Volume I, Chapter 11, "Laboratory Procedures," for details on the equipment and procedures required to run jar tests.

Chemical reagents may be made up by adding 1 gram of reagent to a liter of water.[14] This will produce a 0.1 percent chemical solution. When 1.0 g/L of lime, soda ash, or coagulant is made up as the stock solution, 1 mL of the stock solution in a 1-liter sample of water equals 1 mg/L dosage. If large doses of lime are required, add 10 grams of lime per liter. With this stock solution, 1 mL of stock solution in a 1-liter sample of water equals 10 mg/L dosage.

Set up six samples and estimate the dosage required by adding varying amounts to each sample. Trial and error will put you in the ballpark. To refine the dosage, pick the best looking sample from the settling properties of the floc to establish the optimum lime dose. Then, do the same with varying amounts of soda ash, leaving the lime dosage constant. By running pH, alkalinity, and hardness tests, you can find the optimum dosages that give the desired softened water results.

The exact procedures used to soften water by chemical precipitation using the lime–soda ash process depend on the hardness and other chemical characteristics of the water being treated. A series of jar tests is commonly used to determine optimum dosages. In many cases, the feed rates determined by jar tests do not produce the exact same results in an actual plant. This is because of differences in water temperature, size and shape of jars as compared with plant basins, mixing equipment, and influence of coagulant (a heavy alum feed will neutralize more of the lime). You must remember that jar test results are a starting point. You may have to make additional adjustments to the chemical feeders in your plant based on actual analyses of the treated water.

Convert the jar test results to plant feed rates. Always feed enough chemical to achieve the desired results, but do not overfeed. Overtreating water is a waste of money and quality control will suffer.

14.91 Examples

Let us set up some jar tests to determine the optimum dosages for lime or lime–soda treatment to remove hardness from a municipal water supply. To get started, add 10.0 grams of hydrated lime to a 1-liter graduated cylinder or flask and fill to the 1-liter mark with tap water. Thoroughly mix this stock solution in order to thoroughly suspend all of the lime. One mL of this solution (which has been thoroughly mixed) in a liter of water is the same as a lime dose of 10 mg/L, or 0.5 mL in 500 mL is still the same as a 10 mg/L lime dose.

Set up a series of hardness tests by adding 5.0 mL, 10.0 mL, 15.0 mL, 20.0 mL, 25.0 mL, 30.0 mL, 35.0 mL, and 40.0 mL to 1-liter (1,000 mL) beakers or jars. Fill the beakers to the 1,000 mL mark with the water being tested. Mix thoroughly for as long as normal mixing will occur in your plant. Allow the precipitate to settle (20 minutes if this is the settling time in your plant) and measure the hardness remaining in the water above the precipitate. A plot of the hardness remaining against the lime dosage will reveal the optimum dosage. Examination of Figures 14.8, 14.9, and 14.10[15] reveal that the water of all three cities responded differently to the increasing lime dosage. City 1 (Figure 14.8) should be providing a lime dose of 100 mg/L. The cost of increasing the dosage to 150 mg/L is not worth the slight reduction in hardness from 110 to 100 mg/L as $CaCO_3$. Note that an overfeed of lime will actually increase the hardness.

City 2 (Figure 14.9) should be providing a lime dose of 200 mg/L. A dose of 300 mg/L will reduce hardness, but the increase in lime costs is too great. City 3 (Figure 14.10) should be dosing lime between 200 and 250 mg/L. Note that the greater the lime dose, the less the hardness, but the greater the quantities of sludge that must be handled and disposed.

14. Some operators add 10 grams of reagent to a liter of water. This will produce a 1 percent solution. One mL of the stock solution in one liter will produce a 10 mg/L dosage.

15. These figures were adapted from an article titled, "Use of Softening Curve for Lime Dosage Control," by Michael D. Curry, P.E., which appeared in *The Digester/Over the Spillway*, published by the Illinois Environmental Protection Agency.

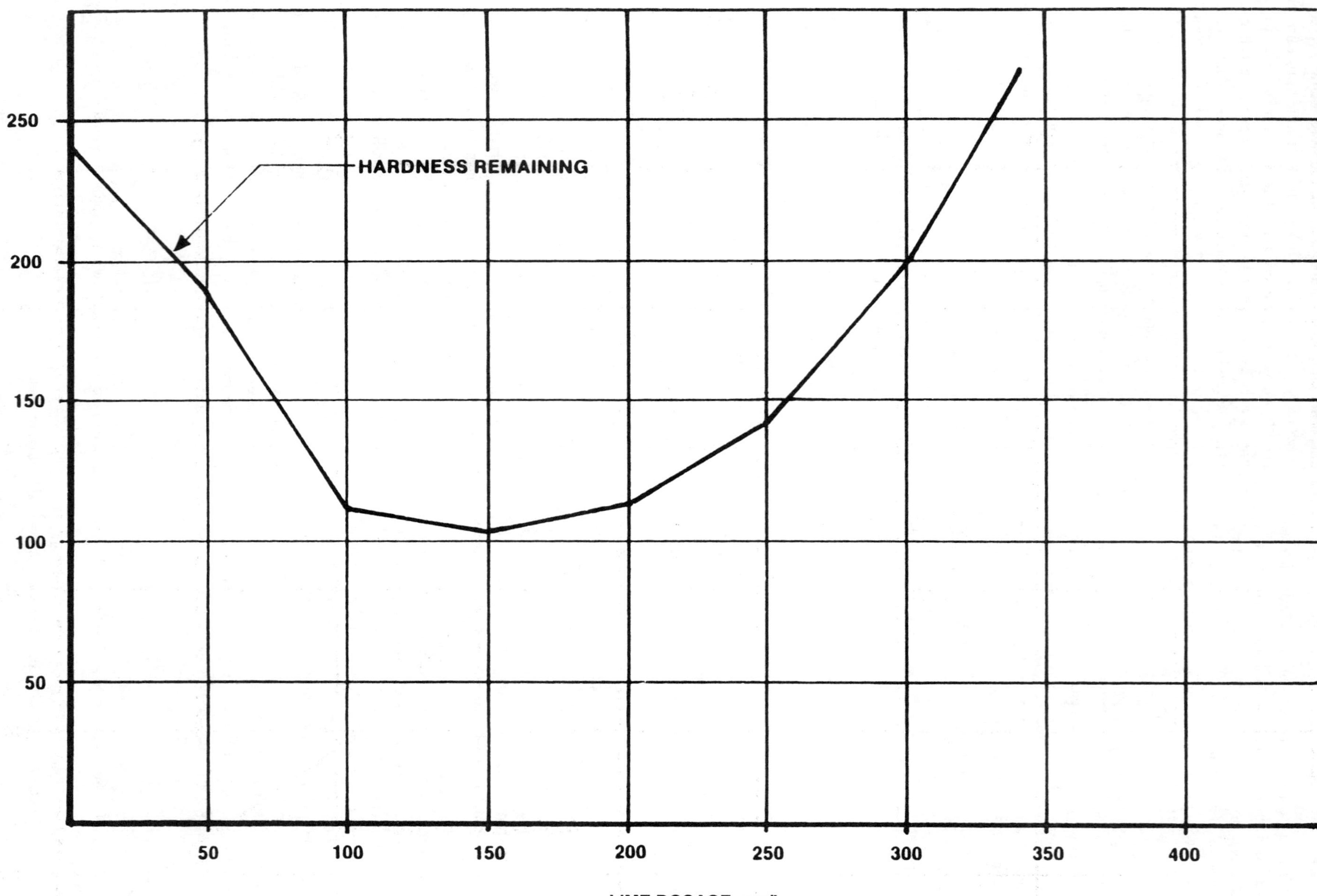

Fig. 14.8 Softening curve for City 1

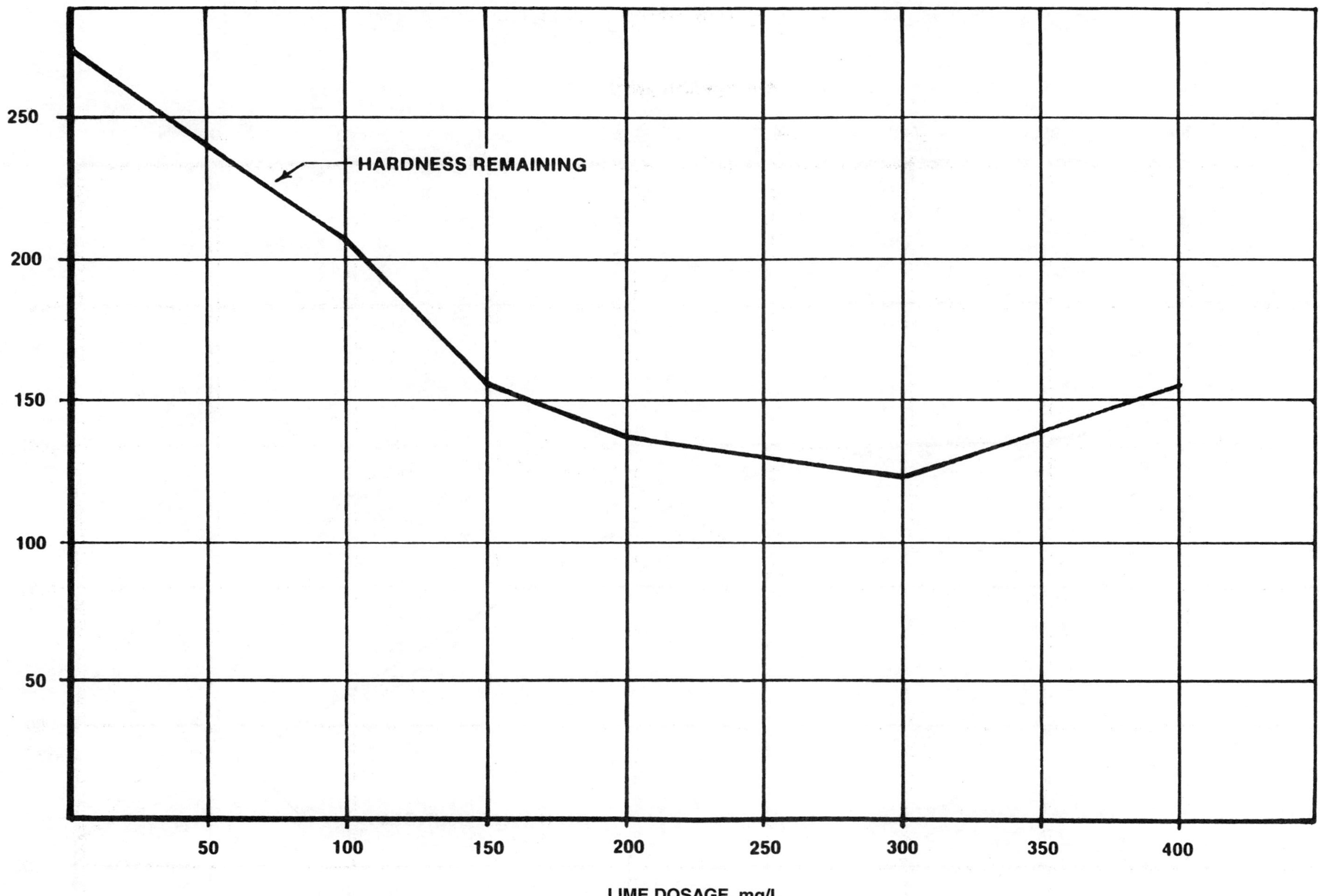

Fig. 14.9 Softening curve for City 2

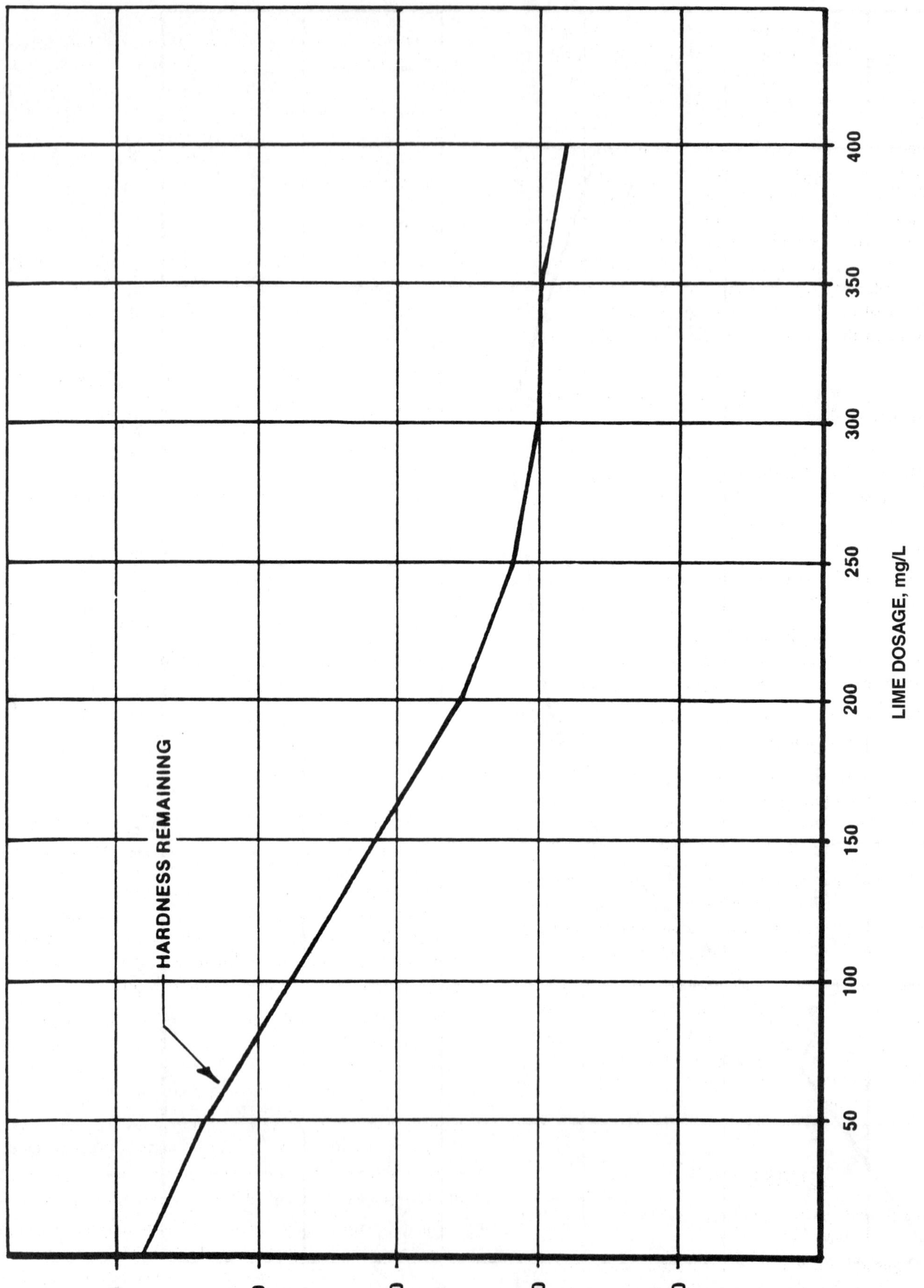

Fig. 14.10 Softening curve for City 3

If lime added to the water does not remove sufficient hardness, select the optimum lime dose and then add varying amounts of soda ash. From Figure 14.9, we found that the optimum lime dose was 200 mg/L (300 mg/L would have reduced the hardness slightly). Let us take six 1-liter containers and add 20 mL of our lime stock solution (a dosage of 200 mg/L). Prepare a stock solution of soda ash similar to our lime solution by adding 10 grams of soda ash to a 1-liter container, fill with distilled water and mix thoroughly. Add zero, 2.5 mL (25 mg/L dose), 5 mL, 7.5 mL, 10 mL, and 12.5 mL to the 1-liter containers. Mix thoroughly, allow the precipitate to settle, and measure the hardness remaining in the water above the precipitate. A plot of hardness remaining against the soda ash dosage will reveal the desired dosage. We would like the final hardness to be in the 80 to 90 mg/L as $CaCO_3$ range in this example.

To select the optimum doses of lime and soda ash, consider the items discussed below.

1. Optimum dosage of lime was based on increments of 50 mg/L. You should refine this test by trying at least two 10 mg/L increments above and below the optimum dose. From Figure 14.8 we found that 100 mg/L was the optimum dose. Try lime doses of 80, 90, 100, 110, and 120 mg/L.
2. Optimum dosage of soda ash can be refined by trying smaller increments also.
3. Try slightly increasing the actual lime dose in your plant to see if there is any decrease in the remaining hardness. Is the decrease in hardness worth the increase in lime costs?
4. Try slightly increasing and decreasing both lime and soda ash dosages at your plant one at a time, and evaluate the results.
5. If you are treating well water or a water of constant quality, all you have to do to maintain proper treatment is to make minor adjustments to keep the system fine-tuned.
6. If you are treating water from a lake or a river and the water quality (including temperature) changes, you will have to repeat these procedures whenever the raw water quality changes. Water quality changes of concern include raw water hardness, alkalinity, pH, turbidity, and temperature.
7. Remember, you do not want to produce water of zero hardness. If you can get the hardness down to around 80 to 90 mg/L, that usually will be low enough for most domestic consumers. When selecting a target hardness level for a water softening plant, consider the uses of the softened water and the cost of softening.

QUESTIONS

Please write your answers to the following questions and compare them with those on page 118.

14.9A If lime added to water does not reduce the hardness of a water sufficiently, what would you do?

14.9B What items should be considered when selecting a target hardness level for a water softening plant?

14.92 Calculation of Chemical Feeder Settings

After chemical doses have been calculated or determined from jar tests, convert the results to plant chemical feed rates. Depending on the type of chemical feeder, you may have to calculate the feed rates in pounds per day, pounds per hour, or pounds per minute. Always feed enough chemical to achieve the desired results, but do not overfeed. Overtreating is a waste of money and quality control will suffer.

Example 4

The optimum lime dosage from the jar tests is 230 mg/L. If the flow to be treated is 6 MGD, what is the feeder setting in pounds per day and the feed rate in pounds per minute?

Known		Unknown
Lime Dose, mg/L	= 230 mg/L	1. Feeder Setting, lb/day
Flow, MGD	= 6 MGD	2. Feed Rate, lb/min

1. Calculate the feeder setting in pounds per day.

$$\text{Feeder Setting, lb/day} = (\text{Flow, MGD})(\text{Lime, mg/L})(8.34\ \text{lb/gal})$$

$$= (6\ \text{MGD})(230\ \text{mg/L})(8.34\ \text{lb/gal})$$

$$= 11{,}509\ \text{lb/day}$$

2. Calculate the feed rate in pounds per minute.

$$\text{Feed Rate, lb/min} = \frac{\text{Feeder Setting, lb/day}}{(60\ \text{min/hr})(24\ \text{hr/day})}$$

$$= \frac{11{,}509\ \text{lb/day}}{(60\ \text{min/hr})(24\ \text{hr/day})}$$

$$= 8.0\ \text{lb/min}$$

When the calculated feed rate of eight pounds of lime per minute is put into the plant process, observations and tests will determine if optimum levels are met. In many instances, jar tests and actual plant feed rates do not agree exactly. This is because of temperature, size and shape of jars versus size and shape of plant facilities, mixing time, and influence of the coagulant (a heavy alum feed would neutralize more of the lime). Jar tests are merely indicators or a point at which to begin.

If underfeeding results, reactions will not be complete and the results will be undertreated water having a hardness higher than desired.

If overfeeding results, chemicals are being wasted. Also, it is quite possible to have excessive calcium in the water. This results in unstable conditions, which cause buildup on filter media sand grains and the interior of the water mains. This is where the stability test enters the picture (refer to Section 14.5, "Stability").

The above discussion has dealt with establishing the proper lime feed. The same process would be used to determine the soda ash requirements if you are removing noncarbonate hardness. Set up the lime feed jar tests as discussed. Pick the optimum dosage. Then set up another series of jars using the same lime feed rate in all jars. Now, vary the soda ash feed rate.

Example 5

How much soda ash is required (pounds per day and pounds per minute) to remove 50 mg/L of noncarbonate hardness as $CaCO_3$ from a flow of 6 MGD?

Known		Unknown
Noncarbonate Hardness Removed, mg/L as $CaCO_3$	= 50 mg/L	1. Feeder Setting, lb/day
Flow, MGD	= 6 MGD	2. Feed Rate, lb/min

1. Calculate the soda ash dose in milligrams per liter. (See Section 14.316, "Calculation of Chemical Dosages," for an explanation of the following formula.)

$$\text{Soda Ash, mg/L} = \left(\begin{array}{c}\text{Noncarbonate Hardness,}\\ \text{mg/L as } CaCO_3\end{array}\right)(106/100)$$

$$= (50 \text{ mg/L})(106/100)$$

$$= 53 \text{ mg/L}$$

2. Determine the feeder setting in pounds per day.

$$\text{Feeder Setting, lb/day} = (\text{Flow, MGD})(\text{Soda Ash, mg/L})(8.34 \text{ lb/gal})$$

$$= (6 \text{ MGD})(53 \text{ mg/L})(8.34 \text{ lb/gal})$$

$$= 2{,}652 \text{ lb/day}$$

3. Calculate the soda ash feed rate in pounds per minute.

$$\text{Feed Rate, lb/min} = \frac{\text{Feeder Setting, lb/day}}{(60 \text{ min/hr})(24 \text{ hr/day})}$$

$$= \frac{2{,}652 \text{ lb/day}}{(60 \text{ min/hr})(24 \text{ hr/day})}$$

$$= 1.8 \text{ lb/min}$$

After you determine the proper feed rates and implement them, if you are treating well water or other water of constant quality, all you have to do to maintain proper treatment is make minor adjustments to keep the system fine-tuned.

On the other hand, if you treat a river or lake supply subject to constant or frequent changes in water quality, it is an entirely different set of circumstances. Until you learn from experience to judge the chemical changes necessary by the fluctuations in raw water hardness, alkalinity, and turbidity, you almost have to check yourself daily by the jar test method.

When your treatment process does not work properly, the first thing to check is whether or not the feeder is feeding properly. If it is, the next step is to check your source water quality. Generally, one of these two will be the cause of your problem.

QUESTIONS

Please write your answers to the following questions and compare them with those on pages 118 and 119.

14.9C Why should overfeeding of chemicals be avoided?

14.9D What should be the lime feeder setting in pounds per day to treat a flow of 2 MGD when the optimum lime dose is 160 mg/L?

14.9E How much soda ash is required in pounds per day to remove 40 mg/L of noncarbonate hardness from a flow of 2 MGD?

END OF LESSON 1 OF 2 LESSONS

on

SOFTENING

Please answer the discussion and review questions next.

DISCUSSION AND REVIEW QUESTIONS

Chapter 14. SOFTENING

(Lesson 1 of 2 Lessons)

At the end of each lesson in this chapter, you will find discussion and review questions. Please write your answers to these questions to determine how well you understand the material in the lesson.

1. Why should water be softened?
2. What are the benefits of softening water in addition to hardness removal?
3. In the two-step recarbonation process, what is accomplished by a second addition of carbon dioxide after treatment?
4. What are the advantages of using liquid caustic soda to soften water?
5. Why should quicklime never be stored close to combustible materials?
6. How can operators protect themselves from lime?
7. Why is the stability of a water important?
8. Why is a paste-type slaker dangerous?
9. Why should you avoid using the same conveyor or bin for alternately handling both quicklime and one of the coagulants containing water, such as alum?
10. Why is sludge sometimes recirculated back into the primary mix area of conventional plants?
11. When running jar tests, how would you determine the optimum chemical dosages for the lime–soda softening process?
12. When your lime–soda ash softening treatment process does not work properly, what is the first thing you should check?

CHAPTER 14. SOFTENING

Ion Exchange Softening by Marty Reynolds

(Lesson 2 of 2 Lessons)

14.10 DESCRIPTION OF ION EXCHANGE SOFTENING PROCESS

The term "zeolite" is most often associated with sodium ion exchangers and should be considered to mean the same as the term ion exchange. Most ion exchange units in use today use sulfonated polystyrene resins as the exchange media. Ion exchange softening can be defined as exchanging hardness-causing ions (calcium and magnesium) for the sodium ions that are attached to the ion exchange resins to create a soft water.

The treatment plant operator should be aware of the three basic types of softeners on the market.

1. An upflow unit in which the water enters from the bottom and flows up through the ion exchange bed and out the top.
2. A unit that is constructed and operated like a gravity rapid sand filter. The water enters the top, flows down through the ion exchange bed, and out the bottom.
3. The pressure downflow ion exchange softener, which is the most common, will be discussed in this lesson (see Figures 14.11 and 14.12). Pressure filters may be either horizontal or vertical units. Vertical units are preferred because there is less chance of short-circuiting.

To help explain the construction and activity that occurs in an ion exchanger, let us compare it to a pressure filter. The water enters the unit through an inlet distributor located in the top; it is forced (usually pumped) down through a bed of some type of media into an underdrain structure. From the underdrain structure, the treated water flows out of the unit and into storage or into the distribution system.

The flow pattern through a filter and softener are similar, the key difference being the action that takes place in the media or bed of each unit. The filter bed may be considered an adsorption and mechanical straining device used to remove suspended solids from the water. The bed usually consists of sand, anthracite (crushed coal), or a combination of both. Once the bed becomes saturated with the insoluble material (usually clay, suspended solids and iron, manganese hydroxide), the filter is taken out of service, backwashed, and returned to service. This pressure filter will continue to operate until the condition reoccurs and the procedure is repeated.

The bed, media, or resin in an ion exchange softener, however, is much more complex. This resin serves as a medium in which an ion exchange takes place. As hard water is passed through the resin, the sodium ions on the resin are exchanged for the calcium and magnesium ions (in the case of sodium exchange resins). The sodium ions are released from the exchange resin and remain in the water, which flows out of the softener. The calcium and magnesium ions, however, are retained by the resin. The softener effluent is free from calcium and magnesium ions and, therefore, is softened (Figure 14.13).

Once a softener has exchanged all of its sodium ions and the resin is saturated with calcium and magnesium, it will no longer produce soft water. At this time, the unit must be taken out of service and the calcium and magnesium must be removed from the resin by exchange with sodium ions. This process is referred to as a regeneration cycle.

In a regeneration cycle, the calcium and magnesium ions that have been retained by the resin must be removed and the sodium ions restored. In order for the exchange to take place, the resin must hold all ions loosely. If the calcium and magnesium ions cannot be removed, the resin will not accept the addition of new sodium ions that are necessary for additional softening.

Salt, in the form of a concentrated brine solution, is used to regenerate (recharge) the ion exchange resin. When salt is added to water it changes into or ionizes to form sodium cation (Na^+) and chloride anions (Cl^-). When the brine solution is fed into the resin, the sodium cations are exchanged for calcium and magnesium cations. As the brine solution travels down through the resin, the sodium cations are attached to the resin while the calcium, magnesium, and chloride (from the salt) ions flow to waste. After the regeneration has taken place, the bed is ready to be placed in service again to remove calcium and magnesium by ion exchange.

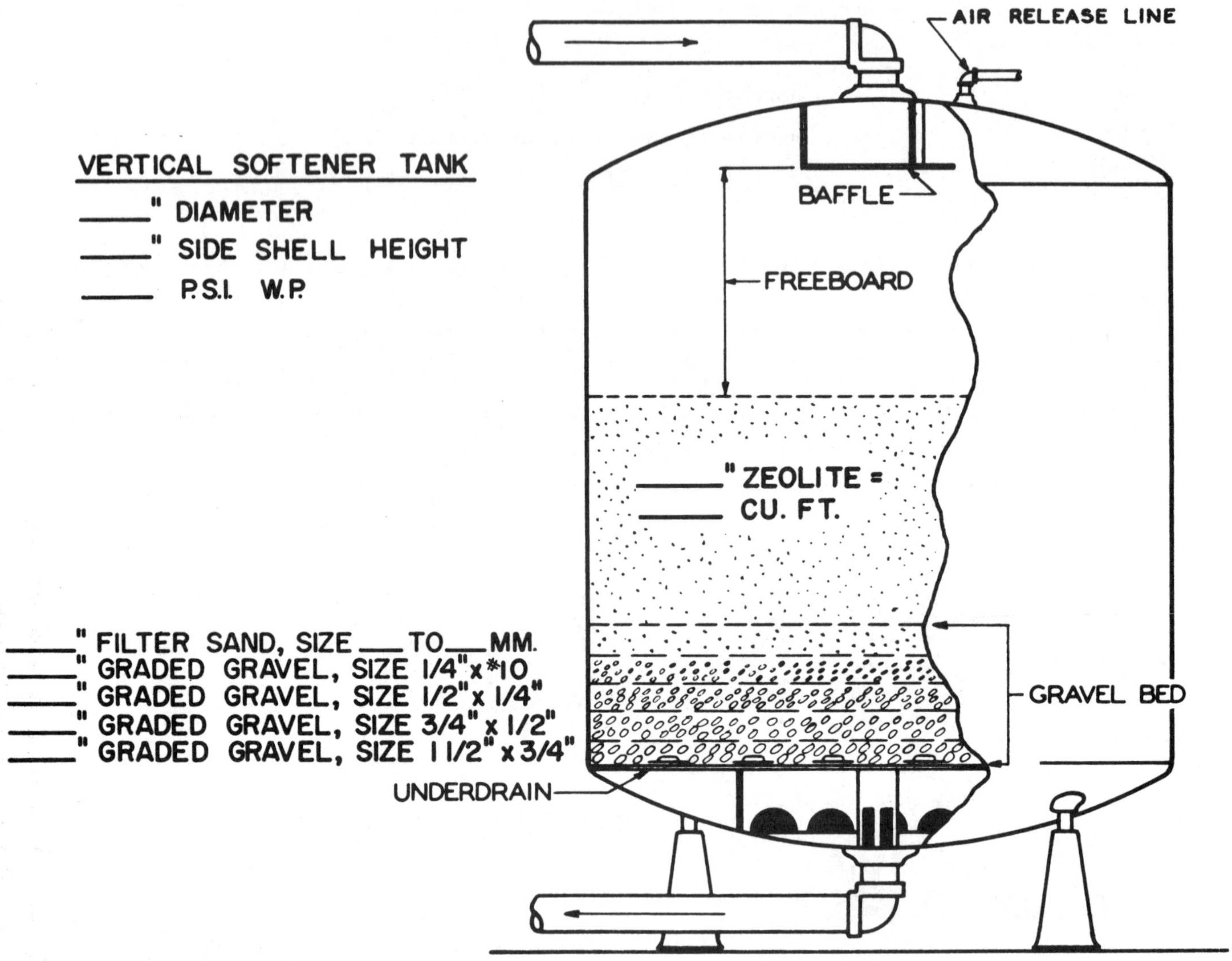

AIR RELEASE LINE

VERTICAL SOFTENER TANK

____" DIAMETER
____" SIDE SHELL HEIGHT
____ P.S.I. W.P.

BAFFLE

FREEBOARD

____" ZEOLITE =
____ CU. FT.

____" FILTER SAND, SIZE ___ TO ___ MM.
____" GRADED GRAVEL, SIZE 1/4" x *10
____" GRADED GRAVEL, SIZE 1/2" x 1/4"
____" GRADED GRAVEL, SIZE 3/4" x 1/2"
____" GRADED GRAVEL, SIZE 1 1/2" x 3/4"

GRAVEL BED

UNDERDRAIN

STANDARD SOFTENER

UNDERDRAIN:-
RIGIDLY SUPPORTED PLATE OVER 100 % FILTER AREA WITH STAINLESS STEEL BAFFLE ASSEMBLIES.

ZEOLITE:- ____ CUBIC FEET.

SUPPORTING GRAVEL:- ____ " GRADED.

STAINLESS STEEL BAFFLE ASSEMBLY

Fig. 14.11 Pressure downflow ion exchange softener

(Permission of General Filter Company)

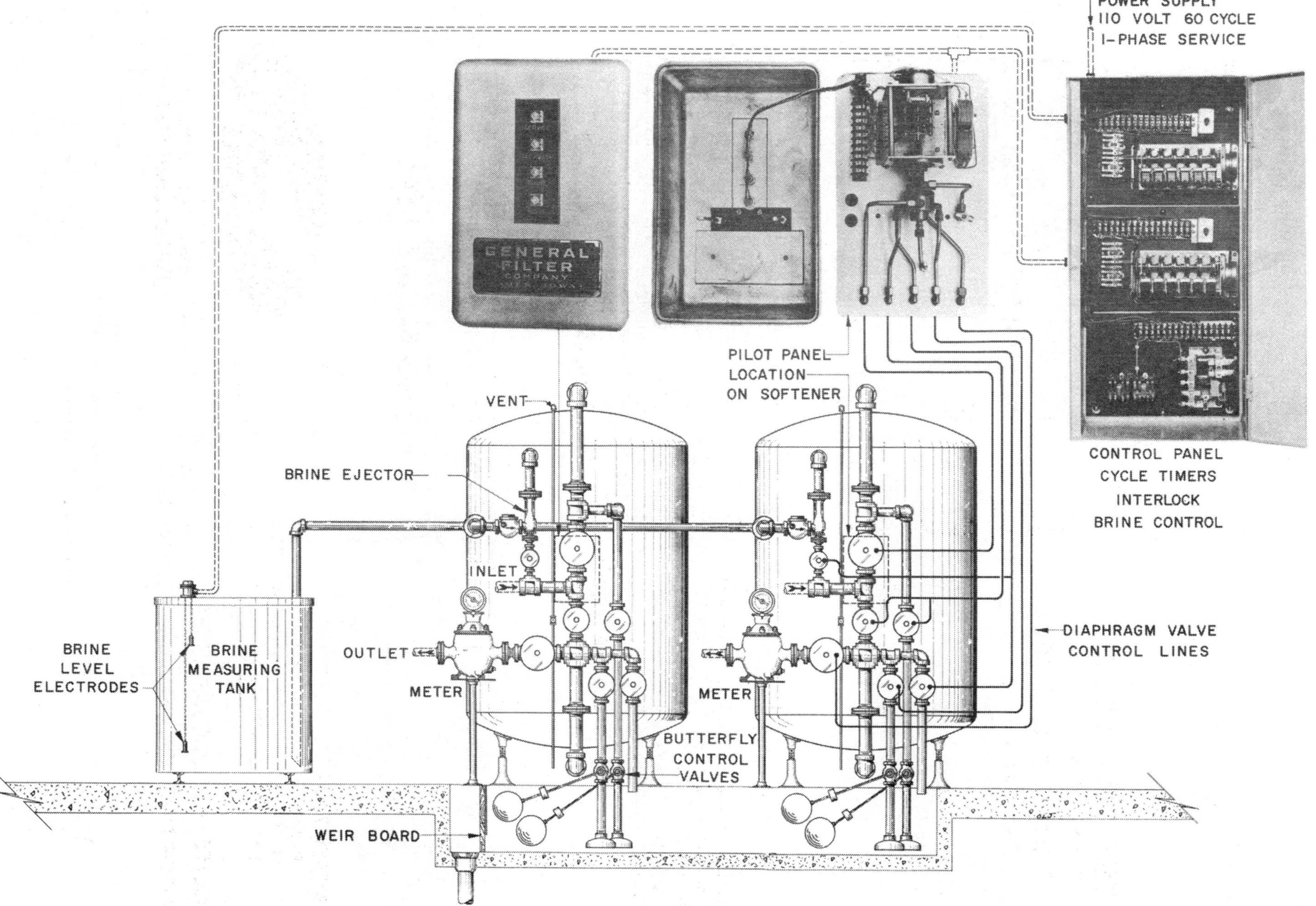

Fig. 14.12 Semiautomatic controls for ion exchanger
(Permission of General Filter Company)

Na^+	Na^+	Na^+
Na^+	Na^+	Na^+
Na^+	Na^+	Na^+
Na^+	Na^+	Na^+
Na^+	Na^+	Na^+

[1]
EXCHANGE RESIN
PRIOR TO SOFTENING
(AFTER REGENERATION)

PLUG FLOW-PISTON EFFECT
HARDNESS APPLIED

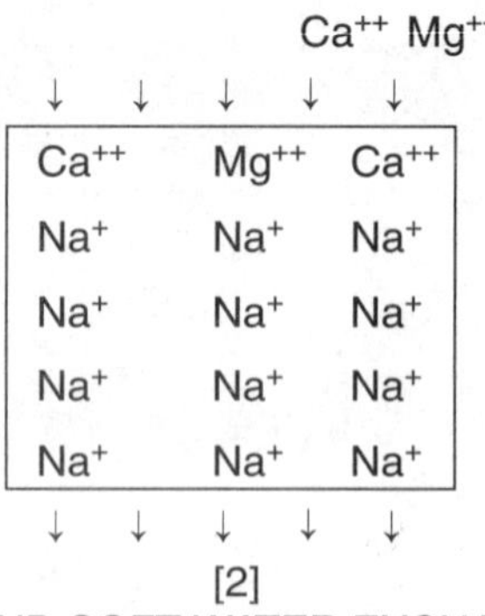

[2]
Na^+ AND SOFT WATER EXCHANGE
RESIN AFTER START OF SOFTENING CYCLE

PLUG FLOW-PISTON EFFECT
HARDNESS APPLIED

Ca^{++} Mg^{++}

Ca^{++}	Mg^{++}	Ca^{++}
Ca^{++}	Ca^{++}	Mg^{++}
Mg^{++}	Ca^{++}	Ca^{++}
Ca^{++}	Na^+	Ca^{++}
Na^+	Na^+	Na^+

[3]
Na^+ AND SOFT WATER EXCHANGE
RESIN DURING SOFTENING CYCLE

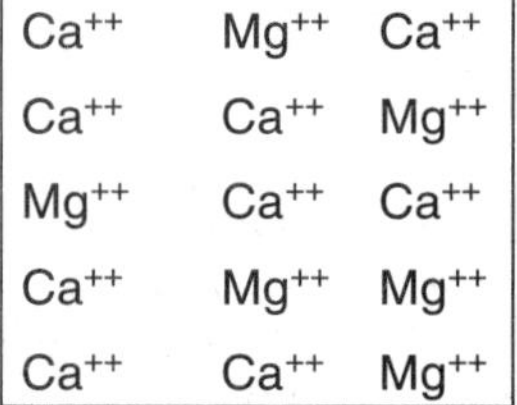

[4]
EXCHANGE RESIN EXHAUSTED
(ALL SOFTENING CAPACITY LOST-
READY TO REGENERATE, SEE [1])

Fig. 14.13 Ion exchange resin condition during softening cycle

QUESTIONS

Please write your answers to the following questions and compare them with those on page 119.

14.10A List the three basic types of softeners on the market.

14.10B What happens during the regeneration cycle of an ion exchange softener?

14.11 OPERATIONS

Many factors influence the procedures used to operate an ion exchange unit and the efficiency of the softening process. These factors include:

1. Characteristics of the ion exchange resin
2. Quality of the source water
3. Rate of flow applied to the softener
4. Salt dosage during regeneration
5. Brine concentration
6. Brine contact time

Each ion exchange softener, regardless of manufacturer, will have at least four common stages of operation. These stages are listed below and will be explained as each occurs in the softener operation (see Figure 14.14).

1. Service
2. Backwash
3. Brine
4. Rinse

14.110 Service

The service stage of each unit is where the actual softening of the water occurs. Hard water is forced into the top of the unit and allowed to flow down through the exchange resin. As this takes place, the calcium and magnesium ions exchange with sodium on the resin. The sodium ions are released into the water and the exchange capacity of the unit is slowly exhausted.

The length of each service stage is dependent on several factors; source water hardness is a main consideration. The harder the water, the more calcium and magnesium must be removed to

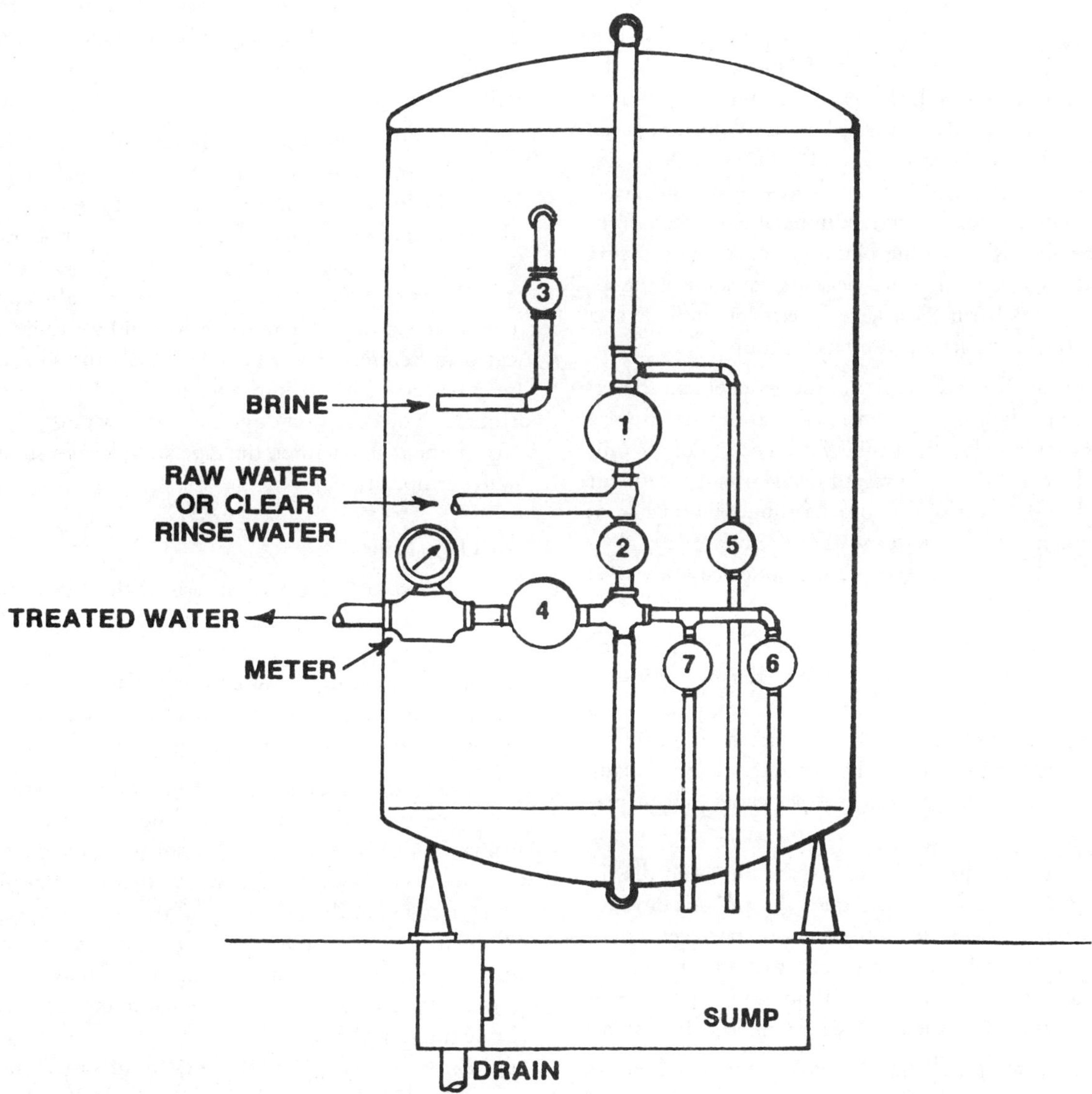

OPERATION	VALVE NUMBER						
	1	2	3	4	5	6	7
SERVICE	OPEN	CLOSE	CLOSE	OPEN	CLOSE	CLOSE	CLOSE
BACKWASH	CLOSE	OPEN	CLOSE	CLOSE	OPEN	CLOSE	CLOSE
BRINE	CLOSE	CLOSE	OPEN	CLOSE	CLOSE	OPEN	CLOSE
RINSE	OPEN	CLOSE	CLOSE	CLOSE	CLOSE	CLOSE	OPEN

Fig. 14.14 Valve positions for each stage of ion exchange softener operation
(Permission of General Filter Company)

reach a level of zero hardness. Simply stated, the harder the water, the less water you can treat before the resin becomes exhausted. As long as the design flow for the ion exchange unit is not exceeded, changes in the hardness of the source water may be automatically adjusted for in the ion exchange unit. The effluent from the unit usually will have zero hardness until the unit needs regeneration. If the total dissolved solids (TDS) in the water supply is fairly high (above 500 mg/L), there may be some leakage. If a high TDS water has a high sodium content, the sodium may hinder the process by causing a local exchange on the media of calcium and magnesium (hardness leakage) for some of the sodium. The amount of hardness leakage depends on the TDS and the salt dosage (percent salt) used for regeneration.

Other factors involved are the size of the softener and the exchange capacity of the resin. Softeners can vary in size from a few cubic feet to several hundred cubic feet. The size of the unit will generally be consistent with regard to the overall treatment plant design. In other words, the softener should be capable of producing enough softened water so that the mix or blend of softened and unsoftened water will produce a treated water with the desired level of hardness.

The exchange resin will also vary in its removal capacity. There are many types of strong acid cation exchange resins on the market today. Most will range in capacity from 20,000 to 30,000 grains of hardness removal per cubic foot (0.011 to 0.016 kg/m^3) of resin. The removal ability of the resin is usually expressed in grains of hardness removal per cubic foot of material or resin.

The source water hardness, the size of the unit, and the removal capacity of the resin will determine the amount of water that can be treated before the softener must be regenerated. With a few simple calculations, an operator can determine the softening capacity of the units. Calculations and examples will be given at the end of the lesson. See Example 8 in Section 14.18, "Ion Exchange Arithmetic."

14.111 Backwash

The second stage of the ion exchange softener process is the backwash. In this stage, the unit is taken out of service and the flow pattern through the unit is reversed. The purpose of this is to expand and clean the resin particles and to free any material such as iron, manganese, and particulates that might have been removed during the softening stage. The backwash water entering the softener at the beginning of this stage should be applied at a slow, steady rate. If the water enters the unit too quickly, it could create a surge in the resin and wash it out of the unit with the water going to waste.

Ideal bed expansion during the softener backwash should be 75 to 100 percent. In other words, when the unit is backwashed, the resin should expand to occupy a volume from 75 to 100 percent greater than when in normal service. An example of this would be an ion exchange softener with 24 inches (60 cm) of resin while in service. When the unit is backwashed, the resin should expand to 48 inches (120 cm) for a 100 percent expansion of the bed. As the bed expands, a shearing action due to the backwash water and some scrubbing action will free any material that might have formed on the resin particles during the softening stage.

During the backwash, a small amount of resin could be lost. This amount, however, should be minimal and you should check the backwash effluent at different intervals to ensure that the resin is not being lost. A glass beaker can be used to catch a sample of the effluent while the unit is backwashing. A trace amount of resin should cause no alarm, but a steady loss of resin could indicate a problem in the unit and the cause should be located and corrected as soon as possible. Too much loss of resin may be caused by an improper freeboard on the tank or wash troughs. The backwash duration and flow rate will vary depending on the manufacturer, the type and size of resin used, and the water temperature.

14.112 Brine

The third stage is most often termed the regeneration or brine stage. At this point, the sodium ion concentration of the resin is recharged by pumping a concentrated brine solution onto the resin. The solution is allowed to circulate through the unit and displace all water from the resin in order to provide full contact between the brine solution and the resin.

Most treatment plants use a brine solution to regenerate their softening units. The optimum brine concentration coming in contact with the ion exchange resin is around 10 to 14 percent sodium chloride solution. Concentrated brine is only used when the water within the softener tank serves as the dilution water. A 26 percent brine solution (fully concentrated or saturated) causes too great of an osmotic shock on the ion exchange resin and can cause it to break up. The salt dosage used to prepare the brine solution is one of the most important factors affecting the ion exchange capacity and ranges from 5 to 15 pounds of salt per cubic foot (80 to 240 kg/m^3) of resin. See Example 10 on page 112 for procedures on how to calculate the salt dosage and gallons of brine solution required. Brine concentrations less than saturated require longer contact time and more solution must be applied to the unit to achieve a successful regeneration.

The regeneration stage of the softener is very important and the operator should be certain it is properly carried out. In the regeneration stage, the sodium ions present in the brine solution are exchanged with the calcium and magnesium ions on the resin. The ions on the resin were exchanged during the service or softening stage. The regeneration rate is usually 1 to 2 GPM per cubic foot (2.2 to 4.4 liters per second per cubic meter) of resin for the first 55 minutes and then 3 to 5 GPM per cubic foot (6.6 to 11 liters per second per cubic meter) for the last 5 minutes of fast drain. If the regeneration process is performed correctly, the result is a bed that is completely recharged with sodium ions and will again soften water when the unit is returned to service.

14.113 Rinse

The fourth and final stage of softener operations is the rinse stage. After adequate contact time has been allowed between the brine solution and resin, a clear rinse is applied from the top of the unit to remove the waste products and excess brine solution from the softener. The flow pattern is very similar to the service stage except that the softener effluent goes to waste instead of storage. The waste discharge contains high concentrations of calcium and magnesium chloride. Most rinse stages will last between 20 and 40 minutes, depending on the size of the unit and the manufacturer. See Section 14.14, "Disposal of Spent Brine," for procedures on how to dispose of the waste discharge.

Again, the operator should pay close attention to the softener while it rinses. The rinse must be long enough to remove the heavy concentration of waste from the unit. If the rinse is not of the correct length and the unit returns to service, a salty taste will be very noticeable in the softener effluent. Taste the waste effluent near the end of the rinse stage to determine if the majority of chloride ions have been removed. The chloride ion concentration may also be measured by titration as outlined in Chapter 21, "Advanced Laboratory Procedures," or by measuring the conductivity of the water. If the water still has a strong salty taste or excessive chloride ions are present, check the rinse rate and timer settings. The unit may need adjustment to increase the duration of the rinse stage.

QUESTIONS

Please write your answers to the following questions and compare them with those on page 119.

14.11A What is the main consideration in determining the length of the service stage of an ion exchange softener?

14.11B What is the purpose of the backwash stage of an ion exchange softener?

14.11C How are ion exchange softeners regenerated?

14.11D Where does the softener effluent go during the rinse stage?

14.12 CONTROL TESTING OF ION EXCHANGE SOFTENERS

In most small treatment plants, the operator has to perform many jobs and may not always have time to monitor the softening units as they should be monitored. If a few simple test procedures are learned and carried out on a regular basis, the operator can feel confident the ion exchange units are operating properly. Control tests the operator should perform are listed in this section.

1. Softener Influent

   Be aware of the iron and manganese levels entering the softener. These levels should be kept to a minimum to prevent fouling of the media bed as the unit will remove a certain amount of iron and manganese before becoming plugged. Insoluble particles of iron and manganese will plug the filter media. Soluble ionic iron (Fe^{2+}) and manganese (Mn^{2+}) will exchange onto the media and will not be fully removed by regeneration. If the source water entering the plant is high in iron and manganese, proper oxidation and filtration of the water before the softener should reduce the levels and prevent problems from developing in the softeners.

   Monitor source water hardness on a routine basis. Generally, hardness will not vary, but if it changes, you will need to adjust the amount of water treated by each softener before the media becomes exhausted and the unit must be regenerated.

2. Softener Effluent

   At the end of a regeneration stage, as the unit goes back into service, check the effluent for hardness. This one test will tell you if the regeneration of the softener has been properly conducted. Allow a few minutes to ensure that all of the rinse water in the unit has been purged (removed). Run a hardness test on the effluent side of the unit. The results should indicate a water of zero hardness. Several test kits are available on the market today that are fairly quick and simple to use to measure water hardness.

14.13 LIMITATIONS CAUSED BY IRON AND MANGANESE

Ion exchange units are very versatile. The primary purpose of the unit is to remove calcium and magnesium from the water thus making the water soft. Ion exchange softeners, however, will also remove iron and manganese in either the soluble or precipitated form. If this occurs, the iron and manganese will seriously affect the life of the exchange capacity of the resin. If water high in iron and manganese is applied to the ion exchange resin for very long, iron fouling or the loss of exchange capacity will result.

When the softeners remove iron in the ferrous (soluble) or ferric (solid) form, two problems could result:

1. If water with iron in the ferrous (soluble) form is applied to the softener, the resin will remove the iron from the water. The iron can be retained on the surface of the resin or is sometimes captured deep inside the resin itself. As this happens, the resin bed will develop an orange or rusty appearance. If the resin becomes iron coated, the efficiency of the softener will be reduced greatly.
2. When water containing iron in the ferric (solid) form is applied to the unit, it will act like a filter and strain the iron from the water, leaving the iron trapped in the bed. This may cause plugging or clogging of the resin bed. If high iron loadings continue, the upper layer of the bed could become plugged, forcing the water to channel or short-circuit through the bed. The result is incomplete contact between the water and media, thus creating hardness leakage and loss of softening efficiency.

Iron and manganese must be reduced to their lowest possible limits before applying water to the softener. Oxidation of iron and manganese (see Chapter 12, "Iron and Manganese Control") before applying water to ion exchange units is very helpful. You should also be aware of the chlorine levels applied to the softening units. Normal chlorine dosages will not present a problem, but high residuals could damage the resin and reduce its life span.

14.14 DISPOSAL OF SPENT BRINE

One of the largest problems associated with the design and operation of ion exchange softening plants is the disposal of the softener waste.

The waste discharge from softeners consists mostly of calcium, magnesium, and sodium chlorides. These byproducts are corrosive to material they contact and possess varying toxic levels in relationship to the environment.

Many water treatment plants discharge spent brine into nearby sewers. This procedure may be approved if the downstream wastewater treatment plant and receiving waters can handle the brine. Usually, the water treatment plant must have some type of holding tank to store the spent brine. The brine is slowly discharged into the sewer at a rate that will not upset (or be toxic to) the biological treatment processes at the wastewater treatment plant. Also, the salt level in the effluent from the wastewater treatment plant must not adversely impact the aquatic life in the receiving waters or cause a violation of the wastewater treatment plant's *NPDES PERMIT.*[16]

Some water treatment plants may be issued an NPDES permit to discharge spent brine into receiving waters. This could happen only if the flow in the receiving waters was very high (plenty of dilution) and the flow of spent brine was very low. A holding tank would be needed for the spent brine and the brine could be discharged very slowly. The receiving waters would have to be monitored to be sure that the discharge of brine will not cause a significant increase in the level of brine.

Sanitary landfills also may be an acceptable means of disposing of spent brine. See Chapter 17, "Handling and Disposal of Process Wastes," for additional information.

Each ion exchange treatment plant probably has only one approved method of waste disposal. Very few options are available to plants discharging this type of waste. Alternate waste disposal methods available for spent brine are discussed in Chapter 17, "Handling and Disposal of Process Wastes."

The operator needs to be aware of the seriousness involved with softener waste. If a problem develops at the treatment plant, the operator should be working with the agency in the area that governs waste disposal as several considerations must be studied when changing a disposal method.

QUESTIONS

Please write your answers to the following questions and compare them with those on page 119.

14.12A Which water quality indicator should be monitored in the effluent of an ion exchange softener?

14.13A What happens when high chlorine residual levels are applied to the softening units?

14.14A Why is the disposal of spent brine a problem?

14.15 MAINTENANCE

Most of the ion exchange water softening equipment on the market today is fully automated (Figure 14.12). The reason for most of this automation is to reduce the time an operator must spend with each unit. Automation is fine for operational control, but it does not mean a unit is maintenance free. Systems like this have a tendency to lead operators astray. A small routine maintenance item can go unnoticed until it becomes a full-scale problem if the operator does not run a regular maintenance schedule on the equipment. For example, most valves on ion exchange units are pneumatic or are equipped with some type of self-operating device (valve actuators). This does not mean, however, the valve will operate each time it is required to do so without a regular examination and overhaul. The operator must check the equipment to ensure it is always in proper working order. One valve that fails to open or to close during a regeneration stage could mean trouble (a storage tank full of salty water or no brine at all).

The components of an ion exchange softening system that should receive constant attention are the brine pumps and piping. A saturated brine solution is very corrosive and will attack any unprotected metallic surface it comes in contact with. Try to keep the system as tight as possible. An uncontained brine leak will only get worse.

If you must change the pipe work in the brine system, give serious consideration to installing PVC pipe. The material is much cheaper than bronze and will outlast steel or galvanized pipe when properly installed and supported. Future repairs are also much easier to make if PVC pipe is used.

The pump on the brine system is most often made of brass, which offers some additional protection from the brine solution. The impeller should be bronze and the shaft stainless steel. A strainer or screen device should be installed ahead of the pump on the suction side.

---

16. *NPDES Permit.* National Pollutant Discharge Elimination System permit is the regulatory agency document issued by either a federal or state agency that is designed to control all discharges of potential pollutants from point sources and stormwater runoff into US waterways. NPDES permits regulate discharges into US waterways from all point sources of pollution, including industries, municipal wastewater treatment plants, sanitary landfills, large animal feedlots, and return irrigation flows.

Most treatment plants buy salt in bulk to make their brine solution. Regardless of the salt supplier, a certain amount of insoluble material will accompany each bulk delivery. This insoluble material will consist of rocks, coal, sand, and other particles that can clog or destroy an impeller if they reach the pump. Check the strainer assembly on the brine pumps quite often and keep spare parts on hand in case replacement is required.

The use of packing on the pumps is recommended over mechanical seals. Regardless of how well the strainers perform, a small amount of sand will usually end up in the pump. The combination of sand and mechanical seals will most often result in high repair and maintenance costs. Packing is cheaper and easier to install and maintain than mechanical seals and packing will usually outlast mechanical seals in this type of installation.

The brine pump motor should have a heavy-duty rating and a body made of cast iron is preferable. Aluminum or mild steel motor housings do not hold up as well as cast iron when subjected to the corrosive environment around the brine pumping station.

An area most often neglected until problems arise is the bulk brine storage area of the treatment plant. Most storage areas are underground pits equipped with rock or gravel strainers above some type of underdrain collection system. Over a period of time, the strainers will silt in with sand and impurities received with salt deliveries. The best way to prevent this from occurring is to regularly shut down, drain, and replace the strainer systems in the pit. This is a great deal of work, but it is a necessity if the brine system is to stay in operation.

Some brine storage areas have become so clogged that the brine solution could not penetrate the strainer media and reach the underdrain system. Like the head loss on a filter, the strainers can become so clogged that the solution cannot seep through to the underdrain system. If this happens and the system cannot be shut down for cleaning, a pipe can be driven down through the sand and impurities into the gravel layers. If enough holes are driven through the zones of impurities, the solution will eventually seep into the underdrains and can be pumped into the softeners. This is a temporary repair measure only and the storage area should be cleaned as soon as possible.

Inspect the brine solution makeup water line while the storage area is shut down. This line must be kept in good working order because it provides potable water to the salt supply. This water makes the saturated brine solution that is used to regenerate the softener. PVC pipe would provide excellent service in a corrosive environment such as a storage area.

Wet salt brine storage tanks are another location at a water treatment plant where sanitary defects may develop. The makeup water line must have a free fall or air gap someplace in the system to prevent the backflow of a brine solution into the potable water supply. The brine tanks must be protected from contamination just like any other water storage facility. The cover and access hatches must be of the raised-lip, overlapping-cover type. All vents must be properly screened to keep out insects, birds, and rodents.

Not every area of maintenance in an ion exchange softening plant can be covered here because each plant will differ with the type of equipment used and its method of operation. Set up a maintenance routine that is appropriate for your treatment plant. The objective of the maintenance routine must be to keep the plant operating and hold repair costs to a minimum.

QUESTIONS

Please write your answers to the following questions and compare them with those on page 119.

14.15A What could happen if a valve fails to open or close during a regeneration stage?

14.15B Why should the brine pumps and piping receive constant attention?

14.15C Why is packing on brine pumps recommended over mechanical seals?

14.16 TROUBLESHOOTING

14.160 Test Units

Ion exchange softeners, if properly operated and maintained, will usually provide years of trouble-free service. If problems arise, however, the operator should be able to identify and correct the situation without a great deal of difficulty.

The best way to ensure that a softener will continue to operate properly is to occasionally test the unit during various stages of operation. Learn to recognize minor problems as they develop and make the necessary repairs before a full-scale problem exists.

Items an operator should check on in each stage are discussed in this section.

14.161 Service Stage

In the service stage, while the unit is producing soft water, hardness tests should be run on the softener effluent to ensure the water has a hardness of zero. One grain hardness per gallon of water (17.1 mg/L) showing up in the effluent should not cause alarm, but concentrations of hardness higher than 1 grain hardness per gallon signal the need to investigate the softener's operation more closely.

14.162 Backwash Stage

Check the backwash stage for adequate flow rates and for sufficient time to complete the stage. Unless the flow rate is high enough to remove trapped turbidity particles and other insoluble material that is trapped in the resin, a loss of softener efficiency could result. Check the timer on the backwash stage to make sure the unit is washed for the required length of time.

If high iron concentrations have been applied to the softener, check the condition of the resin by visually inspecting the top layer of the bed. Color is a key factor to watch for in units beginning to show signs of an iron-fouled resin. The resin will be an orange, rusty color, while the backwash effluent will appear a light orange at the end of the backwash stage. Also, the head loss on the unit will run higher than normal as the bed becomes plugged with iron.

If iron fouling appears to be a problem, the length of the backwash stage should be increased to wash as much of the particulate matter from the resin as possible. A means of surface washing the resin must be provided for this procedure to be effective. Avoid exceeding recommended manufacturer's flow rates to prevent washing resin from the unit.

A chemical cleaner can be used to remove heavy iron coatings from the resin itself. These cleaners are mostly sodium hydrosulfite and can be mixed in solution form and poured into the softener. The hydrosulfite could also be added to the resin during the regeneration stage by dumping a concentrated powder form in with the brine solution. Consult the resin supplier or manufacturer before using any cleaner on the resin.

14.163 Brine Injection Stage

The brine injection stage of the softener sequence must be correctly applied or the unit will not perform satisfactorily when it returns to service. If the resin is not regenerated during this stage, there will be no sodium ions to exchange with the hardness ions in the water when the water is applied to the unit.

The brine storage area should always contain enough salt to provide a brine solution when makeup water is added.

Also, check the amount of salt solution that is pumped into the softener. This is usually done with a meter that is preset to deliver the exact amount of brine solution required to regenerate the softener. If a brine solution is less than saturated, longer contact time is required between the media and the solution.

The required amount of solution must be delivered consistently to achieve a successful regeneration of the unit. If hardness leakage appears early in the service or softening stage, check the amount and saturation of brine solution in the brine system since these are the main reasons for hardness leakage.

If hardness leakage is excessive immediately following a regeneration stage, shut the unit down and check the media level. The bed could be disrupted from excessive backwash or rinse rates. Iron fouling could also cause a channeling condition to occur and cause the water to short-circuit through the media without contacting the complete bed volume.

14.164 Rinse Stage

The rinse stage of the softener should be checked when tests indicate problems are developing. The rinse rate is a key factor in keeping the softener functioning properly. If the rinse starts too soon, the brine solution could be forced out of the unit before adequate contact time has elapsed. If the rinse rate is too low, all the waste material might not be removed from the unit before it goes into the service stage.

The rate settings on the unit should be compared, on all stages, against the actual manufacturer's recommended settings. As equipment ages, it wears. Over a period of time, valves might need adjustment to keep the unit operating within the manufacturer's guidelines.

QUESTIONS

Please write your answers to the following questions and compare them with those on page 119.

14.16A What is the maximum hardness level expected in the effluent of a properly operating ion exchange softener before the operator should investigate the operation more closely?

14.16B What is the purpose of the backwash stage of an ion exchange softener operation?

14.16C What would you do if hardness leakage is excessive immediately following a regeneration stage?

14.16D What problems may occur if the rinse stage starts too soon or the rinse rate is too low?

14.17 STARTUP AND SHUTDOWN OF UNIT

At times, it becomes necessary to shut a unit down and take it out of service for repairs or inspection. If the operator will follow common-sense guidelines, no problems should arise from unit shutdowns.

When dealing with these units, drain and fill them slowly. This will prevent surging of the media, which will either wash it out of the unit or disrupt it, thus making the media uneven and creating channeling problems. If you suspect problems with a unit, the last thing you want to do is make the situation worse by rapidly backfilling or backwashing the softener.

Most units are equipped with automatic air release valves (Figure 14.15). Be certain these valves are operating properly, because the venting of trapped air is an important step in filling a unit after it has been shut down.

During unit shutdown, make a complete visual inspection. Now is the time to detect and correct minor problems that might otherwise develop into bigger ones. Check the brine inlet distributors while the unit is down. They should be visible from the top hatch on most softeners.

Make sure the pipe work is level and the nozzles or openings are not clogged. The distributors play an important role in the regeneration stage by applying the brine solution evenly to the top of the resin bed. If the pipe work is deteriorating from the brine solution, PVC pipe should be used as a replacement.

If the resin and gravel support material are removed from the softener, check the underdrain structures in the unit and repair any problems you discover. In filling the unit with gravel and resin, each zone of the bed should be leveled and sized according to manufacturer's specifications.

The procedure for filling a unit with water after a total shutdown is very important. The flow into the unit should be from the bottom at a slow, controlled rate. This is done by putting the unit in the backwash position, running water into the unit from the bottom and out the backwash effluent valve. The purpose of this procedure is to fill the unit with water and purge the air that was trapped in the resin and softener during the replacement process.

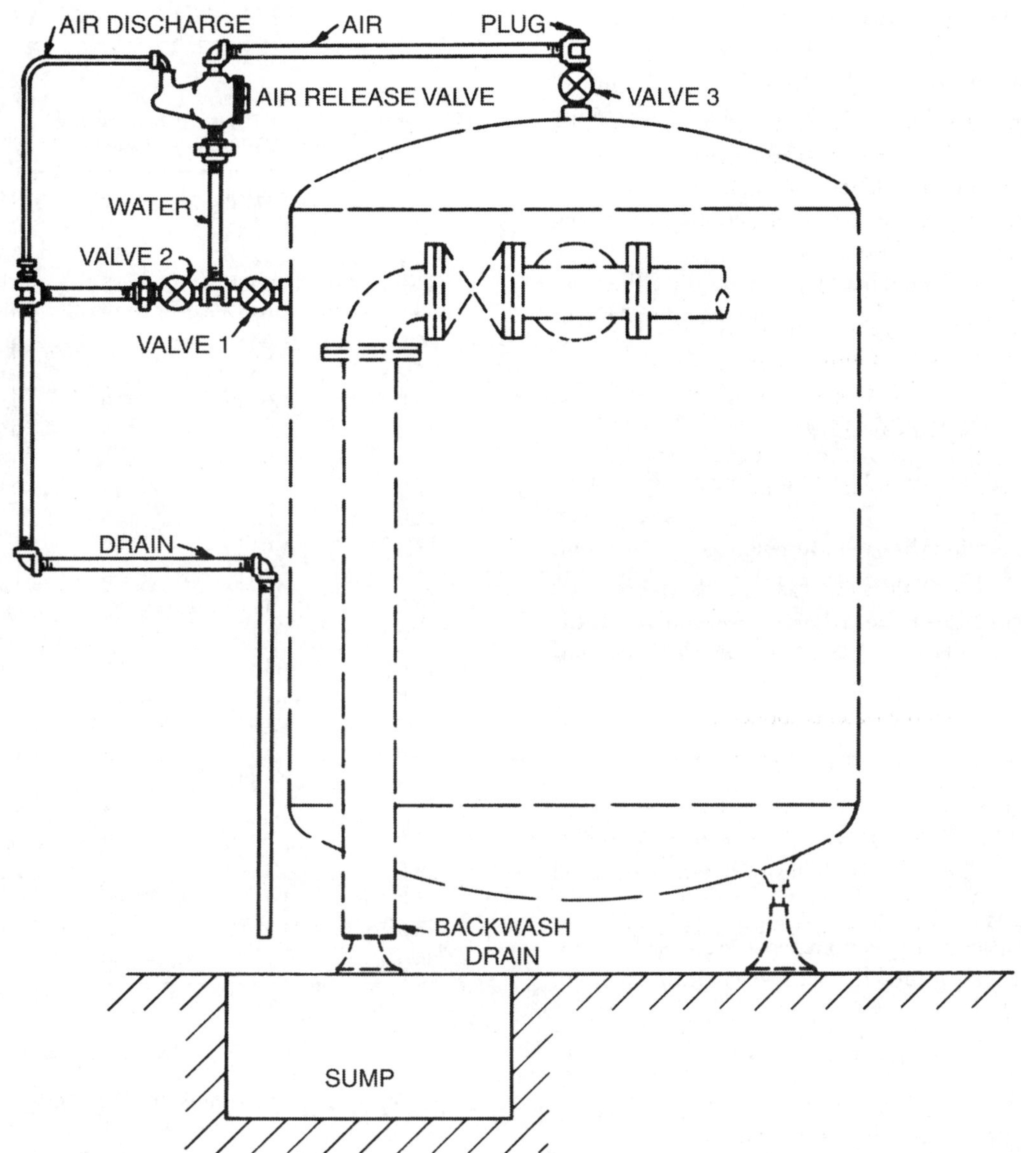

Pressure aeration and pressure filtration type filter plants require an automatic air release assembly to prevent accumulation of an excessive volume of air in the pressure filter tanks. This air release assembly consists of an automatic air release valve and necessary pipe, valves, and fittings to install on filter tank. The air release valve is a float operated type and must be installed with center line of valve level with or above top of filter tank. Air from top of filter enters air release valve at top connection and water from filter inlet pipe enters valve at bottom connection. Excessive air from filter fills valve body with air forcing water level down and thereby allowing float to drop. Downward movement of float allows excessive air to escape through the needle valve until an air–water pressure balance is restored.

In normal operation, valves 1 and 3 are open and valve 2 is closed. To flush air release valve, close valve 1 and open valve 2, which allows water from top of filter to flush down through the valve to the drain. Valve 3 is left open at all times unless it is necessary to remove air release valve.

Fig. 14.15 Air release assembly
(Permission of General Filter Company)

After the unit is filled, the backwash rate should be increased to normal and continued until the effluent is clear. Again, care should be taken when bringing the rates up to the manufacturer's recommendations to prevent disrupting or displacing resin from the bed.

Once the unit has been satisfactorily backwashed, the bed should be regenerated. This can be accomplished by running the softener through a normal brine and rinse procedure before it is returned to service. Run a hardness test on the effluent to ensure all stages have performed correctly and the unit is softening water.

QUESTIONS

Please write your answers to the following questions and compare them with those on page 119.

14.17A Why must ion exchange softeners be drained and filled slowly during startup and shutdown?

14.17B What should be done if the pipe work in an ion exchange softener is deteriorating from the brine solution?

14.18 ION EXCHANGE ARITHMETIC

Hardness is usually expressed as mg/L of $CaCO_3$. In ion exchange softening, however, hardness is most often expressed in terms of grains per gallon (gpg) or grains of hardness removed from the water being treated.

The exchange capacity of most softeners is expressed as kilograins (1,000 grains) of hardness removed per cubic foot of resin.

Salt, in solution form, is used to regenerate ion exchange softeners. The theoretical salt requirement is 0.17 pound (0.08 kg) of salt for 1,000 grains of hardness removed. Most regenerations, however, require 0.3 to 0.5 pound (0.14 to 0.23 kg) of salt per 1,000 grains of hardness removal.

In this section, you will learn how to calculate the volume of the brine solution required to regenerate the softening unit as well as the pounds of salt required for regeneration. The concentration of brine solution used at each treatment plant may vary. Table 14.6 lists the pounds of salt present in the percentage of brine solution being used.

TABLE 14.6 SALT SOLUTION CHARACTERISTICS

Percent NaCl or grams per 100 grams of solution	Specific Gravity at 15°C or 59°F	Salameter Degree	Lb NaCl per US Gal	Lb NaCl per Cubic Feet
1.0	1.0073	4	0.084	0.63
2.0	1.0145	8	0.169	1.27
3.0	1.0217	11	0.255	1.91
4.0	1.0290	15	0.343	2.57
5.0	1.0362	19	0.432	3.23
6.0	1.0437	23	0.522	3.90
7.0	1.0511	27	0.612	4.59
8.0	1.0585	30	0.705	5.28
9.0	1.0659	34	0.799	5.98
10.0	1.0734	38	0.874	6.69
11.0	1.0810	42	0.990	7.41
12.0	1.0885	45	1.09	8.14
13.0	1.0962	49	1.19	8.83
14.0	1.1038	53	1.29	9.63
15.0	1.1115	57	1.39	10.4
16.0	1.1194	60	1.49	11.2
17.0	1.1273	65	1.60	12.0
18.0	1.1352	68	1.70	12.7
19.0	1.1432	72	1.81	13.5
20.0	1.1511	76	1.92	14.4
21.0	1.1593	80	2.03	15.2
22.0	1.1676	84	2.14	16.0
23.0	1.1758	87	2.25	16.9
24.0	1.1840	91	2.37	17.7
25.0	1.1923	95	2.48	18.6
26.0	1.2010	99	2.60	19.5
26.4	1.2040	100	2.65	19.8

FORMULAS AND CONVERSION FACTORS

Hardness is usually expressed as milligrams of hardness per liter of water as $CaCO_3$.

Treatment for hardness is often discussed as grains of hardness per gallon of water.

1 grain per gallon = 17.1 milligrams per liter

or 1 gpg = 17.1 mg/L

7,000 grains = 1 pound

To convert grains per gallon to milligrams per liter.

$$\text{Hardness, mg/L} = \frac{(\text{Hardness, gpg})(17.1\text{ mg/L})}{1\text{ gpg}}$$

To convert milligrams per liter to grains per gallon.

$$\text{Hardness, gpg} = \frac{(\text{Hardness, mg/L})(1\text{ gpg})}{17.1\text{ mg/L}}$$

To find the exchange capacity (removal capacity) of a softener, you need to know the removal capacity of the softener in grains per cubic foot of resin or in kilograins per cubic foot of resin and the volume of the resin in cubic feet.

$$\text{Exchange Capacity, kilograins} = (\text{Removal Capacity, kilograins/ft}^3)(\text{Media Volume, ft}^3)$$

$$\text{Water Treated, gal} = \frac{\text{Exchange Capacity, kilograins}}{\text{Hardness Removed, kilograins/gal}}$$

$$\text{Operating Time, hr (At a Given Flow Rate Before Regeneration)} = \frac{(\text{Water Treated, gal})(24\text{ hr/day})}{\text{Avg Daily Flow, gal/day}}$$

To determine the amount of salt required for regeneration, you need to know the pounds of salt per 1,000 grains required for regeneration. To calculate the gallons of brine required for regeneration, you need to know the percent brine solution or the pounds of salt per gallon of brine.

$$\text{Salt Needed, lb} = (\text{Salt Required, lb/1,000 gr})(\text{Hardness Removed, gr})$$

$$\text{Brine, gallons} = \frac{\text{Salt Needed, lb}}{\text{Salt Solution, lb Salt/gallon of Brine}}$$

Example 6

How many milligrams of hardness per liter are there in a water with 16 grains of hardness per gallon of water?

Known	Unknown
Hardness, gpg = 16 gpg	Hardness, mg/L

Calculate the hardness of the water in milligrams per liter.

$$\text{Hardness, mg/L} = \frac{(\text{Hardness, gpg})(17.1\ \text{mg/L})}{1\ \text{gpg}}$$

$$= \frac{(16\ \text{gpg})(17.1\ \text{mg/L})}{1\ \text{gpg}}$$

$$= 274\ \text{mg/L}$$

Example 7

Convert the hardness of a water at 290 mg/L to grains per gallon.

Known	Unknown
Hardness, mg/L = 290 mg/L	Hardness, gpg

Convert the hardness from milligrams per liter to grains per gallon.

$$\text{Hardness, gpg} = \frac{(\text{Hardness, mg/L})(1\ \text{gpg})}{17.1\ \text{mg/L}}$$

$$= \frac{(290\ \text{mg/L})(1\ \text{gpg})}{17.1\ \text{mg/L}}$$

$$= 17\ \text{gpg}$$

Example 8

An ion exchange softener contains 50 cubic feet of resin with a hardness removal capacity of 20 kilograins per cubic foot of resin. The water being treated has a hardness of 300 mg/L as $CaCO_3$. How many gallons of water can be softened before the softener will require regeneration?

Known	Unknown
Resin Volume, ft^3 = 50 ft^3	Water Treated, gal
Removal Capacity, gr/ft^3 = 20,000 grains/ft^3	
Hardness, mg/L = 300 mg/L	

1. Convert the hardness from mg/L to grains per gallon.

$$\text{Hardness, gpg} = \frac{(\text{Hardness, mg/L})(1\ \text{gpg})}{17.1\ \text{mg/L}}$$

$$= \frac{(300\ \text{mg/L})(1\ \text{gpg})}{17.1\ \text{mg/L}}$$

$$= 17.5\ \text{gpg}$$

2. Calculate the exchange capacity of the softener in grains.

$$\text{Exchange Capacity, grains} = (\text{Resin Vol, ft}^3)(\text{Removal Capacity, grains/ft}^3)$$

$$= (50\ \text{ft}^3)(20{,}000\ \text{grains/ft}^3)$$

$$= 1{,}000{,}000\ \text{grains of Hardness Removal Capacity}$$

3. Calculate the volume of water in gallons that may be treated before regeneration.

$$\text{Water Treated, gal} = \frac{\text{Exchange Capacity, grains}}{\text{Hardness, gpg}}$$

$$= \frac{1{,}000{,}000\ \text{grains}}{17.5\ \text{gpg}}$$

$$= 57{,}143\ \text{gallons}$$

Therefore, 57,000 gallons of water with 17.5 grains of hardness per gallon of water can be treated before the resin becomes exhausted.

Example 9

An ion exchange softening plant has two softeners that are eight feet in diameter and the units have a resin depth of 6 feet. The resin has a 20-kilograin removal ability. How many gallons of water can be treated if the hardness is 14 grains per gallon? If the flow rate to the softeners is 500 gallons per minute, how long will they operate before regeneration is required?

Known	Unknown
Number of Softeners = 2 Softeners	1. Water Treated, gal
Diameter, ft = 8 ft	2. Operating Time, hr
Resin Depth, ft = 6 ft	
Removal Capacity, grains/ft^3 = 20,000 grains/ft^3	
Hardness, gpg = 14 gpg	
Flow, gallons/min = 500 gallons/min	

1. Calculate the total volume of softener media.

$$\text{Resin Vol, ft}^3 = (0.785)(\text{Diameter, ft})^2(\text{Depth, ft})(\text{No. Softeners})$$

$$= (0.785)(8\ \text{ft})^2(6\ \text{ft})(2\ \text{Softeners})$$

$$= 603\ \text{ft}^3$$

2. Calculate the total exchange capacity of the two softeners in grains.

$$\text{Exchange Capacity, grains} = (\text{Resin Vol, ft}^3)(\text{Removal Capacity, grains/ft}^3)$$

$$= (603\ \text{ft}^3)(20{,}000\ \text{grains/ft}^3)$$

$$= 12{,}060{,}000\ \text{grains of Exchange Capacity of the Beds}$$

3. Calculate the volume of water in gallons that may be treated before the resin is exhausted.

$$\text{Water Treated, gal} = \frac{\text{Exchange Capacity, grains}}{\text{Hardness, gpg}}$$

$$= \frac{12{,}060{,}000 \text{ grains}}{14 \text{ gpg}}$$

$$= 861{,}429 \text{ gallons Can Be Treated Before Resin Is Exhausted}$$

4. Find the length of time the softeners can run before requiring regeneration.

$$\text{Operating Time, hr} = \frac{\text{Water Treated, gal}}{(\text{Avg Daily Flow, gal/min})(60 \text{ min/hr})}$$

$$= \frac{861{,}429 \text{ gal}}{(500 \text{ gal/min})(60 \text{ min/hr})}$$

$$= 28.7 \text{ hours of Operation Before Regeneration Is Required}$$

Example 10

An ion exchange softener will remove 1,000,000 grains of hardness before the resin becomes exhausted. If 0.3 pounds of salt are required per 1,000 grains of hardness, how many pounds of salt are needed? If a 15 percent salt solution is used to regenerate the unit, how many gallons of brine are required? Table 14.6 indicates that 1.39 pounds of salt are present in each gallon of 15 percent brine solution.

Known		Unknown
Hardness Removal, grains	= 1,000,000 grains	1. Salt Needed, lb
Salt Required, lb/1,000 gr	= 0.3 lb/1,000 gr	2. Brine, gal
Salt Solution, lb/gal	= 1.39 lb/gal	

1. Determine the pounds of salt needed for regeneration.

$$\text{Salt Needed, lb} = (\text{Salt Required, lb/1,000 gr})(\text{Hardness Removal, gr})$$

$$= \frac{(0.3 \text{ lb Salt})(1{,}000{,}000 \text{ grains})}{(1{,}000 \text{ grains})}$$

$$= 300 \text{ lb of Salt}$$

2. Find the gallons of brine solution required.

$$\text{Brine, gal} = \frac{\text{Salt Needed, lb}}{\text{Salt Solution, lb/gallon of Brine}}$$

$$= \frac{300 \text{ lb of Salt}}{1.39 \text{ lb of Salt/gallon of Brine}}$$

$$= 216 \text{ gallons of Brine (15 percent Salt Solution)}$$

Example 11

Use the same information as in Example 10, except use a 12 percent brine solution. Table 14.6 indicates that 1.09 pounds of salt are present in each gallon of 12 percent brine solution. Three hundred pounds of salt are needed for regeneration. How many gallons of 12 percent brine solution are required?

Known		Unknown
Salt Needed, lb	= 300 lb	Brine, gal
Salt Solution, lb/gal	= 1.09 lb/gal	

Find the gallons of brine solution required.

$$\text{Brine, gal} = \frac{\text{Salt Needed, lb}}{\text{Salt Solution, lb/gal of Brine}}$$

$$= \frac{300 \text{ lb}}{1.09 \text{ lb/gal}}$$

$$= 275 \text{ gallons of 12-percent Brine Solution}$$

NOTE: More gallons of brine solution are required when using a 12 percent brine solution than when using a 15 percent solution. The weaker concentration requires more gallons to achieve the same results.

QUESTIONS

Please write your answers to the following questions and compare them with those on page 120.

14.18A A source water has a hardness of 150 mg/L as $CaCO_3$. What is the hardness in grains per gallon?

14.18B An ion exchange softener contains 60 cubic feet of resin with a hardness removal capacity of 25 kilograins per cubic foot of resin. The water being treated has a hardness of 250 mg/L as $CaCO_3$. How many gallons of water can be softened before the softener will require regeneration?

14.19 BLENDING

Ion exchange softeners will produce a water with zero hardness. Water with zero hardness must not be sent into a distribution system. Water with zero hardness is very corrosive and, over a period of time, will attack steel pipes in the system and create red water problems. Also, providing a water supply with zero hardness water would be very expensive.

At most softening plants, the zero hardness effluent from the softeners is mixed with filtered water having a known hardness concentration. In other words, a certain amount of water the treatment plant produces will bypass the softening units (split treatment). This water has a known hardness concentration and is mixed in various proportions with the softener effluent to arrive at a desired level of hardness in the finished water (Figure 14.16).

An example would be a treatment plant that has a filtered water hardness of 16 grains per gallon. If the desired plant effluent hardness is 8 grains per gallon, 50 percent of the plant influent must be softened and the other 50 percent would be filtered water mixed together with the softener effluent. The result would be water that has a hardness of 8 grains per gallon.

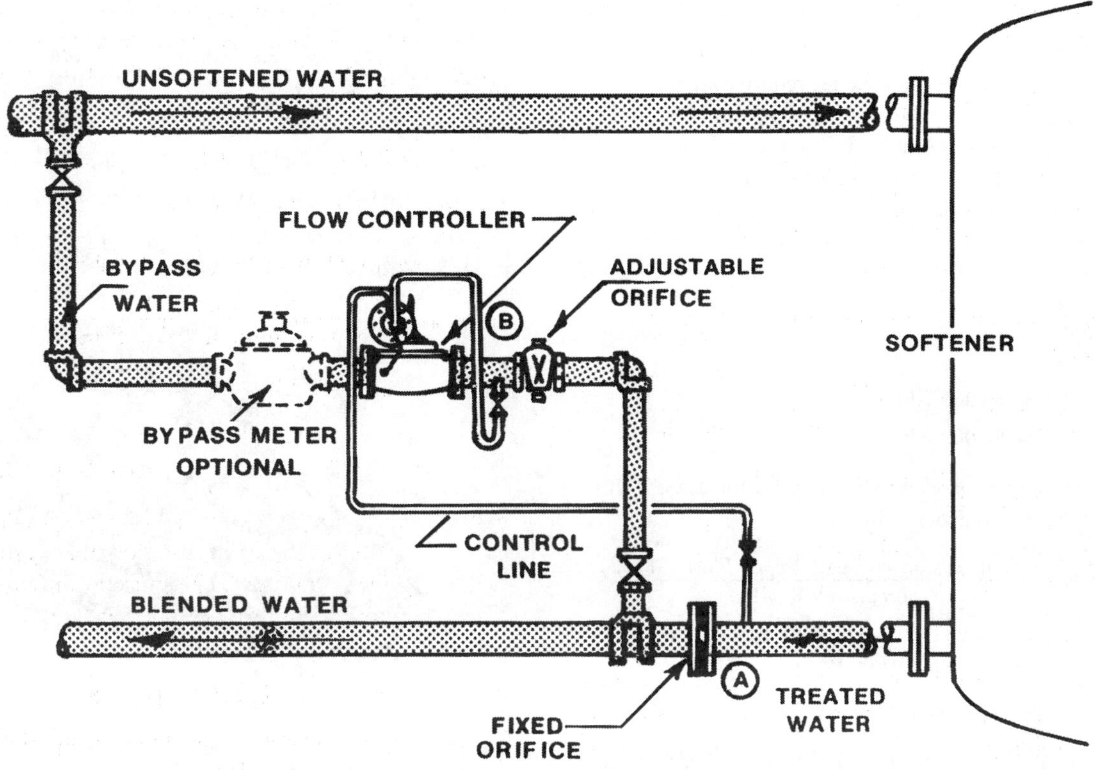

Fig. 14.16 Automatic softener bypass
(Permission of General Filter Company)

The blending of water is very simple and is usually controlled by a valve and meter. The operator adjusts the exact gallons per minute bypassing the softener to produce the desired hardness.

FORMULAS

To calculate the bypass flow in gallons per day to blend water, determine the total flow, the source water hardness, and the desired effluent hardness.

The softener capacity in gallons and both the softener and bypass flows in gallons per day are needed to determine the volume of bypass water.

The total flow produced by the plant before regeneration is the sum of the flows through the softener and the bypass flow.

$$\text{Bypass Flow, GPD} = \frac{(\text{Total Flow, GPD})(\text{Plant Effl Hardness, gpg})}{\text{Source Water Hardness, gpg}}$$

$$\text{Bypass Water, gal} = \frac{(\text{Softener Capacity, gal})(\text{Bypass Flow, GPD})}{\text{Softener Flow, GPD}}$$

$$\text{Total Flow, gal} = \text{Softener Capacity, gal} + \text{Bypass Water, gal}$$

Example 12

A softener plant treats 120,000 gallons per day. The source water has a hardness of 15 grains per gallon (256 mg/L) and the desired hardness in the plant effluent is 5 grains per gallon (86 mg/L). How much water in gallons per day must bypass the softener to produce the desired level of hardness?

Known		Unknown
Total Flow, GPD	= 120,000 GPD	Bypass Flow, GPD
Source Hardness, gpg	= 15 gpg	
Effl Hardness, gpg	= 5 gpg	

Calculate the bypass flow in gallons per day.

$$\text{Bypass Flow, GPD} = \frac{(\text{Total Flow, GPD})(\text{Plant Effl Hardness, gpg})}{\text{Source Water Hardness, gpg}}$$

$$= \frac{(120{,}000\ \text{GPD})(5\ \text{gpg})}{15\ \text{gpg}}$$

$$= 40{,}000\ \text{GPD}$$

Example 13

Using the information in Example 12, how many gallons of water will be bypassed before the softener requires regeneration? The softener has the capacity to treat 105,000 gallons. From Example 12, the bypass flow is 40,000 GPD and the total flow is 120,000 GPD. Therefore, the softener flow is 80,000 GPD (120,000 GPD – 40,000 GPD). What is the total flow produced by the plant per regeneration?

Known		Unknown
Softener Capacity, gal	= 105,000 gal	1. Bypass Water, gal
Softener Flow, GPD	= 80,000 GPD	2. Total Flow, gal
Bypass Flow, GPD	= 40,000 GPD	

1. Calculate the gallons of water that will be bypassed before the softener requires regeneration.

$$\text{Bypass Water, gal} = \frac{(\text{Softener Capacity, gal})(\text{Bypass Flow, GPD})}{\text{Softener Flow, GPD}}$$

$$= \frac{(105{,}000 \text{ gal})(40{,}000 \text{ GPD})}{80{,}000 \text{ GPD}}$$

$$= 52{,}500 \text{ gallons}$$

2. Determine the total flow produced by the plant per regeneration.

$$\text{Total Flow, gal} = \text{Softener Capacity, gal} + \text{Bypass Water, gal}$$

$$= 105{,}000 \text{ gal} + 52{,}500 \text{ gal}$$

$$= 157{,}500 \text{ gallons}$$

14.20 RECORDKEEPING

Keeping correct and up-to-date records is as important as performing scheduled maintenance on a regular routine. The recordkeeping system should be set up to record data on a daily basis. Record the total flow through the softener each day, along with the blend rates and gallons that have bypassed the unit. The total gallons of brine used each day, along with the pounds of salt used to keep the ion exchange softener in good working order should be recorded. Records of the results of tests performed on the softeners (source water, softener effluent, and blended water) should be kept up to date in order to warn the operator of any problems that might be developing with the softening unit. Good records are an important part of a successful treatment plant operation. Many problems can be avoided or solved with an adequate recordkeeping system if you review your daily records and compare them with the normal records to determine operating problems.

QUESTIONS

Please write your answers to the following questions and compare them with those on page 120.

14.19A Why is the zero hardness effluent from the ion exchange softener mixed (blended) with filtered water having a known hardness concentration?

14.20A What records should be kept by the operator of an ion exchange softening plant?

14.21 ARITHMETIC ASSIGNMENT

Turn to the Arithmetic Appendix at the back of this manual. Read and work the problems in Section A.32, "Softening." Check the arithmetic in this section using a calculator. You should be able to get the same answers. Section A.52 contains similar problems using metric units.

14.22 ADDITIONAL READING

Texas Manual, chapter on "Special Water Treatment (Softening and Ion Exchange)."

14.23 ACKNOWLEDGMENTS

Portions of the material discussed on ion exchange softening came from the following sources:

1. Bowers, Eugene, "Ion Exchange Softening" in *Water Quality and Treatment*, 4th Ed., published by American Water Works Association (AWWA), 6666 West Quincy Avenue, Denver, CO 80235.
2. Lipe, L. A. and M. D. Curry, "Ion Exchange Water Softening," a discussion for water treatment plant operators, 1974–75 seminar series sponsored by Illinois Environmental Protection Agency.
3. Hoover, Charles P., *Hoover's Water Supply and Treatment*, revised by Merrill L. Riehl, 1994. Available from the National Lime Association at lime.org or phone (703) 243-5463. Order No. 211.

END OF LESSON 2 OF 2 LESSONS

on

SOFTENING

Please answer the discussion and review questions next.

DISCUSSION AND REVIEW QUESTIONS

Chapter 14. SOFTENING

(Lesson 2 of 2 Lessons)

Please write your answers to the following questions to determine how well you understand the material in the lesson. The question numbering continues from Lesson 1.

13. What happens in the resin or media in an ion exchange softener during the softening stage?
14. How would you ensure that large amounts of resin are not being lost during the backwash stage?
15. How would you determine if an ion exchange softener rinse stage has been successful?
16. What happens if an ion exchange softener removes iron in the ferrous (soluble) or ferric (solid) form?
17. What types of insoluble material may be found in salt? What problems can be caused by this material and how can these problems be prevented?
18. How would you prevent the strainers under the bulk brine storage area from silting in with sand and impurities?
19. How would you determine if iron has fouled the resin of an ion exchange softener?
20. How are ion exchange units filled with water after total shutdown?
21. Why is water with zero hardness not delivered to consumers?

SUGGESTED ANSWERS

Chapter 14. SOFTENING

ANSWERS TO QUESTIONS IN LESSON 1

Answers to questions on page 78.

14.0A Hardness is caused mainly by the calcium and magnesium ions in water.

14.0B Excessive hardness is undesirable because it causes the formation of soap curds, increased use of soap, deposition of scale in boilers, damage in some industrial processes, and sometimes causes objectionable tastes in drinking water.

14.1A Limitations of the ion exchange softening process include an increase in the sodium content of the softened water and the ultimate disposal of spent brine and rinse waters from softeners can be a major problem for many installations.

Answers to questions on page 82.

14.2A Hardness is commonly measured by titration. Individual divalent cations may be measured in the laboratory using an atomic absorption (AA) spectrophotometer.

14.2B Determine the total hardness as $CaCO_3$ for a sample of water with a calcium content of 25 mg/L and a magnesium content of 14 mg/L.

Known		Unknown
Calcium, mg/L	= 25 mg/L	Total Hardness, mg/L as $CaCO_3$
Magnesium, mg/L	= 14 mg/L	

Calculate the total hardness as milligrams per liter of calcium carbonate equivalent.

$$\text{Total Hardness, mg/L as } CaCO_3 = \text{Calcium Hardness, mg/L as } CaCO_3 + \text{Magnesium Hardness, mg/L as } CaCO_3$$

$$= 2.50(\text{Ca, mg/L}) + 4.12(\text{Mg, mg/L})$$

$$= 2.50(25 \text{ mg/L}) + 4.12(14 \text{ mg/L})$$

$$= 62.5 \text{ mg/L} + 57.7 \text{ mg/L}$$

$$= 120.2 \text{ mg/L as } CaCO_3$$

14.2C Water treatment chemicals that lower the pH when added to water include alum, carbon dioxide, chlorine (Cl_2), ferric chloride, hydrofluosilicic acid, and sulfuric acid.

14.2D Results from alkalinity titrations on a sample of water were as follows: sample size, 100 mL; mL titrant used to pH 8.3, 1.2 mL; total mL of titrant used, 5.6 mL; and the acid normality was 0.02 N H_2SO_4. Calculate the total, bicarbonate, carbonate, and hydroxide alkalinity as $CaCO_3$.

Known

Sample Size, mL	= 100 mL
mL Titrant Used to pH 8.3, A	= 1.2 mL
Total mL of Titrant Used, B	= 5.6 mL
Acid Normality, N	= 0.02 N H_2SO_4

Unknown

1. Total Alkalinity, mg/L as $CaCO_3$
2. Bicarbonate Alkalinity, mg/L as $CaCO_3$
3. Carbonate Alkalinity, mg/L as $CaCO_3$
4. Hydroxide Alkalinity, mg/L as $CaCO_3$

1. Calculate the phenolphthalein alkalinity in mg/L as $CaCO_3$.

$$\text{Phenolphthalein Alkalinity, mg/L as } CaCO_3 = \frac{A \times N \times 50{,}000}{\text{mL of Sample}}$$

$$= \frac{(1.2 \text{ mL}) \times (0.02\ N) \times 50{,}000}{100 \text{ mL}}$$

$$= 12 \text{ mg/L as } CaCO_3$$

2. Calculate the total alkalinity in mg/L as $CaCO_3$.

$$\text{Total Alkalinity, mg/L as } CaCO_3 = \frac{B \times N \times 50{,}000}{\text{mL of Sample}}$$

$$= \frac{(5.6 \text{ mL}) \times (0.02\ N) \times 50{,}000}{100 \text{ mL}}$$

$$= 56 \text{ mg/L as } CaCO_3$$

3. Refer to Table 14.4 for alkalinity constituents.

From Table 14.4 we want the second row because P = 12 mg/L, which is less than ½T[(½)(56 mg/L) = 28 mg/L]. Therefore,

a. $$\text{Bicarbonate Alkalinity, mg/L as } CaCO_3 = T - 2P$$

$$= 56 \text{ mg/L} - 2(12 \text{ mg/L})$$

$$= 32 \text{ mg/L as } CaCO_3$$

b. $$\text{Carbonate Alkalinity, mg/L as } CaCO_3 = 2P$$

$$= 2(12 \text{ mg/L})$$

$$= 24 \text{ mg/L as } CaCO_3$$

c. $$\text{Hydroxide Alkalinity, mg/L as } CaCO_3 = 0 \text{ mg/L as } CaCO_3$$

Answers to questions on page 83.

14.3A The minimum hardness that can be achieved by the lime–soda ash process is around 30 to 40 mg/L as $CaCO_3$.

14.3B Benefits that could result from the lime–soda softening process in addition to softening the water include:

1. Removal of iron and manganese
2. Reduction of solids
3. Removal and inactivation of bacteria and viruses due to high pH
4. Control of corrosion and scale formation with proper stabilization of treated water
5. Removal of excess fluoride

Answers to questions on pages 84 and 85.

14.3C The pH increases during the lime–soda softening process when lime is added to water, which increases the hydroxide concentrations.

14.3D After the chemical softening process, the scale-forming tendencies of water can be decreased by the use of carbon dioxide.

14.3E Recarbonation is a process in which carbon dioxide is bubbled into the water being treated to lower the pH. Recarbonation may be carried out in two steps. The first addition of carbon dioxide would follow excess lime addition to lower the pH to about 10.4 and encourage the precipitation of calcium carbonate and magnesium hydroxide. A second addition of carbon dioxide after treatment removes noncarbonate hardness. By carrying out recarbonation prior to filtration, the buildup of excess lime and calcium carbonate and magnesium hydroxide precipitates in the filters will be prevented or minimized.

14.3F Caustic soda softening might be used in place of soda ash. The decision to use caustic soda rather than soda ash depends on the quality of the source water and the delivered costs of various chemicals.

Answer to question on page 86.

14.3G Calculate the hydrated lime ($Ca(OH)_2$) with 90 percent purity, soda ash, and carbon dioxide dose requirements in milligrams per liter for the water shown below.

Known

Constituents	Source Water	Softened Water After Recarbonation and Filtration
CO_2, mg/L	= 5 mg/L	= 0 mg/L
Total Alkalinity, mg/L	= 150 mg/L as $CaCO_3$	= 20 mg/L as $CaCO_3$
Total Hardness, mg/L	= 240 mg/L as $CaCO_3$	= 50 mg/L as $CaCO_3$
Mg^{2+}, mg/L	= 16 mg/L	= 2 mg/L
pH	= 7.4	= 8.8
Lime Purity, %	= 90%	

Unknown

1. Hydrated Lime, mg/L
2. Soda Ash, mg/L
3. Carbon Dioxide, mg/L

1. Calculate the hydrated lime ($Ca(OH)_2$) required in milligrams per liter.

$$A = (CO_2, \text{mg/L})(74/44)$$
$$= (5 \text{ mg/L})(74/44)$$
$$= 8 \text{ mg/L}$$

$$B = (\text{Alkalinity, mg/L})(74/100)$$
$$= (150 \text{ mg/L} - 20 \text{ mg/L})(74/100)$$
$$= 96 \text{ mg/L}$$

$$C = 0 \qquad \text{Hydroxide Alkalinity} = 0$$

$$D = (Mg^{2+}, \text{mg/L})(74/24.3)$$
$$= (16 \text{ mg/L} - 2 \text{ mg/L})(74/24.3)$$
$$= 43 \text{ mg/L}$$

$$\text{Hydrated Lime } (Ca(OH)_2) \text{ Feed, mg/L} = \frac{(A + B + C + D)1.15}{\text{Purity of Lime, as a decimal}}$$

$$= \frac{(8 \text{ mg/L} + 96 \text{ mg/L} + 0 + 43 \text{ mg/L})1.15}{0.90}$$

$$= \frac{(147 \text{ mg/L})(1.15)}{0.90}$$

$$= 188 \text{ mg/L}$$

2. Calculate the soda ash required in milligrams per liter.

$$\text{Total Hardness Removed, mg/L as } CaCO_3 = \text{Total Hardness, mg/L as } CaCO_3 - \text{Total Hardness Remaining, mg/L as } CaCO_3$$
$$= 240 \text{ mg/L} - 50 \text{ mg/L}$$
$$= 190 \text{ mg/L as } CaCO_3$$

$$\text{Noncarbonate Hardness, mg/L as } CaCO_3 = \text{Total Hardness Removed, mg/L as } CaCO_3 - \left(\text{Carbonate Hardness, mg/L as } CaCO_3 - \text{Carbonate Hardness Remaining, mg/L as } CaCO_3\right)$$
$$= 190 \text{ mg/L} - (150 \text{ mg/L} - 20 \text{ mg/L})$$
$$= 60 \text{ mg/L as } CaCO_3$$

$$\text{Soda Ash } (Na_2CO_3) \text{ Feed, mg/L} = \left(\text{Noncarbonate Hardness, mg/L as } CaCO_3\right)(106/100)$$
$$= (60 \text{ mg/L})(106/100)$$
$$= 63.6 \text{ mg/L}$$

3. Calculate the dosage of carbon dioxide required for recarbonation.

$$\text{Excess Lime, mg/L} = (A + B + C + D)(0.15)$$
$$= (8 \text{ mg/L} + 96 \text{ mg/L} + 0 + 43 \text{ mg/L})(0.15)$$
$$= (147 \text{ mg/L})(0.15)$$
$$= 22 \text{ mg/L}$$

$$\text{Total } CO_2 \text{ Feed, mg/L} = (Ca(OH)_2 \text{ excess, mg/L})(44/74) + (Mg^{2+} \text{ residual, mg/L})(44/24.3)$$
$$= (22 \text{ mg/L})(44/74) + (2 \text{ mg/L})(44/24.3)$$
$$= 13 \text{ mg/L} + 4 \text{ mg/L}$$
$$= 17 \text{ mg/L}$$

Answers to questions on page 87.

14.3H In the lime softening process, calcium is precipitated out as calcium carbonate and magnesium as magnesium hydroxide.

14.3I Partial lime softening (no magnesium removal) removes hardness caused by calcium ions. This may be referred to as calcium hardness.

14.3J In split lime treatment, a portion of the water is treated with excess lime to remove the magnesium at a high pH. Then, source water (the remaining portion) is added in the next basin to neutralize (lower the pH) the excess-lime-treated portion.

Answers to questions on page 89.

14.3K Lime–soda ash softening is used when lime alone will not remove enough hardness.

14.3L Noncarbonate hardness is removed by the addition of soda ash (sodium carbonate) in the chemical precipitation softening process.

Answers to questions on page 90.

14.3M Where the daily requirements for lime are small, lime is usually delivered to the water treatment plant in bags.

14.3N Considerable heat is generated if quicklime accidentally gets wet.

14.3O Lime may be applied by dry feeding techniques using volumetric or gravimetric feeders.

Answers to questions on page 90.

14.4A When the alum dose increases for coagulation, the lime dose must be increased also.

14.4B Color can be removed from water by coagulation with alum at low pH values. The high pH values required during softening tend to set the color, which then becomes very difficult to remove.

Answers to questions on page 91.

14.5A A slight excess of lime can cause a scale to form on the filter sand, distribution mains, and household plumbing.

14.5B Excess caustic and unprecipitated carbonate ions (pin floc) can be removed from softened water by recarbonation. Recarbonation is the bubbling of carbon dioxide through the water being treated to lower the pH. Recarbonation can be accomplished, to a degree, by using source water in the split treatment mode.

14.5C The marble test is used to determine if a water is stable. The Langelier Index is also used to determine the corrosivity of water.

14.5D Suspending a couple of nails on strings in a filter can indicate if the water is stable. If the nails are rusting, the water is corrosive. If a scale forms on the nails, then scale is forming on your filter media and in your distribution system.

Answers to questions on page 92.

14.6A Wooden paddles should be used as cleaning tools on any slaker in operation. A metal tool will damage the slaker and could even injure the operator if dropped by accident. However, a wooden paddle will likely be broken up with no damage to the equipment or the operator.

14.6B Information on how to safely maintain equipment may be found in equipment manuals provided by equipment suppliers and manufacturers.

Answers to questions on page 93.

14.7A A disadvantage of recirculating sludge back to the primary mix area is that an increase in magnesium could result.

14.7B Only trial and error will really determine if sludge recirculation will serve a useful purpose in your plant.

14.8A Records should be kept on the amounts of treatment plant chemicals ordered and the amounts fed.

Answers to questions on page 97.

14.9A If lime added to water does not reduce the hardness of a water sufficiently, use the optimum lime dose and run jar tests with varying soda ash doses. Select the soda ash dose that will produce a water with a hardness of around 80 to 90 mg/L.

14.9B When selecting a target hardness level for a water softening plant, consider the uses of the softened water and the cost of softening.

Answers to questions on page 98.

14.9C The overfeeding of chemicals is a waste of money and quality control will suffer.

14.9D What should be the lime feeder setting in pounds per day to treat a flow of 2 MGD when the optimum lime dose is 160 mg/L?

Known		Unknown
Flow, MGD	= 2 MGD	Feeder Setting, lb/day
Lime Dose, mg/L	= 160 mg/L	

Calculate the lime feeder setting in pounds per day.

$$\text{Feeder Setting, lb/day} = (\text{Flow, MGD})(\text{Lime, mg/L})(8.34\ \text{lb/gal})$$
$$= (2\ \text{MGD})(160\ \text{mg/L})(8.34\ \text{lb/gal})$$
$$= 2{,}669\ \text{lb/day}$$

14.9E How much soda ash is required in pounds per day to remove 40 mg/L of noncarbonate hardness from a flow of 2 MGD?

Known		Unknown
Flow, MGD	= 2 MGD	Feeder Setting, lb/day
Hardness Removed, mg/L	= 40 mg/L	

1. Calculate the soda ash dose in milligrams per liter.

$$\text{Soda Ash, mg/L} = (1.06)(\text{Hardness Removed, mg/L})$$
$$= (1.06)(40\ \text{mg/L})$$
$$= 42.4\ \text{mg/L}$$

2. Determine the soda ash feeder setting in pounds per day.

$$\text{Feeder Setting, lb/day} = (\text{Flow, MGD})(\text{Soda Ash, mg/L})(8.34\ \text{lb/gal})$$

$$= (2\ \text{MGD})(42.4\ \text{mg/L})(8.34\ \text{lb/gal})$$

$$= 707\ \text{lb/day}$$

ANSWERS TO QUESTIONS IN LESSON 2

Answers to questions on page 102.

14.10A There are three basic types of softeners on the market:

1. An upflow unit
2. A gravity flow unit
3. A pressure downflow unit (the most common)

14.10B During the regeneration cycle, the softener is taken out of service. Salt, in the form of a concentrated brine solution, is used to regenerate (recharge) the ion exchange resin. When the brine solution is fed into the resin, the sodium cations are exchanged for calcium and magnesium cations. As the brine solution travels down through the resin, the sodium cations are attached to the resin while the calcium, magnesium, and chloride (from the salt) ions flow to waste. After the regeneration has taken place, the bed is ready to be placed in service again to remove calcium and magnesium by ion exchange.

Answers to questions on page 105.

14.11A The source water hardness is the main consideration in determining the length of the service stage of an ion exchange softener.

14.11B The purpose of the backwash stage is to expand and clean the media or resin particles and to free any material such as iron, manganese, and particulates that might have been removed during the softening stage.

14.11C Ion exchange softeners are regenerated by the use of a concentrated brine solution.

14.11D During the rinse stage, the softener effluent goes to waste.

Answers to questions on page 106.

14.12A Hardness should be monitored in the effluent of an ion exchange softener.

14.13A High chlorine residual levels applied to softening units could damage the resin and reduce its life span.

14.14A The disposal of spent brine is a problem because the brine is very corrosive and toxic to many living things in the environment.

Answers to questions on page 107.

14.15A One valve that fails to open or close during a regeneration stage could mean a storage tank full of salty water or no brine.

14.15B Brine pumps and piping must receive constant attention because a saturated brine solution is very corrosive and will attack any unprotected metallic surface it comes in contact with. Try to keep the system as tight as possible. An uncontained brine leak will only get worse.

14.15C Packing is recommended over mechanical seals on brine pumps because, regardless of how well the strainers perform, a small amount of sand will usually end up in the pump. The combination of sand and mechanical seals will most often result in high repair and maintenance costs. Packing is cheaper and easier to install and maintain than mechanical seals and packing will usually outlast mechanical seals in this type of installation.

Answers to questions on page 108.

14.16A The maximum expected hardness level in the effluent from an ion exchange softener should not exceed 1 grain hardness per gallon (17.1 mg/L). Concentrations of hardness higher than 1 grain hardness per gallon signal the need to investigate the softener's operation more closely.

14.16B The purpose of the backwash stage is to remove trapped turbidity particles and other insoluble material that is trapped in the resin.

14.16C If hardness leakage is excessive immediately following a regeneration stage, shut the unit down and check the media level. The bed could be disrupted from excessive backwash or rinse rates. Iron fouling could also cause a channeling condition to occur and cause the water to short-circuit through the media without contacting the complete bed volume.

14.16D If the rinse stage starts too soon, the brine solution could be forced out of the unit before adequate contact time has elapsed. If the rinse rate is too low, all the waste material might not be removed from the unit before it goes into the service stage.

Answers to questions on page 110.

14.17A Ion exchange softeners must be drained and filled slowly during startup and shutdown to prevent surging of the media, which will either wash it out of the unit or disrupt it, thus making the media uneven and creating channeling problems.

14.17B If the pipe work in an ion exchange softener is deteriorating from the brine solution, PVC pipe should be used as a replacement.

Answers to questions on page 112.

14.18A A source water has a hardness of 150 mg/L as $CaCO_3$. What is the hardness in grains per gallon?

Known	Unknown
Hardness, mg/L = 150 mg/L	Hardness, gpg

Calculate the source water hardness in grains per gallon.

$$\text{Hardness, gpg} = \frac{(\text{Hardness, mg/L})(1\text{ gpg})}{17.1\text{ mg/L}}$$

$$= \frac{(150\text{ mg/L})(1\text{ gpg})}{17.1\text{ mg/L}}$$

$$= 8.8\text{ grains/gal}$$

14.18B An ion exchange softener contains 60 cubic feet of resin with a hardness removal capacity of 25 kilograins per cubic foot of resin. The water being treated has a hardness of 250 mg/L as $CaCO_3$. How many gallons of water can be softened before the softener will require regeneration?

Known	Unknown
Resin Volume, ft^3 = 60 ft^3	Water Treated, gal
Removal Capacity, gr/ft^3 = 25,000 grains/ft^3	
Hardness, mg/L = 250 mg/L	

1. Convert the hardness from mg/L to grains per gallon.

$$\text{Hardness, gpg} = \frac{(\text{Hardness, mg/L})(1\text{ grain/gal})}{17.1\text{ mg/L}}$$

$$= \frac{(250\text{ mg/L})(1\text{ grain/gal})}{17.1\text{ mg/L}}$$

$$= 14.6\text{ gpg}$$

2. Calculate the exchange capacity of the softener in grains.

$$\text{Exchange Capacity, grains} = (\text{Resin Vol, ft}^3)(\text{Removal Capacity, grains/ft}^3)$$

$$= (60\text{ ft}^3)(25{,}000\text{ grains/ft}^3)$$

$$= 1{,}500{,}000\text{ grains of Hardness Removal Capacity}$$

3. Calculate the volume of water in gallons that may be treated before regeneration.

$$\text{Water Treated, gal} = \frac{\text{Exchange Capacity, grains}}{\text{Hardness, grains/gallon}}$$

$$= \frac{1{,}500{,}000\text{ grains}}{14.6\text{ grains/gal}}$$

$$= 102{,}740\text{ gal}$$

Therefore, 102,700 gallons of water with 14.6 grains of hardness per gallon of water can be treated before the resin becomes exhausted.

Answers to questions on page 114.

14.19A Zero hardness effluent from the ion exchange softener is mixed (blended) with filtered water having a known hardness concentration to achieve a desired level of hardness in the finished water. Water with zero hardness must not be sent into a distribution system. Water with zero hardness is very corrosive and, over a period of time, will attack steel pipes in the system and create red water problems. Also, providing a water supply with zero hardness water would be very expensive.

14.20A Records that should be kept by the operator of an ion exchange softening plant include:

1. Total daily flow through unit
2. Blend rates
3. Total daily gallons that have bypassed unit
4. Gallons of brine used each day
5. Pounds of salt used each day
6. Results of tests performed on source water, softener effluent, and blended water

CHAPTER 15

SPECIALIZED TREATMENT PROCESSES

by

Mike McGuire

and

Chet Anderson

NOTICE

Contact your state drinking water agency to obtain current information about rules and regulations that apply to your water utility. For information about federal drinking water regulations, visit the US EPA online at water.epa.gov/drink/hotline/ or call the Safe Drinking Water Hotline at (800) 426-4791.

TABLE OF CONTENTS

Chapter 15. SPECIALIZED TREATMENT PROCESSES

LEARNING OBJECTIVES

Chapter 15. SPECIALIZED TREATMENT PROCESSES

Following completion of Chapter 15, you should be able to:

1. Describe how trihalomethanes are formed.
2. Explain why trihalomethanes are a problem in drinking water.
3. Collect samples for trihalomethane analysis.
4. Identify control strategies for trihalomethanes.
5. Describe treatment processes capable of controlling trihalomethanes.
6. Select and implement a cost-effective means of controlling trihalomethanes.
7. Describe why we are concerned about arsenic.
8. Review the various water treatment processes used for reduction or removal of arsenic.
9. Safely operate, maintain, and troubleshoot a typical arsenic treatment plant.
10. Review the plans and specifications for an arsenic treatment plant.
11. Properly dispose of wastewater and residuals from an arsenic treatment plant.
12. Monitor an arsenic treatment plant for process control and compliance.
13. Properly perform recordkeeping and reporting tasks for an arsenic treatment plant.

WORDS

Chapter 15. SPECIALIZED TREATMENT PROCESSES

ABSORPTION (ab-SORP-shun) ABSORPTION

The taking in or soaking up of one substance into the body of another by molecular or chemical action (as tree roots absorb dissolved nutrients in the soil).

ACTIVATED ALUMINA ACTIVATED ALUMINA

A charged form of aluminum, used with a synthetic, porous media in an ion exchange adsorption process to remove charged contaminants.

ADSORPTION (add-SORP-shun) ADSORPTION

The gathering of a gas, liquid, or dissolved substance on the surface or interface zone of another material.

CARCINOGEN (kar-SIN-o-jen) CARCINOGEN

Any substance that tends to produce cancer in an organism.

COPRECIPITATION COPRECIPITATION

A treatment process that occurs when ferrous iron is added to water or metallic wastestreams and subsequently oxidized in an aerator. The oxidized iron, which is insoluble, precipitates along with other metallic contaminants present in the water or wastestream, thereby enhancing metals removal.

EMPTY BED CONTACT TIME (EBCT) EMPTY BED CONTACT TIME (EBCT)

A measure of the time during which a water to be treated is in contact with the treatment medium in a contact vessel, assuming that all liquid passes through the vessel at the same velocity. EBCT is equal to the volume of the empty bed divided by the flow rate.

ENDOCRINE (EN-doe-krin) EFFECTS ENDOCRINE EFFECTS

An altering of the organs in the human body responsible for secreting hormones into the bloodstream. Endocrine glands include the thyroid gland, the pancreas, and the adrenal glands.

HALOACETIC (HAL-o-uh-SEE-tick) ACID (HAA) HALOACETIC ACID (HAA)

A class of disinfection byproducts, formed mainly during the chlorination of water, containing natural organic matter. HAA5 is the sum of the concentrations, in milligrams per liter, of five haloacetic acid compounds.

MCL MCL

Maximum Contaminant Level. The largest allowable amount. MCLs for various water quality indicators are specified in the National Primary Drinking Water Regulations (NPDWR).

NONVOLATILE MATTER NONVOLATILE MATTER

Material such as sand, salt, iron, calcium, and other mineral materials that are only slightly affected by the actions of organisms and are not lost on ignition of the dry solids at 1,022°F (550°C). Volatile materials are chemical substances usually of animal or plant origin. Also see INORGANIC WASTE and VOLATILE SOLIDS.

NSF NSF

NSF International is a noncommercial, not-for-profit organization concerned with public health safety and environmental protection. NSF Standard 60 lists certified drinking water chemicals and NSF Standard 61 lists certified drinking water system components.

PRECURSOR, THM (PRE-curse-or) PRECURSOR, THM

Natural, organic compounds found in all surface and groundwaters, which may react with halogens (such as chlorine) to form trihalomethanes (THMs); they must be present in order for THMs to form.

REPRESENTATIVE SAMPLE REPRESENTATIVE SAMPLE

A sample portion of material, water, or wastestream that is as nearly identical in content and consistency as possible to that in the larger body being sampled.

SPLIT SAMPLE SPLIT SAMPLE

A single grab sample that is separated into at least two parts such that each part is representative of the original sample. Often used to compare test results between field kits and laboratories or between two laboratories.

TRIHALOMETHANES (THMs) (tri-HAL-o-METH-hanes) TRIHALOMETHANES (THMs)

Derivatives of methane, CH_4, in which three halogen atoms (chlorine or bromine) are substituted for three of the hydrogen atoms. Often formed during chlorination by reactions with natural organic materials in the water. The resulting compounds (THMs) are suspected of causing cancer.

VISCOSITY (vis-KOSS-uh-tee) VISCOSITY

A property of water, or any other fluid, that resists efforts to change its shape or flow. Syrup is more viscous (has a higher viscosity) than water. The viscosity of water increases significantly as temperatures decrease. Motor oil is rated by how thick (viscous) it is; 20-weight oil is considered relatively thin while 50-weight oil is relatively thick or viscous.

VOLATILE (VOL-uh-tull) VOLATILE

(1) A volatile substance is one that is capable of being evaporated or changed to a vapor at relatively low temperatures. Volatile substances can be partially removed from water or wastewater by the air stripping process.

(2) In terms of solids analysis, volatile refers to materials lost (including most organic matter) upon ignition in a muffle furnace for 60 minutes at 1,022°F (550°C). Natural volatile materials are chemical substances usually of animal or plant origin. Manufactured or synthetic volatile materials, such as plastics, ether, acetone, and carbon tetrachloride, are highly volatile and not of plant or animal origin. Also see NONVOLATILE MATTER.

CHAPTER 15. SPECIALIZED TREATMENT PROCESSES

Trihalomethanes by Mike McGuire

(Lesson 1 of 2 Lessons)

15.0 THE TRIHALOMETHANE (THM) PROBLEM

For the past few decades, water utilities have been concerned about the presence of organic compounds in drinking water. The analytical methods for detecting inorganic compounds such as calcium, magnesium, and iron have been known for many decades. However, the ability to analyze for organic compounds in water has been developed only recently. What are organic compounds? Organic compounds are defined as those compounds that contain a carbon atom. Carbon is one of the basic chemical elements. Examples of organic compounds include proteins, carbohydrates, fats, vitamins, and a wide variety of compounds that modern technology has created.

In 1974, researchers with the US Environmental Protection Agency (EPA) and in the Netherlands published their findings that trihalomethanes are formed in drinking water when free chlorine comes in contact with naturally occurring organic compounds (*THM PRECURSORS* [1]). *TRIHALOMETHANES* are a class of organic compounds in which there has been a replacement of the three hydrogen atoms in the methane molecule with three halogen atoms (chlorine or bromine). The four most commonly found THMs are chloroform, bromodichloromethane, dibromochloromethane, and bromoform (Figure 15.1). While it is theoretically possible to form iodine-substituted THMs, they are rarely found in treated water and they are not regulated at this time. In general, methane is not involved in the THM reaction. The production of THMs can generally be shown as:

EQUATION 1

$$\text{Free Chlorine} + \text{Natural Organics (precursor)} + \text{Bromide} \rightarrow \text{THMs} + \text{Other Products}$$

Free chlorine is added to drinking water as a disinfectant. The naturally occurring organics get into water when the water partially dissolves organic materials from algae, leaves, bark, wood, soil, and other similar materials. This dissolving action is similar to what happens when a teabag is placed in hot water; the water dissolves those parts of the tea leaves that are soluble organic and inorganic compounds. While it is possible to form THMs by reactions between chlorine and industrial organic chemicals, the overwhelming bulk of THM precursors in water are from natural organic compounds.

One source of bromide is seawater. Water agencies whose supplies are subject to seawater intrusion can expect THMs in their treated water to have high levels of bromide. Bromide reaction products can be found in most surface waters, even where bromide concentrations are low. The other products formed in this reaction are very poorly understood and are not regulated at this time.

After THMs were discovered in drinking waters around the country, several studies were made of their possible health effects in general and of the THM chloroform in particular. These tests indicated that chloroform caused cancer in laboratory animals (rats and mice) and was suspected of causing cancer in humans. Further studies comparing people who used different sources of drinking water suggested that there may be a link between the presence of manufactured organic compounds such as THMs and increased levels of cancer. Animal feeding experiments and population studies are not definite proof that THMs in drinking water cause cancer. Under the Safe Drinking Water Act, EPA may pass a regulation for any contaminant that may have any adverse health effect.

On November 29, 1979, the THM regulations were published in the *Federal Register;* they were amended on February 28, 1983 (see Section 15.7, "Regulatory Update"). The general aspects of the regulation are outlined below:

MAXIMUM CONTAMINANT LEVEL (MCL): 0.080 mg/L total trihalomethanes (TTHMs)—sum of the concentrations of chloroform, bromodichloromethane, dibromochloromethane, and bromoform (see the *NOTICE* on page 131).

APPLIES TO: All community water systems that add a disinfectant to their water supply and serve a population greater than 10,000 persons.

MONITORING REQUIREMENTS: Monitoring compliance is based on an annual running TTHM (total trihalomethane) average of four quarters of data. Schedule, locations, and numbers of samples depend on system size and should be worked out with the state or EPA.

ENSURING MICROBIOLOGICAL QUALITY: State or EPA must be notified of significant modifications to treatment processes to remove TTHMs in order to ensure microbiological quality of the treated water.

1. *Precursor, THM* (PRE-curse-or). Natural, organic compounds found in all surface and groundwaters, which may react with halogens (such as chlorine) to form trihalomethanes (THMs); they must be present in order for THMs to form.

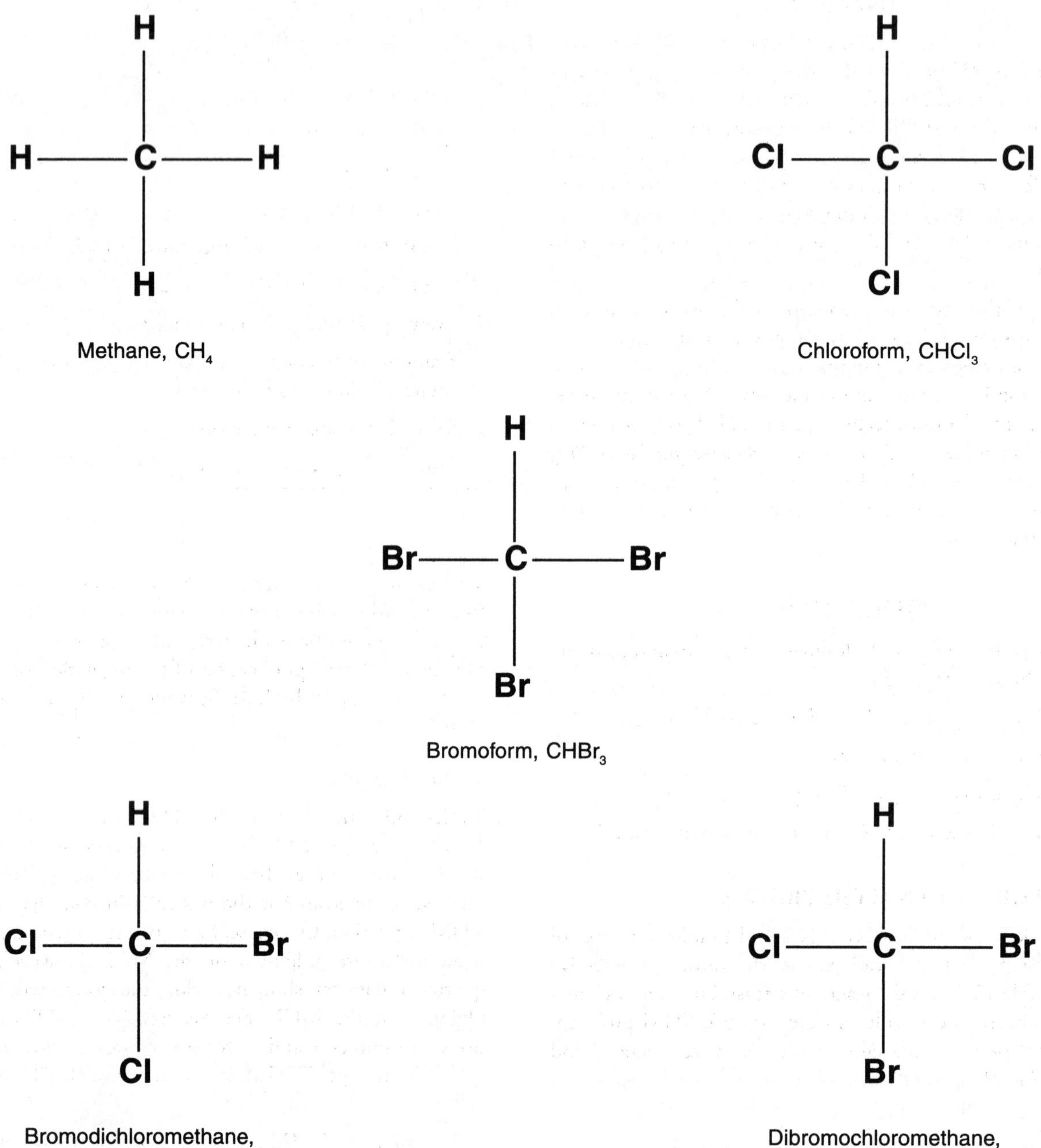

Fig. 15.1 Methane and THMs

The MCL for TTHMs was not established on the basis of the health-effects data, but was set as a feasible level for compliance.

NOTICE

Operators should be aware that the Disinfectants/Disinfection Byproducts (D/DBP) Rule, which was promulgated on December 16, 1998, significantly affected both large and small drinking water utilities. The D/DBP Rule lowered the MCL for total trihalomethanes (TTHMs) from 0.10 mg/L to 0.080 mg/L and set new limits for various disinfectants and disinfection byproducts. See Chapter 22, "Drinking Water Regulations," Section 22.23, "Disinfectants and Disinfection Byproducts (DBP)," for additional details.

The rest of this lesson is devoted to a discussion of how to collect samples for THM analysis and how a utility can evaluate the many alternatives available to control THMs in its system. This discussion is presented in outline form. A much more detailed treatment of control techniques for THMs is presented in an EPA publication entitled *Treatment Techniques for Controlling Trihalomethanes in Drinking Water*, by J. M. Symons, et al., September 1981.[2] A large part of this lesson is a summary of material from that source.

QUESTIONS

Please write your answers to the following questions and compare them with those on page 153.

15.0A How are trihalomethanes formed in drinking water?

15.0B What are THM precursors?

15.0C Why is free chlorine added to drinking water?

15.0D What is one source of bromide in drinking water?

15.1 FEASIBILITY ANALYSIS PROCESS

In any problem-solving process, it is useful to follow a series of prescribed steps that will lead you to the most cost-effective solution. Table 15.1 lists the stages of a feasibility analysis process that has been used to solve a water utility's THM problem. However, the process outlined in Table 15.1 is very general and can also be applied to solving other treatment or operational problems.

TABLE 15.1 FEASIBILITY ANALYSIS PROCESS

1. Determine the extent of the THM problem
   a. Monitor THM levels
   b. THM chemistry (time of formation)
2. Evaluate control strategies
   a. Change sources of supply
   b. Treatment options
      (1) Remove THMs
      (2) Remove precursors
      (3) Adjust or modify the chlorine application points
      (4) Use alternative disinfectants
3. Evaluate existing treatment processes
4. Examine studies of proposed treatment processes (Bench-, pilot-, and full-scale)
5. Select a cost-effective option
6. Implement the chosen option

15.2 PROBLEM DEFINITION

In order to determine the extent of a THM problem in a system, a reliable analytical technique must be used. THM analytical services may be developed by the utility or may be purchased from a contract laboratory. A discussion of the THM analytical methods will not be made here; rather, the reader is referred to the *Federal Register* publication of the regulation or to the EPA document prepared by J. M. Symons, which was previously discussed.

15.20 Sampling

To determine the extent of the THM problem, collect *REPRESENTATIVE SAMPLES*[3] from the distribution system of the water utility and analyze them according to an approved method to determine if the utility is in compliance with the THM regulation. Of course, four quarters of data are needed to make a definite judgment on the MCL. However, even one quarter of data can show how close the system will be to complying with the MCL. See Section 15.21, "THM Calculations," Examples 1 and 2, for procedures on how to calculate quarterly average TTHM levels and annual TTHM running averages.

The following THM sampling protocol was developed from 141.132 CFR and applies to drinking water distribution systems serving at least 10,000 people. Refer to 141.132 CFR for THM sampling protocols for smaller distribution systems.

2. EPA No. 600-2-81-156. This EPA document is available online at epa.gov/nscep/.
3. *Representative Sample.* A sample portion of material, water, or wastestream that is as nearly identical in content and consistency as possible to that in the larger body being sampled.

To collect samples to determine THM levels, use the following procedures:

1. A minimum of four samples per quarter (every 3 months) must be taken on the same day for each treatment plant in the distribution system.
2. Twenty-five percent of these samples must be collected from the extremities of the distribution system (the points farthest from each treatment plant).
3. Seventy-five percent of all the four samples must be representative of the population served by the distribution system.

Do not collect samples from swivel faucets, faucets with aerators, or faucets with hoses because of the possibility of contaminating the sample or loss of THMs.

To collect samples for THM analysis, use a narrow-mouth, screw-cap, glass sample bottle that can hold at least 25 mL of water. Use a polytetrafluorethylene (PTFE)-faced silicon-septia bottle-cap liner to provide an airtight seal over the sample bottle. The bottle caps must screw tightly on the sample bottles.

Some sample bottles will contain a small amount of a chemical reducing agent (usually sodium thiosulfate or sodium sulfite). The reducing agent will stop the chemical reaction that occurs between chlorine and the THM precursors (humic and fluvic acids). By stopping this chemical reaction, THMs will not continue to form in the sample after it has been collected from the distribution system. When using sample bottles that contain a reducing agent, do not rinse out the reducing agent before collecting the sample.

Some sample bottles will not contain a reducing agent. Water samples from these bottles will be tested for the maximum concentration of TTHMs that can form over an extended period of time. These tests cannot be performed if a reducing agent has been added to the sample. When using sample bottles that do not contain a reducing agent, do not add any chemicals to the bottles.

When collecting water samples for THM analysis, use the following procedures:

1. Turn on the sampling tap.
2. Allow sufficient time (about 5 minutes) for the water temperature to become constant.
3. Fill the sample bottle until it begins to overflow.
4. Set the bottle on a level surface and place the bottle-cap liner on top of the bottle.
5. Screw the bottle cap tightly on the bottle and turn the bottle upside down.
6. The sample is properly sealed if no air bubbles are present.
7. If air bubbles are present, remove the bottle cap and bottle-cap liner, turn on the sampling tap and add a small amount of water to the sample in the bottle. Repeat steps 4 through 6.

A good practice is to collect two samples at each location. This procedure allows the laboratory to double-check test results and, if a sample bottle is broken, there will be another sample available for testing.

Each sample bottle must include a label on which important information is recorded. Be sure to write on the label the sample location, date, and name of the person collecting the sample.

Samples should be sent to the laboratory immediately after they are collected and should be analyzed within 14 days. When sending samples to the laboratory, be sure to include the complete name and address of the person to whom the test results are to be returned. Samples must be chilled to 39°F (4°C) immediately after collection and kept at the same temperature during storage. Do not use dry ice when shipping or storing samples because the water in the bottles may freeze and break the sample bottles.

15.21 THM Calculations

FORMULAS

To calculate the average of a group of measurements, sum up the measurements and divide the total by the number of measurements.

$$\text{Average} = \frac{\text{Sum of Measurements}}{\text{Number of Measurements}}$$

To calculate the running annual average, sum up the average measurements for each quarter and divide the total by the number of quarters.

$$\text{Running Average} = \frac{\text{Sum of Averages for Each Quarter}}{\text{Number of Quarters}}$$

Whenever data for a new quarter become available, the newest quarterly average replaces the oldest quarterly average and the running annual average is recalculated.

Example 1

A water utility collected and analyzed eight samples from a water distribution system on the same day for TTHMs. The results are shown below.

Sample No.	1	2	3	4	5	6	7	8
TTHM, μg/L	80	50	70	110	90	120	80	90

What was the average TTHM for the day?

Known	Unknown
Results from analyses of 8 TTHM samples	Average TTHM level for the day, μg/L

Calculate the average TTHM level in micrograms per liter.

$$\text{Avg TTHM, } \mu g/L = \frac{\text{Sum of Measurements, } \mu g/L}{\text{Number of Measurements}}$$

$$= \frac{80\ \mu g/L + 50\ \mu g/L + 70\ \mu g/L + 110\ \mu g/L + 90\ \mu g/L + 120\ \mu g/L + 80\ \mu g/L + 90\ \mu g/L}{8\ \text{Measurements}}$$

$$= \frac{690\ \mu g/L}{8}$$

$$= 86\ \mu g/L$$

Example 2

The results of the quarterly average TTHM measurements for two years are given below. Calculate the running annual average of the four quarterly measurements in micrograms per liter.

Quarter	1	2	3	4	1	2	3	4
Avg Quarterly TTHM, μg/L	87	72	99	82	62	111	138	89

Known	Unknown
Results from analyses of 2 years of TTHM samples	Running annual average of quarterly TTHM measurements, μg/L

Calculate the running annual average of the quarterly TTHM measurements.

$$\text{Annual Running TTHM Average, } \mu g/L = \frac{\text{Sum of Average TTHM for Four Quarters}}{\text{Number of Quarters}}$$

Quarters 1, 2, 3, and 4

$$\text{Annual Running TTHM Average, } \mu g/L = \frac{87\ \mu g/L + 72\ \mu g/L + 99\ \mu g/L + 82\ \mu g/L}{4}$$

$$= \frac{340\ \mu g/L}{4}$$

$$= 85\ \mu g/L$$

Quarters 2, 3, 4, and 1

$$\text{Annual Running TTHM Average, } \mu g/L = \frac{72\ \mu g/L + 99\ \mu g/L + 82\ \mu g/L + 62\ \mu g/L}{4}$$

$$= \frac{315\ \mu g/L}{4}$$

$$= 79\ \mu g/L$$

Quarters 3, 4, 1, and 2

$$\text{Annual Running TTHM Average, } \mu g/L = \frac{99\ \mu g/L + 82\ \mu g/L + 62\ \mu g/L + 111\ \mu g/L}{4}$$

$$= \frac{354\ \mu g/L}{4}$$

$$= 89\ \mu g/L$$

Quarters 4, 1, 2, and 3

$$\text{Annual Running TTHM Average, } \mu g/L = \frac{82\ \mu g/L + 62\ \mu g/L + 111\ \mu g/L + 138\ \mu g/L}{4}$$

$$= \frac{393\ \mu g/L}{4}$$

$$= 98\ \mu g/L$$

Quarters 1, 2, 3, and 4

$$\text{Annual Running TTHM Average, } \mu g/L = \frac{62\ \mu g/L + 111\ \mu g/L + 138\ \mu g/L + 89\ \mu g/L}{4}$$

$$= \frac{400\ \mu g/L}{4}$$

$$= 100\ \mu g/L$$

SUMMARY OF RESULTS

Quarter	1	2	3	4	1	2	3	4
Avg Quarterly TTHM, μg/L	87	72	99	82	62	111	138	89
Annual Running TTHM Avg, μg/L				85	79	89	98	100

15.22 Chemistry of THM Formation

An understanding of the chemistry of THM formation is crucial if a water utility is to solve a THM problem. Equation 1 shown in Section 15.0, "The Trihalomethane (THM) Problem," describes the overall mechanism. Very little is known about the specific reactions that free chlorine and natural organics (precursors) undergo. In general, the effects of time, temperature, pH, and concentrations of the chemicals on the production of THMs have been studied by various investigators and are fairly well understood.

Depending on the type of natural organics present in the water, the time it takes for 0.080 mg/L (80 μg/L) of THMs to form may range from minutes to days. Set up a THM monitoring program on the source water(s) of the utility to measure the production of THMs over an appropriate time period (time from when chlorine is first added to water until water is consumed). A plot of the THMs produced against time will give you an idea of the TTHM (Total Trihalomethanes) formation potential (TTHMFP) of each source water. For many systems, a large part of the production of THMs will take place after the water leaves the treatment plant.

The higher the temperature, the faster the THMs will be produced. As might be expected, a dependence on temperature will probably show up as a seasonal effect—higher THM levels in the summer than in the winter. Temperature may not be the only controlling factor, however; higher THM levels may show up in the winter, as they have in California.

The higher the pH of the water, the faster the production of THMs. For most water utilities this will not be a concern; however, utilities raising the pH of treated water by caustic soda or by lime for corrosion control (Langelier Index) or using lime softening should be aware that free chlorine in contact with natural organics at a pH of 10.5 or higher will produce THMs much faster than if the pH were near 7.0.

The higher the concentrations of free chlorine and natural organics in the water, the more THMs will be produced. In the past, the amount of free chlorine that utilities used was only limited by economics and possible taste and odor complaints from consumers. Careful use of chlorine may help a utility to lower the THMs in its system. However, because of the danger of using too little chlorine (inadequate disinfection) in a system, the THM regulation specifically requires state or EPA approval of major treatment changes to meet the regulation.

The concentration of precursors in water is as important as the types of precursors that are found in water. Some naturally occurring organic compounds can produce 10 or 100 times more THMs on an equivalent basis than organics from another source. Also, some types of precursors will produce THMs faster than others. For these reasons it is important to evaluate the TTHMFP of each source of supply as a possible THM control measure.

The effect of higher bromide concentrations on THM production is not as clear as the effects of temperature and pH. The more bromide present, the more bromide-containing THMs will be formed. Free chlorine selectively attacks the bromide ion and changes it to bromine, which reacts quickly with precursors to form bromoform, dibromochloromethane, and bromodichloromethane. The usual result of high bromide levels is higher THM levels because more molecules are available (chlorine + bromine) to participate in THM-forming chemical reactions.

Now that some of the basics of THM chemistry are understood and a THM problem can be properly defined, it is time to look at some of the possible control strategies.

QUESTIONS

Please write your answers to the following questions and compare them with those on page 153.

15.1A List the major steps that a water utility could take to solve a THM problem.

15.1B List the possible control strategies that could be evaluated to control a THM problem.

15.2A What important factors influence the production of trihalomethanes?

15.2B How does lime used for softening influence the production of THMs?

15.3 CONTROL STRATEGIES

Assuming that a utility discovers a THM problem in its system, there are two ways to control it: change the source of supply or provide some type of treatment. Changing the source of supply can consist of an entire range of alternatives such as shifting between wells of different water quality, drawing water from different levels in a reservoir, or abandoning a surface water supply altogether during part of the year. Since most utilities do not have the flexibility to abandon a source of supply, this alternative will have limited application.

The three treatment options available to control THMs are as follows:

1. Remove THMs after they are formed.
2. Remove THM precursors before chlorine is added.
3. Use a disinfectant other than free chlorine.

A later section will examine each of these options and the processes associated with them. At this point, it is useful to discuss overall treatment strategies. The general equation for forming THMs illustrates how each of the three options can work.

EQUATION 1

$$\text{Free Chlorine} + \text{Natural Organics (precursor)} + \text{Bromide} \rightarrow \text{THMs} + \text{Other Products}$$

Removing THMs after they are formed is generally not the strategy of choice unless there is a particular circumstance at the utility that warrants its evaluation. Since precursors are not necessarily removed when THMs are removed, there is the problem of continued THM formation, especially in the distribution system.

Removing precursors before free chlorine is added has some major advantages, particularly if the precursors can be removed by a fairly inexpensive process. Removing precursors allows the continued use of free chlorine as a disinfectant, which has been proven to be an effective barrier against disease for many decades. As Equation 1 shows, fewer precursors also means the formation of fewer other products. These other products consist of high-molecular-weight organic compounds that contain chlorine and bromine. The health significance of these other products is not always known, but concern has been raised by regulatory agencies.

Using a disinfectant other than free chlorine has a number of advantages and disadvantages that must be evaluated case by case. Abandoning free chlorine is a serious move in view of its superior performance as a disinfectant. However, if any of the alternative disinfectants are lower cost alternatives, they must be given careful consideration.

15.4 EXISTING TREATMENT PROCESSES

Before beginning a complex, expensive research program, it is valuable to examine how well existing treatment processes can control the formation of THMs. The following sections cover the potential of individual processes for THM control; however, some generalizations can be made with regard to existing unit processes. Aeration unit processes are sometimes available in water treatment plants to control tastes and odors. The same process may show measurable removals of THMs after they are formed. Oxidation of tastes and odors with chlorine dioxide (ClO_2) and potassium permanganate are common unit processes available in water treatment plants. Chlorine dioxide does not form THMs. Permanganate sometimes can be used to oxidize THM precursors if they are affected by this kind of treatment.

Coagulation/sedimentation/filtration and softening processes can remove THM precursors depending on the types that are present in the water supply. Powdered activated carbon and granular activated carbon used for taste and odor control can have a limited impact on the removal of both THMs and THM precursors.

QUESTIONS

Please write your answers to the following questions and compare them with those on page 153.

15.3A If a utility discovers a THM problem, what are two ways to control the problem?

15.3B Why is abandoning the use of free chlorine considered a serious move?

15.4A List the water treatment processes that can be used to control THMs.

15.5 TREATMENT PROCESS RESEARCH STUDY RESULTS

15.50 Consider Options

There is a long list of treatment options that can be investigated for the control of THMs. Since a large number of them have already been studied and reported on, it is not necessary for every utility to repeat this work. A careful evaluation of the results published by the US EPA will help a utility focus on the treatment processes that should be looked at on a bench-, pilot-, or full-scale basis.

Most feasible treatment options include the removal of precursor materials before the formation of a THM, the avoidance of generation of a THM by use of an alternate disinfectant, or the actual removal of a THM by means of aeration or carbon adsorption. Also, the geographic and climatological conditions can have a very important influence on the choice of a process. For example, aeration is not a desirable method of treatment where severe cold weather is common.

15.51 Remove THMs After They Are Formed

There are three treatment processes available to remove THMs after they have been formed:

1. Oxidation
   a. Ozone
   b. Chlorine dioxide
   c. Ozone/ultraviolet light
2. Aeration
   a. Open storage
   b. Diffused air
   c. Towers
3. Adsorption
   a. Powdered activated carbon
   b. Synthetic resins

OXIDATION. Oxidation of THMs using any one of the three oxidants listed above has not been very successful. The combination of ozone/ultraviolet light showed some promise; however, the cost effectiveness of the process has yet to be demonstrated.

AERATION. In contrast, aeration is an effective process for removing THMs from water, although the individual THMs are removed at different efficiencies. THM removal efficiencies by aeration, ranging from the easiest to most difficult, are from chloroform to bromodichloromethane to dibromochloromethane and to bromoform. Allowing water containing THMs to stand uncovered will ultimately result in the THMs leaving the water, since they are *VOLATILE*[4] compounds that are poorly soluble in water. In other words, THMs have a natural tendency to migrate from water into the atmosphere if given the chance. Because of this tendency, THM reductions may be noticeable in effluents from uncovered finished water reservoirs after a significant detention time (days). *NOTE:* The Interim Enhanced Surface Water Treatment Rule (IESWTR) prohibited construction of new uncovered finished water facilities after February 1999.

More efficient removal of THMs can be accomplished if energy is put into the aeration process. A convenient way to put energy into aeration is by bubbling air into water. Many water treatment plants currently have an aeration process of some kind to help control tastes and odors in the source water. The

4. *Volatile* (VOL-uh-tull). (1) A volatile substance is one that is capable of being evaporated or changed to a vapor at relatively low temperatures. Volatile substances can be partially removed from water or wastewater by the air stripping process. (2) In terms of solids analysis, volatile refers to materials lost (including most organic matter) upon ignition in a muffle furnace for 60 minutes at 1,022°F (550°C). Natural volatile materials are chemical substances usually of animal or plant origin. Manufactured or synthetic volatile materials, such as plastics, ether, acetone, and carbon tetrachloride, are highly volatile and not of plant or animal origin. Also see NONVOLATILE MATTER.

efficiencies of these existing processes would not be expected to be very great for THM removal.

Operators should realize that aeration of treated water can cause a significant amount of contamination. Air in many areas may contain large amounts of dust, dirt, bacteria, and other pollutants that can contaminate treated water and lead to operation and maintenance problems.

In the research results that are currently available, countercurrent tower aeration (Figure 15.2) has produced the highest removals of THMs with air-to-water ratios (the ratio of the volume of air added to the volume of water treated) in the 20 to 1 to 50 to 1 range. Treatment efficiencies greater than 90 percent removal have been demonstrated with some aeration towers on some types of water.

Countercurrent aeration towers are designed so that the water and air pass over a packing material countercurrent to each other (in opposite directions). A significant amount of theoretical work has been done on the possible tower designs for any given set of treatment conditions. Pilot-scale testing is usually recommended before a full-scale plant is constructed.

Aeration is most effective in removing the more volatile chemicals. Chloroform is the most volatile of the THMs and is generally the most easily removed by aeration. Bromoform, on the other hand, is the least volatile THM and, consequently, is the hardest to remove by aeration. If the TTHM content of the water contains a significant amount of bromoform, aeration may not be the most desirable technique to investigate.

ADSORPTION. THMs can be removed by a wide variety of activated carbons and synthetic resins. The adsorption process involves the individual THM compounds leaving the water and becoming attached to the surface of the carbon or resin. THMs are generally considered difficult to adsorb on any surface. The efficiency by adsorption from easiest to most difficult is bromoform, dibromochloromethane, bromodichloromethane, and chloroform.

Powdered activated carbon (PAC) is usually added as a treatment chemical in the rapid-mix process or in the sedimentation basin effluent. PAC is normally used in water treatment for taste and odor control at dosages of less than 20 mg/L. Studies have shown that PAC dosages of 100 mg/L or more are necessary to get significant removals of THMs. Chloroform is particularly difficult to remove with PAC.

Synthetic resins such as XE-340 have been demonstrated to be effective in removing THMs from water; however, economics must be taken into consideration, since the cost of the resins is high in comparison with other alternatives. Regeneration of the resins has not been worked out satisfactorily. Pilot-scale studies show some promise. Resin manufacturers are continually developing new processes to improve the performance of their products.

QUESTIONS

Please write your answers to the following questions and compare them with those on page 153.

15.5A Which is the better process for removing THMs after they are formed, oxidation or aeration?

15.5B How does the storage of water in uncovered reservoirs affect THM levels?

15.5C How does the adsorption process work?

15.52 Remove THM Precursors

A variety of treatment processes have been investigated to remove THM precursors before they come in contact with chlorine:

1. Aeration
2. Oxidation
   a. Ozone (before coagulation and clarification)
   b. Chlorine dioxide
   c. Permanganate
   d. Ozone/ultraviolet light
   e. Hydrogen peroxide
3. Clarification
   a. Coagulation/sedimentation/filtration
   b. Softening
4. Adsorption
   a. Powdered activated carbon
   b. Granular activated carbon
   c. Synthetic resins
5. Ion Exchange

AERATION. Since THM precursors are not volatile compounds, it is not surprising that aeration is ineffective in removing them from water.

OXIDATION. All of the oxidants listed above have some effect on removing or modifying THM precursors. Since THM precursors vary so much between locations, it is difficult to generalize on the effectiveness of any of the oxidants. In fact, some studies have demonstrated that the formation potential for THMs can *increase* with the application of certain dosages of ozone and potassium permanganate. In general, it is necessary to perform bench- or pilot-scale studies on the water in question before the usefulness of any of these oxidants can be considered. The US EPA is also concerned with the production of potentially harmful byproducts that could result from the use of any of these oxidants. Once again, studies on the water to be treated are necessary to determine whether or not this is a problem.

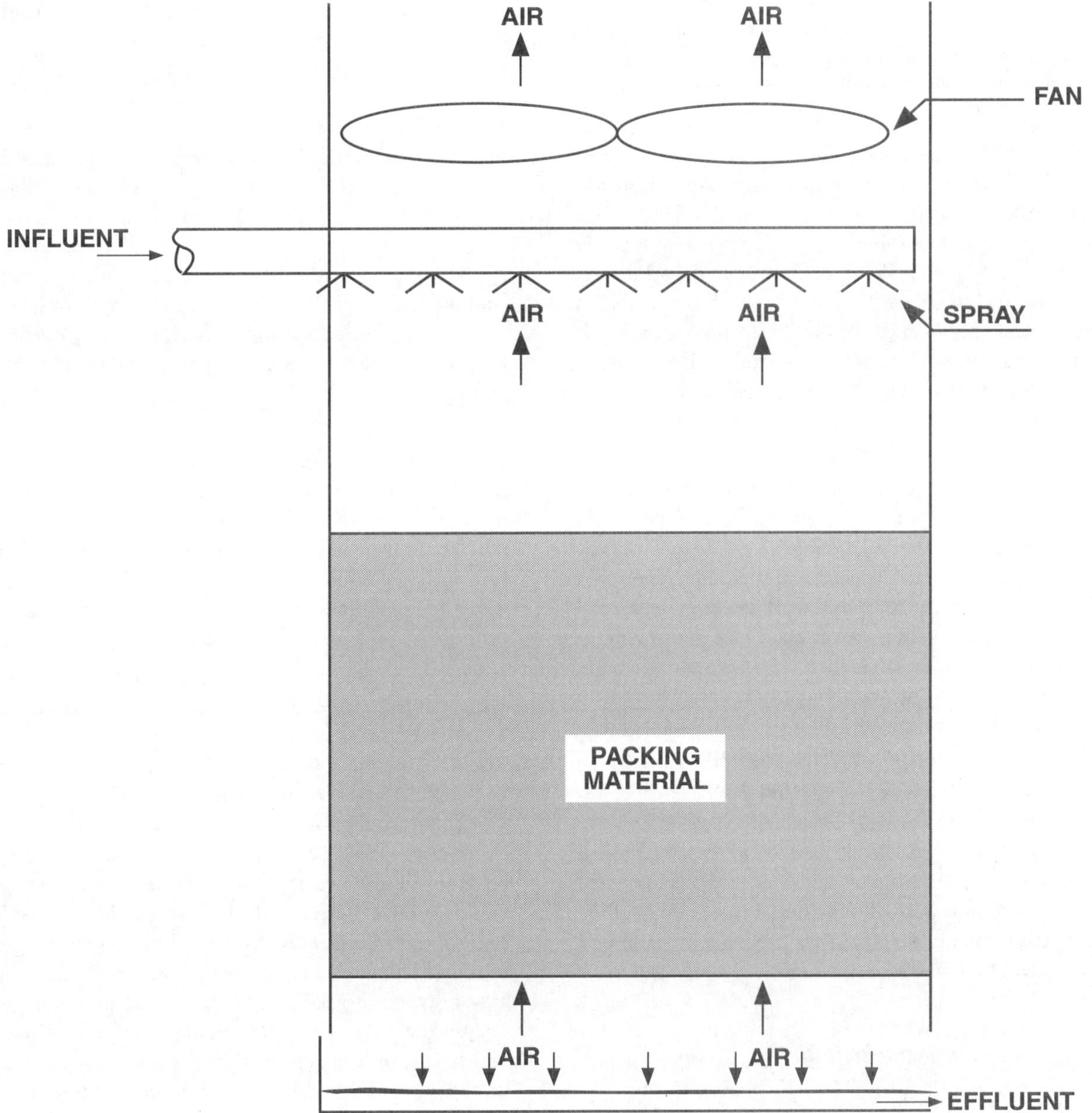

Fig. 15.2 Countercurrent aeration tower

In the southeastern United States and in other locations, water supplies are relatively warm and contain high levels of organic matter. Under these conditions, controlled oxidation with small doses of ozone can actually coagulate organic material and make conventional sedimentation more effective. Too much ozone can break the organics down and make them more reactive with chlorine. However, small, controlled doses of ozone may be an effective microflocculant and may be added to conventional water treatment plants to improve the physical removal of THM precursors to the point that prechlorination disinfection is possible.

CLARIFICATION. The clarification process used in water treatment plants has the potential for removing significant amounts of THM precursors. Dozens of studies by the US EPA have demonstrated widely varying removal efficiencies (0 to 100 percent) because of the highly variable nature of THM precursors from place to place. The use of this process, which is available in most water treatment plants, to remove THM precursors holds great promise for an economical solution to any THM problem. Moving the addition of free chlorine to a point following the clarification process is the key to success for this approach. Many water utilities have adopted this approach to solve their problem.

ADSORPTION. The use of powdered activated carbon (PAC) and granular activated carbon (GAC) are effective in removing THM precursors; however, the economics of these processes must be carefully evaluated. Dozens of studies have reported a wide variety of THM precursor removal efficiencies. Because of the high cost of PAC and GAC, their use as THM control methods will be restricted to those cases where no other alternatives are available. Synthetic resins showed limited removal potential for THM precursors. Effective regeneration of the resins for additional precursor removal has not been demonstrated.

ION EXCHANGE. Anion exchange resins can be effective for removing THM precursors, which generally have a negative charge. Both strong-base and weak-base anion exchange resins have been investigated. As with the activated carbons discussed above, anion exchange resins will only find a role in controlling THMs if the economics of the treatment process for a particular site are favorable. Disposal of the spent regenerant liquid may be a problem.

QUESTIONS

Please write your answers to the following questions and compare them with those on pages 153 and 154.

15.5D List the major treatment processes that have been investigated to remove THM precursors before they come in contact with chlorine.

15.5E What is the key to success in using clarification to remove THM precursors from the water being treated?

15.53 Alternative Disinfectants

Removing free chlorine from the chlorine/bromide/precursor reaction will stop the formation of a significant amount of THMs. However, free chlorine has been an effective barrier between people and disease-causing bacteria since the beginning of the 20th century, and abandoning its use is a very serious step. There are other disinfectants that can be used instead of free chlorine, but the advantages and disadvantages of each alternative must be carefully evaluated.

The most commonly considered alternative disinfectants are ozone, chlorine dioxide, and chloramines. Ozone is a gas that is produced by passing oxygen through an electrical discharge. While ozone is a highly effective disinfectant, it is very expensive, it must be generated on site, and it does not leave a residual in the treated water. Chlorine dioxide is a gas produced by the reaction of free chlorine and sodium chlorite. Chlorine dioxide is a very effective disinfectant that does leave a residual in the treated water; however, there are some concerns regarding the health implications of the inorganic breakdown products, chlorite and chlorate. The THM regulation recommended a 0.5 mg/L limit for the total concentration of chlorine dioxide, chlorite, and chlorate in water after chlorine-dioxide treatment.

Chloramines are produced in water by the reaction between free chlorine and ammonia. Chloramines are weaker disinfectants than free chlorine, ozone, or chlorine dioxide, but the residuals remain much longer than free chlorine and they have been used successfully by dozens of water utilities. For example, the City of Denver has used chloramines for many years. The effectiveness of monochloramines as a disinfectant depends on water temperature, pH, and biological quality, as well as the proper ratio of ammonia to chlorine.

The use of chloramines can also cause problems in a utility's system unless proper precautions are taken. Chloramines must be removed from the water before it is used in kidney dialysis machines. Chloramines in water can pass through kidney dialysis machines and into a patient's blood where the ammonia will decrease the oxygen-carrying capacity of the blood. In addition, chloramines are toxic to fish in home aquariums, and they must be removed from water before it comes in contact with fish. Dechlorination of water with activated carbon, ascorbic acid, or sodium thiosulfate will prevent any of these problems if the removal of chloramines is properly controlled. Any oxidants that are present in drinking water can cause problems with kidney dialysis machines and fish in home aquariums. However, chloramines are somewhat more difficult to remove than the other alternative disinfectants. For additional information, see *Water Treatment Plant Operation*, Volume I, Chapter 7, Section 7.26, "Chloramination."

Before an alternative disinfectant is applied to any system, the source of the water supply, water quality, and treatment effectiveness for bacteriological control must be evaluated. For example, the use of a weaker disinfectant such as chloramines may not be appropriate for a surface water supply that is highly contaminated with discharges from municipal and industrial wastewater treatment plants unless an extra high dosage and a long contact time are provided. Also, many of the conventional water treatment processes are capable of removing bacteria, viruses, and protozoa from the water (for example, softening and coagulation/sedimentation/filtration). These conventional processes may help to provide the required disinfection barrier between a contaminated supply and the population served, which could allow the use of a less potent disinfectant in the distribution system.

Upgraded monitoring (more samples and tests) of the distribution system before and after a disinfectant change must be provided by the water utility. The THM regulation specifies guidelines that the states must use in establishing such a monitoring program. Guidelines describing coliforms, standard plate count, turbidity, and nutrients are included in the suggested monitoring list. With before-and-after monitoring by the water utility, it will be possible to determine if there is any significant degradation of the bacteriological quality in the distribution system. Control of THMs must not be accomplished at the expense of a higher risk of bacterial and viral diseases among the population that is being served. Therefore, a decision to use a disinfectant other than free chlorine must be based on a carefully considered plan. A utility that rushes into the use of an alternative disinfectant without the required studies is likely to experience many problems that are easily avoided with proper planning.

QUESTIONS

Please write your answers to the following questions and compare them with those on page 154.

15.5F What items must be considered before an alternative disinfectant is applied to any system?

15.5G What type of distribution system monitoring must be provided by a utility before and after a disinfectant change?

15.5H What water quality indicators should be monitored before and after a disinfectant change?

15.6 SELECTION AND IMPLEMENTATION OF A COST-EFFECTIVE ALTERNATIVE

A detailed evaluation of the comparative economics of the many treatment processes described above is outside the scope of this lesson. In many cases, a utility will commission a special cost-effectiveness study that will be accomplished in-house or by an outside consultant. The US EPA THM treatment manual presents a detailed look at cost estimates for various alternatives with equivalent THM control levels. However, these data are not current and should be updated to reflect current economic conditions whenever cost studies are conducted.

If existing processes are not capable of solving a utility's THM problem, the least-cost solution will probably be an alternative disinfectant. While no statistics are currently available, evidence from discussions with consultants and utility managers suggests that alternative disinfectants, especially chloramines, are the overwhelming least-cost solution for water utilities with a THM problem. However, the use of chloramines may cause problems for persons using kidney dialysis machines.

Implementation of a THM control strategy requires a number of well-defined steps:

1. Full-scale design
2. Construction
3. Start-up
4. Operation

The length of time required to complete these steps will depend on the complexity of the control strategy chosen and the availability of engineering services to complete the assigned tasks. Throughout the implementation phase, it is important that the bench-, pilot-, and full-scale tests initiated in the feasibility analysis phase be continued so that the chosen strategy can be refined and optimized. For example, a pilot plant can be used to train treatment plant operators to use the new technology that will soon be on line.

QUESTIONS

Please write your answers to the following questions and compare them with those on page 154.

15.6A If existing water treatment processes are not capable of solving a THM problem, what is the most likely least-cost solution?

15.6B What are the most popular alternative disinfectant?

15.7 REGULATORY UPDATE

On February 28, 1983, the US EPA published in the *Federal Register* (page 8406) an amendment to the THM regulation originally published on November 29, 1979. The amendment specifies the treatment alternatives that a utility must consider or investigate in detail before it can apply for and receive a variance from meeting the MCL as defined under the regulation. The US EPA or the state may require a community water system to use a "generally available" technology before granting a variance. "Generally available" or Group 1 treatment techniques are:

1. Use of chloramines or chlorine dioxide as an alternative or supplement to chlorine for oxidation and disinfection.
2. Use of chloramines, chlorine dioxide, or potassium permanganate as an alternative to chlorine for preoxidation.
3. Moving the point of chlorination in order to reduce THM formation.
4. Improvement of existing clarification.
5. Use of powdered activated carbon (PAC), intermittently, as necessary, to reduce TTHM or THM precursors. The dosage of PAC is not to exceed an annual average of 10 mg/L.

Any of these technologies may be required in the variance unless the regulatory agency, US EPA, or the state determines that "such treatment method … is not available and effective for TTHM control for the system." The rule allows exemption from the use of a technique if the method would not be technically appropriate and technically feasible for the system or if the method would result in only a marginal reduction in TTHM.

The rule also allows the regulator to require the study of Group 2 technologies by water systems where Group 1 technologies are not appropriate or sufficient in meeting the MCL. If a Group 2 technology would be technically feasible and economically reasonable and result in significant TTHM reductions in line with the cost of treatment, then the regulator can require the use of a Group 2 technology.

The listed Group 2 technologies are introduction of offline water storage, aeration, introduction of clarification, alternative sources of raw water, and the use of ozone as an alternative or supplement to chlorine for disinfection or oxidation.

The February 28, 1983, amendment to the THM regulations did not mention granular activated carbon (GAC) or biological activated carbon (BAC) as treatment alternatives that must be considered. These two treatment methods were judged to be too expensive and are not widely enough used in the United States to warrant their evaluation for THM control. In general, the amendment was designed to reduce the economic impact of the THM regulation on those utilities that have THM problems and limited resources to drastically modify their treatment procedures.

The Disinfectants/Disinfection Byproducts (D/DBP) Rule, promulgated by the EPA on December 16, 1998, further regulates not only THMs but also other disinfection byproducts and residual concentrations of some disinfectants. The goal of the rule is a balance between projected microbiological and chemical risks. See Chapter 22, "Drinking Water Regulations," Section 22.23, "Disinfectants and Disinfection Byproducts (DBP)," for additional information about specific limits, effective dates of the regulations, and types of systems required to comply with the D/DBP regulations.

Utilities that may be affected by THM regulations are advised to follow future developments in the *Federal Register*.

15.8 SUMMARY AND CONCLUSIONS

1. Trihalomethanes are produced when free chlorine, which is added as a disinfectant, reacts with naturally occurring bromide and organic compounds.
2. A trihalomethane regulation is now in effect that has established a 0.080 mg/L maximum contaminant level and monitoring requirements.
3. A feasibility analysis process is a series of logical steps to arrive at a cost-effective solution to a THM problem:
   a. Determine the extent of THM problem
      (1) Monitor
      (2) Examine THM chemistry
   b. Evaluate control strategies
      (1) Change sources of supply
      (2) Treatment options
         (A) Remove THMs
         (B) Remove precursors
         (C) Use alternative disinfectants
   c. Evaluate existing treatment processes
   d. Research studies of treatment processes
      (1) Bench-, pilot-, and full-scale
   e. Select a cost-effective option
   f. Implement chosen option
4. There are three treatment options available to control THMs:
   a. Remove THMs after they are formed
   b. Remove THM precursors before chlorine is added
   c. Use a disinfectant other than free chlorine
5. A water utility must not create a possible health problem by ignoring bacteriological safeguards in an attempt to solve a THM problem.
6. An amendment to the THM regulation specifies treatment techniques that must be evaluated before a utility may receive a variance. Since this amendment affects a utility's feasibility analysis procedure, the steps outlined in the amendment should be followed.

QUESTIONS

Please write your answers to the following questions and compare them with those on page 154.

15.7A What treatment processes must utilities evaluate before applying for and receiving a variance?

15.7B If treatment processes are not technically feasible and economically reasonable, then what should utilities consider?

15.9 ADDITIONAL READING

Treatment Techniques for Controlling Trihalomethanes in Drinking Water, US Environmental Protection Agency. EPA No. 600-2-81-156. This document is available online at epa.gov/nscep/.

END OF LESSON 1 OF 2 LESSONS

on

SPECIALIZED TREATMENT PROCESSES

Please answer the discussion and review questions next.

DISCUSSION AND REVIEW QUESTIONS

Chapter 15. SPECIALIZED TREATMENT PROCESSES

(Lesson 1 of 2 Lessons)

At the end of each lesson in this chapter, you will find discussion and review questions. Please write your answers to these questions to determine how well you understand the material in the lesson.

1. How are trihalomethanes formed in drinking water?
2. On what basis was the maximum contaminant level (MCL) for total trihalomethanes (TTHMs) established?
3. What is the influence of higher temperatures and pH on the production of trihalomethanes (THMs)?
4. What are some of the options if a water utility decides to investigate changing the source of the water supply?
5. What are the advantages of removing precursors before free chlorine is added?
6. Under what conditions might ozone be used before clarification and filtration?
7. List the three alternative disinfectants to free chlorine and the advantages and limitations of each one.
8. What items must be considered before an alternative disinfectant is applied to any system?
9. What are the advantages of using a pilot plant in the implementation of a THM control strategy?

CHAPTER 15. SPECIALIZED TREATMENT PROCESSES

Arsenic by Chet Anderson

(Lesson 2 of 2 Lessons)

15.10 THE ARSENIC PROBLEM

15.100 Why Are We Concerned About Arsenic?

Arsenic is a natural element that is found in water at low concentrations. Ingestion of very high levels of arsenic can cause acute (short-term) adverse health effects. High levels of arsenic are typically not found in drinking water sources. Low levels of arsenic consumed over a long period of time can cause cancer and noncancerous toxic effects on humans. The types of cancer related to chronic (long-term) exposure to arsenic in drinking water include skin, bladder, lung, kidney, nasal passages, liver, and prostate cancer. Noncarcinogenic (not cancer-causing) effects include cardiovascular, pulmonary, immunological, neurological, and *ENDOCRINE EFFECTS*.[5]

The US EPA MCL for arsenic is 0.010 mg/L (or 10 ppb). States must adopt an MCL at least as stringent as 10 ppb.

15.101 What Are the Sources of Arsenic?

Arsenic in drinking water sources can come from either natural occurrences (such as volcanic action, erosion of rock, or a result of forest fires) or human activities. Arsenic is most commonly found in water sources as a result of natural leaching from soils. Arsenic in water sources also can result from mining, agricultural, or manufacturing operations. Approximately 90 percent of the arsenic used by industry in the US is currently used for wood-preservative purposes. Arsenic is also found in paints, dyes, metals, and other manufactured products.

The occurrence of high levels of arsenic in drinking water is mainly in groundwater, although there are a few exceptions in surface waters. Western states in the US have a disproportionate number of drinking water systems with arsenic problems. The source of this problem is primarily related to naturally occurring arsenic in soil and is normally not related to mining wastes.

15.102 Chemistry of Arsenic

Arsenic is an element with the chemical symbol As. Arsenic is normally a cation with a valence (charge) of either +3 or +5. However, when arsenic is dissolved in water, it will be present as part of a complex that may have either a negative charge or a neutral charge. Arsenic can be either in the trivalent (+3 or III) arsenite form or the pentavalent (+5 or V) arsenate form, depending on the level of oxidation. The oxidation state of arsenic is very important when treating arsenic because the As(III) form is not removed by common treatment processes. In water, the As(III) is part of a neutrally charged complex while As(V) is part of a negatively charged ion. As(V) can be removed by ion exchange, activated alumina, or by the coagulation-filtration systems using ferric salt coagulants. The pH of the water is important in determining the arsenic speciation (As(III) or As(V) form).

Both forms of arsenic commonly occur in groundwater: As(III) and As(V). NP4, an arsenic-reducing microbe from the genus *Sulfurospirillum*, can transform As(V) to As(III), the more toxic form of arsenic. Also, naturally occurring bacteria can increase arsenic concentrations in groundwater. Members of the genus *Geobacter* and other iron-reducing bacteria grow in the absence of oxygen. They can transform solid-phase iron (Fe(III)) into water-soluble Fe(II). Solid-phase Fe(III) can trap arsenic on its surface, but if transformed into Fe(II), it would release the arsenic into groundwater.

QUESTIONS

Please write your answers to the following questions and compare them with those on page 154.

15.10A What are the potential sources of arsenic in drinking water?

15.10B Why is the oxidation state of arsenic very important when treating arsenic?

15.11 TREATMENT FOR REDUCTION OR REMOVAL OF ARSENIC

15.110 New Source Alternative to Treatment

Before considering treatment of a water source to reduce or remove arsenic, water utilities should investigate a new source as an alternative water supply. Construction of new, low-arsenic water sources may be the most economical solution for some systems. However, this alternative will likely be restricted to very

5. *Endocrine* (EN-doe-krin) *Effects*. An altering of the organs in the human body responsible for secreting hormones into the bloodstream. Endocrine glands include the thyroid gland, the pancreas, and the adrenal glands.

small systems where drilling of a single well may create the desired solution. This solution assumes that low-arsenic groundwater is available and can be recovered for drinking purposes. There are costs and risks associated with verifying that an alternative groundwater of acceptable water quality is available. Changing to a surface water supply means a tradeoff between building an arsenic removal plant or a surface water treatment plant.

15.111 Summary of Arsenic Treatment Options

There are a number of treatment methods that can be used to remove arsenic from drinking water sources. These methods can be classified into three general treatment categories: *ADSORPTION*[6] (A), *COPRECIPITATION*[7] (C), and physical separation (P). Those treatment technologies that are considered feasible from a technical and economic viewpoint are described as best available technology (BAT).

BAT processes available for treating arsenic include the following technologies:

1. Engineered blending
2. *ACTIVATED ALUMINA*[8] (A)
3. Coagulation/filtration (C)
4. Ion exchange (A)
5. Lime softening (C)
6. Reverse osmosis (RO) (P)
7. Electrodialysis removal (EDR) (P)
8. Oxidation/filtration (P)
9. Point-of-use (POU) and point-of-entry (POE) devices (A), (C), or (P)

Almost all arsenic treatment processes (except reverse osmosis and blending) depend on the electrostatic interactions of a negatively charged As(V) ion with the treatment media (or chemical). For this reason, it is necessary to add an oxidation agent to ensure that all of the neutral (uncharged) As(III) is changed (oxidized) to the negatively charged anionic As(V) state. This change can be achieved by using chlorine or potassium permanganate ($KMnO_4$). Chlorine is the most common and readily available oxidant. Ozone and hydrogen peroxide could also be effective.

15.112 Engineered Blending

Blending of higher arsenic level sources with lower arsenic level sources is an acceptable arsenic treatment option as long as the blending is properly engineered, operated, and controlled. Because the arsenic levels in both surface and groundwaters tend to be quite stable, blending can be effectively controlled. Variations in source water arsenic levels must be thoroughly evaluated before considering blending as an option.

The following key elements must be provided for in order for blending to be successful:

1. Low-arsenic source(s) with sufficient quantity and reliability.
2. The high- and low-arsenic sources must be located in such proximity that they can be mixed before entering the distribution system.
3. Mixing must be thorough and proven (for example, the use of a storage or detention tank, if available, provides for adequate mixing and a buffer in the event of unexpected changes in raw water arsenic levels).
4. The system must be able to function with both types of arsenic sources out of operation since failure of the low-arsenic source will render both sources unusable.
5. There must be adequate monitoring, operation, and recordkeeping.

Before a blending operation can be implemented, theoretical calculations must be made to determine if blending is feasible and will be an effective solution. Factors that must be known in order to make the required blending calculations include the following:

Q_1 = Pumping capacity of Source #1 (usually in gallons per minute (GPM))

Q_2 = Pumping capacity of Source #2 (GPM)

Q_b = Combined flow of the two sources (GPM)

C_1 = Arsenic level of Source #1 (in mg/L or ppm)

C_2 = Arsenic level of Source #2 (mg/L)

C_b = Arsenic level in blended water (mg/L)

The theoretical resulting arsenic level in the blended water (C_b) is calculated using the following equation:

$$Q_1 \times C_1 + Q_2 \times C_2 = C_b (Q_1 + Q_2)$$

or

$$C_b = \frac{Q_1 C_1 + Q_2 C_2}{Q_b}$$

Where

$$Q_b = Q_1 + Q_2$$

6. *Adsorption* (add-SORP-shun). The gathering of a gas, liquid, or dissolved substance on the surface or interface zone of another material.
7. *Coprecipitation.* A treatment process that occurs when ferrous iron is added to water or metallic wastestreams and subsequently oxidized in an aerator. The oxidized iron, which is insoluble, precipitates along with other metallic contaminants present in the water or wastestream, thereby enhancing metals removal.
8. *Activated Alumina.* A charged form of aluminum, used with a synthetic, porous media in an ion exchange adsorption process to remove charged contaminants.

Example 1

$Q_1 = 200 \text{ GPM}$

$Q_2 = 300 \text{ GPM}$

$C_1 = 0.020 \text{ mg/L}$

$C_2 = 0.000 \text{ mg/L}$

$Q_b = Q_1 + Q_2 = 200 \text{ GPM} + 300 \text{ GPM} = 500 \text{ GPM}$

$$C_b = \frac{Q_1C_1 + Q_2C_2}{Q_b}$$

$$= \frac{200 \text{ GPM} \times 0.020 \text{ mg/L} + 300 \text{ GPM} \times 0.000 \text{ mg/L}}{500 \text{ GPM}}$$

$$= \frac{4 \text{ mg/L}}{500 \text{ GPM}}$$

$$= 0.008 \text{ mg/L}$$

As part of the operational requirements for approval of blending operations, the regulatory agency will probably require frequent, perhaps daily, calculation of the theoretical blended water arsenic levels to provide verification of laboratory analyses and operational procedures.

Some issues that are related to blending that require consideration include:

1. Blending processes may be complex in some situations.
2. Well failure/maintenance failure will affect the availability of low-arsenic sources.
3. Water quality could potentially change over time.
4. System may need water distribution system modeling to ensure reliable blended arsenic levels in the drinking water.

15.113 Ion Exchange (IX)

The basics for using ion exchange for removing arsenic from drinking water sources (Figure 15.3) are similar to any ion exchange treatment process using an anion-based resin.

Some specific concerns related to the use of ion exchange to remove arsenic include:

1. Effect of pH. The ion exchange of the arsenate ion, with chloride on the chloride-form, strong-base resins that are most likely to be used, occurs effectively in the pH range of about 6.5 to 9.0. Typically, most groundwaters are in this pH range. Therefore, adjustment of pH before ion exchange is rarely needed.
2. Effect of competing ions. Because of the relative affinity that the strong-base resins have for different anions, the source raw water quality is very important. Sulfate and nitrate are both preferentially adsorbed instead of arsenic on the ion exchange resin, so the levels of these water quality constituents are critical. If nitrate removal is also needed and is being done, sulfate is especially critical. An effect known as "chromatographic peaking" can occur when sulfate can displace previously adsorbed arsenate ions on the resin and cause higher levels of arsenic in the effluent than in the raw water. The resin bed must be monitored and regenerated well before this peaking situation occurs. Also, if the sulfate levels are above 120 to 150 mg/L, ion exchange costs are prohibitive.
3. Effect of arsenic concentration. Because arsenic is present in groundwater in very low concentrations relative to other competing ions, such as sulfate, and is not the preferred ion by the exchange resin, the concentration of arsenic (whether at 10 ppb or 100 ppb) will not dramatically affect the length of ion exchange run before leakage occurs. For example, the time for the length of ion exchange run could be half the run length time if the arsenic concentration is increased ten times.
4. Resin type. Strong-base resins are normally used for arsenic removal. However, resin selection should be a site-specific decision based on the results of pilot tests.
5. Configuration of ion exchange columns or vessels. Use of multiple ion exchange beds can help address issues of arsenic breakthrough and the chromatographic effect.
6. Direction of flow. The usual procedure is to have the raw water and the regenerant brine both flow through the resin bed in the downward direction (also called cocurrent flow). If the regenerant flow is in the opposite direction (upward flow or countercurrent flow), potential advantages include minimizing the leakage of arsenate ions from the vessel into the treated water and, possibly, the reduction of volume and concentration of brine required.
7. Number of ion exchange units. The use of multiple ion exchange units may have very significant benefits in terms of the removal of arsenic because of the breakthrough potential and the competing water quality indicator issues.

TYPICAL ION EXCHANGE ELEMENTS FOR ARSENIC REMOVAL

- Strong-base anion exchange resin
- Empty bed contact time (EBCT)[9] = 1.5 minutes
- Typical bed volumes treated = 400
- Regenerant, salt
- Salt use, 1,850 lb salt per million gallons treated (222 kilograms salt per million liters treated)
- Recycle brine, up to 36 times
- Precipitate arsenic in brine with ferric chloride before reuse

NOTE: Each treatment situation will be different because of unique water quality factors and local conditions.

9. *Empty Bed Contact Time (EBCT).* A measure of the time during which a water to be treated is in contact with the treatment medium in a contact vessel, assuming that all liquid passes through the vessel at the same velocity. EBCT is equal to the volume of the empty bed divided by the flow rate.

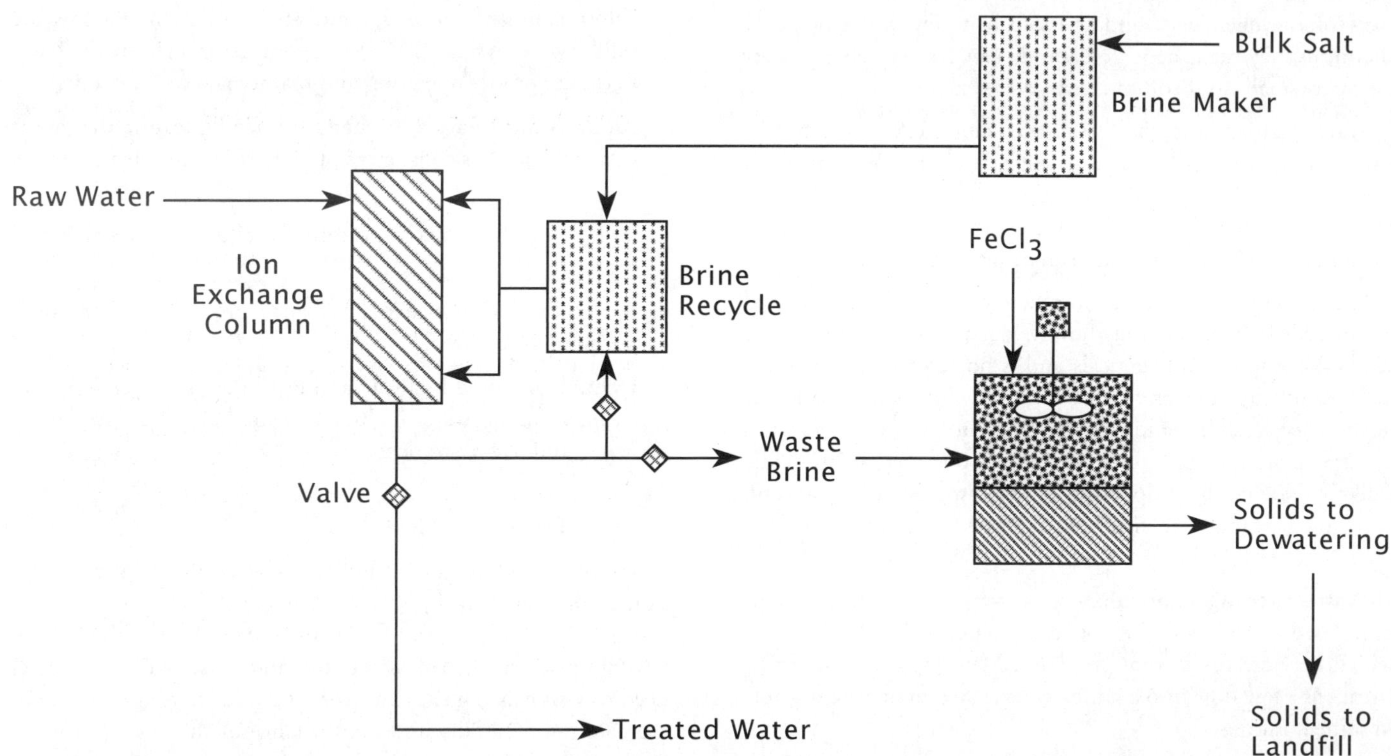

Fig. 15.3 Ion exchange arsenic removal treatment process

15.114 Activated Alumina (AA)

Activated alumina is an adsorption process used to remove arsenic from drinking water (Figure 15.4). Critical elements of an activated alumina treatment process are described in this section.

Importance of pH. The activated alumina adsorption process for removal of arsenic from drinking water is highly dependent on pH. This adsorption process may not perform efficiently unless the pH is lowered to the range of about 5.5 to 6.0. pH reduction requires the addition of acid (usually sulfuric acid) to lower the pH of most groundwaters. After arsenic has been adsorbed on the activated alumina, the treated water will require the pH to be raised to 7.5 or higher by adding caustic (sodium hydroxide) to control corrosion. Operation of this process will require operators to have advanced skills. The use and storage requirements for pH adjustment chemicals, the costs of the chemicals, and the safe handling and storage of chemicals may make activated alumina an unacceptable process for many small- to medium-sized water systems.

Water quality concerns. Activated alumina can be effective when the groundwater contains high levels of total dissolved solids (TDS). Water quality indicators that compete with arsenic ions (arsenate) for adsorptions sites on the media include selenium, fluoride, chloride, and sulfate (at high levels) ions.

Spent media. Activated alumina media may be regenerated or discarded. Regeneration, however, requires the addition of strong-base and acid chemicals and is not complete. For these reasons, on-site regeneration of activated alumina media is not likely to be feasible. Shipping the media to an off-site location for regeneration is typically not feasible either. The spent media is likely to be nonhazardous and can be disposed of in a landfill. If the media cannot be discarded, activated alumina will probably not be feasible for smaller systems.

Wastewater disposal. Regeneration of the activated alumina media produces a highly concentrated brine wastestream, which may create disposal problems. Metering the waste brine stream as a low flow into a sanitary sewer system may be possible in some locations.

Media availability. Activated alumina production is limited at this time and media may not be readily available in some locations and at some points in time in the future.

Configuration. Typically, two activated alumina media columns are operated in sequence to allow efficient operation, process control, and safety benefits. The media in the first column is run to exhaustion (unable to remove any more arsenic) and the second column provides a polishing step. Then, the media in the first column is replaced and the sequence is reversed (the second column removes arsenic and the first column provides the polishing step).

15.115 Oxidation-Filtration and Iron-Based Adsorption

There are several treatment systems that have been used successfully for removing iron and manganese (such as greensand filters) that may provide for co-removal of arsenic. The effectiveness of these processes in removing arsenic depends on the presence of sufficient iron in the raw water. To be effective, the iron/arsenic ratio must be at least 20 parts iron to 1 part arsenic in order to obtain 80 percent arsenic removal efficiency.

15.116 Point-of-Use (POU) and Point-of-Entry (POE) Devices

Point-of-use (POU) and point-of-entry (POE) devices are technologies that could be effective in removing arsenic from drinking water, especially for small systems. The devices generally use one or more BAT treatment processes.

Critical elements of any POU or POE system should include the following items:

1. All units must be owned, controlled, and maintained by the public water system (PWS) or a contractor hired by the PWS to ensure proper operation and maintenance of the units.
2. All units must have a mechanical warning feature that automatically notifies consumers of operational problems.
3. The public water system (PWS) must be able to obtain regular access to the units to perform necessary maintenance and monitoring.
4. A rigorous preventive maintenance program and frequent sampling is required.

Technologies used by POU and POE devices to remove arsenic from drinking water include activated alumina, iron-based sorbents, and reverse osmosis.

15.117 Proprietary Media

There are several other iron-based technologies that use adsorption as the mechanism for arsenic removal. These newer technologies include granular ferric hydroxide (GFH) and several new proprietary media. Operators and water utility agencies need to visit existing sites and conduct pilot tests at their site to evaluate the feasibility and performance of the newer, proprietary media, which are typically continually being improved by the vendor.

The GFH media systems remove arsenic, chromium, lead, selenium, antimony, uranium, and other heavy metals from groundwater. The system is operated as a fixed-bed adsorber. Typical installation is in pressure vessels to allow a single pumping stage for the treatment system. Water continuously passes through the media bed where arsenic is adsorbed from the water onto the media. The vessels are arranged in parallel or in series arrangement with an empty bed contact time of 5 minutes. The pressure vessels are backwashed once every 2 to 6 weeks to prevent bed compaction and remove trapped particulates.

Once the GFH media is exhausted, it is removed from the vessels and new media is installed. Exhausted media in most instances can be disposed of in a nonhazardous landfill.

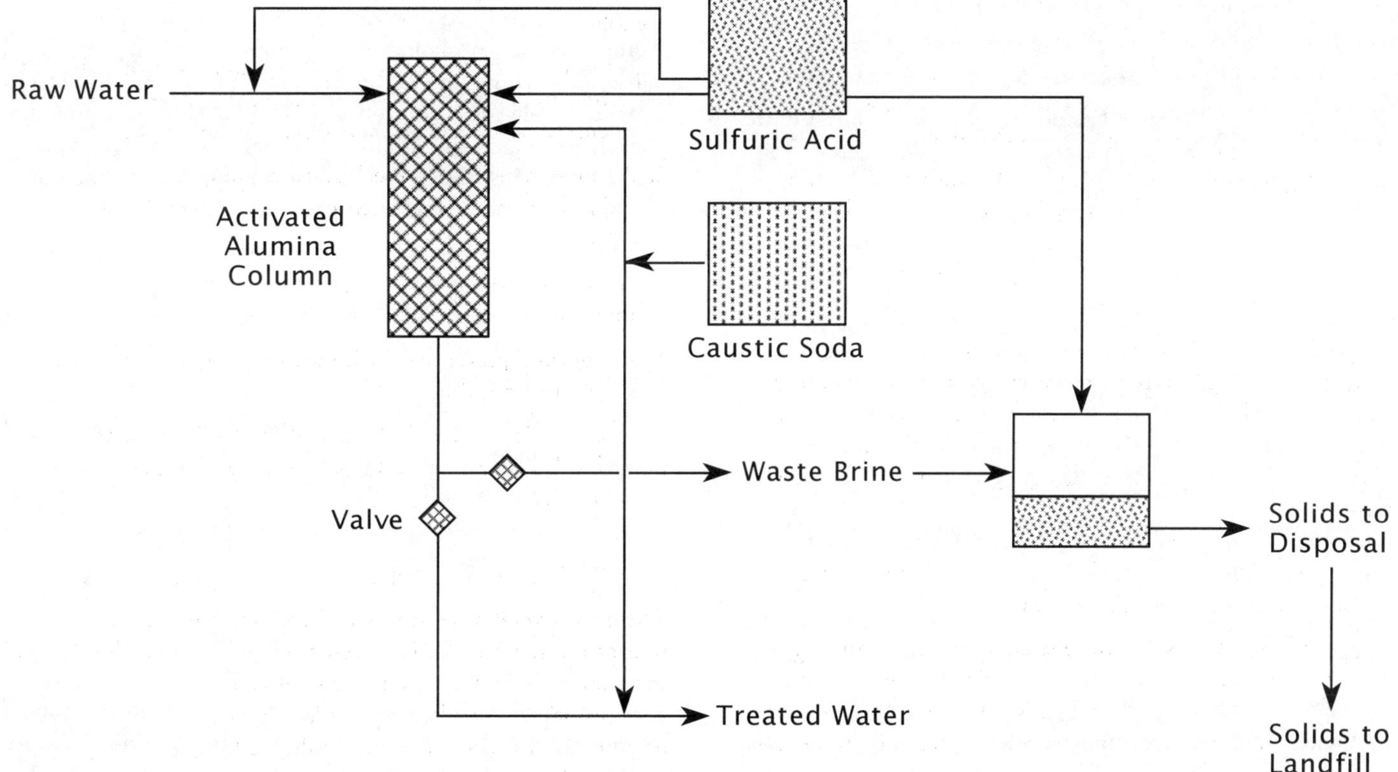

Fig. 15.4 Activated alumina arsenic removal treatment process

Parallel or series operation depends on the required removal concentrations. If a consistent 90 percent reduction of arsenic is needed across the system, the series design is used. However, if the required percentage removal is less than 90 percent, then the parallel design is typically used.

QUESTIONS

Please write your answers to the following questions and compare them with those on page 154.

15.11A What are the three general treatment categories used to remove arsenic from drinking sources?

15.11B Which tasks must be performed for blending to be an acceptable arsenic treatment option?

15.11C Which ions in groundwater are preferentially adsorbed instead of arsenic on an ion exchange resin?

15.11D Which technologies are used by POU and POE devices to remove arsenic from drinking water?

15.12 TYPICAL ARSENIC TREATMENT PLANT

Most arsenic removal treatment plants use either ion exchange or some type of adsorption media. The operation and maintenance (O&M) of these processes are similar. This section will emphasize the O&M of an ion exchange unit. If your plant uses another adsorption media, the O&M procedures will be similar.

15.120 Plant Startup and Shutdown

See Chapter 14, Section 14.17, "Startup and Shutdown of Unit," for an explanation of these procedures for an ion exchange unit.

15.121 Operation

Many factors influence the procedures used to operate an ion exchange unit and the efficiency of the arsenic removal process:

1. Characteristics of the ion exchange resin
2. Quality of the source water
3. Rate of flow applied to the ion exchange unit
4. Salt dosage during regeneration
5. Brine concentration
6. Brine contact time

Every water treatment plant should have an approved operating plan. The plant should always be operated in accordance with the approved plan. The detailed steps for normal operation, abnormal operation, troubleshooting, maintenance, monitoring, records, and reporting events are spelled out in the operating plan. The operating plan is the operator's bible for the plant.

Initial operating plans are frequently prepared by the design engineer who cannot anticipate all of the actual operating conditions. Therefore, it is essential that the operator modify the plan routinely, as needed, to reflect real-world conditions. The operator is responsible for working with others in the agency to make sure the operating plan is effective and up to date.

Each ion exchange unit, regardless of manufacturer, will have at least four common stages of operation. These stages are listed below and explained in previous sections (see Chapter 14, Section 14.11, Figure 14.14).

1. "Service," see Chapter 14, Section 14.110
2. "Backwash," see Chapter 14, Section 14.111
3. "Brine," see Chapter 14, Section 14.112
4. "Rinse," see Chapter 14, Section 14.113

Most ion exchange arsenic removal plants are fully automated, but must be designed to permit manual override upon demand. Automated valves should be hydropneumatically controlled valves actuated by signals through a computer system such as a programmable logic controller (PLC). Each vessel containing the ion exchange resin may have six or more valves to control raw water, brine, rinse water, and backwash water flows in and out of the vessel. In addition, there will be main line raw water, backwash water, wastewater, and other valves servicing all the vessels.

A detailed valve sequencing chart or list is needed showing the exact sequence and timing of operation of each valve or valves on an ion exchange vessel. This chart or list will be specific for a given water treatment plant.

Daily operation of an arsenic removal water treatment plant typically includes:

1. Reading and recording all flowmeter values and recording instantaneous rate-of-flow and cumulative volumes. Unusual readings need to be investigated and corrective action taken, if necessary.
2. Checking salt system and brine levels and verifying flow rates and salt usage.
3. Reading all online instruments for water quality indicators; noting set points for alarms or automatic shutdown; and recording and taking any corrective action needed. Calibrating the instruments, as needed.
4. Collecting grab samples of raw and treated water, as required, and performing analyses or arranging shipment to an approved laboratory. Conducting any calibrations needed based on these results.
5. Checking air compressor and automatic valves.
6. Checking all valve control positions by computer or visually.
7. Logging all data in records.
8. Reporting to superintendent, manager, or regulatory agency, as needed.

Operators must be sure to use only *NSF*[10] certified media and chemicals.

10. *NSF.* NSF International is a noncommercial, not-for-profit organization concerned with public health safety and environmental protection. NSF Standard 60 lists certified drinking water chemicals and NSF Standard 61 lists certified drinking water system components.

15.122 Maintenance

See Chapter 14, Section 14.15, "Maintenance," which contains information on how to maintain ion exchange units.

15.123 Troubleshooting

See Chapter 14, Section 14.16, "Troubleshooting," which contains information on how to troubleshoot the ion exchange units, the service stage, the backwash stage, the brine injection stage, and the rinse stage.

15.124 Safety and Security Issues

Operation of ion exchange and other arsenic removal treatment systems does not present any unique or unusual safety or security issues that are not present in any typical water treatment plant. Safety issues of concern for operators include working with electrical systems and mechanical equipment, and the storage and handling of chemicals and water. Operators are expected to be properly trained to be familiar with and adhere to good safety and security procedures.

15.125 Review of Plans and Specifications

For most treatment plants, it is very beneficial to the design engineer, to the operator, and to the facility for the operator to review the plans and specifications of an expanded or new arsenic removal plant before the completion of the plans and specifications. The design review helps the design engineer know what details to look for to make operation and maintenance easier and to anticipate problems that might otherwise require design modifications after construction is completed. Without the operator's assistance, modification might later be necessary because someone forgot or did not have the knowledge to recommend specific details for better operational control of the arsenic removal process. Operators should very carefully review the location of valves and the ease of adjusting and maintaining valves on the ion exchange unit.

The chemical storage and supply building should be equipped with emergency eye wash stations and deluge showers and floor washdown equipment. Electrical control panels and chemical feed pumps should be protected from accidental spills or splashing during washdown. If polymers are used, all possible locations of polymer spillage should have a false grated floor, with washdown drain, to allow the spilled material to pass to a point where its *VISCOUS,*[11] slippery nature is not a threat to operator safety. The ultimate destination of floor drains should always comply with sewer-use ordinances or discharge permits.

15.13 WASTEWATER AND RESIDUALS

Operators of arsenic removal treatment plants must have facilities capable of handling various types of wastestream liquids, sludges, and solids. The first step is to identify the sources of the liquids, sludges, and solids and then ensure that the treatment plant properly collects, treats, and disposes of these wastestreams. The sources of liquids, sludges, and solids include:

1. Liquids: Backwash water, regenerant solutions (brines, caustic), and reject water
2. Sludges (semiliquid residuals): Sludge from coagulation/precipitation, and iron/manganese sludge (sludge may be subjected to added treatment to concentrate and dewater)
3. Solids: Spent filter/adsorption media, resins, and concentrated salts from evaporation ponds

Alternative methods of ultimately disposing of liquids, sludges, and solids from arsenic removal treatment processes include the following:

1. Discharge to receiving stream (if allowed, pretreatment will be required)
2. Discharge to sanitary sewer that flows to wastewater treatment plant (POTW) (may require pretreatment)
3. Land disposal
4. Evaporation pond, with solids to landfill
5. Sanitary landfill (sludge or solids to landfill and brine to a POTW)
6. Hazardous waste facility

Contaminated brine streams may need pretreatment before discharge either to a receiving stream, sanitary sewer, or land disposal. In addition to the presence of arsenic in the brine stream, total dissolved solids (TDS) may cause a disposal problem by increasing the salinity in the receiving water (surface water or groundwater).

Another concern is the possibility that the contaminated brine stream may be classified as a hazardous waste. The determination of whether the wastestream is hazardous or not may depend on whether federal procedures or local procedures are used to make the hazardous waste determination. This determination may be a critical factor in both the selection of a treatment process and the operation of the facility. If a treatment residual is deemed hazardous, the cost of treating the hazardous waste may be prohibitive.

11. *Viscosity* (vis-KOSS-uh-tee). A property of water, or any other fluid, that resists efforts to change its shape or flow. Syrup is more viscous (has a higher viscosity) than water. The viscosity of water increases significantly as temperatures decrease. Motor oil is rated by how thick (viscous) it is; 20-weight oil is considered relatively thin while 50-weight oil is relatively thick or viscous.

QUESTIONS

Please write your answers to the following questions and compare them with those on page 154.

15.12A What are the common stages of operation of an ion exchange unit?

15.12B What are the safety issues of concern for operators of an ion exchange unit?

15.13A What are the sources of waste liquids, sludges, and solids from an arsenic removal treatment plant?

15.14 MONITORING

15.140 Analysis of Arsenic

Currently, there are four US EPA-approved laboratory methods of analyzing for arsenic in drinking water at levels below the maximum contaminant level (MCL). These methods use mass spectrometry (MS) or atomic adsorption (AA). Most commercial water laboratories can do these analyses.

For arsenic removal process control, there are a number of commercial test kits available that can measure arsenic levels below the MCL. The accuracy and precision of these test kits vary and their reliability in producing data for controlling treatment operations must be verified regularly by comparing results of *SPLIT SAMPLES*[12] obtained by a certified/approved laboratory. The frequency of analyzing verification test samples should depend on experience with previous results.

Currently, there are no readily available continuous, online, arsenic measuring systems to monitor arsenic removal treatment processes. Potential indicator (surrogate) measurements such as pH or conductivity may provide an online indication of the process efficiency in removing arsenic.

Field test kits for arsenic determination are available from the following companies:

1. Hach Company, www.hach.com or phone (800) 227-4224.
2. Industrial Test Systems, Inc., www.sensafe.com or phone (800) 861-9712.

15.141 Types of Arsenic Sampling/Monitoring

There are several types of arsenic sampling/monitoring that are required.

1. Compliance: Compliance monitoring samples are the mandated samples that must be analyzed by an approved laboratory and reported to the regulatory agency. The results include all samples required and used to calculate conformance with the MCL.
2. Process Control: These samples are the essential samples needed on an ongoing basis to verify proper operation of the arsenic removal treatment plant. Many of theses samples are mandated and reportable to the regulatory agency as part of an approved operations plan.
3. Calibration/Verification: These samples are the essential samples taken to calibrate instruments, compare results with laboratory data, and verify settings on process controls.
4. Special: Additional samples may be needed to troubleshoot or investigate problems.

Arsenic at the relatively low levels likely to be encountered in drinking water sources does not pose an acute health effect, if the human exposure is short term (hours or days rather than years or a lifetime). The lack of acute health effects minimizes the need for continuous, online monitoring to prevent any instantaneous exceeding of the MCL or of a treatment target. This fact is recognized in the regulatory scheme for arsenic that allows the averaging of monitoring data over quarters for a year.

15.142 Monitoring for Compliance

Monitoring for determining compliance with the arsenic MCL is mandated by the US EPA regulation and is usually also spelled out by the state/province regulatory agency. The compliance monitoring is based on sampling each point of entry (POE) of water into the distribution system. This means that each source that pumps directly to the system will need to be sampled separately. Similarly, if multiple sources are mixed before their mutual POE, the sampling is performed on the mixed waters. Also, if one or more wells are treated before delivery to the system, the combined treated water will need to be monitored for compliance.

Each POE must be sampled initially once after the effective date (January 23, 2006) and then every three years for wells and every year for surface waters. However, if the initial sample exceeds the MCL, then quarterly monitoring must be conducted thereafter until the system is reliably and consistently below the MCL. The quarterly results are averaged (on a running annual average basis) to determine compliance with the MCL. If the MCL were exceeded, treatment requirements would be imposed on the water system.

Water utilities must do intensive monitoring to determine which sources exceed the arsenic MCL. If sources exceed the MCL, the water utilities will need to plan, fund, design, and install treatment facilities to remove arsenic or to blend waters to meet the MCL. With the approval and permitting of each water treatment plant, the monitoring for determining compliance with the arsenic MCL will be specified. Water utilities operating arsenic removal and blending treatment facilities have stringent requirements for meeting the arsenic MCL and some will have even lower operational targets on an ongoing basis.

12. *Split Sample.* A single grab sample that is separated into at least two parts such that each part is representative of the original sample. Often used to compare test results between field kits and laboratories or between two laboratories.

15.143 Monitoring for Process Control

Water utilities should have a detailed operations plan and a monitoring program that will greatly minimize the likelihood of MCL exceedances occurring. Continuous monitoring of some indicators of arsenic treatment efficiency such as pH or conductivity may be desirable to provide better operational oversight and control.

Regular and frequent samples must be taken of raw water, plant effluent, and, if applicable, blended water. A minimum of daily grab samples from each source location may be required, at least initially. Test kits are useful for analyzing samples for arsenic, provided regular split-sample analyses are performed by a certified laboratory. A frequency of performing at least one split sample per week is typical. If the raw water is demonstrated to be very consistent, less frequent split samples may be acceptable. In most groundwater sources, arsenic levels do not vary dramatically either seasonally or over time, but regular monitoring is needed to document the lack of variation for each source.

Measuring electrical conductivity (EC), a surrogate for total dissolved solids (TDS), and pH can be very helpful process-control guidelines in processes such as ion exchange. However, operators must realize that EC cannot usually be correlated directly with arsenic reduction since other commonly present ions such as sulfate are also reflected by EC tests and are removed at different rates and preferences than arsenic.

Other water quality sampling and analyses may be needed depending on the specific site situation. These other analyses may include pH, sulfate, and bicarbonate. Control testing of the chlorination or other disinfection processes used is essential.

Special sampling of residuals and wastestreams is required for arsenic removal facilities. This special sampling may include determination of whether the waste is hazardous under federal and state/province regulatory definitions. Monitoring of waste discharges is required by water pollution control agencies or local public sewer agencies.

15.15 RECORDKEEPING AND REPORTING

Each water treatment plant should have an operations plan that will list the required records and reporting.

15.150 Records

Detailed plant operating records are essential to enable operators to effectively troubleshoot problems, evaluate long-term trends and subtle changes in treatment capacity or efficiency, assess operating costs, plan for major maintenance, and demonstrate compliance and treatment effectiveness. The records may be very important in the unlikely case of litigation against the water utility.

There are no standard recordkeeping forms that work for all water treatment plants; however, the following records need to be kept for all plants on some type of form.

1. Daily operations log: This information could be kept in a book in which the operator notes the times and types of inspections and records any unusual events and information.
2. Daily inspection form: This form will provide a consistent, organized form for recording all meter readings, laboratory results, notations of valve operation, tank levels, and other important information. The form is essential to record critical information.
3. Monthly summary form: This form will provide an overview for management and may be a required regulatory report summarizing the overall operation and the actual performance of the plant.
4. Blending calculations record: This form simplifies the regular calculation of theoretical blends and enables the operator to compare and verify test results. Detailed records are required if engineered blending is used for arsenic compliance.

15.151 Reporting

There are at least four types of reports that are required in the operation of ion exchange arsenic removal water treatment plants. These types of reports are similar to the types of reports required for any of the arsenic removal treatment plants.

1. Routine: Monthly summary reports are generally required to be submitted to the regulatory agency. These reports are due by the tenth of the following month and must be signed by the responsible utility official. The signature carries with it the responsibility and liability for ensuring that the report is accurate. Annual summary reports also may be required by either management or the regulatory agency.
2. Special: These special reports include appropriate notification of other officials in the organization (other operators, superintendent, manager) of special or unusual conditions that may affect the overall system operation or to alert them of the need for materials, supplies, maintenance, or equipment replacement.
3. Violations: Regulatory law typically has provisions for special notification of the regulatory agency if a violation occurs. The nature and extent of the violation dictates the immediacy and method of the notification required. Often the operator will have taken the necessary corrective action immediately and no further action is required by the utility agency beyond making the required report.
4. Emergency: *NOTE:* Any failure of an arsenic removal treatment plant or exceeding the arsenic MCL probably will not necessitate an emergency warning. This is because the arsenic levels in the water supplied to the consumers probably will not reach the arsenic level that a short-term exposure would cause adverse health effects. However, if the water quality test results of the situation indicate an emergency, an emergency notice may be necessary. This could involve an emergency notice to all consumers using the most expeditious means available (radio, TV, or door-to-door notification). Usually, the decision for such a notice is made by management in cooperation with the regulatory agency. The initial report of an emergency by the operator must be done without delay. In a small water system, the operator may have the responsibility, in the absence of others, to make the decision that there is an emergency and to initiate an immediate notice.

QUESTIONS

Please write your answers to the following questions and compare them with those on page 154.

15.14A How can an operator verify the precision and accuracy of commercial field arsenic test kits?

15.15A What types of records should be kept by the operator of an ion exchange arsenic removal water treatment plant?

15.15B What types of reports are required in the operation of an ion exchange arsenic removal water treatment plant?

15.16 ARITHMETIC ASSIGNMENT

Turn to the Arithmetic Appendix at the back of this manual. Read and work the problems in Section A.33, "Specialized Treatment Processes." Check the arithmetic in this section using a calculator. You should be able to get the same answers.

15.17 ADDITIONAL READING

1. *Arsenic Treatment Technology Evaluation Handbook for Small Systems,* US Environmental Protection Agency. EPA No. 816-R-03-014. This EPA document is available online at epa.gov/nscep/.
2. *Arsenic Removal from Drinking Water by Ion Exchange and Activated Alumina Plants,* US Environmental Protection Agency. EPA No. 600-R-00-088. This EPA document is available online at epa.gov/nscep/.

END OF LESSON 2 OF 2 LESSONS

on

SPECIALIZED TREATMENT PROCESSES

Please answer the discussion and review questions next.

DISCUSSION AND REVIEW QUESTIONS

Chapter 15. SPECIALIZED TREATMENT PROCESSES

(Lesson 2 of 2 Lessons)

Please write your answers to the following questions to determine how well you understand the material in the lesson. The question numbering continues from Lesson 1.

10. What are the potential health effects of the consumption of low levels of arsenic over a long period of time?
11. What should a water utility do before considering treatment of a water source to reduce or remove arsenic?
12. Which issues related to blending must be considered before a blending program is implemented?
13. Which factors influence the procedures used to operate an ion exchange unit and the efficiency of the arsenic removal process?
14. Which items should an operator consider when reviewing the plans and specifications for an ion exchange unit to remove arsenic from drinking water?
15. What are the required types of arsenic sampling/monitoring?
16. Why are detailed arsenic removal treatment plant operating records essential?

SUGGESTED ANSWERS

Chapter 15. SPECIALIZED TREATMENT PROCESSES

ANSWERS TO QUESTIONS IN LESSON 1

Answers to questions on page 131.

15.0A Trihalomethanes are formed in drinking water when free chlorine comes in contact with naturally occurring organic compounds (THM precursors). Trihalomethanes are a class of organic compounds in which there has been a replacement of the three hydrogen atoms in the methane molecule with three halogen atoms (chlorine or bromine).

15.0B THM precursors are defined as natural, organic compounds found in all surface and groundwaters, which may react with halogens (such as chlorine) to form trihalomethanes (THMs); they must be present in order for THMs to form.

15.0C Free chlorine is added to drinking water as a disinfectant.

15.0D One source of bromide in drinking water is seawater.

Answers to questions on page 134.

15.1A The major steps that a water utility could take to solve a THM problem include:

1. Determine the extent of the THM problem
2. Evaluate control strategies
3. Evaluate existing treatment processes
4. Examine studies of proposed treatment processes
5. Select a cost-effective option
6. Implement the chosen option

15.1B Control strategies that could be evaluated to control a THM problem include:

1. Change sources of supply
2. Treatment options
   a. Remove THMs
   b. Remove precursors
   c. Adjust or modify the chlorine application points
   d. Use alternative disinfectants

15.2A Important factors that influence the production of trihalomethanes include time, temperature, pH, and concentrations of chemicals.

15.2B Those utilities that use lime softening should be aware that free chlorine in contact with natural organics at a pH of 10.5 or higher will produce THMs faster than if the pH were near 7.0.

Answers to questions on page 135.

15.3A The two methods of controlling a THM problem are to change the source of supply or provide some type of treatment.

15.3B Abandoning the use of free chlorine is a serious move in view of its superior performance as a disinfectant.

15.4A Water treatment processes that can be used to control THMs include:

1. Aeration
2. Oxidation with chlorine dioxide and potassium permanganate
3. Coagulation, sedimentation, and filtration
4. Softening processes
5. Powdered activated carbon and granular activated carbon applications

Answers to questions on page 136.

15.5A Aeration is a much more effective process than oxidation for removing THMs after they have been formed.

15.5B THM concentrations should be lower in waters that have been stored in uncovered reservoirs because of loss to the atmosphere.

15.5C The adsorption process involves the individual THM compounds leaving the water and becoming attached to the surface of the carbon or resin.

Answers to questions on page 138.

15.5D The major treatment processes that have been investigated to remove THM precursors before they come in contact with chlorine include:

1. Aeration
2. Oxidation (including ozone oxidation before coagulation and clarification)
3. Clarification
4. Adsorption
5. Ion exchange

15.5E Moving the addition of free chlorine to a point following the clarification process is the key to success when using clarification to remove THM precursors.

Answers to questions on page 139.

15.5F Before an alternative disinfectant is applied to any system, the source of the water supply, water quality, and treatment effectiveness for bacteriological control must be evaluated.

15.5G Upgraded monitoring (more samples and tests) of the distribution system before and after a disinfectant change must be provided by a utility.

15.5H Before and after a disinfectant change, the distribution system monitoring program should include coliforms, standard plate count, turbidity, and nutrients.

Answers to questions on page 139.

15.6A If existing water treatment processes are not capable of solving a THM problem, the least-cost solution will probably be an alternative disinfectant.

15.6B The most popular alternative disinfectants are chloramines.

Answers to questions on page 140.

15.7A Before applying for and receiving a variance, utilities must evaluate the following treatment processes:

1. Use of chloramines or chlorine dioxide as an alternative or supplement to chlorine for oxidation and disinfection.
2. Use of chloramines, chlorine dioxide, or potassium permanganate as an alternative to chlorine for preoxidation.
3. Moving the point of chlorination in order to reduce THM formation.
4. Improvement of existing clarification.
5. Use of powdered activated carbon (PAC), intermittently, as necessary, to reduce TTHM or THM precursors. The dosage of PAC is not to exceed an annual average of 10 mg/L.

15.7B If treatment processes are not technically feasible and economically reasonable, then utilities must consider:

1. Offline water storage
2. Aeration
3. Introduction of clarification
4. Alternative sources of raw water
5. Use of ozone as an alternative or supplement to chlorine for disinfection or oxidation

ANSWERS TO QUESTIONS IN LESSON 2

Answers to questions on page 142.

15.10A Potential sources of arsenic in drinking water include either natural occurrences (such as volcanic action, erosion of rock, or a result of forest fires) or from human activities.

15.10B The oxidation state of arsenic is very important when treating arsenic because the arsenic (III) form is not removed by common treatment processes.

Answers to questions on page 148.

15.11A The three general treatment categories used to remove arsenic from drinking water sources are adsorption, coprecipitation, and physical separation.

15.11B Blending of higher arsenic level sources with low-arsenic level sources is an acceptable arsenic treatment option as long as the blending is properly engineered, operated, and controlled.

15.11C Sulfate and nitrate are both preferentially adsorbed instead of arsenic on an ion exchange resin.

15.11D Technologies used by POU and POE devices to remove arsenic from drinking water include activated alumina, iron-based sorbents, and reverse osmosis.

Answers to questions on page 150.

15.12A The common stages of operation of an ion exchange unit are service, backwash, brine, and rinse.

15.12B Safety issues of concern for operators of ion exchange units include working with electrical systems and mechanical equipment, and the storage and handling of chemicals and water.

15.13A Sources of waste liquids, sludges, and solids from an arsenic removal treatment plant include:

1. Liquids: Backwash water, regenerant solutions (brines, caustic), and reject water
2. Sludges (semiliquid residuals): Sludge from coagulation/precipitation; and iron/manganese sludge (sludge may be subjected to added treatment to concentrate and dewater)
3. Solids: Spent filter/adsorption media, resins, and concentrated salts from evaporation ponds

Answers to questions on page 152.

15.14A Operators can verify the precision and accuracy of commercial field arsenic test kits by comparing results of split samples obtained by a certified/approved laboratory.

15.15A Types of records that should be kept by the operator of an ion exchange arsenic removal water treatment plant include a daily operations log, daily inspection form, monthly summary form, and a blending calculations record, if appropriate.

15.15B Types of reports that are required in the operation of an ion exchange arsenic removal water treatment plant include routine, special, violations, and emergency.

CHAPTER 16

MEMBRANE TREATMENT PROCESSES

(MEMBRANE FILTRATION AND DEMINERALIZATION)

by

Dave Argo

and

Ken Kerri

Revised by

Warren Casey

TABLE OF CONTENTS

Chapter 16. MEMBRANE TREATMENT PROCESSES
(MEMBRANE FILTRATION AND DEMINERALIZATION)

LEARNING OBJECTIVES

Chapter 16. MEMBRANE TREATMENT PROCESSES (MEMBRANE FILTRATION AND DEMINERALIZATION)

Following completion of Chapter 16, you should be able to:

1. Describe various membrane technologies used to treat water.
2. Explain various types of membrane filtration processes.
3. Evaluate potential membrane pretreatment alternatives.
4. Operate and maintain a membrane filtration treatment plant.
5. Keep accurate and appropriate records for a membrane filtration plant.
6. Develop, implement, and maintain a membrane filtration plant maintenance program.
7. Describe the various demineralizing processes.
8. Explain how the reverse osmosis process works.
9. Operate and maintain a reverse osmosis demineralization plant.
10. Explain the principles of electrodialysis.
11. Identify and describe the parts of an electrodialysis plant.
12. Operate and maintain an electrodialysis plant.
13. Safely perform your duties around reverse osmosis and electrodialysis plants.

WORDS

Chapter 16. MEMBRANE TREATMENT PROCESSES (MEMBRANE FILTRATION AND DEMINERALIZATION)

ANGSTROM (ANG-strem) ANGSTROM

A unit of length equal to one-tenth of a nanometer or one ten-billionth of a meter (1 Angstrom = 0.000 000 000 1 meter). One Angstrom is the approximate diameter of an atom.

atm atm

The abbreviation for atmosphere. One atmosphere is equal to 14.7 psi or 100 kPa.

BENCH-SCALE ANALYSIS (TEST) BENCH-SCALE ANALYSIS (TEST)

A method of studying different ways or chemical doses for treating water or wastewater and solids on a small scale in a laboratory. Also see JAR TEST.

CHELATION (key-LAY-shun) CHELATION

A chemical complexing (forming or joining together) of metallic cations (such as copper) with certain organic compounds, such as EDTA (ethylene diamine tetracetic acid). Chelation is used to prevent the precipitation of metals (copper). Also see SEQUESTRATION.

COLLOIDS (KALL-loids) COLLOIDS

Very small, finely divided solids (particles that do not dissolve) that remain dispersed in a liquid for a long time due to their small size and electrical charge. When most of the particles in water have a negative electrical charge, they tend to repel each other. This repulsion prevents the particles from clumping together, becoming heavier, and settling out.

CONCENTRATION POLARIZATION CONCENTRATION POLARIZATION

(1) A buildup of retained particles on the membrane surface due to dewatering of the feed closest to the membrane. The thickness of the concentration polarization layer is controlled by the flow velocity across the membrane.

(2) Used in corrosion studies to indicate a depletion of ions near an electrode.

(3) The basis for chemical analysis by a polarograph.

CONDUCTANCE CONDUCTANCE

A rapid method of estimating the total dissolved solids content of a water supply. The measurement indicates the capacity of a sample of water to carry an electric current, which is related to the concentration of ionized substances in the water. Also called specific conductance.

CROSS-FLOW FILTRATION CROSS-FLOW FILTRATION

A type of membrane filtration where the water being filtered flows across the surface of the membrane to keep the particle buildup and fouling to a minimum. The flow that is not filtered becomes concentrated and flows out the end of the membrane fiber as a wastestream.

DEAD-END FILTRATION DEAD-END FILTRATION

A type of membrane filtration where the water being filtered flows through the membrane, but there is no wastestream from the system. All solids accumulate on the membrane during filtration and are removed during backwash.

DEMINERALIZATION (DEE-min-er-al-uh-ZAY-shun) DEMINERALIZATION

A treatment process that removes dissolved minerals (salts) from water.

DISINFECTION BYPRODUCT (DBP) DISINFECTION BYPRODUCT (DBP)

A contaminant formed by the reaction of disinfection chemicals (such as chlorine) with other substances in the water being disinfected.

DISSOLVED ORGANIC CARBON (DOC) DISSOLVED ORGANIC CARBON (DOC)

That portion of the organic carbon in water that passes through a 0.45 µm pore-diameter filter.

DISSOLVED ORGANIC MATTER (DOM) DISSOLVED ORGANIC MATTER (DOM)

That portion of the organic matter in water that passes through a 0.45 µm pore-diameter filter.

ENZYMES (EN-zimes) ENZYMES

Organic substances (produced by living organisms) that cause or speed up chemical reactions. Organic catalysts or biochemical catalysts.

FEEDWATER FEEDWATER

The water that is fed to a treatment process; the water that is going to be treated.

FLUX FLUX

A flowing or flow.

FULVIC ACID FULVIC ACID

A complex organic compound that can be derived either from soil or water. Aquatic fulvic acids are major precursors of disinfection byproducts.

HUMIC ACID HUMIC ACID

A polymeric constituent of soils, lignite, and peat. Aquatic humic acids are major precursors of disinfection byproducts.

HUMIC SUBSTANCES HUMIC SUBSTANCES

Natural organic matter resulting from partial decomposition of plant or animal matter and forming the organic portion of soil.

HYDROLYSIS (hi-DROLL-uh-sis) HYDROLYSIS

(1) A chemical reaction in which a compound is converted into another compound by taking up water.

(2) Usually a chemical degradation of organic matter.

MATERIAL SAFETY DATA SHEET (MSDS) MATERIAL SAFETY DATA SHEET (MSDS)

A document that provides pertinent information and a profile of a particular hazardous substance or mixture. An MSDS is normally developed by the manufacturer or formulator of the hazardous substance or mixture. The MSDS is required to be made available to employees and operators or inspectors whenever there is the likelihood of the hazardous substance or mixture being introduced into the workplace. Some manufacturers are preparing MSDSs for products that are not considered to be hazardous to show that the product or substance is not hazardous. Also see SAFETY DATA SHEET (SDS).

MEMBRANE FOULING MEMBRANE FOULING

The cause of a loss of flow through the membrane as a result of material being retained on the surface of the membrane or within the membrane pores. Membrane fouling may be reversible or irreversible. Loss of flow through the membrane by reversible fouling can be recovered by regular backwashing of the membrane surface. Flow loss by irreversible fouling cannot be recovered.

MICROFILTRATION (MF) MICROFILTRATION (MF)

A pressure-driven membrane filtration process that separates particles down to approximately 0.1 µm diameter from influent water using a sieving process.

NANOFILTRATION (NF) NANOFILTRATION (NF)

A pressure-driven membrane filtration process that separates particles down to approximately 0.002 to 0.005 µm diameter from influent water using a sieving process.

NATURAL ORGANIC MATTER (NOM) NATURAL ORGANIC MATTER (NOM)

Humic substances composed of humic acid and fulvic acid that come from decayed vegetation. Also see HUMIC SUBSTANCES, HUMIC ACID, and FULVIC ACID.

OSMOSIS (oz-MOE-sis) OSMOSIS

The passage of a liquid from a weak solution to a more concentrated solution across a semipermeable membrane. The membrane allows the passage of the water (solvent) but not the dissolved solids (solutes). This process tends to equalize the conditions on either side of the membrane.

PERMEATE (PURR-me-ate) PERMEATE

(1) To penetrate and pass through, as water penetrates and passes through soil and other porous materials.

(2) The liquid (demineralized water) produced from the reverse osmosis process that contains a low concentration of dissolved solids.

PILOT-SCALE STUDY PILOT-SCALE STUDY

A method of studying different ways of treating water or wastewater and solids or to obtain design criteria on a small scale in the field.

POINT-OF-USE/POINT-OF-ENTRY (POU/POE) POINT-OF-USE/POINT-OF-ENTRY (POU/POE)

Point-of-use applications refer to a water treatment device that treats water at the location of the end user. Point-of-entry application refers to a water treatment device that is located at the inlet to an entire building or facility.

REVERSE OSMOSIS (oz-MOE-sis) (RO) REVERSE OSMOSIS (RO)

The application of pressure to a concentrated solution, which causes the passage of a liquid from the concentrated solution to a weaker solution across a semipermeable membrane. The membrane allows the passage of the water (solvent) but not the dissolved solids (solutes). In the reverse osmosis process, two liquids are produced: (1) the reject (containing high concentrations of dissolved solids) and (2) the permeate (containing low concentrations). The clean water (permeate) is not always considered to be demineralized. Also see OSMOSIS.

SAFETY DATA SHEET (SDS) SAFETY DATA SHEET (SDS)

Safety data sheets (SDSs) are an essential component of the Globally Harmonized System of Classification and Labeling of Chemicals (GHS) and are intended to provide comprehensive information about a substance or mixture for use in workplace chemical management. They are used as a source of information about hazards, including environmental hazards, and to obtain advice on safety precautions. In the GHS, they serve the same function that the material safety data sheet (MSDS) does in OSHA's Hazard Communication Standard. The SDS is normally product related and not specific to the workplace; nevertheless, the information on an SDS enables the employer to develop an active program of worker protection measures, including training, which is specific to the workplace, and to consider measures necessary to protect the environment.

SALINITY SALINITY

(1) The relative concentration of dissolved salts, usually sodium chloride, in a given water.

(2) A measure of the concentration of dissolved mineral substances in water.

SCADA (SKAY-dah) SYSTEM SCADA SYSTEM

Supervisory Control And Data Acquisition system. A computer-monitored alarm, response, control, and data acquisition system used to monitor and adjust treatment processes and facilities.

SEQUESTRATION (SEE-kwes-TRAY-shun) SEQUESTRATION

A chemical complexing (forming or joining together) of metallic cations (such as iron) with certain inorganic compounds, such as phosphate. Sequestration prevents the precipitation of the metals (iron). Also see CHELATION.

SPECIFIC CONDUCTANCE SPECIFIC CONDUCTANCE

A rapid method of estimating the total dissolved solids content of a water supply. The measurement indicates the capacity of a sample of water to carry an electric current, which is related to the concentration of ionized substances in the water. Also called conductance.

TOTAL DISSOLVED SOLIDS (TDS) TOTAL DISSOLVED SOLIDS (TDS)

All of the dissolved solids in a water. TDS is measured on a sample of water that has passed through a very fine mesh filter to remove suspended solids. The water passing through the filter is evaporated and the residue represents the total dissolved solids. Also see SPECIFIC CONDUCTANCE.

ULTRAFILTRATION (UF) ULTRAFILTRATION (UF)

A pressure-driven membrane filtration process that separates particles down to approximately 0.01 μm diameter from influent water using a sieving process.

CHAPTER 16. MEMBRANE TREATMENT PROCESSES

(MEMBRANE FILTRATION AND DEMINERALIZATION)

(Lesson 1 of 4 Lessons)

16.0 MEMBRANE TREATMENT TECHNOLOGIES

Membrane treatment technologies are advancing rapidly in the water treatment field. Membranes are being used in municipal water treatment plants, in home *POINT-OF-USE OR POINT-OF-ENTRY (POU/POE)*[1] applications, in reclamation facilities, and wastewater treatment plants to remove both suspended and dissolved minerals from water.

Membranes contain very fine pore openings that allow water to pass through and block the passage of any contaminant larger than the pore diameter. Membranes used in water treatment are classified by their pore diameter. Classifications, from largest pore diameters to smallest, are *MICROFILTRATION (MF),*[2] *ULTRAFILTRATION (UF),*[3] *NANOFILTRATION (NF),*[4] and *REVERSE OSMOSIS (RO).*[5] The size ranges for each classification are presented in Figure 16.1, along with the relative sizes of materials commonly present in water.

Microfiltration (MF) and ultrafiltration (UF) are used in water treatment for particle, sediment, algae, bacteria, and virus removal. MF and UF are effective in the removal of *Giardia* and *Cryptosporidium*. Reverse osmosis (RO) membranes are used for desalination/demineralization and in home drinking water units. Reverse osmosis (RO) and nanofiltration (NF) membranes are used to remove *DISSOLVED ORGANIC MATTER (DOM)*[6] and dissolved contaminants such as arsenic, nitrate, pesticides, and radionuclides. Also, these membranes can remove ions such as calcium and magnesium (for softening) and sodium and chloride (for deionization, desalination, or demineralization of brackish waters). NF can be used to reduce the concentration of *NATURAL ORGANIC MATTER (NOM)*[7] to control the formation of *DISINFECTION BYPRODUCTS (DBPs).*[8]

The type of membrane used depends on the constituents to be removed from the water being treated. During water treatment, water is typically pumped against the surface of the membrane, but may be pulled through the membrane by a vacuum (submerged membrane). The water pressure forces water through the membrane and the constituents that do not pass through form a wastestream that may require treatment and proper disposal.

QUESTIONS

Please write your answers to the following questions and compare them with those on page 224.

16.0A Microfiltration (MF) and ultrafiltration (UF) are used in water treatment for the removal of which constituents?

16.0B Reverse osmosis (RO) and nanofiltration (NF) membranes are used to remove which constituents from water?

---

1. *Point-Of-Use/Point-Of-Entry (POU/POE).* Point-of-use applications refer to a water treatment device that treats water at the location of the end user. Point-of-entry application refers to a water treatment device that is located at the inlet to an entire building or facility.
2. *Microfiltration (MF).* A pressure-driven membrane filtration process that separates particles down to approximately 0.1 µm diameter from influent water using a sieving process.
3. *Ultrafiltration (UF).* A pressure-driven membrane filtration process that separates particles down to approximately 0.01 µm diameter from influent water using a sieving process.
4. *Nanofiltration (NF).* A pressure-driven membrane filtration process that separates particles down to approximately 0.002 to 0.005 µm diameter from influent water using a sieving process.
5. *Reverse Osmosis* (oz-MOE-sis) *(RO).* The application of pressure to a concentrated solution, which causes the passage of a liquid from the concentrated solution to a weaker solution across a semipermeable membrane. The membrane allows the passage of the water (solvent) but not the dissolved solids (solutes). In the reverse osmosis process, two liquids are produced: (1) the reject (containing high concentrations of dissolved solids) and (2) the permeate (containing low concentrations). The clean water (permeate) is not always considered to be demineralized. Also see OSMOSIS.
6. *Dissolved Organic Matter (DOM).* That portion of the organic matter in water that passes through a 0.45 µm pore-diameter filter.
7. *Natural Organic Matter (NOM).* Humic substances composed of humic acid and fulvic acid that come from decayed vegetation.
8. *Disinfection Byproduct (DBP).* A contaminant formed by the reaction of disinfection chemicals (such as chlorine) with other substances in the water being disinfected.

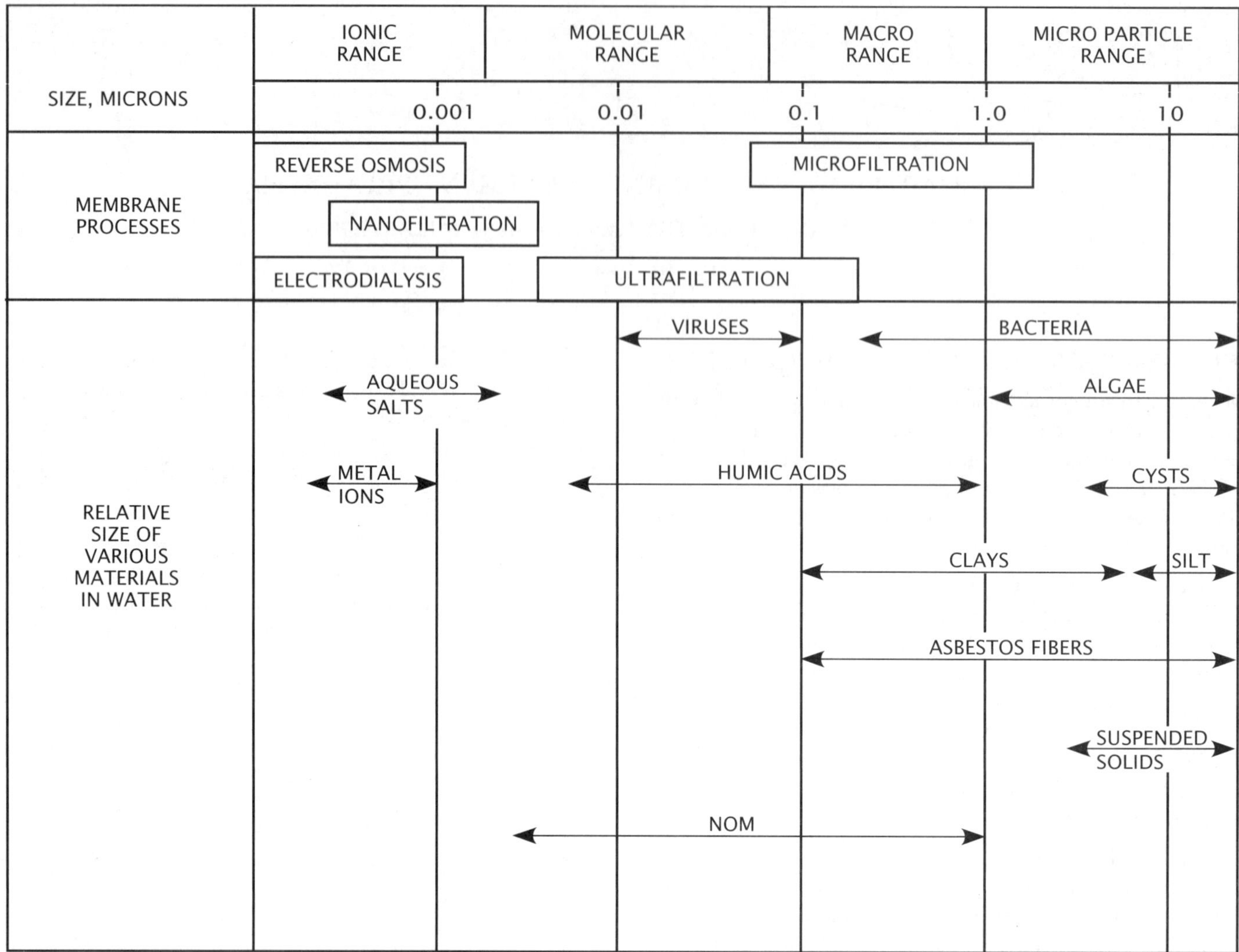

Fig. 16.1 Size ranges of membrane processes and contaminant

(Adapted from AWWA, *Water Quality and Treatment: A Handbook of Community Water Supplies*, 5th edition, McGraw-Hill, 1999)

16.1 DESCRIPTION OF MEMBRANE FILTRATION UNITS

16.10 Pressure Vessels or Submerged Flow

Typical water treatment membrane filtration units are installed either in pressure vessels or submerged in tanks. The membranes are hollow fibers or threads (Figures 16.2 and 16.3) with an outside diameter ranging from 0.5 to 2 mm and a wall thickness (membrane thickness) of 0.07 to 0.6 mm. In the pressure vessel installation, thousands of membrane fibers or threads (Figure 16.3) are arranged in racks or skids, with each pressure vessel from 4 to 12 inches (100 to 300 mm) in diameter and 3 to 18 feet (0.9 to 5.5 m) long (Figure 16.4). Filtration flow may be from outside to inside or from inside to outside of the hollow fiber membrane (Figures 16.5, 16.6, and 16.7).

Submerged or immersed membrane filtration systems (Figures 16.8 and 16.9) are membrane modules suspended in basins containing the water to be treated. The treated water may be pulled through the membrane by vacuum. Some treatment plants have removed sand from the sand filters and installed submerged membrane modules in the old filter basin.

16.11 Membrane Flow Types

There are two different types of membrane filtration feedwater flows: the *CROSS-FLOW FILTRATION*[9] system (Figure 16.7)

9. *Cross-Flow Filtration.* A type of membrane filtration where the water being filtered flows across the surface of the membrane to keep the particle buildup and fouling to a minimum. The flow that is not filtered becomes concentrated and flows out the end of the membrane fiber as a wastestream.

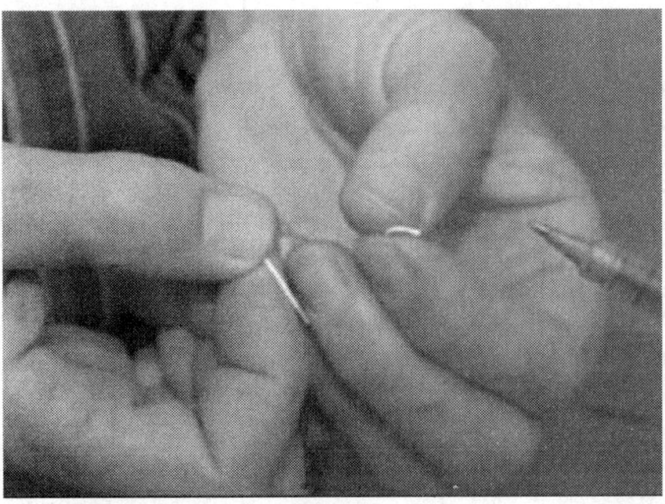

Fig. 16.2 One hollow fiber membrane

Fig. 16.3 Broken membrane tube with fibers showing

and the *DEAD-END FILTRATION*[10] system (Figure 16.6). In the cross-flow filtration system, the flow is from the inside of the membrane, through the membrane, and the filtered water flows out of the system. The flow inside the membrane flows along the inside surface of the membrane, becomes concentrated, and flows out of the end of the membrane fiber as a wastestream. In the dead-end filtration system, the water being filtered may flow either from the outside into the hollow fiber or from the inside to the outside, but there is no wastestream from the system. All solids accumulate on the membrane during filtration and are removed during backwash.

16.12 Membrane Fouling

MEMBRANE FOULING[11] can be a serious problem when operating membrane filtration processes. Membrane fouling can be described by whether the cause of the fouling can be removed (reversible or irreversible), by the material causing the fouling (biological, organic, particulate, or dissolved), and by the means of fouling (cake formation or membrane pore blockage).

Whether the cause of the fouling can be removed (reversible or irreversible) depends on the type of membrane used and the constituents in the source water. During continued operation, the flow through the membrane may decrease, but the flow may be recovered by backwashing and cleaning.

The constituents in the water being filtered may cause fouling. During membrane filtration, microorganisms are transported to the membrane surface where biofouling may occur. These microorganisms may not be removed by backwashing, but may be controlled by chlorine. However, some membrane materials (cellulose acetate [CA] or polypropylene [PP]) may be damaged by chlorine. Membrane manufacturers are now tending to use materials that are not damaged by chlorine.

Dissolved organic matter (DOM) may cause membrane fouling. The extent of the fouling problem depends on the characteristics of the dissolved organic matter, the membrane material, and the characteristics of the water being filtered.

During membrane filtration, particulate matter from the water being filtered collects on the membrane surface in a porous mat called a filter cake. Particulate fouling is usually reversible during periodic backwashing.

Natural organic matter (NOM) can be the most common form of membrane fouling. DOM includes wastes and portions of aquatic plants and animals as well as organic matter washed into surface water from land. Sources of DOM include organic chemicals found in biological systems and dissolved organic chemicals from industrial and commercial wastes. Fouling depends on the characteristics of the DOM, the membrane material, and the source water.

16.13 Pretreatment

16.130 Operation Experience

When operators first started using membrane filtration (microfiltration [MF] and ultrafiltration [UF]), most water sources

10. *Dead-End Filtration.* A type of membrane filtration where the water being filtered flows through the membrane, but there is no wastestream from the system. All solids accumulate on the membrane during filtration and are removed during backwash.
11. *Membrane Fouling.* The cause of a loss of flow through the membrane as a result of material being retained on the surface of the membrane or within the membrane pores. Membrane fouling may be reversible or irreversible. Loss of flow through the membrane by reversible fouling can be recovered by regular backwashing of the membrane surface. Flow loss by irreversible fouling cannot be recovered.

Fig. 16.4 Racks of membrane pressure vessels

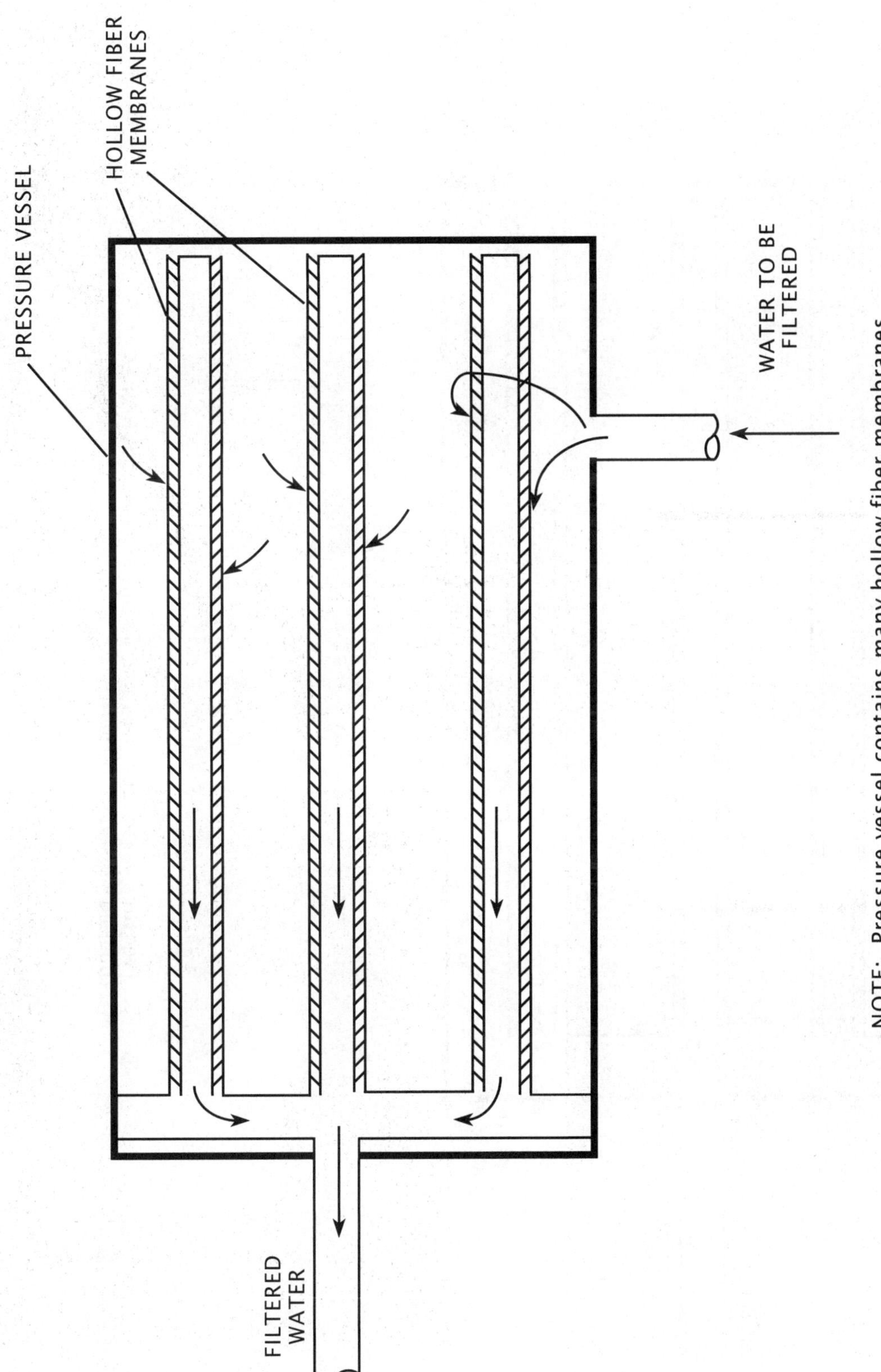

Fig. 16.5 Membrane filtration—outside into membrane flow

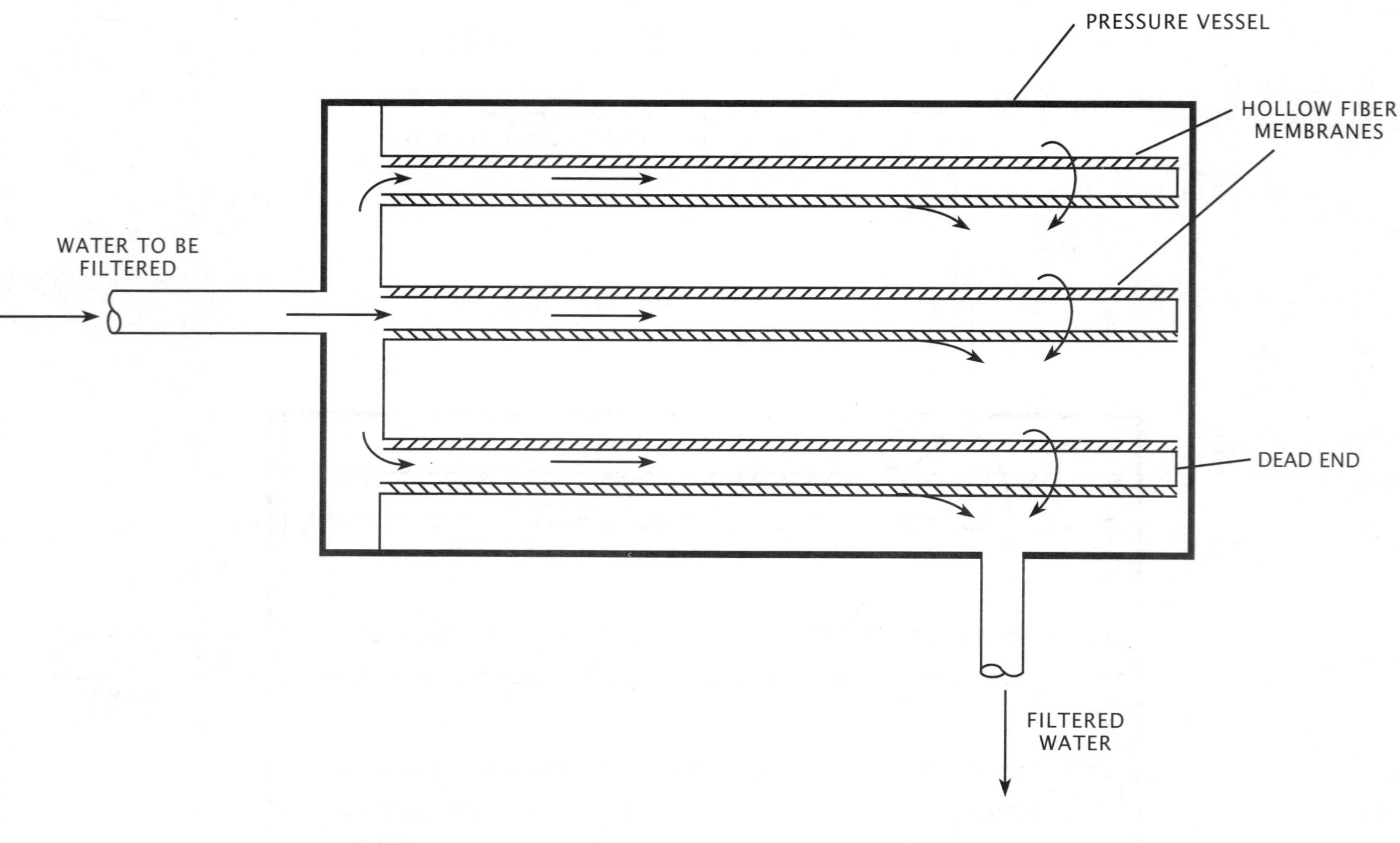

Fig. 16.6 Dead-end membrane filtration—inside membrane flow to outside of membrane

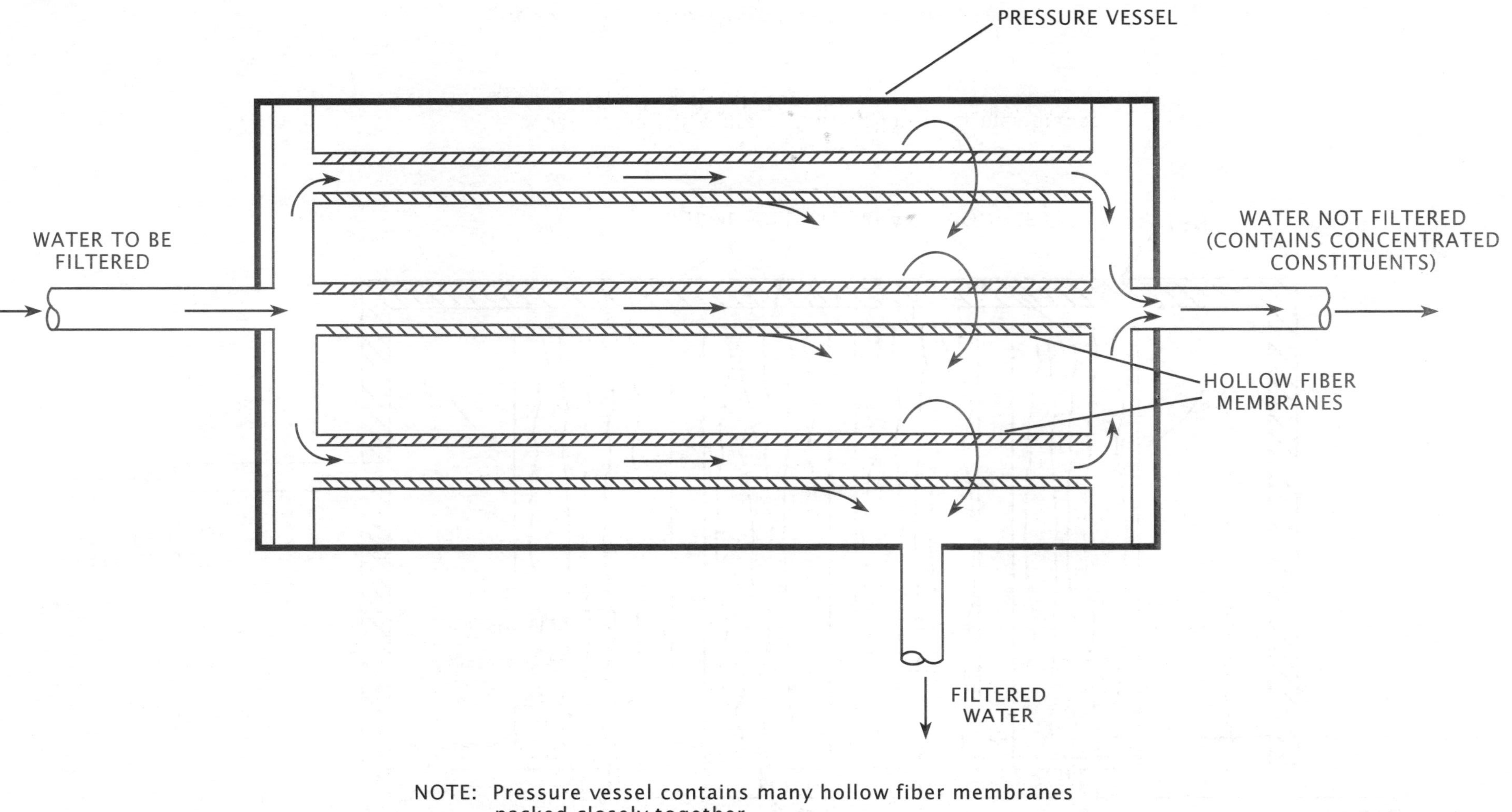

Fig. 16.7 Cross-flow membrane filtration—inside membrane flow to outside of membrane, across membrane surface

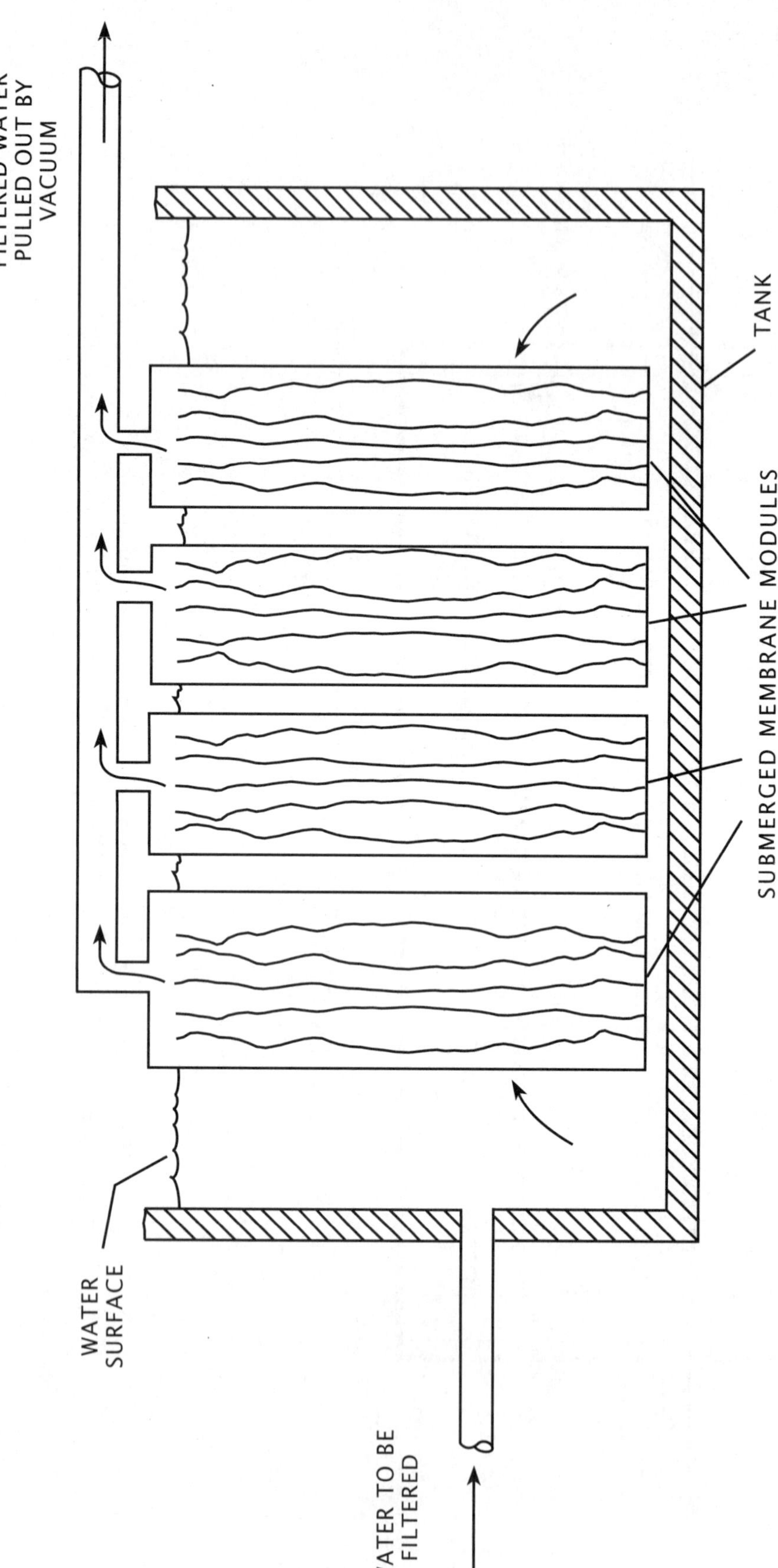

Fig. 16.8 Schematic of submerged membrane filtration system

Fig. 16.9 Submerged membrane filtration system

used were of high-quality, low-turbidity surface waters that required minimum treatment. Membrane filtration effectively removed any turbidity present in the water. The effectiveness of membrane filtration in treating low-turbidity water encouraged operators to consider using coagulation pretreatment before membrane filtration.

Operators have learned that the interactions among coagulants, the various constituents in natural waters, and the membrane materials are very complex, making the effect of coagulation on membrane performance difficult to predict. In some treatment plants, the flow through the membrane increases with coagulation pretreatment and in other plants there is a reduction in flow.

When using coagulation pretreatment before membrane filtration, operators have experienced all of the following, sometimes contradictory, situations:

- When using river water, cellulose acetate (CA) ultrafiltration membranes fouled in 2 days, but pretreatment with 21 mg/L of ferric chloride essentially eliminated fouling.
- After several months of operation, a plant using lake water and CA ultrafiltration membranes experienced rapid fouling during a period of lake turnover. Pretreatment with 14 mg/L ferric chloride restored the system to normal operation.
- Some water treatment plants have experienced improved membrane performance (flows through the membranes) after pretreatment with alum or ferric salts.
- Coagulation with either ferric chloride or a commercial polyaluminum coagulant slowed short-term reversible fouling, but did not reduce the rate or extent of irreversible fouling.
- Coagulation can be effective for removing particulate matter, but may have no effect on fouling by *DISSOLVED ORGANIC CARBON (DOC)*.[12]
- Some plants have found that coagulation makes fouling worse.
- Coagulation with ferric chloride significantly increased fouling of microfiltration and ultrafiltration membranes during

12. *Dissolved Organic Carbon (DOC).* That portion of the organic carbon in water that passes through a 0.45 µm pore-diameter filter.

filtration of river water containing humic and fulvic acid solutions.

- Low alum doses increased the rate of fouling, but high doses decreased the fouling of 10 mg/L solutions of fulvic acids from swamp waters filtered through polysulfone membranes. Also, a similar reduction in membrane performance was observed at low ferric chloride doses, but overall membrane performance improved at higher doses.

Membrane configuration (layout), membrane materials, and coagulation conditions may also affect whether coagulation improves or decreases membrane performance:

- Coagulation pretreatment caused a decline in performance of a pressurized membrane system, but an improvement in performance of a submerged membrane system.
- Differences in membrane performance can be caused by changes in cake resistance. A reduction in fouling can be obtained by the removal of dissolved organic carbon (DOC).

16.131 Summary of Experiences

A summary of these operator experiences indicates that pretreatment coagulation before membrane filtration has produced inconsistent results regarding membrane fouling and a decline of flows through the membrane. These inconsistencies apparently result from the fact that both coagulation and membrane filtration may be performed under a wide variety of operating conditions. For example, coagulation may be performed with or without flocculation and with or without sedimentation before membrane filtration. Membrane filtration may be performed with submerged or pressurized membranes and with constant pressure or constant flow conditions. Studies conducted on a specific source water with an individual coagulant and a set coagulant dose cannot be easily compared with other source waters, coagulants, and doses.

Another important consideration in the relationship between coagulation and membrane performance is that coagulation affects both particulate matter and DOC, both of which can affect membrane performance. Coagulation collects particles into layer masses and, if settling is practiced, removes particles from solution, which may alter membrane fouling because of cake resistance. Coagulation also removes DOC from the water, which may alter membrane fouling caused by adsorption.

16.132 Need for Bench- or Pilot-Scale Studies

In view of these conflicting experiences with pretreatment and membrane filtration, what can an operator do to optimize membrane performance? A suggested approach is to work with the membrane manufacturers. Invite the manufacturers to your facility and ask them to perform *BENCH- OR PILOT-SCALE STUDIES*.[13, 14] Indicate that the results of the studies should provide the following information:

1. The selection of the most appropriate coagulant for your specific application. This selection is dependent on interactions between the coagulant, source water, and membrane material.
2. The selection of the most appropriate coagulant dose. Coagulant dose is an important factor in determining whether membrane performance improves or decreases. For example, at low alum doses, membrane fouling is frequently worse than with no coagulant, but at the dose for enhanced coagulation, membrane performance is usually better.
3. The coagulant selection should include consideration of removing DOC. The coagulant type (for example, alum, ferric sulfate, or PACL) that tends to be most effective at removing DOC in a particular source water is typically the most effective at improving membrane performance.
4. A cost comparison for each alternative investigated.

QUESTIONS

Please write your answers to the following questions and compare them with those on page 224.

16.1A Where are typical water treatment membrane filtration units installed?

16.1B How can a decrease in flow through a membrane caused by fouling be recovered?

16.1C Why do operators consider using coagulation pretreatment before membrane filtration?

16.1D What can an operator do to optimize membrane performance?

16.2 OPERATION AND MAINTENANCE

16.20 SCADA System

The operation of membrane filtration water treatment processes is typically almost fully automated. These systems usually have a *SUPERVISORY CONTROL AND DATA ACQUISITION (SCADA) SYSTEM*[15] and are designed to run unattended (Figures 16.10 and 16.11). The SCADA system is used to monitor operational conditions and to make process adjustments within selected set points. The membrane filtration SCADA system monitors alarm conditions and follows protocols for alarm notification and plant shutdown.

Typically, operators maintain tablets, laptops, or computers with full plant operation compatibility. When the plant is

13. *Bench-Scale Analysis (Test).* A method of studying different ways or chemical doses for treating water or wastewater and solids on a small scale in a laboratory. Also see JAR TEST.
14. *Pilot-Scale Study.* A method of studying different ways of treating water or wastewater and solids or to obtain design criteria on a small scale in the field.
15. *SCADA* (SKAY-dah) *System.* Supervisory Control And Data Acquisition system. A computer-monitored alarm, response, control, and data acquisition system used to monitor and adjust treatment processes and facilities.

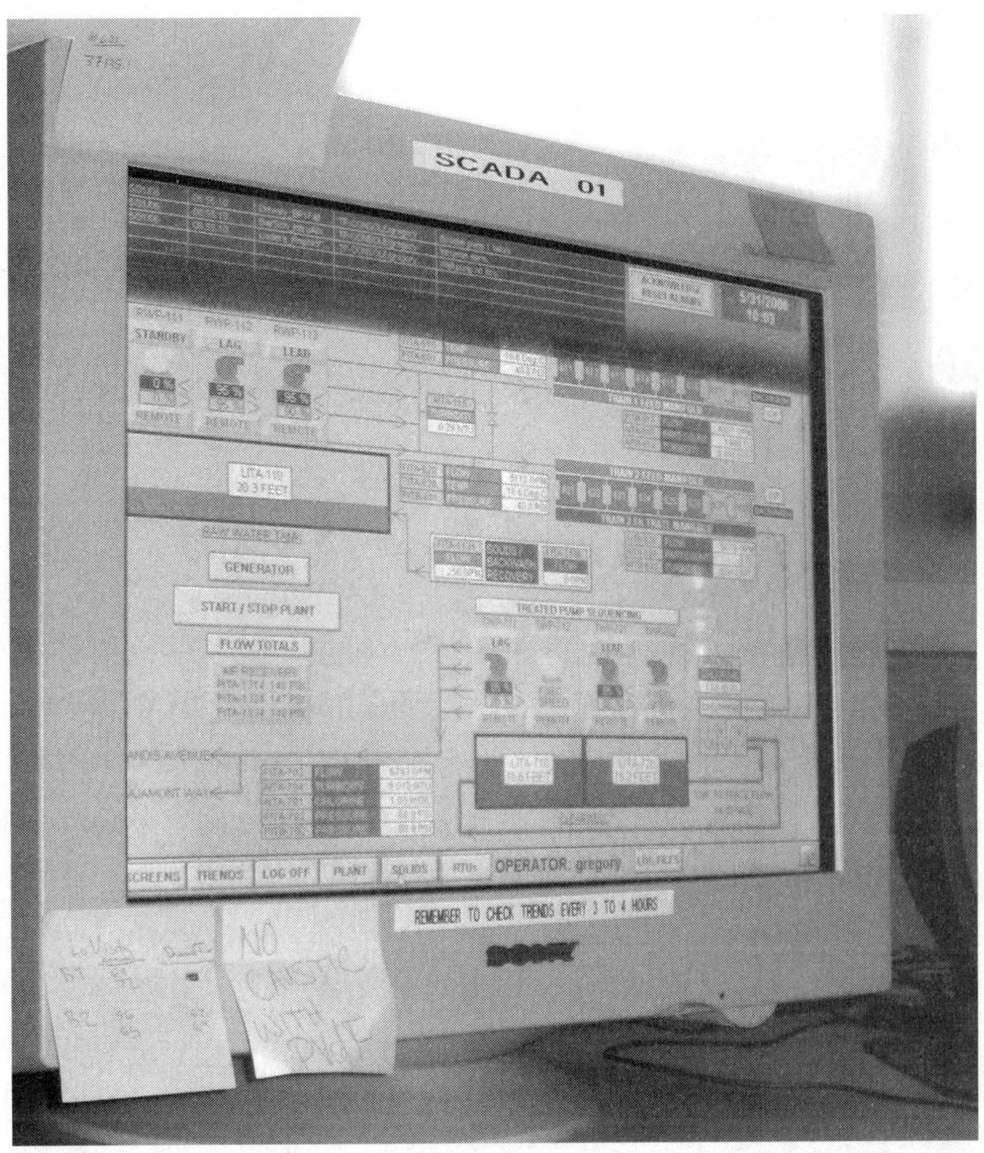

Fig. 16.10 SCADA system (plant control room)

unattended, the on-call operator can use a computer off-site to confirm plant process performance and make process and equipment adjustments before returning to the plant.

At the beginning and end of a shift, the operator should review the SCADA information and walk through the water treatment plant. The plant walk-through includes looking and listening for anything unusual.

16.21 Components of a Typical Plant

The operation and maintenance of a typical membrane filtration water treatment plant requires the operator to be familiar with the purposes of the following units.

1. Raw water source. The raw water sources for most membrane plants are surface waters, including high-quality streams and lakes.
2. Raw water pump station. The raw water pump station diverts water from the source into the treatment units.
3. Pretreatment facilities, if applicable. Pretreatment facilities consist of chemical treatment processes to remove turbidity before filtration.
4. Membrane filters. The membrane filters are used to remove particles, sediment, algae, bacteria, and viruses from the water being treated by water filtration through membranes.
5. Backwashing. Backwashing may occur on the basis of three possible conditions:
   a. Elapsed time since the last backwash
   b. Total water produced since the last backwash
   c. Pressure increase since the last backwash

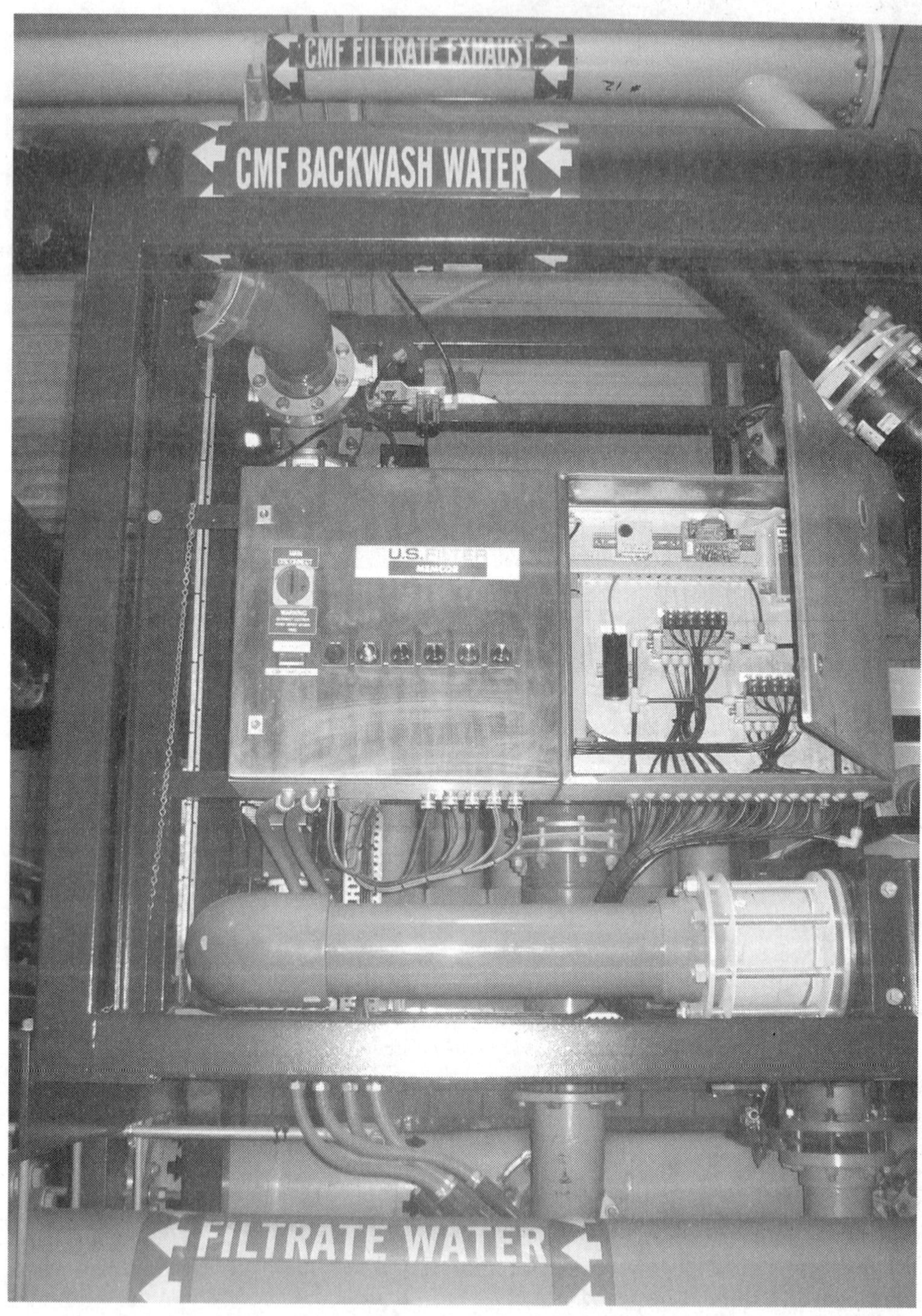

Fig. 16.11 SCADA system (controls at membrane rack)

Backwashing may be by air or by water, depending on the plant. Over time, the membranes require a more aggressive cleaning to remove accumulations and this is accomplished with the use of a proprietary chemical and caustic solution through a clean-in-place (CIP) process. This process results in the discharge of a wastestream that requires proper disposal.

6. Backwash water treatment (Figure 16.12). Water from backwash cleaning requires treatment and, depending on treatment, may be disinfected and discharged to the distribution or storage system or returned to the raw water intake, blended with raw water, and re-treated by pretreatment (if necessary) and membrane filtration.
7. CIP membrane regeneration system (Figure 16.13). The CIP membrane regeneration system uses aggressive chemicals based on the material of the membrane.
8. Chemical feed systems. Chemical feed systems feed chlorine for the disinfection of the treated water. Caustic is fed after chlorination to increase the pH for corrosion control.
9. CT chamber. The chlorine contact chamber (CT chamber) provides the required contact time to achieve pathogen inactivation.
10. Clear well. The clear well provides storage for treated water before the water is pumped into the distribution system.
11. Treated water pump station. Treated water is pumped from the clear well into the distribution system. The pump station discharge is monitored for flow, pressure, turbidity, chlorine residual, and particle count.
12. Standby engine generator. The plant should be equipped with a standby emergency engine generator to provide power in the event of a power failure.

16.22 Operational Procedures

Daily Operations

1. Inspect intake structure and security
2. Inspect building security
3. Inspect and maintain:
   - Pumps, raw water, and treated water
   - Membrane filters, flows, turbidity, and particle count monitoring
   - Chemical feed system, disinfection, and pH (corrosion control)
   - CT chamber or tank
   - Clear well
   - Treated water pump station
   - Standby engine generator
4. Pretreatment facilities (if applicable), inspect and maintain:
   - Chemicals, storage, leaks, quantities (need to order more?)
   - Chemical feeders
   - Chemical doses

Weekly Operations

1. Particle count measurement on each membrane skid effluent

Monthly Operations

1. Raw water and treated water pump stations
   - Check mechanical seals
   - Turbidimeter calibration
2. Membrane filters
   - Turbidimeter calibration
3. Standby engine generator
   - Run/Test

Yearly Operations

1. Raw water and treated water pump stations
   - Check motor efficiency, electrical phase balance
   - Calibrate flowmeters
   - Calibrate pressure transducers
   - Change packing and oil on top bearing and grease bearings of pumps
2. Membrane filters
   - Calibrate flowmeters
   - Calibrate pressure transducers
3. Chemical feed system
   - Change hydraulic fluid in feed pumps
4. CT chamber and clear well
   - Dive inspection

16.23 Membrane Performance Monitoring

The operation of membranes includes monitoring and testing for membrane filtration rate and membrane integrity. This is accomplished by pressure decay testing and sonic testing. When broken or damaged membrane fibers are discovered, they are repaired or replaced.

16.24 Troubleshooting Equipment and Process Failures

When the membrane filtration water treatment plant experiences an equipment or process failure, the operators troubleshoot to determine the cause of the problem. When the cause is determined, they will complete any repairs they are capable of performing. Any parts or outside services that are required will

Fig. 16.12 Backwash water treatment

Fig. 16.13 Clean-in-place chemical preparation

TABLE 16.1 TROUBLESHOOTING MEMBRANE EQUIPMENT AND PROCESS PROBLEMS

Problem	Cause	Corrective Action
Air compressor failure	Mechanical failure inside unit.	Rebuild failed compressor.
	Belt failure.	Replace belt.
	Oil pump leak.	Repair leak.
		Use a portable air compressor.
Raw water pump station failure	Electronic failure of the uninterrupted power supply (UPS).	Bypass the UPS.
		Conduct routine monthly checks of all plant UPS units.
Computer lockup	Central SCADA computer failure.	Operate plant with only local control over plant operation.
		Hire an outside contractor to bring computer back to normal operating conditions.

be ordered or scheduled. The incident is recorded in the operator logbook. This logbook is maintained daily and includes daily and critical events.

Typical problems that may be encountered when operating a membrane filtration water treatment plant are described in Table 16.1 along with the cause and the corrective action.

QUESTIONS

Please write your answers to the following questions and compare them with those on page 224.

16.2A How can an on-call operator use a computer?

16.2B Backwashing membrane filters may occur under what kinds of possible conditions?

16.2C How is membrane performance monitored?

16.2D What kinds of problems may an operator encounter when operating a membrane filtration water treatment plant?

16.3 RECORDKEEPING

16.30 Records

Membrane filtration water treatment plants typically use a SCADA system for water quality and water production monitoring and recordkeeping. These records include influent and effluent water quality, the hours of plant operation, plant flow rates, water production and distribution, process monitoring records, chemical dose rates and residual trends, alarms, and many other important conditions. In addition, the operations staff maintains handwritten records of maintenance, meter readings, chemical storage levels, and other process data that is read during the operator's walk-throughs at the plant.

16.31 Summary of Records

Membrane filtration water treatment plant records are usually maintained both electronically by the SCADA system and manually on hard copies. The records include water quality monitoring data, water production data, maintenance information and tasks performed, calibration dates and procedures, cleaning dates and procedures, testing methods, and *SAFETY DATA SHEETS (SDS)*.[16]

16.310 Water Quality Monitoring Records

The membrane filtration water treatment plant uses the SCADA system to continuously monitor turbidity and chlorine residual. The turbidity monitoring is of the raw, filtered, treated, and backwash returned water. Required disinfection levels are achieved by providing enough contact time and residual chlorine concentration.

16.311 Plant Operation and Water Production

The membrane filtration water treatment plant operation and water production records are maintained by the SCADA system.

16. *Safety Data Sheet (SDS).* Safety data sheets (SDSs) are an essential component of the Globally Harmonized System of Classification and Labeling of Chemicals (GHS) and are intended to provide comprehensive information about a substance or mixture for use in workplace chemical management. They are used as a source of information about hazards, including environmental hazards, and to obtain advice on safety precautions. In the GHS, they serve the same function that the material safety data sheet (MSDS) does in OSHA's Hazard Communication Standard. The SDS is normally product related and not specific to the workplace; nevertheless, the information on an SDS enables the employer to develop an active program of worker protection measures, including training, which is specific to the workplace, and to consider measures necessary to protect the environment.

This includes water levels, flow rates, hours of operations, and quality of water produced.

16.312 Troubleshooting Records

Always record the conditions and causes of problems and failures. Also, record the corrective action or solution taken and the effectiveness of the corrective action. This information is very helpful in preventing future problems and failures and in evaluating potential solutions to failures.

16.313 Membrane Filter Maintenance and Inspection

Filter maintenance and inspection records for the membranes consist of the following:

- Transmembrane pressure (as indication of need for CIP)
- Filtration production by skid
- Backwash occurrence, including water production, transmembrane pressure, and run time for each period before backwash
- Membrane integrity testing using pressure decay testing
- Membrane sonic testing
- Membrane repair testing and membrane fiber repairs

QUESTIONS

Please write your answers to the following questions and compare them with those on page 224.

16.3A Membrane filtration water treatment plants continuously monitor which water quality indicators?

16.3B Which membrane filtration water treatment plant operation and water production records are maintained in a SCADA system?

16.4 MAINTENANCE PROGRAM

16.40 Types of Maintenance

The maintenance program at a membrane filtration water treatment plant includes different levels of activity performed throughout the year. There are three main types of equipment maintenance activities:

1. Routine maintenance
2. Preventive maintenance
3. Breakdown or corrective maintenance

16.41 Routine Maintenance

Routine maintenance is the first level of maintenance performed at a water treatment plant. The operator conducts routine maintenance on a daily basis. The operator observes each piece of equipment and checks for noise, heat, odors, vibration, and leaks. Depending on the skill level required, the operator may investigate the problems encountered and conduct on-site repair or schedule a contract service to maintain the equipment. A daily checklist is filled out to confirm that all equipment is running properly or to report problems encountered.

16.42 Preventive Maintenance

The preventive maintenance program (PMP) (also known as preventive maintenance [PM] program) includes scheduled maintenance based on operating time. To prevent catastrophic equipment failure events and to eliminate or minimize conditions that would result in plant downtime requires an effective PMP. The frequency of PMP activity is set to comply with the equipment manufacturer's recommendations and historical operations and maintenance experience.

Typical PMP activities at a water treatment plant include replacing lubricants, inspecting impellers, replacing bearings, and tensioning drive belts. Equipment at the plant should be monitored continuously by the SCADA system to ensure an effective PMP. The information maintained in the SCADA system associated with the PMP log includes the following items:

- Description of the equipment, equipment number, and location within the plant
- Manufacturer's address, phone number, and person to contact
- Supplier or local representative with address, phone number, and date of purchase
- Size, model type, and serial numbers
- Electrical, mechanical, and other pertinent performance data
- Type and frequency of preventive maintenance tasks to be accomplished
- Space to note when preventive maintenance was performed and by whom
- Lubrication schedule, denoting frequency of oil changes, greasing and general lubrication, types of lubricants to be used, and guidance on disposal of old oil
- Location of drawings and operation manuals
- Spare parts available from stock
- Safety considerations
- Work hours required, costs, and material or supplies consumed

The preventive maintenance activity is planned such that replacement parts are available in order to reduce equipment downtime. The skills required to perform preventive maintenance are different than those required for routine maintenance. Preventive maintenance is often carried out by outside contractors.

16.43 Breakdown or Corrective Maintenance

Breakdown or corrective maintenance relates to repairs to equipment that has failed in service. Due to normal redundancy built into the plant design, breakdown maintenance generally does not interrupt treatment and allows time for the plant repairs or replacement to be undertaken and completed. On completion of the repair task, the action is recorded in the breakdown maintenance log.

16.44 Acknowledgment

Material in this lesson is from the Carmichael Water District's "Bajamont Way Membrane Filtration Plant Operations Plan" prepared by Alex Peterson, P.E., Kennedy/Jenks Consultants. Steven Nugent is the General Manager of the Carmichael Water District. The cooperation of Alex Peterson and Steven Nugent is greatly appreciated.

QUESTIONS

Please write your answers to the following questions and compare them with those on page 224.

16.4A How frequently does an operator conduct routine maintenance?

16.4B What is the preventive maintenance program (PMP) scheduled maintenance based on?

16.4C What does breakdown maintenance relate to?

END OF LESSON 1 OF 4 LESSONS

on

MEMBRANE TREATMENT PROCESSES

Please answer the discussion and review questions next.

DISCUSSION AND REVIEW QUESTIONS

Chapter 16. MEMBRANE TREATMENT PROCESSES

(MEMBRANE FILTRATION AND DEMINERALIZATION)

(Lesson 1 of 4 Lessons)

At the end of each lesson in this chapter, you will find discussion and review questions. Please write your answers to these questions to determine how well you understand the material in the lesson.

1. How can membrane fouling be described?
2. How frequently should operational procedures be performed at a membrane filtration plant?
3. How are membrane filtration water treatment plant records usually maintained?
4. What are the main types of equipment maintenance activities?

CHAPTER 16. MEMBRANE TREATMENT PROCESSES

(MEMBRANE FILTRATION AND DEMINERALIZATION)

(Lesson 2 of 4 Lessons)

16.5 SOURCES OF MINERALIZED WATER

As our country's population continues to grow, so does our demand for more water resources. Traditionally, water supplies have been obtained from fresh water sources. This constantly increasing need for water has started to deplete the available fresh water supplies in some areas of the country.

Faced with potential shortages, water planners must now consider new treatment technologies that, until recently, were not considered to be economically feasible. Since most of the earth's water supplies are saline (the ocean is high in dissolved minerals) rather than fresh, these impurities must be removed. One process receiving considerable attention is demineralization. Demineralization is the process that removes dissolved minerals (salts) from water.

All available water supplies can be classified according to their mineral quality. All waters contain various amounts of *TOTAL DISSOLVED SOLIDS (TDS)*,[17] including fresh water. A majority of the dissolved materials are inorganic minerals (salts). Minerals are compounds commonly found in nature that consist of positive metallic ions (such as calcium, sodium) bonded to negative ions (such as chloride, sulfate, carbonate). Many of these compounds are soluble in water and come from the weathering and erosion of the earth's surface.

Fresh water supplies, which have been the major sources of water developed in the past, usually contain less than 1,000 mg/L of total dissolved solids. Secondary drinking water standards recommend 500 mg/L TDS as the limit. Waters containing slightly higher concentrations can be used without adverse effects.

Brackish water contains from 1,000 to 10,000 mg/L TDS (seawater has 35,000 mg/L TDS). Most brackish water is found in groundwater. Figure 16.14 shows that over one-half of the United States overlays groundwater containing TDS levels ranging from 1,000 to 3,000 mg/L. To date, brackish water has not been widely used for municipal drinking water supplies because of its highly mineralized taste and associated problems such as scaling in pipes. With the advent of new treatment technologies, however, demineralization of brackish waters (including reuse of wastewater) has great potential for further development.

The largest available source of water in terms of quantity is classified as seawater, which usually contains more than 35,000 mg/L TDS. While seawater is becoming an important future water resource because of its seemingly unlimited availability in coastal areas, it is more expensive to treat than brackish water because of its greater TDS concentration.

The purpose of this chapter is to introduce and familiarize the water treatment plant operator with the newer treatment processes that have been developed to remove the dissolved minerals (TDS) from water. The development of the membrane demineralization processes has significantly reduced the cost of demineralization. This savings, combined with diminished fresh water supplies, will increase the use of demineralization treatment processes. The large quantities of mineralized groundwater and the increased *SALINITY*[18] of many rivers and lakes due to waste discharges, agricultural runoff, and other uses will increase the need for demineralization.

Some areas of the United States, such as the Florida Gulf Coast, are already turning to demineralization. The worldwide capacity of brackish water demineralization plants is increasing every year. Thus, it is important that the water treatment plant operator become more knowledgeable concerning the methods used to demineralize water.

QUESTIONS

Please write your answers to the following questions and compare them with those on page 224.

16.5A What is demineralization?

16.5B Why is seawater more expensive to treat than brackish water?

---

17. *Total Dissolved Solids (TDS).* All of the dissolved solids in a water. TDS is measured on a sample of water that has passed through a very fine mesh filter to remove suspended solids. The water passing through the filter is evaporated and the residue represents the total dissolved solids. Also see SPECIFIC CONDUCTANCE.

18. *Salinity.* (1) The relative concentration of dissolved salts, usually sodium chloride, in a given water. (2) A measure of the concentration of dissolved mineral substances in water.

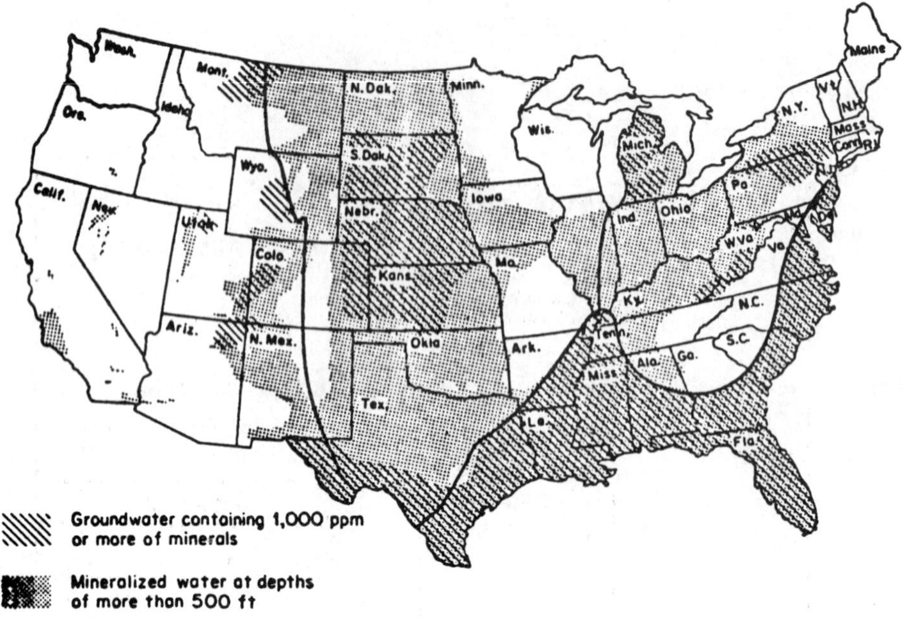

Fig. 16.14 Map of the conterminous United States showing depth to and quality of shallowest groundwater containing more than 1,000 mg/L dissolved solids

(From a paper by Bill Katz, "Treating Brackish Water for Community Supplies," published in Proceedings in "Role of Desalting Technology," a series of Technology Transfer Workshops presented by the Office of Water Research and Technology)

16.6 DEMINERALIZING PROCESSES

Methods of removing minerals from water can be divided into two classes: (1) those that use a phase change such as freezing or distillation, and (2) nonphase change methods such as reverse osmosis, electrodialysis, and ion exchange.

Demineralizing processes have primarily been used to remove dissolved inorganic material (TDS) from industrial water and wastewater, municipal water and wastewater, and seawater. However, some processes will also remove suspended material, organic material, bacteria, and viruses. Application of the various demineralizing processes is partially dependent upon the TDS concentration of the water to be treated. Figure 16.15 illustrates the approximate TDS range for use of two phase change processes (distillation and freezing) and three nonphase change processes (reverse osmosis, electrodialysis, and ion exchange).

The selection of a demineralizing process for a particular application depends on several factors:

1. Mineral concentration in *FEEDWATER*[19] (brackish water supply)
2. Product water quality required
3. Brine disposal facilities
4. Pretreatment required
5. Need to remove other material such as bacteria and viruses
6. Availability of energy and chemicals required for the process
7. Cost

The basic system is similar for all demineralizing processes and includes the processes shown in Figure 16.16.

19. *Feedwater.* The water that is fed to a treatment process; the water that is going to be treated.

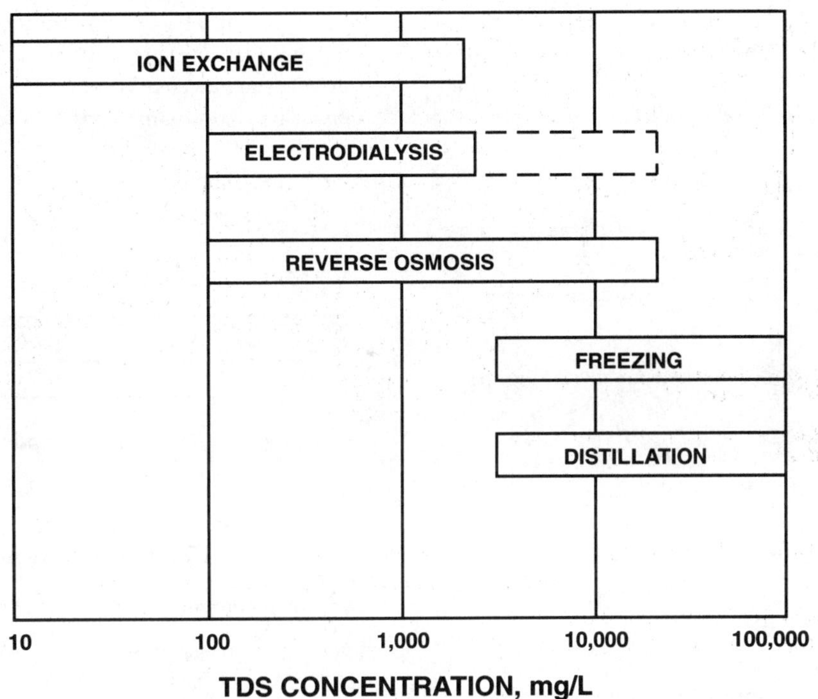

NOTE: The dashed lines indicate a feasible range of operation, but not typical range.

Fig. 16.15 Demineralization processes versus feedwater TDS concentrations

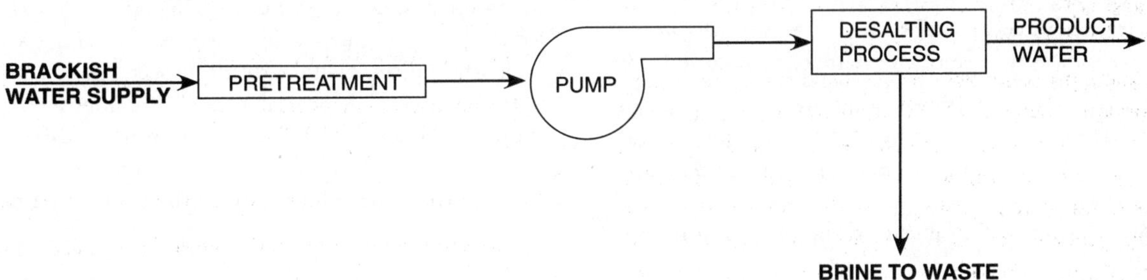

Fig. 16.16 Basic system for demineralizing processes

Since freezing and distillation apply primarily to seawater demineralization and their widespread use seems unlikely, these processes will not be discussed. Neither will ion exchange because of its limited use for brackish water (1,000 to 3,000 mg/L TDS). Currently, activity within the water industry is focused primarily on the membrane demineralizing processes known as reverse osmosis, electrodialysis, and nanofiltration. Also see *Advanced Waste Treatment*, Chapter 4, "Solids Removal From Secondary Effluents," Section 4.5, "Cross Flow Membrane Filtration."

QUESTIONS

Please write your answers to the following questions and compare them with those on pages 224 and 225.

16.6A List the two classes of methods of removing minerals from water.

16.6B List the common membrane demineralizing processes.

16.7 REVERSE OSMOSIS

16.70 What Is Reverse Osmosis?

Osmosis can be defined as the passage of a liquid from a weak solution to a more concentrated solution across a semipermeable membrane. The membrane allows the passage of the water (solvent) but not the dissolved solids (solutes).

Osmosis plays a vital role in many biological processes. Nutrient and waste minerals are transported by osmosis through the cells of animal tissues, which show varying degrees of permeability to different dissolved solids. A striking example of a natural osmotic process is the behavior of blood cells placed in pure water. Water passes through the cell walls to dilute the solution inside the cell. The cell swells and eventually bursts, releasing its red pigment. If the blood cells are placed in a concentrated sugar solution, the reverse process occurs; the cells shrink and shrivel up as water moves out into the sugar solution.

The top half of Figure 16.17 illustrates osmosis. The transfer of the water (solvent) from the freshwater side of the membrane continues until the level (shown in shaded area) rises and the head or pressure is large enough to prevent any net transfer of the solvent (water) to the more concentrated solution. At equilibrium, the quantity of water passing in either direction is equal; the difference in water level between the two sides of the membrane is defined as the osmotic pressure of the solution.

If a piston is placed on the more concentrated solution side of the semipermeable membrane (Figure 16.17) and a pressure, P, is applied that is greater than the osmotic pressure, water flows from the more concentrated solution to the freshwater side of the membrane. This condition illustrates the process of reverse osmosis.

16.71 Reverse Osmosis Membrane Structure and Composition

The two types of semipermeable membranes that are used most often for demineralization are cellulose acetate and thin film composites. Cellulose acetate, the first commercially available membrane (since the mid-1960s) is made by casting a cellulose acetate/solvent solution onto a porous support material. After quenching and annealing operations, the cellulose acetate membrane, often referred to as CA, exhibits the structure shown in Figure 16.18. The cellulose acetate membrane is asymmetric, meaning that one side is different from the other. The total cellulose acetate layer is 50 to 100 microns thick; however, a thin, dense layer approximately 0.2 microns thick exists at the surface. This thin, dense layer serves as the rejecting barrier of the membrane. The characteristics of a CA membrane are given in Table 16.2.

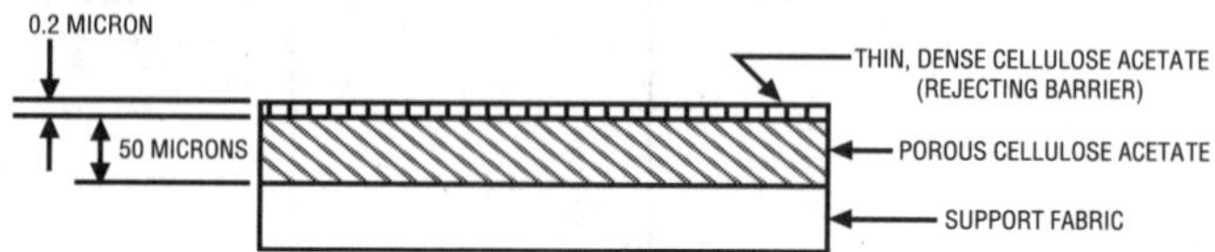

Fig. 16.18 Cellulose acetate membrane cross section

TABLE 16.2 CELLULOSE ACETATE MEMBRANE TYPICAL CHARACTERISTICS

Net Driving Pressure:	400 psi
NaCl Rejection:	92–97%
Flux Rate at 400 psi, 77°F:	25 GFD[a]
Operating pH range:	4.0–6.0
Cleaning pH range:	3.0–6.0
Cost relative to thin film composite membrane:	lower
Allowable feedwater chlorine concentration:	1.0 mg/L
Maximum operating temperature:	104°F (40°C)
Subject to biological attack	
Subject to hydrolysis	
Salt passage rate typically doubles after three years	
Most suitable for treatment of municipal wastes and some heavily pretreated surface supplies (due to lower fouling rate vs. thin film)	

a Gallons of flux per square foot per day.

During the 1970s, researchers realized the need for a membrane with better *FLUX* [20] and rejection characteristics than those of cellulose acetate. The basic approach to developing a better membrane was to improve the efficiencies of the thin rejecting layer and the porous substrate by casting each layer separately during the manufacturing process, hence the term "thin film composite." A typical thin film composite membrane is shown in Figure 16.19. The first step in preparing a thin film composite membrane is the casting of the porous support, usually a polysulfone solution, onto the support fabric. The next

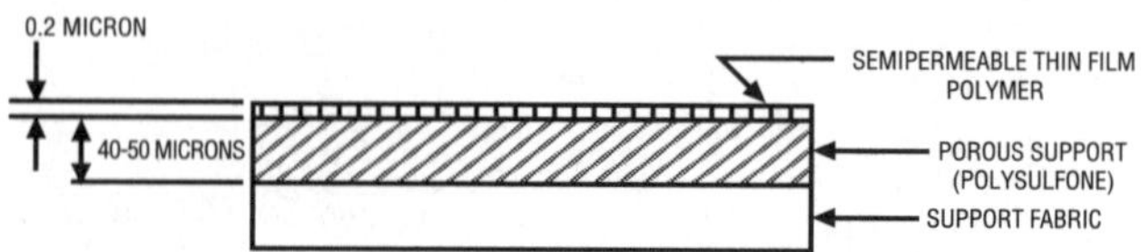

Fig. 16.19 Thin film composite membrane cross section

20. *Flux.* A flowing or flow.

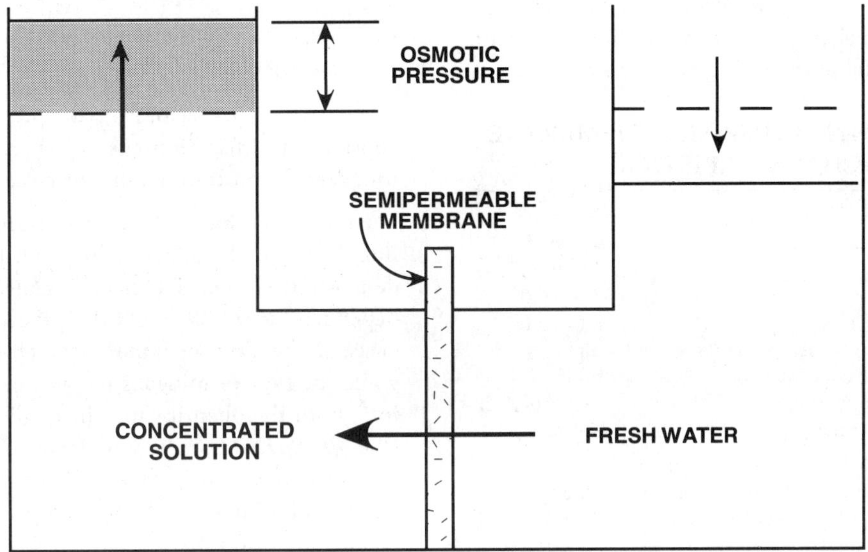

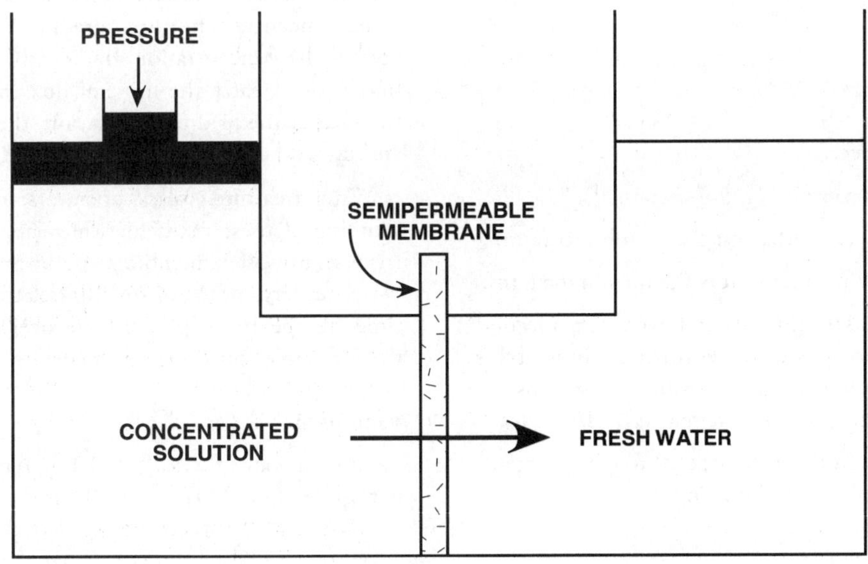

Fig. 16.17 Flows through a semipermeable membrane

step is to contact the composite support with the polymers that will actually form the semipermeable thin film rejecting barrier. The thin film rejecting barrier is formed *in situ*, or in position, by interfacial polymerization. It is the ability to form the semipermeable membrane separate from the support layers that has enabled membrane manufacturers to select polymers that will produce membranes with optimum dissolved solids rejection and water flux rates. The characteristics of thin film composite membranes are given in Table 16.3.

TABLE 16.3 THIN FILM COMPOSITE MEMBRANE TYPICAL CHARACTERISTICS

Net Driving Pressure:	200 psi
NaCl Rejection:	98–99%
Flux Rate at 200 psi, 77°F:	25–30 GFD[a]
Operating pH range:	3.0–10.0
Cleaning pH range:	2.0–12.0
Cost relative to cellulose acetate membrane:	higher
Allowable feedwater chlorine concentration:	none
Maximum operating temperature:	113°F (45°C)
Salt passage increase after 3 years:	≤30%

Not subject to biological attack, hydrolysis, or compaction
Higher rejection and flux rates than cellulose acetate membrane
Higher fouling rates than cellulose acetate on certain waste and surface water supplies
Sensitive to oxidants in feedwater

a Gallons of flux per square foot per day.

16.72 Membrane Performance and Properties

The basic behavior of semipermeable cellulose acetate reverse osmosis membranes can be described by two equations. The product water flow through a semipermeable membrane can be expressed as shown in Equation 1.

EQUATION 1

$$F_w = A(\Delta P - \Delta\pi)$$

Where

F_w = Water flux (g/cm^2-sec)

A = Water permeability constant (g/cm^2-sec atm[21])

ΔP = Pressure differential applied across the membrane (atm)

$\Delta\pi$ = Osmotic pressure differential across the membrane (atm)

Note that the water flux is the flow of water in grams per second through a membrane area of one square centimeter. Think of this as similar to the flow through a rapid sand filter in gallons per minute through a filter area of one square foot (GPM/ft^2).

The mineral (salt) flux (mineral passage) through the membrane can be expressed as shown in Equation 2.

EQUATION 2

$$F_s = B(C_1 - C_2)$$

Where

F_s = Mineral flux (g/cm^2-sec)

B = Mineral permeability constant (cm/sec)

$C_1 - C_2$ = Concentration gradient across the membrane (g/cm^3)

The water permeability (A) and mineral permeability (B) constants are characteristics of the particular membrane that is used and the processing it has received.

An examination of Equations 1 and 2 shows that the water flux (the rate of water flow through the membrane) is dependent on the applied pressure, while the mineral flux is not dependent on pressure. As the pressure of the feedwater is increased, the flow of water through the membrane increases while the flow of minerals remains essentially constant. Therefore, both the quantity and the quality of the purified product (the *PERMEATE*[22]) should increase with increased pressure. This occurs because there is more water to dilute the same amount of mineral.

The water flux (F_w) decreases as the mineral content of the feed increases because the osmotic pressure contribution ($\Delta\pi$) increases with increasing mineral content. In other words, since $\Delta\pi$ increases, the term ($\Delta P - \Delta\pi$) decreases, which results in a decrease in F_w, the water flux. Further, as more and more feedwater passes through the membrane, the mineral content of the feedwater becomes higher and higher (more concentrated). The osmotic pressure contribution ($\Delta\pi$) of the concentrate increases, resulting in a lower water flux.

Finally, since the membrane rejects a constant percentage of mineral, product water quality decreases with increased feedwater concentration. Also, note that Equation 2 reveals that the greater the concentration gradient ($C_1 - C_2$) across the membrane, the greater the mineral flux (mineral flow). Therefore, the greater the feed concentration, the greater the mineral flux and mineral concentration in the product water.

Water treatment plant operators must have a basic understanding of these mathematical relationships that describe RO (reverse osmosis) membrane performance. To help develop a better understanding of the interrelationships of flux, rejection, time, temperature, pH, and recovery, further explanation of these variables continues in the next section.

Example 1

Convert a water flux of 5×10^{-4} g/cm^2-sec to gallons per day per square foot. *NOTE:* 5×10^{-4} is the same as 0.0005.

21. *atm.* The abbreviation for atmosphere. One atmosphere is equal to 14.7 psi or 100 kPa.
22. *Permeate* (PURR-me-ate). (1) To penetrate and pass through, as water penetrates and passes through soil and other porous materials. (2) The liquid (demineralized water) produced from the reverse osmosis process that contains a low concentration of dissolved solids.

Known		Unknown
Water Flux, g/cm²-sec	= 5×10^{-4} g/cm²-sec	Flow, GPD/ft²

Convert the water flux from g/cm²-sec to flow in GPD/ft².

$$\text{Flow, GPD/ft}^2 = \frac{\left(\text{Water Flux, g/cm}^2\text{-sec}\right)(2.54\text{ cm/in})^2(12\text{ in/ft})^2(60\text{ sec/min})(60\text{ min/hr})(24\text{ hr/day})}{(1{,}000\text{ g/L})(3.785\text{ L/gal})}$$

$$= \frac{\left(\frac{0.0005}{\text{g/cm}^2\text{-sec}}\right)(2.54\text{ cm/in})^2(12\text{ in/ft})^2(60\text{ sec/min})(60\text{ min/hr})(24\text{ hr/day})}{(1{,}000\text{ g/L})(3.785\text{ L/gal})}$$

$$= 10.6\text{ GPD/ft}^2$$

QUESTIONS

Please write your answers to the following questions and compare them with those on page 225.

16.7A What is the osmotic pressure of a solution?

16.7B What are the two types of semipermeable membranes that are used most often for demineralization?

16.7C What is the meaning of water flux and of mineral flux? What units are used to express measurement of these quantities?

16.7D When additional pressure is applied to the side of a membrane with a concentrated solution, what happens?

16.7E When higher mineral concentrations occur in the feedwater, what happens to the permeate (or product water)?

16.73 Definition of Flux

The term "flux" is used to describe the rate of water flow through the semipermeable membrane. Flux is usually expressed in gallons per day per square foot of membrane surface or in grams per second per square centimeter.

The average membrane flux rate of a reverse osmosis system is an important operating guideline. In practice, most reverse osmosis systems will require periodic cleaning. It has been demonstrated that the cleaning frequency can be dependent on the average membrane flux rate of the system. Too high a flux rate may result in excessive fouling rates requiring frequent cleaning. Some general industry guidelines for acceptable flux rates are:

Feedwater Source	Flux Rate, GFD
Industrial/Municipal Waste	8–12
Surface (river, lake, ocean)	8–14
Well	14–20

Example 2

The permeate flow through an arrangement (or array) of RO membrane pressure vessels is 1,330,000 gallons per day (GPD). Feedwater first flows to 33 vessels operating in parallel. The concentrate from the 33 first-pass vessels is combined and sent to a set of 11 second-pass vessels. Each vessel (or tube) contains six membrane elements. Each element contains many fiber membranes, which provide a total of 325 square feet of membrane surface area per element. Calculate the average membrane flux rate for the system in GPD per square foot.

Known		Unknown
Permeate Flow, GPD	= 1,330,000 GPD	Average Flux Rate, GFD
No. of Vessels	= 44 (33 + 11)	
No. of Elements per Vessel	= 6	
Membrane Area per Element, ft²	= 325 ft²	

1. Determine the total membrane area in the system.

$$\text{Membrane Area, ft}^2 = (\text{No. Vessels})(\text{No. Elements/Vessel})(\text{Surface Area/Element})$$

$$= (44\text{ Vessels})(6\text{ Elements/Vessel})(325\text{ ft}^2\text{/Element})$$

$$= 85{,}800\text{ ft}^2$$

2. Calculate the average membrane flux rate for the system.

$$\text{Average Flux Rate, GFD} = \frac{\text{Permeate Flow, GPD}}{\text{Membrane Area, ft}^2}$$

$$= \frac{1{,}330{,}000\text{ GPD}}{85{,}800\text{ ft}^2}$$

$$= 15.5\text{ GFD}$$

Even under ideal conditions (pure feedwater and no fouling of the membrane surface), there is a decline in water flux with time. This decrease in flux is due to membrane compaction. This phenomenon is considered comparable to creep observed in other plastics or even metals when subjected to compressing stresses (pressure).

The term "flux decline" is used to describe the loss of water flow through the membrane due to compaction plus fouling. In the real world, feedwaters are never pure and contain suspended solids, dissolved organics and inorganics, bacteria, algae, and other potential foulants. These impurities can be deposited or grow on the membrane surface, thus hindering the flow of water through the membrane.

16.74 Mineral Rejection

The purpose of demineralization is to separate minerals from water; the ability of the membrane to reject minerals is called the mineral rejection. Mineral rejection is defined as:

EQUATION 3

$$\text{Rejection, \%} = \left(1 - \frac{\text{Product Concentration}}{\text{Feedwater Concentration}}\right) \times 100\%$$

Mineral rejections can be determined by measuring the TDS and using Equation 3. Rejections also may be calculated for individual constituents in the solution by using their concentrations.

The basic equations that describe the performance of a reverse osmosis membrane indicate that rejection decreases as the feedwater mineral concentration increases. Remember, this is because the higher mineral concentration increases the osmotic pressure. Figure 16.20 illustrates the rejection performance for a typical RO (reverse osmosis) membrane operating on three different feedwater solutions. This figure shows that as feed mineral concentration increases (TDS in mg/L), rejection decreases at a given feed pressure. Notice also that rejection improves as feed pressure increases.

Typical rejection for most commonly encountered dissolved inorganics is usually between 92 and 99 percent. Divalent ions such as calcium and sulfate are better rejected than monovalent ions such as sodium or chloride. Table 16.4 lists the typical rejections of an RO membrane operating on a brackish feedwater.

TABLE 16.4 TYPICAL REVERSE OSMOSIS REJECTIONS OF COMMON CONSTITUENTS FOUND IN BRACKISH WATER

Contaminant	Units	Feedwater Concentration	Percent Removal
EC[a]	µmhos	1,400	92
TDS[a]	mg/L	900	92
Calcium	mg/L	100	99
Chloride	mg/L	120	92
Sulfate	mg/L	338	99
Sodium	mg/L	158	92
Ammonia	mg/L	22.5	94
Nitrate	mg/L	2.9	55
COD[a]	mg/L	12.5	95
TOC[a]	mg/L	6.0	88
Silver	µg/L	1.2	88
Arsenic	µg/L	<5.0	—
Aluminum	µg/L	71.0	93
Barium	µg/L	24.0	96
Beryllium	µg/L	<1.0	—
Cadmium	µg/L	3.4	98
Cobalt	µg/L	4.6	>90
Chromium	µg/L	3.6	80
Copper	µg/L	12.7	63
Iron	µg/L	24.0	91
Mercury	µg/L	0.8	41
Manganese	µg/L	1.0	85
Nickel	µg/L	2.5	88
Lead	µg/L	<1.0	—
Selenium	µg/L	<5.0	—
Zinc	µg/L	<100.0	—

a EC, Electrical Conductivity; TDS, Total Dissolved Solids; COD, Chemical Oxygen Demand; and TOC, Total Organic Carbon.

Example 3

Estimate the ability of a reverse osmosis plant to reject minerals by calculating the mineral rejection as a percent. The feedwater contains 1,500 mg/L TDS and the product water TDS is 150 mg/L.

Known		Unknown
Feedwater TDS, mg/L	= 1,500 mg/L	Mineral Rejection, %
Product Water TDS, mg/L	= 150 mg/L	

Calculate the mineral rejection as a percent.

$$\text{Mineral Rejection, \%} = \left(1 - \frac{\text{Product TDS, mg/L}}{\text{Feed TDS, mg/L}}\right)(100\%)$$

$$= \left(1 - \frac{150 \text{ mg/L}}{1{,}500 \text{ mg/L}}\right)(100\%)$$

$$= (1 - 0.1)(100\%)$$

$$= 90\%$$

While most demineralization applications require the use of a membrane with high rejection rates (greater than 95 percent), some applications can use a membrane with lower rejection rates (80 percent) and lower operating pressures (less than 150 psi). The membranes that fit this classification are commonly referred to as softening or nanofiltration membranes. These membranes produce the same quantity of water as standard RO membranes but at lower operating pressures. A comparison of standard and softening RO membranes is given below:

	Standard	Softening (Nanofiltration)
Flux	25–30 GFD	25–30 GFD
Applied Pressure	225 psi	150 psi
Minimum Salt Rejection	97–98%	75–80%
Hardness Rejection	>99%	>95%

Softening or nanofiltration membranes are seeing widespread use for demineralization of municipal water supplies that require high rejection rates for hardness and THM formation potential, and moderate TDS rejection.

Reverse osmosis membrane manufacturers provide computer software to project the permeate water quality of a reverse osmosis system. Figures 16.21 and 16.22 are examples of computer printouts of permeate projections showing the expected initial performance of a thin film composite membrane (Figure 16.21) and a cellulose acetate membrane (Figure 16.22). The same feedwater was used for both projections.

QUESTIONS

Please write your answers to the following questions and compare them with those on page 225.

16.7F Water flux is usually expressed in what units?

16.7G What is "flux decline"?

16.7H How is mineral rejection defined?

16.7I What are the two major differences between standard RO membranes and softening membranes?

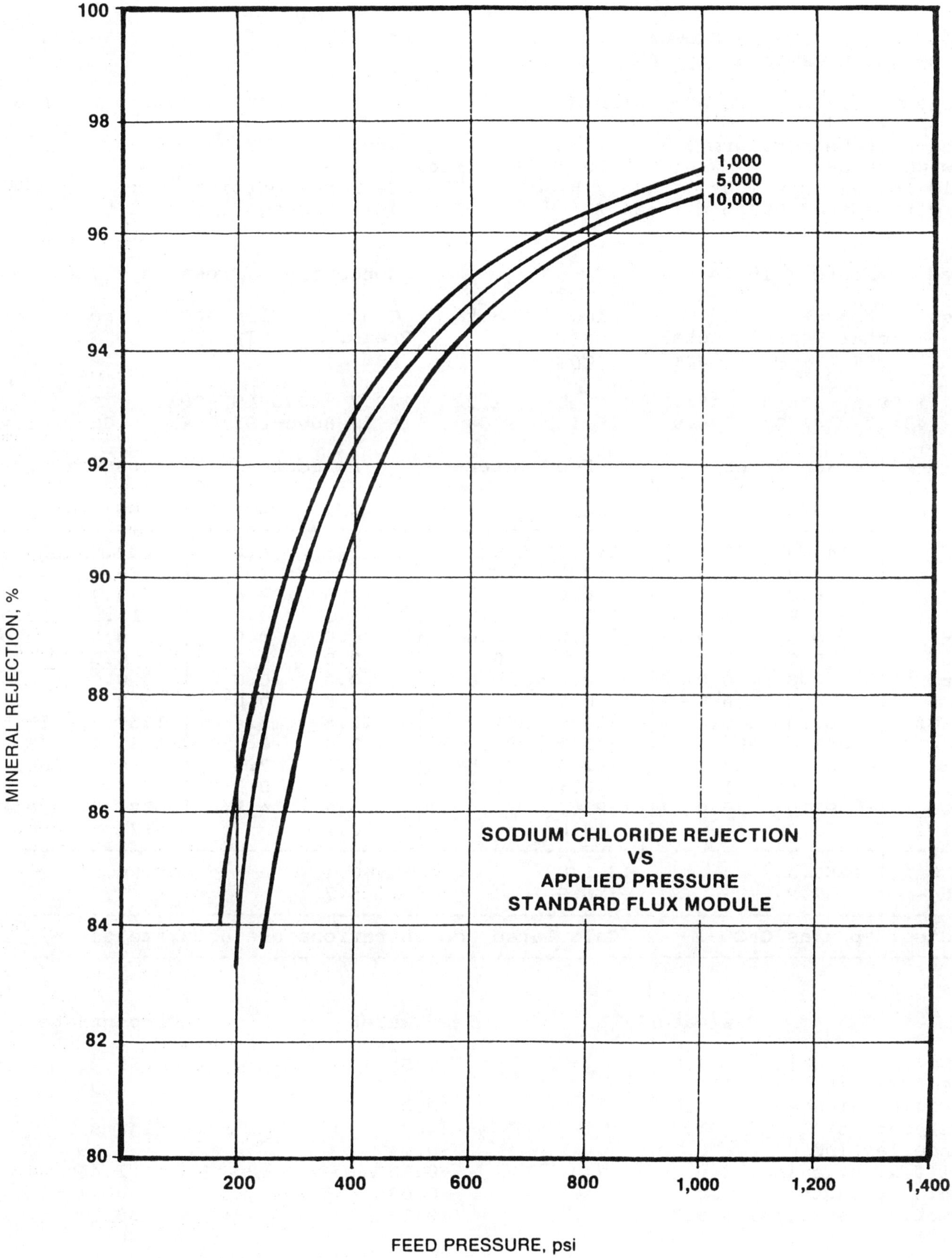

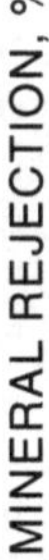

Fig. 16.20 Typical RO rejection for three different feedwater concentrations of TDS in mg/L

(Source: *Reverse Osmosis Principles and Applications*,
by Fluids Systems, Division of UOP, October 1970)

HYDRANAUTICS DESIGN PROGRAM - VERSION 4.01 (1991) 05-26-14
Calculation was made by: CWC

Project name : SANTA ANA WATERSHED Permeate flow : 1330000 GPD

Feedwater temperature : 21.0 C
Raw water pH : 7.20
Acid dosage, ppm(100%): 32.7 H2SO4
Acidified feed CO2,ppm : 66.7

Recovery : 77.0%
Element age : 0.0 years
Flux decline coefficient : -0.025
3-yr salt passage increase factor :1.2

Feed Pressure : 157.4 psi Concentrate Pressure : 118.1 psi

Pass	Feed Flow Total gpm	Feed Flow Vessel gpm	Conc. Flow Total gpm	Conc. Flow Vessel gpm	Beta	Conc. Press. psi	Element Type	Element No.	Array
1	1199.5	36.3	454.7	13.8	1.22	142.7	8040-LSY-CPA2	198	33x6
2	454.7	41.3	275.9	25.1	1.08	118.1	8040-LSY-CPA2	66	11x6

------	---Raw water-----		----Feed water---		----Permeate-----		--Concentrate----	
Ion	mg/l	ppm*	mg/l	ppm*	mg/l	ppm*	mg/l	ppm*
Ca	140.0	349.1	140.0	349.1	1.8	4.6	602.5	1502.5
Mg	42.0	172.8	42.0	172.8	0.6	2.3	180.8	743.8
Na	168.0	365.2	168.0	365.2	10.5	22.8	695.4	1511.7
K	3.8	4.9	3.8	4.9	0.3	0.4	15.5	19.9
NH4	0.0	0.0	0.0	0.0	0.0	0.0	0.0	0.0
Ba	0.0	0.0	0.0	0.0	0.0	0.0	0.0	0.0
Sr	0.0	0.0	0.0	0.0	0.0	0.0	0.0	0.0
CO3	0.3	0.5	0.1	0.2	0.0	0.0	2.1	3.5
HCO3	367.0	300.8	326.6	267.7	18.8	15.4	1357.2	1112.5
SO4	243.0	253.1	275.1	286.5	2.3	2.3	1188.4	1237.9
Cl	162.0	228.5	162.0	228.5	5.2	7.4	686.8	968.7
F	0.4	1.1	0.4	1.1	0.0	0.1	1.7	4.4
NO3	93.0	75.0	93.0	75.0	5.9	4.8	384.5	310.1
SiO2	40.0		40.0		0.7		171.7	
TDS	1259.5		1251.0		46.1		5286.5	
pH	7.2		6.9		5.7		7.5	

Notes: *ppm as CaCO3. Calculated concentrations are accurate to +/- 10%

	Raw water	Feed water	Concentrate
CaSO4/Ksp*100,%	7.9	8.8	55.2
SrSO4/Ksp*100,%	0.0	0.0	0.0
BaSO4/Ksp*100,%	0.0	0.0	0.0
SiO2 sat.,%	34.2	34.2	146.8
Langelier ind.	0.21	-0.14	1.69
Stiff & Davis ind.	0.20	-0.15	1.44
Ionic strength	0.03	0.03	0.11
Osmotic press.,psi	9.2	9.1	38.5

Fig. 16.21 Computer printout of permeate projections for a thin film composite membrane

(Source: Hydranautics)

HYDRANAUTICS DESIGN PROGRAM - VERSION 4.01 (1991) 05-26-14
Calculation was made by: CWC

Project name : SANTA ANA WATERSHED Permeate flow : 1330000 GPD

Feedwater temperature : 21.0 C
Raw water pH : 7.20
Acid dosage, ppm(100%): 0.0 H2SO4
Acidified feed CO2,ppm : 37.9

Recovery : 77.0%
Element age : 0.0 years
Flux decline coefficient : -0.035
3-yr salt passage increase factor :2.0

Feed Pressure : 297.9 psi Concentrate Pressure : 257.5 psi

Pass	Feed Flow Total gpm	Feed Flow Vessel gpm	Conc. Flow Total gpm	Conc. Flow Vessel gpm	Beta	Conc. Press. psi	Element Type	Element No.	Array
1	1199.5	36.3	480.9	14.6	1.21	282.8	8040-MSY-CAB1	198	33x6
2	480.9	43.7	275.9	25.1	1.11	257.5	8040-MSY-CAB1	66	11x6

Ion	Raw water mg/l	Raw water ppm*	Feed water mg/l	Feed water ppm*	Permeate mg/l	Permeate ppm*	Concentrate mg/l	Concentrate ppm*
Ca	140.0	349.1	140.0	349.1	2.5	6.2	600.4	1497.3
Mg	42.0	172.8	42.0	172.8	0.7	3.1	180.1	741.2
Na	168.0	365.2	168.0	365.2	27.9	60.6	637.2	1385.1
K	3.8	4.9	3.8	4.9	0.8	1.0	13.8	17.7
NH4	0.0	0.0	0.0	0.0	0.0	0.0	0.0	0.0
Ba	0.0	0.0	0.0	0.0	0.0	0.0	0.0	0.0
Sr	0.0	0.0	0.0	0.0	0.0	0.0	0.0	0.0
CO3	0.3	0.5	0.3	0.5	0.0	0.0	4.6	7.7
HCO3	367.0	300.8	367.0	300.8	21.3	17.5	1524.3	1249.4
SO4	243.0	253.1	243.0	253.1	1.8	1.9	1050.5	1094.2
Cl	162.0	228.5	162.0	228.5	22.6	31.8	628.8	886.9
F	0.4	1.1	0.4	1.1	0.0	0.1	1.6	4.3
NO3	93.0	75.0	93.0	75.0	24.3	19.6	323.1	260.6
SiO2	40.0		40.0		10.4		139.0	
TDS	1259.5		1259.5		112.3		5103.5	
pH	7.2		7.2		6.0		7.8	

Notes: *ppm as CaCO3. Calculated concentrations are accurate to +/- 10%

	Raw water	Feed water	Concentrate
CaSO4/Ksp*100,%	7.9	7.9	50.1
SrSO4/Ksp*100,%	0.0	0.0	0.0
BaSO4/Ksp*100,%	0.0	0.0	0.0
SiO2 sat.,%	34.2	34.2	118.8
Langelier ind.	0.21	0.21	2.04
Stiff & Davis ind.	0.20	0.20	1.79
Ionic strength	0.03	0.03	0.11
Osmotic press.,psi	9.2	9.2	37.0

Fig. 16.22 Computer printout of permeate projections for a cellulose acetate membrane

(Source: Hydranautics)

16.75 Effects of Feedwater Temperature and pH on Membrane Performance

In reverse osmosis operation, feedwater temperature has a significant effect on membrane performance and must therefore be taken into account in system design and operation. Essentially, the value of the water permeation constant is only constant for a given temperature. As the temperature of the feedwater increases, flux increases. Usually, flux is reported at some standard temperature reference condition, such as 25°C. Figure 16.23 illustrates the increase in flux for a standard RO module over a range of operating temperatures when 400 psi (2,758 kPa or 28 kg/cm^2) net operating pressure is applied.

Cellulose acetate membrane is subject to long-term *HYDROLYSIS.*[23] Hydrolysis results in a lessening of mineral rejection capability. The rate of hydrolysis is accelerated by increased temperature, and is a function of feed pH (Figure 16.24). Slightly acidic pH values (5 to 6) ensure a lower hydrolysis rate, as do cooler temperatures. Therefore, to ensure the longest possible lifetime of the cellulose acetate membrane and to slow hydrolysis, acid is added as a pretreatment step before demineralization. Table 16.5 indicates the relative time for mineral passage to increase 200 percent at different feedwater pH levels. Thin film composite membranes are not subject to hydrolysis but pH adjustment of the feedwater may be required for scale control.

TABLE 16.5 TIME REQUIRED TO ACHIEVE A 200 PERCENT INCREASE IN MINERAL PASSAGE AT 23°C AT VARIOUS pH LEVELS

pH	Time
pH 5.0	6 years
6.0	3.8 years
7.0	1 year
8.0	0.14 year = 51 days
9.0	0.01 year = 3.6 days

16.76 Recovery

Recovery is defined as the percentage of feed flow that is recovered as product water. Expressed mathematically, recovery can be determined by Equation 4.

EQUATION 4

$$\text{Recovery, \%} = \left(\frac{\text{Product Flow}}{\text{Feed Flow}}\right)(100\%)$$

The recovery rate is usually determined or limited by two considerations. The first is the desired product water quality. Since the amount of mineral passing through the membrane is influenced by the concentration differential between the brine and product, there is a possibility of exceeding product quality criteria with excessive recovery. The second consideration concerns the solubility limits of minerals in the brine. One should not concentrate the brine to a degree that would precipitate minerals on the membrane. This effect is commonly referred to as *CONCENTRATION POLARIZATION.*[24]

The most common and serious problem resulting from concentration polarization is the increasing tendency for precipitation of sparingly soluble salts and the deposition of particulate matter on the membrane surface.

In any flowing hydraulic system, the fluid near a solid surface travels more slowly than the main stream of the fluid. In other words, there is a liquid boundary layer at the solid surface. This is also true at the surface of the membrane in a spiral-wound element or in any other membrane packaging configuration. Since water is transmitted through the membrane at a much more rapid rate than minerals, the concentration of the minerals builds up in the boundary layer (concentration polarization) and it is necessary for the minerals to diffuse back into the flowing stream. Polarization will reduce both the flux and rejection of a reverse osmosis system. Since it is impractical to totally eliminate the polarization effect, it is necessary to minimize it by good design and operation.

The boundary layer effect can be minimized by increased water flow velocity and by promoting turbulence within the RO elements. Brine flow rates can be kept high as product water is removed by staging (reducing) the module pressure vessels. This is popularly referred to as a "Christmas Tree" arrangement. Typical flow arrangements such as 4 units - 2 units - 1 unit (85 percent recovery) or 2 units - 1 unit (75 percent recovery) are used most often (Figure 16.25).

These configurations consist of feeding water to a series of pressure vessels in parallel where about 50 percent of the water is separated by the membrane as product water and 50 percent of the water is rejected. The reject is then fed to half as many vessels in parallel where again about 50 percent is product water and 50 percent rejected. This reject becomes the feed for the next set of vessels. By arranging the pressure vessels in the 4-2-1 arrangement, it is possible to recover over 85 percent of the feedwater as product water and to maintain adequate flow rates across the membrane surface to minimize polarization. For example, a

23. *Hydrolysis* (hi-DROLL-uh-sis). (1) A chemical reaction in which a compound is converted into another compound by taking up water. (2) Usually a chemical degradation of organic matter.
24. *Concentration Polarization.* (1) A buildup of retained particles on the membrane surface due to dewatering of the feed closest to the membrane. The thickness of the concentration polarization layer is controlled by the flow velocity across the membrane. (2) Used in corrosion studies to indicate a depletion of ions near an electrode. (3) The basis for chemical analysis by a polarograph.

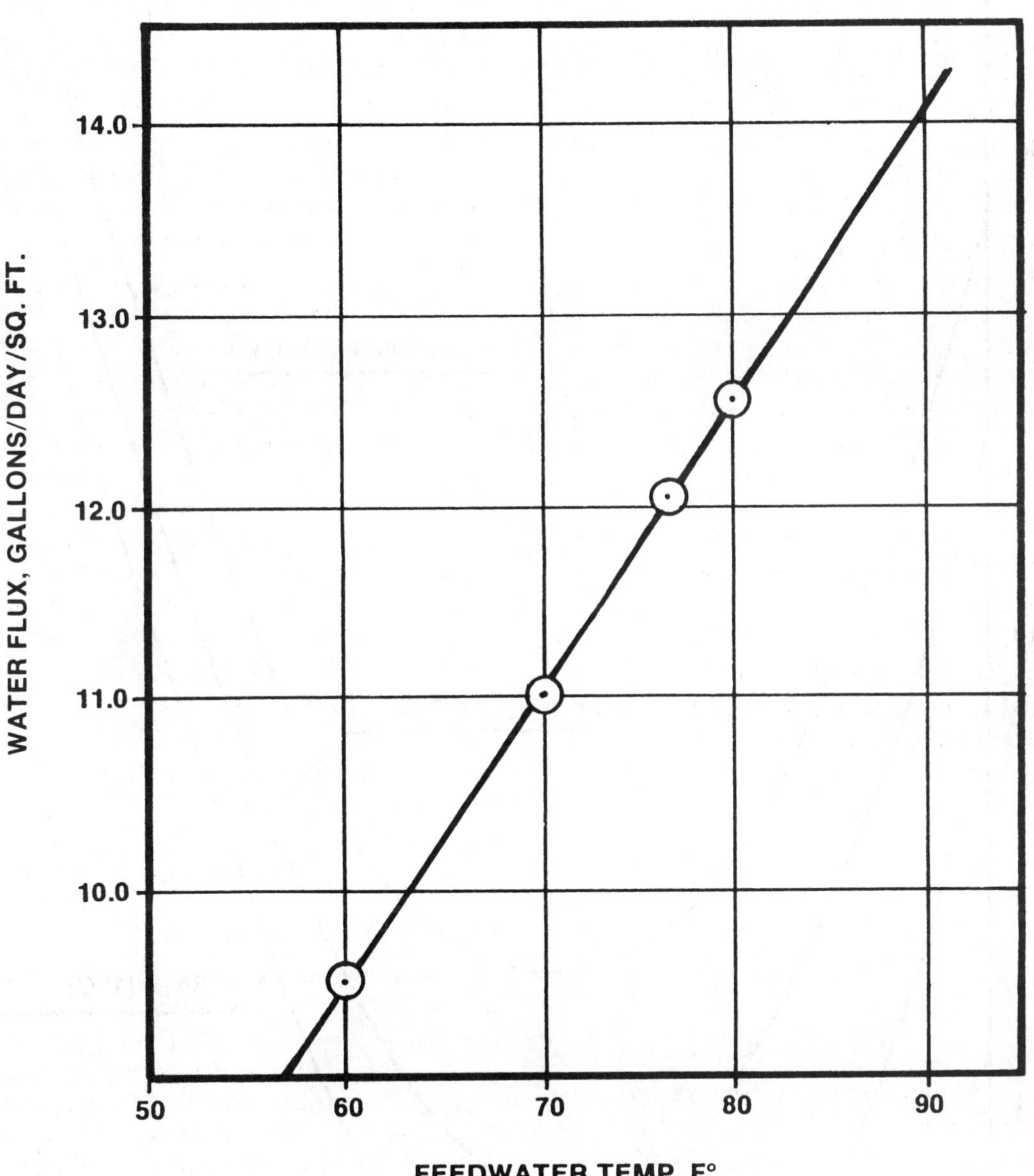

Fig. 16.23 Effect of temperature on water flux rate, cellulose acetate membrane operating pressure at 400 psi (2,758 kPa or 28 kg/cm²) net

(Source: *Reverse Osmosis Principles and Applications*, by Fluids Systems, Division of UOP, October 1970)

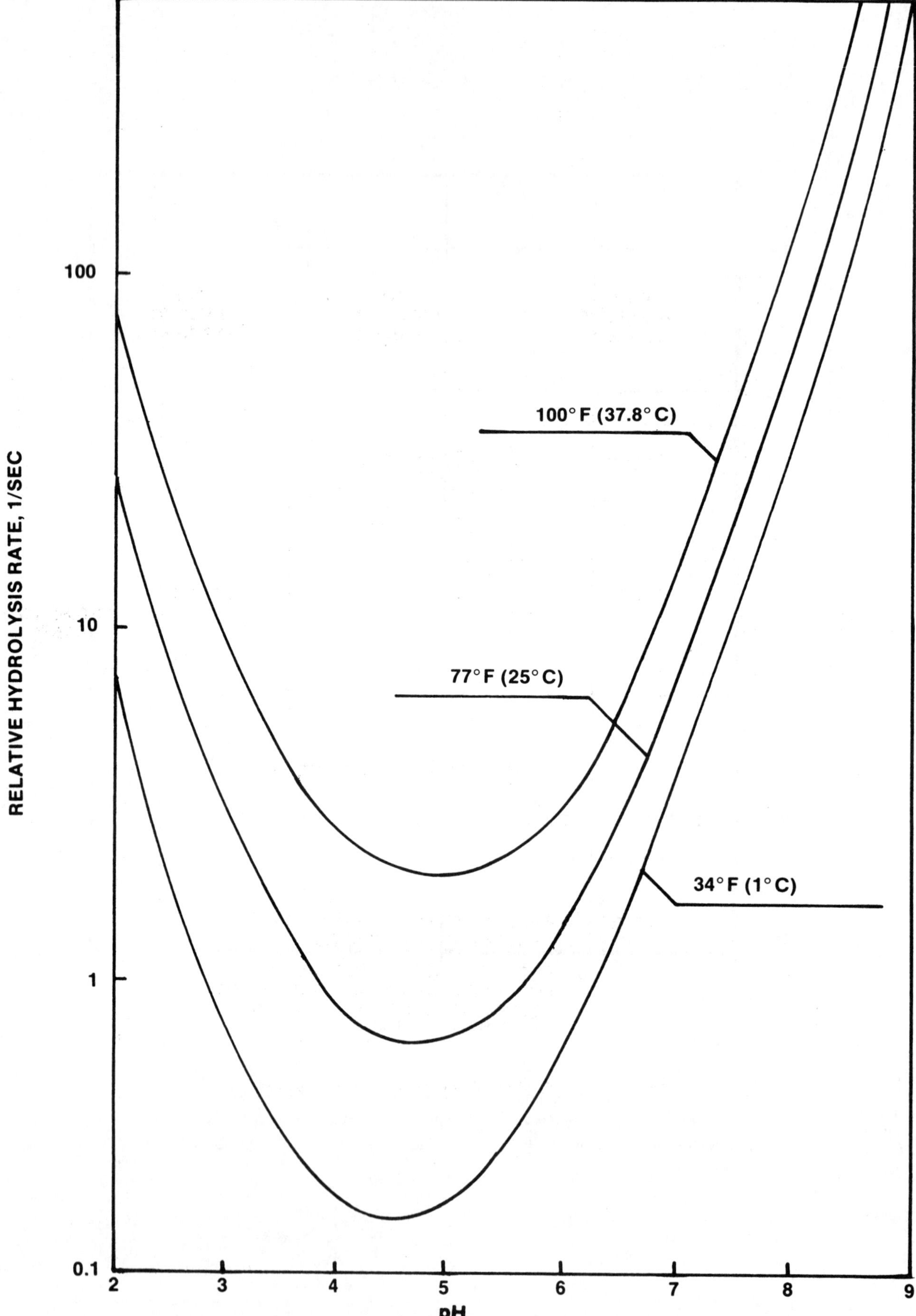

Fig. 16.24 Effect of temperature and pH on hydrolysis rate for cellulose acetate membrane

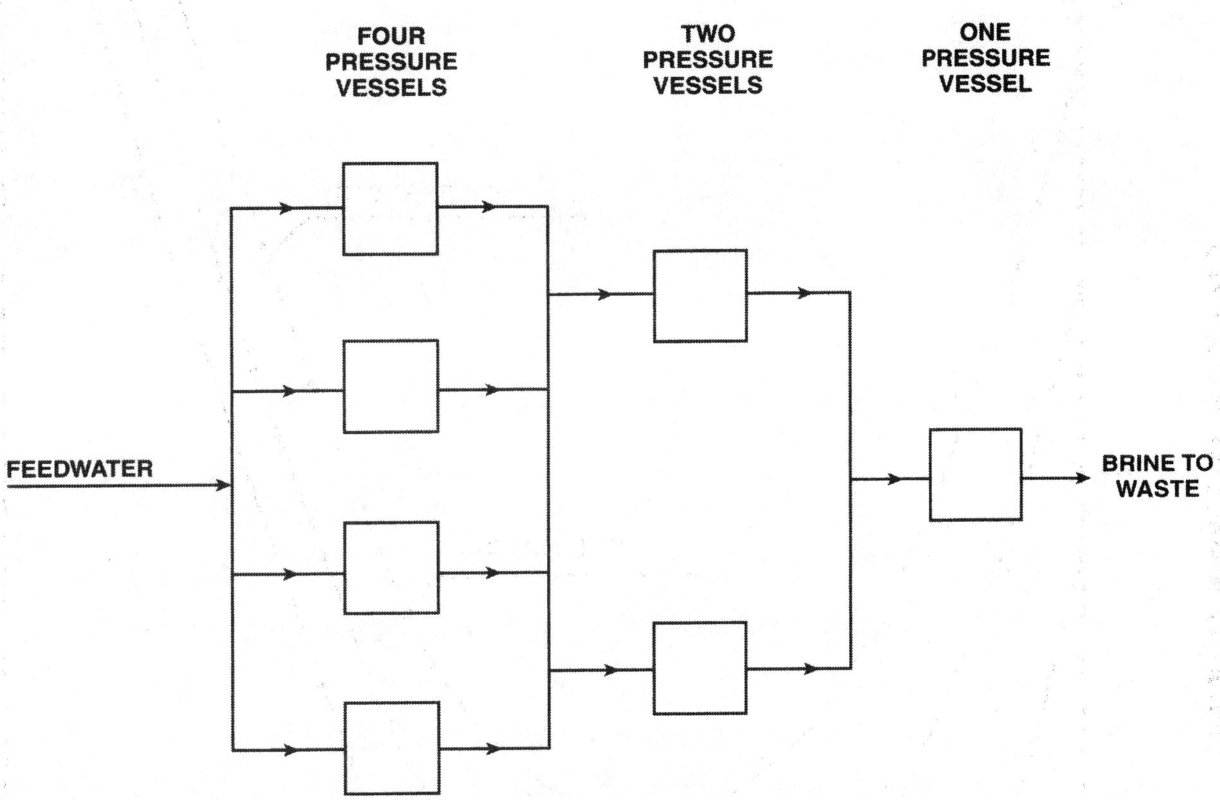

NOTES: 1. BRINE FLOWS OUT OF PRESSURE VESSELS TO NEXT VESSEL.

2. PRODUCT WATER IS NOT SHOWN. PRODUCT WATER FLOWS OUT OF EACH VESSEL INTO A COMMON HEADER.

Fig. 16.25 Typical 4-2-1 "Christmas Tree" arrangement

system consisting of a total of 35 vessels would have a 20-10-5 pressure vessel arrangement for an 85 percent recovery.

Example 4

Estimate the percent recovery of a reverse osmosis unit with a 4-2-1 arrangement if the feed flow is 5.88 MGD and the product flow is 5.0 MGD.

Known		Unknown
Product Flow, MGD	= 5.0 MGD	Recovery, %
Feed Flow, MGD	= 5.88 MGD	

Calculate the recovery as a percent.

$$\text{Recovery, \%} = \frac{(\text{Product Flow, MGD})(100\%)}{\text{Feed Flow, MGD}}$$

$$= \frac{(5.0\text{ MGD})(100\%)}{5.88\text{ MGD}}$$

$$= 85\%$$

QUESTIONS

Please write your answers to the following questions and compare them with those on page 225.

16.7J How will an increase in feedwater temperature influence the water flux?

16.7K How does hydrolysis influence the mineral rejection capability of a membrane?

16.7L How is recovery defined?

16.7M Recovery rate is usually limited by what two considerations?

16.7N Define concentration polarization.

END OF LESSON 2 OF 4 LESSONS

on

MEMBRANE TREATMENT PROCESSES

Please answer the discussion and review questions next.

DISCUSSION AND REVIEW QUESTIONS

CHAPTER 16. MEMBRANE TREATMENT PROCESSES

(MEMBRANE FILTRATION AND DEMINERALIZATION)

(Lesson 2 of 4 Lessons)

Please write your answers to the following questions to determine how well you understand the material in the lesson. The question numbering continues from Lesson 1.

5. Why has brackish water not been widely used for municipal drinking water supplies?
6. What is reverse osmosis?
7. Indicate what will happen to both the water flux and mineral flux when:
   a. Pressure differential applied across the membrane (ΔP) increases
   b. Osmotic pressure differential across the membrane ($\Delta\pi$) increases
   c. Concentration gradient across the membrane ($C_1 - C_2$) increases
8. What usually happens to water flux with time? Explain.
9. How does fouling develop on membranes?
10. What factors influence the rate of hydrolysis of a membrane? Explain.
11. What is the most common and serious problem resulting from concentration polarization?
12. Why do demineralization plants use a pressure vessel Christmas Tree configuration?

CHAPTER 16. MEMBRANE TREATMENT PROCESSES

(MEMBRANE FILTRATION AND DEMINERALIZATION)

(Lesson 3 of 4 Lessons)

16.8 COMPONENTS OF A REVERSE OSMOSIS UNIT

An RO unit is shown in Figure 16.26. Several RO units, along with the necessary auxiliary equipment for a large reverse osmosis demineralization plant, are shown in Figure 16.27. The major components are shown in the schematic, Figure 16.28.

16.80 Pressurization Pump

The pressures required for RO can range from 100 to 1,200 psi. Typically, the pressure ranges can be broken down as follows:

Application	Pressure Range (psi)
Softening	100–200
Brackish	200–500
Brackish/Seawater Mixture, Industrial Concentrating	500–800
Seawater	800–1,200

Two basic types of pumps are used for pressurizing the feedwater: centrifugal and positive displacement. Important characteristics of each type of pump relating to its use in RO applications are listed below.

Centrifugal Pumps

1. Typically used for applications less than 500 psi.
2. Most cost effective for applications below 500 psi.
3. Single impellers geared to operate higher than motor speed create excessive noise.
4. Multistage centrifugal pumps are more costly than single-stage but more efficient.

Positive Displacement Pumps

1. Typically used for applications greater than 500 psi.
2. Very efficient for seawater (800 to 1,200 psi).
3. Flow pulsations require use of pulsation dampener for velocities greater than 2 FPS (feet per second).

The output of a centrifugal pump may be throttled by use of a multiturn throttling valve. This is often done for new systems, or after a successful membrane cleaning.

The output of a positive displacement pump may not be throttled. The pump discharge line should contain a pressure relief mechanism. Optional items would be a bypass valve to control flow to the membrane section and a pulsation dampener.

16.81 Piping

The selection of piping material depends on the water salinity and pressure. Seawater reverse osmosis requires the use of high-grade stainless steel for high-pressure lines. The most common types of materials used currently are 316L and 317L, due to their high molybdenum content. Brackish water plants typically use 304 and 316 stainless steel.

Low-pressure piping is typically made of polyvinyl chloride (PVC) or fiber-reinforced plastic/polymer (FRP). Some exotic materials such as 316SS and polyvinylidene fluoride (PVDF) are used in high-purity applications such as for semiconductor rinse water.

16.82 Pressure Vessel Housings

Several spiral-wound membrane elements (described later in Section 16.810) are connected in series and contained in pressure vessels. For most applications, a maximum of six 40-inch-long spiral-wound elements are contained in a single vessel. Due to improvements in the hydraulics of the spiral-wound design, seven 40-inch-long elements have been placed in one vessel. The standard material of construction is fiber-reinforced plastic/polymer (FRP). The pressure vessels are available in 200, 400, 600, 1,000, and 1,200 psi ratings. Some manufacturers can provide vessels constructed and stamped according to ASME Code—Section X.

Hollow fiber bundles (described later in Section 16.810) are packaged in individual fiberglass housings. For seawater desalination, these housings can be rated up to 1,200 psi.

16.83 Concentrate Control Valve

A regulating valve located in the concentrate line provides a means of applying a back pressure to the membrane. Positioning this valve in conjunction with the pump discharge valve (bypass valve for positive displacement pumps) will set the concentrate and permeate flow rates.

Fig. 16.26 Reverse osmosis unit

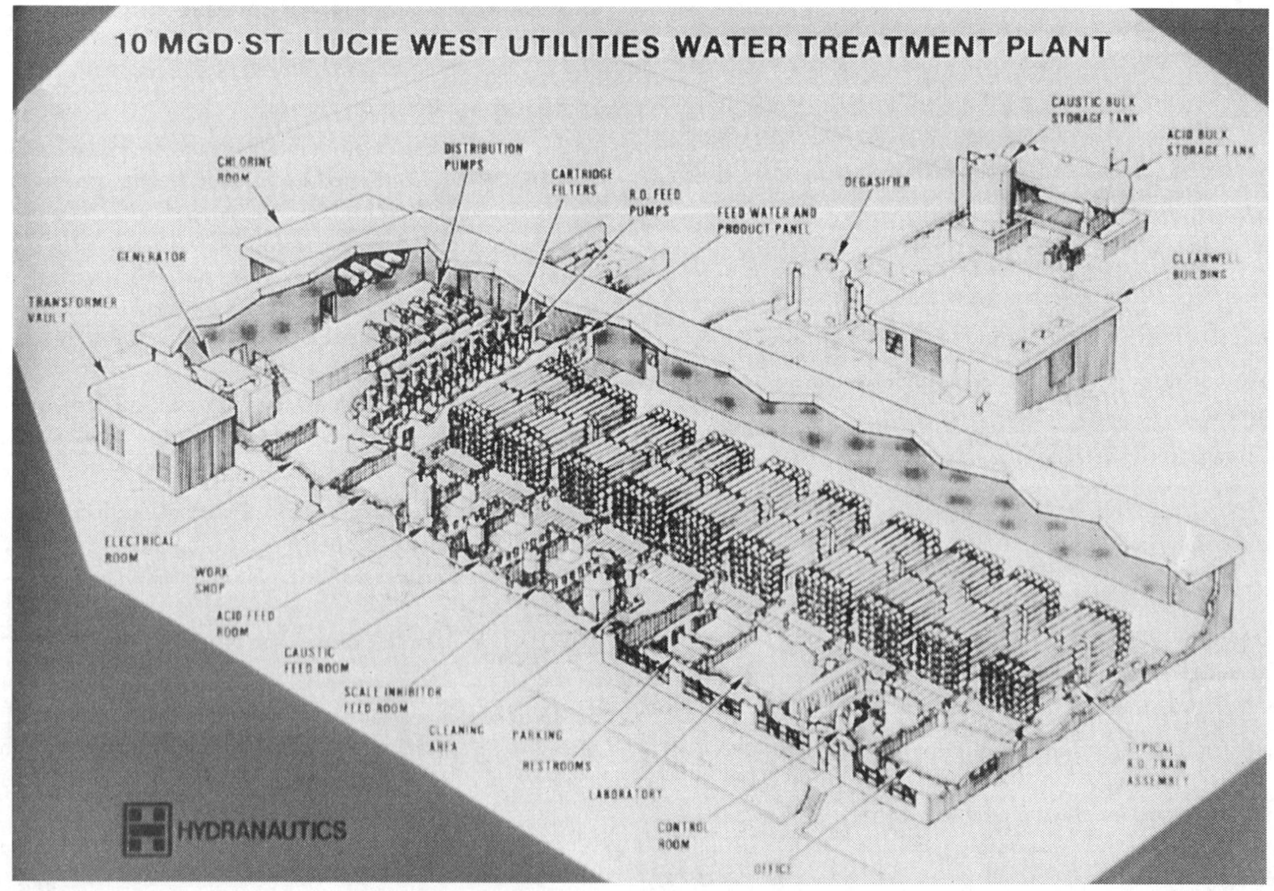

Fig. 16.27 RO units with necessary auxiliary equipment

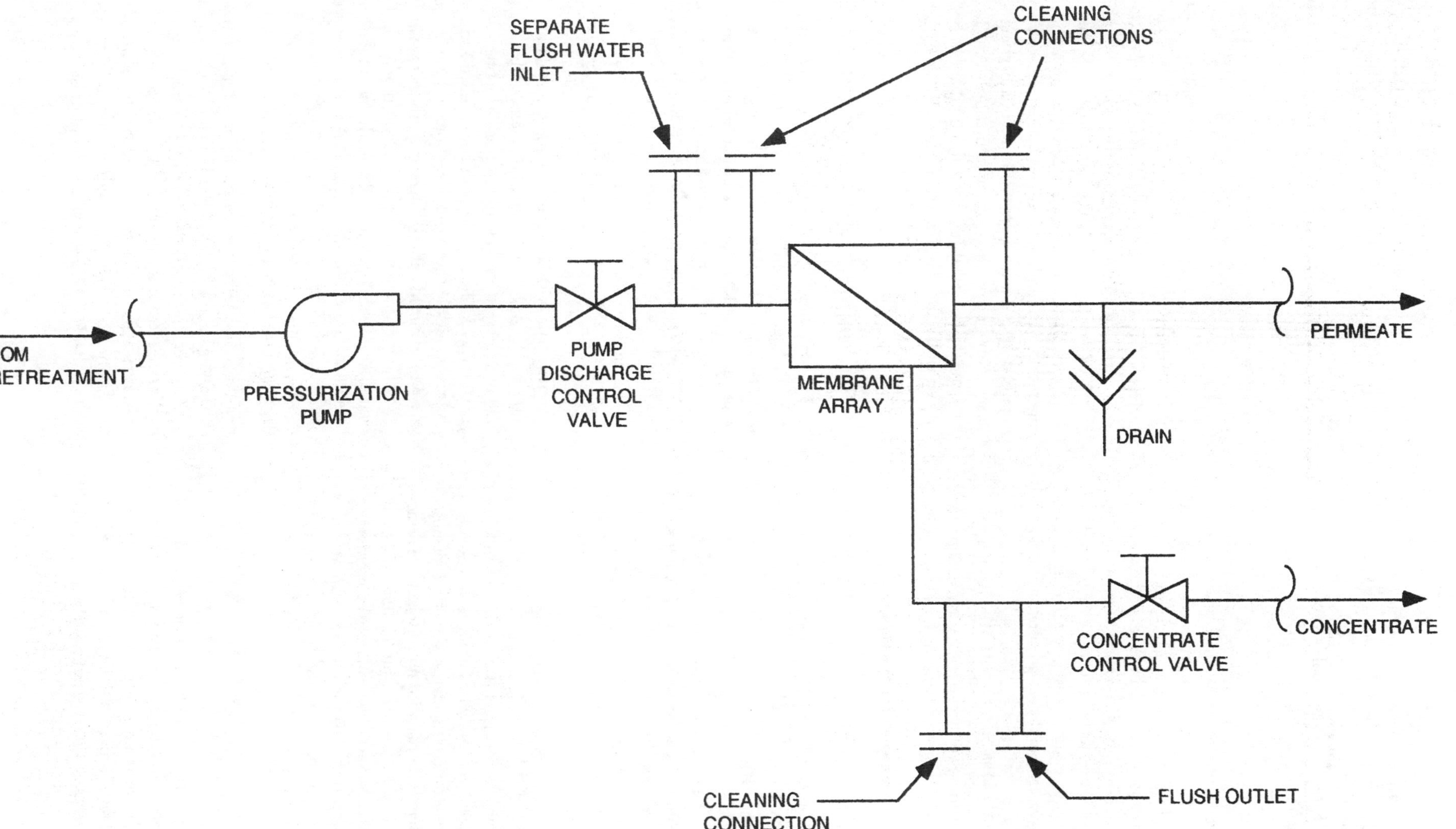

Fig. 16.28 Schematic of reverse osmosis unit

16.84 Sample Valves

Sample valves should be located on the feed, permeate, and concentrate lines. Locations should be such that samples can be taken during all modes of operation such as servicing, flushing, cleaning, and rinsing. Sample valves should also be located in the permeate line of each permeator or pressure vessel.

16.85 Flush Connections

Provisions should be made for flushing the unit for certain applications. Examples would be seawater or brackish waters with high organic content. The flush water could be acidified feed or permeate. If permeate is used, a separate inlet would be required. For all units that require flushing, a separate outlet in the concentrate line upstream of the concentrate control valve should be provided. The concentrate control valve would restrict flush flow, which is usually greater than the design concentrate flow.

16.86 Cleaning Connections

All units should have cleaning connections for each bank of permeators or pressure vessels connected in parallel. Isolation valves for each bank would allow for one bank to soak while the upstream or downstream bank is being cleaned. On large systems with many vessels or permeators connected in parallel, the cleaning system is sized for economic reasons to clean only a portion of the bank. In this case, valves are required to isolate the specific amount of vessels that can be cleaned at one time.

16.87 Permeate Rinse

It is useful to have provisions for sending the permeate from one bank or unit to drain. Some processes require that the permeate achieve quality by rinsing to drain after a shutdown period. Also, for certain troubleshooting procedures, poor quality permeate can be directed to drain while individual vessels are checked for poor quality.

16.88 Permeate Drawback Tank

For seawater applications, a permeate drawback tank may be provided. The purpose of the drawback tank is to provide a supply of water for an offline unit that is subject to osmosis. Upon shutdown with removal of applied pressure, reverse osmosis ceases and osmosis begins. During osmosis, flow will occur from permeate to the feed-concentrate side of the membrane. Flushing the feed-concentrate channels with permeate after shutdown should prevent natural osmosis but, as a precaution, drawback banks may be provided to prevent dehydration of the membrane.

16.89 Energy Recovery Devices

Energy recovery devices installed in the high-pressure concentrate line are used in some seawater reverse osmosis plants. The basic principle of operation is the conversion of the potential energy of the high-pressure concentrate into kinetic energy. A nozzle directs the concentrate flow toward a rotor with dished vanes. The flow strikes the vanes and turns the rotor. The shaft of the rotor is connected to a pump that prepressurizes the seawater feed. Due to lack of cost effectiveness, corrosion problems, and size restrictions, energy recovery turbines of this type have seen limited use.

16.810 Membranes

Operating plants use the RO principle in several different membrane configurations. Three types of commercially available membrane configurations used in operating plants include spiral wound, hollow fine fiber, and tubular.

The spiral-wound RO module was developed by Gulf Environmental Systems Company (now Fluid Systems Division, UOP) under contract to the US Office of Saline Water. This RO unit was conceived as a method of obtaining a relatively high ratio of membrane area to pressure vessel volume. The membrane is supported on both sides of a backing material and sealed with glue on three of the four edges of the laminate. The laminate is also sealed to a central tube that has been drilled to allow the demineralized water to enter. The membrane surfaces are separated by a screen material that acts as a brine spacer. The entire package is then rolled into a spiral configuration and wrapped in a cylindrical form. The membrane modules are loaded, end to end, into a pressure vessel as shown in Figure 16.29. Feed flow is parallel to the central tube while permeate flows through the membrane toward the central tube. Plants using this type of system include the brackish water demineralizing plants at Key Largo, Florida, and Kashima, Japan; the wastewater demineralizing plants in California; and the seawater demineralizing plant at Jeddah in Saudi Arabia. Spiral-wound membrane elements are pictured in Figure 16.29.

The hollow fiber type of membrane was developed by DuPont and Dow Chemical. The membranes manufactured by DuPont are made of aromatic polyamide fibers about the size of a human hair with an inside diameter of about 0.0016 inch (0.04 mm). In these very small diameters, fibers can withstand high pressures. In an operating process, the fibers are placed in a pressure vessel; one end of each fiber is sealed and the other end protrudes outside the vessel. The brackish water is under pressure on the outside of the fibers and product water flows inside of the fiber to the open end. A DuPont module is illustrated in Figure 16.30. For operating plants, the membrane modules are assembled in a configuration similar to the spiral-wound unit. Municipal demineralizing plants (manufactured by DuPont) in Greenfield, Iowa, and in Florida and seawater demineralizing plants in the Middle East use this type of membrane.

Tubular membrane processes operate on much the same principle as the hollow fine fiber except that the tubes are much larger in diameter, on the order of 0.5 inch (12 mm). Use of this type of membrane system is usually limited to special situations such as for wastewater with a high suspended solids concentration. The tubular membrane process is not economically competitive with other available systems for treatment of most waters.

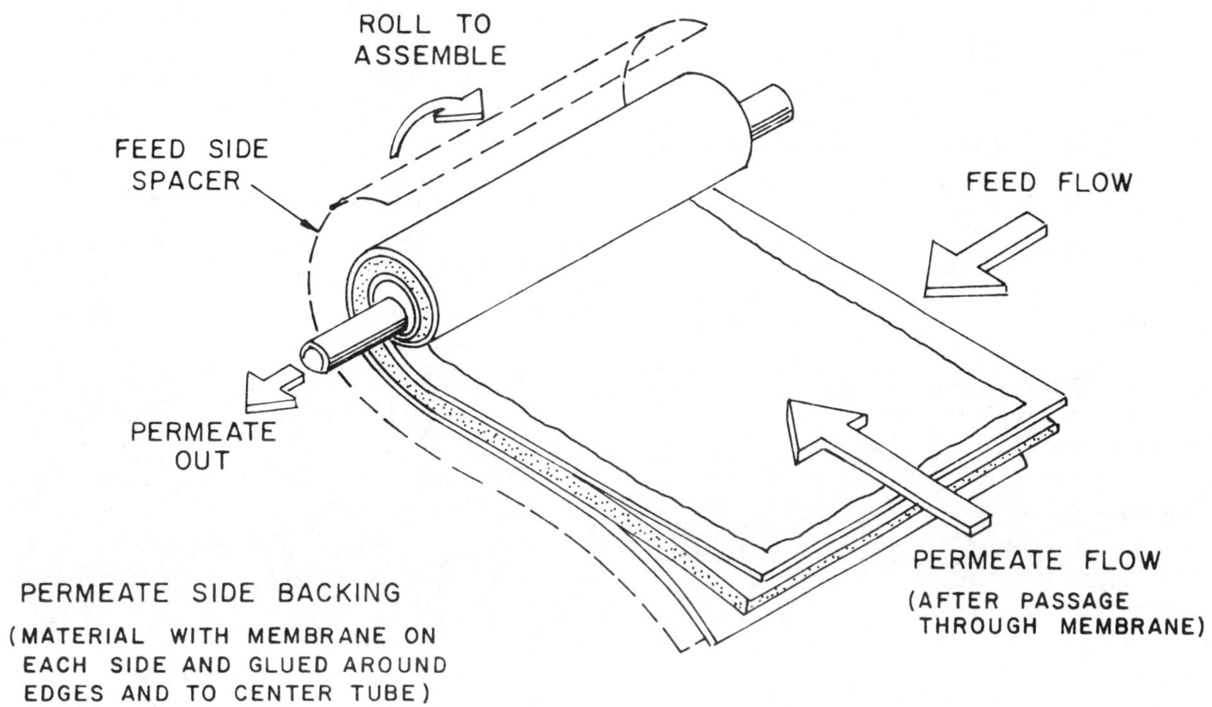

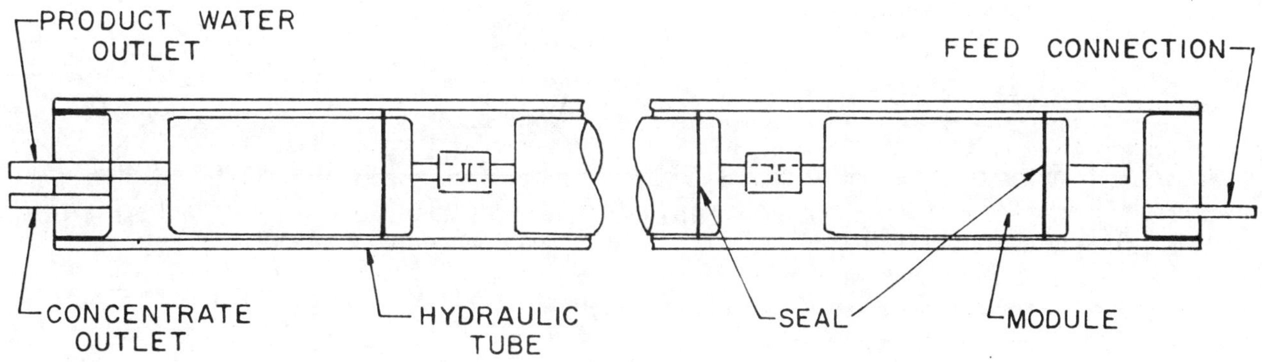

Fig. 16.29 Spiral-wound reverse osmosis module (as manufactured by UOP)

(From a paper by Mack Wesner, "Desalting Process and Pretreatment," published in Proceedings on "Role of Desalting Technology," a series of Technology Transfer Workshops presented by the Office of Water Research and Technology)

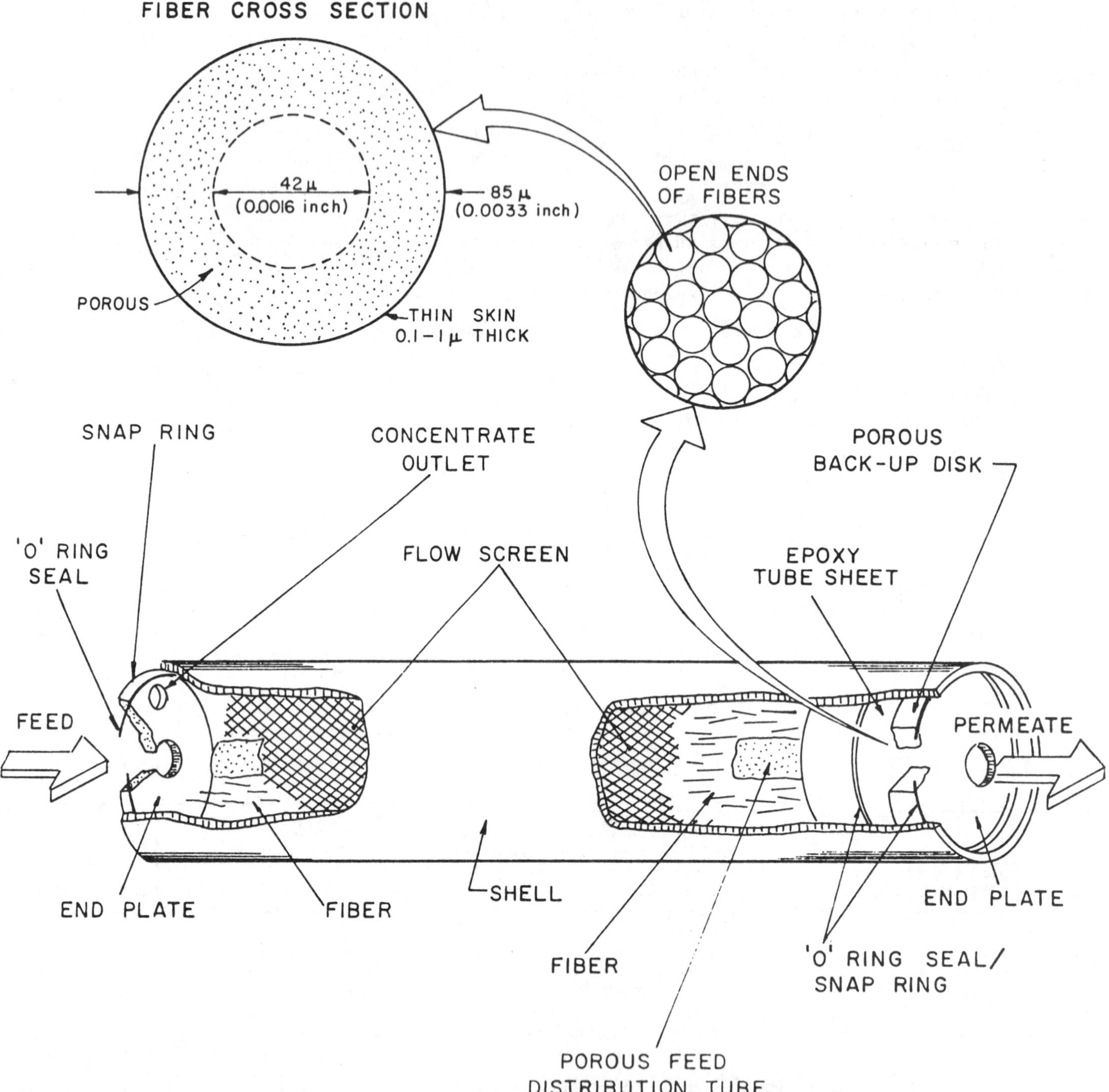

Fig. 16.30 Hollow fiber reverse osmosis module (as manufactured by DuPont)

(From a paper by Mack Wesner, "Desalting Process and Pretreatment," published in Proceedings on "Role of Desalting Technology," a series of Technology Transfer Workshops presented by the Office of Water Research and Technology)

QUESTIONS

Please write your answers to the following questions and compare them with those on page 225.

16.8A List the three types of commercially available membrane configurations that have been used in operating plants.

16.8B What type of membrane process is used to treat wastewater with a high suspended solids concentration?

16.9 OPERATION

16.90 Pretreatment

Water to be demineralized always contains impurities that should be removed by pretreatment to protect the membrane and to ensure maximum efficiency of the reverse osmosis process. Depending on the water to be demineralized, it is usually necessary to treat the feedwater to remove materials and conditions potentially harmful to the RO process:

1. Remove turbidity/suspended solids
2. Adjust pH and temperature
3. Remove materials to prevent scaling or fouling
4. Disinfect to prevent biological growth

Table 16.6 provides a summary of pretreatment requirements for various feedwater constituents.

16.91 Removal of Turbidity and Suspended Solids

In general, the feedwater should be filtered to protect the reverse osmosis system and its accessory equipment. When the water source is a groundwater or previously treated municipal or industrial supply, this may be accomplished by a simple screening procedure. However, such a procedure may not be adequate

TABLE 16.6 RO PRETREATMENT REQUIREMENTS

Constituent	Problem	Treatment
Fine suspended solids (that is, causing turbidity)	Fouling. Collects on face of element and on the feed concentrate mesh spacer.	Clarification or coagulant addition followed by media filtration.
Gross suspended solids and particulates (that is, sand, debris)	Blockage of lead element face and entrapment within feed/concentrate channels.	Media filtration or cartridge filtration. Cartridge filters usually required to protect high-pressure pumps as well.
Oil and Grease	Fouls membrane surface. Difficult to remove by cleaning.	Ultrafiltration. Batch separators, skimmers.
Microbiological matter	Fouling of the feed/concentrate mesh spacer. Adheres to some membrane surfaces.	Addition of oxidizing agent. In limited cases, ultraviolet sterilization.
Oxidizing agents (that is, chlorine)	Causes membrane degradation for some types of membranes.	Activated carbon or addition of reducing agents (sodium bisulfite).
Carbonates (CO_3, HCO_3)	Scaling. Combines with excess calcium to form calcium carbonate scale on membranes' surfaces.	Acid addition to convert alkalinity to carbon dioxide. Softening to remove calcium. Scale inhibitor in some cases.
Sulfate (SO_4)	Scaling. Combines with calcium, barium, strontium, or aluminum to form scale on membranes' surfaces.	Scale inhibitor addition. Softening to remove cations that could combine with sulfate. Reduce recovery of RO unit.
Silica	Scaling. Forms insoluble gel on membrane surface when supersaturated.	Lime softening. Reduce recovery of RO unit.
Iron (Ferric, +3)	Scaling and fouling. Forms iron oxides. Iron in the ferrous (+2) form is acceptable.	For groundwaters, no aeration or addition of oxidizing agents before RO unit. Greensand filters for removal of all forms of iron before RO unit.
Hydrogen sulfide (H_2S)	Scaling and fouling (due to formation of elemental sulfur) when H_2S is aerated or oxidized prior to RO unit.	Best treatment technique is to avoid aeration or oxidation prior to RO unit and to remove H_2S from permeate by degasification.

when the source is an untreated surface water. The amount of suspended matter in surface waters may vary by several orders of magnitude and may change radically in character and composition in a very short time. In such cases, in addition to the mechanical action of the filter, the operator may have to introduce chemicals for coagulation and flocculation and use filtration equipment in which the media can be washed or renewed at low cost. Pressure and gravity sand filters and diatomaceous earth filters may be required, particularly for large installations. When the particulates approach or are *COLLOIDAL*,[25] chemical treatment and filtration are almost essential.

Cartridge filters function as a particle safeguard and not as a primary particle removal device. In general, the influent turbidity to a cartridge filter should be less than 1 turbidity unit (TU). Typical cartridge filter sizes range from 5 to 20 microns and loading rates vary from 2 to 4 GPM/ft^2 (1.4 to 2.8 mm/sec).

16.92 pH and Temperature Control

As previously discussed, an important limiting factor in the life of cellulose acetate membranes in reverse osmosis is the rate of membrane hydrolysis. Cellulose acetate will break down (hydrolyze) to cellulose and acetic acid. The rate at which this hydrolysis occurs is a function of feedwater or source water pH and temperature. As the membrane hydrolyzes, both the amount of water and the amount of solute that permeate the membrane increase and the quality of the product water deteriorates. The rate of hydrolysis is at a minimum at a pH of about 4.7, and it increases with both increasing and decreasing pH. Thus, it is standard practice to inject acid, usually sulfuric acid, to adjust feedwater pH to 5.5. Not only does pH adjustment minimize the effect of hydrolysis, but it is also essential in controlling precipitation of scale-forming or membrane-fouling minerals.

Calcium carbonate and calcium sulfate are probably the most common scaling salts encountered in natural waters and are certainly the most common cause of scale in reverse osmosis systems. The addition of a small amount of acid can reduce the pH to a point where the alkalinity is reduced; this shifts the equilibrium to the point where calcium bicarbonate, which is much more soluble, is present at all points within the reverse osmosis loop. Neutralization of 75 percent of the total alkalinity usually provides sufficient pH adjustment to achieve calcium carbonate scale control and bring the membrane into a reasonable part of the hydrolysis curve. The pH reached by 75 percent neutralization is about 5.7. Calcium carbonate precipitation is also inhibited by the control procedure used for calcium sulfate.

Calcium sulfate is relatively soluble in water in comparison to calcium carbonate. Again, however, as pure or product water is removed from a feed solution containing calcium and sulfate, these chemicals become further concentrated in the feedwater. When the limits of saturation are eventually exceeded, precipitation of calcium sulfate will occur. Since calcium sulfate solubility occurs over a wide pH range, the scale control method used to inhibit calcium sulfate precipitation is a threshold treatment[26] with sodium hexametaphosphate (SHMP). This precipitation inhibitor represses both calcium carbonate and calcium sulfate by interfering with the crystal formation process. Other polyphosphates may also be used but are not as effective as the hexametaphosphate. Generally, 2 to 5 mg/L of SHMP are added to the feedwater to decrease precipitation of calcium sulfate.

16.93 Other Potential Scalants

The oxides or hydroxides most commonly found in water are those of iron, manganese, and silica. The oxidized and precipitated forms of iron, manganese, and silica can be a serious problem to any demineralization scheme because they can coat the reverse osmosis membrane with a tenacious (difficult to remove) film, which will affect performance. The scale inhibitor most frequently used is sodium hexametaphosphate.

16.94 Microbiological Organisms

Reverse osmosis modules provide a large surface area for the attachment and growth of bacterial slimes and molds. These organisms may cause membrane fouling or even module plugging. There is also some evidence that occasionally the enzyme systems of some organisms will attack the cellulose acetate membrane. The continuous application of chlorine to produce a 1 to 2 mg/L chlorine residual will help inhibit or retard the growth of most of the organisms encountered. However, caution must be exercised since continuous exposure of the membrane to higher chlorine residuals will impair membrane efficiency. Shock concentrations of up to 10 mg/L of chlorine are applied from time to time. When an oxidant-intolerant polyamide-type membrane is used, chlorination must be followed with dechlorination. One of the dechlorination agents, sodium bisulfite, is also known to be a disinfectant. Another disinfection option is the use of ultraviolet disinfection, which leaves no oxidant residual in the water.

QUESTIONS

Please write your answers to the following questions and compare them with those on page 225.

16.9A How are turbidity and suspended solids removed from feedwater to the reverse osmosis system?

25. *Colloids* (KALL-loids). Very small, finely divided solids (particles that do not dissolve) that remain dispersed in a liquid for a long time due to their small size and electrical charge. When most of the particles in water have a negative electrical charge, they tend to repel each other. This repulsion prevents the particles from clumping together, becoming heavier, and settling out.
26. Threshold treatment refers to the practice of using the least amount of chemical to produce the desired effect.

16.9B How are colloidal particulates removed from feedwater to the reverse osmosis system?

16.9C What happens to the product water as an acetate membrane hydrolyzes?

16.9D How is the precipitation of calcium sulfate prevented?

16.9E How is biological fouling on membranes controlled?

16.95 RO Plant Operation

Following proper pretreatment, the water to be demineralized is pressurized by high-pressure feed pumps and delivered to the RO pressure vessel membrane assemblies. An example of a typical RO plant layout is given in Figure 16.31. The membrane assemblies consist of a series of pressure vessels (usually fiber-reinforced plastic/polymer) arranged in the Christmas Tree layout depending on the desired recovery. Typical operating pressure for brackish water demineralizing varies from 150 to 400 psi (1,050 to 2,760 kPa or 10.5 to 28 kg/cm^2). A control valve on the influent manifold regulates the operating pressure. The volumes of feed flow and of product water are also monitored. The demineralized water is usually called permeate, and the reject, concentrate (brine). The recovery rate is controlled by increasing feed flow (increase operating pressure) and controlling the concentrate (brine) or reject with a preset brine control valve.

The operator must properly maintain and control all flows and recovery rates to avoid possible damage to the membranes from scaling.

You must remember that the brine flow valves are never to be fully closed. Should they be accidentally closed during operation, 100 percent recovery will result in almost certain damage to the membranes due to the precipitation of inorganic salts ($CaSO_4$). Product or permeate flow is not regulated and varies as feedwater pressure and temperature change as previously discussed.

Most RO systems are designed to operate automatically and require a minimum of operator attention. However, the continuous monitoring of system performance is an important aspect of operation. An example of a typical operation log for monitoring the Orange County Water District's 5 MGD (19 million liters per day) RO plant is given in Figure 16.32.

QUESTIONS

Please write your answers to the following questions and compare them with those on page 225.

16.9F How is the operating pressure on a reverse osmosis unit regulated?

16.9G The demineralized water is usually called ______, the reject, ______ .

16.9H How does the product or permeate flow vary or change?

16.96 Typical RO Plant Operations Checklist

1. Check cartridge filters. Properly installed filters ensure additional removal of suspended solids that could damage either the high-pressure feed pumps or foul the membrane elements. Cartridge filters should be replaced whenever the head loss exceeds the manufacturer's recommendation or effluent turbidity exceeds 1 TU.
2. Start up and check scale inhibitor feeding equipment and adjust feed rate to desired dose (2 to 5 mg/L). Most RO systems should not be operated without the addition of a scale inhibitor to protect membranes from precipitation of calcium sulfate or other inorganics. The scale inhibitor most frequently used is sodium hexametaphosphate.
3. If chlorine is used to prevent biological fouling, start chlorine feed and adjust dose to produce a chlorine residual of between 1 and 2 mg/L.
4. Start up and adjust acid feed system to correct feedwater pH to a level between 5.0 and 6.0 to protect membranes from possible damage due to hydrolysis. Note, feedwater should always be bypassed until the pH is properly adjusted.
5. Most RO systems are designed with automatic controls and various shutdown alarms. These alarms prevent startup or running of the unit until proper operating conditions are reached. After satisfying these conditions, high-pressure feed pumps can be started and water delivered to the RO units. A control valve is used to regulate feedwater pressure. Typical operating pressures vary from 150 to 400 psi (1,050 to 2,760 kPa or 10.5 to 28 kg/cm^2).
6. Adjust permeate and concentrate flow to establish the desired recovery rate.
7. Once flow has been established, check the differential pressure (ΔP) across the RO unit (ΔP = feed pressure – concentrate pressure), which is usually indicated by a meter and recorded. The importance of ΔP relates to cleaning. When the elements become fouled, ΔP usually increases, thus indicating the need for cleaning. The ΔP should not exceed 60 psi per 6-element pressure vessel (414 kPa or 4.1 kg/cm^2) because of possible damage to the RO modules.
8. With the system on line, monitor the performance. Rely on flow measurements, product water quality, and various pressure indications. A sample of a typical log sheet is shown in Figure 16.32.

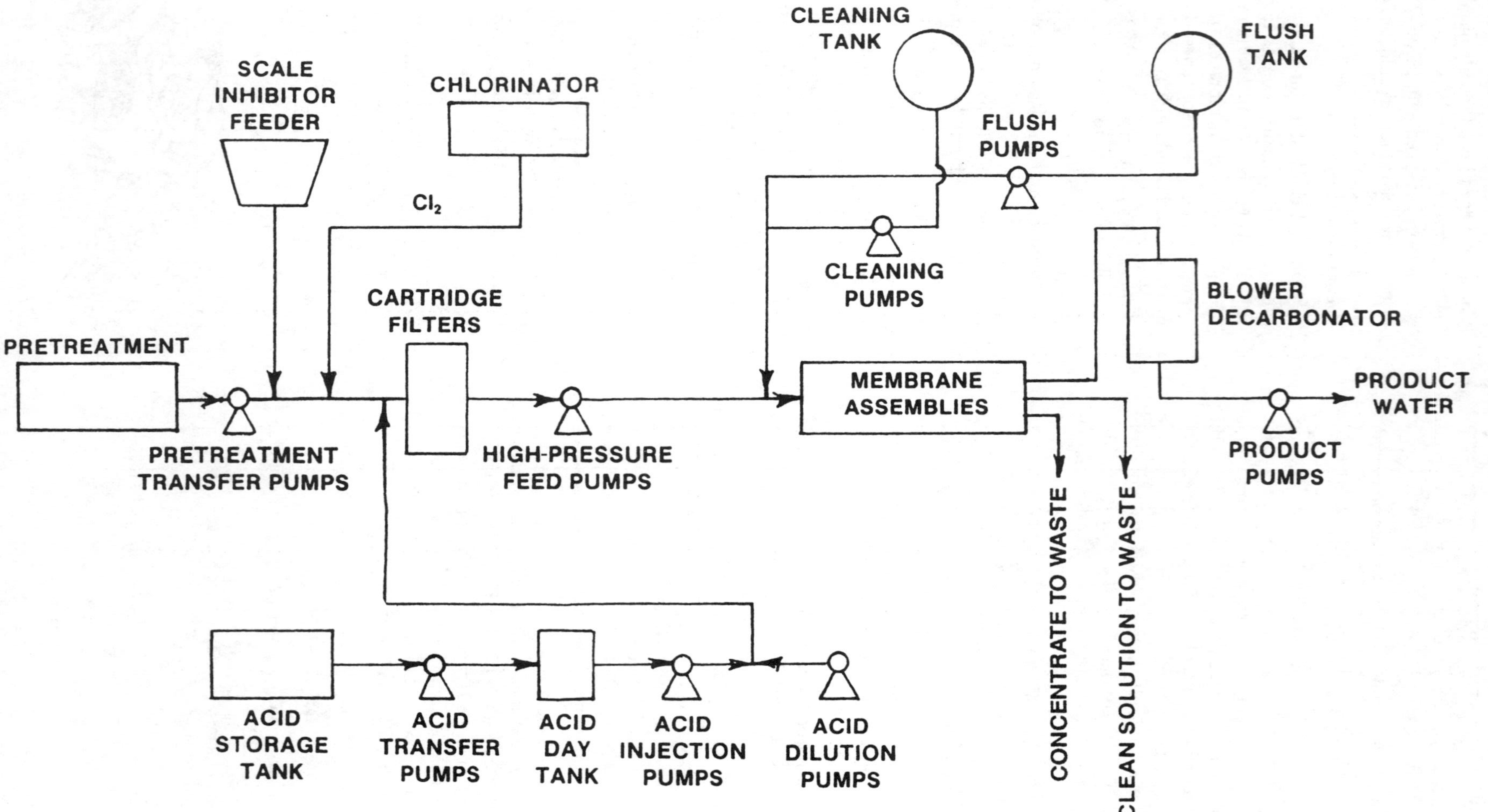

Fig. 16.31 RO flow diagram

TIME	Pretreatment											
	Cl mg/ℓ	Cartridge Filters				Turb NTU	Temp °F	Feed Cond µmhos/cm	pH	Pump-2 Disch.		
		#1 psig	#2 psig	#3 psig	#4 psig					A psig	B psig	C psig
0400												
1200												
2000												

TIME	RO Unit 1			RO Unit 2			Total Brine				Total Product		
	Feed psig	Feed Flow MGD	Product Cond. (µ)	Feed psig	Feed Flow MGD	Product Cond. (µ)	Flow MGD	pH	Conductivity µmhos/cm	KWH Ton	pH	Conductivity µmhos/cm	Flow MGD
0400													
1200													
2000													

TIME	RO Unit 1																	
	Section 1A-Product EC						Section 1B-Product EC						Section 1C-Product EC					
	Feed psig	ΔP 1st	ΔP 2nd	ΔP 3rd	Product gpm	Brine MGD	Feed psig	ΔP 1st	ΔP 2nd	ΔP 3rd	Product gpm	Brine MGD	Feed psig	ΔP 1st	ΔP 2nd	ΔP 3rd	Product gpm	Brine MGD
0400																		
1200																		
2000																		

TIME	RO Unit 2																	
	Section 2A-Product EC						Section 2B-Product EC						Section 2C-Product EC					
	Feed psig	ΔP 1st	ΔP 2nd	ΔP 3rd	Product gpm	Brine MGD	Feed psig	ΔP 1st	ΔP 2nd	ΔP 3rd	Product gpm	Brine MGD	Feed psig	ΔP 1st	ΔP 2nd	ΔP 3rd	Product gpm	Brine MGD
0400																		
1200																		
2000																		

Fig. 16.32 Data sheet, Orange County Water District 5.0 MGD (19 MLD) reverse osmosis system

Shift	Operator

Date

24 Hour Totalizer				
TIME	Feed Flow MGD	Bypass Flow MGD	Total Product MGD	Total Brine MGD
(A)2400 (II)				
(B)2400 (I)				

Cl_2	SHMP	ACID	ELAPSED TIME		POWER
2400 (11)			RO #1	RO #2	2400 (11)
2400 (1)	lbs	gal			2400 (1)

REMARKS:

Fig. 16.32 Data sheet, Orange County Water District 5.0 MGD (19 MLD) reverse osmosis system (continued)

16.97 Membrane Cleaning

Periodically, the performance of the RO system will decline. This is usually observed when either the product water flow rate (flux) decreases or salt removal (rejection) decreases. Table 16.7 summarizes common causes of membrane damage or loss of performance. Note that in Cases III and IV the corrective action requires cleaning of the element. Provisions for the periodic cleaning of the reverse osmosis elements are usually included in the system design. This makes it possible to clean impurities off the membrane surfaces and restore normal flow rates without removing the elements from the pressure vessels. Element cleaning should be performed at regular intervals to ensure as low an operating pressure as practical. The elements should be cleaned when the pressure required to maintain the rated capacity has either been increased by 15 percent (or a 15 percent decrease in product water flow has occurred at constant pressure), or a rise of 15 percent in the system differential pressure has been observed.

Most RO systems are provided with in-place cleaning systems. This includes tanks, pumps, valves, and piping for mixing and pumping cleaning solutions through the membrane elements. For cleaning, the unit is shut down and cleaning solutions are pumped through the vessels in a manner similar to feedwater. Typically, cleaning solutions are passed through the pressure vessels at low pressure and at flow rates where the ΔP does not exceed 60 psi (414 kPa or 4.1 kg/cm^2) to avoid damaging the elements. The cleaning solutions are returned to clean tanks at the end of a cleaning cycle, which usually lasts about one hour. Different cleaning solutions are available for use depending upon the type of fouling. Membranes are typically cleaned for approximately 45 minutes after which the cleaning solution is spent.

To remove inorganic precipitates, use an acid flush of citric acid. For biological or organic fouling, various solutions of detergents, sequestrants, chelating agents, bactericides, and enzymes are available. Examples include sodium tripolyphosphate, B13, Triton X-100, and EDTA.

To improve the long-term performance of an RO system, the membranes should be flushed with flush water during periods of shutdown to remove raw feedwater and concentrate. If raw water is allowed to remain in the unit, precipitation may occur. Flushing is also done after cleaning to remove the cleaning solution prior to system startup. In some cases, where the system is shut down for long periods of time, formaldehyde may be added to the flush water to inhibit biological growth.

QUESTIONS

Please write your answers to the following questions and compare them with those on page 225.

16.9I Why is chlorine added to the feedwater to a reverse osmosis unit?

16.9J Why must the operator check the differential pressure (ΔP) across the RO unit?

16.9K When should the reverse osmosis elements be cleaned?

16.98 Safety

16.980 Use of Proper Procedures

As in any water treatment plant, there are forces and chemicals used in a reverse osmosis plant that must be handled properly to ensure the safety and protection of personnel. Safety needs for demineralization plants can be divided into three general groups consisting of chemicals, electrical, and hydraulics.

16.981 Chemical Safety

Operation of an RO plant requires the use of a wide variety of chemicals. Whenever you must handle chemicals, follow the

TABLE 16.7 SUMMARY OF COMMON CAUSES OF MEMBRANE DAMAGE

	Symptom	Cause	Restoration Procedure
Case I[a]	1. Lower product water flow rate 2. Higher salt rejection	Membrane compaction[b] accelerated by operating pressure greater than 500 psi (3,450 kPa or 35 kg/cm^2).	None. Required element replacement when product water flow rate reaches an unacceptable level.
Case II[a]	1. Higher product water flow rate 2. Lower salt rejection	Membrane hydrolysis 1. pH outside operating limits. 2. Bacterial degradation. 3. Temperature outside operating limits.	Injection of chemical or element replacement.
Case III[c]	1. Lower product water flow rate 2. Lower salt rejection	Membrane fouling.	Element cleaning.
Case IV[c]	1. Lower product water flow rate 2. High ΔP	Membrane fouling.	Element cleaning.

a CA–cellulose acetate membranes.

b Membrane Compaction. Product water flow rate declines with operational time in addition to fouling of the membrane surface due to other factors. Water flow rate plotted versus time on log-log paper will yield a straight line (flow rate decline).

c All types of membranes.

proper procedures for each chemical. The manufacturer's recommendations for use of each chemical must be observed. Chemicals that are commonly used in an RO plant operation and that require special handling include:

1. Acid
2. Chlorine
3. Sodium hexametaphosphate
4. Formaldehyde
5. Citric acid
6. Numerous cleaning agents

See Chapter 20, "Safety," for more detailed procedures on the safe use of hazardous chemicals.

16.982 Hydraulic Safety

For the reverse osmosis process to function properly, hydraulic pressure in excess of the solution's average osmotic pressure (π) is required. Within the plant, therefore, most of the pipes, tubing, vessels, and their associated equipment, along with the substances inside these items, operate under varying levels of hydraulic pressure (150 to 400 psi or 1,050 to 2,760 kPa or 10.5 to 28 kg/cm^2). Therefore, before starting any repairs, modifications, or work of any kind, no matter how minor, know the substances contained, isolate the piece of equipment, and equalize pressure levels to atmospheric pressure.

After being repaired, any piece of equipment should be purged of all foreign substances before being restarted. When bringing a piece or pieces of equipment on line, increase the hydraulic pressures slowly. Keep all personnel in a safe area to maximize their personal safety.

16.983 Electrical Safety

An RO plant consists of a series of electrically powered pumps and mechanical equipment. Electric shocks due to the use of electrical equipment occur without warning and are usually serious. The average individual thinks of the hazards of electric shock in terms of high voltage and does not always realize that it is primarily the electric current that kills, not the voltage. Consequently, persons who work around low-voltage equipment do not always have the same healthy respect for current as they do for high voltage. Whenever working around electrically operated equipment, strictly observe all applicable rules of the current edition of the *National Electrical Safety Code*.[27]

QUESTIONS

Please write your answers to the following questions and compare them with those on page 226.

16.9L List the three general groups of safety needs for a demineralization plant.

16.9M What type of electrical equipment is used around reverse osmosis plants?

END OF LESSON 3 OF 4 LESSONS

on

MEMBRANE TREATMENT PROCESSES

Please answer the discussion and review questions next.

DISCUSSION AND REVIEW QUESTIONS

CHAPTER 16. MEMBRANE TREATMENT PROCESSES

(MEMBRANE FILTRATION AND DEMINERALIZATION)

(Lesson 3 of 4 Lessons)

Please write your answers to the following questions to determine how well you understand the material in the lesson. The question numbering continues from Lesson 2.

13. Why does water to be demineralized require pretreatment?
14. What problems are created for demineralization processes by the oxidized and precipitated forms of iron, manganese, and silica?
15. How does the operator of a reverse osmosis plant avoid possible damage to the membrane from scaling?
16. What will happen in a reverse osmosis plant if the brine flow valves are accidentally closed during operation?
17. What is the purpose of cartridge filters and when should they be replaced?
18. How can the operator determine if the performance of the RO system is declining?
19. What does hydraulic safety consist of around a reverse osmosis process?

27. *National Electrical Safety Code.* Obtain from the Institute of Electrical and Electronic Engineers (IEEE) at www.ieee.org or phone (800) 701-4333.

CHAPTER 16. MEMBRANE TREATMENT PROCESSES

(MEMBRANE FILTRATION AND DEMINERALIZATION)

(Lesson 4 of 4 Lessons)

16.10 ELECTRODIALYSIS

Electrodialysis (ED) is a well-developed process with a history of many years of operation on brackish well water supplies. A 650,000 GPD (2.5 MLD) ED plant, manufactured by Ionics, Inc., Watertown, Massachusetts, began operation on well water at Buckeye, Arizona, in September 1972 and has been in continuous operation to date. ED plants are also in operation demineralizing municipal water supplies in Siesta Key, Florida; Sanibel Island, Florida; Sorrento Shores, Florida; and at the Foss Reservoir in Oklahoma. The process is also used for industrial water demineralizing.

Typical removals of inorganic salts from brackish water by ED range from 25 to 40 percent of dissolved solids per stage of treatment. Higher removals require treatment by multiple stages in series. Less than 20 percent of the organics remaining in activated carbon-treated secondary effluent are removed by electrodialysis. Energy required for ED is about 0.2 to 0.4 kilowatt-hour per 1,000 gallons (kWh/1,000 gal) for each 100 mg/L dissolved solids removed, plus 2 to 3 kWh/1,000 gal for pumping feedwater and brine. Advantages of the ED process include: (1) well-developed technology, including equipment and membranes; (2) efficient removal of most inorganic constituents; and (3) waste brine contains only salts removed plus a small amount of acid used for pH control in some ED applications.

In the ED process, brackish water flows between alternating cation-permeable and anion-permeable membranes as illustrated in Figure 16.33. A direct electric current provides the motive force to cause ions to migrate through the membranes. Many alternating cation- and anion-permeable membranes, each separated by a plastic spacer, are assembled into membrane stacks. The spacers (about 0.04-inch thick [1 mm]) contain the water streams within the stack and direct the flow of water through a tortuous path across the exposed faces of the membranes. Membrane thicknesses generally range between 0.005 and 0.025 inch (0.125 and 0.625 mm).

Physically, the equipment takes the form of a plate-and-frame assembly similar to that of a filter press. The spacers determine the thickness of the solution compartments and define the flow paths of the water over the membrane surface. Several hundred membranes and their separating spacers are usually assembled between a single set of electrodes to form a membrane stack. End plates and tie rods complete the assembly. When a membrane is placed between two salt solutions and subjected to the passage of a direct electric current, most of the current will be carried through the membrane by ions, hence the membrane is said to be ion selective. Typical selectivities are greater than 90 percent. When the passage of current is continued for a sufficient length of time, the solution on the side of the membrane that is furnishing the ions becomes partially desalted, and the solution adjacent to the other side of the membrane becomes more concentrated. These desalting and concentrating phenomena occur in thin layers of solution immediately adjacent to the membrane, resulting in the desalting of the bulk of the solution.

Passage of water between the membranes of a single stack, or stage, usually requires 10 to 20 seconds, during which time the entering minerals in the feedwater are removed. The actual percentage removal that is achieved varies with water temperature, type and amounts of ions present, flow rate of the water, and stack design. Typical removals per stage range from 25 to 40 percent and systems use one to six stages. An ED system will operate at temperatures up to 110°F (43°C) and the removal efficiency increases with increasing temperature. Ion-selective membranes in commercial electrodialysis equipment are commonly guaranteed for as long as 5 years and experience has demonstrated an effective life of over 10 years.

The most commonly encountered problem in ED operation is scaling (or fouling) of the membranes by both organic and inorganic materials. Alkaline scales are troublesome in the concentrating compartments when the diffusion of ions to the surface of the anion membrane in the diluting cell is insufficient to carry the current. Water is then electrolyzed and hydroxide ions pass through the membrane and raise the pH in the cell. This increase is often sufficient to cause precipitation of materials such as magnesium hydroxide or calcium carbonate. The accumulation of particulate matter increases the electrical resistance of the membrane; this may damage or destroy the membranes. This condition can be offset by feeding acid to the concentrate water stream to maintain a negative Langelier Index to ensure scale-free operation.

Ionics, Inc., has developed a type of ED unit that does not require the addition of acid or other chemicals for scale control. This system reverses the DC current direction and the flow path of the dilution and concentrating streams every 15 minutes. The electrodes reverse by switching the polarity of the cathodes and anodes. The stream flow paths also exchange their source every 15 minutes. Motor-operated valves controlled by timers switch the streams so that the flow path that was previously the

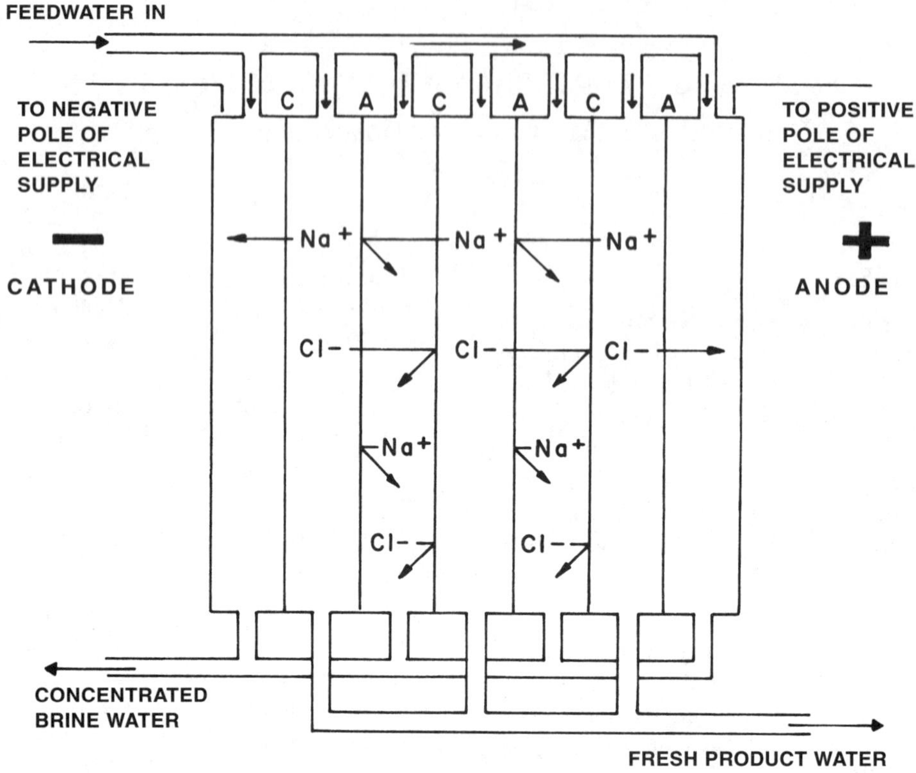

Fig. 16.33 Electrodialysis demineralization process
(From *Standard Operation Instruction Plan for Electrodialysis*, prepared by Ionics, Inc.)

diluting stream becomes the concentrating stream and the flow path that was previously the concentrating stream becomes the diluting stream. This reversing polarity system is commonly referred to as electrodialysis polarity reversal (EDR).

QUESTIONS

Please write your answers to the following questions and compare them with those on page 226.

16.10A What are the typical removals of inorganic salts from brackish water by electrodialysis (ED) per stage of treatment?

16.10B What is a membrane stack in an electrodialysis unit?

16.10C What is the most commonly encountered problem in ED operation?

16.11 PRINCIPLES OF ELECTRODIALYSIS

16.110 Anions and Cations in Water

When most common salts, minerals, acids, and alkalis are dissolved in water, each molecule splits into two oppositely charged particles called ions. All positively charged ions are known as cations and all negatively charged ions, as anions. For instance, when common table salt (sodium chloride or NaCl) is dissolved in water, it separates into positive sodium ions (Na^+) and negative chloride ions (Cl^-). The following ions are found in seawater or brackish water in appreciable quantities.

Cations		Anions	
Sodium	Na^+	Chloride	Cl^-
Calcium	Ca^{2+}	Bicarbonate	HCO_3^-
Magnesium	Mg^{2+}	Sulfate	SO_4^{2-}

16.111 Effect of Direct Current (DC) Potential on Ions

If a DC potential is applied across a solution of salt in water by means of insertion of two electrodes in the solution, the cations will move toward a negative electrode, which is known as the cathode, and the anions will move toward the positive electrode, which is known as the anode. In Figure 16.34 (A) we have a solution of sodium chloride in water. The cations (Na^+) and anions (Cl^-) are moving about at random. In Figure 16.34 (B) a DC potential has been introduced in the solution and the anions move toward the positive electrode and the cations move toward the negative electrode.

16.112 Anion- and Cation-Permeable Membranes and Three-Cell Unit

Advantage could be taken of this movement of ions if proper barriers were available to isolate the purified zone in Figure 16.34 (B) so as to prevent remixing. There are two types of membranes that can be used as such barriers:

1. Cation-Permeable Membranes—Permit only the passage of cations (positively charged ions).
2. Anion-Permeable Membranes—Permit only the passage of anions (negatively charged ions).

Introduction of a cation-permeable membrane and anion-permeable membrane into a salt solution to form three watertight compartments (Figure 16.34 [C]) followed by a direct electric current into the water (Figure 16.34 [D]) will result in the demineralization of the central compartment.

In the three-cell unit shown in Figure 16.34 (C) and (D), "1" is the anode (positive electrode), "2" is the anion-permeable membrane, "3" is the cation-permeable membrane, and "4" is the cathode (negative electrode). In Figure 16.34 (C) there is no electric flow so the ions move at random in their respective compartments. In Figure 16.34 (D) the introduction of a DC potential gives these ions direction: the cations (Na^+) move toward the cathode and the anions (Cl^-) toward the anode. The following occurs:

1. Na^+ from compartment A cannot pass through anion-permeable membrane (2) into compartment B
2. Cl^- from compartment A reacts at the anode (1) to give off chlorine gas
3. Na^+ from compartment B passes through cation-permeable membrane (3) into compartment C
4. Cl^- from compartment B passes through anion-permeable membrane (2) into compartment A
5. Na^+ from compartment C reacts at the cathode to give off hydrogen gas and hydroxyl ions (OH^-)
6. Cl^- from compartment C cannot pass through cation-permeable membrane (3) into compartment B

This description indicates how the overall effect has produced a demineralization of the central compartment.

16.113 Multicompartment Unit

Figure 16.35 presents a multicompartment unit similar in principle to a stack. Letter "A" designates the anion-permeable membranes; letter "C," the cation-permeable membranes; the "+" sign, the anode; the "–" sign, the cathode. A salt solution of Na^+ and Cl^- ions flows between the membranes. On application of the DC potential, the overall effect will be as shown, a movement of ions from the compartments bounded by an anion-permeable membrane on the left and a cation-permeable membrane on the right into the adjacent compartments. The compartments losing salt are labeled "dilute" and those receiving the transferred salt are labeled "brine." Two electrode compartments are also found in the drawing. Each is bordered by a cation-permeable membrane and the electrode. At the anode, a reaction takes place evolving chlorine and oxygen gases; at the cathode, hydrogen gas is produced and hydroxyl ions (OH^-) are left in the solution. Hydroxyl ions are alkaline.

QUESTIONS

Please write your answers to the following questions and compare them with those on page 226.

16.11A What happens if a DC potential is applied across a solution of salt in water by means of insertion of two electrodes in the solution?

16.11B What type of ions can pass through cation-permeable membranes?

16.11C In a multicompartment ED unit, the compartments losing salt are labeled _______ and those receiving the transferred salt, _______ .

16.12 PARTS OF AN ELECTRODIALYSIS UNIT

16.120 Flow Diagram

The basic electrodialysis unit consists of:

1. Pretreatment equipment
2. Pumping equipment (feed, brine, and recirculation)
3. DC power supply
4. Membrane stack and electrodes
5. In-place cleaning system

Figure 16.36 shows a typical flow diagram and Figure 16.37 a photo of an electrodialysis unit.

16.121 Pretreatment

A certain degree of pretreatment of the feedwater supply is necessary in order to prepare it for demineralization in the stacks. Pretreatment depends on the specific water being treated, but it usually includes the removal of suspended or dissolved solids that could adversely affect the surface of the membranes or mechanically block the narrow passageways in the individual cells. Cartridge filters are used as a particle safeguard before the ED

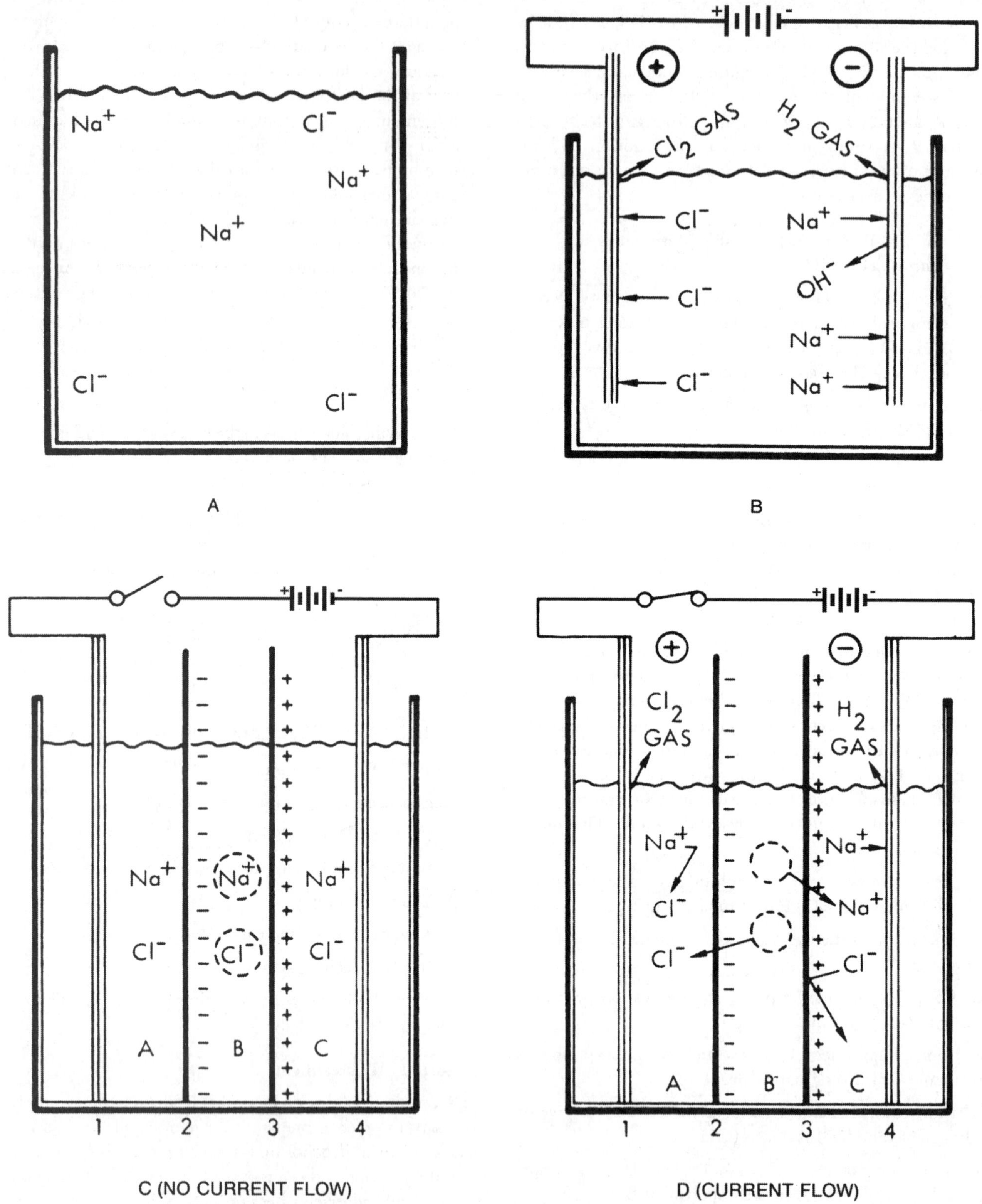

Fig. 16.34 Influence of current flow

(Adapted from *Standard Operation Instruction Plan for Electrodialysis*, prepared by Ionics, Inc.)

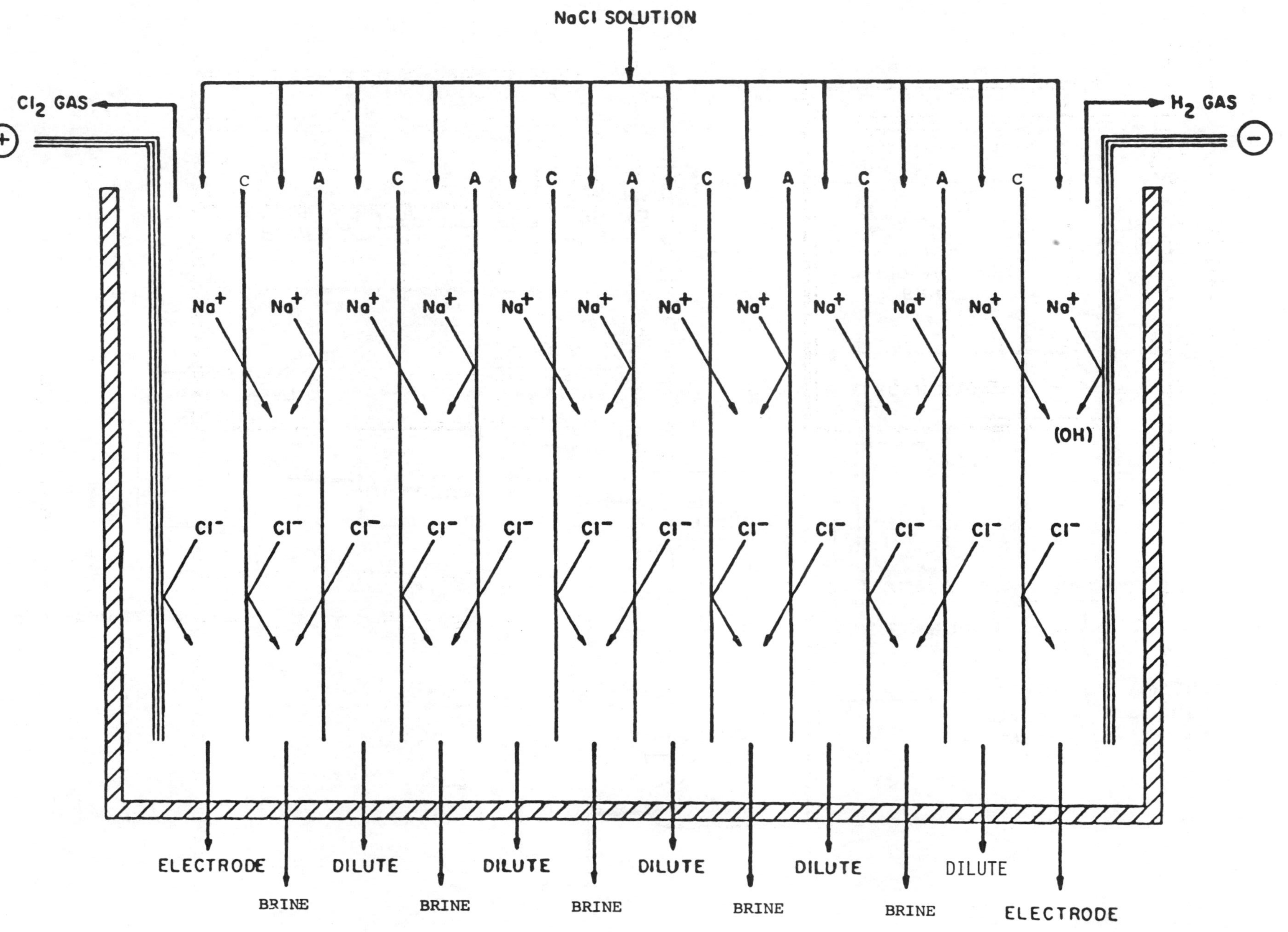

Fig. 16.35 Multicompartment ED stack

(From *Standard Instruction Plan for Electrodialysis*, prepared by Ionics, Inc.)

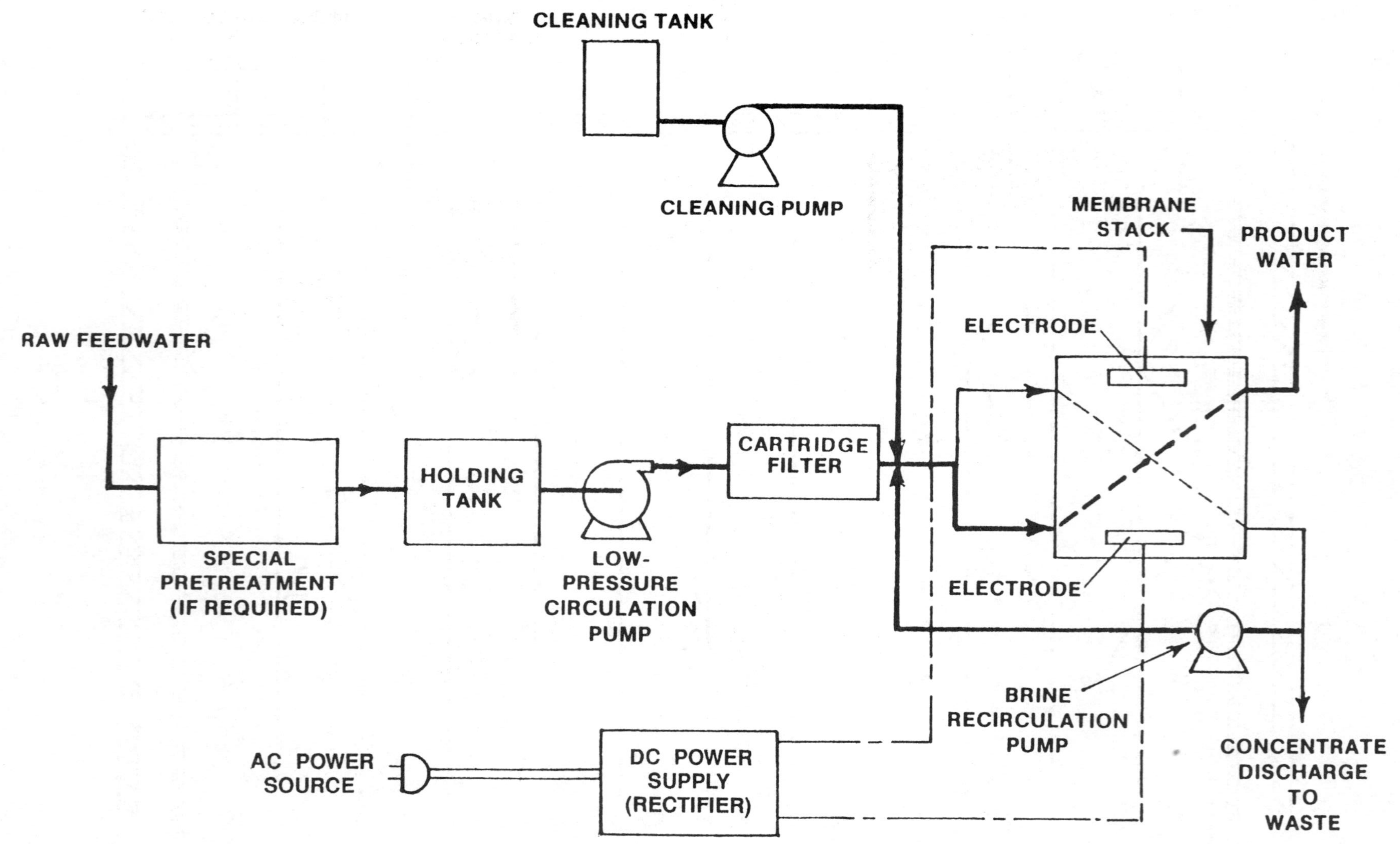

Fig. 16.36 Typical electrodialysis flow diagram

Fig. 16.37 Basic parts of an electrodialysis unit
(Courtesy of Ionics, Inc.)

unit. Before development of the EDR unit, acid addition to prevent carbonate scaling was always practiced. With the electrodialysis reversal process, the requirement for acid addition is reduced or eliminated. Removal of specific materials such as iron, manganese, or chlorine residual, if required, is included in pretreatment.

16.122 Pumping Equipment and Piping

In the electrodialysis process, the water pump(s) is used only for circulation of the water through the stack. The head loss for this circulation varies with the construction of the stacks, number of stages, stacks, and piping, but generally a pumping pressure of only about 50 to 75 psi (345 to 517 kPa or 3.5 to 5.2 kg/cm^2) is needed.

Electrodialysis (ED) systems can be constructed with common materials found in most water treatment applications because only low operating pressures are required, as compared to reverse osmosis (RO) systems. This has allowed the use of a great deal of standard plastic pipe and fittings. The use of plastic pipe produces benefits, including lower cost (compared to stainless steel), high resistance to corrosion in a saline environment, and ease of construction.

16.123 DC Power Supply

The rectifier provides the DC power to the membrane stack assembly. The input (alternating current [AC]) is converted by the rectifier to direct current, which is applied to the electrodes on each side of the membrane stack to remove the ions from the feed stream. This equipment also includes a control module for periodic reversal of the current every 15 to 30 minutes on all new EDR models.

16.124 Membrane Stack

The membrane "stack" is so called because it is composed of a large number of stacked pieces, like a deck of cards. Half of these pieces are spacers and half are membranes, which alternate from the bottom to the top of the stack. In other words, in examining any portion of the stack, you will find a membrane above and below every spacer (except at the electrodes) and a spacer above and below every membrane. Two membranes or two spacers should never occur together.

Each membrane stack constitutes one stage of demineralization and is a separate hydraulic and electrical stage. The total number of stacks in the unit will be arranged in either one line or two lines running in parallel (each with an equal number of stacks). Since all the stacks in a line are connected in a series, the number of stacks per line will equal the number of stages of demineralization.

The membranes and spacers in the main section of the stack make up the number of cell pairs noted in the stack specifications. A cell pair consists of one anion-permeable membrane, one cation-permeable membrane, and two intermembrane spacers and is the basic demineralizing element. The metal electrodes located at the ends of the stack apply the DC electric power required for demineralization.

16.125 Chemical Flush System

ED units are equipped with a clean-in-place (CIP) flush system to allow periodic flushing of the membrane stacks and associated piping with acid solutions down to pH 1 or with brine solutions up to 10 percent sodium chloride.

The two chemical solutions that are used most often for stack cleaning are a 5 percent solution of hydrochloric acid (for removal of scale and normal cleaning), and a 5 percent salt solution that has caustic soda added to adjust the pH to between 12 and 13 (for removal of organic fouling or slime).

QUESTIONS

Please write your answers to the following questions and compare them with those on page 226.

16.12A What must be removed by pretreatment of the feedwater supply to the electrodialysis unit?

16.12B What is the purpose of the rectifier in an electrodialysis unit?

16.13 ROUTINE OPERATING PROCEDURES

16.130 Design Specifications for Feedwater

The electrodialysis desalting unit will produce demineralized water at a rate dependent on water temperature and mineral composition of feedwater. The quality of the feedwater and its ionic composition are extremely important and the design of the ED unit is based on these conditions.

The ions most often encountered in feedwater are:

Cations	Anions
1. Calcium	1. Bicarbonate
2. Iron	2. Chloride
3. Magnesium	3. Sulfate
4. Silica	
5. Sodium	

An excessive concentration of any of these constituents could lead to chemical fouling due to scaling. Iron in the feedwater will cause certain process problems; above 0.1 mg/L certain precautions have to be taken. One of the effects of excess iron in feedwater is the deposit of an orange film onto the membrane surface, which increases the electrical resistance of the membrane stack. Concentrations of iron in excess of 0.3 mg/L should be removed by pretreatment.

There are other important considerations regarding feed-water quality. These include pH, biological quality, and bacteriological quality of the feed. To prevent biological fouling of the cation- and anion-permeable membranes, the feedwater should be free of bacteria. Proper control of feedwater pH is also important, particularly in terms of corrosion control in piping and plumbing equipment. Because chlorine attacks the ED membrane, the feedwater cannot contain any chlorine residual. If prechlorination is practiced, the feedwater must be dechlorinated before entering the ED unit. Generally, the unit should not be operated when the feedwater contains any of the following:

1. Chlorine residual of any concentration
2. Hydrogen sulfide of any concentration
3. Calgon or other hexametaphosphates in excess of 10 mg/L
4. Manganese in excess of 0.1 mg/L
5. Iron in excess of 0.3 mg/L

16.131 Detailed Operating Procedures

Detailed operating procedures vary from one system to the next. Most ED or EDR units come designed with fully automatic control systems. A typical operating log used to monitor an ED system is given in Figure 16.38.

The detailed specifications for any plant will give the proper settings for the various controls on the unit. These control settings should be checked and recorded at least once every 24 hours using the sample log sheet given in Figure 16.38. Any action needed to keep the plant running according to the specifications should be taken immediately.

In addition to checking the specifications, the routine maintenance schedule outlined below should be followed closely in order to reduce the risk of lengthy and expensive downtimes. Any process problems discovered must be acted upon immediately.

Daily

1. Fill out log sheet.
2. Verify that electrodes are bumping and flowing properly.
3. Inspect stacks for excess external leakage (greater than 10 gallons per hour or 38 liters per hour per stack).
4. Check the pressure drop across the cartridge filter and change the cartridges whenever the pressure drop reaches 10 psi (69 kPa or 0.7 kg/cm^2).

Weekly

1. Voltage probe the membrane stacks.
2. Check the oil level on pumps fitted with automatic oilers.
3. Inspect all piping and skid components for leaks.
4. Twice per week, measure all electrode waste flows.

QUESTIONS

Please write your answers to the following questions and compare them with those on page 226.

16.13A List the ions most often encountered in the feedwater to an electrodialysis unit.

16.13B What items must be considered to prevent biological fouling of the cation- and anion-permeable membranes?

16.13C Generally, the electrodialysis unit should not be operated when the feedwater contains ____. (List the appropriate water quality constituents.)

16.13D List the recommended daily activities for the operator of an electrodialysis unit.

16.14 SAFETY PRECAUTIONS

1. Grounding. The entire unit, including the stacks, must be connected to an electrical ground of each potential. At the time of installation, it is necessary to ground the skid or the control panel cabinet, either by a metal conduit or a separate grounding wire.

   Each time the unit is moved or dismantled, check the ground connections before turning on the power. The skid, power supply cabinet, and stack(s) must always be firmly connected to the building ground or other suitable ground.
2. Check the electrode tab connecting bolts and be sure these are tight and there is no corrosion. Loose connections at these points will cause overheating, which could result in serious damage to the membrane stack.
3. Do not touch wet stack sides or electrode tabs when the DC power is on.
4. Always wear rubber gloves when voltage probing the membrane stack.
5. When washing down the area, never direct a hose on the membrane stack when the DC power is on.
6. Never operate a dry centrifugal pump, even when checking rotation.
7. Never apply DC voltage to the membrane stack without water flowing through the stack.
8. Expect the DC amperage to drop when the feedwater temperature drops. Never increase the DC stack voltage as the water temperature drops in an attempt to raise currents to those recorded at the higher temperatures unless you have received specific instructions to do so from the manufacturer.
9. Expect the DC amperage to rise when the feedwater temperature rises. As this happens, the DC stack voltages must be lowered until the DC amperage returns to the normal setting. This conserves power and prevents damage to the stack.
10. Never allow oil, organic solutions, solvents, detergents, wastewater, chlorine, nitric acid, strong bleach, or other oxidizing agents to come in contact with the membranes and spacers unless directed to do so by the manufacturer. Membranes can be damaged by a feedwater containing even 0.1 mg/L free chlorine.
11. Always keep the membranes wet. Store in the membrane tube supplied or in the original plastic bags provided the seals are not broken.

Date										
Polarity										
Feed Temp (°F)										
Feed TDS (mg/L)										
Product TDS (mg/L)										
Product Conductivity										
Dilute Flowrate (GPM)										
Brine Make-up (GPM)										
PRESSURES	Stack Inlet									
	Stack Outlet									
	Differential In									
	Differential Out									
	Before Filter									
	After Filter									
	Electrode Inlet									
Stage 1	Volts	Line 1								
		Line 2								
Stage 1	Amps	Line 1								
		Line 2								
Stage 2	Volts	Line 1								
		Line 2								
Stage 2	Amps	Line 1								
		Line 2								
Stage 3	Volts	Line 1								
		Line 2								
Stage 3	Amps	Line 1								
		Line 2								
Stage 4	Volts	Line 1								
		Line 2								
Stage 4	Amps	Line 1								
		Line 2								
Stage 5	Volts	Line 1								
		Line 2								
Stage 5	Amps	Line 1								
		Line 2								
Stage 6	Volts	Line 1								
		Line 2								
Stage 6	Amps	Line 1								
		Line 2								

Fig. 16.38 Typical operating log sheet for ED unit

12. Do not smoke or use exposed flames or sparks in the gas separator tank area due to the presence of potentially explosive gases.
13. Do not service the gas separator tank when the unit is in operation. Especially avoid the vent lines where toxic and explosive gases can be present. If it is necessary to service the tank, operate the unit for 30 minutes without DC power, then wait an additional hour before beginning work or ventilate with fans to ensure complete dispersion of dangerous gases.
14. If it is necessary to troubleshoot any of the electrical panels, be extremely careful of the live panel voltages. This maintenance should be done only by someone familiar with the circuits and wiring. The unit should never be operated with the panel doors open, except for maintenance purposes, and only by experienced personnel.
15. Should shorting occur from a metal end plate across the plastic end block to the electrode, immediately turn off the rectifier. Try to eliminate the cause of the shorting by wiping excess moisture off the block. Also, be sure to completely remove the black carbon that has formed at the point of shorting. If this is not effectively done, the shorting will recur when the rectifier is turned back on.
16. Feedwater containing Calgon or other hexametaphosphates will cause high membrane stack resistance. Avoid operation when these are present.
17. Red warning lamps are mounted on the wire way for the stack power connections. The lamps are lit when the DC power is applied to the stacks.
18. When the plant is on automatic, the plant is controlled by the product water tank's level switch. Therefore, when working on the equipment, the plant should be switched to manual operation and locked out, thus avoiding the possibility of an unexpected startup.
19. Use of the "STOP" switch or "STOP/START" switch activates an automatic flushing cycle and therefore does not immediately stop operation of all components of the unit. If the operation of the entire unit must be stopped immediately, the main breaker should be switched off.

QUESTIONS

Please write your answers to the following questions and compare them with those on page 226.

16.14A What problems can be created by loose connections at the electrode tab connecting bolts?

16.14B What happens to the DC amperage when the feedwater temperature drops?

16.14C How can shorting be prevented from the metal end plate across the plastic end block to the electrode?

16.14D How can the operation of the entire electrodialysis unit be stopped immediately?

16.15 ARITHMETIC ASSIGNMENT

Turn to the Arithmetic Appendix at the back of this manual. Read and work the problems in Section A.34, "Membrane Treatment Processes." Check the arithmetic in this section using an electronic calculator. You should be able to get the same answers. Section A.54 contains similar problems using metric units.

16.16 ADDITIONAL READING

Texas Manual, chapter on "Special Water Treatment (Desalting)."

END OF LESSON 4 OF 4 LESSONS

on

MEMBRANE TREATMENT PROCESSES

Please answer the discussion and review questions next.

DISCUSSION AND REVIEW QUESTIONS

CHAPTER 16. MEMBRANE TREATMENT PROCESSES

(MEMBRANE FILTRATION AND DEMINERALIZATION)

(Lesson 4 of 4 Lessons)

Please write your answers to the following questions to determine how well you understand the material in the lesson. The question numbering continues from Lesson 3.

20. How does an electrodialysis unit demineralize (desalt) brackish water?
21. What are the parts of a basic electrodialysis unit?
22. What are the benefits of using plastic pipe in an electrodialysis plant?
23. What is the purpose of the chemical flush system in an electrodialysis unit?
24. An excessive concentration of any specific ion in the feedwater to an electrodialysis unit can cause what problem?
25. When should you check to be sure that an electrodialysis unit is properly grounded?

SUGGESTED ANSWERS

Chapter 16. MEMBRANE TREATMENT PROCESSES

(MEMBRANE FILTRATION AND DEMINERALIZATION)

ANSWERS TO QUESTIONS IN LESSON 1

Answers to questions on page 165.

16.0A Microfiltration (MF) and ultrafiltration (UF) are used in water treatment for particle, sediment, algae, bacteria, and virus removal as well as in the removal of *Giardia* and *Cryptosporidium.*

16.0B Reverse osmosis (RO) and nanofiltration (NF) membranes are used to remove dissolved organic matter and dissolved contaminants such as arsenic, nitrate, pesticides, and radionuclides. Also, these membranes can remove ions such as calcium and magnesium (for softening) and sodium and chloride (for deionization, desalination, or demineralization of brackish waters).

Answers to questions on page 174.

16.1A Typical water treatment membrane filtration units are installed either in pressure vessels or submerged in tanks.

16.1B A decrease in flow through a membrane can be recovered by backwashing and cleaning.

16.1C Operators consider using coagulation pretreatment before membrane filtration because of the effectiveness of membrane filtration in treating low-turbidity water.

16.1D To optimize membrane performance, a suggested approach is to work with membrane manufacturers and ask them to perform bench- or pilot-scale studies. The results of the studies should provide the following information:

1. The selection of the most appropriate coagulant for specific applications
2. The selection of the most appropriate coagulant dose
3. The coagulant selection should include consideration of removing DOC
4. A cost comparison for each alternative investigated

Answers to questions on page 180.

16.2A When the plant is unattended, the on-call operator can use a computer off-site to confirm plant process performance and make process and equipment adjustments before to returning to the plant.

16.2B Backwashing membrane filters may occur on the basis of three possible conditions: elapsed time since the last backwash, total water produced since the last backwash, and pressure increase since the last backwash.

16.2C The operation of membranes includes monitoring and testing for membrane filtration rate and membrane integrity.

16.2D Problems that an operator may encounter when operating a membrane filtration water treatment plant include air compressor failure, raw water pump station failure (electronic failure of the uninterrupted power supply [UPS]), and computer lockup.

Answers to questions on page 181.

16.3A Membrane filtration water treatment plants continuously monitor turbidity and chlorine residual.

16.3B Membrane filtration water treatment plant operation and water production records maintained in a SCADA system include water levels, flow rates, hours of operations, and quality of water produced.

Answers to questions on page 182.

16.4A The operator conducts routine maintenance on a daily basis.

16.4B The preventive maintenance program (PMP) includes scheduled maintenance based on operating time.

16.4C Breakdown maintenance relates to repairs to equipment that has failed in service.

ANSWERS TO QUESTIONS IN LESSON 2

Answers to questions on page 183.

16.5A Demineralization is the process that removes dissolved minerals (salts) from water.

16.5B Seawater is more expensive to treat than brackish water because of its greater TDS concentration.

Answers to questions on page 186.

16.6A Methods of removing minerals from water can be divided into two classes:

1. Those that use a phase change such as freezing or distillation
2. Nonphase change methods such as reverse osmosis, electrodialysis, and ion exchange

16.6B The common membrane demineralizing processes are reverse osmosis, electrodialysis, and nanofiltration.

Answers to questions on page 189.

16.7A The osmotic pressure of a solution is the difference in water level between the two sides of a membrane.

16.7B The two types of semipermeable membranes that are used most often for demineralization are cellulose acetate and thin film composites.

16.7C The water flux is the flow of water in grams per second through a membrane area of one square centimeter (or gallons per day per square foot) while the mineral flux is the flow of minerals in grams per second through a membrane area of one square centimeter.

16.7D When additional pressure is applied to the side of a membrane with a concentrated solution, the water flux (rate of water flow through the membrane) will increase, but the mineral flux (rate of flow of minerals) will remain constant.

16.7E When higher mineral concentrations occur in the feedwater, the mineral concentrations will increase in the permeate (or product water).

Answers to questions on page 190.

16.7F Water flux is usually expressed in gallons per day per square foot (or grams per second per square centimeter) of membrane surface.

16.7G The term "flux decline" is used to describe the loss of water flow through the membrane due to compaction plus fouling.

16.7H Mineral rejection is defined as:

$$\text{Rejection, \%} = \left(1 - \frac{\text{Product Concentration}}{\text{Feedwater Concentration}}\right) \times 100\%$$

Mineral rejections can be determined by measuring the TDS and using the above equation. Rejections also may be calculated for individual constituents in the solution by using their concentrations.

16.7I Softening membranes operate at lower pressures and lower rejection rates than standard reverse osmosis membranes.

Answers to questions on page 198.

16.7J An increase in feedwater temperature will increase the water flux.

16.7K Hydrolysis of a membrane results in a lessening of mineral rejection capability.

16.7L Recovery is defined as the percentage of feed flow that is recovered as product water.

$$\text{Recovery, \%} = \left(\frac{\text{Product Flow}}{\text{Feed Flow}}\right)(100\%)$$

16.7M Recovery rate is usually limited by the desired product water quality and the solubility limits of minerals in the brine.

16.7N Concentration polarization is a buildup of retained particles on the membrane surface due to dewatering of the feed closest to the membrane. The thickness of the concentration polarization layer is controlled by the flow velocity across the membrane.

ANSWERS TO QUESTIONS IN LESSON 3

Answers to questions on page 205.

16.8A The three types of commercially available membrane configurations that have been used in operating plants are spiral wound, hollow fine fiber, and tubular.

16.8B The tubular membrane process is used to treat wastewater with a high suspended solids concentration.

Answers to questions on pages 206 and 207.

16.9A To protect the reverse osmosis system and its accessory equipment, the feedwater should be filtered. When the water source is a groundwater or a previously treated municipal or industrial supply, filtration may be accomplished by a simple screening procedure. An untreated surface water will probably require coagulation, flocculation, sedimentation, and filtration.

16.9B Colloidal particulates are removed from feedwater by chemical treatment and filtration.

16.9C As an acetate membrane hydrolyzes, both the amount of water and the amount of solute that permeate the membrane increase and the quality of the product water deteriorates.

16.9D The scale control method that is used to inhibit calcium sulfate precipitation is a threshold treatment with 2 to 5 mg/L of sodium hexametaphosphate (SHMP).

16.9E A 1 to 2 mg/L chlorine residual is maintained to control biological fouling.

Answers to questions on page 207.

16.9F Operating pressure on a reverse osmosis unit is regulated by a control valve on the influent manifold.

16.9G The demineralized water is usually called permeate, the reject, concentrate.

16.9H Product or permeate flow is not regulated and varies as feedwater pressure and temperature change.

Answers to questions on page 211.

16.9I Chlorine is added to the feedwater to prevent biological fouling.

16.9J The operator must check the differential pressure across the RO unit to know when to clean the elements. When the elements become fouled, ΔP usually increases, thus indicating the need for cleaning.

16.9K The reverse osmosis elements should be cleaned when the operator observes lower product water flow rate, lower salt rejection, higher differential pressure (ΔP), and higher operating pressure.

Answers to questions on page 212.

16.9L Safety needs for demineralization plants can be divided into three general groups consisting of chemicals, electrical, and hydraulics.

16.9M Electrical equipment used around reverse osmosis plants consists of a series of electrically powered pumps.

ANSWERS TO QUESTIONS IN LESSON 4

Answers to questions on page 214.

16.10A Typical removals of inorganic salts from brackish water by ED range from 25 to 40 percent of dissolved solids per stage of treatment.

16.10B A membrane stack in an electrodialysis unit consists of several hundred membranes and their separating spacers assembled between a single set of electrodes. End plates and tie rods complete the assembly.

16.10C The most commonly encountered problem in ED operation is scaling (or fouling) of the membranes by both organic and inorganic materials. Alkaline scales are troublesome in the concentrating compartments when the diffusion of ions to the surface of the anion membrane in the diluting cell is insufficient to carry the current.

Answers to questions on page 215.

16.11A If a DC potential is applied across a solution of salt in water by means of insertion of two electrodes in the solution, the cations will move toward a negative electrode, which is known as the cathode, and the anions will move toward the positive electrode, which is known as the anode.

16.11B Only cations (positively charged ions) can pass through cation-permeable membranes.

16.11C In a multicompartment ED unit, the compartments losing salt are labeled "dilute" and those receiving the transferred salt, "brine."

Answers to questions on page 220.

16.12A Iron, manganese, and chlorine residual must be removed from the feedwater supply to the electrodialysis unit.

16.12B The rectifier provides the DC power to the membrane stack assembly. The input (alternating current [AC]) is converted by the rectifier to DC, which is applied to the electrodes on each side of the membrane stack to remove the ions from the feed stream.

Answers to questions on page 221.

16.13A The ions most often encountered in the feedwater to an electrodialysis unit include:

Cations	Anions
1. Calcium	1. Bicarbonate
2. Iron	2. Chloride
3. Magnesium	3. Sulfate
4. Silica	
5. Sodium	

16.13B To prevent biological fouling of the cation- and anion-permeable membranes, the operator must control feedwater pH, biological quality, and bacteriological quality.

16.13C Generally, the electrodialysis unit should not be operated when the feedwater contains any of the following:

1. Chlorine residual of any concentration
2. Hydrogen sulfide of any concentration
3. Calgon or other hexametaphosphates in excess of 10 mg/L
4. Manganese in excess of 0.1 mg/L
5. Iron in excess of 0.3 mg/L

16.13D The recommended daily activities for the operator of an electrodialysis unit include:

1. Fill out log sheet.
2. Verify that electrodes are bumping and flowing properly.
3. Inspect stacks for excess external leakage.
4. Check the pressure drop across the cartridge filter and change the cartridges whenever the pressure drop reaches 10 psi.

Answers to questions on page 223.

16.14A Loose connections at the electrode tab connecting bolts will cause overheating, which could result in serious damage to the membrane stack.

16.14B Expect the DC amperage to drop when the feedwater temperature drops.

16.14C Should shorting occur from the metal end plate across the plastic end block to the electrode, immediately turn off the rectifier. Try to eliminate the cause of the shorting by wiping excess moisture off the block. Also, be sure to completely remove the black carbon that has formed at the point of shorting.

16.14D If the operation of the entire unit must be stopped immediately, the main breaker should be switched off.

CHAPTER 17

HANDLING AND DISPOSAL OF PROCESS WASTES

by

George Uyeno

TABLE OF CONTENTS

Chapter 17. HANDLING AND DISPOSAL OF PROCESS WASTES

LEARNING OBJECTIVES

Chapter 17. HANDLING AND DISPOSAL OF PROCESS WASTES

Following completion of Chapter 17, you should be able to:

1. Explain why the operators of water treatment plants should be concerned about the handling and disposal of process wastes.
2. Identify the sources of water treatment plant wastes.
3. Drain and clean sedimentation tanks.
4. Discharge process wastes to collection systems (sewers).
5. Operate and maintain backwash recovery ponds (lagoons) and sludge drying beds.
6. Dispose of process wastes.
7. Safely operate and maintain sludge handling and disposal equipment.
8. Monitor and report on the disposal of process wastes.

WORDS

Chapter 17. HANDLING AND DISPOSAL OF PROCESS WASTES

CENTRIFUGE — CENTRIFUGE

A mechanical device that uses centrifugal or rotational forces to separate solids from liquids.

CONDITIONING — CONDITIONING

Pretreatment of sludge to facilitate removal of water in subsequent treatment processes.

DECANT (de-KANT) — DECANT

To draw off the upper layer of liquid (water) after the heavier material (a solid or another liquid) has settled.

DEWATER — DEWATER

(1) To remove or separate a portion of the water present in a sludge or slurry. To dry sludge so it can be handled and disposed of.

(2) To remove or drain the water from a tank or a trench. A structure may be dewatered so that it can be inspected or repaired.

SEPTIC (SEP-tick) — SEPTIC

A condition produced by bacteria when all oxygen supplies are depleted. If severe, the bottom deposits produce hydrogen sulfide, the deposits and water turn black, give off foul odors, and the water has a greatly increased chlorine demand.

SLUDGE (SLUJ) — SLUDGE

(1) The settleable solids separated from liquids during processing.

(2) The deposits of foreign materials on the bottoms of streams or other bodies of water or on the bottoms and edges of wastewater collection lines and appurtenances.

SUPERNATANT (soo-per-NAY-tent) — SUPERNATANT

The relatively clear water layer between the sludge on the bottom and the scum on the surface of a basin or container. Also called clear zone.

THICKENING — THICKENING

Treatment to remove water from the sludge mass to reduce the volume that must be handled.

CHAPTER 17. HANDLING AND DISPOSAL OF PROCESS WASTES

17.0 NEED FOR HANDLING AND DISPOSAL OF PROCESS WASTES

The need for handling and disposal of potable water treatment plant wastes is a problem that must be faced by all plant operators. Many articles and books have been published on potable water treatment processes. Their emphasis is usually on producing wholesome and pure water for human consumption in compliance with EPA and state and local health department regulations, but very few mention sludge handling and disposal in any great detail. In response to a growing population and increasing concern about pollution of natural water sources, pollution control agencies, health departments, and fish and game departments established programs to enforce rules to prevent any waste discharge that would tend to discolor, pollute, or generally be harmful to aquatic or plant life or the environment.

The law that restricts or prohibits the discharge of process wastes from water treatment plants is Public Law 92-500, the Water Pollution Control Act Amendments of 1972. This act clearly includes treatment plant wastes such as sludge from a water treatment plant. These wastes are considered an industrial waste that requires compliance with the provisions of the act. Under the National Pollutant Discharge Elimination System (NPDES) provisions, a permit must be obtained in order to discharge wastes from a water treatment plant. The NPDES permit is typically a license for a facility to discharge a specified amount of a pollutant into a receiving water under certain conditions. Permits may also authorize facilities to process, incinerate, landfill, or beneficially use sewage sludge. The two basic types of NPDES permits that can be issued are *individual* and *general* permits.

An *individual permit* is a permit developed for an individual facility. After a facility submits the appropriate applications, the permitting authority develops a permit for that particular facility based on the information contained in the permit application (for example, type of activity, nature of discharge, receiving water quality, and beneficial uses).

A *general permit* is developed and issued by a permitting authority to cover multiple facilities within a specific category. General permits may offer a cost-effective option for utility agencies because of the large number of facilities that can be covered under a single permit. General permits may be written to cover categories of point sources having common elements:

- Stormwater point sources
- Facilities that involve the same or substantially similar types of operations
- Facilities that discharge the same types of wastes or engage in the same types of sludge use or disposal

For additional information on permits, see the EPA's *NPDES Permit Writers' Manual* (EPA No. 833-K-10-001). The manual is also available for download at www.epa.gov/npdes/pubs/pwm_2010.pdf.

Water treatment plants are classified into three categories.

Category 1 Plants that use one of the following three processes: (1) coagulation, (2) oxidative iron and manganese removal, or (3) direct filtration.

Category 2 Plants that use only chemical softening processes.

Category 3 Plants that use combinations of coagulation and chemical softening, or oxidative iron and manganese removal and chemical softening.

Enforcement of Public Law 92-500 is the responsibility of each state. Many NPDES permits have been issued by the states to water treatment plants using state standards applicable to the local conditions at the time the permits were issued. Water quality indicators for which waste discharge limitations have been issued include pH, total suspended solids, settleable solids, total iron and manganese, flow rate, total dissolved solids (TDS), BOD, turbidity, total residual chlorine, temperature, floating solids, and visible forms of waste.

Water treatment plants can no longer simply discharge dirty backwash water or settled sludge into lakes, rivers, streams, or tributaries as was done in the past. Current regulations require daily monitoring of any discharge and analysis of such water quality indicators as pH, turbidity, TDS, settleable solids, or other harmful materials. The results of the analyses must be logged and reported frequently to the proper authorities and must conform to their rigid standards. For these reasons, it is absolutely necessary to make provisions for facilities to handle these wastes on a routine basis. The most important part of an operator's job is still the end product, good potable water, but an operator's duties are not complete until all byproducts and wastes are disposed of in an acceptable and documented manner.

QUESTIONS

Please write your answers to the following questions and compare them with those on page 253.

17.0A Why are strict laws needed regarding the disposal of process wastes?

17.0B If a discharge results from the disposal of process wastes, what water quality indicators require daily monitoring?

17.1 SOURCES OF TREATMENT PROCESS WASTES

Although there are many types of water treatment plants and methods for treating water, most of them probably operate in the following general manner. Alum or polymers are applied to the water in a rapid-mix chamber. The mixture is agitated by mechanical means or through a cylinder designed for hydraulic flash mixing for coagulation.

Following coagulation, the water passes through mechanical flocculators or a series of baffles for flocculation. The water then moves into the sedimentation tank where the floc is allowed to settle out before the water moves to the filters. The sedimentation tanks may be of various shapes and depths; however, they are most commonly rectangular or circular. Many large plants are equipped with either mechanical rakes or scrapers that periodically remove sludge from a hopper, or with a vacuum-type sludge removal device. The sludge is continuously scraped into the hopper. The hopper is emptied from one to three times per day for 20 to 30 minutes each time depending on the size of the hopper and the density of the sludge. Sludge is then usually moved to drying beds. The smaller and older plants may not have these sludge handling facilities available. Many new water treatment plants are equipped with sludge collection headers with squeegees. This system does not need any sludge hoppers. The collection headers are supported by a travelling bridge or floats. The sludge is pumped out of the bottom of the basin and into a sludge channel on the walkway level. This system is described in Volume I, Chapter 5, "Sedimentation."

Another type of plant similar to the one above contains an upflow solids-contact unit with clarifiers. The clarifiers are usually circular and have sludge draw-off lines that must be monitored; the solids are then drawn off periodically as sludge.

For small plants and in areas where water must be pumped, pressure filters may be used and the coagulant is applied directly to the filter. Sedimentation tanks or clarifiers may occasionally accompany the use of pressure filters but this is not usually the case if the quality of the source water is good.

In another type of plant layout (not too commonly used), the sedimentation tank also functions as the backwash recovery area. In this case, the backwash wastewater is pumped back to the head of the plant. Most of the solids will settle out when the water flows through the sedimentation basin. This method does eliminate the need for backwash recovery ponds or lagoons.

Diatomaceous earth filtration is different from all other types of filtration in its method of operation. There is usually no pretreatment of the water. Disposal of backwash wastes is still a problem. Table 17.1 summarizes the various sources of treatment process wastes and the methods of collecting, handling, and disposing of these wastes.

QUESTIONS

Please write your answers to the following questions and compare them with those on page 253.

17.1A How is sludge removed from sedimentation tanks?

17.1B How is sludge removed from an upflow solids-contact type of unit?

TABLE 17.1 COLLECTION, HANDLING, AND DISPOSAL OF PROCESS WASTES

SOURCES OF WASTES

1. Trash racks
2. Grit basins
3. Alum, ferric hydroxide, or polymer sludges from sedimentation basins
4. Filter backwash
5. Lime–soda softening
6. Ion exchange brine

COLLECTION OF SLUDGES

1. Mechanical scrapers or vacuum devices
2. Manual (hoses and squeegees)
3. Pumps (into tank trucks or dewatering facilities)
4. Floating headers

DEWATERING OF SLUDGES

1. Solar drying lagoons
2. Sand drying beds
3. Centrifuges[a]
4. Belt presses
5. Filter presses
6. Vacuum filters

DISPOSAL OF SLUDGES AND BRINES

1. Wastewater collection systems (sewers)
2. Landfills (usually dewatered sludges)
3. Spread on land

a Mechanical devices that use centrifugal or rotational forces to separate solids from liquids (sludge from water).

17.2 PROCESS SLUDGE VOLUMES

All of the different methods of sludge collection require some type of facilities for handling processed waste. The possibilities here include sedimentation tanks, backwash recovery ponds, drying beds, lagoons, ponds, holding tanks, adequate land, access to sewer systems, or vacuum trucks or similar equipment for removal and disposal of the waste material.

The amount of sludge accumulation at a typical water treatment plant depends on the type and amount of suspended matter in the source water being treated as well as on the level of dosage and the type of coagulant used. As an example, let us examine sludge production at two 5 MGD (19 megaliters per day [ML/day]) plants. The annual water production at each was 800 to 900 million gallons (3,000 to 3,400 ML). Plant One had no source water stabilizing reservoir and the raw water turbidity ranged from 3 units during the summer months to over 100 units during the winter, with an annual average alum dosage of 11 mg/L. Yearly sludge accumulation was approximately 500,000 gallons (1.9 ML). Plant Two, with a 15 million-gallon (56.8 ML) source water stabilizing reservoir, treated raw water that never exceeded 20 turbidity units at the intake with an average alum dosage of 8 mg/L. The annual sludge accumulation was approximately 300,000 gallons (1.14 ML). In both cases, nonionic polymer was used for filter aid at approximately 15 ppb. The source water stabilizing reservoir reduced the turbidity level of the water to be treated and provided a more constant quality of water that required less alum and produced less sludge.

Organic polymers may be used instead of alum to reduce the quantity of sludge produced. Polymer sludges are relatively denser and easier to dewater for subsequent handling and disposal. Not all waters can be treated by using polymers instead of alum.

For plants without sludge collection devices, the volume of sludge produced and the frequency of cleaning the sedimentation tank are affected by several factors. Items to consider include:

1. Water demand
2. Suspended solids loads and when peak demands occur
3. Water temperature (as the temperature of the water increases, the settling rate of the solids will increase)
4. Detention time (as the detention time increases, the amount of solids that settle out will increase)
5. Volume of sludge deposited in the basin (as the volume of sludge increases, the detention time decreases as well as the efficiency of the basin)
6. Volume of treated water storage for the system (the greater the volume of treated water storage, the more time is available for sludge removal)
7. Time required to clean and make any necessary repairs during the shutdown
8. Availability of adequate drying beds, lagoons, landfill, a vacuum tank truck, pumps or equipment, and adequate help with all necessary safety equipment and procedures

Sedimentation tanks should be drained and cleaned at least twice a year and more often if the sludge buildup interferes with the treatment processes (filtration and disinfection). Alum or polymer sludge solids content is only 0.5 to 1 percent for continuous sludge removal and 2 to 4 percent when the sludge is allowed to accumulate and compact. Therefore, the sludge can flow readily in pipes or be pumped, especially with wastewater-type pumps.

QUESTIONS

Please write your answers to the following questions and compare them with those on page 253.

17.2A How can a source water stabilizing reservoir reduce the volume of sludge handled?

17.2B How often should sedimentation tanks be drained and cleaned?

17.3 METHODS OF HANDLING AND DISPOSING OF PROCESS WASTES

Various methods are used to handle and dispose of process wastes (Figure 17.1). The disposal facilities at your plant will depend on when the plant was built, the region where the plant is located (topography and climate), the sources of sludge, and the methods of ultimate disposal.

An effective method of handling sludge is to regularly during the day remove sludge from sedimentation tanks to a drying bed. When one drying bed is full of sludge, the sludge is allowed to dry while the other drying beds are being filled. A key to speedy drying is the regular removal of the water on top of the sludge.

Some plants require that portions of the facilities be shut down twice a year, the tanks drained, and the sludge removed. This is an excellent time to inspect the tanks and equipment and perform any necessary maintenance and repairs.

Backwash recovery ponds or lagoons are used to separate the water from the solids after the filters have been backwashed. The water is usually returned or recycled to the plant headworks for treatment with the source water. These ponds also may be used to concentrate or thicken sludges from sedimentation tanks. Sludges from the lime–soda softening process are usually stored in lagoons. The drainage water is removed and the sludge may be covered or hauled off to a disposal site. Lime softening sludges may be applied to agricultural lands to achieve the best soil pH for optimum crop yields.

Larger plants or plants that produce large volumes of sludge may use *THICKENING,*[1] *CONDITIONING,*[2] and *DEWATERING*[3] processes to reduce the volume of sludge that must be handled and ultimately disposed of. Sometimes, polymers are added to sludges for conditioning before dewatering. Belt filter presses, centrifuges, filter presses, vacuum filters, solar lagoons, and sand drying beds are some of the processes used to dewater sludges.

Ultimately, process wastes such as trash, grit, sludge, and brine must be disposed of in a manner that will not harm the environment. Trash and grit may be disposed of in landfills. Sometimes, sludge and brine are discharged into wastewater collection systems (sewers); however, this procedure may cause operational problems for the wastewater treatment plant operator. To avoid upsetting wastewater treatment plants, discharges to sewers must be made very slowly to take advantage of the dilution provided by the wastewater. Sludges are commonly disposed of by spreading on land or dumping in landfills. The method used will depend on the volume of the sludge, sludge moisture content, land available, and distance from the plant to the ultimate disposal site.

1. *Thickening.* Treatment to remove water from the sludge mass to reduce the volume that must be handled.
2. *Conditioning.* Pretreatment of sludge to facilitate removal of water in subsequent treatment processes.
3. *Dewater.* (1) To remove or separate a portion of the water present in a sludge or slurry. To dry sludge so it can be handled and disposed of. (2) To remove or drain the water from a tank or a trench. A structure may be dewatered so that it can be inspected or repaired.

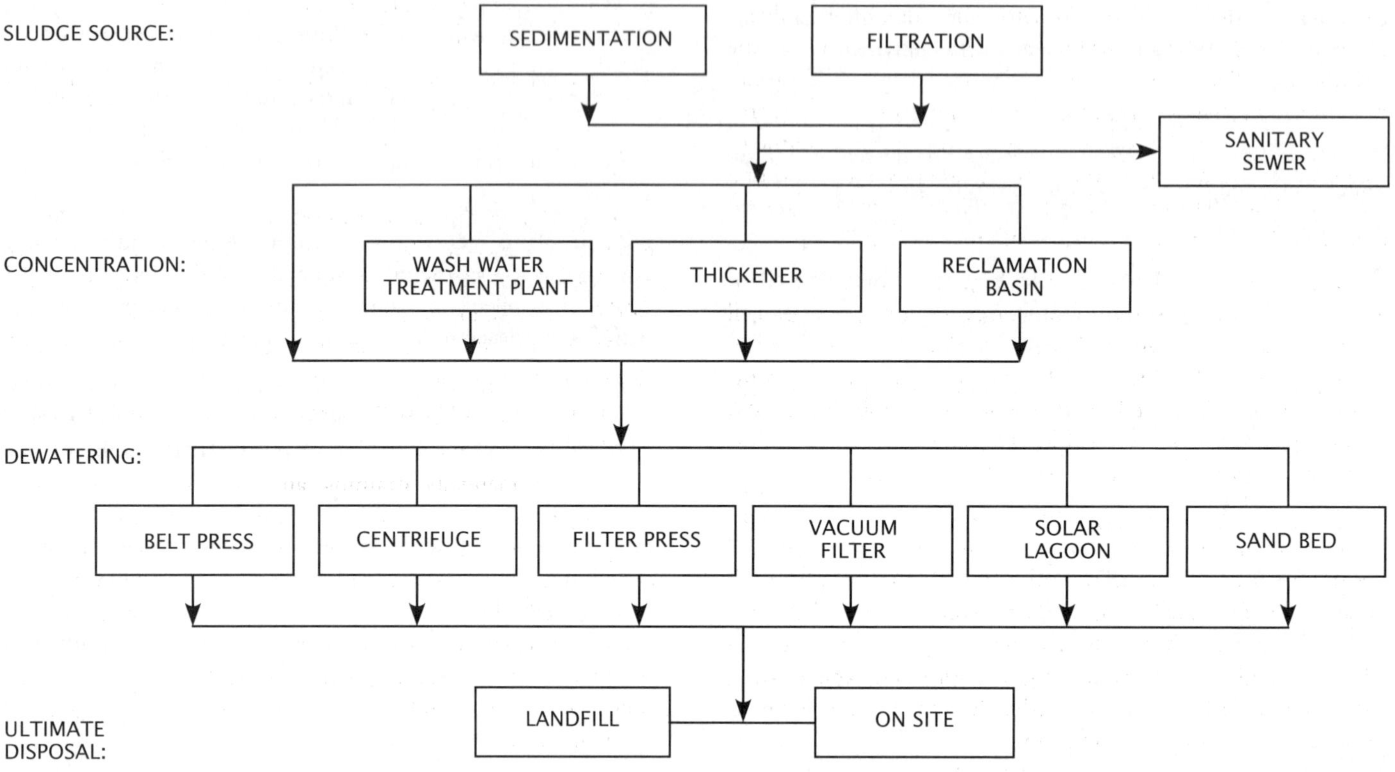

Fig. 17.1 Sludge processing alternatives

The remainder of this chapter will discuss the detailed operational procedures that an operator must consider when handling and disposing of process wastes. Sections are also provided on equipment operation and maintenance as well as on monitoring and reporting.

QUESTIONS

Please write your answers to the following questions and compare them with those on page 253.

17.3A When should sedimentation tanks be inspected and repaired?

17.3B List the methods that may be used to dewater sludge.

17.4 DRAINING AND CLEANING OF TANKS

Plants without mechanical or vacuum-type sludge collectors will require the use of manual labor to remove the sludge once or twice a year. When two or more sedimentation tanks are designed into a plant, the job of cleaning is made easier. While one sedimentation tank is down, the other(s) can remain in operation. The cleaning of sedimentation tanks should be done before or after peak demand months. Generally, early spring and the fall of the year are the better times to take some facilities out of service for cleaning.

Before draining any in-ground tank, always determine the level of the water table. If the water table is high, an empty tank could float like a cork on the water surface and cause considerable damage to the tank and piping. A properly designed tank will have provisions to drain high water tables or will contain other protective features (bottom pressure-relief disks to let groundwater into the tank to prevent damage).

After any necessary intake valve(s) or gate(s) changes are made, drain the water down to the sludge blanket by partially opening the drain valve from the sedimentation tank into the lagoon or drying bed. After the first few minutes (if the valve is not wide open), the water will become clear. This portion of the water can be diverted from the lagoon or drying bed, if proper plumbing is available, and returned to the source to be reprocessed. Most drying beds will not handle this great a volume of water unless this draining process is extended over a period of a few days. Pump(s) can be used to transfer the settled water to another sedimentation tank that is still in operation.

As the water gets down to the sludge, fully open the drain valve into the drying bed(s). A large quantity of the sludge will drain by itself. As shown in Figure 17.2, the tank wall is 10 feet (3 m) high with the sludge level showing about 5 feet (1.5 m) from the top. When the level drops down to about 2 feet (0.6 m) of depth by the drain opening, the sludge will have to be assisted by an operator with a squeegee (Figure 17.3).

During the draining stages, the walls and all the equipment should be completely hosed down and inspected for damage. All necessary repairs should be made at this time. Once sludge dries on any coated surface, it is difficult to remove so it is important to hose everything down during the process of draining and while the sludge is still wet. All gears, sprockets, and moving parts should be lubricated immediately after hosing down to prevent freeze-up resulting from exposure to the air during inspection and repair. By using drying beds and drying bed *DECANT*[4] pumps, ample amounts of water may be used for cleaning and assisting draining of the sludge. Under these conditions, two to three operators can clean out one sedimentation tank for a plant of 5 to 10 MGD (19 to 38 MLD) in 1 day. Additional time is required for initial drawdown, gathering up of tools and equipment, final cleanup, and any repairs that may be needed.

Sludge that settles out near the entrance to the sedimentation tank is more dense, especially when a polymer is used for flocculation aid. Therefore, the drain should be located in the headworks area. Once the sludge ceases to flow freely, even with the dilution water, then operators will have to push it toward the drain with squeegees (Figure 17.4).

The volume of sludge can vary with the size of the basin or clarifier, the quality of the source water being treated, the use of alum, polymer or combinations of both, and the frequency of cleaning. This volume may range from 100,000 to 200,000 gallons (0.38 to 0.76 ML), depending on the size of the basin and how long the sludge has accumulated in the basin.

The following precautions must be exercised whenever operators are in any closed tank (confined space):

1. Do not operate gasoline engines in the tank.
2. Test the atmosphere in the tank for oxygen deficiency/enrichment, flammable or explosive gases, and toxic gases, and provide adequate ventilation of clean air at all times.
3. Provide a source of water to clean off boots and tools where the operators come out of the tank.
4. Use the buddy system. At least one person must be outside the tank and watching anyone inside the tank, and another person must be nearby and available to help in an emergency.

Before filling the tank, thoroughly inspect and repair all equipment and valves. Wash everything down with clean water or a solution of 200 mg/L chlorine to disinfect the basin. If a chlorine wash solution is not used, fill the tank 10 percent full with a 50 mg/L chlorine solution and then finish filling it with clean water from the plant. The final free chlorine residual should not be so high that water with a free chlorine residual greater than 0.5 mg/L reaches the consumers.

Although manually draining and cleaning tanks requires more operator hours and plant downtime than mechanical sludge removal, it does have its advantages. A more sanitary condition in the tank is obtained by cleaning up algae buildups or other deposits that are not picked up by mechanical collectors and regular inspection of equipment can eliminate many potential breakdown conditions. Even basins with continuous sludge collection systems should be drained once a year for inspection and maintenance.

QUESTIONS

Please write your answers to the following questions and compare them with those on page 253.

17.4A How can sludge be removed from tanks without mechanical or vacuum-type sludge collectors?

17.4B When draining a sedimentation tank, what should be done with the settled water above the sludge?

17.4C What precautions must be exercised whenever an operator enters a closed tank (confined space)?

17.5 BACKWASH RECOVERY PONDS (SOLAR LAGOONS)

Because of water pollution control legislation enacted since the 1960s, many water treatment plants have backwash recovery ponds (Figure 17.5). In many instances, these ponds can serve a dual purpose. In addition to their primary function as backwash recovery ponds, they can also be used to collect the sludge from

4. *Decant* (de-KANT). To draw off the upper layer of liquid (water) after the heavier material (a solid or another liquid) has settled.

Fig. 17.2 Sludge being drained from a clarifier

Fig. 17.3 Operator with a squeegee assisting sludge out of clarifier

Fig. 17.4 Operators pushing sludge toward drain with squeegees and vacuum truck removing sludge

Fig. 17.5 Backwash recovery pond

sedimentation tanks and clarifiers with a few modifications. While these modified ponds are capable of performing both functions at the same time, it is critical to time these operations so that they do not overlap. You want to avoid having to backwash the filters while you are draining a sedimentation tank. Water for hosing down the sedimentation basin and assisting the flow of sludge should be used sparingly. Also, the backwash recovery pump suction pipe intake should be floated near the surface, by use of a flexible hose and tire tube or any similar float, so that any excess water can be recycled without also drawing out sludge. This will be very important if the filters must be backwashed at the same time sludge is being cleaned out of the backwash recovery ponds.

A vacuum tank truck will be needed to move the wet sludge (a vacuum tank truck is shown in Figure 17.4). The capacity of the vacuum truck's tank in the picture is 5,000 gallons (19 m^3) and it has a 6-inch (150 mm) suction hose. About 15 minutes are required to fill the tank if sludge is fed to the suction end constantly without breaking the vacuum. A lift of 12 feet (3.6 m) can be obtained without too much difficulty.

Sludge is sometimes applied to land as a soil conditioner. Polymer sludges are suitable as a soil conditioner. Sludges produced by direct filtration, without coagulants, usually make excellent soil conditioners both with and without polymers. The sludge may be applied either wet or dry. Because commercial soil conditioner is becoming more expensive, sale of sludge as a soil conditioner can help to offset sludge handling and disposal costs.

Most plants that use the lime–soda ash softening process collect the sludge and dewater it in a lagoon. A variable length riser or discharge pipe is used to draw off the water that is separated from the sludge. When the lagoon is full and the sludge is dried, the surface may be covered with soil as in a landfill operation. In some plants where space is scarce, the dried sludge is hauled off to a landfill and the lagoon refilled. Lime softening sludges also can be disposed of by application to agricultural soils to adjust the pH for optimum crop yields.

Discharge of lime–soda sludges to the wastewater collection system (sewers) is a poor practice because the sewers could become plugged regularly and the operator at the wastewater treatment plant will have to handle and dispose of the sludge. However, the lime–soda sludge may help the wastewater treatment plant operator by adjusting the pH or serving as a coagulant aid in treating the incoming wastewater.

QUESTIONS

Please write your answers to the following questions and compare them with those on page 253.

17.5A Why is timing critical if backwash recovery ponds are used to handle sludge from sedimentation basins?

17.5B Why should the suction pipe for the backwash recovery pump be floated near the surface of the pond?

17.5C How can lime–soda softening sludge be disposed of ultimately?

17.6 SLUDGE DEWATERING PROCESSES[5]

17.60 Solar Drying Lagoons

Solar drying lagoons are shallow, small-volume storage ponds in which treatment process sludge (sometimes concentrated) is stored for extended time periods. Sludge solids settle to the bottom of the lagoon by plain sedimentation (gravity settling) and the clear *SUPERNATANT*[6] water is skimmed off the top with the aid of an outlet structure that drains the clear surface waters. Evaporation removes additional water and the solar drying process proceeds until the sludge reaches a concentration of from 30 to 50 percent solids. At this point, the sludge can be disposed of on site or at a sanitary landfill. Obviously, the solar drying process is dependent on environmental conditions (weather) and may take many months to complete. For this reason, several lagoons should be provided (a minimum of three) so that sludge loading and drying can be rotated from one lagoon to another.

17.61 Sand Drying Beds

Sand drying beds have been used extensively in municipal wastewater treatment where high solids volumes are handled. Sand drying beds are similar in construction to a sand filter, and consist of a layer of sand, a support gravel layer, an underdrain system, and some means for manual or mechanical removal of the sludge (see Figures 17.6, 17.7, and 17.8). They are built with underdrains covered with gradations of aggregate and sand. The drains discharge into a sump where recovery pumps can return the water drained from the sludge back to the plant to be reprocessed. Frequently, three beds are used so one can be dried out while one is being filled from the drawdown of wet sludge from the sedimentation tanks. The third bed contains dried sludge that is being hauled out.

The efficiency of the sand drying bed dewatering process can be greatly improved by preconditioning the process sludge with chemical coagulants. The drying time can vary from days to weeks, depending on weather conditions and the degree of

5. Portions of this section were prepared by Jim Beard.
6. *Supernatant* (soo-per-NAY-tent). The relatively clear water layer between the sludge on the bottom and the scum on the surface of a basin or container. Also called clear zone.

Fig. 17.6 Sludge drying beds

preconditioning of the sludge. The frequency of removal of dried sludge will vary with different plants depending on the volume of sludge produced, size of drying beds, and drying conditions (weather).

Sludge has a unique characteristic about it that once it has even partially dried, it will not expand; therefore, layer after layer of wet sludge can be added over a period of time. This procedure will work as long as the solids content of the applied sludge is at least 2 to 3 percent. When drying this type of sludge, large cracks and checks will develop on the surface and extend down through the sludge to the sand. The proper time to remove this dried sludge is when no more than 1 foot (0.3 m) has accumulated and dried into a checkered pattern. A piece of dried sludge can then be picked up off the sand. The dried sludge can easily be removed with a front-end loader onto a dump truck and be hauled off to a landfill. However, the operator must exercise extreme caution so that only the dried sludge is picked up with the minimum possible disturbance to the sand and aggregate. The loader bucket should be operated carefully since there may be only about 10 inches (36 cm) of sand cover over the underdrains. The loader bucket capacity should be limited to 1 or 2 cubic yards of sludge. Concrete tracks should be provided for larger equipment to collect the dried sludge.

QUESTIONS

Please write your answers to the following questions and compare them with those on page 254.

17.6A What is the minimum recommended number of solar drying lagoons?

17.6B Describe a typical sludge drying bed.

17.6C When is the proper time to remove dried sludge from the drying bed?

17.6D What precautions must be exercised when operating a front-end loader to remove sludge from a drying bed?

17.62 Belt Filter Presses

Continuous-belt filter presses are popular because of their relative ease of operation, low energy consumption, small land requirements, and their ability to produce a relatively dry filter cake (material removed from the filter press has about 35 to 40 percent solids). There are two primary mechanisms by which free water is separated from the sludge solids in a belt press:

1. Gravity drainage
2. Pressure dewatering

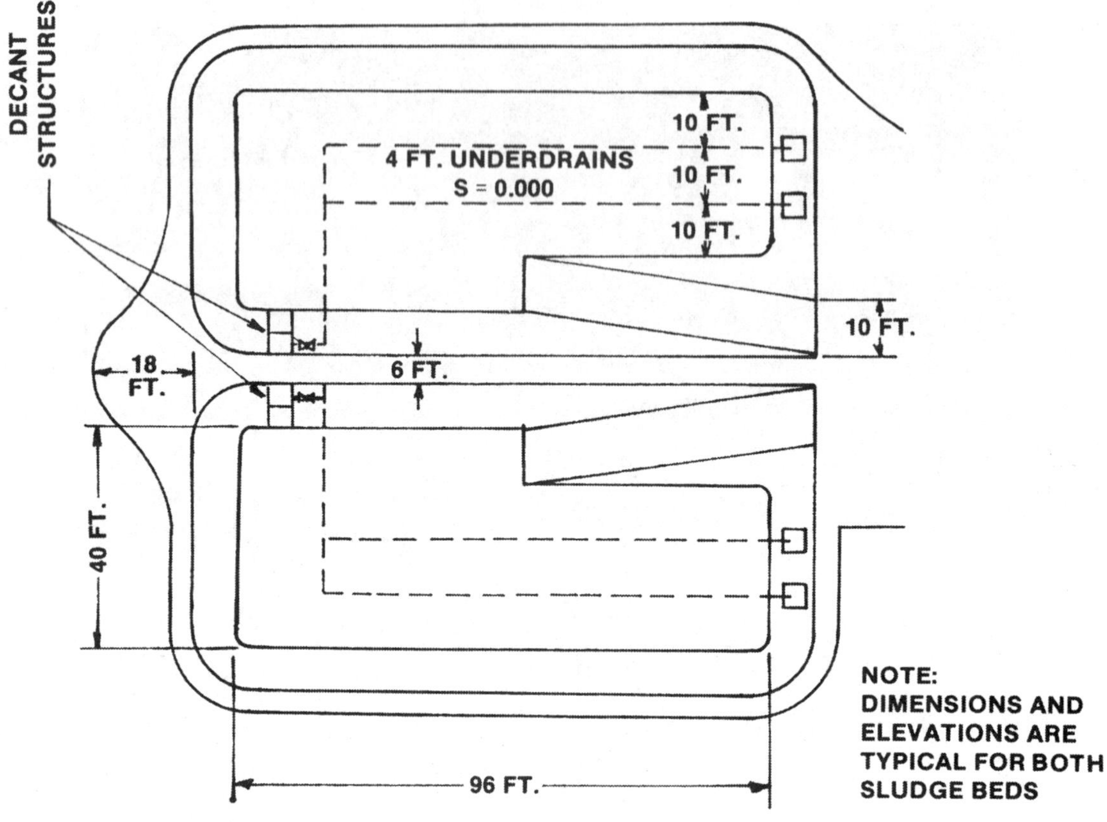

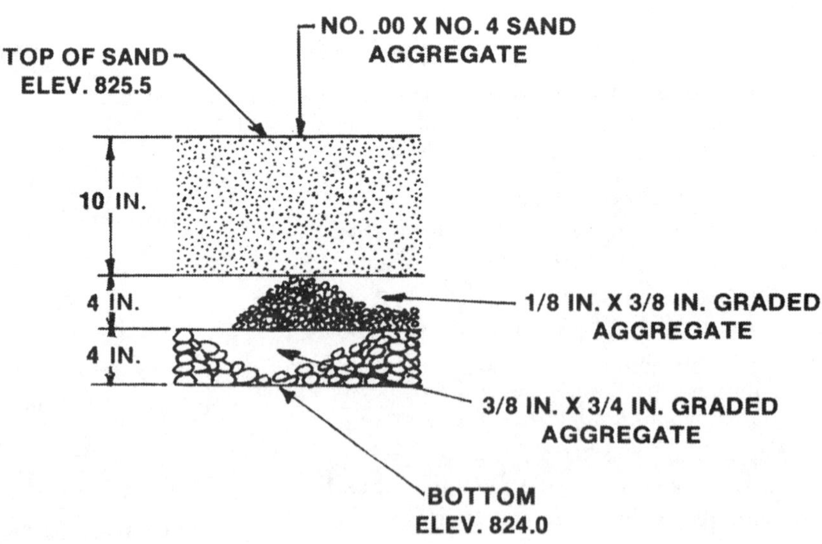

Fig. 17.7 Sludge drying beds

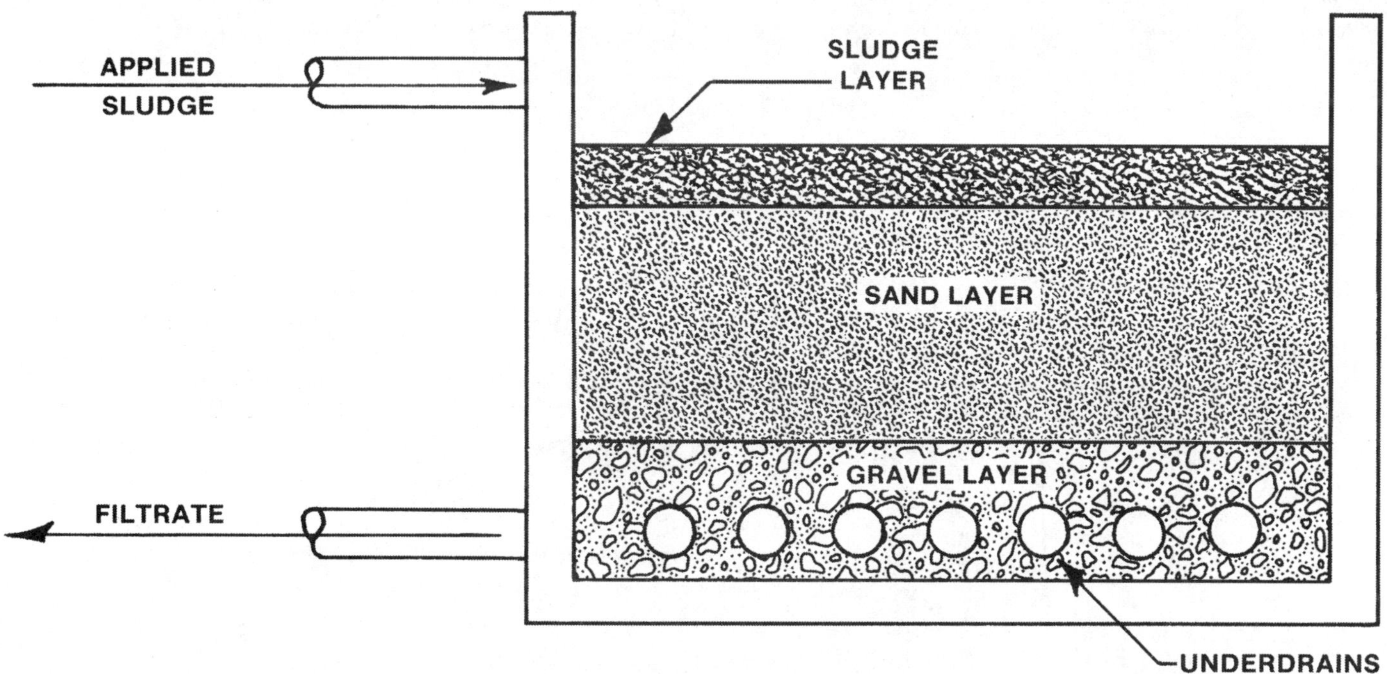

Fig. 17.8 Sectional view of sludge drying bed

Sludge is conveyed and dewatered between two endless belts (Figure 17.9). After the sludge is initially mixed with a polymer in a rotary drum conditioner, it is dewatered in three distinct zones:

1. A horizontal zone for gravity drainage
2. A vertical sandwich draining zone
3. A final dewatering zone containing an arrangement of staggered rollers that produce a multiple-shear force action that squeezes out the remaining free water

Each belt is washed with a high-pressure/low-volume water spray.

17.63 Centrifuges

A *CENTRIFUGE* is a mechanical device that uses centrifugal or rotational forces to separate solids from liquids. Centrifuges have been used to dewater municipal sludges for some time. Problems with the earlier units included erosion of surfaces hit by high-speed particles, and poor performance capacity. Design improvements and the use of polymers have generally eliminated these problems. The principal advantage of this dewatering technique is that the density of the sludge cake can be varied from a thickened liquid slurry to a dry cake. The major limitation of using centrifuges is high energy consumption.

There are two basic types of centrifuges, the scroll type and the basket type. The scroll centrifuge operates continuously, while the basket centrifuge is a batch-type unit. Solids capture is generally greater with the basket centrifuge.

In the scroll centrifuge (Figure 17.10), solids are introduced horizontally into the center of the unit. The spinning action forces the solids against the outer wall of the bowl, where they are transported to the discharge end by a rotating screw conveyor. Clear supernatant liquid is discharged over an adjustable weir on the opposite end of the unit.

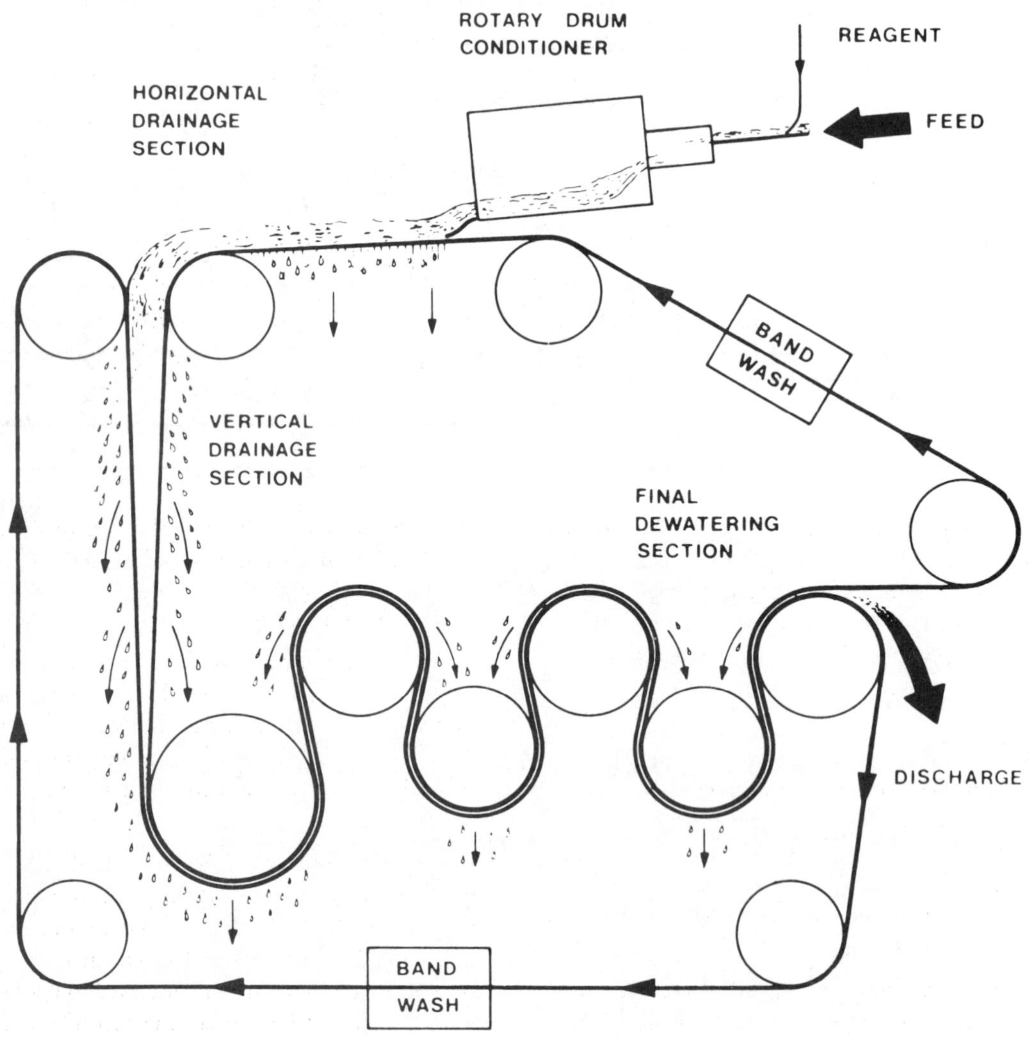

Fig. 17.9 Flow diagram of Winklepress (a continuous-belt filter press)
(Permission of Ashbrook-Simon-Hartley)

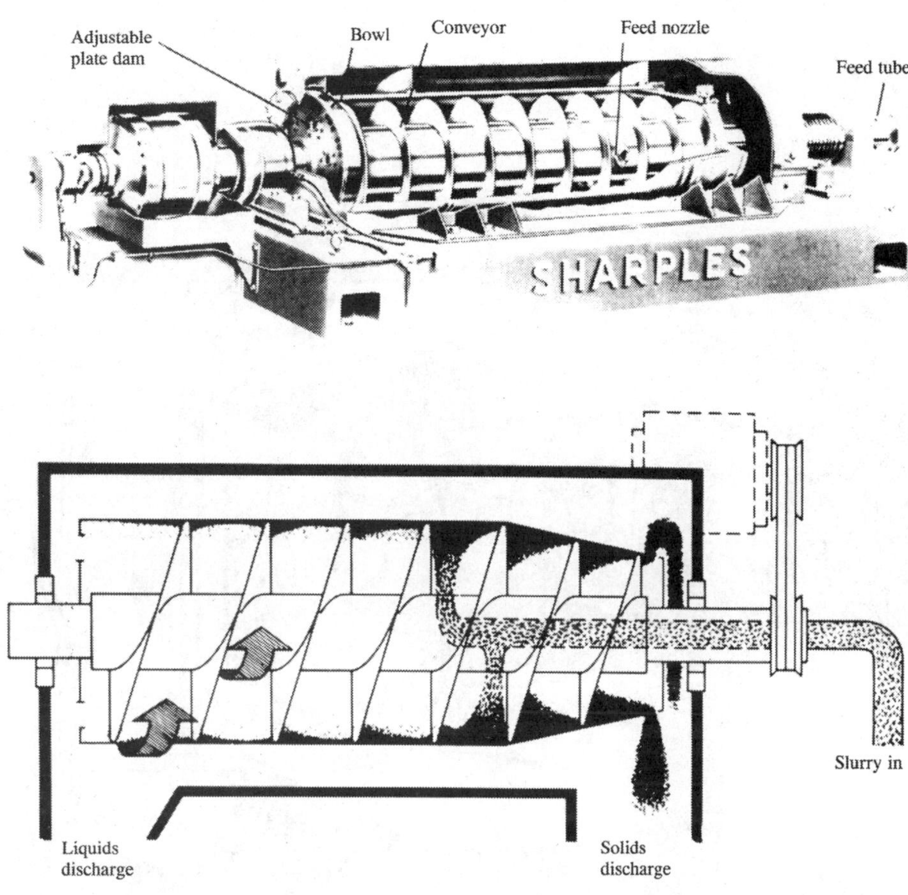

Fig. 17.10 Solid-bowl scroll centrifuge
(Permission of Sharples-Stokes Division, Penwalt Corporation)

In the basket centrifuge (Figure 17.11), sludge is introduced vertically into the bottom of the bowl and the supernatant is discharged over a weir at the top of the bowl. When the solids concentration in the supernatant becomes too high, the operation is stopped and the dense solids cake is removed by a knife unloader.

17.64 Filter Presses

Filter presses have been successfully used to process difficult-to-dewater sludges (alum sludges). These machines are best suited for sludges with a high specific resistance (the internal resistance of a sludge cake to the passage of water). Filter presses produce very dry cakes, a clear filtrate, and have a very high solids capture.

A filter press consists of a series of vertical plates covered with cloth that supports and retains the filter cake (Figure 17.12). These plates are rigidly held in a frame. Sludge is fed into the press at increasing pressures for about half an hour. The plates are then pressed together for 1 to 4 hours at pressures as high as 225 psi (15.8 kg/cm^2 or 1,551 kiloPascals). Water passes through the cloth while the solids are retained, forming a cake that is removed when the press is depressurized.

17.65 Vacuum Filters

Vacuum filtration was once the main chemical sludge dewatering process. However, its use has declined due to development of devices such as the belt press, which consumes less energy, is less sensitive to polymer dosage, and does not require use of a precoat (a substance applied to the filter before applying sludge for dewatering).

A vacuum filter consists of a cylindrical drum that rotates, partially submerged, in a tank of chemically conditioned sludge (Figure 17.13). As the drum slowly rotates, a vacuum is applied under the filter medium (belt) to form a cake on the surface. As the belt rotates, suction is maintained to promote additional dewatering. As the belt passes the top of the drum, it separates from the drum and passes over a small-diameter roller for discharge of the cake. The belt is then washed before it reenters the vat. A precoat of diatomaceous earth is required to dewater gelatinous (jelly-like) alum sludge.

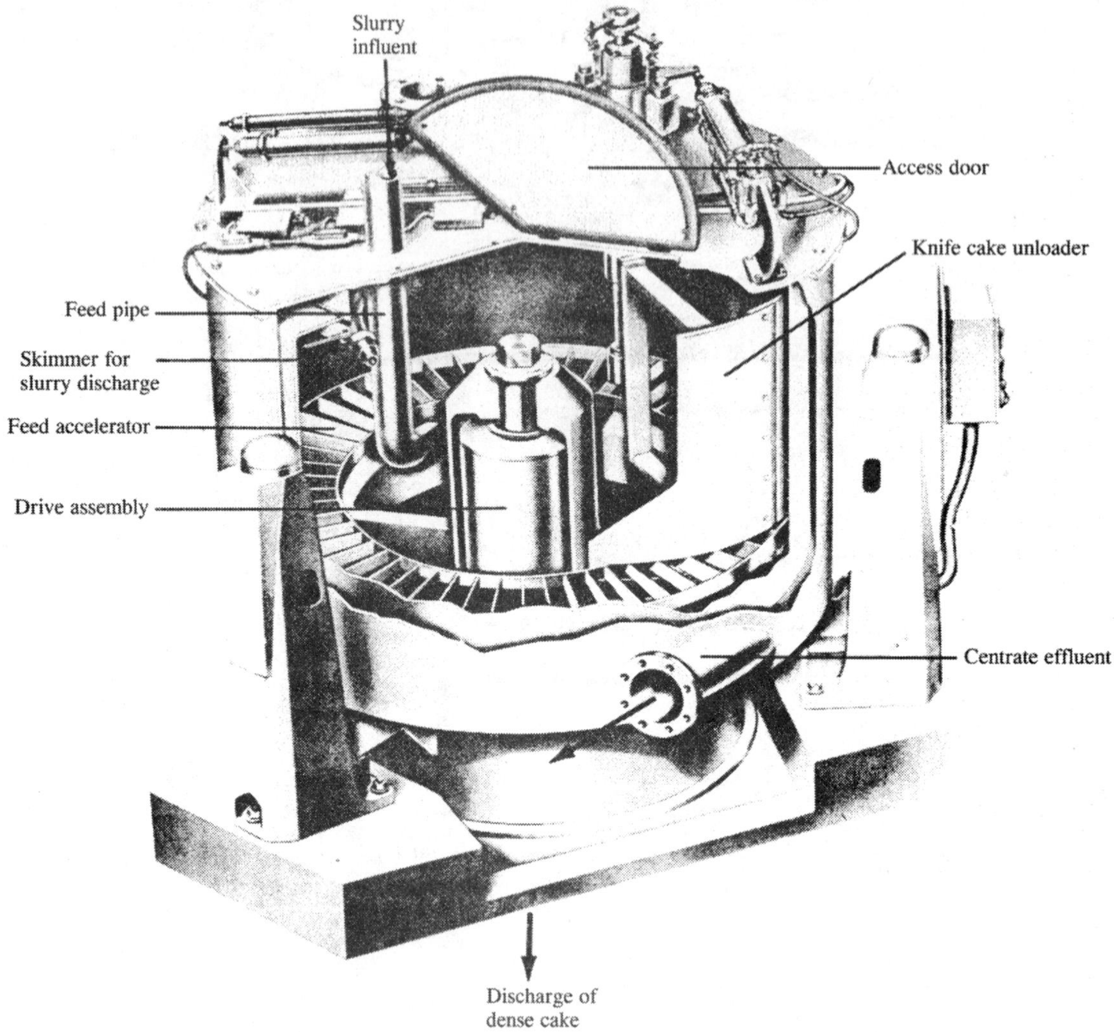

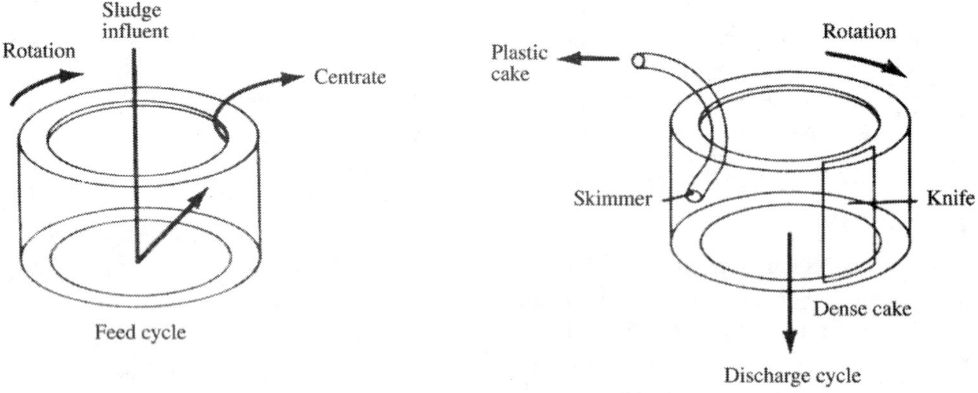

Fig. 17.11 Imperforate basket centrifuge

(Permission of Sharples-Stokes Division, Penwalt Corporation)

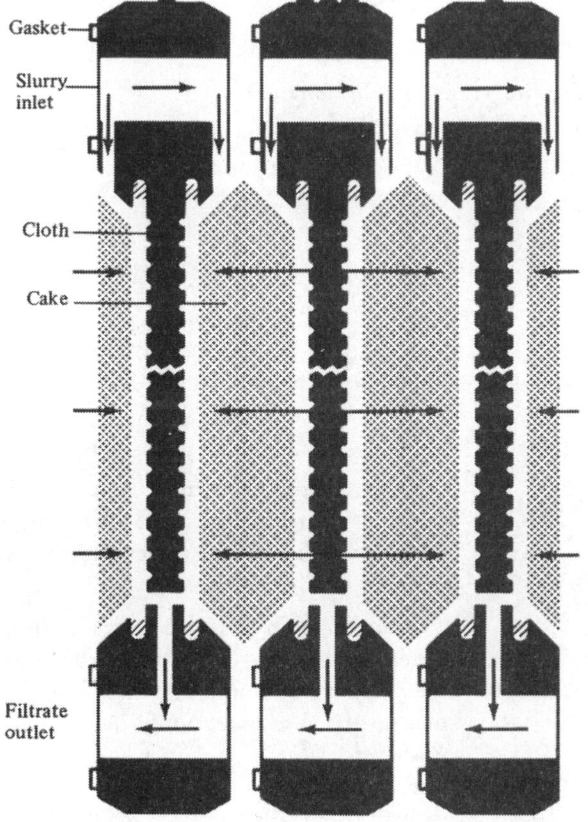

FILTER CHAMBER ASSEMBLY

Fig. 17.12 Filter press

(Permission of Metropolitan Water District of Southern California)

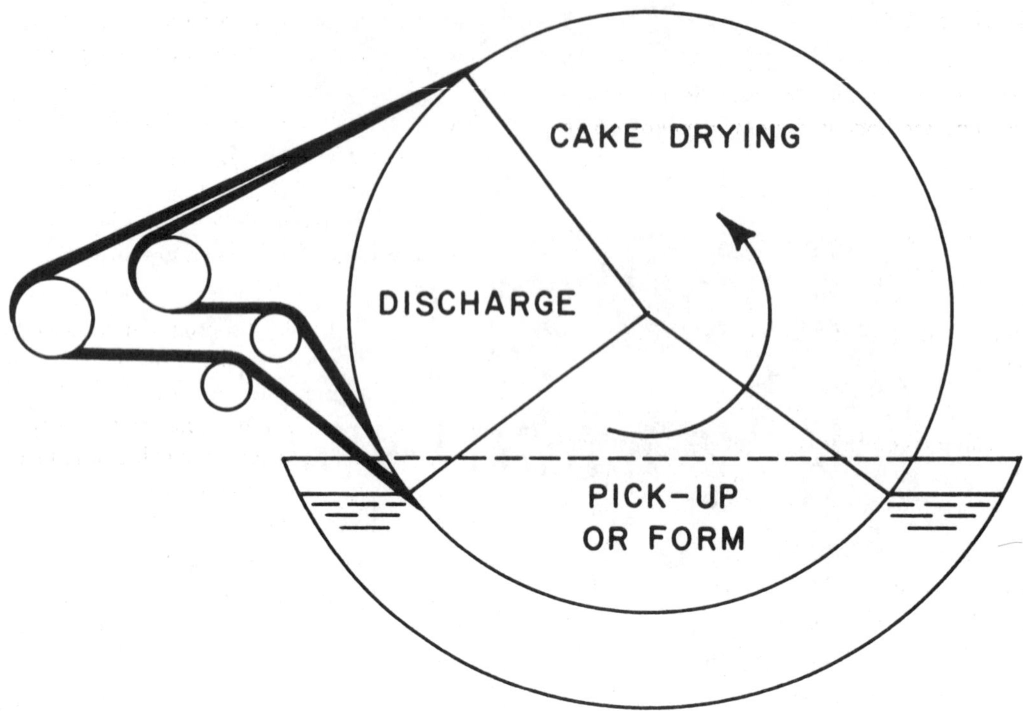

Fig. 17.13 Operating zones of a vacuum filter
(Permission of Metropolitan Water District of Southern California)

QUESTIONS

Please write your answers to the following questions and compare them with those on page 254.

17.6E List the major advantages and limitations of using centrifuges to dewater sludges.

17.6F When is a precoat of diatomaceous earth required on vacuum filters?

17.7 DISCHARGE INTO COLLECTION SYSTEMS (SEWERS)

The easiest method of sludge disposal would be to send the sludge down the wastewater collection (sewer) system. This does create some complications even if the wastewater treatment plant has the capacity to handle the load. Sludge discharged to sewers could cause sewer blockages. Also, the fees charged by the wastewater treatment plant to process and dispose of the sludge could be prohibitive. The charges are usually based upon annual flow, chemical or biochemical oxygen demand, suspended solids, and peak and average discharge. There are also increased monitoring requirements and costs associated with a sewer discharge. The water treatment plant must have a holding tank so that the sludge can be released at a uniform rate throughout the day or released only during the wastewater treatment plant's low-flow period.

Brine from ion exchange units may be discharged into wastewater collection systems. Usually, the brine is discharged during the day to take advantage of high flows for dilution. When you plan such a discharge, notify the operator of the downstream wastewater treatment plant to be sure you will not create any unnecessary problems.

QUESTIONS

Please write your answers to the following questions and compare them with those on page 254.

17.7A What are the complications of discharging sludge to sewers?

17.7B When is brine from ion exchange units usually discharged into wastewater collection systems?

17.8 DISPOSAL OF SLUDGE[7]

Sludge is commonly disposed of in sanitary landfills. Other methods of ultimate disposal include lagoons, land application, and sanitary sewers. The method of disposal depends on the source and type of sludge, as well as economic, environmental, and community awareness/sensitivity considerations.

Water treatment plant lime sludge is an excellent liming agent for agricultural purposes. Lime sludge must be applied at a rate to achieve the best soil pH for optimum crop yield. Optimum levels of nitrogen and phosphorus are also important to achieve high crop yields.

The application of nitrogen fertilizers causes a reduction in soil pH. If optimum soil pH conditions do not exist, crop yields will be reduced. Therefore, sufficient quantities of lime must be applied as a means of counteracting the fertilizer applications.

Lime softening sludges can also aid in the reclamation of spoiled lands by neutralizing acid soils. Disposal of lime softening sludge on strip mine land will help minimize the discharge of acidic compounds and low pH drainage waters.

Although most lime softening sludges are an excellent liming agent for agricultural and land reclamation purposes, some lime softening sludges must be disposed of in a sanitary landfill due to the lack of availability of agricultural land or excessive costs. Landfilling of lime softening sludge is a practical alternative where this method is cost effective (minimum cost of disposal).

Alum sludge has a tendency to cause soils to harden and does not provide any beneficial value. For this reason, water treatment plant alum sludge must not be applied to agricultural land. The sludge may be applied to a dedicated land disposal site. The sludge is applied to the land and plowed into the soil. Landfilling is another method of ultimate disposal of alum sludge.

Some water treatment plants use a slow sand filter or settling pond for the treatment of iron filter backwash wastewater. The slow sand filter must be cleaned occasionally by removing the top 2 to 3 inches (50 to 75 mm) of sand and iron sludge. The material removed must be disposed of in a sanitary landfill.

Water treatment plants that soften water by the ion exchange (zeolite) softening method produce a wastewater that has high concentrations of total dissolved solids and chloride compounds. Ion exchange softening wastes should not be discharged untreated into low-flow streams. The wastewater treatment processes capable of reducing the total dissolved solids and chloride concentrations to acceptable levels are very energy consumptive and expensive. Ion exchange softener regeneration wastewater may be very carefully and slowly discharged into a sanitary sewer system. The operator at the wastewater treatment plant must be notified in advance.

If a sanitary sewer system is not available for the disposal of ion exchange softening wastes, holding tanks should be installed to store the liquid from the regeneration and rinse cycles. This liquid should be ultimately disposed of in a sanitary landfill.

Filter backwash wastewater must be recycled through the water treatment plant or placed in wastewater storage ponds for additional treatment and disposal. The remainder of this section discusses some of the procedures used by operators to dispose of sludges.

Wet sludge can be disposed of in an open field by use of spray bars or dumped out for landfill (Figure 17.14). When releasing the wet sludge at one spot, the back of the truck should face downhill so it will drain faster and be emptied out completely (Figure 17.15). This practice will also reduce the chances of the truck getting stuck in the sludge. Usually, it takes about 10 minutes to empty a truck.

Sometimes, individuals will request the sludge for fill and some contractors have used the sludge to mix with decomposed granite (DG) (a type of rock found in some regions) for fill purposes.

Sludge drying beds for sedimentation tank wastes also can be used to dry sludge from nearby backwash recovery ponds, but this requires the sludge to be handled for a second time. An open field spray bar application is one method for disposing of backwash recovery sludges because after a few weeks, the residual is hardly noticeable. PVC pipe may be used to dispose of backwash recovery sludges instead of using a spray bar.

Where sludge is repeatedly spread in a single landfill site, be prepared to mix the sludge in with the native soil because it is unsightly. Sludge should be disposed of as close as possible to the water treatment plant to reduce hauling costs. Process wastes also can be dried in lagoons and covered over permanently.

In plants with a size range from 5 to 10 MGD (19 to 38 MLD), it will take four operators approximately 2 days to complete the job of draining and cleaning a sedimentation tank or a backwash recovery pond and disposing of the sludge.

In plants where a backwash pond is not available, the sludge can be moved to a sump. A smaller suction hose must be used to empty small sumps; instead of the 6-inch (150 mm) hose, use either a 3- or 4-inch (75 or 100 mm) hose. With a smaller suction hose, it will take 25 minutes or more to fill the vacuum tank. Five or six operators will be needed to keep the sump filled. Of course, during the time that the truck is on the road to the dump site and back again, the operators are standing by and the sludge cannot be moved. If two trucks were used, this disadvantage could be overcome, but the cost may also increase. Therefore, the larger the sump and the truck's tank, the cheaper the operation from the standpoint of labor costs.

7. Portions of this section were obtained from *Illinois EPA Sludge Regulation Guidance Document*, Illinois Environmental Protection Agency, 1021 North Grand Avenue East, PO Box 19276, Springfield, IL 62794-9276.

Fig. 17.14 Use of sludge for a landfill

Fig. 17.15 Emptying sludge from a truck

Plants of 1 MGD (3.8 MLD) capacity may be able to use some of the local septic tank pumpers or vacuum trucks to an advantage. Such companies usually have made arrangements for the use of dump sites that may be used to dispose of the sludge they pump.

QUESTIONS

Please write your answers to the following questions and compare them with those on page 254.

17.8A List the methods of ultimate sludge disposal.

17.8B Why should sludge be disposed of as close as possible to the water treatment plant?

17.9 EQUIPMENT

17.90 Vacuum Trucks

The use of a vacuum truck is highly recommended because these trucks develop the fewest problems with clogging. The larger the suction pipe the better. Any object smaller than the hose size can be readily sucked through unless there are too many objects in the sludge. Wedging of several rocks or sticks can occur, slowing up the process of sludge removal. Any object that can be seen should be removed by hand and discarded. Any well-operated plant should be void of these solids, but they are sometimes accidentally dropped in or thrown in by vandals. Leaves and other small objects that may get by the plant inlet screens are not too much of a problem unless an excessive amount exists. Under these conditions, the heavily accumulated portions should be scooped out to prevent any clogging of drain pipes or suction hoses.

17.91 Sludge Pumps

Many small treatment plants exist, especially in rural areas, where disposal of sludge would appear less troublesome. A couple of thousand gallons of sludge can be moved by gravity or pumped out to an open field and mixed (disked) into the soil. Even a quarter of an acre can handle many years of dried sludge from a small plant. Unfortunately, most of these plants do not have the land available, so the sludge in its wet form must be hauled out to a disposal site or a small lagoon or pond must be excavated for sludge collection. If sludge must be moved wet, use a sludge pump[8] to pump it into a tank or hire a septic tank pumper. Septic tank pumps, especially the suction hose, must be thoroughly rinsed and disinfected before use in any water treatment plant facilities. This applies also to all tools and equipment used.

A sludge pump can be an advantage over a self-priming centrifugal pump because of its large suction and discharge hose (usually 3 inches [75 mm] in diameter). These pumps are rated at about 60 GPM (3.8 L/sec) and can handle more solids. However, there are also self-priming types of wastewater pumps available that are designed so that solids do not actually pass through the impellers. These pumps may be used to pump sludges from water treatment plants. In either case, an excessive amount of foreign solids other than sludge itself can cause some pumping problems.

Wet sludge from a sedimentation tank will flow through pipes by gravity even if there are some ups and downs in the line provided the sludge is under some head. If difficulty arises, add some water for dilution or raise the end of the suction hose closer to the surface of the sump where the sludge is more diluted.

Because of the differences in volume between wet and dried sludge, it is always preferable to contain the wet sludge at the plant for drying. If the plant water source is a canal or reservoir close to the site, construct a small pond parallel to it and return the supernatant to the source for recycling by removing baffle boards. Only fresh backwash wastewater should be recycled. Water separated or drained from old sludge can cause serious problems if recycled. This water is likely to be *SEPTIC*[9] and could cause taste and odor problems. Also, this water will contain millions of bacteria and microorganisms that should not be recycled through a water treatment plant.

If necessary, semidried or dried sludge can be hauled out of the pond.

17.10 PLANT DRAINAGE WATERS

There are several sources of drainage waters in a water treatment plant that must be properly handled and disposed of. These sources include the laboratory, shops, and plant drainage water from leaks and spills. If continuous sampling pumps provide the lab with continuously flowing water from various plant processes, this water could be discharged to a sewer. Any reagents, toxics, or potentially pathogenic wastes from the lab must be properly treated and packaged before ultimate disposal in landfills. Drainage waters from leaks and other sources in the plant

8. Types of sludge pumps include diaphragm, nonclog, and progressive cavity pumps.
9. *Septic* (SEP-tick). A condition produced by bacteria when all oxygen supplies are depleted. If severe, the bottom deposits produce hydrogen sulfide, the deposits and water turn black, give off foul odors, and the water has a greatly increased chlorine demand.

may be discharged to sewers. These drainage waters may be recycled through the plant; however, extreme caution must be exercised at all times to avoid contributing to taste and odor problems, health hazards, or operational problems.

17.11 MONITORING AND REPORTING

The location of the water treatment plant and the methods used to ultimately dispose of the process wastes will dictate the monitoring and reporting requirements. These reporting requirements may be established by local or state health or pollution control agencies. Monitoring and reporting will usually involve measuring and recording volumes of sludges or brines, percent solids, and other measurements that will prove that these processes are not creating any adverse environmental impacts.

QUESTIONS

Please write your answers to the following questions and compare them with those on page 254.

17.9A How do objects that plug sludge suction hoses get into water treatment plants?

17.9B What types of pumps can be used to pump wet sludge into a tank on a truck?

17.10A List the sources of plant drainage waters.

17.11A What factors will dictate the monitoring and reporting requirements for your sludge disposal program?

Please answer the discussion and review questions next.

DISCUSSION AND REVIEW QUESTIONS

Chapter 17. HANDLING AND DISPOSAL OF PROCESS WASTES

Please write your answers to the following discussion and review questions to determine how well you understand the material in the chapter.

1. The amount of sludge accumulation at a typical water treatment plant depends on what factors?
2. What items should be considered when determining the frequency of cleaning a sedimentation basin?
3. What happens to the water separated from sludges in backwash recovery ponds or lagoons?
4. What precaution must be taken before draining any inground tank?
5. What duties should be performed by operators as the sludge is being drained from a sedimentation tank?
6. How would you fill a sedimentation tank after the tank has been emptied, inspected, and the necessary repairs completed?
7. How is sludge from the lime–soda ash softening process dewatered?
8. Why are three sand drying beds frequently installed at water treatment plants?
9. Why should the back of the sludge truck face downhill when releasing wet sludge at one point?
10. What is the purpose of monitoring and reporting for a sludge disposal program?

SUGGESTED ANSWERS

Chapter 17. HANDLING AND DISPOSAL OF PROCESS WASTES

Answers to questions on page 233.

17.0A Strict laws are needed regarding the disposal of process wastes to prevent rivers and streams from becoming more polluted. These laws are designed to prevent any waste discharge that could discolor, pollute, or generally be harmful to aquatic or plant life or the environment.

17.0B Current regulations require daily monitoring of any discharge and analysis of such water quality indicators as pH, turbidity, TDS, settleable solids, or other harmful materials.

Answers to questions on page 234.

17.1A Sludge is removed from sedimentation tanks by mechanical rakes or scrapers that periodically remove sludge from a hopper, or a vacuum-type sludge removal device may be used.

17.1B Sludge is removed from upflow solids-contact units through sludge draw-off lines that must be monitored.

Answers to questions on page 235.

17.2A A source water stabilizing reservoir can reduce the volume of sludge handled by reducing the turbidity in the water being treated. Lower turbidities reduce the amount of alum required and thus the volume of sludge that settles out.

17.2B Sedimentation tanks should be drained and cleaned at least twice a year and more often if the sludge buildup interferes with the treatment processes.

Answers to questions on page 236.

17.3A Sedimentation tanks should be inspected and repaired (if necessary) when the tanks are drained and cleaned.

17.3B Sludge may be dewatered by the use of belt filter presses, centrifuges, filter presses, vacuum filters, solar lagoons, and sand drying beds.

Answers to questions on page 237.

17.4A Sludge must be removed manually from tanks without mechanical or vacuum-type sludge collectors. The tanks must be drained and then operators must push the sludge to the drain lines with squeegees or the sludge must be pumped out into tank trucks.

17.4B When draining a sedimentation tank, after any necessary intake valve(s) or gate(s) changes are made, drain the water down to the sludge blanket by partially opening the drain valve from the sedimentation tank into the lagoon or drying bed. After the first few minutes (if the valve is not wide open), the water will become clear. This portion of the water can be diverted from the lagoon or drying bed, if proper plumbing is available, and returned to the source to be reprocessed. Most drying beds will not handle this great a volume of water unless this draining process is extended over a period of a few days. Pump(s) can be used to transfer the settled water to another sedimentation tank that is still in operation.

17.4C The following precautions must be exercised whenever operators are in any closed tank (confined space):

1. Do not operate gasoline engines in the tank.
2. Test the atmosphere in the tank for oxygen deficiency/enrichment, flammable or explosive gases, and toxic gases, and provide adequate ventilation of clean air at all times.
3. Provide a source of water to clean off boots and tools where the operators come out of the tank.
4. Use the buddy system. At least one person must be outside the tank and watching anyone inside the tank, and another person must be nearby and available to help in an emergency.

Answers to questions on page 240.

17.5A Timing is critical if backwash recovery ponds are used to handle sludge from sedimentation basins because you want to avoid having to backwash the filters while you are draining a sedimentation tank.

17.5B The suction pipe for the backwash recovery pump should be floated near the surface of the pond so that any excess water can be recycled without also drawing out sludge. This will be very important if the filters must be backwashed at the same time sludge is being cleaned out of the backwash recovery ponds.

17.5C Lime–soda softening sludge can be disposed of ultimately by:

1. Covering the lagoon with soil
2. Hauling the dried sludge to a landfill
3. Spreading on agricultural soils to adjust the pH for optimum crop yields

Answers to questions on page 241.

17.6A The minimum recommended number of solar drying lagoons is three.

17.6B Sludge drying beds are made with underdrains covered with gradations of aggregate and sand. The drains discharge into a sump where recovery pumps can return the water drained from the sludge back to the plant to be reprocessed.

17.6C The proper time to remove dried sludge from the drying bed is when no more than 1 foot of sludge has accumulated and dried into a checkered pattern. A piece of dried sludge can then be picked up off the sand.

17.6D When operating a front-end loader to remove sludge from a drying bed, be careful so only the dried sludge is picked up with the minimum possible disturbance to the sand and aggregate. The loader bucket capacity should be limited to 1 or 2 cubic yards of sludge.

Answers to questions on page 248.

17.6E The principal advantage of using centrifuges to dewater sludges is that the density of the sludge cake can be varied from a thickened liquid slurry to a dry cake. The major limitation of using centrifuges is high energy consumption.

17.6F A precoat of diatomaceous earth is required to dewater gelatinous alum sludge.

Answers to questions on page 248.

17.7A Discharging sludge to sewers creates some complications even if the wastewater treatment plant has the capacity to handle the load. Sludge discharged to sewers could cause sewer blockages. Also, the fees charged by the wastewater treatment plant to process and dispose of the sludge could be prohibitive. There are also increased monitoring requirements and costs associated with a sewer discharge. The water treatment plant must have a holding tank so that the sludge can be released at a uniform rate throughout the day or released only during the wastewater treatment plant's low-flow period.

17.7B Brine from ion exchange units is usually discharged into wastewater collection systems during the day to take advantage of high flows for dilution.

Answers to questions on page 251.

17.8A Methods of ultimate sludge disposal include disposal in sanitary landfills, lagoons, land application, and discharge to sanitary sewers.

17.8B Sludge should be disposed of as close as possible to the water treatment plant to reduce hauling costs.

Answers to questions on page 252.

17.9A Objects that plug sludge suction hoses get into water treatment plants by being accidentally dropped in or thrown in by vandals.

17.9B If sludge must be moved wet, use a sludge pump to pump it into a tank. Types of sludge pumps include diaphragm, nonclog, and progressive cavity pumps.

17.10A Sources of plant drainage waters include the laboratory, shops, and plant drainage water from leaks and spills.

17.11A Monitoring and reporting requirements for a sludge disposal program are dictated by the location of the water treatment plant and the methods used to ultimately dispose of the process wastes.

CHAPTER 18

MAINTENANCE

by

Parker Robinson

TABLE OF CONTENTS

Chapter 18. MAINTENANCE

LESSON 5

LEARNING OBJECTIVES

Chapter 18. MAINTENANCE

Following completion of Chapter 18, you should be able to:

1. Develop a maintenance program for your plant, including equipment, buildings, grounds, channels, and tanks.
2. Start a maintenance recordkeeping system that will provide you with information to protect equipment warranties, to prepare budgets, and to satisfy regulatory agencies.
3. Schedule maintenance of equipment at proper time intervals.
4. Perform maintenance as directed by manufacturers.
5. Recognize symptoms that indicate equipment is not performing properly, identify the source of the problem, and take corrective action.
6. Recognize the serious consequences that could occur when inexperienced, unqualified, or unauthorized persons attempt to troubleshoot or repair electrical panels, controls, circuits, wiring, or equipment.
7. Communicate with electricians by indicating possible causes of problems in electrical panels, controls, circuits, wiring, and motors.
8. Properly select and use the following pieces of equipment (if qualified and authorized):
   a. Multimeter
   b. Ammeter
   c. Megger
   d. Ohmmeter
9. Safely operate and maintain auxiliary electrical equipment, including during standby and emergency situations.
10. Describe how a pump is put together.
11. Discuss the application or use of different types of pumps.
12. Start and stop pumps.
13. Maintain the various types of pumps.
14. Operate and maintain a compressor.
15. Develop and conduct an equipment lubrication program.
16. Start up, operate, maintain, and shut down gasoline engines; diesel engines; and heating, ventilating, and air conditioning systems.

NOTE: Special maintenance information is given in the previous chapters on treatment processes, where appropriate.

WORDS

Chapter 18. Maintenance

AIR GAP | AIR GAP

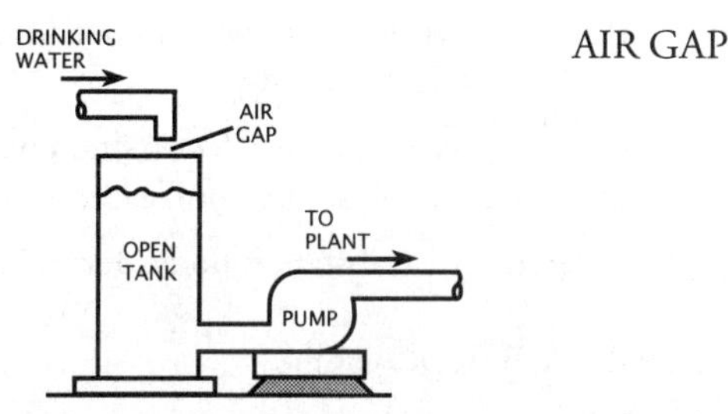

An open, vertical drop, or vertical empty space, between a drinking (potable) water supply and the nonpotable point of use. This gap prevents the contamination of drinking water by backsiphonage because there is no way wastewater can reach the drinking water supply. Air gap devices are also used to provide adequate space above the top of a manhole and the end of the hose from the fire hydrant.

ALTERNATING CURRENT (AC) | ALTERNATING CURRENT (AC)

An electric current that reverses its direction (positive/negative values) at regular intervals.

AMPERAGE (AM-purr-age) | AMPERAGE

The strength of an electric current measured in amperes. The amount of electric current flow, similar to the flow of water in gallons per minute.

AMPERE (AM-peer) | AMPERE

The unit used to measure current strength. The current produced by an electromotive force of 1 volt acting through a resistance of 1 ohm.

AMPLITUDE | AMPLITUDE

The maximum strength of an alternating current during its cycle, as distinguished from the mean or effective strength.

AXIAL TO IMPELLER | AXIAL TO IMPELLER

The direction in which material being pumped flows around the impeller or flows parallel to the impeller shaft.

AXIS OF IMPELLER | AXIS OF IMPELLER

An imaginary line running along the center of a shaft (such as an impeller shaft).

BRINELLING (bruh-NEL-ing) | BRINELLING

Tiny indentations (dents) high on the shoulder of the bearing race or bearing. A type of bearing failure.

CATHODIC (kath-ODD-ick) PROTECTION | CATHODIC PROTECTION

An electrical system for prevention of rust, corrosion, and pitting of metal surfaces that are in contact with water, wastewater, or soil. A low-voltage current is made to flow through a liquid (water) or a soil in contact with the metal in such a manner that the external electromotive force renders the metal structure cathodic. This concentrates corrosion on auxiliary anodic parts, which are deliberately allowed to corrode instead of letting the structure corrode.

CAVITATION (kav-uh-TAY-shun) | CAVITATION

The formation and collapse of a gas pocket or bubble on the blade of an impeller or the gate of a valve. The collapse of this gas pocket or bubble drives water into the impeller or gate with a terrific force that can knock metal particles off and cause pitting on the impeller or gate surface. Cavitation is accompanied by loud noises that sound like someone is pounding on the impeller or gate with a hammer.

CIRCUIT CIRCUIT

The complete path of an electric current, including the generating apparatus or other source; or, a specific segment or section of the complete path.

CIRCUIT BREAKER CIRCUIT BREAKER

A safety device in an electric circuit that automatically shuts off the circuit when it becomes overloaded. The device can be manually reset.

CONDUCTOR CONDUCTOR

(1) A pipe that carries a liquid load from one point to another point. In a wastewater collection system, a conductor is often a large pipe with no service connections. Also called a conduit. Also see INTERCEPTOR (INTERCEPTING) SEWER or INTERCONNECTOR.

(2) In plumbing, a pipe installed to drain water from the roof gutters or roof catchment to the storm drain or other means of disposal. Also called a downspout, roof leader, or roof drain.

(3) In electricity, a substance, body, device, or wire that readily conducts or carries electric current. Also called a conduit.

COULOMB (KOO-lahm) COULOMB

A measurement of the amount of electrical charge carried by an electric current of 1 ampere in 1 second. One coulomb equals about 6.25×10^{18} electrons (6,250,000,000,000,000,000 electrons).

CROSS-CONNECTION CROSS-CONNECTION

(1) A connection between a drinking (potable) water system and an unapproved water supply.

(2) A connection between a storm drain system and a sanitary collection system.

(3) Less frequently used to mean a connection between two sections of a collection system to handle anticipated overloads of one system.

CURRENT CURRENT

A movement or flow of electricity. Electric current is measured by the number of coulombs per second flowing past a certain point in a conductor. A coulomb is equal to about 6.25×10^{18} electrons (6,250,000,000,000,000,000 electrons). A flow of 1 coulomb per second is called 1 ampere, the unit of the rate of flow of current.

CYCLE CYCLE

A complete alternation of voltage or current in an alternating current (AC) circuit.

DATEOMETER (day-TOM-uh-ter) DATEOMETER

A small calendar disk attached to motors and equipment to indicate the year in which the last maintenance service was performed.

DIRECT CURRENT (DC) DIRECT CURRENT (DC)

Electric current flowing in one direction only and essentially free from pulsation. Also see ALTERNATING CURRENT.

ELECTROLYTE (ee-LECK-tro-lite) ELECTROLYTE

A substance that dissociates (separates) into two or more ions when it is dissolved in water.

ELECTROMOTIVE FORCE (EMF) ELECTROMOTIVE FORCE (EMF)

The electrical pressure available to cause a flow of current (amperage) when an electric circuit is closed. Also called voltage.

ELECTRON ELECTRON

(1) A very small, negatively charged particle that is practically weightless. According to the electron theory, all electrical and electronic effects are caused either by the movement of electrons from place to place or because there is an excess or lack of electrons at a particular place.

(2) The part of an atom that determines its chemical properties.

END BELLS END BELLS

Devices used to hold the rotor and stator of a motor in position.

FUSE | FUSE

A protective device having a strip or wire of fusible metal that, when placed in a circuit, will melt and break the electric circuit if heated too much. High temperatures will develop in the fuse when a current flows through the fuse in excess of that which the circuit will carry safely.

GROUND | GROUND

An expression representing an electrical connection to earth or a large conductor that is at the earth's potential or neutral voltage.

HERTZ (Hz) | HERTZ (Hz)

The number of complete electromagnetic cycles or waves in one second of an electric or electronic circuit. Also called the frequency of the current.

HYGROSCOPIC (hi-grow-SKAWP-ick) | HYGROSCOPIC

Absorbing or attracting moisture from the air.

JOGGING | JOGGING

The frequent starting and stopping of an electric motor.

LEAD (LEED) | LEAD

A wire or conductor that can carry electric current.

MANDREL (MAN-drill) | MANDREL

(1) A special tool used to push bearings in or to pull sleeves out.

(2) A testing device used to measure for excessive deflection in a flexible conduit.

MEG | MEG

(1) Abbreviation of MEGOHM.

(2) A procedure used for checking the insulation resistance on motors, feeders, bus bar systems, grounds, and branch circuit wiring. Also see MEGGER.

MEGGER (from megohm) | MEGGER

An instrument used for checking the insulation resistance on motors, feeders, bus bar systems, grounds, and branch circuit wiring. A megger reads in millions of ohms. Also see MEG.

MEGOHM (MEG-ome) | MEGOHM

Millions of ohms. Mega- is a prefix meaning 1 million, so 5 megohms means 5 million ohms.

MULTISTAGE PUMP | MULTISTAGE PUMP

A pump that has more than one impeller. A single-stage pump has one impeller.

NAMEPLATE | NAMEPLATE

A durable, metal plate found on equipment that lists critical installation and operating conditions for the equipment.

OSHA (O-shuh) | OSHA

The Williams-Steiger Occupational Safety and Health Act of 1970 (OSHA) is a federal law designed to protect the health and safety of industrial workers and the operators of water supply systems and treatment plants. The act regulates the design, construction, operation, and maintenance of water supply systems and water treatment plants. OSHA also refers to the federal and state agencies that administer the OSHA regulations.

OHM | OHM

The unit of electrical resistance. The resistance of a conductor in which 1 volt produces a current of 1 ampere.

POLE SHADER POLE SHADER

A copper bar circling the laminated iron core inside the coil of a magnetic starter.

POWER FACTOR POWER FACTOR

The ratio of the true power passing through an electric circuit to the product of the voltage and amperage in the circuit. This is a measure of the lag or lead of the current with respect to the voltage. In alternating current, the voltage and amperes are not always in phase; therefore, the true power may be slightly less than that determined by the direct product.

PRUSSIAN BLUE PRUSSIAN BLUE

A blue paste or liquid (often on a paper like carbon paper) used to show a contact area. Used to determine if gate valve seats fit properly.

RADIAL TO IMPELLER RADIAL TO IMPELLER

Perpendicular to the impeller shaft. Material being pumped flows at a right angle to the impeller.

RESISTANCE RESISTANCE

That property of a conductor or wire that opposes the passage of a current, thus causing electric energy to be transformed into heat.

ROTOR ROTOR

The rotating part of a machine. The rotor is surrounded by the stationary (nonmoving) parts (stator) of the machine.

SEIZING or SEIZE UP SEIZING or SEIZE UP

Seizing occurs when an engine overheats and a part expands to the point where the engine will not run. Also called freezing.

SHEAVE SHEAVE

V-belt drive pulley, which is commonly made of cast iron or steel.

SHIM SHIM

Thin metal sheets that are inserted between two surfaces to align or space the surfaces correctly. Shims can be used anywhere a spacer is needed. Usually shims are 0.001 to 0.020 inch (0.025 to 0.50 mm) thick.

SINGLE-STAGE PUMP SINGLE-STAGE PUMP

A pump that has only one impeller. A multistage pump has more than one impeller.

STATOR STATOR

That portion of a machine that contains the stationary (nonmoving) parts that surround the moving parts (rotor).

STETHOSCOPE STETHOSCOPE

An instrument used to magnify sounds and carry them to the ear.

VOLTAGE VOLTAGE

The electrical pressure available to cause a flow of current (amperage) when an electric circuit is closed. Also called electromotive force (EMF).

WATER HAMMER WATER HAMMER

The sound like someone hammering on a pipe that occurs when a valve is opened or closed very rapidly. When a valve position is changed quickly, the water pressure in a pipe will increase and decrease back and forth very quickly. This rise and fall in pressures can cause serious damage to the system.

CHAPTER 18. MAINTENANCE

(Lesson 1 of 5 Lessons)

18.0 TREATMENT PLANT MAINTENANCE—GENERAL PROGRAM

A water treatment plant operator has many duties—most having to do with the efficient operation of the plant. An operator has the responsibility to produce a water that will meet all the requirements established for the plant. By doing this, the operator develops a good working relationship with the regulatory agencies, water users, and plant neighbors.

Another duty an operator has is that of plant maintenance. A good maintenance program is a must in order to maintain successful operation of the plant. A successful maintenance program will include everything from mechanical equipment to the care of the plant grounds, buildings, and structures.

Mechanical maintenance is of prime importance as the equipment must be kept in good operating condition in order for the plant to maintain peak performance. Manufacturers provide information on the mechanical maintenance of their equipment. You should thoroughly read their literature on your plant equipment and understand the procedures. Contact the manufacturer or the local representative if you have any questions. Follow the instructions very carefully when performing maintenance on equipment. You also must recognize tasks that may be beyond your capabilities or repair facilities, and you should request assistance when needed.

For a successful maintenance program, your supervisors must understand the need for and benefits from equipment that operates continuously as intended. Disabled or improperly working equipment is a threat to the quality of the plant output, and repair costs for poorly maintained equipment usually exceed the cost of maintenance.

18.00 Preventive Maintenance Records

Preventive maintenance programs help operating personnel keep equipment in satisfactory operating condition and aid in detecting and correcting malfunctions before they develop into major problems.

A frequent occurrence in a preventive maintenance program is the failure of the operator to record the work after it is completed. When this happens, the operator must rely on memory to know when to perform each preventive maintenance function. As days pass into weeks and months, the preventive maintenance program is lost in the turmoil of everyday operation.

The only way an operator can keep track of a preventive maintenance program is by good recordkeeping. A good recordkeeping system tells when maintenance is due and provides a record of equipment performance. Poor performance is a good justification for replacement or new equipment. Good records also help keep your warranty in force. Whatever recordkeeping system is used, it should be kept up to date on a daily basis and not left to memory for some other time. Equipment service cards and service record cards are easy to set up and require little time to keep up to date (Figure 18.1). The equipment service card tells what should be done and when, while the service record card is a record of what the operator did and when it was done. Many plants use a computerized maintenance records systems to track this and other types of information.

An *EQUIPMENT SERVICE CARD* (master card) should be filled out for each piece of equipment in the plant. Each card should have the equipment name on it, such as "Raw Water Intake Pump No. 1."

1. List each required maintenance service with an item number.
2. List maintenance services in order of frequency of performance. For instance, show daily service as items 1, 2, and 3 on the card; weekly items as 4 and 5; monthly items as 6, 7, 8, and 9; and so on.
3. Describe each type of service under work to be done.

Make sure all necessary inspections and services are listed. For reference data, list paragraph or section numbers as shown in the pump maintenance section of this lesson (Section 18.23, page 312). Also, list frequency of service as shown in the time schedule columns of the same section. Under time, enter the day or month service is due. Service card information may be changed to fit the needs of your plant or a particular piece of equipment as recommended by the manufacturer. Be sure the information on the cards is complete and correct.

The *SERVICE RECORD CARD* should include the date and a brief description of the work performed, listed by item number and signed by the operator who performed the service. When a service record is completed, it should be retained for reference and a new card should be used. Some operators prefer to keep the master card and service record card clipped together, while others place only the service record card near the equipment.

EQUIPMENT SERVICE CARD				
EQUIPMENT: #1 Raw Water Intake Pump				
Item No.	**Work to Be Done**	**Reference[a]**	**Frequency**	**Time**
1	Check water seal and packing gland	Par. 1	Daily	
2	Listen for unusual noises	Par. 6	Daily	
3	Operate pump alternately	Par. 1	Weekly	Monday
4	Inspect pump assembly	Par. 1	Weekly	Wednesday
5	Inspect and lube bearings	Par. 1	Quarterly	1-4-7-10[b]
6	Check operating temperature of bearings	Par. 1	Quarterly	1-4-7-10[b]
7	Check alignment of pump and motor	Par. 1	Semiann.	4 & 10
8	Inspect and service pumps	Par. 1	Semiann.	4 & 10
9	Drain pump before shutdown	Par. 1		

SERVICE RECORD CARD					
EQUIPMENT: #1 Raw Water Intake Pump					
Date	**Work Done (Item No.)**	**Signed**	**Date**	**Work Done (Item No.)**	**Signed**
1-5-14	1-2-3	J.B.			
1-6-14	1-2	J.B.			
1-7-14	1-2-4-5-6	R.W.			

a Par. 1 refers to Paragraph 1 in Section 18.23 of this manual. Par. 6 (Paragraph 6) is also in Section 18.23.
b 1-4-7-10 represent the months of the year when the equipment should be serviced—1-January, 4-April, 7-July, and 10-October.

Fig. 18.1 Equipment service card and service record card

18.01 Plant Library

A plant library can contain helpful information to assist in plant operation. Material in the library should be organized for easy use. The internet provides fast access to current information from government agencies, educational resources, and industry professionals. Some plants maintain an electronic library of reference materials. Items in a plant library should include:

1. Plant operation and maintenance instruction manuals
2. Plant plans and specifications
3. Manufacturer instructions
4. Reference materials on water treatment
5. Operator training manuals on water treatment
6. Professional journals and publications
7. First-aid information
8. Reports from other plants
9. A dictionary

18.02 Emergencies

If your plant has not developed procedures for handling potential emergencies, do it now. Emergency procedures must be established for operators to follow when emergencies are caused by the release of chlorine or hazardous or toxic chemicals into the raw water supply, power outages, broken transmission lines or distribution mains, and attacks by terrorists. These procedures should include a list of emergency phone numbers located near all telephones, especially one that is unlikely to be affected by the emergency. Include the following agencies and people and update the phone numbers annually.

1. Police
2. Fire
3. Hospital or Physician
4. Responsible Plant Officials
5. Local Emergency Disaster Office
6. *CHEMTREC*, (800) 424-9300
7. Emergency Team (if your plant has one)

The *CHEMTREC* toll-free number may be called at any time. Personnel at this number will give information on how to handle emergencies created by hazardous materials and will notify appropriate emergency personnel.

An emergency team for your plant may be trained and assigned the task of responding to specific emergencies such as chlorine leaks. This emergency team must meet the following strict specifications at all times.

1. Team personnel must be physically and mentally qualified.
2. Proper equipment must be available at all times:
   a. Protective equipment, including self-contained breathing apparatus
   b. Repair kits
   c. Repair tools
   d. First-aid supplies
3. Proper training must take place on a regular basis and include instruction about:
   a. Properties and detection of hazardous chemicals
   b. Safe procedures for handling and storage of chemicals
   c. Types of containers, safe procedures for shipping containers, and container safety devices
   d. Installation of repair devices
   e. First-aid procedures
4. Team members must be exposed regularly to simulated field emergencies or practice drills. Team response must be carefully evaluated and any errors or weaknesses corrected.
5. Emergency team performance must be reviewed annually on a specified date. The review must include:
   a. Training program
   b. Response to actual emergencies
   c. Team physical and mental examinations

WARNING

One person should never be permitted to attempt an emergency repair alone. Always wait for trained assistance. Valuable time, needed to correct a serious emergency, could be lost rescuing a foolish individual.

For additional information on emergencies, see Volume I, Chapter 7, "Disinfection," Section 7.52, "Chlorine Leaks;" Chapter 10, "Plant Operation," Section 10.9, "Emergency Conditions and Procedures;" and, in this volume, Chapter 23, "Administration," Section 23.10, "Emergency Response." Chapter 23 contains information on what to do if a toxic substance gets into your water supply.

18.03 Lockout/Tagout Procedure

Occupational Safety and Health Administration (OSHA) standards require that all equipment that could unexpectedly start up or release stored energy must be locked out and tagged whenever it is being worked on. Common forms of stored energy are electrical energy, spring-loaded equipment, hydraulic pressure, and compressed gases that are stored under pressure.

The operator who will be performing the work on the equipment is the person who installs a lock (either key or combination) on the energy isolating device (switch, valve) to positively lock the switch or valve in a safe position, preventing the equipment from starting or moving. At the same time, a tag such as the one shown in Figure 18.2 is installed at the lock indicating why the equipment is locked out and who is working on it. No other person is authorized to remove the tag or lockout device unless the employer has provided specific procedures and training for removal by others.

DANGER

OPERATOR WORKING ON LINE

DO NOT CLOSE THIS SWITCH WHILE THIS TAG IS DISPLAYED

TIME OF: ____________

DATE: ____________

SIGNATURE: ____________

This is the ONLY person authorized to remove this tag.

INDUSTRIAL INDEMNITY/INDUSTRIAL UNDERWRITERS/ INSURANCE COMPANIES

4E210—R66

Fig. 18.2 Typical warning tag

(Source: Industrial Indemnity/Industrial Underwriters/Insurance Companies)

Even after equipment is locked out and tagged, it still may not be safe to work on. Stored energy in gas, air, water, steam, and hydraulic systems (such as water pumping systems) must be drained or bled down to prevent its sudden release, which could injure an operator working on the system or equipment. Elevated machine members, flywheels, and springs should be physically blocked or secured in place to prevent movement.

Before bleeding down pressurized systems, think about the pressures involved and what will be discharged to the atmosphere and work area (toxic or explosive gases, corrosive chemicals). Will other safety precautions have to be taken? What volume of bleed-down material will there be and where will it go? If dealing with chemicals, greases, or lubricants, how can they be cleaned up or contained? Many times, an operator has removed a pump volute cleanout only to find that the discharge check valve was not seated or one of the isolation valves was not fully seated. Once the pump was open, however, the pump intake structure drained into the pump room through the opened pump.

All rotating mechanical equipment must have guards installed to protect operators from becoming entangled or caught up in belts, pulleys, drive lines, flywheels, and couplings. Keep the guards installed even when no one is working on the equipment.

Various pieces of machinery are equipped with travel limit switches, pressure sensors, pressure reliefs, shear pins, and stall or torque switches to ensure proper and safe operation of the equipment. Never disconnect a device, or install larger shear pins than those specified on the original design, or modify pressure or temperature settings.

The basic elements of a proper lockout/tagout procedure are as follows:

1. Notify all affected employees that a lockout or tagged system is going to be used and the reason why. The authorized employee shall know the type and level of energy that the equipment uses and shall understand the hazard presented by the equipment.
2. If the equipment is operating, shut it down using the procedures established for the equipment.
3. Operate the switch, valve, or other energy isolating device(s) so that the equipment is isolated from its energy source(s). Stored energy such as that in springs; elevated machine parts; rotating flywheels; hydraulic systems; and air, gas, steam, or water pressure must be released or restrained by methods such as repositioning, blocking, or bleeding down.
4. Lock out or tag out the energy isolating device with your assigned individual lock or tag.
5. After ensuring that no personnel are exposed, and as a check that the energy source is disconnected, operate the pushbutton or other normal operating controls to make certain the

equipment will not operate. A common problem when motor control centers (MCCs) contain many breakers is to lock out the wrong equipment (locking pump #3 and thinking it is #2). Always confirm the dead circuit. *CAUTION: Return operating controls to the neutral or OFF position after the test.*

6. The equipment is now locked out or tagged out and work on the equipment may begin.
7. After the work on the equipment is complete, all tools have been removed, guards have been reinstalled, and employees are in the clear, remove all lockout or tagout devices. Operate the energy isolating devices to restore energy to the equipment.
8. Notify affected employees that the lockout or tagout devices have been removed before starting the equipment.

QUESTIONS

Please write your answers to the following questions and compare them with those on page 368.

18.0A Why should you plan a good maintenance program for your treatment plant?

18.0B What general items would you include in your maintenance program?

18.0C Why should you have a good recordkeeping system for your maintenance program?

18.0D What is the difference between an equipment service card and a service record card?

18.0E Prepare a list of emergency phone numbers for your treatment plant.

18.0F Why is it necessary to bleed down pressurized systems before making repairs?

18.1 ELECTRICAL EQUIPMENT

18.10 Beware of Electricity

Do not attempt to install, troubleshoot, maintain, repair, or replace electrical equipment, panels, controls, wiring, or circuits unless you know what you are doing, are qualified, and are authorized.

Section 18.11, "Understanding Electricity," is presented to provide you with an understanding and awareness of electricity. The purpose of the section is to help you provide electricians with the information they will need when you contact them and request their assistance. You must be extremely familiar with electricity before attempting any major repairs.

Due to the wide variety of equipment and manufacturers in the water treatment field, detailed procedures for the maintenance of some types of equipment were very difficult to include in this chapter. Also, manufacturers are continually improving their products and some details would soon be out of date. For details concerning the operation, maintenance, and repair of a particular piece of equipment, refer to the O&M instruction manual or contact the manufacturer.

Effective equipment maintenance is the key to successful system performance. The better your maintenance, the better your facilities will perform. Abuse your equipment and facilities and they will abuse you. Everyone must realize that if the equipment cannot work, no one can work.

18.100 Recognize Your Limitations

In the water departments of all cities, there is a need for maintenance operators to know something about electricity. Duties could range from repairing a taillight on a trailer or vehicle to repairing complex pump controls and motors. Very few maintenance operators do the actual electrical repairs or troubleshooting because this is a highly specialized field and unqualified people can seriously injure themselves and damage costly equipment. For these reasons, you must be familiar with electricity, know the hazards, and recognize your own limitations when you must work with electrical equipment.

Most municipalities employ electricians or contract with a commercial electrical company that they call when major problems occur. However, the maintenance operator should be able to explain how the equipment is supposed to work and what it is doing or is not doing when it fails. After studying this section, you should be able to tell an electrician what appears to be the problem with electrical panels, controls, circuits, and equipment.

The need for safety should be apparent. If proper safe procedures are not followed in operating and maintaining the various electrical equipment used in water treatment facilities, accidents can happen that cause injuries, permanent disability, or loss of life. Some of the serious accidents that have happened and could have been avoided occurred when machinery was not shut off, locked out, and tagged properly (Figure 18.2). Possible accidents include:

1. A maintenance operator could be cleaning a pump when it starts suddenly without warning, causing the operator to lose an arm, hand, or finger.
2. Electric motors or controls that are not properly grounded could cause severe shock, paralysis, or death.
3. Improper circuit connections, jumped safety devices, wrong fuses, or improper wiring can cause fires or injuries due to incorrect operation of machinery.

Another consideration for having a basic working knowledge of electricity is to prevent financial losses resulting from motors burning out and from damage to equipment, machinery, and control circuits. Additional costs result when damages have to be repaired, including payments for outside labor.

> **WARNING**
>
> Never work in electrical panels or on electrical controls, circuits, wiring, or equipment unless you are qualified and authorized. By the time you find out what you do not know about electricity, you could find yourself too dead to use the knowledge.

QUESTIONS

Please write your answers to the following questions and compare them with those on page 368.

18.10A Why must unqualified or inexperienced people be extremely careful when attempting to troubleshoot or repair electrical equipment?

18.10B What could happen when machinery is not shut off, locked out, and tagged properly?

18.11 Understanding Electricity

Most electrical equipment used in water treatment plants is labeled by the manufacturer with a *NAMEPLATE.*[1] All of the information on the nameplate should be recorded and placed in a file for future reference. Many times, the nameplate is painted, corroded, or missing from the unit when the information is needed to repair the equipment or replace parts. Record the serial or model number, the date of installation, and the date the equipment was placed in service on the equipment service card or service record card or file this information in a safe and convenient location.

The nameplate on a piece of electrical equipment will always indicate the proper voltage and allowable current in amps for that particular piece of equipment. The following sections describe the meaning of the terms "voltage" and "amps" and the basic differences between direct current and alternating current.

18.110 Volts

Voltage (E) is also known as electromotive force (EMF). It is the electrical pressure available to cause a flow of current (amperage) when an electric circuit is closed.[2] This pressure can be compared with the pressure or force that causes water to flow in a pipe. Some pressure in a water pipe is required to make the water move. The same is true of electricity. A force is necessary to push electricity or electric current through a wire. This force is called voltage. There are two types of current: direct current (DC) and alternating current (AC).

18.111 Direct Current (DC)

Direct current flows in one direction only and is essentially free from pulsation, that is, the strength of the current remains steady. Direct current is seldom used in water treatment plants except possibly in electronic equipment, some control components of pump drives, and standby lighting. Direct current is used exclusively in automotive equipment, certain types of welding equipment, and a variety of portable equipment. Direct current is found in various voltages such as 6 volts, 12 volts, 24 volts, 48 volts, and 110 volts. All batteries are direct current. DC voltage can be measured by holding the positive and negative leads of a DC multimeter on the corresponding terminals of the DC device such as a battery. Direct current usually is not found in higher voltages (over 24 volts) around plants except in motor generator sets. Care must be taken when installing battery cables and wiring that positive (+) and negative (–) poles are connected properly to wires marked (+) and (–). If not properly connected, there could be an arc of electricity across the unit that could cause an explosion.

18.112 Alternating Current (AC)

An alternating current circuit is one in which the voltage and current periodically change direction and *AMPLITUDE.*[3] In other words, the current goes from zero to maximum strength, back to zero, and to the same strength in the opposite direction. Most AC circuits have a frequency of 60 *CYCLES*[4] per second. "Hertz" is the term we use to describe the frequency of cycles completed per second so our AC voltage would be 60 Hertz (Hz).

There are three classifications of alternating current:

1. Single phase
2. Two phase
3. Three phase or polyphase

The most common of these are single phase and three phase. The various voltages you probably will find on your job are 110 volts, 120 volts, 208 volts, 220 volts, 240 volts, 277 volts, 440 volts, 480 volts, 2,400 volts, and 4,160 volts.

Single-phase power is found in lighting systems, small pump motors, various portable tools, and throughout our homes. This

1. *Nameplate.* A durable, metal plate found on equipment that lists critical installation and operating conditions for the equipment.
2. Electricians often talk about closing an electric circuit. This means they are closing a switch that actually connects circuits together so electricity can flow through the circuit. Closing an electric circuit is like opening a valve on a water pipe.
3. *Amplitude.* The maximum strength of an alternating current during its cycle, as distinguished from the mean or effective strength.
4. *Cycle.* A complete alternation of voltage or current in an alternating current (AC) circuit.

power is usually 120 volts and sometimes 240 volts. Single phase means that only one phase of power is supplied to the main electrical panel at 240 volts and it has three wires or leads. Two of these leads have 120 volts each; the other lead is neutral and usually is coded white. The neutral lead is grounded. Many appliances and power tools have an extra ground (commonly a green wire) on the case for additional protection.

Three-phase power is generally used with motors and transformers found in water treatment plants, but is generally not available in rural areas. Phase converters allow three-phase electric motors to be operated on single-phase power. A converter is wired in between the motor and the power line. The advantages of three-phase motors are lower current draw, simple reversing, and lighter weight. Heavy-duty converters are suitable for long starting, frequent starting, and reversing.

Circuit breakers (Figure 18.3) are used to protect electric circuits from overloads. Most circuit breakers are metal conductors that de-energize the main circuit when excess current passes through a metal strip causing it to overheat and open the main circuit.

Two-phase systems will not be discussed because they are seldom found in water treatment facilities.

Fig. 18.3 Circuit breakers

(Courtesy of Consolidated Electrical Distributors, Inc.)

18.113 Amps

An ampere (A) is the SI (*Le Système International d'Unités* or International System of Units)[5] unit of electric current (I). This is the current produced by a pressure of 1 volt in a circuit having a resistance of 1 ohm. Amperage is the measurement of current or electron flow and is an indication of work being done or how hard the electricity is working.

To understand amperage, one more term must be explained. The ohm is the practical unit of electrical resistance (R). "Ohm's Law" states that in a given electric circuit, the amount of current (I) in amperes is equal to the pressure in volts (E) divided by the resistance (R) in ohms. The following three formulas are given to provide you with an indication of the relationships among current, resistance, and EMF (electromotive force).

$$\text{Current, amps} = \frac{\text{EMF, volts}}{\text{Resistance, ohms}} \qquad I = \frac{E}{R}$$

$$\text{EMF, volts} = (\text{Current, amps})(\text{Resistance, ohms}) \qquad E = IR$$

$$\text{Resistance, ohms} = \frac{\text{EMF, volts}}{\text{Current, amps}} \qquad R = \frac{E}{I}$$

These equations are used by electrical engineers for calculating circuit characteristics. If you memorize the following relationship, you can always figure out the correct formula.

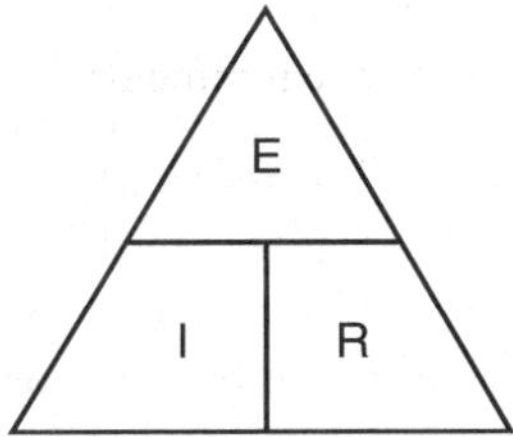

To use the above triangle, you cover up with your finger the term you do not know or are trying to find out. The relationship between the other two known terms will indicate how to calculate the unknown. For example, if you are trying to calculate the current, cover up I. The two knowns (E and R) are shown in the triangle as E/R. Therefore, I = E/R. The same procedure can be used to find E when I and R are known or to find R when E and I are known.

18.114 Watts

Watts (W) and kilowatts (kW) are the units of measurement of the rate at which power is being used or generated. In DC circuits, watts (W) equal the voltage (E) multiplied by the current (I).

$$\text{Power, watts} = (\text{Electromotive Force, volts})(\text{Current, amps})$$

$$\text{or P, watts} = (\text{E, volts})(\text{I, amps})$$

In AC polyphase circuits, the formula becomes more complicated because of the inclusion of two additional factors. First, there is the square root of 3, for three-phase circuits, which is equal to 1.73. Second, there is the power factor, which is the ratio of the true or actual power passing through an electric circuit to the product of the voltage times the amperage in the circuit. For

5. SI (*Le Système International d'Unités* or International System of Units) is a system of physical units (SI Units) based on the meter, kilogram, second, ampere, kelvin, candela, and mole, together with a set of prefixes to indicate multiplication or division by a power of ten.

standard three-phase induction motors, the power factor will be somewhere near 0.9. The formula for power input to a three-phase motor is:

$$\text{Power, kilowatts} = \frac{(\text{E, volts})(\text{I, amps})(\text{Power Factor})(1.73)}{1{,}000 \text{ watts/kilowatt}}$$

Since 0.746 kilowatt equals 1.0 horsepower, then the formula for power output of a motor is:

$$\text{Power Output, horsepower} = \frac{(\text{Power Input, kilowatts})(\text{Efficiency, \%})}{(0.746 \text{ kilowatt/horsepower})(100\%)}$$

18.115 Power Requirements

Power requirements (PR) are expressed in kilowatt hours. One watt for 1,000 hours or 500 watts for 2 hours equals 1 kilowatt hour. The power company charges so many cents per kilowatt hour.

$$\text{Power Req, kW-hr} = (\text{Power, kilowatts})(\text{Time, hours})$$

$$\text{PR, kW-hr} = (\text{P, kW})(\text{T, hr})$$

18.116 Conductors and Insulators

A material, like copper, that permits the flow of electric current is called a conductor. Material that will not permit the flow of electricity, like rubber, is called an insulator. Such material when wrapped or cast around a wire is called insulation. Insulation is commonly used to prevent the loss of electrical flow by two conductors coming into contact with each other.

QUESTIONS

Please write your answers to the following questions and compare them with those on page 368.

18.11A How can you determine the proper voltage and allowable current in amps for a piece of equipment?

18.11B What are two types of current?

18.11C Amperage is the measurement of what?

18.12 Tools, Meters, and Testers

WARNING

Never enter any electrical panel or attempt to troubleshoot or repair any piece of electrical equipment or any electric circuit unless you are qualified and authorized.

18.120 Voltage Testing

To maintain, repair, and troubleshoot electrical equipment and circuits, the proper tools are required. You will need a *MULTIMETER* to check for voltage (Figures 18.4 and 18.5). There are several types on the market and all of them work. They are designed to be used on energized circuits and care must be exercised when testing. By holding one lead on ground and the other on a power lead, you can determine if the circuit is energized.

Be sure the multimeter that you are using has sufficient range to measure the voltage you would expect to find. In other words, do not use a multimeter with a limit of 600 volts on a circuit that normally is energized at 2,400 volts. With the multimeter, you can tell if the current is AC or DC and the intensity or voltage, which will probably be one of the following: 120, 208, 240, 480, 2,400, or 4,160.

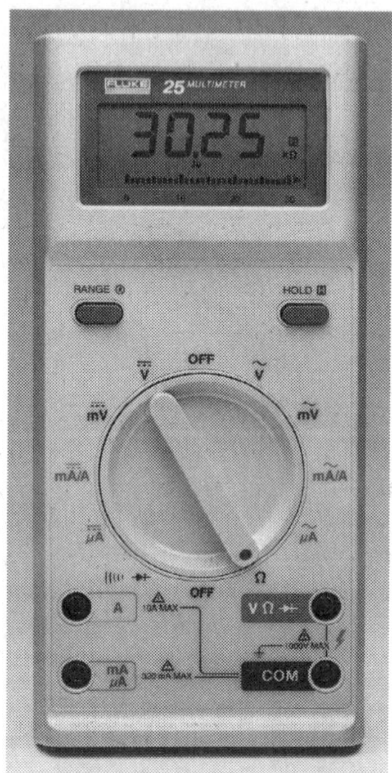

Fig. 18.4 Digital multimeter
(Reproduced with permission of Fluke Corporation)

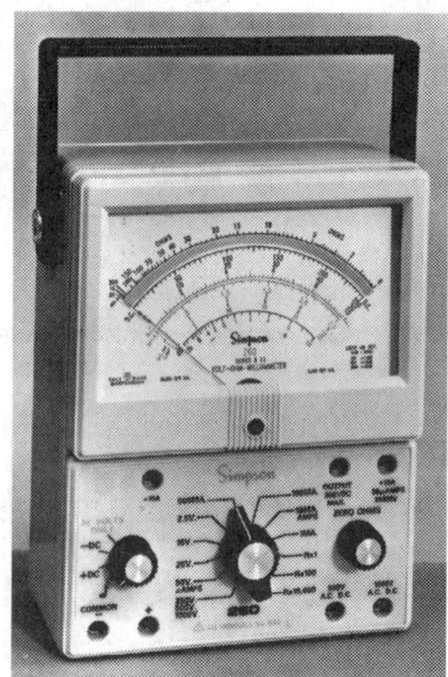

Fig. 18.5 Analog clamp-on multimeter
(Permission of Simpson Electric)

Do not work on any electric circuits unless you are qualified and authorized. Use a multimeter and other circuit testers to determine if a circuit is energized, or if all voltage is off. This should be done after the main switch is turned off to make sure it is safe to work inside the electrical panel. Always be aware of the possibility that even if the disconnect to the unit you are working on is off, the control circuit may still be energized if the circuit originates at a different distribution panel. Also, a capacitor in the unit such as a power factor correction capacitor on a motor may have sufficient energy stored to cause considerable harm to an operator. Test for voltage both before and during the time the switch is pulled off to have a double-check. This procedure ensures that the multimeter is working. Use circuit testers to measure voltage or current characteristics to a given piece of equipment and to make sure that you have or do not have a live circuit.

In addition to checking for power, a multimeter can be used to test for open circuits, blown fuses, single phasing of motors, grounding, and many other uses. Some examples are illustrated in the following paragraphs.

In the single-phase circuit shown in Figure 18.6, test for power by first opening the switch as shown. Using a multimeter with clamp-on leads, clamp one wire (lead) between the switch and the light and the other lead between the light and ground. Now, close the switch. In general, the multimeter should register at 110 volts. If there is no reading, then the switch is probably faulty. Try replacing the switch.

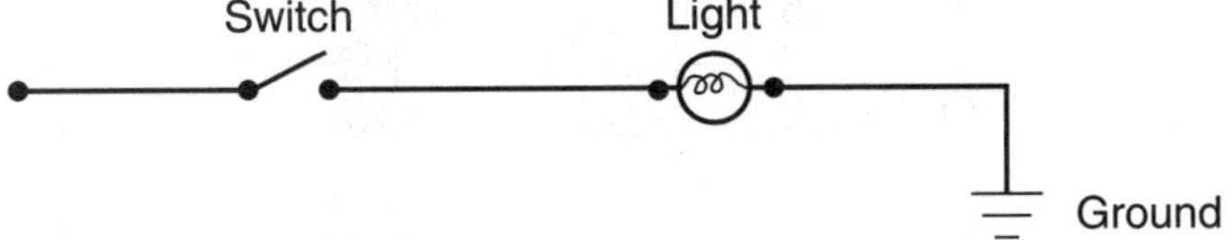

Fig. 18.6 Single-phase circuit (switch in open position)

In the single-phase, three-lead circuit shown in Figure 18.7, test for power by first opening the switches as shown.

1. Using a multimeter with clamp-on leads, clamp one lead on L1 and the other lead on L2 between the fuses and the load (equipment). With the connections made, bring the multimeter and attached leads out of the panel and close the panel door as far as possible without cutting or damaging the meter leads. Now, close the switches. (*NOTE:* Some switches cannot be closed if the panel door is open. The panel door is closed when testing because hot copper sparks could seriously injure you when the circuit is energized and the voltage is high.)
2. The multimeter should register at 220 volts. If there is no reading, either a switch or both switches could be faulty, or a fuse or both fuses could be blown. Let us first determine if the problem is with the fuses.
3. To test the fuses, first open the switches. Clamp a lead on L1 and a lead on L2 between the switches and the line (panel). Close the panel door, then close the switches.

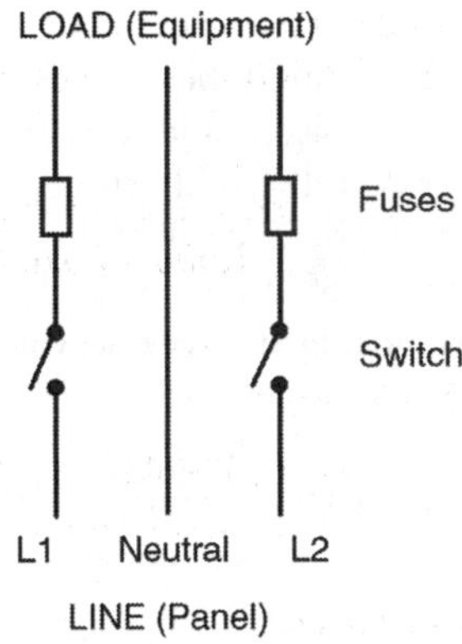

Fig. 18.7 Single-phase, three-lead circuit

4. If no reading is registered, test the fuse on L1 by first opening the switches. Move the test lead on L1 to a position between the fuse and the load; move the test lead on L2 to a position between line and load on the neutral lead. Close the panel door, then close the switches. If a reading of 110 volts is observed, then the fuse on L1 is OK; we also know that the switch on L1 is OK. If there is no reading, the switch on L1 could be faulty or the fuse on L1 could be blown. Open the switches and try replacing the fuse on L1. Retest for power as described in Step 1.
5. The multimeter should register at 220 volts. If there is still no reading, a switch or both switches could be faulty or the fuse on L2 could be blown.
6. Test the fuse on L2 by first opening the switches. Move the test lead on L1 to a position between the fuse and the load on L2. (The other test lead should remain on the neutral lead.) Close the panel door, then close the switches. If a reading of 110 volts is observed, then the fuse on L2 is OK; we also know that the switch on L2 is OK. If there is no voltage reading, the switch on L2 could be faulty or the fuse on L2 could be blown. Open the switches and try replacing the fuse on L2. Retest for power as described in Step 1.
7. The multimeter should register at 220 volts. If there is still no reading, the problem could be faulty switches, a broken neutral line, an open switch in another location, or a power outage.

WARNING

Turn off power and be sure that there is no voltage in either power line before replacing fuses. Use a fuse puller to remove and replace fuses. Test the circuit again in the same manner to make sure fuses or circuit breakers work properly. There should be 220 volts of power or voltage present between L1 and L2. If a fuse or circuit breaker trips again, shut off the power and determine the source of the problem.

Test for power at points A, B, and C in a three-phase circuit (Figure 18.8). Place the multimeter leads on lines A and B. Close all switches. 220 volts should register on the multimeter. Check between lines A and C, and between lines B and C. 220 volts should be recorded between all of these points. If voltage is not present, one or all of the fuses are blown or the circuit breaker has been tripped. First, check for voltage above the fuses at all of these points, A to B, A to C, and B to C, to make sure power is available (see 220 readings in Figure 18.8). If voltage is recorded, move leads back down to the bottom of the fuses. If voltage is present from A to B, but not at A to C and B to C, the fuse on line C is blown. If there were no voltage readings at any of the test points, all the fuses could be blown.

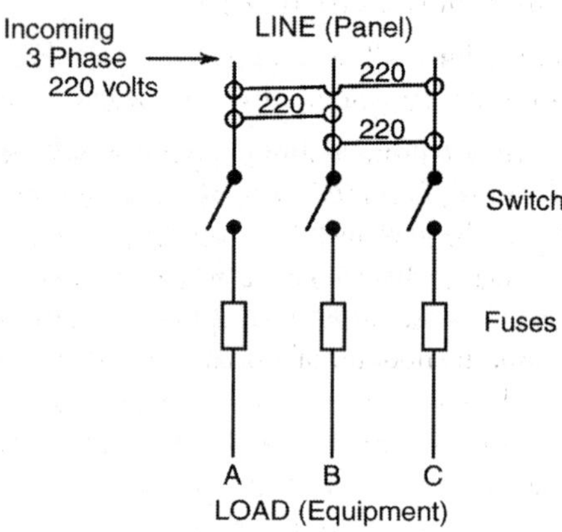

Fig. 18.8 Three-phase circuit, 220 volts

Another way of checking the fuses with the load connected on this three-phase circuit would be to take your multimeter and place one lead on the bottom, and one lead on the top of each fuse. You should not get a voltage reading on the multimeter. This is because electricity takes the path of least resistance. If you get a reading across any of the fuses (top to bottom), that fuse is bad.

Always make sure that when you use a multimeter it is set for the proper voltage. If voltage is unknown and the meter has different scales that are manually set, always start with the highest voltage range and work down. Otherwise, the multimeter could be damaged. Look at the equipment instruction manual or nameplate for the expected voltage. Actual voltage should not be much higher than given unless someone made a mistake when the equipment was wired and inspected.

18.121 Ammeter

Another meter used in electrical maintenance and testing is the *AMMETER*. The ammeter records the current or amps flowing in the circuit. There are several types of ammeters, but only two will be discussed in this section. The ammeter generally used for testing is called a clamp-on type. The term "clamp-on" means that it can be clamped around a wire supplying a motor, and no direct electrical connection needs to be made. Each "leg" or lead on a three-phase motor must be individually checked.

The first step should be to read the motor nameplate data and find what the amperage reading should be for the particular motor or device you are testing. After you have this information, set the ammeter to the proper scale. Set it on a higher scale than necessary if the expected reading is close to the top of the meter scale. Place the clamp around one lead at a time. Record each reading and compare with the nameplate rating. If the readings are not similar to the nameplate rating, find the cause, such as low voltage, bad bearings, poor connections, or excessive load. If the ammeter readings are higher than expected, the high current could produce overheating and damage to the equipment. Try to find the problem and correct it.

Current imbalance is undesirable because it causes uneven heating in a motor that can shorten the life expectancy of the insulation. However, a small amount of current imbalance is to be expected in the leads to a three-phase motor. This imbalance can be caused by either peculiarities in the motor or by a power company imbalance. To isolate the cause, make the following test. Note that this test should be done by a qualified electrician. Refer to Figure 18.9.

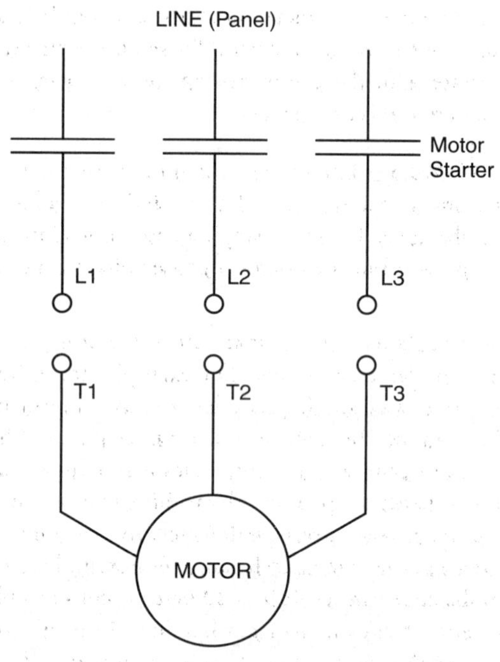

1. With the motor wired to its starter, L1 to T1, L2 to T2, and L3 to T3, measure and record the amperage on L1, L2, and L3.
2. De-energize the circuit and reconnect the motor as follows: L1–T3, L2–T1, L3–T2. This wiring change will not change the direction of the rotation of the motor.
3. Start up the motor and again measure and record the amperage on L1, L2, and L3.

Fig. 18.9 Determination of current imbalance

If the current readings on lines L1, L2, and L3 are about the same both before and after the wiring change, this is an indication that the imbalance is being caused by the power company and they should be asked to make adjustments to correct the condition. However, if the current readings followed the motor terminal (T) numbers rather than the power line (L) numbers, the problem is within the motor and there is not much that can be done except contact the motor manufacturer for a possible exchange.

When using a clamp-on ammeter, be sure to set the meter on a high enough range or scale for the starting current if you are testing during startup. Starting currents range from 500 to 700 percent higher than running currents and using too low a range can ruin an expensive and delicate instrument. Newer clamp-on ammeters automatically adjust to the proper range and can measure both starting or peak current and normal running current.

Another type of ammeter is one that is connected in line with the power lead or leads. Generally, they are not portable and are usually installed in a panel or piece of equipment. They require physical connections to put them in series with the motor or apparatus being tested. Current transformers (CT) are commonly used with this type of ammeter so that the meter does not have to conduct the full motor current. These ammeters are usually more accurate than the clamp-on type and are used in motor control centers and pump panels.

PROBLEM: Voltage imbalance. A common problem found in pump stations with a high rate of motor failure is voltage imbalance or unbalance. Unlike a single-phase condition, all three phases are present but the phase-to-phase voltage is not equal in each phase.

Voltage imbalance can occur in either the utility side or the pump station electrical system. For example, the utility company may have large single-phase loads (such as residential services) that reduce the voltage on a single phase. This same condition can occur in the pump station if a large number of 120/220 volt loads are present. Slight differences in voltage can cause disproportional current imbalance; this may be six to ten times as large as the voltage imbalance. For example, a 2 percent voltage imbalance can result in a 20 percent current imbalance. A 4.5 percent voltage imbalance will reduce the insulation life to 50 percent of the normal life. This is the reason a dependable voltage supply at the motor terminals is critical. Even relatively slight variations can greatly increase the motor operating temperatures and burn out the insulation.

It is common practice for electrical utility companies to furnish power to three-phase customers in open delta or wye configurations. An open delta or wye system is a two-transformer bank that is a suitable configuration where lighting loads are large and three-phase loads are light. This is the exact opposite of the configuration needed by most pumping facilities where three-phase loads are large. (Examples of three-transformer banks include Y-delta, delta-Y, and Y-Y.) In most cases, three-phase motors should be fed from three-transformer banks for proper balance. The capacity of a two-transformer bank is only 57 percent of the capacity of a three-transformer bank. The two-transformer configuration can cause one leg of the three-phase current to furnish higher amperage to one leg of the motor, which will greatly shorten its life.

Operators should acquaint themselves with the configuration of their electric power supply. When an open delta or wye configuration is used, operators should calculate the degree of current imbalance existing between legs of their polyphase motors. If you are unsure about how to determine the configuration of your system or how to calculate the percentage of current imbalance, always consult a qualified electrician. Current imbalance between legs should never exceed 5 percent under normal operating conditions (National Electrical Manufacturers Association [NEMA] Standard MG 1-14.35).

Loose connections will also cause voltage imbalance as will high-resistance contacts, circuit breakers, or motor starters.

Another serious consideration for operators is voltage fluctuation caused by neighborhood demands. A pump motor in near perfect balance (for example, 3 percent imbalance) at 9:00 a.m. could be as much as 17 percent imbalanced by 4:00 p.m. on a hot day due to the use of air conditioners by customers on the same grid. Also, the hookup of a small market or a new home to the power grid can cause a significant change in the degree of current imbalance in other parts of the power grid. Because energy demands are constantly changing, water system operators should have a qualified electrician check the current balances between legs of their three-phase motors at least once a year.

SOLUTION: Motor connections at the circuit box should be checked frequently (semiannually or annually) to ensure that the connections are tight and that vibration has not caused the insulation on the conductors to wear away. Measure the voltage at the motor terminals and calculate the percentage imbalance (if any) using the procedures below.

Do not rely entirely on the power company to detect imbalanced current. Complaints of suspected power problems are frequently met with the explanation that all voltages are within the percentages allowed by law and no mention is made of the percentage of current imbalance, which can be a major source of problems with three-phase motors. A little research of your own can pay big dividends. For example, a small water company in central California configured with an open delta system (and running three-phase imbalances as high as 17 percent as a result) was routinely spending $14,000 a year for energy and burning out one 10 HP motor every 1.5 years, on average. This equals the loss of six 10 HP motors in 9 years. After consultation, the local power utility agreed to add a third transformer to each power board to bring the system into better balance. Pump drop leads were then rotated, bringing overall current imbalances down to an average of 3 percent; heavy-duty three-phase capacitors were added to absorb the voltage surges common in the area; and computerized controls were added to the pumps to shut them off when pumping volumes got too low. These modifications resulted in a saving in energy costs the first year alone of $5,500.

FORMULAS

Percentage of current imbalance can be calculated by using the following formulas and procedures:

$$\text{Average Current} = \frac{\text{Total of Current Value Measured on Each Leg}}{3}$$

$$\text{\% Current Imbalance} = \frac{\text{Greatest Amp Difference from the Average}}{\text{Average Current}} \times 100\%$$

PROCEDURES

1. Measure and record current readings in amps for each leg. (Figure 18.10, Hookup 1.) Disconnect power.
2. Shift or roll the motor leads from left to right so the drop cable lead that was on terminal 1 is now on 2, the lead on 2 is now on 3, and the lead on 3 is now on 1. (Figure 18.10, Hookup 2.) Rolling the motor leads in this manner will not reverse the motor rotation. Start the motor and measure and record current reading on each leg. Disconnect power.
3. Again, shift drop cable leads from left to right so the lead on terminal 1 goes to 2, 2 goes to 3, and 3 to 1. (Figure 18.10, Hookup 3.) Start pump and measure and record current reading on each leg. Disconnect power.
4. Add the values for each hookup.
5. Divide the total by 3 to obtain the average.
6. Compare each single leg reading to the average current amount to obtain the greatest amp difference from the average.
7. Divide this difference by the average to obtain the percentage of imbalance.
8. Use the wiring hookup that provides the lowest percentage of imbalance.

CORRECTING THE THREE-PHASE POWER IMBALANCE

Example: Check for current imbalance for a 230-volt, three-phase, 60 Hz submersible pump motor, 18.6 full load amps.

Solution: Steps 1 to 3 measure and record amps on each motor drop lead for Hookups 1, 2, and 3 (Figure 18.10).

	Step 1 (Hookup 1)	Step 2 (Hookup 2)	Step 3 (Hookup 3)
(T_1)	DL_1 = 25.5 amps	DL_3 = 25 amps	DL_2 = 25.0 amps
(T_2)	DL_2 = 23.0 amps	DL_1 = 24 amps	DL_3 = 24.5 amps
(T_3)	DL_3 = 26.5 amps	DL_2 = 26 amps	DL_1 = 25.5 amps
Step 4	Total = 75 amps	Total = 75 amps	Total = 75 amps

Step 5 Average Current = $\frac{\text{Total Current}}{\text{3 readings}} = \frac{75}{3}$ = 25 amps

Step 6	Greatest amp difference from the average:	(Hookup 1) = 25 – 23	= 2
		(Hookup 2) = 26 – 25	= 1
		(Hookup 3) = 25.5 – 25	= .5
Step 7	% Imbalance	(Hookup 1) = 2/25 × 100	= 8
		(Hookup 2) = 1/25 × 100	= 4
		(Hookup 3) = 0.5/25 × 100	= 2

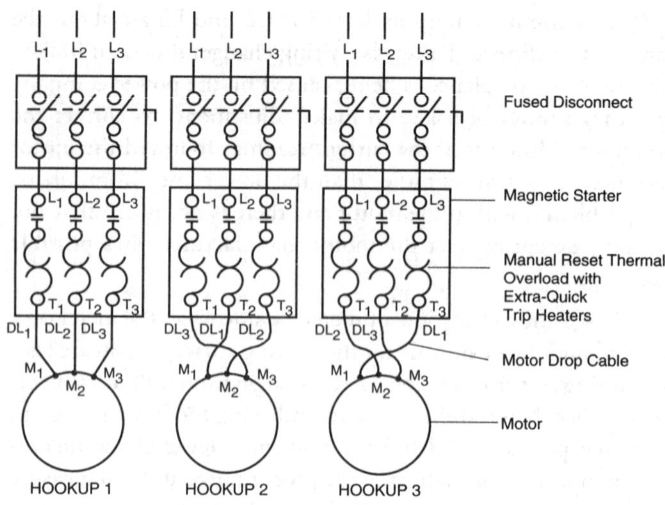

Fig. 18.10 Three hookups used to check for current imbalance

As can be seen, Hookup 3 should be used since it shows the least amount of current imbalance. Therefore, the motor will operate at maximum efficiency and reliability on Hookup 3.

By comparing the current values recorded on each leg, you will note the highest value was always on the same leg, L_3. This indicates the imbalance is in the power source. If the high current values were on a different leg each time the leads were changed, the imbalance would be caused by the motor or a poor connection.

If the current imbalance is greater than 5 percent, contact your power company for help.

Acknowledgment

Material on imbalanced current was provided by James W. Cannell, President, Canyon Meadows Mutual Water Company, Inc., Bodfish, California. His contribution is greatly appreciated.

QUESTIONS

Please write your answers to the following questions and compare them with those on page 368.

18.12A How can you determine if there is voltage in a circuit?

18.12B What are some of the uses of a multimeter?

18.12C What precautions should be taken before attempting to change fuses?

18.12D How do you test for voltage with a multimeter when the voltage is unknown?

18.12E What could be the cause of amp readings different from the nameplate rating?

18.122 Megger

A *MEGGER* is a device used for checking the insulation resistance on motors, feeders, bus bar systems, grounds, and branch circuit wiring.

> **WARNING**
>
> Use a megger only on de-energized circuits or motors.

There are three general types of meggers: crank operated, battery operated, and instrument. There are two leads to connect. One lead is clamped to a ground lead and the other to the lead you are testing. The readings on the megger will range from "0" (ground) to infinity (perfect), depending on the condition of your circuit.

The megger is usually connected to a motor terminal at the starter, and the other lead to the ground lead. Results of this test indicate if the insulation is deteriorating or cut.

Insulation resistance of electrical equipment is affected by many variables such as the equipment design, the type of insulating material used, including binders and impregnating compounds, the thickness of the insulation and its area, cleanliness (or uncleanliness), moisture, and temperature. For insulation resistance measurements to be conclusive in analyzing the condition of equipment being tested, these variables must be taken into consideration.

Such factors as the design of the equipment, the kind of insulating material used, and its thickness and area cease to be variables after the equipment has been put into service, and minimum insulation resistance values can be established within reasonable tolerances. The variables that must be considered after the equipment has been put into service, and at the time that the insulation resistance measurements are being made, are uncleanliness, moisture, temperature, and damage such as fractures.

The most important steps an operator can take to ensure reliable operation of electrical equipment are keeping the equipment clean, preventing moisture from getting into equipment insulation, and keeping control panels and cabinets closed and well sealed with weatherstripping. This is merely good housekeeping but it is essential in the maintenance of all types of electrical equipment. The very fact that insulation resistance is affected by moisture and dirt, with due allowances for temperature, makes the megger insulation test the valuable tool that it is in electrical maintenance. The test is an indication of cleanliness and good housekeeping as well as a detector of deterioration and impending trouble.

Suggested safe levels or minimum values of insulation resistance have been developed. These values should be provided by the equipment manufacturer and should serve as a guide for equipment in service. However, periodic tests on equipment in service will usually reveal readings considerably higher than the suggested minimum safe values. Records of periodic tests must be kept because persistent downward trends in insulation resistance usually give fair warning of impending trouble, even though the actual values may be higher than the suggested minimum safe values.

Also, allowances must be made for equipment in service showing periodic test values lower than the suggested minimum safe values, so long as the values remain stable or consistent. In such cases, after due consideration has been given to temperature and humidity conditions at the time of the test, there may be no need for concern. This condition may be caused by uniformly distributed leakages of a harmless nature, and may not be the result of a dangerous localized weakness. Here again, records of insulation resistance tests over a period of time reveal changes that may justify investigation. The trend of the curve may be more significant than the numerical values themselves.

For many years, one *MEGOHM* [6] has been widely used as a fair allowable lower limit for insulation resistance of ordinary industrial electrical equipment rated up to 1,000 volts. This value is still recommended for those who may not be too familiar with insulation resistance testing practices, or who may not wish to approach the problem from a more technical point of view.

For equipment rated above 1,000 volts, the "one megohm" rule is usually stated: "A minimum of one megohm per thousand volts." Although this rule is somewhat arbitrary, and may be criticized as lacking an engineering foundation, it has stood the test of a good many years of practical experience. This rule gives some assurance that equipment is not too wet or not too dry and has saved many an unnecessary breakdown.

More recent studies of the problem, however, have resulted in formulas for minimum values of insulation resistance that are based on the kind of insulating material used and the electrical and physical dimensions of the types of equipment under consideration.[7]

Motors and wiring should be megged at least once a year, and twice a year, if possible. The readings taken should be recorded and plotted in some manner so that you can determine when insulation is breaking down. Meg motors and wirings after a pump station has been flooded. If insulation is wet, excessive current could be drawn and cause pump motors to kick out.

18.123 Ohmmeter

OHMMETERS, sometimes called circuit testers, are valuable tools used for checking electric circuits. An ohmmeter is used only when the electric circuit is OFF, or de-energized. The ohmmeter supplies its own power by using batteries. An ohmmeter

---

6. *Megohm* (MEG-ome). Millions of ohms. Mega- is a prefix meaning 1 million, so 5 megohms means 5 million ohms.
7. Portions of the preceding paragraphs were taken from *Instruction Manual For Megger Insulation Testers*, No. 21-J, pages 42 and 43, published by Biddle Instruments, no longer in print. For additional information, see *A Stitch in Time: The Complete Guide to Electrical Insulation Testing* (Revised 2006). A PDF of the guide is available from the Megger© division of MeterCenter© at www.biddlemegger.com/biddle/Stitch-new.pdf.

is used to measure the resistance (ohms) in a circuit. These are most often used in testing the control circuit components such as coils, fuses, relays, resistors, and switches. They are used also to check for continuity.

An ohmmeter has several scales that can be used. Typical scales are: R × 1, R × 10, R × 1,000, and R × 10,000. Each scale has a level of sensitivity for measuring different resistances. To use an ohmmeter, set the scale (start at the low point—R × 1), put the two leads across the part of the circuit to be tested, such as a coil or resistor, and read the resistance in ohms. A reading of infinity would indicate an open circuit, and a zero would read no resistance. Ohmmeters usually would be used only by skilled technicians because they are very delicate instruments.

All meters should be kept in good working order and calibrated periodically. They are very delicate, susceptible to damage, and should be well protected during transportation. When readings are taken, they should always be recorded on a machinery history card for future reference. Meters are a good way to determine pump and equipment performance. *CAUTION: Never use a meter for testing electrical equipment unless you are qualified and authorized.*

QUESTIONS

Please write your answers to the following questions and compare them with those on page 368.

18.12F How often should motors and wirings be megged?

18.12G An ohmmeter is used to measure the resistance (ohms) in what control circuit components?

18.13 Switch Gear

18.130 Equipment Protective Devices

Operators need safety devices to protect themselves and plant equipment from the destructive effects of electricity. Water systems have pressure valves, pop-offs, and different safety equipment to protect the pipes and equipment. Similarly, electricity must have safety devices to contain the voltage and amperage that come in contact with the wiring and equipment. The first piece of equipment that must be protected is the main electrical panel or control unit where the power enters. This protection is provided by either fuses or a circuit breaker.

18.131 Fuses

The power company has installed fuses on their power poles to protect their equipment from damage. We also must install fuses to protect the main control panel and wiring from damage due to excessive voltage or amperage.

A fuse is a protective device having a strip or wire of fusible metal that, when placed in a circuit, will melt and break the electrical circuit when subjected to an excessive temperature. This temperature will develop in the fuse when a current flows through the fuse in excess of what the circuit will carry safely. This means that the fuse must be capable of de-energizing the circuit before any damage is done to the wiring it is safely protecting. Fuses are used to protect operators, main circuits, branch circuits, heaters, motors, and various other electrical equipment.

There are several types of fuses, each being used for a certain type of protection. Two types are:

1. Current-Limiting Fuses: These fuses open so quickly while clearing a short-circuit current that the potential fault current is not allowed to reach its peak. They are used to protect power distribution circuits.
2. Dual-Element Fuses: These fuses provide a time delay in the low overload range and a fast-acting element for short-circuit protection. These fuses are used for motor protection circuits.

There are many other types of fuses used for special application, but the above are the most common.

A fuse must never be bypassed or jumped. This is the only protection the circuit has; without it, serious damage to equipment and possible injury to operators can occur. Make sure that all fuses are replaced with the proper size and type indicated for that circuit. If you have any doubt, check the electrical blueprints or contact your electrical engineer.

18.132 Circuit Breakers

The circuit breaker (Figure 18.3) is another safety device and is used in the same place as a fuse. Most circuit breakers consist of a switch that opens automatically when the current or the voltage exceeds or falls below a certain limit. Unlike a fuse that has to be replaced each time it blows, a circuit breaker can be reset after a short delay to allow time for cooling. This is done by moving the handle to the OFF position or slightly past, and then moving it back to the ON position. Also, unlike a fuse, a circuit breaker can be visually inspected to find out if it has been tripped. The handle will be at the mid position between ON and OFF. Several different types of circuit breakers are being used today and each one is selected for a special protective purpose.

18.133 Overload Relays

Three-phase motors are usually protected by overload relays. This is accomplished by having heater strips, bimetal, or solder pots that open on current rise (overheating), and open the control circuit. This, in turn, opens the power control circuit, which de-energizes the starter and stops power to the motor. Such relays are also known as heaters or "thermal overloads." Sizing of these overloads is very critical and should coincide with the nameplate rating on the motor. Sizing depends on the service factor of the electric motor. Usually, they range from 100 to 110 percent of the motor nameplate ratings and should never exceed 125 percent (usually 115 percent) of the motor rating. For example, if the motor is rated for 10 amps, the overloads should be sized from 10 to 11 amps.

Again, never increase the rating of the overload heaters because of tripping. You should find the problem and repair it. There are many other protective devices for electricity such as motor winding thermostats, phase protectors, low-voltage protectors, and ground-fault protectors. Each has its own special applications and should never be tampered with or jammed.

"Ground" is an expression representing an electrical connection to earth or a large conductor that is at the earth's potential or neutral voltage. Motor frames and all electrical tools and equipment enclosures should be connected to ground. This is generally referred to simply as grounding, or equipment ground.

The third prong on cords from electric hand tools is the equipment ground and must never be removed. When an adapter is used with a two-prong receptacle, the green wire on the adapter should be connected under the center screw on the receptacle cover plate. Many times, equipment grounding, especially at home, is achieved by connecting onto a water pipe or drain rather than a rod driven into the ground. This practice generally is not recommended when plastic pipes and other nonconducting pipe materials are used unless it is known that the piping is all metal and not interrupted. Also, corrosion can be accelerated if pipes of different metals are used. A rod driven into dry ground is not very effective as a ground.

18.134 Motor Starters

A motor starter is a device or group of devices that are used to connect the electrical power to a motor. These starters can be either manually or automatically controlled.

Manual and magnetic starters range in complexity from a single ON/OFF switch to a sophisticated automatic device using timers and coils. The simplest motor starter is used on single-phase motors where a circuit breaker is manually turned on and the motor starts. This type of starter also is used on three-phase motors of smaller horsepower and on fan motors, machinery motors, and several other applications where it is not necessary to have automatic control.

Magnetic starters (Figures 18.11 and 18.12) are commonly used to start pumps, compressors, blowers, and anything where automatic or remote control is desired. They permit low-power circuits to energize the starter of equipment at a remote location or to start larger starters (Figure 18.13). A magnetic starter is operated by electromagnetic action. This starter has contactors and they operate by energizing a coil that closes the contact, thus starting the motor. The circuit that energizes the starter is called the control circuit and it may operate on a lower voltage (115 volts) than the motor. Whenever a starter is used as a part of an integrated circuit (such as for flow, pressure, or temperature control), a magnetic starter or controller is necessary.

Magnetic starters are sized for their voltage and horsepower ratings. These are divided into classes. The most common starter is Class A. A Class A starter is an alternating current, air-break and oil-immersed, manual or magnetic controller for service on 600 volts or less. It is capable of interrupting operating overloads up to and including 10 times their normal motor rating, but not short circuits or faults beyond operating overloads. Additional class information can be found in electrical catalogs, manuals, and manufacturers' brochures.

There are a number of different types of three-phase magnetic motor starters available. The simplest and most common is the across-the-line full-voltage starter. This starter consists of three contacts, a magnetic actuating device, and overload detection.

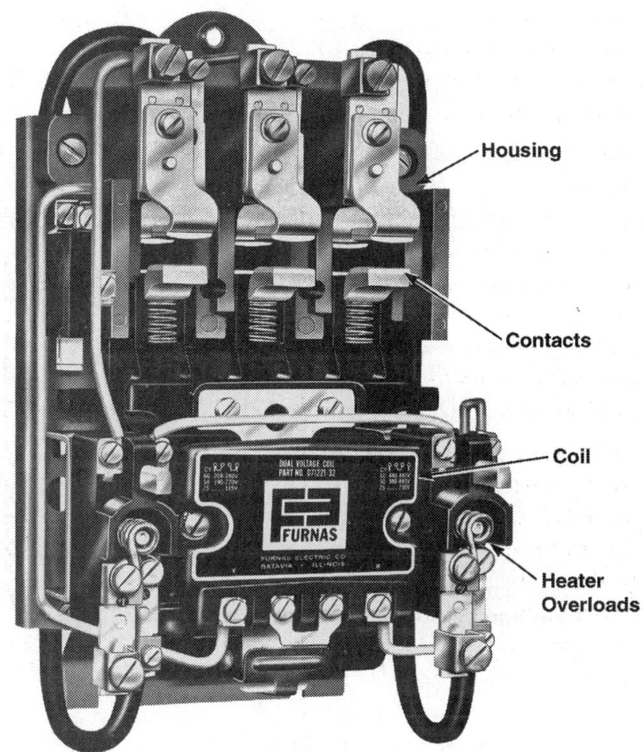

Fig. 18.11 Three-phase magnetic starter

(Courtesy of Furnas Electric Company)

This starter subjects the power system to the full surge current on startup and may cause the lights in the treatment plant to dim momentarily.

To reduce the inrush current when starting polyphase motors, a number of other types of starters are available.

1. Auto-Transformer-Type Reduced-Voltage Starters. These begin the motor start sequence by applying a reduced voltage to the motor for a few seconds. The voltage is controlled by a time-delay relay within the starter. The reduced voltage is obtained from transformers that are a part of the starter. These transformers are designed to operate for only a few seconds at a time and can easily be burned out if the motor is started too frequently.
2. Solid-State Reduced-Voltage Starters. These starters do the same job as the auto-transformer-type reduced-voltage starters but they do not need transformers because the voltage and current are electrically controlled.
3. Part-Winding Starters. These starters are used with special motors that have two separate sets of windings on the same motor frame. By energizing the windings about 1 second apart, the inrush current is limited to about half that of a normal motor with a full-voltage starter.
4. Wye-Delta Starters. These starters are used with motors that have all leads brought out to the terminal box. The motor is first started with wye-connected coils and switched over to a delta connection for running. The result is the same as if you used a reduced-voltage starter.

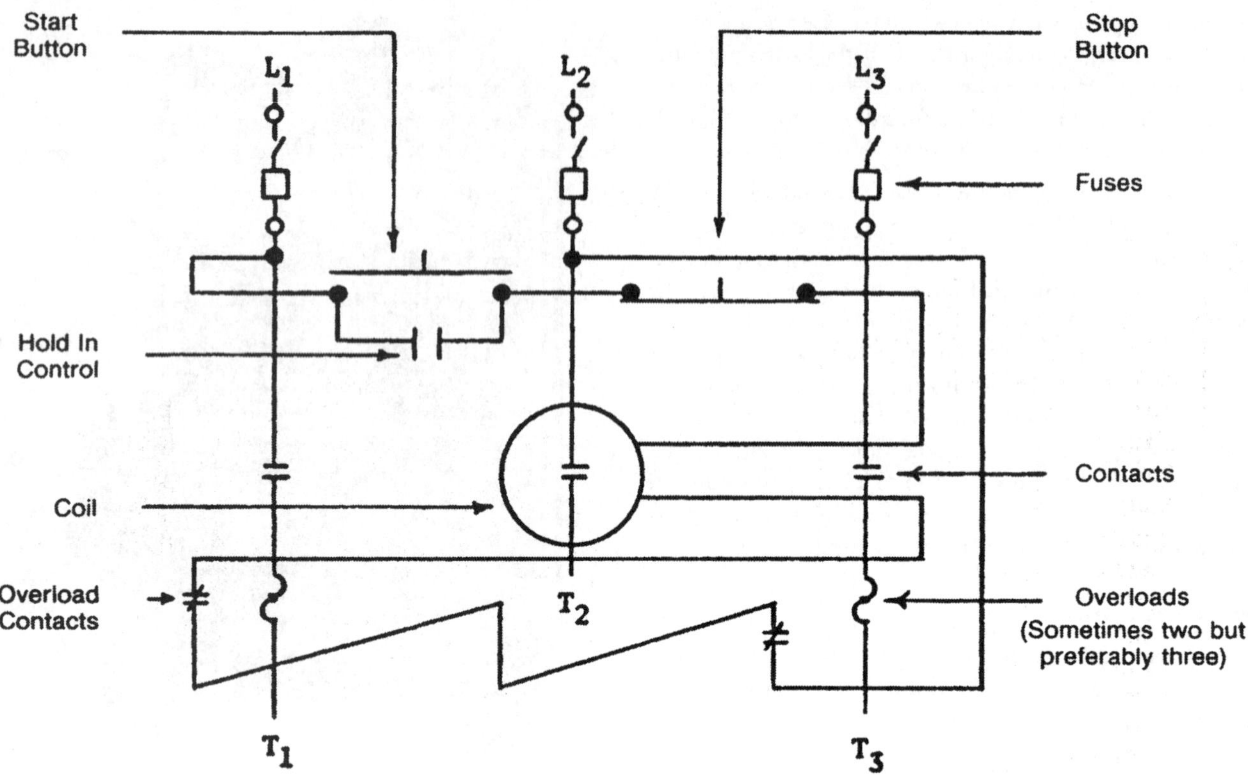

Fig. 18.12 Wiring diagram of three-phase magnetic starter

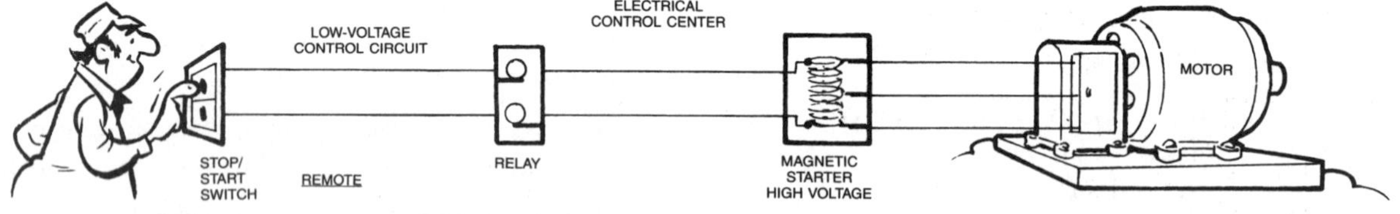

Fig. 18.13 Application of magnetic starter

QUESTIONS

Please write your answers to the following questions and compare them with those on page 368.

18.13A What are two types of safety devices found in main electrical panels or control units?

18.13B What are fuses used to protect?

18.13C Why must a fuse never be bypassed or jumped?

18.13D How does a circuit breaker work?

18.13E How are motor starters controlled?

18.13F When are magnetic starters used?

18.14 Electric Motors

18.140 Classifications

Electric motors are the machines most commonly used to convert electrical energy into mechanical energy. A motor usually consists of a *STATOR*,[8] *ROTOR*,[9] *END BELLS*,[10] and windings. The rotor has an extended shaft that allows a machine to be coupled to it.

There are many different types of electric motors, including the squirrel cage induction motor, wound rotor motor, synchronous motor, and others. Two examples are shown in Figure 18.14. The most common type is the squirrel cage induction motor. Some pumping stations use wound rotor induction motors when speed control is needed.

8. *Stator.* That portion of a machine that contains the stationary (nonmoving) parts that surround the moving parts (rotor).
9. *Rotor.* The rotating part of a machine. The rotor is surrounded by the stationary (nonmoving) parts (stator) of the machine.
10. *End Bells.* Devices used to hold the rotor and stator of a motor in position.

DRIP PROOF

ITEM NO.	PART NAME
1	Wound Stator w/ Frame
2	Rotor Assembly
3	Rotor Core
4	Shaft
5	Bracket
6	Bearing Cap
7	Bearings
8	Seal, Labyrinth
9	Thru Bolts/Caps
10	Seal, Lead Wire
11	Terminal Box
12	Terminal Box Cover
13	Fan
14	Deflector
15	Lifting Lug

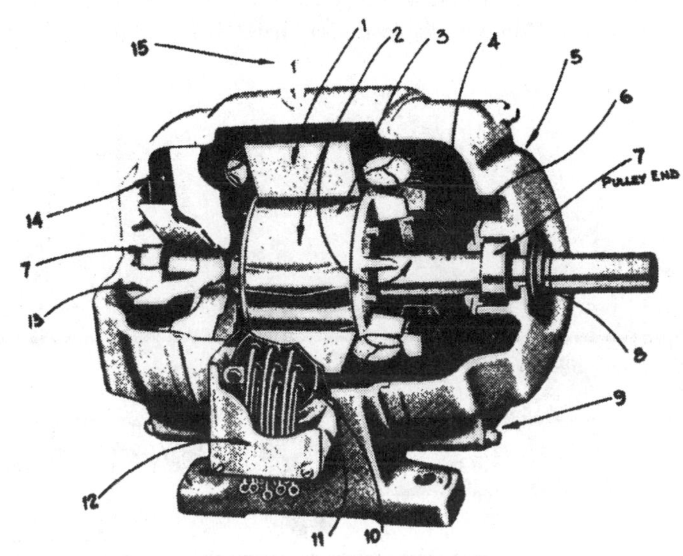

TOTALLY ENCLOSED FAN COOLED

ITEM NO.	PART NAME
1	Wound Stator w/ Frame
2	Rotor Assembly
3	Rotor Core
4	Shaft
5	Brackets
6	Bearings
7	Seal, Labyrinth
8	Thru Bolts/Caps
9	Seal, Lead Wire
10	Terminal Box
11	Terminal Box Cover
12	Fan, Inside
13	Fan, Outside
14	Fan Grill
15	Fan Cover
16	Fan Cover Bolts
17	Lifting Lug

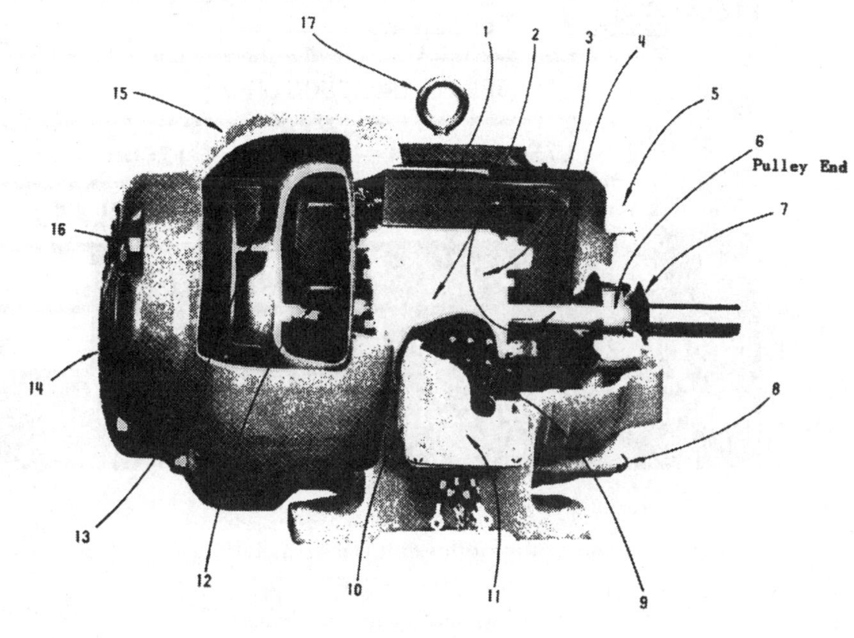

Fig. 18.14 Typical motors
(Courtesy of Sterling Power Systems, Inc.)

Three-phase electric motors are used for operating pumps, compressors, fans, and other machinery. Motors are generally trouble free and, when lubricated properly, cause very few problems. The amperage and voltage readings on motors should be taken periodically to ensure proper operation.

Motors are classified by NEMA with code letters from A through V with A having the lowest starting torque and inrush current and V having the highest starting torque and inrush current. The most commonly available motors have code letters from F through L, which have inrush currents on start of from 500 to 1,000 percent of full load.

Another important consideration in selecting a motor is the class of insulation. The class of insulation designates how high the temperature of the operating motor surface may rise above ambient temperature. The insulation class is indicated on the motor nameplate (Figure 18.15). Temperature ratings listed below are the allowable temperature increases above 104°F (40°C) ambient temperature. Motor insulation classes are as follows:

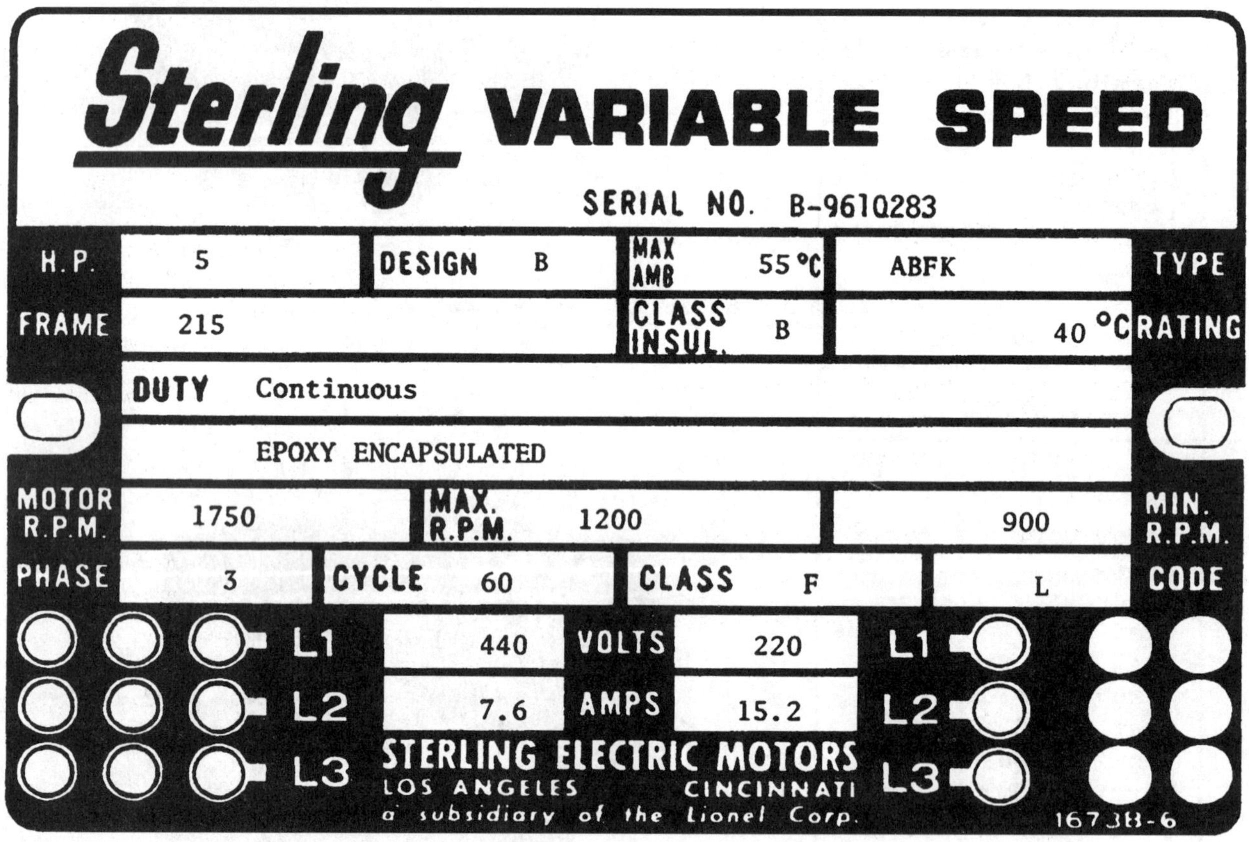

NOTE: 1. The motor for this unit is rated at 1,750 rpm and the maximum speed for the variable drive unit is 1,200 rpm.

2. The insulation is Class B.

Fig. 18.15 Typical nameplate

(Courtesy of Sterling Power Systems, Inc.)

Class	Temperature Rating
A	221°F (105°C)
B	266°F (130°C)
F	311°F (155°C)
H	356°F (180°C)

At present, most motors are Class B insulated. Try to keep the actual operating temperature below the temperature rating or limit to prolong the life of the insulation.

All of this information can be found on the motor nameplate and should be taken into consideration when evaluating a motor. Most of the trouble encountered with electric motors results from bad bearings, shorted windings due to insulation breakdown, or excessive moisture.

All of the information on the motor nameplate (Figure 18.15) should be recorded and placed in a file for future reference. Many times, the nameplate is painted, corroded, or missing from the unit when the information is needed to repair the motor or replace parts. Also record the date of installation and service startup. See Section 18.142, "Recordkeeping," for a typical data sheet for recording the essential information. This information also should be on the manufacturer's data sheet and in the instruction manual. Compare the information for consistency and file in an appropriate location. Be sure you have the correct serial or model numbers.

QUESTIONS

Please write your answers to the following questions and compare them with those on page 369.

18.14A How is electrical energy converted into mechanical energy?

18.14B What are the important parts of an electric motor?

18.14C How can motors be kept trouble free?

18.14D What should be done with motor nameplate data?

18.141 Troubleshooting

18.1410 STEP-BY-STEP PROCEDURES

The key to effective troubleshooting is the use of practical, step-by-step procedures combined with a common-sense approach. Never take anything for granted.

1. Gather preliminary information. The first step in troubleshooting any motor control that has developed trouble is to understand the circuit operation and other related functions. In other words, what is supposed to happen, operate, and so forth when it is working properly? Also, what is it doing now? The qualified maintenance operator should be able to do the following:
   a. *KNOW WHAT SHOULD HAPPEN WHEN A SWITCH IS PUSHED:* When switches are pushed or tripped, what coils go in, contacts close, relays operate, and motors run?
   b. *EXAMINE ALL OTHER FACTORS:* What other unusual things are happening in the pump station (facility) now that this circuit does not work properly? Lights dimmed, other pumps ran faster, lights went out when it broke, everything was flooded, operators were hosing down area, and many other possible factors.
   c. *ANALYZE WHAT YOU KNOW:* What part of it is working correctly? Is switch arm tripped? Everything but this is all right, except pump gets plugged with rags frequently. Is it a mechanical failure or an electrical problem caused by a mechanical failure?
   d. *SELECT SIMPLE PROCEDURES:* To localize the problem, select logical ways that can be simply and quickly accomplished.
   e. *MAKE A VISUAL INSPECTION:* Check for burned wires, loose wires or contacts, pooled or standing water, burned coils, or strange odors.
   f. *CONVERGE ON SOURCE OF TROUBLE:* Mechanical or electrical. Motor or control, whatever it might be. Electrical problems result from some type of mechanical failure.
   g. *PINPOINT THE PROBLEM:* Exactly where is the problem and what do you need for repair?
   h. *FIND THE CAUSE:* What caused the problem? Consider moisture, wear, poor design, voltage, or overloading.
   i. *REPAIR THE PROBLEM AND ELIMINATE THE CAUSE, IF POSSIBLE:* If the problem is inside switch gear or motors, call an electrician. Give the electrician the information you have regarding the equipment. Do not attempt electrical repairs unless you are qualified and authorized, otherwise you could cause excessive damage to yourself and to the equipment.
2. Some of the things to look for when troubleshooting are given in the remainder of this section.

18.1411 TROUBLESHOOTING GUIDE FOR ELECTRIC MOTORS

Symptom	Cause	Result[a]	Remedy
1. Motor does not start (switch is on and not defective)	a. Incorrectly connected	a. Burnout	a. Connect correctly per diagram on motor.
	b. Incorrect power supply	b. Burnout	b. Use only with correctly rated power supply.
	c. Fuse out, loose or open connection	c. Burnout	c. Correct open circuit condition.
	d. Rotating parts of motor may be jammed mechanically	d. Burnout	d. Check and correct: 1. Bent shaft. 2. Broken housing. 3. Damaged bearing. 4. Foreign material in motor.
	e. Driven machine may be jammed	e. Burnout	e. Correct jammed condition.
	f. No power supply	f. None	f. Check for voltage at motor and work back to power supply.
	g. Internal circuitry open	g. Burnout	g. Correct open circuit condition.
2. Motor starts but does not come up to speed	a. Same as 1-a, -b, -c above	a. Burnout	a. Same as 1-a, -b, -c above.
	b. Overload	b. Burnout	b. Reduce load to bring current to rated limit. Use proper fuses and overload protection.
	c. One or more phases out on a three-phase motor	c. Burnout	c. Look for open circuits.
3. Motor noisy (electrically)	a. Same as 1-a, -b, -c above	a. Burnout	a. Same as 1-a, -b, -c above.
4. Motor runs hot (exceeds rating)	a. Same as 1-a, -b, -c above	a. Burnout	a. Same as 1-a, -b, -c above.
	b. Overload	b. Burnout	b. Reduce load.
	c. Impaired ventilation	c. Burnout	c. Remove obstruction.
	d. Frequent starts or stops	d. Burnout	d. 1. Reduce number of starts or reversals. 2. Secure proper motor for this duty.
	e. Misalignment between rotor and stator laminations	e. Burnout	e. Realign.
5. Motor noisy (mechanically)	a. Misalignment of coupling or sprocket	a. Bearing failure, broken shaft, stator burnout due to motor drag	a. Correct misalignment.
	b. Mechanical imbalance of rotating parts	b. Same as 5-a	b. Find imbalanced part, then balance.
	c. Lack of or improper lubricant	c. Bearing failure	c. Use correct lubricant, replace parts as necessary.

18.1411 TROUBLESHOOTING GUIDE FOR ELECTRIC MOTORS (continued)

Symptom	Cause	Result[a]	Remedy
5. Motor noisy (mechanically) *(continued)*	d. Foreign material in lubricant	d. Bearing failure	d. Clean out and replace bearings.
	e. Overload	e. Bearing failure	e. Remove overload condition. Replace damaged parts.
	f. Shock loading	f. Bearing failure	f. Correct causes and replace damaged parts.
	g. Mounting acts as amplifier of normal noise	g. Annoying	g. Isolate motor from base.
	h. Rotor dragging due to worn bearings, shaft, or bracket	h. Burnout	h. Replace bearings, shaft, or bracket as needed.
6. Bearing failure	a. Same as 5-a, -b, -c, -d, -e	a. Burnout, damaged shaft, damaged housing	a. Replace bearings and follow 5-a, -b, -c, -d, -e.
	b. Entry of water or foreign material into bearing housing	b. Burnout, damaged shaft, damaged housing	b. Replace bearings and seals and shield against entry of foreign material (water, dust, etc.). Use proper motor.

Symptom	Caused By	Appearance
1. Shorted motor winding	a. Moisture, chemicals, foreign material in motor, damaged winding	a. Black or burned coil with remainder of winding good.
2. All windings completely burned	a. Overload	a. Burned equally all around winding.
	b. Stalling	b. Burned equally all around winding.
	c. Impaired ventilation	c. Burned equally all around winding.
	d. Frequent reversal or starting	d. Burned equally all around winding.
	e. Incorrect power	e. Burned equally all around winding.
3. Single-phase condition	a. Open circuit in one line. The most common causes are loose connection, one fuse out, loose contact in switch.	a. If 1,800 rpm motor—four equally burned groups at 90° intervals.
		b. If 1,200 rpm motor—six equally burned groups at 60° intervals.
		c. If 3,600 rpm motor—two equally burned groups at 180° intervals.
		NOTE: If wye connected, each burned group will consist of two adjacent phase groups. If delta connected, each burned group will consist of one-phase group.
4. Other	a. Improper connection	a. Irregularly burned groups or spot burns.
	b. Ground	

a Many of these conditions should trip protective devices rather than burn out motors. Also, many burnouts occur within a short period of time after motor is started up. This does not necessarily indicate that the motor was defective, but usually is due to one or more of the above-mentioned causes. The most common causes of failure shortly after startup are improper connections, open circuits in one line, incorrect power supply, or overload.

18.1412 TROUBLESHOOTING GUIDE FOR MAGNETIC STARTERS

Trouble	Possible Cause	Remedy
CONTACTS		
Contact chatter	1. Broken *POLE SHADER*[11]	1. Replace.
	2. Poor contact in control circuit	2. Improve contact or use holding circuit interlock.
	3. Low voltage	3. Correct voltage condition. Check momentary voltage drop.
Welding or freezing	1. Abnormal surge of current	1. Use larger contactor and check for grounds, shorts, or excessive motor load current.
	2. Frequent *JOGGING*[12]	2. Install larger device rated for jogging service or caution operators.
	3. Insufficient contact pressure	3. Replace contact spring; check contact carrier for damage.
	4. Contacts not positioning properly	4. Check for voltage drop during startup.
	5. Foreign matter preventing magnet from seating	5. Clean contacts.
	6. Short circuit	6. Remove short fault and check that fuse and breaker are right.
Short contact life or overheating of tips	1. Contacts poorly aligned, poorly spaced, or damaged	1. Do not file silver-faced contacts. Rough spots or discoloration will not harm contacts. Replace.
	2. Excessively high currents	2. Install larger device. Check for grounds, shorts, or excessive motor currents.
	3. Excessive starting and stopping of motor	3. Caution operators. Check operating controls.
	4. Weak contact pressure	4. Adjust or replace contact springs.
	5. Dirty contacts	5. Clean with approved solvent.
	6. Loose connections	6. Check terminals and tighten.
Coil overheated	1. Starting coil may not kick out	1. Repair coil.
	2. Overload will not let motor reach minimum speed	2. Remove overload.
	3. Overvoltage or high ambient temperature	3. Check application and circuit.
	4. Incorrect coil	4. Check rating; if incorrect, replace with proper coil.
	5. Shorted turns caused by mechanical damage or corrosion	5. Replace coil.

11. *Pole Shader.* A copper bar circling the laminated iron core inside the coil of a magnetic starter.
12. *Jogging.* The frequent starting and stopping of an electric motor.

18.1412 TROUBLESHOOTING GUIDE FOR MAGNETIC STARTERS (continued)

Trouble	Possible Cause	Remedy
CONTACTS (continued)		
Coil overheated *(continued)*	6. Undervoltage, failure of magnet to seal it	6. Correct system voltage.
	7. Dirt or rust on pole faces increasing air gap	7. Clean pole faces.
Overload relays tripping	1. Sustained overload	1. Check for grounds, shorts, or excessive motor currents. Mechanical overload.
	2. Loose connection on all or any load wires	2. Check, clean, and tighten.
	3. Incorrect heater	3. Replace with correct size heater unit.
	4. Fatigued heater blocks	4. Inspect and replace.
Failure to trip	1. Mechanical binding, dirt, or corrosion	1. Clean or replace.
	2. Wrong heater, or heaters omitted and jumper wires used	2. Check ratings. Apply heaters of proper rating.
	3. Motor and relay in different temperatures	3. Adjust relay rating accordingly, or install temperature compensating relays.
MAGNETIC AND MECHANICAL PARTS		
Noisy magnet (humming)	1. Broken shading coil	1. Replace shading coil.
	2. Magnet faces not mating	2. Replace magnet assembly or realign.
	3. Dirt or rust on magnet faces	3. Clean and realign.
	4. Low voltage	4. Inspect system voltage and voltage dips or drops during startup.
Failure to pick up and seal	1. Low voltage	1. Inspect system voltage and correct.
	2. Coil open or shorted	2. Replace.
	3. Wrong coil	3. Check coil number and voltage rating.
	4. Mechanical obstruction	4. With power off, check for free movement of contact and armature assembly. Repair.
Failure to drop out	1. Gummy substance on pole	1. Clean with solvent.
	2. Voltage not removed from coil	2. Check coil circuit.
	3. Worn or rusted parts causing binding	3. Replace or clean parts as necessary.
	4. Residual magnetism due to lack of air gap in magnet path	4. Replace worn magnet parts or align, if possible.
	5. Welded contacts	5. Shorted circuit, grounded, overloaded.

18.1413 TROUBLE/REMEDY PROCEDURES FOR INDUCTION MOTORS

1. Motor will not start.

   Overload control tripped. Wait for overload to cool, then try to start again. If motor still does not start, check for the causes outlined below.

   a. Open fuses: test fuses.
   b. Low voltage: check nameplate values against power supply characteristics. Also check voltage at motor terminals when starting motor under load to check for allowable voltage drop.
   c. Wrong control connections: check connections with control wiring diagram.
   d. Loose terminal-lead connection: turn power off and tighten connections.
   e. Drive machine locked: disconnect motor from load. If motor starts satisfactorily, check driven machine.
   f. Open circuit in stator or rotor winding: check for open circuits.
   g. Short circuit in stator winding: check for short.
   h. Winding grounded: test for grounded wiring.
   i. Bearing stiff: free bearing or replace.
   j. Overload: reduce load.

2. Motor noisy.

   a. Three-phase motor running on single phase: stop motor, then try to start. It will not start on single phase. Check for open circuit in one of the lines.
   b. Electrical load imbalance: check current balance.
   c. Shaft bumping (sleeve-bearing motor): check alignment and conditions of belt. On pedestal-mounted bearing, check for play and axial centering of rotor.
   d. Vibration: driven machine may be imbalanced. Remove motor from load. If motor is still noisy, rebalance.
   e. Air gap not uniform: center the rotor and, if necessary, replace bearings.
   f. Noisy ball bearing: check lubrication. Replace bearings if noise is excessive and persistent.
   g. Rotor rubbing on stator: center the rotor and replace bearings, if necessary.
   h. Motor loose on foundation: tighten hold-down bolts. Motor may possibly have to be realigned.
   i. Coupling loose: insert feelers at four places in coupling joint before pulling up bolts to check alignment. Tighten coupling bolts securely.

3. Motor at higher than normal temperature or smoking. (Measure temperature with thermometer or thermister and compare with nameplate value.)

   a. Overload: measure motor loading with ammeter. Reduce load.
   b. Electrical load imbalance: check for voltage imbalance or single phasing.
   c. Restricted ventilation: clean air passage and windings.
   d. Incorrect voltage and frequency: check nameplate values with power supply. Also check voltage at motor terminals with motor under full load.
   e. Motor stalled by driven tight bearings: remove power from motor. Check machine for cause of stalling.
   f. Stator winding shorted or grounded: test windings by standard method.
   g. Rotor winding with loose connection: tighten, if possible, or replace with another rotor.
   h. Belt too tight: remove excessive pressure on bearings.
   i. Motor used for rapid reversing service: replace with motor designed for this service.

4. Bearings hot.

   a. End shields loose or not replaced properly: make sure end shields fit squarely and are properly tightened.
   b. Excessive belt tension or excessive gear side thrust: reduce belt tension or gear pressure and realign shafts. See that thrust is not being transferred to motor bearing.
   c. Bent shaft: straighten shaft or send to motor repair shop.

5. Sleeve bearings.

   a. Insufficient oil: add oil. If supply is very low, drain, flush, and refill.
   b. Foreign material in oil or poor grade of oil: drain oil, flush, and relubricate using industrial lubricant recommended by a reliable oil manufacturer.
   c. Oil rings rotating slowly or not rotating at all: oil too heavy; drain and replace. If oil ring has worn spot, replace with new ring.
   d. Motor tilted too far: level motor or reduce tilt and realign, if necessary.
   e. Rings bent or otherwise damaged in reassembling: replace rings.
   f. Rings out of slot (oil-ring retaining clip out of place): adjust or replace retaining clip.
   g. Defective bearings or rough shaft: replace bearings. Resurface shaft.

6. Ball bearings.

   a. Too much grease: remove relief plug and let motor run. If excess grease does not come out, flush and relubricate.
   b. Wrong grade of grease: flush bearing and relubricate with correct amount of proper grease.
   c. Insufficient grease: remove relief plug and grease bearing.
   d. Foreign material in grease: flush bearing, relubricate; make sure grease supply is clean (keep can covered when not in use).

18.142 Recordkeeping

Records are a very important part of electrical maintenance. They must be accurate and complete. Whenever something is changed, repaired, or tested, it should be recorded on a material history card of some type. Figures 18.16 and 18.17 are examples of typical record sheets.

PUMP RECORD CARD

NAME__________ MAKE__________ MODEL__________

TYPE__________ SIZE__________ SERIAL #__________

ORDER NUMBER______ SUPPLIER__________ DATE PURCHASED__________

DATE INSTALLED______ APPLICATION__________ PLANT #__________

Name Plate Data and Pump Info	Stuffing Box Data	Motor Data
GPM ______	Diameter____ Depth____	Name______ Serial #______
TDH ______	Pack. Size___ Type____	H.P.______ Speed______
RPM ______	Length____ No. Rings____	Ambient° __________
Gage Press Disc___ ______	Lantern Ring____ Flushed____	RPM____ Frame______
Gage Press Suc ___ ______	Mech. Seal Name____ Size___	Volts____ Amps______
Shut off Press ___ ______	Type__________	Phase____ Cycle______
Suction Head ___ ______	**Pump Materials**	Shaft Size____ Key____
Rotation ___ ______	Casing __________	Bearing Front______
Impeller Type__________	Shaft__________	Rear______
Impeller Dia.__________	Wearing Rings Casing ______	Code______ Type______
Impeller Clear__________	Wearing Rings Impeller______	Amps @ Max. Speed______
Coupl Type & Size__________	Shaft Sleeve__________	Amps @ Shut Off______
Front Brg #__________	Slinger__________	Control Data Info
Rear Brg #__________	Shims__________	Starter__________
Lub Interval__________	Gaskets__________	MENA Size__________
Lubricant__________	"O" Rings__________	Cat. #__________
Wearing Rings__________	Brg. Seals Front__________	Heater Size__________
Shaft Sleeve Size__________	Rear__________	Rated @__________
Pump Shaft Size__________	Casing Wear Ring Size ID______	Control Voltage__________
Pump Keyway__________	OD______	Variable Speed Type__________
__________	Width______	Speed Max__________
Other Related Information:	Impeller Wear Ring ID______	Speed Min__________
	OD______	
	Width______	

Fig. 18.16 Pump record card

MOTOR STARTERS Number______

Title:______

Mfg.:______ Address______

Style:______ Class______ Size______

Type:______ ______ ______

O.L. HEATERS

Style______ Code______

Amps______ ______

______ ______

O.L. TRIP UNITS

Mfg:______ Style:______

Type:______ ______

Amps Range:______ ______

CIRCUIT BREAKER

Mfg:______ Address______

Style:______ Frame:______ Volts______ Amps Setting______

Cat. No.______ ______ ______ ______

MOTOR Number______

TITLE______

Mfg:______ Address______

HP:______ Volts:______ Ser. No.______ Duty:______

Phase:______ Amps:______ Frame:______ Temp:______

Cycles:______ RPM:______ Type______ Class:______

Code:______ S.F.:______ Model______ Spec.:______

SO#______ S#______ Style:______ CSA App:______

Form______ Spec.______ Shft. Brg.______ Rear Brg.______

50 Cycle Data______

Suitable for 208V Network:______

Additional data______

______

______

______

Connection Diagram

(6) (5) (4) (6) (5) (4)

(7) (8) (9) (7) (8) (9)

(1) (2) (3) (1) (2) (3)

Fig. 18.17 Motor record card

QUESTIONS

Please write your answers to the following questions and compare them with those on page 369.

18.14E What is the key to effective troubleshooting?

18.14F What are some of the steps that should be taken when troubleshooting magnetic starters?

18.14G What kind of information should be recorded regarding electrical equipment?

18.15 Auxiliary Electrical Power

18.150 Safety First

Always remember that a qualified electrician should perform most of the necessary maintenance and repair of electrical equipment. If you do not know the how, why, and when of the job, do not do it. You could endanger your life as well as your fellow operators or damage equipment. Never attempt work that you are not qualified to do or are not authorized to perform.

18.151 Standby Power Generation

There are three ways of providing standby power. One is by providing the treatment plant with an engine-driven generator set. The limit of how much power can be produced is determined only by the size of the generator. The second possibility for standby power is batteries. Batteries should only be considered for low power consumption uses such as emergency lighting, communication, and possibly some control and instrumentation functions. The other possibility for standby power is a connection to an alternate power source such as a different substation or another power company.

Because the treatment of water is considered a critical service, it is important to be able to provide drinking water even with the loss of commercial power. A power outage of a short duration probably will not have adverse effects on plant operation. The question you must ask yourself is, "Can my plant meet the needs of the public if a brownout or catastrophic event eliminates commercial power for an extended length of time?" If the answer is "no," then perhaps a form of standby power generation should be considered.

The following six conditions must be analyzed to determine the need for and size of standby power generation:

1. Frequency of power outages in the last 10 to 15 years
2. Duration of the power outage in each occurrence
3. Availability of an additional source of power supply from a different substation in the vicinity
4. Method by which raw water reaches the plant (by gravity or by a raw water pumping station?)
5. Total storage capacity of reservoirs in the distribution system
6. Possibility of obtaining a potable water supply from adjacent cities (for example, a reasonably sized pipe connection between your system and the distribution system of an adjacent city)

If the frequency of power outages is once or twice a year with a 10- to 30-minute duration, the capacity of a standby power generator can be relatively small. The minimum size of a standby power generator should provide sufficient capacity to operate essential equipment:

1. Coagulant and chlorine feeders
2. One-third of flocculators
3. Major electric valve operators and plant control system
4. One-third of pumping capacity (if necessary)
5. Minimum lighting

Where do you begin? You have to consider whether you would like to have all of your facility operating or whether just the vital or key equipment would be sufficient. Since the characteristics and operating conditions of every plant are different, it is difficult to make specific suggestions.

For the sake of illustration, let us pose a hypothetical situation. Consider a 10 MGD (38 MLD) capacity plant with an average flow rate of 6 MGD (23 MLD). Prepare a list of needs that must be met to ensure minimal operation:

1. Raw water pumping
2. Clarification
3. Clear water pumping
4. Chlorination
5. Minimal lighting

Calculate the maximum horsepower or total kilowatts necessary to maintain the limited operation:

1. Raw Water Pump—75 horsepower	56.00 kW
2. Clarification—2½ horsepower	2.24 kW
3. Clear Water Pump—40 horsepower	30.00 kW
4. Chlorination—15 horsepower	11.20 kW
5. Lighting	5.00 kW
	104.44 kW

The minimum power required is 104.44 kW. When sizing a generator for emergency power, you have to make sure that the operator will be able to start the needed motors. Since the locked rotor current of the 75-horsepower induction motor on the raw water pump is approximately four times running current, the generator must be able to handle 224 kW at that instant. Size the generator not only by total load, but also for the highest horsepower motor being started. Consider the sequence in which motors will be started. The starting of all the motors simultaneously (without sequence starting) would be nearly impossible. Consult experts in power generation for answers to your specific questions regarding your plant because each plant has different needs. If you are considering standby power, shop around and get ideas from the equipment manufacturers. You may be able to reduce the size of the generator by using reduced-voltage starters on the larger motors.

After you have determined the size of generator needed, you must be able to connect it to your power distribution system. This may require some sophisticated switch gear. Besides the mechanical functions necessary in connecting the emergency power with your normal system, it is important that the two systems cannot be electrically coupled. (Two electrical systems must be in phase with each other before parallel coupling.) For this reason, mechanical interlocks are used to ensure that one circuit is always open. A kirk-key system, where one key is used for two locks, locking one switch open before the other can be closed, is sometimes used. The manufacturers of most packaged motor-generator systems can provide automatic transfer switches that will automatically start the generator when a power failure occurs and connect the generated power into the plant power distribution wiring.

Looking back at the plant described, a generator of 125 kW with intermittent overload capabilities should handle the load. (*NOTE:* This is an assumption. Actual calculation may indicate a different size.) An engine-generation system of this size could handle your minimal power needs. If your water distribution system has ample capacity, it may be possible to cut the plant production rate to reduce power requirements to what can be handled with a smaller capacity generator.

If you do not have standby power generation at your facility, talk to others in the water treatment field who do and obtain ideas and information. After due consideration, take the necessary steps to ensure your plant against interrupted power.

Standby power generators should be operated once a week to be sure they will operate properly, when needed. Be sure to operate your generator at full load for at least an hour. Commercial power to your plant must be shut off to operate standby power at full load.

18.152 Emergency Lighting

Your treatment plant is required to have an emergency lighting system that has an uninterruptible power supply, responds instantaneously, and provides 1.5 hours of continuous lighting when normal lighting sources fail. The most practical form of emergency lighting, in most instances, is provided by battery-powered, self-contained lighting units. Because they are used primarily for exit lighting, these units are more economical than electrically powered units. All emergency lighting equipment has basically the same design, consisting of a rechargeable battery, battery charger, low-voltage flood lights, and test monitoring and control features. Selecting an appropriate unit for a particular location requires careful consideration of your lighting requirements and the unit's initial cost, battery type, and maintenance requirements.

The three types of batteries most commonly used are lead acid, lead calcium, and nickel cadmium. Because poor battery maintenance is common in emergency lighting systems, maintenance-free batteries are frequently used. These batteries can have a gelatin or acid (wet) *ELECTROLYTE.*[13] The gelatin type is spillproof and can be handled safely without the danger of acid spills. These batteries have a shorter life span than the wet type. Since all batteries undergo evaporation, the gelatin electrolyte will be exhausted before a battery containing liquid. Wet-type, maintenance-free batteries require no refilling and, when handled properly, acid spillage is minimal.

In terms of initial cost, maintenance-free batteries are more expensive; but when you consider the human factor, they may be more reliable and cheaper in the long run. Most systems use a battery charger that monitors the battery voltage. When required, the charger then charges the batteries. In older units, a trickle charger was used, providing constant charging that caused the overcharging of batteries in a short period. This reduced battery life and required more frequent replacement.

The lamps used in emergency lighting systems are normally 6- to 12-volt, sealed-beam, 25-watt lamps. The light pattern provided is most effective when illuminating a work area. A rule of thumb is that one lamp will be sufficient for about 1,000 square feet, providing that the full light pattern can be used.

Before purchasing or installing an emergency lighting system, consult emergency light level code requirements for your facility. These codes vary widely throughout the United States. The National Fire Protection Association (NFPA), a recognized authority on fire, electrical, and building safety standards, details emergency lighting (illumination) standards in its publication, *NFPA 101: Life Safety Code.* The following list is adapted from these standards to provide you with an overview of emergency lighting system requirements.

1. Emergency lighting must activate instantaneously and automatically when the normal power source of lighting fails.
2. Emergency lighting must last for at least 1.5 hours initially following a power outage.
3. At the end of the initial emergency lighting period following a power outage (1.5 hours), it is permissible for illumination to fade to 0.6 foot-candle of light at any location in the building and 0.06 foot-candle of light at floor level along the emergency exit route.
4. Emergency lighting must emit 1 foot-candle of light at any point in the building and 0.1 foot-candle of light at floor level along the emergency exit route.
5. Maximum to minimum illumination uniformity cannot exceed a ratio of 40 to 1.
6. Exit routes, including stairs, aisles, corridors, and ramps are required to have emergency lighting.
7. Exits must be marked with approved signs that are easily visible along the entire evacuation route.
8. The word "EXIT" on illuminated signs must have letters that are at least 6 inches high and ¾ inch wide.
9. Exit signs must be illuminated by an uninterruptible power supply, preferably one in a self-contained, battery-powered unit.
10. When servicing the organization's regular lighting system, there must be an uninterruptible power supply available for the emergency lighting system.
11. Functional testing of an emergency lighting system must be conducted on a monthly basis, with a minimum of 3 weeks and a maximum of 5 weeks between tests.

---

13. *Electrolyte* (ee-LECK-tro-lite). A substance that dissociates (separates) into two or more ions when it is dissolved in water.

18.153 Batteries

This section will discuss wet storage batteries since they are the most prevalent. Automotive and equipment batteries are usually of the lead-acid type in which the dissimilar plates are made of two types of lead and the electrolyte is sulfuric acid. Wet-type batteries can also be nickel cadmium or nickel iron.

Most batteries are a series of cells enclosed in a common case. Each of these cells develops a potential (voltage) of 2.3 volts per cell when fully charged. Hence, a 6-volt battery contains three cells and a 12-volt battery has six cells. The voltage output of a 12-volt battery is 13.8 volts when fully charged. Once a lead-acid battery has been placed in service, the addition of sulfuric acid is not necessary. The water portion of the electrolyte solution evaporates as the battery is charged and discharged. Lost water must be replaced. Deionized or distilled water should be used. Tap water contains impurities that shorten the lifetime of a battery if used to replace lost water. These minute particles become attached to the lead plates and do not allow the battery to rejuvenate itself fully when charged.

When batteries are placed on charge, remove the cell covers to allow the gas (hydrogen) caused by charging to escape and not to build excessive pressure in the battery. A battery on charge is as lethal as a small bomb if you ignite the gas. Do not smoke or cause electrical arcing near the battery. Do not breathe the gas. Make sure that the area where a battery is being charged is well ventilated.

The keys to prolonged life of a battery are to keep the electrolyte level above the cell plates, to keep the battery fully charged, and, above all, to keep the terminals and top clean. When dirt and residue accumulate on the top of a battery, it forms a path for current to flow between the negative and positive posts. Take a multimeter, connect one lead to the proper post (it will cause up-scale deflection), and slowly slide the other lead across the top of the battery toward the other post. If the top is dirty, the meter will deflect more as you proceed across the top.

To clean the battery, use a stiff-bristled brush (not a wire brush) and remove the heavy material. Then, wash with a solution of baking soda and water (4 teaspoons of baking soda to 1 quart of water). This will remove the acid film from the top and neutralize corrosion on the battery terminals. Rinse with fresh water and dry the top with a dry, lintless cloth. Remove cell caps and wipe between them, then replace. At this time, check to be sure that the battery terminals are clean and tight. If a battery is charged, but the terminals are loose, proper voltage and current cannot be delivered.

QUESTIONS

Please write your answers to the following questions and compare them with those on page 369.

18.15A Why should a qualified electrician perform most of the necessary maintenance and repair of electrical equipment?

18.15B What is the purpose of a kirk-key system?

18.15C Why are battery-powered backup lighting units considered better than engine-driven power sources?

18.15D Why should the water lost from a lead-acid battery be replaced with deionized or distilled water?

18.16 High Voltage

18.160 Transmission

In general terms, high voltage is the voltage transmitted to the plant site by the utility company. The voltage level can vary, but 12,000 volts is quite common. After the power reaches the plant, it is transformed down to a usable voltage (460 to 480 volts) either through utility-owned or customer-owned transformers. The National Electrical Code (NEC) denotes high voltages as those over 600 volts.

Why have high voltage? Since current (amperes) varies inversely with voltage, a load of 500 amps on the low-voltage side of the transformer would create a 20-amp load on the high-voltage side of a 12,000-volts/480-volt transformer. Transmission lines would have to be enormous in order to carry the load if a lower voltage were used. Where high-voltage cables terminate at a transformer or switch gear, certain conditions must be adhered to. If outdoor transformers are used that have high-voltage wires exposed, an 8-foot-high (2.4 m) fence is required to prevent accessibility by unqualified or unauthorized persons. Signs attached to the fence must indicate "High Voltage." Specifications for clearances, grounding, access, and enclosures vary with installations. Any modification or repair work must be completed by qualified people only.

18.161 Switch Gear

When we see the term "switch gear," it is usually associated with the equipment used in the interruption, transfer, or disconnecting of voltages over 600 volts. The enclosure is designed and manufactured to safely control high-voltage switching. Most distribution systems have a load-interrupting switch that is capable of disconnecting high-voltage lines that are under load. Because of the arc that is caused in breaking the circuit, special "arc shoes" (arc-suppressant devices) are used to ensure that the contact points are not pitted. A keyed lock system is used to prevent opening of the enclosure in the energized state.

Probably the best preventive maintenance that a treatment plant operator can provide for the switch gear is to keep the exterior and its surroundings clean. If you encounter difficulties in the course of operating the switches, please obtain qualified help to do the inspection or repairs needed. Check with your particular manufacturer to determine what is needed and when this has to be done to keep your system functioning as designed. If your equipment is in a corrosive atmosphere, it may be necessary to remove it from service and epoxy paint the internal buses. All pivoting points should be lubricated with a lubricant specified by the manufacturer.

18.162 Power Distribution Transformers

If the high-voltage transformers are owned by the utility, the inspection and maintenance is carried out by the utility. Any peculiar changes, smells, or noises should be reported to the utility. When transformers are customer owned, a regular inspection program should be established.

Most transformers use an oil to insulate as well as to cool the windings. As heat is generated in the windings, it is transferred to the oil. The oil is then cooled by air passing the cooling fins of the transformer. The primary requirements of the oil are:

1. High dielectric strength
2. Freedom from inorganic acid, alkali, and sulfur to prevent damage to insulation and conductors
3. Low viscosity to provide good heat transfer
4. Freedom from sludging under normal operation conditions

The principal causes of deterioration of insulating oil are water and oxidation. The oil may be exposed to moisture through condensation of moist air due to breathing of the transformer, especially when the transformer is not continuously in service. The moist air condenses on the surface of the oil and on the inside of the tank. Oxidation causes sludging. The amount of sludge formed in a given oil depends upon the temperature and the time of exposure of the oil to the air. Excessive operating temperatures may cause sludging of any transformer oil. Check with the manufacturer to determine how often the oil should be tested. Oil can be revitalized by a cleaning procedure that is accomplished at the transformer site.

Any symptoms such as unusual noises, high or low oil levels, oil leaks, or high operating temperatures should be investigated at once. If your transformer has a thermometer, it is of the alcohol type and should be replaced with that type only. A mercury-type thermometer could cause insulation failures due to the proximity of a metallic substance, regardless of whether the thermometer is intact or broken.

The tank of every power transformer should be grounded to eliminate the possibility of obtaining static shocks from it or from being injured by accidental grounding of the winding to the case.

If repairs are indicated, use the expertise of a qualified person to ensure that the repairs are made safely as well as correctly. Your life and the lives of others may depend on the use of qualified people.

QUESTIONS

Please write your answers to the following questions and compare them with those on page 369.

18.16A Why is electricity transmitted at high voltage?

18.16B What precautions must be taken if outdoor transformers have exposed high-voltage wires?

18.16C What kind of maintenance should a treatment plant operator perform on the switch gear?

18.16D What symptoms indicate that a power distribution transformer may be in need of maintenance or repair?

18.17 Electrical Safety Checklist

Throughout this manual and throughout this chapter the need for electrical safety is always being stressed. This section contains an electrical safety checklist, which is provided to help you ensure that you have minimized electrical hazards in your plant. This list is provided to make you aware of potential electrical hazards. You should add to the list additional electrical hazards that could injure someone at your water treatment plant.

1. Are there any conduits rusted to the point where they might have lost their explosion-proof integrity?
2. Are there any electrical conduit hangers that are so rusted that they are allowing the conduit to sag?
3. Are there any fasteners on the conduit hangers that are rusted and allowing the conduit to hang by the wires?
4. Do all of the extension cords and power tools meet code requirements for use in wet areas?
5. Does the agency or utility have a policy covering the proper placement of portable ventilation equipment when operators work inside enclosed tanks, vaults, and other confined spaces?
6. Does the agency use proper grounding units (ground-fault circuit interrupters) when working in wet areas?
7. Is the grounding of electrical equipment and systems inspected regularly?
8. Are electrical breakers and controls clearly marked?
9. Is there a formal program for locking out, tagging, and blocking out of electrical devices?

If you can answer these questions properly, you are working in the right direction to minimize electrical hazards in your water treatment plant.

QUESTIONS

Please write your answers to the following questions and compare them with those on page 369.

18.17A What is the purpose of an electrical safety checklist?

18.17B Why are rusted conduits of concern to a water treatment plant operator?

18.18 Additional Reading

1. *Basic Electricity* by Van Valkenburgh, Nooger & Neville, Inc. Published by Prompt Publications. ISBN 978-0-7906-1041-2.
2. *Operation of Wastewater Treatment Plants*, Volume II, Chapter 15, "Maintenance." Obtain from the Office of Water Programs, California State University, Sacramento, at www.owp.csus.edu or phone (916) 278-6142. ISBN 978-1-59371-038-5.
3. *Advanced Waste Treatment*, Chapter 9, "Instrumentation and Control Systems." Obtain from the Office of Water Programs, California State University, Sacramento, at www.owp.csus.edu or phone (916) 278-6142. ISBN 1-59371-035-6.
4. *Water Distribution Operator Training Handbook*, chapter on "Instrumentation and Control." Obtain from American Water Works Association (AWWA) at www.awwa.org or phone (800) 926-7337. ISBN 978-1-58321-954-6.
5. *Maintenance Engineering Handbook* by R. Keith Mobley. Published by the McGraw-Hill Companies. ISBN 978-0-07-182661-7.

END OF LESSON 1 OF 5 LESSONS

on

MAINTENANCE

Please answer the discussion and review questions next.

DISCUSSION AND REVIEW QUESTIONS

Chapter 18. MAINTENANCE

(Lesson 1 of 5 Lessons)

At the end of each lesson in this chapter, you will find discussion and review questions. Please write your answers to these questions to determine how well you understand the material in the lesson.

1. Why should operators thoroughly read and understand manufacturers' literature before attempting to maintain plant equipment?
2. Why must administrators or supervisors be made aware of the need for an adequate maintenance program?
3. What is the purpose of a maintenance recordkeeping program?
4. What items should be included in a plant library?
5. Why should your plant have an emergency team to repair chlorine leaks?
6. Why should one person never be permitted to repair a chlorine leak alone?
7. Why should inexperienced, unqualified, or unauthorized persons and even qualified and authorized persons be extremely careful around electrical panels, circuits, wiring, and equipment?
8. What protective or safety devices are used to protect operators and equipment from being harmed by electricity?
9. Why must motor nameplate data be recorded and filed?
10. What might be the cause of a pump motor failing to start?
11. Why should a water treatment plant have standby power?
12. How would you determine the capacity of standby generation equipment?

CHAPTER 18. MAINTENANCE

(Lesson 2 of 5 Lessons)

18.2 MECHANICAL EQUIPMENT

Mechanical equipment commonly used in water treatment plants is described and discussed in this section. Equipment used with specific treatment processes such as flocculation and filtration is not discussed. You must be familiar with equipment and understand what it is intended to do before developing a preventive maintenance program and maintaining equipment.

18.20 Repair Shop

Many large plants have fully equipped machine shops staffed with competent mechanics. But for smaller plants, adequate machine shop facilities often can be found in the community. In addition, most pump manufacturers maintain pump repair departments where pumps can be fully reconditioned.

The pump repair shop in a large plant commonly includes such items as welding equipment, lathes, drill press and drills, power hacksaw, flame-cutting equipment, micrometers, calipers, gauges, portable electric tools, grinders, a forcing press, metal-spray equipment, and sandblasting equipment. You must determine what repair work you can and should do and when you need to request assistance from an expert.

Some agencies have their own repair shops or local machine shops rebuild parts rather than buying direct from manufacturers. Many agencies try to select equipment on the basis of the reputations of distributors for supplying repair parts when needed. A parts inventory is essential for key pieces of equipment.

18.21 Pumps

Pumps serve many purposes in water treatment plants. They may be classified by the character of the material handled, such as raw or filtered water. Or, they may relate to the conditions of pumping: high lift, low lift, or high capacity. They may be further classified by principle of operation, such as centrifugal, propeller, reciprocating, and turbine (Figure 18.18).

The type of material to be handled and the function or required performance of the pump vary so widely that the designing engineer must use great care in preparing specifications for the pump and its controls. Similarly, the operator must conduct a maintenance and management program adapted to the peculiar characteristics of the equipment.

18.210 Centrifugal Pumps

A centrifugal pump is basically a very simple device: an impeller rotating in a casing called the volute. The impeller is supported on a shaft, which is, in turn, supported by bearings. Liquid coming in at the center (eye) of the impeller (Figure 18.19) is picked up by the vanes and by the rotation of the impeller and then is thrown out by centrifugal force into the discharge.

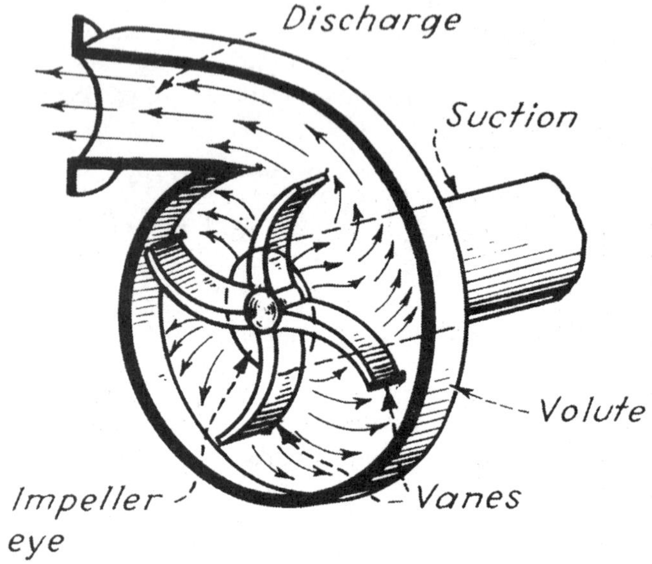

(Refer to Figure 18.24, page 307, for location of impeller in pump.)

Fig. 18.19 Diagram showing details of centrifugal pump impeller
(Source: *Centrifugal Pumps* by Karassik and Carter of Worthington Corporation)

To help you understand how pumps work and the purpose of the various parts, a section titled "Let's Build a Pump!" has been included on the following pages.

18.211 Let's Build a Pump

This section, reproduced with the permission of the Allis-Chalmers Corporation, has been edited for style. Originally, the material was printed in Allis-Chalmers Bulletin No. OBX62568.

Let's Build a Pump!

A student of medicine spends long years learning exactly how the human body is built before attempting to prescribe for its care. Knowledge of pump anatomy is equally basic in caring for centrifugal pumps.

But, whereas the medical student must take a body apart to learn its secrets, it will be far more instructive to us if we put a pump together (on paper, of course). Then, we can start at the

PUMPS
DYNAMIC
DISPLACEMENT
CENTRIFUGAL
AXIAL FLOW
SINGLE STAGE
MULTISTAGE
CLOSED IMPELLER
OPEN IMPELLER
FIXED PITCH
VARIABLE PITCH
MIXED FLOW, RADIAL FLOW
SINGLE SUCTION
DOUBLE SUCTION
SELF-PRIMING
NONPRIMING
SINGLE STAGE
MULTISTAGE
OPEN IMPELLER
SEMI-OPEN IMPELLER
CLOSED IMPELLER
PERIPHERAL
SINGLE STAGE
MULTISTAGE
SELF-PRIMING
NONPRIMING
SPECIAL EFFECT
JET (EDUCTOR)
GAS LIFT
HYDRAULIC RAM
ELECTROMAGNETIC

Dynamic types of pumps

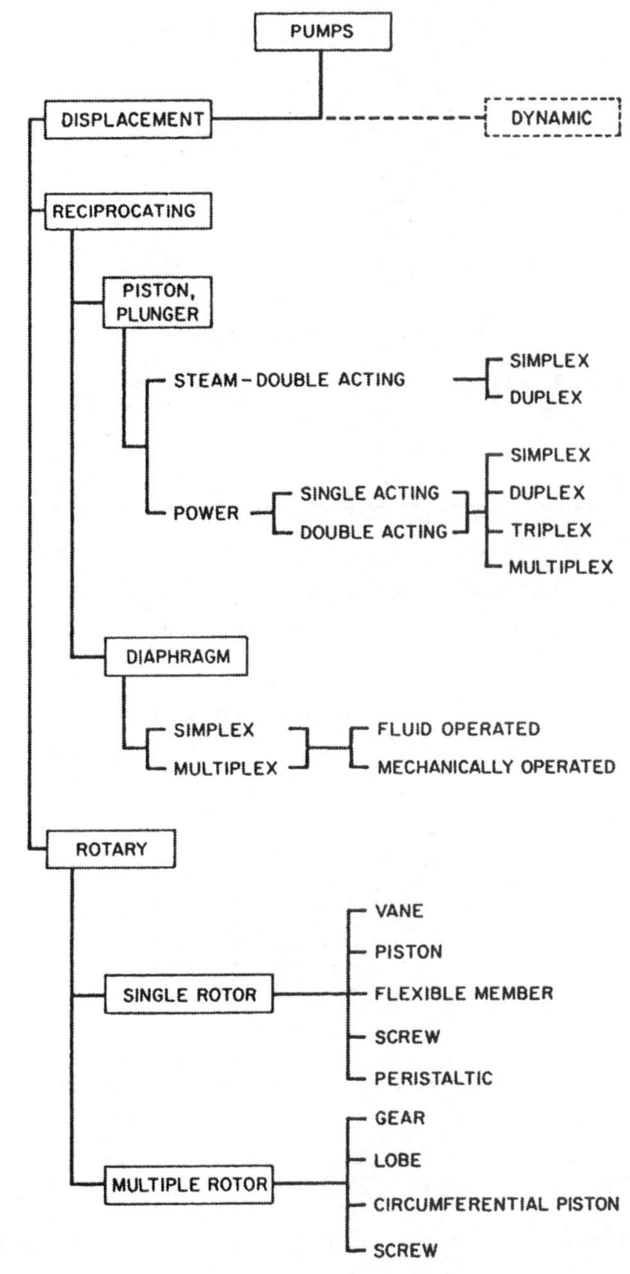

Displacement types of pumps

Fig. 18.18 Classification of pumps

beginning—adding each new part as we need it in logical sequence.

As we see what each part does and how it does it, we will see how it must be cared for.

Another analogy between medicine and maintenance: there are various types of human bodies, but if you know basic anatomy, you understand them all. The same is true of centrifugal pumps. In building one basic type, we will learn about all types.

Part of this will be elementary to some maintenance people, but they will find it a valuable refresher course. After all, maintenance just cannot be too good.

FIRST, WE REQUIRE A DEVICE TO SPIN LIQUID AT HIGH SPEED

That paddle wheel device is called the "impeller," and it is the heart of our pump (Figure 18.19).

Note that the blades curve out from its hub. As the impeller spins, liquid between the blades is impelled outward by centrifugal force (Figure 18.20).

Note, too, that our impeller is open at the center—the "eye." As liquid in the impeller moves outward, it will suck more liquid in behind it through this eye, provided it is not clogged.

Maintenance Rule No. 1: If there is any danger that foreign matter (sticks, refuse, etc.) may be sucked into the pump—clogging or wearing the impeller unduly—provide the intake end of the suction piping with a suitable screen.

NOW WE NEED A SHAFT TO SUPPORT AND TURN THE IMPELLER

Our shaft looks heavy—and it is. It must maintain the impeller in precisely the right place.

But that ruggedness does not protect the shaft from the corrosive or abrasive effects of the liquid pumped, so we must protect it with sleeves slid on from either end.

What these sleeves—and the impeller, too—are made of depends on the nature of the liquid we are to pump. Generally, they are bronze, but various other alloys, ceramics, glass, or even rubber-coating are sometimes required.

Maintenance Rule No. 2: Never pump a liquid for which the pump was not designed.

Whenever a change in pump application is contemplated and there is any doubt as to the pump's ability to resist the different liquid, check with your pump manufacturer.

WE MOUNT THE SHAFT ON SLEEVE, BALL, OR ROLLER BEARINGS

As we will see later, clearances between moving parts of our pump are quite small.

If bearings supporting the turning shaft and impeller are allowed to wear excessively and lower the turning units within a pump's closely fitted mechanism, the life and efficiency of that pump will be seriously threatened.

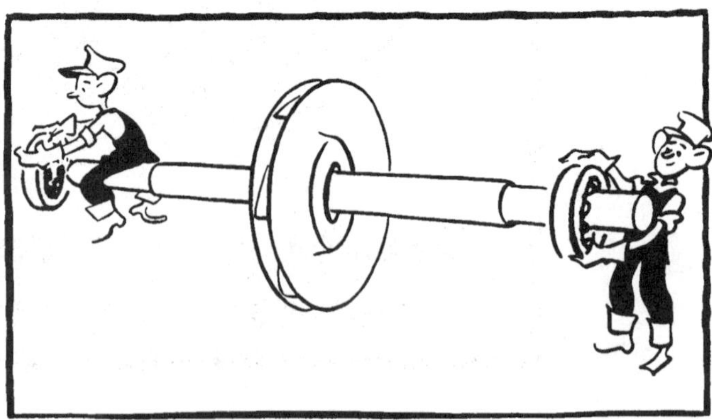

Maintenance Rule No. 3: Keep the right amount of the right lubricant in bearings at all times. Follow your pump manufacturer's lubrication instructions to the letter.

CENTRIFUGAL FORCE IN ACTION--

ALL MOVING BODIES TEND TO TRAVEL IN A STRAIGHT LINE. WHEN FORCED TO TRAVEL IN A CURVE, THEY CONSTANTLY *TRY* TO TRAVEL ON A TANGENT...

Centrifugal force pushes dummy planes swung in a circle *away* from center of rotation.

Centrifugal force tends to push swirling water *outward*... forming vortex in center.

Fig. 18.20 Centrifugal force

The main points to keep in mind are:

1. Although too much oil will not harm sleeve bearings, too much grease in antifriction type bearings (ball or roller) will promote friction and heat. The main job of grease in antifriction bearings is to protect steel elements against corrosion, not friction.
2. Operating conditions vary so widely that no one rule as to frequency of changing lubricant will fit all pumps. So, play it safe: if anything, change lubricant before it is too worn or too dirty.

TO CONNECT WITH THE MOTOR, WE ADD A COUPLING FLANGE

Some pumps are built with pump and motor on one shaft and, of course, offer no alignment problem.

But our pump is to be driven by a separate motor, and we attach a flange to one end of the shaft through which bolts will connect with the motor flange.

Use a straightedge or dial indicator to ensure shaft alignment (see pages 324 and 326, Figures 18.33 and 18.34).

Maintenance Rule No. 4: See that pump and motor flanges are parallel vertically and axially and that they are kept that way.

If shafts are eccentric or meet at an angle, every revolution throws tremendous extra load on bearings of both pump and motor. Flexible couplings will not correct this condition, if excessive.

Checking alignment should be a regular procedure in pump maintenance. Foundations can settle unevenly, piping can change the pump position, bolts can loosen. Misalignment is a major cause of pump and coupling wear.

NOW WE NEED A STRAW THROUGH WHICH LIQUID CAN BE SUCKED

Notice two things about the suction piping: (1) the horizontal piping slopes upward toward the pump; and (2) any reducer that connects between the pipe and pump intake nozzle should be horizontal at the top (eccentric, not concentric).

This up-sloping prevents air pocketing in the top of the pipe where trapped air might be drawn into the pump and cause loss of suction.

Maintenance Rule No. 5: Any down-sloping toward the pump in suction piping (as exaggerated in the previous diagrams) should be corrected.

This rule is very important. Loss of suction greatly endangers a pump, as we will see shortly.

WE CONTAIN AND DIRECT THE SPINNING LIQUID WITH A CASING

We got a little ahead of our story in the previous paragraphs because we did not yet have the casing to which the suction piping bolts. And, the manner in which it is attached is of great importance.

Maintenance Rule No. 6: See that piping puts absolutely no strain on the pump casing.

THE WEIGHT OF PIPING CAN EASILY RUIN A PUMP!

When the original installation is made, all piping should be in place and self-supporting before connection. Openings should meet with no force. Otherwise, the casing is apt to be cracked or sprung enough to allow closely fitted pump parts to rub.

It is good practice to check the piping supports regularly to see that loosening, or settling of the building, has not put strains on the casing.

NOW OUR PUMP IS ALMOST COMPLETE, BUT IT WOULD LEAK LIKE A SIEVE

We are far enough along now to trace the flow of water through our pump. It is not easy to show suction piping in the cross-section view above, so imagine it stretching from your eye to the lower center of the pump.

Our pump happens to be a "double suction" pump, which means that water flow is divided inside the pump casing, reaching the eye of the impeller from either side.

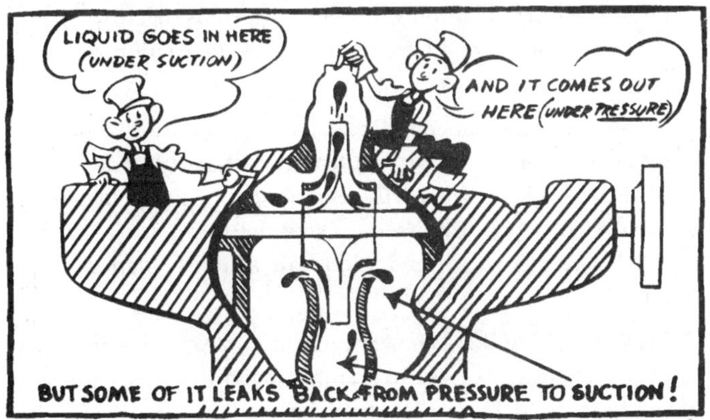

As water is drawn into the spinning impeller, centrifugal force causes it to flow outward, building up high pressure at the outside of the pump (which will force water out) and creating low pressure at the center of the pump (which will draw water in). This situation is shown in the upper half of the pump, above.

So far so good, except that water tends to be drawn back from pressure to suction through the space between impeller and casing—as diagrammed in the lower half of the pump, above—and our next step must be to plug this leak, if our pump is to be very efficient.

SO WE ADD WEARING RINGS TO PLUG INTERNAL LIQUID LEAKAGE

You might ask why we did not build our parts closer fitting in the first place—instead of narrowing the gap between them by inserting wearing rings (see Figure 18.24, page 307, item 16).

The answer is that those rings are removable and replaceable when wear enlarges the tiny gap between them and the impeller. (Sometimes rings are attached to the impeller rather than the casing—or rings are attached to both so they face each other.)

Maintenance Rule No. 7: Never allow a pump to run dry (either through lack of proper priming when starting or through loss of suction when operating). Water is a lubricant between rings and impeller.

Maintenance Rule No. 8: Examine wearing rings at regular intervals. When seriously worn, their replacement will greatly improve pump efficiency.

TO KEEP AIR FROM BEING DRAWN IN, WE USE STUFFING BOXES

We have two good reasons for wanting to keep air out of our pump: (1) we want to pump water, not air; and (2) air leakage is apt to cause our pump to lose suction.

Each stuffing box we use consists of a casing, rings of packing, and a gland at the outside end. A mechanical seal may be used instead.

Maintenance Rule No. 9: Packing should be replaced periodically—depending on conditions—using the packing recommended by your pump manufacturer. Forcing in a ring or two of new packing instead of replacing worn packing is bad practice. It is apt to displace the seal cage.

Put each ring of packing in separately, seating it firmly before adding the next. Stagger adjacent rings so the points where their ends meet do not coincide. See Figure 18.30, "How to pack a pump," on pages 314 and 315.

Maintenance Rule No. 10: Never tighten a gland more than necessary as excessive pressure will wear shaft sleeves unduly.

Maintenance Rule No. 11: If shaft sleeves are badly scored, replace them immediately or packing life will be entirely too short.

TO MAKE PACKING MORE AIRTIGHT, WE ADD WATER SEAL PIPING

In the center of each stuffing box is a "seal cage." By connecting it with piping to a point near the impeller rim, we bring liquid under pressure to the stuffing box.

This liquid acts both to block out air intake and to lubricate the packing. It makes both packing and shaft sleeves wear longer, providing it is clean liquid.

Maintenance Rule No. 12: If the liquid being pumped contains grit, a separate source of sealing liquid should be obtained (for example, it may be possible to direct some of the pumped liquid into a container and allow the grit to settle out).

To control liquid flow, draw up the gland just tight enough so a thin stream (approximately one drop per second) flows from the stuffing box during pump operation.

DISCHARGE PIPING COMPLETES THE PUMP INSTALLATION—AND NOW WE CAN ANALYZE THE VARIOUS FORCES WE ARE DEALING WITH

SUCTION. At least 75 percent of centrifugal pump troubles trace to the suction side. To minimize them, take the following precautions:

1. Total suction lift (distance between centerline of pump and liquid level when pumping, plus friction losses) generally should not exceed 15 feet (4.5 meters).
2. Piping should be at least a size (diameter) larger than pump suction nozzle.
3. Friction in piping should be minimized. Use as few and as easy bends as possible and avoid scaled or corroded pipe.

DISCHARGE lift, plus suction lift, plus friction in the piping from the point where liquid enters the suction piping to the end of the discharge piping equals total head.

Pumps should be operated near their rated heads. Otherwise, the pump is apt to operate under unsatisfactory and unstable conditions, which reduce efficiency and operating life of the unit, or "cavitation" could occur. Note the description of cavitation in Figure 18.21. Cavitation can seriously damage your pump.

PUMP CAPACITY generally is measured in gallons per minute (liters per second or m^3 per second). A new pump is guaranteed to deliver its rating in capacity and head. But whether a pump retains its actual capacity depends to a great extent on its maintenance.

Wearing rings must be replaced when necessary—to keep internal leakage losses down.

Friction must be minimized in bearings and stuffing boxes by proper lubrication, and misalignment must not be allowed to force scraping between closely fitted pump parts.

POWER of the driving motor, like capacity of the pump, will not remain at a constant level without proper motor maintenance.

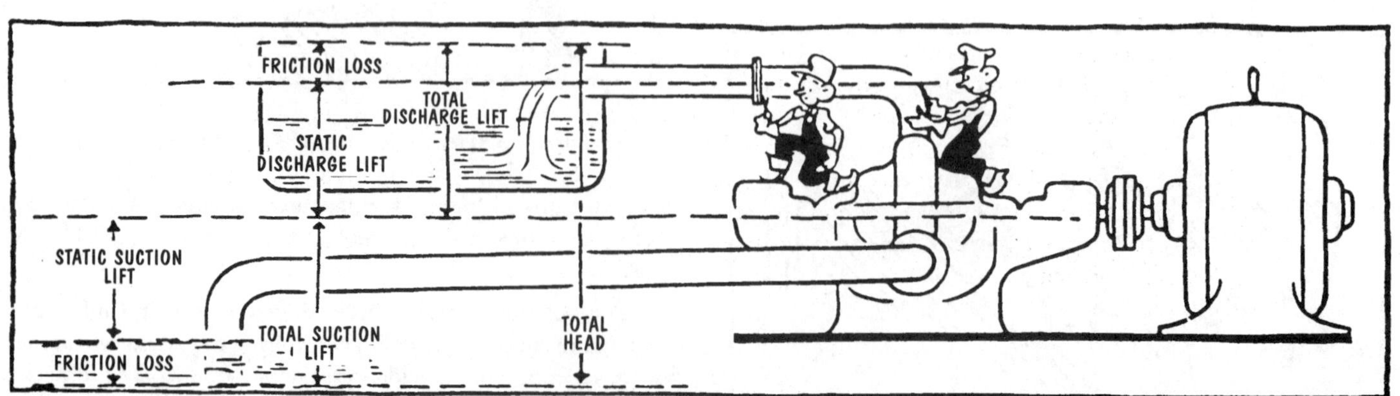

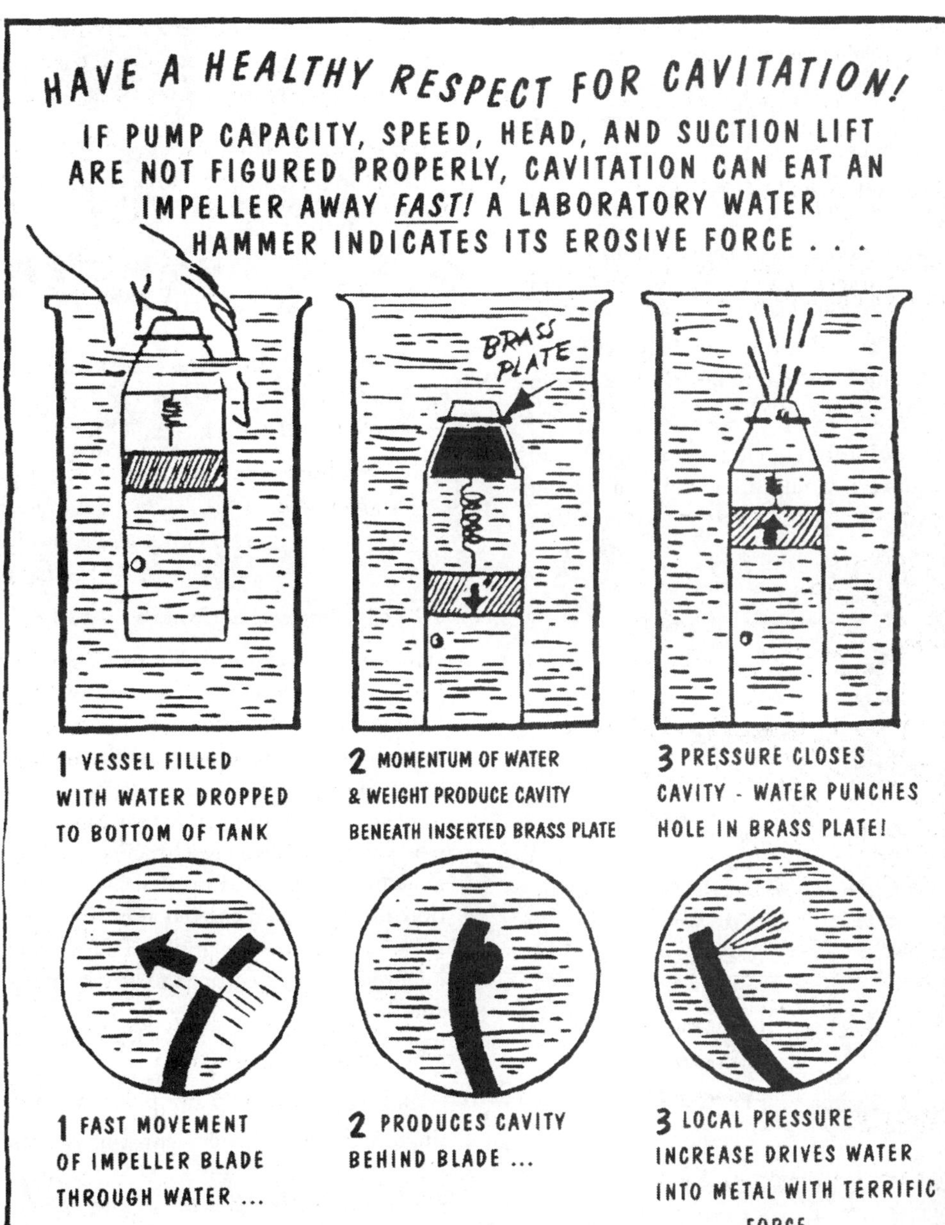

Cavitation is a condition that can cause a drop in pump efficiency, vibration, noise, and rapid damage to the impeller of a pump. Cavitation occurs due to unusually low pressures within a pump. These low pressures can develop when pump inlet pressures drop below the design inlet pressures or when the pump is operated at flow rates considerably higher than design flows. When the pressure within the flowing water drops very low, the water starts to boil and gas pockets or bubbles form on the blade of the impeller. The collapse of these gas pockets or bubbles drives water into the impeller with a terrific force that can knock metal particles off the impeller and cause pitting on the impeller surface. This same action can and does occur on pressure reducing valves and partially closed gate and butterfly valves. Cavitation is accompanied by loud noises that sound like someone is pounding on the impeller with a hammer.

Fig. 18.21 Cavitation

The starting load on motors can be reduced by throttling or closing the pump discharge valve (never the suction valve), but the pump must not be operated for long with the discharge valve closed. Power then is converted into friction, overheating the water with serious consequences.

18.212 Vertical Centrifugal Pumps

Another common configuration for centrifugal pumps is the vertical-suction cased centrifugal pump (Figures 18.22 and 18.23). This is an adaptation of the deep-well turbine pump for booster pump service. They are very flexible in design as the engineers can specify either single- or multistage in a wide variety of sizes and characteristics.

Besides the usual lubrication of the electric motor, the only routine maintenance required is to adjust and repair, as needed, the single packing gland.

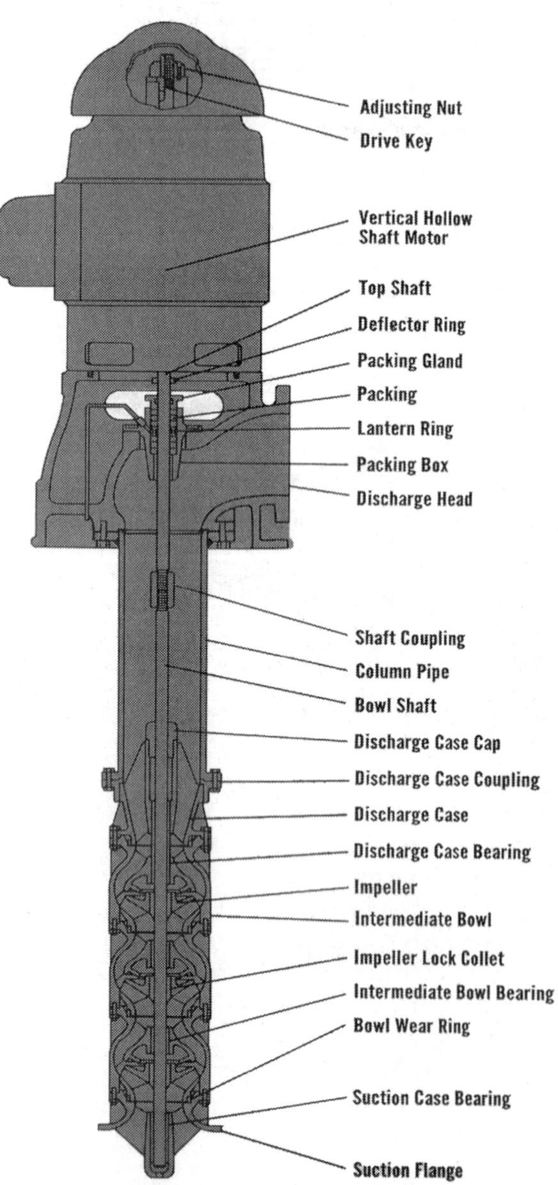

Fig. 18.22 Vertical centrifugal pump (multistage)
(Permission of Aurora Pump Company)

Fig. 18.23 Vertical centrifugal pump (single stage)
(Permission of Aurora Pump Company)

18.213 Horizontal Centrifugal Pumps

Horizontal centrifugal pumps (Figure 18.24), like the one we just constructed on paper in Section 18.211, are available in a number of configurations. The one we built is best described as a single-stage, horizontal, double-suction, split-case centrifugal pump. The pump is a single-stage pump because it has only one impeller. Some horizontal pumps have two impellers that are working in series to create higher heads than can readily be obtained with only one impeller. Our paper pump was double suction in that water entered the impeller from both sides. The advantage of this design over a single-suction design is that the longitudinal thrust from the water entering the impeller is balanced. This greatly reduces the thrust load that the pump's bearings must carry. The split-case designation indicates that the pump case is made in two halves. Some centrifugal pumps have a single suction in line with the shaft. These are described as single-stage, end-suction centrifugal pumps.

18.214 Reciprocating or Piston Pumps

The word "reciprocating" means moving back and forth, so a reciprocating pump is one that moves a liquid by a piston that moves back and forth. A simple reciprocating pump is shown in Figure 18.25. If the piston is pulled to the left, Check Valve A will be open and the liquid will enter the pump and fill the casing. When the piston reaches the end of its travel to the left and is pushed back to the right, Check Valve A will close, Check Valve B will open, and the liquid will be forced out the exit line.

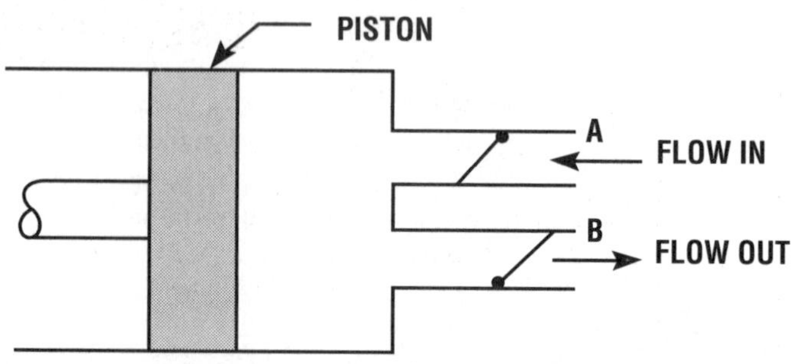

Fig. 18.25 Simple reciprocating pump

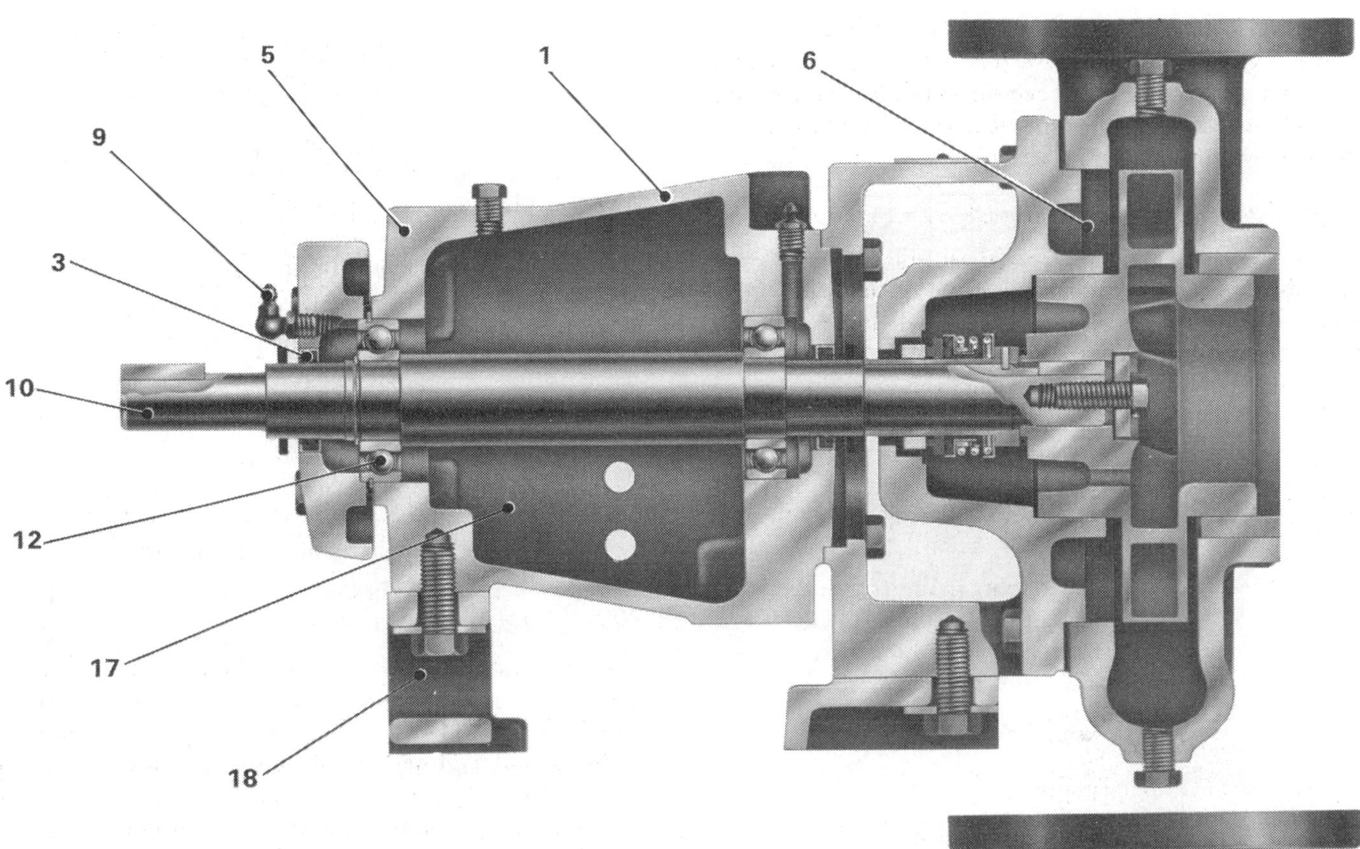

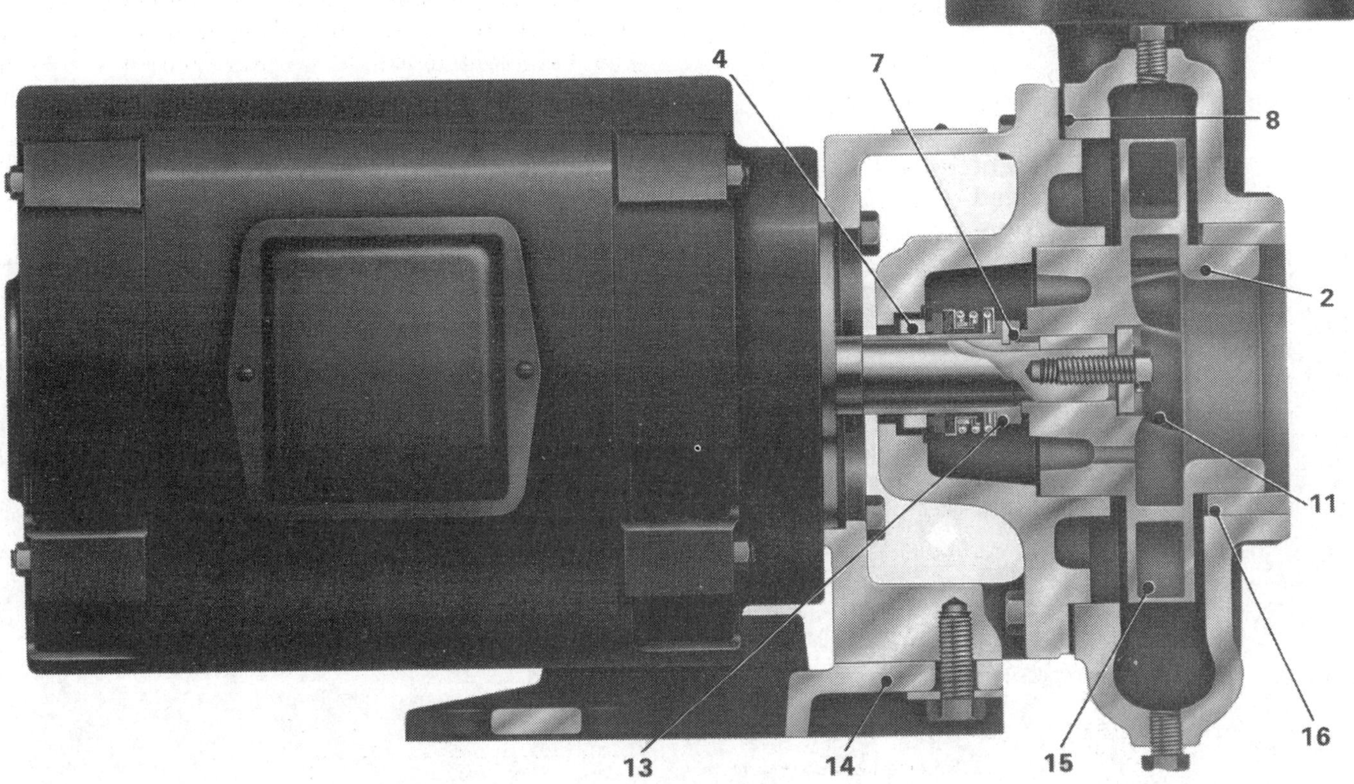

1. Motor frame
2. Impeller
3. Oil seal
4. Mechanical seal
5. Frame
6. Casing
7. Shaft sleeve
8. Back pull-out
9. Lubrication fittings
10. Shaft
11. Impeller
12. Bearings
13. Shaft sleeve
14. Close coupled motor support
15. Impeller
16. Wearing ring
17. Oil reservoir
18. Rear support foot

Fig. 18.24 Centrifugal pump parts

(Permission of Aurora Pump Company)

A piston pump is a positive-displacement pump. Never operate it against a closed discharge valve or the pump, valve, or pipe could be damaged by excessive pressures. Also, the suction valve should be open when the pump is started. Otherwise, an excessive suction or vacuum could develop and cause problems.

18.215 Progressive Cavity (Screw-Flow) Pumps

The progressive cavity pump consists of a screw-shaped rotor snugly enclosed in a nonmoving stator or housing (Figure 18.26). The threads of the screw-like rotor (commonly manufactured of chromed steel) make contact along the walls of the stator (usually made of synthetic rubber). The gaps between the rotor threads are called "cavities." When water is pumped through an inlet valve, it enters the cavity. As the rotor turns, the material is moved along until it leaves the conveyor (rotor) at the discharge end of the pump. The size of the cavities along the rotor determines the capacity of the pump.

All progressive cavity pumps operate on the basic principle described above. To further increase capacity, some models have a shaped inside surface of the stator (housing) with a similarly shaped rotor. In addition, some models use a rotor that moves up and down inside the stator as well as turning on its axis (Figure 18.27). This allows a further increase in the capacity of the pump.

Progressive cavity pumps are recommended for materials containing higher concentrations of suspended solids. They are commonly used to pump sludges. Progressive cavity pumps should never be operated dry (without liquid in the cavities), and should never be run against a closed discharge valve.

18.216 Chemical Metering Pumps

Many chemical metering pumps are a type of positive-displacement pump. For information on chemical metering pumps, see Chapter 13, "Fluoridation," Section 13.30, "Chemical Feeders," and this chapter, Section 18.4, "Chemical Storage and Feeders."

QUESTIONS

Please write your answers to the following questions and compare them with those on pages 369 and 370.

18.20A List the pieces of equipment and special tools commonly found in a pump repair shop.

18.21A What is the purpose of a pump impeller?

18.21B Why should the intake end of suction piping have a suitable screen?

18.21C Why must suction piping always be up-sloping?

18.21D What is cavitation?

18.21E What is an advantage of a double-suction pump over a single-suction pump?

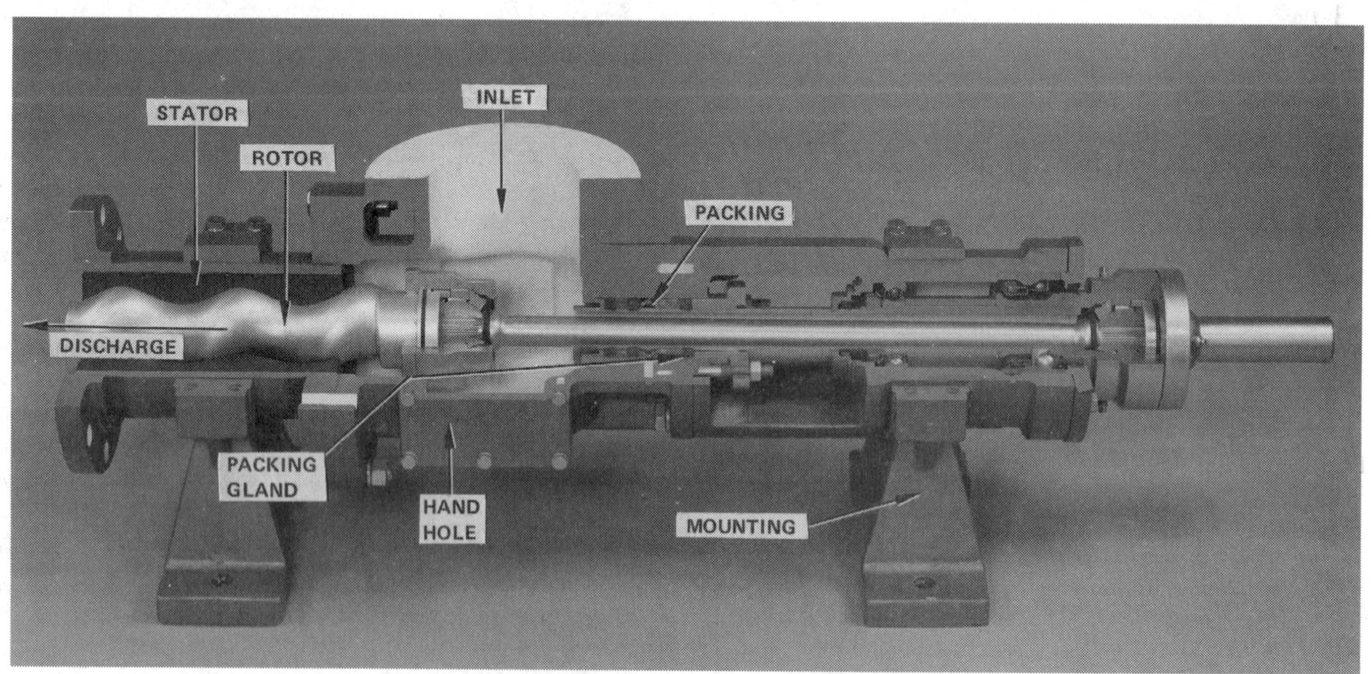

Fig. 18.26 Progressive cavity (screw-flow) pump
(Permission of Moyno Pump Division, Robbins & Meyer, Inc.)

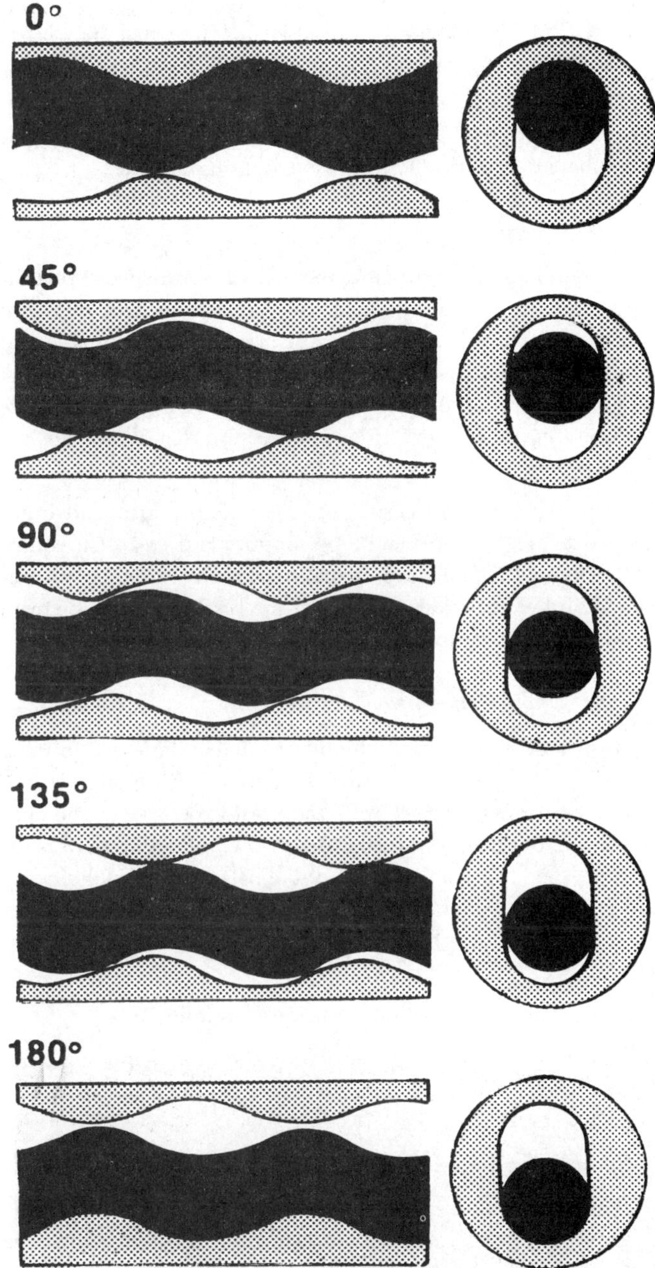

Fig. 18.27 Pumping principle of a progressive cavity pump

(Permission of Allweiler Pumps, Inc.)

18.22 Lubrication

18.220 Purpose of Lubrication

Lubrication of equipment is one of the most important phases of a maintenance operator's job. Without proper lubrication, the tools and equipment used for operating and maintaining water treatment plants would fail. Proper lubrication of tools and equipment is one of the maintenance operator's easiest jobs, but often it is the most neglected.

The purpose of lubrication is to reduce friction between two surfaces. Lubrication also removes heat that is caused by friction. Solid friction of two dry surfaces in contact is changed to a fluid friction of a separating layer of liquid or liquid lubricant. Actually, water is a lubricant, although not a good lubricant.

18.221 Properties of Lubricants

A good lubricant must have the following properties:

1. Form a slippery coating on contacting surfaces so they can slide freely past each other
2. Exert sufficient pressure to keep the surfaces apart when running

To be a good lubricant for a particular job, the lubricant used must have the following qualities:

1. Thickness of the lubricant layer must be sufficient to keep the roughness of the metal parts from touching.
2. Lubricity (slipperiness) must be sufficient to allow molecules to slide freely past each other.
3. Viscosity (resistance to flow) must be sufficient to build up a pressure necessary to keep the surfaces apart. If viscosity alone cannot provide enough pressure, an external pressure must be supplied by a pump.

Viscosity in the United States is the number of seconds it takes 60 cubic centimeters (cc) of an oil to flow through the standard orifice of a Saybolt Universal Viscometer at 100, 130, or 210 degrees Fahrenheit. A 300 - *SSU*[14] @ 130 oil means that it took 300 seconds for 60 cc to flow through a Saybolt Universal Viscometer at 130°F (54.4°C). Viscosity decreases with temperature rise because oil becomes thinner. The specific gravity of an oil is measured by comparing the weight of oil with an equal volume of water, both at 60°F (15.6°C).

Some other important information to know about lubricants is their "pour point," "flash point," and "fire point." Pour point is the temperature at which a lubricant refuses to run. This is important in low-temperature work. Flash point is the temperature at which oil vaporizes enough to ignite momentarily when near a flame. A low flash point means that oil evaporates more readily in service. Fire point is the temperature at which oil vaporizes enough to keep on burning. Oils in service tend to become acid (contaminated) and may cause corrosion, deposits, sludging, and other problems. This condition may not be visible when you look at the oil. Therefore, do not extend the time for an oil change because the oil looks clean.

To detect acid conditions in oils, the neutralization number of an oil is used. The neutralization number is the weight in milligrams of potassium hydroxide required to neutralize one

14. *SSU.* Standard Saybolt Units.

gram of oil. This is used by laboratories that test the oil on large engines, turbines, compressors, and other equipment that have large-volume oil reservoirs to determine when oil changes or additives are needed.

Most lubricants in general use are fluid at room temperature. Mostly, these are petroleum based, but others are used. Greases are mixtures of petroleum products with soaps such as lime, soda, aluminum, and metallic. Metallic soaps, forms of calcium, sodium, potassium, and lithium, have good retention in bearings and can withstand high temperatures and pressures. A sodium-base grease has sodium as the soap mixed with the petroleum.

Solid materials such as graphite and finely ground mica are sometimes used as lubricants. Some recently developed silicon compounds (silicones) work very well under heavy loads and widely varying temperatures.

There are many oil additives on the market today and they are worth investigating. Oil additives are chemical compounds added to an oil to improve certain chemical or physical properties such as stability, lubricity, and foaming. They are used to prevent rust or deposits and many other items that could cause problems.

18.222 Lubrication Schedule

To have proper lubrication, you must first set up a lubrication schedule. This can be a simple check-off sheet or card system or an elaborate computer system. The first thing to do is make a list of everything that needs lubrication down to the smallest item, including chains, rollers, and sprockets. After you have listed every item on paper, go through the manufacturer's instruction books to determine the frequency and type of lubrication required. Is the frequency daily, weekly, monthly, semiannually, or annually? The manufacturer's literature usually lists several different name brands of lubricants that are equal. If you need help determining the type of lubricant or cross-referencing it to your particular brand, contact your supplier. Most oil distributors have a service representative who will come to your facility and go over the individual equipment and specify which lubricants you should use. Next, determine the amount of each lubricant required. This is achieved by counting the number of grease fittings. Determine the locations of fill plugs, drain plugs, oil levels, sight glasses, dipsticks, and other important items. To find these locations, physically inspect each piece of equipment thoroughly and look for all lubrication points. Also, the manufacturer's maintenance manual should show the lubrication points for each piece of equipment.

When you have gathered all this information, transfer it to the equipment history cards for future reference. From this information, you can make up a lubrication chart or form.

As stated earlier, use whatever type of lubrication form you prepare, but follow it. Always record each lubrication job when completed and have the operator who did the job initial the record card. Always keep your lubrication schedules up to date. If there are failures due to the wrong or insufficient lubricant, change or increase the lubrication frequency on the schedule. Also, new equipment must be added and discarded equipment removed from the schedule. Someone must be assigned to take care of the lubrication and records. Assign more than one operator or rotate this job so if an individual is off work or leaves the crew, there is a continuity in the lubrication schedule.

18.223 Precautions

When handling or storing oils and greases, some special precautions must be followed. Make sure the storage area does not create a fire hazard. Most lubricants are highly flammable and should not be stored where there is an open flame. "NO SMOKING" signs must be posted outside the building. Be sure to keep any spills wiped up and make sure that all the lids are tight on their containers.

Keep materials and containers clean. Sand, grit, and other substances can contaminate lube supplies and create an equipment failure that lubrication maintenance is intended to prevent. Another good idea is to direct the first shot of grease from a gun into a waste can.

18.224 Pump Lubrication

Pumps, motors, and drives should be oiled and greased in strict accordance with the recommendations of the manufacturer. Cheap lubricants may often be the most expensive in the end. Oil should not be put in the housing while the pump shaft is rotating because the rotary action of the ball bearings will pick up and retain a considerable amount of oil. When the unit comes to rest, an overflow of oil around the shaft or out of the oil cup will result.

Greased bearings should be lubricated as follows:

1. Shut off, lock out, tag, and block the unit if moving parts that might be a safety hazard are close to the grease fitting or drain plugs.
2. Remove the drain plug from the bearing housing.
3. Remove the grease fitting protective cap and wipe off the grease fitting. Be sure that you do not force dirt into the bearing housing along with the clean grease.
4. Pump in clean grease until the grease coming out of the drain hole is clean. Do not pump grease into a bearing with the drain plug in place. This could easily build up enough pressure to blow out the seals.
5. Put the protective cap back on the grease fitting.
6. With the drain plug still removed, put the unit back in service. As the bearing warms up, excess grease will be expelled from the drain hole. After the unit has been running for a few hours, the drain plug may be put back in place. Special drain plugs with spring-loaded check valves are recommended because they will protect against further buildup.
7. Unless you intend to be very careful, we recommend that bearing grease be purchased in cartridge form to minimize the chance of getting dirt into the lubricant.

QUESTIONS

Please write your answers to the following questions and compare them with those on page 370.

18.22A What is the purpose of lubrication?

18.22B What happens to oils in service?

18.22C What should be done to ensure proper lubrication of equipment?

18.225 Equipment Lubrication

Different authorities may make conflicting lube recommendations for essentially the same item; however, general reference material is available to help select the correct lubricant for a specific application.

Grease is graded on a number scale, or viscosity index, by the National Lubricating Grease Institute. For example, No. 0 is very soft; No. 6 is quite stiff. A typical grease for most treatment plant applications might be a No. 2 lithium or sodium compound grease, which is used for operating temperatures up to 250°F (120°C).

Generally, the time between flushing and repacking for greased bearings should be divided by 2 for every 25°F (14°C) above 150°F (65°C) operating temperature. Also, generally, the time between lubrications should not be allowed to exceed 48 months, since lube component separation and oxidation can become significant after this period of time, regardless of amount of use.

Another point worth noting is that grease is normally not suitable for moving elements with speeds exceeding 12,000 in/min (5 m/sec). Usually, oil lubricating systems are used for higher speeds. Lighter viscosity oils are recommended for high speeds, and, within the same speed and temperature range, a roller bearing will normally require one grade heavier viscosity than a ball bearing.

A good rule of thumb is to change and flush oil completely at the end of 600 hours of operation, or every 3 months, whichever occurs first. More specific procedures for flushing and changing lubricants are outlined by most equipment manufacturers.

Every operator should be aware of the dangers of overfilling with either grease or oil. Overfilling can result in high pressures and temperatures, and ruined seals or other components. It has been observed that more antifriction bearings are ruined by overgreasing than by neglect.

A thermometer can tell a great deal about the condition of a bearing. Ball bearings are generally in trouble above 180°F (82°C). Grease-packed bearings typically run 10 to 50 degrees above ambient temperature.

For clarifier drive units, which are almost always located outdoors, condensation presents a dangerous problem for the lubrication system. Most units of current design have a condensate bailing system to remove water from the gear housing by displacement. These units should be checked often for proper operation, particularly during seasons of wide air temperature fluctuation.

Pumps incorporate many types of seals and gaskets constructed of combinations of elastomers and metals. As for lubricants, conflicting advice can be obtained. A file containing data on general properties of materials used can help in the choice of lubricant.

QUESTIONS

Please write your answers to the following questions and compare them with those on page 370.

18.22D Does a soft grease have a high or low viscosity index as compared with a hard grease?

18.22E Is oil or grease used with higher speeds?

18.22F What problems can result from overfilling with oil or grease?

END OF LESSON 2 OF 5 LESSONS

on

MAINTENANCE

Please answer the discussion and review questions next.

DISCUSSION AND REVIEW QUESTIONS

Chapter 18. MAINTENANCE

(Lesson 2 of 5 Lessons)

Please write your answers to the following questions to determine how well you understand the material in the lesson. The question numbering continues from Lesson 1.

13. What is the purpose of a pump shaft?
14. What is the purpose of pump sleeves?
15. Why should a pump never be allowed to run dry?
16. How would you develop a lubrication schedule for a pump?
17. Why is cleanliness important in the storage and use of lubricants?

CHAPTER 18. MAINTENANCE

(Lesson 3 of 5 Lessons)

18.23 Pump Maintenance

18.230 Section Format

The format of this section differs from the other chapters. This format was designed specifically to assist you in planning an effective preventive maintenance program. The paragraphs are numbered for easy reference when you use the Equipment Service Cards and Service Record Cards mentioned in Section 18.00, page 267, and shown in Figure 18.1.

An entire book could be written on the topics covered in this section. Step-by-step details for maintaining equipment are not provided because manufacturers are continually improving their products and these details could soon be out of date. You are assumed to have some familiarity with the equipment being discussed. For details concerning a particular piece of equipment, you should contact the manufacturer. This section indicates to you the kinds of maintenance you should include in your program and how you could schedule your work. Carefully read the manufacturer's instructions and be sure you clearly understand the material before attempting to maintain and repair equipment. If you have any questions or need any help, do not hesitate to contact the manufacturer or your local representative.

A glossary is not provided in this section because of the large number of technical words that require familiarization with the equipment being discussed. The best way to learn the meaning of these new words is from manufacturers' literature or from their representatives. Some new words are described in the lessons, where necessary.

18.231 Preventive Maintenance

The following paragraphs list some general preventive maintenance services and indicate frequency of performance. There are many makes and types of equipment and the wide variation of functions cannot be included; therefore, you will have to use some judgment as to whether the services and frequencies will apply to your equipment. If something goes wrong or breaks in your plant, you may have to disregard your maintenance schedule and fix the problem now.

NOTE: If you need to shut a unit down, make sure it is also locked out and tagged properly. See Section 18.03, "Lockout/Tagout Procedure."

Paragraph 1: Pumps, General

This paragraph lists some general preventive maintenance services and indicates frequency of performance. Typical centrifugal pump sections are shown in Figure 18.24 on page 307.

Frequency of Service

D 1. CHECK WATER-SEAL PACKING GLANDS FOR LEAKAGE. See that the packing box is protected with a clear water supply from an outside source, make sure that water seal pressure is at least 5 psi (35 kPa or 0.35 kg/cm^2) greater than maximum pump suction pressure. See that there are no *CROSS-CONNECTIONS.*[15] Check packing glands for leakage during operation. Allow a slight seal leakage when pumps are running to keep packing cool and in good condition. The proper amount of leakage depends on equipment and operating conditions. Sixty drops of water per minute is a good rule of thumb. If excessive leakage is found, hand tighten gland nuts evenly, but not too tight. After adjusting packing glands, be sure shaft turns freely by hand. If serious leakage continues, renew packing, shaft, or shaft sleeve.

D 2. CHECK GREASE-SEALED PACKING GLANDS. When grease is used as a packing gland seal, maintain constant grease pressure on packing during operation. When a spring-loaded grease cup is used, keep it loaded with grease. Force grease through packing at a rate of about 1 ounce (30 g) per day. When water is used, adjust seal pressure to 5 psi (35 kPa or 0.35 kg/cm^2) above maximum pump suction pressure. Never allow the seal to run dry.

15. *Cross-Connection.* (1) A connection between a drinking (potable) water system and an unapproved water supply. (2) A connection between a storm drain system and a sanitary collection system. (3) Less frequently used to mean a connection between two sections of a collection system to handle anticipated overloads of one system.

Frequency of Service	
W	3. OPERATE PUMPS ALTERNATELY. If two or more pumps of the same size are installed, alternate their use to equalize wear, keep motor windings dry, and distribute lubricant in bearings.
W	4. INSPECT PUMP CONTROL. Inspect the pump controls to see that the pump responds properly to changes in the controlling variable. This variable may be either a pressure or a water level. This check could be done physically or by analyzing recording gauge records.
D	5. CHECK MOTOR CONDITION. See "Paragraph 6: Electric Motors," on page 319.
W	6. CHECK PACKING GLAND ASSEMBLY. Check packing gland, the unit's most abused and troublesome part. If stuffing box leaks excessively when gland is pulled up with mild pressure, remove packing and examine shaft sleeve carefully. Replace or repair grooved or scored shaft sleeve because packing cannot be held in stuffing box with roughened shaft or shaft sleeve. Replace the packing a strip at a time, tamping each strip thoroughly and staggering joints. (See Figure 18.28.) Position lantern ring (water-seal ring) properly. If grease sealing is used, completely fill lantern ring with grease before putting remaining rings of packing in place. The type of packing used (Figure 18.29) is less important than the manner in which packing is placed. Never use a continuous strip of packing. This type of packing wraps around and scores the shaft sleeve or is thrown out against the outer wall of the stuffing box, allowing water to leak through and score the shaft. The proper size of packing should be available in your plant's equipment files. See Figure 18.30 for illustrated steps on how to pack a pump.

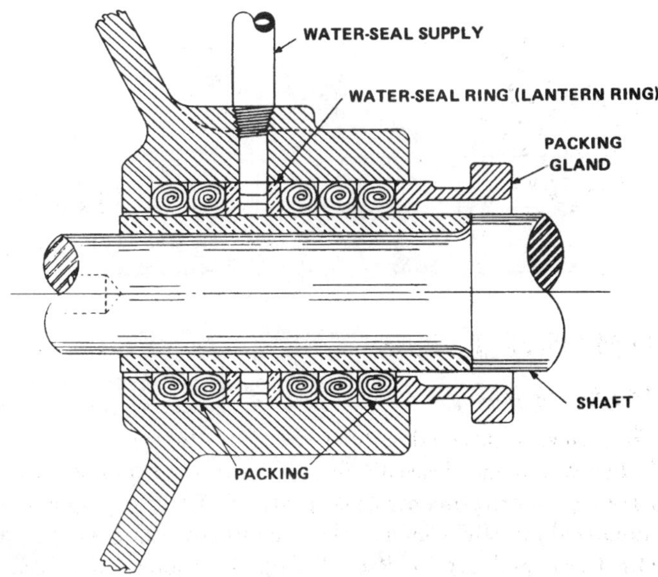

Fig. 18.28 Method of packing shaft
(Source: War Department Technical Manual TM5-666)

Teflon Packing

Graphite Packing

Fig. 18.29 Packing
(Courtesy A. W. Chesteron Co.)

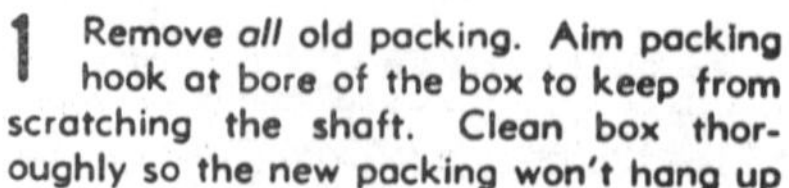

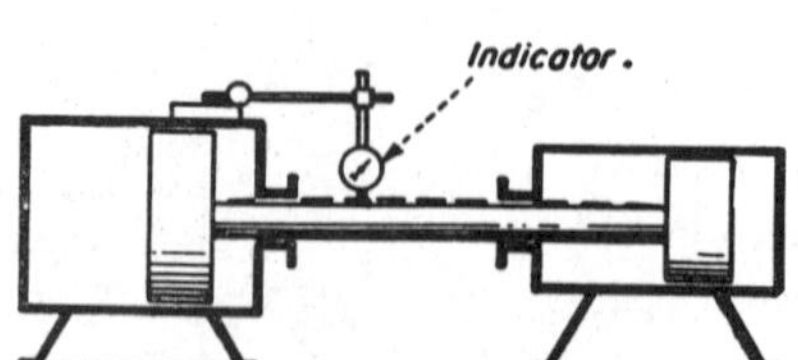

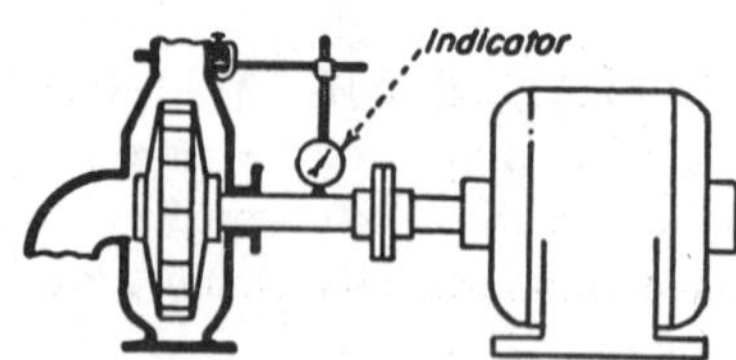

1 Remove *all* old packing. Aim packing hook at bore of the box to keep from scratching the shaft. Clean box thoroughly so the new packing won't hang up

2 Check for bent shaft, grooves or shoulders. If the neck bushing clearance in bottom of box is great, use stiffer bottom ring or replace the neck bushing

3 Revolve rotary shaft. If the indicator runs out over 0.003-in., straighten shaft, or check bearings, or balance rotor. Gyrating shaft beats out packing

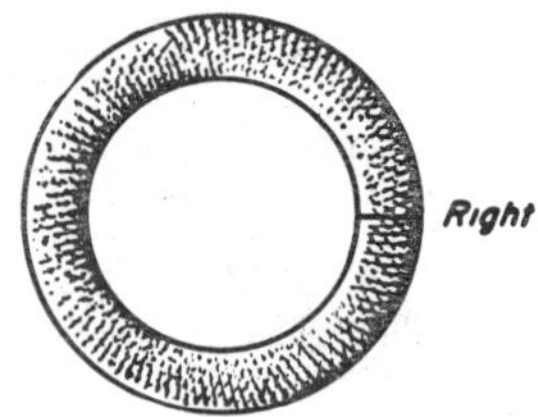

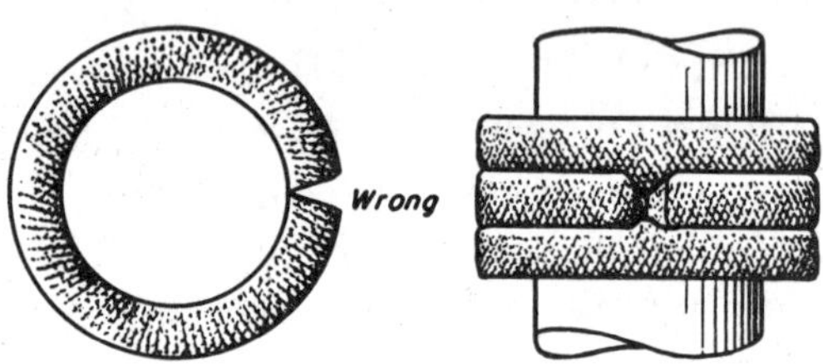

6 Cutting off rings while packing is wrapped around shaft will give you rings with parallel ends. This is very important if packing is to do job

7 If you cut packing while stretched out straight, the ends will be at an angle. With gap at angle, packing on either side squeezes into top of gap and ring cannot close. This brings up the question about gap for expansion. Most packings need none. Channel-type packing with lead core may need slight gap for expansion

HOW TO PACK A PUMP

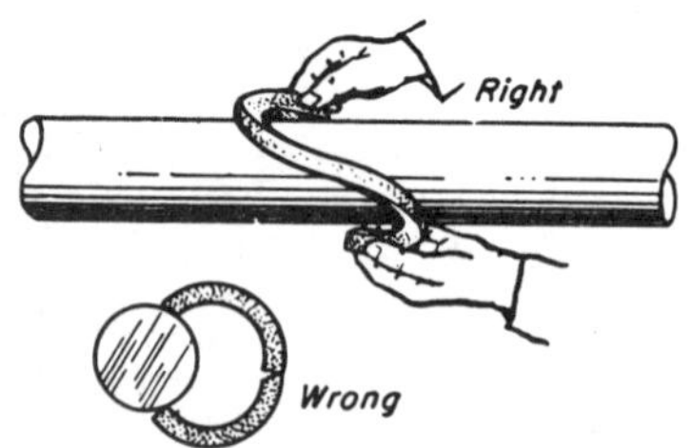

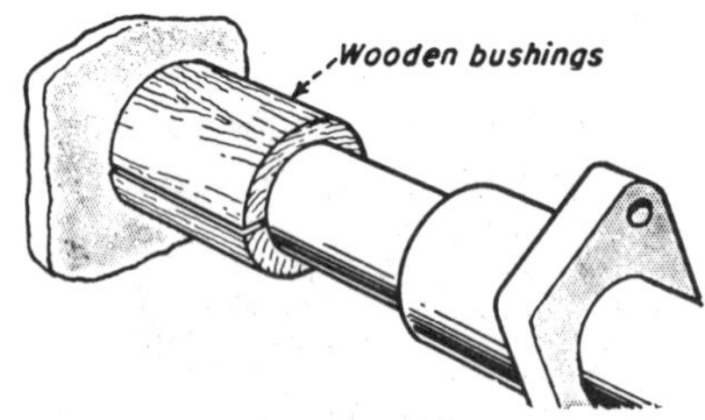

11 Open ring joint sidewise, especially lead-filled and metallic types. This prevents distorting molded circumference—breaking the ring opposite gap

12 Use split wooden bushing. Install first turn of packing, then force into bottom of box by tightening gland against bushing. Seat each turn this way

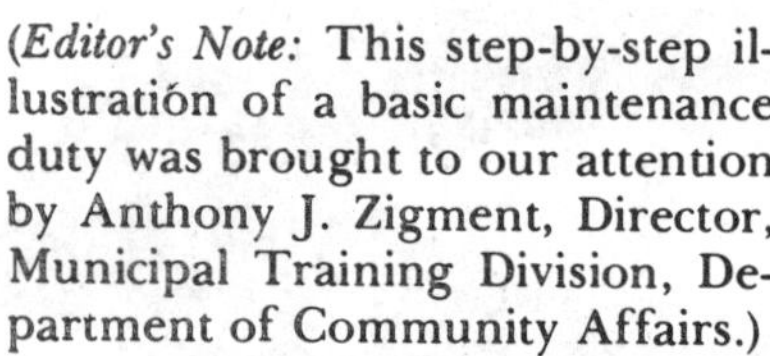

(*Editor's Note:* This step-by-step illustratión of a basic maintenance duty was brought to our attention by Anthony J. Zigment, Director, Municipal Training Division, Department of Community Affairs.)

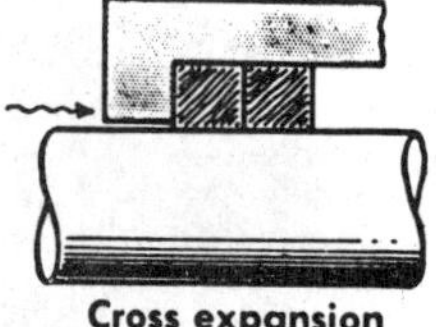

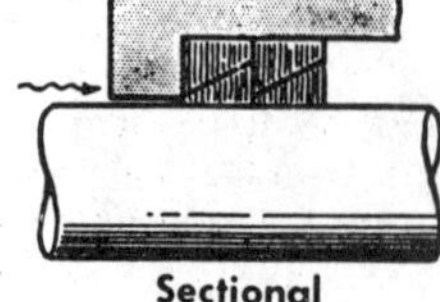

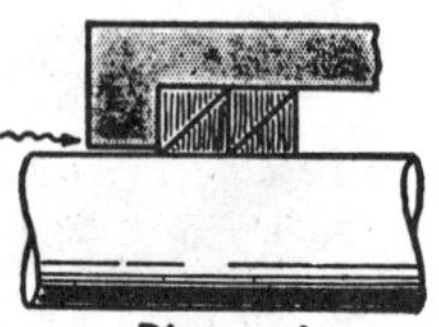

15 Always install cross-expansion packing so plies slope toward the fluid pressure from housing. Place sectional rings so slope between inside and outside ring is toward the pressure. Diagonal rings must also have slope toward the fluid pressure. Watch these details for best results when installing new packing in a box

Fig. 18.30 How to pack a pump

(Source: Water Pollution Control Association of Pennsylvania Magazine)

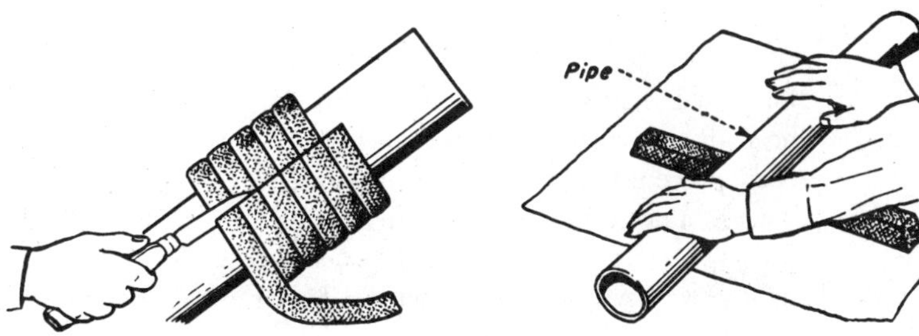

4 To find the right size of packing to install, measure stuffing-box bore and subtract shaft diameter, divide by 2. Packing is too critical for guesswork.

5 Wind packing, needed for filling stuffing box, snugly around shaft (or same size shaft held in vise) and cut through each turn while coiled, as shown. If the packing is slightly too large, never flatten with a hammer. Place each turn on a clean newspaper and then roll out with pipe as you would with a rolling pin

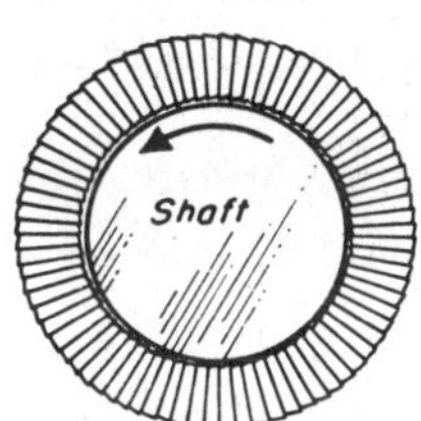

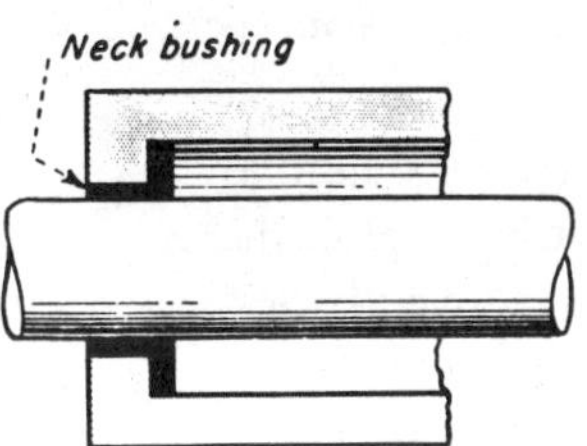

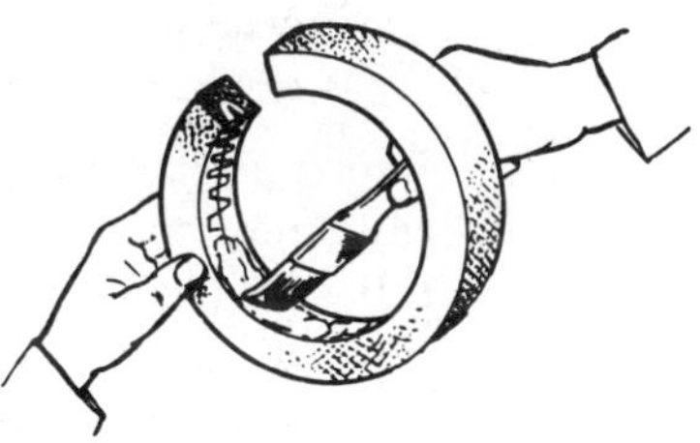

8 Install foil-wrapped packing so edges on inside will face direction of shaft rotation. This is a must; otherwise, thin edges flake off, reduce packing life

9 Neck bushing slides into stuffing box. Quick way to make it is to pour soft bearing metal into tin can, turn and bore for sliding fit into place

10 Swabbing new metallic packings with lubricant supplied by packing maker is OK. These include foil types, leadcore, etc. If the shaft is oily, don't swab it

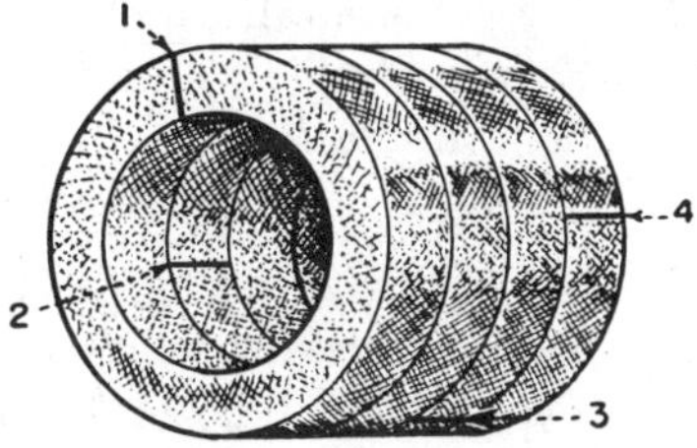

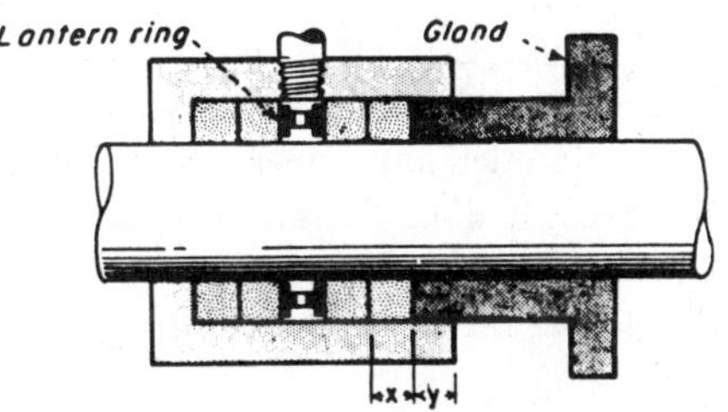

13 Stagger joints 180 degrees if only two rings are in stuffing box. Space at 120 degrees for three rings, or 90 degrees if four rings or more are in set

14 Install packing so lantern ring lines up with cooling-liquid opening. Also, remember that this ring moves back into box as packing is compressed. Leave space for gland to enter as shown. Tighten gland with wrench—back off finger-tight. Allow the packing to leak until it seats itself, then allow a slight operating leakage.

Hydraulic-packing pointers

First, clean stuffing box, examine ram or shaft. Next, measure stuffing box depth and packing set—find difference. Place 1/8-in. washers over gland studs as shown. Lubricate ram and packing set (if for water). If you can use them, endless rings give about 17% more wear than cut rings. Place male adapter in bottom, then carefully slide each packing turn home—don't harm lips. Stagger joints for cut rings. Measure from top of packing to top of washers, then compare with gland. Never tighten down new packing set until all air has chance to work out. As packing wears, remove one set of washers after more wear, remove other washer.

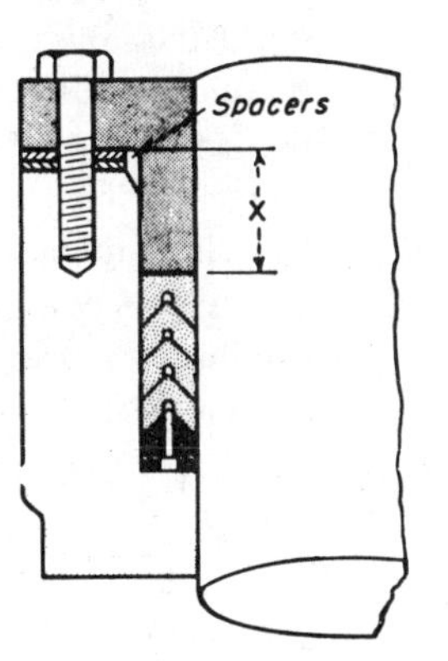

Fig. 18.30 How to pack a pump (continued)

Frequency of Service	

If a bronze shaft sleeve is not too badly scored, the shaft sleeve can be restored to service. The repair procedure consists of turning the sleeve down to a uniform diameter with a rough cut. Then spray the sleeve with stainless steel to a slightly oversized outside diameter followed by machining and polishing to bring the sleeve back to its original diameter. You will probably find that these reworked sleeves will outlast the originals.

W 7. CHECK MECHANICAL SEALS. Mechanical seals usually consist of two subassemblies: (1) a rotating ring assembly, and (2) a stationary assembly.

Inspect seal for leakage and excessive heat. If any part of the seal needs replacing, replace the entire seal (both subassemblies) with a new seal that has been provided by the manufacturer. Before installing a new seal, be sure that there are no chips or cracks on the carbide sealing surface. Keep a new mechanical seal clean at all times.

Always be sure that a mechanical seal is surrounded with water before starting and running the pump.

Q 8. INSPECT AND LUBRICATE BEARINGS. Unless otherwise specifically directed for a particular pump model, lubricate according to the procedures covered in Section 18.224, page 310. Check sleeve bearings to see that oil rings turn freely with the shaft. Repair or replace if defective.

Measure sleeve bearings and replace those worn excessively. Generally, allow clearance of 0.002 inch plus 0.001 inch for each inch or fraction of inch of shaft-journal diameter.

Q 9. CHECK OPERATING TEMPERATURE OF BEARINGS. Check bearing temperature with thermometer, not by hand. If antifriction bearings are running hot, check for overlubrication and relieve if necessary. If sleeve bearings run too hot, check for lack of lubricant. If proper lubrication does not correct condition, disassemble and inspect bearings. Check alignment of pump and motor if high temperatures continue.

S 10. CHECK ALIGNMENT OF PUMP AND MOTOR. For method of aligning pump and motor, see Paragraph 10: Couplings. If misalignment recurs frequently, inspect entire piping system. Unbolt piping at suction and discharge nozzles to see if it springs away, indicating strain on casing. Check all piping supports for soundness and effective support of load.

Vertical pumps usually have flexible shafting, which permits slight angular misalignment; however, if solid shafting is used, align exactly. If beams carrying intermediate bearings are too light or are subject to contraction or expansion, replace beams and realign intermediate bearings carefully.

S 11. INSPECT AND SERVICE PUMPS.

a. Remove rotating element of pump and inspect thoroughly for wear. Order replacement parts where necessary. Check impeller clearance between volute.

b. Remove any deposit or scaling. Clean out water-seal piping.

c. Determine pump capacity by pumping into empty tank of known size or by timing the draining of pit or sump.

$$\text{Pump Capacity, GPM} = \frac{\text{Volume, gallons}}{\text{Time, minutes}}$$

or

$$\text{Pump Capacity, } \frac{\text{liters}}{\text{sec}} = \frac{\text{Volume, liters}}{\text{Time, seconds}}$$

See Example 1 (page 317) for procedures on how to calculate pump capacity.

d. Test pump efficiency. Refer to pump manufacturer's instructions on how to collect data and perform calculations. Also see Chapter 3, Section 3.17, "Pump Testing and Evaluation," in *Small Water System Operation and Maintenance* in this series of operator training manuals.

e. Measure total dynamic suction head or lift and discharge head to test pump and pipe condition. Record figures for comparison with later tests.

f. Inspect foot and check valves, paying particular attention to check valves, which can cause water hammer when pump stops. (Also see "Paragraph 13: Check Valves," on page 341.) Foot valves, which are a type of check valve, are used when pumping raw water.

g. Examine wearing rings. Replace seriously worn wearing rings to improve efficiency. Check wearing ring clearances, which generally should be no more than 0.003 inch per inch of wearing ring diameter.

CAUTION: To protect rings and casings, never allow pump to run dry through lack of proper priming when starting or loss of suction when operating.

A 12. DRAIN PUMP FOR LONG-TERM SHUTDOWN. When shutting down pump for a long period, open motor disconnect switch to disconnect the motor, and, if so equipped, turn on the electric motor winding heaters. Shut all valves on suction, discharge, water-seal, and priming lines; drain pump completely by removing vent and drain plugs. This procedure protects pump against corrosion, sedimentation, and freezing. Inspect pump and bearings thoroughly and perform all necessary servicing. Drain bearing housings and replenish with fresh oil; purge old grease and replace. When a pump is out of service, run it monthly to warm it up and to distribute lubrication so the packing will not freeze to the shaft. Resume periodic checks after pump is put back in service.

FORMULAS

To find the volume of a rectangle in cubic feet, multiply the length times width times depth.

$$\text{Volume, ft}^3 = (\text{Length, ft})(\text{Width, ft})(\text{Depth, ft})$$

To find the volume of a cylinder in cubic feet, multiply 0.785 times the diameter squared times the depth.

$$\text{Volume, ft}^3 = (0.785)(\text{Diameter, ft})^2(\text{Depth, ft})$$

To convert a volume from cubic feet to gallons, multiply the volume in cubic feet times 7.48 gallons per cubic foot.

$$\text{Volume, gal} = (\text{Volume, ft}^3)(7.48\ \text{gal/ft}^3)$$

To calculate the output or capacity of a pump in gallons per minute, divide the volume pumped in gallons by the pumping time in minutes.

$$\text{Pump Capacity, GPM} = \frac{\text{Volume Pumped, gallons}}{\text{Pumping Time, minutes}}$$

EXAMPLE 1

A pump's capacity is measured by recording the time in minutes for water to rise 3 feet in an 8-foot-diameter tank. What is the pumping rate or capacity in gallons per minute when the pumping time is 9 minutes?

Known		Unknown
Diameter, ft	= 8 ft	Pump Capacity, GPM
Depth, ft	= 3 ft	
Time, min	= 9 min	

Calculate the tank volume in cubic feet.

$$\begin{aligned}\text{Volume, ft}^3 &= (0.785)(\text{Diameter, ft})^2(\text{Depth, ft})\\ &= (0.785)(8\ \text{ft})^2(3\ \text{ft})\\ &= 151\ \text{ft}^3\end{aligned}$$

Convert the tank volume from cubic feet to gallons.

$$\begin{aligned}\text{Volume, gal} &= (\text{Volume, ft}^3)(7.48\ \text{gal/ft}^3)\\ &= (151\ \text{ft}^3)(7.48\ \text{gal/ft}^3)\\ &= 1{,}129\ \text{gallons}\end{aligned}$$

Calculate the pump capacity in gallons per minute.

$$\begin{aligned}\text{Pump Capacity, GPM} &= \frac{\text{Volume Pumped, gal}}{\text{Pumping Time, min}}\\ &= \frac{1{,}129\ \text{gal}}{9\ \text{min}}\\ &= 125\ \text{GPM}\end{aligned}$$

QUESTIONS

Please write your answers to the following questions and compare them with those on page 370.

18.23A What is a cross-connection?

18.23B Is a slight water-seal leakage desirable when a pump is running? If so, why?

18.23C How would you measure the capacity of a pump?

18.23D What should be done to a pump before it is shut down for a long time, and why?

18.23E Estimate the capacity of a pump (in GPM) if it lowers the water in a 10-foot-wide by 15-foot-long wet well 1.7 feet in 5 minutes.

Paragraph 2: Reciprocating Pumps, General

The general procedures in this paragraph apply to all reciprocating pumps described in this section.

Frequency of Service	
W	1. CHECK SHEAR PIN ADJUSTMENT. Set eccentric by placing shear pin through proper hole in eccentric flanges to give required stroke. Tighten the two ⅝- or ⅞-inch hexagonal nuts on connecting rods just enough to take spring out of lock washers. (See "Paragraph 11: Shear Pins," on page 325.) When a shear pin fails, eccentric moves toward neutral position, preventing damage to the pump. Remove cause of obstruction and insert new shear pin. Shear pins fail because of one of three common causes: a. Solid object lodged under piston b. Clogged discharge line c. Stuck or wedged valve
D	2. CHECK PACKING ADJUSTMENT. Give special attention to packing adjustment. If packing is too tight, it reduces efficiency and scores piston walls. Keep packing just tight enough to keep sludge from leaking through gland. Before pump is installed or after it has been idle for a time, loosen all nuts on packing gland. Run pump with sludge suction line closed and valve covers open for a few minutes to break in the packing. Turn down gland nuts no more than necessary to prevent sludge from getting past packing. Tighten all packing nuts uniformly.

Frequency of Service

When packing gland bolts cannot be taken up farther, remove packing. Remove old packing and thoroughly clean cylinder and piston walls. Place new packing into cylinder, staggering packing-ring joints, and tamp each ring into place. Break in and adjust packing as explained above. When chevron type packing is used, tighten gland nuts only finger tight because excessive pressure ruins packing and scores plunger.

Q 3. CHECK BALL VALVES. When valve balls are so worn that diameter is ⅝ inch (1.5 cm) smaller than original size, they may jam into guides in valve chamber. Check size of valve balls and replace if badly worn.

Q 4. CHECK VALVE-CHAMBER GASKETS. Valve-chamber gaskets on most pumps serve as a safety device and blow out under excessive pressure. Check gaskets and replace if necessary. Keep additional gaskets on hand for replacement.

A 5. CHECK ECCENTRIC ADJUSTMENT. To take up babbitt bearing, remove brass shims provided on connecting rod. After removing shims, operate pump for at least 1 hour and check to see that eccentric does not run hot.

D 6. NOTE UNUSUAL NOISES. Check for noticeable water hammer when pump is operating. This noise is most pronounced when pumping water or very thin sludge; it decreases or disappears when pumping heavy sludge. Eliminate noise by opening the ¼-inch (0.6 cm) petcock on pump body slightly; this draws in a small amount of air, keeping discharge air chamber full at all times.

D 7. CHECK CONTROL VALVE POSITIONS. Because any plunger pump may be damaged if operated against closed valves in the pipeline, especially the discharge line, make all valve setting changes with pump shut down; otherwise, pumps that are installed to pump from two sources or to deliver to separate tanks at different times may be broken if all discharge line valves are closed simultaneously for a few seconds or discharge valve directly above pump is closed.

W 8. GEAR REDUCER. Check oil level by removing plug on the side of the gear case. Unit should not be in operation.

Q 9. CHANGE OIL AND CLEAN MAGNETIC DRAIN PLUG.

W 10. CONNECTING RODS. Set oilers to disperse two drops per minute.

W 11. PLUNGER CROSSHEAD. Fill plunger as required to half cover the wrist pin with oil.

D 12. PLUNGER TROUGH. Keep small quantity of oil in trough to lubricate the plunger.

M 13. MAIN SHAFT BEARING. Grease bearings monthly. Pump should be in operation when lubricating to avoid excessive pressure on seals.

14. CHECK ELECTRIC MOTOR. See Paragraph 6: Electric Motors.

Paragraph 3: Propeller Pumps, General

D 1. CHECK MOTOR CONDITION. See Paragraphs 6.1 and 6.2.

D 2. CHECK PACKING GLAND ASSEMBLY. See Paragraph 1.6

W 3. INSPECT PUMP CONTROL. See Paragraph 1.4.

W 4. LUBE LINE SHAFT AND DISCHARGE BOWL BEARING. Maintain oil in oiler at all times. Adjust feed rate to approximately four drops per minute.

W 5. LUBE SUCTION BOWL BEARING. Lube through pressure fitting. Usually three or four strokes of gun are enough.

W 6. OPERATE PUMPS ALTERNATELY. See Paragraph 1.3.

A 7. LUBE MOTOR BEARINGS. See Paragraph 6.3.

Paragraph 4: Progressive Cavity Pumps, General (Figure 18.26, page 308).

D 1. CHECK MOTOR CONDITION. See Paragraphs 6.1 and 6.2.

D 2. CHECK PACKING GLAND ASSEMBLY. See Paragraph 1.6.

D 3. CHECK DISCHARGE PRESSURE. A higher than normal discharge pressure may indicate a line blockage or a closed valve downstream. An abnormally low discharge pressure can mean reduced rate of discharge.

S 4. INSPECT AND LUBRICATE BEARING GREASE. If possible, remove bearing cover and visually inspect grease. When greasing, remove relief plug and cautiously add 5 or 6 strokes of the grease gun. Afterward, check bearing temperature with thermometer. If over 220°F (104°C), remove some grease.

Frequency of Service

S 5. LUBEFLUSH MOTOR BEARINGS. See Paragraph 6.3.

S 6. CHECK PUMP OUTPUT. Check how long it takes to fill a vessel of known volume or quantity; or check performance against a meter, if available. See Paragraph 1.11 c.

A 7. SCOPE MOTOR BEARINGS. See Paragraph 6.4.

A 8. SCOPE PUMP BEARINGS. See Paragraph 6.4.

Paragraph 5: Pump Controls

To ensure the best operation of the pump, a systematic inspection of the controls should be made at least once a week.

W 1. CHECK CONTROLS. Controls respond to the control variable.

W 2. STARTUP. The unit starts when the control system makes contact, and the pump stops at the prescribed control setting.

W 3. MOTOR SPEED. The motor comes up to speed quickly and is maintained.

W 4. SPARKING. A brush-type motor does not spark profusely in starting or running.

W 5. INTERFERENCE WITH CONTROLS. Grease and dirt are not interfering with controls.

W 6. ADJUSTMENTS. Any necessary adjustments are properly completed.

QUESTIONS

Please write your answers to the following questions and compare them with those on page 370.

18.23F What are some of the common causes of shear pin failure in reciprocating pumps?

18.23G What may happen when water or a thin sludge is being pumped by a reciprocating pump?

18.23H What could be the causes of a higher than normal discharge pressure in a progressive cavity pump?

Paragraph 6: Electric Motors

To ensure the proper and continuous function of electric motors, the items listed in this paragraph must be performed at the designated intervals. If operational checks indicate a motor is not functioning properly, these items will have to be checked to locate the problem.

D 1. CHECK MOTOR CONDITIONS.

a. Keep motors free from dirt, dust, and moisture.

b. Keep operating space free from articles that may obstruct air circulation.

c. Check for excessive grease leakage from bearings.

D 2. NOTE ALL UNUSUAL CONDITIONS.

a. Unusual noises in operation.

b. Motor failing to start or come to speed normally, sluggish operation.

c. Motor or bearings that feel or smell hot.

d. Continuous or excessive sparking at commutator or brushes. Blackened commutator.

e. Intermittent sparking at brushes.

f. Fine dust under coupling having rubber buffers or pins.

g. Smoke, charred insulation, or solder whiskers extending from armature.

h. Excessive humming.

I. Regular clicking.

j. Rapid knocking.

k. Brush chatter.

l. Vibration.

m. Hot commutator.

A 3. LUBRICATE BEARINGS (Figure 18.31). Check grease in ball bearing and relubricate when necessary.

Follow instructions in Section 18.224, "Pump Lubrication," when lubricating greased bearings.

A 4. USING A *STETHOSCOPE*,[16] CHECK BOTH BEARINGS.

Listen for whines, gratings, or uneven noises. Listen all around the bearing and as near as possible to the bearing. Listen while the motor is being started and shut off. If unusual noises are heard, pinpoint the location.

5. IF YOU THINK THE MOTOR is running unusually hot, check with a thermometer.

16. *Stethoscope.* An instrument used to magnify sounds and carry them to the ear.

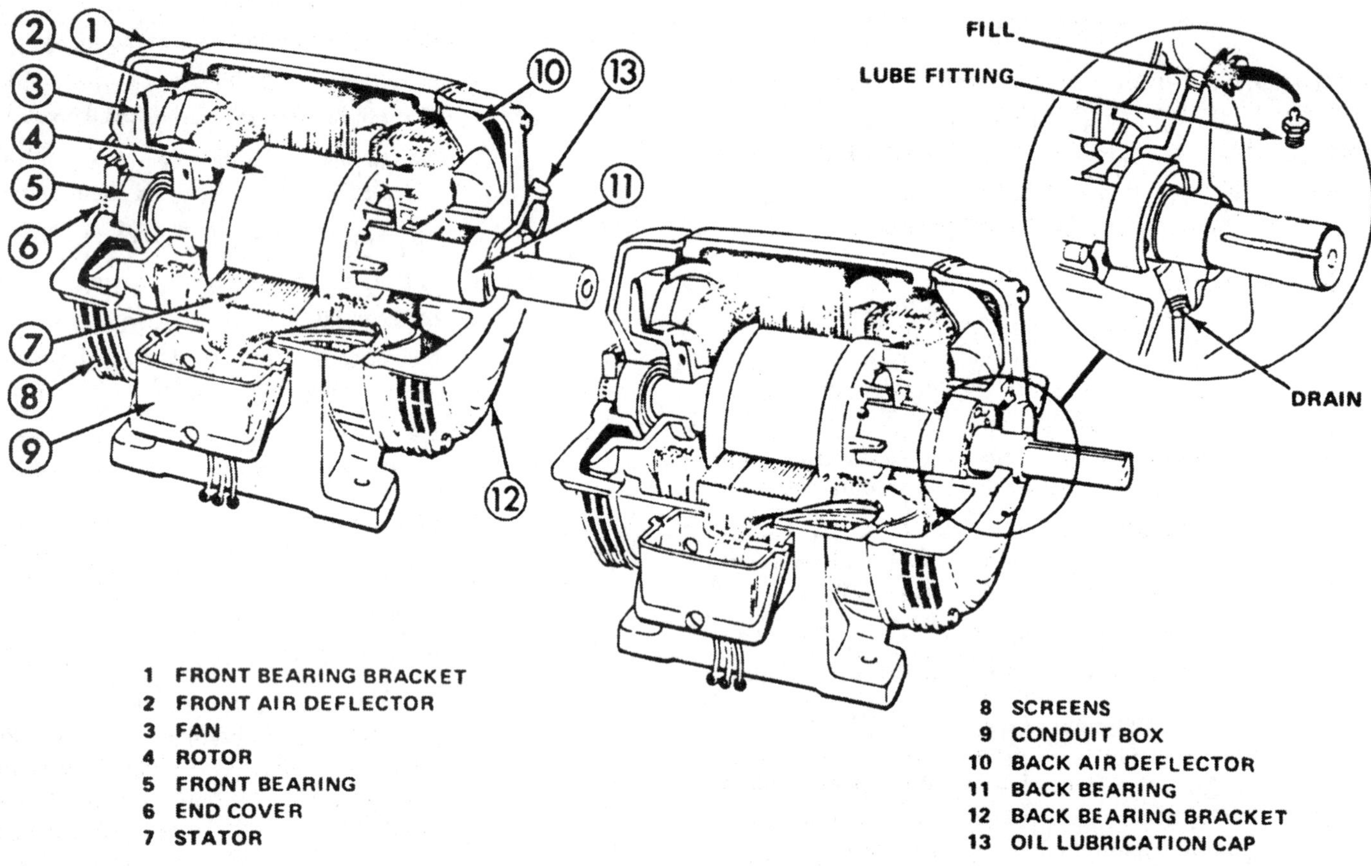

Fig. 18.31 Electric motor lubrication

Frequency of Service

Place the thermometer on the casing near the bearing, holding it there with putty or clay. Check the current on each leg to determine if the currents are balanced and within the motor nameplate limits.

A — 6. *DATEOMETER.*[17] If there is a dateometer on the motor, after changing the oil in the motor, loosen the dateometer screw and set to the corresponding year.

QUESTIONS

Please write your answers to the following questions and compare them with those on page 370.

18.23I What are the major items you would include when checking an electric motor?

18.23J What is the purpose of a stethoscope?

Paragraph 7: Belt Drives

1. GENERAL. Maintaining a proper tension and alignment of belt drives ensures long life of belts and sheaves (pulleys). Incorrect alignment causes poor operation and excessive belt wear. Inadequate tension reduces the belt grip and causes high belt loads, snapping, and unusual wear.

   a. Cleaning belts. Keep belts and sheaves clean and free of oil, which causes belts to deteriorate. To remove oil, take belts off sheaves and wipe belts and sheaves with a rag moistened in a nonoil-based solvent. Carbon tetrachloride is not recommended because exposure to its fumes has many toxic effects on humans. Carbon tetrachloride also is absorbed into the skin on contact and its effects become stronger with each contact.

   b. Installing belts. Before installing belts, replace worn or damaged sheaves, then slack off on adjustments. Do not try to force belts into position. Never use a screwdriver or similar lever to get belts onto sheaves. After belts are installed, adjust tension; recheck tension after eight hours of operation. (See Table 18.1.)

   c. Replacing belts. Replace belts as soon as they become frayed, worn, or cracked. Never replace only one V-belt on a multiple drive. Replace the complete set with a set of matched belts, which can be obtained from any supplier. All belts in a matched set are machine checked to ensure equal size and tension.

   d. Storing spare belts. Store spare belts in a cool, dark place. Tag all belts in storage to identify them with the equipment on which they can be used.

2. V-BELTS. A properly adjusted V-belt has a slight bow in the slack side when running; when idle, it has an alive springiness when thumped with the hand. An improperly tightened belt feels dead when thumped.

   If the slack side of the drive is less than 45 degrees from the horizontal, vertical sag at the center of the span may be adjusted in accordance with Table 18.1.

TABLE 18.1 HORIZONTAL BELT TENSION

Span (inches)		10	20	50	100	150	200
Vertical Sag (inches)	From	.01	.03	.20	.80	1.80	3.30
	To	.03	.09	.58	2.30	4.90	8.60
Span (millimeters)		250	500	1,250	2,500	3,750	5,000
Vertical Sag (millimeters)	From	0.25	0.75	5.00	20.0	45.0	82.5
	To	0.75	2.25	14.50	57.5	122.5	215.0

M — a. Check tension. If tightening belt to proper tension does not correct slipping, check for overload, oil on belts, or other possible causes. Never use belt dressing to stop belt slippage. Worn rubber near the drive is a sign of improper tension, incorrect alignment, or damaged sheaves.

M — b. Check sheave (pulley) alignment. Lay a long straightedge or string across outside faces of pulley, and allow for differences in dimensions from centerlines of grooves to outside faces of the pulleys being aligned. Be very careful in aligning drives with more than one V-belt on a sheave, as misalignment can cause unequal tension.

17. *Dateometer* (day-TOM-uh-ter). A small calendar disk attached to motors and equipment to indicate the year in which the last maintenance service was performed.

Paragraph 8: Chain Drives

Frequency of Service

1. GENERAL. Chain drives may be designated for slow, medium, or high speeds.
   a. Slow-speed drives. Because slow-speed drives are usually enclosed, adequate lubrication is difficult. Heavy oil applied to the outside of the chain seldom reaches the working parts; in addition, the oil catches dirt and grit and becomes abrasive. For lubricating and cleaning methods, see 5 and 6 below.
   b. Medium- and high-speed drives. Medium-speed drives should be continuously lubricated with a device similar to a sight-feed oiler. High-speed drives should be completely enclosed in an oil-tight case and the oil maintained at proper level.

D — 2. CHECK OPERATION. Check general operating condition during regular tours of duty.

Q — 3. CHECK CHAIN SLACK. The correct amount of slack is essential to proper operation of chain drives. Unlike other belts, chain belts should not be tight around the sprocket; when chains are tight, working parts carry a much heavier load than necessary. Too much slack is also harmful; on long centers particularly, too much slack causes vibrations and chain whip, reducing life of both chain and sprocket. A properly installed chain has a slight sag or looseness on the return run.

S — 4. CHECK ALIGNMENT. If sprockets are not in line or if shafts are not parallel, excessive sprocket and chain wear and early chain failure result. Wear on inside of chain, side walls, and sides of sprocket teeth are signs of misalignment. To check alignment, remove chain and place a straightedge against sides of sprocket teeth.

S — 5. CLEAN. On enclosed types, flush chain and enclosure with a petroleum solvent (kerosene). On exposed types, remove chain and soak and wash it in solvent. Clean sprockets, install chain, and adjust tension.

S — 6. CHECK LUBRICATION. Soak exposed-type chains in oil to restore lubricating film. Remove excess lubricant by hanging chains up to drain.

   Do not lubricate underwater chains that operate in contact with considerable grit. If water is clean, lubricate by applying waterproof grease with brush while chain is running.

   Do not lubricate chains on elevators or on conveyors of feeders that handle dirty or gritty materials. Dust and grit combine with lubricants to form a cutting compound that reduces chain life.

S — 7. CHANGE OIL. On enclosed types only, drain oil and refill case to proper level.

S — 8. INSPECT. Note and correct abnormal conditions before serious damage results. Do not put a new chain on worn sprockets. Always replace worn sprockets when replacing a chain because out-of-pitch sprockets cause as much chain wear in a few hours as years of normal operation.

9. TROUBLESHOOTING. Some common symptoms of improper chain drive operation and their remedies follow:
   a. Excessive noise. Correct alignment, if misaligned. Adjust centers for proper chain slack. Lubricate in accordance with aforementioned methods. Be sure all bolts are tight. If chain or sprockets are worn, reverse or renew if necessary.
   b. Wear on chain, side walls, and sides of teeth. Remove chain and correct alignment.
   c. Chain climbs sprockets. Check for poorly fitting sprockets and replace if necessary. Make sure tightener is installed on drive chain.
   d. Broken pins and rollers. Check for chain speed that may be too high for the pitch, and substitute chain and sprockets with shorter pitch, if necessary. Breakage also may be caused by shock loads.
   e. Chain clings to sprockets. Check for incorrect or worn sprockets or heavy, tacky lubricants. Replace sprockets or lubricants if necessary.
   f. Chain whip. Check for too-long centers or high, pulsating loads and correct cause.
   g. Chains get stiff. Check for misalignment, improper lubrication, or excessive overloads. Make necessary corrections or adjustments.

Paragraph 9: Variable-Speed Belt Drives (Figure 18.32)

Frequency of Service	
D	1. CLEAN DISKS. Remove grease, acid, and water from disk faces.
D	2. CHECK SPEED-CHANGE MECHANISM. Shift drive through entire speed range to make sure shafts and bearings are lubricated and disks move freely in lateral direction on shafts.
W	3. CHECK V-BELT. Make sure it runs level and true. If one side rides high, a disk is sticking on shaft because of insufficient lubrication or wrong lubricant. In this case, stop the drive at once, remove V-belt, and clean disk hub and shaft thoroughly with petroleum solvent until disk moves freely. Relubricate with soft ball-bearing grease and replace V-belt in opposite direction from that in which it formerly ran.
M	If drive is not operated for 30 days or more, shift unit to minimum speed position, placing spring on variable-speed shaft at minimum tension and relieving belt of excessive pressure.
	4. LUBRICATE DRIVE. Make sure to apply lubricant at all the six force-feed lubrication fittings (Figure 18.32: A, B, D, E, G, and H) and the one cup-type fitting (C). *NOTE:* If the drive is used with a reducer, fitting E is not provided.
W	a. Once every 10–14 days, use 2–3 strokes of a grease gun through fittings A and B at ends of shifting screw and variable-speed shaft, respectively, to lubricate bearings of movable disks. Then, with unit running, shift drive from one extreme speed position to the other to ensure thorough distribution of lubricant over disk-hub bearings.
Q	b. Add two or three shots of grease through fittings D and E to lubricate frame bearing on variable-speed shaft.
Q	c. Every 90 days, add 2–3 cups of grease to Cup C, which lubricates thrust bearing on constant-speed shaft.
Q	d. Every 90 days, use 2–3 strokes of grease gun through fittings G and H to lubricate motor frame bearings.
	CAUTION: Be sure to follow manufacturer's recommendation on type of grease. After lubricating, wipe excessive grease from sheaves and belt.

QUESTIONS

Please write your answers to the following questions and compare them with those on page 370.

18.23K How can you tell if a V-belt on belt-drive equipment has proper tension and alignment?

18.23L Why should worn sprockets be replaced when replacing a chain in a chain drive unit?

Paragraph 10: Couplings

	1. GENERAL. Unless couplings between the driving and driven elements of a pump or any other piece of equipment are kept in proper alignment, breaking and excessive wear results in either the driven machinery and the driver or both. Burned-out bearings, sprung or broken shaft, and excessively worn or ruined gears are some of the damages caused by misalignment. To prevent outages and the expense of installing replacement parts, check the alignment of all equipment before damage occurs.
	a. Improper original installation of the equipment may not necessarily be the cause of the trouble. Settling of foundations, heavy floor loadings, warping of bases, excessive bearing wear, and many other factors cause misalignment. A rigid base is not always security against misalignment. The base may have been mounted off level, which could cause it to warp.
	b. Flexible couplings permit easy assembly of equipment, but they must be aligned as exactly as flanged couplings if maintenance and repair are to be kept to a minimum. Rubber-bushed types cannot function properly if the bolts cannot move in their bushings.
S	2. CHECK COUPLING ALIGNMENT (straightedge method). Excessive bearing and motor temperatures caused by overload, noticeable vibration, or unusual noises may all be warnings of misalignment. Realign when necessary (Figure 18.33) using a straightedge and thickness gauge or wedge. To ensure satisfactory operation, level up to within 0.005 inch (0.13 mm) as follows:
	a. Remove coupling pins.
	b. Rigidly tighten driven equipment; slightly tighten bolts holding drive.
	c. To correct horizontal and vertical misalignment, shift or shim drive to bring coupling halves into position so no light can be seen

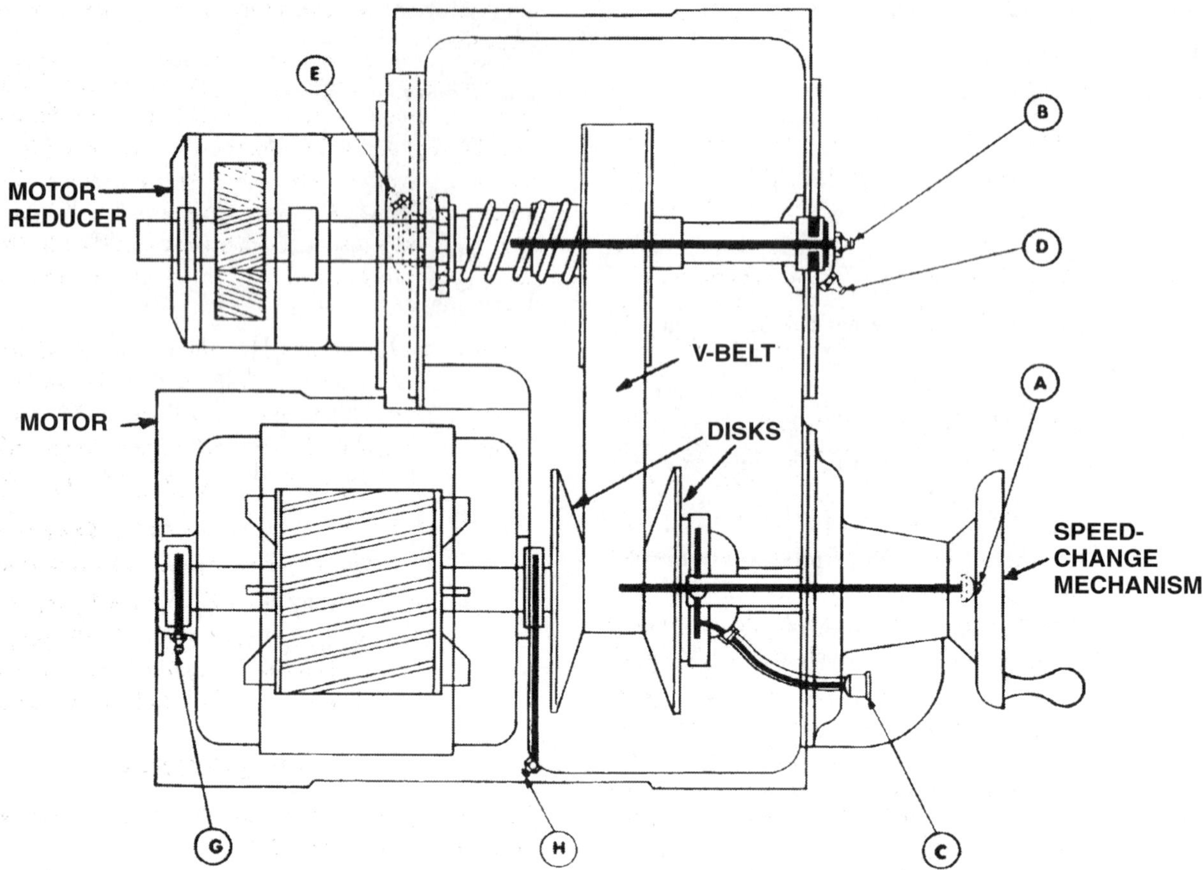

NOTE: A, B, D, E, G, and H are force-feed lubrication fittings. C is a cup-type lubrication fitting.

Fig. 18.32 Reeves varidrive
(Source: War Department Technical Manual TM5-666)

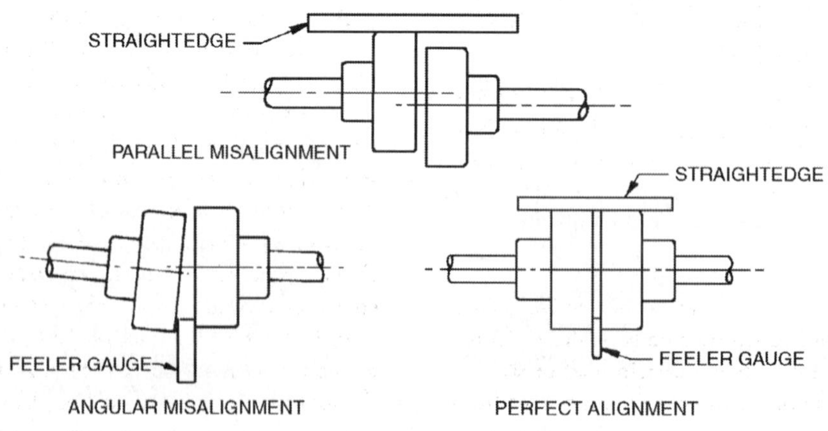

Fig. 18.33 Testing alignment, straightedge

Frequency of Service

under a straightedge laid across them. Place straightedge in four positions, holding a light in back of straightedge to help ensure accuracy.

d. Check for angular misalignment with a thickness or feeler gauge inserted at four places to make certain space between coupling halves is equal.

e. If proper alignment has been secured, coupling pins can be put in place easily using only finger pressure. Never hammer pins into place.

f. If equipment is still out of alignment, repeat the procedure.

S 3. CHECK COUPLING ALIGNMENT (dial indicator method). Dial indicators also are used to measure coupling alignment. This method produces better results than the straightedge method. The dial indicates very small movements or distances, which are measured in mils (1 mil equals 1/1,000 of an inch). The indicator consists of a dial with a graduated face (with plus and minus readings), a pedestal, and a rigid indicator bar (or fixture) as shown in Figure 18.34.

The dial indicator is attached to one coupling via the fixture and adjusted to the zero position or reading. When the shaft of the machine is rotated, misalignment will cause the pedestal to compress (a plus reading), or extend (a minus reading). Literature provided by the manufacturer of machinery usually will indicate maximum allowable tolerances or movement.

Carefully study the manufacturer's literature provided with your dial indicator before attempting to use the device.

A 4. CHANGE OIL IN FAST COUPLINGS. Drain out old oil and add oil to proper level. Correct quantity is given on instruction card supplied with each coupling.

Paragraph 11: Shear Pins

Some water treatment units use shear pins as protective devices to prevent damage in case of sudden overloads. A shear pin is a straight pin that will fail (break) when a certain load or stress is exceeded. The purpose of the pin is to protect equipment from damage due to excessive loads or stresses. To serve this purpose, these devices must be in operational condition at all times. Under some operating conditions, shearing surfaces of a shear pin device may freeze together so solidly that an overload fails to break them.

Manufacturer drawings for particular installations usually specify shear pin material and size. If this information is not available, obtain the information from the manufacturer, giving the model, serial number, and load conditions of unit. When necessary to determine shear pin size, select the lowest strength that does not break under the unit's usual loads. When proper size is determined, never use a pin of greater strength, such as a bolt or a nail.

If necked pins are used, be sure the necked-down portion is properly positioned with respect to shearing surfaces. When a shear pin breaks, determine and remedy the cause of failure before inserting new pin and starting drive in operation.

M 1. GREASE SHEARING SURFACES.

Q 2. REMOVE SHEAR PIN. Operate motor for a short time to smooth out any corroded spots.

A 3. CHECK SPARE INVENTORY. Make sure an adequate supply is on hand, properly identified and with record of proper pin size, necked diameter, and longitudinal dimensions.

QUESTIONS

Please write your answers to the following questions and compare them with those on page 371.

18.23M What factors could cause couplings to become out of alignment?

18.23N What is the purpose of shear pins?

18.24 Pump Operation

18.240 Starting a New Pump

The initial startup work described in this section should be done by a competent and trained person, such as a manufacturer's representative, consulting engineer, or an experienced operator. The operator can learn a lot about pumps and motors by accompanying and helping a competent person put new equipment into operation.

Before starting a pump, lubricate it according to manufacturer's recommendations. Quality lubricants should be used. Turn the shaft by hand to see that it rotates freely. Then check to see that the shafts of the pump and motor are aligned and the flexible coupling adjusted. (Refer to Paragraph 10: Couplings, page 323; also see Section 18.23, "Pump Maintenance," page 312.) If the unit is belt driven, sheave (pulley) alignment and belt adjustment should be checked. (Refer to Paragraph 7: Belt Drives, page 321.) Check the electric voltage with the motor characteristics and inspect the wiring. See that thermal overload units in the starter are set properly. Turn on the motor just long enough to see that it turns the pump in the direction indicated by the rotational arrows marked on the pump. If separate water seal units or vacuum primer systems are used, these should be started. Finally, make sure lines are open. Sometimes, there is an exception (see following paragraph) in the case of the discharge valve.

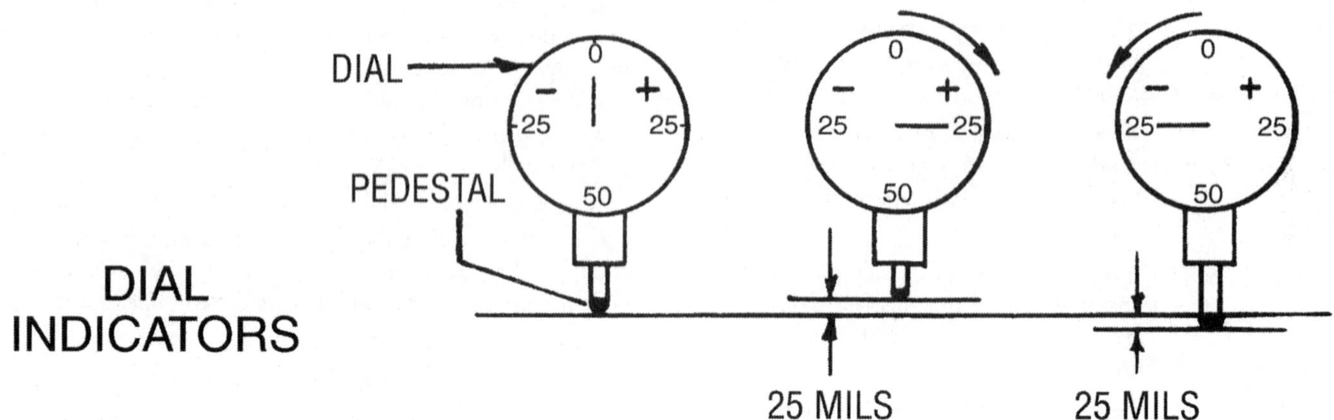

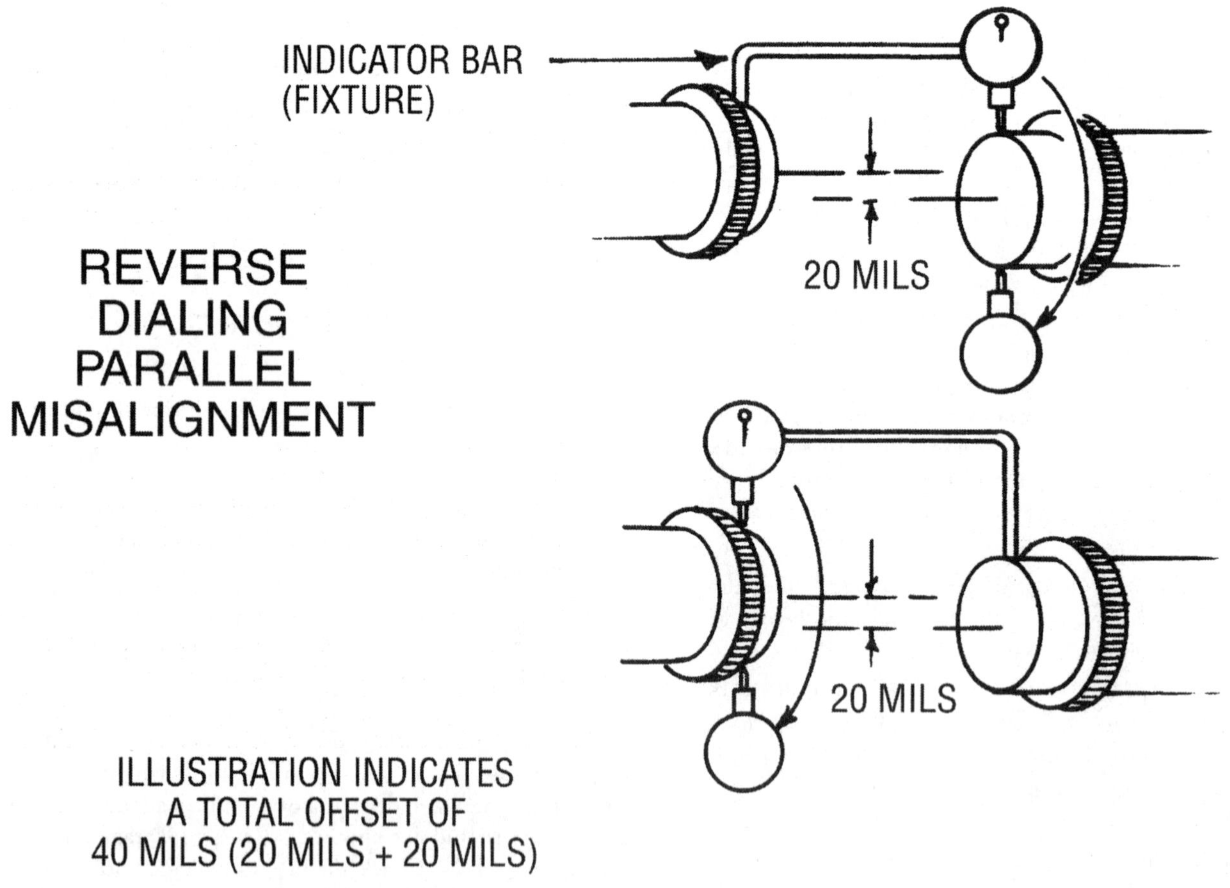

Fig. 18.34 Use of a dial indicator

(Permission of DYMAC, a Division of Spectral Dynamics Corporation)

A pump should not be run without first having been primed. To prime a pump, the operator should completely fill it with water. In some cases, automatic primers are provided. If they are not, it is necessary to vent the casing. Most pumps are provided with a valve to accomplish this. Allow the trapped air to escape until water flows from the vent; then replace the vent cap. In the case of suction lift applications, the pump must be filled with water unless a self-primer is provided. In nearly every case, you may start a pump with the discharge valve open. Exceptions to this, however, are where water hammer or pressure surges might result, or where the motor does not have sufficient margin of safety or power. Sometimes, there are no check valves in the discharge line. In this case (with the exception of positive-displacement pumps), it is necessary to start the pump and then open the discharge lines. Where there are common discharge headers, it is essential to start the pump and then open the discharge valve. A positive-displacement pump (reciprocating or piston types) should never be operated against a closed discharge line.

After starting the pump, again check to see that the direction of rotation is correct. Packing gland boxes (stuffing boxes) should be observed for slight leakage (approximately 60 drops per minute) as described in Paragraph 1: Pumps, General. Check to see that the bearings do not overheat from over- or underlubrication. The flexible coupling should not be noisy; if it is, the noise may be caused by misalignment or improper clearance or adjustment. Check to be sure pump anchorage is tight. Compare delivered pump flows and pressures with pump performance curves. If pump delivery falls below performance curves, look for obstructions in the pipelines and inspect piping for leaks.

18.241 Pump Shutdown

When shutting down a pump for a long period, the motor disconnect switch should be opened, locked out, and tagged with the reason for the tag noted. If the electric motor is equipped with winding heaters, check to be sure they are turned on. This helps to prevent condensation from forming, which can weaken the insulation on the windings. All valves on the suction, discharge, and water-seal lines should be shut tightly. Completely drain the pump by removing the vent and drain plugs.

Inspect the pump and bearings thoroughly so that all necessary servicing may be done during the inactive period. Drain the bearing housing and then add fresh lubricant. Follow any additional manufacturer's recommendations.

18.242 Pump-Driving Equipment

Driving equipment used to operate pumps includes electric motors and internal combustion engines. In rare instances, pumps are driven with steam turbines, steam engines, or air and hydraulic motors.

In all except the large installations, electric motors are used almost exclusively, with synchronous and induction types being the most commonly used. Synchronous motors operate at constant speeds and are used chiefly in large sizes. Three-phase, squirrel-cage induction motors are most often used in water treatment plants. These motors require little attention and, under average operating conditions, the factory lubrication of the bearing will last approximately 1 year. (Check with the manufacturer for average number of operating hours for bearings.) When lubricating motors, remember that too much grease may cause bearing trouble or damage the winding.

Clean and dry all electrical contacts. Inspect for loose electrical contacts. Make sure that hold-down bolts on motors are secure. Check voltage while the motor is starting and running. Examine bearings and couplings.

18.243 Electrical Controls

A variety of electrical equipment is used to control the operation of pumps or to protect electric motors. If starters, disconnect switches, and cutouts are used, they should be installed in accordance with the local regulations (city or county codes) regarding this equipment. In the case of larger motors, the power company often requires starters that do not overload the power lines.

The electrode-type, bubbler-type, and diaphragm-type water-level control systems are all similar in effect to the floatswitch system. Scum is a problem with most water-level controls that operate pumps, and it must be removed on a regular basis.

QUESTIONS

Please write your answers to the following questions and compare them with those on page 371.

18.24A Where would you find out how to lubricate a pump?

18.24B What problems can develop if too much grease is used in lubricating a motor?

18.244 Operating Troubles

The following list of operating troubles includes most of the causes of pump failure or reduced operating efficiency. The remedy or cure is either obvious or may be identified from the description of the cause.

SYMPTOM A—PUMP WILL NOT START

CAUSES:

1. Blown fuses or tripped circuit breakers due to:
   a. Rating of fuses or circuit breakers not correct
   b. Switch (breakers) contacts corroded or shorted
   c. Terminal connections loose or broken somewhere in the circuit
   d. Automatic control mechanism not functioning properly
   e. Motor shorted or burned out
   f. Wiring hookup or service not correct
   g. Switches not set for operation
   h. Contacts of the control relays dirty and arcing
   i. Fuses or thermal units too warm
   j. Wiring short-circuited
   k. Shaft binding or sticking due to rubbing impeller, tight packing glands, or clogging of pump
2. Loose connections, fuse, or thermal unit

SYMPTOM B—REDUCED RATE OF DISCHARGE

CAUSES:

1. Pump not primed
2. Air in the water
3. Speed of motor too low
4. Improper wiring
5. Defective motor
6. Discharge head too high
7. Suction lift greater than anticipated
8. Impeller clogged
9. Discharge line clogged
10. Pump rotating in wrong direction
11. Air leaks in suction line or packing box
12. Inlet to suction line too high, permitting air to enter
13. Valves partially or entirely closed
14. Check valves stuck or clogged
15. Incorrect impeller adjustment
16. Impeller damaged or worn
17. Packing worn or defective
18. Impeller turning on shaft because of broken key
19. Flexible coupling broken
20. Loss of suction during pumping may be caused by leaky suction line or ineffective water or grease seal
21. Belts slipping
22. Worn wearing ring

SYMPTOM C—HIGH POWER REQUIREMENTS

CAUSES:

1. Speed of rotation too high
2. Operating heads lower than rating for which pump was designed, resulting in excess pumping rates
3. Sheaves on belt drive misaligned or maladjusted
4. Pump shaft bent
5. Rotating elements binding
6. Packing too tight
7. Wearing rings worn or binding
8. Impeller rubbing

SYMPTOM D—NOISY PUMP

CAUSES:

1. Pump not completely primed
2. Inlet clogged
3. Inlet not submerged
4. Pump not lubricated properly
5. Worn impellers
6. Strain on pumps caused by unsupported piping fastened to the pump
7. Foundation insecure
8. Mechanical defects in pump
9. Misalignment of motor and pump where connected by flexible shaft
10. Rocks in the impeller
11. Cavitation

QUESTIONS

Please write your answers to the following questions and compare them with those on page 371.

18.24C What items would you check if a pump will not start?

18.24D How would you attempt to increase the discharge from a pump if the flow rate is lower than expected?

18.245 Starting and Stopping Pumps

The operator must determine what treatment processes will be affected by either starting or stopping a pump. The pump discharge point must be known and valves either opened or closed to direct flows as desired by the operator when a pump is started or stopped.

18.2450 CENTRIFUGAL PUMPS

Basic rules for the operation of centrifugal pumps include the following items.

1. Do not operate the pump when safety guards are not installed over or around moving parts.
2. Do not start a pump that has been locked or tagged out for maintenance or repairs. An operator working on the pump could be injured or the equipment could be damaged.
3. Never run a centrifugal pump when the impeller is dry. Always be sure the pump is primed.
4. Never attempt to start a centrifugal pump whose impeller or shaft is spinning backward.
5. Do not operate a centrifugal pump that is vibrating excessively after startup. Shut unit down and isolate pump from system by closing the pump suction and discharge valves. Look for a blockage in the suction line and the pump impeller.

There are several situations in which it may be necessary to start a centrifugal pump against a closed discharge valve. Once the pump is primed, running, and indicating a discharge pressure, slowly open the pump discharge valve until the pump is fully on line. This procedure is used with treatment processes or piping systems with vacuums or pressures that cannot be dropped or allowed to fluctuate greatly while an alternate pump is put on the line.

Most centrifugal pumps used in water treatment plants are designed so that they can be easily started even if they have not been primed. This is accomplished with a positive static suction head or a low suction lift. On most of these arrangements, the pump will not require priming as long as the pump and the piping system do not leak. Leaks would allow the water to drain out of the pump volute. When pumps in water systems lose their prime, the cause is often a faulty check valve on the pump discharge line. When the pump stops, the discharge check valve will not seal (close) properly. Water previously pumped then flows back through the check valve and through the pump. The pump is drained and has lost its prime.

About 95 percent of the time, the centrifugal pumps in water treatment plants are ready to operate with suction and discharge valves open and seal water turned on. When the automatic start or stop command is received by the pump from the controller, the pump is ready to respond properly.

When the pumping equipment must be serviced, take it off the line by locking and tagging out the pump controls until all service work is completed.

QUESTIONS

Please write your answers to the following questions and compare them with those on page 371.

18.24E Why should a pump that has been locked or tagged out for maintenance or repairs not be started?

18.24F Under what conditions might a centrifugal pump be started against a closed discharge valve?

STOPPING PROCEDURES

This section contains a typical sequence of procedures to follow to stop a centrifugal pump. Exact stopping procedures for any pumping system depend upon the condition of the discharge system. The sudden stoppage of a pump could cause severe *WATER HAMMER*[18] problems in the piping system.

1. Inspect process system affected by pump, start alternate pump if required, and notify supervisor or log action.
2. Before stopping an operating pump, check its operation. This will give an indication of any developing problems, required adjustments, or problem conditions of the unit. This procedure only requires a few minutes. Items to be inspected include:
   a. Pump packing gland
      (1) Seal water pressure
      (2) Seal leakage (too much, sufficient, or too little leakage)
      (3) Seal leakage drain flowing clear
      (4) Mechanical seal leakage (if equipped)
   b. Pump operating pressures
      (1) Pump suction (pressure, vacuum)
      A higher vacuum than normal may indicate a partially plugged or restricted suction line. A lower vacuum may indicate a higher suction water level or a worn pump impeller or wearing rings.
      (2) Pump discharge pressure
      System pressure is indicated by the pump discharge pressure. Lower than normal discharge pressures can be caused by:
      (a) Worn impeller or wearing rings in the pump
      (b) A different point of discharge can change discharge pressure conditions
      (c) A broken discharge pipe can change the discharge head

   NOTE: To determine the maximum head a centrifugal pump can develop, slowly close the discharge valve at the pump. Read the pressure gauge between the pump and the discharge valve when the valve is fully closed. This is the maximum pressure the pump is capable of developing. Do not operate the pump longer than a few minutes with the discharge valve closed completely because the energy from the pump is converted to heat and water in the pump can become hot enough to damage the pump.

   c. Motor temperature and pump bearing temperature
   If motor or bearings are too hot to touch, further checking is necessary to determine if a problem has developed or if the temperature is normal. High temperatures may be measured with a thermometer.
   d. Unusual noises, vibrations, or conditions about the equipment
   If any of the above items indicate a change from the pump's previous operating condition, additional service or maintenance may be required during shutdown.
3. Actuate stop switch for pump motor and lockout switch. If possible, use switch next to equipment so that you may observe the equipment stop. Observe the following items:
   a. Check valve closes and seats.
   Valve should not slam shut, or discharge piping will jump or move in their supports. There should not be any leakage around the check valve shaft. If check valve is operated automatically, it should close smoothly and firmly to the fully closed position.
   NOTE: If the pump is not equipped with a check valve, close discharge valve before stopping pump.
   b. Motor and pump should wind down slowly and not make sudden stops or noises during shutdown.
   c. After equipment has completely stopped, pump shaft and motor should not start backspinning. A pump shaft or motor will spin backward if water being pumped flows

18. *Water Hammer.* The sound like someone hammering on a pipe that occurs when a valve is opened or closed very rapidly. When a valve position is changed quickly, the water pressure in a pipe will increase and decrease back and forth very quickly. This rise and fall in pressures can cause serious damage to the system.

back through the pump when the pump is shut off. This will occur if there is a faulty check valve or foot valve in the system. If backspinning is observed in a pump with a check valve or foot valve, close the pump discharge valve slowly. Be extra careful if there is a plug valve on a line with a high head because when the discharge valve is partway closed, the plug valve could slam closed and damage the pump or piping.

4. Go to power control panel containing the pump motor starters just shut down and open motor breaker switch, lock out, and tag.
5. Return to pump and close:
   a. Discharge valve
   b. Suction valve
   c. Seal water supply valve
   d. Pump volute bleed line (if so equipped)
6. If required, close and open appropriate valves along piping system through which pump was discharging.

STARTING PROCEDURES

This section contains a typical sequence of procedures to follow to start a centrifugal pump.

1. Check motor control panel for lock and tags. Examine tags to be sure that no item is preventing startup of equipment.
2. Inspect equipment.
   a. Be sure stop switch is locked out at equipment location.
   b. Guards over moving parts must be in place.
   c. Cleanout on pump volute and drain plugs should be installed and secure.
   d. Valves should be in closed position.
   e. Pump shaft must rotate freely.
   f. Pump motor should be clean and air vents clear.
   g. Pump, motor, and auxiliary equipment lubricant levels must be at proper elevations.
   h. Determine if any special considerations or precautions are to be taken during startup.
3. Follow pump discharge piping route. Be sure all valves are in the proper position and that the pump flow will discharge where intended.
4. Return to motor control panel.
   a. Remove tag.
   b. Remove padlock.
   c. Close motor main breaker.
   d. Place selector switch to manual (if you have automatic equipment).
5. Return to pump equipment.
   a. Open seal water supply line to packing gland. Be sure seal water supply pressure is adequate.
   b. Open pump suction valve slowly.
   c. Bleed air out of top of pump volute in order to prime pump. Some pumps are equipped with air release valves or bleed lines back to the wet well for this purpose.
   d. When pump is primed, slowly open pump discharge valve and recheck prime of pump. Be sure no air is escaping from volute.
   e. Unlock stop switch and actuate start switch. Pump should start.
6. Inspect equipment.
   a. Motor should come up to speed promptly. If ammeter is available, test for excessive draw of power (amps) during startup and normal operation. Most three-phase induction motors used in water treatment plants will draw 5 to 7 times their normal running current during the brief period when they are coming up to speed.
   b. Unusual noise or vibrations should not be observed during startup.
   c. Check valve should be open and no chatter or pulsation should be observed.
   d. Pump suction and discharge pressure readings should be within normal operating range for this pump.
   e. Packing gland leakage should be normal.
   f. If a flowmeter is on the pump discharge, record pump output.
7. If the unit is operating properly, return to the motor control panel and place the motor mode of operation selector in the proper operating position (MANUAL/AUTO/OFF).
8. The pump and auxiliary equipment should be inspected routinely after the pump has been placed back into service.

QUESTIONS

Please write your answers to the following questions and compare them with those on page 371.

18.24G What should be done before stopping an operating pump?

18.24H What could cause a pump shaft or motor to spin backward?

18.24I Why should the position (open or closed) of all valves be checked before starting a pump?

18.2451 POSITIVE-DISPLACEMENT PUMPS

Steps for starting and stopping positive-displacement pumps are outlined in this section. There are two basic differences in the operation of positive-displacement pumps as compared with centrifugal pumps. Centrifugal pumps (due to their design) will permit an operator error, but a positive-displacement pump will not and someone will have to pay for correcting the damages.

Important rules for operating positive-displacement pumps include:

1. **Never operate a positive-displacement pump against a closed valve, especially a discharge valve.** The pipe, valve, or pump could rupture from excessive pressure. The rupture will damage equipment and possibly seriously injure or kill someone standing nearby.
2. Positive-displacement pumps are used to pump solids (sludge). Certain precautions must be taken to prevent injury or damage. If the valves on both ends of a sludge line are closed tightly, the line becomes a closed vessel. Gas from decomposition of the sludge can build up and rupture pipes or valves.
3. Positive-displacement pumps also are used to meter and pump chemicals. Care must be exercised to avoid venting chemicals to the atmosphere.
4. Never operate a positive-displacement pump when it is dry or empty, especially the progressive-cavity types that use rubber stators. A small amount of liquid is needed for lubrication in the pump cavity between the rotor and the stator.

In addition to never closing a discharge valve on an operating positive-displacement pump, the only other difference (when compared with a centrifugal pump) may be that the positive-displacement pump system may or may not have a check valve in the discharge piping after the pump. Installation of a check valve depends upon the designer and the material being pumped.

Other than the specific differences mentioned in this section, the starting and stopping procedures for positive-displacement pumps are similar to the procedures for centrifugal pumps.

QUESTIONS

Please write your answers to the following questions and compare them with those on page 371.

18.24J What is the most important rule regarding the operation of positive-displacement pumps?

18.24K What could happen if a positive-displacement pump is operated against a closed discharge valve?

18.24L Why should both ends of a sludge line never be closed tight at the same time?

END OF LESSON 3 OF 5 LESSONS

on

MAINTENANCE

Please answer the discussion and review questions next.

DISCUSSION AND REVIEW QUESTIONS

Chapter 18. MAINTENANCE

(Lesson 3 of 5 Lessons)

Please write your answers to the following questions to determine how well you understand the material in the lesson. The question numbering continues from Lesson 2.

18. When two or more pumps of the same size are installed, why should they be operated alternately?
19. What should be checked if pump bearings are running hot?
20. What happens when the packing is too tight on a reciprocating pump?
21. Why should adjustments in control valves for reciprocating pumps be made when the pump is shut down?
22. Why would you use a stethoscope to check an electric motor?
23. How would you determine if a motor is running unusually hot?
24. How would you clean belts on a belt drive?
25. Why should you never replace only one belt on a multiple-drive unit?
26. What do rubber wearings near a belt drive indicate?
27. How can you determine if a chain in a chain-drive unit has the proper slack?
28. What happens when couplings are not in proper alignment?
29. How can you determine if a new pump will turn in the direction intended?
30. How can you determine if a new pump is delivering design flows and pressures?
31. When shutting down a pump for a long period, what precautions should be taken with the motor disconnect switch?

CHAPTER 18. MAINTENANCE

(Lesson 4 of 5 Lessons)

18.25 Compressors

Compressors (Figure 18.35) are commonly used in the operation and maintenance of water treatment plants. They are used to activate and control pump control systems (bubblers), valve operators, and water pressure systems. They are also used to operate portable pneumatic tools, such as jackhammers, compactors, air drills, sandblasters, tapping machines, and air pumps.

A compressor is a device used to increase the pressure of air or gas. It can be of a very simple diaphragm or bellows type such as are found in aquarium pumps, or an extremely complex rotary, piston, or sliding vane type compressor. A compressor usually has a suction pipe with a filter and a discharge pipe that connects to an air receiver or storage tank. The compressed air or gas is then used from the air receiver.

Due to the complexity of compressors, the water treatment plant operator usually will not be repairing them. You will, however, be required to maintain these compressors. With proper maintenance, a compressor should give years of trouble-free service.

The first step for compressor maintenance, and this pertains to any mechanical equipment, is to get the manufacturer's instruction book and read it completely. Each compressor is different and the particular manufacturer will provide its recommended maintenance schedules and procedures. Some of the maintenance procedures are discussed in the following paragraphs.

1. Inspect the suction filter of the compressor regularly. The frequency of cleaning depends upon the use of the compressor and the atmosphere around it. Under normal operations, the filter should be inspected at least monthly and cleaned or replaced every 3–6 months. Inspect and replace the filter more frequently in areas with excavation and dust. When breaking up concrete, inspect the filters daily.

   There are several types of filters such as paper, cloth, wire screen, and oil bath. The impregnated paper filters must be replaced when dirty. The cloth filters can be washed with soap and water, dried, and reinstalled. If a cloth filter is used, it is recommended that a spare be kept so one can be cleaned while the other is being used. Wire mesh and oil-bath filters can be cleaned with a standard solvent, reoiled or oil-bath filled, and used again. Never operate a compressor without the suction filter because dirt and foreign materials will collect on the rotors, pistons, or blades and cause excessive wear.

2. Lubrication. Improper or lack of lubrication is probably the biggest cause of compressor failures. Most compressors require oiling of the bearings. They can have crankcase reservoirs, oil cups, grease fittings, a pressure system, or separate pump. Whatever type, it must be inspected daily. Examine the reservoir dipstick or sight glass. Make sure that drip-feed oilers are dripping at the proper rate, force-feed oilers have the proper pressure, and grease fittings are greased at the proper interval. Compressors use a certain amount of oil in their operation and special attention is needed to keep the reservoirs full. Care also must be used to not overfill the crankcase. On some compressors, it is possible for the oil to get into the compression side and lock up the compressor or damage it. Remember, a liquid cannot be compressed.

   When air or gas is compressed, it gives off heat and the compressor becomes very hot. This tends to break down oil faster, so most compressor manufacturers have special oils recommended for their particular compressor. Also, due to the heat and contamination, it is necessary to change oil quite frequently. Compressor oil should be changed at least every 3 months, unless the manufacturer states differently. If there are filters in the oil system, these also should be changed.

3. Cylinder or casing fins should be cleaned weekly with compressed air or vacuumed off. The fins must be clean to ensure proper cooling of the compressor.

4. Unloader. Many compressors have unloaders that allow the compressor to start under a no-load condition. These can be inspected by observing the compressor. When the compressor starts, it should come up to speed and the unloader will change, starting the compression cycle. This can usually be heard by a change in sound. When it stops, you can hear a small pop and hear the air bleed off the cylinders. If the unloader is not working properly, the compressor will stall when starting, not start, or, if belt-driven, burn off the belts.

5. Test the safety valves weekly. The pop-off or safety valves are located on the air receiver or storage tank. They prevent the pressure from building up above a specified pressure by

Fig. 18.35 Two-stage piston compressor
(Courtesy Worthington Corporation)

opening and venting to the atmosphere. In gas compressors, they vent to the suction side of the compressor. Some compressors have high-pressure cutoff switches, low oil pressure switches, and high-temperature cutoff switches. These switches have preset cutoff settings and must not be changed without proper authorization. If for any reason any of the safety switches are not functioning properly, the problem must be corrected before starting the compressor again. The safety switch settings should be recorded and kept in the equipment file.

6. Drain the condensate (condensed water) from the air receiver daily. Due to temperature changes, the air receiver will fill with condensate. Each day the condensate should be drained from the bottom of the tank. There is usually a small valve at the bottom of the air receiver for this purpose. Some air receivers are equipped with automatic drain valves. These must be inspected periodically to ensure they are operating satisfactorily.

7. Inspect belt tension on compressors. Usually, you should be able to press the belt down approximately ¾ of an inch using hand pressure. This is done at the center between the two pulleys. Make sure the compressor is locked off before making this test. Do not overtighten belts because it will cause overheating and excessive wear on bearings and motor overloading.

8. Examine operating controls. Make sure the compressor is starting and stopping at the proper settings. If it is a dual installation, make sure they are alternating if so designed; inspect gauge for accuracy. Compare readings with recorded startup values or other known, accurate readings.

9. Many portable compressors are equipped with tool oilers on the receivers. These are used for mixing a small quantity of oil with the compressed air for lubrication of the tools being used. Tool oilers are located on the discharge side of the air receiver. They have a reservoir that must be filled with rock drill oil.

10. All compressors should be thoroughly cleaned at least monthly. Dirt, oil, grease, and other material must be thoroughly cleaned off the compressor and surrounding area. Compressors have a tendency to lose oil around piping, fittings, and shafts; therefore, constant cleaning is required by the maintenance operator to ensure proper and safe operation.

QUESTIONS

Please write your answers to the following questions and compare them with those on page 371.

18.25A List some of the uses of a compressor in connection with operation and maintenance of a water treatment plant.

18.25B How often should the suction filter of a compressor be cleaned?

18.25C How often should compressor oil be changed?

18.25D How often should the condensate from the air receiver be drained?

18.25E What must be done before testing belt tension on compressors with your hands?

18.26 Valves[19]

18.260 Uses of Valves

Valves are the controlling devices placed in piping systems to stop, regulate, check, divert, or otherwise modify the flow of liquids or gases. There are specific valves that are more suitable for certain jobs than others. The six most common valves that you will find in a water treatment facility are discussed in this section.

18.261 Gate Valves

The basic parts of a gate valve are: the operator (handle), the shaft packing assembly, the bonnet, the valve body with seats, the stem, and the disc (Figures 18.36 and 18.37). Gate valves come in a large number of sizes, but the principle of operation is quite similar for all sizes. One could associate the action of a gate valve to that of a guillotine having a screw shaft instead of the rope. The valve disc is raised or lowered by a threaded shaft and is guided on each side to ensure that it will not hang up in the operation. The disc is screwed down until it wedges itself between two machined valve seats. This makes a leakproof seat on both sides of the disc. The discs are replaceable. Some gate valves have discs with wedges inside. As more force is applied to the screwed stem, the wedges force the discs into tighter contact with the valve seats.

Gate valves are either of the rising (Figure 18.36) or nonrising stem (Figure 18.37) type. The rising stem has companion threads in the valve bonnet. As the valve is opened, the stem is threaded out, lifting the wedged disc. In the nonrising type, the stem is held in place in the bonnet by a collar. The stem is threaded with companion threads in the wedged disc. As the valve opens, the disc rises on the stem. Consequently, the handwheel stays on the same plane.

Gate valves are not commonly used to control flows. With the valve partially open, the water velocity is increased through the valve and minute particles transported in the water can cause undue seat wear. However, the V-ported gate valve can be used in controlling flows. As the valve is opened, the V is widened to allow more flow. Because of the valve design, little damage is done to the valve seats in the V-ported type of gate valve.

Suggested gate valve operation and maintenance procedures are listed below:

1. Open valve fully. When at stop, reverse and close valve one-half turn.
2. Operate all large valves at least yearly to ensure proper operation.
3. Inspect valve stem packing for leaks. Tighten, as needed.
4. If the valve has a rising stem, keep stem threads clean and lubricated.
5. Close valves slowly in pressure lines to prevent water hammer.
6. If a valve will not close by using the handwheel, check for the cause. Using a "cheater" (bar-pipe wrench) will only aggravate the problem.

18.262 Maintenance of Gate Valves

Paragraph 12: Gate Valves

The most common maintenance required by gate valves is oiling, tightening, or replacing the stem stuffing box packing.

Frequency of Service	
A	1. REPLACE PACKING. Modern gate valves can be repacked without removing them from service. Before repacking, open valve wide. This prevents excessive leakage when the packing or the entire stuffing box is removed. It draws the stem collar tightly against the bonnet on a nonrising stem valve, and tightly against the bonnet bushing on a rising stem valve. a. Stuffing box. Remove all old packing from stuffing box with a packing hook or a rattail file with bent end. Clean valve stem of all adhering particles and polish it with fine emery cloth. After polishing, remove the fine grit with a clean cloth to which a few drops of oil have been added. b. Insert packing. Insert new split-ring packing in stuffing box and tamp it into place with packing hook. Stagger ring splits. After stuffing box is filled, place a few drops of oil on stem, assemble gland, and tighten it down on packing.

19. For additional information on valves, see *Water Distribution System Operation and Maintenance*, Chapter 3, "Distribution System Facilities," Section 3.670, "Valves," in this series of operator training manuals.

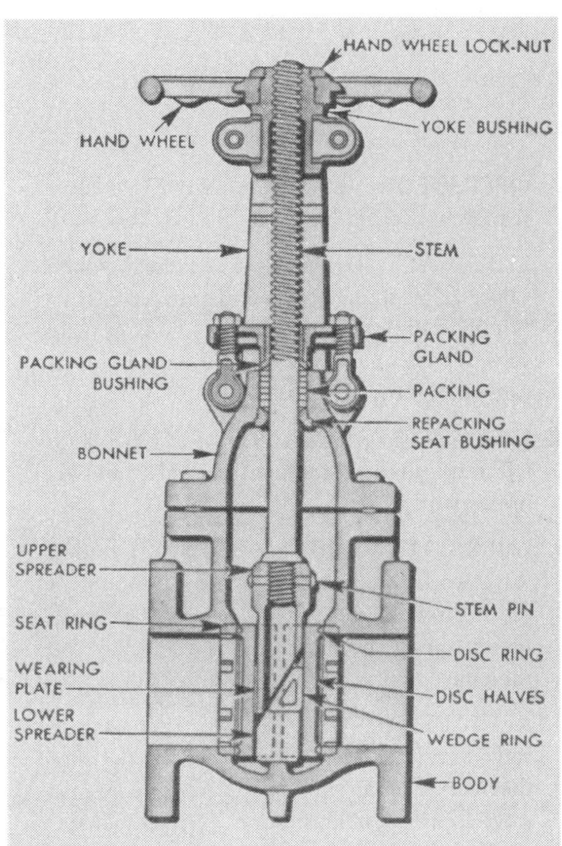

Fig. 18.36 Rising stem gate valve
(Permission of Stockham Valves & Fittings, Copyright, 1976)

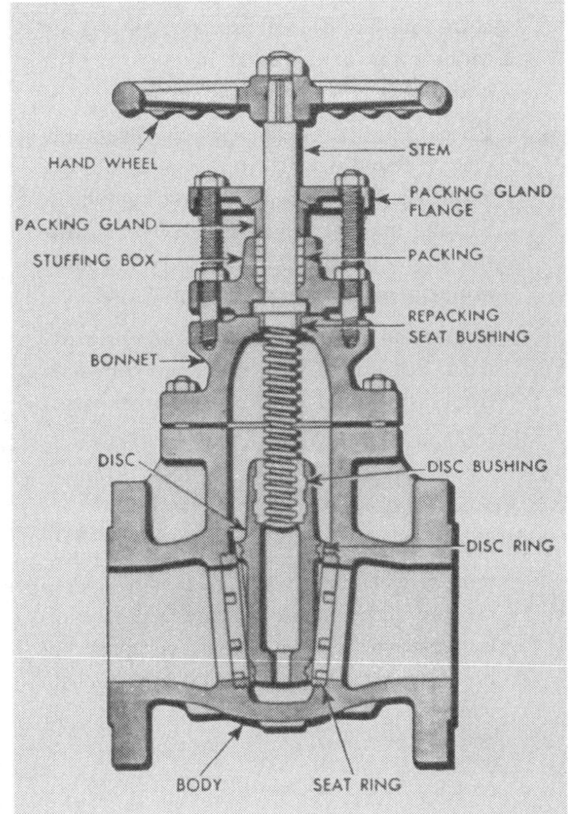

Fig. 18.37 Nonrising stem gate valve
(Permission of Stockham Valves & Fittings, Copyright, 1976)

Frequency of Service	
S	2. OPERATE VALVE. Operate inactive gate valves to prevent sticking.
A	3. LUBRICATE GEARING. Lubricate gate valves as recommended by manufacturer. Lubricate thoroughly any gearing in large gate valves. Wash open gears with solvent and lubricate with grease.
S	4. LUBRICATE RISING STEM THREADS. Clean threads on rising stem gate valves and lubricate with grease.
	5. LUBRICATE BURIED VALVES. If a buried valve works hard, lubricate it by pouring oil down through a pipe that is bent at the end to permit oiling the packing follower below the valve nut.
A	6. REFACE LEAKY GATE VALVE SEATS. If gate valve seats leak, reface them immediately, using the method discussed below. A solid wedge disc valve is used for illustration, but the general method also applies to other types of repairable gate valves. Proceed as follows:

a. Remove bonnet and clean and examine disc and body thoroughly. Carefully determine extent of damage to body rings and disc. If corrosion has caused excessive pitting or eating away of metal, as in guide ribs in body, repairs may be impractical.

b. Check and service all parts of valve completely. Remove stem from bonnet and examine it for scoring and pitting where packing makes contact. Polish lightly with fine emery cloth to put stem in good condition. Use soft jaws if stem is put in vise.

c. Remove all old packing and clean out stuffing box. Clean all dirt, scale, and corrosion from inside of valve bonnet and other parts.

d. Do not salvage an old gasket. Remove it completely and replace with one of proper quality and size.

e. After cleaning and examining all parts, determine whether valve can be repaired by removing cuts from disc and body seat faces or by replacement of body seats. If repair can be made, set disc in vise with face leveled, wrap fine emery cloth around a flat tool, and rub or lap off entire bearing surface on both sides to a smooth, even finish. Remove as little metal as possible.

f. Repair cuts and scratches on body rings, lapping with an emery block small enough to permit convenient rubbing all around rings. Work carefully to avoid removing so much metal that disc will seat too low. When seating surfaces of disc and seat rings are properly lapped in, coat faces of disc with *PRUSSIAN BLUE* [20] and drop disc in body to check contact. When good, continuous contact is obtained, the valve is tight and ready for assembly. Insert stem in bonnet, install new packing, assemble other parts, attach disc to stem, and place assembly in body. Raise stem to prevent contact with seats so bonnet can be properly seated on body before tightening the joint.

g. Test repaired valve before putting it back in line to ensure that repairs have been properly made.

h. If leaky gate valve seats cannot be refaced, remove and replace seat rings with a power lathe. Chuck up body with rings vertical to lathe and use a strong steel bar across ring lugs to unscrew them. They can be removed by hand with a diamond point chisel if care is taken to avoid damaging threads. Drive new rings home tightly. Use a wrench on a steel bar across lugs when putting in rings by hand. Always coat threads with a good lubricant before putting threads into the valve body. This helps to make the threads easier to remove the next time the seats have to be replaced. Lap in rings to fit disc perfectly.

18.263 Globe Valves (Figure 18.38)

The globe valve seating configuration is quite different from the gate valve. Globe valves use a circular disc to make a flat surface contact with a ground-fitted valve seat. This is similar to placing your thumb over the end of a tube. The parts of the valve are similar in name and function to the gate valve. They can be of the rising or nonrising stem type.

What is unique about the globe valve is its internal design (Figure 18.38). This design enables the valve to be used in a controlling mode. The valve seats are not subject to excessive wear when partially opened like the gate valve. After extended use, the valve may not have a positive shutoff but it will still be effective in throttling flows. Procedures for operating and maintaining globe valves are similar to the procedures outlined for gate valves in Section 18.262.

20. *Prussian Blue.* A blue paste or liquid (often on a paper like carbon paper) used to show a contact area. Used to determine if gate valve seats fit properly.

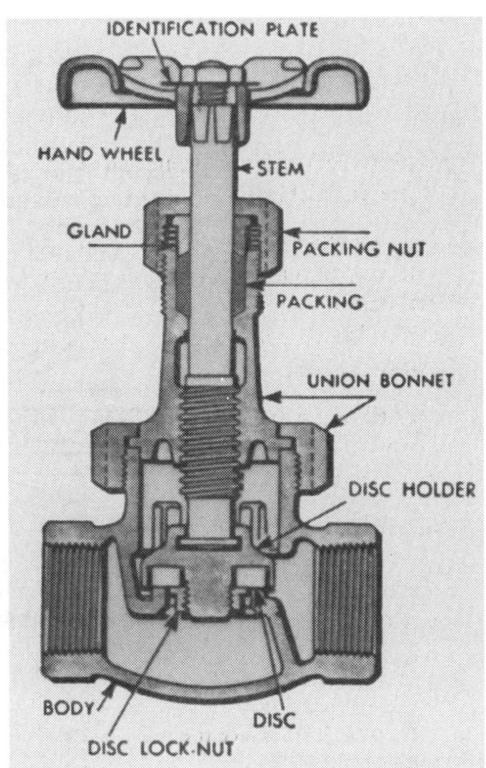

Fig. 18.38 Globe valve
(Permission of Stockham Valves & Fittings)

18.264 *Eccentric Valves (Figures 18.39 and 18.40)*

The eccentric valve has many desirable features. These features include allowance for high flow capacity, quarter-turn operation, no lubrication, excellent resistance to wear, and good throttling characteristics. The eccentric valve uses a cam-shaped plug to match an eccentric valve seat. As the valve is closed, the plug throttles the flow yet maintains a smooth flow rate. The plug does not come into contact with the valve seat until it is in the closed position.

Because the plug has a resilient coating, it ensures a leak-tight seal at the valve seats. The Buna-N, neoprene, or viton plug coating is a very wear-resistant compound and can function well under a wide temperature range. This valve is excellent for controlling the flows of slurries and sludges found in water treatment facilities.

18.265 *Butterfly Valves (Figure 18.41)*

The butterfly valve is used primarily as a control valve. The flow characteristics allow the water to move in straight lines with little turbulence in the area of the valve disc (butterfly). Complete flow shutoff can be accomplished but the psi rating is relatively low in comparison to eccentric or gate valves.

The butterfly valve uses a machined disc that can be opened to 90 degrees to allow full flow through the valve. Quarter-turn operation moves the valve from the closed to open position. The disc is mounted on a shaft eccentric that allows the disc to come into its seat with minimum seating torque and scuffing of the rubber seat. There is no contact between the disc and the seat until the last few degrees of valve closure.

A resilient rubber is used as the seat and is of a continuous form that is not interrupted by a shaft connection. Wear resistance characteristics are good when used in slurry and sludge applications.

When the valve is closed, the disc is forced against the rubber seat. Wedges with jacking screws compress the rubber seat via a jack ring. The rubber seat then conforms to the entire disc circumference. The rubber can be readily replaced, when necessary, without complete valve dismantling. Large valves do not need to be removed from the line for seat replacement.

18.266 *Check Valves (Figure 18.42)*

The term "check valve" describes its function. A check valve allows water to flow in one direction only. If the water attempts to flow in the opposite direction, an internal mechanism closes the valve and "checks" (prevents) the flow. Three types of check mechanisms may be used—the swing check, the wafer check, or the lift check. In the swing check, a movable disc rests at a right angle to the flow and seats against a

ECCENTRIC ACTION

The DeZurik design matches a single-faced eccentric or cam-shaped plug with an eccentric raised body seat. With rotary motion only, the plug advances against the seat as it closes. Here's how it works:

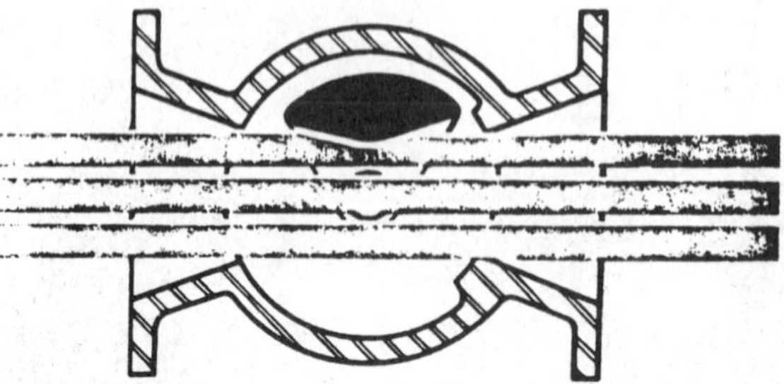

OPEN—The plug is out of the flow path. There is no bonnet or other cavity to fill with slurry material. Flow is straight-through with minimum pressure drop.

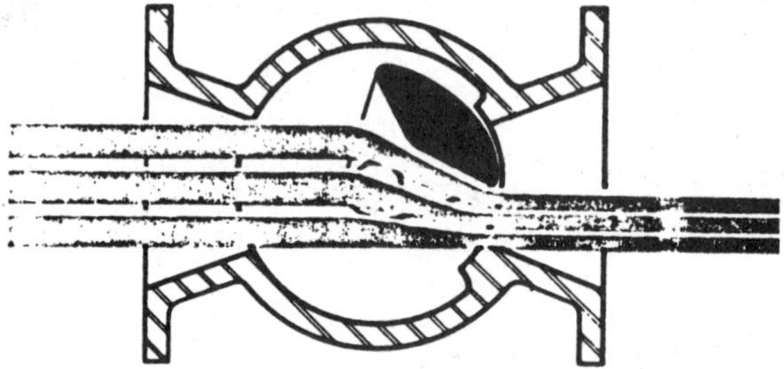

CLOSING—At any position between open and closed, the eccentric plug still has not touched the seat. There is no friction to cause wear or binding. Flow is still smooth and straight. Throttling action is excellent on all types of services from slurries to gas.

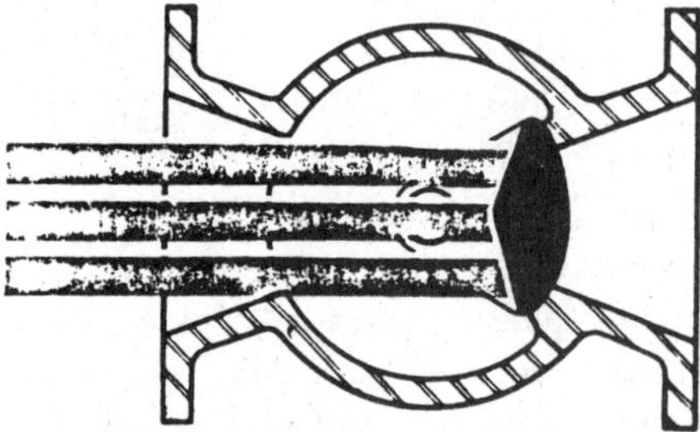

CLOSED—The eccentric plug makes contact with the eccentric seat only in the fully closed position. Action is easy, without binding or scraping. There is no continual seat wear. The plug is moved firmly into the seat to provide a positive, drip-tight, long-lasting seal.

Fig. 18.39 How eccentric valves work

(Permission of DeZurik Corporation, Sartell, MN)

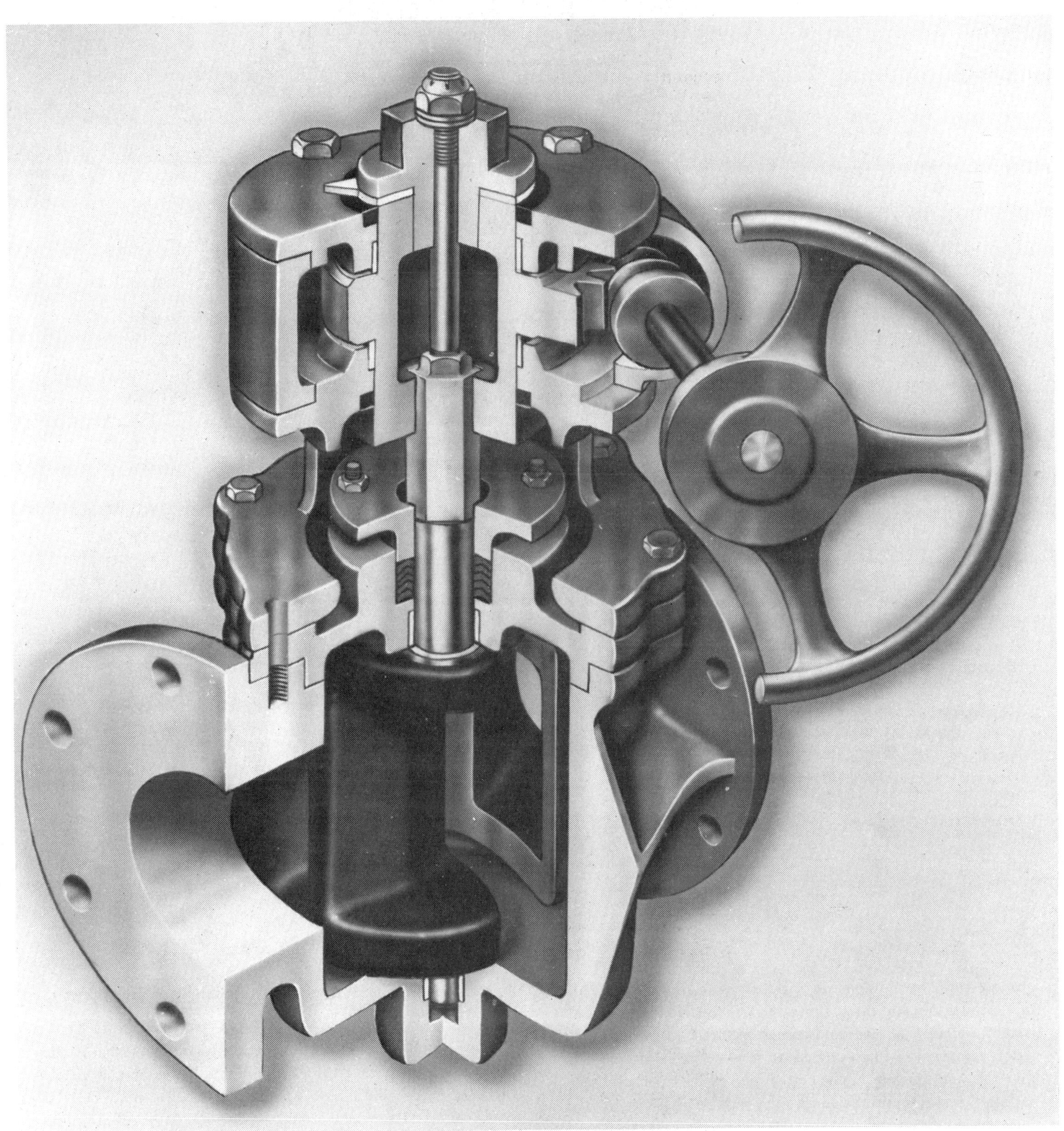

Fig. 18.40 Eccentric valve
(Permission of DeZurik Corporation, Sartell, MN)

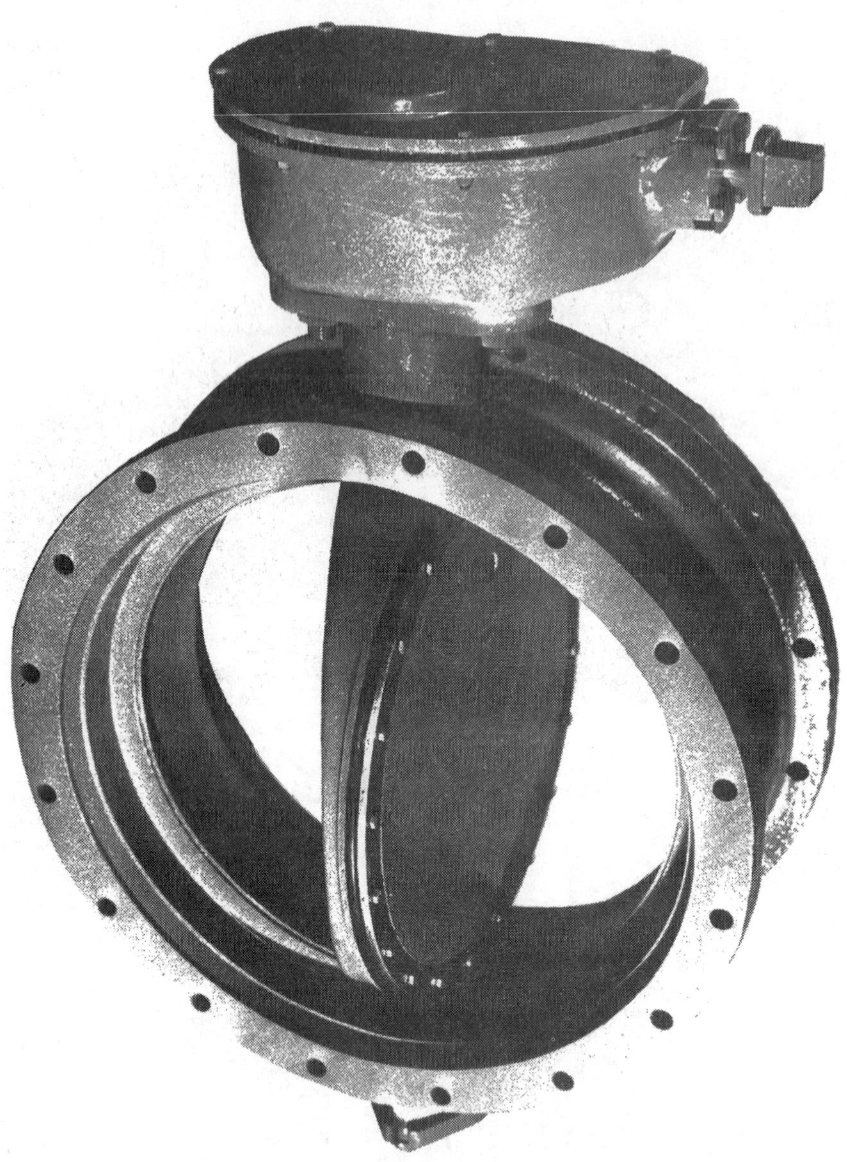

Fig. 18.41 Butterfly valve
(Permission of American-Darling Valve, Birmingham, AL)

ground seat. The movable disc is called the "clapper." The clapper can be one of three types: gravity operated, lever and weight operated, or lever and spring operated. In many installations, the water being pumped must be delivered at a desired flow rate and pressure. A clapper with an external means of adjusting the opening in the check valve may be necessary to produce desired flows and pressures. By positioning the weight on the lever or adjusting the spring tension, a check valve can be made to operate either partially or fully open at various pressures and flows. The spring or counterweight also ensures that the check valve closes at no flow. This is very helpful if the valve is not in a position that will enable gravity alone to operate the clapper. The gravity-operated clapper does not have an external adjustment and relies on the weight of the clapper to close the valve at no-flow conditions.

Most swing check valves provide for full opening, that is, the clapper can move up into the bonnet and thus be completely out of the flow. Head loss in swing check valves may be relatively high and this factor must be considered in selecting the device for a particular application. This type of check valve is quite common in pump installations and often has a dampening feature to cushion the closing of the clapper.

The wafer check has a circular disc that hinges in the center (diameter) of the disc. Water passing through collapses the disc and the stoppage of flow allows the disc to return to its circular

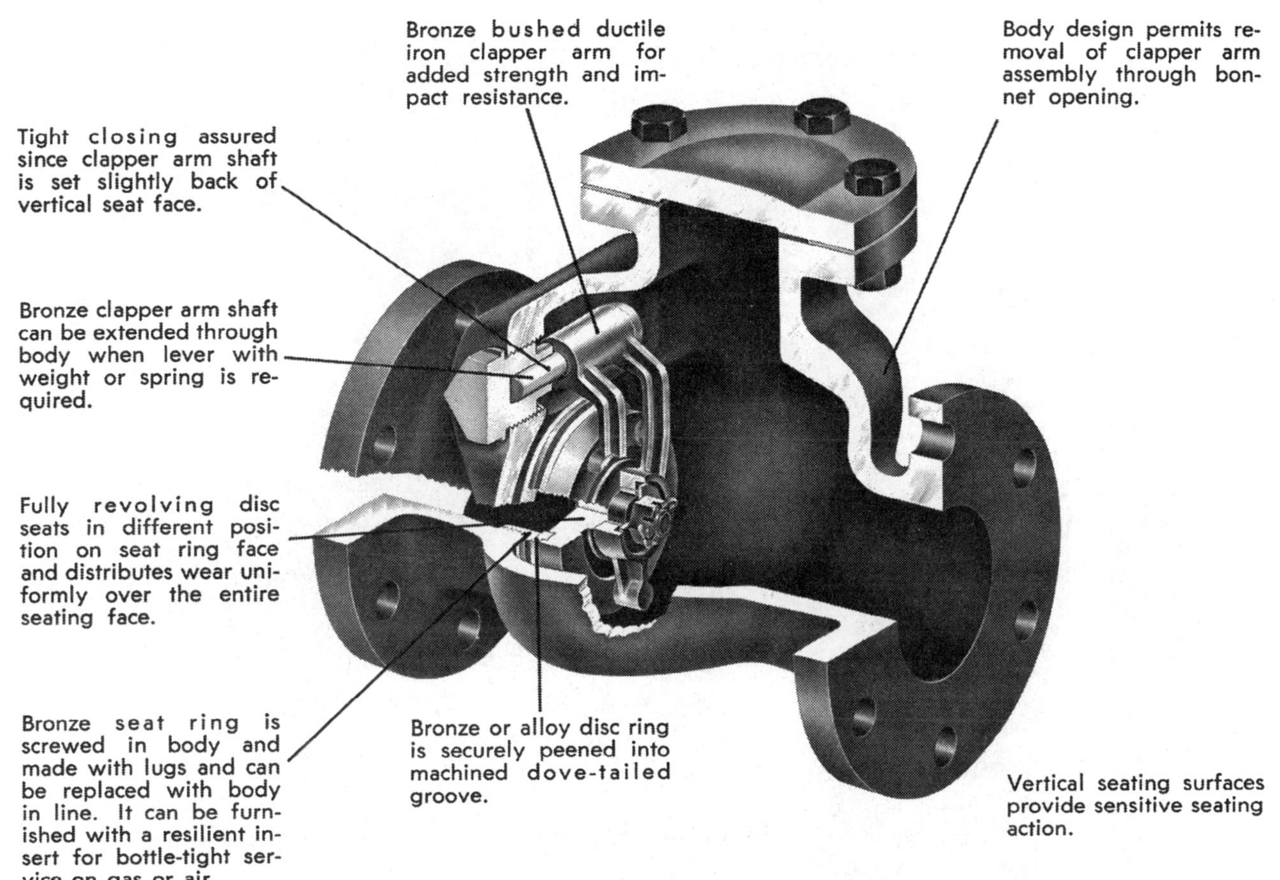

Fig. 18.42 Check valve

(Permission of American-Darling Valve, Birmingham, AL)

form. Because the valve has a tendency to be fouled up by stringy material, it is not commonly used in handling raw water. Wafer check valves are very effective when used with clean water.

The lift check uses a vertical lift disc or ball. When there is flow, the disc or ball is lifted from its ground seat and fluid passes through the valve. As flow stops, the check realigns itself with its seat and checks or prevents water backflow. The movable portion can be a spring or gravity return.

The foot valves used in pump suctions are nearly always of the vertical lift disc design. A check valve of this type is usually applied to handle clean water.

Backflow prevention by check valves is essential in many applications:

1. To prevent pumps from reversing when power is removed
2. To protect water systems from being cross-connected
3. To aid in pump operation as a dampener
4. To ensure full-pipe operation (pipe is full of water)

Table 18.2 provides a comparison of various types of check valves with features of these valves. Figures 18.43 through 18.49 provide drawings and photographs of the different types of check valves listed in Table 18.2.

18.267 Maintenance of Check Valves

Paragraph 13: Check Valves

Frequency of Service	
A	1. INSPECT DISC FACING. Open valves to observe condition of facing on swing check valves equipped with leather or rubber seats on disc. If metal seat ring is scarred, dress it with a fine file and lap with fine emery paper wrapped around a flat tool.
A	2. CHECK PIN WEAR. Check pin wear on balanced disc check valve since disc must be accurately positioned in seat to prevent leakage.

TABLE 18.2 COMPARISON OF FEATURES OF DIFFERENT TYPES OF CHECK VALVES

(Permission of APCO/Valve and Primer Corporation)

Feature	Globe Silent Check Valve	Wafer Silent Check Valve	Double Door Check Valve	Rubber Flapper Check Valve	Cushion Swing Check Valve	Slanting Disc Check Valve	Automatic Control Check Valve
1. Lowest initial cost			X				
2. Shortest laying length			X				
3. Highest head loss (see head loss curves)		X					
4. Resilient seat (optional)	X	X	X		X		
5. For waste and raw sewage				X	X		
6. Clean water only	X	X	X			X	X
7. Cushion closing			X	X	X	X	
8. Silent closing (positively silent)	X	X					X
9. Free open – Free close				X		X	
10. Control open or close or both (optional)					X	X	X
11. Vertical installation flow up or down	X	X					X
12. Can be rubber lined				X	X		X
13. Disc position indicator					X	X	X
14. Buried service				X			
15. Outside lever					X		
16. Surge pressure control						X	X
17. Reverse flow				X	X		X
18. Up to 600# class	X	X				X	X
19. Up to 1500# class		X					
20. Lowest head loss (see head loss curves)						X	
21. Up to 2500# class			X				
22. Control open and close standard							X
23. Shutoff valve							X
24. Throttling valve							X
25. Vertical installation flow up only			X	X	X	X	
26. Electric motor operated							X
27. Remote control							X
28. Control closure upon power failure							X
29. Resilient seat standard			X	X	X		X
30. Velocities in excess of 15 FPS			X			X	X
31. Velocities up to 5 FPS					X		
32. Velocities up to 10 FPS	X	X		X			

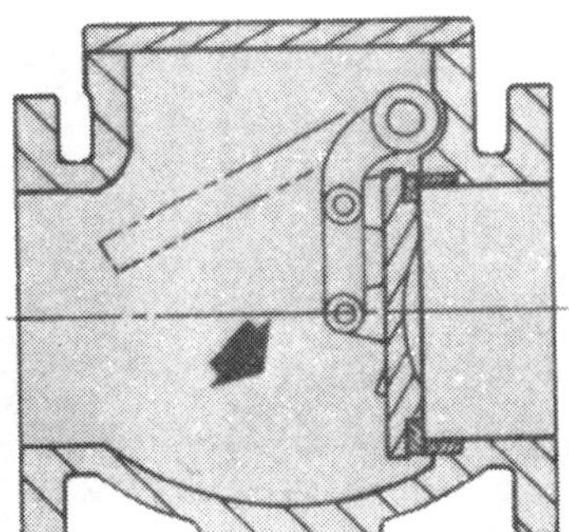

Fig. 18.43 Swing check valves (single disc)
(Permission of APCO/Valve and Primer Corporation)

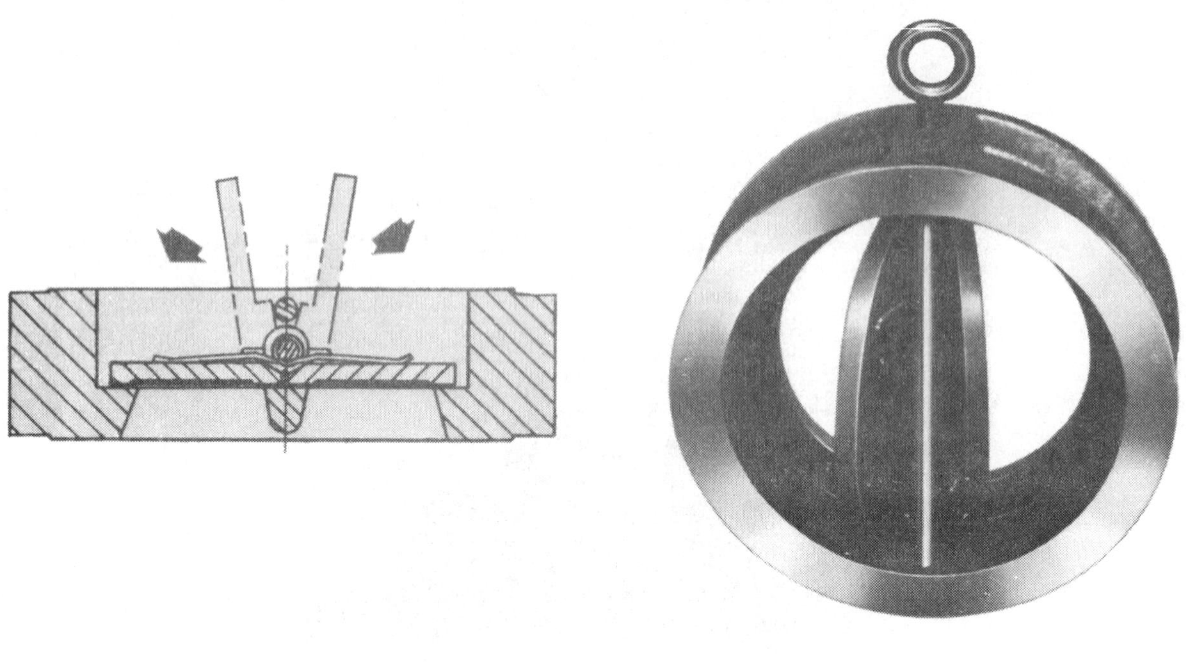

Fig. 18.44 Double disc swing check valves (split swing discs)
(Permission of APCO/Valve and Primer Corporation)

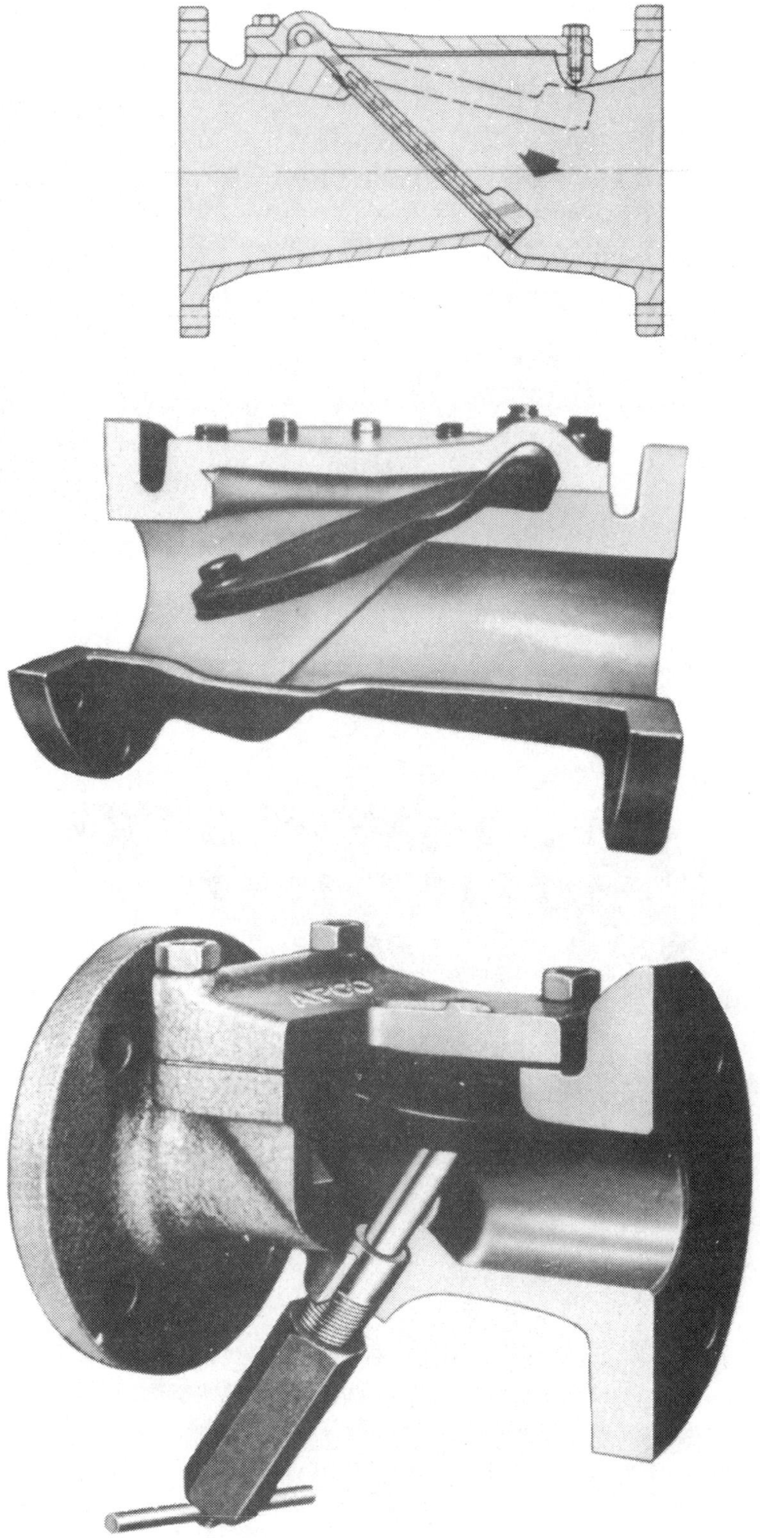

Fig. 18.45 Rubber flapper check valves (angle seating)
(Permission of APCO/Valve and Primer Corporation)

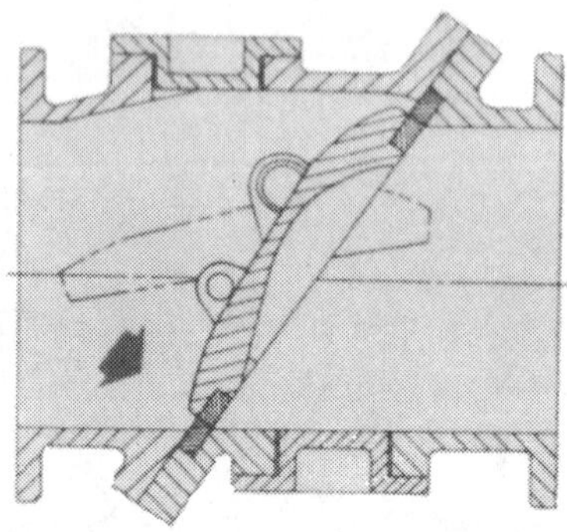

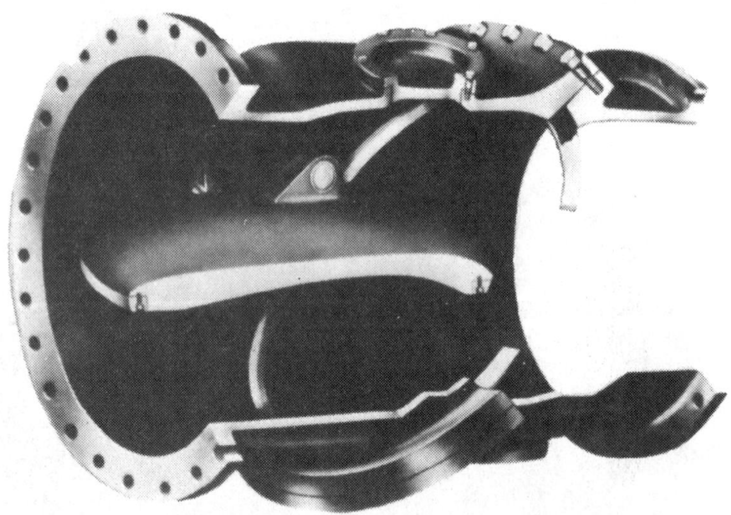

Fig. 18.46 Slanting disc check valves (pivot off center)
(Permission of APCO/Valve and Primer Corporation)

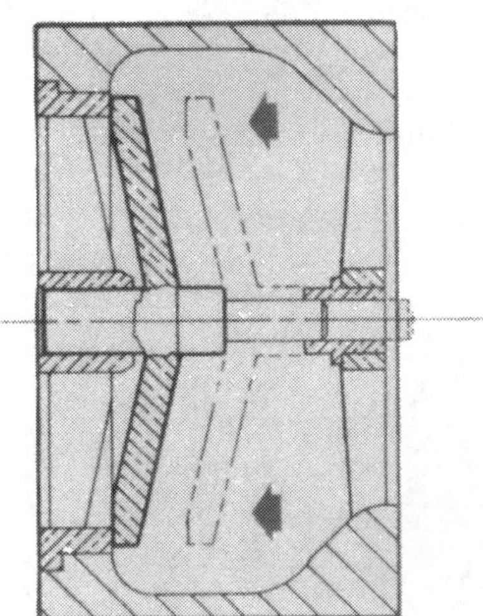

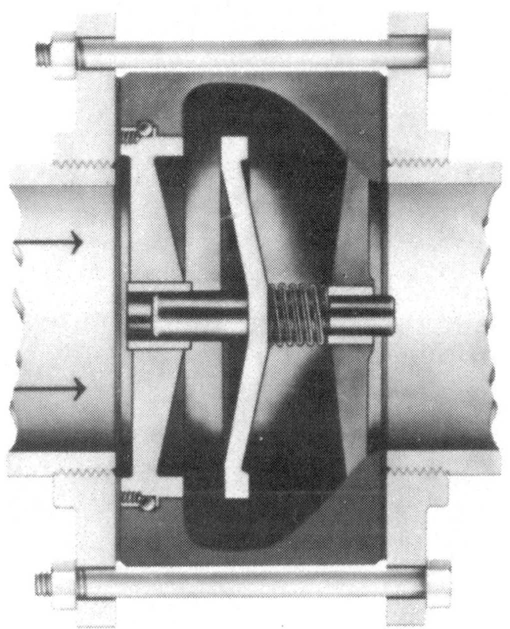

Fig. 18.47 Silent check valves (wafer)
(Permission of APCO/Valve and Primer Corporation)

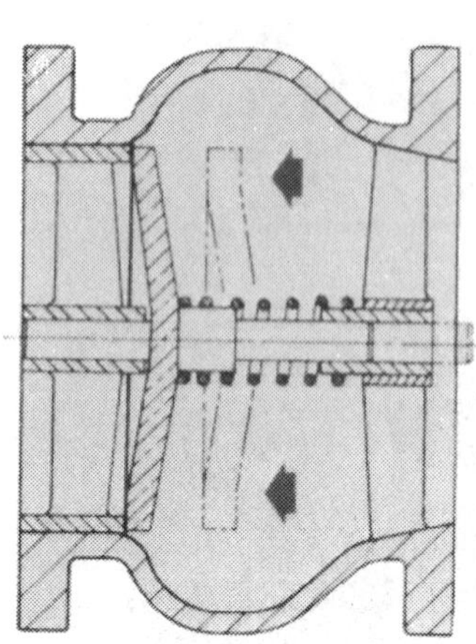

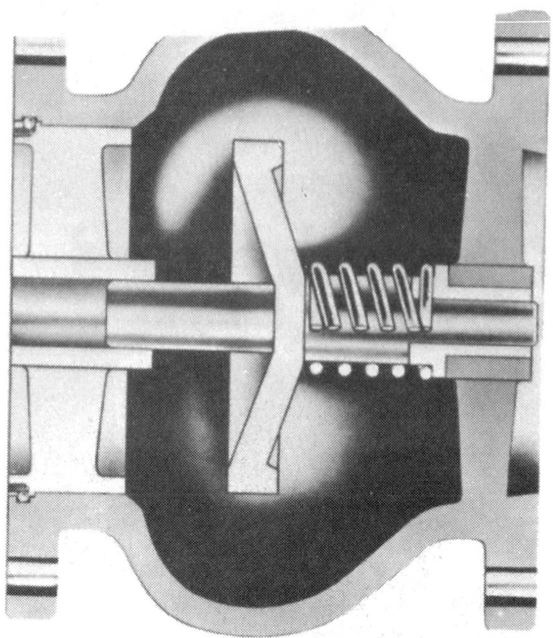

Fig. 18.48 Silent check valves (globe)
(Permission of APCO/Valve and Primer Corporation)

Fig. 18.49 Automatic control check valves
(Permission of APCO/Valve and Primer Corporation)

18.268 Automatic Valves

Water treatment plants usually have a number of automatically operated valves. The simplest type is either open or closed and is not required to operate in an intermediate position. Frequently, these valves are similar to gate valves that have had their threaded stems and handwheels replaced by a smooth shaft and hydraulic piston. Maintenance on these valves is essentially the same as for gate valves.

Other automatically operated valves are used to control flow in water treatment plants and are usually located at some point between tight shut and wide open. These are commonly called modulating valves. A butterfly valve with a hydraulic cylinder operator can be used for this type of service.

The diaphragm-operated globe valve (Figure 18.50) is also used for modulating service. These valves can be equipped with pilot control devices to control pressure, flow, or level either singly or in combination. Maintenance on these valves consists of the following:

1. Periodically clean any strainers in the pilot control system. Scheduling should be adjusted to accommodate the rate at which the strainer collects foreign material.
2. Check the operation of the valve to see that the controls are, in fact, correctly positioning the valve to accomplish the job.
3. If the valve is used in an application where it seldom or never is wide open, it should periodically be exercised manually to cycle from tight shut to wide open. This is to ensure that there is no buildup on the stem that could jam the valve. These valves can be opened wide by drawing all the pressure from the cover chamber. If water does not stop flowing out of the cover chamber when the valve is wide open, this is an indication that the diaphragm is leaking and should be replaced.

QUESTIONS

Please write your answers to the following questions and compare them with those on page 371.

18.26A What is the purpose of valves?

18.26B List six common types of valves found in water treatment facilities.

18.26C What maintenance is required by gate valves?

18.26D What is the purpose of a check valve?

18.26E Why is backflow prevention by check valves essential in many applications?

END OF LESSON 4 OF 5 LESSONS

on

MAINTENANCE

Please answer the discussion and review questions next.

DISCUSSION AND REVIEW QUESTIONS

Chapter 18. MAINTENANCE

(Lesson 4 of 5 Lessons)

Please write your answers to the following questions to determine how well you understand the material in the lesson. The question numbering continues from Lesson 3.

32. What are the uses of a compressor?
33. What items should be maintained on a compressor?
34. What factors can cause wear on gate valve seats?
35. Why should inactive gate valves be operated periodically?

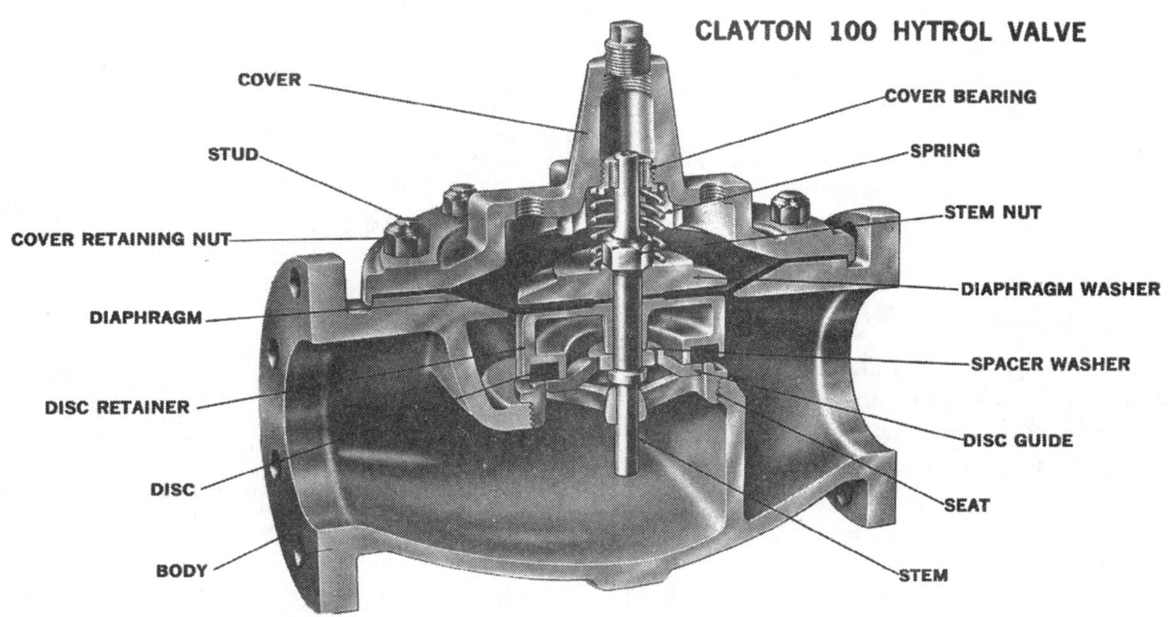

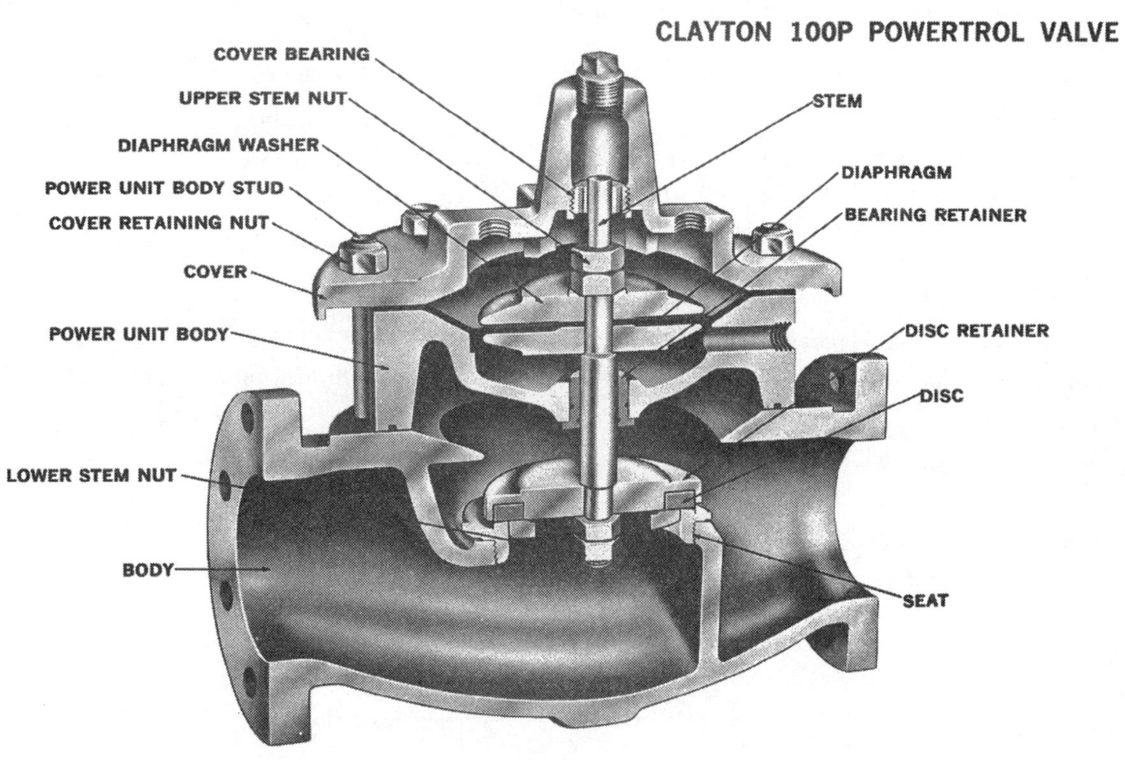

Fig. 18.50 Diaphragm-operated globe valve
(Permission of CLA-VAL Co.)

CHAPTER 18. MAINTENANCE

(Lesson 5 of 5 Lessons)

18.3 INTERNAL COMBUSTION ENGINES

18.30 Gasoline Engines

18.300 Need to Maintain Gasoline Engines

In the water treatment departments of all cities, there is occasion to use gasoline-powered engines that drive pumps, generators, tractors, and vehicles. Although we all drive automobiles that are powered by internal combustion engines, are you aware of the fundamentals?

Very few operators actually do the repair of gasoline-powered engines. Although you may not be able to perform the duties of an engine mechanic, there are a number of steps you can take to ensure that your particular engine is well maintained.

The goal of this section is to provide you with adequate knowledge of how a gasoline engine operates so that you can maintain this type of engine and help it deliver many hours of use at its optimum performance level.

18.301 Maintenance

To have an engine that will provide you with many hours of trouble-free operation, it must be well cared for. Please refer to the owner/operator manual for your particular engine. Typical maintenance procedures are as follows:

1. Change engine oil regularly every 25 hours
2. Clean carburetor air filter every 25 hours
3. Blow dust and chaff from louvered engine vanes regularly
4. Clean carburetor fuel filter/screen every 100 hours
5. Lubricate generator or starter motor, as recommended, every 100 hours
6. Lubricate throttle linkage every 100 hours
7. Clean, gap, or replace spark plug every 100 hours
8. Remove carbon deposits from top of piston and valves every 100 to 300 hours

18.302 Starting Problems

Listed below are some items to check if you have problems starting a gasoline engine.

1. No fuel in tank, valve closed
2. Carburetor not choked
3. Water or dirt in fuel lines or carburetor
4. Carburetor flooded
5. Low compression
6. Loose spark plug
7. No spark at plug
   a. Dirty or improperly gapped spark plug
   b. Broken or wet ignition cables
   c. Breaker points not opening or closing
   d. Magneto grounded

18.303 Running Problems

Check the following items if a gasoline engine does not run properly.

1. Engine misses
   a. Faulty spark plug/gapping
   b. Weak ignition spark
   c. Loose ignition cable
   d. Worn breaker points
   e. Water in fuel
   f. Poor compression
2. Engine surges
   a. Carburetor flooding
   b. Governor spring connected improperly
3. Engine stops
   a. Fuel tank empty
   b. Vapor lock
   c. Tank air vent plugged
4. Engine overheats
   a. Low crankcase oil
   b. Ignition timing wrong
   c. Engine overloaded
   d. Restricted air circulation/high ambient temperature
   e. Poor grade of gasoline
5. Engine knocks
   a. Poor grade of gasoline
   b. Engine under heavy load at low speed

c. Carbon deposits in cylinder head
d. Spark advanced too far
e. Loose connecting rod bearing
f. Worn or loose piston pin

6. Engine backfires through carburetor
   a. Water or dirt in fuel
   b. Cold engine
   c. Poor grade of gasoline
   d. Sticking inlet valves
   e. Spark plug heat range too hot

QUESTIONS

Please write your answers to the following questions and compare them with those on page 372.

18.30A List some possible uses of gasoline engines in water treatment plants.

18.30B What items would you check if you had problems starting a gasoline engine?

18.30C What items could cause a gasoline engine to not run properly?

18.304 How to Start a Gasoline Engine

Because of the wide variety of uses for gasoline-powered engines, no one starting sequence will apply to all engines. In general, gasoline engines can be divided into two groups. In the first group are small engines with magneto ignition and recoil start. Larger engines, with battery-powered ignition and electric start, are in the second group.

18.3040 SMALL ENGINES

The procedure for starting small engines is as follows:

1. Check fuel tank for adequate fuel
2. Ensure fuel shutoff valve from the tank to the carburetor is open
3. Disengage ignition ground (kill switch or mechanism that grounds the spark plug)
4. Check crankcase lubricating oil
5. Set throttle to start position or three-fourths-full throttle
6. Set choke lever or pull out choke on carburetor
7. Pull recoil starter twice
8. If engine has started, push choke to OFF
9. If engine does not start after two pulls, disengage the choke and try three or four more times

If repeated efforts at starting have been unsuccessful, remove the high-tension voltage wire from the spark. Hold the end of the wire (grasp the insulated portion, not the connector) ⅛ inch (3 mm) from the spark plug. Pull the recoil starter. You should see a small blue spark. This will indicate that the points are opening and closing and providing ignition voltage.

Next, using a 13⁄16-inch-deep (20.6 mm) socket, remove the spark plug from the cylinder head. Check for a carbon buildup on the electrode. A piece of carbon may have lodged between the center electrode and the side electrode. Also, check to see if the plug is wet with fuel or oil. This could indicate that you have flooded the cylinder with fuel by having the choke on too long. If there is oil residue, it could indicate worn piston rings.

Replace the spark plug with a new one if in doubt. If you must use the one you have, clean it by buffing with a wire brush. Check the gap between the center electrode and side electrode; it should be approximately 0.03 inch (0.76 mm).

Try starting the engine as previously described. If the engine does not sputter or pop, close the fuel shutoff valve, remove fuel sedimentation bowl, and clean it. Open fuel valve. Catch a small amount of fuel in the palm of your hand and examine the fuel for grit or water. If everything looks OK, replace sedimentation bowl and open fuel valve.

Try to start the engine. If you still cannot achieve ignition, you may have other problems that will require further checking by a small-engine mechanic. Do not feel disgruntled; you have checked for the most common problems.

18.3041 LARGE ENGINES

The procedure for starting large engines is as follows:

1. Check fuel tank for fuel
2. Check crankcase for oil
3. Check radiator for coolant (if water-cooled)
4. Set throttle to one-half-full position
5. Pull out choke
6. Turn on ignition switch and press start button
7. After four or five engine revolutions, push in the choke
8. Engine should start

After repeated tries, further investigation by a mechanic may be needed.

NOTE: Do not crank engine with the starter motor for more than one minute initially. Wait 2 minutes and try again for 45 seconds. After three tries, let starter motor cool for 5 minutes before trying again. This will avoid starter motor damage.

Preliminary checks for a large engine that will not start are similar to procedures for small engines. Remove spark plug wires. Test each one by holding it ⅛ inch (3 mm) from the spark plug or ground, and turn engine over with the starter. You should see a small blue spark. If you have no spark, the points are not opening or high-tension voltage is not present from the ignition coil. Check further, as needed. If spark is present, inspect spark plugs. Clean or replace, if needed.

After checking the ignition system, make sure fuel is present at the carburetor. Remove the fuel line at the carburetor and direct it away from you and the engine. Engage starter motor for two revolutions. Fuel should spurt from the line if the fuel pump is working satisfactorily. Replace fuel line and wipe away any fuel that may be present on the engine.

With fuel and ignition voltage present, the engine should start. Repeat start procedure. If you still cannot start the engine, call on your mechanic to look for the problem.

NOTE: Some engines have a low-oil pressure switch that must be manually held in until sufficient oil pressure is present.

Do not use a starting fluid on gasoline engines unless it is a last resort effort to get a critical piece of equipment running. Hard-starting engines should be inspected and repaired by a reliable mechanic.

After an engine has been started, give it an opportunity to warm up before applying the load. Follow manufacturer's recommendations for the starting procedure since there is some variation between different makes of engines.

QUESTIONS

Please write your answers to the following questions and compare them with those on page 372.

18.30D If a gasoline engine will not start and the spark plug is wet with fuel or oil, what has happened?

18.30E If a gasoline engine will not start and there is an oil residue on the spark plug, what has happened?

18.30F After an engine has started, what should be done before applying the load?

18.31 Diesel Engines

18.310 How Diesel Engines Work (Figure 18.51)

Diesel engines are similar to gasoline engines and are either two or four cycle. They can be air or water cooled. In general, the diesel engine is of heavier construction to withstand the higher pressures resulting from higher compression ratios.

The diesel does not use spark plugs, but instead relies on heat generated by air compressed in the cylinder (1,000°F/540°C) to ignite the fuel mixture. The fuel is a petroleum product that is heavier than gasoline and has a higher flash point. Gasoline cannot be used in a diesel because it would start to burn from the heat generated by compression before the piston reached the top of the stroke.

A diesel has no carburetor. The fuel is sprayed (injected) into the cylinder while the cylinder is compressing air. The heat of compression ignites the fuel-air mixture and burns, producing power similar to a gasoline engine. The introduction of fuel into the cylinder must be timed in the same manner as spark to the plug in a gasoline engine. Fuel is pumped by a pumping device that is geared to the crankshaft.

Diesel fuel, unlike gasoline, does not vaporize readily. The fuel must be broken up into fine particles and sprayed into the cylinder. The atomization of fuel is accomplished by forcing the fuel through a nozzle at the top of the combustion chamber. As the fuel combines with the air in the cylinder, it becomes a combustible mixture. Since the diesel engine depends upon the heat of compressed air to ignite the fuel-air mixture, compression pressure must be maintained. Leaking valves or piston rings (causing blowby) cannot be tolerated.

The fuel is also important. The automotive-type diesel is designed to run on a specific type or grade of fuel. Trouble can be expected if an attempt is made to use other than the proper type.

18.311 Operation

In the two-cycle engine, intake and exhaust take place during part of the compression and power strokes; whereas, the four-cycle engine requires four strokes to complete the operating cycle. During one-half of the cycle, the four-stroke acts as an air pump. The two-stroke must have a blower (air pump) to provide the necessary air to expel the exhaust gases and recharge the cylinder with fresh air.

In the two-cycle engine, a series of ports surround the cylinder at a point higher than the lowest position of the piston. These are the intake ports that allow air into the cylinder. The four-cycle engine uses intake valves. The incoming air forces the expended gases out the exhaust valve, leaving the cylinder full of clean air.

As the piston starts its upward stroke, the exhaust valve closes, the intake ports are sealed off by the piston, and the air in the cylinder is compressed. Shortly before the piston reaches the top of the stroke, the required amount of fuel is sprayed into the combustion chamber by the fuel injector. The intense heat of compression ignites the fuel-air mixture with the resulting combustion driving the piston down on its power stroke.

As the piston nears the bottom of the stroke, the exhaust valve opens and the spent gases are released, assisted by the incoming fresh air. The cycle is complete.

18.312 Fuel System (Figure 18.52)

The basic parts of the fuel system are:

1. Primary fuel filter
2. Secondary fuel filter
3. Fuel injection pump
4. Fuel injector

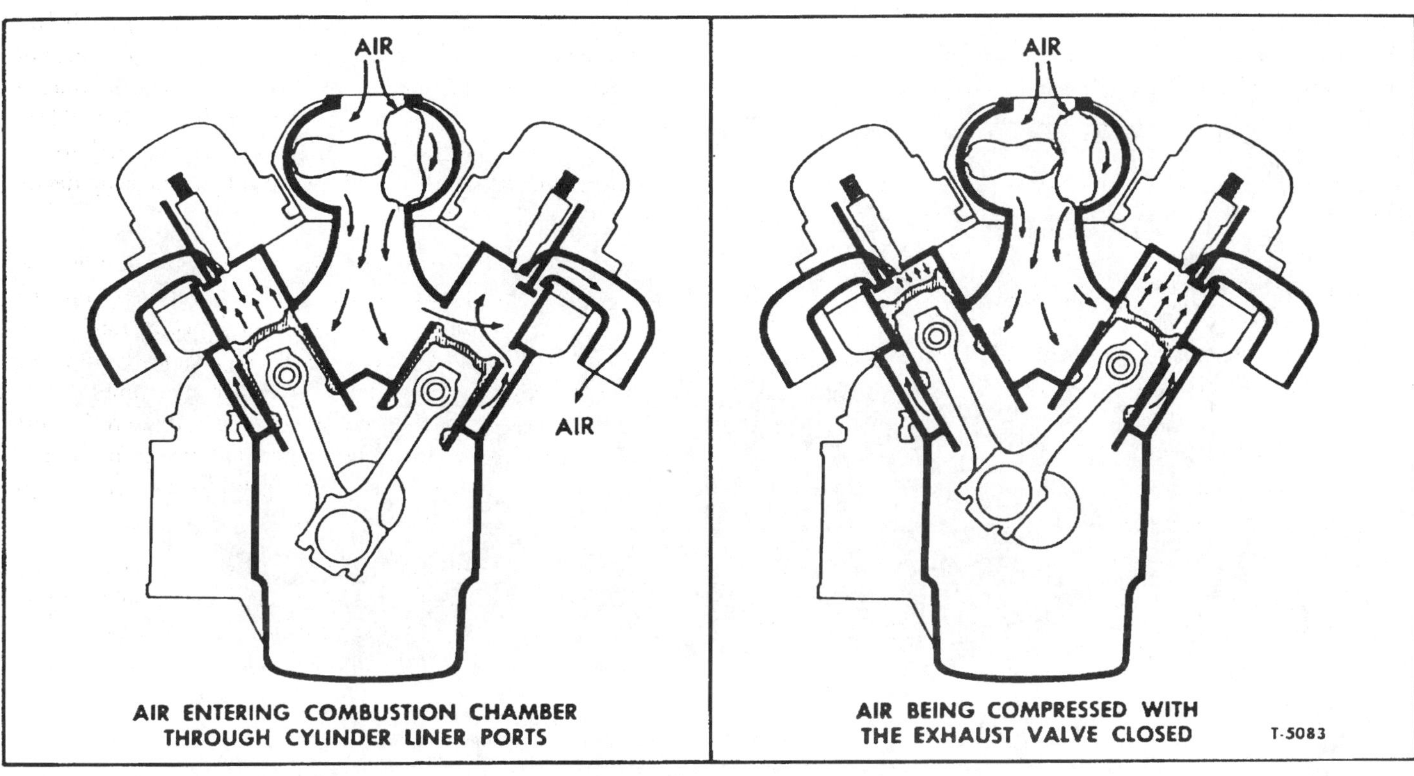

Scavenging and Compression

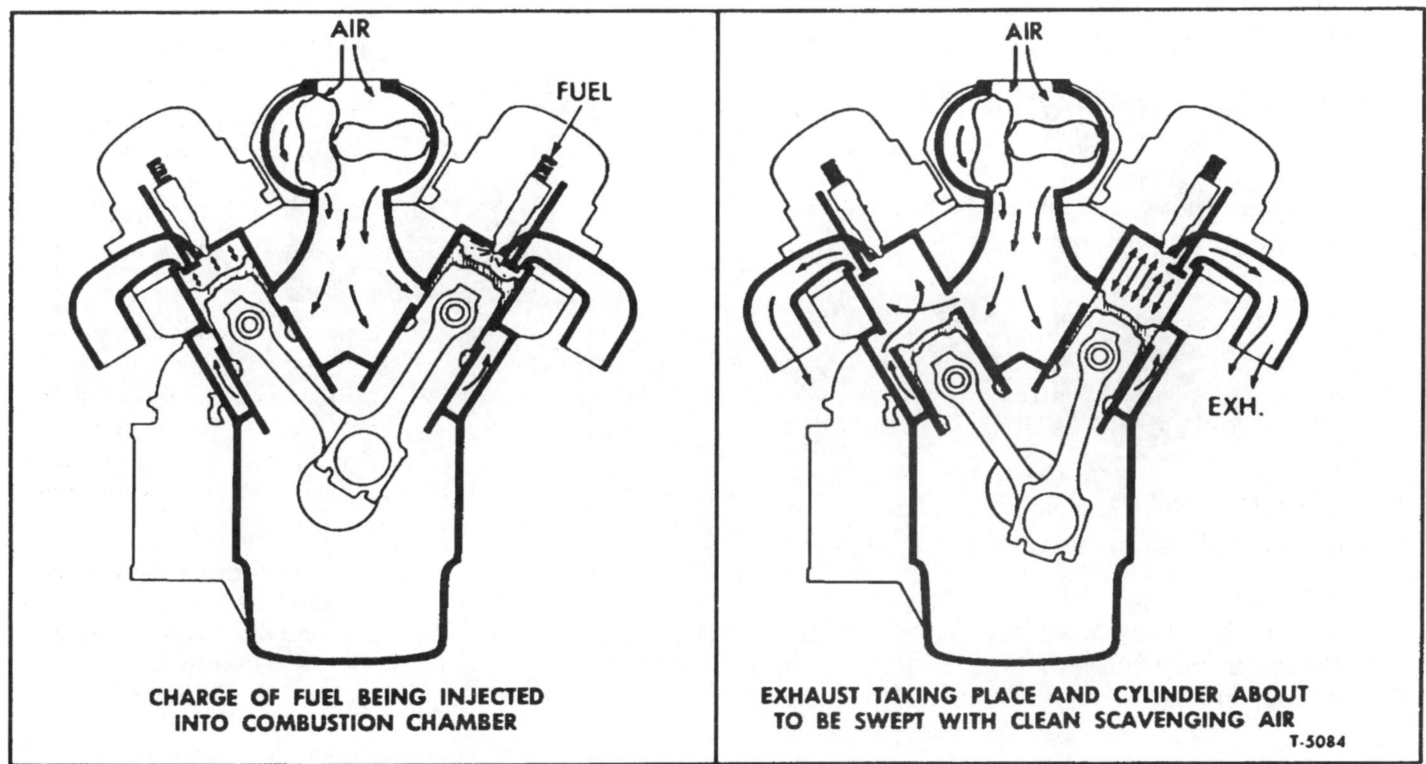

Power and Exhaust

Fig. 18.51 How diesel engines work

(Source: GMC Truck Overhaul Manual Series 53, permission of General Motors Corp.)

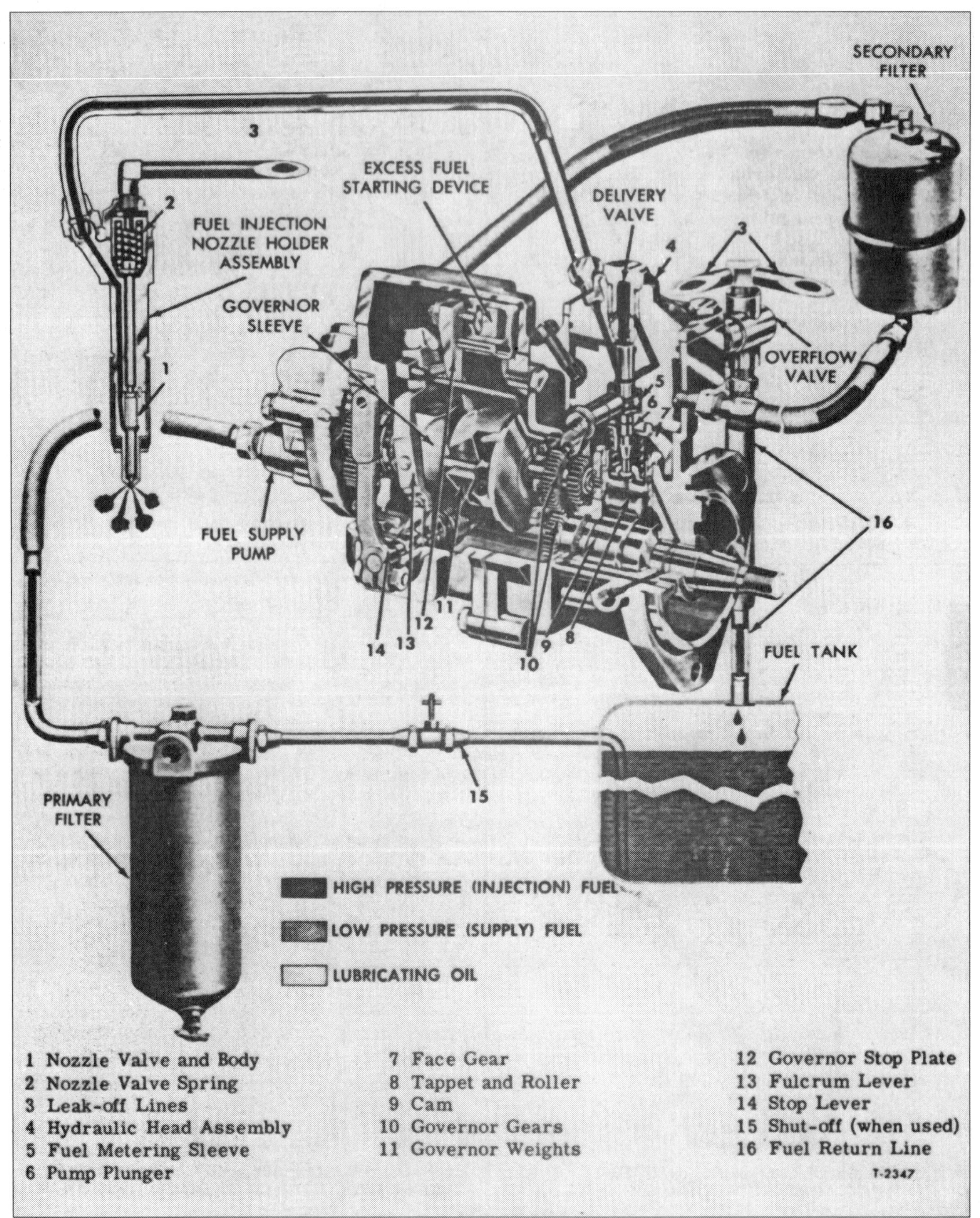

Fig. 18.52 Diesel engine fuel system

(Source: Maintenance Manual, permission of General Motors Corp.)

The primary filter removes all coarse particles from the fuel and the secondary filter removes any minute particles that remain. This ensures a clean fuel that will not clog the injector pump or fuel injectors. The heart of the fuel system is the injection pump (Figure 18.53). This pump is a gear-type positive-displacement pump that can deliver fuel to the injector at a very high pressure. Incorporated into the pump is a timing advance mechanism to advance or retard the instant when fuel is injected into the cylinder. At a high engine speed, injection would take place sooner in the cycle. The reverse happens for lower speeds.

A governor, which uses centrifugal weights and is driven by the pump shaft, activates a fuel control unit. When the engine speed increases, the weights are thrown toward their outer limit. Geared to the assembly, the fuel control valve is opened wider allowing more fuel to flow to the injector.

We now have higher engine speed, advanced timing of injection, and the necessary fuel to sustain the faster operation. When the engine is slowed, the reverse takes place.

Fuel under pressure is fed from the injection pump to the appropriate fuel nozzles. When the pressure reaches approximately 3,000 psi (20,700 kPa or 207 kg/cm^2), the valve in the injector opens allowing fuel to be injected into the combustion chambers. As line pressure drops, the return spring closes the nozzle valve. Fuel left in the line is fed back to the pump through leak-off lines.

18.313 Water-Cooled Diesel Engines

Usually, the larger diesel engines are of the water-cooled type, similar to gasoline engines. To deliver a sustained amount of high

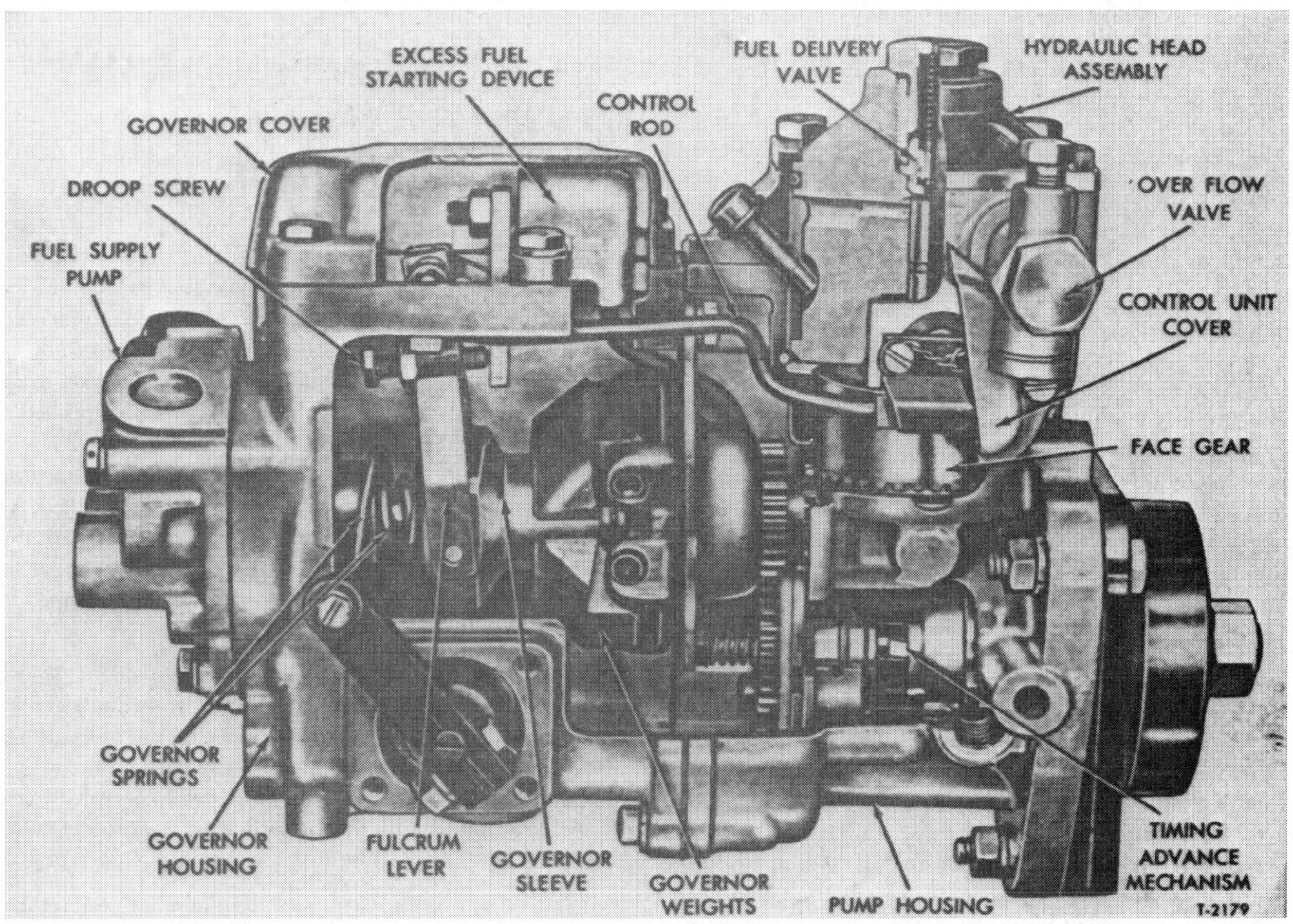

Fig. 18.53 Cutaway view of fuel injection pump for 6-cylinder engine

(Source: Maintenance Manual, permission of General Motors Corp.)

horsepower, an effective cooling system is necessary to dissipate the extreme heat of combustion. Because of this fact, a water-cooled engine of comparative horsepower to the air-cooled will cost more to manufacture, and, subsequently, to maintain.

18.314 Air-Cooled Diesel Engines

When a lighter weight, lower horsepower, and more compact engine is desired, the air-cooled engine will serve your needs. You get the benefits of a diesel engine in a smaller package.

The initial cost is greater for a diesel engine; however, there are some advantages to the diesel engine over the gasoline engine, including:

1. Requires less maintenance because there are no plugs, contact points to pit, ignition coils, or high-tension wires, which means fewer tune-ups
2. Cheaper to operate because diesel fuel may be cheaper and fuel efficiency is better.

Perhaps the biggest drawbacks of diesel engines are initial investment costs and repair costs.

The pros and cons must be weighed to provide you with an engine that will meet your particular needs. Whichever engine you select, remember that a well-cared-for engine will be there to serve you when it is needed and will provide trouble-free operation that is essential to most users.

18.315 How to Start Diesel Engines

Diesel engines vary in size and use and have varied starting procedures. Follow the manufacturer's suggested procedures for your particular engine. As with the gasoline engine, check fuel, oil, and coolant.

To start a diesel engine, the procedures are as follows:

1. Push in "stop" control
2. Set throttle to one-third-full
3. Turn on switch and engage starter
4. Engine should start

Some engines have glow plugs that are energized when the switch is placed in the start position. They preheat the air-fuel mixture in the cylinder to aid in starting. After the engine is started, maintain the lower rpm on the engine tachometer and allow the engine to warm up. The warm-up period is vital to the diesel engine for efficient engine performance. When operating the engine, maintain adequate engine rpm as recommended by the manufacturer.

When a diesel engine will not start after repeated tries, a small amount of starting fluid sprayed into the air intake may be needed to start the engine. If you use starting fluid, do not get carried away with its use; a little goes a long way. Use it only as a last resort or as specified by the manufacturer. If your efforts have failed to start the engine, have a mechanic who is familiar with diesel engines determine the cause of the problem.

18.316 Maintenance and Troubleshooting

For detailed maintenance procedures for your diesel engine, see your diesel manufacturer's service manual.

TROUBLESHOOTING

Certain abnormal conditions that sometimes interfere with satisfactory engine operation are listed in this section.

Satisfactory engine operation depends primarily on an adequate supply of air compressed to a sufficiently high compression pressure and the injection of the proper amount of fuel at the right time.

Lack of power, uneven running, excessive vibration, stalling at idle speed, and hard starting may be caused by low compression, faulty injection in one or more cylinders, or lack of sufficient air. Several problems that can reduce engine performance are listed below:

1. Misfiring cylinders
2. Improper compression pressure
3. Engine out of fuel
4. Inadequate fuel flow
5. Excessive crankcase pressure
6. Excessive back pressure
7. Improper air box pressure
8. Restricted air inlet
9. Low oil pressure
10. Improper engine coolant operating temperature

Solutions to these problems can be found in the operation and maintenance instructions for the engines.

18.32 Cooling Systems (Figure 18.54)

In an air-cooled engine, the heat generated by combustion is dissipated by the air circulating past the louvered cylinder block. With a water-cooled system, the same effect is achieved by using water. Each cylinder is surrounded with a water jacket through which the coolant (water) circulates. This is accomplished by a water pump that is belt driven from the crankshaft. The heat transfers from the cylinder wall to the water which, in turn, is pumped back to the radiator where the heat is dissipated. A fan mounted on the same shaft as the water pump ensures that a large volume of air is blown across the radiator coils to facilitate rapid dispersal of heat. The cooled water is then pumped back into the engine.

Internal combustion engines operate more efficiently when their temperature is maintained within narrow limits. This objective is achieved with the insertion of a thermostatic valve, which is opened by a temperature-activated thermostat, in the cooling system. When the engine is cold, the valve remains closed, not allowing the water to circulate back to the radiator. As the engine temperature increases to normal operating temperature, the valve opens.

The radiator cap provides a function other than preventing coolant from splashing out the filter opening. The cap is designed to seal the cooling system so that it operates under pressure. This improves cooling efficiency and prevents evaporation of coolant. The boiling point of water is 212°F (100°C). However, for every pound of pressure applied to the system, the boiling point rises 3.25°F (1.8°C). If your cooling system had a

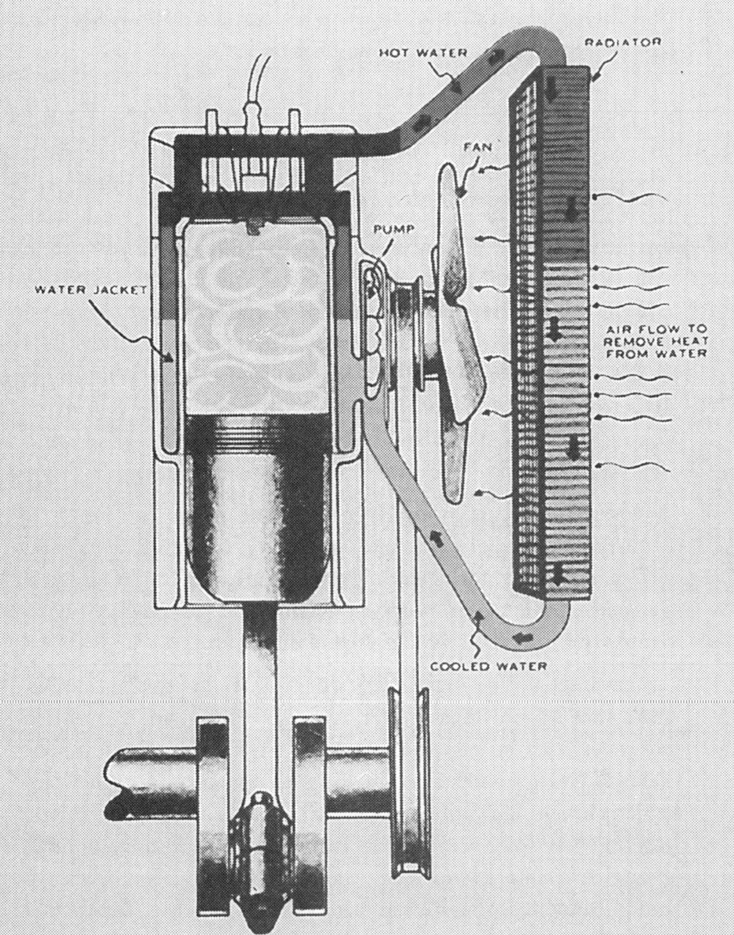

Engine temperatures are regulated by transferring excess heat to surrounding air.

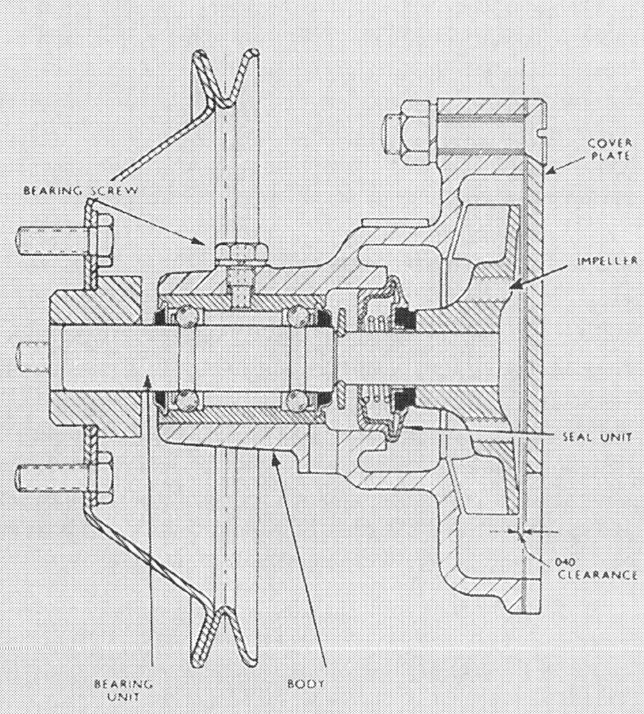

Shaft ball bearings are sealed at each end to keep lubricant in and water out of bearings. A spring-loaded seal (in color) is used to avoid water leakage around pump shaft. Note clearance between impeller and cover plate.

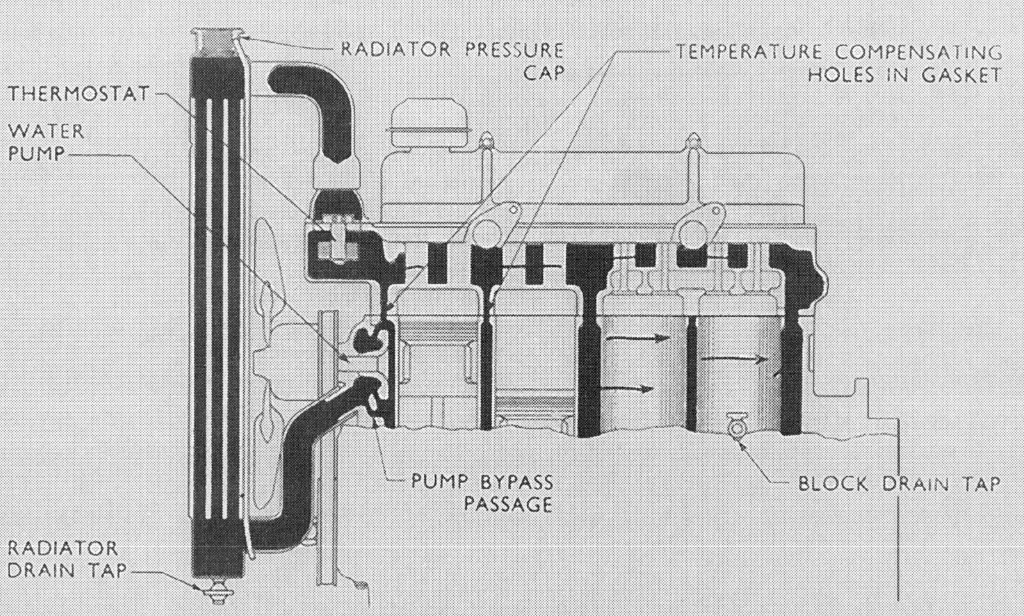

With water jackets entirely around each cylinder and valve, there is a great amount of area exposed to the circulating coolant.

Fig. 18.54 Water cooling system

(Source: Automotive Encyclopedia, permission of the Goodheart-Wilcox Co., Inc.)

15 psi (100 kPa or 1 kg/cm^2) radiator cap and used water for coolant, it would have a boiling point near 260°F (127°C).

The use of coolant/antifreeze provides protection against the radiator coolant freezing and rupturing the system and provides better heat transfer and heat dissipation characteristics than water. Most of the name-brand coolants contain rust inhibitors. Rust buildup in the cooling system interferes with good heat transfer and the sloughing of rust scale can block narrow passages.

Stationary internal combustion engines, such as those that are used to drive pumps and generators at water treatment plants, are often installed in a building where free circulation of air for radiator cooling may not be possible. For these installations, a liquid-to-liquid heat exchanger often replaces the radiator (which is a liquid-to-air heat exchanger). In this case, instead of the heat in the cooling jacket water being transferred to the surrounding air, it is transferred to another liquid, usually tap water. This water may be wasted if the engine is a standby unit and not operated very much, or the cooling water may be recovered in a cooling tower if the engine is in regular use.

In liquid-to-liquid heat exchanger systems, a thermostatically controlled valve is usually installed to regulate the flow of cooling water through the exchanger. This valve should be checked periodically to see that it provides sufficient water flow when the engine is running and closes off tight when the engine is shut down to prevent waste.

Cooling water from a heat exchanger should not be put back into a potable water system. Any leakage in the heat exchanger could result in engine jacket coolant contaminating the potable water supply.

QUESTIONS

Please write your answers to the following questions and compare them with those on page 372.

18.31A Why is gasoline not used as a fuel in diesel engines?

18.31B List the four basic parts of a diesel fuel system.

18.31C What is the purpose of the fuel injection pump?

18.32A How is heat removed from the cylinders in a water-cooled engine?

18.33 Fuel Storage

18.330 Code Requirements

Storage and use of fuels for internal combustion engines must always be in accordance with local building and fire codes. In addition, the water treatment plant operator should be familiar with the particular problems associated with each of the commonly used fuels.

18.331 Diesel Fuel

This fuel comes in two grades known as #1 and #2. Be sure to use the grade recommended by the engine manufacturer. Be aware that the fuel grade recommendation may vary with the season. Diesel fuel is often stored in aboveground tanks. The fuel may be kept in storage for years without deteriorating. To protect stored diesel fuel from water contamination, keep the storage tanks full and use special additives.

18.332 Gasoline

Except for very small quantities, gasoline is stored in underground tanks. This can result in problems for the operator. If the storage tank develops a leak, either fuel can leak out or water can leak in. Either condition is undesirable. Fuel loss cannot only be an unwanted operating expense, but can be a danger to underground plant piping. Gasoline deteriorates rubber and, if the piping is put together with rubber gaskets or rings, the deterioration can result in major leaks and broken couplings. The owner of a leaking underground tank can be liable for any resulting environmental damage and responsible for the cleanup. Fuel loss can best be monitored by careful accounting.

Water leakage into underground fuel tanks can result in engine stoppages and possible damage to the engine. Special devices are available for detecting water in gasoline tanks and such a test should be run routinely. These devices can be obtained by contacting your local wholesale fuel distributor.

Gasoline, unlike diesel fuel, deteriorates in storage. For engines that are in normal everyday use, this is not a problem. However, fuel storage for standby, engine-driven equipment requires further consideration. Engine operation and fuel tank replenishment should be scheduled so that at least half of the gasoline in storage is used each year. Failure to do this can result in engines that are hard to start and in the formation of varnish and gummy deposits that can cause malfunctions in the parts of the fuel system.

18.333 Liquified Petroleum Gas (LPG)

LPG is usually a mixture of propane and butane. The proportion of each is varied according to the weather temperature. The cooler the weather, the greater the proportion of propane.

This fuel is always stored under pressure in aboveground tanks that are located out in the open. LPG does not deteriorate in storage and, therefore, can be kept for many years.

LPG is heavier than air and will collect in low areas if there is any leakage. This poses an extremely dangerous explosive threat that treatment plant operators must constantly guard against.

18.334 Natural Gas

This fuel is usually obtained from the local gas company through a metered connection from their distribution system. There is no on-site storage.

Natural gas, being lighter than air, tends to rise and dissipate from leaks and therefore is less dangerous to handle than LPG. Explosions can occur, however, if the leakage is confined inside a building.

18.34 Standby Engines

Internal combustion engines must be run periodically to ensure that, when needed, they will function properly. An engine that is not in regular service should be started up and test run at least once a week. The test run should be long enough for the engine to come up to its normal operating temperature before the engine is shut down. If at all possible, run the engine under its normal load. Just idling an engine for 20 minutes does not give you much of an indication whether it can handle a load. Check and make note of the engine instruments. Look for changes that may indicate a need for repairs. Lube oil pressure and intake manifold pressure (on spark ignition engines without supercharging or fuel injection) are two key indicators of engine condition.

QUESTIONS

Please write your answers to the following questions and compare them with those on page 372.

18.33A The storage and use of fuels for internal combustion engines must be in accordance with what codes?

18.33B List four types of fuels commonly used by internal combustion engines.

18.34A How often should standby internal combustion engines be test run when not in regular service?

18.34B Under what conditions should standby engines be test run?

18.4 CHEMICAL STORAGE AND FEEDERS [21]

18.40 Chemical Storage

Certain dry chemicals such as alum, ferric chloride, and soda ash are *HYGROSCOPIC*.[22] These chemicals require special considerations to protect them from moisture during storage. Dry quicklime should be kept dry because of the tremendous heat that is generated when it comes in contact with water. This heat is sufficient to cause a fire.

Some liquid chemicals such as sodium hydroxide (caustic soda) should not be exposed to air because of the formation of calcium carbonate (a solid) due to the carbon dioxide in the air. Also, some liquid chemicals may freeze. A 50 percent sodium hydroxide solution becomes crystalized (forms a solid) at temperatures below 55°F (13°C). Therefore, a heater may be required to keep the storage area warm or the solution may have to be diluted down to a 25 percent solution.

Potassium permanganate ($KMnO_4$) can be kept indefinitely if stored in a cool, dry area in closed containers. The drums should be protected from damage that could cause leakage. Potassium permanganate should be stored in fire-resistant buildings, having concrete floors as opposed to wooden floors. It should not be exposed to intense heat, or stored next to heated pipes. Any organic solvent, such as greases and oils in general, should be kept away from stored $KMnO_4$.

Potassium permanganate spills should be swept up and removed immediately. Flushing with water is an effective way to eliminate spillage on floors. Potassium permanganate fires should be extinguished with water.

Carbon should be stored in a clean, dry place, in single or double rows, and with access aisles around every stack for frequent fire inspections. The removal of burning carbon will thus be facilitated. Carbon should never be stored in large stacks.

The storage area should be of fireproof construction, with self-closing fire doors separating the carbon room from other sections. Storage bins for dry bulk carbon should be of fireproof construction equipped for fire control by the installation of carbon dioxide equipment, or should be so arranged that they can be flooded with a fine spray of water.

Carbon storage areas should be protected from contact with flammable materials. (Carbon dust mixed with oily rags or chlorine compounds can ignite in spontaneous combustion.) Smoking is prohibited at all times during the handling and unloading of carbon and in the storage area. Carbon should not be stored where a spark from overhead electric equipment could start a fire. If a fire occurs, the carbon monoxide hazard should be taken into account.

Electric equipment should be protected from carbon dust and cleaned frequently or, better, explosion-proof electric wiring and equipment should be used. (The heat from a motor may ignite the accumulated carbon dust; this material, especially when damp, is a good conductor of electricity and could short-circuit the mechanism.)

Polymer solutions will be degraded (lose their strength) by biological contamination. A good cleaning of polymer storage tanks is recommended before a new shipment is delivered to the plant.

Liquid chemical storage tanks should have a berm or earth bank around the tanks to contain any chemicals released if the tank fails due to an earthquake, corrosion, or any other reason.

Some chemicals such as chlorine and fluoride compounds are harmful to the human body when they are released as the result of a leak. Continual surveillance and maintenance of the storage and feeding systems are required.

18.41 Drainage from Chemical Storage and Feeders

Safety regulations prohibit a single drainage pit that can accept and contain both acid and alkali chemicals because of the possibility of an explosion whenever these two types of chemicals come in contact. Also, any organic chemical waste such as a polymer solution should not be allowed to be discharged into a pit or sump that could also receive a waste from oxidizing chemicals such as potassium permanganate ($KMnO_4$) because of the possibility of a fire. Therefore, separate drainage systems or a high dilution of certain chemicals is necessary for a safe drainage system.

21. For additional information on chemical feeders, see Chapter 13, "Fluoridation," Section 13.30, "Chemical Feeders."
22. *Hygroscopic* (hi-grow-SKAWP-ick). Absorbing or attracting moisture from the air.

18.42 Use of Feeder Manufacturer's Manual

Water treatment plants will have a number of chemical feeders to accurately control the rate at which chemicals are fed into the water as a part of the treatment processes. There are many types of feeders and they work on many different principles. Study the feeder manufacturer's manual that you should find in the treatment plant library for details on maintaining the equipment. Additional information on chemical feeders is contained in specific chapters on treatment processes that require the use of chemical feeders.

18.43 Solid Feeders

Solid feeders usually handle powdered material and usually have many moving mechanical parts that need adjustment, lubrication, and replacement when worn. The chemical supply is usually stored in a hopper. Keep the hopper and feeder clean and dry to prevent "bridging" (a hardened layer that can form an arch and prevent flow) of the chemical in the hopper and clogging in the feeder.

18.44 Liquid Feeders

Liquid feeders handle many types of chemicals, some of which may be corrosive or have a tendency to plug up the mechanism. The key to reliable operations is constant vigilance and cleaning, as needed.

18.45 Gas Feeders

The principal chemical found in gaseous form at water treatment plants is chlorine. Chlorine is quite poisonous to humans and must be handled with great caution.

18.46 Calibration of Chemical Feeders

To ensure the accuracy of chemical feed rates, liquid-chemical metering pumps and dry-chemical feed systems should be tested and calibrated when first installed and at regular intervals thereafter. This section presents general procedures for calibrating several types of liquid- and dry-chemical feeders. For additional information on calibration of chemical feeders, see Volume I, Appendix, "How to Solve Water Treatment Plant Arithmetic Problems," Section A.131, "Chemical Doses."

18.460 Large-Volume Metering Pumps

Pumps metering chemicals such as liquid alum deliver a relatively large volume of chemical in a short time period. These pumps can be accurately calibrated with a clear plastic sight tube and a stopwatch (Figure 18.55).

To calibrate the pump, fill the sight tube from the chemical solution tank, then set the valve so the tube is the only source of liquid chemical entering the pump. Run the pump for exactly one minute (use the stopwatch) at each of five or six representative settings of the pump-control scale. Record the amount

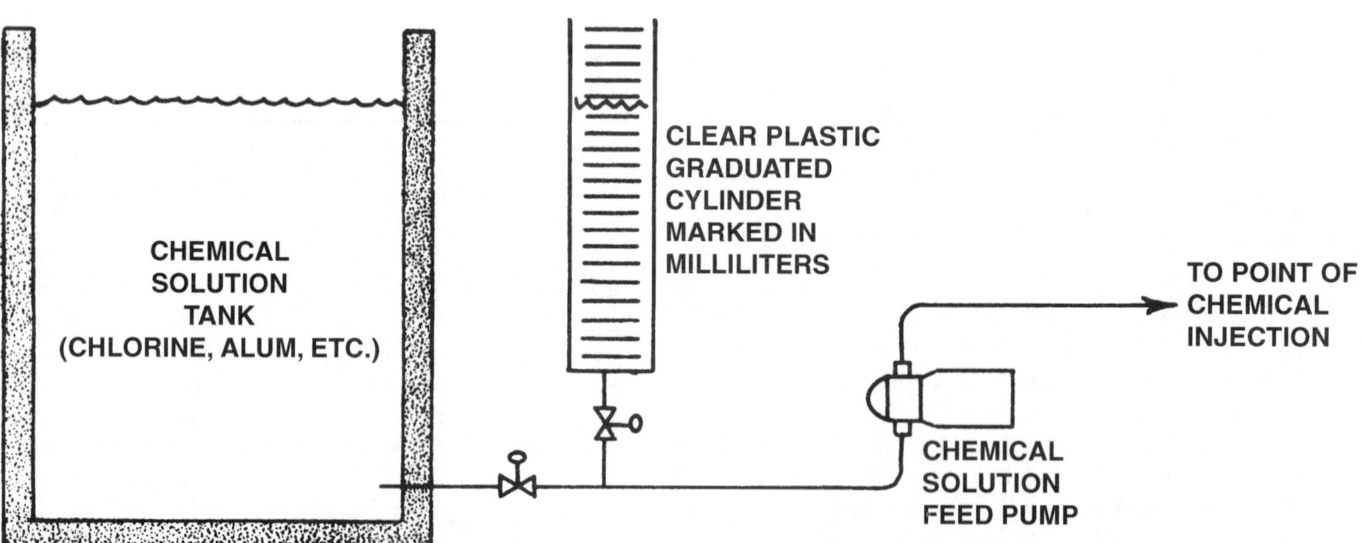

The feed rate of a chemical solution feed pump can be determined by measuring the amount of solution withdrawn from a graduated cylinder in a given time period. Allow the cylinder to fill with solution. Then close the valve on the line from the tank so the feed pump takes suction from the cylinder only. Observe the milliliters of solution used in 1 minute. Compare this result with the desired feed rate and adjust the feed pump accordingly.

Fig. 18.55 Calibration of a chemical feed pump

pumped at each setting as observed in the sight tube. Use this information to develop curves of pump setting versus chemical dose in mg/L or chemical feed in gallons per day for your plant (Figure 18.56).

The graph developed by this process is called a calibration curve. It can be used to determine the pump setting needed to deliver a required chemical feed rate, or the commonly used range of feed rates can be marked in gallons per day directly on the pump control panel.

18.461 Small-Volume Metering Pumps

Pumps metering a chemical such as sodium hexametaphosphate, a lime feed solubility enhancer, feed a very small volume per day. The procedure for calibration of these pumps is similar to the procedure for large-volume units. For very low feed rates, pumping times of longer than one minute may be required to give accurately measurable results.

Once the test data have been recorded, convert the test results to appropriate units and draw a calibration curve to be used as for the larger pumps.

18.462 Dry-Chemical Systems

Dry-chemical feed systems are used for chemicals such as activated carbon, fluoride, and lime. Two types of systems are common, the rocker-dump type and the helix-feed type. The rocker-dump chemical feed system uses a scraper moving back and forth on a platform located at the bottom of a hopper filled with dry chemical. The platform may be adjusted up and down to regulate the thickness of the ribbon of chemical, and the length of stroke for the scraper can be adjusted, usually by means of an indicator on an exterior arm.

The helix-type feeder feeds the dry chemical with a rotating screw (helix). The feed rate is adjusted by varying the drive-motor speed. The speed can usually be varied from 0 to 100 percent.

To calibrate either type of feed system, choose five or six representative settings of the arm (rocker-dump) or of the motor speed control (helix type), and at each of the settings catch the amount of chemical fed during a precisely measured time interval. Next, weigh each volume of chemical as accurately as possible and convert the information into pounds per day. Use the data to construct a calibration curve with one axis representing feeder

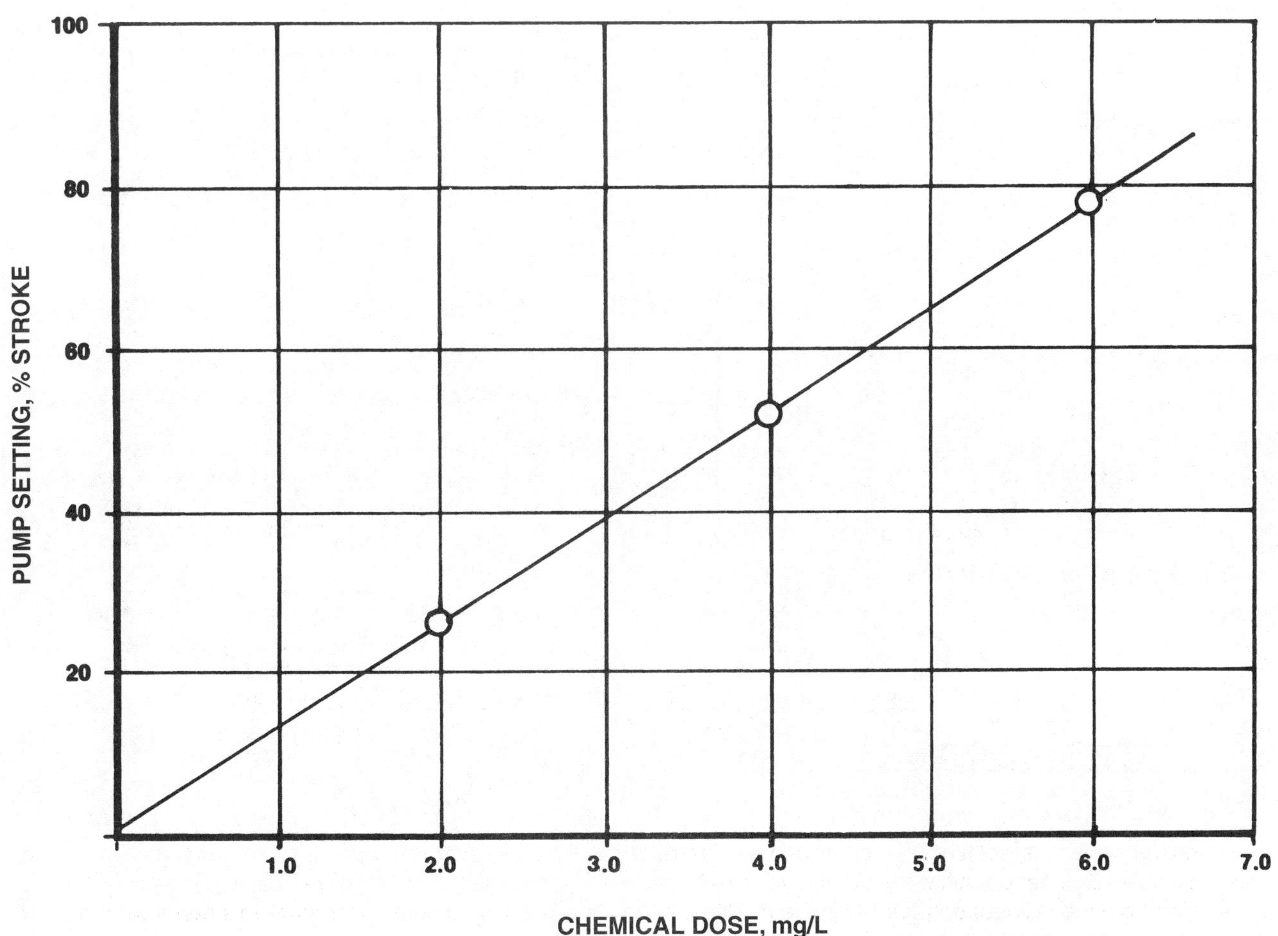

Fig. 18.56 Chemical feed pump settings for various chemical doses

settings and the other representing pounds per day. The curve is used in the same manner as the curves for liquid-feed pumps.

FORMULAS

To determine the chemical feed rate or flow from a chemical feeder, we need to know the amount or volume fed during a known time period. The flow from a chemical feeder can be calculated by knowing the volume pumped from a chemical storage tank and the time period.

$$\text{Flow, GPM} = \frac{\text{Volume Pumped, gal}}{\text{Pumping Time, minutes}}$$

or

$$\text{Flow, GPD} = \frac{(\text{Volume Pumped, gal})(24\text{ hr/day})}{\text{Pumping Time, hour}}$$

Liquid polymer feed rates are often measured in pounds per day. To calculate this feed rate, we need to know the strength of the polymer solution as a percent or as milligrams per liter, the specific gravity of the solution, the volume pumped, and the time period.

$$\text{Polymer Feed, lb/day} = \frac{(\text{Poly Conc, mg/L})(\text{Vol Pumped, mL})(60\text{ min/hr})(24\text{ hr/day})}{(\text{Time Pumped, min})(1{,}000\text{ mL/L})(1{,}000\text{ mg/g})(454\text{ g/lb})}$$

To determine the actual feed from a dry chemical feeder, we need to know the pounds of chemical fed and the time period.

$$\text{Chemical Feed, lb/day} = \frac{(\text{Chemical Fed, lb})(60\text{ min/hr})(24\text{ hr/day})}{\text{Time, minutes}}$$

EXAMPLE 2

A chemical feed pump lowered the chemical solution in a 4-foot-diameter chemical storage tank 2 feet during a 7-hour period. Estimate the flow delivered by the pump in gallons per minute and gallons per day.

Known		Unknown
Tank Diameter, ft	= 4 ft	1. Flow, GPM
Chemical Drop, ft	= 2 ft	2. Flow, GPD
Time, hr	= 7 hr	

1. Determine the volume of water pumped in gallons.

$$\begin{aligned}\text{Volume, gal} &= (0.785)(\text{Diameter, ft})^2(\text{Drop, ft})(7.48\text{ gal/ft}^3)\\ &= (0.785)(4\text{ ft})^2(2\text{ ft})(7.48\text{ gal/ft}^3)\\ &= 188\text{ gal}\end{aligned}$$

2. Calculate the flow from the chemical feed pump in gallons per minute.

$$\begin{aligned}\text{Flow, GPM} &= \frac{\text{Volume Pumped, gal}}{(\text{Pumping Time, hr})(60\text{ min/hr})}\\ &= \frac{188\text{ gal}}{(7\text{ hr})(60\text{ min/hr})}\\ &= 0.45\text{ GPM}\end{aligned}$$

3. Calculate the flow from the chemical feed pump in gallons per day.

$$\begin{aligned}\text{Flow, GPD} &= \frac{\text{Volume Pumped, gal}}{\text{Pumping Time, hr}}\left(\frac{24\text{ hr}}{\text{day}}\right)\\ &= \frac{188\text{ gal}}{7\text{ hr}}\left(\frac{24\text{ hr}}{\text{day}}\right)\\ &= 645\text{ GPD}\end{aligned}$$

EXAMPLE 3

Determine the chemical feed in pounds of polymer per day from a chemical feed pump. The polymer solution is 2.0 percent or 20,000 mg polymer per liter. Assume a specific gravity of the polymer solution of 1.0. During a test run, the chemical feed pump delivered 750 mL of polymer solution during 6 minutes.

Known		Unknown
Polymer Solution, %	= 2.0%	Polymer Feed, lb/day
Polymer Conc, mg/L	= 20,000 mg/L	
Polymer Sp Gr	= 1.0	
Volume Pumped, mL	= 750 mL	
Time Pumped, min	= 6 min	

Calculate the polymer feed by the chemical feed pump in pounds of polymer per day.

$$\begin{aligned}\text{Polymer Feed, lb/day} &= \frac{(\text{Poly Conc, mg/L})(\text{Vol Pumped, mL})(60\text{ min/hr})(24\text{ hr/day})}{(\text{Time Pumped, min})(1{,}000\text{ mL/L})(1{,}000\text{ mg/g})(454\text{ g/lb})}\\ &= \frac{(20{,}000\text{ mg/L})(750\text{ mL})(60\text{ min/hr})(24\text{ hr/day})}{(6\text{ min})(1{,}000\text{ mL/L})(1{,}000\text{ mg/g})(454\text{ g/lb})}\\ &= 7.9\text{ lb Polymer/day}\end{aligned}$$

EXAMPLE 4

Determine the actual chemical fed in pounds per day from a dry chemical feeder. A pie tin placed under a chemical feeder collected 1,000 grams of chemical in 5 minutes.

Known		Unknown
Chemical Fed, g	= 1,000 g	Chemical Feed, lb/day
Time, min	= 5 min	

Determine the chemical feed in pounds of chemical applied per day.

$$\begin{aligned}\text{Chemical Feed, lb/day} &= \frac{(\text{Chemical Fed, g})(60\text{ min/hr})(24\text{ hr/day})}{(454\text{ g/lb})(\text{Time, min})}\\ &= \frac{(1{,}000\text{ g})(60\text{ min/hr})(24\text{ hr/day})}{(454\text{ g/lb})(5\text{ min})}\\ &= 634\text{ lb/day}\end{aligned}$$

18.47 Chlorinators

Chlorine gas leaks around chlorinators or containers of chlorine will cause corrosion of equipment. Check every day for leaks. Large leaks will be detected by odor; small leaks may go unnoticed until damage results. A green or reddish deposit on metal indicates a chlorine leak. Any chlorine gas leakage in the presence of moisture will cause corrosion. Always plug the ends of any open connection to prevent moisture from entering the lines. Never pour water on a chlorine leak because this will only create a bigger problem by enlarging the leak. Chlorine gas reacts with water to form hydrochloric acid.

WARNING

Be proactive in preventing chlorine leaks because chlorine gas is toxic to humans.

Ammonia water will detect any chlorine leak. A small piece of cloth, soaked with ammonia water[23] and wrapped around the end of a short stick, makes a good leak detector. Wave this stick in the general area of the suspected leak (do not touch the equipment with it). If chlorine gas leakage is occurring, a white cloud of ammonium chloride will form. Perform this test at all gas pipe joints, both inside and outside the chlorinators, at regular intervals. Bottles of ammonia water should be kept tightly capped to avoid loss of strength. All pipe fittings must be kept tight to avoid leaks. New gaskets should be used for each new connection.

Do not use a spray bottle filled with ammonia water in a room where a large amount of chlorine gas has already leaked into the air. After one squeeze, the entire area may fill with white vapor, making it hard for you to locate the leak. Under these conditions, use a cloth soaked in ammonia water to look for leaks.

CAUTION

Do not spray or swab equipment with ammonia water to locate a chlorine leak! Wave an ammonia-soaked rag or paintbrush in the general area and you can detect the presence of many leaks. Some operators prefer to wave a stick with a cloth on the end in front of them when they are looking for chlorine leaks.

The exterior casing of chlorinators should be painted, as required; however, most chlorinators manufactured recently have plastic cases that do not require protective coatings. A clean machine is a better operating machine. Parts of a chlorinator handling chlorine gas must be kept dry to prevent the chlorine and moisture from forming hydrochloric acid. Some parts may be cleaned, when required, first with water to remove water-soluble material, then with wood alcohol, followed by drying. The above chemicals leave no moisture residue. Another method would be to wash them with water and dry them over a pan or heater to remove all traces of moisture.

Water strainers on chlorinators frequently clog and require attention. They may be cleaned by flushing with water or, if badly fouled, they may be cleaned with diluted hydrochloric acid, followed with a water rinse.

The atmosphere vent lines from chlorinators must be open and free. These vent lines evacuate the chlorine to the outside atmosphere when the chlorinator is being shut down. Place a screen over the end of the pipe to keep insects from building a nest in the pipe and clogging it.

When chlorinators are removed from service, as much chlorine gas as possible should be removed from the supply lines and machines. The chlorine valves at the containers are shut off and the chlorinator injector is operated for a period to remove the chlorine gas. In V-notch chlorinators, the rotameter goes to the bottom of the manometer tube when the chlorine gas has been expelled.

All chlorinators will give continuous, trouble-free operation if properly maintained and operated. Each chlorinator manufacturer provides with each machine a maintenance and operations instruction booklet with line diagrams showing the operation of the component parts of the machine. Manufacturer's instructions should be followed for maintenance and lubrication of your particular chlorinator. If you do not have an instruction booklet, you may obtain one by contacting the manufacturer's representative in your area.

QUESTIONS

Please write your answers to the following questions and compare them with those on page 372.

18.4A How can an operator locate information on how to operate, control, and maintain chemical feeders?

18.4B List three common types of chemical feeders.

18.4C Why should chlorine leaks be detected and repaired?

18.4D How would you search for chlorine leaks?

23. Use a concentrated ammonia solution containing 28 to 30 percent ammonia as NH_3 (this is the same as 58 percent ammonium hydroxide, NH_4OH, or commercial 26° Baumé).

18.5 TANKS AND RESERVOIRS[24]

18.50 Scheduling Inspections

Plant tanks should be drained and inspected at regular intervals. If the interior is well protected, 5-year intervals between inspections may be sufficient. If the tank is below the surface of the ground, be sure the groundwater level is down far enough (below the bottom of the tank) so the tanks will not float on the groundwater when empty or develop cracks from groundwater pressure.

Schedule inspections of tanks and channels during periods of low plant demand so that plant operation will not be disrupted.

18.51 Steel Tanks

All steel tanks must be protected from rusting. Once metal is lost because of rusting, it cannot be recovered. The exteriors of the tanks are easily inspected—do not forget the roof—and should be repainted, as needed, not only to protect the steel surface but to provide a pleasing appearance. The interiors of steel tanks are exposed to a much harsher environment due either to being constantly submerged or to constant high humidity.

Protective coatings for steel tank interiors must be carefully selected to provide superior protection and, at the same time, impart neither taste nor odors to the water. Proper surface preparation and application is as important as the coating materials in getting interior protection that will last a reasonable period of time.

When interior tank recoating is required, schedule the work when plant demand is low but not during rainy weather when it may be impossible to maintain a dry steel surface warm enough to ensure proper curing of the coating. This type of work is usually done by outside contractors. Constant inspection is a must if the work is going to be completed according to the specifications.

18.52 Cathodic Protection[25]

An alternative to repainting the submerged interior surfaces of a steel water tank is installation of a cathodic protection system. The rusting of steel is accompanied by the flow of small electrical currents. Cathodic protection systems prevent rusting of bare steel surfaces by causing an electrical current to flow from anodes hung in the water to the tank surface. The polarity of this current is opposite to what it would be if rust were forming. The current can be obtained from sacrificial anodes that make the tank into a giant low-voltage battery, or from electronic rectifiers that are powered from the commercial power lines.

Cathodic protection systems provide protection only so long as they are operating and properly adjusted. Systems with rectifiers should be checked weekly. The inspection consists of reading and recording the DC multimeter and ammeter readings. Compare the readings with previous readings and with the readings recommended by the corrosion engineer or technician. Deviations from normal should be investigated without delay.

Once a year, a corrosion specialist should be called in to take potential profile readings on the inside of the tank and to set the rectifier and recommend new normal current settings. Over the years, as more and more of the interior tank coating fails, the bare surface area to be protected by the cathodic protection system will increase. For this reason, it is to be expected that the current required to provide protection will increase in small amounts or increments each year.

18.53 Concrete Tanks

Concrete tanks are not usually coated on the inside and are painted on the exterior for appearance purposes only. This would seem to indicate that maintenance on concrete tanks is minimal, but this may not be true. Concrete tanks are all reinforced with steel. Steel can rust. If too much steel is lost to rust, the structural strength of the tank can be threatened. Periodically inspect the tank for signs of rusting. This is particularly important for prestressed concrete tanks that have a tensioned wire wrap on the exterior. The wires are small in diameter and even a small amount of rust could reduce the size of the wire to the point where it might fail.

QUESTIONS

Please write your answers to the following questions and compare them with those on page 372.

18.5A How often should tanks and reservoirs be drained and inspected?

18.5B Why must the groundwater level be below the bottom of a tank before it is drained?

18.5C What is an alternative to applying a protective coating to prevent corrosion of a steel tank?

18.6 BUILDING MAINTENANCE

Building maintenance is another program that should receive attention on a regular schedule. Buildings in a treatment plant are usually built of sturdy materials to last for many years, if they are kept in good repair. In selecting paint for a treatment plant, it is always a good idea to have a painting expert help the operator select the types of paint needed to protect the buildings from deterioration. The expert also will have some good ideas as to color schemes to help blend the plant in with the surrounding area. Consideration should also be given to the quality of paint. A good quality, more expensive material will usually give better service over a longer period of time than the economy-type products.

Building maintenance programs depend on the age, type, and use of a building. New buildings require a thorough check to be

24. Also see *Water Distribution System Operation and Maintenance*, Chapter 2, "Storage Facilities," Section 2.4, "Maintenance," in this series of operator training manuals.
25. Also see *Water Treatment Plant Operation*, Volume I, Chapter 8, "Corrosion Control," Section 8.45, "Cathodic Protection," in this series of operator training manuals.

certain essential items are available and working properly. Older buildings require careful watching and prompt attention to keep ahead of leaks, breakdowns, replacements when needed, and changing uses of the building. Attention must be given to the maintenance requirements of many items in all plant buildings such as electrical systems, plumbing, heating, cooling, ventilating, floors, windows, roofs, and drainage around the buildings. Regularly scheduled examinations and necessary maintenance of these items can prevent many costly and time-consuming problems in the future.

In each plant building, periodically check all stairways, ladders, catwalks, and platforms for adequate lighting, head clearance, and sturdy and convenient guardrails. Protective devices should surround all moving equipment. Whenever any repairs, alterations, or additions are built, avoid building accident traps such as pipes laid on top of floors or hung from the ceiling at head height, which could create serious safety hazards.

Organized storage areas should be provided and maintained in an accessible and neat manner.

Keep all buildings clean and orderly. Janitorial work should be done on a regular schedule. All tools and plant equipment should be kept clean and in their proper place. Floors, walls, and windows should be cleaned at regular intervals to maintain a neat appearance. A treatment plant kept in a clean, orderly condition makes a safe place to work and aids in building good public and employee relations.

QUESTIONS

Please write your answers to the following questions and compare them with those on page 372.

18.6A What factors influence the type of building maintenance program that might be needed for your water treatment plant?

18.6B What items should be included in a building maintenance program?

18.7 ARITHMETIC ASSIGNMENT

Turn to the Arithmetic Appendix at the back of this manual. Read and work the problems in Section A.35, "Maintenance." Check the arithmetic in this section using an electronic calculator. You should be able to get the same answers. Section A.55 contains similar problems using metric units.

18.8 ADDITIONAL READING

1. *Texas Manual*, chapter on "Pumps and Measurement of Pumps."
2. *Pump Handbook*, Fourth Edition, edited by Igor Karassik, Joseph Messina, Paul Cooper, and Charles Heald. Published by the McGraw-Hill Companies. ISBN 978-0-07-146044-6.

18.9 ACKNOWLEDGMENTS

Major portions of this chapter were taken from the following California State University, Sacramento, operator training manuals:

1. *Operation of Wastewater Treatment Plants*, Volume II, Chapter 15, "Maintenance," by Norman Farnum, Stan Walton, John Brady, Roger Peterson, Malcolm Carpenter, and Rick Arbour.
2. *Operation and Maintenance of Wastewater Collection Systems*, Volume II, Chapter 9, "Equipment Maintenance," by Lee Doty and Rick Arbour.
3. *Industrial Waste Treatment*, Volume II, Chapter 8, "Maintenance," by Norman Farnum, Stan Walton, John Brady, Roger Peterson, Malcolm Carpenter, and Rick Arbour.

END OF LESSON 5 OF 5 LESSONS

on

MAINTENANCE

Please answer the discussion and review questions next.

DISCUSSION AND REVIEW QUESTIONS

Chapter 18. MAINTENANCE

(Lesson 5 of 5 Lessons)

Please write your answers to the following questions to determine how well you understand the material in the lesson. The question numbering continues from Lesson 4.

36. What factors could cause gasoline engine starting problems?
37. What is the purpose of the filters in the diesel fuel system?
38. What are the advantages of air-cooled diesel engines as compared with water-cooled types?
39. Why is rust a problem in water-cooled systems?
40. How should large quantities of gasoline be stored?
41. Why is idling not a satisfactory method of testing standby engines?
42. Why should solid feeders and hoppers be kept clean and dry?
43. What problems can be caused by chlorine gas leaks around chlorinators or containers of chlorine?
44. Why should a treatment plant be kept in a clean and orderly condition?

SUGGESTED ANSWERS

Chapter 18. MAINTENANCE

ANSWERS TO QUESTIONS IN LESSON 1

Answers to questions on page 271.

18.0A A good maintenance program is essential to maintain successful operation of the treatment plant.

18.0B A successful maintenance program will include everything from mechanical equipment to the care of the plant grounds, buildings, and structures.

18.0C A good recordkeeping system tells when maintenance is due and provides a record of equipment performance. Poor performance is a good justification for replacement or new equipment. Good records also help keep your warranty in force.

18.0D An equipment service card tells what service or inspection work should be done on a piece of equipment and when, while the service record card is a record of what was done and when.

18.0E Emergency phone numbers for a treatment plant should include the phone numbers for police, fire, hospital or physician, responsible plant officials, local emergency disaster office, emergency team, and *CHEMTREC*, (800) 424-9300.

18.0F Stored energy must be bled down to prevent its sudden release, which could injure an operator working on the system or equipment.

Answers to questions on page 272.

18.10A Unqualified or inexperienced people must be extremely careful when attempting to troubleshoot or repair electrical equipment because they can be seriously injured and damage costly equipment if a mistake is made.

18.10B When machinery is not shut off, locked out, and tagged properly, the following accidents could occur:

1. Maintenance operator could be cleaning pump and have it start, thus losing an arm, hand, or finger.
2. Electric motors or controls not properly grounded could lead to possible severe shock, paralysis, or death.
3. Improper circuits such as a wrong connection, safety devices jumped, wrong fuses, or improper wiring can cause fires or injuries due to incorrect operation of machinery.

Answers to questions on page 274.

18.11A The proper voltage and allowable current in amps for a piece of equipment can be determined by reading the nameplate information or the instruction manual for the equipment.

18.11B The two types of current are direct current (DC) and alternating current (AC).

18.11C Amperage is the measurement of current or electron flow and is an indication of work being done or how hard the electricity is working.

Answers to questions on page 278.

18.12A You test for voltage by using a multimeter.

18.12B A multimeter can be used to test for voltage, open circuits, blown fuses, single phasing of motors, and grounding.

18.12C Before attempting to change fuses, turn off power and check both power lines for voltage. Use a fuse puller.

18.12D If the voltage is unknown and the multimeter has different scales that are manually set, always start with the highest voltage range and work down. Otherwise, the multimeter could be damaged.

18.12E Amp readings different from the nameplate rating could be caused by low voltage, bad bearings, poor connections, or excessive load.

Answers to questions on page 280.

18.12F Motors and wiring should be megged at least once a year, and twice a year, if possible.

18.12G An ohmmeter is used to measure the resistance (ohms) in control circuit components such as coils, fuses, relays, resistors, and switches.

Answers to questions on page 282.

18.13A The two types of safety devices in main electrical panels or control units are fuses or circuit breakers.

18.13B Fuses are used to protect operators, wiring, circuits, heaters, motors, and various other electrical equipment.

18.13C A fuse must never be bypassed or jumped because the fuse may be the only protection the circuit has; without it, serious damage to equipment and possible injury to operators can occur.

18.13D A circuit breaker is a switch that is opened automatically when the current or the voltage exceeds or falls below a certain limit. Unlike a fuse that has to be replaced each time it blows, a circuit breaker can be reset after a short delay to allow time for cooling.

18.13E Motor starters can be either manually or automatically controlled.

18.13F Magnetic starters are commonly used to start pumps, compressors, blowers, and anything where automatic or remote control is desired.

Answers to questions on page 285.

18.14A Electrical energy is commonly converted into mechanical energy by electric motors.

18.14B An electric motor usually consists of a stator, rotor, end bells, and windings.

18.14C Motors can be kept trouble free with proper lubrication and maintenance.

18.14D Motor nameplate data should be recorded, compared with manufacturer's data sheet and instruction manual for consistency, and placed in a file for future reference. Many times the nameplate is painted, corroded, or missing from the unit when the information is needed to repair the motor or replace parts.

Answers to questions on page 293.

18.14E The key to effective troubleshooting is practical, step-by-step procedures combined with a common-sense approach.

18.14F When troubleshooting magnetic starters:

1. Gather preliminary information.
2. Inspect:
   a. Contacts
   b. Mechanical parts
   c. Magnetic parts

18.14G Information that should be recorded regarding electrical equipment includes nameplate data and every change, repair, and test.

Answers to questions on page 295.

18.15A A qualified electrician should perform most of the necessary maintenance and repair of electrical equipment to avoid endangering lives and to avoid damage to equipment.

18.15B The purpose of a kirk-key system (one key is used for two locks) is to ensure proper connection of standby power into your power distribution system. The commercial power system must be locked out by the use of switch gear before the standby power is connected to your power distribution system.

18.15C Battery-powered backup lighting units are considered better than engine-driven power sources because they are more economical. If you have a momentary power outage, the system responds without an engine-generator startup.

18.15D If water lost from a lead-acid battery is replaced with tap water, the impurities in the water will become attached to the lead plates and shorten the life of the battery.

Answers to questions on page 296.

18.16A Electricity is transmitted at high voltage to reduce the size of transmission lines.

18.16B If outdoor transformers have exposed high-voltage wires, the following precautions must be taken:

1. An 8-foot-high (2.4 m) fence is required to prevent accessibility by unqualified or unauthorized persons.
2. Signs attached to the fence must indicate "High Voltage."

18.16C The treatment plant operator must keep the exterior and surroundings of the switch gear clean.

18.16D Symptoms that a power distribution transformer may be in need of maintenance or repair include unusual noises, high or low oil levels, oil leaks, or high operating temperatures.

Answers to questions on page 296.

18.17A Electrical safety checklists are used to make operators aware of potential electrical hazards in their water treatment plant.

18.17B Rusted conduits are of concern because they could become the source of a spark that could cause an explosion.

ANSWERS TO QUESTIONS IN LESSON 2

Answers to questions on page 308.

18.20A Pieces of equipment and special tools commonly found in a pump repair shop include welding equipment, lathes, drill press and drills, power hacksaw, flame-cutting equipment, micrometers, calipers, gauges, portable electric tools, grinders, a forcing press, metal-spray equipment, and sandblasting equipment.

18.21A The purpose of a pump impeller is to suck water in the suction piping and to throw water out between the impeller blades.

18.21B A suitable screen should be installed on the intake end of suction piping to prevent foreign matter (sticks, refuse) from being sucked into the pump and clogging or wearing the impeller.

18.21C Suction piping must be up-sloping to prevent air pockets from forming in the top of a pipe where air could be drawn into the pump and cause the loss of suction.

18.21D Cavitation is a condition that occurs when the pressure within the flowing water drops very low, the water starts to boil, and gas pockets or bubbles form on the blade of the impeller. The collapse of these gas pockets or bubbles drives water into the impeller with a terrific force that can knock metal particles off the impeller and cause pitting on the impeller surface. This same action can and does occur on pressure reducing valves and partially closed gate and butterfly valves. Cavitation is accompanied by loud noises that sound like someone is pounding on the impeller with a hammer.

18.21E An advantage of a double-suction pump over a single-suction design is that the longitudinal thrust from the water entering the impeller is balanced.

Answers to questions on page 311.

18.22A The purpose of lubrication is to reduce friction between two surfaces and to remove heat caused by friction.

18.22B Oils in service tend to become acid (contaminated) and may cause corrosion, deposits, sludging, and other problems.

18.22C To ensure proper lubrication of equipment, determine the proper lubrication schedule, lubricant, and amount of lubricant, and prepare a lubrication chart.

Answers to questions on page 311.

18.22D A soft grease has a low viscosity index as compared with a hard grease.

18.22E Oil is used with higher speeds.

18.22F Overfilling with oil or grease can result in high pressures and temperatures, and ruined seals or other components.

ANSWERS TO QUESTIONS IN LESSON 3

Answers to questions on page 317.

18.23A A cross-connection is a connection between a drinking (potable) water system and an unapproved water supply.

18.23B A slight water-seal leakage is desirable when a pump is running to keep the packing cool and in good condition.

18.23C To measure the capacity of a pump, measure the volume pumped during a specific time period.

$$\text{Pump Capacity, GPM} = \frac{\text{Volume, gallons}}{\text{Time, minutes}}$$

or

$$\text{Pump Capacity, } \frac{\text{liters}}{\text{sec}} = \frac{\text{Volume, liters}}{\text{Time, seconds}}$$

18.23D Before a prolonged shutdown, the pump should be drained to prevent damage from corrosion, sedimentation, and freezing. Also, the motor disconnect switch should be opened to disconnect the motor.

18.23E Estimate the capacity of a pump (in GPM) if it lowers the water in a 10-foot-wide by 15-foot-long wet well 1.7 feet in 5 minutes.

Known	Unknown
Width, ft = 10 ft	Pump Capacity, GPM
Length, ft = 15 ft	
Depth, ft = 1.7 ft	
Time, min = 5 min	

$$\text{Capacity, GPM} = \frac{\text{Volume, gallons}}{\text{Time, minutes}}$$

$$= \frac{(10\text{ ft})(15\text{ ft})(1.7\text{ ft})(7.48\text{ gal/ft}^3)}{5\text{ minutes}}$$

$$= 381.5\text{ GPM}$$

or

$$\text{Capacity, } \frac{\text{liters}}{\text{sec}} = \frac{\text{Volume, liters}}{\text{Time, sec}}$$

$$= \frac{(3\text{ m})(5\text{ m})(0.5\text{ m})(1{,}000\text{ L/m}^3)}{(5\text{ minutes})(60\text{ sec/min})}$$

$$= 25\text{ liters/sec}$$

Answers to questions on page 319.

18.23F Shear pins commonly fail in reciprocating pumps because of a solid object lodged under piston, a clogged discharge line, or a stuck or wedged valve.

18.23G A noise may develop when a reciprocating pump is pumping thin sludge due to water hammer, but it will disappear when heavy sludge is pumped.

18.23H Higher than normal discharge pressures in a progressive cavity pump may indicate a line blockage or a closed valve downstream.

Answers to questions on page 321.

18.23I When checking an electric motor, the following items should be checked periodically, as well as when trouble develops:

1. Motor conditions
2. Note all unusual conditions
3. Lubricate bearings
4. Listen to motor
5. Check temperature

18.23J The purpose of a stethoscope is to magnify sounds and carry them to the ear. This instrument is used to detect unusual sounds in electric motors such as whines, gratings, or uneven noises.

Answers to questions on page 323.

18.23K A properly adjusted V-belt has a slight bow in the slack side when running. When idle, it has an alive springiness when thumped with the hand. To check for proper alignment, lay a long straightedge or string across outside faces of the pulley, and allow for differences in dimensions from centerlines of grooves to outside faces of the pulleys being aligned. Be very careful in aligning drives with more than one V-belt on a sheave, as misalignment can cause unequal tension.

18.23L Always replace worn sprockets when replacing a chain in a chain drive unit because out-of-pitch sprockets cause as much chain wear in a few hours as years of normal operation.

Answers to questions on page 325.

18.23M Improper original installation of equipment, settling of foundations, heavy floor loadings, warping of bases, and excessive bearing wear could cause couplings to become out of alignment.

18.23N A shear pin is a straight pin that will fail (break) when a certain load or stress is exceeded. The purpose of the pin is to protect equipment from damage due to excessive loads or stresses.

Answers to questions on page 327.

18.24A Pumps must be lubricated in accordance with the manufacturer's recommendations. Quality lubricants should be used.

18.24B In lubricating motors, too much grease may cause bearing trouble or damage the winding.

Answers to questions on page 328.

18.24C If a pump will not start, check for blown fuses or tripped circuit breakers and the cause. Also check for a loose connection, fuse, or thermal unit.

18.24D To increase the rate of discharge from a pump, you should look for something causing the reduced rate of discharge, such as pumping air, motor malfunction, plugged lines or valves, impeller problems, or other factors.

Answers to questions on page 329.

18.24E If a pump that has been locked or tagged out for maintenance or repairs is started, an operator working on the pump could be seriously injured and equipment could be damaged.

18.24F Normally, a centrifugal pump should be started after the discharge valve is opened. Exceptions are treatment processes or piping systems with vacuums or pressures that cannot be dropped or allowed to fluctuate greatly while an alternate pump is put on the line.

Answers to questions on page 330.

18.24G Before stopping an operating pump, start another pump (if appropriate). Inspect the operating pump by looking for developing problems, required adjustments, and problem conditions of the unit.

18.24H A pump shaft or motor will spin backward if water being pumped flows back through the pump when the pump is shut off. This will occur if there is a faulty check valve or foot valve in the system.

18.24I The position of all valves should be checked before starting a pump to ensure that the water being pumped will go where intended.

Answers to questions on page 331.

18.24J The most important rule regarding the operation of positive-displacement pumps is to never operate the pump against a closed valve, especially a discharge valve.

18.24K If a positive-displacement pump is operated against a closed discharge valve, the pipe, valve, or pump could rupture from excessive pressure. The rupture will damage equipment and possibly seriously injure or kill someone standing nearby.

18.24L Both ends of a sludge line should never be closed tight at the same time because gas from decomposition of the sludge can build up and rupture pipes or valves.

ANSWERS TO QUESTIONS IN LESSON 4

Answers to questions on page 334.

18.25A Compressors are used to activate and control pump control systems (bubblers), valve operators, and water pressure systems. They are also used to operate portable pneumatic tools, such as jackhammers, compactors, air drills, sandblasters, tapping machines, and air pumps.

18.25B The frequency of cleaning a suction filter on a compressor depends on the use of the compressor and the atmosphere around it. The filter should be inspected at least monthly and cleaned or replaced every 3–6 months. More frequent inspection, cleaning, and replacement are required under dusty conditions such as near jackhammer operation on a street.

18.25C Compressor oil should be changed at least every 3 months, unless the manufacturer states differently. If there are filters in the oil system, these also should be changed.

18.25D Drain the condensate from the air receiver daily.

18.25E Before testing belt tension on a compressor with your hands, make sure the compressor is locked off.

Answers to questions on page 350.

18.26A Valves are the controlling devices placed in piping systems to stop, regulate, check, divert, or otherwise modify the flow of liquids or gases.

18.26B Six common types of valves found in water treatment facilities include gate valves, globe valves, eccentric valves, butterfly valves, check valves, and automatic valves.

18.26C The most common maintenance required by gate valves is oiling, tightening, or replacing the stem stuffing box packing.

18.26D The purpose of the check valve is to allow water to flow in one direction only.

18.26E Backflow prevention by check valves is essential in many applications to prevent pumps from reversing when power is removed, to protect water systems from being cross-connected, to aid in pump operation as a dampener, and to ensure full-pipe operation (pipe is full of water).

ANSWERS TO QUESTIONS IN LESSON 5

Answers to questions on page 353.

18.30A Gasoline engines may be used in water treatment plants to drive pumps, generators, tractors, and vehicles.

18.30B If a gasoline engine will not start, check the following items:

1. No fuel in tank, valve closed
2. Carburetor not choked
3. Water or dirt in fuel lines or carburetor
4. Carburetor flooded
5. Low compression
6. Loose spark plug
7. No spark at plug

18.30C A gasoline engine may not run properly due to:

1. Engine misses
2. Engine surges
3. Engine stops
4. Engine overheats
5. Engine knocks
6. Engine backfires through carburetor

Answers to questions on page 354.

18.30D If a gasoline engine will not start and the spark plug is wet with fuel or oil, this could indicate that the cylinder is flooded with fuel from having the choke on too long.

18.30E If a gasoline engine will not start and there is an oil residue on the spark plug, this could indicate worn piston rings.

18.30F After an engine has started, give it an opportunity to warm up before applying the load.

Answers to questions on page 360.

18.31A Gasoline is not used as a fuel in diesel engines because it would start to burn from the heat generated by compression before the piston reached the top of the stroke.

18.31B The four basic parts of a diesel fuel system are:

1. Primary fuel filter
2. Secondary fuel filter
3. Fuel injection pump
4. Fuel injector

18.31C The purpose of the fuel injection pump is to deliver fuel to the injector at a very high pressure.

18.32A Heat is removed from the cylinders by a water cooling system. Each cylinder is surrounded with a water jacket through which the coolant (water) circulates. This is accomplished by a water pump that is belt driven from the crankshaft.

Answers to questions on page 361.

18.33A The storage and use of fuels for internal combustion engines must be in accordance with local building and fire codes.

18.33B Four types of fuels commonly used by internal combustion engines are diesel, gasoline, liquified petroleum gas (LPG), and natural gas.

18.34A Standby internal combustion engines not in regular service should be started up and test run at least once a week.

18.34B Standby engines should be test run long enough for the engine to come up to its normal operating temperature. If at all possible, the engine should be run under its normal load.

Answers to questions on page 365.

18.4A Information on how to operate, control, and maintain chemical feeders may be found in the feeder manufacturer's literature.

18.4B The three common types of chemical feeders are solid feeders, liquid feeders, and gas feeders.

18.4C Chlorine is toxic to humans and will cause corrosion damage to equipment.

18.4D Large chlorine leaks can be detected by smell. Small leaks are detected by soaking a cloth with ammonia water and holding the cloth near areas where leaks might develop. A white cloud will indicate the presence of a leak.

Answers to questions on page 366.

18.5A Tanks and reservoirs should be drained and inspected at least once every 5 years if the interior is well protected, more often if it is not well protected.

18.5B The groundwater level should be below the bottom of a tank before it is drained so the tank will not float on the groundwater when empty or develop cracks from groundwater pressure.

18.5C Cathodic protection is an alternative to applying a protective coating to prevent corrosion of a steel tank.

Answers to questions on page 367.

18.6A Factors that influence the type of building maintenance program needed by a water treatment plant include the age, type, and use of each building.

18.6B A building maintenance program will keep the building in good shape and includes painting, when necessary. Attention also must be given to electrical systems, plumbing, heating, cooling, ventilating, floors, windows, roofs, and drainage around the buildings. The building should be kept clean, tools should be stored in their proper place, and essential storage should be available.

CHAPTER 19

INSTRUMENTATION AND CONTROL SYSTEMS

by

Leonard Ainsworth

Revised by

William H. Hendrix

TABLE OF CONTENTS

Chapter 19. INSTRUMENTATION AND CONTROL SYSTEMS

LESSON 2

LEARNING OBJECTIVES

Chapter 19. INSTRUMENTATION AND CONTROL SYSTEMS

Following completion of Chapter 19, you should be able to:

1. Explain the purpose and nature of instrumentation and control systems.
2. Identify, avoid, and correct safety hazards associated with instrumentation work.
3. Recognize various types of sensors and transducers.
4. Read instruments and make proper adjustments in the operation of water treatment facilities.
5. Identify symptoms of measurement and control system problems.

WORDS

Chapter 19. INSTRUMENTATION AND CONTROL SYSTEMS

ACCURACY ACCURACY

How closely an instrument measures the true or actual value of the process variable being measured or sensed.

ALARM CONTACT ALARM CONTACT

A switch that operates when some preset low, high, or abnormal condition exists.

ANALOG ANALOG

The continuously variable signal type sent to an analog instrument (for example, 4–20 mA).

ANALOG READOUT ANALOG READOUT

The readout of an instrument by a pointer (or other indicating means) against a dial or scale. Also see DIGITAL READOUT.

ANALYZER ANALYZER

A device that conducts a periodic or continuous measurement of turbidity or some factor such as chlorine or fluoride concentration. Analyzers operate by any of several methods, including photocells, conductivity, or complex instrumentation.

CALIBRATION CALIBRATION

A procedure that checks or adjusts an instrument's accuracy by comparison with a standard or reference.

CONTACTOR CONTACTOR

An electric switch, usually magnetically operated.

CONTROLLER CONTROLLER

A device that controls the starting, stopping, or operation of a device or piece of equipment.

CONTROL LOOP CONTROL LOOP

The combination of one or more interconnected instrumentation devices that are arranged to measure, display, and control a process variable. Also called a loop.

CONTROL SYSTEM — CONTROL SYSTEM

An instrumentation system that senses and controls its own operation on a close, continuous basis in what is called proportional (or modulating) control.

DANGEROUS AIR CONTAMINATION — DANGEROUS AIR CONTAMINATION

An atmosphere presenting a threat of causing death, injury, acute illness, or disablement due to the presence of flammable or explosive, toxic, or otherwise injurious or incapacitating substances.

(1) Dangerous air contamination due to the flammability of a gas, vapor, or mist is defined as an atmosphere containing the gas, vapor, or mist at a concentration greater than 10 percent of its lower explosive (lower flammable) limit (LEL).

(2) Dangerous air contamination due to a combustible particulate is defined as a concentration that meets or exceeds the particulate's lower explosive limit (LEL).

(3) Dangerous air contamination due to the toxicity of a substance is defined as the atmospheric concentration that could result in employee exposure in excess of the substance's permissible exposure limit (PEL).

NOTE: A dangerous situation also occurs when the oxygen level is less than 19.5 percent by volume (OXYGEN DEFICIENCY) or more than 23.5 percent by volume (OXYGEN ENRICHMENT).

DESICCANT (DESS-uh-kant) — DESICCANT

A drying agent that is capable of removing or absorbing moisture from the atmosphere in a small enclosure.

DIGITAL — DIGITAL

The encoding of information that uses binary numbers (ones and zeros) for input, processing, transmission, storage, or display, rather than a continuous spectrum of values (an analog system) or non-numeric symbols such as letters or icons.

DIGITAL READOUT — DIGITAL READOUT

The readout of an instrument by a direct, numerical reading of the measured value or variable.

DISCRETE CONTROL — DISCRETE CONTROL

ON/OFF control; one of the two output values is equal to zero.

DISCRETE I/O (INPUT/OUTPUT) — DISCRETE I/O (INPUT/OUTPUT)

A digital signal that senses or sends either ON or OFF signals. For example, a discrete input would sense the position of a switch; a discrete output would turn on a pump or light.

DISTRIBUTED CONTROL SYSTEM (DCS) — DISTRIBUTED CONTROL SYSTEM (DCS)

A computer control system having multiple microprocessors to distribute the functions performing process control, thereby distributing the risk from component failure. The distributed components (input/output devices, control devices, and operator interface devices) are all connected by communications links and permit the transmission of control, measurement, and operating information to and from many locations.

EFFECTIVE RANGE — EFFECTIVE RANGE

That portion of the design range (usually from 10 to 90+ percent) in which an instrument has acceptable accuracy. Also see RANGE and SPAN.

FAIL-SAFE — FAIL-SAFE

Design and operation of a process control system whereby failure of the power system or any component does not result in process failure or equipment damage.

FEEDBACK — FEEDBACK

The circulating action between a sensor measuring a process variable and the controller that controls or adjusts the process variable.

HERTZ (Hz) — HERTZ (Hz)

The number of complete electromagnetic cycles or waves in one second of an electric or electronic circuit. Also called the frequency of the current.

HUMAN MACHINE INTERFACE (HMI) | HUMAN MACHINE INTERFACE (HMI)

The device at which the operator interacts with the control system. This may be an individual instrumentation and control device or the graphic screen of a computer control system. Also see MAN MACHINE INTERFACE (MMI) and OPERATOR INTERFACE.

INTEGRATOR | INTEGRATOR

A device or meter that continuously measures and sums a process rate variable in cumulative fashion over a given time period. For example, total flows displayed in gallons per minute, million gallons per day, cubic feet per second, or some other unit of volume per time period. Also see TOTALIZER.

INTERLOCK | INTERLOCK

A physical device, equipment, or software routine that prevents an operation from beginning or changing function until some condition or set of conditions is fulfilled. An example would be a switch that prevents a piece of equipment from operating when a hazard exists.

LAG TIME | LAG TIME

The time period between the moment a process change is made and the moment such a change is finally sensed by the associated measuring instrument.

LINEARITY (lin-ee-AIR-it-ee) | LINEARITY

How closely an instrument measures actual values of a variable through its effective range.

MAN MACHINE INTERFACE (MMI) | MAN MACHINE INTERFACE (MMI)

The device at which the operator interacts with the control system. This may be an individual instrumentation and control device or the graphic screen of a computer control system. Also see HUMAN MACHINE INTERFACE (HMI) and OPERATOR INTERFACE.

MEASURED VARIABLE | MEASURED VARIABLE

A factor (flow, temperature) that is sensed and quantified (reduced to a reading of some kind) by a primary element or sensor.

OPERATOR INTERFACE | OPERATOR INTERFACE

The device at which the operator interacts with the control system. This may be an individual instrumentation and control device or the graphic screen of a computer control system. Also see HUMAN MACHINE INTERFACE (HMI) and MAN MACHINE INTERFACE (MMI).

ORIFICE (OR-uh-fiss) | ORIFICE

An opening (hole) in a plate, wall, or partition. An orifice flange or plate placed in a pipe consists of a slot or a calibrated circular hole smaller than the pipe diameter. The difference in pressure in the pipe above and at the orifice may be used to determine the flow in the pipe. In a trickling filter distributor, the wastewater passes through an orifice to the surface of the filter media.

PRECISION | PRECISION

The ability of an instrument to measure a process variable and repeatedly obtain the same result. The ability of an instrument to reproduce the same results.

PRIMARY ELEMENT | PRIMARY ELEMENT

(1) A device that measures (senses) a physical condition or variable of interest. Floats and thermocouples are examples of primary elements. Also called a sensor.

(2) The hydraulic structure used to measure flows. In open channels, weirs and flumes are primary elements or devices. Venturi meters and orifice plates are the primary elements in pipes or pressure conduits.

PROCESS VARIABLE | PROCESS VARIABLE

A physical or chemical quantity that is usually measured and controlled in the operation of a water, wastewater, or industrial treatment plant. Common process variables are flow, level, pressure, temperature, turbidity, chlorine, and oxygen levels.

PROGRAMMABLE LOGIC CONTROLLER (PLC) | PROGRAMMABLE LOGIC CONTROLLER (PLC)

A microcomputer-based control device containing programmable software; used to control process variables.

RANGE RANGE

The spread from minimum to maximum values that an instrument is designed to measure. Also see EFFECTIVE RANGE and SPAN.

READOUT READOUT

The reading of the value of a process variable from an indicator or recorder or on a computer screen.

RECEIVER RECEIVER

A device that indicates the result of a measurement, usually using either a fixed scale and movable indicator (pointer), such as a pressure gauge, or a moving chart with a movable pen like those used on a circular flow-recording chart. Also called an indicator.

RECORDER RECORDER

A device that creates a permanent record, on a paper chart, magnetic tape, or in a computer, of the changes in a measured variable.

REFERENCE REFERENCE

A physical or chemical quantity whose value is known exactly, and thus is used to calibrate instruments or standardize measurements. Also called a standard.

ROTAMETER (ROTE-uh-ME-ter) ROTAMETER

A device used to measure the flow rate of gases and liquids. The gas or liquid being measured flows vertically up a tapered, calibrated tube. Inside the tube is a small ball or bullet-shaped float (it may rotate) that rises or falls depending on the flow rate. The flow rate may be read on a scale behind or on the tube by looking at the middle of the ball or at the widest part or top of the float.

SCADA (SKAY-dah) SYSTEM SCADA SYSTEM

Supervisory Control And Data Acquisition system. A computer-monitored alarm, response, control, and data acquisition system used to monitor and adjust treatment processes and facilities.

SCALE SCALE

(1) A combination of mineral salts and bacterial accumulation that sticks to the inside of a collection pipe under certain conditions. Scale, in extreme growth circumstances, creates additional friction loss to the flow of water. Scale may also accumulate on surfaces other than pipes.
(2) The marked plate against which an indicator or recorder reads, usually the same as the range of the measuring system. Also see RANGE.

SENSITIVITY SENSITIVITY

The smallest change in a process variable that an instrument can sense.

SENSOR SENSOR

A device that measures (senses) a physical condition or variable of interest. Floats and thermocouples are examples of sensors. Also called a primary element.

SET POINT SET POINT

The position at which the control or controller is set. This is the same as the desired value of the process variable. For example, a thermostat is set to maintain a desired temperature.

SOFTWARE PROGRAM SOFTWARE PROGRAM

Computer program; the list of instructions that tell a computer how to perform a given task or tasks. Some software programs are designed and written to monitor and control treatment processes.

SOLENOID (SO-luh-noid) SOLENOID

A magnetically operated mechanical device (electric coil). Solenoids can operate small valves or electric switches.

SPAN SPAN

The scale or range of values an instrument is designed to measure. Also see RANGE.

STANDARD STANDARD

A physical or chemical quantity whose value is known exactly, and thus is used to calibrate instruments or standardize measurements. Also called a reference.

STANDARDIZE STANDARDIZE

To compare with a standard.

(1) In wet chemistry, to find out the exact strength of a solution by comparing it with a standard of known strength. This information is used to adjust the strength by adding more water or more of the substance dissolved.

(2) To set up an instrument or device to read a standard. This allows you to adjust the instrument so that it reads accurately, or enables you to apply a correction factor to the readings.

STARTERS (MOTOR) STARTERS (MOTOR)

Devices used to start up large motors gradually to avoid severe mechanical shock to a driven machine and to prevent disturbance to the electrical lines (causing dimming and flickering of lights).

TELEMETRY (tel-LEM-uh-tree) TELEMETRY

The electrical link between a field transmitter and the receiver. Telephone lines are commonly used to serve as the electrical line.

THERMOCOUPLE THERMOCOUPLE

A heat-sensing device made of two conductors of different metals joined together. An electric current is produced when there is a difference in temperature between the ends.

TIMER TIMER

A device for automatically starting or stopping a machine or other device at a given time.

TOTALIZER TOTALIZER

A device or meter that continuously measures and sums a process rate variable in cumulative fashion over a given time period. For example, total flows displayed in gallons per minute, million gallons per day, cubic feet per second, or some other unit of volume per time period. Also see INTEGRATOR.

TRANSDUCER (trans-DUE-sir) TRANSDUCER

A device that senses some varying condition measured by a primary sensor and converts it to an electrical or other signal for transmission to some other device (a receiver) for processing or decision making.

VARIABLE, MEASURED VARIABLE, MEASURED

A factor (flow, temperature) that is sensed and quantified (reduced to a reading of some kind) by a primary element or sensor.

VARIABLE, PROCESS VARIABLE, PROCESS

A physical or chemical quantity that is usually measured and controlled in the operation of a water, wastewater, or industrial treatment plant.

ABBREVIATIONS AND SYMBOLS

Chapter 19. INSTRUMENTATION AND CONTROL SYSTEMS

Special symbols are used for simplicity and clarity on circuit drawings for instruments. Usually, instrument manufacturers and design engineers provide lists of symbols they use with an explanation of the meaning of each symbol. This section contains a list of typical instrumentation abbreviations and symbols used in this chapter and used by the waterworks profession.

ABBREVIATIONS

A — Analyzer, such as a device used to measure a water quality indicator (pH, temperature).

C — Controller, such as a device used to start, operate, or stop a pump.

D — Differential, such as a "differential pressure" (DP) cell used with a flowmeter.

E — Electrical or Voltage.
Element, such as a primary element.

F — Flow rate (not total flow).

H — Hand (manual operation).
High as in high-level switch.

I — Indicator, such as the indicator on a flow recording chart.
I = E/R where I is the electric current in amps.

L — Level, such as the level of water in a tank.
Low, as in a low-level switch.
Light, as in indicator light.

M — Motor.
Middle, as in a mid-level switch.

P — Pressure (or vacuum).
Pump.
Program, as in a software program.

Q — Quantity, such as a totalized volume (Σ for summation is also used).

R — Recorder (or printer), such as a chart recorder.
Receiver.
Relay.

S — Switch.
Speed, such as the revolutions per minute (rpm) of a motor.
Starter, such as a motor starter.
Solenoid.

T — Transmitter.
Temperature.
Tone.

V — Valve.
Voltage.

W — Weight.
Watt.

X — Special or unclassified variable.

Y — Computing function, such as a square root ($\sqrt{\ }$) extraction.

Z — Position, such as a percent valve opening.

TYPICAL PROCESS AND ELECTRICAL SYMBOLS

1. 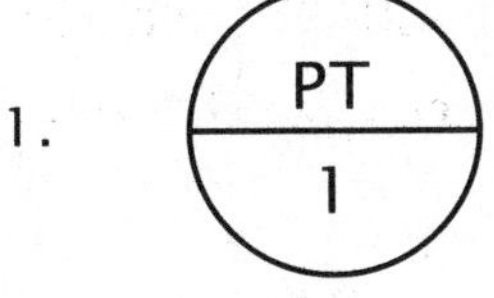

Pressure transmitter #1

2. 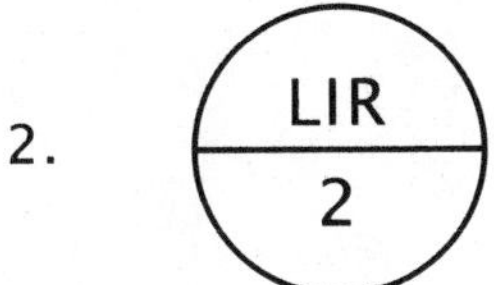

Level indicator/recorder #2

3. 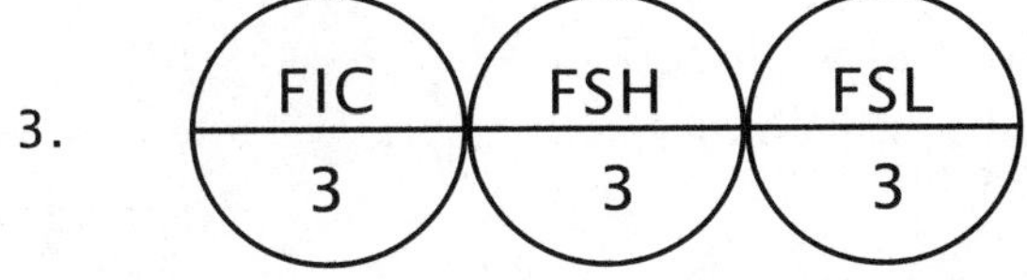

Flow indicator/controller #3 with high-low control switches in the same instrument

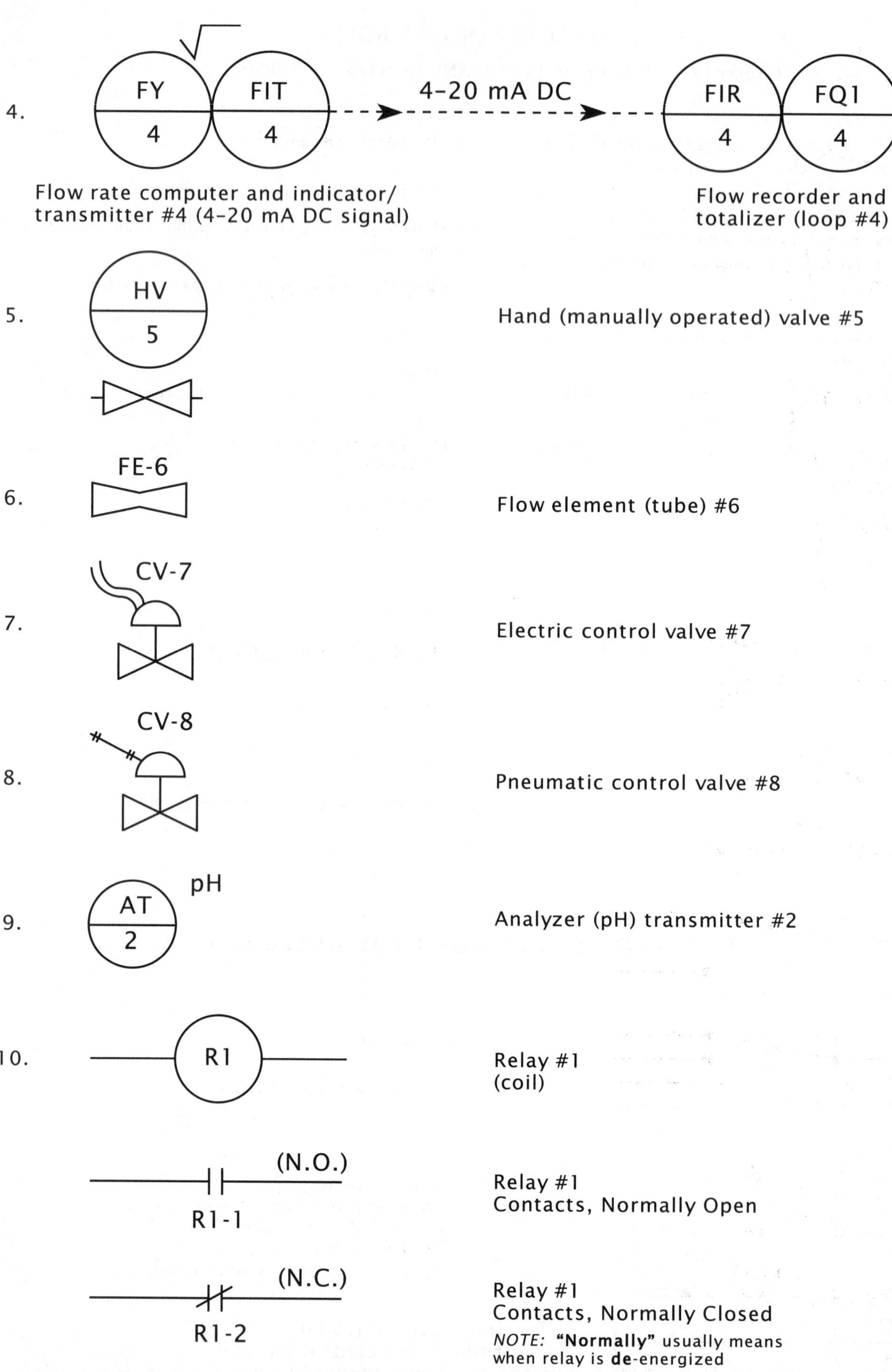

4.
FY
4
FIT
4
4–20 mA DC
FIR
4
FQ1
4
Flow rate computer and indicator/
transmitter #4 (4–20 mA DC signal)
Flow recorder and
totalizer (loop #4)
5.
HV
5
Hand (manually operated) valve #5
6.
FE-6
Flow element (tube) #6
7.
CV-7
Electric control valve #7
8.
CV-8
Pneumatic control valve #8
9.
AT
2
pH
Analyzer (pH) transmitter #2
10.
R1
Relay #1
(coil)
(N.O.)
R1-1
Relay #1
Contacts, Normally Open
(N.C.)
R1-2
Relay #1
Contacts, Normally Closed
NOTE: "Normally" usually means
when relay is de-energized

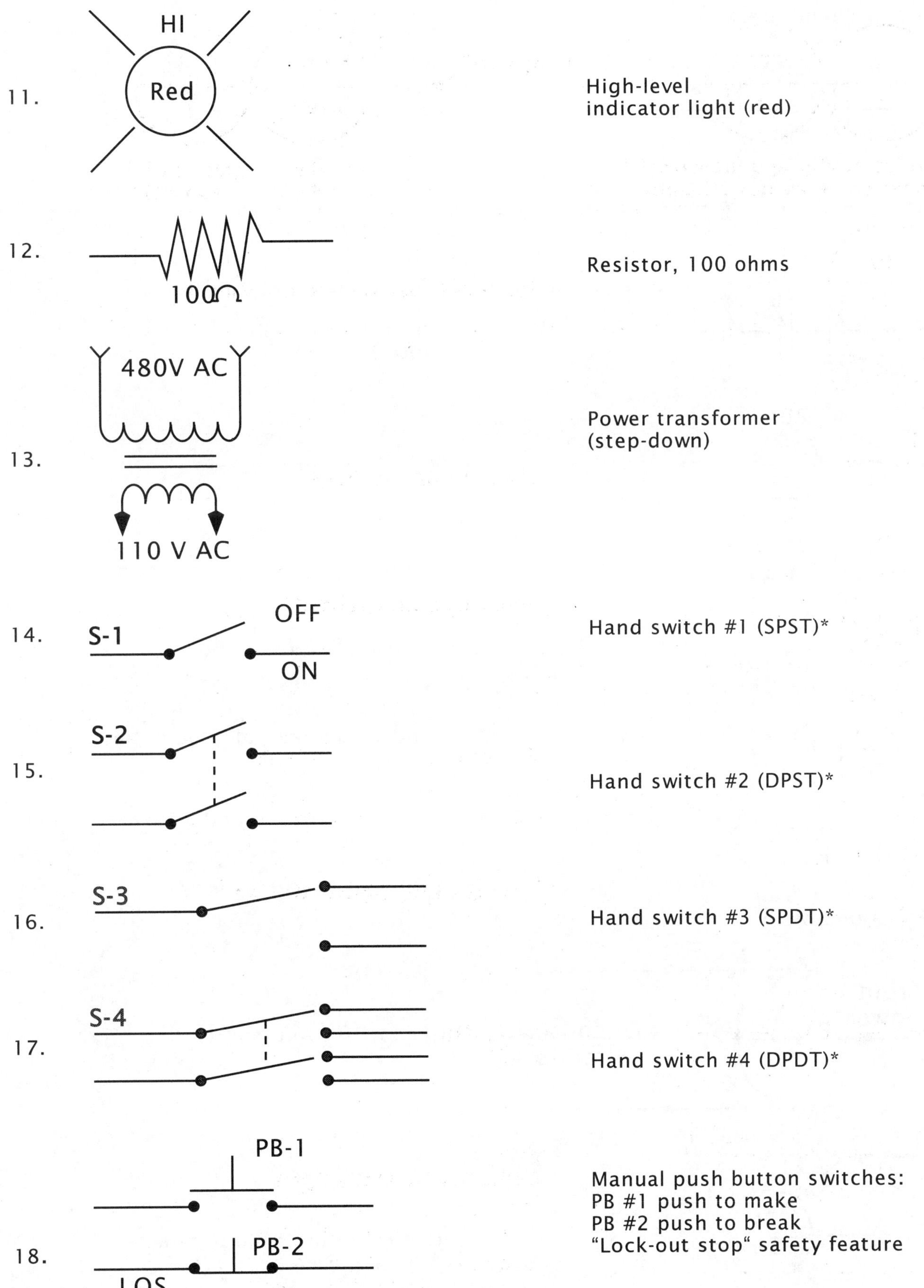

* SPST means single-pole, single-throw; DPST means double-pole, single-throw; SPDT means single-pole, double-throw; DPDT means double-pole, double throw

19. Fuses and Circuit Breakers

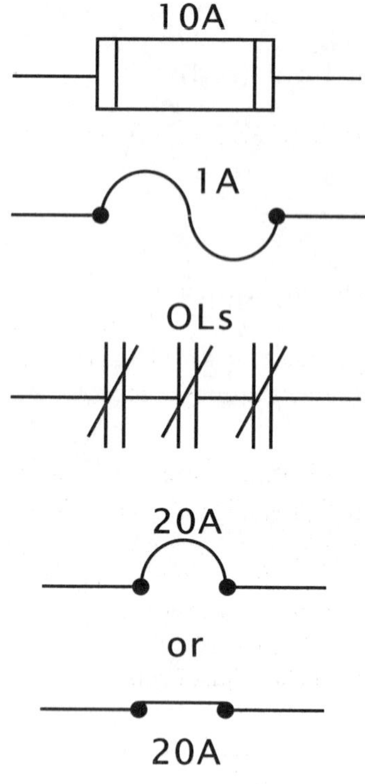

10-amp cartridge fuse

1-amp in-line fuse

Thermal overload contacts (motor)

20-amp circuit breaker

20.

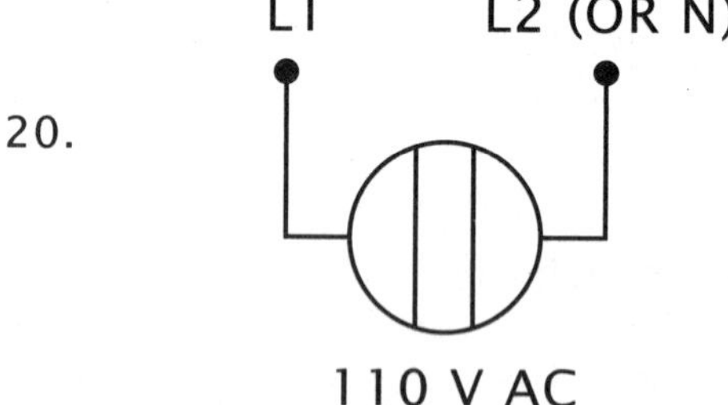

Line 1 and Line 2 (neutral) to standard duplex wall plug outlet

21. HOA Function Switch

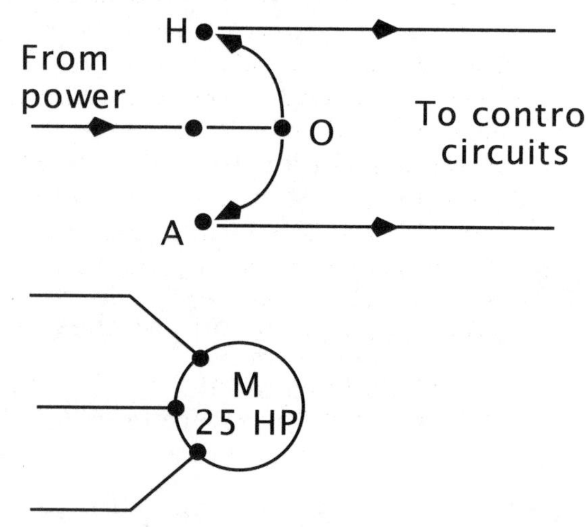

(Hand - Off - Automatic)

22.

Electric motor, 3-phase power, 25 horsepower

CHAPTER 19. INSTRUMENTATION AND CONTROL SYSTEMS

(Lesson 1 of 2 Lessons)

19.0 INSTRUMENTATION AND CONTROL SYSTEMS

19.00 Importance and Nature of Instrumentation and Control Systems

The plant process instrumentation and control system can include manual controls, remote controls, and central computer processing units in any combination. The instrumentation and control system may be computerized or manual: both types provide similar functionality in operating the plant. Today, most facilities are operated through computer control systems. Classic instrumentation *LOOP*[1] controllers are still used in new facilities for specific purposes and are in general use in older process facilities. The operator's interface with the control system is a small window into the system. In this chapter, you will learn instrumentation and control system concepts and practices that apply to computerized and manual control systems.

Instrumentation is used to sense process variables that can and should be measured. The treatment plant operator must have a good working familiarity with instrumentation and control systems to properly monitor and control the treatment processes. Operators' primary, and sometimes only, links to the process are the instruments and automatic controls in the treatment plant. The range of instrumentation and control system capabilities is extremely large and is beyond what you will cover in this chapter. Here, you will learn general instrumentation functions and control system capabilities. Once the treatment plant is built, the main processes will be permanently in place. However, by using instrumentation and controls, these processes can be manipulated to achieve and maintain maximum process efficiency. The operator's knowledge and use of instrumentation and control will benefit plant operations.

Your knowledge about instrumentation and control systems can only enhance the effective and efficient operation of your facility. Specifically, if you can recognize that instrumentation and control equipment is not operating properly, your process decisions can be based on that knowledge rather than blind faith. This is especially true for computer automation systems, which tend to be trusted implicitly. The automated instrumentation and control is your main link to the process. Some systems provide backup controls to continue critical operation when the computer control system is out of service. Familiarize yourself with these backup controls and how to operate them for your facility. Instrumentation and control systems have self-diagnostic capabilities that will inform the operator of failures and malfunctions. Again, you will want to familiarize yourself with the control systems to understand the capabilities of their diagnostic features.

Basic process variable measurement functions and concepts are the same, whether applied as a single-loop instrumentation and control system or a fully automated computer control system. Even the most modern plants have many instruments that require the operator to perform some minor preventive maintenance. The operator who understands how to recognize failures, troubleshoot, and make adjustments and minor repairs will be more comfortable in operating the plant without the need to call out maintenance personnel. Thus, the more you know about your plant's instrumentation and control systems, the better operator you become.

19.01 Importance to the Water Treatment Operator

Instrumentation and control systems are essential components in treatment facility operation, providing the operator the ability to monitor and manipulate plant processes remotely.

In a very real sense, measurement instruments can be considered extensions of and improvements to your senses of vision, touch, hearing, and even smell. Not only can instruments provide continuous and simultaneous monitoring of the many process variables throughout the plant, they do so in a more precise and consistent manner than the human senses. In addition, instruments often provide a permanent record of measurements taken. The automated control systems, in turn, provide operators with far-reaching and powerful hands to manipulate switches, valves, motors, and pumps in specific ways. In effect, the control and instrumentation system provides you with a staff of obedient and hard-working assistants, always on the job to help you operate your process more easily and efficiently. To appreciate these great advantages of automation, consider what a treatment plant would be like without modern controls. In the past, and even in some treatment plants today, the situation described in the following paragraph is the standard operating procedure for some operations.

The plant experiences a power failure in the main and backup electrical systems supplying the main automated control system.

1. *Control Loop.* The combination of one or more interconnected instrumentation devices that are arranged to measure, display, and control a process variable. Also called a loop.

However, other plant power and process equipment continue operating. The equipment failure is such that the operator cannot repair or reset to restore power. As the operator, you must try to keep the plant on-line manually by attempting to control all process variables without any of the normal data indications given by the control system. Flows and levels have to be estimated visually, valves turned manually, and pumps started and stopped by hand. Additionally, you are forced to observe and try to regulate other process variables using only your eyes, ears, sense of touch, and probably, in time, even your sense of smell.

Think that it's impossible to operate a plant under such conditions? It has been done in old or poorly maintained plants, but even if you could do it at your plant temporarily, you could not exercise close process control or maintain the plant under such conditions for any extended period. Operating a larger or sophisticated plant would be impossible without its process instrumentation and controls, even with a sizeable crew. Your plant emergency systems and operator knowledge and skills are the most important skills to have in such a situation.

The plant instrumentation and control system is designed with safeguards that nearly eliminate such failures. However, infrequent failures do occur, even in the best-designed instrumentation and control system. Operators who develop their knowledge of the benefits and limitations of the instrumentation and control systems will have enhanced skills during such upset conditions of operation.

QUESTIONS

Please write your answers to the following questions and compare them with those on page 435.

19.0A Why must a treatment plant operator have a good working familiarity with instrumentation and control systems?

19.0B What kind of information is presented in this chapter with regard to instrumentation?

19.0C Measurement instruments can serve as an extension of and improvement to which of the operator's physical senses?

19.02 Nature of the Measurement Process

Our senses provide us with qualitative (for example, color, sound, odor) and relatively short-term quantitative (for example, brighter, louder, stronger) information; instrumentation measurements provide us with exact quantitative data. That is to say, measurements give us numbers. Without this objective quantification of our environment (that is, the numbers we use to describe it), we could accomplish very little—just as primitive tribes with no real number systems have no technology worthy of the term. Their processing of natural materials is very limited because the information they use consists only of the relative terms "more than" or "less than." Without objective measurements, they have little control over their environment. Modern water treatment plants, on the other hand, depend absolutely upon accurate and reproducible measurements of chemical and physical processes, which is what good instrumentation provides.

A measurement is, by definition, the comparison of a quantity with a standard unit of measure. Thus, a tape measure is marked in feet or meters, standard units of length; clocks are marked in hours, minutes, and seconds as the standard units of time; and a small container may have fluid ounces marked on its side. These units, usually of a convenient magnitude (size), have been agreed upon internationally to serve as the accepted standard units. The primary standard for mass (weight) is the weight of an actual physical object, which is located in Paris, France; a standard kilogram is defined as the weight of this one specific object. The primary standards for time are defined in wavelengths of light, and the primary standard for length is the wavelength of a certain spectrum line of the element cesium. Other primary standards can be set up for certain quantities—for example, solution strengths—but generally, industrial measurements depend on secondary standards that are based on more or less exact comparisons to a primary standard. A machinist's caliper is calibrated against precision measuring blocks, which are such secondary standards.

Of course, all the measurements used in water and wastewater treatment ultimately refer back to a few primary and secondary standards. Measurements of length, and the related calculations for area and volume, are in effect comparisons to the standard foot (meter). Weights of chemicals are traceable to the standard pound (kilogram). Timepieces are designed around the standard second. All other measurements are derived from the fundamental units of measure. These include flow rate (volume per unit of time), pressure (weight per unit of area), chemical concentration/dosage (weight of chemical per unit of liquid volume), and chemical feed (weight of chemical per unit of time).

Some important terms directly relating to the measurement process also need to be defined: a *PROCESS VARIABLE* is a physical or chemical quantity, such as flow rate or pH, that is measured or controlled in a water treatment process. *ACCURACY* refers to how closely an instrument measures the true or actual value of a process variable. Accuracy is usually expressed as within plus or minus a given small percent of the true value, for example, ±2 percent is typical instrument accuracy of measurement. Accuracy depends partly upon the *PRECISION* of an instrument, which amounts to how closely the device can reproduce a given reading (or measurement) time after time (a good example of a precise but inaccurate instrument is a precision electrical gauge—with a bent needle). *SENSITIVITY* refers to the smallest change in a process variable that an instrument can sense. It is usually expressed as a percent of the full-scale value of the instrument. Typical values are ±0.1 percent of full scale for a high-sensitivity instrument and ±2 percent of full scale for an average-sensitivity instrument. *CALIBRATION* is the complete test of an instrument by measuring several (secondary) standards and its complete adjustment (if required) to read the standard values at several points in its range. *STANDARDIZATION* is an abbreviated calibration where only one standard is measured and the instrument is adjusted to read the proper value, such as standardizing a pH meter with pH 7 buffer. The *RANGE* of an instrument is the spread between the minimum and maximum values it is designed to measure, most often from zero to a maximum value, as with, say, a 0–100 psi pressure gauge. And the *EFFECTIVE RANGE* is that portion of an instrument's complete range within which the instrument has acceptable accuracy, commonly between 10 percent and 90+ percent of its design range. *SPAN* is a closely related term to express that procedure in an instrument's calibration that adjusts it to read both low and high standards accurately; an improperly spanned 0–100 psi pressure gauge might read 10 psi, but then indicate a true 80 psi as 70 psi. *LINEARITY* is the term to describe how well a device tracks a series of standards throughout its effective range—the preceding poorly spanned gauge is off in linearity by −12.5 percent at 80 percent of scale (off 10 psi at 80 psi). It is evident that to be truly accurate, an instrument must have acceptable precision, be in calibration, and be designed with an effective range that is wide enough to measure the range of the treatment plant's process variables.

Finally, two other common terms should be defined. *ANALOG* displays show a reading as a pointer (or other cursor) against a marked scale such as simple pressure or level gauges. Numeric *DIGITAL*[2] displays provide direct, numerical readouts, as with most modern instrumentation. The best example of both is a watch that has both readouts; the hands give a quick analog reference and the digital time display gives a more precise reading. Analog process variable values may be transmitted by a continuously variable 4–20 mA (milliamp) signal or a digital-data signal (not to be confused with a numeric digital display). Figure 19.1 illustrates the difference between the connections required for each. Computer readouts are analog input signals that have been encoded as digital data. The *OPERATOR INTERFACE*[3] on the computer graphic may be programmed to display the value as an analog or digital readout.

An instrument loop encompasses all devices from the process sensing element to the indicators on a control panel or control system computer screen. Typically, loop components are connected together by a 4–20 mA signal or digital-data link (Figure 19.2).

QUESTIONS

Please write your answers to the following questions and compare them with those on page 435.

19.0D What is the main difference between measurements by our senses and measurements by instruments?

19.0E What is the definition of a measurement?

19.0F Explain the difference between accuracy and precision.

19.0G What is the difference between analog and digital displays?

19.03 Explanation of Control Systems

Control systems are the means by which process variables such as pressure, level, weight, or flow are controlled. The terms *CONTROLLER* and *CONTROL SYSTEMS* are used to refer to two different types of instrumentation control systems: (1) modulating systems, which sense and control their own operation on a close, continuous basis; and (2) motor control stations, which control only the ON/OFF operation of motors and other devices.

19.030 Modulating Control Systems

The technical use of the terms controller and control system refers to those systems that provide modulating control of process variables. Examples of modulating control systems include chlorine residual analyzers/controllers, flow-paced (open-loop) chemical feeders, pressure- or flow-regulating valves, continuous level control of process basins, and variable-speed pumping systems for flow/level control.

In order for a process variable, whether pressure, level, weight, or flow, to be closely controlled, it must be measured precisely and continuously. In a modulating control system, the measuring device or primary element sends an electrical or pneumatic signal proportional to the value of the variable to the actual system controller.

Within the controller, the signal is compared to the set-point value. A difference between the actual and set-point values results in the controller sending a command signal to the controlled element, usually a valve, pump, or chemical feeder. Such an error signal produces an adjustment in the system that causes a corresponding change in the variable measured originally, making it more closely match the set point value. This continuous, fine-tuning of variables using process controls provides a way to maintain variables, such as flow rates, pressures, levels, or chemical feed rates, at constant values. The term applied to this circulating action of the variable in such a controller is *FEEDBACK*. The path through the control system is the control loop. A diagram of such a control loop, measuring the process variable of water flow, is shown in Figure 19.3.

2. *Digital.* The encoding of information that uses binary numbers (ones and zeros) for input, processing, transmission, storage, or display, rather than a continuous spectrum of values (an analog system) or non-numeric symbols such as letters or icons.
3. *Operator Interface.* The device at which the operator interacts with the control system. This may be an individual instrumentation and control device or the graphic screen of a computer control system. Also see HUMAN MACHINE INTERFACE (HMI) and MAN MACHINE INTERFACE (MMI).

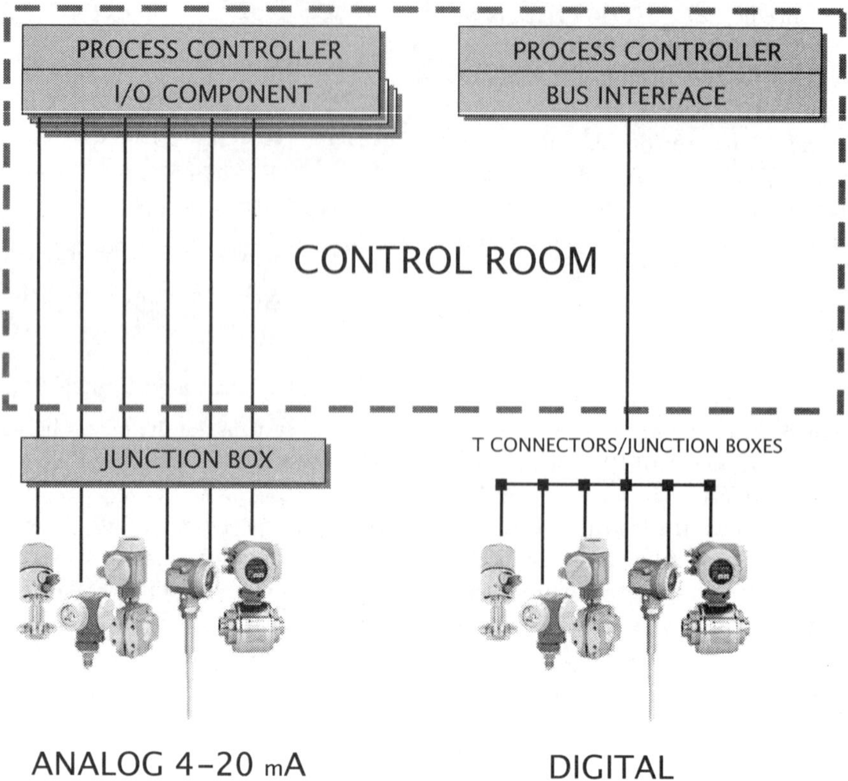

Fig. 19.1 Connections required for analog and digital transmissions
(Adapted from figure provided by Endress+Hauser)

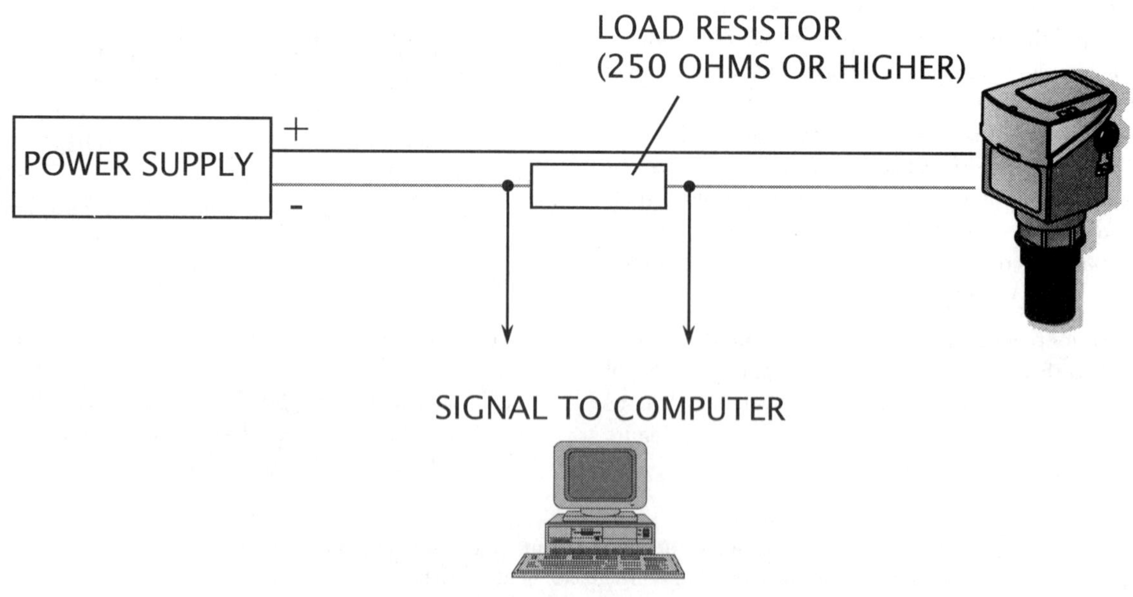

Fig. 19.2 Typical 4–20 mA instrument loop
(Adapted from figure provided by Endress+Hauser)

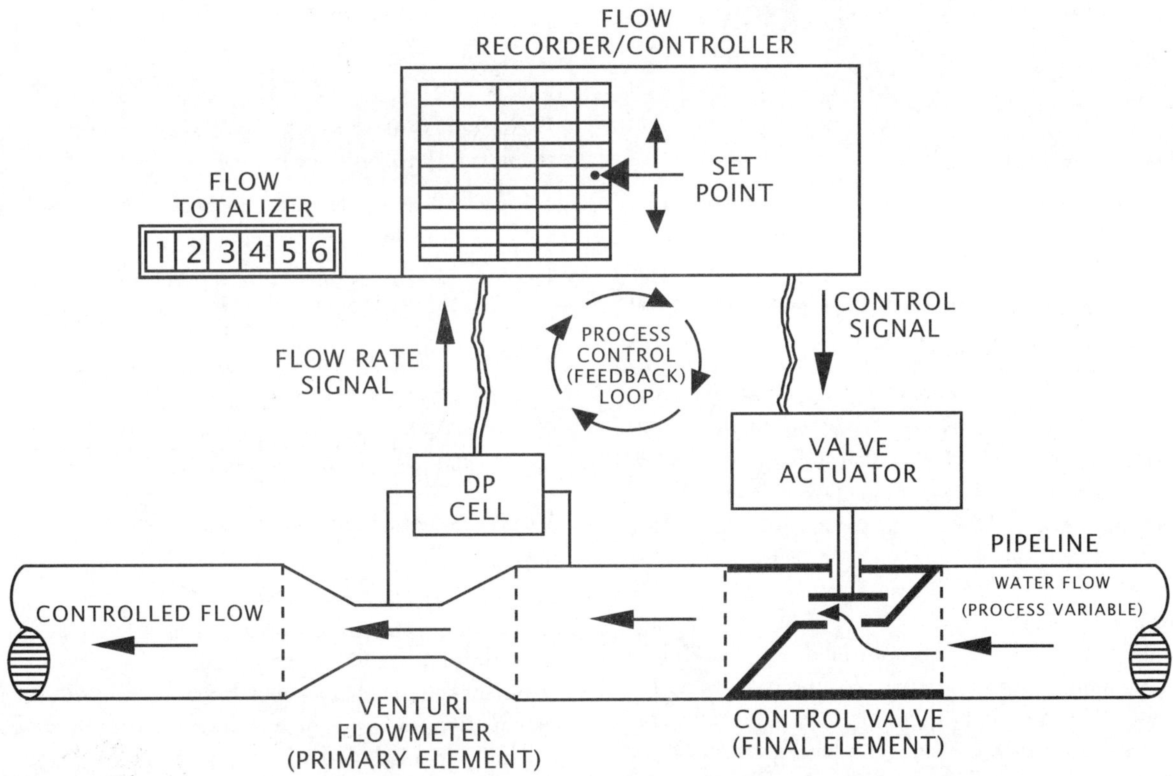

NOTE: Measurement and control system instruments may be electric or pneumatic.

Fig. 19.3 Automatic control system diagram (closed-loop modulating control)

The internal settings of the controller can be quite critical, since close control depends on sensitive adjustments. Thus, you should not try to adjust any such control system unless you know exactly what you are doing. Many plant and system operations have been drastically upset due to such efforts, however well intentioned, by unqualified personnel.

19.031 *Motor Control Stations*

A motor control station or panel (Figure 19.4) and the related circuitry (Figure 19.5) essentially provide only ON/OFF operation of an electric motor or other device. The electric motor or device, in turn, might power a pump, control valve, or chemical feeder. With this system, control could be manual or automatic, with a switch responding to a preset value of level, pressure, or some other variable. Motor control stations are typically made up of a standard electric power control panel with manual push buttons, overload relays, and Hand-Off-Automatic (HOA) or similar On-Off-Remote switch. Additionally, they may include, in good electrical *FAIL-SAFE*[4] design practice, provisions for power failure or loss-of-phase, and such protective devices as high- or low-pressure/temperature/level cutoff switches. For this type of panel to be considered a controller (within our secondary meaning of the term), its operation must be directly controlled by the value or values of some variable, not merely by a device such as an ON/OFF timer. In other words, it must be turned on and off as a result of a measurement of a level, pressure, flow, chemical concentration, or other variable as it reaches a predetermined setting or settings. In the automatic mode (A on the HOA switch), its operation thus is, in fact, automatic in the sense that the variable is automatically controlled, even though the limits of its value may be quite wide compared to those attainable with a modulating controller as previously described. While a basin level modulating controller may allow only an inch (centimeter) or so of water level change, an ON/OFF system might operate within a few feet (meters) of level difference. In many applications, however, such wide control is of no particular disadvantage, and sometimes is even desirable, as with a tank level where regular exchange of contents is desirable.

Terms used in control practice can now be defined operationally: feedback and control loop have been mentioned previously, however the term control loop needs some qualification. An open-loop control system controls one variable on the basis of another. A good example of this is a chlorinator paced by process flow signals (rather than by a chlorine residual analyzer). Closed-loop control remains as discussed previously, the true control system with measurement and control of the same variable (and feedback). Proportional-band, reset, and derivative actions are adjustments of a controller that relate to the effectiveness and speed of its control action. Offset is the difference between the desired value of the variable (the set point) and the controlled (actual) value. Lag time, common to chlorinator control systems, refers to the time period between the moment a process change is made and the moment such a change is finally sensed by the associated measuring instrument. Long lag times can result in unstable processes and poor control.

QUESTIONS

Please write your answers to the following questions and compare them with those on page 435.

19.0H What kinds of water treatment process variables do control systems control?

19.0I List three examples of modulating control systems.

19.0J What is the essential purpose of a motor control station or panel?

19.1 SAFETY HAZARDS OF INSTRUMENTATION AND CONTROL SYSTEMS

19.10 General Precautions

The general principles for safe performance on the job, summed up as always avoiding unsafe acts and correcting unsafe conditions, apply as much to instrumentation work as to other plant operations. There are several specific dangers associated with instrument systems that are worth repeating in this section for the sake of safe practice. These include electrical hazards, mechanical and pneumatic hazards, confined spaces, oxygen deficiency and enrichment, explosive gas mixtures, and falls and associated hazards.

19.11 Electrical Hazards

A hidden aspect of energized electrical equipment is that it looks normal; that is, there are no moving parts or other obvious signs that would discourage someone from touching it. In fact, there seems to be a peculiar fascination with seeing if circuit components are live (energized) by quickly touching them with a tool, often a screwdriver (hopefully with an insulated handle) but sometimes even with a finger. Only proper safety training (coupled with bad experience, at times) is effective in changing this behavior. Even so, many electricians' have tools with an "arc-mark trademark," evidence of the need for continuing

4. *Fail-Safe.* Design and operation of a process control system whereby failure of the power system or any component does not result in process failure or equipment damage.

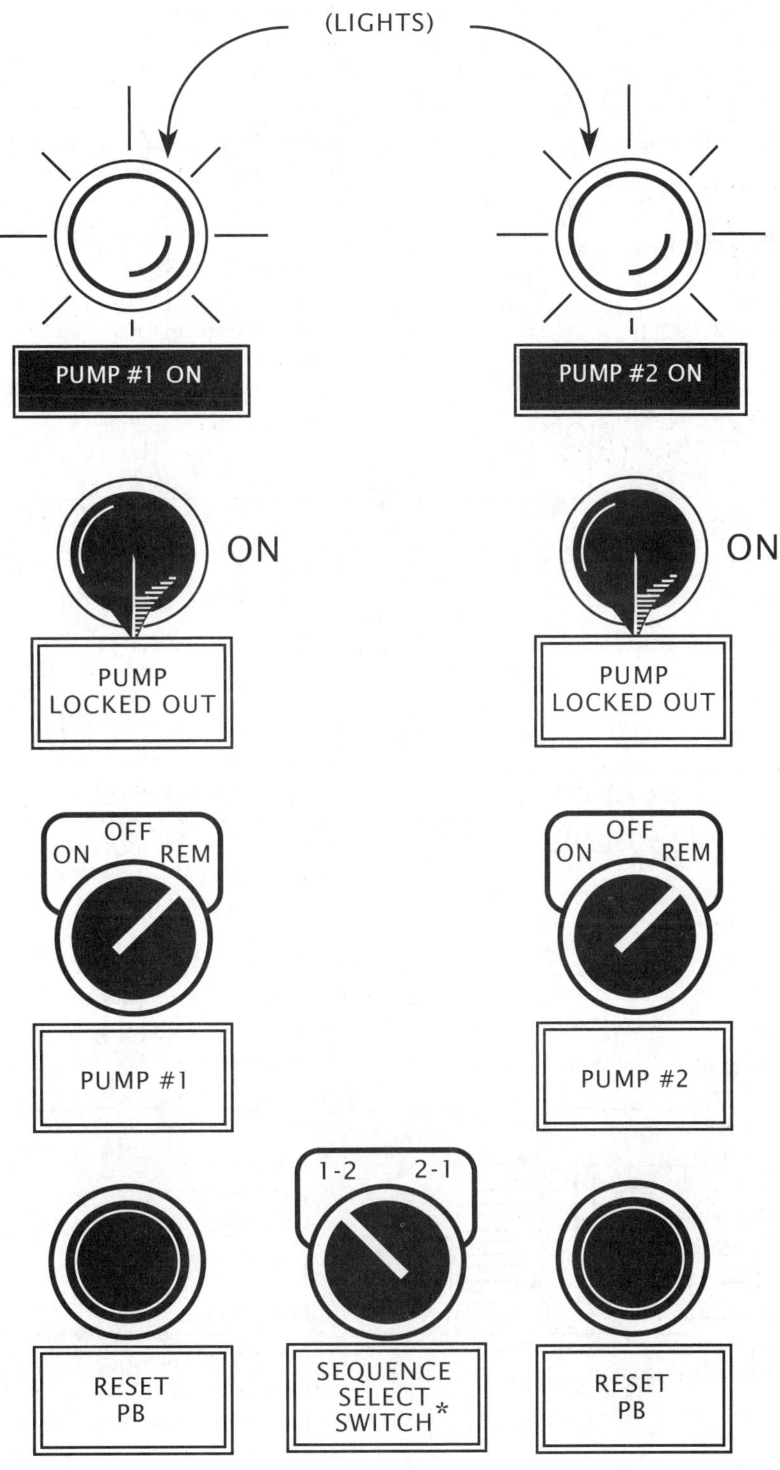

Note: "REM" on switch above means "remote," which is automatic operation in this application

*Determines which is "lead" (first to come on) and "lag" (next to come on) pump in automatic operation

Fig. 19.4 Typical duplex (two-pump) motor control panel

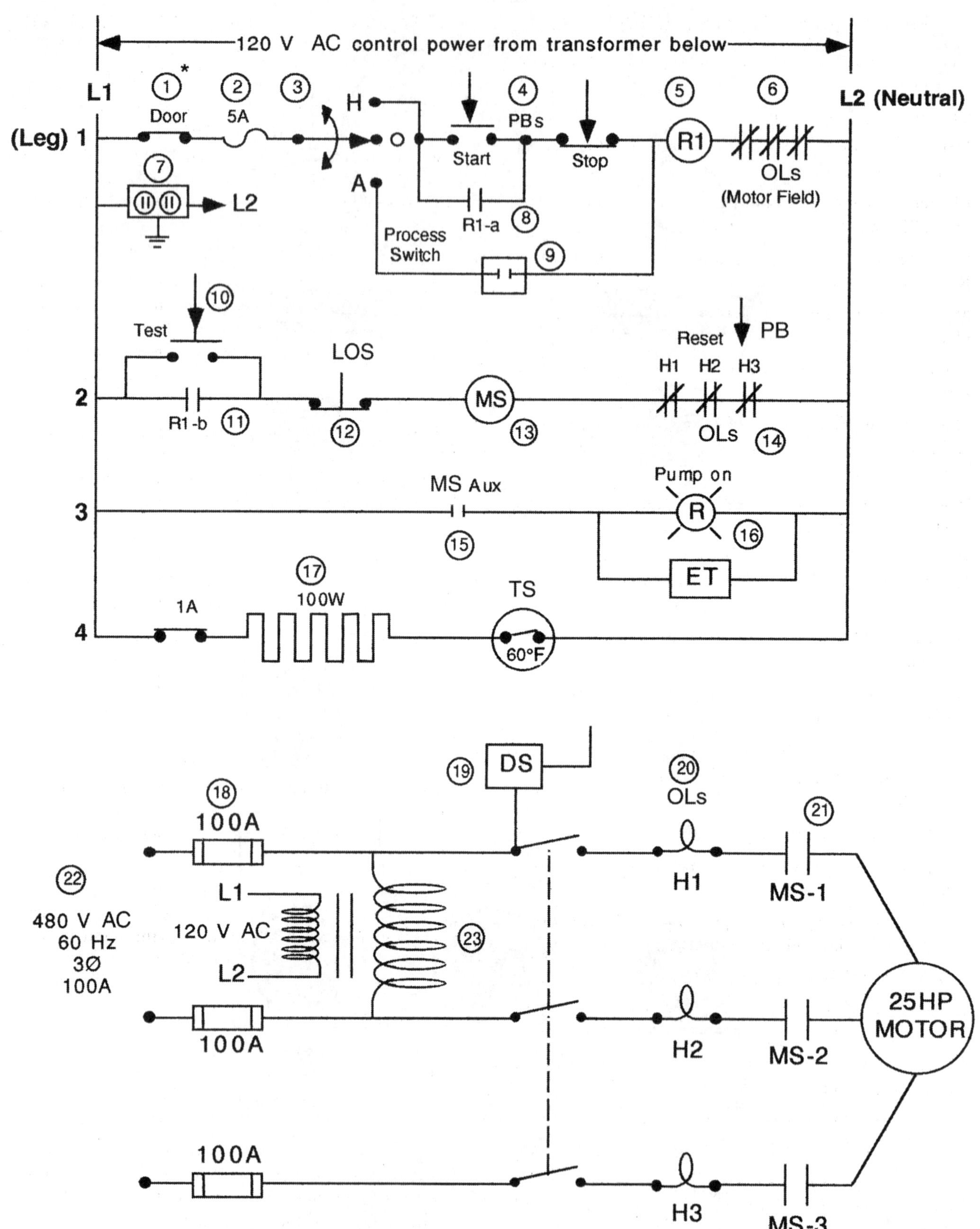

Fig. 19.5 Typical motor control elementary diagram circuitry (do not use for design) (simplified, self-explanatory ladder diagram) *(see page 395 for parts legend)*

FIGURE 19.5 PARTS LEGEND

1. Door or interlock switch to kill control power if panel door is opened.
2. Control circuit fuse, 5 amp.
3. Hand-Off-Automatic (HOA) function switch for manual or automatic operation (shown Off).
4. Momentary contact (spring-loaded) push buttons (Start-Stop), PB station.
5. Control relay coil (other relays, if used, would be shown as R2, R3, etc.)
6. Motor thermal protection elements embedded in three-phase motor field (stator) coils.
7. Duplex 120 V AC outlet within panel (for test equipment usage).
8. Holding contacts of control relay; start PB energizes relay coil to close contacts, which remain closed until coil is de-energized.
9. Process control contacts (ON/OFF control); for example, level, pressure, or analytical process variable.
10. Push-to-test motor starter relay circuit leg (if MS coil and OLs OK, pump will start); used for troubleshooting purposes.
11. Main control contacts of relay R1, energizes MS coil to start pump.
12. Lockout Stop switch to prevent motor operation while being serviced, key lockable in open position.
13. Motor Starter (Mag starter) relay coil.
14. Manually (push button) resettable overload relay contacts, opened by heaters (see 20).
15. Auxiliary contacts in MS relay.
16. Pump running light (red) and elapsed time (total hours run) meter.
17. Enclosure heater in panel (keeps components dry) with 1 amp circuit breaker and thermostat switch.
18. 100-amp 480 V AC cartridge-type fuses.
19. Main 480 V AC manual disconnect switch (Main switch) local at the motor.
20. Motor circuit thermal overload heaters (set for maximum load motor current).
21. Motor starter relay contactors (large contacts).
22. Motor power, 480 V AC, 60 cycle, three-phase, 100-amp service.
23. Control power transformer (for control circuit above), 120 volts AC.

DESCRIPTION OF TYPICAL MOTOR CONTROL CIRCUIT OPERATION

(Refer to Figure 19.5)

Control circuit is shown Off; the HOA function switch is at O. To start motor manually, place HOA in H and depress Start push button (PB) to energize main control relay R1. With power through coil, holding contacts R1-a close around the start PB to keep R1 energized (and motor running) when the operator's finger is removed from the button (such spring-loaded PBs are termed "momentary" contacts).

When R1 is energized, it also pulls-in (closes) contacts R1-b to energize motor starter relay MS in turn. With control power to the MS coil, its contactors MS-1, 2, and 3 make (close) to supply 480 V AC to the motor—assuming resettable thermal overloads H1, H2, and H3 are all closed (and the main disconnect is closed and 100A fuses are good).

To stop the motor manually, with the HOA still in the H position, one pushes the Stop PB to break the circuit to the R1 coil, allowing contacts R1-a to fall out (open) so the control relay is de-energized (even after this momentary open PB is released).

For automatic operation, HOA switch is placed in A so motor will start and stop with the open and close, respectively, of the process controller contacts. Note that the PBs cannot start or stop the motor with HOA switch in the Auto position.

The Test PB, in Leg 1 of the schematic, will start stopped pump as long as it is depressed (being a momentary close PB), by bypassing the control contacts R1-b. This feature permits isolation of a no-start problem to Leg 1 circuitry, in that the motor will start if Leg 2 is OK.

In Leg 3, the motor-on light and elapsed-time meter are turned on when the MS relay is energized, through its auxiliary contacts (integral to the relay). Leg 4 consists of a small fused and thermostated space heater to keep all electrical panel components dry.

120 V AC control power for the control circuit (L1 and L2) is transformed from the main 480 V AC 3Ø motor power service. A main disconnect switch kills all power to the motor, for servicing, and all three phases are protected with 100-amp cartridge fuses.

This typical motor control circuitry has two types of motor overheating protection:

1. In-line heater elements H1, H2, and H3, which open any or all of their respective contacts above when the motor draws higher currents for a longer time than it is designed for. These interrupt the power to the MS relay (Leg 2).
2. Small, heat-sensitive contacts embedded in all three stator motor windings, which open at a predetermined maximum temperature to interrupt the control circuit in Leg 1, stopping the motor before it gets too hot.

Only the former (overload heaters) are manually resettable using the PB shown; the latter (stator-imbedded) must be allowed to cool before they remake contact.

self-discipline in this area. Though such mention may conjure up humorous images of the maintenance person's surprise and shock upon such an incident, one only need consider that electric shock can and does regularly cause serious burns and even death (by asphyxiation due to paralysis of the muscles used in breathing) to bring the problem into perspective. Also, the expense and effort caused by a needless shorting-out of an electrical device could be significant. The point is, resist the urge to test any electrical device with a hand tool or any part of your body.

If you have any doubt about whether all sources of voltage to a device have been switched off or disconnected (not merely the local switch), then do not touch it, except possibly with the insulated probes of a test meter. Remember, you cannot see even the highest voltage, and assuming that a circuit is dead can be hazardous or even deadly.

Do not simulate an electrical action, for example by pressing down a relay armature, within an electrical panel without a clear understanding of the circuitry. Your action may cause an electrical explosion to shower you with molten metal or cause another type of injury. Remember, when it comes to electricity: When in doubt, do not touch.

Usually, a plant operator does not have the test equipment or the technical knowledge to correct an electrical malfunction, other than possibly resetting a circuit breaker, regardless of how critical the device's function is to plant operations. In addition to the shock hazards of motor control centers, there also may be a shock hazard within measurement instrument cases. Expensive instrument components can easily be damaged or destroyed when a foolhardy operator uses the tool-touch-test method.

Most panels have an *INTERLOCK* on the door that interrupts all (local) power to a panel or device when the door is opened or the circuit is exposed for service. Never disconnect or disable interlocks that interrupt control power. Warning labels, insulating covers (over hot terminals), safety switches, lockouts, and other safety provisions on electrical equipment must remain functional at all times. Your attention to this crucial aspect of your workplace may save a life, and, as the saying goes, it could be your own.

Operators often use hand power tools around electrical equipment. All such power tools (even double-insulated ones) can present a shock hazard, treatment plants being damp places, at times. Power tools can be a mechanical hazard as well. For your own safety, use power tools only when you can have an observer on hand in case of an accident. Never stand in water with a power tool, even when it is turned off. Brace yourself, if necessary, in such a way that electric current cannot flow from arm to arm in case of a faulty power tool. Shocks through the upper body could involve your heart or your head, whose importance to you is self-evident.

QUESTIONS

Please write your answers to the following questions and compare them with those on page 435.

19.1A What are the general principles for safe performance on the job?

19.1B How can electric shock cause death?

19.1C What could happen to you as a result of an electrical explosion?

19.1D Why should you brace yourself when operating power equipment so that electric current cannot flow from arm to arm in case of a faulty tool?

19.12 Mechanical and Pneumatic Hazards

There exists a special danger when working around powered mechanical equipment, such as electric motors, valve operators, and chemical feeders that are operated remotely or by an automatic control system. Directly stated, the machinery may suddenly start or move when you are not expecting it. Most devices are powered by motors with enough torque or rpm to severely injure anyone in contact with a moving part. Even when the exposed rotating or meshing elements are fitted with guards in compliance with safety regulations, a danger may exist. A motor started remotely may catch a shirt sleeve, finger, or tool hanging near a loose or poorly fitted shaft or gear train guard.

The sudden, automatic operation of equipment, even when half-expected, may startle a nearby operator into a fall or slip. Signs indicating that "This Equipment May Start At Any Time" tend to be ignored after a while. Accordingly, you must stay alert to the fact that any automatic device may begin to operate at any time. Thus, you must stay well clear of active automatic equipment, especially when it is not operating.

Lockout devices on electric switches must be respected at all times. The electrician who attaches a lockout device to physically prevent the operation of an electric circuit is, in effect, trusting his or her life and health to the device. Once the lockout device is attached to the switch (whether the switch is tagged off or actually locked with lock and key), the electrician will consider the circuit and its connected equipment deenergized and safe and will feel free to work on it. Consider the potential consequences, then, of a careless operator who removes a lockout to place needed equipment back into service, presuming the electrician is finished (as might occur after several hours). The point cannot be overstressed:

> Respect all lockout devices and all tagged-off equipment as if a life is entrusted to you ... it may well be.

PNEUMATIC HAZARDS

Working with and around pneumatic instrumentation presents an additional hazard associated with high-pressure air. Pneumatic devices are powered by air at supply pressures from about 50 to 100 psi (350 to 700 kPa). Serious injury can be caused by the high air (and trapped particle) velocities that can be produced. A pressure regulator normally reduces the high air pressures to a safer 30 psi (210 kPa) or so for each device, but even these lower pressures can cause injury if directed toward the eye or other delicate tissues.

Before disconnecting any air line, put on your safety goggles (not just glasses, but wraparound goggles). Then valve the line off, from both directions ideally, and crack (open slightly) a fitting on the pipe or tubing to permit the pressurized air to bleed out slowly. If it is necessary to purge a line of moisture after inspection or repair, do so through the filter trap valve (pointing down, if standard) on each of a supply line's connected components. If a very dirty, oily, or moist line must be purged with supply air, set up a temporary purge line to a low area and tape a sock or closed rag to the open end to trap particles and minimize dirt pickup by the exhausting air. Never direct a pressurized air stream toward any part of your body or anyone else's body. High-velocity air can easily penetrate the body's tissues, even the skin of the hands. The tiny particles that are always present, even in filtered air (picked up from the inside of piping), can enter the eye's delicate tissues and cause an irritation at the least, or corneal damage and infection at worst. Breathing oil-laden air (many compressors seal with oil) from extensive line and filter purging can cause chemical pneumonia or even bacterial lung infections in sensitive individuals. Wearing a simple painter's (filter) mask will keep the suspended oil droplets and other particulates out of your respiratory system.

Though it is recognized that plant operators do little, if any, corrective maintenance around instrumentation and control components, it seems there is always some exposure to electrical, mechanical, and pneumatic hazards. Operators often perform preventive maintenance tasks and even minor repairs on many devices. Therefore, always play it smart and safe around all instrumentation.

19.13 Confined Spaces

Many measurement and control systems include remotely installed sensors and control valves. Quite often, these are found in vaults or other closed concrete structures. This section outlines procedures for preventing personal exposure to *DANGEROUS AIR CONTAMINATION*[5] or oxygen deficiency/enrichment when working within such spaces. If you enter confined spaces, you must develop and implement written, understandable procedures in compliance with OSHA standards and you must provide training in the use of these procedures for all persons whose duties may involve confined space entry. The procedures presented here are intended as guidelines. Exact procedures for work in confined spaces may vary with different agencies and geographical locations and must be confirmed with the appropriate regulatory safety agency.

A confined space may be defined as any space that (1) is large enough and so configured that an employee can bodily enter and perform assigned work; (2) has limited or restricted means for entry or exit (for example, manholes, tanks, vessels, silos, storage bins, hoppers, vaults, and pits are spaces that may have limited means of entry); and (3) is not designed for continuous employee occupancy. One easy way to identify a confined space is by whether or not you can enter it by simply walking while standing fully upright. In general, if you must duck, crawl, climb, or squeeze into the space, it is considered a confined space.

A major concern in confined spaces is whether the existing ventilation is capable of removing dangerous air contamination or oxygen deficiency/enrichment that may exist or develop. In water treatment, we are concerned primarily with oxygen deficiency (less than 19.5 percent oxygen by volume), oxygen enrichment (greater than 23.5 percent oxygen by volume), methane (explosive), and hydrogen sulfide (toxic).

5. *Dangerous Air Contamination.* An atmosphere presenting a threat of causing death, injury, acute illness, or disablement due to the presence of flammable or explosive, toxic, or otherwise injurious or incapacitating substances.
   (1) Dangerous air contamination due to the flammability of a gas, vapor, or mist is defined as an atmosphere containing the gas, vapor, or mist at a concentration greater than 10 percent of its lower explosive (lower flammable) limit (LEL).
   (2) Dangerous air contamination due to a combustible particulate is defined as a concentration that meets or exceeds the particulate's lower explosive limit (LEL).
   (3) Dangerous air contamination due to the toxicity of a substance is defined as the atmospheric concentration that could result in employee exposure in excess of the substance's permissible exposure limit (PEL).
   NOTE: A dangerous situation also occurs when the oxygen level is less than 19.5 percent by volume (OXYGEN DEFICIENCY) or more than 23.5 percent by volume (OXYGEN ENRICHMENT).

The potential for buildup of toxic or explosive gas mixtures or oxygen deficiency/enrichment exists in all confined spaces. The atmosphere must be checked with reliable, calibrated instruments before every entry. When testing the atmosphere, first test for oxygen deficiency/enrichment, then combustible gases and vapors, and then toxic gases and vapors. The oxygen concentration in normal breathing air is 20.9 percent. The atmosphere in the confined space must not fall below 19.5 percent or exceed 23.5 percent oxygen. Engineering controls are required to prevent low or high oxygen levels. However, personal protective equipment is necessary if engineering controls are not possible. In atmospheres where the oxygen content is less than 19.5 percent, supplied air or a self-contained breathing apparatus (SCBA) is required. SCBAs are sometimes referred to as scuba gear because they look and work much like the air tanks used by divers.

Entry into confined spaces is never permitted until the space has been properly ventilated using specially designed forced-air ventilators. These blowers force all the existing air out of the space, replacing it with fresh air from outside. This crucial step must always be taken even if atmospheric monitoring instruments show the atmosphere to be safe. Because some of the gases likely to be encountered in a confined space are combustible or explosive, the blowers must be specially designed so that the blower itself will not create a source of ignition that could cause an explosion.

There are two general classifications of confined spaces: (1) non-permit confined spaces, and (2) permit-required confined spaces (permit spaces).

A non-permit confined space is a confined space that does not contain or, with respect to atmospheric hazards, have the potential to contain any hazard capable of causing death or serious physical harm. The following steps are recommended before entry into a confined space:

1. Ensure that all employees involved in confined space work have been effectively trained.
2. Identify and close off or reroute any lines that may carry harmful substance(s) to, or through, the work area.
3. Empty, flush, or purge the space of any harmful substance(s) to the extent possible.
4. Monitor the atmosphere at the work site and within the space to determine if dangerous air contamination or oxygen deficiency/enrichment exists.
5. Record the atmospheric test results and keep them at the site throughout the work period.
6. If the space is interconnected with another space, each space must be tested and the most hazardous conditions found must govern subsequent steps for entry into the space.
7. If an atmospheric hazard is noted, use portable blowers to further ventilate the area; retest the atmosphere after a suitable period of time. Do not place the blowers inside the confined space.
8. If the only hazard posed by the space is an actual or potential hazardous atmosphere and the preliminary ventilation has eliminated the atmospheric hazard or continuous forced ventilation alone can maintain the space safe for entry, entry into the area may proceed.

A permit-required confined space (permit space) is a confined space that has one or more of the following characteristics:

1. Contains or has the potential to contain a hazardous atmosphere
2. Contains a material that has the potential for engulfing an entrant
3. Has an internal configuration such that an entrant could be trapped or asphyxiated by inwardly converging walls or by a floor that slopes downward and tapers to a smaller cross section
4. Contains any other recognized serious safety or health hazard

OSHA regulations require that a confined space entry permit be completed for each permit-required confined space entry (Figure 19.6). The permit must be renewed each time the space is left and reentered, even if only for a break or lunch, or to go get a tool. The confined space entry permit is "an authorization and approval in writing that specifies the location and type of work to be done, certifies that all existing hazards have been evaluated by a competent person, and that necessary protective measures have been taken to ensure the safety of each worker." A competent person, in this case, is a person designated in writing as capable, either through education or specialized training, of anticipating, recognizing, and evaluating employee exposure to hazardous substances or other unsafe conditions in a confined space. This person is authorized to specify control procedures and protective actions necessary to ensure worker safety.

The following procedures must be observed before entry into a permit-required confined space:

1. Ensure that personnel are effectively trained.
2. If the confined space has both side and top openings, enter through the side opening if it is within 3½ feet (1.1 meters) of the bottom.
3. Wear appropriate, approved, respiratory protective equipment.
4. Ensure that written operating and rescue procedures are at the entry site.
5. Wear an approved harness with an attached line. The free end of the line must be secured outside the entry point.
6. Test for atmospheric hazards as often as necessary to determine that acceptable entry conditions are being maintained.
7. Station at least one person to stand by on the outside of the confined space and at least one additional person within sight or call of the standby person.
8. Maintain effective communication between the standby person and the entry person.
9. The standby person, equipped with appropriate respiratory protection, should only enter the confined space in case of emergency.

Confined Space Pre-Entry Checklist/Confined Space Entry Permit

Date and Time Issued: ____________ Date and Time Expires: ____________ Job Site/Space I.D.: ____________

Job Supervisor: ____________ Equipment to be worked on: ____________ Work to be performed: ____________

Standby personnel: ____________ ____________ ____________

1. Atmospheric Checks: Time ______ Oxygen ______ % Toxic ______ ppm
   Explosive ______ % LEL Carbon Monoxide ______ ppm

2. Tester's signature: ____________

3. Source isolation: (No Entry)

	N/A	Yes	No
Pumps or lines blinded, disconnected, or blocked	( )	( )	( )

4. Ventilation Modification:

	N/A	Yes	No
Mechanical	( )	( )	( )
Natural ventilation only	( )	( )	( )

5. Atmospheric check after isolation and ventilation: Time ______

   Oxygen ______ $\% > 19.5\%$ $< 23.5\%$ Toxic ______ ppm < 10 ppm H_2S
   Explosive ______ % LEL $< 10\%$ Carbon Monoxide ______ ppm < 35 ppm CO

Tester's signature: ____________

6. Communication procedures: ____________
7. Rescue procedures: ____________
   ____

8. Entry, standby, and backup persons

	Yes	No
Successfully completed required training?	( )	( )
Is training current?	( )	( )

9. Equipment:

	N/A	Yes	No
Direct reading gas monitor tested	( )	( )	( )
Safety harnesses and lifelines for entry and standby persons	( )	( )	( )
Hoisting equipment	( )	( )	( )
Powered communications	( )	( )	( )
SCBAs for entry and standby persons	( )	( )	( )
Protective clothing	( )	( )	( )
All electric equipment listed for Class I, Division I, Groups A, B, C, and D, and nonsparking tools	( )	( )	( )

10. Periodic atmospheric tests:

Oxygen:	____% Time ____;	____% Time ____;	____% Time ____;	____% Time ____;
Explosive:	____% Time ____;	____% Time ____;	____% Time ____;	____% Time ____;
Toxic:	____ppm Time ____;	____ppm Time ____;	____ppm Time ____;	____ppm Time ____;
Carbon Monoxide:	____ppm Time ____;	____ppm Time ____;	____ppm Time ____;	____ppm Time ____;

We have reviewed the work authorized by this permit and the information contained herein. Written instructions and safety procedures have been received and are understood. Entry cannot be approved if any brackets ( ) are marked in the "No" column. This permit is not valid unless all appropriate items are completed.

Permit Prepared By: (Supervisor) ____________ Approved By: (Unit Supervisor) ____________

Reviewed By: (CS Operations Personnel) ____________
(Entrant) (Attendant) (Entry Supervisor)

This permit to be kept at job site. Return job site copy to Safety Office following job completion.

Fig. 19.6 Confined space pre-entry checklist/confined space entry permit

10. If the entry is made through a top opening, use a hoisting device with a harness that suspends a person in an upright position. A mechanical device must be available to retrieve personnel from vertical spaces more than 5 feet (1.5 meters) deep.
11. If the space contains, or is likely to develop, flammable or explosive atmospheric conditions, do not use any tools or equipment (including electrical) that may provide a source of ignition.
12. Wear appropriate protective clothing when entering a confined space that contains corrosive substances or other substances harmful to the skin.
13. At least one person trained in first aid and cardiopulmonary resuscitation (CPR) should be immediately available during any confined space job.

    Individuals designated to provide first aid or CPR should be included in a bloodborne pathogens (BBP) program. These employees may be exposed to contact with blood or other potentially infectious materials from the performance of their duties. The BBP program includes training in exposure potential determination, engineering and work practice controls, personal protective equipment (PPE), and the availability of the hepatitis B vaccination series (29 CFR 1910.1030). If the operator(s) must enter confined spaces to perform rescue services, they must be trained specifically to perform the assigned rescue duty and to use required PPE and rescue equipment. Rescue practice sessions must be held at least once every 12 months.

If you arrange to have a contractor perform work in confined spaces at your facility or within your collection system, you must inform the contractor:

- That the contractor must comply with confined space regulations
- Of hazards that you have identified and your experience with the space(s)
- Of precautions or procedures you have implemented for the protection of employees in or near the space where the contractor's personnel will be working
- That a debriefing must occur at the conclusion of the entry operations regarding the confined space program followed and any hazards encountered or created during the entry operations

To enhance safety, communications, and coordination of confined space activities, the contractor is also required to obtain available information from you and to inform you of the confined space program the contractor will follow. This exchange of information must occur before the confined space is entered by any operator.

Confined space work can present serious hazards if you are uninformed or untrained. The procedures presented are only guidelines and exact requirements for confined space work for your locale may vary. Contact your local regulatory safety agency for specific requirements.

19.14 Oxygen Deficiency or Enrichment

Low oxygen levels may exist in any poorly ventilated, low-lying structure where gases such as hydrogen sulfide, gasoline vapor, carbon dioxide, or chlorine may be produced or may accumulate. Oxygen in a concentration above 23.5 percent (oxygen enrichment) also can be dangerous because it speeds up combustion.

Oxygen deficiency is most likely to occur when structures or channels are installed below grade (ground level). Several gases (including hydrogen sulfide and chlorine) have a tendency to collect in low places because they are heavier than air. The specific gravity of a gas indicates its weight as compared to an equal volume of air. Since air has a specific gravity of exactly 1.0, any gas with a specific gravity greater than 1.0 may sink to low-lying areas and displace the air from that area or structure. (On the other hand, methane may rise out of a manhole because it has a specific gravity of less than 1.0, which means that it is lighter than air.) You should never rely solely on the specific gravity of a gas to tell you where it is. Air movement or temperature differences within a confined space may affect the location of atmospheric hazards. The only effective way of ensuring safe atmospheric conditions before entering a confined space is to test the atmosphere with an appropriate monitor(s) at various levels and locations throughout the space.

When oxygen deficiency or enrichment is discovered, the area should be ventilated with fans or blowers and checked again for oxygen deficiency/enrichment before anyone enters the area to work. Ventilation may be provided by fans or blowers. Follow confined space procedures before entering and during occupancy of any suspect area. Always get air into the confined space before you enter to work and maintain the ventilation until you have left the space. Equipment is available to measure oxygen concentration as well as toxic and combustible atmospheric conditions. You must use this equipment whenever you encounter a potential confined space situation. Ask your local safety regulatory agency or water association about sources of this type of equipment in your area.

19.15 Explosive Gas Mixtures

Explosive gas mixtures may develop in many areas of a treatment facility from mixtures of air and methane, natural gas, manufactured fuel gas, hydrogen, or gasoline vapors. The upper explosive limit (UEL) and lower explosive limit (LEL) indicate the range of concentrations at which combustible or explosive gases will ignite when an ignition source is present at ambient temperature. No explosion or ignition occurs when the concentration is outside these ranges. Gas concentrations below the LEL are too lean to ignite; there is not enough flammable gas or vapor to support combustion. Gas concentrations higher than the UEL are too rich to ignite; there is too much flammable gas or vapor and not enough oxygen to support combustion (Figure 19.7).

Explosive gas ranges can be measured by using a combustible gas detector calibrated for the gas of concern. Do not rely on your nose to detect gases. The sense of smell is absolutely unreliable for evaluating the presence of dangerous gases. Some gases

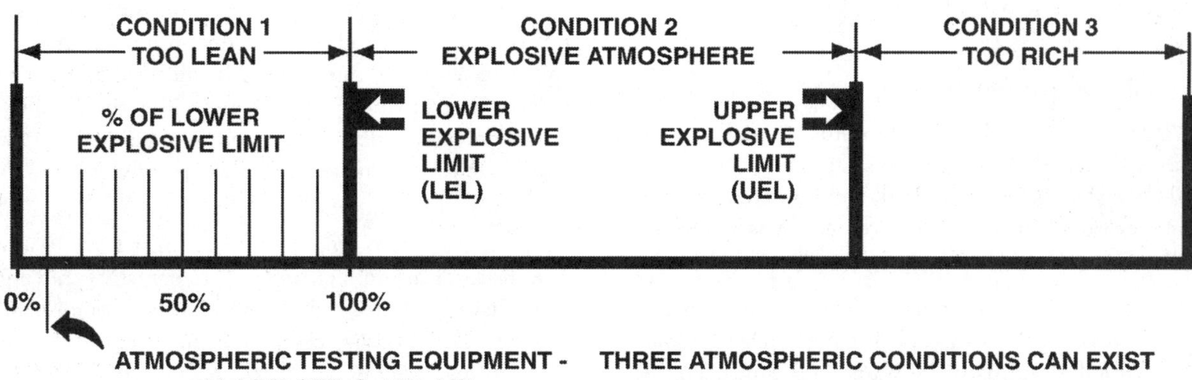

Fig. 19.7 Relationship between the lower explosive limit (LEL) and the upper explosive limit (UEL) of a mixture of air and gas

have no smell and hydrogen sulfide can paralyze the sense of smell.

Avoid explosions by eliminating all sources of ignition in areas potentially capable of developing explosive mixtures. Only explosion-proof electrical equipment and fixtures should be used in these areas (inflow/screen rooms, gas compressor areas, battery charging stations). Provide adequate ventilation in all areas that have the potential to develop an explosive atmosphere.

The National Fire Protection Association Standard 820 (NFPA 820), *Fire Protection in Wastewater Treatment and Collection Facilities*, lists requirements for electrical classifications, ventilation, gas detection, and fire control methods in various wastewater treatment and collection system areas. (For ordering information, see Section 19.5, "Additional Reading," item 3.) This information is also applicable to water treatment facilities. Comparing these requirements with your plant's existing design and equipment may indicate deficiencies, which should be remedied to minimize potential hazards.

19.16 Falls and Associated Hazards

Falls from or into tanks, wet wells, catwalks, or conveyors can be disabling or deadly. Most of these can be avoided by the proper use of ladders, hand tools, and safety equipment, and by following established safety procedures. Strains and sprains are probably the most frequent injuries in water treatment facilities. Keep in mind that an electric shock of even minor intensity can result in a serious fall. When working above ground on a ladder, use the proper, nonconductive type of ladder (fiberglass rails) and position it safely. Even an alert, safety-conscious operator could be knocked off balance by a slight electric shock. When required to do preventive maintenance from a ladder, turn off the power to the equipment being serviced, if at all possible. If this is not feasible, take special care to stay out of contact with any component inside the enclosure of an operating mechanism, and well away from terminal strips, unconduited wiring, and electrical black boxes. Though not commonly considered essential, wearing thin rubber or plastic gloves can greatly reduce your chances of electric shock (whether on a ladder or off).

Make provisions for carrying tools or other required objects on an electrician's belt rather than in your hands when climbing up or down ladders. Finally, never leave tools or any object on a step or platform of the ladder when you climb down, even temporarily. You might be the one upon whom they fall if the ladder is moved or even steadied from below. In this regard, it is always a good idea (even if not required) for preventive maintenance personnel to wear a hard hat whenever working on or near equipment, especially when a ladder must be used.

QUESTIONS

Please write your answers to the following questions and compare them with those on page 435.

19.1E Why should operators be especially careful when working around powered mechanical equipment?

19.1F What is the purpose of an electrical lockout device?

19.1G When testing the atmosphere before entry in any confined space, what procedure should be used?

19.1H What do the initials UEL and LEL stand for?

19.1I What kind of specific protective clothing could be worn to protect you from electric shock?

19.2 MEASURED VARIABLES AND TYPES OF SENSORS

19.20 General Principles of Sensors

A *MEASURED VARIABLE* is any factor that is sensed and quantified by the *PRIMARY ELEMENT* or sensor. In water treatment, the variables of pressure, level, and flow are the most common ones measured; at times, chemical feed rates and some physical or chemical process characteristics are also sensed.

The primary element performs the initial conversion of measurement energy to a signal. An example is a pressure gauge that converts the hydraulic action of the water into a mechanical motion to drive a meter indicator, or produces an electrical signal for a remote readout device. For transmitters not having a local readout, the sensing portion is the primary element. If such a signal is produced, be it electric or pneumatic, the sensor is also then considered a transmitter.

The signal produced is not necessarily a continuous one proportional to the variable (that is, an analog signal), but often merely a switch set to detect when the variable goes above or below preset limits. In this type of ON/OFF control, the predetermined setting is called a *DISCRETE*[6] point. This distinction between continuous and ON/OFF operation relates to the two types of controllers discussed previously.

Process measuring instruments have a common goal: to detect a variable and create an output for use by the control system or the operator. There are two basic types of instrument signals: continuously variable analog and discrete ON/OFF. Signals to and from a control device are termed as input/output (I/O). The signals can further be classified as analog input, analog output, *DISCRETE INPUT*, and *DISCRETE OUTPUT*.[7] Computer control system inputs are converted from the analog and discrete signals and coded into digital-data representations of the inputs. Digital-data output from the computer control system is converted to analog and discrete outputs.

19.21 Pressure Measurements

Since pressure is defined as a force per unit of area (pound per square inch, or kiloPascal), you might expect that sensing pressure would thus entail the small movement of some flexible element subjected to a force. In fact, that is how pressure is measured in practice. There are many classes and brands of sensors, but the most common types contain mechanically deformable components such as the Bourdon tube (Figures 19.8), bellows, or diaphragm arrangements (Figure 19.9). The slight motion each exhibits, directly proportional to the applied force, is then amplified mechanically by levers or gears to position a pointer on a scale or provide an input for an associated transmitter. A blind transmitter for pressure, or any variable, has no local readout.

Some pressure sensors are fitted with surge or overrange protection (snubbers) to limit the effect water hammer or pressure spikes have on the instrument. In most cases, such protection devices function by restricting flow into the sensing element. Surge protection devices thus prevent sudden pressure surges, which can easily damage most pressure sensors.

A snubber (Figure 19.10) consists of a restriction through which the pressure-producing fluid must flow. A more elaborate mechanical snubber responds to surges by moving a piston or plunger that effectively controls the size of an orifice. Some snubbers are subject to clogging or being adjusted so tight as to prevent any response at all to pressure changes. If a pressure sensor is not performing properly, look first for such clogging or adjustment that has become too restrictive.

Another dampening device is an air cushion chamber (Figure 19.11), which is simply constructed yet very effective. The top part of the chamber contains air; water flows into the bottom part. A sudden change in water pressure compresses the air within the chamber, taking the shock. The rate of response can be further dampened by also placing a snubber into the air chamber line.

QUESTIONS

Please write your answers to the following questions and compare them with those on page 435.

19.2A What is a primary element or sensor?

19.2B How is pressure measured?

19.2C Why are some pressure sensors fitted with surge or overrange protection?

19.22 Level Measurements

Systems for sensing the level of water or any other liquid, either continuously or at a single point, are very common in municipal and water treatment plants. Pumps are controlled, filters operated, basins and tanks filled and emptied, chemicals fed and ordered, sumps emptied, and many other variables controlled on the basis of liquid levels. Fortunately, level sensors usually are simple devices. A float, for example, can be a very reliable liquid level sensor both for single-point and continuous-level sensing. Other types of liquid level-sensing devices include displacers, electrical probes, direct hydrostatic pressure, pneumatic bubbler tubes, and ultrasonic (sonar) devices.

Single-point detection of level is very common for levels controlled by pumps or valves within fairly wide limits. Single-point ball float levels are very simple devices and have proven reliable in applications where set-point changes are rarely necessary. Single-point detectors mount in a sump or tank to sense a fixed elevation. When the water level rises, the ball floats and rolls onto its side, actuating an internal mercury switch. The switch

6. *Discrete Control.* ON/OFF control; one of the two output values is equal to zero.

7. *Discrete I/O (Input/Output).* A digital signal that senses or sends either ON or OFF signals. For example, a discrete input would sense the position of a switch; a discrete output would turn on a pump or light.

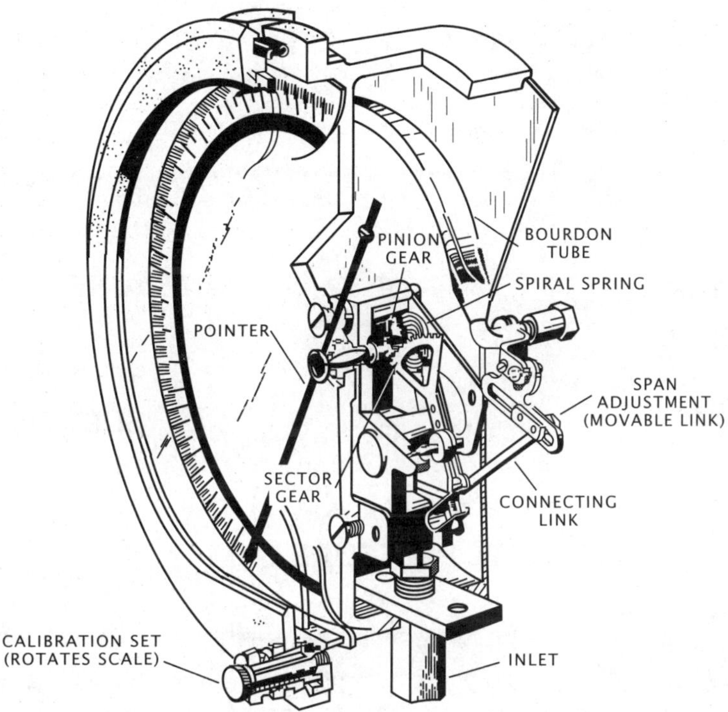

A precision industrial pressure gauge with a Bourdon-tube pressure element.

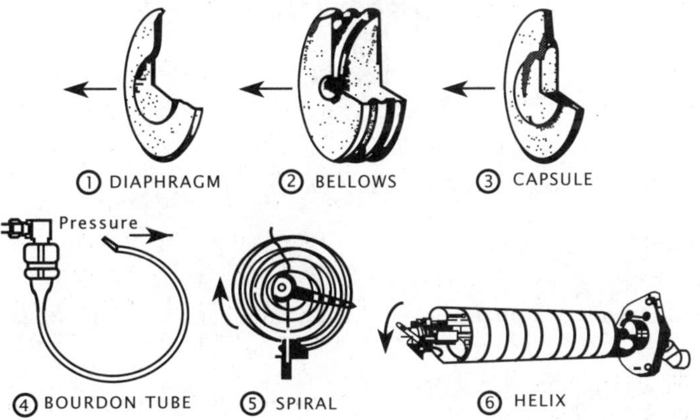

Elastic deformation elements. Pressure tends to expand or unroll elements to indicate pressure as shown by arrows.

Fig. 19.8 Bourdon tube and other pressure-sensing elements

(Permission of Heise Gauge, Dresser Industries)

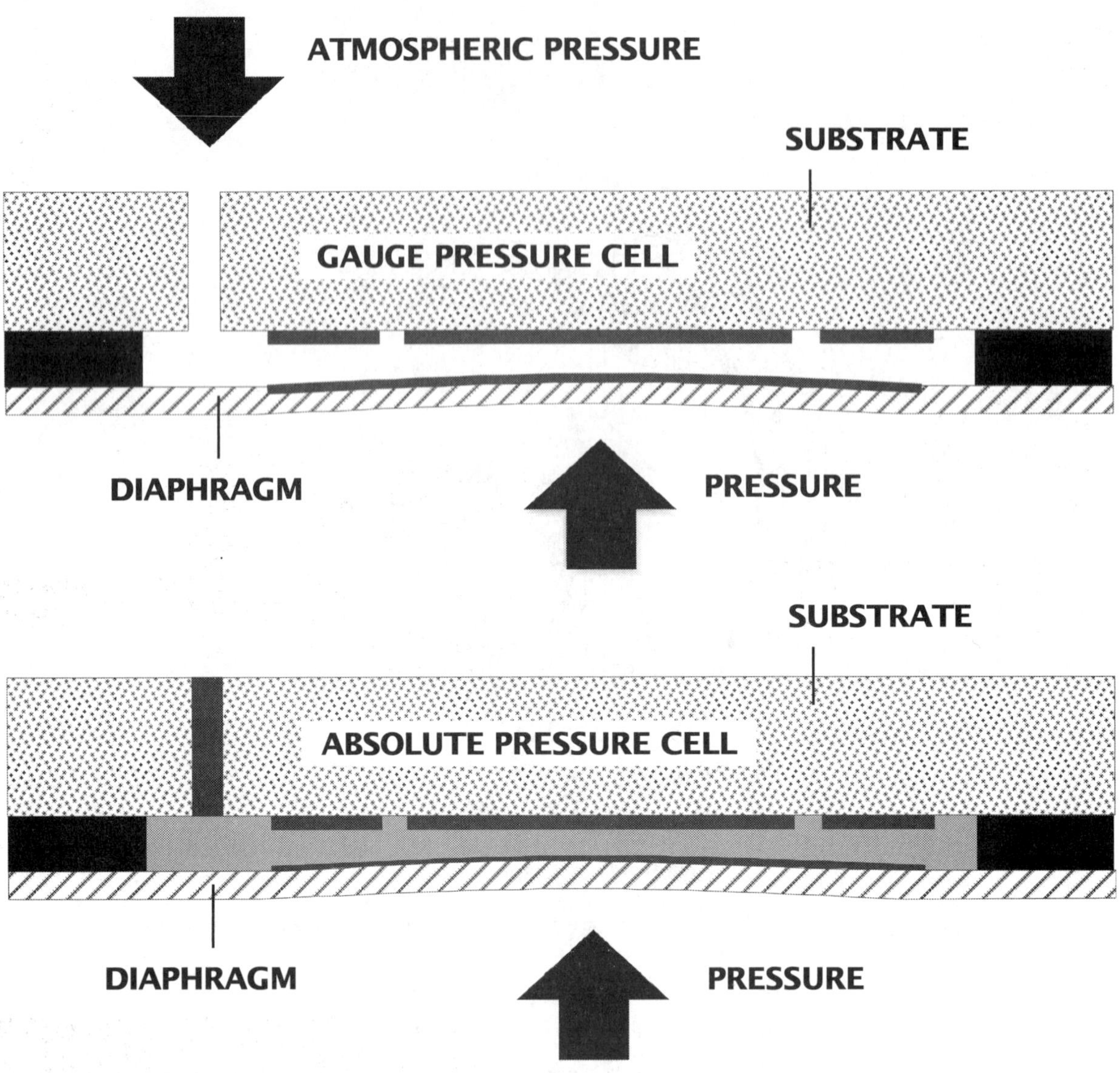

Fig. 19.9 Ceramic diaphragms to measure atmospheric pressure

provides a discrete ON/OFF signal used to control equipment or send level data to the control system.

An alternative to a float for single-point sensing, is the use of a probe to sense liquid level (Figure 19.12). Only single-point determinations can be made this way, though several probes or one with several sensors along its length can be set up to detect several different levels. Level probes are used where a mechanical system is impractical, such as within sealed or pressurized tanks or with chemically active liquids.

The probes can be small-diameter rods of electrically conductive material inserted into a tank through a fitting, usually through the top but at times in the side of a vessel (Figure 19.13). Each rod is cut to length corresponding to a specific liquid level in the case of the top-entering probes; in the side-entering setup, a short rod merely enters the vessel at the appropriate height or depth. One problem encountered with probes is the accumulation of scum or caking (such as grease) on the surface of the rods.

A number of methods are used to detect whether the probe is immersed in liquid. A simple method for conductive liquids is to apply a small voltage to the probe(s) by the system's power supply, with current flowing only when the probe just becomes immersed in the liquid. When immersion is detected, a switch activates a pump/valve control or alarm at as many places as necessary through a discrete output from the control system. Though at times only a single probe is used, with the metal tank completing the circuit as a ground, usually at least two probes are found. The ground probe extends to within a short distance of the bottom of the tank so as to be in constant contact with the electrically conducting liquid (a liquid ground, as it were). Nonconductive liquids must use other methods to detect immersion of the probe in the liquid.

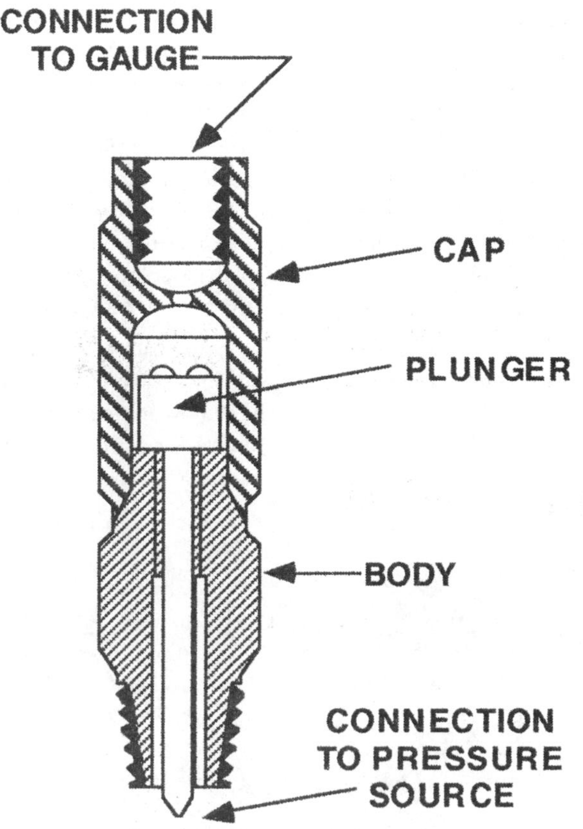

Fig. 19.10 Internal parts of a plunger-type snubber for surge protection

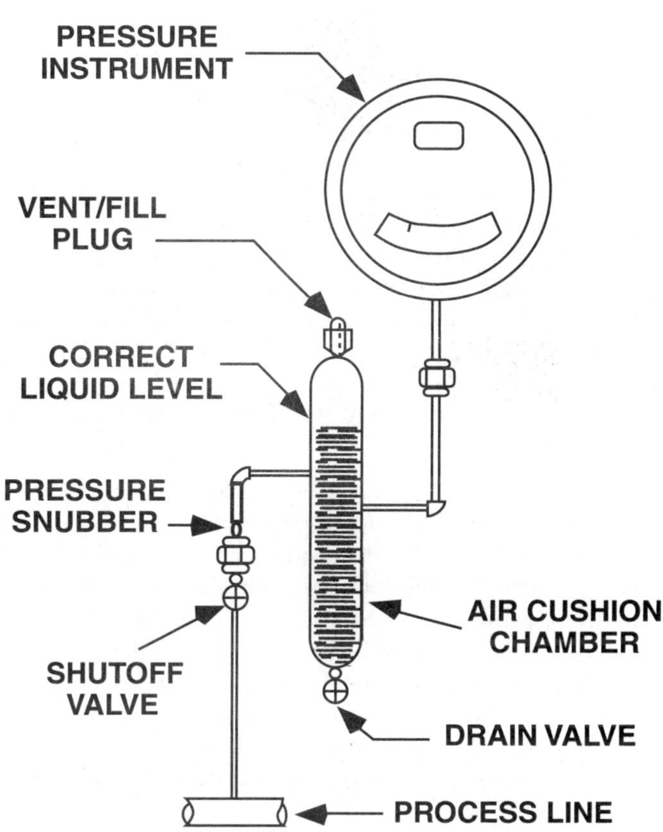

Fig. 19.11 Air cushion chamber and snubber for surge protection of pressure gauge

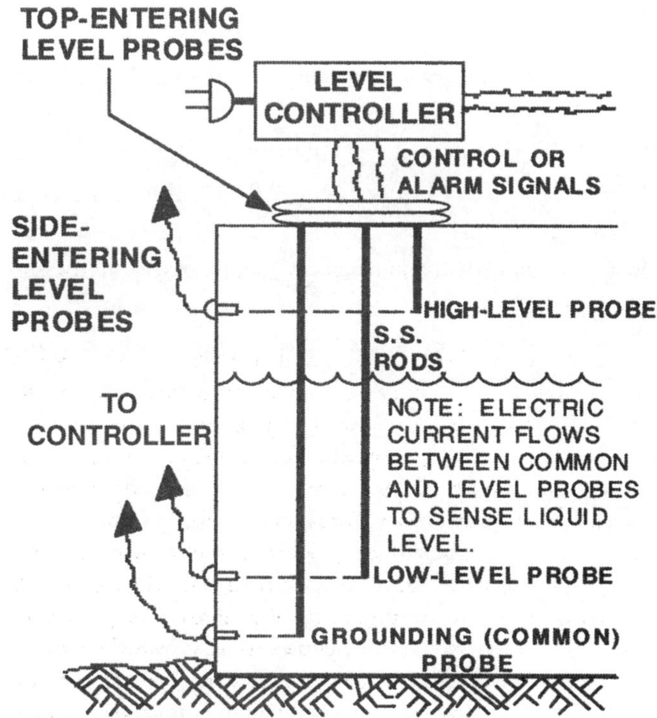

Fig. 19.12 Electrical probes (multi-point)

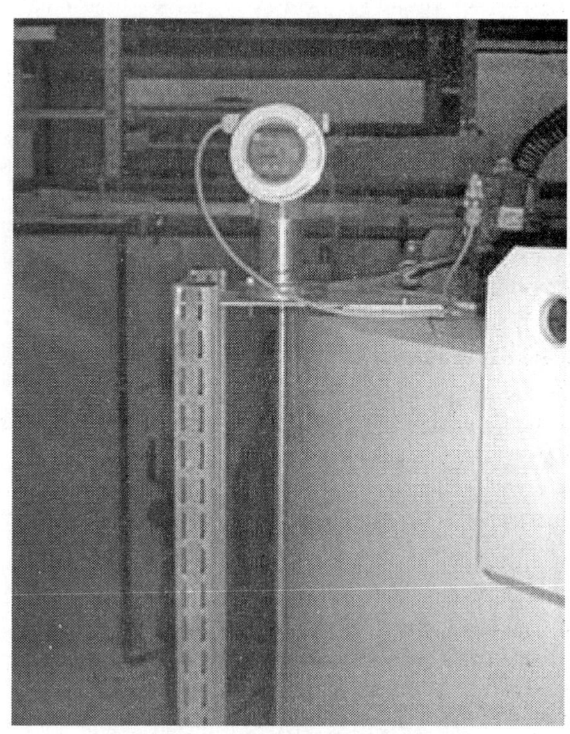

Fig. 19.13 Single-point level probe

(Courtesy Endress+Hauser)

Levels can be sensed continuously by measurement of liquid hydrostatic pressure near the bottom of a vessel or basin. The pressure elements used for level sensing must be quite sensitive to the low pressures created by liquid level (23 feet of water column equals only 10 psi, or 70 kPa). Therefore, simple off-the-shelf pressure gauges such as those found on pumps are not used to measure water levels. Instead, very sensitive water level sensors are used to measure levels of filter basins or in process tanks where control or monitoring must be close, continuous, and positive. Rather than being calibrated in units of pressure (psi or kPa), these gauges read directly in units of liquid level (feet or meters). One or more single-point control/alarm contacts can be made a part of this, or any, continuous type of level-sensing system.

Ultrasonic, microwave, and time domain reflectometry (TDR) level sensors have proven very reliable for a wide variety of applications. Typically, the instruments in this group detect level by emitting a pulse that is reflected back to the receiver. The sensor periodically generates a pulse of waves that bounce off the liquid surface and reflect back. The sensor detects the reflected pulse. Based on the speed of the pulse, the travel time between sending and receiving is measured and the distance is calculated and converted into a level measurement. The instruments have many display options: digital readout, analog 4–20 mA, discrete ON/OFF, and digital data. Any combination of one or more may be available on a single device. They are very versatile and reliable and can be used to detect levels of liquid and solid materials.

A very precise method of measuring liquid level is the bubbler tube with its associated pneumatic instrumentation (Figure 19.14). The pressure created by the liquid level is sensed, but not directly as with a liquid pressure element. A bubbler measures the level of a liquid by sensing the air pressure necessary to cause bubbles to just flow out the end of a submerged tube. Air pressure is created in a bubbler tube to just match the pressure applied by the liquid above the open end of the tube when it is immersed to a precisely determined depth in a tank or basin. The air pressure in the tube is then measured as proportional to the liquid level above the end of the tube. This indirect determination of level using air permits the placement of the instrumentation anywhere above or below the liquid's surface, whereas direct pressure-to-level gauges must be installed at or very close to the point where liquid pressure must be sensed.

Bubbler systems are adjusted so air just begins to bubble slowly out of the submerged end of the sensing tube. They automatically compensate for changes in liquid level by providing a small, constant flow of air through the bubbler tube by means of the constant air flow (also called constant differential) regulator. There is no advantage to turning up the amount of air to create more intense bubbling because the back pressure will still mainly depend on the water level. In fact, increasing the air flow may create a sizeable measuring error in the system, so any air flow changes should be left to qualified instrument service personnel.

Bubbler tube systems are common in basin-level controllers that must maintain water levels within a range of a few inches (centimeters) and for measurement of flow in open channels or over weirs. Usually, the level transmitter for basin-level controllers is blind since it only controls liquid level and need not provide a local readout of the level.

QUESTIONS

Please write your answers to the following questions and compare them with those on page 436.

19.2D List the different types of liquid level-sensing devices.

19.2E How does a single-point ball float level generate a level signal?

19.2F Under what circumstances are probes used to measure liquid level?

19.2G How does an ultrasonic sensor measure the level of a liquid?

19.23 Flow (Rate of Flow and Total Flow)

The two basic types of flow readings are rate of flow, such as MGD, ft^3/sec, gpm, m^3/sec, or liters/sec (volume per unit of time), or total flow, in simple units of volume, such as the corresponding million gallons, cubic feet, gallons, liters, or cubic meters. Total flow volumes are usually obtained as a running total, with a comparatively long time period for the flow delivery, such as a day or month. The flow instrument may provide both the rate of flow and total flow values. Flow instruments applied to computer control systems typically only provide flow rate; total

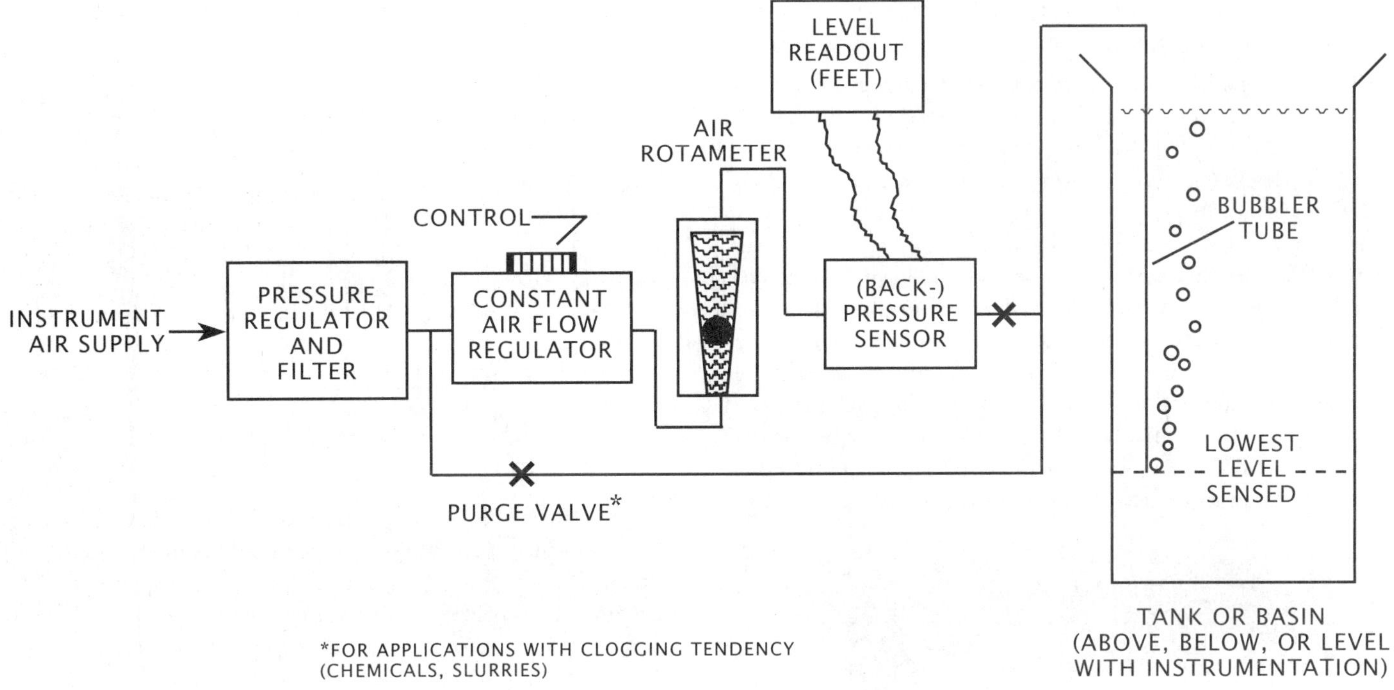

Constant-flow regulator and rotameter on left, back-pressure sensor (DP cell) to right.

Fig. 19.14 Diagram and photo of a bubbler tube system for measuring liquid level flow is calculated by accumulating the flow rate over some period, such as a day, month, or year.

While it is possible, in principle, to measure process flows directly, it is quite impractical. Direct measurement would involve the constant filling and emptying of, say, a gallon container with water flowing from a pipe on a timed basis. This method is obviously not practical. Therefore, sensing of process flows in waterworks practice is done inferentially, that is by inferring what the flow is from the observation of some associated hydraulic action or effect of the water. Figure 19.15 illustrates measuring level in a flume as a process flow instrument, inferring flow from the level measured. The inferential flow-sensing techniques that are used in flow measurement are velocity, differential pressure, magnetic, and ultrasonic. First, let us look at one device, the rotameter, used in flow sensing for some specialized applications before studying the devices for process stream flows.

Fig. 19.15 Inferring flow ultrasonically in a flume
(Courtesy Endress+Hauser)

ROTAMETERS (Figures 19.16 and 19.17) are transparent (usually) tubes with a tapered bore containing a ball or float. The float rises up within the tube to a point corresponding to a particular rate of flow. The rotameter tube is set against, or has etched upon it, marks calibrated in whatever flow rate unit is appropriate. Rotameters are used to indicate approximate liquid or gas flow. For example, the readout for a gas chlorinator could be a rotameter. Sometimes a simple rotameter is installed merely to indicate a flow or no-flow condition in a pipe, for example, on a chlorinator injector supply line.

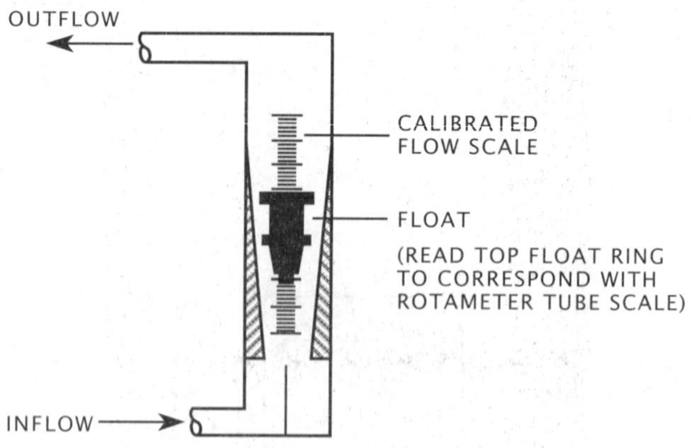

Fig. 19.16 Rotameter

For process flows, as mentioned, velocity-sensing meters measure water speed within a pipeline or channel. One way of doing this is by sensing the rate of rotation of a special impeller (Figure 19.18) placed within the flowing stream; the rate of flow is directly proportional to impeller rpm (within certain limits). Since normal water velocities in pipes and channels are under 10 feet per second (about 7 mph or 3 m/sec), the impeller turns rather slowly. This rotary motion drives a train of gears, which indicates rate of flow, in the same way a speedometer indicates the rate of travel for a car. Total flow then appears as the cumulative number of revolutions, denoting total volume flowed, like the odometer on your car's speedometer indicates total mileage traveled.

Rotation of the velocity-sensing element is not always transferred by gears, but may be picked up as magnetic or electric pulses by a transducer. Nor is velocity always sensed mechanically; it may also be detected or measured purely electrically (the thermistor type), or hydraulically (the Pitot tube). In each case, the principle of equating water velocity with rate of flow (within a constant flow area) is the same. Of course, all such flowmeters are calibrated to read out in an appropriate unit of flow rate, rather than velocity units.

Typically, a velocity-sensing element transmits its reading to a remote site as electric pulses, although other devices can be used to convert to any standard electrical or pneumatic signal. Figure 19.19 illustrates an electrical method of sensing velocity to measure flow.

Preventive maintenance of impeller-type flowmeters centers around regular lubrication of rotating parts, at least for the older types. Propeller meters, as they are called, have a long history of reliability and acceptable accuracy in both municipal and industrial applications. When propeller meters become old, they can become susceptible to underregistration (read low) due to bearing wear and gear train friction. Accordingly, annual teardown for inspection is indicated. Overregistration is rare, but a partially full pipeline, wrong gears installed, or a malfunctioning transmitter or receiver can cause high readings.

Rotameters (gas rate-of-flow)

Fig. 19.17 Flow-sensing devices for fluids

Differential pressure-sensing tubes (Figures 19.20 and 19.21), also called Venturi or differential meters (or flow tubes), depend for their operation upon a basic principle of hydraulics, the Bernoulli effect: When a liquid is forced to go faster in a pipe or channel, its internal pressure drops. If a carefully sized restriction is placed within the pipe or flow channel, the flowing water must speed up to get through it. In doing so, its pressure drops a little, and it drops an exact amount for a given flow rate. This small pressure drop, the "pressure differential," is the difference between the water pressure before the restriction and within the restriction. This difference is proportional in a certain way (but not directly proportional) to the rate of flow. The difference in pressure is measured very precisely by the instrumentation associated with the particular flow tube installed. Typically, a difference of only a few psi (kPa) is required. This small value of pressure difference is often described in inches (centimeters) of water (head).

Measuring flow by the differential pressure method removes a little hydraulic energy from the water. However, the modern flow tube, with its carefully tapered form, allows recovery of well over 95 percent of the original pressure throughout its range of flows. Other ways of constricting the flow, such as installation of orifice plates, do not allow such high recoveries of pressure or the accuracy possible with other modern flow tubes.

An orifice plate (Figures 19.20 and 19.21) is a steel plate with a precisely sized hole (orifice) in it. The plate is inserted between flanges in a pipe. The pressure drop is sensed right at the orifice, or immediately downstream, to yield a less accurate flow indication than a Venturi meter. This drop in pressure is not recovered; that is, a permanent pressure loss occurs with orifice plate installations, unlike Venturi flow tubes.

Differential devices require little, if any, preventive maintenance by the operator since there are no moving parts. Occasional

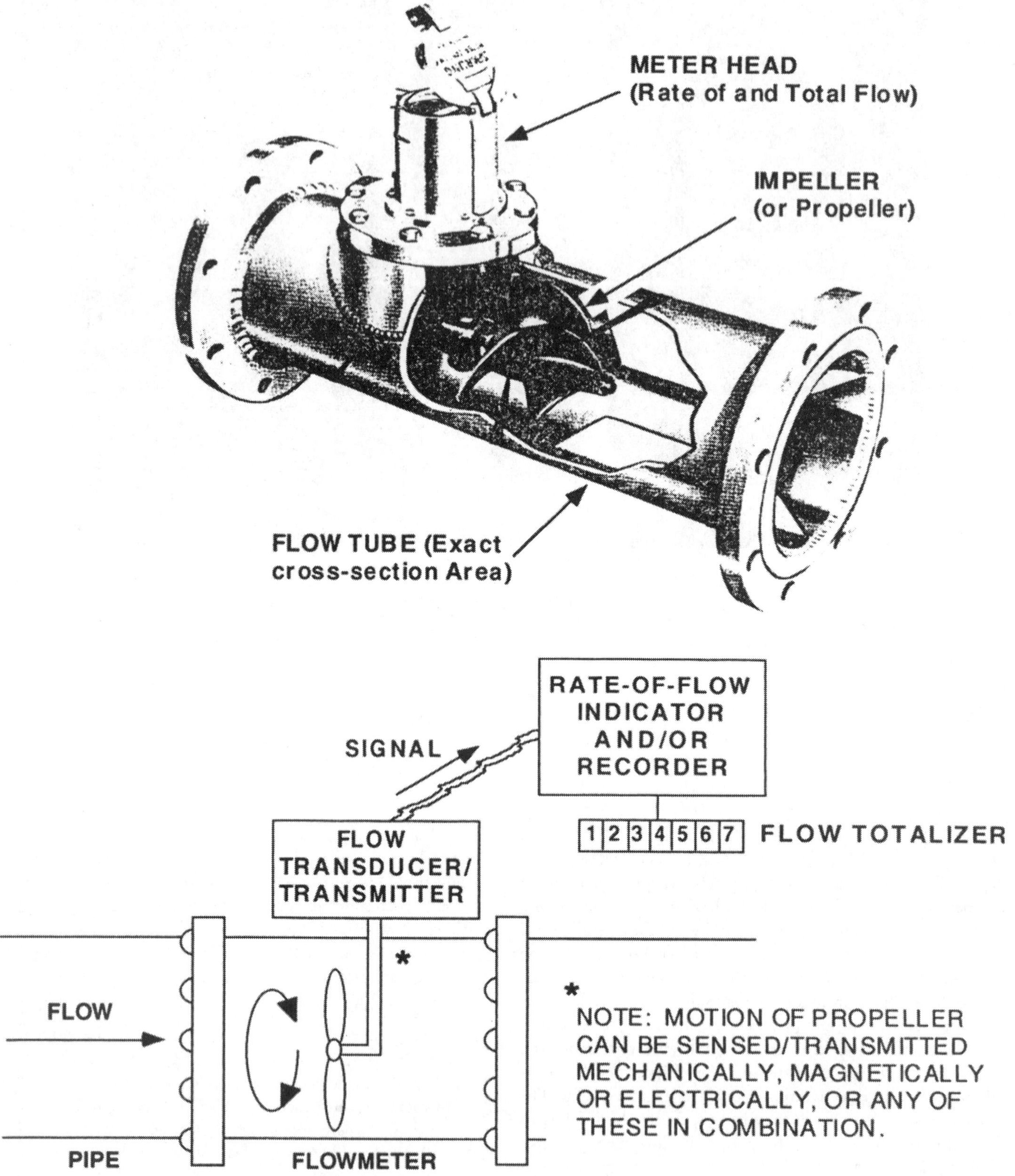

Fig. 19.18 Propeller meter (a type of velocity meter)

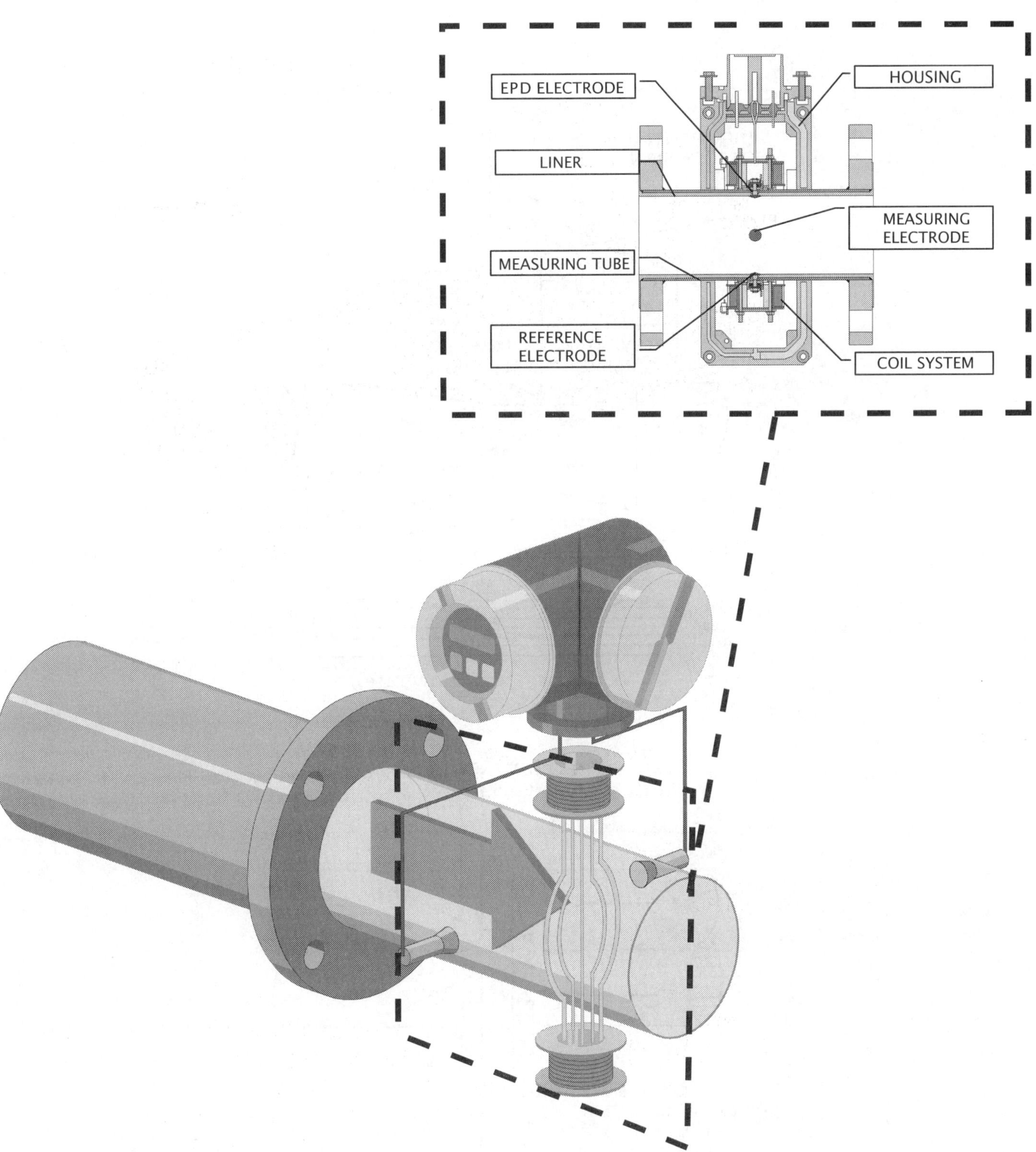

Fig. 19.19 Magnetic flowmeter

(Adapted from figure provided by Endress+Hauser)

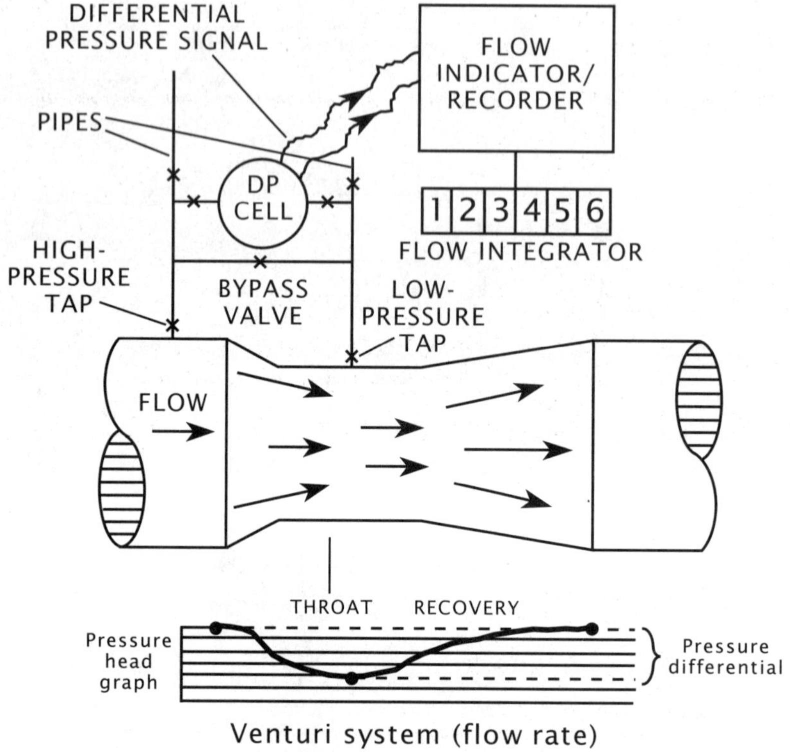

Venturi system (flow rate)

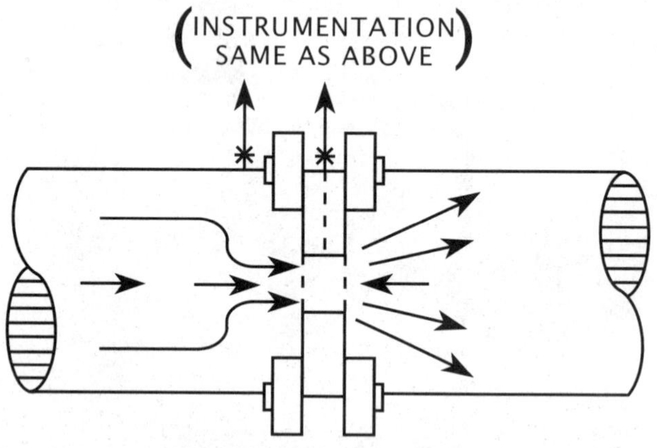

Orifice plate installation (flow rate)

Fig. 19.20 Schematic diagrams of differential pressure flow-measuring devices

54-inch Venturi tube

24-inch orifice plate

Fig. 19.21 Photos of differential pressure flow tubes

flushing of the hydraulic sensing lines is good practice. However, flushing should only be done by a qualified person. When dealing with an instrument sensitive to fractions of a psi or Pascals, opening the wrong valve can instantly damage the internal parts severely. Also, if an older differential pressure (DP) cell containing mercury is used, this toxic (and expensive) metal can easily be blown out of the device and into the process pipeline. Thus, all valve manipulations must be understood and done deliberately after careful planning by a qualified person.

In nearly all cases, the signals from larger flow tubes are transmitted to a remote readout station. Local readout is also provided (sometimes inside the case only) for purposes of calibration. Differential pressure transmitters may be electrical or pneumatic types. The signal transmitted is proportional to the square root of the differential pressure.

Venturi meters have been in use for many decades and can produce very close accuracies year after year. Older flow tubes are quite long physically (to yield maximum accuracy and pressure recovery). Newer units are much shorter but have very good pressure recovery and even better accuracy. With no moving parts, the Venturi-type meter is not subject to mechanical failure as is the comparable propeller meter. Flow tubes, however, must be kept internally clean and without obstructions upstream (and even downstream) to provide the designed accuracies.

The piping design for flowmeters must provide for adequate lengths of straight pipe runs upstream and downstream from the meter. Flowmeters in pipes will produce accurate flowmeter readings when the meter is located at least five pipe diameters distance downstream from any pipe bends, elbows, or valves and at least two pipe diameters distance upstream from any pipe bends, elbows, or valves. Flowmeters also should be calibrated in place to ensure accurate flow measurements.

In summary, all of the flowmeters described in this section provide rate of flow indication. The rate of flow can also be (and usually is) continuously totalized to give a reading of total flow past the measuring point. Total flow is usually indicated at the readout instrument in units of gallons or cubic feet (liters or cubic meters).

QUESTIONS

Please write your answers to the following questions and compare them with those on page 436.

19.2H What are two basic types of flow readings?

19.2I List the inferential flow-sensing techniques that are used in flow measurement.

19.2J How do velocity-sensing meters measure process flows?

19.2K Flows measured with Venturi meters take advantage of what hydraulic principle?

19.24 Chemical Feed Rate

Chemical feed rate indicators are often an integral part of a particular chemical feed system and thus are usually not considered instrumentation as such. For example, a dry feeder for lime may be provided with an indicator for feed rate in units of weight per time, such as pounds/hour or grams/minute. In a fluid (liquid or gas) feeder, the indication of quantity per unit of time, such as gallons/hour or pounds/day (liters/hour or kilograms/day), may be provided by use of a rotameter (Figure 19.16) or built-in calibrated pump with indicated output settings.

19.25 Process Instrumentation

Process instrumentation, by definition, provides for continuous analysis of physical or chemical indicators of process variables in a municipal or industrial plant. This does not include laboratory bench instruments (unless set up to measure sample water continuously), although the operating principles are usually quite similar. The process variables of turbidity and pH are often monitored closely in a water treatment plant (Figure 19.22). Very frequently, chlorine residuals are also continuously measured and controlled. These variables are usually measured at several locations. Additionally, other indicators of process status may be sensed on a continuous basis, such as electrical conductivity (total dissolved solids, TDS), oxidation-reduction potential (ORP), water alkalinity, and temperature. In every case, the instrumentation is specific as to operating principle, standardization procedures, preventive maintenance, and operational checks. The manufacturer's technical manual describes routine procedures to check and operate this sensitive type of equipment.

Operators must realize that most process instrumentation is quite delicate and thus requires careful handling and special training to service. No adjustments should be made without a true understanding of the specific device. Generally speaking, this category of instrumentation must be maintained by the plant's instrument specialist, or the factory representative, rather than by an operator (unless specially instructed).

19.26 Signal Transmitters/Transducers

Common system installations measure a variable at one location and provide a readout of the value at a remote location, such as a main control room. Except in the case of a blind transmitter, local indication is provided at the field site as well as being presented at the remote site.

In order to transmit a measured value to a remote location for readout, it is necessary to generate an analog or digital-data signal directly proportional to the value measured. This signal is then transmitted to a remote receiver, which provides a reading based on the signal. Also, a controller may use the signal to control the measured variable and a totalizer to integrate it.

Presently, two general systems for transmission of signals are used in most water treatment facilities: analog or digital data. Analog 4–20 mA transmission has a limited range, typically a few hundred feet or meters, and is being replaced by computer digital-data transmission. For transmission over longer distances, the transmitter output is converted to a digital-data representation of the process variable's measured value. The digital-data transmission simplifies the wired connections that are required for signal transmission.

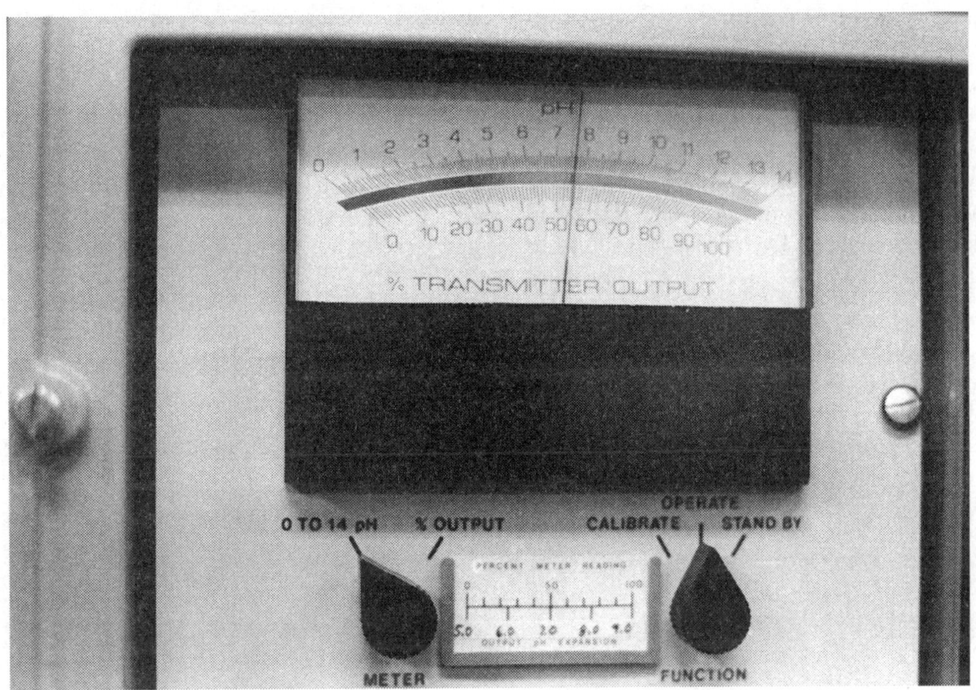

pH meter

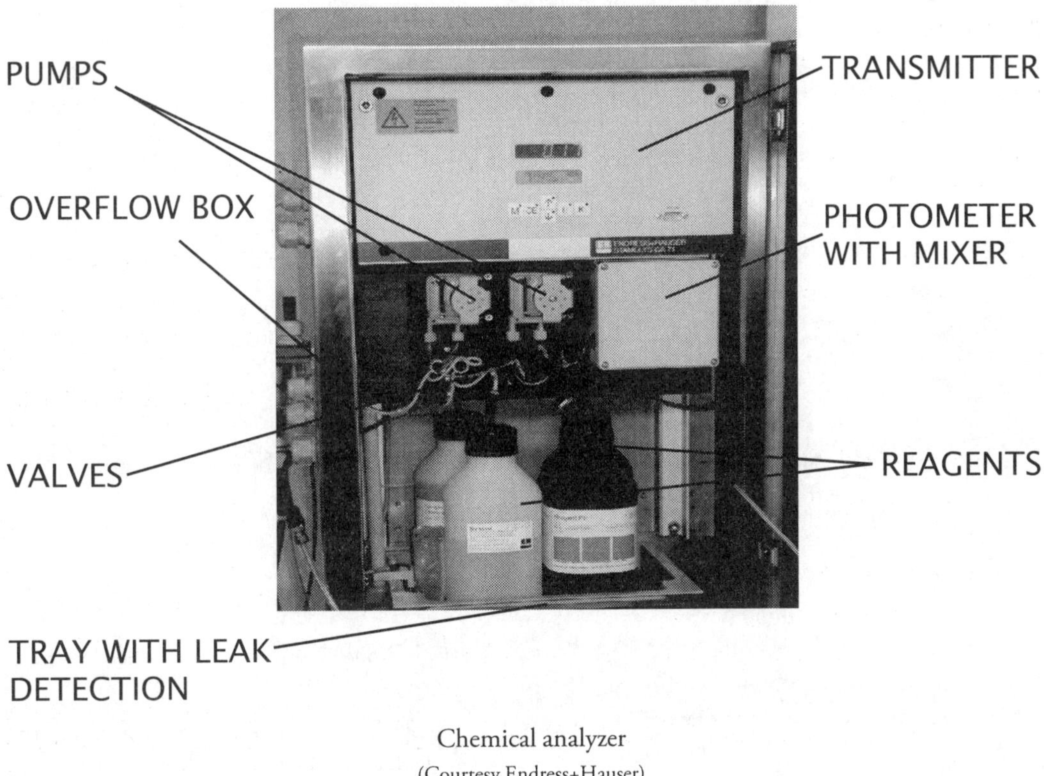

Chemical analyzer
(Courtesy Endress+Hauser)

Fig. 19.22 Water treatment plant analytical process instrumentation

A power supply to generate the required electrical energy may be located at the analog 4–20 mA transmitter, at the receiver, or at another location in the control instrument loop. The transmitter may be an integral part of the measurement or readout transducer, or it may be housed separately. In either case, the transmitter adjusts the signal to a corresponding value of the measured variable, and the receiver, in turn, converts this signal to a visible indication, the readout. Digital-data transmission distance is limited only by the computer network capability and reliability.

Instrumentation power is required for all components in the instrument loop, for both analog and digital devices. A reliable source of power such as a battery-backed power supply is used to ensure availability during power interruptions. Digital devices have lower power requirements, allowing for smaller and simpler backup power systems.

QUESTIONS

Please write your answers to the following questions and compare them with those on page 436.

19.2L What is one way to measure liquid chemical feed rates?

19.2M What process variables are commonly monitored or controlled by process instrumentation?

19.2N What are the two general systems for transmission of measurement signals?

END OF LESSON 1 OF 2 LESSONS

on

INSTRUMENTATION AND CONTROL SYSTEMS

Please answer the discussion and review questions next.

DISCUSSION AND REVIEW QUESTIONS

Chapter 19. INSTRUMENTATION AND CONTROL SYSTEMS

(Lesson 1 of 2 Lessons)

At the end of each lesson in this chapter, you will find discussion and review questions. Please write your answers to these questions to determine how well you understand the material in the lesson.

1. Why should operators understand instrumentation and control systems?
2. How can control and instrumentation systems make an operator's job easier?
3. What is the difference between accuracy and precision?
4. Why should a screwdriver not be used to test an electric circuit?
5. What should you do when you discover an area with an oxygen deficiency or enrichment?
6. How can water levels be measured?
7. What problems can develop with propeller meters when they become old or worn?

CHAPTER 19. INSTRUMENTATION AND CONTROL SYSTEMS

(Lesson 2 of 2 Lessons)

19.3 CATEGORIES OF INSTRUMENTATION

19.30 Primary Elements

The first system element that responds quantitatively to the measured variable is the primary element or sensor. *TRANSDUCERS* convert the sensor's minute actions to a usable indication or a signal. If remote transmission of the value is required, a transmitter may be part of the transducer. An illustrative example of these three components is the typical Venturi meter (shown in Figure 19.20): the flow tube is the primary element, the differential pressure-sensing device (DP cell) is the transducer, and the signal-producing component is the transmitter. An understanding of the separate functions of each section of such a flowmeter is important to the proper understanding of equipment problems.

19.31 Panel Instruments

Process variable readouts are provided locally with indicators and repeated in the control panel and the process computer. These particular components are important to the operator and, hence, to plant operation itself, because they display or control the variable directly. The main control panel devices can also produce alarm signals to indicate if a variable is outside its range of expected values. The controllers are often installed on (or behind) the main panel along with the operating buttons, switches, and indicator lights for the plant's equipment. The controller in an instrument loop control system produces the alarm and control signals.

19.310 Indicators

Indicators give a visual representation of a variable's present value, either as an analog or digital display (Figures 19.23 and 19.24). The analog display uses some type of pointer or graphical display against a scale. A digital display is a direct numerical readout. Chart recorders, which by nature also serve as indicators, give a permanent record of how the variable changes with time by way of a moving chart. These historical records are now largely stored electronically as part of the computerized control system. They may be displayed on a computer screen or printed out. Indicators out in the plant or field provide operators with local readouts of the process variables.

Indicators with digital readouts may be devices in a 4–20 mA instrument loop or may receive process variable data from a control system communication network. Digital readouts may be read more quickly and precisely from a longer distance, and can respond virtually instantly to variable changes. But analog indicators are cheaper, more rugged, and may not even require

Fig. 19.23 Analog chlorine residual indicator

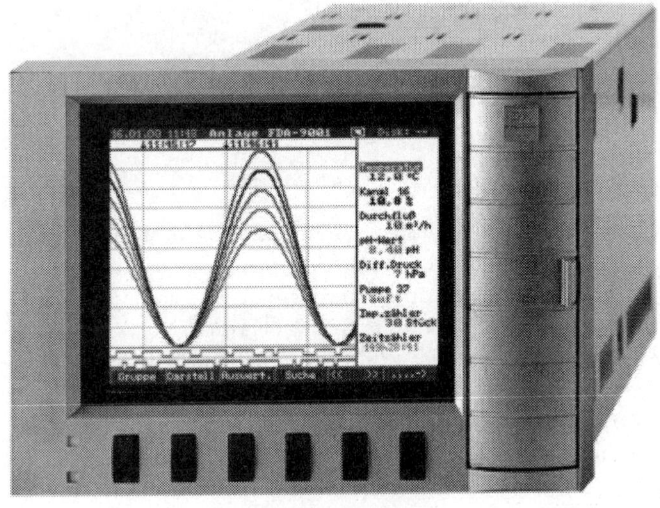

Fig. 19.24 Paperless chart recorder

(Courtesy Endress+Hauser)

electrical power, an advantage during a power failure or in hazardous environments. Another advantage of the analog display is that an incorrect readout may be more recognizable than with a digital readout or computer control system, and is more easily corrected by the operator. For example, the pointer on a flowmeter gauge may merely be stuck, as evidenced by a perfectly constant reading.

Computer control systems provide diagnostic tools to monitor the functionality of instrumentation and troubleshoot process instrumentation loops. Because the capabilities of these systems vary widely, operators must learn the specifics of the system installed. Learning to use the diagnostic tools allows the operator to determine the reliability of the control system.

With all-electronic instrumentation, as advantageous as it may seem from a technical and economic standpoint, the operator has little recourse in case of malfunction of critical instrumentation. Temporary power failures, tripped panel circuit breakers, voltage surges resulting in blown fuses, static electricity, and excessive heat can disrupt the instrument and control systems. Electromechanical or pneumatic instruments may keep operating, or recover operation readily, after such power or heat problems. Computer control systems are typically designed so that the operator can, with training, easily restart the control system and initiate plant operation. Accordingly, the operator should insist upon some input into the design phase of instrument and control systems to ensure that the plant is still operable during power outages, hot weather, and other contingencies. Standby power generators and battery-backed power supplies are used to keep plants operating during commercial power outages, but even they have shortcomings if plant operations depend entirely on electrical power.

QUESTIONS

Please write your answers to the following questions and compare them with those on page 436.

19.3A What is the purpose of instrumentation indicators?

19.3B Describe an analog display.

19.3C What is the purpose of chart recorders?

19.3D What factors can disrupt instrument and control systems?

19.311 Recorders

Chart recorder functions have been incorporated into computer control systems (Figure 19.24). However, local recorders are still necessary at some locations throughout a treatment plant to ensure reliable, continuous recording of critical process variables. *RECORDERS* are indicators designed to show how the value of the variable has changed with time (Figure 19.25). Usually, this is done by attaching a pen (or stylus) to an indicator's arm, which then marks or scribes the value of the variable onto a continuously moving chart. The chart is marked on a horizontal, vertical, or circular scale in time units. Chart records are also stored on a computer disk or in memory for download from the recorders.

There are two main types of chart recorders: the strip-chart type and the circular-chart type. The strip-chart type carries its chart on a roll or as folded stock, with typically several weeks' supply of chart available. Several hours of charted data are usually visible, or easily available, for the operator to read. On a circular recorder, the chart makes one revolution every day, week, or month, with the advantage that the record of the entire elapsed time period is visible at any time.

Changing of charts is usually the operator's duty. It is easier with circular recorders, though not that difficult with most strip-chart units with some practice.

Chart recorders are typically electrical, powered by the instrumentation and control system electrical supply. Battery-backed power for the recorders may be used, if required, as a backup power source. Recorders are most commonly described by the nominal size of the strip-chart width or circular-chart diameter (for example, a 4-inch [100 mm] strip-chart, or a 10-inch [250 mm] circular-chart recorder). Figure 19.25 shows a combination indicator/recorder and Figure 19.26 presents two models of recorders.

19.312 Totalizers

Rate of flow, as a variable, is a time rate; that is, it involves time directly, such as in gallons per minute, or million gallons per day, or cubic meters per second. Flow rate units become units of volume with the passage of time. For example, flow in gallons per minute accumulates as total gallons during an hour or day. The process of calculating and presenting an ongoing running total of flow volumes passing through a meter is termed "integration" or totalizing.

The totalized flow functions are commonly incorporated into a computer control system, which performs the calculations and stores the historical record of the totals. The historical data may be displayed on the operator interface in trend displays, which are graphical displays similar to strip-chart recorder outputs. The data can be displayed on the computer screen, printed, or sent electronically for analysis by charting and graphing programs.

Large quantities of water (or liquid chemical) are commonly read out in units of hundreds or thousands of gallons (liters). On the face of a totalizer you may find a multiplier such as × 100 or × 1,000. This indicates that the reading is to be multiplied by this factor to yield the full amount of gallons or cubic meters. If the readout uses a large unit, such as mil gal, a decimal will appear between appropriate numbers on the display, or a fractional multiplier (× 0.001, for example) may appear on the face of the totalizer.

Every operator should be able to calculate total flow for a given time period in order to verify that the totalizer is actually producing the correct value. Accuracy to one or two parts in a hundred (1 or 2 percent) is usually acceptable in a totalizer. There are methods to integrate (add up) the area under the flow-rate curve on a recorder chart, to check for long-term accuracy of total flow calculations, but it is cumbersome and rarely necessary.

Fig. 19.25 Chart recorder with digital indicator

Strip-chart recorders

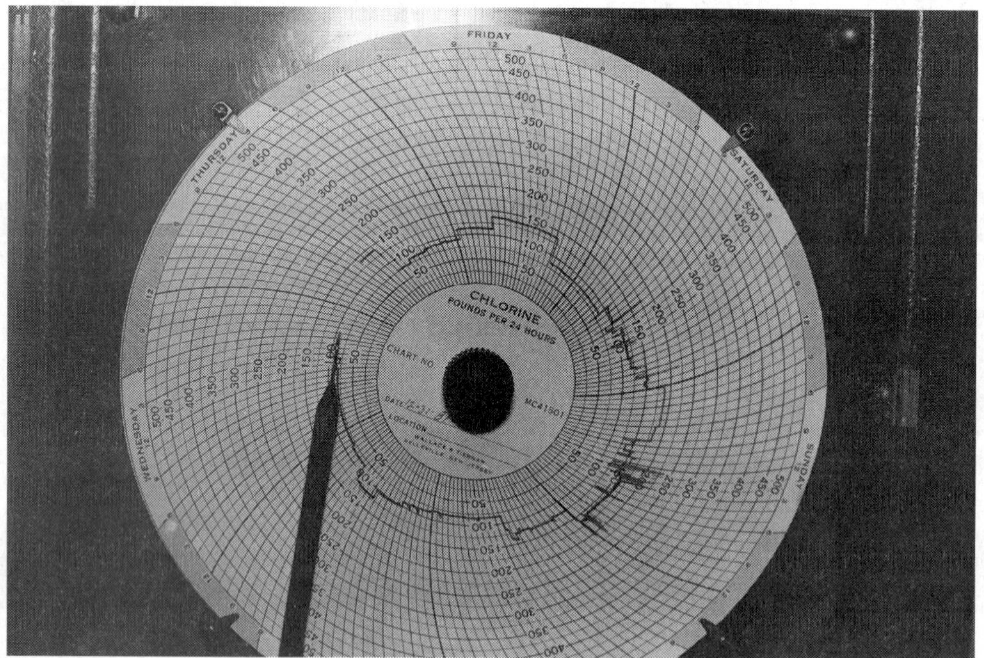

Seven-day circular-chart recorder

Fig. 19.26 Recorders, strip-chart and circular-chart

19.313 Alarms

Alarms are visual or audible signals that a variable is out of bounds, or that a condition exists in the plant requiring the operator's attention. Computer control systems have incorporated the alarm annunciator functions into the human machine interface (HMI). Only conditions that require operator attention need alarm signals. For noncritical conditions, a change in color on the operator interface graphic is sufficient notice. For more important variables or conditions, and especially when no computer system is available, an attention-getting annunciator panel (Figure 19.27) with flashing lights and an unmistakable and penetrating alarm horn is commonly used.

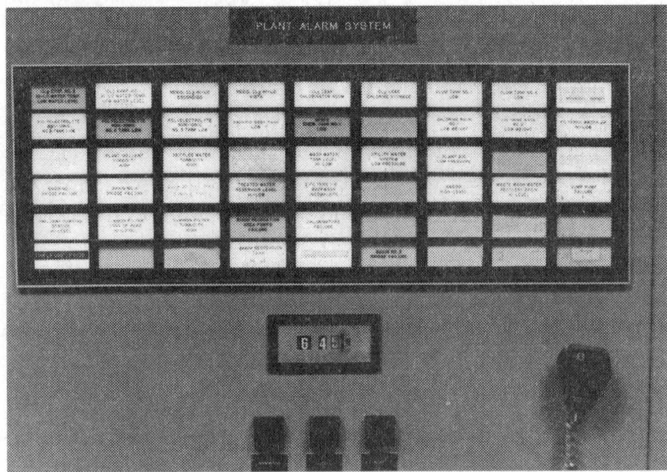

Fig. 19.27 Annunciator (alarm) panel (each rectangle represents a monitored location)

Operator interface and annunciator panels should all have acknowledge and reset features to allow the operator to squelch the alarm sound (leaving the visible indication alone), and then to reset the system after the alarm condition is corrected. Annunciator panels should also have a test button so that an operator can confirm that no alarm lamps have burned out. The alarm contacts that activate annunciator panel alarms are within the field instrumentation devices. Computer control system alarms are activated by the software's alarm set points and may be programmable from the operator interface. The operator may be responsible for setting the alarm set point and must use judgment as to the actual limits of the particular variables that will ensure meeting proper operational goals. Each system is different, so no attempt will be made here to instruct operating personnel in alarm resetting procedures.

Sometimes, operators fail to reset alarm limits as conditions and judgments change in the plant. It is not uncommon to see a plant's operator ignoring alarm conditions on an annunciator panel as normal practice. Such practice is not advised because a true alarm condition requiring immediate operator attention may be lost in the resulting general indifference to the alarm system. For some operators, acknowledging an alarm sound to get rid of the noise is second nature, without due attention to each and every activating condition. All alarm contact limits should be reset (or deactivated), as necessary, to ensure that the operator is as attentive to the alarm system as necessary to handle real emergencies. The system design should include provisions for disabling alarms when the associated instrument or device is out of service for long periods of time.

QUESTIONS

Please write your answers to the following questions and compare them with those on page 436.

19.3E What is the purpose of recorders?

19.3F What are the two main types of chart recorders?

19.3G What are typical power sources for chart recorders?

19.3H List the two kinds of warning signals that are produced by alarms.

19.32 Automatic Controllers

Section 19.0 explains the nature of control systems as they are used in water treatment plant operations. Automatic controllers may be individual instrumentation devices or a function programmed into a computer control system. Indications of proper and improper control need to be recognized by operators, but adjustment of the controller is left to a qualified instrument technician. By shifting to the manual mode, the operator can bypass the operation of any controller, whether electric or pneumatic. Learn how to shift all your controllers to manual operation. This will allow you to take over control of a critical system, when necessary, in an emergency, as well as at any other time it suits your purposes. For example, you may be able to quickly correct a cycling or sluggish variable by using manual control rather than waiting for the controller to correct the condition in time (if it ever does).

A controller is limited in its capability. It can only do what it has been programmed to do. You, as the operator, can exercise judgment based on your experience and observations, so do not hesitate to intervene if a controller is not exercising control within sensible limits. Of course, you must be sure of your conclusions and competent to take over control if you decide to operate manually.

To repeat a few of the more important operational control considerations, remember that ON/OFF control is quite different in operation from proportional control. Both methods can exercise close control of a variable; however, proportional control is better suited for close control. Attempting to set up an ON/OFF control system to maintain a variable within too close a tolerance may result in rapid ON/OFF operation of equipment. Such operation can damage both the equipment and the switching devices. Therefore, the basic principle to guide you in setting up the frequency of ON/OFF operation of a piece of equipment is to set the ON/OFF controls to operate or cycle associated equipment on and off no more often than actually necessary for plant operation. A level controller, for instance, should be set to cycle a pump or valve only as often as needed.

In the case of a modulating controller, it too may begin to cycle its final control element (pump or valve) through a wide range if any of the internal settings, namely proportional band or reset, are adjusted so as to attempt closer control of the variable than is reasonable. Accordingly, it may be better to accommodate to a small offset (difference between set point and control point) than risk an upset in control by attempting too close control.

19.33 Pump Controllers

Control of pumping systems can be achieved, as we have seen, by an ON/OFF type of controller starting and stopping pumps according to a level, pressure, or flow measurement.

Usually, an ON/OFF pump control system responds to level changes in a tank of some type. Water level can be sensed directly with a float or by a pressure change at the tank or pump site. The pump is thus turned off or on as the tank level rises above or falls below predetermined level or pressure limits. Control is rather simple in this case.

However, such systems may include several extra electrical control features to ensure fail-safe operation. To prevent the pump from running after a loss of level signal, electrical circuitry should be designed so the pump will turn off on an open signal circuit and on only with a closed circuit. (Ideally, the controller would be able to distinguish between an open or closed remote level/pressure contact and an open or shorted signal line.) Larger pump systems also will often have a low-pressure cutoff switch on the suction side to prevent the pump from running when no water is available, such as with an empty tank or closed suction valve.

Controllers may also protect against overheating a pump (as happens when continuing to pump against a closed discharge valve) by a high-pressure (or low-flow) cutoff switch on the discharge piping. Both the high- and low-pressure switches should shut off a pump through a time delay circuit so that short-term pressure surges (dips and spikes) in the pump's piping can be tolerated. Ideally, the low- or high-pressure switches also key alarms to notify the operator of the condition. For remote stations, a plant's main panel may include indicator lights to show the pumps' operating conditions. Figure 19.28 shows a simplified diagram of pump control circuitry.

Pump control panels (Figure 19.29) may also include automatic or manual alternators (two pumps) or sequencers (more than two pumps). This provision allows the total pump operating time required for the particular system to be distributed equally among all the pumps at a pump station. A manual switch for a two-pump station, for example, may read 1-2 in one position and 2-1 in the other position. In the first position, pump #1 is the lead pump (which runs most of the time) and #2, the lag pump (which runs less). When the operator changes the switch to 2-1, the lead-lag order of pump operation is reversed, as it should be periodically, to keep the running time (as read on the elapsed time meters, or as estimated) of both pumps to about the same number of total hours. In a station with multiple pumping units, an automatic alternator or sequencer regularly changes the order of the pumps' startup to maintain similar operating times for all pumps. To protect a pump's electric motor from overheating, level controls should be set so that the pump starts no more often than six times per hour (average one start every 10 minutes).

QUESTIONS

Please write your answers to the following questions and compare them with those on page 436.

19.3I Under what conditions might an operator decide to bypass the operation of a controller? How could this be done?

19.3J What basic principle should guide you in setting up the frequency of ON/OFF operation of a piece of equipment?

19.3K How can pumps be prevented from running after a loss of level signal?

19.3L What provision allows the total pump operating time required for a particular system to be distributed equally among all the pumps at a pump station?

19.34 Air Supply Systems

Pneumatic instrumentation depends on a constant source of clean, dry, pressurized air for reliable operation. Given a quality air supply, pneumatic devices can operate for long periods without significant problems. Without a quality air supply, operational problems can be frequent. The operator of a plant is usually assigned the task of ensuring that the instrument air is always available and dry, so operators must learn how to accomplish this; it cannot be assumed that clean air is there automatically.

The plant's instrument air supply system consists of a compressor with its own controls, master air pressure regulator, air filter, and air dryer, as well as the individual pressure regulator/filters in the line at each pneumatic plant instrument (Figure 19.30). Only the instrument air is filtered and dried; the plant air usually does not require such measures since it is being used only for other purposes.

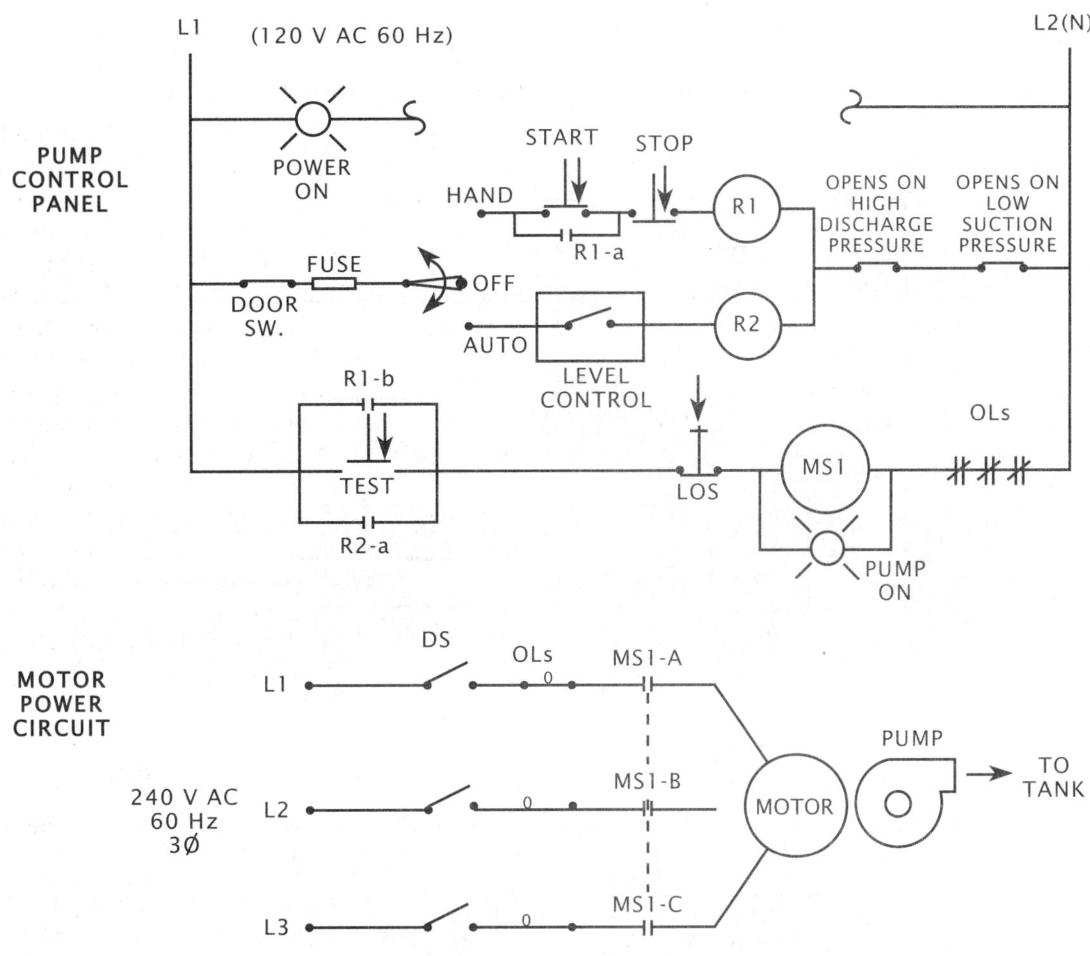

REFER TO FIGURE 19.5 (PAGE 394), AND PAGE 395 FOR PARTS LEGEND

Fig. 19.28 Pump control station ladder diagram (ON/OFF control) (simplified schematic)

Fig. 19.29 Photo of pump control station

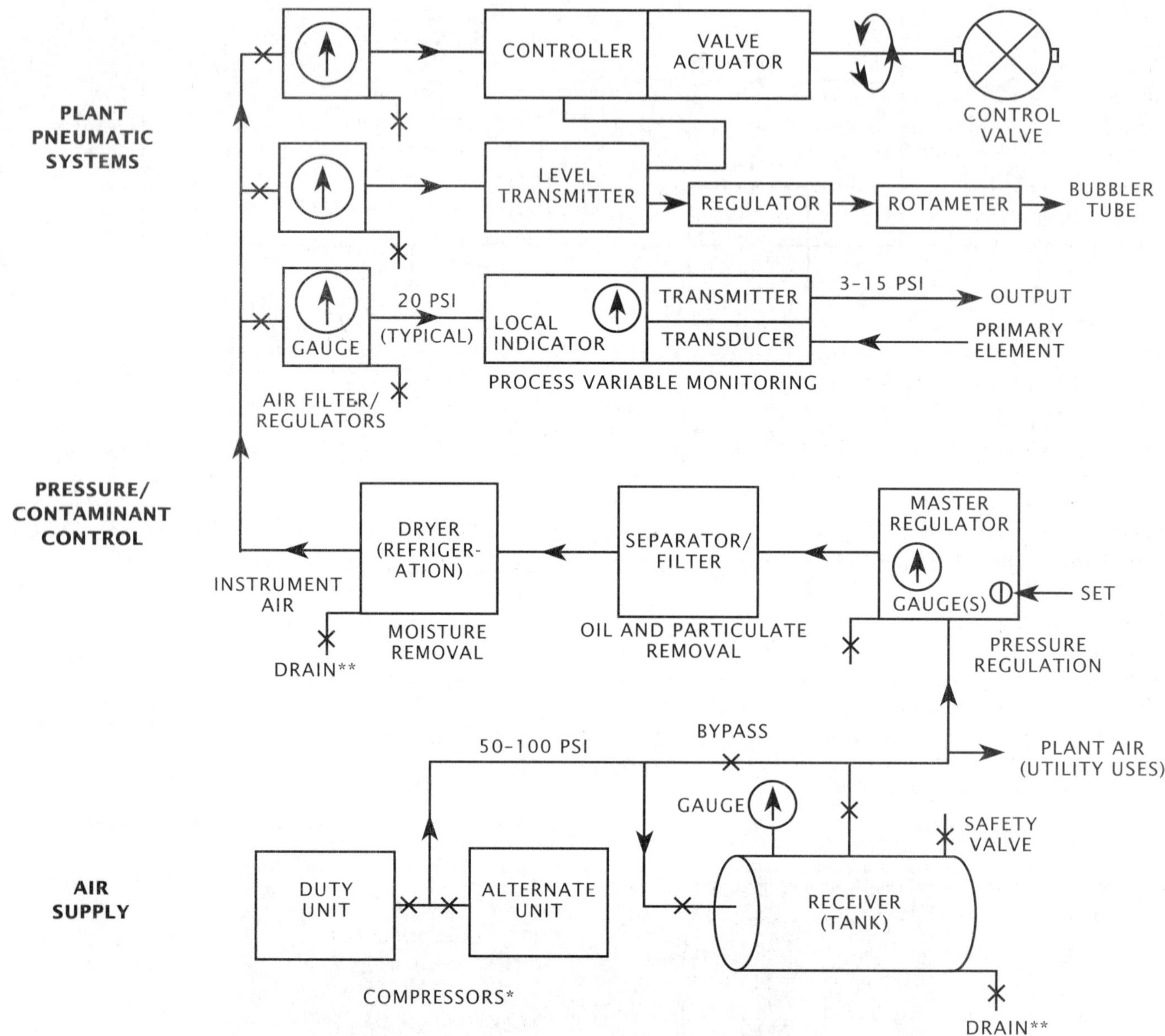

* Compressor and receiver (tank) are often an integral unit in small installations.
** Drains may be automatic type (on timer).

Fig. 19.30 Typical plant instrument air system functional diagram (simplified, not all valves and piping shown)

As air passes through a compressor, it not only can pick up some oil but the air's moisture content is concentrated by the compression process. Special measures must be taken to remove both of the liquids. Oil is removed by filtering the air through special oil-absorbent elements. Water is removed by passing the moisture-laden air through an air-drying system consisting of a moisture separation column and air dryers. Air-drying systems are typically provided in pairs to allow one dryer to regenerate while the other is in service. You must recognize that the capacity of any of these systems of oil or water removal is limited to amounts of liquid encountered under normal conditions.

If the compressor is worn so as to pass more oil than usual, the oil separation process may permit troublesome amounts of oil to pass into the air supply. If the air source contains excessive humidity (due to a rainy day, perhaps, or a wet compressor room), the air-drying system may not handle the excess moisture. Learn enough about the instrument air system to be able to open the drain valves, cycle the air dryer, or even bypass the tank in order to prevent instrumentation problems due to an oily or moisture-saturated air supply.

Operators should regularly crack the regulator/filter drain valves at each of the plant's process instruments. An unusual quantity of liquid drainage may indicate an overloading or failure in the instrument air filter/drying parts. Also, pneumatic indicators/recorders should be watched for erratic pointer/pen movements, which usually are indicative of air quality problems.

The seriousness of plant power or compressor failures can be lessened if you temporarily turn off all nonessential usages of compressed air in the plant. The air storage tank is usually sized so that there is enough air on hand to last for several hours, if conserved. Knowing this, you may be able to wait out a power failure without undue drastic action by conserving the remaining available pressure of the air supply.

QUESTIONS

Please write your answers to the following questions and compare them with those on pages 436 and 437.

19.3M What are the essential qualities of the air supply needed for reliable operation of pneumatic instrumentation?

19.3N How are moisture and oil removed from instrument air?

19.35 Laboratory Instruments

This category of instrumentation includes those analytical units typically found in water treatment plant laboratories (Figure 19.31). Some examples are turbidimeters, colorimeters and comparators, pH, and conductivity (TDS) meters. We have already seen that process models of each of the units monitor these same variables out in the plant. The models used in the laboratory are usually referred to as "bench" models rather than "process" instrumentation.

Operators often are required to make periodic readings from laboratory instruments, and periodic standardizing of particular instruments is often required before determinations are made. Preventive, and certainly corrective, maintenance is handled by the laboratory staff, factory representative, or instrument technician since each unit can be quite complex. Some of these countertop instruments or devices are very delicate and replacement parts, such as glass vessels or pH electrodes, are quite expensive. Moreover, the use of some of these instruments requires the regular handling of laboratory glassware and other breakable items. The operator who, through carelessness, lack of knowledge, or simple hurrying, consistently breaks glassware or "finds the darn meter broken again" does not become popular with the chemist, supervisor, or other operators. The byword in the laboratory is caution. Protect valuable and essential instrumentation and supplies.

19.36 Test and Calibration Equipment

Plant measuring systems must be periodically calibrated to ensure accurate measurements. In most larger water treatment facility operations, the plant operating staff has little occasion to use testing and calibration devices on the plant instrumentation systems. A trained technician will usually be responsible for using such equipment. There are, however, some general considerations the operator should understand concerning the testing and calibration of plant measuring and control systems. With this basic knowledge, you may be able to discuss needed repairs or adjustments with an instrument technician and perhaps assist with that work. A better understanding of your plant's instrument systems may also enable you to analyze the effects of instrument problems on continued plant operation, and to handle emergency situations created by instrument failure. Your skills in instrument testing and calibration may even eventually result in a job promotion or pay raise.

The most useful piece of general electrical test equipment is the volt-ohm-milliammeter (VOM), commonly referred to as a multimeter (Figure 19.32). To use this instrument, you will need a basic understanding of electricity, but once you learn to use it, the VOM has potential for universal usage in instrument and general electrical work. Local colleges and other educational institutions may offer courses in basic electricity, which undoubtedly include practice with a VOM. You, as a professional plant operator, are unlikely to find technical training of

Fig. 19.31 Water treatment plant laboratory

SPECIFICATIONS

Voltage Ranges

0–199.9/750 V AC 15 kV AC
0–1.999/19.99/199.9/1,000 V DC 15 kV DC
0–1,999 mV DC

Resistance Ranges

0–199.9/1,999 ohms
0–19.99/199.9/1,999 K ohms

Current Ranges

0–1,999 μA DC
0–19.99/199.9/1,999 mA DC
0–10 Amps DC
0–19.99/199.9/1,999 mA AC
0–10 Amps AC

NOTE: AC accuracy may be affected by outside interference.

Accuracy

DC V: ± 0.5% of rdg ± 2LSD
AC V: ± 1.5% of rdg ± 2LSD
DC Amps: All ranges ± 1.0% of rdg ± 2LSD except 10 Amp range, which is ± 1.5% of rdg ± 3LSD.
AC Amps: All ranges ± 1.5% of rdg ± 2LSD except 10 Amp range, which is ± 2.0% of rdg ± 3LSD.
Ohms: All ranges ± 0.75% of rdg ± 2LSD except 2 megohm range, which is ± 1% of rdg ± 2LSD.
15 kV AC/DC high voltage probe: add up to ± 2% of rdg.

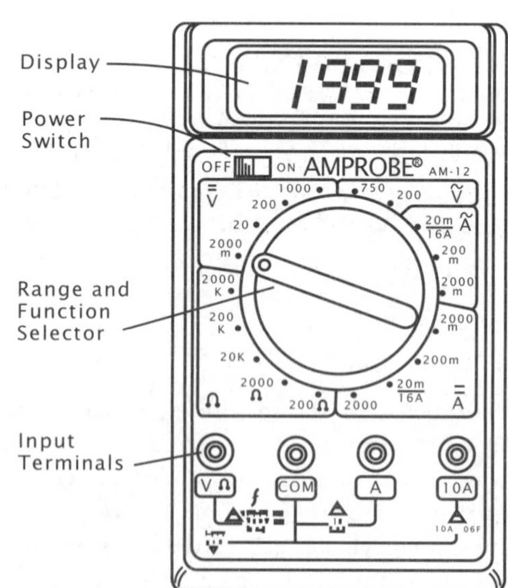

Fig. 19.32 Digital multimeter (VOM)

greater practical value than this type of course or program. Your future use of test and calibration equipment, in general, certainly should be preceded by instruction in the fundamentals of electricity.

QUESTIONS

Please write your answers to the following questions and compare them with those on page 437.

19.3O Why should an operator be especially careful when working with laboratory instruments?

19.3P Why should an operator become familiar with the testing and calibration of plant measuring and control systems?

19.3Q What is a VOM?

19.37 Process Computer Control Systems

19.370 Computer Control Systems

The computer control system is a computer-monitored alarm, response, control, and data acquisition system used by operators to monitor and adjust their treatment processes and facilities. Computer control systems used for process control may be classified by two commonly used terms: the *DISTRIBUTED CONTROL SYSTEM (DCS)*[8] and the *SUPERVISORY CONTROL AND DATA ACQUISITION (SCADA) SYSTEM.*[9] The DCS and the SCADA system perform the same functions in different settings. DCSs are typically used to control and monitor processes in treatment plants. SCADA systems are most commonly used to control and monitor distribution system facilities that are widely separated geographically. Larger water facilities may have both DCSs and SCADA systems to provide treatment process control and distribution system controls. Smaller utilities often combine all the controls necessary into a SCADA system. In large and small utilities, the operator interface for both distribution system and plant processes is provided in a single control room.

The computer control system collects, stores, and analyzes information about all aspects of operation and maintenance, transmits alarm signals, when necessary, and allows fingertip control of alarms, equipment, and processes. The computer control system provides the information that operators need to solve minor problems before they become major incidents. As the nerve center at the treatment plant, the system allows operators to enhance the efficiency of their facility by keeping them fully informed and fully in control. Figure 19.33 shows the basic interaction cycle of the operator with the control systems and plant treatment processes.

The five basic components of the computer control system (Figure 19.34) are as follows:

- Process instrumentation and control devices sense process variables in the field and actuate equipment.
- The input/output (I/O) interface sends and receives data with the process instrumentation and control devices.
- The central processing unit (CPU) is the system component that contains the program instructions for the control system. These instructions are programmed to react based on a control strategy. The CPU gathers data from the various interfaces and sends commands to field devices to operate the plant processes.
- The communication interfaces provide the means for the computer control system to send data to and from outside computer systems, business systems, other process control systems, and equipment.
- The *HUMAN MACHINE INTERFACE (HMI)*[10] is commonly a computer workstation that is running the computer control system software that provides the plant data to the operator on the workstation screen.

These components are the means by which the control system gathers and distributes information for the human operator and the process instrumentation and other equipment.

The computer control systems may be used in various capacities, from data collection and storage only, to total data analysis, interpretation, and process control.

---

8. *Distributed Control System (DCS).* A computer control system having multiple microprocessors to distribute the functions performing process control, thereby distributing the risk from component failure. The distributed components (input/output devices, control devices, and operator interface devices) are all connected by communications links and permit the transmission of control, measurement, and operating information to and from many locations.
9. *SCADA* (SKAY-dah) *System.* Supervisory Control And Data Acquisition system. A computer-monitored alarm, response, control, and data acquisition system used to monitor and adjust treatment processes and facilities.
10. *Human Machine Interface (HMI).* The device at which the operator interacts with the control system. This may be an individual instrumentation and control device or the graphic screen of a computer control system. Also see MAN MACHINE INTERFACE (MMI) and OPERATOR INTERFACE.

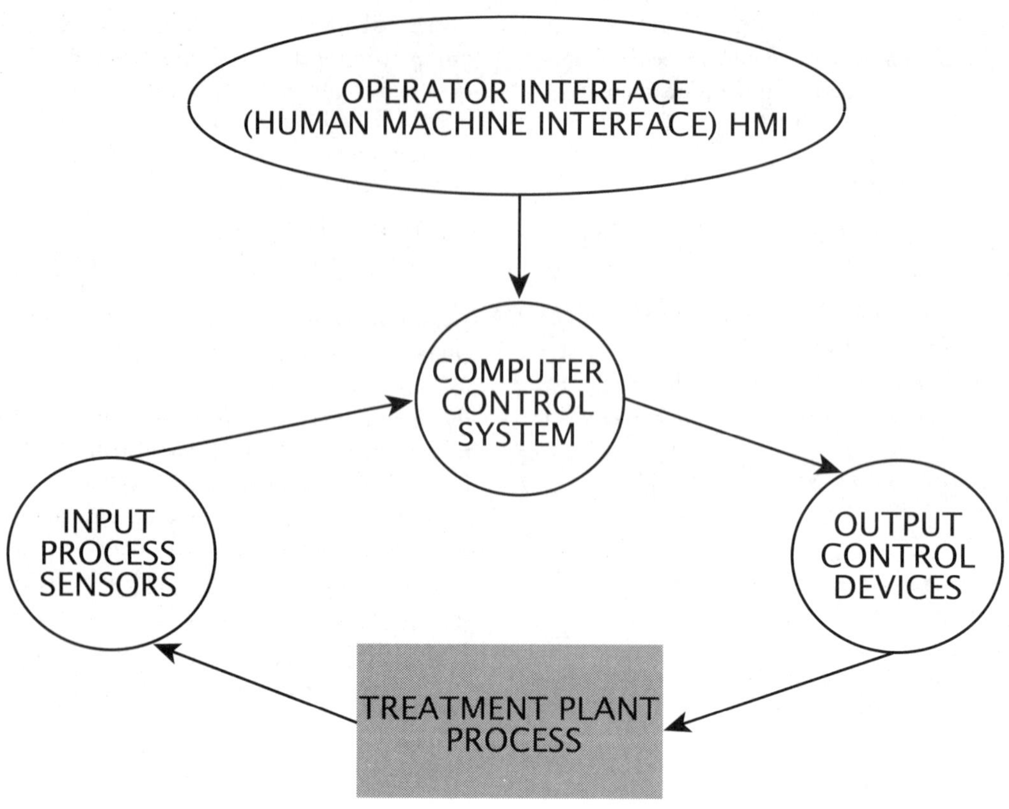

Fig. 19.33 Operator interaction with plant processes

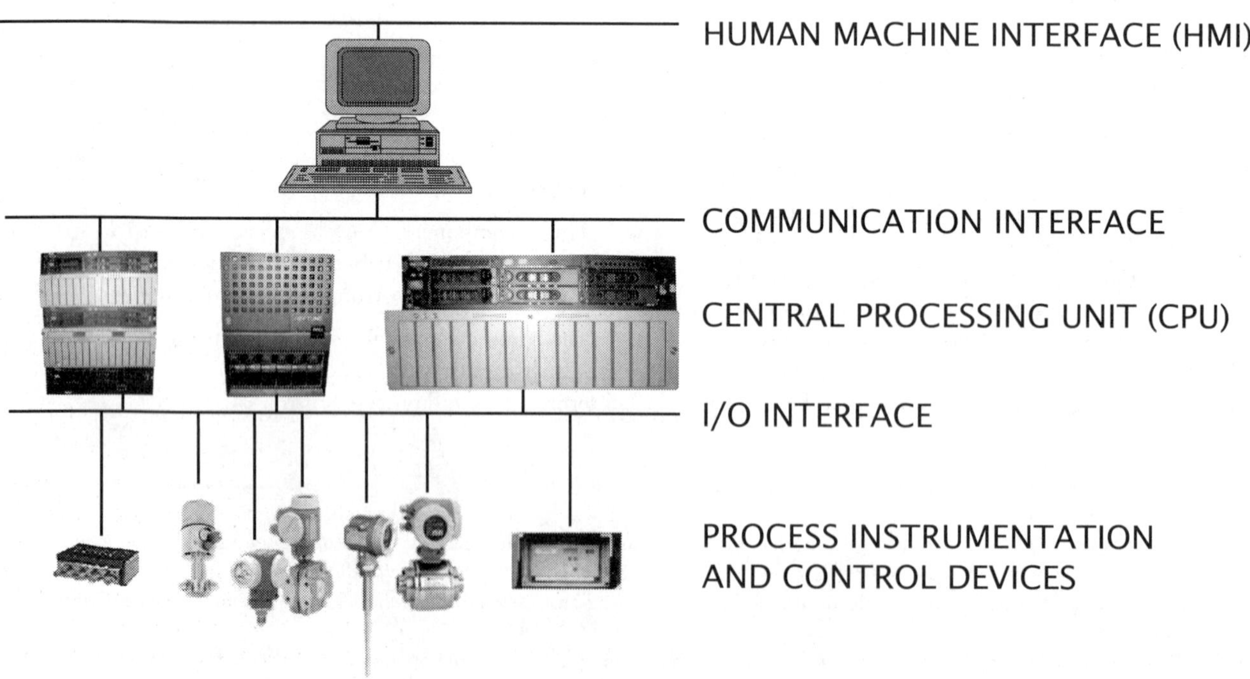

Fig. 19.34 Computer control system components

Computer control systems monitor levels, pressures, and flows and operate pumps, valves, and alarms. They monitor temperatures, speeds, motor currents, pH, turbidity, and other operating parameters, and provide control, as necessary. Computer control systems provide a log of historical data for events, analog signal trends, and equipment operating time for maintenance purposes. The information collected may be read by an operator on computer screen readouts or analyzed and plotted by the computer as trend charts.

Computer control systems provide a picture of the plant's overall status on a computer screen. In addition, detailed pictures of specific portions of the system can be examined by the operator through the computer workstation. The graphical displays on the computer screens can include current operating information, which the operator can use to determine if the guidelines are within acceptable operating ranges or if any adjustments are necessary.

Computer control systems are capable of analyzing data and providing operating, maintenance, regulatory, and annual reports. Operation and maintenance personnel rely on a computer control system to help them prepare daily, weekly, and monthly maintenance schedules, monitor the spare parts inventory status, order additional spare parts, print out work orders, and record completed work assignments.

Computer control systems can also be used to enhance energy conservation programs. For example, operators can develop energy management control strategies that allow for both maximum energy savings and maximum treatment flow before peak-flow periods. In this type of system, power meters are used to accurately measure and record power consumption. The information can then be reviewed by operators to watch for changes that may indicate equipment problems.

Emergency response procedures can also be programmed into a computer control system. Operator responses can be provided for different operational scenarios that might be encountered as a result of adverse weather changes, fires, earthquakes, or other emergency situations.

QUESTIONS

Please write your answers to the following questions and compare them with those on page 437.

19.3R Computer control systems used for process control are classified by which two commonly used terms?

19.3S What are the five basic components of the computer control system?

19.3T How do computer control systems help operation and maintenance personnel?

19.371 Typical Computer Control System Functions

Computer control systems for water treatment plants and distribution systems are usually operated together, with the controls located at the treatment plant. Information that historically was recorded on paper strip-charts is now being recorded and stored (archived) by computers. This information can be retrieved and reviewed easily by the operator, whereas before, years of strip-chart records would need to be examined to find needed information. Therefore, computer control systems are more efficient in providing operators with the information they need to make informed and timely decisions. Figure 19.35 is a typical modern water treatment plant control room.

Computer control systems give the treatment plant operator the tools to optimize plant processes based on current and historical operating information. The treatment plant influent and effluent are monitored continuously for many process variables such as flow, turbidity, pH, ammonium, chlorine, and nitrogen. If these indicators change significantly or exceed predetermined levels, the computer control system alerts the operator or changes the process based on a preprogrammed control strategy defined by the operator.

Historical operating data stored in a computer control system is readily available at any time. The computer control system can be queried to identify, for instance, when peak plant influent flow was greater than 50 MGD during the previous 2 years. Plant performance under these conditions can be recalled using the computer control system, analyzed by the operators, and the results used to operate the plant accordingly.

Electrical energy consumption can be optimized by the use of computer control systems. Most power companies are eager to help operators save money by structuring their rates to encourage electrical energy consumption when demands for power are low and to discourage consumption when demands for power are high. Computer controls can be programmed with a control strategy to reduce energy costs by automatically operating equipment when demands for power are low.

Computer control systems are being continually improved to help operators do a better job. Operators can create their own display screens, their own graphics, and show whatever operating characteristics they wish to display. The main screen could be a flow diagram from influent to effluent showing the main treatment and auxiliary process areas. Critical operating information could be displayed for the main treatment flow path and process area, with navigation capabilities to easily access detailed screens for each piece of equipment.

Information on the screen should be color coded to indicate if a pump is running, ready, unavailable, or failed, or if a valve is open, closed, moving, unavailable, or failed. The computer uses a failed signal to inform the operator that something is wrong with the information or the signal it is receiving or is being instructed to display. For example, if there is no power to a motor, then the motor cannot be running even though the computer is receiving a signal that indicates it is running. In this case, the computer would send a failed signal, indicating that the information it is receiving is not consistent with the rest of the information available.

The operator can request a computer to display a summary of all alarm conditions in a plant, a particular plant area, or a process system. A blinking alarm signal indicates that the alarm condition has not yet been acknowledged by the operator. A steady alarm signal, one that is not blinking, indicates that the alarm

Fig. 19.35 Control room

has been acknowledged but the condition causing it has not yet been fixed. Also, the screen could be set up to automatically designate certain alarm conditions as priority alarms, requiring immediate operator attention.

With proper security implementation, computer control systems allow operators to have remote access to plant controls from anywhere using a laptop or remote workstation. This provides the flexibility for off-duty staff to help on-duty operators solve operational problems. Computer networking systems allow operators at terminals in offices, in plants, and in the field to work together and use the same information or whatever information they need from one central computer database.

A drawback of some computer control systems is that when the system goes down due to a power failure, the numbers displayed will be the numbers that were registered immediately before the failure, not the current numbers. The operator may therefore experience a period of time when accurate, current information about the system is not immediately available.

Customer satisfaction with the performance of a water utility can be enhanced by the use of an effective computer control system. Coordination of the treatment facility control and the distribution system control is used to avoid water shortages and low pressures.

When operators decide to initiate or expand a computer control system for their plant process system, the first step is to decide what the computer control system should do to make the operators' jobs easier, more efficient, and safer, and to make their facilities' performance more reliable and cost effective. Cost savings associated with the use of a computer control system frequently include reduced labor costs for operation, maintenance, and monitoring functions that were formerly performed manually. Precise control of chemical feed rates by a computer control system eliminates wasteful overdosing. Preventive maintenance monitoring can save on equipment and repair costs, and, as previously noted, energy savings may result from off-peak (reduced) electrical power rates. Operators should visit facilities with computer control systems and talk to the operators about what they find beneficial and detrimental with regard to computer control systems and how the systems contribute to their performance as operators.

The greatest challenge for operators using computer control systems is to realize that just because a computer says something (a pump is operating as expected), this does not mean that the computer is always correct. Also, when the system fails due to a power failure or for any other reason, operators will be required to operate the plant manually and without critical information. Could you do this?

Operators will always be needed to question and analyze the results from computer control systems. They will be needed to see if the effluent looks OK, to listen to a pump to be sure it sounds right, and to smell the process and the equipment to determine if unexpected or unidentified changes are occurring. Treatment plants and distribution systems will always need alert, knowledgeable, and experienced operators who have a feel for their plant and their distribution system.

QUESTIONS

Please write your answers to the following questions and compare them with those on page 437.

19.3U What do computer control systems do with information that historically was recorded on paper strip-charts?

19.3V Cost savings associated with the use of a computer control system frequently include which items?

19.3W What is the greatest challenge for operators using computer control systems?

19.4 OPERATION AND PREVENTIVE MAINTENANCE

19.40 Proper Care of Instruments

Usually, instrumentation systems are remarkably reliable year after year, assuming proper application, setup, operation, and maintenance. Reliable measurement systems, though outdated by today's standards, are still found in regular service at some plants up to 50 years after installation. To a certain extent, good design and application account for such long service life, but most important is the careful operation and regular maintenance of the instruments' components. The key to proper operation and maintenance (O&M) is the operator's practical understanding of the system. Operators must know how to recognize malfunctioning instruments so as to prevent prolonged damaging operation, shut down and prepare devices for seasonal or other long-term nonoperation, and perform preventive (and minor corrective) maintenance tasks to ensure proper operation in the long term. A sensitive instrumentation system can be ruined in short order with neglect in any one of these three areas.

Operators should be familiar with the technical manual (also called the instruction book or operating manual) of each piece of equipment and instrument encountered in a plant. Each manual will have a section devoted to the operation of a certain component of a complete measuring or control system (although frequently not for the entire system). Detailed descriptions of maintenance tasks and operating checks will usually be found in the manual. Depending on the general type of instrument (electromechanical, pneumatic, or electronic), the suggested frequency of the operation and maintenance/checking tasks can range from none to monthly. Accordingly, this section of the course only describes those common and general tasks an operator might be expected to perform to operate and maintain instrumentation systems. From an operations standpoint, these tasks include learning, and constant attention to, what constitutes normal function. From a maintenance standpoint, they include ensuring proper and continuing protection and care of each component.

19.41 Indications of Proper Function

The usual pattern of day-to-day operation of every measuring and control system in a plant should become so familiar to operators that they almost unconsciously sense any significant change. This will be especially evident and true for systems with recorders where the pen trace is visible. An operator should thus watch indicators and controllers for their characteristic actions and pay close attention to trend records. Using the trend capability of the computer control system provides a method to analyze the reaction of one process variable to a change in another or other process variables.

Two of the surest indications of a serious electrical problem in instruments or power circuits are, of course, smoke or a burning odor. Such signs of a problem should never be ignored. Smoke/odor means heat, and no device can operate long at unduly high temperatures. Any electrical equipment that begins to show signs of excessive heating must be shut down immediately, regardless of how critical it is to plant operation. Overheated equipment will very likely fail very soon anyway, with the damage being aggravated by continued usage. Fuses and circuit breakers do not always de-energize circuits before damage occurs, so they cannot be relied upon to do so.

Finally, operators frequently forget to reset an individual alarm, either after an actual occurrence or after a system test. This is especially prevalent when an annunciator panel is allowed to operate day after day with lit-up alarm indicators (contrary to good practice) and one light (or more) is not easily noticeable. Also, when a plant operator must be away from the main duty station, the system may be set so the audible part of the alarm system is temporarily squelched. When the operator returns, the audible system may inadvertently not be reactivated. In both instances (individual or collective loss of audible alarm), the consequences of such inattention can be serious. Therefore, develop the habit of checking your annunciator system often.

QUESTIONS

Please write your answers to the following questions and compare them with those on page 437.

19.4A List the three areas of operator responsibility that are key to proper instrument operation and maintenance (O&M).

19.4B What general tasks are expected of an operator of instrumentation systems from an operations standpoint and from a maintenance standpoint?

19.4C What are two of the surest indications of a serious electrical problem in instruments or power circuits?

.2 Startup/Shutdown Considerations

The startup and periodic or prolonged shutdown of instrumentation equipment usually require very little extra work by the operator. Startup is limited mainly to undoing or reversing the shutdown measures taken.

When shutting down any pressure-, flow-, or level-measuring system, valve off the liquid to the measuring element. Exercise particular care, as explained previously, regarding the order in which the valves are manipulated for any flow tube installation. Also, the power source of some instruments may be shut off, unless the judgment is made that keeping an instrument case warm (and thus dry) is in order. Constantly moving parts, such as chart drives, should be turned off. With an electrical panel room containing instrumentation, it is good practice to leave some power components on (such as a power transformer) to provide space heat for moisture control. In a known moist environment, sealed instrument cases may be protected for a while with a container of *DESICCANT*[11] (indicating silica gel, which is blue if OK and pink when the moisture-absorbent capacity is exhausted).

Although preventing the access of insects and rodents into any area is difficult, general cleanliness seems to help considerably. Rodenticides are available to control rodents; this is good preventive maintenance practice in any electrical space. Rats and mice will chew off wire and transformer insulation, and may urinate on other insulator material, leading to serious problems for equipment (not to mention the rodent).

Nest-building activities of some birds can also be a problem. Screening some buildings and equipment against entry by birds has become a design practice of necessity. Insects and spiders are not generally known to cause specific functional problems, but startup and operation of systems invaded by ants, bees, or spiders should await cleanup of each such component of the system. All of these pests can bite or sting, so take care.

With pneumatic instrumentation, it is desirable to purge each device with dry air before shutdown. This measure helps rid the individual parts of residual oil and moisture to minimize internal sticking and corrosion while standing idle. As before, periodic blow-off of air receivers and filters keeps these liquids out of the instruments to a large degree. Before shutdown, however, extra attention should be paid to instrument air quality for purging. Before startup, each filter/receiver should again be purged.

Finally, pay attention to the pens and chart drives of recorders upon shutdown. Ink containers (capsules) may be removed if deemed necessary, and chart drives turned off. A dry pen bearing against one track (such as zero) of a chart for weeks on end is an invitation to startup problems. Re-inking and chart replacement at startup are easy if the proper shutdown procedures were followed.

QUESTIONS

Please write your answers to the following questions and compare them with those on page 437.

19.4D How can moisture be controlled in an electrical panel room containing instrumentation or in sealed instrument cases?

19.4E Why should pneumatic instrumentation be purged with dry air before shutdown?

19.43 Preventive Maintenance

Preventive maintenance (PM) means that attention is given periodically to equipment in order to prevent future malfunctions. Corrective maintenance involves actual, significant repairs, which are beyond the scope of this chapter and, in most cases, are not the responsibility of the operator. Routine operational checks are part of all PM programs in that a potential problem may be discovered and thereby corrected before it becomes serious.

PM duties for instrumentation should be included in the plant's general PM program. If your plant has no formal, routine PM program, it should have. Such a program must be set up on paper (or on a computer program). That is, the regular duties required are printed on forms or cards (or appear on a computer screen) that the operator (or technician) uses as a reminder, guide, and record of PM tasks performed. Without such explicit measures, experience shows that preventive maintenance will almost surely be put off indefinitely. Eventually, the press of critical corrective maintenance (often due to lack of preventive maintenance) and even equipment replacement projects may well eliminate forever any hope of a regular PM program. The fact that instrumentation is usually very reliable (being of quality design) may keep it running long after pumps and other equipment have failed. Nevertheless, instrumentation does require proper attention periodically to maximize its effective life. PM tasks and checks on modern instrument systems are quite minimal (even virtually nonexistent on some), so there are no valid reasons for ever failing to perform these tasks.

The technical manual for each item of instrumentation in your plant should be available so you can refer to it for O&M purposes. When a manual cannot be located, contact the manufacturer of the unit. Be sure to give all relevant serial/model numbers in your request for the manual. Request two manuals, one to use and one to put in reserve. All equipment manuals should be kept in one protected location and signed out, as needed. Become familiar with the sections of these manuals related to O&M, and follow their procedures and recommendations closely.

A good practice is to have on hand any supplies and spare parts that are or may be necessary for instrument operation (such as charts) or service (such as pens, pen cleaners, and ink). Some technical manuals contain a list of recommended spare parts that you could use as a guide. Try to obtain these supplies/parts for

11. *Desiccant* (DESS-uh-kant). A drying agent that is capable of removing or absorbing moisture from the atmosphere in a small enclosure.

your equipment. A new pen on hand for a critical recorder can be a lifesaver, at times.

Since PM measures can be so diverse for different types, brands, and ages of instrumentation, only the few general considerations applicable to all will be covered in this section.

1. Protect all instrumentation from moisture (except as needed by design), vibration, mechanical shock, vandalism (a very real problem in the field), and unauthorized access.
2. Keep instrument components clean on the outside, and closed/sealed against inside contamination (for example, spider webs and rodent wastes).
3. Do not presume to lubricate, tweak, fix, calibrate, free up, or modify any component of a system arbitrarily. If you are not qualified to take any of these measures, then do not do it.
4. Do keep recorder pens and charts functioning as designed by frequent checking and service, bleed pneumatic systems regularly, as instructed, ensure continuity of power for electrical devices, and do not neglect routine analytical instrument cleanings and standardizing duties as required by your plant's established procedures.

As a final note, it is a good idea to get to know and cooperate fully with your plant's instrument service person. Good communication between this person and the operating staff can only result in better all-around operation. If your agency is too small to staff such a specialist, it may be a good idea to enter into an instrumentation service contract with an established company or possibly even with the manufacturer of the majority of the components. With rare exceptions, general maintenance persons (even journeyman electricians) are not qualified to perform extensive maintenance on modern instrumentation. Be sure that someone takes good care of your instruments and they will take good care of you.

QUESTIONS

Please write your answers to the following questions and compare them with those on page 438.

19.4F Why should regular preventive maintenance duties be printed on forms or cards or appear on a computer screen?

19.4G How can the technical manual for an instrument be obtained if the only copy in a plant is lost?

19.4H What instrument supplies and spare parts should always be available at your plant?

19.44 Operational Checks

Computer control systems provide the best tools for observing the operating functions of plant process systems. The operator interface (also called the human machine interface [HMI]) provides the ability to view all areas of plant operation. Most systems provide the ability to display trends of multiple process variables on the same graphical display. These trends are tools you will learn to use extensively to monitor and control your facility's processes.

Operational checks are most efficiently made by always observing each system for its continuing signs of normal operation. However, some measuring systems may be cycled within their range of action as a check on the responsiveness of components. For instance, if a pressure-sensing system indicates only one pressure for months on end, and some doubt arises as to whether it is working or not, the operator may bleed off a little pressure at the primary element to produce a small fluctuation. Or, if a flow has appeared constant for an overly long period, the bypass valve in the DP (differential pressure) cell piping may be cracked open briefly to cause a drop in the reading.[12] Be sure you crack the bypass valve, not one of the others on the piping. If you open the wrong valve, the pressure may be excessive and be beyond the range of the DP cell, which could cause problems. A float for a level recorder suspected of being stuck (very constant level indication) may be freed by jiggling its cable, or by taking other measures to cause a slight fluctuation in the reading.

Whenever an operator or a technician disturbs normal operation during checking or for any reason, plant process operating personnel must be informed—ideally before the disturbance. If a recorder trace is altered from its usual pattern in the process, the person causing the upset should initial the chart appropriately and note the time. Some plants require operators to mark or date each chart at midnight (or noon) of each day for easy reference and filing.

In the case where a pen or pointer is thought to be stuck mechanically, that is, it does not respond at all to simulated or actual change in the measured variable, it is normally permissible to open an instrument's case and try to move the pointer or pen, but only to the minimum extent necessary to free it. Further deflection may well bend or break the device's linkage. A dead pen often is due only to loss of power or air to the readout mechanism. Any hard or repeated striking of an instrument to make it work identifies the striker as ignorant of good operational practice and can ruin the equipment. Insertion of tools into an instrument case in a random fix-it attempt can easily damage an instrument. Generally speaking, any extensive operating check of instrumentation should be performed by the instrument technician during routine PM program activities.

QUESTIONS

Please write your answers to the following questions and compare them with those on page 438.

19.4I How are operational checks performed on instrumentation equipment?

19.4J What should be done if a recorder trace is altered from its usual pattern during the process of checking an instrument?

12. There is no similar easy way to check a propeller meter's response.

19.5 ADDITIONAL READING

1. *Instrumentation: Handbook for Water and Wastewater Treatment Plants*, Robert G. Skrentner. Obtain from CRC Press at www.crcpress.com or phone (800) 272-7737. ISBN 978-0-87371-126-5. Order No. L126.
2. *Hands-on Water/Wastewater Equipment Maintenance*. Obtain from CRC Press at www.crcpress.com or phone (800) 272-7737. Order No. TX68179.
3. National Fire Protection Association Standard 820 (NFPA 820), *Fire Protection in Wastewater Treatment and Collection Facilities*. Obtain from National Fire Protection Association at www.nfpa.org or phone (800) 344-3555. Item No. NFPA_820.
4. *Instrumentation and Control* (M2). Obtain from American Water Works Association (AWWA) at www.awwa.org or phone (800) 926-7337. Order No. 30002. ISBN 978-1-58321-125-0.
5. *Successful Instrumentation and Control Systems Design*, Michael D. Whitt. Obtain from the International Society of Automation (ISA) at www.isa.org or phone (919) 549-8411. ISBN 978-1-936007-45-5.
6. *SCADA: Supervisory Control and Data Acquisition*, Stuart A. Boyer. Obtain from the International Society of Automation (ISA) at www.isa.org or phone (919) 549-8411. ISBN 978-1-936007-09-7.

END OF LESSON 2 OF 2 LESSONS

on

INSTRUMENTATION AND CONTROL SYSTEMS

Please answer the discussion and review questions next.

DISCUSSION AND REVIEW QUESTIONS

Chapter 19. INSTRUMENTATION AND CONTROL SYSTEMS

(Lesson 2 of 2 Lessons)

Please write your answers to the following questions to determine how well you understand the material in the lesson. The question numbering continues from Lesson 1.

8. What are the advantages and limitations of analog versus digital indicators?
9. Why is it poor practice to ignore the lamps that are lit up (alarm conditions) on an annunciator panel?
10. How should the constantly lit-up lamps (alarm conditions) on an annunciator panel be handled?
11. What electrical control features are available to protect pumps from damage?
12. What problems are created by oil and moisture in instrument air, and how can these contaminants be removed?
13. Why should plant measuring systems be periodically calibrated?
14. Why should rodents be kept out of instrumentation equipment?
15. How could you tell if a float for a level recorder might be stuck, and how would you determine if it was actually stuck?

SUGGESTED ANSWERS

Chapter 19. INSTRUMENTATION AND CONTROL SYSTEMS

ANSWERS TO QUESTIONS IN LESSON 1

Answers to questions on page 388.

19.0A A treatment plant operator must have a good working familiarity with instrumentation and control systems to properly monitor and control the treatment processes.

19.0B This chapter contains information on general instrumentation functions and control system capabilities.

19.0C Measurement instruments can serve as an extension of and improvement to an operator's senses of vision, touch, hearing, and even smell.

Answers to questions on page 389.

19.0D Our senses provide us with qualitative and relatively short-term quantitative information; instrumentation measurements provide us with exact quantitative data.

19.0E A measurement is, by definition, the comparison of a quantity with a standard unit of measure.

19.0F Accuracy refers to how closely an instrument measures the true or actual value of a process variable, while precision refers to how closely the device can reproduce a given reading (or measurement) time after time.

19.0G Analog displays show a reading as a pointer against a marked scale and digital displays provide direct, numerical readouts.

Answers to questions on page 392.

19.0H Control systems are the means by which such process variables as pressure, level, weight, or flow are controlled.

19.0I Examples of modulating control systems include chlorine residual analyzers/controllers, flow-paced (open-loop) chemical feeders, pressure- or flow-regulating valves, continuous level control of process basins, and variable-speed pumping systems for flow/level control.

19.0J A motor control station or panel and the related circuitry essentially provide only ON/OFF operation of an electric motor or other device.

Answers to questions on page 396.

19.1A The general principles for safe performance on the job are to always avoid unsafe acts and correct unsafe conditions immediately.

19.1B Electric shock can cause serious burns and even death (by asphyxiation due to paralysis of the muscles used in breathing).

19.1C An electrical explosion could shower you with molten metal or cause another type of injury.

19.1D If electric current flows through your upper body, electric shock could harm your heart or your head.

Answers to questions on page 401.

19.1E Operators should be especially careful when working around powered mechanical equipment because the equipment could start unexpectedly and cause serious injury.

19.1F The purpose of an electrical lockout device is to physically prevent the operation of an electric circuit, or to de-energize the circuit temporarily.

19.1G When testing the atmosphere before entry in any confined space, first test for oxygen deficiency/enrichment, then combustible gases and vapors, and then toxic gases and vapors.

19.1H UEL stands for upper explosive limit and LEL stands for lower explosive limit. UEL and LEL indicate the range of concentrations at which combustible or explosive gases will ignite when an ignition source is present at ambient temperature.

19.1I Wearing thin rubber or plastic gloves can greatly reduce your chances of electric shock.

Answers to questions on page 402.

19.2A A primary element or sensor senses and quantifies measured variables such as pressure, level, and flow; at times, chemical feed rates and some physical or chemical process characteristics are also sensed. The primary element performs the initial conversion of measurement energy to a signal.

19.2B Pressure is measured by the movement of a flexible element or a mechanically deformable device subjected to the force of the pressure being measured.

19.2C Some pressure sensors are fitted with surge or overrange protection to limit the effect of pressure spikes or water hammer on the device.

Answers to questions on page 406.

19.2D Different types of liquid level-sensing devices include floats, displacers, electrical probes, direct hydrostatic pressure, pneumatic bubbler tubes, and ultrasonic (sonar) devices.

19.2E Single-point ball float levels mount in a sump or tank to sense a fixed elevation. When the water level rises, the ball floats and rolls onto its side, actuating an internal mercury switch. The switch provides a discrete ON/OFF signal used to control equipment or send level data to the control system.

19.2F Probes are used to measure liquid levels where mechanical systems are impractical such as within sealed or pressurized tanks or with chemically active liquids.

19.2G Ultrasonic sensors measure the level of a liquid by emitting a pulse that is reflected back to the receiver. The sensor periodically generates a pulse of waves that bounce off the liquid surface and reflect back. The sensor detects the reflected pulse. Based on the speed of the pulse, the travel time between sending and receiving is measured and the distance is calculated and converted into a level measurement.

Answers to questions on page 414.

19.2H The two basic types of flow readings are rate of flow (volume per unit of time) and total flow (volume units).

19.2I The inferential flow-sensing techniques that are used in flow measurement are velocity, differential pressure, magnetic, and ultrasonic.

19.2J Velocity-sensing meters measure process flows by measuring water speed within a pipeline or channel. One way of doing this is by sensing the rate of rotation of a special impeller placed within the flowing stream; the rate of flow is directly proportional to impeller rpm (within certain limits).

19.2K Flows measured with Venturi meters take advantage of a basic principle of hydraulics, the Bernoulli effect: When a liquid is forced to go faster in a pipe or channel, its internal pressure drops.

Answers to questions on page 416.

19.2L In a fluid (liquid or gas) chemical feeder, the indication of quantity per unit of time, or liquid chemical feed rate, may be provided by use of a rotameter or built-in calibrated pump with indicated output settings.

19.2M Process variables commonly monitored or controlled by process instrumentation include turbidity, pH, chlorine residuals, electrical conductivity, ORP, alkalinity, and temperature.

19.2N The two general systems for transmission of measurement signals are analog (4–20 mA) or digital data.

ANSWERS TO QUESTIONS IN LESSON 2

Answers to questions on page 418.

19.3A The purpose of instrumentation indicators is to give a visual representation of a variable's present value, either as an analog or digital display.

19.3B An analog display uses some type of pointer or graphical display against a scale.

19.3C Chart recorders, which by nature also serve as indicators, give a permanent record of how a variable changes with time by way of a moving chart.

19.3D Factors that can disrupt instrument and control systems include temporary power failures, tripped panel circuit breakers, voltage surges resulting in blown fuses, static electricity, and excessive heat.

Answers to questions on page 421.

19.3E Recorders are indicators designed to show how the value of the variable has changed with time.

19.3F The two main types of chart recorders are the strip-chart type and the circular-chart type.

19.3G Chart recorders are typically electrical, powered by the instrumentation and control system electrical supply. Battery-backed power for the recorders may be used, if required, as a backup power source.

19.3H Alarms may produce visual or audible warning signals.

Answers to questions on page 422.

19.3I An operator might bypass the operation of a controller in an emergency or when, in the judgment of the operator, the controller is not exercising control within sensible limits. To bypass a controller, switch to the manual mode of operation.

19.3J The basic principle to guide you in setting up the frequency of ON/OFF operation of a piece of equipment is to set the ON/OFF controls to operate or cycle associated equipment on and off no more often than actually necessary for plant operation.

19.3K Pumps can be prevented from running after a loss of level signal by electrical circuitry designed so the pump will turn off on an open signal circuit and on only with a closed circuit.

19.3L Pump control panels may include automatic or manual alternators (two pumps) or sequencers (more than two pumps). This provision allows the total pump operating time required for a particular system to be distributed equally among all the pumps at a pump station.

Answers to questions on page 425.

19.3M Pneumatic instrumentation depends on a constant source of clean, dry, pressurized air for reliable operation.

19.3N Oil is removed from instrument air by filtering the air through special oil-absorbent elements. Water is removed from the air by passing the moisture-laden air through an air-drying system consisting of a moisture separation column and air dryers.

Answers to questions on page 427.

19.3O An operator should be especially careful when working with laboratory instruments because some of these instruments are very delicate and replacement parts, such as glass vessels or pH electrodes, are quite expensive. Moreover, the use of some of these instruments requires the regular handling of laboratory glassware and other breakable items.

19.3P Operators should become familiar with the testing and calibration of plant measuring and control systems in order to discuss needed repairs or adjustments with an instrument technician and perhaps assist with that work. This knowledge also enables operators to analyze the effects of instrument problems on continued plant operation and to handle emergency situations created by instrument failure.

19.3Q The most useful piece of general electrical test equipment is the VOM, that is the volt-ohm-milliammeter, commonly referred to as a multimeter.

Answers to questions on page 429.

19.3R Computer control systems used for process control are classified by two commonly used terms: the distributed control system (DCS) and the supervisory control and data acquisition (SCADA) system.

19.3S The five basic components of the computer control system are the process instrumentation and control devices, the input/output (I/O) interface, the central processing unit (CPU), the communication interfaces, and the human machine interface (HMI). These components are the means by which the control system gathers and distributes information for the human operator and the process instrumentation and equipment.

19.3T Computer control systems help operation and maintenance personnel in preparing daily, weekly, and monthly maintenance schedules, monitoring the spare parts inventory status, ordering additional spare parts, printing out work orders, and recording completed work assignments.

Answers to questions on page 431.

19.3U Computer control systems record and store (archive) information that historically was recorded on paper strip-charts.

19.3V Cost savings associated with the use of a computer control system frequently include reduced labor costs for operation, maintenance, and monitoring functions that were formerly performed manually. Precise control of chemical feed rates by a computer control system eliminates wasteful overdosing. Preventive maintenance monitoring can save on equipment and repair costs, and, as previously noted, energy savings may result from off-peak (reduced) electrical power rates.

19.3W The greatest challenge for operators using computer control systems is to realize that just because a computer says something, this does not mean that the computer is always correct. Operators will always be needed to question and analyze the results from computer control systems.

Answers to questions on page 431.

19.4A The key to proper instrument operation and maintenance (O&M) is the operator's practical understanding of the system. Operators must know how to recognize malfunctioning instruments so as to prevent prolonged damaging operation, shut down and prepare devices for seasonal or other long-term nonoperation, and perform preventive (and minor corrective) maintenance tasks to ensure proper operation in the long term.

19.4B General tasks expected of operators of instrumentation systems can be summed up, from an operations standpoint, as learning, and constant attention to, what constitutes normal function; and, from a maintenance standpoint, as ensuring proper and continuing protection and care of each component.

19.4C Two of the surest indications of a serious electrical problem in instruments or power circuits are smoke or a burning odor.

Answers to questions on page 432.

19.4D Moisture can be controlled in an electrical panel room containing instrumentation by leaving some power components on (such as a power transformer) to provide space heat. In a known moist environment, sealed instrument cases may be protected for a while with a container of desiccant.

19.4E Pneumatic instrumentation should be purged with dry air before shutdown to rid the individual parts of residual oil and moisture to minimize internal sticking and corrosion while standing idle.

Answers to questions on page 433.

19.4F Regular preventive maintenance duties should be printed on forms or cards (or appear on a computer screen) for use by operators as a reminder, guide, and record of preventive maintenance tasks performed.

19.4G To obtain a technical manual for an instrument, contact the manufacturer. Be sure to provide all relevant serial/model numbers in your request to the manufacturer for a manual.

19.4H Instrument supplies and spare parts that should always be available include charts, pens, pen cleaners, and ink, and any other parts listed in the technical manual as necessary for instrument operation or service.

Answers to questions on page 433.

19.4I Operational checks on instrumentation equipment are performed by always observing each system for its continuing signs of normal operation, and cycling some indicators by certain test methods.

19.4J If a recorder trace is altered from its usual pattern during the process of checking an instrument, the operator causing the upset should initial the chart appropriately and note the time.

CHAPTER 20

SAFETY

by

Joe Monscvitz

TABLE OF CONTENTS

Chapter 20. SAFETY

LESSON 3

LEARNING OBJECTIVES

Chapter 20. SAFETY

Following completion of Chapter 20, you should be able to:

1. List the responsibilities of operators and management for waterworks safety.
2. Identify and safely handle hazardous chemicals.
3. Recognize fire hazards and properly extinguish various types of fires.
4. Safely maintain waterworks equipment and facilities.
5. Properly operate and maintain vehicles.
6. Recognize electrical hazards.
7. Safely perform duties in a laboratory.
8. Protect other operators and yourself while working in and around waterworks facilities.

WORDS

Chapter 20. SAFETY

ACUTE HEALTH EFFECT ACUTE HEALTH EFFECT

An adverse effect on a human or animal body, with symptoms developing rapidly.

ARCH ARCH

(1) The curved top of a sewer pipe or conduit.

(2) A bridge or arch of hardened or caked chemical that will prevent the flow of the chemical.

CONFINED SPACE CONFINED SPACE

A space that has the following characteristics:

(1) Is large enough and so configured that an employee can bodily enter and perform assigned work

(2) Has limited or restricted means for entry or exit (for example, manholes, tanks, vessels, silos, storage bins, hoppers, vaults, and pits are spaces that may have limited means of entry)

(3) Is not designed for continuous employee occupancy

Also see DANGEROUS AIR CONTAMINATION and OXYGEN DEFICIENCY.

CONFINED SPACE, CLASS A CONFINED SPACE, CLASS A

A confined space that presents a situation that is immediately dangerous to life or health (IDLH). These include but are not limited to oxygen deficiency, explosive or flammable atmospheres, and concentrations of toxic substances.

CONFINED SPACE, CLASS B CONFINED SPACE, CLASS B

A confined space that has the potential for causing injury and illness, if preventive measures are not used, but is not immediately dangerous to life and health.

CONFINED SPACE, CLASS C CONFINED SPACE, CLASS C

A confined space in which the potential hazard would not require any special modification of the work procedure.

CONFINED SPACE, PERMIT-REQUIRED (PERMIT SPACE) CONFINED SPACE, PERMIT-REQUIRED (PERMIT SPACE)

A confined space that has one or more of the following characteristics:

(1) Contains or has a potential to contain a hazardous atmosphere

(2) Contains a material that has the potential for engulfing an entrant

(3) Has an internal configuration such that an entrant could be trapped or asphyxiated by inwardly converging walls or by a floor that slopes downward and tapers to a smaller cross section

(4) Contains any other recognized serious safety or health hazard

DANGEROUS AIR CONTAMINATION DANGEROUS AIR CONTAMINATION

An atmosphere presenting a threat of causing death, injury, acute illness, or disablement due to the presence of flammable or explosive, toxic, or otherwise injurious or incapacitating substances.

(1) Dangerous air contamination due to the flammability of a gas, vapor, or mist is defined as an atmosphere containing the gas, vapor, or mist at a concentration greater than 10 percent of its lower explosive (lower flammable) limit (LEL).

(2) Dangerous air contamination due to a combustible particulate is defined as a concentration that meets or exceeds the particulate's lower explosive limit (LEL).

(3) Dangerous air contamination due to the toxicity of a substance is defined as the atmospheric concentration that could result in employee exposure in excess of the substance's permissible exposure limit (PEL).

NOTE: A dangerous situation also occurs when the oxygen level is less than 19.5 percent by volume (OXYGEN DEFICIENCY) or more than 23.5 percent by volume (OXYGEN ENRICHMENT).

DECIBEL (DES-uh-bull) DECIBEL

A unit for expressing the relative intensity of sounds on a scale from zero for the average least perceptible sound to about 130 for the average level at which sound causes pain to humans. Abbreviated dB.

MATERIAL SAFETY DATA SHEET (MSDS) MATERIAL SAFETY DATA SHEET (MSDS)

A document that provides pertinent information and a profile of a particular hazardous substance or mixture. An MSDS is normally developed by the manufacturer or formulator of the hazardous substance or mixture. The MSDS is required to be made available to employees and operators or inspectors whenever there is the likelihood of the hazardous substance or mixture being introduced into the workplace. Some manufacturers are preparing MSDSs for products that are not considered to be hazardous to show that the product or substance is not hazardous. Also see SAFETY DATA SHEET (SDS).

OLFACTORY (all-FAK-tore-ee) FATIGUE OLFACTORY FATIGUE

A condition in which a person's nose, after exposure to certain odors, is no longer able to detect the odor.

OSHA (O-shuh) OSHA

The Williams-Steiger Occupational Safety and Health Act of 1970 (OSHA) is a federal law designed to protect the health and safety of industrial workers and the operators of water supply systems and treatment plants. The act regulates the design, construction, operation, and maintenance of water supply systems and water treatment plants. OSHA also refers to the federal and state agencies that administer the OSHA regulations.

OXYGEN DEFICIENCY OXYGEN DEFICIENCY

An atmosphere containing oxygen at a concentration of less than 19.5 percent by volume.

OXYGEN ENRICHMENT OXYGEN ENRICHMENT

An atmosphere containing oxygen at a concentration of more than 23.5 percent by volume.

PERMIT-REQUIRED CONFINED SPACE (PERMIT SPACE) PERMIT-REQUIRED CONFINED SPACE (PERMIT SPACE)

See CONFINED SPACE, PERMIT-REQUIRED (PERMIT SPACE).

SAFETY DATA SHEET (SDS) SAFETY DATA SHEET (SDS)

Safety data sheets (SDSs) are an essential component of the Globally Harmonized System of Classification and Labeling of Chemicals (GHS) and are intended to provide comprehensive information about a substance or mixture for use in workplace chemical management. They are used as a source of information about hazards, including environmental hazards, and to obtain advice on safety precautions. In the GHS, they serve the same function that the material safety data sheet (MSDS) does in OSHA's Hazard Communication Standard. The SDS is normally product related and not specific to the workplace; nevertheless, the information on an SDS enables the employer to develop an active program of worker protection measures, including training, which is specific to the workplace, and to consider measures necessary to protect the environment.

TAILGATE SAFETY MEETING TAILGATE SAFETY MEETING

Brief (10 to 20 minutes) safety meetings held every 7 to 10 working days. The term comes from the safety meetings regularly held by the construction industry around the tailgate of a truck.

CHAPTER 20. SAFETY

(Lesson 1 of 3 Lessons)

20.0 IMPORTANCE OF SAFETY

20.00 What Is Safety?

Webster states that safety is the following:

> "The condition of being safe; freedom from exposure to danger; exemption from hurt, injury, or loss; to protect against failure, breakage, or other accidents; knowledge of or skill in methods of avoiding accident or disease."

Safety is more than words. Safety is the action of Webster's definition. Safety is using one's knowledge or skill to avoid accidents or to protect oneself and others from accidents. Safety is a form of preventive maintenance that includes equipment and machinery and its proper handling. Who is responsible for safety? Everyone should be responsible, from top management to all employees.

Management has its responsibilities for safety; its main function is to set the tone and provide the training and monies for an effective safety program. (See Chapter 23, "Administration," for a description of the elements of an effective safety program and a detailed discussion of management's responsibilities for safety.) Management should be responsible for the following:

1. Establish a safety policy
2. Assign responsibility for accident prevention, which includes descriptions of the duties and responsibilities of a safety officer, department heads, line supervisors, safety committees, and operators
3. Appoint a safety officer or coordinator
4. Establish realistic goals and periodically revise them to ensure continuous and maximum effort
5. Evaluate the results of the safety program

While management has its responsibilities, the operators also have their particular responsibilities. Operators should:

1. Perform their jobs in accordance with established safe procedures
2. Recognize the responsibility for their own safety and that of fellow operators
3. Report all injuries
4. Report all observed hazards
5. Actively participate in the safety program

QUESTIONS

Please write your answers to the following questions and compare them with those on page 487.

20.0A What are management's responsibilities for safety?

20.0B What are the operators' responsibilities for safety?

20.01 Causes of Accidents

A safety program has one objective: To prevent accidents. Accidents do not happen, they are caused. They may be caused by unsafe acts of operators or result from hazardous conditions or may be a combination of both. An analysis of accident statistics reveals that operator negligence and carelessness are the causes of most accidents. We know how to do our job safely, but we just do not do it safely.

Accidents reduce efficiency and effectiveness. An accident is that occurrence in a sequence of events that usually produces unintended injury, death, or property damage. Accidents affect the lives and morale of operators. They raise the costs not only to management but also to operators.

Safety authorities tell us that nine out of ten injuries are the result of unsafe acts of either the person injured or someone else. Here are some of the principal reasons for unsafe acts.

IGNORANCE. This may be due to lack of experience or training, or to a temporary condition that prevents the recognition of a hazard.

INDIFFERENCE. Some people know better but do not care. They take unnecessary risks and disregard the rules or instructions.

POOR WORK HABITS. Some people either do not learn the right way of doing things, or they develop a wrong way. Supervisors or fellow operators who see things done unsafely must speak up.

LAZINESS. Laziness affects speed and quality of work. Laziness also affects safety because safety requires an effort. In most jobs, you cannot reduce your safety effort and still maintain the same level of safety, even if you slow down or lower your quality standards.

HASTE. When we rush, we work too fast to think about what we are doing, we take dangerous shortcuts, and we are more likely to be injured.

POOR PHYSICAL CONDITION. Some of us just will not take reasonable care of ourselves. We ignore our bodily needs in regard to exercise, rest, and diet, lessening our endurance and alertness.

TEMPER. Impatience and anger cause many accidents. Again, our thinking is interfered with and the way prepared for an accident.

Every one of these reasons for unsafe acts could be considered operator negligence or carelessness.

20.02 Steps to Avoid Accidents

Some specific steps you can take to avoid accidents are the following:

1. Ask about the safety program established by management and then actively participate in safety training sessions and *TAILGATE SAFETY MEETINGS.*[1]
2. Routinely use the safety equipment described in Section 20.7 (for example, self-contained breathing apparatus) and personal protective gear (hard hat, safety goggles, gloves) provided by your employer.
3. Follow safe procedures at all times. This means taking the time needed to do every job safely.

QUESTIONS

Please write your answers to the following questions and compare them with those on page 487.

20.0C What is an accident?

20.0D List the principal reasons for unsafe acts.

20.0E What problems are caused by laziness?

20.1 CHEMICAL HANDLING

20.10 Safe Handling of Chemicals

The water treatment plant operator may be required to handle a wide variety of chemicals, depending on the type of plant. In a simple well system, chlorine may be the only chemical used. In more complicated plants, there may be chlorine, other gases, sulfuric acid, lime, alum, powdered activated carbon, and anhydrous ammonia. All of these chemicals fall into groups: acids, hydroxides, gases, salts, organics, and solvents. Each group requires you, the operator, to have a good understanding of all types of chemicals. You must know how to handle the many problems associated with each of these elements or compounds. For example, you must know how to store chemicals; understand the fire problem; be aware of the tendency of chemicals to *ARCH*[2] in a storage bin; and know how to feed dry, how to feed liquid, and how to make up solutions. All of these situations may cause an unsafe condition. Overheating gas containers, dust problems with powdered carbon, burns caused by acid, and reactivity of each chemical under a variety of conditions that may cause fire and explosions are other safety hazards. You will need to know the usable limits of each chemical because of toxicity, the protective equipment required for each chemical, each chemical's antidote, and how to control fires caused by each chemical.

Although you may not regularly handle all of the chemicals listed here, you may come into contact with them from time to time. Try at least to learn all of the characteristics of the chemicals you will use regularly. For example, learn the boiling point, explosive limits, reactivity, flammability, first aid used for each chemical, and other characteristics that may prove helpful in preventing a safety hazard. Study the chemistry of each chemical used in the plant in order to have a safe plant in which to work. The following discussions concentrate on the characteristics of each substance and point out its hazards, reactivity, and the information needed to avoid conditions that may cause a safety problem.

20.11 Acids

Acids are used extensively in water treatment. For example, hydrofluoric acid is used in fluoride addition and hydrochloric acid is often used in cleaning. Table 20.1 lists many of the acids used in water treatment and gives their characteristics. This section can serve as a quick reference and a guide for learning some of each acid's limitations and its reactivity with other compounds.

The antidote to all acids is neutralization. However, one must be careful in how this is performed. Most often, large amounts of water will serve the purpose, but if the acid is ingested (swallowed), then milk of magnesia may be needed. If vapors are inhaled, first aid usually consists of providing fresh air, artificially restoring breathing (CPR), or supplying oxygen. In general, acids are neutralized by a base or alkaline substance. Baking soda is often used to neutralize acids on skin because it is not harmful on contact with your skin. To understand these reactions, you will need to know some acid/base chemistry. The knowledge of acid/base chemistry and fast reactions on your part may reduce the safety hazards involved in handling acids in water treatment.

20.110 Acetic Acid (Glacial) (CH_3COOH)

This chemical is stable when stored and handled properly. However, it may react violently with certain compounds such as ammonium nitrate, potassium hydroxide, and other alkaline materials. Strong oxidizing glacial acetic acid is a combustible material. Fires involving the acid may be extinguished with water, dry chemical, or carbon dioxide. Under such conditions as adding water, the diluted acid may produce hydrogen gas when it comes in contact with metals. When the chemical is

1. *Tailgate Safety Meeting.* Brief (10 to 20 minutes) safety meetings held every 7 to 10 working days. The term comes from the safety meetings regularly held by the construction industry around the tailgate of a truck.
2. *Arch.* (1) The curved top of a sewer pipe or conduit. (2) A bridge or arch of hardened or caked chemical that will prevent the flow of the chemical.

TABLE 20.1 ACIDS USED IN WATER TREATMENT

Name, Formula	Common Name	Available Forms	Specific Gravity	Flam-mability	Color	Odor	Containers
Acetic Acid, CH_3COOH	Ethanoic Acid	Solution	1.05	Yes	Clear	Sharp, pungent	Carboys, drums
Hydrofluosilicic Acid, H_2SiF_6	Fluosilicic Acid	Solution	1.4634	N/A[a]	Clear	Pungent fumes	Drums, trucks, R.R. tank cars
Hydrogen Fluoride Acid, HF	Hydrofluoric Acid	Liquid	0.987	N/A	Clear	Fumes, toxic	Drums, tank cars
Hydrochloric Acid, HCl	Muriatic Acid	Solution	1.16	N/A	Clear to yellow	Pungent, suffocating	Drums, carboys
Nitric Acid, HNO_3	—	Liquid	1.5027	N/A	Colorless, yellowish	Toxic fumes in presence of light	Drums, carboys, bottles
Sulfuric Acid, H_2SO_4	Oil of Vitriol; Vitriol	Solution	(60–66° Bé [b]) 1.841	N/A	Clear	Odorless	Bottles, carboys, drums, trucks, tank cars

a N/A = Not Applicable.
b Bé = Baumé scale, a scale for measuring specific gravity of liquids.

involved in a fire situation, a self-contained breathing apparatus must be used to protect the operator against suffocation and problems caused by corrosive vapors.

Most people find inhalation of acetic acid vapors in concentrations over 50 parts per million (ppm) intolerable, resulting in nose and throat irritations. Repeated exposure to high concentrations may produce congestion of the larynx. Skin contact with concentrated acetic acid can produce deep burns with skin destruction. High vapor concentration may blacken the skin and produce allergic reactions and eye irritation. Possible permanent damage or immediate burns are caused when the acid comes into contact with the eyes. If the acid is ingested, severe intestinal irritation will result as well as burns to the mouth and upper respiratory tract.

Operators should be protected by adequate exhaust facilities to ensure ventilation when working with acetic acid. At a minimum, exhaust hoods should have air velocity of 100 feet per minute (fpm; 30 m/min). Wear rubber gloves and an apron to prevent skin contact. Wear splash-proof goggles or a face shield to prevent any eye contact. Gas-tight goggles may also be needed to prevent vapors from irritating your eyes. An eye wash station must be readily available where this chemical is being handled. Also, respiratory equipment should be available for emergency use. Acetic acid can be handled safely by using adequate ventilation and safety equipment to prevent skin and eye contact. Remember also that acetic acid vapors can cause *OLFACTORY* (all-FAK-tore-ee) *FATIGUE*. This is a condition in which a person's nose, after exposure to certain odors, is no longer able to detect the odor. Acetic acid is detectable by your nose at 1 ppm, but documentation has shown operators tolerating up to 200 ppm.

If a leak or spill should occur, notify safety personnel and provide adequate ventilation. When cleaning up large spills, wear a self-contained breathing apparatus and equipment to prevent contact with eyes and skin. To clean up spill areas and remove chemical residue, cover the area with sodium bicarbonate and flush away with large quantities of water.

First aid for acetic acid exposure calls for removal of the victim to fresh air, rinsing the mouth and nasal passages with water, and checking for inhalation problems. If the acid made contact with the eyes, immediately irrigate with water for at least 15 minutes. Obtain medical attention. For skin contact problems, wash with water immediately; if the acid was swallowed, give three glasses of milk or water and obtain medical attention quickly. Acetic acid exposure, like all other acids, must be treated immediately to prevent damage to the victim.

20.111 Hydrofluosilicic Acid (H_2SiF_6)

This chemical is hazardous to handle under any conditions. Be extremely careful using this acid. The acid is a colorless, transparent, fuming, corrosive liquid. A pungent odor is created by the acid and contact causes skin irritation. When the acid vaporizes, it decomposes into hydrofluoric acid and silicon tetrafluoride. Hydrofluoric acid can attack glass.

When handling hydrofluosilicic acid, always use complete protective equipment, rubber gloves, goggles or face shield, rubber apron, rubber boots, and have lime slurry barrels, Epsom salt solution, and safety showers (Figure 20.1) available. Always provide adequate ventilation because its vapor can cause irritation to the respiratory system. Careful maintenance of protective equipment is essential because the fumes of the acid corrode metal or etch glass on the protective equipment.

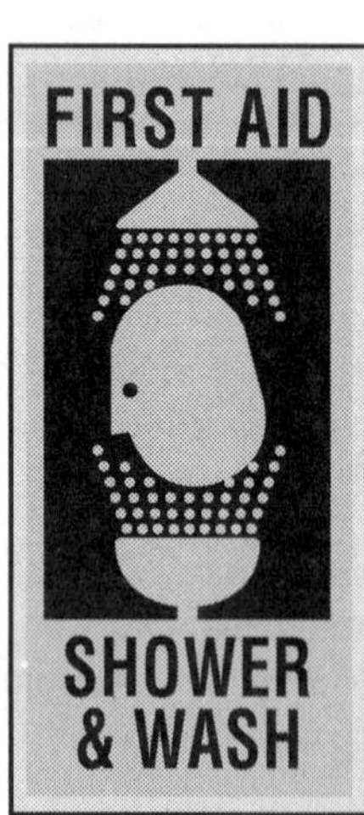

Fig. 20.1 Safety shower with face/eye wash facility
(Permission of Nevada Safety & Supply)

First aid for eye contact is to thoroughly flush with water for 15 minutes and get medical aid as soon as possible. For skin contact, wash the affected areas with water. For gross (large) contact, remove contaminated clothing under a safety shower and thoroughly wash entire body for 15 minutes or longer. In case of inhalation, remove operator to fresh air, restore breathing, if required, and get medical aid.

20.112 Hydrogen Fluoride (HF)

This acid is extremely poisonous, and it produces terrible sores when allowed to come into contact with the skin. The acid is a clear, corrosive liquid that has a pungent odor. All of the precautions discussed for hydrofluosilicic acid apply to this acid also.

20.113 Hydrochloric Acid (HCl)

This acid is used most often for cleaning in and around the treatment plant and is known as muriatic acid. The acid is also used very frequently in the laboratory. Hydrochloric acid is stable when properly contained and handled. This acid is one of the strong mineral acids and, therefore, is highly reactive with metals and these oxides: hydrocarbon, amine, and carbonate compounds. The acid liberates significant levels of hydrogen chloride gas (HCl) because of its vapor pressure at room temperature, and gives off large amounts of gas when heated. In reactions with most metals, hydrochloric acid will produce hydrogen gas.

Inhalation of HCl vapors or mists for long periods can cause damage to teeth and irritation to the nasal passages. Concentrations of 750 ppm or more will cause coughing, choking, and produce severe damage to the mucous membranes of the respiratory tract. In concentrations of 1,300 ppm, HCl is dangerous to life. Ingestion can cause burns of the mouth and digestive tract.

When handling HCl, provide adequate exhaust facilities to ensure ventilation and wear protective clothing and equipment to prevent body contact with the acid. Use rubber gloves, rubber apron, rubber boots, and wear a long-sleeved shirt when handling hydrochloric acid. To protect your eyes against splashing of the acid, you must wear safety goggles or a face shield. There should always be an eye wash station and safety shower located near areas where this acid is to be used.

First aid consists of thoroughly flushing the eyes with running water for 15 minutes and securing medical aid. If hydrochloric acid comes in contact with skin, wash the affected areas with water. For gross contact, remove clothing under the safety shower and continue showering for 15 minutes or longer. Should the acid be ingested, give water and milk of magnesia. Do not induce vomiting; get medical aid. In case of inhalation, remove the victim to fresh air, restore breathing if required, and get medical aid.

Store acid containers closed in a clean, cool, open, and well-ventilated area. Keep out of the sun. Keep the acid away from oxidizing agents or alkaline materials. Provide emergency neutralization materials in use areas.

20.114 Nitric Acid (HNO_3)

Like hydrochloric acid, nitric acid is one of the most commonly used acids in the water treatment laboratory and plant. The acid is a powerful oxidizing agent and attacks most metals. Nitric acid is stable when properly handled and placed into a proper container. The acid is one of the strong mineral acids, and is highly reactive with materials such as metals. When handling nitric acid, use protective clothing and equipment to prevent body contact with the liquid. Such equipment includes rubber gloves, rubber apron, and safety goggles or a face shield for eye protection against splashing of the acid. Nitric acid is a strong, poisonous, and highly corrosive liquid and must be handled carefully. The acid forms toxic fumes in the presence of light; therefore, it should be kept out of the sun.

Like other acids, nitric acid should be stored in clean, cool, well-ventilated areas. The areas should have an acid-resistant floor and adequate drainage. Keep it away from oxidizing agents and alkaline materials. Protect containers from damage or breakage. Avoid contact with skin and provide emergency neutralization materials and safety equipment in use areas.

First aid for skin contact is to flush thoroughly with water for 15 minutes. Get medical aid if needed. This acid will cause burns, but these can be greatly reduced if the contact area is immediately flushed. For gross (large) contact, remove contaminated clothing under a safety shower. For eye contact, flush with water for 15 minutes and get medical aid. If acid is ingested, give water with milk of magnesia; get medical aid. For inhalation, remove to fresh air, restore breathing if required, and get medical aid.

20.115 Sulfuric Acid (H_2SO_4)

This mineral acid is highly corrosive and will attack most metals. Sulfuric acid is also very reactive to the skin and must be handled with extreme care or you will suffer severe burns. Even when the acid is diluted, it is highly corrosive and must be contained in rubber-, glass-, or plastic-lined equipment. The acid will decompose clothing and shoes. Sulfuric acid should not come in contact with potassium permanganate or similar compounds. Sulfuric acid reacts violently with water. Always pour acid into water while stirring to prevent the generation of steam and hot water, which could boil over the container and cause serious acid burns.

As with other mineral acids, this material is stable when properly contained and handled. When you are handling sulfuric acid, you must use protective clothing and equipment to prevent body contact with the acid. Wear rubber gloves and safety goggles or a face shield for eye protection against splashing. Also, wear a rubber apron, rubber boots, and long-sleeved shirt. The eye wash station and safety shower must be located nearby where the acid is being handled. The area should be well ventilated, the acid should be stored in closed containers in a clean, cool, open area. The area should have an acid-resistant floor that drains well. Keep away from oxidizing agents and alkaline materials and protect the containers from damage or breakage.

Because the acid is highly corrosive and causes severe burns, first aid for sulfuric acid exposure must be immediate to avoid substantial damage to human tissue. For eye contact, flush thoroughly with running water for 15 minutes and get medical aid—the first seconds are important. For skin contact, wash affected areas thoroughly with water. For gross (large) contact, remove contaminated clothing under the safety shower with prolonged washing for at least 15 minutes. In cases of ingestion, give water and milk of magnesia; get medical aid. When working with sulfuric acid, avoid skin contact and always provide emergency neutralization materials and a safety shower near the work areas.

QUESTIONS

Please write your answers to the following questions and compare them with those on pages 487 and 488.

20.10A What does an operator need to know about chemicals used in a water treatment plant?

20.11A What should be done if an operator inhales acid vapors?

20.11B Acetic acid will react violently with which compounds?

20.11C Under what conditions can acetic acid be handled safely?

20.11D What protective equipment is necessary for handling hydrofluosilicic acid?

20.11E How can the inhalation of hydrochloric acid (HCl) vapors or mists cause damage to operators?

20.11F How should nitric acid be stored?

20.12 Bases

The bases that are used in water treatment are known as hydroxides. From a functional standpoint, they are used to raise pH. Most common bases are compounds of sodium, calcium, and ammonium, which are strong bases. However, there are other weak bases, such as silicate, carbonate, and hypochlorite. From the standpoint of safety, both weak and strong bases must be given the same consideration when being handled. Some are very toxic and will attack human tissue very rapidly and cause burns. Explosive reactions will occur when bases come in contact with an acid and hazardous decomposition products are created under certain conditions.

Bases must be neutralized with dilute acids. However, the operator must work carefully because, under some conditions, there may be other reactions, such as with hypochlorite compounds. Therefore, you should understand acid/base chemistry before handling any of the basic compounds used in water treatment. Table 20.2 gives some of the common basic compounds used in water treatment. The following sections will discuss some of their characteristics and the precautions the operator must use to safely handle such compounds.

20.120 Ammonia (NH_3)

The operator may use one of two forms of ammonia: anhydrous or hydroxide. The first (anhydrous) is in a gas form and requires one type of consideration. The hydroxide is a liquid and requires another type of consideration. Anhydrous ammonia in the gaseous state is colorless; it is about 0.6 times as heavy as air. In a liquid state, ammonia is also colorless, 0.68 times as heavy as water, and it vaporizes rapidly. Ammonia gas is capable of

TABLE 20.2 BASES USED IN WATER TREATMENT

Name, Formula	Common Name	Available Forms	Spec. Grav. or lb/ft³	Flammability	Color	Odor	Containers
Ammonia, NH_3	Ammonia	Liquid, gas	0.04813 lb/ft³ @ 0°C	Explosive with air; will burn in air	Colorless	Irritating	Cylinders, tanks, trucks
Calcium Hydroxide and Oxide, $Ca(OH_2)$ and CaO	Hydrated Lime and Quicklime	Dry powder, lump	50–70 lb/ft³	N/A[a]	White	Dust	Bags, bulk, trucks
Sodium Hydroxide, $NaOH$	Caustic Soda, Lye	Lump, liquid, flake	1.524	May cause flammable condition	Opaque white	Toxic, pungent	Drums, bulk, trucks
Sodium Silicate, Na_2SiO_2	Water Glass	Liquid	1.35–1.42	N/A	Opaque	N/A	Drums, bulk, trucks
Hypochlorite Compounds, $NaOCl$, $Ca(OCl)_2$	HTH	Powder, liquid	—	Explosive with antifreeze	White	Toxic Cl_2, pungent	Cans, drums
Sodium Carbonate, Na_2CO_3	Soda Ash	Powder	23, 35, and 65 lb/ft³	N/A	White	Dust	Bags, bulk, trucks

a N/A = Not Applicable.

forming explosive mixtures with air. For your own safety, be aware of the possibility of suffocation since the gas can displace air, which contains oxygen. Although the vapors are not poisonous, they can and will irritate the mucous membranes of the eyes, nose, throat, and lungs. Irritation will be detected in concentrations of 5.0 ppm and when human tissue comes in contact with the liquid, it will cause severe burns.

When handling ammonia or working in an ammonia environment, respiratory protection is a requirement. For entry into emergency areas, use only a self-contained breathing apparatus. Install a good ventilation system to control vapors in the application room. Use protective clothing, rubber gloves, apron, boots, and face and eye protection if you are going to work with ammonia for long periods.

Care must be used when storing or transporting containers. Always keep cylinder caps in place when cylinders are not in use. Store cylinders in a cool, dry location away from heat, and protect from direct sunlight. Storage near radiators, steam pipes, or other sources of heat may raise the pressure to a dangerous point, whereas dampness may cause excessive corrosion. Do not store in the same room with chlorine. Always use lifting clamps or cradles. Avoid hoisting the cylinders using ropes, cables, or slings, and never drop the containers. Control ammonia leaks. They can be detected by odor or by using a cloth swab soaked with hydrochloric acid. This will form a white cloud of ammonium chloride.

Ammonia gas will burn if it is blended with air in a mixture containing 15 to 28 percent ammonia by volume. Check cylinder valve stems for leaks; tighten the packing gland nut only with a special wrench provided for such purposes. If a serious leak in a cylinder cannot be controlled, place the container in a vat of water. Fifty-three pounds of ammonia will dissolve in 100 pounds of water at 68°F (20°C). Never neutralize liquid ammonia with an acid. The reaction generates a lot of heat, which may speed up the release of ammonia gas.

First aid for skin contact with ammonia is to flush with large amounts of water for 5 to 10 minutes, and get medical aid. Remove contaminated clothing under a safety shower. For eye contact, flush thoroughly with water for 15 minutes immediately, and get medical aid. In the case of inhalation, remove to fresh air and restore breathing. If required, get medical aid. Throat burns should be rinsed with 2 percent boric acid solution. Urge the patient to drink large amounts of milk. For irritation of nose and throat because of exposure, see a physician.

Ammonium hydroxide is an aqueous (watery) solution of anhydrous ammonia and is quite volatile (will evaporate) at atmospheric temperatures and pressures. This solution can cause local skin irritations. A strong solution will cause human tissue destruction on contact with eyes, skin, and mucous membranes of the respiratory system, so avoid contact with the compound. The solution will cause severe burns depending upon solution concentration and length of contact time. The solution's vapor causes the same effects as the gas. First aid should be the same as for anhydrous ammonia.

20.121 Calcium Hydroxide ($Ca(OH)_2$)

Hydrated lime (calcium hydroxide) and quicklime (calcium oxide) are two forms of lime commonly used in water treatment plants. Hydrated lime is the least troublesome of the two forms. Hydrated lime is less caustic and is, therefore, less irritating to the skin, but can cause injury to eyes. However, as a dust it is just as hazardous as quicklime. Quicklime is a strong caustic and irritating to personnel exposed to the compound. Although quicklime is not explosive, when mixed with water, a great deal of heat is generated, which can cause explosions.

Both quicklime and hydrated lime should be stored in cool, dry areas. Care must be taken to avoid mixtures of alum and quicklime, since quicklime tends to absorb the water away from the alum that forms as alum crystallizes (water of crystallization). In a closed container, this could lead to a violent explosion. Equal care should be taken to avoid mixtures of ferric sulfate and lime.

When handling both forms of lime, the operator should use chemical goggles and a suitable dust mask to protect the eyes and mucous membranes. Also, wear proper clothing to protect the skin because, with long contact, the lime can cause dermatitis (skin irritation) or burns, particularly at perspiration points. Always shower after handling quicklime. All operators should wear a face shield when inspecting lime slakers. Hot lime suspension that splatters on the operator may cause severe burns of the eyes or skin. The hot mist coming from the slakers is also dangerous. The loss of water supply to a lime slaker can create explosive temperatures.

First aid for lime burns, which are like burns from other caustics, consists of alternately washing with water and a mild acetic acid solution. One may also use large amounts of soap and water. For eye contact, wash immediately with large amounts of warm water and rinse with a boric acid solution. Get medical aid. For irritation of nose and throat because of exposure, see a physician.

20.122 Sodium Hydroxide (Caustic Soda) ($NaOH$)

Sodium hydroxide is available in pellet, liquid, and flake forms. Caustic soda usually comes as a 50 percent solution of sodium hydroxide. This base is a strong caustic alkali and very hazardous to the operator. This compound is extremely reactive. Sodium hydroxide absorbs carbon dioxide from the air, reacting violently or explosively with acid and a number of organic compounds. Caustic soda dissolves human skin; generates heat when mixed with water; and reacts with amphoteric metals (such as aluminum) generating hydrogen gas, which is flammable and may explode if ignited. Sodium hydroxide can be dissolved in water; the resulting solution is often used for the adjustment of pH because it is a liquid and easy to feed. This base is extensively used in water treatment. Because of its everyday use, you may forget just how hazardous this compound is, and through neglect may injure yourself or another operator. Only trained and protected operators should undertake spill cleanup. The operator must act cautiously, dilute the spill with water, and neutralize with a dilute acid, preferably acetic.

When handling caustic soda, control the mists with good ventilation. Protect your nose and throat with an approved respiratory system. For eye protection, you must wear chemical worker's goggles or a full face shield to protect your eyes. There must be an eye wash station and safety shower at or near the work station for this chemical. Protect your body by being fully clothed, and by using impervious gloves, boots, apron, and face shield.

Special precautions to be taken when handling or storing caustic soda include: prevent eye and skin contact, do not breathe dusts or mists, and avoid storing this chemical next to strong acids. Dissolving sodium hydroxide in water or other substances generates excessive heat and causes splattering and mists. Solutions of sodium hydroxide are viscous (thick) and slippery. Caustic can quickly and permanently cloud vision if not immediately flushed out of eyes. Determine location of safety showers and eye wash stations before starting to work with caustic soda.

First aid for the eyes consists of irrigating the eyes immediately and continuously with flowing water for at least 30 minutes. Prompt medical attention is essential. For skin burns, immediate and continuous, thorough washing in flowing water for 30 minutes is important to prevent damage to the skin. Consult a physician, if required. In case of inhalation, remove victim to fresh air, call physician, or transport injured person to a medical facility. For ingestion, give large amounts of water or milk and immediately transport injured person to a medical facility; do not induce vomiting.

You may also have occasion to use sodium hydroxide as flakes or pellets. All of the precautions stated for liquid caustic also apply for the flake form.

20.123 Sodium Silicate (Na_2SiO_2)

This chemical is a liquid as used in water treatment. However, it is nontoxic, nonflammable, and nonexplosive, but presents the same hazards to the eyes and skin as any other base compounds. Sodium silicate is a strong alkali and should be handled with care by using goggles or face shield and wearing gloves and protective clothing. The chemical will cause damage to the eyes and skin, but it is less dangerous than other alkaline compounds used in water treatment.

First aid for the eyes is to flush immediately and thoroughly with flowing water for at least 15 minutes. Get medical attention. If sodium silicate makes contact with skin, wash thoroughly with water, particularly if the solution is hot. Then, wash the skin with a 10 percent solution of ammonium chloride or 10 percent acetic acid. For ingestion, give plenty of water and dilute vinegar, lemon, or orange juice. Follow this with milk, white of eggs beaten with water, or olive oil. Call a physician.

20.124 Hypochlorite (OCl^-)

A number of hypochlorite compounds are commercially available for use in water treatment. If you understand the precautions for one such compound, you will know what steps must be taken with other hypochlorite compounds such as calcium, sodium, or lithium. These chemicals may be used in either a liquid or dry form. There are several grades of hypochlorite compounds, but all are good oxidizers and are used for disinfection. When these compounds come into contact with organic materials, their decomposition releases heat very rapidly and produces oxygen and chlorine. Although hypochlorite compounds are nonflammable, they may cause fires when they come in contact with heat, acids, organic matter, or other oxidizable substances.

All solutions of hypochlorite compounds attack the skin, eyes, or other body tissues with which they come into contact. When handling hypochlorite, liquid or dry, use suitable protective clothing such as rubber gloves, aprons, and goggles or a face shield. Be aware that many times these compounds are stored in containers and give off chlorine gas when opened. Store these compounds in a cool, dry, dark area.

First aid for eyes is to flush with plenty of water for at least 15 minutes and see a physician. If hypochlorite compounds come in contact with the skin, flush thoroughly with water for at least 15 minutes, and get medical attention, as needed. In case of ingestion, wash out mouth thoroughly with water, give plenty of water to drink, and get medical attention. For inhalation, move the victim into fresh air and get medical attention.

Overexposure to any of the hypochlorite compounds may produce severe burns, so avoid contact with these compounds. They are hazardous and can attack skin, eyes, mucous membranes, and clothing.

20.125 Sodium Carbonate (Na_2CO_3)

Soda ash (sodium carbonate) is a mild alkaline compound, but requires safety precautions to minimize hazards when handling the chemical. An adequate ventilation system is needed to control the dust generated by the compound. Wear protective gear such as chemical safety goggles or a face shield, a well-fitting dust respirator, and protective clothing to avoid skin contact. You should protect yourself by using a suitable cream or petroleum jelly on exposed skin surfaces such as neck and hands. This compound's dust irritates the mucous membranes and prolonged exposure can cause sores in your nasal passage.

First aid for exposure to eyes (dust or solution) requires irrigation with water immediately for at least 15 minutes. Consult a physician if the exposure has been severe. For skin exposure, wash with large amounts of water; for contaminated clothing, wash before reusing. For inhalation or irritation of the respiratory tract, gargle or spray with warm water, and consult a physician, as needed.

QUESTIONS

Please write your answers to the following questions and compare them with those on page 488.

20.12A What are the two forms of ammonia used by operators?

20.12B How should ammonia be stored?

20.12C What are the two forms of lime used in water treatment plants?

20.12D What would you do if someone swallowed sodium hydroxide?

20.12E What would you do if sodium silicate came in contact with your skin?

20.13 Gases

There are a number of gases used in water treatment (Table 20.3). Most are supplied in steel drum containers; others must be generated on site. Some gases can be seen; others call the operator's attention by odor; and still others cannot be seen or detected by odor, yet are deadly. In this section, we will only discuss those gases that are supplied in containers that the operator must connect, disconnect, handle, or store. (See Section 20.120 for a discussion of ammonia.)

Exposure to the liquid forms of these gases usually will cause damage to human tissue, such as skin burns, but the most important factor to remember is the displacement of oxygen. Most gases are heavier than air and remove air from a room by displacement. Therefore, it is very important to have the right type of ventilation and respiratory protection. Use only the self-contained breathing apparatus when working in emergency areas.

20.130 Chlorine (Cl_2)

Safety is of the utmost importance when handling chlorine. Do not treat chlorine cylinders roughly; never drop them or permit collision of two or more cylinders. Never hoist chlorine cylinders by the neck. Always use lifting clamps or cradles—do not use ropes, cables, or chains. Store the cylinders in such a way that they cannot fall. Do not store chlorine cylinders below ground level and always keep the protective cap on the cylinder when it is not in use. Mark the empty containers and store them apart from full cylinders. Always store containers in an upright position in a clean, dry location free of flammable materials. The storage area must be equipped with forced-exhaust ventilation with starting switches located on the outside of the storage room. Some systems have a microswitch on the door so the ventilation will start automatically whenever the door is opened. Ventilation must provide at least one complete air change per minute. The temperature of the storage room should never be permitted to approach 140°F (60°C). Protect the chlorine cylinder from heat sources and never use an open flame on cylinders or pipes carrying chlorine. If chlorine is heated, the increase in temperature will cause an expansion of the gas, which results in an increase in pressure inside of the cylinders or piping, resulting in rupture of the containers.

When working with chlorine, be equipped to control chlorine leaks, which are most often found in the control valve. Repair kits are available for the 100- and 150-pound (45 and 68 kg) cylinders, as is emergency equipment for the 1-ton (909 kg) tanks for controlling leaks. Each operator must be trained in the use of these emergency kits and must practice with the equipment at least once a year. Always check out even the slightest odor of chlorine; it may indicate a leak. Chlorine leaks only get worse. Small leaks can grow very rapidly causing serious problems that could have been easily solved as a small leak. There should always be two operators attending a chlorine leak, one to do the repairs and the other to act as safety observer. Some repairs require two operators to do the job (depends on the leak and the repair kit). Once again, use only the self-contained breathing apparatus when repairing a chlorine leak.

When connecting chlorine cylinders, be very careful with the threaded connections; never use two washers, use only one. If it does not work well, remove the washer and use another one. Do not reuse an old or used washer; always use a new washer. By taking this precaution, by cleaning the threads and washer, and by being careful with the thread setting, many chlorine leaks will be prevented. You must be aware whether you are using gas or liquid when connecting the container. On the 1-ton (909 kg) tank, the top valve is for gas, the bottom valve is for liquid. If liquid chlorine is allowed into a gas feed system it will cause

TABLE 20.3 GASES USED IN WATER TREATMENT

Name, Formula	Common Name	Available Forms	Spec. Grav. or lb/ft^3	Flam-mability	Color	Odor	Containers
Ammonia, NH_3	Ammonia	Liquid, gas	0.04813 lb/ft^3 @ 0°C	Explosive with air; will burn in air	Colorless	Irritating	Cylinders, tanks, trucks
Chlorine, Cl_2	Liquid Chlorine	Liquid, gas	1.468 @ 0°C	None	Greenish yellow	Irritating	Cylinders, 1-ton units, tank cars
Carbon Dioxide, CO_2	Dry Ice	Liquid, gas	0.914	None	Colorless	Odorless	Bulk liquid under pressure
Sulfur Dioxide, SO_2	Sulfurous Acid Anhydride	Liquid, gas	1.436 @ 0°C	None	Colorless	Suffocating, pungent	Cylinders, 1-ton units, tank cars

"freezing"[3] and shut down (plug) the system. Similarly, if liquid gets into the gas outlet, it will cause a problem by freezing. You must not panic in this situation. Do not do anything as foolish as adding heat by open flame or electrical heaters to clear a frozen (plugged) gas line. Get help from someone experienced with chlorine cylinders.

Never make repairs to the valve or chlorine container. Just stop the leak, perhaps by tightening the packing on the valve stem or placing the safety device onto the cylinder. Let the chlorine supplier repair the container. Never use a wrench longer than 6 inches (15 cm) to open the cylinder valve; make only one complete turn of the valve stem in a counterclockwise direction. The one turn will open the valve sufficiently for the maximum discharge. As a safety feature, cylinders are equipped with fusible metal plugs that are designed to melt at 158 to 168°F (70 to 76°C). This will allow the cylinder contents to discharge and prevent rupturing of the tank. On 100- to 150-pound (45 to 68 kg) cylinders, the plug is located just below the valve seat. The 1-ton (909 kg) tanks have six such plugs, three on each end. Should one of these plugs melt, permitting liquid chlorine to discharge, place the tank in a position with the leak at the top of the tank so that it permits the chlorine gas (rather than the liquid) to discharge. This action will reduce the amount of chlorine being discharged because the liquid will change to a gas to escape. In doing so, it will lower the temperature of the container, reducing the discharge rate.

All employees, maintenance personnel, and operators who handle chlorine must have access to an approved chlorine gas mask (Figure 20.2). They must be instructed in the use and maintenance of this equipment. A monthly program should be conducted to familiarize and train each user of the safety equipment. Those employees who are to use the chlorine emergency equipment should practice with this equipment at least once a year while wearing a self-contained breathing apparatus. The emergency kit consists of clamps, gaskets, drift pins, hammers, wrenches, and other tools needed for repairing leaks. The operator may not be able to practice with all of the tools, but inspection and practice gives the operator an opportunity to do maintenance on the emergency equipment.

All operators working with chlorine should be familiar with methods of detecting chlorine leaks. When testing for leaks, use ammonia water on a small cloth or swab on a stick or use an aspirator containing ammonia water. This will form a white cloud of ammonium chloride. Leaks should be repaired immediately. Do not apply the ammonia swab directly to the equipment surface. Also, do not spray ammonia into a room full of chlorine because a white cloud will form and you will not be able to see anything.

Many plants are equipped with chlorine gas detectors. This equipment must be maintained weekly. If not properly maintained, it may not be operable when you need it. Change the electrolyte regularly, test the alarm, and keep the detectors clean and in good repair.

Someone must be assigned the responsibility for maintenance of the self-contained breathing apparatus. That operator must keep records of the maintenance problems and of monthly drills using the gear. The assigned operator should check the masks for leaks, loose eyepieces, faulty tubing, or other worn or defective spots. If inspection indicates any defective parts, they should be discarded or repaired by a properly trained employee. Remember, in high concentrations of chlorine within a confined space where oxygen can be displaced, do not use the canister-type mask. No canister can provide oxygen. Therefore, use only a self-contained breathing apparatus or a hose-type mask supplied with air (Figure 20.3).

If you are caught in an area containing chlorine, do not panic, but leave immediately. Do not breathe or cough, and keep your head high until you are out of the affected area.

The first safety measure you can take when entering a chlorination room is to make sure the ventilating system is working. The ventilating system for the chlorination room should be working all the time. Doors of chlorination rooms should have panic bars as door openers so that in an emergency you will not have to search for the door opener. All safety equipment should be located outside of the chlorination room, but close enough so you can find the equipment, when needed.

First aid for eyes exposed to liquid chlorine is immediate irrigation with flowing water for at least 30 minutes. Medical attention is essential. If the eyes are exposed to chlorine gas, immediately irrigate with flowing water for a period of 15 minutes. Get medical aid. If skin is exposed to liquid chlorine, it will most likely cause burns. The skin should be washed with flowing water for 30 minutes. If the skin is burned, get medical attention. Chlorine gas can become trapped in the clothing and react with body moisture to form hydrochloric acid, which could burn the skin. Remove the clothing of the victim and wash the body down with water. In case of inhalation, remove the victim to fresh air, administer oxygen, if available, call a physician, or transport the injured person to a medical facility. Ingestion is not a problem because chlorine is a gas at room temperature.

Each operator should have a copy of the Chlorine Institute's *Chlorine Basics*.[4] You should read this manual and review it at least once every year. The manual gives data concerning chlorine as an element and gives you suggestions for safely handling this hazardous liquid or gas.

---

3. Liquid chlorine becomes a solid around –103 to –105°C. The liquid can plug a chlorine gas line, which operators refer to as a "frozen" line.
4. *Chlorine Basics*, Seventh Edition. Available as a free download from the Chlorine Institute at www.chlorineinstitute.org. Pamphlet 1.

SCOTT PRESUR-PAK IIa

SPECIFICATIONS	BACK-PAK STYLE
AIR SUPPLY	
Rated Duration at moderate exertion (MESA/NIOSH test procedure)*	30 min.
Cylinder Capacity at 2216 psi	45 cu. ft.
USE FACTORS	
Weight, as worn, fully charged (approx.)	32 lbs.
Donning Speed (trained personnel)	under 30 secs.
Facepiece (Scottoramic w/nose cup)	Wide vision, anti-fogging
Cylinder and Valve Connection	Straight thread, gasket seal
Cylinder Change	Hand disconnect, no tools req'd
Harness Webbing (replaceable without tools or rivets)	Polypropylene
Transport and Storage	Custom Molded High Density Polyethylene Case
SHIPPING WEIGHT	48 lbs.

NOTES: * POSITIVE PRESSURE

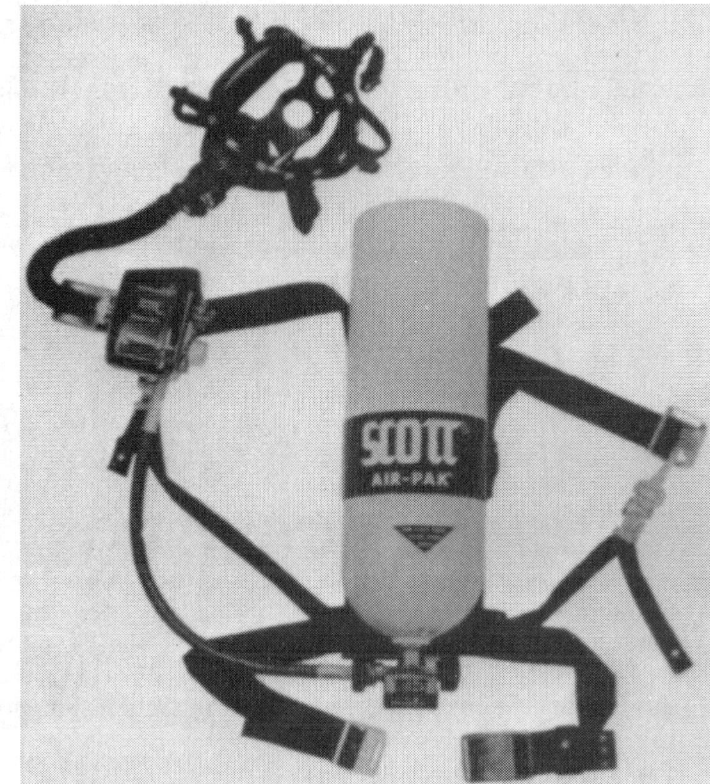

Fig. 20.2 Chlorine gas mask
(Permission of Nevada Safety & Supply)

20.131 Carbon Dioxide (CO_2)

Water treatment plant operators are not often exposed to carbon dioxide because of its limited use, but it is hazardous and can cause suffocation due to the lack of oxygen. Therefore, when using CO_2, keep in mind the carbon dioxide safety considerations. The problem with carbon dioxide is that it is odorless, colorless, and will accumulate at the lowest possible level because it is heavier than air.

Carbon dioxide is obtained in bulk lots, as a liquid (under pressure), which must be vaporized before using. CO_2 is also prepared by generation on site. In either case, good ventilation will reduce the hazards of using CO_2. This will control the accumulating effects of the gas. If you must go into a room filled with CO_2, use a self-contained breathing apparatus, not a canister gas mask. Carbon dioxide displaces oxygen and you may suffocate with the canister type of mask. Exposure to carbon dioxide does not require any protection of the eyes, skin, or other parts of the body, but take precautions when entering rooms, low spots, or manholes that may be filled with carbon dioxide.

First aid involves moving the victim to fresh air, giving resuscitation if the victim has stopped breathing, and getting medical attention.

20.132 Sulfur Dioxide (SO_2)

This gas is about 2.3 times as heavy as air and, therefore, will accumulate in low areas. Sulfur dioxide is colorless in the gaseous form. As a liquid, it is also colorless and, when unconfined, will vaporize rapidly into a gas. The gas is extremely irritating and, like chlorine, will react readily with the respiratory system if inhaled. Sulfur dioxide causes varying degrees of irritation to the mucous membranes—eyes, nose, throat, and lungs. The damage is caused by the formation of sulfurous acid

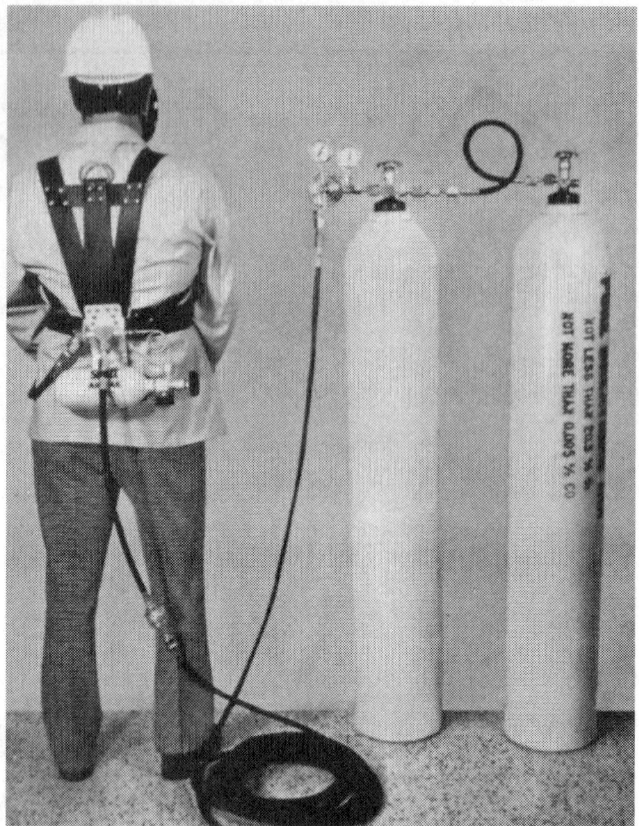

Dual air supply cylinder being used with 900007 series hoseline Air-Pak with Egress.

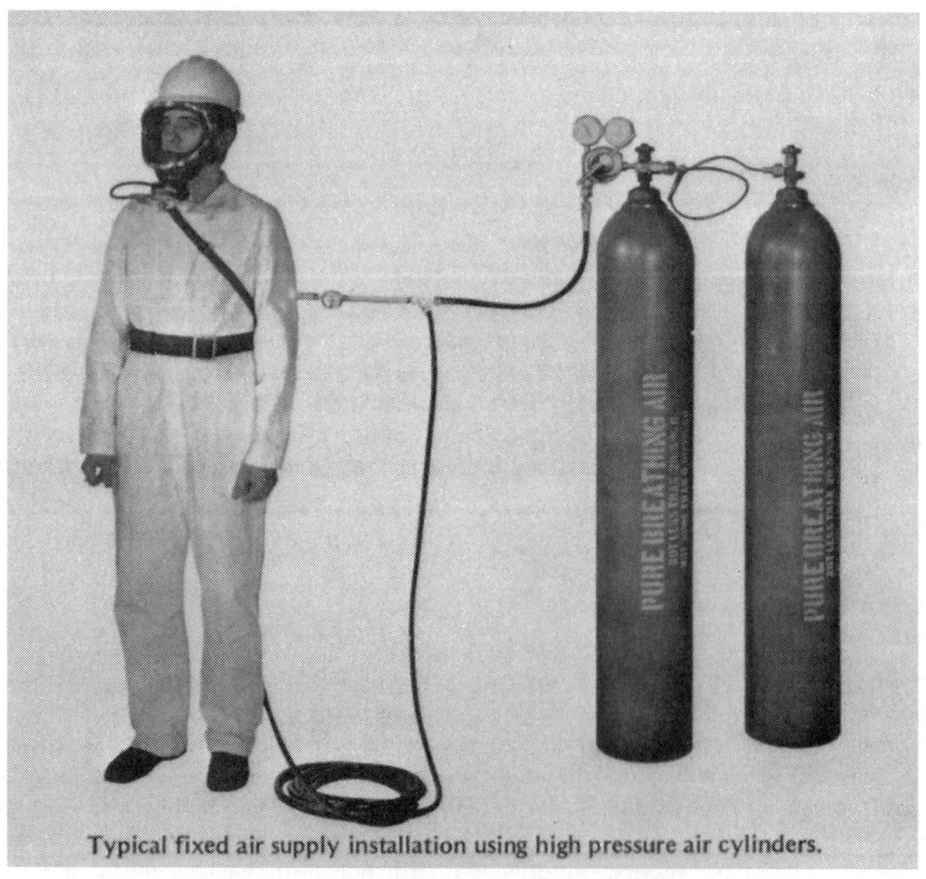

Typical fixed air supply installation using high pressure air cylinders.

Fig. 20.3 Hose-type mask supplied with air

(Permission of Nevada Safety & Supply)

in reaction with moisture in these locations. Sulfur dioxide can be readily detected in concentrations of 3 to 5 ppm. In higher concentrations, it is unlikely that you will remain in the area unless you are unconscious or trapped. If the liquid comes into contact with the skin, it may cause local freezing as the liquid evaporates.

Always use a self-contained breathing apparatus around sulfur dioxide and never use the canister type. As with other gases, good ventilation is essential in a room where sulfur dioxide is being used. The fans should be used to dissipate any gas vapors that may occur. There should always be an eye wash fountain close to the work area where sulfur dioxide is used.

First aid for eyes exposed to or splashed with sulfur dioxide is washing immediately with water for at least 15 minutes, then getting medical attention. In case of inhalation, remove victim to fresh air, give resuscitation if needed, consult a physician, or transport the injured person to a medical facility. Sulfur dioxide leaks and injuries should be treated similarly to chlorine problems.

QUESTIONS

Please write your answers to the following questions and compare them with those on page 488.

20.13A Where are chlorine leaks most often found?

20.13B What is the purpose of the fusible metal plugs on a chlorine cylinder?

20.13C How can chlorine leaks be detected?

20.13D Why is carbon dioxide a safety hazard and what safety precautions must be taken when using it?

20.14 Salts

There are many salts (chemicals) used in water treatment. Table 20.4 lists the salts that will be discussed in this section. If you would like additional information about compounds not listed, request a data sheet about specific compounds from the suppliers of the chemicals. Review the chemistry of the compounds in Table 20.4 and become familiar with each chemical's

TABLE 20.4 SALTS USED IN WATER TREATMENT

Name, Formula	Common Name	Available Forms	Density, lb/ft^3	Flammability	Color	Odor	Containers
Aluminum Sulfate, $Al_2(SO_4)_3 \bullet 14H_2O$	Alum, Filter Alum	Liquids, powder, lump	1.69 (S.G.) 38–67	None	Ivory	N/A[a]	Bags, tank truck, bulk
Ferric Chloride, $FeCl_3$	Ferrichlor, Chloride of Iron	Syrup, liquid, lump	60–90	None	Dark brown, yellow-brown	N/A	Carboys, tank cars
Ferric Sulfate, $Fe_2(SO_4)_3$	Ferrifloc, Ferrisul	Powder, granule	70–72	None	Red-brown	N/A	Bags, drums
Ferrous Sulfate, $FeSO_4 \bullet 7H_2O$	Coppras, Green Vitriol	Crystal, liquid, granule, lump	63–66	None	Green	N/A	Bags, drums, bulk
Sodium Aluminate, $Na_2O \bullet Al_2O_3$	Soda Alum	Dry crystal, liquid	(27° Bé [b])	None	White, green-yellow	N/A	Bags, bulk
Fluoride Compounds, NaF and H_3SiF_6	—	Liquid and powder	50–75	None	Blue, white	Dust	Bags, carboys, tank trucks
Sodium Hexametaphosphate, $(NaPO_3)_6$	Calgon, Glassy Phosphate	Crystal, flake	47	None	White	N/A	Bags, drums
Copper Sulfate, $CuSO_4$	Blue Vitriol, Blue Stone	Crystal, lump, powder	60–90	None	Blue	None	Bags, drums
Sodium Chlorite, NaOCl	Technical Sodium Chlorite	Powder, flake, liquid	70 dry	Oxidizer	Light orange	None	Tank truck, 100 lb drums
Potassium Permanganate, $KMnO_4$	Permanganate	Crystal, powder	90–100	Oxidizer	Purple	None	Drums, bulk

a N/A = Not Applicable.
b Bé = Baumé scale, a scale for measuring specific gravity of liquids.

characteristics. You should become well trained in how to handle each chemical and know and observe the appropriate safety precautions. For most of these salts, ventilation, respiratory protection, and eye protection will prove adequate. Other problems may involve chemical solutions and dust. The solution may attack skin and clothing. The dust may attack the respiratory system or cause an explosion. Even though these chemicals do not normally react violently, use the following procedures when handling such salts to reduce the hazards and to provide a safe working location.

When handling, storing, or preparing solutions of chemicals, treat them all as being hazardous. All chemicals require careful consideration. They may be sources of an explosion, violent reaction, loss of eyesight, burns, and illness.

Do not store acidic or basic compounds with salts. Keep these chemicals in a clean, dry area. When handling dry, bulk materials, store in a fire-safe area. Keep all lids on containers and follow the instructions on the container. Make sure that the operator who is mixing or dispensing these chemicals is well trained and wears proper clothing to meet all safety requirements such as chemical goggles, face shield, rubber gloves, rubber boots, rubber apron, and chemical respirator. When working with chemical salts, be aware that fumes, gases, vapors, dusts, or mists may be given off and this represents a hazard to the safety of the operator. Protect yourself.

20.140 Aluminum Sulfate (Alum) ($Al_2(SO_4)_3 \cdot 14H_2O$)

There are two forms of alum: dry and liquid. Both have to be handled with care. Dry alum is available in lump, ground, or powdered forms and should be stored in a dry location because moisture can cause caking. Liquid alum is acidic and very corrosive. Store liquid alum in corrosion-resistant storage tanks:

1. Steel, wood (Douglas fir), or concrete-lined, all lined with a material corrosion-resistant to alum
2. Steel, lined with 3⁄16-inch (5 mm) soft rubber
3. Stainless steel
4. Steel, lined with plastic if temperature remains below 150°F (65°C)
5. Glass-reinforced epoxy or polyester plastic

When working with dry alum, use respiratory protection and ensure adequate ventilation of the work area. There should be a good mechanical dust collection system to minimize any dust accumulation. Exposure to alum dust concentrations greater than 15 milligrams per cubic meter of air for more than an 8-hour period is dangerous. Avoid skin exposure to this chemical by using long-sleeved, loose-fitting, dust-proof clothing. Never use the same conveyor for quicklime and alum. This mixture may explode under certain conditions.

Liquid alum is an acidic solution and should be handled as you would handle a weak acid. Reduce exposures to the skin and eyes. Avoid ingestion. Although the chemical will not cause any lasting internal damage, it will cause discomfort. Use good ventilation for removing any mists. Rubber gloves and protective clothing are recommended.

As a general precaution, avoid prolonged exposure to dry or liquid forms of alum. If used dry, a dust mask and goggles are desirable for the comfort of the operator. Alum dust can be extremely irritating to the eyes. When handling the liquid, normal precautions should be used to prevent splashing of the compound onto the operator, particularly if the liquid is hot. Wear a face shield to protect your eyes and a rubber apron to protect clothing.

First aid for liquid or dry alum is immediate flushing of the eyes for 15 minutes with large amounts of water. Alum should also be washed off the skin with water because prolonged contact will cause irritation.

20.141 Ferric Chloride ($FeCl_3$)

This is a very corrosive compound and should be treated as you would treat any acid. The salt is highly soluble in water, but in the presence of moist air or light, it decomposes to give off hydrochloric acid, which may cause other safety hazards. Avoid prolonged exposure to this liquid (there is a dry form but it is not often used). When handling liquid ferric chloride, normal precautions should be taken to prevent splashing, particularly if the liquid is hot. Use a face shield to protect your eyes and rubber aprons to protect clothing. This compound will not only attack the clothing, but also stain it. First aid for eyes exposed to the liquid is that the eyes must be flushed out immediately for 15 minutes with large amounts of water. Ferric chloride should also be washed off the skin with water because prolonged contact will cause irritation and staining of the skin.

20.142 Ferric Sulfate ($Fe_2(SO_4)_3$)

This compound produces an acidic solution when mixed with water. Because of its acidic nature, operators using this compound should be provided with protection suitable for dry or liquid alum. The hazards associated with the use of dry ferric sulfate are those usually connected with an acid. Use protective clothing, neck cloths, gloves, goggles or face shield, and a respirator. Avoid prolonged exposure to the dry form because of its acidic reaction with moisture on the skin, eyes, and throat. The normal precautions should be used including a dust mask and protective clothing. First aid for exposure to the eyes requires the eyes to be flushed immediately with lots of water. The skin should also be flushed with large amounts of water. Prolonged contact may cause irritation.

20.143 Ferrous Sulfate ($FeSO_4 \cdot 7H_2O$)

This chemical may be obtained in liquid or dry form. The safety hazards are some of those for dry or liquid forms of alum. The operator should be provided with adequate ventilation and respiratory protection. The material should be stored in a clean, dry location. Mechanical dust collecting equipment must be used to minimize the dust. Wear chemical goggles or a face shield, loose-fitting, long-sleeved clothing, and make an effort to minimize all skin exposure.

First aid for ferrous sulfate in the eyes is to flush out immediately with large amounts of water for 15 minutes. The chemical should be washed off the skin to reduce irritations.

20.144 Sodium Aluminate ($Na_2O \cdot Al_2O_3$)

Sodium aluminate dissolved in water produces a noncorrosive solution. In the dry form, its powder consistency raises the usual dust problems. There are few hazards with this compound, but as with other chemicals, you should use precautions when handling it. Use respiratory protection when handling the dry compound to prevent the inhalation of dust. First aid for eyes that are exposed is to flush with water; keep the skin clean with water.

20.145 Fluoride Compounds (NaF and H_3SiF_6)

All fluoride compounds should be treated with care when you are handling them because of their long-term accumulative effects. Provide good ventilation; always wear respiratory protection; and be careful not to expose any open cuts, lesions (wounds), or sores to fluoride compounds. Clean up any spills promptly and wash immediately after handling such compounds.

When handling acid compounds of fluoride, always wear a face shield or chemical goggles, rubber gloves, rubber apron, and rubber boots. Your wearing apparel should always be washed after working around fluoride, and the respirator should be kept clean and sanitary. Keep the acid feeder for fluoride in good repair. Use plastic guards to prevent acid spray from glands or other parts of the chemical feeder. This prevents attack upon the equipment and protects operators. All fluoride compounds must be regarded as hazardous chemicals that are toxic to operators. Every means possible must be taken to prevent exposure to these compounds by use of a respirator and protective clothing.

Once it is established that fluoride is the cause of the poisoning, first aid should be started while waiting for medical help. The following are recommended first-aid procedures:

1. Move the person away from any contact with fluoride and keep warm.
2. Give the person 3 teaspoons (1 tablespoon) of table salt in a glass of warm water.
3. If the person is conscious, and has swallowed sodium fluoride, induce vomiting by rubbing the back of the throat with a spoon or your finger; if available, use syrup of ipecac. Do not induce vomiting if hydrofluosilicic acid is swallowed. Vomiting may cause the acid to go the wrong way into the lungs and cause more serious problems.
4. Give the person a glass of milk.
5. Repeat the salt and vomiting several times.
6. Take the person to the hospital as soon as possible.

First aid for a person with a nosebleed from inhaling a high concentration of fluoride is:

1. Take the person away from the source of the fluoride.
2. Tip the person's head back while placing cotton, cloth, or tissues inside the nostrils (change these often).
3. Take the person to a doctor if you cannot stop the bleeding.

QUESTIONS

Please write your answers to the following questions and compare them with those on page 488.

20.14A What kind of protection does an operator need when handling salts?

20.14B What is the recommended first aid when either liquid or dry alum comes in contact with your skin or your eyes?

20.14C What happens when ferric chloride is exposed to moist air or light?

20.15 Powders

20.150 Potassium Permanganate ($KMnO_4$)

Under normal conditions in a water treatment plant, potassium permanganate is considered to be a safe chemical (see Table 20.4). However, potassium permanganate is a strong oxidizing agent and will react with certain easily oxidizable substances. Keep potassium permanganate away from the possibility of reacting with sulfuric acid, hydrogen peroxide, metallic powders, elemental sulfur, phosphorus, carbon, hydrochloric acid, hydrazine, hydroxylamine, and metal hydrides. When in contact with potassium permanganate, the following compounds may ignite: ethylene glycol (antifreeze), glycerine, sawdust compounds, propylene glycol, and sulfuric oxide.

Potassium permanganate is available as pellets, as a powder, or in crystal form. This chemical can be kept indefinitely if stored in a cool, dry area in closed containers. The drums should be protected from damage that could cause leaks or spills. Potassium permanganate should be stored in fire-resistant buildings having concrete floors instead of wooden floors. The chemical must not be exposed to intense heat, or stored next to heated pipes. Organic solvents, such as greases and oils, should be kept away from stored potassium permanganate.

Potassium permanganate spills should be swept up and removed immediately. Flushing with water is an effective way to eliminate any residue remaining on floors. Potassium permanganate fires should be extinguished with water.

To avoid inhalation of potassium permanganate dust, use an approved mask that is an air-purifying, half-mask respirator with an outblower. Wear safety glasses or a full face shield to protect your eyes. Protective clothing that should be worn includes rubber or plastic gloves and apron, and a long-sleeved shirt for handling both dry and dissolved potassium permanganate.

Mild exposure will cause sneezing and mild irritation of the mucous membranes. Prolonged inhalation of potassium permanganate should be avoided. If potassium permanganate gets on your skin, flood the contacted skin with water. If it gets in your eyes, flush with plenty of water and call a physician immediately.

20.151 Powdered Activated Carbon

Powdered activated carbon is the most dangerous powder that you will be exposed to as a treatment plant operator. If you understand how to handle activated carbon properly, other dust problems or powdered chemicals will not be very difficult for you to handle.

There are two potential problems when handling activated carbon. One is dust and the second is fire. The two may or may not be related. Dust can cause uncomfortable working conditions and fire can cause damage to equipment and is a hazard to personnel. If the two problems are treated together, it will reduce the hazards to operators. Left unattended they may cause loss of life and property. If you will use the following safety precautions, you can minimize the hazards of handling activated carbon and aid the other operators in handling other powders.

Store activated carbon in a clean, dry, fireproof location. Keep the area free of dust, protect it from flammable materials, and do not permit smoking in the area at any time when handling or unloading activated carbon. Install carbon dioxide fire extinguishers. Store bagged carbon in single rows. Keep access aisles clear to prevent damage to the bags and thus reduce the dust and fire potential.

Electrical equipment in and around activated carbon storage should be explosion-proof and protected from the carbon dust. Keep the equipment clean and dry. Wet or damp carbon is a good conductor of electrical current and can cause short-circuit fires. Heat can also build up from the motors if covered with carbon dust, causing fires. The key to preventing fires with activated carbon is keeping the storage area clean and dust free.

Next to electrical fires, activated carbon gives the operator the most difficult fire to control. The carbon gives off an intense heat; it burns without smoke or visible flame. The fires are difficult to locate and are very hard to control. They cannot readily be detected in a large storage bin or in large stacks of bags. Bags of activated carbon should be stored in single rows.

You will detect the indications of the fire, such as the smell of charred paper, burned paint, or other odor, before seeing any evidence of flames.

Do not douse a carbon fire with a stream of water. The water may cause burning carbon particles to fly, resulting in a greater fire problem. The carbon fire should be controlled with carbon dioxide (CO_2) extinguishers or hoses equipped with fog nozzles. However, when using CO_2, be aware that there is a potential of carbon monoxide formation and take the precaution of using a self-contained breathing apparatus.

Activated carbon supports fire without atmospheric oxygen because it may have adsorbed sufficient oxygen for combustion. The best means of controlling a carbon fire is to reduce its temperature below the ignition point. This can be done by applying cold water with fog or spray nozzles and soaking the burning carbon, but do not hit the carbon with a stream of water. For a small fire involving just a few bags, move the bags to a safe location and use CO_2 or spray nozzles to extinguish the fire. Blocks of dry ice may help control fires in storage bins or other confined areas, but do not expect this method to be very effective.

A final word about fire and explosions involving activated carbon. Tests performed by carbon manufacturers have not shown that dust mixtures of carbon have explosive tendencies. Activated carbon is a charcoal and performs in a like manner; the carbon burns without smoke or visible flame, burns very hot, and will spread if doused with a large stream of water.

There are no specific first-aid methods for carbon exposure because the carbon will not attack the human body. Carbon does sometimes cause problems with the nasal passages, however, and may be difficult to wash off your hands and body. Therefore, methods here are those of prevention. Provide good dust collection at the point where carbon is being unloaded, in storage bins for liquid preparation, and in dry storage bins. Wear an approved dust mask, loose-fitting and dust-proof clothing. If there is an excess of dust, you should use chemical goggles, close your shirt collar, and tape your trousers to cover your ankles. You will need adequate shower facilities and should use mild liquid soap. Most, if not all, hard soap bars are ineffective in removing activated carbon dust from pores of the human body.

20.152 Other Powders

You may come in contact with other powdered compounds in the water treatment plant, but they do not present the problem in handling that activated carbon does. Bentonite creates no significant hazard other than dust, and this can be controlled by using dust collection systems.

Similarly, calcium carbonate presents a slight dust hazard that can be controlled by a dust collection system. Also, when handling calcium carbonate in bags, use the same methods of control that you would use with bags of carbon.

Of the many organic coagulant aids used in water treatment, only a few are applied in powder form. Most of these compounds are used in the liquid form, which reduces the danger to operators. The dry compounds present a slight hazard in dust irritation to the nasal passages; this can be prevented by use of approved dust masks. The liquid compounds can and will attack the skin, but can be treated by washing with ample amounts of water or soap and water. Organic coagulant aids (polymers) are extremely slippery when wet. Floors and walkways should be clean and dry to prevent slips and falls. Spilled polymers may be cleaned up using inert, absorbent materials such as sand or earth.

As new compounds are introduced to the waterworks field, ask for training in their use. Such a training program should provide information about detailed safety precautions, the toxicity of the compound, and the appropriate first-aid methods. Supervisors have the responsibility to provide this type of training, either in conjunction with the supplier or sponsored entirely by the utility. The training must be reinforced periodically for compounds that are not often used by the operator. Reviewing and updating information regarding safely handling powdered compounds will help prevent accidents.

QUESTIONS

Please write your answers to the following questions and compare them with those on page 488.

20.15A How can potassium permanganate spills be cleaned up?

20.15B What is the most dangerous powder the water treatment plant operator will be exposed to?

20.15C How should activated carbon be stored?

20.15D How can fires caused by activated carbon be prevented?

20.15E How should an activated carbon fire be extinguished?

20.16 Labeling of Chemical Containers

For information on proper labeling of chemical containers, refer to the section on "Hazard Communication Standard (HCS) and Worker Right-To-Know (RTK) Laws" in *Operation and Maintenance of Wastewater Collection Systems*, Volume I, in this series of operator training manuals.

20.17 Chemical Storage Drains

Safety regulations prohibit the use of a common drain and sump for acid and alkali chemicals, oxidizing chemicals, and organic chemicals because of the possibility of the release of toxic gases, explosions, and fires. If an acid and an alkali chemical come in contact, an explosion could occur. If an organic chemical such as a polymer solution comes in contact with an oxidizing chemical, such as potassium permanganate, a fire could develop. Be proactive if a leak develops from any chemical container or storage facility by verifying that the chemical will not be able to reach, mix, or react with another chemical.

QUESTIONS

Please write your answers to the following questions and compare them with those on page 488.

20.17A Why should drains from chemical storage areas not use common drains and sumps?

20.17B What could happen if a leak from a polymer storage container comes in contact with potassium permanganate?

END OF LESSON 1 OF 3 LESSONS

on

SAFETY

Please answer the discussion and review questions next.

DISCUSSION AND REVIEW QUESTIONS

Chapter 20. SAFETY

(Lesson 1 of 3 Lessons)

At the end of each lesson in this chapter, you will find discussion and review questions. Please write your answers to these questions to determine how well you understand the material in the lesson.

1. What is safety?
2. Who is responsible for safety?
3. What are the causes of accidents?
4. What does an operator need to know about chemicals used in a water treatment plant?
5. How can hydrochloric acid be handled safely?
6. How should hydrochloric acid be stored?
7. How can ammonia leaks be detected?
8. What is the first-aid treatment for lime burns?
9. What special precautions should be taken when handling and storing caustic soda?
10. How should hypochlorite be handled and stored?
11. What first aid is required for a person overcome by carbon dioxide?
12. What safety hazards may be caused by salt dust?
13. What safety precautions should an operator take when handling alum?
14. What are the two major problems encountered when handling activated carbon?
15. How can an operator detect an activated carbon fire?

CHAPTER 20. SAFETY

(Lesson 2 of 3 Lessons)

20.2 FIRE PROTECTION

20.20 Fire Prevention

Fire prevention is the best fire protection the plant operator can afford. Fire protection is just good housekeeping. The word "housekeeping" best describes the action any water plant operator can take to protect from or prevent fires. This means a well-kept, neat, and orderly plant represents a good fire safety policy.

Fire hazards can be easily removed. The prompt disposal of cartons, crates, and other packing materials, a system of waste-paper collection, and the removal of other debris can greatly reduce fire hazards. Provide suitable containers for used wiping cloths and have fire extinguishers conspicuously located in hallways, near work areas, and near potential fire problem areas. All of these housekeeping activities are low-cost measures that also improve the appearance of the plant and create a better work environment.

You can call upon the local fire department for advice on fire prevention in and around the treatment plant. You may also ask the utility insurance underwriter for cooperation in your fire prevention program. All operators should be trained in the proper use and maintenance of fire control equipment. These simple steps can reduce fire losses to a minimum and prevent most fires from happening at very low cost to the utility.

You should make a fire analysis of your plant once a year to determine what new measures should be taken to prevent fires. As activity changes occur, there may be a need to change the location of hoses and extinguishers or it may be necessary to add fire control equipment. Fire and police departments' telephone numbers must be posted in a conspicuous location along with escape routes. Post emergency numbers near all telephones throughout the plant. In hazardous locations, the means of exit should be lighted and all doors should be equipped with panic bars. As indicated above, your best fire protection or prevention is good housekeeping.

20.21 Classification of Fires and Extinguishers

Fire classifications are important for determining the type of fire extinguisher needed to control the fire. Classifications also aid in recordkeeping. Fires are classified as A, B, C, or D fires based on the type of material being consumed: A, ordinary combustibles; B, flammable liquids and vapors; C, energized electrical equipment; and D, combustible metals. Fire extinguishers are also classified as A, B, C, or D to correspond with the class of fire each will extinguish.

Class A fires: ordinary combustibles such as wood, paper, cloth, rubber, many plastics, dried grass, hay, and stubble. Use foam, water, soda-acid, carbon dioxide gas, or almost any type of extinguisher.

Class B fires: flammable and combustible liquids such as gasoline, oil, grease, tar, oil-based paint, lacquer, and solvents, and flammable gases. Use foam, carbon dioxide, or dry chemical extinguishers.

Class C fires: energized electrical equipment, such as starters, breakers, and motors. Use carbon dioxide or dry chemical extinguishers to smother the fire; both types are nonconductors of electricity.

Class D fires: combustible metals such as magnesium, sodium, zinc, and potassium. Operators rarely encounter this type of fire. Use a Class D extinguisher or use fine, dry soda ash, sand, or graphite to smother the fire. Consult with your local fire department about the best methods to use for specific hazards that exist at your facility.

Multipurpose extinguishers are also available such as a Class BC carbon dioxide extinguisher that can be used to smother Class B and Class C fires. A multipurpose ABC carbon dioxide extinguisher will handle most laboratory fire situations. (When using carbon dioxide extinguishers, remember that the carbon dioxide can displace oxygen—take appropriate precautions.)

There is no single type of fire extinguisher that is effective for all fires so it is important that you understand the class of fire you are trying to control. You must be trained in the use of the different types of extinguishers, and the proper type should be located near the area where that class of fire may occur.

20.22 Fire Extinguisher Operation and Maintenance

There are many types of hand-held fire extinguishers. Each type of extinguisher has different operation and maintenance requirements.

Water Extinguishers

There are four types of water extinguishers: stored pressure, cartridge operated, water pump tank, and soda-acid. All of these perform well in Class A fires, but they do require maintenance. A preventive maintenance schedule on all water extinguishers should include a monthly check by the operator responsible for the maintenance and completion of appropriate maintenance records. Some agencies make the safety officer responsible for ensuring that an operator checks the fire extinguishers and records each inspection.

1. The method of operation for a stored pressure extinguisher is simply to squeeze the handle or turn a valve. The maintenance is also simple: check air pressure and recharge the extinguisher, as needed.
2. For the cartridge type, the maintenance consists of weighing the gas cartridge and adding water, as required. To operate, turn upside down and pump.
3. To use the water pump tank type of extinguisher, simply operate the pump handle. For maintenance, one has only to discharge the contents and refill with water annually or as needed.
4. The soda-acid type must be turned upside down to operate; it also requires annual recharging.

Foam Extinguishers

The foam type of extinguishers will control Class A and Class B fires well. They, like soda-acid types, operate by turning upside down, and require annual recharging.

The foam and water type extinguishers should not be used for fires involving electrical equipment. However, they can be used in controlling flammable liquids such as gasoline, oil, paints, grease, and other Class B fires.

Carbon Dioxide Extinguishers

The carbon dioxide (CO_2) extinguishers are common (Figures 20.4 and 20.5). They are easy to operate, just pull the pin and squeeze the lever. For maintenance, they must be weighed at least semiannually. Many of these extinguishers will discharge with age. They can be used on a Class C (electrical) fire. All electric circuits should be killed, if possible, before trying to control this type of fire. A carbon dioxide extinguisher is also satisfactory for Class B fires, such as gasoline, oil, and paint, and may be used on surface fires of the Class A type.

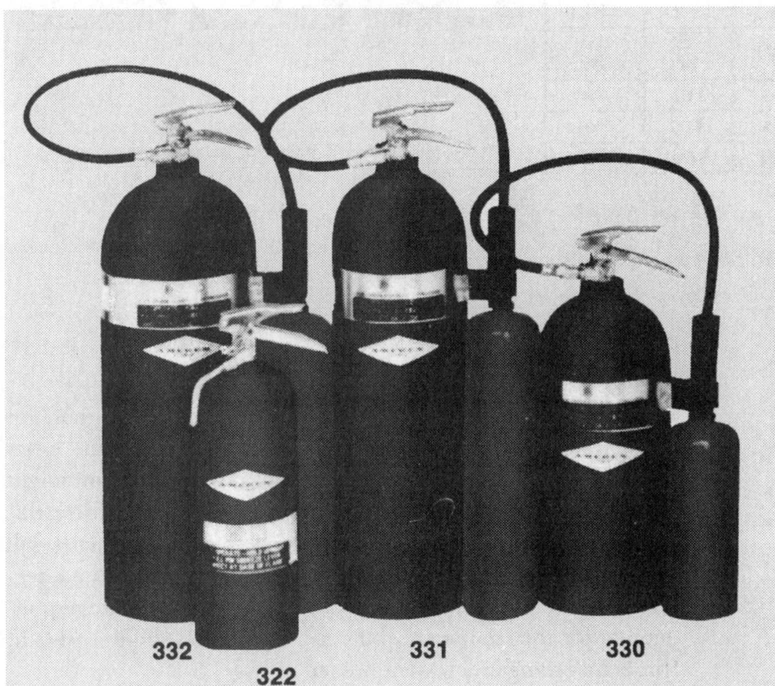

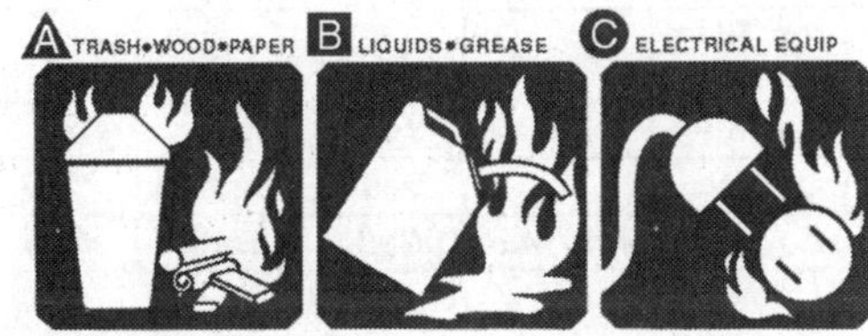

SPECIFICATIONS	CARBON DIOXIDE			
Size/Type	5 Hose	10 Hose	15 Hose	20 Hose
Model Number	322	330	331	332
U/L Rating	5B:C	10B:C	10B:C	10B:C
Capacity (lbs.)	5	10	15	20
Shipping Wt. (lbs.)	15	29½	39½	51½
Height	17¾"	24"	30"	30"
Width	8¼"	12"	12"	13"
Depth (Diam.)	5¼"	7"	7"	8"
Range (Ft.)	3-8	3-8	3-8	3-8
Discharge Time—Seconds	10	10	12.5	19
Coast Guard App.	Yes	Yes	Yes	Yes
FM Approved	Yes	Yes	Yes	Yes
Bracket	Wall	Wall	Wall	Wall

Fig. 20.4 Carbon dioxide extinguishers

(Permission of Nevada Safety & Supply)

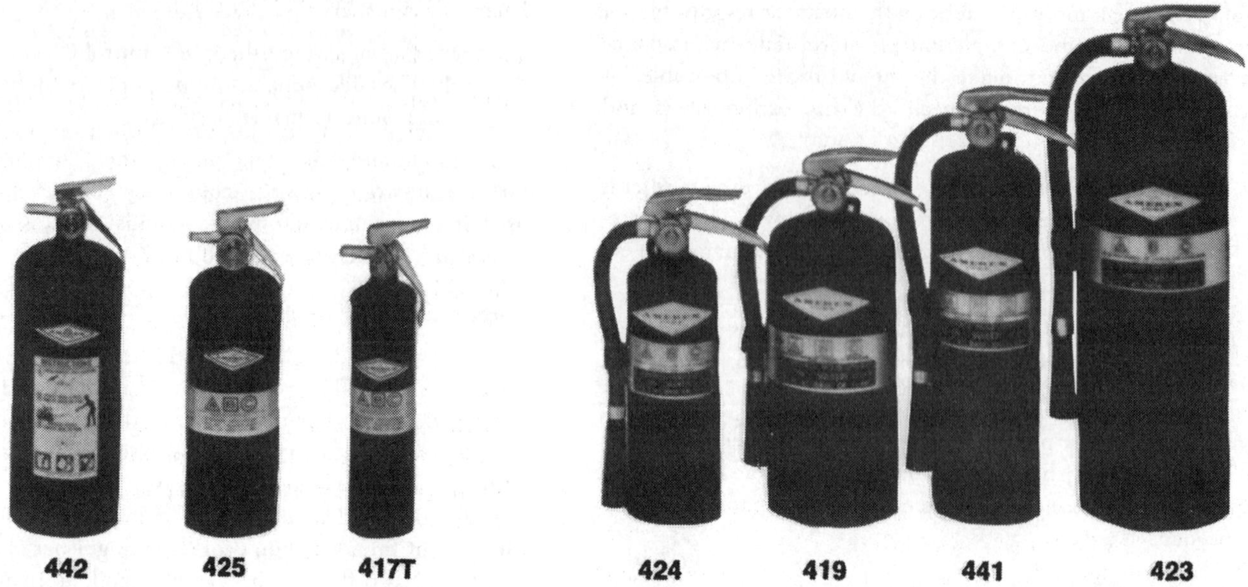

442 425 417T 424 419 441 423

SPECIFICATIONS	ABC						
Size/Type	2½ Nozzle	5 Nozzle	6 Nozzle	5 Hose	10 Short Hose	10 Tall Hose	20 Hose
Model Number	417T	425	442	424	419	441	423
U/L Rating	1A:10 B:C	2A:10 B:C	3A:40 B:C	2A:10 B:C	4A:60 B:C	4A:60 B:C	20A:120 B:C
Capacity (lbs.)	2½	5	6	5	10	10	20
Ship. Wt. (lbs.)	5½	8½	11	10½	19½	18	40
Height	14⅛"	14⅝"	15½"	14⅝"	17"	20½"	24"
Width	3⅞"	5"	5"	8"	9½"	9"	10"
Depth (Diam.)	3"	4¼"	5"	4¼"	6"	5"	7"
Range (Ft.)	9-15	12-18	12-18	12-18	15-21	15-21	15-21
Discharge Time-Seconds	10	10	14	10	17	17	30
Coast Guard Ap.	Yes	Yes	Yes	Yes	Yes	Yes	Yes
FM Approved	Yes	Yes	Pending	Yes	Yes	Yes	Yes
Bracket	Veh/Mar	Wall	Wall	Wall	Wall	Wall	Wall

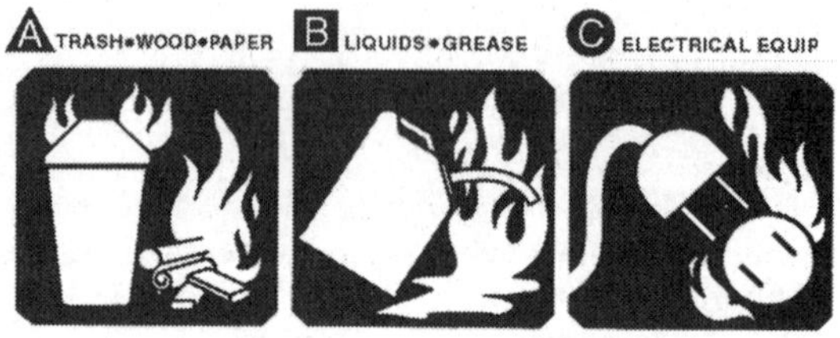

A TRASH•WOOD•PAPER B LIQUIDS•GREASE C ELECTRICAL EQUIP

Fig. 20.5 Typical carbon dioxide extinguishers
(Permission of Nevada Safety & Supply)

Dry Chemical Extinguishers

There are two types of dry chemical extinguishers. These extinguishers are either cartridge operated or stored pressure. These are recommended for Class B and C fires and may work on small surface Class A fires.

1. The cartridge-operated extinguishers only require you to rupture the cartridge, usually by squeezing the lever. The maintenance is a bit more difficult, requiring weighing of the gas cartridge and checking the condition of the dry chemical.

2. For the stored-pressure extinguishers, the operation is the same as the CO_2 extinguisher. Just pull the pin and squeeze the lever. The maintenance requires a check of the pressure gauges and condition of the dry chemical.

As suggested above, a preventive maintenance program for fire extinguishers requires a considerable amount of time from the operator and requires a system of recordkeeping. You might consider hiring a local fire prevention agency to perform this part of your maintenance program. These service agencies will check and maintain the plant's firefighting equipment on a regular basis. This does not relieve the operator of ultimate responsibility for the equipment, but ensures that the equipment is in proper working order when needed.

20.23 Fire Hoses

Fire hoses are usually stationed throughout the treatment and pumping plants. These are the type of firefighting equipment that an operator may see every day, but never give due consideration as to their maintenance. Without proper maintenance, the hoses may develop dry rot and be untrustworthy at the time they are needed. Under some conditions, you may be tempted to use these hoses for cleaning settling basins or filters. The fire hoses should only be used for fighting fires and, after their use, they must be cleaned and stored properly. The hose should be tested periodically and replaced as required, or at regular time intervals. Check with the local fire department for recommendations.

20.24 Storage of Flammables

The storage of flammable material should be isolated, if possible, from other plant structures. Ideally, these storage areas should have explosion-proof lighting. The floor should be grounded and the operator should only use spark-proof tools when working near or handling flammable materials. The room should have an alarm system, be equipped with automatic extinguishers, and have supplementary equipment located outside of the room. In and around the storage area, smoking or welding must be prohibited. The flammable storage areas must be clearly marked with distinctive signs and all entrances should be lighted.

More often than not, however, you will be compelled to use rooms within the plant for storage of flammable material. Here you must make the room fireproof, equip the room with a fire door, automatic extinguishers, and alarms. Keep passageways free from obstructions. Station firefighting equipment at a suitable location, readily accessible and with plainly labeled operating instructions. The room must be equipped with explosion-proof lights, grounded floor, no-smoking signs, and other distinctive signs indicating that the room is a flammable storage area.

20.25 Exits

Well-placed and signed exits are very important for plant operator safety. All exit signs should be well lighted and easily visible. All doors should open outward and, in hazardous areas, there should be panic bars on the doors. To provide positive protection around filter and sedimentation basins, install handrails or enclosures for the protection of operating personnel as well as visitors.

In high-fire-hazard occupied areas, there should be at least two means of emergency exit located, if possible, at opposite ends of the room or building. These would include areas containing woodworking and paint-spraying residues that burn rapidly or give off poisonous fumes.

QUESTIONS

Please write your answers to the following questions and compare them with those on page 489.

20.2A Class A fires involve what types of materials?

20.2B What kinds of fires can be controlled by a foam type of extinguisher?

20.2C How can an electrical fire be extinguished?

20.3 PLANT MAINTENANCE

20.30 Maintenance Hazards

Plant maintenance, housekeeping, cleaning up, or whatever you wish to call it, is a very important function of the treatment plant and essential for plant equipment. This function requires the use of cleaning materials and hand tools. Maintenance may require you to go into a manhole, repair electric motors, lift boxes, and use power tools. All of these functions may in some way be hazardous and may cause injury, fire, disease, or even death unless proper safety precautions are taken.

20.31 Cleaning

Any effort spent keeping the entire plant clean and sanitary will provide a much nicer place for you to work and will also make visitors feel as if the water being produced is safe. Even if you can just keep all working areas free of tripping hazards, this will add greatly to the safety in the plant.

Cleaning duties should be performed at such times of day or night as to cause a minimum of exposure to other operators. For example, floors become slippery when wet so give some consideration to the time of day and the type of floor wax to be used. When cleaning floors, there are problems of exposure to others of cleaning equipment, mops, mop and broom handles, other tools, cleaning compounds and, most of all, wet floors. When cleaning, try to keep others out of the area. Warn others about newly waxed floors. Use wax compounds containing nonslip ingredients. Try to do such cleaning and waxing on weekends or at night when foot traffic through the plant is minimal.

As part of your maintenance program, provide trash containers for collecting wastepaper and for separating used, oily rags. Dispose of garbage and flammable refuse on a routine, frequent basis. Hazardous waste, acids, and caustics should be cleaned up immediately. These steps will add to the safety in the plant and to the safety of operators in the plant.

Keeping aisles, doorways, stairs, and work areas free of refuse reduces hazards of tripping and other injuries, as well as reducing the possibility of fires.

Cleaning windows is a hazardous occupation, but the operator who gives due consideration to the task can perform it safely. If windows are high, time of day is important. Cleaning tools may be dropped and fall on pedestrians or vehicles. There may be a need for safety harnesses; check the harness each time it is used. Make sure all parts of the harness are in good working condition. Cleaning compounds that are acidic or alkaline may attack the harness or the safety rope. Also, such compounds may attack human skin; therefore, use rubber gloves, when appropriate.

20.32 Painting

There are a number of considerations when painting in a treatment plant. First, is the paint exposed to the drinking water being treated and are there toxic compounds in the paint? Next, is the paint being applied by brush or spray? In either case, is there sufficient ventilation if the operator is painting indoors or in a closed area?

When working with toxic paints, for example, those containing zinc or organics, be sure to clean your hands before eating or handling food. Also, avoid exposing your skin to solvents and thinners and try not to use compounds such as carbon tetrachloride. When spray painting, always use a respirator to avoid inhaling fumes. Do not allow smoking or open flames of any kind around areas being painted. Also, when painting or cleaning the spraying equipment, avoid closed containers where heat is involved. At a certain temperature, called the flash point, spray or vapors could ignite and burn the operator or start fires. Always clean the spray equipment in an area having sufficient ventilation. If the painting operation is taking place in a paint booth, use only explosion-proof lighting and permit no open flame and no electric switch that may cause a spark.

Other considerations when painting involve the safe use of scaffolding and ladders, proper disposal of solvent-soaked rags, and protection of your personal health. Be very careful when using scaffolding and ladders. The scaffolding must be in good repair and conform to current safety regulations. Ladders must also be in good repair. If they are broken or badly worn, they should be replaced with new ones. Rags are always a problem if they contain oils, paint, or other cleaning compounds; there is always the possibility of fire. The rags should be placed into a closed metal container to reduce the fire hazard. To protect your skin, consider using creams to help reduce direct exposure to paint and solvents. Always use an approved respirator to reduce inhalation of fumes and paint mists.

20.33 Cranes

Overhead traveling cranes require safety considerations. First, only authorized personnel should be allowed to operate them. Inspections should be made to check out the circuit breaker, limit switches, the condition of the hook, the wire rope, and other safety devices. The load limits should be posted on the crane and you should never overload the unit. Always check out each lift for proper balance. Use only a standard set of hand signals, and make sure that each operator involved with the crane knows all of the signals. Personnel in and around an overhead crane should be required to wear hard hats. When making repairs to the crane, lock out the main power switch, physically block any movable parts, and allow only authorized personnel to make repairs.

Never move loads over areas where operators or other people are working. Do not let the load remain over the heads of operators or other workers or allow anyone to work under loaded cranes. If loads must be moved over populated areas, give a warning signal and make sure everyone is in a safe location. Set up monthly safety inspection forms to be filled out and placed into the maintenance file. The plant supervisor should review the forms and authorize any maintenance necessary on the crane in addition to following a good preventive maintenance program.

20.34 Confined Spaces [5]

Special safety rules and regulations have been developed to protect the life and health of persons who must enter confined spaces. A confined space may be defined as a space that has limited or restricted means for entry and exit and that is not designed for continuous occupancy. One easy way to identify a confined space is by whether or not you can enter it by simply walking while standing fully upright. In general, if you must duck, crawl, climb, or squeeze into the space, it is considered a confined space. Typical examples of confined spaces found in water treatment plants and within water systems are pits, tanks, basins, and manholes.

Dangerous air contamination presents a threat of causing death, injury, *ACUTE*[6] illness, or disablement due to the presence of flammable, explosive, toxic, or otherwise injurious or incapacitating substances. Accumulations of gases or vapors, for example, may produce an explosive or flammable atmosphere. Similarly, if particles of chemical dust become airborne when filling hoppers or during operation of chemical feed equipment, the dust concentration in a poorly ventilated space may reach a level where a spark could cause the dust to ignite or explode.

An oxygen-deficient atmosphere is one in which the air with its oxygen content has been reduced by another gas or the oxygen has been removed. Normal air contains approximately 20.9 percent oxygen at sea level with the remaining constituents being primarily nitrogen (about 78.1 percent), argon (about 1 percent), and a few traces of other inert gases. When the oxygen content of air drops below 19.5 percent by volume, a potentially dangerous condition exists. A person breathing air with only 17 percent oxygen will experience shortness of breath. Loss of consciousness occurs rapidly when the oxygen content drops to 6 to 10 percent, and death occurs rapidly (within minutes) at less than 6 percent.

Oxygen deficiency in an enclosed atmosphere can occur for several reasons. Some of the more common causes include bacterial action that uses up the oxygen, displacement by other gases (can be both toxic and inert gases like nitrogen or carbon dioxide), oxidation of metals or other materials that deplete the oxygen level, adsorption of the oxygen onto surfaces, and combustion. Since the oxygen content of an enclosed atmosphere can change quickly, it must be continually monitored to ensure that the oxygen content does not drop below an acceptable level. High levels of oxygen enrichment (above 22.5 percent) are also dangerous because they increase the possibility of an explosion.

---

5. *Confined Space.* A space that has the following characteristics:
   (1) Is large enough and so configured that an employee can bodily enter and perform assigned work
   (2) Has limited or restricted means for entry or exit (for example, manholes, tanks, vessels, silos, storage bins, hoppers, vaults, and pits are spaces that may have limited means of entry)
   (3) Is not designed for continuous employee occupancy
   Also see DANGEROUS AIR CONTAMINATION and OXYGEN DEFICIENCY.
6. *Acute Health Effect.* An adverse effect on a human or animal body, with symptoms developing rapidly.

To check the safety of the atmosphere in a confined space, use a gas-detection instrument (Figure 20.6). These devices can detect explosive gases, oxygen deficiency, and toxic conditions. Remember, just because there are no toxic or explosive gases present does not mean that you may not lose your life because of a deficiency of oxygen.

According to OSHA regulations, each entry into a confined space requires a confined space entry permit (Figure 20.7). The purpose of the permit is to ensure that a supervisor or other qualified person has taken the necessary steps to identify any hazards associated with entry into the confined space and has taken appropriate safety measures to protect operators from injury. The permit should be renewed each time the space is left and re-entered, even for a break or lunch, or to go get a tool.

A comprehensive set of written, understandable operating procedures should be developed and provided to all persons whose duties may involve work in *PERMIT-REQUIRED* [7] or non-permit confined spaces. Training in the use of procedures should be provided, along with training in the use of safety equipment, rescue techniques, and cardiopulmonary resuscitation (CPR). Exact procedures for work in confined spaces may vary with different agencies and geographical locations and must be confirmed with the appropriate regulatory safety agency.

The following steps are recommended before entry into a confined space.

1. Identify and close off or reroute any lines that may convey harmful substance(s) to, or through, the work area.
2. Empty, flush, or purge the space of any harmful substance(s) to the greatest extent possible.
3. Monitor the atmosphere at the work site and within the space to determine if dangerous air contamination or oxygen deficiency exists.
4. Record the atmospheric test results and keep them at the site throughout the work period.
5. If the space is interconnected with another space, each space should be tested and the most hazardous conditions found should govern subsequent steps for entry into the space.
6. If an atmospheric hazard is noted, use portable blowers to further ventilate the area; retest the atmosphere after a suitable period of time.
7. If dangerous air contamination and oxygen deficiency do not exist before or after ventilation, entry into the area may proceed.
8. Provide appropriate, approved respiratory protective equipment for the standby person and place it where it will be readily available for immediate use in case of emergency.

Whenever an atmosphere free of dangerous air contamination or oxygen deficiency cannot be ensured through source isolation, ventilation, flushing, or purging, observe the following procedures.

1. If the confined space has both side and top openings, enter through the side opening, whenever possible.
2. Wear appropriate, approved respiratory protective equipment.
3. Wear an approved safety belt with an attached line. The free end of the line should be secured outside the entry point.
4. Station at least one person to stand by on the outside of the confined space and at least one additional person within sight or call of the standby person.
5. Maintain frequent, regular communication between the standby person and the entry person.
6. The standby person, equipped with appropriate respiratory protection, should only enter the confined space in case of emergency.

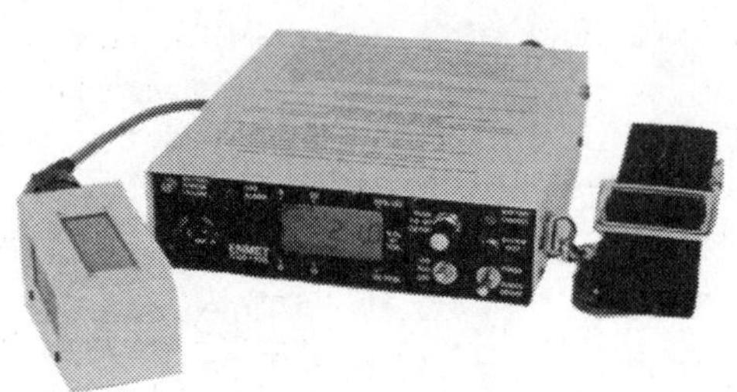
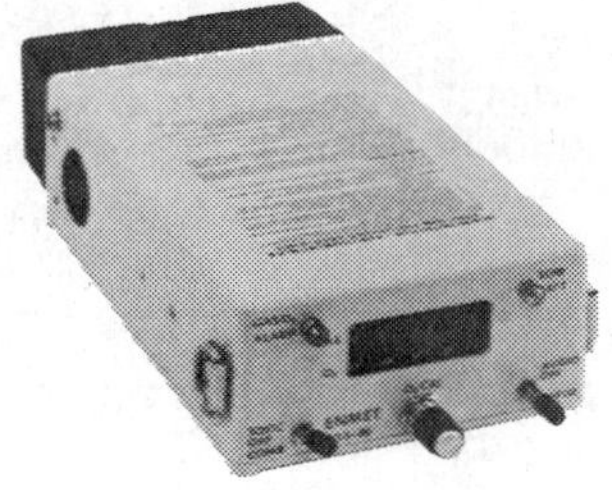

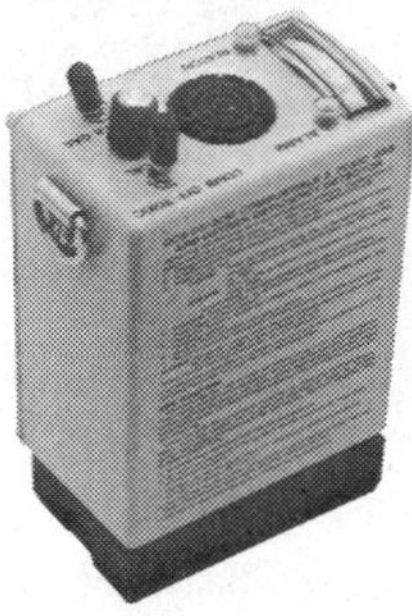

Fig. 20.6 Gas-detection instruments (toxic gas, combustible gas, and oxygen deficiency)
(Permission of ENMET Corporation)

7. *Confined Space, Permit-Required (Permit Space).* A confined space that has one or more of the following characteristics:
   (1) Contains or has a potential to contain a hazardous atmosphere
   (2) Contains a material that has the potential for engulfing an entrant
   (3) Has an internal configuration such that an entrant could be trapped or asphyxiated by inwardly converging walls or by a floor that slopes downward and tapers to a smaller cross section
   (4) Contains any other recognized serious safety or health hazard

Confined Space Pre-Entry Checklist/Confined Space Entry Permit

Date and Time Issued: ____________ Date and Time Expires: ____________ Job Site/Space I.D.: ____________

Job Supervisor: ____________ Equipment to be worked on: ____________ Work to be performed: ____________

Standby personnel: ____________ ____________ ____________

1. Atmospheric Checks: Time ______ Oxygen ______ % Toxic ______ ppm
   Explosive ______ % LEL Carbon Monoxide ______ ppm

2. Tester's signature: ____________

3. Source isolation: (No Entry) N/A Yes No
   Pumps or lines blinded, disconnected, or blocked ( ) ( ) ( )

4. Ventilation Modification: N/A Yes No
   Mechanical ( ) ( ) ( )
   Natural ventilation only ( ) ( ) ( )

5. Atmospheric check after isolation and ventilation: Time ______
   Oxygen ______ % > 19.5% < 23.5% Toxic ______ ppm < 10 ppm H_2S
   Explosive ______ % LEL < 10% Carbon Monoxide ______ ppm < 35 ppm CO

Tester's signature: ____________

6. Communication procedures: ____________
7. Rescue procedures: ____________

8. Entry, standby, and backup persons Yes No
   Successfully completed required training? ( ) ( )
   Is training current? ( ) ( )

9. Equipment:

	N/A	Yes	No
Direct-reading gas monitor tested	( )	( )	( )
Safety harnesses and lifelines for entry and standby persons	( )	( )	( )
Hoisting equipment	( )	( )	( )
Powered communications	( )	( )	( )
SCBAs for entry and standby persons	( )	( )	( )
Protective clothing	( )	( )	( )
All electric equipment listed for Class I, Division I, Groups A, B, C, and D and nonsparking tools	( )	( )	( )

10. Periodic atmospheric tests:

Oxygen:	___% Time ___;	___% Time ___;	___% Time ___;	___% Time ___;
Explosive:	___% Time ___;	___% Time ___;	___% Time ___;	___% Time ___;
Toxic:	___ppm Time ___;	___ppm Time ___;	___ppm Time ___;	___ppm Time ___;
Carbon Monoxide:	___ppm Time ___;	___ppm Time ___;	___ppm Time ___;	___ppm Time ___;

We have reviewed the work authorized by this permit and the information contained herein. Written instructions and safety procedures have been received and are understood. Entry cannot be approved if any brackets ( ) are marked in the "No" column. This permit is not valid unless all appropriate items are completed.

Permit Prepared By: (Supervisor) ____________ Approved By: (Unit Supervisor) ____________

Reviewed By: (CS Operations Personnel) ____________
(Entrant) (Attendant) (Entry Supervisor)

This permit to be kept at job site. Return job site copy to Safety Office following job completion.

Fig. 20.7 Confined space pre-entry checklist/confined space entry permit

7. If the entry is made through a top opening, use a hoisting device with a harness that suspends a person in an upright position.
8. If the space contains, or is likely to develop, flammable or explosive atmospheric conditions, do not use any tools or equipment (including electrical) that may provide a source of ignition.
9. Wear appropriate protective clothing when entering a confined space that contains corrosive substances, or other substances harmful to the skin.
10. At least one person trained in first aid and cardiopulmonary resuscitation (CPR) should be immediately available during any confined space job.

Confined space work can present serious hazards if you are uninformed or untrained. The procedures presented are only guidelines and exact requirements for confined space work for your locale may vary. Contact your local regulatory safety agency for specific requirements.

20.35 Manholes

There are many hazards involved with manholes and all of them can cause injury to the operator. Just removing the manhole cover can cause the loss of hands or fingers. You should never remove the manhole cover with your hands. Use a manhole hook or special tool such as a pick with a bent point to remove the lid. Be very careful when lifting the lid. Use your legs, not your back, for lifting. This will help prevent back strains. Locate the cover outside the working area to provide adequate working area around the manhole opening.

Next is the problem of traffic around an open manhole. The public, other operators, and vehicles must be protected. Traffic can be warned that operators are working in a manhole by the use of barricades, signs, flags, lights, and other warning devices. Warning devices and procedures must conform to local and state regulations. There also should be a barricade around the manhole to protect the operators. All personnel around manholes should wear hard hats for their safety.

Always inspect and test the ladder rungs in the manhole before using them. They may become loose or corroded and could collapse under your full weight. One should never enter a manhole alone; there should be at least one other person standing by at the top and one or more people within hearing distance in case of injury.

Perhaps the greatest threats to operators working in manholes are air contamination or depletion of oxygen. If possible, test the atmosphere in the manhole before removing the lid. Many operators have lost their lives because of leaking gas mains, gases produced by decaying vegetation, or other gases. Never enter a manhole without checking the atmosphere for sufficient oxygen, presence of toxic gases (hydrogen sulfide), or explosive conditions (methane or natural gas). Always provide adequate ventilation, which will remove any hazardous gases.

Smoking should never be permitted in or within 10 feet (3 meters) of manholes. Always use a mechanical lifting aid (rope and bucket) for raising or lowering tools and equipment into and out of a manhole. The use of a bucket or basket will keep your hands free when climbing down into or out of the manhole.

To review the hazards of underground structures, remember to use the proper tools for opening and closing the manhole. Keep in mind the need for barricades and lights to warn traffic and to prevent endangering other operators. Be sure that operators are trained in CPR methods and in the way to test the manhole for oxygen deficiency, oxygen enrichment, explosive gases, and toxic gases.

QUESTIONS

Please write your answers to the following questions and compare them with those on page 489.

20.3A What safety precautions should be taken when waxing floors?

20.3B How should rags containing oils, paint, or other cleaning compounds be stored?

20.3C What safety precautions should be exercised when operating an overhead crane?

20.3D How can traffic be warned that operators are working in a manhole?

20.3E How should tools and equipment be lowered into and removed from manholes?

20.36 Power Tools

The two general classes of portable power tools are pneumatic and electrical. Safety precautions for handling these types of tools are much the same for both types. Wear eye and ear protection when operating grinding, chipping, buffing, or pavement-breaking equipment. When using grinding or buffing tools, you will sometimes encounter toxic materials and will need respiratory protection. At other times, there is a need for full-face protection because of flying particles; you should use a face shield or at least goggles. In the use of electrical tools, always replace worn extension cords, and never expose cords to oils or chemicals. Extension cords also present a tripping hazard if left in the way. Avoid leaving extension cords in aisles or in work areas. Do not hang extension cords over sharp edges that could cut the cord, and always store the cords in a clean, dry location.

When working in a wet or damp location, consider using rubber mats or insulated platforms. Use only grounded tools. For protection against electric shock, use a ground-fault circuit interrupter (GFCI), especially for outdoor work.

When using pneumatic tools, never use the compressed air to clean off your clothing or parts of your body. Air can enter your tissues or other openings and cause problems. Always check hose clamps. If they are loose or worn, tighten or replace, as needed. Air hoses, like extension cords, are a tripping hazard. Therefore, consider their location when working with pneumatic tools. For the large (¾-inch or 18 mm) hoses, always use an approved safety-type hose connection with a short safety chain or other safety device attached. Air hoses that come apart can cause injuries as they are whipping about. Like electrical cords, keep air hoses away from oils, chemicals, or sharp objects.

Sandblasting with a pneumatic tool requires some special precautions. The operator should protect all skin surfaces with protective clothing, wear eye and face protection, use a respirator, and be very careful of toxic fumes that are discharged from a blasting operation.

The grinding wheel, pneumatic or electric, requires the same safety considerations. Eye and face protection is required. Do not use this tool without safety guards. Be careful of gloves being caught on the grinding wheel. Never operate a wheel with loose nuts on its spindle. When the grinding wheel is badly worn, replace it and use the proper wheel and speed of rotation.

All persons using power tools must be trained in their use and maintenance. Use the manufacturer's operations and maintenance guide for details of proper training. Most injuries by power tools are caused by incorrect setup and operation due to poor training.

Finally, power tools often produce a high level of noise. For example, air drills produce 95 *DECIBELS*[8] and circular saws, 105 dB. Ear protection must be provided when operators are exposed to long periods of high levels of noise. In areas of noise exposure, all operators should be provided with approved ear protective devices.

20.37 Welding

The first safety rule in the operation of gas or electric welding equipment is that the operator must be thoroughly trained in the correct operating procedures. The second rule concerns fire protection. The third rule is personnel protection. None of these rules is first or last—they should all be followed.

If you are not thoroughly trained in the use of the welding equipment, do not use it. If you absolutely must use the equipment, do so only under the supervision of a trained welder. Whenever such work must be performed in or around a water treatment plant, take time to consider the fire problem. For example, welding can be very dangerous in an oxidizing chemical location, near powdered activated carbon storage, and in storage areas for other bagged chemicals. Avoid welding around oil and grease, when possible; when that is not possible, at least provide for ventilation of fumes. When welding or cutting is done in the vicinity of any combustible material, you must take special precautions to prevent sparks or slag from reaching the combustible material and causing a fire. Some insurance companies require a fire watch to be present during welding operations. The fire watch is responsible for making the welder immediately aware of fire and to assist in putting out any fires that develop.

Regarding the safety of other personnel in the welding area, eye protection comes first. The person using the welding equipment must wear protective clothing, gloves, helmet, and goggles. Others in and around the welding operation should be kept at a safe distance. Always be careful of overhead welding because of falling sparks and slag. If other operators are (or must be) working in the vicinity of the welding operation, they too must be protected from the rays of arc welding; never look at the welding operation without eye protection.

Store welding gas cylinders in the same manner as other gas cylinders. They are stored upright, kept out of radiation of heat and sunlight, and stored with protective covers in place when not in use. Store cylinders away from elevators and stairs, and secure them with a chain or other suitable device.

20.38 Safety Valves

There are quite a number of safety valves in a water treatment plant; operators are not always aware of their locations or functions. For example, most operators know of the safety plugs on chlorine cylinders, but there are also large safety valves in any plant that stores large amounts of chlorine on site. These containers take on truckload lots of 17 tons (15,540 kilograms). The safety valves on such containers should be certified at least every 2 years or as often as the state requires. Such relief valves must be maintained on a regular basis. Inspect the inside of these tanks at regular time intervals and keep a record of the findings, for example, evidence of deposits and corrosion.

Water heater safety valves should be checked on an annual basis and maintained or replaced, as needed. If the plant has a boiler room, the steam safety valve should be maintained and checked for proper operating pressures. These valves should not discharge in such a manner as to be a hazard to operating personnel.

There also may be surge relief valves on discharge piping (high lift) of the treatment plant. These valves also act as a safety valve to the pumping equipment and must be maintained on some regular time interval. They should be checked for proper pressure setting, with all pilot valves being reconditioned or replaced, as needed.

There may be other safety valves located in the pumping plant's hydraulic system for opening and closing discharge valves that require maintenance. In the maintenance of water treatment plants, you or your supervisor must set up a maintenance system for all equipment. Hand tools, power tools, and other maintenance equipment must also be kept in safe working condition. Operators must be furnished protection for the eyes, the ears, the hands, the head, the feet and, at other times, the body. Work areas should be well ventilated and noise should be reduced, whenever possible. Each operator should always be on the lookout for additional ways of making the treatment plant a safer place to work.

---

8. *Decibel* (DES-uh-bull). A unit for expressing the relative intensity of sounds on a scale from zero for the average least perceptible sound to about 130 for the average level at which sound causes pain to humans. Abbreviated dB.

QUESTIONS

Please write your answers to the following questions and compare them with those on page 489.

20.3F What type of protection do operators need when operating portable power tools?

20.3G How can operators be protected from high noise levels when operating air drills and circular saws?

20.3H What personal protection should be used when operating welding equipment?

20.4 VEHICLE MAINTENANCE AND OPERATION

20.40 Types of Vehicles

Many types of vehicles are used in the waterworks industry. However, the plant operator may only come into contact with a few. Cars, pickup trucks, forklifts, dump trucks, and some electrically driven cars are the types of vehicles the operator is most likely to be involved with and to need to maintain. In addition to motor vehicle safety, this section will also consider the storage of fuel for these and other engines in the plant.

20.41 Maintenance

To have a safe motor vehicle, there must be a preventive maintenance program. Figure 20.8 gives a checklist for finding potential safety problems and a means of recording the preventive maintenance.

Tire inflation is a good example of proper safety checks. Not only is it unsafe to operate on underinflated tires, but it also causes undue wear on the tire. Therefore, tires should be checked regularly for wear, which may be caused by misalignment or low inflation. If tires are badly worn, they should be replaced. Always maintain the recommended pressure in the tires; when checking, set the hand brake and turn off the motor.

Next, when changing tires, be sure the jack you are using has sure footing. Position the jack at right angles to the direction of the lift. Jacks are a problem in general, and you should make sure that proper jacks are in each vehicle. In other words, select the proper jack for each job and choose only one that is safe and strong enough. If blocking is required, only use safe supports, avoid leaning on the jack, and protect hands. Always stay a safe distance from the jack handle as many injuries are caused by flying jack handles. Also, injuries are caused by overloading jacks. In addition, where needed, use braces or other supports to prevent tipping the vehicle over, another cause of serious injuries.

Fueling motor vehicles also involves some hazards. Always stop the vehicle's engine. Remember to remove the fuel hose immediately after using it. You could start a fire if you carelessly drive off with the hose still attached. Also, make sure the cap is replaced tightly on the tank. Do not permit smoking in the vicinity of gas delivery pumps at any time. Avoid any sparks and skin contact. Use only high-flash-point solvent for cleaning up any gasoline. Some other safety tips when refueling are: do not let the tank overflow, always set the brake, hang nozzle up properly onto the pump, and eliminate any leaks on the hose connections.

Most small water treatment plants do not have hoists or pits for vehicle lubrication. However, most of the following suggestions are applicable. First, keep all walkways, steps, tools, and containers free from grease, oil, and dirt. This will reduce the possibility of accidents caused by these items. As when maintaining any equipment, use the right tools, keep shoes free of all oil or grease, and wear only shoes with nonslip soles.

If the plant is equipped with a hoist, do not permit anyone to remain inside the vehicle when it is on the hoist. When lifting the vehicle, do not permit tools on the hoist or vehicle that may fall onto you or other personnel in the work area. Keep the driveway free of hoses, tools, and cars, and always keep your hand on the operating lever when raising or lowering the vehicle.

Most water treatment plants have assigned specific areas for washing or steam cleaning vehicles. If your plant does not have such an area, you should have one assigned. This area does not have to be elaborate, but should have water hoses and steam-cleaning equipment and be adequately drained. The same safety rules apply to makeshift installations as apply to completely equipped cleaning areas. The most important consideration is the steam cleaner. Keep the nozzle clean, check water level on coils before turning on the flame, and always wear protection for your eyes and face. Make sure that the cleaner is adequately grounded and be careful of cleaning compounds (see Section 20.1, "Chemical Handling," for precautions to prevent burns). Maintain steam hoses, check connections, and never permit horseplay with steam-cleaning equipment. As in other work areas, keep the wash rack free from grease and oil and oily rags. Hoses should be stored on the rack when not in use. Always use scaffolding or platforms when cleaning the tops of vehicles.

20.42 Seat Belts

Many water treatment plant operators have some excuse for not wearing seat belts. The excuses may sound good, but they will not protect you in the event of an accident. Many lives would have been saved if seat belts were used. The water utility should equip all vehicles with seat belts and require every operator to use them.

20.43 Accident Prevention

The best overall means of preventing vehicle accidents is defensive driving. This method requires training and a certain mental outlook on the part of the vehicle operator. Most, if not all, drivers think they are good at what they do and this may be true to some extent. However, if each driver would operate all vehicles as if all other drivers were the world's worst drivers, accidents would be greatly reduced.

Good drivers check out their vehicles each time they use them and have any maintenance performed, when needed. They use proper signals for directional change, always observe traffic regulations, and show courtesy to others. Remember that

Month __________ Vehicle Number __________ Assignment __________

	CHECK	WEEKS				
		1st	2nd	3rd	4th	5th
1	Oil					
2	Water					
3	Tires					
4	Horn					
5	Headlights, High - Low					
6	Tail Lights					
7	Turn Signals					
8	Stop Lights					
9	Battery Water					
10	Fire Extinguisher					
11	First Aid Kits					
12	Windshield Wipers					
13	Visual Inspection - Wire Rope					
14	" " - Hook					
15	" " - Sheaves					
16	" " - Boom					
17	" " - Hydraulic Level					
18	Operational Test Controls					

SERVICE MILEAGE READINGS				FUEL CONSUMPTION	
Week	Present	Last Service	Difference	Start Mileage	
1st				End Mileage	
2nd				Total Miles	
3rd				Fuel Used	
4th				MPG Average	
5th					

Fig. 20.8 Mobile equipment checklist

drivers in an agency vehicle represent the agency. Therefore, good driving skills are good for public relations.

Another way to avoid accidents is not to tailgate. Tailgating is a very unwise practice that is dangerous to the vehicle and hazardous to its operator. A good rule to use when following another vehicle is the old one-car-length for every 10 mph, and if there is limited visibility, increase that distance. Another rule is the "Three Second Rule," which says you must be at least 3 seconds behind the car in front of you. Take precautions when backing up. Always set the brake or shift to Park when parking the vehicle. Be cautious at intersections. As a defensive driver, always be ready to give the right of way. No right of way is worth injuring oneself.

In some cases, even the most defensive and careful driver has an accident. Because of this, each vehicle should carry flashlights, flares, flags, and a fire extinguisher, along with a first-aid kit. In the event of an accident, the driver should know how to fill out all the forms, a supply of which should be provided in the vehicle.

Remember, when operating a vehicle, an accident can be prevented by defensive driving. The plant supervisor should have each member of the staff take a defensive driver training course. Each driver should develop a defensive driver frame of mind. Developing a good attitude and driving skills are the key to accident prevention when operating a vehicle. All new employees should be given road tests in operating the types of vehicles they will be using.

Here are a few reminders when operating a vehicle. During a storm, roadways and pavement are likely to be slippery. Slow down, pump the brakes when stopping, and remember the minimum distance rules. No driver should be required to operate an unsafe vehicle. Keep copies of a suitable form for reporting mechanical problems in each vehicle and encourage operators to use them.

20.44 Forklifts

Most water treatment plants and pumping stations have a forklift. Most, if not all, plant operators use this vehicle to move chemicals, repair parts, and even use it when making repairs to lift heavy objects. Therefore, every plant operator should be trained in the use of the forklift.

Following are a few points regarding safe operation of the forklift with suggestions for operator safety as well as protection of others who may be in the operating area of the forklift. Keep all aisles free of boxes and other debris. Do not permit anyone to ride on the forklift except the operator. Never overload the forklift. Always be sure the warning signals are operational, and never leave the power on when leaving the forklift. Like other vehicles, check out the brakes before operating. Be careful at intersections of aisles and always face the direction of travel. If a loaded forklift is to be placed on an elevator, be sure that the load on the forklift and the weight of the vehicle do not exceed the weight limit of the elevator. Also, make sure the forklift load is stacked properly before lifting or moving. When handling drums, special lifting and retaining devices are needed.

Figure 20.9 is a typical forklift inspection form.

QUESTIONS

Please write your answers to the following questions and compare them with those on page 489.

20.4A What causes tire wear on motor vehicles?

20.4B Motor vehicles should contain what safety devices?

END OF LESSON 2 OF 3 LESSONS

on

SAFETY

Please answer the discussion and review questions next.

DISCUSSION AND REVIEW QUESTIONS

Chapter 20. SAFETY

(Lesson 2 of 3 Lessons)

Please write your answers to the following questions to determine how well you understand the material in the lesson. The question numbering continues from Lesson 1.

16. How can an operator prevent fires in a water treatment plant?
17. How would you maintain water-type fire extinguishers?
18. How would you maintain fire hoses in your water treatment plant?
19. What precautions should be taken in areas where flammable material is stored?
20. How can an operator make visitors feel as if the water being produced is safe to drink?
21. How would you remove a manhole cover?
22. What precautions should be taken when operating power tools in a wet or damp location?
23. What fire hazards should be considered before doing any welding?
24. How would you safely refuel a motor vehicle?
25. What safety precautions should be taken when driving during a storm?

SOUTHERN NEVADA WATER SYSTEM
For Truck Operator Inspection

Operator	Brake	Boom		Tilt		Back-up Horn	Horn	Wheel	Remarks
		Up	Dn.	F	R				

FORM #268 — S.N.W.S.

Fig. 20.9 Forklift inspection form

CHAPTER 20. SAFETY

(Lesson 3 of 3 Lessons)

20.5 ELECTRICAL EQUIPMENT

20.50 Electrical Safety

As a water treatment plant operator, you are not expected to be an expert in electrical equipment, but you must have a working understanding of electricity. This includes an understanding of the safety precautions needed to operate the electrical equipment. After all, electrical energy is required to power most of the treatment plant operations. The objective of this section is to show you how to operate safely and to become involved, to a limited degree, in the maintenance of electrical equipment. Electricity is unforgiving to the careless treatment plant operator. If you have any doubts about your understanding of this section, you may wish to reread Chapter 18, "Maintenance," Section 18.1, "Electrical Equipment."

20.51 Current—Voltage

Many types of electrical current are used in water treatment plants and the associated pumping plant. Each day the plant operator is exposed to this equipment, giving little thought to the potential hazards of the equipment. Current may come into the plant at a high voltage, for example, 69 KVA, reduced to 4,160 volts or 2,300 volts. This current may power pump motors, blowers, and other equipment at lower voltages of 440, 220, or 120 volts and within starters may be reduced to 24 or 12 volts or changed over into DC voltages. Given all of these various voltages, the operator must be careful not to become careless working with equipment. Therefore, become familiar with the types of current and voltage in the plant. By knowing this, you will avoid the mistake of becoming involved with unsafe electric currents or practices for which you are not trained. This will also enable you to know when to ask for a qualified person to perform any necessary repairs.

20.52 Transformers

Electrical power entering a plant is routed through transformers to reduce the voltage, in most cases. There are many types of transformers although the operator may only think of the larger ones that bring the power into the plant. Sometimes, these are owned by the waterworks agency and, therefore, the maintenance is the responsibility of the operator. There are few, if any, plant operators who are qualified to perform such maintenance. Never attempt to work on a high-voltage transformer without the assistance of qualified personnel. Such personnel can be located at the power company or contact an electrical contractor who specializes in the repair and maintenance of electrical transformers. You will, however, need to keep records of the transformer's operation. This information is helpful to repair personnel and is useful to operators who need to know the status of the transformer.

There are many small transformers within the plant's operating gear and it is these you may have to maintain. Most often, these low-voltage transformers become overworked; they overheat and burn out. Any fire in the electrical gear can be hazardous. Be very careful when opening a starter, breaker box, or indicating instrumentation if a fire or overheated transformer is suspected. Operators have been badly burned by not thinking before opening such devices when they smell smoke in pumping stations or treatment plants. When solving problems with hot, overheated, or burning transformers, remember what you learned in Section 20.2 about electrical fires. For the safety of operating personnel and the safety of the plant, regularly inspect, or have someone inspect, both large and small transformers. If you detect any overheating, have a qualified electrician inspect and replace any transformer that is not functioning properly.

There should be a fence around the transformer station with a locked gate and only a limited number of keys issued to plant personnel. The operator may perform routine preventive maintenance such as removing weeds or brush and general cleanup. Replacing fuses and major maintenance or repairs should be made by the power company or qualified electricians. Maintenance must be performed by qualified and well-trained personnel.

20.53 Electric Starters

As a treatment plant operator, your most frequent contact with electrical power will probably be with electric starters on the motor control panels. These devices are used throughout the plant and provide an interface between the operator and the flow of energy. The starter may be located on a switch panel or there may be a switch that is remotely located from the starter. One of the first safety procedures you should take is to use a special insulated mat on the floor at all switchboards. The starter should be provided with adequate lighting and clearly marked Start/Stop buttons. Replace indicating lights as needed without delay. There should always be clear and adequate working space around the starter or switch panels. To reduce the hazards of fire in electric starters, they should be cleaned and maintained on a regular basis. Such maintenance must be performed by trained, qualified personnel. In electrical starting equipment, fires can easily occur because of accumulation of dust and dirt on the contactors. When they become so badly burned that they do not make proper contact, they could overheat and start fires. The key to preventing fires in starting equipment is a good preventive maintenance program.

20.54 Electric Motors

The treatment plant operator is exposed to many types and sizes of electric motors. In some plants, the motors are old and require more attention because of exposed parts. The newer electric motors are enclosed and have all parts protected. For the old motors, you should install safety guards or guardrails to prevent accidental contact with live parts of the motors.

Some of the electric motors may have exposed couplings, pulleys, gears, or sprockets that also require consideration. For these and other moving devices, a wire cloth gear guard may be installed. The gear guard can also be made of sheet metal. However, no matter which type of guard is used, it must be securely fastened onto the floor or some other solid support. The safety guards must be constructed and fitted to prevent material being handled by operators from coming into contact with the moving parts driven by the motors.

Another consideration is projections on couplings, pulley shafts, and other revolving parts on the motor or on the device being driven. These projections can be bolts, keys, set screws, or other projections. The projections should be removed, reduced, or protected by one of the above guards.

Check grounding on all electric motors as part of a routine maintenance program. The motor frames themselves must be grounded if the wires to the motor are not enclosed in an armored conduit or other metallic raceway. Check that all joints are mechanically secure to ensure good grounding. In the case of portable electric motors, the simplest way of grounding is an extra conductor in the cord serving the motor. The best way is to install a ground-fault circuit interrupter (GFCI) receptacle. This device will automatically disconnect the tool from the power supply if the ground is not connected and will supply the greatest protection to the operator and to the equipment.

When using portable electric motors, always check the service cord. If the cord or receptacles are in poor condition or showing signs of wear, they should be replaced. A badly worn cord must never be used in a wet location.

20.55 Instrumentation

In this area of water treatment, the operator is not exposed to a great deal of hazard, but must give some consideration to instrumentation devices since they are operated by electric current. This is also true of all other automatic equipment. Although most instruments protect the operator, there is still a degree of hazard when changing charts, calibrating, or performing other maintenance. First, when calibrating an instrument, you are exposed to at least 12 volts DC to 120 volts AC. If you become grounded with the 120 volts AC, you may be killed or severely injured. Also, when maintaining automatic control equipment, adjustment of one instrument may start another device, exposing another operator to a hazard because of an unexpected start. As mentioned above, electronic devices operate on low current, but do not forget that there is still high voltage located somewhere in the instrument.

20.56 Control Panels

Always provide adequate working space in and around control panels. As with electric motors, the panels must be well grounded. At some locations, there may be a need for special insulating mats, such as in wet locations. Adequate lighting must be available inside the control panel as well as outside for those who do the maintenance. Moisture or corrosive gases must be kept away from the control panels. To reduce fire hazard, never store any hazardous material next to switchboards or control panels. Panels carrying greater than 600 volts must be permanently marked warning of the hazards. Areas of high voltage should be screened off and locked with a limited number of keys given to authorized personnel only.

CAUTION

Unless you are a qualified electrician, stay out of the inside of all electrical panels. If you do not know or are not familiar with the equipment, leave it alone!

20.57 Lockout/Tagout Procedure

Any equipment that could suddenly start up or release stored energy is a potential hazard to operators. Stored energy may be in the form of electrical power, hydraulic pressure, pneumatic pressure, or even the tension in a spring. Two hazards due to the lack of a good lockout procedure are accidentally starting a piece of equipment, exposing a fellow operator to a hazard; and turning electrical power on when someone is still working on the equipment, exposing that person to danger. Every water treatment plant should develop a standard operating procedure (SOP) (Figure 20.10) for lockout of equipment to protect personnel from accidental injury.

A lockout device is a positive means of holding a switch, valve, or other energy isolating device in a safe position to prevent the equipment from becoming energized, either directly or by release of stored energy. Positive control of the energy source is ensured by the use of a keyed lock or a combination lock. Besides physically locking out or isolating the source of energy, it is also important to warn other operators, by means of a tagout device, that the switch or piece of equipment has been shut down. A tagout device is a prominent warning tag that can be securely fastened to the energy isolating device to indicate that both it and the equipment being controlled may not be operated until the tagout device is removed.

Even after equipment is locked out and tagged, it still may not be safe to work on. Stored energy in gas, air, water, steam, and hydraulic systems (such as water pumping systems) must be drained or bled down to release the stored energy. Before bleeding down pressurized systems, think about the pressures involved and what will be discharged to the atmosphere and work area (toxic or explosive gases, corrosive chemicals). Will other

SOUTHERN NEVADA WATER SYSTEM
Standard Operating Procedure

SUBJECT: REMOVING EQUIPMENT FROM SERVICE	Number: 35
DIVISION: SAFETY	Prepared By: JAR 4/22/14

PURPOSE

This procedure prescribes the method for safely removing a piece of equipment from service prior to performing the required maintenance or repairs.

GENERAL

Equipment may be removed from service for the following reasons:

1. To prevent an unsafe or damaging operating condition.
2. To perform maintenance or repairs that would be unsafe or impractical to perform while equipment is in service.

Equipment should only be removed from service when one of the above conditions exists.

METHOD

RESPONSIBILITY: Operations or Maint. Employee

ACTION:

1. Upon receipt of a work order requiring removal of equipment from service or upon discovering a condition that requires removal of equipment from service, checks the following as applicable:
   - motor(s) is off
   - discharge valves are closed
   - selector switches are off
   - alarm system is bypassed
   - etc.
2. Attaches a "lockout" device to the piece of equipment, if possible.

NOTE: Once a "lockout" device is installed, it is not to be removed until repairs are complete, unless removal is authorized by the Director or a Division Head.

3. Prepares a "Do Not Operate" tag, entering the following:
   - date
   - reason for equipment being removed from service
   - signature of employee preparing tag
4. Affixes tag to the piece of equipment.
5. Notifies operations by radio or telephone of the removal from service.
6. If the removal from service was not initiated by a work order, checks to make sure that no work order is already scheduled for the repair.
7. If no active work order is on file for the repair, prepares and processes a work order in accordance with SOP #39, subject: "Work Orders."
8. Removes the "lockout" device, if possible, or requests that the device be removed by the employee who installed it.

NOTE: If the employee who installed the device is unavailable, requests removal or authorization for removal from the Superintendent.

9. Performs scheduled maintenance or repairs and installs own "lockout" device as needed until work is completed.
10. Maintenance Technician, upon completion of scheduled maintenance or repairs, removes the tag and the "lockout" device, if applicable, and returns the piece of equipment to service status.
11. Notifies Operations that the equipment has been returned to service.

Fig. 20.10 Standard operating procedure for locking out electrical equipment

safety precautions have to be taken? What volume of bleed-down material will there be and where will it go? If dealing with chemicals, greases, or lubricants, how can they be cleaned up or contained? Many times, an operator has removed a pump volute cleanout only to find that the discharge check valve was not seated or one of the isolation valves was not fully seated. Once the pump was open, however, the pump intake structure drained into the pump room through the opened pump.

Many types of equipment also require the use of a blockout procedure. Blockout means the insertion of some device that prevents moving parts in equipment from moving. Physically block or secure in place any elevated machine members, flywheels, and springs to prevent movement that could injure an operator working on the equipment.

All rotating mechanical equipment must have guards installed to protect operators from becoming entangled or caught up in belts, pulleys, drive lines, flywheels, and couplings. Keep the guards installed even when no one is working on the equipment.

Various pieces of machinery are equipped with travel limit switches, pressure sensors, pressure reliefs, shear pins, and stall or torque switches to ensure proper and safe operation of the equipment. Never disconnect a device, or install larger shear pins than those specified on the original design, or modify pressure or temperature settings.

In a safe lockout procedure, the switches are locked open and are properly tagged; only the operator who is doing the maintenance should have a key. In fact, all people who perform electrical maintenance should have their own individual lock and key so as to maintain control over the equipment being worked on by each individual. Only the operator who installed the lockout device and warning tag may remove them unless the employer has provided specific procedures and training for removal by others.

The basic elements of a proper lockout/tagout procedure[9] are as follows:

1. Notify all affected employees that a lockout or tagout system is going to be used and the reason why. The authorized employee is responsible for knowing the type and magnitude of energy that the equipment uses and must understand the potential hazards.
2. If the equipment is operating, shut it down by the normal stopping procedure.
3. Operate the switch, valve, or other energy isolating device(s) so that the equipment is isolated from its energy source(s). Stored energy such as that in springs, elevated machine members, rotating flywheels, hydraulic systems, and air, gas, steam, or water pressure must be dissipated or restrained by methods such as repositioning, blocking, or bleeding down.
4. Lock out and tag the energy isolating device with assigned individual lock or tag.
5. After ensuring that no personnel are exposed, and as a check that the energy source is disconnected, operate the push button or other normal operating controls to make certain the equipment will not operate. A common problem when motor control centers (MCCs) contain many breakers is to lock out the wrong equipment (locking pump #3 and thinking it is #2). Always confirm the dead circuit. *CAUTION: Return operating controls to the neutral or OFF position after the test.*
6. The equipment is now locked out or tagged out and work on the equipment may begin.
7. After the work on the equipment is complete, all tools have been removed, guards have been reinstalled, and employees are in the clear, remove all lockout/tagout devices. Operate the energy isolating devices to restore energy to the equipment.
8. Notify affected employees that the lockout or tagout device(s) has been removed before starting the equipment.

QUESTIONS

Please write your answers to the following questions and compare them with those on page 489.

20.5A Why should each plant operator become familiar with the type of current and voltage in the water treatment plant?

20.5B List the moving parts on electric motors that require safety guards.

20.5C What are two hazards created by the lack of a good lockout procedure for control panels and switchboards?

20.5D List four examples of stored energy that could be hazardous to operators.

20.6 LABORATORY SAFETY[10]

20.60 Laboratory Hazards

In general, water plant operators do not experience a great deal of exposure to hazardous laboratory conditions. However, you will be in contact with potentially hazardous glassware, toxic chemicals, flammable chemicals, corrosive acids, and alkalies. There may be times when you will be exposed to hazardous bacteriological agents. The seriousness of the hazards depends mainly upon the size of the plant and the operating procedures in the treatment plant. For your own safety, learn the proper procedures for handling laboratory equipment and chemicals.

20.61 Glassware

An important item in laboratory safety is handling of glassware. Almost all tests performed by an operator will require the use of some glassware. The operator's hands, of course, are exposed to the greatest hazard. To reduce accidents when handling glassware, never used chipped, cracked, or broken glassware in any testing procedure. All such glassware should be disposed of in a container marked "For Broken Glass Only." Never put broken glass in wastebaskets. Although it may not be a hazard to the operator, it is a danger to those who clean out the wastebaskets. Clean up any broken glass or spilled chemicals to reduce hazards to others. Never let broken pieces of glass remain in the sink or in sink drains. This may cause cuts to others who unknowingly try to clean the sink.

Washing glassware is always potentially hazardous. The glassware can be broken while being washed, causing cuts, or cuts can be caused by chipped or cracked glassware. Also, the cleaning compounds themselves can be a problem. Sometimes strong acid cleaners are used to remove stains from the glassware. Without protective gloves, your hands could be seriously burned by these acids.

20.62 Chemicals

When handling liquid chemicals, such as acids and bases, always use safety glasses or face shields. If working with ether or chloroform, avoid inhalation of fumes and always do this type of work under the ventilation hood. Be sure to turn the ventilation fan on. Of course, be careful of open flames when using flammables such as ether. As a general rule, do not permit smoking in the laboratory. All chemicals should be stored in proper locations; do not set chemicals on the laboratory benches where they may cause an accident if spilled or if the container is broken.

When handling laboratory gases, give consideration to their location and potential accident hazards. Gas cylinders must be prevented from falling by using safety retaining devices such as chains. The valve and cylinder regulator should be protected from being struck by stools, ladders, and other objects.

9. Adapted from *Wastewater System Operations*, State of Oklahoma Certification Study Guide.
10. See *Fisher Safety Catalog*. Obtain from Literature Fulfillment at the Fisher Scientific website, www.fishersci.com, or phone (800) 772-6733.

When mixing acid with water, always pour the acid into the water while stirring. Never add water to acid because the water may remain on top of the acid, causing splattering and excess heat generation.

Always use safety goggles, gloves, and protective garments. When cleaning up acid or alkali spills, dilute with lots of water even if you flush them down the sink drain. Baking soda can be used to neutralize acids, and vinegar is used to neutralize bases. Never allow mercury, gasoline, oil, or organic compounds into the laboratory drains. Use only a toxic waste disposal drain system for these items. Pouring such compounds down sink drains can cause an explosion, allow toxic gases and vapors to enter the lab, or destroy the piping.

You must never use your mouth with the pipet for transferring toxic chemicals, acids, or alkalis. Use a suction bulb, aspirator, pump, or vacuum line. If you use your mouth, there is always the danger of getting the toxic solutions into your mouth.

20.63 Biological Considerations

Do not take chances with bacteria. A good policy is to have each operator immunized with antityphoid vaccine and to keep their booster shots current. Always use good sanitary practices, particularly when working with unknown bacteria or known pathogens. Never pipet bacteriological samples by mouth. Always use a pipet bulb.

When exposed to any bacteria, you should make it a habit to always wash your hands before eating or smoking. If you have any cuts or broken skin areas, these wounds should not come in contact with bacterial agents. You should wear protective gloves or cover the wound with a bandage when working with any kind of bacteria.

All work areas should be swabbed down with a good bactericidal disinfectant before and after preparing samples. As a general policy, the preparation or serving of food should never be permitted in the laboratory. Also, give some consideration to proper ventilation, because some bacteria may be transmitted through the air system.

20.64 Radioactivity

There are many laboratory and treatment plant instruments that use radioactive isotopes in laboratory tests and research. A plant operator may be exposed to radioactive compounds when calibrating sludge density meters or using research isotopes. From a safety standpoint, only qualified personnel should be involved in the use of radioactive compounds. If radioactive compounds are present in the laboratory, warning signs should be posted. The disposal of all radioactive compounds must be performed by qualified laboratory personnel in strict accordance with government regulations.

20.65 Laboratory Equipment

20.650 Hot Plates

You will probably use a hot plate in your threshold odor number (TON) tests. You should turn the hot plate off when not in use; never place bare hands on the hot plate to check if it is hot. When using the hot plate, always use the hood and turn on the hood ventilation fan to remove gas or fumes. Never place glassware onto a hot plate if the outside of the glassware has water or moisture on the surface between the glass and the hot plate. Steam will form at this interface and cause the glass to break. When taking hot glassware off the hot plate, always use gloves.

20.651 Water Stills

Most water stills in the laboratory today are the electrical type. To observe good safety practices, check the items described in the electrical safety section of this chapter, such as good grounding. Set up a standard operating procedure (SOP) for proper operation of the still and follow the manufacturer's instructions for proper starting and stopping. The still will require cleaning from time to time. Be very careful when disassembling the still. Parts may be frozen together because of hardness in the water and may require an acid wash to separate. Be sure that the boiler unit is full of water before turning the still on. Never allow cold water into a hot boiler unit because it may cause the unit to break.

20.652 Sterilizers

There are two types of sterilizers: dry electrical sterilizers and wet sterilizers (autoclaves). In the dry electrical sterilizers, check the cords frequently because the high heat may damage the wiring. Always let the unit cool off before removing its contents. Wet sterilizers (autoclaves) are under pressure by steam and should be opened very slowly. Use gloves and protective clothing when unloading the autoclave. Cover the steam exhaust with an insulated covering to prevent burns. Any leakage around the door should be repaired by replacing door gaskets or even the door if it is worn or bent. Always load the autoclave in accordance with the manufacturer's recommendations. Never allow an operator to work with this equipment without proper instruction in its operation.

20.653 Pipet Washers

Cleaning pipets can be hazardous. Many laboratories have a pipet cleaner that contains an acid compound or other cleaning agents. Always use protective clothing and a face shield when working with the washer unit. Try to avoid dripping or spilling the cleaning compound when transferring the pipets. If the acid comes into contact with your skin, use the remedies recommended in Section 20.11, "Acids."

QUESTIONS

Please write your answers to the following questions and compare them with those on pages 489 and 490.

20.6A Why is washing glassware always a potential hazard?

20.6B What precautions should be taken when handling liquid chemicals, such as acids and bases?

20.6C Why should mercury, gasoline, oil, or organic compounds never be allowed into laboratory drains?

20.6D How can an operator be exposed to radioactive compounds?

20.6E Why should cold water never be allowed into the hot boiler unit of a water still?

20.7 OPERATOR PROTECTION

20.70 Operator Safety

So far in this chapter, we have discussed many means by which you can protect yourself and your equipment. In this section, we wish to discuss your own personal protection. Take a look at the means of protecting the eye, the foot, the head, and, most of all, look at water safety. After all, a plant operator is always in contact with water. The water may be found in raw water reservoirs, settling basins, clear wells, or filters. Operators have lost their lives by falling into the backwash gullet. Operators have lost their lives in the finished water reservoirs. As unlikely as it seems, fatal accidents have happened in the past and will happen again in the future. Therefore, it is the responsibility of each operator to watch for any safety problems in the plant's reservoirs, pumping stations, or filters. You, the operator, are responsible for yourself. You should never expose yourself to unsafe conditions.

A major problem confronting many operators is locating reliable safety vendors and equipment. State, regional, and national professional meetings, such as those sponsored by the American Water Works Association, often have displays or exhibits featuring manufacturers of safety equipment. This is an excellent opportunity to meet the representatives of these companies and discuss with them what equipment they would recommend for your situation. Also, other operators who have had experience with safety equipment of interest to you often attend these meetings and are eager to share their experiences with you.

If you are unable to attend these meetings, the program announcements will often have a short description of the types of safety equipment that will be featured by each vendor exhibiting at the conference. You can obtain the vendor's address by looking in professional journals, buyers' guides, or by writing to the sponsor of the conference.

20.71 Respiratory Protection

There are many respiratory hazards in and around the treatment plant that an operator is exposed to daily, including chemical dusts, chemical fumes, and chemical gases, such as chlorine, ammonia, sulfur dioxide, and acid fumes, to name only a few. Whenever working around or handling these and other compounds, you must take adequate precautions.

Two types of conditions for which you should be prepared are an oxygen-deficient atmosphere and one with sufficient oxygen but contaminated by toxic gases or explosive conditions. In either circumstance, you will need an independent oxygen supply. However, an independent oxygen supply will not protect you from an explosion.

> Never enter a confined space with an explosive atmosphere.

Call your local gas company and ask their experts to enter the explosive area, if entry is essential. Your independent oxygen supply should be of the positive-pressure type to protect you if there are any leaks in your mask. Good ventilation can reduce explosive conditions.

20.72 Safety Equipment

All waterworks safety equipment, such as life lines, life buoys, fire extinguishers, fencing, guards, and respiratory apparatus, must be kept in good repair. This and other safety equipment is necessary to protect operators or visitors from injury or death. Safety equipment may fall into disrepair because it is only used occasionally and may deteriorate due to heat, time, and other environmental factors. First-aid equipment should also be provided and kept resupplied as it is used. The operating staff should be given regular instructions in the use and maintenance of the safety equipment.

Provide protective clothing for all operators handling chemicals or dangerous materials. Keep the clothing clean and store it in a protected area when not in use.

The water utility is responsible for providing outward opening doors, remote-controlled ventilation, inspection windows, and similar safety devices where appropriate. This equipment should be exercised and kept clean and well maintained so that it will operate when needed. Respiratory (self-contained breathing) apparatus must be stored in unlocked cabinets outside of chlorination, sulfur dioxide, carbon dioxide, ozone, and ammonia rooms. The storage cabinets must have a controlled environment to prevent deterioration of the equipment.

The operator has the responsibility to inspect each apparatus for deterioration and need for repair. Safety equipment is of no use to the operator if it fails when put to use, and it may cost you your life if it is in poor condition. Some self-contained breathing apparatus (air packs) depend on compressed air to supply the oxygen. Under conditions of deficient oxygen supply, the canister type of respirator is useless. You could lose your life by entering a room containing chlorine gas (which is heavier than air) while depending on a poorly maintained respirator. Although you might have protection from the chlorine, you would not have adequate oxygen. Never take a chance with a toxic gas. In water treatment plants, use only the positive-pressure type of self-contained breathing apparatus.

Many newer plants are being constructed with independent air supplies consisting of a helmet, hose, and compressed air. The helmet is connected by a hose to an uncontaminated air source. The key word here is "uncontaminated." Not only should the operator follow a maintenance program for the hose and mouthpieces of the apparatus, but the operator must maintain the air supply. The air is supplied by mechanical equipment that requires maintenance. The air pressure is controlled by a reducer or regulator that must be kept clean and maintained to be available when needed. Set up a preventive maintenance program for this equipment. It should be checked out on a weekly basis, and records should be kept of each inspection. The record should show conditions of the hoses, regulators, air filters, compressors, helmet, and any other apparatus furnished with the system.

The old standby, of course, has been the air packs or self-breathing apparatus. These units are carried by the user, giving the operator an independent source of air (oxygen). The unit can be used in any concentration of contamination of gases, dust, or anywhere the atmosphere is oxygen deficient. There are two types of units. One type of unit depends on compressed air or oxygen, and the second system generates oxygen by use of chemicals in a canister. The oxygen is generated by the moisture exhaled by the user. Because this equipment is not used daily in the water treatment plant, there must be a preventive maintenance program, with records, inspection, and operator checkout. As with any system, self-contained breathing equipment requires maintenance. This is vital because the operator's life will depend on how well this apparatus performs.

Training is another important consideration. Even though you may have used the breathing apparatus many times in the past, you should be checked out each month on the equipment. There should be a maximum allowable time for putting on the apparatus. The apparatus should be checked out under field conditions, such as using ammonia or some other nontoxic gas. Remember, it is too late to learn how fast an operator can put on a breathing apparatus when a room is filled with chlorine.

Be aware that there have been cases where operators have been saved because they knew how to use the breathing apparatus properly. Only repeated practice will enable you to master this survival skill.

20.73 Eye Protection

The water treatment plant operator has only two eyes. You may think that everyone is aware of this fact. However, some operators behave as though they have many eyes and are very careless about protecting the eyes they do have from hazards.

Because some operators fail to see the value of eye protection, it will take a maximum effort on the part of the supervisory staff to enforce an adequate eye protection program. There must be an intense program of education, persuasion, and appeal to guarantee compliance with an eye protection program.

Most conditions in which a plant operator needs eye protection are not too difficult to understand. Eye protection is needed when handling many of the liquid chemicals, acids, and caustics. Some of the tests performed in the laboratory require eye protection. Only a few moments are required to put on a face shield or safety glasses, and remember—the loss of an eye will last a lifetime.

QUESTIONS

Please write your answers to the following questions and compare them with those on page 490.

20.7A When entering an oxygen-deficient atmosphere, what type of oxygen supply is recommended?

20.7B Where should respiratory apparatus be stored?

20.7C How frequently should independent air supply equipment be checked out and what should be inspected?

20.7D How can compliance with an eye protection program be encouraged?

20.7E Under what conditions does an operator need eye protection?

20.74 Foot Protection

There are few situations under which a water treatment plant operator needs foot protection. In the normal routine of daily operation of the plant, there are not many hazards to the operator's feet. But in some plants the operator also performs plant maintenance. Here the steel-toed safety shoes are useful. The shoe should be able to resist at least a 300-pound (136 kg) impact. An important consideration in any plant under operating conditions is the use of rubber boots. The rubber boots are needed when handling acid or caustic or when the operator is working in wet conditions such as reservoirs, filters, or chemical tanks. Under these circumstances, the agency should have an adequate supply of boots with nonskid soles in various sizes. If these conditions are something that the operator is exposed to daily or weekly, the agency should give the operator a pair of boots for personal use.

20.75 Hand Protection

The treatment plant operator's hands are always exposed to hazards. These include not only minor scratches or cuts but also exposure to chemicals that may not attack immediately. Some compounds such as alum, attack the skin slowly. Because there is no immediate pain, you may think there is no damage. This is not true; the attack on the skin is slow and may cause an infection at a later date. Therefore, when handling chemicals, always use rubber gloves. As part of a safety package, each operator should be issued a pair of rubber gloves and a pair of leather gloves. These gloves should be replaced when they no longer provide the necessary protection.

There are other compounds such as solvents that will absorb through the skin and can cause long-term effects. For such special problems, there is a need for neoprene gloves. Another problem is that of handling hot materials, such as laboratory flasks and beakers. Here you may need insulated fabric gloves. When working around machinery that is revolving, wearing gloves or other hand protection can be dangerous. If a glove gets caught in the machinery, you could become injured. Do not let your protective equipment itself become a hazard.

Be sure that the gloves you are wearing are the right type for the job you are doing. The gloves should allow for quick removal and be in good condition. Always check for cracks and holes, flexibility, and grip. Keep them clean and in good condition. There are many types of gloves and the proper type should be worn for each job.

1. *CLOTH GLOVES* protect from general wear, dirt, chafing, abrasions, wood slivers, and low heat.
2. *LEATHER GLOVES* protect from sparks, chips, rough material, and moderate heat.
3. *RUBBER GLOVES* protect against acids and some chemical burns.
4. *NEOPRENE AND CORK-DIPPED GLOVES* give better grip on slippery or oily jobs.
5. *ALUMINIZED GLOVES* are heat-resistant to protect against sparks, flames, and heat.

6. *METAL MESH GLOVES* protect from cuts, rough materials, and blows from edge tools.
7. *PLASTIC GLOVES* protect from chemicals and corrosive substances.
8. *INVISIBLE GLOVES* (barrier cream, such as petroleum jelly) protect from excessive water contact and from substances that dissolve in skin oil.

20.76 Head Protection

In most areas of a water treatment plant, there is really no need for a hard hat. However, there are certain hazardous conditions under which the operators should be required to wear a metal, plastic-impregnated fabric, or fiberglass hat. The hard hat should have a suspended crown with an adjustable head band, provide good ventilation, and be water resistant. Operators should be required to wear the hard hats when work is being performed overhead or in any location where there is danger of tools or other materials falling, for example, working in filters, settling basins, manholes, or trenches. There has been a long history showing the value of hard hats in reducing injuries and death.

20.77 Water Safety

Every operator in a treatment plant is exposed to situations in which the operator's life can be lost due, either directly or indirectly, to water. Although during your daily activity you may never think in terms of drowning, this hazard is always present in the treatment plant. If you are working at a reservoir or a lake in a boat, you may think of water safety, but still never pay real attention to the danger.

Starting at the treatment plant, you can take simple measures that will reduce hazards. To reduce the hazard of slipping on catwalks or when working around clarifiers or settling basins, use nonskid surfaces on ladders and walkways going into and out of clarifiers or sedimentation basins. Be very cautious during cold or wet weather. Water freezes into ice, which is slippery.

Keep all handrails or other guards in good repair; replace any that become unsafe. Many older plants do not have protective handrails; install rails or chain off the unsafe area to all employees and mark off with warning signs. The unsafe areas can be guarded with ⅜-inch (9 mm) Manila rope, chains, or cables that you may have around the plant.

Filters are an important area of safety consideration because there is always activity in or around each filter, such as washing or maintenance. Make repairs to handrails immediately, when needed. Station emergency gear around the filter areas; equipment such as life jackets are good, but buoys, ⅜-inch (9 mm) Manila line, or a long wooden pole are much more useful. These types of devices can be used to rescue someone who has fallen into the filter. An operator should never work in the filter when it is being backwashed. There is always the danger of falling into the wash-water gullet and being unable to get out before drowning.

Sedimentation basins, flocculation basins, or clarifiers present many of the same problems as filters. Maintain handrails, place warning signs, or put up guard ropes or chains. Also, keep life rings and Manila or nylon lines in good repair. A lift ring, life pole, and lines should be stationed at each basin. A good idea is to shelter the safety gear from the weather, but do not cause the gear to become inaccessible.

In reservoir operation and maintenance, you will encounter two types of water: raw water and treated water. In a raw water reservoir or lake, you have to worry only about personnel safety. In a treated water reservoir, you must also be concerned about the safety of the water going to the customer. If you are working out of a boat, make sure that everyone in the boat is wearing a life jacket. Also, take on board both a safety line and buoys. Cold-weather conditions are an added problem. Even though you may be a good or excellent swimmer, the thermal shock of cold water may quickly paralyze you, making you unable to save yourself. Under such conditions, if a second operator goes into the water to save you, there may be two lives lost.

When taking a boat out on the water, it should first be checked out for safety. Check the bilge pump, ventilation in the compartments, safety cushions, fire extinguishers, battery, and engine. Also, check for safety equipment, oars, life jackets, lights, mooring lines, and fuel. If you are applying copper sulfate powder or solution, other safety equipment will be needed, such as respiratory and eye protection equipment. Prepare a detailed equipment checklist to use each time the boat goes out onto the lake or reservoir. The boat should never be taken out on choppy waters or when the wind is high.

On some occasions, there is a need for underwater examination of valves, intake, or other underwater equipment or apparatus. Such work should only be performed by employees who are trained in underwater diving. If there are no qualified divers on your staff, you should hire such personnel to do the diving and underwater inspections. There are organizations with people who do this type of work and they are well qualified in underwater examinations. If an operator on staff is to do the diving, the operator should be certified by a local diving school or other certifying agency. The operator's certificate should always be kept current and the operator should be required to perform the number of dives necessary to keep this certification current.

In closing, all plant operators should know how to swim. If they do not, they should take a Red Cross class and learn the minimum fundamentals to save their own lives. Any operator working over open water should be required to wear a buoyant vest. All basins should have approved safety vests, buoys, and life lines stationed at outside edges.

QUESTIONS

Please write your answers to the following questions and compare them with those on page 490.

20.7F Under what operating conditions should an operator wear rubber boots?

20.7G Under what specific conditions should an operator be very careful wearing gloves?

20.7H Why should operators never work in a filter when it is being backwashed?

20.7I What items should be checked before an operator takes a boat out on the water?

20.8 PREPARATION FOR EMERGENCIES[11]

Emergencies are very difficult to plan and prepare for because you never know what will happen and when it will occur. Catastrophic events could include floods, tornados, hurricanes, fires, and earthquakes. Serious injury to anyone on the plant grounds is an emergency.

Conduct periodic tours of your facilities with the local fire, police, and emergency response organizations to familiarize them with the site, potentially hazardous locations, and location of fire hydrants. Their familiarity with the plant layout will be very helpful if an emergency ever occurs. Emphasize to these people that, if a disaster occurs, it is important for your plant to be a top priority for assistance because the entire community relies on you for its drinking water.

You should know the names and phone numbers of your local and state civil preparedness coordinators.

If a chemical emergency occurs, such as a chemical spill, leak, fire, exposure, or accident, phone *CHEMTREC*, (800) 424-9300. CHEMTREC (Chemical Transportation Emergency Center) provides immediate advice for those at the scene of a chemical emergency, and then quickly and promptly alerts experts from the manufacturers whose products are involved for more detailed assistance and appropriate followup. Also, notify local and state civil preparedness coordinators.

Prepare a procedure for quick and efficient handling of all accidents or injuries occurring in your treatment facilities or involving your outside crews. All personnel must be familiar with these procedures and must be prepared to carry them out with a minimum amount of delay or confusion.

A copy of these procedures must be posted in all working areas accessible to a phone and in all vehicles containing work crews. Names, addresses, and phone numbers of operators in each working area should be listed in that area and those immediately available (day or night) by telephone.

Everyone must study these procedures carefully and be able to respond properly and quickly. Your health and life may depend on these procedures.

Figure 20.11 is an example of a typical emergency safety procedure and Figure 20.12 is a checklist of what must be done if someone is seriously injured.

QUESTIONS

Please write your answers to the following questions and compare them with those on page 490.

20.8A What types of emergencies should operators be prepared to handle?

20.8B Who should be contacted if a serious chemical emergency occurs, such as a chemical spill, leak, fire, exposure, or accident?

20.8C What is *CHEMTREC*?

20.9 ARITHMETIC ASSIGNMENT

Turn to the Arithmetic Appendix at the back of this manual. Read and work the problems in Section A.38, "Safety." Check the arithmetic in this section using an electronic calculator. You should be able to get the same answers.

20.10 ADDITIONAL READING

This chapter cannot provide you with all of the information you need to respond to the many safety problems that occur daily. You should have other sources of ready references. Because this chapter does not give all of the detail you may need, the following bibliography is given. You should have copies of these documents for your reference or have your employer obtain them.

1. *Chlorine Basics*, Seventh Edition. Available as a free download from the Chlorine Institute at www.chlorineinstitute.org. Pamphlet 1.
2. *Fisher Safety Catalog*. Obtain from Literature Fulfillment at the Fisher Scientific website, www.fishersci.com, or phone (800) 772-6733.
3. The American Red Cross is another source of up-to-date information on safety and first aid. Contact your local Area Chapter for a catalog of materials.
4. *Safety Management for Water Utilities* (M3). Obtain from American Water Works Association (AWWA) at www.awwa.org or phone (800) 926-7337. Order No. 30003-7E. ISBN 978-1-58321-999-7.
5. *Let's Talk Safety: 52 Talks on Common Utility Safety Practices for Water Professionals*. Obtain from American Water Works Association (AWWA) at www.awwa.org or phone (800) 926-7337. Order No. 10123.

END OF LESSON 3 OF 3 LESSONS

on

SAFETY

Please answer the discussion and review questions next.

11. Some of the information in this section was provided by Richard R. Metcalf, Training Officer, County of Onondaga, NY.

EMERGENCY SAFETY PROCEDURE

1. *DO NOT MOVE THE INJURED PERSON*
   except when conditions would cause additional injury, such as a gas leak or a fire.
2. *ADMINISTER ONLY SUCH AID AS NECESSARY TO PRESERVE LIFE—TREAT FOR SHOCK*
   a. clear throat and restore breathing
   b. stop bleeding
   c. closed heart massage
3. *DO NOT ATTEMPT MEDICAL TREATMENT* such as
   a. do not apply splints or attempt to set broken bones
   b. do not remove foreign objects from the body
   c. do not administer liquids or oxygen
4. *NOTIFY YOUR SUPERVISOR*
   IF AN AMBULANCE IS REQUIRED:
   a. CALL AMBULANCE—phone ____________
   b. Give this information carefully and accurately:
      (1) Location of the injured—be specific
         (A) Street location and number, town or city
            **Remember some streets have north or south or east or west designation—use the full street name. Also, many streets in different towns have the same name—specify the town.
         (B) Location within the plant area
      (2) Phone number from which you are calling
      (3) Number of persons injured and nature of the injury
      (4) Post an operator to direct the ambulance to the victim
   c. Upon arrival of the ambulance:
      (1) give name, address, and phone number of the injured person to the ambulance crew
      (2) notify relatives of injury and hospital to which person is being taken
         (Medical treatment cannot be given without the permission of the injured or a relative, if a minor)
   d. Call your supervisor

IF AN AMBULANCE IS NOT REQUIRED:
1. Take injured: (see map) (phone____________, ask for Emergency Room)
   Emergency Room
   St. Joseph's Hospital
   301 Prospect Avenue
2. If possible, call ahead. Give the names of injured and nature of injury
3. Notify relatives of injury and address of hospital
4. Call your supervisor

Fig. 20.11 Emergency safety procedure

INJURED PERSON CHECKLIST

1. *CALL AMBULANCE SERVICE*, phone ____________
   LOCATION OF INJURED____________
   STREET
   ____
   TOWN
   ____
   BUILDING LOCATION
   ____
   PHONE NUMBER
   NUMBER OF PERSONS INJURED ____________
   NATURE OF INJURY____________
2. *POST OPERATOR TO DIRECT THE AMBULANCE*
   NAME, ADDRESS, PHONE OF INJURED PERSON
   NAME____________
   ADDRESS ____________
   ____
   ____
   PHONE ____________
3. *GIVE ABOVE INFORMATION TO AMBULANCE CREW*
   NAME AND LOCATION OF HOSPITAL
   HOSPITAL NAME ____________
   ADDRESS ____________
   ____
   ____
   PHONE ____________
4. *NOTIFY RELATIVES*
5. *NOTIFY SUPERVISORS*
6. *MAKE OUT ACCIDENT REPORT AS SOON AS POSSIBLE*

Fig. 20.12 Injured person checklist

DISCUSSION AND REVIEW QUESTIONS

Chapter 20. SAFETY

(Lesson 3 of 3 Lessons)

Please write your answers to the following questions to determine how well you understand the material in the lesson. The question numbering continues from Lesson 2.

26. How can the hazards of fire be reduced in electric starters?
27. How should gas cylinders be stored in laboratories?
28. What safety precautions should be taken with regard to bacteria while working in the laboratory?
29. What is the most common reason safety equipment falls into disrepair?
30. How can an operator reduce the hazard of slipping when working around clarifiers or settling basins?

SUGGESTED ANSWERS

Chapter 20. SAFETY

ANSWERS TO QUESTIONS IN LESSON 1

Answers to questions on page 447.

20.0A Management's responsibilities for safety include:

1. Establishing a safety policy
2. Assigning responsibility for accident prevention
3. Appointing a safety officer or coordinator
4. Establishing realistic goals and periodically revising them
5. Evaluating the results of the safety program

20.0B Operators' responsibilities for safety include:

1. Performing their jobs in accordance with established safe procedures
2. Recognizing the responsibility for their own safety and that of fellow operators
3. Reporting all injuries
4. Reporting all observed hazards
5. Actively participating in the safety program

Answers to questions on page 448.

20.0C An accident is that occurrence in a sequence of events that usually produces unintended injury, death, or property damage.

20.0D The principal reasons for unsafe acts include ignorance, indifference, poor work habits, laziness, haste, poor physical condition, and temper.

20.0E Laziness affects the speed and quality of work. Laziness also affects safety because safety requires an effort. In most jobs, you cannot reduce your safety effort and still maintain the same level of safety, even if you slow down or lower your quality standards.

Answers to questions on page 452.

20.10A An operator needs to know how to handle the problems associated with the chemicals used in a water treatment plant. The operator needs to know how to store chemicals, understand the fire problem, be aware of the tendency of chemicals to arch in a storage bin, and know how to feed dry, how to feed liquid, and how to make up solutions. Overheating gas containers, dust problems with powdered carbon, burns caused by acid, and reactivity of each chemical under a variety of conditions that may cause fire and explosions are other safety hazards that an operator needs to know about and know how to control. Also, the operator needs to know the usable limits of each chemical because of toxicity, the protective equipment required for each chemical, each chemical's antidote, and how to control fires caused by each chemical.

20.11A To give first aid when acid vapors are inhaled, remove the victim to fresh air, restore breathing, or give oxygen, when necessary.

20.11B Acetic acid will react violently with ammonium nitrate, potassium hydroxide, and other alkaline materials.

20.11C Acetic acid can be handled safely if the operator uses adequate ventilation and prevents skin and eye contact.

20.11D When handling hydrofluosilicic acid, always use complete protective equipment including rubber gloves, goggles or face shield, rubber apron, rubber boots, and have lime slurry barrels, Epsom salt solution, and safety showers available. Always provide adequate ventilation.

20.11E Inhalation of hydrochloric acid (HCl) vapors or mists can cause damage to teeth and irritation to the nasal passages. Concentrations of 750 ppm or more will cause coughing, choking, and produce severe damage to the mucous membranes of the respiratory tract. In concentrations of 1,300 ppm, HCl is dangerous to life.

20.11F Like other acids, nitric acid should be stored in clean, cool, well-ventilated areas. The area should have an acid-resistant floor and adequate drainage. Keep away from oxidizing agents and alkaline materials. Protect containers from damage or breakage. Avoid contact with skin and provide emergency neutralization materials and safety equipment in use areas.

Answers to questions on pages 454 and 455.

20.12A Operators use two forms of ammonia: the gaseous form (anhydrous) and the liquid form (hydroxide).

20.12B Care must be used when storing or transporting ammonia containers. Always keep cylinder caps in place when cylinders are not in use. Store cylinders in a cool, dry location away from heat, and protect from direct sunlight. Do not store in the same room with chlorine.

20.12C The two forms of lime used in water treatment plants are hydrated lime (calcium hydroxide) quicklime (calcium oxide).

20.12D If someone swallowed sodium hydroxide, give large amounts of water or milk and immediately transport to a medical facility. Do not induce vomiting.

20.12E If sodium silicate comes in contact with your skin, wash thoroughly with water, followed by washing with a 10 percent solution of ammonium chloride or 10 percent acetic acid.

Answers to questions on page 459.

20.13A Chlorine leaks are most often found in the control valve.

20.13B If a chlorine cylinder becomes overheated, its fusible metal plugs act as a safety feature, melting at 158 to 168°F (70 to 76°C) and letting the gas escape so the cylinder does not burst.

20.13C Chlorine leaks can be detected by the odor, by the use of ammonia water on a small cloth or swab on a stick, or by the use of an aspirator containing ammonia water. (Remember not to spray ammonia into a room full of chlorine because a white cloud will form and you will not be able to see anything.) Also, a chlorine gas detector may be used.

20.13D Carbon dioxide is a safety hazard because it is odorless, colorless, and will accumulate at the lowest possible level. Carbon dioxide will displace oxygen so you must use a self-contained breathing apparatus, not a canister gas mask.

Answers to questions on page 461.

20.14A For handling most salts, ventilation, respiratory protection, and eye protection will prove adequate.

20.14B First aid when liquid or dry alum gets into the eyes consists of flushing them immediately for 15 minutes with large amounts of water. Alum should be washed off the skin with water because prolonged contact will cause skin irritation.

20.14C When exposed to moist air or light, ferric chloride decomposes and gives off hydrochloric acid.

Answers to questions on page 463.

20.15A Potassium permanganate spills can be swept up. Flushing with water is an effective way to eliminate any residue remaining on floors.

20.15B Powdered activated carbon is the most dangerous powder the water treatment plant operator will be exposed to.

20.15C Activated carbon should be stored in a clean, dry, fireproof location. Keep the area free of dust, protect it from flammable materials, and do not permit smoking in the area at any time when handling or unloading activated carbon.

20.15D The key to preventing activated carbon fires is keeping the storage area clean and free of dust.

20.15E Carbon fires should be controlled by carbon dioxide (CO_2) extinguishers or hoses equipped with fog nozzles. An activated carbon fire should not be doused with a stream of water. The water may cause burning carbon particles to fly, resulting in a greater fire problem.

Answers to questions on page 463.

20.17A Safety regulations prohibit the use of common drains and sumps from chemical storage areas to avoid the possibility of chemicals reacting and producing toxic gases, explosions, and fires.

20.17B If a polymer solution comes in contact with potassium permanganate, a fire could develop.

ANSWERS TO QUESTIONS IN LESSON 2

Answers to questions on page 467.

20.2A Class A fires involve ordinary combustible materials. These include wood, paper, cloth, rubber, many plastics, dried grass, hay, and stubble.

20.2B Foam extinguishers can control Class A and Class B fires. Class A fires are those that consume ordinary combustibles such as wood, paper, cloth, rubber, many plastics, dried grass, hay, and stubble. Class B fires are those that consume flammable and combustible liquids such as gasoline, oil, grease, tar, oil-based paint, lacquer, and solvents, and also flammable gases.

20.2C An electrical fire can be extinguished by the use of carbon dioxide (CO_2) extinguishers or with a dry chemical extinguisher.

Answers to questions on page 471.

20.3A When waxing floors, use compounds containing nonslip ingredients. Warn others about newly waxed floors. Try to do cleaning and waxing during weekends or at night when foot traffic is light.

20.3B Rags are always a problem if they contain oils, paint, or other cleaning compounds; there is always the possibility of fire. The rags should be placed into a closed metal container to reduce the fire hazard.

20.3C When operating an overhead crane, the following safety precautions must be exercised:

1. Allow only trained and authorized personnel to operate the overhead crane.
2. Inspect the circuit breaker, limit switches, the condition of the hook, the wire rope, and other safety devices.
3. Post load limits on the crane and never overload the crane.
4. Check each lift for proper balance.
5. Use a standard set of hand signals.
6. Be sure everyone in the vicinity wears a hard hat.
7. Allow only authorized personnel to make repairs.
8. Lock out the main power switch before repairs begin.
9. Try to avoid moving loads over populated areas.
10. Set up monthly safety inspection forms to be filled out and placed into the maintenance file.

20.3D Traffic can be warned that operators are working in a manhole by the use of barricades, signs, flags, lights, and other warning devices. Warning devices and procedures must conform to local and state regulations.

20.3E Operators should always use a mechanical lifting aid (rope and bucket) for raising or lowering tools and equipment into and out of a manhole. The use of a bucket or basket will keep your hands free when climbing down into or out of the manhole.

Answers to questions on page 473.

20.3F Operators should wear eye and ear protection when operating grinding, chipping, buffing, or pavement-breaking equipment. Sometimes, when using grinding or buffing tools, operators encounter toxic dusts or fumes and therefore need respiratory protection. At other times, there is a need for full-face protection because of flying particles.

20.3G Operators can be protected from high noise levels by wearing approved ear protective devices.

20.3H When operating welding equipment, the operator should wear protective clothing, gloves, helmet, and goggles.

Answers to questions on page 475.

20.4A Tire wear is caused by misalignment and low inflation.

20.4B Motor vehicles should have flashlights, flares, flags, fire extinguisher, and first-aid kit.

ANSWERS TO QUESTIONS IN LESSON 3

Answers to questions on page 480.

20.5A Each operator should become familiar with the type of current and voltage in the plant in order to avoid any mistake of becoming involved with unsafe electric currents or practices for which they are not trained. This will also enable operators to know when to ask for a qualified person to perform any necessary repairs.

20.5B Moving parts on electric motors that require safety guards include exposed couplings, pulleys, gears, and sprockets, as well as projections such as bolts, keys, or set screws.

20.5C Two hazards due to the lack of a good lockout procedure for control panels and switchboards are accidentally starting a piece of equipment, exposing a fellow operator to a hazard; and turning electrical power on when someone is still working on the equipment, exposing that person to danger.

20.5D Examples of stored energy that could be hazardous to operators include electrical power; air, gas, or steam pressure; water (hydraulic) pressure; spring tension; rotational forces in flywheels; and the energy in elevated machine parts if they suddenly swing free.

Answers to questions on page 481.

20.6A Washing glassware is always a potential hazard because the glassware can be broken while being washed, causing cuts, or cuts can be caused by chipped or cracked glassware. If your hands come in contact with strong acid cleaners, the acids may cause serious burns.

20.6B When handling liquid chemicals, such as acids and bases, always use safety glasses or face shields.

20.6C Never allow mercury, gasoline, oil, or organic compounds into the laboratory drains. Use only a toxic waste disposal drain system for these items. Flushing such compounds down sink drains can cause an explosion, allow toxic gases and vapors to enter the lab, or destroy the piping.

20.6D The plant operator may be exposed to radioactive compounds when calibrating sludge density meters or using research isotopes.

20.6E Never allow cold water into the hot boiler unit of a water still because it may cause the unit to break.

Answers to questions on page 483.

20.7A When entering an oxygen-deficient atmosphere, you should have an independent oxygen supply of the positive-pressure type to protect you if there are any leaks in your mask.

20.7B Respiratory apparatus must be stored outside of chlorination, sulfur dioxide, carbon dioxide, ozone, and ammonia rooms in an unlocked cabinet. The storage cabinets must have a controlled environment to prevent deterioration of the equipment.

20.7C Independent air supply equipment should be checked out on a weekly basis, and records kept of each inspection. The record should show conditions of the hoses, regulators, air filters, compressors, helmet, and any other apparatus furnished with the system.

20.7D To obtain compliance with an eye protection program, supervisors should undertake an intense program of education, persuasion, and appeal.

20.7E An operator needs eye protection when handling many of the liquid chemicals, acids, and caustics. Many of the tests performed in the laboratory also require eye protection.

Answers to questions on page 484.

20.7F Rubber boots are needed when handling acid or caustic or when the operator is working in wet conditions such as reservoirs, filters, or chemical tanks.

20.7G An operator should be very careful wearing gloves when working around machinery that is revolving.

20.7H Operators should never work in a filter when it is being backwashed because there is always the danger of falling into the wash-water gullet and being unable to get out before drowning.

20.7I Before taking a boat out on the water, it should be checked out for safety. The operator should check the bilge pump, ventilation in the compartments, the safety cushions, fire extinguishers, battery, and the engine. Also, check for safety equipment, oars, life jackets, lights, mooring lines, and fuel.

Answers to questions on page 485.

20.8A Operators should be prepared for catastrophic events such as floods, tornados, hurricanes, fires, and earthquakes. Serious injury to anyone on the plant grounds is an emergency.

20.8B If a serious chemical emergency occurs, such as a chemical spill, leak, fire, exposure, or accident, phone *CHEMTREC,* (800) 424-9300. Also, local and state civil preparedness coordinators should be notified.

20.8C *CHEMTREC* is an emergency response service that provides immediate advice (by telephone) about what to do in a chemical emergency such as a spill, leak, fire, exposure, or accident.

CHAPTER 21

ADVANCED LABORATORY PROCEDURES

by

Jim Sequeira

Additional test procedures not covered in this chapter, including alkalinity, chlorine residual and demand, coliform bacteria, hardness, jar test, pH, temperature, and turbidity, may be found in *Water Treatment Plant Operations, Volume I*, "Laboratory Procedures."

TABLE OF CONTENTS

Chapter 21. ADVANCED LABORATORY PROCEDURES

LEARNING OBJECTIVES

Chapter 21. ADVANCED LABORATORY PROCEDURES

Following completion of Chapter 21, you should be able to:

1. Explain how a spectrophotometer analyzes samples of water.
2. Perform the following field or laboratory tests: algae counts, calcium, chloride, color, dissolved oxygen, fluoride, iron, manganese, marble test, metals, nitrate, pH, specific conductance, sulfate, taste and odor, trihalomethanes, and total dissolved solids.

CHAPTER 21. ADVANCED LABORATORY PROCEDURES

(Lesson 1 of 2 Lessons)

21.0 USING A SPECTROPHOTOMETER

In the field of water analysis, measuring the intensity of color in a sample at a particular wavelength helps determine many components, including iron, manganese, and phosphorus. The color is formed in the sample by adding a specific developing reagent to it. The intensity of the color formed is directly related to the amount of a substance in the sample. For example, for the analysis of phosphorus present in a water sample, ammonium molybdate is added as the developing reagent. If phosphorus is present, a blue color develops. The more phosphorus there is, the deeper and darker the blue color.

The human eye can detect some differences in color intensity; however, for very precise measurements an instrument called a spectrophotometer (SPEK-tro-fo-TOM-uh-ter) is used. The *SPECTROPHOTOMETER* (Figure 21.1) is an instrument generally used to measure the color intensity of a chemical solution. In its simplest form, it consists of a light source that is focused on a prism or other suitable light dispersion device to separate the light into its separate bands of energy. Each different wavelength or color may be selectively focused through a narrow slit. This beam of light then passes through the sample to be measured. The sample is usually contained in a glass tube called a cuvette (kyoo-VET). Most cuvettes are standardized to have a 1.0 cm light path length, however, other sizes are available.

After the selected beam of light has passed through the sample, it emerges and strikes the photodetector. If the solution in the sample cell has absorbed any of the light, the total energy content will be reduced. If the solution in the sample cell does not absorb the light, then there will be no change in energy. When the transmitted light beam strikes the photodetector, it generates an electric current that is proportional to the intensity of light energy striking it. The intensity of the transmitted beam is measured by connecting the photodetector to a device for measuring electric current that produces a reading either digitally or using a scale. This is a general description of how the spectrophotometer works. Always follow the manufacturer instructions provided with the instrument.

UNITS OF SPECTROSCOPIC MEASUREMENT. The scale on a spectrophotometer is generally graduated in two ways:

1. In units of percent transmittance (%T), an arithmetic scale with units graded from 0 to 100%
2. In units of absorbance (A), a logarithmic scale of nonequal divisions graduated from 0.0 to 2.0

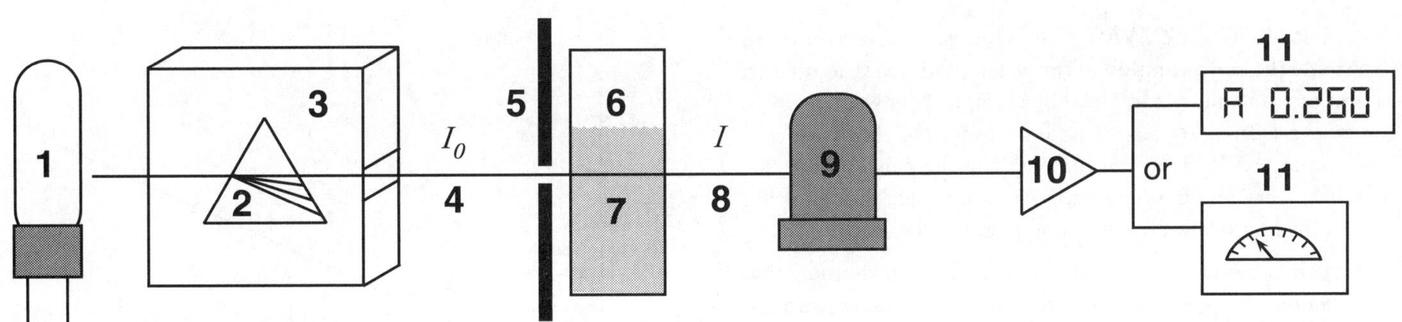

1. Light Source
2. Prism Inside
3. Monochromator
4. Monochromatic Incident Light (I_0)
5. Slit
6. Cuvette containing sample
7. Sample inside cuvette
8. Transmitted Light (I)
9. Photodetector
10. Amplifier
11. Digital Output or Meter

Fig. 21.1 Working parts of a spectrophotometer

(Using a Spectrophotometer)

Both the units, percent transmittance and absorbance, are associated with the term "color intensity," meaning that a sample with a low color intensity will have a high percent transmittance but a low absorbance.

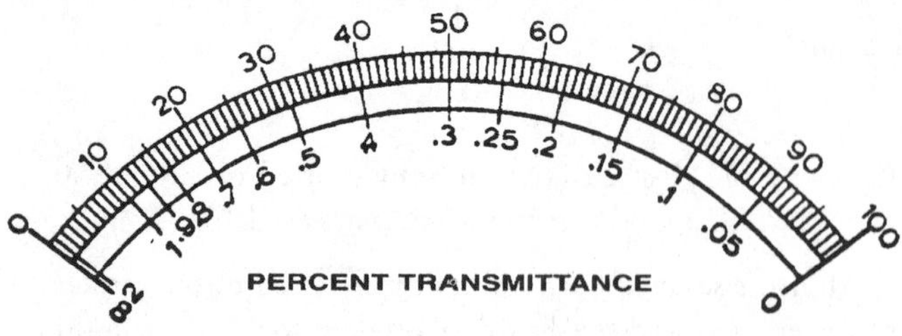

Absorbance Scale

The absorbance scale is ordinarily calibrated on the same scale as the percent transmittance on a spectrophotometer. The chief usefulness of measuring absorbance lies in the fact that it is a logarithmic function rather than linear (arithmetic). Beer's Law states that the concentration of a light-absorbing, colored solution is directly proportional to absorbance over a given range of concentrations. If one were to plot a graph showing %T versus concentration on straight graph or lined paper, and another showing absorbance versus concentration on the same piece of paper, the following curves (graphs) would result:

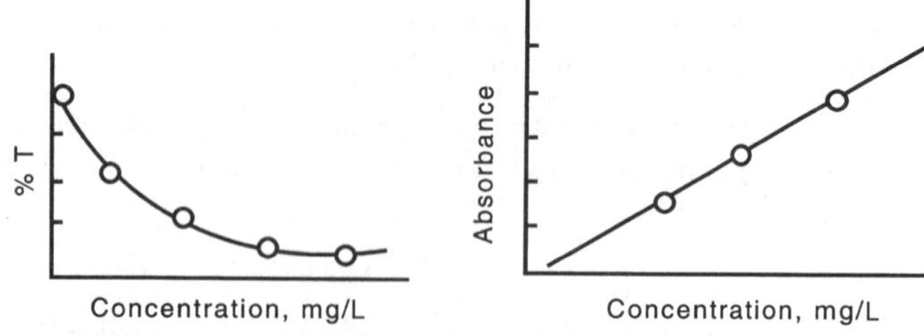

CALIBRATION CURVES. The calibration curve is used to determine the concentration of the water quality indicator (such as iron or manganese) contained in a sample. Many spectrophotometers are capable of constructing an internal electronic calibration curve. If your instrument can do this, follow the manufacturer's instructions. Otherwise, complete the following three steps to prepare a spectrophotometer calibration graph:

1. Prepare a series of standards. A standard is a solution that contains a known amount of the same chemical constituent that is being determined in the sample.
2. Treat the standard solutions and a sample containing none of the constituent being tested for with the developing reagent, in the same manner the sample would be treated. The sample in this step usually consists of distilled water and is generally referred to as a blank.
3. Use a spectrophotometer to determine the absorbance or transmittance of the standards and blank at the specified wavelength. From the values obtained, a calibration curve of absorbance versus concentration can be plotted. After these points are plotted, you can then extend the plotted points by connecting the known points with a straight line. For example, with the data given below you could graph the following calibration curve.

Absorbance	Concentration, mg/L
0.0	0.0
0.30	0.25
0.55	0.50
0.80	0.75

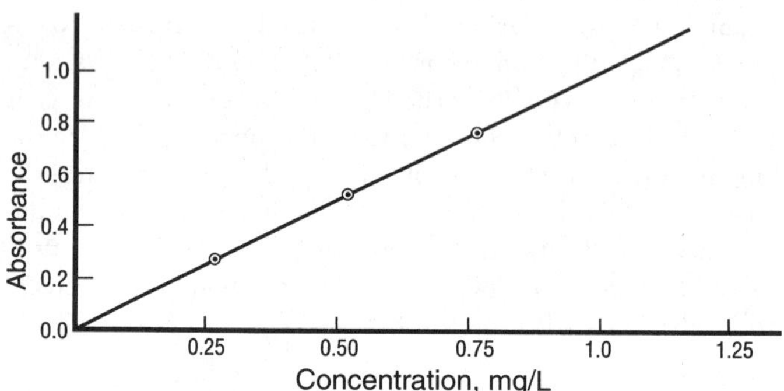

Once you have established a calibration curve for the water quality indicator in question, you can easily determine the amount of that substance contained in a solution of unknown concentration. Do this by taking an absorbance reading on the color developed by the unknown and then locating it on the vertical axis. Next, draw a straight line to the right on the graph until it intersects with the experimental standard curve. Finally, draw a line to the horizontal axis to identify the value of the concentration of the unknown water quality indicator. In the following graph, an absorbance reading of 0.32 was read on the unknown solution or sample, which indicates a concentration of about 0.37 mg/L.

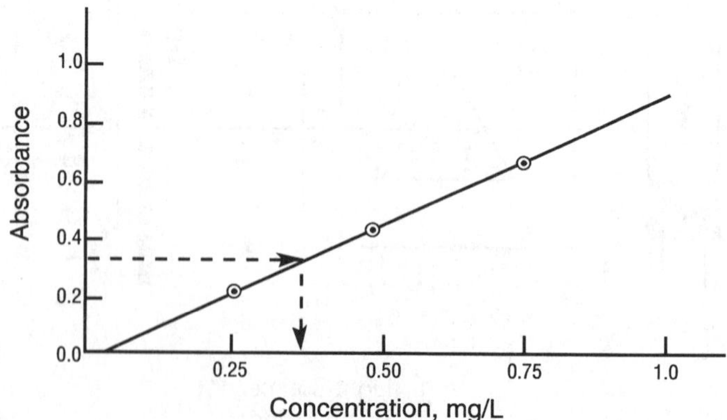

QUESTIONS

Please write your answers to the following questions and compare them with those on page 532.

21.0A When measuring phosphorus concentrations, the intensity of what color is measured?

21.0B What are the units of measurements for a spectrophotometer?

21.0C Using the previous absorbance versus concentration calibration graph, if an absorbance of 0.60 was read on an unknown solution or sample, what would be the concentration of the unknown?

21.1 TEST PROCEDURES

1. Algae Counts

Discussion

The quality of water in any lake, reservoir, or stream has a very direct effect on the abundance and types of aquatic organisms found. By knowing the nature and numbers of these aquatic organisms, one can obtain a good idea of the water quality. A biological method used for measuring water quality is the collection, counting, and identification of algae contained in a particular body of water. Information from algae counts can serve one or more of the following purposes:

1. Help explain the cause of color and turbidity and the presence of flavors and odors in the water
2. Help explain the clogging of screens or filters
3. Help document variability in the water quality

Algae counting and identification may be done very simply or it may be developed into a highly technical operation. The beginner should use great caution applying the results of algae identifications until considerable experience has been gained.

Some operators perform algae counts on both the raw water and treated water. Taking algae counts on treated water is a means of studying the effectiveness of coagulation and the performance of filters. If the filters are performing properly, there should not be any countable algae in the treated water.

Materials and Procedures

See *Standard Methods*, 22nd Edition, Method 10200 (B), page 10-3.[1]

2. Calcium

Discussion

In most natural waters, calcium is the principal cation. The element is widely distributed in the common minerals of rocks and of soil. Calcium in the form of lime or calcium hydroxide may be used to soften water or to control corrosion through pH adjustment.

What Is Tested?

Sample	Common Range, mg/L
Raw and Treated Surface Water	5 to 50
Well Water	10 to 100

Apparatus Required

Buret, 25 mL
Buret support
Graduated cylinder, 100 mL
Beaker, 100 mL
Magnetic stirrer
Magnetic stir-bar

Reagents

Standardized solutions are commercially available for most reagents, including the following:

1. Sodium hydroxide (NaOH), 1 *N*.
2. Eriochrome Blue Black R indicator.
3. Standard EDTA titrant, 0.01 *M*. Standardize against Standard Calcium Solution and store in plastic polyethylene bottle.
4. Standard Calcium Solution. Store in polyethylene plastic bottle. 1 mL of this solution = 1 mg calcium hardness as $CaCO_3$ or 400.8 micrograms (μg) Ca.
5. 1 + 1 HCl: Carefully add 50 mL concentrated HCl to 50 mL distilled water.
6. Ammonium hydroxide, 3 *N*.
7. Methyl orange indicator solution.

Refer to *Standard Methods* if you wish to prepare your own reagents.

Procedure

1. Take a clean beaker and add 50 mL of sample.
2. Add 2.0 mL NaOH solution.
3. Add 0.1 to 0.2 g indicator mixture.
4. Titrate immediately with EDTA titrant until last reddish-purple tinge disappears. Mix with magnetic stirrer during titration.
5. Calculate calcium concentration.

Example

Results from calcium testing of a treated water sample were as follows:

Sample Size = 50 mL
mL EDTA Titrant Used, A = 7.3 mL

Calculation

$$\text{mg Ca/L} = \frac{A \times 400.8^*}{\text{mL of Sample}}$$

$$= \frac{7.3\text{ mL} \times 400.8}{50\text{ mL}}$$

$$= 59\text{ mg/L}$$

*400.8 is a constant for this calculation. It contains the conversion units to produce mg/L.

1. *Standard Methods for the Examination of Water and Wastewater*, 22nd Edition. Obtain from American Water Works Association (AWWA) at www.awwa.org or phone (800) 926-7337. ISBN 978-0-87553-013-0.

(Calcium—Chloride)

OUTLINE OF PROCEDURE FOR CALCIUM

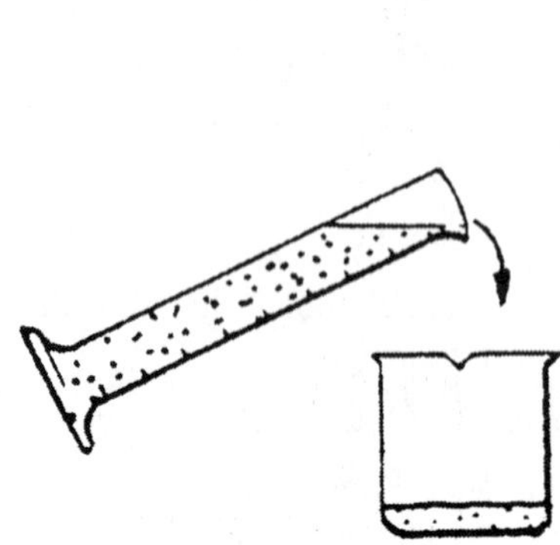

1. Add 50 mL to a clean beaker.

2. Add 2 mL NaOH and 0.2 g indicator mixture.

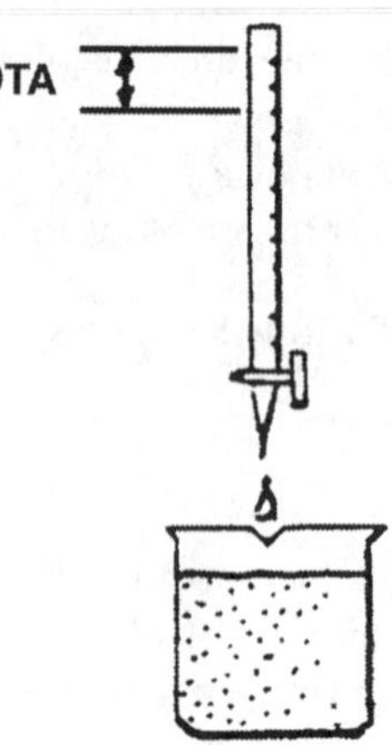

3. Titrate with EDTA. Mix with magnetic stirrer.

Precautions

1. Titrate immediately after adding NaOH solution.
2. Use 50 mL or a smaller portion of sample diluted to 50 mL with distilled water so that the calcium content is about 5 to 10 mg.
3. For hard waters with alkalinity greater than 300 mg $CaCO_3$/L, use a smaller portion or neutralize alkalinity with acid, boiling for one minute, and cooling before beginning the titration.

Reference

See *Standard Methods*, 22nd Edition, Method 3500-Ca (B), page 3-67.

3. Chloride

Discussion

Chloride occurs in all natural waters, usually as a metallic salt. In most cases, the chloride content increases as mineral content increases. Mountain water supplies usually are quite low in chloride while groundwaters and valley rivers often contain a considerable amount. The maximum allowable chloride concentration of 250 mg/L in drinking water has been established for reasons of flavor rather than as a safeguard against a physical or a health hazard. At concentrations above 250 mg/L, chloride may give a salty taste to the water that is objectionable to many people.

What Is Tested?

Sample	Common Range, mg/L
Surface or Groundwater	2 to 100

Apparatus Required

Graduated cylinder, 100 mL
Buret, 50 mL
Erlenmeyer flask, 250 mL
Pipet, 10 mL
Magnetic stirring apparatus

Reagents

Standardized solutions such as the following are commercially available for most reagents:

1. Chloride-free water (distilled or deionized water).
2. Potassium chromate (K_2CrO_4) indicator solution.
3. Standard Silver Nitrate ($AgNO_3$) Titrant, 0.0141 *N*.
4. Standard Sodium Chloride, 0.0141 *N*.

Procedure

1. Place 100 mL or a suitable portion of sample diluted to 100 mL in a 250 mL Erlenmeyer flask.
2. Add 1.0 mL K_2CrO_4 indicator solution.
3. Titrate with standard silver nitrate to a pinkish yellow end point. Be consistent in end point recognition. Compare with known standards of various chloride concentrations.

Calculation

$$\text{Chloride (as Cl), mg/L} = \frac{(A - B) \times N \times 35{,}450^*}{\text{mL of Sample}}$$

*35,450 is a constant for this calculation. It contains the conversion units to produce mg/L.

A = mL $AgNO_3$ used for titration of sample

B = mL $AgNO_3$ used for blank

N = normality of $AgNO_3$

(Chloride)

OUTLINE OF PROCEDURE FOR CHLORIDE

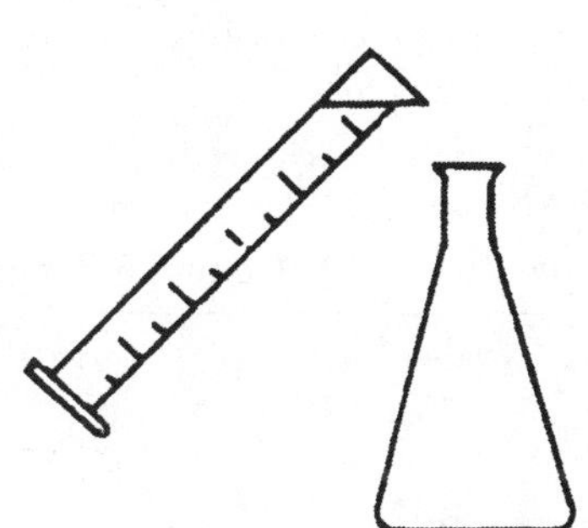

1. Place 100 mL or other measured sample in flask.

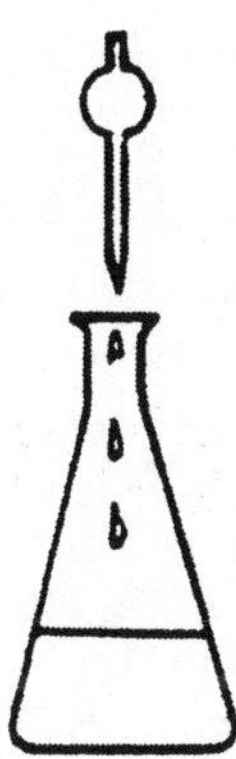

2. Add 1 mL chromate indicator.

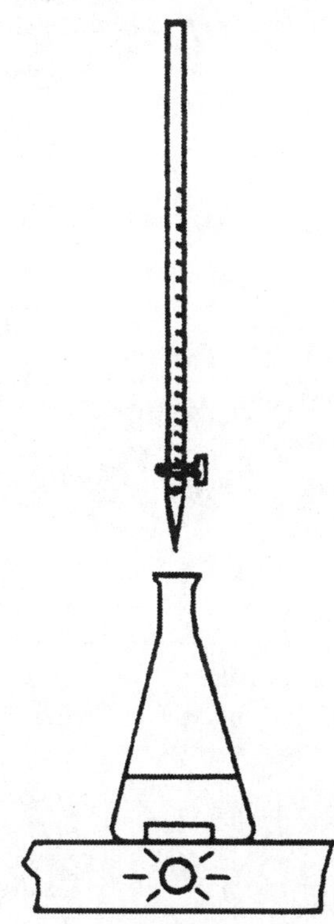

3. Place flask on magnetic stirrer and titrate with standard silver nitrate.

Example

Sample Size = 100 mL

A = mL $AgNO_3$ used for titration of sample = 10.0 mL

B = mL $AgNO_3$ used for blank = 0.4 mL

N = normality of $AgNO_3$ = 0.0141 N

$$\text{Chloride, mg/L} = \frac{(10.0\text{ mL} - 0.4\text{ mL}) \times (0.0141\ N) \times 35{,}450}{100\text{ mL}}$$

$$= 48\text{ mg/L}$$

Special Notes

1. Sulfide, thiosulfate, and sulfite ions interfere, but can be removed by treatment with 1 mL of 30 percent hydrogen peroxide (H_2O_2).
2. Highly colored samples must be treated with an aluminum hydroxide suspension and then filtered.
3. Orthophosphate in excess of 25 mg/L and iron in excess of 10 mg/L also interfere.
4. If the pH of the sample is not between 7 and 10, adjust with 1 N sulfuric acid or 1 N sodium hydroxide.

(Chloride—Color)

5. Procedure for standardization of $AgNO_3$
   a. Add 10 mL (1 mg Cl) standard sodium chloride solution to a clean 250 mL Erlenmeyer flask.
   b. Add 90 mL distilled water.
   c. Titrate using procedure described above.
   d. Calculation

$$\text{Normality, } N, AgNO_3 = \frac{\text{mL CaCl Standard} \times 0.0141}{\text{mL } AgNO_3 \text{ Used in Titration}}$$

   e. Example

   10.0 mL NaCl standard used

   10.0 mL $AgNO_3$ used in titration

   0.0141 N = normality of NaCl standard

$$\text{Normality, } N, AgNO_3 = \frac{10.0 \text{ mL} \times 0.0141}{10 \text{ mL}} = 0.0141$$

Reference

See *Standard Methods*, 22nd Edition, Method 4500-Cl$^-$ (B), page 4-72.

QUESTIONS

Please write your answers to the following questions and compare them with those on page 532.

21.1A Does the quality of water in any lake, reservoir, or stream affect the abundance and types of aquatic organisms found in the water? Yes or No?

21.1B How are calcium compounds used to treat water?

21.1C How soon should a sample be titrated for calcium after the sodium hydroxide (NaOH) solution has been added?

21.1D Why are concentrations of chloride above 250 mg/L objectionable to many people?

4. Color

Discussion

Color in water supplies may result from the presence of metallic ions (iron, manganese, and copper), organic matter of vegetable or soil origin, and industrial wastes. The most common colors that occur in raw water are yellow and brown. There are two general types of color found in water. True color results from the presence of dissolved organic substances or from certain minerals such as copper sulfate dissolved in the water. Suspended materials (including colloidal substances) can add what is called apparent color. True color is normally removed or at least reduced by coagulation and chlorination or ozonation. The method given below is suitable only for the measurement of color in clear treated water supplies having less than one unit of turbidity. When greater amounts of turbidity are present in the sample, some form of pretreatment for turbidity removal must be used before measuring the color.

What Is Tested?

Sample	Common Range, units
Treated Surface Water	1 to 10
Groundwater	0 to 5

Apparatus Required

Nessler tubes, matched, 50 mL tall form
Pipet, 1.0 mL

Reagents

1. Color Standard. Use a stock standard with a color of 500 units.
2. Prepare color standards by adding the following increments of stock color standard to a Nessler tube and diluting to 50 mL.

Color Unit Standard	mL of Stock Color Standard
1	0.1
2	0.2
3	0.3
4	0.4
5	0.5

Protect the standards against evaporation and contamination when not in use.

Procedure

1. Fill a clean matched Nessler tube to the 50 mL mark with the sample.
2. Compare the sample with the various color standards by looking downward vertically through the tubes toward a white surface.
3. Match the sample color as closely as possible with a color standard.

Other Procedures

Color may also be measured by the use of:

1. Color comparator kits
2. Spectrophotometers

(Color)

OUTLINE OF PROCEDURE FOR COLOR

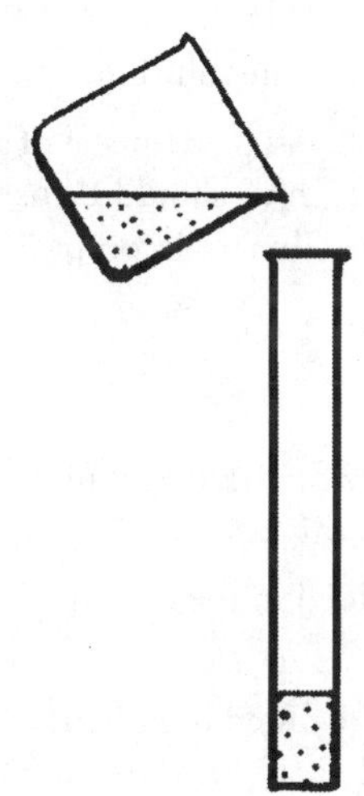

1. Fill Nessler tube to 50 mL.

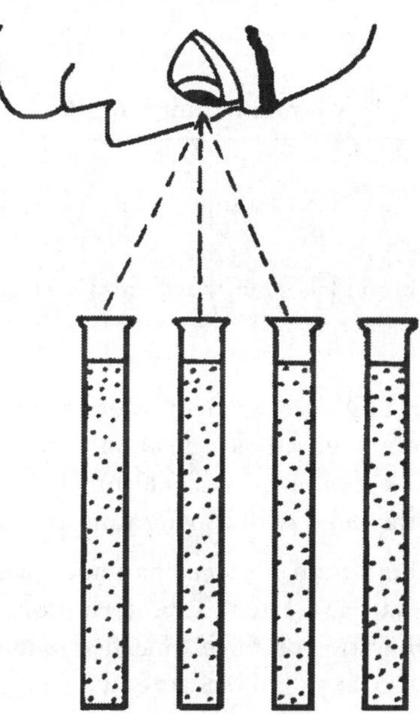

2. Compare the sample to color standards.

Notes

1. If the color exceeds 70 units, dilute sample with distilled water in known proportions until the color is within range of the standards. Calculate color units by the following equation:

$$\text{Color Units} = \frac{A \times 50}{B}$$

where:

A = estimated color of diluted sample

B = mL of sample taken from dilution

2. If turbidity is greater than one unit, consult *Standard Methods* for pretreatment for turbidity removal.

Reference

See *Standard Methods*, 22nd Edition, Method 2120 (B), page 2-6.

QUESTIONS

Please write your answers to the following questions and compare them with those on page 532.

21.1E What are the most common colors that occur in raw water?

21.1F How can true color be removed from water?

21.1G When not in use, stock color standards should be protected against what?

(Dissolved Oxygen)

5. Dissolved Oxygen

Discussion

Dissolved oxygen (DO) is important to the water treatment plant operator for a number of reasons. In surface waters, DO must be present in order for fish and smaller aquatic organisms to survive. The flavor of water is improved by DO. However, the presence of DO in water can contribute to corrosion of piping systems. Low or zero DO levels at the bottom of lakes or reservoirs often cause flavor and odor problems in drinking water.

What Is Tested?

Sample	Common Range, mg/L
Surface Water	5 to 11*
Groundwaters	0 to 2

*Some reservoirs and lakes may have zero DO near the bottom.

Precautions

1. Samples for DO measurements should be collected very carefully. Do not let sample remain in contact with air or be agitated. Collect samples in a 300 mL BOD bottle. Avoid entraining (trapping) or dissolving atmospheric oxygen.
2. When sampling from a water line under pressure, attach a tube to the tap and extend tube to bottom of bottle. Let bottle overflow two or three times its volume and replace glass stopper so no air bubbles are entrapped.
3. Use suitable sampler for streams, reservoirs, or tanks of moderate depth such as that shown in Figure 21.2. Use a Kemmerer-type sampler for samples collected from depths greater than 6½ feet (2 m).
4. Always record temperature of water at time of sampling.
5. When working with a lake or reservoir, examine the temperature and DO profile (measure temperature and DO at surface and at various depths all the way down to the bottom).
6. Measure the DO in the sample as soon as possible.

Sodium Azide Modification of the Winkler Method

Apparatus Required

Buret, graduated to 0.1 mL
Buret support
BOD bottle, 300 mL
Magnetic stirrer
Magnetic stir-bar
Pipets, 10 mL

Reagents

Standardized solutions are commercially available for most reagents, including the following:

1. Manganous Sulfate solution.
2. Alkaline Iodide-Sodium Azide solution.
3. Sulfuric Acid: Use concentrated reagent-grade acid (H_2SO_4). Handle carefully because this material will burn hands and clothes. Rinse affected parts with tap water to prevent injury.

CAUTION

When working with alkaline azide and sulfuric acid, keep a nearby water faucet running for frequent hand rinsing.

4. 0.025 *N* Phenylarsine Oxide (PAO) solution.
5. 0.025 *N* Sodium Thiosulfate solution.

   For preservation, add 0.4 g or 1 pellet of sodium hydroxide (NaOH). Solutions of "thio" should be used within 2 weeks to avoid loss of accuracy due to decomposition of solution.
6. Starch solution.

Procedure

NOTE: The sodium azide destroys nitrate, which would otherwise interfere with this test.

The reagents are to be added in the quantities, order, and methods as follows:

1. Collect a sample to be tested in 300 mL BOD bottle taking special care to avoid aeration of the liquid being collected. Fill bottle completely and stopper.
2. Remove stopper and add 1 mL of manganous sulfate solution below surface of the liquid.
3. Add 1 mL of alkaline iodide-sodium azide solution below the surface of the liquid.
4. Replace the stopper, avoid trapping air bubbles, and mix well by inverting the bottle several times. Repeat this mixing after the floc has settled halfway. Allow the floc to settle halfway a second time.
5. Acidify with 1 mL of concentrated sulfuric acid by allowing the acid to run down the neck of the bottle above the surface of the liquid.
6. Restopper and mix well until the precipitate has dissolved. The solution will then be ready to titrate. Handle the bottle carefully to avoid acid burns.
7. Pour 201 mL from bottle into an Erlenmeyer flask.
8. If the solution is brown in color, titrate with 0.025 *N* PAO until the solution is a pale yellow color. Add a small quantity of starch indicator and proceed with Step 10. (Note: Either PAO or 0.025 *N* sodium thiosulfate can be used.)
9. If the solution has no brown color, or is only slightly colored, add a small quantity of starch indicator. If no blue color develops, there is zero DO. If a blue color does develop, proceed to Step 10.

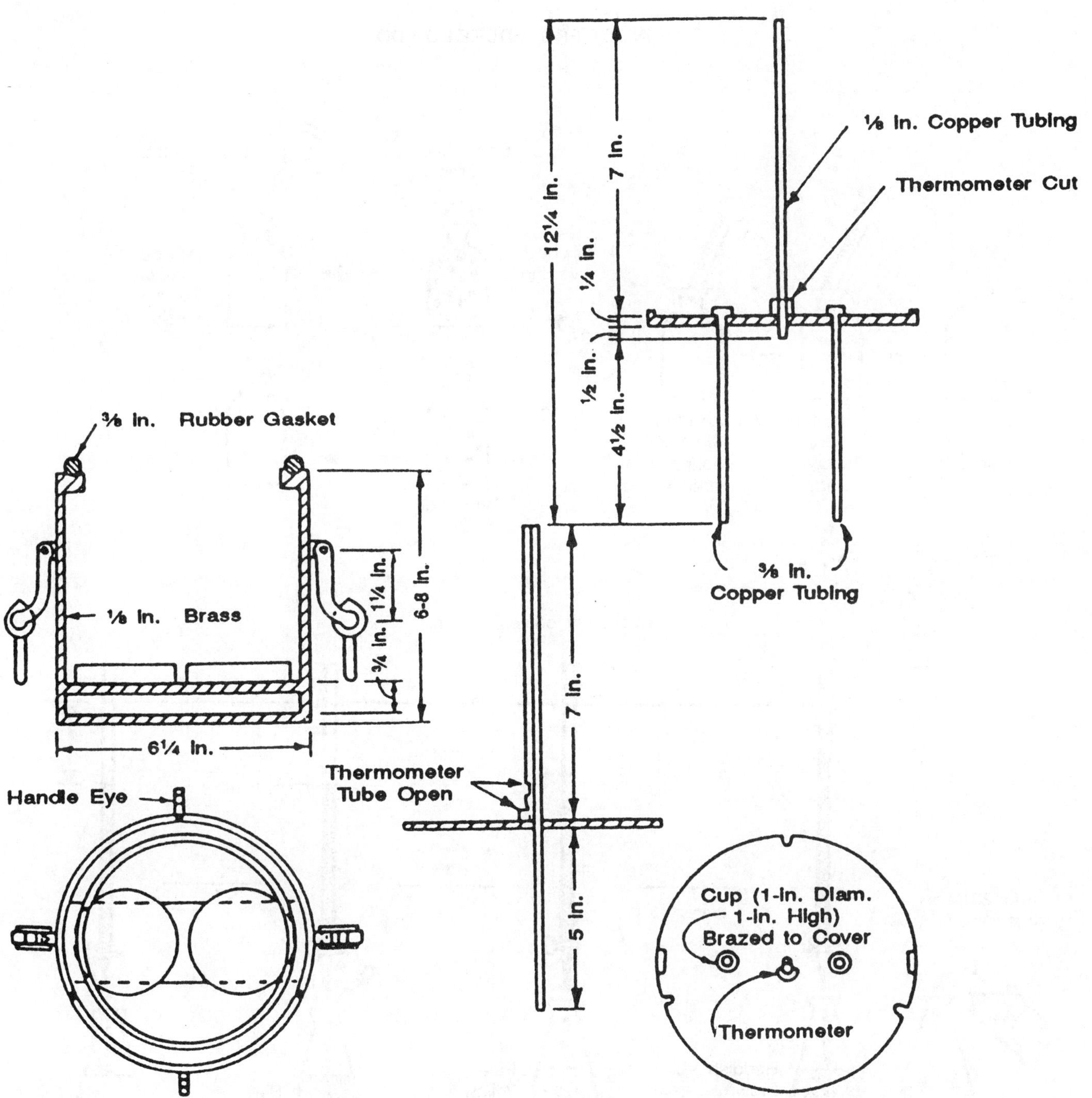

Fig. 21.2 DO Sampler

(Reprinted from *Standard Methods*, 15th Edition with permission from the American Public Health Association)

(Dissolved Oxygen)

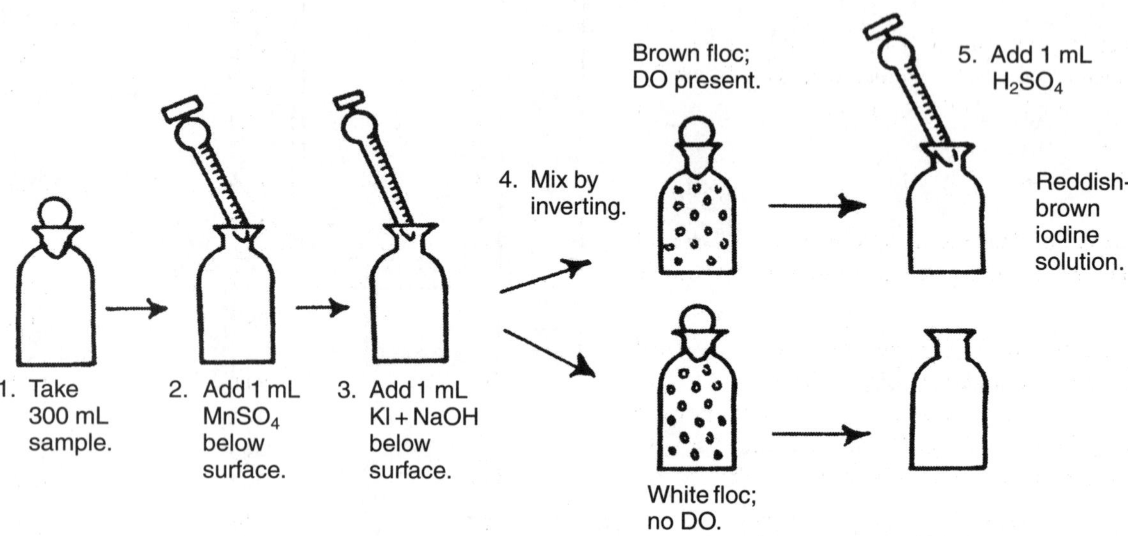

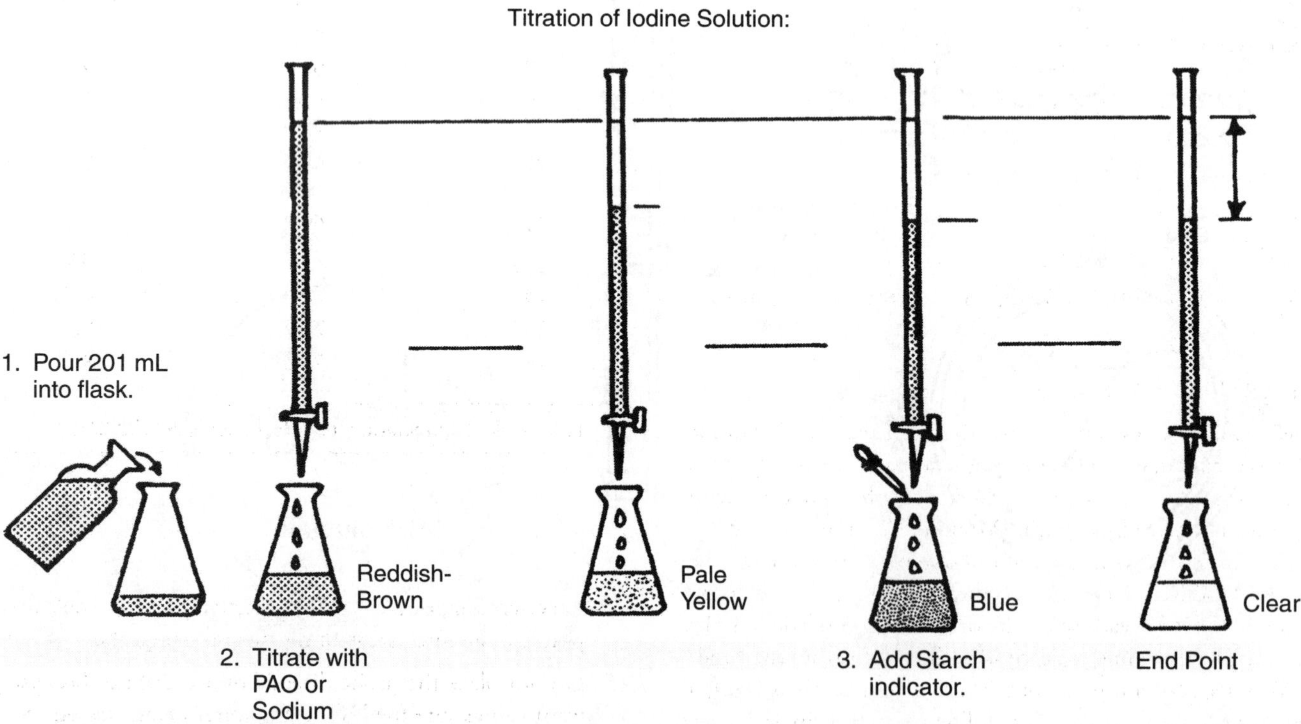

10. Titrate to the first disappearance of the blue color. Record the number of mL of PAO used.
11. The amount of oxygen dissolved in the original solution will be equal to the number of mL of PAO used in the titration provided significant interfering substances are not present.

$$\text{mg DO/L} = \text{mL PAO}$$

Example

A sample is collected from just upstream of a river intake to a water treatment facility. The water temperature is 64°F (18°C). The sample is tested for DO and the operator uses 9.1 mL of 0.025 *N* PAO titrant.

Calculation

The DO titration of the 201 mL sample required 9.1 mL of 0.025 *N* PAO. Therefore, the DO concentration in the sample is 9.1 mg/L.

The percent saturation of DO in the river can be calculated using the DO saturation values given in Table 21.1. Note that as the temperature of water increases, the DO saturation value (saturation column) decreases. Table 21.1 gives 100 percent DO saturation values for temperatures in °C and °F.

$$\text{DO Saturation, \%} = \frac{\text{DO of Sample, mg/L} \times 100\%}{\text{DO at 100\% Saturation, mg/L}}$$

For example, given

$$\text{DO Saturation, \%} = \frac{9.1\text{ mg/L}}{9.5\text{ mg/L}} \times 100\%$$

$$= 0.96 \times 100\%$$

$$= 96\%$$

where

9.1 mg/L = DO of Sample

9.5 mg/L = DO at 100% Saturation at 18°C (river temperature)

TABLE 21.1
EFFECT OF TEMPERATURE ON DISSOLVED OXYGEN SATURATION FOR A CHLORIDE CONCENTRATION OF ZERO mg/L

°C	°F	DO at 100% Saturation, mg/L
0	32.0	14.6
1	33.8	14.2
2	35.6	13.8
3	37.4	13.5
4	39.2	13.1
5	41.0	12.8
6	42.8	12.5
7	44.6	12.2
8	46.4	11.9
9	48.2	11.6
10	50.0	11.3
11	51.8	11.1
12	53.6	10.8
13	55.4	10.6
14	57.2	10.4
15	60.0	10.2
16	61.8	10.0
17	63.6	9.7
18	65.4	9.5
19	67.2	9.4
20[a]	68.0	**9.2**[a]
21	69.8	9.0
22	71.6	8.8
23	73.4	8.7
24	75.2	8.5
25	77.0	8.4

a The standard temperature for a BOD test is 20°C (68°F).

Membrane Electrode Method

Follow manufacturer's instructions. To ensure that the DO probe reading is accurate, the probe must be calibrated frequently. Take a sample that does not contain substances that interfere with either the probe reading or the Modified Winkler procedure. Split the sample. Measure the DO in one portion of the sample using the Modified Winkler procedure and compare this result with the DO probe reading on the other portion of the sample. Adjust the probe reading to agree with the results from the Modified Winkler procedure. To obtain good results when using a probe, you should be aware of the following precautions:

1. Periodically check the calibration of the probe.
2. Keep the membrane in the tip of the probe from drying out.
3. Dissolved inorganic salts, such as found in seawater, can influence the readings from a probe.
4. Reactive compounds, such as reactive gases and sulfur compounds, can interfere with the output of a probe.
5. Do not place the probe directly over a diffuser because you want to measure the DO in the water being treated, not the oxygen in the air supply to the aerator.

Reference

See *Standard Methods*, 22nd Edition, Method 4500-O (C), page 4-139; and Method 4500-O (G), page 4-143.

(Fluoride)

6. Fluoride

Discussion

Fluoride may occur naturally or it may be added in controlled amounts. The concentration of fluoride in most natural waters is less than 1.0 mg/L. There are, however, several areas in the United States that have natural fluoride concentrations as high as 30 mg/L. The importance of fluoride in forming human teeth and the role of fluoride intake from drinking water in controlling the characteristics of tooth structure have been accurately documented. Studies have shown that a fluoride concentration of approximately 1.0 mg/L reduces dental caries of young people without harmful effects on health.

What Is Tested?

Sample	Common Range, mg/L
Fluoridated Water	0.8 to 1.2

Apparatus Required

Spectrophotometer for use at 570 nanometers (nm) wavelength
Pipets, 5 mL
Erlenmeyer flasks, 125 mL

Reagents

1. Stock fluoride solution. 1.0 mL = 0.100 mg F.
2. Standard fluoride solution: Dilute 100 mL stock fluoride solution to 1,000 mL with distilled water; 1.0 mL = 0.010 mg F.
3. SPADNS solution. This solution is stable indefinitely if protected from direct sunlight.
4. Zirconyl-acid reagent.
5. Acid zirconyl-SPADNS reagent: Mix equal volumes of SPADNS solution and zirconyl-acid reagent. The combined reagent is stable for at least 2 years.
6. Reference solution: Add 10 mL SPADNS solution to 100 mL distilled water. Dilute 7 mL concentrated HCl to 10 mL and add to the diluted SPADNS solution. The resulting solution, used for setting the instrument reference point at 0, is stable and may be reused indefinitely. Alternatively, use a prepared standard as a reference.
7. Sodium arsenite solution. (*CAUTION: Toxic—avoid ingestion.*)

Procedure

1. Measure 50 mL of sample and add to a clean 125 mL Erlenmeyer flask. (If sample contains residual chlorine, add one drop $NaAsO_2$ solution per 0.1 mg chlorine residual and mix.)
2. Add 5.0 mL each of SPADNS solution and zirconyl-acid reagent, or 10.0 mL acid zirconyl-SPADNS reagent. Mix.
3. Set spectrophotometer to 0.730 absorbance with reference solution containing 0 mg/L of fluoride (see G. Example).
4. Read absorbance at 570 nm with spectrophotometer and determine the amount of fluoride from standard curve.

NOTE: A colorimeter may also be used to measure fluoride.

OUTLINE OF PROCEDURE FOR FLUORIDE

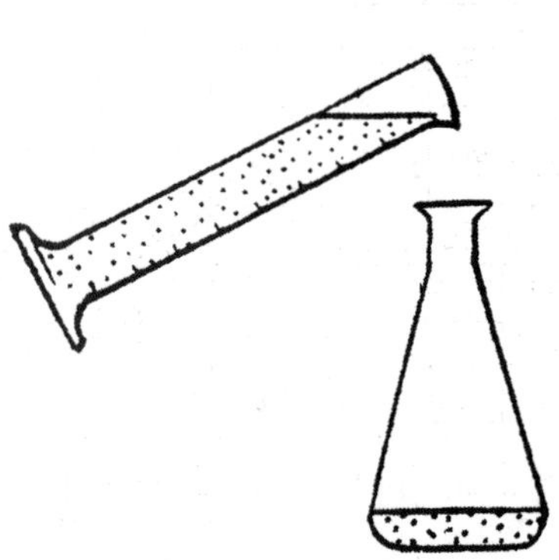

1. Measure 50 mL of sample into flask. Dechlorinate, if necessary.

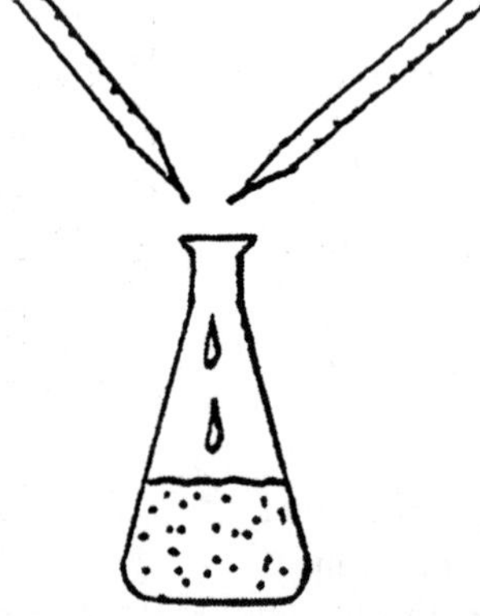

2. Add 5 mL each of SPADNS solution and zirconyl-acid reagent.

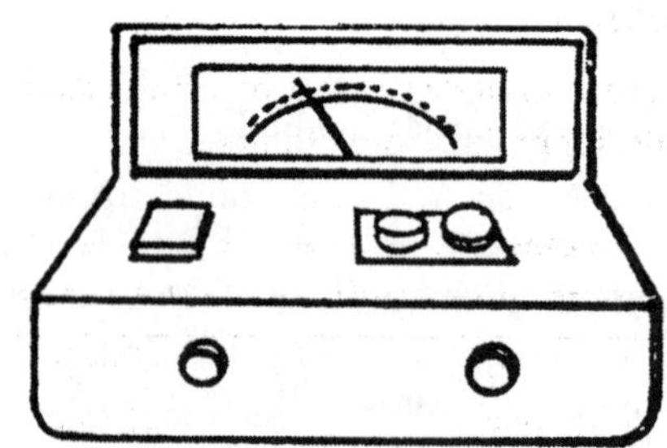

3. Measure absorbance at 570 nm with spectrophotometer.

(Fluoride)

Construction of Standard Calibration Curve

If your instrument can construct an internal electronic calibration curve, follow the manufacturer's instructions. If not, complete the following steps to construct a calibration curve.

1. Using the standard fluoride solution, prepare the following standards in 100 mL volumetric flasks.

mL of Standard Fluoride Solution Placed in 100 mL Volumetric Flask	Fluoride Concentration, mg/L
5.0	0.50
7.5	0.75
10.0	1.00
12.5	1.25

2. Dilute flasks to 100 mL.
3. Transfer 50 mL to 125 mL Erlenmeyer flask.
4. Determine amount of fluoride as outlined previously.
5. Prepare a standard curve by plotting the absorbance values of standards versus the corresponding fluoride concentrations.

Example

Results from a series of tests for fluoride were as follows:

Flask No.	Sample	Volume, mg/L	Absorbance
1	Distilled Water	50	0.730
2	C Street Well	50	0.470
3	Plant Effluent	50	0.520
4	0.5 mg/L Fluoride Standard	50	0.625
5	0.75 mg/L Fluoride Standard	50	0.560
6	1.0 mg/L Fluoride Standard	50	0.500
7	1.25 mg/L Fluoride Standard	50	0.444

Calculation

1. Prepare a standard curve by using data from prepared standards. From the above example:

Fluoride Concentration, mg/L	Absorbance
0.0	0.730
0.5	0.625
0.75	0.560
1.0	0.500
1.25	0.444

The graph below is a result of plotting concentration of fluoride standards versus their corresponding absorbance.

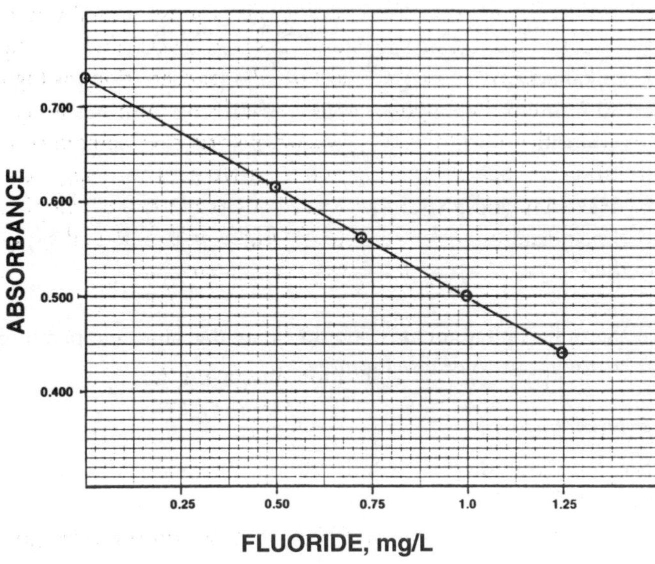

2. Obtain concentration of unknown samples from curve.

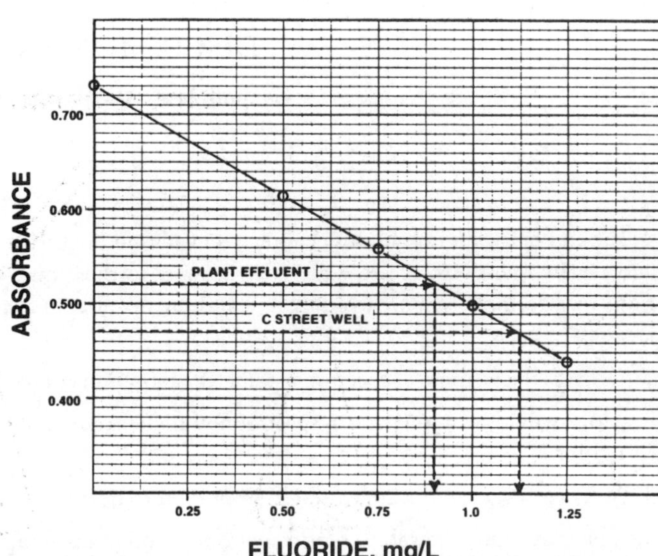

(Fluoride)

Precautions

1. Whenever any of the following substances are present in the listed quantities, the sample must be distilled before analysis.

Substance	Concentration, mg/L
Alkalinity	5,000
Aluminum	0.1
Chloride	7,000
Iron	10
Hexametaphosphate	1.0
Phosphate	16
Sulfate	200

2. Samples and standards should be at the same temperature throughout color development.

Reference

See *Standard Methods*, 22nd Edition, Method 4500-F^- (D), page 4-87.

An alternative to the wet chemistry method described in this section on fluoride is the ion-selective electrode method, which requires a specialized instrument. See *Standard Methods*, 22nd Edition, Method 4500-F^- (C), page 4-85.

QUESTIONS

Please write your answers to the following questions and compare them with those on page 532.

21.1H Why is the presence of dissolved oxygen (DO) in water in piping systems of concern to operators?

21.1I What is the common range of fluoride in fluoridated drinking water?

END OF LESSON 1 OF 2 LESSONS

on

ADVANCED LABORATORY PROCEDURES

Please answer the discussion and review questions next.

DISCUSSION AND REVIEW QUESTIONS

Chapter 21. ADVANCED LABORATORY PROCEDURES

(Lesson 1 of 2 Lessons)

At the end of each lesson in this chapter, you will find discussion and review questions. Please write your answers to these questions to determine how well you understand the material in the lesson.

1. What is the purpose of spectrophotometer calibration curves?
2. How would you prepare a spectrophotometer calibration graph?
3. Why are algae counts in raw waters important to operators?
4. The maximum allowable chloride concentration in drinking water has been established on what basis?
5. What are the two general types of color found in water and what is the cause of each type?
6. Why is dissolved oxygen (DO) in water important to the treatment plant operator?
7. What precautions would you take when collecting a lake sample for a dissolved oxygen measurement?
8. How does fluoride get into drinking waters?

CHAPTER 21. ADVANCED LABORATORY PROCEDURES

(Lesson 2 of 2 Lessons)

7. Iron (Total)

Discussion

Iron is an abundant and widespread constituent of rocks and soils. The most common form of iron in solution in groundwater and in water under anaerobic conditions (bottom of a lake or reservoir) is the ferrous ion, Fe^{2+}. Ferric iron can occur in soils; in aerated water; and in acid solutions as Fe^{3+}, ferric hydroxide, and in polymeric forms depending upon the pH. Above a pH of 4.8, however, the solubility of the ferric species is less than 0.1 mg/L. Colloidal ferric hydroxide is commonly present in surface water and small quantities may persist even in water that appears clear.

Iron in a domestic water supply can stain laundry, concrete, and porcelain. A bitter, astringent flavor can be detected by some people at levels above 0.3 mg/L. When iron reacts with oxygen, a red precipitate (rust) is formed.

What Is Tested?

Source	Common Range, mg/L
Untreated Surface Water	0.10 to 1.0
Treated Surface Water	<0.01 to 0.20
Groundwater	<0.01 to 10

Apparatus Required

Spectrophotometer for use at 510 nm

Acid-washed glassware. Wash all glassware with concentrated HCl and rinse with distilled water to remove deposits of iron oxide, which could give false results.

Erlenmeyer flasks, 125 mL

Pipets, 5 and 10 mL

Volumetric flasks, 100 mL

Hot plate

Reagents

Use reagents low in iron. Use iron-free distilled water. Store reagents in glass-stoppered bottles. The hydrochloric acid and ammonium acetate solutions are stable indefinitely if tightly stoppered. The hydroxylamine, phenanthroline, and stock iron solutions are stable for several months. The standard iron solutions are not stable; prepare daily, as needed, by diluting the stock solution. Visual standards in Nessler tubes are stable for several months if sealed and protected from light.

1. Hydrochloric acid, HCl.
2. Hydroxylamine solution.
3. Ammonium acetate buffer solution. Because even a good grade of $NH_4C_2H_3O_2$ contains a significant amount of iron, prepare new reference standards with each buffer preparation.
4. Sodium acetate solution.
5. Phenanthroline solution. (*NOTE:* One milliliter of this reagent is sufficient for no more than 100 µg Fe.)
6. Stock iron solution: 1.00 mL = 0.200 mg Fe.
7. Standard iron solutions. Prepare daily for use. Pipet 50 mL stock solution into a 1-liter volumetric flask and dilute to mark with iron-free distilled water; 1.00 mL = 0.010 mg Fe.

Procedure

TOTAL IRON

1. Measure 50 mL of thoroughly mixed sample into a 125 mL Erlenmeyer flask.
2. Add 2 mL concentrated HCl and 1 mL hydroxylamine solution.
3. Heat to boiling. Boil sample until volume is reduced to 20 mL. Cool to room temperature.
4. Transfer to 100 mL volumetric flask.
5. Add 10 mL acetate buffer solution and 2 mL phenanthroline solution. Dilute to 100 mL mark with iron-free distilled water. Mix thoroughly.
6. After 15 minutes, measure the absorbance at 510 nm and determine the amount of iron from the standard curve.

(Iron)

OUTLINE OF PROCEDURE FOR IRON

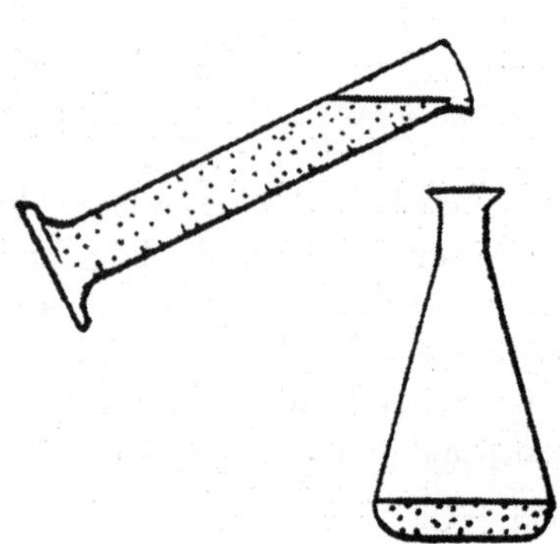

1. Measure 50 mL into flask.

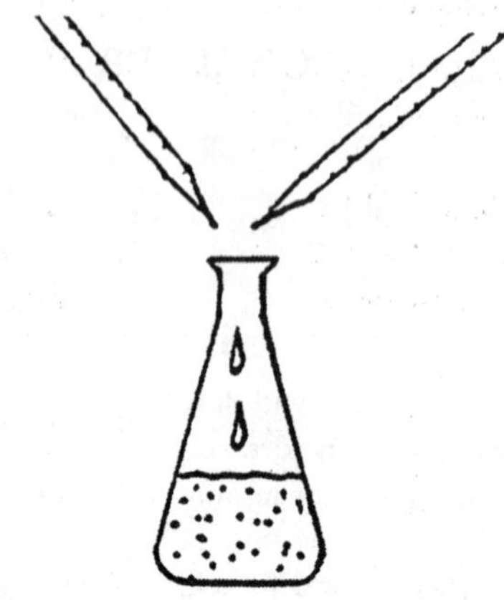

2. Add 2 mL conc. HCl and 1 mL hydroxylamine solution.

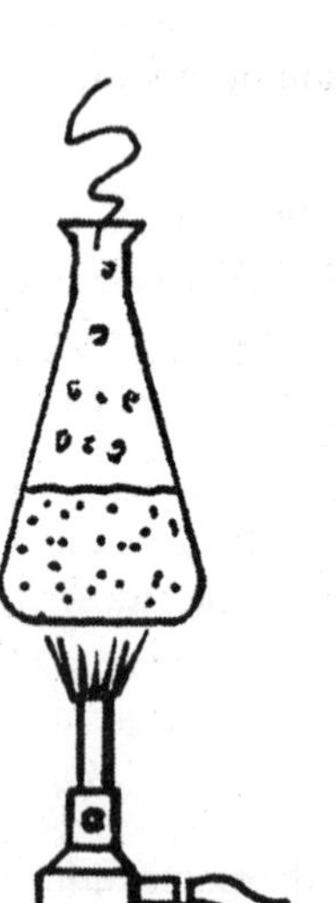

3. Heat to boiling. Reduce volume to 20 mL. Cool.

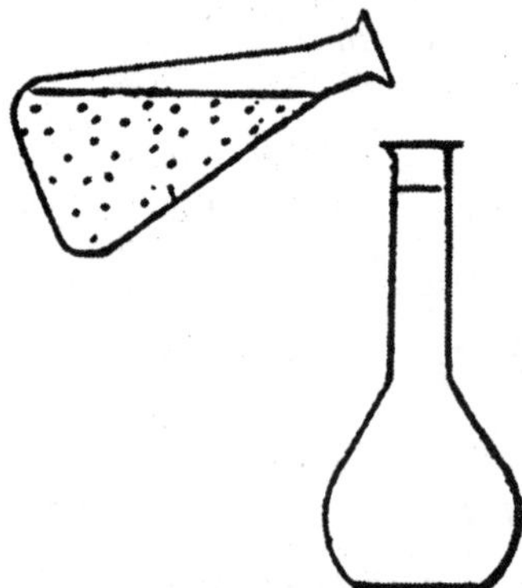

4. Transfer to 100 mL volumetric flask.

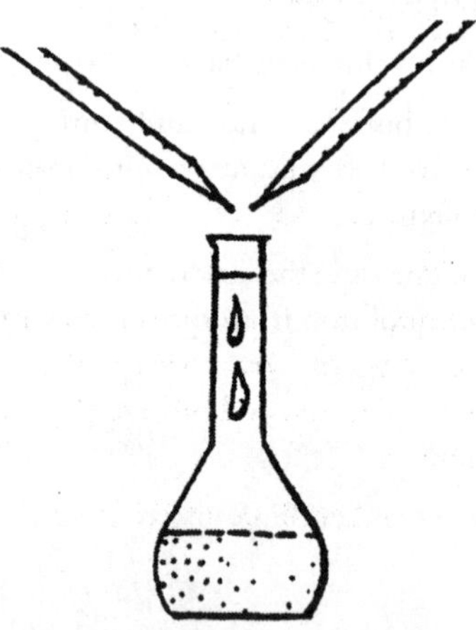

5. Add 10 mL acetate buffer and 2 mL phenanthroline solution. Dilute to 100 mL.

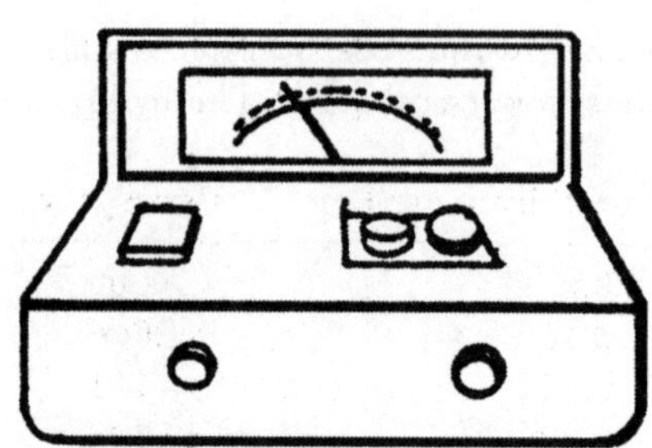

6. Measure absorbance at 510 nm with spectrophotometer.

(Iron)

Construction of Standard Calibration Curve

If your instrument can construct an internal electronic calibration curve, follow the manufacturer's instructions. If not, complete the following steps to construct a calibration curve.

1. Using the standard solution, prepare the following standards in 100 mL volumetric flasks.

mL of Standard Iron Solution Placed in 100 mL Volumetric Flask	Iron Concentration, mg/L
0	0
1.0	0.10
2.5	0.25
5.0	0.50
7.5	0.75
10.0	1.00

2. Dilute flasks to 100 mL.
3. Transfer 50 mL to 100 mL volumetric flask.
4. Add 1 mL hydroxylamine solution and 1 mL acetate solution to each flask.
5. Dilute to about 75 mL, add 10 mL phenanthroline solution, dilute to 100 mL mark. Mix thoroughly.
6. Measure absorbance at 510 nm against the reference blank.
7. Prepare a standard curve by plotting the absorbance values of standards versus the corresponding iron concentrations.

Example

A series of tests for total iron produced these results:

Flask No.	Sample	Absorbance
1	Distilled Water	0.000
2	Plant Clear Well	0.100
3	River Sample	0.420
4	0.10 mg/L Fe Standard	0.066
5	0.25 mg/L Fe Standard	0.161
6	0.50 mg/L Fe Standard	0.328
7	0.75 mg/L Fe Standard	0.495
8	1.00 mg/L Fe Standard	0.658

Calculation

1. Prepare a standard curve by using data from prepared standards. From the above example:

Concentration Iron, mg/L	Absorbance
0.0	0.000
0.10	0.066
0.25	0.161
0.50	0.328
0.75	0.495
1.00	0.658

The graph below is a result of plotting concentration of standards versus their corresponding absorbance.

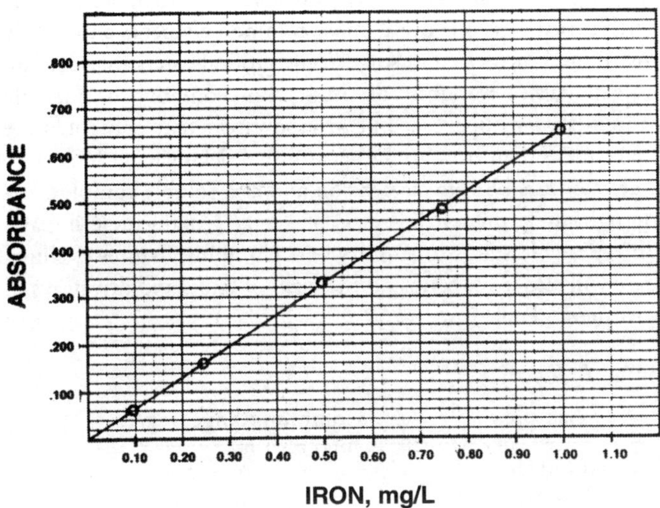

2. Obtain concentration of unknown clear well and river samples from curve.

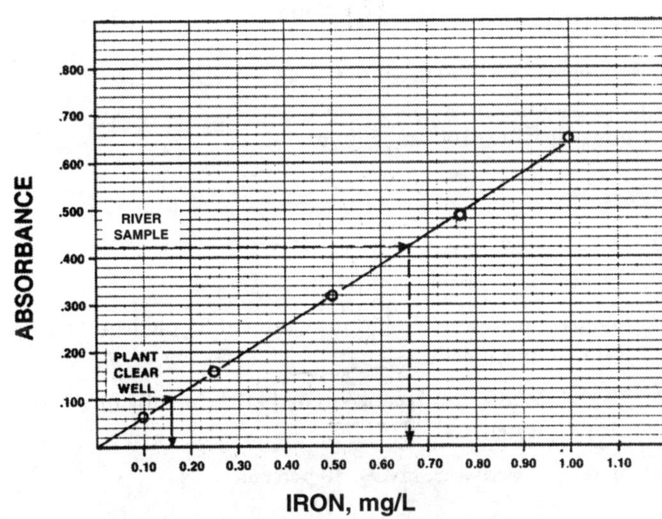

PLANT CLEAR WELL = 0.16 mg/L Fe
RIVER SAMPLE = 0.66 mg/L Fe

Notes

1. Iron in well water or tap samples may vary in concentration and form with the duration and degree of flushing before and during sampling.
2. For precise determination of total iron, use a separate container for sample collection. Treat with acid at time of collection to place iron in solution and prevent deposition on walls of sample container.
3. Exercise caution when handling sulfuric acid.

Reference

See *Standard Methods*, 22nd Edition, Method 3500-Fe (B), page 3-77.

(Manganese)

8. Manganese

Discussion

Although manganese is much less abundant than iron in the earth's crust, it is one of the most common elements and widely distributed in rocks and soils. Some groundwaters that contain objectionable amounts of iron also contain considerable amounts of manganese, but groundwaters that contain more manganese than iron are rather unusual. Manganese in surface waters occurs both in suspension and as a soluble complex. Although rarely present in excess of 1 mg/L, manganese imparts objectionable stains to laundry and plumbing fixtures. Manganese will also cause stains on the walls of tanks and driveways in treatment plants.

What Is Tested?

Sample	Common Range, mg/L
Treated and Untreated Surface Water	<0.01 to 0.10
Groundwater	<0.01 to 1.0

Apparatus Required

Spectrophotometer for use at 525 nm
Hot plate
Erlenmeyer flask, 250 mL
Pipets, 5 and 10 mL
Volumetric flasks, 100 and 500 mL

Reagents

1. Special reagent.
2. Ammonium persulfate, $(NH_4)_2S_2O_8$, solid.
3. Standard manganese solution. 1 mL = 0.01 mg Mn. Prepare dilute solution daily.
4. 1% HCl: Add 10 mL concentrated HCl carefully to 990 mL distilled water.
5. Hydrogen peroxide, H_2O_2, 30 percent.

Procedure

1. Measure 100 mL of thoroughly mixed sample into a 250 mL Erlenmeyer flask that has been marked with a line at the 90 mL level.
2. Add 5 mL special reagent and 1 drop H_2O_2.
3. Concentrate to 90 mL by boiling. Add 1 gram ammonium persulfate. Cool immediately under water tap.
4. Dilute to 100 mL.
5. Measure the absorbance at 525 nm with a spectrophotometer and determine the amount of manganese from the standard curve.

Construction of Calibration Curve

If your instrument can construct an internal electronic calibration curve, follow the manufacturer's instructions. If not, complete the following steps to construct a calibration curve.

1. Using the standard manganese solution, prepare the following standards in 100 mL volumetric flasks.

mL of Standard Manganese Solution Placed in 100 mL Volumetric Flask	Manganese Concentration, mg/L
0	0
1.0	0.10
2.0	0.20
3.0	0.30
4.0	0.40

2. Dilute flasks to 100 mL.
3. Transfer to 250 mL Erlenmeyer flask.
4. Determine amount of manganese as outlined previously.
5. Prepare a standard curve by plotting the absorbance values of standards versus the corresponding manganese concentrations.

Example

Results from a series of tests for manganese were as follows:

Flask No.	Sample	Absorbance
1	Distilled Water	0.000
2	Plant Effluent	0.000
3	Jones St. Well	0.030
4	0.05 mg/L Mn Standard	0.009
5	0.10 mg/L Mn Standard	0.018
6	0.20 mg/L Mn Standard	0.036
7	0.30 mg/L Mn Standard	0.053
8	0.40 mg/L Mn Standard	0.071

Calculation

1. Prepare a standard curve by using data from prepared standards. From the above example:

Concentration Manganese, mg/L	Absorbance
0.0 (distilled water)	0.000
0.05	0.009
0.10	0.018
0.20	0.036
0.30	0.053
0.40	0.071

OUTLINE OF PROCEDURE FOR MANGANESE

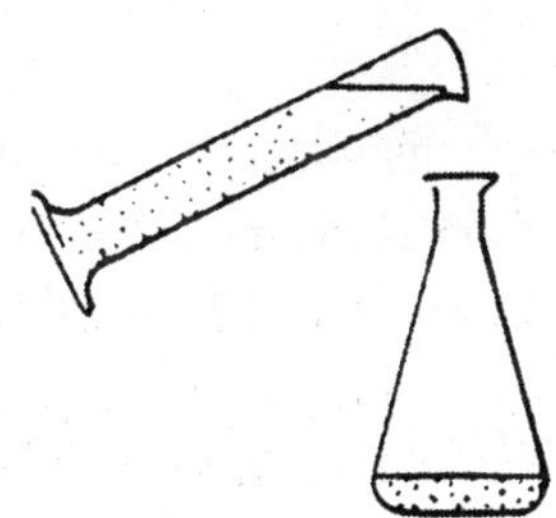

1. Measure 100 mL into flask.

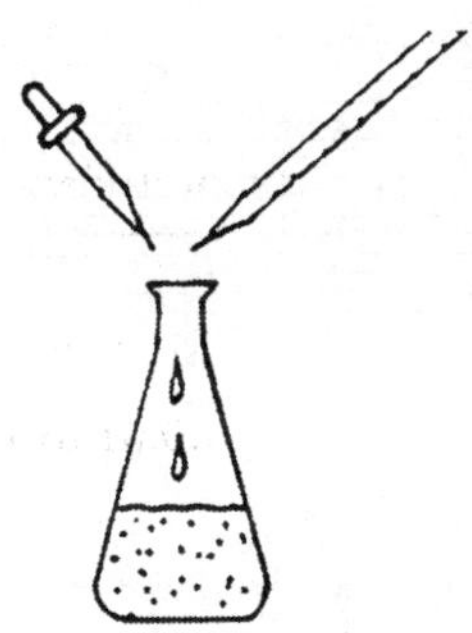

2. Add 5 mL special reagent and 1 drop H_2O_2.

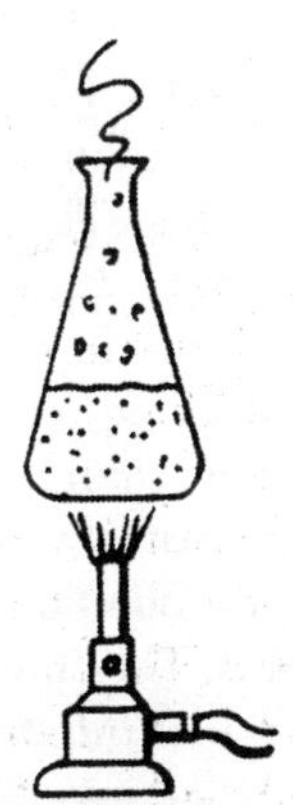

3. Concentrate to 90 mL, then add 1 g ammonium persulfate. Dilute to 100 mL after cooling.

4. Measure absorbance at 525 nm with spectrophotometer.

(Manganese—Marble Test)

The graph below is the result of plotting concentration of standards versus their corresponding absorbance.

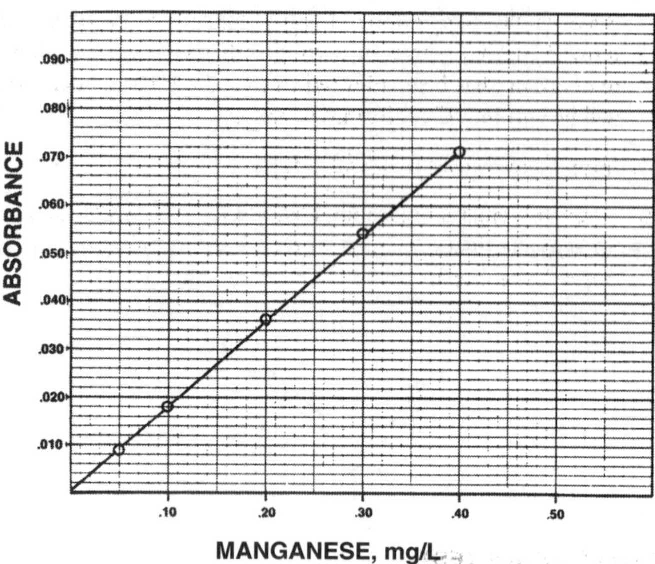

2. Obtain concentration of unknown plant effluent and well sample from curve.

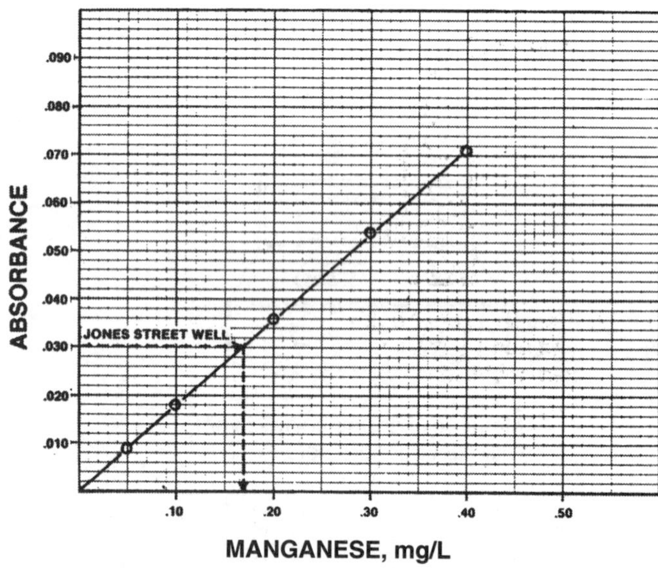

Notes

1. If turbidity or interfering color is present, use the following bleaching method: as soon as the spectrophotometer reading has been made, add 0.05 mL hydrogen peroxide solution directly to the optical cell. Mix and read again as soon as the color has faded. Deduct the absorbance of the bleached solution from the initial absorbance to obtain the absorbance due to manganese.
2. Determine manganese as soon as possible after sample collection. If this is not possible, acidify the sample with nitric acid to a pH less than 2.

Reference

See *Standard Methods*, 22nd Edition, Method 3500-Mn (B), page 3-85.

QUESTIONS

Please write your answers to the following questions and compare them with those on page 532.

21.1J Iron in a domestic water supply may cause what problems?

21.1K Why must all glassware be acid washed when analyzing samples for iron?

21.1L In what forms does manganese occur in surface waters?

21.1M If the manganese concentration in a sample cannot be tested immediately, what would you do?

9. Marble Test (Calcium Carbonate Saturation Test)

Discussion

The Marble Test is used to determine the degree to which a sample of water is saturated with calcium carbonate. Marble Test results are used in the lime–soda softening process and to control corrosion. Water in intimate contact with powdered calcium carbonate (calcite) will approach saturation. The water being tested should not be exposed to atmospheric carbon dioxide. The Marble Test must be conducted at the specific temperature because the solubility of calcium carbonate varies with temperature. However, equipment that will maintain a constant temperature (either lower or higher than room temperature) while mixing the solution is not commonly available in water treatment plants. The only other way to keep a reasonably uniform temperature is to run the test as rapidly as possible.

What Is Tested?

Source	Common Range (Initial pH – Final pH)
Untreated Surface Water	– 1 to +1
Treated Surface Water	– 0.2 to +0.2
Well Water	– 0.1 to +1

Apparatus Required

BOD bottle, 300 mL
Magnetic stirrer
Stir-bar
Thermometer
Glass funnel, 125 mm
Filter paper, Whatman #50 (18.5 inch)
Equipment for determining pH and hardness

Reagents

1. Calcium carbonate, reagent grade.
2. Reagents for determining pH and hardness.

Procedure

1. Measure the temperature of the water to be tested.
2. Measure the pH, hardness, and, if desired, the alkalinity of the sample being tested.
3. Insert the stirring bar in the BOD bottle and fill with the water being tested. Adjust the water temperature to within 1°C of the initial temperature. Add approximately 1 gram of calcium carbonate and stir for 5 minutes at a rate high enough to keep the calcium carbonate in suspension and the sample vigorously agitated.
4. Recheck the temperature. If the temperature has changed more than 1°C, repeat the stirring with a fresh sample whose temperature has been adjusted so that the final temperature will be within 1°C of the initial temperature.
5. Immediately measure the final pH.
6. Filter the remaining sample. Determine the hardness and, if desired, the final alkalinity on the filtrate (water that passed through the filter).

OUTLINE OF PROCEDURE FOR MARBLE TEST

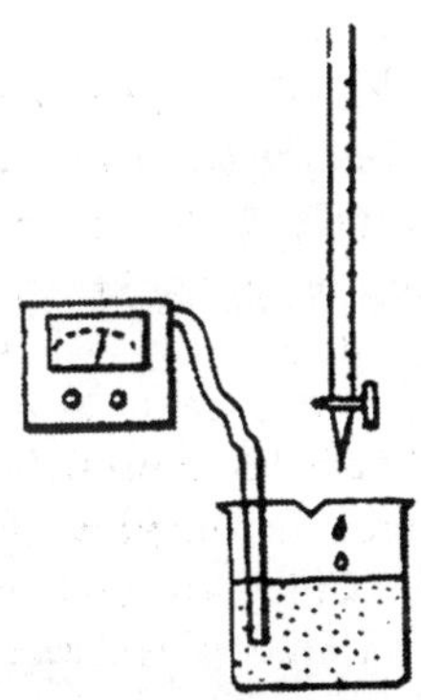

1. Measure temperature, pH, hardness, and alkalinity of sample being tested.

2. Transfer to BOD bottle and add 1 g calcium carbonate. Mix.

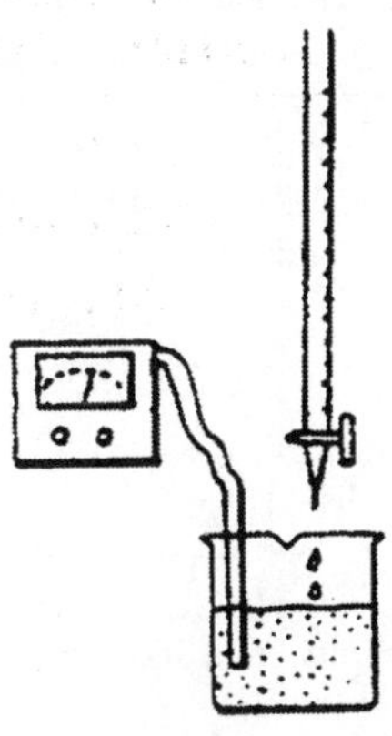

3. Measure final pH and temperature.

4. Filter.

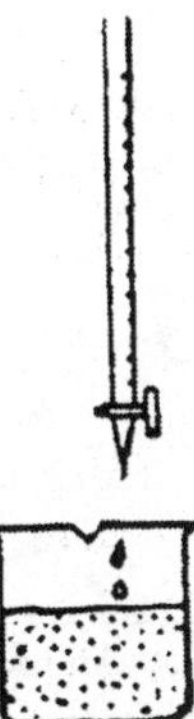

5. Determine hardness and alkalinity of filtrate.

(Marble Test)

Example

Results from a series of tests for the calcium carbonate precipitation potential were as follows:

Filtered Water Sample	
Initial Temperature	14°C
Final Temperature	14°C
Initial pH	8.7
Final pH	9.1
Initial Hardness	34 mg/L
Final Hardness	38 mg/L
Initial Alkalinity	24 mg/L
Final Alkalinity	27 mg/L

Calculation

$$\text{Calcium Carbonate Precipitation, Potential, mg/L} = \text{Initial Hardness, mg/L} - \text{Final Hardness, mg/L}$$

The Langelier Index[2] is approximately equal to the initial pH minus the final pH. If the value of this index is less than 0.2, this value will indicate that the water is very near the saturation level. In any event, the sign of this value will be the same as the sign of the Langelier Index. That is to say, both the Langelier Index and the calcium carbonate precipitation potential will be negative if the water is undersaturated, and positive if the water is supersaturated.

From the example above:

$$\text{Calcium Carbonate Precipitation, Potential, mg/L} = \text{Initial Hardness, mg/L} - \text{Final Hardness, mg/L}$$

$$= 34 \text{ mg/L} - 38 \text{ mg/L}$$

$$= -4 \text{ mg/L}$$

$$\text{Langelier Index} \cong \text{Initial pH} - \text{Final pH}$$

$$\cong 8.7 - 9.1$$

$$\cong -0.4$$

This water is undersaturated (and therefore corrosive) with respect to calcium carbonate.

10. Metals

Discussion

The presence of certain metals in drinking water can be a matter of serious concern because of the toxic properties of these materials. The analyses of these metals are generally done by using atomic absorption spectroscopy or colorimetric methods. The term "metals" includes the following elements:

Aluminum	Cobalt	Potassium
Antimony	Copper	Selenium
Arsenic	Iron	Silver
Barium	Lead	Sodium
Beryllium	Magnesium	Thallium
Cadmium	Manganese	Tin
Calcium	Mercury	Titanium
Chromium	Molybdenum	Vanadium
	Nickel	Zinc

Reference

Marble Test: See *Standard Methods*, 22nd Edition, Method 2330 (C), page 2-41.

Hardness: See *Standard Methods*, 22nd Edition, Method 2340 (C), page 2-44.

Alkalinity: See *Standard Methods*, 22nd Edition, Method 2320 (B), page 2-34.

QUESTIONS

Please write your answers to the following questions and compare them with those on page 532.

21.1N Why is temperature important when running the Marble Test?

21.1O The results from the Marble Test produce an initial pH of 8.9 and a final pH of 8.6. Would this water be considered corrosive?

21.1P How are the concentrations of most metals in water measured?

2. *Langelier Index (LI).* An index reflecting the equilibrium pH of a water with respect to calcium and alkalinity. This index is used in stabilizing water to control both corrosion and the deposition of scale.

$$\text{Langelier Index} = pH - pH_s$$

where pH = actual pH of the water

pH_s = pH at which water having the same alkalinity and calcium content is just saturated with calcium carbonate

(Nitrate)

11. Nitrate

Discussion

Nitrate represents the most completely oxidized form of nitrogen found in water. High levels of nitrate in raw water samples indicate biological wastes in the final stage of stabilization or runoff from fertilized areas. High nitrate levels degrade water quality by stimulating excessive algal growth. Drinking water that contains excessive amounts of nitrate can cause infant methemoglobinemia (blue babies). For this reason, a level of 10 mg/L (as nitrogen) has been established as a maximum level. The procedure given below measures the amount of both nitrate and nitrite nitrogen present in a sample by reducing all nitrate to nitrite through the use of a copper-cadmium column. The total nitrate (any nitrite present originally plus the reduced nitrate) is then measured colorimetrically.

What Is Tested?

Sample	Common Range, mg/L
Treated Surface Water	<0.1 to 5
Groundwater	0.5 to 10

Apparatus

Reduction column. The column in Figure 21.3 was constructed from a 100 mL volumetric pipet by removing the top portion. This column may also be constructed from two pieces of tubing joined end to end. A 10 cm length of 3 cm inside diameter (ID) tubing is joined to a 25 cm length of 3.5 mm ID tubing.
Spectrophotometer for use at 540 nm, providing a light path of 1 cm or longer
Beakers, 125 mL
Glass wool
Glass-fiber filter or 0.45-micron membrane filter
Filter holder assembly
Filter flask
pH meter
Separatory funnel, 250 mL
Volumetric pipets, 1, 2, 5, and 10 mL

Reagents

1. Granulated cadmium: 40 to 60 mesh.
2. Copper-Cadmium: The cadmium granules (new or used) are cleaned with 6 *N* HCl and copperized with a 2 percent solution of copper sulfate in the following manner:
   a. Wash the cadmium with 6 *N* HCl and rinse with distilled water. The color of the cadmium should be silver.
   b. Swirl 25 g cadmium in 100 mg/L portions of a 2 percent solution of copper sulfate for 5 minutes or until the blue color partially fades, decant, and repeat with fresh copper until a brown precipitate forms.
   c. Wash the copper-cadmium with distilled water at least 10 times to remove all the precipitated copper. The color of the cadmium should now be black.
3. Dilute ammonium chloride-EDTA solution. Dilute 300 mL of concentrated ammonium chloride-EDTA solution (reagent 4) to 500 mL with distilled water.
4. Color reagent.
5. Zinc sulfate solution.
6. Sodium hydroxide, 6 *N*.
7. Ammonium hydroxide, concentrated.
8. Hydrochloric acid, 6 *N*. Dilute 50 mL concentrated HCl to 100 mL with distilled water.
9. Copper sulfate solution, 2 percent.
10. Nitrate stock solution. 1.0 mL = 1.00 mg NO_3-N. Preserve with 2 mL of chloroform per liter. This solution is stable for at least 6 months.
11. Nitrate standard solution. 1.0 mL = 0.01 mg NO_3-N. Dilute 10.0 mL of nitrate stock solution (reagent 12) to 1,000 mL with distilled water.
12. Chloroform.

Procedure

REMOVAL OF INTERFERENCES (IF NECESSARY)

1. Turbidity removal. Use one of the following methods to remove suspended matter that can clog the reduction column.
   a. Filter sample through a glass-fiber filter or a 0.45-micron pore size filter as long as the pH is less than 8.
   b. Add 1 mL zinc solution (reagent 5) to 100 mL of sample and mix thoroughly. Add enough (usually 8 to 10 drops) sodium hydroxide solution (reagent 6) to obtain a pH of 10.5. Let treated sample stand a few minutes to allow the heavy flocculant precipitate to settle. Clarify by filtering through a glass-fiber filter.

REDUCTION OF NITRATE TO NITRITE

1. Preparation of reaction column: Insert a glass wool plug into the bottom of the reduction column and fill with distilled water. Add sufficient copper-cadmium granules to produce a column 18.5 cm in length. Maintain a level of distilled water above the copper-cadmium granules to eliminate entrapment of air. Wash the column with 200 mL of dilute ammonium chloride-EDTA solution (reagent 3). The column is then activated by passing through the column 100 mL of solution composed of 25 mL of a 1.0 mg/L NO_2-N standard and 75 mL of concentrated ammonium chloride-EDTA solution. Use a flow rate of 7 to 10 mL per minute. Collect the reduced standard until the level of solution is 0.5 cm above the top of the granules. Close the screw clamp to stop flow. Discard the reduced standard.
2. Measure about 40 mL of concentrated ammonium chloride-EDTA and pass through column at 7 to 10 mL per minute to wash nitrate standard off column. Always leave at least 0.5 cm of liquid above top of granules. The column is now ready for use.

(Nitrate)

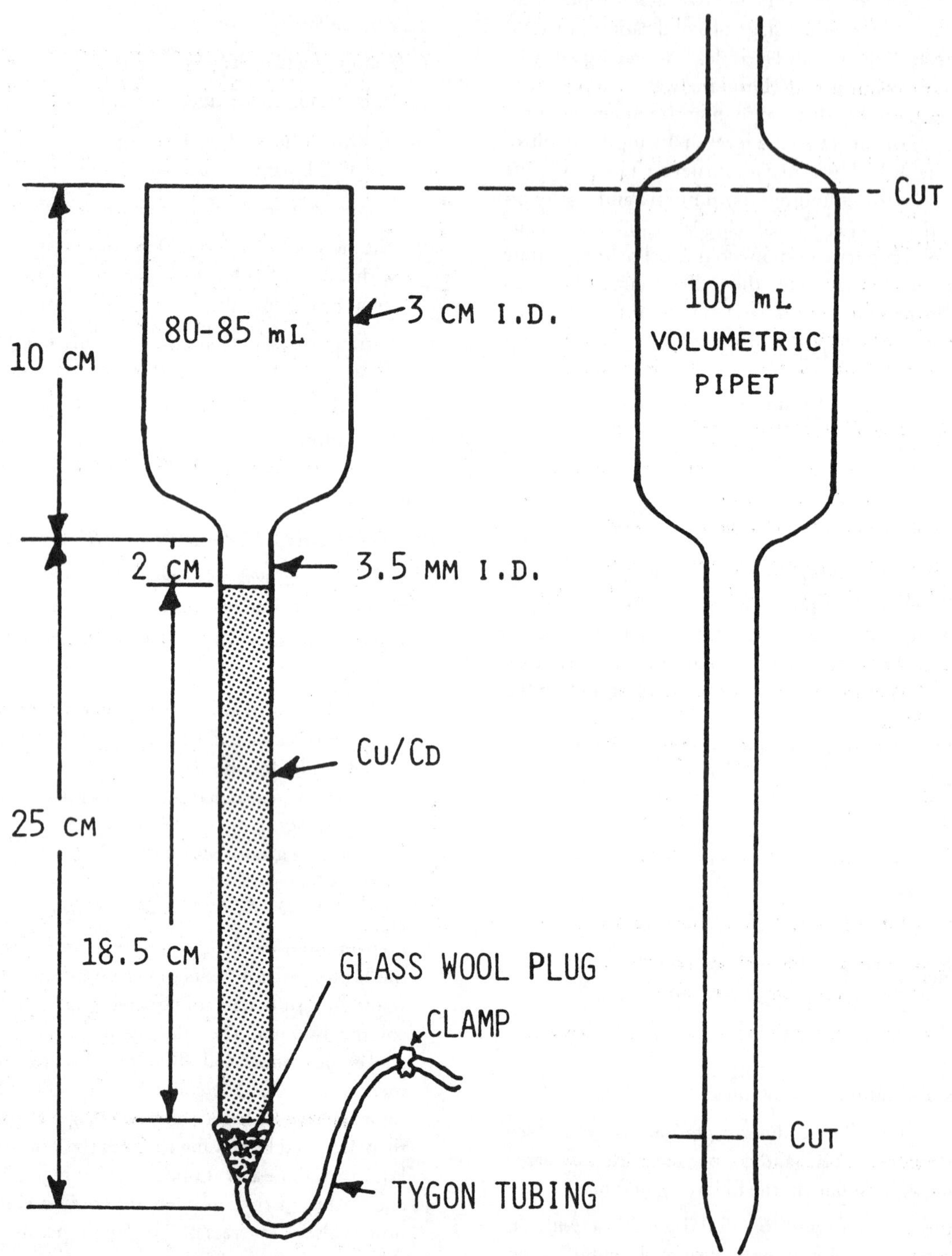

Fig. 21.3 Reduction column

(Nitrate)

3. Using a pH meter, adjust the pH of sample (or standard) to between 5 and 9 either with concentrated HCl or concentrated NH_4OH.
4. To 25 mL of sample (or standard) or aliquot diluted to 25 mL, add 75 mL of concentrated ammonium chloride-EDTA solution and mix.
5. Pour sample into column and collect reduced sample at a rate of 7 to 10 mL per minute.
6. Discard the first 25 mL. Collect the rest of the sample (approximately 70 mL) in the original sample flask. Reduced samples should not be allowed to stand longer than 15 minutes before addition of color reagent.
7. Add 2 mL of color reagent to 50 mL of sample. Allow 10 minutes for color development. Within 2 hours, measure the absorbance at 540 nm against a reagent blank (50 mL distilled water to which 2 mL color reagent has been added).

Construction of Standard Calibration Graph

If your instrument can construct an internal electronic calibration curve, follow the manufacturer's instructions. If not, complete the following steps to construct a calibration curve.

1. Prepare working standards by pipeting the following volumes of nitrate standard solution into each of five 100 mL volumetric flasks.

Add This Volume of Nitrate Standard Solution to 100 mL Flask	Concentration of NO_3-N in mg/L
0.0	0.00
1.0	0.10
2.0	0.20
5.0	0.50
10.0	1.00

Dilute each to 100 mL with distilled water and mix.

2. Determine the amount of nitrate-nitrite as outlined above in the procedure for reduction of nitrate to nitrite.
3. Plot on a sheet of graph paper the absorbance versus concentration.

Example

Results from the analyses of samples and working standards for nitrate-nitrite were as follows:

Flask No.	Sample	Volume	Absorbance
1	Jones St. Well	25 mL	0.440
2	Blank (distilled water)	25 mL	0.00
3	0.10 mg/L NO_3-N Standard	25 mL	0.075
4	0.20 mg/L NO_3-N Standard	25 mL	0.142
5	0.50 mg/L NO_3-N Standard	25 mL	0.355
6	1.00 mg/L NO_3-N Standard	25 mL	0.700

Calculation

1. Using graph paper, plot the absorbance values of working standards versus their known concentrations. For example, from the above data the following graph can be constructed.

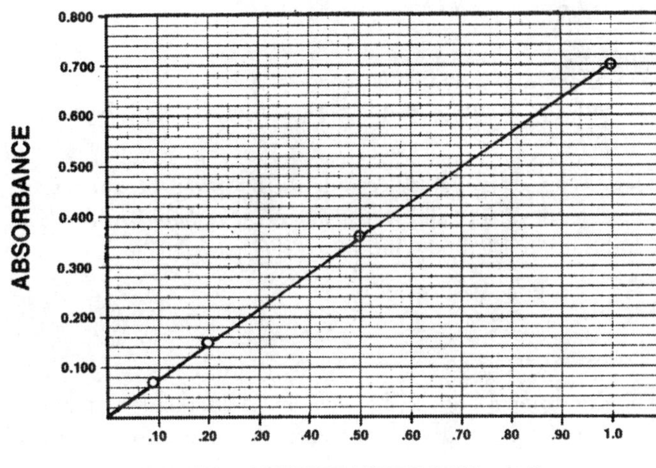

2. Read concentration of NO^-_3 + NO^-_2 nitrogen from graph shown below.

   mg/L Nitrate + Nitrite Nitrogen in Sample = 0.62 mg/L

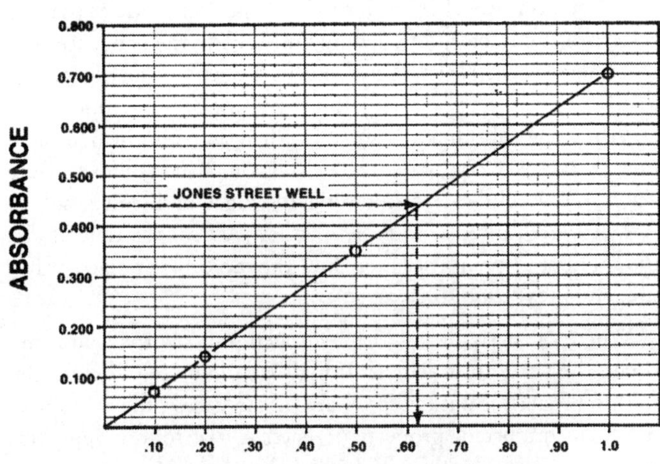

3. Determine concentration of nitrite nitrogen (NO_2-N) in sample using nitrite procedure.
4. Subtract nitrite from NO^-_2 + NO^-_3 nitrogen concentration. The result is the amount of nitrate nitrogen in sample.
5. For example, if the Jones Street Well sample used in the above example contained no nitrite nitrogen then the nitrate nitrogen (NO_3-N) would be 0.62 mg/L.

(Nitrate—pH)

Notes

1. If concentration of nitrate in the sample is greater than 1 mg/L, then the sample must be diluted.
2. Cadmium metal is highly toxic, thus caution must be exercised in its use. Rubber gloves should be used whenever it is handled.

Reference

See *Standard Methods*, 22nd Edition, Method 4500-NO_3^- (E), page 4-125.

There is an ion-specific electrode method available for nitrate, which requires specialized instruments. See *Standard Methods*, 22nd Edition, Method 4500-NO_3^- (D), page 4-124.

12. pH

Discussion

This discussion is presented to give you a better understanding of what a pH value actually represents. Procedures for measuring pH are given in the "Laboratory Procedures" chapter in Volume I of this manual.

Pure water dissociates according to the following reaction:

$$H_2O = H^+ + OH^-$$

At 25°C and a pH of 7, the activity of the hydrogen ion is equal to the activity of the hydroxyl ion at .000 000 1 moles/liter. "Activity" is a term used by chemists to allow real atoms, molecules, and ions to behave as if they were perfect particles (having zero size). Activity is obtained by multiplying the concentration by an activity coefficient. The value of the activity coefficient depends on the electrical charge on the particle, the temperature, and the other substances dissolved in the water. For the hydrogen ion, the activity coefficient at 25°C varies from 0.996 in pure water to 0.900 in water containing 400 mg/L of dissolved solids. Activities are expressed in moles per liter, which is assumed to be the number of grams per liter since the molecular weight of hydrogen ion is 1.008 (or near 1.0).

When the activities of the hydrogen and hydroxyl ions are equal, the solution is neutral. If hydrogen ions are in excess, the solution is acid and if hydroxyl ions are in excess, the solution is basic. An important property of water is that for any temperature, the product of the activities of these ions is a constant. At 25°C, this constant is .000 000 000 000 01.

In a strong solution of hydrochloric acid, the hydrogen ion activity may be as high as 1 mole per liter, while in a strong solution of lye, the hydroxyl ion concentration may be as high as 1 mole per liter. To avoid the inconvenience of writing these very small numbers, hydrogen ion activities are expressed in terms of pH, with

$$pH = \text{Log}\frac{1}{\{H^+\}}$$

The relation between pH, H^+, and OH^- at 25°C is shown in Table 21.2.

TABLE 21.2 RELATION BETWEEN pH, H^+, AND OH^- AT 25°C

pH	Activity of H^+, moles/L	Activity of OH^-, moles/L
0	1.0	0.000 000 000 000 01
1	0.1	0.000 000 000 000 1
2	0.01	0.000 000 000 001
3	0.001	0.000 000 000 01
4	0.000 1	0.000 000 000 1
5	0.000 01	0.000 000 001
6	0.000 001	0.000 000 01
7	0.000 000 1	0.000 000 1
8	0.000 000 01	0.000 001
9	0.000 000 001	0.000 01
10	0.000 000 000 1	0.000 1
11	0.000 000 000 01	0.001
12	0.000 000 000 001	0.01
13	0.000 000 000 000 1	0.1
14	0.000 000 000 000 01	1.0

Most natural waters have pH values between 6.5 and 8.5. The pH of natural waters is controlled by the relative amounts of carbon dioxide, bicarbonate, and carbonate ions. Rainwater usually has a pH of slightly less than 7 because carbon dioxide from the air dissolves to form carbonic acid.

Human blood has a pH of 7.4 and the gastric juices in your stomach have a pH of approximately 0.9 to aid in the digestion of food.

Alum coagulates most effectively at pH values near 6.8.

Reference

See *Standard Methods*, 22nd Edition, Method 4500-H^+ (B), page 4-92.

QUESTIONS

Please write your answers to the following questions and compare them with those on pages 532 and 533.

21.1Q How is nitrate measured in the nitrate test?

21.1R If turbidity is interfering with a nitrate analysis, how can turbidity be removed?

21.1S The pH of natural waters is usually controlled by the relative amounts of what ions?

13. Specific Conductance (Conductivity)

Discussion

Specific conductance or conductivity is a numerical expression (expressed in micromhos per centimeter) of the ability of a water to conduct an electric current. This number depends on the total concentration of the minerals dissolved in the sample (TDS) and the temperature. Changes in conductivity from normal levels may indicate changes in mineral composition of the water, seasonal variations in lakes and reservoirs, or intrusion of pollutants. The custom of reporting conductivity values in micromhos/cm at 25°C requires the accurate determination of each sample's temperature at the time of conductivity measurement.

Specific conductance is measured by the use of a conductivity meter.

What Is Tested?

Sample	Common Range, micromhos/cm
Raw and Treated Surface Waters	30 to 500
Groundwater	100 to 1,000

Materials and Procedure

Follow instrument manufacturer's instructions. Also see *Standard Methods*, 22nd Edition, Method 2510 (B), page 2-54.

14. Sulfate

Discussion

The sulfate ion is one of the major anions occurring in natural waters. Sulfate ions are of importance in water supplies because of the tendency of appreciable amounts to form hard scales in boilers and heat exchangers. EPA is evaluating the need to regulate sulfate in drinking water.

What Is Tested?

Sample	Common Range, mg/L
Raw or Treated Water Supply	5 to 100

Apparatus Required

Turbidimeter or spectrophotometer
Stopwatch or timer
Measuring spoon, 0.3 mL
Magnetic stirrer
Magnetic stir-bar
Pipet, 10 mL
Erlenmeyer flasks, 250 mL
Volumetric flasks, 100 mL

Reagents

Standardized solutions such as the following are commercially available:

1. Conditioning reagent.
2. Barium chloride, $BaCl_2$, crystals: Sized for turbidimetric work (Baker No. 0974 or equivalent). To ensure uniformity of results, construct a standard curve for each batch of $BaCl_2$ crystals.
3. Standard sulfate solution: Prepare a standard sulfate solution as described in a. or b. below; 1.00 mL = 0.10 mg SO_4.
   a. Dilute 10.41 mL standard 0.0200 *N* H_2SO_4 titrant specified in Alkalinity Test (see the "Laboratory Procedures" chapter in Volume I of this manual) to 100 mL with distilled water.
   b. Dissolve 147.9 mg anhydrous Na_2SO_4 in distilled water and dilute to 1,000 mL.

Procedure

1. Place 100 mL of sample or a suitable portion diluted to 100 mL into a clean 250 mL Erlenmeyer flask.
2. Add 5.0 mL of conditioning reagent and mix.
3. While stirring, add a spoonful (0.3 mL) of barium chloride crystals. Stir for exactly 1 minute.
4. Measure turbidity at 30-second intervals for 4 minutes. Consider turbidity to be the maximum reading obtained in the 4-minute interval.

Construction of Standard Calibration Curve

1. Using the standard solution, prepare the following standards in 100 mL volumetric flasks.

mL of Standard Sulfate Solution Placed in 100 mL Volumetric Flask	Sulfate Concentration, mg/L
5.0	5.0
10.0	10.0
15.0	15.0
20.0	20.0
25.0	25.0

2. Dilute flasks to 100 mL.
3. Transfer to 250 mL Erlenmeyer flask.
4. Determine amount of sulfate as outlined previously.
5. Prepare a standard curve by plotting turbidity values of standards versus the corresponding sulfate concentrations. Set nephelometer (or spectrophotometer) at zero sulfate concentration using distilled water as a control.

(Sulfate)

OUTLINE OF PROCEDURE FOR SULFATE

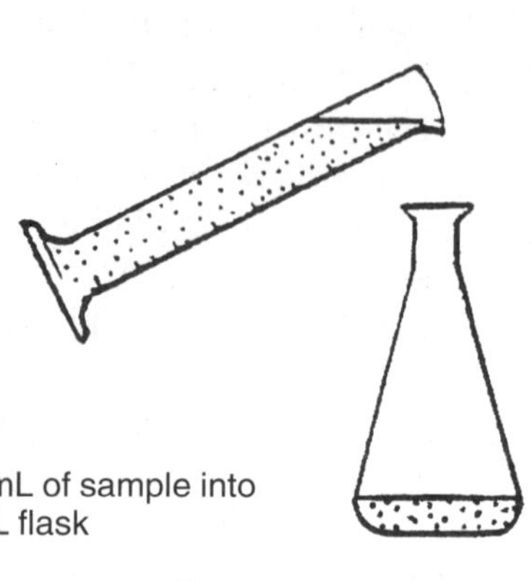

1. Measure 100 mL of sample into a clean 250 mL flask

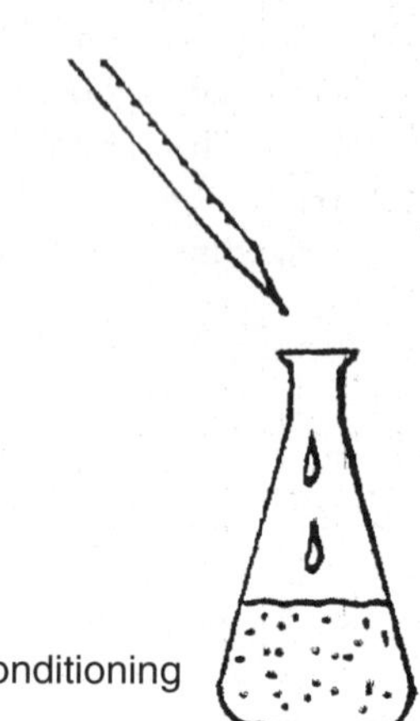

2. Add 5.0 mL conditioning reagent. Mix.

3. Add barium chloride and stir for 1 minute.

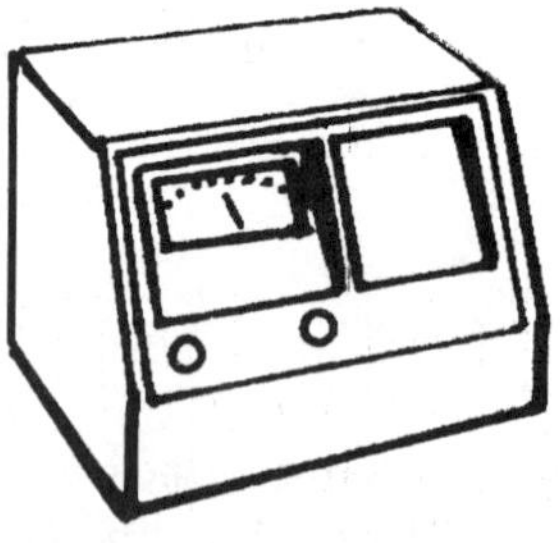

4. Measure turbidity.

Example

Results from a series of tests for sulfate were as follows:

Flask No.	Sample	Volume	Turbidity
1	Distilled Water	100 mL	0
2	Plant Effluent	100 mL	35
3	Jones St. Well	50 mL	45
4	5.0 mg/L SO_4 Standard	100 mL	11
5	10.0 mg/L SO_4 Standard	100 mL	29
6	15.0 mg/L SO_4 Standard	100 mL	40
7	20.0 mg/L SO_4 Standard	100 mL	53

Calculation

1. Prepare a standard curve by using data from prepared standards. From the above example:

Concentration Sulfate, mg/L	Turbidity, TU
0.0	0
5.0	11
10.0	29
15.0	40
20.0	53

(Sulfate)

The graph below is the result of plotting concentration of standards versus their corresponding turbidity.

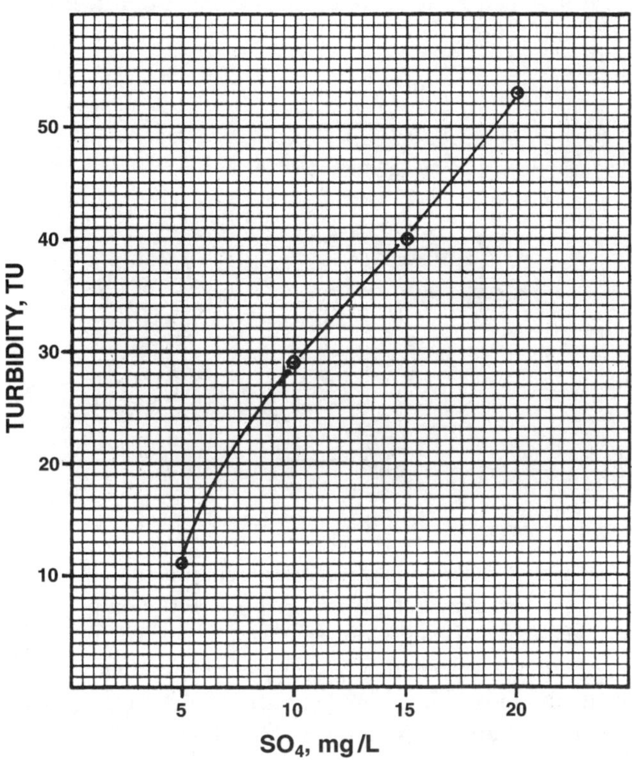

2. Obtain concentration of unknown plant and well samples from curve.

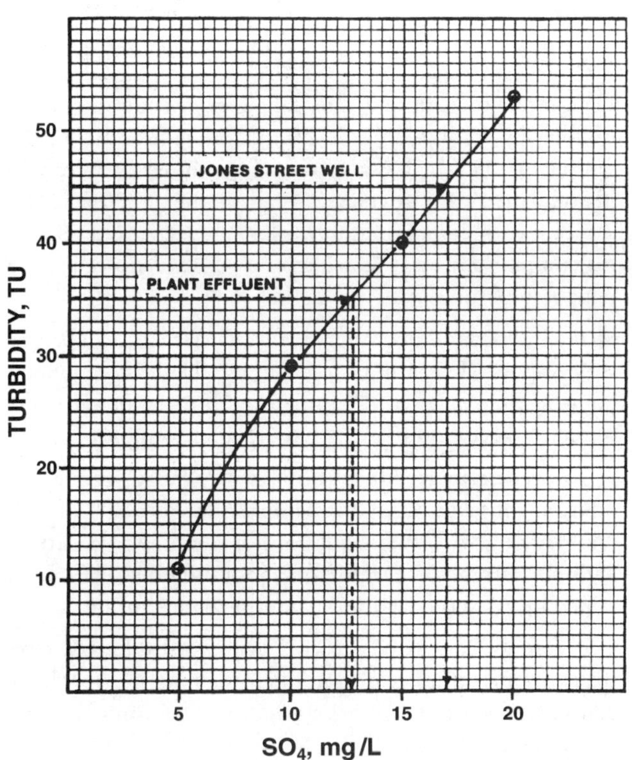

3. Correct (if necessary) for samples of less than 100 mL by using the following formula:

$$\text{Sulfate } (SO_4), \text{ mg/L} = \frac{(\text{Graph Sulfate, mg/L})(100 \text{ mL})}{\text{Sample Size, mL}}$$

Using data from example:

	Sample Volume	Turbidity	Concentration from Graph
Plant Effluent	100 mL	35 TU	13 mg/L

Sulfate, mg/L = 13 mg/L

	Sample Volume	Turbidity	Concentration from Graph
Jones St. Well	50 mL	45 TU	17 mg/L

$$\begin{aligned} \text{Sulfate, mg/L} &= \frac{(\text{Sulfate, mg/L})(100 \text{ mL})}{\text{Sample Size, mL}} \\ &= \frac{(17 \text{ mg/L})(100 \text{ mL})}{50 \text{ mL}} \\ &= 34 \text{ mg/L } SO_4 \end{aligned}$$

Notes

1. A spectrophotometer can be used to measure absorbance of barium sulfate suspension. Use at 420 nanometers (nm) wavelength.
2. Color or suspended matter will interfere when present in large amounts. Correct for these items by testing blanks from which barium chloride is withheld.
3. Analyze samples and standards with their temperatures in the range of 20 to 25°C.

Reference

See *Standard Methods*, 22nd Edition, Method 4500-SO_4^{2-} (E), page 4-190.

QUESTIONS

Please write your answers to the following questions and compare them with those on page 533.

21.1T What is the meaning of specific conductance or conductivity?

21.1U Sulfate ions are of concern in drinking water for what reason?

21.1V A 50 mL sample from a well produced a turbidity reading of 40 TU using a nephelometer (turbidimeter). What was the sulfate concentration in mg/L?

(Taste and Odor)

15. Taste and Odor

Discussion

Taste (flavor) and odor are sensory clues that provide the first warning of potential hazards in the environment. Water, in its pure form, cannot produce odor or taste sensations. However, algae, *Actinomycetes*, bacteria, decaying vegetation, metals, and pollutants can cause tastes and odors in drinking water. Corrective measures designed to reduce unpleasant tastes and odors include aeration or the addition of chlorine, chlorine dioxide, potassium permanganate, or activated carbon. See the "Taste and Odor Control" chapter in Volume I of this manual.

Odor is considered a quality factor affecting acceptability of drinking water (and foods prepared with it), tainting of fish and other aquatic organisms, and aesthetics of recreational waters. Most organic and some inorganic chemicals contribute to taste and odor. These chemicals may originate from municipal and industrial waste discharges, from natural sources (such as decomposition of vegetable matter), or from associated microbial activity.

Some substances, such as certain inorganic salts, produce taste without odor. Many taste sensations actually are odor sensations, even though the sensation is not noticed until the water is in the mouth.

Taste, like odor, is one of the chemical senses. Taste and odor are different in that odors are sensed high up in our nose and tastes are sensed on our tongue. Taste is simpler than odor because there may be only four true taste sensations: sour, sweet, salty, and bitter. Dissolved inorganic salts of copper, iron, manganese, potassium, sodium, and zinc can be detected by taste.

Operators must remember that a flavorless water is not the most desirable water. Distilled water is considered less pleasant to drink than a high-quality drinking water. The taste test must determine the taste intensity by the threshold odor test and evaluate the quality of the drinking water on the basis of desirability to consumers.

Apparatus Required

Sample bottles, glass-stoppered or with TFE-lined closures
Constant temperature bath
Odor flasks, 500 mL glass-stoppered Erlenmeyer flasks
Transfer and volumetric pipets or graduated cylinders, 200, 100, 50, and 25 mL
Measuring pipets, 10 mL, graduated in 0.1 mL increments
Thermometer, 0 to 110°C

Precautions

Use preliminary tests to select the persons to make taste or odor tests. Use only persons who want to participate in the test. Avoid distracting odors such as those caused by smoking, foods, soaps, perfumes, and shaving lotions. The testers should not have colds or allergies that affect odor response. Do not have the testers perform too many tests and allow frequent rests so the testers will not become tired and lose their sensitivity. Keep the room in which the tests are conducted free from distractions, drafts, and odors.

A panel of five or more testers is recommended for precise work. Do not allow the testers to prepare the samples or to know the dilution concentrations being evaluated. Familiarize testers with the procedure before they participate in a panel test. Present the most dilute sample first to avoid tiring the senses with a concentrated sample. Keep temperature of samples during test within 1°C of the specified temperature. Use opaque or darkly colored flasks to avoid biasing the results due to turbid or colored waters being tested.

Procedure

ODOR

1. Determine the approximate range of the threshold odor number by adding 200 mL, 50 mL, 12 mL, and 2.8 mL of sample to 500 mL glass-stoppered Erlenmeyer flasks containing odor-free water to make a total volume of 200 mL. Use a separate flask containing only odor-free water as a reference for comparison. Heat dilutions and reference to desired test temperature (usually 60°C or 140°F).
2. Shake flask containing odor-free water, remove stopper, and sniff vapors. Test sample containing the least amount of odor-bearing water in the same way. If an odor can be detected in this dilution, prepare more dilute samples.

   To prepare more dilute samples, prepare an intermediate dilution consisting of 20 mL sample diluted to 200 mL with odor-free water. Use this dilution for the threshold determination. Multiply the threshold odor number (TON) obtained by 10 to correct for the intermediate dilution.

   If an odor cannot be detected in the first dilution, repeat the above procedure using sample containing the next higher concentration of odor-bearing water and continue this process until odor is detected clearly.

3. Based on the results obtained in the preliminary test, prepare a set of dilutions using Table 21.3 as a guide. Prepare the five dilutions shown in the appropriate column and the three next most concentrated in the next column to the right in Table 21.3. For example, if odor was first noted in the flask containing the 50 mL sample in the preliminary test, prepare flasks containing 50, 35, 25, 17, 12, 8.3, 5.7, and 4.0 mL samples, each diluted to 200 mL with odor-free water. This procedure is necessary to challenge the range of sensitivities of the entire panel of testers.

TABLE 21.3 DILUTIONS FOR VARIOUS ODOR INTENSITIES

PRELIMINARY TEST			
Sample Volume in Which Odor First Noted			
200 mL	50 mL	12 mL	3.8 mL
FINAL TEST			
Volume in mL of Sample to be Diluted to 200 mL			
200	50	12	(Intermediate dilution)
140	35	8.3	
100	25	5.7	
70	17	4.0	
50	12	2.8	

Insert two or more blanks near the expected threshold, but avoid any repeated patterns. Do not let the testers know which dilutions are odorous and which are blanks. Instruct each tester to smell each flask in sequence, beginning with the least concentrated sample, until odor is detected with certainty.

4. Record observations by indicating whether odor is noted in each flask. For example,

mL Sample Diluted to 200 mL	12	0	17	25	0	35	50
Response	−	−	−	+	−	+	+

5. Calculate the threshold odor number (TON) as shown in the "Calculations" section on the following page.

FLAVOR THRESHOLD TEST

1. The flavor threshold test is used when the purpose is quantitative measurement of detectable taste. When odor is the predominant sensation, as in the case of chlorophenols, the threshold odor test is used.
2. Use the dilution and random blank system described for odor tests when preparing taste samples.
3. Present each dilution and blank to the tester in a clean 50 mL plastic container filled to the 30 mL level. Use high-quality clear plastic containers. Discard the plastic containers when finished. Do not use glass containers because the soap used to clean the glass could leave a residue that may affect the results.
4. *Standard Methods* recommends maintaining the sample presentation for taste tests at 59°F (15°C).
5. Present the series of samples to each tester. Pair each sample with a known blank.
6. Have each tester taste the sample by taking into the mouth whatever volume is comfortable, holding it in the mouth for several seconds, and discharging the sample without swallowing the water.

7. Have the tester compare the sample with the blank and record whether a taste or aftertaste is detectable in the sample.
8. Submit samples in an increasing order of concentration until the tester's taste threshold has been passed.
9. Calculate individual threshold and threshold of the panel as shown in the "Calculations" section on the following page.

FLAVOR RATING TEST

1. When the purpose of the test is to estimate the taste acceptability, use the flavor rating test procedure described below.
2. Samples for this test usually represent treated water ready for human consumption. If experimentally treated water is tested, be certain that the water is safe to drink (no pathogens and no toxic chemicals present).
3. Give testers thorough instructions and trial or orientation sessions followed by questions and discussions of procedures.
4. Select panel members on the basis of performance in trial sessions.
5. Place testers in separate locations to test samples.
6. Present samples at a temperature that testers find pleasant for drinking water. Maintain this temperature by the use of a water bath apparatus. A temperature of 59°F (15°C) is recommended. Do not let the test temperature exceed tap water temperatures that are customary at the time of the test. Specify the test temperature in reporting results.

(Taste and Odor)

7. Present each dilution and blank to the tester in a clean 50 mL plastic container filled to the 30 mL level. Use high-quality clear plastic containers. Discard the plastic containers when finished. Do not use glass containers because the soap used to clean the glass could leave a residue that may affect the results.

8. Each tester is presented with a list of nine statements about the water, ranging on a scale from very favorable to very unfavorable (Table 21.4). The tester's task is to select the statement that best expresses the tester's opinion. The scored rating is the scale number of the statement selected. The panel rating is the arithmetic mean (average) of the scale numbers of all testers.

9. Rating involves the following steps:
   a. Initial tasting of about half the sample by taking water into the mouth, holding it for several seconds, and discharging it without swallowing
   b. Forming an initial judgment on the rating scale
   c. A second tasting is made in the same manner as the first
   d. A final rating is made for the sample and the result is recorded on the appropriate data form
   e. Rinse mouth with taste- and odor-free water
   f. Rest 1 minute before repeating the above steps on the next sample

10. Independently randomize sample order for each tester. Allow at least 30 minutes rest between repeated rating sessions. Testers should not know the composition or source of the samples.

Calculations

FORMULAS

1. Odor

   The threshold odor number (TON) for an individual tester is calculated using the following formula:

$$TON = \frac{A + B}{A}$$

where:

A = mL sample

B = mL odor-free water

The threshold odor number for a group is presented as the geometric mean of the individual tester thresholds.

$$\text{Geometric Mean} = (X_1 \times X_2 \times X_3 \ldots X_n)^{1/n}$$

where:

X_1 = threshold odor number for tester number 1

X_2 = threshold odor number for tester number 2

X_n = threshold odor number for the nth tester

n = total number of testers

2. Flavor Threshold

   Calculate the individual tester's threshold taste number and the threshold taste number for a panel using the same formulas that are used for the threshold odor tests.

TABLE 21.4 ACTION TENDENCY RATING SCALE FOR FLAVOR RATING TEST

1. I would be very happy to drink this water as my everyday drinking water.
2. I would be happy to accept this water as my everyday drinking water.
3. I am sure that I could accept this water as my everyday drinking water.
4. I could accept this water as my everyday drinking water.
5. Maybe I could accept this water as my everyday drinking water.
6. I do not think I could accept this water as my everyday drinking water.
7. I could not accept this water as my everyday drinking water.
8. I could never drink this water.
9. I cannot stand this water in my mouth and I could never drink it.

(Taste and Odor)

3. Flavor Rating

Determine the taste rating for a water by calculating the arithmetic mean and *STANDARD DEVIATION*[3] of all ratings given for each sample.

$$\text{Arithmetic Mean, } \bar{X} = \frac{X_1 + X_2 + X_3 + \ldots X_n}{n}$$

where: X_1 = flavor rating for tester number 1

X_2 = flavor rating for tester number 2

X_n = flavor rating for the nth tester

n = total number of testers

$$\text{Standard Deviation} = \left[\frac{(X_1 - \bar{X})^2 + (X_2 - \bar{X})^2 + \ldots (X_n - \bar{X})^2}{n-1}\right]^{0.5}$$

$$\text{or} = \left[\frac{(X_1^2 + X_2^2 + \ldots X_n^2) - (X_1 + X_2 + \ldots X_n)^2/n}{n-1}\right]^{0.5}$$

If the distribution of the results is not reasonably symmetrical, use the median or geometric mean instead (see odor calculations).

EXAMPLE 1

Calculate the threshold odor number (TON) for a sample when the first detectable odor occurred when the 25 mL sample was diluted to 200 mL (175 mL of odor-free water was added to the 25 mL sample.)

Known		Unknown
A or Sample Size, mL	= 25 mL	TON
B or Odor-Free Water, mL	= 175 mL	

Calculate the threshold odor number (TON).

$$\text{TON} = \frac{A + B}{A}$$

$$= \frac{25 \text{ mL} + 175 \text{ mL}}{25 \text{ mL}}$$

$$= 8$$

EXAMPLE 2

Determine the geometric mean threshold odor number for a panel of five testers given the following results.

Known	Unknown
Tester 1, X_1 = 8	Geometric Mean Threshold Odor Number
Tester 2, X_2 = 6	
Tester 3, X_3 = 12	
Tester 4, X_4 = 8	
Tester 5, X_5 = 4	

Calculate the geometric mean.

$$\text{Geometric Mean TON} = (X_1 \times X_2 \times X_3 \times X_4 \times X_5)^{1/n}$$

$$= (8 \times 6 \times 12 \times 8 \times 4)^{1/5}$$

$$= (18{,}432)^{0.2}$$

$$= 7.1$$

EXAMPLE 3

Calculate the flavor threshold number (FTN) for a sample when the first detectable flavor occurred when the 50 mL sample was diluted to 200 mL (150 mL of taste-free water was added to the 50 mL sample).

Known		Unknown
A or Sample Size, mL	= 50 mL	FTN
B or Flavor-Free Water, mL	= 150 mL	

Calculate the flavor threshold number (FTN).

$$\text{FTN} = \frac{A + B}{A}$$

$$= \frac{50 \text{ mL} + 150 \text{ mL}}{50 \text{ mL}}$$

$$= 4$$

EXAMPLE 4

Determine the flavor rating for a water by calculating the arithmetic mean and standard deviation for the panel ratings given below.

Known	Unknown
Tester 1, X_1 = 2	1. Arithmetic Mean, $\bar{X}$
Tester 2, X_2 = 5	2. Standard Deviation, S
Tester 3, X_3 = 3	
Tester 4, X_4 = 6	
Tester 5, X_5 = 2	
Tester 6, X_6 = 6	

1. Calculate the arithmetic mean, $\bar{X}$, flavor rating.

$$\text{Arithmetic Mean, } \bar{X}, \text{ Flavor Rating} = \frac{X_1 + X_2 + X_3 + X_4 + X_5 + X_6}{n}$$

$$= \frac{2 + 5 + 3 + 6 + 2 + 6}{6}$$

$$= \frac{24}{6}$$

$$= 4$$

3. *Standard Deviation.* A measure of the spread or dispersion of data.

(Taste and Odor—Trihalomethanes)

2. Calculate the standard deviation, S, of the flavor rating.

$$\text{Standard Deviation, S} = \left[\frac{(X_1-\bar{X})^2+(X_2-\bar{X})^2+(X_3-\bar{X})^2+(X_4-\bar{X})^2+(X_5-\bar{X})^2+(X_6-\bar{X})^2}{n-1}\right]^{0.5}$$

$$= \left[\frac{(2-4)^2+(5-4)^2+(3-4)^2+(6-4)^2+(2-4)^2+(6-4)^2}{6-1}\right]^{0.5}$$

$$= \left[\frac{(-2)^2+(1)^2+(-1)^2+(2)^2+(-2)^2+(2)^2}{5}\right]^{0.5}$$

$$= \left[\frac{4+1+1+4+4+4}{5}\right]^{0.5}$$

$$= \left[\frac{18}{5}\right]^{0.5}$$

$$= (3.6)^{0.5}$$

$$= 1.9$$

or

$$\text{Standard Deviation, S} = \left[\frac{(X_1^2+X_2^2+X_3^2+X_4^2+X_5^2+X_6^2)-(X_1+X_2+X_3+X_4+X_5+X_6)^2/n}{n-1}\right]^{0.5}$$

$$= \left[\frac{(2^2+5^2+3^2+6^2+2^2+6^2)-(2+5+3+6+2+6)^2/6}{6-1}\right]^{0.5}$$

$$= \left[\frac{(4+25+9+36+4+36)-(24)^2/6}{5}\right]^{0.5}$$

$$= \left[\frac{114-96}{5}\right]^{0.5}$$

$$= \left[\frac{18}{5}\right]^{0.5}$$

$$= (3.6)^{0.5}$$

$$= 1.9$$

Reference

Odor: *Standard Methods*, 22nd Edition, Method 2150 (B), page 2-16. Taste: Method 2160 (B), page 2-20; and Method 2160 (C), page 2-22.

QUESTIONS

Please write your answers to the following questions and compare them with those on page 533.

21.1W List the items that can cause tastes and odors in drinking water.

21.1X Calculate the threshold odor number (TON) for a sample when the first detectable odor occurred when the 12 mL sample was diluted to 200 mL (188 mL of odor-free water was added to the 12 mL sample).

16. Trihalomethanes

Discussion

The trihalomethanes (THMs) are members of the family of organohalogen compounds, which are named as derivatives of methane. Current analytical chemistry applied to drinking water has thus far detected chloroform, bromodichloromethane, dibromochloromethane, bromoform, and dichloroiodomethane.

The principal source of chloroform and other trihalomethanes in drinking water is the chemical interaction of chlorine added for disinfection and other purposes with the commonly present natural humic substances and other precursors produced either by normal organic decomposition or by the metabolism of aquatic organisms. Since these natural organic precursors are more commonly found in surface water, water taken from a surface source is more likely to produce high THM levels than most groundwaters.

Generally, the THM-producing reaction is:

Chlorine + Precursors = Chloroform + Other THMs

Chloroform is the most common THM found in drinking water and it is usually the highest concentration of all the THMs. The presence in drinking water of chloroform and other THMs and synthetic organic chemicals may have an adverse effect on the health of consumers; therefore, human exposure to these chemicals should be reduced.

Reference

For materials and procedures, see *Standard Methods*, 22nd Edition, Method 5710, page 5-64. Testing for THMs requires analysis by gas chromatography. Most treatment plants contract with an external laboratory for this type of analysis.

17. Total Dissolved Solids

Discussion

Total dissolved solids (TDS) refer to material that passes through a standard glass-fiber filter disk and remains after evaporation at 180°C. The amount of dissolved solids present in water is a consideration in its suitability for domestic use. In general, waters with a TDS content of less than 50 mg/L are most desirable for such purposes. The higher the TDS concentration, the greater the likelihood of tastes and odors and scaling problems. As TDS increase, the number of times the water can be recycled and reclaimed before requiring demineralization decreases. In potable waters, TDS consist mainly of inorganic salts, small amounts of organic matter, and dissolved gases.[4]

What Is Tested?

Sample	Common Range, mg/L
Raw and Treated Surface Waters	20 to 700
Groundwater	100 to 1,000

Apparatus Required

Glass-fiber filter disks (Millipore AP40; or Gelman Type A/E)
Suction flask, 500 mL
Filter holder or Gooch crucible adapter
Gooch crucibles (25 mL if 2.2 cm filter used)
Evaporating dishes, 100 mL (high-silica glass)
Drying oven, 180°C
Steam bath
Vacuum source
Desiccator
Analytical balance
Muffle furnace, 550°C

Procedure

PREPARATION OF DISH

1. Ignite a clean evaporating dish at 550±50°C for 1 hour in muffle furnace.
2. Cool in desiccator, then weigh and record weight. Store in desiccator until needed.

PREPARATION OF GLASS-FIBER FILTER DISK

1. Place the disk on the filter apparatus or insert into the bottom of a suitable Gooch crucible. While vacuum is applied, wash the filter disk with three successive 20 mL volumes of distilled water. Continue the suction to remove all traces of water from the disk and discard the washings.

SAMPLE ANALYSIS

1. Shake the sample vigorously and transfer 100 to 150 mL to the funnel or Gooch crucible by means of a 150 mL graduated cylinder.

2. Filter the sample through the glass-fiber filter and continue to apply vacuum for about 3 minutes after filtration is complete to remove as much water as possible.
3. Transfer 100 mL of the filtrate to the weighed evaporating dish and evaporate to dryness on a steam bath.
4. Dry the evaporated sample for at least 1 hour at 180°C. Cool in desiccator and weigh. Repeat drying cycle until constant weight is obtained or until weight loss is less than 0.5 mg.

Example

Results from weighings were:

Clean Dish = 47.0028 grams (47,002.8 mg)
Dissolved Residue + Dish = 47.0453 grams (47,045.3 mg)
Sample Volume = 100 mL

4. Reference. *Chemistry for Environmental Engineering and Science*, Fifth Edition, 2003, by Clair N. Sawyer, Perry L. McCarty, and Gene F. Parkin. Published by the McGraw-Hill Companies. ISBN 978-0-07-248066-5.

(Total Dissolved Solids)

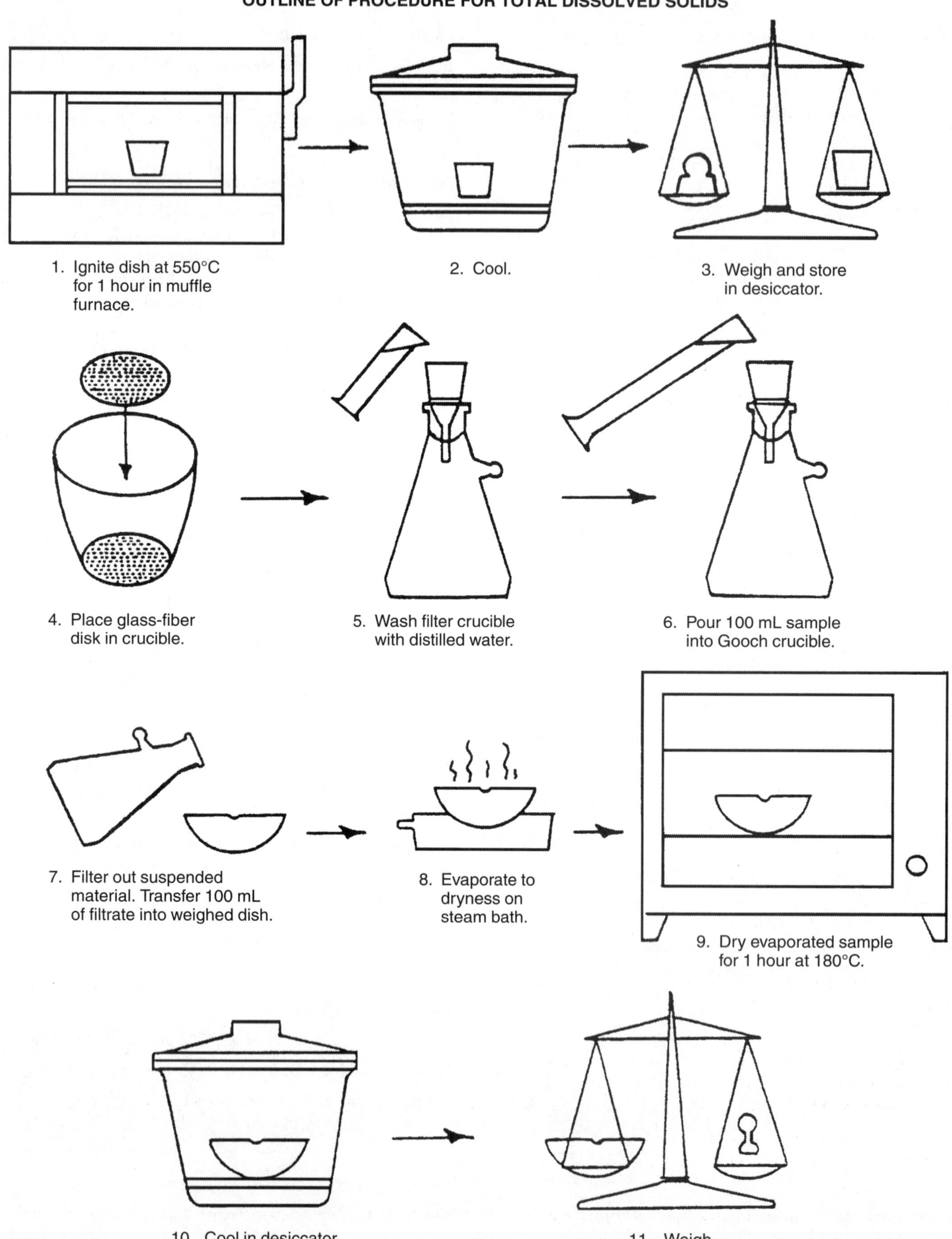

(Total Dissolved Solids)

Calculations

1. Total Dissolved Solids, mg/L $= \dfrac{(A - B) \times 1{,}000}{\text{mL Sample Volume}}$

where: A = weight of dish and dissolved material in milligrams (mg)

B = weight of clean dish in milligrams (mg)

2. From example:

$$\text{Total Dissolved Solids, mg/L} = \frac{(A - B) \times 1{,}000}{\text{mL Sample Volume}}$$

$$= \frac{(47{,}045.3 \text{ mg} - 47{,}002.8 \text{ mg})(1{,}000 \text{ mL/L})}{100 \text{ mL}}$$

$$= 425 \text{ mg/L}$$

Comments

Because excessive residue in the evaporating dish may form a water-entrapping crust, use a sample that yields no more than 200 mg of residue.

Reference

See *Standard Methods*, 22nd Edition, Method 2540 (C), page 2-65.

QUESTIONS

Please write your answers to the following questions and compare them with those on page 533.

21.1Y How are trihalomethanes produced?

21.1Z What are total dissolved solids (TDS)?

END OF LESSON 2 OF 2 LESSONS

on

ADVANCED LABORATORY PROCEDURES

Please answer the discussion and review questions next.

DISCUSSION AND REVIEW QUESTIONS

Chapter 21. ADVANCED LABORATORY PROCEDURES

(Lesson 2 of 2 Lessons)

Please write your answers to the following questions to determine how well you understand the material in the lesson. The question numbering continues from Lesson 1.

9. Why is iron undesirable in a domestic water supply?
10. What precautions must be exercised when collecting samples to be analyzed for iron?
11. How would you obtain the manganese concentration in a sample by using a spectrophotometer if turbidity or color is interfering with the results?
12. What is the purpose of the Marble Test?
13. Why is the presence of certain metals in drinking water of serious concern?
14. How would you interpret the results of lab tests that indicate high levels of nitrate in a raw water sample?
15. When performing the nitrate determination, why should caution be exercised when using cadmium and what precautions should be used?
16. How would you interpret changes (away from normal) in conductivity in water?
17. Why are sulfate ions of concern in water supplies?
18. How would you attempt to reduce unpleasant tastes and odors in drinking water?
19. Why should exposure to trihalomethanes (THMs) be reduced?
20. Why is the amount of dissolved solids present in water a consideration in its suitability for domestic use?

SUGGESTED ANSWERS

Chapter 21. ADVANCED LABORATORY PROCEDURES

ANSWERS TO QUESTIONS IN LESSON 1

Answers to questions on pages 496 and 497.

21.0A The intensity of a blue color is measured when measuring the concentration of phosphorus in water.

21.0B The scale on spectrophotometers is usually graduated in two ways:

1. In units of percent transmittance (%T), an arithmetic scale with units graded from 0 to 100%
2. In units of absorbance (A), a logarithmic scale of nonequal divisions graduated from 0.0 to 2.0

21.0C If an absorbance reading was 0.60, the unknown concentration would be 0.70 mg/L.

Answers to questions on page 500.

21.1A Yes, the quality of water in any lake, reservoir, or stream has a very direct effect on the abundance and types of aquatic organisms found.

21.1B Calcium in the form of lime or calcium hydroxide may be used to soften water or to control corrosion through pH adjustment.

21.1C Titrate a sample for calcium immediately after adding sodium hydroxide (NaOH) solution.

21.1D Chloride concentrations above 250 mg/L are objectionable to many people due to a salty taste.

Answers to questions on page 501.

21.1E The most common colors that occur in raw water are yellow and brown.

21.1F True color is normally removed or at least decreased by coagulation and chlorination or ozonation.

21.1G Stock color standards should be protected against evaporation and contamination when not in use.

Answers to questions on page 508.

21.1H The presence of dissolved oxygen (DO) in water can contribute to corrosion of piping systems.

21.1I The common range of fluoride in fluoridated drinking water is 0.8 to 1.2 mg/L.

ANSWERS TO QUESTIONS IN LESSON 2

Answers to questions on page 514.

21.1J Problems that may be caused by iron in a domestic water supply include staining of laundry, concrete, and porcelain. A bitter, astringent taste can be detected by some people at levels above 0.3 mg/L.

21.1K All glassware must be acid washed when analyzing samples for iron to remove deposits of iron oxide, which could give false results.

21.1L Manganese occurs both in suspension and as a soluble complex in surface waters.

21.1M If the manganese concentration cannot be tested immediately, acidify the sample with nitric acid to a pH less than 2.

Answers to questions on page 516.

21.1N Temperature is important in the Marble Test because the solubility of calcium carbonate varies with temperature. Therefore, the test must be performed immediately after the sample is collected and as rapidly as possible.

21.1O Langelier Index ≅ Initial pH – Final pH

≅ 8.9 – 8.6

≅ 0.3

Since the Langelier Index is positive, the water is supersaturated with calcium carbonate and not considered corrosive.

21.1P The concentrations of most metals in water are determined by using atomic absorption spectroscopy or colorimetric methods.

Answers to questions on page 520.

21.1Q In the nitrate test, all nitrate is reduced to nitrite and then measured colorimetrically.

21.1R Removal of turbidity interfering with nitrate analyses can be accomplished by one of the following methods to remove suspended matter that can clog the reduction column.

1. Filter sample through a glass-fiber filter or a 0.45-micron pore size filter as long as the pH is less than 8.
2. Add 1 mL zinc solution to 100 mL of sample and mix thoroughly. Add enough sodium hydroxide solution to obtain a pH of 10.5. Let treated sample stand a few minutes to allow the heavy flocculant precipitate to settle. Clarify by filtering through a glass-fiber filter.

21.1S The pH of natural waters is controlled by the relative amounts of carbon dioxide, bicarbonate, and carbonate ions.

Answers to questions on page 523.

21.1T Specific conductance or conductivity is a numerical expression (expressed in micromhos per centimeter) of the ability of a water to conduct an electric current. This number depends on the total concentration of the minerals dissolved in the sample (TDS) and the temperature.

21.1U Sulfate ions are of importance in water supplies because of the tendency of appreciable amounts to form hard scales in boilers and heat exchangers.

21.1V A 50 mL sample from a well produced a turbidity reading of 40 TU using a nephelometer. What was the sulfate concentration in mg/L?

Known		Unknown
Sample Size, mL	= 50 mL	Sulfate, mg/L
Turbidity, TU	= 40 TU	

1. Determine the sulfate concentration from the graph.

Sulfate Concentration, mg/L = 15 mg/L

2. Calculate the sulfate concentration in mg/L.

$$\text{Sulfate, mg/L} = \frac{(\text{Graph Sulfate, mg/L})(100\text{ mL})}{\text{Sample Size, mL}}$$

$$= \frac{(15\text{ mg/L})(100\text{ mL})}{50\text{ mL}}$$

$$= 30\text{ mg/L}$$

Answers to questions on page 528.

21.1W Tastes and odors can be caused in drinking water by algae, *Actinomycetes*, bacteria, decaying vegetation, metals, and pollutants (most organic chemicals and some inorganic chemicals). Dissolved inorganic salts of copper, iron, manganese, potassium, sodium, and zinc can be detected by taste.

21.1X Calculate the threshold odor number (TON) for a sample when the first detectable odor occurred when the 12 mL sample was diluted to 200 mL (188 mL of odor-free water was added to the 12 mL sample).

Known		Unknown
A or Sample Size, mL	= 12 mL	TON
B or Odor-Free Water, mL	= 188 mL	

Calculate the threshold odor number (TON).

$$\text{TON} = \frac{A + B}{A}$$

$$= \frac{12\text{ mL} + 188\text{ mL}}{12\text{ mL}}$$

$$= 17$$

Answers to questions on page 531.

21.1Y The principal source of chloroform and other trihalomethanes in drinking water is the chemical interaction of chlorine added for disinfection and other purposes with the commonly present natural humic substances and other precursors produced either by normal organic decomposition or by the metabolism of aquatic organisms.

21.1Z Total dissolved solids (TDS) refers to material that passes through a standard glass-fiber filter disk and remains after evaporation at 180°C.

CHAPTER 22

DRINKING WATER REGULATIONS

by

Tim Gannon

Revised by

Jim Sequeira

and

Ken Kerri

NOTICE

EPA drinking water rules and regulations change regularly and some state rules and regulations may be stricter than their federal equivalents. Obtain current information about rules and regulations that apply to your water utility from your state water agency. For information about federal drinking water regulations, visit the EPA online at water.epa.gov/drink/hotline/ or call the Safe Drinking Water Hotline at (800) 426-4791.

TABLE OF CONTENTS

Chapter 22. DRINKING WATER REGULATIONS

LEARNING OBJECTIVES

Chapter 22. DRINKING WATER REGULATIONS

Following completion of Chapter 22, you should be able to:

1. Identify the two basic types of water systems.
2. List the types of primary contaminants.
3. Explain the Total Coliform Rule.
4. Explain the Surface Water Treatment Rules.
5. Describe the Primary Drinking Water Standards.
6. List the secondary contaminants.
7. Develop and conduct a sampling program.
8. Record and report results.
9. Comply with notification requirements.
10. Prepare a Consumer Confidence Report (CCR).

WORDS

Chapter 22. DRINKING WATER REGULATIONS

ACUTE HEALTH EFFECT ACUTE HEALTH EFFECT

An adverse effect on a human or animal body, with symptoms developing rapidly.

CT VALUE CT VALUE

Residual concentration of a given disinfectant in mg/L times the disinfectant's contact time in minutes.

CHECK SAMPLING CHECK SAMPLING

Whenever an initial or routine sample analysis indicates that a Maximum Contaminant Level (MCL) has been exceeded, check sampling is required to confirm the routine sampling results. Check sampling is in addition to the routine sampling program.

CHLORAMINATION (KLOR-uh-min-NAY-shun) CHLORAMINATION

The application of chlorine and ammonia to water to form chloramines for the purpose of disinfection.

CHRONIC HEALTH EFFECT CHRONIC HEALTH EFFECT

An adverse effect on a human or animal body with symptoms that develop slowly over a long period of time or that recur frequently.

CONSUMER CONFIDENCE REPORT (CCR) CONSUMER CONFIDENCE REPORT (CCR)

An annual report prepared by a water utility to communicate with its consumers. The report provides consumers with information on the source and quality of their drinking water and is an opportunity for positive communication with consumers.

ENTERIC ENTERIC

Of intestinal origin, especially applied to wastes or bacteria.

HETEROTROPHIC (HET-er-o-TROF-ick) HETEROTROPHIC

Describes organisms that use organic matter for energy and growth. Animals, fungi, and most bacteria are heterotrophs.

INFORMATION COLLECTION RULE (ICR) INFORMATION COLLECTION RULE (ICR)

The Information Collection Rule (ICR) was promulgated on May 14, 1996 and approved by the Director of the Federal Register on June 18, 1996. It was to remain effective until December 31, 2000. The rule specified requirements for monitoring microbial contaminants and disinfection byproducts (DBPs) by large public water systems (PWSs). It required large PWSs to conduct either bench- or pilot-scale testing of advanced treatment techniques. The data reported under the ICR were used by EPA to learn more about the occurrence of microbial contamination and disinfection byproducts, the health risks posed, appropriate analytical methods, and effective forms of treatment. The ICR data form the scientific basis for EPA's development of the Enhanced Surface Water Treatment Rule and the Disinfectants and Disinfection Byproducts Rule (DBPR).

INITIAL SAMPLING INITIAL SAMPLING

The very first sampling conducted under the Safe Drinking Water Act (SDWA) for each of the applicable contaminant categories.

MBAS MBAS

Methylene Blue Active Substance. Another name for surfactants or surface active agents. The determination of surfactants is accomplished by measuring the color change in a standard solution of methylene blue dye.

MCL MCL

Maximum Contaminant Level. The largest allowable amount. MCLs for various water quality indicators are specified in the National Primary Drinking Water Regulations (NPDWR).

MCLG MCLG

Maximum Contaminant Level Goal. MCLGs are health goals based entirely on health effects. They are a preliminary standard set but not enforced by EPA. MCLs consider health effects, but also take into consideration the feasibility and cost of analysis and treatment of the regulated MCL. Although often less stringent than the corresponding MCLG, the MCL is set to protect health.

MRDL MRDL

Maximum Residual Disinfectant Level. The highest level of a disinfectant allowed in drinking water without causing an unacceptable possibility of adverse health effects.

NTU NTU

Nephelometric Turbidity Units. See TURBIDITY UNITS (TU).

pCi/L pCi/L

picoCurie per liter. A picoCurie is a measure of radioactivity. One picoCurie of radioactivity is equivalent to 0.037 nuclear disintegrations per second.

POTABLE (POE-tuh-bull) WATER POTABLE WATER

Water that does not contain objectionable pollution, contamination, minerals, or infective agents and is considered satisfactory for drinking.

PRIMACY PRIMACY

Under the Safe Drinking Water Act (SDWA), primacy is the responsibility for ensuring that a law is implemented, and the authority to enforce a law and related regulations (40 CFR 142.2). A primacy agency has the primary responsibility for administrating and enforcing regulations.

ROUTINE SAMPLING ROUTINE SAMPLING

Sampling repeated on a regular basis.

SANITARY SURVEY SANITARY SURVEY

A detailed evaluation or inspection of a source of water supply and all conveyances, storage, treatment, and distribution facilities to ensure protection of the water supply from all pollution sources.

SURFACTANT (sir-FAC-tent) SURFACTANT

Abbreviation for surface-active agent. The active agent in detergents that possesses a high cleaning ability.

THM THM

See TRIHALOMETHANES (THMs).

THRESHOLD ODOR NUMBER (TON) THRESHOLD ODOR NUMBER (TON)

The greatest dilution of a sample with odor-free water that still yields a just-detectable odor.

TRIHALOMETHANES (THMs) (tri-HAL-o-METH-hanes) TRIHALOMETHANES (THMs)

Derivatives of methane, CH_4, in which three halogen atoms (chlorine or bromine) are substituted for three of the hydrogen atoms. Often formed during chlorination by reactions with natural organic materials in the water. The resulting compounds (THMs) are suspected of causing cancer.

TURBIDITY (ter-BID-it-tee) UNITS (TU) TURBIDITY UNITS (TU)

Turbidity units are a measure of the cloudiness of water. If measured by a nephelometric (deflected light) instrumental procedure, turbidity units are expressed in nephelometric turbidity units (NTU) or simply TU. Those turbidity units obtained by visual methods are expressed in Jackson turbidity units (JTU), which are a measure of the cloudiness of water; they are used to indicate the clarity of water. There is no real connection between NTUs and JTUs. The Jackson turbidimeter is a visual method and the nephelometer is an instrumental method based on deflected light.

CHAPTER 22. DRINKING WATER REGULATIONS

(Lesson 1 of 2 Lessons)

All water treatment plant operators need to be thoroughly familiar with the state and federal laws and standards that apply to domestic water supply systems. These regulations are the goals and guideposts for the water supply industry. Their purpose is to ensure the uniform delivery of safe and aesthetically pleasing drinking water to the public.

This chapter will introduce the major drinking water regulations and explain the monitoring and reporting requirements. For more detailed information, you will need to refer to current copies of your state's regulations and the most recent federal standards. These publications should be made readily available to all operators since operators will only know whether their system is in compliance by comparing monitoring test data with the actual current regulations. For the most current and accurate information, visit the US Environmental Protection Agency (EPA) website at epa.gov.

22.0 HISTORY OF DRINKING WATER LAWS AND STANDARDS

Up until shortly after 1900, there were no standards for drinking water. The first standards, established in 1914, were designed in large part to control waterborne bacteria and viruses that cause diseases such as cholera, typhoid, and dysentery. These new standards were overwhelmingly successful in curbing the spread of such diseases. However, with time and technology, other types of contaminants, this time chemicals, again stirred public concern. In 1962, the US Public Health Service (the forerunner of the EPA) revised the national drinking water standards to include limits on selected organic chemicals.

In 1970, the EPA was created to protect human health and safeguard our natural environment. Among its many responsibilities, EPA sets the standards for drinking water quality and oversees the states, localities, and water suppliers who implement those standards.

In 1972, a series of reports detailing organic contamination in the drinking water supplied to the residents of New Orleans from the Mississippi River triggered profound changes in drinking water regulations. A study by the Environmental Defense Fund found that people drinking treated Mississippi River water in New Orleans had a greater chance of developing certain cancers than those in neighboring areas whose drinking water came from groundwater sources. Heightened public awareness and concern regarding cancer became major factors behind the push for legislative action on the issue of drinking water contamination. The finding of suspected carcinogens in drinking water established a widespread sense of urgency that led to the passage and signing into law of the Safe Drinking Water Act (SDWA) in December 1974.

The SDWA gave the federal government, through the EPA, the authority to:

- Set national standards regulating the levels of contaminants in drinking water
- Require public water systems to monitor and report the levels of their identified contaminants
- Establish uniform guidelines specifying the acceptable treatment technologies for removing unsafe levels of pollutants from drinking water

The SDWA is the main federal law that ensures the quality of Americans' drinking water. Under this law the EPA sets legal limits on the levels of certain contaminants in drinking water. The legal limits reflect both the levels that safeguard human health and the degree to which water systems must protect drinking water using the best available technology.

In addition to prescribing these legal limits, EPA also develops rules for setting water-testing schedules and methods that water systems must follow. The rules also list acceptable standards or practices to meet the legal limits. EPA's standards include acceptable techniques for treating contaminated water. The SDWA gives individual states the opportunity to set and enforce their own drinking water standards if the standards are at least as strong as EPA's national standards. Most states and territories directly oversee the water systems within their borders.

While the SDWA gave EPA responsibility for promulgating (passing into law) drinking water regulations, it gave state regulatory agencies the opportunity to assume primary responsibility for enforcing those regulations. This responsibility is referred to as *PRIMACY*.[1] All states, territories, and the Navajo Nation, except Wyoming and the District of Columbia, currently have primary enforcement authority for implementing SDWA regulations. EPA is responsible or implementing the SDWA on all other tribal lands.

1. *Primacy.* Under the Safe Drinking Water Act (SDWA), primacy is the responsibility for ensuring that a law is implemented, and the authority to enforce a law and related regulations (40 CFR 142.2). A primacy agency has the primary responsibility for administrating and enforcing regulations.

Implementation of the SDWA has greatly improved basic drinking water purity across the nation. However, EPA surveys of surface water and groundwater indicate the presence of synthetic organic chemicals in 20 percent of the nation's water sources, with a small percentage at levels of concern. In addition, research studies suggest that some naturally occurring contaminants may pose even greater risks to human health than the synthetic contaminants.

With the advent of increasingly sensitive analytical techniques, it is easier to detect the occurrence of contaminants in water sources. One such occurrence developed in 2008. Research conducted by scientists and reports by the media raised concerns about the presence of pharmaceuticals and personal care products (PPCPs) in both raw and *POTABLE*[2] drinking water. Although PPCPs have been present for more than thirty years, these reports prompted EPA and the states to increase their focus and oversight of PPCPs in drinking water.

In the years following passage of the SDWA, Congress felt that EPA was slow to regulate contaminants and states were lax in enforcing the law. Consequently, in 1986 Congress enacted amendments designed to strengthen the 1974 SDWA. These amendments set deadlines for the establishment of maximum contaminant levels, placed greater emphasis on enforcement, authorized penalties for tampering with drinking water supplies, mandated the complete elimination of lead from drinking water, and placed considerable emphasis on the protection of underground drinking water sources.

The 1986 SDWA amendments set up a timetable under which EPA was required to develop primary standards for 83 contaminants. Other major provisions of the 1986 SDWA amendments required EPA to do the following:

1. Define an approved treatment technique for each regulated contaminant
2. Specify criteria for filtration of surface water supplies
3. Specify criteria for disinfection of surface and groundwater supplies
4. Prohibit the use of lead products in materials used to convey drinking water

To comply with the provisions of the 1986 SDWA amendments, the EPA, the states, and the water supply industry undertook significant new programs to clean up the country's water supplies.

In 1996, the President signed new SDWA amendments into law as Public Law 104–182. These amendments made sweeping changes to the existing SDWA, created several new programs, and included a total authorization of more than $12 billion in federal funds for various drinking water programs and activities from fiscal year (FY) 1997 through FY 2003.

Topics covered in the amendments include arsenic research; financial assistance for water infrastructure; support and assistance for watersheds; assistance to colonias (low-income communities located along the US-Mexico border); backwash water recycling; bottled water; capacity development (technical, managerial, and financial) support for small water systems; conditions that qualify water systems for exemptions from regulations; consumer awareness; contaminant selection and standard-setting authority; definitions of public water system, community water system, and noncommunity water system; disinfectants and disinfection byproducts (DBPs); drinking water studies and research; effective date of regulations; enforcement; environmental finance centers and capacity clearinghouse; estrogenic substances screening program; groundwater disinfection; groundwater protection programs; lead plumbing and pipes; monitoring and information gathering; monitoring relief; monitoring for unregulated contaminants; occurrence of contaminants in drinking water database; operator certification; primacy; public notification; drinking water regulations for radon; review of National Primary Drinking Water Regulations (NPDWRs); risk assessment application to establishing NPDWRs; small systems (technical assistance, treatment technology, variances); source water quality assessment and petition programs; state revolving loan fund; authorization to promulgate an NPDWR for sulfate; Surface Water Treatment Rule (SWTR) compliance; variance treatment technologies; water conservation; and waterborne disease study and training. For additional information and details, see "Overview of the Safe Drinking Water Act Amendments of 1996," by Frederick W. Pontius, *Journal American Water Works Association*, October 1996, pages 22–33.

In 1998, two major regulations were signed into federal law: the Disinfectants and Disinfection Byproducts Rule (DBPR) (see Section 22.23 for details) and the Interim Enhanced Surface Water Treatment Rule (IESWTR) (see Section 22.2226 for details). The DBPR was developed to protect the public from harmful concentrations of disinfectants and from trihalomethanes (THMs), which could form when DBPs combine with organic matter in drinking water. The goal of the IESWTR is to reduce the occurrence of *Cryptosporidium* and other pathogens in drinking water. This rule was developed to ensure that protection against microbial contaminants is not lowered as water systems comply with the DBPR.

In 2010, EPA announced a new approach to implementing the SDWA. This new approach involves four key elements:

1. Addressing contaminants as a group rather than one at a time to be more cost effective in regulating drinking water protection
2. Promoting the development of new drinking water treatment technologies to address health risks
3. Using the authority of multiple statutes to help protect drinking water
4. Partnering with states to share data collected from monitoring drinking water systems

Operators are urged to develop close working relationships with their local regulatory agencies to keep themselves informed of the frequent changes in regulations and requirements.

2. *Potable* (POE-tuh-bull) *Water.* Water that does not contain objectionable pollution, contamination, minerals, or infective agents and is considered satisfactory for drinking.

QUESTIONS

Please write your answers to the following questions and compare them with those on page 605.

22.0A What were the first drinking water standards designed to control?

22.0B Why is it important for operators to establish good working relationships with local regulatory agencies?

22.1 HOW EPA DEVELOPS DRINKING WATER STANDARDS

The process by which EPA sets drinking water standards is both long and complicated. It involves deciding which contaminants may endanger public health, conducting studies of the effects of these contaminants, defining the maximum safe level of each contaminant, estimating the costs and benefits of regulation, proposing a standard or limit, listening to and evaluating public reactions to the proposed standard, revising the standard (if needed), proposing the standard in final form, seeking and evaluating additional public input, and finally, publishing the standard. The entire process by which a standard is proposed and promulgated is governed by strict procedural guidelines. It often takes three or more years to produce a standard. The remainder of this section describes the general types of contaminants that are considered public health threats and how EPA decides which contaminants to regulate first, how it approaches the task of regulating contaminants, and how it defines a standard.

22.10 Types of Contaminants

Five types of primary contaminants are considered to be of public health importance:

1. *INORGANIC CONTAMINANTS*, such as lead and mercury
2. *ORGANIC CONTAMINANTS*, which include pesticides, herbicides, THMs, solvents, and other synthetic organic compounds
3. *TURBIDITY*, such as small particles suspended in water that interfere with light penetration and disinfection
4. *MICROBIAL CONTAMINANTS*, such as bacteria, viruses, and protozoa
5. *RADIOLOGICAL CONTAMINANTS*, which include natural and manmade sources of radiation

22.11 Identifying Contaminants to Be Regulated

The 1986 SDWA required EPA to establish a priority list of contaminants that may have adverse health effects and may require regulation. The 1996 SDWA amendments retained the list of 83 contaminants developed earlier but revised the process by which contaminants are selected for regulation. Before selecting a contaminant for regulation, the EPA must now consult with the scientific community, solicit public comments, and demonstrate that the contaminant actually occurs in public water systems. For this latter purpose, the 1996 amendments required EPA to establish an occurrence database by August 1999. The database contains information on regulated and unregulated contaminants and the information is available to the public.

To regulate a substance, EPA must demonstrate that the contaminant meets three criteria:

1. The contaminant has an adverse effect on human health
2. It occurs, or is likely to occur, in public water systems at a frequency and concentration of significance to public health
3. Regulation of the contaminant offers a meaningful opportunity to reduce health risks for people served by public water systems

In addition to meeting these three criteria, EPA must also weigh the relative health risks of various contaminants being considered for regulation, the risk reduction that regulation would accomplish, and the costs of implementing the regulations.

Working on a 5-year cycle, the EPA is required to select at least five contaminants from its Drinking Water Contaminant Candidate List (DWCCL) and decide whether there is sufficient reason to regulate them. The first DWCCL included 50 chemicals and 10 microbial contaminants. In June 2002, the EPA announced a preliminary finding that their research revealed no need to regulate ten of the listed contaminants. Based on public comments and continued research and evaluation, EPA will make a final determination about regulating any of these contaminants and will begin to develop proposed regulations as necessary.

22.12 Unregulated Contaminants

EPA uses the Unregulated Contaminant Monitoring (UCM) program to collect data for contaminants suspected to be present in drinking water, but that do not have health-based standards set under the SDWA. Every 5 years, EPA reviews the list of contaminants, largely based on the Contaminant Candidate List. The SDWA Amendments of 1996 provide for the following:

- Monitoring no more than 30 contaminants per 5-year cycle
- Monitoring only a representative sample of public water systems serving fewer than 10,000 people
- Storing analytical results in a National Contaminant Occurrence Database (NCOD)

The UCM program progressed in several stages. Currently, EPA manages the program directly as specified in the Unregulated Contaminant Monitoring Regulation (UCMR). EPA has established a timeline and plans for monitoring unregulated contaminants through 2015.

Under the UCMR, community water systems and nontransient noncommunity water systems serving more than 10,000 people are required to monitor for unregulated contaminants; smaller systems may also have to conduct monitoring if required by the state. Transient water systems are not affected by the UCMR.

Always keep in mind that EPA's regulatory process is basically just that—a process. No list is permanent; regulations change frequently as new information becomes available. Water treatment plant operators must continually seek opportunities to learn about the regulations that affect the water industry. Also, operators should feel free to express their concerns regarding the regulations and the regulatory process to EPA. For additional information on EPA's Unregulated Contaminant Monitoring (UCM) program, visit EPA's website at water.epa.gov/lawsregs/rulesregs/sdwa/ucmr/index.cfm.

QUESTIONS

Please write your answers to the following questions and compare them with those on page 605.

22.1A List the five types of primary contaminants considered to be of public health importance.

22.1B What criteria must be met before EPA selects a contaminant for regulation?

22.1C Why are water systems required to monitor some unregulated contaminants?

22.13 Newer and Proposed Regulations

To help operators and managers quickly learn the basics of newer and proposed drinking water regulations, the US EPA has developed a "Compilation of Quick Reference Guides." These guides provide the following information: an overview of the rules, public health-related benefits, critical deadlines and requirements, compliance determination, and monitoring requirements.

The purpose of this section is to provide you with the important information and requirements of these newer and proposed regulations. Additional information and details are provided in sections in the remainder of this chapter.

If you would like to know how these rules and regulations apply to your water supply system, contact your local drinking water regulator or visit the US EPA website at water.epa.gov/drink/standardsriskmanagement.cfm. Also, a phone call to EPA's Safe Drinking Water Hotline at (800) 426-4791 can be very helpful.

The information in this section is based on EPA's "Compilation of Quick Reference Guides" at the time this manual was revised. Each portion of this section also refers to more details in other sections of this chapter.

22.130 Arsenic Rule

The purpose of the Arsenic Rule is to improve public health by reducing exposure to arsenic in drinking water. The Arsenic Rule reduces the arsenic Maximum Contaminant Level (MCL) to 10 µg/L. The rule requires new systems and new drinking water sources to demonstrate compliance as specified by the state. The public health benefits from the Arsenic Rule include the avoidance of bladder and lung cancers and a reduction in the frequency of noncarcinogenic diseases.

All samples must be collected at each entry point to the distribution system, unless otherwise specified by the state. Compliance with the MCLs for IOCs (Inorganic Chemicals), SOCs (Synthetic Organic Chemicals), and VOCs (Volatile Organic Chemicals) is based on a running annual average at each sampling point.

Monitoring requirements for total arsenic include one sample after the effective date of the MCL (January 23, 2006). Surface water systems must take annual samples. Groundwater systems must take one sample during the 2005–2007 compliance period. If the monitoring result is less than the MCL, groundwater systems must collect one sample every three years and surface water systems must continue to collect annual samples. Also see Section 22.201, "Arsenic."

22.131 Lead and Copper Rule

The purpose of the Lead and Copper Rule is to protect the public health by minimizing lead (Pb) and copper (Cu) levels in drinking water, primarily by reducing water corrosivity. Lead and copper enter drinking water mainly from corrosion of lead and copper in plumbing materials.

The Lead and Copper Rule establishes an action level (AL) of 0.015 mg/L for lead and 1.3 mg/L for copper, based on the 90th percentile level of tap-water samples. An AL exceedance is not a violation of the rule, but can trigger other requirements that include water quality parameter (WQP) monitoring, corrosion control treatment (CCT), source water monitoring/treatment, public education, and lead service line replacement (LSLR).

The public health benefits from the Lead and Copper Rule include a reduced risk of exposure to lead, which can cause damage to the brain, red blood cells, and kidneys, especially for young children and pregnant women. Benefits also include a reduction in the risk of exposure to copper, which can cause stomach and intestinal distress, liver and kidney damage, and complications of Wilson's disease in genetically predisposed people.

Treatment technique and sampling requirements if the AL is exceeded include the following:

1. Water quality parameter (WQP) monitoring
2. Public education (PE)
3. Source water monitoring and treatment (SOWT)
4. Corrosion control treatment (CCT)

If the system continues to exceed the action level after installing CCT or SOWT, then requirements include the following:

5. Lead service line (LSL) monitoring
6. Lead service line replacement (LSLR)

Also see Section 22.209, "Copper," and Section 22.2012, "Lead and Copper."

22.132 Total Coliform Rule (TCR)

The purpose of the Total Coliform Rule (TCR) is to improve public health protection by reducing fecal pathogens to minimal levels through control of total coliform bacteria, including fecal coliforms and *Escherichia coli (E. coli)*. The TCR establishes an MCL based on the presence or absence of total coliforms or *E. coli*, requires use of a sample siting plan, and requires sanitary surveys for systems collecting fewer than 5 samples per month.

The public health benefits from the TCR include a reduction in the risk of illness from disease-causing organisms associated with wastewater or animal wastes. Disease symptoms may include diarrhea, cramps, nausea, jaundice, and associated headaches and fatigue.

Also see Section 22.220, "Total Coliform Rule."

22.133 Surface Water Treatment Rules

On January 14, 2002, the EPA promulgated the Long Term 1 Enhanced Surface Water Treatment Rule (LT1ESWTR). The purpose of this rule is to improve public health protection through the control of microbial contaminants, particularly *Cryptosporidium* and to prevent significant increases in microbial risk that might otherwise occur when systems implement the Stage 1 Disinfectants and Disinfection Byproducts Rule (Stage 1 DBPR).

On December 15, 2005, the EPA promulgated the Long Term 2 Enhanced Surface Water Treatment Rule (LT2ESWTR), which increases the monitoring and treatment requirements for water systems that are prone to outbreaks of microorganisms such as *Cryptosporidium.* A purpose of this rule is to reduce the risk of disease-causing microorganisms from entering drinking water systems. This rule also contains provisions to reduce risks from uncovered finished water reservoirs and to ensure that systems maintain microbial protection when they take steps to decrease the formation of DBPs that results from chemical water treatment.

The major provisions of these rules include the control of *Cryptosporidium* and combined filter effluent (CFE) turbidity performance standards. Turbidity monitoring requirements are also included in these rules.

For additional information, see Section 22.222, "Surface Water Treatment Rules."

22.134 Filter Backwash Recycling Rule (FBRR)

The Filter Backwash Recycling Rule was promulgated by the EPA on June 8, 2001. The purpose of this rule is to improve public health protection by assessing and changing, where needed, recycle practices for improved contaminant control, particularly microbial contaminants. The FBRR requires systems that recycle certain flows to return specific recycle flows through all processes of the system's existing conventional or direct filtration system or to recycle to an alternate location approved by the regulatory agency.

The FBRR applies to public water systems that use surface water or groundwater under the direct influence of surface water, practice conventional or direct filtration, and recycle spent filter backwash, thickener supernatant, or liquids from dewatering processes. Conventional filtration is a series of processes including coagulation, flocculation, sedimentation, and filtration resulting in substantial particulate removal. Direct filtration is a series of processes including coagulation and filtration, but excluding sedimentation, and resulting in substantial particulate removal. Typically, direct filtration can be used only with high-quality raw water that has low levels of turbidity and suspended solids.

Recycle flows include spent filter backwash water, which is a stream containing particles that are dislodged from filter media when water is forced back through a filter (backwashed) to clean the filter. Thickener supernatant is a stream containing the decant from a sedimentation basin, clarifier, or other treatment unit that is used to treat water, solids, or semisolids from the primary treatment processes. Liquids from dewater processes include streams containing liquids generated from a treatment unit used to concentrate solids for disposal.

The public health benefits from the FBRR include a reduction in the risk of illness from microbial pathogens in drinking water, particularly *Cryptosporidium.*

For additional information, see Section 22.2221, "Requirements for Filtered Water Systems."

22.135 Disinfectants and Disinfection Byproducts Rules (DBPRs)

EPA promulgated the Stage 1 DBPR on December 16, 1998. The purpose of this rule is to improve public health protection by reducing exposure to DBPs. Some DBPs have been shown to cause cancer and cause reproductive effects in laboratory animals, which suggests the potential to cause bladder cancer and reproductive effects in humans. The Stage 1 DBPR is the first of a staged set of rules that will reduce the allowable levels of DBPs in drinking water. The rule establishes seven new standards and a treatment technique of enhanced coagulation or enhanced softening to further reduce DBP exposure. The rule is designed to limit capital investments and avoid major shifts in disinfection technologies until additional information is available on the occurrence and health effects of DBPs. For additional information, see *Water Treatment Plant Operation*, Volume I, Section 4.8, "Enhanced Coagulation."

The public health benefits from the Stage 1 DBPR include a reduction in THM levels in drinking water, and a reduction in exposure to the major DBPs from the use of ozone (bromate) and chlorine dioxide (chlorite).

On December 15, 2005, EPA promulgated the Stage 2 Disinfectants and Disinfection Byproducts Rule (Stage 2 DBPR), which establishes standards for controlling the harmful byproducts of drinking water disinfection measures. A purpose of this rule is to limit the amount of potentially harmful DBPs that end up in drinking water systems. This rule strengthens public health protection for consumers by tightening compliance monitoring requirements for two groups of DBPs: total trihalomethanes (TTHMs) and haloacetic acids (HAA5). The rule targets systems with the greatest risk. It will reduce potential health risks related to DBP exposure and provide more equitable public health protection.

Also see Section 22.23, "Disinfectants and Disinfection Byproducts (DBP)."

22.136 Ground Water Rule (GWR)

On October 11, 2006, the final Ground Water Rule (GWR) was signed by the EPA Administrator to protect the public from pathogen contamination in water systems that use groundwater. Pathogens are microorganisms, such as bacteria *(E. Coli)*, viruses, and protozoa, such as *Cryptosporidium* and *Giardia lamblia*. Essentially, the GWR aims to identify operating deficiencies in water systems and requires them to do more frequent monitoring.

The GWR applies to all systems using groundwater and it takes a targeted, risk-based approach. There are no mandatory disinfection requirements, but the rule does build on existing state programs and provides flexibility in defining significant deficiencies.

A significant deficiency refers to a problem that could have immediate potential to affect public health. Some possible examples include the following:

- Maximum contaminant level violations
- Not enough chlorine contact time
- Insufficient water for normal demand
- Not enough disinfectant residual to meet the minimum requirement
- No licensed operator for the water system

Under the GWR, primacy agencies must complete a sanitary survey of treatment plants that identifies any problems that could cause contamination. A hydrogeology sensitivity assessment must also be completed to determine the groundwater's vulnerability to contamination. For those water systems that use groundwater and that do not disinfect, the rule requires continuous monitoring for susceptible systems. For systems that disinfect, a 4 log (99.99 percent) inactivation of viruses should be demonstrated.

22.137 Radionuclides Rule

EPA promulgated the Radionuclides Rule on December 7, 2000. The purpose of the Radionuclides Rule is to reduce the exposure to radionuclides in drinking water, which will reduce the risk of cancer. This rule will also improve public health protection by reducing exposure to all radionuclides. The rule applies to all community water systems. The public health benefits from the rule include reduced exposure to uranium, which can cause toxic kidney effects and cancer. For additional information, see Section 22.24, "Radiological Standards."

22.138 Regulation of Microbial Contaminants in Drinking Water

Microbial contamination of drinking water can pose a potential public health risk in terms of acute outbreaks of disease. The illnesses associated with contaminated drinking water are mainly gastrointestinal in nature, although some pathogens are capable of causing severe and life-threatening illnesses.

In many cases, source water from a lake, river, reservoir, or groundwater aquifer needs to be disinfected to inactivate (or kill) microbial pathogens. Microbial pathogens include a few types of bacteria, viruses, protozoa, and other organisms. Some pathogens are often found in water, frequently as a result of the following:

- Fecal matter from sewage discharges
- Leaking septic tanks
- Runoff from animal feedlots into bodies of water

To protect drinking water from these pathogens, water suppliers often add a disinfectant to drinking water such as chlorine. However, disinfection practices can be problematic for the following reasons:

- Certain microbial pathogens, such as *Cryptosporidium*, are highly resistant to traditional disinfection practices.
- Disinfectants themselves can react with naturally occurring materials in the water to form byproducts, such as THMs and haloacetic acids, which may pose health risks.

A major challenge for water suppliers is how to balance the risks from microbial pathogens and DBPs. It is important to provide protection from microbial pathogens while simultaneously minimizing health risks to the population from DBPs.

22.139 Standardized Monitoring Framework (SMF)

The Standardized Monitoring Framework (SMF), promulgated by EPA in the Phase II Rule on January 30, 1991, and revised under Phases IIB and V, includes contaminants regulated under Phases I, II, IIB, V, the revised Arsenic Rule, and the Radionuclides Rule. Monitoring under the SMF began in 1993.

The SMF goal is to standardize, simplify, and consolidate drinking water monitoring requirements across contaminant groups. The SMF reduces the variability within monitoring requirements for chemical and radiological contaminants across system sizes and types.

The practical benefits of the SMF include the following:

- Increased public health protection through consistent monitoring practices
- A reduction in the complexity of water quality monitoring for both primacy agencies and water systems
- Balancing of the resources and expenses for monitoring and vulnerability assessments
- Increased water system compliance with monitoring requirements

22.1310 Consumer Confidence Report (CCR) Rule

The purpose of the Consumer Confidence Report (CCR) Rule is to improve public health protection by providing educational material to allow consumers to make educated decisions regarding any potential health risks pertaining to the quality, treatment, and management of their drinking water supply. The CCR Rule requires all community water systems to prepare and distribute a brief annual water quality report summarizing information regarding source, any detected contaminants, compliance, and educational information.

The public health-related benefits from the CCR Rule include the following:

- Increased consumer knowledge of drinking water quality, sources, susceptibility, treatment, and drinking water supply management
- Increased awareness of consumers to potential health risks, so they may make informed decisions to reduce those risks, including taking steps toward protecting their water supply
- Increased dialog with drinking water utilities and increased understanding of consumers to take steps toward active participation in decisions that affect public health

Information that should be included in the Consumer Confidence Report includes the following:

- Water system information
- Source of water
- Definitions of water terms
- Detected contaminants
- Compliance with drinking water regulations
- Required educational information

For additional information, see Section 22.8, "Consumer Confidence Reports (CCRs)."

QUESTIONS

Please write your answers to the following questions and compare them with those on page 605.

22.1D What are the public health benefits from the Arsenic Rule?

22.1E What is the purpose of the Total Coliform Rule (TCR)?

22.1F The Filter Backwash Recycling Rule (FBRR) refers to which recycle flows?

22.1G What is the purpose of the Consumer Confidence Report (CCR) Rule?

22.14 Setting Standards

A standard is usually the maximum level of a substance that EPA has deemed acceptable in drinking water. The first step in the setting of a standard is to study the human and animal health effects of a given chemical. These studies are normally performed using rats or mice. Based on these studies, EPA establishes a "no observed adverse effect level" (NOAEL). A safety factor is added to the NOAEL and the result is an acceptable daily intake limit of the chemical in question. The limit is adjusted to take into account the average weight and water consumption of the consumer, and the resulting figure is called a maximum contaminant level goal, or MCLG. By policy, the EPA sets MCLGs at zero for known or probable human carcinogens. MCLGs for noncarcinogens are set at a level where no adverse health effects would occur with a margin of safety.

The MCLG represents what EPA believes to be a safe level of consumption based solely on its studies of health effects. It is, however, a goal rather than an immediately achievable constituent limit. To develop more realistic, enforceable limits, EPA further revises the MCLG to take into account existing laboratory detection technology, costs, and reasonableness. After adjusting for these factors, EPA sets the MCL as close to the MCLG as is realistically feasible. The important difference between the two levels is that the MCLG is a nonenforceable goal and the MCL is an enforceable standard.

EPA has established standards (maximum contaminant levels) for chemicals, pesticides, bacteria and viruses, radioactivity, turbidity, and THMs. Most of these substances occur naturally in our environment and in the foods we eat. However, the MCLs apply whether the contaminant is from naturally occurring sources or from manmade pollution. The national drinking water standards set by EPA reflect the levels we can safely consume in our water, taking into account the amounts we are exposed to from these other sources.

Drinking water regulations are sometimes called "interim regulations" because research continues on drinking water contaminants. The existing standards may be strengthened and new standards may be established for other substances based on studies being conducted by the National Academy of Sciences, EPA, and others. The 1996 SDWA amendments gave EPA the authority to issue interim regulations for contaminants that do not appear on the Drinking Water Contaminant Candidate List if the agency determines that the contaminant presents an urgent threat to public health. In making this determination, EPA must consult with the Department of Health and Human Services.

22.15 Types of Water Systems

All of the drinking water regulations apply to all public water systems. It makes no difference whether the water system is publicly or privately owned. A *PUBLIC WATER SYSTEM (PWS)* is defined as any system that has the following characteristics:

1. Has at least 15 service connections
2. Regularly serves an average of at least 25 individuals daily at least 60 days out of the year

Any water system that provides services for fewer connections or persons than this is not covered by the SDWA. Certain other individuals and residences also are excluded, such as those whose water is supplied by an irrigation, mining, or industrial water system. However, regardless of size, all operators must strive to provide consumers with a potable drinking water.

Drinking water regulations also take into account the type of population served by the system and classify water systems as community or noncommunity systems. Therefore, in order to understand what requirements apply to any specific system, it is first necessary to determine whether the system is considered a community system or a noncommunity system. A *COMMUNITY WATER SYSTEM* is defined as a public water system that has the following characteristics:

1. Has at least 15 service connections used by all-year residents
2. Regularly serves at least 25 all-year residents

Any public water system that is not a community water system is classified as a *NONCOMMUNITY WATER SYSTEM*. Restaurants, campgrounds, and hotels could be considered noncommunity systems for purposes of drinking water regulations.

In addition to distinguishing between community and noncommunity water systems, EPA identifies some small systems as *NONTRANSIENT NONCOMMUNITY* systems if they regularly serve at least 25 of the same persons over 6 months per year. This classification applies to water systems for facilities such as schools or factories where the consumers served are nearly the same every day but do not actually live at the facility. In general, nontransient noncommunity systems must meet the same requirements as community systems.

A *TRANSIENT NONCOMMUNITY* water system is a system that does not regularly serve drinking water to at least 25 of the same persons over 6 months per year. This classification is used by EPA only in regulating nitrate levels and total coliform. Examples of a transient noncommunity system might be campgrounds or service stations if those facilities do not meet the definition of a community, noncommunity, or nontransient noncommunity system.

QUESTIONS

Please write your answers to the following questions and compare them with those on page 605.

22.1H What is the MCLG for known or probable human carcinogens?

22.1I What is the difference between a maximum contaminant level (MCL) and a maximum contaminant level goal (MCLG)?

22.1J Define a community water system.

22.2 PRIMARY DRINKING WATER STANDARDS

Primary standards or MCLs are set for substances that are thought to pose a threat to health when present in drinking water at certain levels. Because these substances are of health concern, primary standards are enforceable by law. (In contrast, secondary standards relate to cosmetic factors and are not federally enforceable.) A primary standard is usually expressed as an MCL. Some contaminants, such as pathogenic organisms, are very difficult or expensive to measure so EPA requires the use of specific treatment techniques (disinfection or filtration, for example) that are known to be effective in reducing the health risks of these contaminants. Treatment technique requirements, therefore, are also referred to as primary standards. Table 22.1 lists the primary standards and health concerns associated with the contaminants.

22.20 Inorganic Chemical Standards

Inorganic chemicals are metals, salts, and other chemical compounds that do not contain carbon. The health concerns about inorganic chemicals are not centered on cancer, but rather on their suspected links to several different human disorders. For example, lead is suspected of contributing to mental retardation in children. The following paragraphs briefly discuss each of the inorganic contaminants regulated by the national drinking water standards. Waters exceeding the MCL for these elements for short periods of time will pose no immediate threat to health. However, studies show that these substances must be controlled because consumption of drinking water that exceeds these standards over long periods of time may prove harmful.

22.200 Antimony

Code of Federal Regulations Reference: 40 CFR 721.1930

Antimony is used in the production of ceramics, glassware, and pigments. The major sources of antimony in drinking water are discharged from petroleum refineries, fire retardants, ceramics, electronics, and solder. Most exposures to antimony occur in industrial settings where workers may inhale dust containing particles of the metal.

Purpose of Regulating Antimony: To improve public health protection by reducing exposure to the amount of antimony in drinking water.

Public Health Benefit: Some people who drink water containing antimony well in excess of the MCL for many years could experience increases in blood cholesterol and decreases in blood sugar. Ingestion of antimony can alter cholesterol and glucose levels and may cause chromosome damage.

MCL/MCLG: 0.006 mg/L/0.006 mg/L

Monitoring Requirements: Once a year for surface water; once every 3 years for groundwater. To minimize monitoring costs, use historical data, waivers, susceptibility waivers, or make composite samples.

22.201 Arsenic

Code of Federal Regulations Reference: 40 CFR 141 Subpart B

Arsenic occurs naturally in the environment, especially in the western United States, and it is used in insecticides. Arsenic is found in foods, tobacco, shellfish, drinking water, and in the air in some locations. Anyone who drinks water that continuously exceeds the national standard by a substantial amount over a lifetime may experience fatigue and loss of energy. Extremely high levels of arsenic can cause poisoning.

The regulation of arsenic has been vigorously debated. In 1997, EPA released a health effects study plan. Under this plan, EPA worked with the Water Research Foundation (WRF) and the Association of California Water Agencies to assess the health risks from exposure to low levels of arsenic. Based on the results of these studies, EPA proposed an MCL for arsenic in June 2000, and a final rule was issued January 22, 2001. Technologies available for removing arsenic include ion exchange, iron

TABLE 22.1 NATIONAL PRIMARY DRINKING WATER STANDARDS

Contaminant	MCL	Health Effects
Inorganics		
Antimony	0.006 mg/L	decreases longevity, alters cholesterol and glucose levels
Arsenic	0.01 mg/L	skin/nervous system toxicity; possible cancer
Asbestos	7 million fibers/L	possible cancer
Barium	2.0 mg/L	circulatory system effects
Beryllium	0.004 mg/L	bone/lung damage; possible cancer
Bromate	0.010 mg/L	possible cancer
Cadmium	0.005 mg/L	kidney effects
Chlorite	1.0 mg/L	blood; developing nervous system
Chromium	0.1 mg/L	skin sensitization; liver/kidney/circulatory/nervous system disorders and respiratory problems
Copper	1.3 mg/L[a]	nervous system/kidney/gastrointestinal effects; toxicity
Cyanide	0.2 mg/L	spleen/liver/brain effects
Fluoride	4.0 mg/L	brown staining or pitting of teeth
Lead	0.015 mg/L[a] (at tap)	interference with red blood cell chemistry; developmental delays in children; blood pressure effects in some adults; toxic to infants and pregnant women
Mercury	0.002 mg/L	kidney disorders
Nitrate (as N) and Nitrite (as N)[b]	10.0 mg/L 1.0 mg/L	Methemoglobinemia ("blue baby" syndrome)
Selenium	0.05 mg/L	nervous system damage
Thallium	0.002 mg/L	liver/kidney/intestines/brain effects
Organics		
VOLATILE ORGANICS		
Benzene	0.005 mg/L	possible cancer
Carbon tetrachloride	0.005 mg/L	possible cancer
o-Dichlorobenzene	0.6 mg/L	kidney and liver effects; blood cell damage
p-Dichlorobenzene	0.075 mg/L	kidney and liver effects
1,2-Dichloroethane	0.005 mg/L	possible cancer
1,1-Dichloroethylene	0.007 mg/L	kidney and liver damage
cis-1,2-Dichloroethylene	0.07 mg/L	nervous system/liver/circulatory system damage
trans-1,2-Dichloroethylene	0.1 mg/L	nervous system/liver/circulatory system damage
Dichloromethane	0.005 mg/L	possible cancer
1,2-Dichloropropane	0.005 mg/L	possible cancer
Ethylbenzene	0.7 mg/L	kidney/liver/nervous system damage
Monochlorobenzene	0.1 mg/L	kidney/liver/nervous system damage
Styrene	0.1 mg/L	nervous system/liver damage
Tetrachloroethylene	0.005 mg/L	possible cancer
Toluene	1.0 mg/L	nervous system/kidney/liver damage
1,2,4-Trichlorobenzene	0.07 mg/L	adrenal gland and internal organ damage
1,1,1-Trichloroethane	0.2 mg/L	liver/nervous system/circulatory system damage
1,1,2-Trichloroethane	0.005 mg/L	liver/kidney effects
Trichloroethylene (TCE)	0.005 mg/L	possible cancer
Vinyl chloride	0.002 mg/L	possible cancer
Xylenes (total)	10.0 mg/L	liver/kidney/nervous system damage
PESTICIDES AND SYNTHETIC ORGANICS		
Acrylamide	treatment technique	possible cancer; nervous system effects
Alachlor	0.002 mg/L	possible cancer
Atrazine	0.003 mg/L	cardiac/reproductive system damage
Benzo[a]pyrene	0.0002 mg/L	possible cancer
Carbofuran	0.04 mg/L	nervous system/reproductive system damage
Chlordane	0.002 mg/L	possible cancer
Dalapon	0.2 mg/L	kidney/liver damage
Dibromochloropropane (DBCP)	0.0002 mg/L	possible cancer

TABLE 22.1 NATIONAL PRIMARY DRINKING WATER STANDARDS *(continued)*

Contaminant	MCL	Health Effects
PESTICIDES AND SYNTHETIC ORGANICS (continued)		
Di(2-ethylhexyl)adipate	0.4 mg/L	liver/reproductive system effects
Di(2-ethylhexyl)phthalate	0.006 mg/L	possible cancer
Dinoseb	0.007 mg/L	thyroid/reproductive system damage
Diquat	0.02 mg/L	kidney/gastrointestinal damage; cataract risk
Endothall	0.1 mg/L	liver/kidney/gastrointestinal and reproductive system effects
Endrin	0.002 mg/L	liver/kidney/heart damage
Epichlorohydrin	treatment technique	possible cancer
Ethylene dibromide (EDB)	0.00005 mg/L	possible cancer
Glyphosate	0.7 mg/L	liver/kidney damage
Heptachlor	0.0004 mg/L	possible cancer
Heptachlor epoxide	0.0002 mg/L	possible cancer
Hexachlorobenzene	0.001 mg/L	possible cancer
Hexachlorocyclopentadiene	0.05 mg/L	stomach/kidney damage
Lindane	0.0002 mg/L	nervous system/immune system/liver/kidney effects
Methoxychlor	0.04 mg/L	nervous system/reproductive system/kidney/liver effects
Oxamyl (Vydate)	0.2 mg/L	kidney damage
PCBs	0.0005 mg/L	possible cancer
Pentachlorophenol	0.001 mg/L	liver/kidney damage
Picloram	0.5 mg/L	liver/kidney damage
Simazine	0.004 mg/L	possible cancer
Toxaphene	0.003 mg/L	possible cancer
2,4-D	0.07 mg/L	liver/kidney/nervous system damage
2,3,7,8-TCDD (Dioxin)	0.00000003 mg/L	possible cancer
2,4,5-TP (Silvex)	0.05 mg/L	liver/kidney damage
Microbial		
Total Coliform	1 per 100 mL <40 samples/mo—no more than 1 positive >40 samples/mo—no more than 5% positive	indicators of disease-causing organisms
Giardia lamblia	3 log (99.9%) removal[c]	Giardiasis; gastrointestinal effects
Legionella	treatment technique	Legionnaire's Disease; affects respiratory system
Enteric viruses	4 log (99.99%) removal[c]	gastrointestinal and other viral infections
Heterotrophic bacteria	treatment technique[c]	gastrointestinal infections
Physical		
Turbidity[c]	0.5 to 5 NTU	interferes with disinfection
Turbidity[d]	0.3 to 1 NTU	interferes with disinfection
Radionuclides		
Gross alpha particles	15 pCi/L	cancer risk
Gross beta particles[e]	4 mrem/yr	cancer risk
Radium 226 & 228	5 pCi/L	bone cancer
Uranium	30 μg/L	kidney damage; possible cancer
Disinfection Byproducts		
TTHMs[f]	0.080 mg/L	cancer risk
Haloacetic Acids (HAA5)[f]	0.060 mg/L	nervous system/liver effects

a Action level for treatment.
b Applies to community, nontransient noncommunity, and transient noncommunity water systems.
c Applies to systems using surface water or groundwater under the influence of surface water.
d Applies to conventional and direct filtration systems.
e Applies to surface water systems serving more than 100,000 persons and any system determined by the state to be vulnerable.
f Applies to systems serving more than 10,000 persons and to all surface water systems that meet the criteria for avoiding filtration.

hydroxide coagulation followed by microfiltration, and activated alumina.

Purpose of Regulating Arsenic: To improve public health protection by reducing exposure to arsenic in drinking water.

Public Health Benefit: The avoidance of bladder and lung cancer, dental problems, and a reduction in the frequency of noncarcinogenic diseases.

MCL/MCLG: 0.010 mg/L/zero

Monitoring Requirements: Once a year for surface water; once every 3 years for groundwater. To minimize monitoring costs, use historical data, waivers, susceptibility waivers, or make composite samples.

22.202 Asbestos

Code of Federal Regulations Reference: 40 CFR 141.62

The general term "asbestos" refers to a family of fibrous silicate minerals that have been widely used in the manufacture of commercial products. Some examples of products containing asbestos include floor and ceiling tiles, paper products, paint and caulking, plastics, brake linings, insulation, cement, and filters. Asbestos in cement pipes is a potential source of contamination of drinking water. Deterioration of this material with age and with exposure to corrosive water is thought to release asbestos particles. Mining for asbestos minerals has also contributed to the contamination of some drinking water sources throughout the United States.

Scientific evidence on the harmful effects of asbestos consumed in drinking water is less conclusive than the evidence relating to inhaled asbestos.

Purpose of Regulating Asbestos: To improve public health protection by reducing exposure to asbestos in drinking water.

Public Health Benefit: To reduce the risk of developing benign intestinal polyps.

MCL/MCLG: 7 million fibers per liter (MFL)/7 MFL

Monitoring Requirements: Once every 9 years.

22.203 Barium

Code of Federal Regulations Reference: 40 CFR 141.62

Although not as widespread as arsenic, this element also occurs naturally in the environment in some areas of the US. Barium can also enter water supplies through industrial waste discharges. Small doses of barium are not harmful. However, it is quite dangerous when consumed in large quantities and will bring on increased blood pressure, nerve damage, and even death.

Purpose of Regulating Barium: To improve public health protection by reducing exposure to large quantities of barium in drinking water.

Public Health Benefit: To reduce the risk of high blood pressure.

MCL/MCLG: 2 mg/L/2 mg/L

Monitoring Requirements: Once a year for surface water; once every 3 years for groundwater. To minimize monitoring costs, use historical data, waivers, susceptibility waivers, or make composite samples.

22.204 Beryllium

Code of Federal Regulations Reference: 40 CFR 141.62

The major source of beryllium in the environment is the combustion of coal and oil. Airborne particulates containing beryllium can be inhaled or washed by precipitation into drinking water supplies. Animal studies show that when large doses of beryllium are ingested, it is carried by the bloodstream from the stomach to the liver but it is gradually transferred to the animal's bones.

Purpose of Regulating Beryllium: To improve public health protection by reducing exposure to large quantities of beryllium in drinking water.

Public Health Benefit: To reduce the risk of intestinal lesions.

MCL/MCLG: 0.004 mg/L/0.004 mg/L

Monitoring Requirements: Once a year for surface water; once every 3 years for groundwater. To minimize monitoring costs, use historical data, waivers, susceptibility waivers, or make composite samples.

22.205 Bromate

Code of Federal Regulations Reference: 40 CFR Parts 9, 141, and 142

Bromate is formed as a byproduct of ozone disinfection of drinking water. Ozone reacts with the naturally occurring bromide ion in the water to form bromate. Bromate is an inorganic ion that is tasteless, colorless, and has a low volatility. Bromate dissolves easily in water and is fairly stable. Bromate is not typically derived from natural sources. It is commonly found in raw water supplies such as streams, lakes, and reservoirs.

Purpose of Regulating Bromate: To improve public health protection by reducing exposure to bromate in drinking water.

Public Health Benefit: To reduce the risk of cancer.

MCL/MCLG: 0.010 mg/L/zero

Monitoring Requirements: Monthly (ozone systems only).

22.206 Cadmium

Code of Federal Regulations Reference: 40 CFR 141.62

Only extremely small amounts of cadmium are found in natural waters in the United States. Waste discharges from the electroplating, photography, insecticide, and metallurgy industries can increase cadmium levels. The most common source of cadmium in our drinking water is from galvanized pipes and fixtures.

Purpose of Regulating Cadmium: To improve public health protection by reducing exposure to cadmium in drinking water.

Public Health Benefit: To prevent kidney damage.

MCL/MCLG: 0.005 mg/L/0.005 mg/L

Monitoring Requirements: Once a year for surface water; once every 3 years for groundwater. To minimize monitoring costs, use historical data, waivers, susceptibility waivers, or make composite samples.

22.207 Chlorite

Code of Federal Regulations Reference: 40 CFR Parts 9, 141, and 142

The main source of chlorite in drinking water is chlorine dioxide used for disinfection. Chlorite forms as the chlorine dioxide breaks down during the disinfection process. Chlorite is an inorganic ion that is colorless, odorless, tasteless, and dissolves easily in water. Chlorite is one of the two or three chemicals involved in the process of generating chlorine dioxide for water treatment. Chlorite is not typically derived from natural sources.

Purpose of Regulating Chlorite: To improve public health protection by reducing exposure to chlorite in drinking water, especially for infants and young children.

Public Health Benefit: To prevent adverse effects of the nervous system and anemia.

MCL/MCLG: 1.0 mg/L/0.8 mg/L

Monitoring Requirements: For systems that add chlorine dioxide, daily samples are required at distribution system entry points.

22.208 Chromium

Code of Federal Regulations Reference: 40 CFR 141.62

Chromium is found in cigarettes, some of our foods, and the air. It is also discharged from steel and pulp mills and the erosion of natural deposits. Some studies suggest that in very small amounts, chromium may be essential to human beings, but this has not been proven.

Purpose of Regulating Chromium: To improve public health protection by reducing exposure to chromium in drinking water.

Public Health Benefit: To prevent allergic dermatitis.

MCL/MCLG: 0.1 mg/L/0.1 mg/L

Monitoring Requirements: Once a year for surface water; once every 3 years for groundwater. To minimize monitoring costs, use historical data, waivers, susceptibility waivers, or make composite samples.

22.209 Copper

Code of Federal Regulations Reference: 40 CFR 141.86

Copper in drinking water usually results from the reaction of water on copper plumbing. Treatment of surface water in storage reservoirs to control algae may also cause high levels of copper. Copper is an essential nutrient; adults require 2 mg daily and children of preschool age require about 0.1 mg daily for normal growth. The presence of copper in drinking water is also undesirable because of its aesthetic effects. At low levels (0.5 mg/L in soft waters) blue or blue-green staining of porcelain occurs. At higher levels (4 mg/L), it will stain clothing and blond hair. Concentrations greater than 1 mg/L can produce insoluble green curds when reacting with soap.

The primary standard for copper includes a treatment technique that requires action to be taken when more than 10 percent of the taps sampled exceed the action level (AL) of 1.3 mg/L. Treatment technique requirements are contained in the Lead and Copper Rule, which is described in Section 22.2012, "Lead and Copper."

Purpose of Regulating Copper: To improve public health protection by reducing water corrosivity.

Public Health Benefit: To prevent gastrointestinal, kidney, and liver problems.

MCL/MCLG: A treatment technique (TT) is triggered at AL = 1.3 mg/L/1.3 mg/L

Monitoring Requirements: Residential sample taken at the kitchen or bathroom sink tap. ALs must be met in 90 percent of the samples. Follow-up monitoring every 6 months after corrosion controls initiated or optimized. Reduced monitoring for systems consistently meeting AL.

22.2010 Cyanide

Code of Federal Regulations Reference: 40 CFR 141.62

Cyanide is a common ingredient of rat poisons, silver and metal polishes, and photo processing chemicals. The major source of cyanide in drinking water is discharge from industrial chemical factories.

Ingestion of even very small amounts of sodium cyanide or potassium cyanide (both readily available to the general public) can cause death within a few minutes to a few hours. Typical symptoms of cyanide poisoning include nausea without vomiting, anxiety, vertigo, lower jaw stiffness, convulsions, paralysis, irregular heartbeat, and coma.

Purpose of Regulating Cyanide: To improve public health protection by reducing long-term exposure to cyanide.

Public Health Benefit: To prevent thyroid/neurological damage.

MCL/MCLG: 0.2 mg/L/0.2 mg/L

Monitoring Requirements: Once a year for surface water; once every 3 years for groundwater. To minimize monitoring costs, use historical data, waivers, susceptibility waivers, or make composite samples.

22.2011 Fluoride

Code of Federal Regulations Reference: 40 CFR 141.62

This is a natural mineral and many sources of drinking water contain some fluoride. Fluoride produces two effects, depending on its concentration. EPA has set both primary and secondary limits to regulate it. Many communities add fluoride to their drinking water to promote dental health. The decision to fluoridate a water supply is made by the state or local municipality, and is not mandated by EPA or any other federal entity.

However, EPA has completed and peer-reviewed a quantitative dose-response assessment based on the available data for severe dental fluorosis as recommended by the National Research Council (NRC).

At levels of 6 to 8 mg/L, fluoride may cause brittle bones and stiffening of the joints. On the basis of this health hazard, fluoride has been added to the list of primary standards.

At levels of 2 mg/L and greater, fluoride may cause dental fluorosis, which is discoloration and mottling of the teeth, especially in children. EPA has reclassified dental fluorosis as a cosmetic effect, raised the primary drinking water standard from 1.4–2 mg/L to 4 mg/L, and established a secondary standard of 2 mg/L for fluoride.

Purpose of Regulating Fluoride: To improve public health protection by preventing long-term exposure to fluoride.

Public Health Benefit: To prevent bone disease.

MCL/MCLG: 4.0 mg/L/4.0 mg/L

Monitoring Requirements: Once a year for surface water; once every 3 years for groundwater. To minimize monitoring costs, use historical data, waivers, susceptibility waivers, or make composite samples.

22.2012 Lead and Copper

Code of Federal Regulations Reference: 40 CFR 141 Subpart I: Control of Lead and Copper

To ensure full compliance, please consult the federal regulations in 40 CFR 141 and any approved state requirements.

Basic Requirements of the Lead and Copper Rule (LCR)—The LCR has four basic requirements:

1. Require water suppliers to optimize their treatment system to control corrosion in customers' plumbing.
2. Determine tap-water levels of lead and copper for customers who have lead service lines or lead-based solder in their plumbing system.
3. Rule out the source water as a source of significant lead levels.
4. If lead ALs are exceeded, require the suppliers to educate their customers about lead and suggest actions they can take to reduce their exposure to lead through public notices and public education programs. If a water system, after installing and optimizing corrosion control treatment, continues to fail to meet the lead AL, it must begin replacing the lead service lines under its ownership.

Revisions to the Lead and Copper Rule (LCR)—EPA is promulgating a rule that makes several targeted regulatory revisions to the existing national primary drinking water regulations (NPDWRs) for lead and copper. The revisions to the LCR will do the following:

- Enhance the implementation of the LCR in the areas of monitoring, treatment, customer awareness, and lead service line replacement
- Improve compliance with the public education requirements of the LCR to ensure drinking water consumers receive meaningful, timely, and useful information needed to help them limit their exposure to lead in drinking water

Lead—Lead is a naturally occurring metal that was used regularly in a number of industrial capacities and commonly used in household plumbing materials and water service lines. Lead was used as a component of paint, piping, solder, brass, and as a gasoline additive until the 1980s. And, the greatest exposure to lead is swallowing or breathing in lead paint chips and dust. Research has confirmed that lead is highly toxic.

Lead is rarely found in source water. Mostly, lead contamination occurs after water has left the treatment plant due to corrosion of plumbing materials. Since water is naturally corrosive, it corrodes the pipes and plumbing through which it passes. Homes built before 1986 are more likely to have lead pipes, fixtures, and solder. However, new homes are also at risk because even legally lead-free plumbing may contain up to 8 percent lead. Grounding of electrical circuits in homes to water pipes and galvanic action between two dissimilar metals may increase corrosion that could cause lead to leach into the water.

Purpose of Regulating Lead: To protect public health by minimizing lead levels in drinking water, primarily by reducing water corrosivity.

Public Health Benefit: To prevent kidney problems, high blood pressure, and delays in physical and mental development in infants and children.

MCL/MCLG: A TT is triggered at AL = 0.015 mg/L/zero

Monitoring Requirements/Comments: Residential samples taken at the kitchen or bathroom sink tap. ALs must be met in 90 percent of the samples. Follow-up monitoring every 6 months after corrosion controls initiated or optimized. There are reduced monitoring requirements for systems consistently meeting the AL. An AL exceedance is not a violation but can trigger other requirements that include monitoring for the following water quality indicators: pH, corrosion, control treatment, source water monitoring/treatment, public education, and lead service line replacement.

Copper—Copper is a metal found in natural deposits such as ores containing other elements. Copper is widely used in household plumbing materials.

Purpose of Regulating Copper: To protect public health by minimizing the risk of ingesting copper in drinking water, primarily from the corrosion of pipes.

Public Health Benefit: To prevent gastrointestinal distress due to short-term exposure and to prevent liver or kidney damage from long-term exposure.

MCL/MCLG: A TT is triggered at AL = 1.3 mg/L/1.3 mg/L

Monitoring Requirements/Comments: Residential samples taken at the kitchen or bathroom sink tap. ALs must be met in 90 percent of the samples. Follow-up monitoring every 6 months after corrosion controls initiated or optimized. There are reduced monitoring requirements for systems consistently meeting the AL. An AL exceedance is not a violation but can trigger other requirements that include monitoring for the following water quality indicators: pH, corrosion control treatment, source water monitoring/treatment, public education, and lead service line replacement.

All water systems are required to optimize corrosion control; that is to say, they must do the best they can to control corrosion. Small- and medium-sized systems are considered to have optimized corrosion control if they meet the lead and copper action levels during each of two consecutive six-month sampling periods. Large systems have optimized corrosion control if the difference between the source water lead level and the 90th percentile tap-water lead level is less than 0.005 mg/L, and the 90th percentile copper concentration is less than or equal to half the copper action level (0.65 mg/L) for two consecutive six-month periods. Any system that is achieving optimal corrosion control (and therefore meeting the action limits) is permitted to reduce monitoring and reporting frequencies.

Large water systems that do not meet the guidelines for optimal corrosion control are required to conduct studies of various types of corrosion control treatment and then submit to the state a plan that would adequately control corrosion in their distribution system. Small- and medium-sized systems exceeding the action levels must make a written recommendation to the state for a proposed treatment method, but may also be required to conduct similar studies before the state makes a final determination. Once the state directs a water supplier to install treatment, the agency has 24 months to install the specified treatment and 12 months to collect follow-up samples to determine if the system is working. The state will also determine maximum acceptable levels for pH, alkalinity, calcium (if carbonate stabilization is used), orthophosphate (if an inhibitor with a phosphate compound is used), and silica (if an inhibitor with a silicate compound is used). The system must then continue to operate within these water quality guidelines.

If corrosion control methods fail to reduce lead and copper concentrations below the action levels, the water supplier must collect and analyze source water samples. Data from this sampling is sent to the state along with the water agency's recommended treatment plan for reducing contaminant levels. The state may accept the agency's recommended plan or may require some alternative treatment such as ion exchange, reverse osmosis, lime softening, or coagulation-filtration. When required to install source water treatment, the agency is allowed 24 months for installation and 12 months to collect follow-up samples of specific water quality indicators. Based on the results of the follow-up sampling, the state determines the maximum allowable source water lead and copper concentrations and the water system is required to meet these state limits on an ongoing basis.

Water systems that continue to exceed the lead action level even after installing optimal corrosion control treatment and source water treatment will have to begin a program to replace lead service lines within a maximum of 15 years.

SAMPLING GUIDELINES

The sampling required by the Lead and Copper Rule includes tap-water sampling at the consumer's faucet and distribution system sampling. Tap-water samples for initial and routine lead and copper monitoring must be collected at high-risk locations, which are defined as homes with lead solder installed after 1982, homes with lead pipes, and homes with lead service lines. Tap-water samples for water quality indicators (when required) should be taken from representative taps throughout the distribution system. The representative taps can be the same as total coliform sampling sites. These samples will be analyzed for water quality indicator levels to identify optimal treatment and monitor compliance with the rule.

The rule calls for first-draw or first-flush samples, which are water samples taken after the water stands motionless in the plumbing pipes for at least six hours. This usually means taking a sample early in the day before water is used in the kitchen or bathroom. The rule permits water utilities to instruct consumers in sampling techniques so that they may collect the samples. One-liter samples are needed.

Distribution system sampling conducted in connection with corrosion control studies requires samples from representative taps throughout the system as well as samples from entry points to the distribution system.

INITIAL AND BASE MONITORING SAMPLING FREQUENCIES

Compliance with the Lead and Copper Rule depends on meeting the action levels during two consecutive six-month monitoring periods. All public water systems are required to collect one sample for lead and copper analysis from the following number of sites during each six-month monitoring period.

System Size (Population)	No. of Sampling Sites (Initial Base Monitoring)	No. of Sampling Sites (Reduced Monitoring)
>100,000	100	50
10,001 to 100,000	60	30
3,301 to 10,000	40	20
501 to 3,300	20	10
101 to 500	10	5
≤100	5	5

If a system meets the lead and copper action levels or maintains optimal corrosion control treatment for two consecutive six-month periods, tap-water sampling for lead and copper can be reduced to once per year and half the number of sites (see chart above). After three consecutive years of acceptable performance, tap-water sampling can be reduced to once every three years at the reduced number of sites listed above.

WATER QUALITY INDICATOR SAMPLING FREQUENCIES

All large systems (more than 50,000 persons) and any small- or medium-sized system that exceeds the lead or copper action level will be required to monitor for the following water quality indicators: pH, alkalinity, calcium, conductivity, orthophosphate, silica, and water temperature. Samples must be taken both from representative taps throughout the system and at entry points to the distribution system. Two tap-water samples must be collected for each applicable water quality indicator from the following number of sites during each six-month monitoring period.

System Size (Population)	No. of Sampling Sites (Initial Base Monitoring)	No. of Sampling Sites (Reduced Monitoring)
>100,000	25	10
10,001 to 100,000	10	7
3,301 to 10,000	3	3
501 to 3,300	2	2
101 to 500	1	1
≤100	1	1

In addition to the sampling from representative taps, one sample must be collected every two weeks for each applicable water quality indicator at each entry point to the distribution system.

All large water systems and any small- and medium-sized system that exceed the lead or copper action levels after installing optimal corrosion control treatment must continue to collect samples as follows:

- 2 samples for each applicable water quality indicator at each of the sampling sites (taps) every 6 months.
- 1 sample for each applicable water quality indicator at each entry point to the distribution system every 2 weeks.
- After meeting water quality guidelines for 2 consecutive 6-month monitoring periods, systems may reduce the number of tap samples collected (as indicated in the chart above), and after 3 consecutive years of acceptable performance they will be permitted to collect annual samples at the reduced number of sampling points.

CALCULATION OF 90TH PERCENTILE

The action levels that trigger implementation of treatment techniques to reduce lead and copper concentrations are based on the principle that no more than 10 percent of the samples should exceed the limit. To determine whether a system meets this test, it is necessary to figure out the lead or copper concentration on the sample that falls at the 90th percentile.

To calculate the 90th percentile value, write down the results from a set of samples arranging them in order of increasing or decreasing concentration. Number each sample beginning with the number 1 for the smallest concentration. Multiply the total number of samples (which will be the same as the number assigned to the highest concentration) by 0.9. Then, use the value from this calculation to identify the sample with the same number. The concentration of the sample whose number matches your calculated value is the 90th percentile and is used in determining compliance with the action level. (For systems serving fewer than 100 people, the 90th percentile is the average of the highest concentration and the second highest concentration.)

PUBLIC EDUCATION

A public education program developed by the EPA must be delivered to water system customers within 60 days whenever the system exceeds the lead or copper action levels. The information describes the harmful effects of lead in drinking water and tells customers what they can do to reduce their exposure. The education program should continue as long as the system fails to meet the lead or copper action levels.

In conjunction with the public education program, the water system must offer to sample a customer's water if asked to do so. However, the water system is not required to pay for the sampling and analysis and is not required to collect and analyze the sample itself.

LEAD SERVICE LINE REPLACEMENT

Any water system that continues to exceed the lead and copper action levels even after optimizing corrosion control and installing source water treatment will be required to replace its lead distribution pipes within a maximum of 15 years. A lead pipeline qualifies for replacement if it contributes more than 15 parts per billion to the total tap-water lead levels. Seven percent of the lines meeting this standard must be replaced annually. Customers must be notified 45 days before replacing the lines.

22.2013 Mercury

Code of Federal Regulations Reference: 40 CFR 141.62

Mercury is found naturally throughout the environment. Large increases in mercury levels in water can be caused by industrial and agricultural use. The health risk from mercury is greater from mercury in fish than simply from waterborne mercury. Mercury poisoning may have an *ACUTE HEALTH EFFECT*[3] in large doses, or a *CHRONIC HEALTH EFFECT*[4] in lower doses taken over an extended time period.

Purpose of Regulating Mercury: To improve public health protection and reduce ingestion of mercury.

Public Health Benefit: To prevent kidney damage.

MCL/MCLG: 0.002 mg/L/0.002 mg/L

Monitoring Requirements: Once a year for surface water; once every 3 years for groundwater. To minimize monitoring costs, use historical data, waivers, susceptibility waivers, or make composite samples.

22.2014 Nitrate

Code of Federal Regulations Reference: 40 CFR 141.62

Nitrate (as N) in drinking water above the national standard of 10 mg/L poses an immediate threat to children under three months of age. In some infants, excessive levels of nitrate have

3. *Acute Health Effect.* An adverse effect on a human or animal body, with symptoms developing rapidly.
4. *Chronic Health Effect.* An adverse effect on a human or animal body with symptoms that develop slowly over a long period of time or that recur frequently.

been known to react with intestinal bacteria that change nitrate to nitrite. It is often difficult to pinpoint sources of nitrate because there are so many possibilities. Sources of nitrogen may include runoff or seepage from fertilized agricultural lands, municipal and industrial wastewater, refuse dumps, animal feedlots, septic tanks and private sewage disposal systems, urban drainage, and decaying plant debris.

Purpose of Regulating Nitrate: To improve public health protection.

Public Health Benefit: To prevent methemoglobinemia ("blue baby syndrome")/diuresis.

MCL/MCLG: 10 mg/L/10 mg/L

Monitoring Requirements: Groundwater annually; surface water quarterly initially, then annually.

22.2015 Nitrite

Code of Federal Regulations Reference: 40 CFR 141.62

Nitrite reacts with hemoglobin in the blood. This reaction will reduce the oxygen-carrying ability of the blood and produce an anemic condition commonly known as "blue baby syndrome." Sources of nitrite may include runoff or seepage from fertilized agricultural lands, municipal and industrial wastewater, refuse dumps, animal feedlots, septic tanks and private sewage disposal systems, urban drainage, and decaying plant debris.

Purpose of Regulating Nitrite: To improve public health protection.

Public Health Benefit: To prevent methemoglobinemia ("blue baby syndrome")/diuresis.

MCL/MCLG: 1 mg/L/1 mg/L

Monitoring Requirements: One sample during first 3-year compliance period. Repeat frequency determined by state.

22.2016 Selenium

Code of Federal Regulations Reference: 40 CFR 141.62

This mineral occurs naturally in soil and plants, especially in western states. Selenium is found in meat and other foods. Although it is believed to be essential in the diet, there are indications that excessive amounts of selenium may be toxic. Studies are underway to determine the amount required for good nutrition and the amount that may be harmful.

If a person's intake of selenium came only from drinking water, it would take an amount many times greater than the standard to produce any ill effects.

Purpose of Regulating Selenium: To improve public health protection.

Public Health Benefit: To prevent hair or fingernail loss, numbness of fingers or toes, and circulatory problems.

MCL/MCLG: 0.05 mg/L/0.05 mg/L

Monitoring Requirements: Once a year for surface water; once every 3 years for groundwater. To minimize monitoring costs, use historical data, waivers, susceptibility waivers, or make composite samples.

22.2017 Thallium

Code of Federal Regulations Reference: 40 CFR 141.62

Thallium is a metal found in natural deposits such as ores containing other elements. The major sources of thallium in drinking water are leaching from ore-processing sites and discharge from electronics, glass, and drug factories. Thallium is an extremely toxic metal that is used in refining iron, cadmium, and zinc. The greatest use of thallium is in specialized electronic research equipment. It is also used for production of optical lenses, jewelry, semiconductors, dyes, and pigments. The most common source of thallium poisoning is ingestion of rat poisons and insecticides.

Purpose of Regulating Thallium: To improve public health protection.

Public Health Benefit: To prevent hair loss, changes in blood, and kidney/liver/intestinal problems.

MCL/MCLG: 0.002 mg/L/0.0005 mg/L

Monitoring Requirements: Once a year for surface water; once every 3 years for groundwater. To minimize monitoring costs, use historical data, waivers, susceptibility waivers, or make composite samples.

QUESTIONS

Please write your answers to the following questions and compare them with those on pages 605 and 606.

22.2A Why are primary standards sometimes expressed as treatment techniques rather than MCLs?

22.2B What are inorganic chemicals?

22.2C Why is arsenic listed as a primary contaminant?

22.2D What are two harmful effects of excessive levels of fluoride?

22.2E How can new homes be at risk for lead?

22.2F Why is nitrate in drinking water above the national standard of 10 mg/L considered an immediate threat to public health?

22.21 Organic Chemical Standards

Organic chemicals are either natural or synthetic chemical compounds that contain carbon. Synthetic organic chemicals (SOCs) are manmade compounds and are used throughout the world as pesticides, paints, dyes, solvents, plastics, and food additives. Volatile organic chemicals (VOCs) are a subcategory of organic chemicals. These chemicals are termed "volatile" because they evaporate easily. The most commonly found VOCs are THMs, trichloroethylene (TCE), tetrachloroethylene, and 1,1-dichloroethylene. THMs were the first regulated VOCs when EPA finalized regulations in 1979.

The most common sources of organic contamination of drinking water are pesticides and herbicides, industrial solvents, and DBPs (THMs and haloacetic acids). Millions of pounds of pesticides are used on croplands, forests, lawns, and gardens in the United States each year. They drain off into surface waters or seep into underground water supplies. Spills, poor storage, improper application, and haphazard disposal of organic chemicals have resulted in widespread groundwater contamination. This is a critical problem because once groundwater is contaminated, it may remain that way for a long time.

Many organic chemicals pose health problems if they get into drinking water and the water is not properly treated. The maximum limits for pesticides in drinking water are shown in Table 22.1 on pages 553 and 554. The next several paragraphs describe the sources and maximum contaminant levels of some of the most commonly used organic chemical contaminants.

In 2011, EPA decided to address up to 16 volatile organic compounds (VOCs) as a group that may cause cancer. The agency determined that they represent a near-term opportunity and it is likely that the public health goal for all VOCs would likely be set at zero because they may cause cancer. A preliminary evaluation of occurrence indicates that some of these VOCs may co-occur. This group will include trichloroethylene (TCE) and tetrachloroethylene (PCE). EPA determined in March 2010 that the drinking water standards for TCE and PCE need to be revised. Addressing these VOCs as a group will help reduce exposure to these contaminants.

22.210 Trichloroethylene (TCE)

Code of Federal Regulations Reference: 40 CFR 141.50

Although the use of TCE is declining because of stringent regulations, it was, for many years, a common ingredient in household products (spot removers, rug cleaners, and air fresheners), dry cleaning agents, industrial metal cleaners and polishes, refrigerants, and even anesthetics. Its wide range of use is perhaps why TCE is the organic contaminant most frequently encountered in groundwater.

Purpose of Regulating Trichloroethylene: To improve public health protection.

Public Health Benefit: To prevent cancer risks; liver problems.

MCL/MCLG: 0.005 mg/L/zero

Monitoring Requirements: Four consecutive quarterly samples during first compliance period. Compliance is based on annual average of quarterly samples. If no detections are found during the initial round, two quarterly samples are required each year for systems serving greater than 3,300; one sample is required every 3 years for smaller systems. These requirements apply to both community and noncommunity water systems.

22.211 1,1-Dichloroethylene

Code of Federal Regulations Reference: 40 CFR 141.50

This solvent is used in manufacturing plastics and, more recently, in the production of 1,1,1-trichloroethane.

Purpose of Regulating 1,1-Dichloroethylene: To improve public health protection.

Public Health Benefit: To prevent impacts to the liver, kidneys, heart, and central nervous system

MCL/MCLG: 0.007 mg/L/0.007 mg/L

Monitoring Requirements/Comments: For VOCs: Four consecutive quarterly samples during first compliance period. Compliance is based on annual average of quarterly samples. If no detections are found during the initial round, two quarterly samples are required each year for systems serving greater than 3,300; one sample is required every 3 years for smaller systems. With the completion of source water assessments, primacy agencies are allowed to develop alternative monitoring requirements. Contact your local primacy agency for further information. Applies to community water systems and nontransient noncommunity water systems.

22.212 Vinyl Chloride

Code of Federal Regulations Reference: 40 CFR 141.50

Billions of pounds of this solvent are used annually in the United States to produce polyvinyl chloride (PVC), the most widely used ingredient for manufacturing plastics throughout the world. There is also evidence that vinyl chloride may be a biodegradation end-product of TCE and PCE under certain environmental conditions.

Purpose of Regulating Vinyl Chloride: To improve public health protection.

Public Health Benefit: To prevent cancer in humans.

MCL/MCLG: 0.002 mg/L/zero

Monitoring Requirements/Comments: For VOCs: Four consecutive quarterly samples during first compliance period. Compliance is based on annual average of quarterly samples. If no detections are found during the initial round, two quarterly samples are required each year for systems serving greater than 3,300; one sample is required every 3 years for smaller systems. With the completion of source water assessments, primacy agencies are allowed to develop alternative monitoring requirements. Contact your local primacy agency for further information. Applies to community water systems and nontransient noncommunity water systems.

22.213 1,1,1-Trichloroethane

Code of Federal Regulations Reference: 40 CFR 141.24

This chemical has replaced TCE in many industrial and household products. It is the principal solvent in septic tank degreasers, cutting oils, inks, shoe polishes, and many other products. Among the VOCs found in groundwater, 1,1,1-trichloroethane and TCE are encountered most frequently and in the highest concentrations.

Purpose of Regulating 1,1,1-Trichloroethane: To improve public health protection.

Public Health Benefit: To prevent liver/circulatory and nervous system problems.

MCL/MCLG: 0.2 mg/L/0.2 mg/L

22.214 1,2-Dichloroethane

Code of Federal Regulations Reference: 40 CFR 141.50

This substance is used in making chemicals involved in plastics, rubber, and synthetic textile fibers. Other uses include the following: as a solvent for resins and fats, photography, photocopying, cosmetics, drugs, and as a fumigant for grains and orchards.

Purpose of Regulating 1,2-Dichloroethane: To improve public health protection.

Public Health Benefit: To prevent liver problems and cancer.

MCL/MCLG: 0.005 mg/L/zero

Monitoring Requirements: Four quarterly samples every 3 years and if the chemical is detected at greater than 0.0005 mg/L; the water system may return to the initial frequency after one year of no detection.

22.215 Carbon Tetrachloride

Code of Federal Regulations Reference: 40 CFR 141.50

Carbon tetrachloride was once a popular household solvent, a frequently used dry cleaning agent, and a charging agent for fire extinguishers. Since 1970, however, carbon tetrachloride has been banned from all use in consumer goods in the United States. In 1978, carbon tetrachloride was banned as an aerosol propellant. Currently, its principal use is in the manufacture of fluorocarbons, which are used as refrigerants.

Purpose of Regulating Carbon Tetrachloride: To improve public health protection.

Public Health Benefit: To prevent liver problems and cancer.

MCL/MCLG: 0.005 mg/L/zero

Monitoring Requirements: Four consecutive quarterly samples during first compliance period. Compliance is based on annual average of quarterly samples. If no detections are found during the initial round, two quarterly samples are required each year for systems serving greater than 3,300; one sample is required every 3 years for smaller systems. These requirements apply to both community and noncommunity water systems.

22.216 Benzene

Code of Federal Regulations Reference: 40 CFR 141.50

Benzene is used primarily in the synthesis of styrene (for plastics), phenol (for resins), and cyclohexane (for nylon). Other uses include the production of detergents, drugs, dyes, and insecticides. Benzene is still being used as a solvent and as a component of gasoline. It is also discharged from factories and leaches from gas storage tanks and landfills.

Purpose of Regulating Benzene: To improve public health protection.

Public Health Benefit: To prevent anemia, a decrease in blood platelets, and cancer.

MCL/MCLG: 0.005 mg/L/zero

Monitoring Requirements: Four consecutive quarterly samples during first compliance period. Compliance is based on annual average of quarterly samples. If no detections are found during the initial round, two quarterly samples are required each year for systems serving greater than 3,300; one sample is required every 3 years for smaller systems. These requirements apply to both community and noncommunity water systems.

22.217 Para-Dichlorobenzene (p-Dichlorobenzene)

Code of Federal Regulations Reference: 40 CFR 141.50

The principal uses of p-dichlorobenzene are in moth control (moth balls and powders) and as lavatory deodorants. It is also used as an insecticide and fungicide on crops; in the manufacture of other organic chemicals; and in plastics, dyes, and pharmaceuticals. The major source of p-dichlorobenzene in drinking water is discharge from industrial chemical factories.

Purpose of Regulating Para-Dichlorobenzene (p-Dichlorobenzene): To improve public health protection.

Public Health Benefit: To prevent kidney/liver/spleen/circulatory system problems.

MCL/MCLG: 0.075 mg/L/0.075 mg/L

Monitoring Requirements: Four consecutive quarterly samples during the first compliance period. Compliance is based on annual average of quarterly samples. If no detections are found during the initial round, two quarterly samples are required each year for systems serving greater than 3,300; one sample is required every 3 years for smaller systems. These requirements apply to both community and noncommunity water systems.

QUESTIONS

Please write your answers to the following questions and compare them with those on page 606.

22.2G What are organic chemicals?

22.2H What have been the common uses of trichloroethylene (TCE)?

22.22 Microbial Standards

Microbial contamination of drinking water can pose a potential public health risk in terms of acute outbreaks of disease. Bacteria, viruses, and other organisms have long been recognized as serious contaminants of drinking water. Organisms such as *Giardia*, *Cryptosporidium*, and *E. coli* (a type of coliform bacteria) cause almost immediate gastrointestinal illness when people consume them in water. Waterborne diseases such as typhoid, cholera, infectious hepatitis, and dysentery have been traced to improperly disinfected drinking water.

In most communities, drinking water is treated to remove contaminants before being piped to consumers, and bacterial contamination of municipal water supplies has been largely eliminated by adding chlorine or applying other forms of disinfection to drinking water to prevent waterborne diseases. By treating drinking water and wastewater, diseases such as typhoid and cholera have been virtually eliminated.

Many types of coliform bacteria from human and animal wastes may be found in drinking water if the water is not properly treated. Often, the bacteria themselves do not cause diseases transmitted by water, although certain coliforms have been identified as the cause of "traveler's" diarrhea. In general, however, the presence of coliform bacteria indicates that other harmful organisms may be present in the water.

Throughout the early 1990s, chemical contaminants in drinking water were the focus of EPA's regulatory efforts. With passage of the 1996 SDWA amendments, Congress directed EPA to focus on microbial contaminants and to promulgate a series of new regulations to improve treatment effectiveness. Significant new regulations of the microbial contaminant *Cryptosporidium* were promulgated in December 1998. The new regulations consist of mandatory treatment techniques similar to the current regulations for *Giardia* and coliform bacteria. Filtration and disinfection requirements for surface water systems are currently in place and the final Ground Water Rule was signed in 2006. These regulations significantly affect the water treatment processes of most water suppliers.

For additional information about the new regulations, see Section 22.2226, "Interim Enhanced Surface Water Treatment Rule (IESWTR)," and Section 22.23, "Disinfectants and Disinfection Byproducts (DBP)." For more information about *Cryptosporidium*, see Chapter 23, Section 23.1026, "*Cryptosporidium*."

22.220 Total Coliform Rule

Code of Federal Regulations Reference: 40 CFR 141 Subpart C

Coliforms are bacteria that are always present in the digestive tracts of animals, including humans, and are found in their wastes. They are also found in plant and soil material. The most basic test for bacterial contamination of a water supply is the test for total coliform bacteria. Total coliform counts give a general indication of the sanitary condition of a water supply. Coliforms include the following:

- **Total coliforms**—include bacteria that are found in the soil, in water that has been influenced by surface water, and in human or animal waste.
- **Fecal coliforms**—are the group of the total coliforms that are considered to be present specifically in the gut and feces of warm-blooded animals. Because the origins of fecal coliforms are more specific than the origins of the more general total coliform group of bacteria, fecal coliforms are considered a more accurate indication of animal or human waste than the total coliforms.
- ***Escherichia coli (E. coli)***—is the major species in the fecal coliform group. Of the five general groups of bacteria that comprise the total coliforms, only *E. coli* is generally not found growing and reproducing in the environment. Consequently, *E. coli* is considered to be the species of coliform bacteria that is the best indicator of fecal pollution and the possible presence of pathogens.

22.221 2012 Revised Total Coliform Rule (RTCR)

EPA has revised the Total Coliform Rule so that public water systems in the US take steps to ensure the integrity of the drinking water distribution system and monitoring for the presence of microbial contamination. Public water systems and the state and local agencies that oversee them must comply with the RTCR beginning April 1, 2016. The final RTCR will require the following:

- Public water systems are to notify the public if a test exceeds the MCL for *E. coli* in drinking water
- Public water systems that are at risk and therefore vulnerable to microbial contamination are required to identify and fix the problems, including the potential sources and the pathways of contamination
- Criteria will be used for public water systems that are well operated to qualify for and stay on reduced monitoring, which could reduce water system burden
- Provide incentives for better system operation, in particular small systems
- Require additional monitoring requirements for seasonal systems such as campgrounds and state and national parks

The RTCR establishes an MCLG and an MCL for *E. coli* and eliminates the MCLG and MCL for total coliforms, replacing them with treatment techniques for coliforms that require assessment and corrective action.

EPA has established an MCLG of zero for *E. coli*, which is a more specific indicator of fecal contamination and those pathogens that are potentially more harmful than total coliform. EPA has removed the 1989 MCLG and MCL for total coliforms, although the acute total coliform MCL violation under the 1989 TCR has been maintained as the MCL for *E. coli* under the RTCR.

Under the new treatment technique, total coliforms serve as the indicator of a potential pathway of contamination into the system's distribution system. Public water supplies that exceed a specified frequency of total coliform occurrence must conduct an assessment to determine if any sanitary deficiencies exist; if found, they must be corrected. A public water system that experiences an *E. coli* MCL violation must conduct an assessment and correct any sanitary defects that are identified.

Monthly notification requirements based only on the presence of total coliforms will no longer be required. Instead, the RTCR will require public notification when an *E. coli* MCL violation occurs, indicating a potential health threat or when a water system fails to assess and take corrective action.

22.2210 SANITARY SURVEY

Code of Federal Regulations Reference: 40 CFR 141.2

Sanitary surveys are conducted to identify possible health risks that may not be discovered by routine coliform sampling. The Total Coliform Rule required community water systems serving fewer than 4,100 persons to complete an initial sanitary survey by June 1994 and to conduct subsequent surveys every 5 years. Noncommunity water systems were required to complete an initial sanitary survey by June 1999 and subsequent surveys every 5 years with the exception of those systems with protected and disinfected groundwater sources, which are allowed 10 years for subsequent surveys.

A sanitary survey requires detailed planning, a thorough system survey, and reporting of the results. The planning portion involves a review of water quality records for compliance with applicable microbial, inorganic chemical, organic chemical, and radiological contaminant MCLs as well as the records of compliance with the monitoring requirements for those contaminants. The actual field survey is a detailed evaluation and inspection of the source of the water supply and all conveyances, storage, treatment, and distribution facilities to ensure protection from all pollution sources. The final report includes the date(s) of the survey, who was present during the survey, the survey findings, the recommended improvements to correct identified problems, and the target dates of completion for any improvements.

22.2211 SAMPLING PLAN

The Total Coliform Rule requires each water supply system to develop and follow a written sampling plan. Each plan must specifically identify sampling points throughout the distribution system. Sampling plans must be approved by the state regulatory agency and it will be necessary to check with your state agency to determine the details of their review process and what documents need to be submitted.

22.2212 LABORATORY PROCEDURES

Code of Federal Regulations Reference: 40 CFR 141.21 and 40 CFR 141 Subpart C

With regard to coliform testing methodology, EPA maintains a current list of analytical methods that are approved for monitoring drinking water supplies that is published and updated in 40 CFR 141. Examples of the acceptable analytical methods for determining total coliforms include the following:

- Multiple-tube fermentation technique (MTF)
- Membrane filter technique (MF)
- Presence-absence coliform test (P-A)
- Colilert™ system (ONPG-MUG test)
- Colisure test

Regardless of the method used, the standard sample volume for total coliform testing is now 100 mL. This is an increase over the past testing method using 50 mL for the MTF technique.

The multiple-tube fermentation method of testing for coliforms determines the presence or absence of coliforms by the multiple-tube dilution method. This is a process whereby equal portions of a sample (100 mL) are added to 10 tubes containing a culture medium and an inverted vial. If gas accumulates in the inverted vial, it indicates presumptive evidence of coliform organisms and the sample is considered coliform-positive. If no gas forms in any of the vials, the sample is coliform-negative.

The membrane filter method provides for filtering a 100-mL water sample through a thin, porous, cellulose membrane filter under a partial vacuum. The filter is placed in a sterile container and incubated in contact with a special liquid called a "culture medium," which the bacteria use as a food source. Colonies of bacteria then grow on the media. The coliform colonies are visually identified, counted, and recorded as the number of coliform colonies per 100 mL of sample.

The presence-absence (P-A) test for the coliform group is a simple modification of the multiple-tube procedure. Even though this simplified test uses only one large test portion (100 mL) in a single culture bottle, the results still indicate the presence or absence of coliform bacteria and thus meet the testing requirements.

The Colilert™ system is a privately developed version of the ONPG-MUG test procedure that meets the EPA's laboratory methods requirements. The ONPG-MUG test indicates the presence or absence of coliform bacteria within 24 hours.

Although the Colilert™ method can yield presence-absence results within 24 hours and is easier to perform than the Membrane Filtration (MF) method, operators should be aware of the limitations of these tests in evaluating samples for regulatory purposes. The results for the Colilert™ and MF tests are not always comparable, which may be due to the following:

- Interferences in the sample that may suppress or mask bacterial growth
- Greater sensitivity of the Colilert™ media
- Added stress to organisms related to filtering
- The fact that different media may obtain better growth for some bacteria

Both test methods are approved by the EPA for reporting under the Safe Drinking Water Program. When these tests produce conflicting results, however, the safest course of action is to increase monitoring and treatment efforts until the results for both tests are negative.

The Colisure test is a presence-absence test for coliform bacteria. A sample of water is added to a dehydrated medium and, after 24 to 48 hours, the medium is examined for the presence of coliforms. (The Colisure test is available from IDEXX Laboratories at www.idexx.com.)

Water supply systems may contain various other types of bacteria besides coliforms. These are nonpathogenic bacteria and are sometimes referred to as *HETEROTROPHIC*[5] bacteria. During the analyses of water samples for coliforms, heterotrophic bacteria sometimes interfere with the test causing one of the following reactions:

- MTF technique: a turbid culture with no gas production
- P-A test: a turbid culture in the absence of an acid reaction
- MF technique: confluent growth or a coliform colony number too numerous to count

The sample being tested can be considered invalid if any of these situations occurs and total coliforms are not detected. A replacement sample must be taken from the same location within 24 hours of receiving the laboratory results. If the replacement sample also shows heterotrophic interference but tests positive for total coliforms, the sample is considered valid.

5. *Heterotrophic* (HET-er-o-TROF-ick). Describes organisms that use organic matter for energy and growth. Animals, fungi, and most bacteria are heterotrophs.

22.2213 MONITORING FREQUENCY

The routine monitoring frequency for community water systems is based on the population served, as shown in Table 22.2. Routine monitoring frequencies for noncommunity systems are based on the source of supply and, in some cases, the population served. See Table 22.2, Footnote a, for the specific details about monitoring by noncommunity water systems.

Whenever a routine sample tests positive for total coliform, two provisions of the Total Coliform Rule take effect: repeat samples (formerly referred to as check samples) must be taken and the original coliform-positive sample must be tested for the presence of fecal coliforms or *E. coli* to determine whether there is an actual or potential violation of the coliform MCL. The second provision above will be discussed in the next section, "Determining Compliance."

The reason repeat samples are required is to investigate whether the original coliform-positive sample was caused by a contamination problem that exists throughout the distribution system or if it is a localized problem that exists only at that one sampling point. With this information, appropriate corrective action can be taken to eliminate the problem as quickly as possible.

The Total Coliform Rule specifies how many repeat samples must be taken, when, and from what locations. It also directs the water utility to collect a specific number of samples the following month, based on the number of routine samples ordinarily

TABLE 22.2 TOTAL COLIFORM MONITORING FREQUENCY FOR COMMUNITY WATER SYSTEMS

Population Served	Minimum Number of Routine Samples per Month[a]	Population Served	Minimum Number of Routine Samples per Month
25 to 1,000[b]	1[c]	59,001 to 70,000	70
1,001 to 2,500	2	70,001 to 83,000	80
2,501 to 3,300	3	83,001 to 96,000	90
3,301 to 4,100	4	96,001 to 130,000	100
4,101 to 4,900	5	130,001 to 220,000	120
4,901 to 5,800	6	220,001 to 320,000	150
5,801 to 6,700	7	320,001 to 450,000	180
6,701 to 7,600	8	450,001 to 600,000	210
7,601 to 8,500	9	600,001 to 780,000	240
8,501 to 12,900	10	780,001 to 970,000	270
12,901 to 17,200	15	970,001 to 1,230,000	300
17,201 to 21,500	20	1,230,001 to 1,520,000	330
21,501 to 25,000	25	1,520,001 to 1,850,000	360
25,001 to 33,000	30	1,850,001 to 2,270,000	390
33,001 to 41,000	40	2,270,001 to 3,020,000	420
41,001 to 50,000	50	3,020,001 to 3,960,000	450
50,001 to 59,000	60	3,960,001 or more	480

a A noncommunity water system using groundwater and serving 1,000 persons or fewer may monitor at a lesser frequency specified by the state until a sanitary survey is conducted and the state reviews the results. Thereafter, noncommunity water systems using groundwater and serving 1,000 persons or fewer must monitor in each calendar quarter during which the system provides water to the public, unless the state determines that some other frequency is more appropriate and notifies the system (in writing). In all cases, noncommunity water systems using groundwater and serving 1,000 persons or fewer must monitor at least once/year.
A noncommunity water system using surface water, or groundwater under the direct influence of surface water, regardless of the number of persons served, must monitor at the same frequency as a like-sized community public system. A noncommunity water system using groundwater and serving more than 1,000 persons during any month must monitor at the same frequency as a like-sized community water system, except that the state may reduce the monitoring frequency for any month the system serves 1,000 persons or fewer.

b Includes public water systems that have at least 15 service connections, but serve fewer than 25 persons.

c For a community water system serving 25–1,000 persons, the state may reduce this sampling frequency if a sanitary survey conducted in the last five years indicates that the water system is supplied solely by a protected groundwater source and is free of sanitary defects. However, in no case may the state reduce the sampling frequency to less than once/quarter.

NOTE: These sampling frequencies took effect December 31, 1990.

collected. Table 22.3 lists the repeat sampling frequencies triggered by a coliform-positive routine sample.

TABLE 22.3 COLIFORM MONITORING—ROUTINE AND REPEAT SAMPLING FREQUENCIES

Number of Routine Samples per Month	Number of Repeat Samples[a]	Minimum Samples per Month Next Month[b]
1 or fewer	4	5
2	3	5
3	3	5
4	3	5
5 or more	3	See Table 22.2

a Number of repeat samples in the same month for each total coliform-positive routine sample.
b Except where the state has invalidated the original routine sample, substitutes an on-site evaluation of the problem, or waives the requirement on a case-by-case basis.

Noncommunity systems serving more than 1,000 people have the same requirements as community water systems (Table 22.2). Noncommunity systems serving fewer than 1,000 people are required to collect one routine sample per quarter. When the routine sample tests positive, four repeat samples are required in the same quarter and five samples are required the following quarter.

As Table 22.3 indicates, whenever a routine or repeat sample tests positive for total coliform, the water agency must collect a set of either three or four repeat samples within 24 hours of receiving the laboratory results. At least one of the repeat samples must be taken from the same tap as the original coliform-positive sample; the remaining repeat samples in the set must be collected at nearby taps (within five service connections of the original sampling point), both upstream and downstream of the original. Repeat samples must be taken until no coliforms are detected or until the MCL is exceeded and the state is notified.

For additional information about sampling and compliance procedures, see "Rule Changes Way Systems Will Look at Coliform," *Opflow*, Vol. 16, No. 12, December 1990, pages 1, 4, and 5, and "Corrections," *Opflow*, Vol. 17, No. 2, February 1991, page 2.

22.2214 DETERMINING COMPLIANCE

As discussed earlier, the MCLG for total coliforms (including fecal coliforms and *E. coli*) is zero. The MCL, based on the presence-absence concept, is as follows:

- For water systems analyzing at least 40 samples per month, no more than 5.0 percent of the samples (including routine and repeat samples) may be positive for total coliforms.
- For water systems analyzing fewer than 40 samples per month, no more than one sample per month may be positive for total coliforms.

The Total Coliform Rule makes another significant change in the way compliance is calculated. That is, all valid coliform-positive samples—both routine and repeat samples—must be counted when calculating compliance with the monthly MCL. Under previous regulations, check or repeat samples were not included in the monthly MCL calculation.

On a case-by-case basis, the state may declare a sample invalid for one of several reasons, including interference by heterotrophic bacteria during laboratory analysis, as previously discussed. Total-coliform-positive samples may be invalidated by the state under any of the following conditions:

- The analytical laboratory acknowledges that improper sample analysis caused the positive result
- The state determines that the contamination is a local plumbing problem
- The state has substantial grounds to believe that the positive result was unrelated to the quality of drinking water in the distribution system

A simplified decision tree for determining compliance with the Total Coliform Rule is shown in Table 22.4. It is strongly recommended that operators work closely with their state regulatory agencies in implementing this rule.

22.2215 REPORTING AND NOTIFICATION REQUIREMENTS

Reporting frequencies for coliform test results increase in step with the urgency of the problem. A water agency must report the results of monthly coliform testing to the state regulatory agency within the first 10 days of the following month.

Any time a water agency fails to collect a sample as required, the state must be notified within 10 days after the system learns of the violation. An invalid sample result is considered a failure to monitor and must be reported.

If the MCL is exceeded, the state must be notified no later than the end of the next business day and the public within 14 days. This could occur when the water agency exceeds its monthly coliform-positive limit or when test results show the presence of fecal coliforms or *E. coli* in any sample. (See Section 22.6 for further details about public notification procedures.)

The most critical situation exists when either of two situations occurs:

1. A routine sample tests positive for total coliforms and for fecal coliforms or *E. coli*, and any repeat sample tests positive for total coliforms.
2. A routine sample tests positive for total coliforms and negative for fecal coliforms or *E. coli*, and any repeat sample tests positive for fecal coliforms or *E. coli*.

Either of these situations is considered an acute risk to health. This is a Tier 1 violation, which requires that the state and the public be notified within 24 hours, as described in Section 22.6.

TABLE 22.4 SIMPLIFIED DECISION TREE FOR DETERMINING COMPLIANCE WITH THE TOTAL COLIFORM RULE[a]

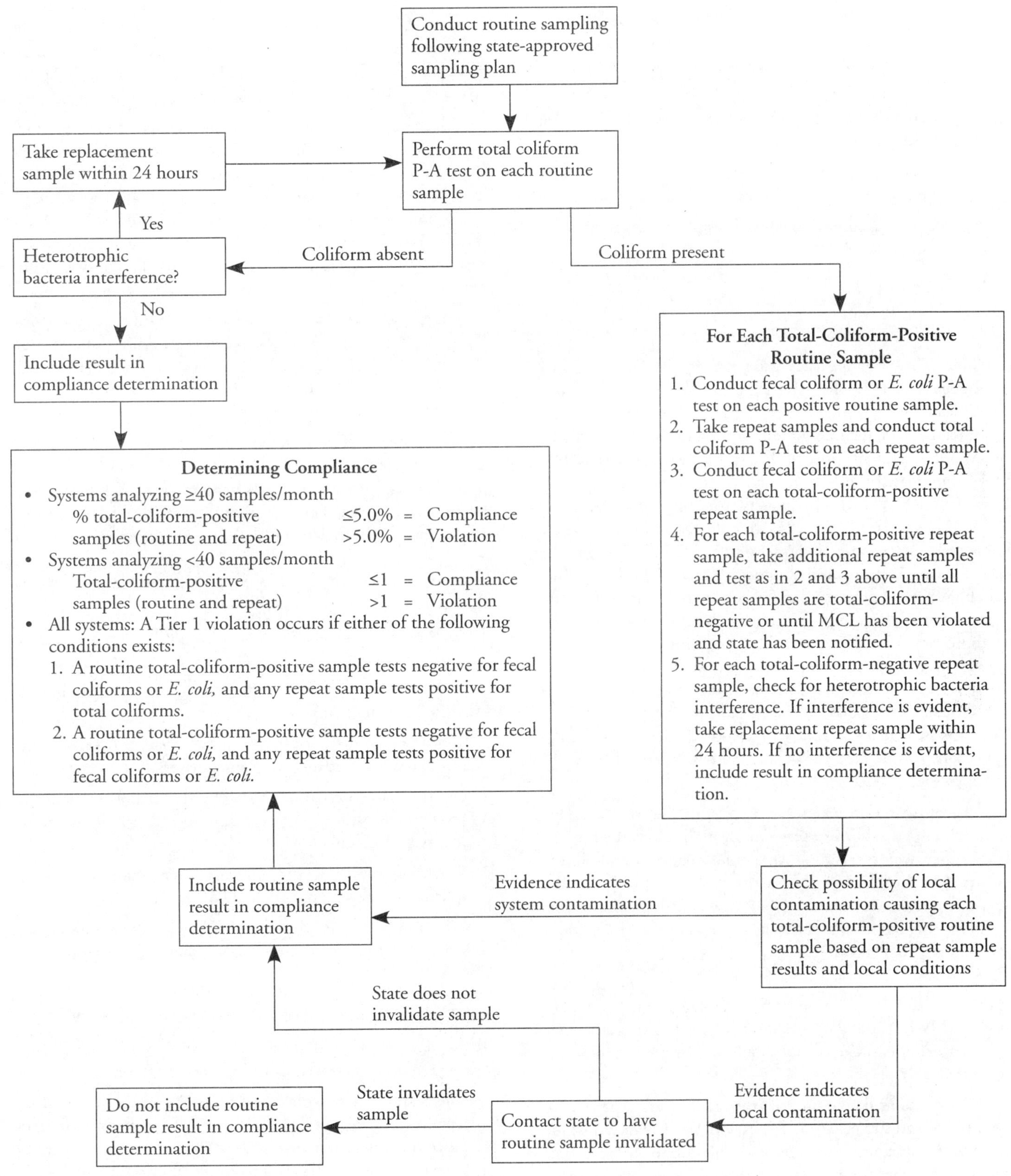

a Reprinted from *Opflow*, Vol. 16, No. 12, December 1990, by permission. Copyright © 1990, American Water Works Association.

QUESTIONS

Please write your answers to the following questions and compare them with those on page 606.

22.2I Why are microbial contaminants a public health concern?

22.2J What is the most basic test for bacterial contamination of a water supply?

22.2K List the five analytical methods EPA will accept for coliform testing.

22.2L What are reasons for collecting repeat coliform samples?

22.2M What sampling locations are used for repeat coliform samples?

22.222 Surface Water Treatment Rules

On June 29, 1989, the EPA promulgated the Surface Water Treatment Rule (SWTR). The 1996 SDWA amendments imposed deadlines by which EPA must complete work on several major regulations, including promulgation of an Enhanced Surface Water Treatment Rule (ESWTR). On December 16, 1998, EPA promulgated the Interim Enhanced Surface Water Treatment Rule (IESWTR). On January 14, 2002, and December 15, 2002, respectively, the EPA promulgated the Long Term 1 Enhanced Surface Water Treatment Rule (LT1ESWTR) and the Long Term 2 Enhanced Surface Water Treatment Rule (LT2ESWTR). These regulations set forth primary drinking water regulations requiring treatment of surface water supplies or groundwater supplies under the direct influence of surface water. These regulations require surface water systems to use a specific water treatment technique (filtration or disinfection) rather than meeting MCLs for *Cryptosporidium*, *Giardia lamblia*, viruses, *Legionella*, and heterotrophic bacteria. Also, the regulations require that all systems must be operated by qualified operators.

The protozoan *Giardia lamblia* is presently the organism most implicated in waterborne disease outbreaks in the United States. These microscopic creatures are found mainly in mountain streams. Once inside the body, they cause a painful and disabling illness. The infection caused by *Giardia* is called Giardiasis. The symptoms of Giardiasis are usually severe diarrhea, gas, cramps, nausea, vomiting, and fatigue.

Giardia and viruses have been added to the traditional coliform and turbidity indicators of microbiological quality. In this case, the MCLGs are zero because the organisms are pathogens, or indicators of pathogens, and should not be present in drinking water.

The regulations define surface water as "all water open to the atmosphere and subject to surface runoff." This would include rivers, lakes, streams, and reservoirs. This surface water definition also includes groundwaters that are directly influenced by surface water. The determination as to whether a supply uses surface water is left up to the state. Generally, if a groundwater source has significant and relatively rapid shifts in water quality such as turbidity, temperature, conductivity, or pH that closely correlate to climatological or nearby surface water conditions, that source can be considered surface water.

At a minimum, the treatment required for surface water would include disinfection systems for very clean and protected source waters. These systems would only be required to disinfect to achieve removal of coliforms to meet the requirements.

Turbidity measurements are a key element in determining compliance with both the filtration and the disinfection requirements of the Surface Water Treatment Rules. The effectiveness of a filtration system in removing particles and organisms is easily obtained by measuring the water's turbidity before and after filtration. High turbidity also reduces the effectiveness of disinfection processes and, therefore, must be controlled and closely monitored.

Water supply systems that are required to filter can use a variety of treatment technologies, including disinfection, to meet the expected performance levels. These technologies include conventional treatment, direct filtration, slow sand filtration, and diatomaceous earth filtration.

The two-pronged treatment technique of the regulations aims to ensure that harmful microorganisms are either removed by filtration processes or inactivated (and thus made harmless) by disinfection processes. Success, therefore, depends heavily on consistently well-operated filtration and disinfection systems. The general performance standards to be met by all surface water systems and any systems using groundwater under the influence of surface water are as follows:

1. At least 99.9 percent (also called 3 log for the 3 nines) removal or inactivation of *Giardia lamblia* cysts
2. At least 99.99 percent (4 log) removal or inactivation of enteric (intestinal) viruses

In general, compliance by the surface water purveyor could be through one of the following alternatives:

1. Meeting the criteria for which filtration is not required and providing disinfection according to the specific requirements in the regulations
2. Providing filtration and meeting disinfection criteria required for those supplies that are filtered

Operators are urged to work closely with their state regulatory agencies to stay informed about changes in water regulations as they occur in the future.

22.2220 CRITERIA FOR AVOIDING FILTRATION

To avoid mandatory filtration, a water utility must meet the following requirements to ensure adequate disinfection:

1. The fecal coliform concentration in the water before disinfection must not exceed 20/100 mL, or total coliform concentration must not exceed 100/100 mL in more than 10 percent of the samples analyzed in the previous 6 months.
2. The system must currently provide at least 99.9 percent (3 log) *Cryptosporidium* inactivation.

3. The turbidity of the water prior to disinfection cannot exceed 5 *NTUs*[6] based on grab samples collected every four hours or by continuous monitoring.
4. Certain site-specific conditions as follows:
   (a) has disinfection that achieves 99.9 percent inactivation of *Giardia* and 99.99 percent inactivation of viruses (CT requirement—see Section 22.2225)
   (b) watershed control or sanitary surveys that satisfy regulatory requirements
   (c) no history of waterborne disease outbreak without making treatment corrections
   (d) compliance with long-term coliform MCL
   (e) compliance with total TTHMs MCL, if the system serves more than 10,000 people
   (f) disinfection system must have an alternative power supply and automatic alarm and start-up to ensure continuous disinfection
   (g) ensure that disinfectant residual at entrance to the distribution system does not drop below 0.2 mg/L for more than four hours

If a system cannot meet the source water quality criteria and site-specific conditions listed above, then the system must install and operate appropriate filtration facilities.

22.2221 REQUIREMENTS FOR FILTERED WATER SYSTEMS

For systems that filter, the primary concern is adequate disinfection and filtration performance. The requirements are as follows:

1. For conventional or direct filtration systems, the filtered water turbidity must be less than or equal to 0.3 NTU for at least 95 percent of each month's measurements. For slow sand or diatomaceous earth filtration, the filtered water turbidity must be less than 1 NTU in at least 95 percent of the measurements taken each month.
2. Filtered water turbidity must never exceed 5 NTUs.
3. A disinfectant residual at the entrance to the distribution system of at least 0.2 mg/L.
4. Filtered water supplies must achieve the same disinfection as required for unfiltered systems (that is, 99.9 percent and 99.99 percent removal/inactivation of *Giardia lamblia* and viruses) through a combination of filtration and application of a disinfectant.

22.2222 CHLORINE RESIDUAL SUBSTITUTION

It is important to remember that this is a two-pronged approach and both elements are considered when determining compliance by systems that are required to install filtration. Filtration alone accomplishes part of the job of removing harmful organisms; disinfection does the rest of the job.

At the discretion of the state and based on a review of the water system, chlorine residual testing may be substituted for some of the bacteriological testing. Chlorine residual testing could give the operator a quicker indication of the condition of the system. However, the following requirements must be met:

1. Samples must be taken at points that are representative of conditions within the distribution system.
2. Chlorine residual testing can replace only up to 75 percent of the bacteriological testing.
3. At least four chlorine residual tests must be taken to substitute for one bacteriological sample.
4. A free chlorine residual of at least 0.2 mg/L must be maintained throughout the distribution system.
5. If free chlorine residual falls below 0.2 mg/L, check samples must be taken for bacteriological testing and a report must be submitted to the state within 48 hours.
6. Chlorine residual must be determined daily.

22.2223 TURBIDITY REQUIREMENTS

Turbidity is undesirable in a finished or treated water because it causes cloudiness resulting in an unattractive water. However, the major reason that turbidity is undesirable is that it causes a health hazard in the following ways:

1. Interfering with disinfection by reducing the ability of the disinfectant to inactivate or kill disease-causing organisms
2. Exerting a chlorine demand that makes it difficult to maintain a residual throughout the distribution system
3. Interfering with the bacteriological examination of the water
4. Preventing satisfactory reduction of tastes and odors and asbestos fibers

The MCLs and monitoring requirements for turbidity (which apply to surface water only) are as follows:

- For unfiltered surface water supplies, the turbidity of the raw water prior to disinfection cannot exceed 5 NTU based on grab samples collected every 4 hours or by continuous turbidity monitoring.

6. *Turbidity* (ter-BID-it-tee) *Units (TU).* Turbidity units are a measure of the cloudiness of water. If measured by a nephelometric (deflected light) instrumental procedure, turbidity units are expressed in nephelometric turbidity units (NTU) or simply TU. Those turbidity units obtained by visual methods are expressed in Jackson turbidity units (JTU), which are a measure of the cloudiness of water; they are used to indicate the clarity of water. There is no real connection between NTUs and JTUs. The Jackson turbidimeter is a visual method and the nephelometer is an instrumental method based on deflected light.

- For surface water supplies using conventional or direct filtration, the process must achieve a turbidity level of less than 1 NTU at all times and not more than 0.3 NTU in more than 5 percent of the samples analyzed during a given month. Turbidity must be sampled every 4 hours or by continuous monitoring.
- Those systems using slow sand or diatomaceous earth filtration must achieve a turbidity level of less than 5 NTU at all times and not more than 1 NTU in more than 5 percent of the samples collected each month. Systems using slow sand filtration must sample at least once a day; diatomaceous earth filtration systems must monitor turbidity every 4 hours or by continuous sampling.

Other filtration technologies may be used if they meet, as a minimum, the turbidity requirements for slow sand filters, the disinfection requirements, and are approved by the state.

Unfiltered systems are required to begin with a clean source water and have a watershed that is protected from human activities that might otherwise have an adverse impact on water quality. Unfiltered systems would have very little, if any, virus contamination. For these systems, the major concern is *Giardia* contamination from animal activities that cannot be prevented by watershed protection. The purpose of the turbidity limit for unfiltered water is to ensure a high probability that turbidity does not interfere with disinfection of *Giardia* cysts. The turbidity limit of 5 NTU serves this purpose.

For filtered water systems, the major burden for *Giardia* removal rests with filtration. With conventional treatment and direct filtration, low turbidity levels (<0.5 NTU) are needed to ensure effective *Giardia* cyst removals. Disinfection of either *Giardia* or viruses will not be hampered at these turbidity levels.

For slow sand filtration and diatomaceous earth filtration, effective *Giardia* removal does not necessarily correlate with low treated water turbidities. However, to ensure effective virus inactivation, a low filtered water turbidity is needed. Viruses are much smaller than *Giardia*, and thus a lower turbidity limit of 1 NTU is needed compared with the turbidity level of 5 NTU for unfiltered supplies to ensure effective disinfection.

22.2224 MONITORING REQUIREMENTS

Unfiltered surface water systems must do the following:

1. Monitor raw water supply for fecal and total coliform. Samples must be collected and analyzed at a minimum frequency based on system size as shown below.

Number of People Served	Samples/Week
<501	1
501–3,300	2
3,301–10,000	3
10,001–25,000	4
>25,000	5

   If the fecal coliform concentration in the water before disinfection exceeds 20 coliforms/100 mL, or the total coliform concentration exceeds 100 coliforms/100 mL in more than 10 percent of the samples analyzed in a 6-month period, the system no longer meets the criteria to avoid filtration.
2. Monitor turbidity every 4 hours or by continuous measurement.
3. Monitor and control activities on the watershed in order to minimize adverse impacts on water quality. Annual on-site inspections are also required.
4. Continuously monitor the disinfectant residual to ensure a minimum of 0.2 mg/L at the entrance to the distribution system. Systems serving fewer than 3,300 persons may take grab samples at the frequencies shown below.

Number of People Served	Samples/Week
<501	1
501–1,000	2
1,001–2,500	3
2,501–3,300	4

   In addition, disinfectant residuals in the distribution system must be monitored at the same frequency and location as total coliform samples.
5. Monitor daily to demonstrate that the level of disinfection achieved is 99.9 percent inactivation of *Giardia* and 99.99 percent inactivation/removal of enteric viruses.

Filtered systems must meet the following requirements:

1. Perform turbidity measurements of representative water every 4 hours (which can be continuous monitoring). For any system using slow sand filtration or any other filtration treatment other than conventional treatment, such as direct filtration or diatomaceous earth filtration, the state may reduce this monitoring to once per day.
2. Continuously monitor the disinfectant residual entering the distribution system to ensure a minimum of 0.2 mg/L at the entrance to the distribution system. Systems serving fewer than 3,300 people may take grab samples at the frequencies shown previously. Disinfectant residuals in the distribution system must be monitored at the same frequency and location as total coliform samples.
3. Monitor daily to demonstrate that a combination of filtration and application of disinfectant achieves 99.9 percent and 99.99 percent removal or inactivation for *Giardia* and viruses, respectively.
4. Measurements for heterotrophic plate count (HPC) bacteria may be substituted for disinfectant residual measurements. If the HPC is less than 500 colonies per mL, then the sample is considered equivalent to a detectable disinfectant residual. For systems serving fewer than 500 people, the regulatory agency may determine the adequacy of the disinfectant residual in place of monitoring.

22.2225 CT VALUES

The purpose of the SWTR is to ensure that pathogenic organisms are removed or inactivated by the treatment process. To meet this goal, all systems are required to disinfect their water supplies. For some water systems using very clean source water and meeting the other criteria to avoid filtration, disinfection alone can achieve the 3 log (99.9 percent) *Giardia* and 4 log (99.99 percent) virus inactivation levels required by the Surface Water Treatment Rule.

Several methods of disinfection are in common use, including free chlorination, chloramination, use of chlorine dioxide, and application of ozone. The concentration of chemical needed and the length of contact time needed to ensure disinfection are different for each disinfectant. Therefore, the effectiveness of the disinfectant is measured by the time (T) in minutes of the disinfectant's contact in the water and the concentration (C) of the disinfectant residual in mg/L measured at the end of the contact time. The product of these two factors (C × T) provides a measure of the degree of pathogenic inactivation. The required CT value to achieve pathogenic inactivation is dependent upon the organism in question, type of disinfectant, pH, and temperature of the water supply.

Time or T is measured from point of application to the point where C is determined. T must be based on peak hour flow rate conditions. In pipelines, T is calculated by dividing the volume of the pipeline in gallons by the flow rate in gallons per minute (GPM). In reservoirs and basins, dye tracer tests must be used to determine T. In this case, T is the time it takes for 10 percent of the tracer to pass the measuring point.

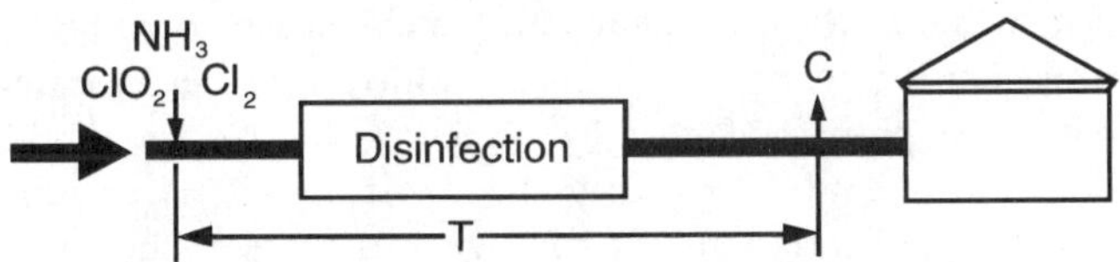

A properly operated filtration system can achieve limited removal or inactivation of microorganisms. Because of this, systems that are required to filter their water are permitted to apply a factor that represents the microorganism removal value of filtration when calculating CT values to meet the disinfection requirements. The factor (removal credit) varies with the type of filtration system. Its purpose is to take into account the combined effect of both disinfection and filtration in meeting the SWTR microbial standards.

Please refer to the Arithmetic Appendix at the end of this manual (Section A.6, "Calculation of CT Values") for instructions on how to perform these calculations for a water treatment plant.

22.2226 INTERIM ENHANCED SURFACE WATER TREATMENT RULE (IESWTR)

Code of Federal Regulations Reference: 40 CFR 141

Purpose of Regulation: To improve public health protection.

Public Health Benefit: To prevent significant increases in microbial contamination, particularly *Cryptosporidium.*

MCL/MCLG for *Cryptosporidium*: 0 mg/L/zero

The IESWTR was promulgated by EPA in 1998 (*Federal Register* 63, No. 241). Most provisions of the IESWTR apply to all public water systems using surface water or groundwater under the direct influence of surface water and serving 10,000 or more people.

Cryptosporidium was included in filtration avoidance criteria and groundwater under the direct influence of surface water determinations. These include tightening of combined filter effluent turbidities to less than 0.3 NTU 95 percent of the time and less than 1 NTU all the time for conventional and direct filtration plants. Monitoring of individual filter effluent turbidities is also required, although exceedances are not violations, but only trigger filter studies. Systems must prepare disinfection profiles and benchmarks to be used in discussions with their primacy agencies. The IESWTR also prohibits the construction of any new uncovered finished water facilities after February 1999.

One provision of the IESWTR applies to all systems using surface water or groundwater under the influence of surface water, regardless of size: the state must conduct a periodic sanitary survey that covers the entire system from sources through treatment plant, distribution, and storage. Significant deficiencies must be addressed in a timely manner.

Large systems (serving more than 10,000 persons) that use surface water or groundwater under the direct influence of surface water were required to comply with the IESWTR by January 2002. Small systems (serving fewer than 10,000) that use surface water or groundwater under the direct influence of surface water should have complied by January 1, 2004. The full text of the rule is available at the EPA home page (epa.gov).

22.2227 LONG TERM 1 ENHANCED SURFACE WATER TREATMENT RULE (LT1ESWTR)

Code of Federal Regulations Reference: 40 CFR 141.500–141.571

Purpose of Regulation: To improve public health protection.

Public Health Benefit: To prevent illness associated with *Cryptosporidium.*

The Long Term 1 Enhanced Surface Water Treatment Rule (LT1ESWTR) was promulgated by EPA in 2002. The provisions of the LT1ESWTR apply to public water systems that use surface water or groundwater under the direct influence (GWUDI) of surface water and serve fewer than 10,000 people. The LT1ESWTR is the smaller system counterpart of the IESWTR.

The major provisions of the LT1ESWTR include control of *Cryptosporidium* and combined filter effluent (CFE) turbidity performance standards. Provisions for the control of *Cryptosporidium* include the MCLG of zero for *Cryptosporidium.* Filtered water systems must physically remove 99 percent (2 log removal) of the *Cryptosporidium.* Unfiltered systems must update their watershed control programs to minimize the potential for contamination of *Cryptosporidium* oocysts. Also, *Cryptosporidium* is included as an indicator that the groundwater is under the direct influence of surface water.

The specific CFE turbidity requirements depend on the type of filtration used by the system. For conventional and direct filtration systems, the turbidity must be less than or equal to 0.3 NTU in at least 95 percent of the measurements taken each month and the maximum level of turbidity is 1 NTU. Slow sand and diatomaceous earth (DE) filtration systems must meet the CFE turbidity limits specified in the SWTR of less than 1 NTU in at least 95 percent of the measurements taken each month and a maximum level of turbidity of 5 NTU. Alternative technologies (other than conventional, direct, slow sand, or DE) must meet turbidity levels established by the drinking water authority based on filter demonstration data submitted by the system, but these limits must not exceed 1 NTU (in at least 95 percent of the measurements) or 5 NTU (maximum).

Turbidity monitoring requirements for CFE include monitoring performed at least every four hours to ensure compliance with CFE turbidity performance standards. Since the CFE may meet regulatory requirements even though one filter is producing high-turbidity water, the individual filter effluent (IFE) is measured to assist conventional and direct filtration treatment plant operators in understanding and assessing individual filter performance. The IFE monitoring requirements include monitoring continuously (recorded at least every 15 minutes). Systems with two or fewer filters may conduct continuous monitoring of CFE turbidity in place of individual filter effluent monitoring. Follow-up actions (i.e., additional reporting, filter self-assessment, or comprehensive performance evaluations (CPEs)) are required if the IFE turbidity (or CFE for systems with two filters) exceeds 1.0 NTU in two consecutive readings or more.

22.2228 LONG TERM 2 ENHANCED SURFACE WATER TREATMENT RULE (LT2ESWTR)

Code of Federal Regulations Reference: 40 CFR 141.700–141.722

Purpose of Regulation: To improve public health protection.

Public Health Benefit: To prevent gastrointestinal illnesses and problems associated with DBPs.

In 2005, the EPA promulgated the Long Term 2 Enhanced Surface Water Treatment Rule (LT2ESWTR or LT2). The purpose of this rule is to reduce illness linked with the contaminant *Cryptosporidium* and other pathogenic microorganisms in drinking water. The LT2ESWTR supplements existing regulations by targeting additional *Cryptosporidium* treatment requirements to higher-risk systems. This rule also contains provisions to reduce risks from uncovered finished water reservoirs and provisions to ensure that systems maintain microbial protection when they take steps to decrease the formation of DBPs that result from chemical water treatment.

Current regulations require filtered water systems to reduce source water *Cryptosporidium* levels by 2 log (99 percent). Data on *Cryptosporidium* infectivity and occurrence indicate that this treatment requirement is sufficient for most systems, but additional treatment is necessary for certain higher-risk systems. These higher-risk systems include filtered water systems with high levels of *Cryptosporidium* in their water sources and all unfiltered water systems that do not treat for *Cryptosporidium*.

The LT2ESWTR was promulgated simultaneously with the Stage 2 DBPR to address concerns about risk tradeoffs between pathogens and DBPs.

Monitoring. Under the LT2ESWTR, systems will monitor their water sources to determine treatment requirements. This monitoring includes an initial two years of monthly sampling for *Cryptosporidium*. To reduce monitoring costs, small filtered water systems will first monitor for *E. coli* (a bacterium that is less expensive to analyze than *Cryptosporidium*) and will monitor for *Cryptosporidium* only if their *E. coli* results exceed specified concentration levels.

Monitoring dates are staggered by system size, with smaller systems beginning monitoring after larger systems. Systems must conduct a second round of monitoring six years after completing the initial round to determine if source water conditions have changed significantly. Systems may use (grandfather) previously collected data in lieu of conducting new monitoring, and systems are not required to monitor if they provide the maximum level of treatment required under the rule.

***Cryptosporidium* Treatment.** Filtered water systems will be classified in one of four treatment categories (bins) based on their monitoring results. The majority of systems will be classified in the lowest treatment bin, which carries no additional treatment requirements. Systems classified in higher treatment bins must provide 90 to 99.7 percent (1.0 to 2.5 log) additional treatment for *Cryptosporidium*. Systems will select from a wide range of treatment and management strategies in the microbial toolbox to meet their additional treatment requirements. All unfiltered water systems must provide at least 99 or 99.9 percent (2 or 3 log) inactivation of *Cryptosporidium* depending on the results of their monitoring. These *Cryptosporidium* treatment requirements reflect consensus recommendations of the Stage 2 Microbial and Disinfection Byproducts Federal Advisory Committee.

Other Requirements. Systems that store treated water in open reservoirs must either cover the reservoir or treat the reservoir discharge to inactivate 4 log virus, 3 log *Giardia lamblia*, and 2 log *Cryptosporidium*. These requirements are necessary to protect against the contamination of water that occurs in open reservoirs. In addition, systems must review their current level of microbial treatment before making a significant change in their disinfection practice. This review will assist systems to reduce the formation of DBPs under the Stage 2 DBPR, which EPA has finalized along with the LT2ESWTR.

QUESTIONS

Please write your answers to the following questions and compare them with those on page 606.

22.2N What do the Surface Water Treatment Rules specifically require?

22.2O How do the Surface Water Treatment Rules define surface water?

22.2P How can a water utility avoid mandatory filtration?

22.2Q What are the MCLs for turbidity for surface waters?

22.2R How is the effectiveness of a disinfectant measured?

22.23 Disinfectants and Disinfection Byproducts (DBP)

Disinfection of drinking water by the addition of chlorine has long been considered a highly effective yet relatively low-cost method of preventing widespread outbreaks of waterborne diseases. In addition to reacting with disease-causing organisms in water, however, chlorine also reacts with many other types of organic materials. There is growing scientific evidence that the byproducts of these chemical reactions can produce adverse health effects in humans.

The highest priority health risk concern in the regulation of drinking water is the potential risk-risk tradeoff between the control of microbiological contamination (bacteria, viruses, and protozoa) on one side and DBPs on the other. This risk-risk tradeoff arises because, typically, the least expensive way for a public water system to increase microbial control is to increase disinfection (which generally increases byproduct formation) and the easiest way to reduce byproducts is to decrease disinfection (which generally increases microbial risk). Microbiological contamination often causes flu-like symptoms, but can also cause serious diseases such as hepatitis, giardiasis, cryptosporidiosis, and Legionnaire's Disease. DBPs may pose the risk of cancer and developmental effects.

THMs are an example of a compound formed by the reaction of chlorine with organic matter in water (also see Chapter 15, "Specialized Treatment Processes"). THMs are suspected of being carcinogenic and have been regulated by EPA in the 1996 SDWA amendments. The MCL for TTHMs is 0.080 milligram per liter or 80 micrograms per liter.

In May 1996, EPA published the Information Collection Rule (ICR). This rule required large public water systems to undertake extensive monitoring of microbial contaminants and DBPs in their water systems. Also, some water systems conducted studies on the use of granular activated carbon and membrane processes. The data reported under the ICR were used by EPA to learn more about the occurrence of microbial contamination and DBPs, the health risks posed, appropriate analytical methods, and effective forms of treatment. The ICR data form the scientific basis for EPA's development of the Enhanced Surface Water Treatment Rule and the Disinfectants and Disinfection Byproducts Rule.

EPA issued the Stage 1 DBPR on December 16, 1998 (*FEDERAL REGISTER* 63, No. 241). This rule set new MCLGs and MCLs for TTHMs, HAA5, bromate, and chlorite. Maximum residual disinfectant level goals (MRDLGs) and maximum residual disinfectant levels (MRDLs) have also been set for chlorine, chloramine, and chlorine dioxide, as follows:

Disinfection Byproducts	MCLG (mg/L)	MCL (mg/L)
TTHM		0.080
Bromoform	0	
Chloroform	0	
Bromodichloromethane	0	
Dibromochloromethane	0.06	
HAA5		0.060
Dichloroacetic acid	0	
Trichloroacetic acid	0.3	
Bromate	0	0.010
Chlorite	0.8	1.0
Disinfectants	**MRDLG (mg/L)**	**MRDL (mg/L)**
Chlorine	4 (as Cl_2)	4.0 (as Cl_2)
Chloramine	4 (as Cl_2)	4.0 (as Cl_2)
Chlorine dioxide	0.8 (as ClO_2)	0.8 (as ClO_2)

The Stage 1 DBPR attempts to further reduce potential formation of harmful DBPs by requiring the removal of THM precursors. A treatment technique of enhanced coagulation, enhanced softening, or use of granular activated carbon (GAC) applies to conventional filtration systems. In most cases, systems must reduce total organic carbon (TOC) levels based on specific source water quality factors.

For large systems (serving more than 10,000 persons) that use surface water or groundwater under the direct influence of surface water, the compliance date for the Stage 1 DBPR was January 1, 2002. Small systems (serving fewer than 10,000) that use surface water or groundwater under the direct influence of surface water and all groundwater systems must have complied by January 1, 2004. The full text of the Stage 1 DBPR is available by visiting EPA's website at water.epa.gov/drink/standardsriskmanagement.cfm.

The Stage 1 DBPR has very specific laboratory and monitoring requirements. The routine monitoring requirements include the following regulated contaminants/disinfectants:

- TTHM/HAA5
- Bromate
- Chlorite
- Chlorine/chloramines
- Chlorine dioxide
- DBP precursors (TOC/alkalinity/specific UV absorbance)

Also, the Stage 1 DBPR specifies the monitoring coverage in terms of surface waters, groundwaters, and groundwater under direct influence (GWUDI), population served, and the type of filtration system and disinfection system. Monitoring frequency depends on the type of source water, population served, and type of treatment and disinfection system.

The routine monitoring requirements are based on the regulated contaminants/disinfectants and include the MCL, MRDL, analytical method, preservation/quenching agent, holding time for sample/extract, and sample container size and type.

On December 15, 2005, EPA promulgated the Stage 2 DBPR. This rule will reduce potential cancer and reproductive and developmental health risks from DBPs in drinking water, which form when disinfectants are used to control microbial pathogens.

This final rule strengthens public health protection for consumers by tightening compliance monitoring requirements for two groups of DBPs: TTHMs and HAA5. The rule targets systems with the greatest risk and builds incrementally on existing rules. This regulation reduces DBP exposure and related potential health risks and provides more equitable public health protection.

The Stage 2 DBPR was promulgated simultaneously with the Long Term 2 Enhanced Surface Water Treatment Rule to address concerns about risk tradeoffs between pathogens and DBPs.

Under the Stage 2 DBPR, systems will conduct an evaluation of their distribution systems, known as an Initial Distribution System Evaluation (IDSE), to identify the locations with high DBP concentrations. These locations will then be used by the systems as the sampling sites for Stage 2 DBPR compliance monitoring.

Compliance with the maximum contaminant levels for two groups of DBPs (TTHMs and HAA5) will be calculated for each monitoring location in the distribution system. This approach, referred to as the locational running annual average (LRAA), differs from previous requirements, which determine compliance by calculating the running annual average of samples from all monitoring locations across the system.

The Stage 2 DBPR also requires each system to determine if they have exceeded an operational evaluation level, which is identified using their compliance monitoring results. The operational evaluation level provides an early warning of possible future MCL violations, which allows the system to take proactive steps to remain in compliance. A system that exceeds an operational evaluation level is required to review their operational practices and submit a report to their state that identifies actions that may be taken to mitigate future high DBP levels, particularly those that may jeopardize their compliance with the DBP MCLs.

22.24 Radiological Standards

Code of Federal Regulations Reference: 40 CFR 141.24

Radioactivity is the only contaminant that has been shown to cause cancer for which standards have been set. Three radioactive elements, radon, radium, and uranium, occur naturally in the ground and dissolve into groundwater supplies. However, the possible exposure to radiation in drinking water is only a fraction of the exposure from all natural sources. The main source of radioactive material in surface water is fallout from nuclear testing. Other sources could be nuclear power plants, nuclear fuel processing plants, and uranium mines. Those sources are monitored constantly and there is no great risk of contamination, barring accidents.

Alpha and radium radioactivity occur naturally in groundwater in parts of the West, Midwest, and Northeast. Standards for those types of radioactivity and for manmade, or beta radiation, have been set at levels of safety comparable to other contaminants.

The MCLs for radiological contaminants are divided into two categories: natural radioactivity that results from well water passing through deposits of naturally occurring radioactive materials and manmade radioactivity such as might result from industrial wastes, hospitals, or research laboratories. Table 22.5 summarizes the MCLs for radioactivity.

TABLE 22.5 MCLs FOR RADIOACTIVITY

Constituent	Maximum Contaminant Level
Combined Radium 226 and Radium 228	5 pCi/L[a]
Gross Alpha Activity (including Radium 226 but excluding Radon and Uranium)	15 pCi/L[a]
Beta/Photon Emitters[b]	4 mrem/yr
Uranium	30 μg/L

a pCi/L. picoCurie per Liter. A picoCurie is a measure of radioactivity. One picoCurie of radioactivity is equivalent to 0.037 nuclear disintegrations per second.

b For surface water systems serving more than 100,000 persons and systems determined by the state to be vulnerable.

Monitoring for natural radioactivity contamination is required every four years for both surface water and groundwater community systems. Routine monitoring procedures to follow are as follows:

1. Test for gross alpha activity; if gross alpha exceeds 5 pCi/L, then
2. Test for radium 226; if radium 226 exceeds 3 pCi/L, then
3. Test for radium 228.

The MCL for gross beta particle activity is 4 mrem/yr for water systems using surface water and serving more than 100,000 people, and for any other community system determined by the state to be vulnerable to this type of contamination. After the initial monitoring period, beta radiation must be monitored every 4 years unless the MCL is exceeded or the state determines that more frequent monitoring is appropriate.

Regulation of radon continues to be a source of considerable debate. EPA proposed a maximum contaminant level in 1991 and then withdrew the proposed MCL in 1997 in response to the 1996 SDWA amendments. EPA published a proposed regulation for radon in 1999, but the final MCL has not been published.

The final Radionuclides Rule was published in 2000. The rule retains the current standards for radium 226 and 228, gross alpha activity, and gross beta particle activity (see Table 22.5), but establishes a new MCL 30 µg/L for uranium. The EPA has indicated that it will continue to review the standard for beta particle activity and may revise the MCL in the future. EPA must base its new rules on a scientific risk assessment and an analysis of both the amount of health risk reduction that will be achieved and the cost of implementing the new rules.

QUESTIONS

Please write your answers to the following questions and compare them with those on page 607.

22.2S What are trihalomethanes (THMs)?

22.2T How does drinking water become contaminated with radioactive elements?

22.2U The MCLs for radiological contaminants are divided into what two categories?

END OF LESSON 1 OF 2 LESSONS

on

DRINKING WATER REGULATIONS

Please answer the discussion and review questions next.

DISCUSSION AND REVIEW QUESTIONS

Chapter 22. DRINKING WATER REGULATIONS

(Lesson 1 of 2 Lessons)

At the end of each lesson in this chapter, you will find discussion and review questions. Please write your answers to these questions to determine how well you understand the material in the lesson.

1. What criteria must be met before EPA selects a contaminant for regulation?
2. What are maximum contaminant levels (MCLs)?
3. What is the definition of a public water system?
4. What is a community water system?
5. Inorganic chemicals pose what type of health concerns?
6. What are the basic requirements of the Lead and Copper Rule?
7. What are the most common sources of organic contamination of drinking water?
8. Why are coliform bacteria found in drinking water and what does their presence indicate?
9. What is the purpose of the Surface Water Treatment Rules?
10. Why is turbidity undesirable in a finished or treated water?
11. Why are trihalomethanes (THMs) regulated?
12. What were the requirements and purposes of the Information Collection Rule?
13. What are the two categories of MCLs for radiological contaminants?

CHAPTER 22. DRINKING WATER REGULATIONS

(Lesson 2 of 2 Lessons)

22.3 SECONDARY DRINKING WATER STANDARDS

22.30 Enforcement of Regulations

The National Secondary Drinking Water Regulations control contaminants in drinking water that primarily affect the aesthetic qualities relating to the public acceptance of drinking water. At considerably higher concentrations of these contaminants, health implications may also exist as well as aesthetic degradation. These regulations are not federally enforceable; however, some states have passed laws requiring the state health agency to enforce the regulations.

22.31 Secondary Maximum Contaminant Levels

Secondary Maximum Contaminant Levels (SMCLs) apply to public water systems and, in the judgment of the EPA Administrator, are necessary to protect the public welfare or for public acceptance of the drinking water. The SMCL means the maximum permissible level of a contaminant that is delivered to the free-flowing outlet of the ultimate user of a public water system. Contaminants added to the water under circumstances controlled by the user, except those resulting from corrosion of piping and plumbing caused by water quality, are excluded by definition. The only exceptions are those resulting from corrosion of piping and plumbing caused by water quality. Currently there are 15 secondary standards (see Table 22.6).

States may establish higher or lower levels depending on local conditions, providing that public health and welfare are adequately protected.

Aesthetic qualities are important factors in public acceptance and confidence in a public water system. States are encouraged to implement SMCLs so that the public will not be driven to obtain drinking water from potentially lower-quality, higher-risk sources. Many states have chosen to enforce both Primary and SMCLs to ensure that the consumer is provided with the best quality water available.

TABLE 22.6 SECONDARY DRINKING WATER STANDARDS

Contaminant	SMCL	Effect
Aluminum	0.05 to 0.2 mg/L	precipitates cause cloudy looking water
Chloride	250 mg/L	taste and corrosion of water pipes
Color	15 Color Units	aesthetic
Copper	1.0 mg/L	staining
Corrosivity	Noncorrosive	corrosion control
Fluoride	2 mg/L	dental fluorosis (a brownish discoloration of the teeth)
Foaming Agents	0.5 mg/L	aesthetic and taste
Iron	0.3 mg/L	taste and staining
Manganese	0.05 mg/L	taste and staining
Odor	3 Threshold Odor Number[a]	aesthetic and taste
pH	6.5 to 8.5	corrosion control and taste
Silver	0.10 mg/L	blue-gray discoloration of the skin, mucous membranes, and eyes
Sulfate	250 mg/L	taste and laxative effects
Total Dissolved Solids (TDS)	500 mg/L	taste
Zinc	5 mg/L	taste

a Threshold Odor Number (TON). The greatest dilution of a sample with odor-free water that still yields a just-detectable odor.

22.32 Monitoring

Collect samples for secondary contaminants at a free-flowing outlet of water being delivered to the consumer. Monitor contaminants in these regulations at least as often as the monitoring performed for inorganic chemical contaminants listed in the Primary Drinking Water Regulations as applicable to community water systems. For surface water systems, this means yearly monitoring is required; for groundwater systems, monitor at least once every three years. Collect monthly distribution system physical water quality monitoring samples for color and odors. More frequent monitoring would be appropriate for specific contaminants such as pH, color, odor, or others, under certain circumstances as directed by the state.

QUESTIONS

Please write your answers to the following questions and compare them with those on page 607.

22.3A Under what conditions are secondary drinking water regulations enforceable?

22.3B List the secondary drinking water contaminants.

22.3C How frequently should the contaminants in the secondary regulations be monitored?

22.33 Secondary Contaminants

22.330 Aluminum

Until relatively recently, most environmental aluminum was found in forms that did not greatly affect humans and other animals. Acid rain, however, has caused a significant increase in aluminum exposure. The average daily intake of aluminum in the general population is about 20 milligrams per day; the SMCL is 0.05 to 0.2 mg/L. Although aluminum is regulated as a secondary contaminant because precipitates of the metal can cause cloudy-looking water, aluminum compounds may also interfere with absorption of fluoride in the gastrointestinal tract and may decrease the absorption of calcium, iron, and cholesterol. Many over-the-counter antacids contain aluminum and use of these products frequently causes constipation, which is thought to be the result of the gastrointestinal effects described above.

22.331 Chloride

The SMCL for chloride is 250 mg/L.

UNDESIRABLE EFFECTS

1. Objectionable salty taste in water
2. Corrosion of the pipes in hot water systems and other pipelines

STUDIES ON THE MINERALIZATION OF WATER INDICATE THE FOLLOWING

1. Major taste effects are produced by anions (where TDS was studied).
2. Chloride produces a taste effect somewhere between the milder sulfate and the stronger carbonate.
3. Laxative effects are caused by highly mineralized waters. Mineralized waters often contain chloride as well as high levels of sodium and magnesium sulfate.

CORROSION EFFECT

1. Studies indicate that corrosion depends on concentration of TDS (TDS may contain 50 percent chloride ions).
2. Domestic plumbing, water heaters, and municipal waterworks equipment will deteriorate when high concentrations of chloride ions are present.

EXAMPLE

Where the TDS = 200 mg/L (Chloride = 100 mg/L), water heater life will range from 10 to 13 years. Water heater life declines uniformly as a function of TDS—1 year shortened life per 200 mg/L additional TDS.

22.332 Color

The SMCL for color is 15 color units. The level of this water quality indicator is not known to be a measure of the safety of water. However, high color content may indicate the following:

1. High organic chemical contamination
2. Inadequate treatment
3. High disinfectant demand and the potential for production of excess amounts of DBPs

Color may be caused by the following:

1. Natural color-causing solids such as aromatic, polyhydroxy, methoxy, and carboxylic acids
2. Fulvic and humic acid fractions
3. Presence of metals such as copper, iron, and manganese

Rapid changes in color levels may provoke more citizen complaints than relatively high, constant color levels.

22.333 Copper

The SMCL for copper is 1.0 mg/L. Soft water containing a low level (0.5 mg/L) of copper may cause blue or blue-green staining of porcelain. Higher levels (4 mg/L) of copper will stain clothing and blond hair. When soap is used with water having a copper concentration greater than 1 mg/L, insoluble green curds will form.

22.334 Corrosivity

The corrosivity of water indicates the rate at which water causes the gradual decomposition or destruction of a material (such as a metal or cement lining). The severity and type of corrosivity are dependent on the chemical and physical characteristics of the water and the material.

Corrosivity causes materials to deteriorate and go into solution (be carried by the water). Corrosion of toxic metal pipe materials such as lead can create a serious health hazard. Corrosion of iron may produce a flood of unpleasant telephone calls from consumers complaining about rusty water, stained laundry, and bad tastes.

Corrosivity can cause the reduction of the carrying capacity of a water main. This reduced carrying capacity can cause an increase in pump energy costs and may reduce distribution system pressures. Leaks in water mains due to corrosivity may eventually require replacement of a water main.

22.335 Fluoride

Fluoride produces two effects, depending on its concentration. At levels of 6 to 8 mg/L, fluoride may cause skeletal fluorosis (the bones become brittle) and stiffening of the joints. For this reason, fluoride has been added to the list of primary standards (those that have health effects).

At levels of 2 mg/L and greater, fluoride may cause dental fluorosis, which is discoloration and mottling of the teeth, especially in children. EPA has recently reclassified dental fluorosis as a cosmetic effect, raised the primary drinking water standard from 1.4 to 2 mg/L to 4 mg/L, and established a secondary standard of 2 mg/L for fluoride.

22.336 Foaming Agents

The SMCL for foaming agents is 0.5 mg/L.

UNDESIRABLE EFFECTS

1. Causes frothing and foaming, which are associated with contamination (greater than 1.0 mg/L)
2. Imparts an unpleasant taste (oily, fishy, perfume-like) (less than 1.0 mg/L)

INFORMATION ITEMS

1. Because no convenient foamability test exists and because *SURFACTANTS*[7] are one major class of substances that cause foaming, this property is determined indirectly by measuring the anionic surfactant (*MBAS*[8]) concentration in the water.
2. Surfactants are synthetic organic chemicals and are the principal ingredient of household detergents.
3. The requirement for biodegradability led to the widespread use of linear alkylbenzene sulfonate (LAS), an anionic surfactant.
4. Concentrations of anionic surfactants found in drinking waters range from 0 to 2.6 mg/L in well supplies and 0 to 5 mg/L in surface water supplies.
5. LAS is essentially odorless. The odor and taste characteristics are likely to arise from the degradation of waste products rather than the detergents.
6. If water contains an average concentration of 10 mg/L surfactants, the water is likely to be entirely of wastewater origin.
7. From a toxicological standpoint, an MCL of 0.5 mg/L, assuming a daily adult human water intake of 2 liters, would give a safety factor of 15,000.

22.337 Iron and Manganese

1. Iron and manganese are frequently found together in natural waters and produce similar adverse environmental effects and color problems. Excessive amounts of iron and manganese are usually found in groundwater and in surface water contaminated by industrial waste discharges.
2. Before 1962, both were covered by a single recommended limit.
3. In 1962, the US Public Health Service recommended separate limits for both iron and manganese to reflect more accurately the levels at which adverse effects occur for each.
4. Both are highly objectionable in large amounts in water supplies for either domestic or industrial use.
5. Both impart color to laundered goods and plumbing fixtures.
6. Taste thresholds in drinking water are considerably higher than the levels that produce staining effects.
7. Both are part of our daily nutritional requirements, but these requirements are not met by the consumption of drinking water.

22.338 Iron

The SMCL for iron is 0.3 mg/L.

UNDESIRABLE EFFECTS

1. At levels greater than 0.05 mg/L some color may develop, staining of fixtures may occur, and precipitates may form.
2. The magnitude of the staining effect is directly proportional to the concentration.
3. Depending on the sensitivity of taste perception, a bitter, astringent taste can be detected from 0.1 mg/L to 1.0 mg/L.
4. Precipitates that are formed create not only color problems but also lead to bacterial growth of slimes and of the iron-loving bacteria, *Crenothrix*, in wells and distribution piping.

NUTRITIONAL REQUIREMENTS

1. Daily requirement is 1 to 2 mg, but intake of larger quantities is required as a result of poor absorption.
2. The limited amount of iron permitted in water (because of objectionable taste or staining effects) constitutes only a small fraction of the amount normally consumed and does not have toxicologic (poisonous) significance.

22.339 Manganese

The SMCL for manganese is 0.05 mg/L.

UNDESIRABLE EFFECTS

1. A concentration of more than 0.02 mg/L may cause buildup of coatings in distribution piping.

---

7. *Surfactant* (sir-FAC-tent). Abbreviation for surface-active agent. The active agent in detergents that possesses a high cleaning ability.
8. *MBAS.* Methylene Blue Active Substance. Another name for surfactants or surface active agents. The determination of surfactants is accomplished by measuring the color change in a standard solution of methylene blue dye.

2. If these coatings slough off, they can cause brown blotches in laundry items and black precipitates.
3. Manganese imparts a taste to water above 0.15 mg/L.
4. The application of chlorine, even at low levels, increases the likelihood of precipitation of manganese at low levels.
5. Unless the precipitate is removed, precipitates reaching pipelines will promote bacterial growth.

TOXIC EFFECTS

1. Toxic effects are reported as a result of inhalation of manganese dust or fumes.
2. Liver cirrhosis has arisen in controlled feeding of rats.
3. Neurological effects have been suggested; however, these effects have not been scientifically confirmed.

NUTRITIONAL REQUIREMENTS

1. Daily intake of manganese from a normal diet is about 10 mg.
2. Manganese is essential for proper nutrition.
3. Diets deficient in manganese will interfere with growth, blood and bone formation, and reproduction.

QUESTIONS

Please write your answers to the following questions and compare them with those on page 607.

22.3D Why is chloride a secondary contaminant?

22.3E What are surfactants?

22.3F What is the impact of chlorine on manganese?

22.3310 Odor

The SMCL for odor is a *THRESHOLD ODOR NUMBER (TON*[9]*)* of 3. Important facts to remember when dealing with odors include the following:

1. Taste and odor go hand-in-hand.
2. Absence of taste and odor helps to maintain the consumers' confidence in the quality of their water, even though it does not guarantee that the water is safe.
3. Research indicates that there are only four true taste sensations: sour, sweet, salty, and bitter.
4. All other sensations ascribed to the sense of taste are actually odors, even though the sensation is not noticed until the material is taken into the mouth.
5. Odor tests are less fatiguing to people than taste tests when testing for tastes and odors.
6. Taste and odor tests are useful:
   a. As a check on the quality of raw[10] and treated water
   b. To help control odor throughout the plant
7. Odor is a useful test:
   a. For determining the effectiveness of different kinds of treatment
   b. As a means for tracing the source of contaminants
8. Hydrogen sulfide is included under the odor SMCL.

22.3311 pH

The SMCL for pH is defined as pH values beyond the acceptable range from 6.5 to 8.5. A wide range of pH values in drinking water can be tolerated by consumers.

UNDESIRABLE EFFECTS

1. When the pH increases, the disinfection activity of chlorine falls significantly.
2. High pH may cause increased halogen reactions, which produce chloroform and other THMs during chlorination.
3. Both excessively high and low pHs may cause increased corrosivity, which can, in turn, create taste problems, staining problems, and significant health hazards.
4. Metallic piping in contact with low-pH water will impart a metallic taste.
5. If the piping is iron or copper, high pH will cause oxide and carbonate compounds to be deposited, leaving red or green stains on fixtures and laundry.
6. At a high pH, drinking water acquires a bitter taste.
7. The high degree of mineralization often associated with basic waters results in encrustation of water pipes and water-using appliances.

22.3312 Silver

Silver is a nonessential element, providing no beneficial effects from its ingestion in trace amounts. Chronic toxicity causes an unsightly blue-gray discoloration of the skin, mucous membranes, and eyes. Apparently, beside cosmetic changes, there are no physiologic effects. Ingestion of trace amounts of silver or silver salts results in its accumulation in the body, particularly the skin and eyes. There is some evidence that changes to the kidneys, liver, and spleen can occur. The SMCL for silver is 0.10 mg/L.

22.3313 Sulfate

The SMCL for sulfate is 250 mg/L.

UNDESIRABLE EFFECTS AT HIGH LEVELS

1. Tends to form hard scales in boilers and heat exchangers.
2. Causes taste effects.
3. Causes a laxative effect. This effect is commonly noted by newcomers or casual or intermittent users of water high in sulfate. Water containing more than 750 mg/L of sulfate usually produces the laxative effect while water with less than

9. *Threshold Odor Number (TON).* The greatest dilution of a sample with odor-free water that still yields a just-detectable odor.
10. When testing raw water, be sure there are no pathogens or toxic chemicals present.

600 mg/L sulfate usually does not. An individual can adjust to sulfate in drinking water.

4. Sodium sulfate and magnesium sulfate are more active as laxatives, whereas calcium sulfate is less active.
5. When the magnesium sulfate content is 200 mg/L, the most sensitive person will feel the laxative effect; however, magnesium sulfate levels between 500 mg/L and 1,000 mg/L will induce diarrhea in most individuals.
6. Tastes may sometimes be detected at 200 mg/L of sulfate, but generally are detected in the range of 300 to 400 mg/L.

22.3314 Total Dissolved Solids (TDS)

The SMCL for total dissolved solids is 500 mg/L.

UNDESIRABLE EFFECTS

1. TDS imparts adverse taste effects at greater than 500 mg/L.
2. Highly mineralized water influences the deterioration of distribution systems as well as domestic plumbing and appliances (the life of a water heater will decrease one year with each additional 200 mg/L of TDS above a typical 200 mg/L value).
3. Mineralization can also cause precipitates to form in boilers and other heating units, sludge in freezing processes, rings on utensils, and precipitates in food being cooked.
4. There may be a great difference between a detectable concentration and an objectionable concentration of the neutral salts. Many people can become accustomed to high levels.
5. Studies show that the temperature of mineralized waters influences their acceptability to the public.

22.3315 Zinc

The SMCL for zinc is 5 mg/L.

UNDESIRABLE EFFECTS

1. High concentrations of zinc produce adverse physiological effects.
2. Zinc imparts a bitter, astringent taste that is distinguishable at 4 mg/L. Also, at 4 mg/L a metallic taste will exist.
3. Zinc will cause a milky appearance in water at 30 mg/L.
4. Zinc may increase lead and cadmium concentrations.
5. The activity of several enzymes is dependent on zinc. Enzymes are proteins that catalyze chemical reactions. Zinc is an essential enzyme growth element in terms of nutrition at low doses (zinc must be present) and is toxic to enzymes at high doses.
6. Cadmium and lead are common contaminants of zinc used in galvanizing steel pipe.

PHYSIOLOGICAL EFFECTS

1. A concentration of 30 mg/L can cause nausea and fainting.
2. Zinc salts act as gastrointestinal irritants. This symptom of illness is acute and transitory.
3. The vomiting concentration range is 675 to 2,280 mg/L.
4. A wide margin of safety exists between normal food intake and concentrations in water high enough to cause oral toxicity.

DIETARY REQUIREMENTS

1. The daily requirement for preschool children is 0.3 mg Zn/kg of weight.
2. Total zinc in an adult human body averages two grams.
3. Zinc most likely concentrates in the retina of the eye and in the prostate.
4. Zinc deficiency in animals leads to growth retardation.

QUESTIONS

Please write your answers to the following questions and compare them with those on page 607.

22.3G What are the undesirable effects of abnormal pH values?

22.3H Why are high levels of sulfate undesirable in drinking water?

22.3I Why are high levels of zinc undesirable in drinking water?

22.4 SAMPLING PROCEDURES

This section discusses the regulations governing sampling, sampling procedures, reporting procedures, public notification, recordkeeping, and consumer confidence reporting requirements for water utilities.

Life cannot exist without water. Water utilities have the important responsibility of making sure that the water provided to their customers is safe to drink. This means that the water is free of bacteria or any substances that might cause disease or affect the human immune system. Making sure the water can be used for public consumption is required under the SDWA. There are specific requirements in the testing of routine water samples to ensure that the water meets the public water quality and bacteriological standards. This section is designed to provide water operators with information about setting up and implementing a public drinking water sampling program.

22.40 Safe Drinking Water Act Regulations

The SDWA and accompanying regulations require that you take the following actions to comply:

1. Sampling
2. Testing
3. Recordkeeping
4. Reporting

Understanding and implementing each of these steps will help ensure the success of your operation.

Operators are urged to become familiar with and stay current on state and federal drinking water rules and regulations. Contact your local health departments and state regulatory agencies for

Fig. 22.1 Water sampling station

(Reprinted with permission from California Water Service Company, San Jose, California)

state requirements. For information about federal drinking water regulations, visit the EPA online at water.epa.gov/drink/hotline/ or call the Safe Drinking Water Hotline at (800) 426-4791.

22.41 Overview of Sampling

What is sampling? Sampling is the selection of a small portion of water to identify the characteristics of a drinking water supply. Sampling is used daily to test water for the purpose of making certain the water complies with public health regulations. Water samples are collected by water utility operators and then submitted to approved water quality laboratories for testing and analysis. These tests detect the absence or the presence of contamination that may affect public health. The value of a laboratory test depends on the quality of the samples collected. Therefore, it is important to understand the steps required to conduct proper water system sampling and analysis.

Water utilities are required to take two types of water samples: raw water and finished (treated) water samples. Raw water samples provide the specific characteristics of the water that will determine the way water is treated. Finished water samples provide the evidence of the effectiveness of the treatment process used by the water utility.

22.42 General Guidelines for Water Sampling

The following guidelines represent an overview of the steps in taking water samples. Water operators should obtain more specific information for delivering water to consumers by contacting their state regulatory agency.

1. **Using Dedicated Sampling Points or Taps**—Using dedicated sampling points will help prevent the contamination of water samples by dust, dirt, or other contaminants. Water utilities should construct sampling taps from source materials that will protect the samples from potential contamination. The use of dedicated sampling taps provides the necessary controls and protection of the water sampling procedures by the water utility. Figure 22.1 shows a typical water sampling station used to collect samples of drinking water for water quality compliance and public health protection.

2. **Labeling Water Sample Containers**—Each water sample should be clearly labeled using a waterproof pen and must include the following information:
   - Sample location
   - Sample code or number
   - Date and time of collection
   - Initials or name of sample collector
   - Type of analysis performed
   - Preservative
3. **Transporting Water Samples**—Samples need to be transported in sanitary coolers or ice chests to the laboratory. Dry ice should not be used in the cooler because it may cause the water samples to freeze. If the water samples freeze, the containers may break or if any microorganisms are indeed present, they may be killed. This would invalidate the sample. Sample containers should be packed securely, preferably in bubble-wrap sleeves to prevent breakage. When packing water samples, it is necessary to identify and separate raw water and finished water samples in different containers or trays. The water utility should provide the staff with trays or baskets to secure the water samples while they are being transported to the water quality laboratory (see Figure 22.2.)

   Water operators should contact the state regulatory agency with primacy over public water systems or personnel from the designated water quality laboratory for specific guidance and recommendations for transporting water quality samples. Figure 22.3 shows the type of container that can be used to carry water samples collected and then transported from the water sampling station to the water quality laboratory.

Fig. 22.2 Water sampling tray

4. **Collecting Representative Samples**—Flushing or running the water for 2 or 3 minutes will allow the water temperature to stabilize and displace any stagnant water to ensure a representative sample. Wells need to run at least 2 hours before sampling to stabilize the flow. After the flow has stabilized, it should not be adjusted while the sample is being taken. This will prevent contamination of the sample.

Fig. 22.3 Water sampling container/carrier

(Reprinted with permission from California Water Service Company, San Jose, California)

5. **Avoid Rinsing Sampling Containers**—Preservatives or quenching agents are used to coat the inside of sampling containers and sampling containers should not be rinsed. If the preserving agent is diluted, the sample may be contaminated and the results may be invalid. Again, do not rinse or overfill the sampling containers.

When collecting water samples, should there be any reason to doubt the loss of the preservative or if potential contamination of the container is suspected, use a new container. Be sure to label the old container to prevent reusing it by accident.

22.43 Selecting Sampling Locations

Water utilities should contact their state regulatory agency or county health department to obtain information about the number of samples the utility is required to collect.

Representative sample points are essential to certify that sampling results provide true indicators of water quality. Having representative samples will help to ensure that localized problems are detected within the scope of a water utility's entire distribution system. This will also help water utility personnel identify other specific problems such as low pressure, cross-connection control hazards, or deteriorating water lines or water mains. The following considerations are important in the development of a sample siting plan:

Acceptable sampling locations

- Pressure zones
- Sources
- Area distribution
- Customer and utility sampling points

Unacceptable sampling locations

- Drinking fountains
- Taps with aerators, strainers, or swivels
- Fire hydrants
- Restroom taps
- Home sites with attached hoses or leaking hose bibs
- Leaking taps

Interactive Sampling Guide for Drinking Water System Operators—EPA has developed an interactive CD and website as part of the agency's ongoing public drinking water sustainable infrastructure efforts. The purpose of this guide is to help drinking water system owners and operators better understand the general procedures involved in the collection of SDWA compliance samples. Additional sample shipping suggestions, sample tips, and sample requirements are made available within the body of this interactive guide. This information may be found on EPA's website water.epa.gov/type/drink/pws/smallsystems/samplingcd.cfm or by calling (800) 490-9198.

22.44 Use of Dedicated Sampling Stations

Dedicated sampling stations (commercially manufactured) are important because they minimize the potential sources of contamination. Although utilities may construct their own sampling stations, there are numerous sampling stations available commercially. There are several factors to consider in selecting a commercially manufactured sampling station:

- Exposure to weather and wind (both at rest and during sampling)
- Protection from vandalism
- Can be located near a water main
- Good drainage
- Aesthetically pleasing
- Narrow outlet port
- Has an even-flowing spigot and smooth operating ball valve

22.45 Sampling Points

Some of the samples required to determine compliance with primary regulations can be taken from the routine sampling points. By coordinating the present sampling points with the sampling required by the regulations, additional monitoring costs can be minimized. The number of sampling points required will depend on the specific size of the population served and the layout of each water system.

Samples for organics, inorganics, and turbidity must be taken at the points where water enters the distribution system, and samples collected for coliform bacteria must be taken at representative points within the distribution system.

At the very minimum, a small system (with a population of 25 to 1,000) must sample for turbidity and coliform bacteria and must have two sampling points:

1. One where the water enters the distribution system
2. One at a consumer faucet at a representative point in the distribution system

QUESTIONS

Please write your answers to the following questions and compare them with those on pages 607 and 608.

22.4A What is sampling?

22.4B What are the two types of water samples water utilities are required to take?

22.4C What topics are covered in the general guidelines for water sampling?

22.4D What are the minimum sampling requirements for a small system with a population of 100 people?

22.46 Sampling Point Selection

There are two major considerations in determining the number and location of sampling points:

1. They should be representative of each different surface water source entering the system.
2. They should be representative of conditions within the system such as dead ends, loops, storage facilities, and pressure zones.

22.47 Sampling Schedule

A sampling schedule should be prepared that indicates all of the samples that will be collected during a yearly period. The schedule should include the following information:

1. Sampling frequency
2. Sampling point designation
3. Location (address)
4. Type of test
5. Sample volume
6. Special handling instructions

In addition to the requirement for routine sampling, water utilities are required to perform bacteriological monitoring in the following situations:

- After construction or repair of wells
- After new main installations
- After construction, repair, or maintenance of storage facilities
- After any system event when the water pressure drops below 5 pounds per square inch (psi); psi is the minimum operating pressure within the distribution system and should not be less than 20 psi

Please note that the results of these samples are not considered part of the calculation of the average MCL for the month.

Each water utility should contact the state regulatory agency to create and review the sampling schedule to determine its adequacy in meeting the SDWA regulations. Figure 22.4 is an example of a sampling site plan worksheet, a form that is used to record the information being documented in the field by water operations personnel.

Illinois Environmental Protection Agency
Public Water Supply Lead and Copper Sample Site Plan

FACILTY NUMBER ___ ___ ___ ___ ___ ___ ___ **DATE** ___ ___/___ ___/___ ___

FACILITY NAME ____________________

AGENCY USE ONLY Sample Site Number	Status	Enter Sample Site Description (Address, Street/City, Rural route + Description, Name of Resident, etc.)	Primary Or Alternate (P or A)	Tier	Tier Type
L	A				
L	A				
L	A				
L	A				
L	A				

PREPARED BY ____________________ **PHONE NUMBER** (___ ___ ___) ___ ___ ___ - ___ ___ ___ ___

Fig. 22.4 Lead and copper sample site plan

(Reprinted with permission from the Compliance Assurance Section, Division of Public Water Supplies, Illinois Environmental Protection Agency)

It is important to note that each water utility is required to prepare a written sampling plan for coliform testing that must be approved by the state regulatory agency. Figures 22.5 and 22.6 are examples of forms used to report the information for sampling and testing for the presence of synthetic organic chemicals and inorganic chemicals in drinking water, used by water operations personnel to record water quality sampling information.

22.48 Sampling Route

After selection of the sampling points and preparation of the sampling schedule, the next step is to select a route. Arrange your route so that samples that must be analyzed immediately are not delayed while other sampling is done. Field data forms must be completed by the person doing the sampling and submitted to the laboratory with the samples.

22.49 Sample Collection

Good sampling techniques are the key to a meaningful and useful sampling program. The following eight steps are necessary to the collection of an acceptable sample:

1. Obtain a sample that is truly representative of the existing conditions.
2. Flush the line before sample collection (except when a first-draw sample is required, such as in testing for lead).
3. Fill the sample bottle without leaving any air pocket.
4. Analyze residual chlorine when the sample is taken.
5. Handle and store the sample so that it does not become contaminated before it reaches the laboratory.
6. Use preservation techniques. These preservation methods are generally pH control and refrigeration.

Illinois Environmental Protection Agency
Public Water Supply Lead and Copper Sample Site Plan

Instructions For Filling Out Lead and Copper Sample Site Data Input Form

1) Facility Number: Enter the seven digit facility number.
2) Facility Name: Print the name of your facility.
3) Sample Site Number and Status: Leave this portion of the form blank.
4) Sample Site Description: Enter the Address of the location where sample will be collected. One letter per box.
5) Primary or Alternate: Enter "P" for a primary sampling site or an "A" for an alternate sampling site.
6) Tier: Enter 1 for Tier 1: 2 for Tier 2: or a 3 for Tier 3. (see explanation below)

Tier 1: a) Includes single and multifamily structures that contain copper pipes with lead solder that was installed after 1982.
b) Lead pipes.
c) Is served by a lead service line.

Tier 2: a) Includes buildings that contain copper pipes with lead solder installed after 1982
b) Lead pipes
c) Is served by a lead service line.

Tier3: a) Includes single family structures that contain copper pipes with lead solder that were installed prior to 1983
b) If not enough Tier 1, 2, or 3 sites are available, then random sites may be chosen.

Tier	Tier Type
Tier 1	A – Single Family, copper pipe with lead solder constructed after 1982 B – Single Family, lead pipes C – Single Family, lead service line D – Multifamily, copper pipe with lead solder constructed after 1982 E – Multifamily, lead pipes F – Multifamily, lead service line
Tier 2	J – Building, copper pipe with lead solder constructed after 1982 K– Building, lead pipes L -Building, lead service line
Tier3	S – Single family, copper pipe with lead solder constructed before 1983 R – Random location

Fig. 22.4 Lead and copper sample site plan (continued)

(Reprinted with permission from the Compliance Assurance Section, Division of Public Water Supplies, Illinois Environmental Protection Agency)

SOC

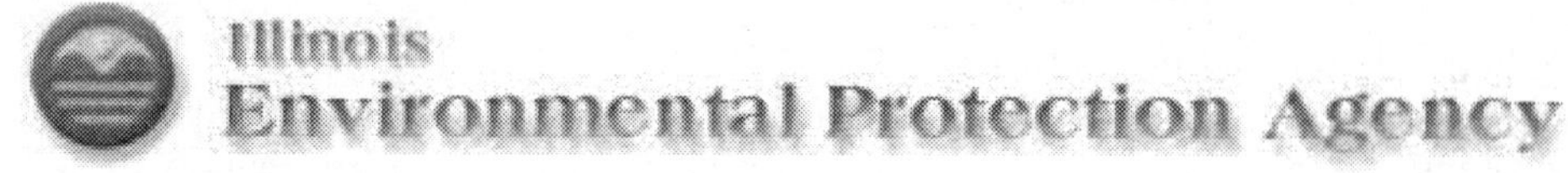

Water System Name: ______________________ Water System Number: __________

Analysis Report Form

-Water System Section-

Water System Name: ______________________

Water System Number: __________

-Sampling Point Section-

WSF State Asgn ID: __________ / Descrpt.: __________

Sampling Point: __________ / Descrpt.: __________

Collection Date (MMDDYYYY): __________

Collection Time: __________

Sample Collector Name Telephone No.: __________

Sample Purpose (Circle One): **Routine (RT)** **Repeat (RP)** **Special (SP)**

Sample Type (Circle One): **Finished (FN)** **Raw (RW)**

Fig. 22.5 Synthetic organic chemicals (SOC) water quality analysis report form

(Reprinted with permission from the Compliance Assurance Section, Division of Public Water Supplies, Illinois Environmental Protection Agency)

SOC

Illinois Environmental Protection Agency

Water System Name: ____________________ Water System Number: __________

- Required Sampling at Sample Point ________ -

Analyte Group Code: **SOC**

Analyte	Analyte Code	Method Code*	w/ Units of Measurement* Lab Reporting Level	Concentration
ENDRIN	2005			
BHC-GAMMA (LINDANE)	2010			
METHOXYCHLOR	2015			
TOXAPHENE	2020			
DALAPON	2031			
DIQUAT	2032			
ENDOTHALL	2033			
DI(2-ETHYLHEXYL) - ADIPATE	2035			
OXAMYL (VYDATE)	2036			
SIMAZINE	2037			
DI(2-ETHYLHEXYL) - PHTHALATE	2039			
PICLORAM	2040			
DINOSEB	2041			
HEXACHLOROCYCLOPENTADIENE	2042			
ALDICARB SULFOXIDE	2043			
ALDICARB SULFONE	2044			
CARBOFURAN	2046			
ALDICARB	2047			
ATRAZINE	2050			
ALACHLOR (LASSO)	2051			
HEPTACHLOR	2065			
HEPTACHLOR EPOXIDE	2067			
DIELDRIN	2070			
2,4-D	2105			
2,4,5-TP (SILVEX)	2110			
HEXACHLOROBENZENE	2274			
BENZO (A) PYRENE	2306			
PENTACHLOROPHENOL	2326			
ALDRIN	2356			
POLYCHLORINATED BIPHENYLS (PCB)	2383			
TOTAL DDT	2775			
DIBROMOCHLOROPROPANE (DBCP)	2931			
ETHYLENE DIBROMIDE (EDB)	2946			
CHLORDANE	2959			

Fig. 22.5 Synthetic organic chemicals (SOC) water quality analysis report form (continued)

(Reprinted with permission from the Compliance Assurance Section, Division of Public Water Supplies, Illinois Environmental Protection Agency)

SOC

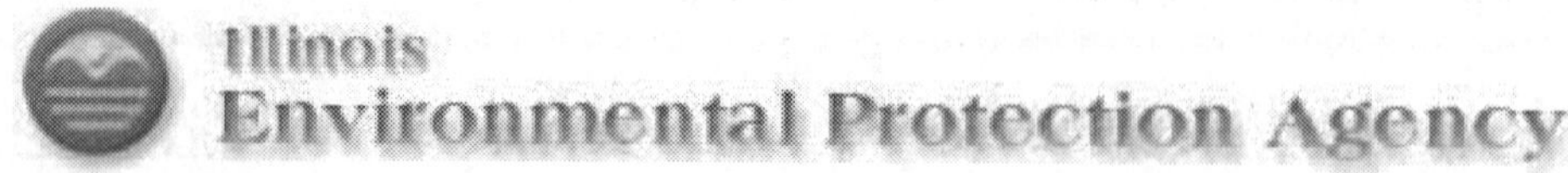

Water System Name: ______________________ Water System Number: __________

-Laboratory Section-

Laboratory State ID Number: ______________

Laboratory Name:

Lab Sample Number: ______________

Date Lab Rcpt.: ______________

Complete Date: ______________

Complete Time: ______________

Comments (Data Quality Issues):

____

Mail Results to: **Illinois Environmental Protection Agency**
Drinking Water Compliance Unit, Mailstop #19
1021 North Grand Avenue East, P.O. 19276
Springfield, IL 62704-9276

Questions Call: (217) 785-0561

Fax: (217) 557-1407

Signature of Analyst or Official ______________________

Date Forwarded

*** See List of Permitted Values**

This Agency is authorized to require this information under Illinois Revised Statutes, 1987, Chapter 111 1/2, Section 1004(H). Disclosure of this information is required. Failure to do so may result in a civil penalty up to $10,000.00 and an additional civil penalty up to $1,000.00 for each day the failure continues. a fine up to $1,000.00 and imprisonment up to one year. This form has been approved by the Forms Management Center.

Fig. 22.5 Synthetic organic chemicals (SOC) water quality analysis report form (continued)
(Reprinted with permission from the Compliance Assurance Section, Division of Public Water Supplies, Illinois Environmental Protection Agency)

IOC

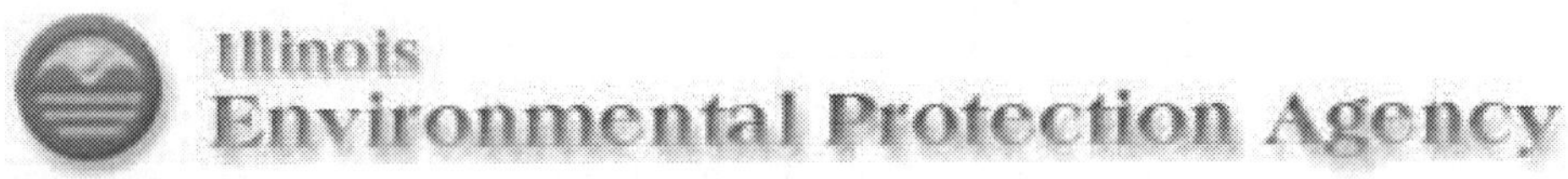

Illinois Environmental Protection Agency

Water System Name: ______________________ Water System Number: ____________

Analysis Report Form

-Water System Section-

Water System Name: ______________________

Water System Number: ______________________

-Sampling Point Section-

WSF State Asgn ID: ______________ / Descrpt.: ______________

Sampling Point: ______________ / Descrpt.: ______________

Collection Date (MMDDYYYY): ______________

Collection Time: ______________

Sample Collector Name _Telephone No.: ______________

Sample Purpose (Circle One): **Routine (RT)** **Repeat (RP)** **Special (SP)**

Sample Type (Circle One): **Finished (FN)** **Raw (RW)**

- Required Sampling at Sample Point ________ -

Analyte Group Code: **IOC**

Analyte	Analyte Code	Method Code*	Lab Reporting Level	Concentration w/ Units of Measurement*
ARSENIC	1005			
BARIUM	1010			
CADMIUM	1015			
CHROMIUM	1020			
CYANIDE	1024			
FLUORIDE	1025			
IRON	1028			
MANGANESE	1032			
MERCURY	1035			
NICKEL	1036			
SELENIUM	1045			
SODIUM	1052			
SULFATE	1055			
ANTIMONY	1074			
BERYLLIUM	1075			
THALLIUM	1085			
ZINC	1095			

Fig. 22.6 Inorganic chemicals (IOC) water quality analysis report form

(Reprinted with permission from the Compliance Assurance Section, Division of Public Water Supplies, Illinois Environmental Protection Agency)

IOC

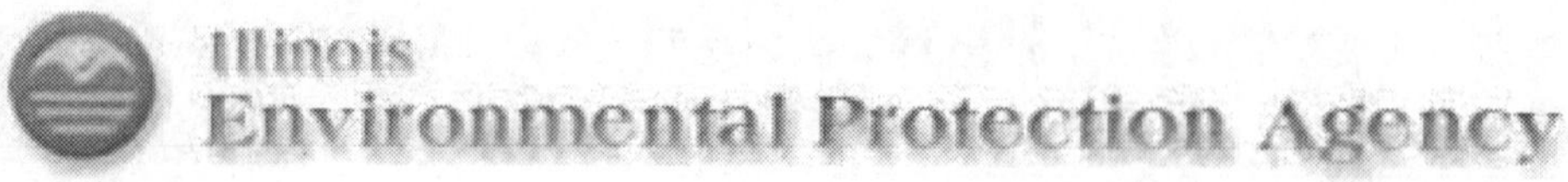

Water System Name: ______________________ Water System Number: ________

-Laboratory Section-

Laboratory State ID Number: ____________

Laboratory Name:

Lab Sample Number: ____________

Date Lab Rcpt.: ____________

Complete Date: ____________

Complete Time: ____________

Comments (Data Quality Issues):

____

Mail Results to: **Illinois Environmental Protection Agency**
Drinking Water Compliance Unit, Mailstop #19
1021 North Grand Avenue East, P.O. 19276
Springfield, IL 62704-9276

Questions Call: (217) 785-0561

Fax: (217) 557-1407

Signature of Analyst or Official ____________________

Date Forwarded

*** See List of Permitted Values**

This Agency is authorized to require this information under Illinois Revised Statutes, 1987, Chapter 111 1/2, Section 1004(H). Disclosure of this information is required. Failure to do so may result in a civil penalty up to $10,000.00 and an additional civil penalty up to $1,000.00 for each day the failure continues. a fine up to $1,000.00 and imprisonment up to one year. This form has been approved by the Forms Management Center.

Fig. 22.6 Inorganic chemicals (IOC) water quality analysis report form (continued)

(Reprinted with permission from the Compliance Assurance Section, Division of Public Water Supplies, Illinois Environmental Protection Agency)

7. Label the sample immediately and keep accurate records of every sample collected including the following:
   a. Date and time sampled.
   b. Location sampled.
   c. Name of sample collected.
   d. Bottle number.
   e. Type of sample.
   f. Name of person collecting sample. Any person collecting samples should be required to complete a form providing the above information at the time of sample collection. This form should be supplied by the laboratory.
8. Keep the time between the collection of the sample and analysis as short as possible.

22.410 Frequency of Sampling

Initial Sampling

Initial sampling refers to the very first sampling conducted as required under the SDWA for each of the applicable contaminant categories. The timing of this sampling depends on the following:

1. The type of contaminant being monitored
2. Whether the system is a community or noncommunity water system
3. Whether the water source is a surface water or groundwater supply

Routine Sampling

Routine sampling refers to sampling repeated on a regular basis. Coliform sampling must be conducted in accordance with a written sampling plan that has been approved by the state regulatory agency.

Repeat Sampling

Whenever an initial or routine sample analysis indicates that an MCL has been exceeded, repeat sampling is required to confirm the routine sampling results. Repeat sampling is in addition to the routine sampling program. Although repeat sampling cannot be scheduled in advance, there are specific repeat sampling procedures to follow. The number of samples, sampling points, and frequency of sampling vary according to the particular contaminant. For example, the regulations specify that wherever coliform bacteria repeat samples are required, at least one of the repeat samples must be taken from the same sampling point as the original and two of the repeat samples must be collected from nearby points upstream and downstream from the original.

22.411 Chain-of-Custody Procedures

Water utilities are responsible for guaranteeing the identity and integrity of a water sample and its data from collection through reporting of the test results. A form documenting the chain of custody (COC) for a water sample is required and may prove useful in the event of litigation. A COC form should contain the following information for each sample:

- The sampler's name
- Location of the sample
- Collection date and time
- Analysis required
- Note preservation times (if any)
- Sampling conditions
- Storage information
- Transportation dates and times
- Proof of laboratory acquisition

There is no standardized COC form that utilities are required to use. Sample COC forms may be found online. The US EPA website (epa.gov) is a useful resource for information about COC procedures and forms.

QUESTIONS

Please write your answers to the following questions and compare them with those on page 608.

22.4E List the information that should be included in a sampling schedule.

22.4F List the elements necessary to the collection of an acceptable sample.

22.5 REPORTING PROCEDURES

The primary purpose of the SDWA is to protect the public's health. Two general categories of reporting are called for by the act:

1. Reporting to the public (public notification)
2. Reporting to the state

There are four types of reports that must be sent to the state:

1. Initial reports
2. Routine sampling reports
3. Repeat sampling reports
4. Violation reports

The SDWA requires water utilities to submit repeat sampling forms to their state primacy agency. Figure 22.7 is an example of a repeat sampling form that can be used to report water quality results to the state primacy agency.

Tables 22.7 through 22.15 outline reporting procedures for various contaminants.

Attention Sample Collector

Immediate Action is needed on all Checked Boxes

Water System Number: ________________ **Water System Name** : ______________________________

<table>
<tr><td>☐</td><td>The coliform sample collected on ________________ from ______________________________ was positive for (circle one) total coliform / fecal colifom / E. coli. Therefore, you must collect three repeat samples. One repeat sample must be collected at the original location. The second repeat sample must be collected within five service connections upstream and the last repeat sample collected within five service connections downstream. You MUST collect all repeat samples and return to the laboratory by ________________. Please be sure to mark sample purpose on the new form as “REPEAT SAMPLE”.

In <u>ADDITION</u> to the above, you MUST <u>READ and UNDERSTAND</u> (#1) and (#2) below to INSURE COMPLIANCE.

1) FOR GROUND WATER SUPPLIES ONLY. If you do not have a triggered source water monitoring (TSWM) Special Exception Permit (SEP), you MUST collect a repeat sample for E. coli <u>from each active ground water source</u> (well) within 24 hours. If you have multiple Routine Distribution total coliform positive samples, you must collect multiple well samples (Example; if you have three TC positive routine distribution samples, you must collect three TSWM samples from each active well. All three well samples may be collected one right after another.)

2) If your water system collects only one routine distribution sample per month repeats are required to be collected as directed above- IN ADDITION another repeat sample must be collected anywhere within your distribution system. This fourth repeat should be marked as “OTHER”</td></tr>
<tr><td>☐</td><td>The coliform sample collected on ________________ from ________________________ was positive for total (circle one) coliform / fecal coliform / E. coli. Therefore, you must collect one repeat sample from the same location. Please be sure to mark sample purpose on the new form as “REPEAT SAMPLE”.</td></tr>
<tr><td>☐</td><td>The coliform sample listed on the enclosed laboratory report form was invalid due to excessive growth, was damaged during shipment, or received more than 30 hours of collection. A replacement sample must be collected and returned to the laboratory as soon as possible. Please be sure to mark sample purpose on the new form as “REPLACEMENT”.</td></tr>
<tr><td>☐</td><td>A triggered source water sample collected on ____________from Well __________ tested positive for E. coli. Therefore, five additional samples must be collected from the well and returned to the laboratory by ____________ - (within 24-hours of notification)</td></tr>
</table>

If any sample in the set of repeat samples is unsatisfactory for any reason, the entire set of three (or four) repeat samples must be collected at the same locations as the first set of repeat samples.

If you are unable to collect repeat samples within 24 hours of notification, you must call the Drinking Water Compliance Unit at 217-785-0561 or your Regional Office and request an extension.

Do NOT discard this document. You are required to keep a copy for your records.

Lab Use Only

Sample ID No.		**Date Repeats mailed and number of bottles**	
Date and Time WS Contacted		**Name of Person Contacted and Phone Number**	
Laboratory		**Lab Contact and Phone Number**	

Available online at http://www.epa.state.il.us/water/compliance/drinking-water/forms/sample-collector.pdf
Rvsd 11/2009 **Illinois EPA DWCU FAX number: 217-557-1407**

Fig. 22.7 Repeat water sampling form

(Reprinted with permission from the Compliance Assurance Section, Division of Public Water Supplies, Illinois Environmental Protection Agency)

TABLE 22.7 REPORTING PROCEDURES—
INORGANIC CHEMICALS (EXCEPT NITRATE) AND ORGANIC CHEMICALS

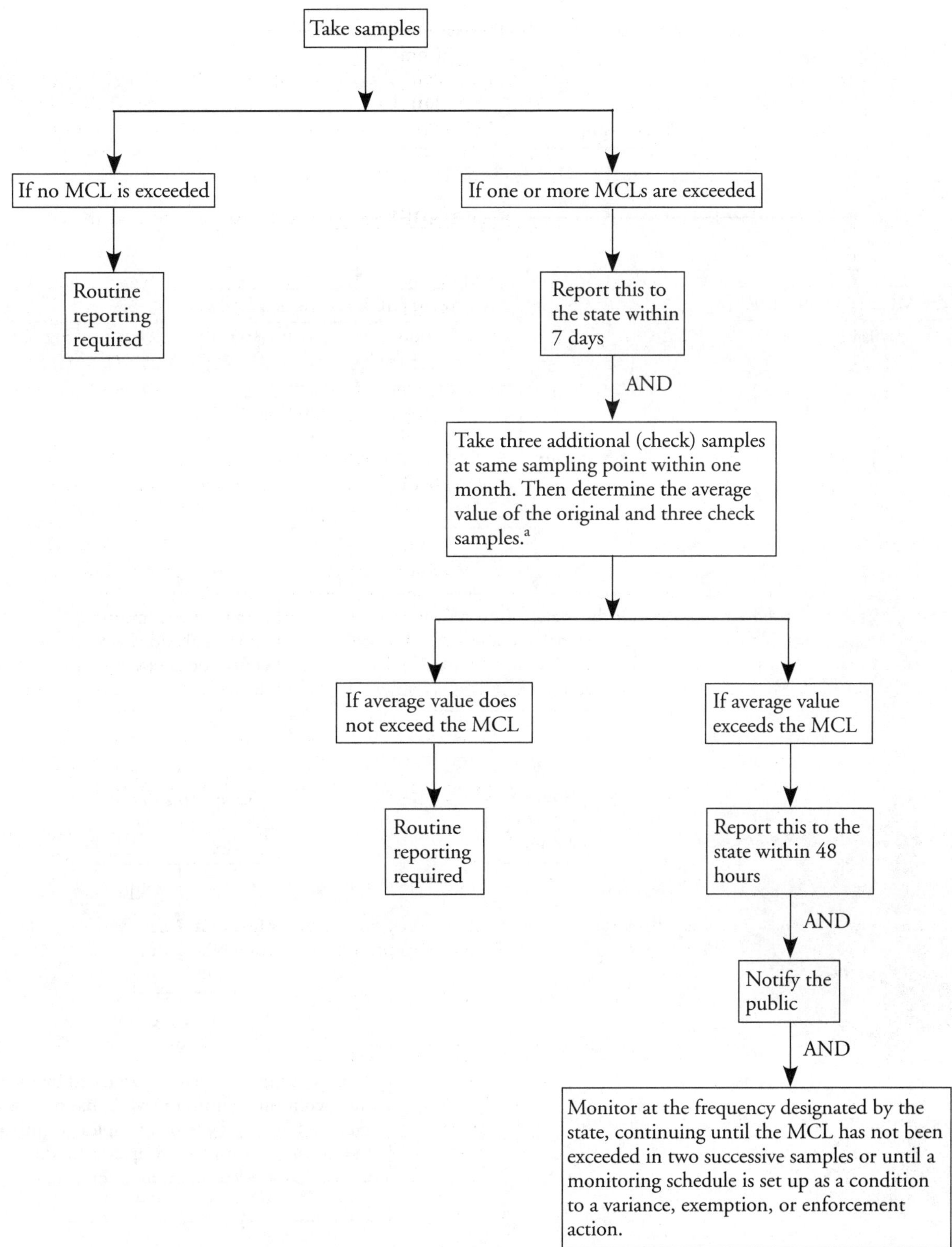

a Average Value = $\frac{\text{Total of Original Sample + 3 Check Samples}}{4}$

TABLE 22.8 REPORTING PROCEDURES—NITRATE

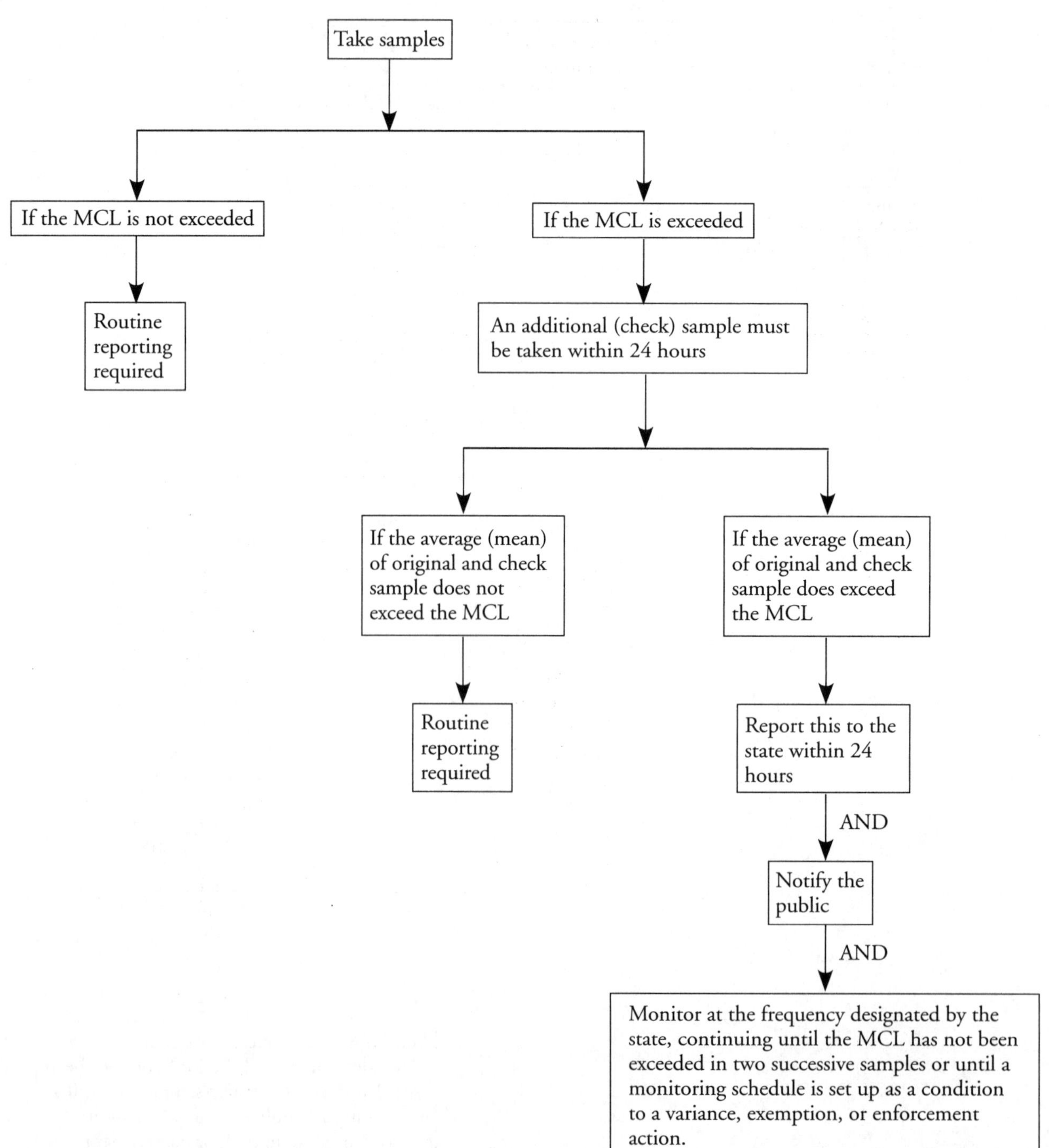

TABLE 22.9 REPORTING PROCEDURES—TURBIDITY MONITORING
(Surface Water Using Conventional or Direct Filtration)

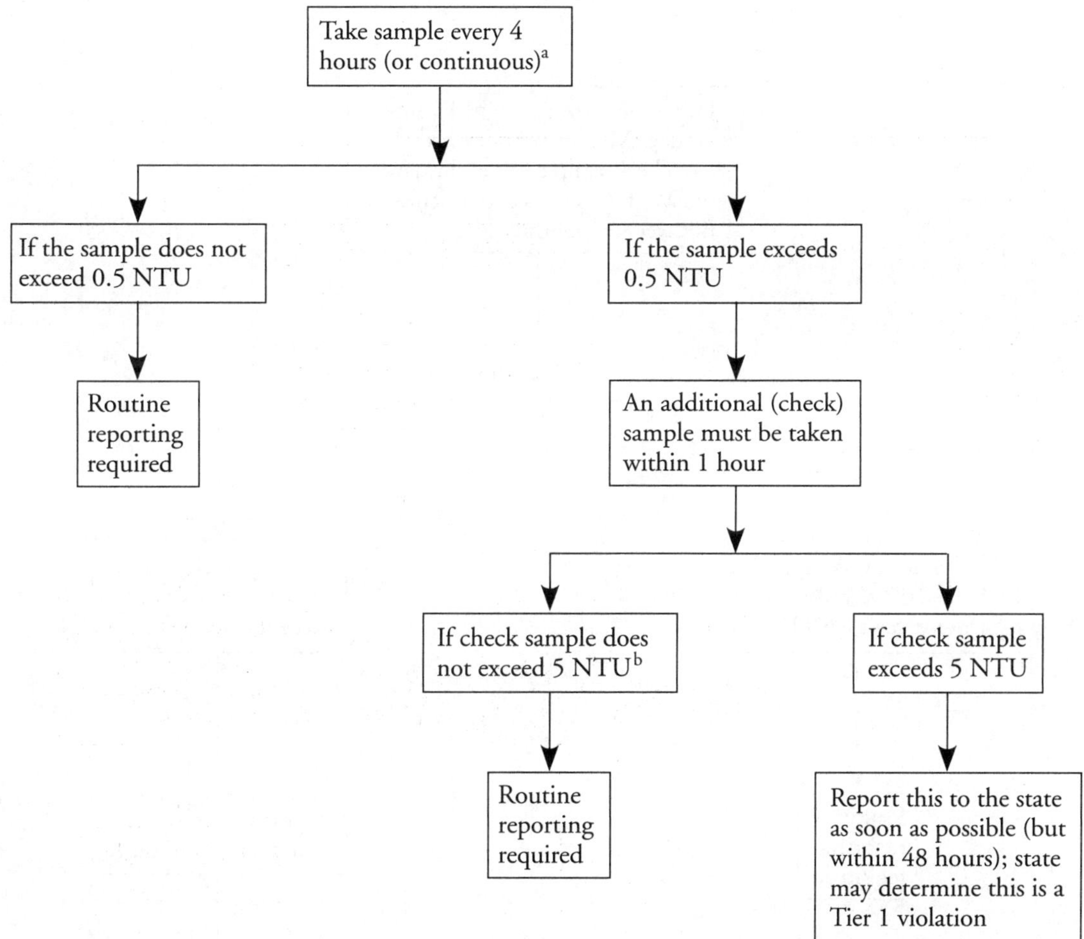

a Less frequent sampling may be established at state option.
b Filter effluent turbidity must be 1 NTU in 95% of samples.

TABLE 22.10 REPORTING PROCEDURES—WHEN CALCULATING MONTHLY AVERAGE TURBIDITY VALUE

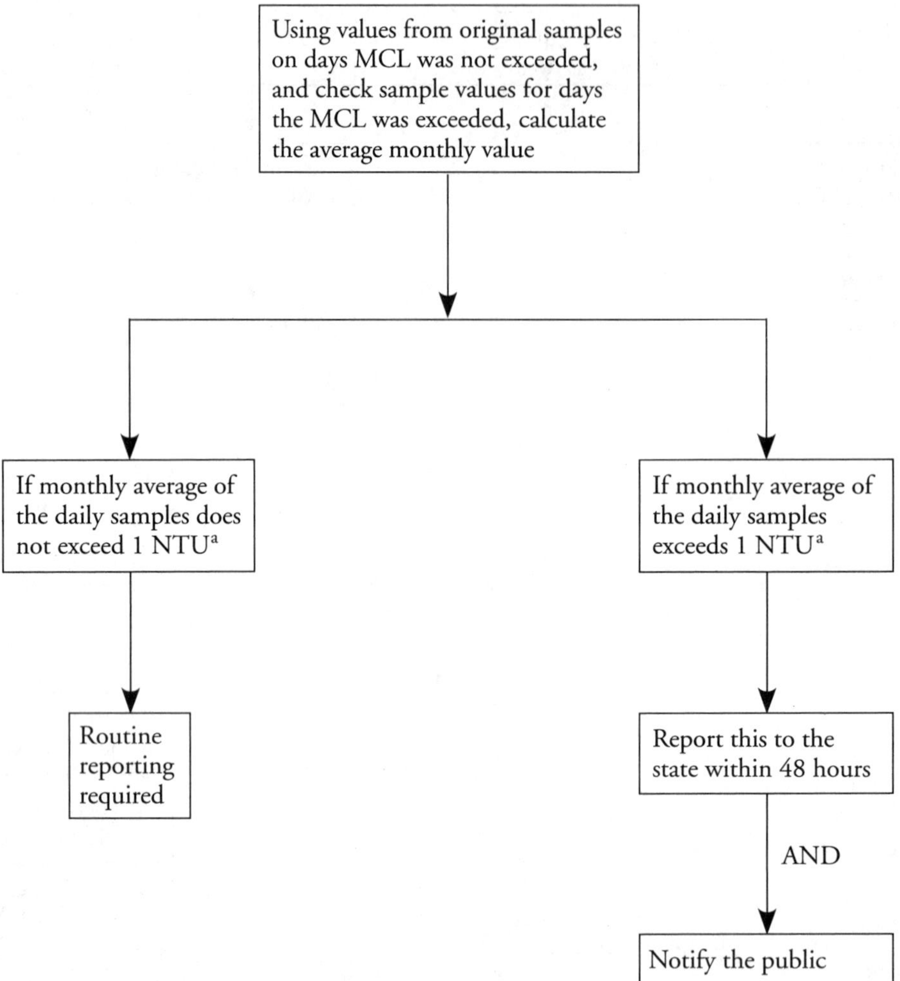

a MCL of 5 NTU may be established at state option.

TABLE 22.11 REPORTING PROCEDURES—
MICROBIAL CONTAMINANTS—TOTAL COLIFORM RULE

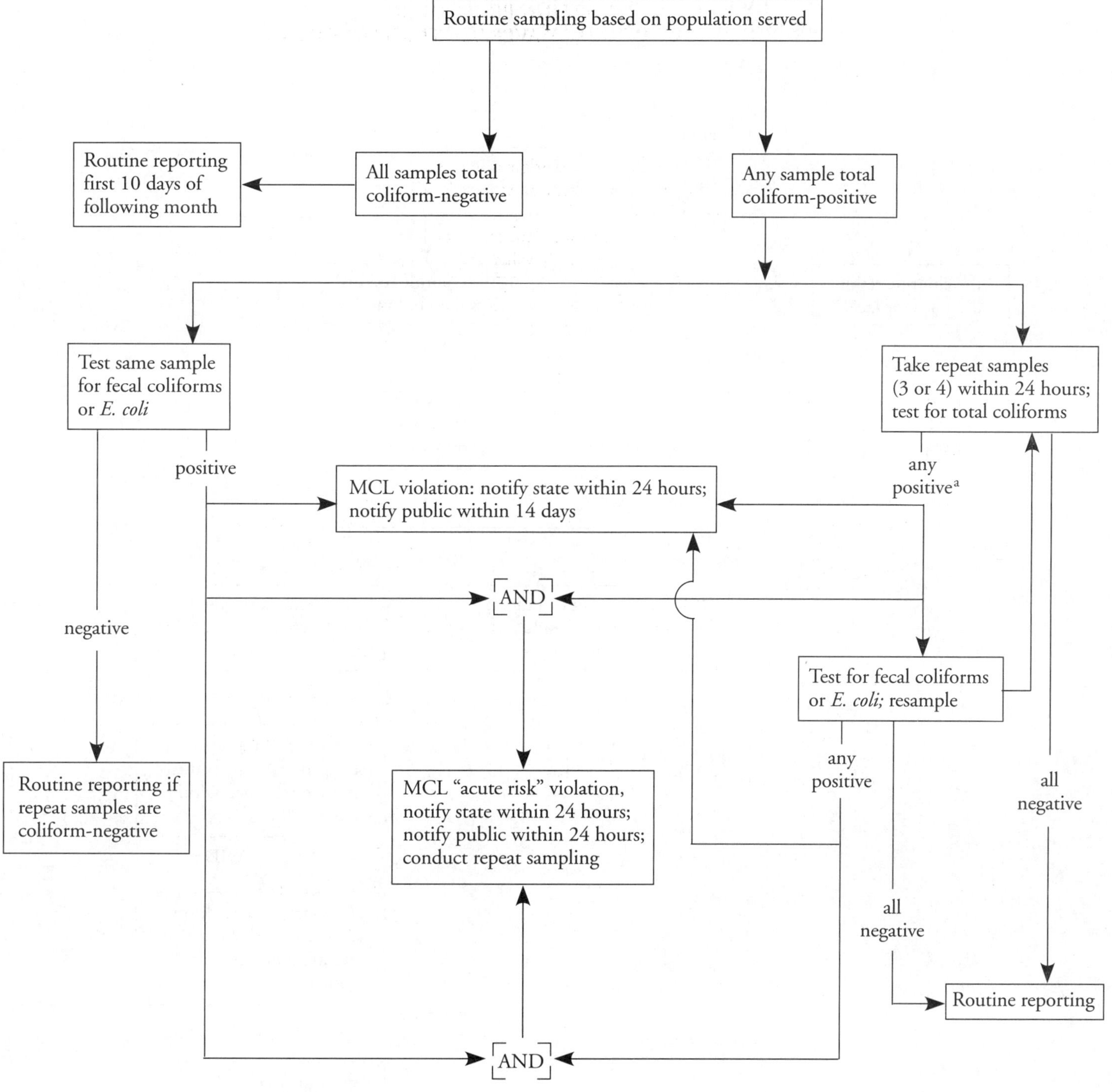

a For systems collecting more than 40 samples per month, MCL is violated when more than 5% are total coliform-positive. For systems collecting fewer than 40 samples per month, MCL is violated when more than one sample is total coliform-positive. Monthly calculations must include repeat samples.

TABLE 22.12 REPORTING PROCEDURES—
MICROBIAL CONTAMINANTS—CHLORINE RESIDUAL

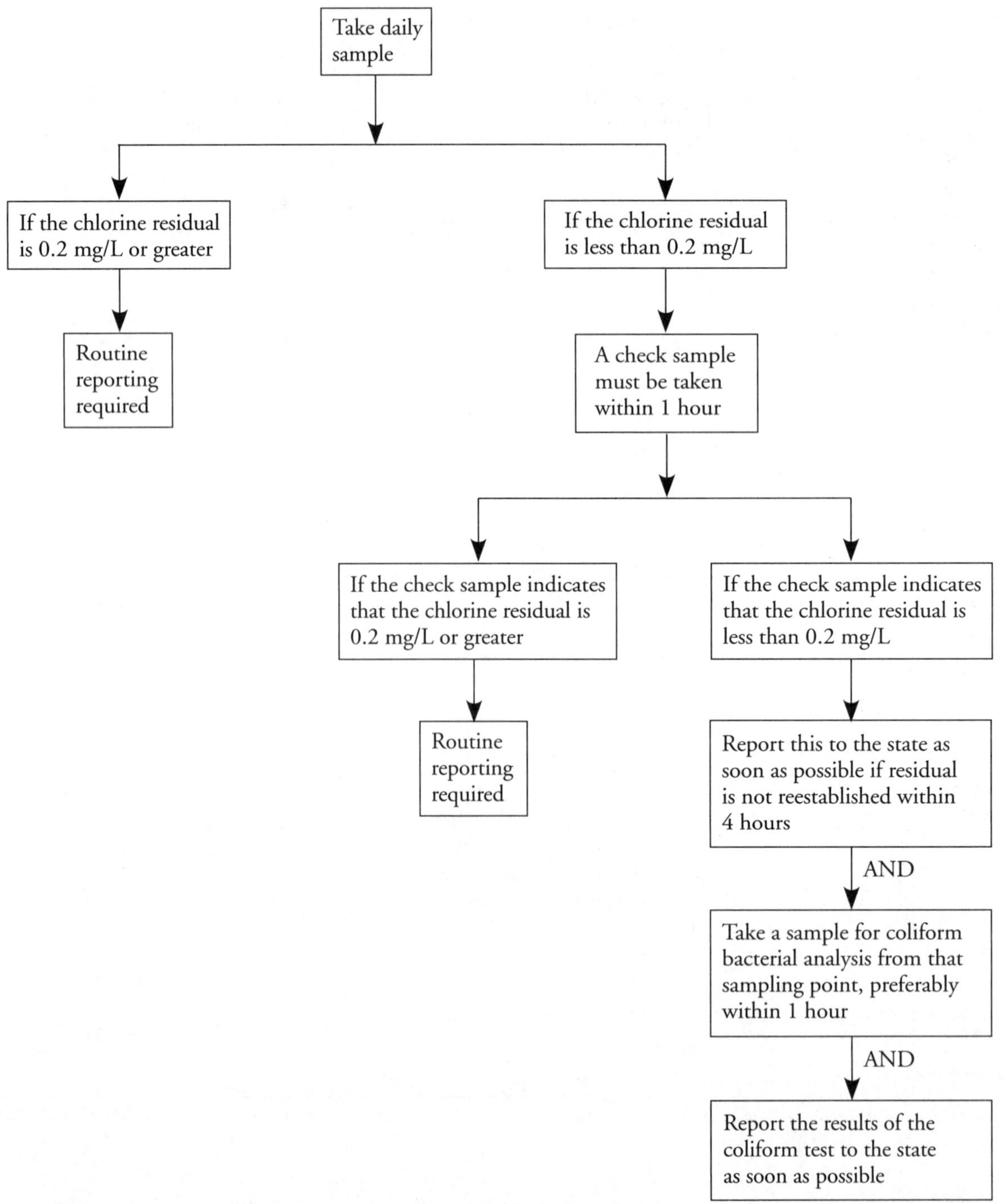

TABLE 22.13 REPORTING PROCEDURES—RADIOLOGICAL CONTAMINANTS—NATURAL (Test for Gross Alpha Activity)

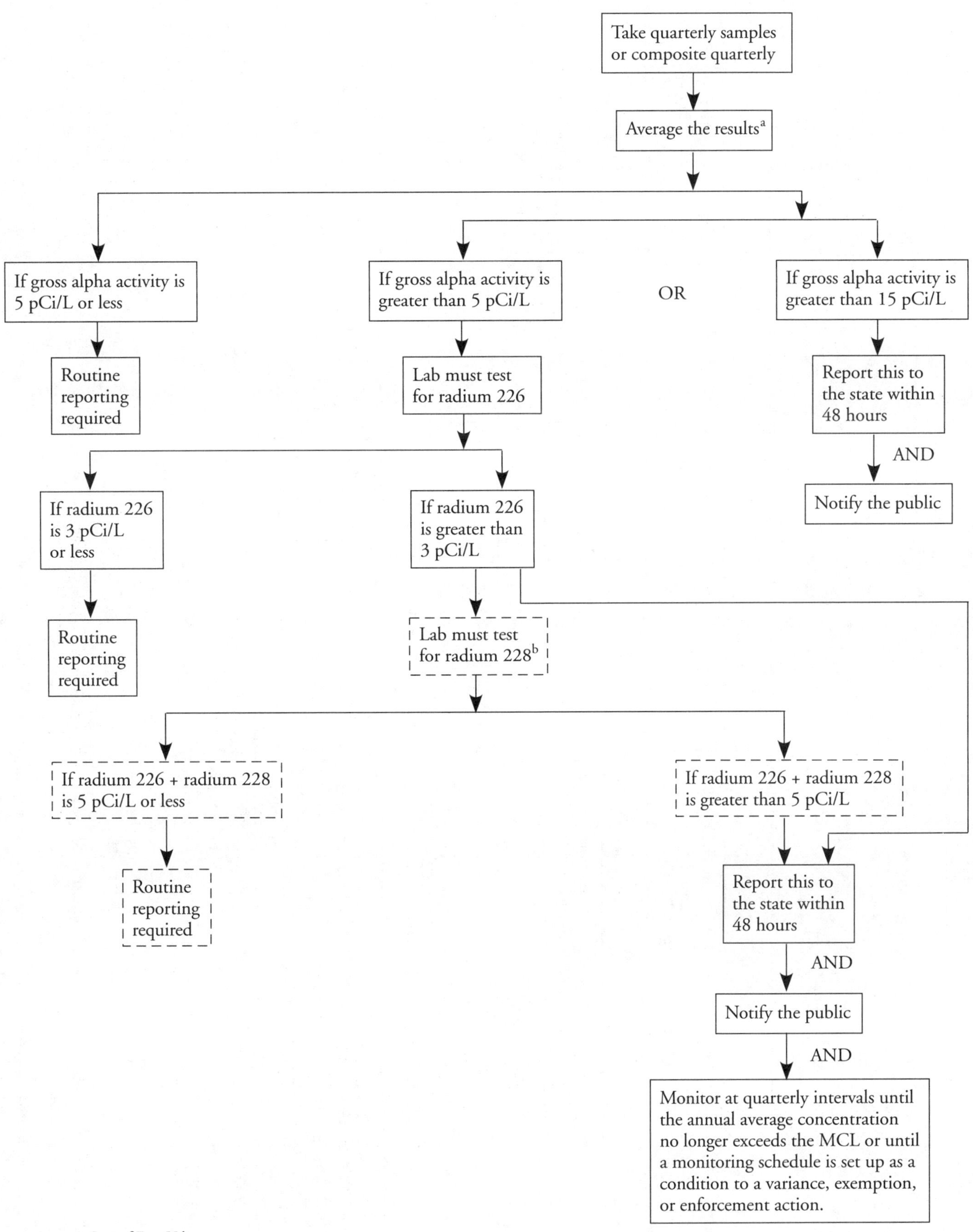

a Average = $\frac{\text{Sum of Four Values}}{4}$

b No averaging is required if the quarterly samples were composited. In that case, use the results of the single sample. This step is required only for the initial monitoring period and not for routine monitoring, except as required by the state.

TABLE 22.14 REPORTING PROCEDURES—RADIOLOGICAL CONTAMINANTS—MANMADE[a]

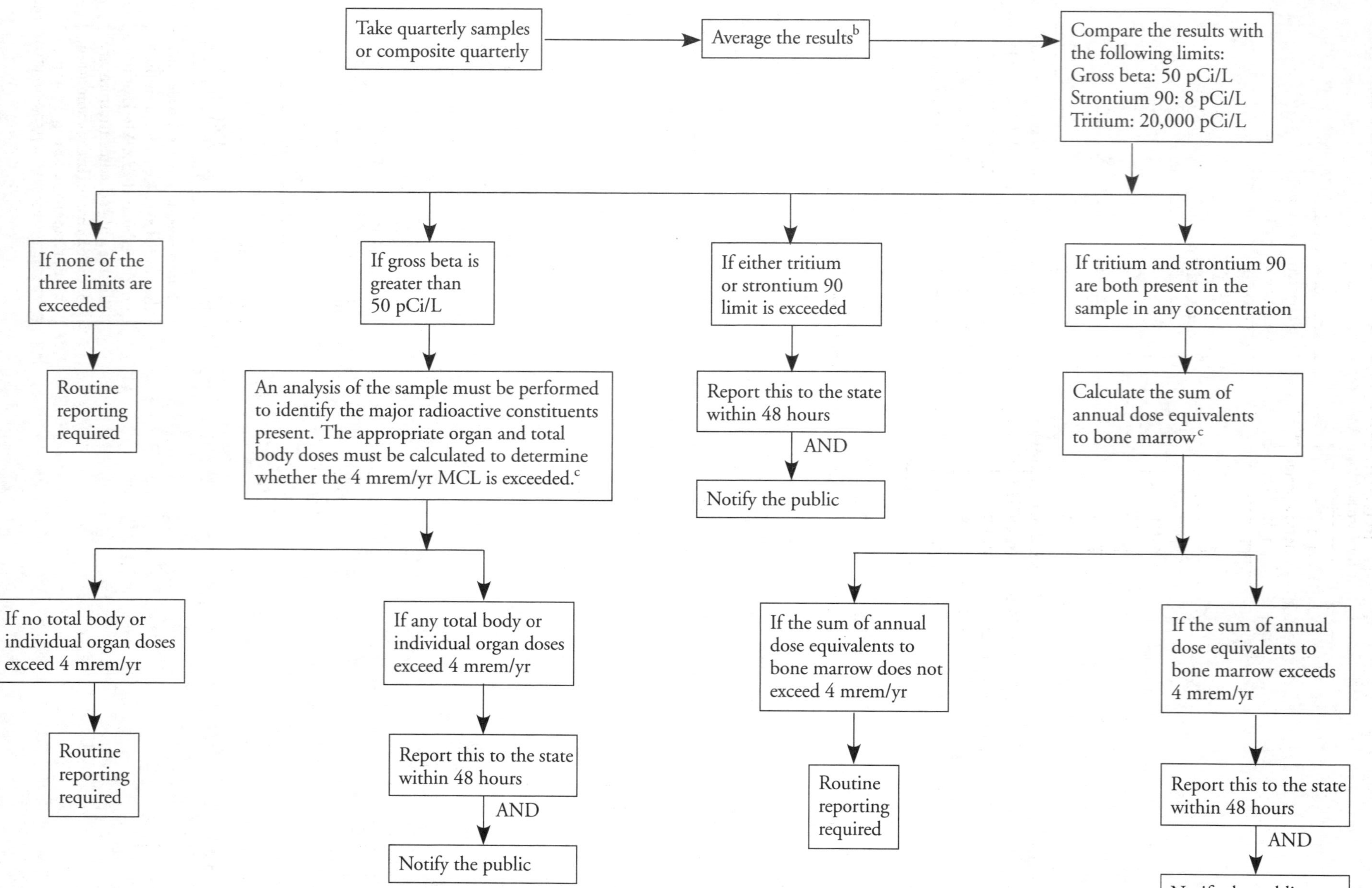

a Applies to surface water systems serving populations of 100,000 or more and systems determined by the state to be vulnerable.

b $\text{Average} = \dfrac{\text{Sum of Four Values}}{4}$

No averaging is required if the quarterly samples were composited. In that case, use the results of the single sample.

c It is likely that the laboratory will not make these calculations. You will probably have to get help from state water supply personnel in making these calculations.

TABLE 22.15 REPORTING PROCEDURES—TOTAL TRIHALOMETHANES[a]

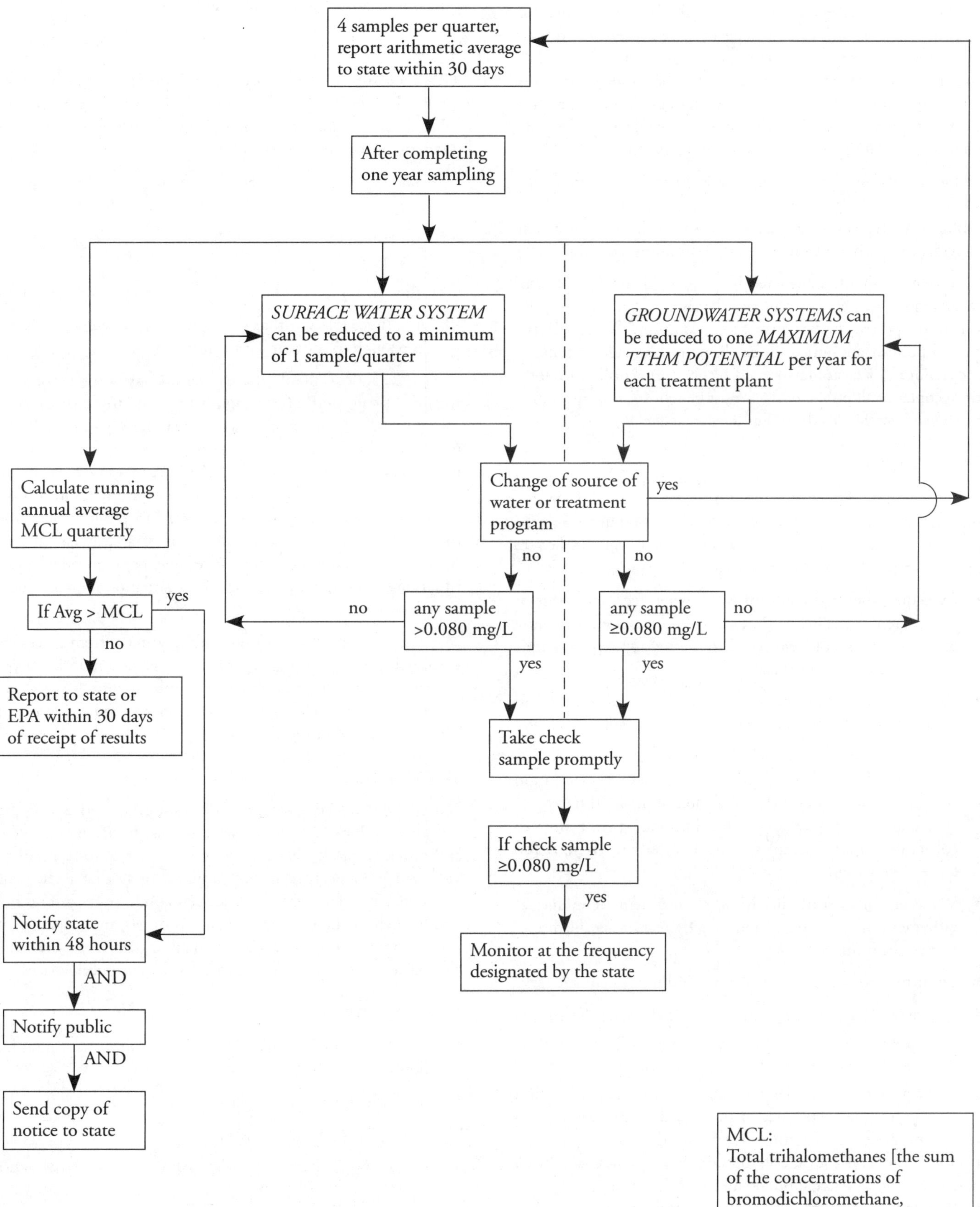

a Applies to community surface water systems serving populations of 10,000 or more and all surface water systems that meet the SWTR criteria for avoiding filtration.

22.6 NOTIFICATION REQUIREMENTS

Public notification regulations were developed to make sure that consumers will be informed whenever there is a serious problem with their drinking water. Operators are urged to contact their state regulatory agency to determine how the public notification regulations affect their specific water system. Also, contact the EPA to request the guidance documents it has developed for implementation of the public notification regulations.

The regulations define three levels (tiers) of violations and specify which groups of consumers must be notified in each instance, what type of notice (electronic broadcast, print) must be provided, and what information the notice must contain.

All notices should be written in language that is free of technical jargon, printed in a type size that is easy to read, and published or posted in a conspicuous location. Water utilities should publish the notices in English and languages specific to the majority of customers served by the water utility. In addition, copies of all public notices must be sent to the state regulatory agency within 10 days after they are issued.

Tier 1 Violations

Tier 1 violations are any violations or situations in which even a brief violation or short-term exposure to a contaminant has the potential to cause a serious, negative effect on human health. Some examples of Tier 1 violations are as follows:

- Exceeding the total coliform MCL when fecal coliforms or *E. coli* are present in the water distribution system or failure to test for fecal coliforms or *E. coli* after any repeat sample tests positive for coliforms.
- Exceeding the MCLs for nitrate, nitrite, or total nitrate and nitrite or failure to take a confirmation sample within 24 hours of learning that the MCL has been exceeded.
- Exceeding the MRDL for chlorine dioxide in a distribution system sample the day after a violation of the MRDL has occurred at the entrance to the distribution system. Failure to collect the required samples in the distribution system is also a Tier 1 violation.
- Violations of the turbidity MCL or treatment technique requirements may be considered Tier 1 violations at the discretion of the primacy agency.
- Occurrence of a waterborne disease outbreak or emergency.
- Any other violations or conditions that the state determines could pose a significant potential to cause adverse health effects upon short-term exposure.

Public notification of Tier 1 violations must take place within 24 hours after the occurrence of the violation. Bill-paying customers of the water system must be notified as well as other people in the service area, even though they may not actually be customers of the water supplier. The precise wording of the notices, called "mandatory health effects language," is provided in the Public Notification Rule. The method of delivery should include at least one of the following methods: appropriate broadcast media (radio, television), newspapers, posting of notices in conspicuous locations, or hand delivery of notices.

Tier 2 Violations

Tier 2 violations include any other situations or violations not included in the Tier 1 notice category that have the potential to cause negative effects on human health. Some examples of Tier 2 violations include the following:

- Violation of the MCL, MRDL, or treatment technique requirements except when Tier 1 notice is required
- Violation of monitoring and testing requirements, if directed by the primacy agency
- Failure to comply with the terms of a variance or exemption
- Any other violation that the primacy agency decides is serious enough to warrant Tier 2 notification

Tier 2 notices should be distributed as soon as practical within 30 days and efforts should be made to reach everyone using the water, not just bill-paying customers. Community systems should mail or deliver the notices to their customers and other people in the area; noncommunity systems may mail or deliver the notices to their customers or post notices in conspicuous public places. If notices are posted in public places, they should remain posted for a minimum of at least seven days. After the first seven days, the notices may be removed when the situation has been corrected, but if the situation remains uncorrected, new notices should be posted every 30 days.

It is the responsibility of the public water system to consult with their primacy agency regarding any violation of the turbidity MCL, treatment techniques, or both to allow the primacy agency to determine if the incident is serious enough to require Tier 1 notification procedures.

Tier 3 Violations

Tier 3 violations consist mainly of monitoring and testing procedure violations. The water system must notify its customers and other people in the area within one year of learning of the violation and repeat the notice annually for as long as the violation continues. As with Tier 2 notices, community systems may mail or deliver the notices to their customers and others in the area; noncommunity systems may mail or deliver the notices to their customers or post notices in conspicuous public places.

22.7 RECORDKEEPING

Water suppliers are required to keep certain information on file, as follows:

- Bacteriological results: 5 years
- Chemical results: 10 years
- Actions taken to correct violations: 3 years after the action was taken
- Sanitary survey reports: 10 years
- Variance or exemption records: 5 years

22.8 CONSUMER CONFIDENCE REPORTS (CCRs)

EPA has developed regulations requiring every community water system to prepare and distribute to its customers an annual Consumer Confidence Report (CCR). The reports are an opportunity for positive communication with the utility's customers and an effective means for a water utility to inform consumers that their water is safe to drink. CCRs also provide an opportunity to convince consumers of the importance of paying for good-quality water and the need for sufficient funds to properly operate and maintain the water supply, treatment, and distribution systems. If higher rates are necessary to fund a capital improvement program, the CCR can explain the importance of having sufficient water with adequate pressure and high quality.

Items that should be covered in CCRs include the following:

1. The name(s), location(s), and type(s) of source water.
2. EPA's definitions of MCLG and MCL.
3. EPA's definitions of "treatment technique" and "action level" (only required if the CCR contains data on a contaminant for which EPA has specified use of a treatment technique or action level).
4. EPA's specific descriptions of health effects for any regulated contaminant for which there was a violation of the MCL during the year covered by the report. The rule lists specific instructions about how the data should be reported and explained for turbidity, lead and copper, total and fecal coliform, *Cryptosporidium*, radon, arsenic, nitrate, and any other contaminants. The CCR must also include EPA's specific warnings to people who may be more vulnerable to contaminants than the general population.
5. The most likely source(s) of detected contaminants. Information from sanitary surveys or source water assessments should be used, if available and appropriate. If this type of site-specific information is not available, the operator should include one or more of the typical sources for that contaminant listed in the CCR Rule.
6. Any violation(s) of the National Primary Drinking Water Regulations that occurred during the year covered by the CCR must be identified in the report. For certain violations, such as a surface water system's failure to install filtration or disinfection, failure to comply with lead and copper requirements, or violation of treatment technique requirements for acrylamide and epichlorohydrin, the EPA-prescribed wording must be used to describe the violation.
7. If the system is operating under a variance or exemption, the CCR must include EPA's definition of "variances and exemptions," the reasons for the variance or exemption, the date it was granted, and a status report on the steps the system is taking to comply with the terms and schedules of the variance or exemption.
8. Notification that the presence of contaminants in drinking water does not necessarily mean the drinking water poses a health risk. Consumers should be advised that they can call the Safe Drinking Water Hotline (800) 426-4791 for more information about contaminants and their potential health effects.
9. Telephone number of the owner, operator, or other authorized person customers can call for additional information about the CCR.
10. If the state primacy agency determines there is a large proportion of non-English-speaking residents in an area, the CCR must use appropriate language(s) to inform consumers about the importance of the report and tell them how to obtain a translated copy or how to get assistance in the appropriate language.
11. Information about how the public may participate in decisions affecting water quality (for example, time and place of regularly scheduled board meetings).
12. Any other information the water utility wishes to communicate to its customers regarding issues covered in the CCR.

22.80 Options for Communicating to the Public

Some water utilities publish the CCR in the local newspaper explaining the report before it is made available to consumers. This advance information helps the consumers understand the report. The report should be a short, concise letter report of one or two pages that can be mailed directly to consumers or mailed in the envelope with the utility bill. All utility staff should be familiar with the contents of the report because consumers who know utility personnel frequently will ask personnel questions about the report. Water utilities should emphasize the good job they are doing as guardians of health in these CCRs.

The CCR regulations not only require water utilities to send annual reports to their direct customers, but also to make a reasonable attempt to reach other people who regularly use water from the system, such as renters and people who work in the area but live elsewhere. A wide variety of methods are available for reaching these individuals, including the following: posting the report on the internet, mailing the report to postal patrons in metropolitan areas, announcing availability of the report in the news media, publication in a local newspaper, posting in public places, delivery of multiple copies to apartment buildings or large private employers, and delivery of copies to community organizations.

Large water systems (serves 100,000 or more) must post their current year's report on the internet, and all systems must make their reports available to the public upon request. The utility should send a copy to the state primacy agency at the same time it distributes copies to its customers and should keep a copy on file for 5 years. The deadline for state primacy agencies to adopt CCR regulations was August 2000.

The CCR regulations give state governors authority to modify some of the notification requirements for small water systems serving fewer than 10,000 persons. Instead of mailing the CCR directly to customers, these systems may be permitted to advise their customers that the CCR will be published in one or more local newspapers and that a copy of the report is available from the utility upon request.

Very small community water systems (500 persons or fewer) that are not required to meet the direct mailing requirements still must prepare a CCR annually but may make it available to customers on request. In such cases, the water utility still must advise its customers (by mail, door-to-door notices, posted signs, or other means authorized by the regulations) that the report is available.

Contact your local or state drinking water supply agency for suggestions and details on how to prepare a CCR. EPA, AWWA, and Rural Water Association all have examples of typical CCRs that can serve as guides for preparing one for your drinking water system.

QUESTIONS

Please write your answers to the following questions and compare them with those on page 608.

22.5A What are the two general categories of reporting called for by the SDWA?

22.5B What are the four types of reports that must be sent to the state?

22.6A If a Tier 1 violation occurs, by what means is the public notified?

22.8A Why should all utility staff be familiar with the contents of the consumer confidence report?

END OF LESSON 2 OF 2 LESSONS

on

DRINKING WATER REGULATIONS

Please answer the discussion and review questions next.

DISCUSSION AND REVIEW QUESTIONS

Chapter 22. DRINKING WATER REGULATIONS

(Lesson 2 of 2 Lessons)

Please write your answers to the following questions to determine how well you understand the material in the lesson. The question numbering continues from Lesson 1.

14. What do secondary drinking water regulations control?
15. Why are secondary drinking water regulations important?
16. How may color be caused in water?
17. Why are iron and manganese undesirable in drinking water?
18. Why is the absence of tastes and odors in drinking water important?
19. Why is hydrogen sulfide not listed under the Secondary Drinking Water Standards?
20. Why are high levels of total dissolved solids undesirable in drinking water?
21. How is the number of sampling points determined?
22. What are the major considerations in determining the number and location of sampling points?
23. How is a sampling route selected?
24. How can samples be preserved?
25. What communication opportunities are provided by Consumer Confidence Reports?

SUGGESTED ANSWERS

Chapter 22. DRINKING WATER REGULATIONS

ANSWERS TO QUESTIONS IN LESSON 1

Answers to questions on page 547.

22.0A The first drinking water standards were designed to control waterborne bacteria and viruses that cause diseases such as cholera, typhoid, and dysentery.

22.0B Operators should develop good working relationships with their local regulatory agencies in order to keep up to date on the latest changes in regulations that apply to their water treatment system.

Answers to questions on page 548.

22.1A Five types of primary contaminants are considered to be of public health importance:

1. Inorganic contaminants
2. Organic contaminants
3. Turbidity
4. Microbial contaminants
5. Radiological contaminants

22.1B To regulate a substance, EPA must demonstrate that the contaminant meets three criteria:

1. The contaminant has an adverse effect on human health
2. It occurs, or is likely to occur, in public water systems at a frequency and concentration of significance to public health
3. Regulation of the contaminant offers a meaningful opportunity to reduce health risks for people served by public water systems

22.1C EPA requires water systems to monitor some unregulated contaminants to assist the agency in determining the extent of contamination by these substances.

Answers to questions on page 551.

22.1D The public health benefits from the Arsenic Rule include the avoidance of bladder and lung cancers and a reduction in the frequency of noncarcinogenic diseases.

22.1E The purpose of the Total Coliform Rule (TCR) is to improve public health protection by reducing fecal pathogens to minimal levels through control of total coliform bacteria, including fecal coliforms and *Escherichia coli (E. coli).*

22.1F Recycle flows referred to in the Filter Backwash Recycling Rule (FBRR) include spent filter backwash water, thickener supernatant, and liquids from dewater processes.

22.1G The purpose of the Consumer Confidence Report (CCR) Rule is to improve public health protection by providing educational material to allow consumers to make educated decisions regarding any potential health risks pertaining to the quality, treatment, and management of their drinking water supply.

Answers to questions on page 552.

22.1H EPA sets MCLGs at zero for known or probable human carcinogens as a matter of policy.

22.1I A maximum contaminant level goal (MCLG) represents what EPA believes to be a safe level of consumption of a contaminant based solely on health effects, but this is not a legally enforceable standard. The maximum contaminant level (MCL) is a legally enforceable standard that protects public health but takes into account factors such as existing laboratory detection technology, costs, and reasonableness.

22.1J A community water system is defined as a public water system that has the following characteristics:

1. Has at least 15 service connections used by all-year residents
2. Regularly serves at least 25 all-year residents

Answers to questions on page 560.

22.2A Primary standards for some contaminants are specified as treatment techniques because the contaminants are too difficult or expensive to detect and measure. The specified treatment technique is known to adequately reduce the public health threat of the contaminant.

22.2B Inorganic chemicals are metals, salts, and other chemical compounds that do not contain carbon.

22.2C Arsenic is listed as a primary contaminant because anyone who drinks water that continuously exceeds the national standard by a substantial amount over a lifetime may experience fatigue and loss of energy. Extremely high levels can cause poisoning.

22.2D At levels of 6 to 8 mg/L, fluoride may cause brittle bones and stiffening of the joints. At levels of 2 mg/L and greater, fluoride may cause dental fluorosis (discoloration and mottling of the teeth), especially in children.

22.2E New homes can be at risk for lead because even legally lead-free plumbing may contain up to 8 percent lead. Grounding of electrical circuits in homes to water pipes and galvanic action between two dissimilar metals may increase corrosion that could cause lead to leach into the water.

22.2F Nitrate (as N) in drinking water above the national standard of 10 mg/L poses an immediate threat to children under 3 months of age. In some infants, excessive levels of nitrate have been known to react with intestinal bacteria that change nitrate to nitrite. Nitrite reacts with hemoglobin in the blood. This reaction will reduce the oxygen-carrying ability of the blood and produce an anemic condition commonly known as "blue baby syndrome."

Answers to questions on page 562.

22.2G Organic chemicals are either natural or synthetic chemical compounds that contain carbon. Synthetic organic chemicals (SOCs) are manmade compounds that are widely used as pesticides, paints, dyes, solvents, plastics, and food additives.

22.2H Trichloroethylene (TCE) has been widely used as an ingredient in many household products (spot removers, rug cleaners, air fresheners), dry cleaning agents, industrial metal cleaners and polishes, refrigerants, and even anesthetics.

Answers to questions on page 568.

22.2I Microbial contaminants are a public health concern because they cause almost immediate gastrointestinal illness when people consume them in water. Waterborne diseases such as typhoid, cholera, infectious hepatitis, and dysentery have been traced to improperly disinfected drinking water.

22.2J The most basic test for bacterial contamination of a water supply is the test for total coliform bacteria. Total coliform counts give a general indication of the sanitary condition of a water supply.

22.2K EPA will accept any one of the five analytical methods for determining total coliforms:

1. Multiple-tube fermentation technique (MTF)
2. Membrane filter technique (MF)
3. Presence-absence coliform test (P-A)
4. Colilert™ system (ONPG-MUG test)
5. Colisure test

22.2L Repeat coliform samples are collected to confirm the presence of coliform bacteria and to help determine whether the contamination is a localized problem or a systemwide problem.

22.2M When repeat samples must be taken, one must be taken from the same tap as the original coliform-positive sample and at least two must be taken from nearby points (within five service connections) upstream and downstream.

Answers to questions on pages 572 and 573.

22.2N The Surface Water Treatment Rules set forth primary drinking water regulations requiring treatment of surface water supplies or groundwater supplies under the direct influence of surface water. These regulations require a specific water treatment technique (filtration or disinfection) in place of the establishment of maximum contaminant levels (MCLs) for *Cryptosporidium, Giardia lamblia,* viruses, *Legionella*, and bacteria. Also, the regulations require that all systems be operated by qualified operators.

22.2O The Surface Water Treatment Rules define surface water as "all water open to the atmosphere and subject to surface runoff." This would include rivers, lakes, streams, reservoirs, and groundwaters directly influenced by surface water.

22.2P A water utility can avoid mandatory filtration by meeting the following requirements:

1. The fecal coliform concentration in the water before disinfection must not exceed 20/100 mL, or total coliform concentration must not exceed 100/100 mL in more than 10 percent of the samples analyzed in the previous 6 months.
2. The system must currently provide at least 99.9 percent (3 log) *Cryptosporidium* inactivation.
3. The turbidity of the water before disinfection cannot exceed 5 NTUs based on grab samples collected every 4 hours or by continuous monitoring.
4. Certain site-specific conditions.

22.2Q The MCLs for turbidity for surface waters are as follows:

1. For unfiltered surface water supplies, the turbidity of the raw water before disinfection cannot exceed 5 NTU based on grab samples collected every four hours or by continuous turbidity monitoring.
2. For filtered surface water supplies, the filtration process must achieve a turbidity level of less than 1 NTU at all times and not more than 0.3 NTU in more than 5 percent of the samples analyzed during a given month.
3. Those systems using slow sand or diatomaceous earth filtration must achieve a turbidity level of less than 5 NTU at all times and not more than 1 NTU in more than 5 percent of the samples collected each month.

22.2R The effectiveness of a disinfectant is measured by the time (T) in minutes of the disinfectant's contact in the water and the concentration (C) of the disinfectant residual in mg/L measured at the end of the contact time. The product of these two factors ($C \times T$) provides a measure of the degree of pathogenic inactivation.

Answers to questions on page 575.

22.2S Trihalomethanes (THMs) are the product of chlorine combining with organic material in the water; they are suspected carcinogens.

22.2T Drinking water can become contaminated with radioactive elements when these naturally occurring elements dissolve and are carried to groundwater basins as water passes through the soil.

22.2U The MCLs for radiological contaminants are divided into two categories: natural radioactivity that results from well water passing through deposits of naturally occurring radioactive materials and manmade radioactivity such as might result from industrial wastes, hospitals, or research laboratories.

ANSWERS TO QUESTIONS IN LESSON 2

Answers to questions on page 577.

22.3A Secondary drinking water regulations are enforceable after a state has passed a law requiring the state health agency to enforce the regulations.

22.3B The secondary drinking water contaminants include the following:

1. Aluminum
2. Chloride
3. Color
4. Copper
5. Corrosivity
6. Fluoride
7. Foaming Agents
8. Iron
9. Manganese
10. Odor
11. pH
12. Silver
13. Sulfate
14. Total Dissolved Solids (TDS)
15. Zinc

22.3C Contaminants in the secondary regulations should be monitored at least as often as the monitoring performed for inorganic chemical contaminants listed in the Primary Drinking Water Regulations as applicable to community water systems. More frequent monitoring would be appropriate for specific contaminants such as pH, color, odor, or others, under certain circumstances as directed by the state.

Answers to questions on page 579.

22.3D Chloride is a secondary contaminant because it affects the aesthetic quality of water by imparting an objectionable salty taste in water and because it causes corrosion of the pipes in hot water systems and other pipelines.

22.3E Surfactants are synthetic organic chemicals and are the principal ingredient of household detergents.

22.3F The application of chlorine to waters containing manganese increases the likelihood of precipitation of manganese at low levels. Unless the precipitate is removed, precipitates reaching pipelines will promote bacterial growth.

Answers to questions on page 580.

22.3G The undesirable effects of abnormal pH values include the following:

1. When the pH increases, the disinfection activity of chlorine falls significantly.
2. High pH may cause increased halogen reactions, which produce chloroform and other trihalomethanes during chlorination.
3. Both excessively high and low pHs may cause increased corrosivity, which can, in turn, create taste problems, staining problems, and significant health hazards.
4. Metallic piping in contact with low-pH water will impart a metallic taste.
5. If piping is iron or copper, high pH will cause oxide and carbonate compounds to be deposited, leaving red or green stains on fixtures and laundry.
6. At a high pH, drinking water acquires a bitter taste.
7. The high degree of mineralization often associated with basic waters results in encrustation of water pipes and water-using appliances.

22.3H High levels of sulfate are undesirable in drinking water for the following reasons:

1. Tend to form hard scales in boilers and heat exchangers
2. Cause taste effects
3. Cause a laxative effect

22.3I High levels of zinc are undesirable in drinking water for the following reasons:

1. Produce adverse physiological effects
2. Impart undesirable tastes
3. Cause a milky appearance in the water
4. May increase lead and cadmium concentrations

Answers to questions on page 583.

22.4A Sampling is the selection of a small portion of water to identify the characteristics of a drinking water supply. Sampling is used daily to test water for the purpose of making certain the water complies with public health regulations.

22.4B Water utilities are required to take two types of water samples: raw water and finished (treated) water samples. Raw water samples provide the specific characteristics of the water that will determine the way water is treated. Finished water samples provide the evidence of the effectiveness of the treatment process used by the water utility.

22.4C General guidelines for water sampling include five main topics: using dedicated sampling points or taps, labeling water sample containers, transporting water samples, collecting representative samples, and avoiding rinsing sampling containers.

22.4D At the very minimum, a small system (with a population of 25 to 1,000 people) must sample for turbidity and coliform bacteria and must have two sampling points:

1. One where the water enters the distribution system
2. One at a consumer faucet at a point representative of the distribution system

Answers to questions on page 591.

22.4E A sampling schedule should include the following information:

1. Sampling frequency
2. Sampling point designation
3. Location
4. Type of test
5. Sample volume
6. Special handling instructions

22.4F The following elements are necessary to the collection of an acceptable sample:

1. Obtain a sample that is truly representative of the existing conditions.
2. Flush the line before sample collection (except for when a first-draw sample is required).
3. Fill the sample bottle without leaving any air pocket.
4. Analyze residual chlorine when the sample is taken.
5. Maintain the sample so that it does not become contaminated before it reaches the laboratory.
6. Use preservation techniques (pH control and refrigeration).
7. Keep accurate records of every sample collected (date, time, location, name of sample, bottle number, type of sample, and name of person collecting sample).
8. Keep the time between the collection of the sample and analysis as short as possible.

Answers to questions on page 604.

22.5A Two general categories of reporting are called for by the SDWA:

1. Reporting to the public (public notification)
2. Reporting to the state

22.5B There are four types of reports that must be sent to the state:

1. Initial reports
2. Routine sampling reports
3. Check or repeat sampling reports
4. Violation reports

22.6A When a Tier 1 violation occurs, the public must be notified by appropriate broadcast media (radio, television), newspapers, posting of notices in conspicuous locations, or hand delivery of notices.

22.8A All utility staff should be familiar with the contents of the consumer confidence report because consumers who know utility personnel frequently will ask personnel questions about the report.

CHAPTER 23

ADMINISTRATION

by

Lorene Lindsay

Tim Gannon

and

Jim Sequeira

TABLE OF CONTENTS

Chapter 23. ADMINISTRATION

LESSON 2

LESSON 3

IN
OUT

LEARNING OBJECTIVES

Chapter 23. ADMINISTRATION

Following completion of Chapter 23, you should be able to:

1. Identify the functions of a manager.
2. Describe the benefits of short-term, long-term, and emergency planning.
3. Define the following terms:
   a. Authority
   b. Responsibility
   c. Delegation
   d. Accountability
   e. Unity of command
4. Read and construct an organizational chart identifying lines of authority and responsibility.
5. Write a job description for a specific position within the utility.
6. Develop good interview questions.
7. Conduct employee evaluations.
8. Describe the steps necessary to provide equal and fair treatment to all employees.
9. Prepare a written or oral report on the utility's operations.
10. Communicate effectively within the organization, with media representatives, and with the community.
11. Describe the financial strength of your utility.
12. Calculate your utility's operating ratio, coverage ratio, and simple payback.
13. Prepare a contingency plan for emergencies.
14. Prepare a plan to strengthen the security of your utility's facilities.
15. Set up a safety program for your utility.
16. Collect, organize, use, and dispose of plant records.
17. Describe the basic elements of water and energy conservation and associated best management practice strategies.

WORDS

Chapter 23. ADMINISTRATION

ACCOUNTABILITY — ACCOUNTABILITY

When a manager gives power/responsibility to an employee, the employee ensures that the manager is informed of results or events.

AUTHORITY — AUTHORITY

The power and resources to do a specific job or to get that job done.

BACKFLOW — BACKFLOW

A reverse flow condition, created by a difference in water pressures, that causes water to flow back into the distribution pipes of a potable water supply from any source or sources other than an intended source. Also see BACKSIPHONAGE.

BACK PRESSURE — BACK PRESSURE

A pressure that can cause water to backflow into the water supply when a user's water system is at a higher pressure than the public water system.

BACKSIPHONAGE — BACKSIPHONAGE

A form of backflow caused by a negative or below atmospheric pressure within a water system. Also see BACKFLOW.

BOND — BOND

(1) A written promise to pay a specified sum of money (called the face value) at a fixed time in the future (called the date of maturity). A bond also carries interest at a fixed rate, payable periodically. The difference between a note and a bond is that a bond usually runs for a longer period of time and requires greater formality. Utility agencies use bonds as a means of obtaining large amounts of money for capital improvements.

(2) A warranty by an underwriting organization, such as an insurance company, guaranteeing honesty, performance, or payment by a contractor.

CALL DATE — CALL DATE

First date a bond can be paid off.

CERTIFICATION EXAMINATION — CERTIFICATION EXAMINATION

An examination administered by a state agency or professional association that operators take to indicate a level of professional competence. In the United States, certification of operators of water treatment plants, wastewater treatment plants, water distribution systems, and small water supply systems is mandatory. In many states, certification of wastewater collection system operators, industrial wastewater treatment plant operators, pretreatment facility inspectors, and small wastewater system operators is voluntary; however, current trends indicate that more states, provinces, and employers will require these operators to be certified in the future. Operator certification is mandatory in the United States for the Chief Operators of water treatment plants, water distribution systems, and wastewater treatment plants.

CODE OF FEDERAL REGULATIONS (CFR) — CODE OF FEDERAL REGULATIONS (CFR)

A publication of the US government that contains all of the proposed and finalized federal regulations, including safety and environmental regulations.

CONFINED SPACE CONFINED SPACE

A space that has the following characteristics:

(1) Is large enough and so configured that an employee can bodily enter and perform assigned work

(2) Has limited or restricted means for entry or exit (for example, manholes, tanks, vessels, silos, storage bins, hoppers, vaults, and pits are spaces that may have limited means of entry)

(3) Is not designed for continuous employee occupancy

Also see DANGEROUS AIR CONTAMINATION and OXYGEN DEFICIENCY.

CONFINED SPACE, PERMIT-REQUIRED (PERMIT SPACE) CONFINED SPACE, PERMIT-REQUIRED (PERMIT SPACE)

A confined space that has one or more of the following characteristics:

(1) Contains or has a potential to contain a hazardous atmosphere

(2) Contains a material that has the potential for engulfing an entrant

(3) Has an internal configuration such that an entrant could be trapped or asphyxiated by inwardly converging walls or by a floor that slopes downward and tapers to a smaller cross section

(4) Contains any other recognized serious safety or health hazard

COVERAGE RATIO COVERAGE RATIO

The coverage ratio is a measure of the ability of the utility to pay the principal and interest on loans and bonds (this is known as debt service) in addition to any unexpected expenses.

DEBT SERVICE DEBT SERVICE

The amount of money required annually to pay the (1) interest on outstanding debts or (2) funds due on a maturing bonded debt or the redemption of bonds.

DELEGATION DELEGATION

The act in which power is given to another person in the organization to accomplish a specific job.

GLOBALLY HARMONIZED SYSTEM OF CLASSIFICATION AND LABELING OF CHEMICALS (GHS) GLOBALLY HARMONIZED SYSTEM OF CLASSIFICATION AND LABELING OF CHEMICALS (GHS)

A worldwide initiative to promote standard criteria for classifying chemicals according to their health, physical, and environmental hazards. It uses harmonized pictograms, hazard statements, precautionary statements, and the signal words “Danger” and “Warning” to communicate hazard information on product labels and safety data sheets (SDSs) in a logical and comprehensive way. The primary goals of the GHS are the following:

(1) To enhance the protection of human health and the environment by providing an internationally comprehensible system for hazard communication

(2) To provide a recognized framework for those countries without an existing system

(3) To reduce the need for testing and evaluation of chemicals

(4) To facilitate international trade in chemicals whose hazards have been properly assessed and identified on an international basis

HAZARD STATEMENT HAZARD STATEMENT

A statement assigned to a hazard class and category that describes the nature of the hazard(s) of a chemical, including, where appropriate, the degree of hazard.

MATERIAL SAFETY DATA SHEET (MSDS) MATERIAL SAFETY DATA SHEET (MSDS)

A document that provides pertinent information and a profile of a particular hazardous substance or mixture. An MSDS is normally developed by the manufacturer or formulator of the hazardous substance or mixture. The MSDS is required to be made available to employees and operators or inspectors whenever there is the likelihood of the hazardous substance or mixture being introduced into the workplace. Some manufacturers are preparing MSDSs for products that are not considered to be hazardous to show that the product or substance is not hazardous. Also see SAFETY DATA SHEET (SDS).

OSHA (O-shuh) OSHA

The Williams-Steiger Occupational Safety and Health Act of 1970 (OSHA) is a federal law designed to protect the health and safety of industrial workers and also the operators of water supply systems and treatment plants. The act regulates the design, construction, operation, and maintenance of water supply systems and water treatment plants. OSHA also refers to the federal and state agencies that administer the OSHA regulations.

OPERATING RATIO OPERATING RATIO

The operating ratio is a measure of the total revenues divided by the total operating expenses.

ORGANIZING ORGANIZING

Deciding who does what work and delegating authority to the appropriate persons.

OUCH PRINCIPLE OUCH PRINCIPLE

This principle says that as a manager when you delegate job tasks you must be **O**bjective, **U**niform in your treatment of employees, and the tasks must be **C**onsistent with utility policies, and **H**ave job relatedness.

PICTOGRAM PICTOGRAM

A graphical composition that may include a symbol plus other graphic elements, such as a border, background pattern, or color, that is intended to convey specific information about the hazards of a chemical. There are nine pictograms under the Globally Harmonized System (GHS) to convey the health, physical, and environmental hazards.

PLANNING PLANNING

Management of utilities to build the resources and financial capability to provide for future needs.

PRECAUTIONARY STATEMENT PRECAUTIONARY STATEMENT

A phrase that describes recommended measures to be taken to minimize or prevent adverse effects resulting from exposure to a hazardous chemical, or improper storage or handling of a hazardous chemical.

RESPONSIBILITY RESPONSIBILITY

Answering to those above in the chain of command to explain how and why you have used your authority.

SAFETY DATA SHEET (SDS) SAFETY DATA SHEET (SDS)

Safety data sheets (SDSs) are an essential component of the Globally Harmonized System of Classification and Labeling of Chemicals (GHS) and are intended to provide comprehensive information about a substance or mixture for use in workplace chemical management. They are used as a source of information about hazards, including environmental hazards, and to obtain advice on safety precautions. In the GHS, they serve the same function that the material safety data sheet (MSDS) does in OSHA's Hazard Communication Standard. The SDS is normally product related and not specific to the workplace; nevertheless, the information on an SDS enables the employer to develop an active program of worker protection measures, including training, which is specific to the workplace, and to consider measures necessary to protect the environment.

SCADA (SKAY-dah) SYSTEM SCADA SYSTEM

Supervisory Control And Data Acquisition system. A computer-monitored alarm, response, control, and data acquisition system used to monitor and adjust treatment processes and facilities.

SIGNAL WORD SIGNAL WORD

A single word used to indicate the relative level of severity of a chemical hazard and alert the reader to a potential hazard on the label. The signal words used are "Danger" and "Warning." "Danger" is used for the more severe hazards, while "Warning" is used for less severe hazards.

TAILGATE SAFETY MEETING TAILGATE SAFETY MEETING

Brief (10 to 20 minutes) safety meetings held every 7 to 10 working days. The term comes from the safety meetings regularly held by the construction industry around the tailgate of a truck.

CHAPTER 23. ADMINISTRATION

(Lesson 1 of 3 Lessons)

23.0 NEED FOR UTILITY MANAGEMENT

The management of a public or private utility, large or small, is a complex and challenging job. Communities are concerned about their drinking water and their wastewater. They are aware of past environmental disasters and they want to protect their communities, but they want this protection with a minimum investment of money. In addition to the local community demands, the utility manager must also keep up with increasingly stringent regulations and monitoring from regulatory agencies. While meeting these external (outside the utility) concerns, the manager faces the normal challenges from within the organization: personnel, resources, equipment, and preparing for the future. For the successful manager, all of these responsibilities combine to create an exciting and rewarding job.

A brief quiz is given in Table 23.1 that asks some basic management questions. This quiz can be used as a guide to management areas that may need some attention in your utility. You should be able to answer yes to most of the questions; however, all utilities have areas that can be improved.

In the environmental field, as well as other fields, the workforce itself is changing. Minorities, women, and people with disabilities provide new opportunities for growth in the utility. For the employee, however, overcoming employment barriers can be difficult, especially when the workload is demanding and physically challenging. The utility manager must provide adequate support services for these workers and learn to deal with organized worker groups.

Changes in the environmental workplace also are created by advances in technology. The environmental field has exploded with new technologies such as computer-controlled water treatment processes and distribution systems. The utility manager must keep up with these changes and provide the leadership to keep everyone at the utility up to speed on new ways of doing things. In addition, the utility manager must provide a safer, cleaner work environment while constantly training and retraining employees to understand new technologies.

QUESTIONS

Please write your answers to the following questions and compare them with those on page 698.

23.0A What are the local community demands on a utility manager?

23.0B What has created changes in the environmental workplace?

23.1 FUNCTIONS OF A MANAGER

The functions of a utility manager are the same as for the chief executive officer (CEO) of any big company: planning, organizing, staffing, directing, and controlling. In small communities, the utility manager may be the only one who has these responsibilities and the community depends on the manager to handle everything.

Planning (see Section 23.2) consists of determining the goals, policies, procedures, and other elements to achieve the goals and objectives of the agency. Planning requires the manager to collect and analyze data, consider alternatives, and then make decisions. Planning must be done before the other managing functions. Planning may be the most difficult in smaller communities, where the future may involve a decline in population instead of growth.

Organizing (see Section 23.3) means that the manager decides who does what work and delegates authority to the appropriate operators. The organizational function in some utilities may be fairly loose while some communities are very tightly controlled.

Staffing (see Section 23.4) is the recruiting of new operators and staff and determining if there are enough qualified operators and staff to fill available positions. The utility manager's staffing responsibilities include selecting and training employees, evaluating their performance, and providing opportunities for advancement for operators and staff in the agency.

Directing includes guiding, teaching, motivating, and supervising operators and utility staff members. Direction includes issuing orders and instructions so that activities at the facilities or in the field are performed safely and are properly completed.

TABLE 23.1 HOW WELL DOES YOUR SYSTEM MANAGE?

The following self-test is designed for water treatment facilities to provide a guide for identifying areas of concern and for improving system management.

1. Is the treatment system budget separate from other accounts so that the true cost of treatment can be determined?
2. Are the funds adequate to cover operating costs, debt service, and future capital improvements?
3. Do operational personnel have input into the budget process?
4. Is there a monthly or quarterly review of the actual operating costs compared to the budgeted costs?
5. Does the user charge system adequately reflect the cost of treatment?
6. Are all users properly metered and does the unaccounted for water not exceed 20 percent of the total flow?
7. Are finished water quality tests representative of plant performance?
8. Are operational control decisions based on process control testing within the plant?
9. Are provisions made for continued training for plant personnel?
10. Are qualified personnel available to fill job vacancies and is job turnover relatively low?
11. Are the energy costs for the system not more than 20 to 30 percent of the total operating costs?
12. Is the ratio of corrective (reactive) maintenance to preventive (proactive) maintenance remaining stable and is it less than 1.0?
13. Are maintenance records available for review?
14. Is the spare parts inventory adequate to prevent long delays in equipment repairs?
15. Are old or outdated pieces of equipment replaced as necessary to prevent excessive equipment downtime, inefficient process performance, or unreliability?
16. Are technical resources and tools available for repairing, maintaining, and installing equipment?
17. Is the utility's pump station equipment providing the expected design performance?
18. Are standby units for key equipment available to maintain process performance during breakdowns or during preventive maintenance activities?
19. Are the plant processes adequate to meet the demand for treatment?
20. Does the facility have an adequate emergency response plan including an alternate water source?

Controlling involves taking the steps necessary to ensure that essential activities are performed so that objectives will be achieved as planned. Controlling means being sure that progress is being made toward objectives and taking corrective action as necessary. The utility manager is directly involved in controlling the treatment process to ensure that water is being properly treated and to make sure that the utility is meeting its short- and long-term goals.

QUESTIONS

Please write your answers to the following questions and compare them with those on page 698.

23.1A What are the functions of a utility manager?

23.1B In small communities, what does the community depend on the utility manager to do?

23.2 PLANNING[1]

A very large portion of any manager's typical work day will be spent on activities that can be described as planning activities since nearly every area of a manager's responsibilities require some type of planning.

Planning is one of the most important functions of utility management and one of the most difficult. Communities must have good, safe drinking water and the management of water utilities must include building the resources and financial capability to provide for future needs. The utility must plan for future growth, including industrial development, and be ready to provide the water that will be needed as the community grows. The most difficult problem for some small communities is recognizing and planning for a decline in population. The utility manager must develop reliable information to plan for growth or decline. Decisions must be made about goals, both short- and long-term. The manager must prepare plans for the next 2 years and the next 10 to 20 years. Remember that utility planning should include operational personnel, local officials (decision makers), and the public. Everyone must understand the importance of planning and be willing to contribute to the process.

Operation and maintenance (O&M) of a utility also involves planning by the utility manager. A preventive maintenance program should be established to keep the system performing as intended and to protect the community's investment in water supply and distribution facilities. (Section 23.9 describes the various types of maintenance and the benefits of establishing maintenance programs.)

The utility also must have an emergency response plan to deal with natural or human disasters. Without adequate planning, your utility will be facing system failures, inability to meet compliance regulations, and inadequate service capacity to meet community needs. Plan today and avoid disaster tomorrow. (Section 23.10, "Emergency Response," describes the basic elements of an emergency operations plan.)

23.3 ORGANIZING[2]

A utility should have a written organizational plan and written policies. In some communities, the organizational plan and policies are part of the overall community plan. In either case, the utility manager and all plant personnel should have a copy of the organizational plan and written policies of the utility.

The purpose of the organizational plan is to show who reports to whom and to identify the lines of authority. The organizational plan should show each person or job position in the organization with a direct line showing to whom each person reports in the organization. Remember, an employee can serve only one supervisor (unity of command) and each supervisor should ideally manage only six or seven employees. The organizational plan should include a job description for each of the positions on the organizational chart. When the organizational plan is in place, employees know who is their immediate boss and confusion about job tasks is eliminated. A sample organizational plan for a water/wastewater utility is shown in Figure 23.1. The basic job duties for some typical utility positions are described in Table 23.2.

To understand organization and its role in management, we need to understand some other terms including authority, responsibility, delegation, and accountability. *AUTHORITY* means the power and resources to do a specific job or to get that job done. Authority may be given to an employee due to their position in the organization (this is formal authority) or authority may be given to the employee informally by their co-workers when the employee has earned their respect. *RESPONSIBILITY* may be described as answering to those above in the chain of command to explain how and why you have used your authority. *DELEGATION* is the act in which power is given to another person in the organization to accomplish a specific job. Finally, when a manager gives power/responsibility to an employee, then the employee is *ACCOUNTABLE*[3] for the results.

Organization and effective delegation are very important to keep any utility operating efficiently. Effective delegation is uncomfortable for many managers since it requires giving up power and responsibility. Many managers believe that they can do the job better than others, they believe that other employees are not well trained or experienced, and they are afraid of mistakes. The utility manager retains some responsibility even after delegating to another employee and, therefore, the manager is often reluctant to delegate or may delegate the responsibility but not the authority to get the job done. For the utility manager, good organization means that employees are ready to accept responsibility and have the power and resources to make sure that the job gets done.

---

1. *Planning.* Management of utilities to build the resources and financial capability to provide for future needs.
2. *Organizing.* Deciding who does what work and delegating authority to the appropriate persons.
3. *Accountability.* When a manager gives power/responsibility to an employee, the employee ensures that the manager is informed of results or events.

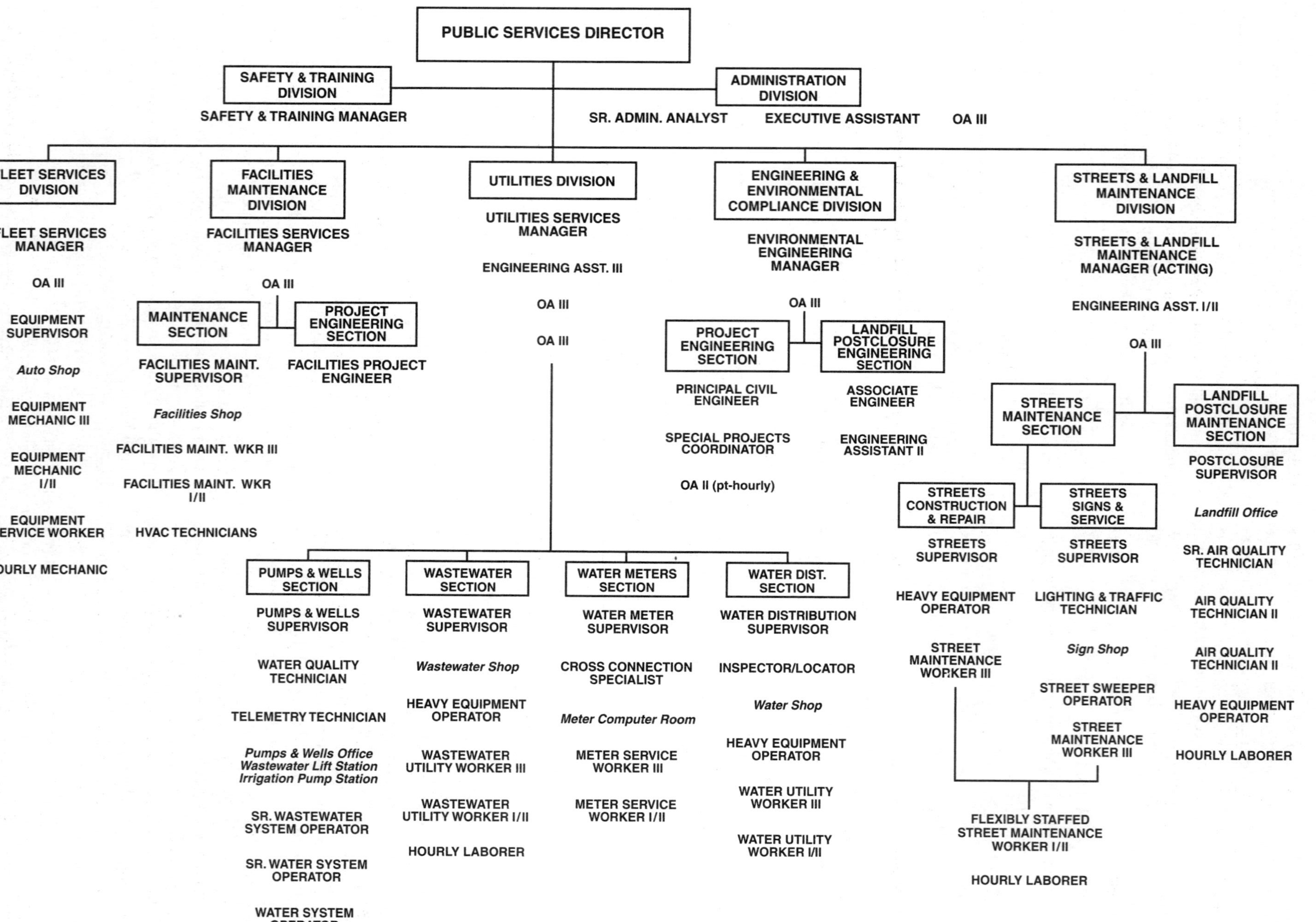

Fig. 23.1 Organizational chart for medium-sized utility
(Courtesy of City of Mountain View, California)

TABLE 23.2 JOB DUTIES FOR STAFF OF A MEDIUM-SIZED UTILITY

Job Title	Job Duties
Superintendent	Responsible for administration, operation, and maintenance of entire facility. Exercises direct authority over all plant functions and personnel.
Assistant Superintendent	Assists superintendent in review of operation and maintenance function, plans special operation and maintenance tasks.
Clerk/Typist	Performs all clerical duties.
Operations Supervisor	Coordinates activities of plant operators and other personnel. Prepares work schedules, inspects plant, and makes note of operational and maintenance requirements.
Lead Utility Worker	Supervises operations and manages all operators.
Utility Worker II (Journey Level)	Controls treatment processes. Collects samples and delivers them to the lab for analysis. Makes operational decisions.
Utility Worker I	Performs assigned job duties.
Maintenance Supervisor	Supervises all maintenance for plant. Plans and schedules all maintenance work. Responsible for all maintenance records.
Maintenance Foreman	Supervises mechanical maintenance crew. Performs inspections and determines repair methods. Schedules all maintenance including preventive maintenance.
Maintenance Mechanic II	Selects proper tools and assigns specific job tasks. Reports any special considerations to foreman.
Maintenance Mechanic I	Performs assigned job duties.
Electrician II	Schedules and coordinates electrical maintenance with other planned maintenance. Plans and selects specific work methods.
Electrician I	Performs assigned job duties.
Chemist	Directs all laboratory activities and makes operational recommendations to operations supervisor. Reports and maintains all required laboratory records. Oversees laboratory quality control.
Laboratory Technician	Performs laboratory tests. Manages day-to-day laboratory operations.

Employees should not be asked to accept responsibilities for job tasks that are beyond their level of authority in the organizational structure. For example, an operator or lead utility worker should not be asked to accept responsibility for additional lab testing. The responsibility for additional lab testing must be delegated to the lab supervisor. Authority and responsibility must be delegated properly to be effective. When these three components—proper job assignments, authority, and responsibility—are all present, the supervisor has successfully delegated. The success of delegation is dependent upon all three components.

An important and often overlooked part of delegation is follow-up by the supervisor. A good manager will delegate and follow up on progress to make sure that the employee has the necessary resources to get the job done. Well-organized managers can delegate effectively and do not try to do all the work themselves, but are responsible for getting good results. The Management Case Study No. 1 that follows describes what can happen when delegation is improperly conducted, and illustrates how an organizational plan can prevent disaster.

Management Case Study No. 1

The City Manager of Pleasantville calls the Director of Public Works and asks for a report on the need for and cost of a new utility truck to be presented at the September 13 meeting of the City Council. The Director of Public Works calls the Plant Manager and asks for a report on the need and cost for a new utility truck with a deadline of September 12. The Plant Manager calls the Lead Utility Worker, an operator, who has been asking for a new utility truck and has been looking into the details. The Plant Manager requests that the Lead Utility Worker provide a report on September 12 about the purchase of the truck. The Lead Utility Worker gathers all the notes and hand writes a report identifying the need for the truck, the features required, and the cost. The Lead Utility Worker takes the report to City Hall to be typed and leaves it with a secretary on September 12. On September 13, the City Manager is preparing for the City Council meeting and does not have the report. Who is responsible? Who is accountable? How could this situation have been avoided?

Responsibility: The Lead Utility Worker's responsibility has been carried out with the authority and resources made available. Both the Director of Public Works and the Plant Manager failed to follow up on the report on September 12. No one informed the Lead Utility Worker that the report must be presented to the City Council on September 13, nor was the Lead Utility Worker supplied with the resources for getting the report in final form. However, the City Manager is ultimately responsible for reporting to the City Council.

Accountability: Starting with the Lead Utility Worker and working upward, each employee is accountable to his or her supervisor and should have communicated the status of the report.

How to avoid this situation: Good communication and follow-up by each of these supervisors could have prevented this situation completely. The City Manager should have asked to see the report on September 12; the Director of Public Works should have asked the Plant Manager to deliver the report no later than September 11; and the Plant Manager should have asked the Lead Utility Worker to submit the typed report (to the Plant Manager) no later than September 10. When delegating this task, the Plant Manager should have arranged for a secretary or clerk to assist the Lead Utility Worker in typing the report. Providing clerical support enables the Lead Utility Manager to complete the assigned task in a timely manner.

At each step in this chain of delegation, setting an early deadline gives the individual receiving the report an opportunity to review the document and make revisions, if necessary, before forwarding it up the chain of authority and ensures that the report reaches the City Manager no later than September 12.

QUESTIONS

Please write your answers to the following questions and compare them with those on page 698.

23.2A Who must be included in utility planning?

23.3A What is the purpose of an organizational plan?

23.3B Why is it sometimes difficult or uncomfortable for supervisors or managers to delegate effectively?

23.3C What is an important and often overlooked part of delegation?

NOTICE

The information provided in this section on staffing should not be viewed as legal advice. The purpose of this section is simply to identify and describe in general terms the major components of a utility manager's responsibilities in the area of staffing. One issue, harassment, is discussed in somewhat greater detail to illustrate the broad scope of a manager's responsibilities within a single policy area. Personnel administration is affected by many federal and state regulations. Legal requirements of legislation such as the Americans with Disabilities Act (ADA), Equal Employment Opportunity (EEO) Act, Family and Medical Leave Act (FMLA), and wages and hours laws are complex and beyond the scope of this manual. If your utility does not have established personnel policies and procedures, consider getting help from a labor law attorney to develop appropriate policies. At the very least, you should get help from a recruitment specialist to develop and document hiring procedures that meet the federal guidelines for Equal Employment Opportunity.

23.4 STAFFING

23.40 The Utility Manager's Responsibilities

The utility manager is also responsible for staffing, which includes hiring new employees, training employees, and evaluating job performance. The utility should have established procedures for job hiring that include requirements for advertising the position, application procedures, and the procedures for conducting interviews.

In the area of staffing, more than any other area of responsibility, a manager must be extremely cautious and consider the consequences before taking action. Personnel management practices are changing dramatically and continue to be redefined almost daily by the courts. A manager who violates an employee's or job applicant's rights can be held both personally and professionally liable in court. Throughout this section on staffing you will repeatedly find references to two terms: "job-related" and "documentation." These are key concepts in personnel management today. Any personnel action you take must be job-related, from the questions you ask during interviews to disciplinary actions or promotions. And while almost no one wants more paperwork, documentation of personnel actions detailing what you did, when you did it, and why you did it (the reasons will be job-related, of course) is absolutely essential. There is no way to predict when you might be called upon to defend your actions in court. Good records not only serve to refresh your memory about past events but can also be used to demonstrate your pattern of lawful personnel behavior over time.

23.41 How Many Employees Are Needed?

There is a common tendency for organizations to add personnel in response to changing conditions without first examining how the existing workforce might be reorganized to achieve greater efficiency and meet the new work demands. In water supply and distribution utilities, aging of the system, changes in use, and expansion of the system often mean changes in the O&M tasks being performed. The manager of a utility should periodically review the agency's work requirements and staffing to ensure that the utility is operating as efficiently as possible. A good time to conduct such a review is during the annual budgeting process or when you are considering hiring a new employee because the workload seems to be greater than the current staff can adequately handle.

The staffing analysis procedure outlined in this section illustrates how to conduct a comprehensive analysis of the type needed for a complete reorganization of the agency. In practice, however, a complete reorganization may not be desirable or even possible. Frequent organizational changes can make employees anxious about their jobs and may interfere with their work performance. Some employees show strong resistance to any change in job responsibilities. Nonetheless, by thoroughly examining the functions and staffing of the utility on a periodic basis, the manager may spot trends or discover inefficiencies that could be corrected over a period of time such as an increase in the amount of time spent maintaining certain equipment or portions of the system.

The first step in analyzing the utility's staffing needs is to prepare a detailed list of all the tasks to be performed to operate and maintain the utility. Next, estimate the number of staff hours per year required to perform each task. Be sure to include the time required for supervision and training.

When you have completed the task analysis, prepare a list of the utility's current employees. Assign tasks to each employee based on the person's skills and abilities. To the extent possible, try to minimize the number of different work activities assigned to each person but also keep in mind the need to provide opportunities for career advancement. One full-time staff year equals 260 days, including vacation and holiday time: (52 wks/yr)(5 days/wk) = 260 days/yr.

You can expect to find that this ideal staffing arrangement does not exactly match up with your current employees' job assignments. Most likely, you will also find that the number of staff hours required does not exactly equal the number of staff hours available. Your responsibility as a manager is to create the best possible fit between the work to be done and the personnel/skills available to do it. In addition to shifting work assignments between employees, other options you might consider are contracting out some types of work, hiring part-time or seasonal staff, or setting up a second shift (to make fuller use of existing equipment). Of course, you may find that it is time to hire another full- or part-time operator.

23.42 Qualifications Profile

Hiring new employees requires careful planning before the personal interview process. In an effort to limit discriminatory hiring practices, the law and administrative policy have carefully defined the hiring methods and guidelines employers may use. The selection method and examination process used to evaluate applicants must be limited to the applicant's knowledge, skills, and abilities to perform relevant job-related activities. In all but rare cases, factors such as age and level of education may not be used to screen candidates in place of performance testing. A description of the duties and qualifications for the job must be clearly defined in writing. The job description may be used to develop a qualifications profile. This qualifications profile clearly and precisely identifies the required job qualifications. All job qualifications must be relevant to the actual job duties that will be performed in that position. The following list of typical job qualifications may be used to help you develop your own qualifications profiles with advice from a recruitment specialist.

1. General Requirements:
   a. Knowledge of methods, tools, equipment, and materials used in water utilities
   b. Knowledge of work hazards and applicable safety precautions
   c. Ability to establish and maintain effective working relations with employees and the general public
   d. Possession of a valid state driver's license for the class of equipment the employee is expected to drive
2. General Educational Development:
   a. Reasoning: Apply common-sense understanding to carry out instructions furnished in oral, written, or diagrammatical form.
   b. Mathematical: Use a pocket calculator to make arithmetic calculations relevant to the utility's O&M processes.
   c. Language: Communicate with fellow employees and train subordinates in work methods. Fill out maintenance report forms.

3. Specific Vocational Preparation: Three years of experience in water utility O&M.
4. Interests: May or may not be relevant to knowledge, skills, and ability; for example, an interest in activities concerned with objects and machines, ecology, or business management.
5. Temperament: Must adjust to a variety of tasks requiring frequent change and must routinely use established standards and procedures.
6. Physical Demands: Medium to heavy work involving lifting, climbing, kneeling, crouching, crawling, reaching, hearing, and seeing. Must be able to lift and carry objects weighing______pounds for a distance of______feet.
7. Working Conditions: The work involves wet conditions, cramped and awkward spaces, noise, risks of bodily injury, and exposure to weather.

QUESTIONS

Please write your answers to the following questions and compare them with those on page 698.

23.4A What do staffing responsibilities include?

23.4B What are two key personnel management concepts a manager should always keep in mind?

23.4C List the steps involved in a staffing analysis.

23.4D What is a qualifications profile?

23.43 Applications and the Selection Process

23.430 Advertising the Position

To advertise a job opening, first prepare a written description of the required job qualifications, compensation, job duties, selection process, and application procedures (with a closing date). The utility should have established procedures about how to advertise the position and conduct the application process. The application procedure may require that the job be posted first within the utility to allow existing personnel first chance at the job opportunity.

23.431 Paper Screening

The next step in the selection process is known as paper screening. The personnel department and the utility manager review each application and eliminate those who are not qualified. The qualified applicants may be given examinations to verify their qualifications. Usually, the top 3 to 12 applicants are selected for an interview, depending on the agency's preference.

23.432 Interviewing Applicants

The purpose of the job interview is to gain additional information about the applicants so that the most qualified person can be selected. The utility manager should prepare for the interview in advance. Review the background information on each applicant. Draw up a list of job-related questions that will be asked of each applicant. During the interviews, briefly note the answers each applicant gives.

It used to be thought that the best way to learn about applicants was to give them plenty of time to talk about themselves because the content and type of information applicants volunteer might provide a deeper insight into the person and what type of employee they will become. Be very careful about open-ended, unstructured conversations with job applicants, even the friendly remarks you make initially to put the applicant at ease during the interview. If the applicant begins to volunteer information that you could not otherwise legally ask for (such as marital status, number of children, religious affiliation, or age), be polite but firm in promptly redirecting the conversation. Even if this information was provided to you voluntarily, an applicant who did not get the job could later allege that you discriminated against them based on age or religion.

The only type of information you may legally request is information about the applicant's job skills, abilities, and experience relating directly to the job for which the person is applying. You must always be sensitive to the civil rights of the applicant and the affirmative action policies of the utility, which is another good reason to prepare a list of questions before the interview process begins. Structure the questions so that you avoid simple yes-and-no answers. Table 23.3 summarizes acceptable and unacceptable pre-employment inquiries to guide you in developing a good list of questions.

If other utility staff members are participating in the interviews, their participation should be confined to the pre-selected questions. Under no circumstances should frontline employees conduct interviews in the absence of the manager or another person knowledgeable about personnel policies and practices.

The interview should be conducted in a quiet room without interruptions. Most applicants will be nervous, so start the interview on a positive note with introductions and some general remarks to put the applicant at ease. Explain the details of the job, working conditions, wages, benefits, and potential for advancement. Allow the applicant a chance to ask questions about the job. Ask each applicant the questions you have prepared and jot down brief notes on their responses. Taking notes while interviewing is awkward for some people but it becomes easier with practice. Notes are important because after interviewing

TABLE 23.3 ACCEPTABLE AND UNACCEPTABLE PRE-EMPLOYMENT INQUIRIES[a]

Acceptable Pre-Employment Inquiries	Subject	Unacceptable Pre-Employment Inquiries
"Have you ever worked for this agency under a different name?"	NAME	Former name of applicant whose name has been changed by court order or otherwise.
Applicant's place of residence. How long applicant has been resident of this state or city.	ADDRESS OR DURATION OF RESIDENCE	
"If hired, can you submit a birth certificate or other proof of US citizenship or age?"	BIRTHPLACE	Birthplace of applicant. Birthplace of applicant's parents, spouse, or other relatives. Requirement that applicant submit a birth certificate, naturalization, or baptismal record.
"If hired, can you furnish proof of age?" or statement that hire is subject to verification that applicant's age meets legal requirements.	AGE	Questions that tend to identify applicants 40 to 64 years of age.
Statement by employer of regular days, hours, or shift to be worked.	RELIGION	Applicant's religious denomination or affiliation, church, parish, pastor, or religious holidays observed. "Do you attend religious services or a house of worship?" Applicant may not be told, "This is a Catholic/Protestant/Jewish/atheist organization."
	RACE OR COLOR	Complexion, color of skin, or other questions directly or indirectly indicating race or color.
Statement that photograph may be required after employment.	PHOTOGRAPH	Requirement that applicant affix a photograph to his/her application form. Request applicant, at his/her option, to submit photograph. Requirement of photograph after interview but before hiring.
Statement by employer that, if hired, applicant may be required to submit proof of eligibility to work in the United States.	CITIZENSHIP	"Are you a United States citizen?" Whether applicant or applicant's parents or spouse are naturalized or native-born US citizens. Date when applicant or parents or spouse acquired US citizenship. Requirement that applicant produce naturalization papers or first papers. Whether applicant's parents or spouse are citizens of the US
Applicant's work experience. Applicant's military experience in armed forces of United States, in a state militia (US), or in a particular branch of the US armed forces.	EXPERIENCE	"Are you currently employed?" Applicant's military experience (general). Type of military discharge.
Applicant's academic, vocational, or professional education; schools attended.	EDUCATION	Date last attended high school.

TABLE 23.3 ACCEPTABLE AND UNACCEPTABLE PRE-EMPLOYMENT INQUIRIES *(continued)*

Acceptable Pre-Employment Inquiries	Subject	Unacceptable Pre-Employment Inquiries
Language applicant reads, speaks, or writes fluently.	NATIONAL ORIGIN OR ANCESTRY	Applicant's nationality, lineage, ancestry, national origin, descent, or parentage.
		Date of arrival in United States or port of entry; how long a resident.
		Nationality of applicant's parents or spouse; maiden name of applicant's wife or mother.
		Language commonly used by applicant. "What is your mother tongue?"
		How applicant acquired ability to read, write, or speak a foreign language.
	CHARACTER	"Have you ever been arrested?"
Names of applicant's relatives already employed by the agency.	RELATIVES	Marital status or number of dependents.
		Name or address of relative, spouse, or children of adult applicant.
		"With whom do you reside?"
		"Do you live with your parents?"
Organizations, clubs, professional societies, or other associations of which applicant is a member, excluding any names the character of which indicate the race, religious creed, color, national origin, or ancestry of its members.	ORGANIZATIONS	"List all organizations, clubs, societies, and lodges to which you belong."
"By whom were you referred for a position here?"	REFERENCES	Requirement of submission of a religious reference.
"Do you have any physical condition that may limit your ability to perform the job applied for?"	PHYSICAL CONDITION	"Do you have any physical disabilities?"
Statement by employer that offer may be made contingent on passing a physical examination.		Questions on general medical condition.
		Inquiries as to receipt of Workers' Compensation.
Notice to applicant that any misstatements or omissions of material facts in his/her application may be cause for dismissal.	MISCELLANEOUS	Any inquiry that is not job-related or necessary for determining an applicant's eligibility for employment.

a Courtesy of Marion B. McCamey, Affirmative Action Officer, California State University, Sacramento, CA.

several candidates you may not be able to remember what each one said. Also, as mentioned earlier, notes taken at the time of an interview can be valuable evidence in court if an unsuccessful applicant files a lawsuit for unfair hiring practices. At the end of the interview, tell the applicant when a decision will be made and how the applicant will be informed of the decision.

If an applicant's responses during the interview indicate that the person clearly is not qualified for this job but may be qualified for another job, briefly describe the other opportunity and how the person can apply for that position. The applicant may ask to be interviewed immediately for the second position. However, do not violate the utility's hiring procedures for the convenience of a job applicant. The same sequence of hiring procedures should be followed each time a position is filled. Tell this applicant that it will still be necessary to apply for the other position and that another interview may or may not follow, depending on the qualifications of the other applicants for the position.

23.433 Selecting the Most Qualified Candidate

Once the interviews are over, the job of evaluating and selecting the successful candidate begins. Review your interview notes and check the candidates' references. Checking references will verify the job experience of the applicant and may provide insight into the applicant's work habits. Questions you might ask previous employers include: Was the employee reliable and punctual? How well did the employee relate to co-workers? Did the employee consistently practice safe work procedures? Would you rehire this employee?

The rights of certain protected groups in the workforce today, such as minorities, women, disabled persons, persons over 40 years of age, and union members, are protected by law. A manager's responsibilities regarding protected groups begins with the hiring process and continues for as long as the employer/employee relationship lasts. The best principle to deal with protected groups (and all other employees, for that matter) is the *OUCH PRINCIPLE,* which says that when you hire new employees or delegate job tasks to current employees, you must be **O**bjective, **U**niform in your treatment of applicants or employees, **C**onsistent with utility policies, and **H**ave job relatedness. If you do not manage using these characteristics, you may find yourself in a questionable position with regard to the treatment of protected workers and other employees in general.

Objectivity is the first hurdle. Often the physical characteristics of a person, such as large or small size, may make a person seem more or less job capable. However, many utility agency jobs are done with power tools or other technology that allows all persons, regardless of size, to manage most tasks. Try to remain objective but reasonable in assessing job applicants and making job assignments.

Uniform treatment of job applicants and employees is necessary to protect yourself and other employees. Nothing will destroy morale more quickly than unequal treatment of employees. Your role as a manager is to consistently apply the policies and procedures that have been adopted by the utility. Often, policies and procedures exist that are not popular and may not even be appropriate. However, the job of the utility manager is to consistently uphold and apply the policies of the utility.

The last part of the OUCH principle is having job relatedness. Any hiring decision must be based on the applicant's qualifications to meet the specific job requirements and any job assignment given to an employee must be related to that employee's job description. Extra assignments, such as buying personal gifts for the boss's family or washing the boss's car, are not appropriate. These types of job assignments will eventually catch up to the manager and can be particularly embarrassing if the public gets involved. So to protect yourself and your utility, remember the OUCH principle as you hire and manage your employees.

Once you have made your selection, the applicant is usually required to pass a medical examination. When this has been successfully completed and the applicant has accepted the position, notify the other applicants that the position has been filled.

23.44 New Employee Orientation

During the first day of work, a new employee should be given all the information available in written and verbal form on the policies and practices of the utility, including compensation, benefits, attendance expectations, alcohol and drug testing (if the utility does this), and employer/employee relations. Answer any questions from the new employee at this time and try to explain the overall structure of the utility as well as identify who can answer employee questions when they arise. Introduce the new employee to co-workers and tour the work area. Every utility should have a safety training session for all new employees and specific safety training for some job categories. Provide safety training (see Section 23.11, "Safety Program") for new employees on the first day of employment or as soon thereafter as possible. Establishing safe work practices is a very important function of management.

QUESTIONS

Please write your answers to the following questions and compare them with those on page 698.

23.4E What is the purpose of a job interview?

23.4F List four protected groups.

23.4G What does the OUCH principle stand for?

23.4H When should a new employee's safety training begin?

23.45 Employment Policies and Procedures

23.450 Probationary Period

Many employers now use a probationary period for all new employees. The probationary period is typically 3 to 6 months but may be as long as 1 year. This period begins on the first day of work. Management may reserve the right to terminate employment of the person with or without cause during this probationary period. The employee must be informed of this probationary period and must understand that successful completion of the probationary period is required in order to move into regular employment status.

The probationary period provides a time during which both the employer and employee can assess the fit between the job and the person. Normally, a performance evaluation is completed near the end of the probationary period. A satisfactory performance evaluation is the mechanism used to move an employee from probationary status into regular employment.

23.451 Compensation

The compensation an employee receives for the work performed includes satisfaction, recognition, security, appropriate pay, and benefits. All are important to keep good employees satisfied. Salaries should be a function of supply and demand. Pay should be high enough to attract and retain qualified employees. Salaries are usually determined by the governing body of the utility in negotiation with employee groups, when appropriate. The salary structure should meet all state and federal regulations and accurately reflect the level of service given by the employee. A survey of salaries from other utilities in the area may provide valuable information in the development of a salary structure.

The benefits provided by an employer are an important part of the compensation package. Benefits generally consist of retirement pay, health insurance, life insurance, employer's portion of social security, holiday and vacation pay, sick leave, personal leave, parental leave, worker's compensation, and protective clothing. Many employers now provide dental and vision insurance, long-term disability insurance, educational bonus or costs and leave, bereavement leave, and release time for jury duty. Some employers also include in their benefit package cash bonus programs and longevity pay. The value of an employee's entire benefit package is often computed and printed on the pay stub as a reminder that salary alone is not the only compensation being provided.

23.452 Training and Certification[4]

Training has become an ongoing process in the workplace. The utility manager must provide new employee training as well as ongoing training for all employees. Safety training is particularly important for all utility operators and staff members and is discussed in detail later in Section 23.11. Certified water supply system and treatment plant operators earn their certificates by knowing how to do their jobs safely. Preparing for certification examinations is one means by which operators learn to identify safety hazards and to follow safe procedures at all times under all circumstances.

Although it is extremely important, safety is not the sole benefit to be derived from a certification program. Other benefits include protection of the public's investment in water supply and treatment facilities and employee pride and recognition.

Vast sums of public funds have been invested in the construction of water supply and treatment facilities. Certification of operators assures utilities that these facilities will be operated and maintained by qualified operators who possess a certain level of competence. These operators should have the knowledge and skills not only to prevent unnecessary deterioration and failure of the facilities, but also to improve O&M techniques.

Achievement of a level of certification is a public acknowledgment of a water supply system or treatment plant operator's skills and knowledge. Presentation of certificates at an official meeting of the governing body will place the operators in a position to receive recognition for their efforts and may even get press coverage and public opinion that is favorable. An improved public image will give the certified operator more credibility in discussions with property owners.

Recognition for their personal efforts raises the self-esteem of all certified operators. Certification also gives water supply system and treatment plant operators an upgraded image. If properly publicized, certification ceremonies give the public a more accurate image of the many dedicated, well-qualified operators working for them.

All states and most Canadian provinces now require that water treatment plant operators be certified. To maintain current certification, these operators must complete additional training classes every 1 to 5 years. In the environmental field, new technologies and regulations require operators to attend training to keep up with their field. The utility manager has the responsibility to provide employees with high-quality training opportunities. Many types of training are available to meet the different training needs of utility operators, for example, in-house training, training conducted by training centers, professional organizations, engineering firms, or regulatory agencies, and also on-line courses and correspondence courses such as this one by the Office of Water Programs.

ABC stands for the Association of Boards of Certification for Operating Personnel in Water Utilities and Pollution Control Systems. If you wish to find out how operators can become certified in your state or province, contact ABC at www.abccert.org or phone (515) 232-3623.

ABC will provide you with the name and address of the appropriate contact person.

One area of training that is frequently overlooked is training for supervisors. Managing people requires a different set of skills than performing the day-to-day work of operating and maintaining a water treatment facility. Supervisors need to know how to communicate effectively and how to motivate others, as well as how to delegate responsibility and hold people accountable for their performance. Supervisors share management's responsibility for fair and equitable treatment of all workers and are required to act in accordance with applicable state and federal personnel regulations. Making the transition from operator to supervisor also requires a change in attitude. A supervisor is part of the management team and is therefore obliged to promote the best interests of the utility at all times. When the interests of the utility conflict with the desires of one or more employees, the supervisor must support management's decisions and policies regardless of the supervisor's own personal opinion about the issue. It is the responsibility of the utility manager to ensure that supervisors receive appropriate training in all of these areas.

Training on how to motivate people, deal with co-workers, and supervise or manage people working for you has become a very highly specialized field of training. These are complex topics that are beyond the scope of this manual. If you have a need for or wish to learn more about how to deal with people,

4. *Certification Examination.* An examination administered by a state agency or professional association that operators take to indicate a level of professional competence. In the United States, certification of operators of water treatment plants, wastewater treatment plants, water distribution systems, and small water supply systems is mandatory. In many states, certification of wastewater collection system operators, industrial wastewater treatment plant operators, pretreatment facility inspectors, and small wastewater system operators is voluntary; however, current trends indicate that more states, provinces, and employers will require these operators to be certified in the future. Operator certification is mandatory in the United States for the Chief Operators of water treatment plants, water distribution systems, and wastewater treatment plants.

consider enrolling in courses or reading books on supervision or personnel management. An excellent book is *Manage for Success* in this series of operator training manuals from the Office of Water Programs.

QUESTIONS

Please write your answers to the following questions and compare them with those on page 698.

23.4I What is the purpose of a probationary period for new employees?

23.4J What kind of compensation does an employee receive for work performed?

23.4K Why should utility managers provide training opportunities for employees?

23.453 Performance Evaluation

Most organizations conduct some type of performance evaluation, usually on an annual basis. The evaluation may be written or oral; however, a written evaluation is strongly recommended because it will provide a record of the employee's performance. Documentation of this type may be needed in the future to support taking disciplinary action if the employee's performance consistently fails to meet expectations. The evaluation of employee performance can be a challenging task, especially when performance has not been acceptable. However, evaluations are also an opportunity to provide employees with positive feedback and let them know their contributions to the organization have been noticed and appreciated.

A formal performance evaluation typically begins with an employee's immediate supervisor filling out the performance evaluation form (a sample evaluation form is shown in Figure 23.2). Complete the entire form and be specific about the employee's achievements as well as areas needing improvement. Next, schedule a private meeting with the employee to discuss the evaluation. Give the employee frequent opportunities to be heard and listen carefully. If some of the employee's accomplishments were overlooked, note them on the evaluation form and consider whether this new information changes your overall rating of performance in one or more categories.

After reviewing the employee's performance for the past year, set performance goals for the next year. Be sure to document the goals you have agreed upon. Setting performance goals is particularly important if an employee's performance has been poor and improvement is needed. If appropriate, develop a written performance improvement plan that includes specific dates when you will again review the employee's progress in meeting the performance goals. Some supervisors find it helpful to schedule an informal mid-year meeting with each employee to review their progress and to avoid surprises during the next performance evaluation.

Many employees and managers dread even the thought of a performance evaluation and see it as an ordeal to be endured. When properly conducted, however, a performance review can strengthen the lines of communication and increase trust between the employee and the manager. Use this opportunity to acknowledge the employee's unique contributions and to seek solutions to any problems the employee may be having in completing work assignments. Ask the employee how you can be of assistance in removing any obstacles to getting the job done. If necessary, provide coaching to help the employee understand both how and why certain tasks are performed. Be generous (but sincere) with praise for the good work the employee does well every day and try to keep the employee's shortcomings in perspective. If the person is doing a good job 95 percent of the time, do not let the entire discussion consist of criticism about the remaining 5 percent of the person's job assignments.

At the end of the meeting, ask the employee to sign the evaluation form to acknowledge having seen and discussed it. Give the employee a copy of the evaluation. If the employee disagrees with any part of the evaluation, invite the person to submit a written statement describing the reasons for their disagreement. The written statement should be filed with the completed performance evaluation form.

23.454 Dealing with Disciplinary Problems

Handling employee discipline problems is difficult, even for an experienced manager. But remember, no discipline problem ever solves itself and the sooner you deal with the problem, the better the outcome will be. If problem behavior is not corrected, then other employees will become dissatisfied and the problems will increase.

Every utility, no matter how small, should have written employment policies enabling the manager to deal effectively with employee problems. It should also provide a formal complaint or grievance procedure by which employees can have their complaint heard and resolved without fear of retaliation by the supervisor.

Dealing with employee discipline requires tact and skill. You will have to find your own style and then try to stay flexible, calm, and open-minded when the situation gets really tough. If you repeatedly find yourself unable to deal successfully with disciplinary problems, consult with the utility's personnel office (if available) or consider enrolling in a management training course designed specifically around strategies and techniques for disciplining employees.

A commonly accepted method for dealing with job-related employee problems is to first discuss the problem with the employee in private. Most employers will give a person two or three verbal warnings; then the warnings should be written with copies given to the employee. Finally, if the written warnings do not produce positive results, the employee may have to be suspended or dismissed. Your job is to make sure that all warnings are documented with specific descriptions of unsatisfactory behaviors and to make sure that all employees are treated fairly.

Start the disciplinary discussion with a positive comment about the employee. Then, identify the problem but keep emotion and blame out of the discussion. The best approach is to state the problem and then ask the employee to suggest a solution. If

EMPLOYEE EVALUATION FORM

Employee Name ______________________________ Date ______________________

Job Title ______________________________ Department ______________________

Evaluate the employee on the job now being performed. Circle the number which most nearly expresses your overall judgment. In the space for comments, consider the employee's performance since their last evaluation and make notes about the progress or specific concerns in that area. The care and accuracy of this appraisal will determine its value to you, the employee, and your employer.

JOB KNOWLEDGE: (Consider knowledge of the job gained through experience, education, and special training)

5. Well informed on all phases of work
4. Knowledge thorough enough to perform well without assistance
3. Adequate grasp of essentials, some assistance required
2. Requires considerable assistance to perform
1. Inadequate knowledge

Comments: ____

____

____

QUALITY OF WORK: (Consider accuracy and dependability of the results)

5. Exceptionally accurate, practically no mistakes
4. Usually accurate, seldom necessary to check results
3. Acceptable, occasional errors
2. Often unacceptable, frequent errors, needs supervision
1. Unacceptable, too many errors

Comments: ____

____

____

INITIATIVE: (Consider the speed with which the employee grasps new job skills)

5. Excellent, grasps new ideas and suggests improvements, is a leader with others
4. Very resourceful, can work unsupervised, manages time well, is reliable
3. Shows initiative on occasion, is reliable
2. Lacks initiative, must be reminded to complete tasks
1. Needs constant prodding to complete job tasks, is unreliable

Comments: ____

____

____

Fig. 23.2 Employee evaluation form

COOPERATION AND RELATIONSHIPS: (Consider manner of handling relationships with co-workers, superiors, and the public)

5. Excellent cooperation and communication with co-workers, supervisors, and others, takes and gives instructions well
4. Gets along well with co-workers
3. Acceptable, usually gets along well, occasionally complains
2. Shows a reluctance to cooperate, complains
1. Very poor cooperation, does not follow instruction, dislikes fellow employees

Comments: ______________________________

____

____

ATTENDANCE: (Consider frequency of absences, reasons for absences or tardiness, and promptness in giving notice about absences)

5. Excellent, absent only for emergencies, illness, civic duties, always on time, gives notice when absent
4. Rarely absent or late, always gives notice and good reason
3. Occasionally absent, less important reasons, usually gives notice, but not always in time
2. Often absent, lack of adequate notice or reasons for absenteeism
1. Unexcusable absenteeism, does not give notice, reasons are unacceptable, cannot be depended upon

Comments: ______________________________

____

____

OVERALL EVALUATION: Superior ____ Good ____ Satisfactory ____ Unsatisfactory ____

Comments: ______________________________

____

____

I hereby certify that this appraisal is my best judgment of the service value of this employee and is based on personal observation and knowledge of the employee's work.

Supervisor's Signature ______________________ Date ______________

I hereby certify that I have personally reviewed this report.

Employee's Signature ______________________ Date ______________

Fig. 23.2 Employee evaluation form (continued)

they respond inappropriately, you must restate the problem and explain that you are trying to find a positive solution that is acceptable to everyone.

Try to keep the discussion focused on solving the problems and do not permit the employee to heap on general complaints, report on what other employees do, or wander from the topic. The following is an example of how you might start the discussion for an employee who is tardy every day. "Joe, you have done a good job in keeping that north side pump station running. You are an asset to this operation. Your tardiness every morning, however, is causing problems. Is there some reason for you to be tardy? We need to find a solution to this problem because your being late creates a bad situation for the night shift. What do you suggest?"

Always remain calm and do not allow yourself to become angry when dealing with an employee about performance issues. If you begin to feel angry and are about to lose control, or if the employee becomes combative or abusive, suggest that the meeting is not producing positive results and schedule an alternative meeting time. Do not let the emotions of the moment carry you into a rage in front of employees. If either you or one of your employees expresses extreme emotions, then the discussion should be postponed until everyone cools down. The following steps may serve as a guide to dealing with confrontation; they apply equally to the employee and the supervisor.

- Maintain an adult approach—positive criticism should be taken/given to improve job skills.
- Create a private environment—job performance issues should be discussed in private between the employee and supervisor.
- Listen very carefully—be sure that both you and the employee understand the situation in the same way. If not, you need to keep talking until both parties are in agreement about the problem and the solution.
- Keep your language appropriate—anger and use of bad language will cause the situation to escalate. Keep your cool and hold your tongue.
- Stay focused in the present—let go of all the past slights, misunderstandings, and dissatisfaction. Problems must be solved one at a time.
- Aim for a permanent solution—changes in job performance need to be permanent to be effective.

Reports of violence in the workplace appear regularly in newspapers and on television. Managers and supervisors should be alert for signs that an employee might become violent and should take any threat of violence seriously.

Violence in the workplace may take the form of physical harm, psychological harm, or property damage. Common warning signs include abusive language, threatening or confrontational behavior, and brandishing a weapon. Some examples of physical harm include pushing, hitting, shoving, or any other form of physical assault. Threats and harassment are forms of psychological harm, and property damage can range from theft or destruction of equipment to sabotage of the employer's computer systems.

The utility's safe workplace policy should be a zero tolerance policy. Any employee who is the target of violent behavior or who witnesses such behavior should be encouraged to immediately report the behavior to a supervisor and the incident should be investigated promptly. If necessary for the immediate safety of other employees, the offending employee should be placed on administrative leave, escorted from the work environment, and permitted to return to work only after the investigation has been completed.

Management Case Study No. 2

Sue has been working for 5 years as a laboratory technician and was recently passed over for promotion. A lab director who has more college experience than Sue was hired from another plant. Since then, Sue's work has not been very good, she has come to work late, and she does not always get all of the lab tests done during her workday. What should be done about Sue? If you were the supervisor, how would you handle this problem with Sue?

Actions: As the supervisor, you should ask Sue to come by your office. In private, you should discuss with Sue why the new lab director was hired. Discuss with her the good work record she has maintained over the past 5 years, and explain the changes you have seen in her work recently. At this point you might ask her to evaluate her own performance or what she would do if she were in your situation. She might need to express her resentment about the new lab director. If she does, let her express herself for a few minutes. You might say, "OK, you are unhappy, but what are we going to do to change this situation? How can I help you to regain your motivation and improve your work habits?" If possible you might help her figure out a way to continue her college education, go to additional training classes, or reorganize the lab so that her job duties change somewhat. There are many other possibilities for helping Sue to become motivated again and she should be part of the process. It is important to clearly communicate and document that her job performance must improve.

QUESTIONS

Please write your answers to the following questions and compare them with those on page 698.

23.4L Who is the appropriate person to conduct an employee's performance evaluation?

23.4M What should be the attitude of a supervisor or manager when dealing with disciplinary problems?

23.4N What are some common warning signs that an employee could become violent?

23.455 Example Policy: Harassment

Harassment is any behavior that is offensive, annoying, or humiliating to an individual and that interferes with a person's ability to do a job. This behavior is uninvited, often repeated, and creates an uncomfortable or even hostile environment in the workplace. Harassment is not limited to physical behavior but also may be verbal or involve the display of offensive pictures or other images. Sexual harassment is legally defined as unwanted sexual advances, or visual, verbal, or physical conduct of a sexual nature. Any type of harassment is inappropriate in the workplace. A manager's responsibilities with regard to harassment include:

- Establish a written policy (such as the one shown in Figure 23.3) that clearly defines and prohibits harassment of any type.
- Distribute copies of the harassment policy to all employees and take whatever steps are necessary (small group discussions, general staff meetings, training sessions) to ensure that all employees understand the policy.
- Encourage employees to report incidents of harassment to their immediate supervisor, a manager, or the personnel department.
- Investigate every reported case of harassment.
- Document all aspects of the complaint investigation, including the procedures followed, statements by witnesses, the complainant, and the accused person, the conclusions reached in the case, and the actions taken (if any).

How do you know when offensive behavior could be considered sexual harassment? Unwelcome sexual advances or other verbal or physical conduct of a sexual nature could be interpreted by an employee as sexual harassment under the following conditions:

- A person is required, or feels they are required, to accept unwelcome sexual conduct in order to get a job or keep a job.
- Decisions about an employee's job or work status are made based upon either the employee's acceptance or rejection of unwelcome sexual conduct.
- The conduct interferes with the employee's work performance or creates an intimidating, hostile, or offensive working environment.

The following is a list of examples of the kind of behavior that is unacceptable and illegal. It is only a partial list to give you an idea of the scope of the requirements.

- Unwanted hugging, patting, kissing, brushing up against someone's body, or other inappropriate sexual touching
- Subtle or open pressure for sexual activity
- Persistent sexually explicit or sexist statements, jokes, or stories
- Repeated leering or staring at a person's body
- Suggestive or obscene notes or phone calls
- Display of sexually explicit pictures or cartoons

The best way to prevent harassment is to set an example by your own behavior and to keep communication open between employees. In most cases, an open discussion with employees about harassment can help everyone understand that innuendo and slurs about a person's race, religion, sex, appearance, or any other personal belief or characteristic are humiliating. The most productive way to control such behavior is by enlisting the help of all employees to feel that they have the right and the responsibility to stop harassment. When you get employees to think about their behavior and how their behavior makes others feel, they will usually realize they should speak up to prevent harassment.

Here is an example of how employees handled a problem of harassment. A group of operators often had coffee in the office of the utility during their morning break. One of the operators often used foul language, which was embarrassing and offensive to one of the operator's co-workers. The manager sent a memo to all employees that mentioned respect for fellow workers and included a reminder about inappropriate language. The next day, all of the other operators had taped their copy of the memo to the mailbox of the one operator who was most vocal. These operators found a way to send their message loud and clear—no one wants to work in an environment that is unpleasant to others. As a manager, you must establish an atmosphere that is open, congenial, and harmonious for all employees.

Occasional flirting, innuendo, or jokes may not meet the legal definition of sexual harassment. Nevertheless, they may be offensive or intimidating to others. Every employee has a right to a workplace free of discrimination and harassment, and every employee has a responsibility to respect the rights of others.

Managers need to be aware of and take action to prevent any type of harassment in the workplace. It is not enough to simply distribute copies of the utility's harassment policy. If legal action is taken against the utility due to harassment or the existence of a hostile work environment, the utility manager may face both personal and professional liability if it can be shown that the manager should have known harassment was occurring or that the manager permitted a hostile environment to continue to exist.

Management Case Study No. 3

The maintenance crew is a group of five men who have been with the utility for many years and are well respected for their work habits. However, the maintenance shed walls are covered by calendars with photos of scantily clad women and the language used in the shed is sometimes inappropriate. One of your operators is a woman who is well respected by her co-workers and is a very good operator. She comes to your office to complain about the situation in the maintenance shed and demands that

SUBJECT: HARASSMENT POLICY AND COMPLAINT PROCEDURE **NO:** ____________

PURPOSE:

To establish a strong commitment to prohibit harassment in employment, to define discrimination harassment and to set forth a procedure for investigating and resolving internal complaints of harassment.

POLICY:

Harassment of an applicant or employee by a supervisor, management, employee, or co-worker on the basis of race, religion, color, national origin, ancestry, handicap, disability, medical condition, marital status, familial status, sex, sexual orientation, or age will not be tolerated. This policy applies to all terms and conditions of employment, including, but not limited to, hiring, placement, promotion, disciplinary action, layoff, recall, transfer, leave of absence, compensation, and training.

Disciplinary action up to and including termination will be instituted for behavior described in the definition of harassment set forth below:

- Any retaliation against a person for filing a harassment charge or making a harassment complaint is prohibited. Employees found to be retaliating against another employee shall be subject to disciplinary action up to and including termination.

DEFINITION:

Harassment includes, but is not limited to:

A. Verbal Harassment—For example, epithets, derogatory comments or slurs on the basis of race, religious creed, color, national origin, ancestry, handicap, disability, medical condition, marital status, familial status, sex, sexual orientation, or age. This might include inappropriate sex-oriented comments on appearance, including dress or physical features or race-oriented stories.

B. Physical Harassment—For example, assault, impeding or blocking movement, with a physical interference with normal work or movement when directed at an individual on the basis of race, religion, color, national origin, ancestry, handicap, disability, medical condition, marital status, familial status, age, sex, or sexual orientation. This could be conduct in the form of pinching, grabbing, patting, propositioning, leering, or making explicit or implied job threats or promises in return for submission to physical acts.

C. Visual Forms of Harassment—For example, derogatory posters, notices, bulletins, cartoons, or drawings on the basis of race, religious creed, color, national origin, ancestry, handicap, disability, medical conditions, marital status, familial status, sex, sexual orientation, or age.

D. Sexual Favors—Unwelcome sexual advances, requests for sexual favors, and other verbal or physical conduct of a sexual nature which is conditioned upon an employment benefit, unreasonably interferes with an individual's work performance, or creates an offensive work environment.

Fig. 23.3 Harassment policy

(Courtesy of City of Mountain View, California)

SUBJECT: HARASSMENT POLICY AND COMPLAINT PROCEDURE (continued)

COMPLAINT PROCEDURE:

A. Filing:

An employee who believes he or she has been harassed may make a complaint orally or in writing with any of the following:

1. Immediate supervisor.
2. Any supervisor or manager within or outside of the department.
3. Department head.
4. Employee Services Director (or his/her designee).

Any supervisor or department head who receives a harassment complaint should notify the Employee Services Director immediately.

B. Upon notification of the harassment complaint, the Employee Services Director shall:

1. Authorize the investigation of the complaint and supervise and/or investigate the complaint. The investigation will include interviews with:
   (a) The complainant;
   (b) The accused harasser; and
   (c) Any other persons the Employee Services Director has reasons to believe has relevant knowledge concerning the complaint. This may include victims of similar conduct.
2. Review factual information gathered through the investigation to determine whether the alleged conduct constitutes harassment; giving consideration to all factual information, the totality of the circumstances, including the nature of the verbal, physical, visual, or sexual conduct and the context in which the alleged incidents occurred.
3. Report the results of the investigation and the determination as to whether harassment occurred to appropriate persons, including to the complainant, the alleged harasser, the supervisor, and the department head. If discipline is imposed, the discipline may or may not be communicated to the complainant.
4. If harassment occurred, take and/or recommend to the appropriate department head or other appropriate authority prompt and effective remedial action against the harasser. The action will be commensurate with the severity of the offense.
5. Take reasonable steps to protect the victim and other potential victims from further harassment.
6. Take reasonable steps to protect the victim from any retaliation as a result of communicating the complaint.

DISSEMINATION OF POLICY:

All employees, supervisors, and managers shall be sent copies of this policy.

Effective Date:
Revision Date:

____
City Manager

Fig. 23.3 Harassment policy (continued)

(Courtesy of City of Mountain View, California)

you remove the pictures from the walls in the shed. She goes on to report that when she went to the shed and requested assistance to check on a pump, which was noisy and running hot, she was told "Sucks to be you, sweetie!" As the manager, what should you do? Should you immediately go to the maintenance shed and rip down the pictures? What should you say to the female operator who comes to your office to complain?

Actions: Your first response should be to reassure your operator that you understand her anger and frustration. Let her know that you will investigate the matter immediately and take action to correct any problems you find. Ask her to write out a complete statement of the facts, including her concerns about the pump, when and how her request for assistance was made, to whom the request was made, who made the offensive remark, the names of any witnesses, and what responses she has gotten from the maintenance crew in the past. Try to establish if this is a one-time incident or if this problem has been going on for some time.

Begin your investigation with a trip to the maintenance shed to observe and evaluate what is hanging on the walls. Discuss the situation with the crew. Try to make this an open discussion so that everybody understands how the photos affect the atmosphere and the image of professionalism of the utility. The best solution is to get the crew to understand how this type of behavior looks to persons outside their own small group and then have them take down the calendars. Be sure to follow up later to confirm that the calendars or other offensive material has not reappeared.

Next, set up private interviews with the person accused of making the offensive remark and each of the witnesses. Ask each person to describe the encounter in the maintenance shed and make detailed notes of their responses. Depending on the complexity of the situation, it may sometimes be appropriate to have each person involved submit a written statement describing what occurred.

After you have thoroughly investigated the incident, discuss with the crew the use of acceptable language in response to other employees. An open discussion and increased awareness of your sexual harassment policy should help change this situation. If it does not, arrange for a training program in sexual harassment awareness. Be sure to establish a policy on the consequences of inappropriate behavior and be prepared to enforce the policy when needed.

Retaliation against an employee for filing a complaint about harassment or a hostile work environment is illegal. Some examples of retaliation are demotion, suspension, failure to hire or consider for hire, failure to make impartial employment recommendations or decisions, adversely changing working conditions, spreading rumors, or denying any employment benefit. Retaliation could be the basis for a lawsuit involving not only the person who is accused of retaliation, but also the immediate supervisor, the manager, and the utility.

Most areas of personnel management, including harassment and retaliation issues, are complex and have significant legal consequences for everyone involved. The discussion in this section is not a complete explanation of harassment or retaliation. If your utility is large enough to employ a personnel specialist or a labor law attorney, ask them to review your staffing policies and procedures and consult with them whenever you have questions about personnel matters. If you manage a small utility and have no in-house sources for technical or legal advice, enroll in appropriate training courses or consider working with an attorney on a contract basis.

23.456 Labor Laws Governing Employer/Employee Relations

Many employers take pride in advertising that they are an equal opportunity employer, and one often sees this claim in newspaper help wanted ads and other forms of job postings. It means an employer's staffing policies and procedures do not discriminate against anyone based on race; religion; national origin; color; citizenship; marital status; gender; age; Vietnam-era or disabled veteran status; or the presence of a physical, mental, or sensory disability. An employer must meet specific requirements of the federal Equal Employment Opportunity Act to be eligible to advertise as an equal opportunity employer. These requirements include adoption of nondiscriminatory personnel policies and procedures and periodic submission of reports of personnel actions for review by the Equal Employment Opportunity Commission.

The Family and Medical Leave Act of 1993 (FMLA) is a federal law that requires all public agencies as well as companies with 50 or more employees to permit eligible employees to take up to 12 weeks of time off in a 12-month period for the employee's own serious health condition; to care for a child following birth or placement for adoption or foster care; or to care for the employee's spouse, child, or parent who has a serious health condition. To be eligible to receive this benefit, an employee must have been employed for at least 1 year before using FMLA leave. The employer is not required to pay the employee's salary during the time off work, but many employers permit (or require) employees to use accrued sick leave and vacation time during the period of unpaid FMLA leave.

The Americans with Disabilities Act of 1990 (ADA) prohibits employment discrimination based on a person's mental or physical disability. The law applies to employers engaged in an industry affecting commerce who have 15 or more employees.

In general, the ADA defines disability as a physical or mental impairment that substantially limits one or more of the major life activities of an individual. The exact meaning of this definition is evolving as the courts settle lawsuits in which individuals allege they were discriminated against because of a physical or mental disability. The original ADA legislation listed more than 40 specific types of impairments and the courts continue to expand the list.

Under the ADA, employers must make reasonable accommodations to enable a disabled person to function successfully in the work environment, for example, installing a ramp to make facilities accessible to someone in a wheelchair, or restructuring an individual's job or work schedule, or providing an interpreter. The requirements of each situation are unique. In each case, the nature and extent of the disability and the reasonableness (including cost factors) of the requested accommodation by the employer must be weighed. Employers are not automatically required to do everything possible to accommodate disabled persons, but rather to take whatever reasonable steps they can to do so.

All of these personnel laws are very complex and managers of any utility that may be covered by them are strongly urged to seek the assistance of an experienced labor law attorney or personnel specialist.

23.457 Personnel Records

A personnel file should be maintained for each utility employee. This file should contain all documents related to the employee's hiring, performance reviews, promotions, disciplinary actions, and any other records of employment-related matters. Since these records often contain sensitive, confidential information, access to personnel records should be closely controlled. (Also see Section 23.12, "Recordkeeping," for more information about what records should be kept and how long they should be kept.)

QUESTIONS

Please write your answers to the following questions and compare them with those on page 699.

23.4O What is harassment?

23.4P List three types of behavior that could be considered sexual harassment.

23.4Q What is the best way to prevent harassment?

23.4R What is the meaning of "disability" under the Americans with Disabilities Act?

23.46 Unions

Whether your utility operators belong to a union now or may join one in the future, a good employee-management relationship is crucial to keeping an agency functioning properly. Managers, supervisors, crew leaders, and operators all have to work together to develop this relationship.

Most of a manager's union contacts are with a shop steward. The shop steward is elected by the union employees and is their official representative to management and the local union. The steward is in an awkward position because the steward is an employee who is expected to do a full-time job like other employees, while also representing all of the employees. The steward must create an effective link between the utility manager or supervisors and the employees.

During contract negotiations between management and the employees' union representatives, management should be in constant consultation with the supervisors. Many employee demands regarding working conditions originate from the supervisor's daily dealings with the employees. An effective supervisor can minimize unreasonable demands. Also, any demands that are agreed upon must be implemented and carried out by a supervisor. A supervisor can help both sides reach an acceptable contractual agreement.

Once a contract has been agreed upon by both the union and the utility, the utility manager and the other supervisors must manage the organization within the framework of the contract. Do not attempt to ignore or get around the contract even if you disagree with some aspects of it. If you do not understand certain contract provisions, ask for clarification before you begin implementation of those provisions.

Union contracts do not change the supervisor's delegated authority or responsibility. Operators must carry out the supervisor's orders and get the work done properly, safely, and within a reasonable amount of time. As a supervisor, you have the right and even the duty to make decisions. However, a contract gives a union the right to protest or challenge your decision. When an operator requires discipline, disciplinary action is a management responsibility.

Handling employee grievances within the framework of a union contract can be a very time-consuming job for the supervisor and the steward. Union contracts usually spell out in great detail the steps and procedures the steward and the supervisor must follow to settle differences. Grievances can develop over disciplinary action, distribution of overtime, transfers, promotions, demotions, and interpretation of labor contracts. The shop steward must communicate complaints and grievances from operators to the supervisor. Then, the supervisor and the steward must work together to settle complaints and adjust grievances. When a shop steward and supervisor can work together, the steward can help the supervisor to be an effective manager.

An effective manager is available to discuss problems. Dealing with grievances as quickly as possible often prevents small problems from growing into large problems. When a shop steward presents a grievance to you, listen carefully and sympathetically to the steward. Discuss the problem with the employee directly with the help of the shop steward. Try to identify the facts and cause of the problem and keep a written record of your findings. Focus on the problem and do not get caught up in irrelevant issues. Make every effort to settle the grievance quickly and to everyone's satisfaction.

The consequences of any solution to a grievance must be considered and solutions must be consistent and fair to other operators. The solution or settlement should be clear and understandable to everyone involved. Once a solution has been agreed upon, prepare a written summary of the agreement. Review this final report with the shop steward to be sure the intent of the solution is understood and properly documented. The entire grievance procedure must be documented and properly filed, from initial presentation to final solution and settlement.

Union activities are governed by the National Labor Relations Act. When a union attempts to organize the utility's employees, your rights and actions as a manager are also governed by this act. Seek competent legal assistance if you have limited or no experience in dealing with unions.

QUESTIONS

Please write your answers to the following questions and compare them with those on page 699.

23.4S What is the role of a shop steward?

23.4T How does a union contract affect a supervisor's authority?

END OF LESSON 1 OF 3 LESSONS

on

ADMINISTRATION

Please answer the discussion and review questions next.

DISCUSSION AND REVIEW QUESTIONS

Chapter 23. ADMINISTRATION

(Lesson 1 of 3 Lessons)

At the end of each lesson in this chapter, you will find discussion and review questions. Please write your answers to these questions to determine how well you understand the material in the lesson.

1. What are the different types of demands on a utility manager?
2. List the basic functions of a manager.
3. What can happen without adequate utility planning?
4. What is the purpose of an organizational plan?
5. Define the following terms:
    1. Authority
    2. Responsibility
    3. Delegation
    4. Accountability
6. When has a supervisor successfully delegated?
7. What two concepts should a manager keep in mind to avoid violating the rights of an employee or job applicant?
8. Why should a manager thoroughly examine the functions and staffing of the utility on a periodic basis?
9. When hiring new employees, the selection method and examination process used to evaluate applicants must be based on what criteria?
10. Why should you make notes of applicants' responses during job interviews?
11. What information should be provided to a new employee during orientation?
12. What type of training should be provided for supervisors?
13. Why is it important to formally document each employee's performance on a regular basis?
14. How should discipline problems be solved?
15. What steps can you take to help reach a successful resolution to a confrontation with an employee?
16. What is a manager's responsibility for preventing harassment in the workplace?
17. How are employee grievances usually handled under a union contract?

CHAPTER 23. ADMINISTRATION

(Lesson 2 of 3 Lessons)

23.5 COMMUNICATION

Good communication is an essential part of good management skills. Both written and oral communication skills are needed to effectively organize and direct the operation of a treatment facility. Remember that communication is a two-part process; information must be given and it must be understood. Good listening skills are as important in communication as the information you need to communicate. As the manager of a water utility, you will need to communicate with employees, with your governing body, and with the public. Your communication style will be slightly different with each of these groups but you should be able to adjust easily to your audience.

23.50 Oral Communication

Oral communication may be informal, such as talking with employees, or it may be formal, such as giving a technical presentation. In both cases, your words should be appropriate to the audience, for example, avoid technical jargon when talking with nontechnical audiences. As you talk, you should be observing your audience to be sure that what you are saying is getting across. If you are talking with an employee, it is a good idea to ask for feedback from the employee, especially if you are giving instructions. When the employee is talking, watch and listen carefully and clarify areas that seem unclear. Likewise, in a more formal presentation, watching your audience will give you feedback about how well your message is being received. Some tips for preparing a formal speech are given in Table 23.4.

23.51 Written Communication

Written communication is more demanding than oral communication and requires more careful preparation. Again, keep your audience in mind and use language that will be understood. Written communication requires more organization since you cannot clarify and explain ideas in response to your audience. Before you begin, you should have a clear idea of exactly what you wish to communicate, then keep your language as concise as possible. Extra words and phrases tend to confuse and clutter your message. Good writing skills develop slowly, but you should be able to find good writing classes in your community if you need help improving your skills. In addition, many publications and computer software programs are available that will assist you in writing the most commonly needed documents such as memos, letters, press releases, résumés, monitoring reports, and the annual report.

Before you can write a report you must first organize your thoughts. Ask yourself, what is the objective of this report? Am I trying to persuade someone of something? What information is important to communicate in this report? For whom is the report being written? How can I make it interesting? What does the reader want to learn from this report?

After you have answered the above questions, the next step is to prepare a general outline of how you intend to proceed with the preparation of the report. List not only key topics, but try to list all of the related topics. Then, arrange the key topics in sequence so there is a workable, smooth flow from one topic to the next. Do not attempt to make your outline perfect. It is just a guide. It should be flexible. As you write, you will find that you need to remove nonessential points and expand on more important points.

You might, for example, outline the following points in preparation for writing a report on a polymer testing program.

- A problem condition of high turbidity was discovered
- Polymer testing offered the best means of reducing turbidity
- Funds, equipment, and material were acquired
- Operators were trained
- Tests were conducted
- Results and conclusions were reached
- Corrective actions were planned and taken
- Conclusion, the tests did or did not produce the anticipated results or correct the problem

Once you are fairly sure you have included all the major topics you will want to discuss, go through the outline and write down facts you want to include on each topic. As you work through it, you may decide to move material from one topic to another. The outline will help you organize your ideas and facts.

TABLE 23.4 TIPS FOR GIVING AN ORAL PRESENTATION

1. Arrive early. Give yourself plenty of time to become familiar with the room, practice using your audiovisuals, and make any necessary changes in room setup.
2. Be ready for mistakes. Number the pages in your presentation and your audiovisuals. Check the order of the pages before the meeting begins.
3. Pace yourself. Do not speak too quickly, speak slowly and carefully. Keep a careful eye on audience reaction to be sure that you are speaking at a pace that can be understood.
4. Project yourself. Speak loudly and look at the audience. Do not talk with your back to the audience. Check that those in the back can hear you.
5. Be natural. Try not to read your presentation. Practice ahead of time so that you can speak normally and keep eye contact with your audience.
6. Connect with the audience. Try to smile and make eye contact with the audience.
7. Involve the audience. Allow for audience questions and invite their comments.
8. Repeat audience questions. Always repeat the question so everyone can hear and to be sure that you hear the question correctly.
9. Know when to stop. Keep your remarks within the time allocated for your presentation and be aware that long, rambling speeches create a negative impression on the audience.
10. Use audiovisuals that can be heard and read. Audiovisuals should enhance and reinforce your words. Be sure that all members of the audience can hear, see, and read your audiovisuals. Normal typewritten text is not readable on overheads; use large type so everyone can see. Use no more than five to seven key ideas per overhead.
11. Organize your presentation. Prepare an introduction, body of the speech, and conclusion. "Tell them what you are going to say, say it, and then tell them what you said." The presentation should have three to five main points presented in some logical order, for example, chronologically or from simple to complex.

When your outline is complete, you will have the essentials of your report. Now you need to tailor it to the audience that will be reading it. Take a few minutes to think about your audience. What information do they want? What aspects of the topics will they be most interested in reading? Each of the following groups may be interested in specific topics in the report. Consider these interests as you write.

1. Management

   Management will have specific interests that relate to the cost effectiveness of the program. A report to management should include a summary that presents the essential information, procedures used, an analysis of the data (including trends), and conclusions. Be sure to include complete cost information. Did the benefits warrant the costs? As a result of the tests, can future expenses be reduced? Backup information and field data can be included in an appendix for those who want more information.

2. Other Utilities

   Other utilities will be interested in costs but will also want more detailed information about how the program was performed. They will also be interested in the results and benefits of the program. Explain how the tests were done, the procedures, size of the crew, equipment used, source and availability of materials, and difficulties encountered and how they were overcome.

3. Citizen Groups

   Citizens' interest will be more general. What is a polymer test and why is it needed? Is the polymer harmless? Will it injure fish or birds? How does the polymer test work? Who pays for the test? How much will it cost to implement results and will they be effective?

Your report may be written to include all of these groups. Adjust the outline to include the topics of interest to each group and identify the topics so readers can find the information most interesting to them. Keep the following information in mind as you write your report:

- Drafts. Good reports are not perfect the first time. Re-read and improve your report several times.
- Facts. Confine your writing to the facts and events that occurred. Include figures and statistics only when they make the report more effective. Include only the relevant facts. Large amounts of data should be put in the appendix. Do not clutter the body of the report with unimportant data.
- Continuity. To be interesting and understood, a report must have continuity. It must make sense to the reader and be organized logically. In the report on the polymer testing, the report should be organized to show you had a problem, you had to find a way to identify where the problem existed, you did the testing, and you identified the problems and the corrective actions.
- Effective. To be effective, a report should achieve the objective for which it was written. In this example, we wanted to justify the costs for the program to management, help other utilities in conducting a similar program, and help citizens to understand what we were doing and why.
- Candid. A good report should be frank and straightforward. Keep the language appropriate for the audience. Do not try to impress your readers with technical terms they do not understand. Your purpose is to communicate information. Keep the information accurate and easy to understand.

The annual report is an important part of the management of the utility. It is one of the most involved writing projects that the utility must put together. The annual report should be a review of what and how the utility operated during the past year and it should also include the goals for the next year. In many small communities, the annual report may be presented orally to the city council rather than written. If this report is well written, it can be used to highlight accomplishments and provide support for future planning.

The first step to organizing the report is to make a list of three or four major accomplishments of the last year, then make a list of the top three goals for next year. These accomplishments and goals should be the focus of the report. The annual report should be a summary of the expenses, treatment services provided, and revenues generated over the last year. As you organize this information, keep those accomplishments in mind and let the data tell the story of how the utility accomplished last year's tasks. The data by itself may seem boring but as you organize the data it becomes a meaningful description of the year's accomplishments. Conclude with projections for next year. The facts and figures should tell the audience how you plan to accomplish your goals for the next year. The annual report may be simple or complex depending on your community needs. A sample table of contents for a medium-sized utility is given in Table 23.5. When you are finished, the annual report will be a valuable planning tool for the utility and can be used to build support for new projects.

TABLE 23.5 EXAMPLE OF TABLE OF CONTENTS FOR THE ANNUAL REPORT OF A UTILITY

Table of Contents	
Executive Summary	
Summary of the Treatment Process, Including Flows and Costs	
Review of Goals and Objectives for the Year	
Special Projects Completed	
Professional Awards or Recognition for the Utility or Its Staff	
General Operating Conditions Including Regulatory Requirements	
Expectations for the Next Year—Goals and Objectives	
Recommended Changes for the Utility in Organizing, Staffing, Equipment, or Resources Summary	
Appendixes:	Operating Data
	Budget
	Information on Special Projects

QUESTIONS

Please write your answers to the following questions and compare them with those on page 699.

23.5A What kinds of communication skills are needed by a manager?

23.5B What are the most common written documents that a utility manager must write?

23.5C What should be included in the annual report?

23.6 CONDUCTING MEETINGS

As a utility manager, you will be asked to conduct meetings. These meetings may be with employees, your governing board, the public, or with other professionals in your field. Many new managers fail to prepare for these meetings and the meetings end up as a terrible waste of time. As a manager, you need to learn to conduct meetings in a way that is productive and guides the participants into an active role. The following steps should be taken to conduct a productive meeting.

Before the meeting:

- Prepare an agenda and distribute it to all participants.
- Find an adequate meeting room.
- Set a beginning and ending time for the meeting.

During the meeting:

- Start the meeting on time.
- Clearly state the purpose and objectives of the meeting.
- Involve all the participants.
- Do not let one or two individuals dominate the meeting.
- Keep the discussion on track and on time with the agenda.

- When the group makes a decision or reaches consensus, restate your understanding of the results.
- Make clear assignments for participants and review them with everyone during the meeting.

After the meeting:

- Send out minutes of the meeting.
- Send out reminders, when appropriate, about any assignments made for participants, and the next meeting time.

QUESTIONS

Please write your answers to the following questions and compare them with those on page 699.

23.6A With whom may a utility manager be asked to conduct meetings?

23.6B What should be done before a meeting?

23.7 PUBLIC RELATIONS

23.70 Establish Objectives

The first step in organizing an effective public relations campaign is to establish objectives. The only way to know whether your program is a success is to have a clear idea of what you expect to achieve—for example, better customer relations, greater water conservation, and enhanced organizational credibility. Each objective must be specific, achievable, and measurable. It is also important to know your audience and tailor various elements of your public relations effort to specific groups you wish to reach, such as community leaders, school children, or the average customer. Your objective may be the same in each case, but what you say and how you say it will depend upon your target audience.

23.71 Utility Operations

Good public relations begin at home. Any time you or a member of your utility comes in contact with the public, you will have an impact on the quality of your public image. Dedicated, service-oriented employees provide for better public relations than paid advertising or complicated public relations campaigns. For most people, contact with an agency employee establishes their first impression of the competence of the organization, and those initial opinions are difficult to change.

In addition to ensuring that employees are adequately trained to do their jobs and knowledgeable about the utility's operations, management has the responsibility to keep employees informed about the organization's plans, practices, and goals. Newsletters, bulletin boards, and regular, open communication between supervisors and subordinates will help build understanding and contribute to a team spirit.

Despite the old adage to the contrary, the customer is not always right. Management should try to instill among its employees the attitude that while the customer may be confused or unclear about the situation, everyone is entitled to courteous treatment and a factual explanation. Whenever possible, employees should phrase responses as positively, or neutrally, as possible, avoiding negative language. For example, "Your complaint" is better stated as "Your question." "You should have ..." is likely to make the customer defensive, while "Will you please ..." is courteous and respectful. "You made a mistake" emphasizes the negative, "What we will do ..." is a positive, problem-solving approach.

23.72 The Mass Media

We live in the age of communications, and one of the most effective and least expensive ways to reach people is through the mass media—radio, television, newspapers, and the internet. Each medium has different needs and deadlines, and obtaining coverage for your issue or event is easier if you are aware of these constraints. Television must have strong visuals, for example. When scheduling a press conference, provide an interesting setting and be prepared to suggest good shots to the reporter. Radio's main advantage over television and newspapers is immediacy, so have a spokesperson available and prepared to give an interview over the telephone, if necessary. Newspapers give more thorough, in-depth coverage to stories than do the broadcast media, so be prepared to spend extra time with print reporters and provide written backup information and additional contacts.

It is not difficult to get press coverage for your event or press conference if a few simple guidelines are followed:

1. Demonstrate that your story is newsworthy, that it involves something unusual or interesting.
2. Make sure your story will fit the targeted format (television, radio, newspaper, or the internet).
3. Provide a spokesperson who is interesting, articulate, and well prepared.

23.73 Being Interviewed

Whether you are preparing for a scheduled interview or are simply contacted by the press on a breaking news story, here are some key hints to keep in mind when being interviewed.

1. Speak in personal terms, free of institutional jargon.
2. Do not argue or show anger if the reporter appears to be rude or overly aggressive.
3. If you do not know an answer, say so and offer to find out. Do not bluff.
4. If you say you will call back by a certain time, do so. Reporters face tight deadlines.
5. State your key points early in the interview, concisely and clearly. If the reporter wants more information, he or she will ask for it.
6. If a question contains language or concepts with which you disagree, do not repeat them, even to deny them.

7. Know your facts.
8. Never ask to see a story before it is printed or broadcast. Doing so indicates that you doubt the reporter's ability and professionalism.

23.74 Public Speaking

Direct contact with people in your community is another effective tool in promoting your utility. Though the audiences tend to be small, a personal, face-to-face presentation generally leaves a strong and long-lasting impact on the listener.

Depending upon the size of the organization, your utility may wish to establish a speaker's bureau and send a list of topics to service clubs in the area. Visits to high schools and college campuses can also be beneficial, and educators are often looking for new and interesting topics to supplement their curriculum.

Effective public speaking takes practice. It is important to be well prepared while retaining a personal, informal style. Find out how long your talk is expected to be, and do not exceed that time frame. Have a definite beginning, middle, and end to your presentation. Visual aids such as charts, slides, or models can assist in conveying your message. The use of humor and anecdotes can help to warm up the audience and build rapport between the speaker and the listener. Just be sure the humor is natural, not forced, and that the point of your story is accessible to the particular audience. Try to keep in mind that audiences only expect you to do your best. They are interested in learning about their water supply and will appreciate that you are making a sincere effort to inform them about an important subject.

23.75 Telephone Contacts

First impressions are extremely important, and frequently a person's first contact with your water utility is over the telephone. A person who answers the phone in a courteous, pleasant, and helpful manner goes a long way toward establishing a friendly, cooperative atmosphere. Be sure anyone answering telephone inquiries receives appropriate training and conveys a positive image for the utility.

Following a few simple guidelines will help to start your utility off on the right note with your customers:

1. *Answer calls promptly.* Your conversation will get off to a better start if the phone is answered by the third or fourth ring.
2. *Identify yourself.* This adds a personal note and lets the caller know whom he or she is talking to.
3. *Pay attention.* Do not conduct side conversations. Minimize distractions so you can give the caller your full attention, avoiding repetitions of names, addresses, and other pertinent facts.
4. *Minimize transfers.* Nobody likes to get the run-around. Few things are more frustrating to a caller than being transferred from office to office, repeating the situation, problem, or concern over and over again. Transfer only those calls that must be transferred, and make certain you are referring the caller to the right person. Then, explain why you are transferring the call. This lets the caller know you are referring him or her to a co-worker for a reason and reassures the customer that the problem or question will be dealt with. In some cases, it may be better to take a message and have someone return the call than to keep transferring the customer's call.

23.76 Consumer Inquiries

No single set of rules can apply to all types of consumer questions or complaints about water quality and service. There are, however, basic principles to follow in responding to inquiries and concerns, including:

1. *Be prepared.* Your employees should be familiar enough with your utility's organization, services, and policies to either respond to the question or complaint or locate the person who can.
2. *Listen.* Ask the customer to describe the problem and listen carefully to the explanation. Take written notes of the facts and addresses.
3. *Do not argue.* Callers often express a great deal of pent-up frustration in their contacts with a utility. Give the caller your full attention. Once you have heard them out, most people will calm down and state their problems in more reasonable terms.
4. *Avoid jargon.* The average consumer lacks the technical knowledge to understand the complexities of water quality. Use plain, nontechnical language and avoid telling the consumer more than he or she wants to know.
5. *Summarize the problem.* Repeat your understanding of the situation back to the caller. This will assure the customer that you understand the problem and offer the opportunity to clear up any confusion or missed communication.
6. *Promise specific action.* Make an effort to give the customer an immediate, clear, and accurate answer to the problem. Be as specific as possible without promising something you cannot deliver.

In some cases, you may wish to have a representative of the utility visit the customer and observe the problem first hand. If the complaint involves water quality, take samples, if necessary, and report back to the customer to be sure the problem has been resolved.

Complaints can be a valuable asset in determining consumer acceptance and pinpointing water quality problems. Customer calls are frequently your first indication that something may be wrong. Responding to complaints and inquiries promptly can save the utility money and staff resources, and minimize the number of customers who are inconvenienced. Still, education can greatly reduce complaints about water quality. Information brochures, utility bill inserts, and other educational tools help to inform customers and avoid future complaints.

23.77 Plant Tours

Tours of water treatment plants can be an excellent way to inform the public about your utility's efforts to provide a safe, high-quality water supply. Political leaders, such as members of the City Council and Board of Supervisors, should be invited and encouraged to tour the facilities, as should school groups and service clubs.

A brochure describing your utility's goals, accomplishments, operations, and processes can be a good supplement to the tour and should be handed out at the end of the visit. The more visually interesting the brochure is, the more likely that it will be read, and the use of color, photographs, graphics, or other design features is encouraged. If you have access to the necessary equipment, production of a videotape program about the utility can also add interest to the facility tour.

The tour itself should be conducted by an employee who is very familiar with plant operations and can answer the types of questions that are likely to arise. Consider including:

1. A description of the sources of water supply
2. History of the plant, the years of operation, modifications and innovations over the years
3. Major plant design features, including plant capacity and safety features
4. Observation of the treatment processes, including filtration, sedimentation, flocculation, and disinfection
5. A visit to the laboratory, including information on the quality of water distributed to consumers
6. Anticipated improvements, expansions, and long-range plans for meeting future service needs

Plant tours can contribute to a water utility's overall program to gain financing for capital improvements. If the City Council or other governing board has seen the treatment process first hand, it is more likely to understand the need for enhancement and support future funding.

As beneficial as plant tours may be for promoting public interest and confidence in the water utility, security precautions should be carefully considered when planning for visitors to the plant. For more information, see Section 23.101, "Homeland Defense."

QUESTIONS

Please write your answers to the following questions and compare them with those on page 699.

23.7A What is the first step in organizing a public relations campaign?

23.7B How can employees be kept informed of the utility's plans, practices, and goals?

23.7C Which news medium is more likely to give a story thorough, in-depth coverage?

23.7D What is the key to effective public speaking?

23.7E How do customer complaints help a utility?

23.8 FINANCIAL MANAGEMENT

Financial management for a utility should include providing financial stability for the utility, careful budgeting, and providing capital improvement funds for future utility expansion. These three areas must be examined on a routine basis to ensure the continued operation of the utility. They may be formally reviewed on an annual basis or more frequently when the utility is changing rapidly. The utility manager should understand what is required for each of the three areas and be able to develop record systems that keep the utility on track and financially prepared for the future.

23.80 Financial Stability

How do you measure financial stability for a utility? Two very simple calculations can be used to help you determine how healthy and stable the finances are for the utility. These two calculations are the *OPERATING RATIO* and the *COVERAGE RATIO*. The operating ratio is a measure of the total revenues divided by the total operating expenses. The coverage ratio is a measure of the ability of the utility to pay the principal and interest on loans and bonds (this is known as *DEBT SERVICE* [5]) in addition to any unexpected expenses. A utility that is in good

5. *Debt Service.* The amount of money required annually to pay the (1) interest on outstanding debts or (2) funds due on a maturing bonded debt or the redemption of bonds.

financial shape will have an operating ratio and coverage ratio above 1.0. In fact, most bonds and loans require the utility to have a coverage ratio of at least 1.25. As state and federal funds for utility improvements have become much more difficult to obtain, these financial indicators have become more important for utilities. Being able to show and document the financial stability of the utility is an important part of getting funding for more capital improvements.

The operating ratio is perhaps the simplest measure of a utility's financial stability. In essence, the utility must be generating enough revenue to pay its operating expenses. The actual ratio is usually computed on a yearly basis, since many utilities may have monthly variations that do not reflect the overall performance. The total revenue is calculated by adding up all revenue generated by user fees, hookup charges, taxes or assessments, interest income, and special income. Next, determine the total operating expenses by adding up the expenses of the utility, including administrative costs, salaries, benefits, energy costs, chemicals, supplies, fuel, equipment costs, equipment replacement fund, principal and interest payments, and other miscellaneous expenses.

Example 1

The total revenues for a utility are $1,686,000 and the operating expenses for the utility are $1,278,899. The debt service expenses are $560,000. What is the operating ratio? What is the coverage ratio?

Known		Unknown
Total Revenue	= $1,686,000	Operating Ratio
Operating Expenses	= $1,278,899	Coverage Ratio
Debt Service Expenses	= $560,000	

1. Calculate the operating ratio.

$$\text{Operating Ratio} = \frac{\text{Total Revenue}}{\text{Operating Expenses}}$$

$$= \frac{\$1,686,000}{\$1,278,899}$$

$$= 1.32$$

2. Calculate nondebt expenses.

$$\text{Nondebt Expenses} = \text{Operating Exp} - \text{Debt Service Exp}$$

$$= \$1,278,899 - \$560,000$$

$$= \$718,899$$

3. Calculate coverage ratio.

$$\text{Coverage Ratio} = \frac{\text{Total Revenue} - \text{Nondebt Expenses}}{\text{Debt Service Expenses}}$$

$$= \frac{\$1,686,000 - \$718,899}{\$560,000}$$

$$= 1.73$$

These calculations provide a good starting point for looking at the financial strength of the utility. Both of these calculations use the total revenue for the utility, which is an important component for any utility budgeting. As managers, we often focus on the expense side and forget to look carefully at the revenue side of utility management. The fees collected by the utility, including hook-up fees and user fees, must accurately reflect the cost of providing service. These fees must be reviewed annually and they must be increased as expenses rise to maintain financial stability. Some other areas to examine on the revenue side include how often and how well user fees are collected, the number of delinquent accounts, and the accuracy of meters. Some small communities have found they can cut their administrative costs significantly by switching to a quarterly billing cycle. The utility must have the support of the community to determine and collect user fees, and the utility must keep track of revenue generation as carefully as resource spending.

23.81 Budgeting

Budgeting for the utility is perhaps the most challenging task of the year for many managers. The list of needs usually is much larger than the possible revenue for the utility. The only way for the manager to prepare a good budget is to have good records from the year before. A system of recording or filing purchase orders (see Section 23.124, "Procurement Records") or a requisition records system must be in place to keep track of expenses and prevent spending money that is not in the budget.

To budget effectively, a manager needs to understand how the money has been spent over the last year, the needs of the utility, and how the needs should be prioritized. The manager also must take into account cost increases that cannot be controlled while trying to minimize the expenses as much as possible. The following problem is an example of the types of decisions a manager must make to keep the budget in line while also improving service from the utility.

Example 2

A pump that has been in operation for 25 years pumps a constant 600 gpm through 47 feet of dynamic head. The pump uses 6,071 kilowatt-hours of electricity per month, at a cost of $0.085 per kilowatt-hour. The old pump efficiency has dropped to 63 percent. Assuming a new pump that operates at 86 percent efficiency is available for $9,730.00, how long would it take to pay for replacing the old pump?

Known		Unknown
Electricity, kW-hr/mo	= 6,071 kW-hr/mo	New Pump Payback Time, yr
Electricity Cost, $/kW-hr	= $0.085/kW-hr	
Old Pump Efficiency, %	= 63%	
New Pump Efficiency, %	= 86%	
New Pump Cost, $	= $9,730	

1. Calculate old pump operating costs in dollars per month.

$$\text{Old Pump Operating Costs, \$/mo} = \left(\text{Electricity, kW-hr/mo}\right)\left(\text{Electricity Cost, \$/kW-hr}\right)$$

$$= (6{,}071 \text{ kW-hr/mo})(\$0.085/\text{kW-hr})$$

$$= \$516/\text{mo}$$

2. Calculate new pump operating electricity requirements.

$$\text{New Pump Electricity, kW-hr/mo} = (\text{Old Pump Elect, kW-hr/mo})\frac{(\text{Old Pump Eff, \%})}{(\text{New Pump Eff, \%})}$$

$$= (6{,}071 \text{ kW-hr/mo})\frac{(63\%)}{(86\%)}$$

$$= 4{,}447 \text{ kW-hr/mo}$$

3. Calculate new pump operating costs in dollars per month.

$$\text{New Pump Operating Costs, \$/mo} = \left(\text{Electricity, kW-hr/mo}\right)\left(\text{Electricity Cost, \$/kW-hr}\right)$$

$$= (4{,}447 \text{ kW-hr/mo})(\$0.085/\text{kW-hr})$$

$$= \$378/\text{mo}$$

4. Calculate annual cost savings of new pump.

$$\text{Cost Savings, \$/yr} = (\text{Old Costs, \$/mo} - \text{New Costs, \$/mo})(12 \text{ mo/yr})$$

$$= (\$516/\text{mo} - \$378/\text{mo})(12 \text{ mo/yr})$$

$$= \$1{,}656/\text{yr}$$

5. Calculate the new pump payback time in years.

$$\text{Payback Time, yr} = \frac{\text{Initial Cost, \$}}{\text{Savings, \$/yr}}$$

$$= \frac{\$9{,}730.00}{\$1{,}656/\text{yr}}$$

$$= 5.9 \text{ years}$$

In this example, a payback time of 5.9 years is acceptable and would probably justify the expense for a new pump. This calculation was a simple payback calculation that did not take into account the maintenance on each pump, depreciation, and inflation. Many excellent references are available from EPA to help utility managers make more complex decisions about purchasing new equipment.

The annual report should be used to help develop the budget so that long-term planning will have its place in the budgeting process. The utility manager must track revenue generation and expenses with adequate records to budget effectively. The manager must also get input from other personnel in the utility as well as community leaders as the budgeting process proceeds. This input from others is invaluable to gain support for the budget and to keep the budget on track once adopted.

23.82 Equipment Repair/Replacement Fund

To adequately plan for the future, every utility must have a repair/replacement fund. The purpose of this fund is to generate additional revenue to pay for the repair and replacement of capital equipment as the equipment wears out. To prepare adequately for this repair/replacement, the manager should make a list of all capital equipment (this is called an asset inventory) and estimate the replacement cost for each item. The expected life span of the equipment must be used to determine how much money should be collected over time. When a treatment plant is new, the balance in repair/replacement fund should be increasing each year. As the plant gets older, the funds will have to be used and the balance may get dangerously low as equipment breakdowns occur. Perhaps the hardest job for the utility manager is to maintain a positive balance in this account with the understanding that this account is not meant to generate a profit for the utility but rather to plan for future equipment needs. In water treatment facilities construction, providing an adequate repair/replacement fund is very important, but if this repair/replacement fund has not been reviewed annually, it must be updated.

To set up a repair/replacement fund for your utility, you should first put together a list of the equipment required for each process in your utility. Once you have this list, you need to estimate the life expectancy of the equipment and the replacement cost. From this list you can predict the amount of money you should set aside each year so that when each piece of equipment wears out, you will have enough money to replace that piece of equipment. The publications listed in Section 23.15,

"Additional Reading," at the end of this chapter are excellent references for utility planning.

23.83 Water Rates

The process of determining the cost of water and establishing a water rate schedule for customers is a subject of much controversy. There is no single set of rules for determining water rates. The establishment of a rate schedule involves many factors including the form of ownership (investor or publicly owned), differences in regulatory control over the water utility (state commission or local authority), and differences in individual viewpoints and preferences concerning the appropriate philosophy to be followed to meet local conditions and requirements.

Generally, the development of water rate schedules involves the following procedures:

- A determination of the total revenue requirements for the period that the rates are to be effective (usually 1 year).
- A determination of all the cost components of system operations. That is, how much does it cost to treat the water? How much does it cost to distribute? How much does it cost to install a water service to a customer? How much are administrative costs?
- Distribution of the various component costs to the various customer classes in accordance with their requirements for service.
- The design of water rates that will recover from each class of customers, within practical limits, the cost to serve that class of customers.

Sales of water to customers may be metered or unmetered. In the case of metered sales, the charge to the customer is based on a rate schedule applied to the amount of water used through each water meter. If meters are not used, the charge per customer is based on a flat rate per period of time per fixture, foot of frontage, number of rooms, or other measurable unit. Although the flat-rate basis still is fairly common, meter-based rates are more widely used.

See *Small Water System Operation and Maintenance*, Chapter 8, "Setting Water Rates and System Security for Small Water Utilities," for an explanation and examples of how to determine and set water rates. This publication is available from the Office of Water Programs, California State University, Sacramento, at www.owp.csus.edu or phone (916) 278-6142.

23.84 Capital Improvements and Funding in the Future

A capital improvements fund must be a part of the utility budget and included in the operating ratio. Your responsibility as the utility manager is to be sure that everyone, your governing body and the public, understands the capital improvement fund is not a profit for the utility but a replacement fund to keep the utility operating in the future.

Capital planning starts with a look at changes in the community. Where are the areas of growth in the community, where are the areas of decline, and what are the anticipated changes in industry within the community? After identifying the changing needs in the community, you should examine the existing utility structure. Identify your weak spots (in the distribution system or with in-plant processes). Make a list of the areas that will be experiencing growth, weak spots in the system, and anticipated new regulatory requirements. The list should include expected capital improvements that will need to be made within one, two, five, and ten years. You can use the information in your annual reports and other operational logs to help compile the list.

Once you have compiled this information, prioritize the list and make a timetable for improving each of the areas. Starting at the top of the priority list, estimate the costs for improvements and incorporate these costs into your capital improvement budget. The calculations you have made previously, including corrective to preventive maintenance ratios, operating ratio, coverage ratio, and payback time will all be useful in prioritizing and streamlining your list of needs.

You may find that some of your capital improvement needs could be met in more than one way. How do you decide which of several options is most cost effective? How do you compare fundamentally different solutions? For example, assume your community's population is growing rapidly and you will need to increase treated water production by 20 percent by the end of the next 10 years. Your existing wells cannot provide that amount of additional water and your filtration equipment is operating at 90 percent of design capacity. Possible solutions might include a combination of the following options, some of which might be implemented immediately while others might be brought on line in 5 or 10 years:

- Rehabilitate some declining wells.
- Drill additional new wells.
- Develop an available surface water source.
- Install another filtration unit.
- Install additional distribution system storage reservoirs.

To compare alternative plans, you will need to calculate the present value (or present worth) of each plan; that is, the cost of each plan in today's dollars. This is done by adding up all of the costs and benefits of the plan over the entire service life of the facility or equipment. Costs should include not only the initial purchase price or construction costs, but also financing costs over the life of the loans or bonds and all O&M costs. Benefits include all of the revenue that would be produced by this facility or equipment, including connection and use fees. With the help of an experienced accountant, apply standard inflation, depreciation, and other factors to calculate the present value of each plan. This will give you the cost of each plan in the equivalent of today's dollars.

Remember to involve all of your local officials and the public in development of this capital improvement budget so they understand what will be needed.

Long-term capital improvements such as a new plant or a new treatment process are usually anticipated in your 10-year or 20-year projection. These long-term capital improvements usually require some additional financing. The basic ways for a utility to finance capital improvements are through general obligation bonds, revenue bonds, or loan funding programs.

General obligation bonds or *ad valorem* (based on value) taxes are assessed based on property taxes. These bonds usually have a lower interest rate and longer payback time, but the total bond limit is determined for the entire community. This means that the water utility will have available only a portion of the total bond capacity of the community. These bonds are not often used for funding water utility improvements today.

The second type of bond, the revenue bond, is commonly used to fund utility improvements. This bond has no limit on the amount of funds available and the user charges provide repayment on the bond. To qualify for these bonds, the utility must show sound financial management and the ability to repay the bond. As the utility manager, you should be aware of the provisions of the bond. Be sure the bond has a call date, which is the first date when you can pay off the bond. The common practice is for a 20-year bond to have a 10-year call date and for a 15-year bond to have an 8-year call date. The bond will also have a call premium, which is the amount of extra funds needed to pay off the debt on the call date. You should try to get your bonds a call premium of no more than 102 percent par (average). This means that for a debt of $200,000 on the call date, the total payoff would be $204,000, which includes the extra 2 percent for the call premium. Consult with a financial advisor to prepare for and issue bonds. An advisor can help you negotiate the best bond structure for your community.

Special assessment bonds may be used to extend services into specific areas. The direct users pay the capital costs and the assessment is usually based on frontage or area of real estate. These special assessments carry a greater risk to investors but may be the best way to extend service to some areas.

The most common way to finance water supply system improvements in the past has been federal and state grant programs. Block grants from Department of Housing and Urban Development (HUD) are still available for some projects and Rural Utilities Service (RUS) loans may also be used as a funding source. In addition, state revolving fund (SRF) programs provide loans (but not direct grants) for improvements. The SRF program has already been implemented with wastewater improvements and the Safe Drinking Water Regulations include an SRF program for funding water treatment improvements. These SRF programs will be very competitive and utilities must provide evidence of sound financial management to qualify for these loans. You should contact your state regulatory agency to find out more about the SRF program in your state.

23.85 Financial Assistance

Many small water treatment and distribution systems need additional funds to repair and upgrade their systems. Potential funding sources include loans and grants from federal and state agencies, banks, foundations, and other sources. Some of the federal funding programs for small public utility systems include:

- Appalachian Regional Commission (ARC)
- Department of Housing and Urban Development (HUD) (provides Community Development Block Grants)
- Economic Development Administration (EDA)
- Indian Health Service (IHS)
- Rural Utilities Service (RUS) (formerly Farmer's Home Administration [FmHA] and Rural Development Administration [RDA])

Another valuable contact is the Environmental Financing Information Network (EFIN), which provides information on financing alternatives for state and local environmental programs and projects in the form of abstracts of publications, case studies, and contacts. EFIN is a component of the US EPA's Center for Environmental Finance. For more information, visit the EPA's website at www2.epa.gov/envirofinance.

Also, many states have one or more special financing mechanisms for small public utility systems. These funds may be in the form of grants, loans, bonds, or revolving loan funds. Contact your state drinking water agency for more information.

QUESTIONS

Please write your answers to the following questions and compare them with those on page 699.

23.8A List the three main areas of financial management for a utility.

23.8B How is a utility's operating ratio calculated?

23.8C Why is it important for a manager to consult with other utility personnel and with community leaders during the budget process?

23.8D How can long-term capital improvements be financed?

23.8E What is a revenue bond?

END OF LESSON 2 OF 3 LESSONS

on

ADMINISTRATION

Please answer the discussion and review questions next.

DISCUSSION AND REVIEW QUESTIONS

Chapter 23. ADMINISTRATION

(Lesson 2 of 3 Lessons)

Please write your answers to the following questions to determine how well you understand the material in the lesson. The question numbering continues from Lesson 1.

18. With whom do managers need to communicate?
19. What information should be included in the utility's annual report?
20. List four steps that can be taken during a meeting to make sure it is a productive meeting.
21. What happens any time you or a member of your utility comes in contact with the public?
22. What attitude should management try to develop among its employees regarding the customer?
23. What is the value of customer complaints?
24. How do you measure financial stability for a utility?
25. How can a manager prepare a good budget?

CHAPTER 23. ADMINISTRATION

(Lesson 3 of 3 Lessons)

23.9 OPERATION AND MAINTENANCE

23.90 The Manager's Responsibilities

A utility manager's specific O&M responsibilities vary depending on the size of the utility. At a small utility, the manager may oversee all utility operations while also serving as chief operator and supervising a small staff of operation and maintenance personnel. In larger utility agencies, the manager may have no direct, day-to-day O&M responsibilities but is ultimately responsible for efficient, cost-effective operation of the entire utility. Whether large or small, every utility needs an effective O&M program.

23.91 Purpose of O&M Programs

The purpose of O&M programs is to maintain design functionality (capacity) or to restore the system components to their original condition and thus functionality. Stated another way, does the system perform as designed and intended? The ability to effectively operate and maintain a water supply utility so it performs as intended depends greatly on proper design (including selection of appropriate materials and equipment), construction and inspection, acceptance, and system startup. Permanent system deficiencies that affect O&M of the system are frequently the result of these phases. O&M staff should be involved at the beginning of each project, including planning, design, construction, acceptance, and startup. When a utility system is designed with future O&M considerations in mind, the result is a more effective O&M program in terms of O&M cost and performance.

Effective O&M programs are based on knowing what components make up the system, where they are located, and the condition of the components. With that information, proactive maintenance can be planned and scheduled, rehabilitation needs identified, and long-term capital improvement programs (CIP) planned and budgeted. High-performing agencies have all developed performance measurements of their O&M program and track the information necessary to evaluate performance.

23.92 Types of Maintenance

Water supply and distribution system maintenance can be either a proactive or a reactive activity. Commonly accepted types of maintenance include three classifications: corrective maintenance, preventive maintenance, and predictive maintenance.

Corrective maintenance, including emergency maintenance, is reactive. For example, a piece of equipment or a system is allowed to operate until it fails, with little or no scheduled maintenance occurring before the failure. Only when the equipment or system fails is maintenance performed. Reliance on reactive maintenance will always result in poor system performance, especially as the system ages. Utility agencies taking a corrective maintenance approach are characterized by:

- The inability to plan and schedule work
- The inability to budget adequately
- Poor use of resources
- A high incidence of equipment and system failures

Emergency maintenance involves two types of emergencies: normal emergencies and extraordinary emergencies. Public utilities are faced with normal emergencies such as water main breaks on a daily basis. Normal emergencies can be reduced by an effective maintenance program. Extraordinary emergencies, such as high-intensity rainstorms, hurricanes, floods, and earthquakes, will always be unpredictable occurrences. However, the effects of extraordinary emergencies on the utility's performance can be minimized by implementation of a planned maintenance program and development of a comprehensive emergency response plan (see Section 23.10).

Preventive maintenance is proactive and is defined as a programmed, systematic approach to maintenance activities. This type of maintenance will always result in improved system performance except in the case where major chronic problems are the result of design or construction flaws that cannot be corrected by O&M activities. Proactive maintenance is performed on a periodic (preventive) basis or an as-needed (predictive) basis. Preventive maintenance can be scheduled on the basis of specific criteria such as equipment operating time since the last maintenance was performed, or passage of a certain amount of time (calendar period). Lubrication of motors, for example, is frequently based on running time.

The major elements of a good preventive maintenance program include:

1. Planning and scheduling
2. Records management
3. Spare parts management
4. Cost and budget control

5. Emergency repair procedures
6. Training program

Some benefits of taking a preventive maintenance approach include:

1. Maintenance can be planned and scheduled
2. Work backlog can be identified
3. Adequate resources necessary to support the maintenance program can be budgeted
4. CIP items can be identified and budgeted for
5. Human and material resources can be used effectively

Predictive maintenance, which is also proactive, is a method of establishing baseline performance data, monitoring performance criteria over a period of time, and observing changes in performance so that failure can be predicted and maintenance can be performed on a planned, scheduled basis. Knowing the condition of the system makes it possible to plan and schedule maintenance as required and thus avoid unnecessary maintenance.

In reality, every agency operates their system with corrective and emergency maintenance, preventive maintenance, and predictive maintenance methods. The goal, however, is to reduce the corrective and emergency maintenance efforts by performing preventive maintenance, which will minimize system failures that result in stoppages and overflows.

System performance is frequently a reliable indicator of how the system is operated and maintained. Agencies that rely primarily on corrective maintenance as their method of operating and maintaining the system are never able to focus on preventive and predictive maintenance. With most of their resources directed at corrective maintenance activities, it is difficult to free up these resources to begin developing preventive maintenance programs. For an agency to develop an effective proactive maintenance program, they must add initial resources over and above those currently existing.

23.93 Benefits of Managing Maintenance

The goal of managing maintenance is to minimize investments of labor, materials, money, and equipment. In other words, we want to manage our human and material resources as effectively as possible, while delivering a high level of service to our customers. The benefits of an effective O&M program include:

- Ensuring the availability of facilities and equipment as intended.
- Maintaining the reliability of the equipment and facilities as designed. Utility systems are required to operate 24 hours per day, 7 days per week, 365 days per year. Reliability is a critical component of the O&M program. If equipment and facilities are not reliable, then the ability of the system to perform as designed is impaired.
- Maintaining the value of the investment. Water supply and distribution systems represent major capital investments for communities and are major capital assets of the community. If maintenance of the system is not managed, equipment and facilities will deteriorate through normal use and age. Maintaining the value of the capital asset is one of the utility manager's major responsibilities. Accomplishing this goal requires ongoing investment to maintain existing facilities and equipment, extend the life of the system, and establish a comprehensive O&M program.
- Obtaining full use of the system throughout its design life.
- Collecting accurate information and data on which to base the O&M of the system and justify requests for the financial resources necessary to support it.

QUESTIONS

Please write your answers to the following questions and compare them with those on pages 699 and 700.

23.9A What is the purpose of an operation and maintenance (O&M) program?

23.9B What are the three common types of maintenance?

23.9C List the major elements of a good preventive maintenance program.

23.94 Computer Control Systems

Computer control systems are used by operators and administrators to monitor and adjust their treatment processes and facilities. One type of computer control system is the SCADA system. *SCADA* stands for Supervisory Control And Data Acquisition system.

23.940 Description of SCADA Systems

A SCADA system collects, stores, and analyzes information about all aspects of O&M, transmits alarm signals, when necessary, and allows fingertip control of alarms, equipment, and processes. SCADA provides the information that operators need to solve minor problems before they become major incidents. As the nerve center of a water utility, the system allows operators to enhance the efficiency of their water facility by keeping them fully informed and fully in control.

A typical SCADA system is made up basically of five groups of components:

1. Process instrumentation and control devices
2. Input/output (I/O) interface
3. Central processing unit (CPU)
4. Communications interface
5. Human machine interface (HMI)

Applications for SCADA systems include water distribution system control and monitoring, water treatment plant control monitoring, and other related applications. SCADA systems can vary from merely data collection and storage to total data analysis, interpretation, and process control.

In water applications, SCADA systems monitor levels, pressures, and flows and operate pumps, valves, and alarms. They monitor temperatures, speeds, motor currents, pH, chlorine residual, dissolved oxygen levels, and other operating guidelines, and provide control as necessary. SCADA also logs event and analog signal trends and monitors equipment operating time for maintenance purposes.

A SCADA system might include liquid level, pressure, and flow sensors. The measured (sensed) information could be transmitted to a computer system, which stores, analyzes, and presents the information. The information may be read by an operator on dials or as digital readouts or analyzed and plotted by the computer as trend charts.

Most SCADA systems present a graphical picture of the overall system on a computer screen. In addition, detailed pictures of specific portions of the system can be examined by the operator following a request and instructions to the computer. The graphical displays on the computer screen can include current operating information. The operator can observe this information, analyze it for trends, or determine if it is within acceptable operating ranges, and then decide if any adjustments or changes are necessary.

SCADA systems are capable of analyzing data and providing operating, maintenance, regulatory, and annual reports. In some plants, operators rely on a SCADA system to help them prepare daily, weekly, and monthly maintenance schedules; monitor the spare parts inventory status; order additional spare parts when necessary; print out work orders; and record completed work assignments. SCADA systems can also be used to enhance energy conservation programs and emergency response procedures.

QUESTIONS

Please write your answers to the following questions and compare them with those on page 700.

23.9D What does SCADA stand for?

23.9E What does a SCADA system do?

23.941 Typical Water Treatment and Distribution SCADA Systems

Water treatment and water distribution SCADA systems are usually linked together with the controls located at the water treatment plant. Information that historically was recorded on paper strip-charts is now being recorded and stored (archived) by computers. This information can be easily retrieved and reviewed by a SCADA system, rather than examining years of strip-chart records to find needed information. Therefore, SCADA systems quickly provide operators with the information they need to make informed decisions.

SCADA systems are also used by operators at water treatment plants to do the following:

- Monitor filtration system influent and effluent turbidity levels, head losses through the filters, filter flows, and filter valve settings
- Control filter backwashing
- Study clear well water levels and fluctuations, as well as expected consumer demands

Distribution system service storage tank levels, system demands, and system pressures are all recorded and plotted by SCADA systems. Operators study this information and decide if adjustments are necessary in booster pump operation or target levels in service storage tanks. SCADA systems can plot system pressures over time against flow demands. A steady pressure in spite of fluctuating flows shows that the operators are in control of the system. If a customer phones and complains about a loss of water pressure, a review of system pressures can indicate if a problem exists or if the problem might be in the customer's home.

Many utilities use smartphones, tablets, and laptop computers to assign work orders to field crews. Crews can be instructed to perform routine preventive maintenance on hydrants and valves or to investigate a drinking water taste or odor complaint while in the field. The field crew uses these devices to record standard tasks performed and to record special comments. The standardized comments allow for field information to be retrieved and analyzed.

Electrical energy consumption can be optimized by the use of SCADA systems. Time-power management describes procedures used to minimize power costs and meet consumer water and pressure demands. Most power companies are eager to work with operators to try to increase water treatment and distribution power consumption during periods when electrical system power demands are low and to decrease water power consumption during periods of peak demands on electrical power supplies. SCADA systems also monitor power consumption and conduct a diagnostic performance of pumps. Operators can review this information and then identify potential pump problems before they become serious.

With proper security implementation, SCADA systems allow operators to have remote access to plant controls from anywhere using a laptop or remote workstation. This provides the flexibility for off-duty staff to help on-duty operators solve operational problems. Computer networking systems allow operators at terminals in offices, in plants, and in the field to work together and use the same information or whatever information they need from one central computer database.

The greatest challenge for operators using SCADA systems is to realize that just because a computer says something (a pump is operating as expected), this does not mean that the computer is always correct. Also, when the system fails due to a power failure or for any other reason (natural disaster), operators will be required to operate the plant manually and without critical information. Could you do this?

Operators will always be needed to question and analyze the results from SCADA systems. They will be needed to see if the floc looks OK, to listen to a pump to be sure it sounds proper, and to smell the water and the equipment to determine if unexpected or unidentified changes are occurring. Water treatment plants and distribution systems will always need alert, knowledgeable, and experienced operators who have a feel for their plant and their distribution system.

For more information on SCADA systems, see Chapter 19, "Instrumentation and Control Systems," Section 19.37, "Process Computer Control Systems."

QUESTIONS

Please write your answers to the following questions and compare them with those on page 700.

23.9F List four items that operators at water treatment plants could use a SCADA system to monitor.

23.9G What can an operator do when a customer phones and complains about a loss of water pressure?

23.9H What are the greatest challenges for operators using SCADA systems?

23.95 Cross-Connection Control Program

23.950 Importance of Cross-Connection Control

BACKFLOW[6] of contaminated water through cross-connections into community water systems is not just a theoretical problem. Contamination through cross-connections has consistently caused more waterborne disease outbreaks in the United States than any other reported factor. Inspections have often disclosed numerous unprotected cross-connections between public water systems and other piped systems on consumers' premises, which might contain wastewater; stormwater; processed waters (containing a wide variety of chemicals); and untreated supplies from private wells, streams, and ocean waters. Therefore, an effective cross-connection control program is essential.

Backflow results from either *BACK PRESSURE*[7] or *BACKSIPHONAGE*[8] situations in the distribution system. Back pressure occurs when the user's water supply is at a higher pressure than the public water supply system (Figure 23.4). Typical locations where back pressure problems could develop include services to premises where wastewater or toxic chemicals are handled under pressure or where there are unapproved auxiliary water supplies such as a private well or the use of surface water or seawater for firefighting. Backsiphonage is caused by the development of negative or below atmospheric pressures in the water supply piping (Figure 23.5). This condition can occur when there are extremely high water demands (firefighting), water main breaks, or the use of on-line booster pumps.

The most common cross-connections are:

1. Connection of a sprinkler system using nonpotable water to a potable water supply.
2. Connection of potable water as a seal supply to a pump delivering unapproved or nonpotable water.
3. A hose left in the swimming pool. When customers draw water from their taps, they suck in the pool water.
4. Customers using a hose connected to the home to deliver pesticides without a vacuum breaker or other backflow device at the house connection.
5. Flushing car radiators using a hose connected to the home without backflow protection. Often the hose is left in the radiator after it has been filled with antifreeze. When customers draw water from their taps, the blueness of the water alarms them.

The best way to prevent backflow is to permanently eliminate the hazard. Back pressure hazards can be eliminated by severing (eliminating) any direct connection at the pump causing the back pressure with the domestic water supply system. Another solution to the problem is to require an air gap separation device (Figure 23.6) where the water supply service line connects to the private system under pressure from the pump. To eliminate or minimize backsiphonage problems, proper enforcement of plumbing codes and improved water distribution and storage facilities will be helpful. As an additional safety factor for certain selected conditions, a double check valve (Figure 23.7) may be required at the meter.

23.951 Program Responsibilities

Responsibilities in the implementation of cross-connection control programs are shared by water suppliers, water users (businesses and industries), health agencies, and plumbing officials.

The water supplier is responsible for preventing contamination of the public water system by backflow. This responsibility begins at the source, includes the entire distribution system, and ends at the user's connection. To meet this responsibility, the water supplier must issue (promulgate) and enforce needed laws, rules, regulations, and policies. Water service should not be provided to premises where the strong possibility of an unprotected cross-connection exists. The essential elements of a water supplier cross-connection control program are discussed in the next section.

The water user is responsible for keeping contaminants out of the potable water system on the user's premises. When backflow prevention devices are required by the health agency or water supplier, the water user must pay for the installation, testing, and maintenance of the approved devices. The user is also responsible for preventing the creation of cross-connections through modifications of the plumbing system on the premises. The health agency or water supplier may, when necessary, require a water user to designate a water supervisor or foreman to be responsible for the cross-connection control program within the water user's premises.

6. *Backflow.* A reverse flow condition, created by a difference in water pressures, that causes water to flow back into the distribution pipes of a potable water supply from any source or sources other than an intended source. Also see BACKSIPHONAGE.
7. *Back Pressure.* A pressure that can cause water to backflow into the water supply when a user's water system is at a higher pressure than the public water system.
8. *Backsiphonage.* A form of backflow caused by a negative or below atmospheric pressure within a water system. Also see BACKFLOW.

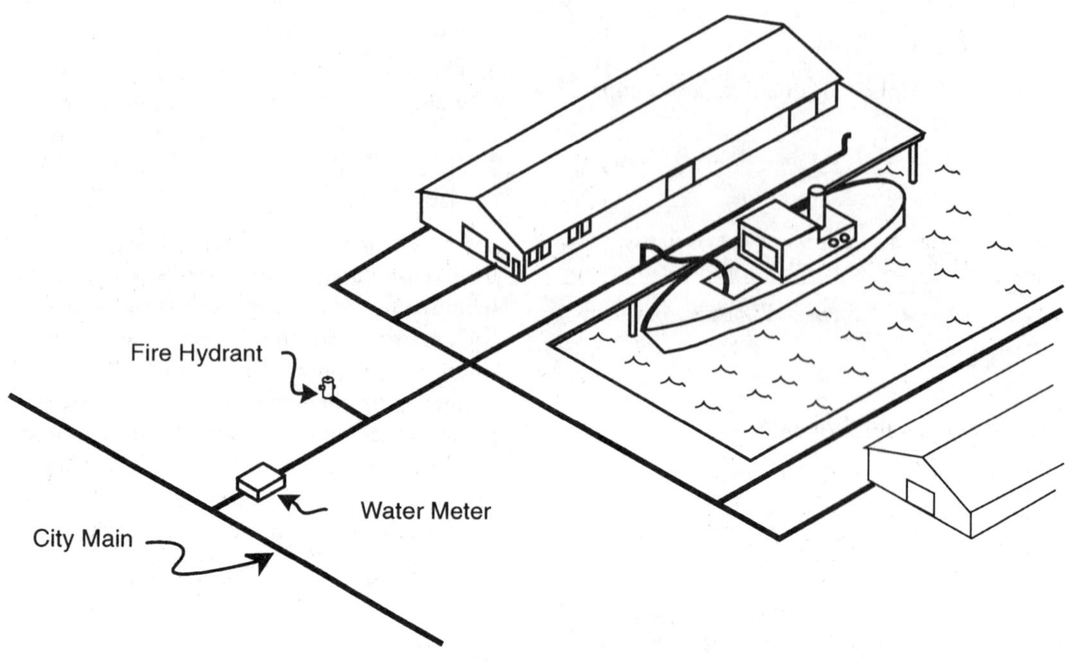

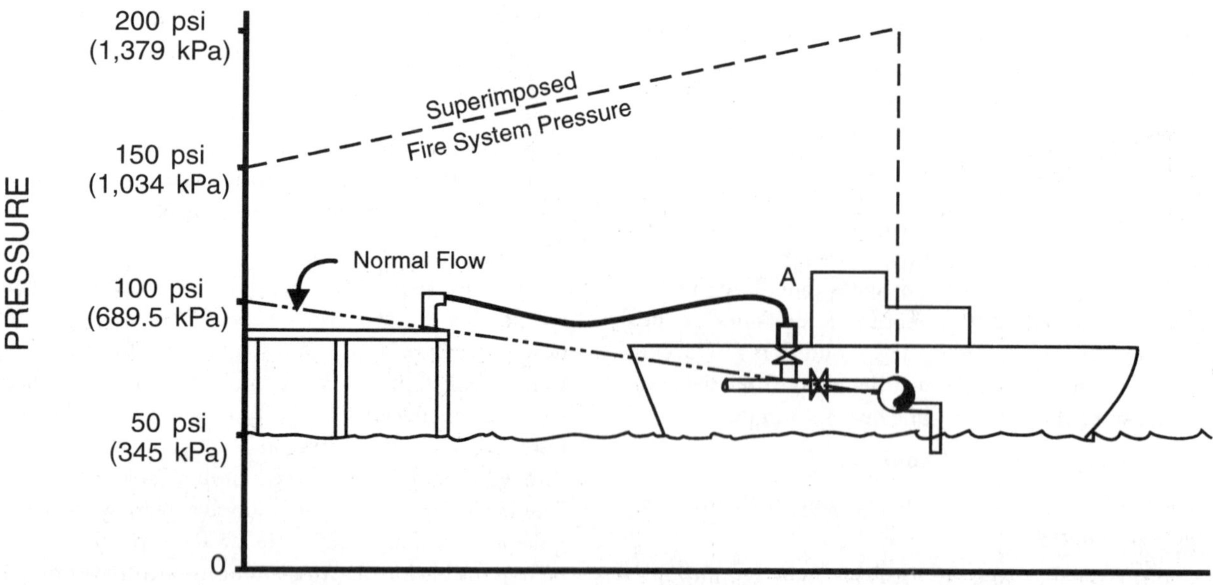

Fig. 23.4 Backflow due to back pressure

(Source: *Manual of Cross-Connection Control Procedures and Practices*, Sanitary Engineering Branch, California Department of Health Services [now California Department of Public Health])

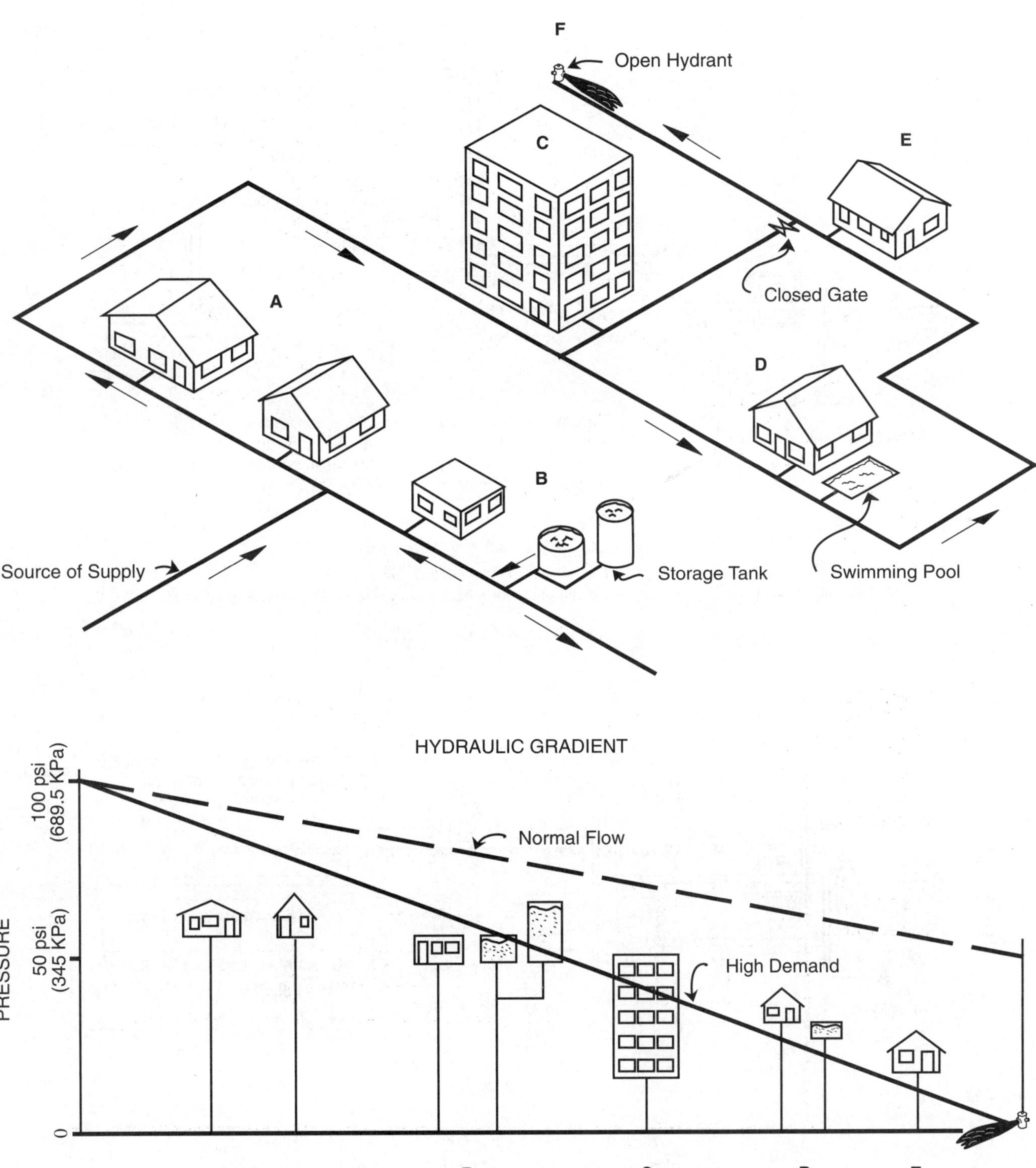

Fig. 23.5 Backsiphonage due to extremely high water demand

(Source: *Manual of Cross-Connection Control Procedures and Practices*, Sanitary Engineering Branch, California Department of Health Services [now California Department of Public Health])

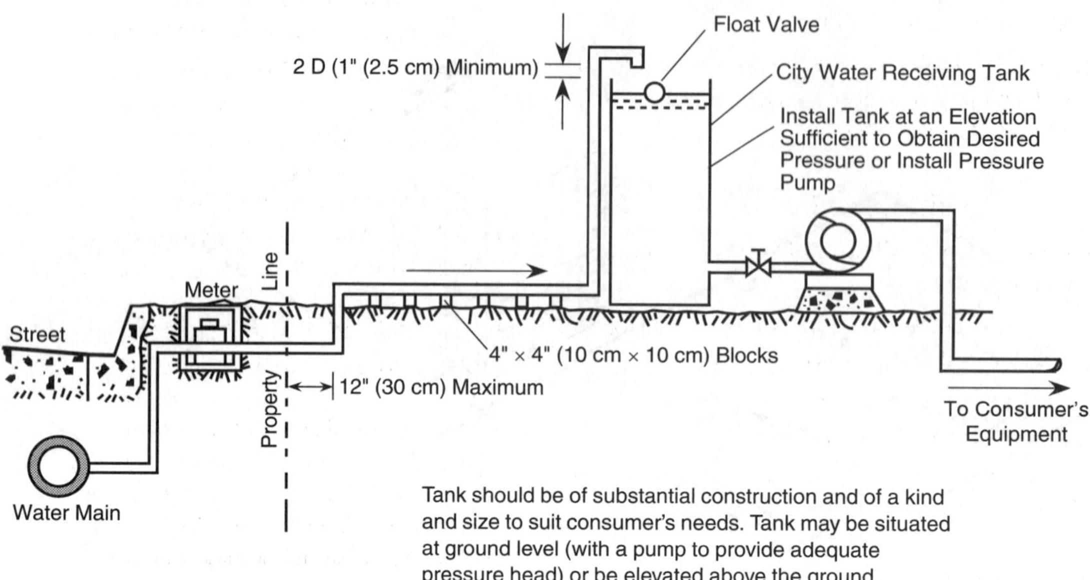

Fig. 23.6 Typical air gap separation

(Source: *Manual of Cross-Connection Control Procedures and Practices*, Sanitary Engineering Branch, California Department of Health Services [now California Department of Public Health])

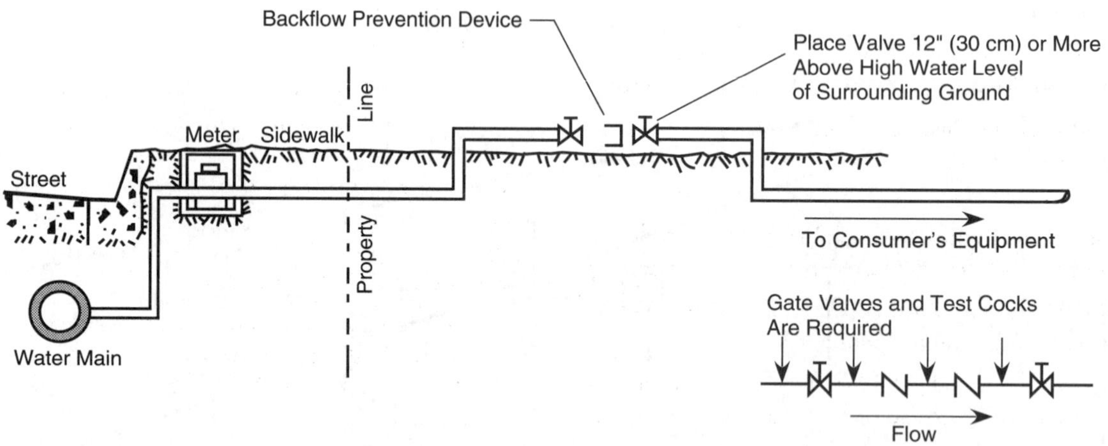

Fig. 23.7 Typical double check valve backflow prevention device

(Source: *Manual of Cross-Connection Control Procedures and Practices*, Sanitary Engineering Branch, California Department of Health Services [now California Department of Public Health])

The local or state health agency is responsible for issuing and enforcing laws, rules, regulations, and policies needed to control cross-connections. Also, this agency must have a program that ensures maintenance of an adequate cross-connection control program. Protection of the system on the user's premises is provided, where needed, by the water utilities.

The plumbing agency (building inspector) is responsible for the enforcement of building regulations relating to prevention of cross-connections on the user's premises.

QUESTIONS

Please write your answers to the following questions and compare them with those on page 700.

23.9I What has caused more waterborne disease outbreaks in the United States than any other reported factor?

23.9J Who is usually responsible for the implementation of cross-connection control programs?

23.9K What is the water user's cross-connection control responsibility?

23.952 Water Supplier Program

The following elements should be included in each water supplier's cross-connection control program:

1. Enactment of an ordinance providing enforcement authority if the supplier is a government agency, or enactment of appropriate rules of service if the system is investor-owned.[9]
2. Training of personnel on the causes of and hazards from cross-connections and procedures to follow for effective cross-connection control.
3. Listing and inspection or reinspection on a priority basis of all existing facilities where cross-connections are of concern.
4. Review and screening of all applications for new services or modification of existing services for cross-connection hazards to determine if backflow protection is needed.
5. Obtaining a list of approved backflow prevention devices and a list of certified testers, if available.
6. Acceptable installation of the proper type of device needed for the specific hazard on the premises.
7. Routine testing of installed backflow prevention devices as required by the health agency or the water supplier. Contact the health agency for approved procedures.
8. Maintenance of adequate records for each backflow prevention device installed, including records of inspection and testing.
9. Notification of each water user when a backflow prevention device has to be tested. This should be done after installation or repair of the device and at least once a year.
10. Maintenance of adequate pressures throughout the distribution system at all times to minimize the hazards from any undetected cross-connections that may exist.

All field personnel should be constantly alert for situations where cross-connections are likely to exist, whether protection has been installed or not. An example is a contractor using a fire hose from a hydrant to fill a tank truck for dust control or the jetting (for compaction) of pipe trenches. Operators should especially be on the lookout for illegal bypassing of installed backflow prevention devices.

23.953 Types of Backflow Prevention Devices

Different types of backflow prevention devices are available. The particular type of device most suitable for a given situation depends on the degree of health hazard, the probability of backflow occurring, the complexity of the piping on the premises, and the probability of the piping being modified. The higher the assessed risk due to these factors, the more reliable and positive the type of device needed. The types of backflow prevention devices normally approved are listed below according to the degree of assessed risk, with the type of device providing the greatest protection listed first. Only the first three devices are approved for use at service connections.

1. Air gap separation
2. Reduced pressure principle (RPP) device
3. Double check valve
4. Pressure vacuum breaker (only used for internal protection on the premises)
5. Atmospheric (nonpressure) vacuum breaker

Figure 23.6 shows a typical air gap separation device and its recommended location. Figure 23.7 shows the installation of a typical double check valve backflow prevention device. These devices are normally installed on the water user's side of the connection to the utility's system and as close to the connection as practical. Figure 23.8 shows typical installations of atmospheric and pressure vacuum breakers.

Only backflow prevention devices that have passed both laboratory and field evaluations by a recognized testing agency and that have been accepted by the health agency and the water supplier should be used.

23.954 Devices Required for Various Types of Situations

The state or local health agency should be contacted to determine the actual types of devices acceptable for various situations

9. A typical ordinance is available in *Cross-Connection Control Manual*, EPA No. 570-9-89-007. This EPA document is available online at epa.gov/nscep/.

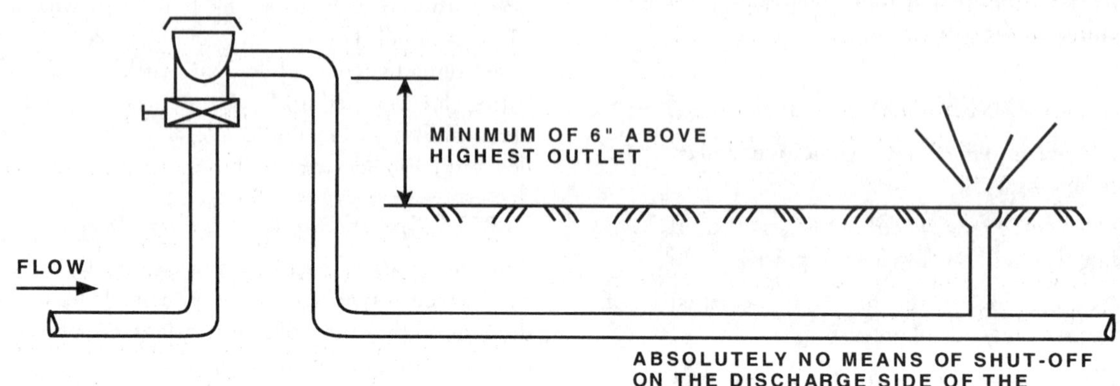

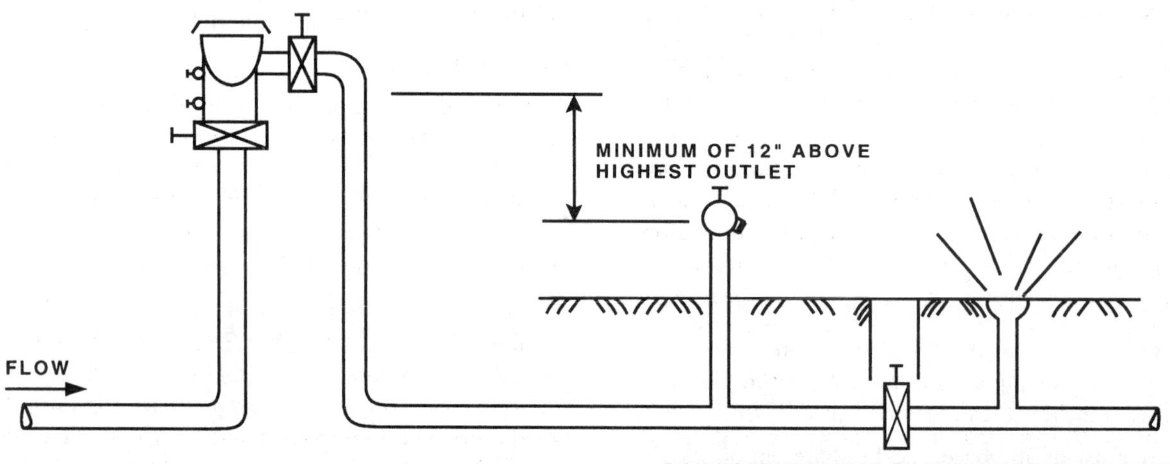

Fig. 23.8 Typical installations of atmospheric (top) and pressure (bottom) vacuum breakers

(Source: *Manual of Cross-Connection Control Procedures and Practices*,
Sanitary Engineering Branch, California Department of Health Services [now California Department of Public Health])

inside the consumer's premises. However, the types of devices generally acceptable for particular situations can be mentioned.

An air gap separation or an RPP device is normally required at services to wastewater treatment plants, wastewater pumping stations, reclaimed water reuse areas, areas where toxic substances in toxic concentrations are handled under pressure, and premises having an auxiliary water supply that is or may be contaminated. The ultimate degree of protection is also needed in cases where fertilizer, herbicides, or pesticides are injected into a sprinkler system.

A double check valve device should be required when a moderate hazard exists on the premises or where an auxiliary supply exists, but adequate protection on the premises is provided.

Atmospheric and pressure vacuum breakers are usually required for irrigation systems; however, they are not adequate in situations where they may be subject to back pressure. If there is a possibility of back pressure, a reduced pressure principle device is needed.

23.96 Geographic Information System (GIS)

The geographic information system (GIS) is a computer program that combines mapping with detailed information about the physical structures within geographic areas. To create the database of information, entities within a mapped area, such as streets, fire hydrants, distribution system line segments, storage facilities, and wells, are given "attributes." Attributes are simply the pieces of information about a particular feature or structure that are stored in a database. The attributes can be as basic as an address, tank storage capacity, or line segment length, or they may be as specific as diameter, tank dimensions, and quadrant (coordinate) location. Attributes of a main line segment might include engineering information, maintenance information, and inspection information. Thus, an inventory of entities and their properties is created. The system allows the operator to periodically update the map entities and their corresponding attributes.

The power of a GIS is that information can be retrieved geographically. An operator can choose an area to look at by pointing to a specific place on the map or outlining (windowing) an area of the map. The system will display the requested section on the screen and show the attributes of the entities located on the map. A printed copy may also be requested. In most cases, CMMS (computer-based maintenance management systems) software has the ability to communicate with geographic information systems so that attribute information from the distribution system can be copied into the GIS.

A GIS can generate work orders in the form of a map with the work to be performed outlined on the map. This minimizes paper work and gives the work crew precise information about where the work is to be performed. Completion of the work is recorded in the GIS to keep the work history for the area and entity up to date. Reports and other inquiries can be requested, as needed, for example, a listing of all line segments in a specific area could be generated for a report.

In many areas, GISs are being developed on an area-wide basis with many agencies, utilities, counties, cities, and state agencies participating. Usually, a county-wide base map is developed and then all participants provide attributes for their particular systems. For example, information on the sanitary sewer collection system might be one map layer, the second map layer might be the water distribution system, and the third layer might be the electric utility distribution system. In addition to sharing databases with CMMSs, GISs have the ability to operate smoothly with computer-aided design (CAD) systems.

QUESTIONS

Please write your answers to the following questions and compare them with those on page 700.

23.9L What factors should be considered when selecting a suitable backflow prevention device for a given situation?

23.9M Air gap separation or reduced pressure principle backflow prevention devices are installed under what conditions?

23.9N What makes a geographic information system (GIS) a potentially powerful tool for a water treatment or distribution system operator?

23.10 EMERGENCY RESPONSE

23.100 Planning

Contingency planning is an essential facet of utility management and one that is often overlooked. Although utilities in various locations will be vulnerable to somewhat different kinds of natural disasters, the effects of these disasters often will be similar. As a first step toward an effective contingency plan, each utility should make an assessment of its own vulnerability and then develop and implement a comprehensive plan of action.

All utilities suffer from common problems such as equipment breakdowns and leaking pipes. During the past few years, there has also been an increasing amount of vandalism, civil disorder, toxic spills, and employee strikes, which have threatened to disrupt utility operations. In observing today's international tension and the potential for nuclear war or the effects of terrorist-induced chemical or biological warfare, water utilities must seriously consider how to respond. Natural disasters such as floods, earthquakes, hurricanes, forest fires, avalanches, and blizzards are a more or less routine occurrence for some utilities. When such catastrophic emergencies occur, the utility must be prepared to minimize the effects of the event and have a plan for rapid recovery. Such preparation should be a specific obligation of every utility manager.

Start by assessing the vulnerability of the utility during various types of emergency situations. If the extent of damage can be estimated for a series of most probable events, the weak elements can be studied and protection and recovery operations can center on these elements. Although all elements are important for the utility to function, experience with disasters points out elements that are most subject to disruption. These elements are:

1. The absence of trained personnel to make critical decisions and carry out orders
2. The loss of power to the utility's facilities
3. An inadequate amount of supplies and materials
4. Inadequate communication equipment

The following steps should be taken in assessing the vulnerability of a system:

1. Identify and describe the system components.
2. Assign assumed disaster characteristics.
3. Estimate disaster effects on system components.
4. Estimate customer demand for service following a potential disaster.
5. Identify key system components that would be primarily responsible for system failure.

If the assessment shows a system is unable to meet estimated requirements because of the failure of one or more critical components, the vulnerable elements have been identified. Repeating this procedure using several typical disasters will usually point out system weaknesses. Frequently, the same vulnerable element appears for a variety of assumed disaster events.

You might consider, for example, the case of the addition of toxic pollutants to water supplies. The list of toxic agents that may have a harmful effect on humans is almost endless. However, it is recognized that there is a relationship between the quantity of toxic agents added to the treatment provided for the supply. Adequate chlorination is effective against most biological agents. Other considerations are the amount of dilution water and the solubility of the chemical agents. There is the possibility that during normal detention times many of the biological agents will die off with adequate chlorination.

Although the drafting of an emergency plan for a water system may be a difficult job, the existence of such a plan can be of critical importance during an emergency situation.

An emergency operations plan need not be too detailed, since all types of emergencies cannot be anticipated and a complex response program can be more confusing than helpful. Supervisory personnel must have a detailed description of their responsibilities during emergencies. They will need information, supplies, equipment, and the assistance of trained personnel. All these can be provided through a properly constructed emergency operations plan that is not extremely detailed.

The following outline can be used as the basis for developing an emergency operations plan:

1. Make a vulnerability assessment.
2. Inventory organizational personnel.
3. Provide for a recovery operation (plan).
4. Provide training programs for operators in carrying out the plan.
5. Coordinate with local and regional agencies such as the health, police, and fire departments to develop procedures for carrying out the plan.
6. Establish a communications procedure.
7. Provide protection for personnel, plant equipment, records, and maps.

By following these steps, an emergency plan can be developed and maintained even though changes in personnel may occur. "Emergency simulation" training sessions, including the use of standby power, equipment, and field test equipment will ensure that equipment and personnel are ready at times of emergency.

A list of phone numbers for operators to call in an emergency should be prepared and posted by a phone for emergency use. The list should include:

1. Plant supervisor
2. Director of public works or head of utility agency
3. Police
4. Fire
5. Doctor (2 or more)
6. Ambulance (2 or more)
7. Hospital (2 or more)

If appropriate for your utility, also include the following phone numbers on the emergency list:

8. Chlorine supplier and manufacturer
9. *CHEMTREC*, (800) 424-9300, for hazardous chemical spills; sponsored by the Manufacturing Chemists Association
10. US Coast Guard's National Response Center, (800) 424-8802
11. Local and state poison control centers
12. Local hazardous materials spill response team

You should prepare a list for your plant now, if you have not already done so, and update the numbers annually.

For additional information on emergencies, see Volume I, Chapter 7, "Disinfection," Section 7.52, "Chlorine Leaks"; Chapter 10, "Plant Operation," Section 10.9, "Emergency Conditions and Procedures"; and this volume, Chapter 18, "Maintenance," Section 18.02, "Emergencies."

QUESTIONS

Please write your answers to the following questions and compare them with those on page 700.

23.10A What is the first step toward an effective contingency plan for emergencies?

23.10B Why is too detailed an emergency operations plan not needed or even desirable?

23.10C An emergency operations plan should include what specific information?

23.101 Homeland Defense

World events in recent years have heightened concern in the United States over the security of one of America's most valuable resources, the critical drinking water treatment, storage, and distribution infrastructure. Water distribution pipelines form an extensive network that runs near or beneath key buildings and roads and is physically close to many communication and transportation networks. Significant damage to the nation's water supply or distribution facilities could result in loss of life; contamination of drinking water supplies; long-term public health impacts; catastrophic environmental damage to rivers, lakes, and wetlands; and disruption to commerce, the economy, and our normal way of life.

Water treatment and distribution facilities have been identified as a target for international and domestic terrorism. This knowledge, coupled with the responsibility of the facility to provide a safe and healthful workplace, requires that management establish rules to protect the workers as well as the facilities. Emergency action and fire prevention plans must identify what steps need to be taken when the threat analysis indicates a potential for attack. These plans must be in writing and be practiced periodically so that all workers know what actions to take.

Some actions that should be taken at all times to reduce the possibility of a terrorist attack are:

- Ensure that all visitors sign in and out of the facilities with a positive ID check.
- Reduce the number of visitors to a minimum.
- Discourage parking by the public near critical buildings to eliminate the chances of car bombs.
- Be cautious with suspicious packages that arrive.
- Be aware of the hazardous chemicals used and how to defend against spills.
- Keep emergency numbers posted near telephones and radios.
- Patrol the facilities frequently, looking for suspicious activity or behavior.
- Maintain, inspect, and use your personal protective equipment (PPE) (hard hats, respirators).

The following recommendations by the EPA[10] include many straightforward, common-sense actions a utility can take to increase security and reduce threats from terrorism.

Guarding Against Unplanned Physical Intrusion

- Lock all doors and set alarms at your office, booster pump stations, treatment plants, and vaults, and make it a rule that doors are locked and alarms are set.
- Limit access to facilities and control access to booster pump stations, chemical and fuel storage areas, giving close scrutiny to visitors and contractors.
- Post guards at treatment plants and post "Employee Only" signs in restricted areas.
- Increase lighting in parking lots, treatment bays, and other areas with limited staffing.
- Control access to computer networks and control systems and change the passwords frequently.
- Do not leave keys in equipment or vehicles at any time.

Making Security a Priority for Employees

- Conduct background security checks on employees at hiring and periodically thereafter.
- Develop a security program with written plans and train employees frequently.
- Ensure all employees are aware of established procedures for communicating with law enforcement, public health, environmental protection, and emergency response organizations.

10. Adapted from "What Wastewater Utilities Can Do Now to Guard Against Terrorist and Security Threats," US Environmental Protection Agency, Office of Wastewater Management, October 2001.

- Ensure that employees are fully aware of the importance of vigilance and the seriousness of breaches in security.
- Make note of unaccompanied strangers on the site and immediately notify designated security officers or local law enforcement agencies.
- If possible, consider varying the timing of operational procedures so that, to anyone watching for patterns, the pattern changes.
- Upon the dismissal of an employee, change pass codes and make sure keys and access cards are returned.
- Provide customer service staff with training and checklists of how to handle a threat if it is called in.

Coordinating Actions for Effective Emergency Response

- Review existing emergency response plans and ensure that they are current and relevant.
- Make sure employees have the necessary training in emergency operating procedures.
- Develop clear procedures and chains of command for reporting and responding to threats and for coordinating with emergency management agencies, law enforcement personnel, environmental and public health officials, consumers, and the media. Practice the emergency procedures regularly.
- Ensure that key utility personnel (both on and off duty) have access to critical telephone numbers and contact information at all times. Keep the call list up to date.
- Develop close relationships with local law enforcement agencies and make sure they know where critical assets are located. Ask them to add your facilities to their routine rounds.
- Report to county or state health officials any illness among the employees that might be associated with water contamination.
- Immediately report criminal threats, suspicious behavior, or attacks on water utilities to law enforcement officials and the nearest field office of the Federal Bureau of Investigation.

Investing in Security and Infrastructure Improvements

- Assess the vulnerability of the distribution system, water storage facilities, major pumping stations, water treatment plants, chemical and fuel storage areas, and other key infrastructure elements.
- Improve computer system and remote operational security.
- Use local citizen watches.
- Seek financing for more expensive and comprehensive system improvements.

Following 9/11, the Department of Homeland Security (DHS) established a five-tiered Homeland Security Advisory System to provide a national framework for notification about the nature and degree of terrorist threats. The system identified 5 threat levels, which were color coded, beginning with green and increasing in severity through blue, yellow, orange, and red. In a coordinating effort, the EPA detailed the steps drinking water and wastewater utilities (water utilities) should consider implementing for each of the 5 threat levels to guard against terrorist and security threats. For each level, the measures focus on detection, preparedness, prevention, and protection. While the color-coded threat levels have been replaced, EPA's suggested protective measures are still valid. Terrorist attacks are summarized in the following list from lowest risk to the highest risk:

- Ongoing facility assessments; and the development, testing, and implementation of emergency plans
- Activating employee and public information plans; exercising communication channels with response teams and local agencies; and reviewing and exercising emergency plans
- Increasing surveillance of critical facilities; coordinating response plans with allied utilities and response teams and local agencies; and implementing emergency plans, as appropriate
- Limiting facility access to essential staff and contractors, and coordinating security efforts with local law enforcement officials and the armed forces, as appropriate
- The decision to close specific facilities and the redirection of staff resources to critical operations

In 2011, the color coded Homeland Security Advisory System was replaced with the National Terrorism Advisory System (NTAS). This new alert system will more effectively communicate information about terrorist threats by providing timely, detailed information to the public, government agencies, first responders, airports and other transportation hubs, and the private sector.

Under the new system, the DHS will coordinate with other federal entities to issue formal, detailed alerts when the federal government receives information about a specific or credible terrorist threat. These alerts will include a clear statement that there is an "imminent threat" or "elevated threat." The alerts also will provide a concise summary of the potential threat, information about actions being taken to ensure public safety, and recommended steps that individuals and communities, businesses, and governments can take.

The NTAS alerts will be based on the nature of the threat: in some cases, alerts will be sent directly to law enforcement or affected areas of the private sector, while in others, alerts will be issued more broadly to the American people through both official and media channels—including a designated DHS webpage (www.dhs.gov/alerts), as well as social media channels, including Facebook and Twitter (@NTASAlerts). Additionally, NTAS will have a "sunset provision," meaning that individual threat alerts will be issued with a specified end date. Alerts may be extended if new information becomes available or if the threat evolves significantly. For more information on the National Terrorism Advisory System, visit www.dhs.gov/alerts.

To address the security of our nation's critical infrastructure, including its drinking water and wastewater systems, the EPA has developed a series of Security Product Guides to assist treatment plant operators and utility managers in reducing risks from, and providing protection against, possible natural disasters and intentional terrorist attacks. These guides provide information on a variety of products:

- Physical security (such as walls, gates, and manhole locks to delay unauthorized entry into buildings or pipe systems)
- Electronic or cyber security (such as computer firewalls and remote monitoring systems that can report on outlying processes)
- Monitoring tools that can be used to identify anomalies in process streams or finished water that may represent potential threats

Individual products evaluated in these guides will be applicable to distribution systems, wastewater collection systems, pumping stations, treatment processes, main plant and remote sites, personnel entry, chemical delivery and storage, SCADA, and control systems for water and wastewater treatment systems. These EPA Security Product Guides are available online at http://cfpub.epa.gov/safewater/watersecurity/guide/.

23.102 Handling the Threat of Contaminated Water Supplies[11]

23.1020 Importance

More than 50 water utilities in southern Louisiana were threatened with cyanide poisoning in their water supplies in 1 year. Such threats can occur anywhere, and every water utility should be prepared to handle this type of emergency.

23.1021 Toxicity

The term "toxicity" is often used when discussing contamination of a water supply with the intention of creating a serious health hazard. Toxicity is the ability of a contaminant (chemical or biological) to cause injury when introduced into the body. The degree of toxicity varies with the concentration of a contaminant required to cause injury, the speed with which the injury takes place, and the severity of the injury.

The effect of a toxic contaminant, once added to a water supply, depends on several things. First, the amount of contaminant added can vary, as can the size of the water supply. In general, it takes larger quantities of a contaminant to be toxic in a larger water supply. Second, the solubility of the contaminant can vary. The more soluble the substance is in water, the more likely it is to cause problems. Finally, the detention time of the contaminant in the water can vary. For example, many biological agents will die before they can cause a problem in the water supply.

Generally, the terms "acute" and "chronic" are used to describe toxic agents and their effects. An acute toxic agent causes injury quickly. When the contaminant causes illness in seconds, minutes, or hours after a single exposure or a single dose, it is considered an acute toxic agent. A chronic toxic agent causes injury to occur over an extended period of exposure. Generally, the contaminant is ingested in repeated doses over a period of days, months, or years.

23.1022 Emergency Contaminant Limits

When determining the emergency limits of a contaminant (the amount of that contaminant necessary to cause injury), the following factors should be considered:

1. Quantity or concentration of the contaminant
2. Duration of exposure to the contaminant
3. Physical form of the contaminant (size of particle; physical state—solid, liquid, gas)
4. Attraction of the contaminant to the organism being contaminated
5. Solubility of the contaminant in the organism
6. Sensitivity of the organism to the contaminant

Table 23.6 lists emergency limits for some chemical pollutants in drinking water.

TABLE 23.6 EMERGENCY LIMITS OF SOME CHEMICAL POLLUTANTS IN DRINKING WATER[a]

	Concentration Limits, mg/L	
Chemical	Emergency Short Term (Three Days)	Long Term
Cyanide (CN)	5.0	0.01
Aldrin	0.05	0.032
Chlordane	0.06	0.003
DDT	1.4	0.042
Dieldrin	0.05	0.017
Endrin	0.01	0.001
Heptachlor	0.1	0.018
Heptachlor epoxide	0.05	0.018
Lindane	2.0	0.056
Methoxychlor	2.8	0.035
Toxaphene	1.4	0.005
Beryllium	0.1	0.000
Boron	25.0	1.000
2,4-D	2.0	0.1
Ethylene chlorohydrin	2.0	
Organophosphorus and carbamate pesticides	2.0	0.100
Trinitrotoluene $(NO_2)(C_6H_2CH_3)$	0.75	0.005

a These limits, based on current knowledge and informed judgment, have been recommended by knowledgeable persons in the field of toxicology. They are subject to change should new information indicate the need. Additional information on some of the chemicals listed can be found in "Report of the Secretary's Commission on Pesticides and Their Relationship to Environmental Health," Parts I and II, USDHHS, Washington, DC.

11. This section was reprinted from *Opflow*, Vol. 9, No. 3, March 1983, by permission. Copyright 1983, the American Water Works Association.

The concentration of a contaminant can be expressed in two ways. The maximum allowable concentration (MAC) is the maximum concentration of the contaminant allowed in drinking water. Table 23.6 lists several contaminants and their MACs, specifically for short-term emergencies ranging up to 3 days. The MACs should not be confused with concentrations required to have an acute effect on the population. Lethal dose 50 (LD 50) is used to express the concentration of a contaminant that will produce 50 percent fatalities from an average exposure.

23.1023 Protective Measures

A utility can take three approaches to protect its water supply from contamination. First, the utility can isolate those reservoirs that offer easy access to the general public. These reservoirs can be fenced off and patrolled, or they can be covered. If access to on-line reservoirs is limited, persons attempting to contaminate the water supply will generally be forced to look to larger bodies of water. Contamination of these large water bodies requires larger quantities of contaminant, increases the detention time of the contaminant, and increases the likelihood of its detection.

As a second means of protection, the water utility can develop an extensive detection and monitoring program. Detecting any contaminant that might be added to a water supply is difficult and expensive. However, because most contaminants cause secondary effects in a water supply, such as taste, color, odor, or chlorine demand, detection is easier.

Because utility operators know their water supply (they know its characteristics), any subtle changes in taste, odor, color, and chlorine demand are instantly recognized. Once it has been determined that the water supply may be contaminated, water samples can be tested. Tests can either be done at the utility's laboratory, if it is a large utility, or the samples can be sent to the state health department.

Finally, the utility can maintain a high chlorine residual. Generally, chlorine residuals of 1 mg/L or higher effectively oxidize or destroy most contaminants. For example, infectious hepatitis virus will not survive a free residual chlorine level of 0.7 mg/L.

23.1024 Emergency Countermeasures

Following is a list of emergency countermeasures that, when used over a short time period, can increase protection of a water supply:

1. Maintaining a high chlorine residual in the system
2. Having engineers, chemists, and medical personnel on 24-hour alert
3. Continuously monitoring key points in the distribution system (monitoring chlorine residual is mandatory)
4. Increasing security around exposed on-line reservoirs
5. Sealing off access to manholes within a three- to six-block radius of highly populated areas
6. Setting up emergency crews who can isolate sections of the distribution system
7. Staffing the treatment facility on a 24-hour basis

23.1025 In Case of Contamination

If contamination of the water supply is discovered, the immediate concern must be the safety of the public. If the contaminated water has already entered the distribution system, immediate public notification is the highest priority. The local police chief, sheriff, or other responsible governmental authority will help with notification. Alternative sources of water may need to be provided.

If the contaminated water has not entered the distribution system, it may be possible to isolate the contaminated source and continue to supply water from other, unaffected sources. If the contaminated water is the only source for the community, treatment measures may be available that will remove the contaminant or reduce its toxicity.

Table 23.7 lists a series of emergency treatment steps that can be taken when identified chemicals are found in the system. These emergency treatment methods are effective only if the contaminant has been identified.

QUESTIONS

Please write your answers to the following questions and compare them with those on page 700.

23.10D What does the word "toxicity" mean?

23.10E The degree of toxicity varies with what factors?

23.10F List possible secondary effects in a water supply that may allow detection of a contaminant without specific testing.

23.1026 Cryptosporidium

Cryptosporidium is a parasite that has become a significant public health concern for drinking water utilities. Even when drinking water meets or exceeds all current state and federal standards, it may contain sufficient *Cryptosporidia* to cause serious illness in sensitive individuals. *Cryptosporidium* oocyst contamination is widespread in the water environment. Potential sources of the parasite in the watersheds of your water supply need to be identified and controlled. The most serious threat to a public water supply occurs during periods of heavy rains or snow melts that flush areas that are sources of high concentrations of oocysts into waters upstream of a plant intake. In the watershed, these potentially contaminated areas include wastewater treatment plants, cattle feedlots, and pastures where livestock graze.

TABLE 23.7 EMERGENCY TREATMENT FOR REDUCING CONCENTRATION OF SPECIFIC CHEMICALS IN COMMUNITY WATER SUPPLIES[a]

Concentration	Treatment
Arsenicals	
Unknown organic and inorganic arsenicals in groundwater at concentrations of 100 mg/L	Precipitation with ferric sulfate and liming to pH 6.8, followed by sedimentation and filtration.
Cyanides	
Hydrogen cyanide	Prechlorination to free residual with pH 7, followed by coagulation, sedimentation, and filtration. *CAUTION: Housed facilities must be adequately ventilated.*
	Precipitation with ferrous or ferric salts to form Prussian blue (iron ferric cyanide) followed by coagulation, sedimentation, and clarification. As long as an excess of iron coagulant is applied, the filtered water should be nontoxic even though it is blue.
Acetone cyanohydrin	Same as for hydrogen cyanide.
Cyanogen chloride	Same as for hydrogen cyanide.
Hydrocarbons	
Kerosene, peak concentrations of 140 mg/L	Preapplications of bleaching clay and activated carbon, plus some increase in normal dosages of alum, chlorine dioxide, lime, and carbon, to provide treatment enabling continued production of water.
Miscellaneous Organic Chemicals	
LSD (lysergic acid derivative)	Chlorination in alkaline water, or water made alkaline by addition of lime or soda ash, to provide a free chlorine residual. Two parts free chlorine are required to react with each part LSD.
Nerve Agents	
(Organophosphorus compounds)	Superchlorination at pH 7 to provide at least 40 mg/L residual after 30-min chlorine contact time, followed by dechlorination and conventional clarification processes.
Pesticides	
2,4-DCP (2,4-Dichlorophenol), an impurity in commercial 2,4-D herbicides	Adsorption on activated carbon followed by coagulation, sedimentation, and filtration.
DDT (dichlorodiphenyltrichloroethane), concentrations of 10 g/L	Chemical coagulation, sedimentation, and filtration.
Dieldrin, concentrations of 10 g/L	Chemical coagulation, sedimentation, and filtration. Supplemental treatment with activated carbon may be necessary.
Endrin, concentrations of 10 g/L	Chemical coagulation, sedimentation, and filtration. Supplemental treatment with activated carbon may be necessary.
Lindane, concentrations of 10 g/L	Application of activated carbon followed by chemical coagulation, sedimentation, and filtration.
Parathion, concentrations of 10 g/L	Chemical coagulation, sedimentation, and filtration. Supplemental treatment with activated carbon may be necessary. Omit prechlorination because chlorine reacts with parathion to form paraoxon, which is more toxic than parathion.

a Source: National Water Supply Research Laboratory, Technical Services, USSR Program.

A water treatment plant can be operated to protect the public from *Cryptosporidium* by developing a strategy for protecting their source water supplies and optimizing the treatment processes. The operator must continually verify optimum coagulant doses, be sure filtered water turbidity levels are low, and provide adequate disinfection chemicals and contact times. This section contains suggestions that operators should consider to protect consumers from "*crypto*."

Conventional water treatment plants using coagulation, flocculation, sedimentation, filtration, and disinfection can provide effective treatment to protect drinking water from *Cryptosporidium*. Be especially alert during periods when the intake water has high turbidity levels resulting from stormwater runoff, snow melt runoff, or lake overturns. During these periods, all treatment processes must operate effectively. Try to achieve turbidity levels of 0.1 nephelometric turbidity unit (NTU) or lower at all times. Run frequent jar tests to determine or simply to verify that you are using the optimum coagulant doses as intake turbidity and other water quality indicators change (pH, temperature, alkalinity). Also, monitor your filtration processes continuously to avoid any increases in turbidity in the treated water. If the water used for backwashing filters is routinely returned to the headworks for conservation purposes, avoid recycling the backwash water during periods of high intake water turbidity. Recycling backwash water may tend to concentrate oocysts in the filter media. Instead, consider wasting the backwash water until the high turbidity levels drop back to normal levels.

Chemical inactivation of *Cryptosporidium* oocysts by disinfection is influenced by several factors. The effectiveness of chemical disinfectants such as chlorine, chlorine dioxide, and ozone is reduced by the presence of high levels of total organic carbon (TOC) (caused by color and turbidity), lower water temperatures, and shorter disinfection contact times. Therefore, effective disinfection can be difficult during periods of high turbidity caused by high stormwater or snow melt runoff flows.

The best approach to evaluating the potential threat of the drinking water to the public is the analysis for *Cryptosporidium* oocysts in the treated drinking water. However, current sampling and analytical methods make it difficult for operators to use the detection of *Cryptosporidium* oocysts for determining the efficiency and effectiveness of filtration and chemical inactivation treatment processes. Today, operators are using turbidity and particle counting measurements as a means of determining the treatment processes' ability to remove or inactivate oocysts. Test methods for identifying the presence or potential presence of *Cryptosporidium* are evolving and should be used to analyze both source and treated (finished) waters for *crypto*.

Keeping the public informed is another good strategy for preventing outbreaks of disease due to *Cryptosporidium*. Operators need to educate the media and consumers regarding the sources of the parasite, possible health risks, monitoring efforts, and treatment processes. Sensitive populations must be informed that even properly treated municipal drinking water, bottled water, and water treated by a home water treatment device still may not be free of *Cryptosporidia*. The Centers for Disease Control and Prevention suggests, "Immunocompromised persons who wish to take independent action to reduce the risk of waterborne *Cryptosporidium* may choose to take precautions similar to those recommended during outbreaks (such as boiling tap water for 1 minute). Such decisions should be made in conjunction with their health care provider."

There are alternatives to boiling water that may be effective when used with proper precautions. Point-of-use filters that remove particles 1 micrometer (1 μm) in diameter or smaller are effective. One-micrometer filters for cyst removal and reverse osmosis units are also acceptable. Bottled water from protected springs and wells may be a safe drinking water, especially if treated by reverse osmosis or distillation to remove *Cryptosporidium* before bottling.

Operators must stay current with the efforts of our profession to prevent outbreaks of waterborne *Cryptosporidium*.

QUESTIONS

Please write your answers to the following questions and compare them with those on page 701.

23.10G Why is *Cryptosporidium* a significant public health concern for drinking water utilities?

23.10H When does the most serious threat of *Cryptosporidium* occur to a public water supply?

23.10I How can a water treatment plant be operated to protect the public from *Cryptosporidium*?

23.11 SAFETY PROGRAM

23.110 Responsibilities

23.1100 Everyone Is Responsible for Safety

Waterworks utilities, regardless of size, must have a safety program if they are to realize a low frequency of accident occurrence. A safety program also provides a means of comparing frequency, disability, and severity of injuries with other utilities. The utility should identify the causes of accidents and injuries, provide safety training, implement an accident reporting system, and hold supervisors responsible for implementing the safety program. Each utility should have a safety officer or supervisor evaluate every accident, offer recommendations, and keep and apply statistics.

The effectiveness of any safety program will depend upon how the utility holds its supervisors responsible. If the utility holds only the safety officer or the employees responsible, the program will fail. The supervisors are key in any organization. They should be responsible for the implementation of a safety program. If they disregard safety measures, essential parts of the program will not work. The results will be an overall poor safety record. After all, the first line supervisor is where the work is being performed, and some may take advantage of an unsafe situation in order to get the job completed. The organization must discipline such supervisors and make them aware of their responsibility for their own and their operators' safety.

Safety is good business, both for the operator and the agency. For a good safety record to be accomplished, all individuals must be educated and must believe in the program. All individuals involved must have the conviction that accidents can be prevented. The operations should be studied to determine the safe way of performing each job. Safety pays, both in monetary savings and in the health and well-being of the operating staff.

23.1101 Regulatory Agencies

Many state and federal agencies are involved in ensuring safe working conditions. The one law that has had the greatest impact is the *OCCUPATIONAL SAFETY AND HEALTH ACT OF 1970 (OSHA)*,[12] Public Law 91-596, which took effect on December 29, 1970. This legislation affects more than 75,000,000 employees and is the basis for most of the current state laws covering employees. Also, many state regulatory agencies enforce the OSHA requirements.

The OSHA regulations provide for safety inspections, penalties, recordkeeping, and variances. Managers and supervisors must understand the OSHA regulations and must furnish each operator with the rules of conduct in order to comply with occupational safety and health standards. The intent of the regulations is to create a place of employment that is free from recognized hazards that could cause serious physical harm or death to an operator.

Civil and criminal penalties are allowed under the OSHA law, depending upon the size of the business and the seriousness of the violation. A routine violation could cost an employer or supervisor up to $1,000 for each violation. A serious, willful, or repeated violation could cause the employer or supervisor to be assessed a penalty of not more than $10,000 for each violation. Penalties are assessed against the supervisor responsible for the injured operator. Operators should become familiar with the OSHA regulations as they apply to their organizations. Managers and supervisors must correct violations and prevent others from occurring.

23.1102 Managers

The utility manager is responsible for the safety of the agency's personnel and the public exposed to the water utility's operations. Therefore, the manager must develop and administer an effective safety program and must provide new employee safety training as well as ongoing training for all employees. The basic elements of a safety program include a safety policy statement, safety training and promotion, and accident investigation and reporting.

A safety policy statement should be prepared by the top management of the utility. The purpose of the statement is to let employees know that the safety program has the full support of the agency and its management. The statement should:

1. Define the goals and objectives of the program.
2. Identify the persons responsible for each element of the program.
3. Affirm management's intent to enforce safety regulations.
4. Describe the disciplinary actions that will be taken to enforce safe work practices.

Give a copy of the safety policy statement to every current employee and each new employee during orientation. Figure 23.9 is an example of a safety policy statement for a water supply utility.

The following list of responsibilities for safety is from the *Plant Manager's Handbook*.[13] These responsibilities represent a typical list but may be incomplete if your agency is subject to stricter local, state, or federal regulations than what is shown here. Check with your safety professional.

Management has the responsibility to:

1. Formulate a written safety policy.
2. Provide a safe workplace.
3. Set achievable safety goals.
4. Provide adequate training.
5. Delegate authority to ensure that the program is properly implemented.

The manager is the key to any safety program. Implementation and enforcement of the program is the responsibility of the manager. The manager also has the responsibility to:

1. Ensure that all employees are trained and periodically retrained in proper safe work practices.
2. Verify that proper safety practices are implemented and continued as long as the policy is in effect.
3. Investigate all accidents and injuries to determine their cause.
4. Institute corrective measures where unsafe conditions or work methods exist.
5. Ensure that equipment, tools, and the work are maintained to comply with established safety standards.

---

12. *OSHA* (O-shuh). The Williams-Steiger Occupational Safety and Health Act of 1970 (OSHA) is a federal law designed to protect the health and safety of industrial workers and also the operators of water supply systems and treatment plants. The Act regulates the design, construction, operation, and maintenance of water supply systems and water treatment plants. OSHA also refers to the federal and state agencies that administer the OSHA regulations.
13. *Plant Manager's Handbook* (MOP SM-4), Water Environment Federation (WEF), no longer in print.

SAFETY POLICY STATEMENT

It is the policy of the Las Vegas Valley Water District that every employee shall have a safe and healthy place to work. It is the District's responsibility; its greatest asset, the employees and their safety.

When a person enters the employ of the District, he or she has a right to expect to be provided a proper work environment, as well as proper equipment and tools, so that they will be able to devote their energies to the work without undue danger. Only under such circumstances can the association between employer and employee be mutually profitable and harmonious. It is the District's desire and intention to provide a safe workplace, safe equipment, proper materials, and to establish and insist on safe work methods and practices at all times. It is a basic responsibility of all District employees to make the SAFETY of human beings a matter for their daily and hourly concern. This responsibility must be accepted by everyone who works at the District, regardless of whether he or she functions in a management, supervisory, staff, or the operative capacity. Employees must use the SAFETY equipment provided; Rules of Conduct and SAFETY shall be observed; and, SAFETY equipment must not be destroyed or abused. Further, it is the policy of the Water District to be concerned with the safety of the general public. Accordingly, District employees have the responsibility of performing their duties in such a manner that the public's safety will not be jeopardized.

The joint cooperation of employees and management in the implementation and continuing observance of this policy will provide safe working conditions and relatively accident-free performance to the mutual benefit of all involved. The Water District considers the SAFETY of its personnel to be of primary importance, and asks each employee's full cooperation in making this policy effective.

Fig. 23.9 Safety policy statement
(Permission of Las Vegas Valley Water District)

QUESTIONS

Please write your answers to the following questions and compare them with those on page 701.

23.11A What should be the duties of a safety officer?

23.11B Who should be responsible for the implementation of a safety program?

23.11C Who enforces the OSHA requirements?

23.11D What are the utility manager's responsibilities with regard to safety?

23.11E What should be included in a utility's policy statement on safety?

23.1103 Supervisors

The success of any safety program will depend upon how the supervisors of the utility view their responsibility. The supervisor who has the responsibility for directing work activities must be safety conscious. This supervisor controls the operators' general environment and work habits and influences whether or not the operators comply with safety regulations. The supervisor is in the best position to counsel, instruct, and review the operators' working methods and thereby effectively ensure compliance with all aspects of the utility's safety program.

The problem, however, is one of the supervisor accepting this responsibility. The supervisor who wishes to complete the job and go on to the next one without taking time to be concerned about working conditions, the welfare of operators, or considering any aspects of safety is a poor supervisor. Only after an accident occurs will a careless supervisor question the need for a work program based on safety. At this point, however, it is too late, and the supervisor may be tempted to simply cover up past mistakes. As sometimes happens, the supervisor may even be partially or fully responsible for the accident by causing unsafe acts to take place, by requiring work to be performed in haste, by disregarding an unsafe work environment, or by overlooking or failing to consider any number of safety hazards. This negligent supervisor could be fined, sentenced to a jail term, or even be barred from working in the profession.

All utilities should make their supervisors bear the greatest responsibility for safety and hold them accountable for planning, implementing, and controlling the safety program. If most accidents are caused and do not just happen, then it is the supervisor who can help prevent most accidents.

Equally important are the officials above the supervisor. These officials include commissioners, managers, public works directors, chief engineers, superintendents, and chief operators. The person in responsible charge for the entire agency or operation must believe in the safety program. This person must budget, promote, support, and enforce the safety program by vocal and visible examples and actions. The top person's support is absolutely essential for an effective safety program.

23.1104 Operators

Each operator also shares in the responsibility for an effective safety program. After all, operators have the most to gain since they are the most likely victims of accidents. A review of accident causes shows that the accident victim often has not acted responsibly. In some way, the victim has not complied with the safety regulations, has not been fully aware of the working conditions, has not been concerned about fellow employees, or just has not accepted any responsibility for the utility's safety program.

Each operator must accept, at least in part, responsibility for fellow operators, for the utility's equipment, for the operator's own welfare, and even for seeing that the supervisor complies with established safety regulations. As pointed out above, the operator has the most to gain. If the operator accepts and uses unsafe equipment, it is the operator who is in danger if something goes wrong. If the operator fails to protect the other operators, it is the operator who must make up the work lost because of injury. If operators fail to consider their own welfare, it is they who suffer the pain of any injury, the loss of income, and maybe even the loss of life.

The operator must accept responsibility for an active role in the safety program by becoming aware of the utility's safety policy and conforming to established regulations. The operator should always call to the supervisor's attention unsafe conditions, environment, equipment, or other concerns operators may have about the work they are performing. Safety should be an essential part of the operator's responsibility.

23.111 First Aid

By definition, first aid means emergency treatment for injury or sudden illness, before regular medical treatment is available. Everyone in an organization should be able to give some degree of prompt treatment and attention to an injury.

First-aid training in the basic principles and practices of life-saving steps that can be taken in the early stages of an injury are available through the local Red Cross, Heart Association, local fire departments, and other organizations. Such training should periodically be reinforced so that the operator has a complete understanding of water safety, cardiopulmonary resuscitation (CPR), and other life-saving techniques. All operators need training in first aid, but it is especially important for those who regularly work with electrical equipment or must handle chlorine and other dangerous chemicals. See Chapter 20, "Safety," for specific first-aid procedures for exposure to a variety of water treatment chemicals.

First aid has little to do with preventing accidents, but it has an important bearing upon the survival of the injured patient. A well-equipped first-aid chest or kit is essential for proper treatment. The kit should be inspected regularly by the safety officer to ensure that supplies are available when needed. First-aid kits should be prominently displayed throughout the treatment plant and in company vehicles. Special consideration must be given to the most hazardous areas of the plant such as shops, laboratories, and chemical handling facilities.

Regardless of size, each utility should establish standard operating procedures (SOPs) for first-aid treatment of injured personnel. All new operators should be instructed in the utility's first-aid program.

QUESTIONS

Please write your answers to the following questions and compare them with those on page 701.

23.11F How could a supervisor be responsible for an accident?

23.11G What types of safety-related responsibilities must each operator accept?

23.11H What is first aid?

23.11I First-aid training is most important for operators involved in what types of activities?

23.112 Hazard Communication Standard (HCS) and Worker Right-To-Know (RTK) Laws

OSHA's Hazard Communication Standard (HCS), 29 *CFR*[14] 1910.1200, and worker Right-To-Know (RTK) laws are now aligned with the UN *GLOBALLY HARMONIZED SYSTEM OF CLASSIFICATION AND LABELING OF CHEMICALS (GHS)*,[15] Revision 3, issued in the Federal Register, March 26, 2012. This update to the HCS will provide a common and coherent approach to classifying chemicals and communicating hazard information on labels and *SAFETY DATA SHEETS*.[16] Once implemented, the revised standard will improve the quality and consistency of hazard information in the workplace, making it safer for workers by providing easily understandable information on appropriate handling and safe use of hazardous

14. *Code of Federal Regulations (CFR).* A publication of the US government that contains all of the proposed and finalized federal regulations, including safety and environmental regulations.
15. *Globally Harmonized System of Classification and Labeling of Chemicals (GHS).* A worldwide initiative to promote standard criteria for classifying chemicals according to their health, physical, and environmental hazards. It uses harmonized pictograms, hazard statements, precautionary statements, and the signal words "Danger" and "Warning" to communicate hazard information on product labels and safety data sheets (SDSs) in a logical and comprehensive way. The primary goals of the GHS are the following:
    (1) To enhance the protection of human health and the environment by providing an internationally comprehensible system for hazard communication
    (2) To provide a recognized framework for those countries without an existing system
    (3) To reduce the need for testing and evaluation of chemicals
    (4) To facilitate international trade in chemicals whose hazards have been properly assessed and identified on an international basis
16. *Safety Data Sheet (SDS).* Safety data sheets (SDSs) are an essential component of the Globally Harmonized System of Classification and Labeling of Chemicals (GHS) and are intended to provide comprehensive information about a substance or mixture for use in workplace chemical management. They are used as a source of information about hazards, including environmental hazards, and to obtain advice on safety precautions. In the GHS, they serve the same function that the material safety data sheet (MSDS) does in OSHA's Hazard Communication Standard. The SDS is normally product related and not specific to workplace; nevertheless, the information on an SDS enables the employer to develop an active program of worker protection measures, including training, which is specific to the workplace, and to consider measures necessary to protect the environment.

chemicals. This update will also help reduce trade barriers and result in productivity improvements for American businesses that regularly handle, store, and use hazardous chemicals while providing cost savings for American businesses that periodically update safety data sheets and labels for chemicals covered under the hazard communication standard.

Major changes to the Hazard Communication Standard include:

1. Hazard classification: The new standard provides specific criteria for classification of health and physical hazards, as well as classification of mixtures.
2. Labels: Chemical manufacturers and importers will be required to provide a label that includes a harmonized *SIGNAL WORD,*[17] *PICTOGRAM,*[18] *HAZARD STATEMENT,*[19] and *PRECAUTIONARY STATEMENT*[20] for each hazard class and category.
3. Safety Data Sheets: The new format will have 16 specific sections.
4. Information and training: Employers are required to train workers on the new label elements and safety data sheets format to facilitate recognition and understanding.

For more information on the revised standard and the effective completion dates for changes, see www.osha.gov/dsg/hazcom/HCSFactsheet.html.

Operators must be concerned with the increased emphasis nationally and globally on hazardous materials and wastes. Much of this attention has focused on hazardous and toxic waste dumps and the efforts to clean them up after the long-term effects on human health were recognized. Each year, thousands of new chemical compounds are produced for industrial, commercial, and household use. Frequently, the long-term effects of these chemicals are unknown.

Federal and state laws have been enacted to control all aspects of hazardous materials handling and use. These laws are more commonly known as worker Right-To-Know (RTK) laws. Every state is covered by one or more RTK laws. OSHA's Hazard Communication Standard (HCS), 29 CFR 1910.1200, forms the basis of most laws. Although the federal standards were originally directed at the manufacturing sector, they now include all industries including the public sector and, therefore, water treatment plants.

In many cases, the individual states have the authority under the OSHA standard to develop their own state RTK laws and most states have adopted their own laws. Unfortunately, these laws vary significantly from state to state. The state laws that have been passed are at least as stringent as the federal standard and, in most cases, are even more stringent and already apply to water treatment plant operators. State laws are also under continuous revision and, because a strong emphasis is being placed on hazardous materials and worker exposure, state laws can be expected to be amended in the future to apply to virtually everybody in the workplace.

The purpose of this section is to familiarize you with general requirements of worker RTK elements so that you are better prepared as a water treatment plant operator to minimize the risk to yourself and your co-workers from hazardous materials and to comply with state and federal laws as well. Because of the wide diversity of existing laws, these guidelines may or may not meet your state's requirements. This section will give you an overview of the basic elements of a hazard communication program, which can be particularly useful if your agency currently does not have one. An effective program can be developed in house through your safety program and committee. By cutting through the bureaucratic/legal language and applying some common sense to what the law is trying to accomplish (protection of the worker in the workplace from hazardous materials), an effective program can be developed.

The basic elements of a hazard communication program are listed in Table 23.8 and described in the following paragraphs.

1. Identify Hazardous Materials

   While there are thousands of chemical compounds that would fall under this definition in a technical sense, water treatment plant operators should be concerned, first of all, with the materials they use in their everyday maintenance activities. Hazardous materials can be broken down into general categories as:

   a. Corrosives
   b. Toxics
   c. Flammables and explosives
   d. Asphyxiants
   e. Harmful physical agents
   f. Infectious agents

---

17. *Signal Word.* A single word used to indicate the relative level of severity of a chemical hazard and alert the reader to a potential hazard on the label. The signal words used are "Danger" and "Warning." "Danger" is used for the more severe hazards, while "Warning" is used for less severe hazards.
18. *Pictogram.* A graphical composition that may include a symbol plus other graphic elements, such as a border, background pattern, or color, that is intended to convey specific information about the hazards of a chemical. There are nine pictograms under the Globally Harmonized System (GHS) to convey the health, physical, and environmental hazards.
19. *Hazard Statement.* A statement assigned to a hazard class and category that describes the nature of the hazard(s) of a chemical, including, where appropriate, the degree of hazard.
20. *Precautionary Statement.* A phrase that describes recommended measures to be taken to minimize or prevent adverse effects resulting from exposure to a hazardous chemical, or improper storage or handling of a hazardous chemical.

TABLE 23.8 ELEMENTS OF A HAZARD COMMUNICATION PROGRAM

I. *WRITTEN HAZARD COMMUNICATION PROGRAM*
  A. Hazard Determination (Chemical manufacturers and importers only)
    1. Person(s) responsible for evaluating the chemical(s)
    2. List of sources to be consulted
    3. Criteria to be used to evaluate studies
    4. A plan for reviewing information to update safety data sheets (SDSs) if new and significant information is found
  B. Labels and Other Forms of Warning
    1. Person(s) responsible for labeling in-plant containers, if used
    2. Person(s) responsible for labeling shipped containers
    3. Description of labeling system
    4. Description of written alternatives to labeling of in-plant containers, if used
    5. Procedure to review and update label information
  C. Safety Data Sheets (SDSs)
    1. Person(s) responsible for obtaining/maintaining the SDSs
    2. Description of how the SDSs will be made available to employees
    3. Procedure to follow when the SDS is not received at time of first shipment
    4. Procedure to review and update SDS information
    5. Description of alternatives to actual data sheets in the workplace, if used
  D. Training
    1. Person(s) responsible for conducting training
    2. Format of the program to be used
    3. Elements of the training program
      a. Requirements of the OSHA standard
      b. Operations, where hazardous chemicals are present in routine tasks, nonroutine tasks, and foreseeable emergencies
      c. Location and availability of
        i. Written Hazard Communication Program
        ii. List of hazardous chemicals
        iii. SDSs
      d. Methods and observations that may be used to detect the presence or release of hazardous chemicals in the work area
      e. Physical and health hazards of chemicals in the work area
      f. Measures employees can take to protect themselves from hazards
      g. Details of the Hazard Communication Program developed
    4. Procedure to train new employees, as well as current employees, when a new chemical hazard is introduced into the workplace
  E. List of hazardous chemicals
  F. Procedure to inform employees of the hazards of chemicals in unlabeled pipes
  G. Procedure to inform on-site contractors

II. *LABELS AND OTHER FORMS OF WARNING*
  A. Labels or other markings on each container of hazardous chemicals
    1. Process vessels
    2. Storage tanks
    3. Compressed gas cylinders
    4. Product containers
    5. Tank truck/tank car labels

III. *SAFETY DATA SHEETS*
  A. SDSs developed or obtained for all hazardous chemicals
  B. Employees have access on each shift
  C. SDSs completed appropriately
  D. When no SDS is available, the documentation requesting an SDS from supplier is maintained

IV. *TRAINING*
  A. Employee training files
  B. Employee questioning:
    1. Awareness of the Hazard Communication Program and its requirements
    2. Have they received training
    3. Ability to locate the SDSs
    4. General familiarity with the hazardous properties of the chemicals in their workplace

A complete inventory of materials in use will produce a list similar to the following:

a. Corrosives
   (1) Sodium hydroxide
   (2) Calcium oxide (lime)
   (3) Hydrochloric acid
   (4) Ferric chloride
b. Toxics
   (1) Hydrogen sulfide
   (2) Chlorine
   (3) Carbon monoxide
c. Flammables and Explosives
   (1) Methane
   (2) Acetylene
   (3) Gasoline
   (4) Solvents
d. Asphyxiants
   (1) Carbon dioxide
   (2) Nitrogen
e. Harmful Physical Agents
   (1) Noise
   (2) Temperature
   (3) Radiation
f. Infectious Agents (see Table 23.9)

2. Obtain Chemical Information and Define Hazardous Conditions

   Once the inventory is complete, the next step is to obtain specific information on each of the chemicals and hazardous materials. This information is commonly available from manufacturers and is generally incorporated into a safety data sheet (SDS), also known as a material safety data sheet (MSDS). The purpose of the SDS is to have a comprehensive, readily available reference document that contains critical information on substance or mixture identification and supplier, hazard(s) identification, composition/information on ingredients, first-aid measures, firefighting measures, accidental release measures, handling and storage, and other information about every hazardous substance.

   OSHA's revised HCS requires that the information on the SDS is presented using consistent headings in a specified sequence. Paragraph (g) of the final rule (HazCom 2012) indicates the headings of information to be included on the SDS and the order in which they are to be provided. Table 23.10 outlines the minimum information required for an SDS and the specified format. The SDS format is the same as the American National Standards Institute (ANSI) standard format, which is widely used in the US and is already familiar to many operators.

TABLE 23.9 INFECTIOUS AGENTS POTENTIALLY PRESENT IN UNTREATED DOMESTIC WASTEWATER (EPA, 1999)

	Organism	Disease Caused
Bacteria	*Escherichia coli*	Gastroenteritis
	Leptospira (spp.)	Leptospirosis
	Salmonella typhi	Typhoid fever
	Salmonella (=2,100 serotypes)	Salmonellosis
	Shigella (4 spp.)	Shigellosis (bacillary dysentery)
	Vibrio cholerae	Cholera
Protozoa	*Balantidium coli*	Balantidiasis
	Cryptosporidium parvum	Cryptosporidiosis
	Entamoeba histolytica	Amebiasis (amoebic dysentery)
	Giardia lamblia	Giardiasis
Helminths	*Ascaris lumbricoides*	Ascariasis
	T. solium	Taeniasis
	Trichuris trichiura	Trichuriasis
Viruses	*Enteroviruses (72 types, e.g., polio, echo, and coxsackie viruses)*	Gastroenteritis, heart anomalies, meningitis
	Hepatitis A virus	Infectious hepatitis
	Norwalk agent	Gastroenteritis
	Rotavirus	Gastroenteritis

   Many agencies request an SDS when purchase orders are generated and will refuse to accept delivery of a shipment if the SDS is not included. Operators must be trained to read and understand the SDS forms. The forms themselves must be stored in a convenient location where they are readily available for reference.

3. Properly Label Hazards

   Once the physical, chemical, and health hazards have been identified and listed, a labeling system must be implemented. One of the major changes to OSHA's HCS is that chemical manufacturers and importers will be required to provide a label that includes a harmonized signal word, pictogram, and hazard statement for each hazard class and category. Precautionary statements must also be provided. Under the current HCS (HazCom 1994), the label preparer must provide the identity of the chemical and the appropriate hazard warnings. This may be done in a variety of ways, and the method to convey the information is left to the preparer. Under the revised HCS (HazCom 2012), once the hazard classification is completed, the standard specifies what information is to be provided for each hazard class and category.

TABLE 23.10 MINIMUM INFORMATION FOR A SAFETY DATA SHEET (SDS)

Heading	Subheading
1. Identification	(a) Product identifier used on the label (b) Other means of identification (c) Recommended use of the chemical and restrictions on use (d) Name, address, and telephone number of the chemical manufacturer, importer, or other responsible party (e) Emergency phone number
2. Hazard(s) identification	(a) Classification of the chemical in accordance with paragraph (d) of §1910.1200 (b) Signal word, hazard statement(s), symbol(s), and precautionary statement(s) in accordance with paragraph (f) of §1910.1200. (Hazard symbols may be provided as graphical reproductions in black and white or the name of the symbol, e.g., flame or skull and crossbones) (c) Describe any hazards not otherwise classified that have been identified during the classification process (d) Where an ingredient with unknown acute toxicity is used in a mixture at a concentration of 1% and the mixture is not classified based on testing of the mixture as a whole, a statement that X% of the mixture consists of ingredient(s) of unknown acute toxicity is required
3. Composition/ information on ingredients	Except as provided for in paragraph (i) of §1910.1200 on trade secrets: **For Substances** (a) Chemical name (b) Common name and synonyms (c) CAS number and other unique identifiers (d) Impurities and stabilizing additives which are themselves classified and which contribute to the classification of the substance **For Mixtures** In addition to the information required for substances: (a) The chemical name and concentration (exact percentage) or concentration ranges of all ingredients which are classified as health hazards in accordance with paragraph (d) of §1910.1200 and (1) are present above their cutoff/concentration limits, or (2) present a health risk below the cutoff/concentration limits (b) The concentration (exact percentage) shall be specified unless a trade secret claim is made in accordance with paragraph (i) of §1910.1200, when there is batch-to-batch variability in the production of a mixture, or for a group of substantially similar mixtures (See A.0.5.1.2) with similar chemical composition. In these cases, concentration ranges may be used. **For All Chemicals Where a Trade Secret is Claimed** Where a trade secret is claimed in accordance with paragraph (i) of §1910.1200, a statement that the specific chemical identity and/or exact percentage (concentration) of composition has been withheld as a trade secret is required.
4. First-aid measures	(a) Description of necessary measures, subdivided according to the different routes of exposure, i.e., inhalation, skin and eye contact, and ingestion (b) Most important symptoms/effects, acute and delayed (c) Indication of immediate medical attention and special treatment needed, if necessary
5. Firefighting measures	(a) Suitable (and unsuitable) extinguishing media (b) Specific hazards arising from the chemical (e.g., nature of any hazardous combustion products) (c) Special protective equipment and precautions for firefighters

TABLE 23.10 MINIMUM INFORMATION FOR A SAFETY DATA SHEET (SDS) *(continued)*

Heading	Subheading
6. Accidental release measures	(a) Personal precautions, protective equipment, and emergency procedures (b) Methods and materials for containment and cleaning up
7. Handling and storage	(a) Precautions for safe handling (b) Conditions for safe storage, including any incompatibilities
8. Exposure controls/ personal protection	(a) OSHA permissible exposure limit (PEL), American Conference of Governmental Industrial Hygienists (ACGIH) Threshold Limit Value (TLV), and any other exposure limit used or recommended by the chemical manufacturer, importer, or employer preparing the safety data sheet (b) Appropriate engineering controls (c) Individual protection measures, such as personal protective equipment
9. Physical and chemical properties	(a) Appearance (physical state, color, etc.) (b) Odor (c) Odor threshold (d) pH (e) Melting point/freezing point (f) Initial boiling point and boiling range (g) Flash point (h) Evaporation rate (i) Flammability (solid, gas) (j) Upper/lower flammability or explosive limits (k) Vapor pressure (l) Vapor density (m) Relative density (n) Solubility(ies) (o) Partition coefficient: n-octanol/water (p) Auto-ignition temperature (q) Decomposition temperature (r) Viscosity
10. Stability and reactivity	(a) Reactivity (b) Chemical stability (c) Possibility of hazardous reactions (d) Conditions to avoid (e.g., static discharge, shock, or vibration) (e) Incompatible materials (f) Hazardous decomposition products
11. Toxicological information	Description of the various toxicological (health) effects and the available data used to identify those effects, including: (a) Information on the likely routes of exposure (inhalation, ingestion, skin and eye contact) (b) Symptoms related to the physical, chemical, and toxicological characteristics (c) Delayed and immediate effects as well as chronic effects from short- and long-term exposure (d) Numerical measures of toxicity (such as acute toxicity estimates) (e) Whether the hazardous chemical is listed in the National Toxicology Program (NTP) Report on Carcinogens (latest edition) or reported to be a potential carcinogen in the International Agency for Research on Cancer (IARC) Monographs (latest edition), or by OSHA

TABLE 23.10 MINIMUM INFORMATION FOR A SAFETY DATA SHEET (SDS) *(continued)*

Heading	Subheading
12. Ecological information (Non-mandatory)	(a) Ecotoxicity (aquatic and terrestrial, where available) (b) Persistence and degradability (c) Bioaccumulative potential (d) Mobility in soil (e) Other adverse effects (such as hazardous to the ozone layer)
13. Disposal considerations (Non-mandatory)	Description of waste residues and information on their safe handling and methods of disposal, including the disposal of any contaminated packaging
14. Transport information (Non-mandatory)	(a) UN number[a] (b) UN proper shipping name (c) Transport hazard class(es) (d) Packing group, if applicable (e) Environmental hazards (e.g., Marine pollutant [Yes/No]) (f) Transport in bulk (according to Annex II of MARPOL 73/78 and the IBC Code) (g) Special precautions which a user needs to be aware of, or needs to comply with, in connection with transport or conveyance either within or outside their premises
15. Regulatory information (Non-mandatory)	Safety, health, and environmental regulations specific for the product in question
16. Other information, including date of preparation or last revision	The date of preparation of the SDS or the last change to it

a UN numbers or UN IDs are four-digit numbers that identify hazardous substances such as explosives, flammable liquids, and toxic substances in the framework of international transport. They are assigned by the United Nations Committee of Experts on the Transport of Dangerous Goods.

What pictograms are required in the revised Hazard Communication Standard? What hazard does each identify?

There are nine pictograms under the GHS to convey health, physical, and environmental hazards. The final HCS requires eight of these pictograms, the exception being the environmental pictogram, as environmental hazards are not within OSHA's jurisdiction. The hazard pictograms and their corresponding hazards are shown in Figure 23.10.

When must label information be updated?

In the revised HCS, OSHA is lifting the stay on enforcement regarding the provision to update labels when new information on hazards becomes available. Chemical manufacturers, importers, distributors, or employers who become newly aware of any significant information regarding the hazards of a chemical must revise the labels for the chemical within 6 months of becoming aware of the new information, and must ensure that labels on containers of hazardous chemicals shipped after that time contain the new information. If the chemical is not currently produced or imported, the chemical manufacturer, importer, distributor, or employer must add the information to the label before the chemical is shipped or introduced into the workplace again.

How will workplace labeling provisions be changing under the revised Hazard Communication Standard?

The current standard provides employers with flexibility regarding the type of system to be used in their workplaces and OSHA has retained that flexibility in the revised HCS. Employers may choose to label workplace containers either with the same label that would be on shipped containers for the chemical under the revised rule, or with label alternatives

HAZARD COMMUNICATION STANDARD (HCS)

Health Hazard	Flame	Exclamation Mark

• Carcinogen • Mutagenicity • Reproductive Toxicity • Respiratory Sensitizer • Target Organ Toxicity • Aspiration Toxicity	• Flammables • Pyrophorics • Self-Heating • Emits Flammable Gas • Self-Reactives • Organic Peroxides	• Irritant (skin and eye) • Skin Sensitizer • Acute Toxicity (harmful) • Narcotic Effects • Respiratory Tract Irritant • Hazardous to Ozone Layer (Non-Mandatory)
Gas Cylinder	**Corrosion**	**Exploding Bomb**

• Gases Under Pressure	• Skin Corrosion/Burns • Eye Damage • Corrosive to Metals	• Explosives • Self-Reactives • Organic Peroxides
Flame Over Circle	**Environment** (Non-Mandatory)	**Skull and Crossbones**

• Oxidizers	• Aquatic Toxicity	• Acute Toxicity (fatal or toxic)

Fig. 23.10 Hazard Communication Standard (HCS) pictograms and hazards

that meet the requirements for the standard. Alternative labeling systems such as the National Fire Protection Association (NFPA) 704 Hazard Rating and the Hazardous Material Information System (HMIS) are permitted for workplace containers. However, the information supplied on these labels must be consistent with the revised HCS; that is, no conflicting hazard warnings or pictograms.

The NFPA system (Figure 23.11) is characterized by a diamond shaped symbol. It identifies the hazards of a material and the degree of severity of the health, flammability, and instability hazards. Hazard severity is indicated by a numerical rating that ranges from 0 indicating a minimal hazard to 4 indicating a severe hazard. The hazards are arranged spatially as follows on the symbol: health at the 9 o'clock position, flammability at the 12 o'clock position, and instability at the 3 o'clock position. Hazards are also color-coded: blue for health, red for flammability, and yellow for instability. The 6 o'clock position on the symbol represents special hazards and has a white background. The special hazards in use are white, indicating unusual reactivity with water and is a caution about the use of water in either firefighting or spill control response; and OX, indicating that the material is an oxidizer.

By contrast, the HMIS (Figure 23.12) is a numerical hazard rating that incorporates the use of labels with color-coded bars as well as training materials. It was developed by the American Coatings Association as a compliance aid for the HCS. The four bars are color coded, with blue indicating the level of health hazard, red for flammability, orange for a physical hazard, and white for personal protection. The number ratings range from 0 to 4.

In some cases, the SDS can be incorporated into the labeling requirements by locating the appropriate SDS in close proximity to chemical drums or storage areas. The labeling requirements offer the water treatment plant operator virtually instant recognition of the hazards in dealing with specific substances, protective equipment required, and other information.

4. Train Operators

   The last element in a hazard communication program is training and making information available to the water treatment plant operator. The GHS does not include harmonized training provisions, but recognizes that training is essential to an effective hazard communication approach. OSHA's revised HCS requires that workers be retrained within 2 years of the publication of the final rule to facilitate recognition and understanding of the new labels and safety data sheets. OSHA requires that employees be trained on the new label elements (pictograms and signal words) and SDS format. While many countries are in various stages of implementing the GHS, OSHA believes that it is possible that American workplaces may begin to receive labels and SDSs that are consistent with the GHS shortly after publication. It is important to ensure that when employees begin to see the new labels and SDSs in their workplaces, they will be familiar with them, understand how to use them, and access the information effectively.

A common sense approach eliminates the confusion about which of the thousands of substances operators should be trained for, and concentrates on those that they will be exposed to or use in everyday maintenance routines.

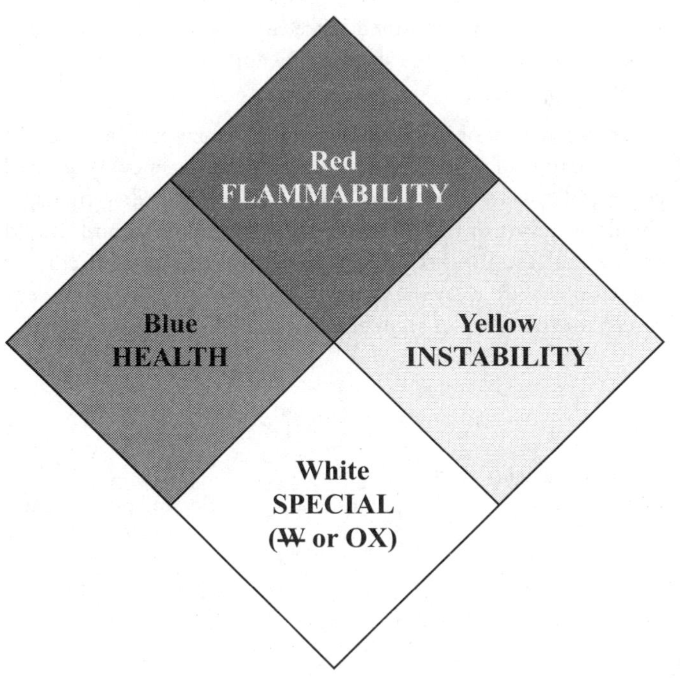

Fig. 23.11 National Fire Protection Association (NFPA) 704 Hazard Rating

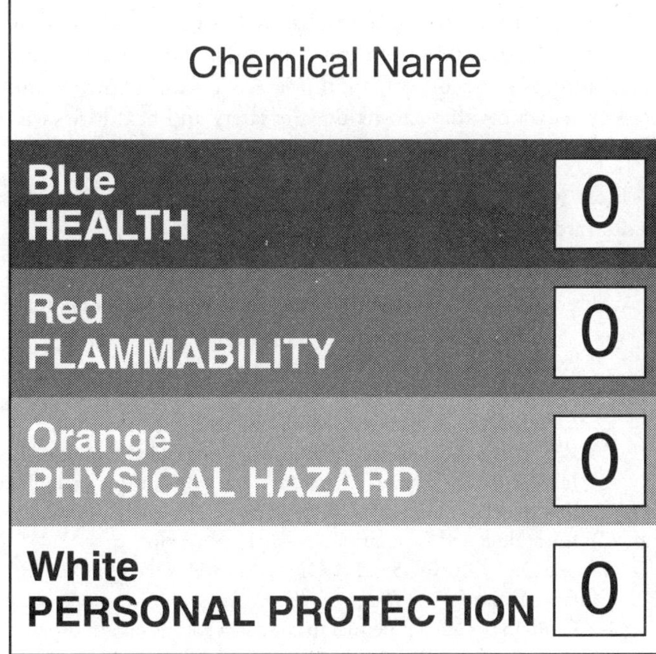

Fig. 23.12 The Hazardous Materials Identification System (HMIS)

The last element is a formal training program provided by qualified personnel designed to accomplish the following objectives:

1. Make operators and other employees aware of the HCS and its requirements.
2. Familiarize operators with potentially dangerous operations and hazardous substances that are present in their routine and nonroutine maintenance tasks and how to deal with emergency situations involving those materials and conditions.
3. Inform employees of the location and availability of the written hazard communication program.
4. Train employees in the methods and observations they may use to detect the presence or release of a hazardous substance in the work area, such as visual appearance, odor, and monitoring.
5. Inform employees of the measures they can take to protect themselves from the physical and health hazards of the chemicals in the work area.
6. Explain how to read and interpret SDS forms and have an SDS file readily available for reference.
7. Train operators and other employees to use the required labeling format.
8. Document what training has been performed.

OSHA expects that the modifications to the HCS will result in increased safety and health for the affected employees and reduce the numbers of accidents, fatalities, injuries, and illnesses associated with exposures to hazardous chemicals. The GHS revisions to the HCS for labeling and safety data sheets would enable employees exposed to workplace chemicals to more quickly obtain and to more easily understand information about the hazards associated with those chemicals. In addition, the revisions to the HCS are expected to improve the use of appropriate exposure controls and work practices that can reduce the safety and health risks associated with exposure to hazardous chemicals.

How will OSHA manage and communicate changes to the Hazard Communication Standard?

It is expected that the GHS will be a living document that will remain up to date and relevant; therefore, further changes may be adopted on a 2-year cycle. Presently, most of the recent updates have been clarifications of text. However, OSHA anticipates that future updates of the HCS may be necessary and can be done through various rulemaking options, including the following:

1. Technical updates for minor terminology changes
2. Direct Final Rules for text clarification
3. Notice and Comment rulemaking for more substantive or controversial updates such as additional criteria or changes in health or safety hazard classes or categories

OHSA's Hazard Communication page includes downloadable versions of the revised 1910.1200 Final Rule and appendices, updated to align with the GHS; a comparison of the HCS, issued in 1994 (HazCom 1994), with the revised Hazard Communication Final Rule issued in 2012 (HazCom 2012); frequently asked questions on the revisions; and new guidance materials on the revisions. The page also contains the full regulatory text and appendices of HazCom 1994. See http://www.osha.gov/dsg/hazcom/index.html for more information and to keep pace with the latest developments.

23.113 Confined Space Entry Procedures

CONFINED SPACES[21,22] pose significant risks for a large number of workers, including many utility operators. OSHA has therefore defined very specific procedures to protect the health and safety of operators whose jobs require them to enter or work in a confined space. The regulations (which can be found in the Code of Federal Regulations at 29 CFR 1910.146) require conditions in the confined space to be tested and evaluated before anyone enters the space. If conditions exceed OSHA's limits for safe exposure, additional safety precautions must be taken and a confined space entry permit (Figure 23.13) must be approved by the appropriate authorities before anyone enters the space.

The managers of water utilities may or may not be involved in the day-to-day details of enforcing the agency's confined space policy and procedures. However, every utility manager should be aware of the current OSHA requirements and should ensure that the utility's policies not only comply with current regulations, but that the agency's policies are vigorously enforced for the safety of all operators.

21. *Confined Space.* A space that has the following characteristics:
   (1) Is large enough and so configured that an employee can bodily enter and perform assigned work
   (2) Has limited or restricted means for entry or exit (for example, manholes, tanks, vessels, silos, storage bins, hoppers, vaults, and pits are spaces that may have limited means of entry)
   (3) Is not designed for continuous employee occupancy
   Also see DANGEROUS AIR CONTAMINATION and OXYGEN DEFICIENCY.
22. *Confined Space, Permit-Required (Permit Space).* A confined space that has one or more of the following characteristics:
   (1) Contains or has a potential to contain a hazardous atmosphere
   (2) Contains a material that has the potential for engulfing an entrant
   (3) Has an internal configuration such that an entrant could be trapped or asphyxiated by inwardly converging walls or by a floor that slopes downward and tapers to a smaller cross section
   (4) Contains any other recognized serious safety or health hazard

Confined Space Pre-Entry Checklist/Confined Space Entry Permit

Date and Time Issued: ______________ Date and Time Expires: __________ Job Site/Space I.D.: __________

Job Supervisor: ________________ Equipment to be worked on: ________ Work to be performed: __________

Standby personnel: ________________ ________________ ________________

1. Atmospheric Checks: Time ________ Oxygen ________ % Toxic ________ ppm
   Explosive ________ % LEL Carbon Monoxide ________ ppm

2. Tester's signature: ________________________
3. Source isolation: (No Entry) N/A Yes No
   Pumps or lines blinded, disconnected, or blocked ( ) ( ) ( )
4. Ventilation Modification: N/A Yes No
   Mechanical ( ) ( ) ( )
   Natural ventilation only ( ) ( ) ( )
5. Atmospheric check after isolation and ventilation: Time ________
   Oxygen ________ % > 19.5% < 23.5% Toxic ________ ppm < 10 ppm H_2S
   Explosive ________ % LEL < 10% Carbon Monoxide ________ ppm < 35 ppm CO

Tester's signature: ____

6. Communication procedures: ____
7. Rescue procedures: ____
   ____
8. Entry, standby, and backup persons Yes No
   Successfully completed required training? ( ) ( )
   Is training current? ( ) ( )
9. Equipment: N/A Yes No
   Direct reading gas monitor tested ( ) ( ) ( )
   Safety harnesses and lifelines for entry and standby persons ( ) ( ) ( )
   Hoisting equipment ( ) ( ) ( )
   Powered communications ( ) ( ) ( )
   SCBAs for entry and standby persons ( ) ( ) ( )
   Protective clothing ( ) ( ) ( )
   All electric equipment listed for Class I, Division I, Groups A, B, C, and D and nonsparking tools ( ) ( ) ( )
10. Periodic atmospheric tests:
    Oxygen: ____% Time ____; ____% Time ____; ____% Time ____; ____% Time ____;
    Explosive: ____% Time ____; ____% Time ____; ____% Time ____; ____% Time ____;
    Toxic: ____ppm Time ____; ____ppm Time ____; ____ppm Time ____; ____ppm Time ____;
    Carbon Monoxide: ____ppm Time ____; ____ppm Time ____; ____ppm Time ____; ____ppm Time ____;

We have reviewed the work authorized by this permit and the information contained herein. Written instructions and safety procedures have been received and are understood. Entry cannot be approved if any brackets ( ) are marked in the "No" column. This permit is not valid unless all appropriate items are completed.

Permit Prepared By: (Supervisor) ________________ Approved By: (Unit Supervisor) ________________

Reviewed By: (CS Operations Personnel) ________________________________
(Entrant) (Attendant) (Entry Supervisor)

This permit to be kept at job site. Return job site copy to Safety Office following job completion.

Fig. 23.13 Confined space pre-entry checklist/confined space entry permit

QUESTIONS

Please write your answers to the following questions and compare them with those on page 701.

23.11J List the four basic elements of a hazard communication program.

23.11K List the general categories of hazardous materials.

23.11L What is the purpose of the safety data sheet (SDS)?

23.11M What pictograms are required on a chemical label in OSHA's revised Hazard Communication Standard (HCS)?

23.11N What are a manager's responsibilities for ensuring the safety of operators entering or working in confined spaces?

23.114 Reporting

The mainstay of a safety program is the method of reporting and keeping of statistics. These records are needed regardless of the size of the utility because they provide a means of identifying accident frequencies and causes as well as the personnel involved. The records can be looked upon as the operator's safety report card. Therefore, it becomes the responsibility of each injured operator to fill out the utility's accident report.

All injuries should be reported, even if they are minor in nature, so as to establish a record in case the injury develops into a serious injury. It may be difficult at a later date to prove the accident did occur on the job and have the utility accept the responsibility for costs. The responsibility for reporting accidents affects several levels of personnel. First, of course, is the injured person. Next, it is the responsibility of the supervisor, and finally, the responsibility of management to review the causes and take steps to prevent such accidents from happening in the future.

Accident report forms may be very simple. However, they must record all details required by law and all data needed for statistical purposes. The forms shown in Figures 23.14 and 23.15 are examples for you to consider for use in your plant. The report must show the name of the injured, employee number, division, time of accident, nature of injury, cause of accident, first aid administered, and remarks for items not covered elsewhere. There should be a review process by foreman, supervisor, safety officer, and management. Recommendations are needed as well as a follow-up review to be sure that proper action has been taken to prevent recurrence.

In addition to reports needed by the utility, other reports may be required by state or federal agencies. For example, vehicle accident reports must be submitted to local police departments. If a member of the public is injured, additional forms are needed because of possible subsequent claims for damages. If the accident is one of occupational injury, causing lost time, other reports may be required. Follow-up investigations to identify causes and responsibility may require the development of other specific types of record forms.

In the preparation of accident reports, it is the operator's responsibility to correctly fill out each form, giving complete details. The supervisor must be sure no information is overlooked that may be helpful in preventing recurrence. The safety officer must review the reports, determine corrective actions, and make recommendations.

In day-to-day actions, operators, supervisors, and management often overlook opportunities to counsel individual operators in safety matters. Then, when an accident occurs, they are not inclined to look too closely at accident reports. First, the accident is a series of embarrassments, to the injured person, to the supervisor, and to management. Therefore, there is a reluctance to give detailed consideration to accident reports. However, if a safety program is to function well, it will require a thorough effort on the part of the operator, supervisor, and management in accepting their responsibility for the accident and making a greater effort through good reporting to prevent future similar accidents. Accident reports must be analyzed, discussed, and the real cause of the accident identified and corrected.

Emphasis on the prevention of future accidents cannot be overstressed. We must identify the causes of accidents and implement whatever measures are necessary to protect operators from becoming injured.

23.115 Training

If a safety program is to work well, management will have to accept responsibility for the following three components of training:

1. Safety education of all employees
2. Reinforced education in safety
3. Safety education in the use of tools and equipment

Or to put it another way, the three most important controlling factors in safety are education, education, and more education.

Date ______________

Name of injured employee ______________ Employee # ______________ Area ______________

Date of accident ______________ Time ______________ Employee's Occupation ______________

Location of accident ______________ Nature of injury ______________

First aid administered ______________

Name of doctor ______________ Address ______________

Name of hospital ______________ Address ______________

Witnesses (name & address) ______________

____

PHYSICAL CAUSES

Indicate below by an "X" whether, in your opinion, the accident was caused by:

_____ Improper guarding

_____ Defective substances or equipment

_____ Hazardous arrangement

_____ Improper illumination

_____ Improper dress or apparel

_____ Not listed (describe briefly) ______________

_____ No mechanical cause

_____ Working methods

_____ Lack of knowledge or skill

_____ Wrong attitude

_____ Physical defect

UNSAFE ACTS

Sometimes the injured person is not directly associated with the causes of an accident. Using an "X" to represent the injured worker and an "O" to represent any other person involved, indicate whether, in your opinion, the accident was caused by:

_____ Operating without authority

_____ Failure to secure or warn

_____ Working at unsafe speed

_____ Made safety device inoperative

_____ Unsafe equipment or hands instead of equipment

_____ No unsafe act

_____ Not listed (describe briefly) ______________

_____ Unsafe loading, placement & etc.

_____ Took unsafe position

_____ Worked on moving equipment

_____ Teased, abused, distracted & etc.

_____ Did not use safe clothing or personal protective equipment

____

What job was the employee doing? ______________

What specific action caused the accident? ______________

What steps will be taken to prevent recurrence? ______________

____

____

Date of Report ______________ Immediate Supervisor ______________

REVIEWING AUTHORITY

Comments:

Safety Officer Date

Comments:

Department Director Date

Fig. 23.14 Supervisor's accident report

INJURED: COMPLETE THIS SECTION

Name ______________________________ Age_____ Sex _____

Address ______________________________

Title ______________________ Dept. Assigned ______________________

Place of Accident ______________________________

Street or Intersection ______________________________

Date ______________________ Hour______ A.M. ______ P.M. ______

Type of Job You Were Doing When Injured

____

Object Which Directly Injured You | Part of Body Injured

____

How Did Accident Happen? (Be specific and give details; use back of sheet if necessary.)

____

First Aid Administered

____

Did You Report Accident or Exposure at Once? (Explain "No") Yes ☐ No ☐

Did You Report Accident or Exposure to Supervisor?
Give Name Yes ☐ No ☐

Were There Witnesses to Accident or Exposure?
Give Names Yes ☐ No ☐

Did You See a Doctor? (If Yes, Give Name) Yes ☐ No ☐

Are You Going to See a Doctor? (Give Name) Yes ☐ No ☐

______________________ ______________________

Date Signature

SUPERVISOR: COMPLETE THIS SECTION — (Return to Personnel as Soon as Possible)

Was an Investigation of Unsafe Conditions and/or
Unsafe Acts Made? If Yes, Please Submit Copy. Yes ☐ No ☐

Was Injured Intoxicated or Behaving Inappropriately
at Time of Accident? (Explain "Yes") Yes ☐ No ☐

Date Disability
Commenced ______________ Last Day
Wages Earned ______________ Date Back
on Job ______________

Date Report Completed ______________ 20 ____ Signed By ______________

Title ______________

Distribution: Canary - Department Head, Pink - Supervisor, White - Personnel

Fig. 23.15 Accident report

Responsibility for overall training must be that of upper management. A program that will educate operators and then reinforce this education in safety must be planned systematically and promoted on a continuous basis. There are many avenues to achieving this goal.

The safety education program should start with the new operator. Even before employment, verify the operator's past record and qualifications and review the pre-employment physical examination. In the new operator's orientation, include instruction in the importance of safety at your utility or plant. Also, discuss the matter of proper reporting of accidents as well as the organization's policies and practices. Give new operators copies of all safety SOPs and direct their attention to parts that directly involve them. Ask the safety officer to give a talk about utility policy, safety reports, and past accidents, and to orient the new operator toward the importance of safety to operators and to the organization.

The next consideration must be one of training the new operator in how to perform assigned work. Most supervisors think in terms of on-the-job training (OJT). However, OJT is not a good way of preventing accidents with an inexperienced operator. The idea is all right if the operator comes to the organization trained in how to perform the work, such as a treatment operator from another plant. Then you only need to explain your safety program and how your policies affect the new operator. For a new operator who is inexperienced in water treatment or in utility operation, the supervisor must give detailed consideration to the operator's welfare. In this instance, the training should include not only a safety talk, but the foreman (supervisor) must train the inexperienced operator in all aspects of treatment plant safety. This training includes instruction in the handling of chemicals, the dangers of electrical apparatus, fire hazards, and proper maintenance of equipment to prevent accidents. Special instructions will also be needed for specific work environments such as manholes, gases (chlorine and hydrogen sulfide [H_2S]), water safety, and any specific hazards that are unique to your facility. The new operator must be checked out on any equipment personnel may operate such as vehicles, forklifts, valve operators, and radios. All new operators should be required to participate in a safety orientation program during the first few days of their employment, and an overall training program in the first few months.

The next step in safety education is reinforcement. Even if the operator is well trained, mistakes can occur; therefore, safety education must be continual. Many organizations use the *TAILGATE*[23] method as a means of maintaining the operator's interest in safety. The program should be conducted by the first line supervisor. Schedule the informal tailgate meeting for a suitable location, keep it short, avoid distractions, and be sure that everyone can hear. Hand out literature, if available. Tailgate talks should communicate to the operator specific considerations, new problems, and accident information. These topics should be published. One resource for such meetings can be those operators who have been involved in an accident. Although it is sometimes embarrassing to the injured, the victim is now the expert on how the accident occurred, what could have been done to prevent it, and how it felt to have the injury. Encourage all operators, new and old, to participate in tailgate safety sessions.

Use safety posters to reinforce safety training and to make operators aware of the location of dangerous areas or show the importance of good work habits. Such posters are available through the National Safety Council's catalog.[24]

Awards for good safety records are another means of keeping operators aware of the importance of safety. The awards could be given to individuals in recognition of a good safety record. Publicity about the awards may provide an incentive to the operators and demonstrates the organization's determination to maintain a good safety record. The awards may include safety lapel pins, certificates, or plaques showing number of years without an accident. Consider publishing a utility newsletter on safety tips or giving details concerning accidents that may be helpful to other operators in the organization. Awards may be given to the organization in recognition of its effort in preventing accidents or for its overall safety program. A suggestion program concerning safety will promote and reinforce the program and give recognition to the best suggestions. The goal of all these efforts is to reinforce concerns for the safety of all operators. If safety, as an idea, is present, then accidents can be prevented.

Education of the operator in the use of tools and equipment is necessary. As pointed out above, OJT is not the answer to a good safety record. A good safety record will be achieved only with good work habits and safe equipment. If the operator is trained in the proper use of equipment (hand tools or vehicles), the operator is less likely to misuse them. However, if the supervisor finds an operator misusing tools or equipment, then it is the supervisor's responsibility to reprimand the operator as a means of reinforcing utility policies. The careless operator who misuses equipment is a hazard to other operators. Careless operators will also be the cause of a poor safety record in the operator's division or department.

An important part of every job should be the consideration of its safety aspects by the supervisor. The supervisor should instruct the foreman or operators about any dangers involved in job assignments. If a job is particularly dangerous, then the supervisor must bring that fact to everyone's attention and clarify utility policy in regard to unsafe acts and conditions.

If the operator is unsure of how to perform a job, then it is the operator's responsibility to ask for the training needed. Each operator must think, act, and promote safety if the organization is to achieve a good safety record. Training is the key to achieving this objective and training is everyone's responsibility—management, the supervisors, foremen, and operators.

23. *Tailgate Safety Meeting.* Brief (10 to 20 minutes) safety meetings held every 7 to 10 working days. The term comes from the safety meetings regularly held by the construction industry around the tailgate of a truck.
24. Contact your local safety council or the National Safety Council online at www.nsc.org or by phone at (800) 621-7615.

QUESTIONS

Please write your answers to the following questions and compare them with those on pages 701 and 702.

23.11O What is the mainstay of a safety program?

23.11P Why should you report even a minor injury?

23.11Q Why should a safety officer review an accident report form?

23.11R A new, inexperienced operator must receive instruction on what aspects of treatment plant safety?

23.11S What should an operator do if unsure of how to perform a job?

23.116 Measuring

To be complete, a safety program must also include some means of identifying, measuring, and analyzing the effects of the program. The systematic classification of accidents, injuries, and lost time is the responsibility of the safety officer. This person should use an analytical method that would refer to types and classes of accidents. Reports should be prepared using statistics showing lost time, costs, type of injuries, and other data, based on a specific time interval.

Such data call attention to the effectiveness of the program and promote awareness of the types of accidents that are happening. Management can use this information to decide where the emphasis should be placed to avoid accidents. However, statistical data are of little value if a report is prepared and then set on the bookshelf or placed in a supervisor's desk drawer. The data must be distributed and read by all operating and maintenance personnel.

As an example, injuries can be classified as fractures, burns, bites, eye injuries, cuts, and bruises. Causes can be classified as related to heat, machinery, falls, handling objects, chemicals, unsafe acts, and miscellaneous. Cost can be classified as lost time, lost dollars, lost production, contaminated water, or any other means of showing the effects of the accidents.

Good analytical reporting will provide a great deal of detail without a lot of paper to read and comprehend. Keep the method of reporting simple and easy to understand by all operators, so they can identify with the causes and be aware of how to prevent the accident happening to themselves or other operators. Table 23.11 gives one method of summarizing the causes of various types of injuries.

There are many other methods of compiling data. Table 23.11 could reflect cost in dollars or in work hours lost. Not all accidents mean time lost, but there can be other cost factors. The data analysis should also indicate if the accidents involve vehicles, company personnel, the public, company equipment, loss of chemical, or other factors. Results also should show direct and indirect costs to the agency, operator, and the public.

Once the statistical data have been compiled, someone must be responsible for reviewing it in order to take preventive actions. Frequently, such responsibility rests with the safety committee. In fact, safety committees may operate at several levels, for example management committee, working committee, or an accident review board. In any event, the committee must be active, be serious, and be fully supported by management.

TABLE 23.11 SUMMARY OF TYPES AND CAUSES OF INJURIES

	Primary Cause of Injury										
Type of Injury	Unsafe Act	Chemical	Falls	Handling Objects	Heat	Machinery	Falling Objects	Stepping	Striking	Miscellaneous	TOTAL
Fractures											
Sprains											
Eye Injuries											
Bites											
Cuts											
Bruises											
Burns											
Miscellaneous											

Another means of determining the effectiveness of your safety program is by reviewing your agency's injury statistics and calculating injury rates during a given time period, typically a year. Injury rates are based on 200,000 operator work hours as a standardized conversion factor representing a workforce of 100 full-time workers who work 40 hours per week for 50 weeks per year (assuming 2 weeks for vacation and holidays). Typically, rates that measure total injuries (injury frequency rate) and lost work days (injury severity rate) are the most useful. The formulas for calculating these rates are:

$$\text{Injury Frequency Rate} = \left(\frac{\text{Number of Injuries}}{\text{Number of Hours Worked During the Period Covered}}\right) \times 200{,}000$$

and

$$\text{Injury Severity Rate} = \left(\frac{\text{Number of Lost Work Days}}{\text{Number of Hours Worked During the Period Covered}}\right) \times 200{,}000$$

These standard formulas were developed by the Department of Labor's Bureau of Labor Statistics. They allow apples-to-apples comparisons because they take into account operator variations such as headcount and individual work hours, as well as unequal time periods. You can use these injury rates to identify trends and spot safety needs or concerns. You can also compare your rates to those of similar agencies and established sources such as the Bureau of Labor Statistics or the National Safety Council.

The injury severity rate is considered the better indicator of safety performance because it measures the more serious injuries involving lost work days. The injury frequency rate is a good general indicator, but is limited because it gives equal weighting to both major and minor injuries.

Calculating injury frequency rates and injury severity rates involves a three-step process: (1) compile the number of hours worked during the period covered, (2) compile the number of injuries or lost work days, and (3) apply the formula as shown in the following examples.

Example 3

A rural water company had a total of four operator injuries during the previous year. The total work hours were 71,856. Calculate the injury frequency rate.

Known		Unknown
Number of Injuries/yr	= 4 injuries/yr	Injury Frequency Rate
Number of Hours Worked/yr	= 71,856 hr/yr	

Calculate the injury frequency rate.

$$\text{Injury Frequency Rate} = \frac{(\text{Number of Injuries/yr})(200{,}000)}{\text{Number of Hours Worked/yr}}$$

$$= \frac{(4 \text{ injuries/yr})(200{,}000)}{71{,}856 \text{ hr/yr}}$$

$$= 11.1$$

This can be expressed as 11.1 injuries per 100 full-time employees (or 200,000 hours worked).

Example 4

Of the four injuries suffered by the operators in Example 3, one was a disabling injury. Calculate the injury frequency rate for the disabling injuries.

Known		Unknown
Number of Disabling Injuries/yr	= 1 injury/yr	Injury Frequency Rate (Disabling Injuries)
Number of Hours Worked/yr	= 71,856 hr/yr	

Calculate the injury frequency rate for disabling injuries.

$$\text{Injury Frequency Rate (Disabling Injuries)} = \frac{(\text{No. of Disabling Injuries/yr})(200{,}000)}{\text{No. of Hours Worked/yr}}$$

$$= \frac{(1 \text{ injury/yr})(200{,}000)}{71{,}856 \text{ hr/yr}}$$

$$= 2.8$$

This can be expressed as 2.8 disabling injuries per 100 full-time employees (or 200,000 hours worked).

Example 5

The water company described in Examples 3 and 4 experienced 5 lost work days in 1 year due to injuries while the operators worked 71,856 hours. Calculate the injury severity rate.

Known		Unknown
Number of Lost Work Days/yr	= 5 days/yr	Injury Severity Rate
Number of Hours Worked/yr	= 71,856 hr/yr	

Calculate the injury severity rate.

$$\text{Injury Severity Rate} = \frac{(\text{Number of Lost Work Days/yr})(200{,}000)}{\text{Number of Hours Worked/yr}}$$

$$= \frac{(5 \text{ Days/yr})(200{,}000)}{71{,}856 \text{ hr/yr}}$$

$$= 13.9$$

This can be expressed as 13.9 lost work days per 100 full-time employees (or 200,000 hours worked).

Note that by using a 1-year time interval in the previous injury rate calculations, the results may be used by the safety officer in preparing an annual report.

23.117 Human Factors

First, you may ask, what is a human factor? Well, it is not too often that a safety text considers human factors as part of the safety program. However, if these factors are understood and emphasis is given to their practical application, then many accidents can be prevented. Human factors engineering is the specialized study of technology relating to the design of operator-machine interfaces. That is to say, it examines ways in which machinery might be designed or altered to make it easier to use, safer, and more efficient for the operator. We hear a lot about making computers more user friendly, but human factors engineering is just as important to everyday operation of other machinery in the everyday plant.

Many accidents occur because the operator forgets the human factors. The ultimate responsibility for accidents due to human factors belongs to the management group. However, this does not relieve the operator of the responsibility to point out the human factors as they relate to safety. After all, it is the operator using the equipment who can best tell if it meets all the needs for an interrelationship between an operator and a machine.

The first step in the prevention of accidents takes place in the plant design. Even with excellent designs, accidents can and do happen. However, every step possible must be taken during design to ensure a maximum effort of providing a safe plant environment. Most often, the operator has little to do with design, and, therefore, needs to understand human factors engineering to be able to evaluate these factors as the plant is being operated. As newer plants become automated, this type of understanding may even be more important.

Other contributing human factors are the operator's mental and physical characteristics. The operator's decision-making abilities and general behavior (response time, sense of alarm, and perception of problems and danger) are all important factors. Ideally, tools and machines should function as intuitive extensions of the operator's natural senses and actions. Any factors disrupting this flow of action can cause an accident. Therefore, be on the lookout for such factors. When you find a system that cannot be operated in a smooth, logical sequence of steps, change it. You may prevent an accident. If the everyday behavior of an operator is inappropriate with regard to a specific job, reconsider the assignment to prevent an accident.

The human factor in safety is the responsibility of design engineers, supervisors, and operators. However, the operator who is doing the work will have a greater understanding of the operator-machine interface. For this reason, the operator is the appropriate person to evaluate the means of reducing the human factor's contribution to the cause of accidents, thereby improving the plant's safety record.

QUESTIONS

Please write your answers to the following questions and compare them with those on page 702.

23.11T Statistical accident reports should contain what types of accident data?

23.11U How can injuries be classified?

23.11V How can causes of injuries be classified?

23.11W How can costs of accidents be classified?

23.12 RECORDKEEPING

23.120 Purpose of Records

Accurate records are a very important part of effective operation of a water treatment plant and distribution system facilities. Records are a valuable source of information. They can save time when trouble develops and provide proof that problems were identified and solved. Pertinent and complete records should be used as a basis for plant operation, interpreting results of water treatment, preparing preventive maintenance programs, and preparing budget requests. When accurately kept, records provide an essential basis for design of future changes or expansions of the treatment plant, and can be used to aid in the design of other water treatment plants where similar water may be treated and similar problems may develop.

If legal questions or problems occur in connection with the treatment of the water or the operation of the plant, accurate and complete records will provide evidence of what actually occurred and what procedures were followed.

Records are essential for effective management of water treatment facilities and to satisfy legal requirements. Some of the important uses of records include:

1. Aiding operators in solving treatment and water quality problems
2. Providing a method of alerting operators to changes in source water quality
3. Showing that the treated water is acceptable to the consumer
4. Documenting that the final product meets plant performance standards, as well as the standards of the regulatory agencies
5. Determining performance of treatment processes, equipment, and the plant
6. Satisfying legal requirements
7. Aiding in answering complaints
8. Anticipating routine maintenance
9. Providing data for cost analysis and preparation of budgets
10. Providing data for future engineering designs
11. Providing information for monthly and annual reports

23.121 Types of Records

There are many different types of records that are required for effective management and operation of water supply, treatment, and distribution system facilities. Essential records include:

1. Source of supply
2. Operation
3. Laboratory
4. Maintenance
5. Chemical inventory and usage
6. Purchases
7. Chlorination station
8. Main disinfection
9. Cross-connection control
10. Personnel
11. Accidents
12. Customer complaints

23.122 Types of Plant Operations Data [25]

Plant operations logs can be as different as the plants and water systems whose information they record. The differences in amount, nature, and format of data are so significant that any attempt to prepare a typical log would be very difficult. This section will outline the kinds of data that are usually required to help you develop a useful log for your facilities.

Treatment plant data such as total flows, chemical use, chemical doses, filter performance, reservoir levels, quality control tests, and rainfall and runoff information represent the bulk of the data required for proper plant operation. Frequently, however, source and distribution system data such as reservoir storage and water quality data are included because of the impact of this information on plant operation and operator responsibilities. Typical plant operations data include:

1. Plant title, agency, and location
2. Date
3. Names of operators and supervisors on duty
4. Source of supply
   a. Reservoir elevation and volume of storage
   b. Reservoir inflow and outflow

25. Also see Volume I, Chapter 10, "Plant Operation," Section 10.6, "Operating Records and Reports," for additional details and recordkeeping forms.

   c. Evaporation and precipitation
   d. Apparent runoff, seepage loss, or infiltration gain
   e. Production figures from wells
5. Water treatment plant
   a. Plant inflow
   b. Treated water flow
   c. Plant operating water (backwash)
   d. Clear well level
6. Distribution system
   a. Flows to system (system demand)
   b. Distribution system reservoir levels and changes
   c. Comparison of production with deliveries (unaccounted-for water)
7. Chemical inventory and usage
   a. Chemical inventory/storage (measured use and deliveries)
   b. Metered or estimated plant usages
   c. Calculated usage of chemicals (compare with actual use)
8. Quality control tests
   a. Turbidity
   b. Chlorine residual
   c. Coliforms
   d. Odor
   e. Color
   f. Other
9. Filter performance
   a. Operation
      (1) Total hours, all units
      (2) Filtered water turbidities
      (3) Head losses
      (4) Levels
      (5) Flow rates
   b. Backwash
      (1) Total hours
      (2) Head losses
      (3) Total wash water used
      (4) Duration and rate of back/surface wash
10. Meteorologic
    a. Rainfall, evaporation, and temperature of both water and air
    b. Weather (clear, cloudy, windy)
11. Remarks

    Space should be provided to describe or explain unusual data or events. Extensive notes should be entered on a daily worksheet or journal.

23.123 Maintenance Records

A good plant maintenance effort depends heavily upon good recordkeeping. There are several areas where proper records and documentation can definitely improve overall plant performance.

23.124 Procurement Records

Ordering repair parts and supplies usually is done when the on-hand quantity of a stocked part or chemical falls below the reorder point, a new item is added to stock, or an item has been requested that is not stocked. Most organizations require employees to submit a requisition (similar to the one shown in Figure 23.16) when they need to purchase equipment or supplies. When the requisition has been approved by the authorized person (a supervisor or purchasing agent, in most cases), the items are ordered using a form called a purchase order. A purchase order contains a number of important items. These items include the date, a complete description of each item and quantity needed, prices, the name of the vendor, and a purchase order number.

A copy of the purchase order should be retained in a suspense file or on a clipboard until the ordered items arrive. This procedure helps keep track of the items that have been ordered but have not yet been received.

All supplies should be processed through the storeroom immediately upon arrival. When an item is received, it should be so recorded on an inventory card. The inventory card will keep track of the numbers of an item in stock, when last ordered, cost, and other information. Furthermore, by always logging in supplies immediately upon receipt, you are in a position to reject defective or damaged shipments and control shortages or errors in billing. Some utilities use personal computers to keep track of orders and deliveries.

23.125 Inventory Records

An inventory consists of the supplies the treatment plant needs to keep on hand to operate the facility. These maintenance supplies may include repair parts, spare valves, electrical supplies, tools, and lubricants. The purpose of maintaining an inventory is to provide needed parts and supplies quickly, thereby reducing equipment downtime and work delays.

In deciding what supplies to stock, keep in mind the economics involved in buying and stocking an item as opposed to depending upon outside availability to provide needed supplies. Is the item critical to continued plant or process operation? Should certain frequently used repair parts be kept on hand? Does the item have a shelf life?

Inventory costs can be held to a minimum by keeping on hand only those parts and supplies for which a definite need exists or that would take too long to obtain from an outside vendor. A definite need for an item is usually demonstrated by a history of regular use. Some items may be infrequently used but may be vital in the event of an emergency; these items should also be stocked. Take care to exclude any parts and supplies that may become obsolete, and do not stock parts for equipment scheduled for replacement.

P - 1

CITY OF SACRAMENTO

REQUISITION

1 BID NO. | DUE DATE

2 PURCHASE ORDER NO. | DATE

3 DELIVER TO:

4 FUND GP.	5 FUND	6 ORGAN.	7 REV. COST CTR.	8 OBJECT

9 REQUISITION NO. | DATE

10 REFER QUESTIONS TO:

VENDORS: USE THESE COLS. ONLY

11 ITEM NO.	12 COMMODITY CODE NO.	13 DESCRIPTION	14 QUANTITY / UNIT	15 PRICE / UNIT	16 BRAND OFFERED	17 LOT PRICE FOR EACH ITEM

18 DEPT.

19 ORIG. BY

TEL. NO.

THIS REQUISITION WILL BE REPRODUCED TO CREATE A PURCHASE ORDER QUICKLY AND PROVIDE FASTER DELIVERY OF THESE ITEMS TO YOUR DEPARTMENT. PLEASE FOLLOW INSTRUCTIONS CAREFULLY.

SUB TOTAL | SALES TAX

INVOICE AMOUNT

20 EST. OF COST

22 SPECIAL REQUIREMENTS OR INSTRUCTIONS. PREVIOUS P. O. NO. OR BID NO., OR RECOMMENDED VENDORS

21 NUMBER OF REQUISITION ATTACHMENT SHEETS

23 CERTIFICATION IS HEREBY MADE THAT THE ABOVE IS A LEGAL CHARGE AGAINST THE APPROPRIATION INDICATED

DEPT. HEAD ______ BY ______

OFFICIAL TITLE

24

VENDOR ______ (A)

ADDRESS ______ (B)

CITY, STATE, AND ZIP ______ (C)

TERMS ______ TEL. NO. ______ (D)

DELIVERY ______ FOB DESTINATION ______ (E)

BY ______ TITLE ______

Fig. 23.16 Requisition/purchase order form

Tools should be inventoried. Tools that are used by operators on a daily basis should be permanently signed out to them. More expensive tools and tools that are only occasionally used, however, should be kept in a storeroom. These tools should be signed out only when needed and signed back in immediately after use.

23.126 Equipment Records

You will need to keep accurate records to monitor the O&M of plant equipment. Equipment control cards and work orders can be used to do the following:

- Record important equipment data such as make, model, serial number, and date purchased.
- Record maintenance and repair work performed to date.
- Anticipate preventive maintenance needs.
- Schedule future maintenance work.

See Chapter 18, Section 18.00, "Preventive Maintenance Records," for additional information.

23.127 Computer Recordkeeping Systems

Most water utilities use computer-based recordkeeping systems for the administration of their water supply systems. There are numerous utility software products available, both computer- and web-based, that enable utilities to track plant data, produce reports and spreadsheets, and to manage customer accounts and billing. Small drinking water systems can purchase software that is designed to help with developing budgets, setting user rates, and tracking expenses.

23.128 Disposition of Plant Records

Good recordkeeping is very important because records indicate potential problems, adequate operation, and are a good waterworks practice. Usually, the only record required by the health agency is the summary of the daily turbidity of the treated surface water as it enters the distribution system. Chlorine residual and bacterial counts are often required. Other records that may also be required include:

1. Total trihalomethane (TTHM) data (frequency of this report is based on the number of people served).
2. The daily log and records of the analyses to control the treatment process may be required when there are chronic treatment problems.
3. Chlorination, constituent removal, and sequestering records may be required from small systems (especially those demonstrating little understanding of the processes).
4. Records showing the quantity of water from each source in use may be required from systems with sources producing water not meeting state or local health department water quality standards.

An important question is how long records should be kept. Records should be kept as long as they may be useful. Some information will become useless after a short time, while other data may be valuable for many years. Data that might be used for future design or expansion should be kept indefinitely. Laboratory data will always be useful and should be kept indefinitely. Regulatory agencies may require you to keep certain water quality analyses (bacteriological test results) and customer complaint records on file for specified time periods (10 years for chemical analyses and bacteriological tests).

Even if old records are not consulted every day, this does not lessen their potential value. For orderly records handling and storage, set up a schedule to periodically review old records and to dispose of those records that are no longer needed. A decision can be made when a record is established regarding the time period for which it must be retained.

QUESTIONS

Please write your answers to the following questions and compare them with those on page 702.

23.12A List some of the important uses of records.

23.12B What is "unaccounted-for water"?

23.12C What chemical inventory and usage records should be kept?

23.12D List the important items usually contained on a purchase order.

23.13 WATER AND ENERGY CONSERVATION

23.130 Need for Conservation

An effective water and energy conservation program is an excellent means of saving money. Water conservation can reduce energy costs. This effort can reduce the flows to wastewater treatment plants, thus saving energy and possibly postponing the need for capacity expansions. Reducing water and energy consumption results in conservation of natural resources.

This section emphasizes possible approaches to water conservation programs. However, many of the concepts used to conserve water also can be applied to an energy conservation program.

23.131 What Is Water Conservation?

When developing a water conservation program, the water utility manager must realize that consumers have different perspectives regarding water conservation. A preservation perspective emphasizes aesthetic and naturalistic goals over all else, including economic development, growth, and efficiency. An ecological perspective emphasizes avoiding the tragedy that occurs when a common resource (water) is overconsumed. A hydrological perspective emphasizes the water cycle and maintaining water quality through sufficient levels of water supply. A traditional economics perspective emphasizes allocating and using water efficiently. A resource economics perspective goes a step further by emphasizing the need to control rates of water use to ensure a sustainable future.

One simple definition equates water conservation with reduced water use. This definition is too simple for practical purposes. Some practices that decrease water use can be undesirable, particularly if the water-use reductions unnecessarily impair consumer lifestyles or come at the expense of other valued resources. The concept of water conservation should take into account the impact of conservation on the mix of all resources.

Developing conservation programs, including conservation-oriented rate structures, requires community involvement through both public participation and consumer education. Community involvement and consumer education can improve the effectiveness of conservation efforts and help mitigate many of the problems associated with implementing conservation programs and rate structures.

The key practical long-term benefit of water conservation is the postponement or deferral of additional treatment and source development capacity. Avoided costs associated with new capacity or sources include not only the capital and operating costs associated with facilities, but also the procedural costs associated with obtaining regulatory approvals, the political costs associated with securing support for projects and necessary revenues, and the environmental costs associated with developing water supplies. By inducing consumers to control their usage, conservation practices and rate structures can help extend the useful service life of existing capacity.

Reductions in water usage can lead to reductions in wastewater flows. These reduced wastewater flows can result in the postponement or deferral of expanded wastewater treatment capacity.

The most important effect of water conservation pricing is improved economic behavior on the part of consumers and improved economic performance on the part of water-supply agencies. Cost-based conservation rates send important economic signals about the value of water. The resulting efficiency benefits water consumers, the water agency, the environment, and society.

23.132 Elements of Conservation Programs

Water utility agencies should develop a water conservation strategy. The basic elements of this strategy consist of water conservation best management practice (BMP) strategies. A BMP is described as a policy, program, practice, rule, regulation, or ordinance or the use of devices, equipment, or facilities that meet either of the following criteria:

1. An established and generally accepted practice among water suppliers that results in more efficient use or conservation of water.
2. A practice for which sufficient data are available from existing water conservation projects to indicate that significant conservation or conservation-related benefits can be achieved; that the practice is technically and economically reasonable and not environmentally or socially unacceptable; and that the practice is not otherwise unreasonable for most water suppliers to carry out.

The BMPs described in this section were developed by the California Urban Water Conservation Council. Water utility managers need to evaluate the BMPs outlined in terms of effort to develop and implement the BMP, including costs, with the anticipated results and benefits.

23.1320 Residential Water Surveys

Residential water survey BMP includes indoor and outdoor audits of residential water use and distribution of water-saving devices. The surveys should include the following elements:

1. Indoor
   a. Check for leaks, including toilets, faucets, and meter check.
   b. Check shower head flow rates, check aerator flow rates, and offer to replace or recommend replacement, as necessary.
   c. Check toilet flow rates and offer to install or recommend installation of displacement device or direct customer to ULFT (ultra-low-flush [flow] toilet) replacement program, as necessary; replace leaking toilet flapper, as necessary.
2. Outdoor
   a. Check irrigation system and timers.
   b. Review or develop customer irrigation schedule.
3. Recommended
   a. Measure currently landscaped area.
   b. Measure total irrigable area.

Following completion of a residential water survey, the customer should be provided with an evaluation of the results and water-saving recommendations. The customer should be provided a water conservation information packet.

23.1321 Residential Plumbing Retrofits

The residential plumbing retrofit BMP includes distribution or installation of water-saving devices in pre-1992 residences. The retrofit program should include at least the following actions:

1. Identify single-family and multifamily residences constructed before 1992. Develop a targeting and marketing strategy to distribute or directly install high-quality, low-flow shower heads (rated 2.5 gpm [10.5 liters per minute] or less), toilet displacement devices (as needed), toilet flappers (as needed), and faucet aerators (rated 2.2 gpm [8.3 liters per minute] or less) as practical to residences requiring them.
2. Maintain distribution or direct installation programs until most single-family residences and multifamily units are fitted with high-quality, low-flow shower heads.
3. Track the type and number of retrofits completed, devices distributed, and program costs.

23.1322 System Water Audits, Leak Detection, and Repair

The system water audits BMP consists of annually calculating unaccounted-for water and performing distribution system water audits, as required. Elements of this BMP include:

1. Complete a prescreening system audit annually to determine the need for a full-scale system audit. The prescreening system audit should be calculated by:
   a. Determining metered sales.
   b. Determining other system verifiable uses.
   c. Determining total supply into the system.
   d. Dividing metered sales plus other verifiable uses by total supply into the system. If this quantity is less than 0.9, a full-scale system audit is indicated.

2. When indicated, water agencies should complete water audits of their distribution systems using methodology consistent with that described in AWWA's guidebook, *Water Audits and Loss Control Programs*.[26]

3. Water agencies should advise customers whenever it appears that leaks exist on the customer's side of the meter; perform distribution system leak detection when warranted and cost-effective; and repair leaks when found.

23.1323 Metering with Commodity Rates

This BMP requires the metering of all new connections and retrofitting of existing connections to ensure metering of all consumptions and billing on the basis of volume of water consumed. Elements of this BMP include:

1. Requiring meters for all new connections and billing by volume of water used.

2. Establishing a program for retrofitting existing unmetered connections and billing by volume of water consumed.

3. Identifying intra- and interagency disincentives or barriers to retrofitting mixed-use commercial accounts with dedicated landscape meters, and conducting a feasibility study to assess the merits of a program to provide incentives to switch mixed-use accounts to dedicated landscape meters.

23.1324 Large Landscape Conservation Programs

Large landscape conservation programs and incentives can be a very effective BMP. A major element of this program is evapotranspiration-based (ETo-based)[27] water-use budgets for large landscape irrigators. This BMP includes provisions for customer support, education, and assistance. Water agencies should provide nonresidential customers with support and incentives to improve their landscape water-use efficiency. This support should include the following elements:

1. Accounts with Dedicated Irrigation Meters

   a. Identify accounts with dedicated irrigation meters and assign ETo-based water-use budgets.

   b. Provide notices with each billing cycle to accounts with water-use budgets showing the relationship between the budget and actual consumption; the water agency may choose not to notify customers whose use is less than their water-use budget.

2. Commercial/Industrial/Institutional (CII) Accounts with Mixed-Use Meters or Not Metered

   a. Develop and implement a strategy targeting and marketing large landscape water-use surveys to commercial/industrial/institutional accounts with mixed-use meters.

   b. Unmetered service areas will actively market landscape surveys to existing accounts with large landscapes, or accounts with landscapes that have been determined by the water purveyor not to be water efficient.

   c. Offer the following measures when cost-effective:

      (1) Landscape water-use analysis/surveys
      (2) Voluntary water-use budgets
      (3) Installation of dedicated landscape meters
      (4) Training (multilingual, where appropriate) in landscape maintenance, irrigation system maintenance, and irrigation system design
      (5) Financial incentives to improve irrigation system efficiency such as loans, rebates, and grants for the purchase or installation of water-efficient irrigation systems
      (6) Follow-up water-use analyses/surveys consisting of a letter, phone call, or site visit, where appropriate

   d. Survey elements will include measurement of landscape area; measurement of total irrigable area; irrigation system check and distribution uniformity analysis; review or development of irrigation schedules, as appropriate; and provision of a customer survey report and information packet.

   e. Track survey offers, acceptance, findings, devices installed, savings potential, and survey cost.

3. New or Change-of-Service Accounts

   Provide information on climate-appropriate landscape design and efficient irrigation equipment/management to new customers and change-of-service customer accounts.

4. Recommended

   a. Install climate-appropriate, water-efficient landscaping at water agency facilities, and dual metering, where appropriate.

   b. Provide customer notices before the start of the irrigation season alerting them to check their irrigation systems and to make repairs, as necessary. Provide customer notices at the end of the irrigation season advising them to adjust their irrigation system timers and irrigation schedules.

23.1325 High-Efficiency Clothes Washers

This BMP provides rebates for high-efficiency washing machines. Important elements of this BMP include the water agency determining that significant water savings are available from high-efficiency washing machines and that there is widespread product availability. Also, financial incentive programs offered by energy service providers may create an effective partnership with the water utility.

---

26. Available from American Water Works Association (AWWA) at www.awwa.org or phone (800) 926-7337. Order No. 30036. ISBN 978-1-58321-631-6.

27. ETo is an estimate of the depth of water evaporated and transpired from a reference crop if water is not a limiting factor. The reference crop for landscape is tall fescue grass that is actively growing, completely shading the soil, and cut 4- to 6-inches (9- to 15-cm) high.

23.1326 Public Information Programs

Public information programs are an effective means of promoting water conservation. Important elements of a water agency's public information programs should include:

1. Information to promote water conservation and water conservation-related benefits.
2. The program should include providing speakers to employees, community groups, and media; using paid and public service advertising; using billing inserts; providing information on customers' bills showing use in gallons per day (liters per day) for the last billing period compared to the same period the year before; providing public information to promote water conservation practices; and coordinating with other government agencies, industry groups, public interest groups, and the media.

23.1327 School Education Programs

School education programs include the provision of educational materials and services to schools. Educating children of the need to conserve water is also a way for the water agency to communicate to adults by information children bring home from school. A school education water conservation program should include the following elements:

1. Information to promote water conservation and water conservation-related benefits.
2. Programs should include working with school districts and private schools in the water supplier's service area to provide instructional assistance, educational materials, and classroom presentations that identify urban, agricultural, and environmental issues and conditions in the local watershed. Educational materials should meet the local education framework requirements, and grade-appropriate materials should be distributed to grade levels K–3, 4–6, 7–8, and high school.

23.1328 Conservation Programs for Commercial, Industrial, and Institutional (CII) Sectors

The purpose of this BMP is to increase water-use efficiency in the commercial, industrial, and institutional (CII) sectors. Activities for this BMP include:

1. Identify and rank commercial, industrial, and institutional (CII) accounts according to water use. For the purposes of this BMP, CII accounts are defined as:
   a. Commercial Accounts: Any water use that provides or distributes a product or service, such as hotels, restaurants, office buildings, commercial businesses, or other places of commerce. These do not include multifamily residences, agricultural users, or customers that fall within the industrial or institutional classifications.
   b. Industrial Accounts: Any water users that are primarily manufacturers or processors of materials as defined by the Standard Industrial Classifications (SIC) code numbers 2000 through 3999 (or by the North American Industry Classification System [NAICS] code numbers 311611 through 339999).
   c. Institutional Accounts: Any water-using establishment dedicated to public service. This includes schools, courts, hospitals, and government facilities. All facilities serving these functions are to be considered institutions regardless of ownership.
2. A program to accelerate replacement of existing high-water-using toilets with ultra-low-flush (1.6 gallons [6 liters] or less) toilets in commercial, industrial, and institutional facilities.
3. Implementation of a CII water-use survey and customer incentives program.
4. Implementation of programs to achieve annual water-use savings.

23.1329 Wholesale Agency Assistance Programs

The purpose of this BMP is for the wholesaler to provide support for the water conservation programs of retail water suppliers. Important elements of this program include:

1. Financial Support
   a. Wholesale water supplier should provide financial incentives or equivalent resources, as appropriate, beneficial, and mutually agreeable, to the retail water agency customers to advance water conservation efforts and effectiveness.
   b. All BMPs implemented by retail water agency customers that can be shown to be cost-effective in terms of avoided cost of water from the wholesaler's perspective, using cost-effectiveness analysis procedures, should be supported.
2. Technical Support

   Wholesale water agencies should provide conservation-related technical support and information to all retail agencies for whom they serve as a wholesale supplier. At a minimum, this requires:

   a. Conducting, funding, or promoting workshops addressing the following topics:
      (1) Procedures for calculating program savings, costs, and cost-effectiveness.
      (2) Retail agencies' BMP implementation reporting requirements.

(3) The technical, programmatic, strategic, or other pertinent issues and developments associated with water-conservation activities in each of the following areas: ultra-low-flush toilet (ULFT) replacement; residential retrofits; commercial, industrial, and institutional surveys; residential and large turf irrigation; and conservation-related rates and pricing.

b. Having the necessary staff or equivalent resources available to respond to retail agencies' technical and programmatic questions involving the BMPs and their associated reporting requirements.

3. Program Management

   Wholesale and retail agencies should retain maximum local flexibility in designing and implementing locally cost-effective BMP conservation programs. Cooperatively designed regional programs should be encouraged. When mutually agreeable and beneficial, the wholesaler may operate all or any part of the conservation-related activities that a given retail supplier is obligated to implement under the BMP's cost-effectiveness test.

4. Water Shortage Allocations

   Wholesale agencies should work in cooperation with their customers to identify and remove potential disincentives to long-term conservation created by water shortage allocation policies and to identify opportunities to encourage and reward cost-effective investments in long-term conservation shown to advance regional water supply reliability and sufficiency.

23.13210 Conservation Pricing

Conservation pricing includes uniform or increasing block rate structure, volume-related sewer charges, and service cost recovery. Implementation methods should be at least as effective as eliminating nonconserving pricing and adopting conserving pricing. For agencies supplying both water and sewer service, this BMP applies to pricing of both water and sewer service. Agencies that supply water but not sewer service should make good-faith efforts to work with sewer agencies so that those sewer agencies adopt conservation pricing for sewer service.

1. Nonconserving pricing provides no incentives to customers to reduce use. Such pricing is characterized by one or more of the following components: rates in which the unit price decreases as the quantity used increases (declining block rates); rates that involve charging customers a fixed amount per billing cycle regardless of the quantity used; pricing in which the typical bill is determined by high fixed charges and low commodity (water-use) charges.
2. Conservation pricing provides incentives to customers to reduce average or peak use, or both. Such pricing includes rates designed to cover the cost of providing service and billing for water and sewer service based on metered water use. Conservation pricing is also characterized by one or more of the following components: rates in which the unit rate is constant regardless of the quantity used (uniform rates) or increases as the quantity used increases (increasing block rates); seasonal rates or excess-use surcharges to reduce peak demands during summer months; or rates based upon the long-run marginal cost or the cost of adding the next unit of capacity to the system.
3. Adoption of lifeline rates for low-income customers should neither qualify nor disqualify a rate structure as meeting the requirements of this BMP.

23.13211 Conservation Coordinator

This BMP consists of the designation of a staff coordinator for the agency conservation programs. This BMP includes the following actions:

1. Designation of a water conservation coordinator and support staff (if necessary), whose duties should include:
   a. Coordination and oversight of conservation programs and BMP implementation.
   b. Preparation and submittal of the BMP implementation report.
   c. Communication and promotion of water conservation issues to agency senior management; coordination of agency conservation programs with operations and planning staff; preparation of annual conservation budget; and preparation of the conservation elements of the agency's urban water management plan.
2. Agencies jointly operating regional conservation programs are not expected to staff duplicative and redundant conservation coordinator positions.

23.13212 Water Waste Prohibition

The purpose of this BMP is to enforce prohibition of wasteful use of water. Elements of this BMP include the enacting and enforcing of measures prohibiting gutter flooding, single-pass cooling systems in new connections, nonrecirculating systems in all new conveyer car wash and commercial laundry systems, and nonrecycling decorative water fountains.

Water agencies should support efforts to develop laws regarding exchange-type water softeners that would allow the sale of only more efficient, demand-initiated regenerating (DIR) models; develop minimum appliance efficiency standards that increase the regeneration efficiency standard to at least 3,350 grains of hardness removed per pound (477 grams/kg) of common salt used and implement an identified maximum number of gallons discharged per gallon of soft water produced; and allow local agencies, including municipalities and special districts, to set more stringent standards or to ban on-site regeneration of water softeners if it is demonstrated and found by the agency governing board that there is an adverse effect on the reclaimed water or groundwater supply.

Water agencies should also include water softener checks in home water audit programs and include information about DIR and exchange-type water softeners in their educational efforts to encourage replacement of less efficient timer models.

23.13213 Residential ULFT Replacement Programs

This BMP encourages programs that promote the replacement of high-water-using toilets with ultra-low-flow toilets (ULFTs). Important elements of this BMP include:

1. Implementation of programs for replacing existing high-water-using toilets with ultra-low-flush (1.6 gallons [6 liters] or less) toilets in single-family and multifamily residences.
2. Programs should be at least as effective as those programs requiring toilet replacement at time of resale with program effectiveness determined using a methodology for calculating water savings.

23.13214 Potential Best Management Practices

This section lists other potential best management practices that a water utility might wish to consider when developing and implementing a water conservation program.

- Graywater use
- Efficiency standards for water-using appliances and irrigation devices
- Retrofit of existing car washes
- Distribution system pressure regulation
- Swimming pool and spa conservation including covers to reduce evaporation
- Point-of-use water heaters, recirculating hot water systems, and hot water pipe insulation
- Efficiency standards for new industrial and commercial processes

23.133 EPA's WaterSense: Efficiency Made Easy

The US EPA has implemented a water efficiency program to educate consumers about making smart water choices that save money and maintain high environmental standards. Manufacturers can use the EPA Water-Sense label on products and services that perform at least 20 percent more efficiently than less efficient counterparts. To learn more about EPA's WaterSense program, visit the website at epa.gov.

23.134 Acknowledgment

Information presented in this section on water and energy conservation was obtained from the many excellent publications of the California Urban Water Conservation Council. Permission to use this information is greatly appreciated.

23.135 Additional Reading

1. *Water Conservation Programs—A Planning Manual* (M52). Obtain from American Water Works Association (AWWA) at www.awwa.org or phone (800) 926-7337. Order No. 30052. ISBN 978-1-58321-391-9.
2. *Water Audits and Loss Control Programs* (M36). Obtain from American Water Works Association (AWWA) at www.awwa.org or phone (800) 926-7337. Order No. 30036. ISBN 978-1-58321-631-6.

Many valuable publications are available on water conservation. For additional reading on this subject, contact:

- California Urban Water Conservation Council (CUWCC) at www.cuwcc.org or phone (916) 552-5885.
- American Water Works Association (AWWA) at www.awwa.org or phone (800) 926-7337.

QUESTIONS

Please write your answers to the following questions and compare them with those on page 702.

23.13A What are the different perspectives that consumers have regarding water conservation?

23.13B What are the key elements of a residential water survey BMP?

23.13C What is a major element of a large landscape water conservation program BMP?

23.13D What are the important elements of wholesale agency assistance programs?

23.14 ACKNOWLEDGMENTS

We appreciate the contributions made during the revision of this chapter by Lynn Scarpa, Phil Scott, Chris Smith, and Rich von Langen, all members of the California Water Environment Association.

23.15 ADDITIONAL READING

1. *Manage for Success*, especially Chapter 11, "Emergency Planning." Obtain from the Office of Water Programs, California State University, Sacramento, at www.owp.csus.edu or phone (916) 278-6142.
2. *Water Utility Management* (M5). Obtain from American Water Works Association (AWWA) at www.awwa.org or phone (800) 926-7337. Order No. 30005. ISBN 978-1-58321-361-2.
3. *A Water and Wastewater Manager's Guide for Staying Financially Healthy*, US Environmental Protection Agency. EPA No. 430-9-89-004. This EPA document is available online at epa.gov/nscep/.

4. *Wastewater Utility Recordkeeping, Reporting and Management Information Systems*, US Environmental Protection Agency. EPA No. 430-9-82-006. This EPA document is available online at epa.gov/nscep/.

5. *Supervision: Concepts and Practices of Management*, 12th Edition, Edwin Leonard, Jr. Obtain from Cengage Learning at www.cengage.com. ISBN 978-1-111-96979-0.

6. *Texas Manual*, chapters on "Effective Public Relations in Water Works Operations" and "Planning and Financing."

7. *Principles of Water Rates, Fees, and Charges* (M1). Obtain from American Water Works Association (AWWA) at www.awwa.org or phone (800) 926-7337. Order No. 30001. ISBN 978-1-58321-863-1.

8. *Emergency Planning for Water Utilities* (M19). Obtain from American Water Works Association (AWWA) at www.awwa.org or phone (800) 926-7337. Order No. 30019. ISBN 978-1-58321-135-9.

9. *GIS Implementation for Water and Wastewater Treatment Facilities* (MOP 26). Obtain from Water Environment Federation (WEF) at wef.org or phone (800) 666-0206. Order No. WPM401. ISBN 978-0-07-145305-9.

10. To learn more about homeland defense, visit EPA's website at water.epa.gov/infrastructure/watersecurity/index.cfm.

END OF LESSON 3 OF 3 LESSONS

on

ADMINISTRATION

Please answer the discussion and review questions next.

DISCUSSION AND REVIEW QUESTIONS

Chapter 23. ADMINISTRATION

(Lesson 3 of 3 Lessons)

Please write your answers to the following questions to determine how well you understand the material in the lesson. The question numbering continues from Lesson 2.

26. What can happen when agencies rely primarily on corrective maintenance to keep the system running?
27. What does a SCADA system do?
28. What items should be included in a water supplier's cross-connection control program?
29. What factors should be considered when selecting a backflow prevention device for a particular situation?
30. How would you assess the vulnerability of a water supply system?
31. How can a utility protect its water supply from contamination?
32. What would you do if you discovered that contaminated water has entered your distribution system?
33. Why must waterworks utilities have a safety program?
34. How can a good safety record be accomplished?
35. What is the intent of the OSHA regulations?
36. What are the four main elements of a hazard communication program?
37. What topics should be included in a safety officer's talk to new operators?
38. Why are records important?
39. Why do water utilities need an effective water and energy conservation program?
40. What is the most important effect of water conservation pricing?

SUGGESTED ANSWERS

Chapter 23. ADMINISTRATION

ANSWERS TO QUESTIONS IN LESSON 1

Answers to questions on page 619.

23.0A Local community demands on a utility manager include protection from environmental disasters with a minimum investment of money.

23.0B Changes in the environmental workplace are created by changes in the workforce and advances in technology.

Answers to questions on page 621.

23.1A The functions of a utility manager include planning, organizing, staffing, directing, and controlling.

23.1B In small communities, the community depends on the manager to handle everything.

Answers to questions on page 624.

23.2A Utility planning must include operational personnel, local officials (decision makers), and the public.

23.3A The purpose of an organizational plan is to show who reports to whom and to identify the lines of authority.

23.3B Effective delegation is uncomfortable for many managers since it requires giving up power and responsibility. Many managers believe that they can do the job better than others, they believe that other employees are not well trained or experienced, and they are afraid of mistakes. The utility manager retains some responsibility even after delegating to another employee and, therefore, the manager is often reluctant to delegate or may delegate the responsibility but not the authority to get the job done.

23.3C An important and often overlooked part of delegation is follow-up by the supervisor.

Answers to questions on page 626.

23.4A Staffing responsibilities include hiring new employees, training employees, and evaluating job performance.

23.4B The two personnel management concepts a manager should always keep in mind are "job-related" and "documentation."

23.4C The steps involved in a staffing analysis include:

1. List the tasks to be performed.
2. Estimate the number of staff hours per year required to perform each task.
3. List the utility's current employees.
4. Assign tasks based on each employee's skills and abilities.
5. Adjust the work assignments as necessary to achieve the best possible fit between the work to be done and the personnel/skills available to do it.

23.4D A qualifications profile is a clear statement of the knowledge, skills, and abilities a person must possess to perform the essential job duties of a particular position.

Answers to questions on page 629.

23.4E The purpose of a job interview is to gain additional information about the applicants so that the most qualified person can be selected to fill a job opening.

23.4F Protected groups include minorities, women, disabled persons, persons over 40 years of age, and union members.

23.4G The OUCH principle says that when you hire new employees or delegate job tasks to current employees, you must be Objective, Uniform in your treatment of applicants or employees, Consistent with utility policies, and Have job relatedness. If you do not manage with all of these characteristics, you may find yourself in a questionable position with regard to the treatment of protected workers and other employees in general.

23.4H A new employee's safety training should begin on the first day of employment or as soon thereafter as possible.

Answers to questions on page 631.

23.4I The purpose of a probationary period for new employees is to provide a time during which both the employer and employee can assess the fit between the job and the person.

23.4J The compensation an employee receives for the work performed includes satisfaction, recognition, security, appropriate pay, and benefits.

23.4K Utility managers should provide training opportunities for employees so they can keep informed of new technologies and regulations. Training for supervisors is also important to ensure that supervisors have the knowledge, skills, and attitude that will enable them to be effective supervisors.

Answers to questions on page 634.

23.4L An employee's immediate supervisor should conduct the employee's performance evaluation.

23.4M Dealing with employee discipline requires tact and skill. The manager or supervisor should stay flexible, calm, and open-minded.

23.4N Common warning signs of potential violence include abusive language, threatening or confrontational behavior, and brandishing a weapon.

Answers to questions on page 639.

23.4O Harassment is any behavior that is offensive, annoying, or humiliating to an individual and that interferes with a person's ability to do a job. This behavior is uninvited, often repeated, and creates an uncomfortable or even hostile environment in the workplace.

23.4P Types of behavior that could be considered sexual harassment include:

- Unwanted hugging, patting, kissing, brushing up against someone's body, or other inappropriate sexual touching
- Subtle or open pressure for sexual activity
- Persistent sexually explicit or sexist statements, jokes, or stories
- Repeated leering or staring at a person's body
- Suggestive or obscene notes or phone calls
- Display of sexually explicit pictures or cartoons

23.4Q The best way to prevent harassment is to set an example by your own behavior and to keep communication open between employees. A manager must also be aware of and take action to prevent any type of harassment in the workplace.

23.4R In general, the ADA defines disability as a physical or mental impairment that substantially limits one or more of the major life activities of an individual.

Answers to questions on page 640.

23.4S The shop steward is elected by the union employees and is their official representative to management and the local union.

23.4T Union contracts do not change the supervisor's delegated authority or responsibility. Operators must carry out the supervisor's orders and get the work done properly, safely, and within a reasonable amount of time. However, a contract gives a union the right to protest or challenge a supervisor's decision.

ANSWERS TO QUESTIONS IN LESSON 2

Answers to questions on page 643.

23.5A A manager needs both written and oral communication skills.

23.5B The most common written documents that a utility manager must write include memos, business letters, press releases, résumés, monitoring reports, and the annual report.

23.5C The annual report should be a review of what and how the utility operated during the past year and the goals for the next year.

Answers to questions on page 644.

23.6A A utility manager may be asked to conduct meetings with employees, the governing board, the public, and with other professionals in your field.

23.6B Before a meeting, prepare and distribute an agenda, find an adequate meeting room, and set a beginning and ending time.

Answers to questions on page 646.

23.7A The first step in organizing a public relations campaign is to establish objectives so you will have a clear idea of what you expect to achieve.

23.7B Employees can be informed about the utility's plans, practices, and goals through newsletters, bulletin boards, and regular, open communication between supervisors and subordinates.

23.7C Newspapers give more thorough, in-depth coverage to stories than do the broadcast media.

23.7D Practice is the key to effective public speaking.

23.7E Complaints can be a valuable asset in determining consumer acceptance and pinpointing water quality problems. Customer calls are frequently the first indication that something may be wrong. Responding to complaints and inquiries promptly can save the utility money and staff resources, and minimize customer inconvenience.

Answers to questions on page 650.

23.8A The three main areas of financial management for a utility include providing financial stability for the utility, careful budgeting, and providing capital improvement funds for future utility expansion.

23.8B The operating ratio for a utility is calculated by dividing total revenues by total operating expenses.

23.8C It is important for a manager to get input from other personnel in the utility as well as community leaders as the budgeting process proceeds in order to gain support for the budget and to keep the budget on track once adopted.

23.8D The basic ways for a utility to finance capital improvements are through general obligation bonds, revenue bonds, or loan funding programs.

23.8E A revenue bond is commonly used to fund utility improvements. This bond has no limit on the amount of funds available and the user charges provide repayment on the bond. To qualify for these bonds, the utility must show sound financial management and the ability to repay the bond.

ANSWERS TO QUESTIONS IN LESSON 3

Answers to questions on page 653.

23.9A The purpose of an O&M program is to maintain design functionality or to restore the system components to their original condition and thus functionality, that is, to ensure that the system performs as designed and intended.

23.9B Commonly accepted types of maintenance include corrective maintenance, preventive maintenance, and predictive maintenance.

23.9C The major elements of a good preventive maintenance program include:

1. Planning and scheduling
2. Records management
3. Spare parts management
4. Cost and budget control
5. Emergency repair procedures
6. Training program

Answers to questions on page 654.

23.9D SCADA stands for Supervisory Control And Data Acquisition system.

23.9E A SCADA system collects, stores, and analyzes information about all aspects of O&M, transmits alarm signals, when necessary, and allows fingertip control of alarms, equipment, and processes.

Answers to questions on page 655.

23.9F SCADA systems are used by operators at water treatment plants to monitor filtration system influent and effluent turbidity levels, head losses through the filters, filter flows, and filter valve settings.

23.9G When a customer phones and complains about a loss of water pressure, the operator can review system pressures to determine if the problem is in the system or in the customer's home.

23.9H The greatest challenges for operators using SCADA systems are to realize that computers may not be correct and to have the ability to operate when the SCADA system fails.

Answers to questions on page 659.

23.9I Contamination through cross-connections has consistently caused more waterborne disease outbreaks in the United States than any other reported factor.

23.9J Responsibilities for the implementation of cross-connection control programs are shared by water suppliers, water users, health agencies, and plumbing officials.

23.9K The water user's cross-connection control responsibility is to keep contaminants out of the potable water system on the user's premises. The user also has the responsibility to prevent the creation of cross-connections through modifications of the plumbing system on the premises.

Answers to questions on page 661.

23.9L The particular type of backflow prevention device most suitable for a given situation depends on the degree of health hazard, the probability of backflow occurring, the complexity of the piping on the premises, and the probability of the piping being modified. The higher the assessed risk due to these factors, the more reliable and positive the type of device needed.

23.9M Air gap separation or reduced pressure principle backflow prevention devices are installed at wastewater treatment plants; wastewater pumping stations; reclaimed water reuse areas; areas where toxic substances in toxic concentrations are handled under pressure; premises having an auxiliary water supply that is or may be contaminated; and in locations where fertilizer, herbicides, or pesticides are injected into a sprinkler system.

23.9N The power of a geographic information system (GIS) is that information can be retrieved geographically. An operator can easily look at or print out a specific area of a map. The map will contain an inventory of the distribution system pipes and structures within the selected area and it will provide detailed information about each of the pipes and structures (or entities).

Answers to questions on page 663.

23.10A The first step toward an effective contingency plan for emergencies is to make an assessment of vulnerability. Then, a comprehensive plan of action can be developed and implemented.

23.10B An emergency operations plan need not be too detailed, since all types of emergencies cannot be anticipated and a complex response program can be more confusing than helpful.

23.10C The following outline can be used as the basis for developing an emergency operations plan:

1. Make a vulnerability assessment.
2. Inventory organizational personnel.
3. Provide for a recovery operation (plan).
4. Provide training programs for operators in carrying out the plan.
5. Coordinate with local and regional agencies such as the health, police, and fire departments to develop procedures for carrying out the plan.
6. Establish a communications procedure.
7. Provide protection for personnel, plant equipment, records, and maps.

Answers to questions on page 666.

23.10D Toxicity is the ability of a contaminant (chemical or biological) to cause injury when introduced into the body.

23.10E The degree of toxicity varies with the concentration of a contaminant required to cause injury, the speed with which the injury takes place, and the severity of the injury.

23.10F Possible secondary effects in a water supply that may allow detection of a contaminant without specific testing include taste, color, odor, and chlorine demand.

Answers to questions on page 668.

23.10G *Cryptosporidium* is a significant public health concern because sensitive populations may become seriously ill even when the drinking water meets or exceeds all current state and federal standards.

23.10H The most serious threat of *Cryptosporidium* to a public water supply occurs during periods of heavy rains or snow melts that flush areas that are sources of high concentrations of oocysts into waters upstream of a plant intake.

23.10I A water treatment plant can be operated to protect the public from *Cryptosporidium* by optimizing the treatment processes. The operator must continually verify optimum coagulant doses, be sure filtered water turbidity levels are low, and provide adequate disinfection chemicals and contact times.

Answers to questions on page 670.

23.11A A safety officer should evaluate every accident, offer recommendations, and keep and apply statistics.

23.11B The supervisors should be responsible for the implementation of a safety program.

23.11C Both state and federal regulatory agencies enforce the OSHA requirements.

23.11D The utility manager is responsible for the safety of the agency's personnel and the public exposed to the water utility's operations. Therefore, the manager must develop and administer an effective safety program and must provide new employee safety training as well as ongoing training for all employees.

23.11E A safety policy statement should do the following:

1. Define the goals and objectives of the program.
2. Identify the persons responsible for each element of the program.
3. Affirm management's intent to enforce safety regulations.
4. Describe the disciplinary actions that will be taken to enforce safe work practices.

Answers to questions on page 671.

23.11F A supervisor may be responsible, in part or completely, for an accident by causing unsafe acts to take place, by requiring that work be performed in haste, by disregarding an unsafe environment of the workplace, or by failing to consider any number of safety hazards.

23.11G Each operator must accept, at least in part, responsibility for fellow operators, for the utility's equipment, for the operator's own welfare, and even for seeing that the supervisor complies with established safety regulations.

23.11H First aid means emergency treatment for injury or sudden illness, before regular medical treatment is available.

23.11I First-aid training is most important for operators who regularly work with electrical equipment and those who must handle chlorine and other dangerous chemicals.

Answers to questions on page 682.

23.11J The four basic elements of a hazard communication program are:

1. Identify hazardous materials.
2. Obtain chemical information and define hazardous conditions.
3. Properly label hazards.
4. Train operators.

23.11K The general categories of hazardous materials are:

1. Corrosives
2. Toxics
3. Flammables and explosives
4. Asphyxiants
5. Harmful physical agents
6. Infectious agents

23.11L The purpose of the SDS is to have a comprehensive, readily available reference document that contains critical information on substance or mixture identification and supplier, hazard(s) identification, composition/information on ingredients, first-aid measures, firefighting measures, accidental release measures, handling and storage, and other information about every hazardous substance.

23.11M There are nine pictograms under the Globally Harmonized System of Classification and Labeling of Chemicals (GHS) to convey the health, physical, and environmental hazards. OSHA's final Hazard Communication Standard (HCS) requires eight of these pictograms, the exception being the environmental pictogram because environmental hazards are not within OSHA's jurisdiction.

23.11N A utility manager may or may not be involved in the day-to-day details of enforcing the agency's confined space policy and procedures. However, every utility manager should be aware of the current OSHA requirements and should ensure that the utility's policies not only comply with current regulations, but that the agency's policies are vigorously enforced for the safety of all operators.

Answers to questions on page 686.

23.11O The mainstay of a safety program is the method of reporting and keeping statistics.

23.11P All injuries should be reported, even if they are minor in nature, so as to establish a record in case the injury develops into a serious injury. It may be difficult at a later date to prove whether the accident occurred on or off the job and this information may determine who is responsible for the costs.

23.11Q A safety officer should review an accident report form to make recommendations as well as to follow up to be sure that proper action has been taken to prevent recurrence.

23.11R A new, inexperienced operator must receive instruction on all aspects of plant safety. This training includes instruction in the handling of chemicals, the dangers of electrical apparatus, fire hazards, and proper maintenance of equipment to prevent accidents. Special instructions are required for specific work environments such as manholes, gases (chlorine and hydrogen sulfide [H_2S]), water safety, and any specific hazards that are unique to your facility. All new operators should be required to participate in a safety orientation program during the first few days of employment, and an overall training program in the first few months.

23.11S If an operator is unsure of how to perform a job, then it is the operator's responsibility to ask for the training needed.

Answers to questions on page 688.

23.11T Statistical accident reports should contain accident statistics showing lost time, costs, type of injuries, and other data, based on some time interval.

23.11U Injuries can be classified as fractures, burns, bites, eye injuries, cuts, and bruises.

23.11V Causes of injuries can be classified as heat, machinery, falls, handling objects, chemicals, unsafe acts, and miscellaneous.

23.11W Costs of accidents can be classified as lost time, lost dollars, lost production, contaminated water, or any other means of showing the effects of the accidents.

Answers to questions on page 691.

23.12A Some of the important uses of records include:

1. Aiding operators in solving treatment and water quality problems
2. Providing a method of alerting operators to changes in source water quality
3. Showing that the treated water is acceptable to the consumer
4. Documenting that the final product meets plant performance standards, as well as the standards of the regulatory agencies
5. Determining performance of treatment processes, equipment, and the plant
6. Satisfying legal requirements
7. Aiding in answering complaints
8. Anticipating routine maintenance
9. Providing data for cost analysis and preparation of budgets
10. Providing data for future engineering designs
11. Providing information for monthly and annual reports

23.12B "Unaccounted-for water" is the difference between the amount of treated water that enters the distribution system and water delivered to consumers.

23.12C Chemical inventory and usage records that should be kept include chemical inventory/storage (measured use and deliveries); metered or estimated plant usages; and calculated usage of chemicals (compare with actual use).

23.12D Important items usually contained on a purchase order include the date, a complete description of each item and quantity needed, prices, the name of the vendor, and a purchase order number.

Answers to questions on page 696.

23.13A The different perspectives that consumers have regarding water conservation include preservation, ecological, hydrological, traditional economics, and resource economics.

23.13B The key elements of a residential water survey BMP include indoor and outdoor audits of residential water use and distribution of water-saving devices.

23.13C A major element of a large landscape water conservation program BMP is evapotranspiration-based (ETo-based) water-use budgets for large landscape irrigators.

23.13D The important elements of wholesale agency assistance programs include financial support, technical support, program management, and water shortage allocations.

APPENDIX

WATER TREATMENT PLANT OPERATION

(VOLUME II)

Comprehensive Review Questions and Suggested Answers

How To Solve Water Treatment Plant Arithmetic Problems

Abbreviations

Water Words

Subject Index

COMPREHENSIVE REVIEW QUESTIONS

VOLUME II

This section was prepared to help you review the material in Volume II for civil service and certification examinations. The questions are divided into five types:

1. True-False
2. Best Answer
3. Multiple Choice
4. Short Answer
5. Problems

Please *do not send* your answers to California State University, Sacramento.

True-False

1. Both iron and manganese are essential to the growth of many plants and animals, including humans.
   1. True
   2. False
2. Only one cell of iron bacteria is needed to start an infestation of iron bacteria in a well or distribution system.
   1. True
   2. False
3. pH is decreased by the removal of carbon dioxide.
   1. True
   2. False
4. Fluoride saturators are a special application of a solution feeder.
   1. True
   2. False
5. Fluoride solutions are extremely corrosive.
   1. True
   2. False
6. Excessively hard water can cause scaling on the inside of pipes and thereby restrict flow.
   1. True
   2. False
7. The stability of treated water is determined by measuring the pH and calculating the Langelier Index.
   1. True
   2. False
8. The addition of acid to prevent calcium carbonate scale will accomplish the same things as recarbonation and the addition of polyphosphate.
   1. True
   2. False
9. The feed rates determined by jar tests always produce the exact same results in the actual plant.
   1. True
   2. False
10. The effluent from an ion exchange unit usually will have zero hardness until the unit needs regeneration.
    1. True
    2. False
11. The overwhelming bulk of THM precursors in water are from industrial organic chemicals.
    1. True
    2. False
12. When collecting water samples for THM analysis, be sure to collect the sample before the water temperature becomes constant.
    1. True
    2. False
13. The higher the pH of the water, the slower the production of THMs.
    1. True
    2. False
14. Low levels of arsenic consumed over a long period of time can cause cancer and noncancerous toxic effects on humans.
    1. True
    2. False
15. Electrical conductivity (EC) can usually be correlated directly with arsenic reduction.
    1. True
    2. False
16. All waters contain various amounts of total dissolved solids (TDS), including fresh water.
    1. True
    2. False

17. The rate of hydrolysis is accelerated by decreased temperature.

    1. True
    2. False

18. The output of a positive-displacement pump may be throttled.

    1. True
    2. False

19. At a minimum, the rate of membrane hydrolysis is at a pH of about 4.7, and it increases with both increasing and decreasing pH.

    1. True
    2. False

20. Trash and grit may be disposed of in landfills.

    1. True
    2. False

21. Before filling a drained tank after removing sludge, thoroughly disinfect the tank.

    1. True
    2. False

22. The application of nitrogen fertilizers causes an increase in soil pH.

    1. True
    2. False

23. One person may be permitted to attempt an emergency repair alone.

    1. True
    2. False

24. Operators must be familiar with equipment and understand what it is intended to do before developing a preventive maintenance program and maintaining equipment.

    1. True
    2. False

25. Check the temperature of a centrifugal pump bearing temperature by hand.

    1. True
    2. False

26. Always be sure a centrifugal pump is primed before starting the pump.

    1. True
    2. False

27. Diesel engines can be air or water cooled.

    1. True
    2. False

28. Never pour water on a chlorine leak because this will only create a bigger problem by enlarging the leak.

    1. True
    2. False

29. Analog instruments provide a direct numerical readout.

    1. True
    2. False

30. Long detection lags can result in stable processes and good control.

    1. True
    2. False

31. Entry into confined spaces is never permitted until the space has been properly ventilated using specially designed forced-air ventilators.

    1. True
    2. False

32. Pneumatic instrumentation depends upon a constant source of clean, dry, pressurized air for reliable operation.

    1. True
    2. False

33. The fumes from hydrofluosilicic acid can corrode metal and etch glass on protective equipment.

    1. True
    2. False

34. In reactions with most metals, hydrochloric acid will produce hydrogen gas.

    1. True
    2. False

35. Anhydrous ammonia (gas) may be stored in the same room with chlorine (gas).

    1. True
    2. False

36. When working with chlorine, be equipped to control chlorine leaks, which are most often found in the control valve.

    1. True
    2. False

37. Ferric chloride is a very corrosive compound and should be treated as you would treat any acid.

    1. True
    2. False

38. The most common form of iron in solution in groundwater and in water under anaerobic conditions (bottom of a lake or reservoir) is ferric ion.

    1. True
    2. False

39. Nitrate represents the least completely oxidized form of nitrogen found in water.
    1. True
    2. False
40. Chloroform is the most common THM found in drinking water.
    1. True
    2. False
41. EPA regulations change frequently as new information becomes available.
    1. True
    2. False
42. MCLs apply whether the contaminant is from naturally occurring sources or from man-made pollution.
    1. True
    2. False
43. The health risk from mercury is less from mercury in fish than simply from waterborne mercury.
    1. True
    2. False
44. The presence of coliform bacteria indicates that other harmful organisms may be present in the water.
    1. True
    2. False
45. Tier 1 violations are any violations or situations in which even a brief violation or short-term exposure to a contaminant has the potential to cause a serious, negative effect on human health.
    1. True
    2. False
46. No discipline problem ever solves itself and the sooner you deal with the problem, the better will be the outcome.
    1. True
    2. False
47. Written communication is less demanding than oral communication and requires less careful preparation.
    1. True
    2. False
48. Customer calls are frequently your first indication that something may be wrong.
    1. True
    2. False
49. Corrective maintenance, including emergency maintenance, is considered reactive maintenance.
    1. True
    2. False
50. The SCADA system allows operators to enhance the efficiency of their water facility by keeping them fully informed and fully in control.
    1. True
    2. False
51. The operator should always call to the supervisor's attention unsafe conditions, environment, and equipment or other concerns about the work they are performing.
    1. True
    2. False
52. On-the-Job Training (OJT) is a good way of preventing accidents with an inexperienced operator.
    1. True
    2. False

Best Answer (Select only the closest or best answer.)

1. How can the growth of iron bacteria be controlled?
    1. By caustic soda
    2. By chlorination
    3. By permanganate
    4. By sulfuric acid
2. What is the best way to determine if there is an iron and manganese problem in a water supply?
    1. Collect some samples of water and analyze them for iron and manganese
    2. Consult with the local health authorities
    3. Interview concerned homeowners
    4. Look for stained plumbing fixtures in a couple of houses
3. What should an operator do when operating an ion exchange unit and iron and manganese start to appear in the treated water?
    1. Inform consumers of the presence of iron and manganese
    2. Keep the unit operating until breakthrough becomes excessive
    3. Notify the regulatory agency or the health department
    4. Regenerate with a brine solution
4. What happens if children drink a recommended dose of fluoride?
    1. The children's bones will become stronger
    2. The children's growth will be stunted
    3. The children will have fewer dental caries (decay or cavities)
    4. The children will perform better in school
5. When working with fluoridation systems using a sodium fluoride solution, why is the hardness of the water very important?
    1. Corrosion can be caused by hardness
    2. Pump cavitation is a problem when hardness is present
    3. Scaling can be caused by hardness
    4. Water hammer problems are caused by hardness

6. Water hardness is a measure of which item?
   1. The corrosivity of water
   2. The density of water
   3. The penetration power of water
   4. The soap- or detergent-consuming power of water

7. Why should bagged quicklime never be stored close to combustible materials?
   1. Because combustible materials can use lime for fuel when burning
   2. Because considerable heat will be generated if the lime accidentally gets wet
   3. Because heat from a fire could melt the lime
   4. Because lime easily catches on fire

8. How can an operator prevent scale formation on the filter sand, distribution mains, and household plumbing?
   1. By converting the excess caustic and unprecipitated carbonate ions to soluble forms
   2. By implementing a flushing program
   3. By maintaining high water velocities throughout the system
   4. By regular backwashing

9. Why are restrictions placed on the locations where THM samples are collected?
   1. To allow easy access for sample containers
   2. To avoid the possibility of contaminating samples or losing THMs
   3. To protect homeowners from intrusive sample collectors
   4. To provide operators with safe access for sample collection

10. Why is the reducing agent left out of some THM sample bottles?
    1. To avoid introducing THM precursors into the sample
    2. To determine the maximum concentration of TTHMs that can form over an extended period of time
    3. To estimate the hourly rate of TTHM formation
    4. To prevent the formation of TTHMs until the sample is delivered to the lab

11. What is a major advantage of removing precursors before chlorine is added?
    1. Allows the continued use of free chlorine as a disinfectant
    2. Facilitates the ultimate disposal of the precursors
    3. Is the option preferred by most water consumers
    4. Promotes the use of chloramination

12. Why are arsenic split-sample analyses performed by a certified laboratory?
    1. To compare the efficiencies of field test kits
    2. To confirm results from arsenic test kit analyses
    3. To evaluate the time required to perform the analyses
    4. To test the operator's ability to accurately split samples

13. What is the most common pretreatment process to remove turbidity and suspended solids?
    1. Chlorination
    2. Coagulation/flocculation
    3. Filtration
    4. Sedimentation

14. What can be applied to inhibit or retard the growth of bacterial slimes and molds on membrane surfaces?
    1. Acetone
    2. Chlorine
    3. Soda ash
    4. Sulfuric acid

15. Why is acid addition a part of the pretreatment of the feedwater supply to prepare for electrodialysis demineralization?
    1. To prevent carbonate scaling
    2. To prevent membrane compaction
    3. To prevent membrane corrosion
    4. To prevent polarization

16. What is the key to speedy sludge drying in a sludge drying bed?
    1. The regular application of a thick layer of sludge
    2. The regular application of a thin layer of sludge
    3. The regular removal of the dried portion of the sludge
    4. The regular removal of the water on top of the sludge

17. What is the principal advantage of using centrifuges for sludge dewatering?
    1. The density of the sludge cake can be varied
    2. The erosion of surfaces hit by high-speed particles
    3. The low energy consumption
    4. The poor performance capacity

18. What is an ampere?
    1. The electrical pressure available to cause a flow of current (amperage) when an electric circuit is closed
    2. The level of shock provided by a circuit
    3. The measure of the number of phases provided by the current
    4. The unit of electric current or an indication of how hard the electricity is working

19. How can an operator determine if a circuit breaker has been tripped?
    1. The handle will be at the mid position between on and off
    2. The lockout/tagout device is secure
    3. The overload relay is overheated
    4. The strip or wire of fusible metal is melted

20. Why should bearing grease be purchased in cartridge form?
    1. To facilitate applying grease
    2. To minimize the chance of getting dirt into the lubricant
    3. To optimize grease storage space
    4. To reduce the costs of purchasing grease

21. How can an operator determine if a pump is primed?
    1. The pump is in good shape
    2. The pump is ready to start
    3. The pump must be completely filled with water
    4. The pump must be properly aligned
22. What does a compressor unloader allow the compressor to do?
    1. Control compressed air
    2. Discharge used oils and grease
    3. Release stored air pressure
    4. Start under a no-load condition
23. What is the purpose of cathodic protection?
    1. To prevent operators from being exposed to toxic gases
    2. To prevent the formation of slimes on walkways
    3. To prevent the rusting of base metal surfaces
    4. To prevent the scaling and deposition of chemicals on metal surfaces
24. What is accuracy?
    1. A physical or chemical quantity which is measured and/or controlled in a wastewater treatment process
    2. How closely an instrument measures the true or actual value of a process variable
    3. How closely the device can reproduce a given reading (or measurement) time after time
    4. The smallest change in a process variable able to produce an appropriate response from an instrument
25. What is a very precise method of measuring fluctuating water levels?
    1. Bubbler tube
    2. Displacer
    3. Electrical probe
    4. Pump pressure gauge
26. What is a "test" button on an annunciator panel used for?
    1. To confirm that no alarm lamps have burned out
    2. To initiate an alarm condition at each alarm site
    3. To perform alarm drills throughout the plant
    4. To verify the noise level of each alarm
27. What is the most useful piece of general electrical test equipment?
    1. Battery charger tester
    2. Ground fault tester
    3. Live circuit tester
    4. Multimeter tester
28. How can bases be neutralized?
    1. With ammonia
    2. With deionized water
    3. With dilute acids
    4. With strong acids
29. Why is carbon dioxide gas considered hazardous?
    1. Can cause acidic types of skin burns
    2. Can cause cancer
    3. Can cause injury to the eyes
    4. Can cause suffocation due to the lack of oxygen
30. Why are high levels of oxygen (enrichment above 22.5 percent) dangerous?
    1. Because high levels can cause excessive exertion by operators
    2. Because high levels can confuse an operator's thinking ability
    3. Because high levels can increase the possibility of an explosion
    4. Because high levels can mask the odor of toxic gases
31. What type of protection must be used when operating power tools outdoors for protection against electric shock?
    1. Ground-fault circuit interrupter (GFCI)
    2. Insulated screwdrivers
    3. Lockout/tagout
    4. Rubber gloves
32. What is the purpose of a spectrophotometer calibration curve?
    1. To determine if the sample has exceeded its holding time
    2. To determine if the spectrophotometer needs recalibration
    3. To determine if the spectrophotometer reagents are deteriorating
    4. To determine the concentration of the water quality indicator contained in a sample
33. What causes apparent color in water?
    1. Dissolved organic substances or certain minerals
    2. Hardness ions
    3. Iron and manganese
    4. Suspended materials (including colloidal substances)
34. Specific conductance or conductivity is a numerical expression of which item?
    1. Ability of water to conduct an electric current
    2. Ease with which an electric current heats the water sample
    3. Number of total dissolved solids in a water sample
    4. Transmissibility of a sample of water in a pipeline
35. The Disinfectants/Disinfection Byproducts (D/DBP) Rule is designed to protect the public from which of the following?
    1. Chlorine
    2. Disease
    3. Disinfectant flavors and odors
    4. Harmful concentrations of disinfectants
36. What is the purpose of the Total Coliform Rule (TCR)?
    1. To avoid public exposure to disease-causing organisms associated with wastewater or animal wastes
    2. To eliminate the need to test for protozoa in drinking water
    3. To improve public health protection by reducing fecal pathogens to minimal levels through control of total coliform bacteria
    4. To prevent the risk of illness from disease-causing organisms

37. What is the purpose of the Filter Backwash Recycling Rule (FBRR)?
    1. To improve filter backwash pumping practices to conserve energy and enhance pumping efficiencies
    2. To improve public health protection by assessing and changing, where needed, recycling practices for improved contaminant control, particularly microbial contaminants
    3. To improve water filtration plant practices for disposing of filter backwash wastewaters
    4. To improve water treatment plant filtration practices to conserve, recycle, and reclaim filter backwash practices

38. What is the important difference between MCL and MCLG?
    1. MCLG is always established before the MCL
    2. MCLG is nonenforceable and MCL is an enforceable standard
    3. MCLG is preferred by regulatory agencies in lieu of the MCL
    4. MCL is always lower than the MCLG

39. What causes backflow situations in distribution systems?
    1. Defective clear well check valves
    2. Either back pressure or backsiphonage
    3. Failure of filter backwash system
    4. Malfunctioning distribution system sampling stations

40. What is the National Terrorism Advisory System?
    1. A media group with instructions to broadcast warnings of terrorist activity
    2. A network of agents throughout the world that monitors the potential for terrorist activity
    3. A system to issue formal, detailed alerts about specific or credible terrorist threats
    4. A terrorist detection system programmed to alert water utility agencies

41. What is an acute toxic agent?
    1. An agent that bioaccumlates in the system of a susceptible individual
    2. An agent that causes injury quickly
    3. An agent that causes injury to occur over an extended period of exposure
    4. An agent that is ingested by an individual

42. If contamination of the water supply is discovered, what is the immediate concern?
    1. Flushing the contaminant out of the system
    2. Protection of source water
    3. Safety of the public
    4. Source of contaminant

43. What is the purpose of the Safety Data Sheet (SDS)?
    1. To have a readily available reference document that includes complete information on the hazardous substance
    2. To protect operators from exposure to hazardous atmospheres
    3. To protect operators from exposure to hazardous substances
    4. To require manufacturers to submit safety information upon delivery of a hazardous substance

Multiple Choice (Select all correct answers.)

1. Why is the presence of iron and manganese in drinking water objectionable?
    1. Clothes laundered in the water may come out stained
    2. Iron and manganese promote the growth of iron bacteria
    3. Plumbing fixtures, bathtubs, and sinks may become stained
    4. The water is very corrosive
    5. They create slimes that cause foul flavors and odors

2. What can be added to increase the pH of water being treated with a continuous regeneration (CR) manganese greensand plant?
    1. Muriatic acid
    2. Sodium carbonate
    3. Sodium chloride
    4. Sodium hydroxide
    5. Sulfuric acid

3. How can an operator minimize red water or dirty water problems and complaints?
    1. Deliver stable (noncorrosive) water to the distribution system
    2. Ensure that water pumped to the distribution system contains little or no iron and manganese
    3. Keep the health department informed of the water quality conditions
    4. Maintain a free chlorine residual in all water throughout the distribution system
    5. Provide adequate treatment to control iron and manganese

4. What should warning system alarms at fluoridation plants indicate?
    1. Downtime measured by a time meter
    2. High or low fluoride levels
    3. Low injection water pressure
    4. Low water flow in the main pipeline
    5. Power outages

5. Which items should be checked before startup of a chemical feeder system?
   1. Be sure that safety equipment, such as eyewash, drench showers, dust masks, face shields, gloves, and vent fans, are in place and functional
   2. Check the operation of alarms and safety shutoffs
   3. Confirm that safety guards are in place
   4. Examine the operation of all auxiliary equipment, including the dust collectors, fans, cooling water, mixing water, and safety equipment
   5. Record all important nameplate data and place it in the plant files for future reference
6. How can an operator come in contact with fluoride?
   1. Eye contact
   2. Ingestion
   3. Inhalation
   4. Noise exposure
   5. Skin contact
7. What undesirable effects are caused by excessive hardness in water?
   1. Damage in some industrial processes
   2. Deposition of scale in boilers
   3. Formation of soap curds
   4. Increased use of soap
   5. Objectionable flavors
8. Which items are limitations of the ion exchange softening process?
   1. Formation of hardness scale in water mains
   2. Increase in sodium content of softened water
   3. Product water with zero hardness
   4. Ultimate disposal of rinse water
   5. Ultimate disposal of spent brine
9. Which items should an operator check when troubleshooting the operation of the backwash stage of an ion exchange softener?
   1. Check for adequate flow rates
   2. Check for excessive flow rates to prevent washing resin from the unit
   3. Check for higher than normal head loss
   4. Check the condition of the resin by visually inspecting the top layer of the bed
   5. Check the timer to make sure the unit is washed for the required length of time
10. Which treatment processes are available to remove THMs after they have been formed?
    1. Adsorption
    2. Aeration
    3. Filtration
    4. Oxidation
    5. Sedimentation
11. Which oxidation processes may remove or modify THM precursors?
    1. Chlorine dioxide
    2. Hydrogen peroxide
    3. Ozone
    4. Ozone/ultraviolet light
    5. Permanganate
12. Which Best Available Technology (BAT) processes are available for treating arsenic?
    1. Activated alumina
    2. Coagulation/filtration
    3. Ion exchange
    4. Oxidation filtration
    5. Reverse osmosis
13. Which technologies are used by POU and POE devices to remove arsenic from drinking water?
    1. Activated alumina
    2. Aeration
    3. Blending
    4. Iron-based sorbents
    5. Reverse osmosis
14. Which methods can be used to remove minerals from water?
    1. Distillation
    2. Electrodialysis
    3. Ion exchange
    4. Reverse osmosis
    5. Thawing
15. Nanofiltration membranes are used for demineralization to reject which items?
    1. Alkalinity
    2. Corrosivity
    3. Hardness
    4. TDS
    5. THM formation potential
16. Which constituents in feedwater to a reverse osmosis unit require pretreatment to prevent scaling?
    1. Carbonates
    2. Ferric iron
    3. Hydrogen sulfide
    4. Silica
    5. Sulfate
17. Which commonly used chemicals require special handling in a reverse osmosis plant operation?
    1. Acids
    2. Chlorine
    3. Citric acid
    4. Formaldehyde
    5. Sodium hexametaphosphate

18. What methods are used to dispose of sludges and brines produced at water treatment plants?
    1. Deposit in landfills (usually dewatered sludges)
    2. Spread on land
    3. Storm drains
    4. Storm sewers
    5. Wastewater collection systems (sewers)
19. When recycling drainage waters through a water treatment plant, exercising extreme caution will help avoid contributing to which conditions?
    1. Excessive solids loadings
    2. Flavor and odor problems
    3. Health hazards
    4. Operational problems
    5. Plant flow overloads
20. Which factors could cause voltage imbalance?
    1. Changing energy demands
    2. Circuit breakers
    3. High-resistance contacts
    4. Loose connections
    5. Motor starters
21. Which items could be the cause of an electric motor control problem?
    1. Moisture
    2. Overload
    3. Poor design
    4. Voltage
    5. Wear
22. Which items are possible causes of pump misalignment?
    1. Bolts loosening
    2. Change in suction conditions
    3. Foundations settling unevenly
    4. Piping changing position
    5. Water cavitation
23. Which items are types of pump operating troubles?
    1. Excessive discharge flows
    2. High power requirements
    3. Noisy pump
    4. Pump will not start
    5. Reduced rate of discharge
24. What could cause a reduced rate of discharge from a pump?
    1. Air in the water
    2. Stuck or clogged check valves
    3. Discharge head too high
    4. Clogged impeller
    5. Greater than anticipated suction lift
25. Which maintenance procedures should be performed on compressors?
    1. Clean the cylinder or casing fins weekly
    2. Drain the condensate (condensed water) from the air receiver
    3. Inspect the suction filter of the compressor regularly
    4. Lubricate the compressor regularly
    5. Test the safety valves
26. Which items are examples of modulating control systems for water treatment process variables?
    1. Chlorine residual analyzer/controller
    2. Continuous level control of process basins
    3. Flow-paced chemical feeders
    4. Pressure- or flow-regulating valves
    5. Variable-speed pumping systems for flow/level control
27. Which gases may develop explosive gas mixtures when combined with air?
    1. Gasoline vapors
    2. Hydrogen
    3. Manufactured fuel gas
    4. Methane
    5. Natural gas
28. Which items can cause electronic instrument problems?
    1. Excessive heat
    2. Static electricity
    3. Temporary power failures
    4. Tripped panel circuit breakers
    5. Voltage surges (e.g., lightning) resulting in blown fuses
29. Which items are critical preventive maintenance measures for instrumentation systems?
    1. Bleed pneumatic systems regularly as instructed
    2. Do not lubricate, "tweak," fix, calibrate, free up, or modify any component of a system arbitrarily
    3. Keep instrument components clean on the outside and closed/sealed against inside contamination
    4. Keep recorder pens and charts functioning as designed by frequent checking and service
    5. Protect all instrumentation from moisture, vibration, and mechanical shock
30. Which items must an operator know in order to safely handle chemicals?
    1. Be aware of the tendency of chemicals to form an arch in a storage bin
    2. Know how to feed the chemical
    3. Know how to make up solutions
    4. Know how to store chemicals
    5. Understand the fire problem associated with each chemical

31. Which bases used in water treatment are flammable or explosive?
    1. Ammonia
    2. Caustic soda
    3. Hypochlorite compounds
    4. Quicklime
    5. Soda ash
32. Which safety precautions should be taken when handling most salts used in water treatment plants?
    1. Eye protection
    2. Hard hat
    3. Respiratory protection
    4. Steel-toed shoes
    5. Ventilation
33. Which steps or tasks are recommended before entry into a confined space?
    1. Empty, flush, or purge the space of any harmful substance(s) to the greatest extent possible
    2. Identify and close off or reroute any lines that may convey harmful substance(s) to, or through, the work area
    3. If an atmospheric hazard is noted, use portable blowers to further ventilate the area
    4. Monitor the atmosphere at the work site and within the space to determine if dangerous air contamination and/or oxygen deficiency exists
    5. Record the atmospheric test results and keep them at the site throughout the work period
34. Which steps must be completed in order to prepare a spectrophotometer calibration graph?
    1. Develop a laboratory bench sheet for recording the test results
    2. Prepare a series of standards
    3. Read the spectrophotometer manufacturer's operation manual
    4. Treat the standard solution and a blank with the color-developing reagent
    5. Use the spectrophotometer to determine the absorbance or transmittance at the specified wave length
35. Which items can be used to measure color in water?
    1. Color comparator kits
    2. Colorimeters
    3. Colorimetric meters
    4. Nessler tubes
    5. Spectrophotometers
36. What do high levels of nitrate in raw water samples indicate?
    1. Biological wastes in the final stage of stabilization
    2. Melted snow water
    3. Runoff from fertilized areas
    4. Water from above the timberline
    5. Water of groundwater origin
37. Which topics were covered in the 1996 amendments to the SDWA?
    1. Backwash water recycling
    2. Bottled water
    3. Capacity development (technical, financial, and managerial)
    4. Consumer awareness
    5. Disinfectants and disinfection byproducts
38. Which items are considered microbial contaminants?
    1. Bacteria
    2. Mercury
    3. Pesticides
    4. Protozoa
    5. Viruses
39. Before selecting a contaminant for regulation, the EPA must perform which tasks?
    1. Consult with the scientific community
    2. Demonstrate that the contaminant actually occurs in public water systems
    3. Evaluate best possible treatment processes
    4. Examine costs of alternative treatment systems
    5. Solicit public comments
40. Which items are synthetic organic chemicals (SOCs)?
    1. Food additives
    2. Man-made compounds
    3. Paints
    4. Pesticides
    5. Plastics
41. Which methods of disinfection are in use today?
    1. Application of ozone
    2. Chloramination
    3. Free chlorination
    4. Freezing
    5. Use of chlorine dioxide
42. The Stage 1 Disinfectants and Disinfection Byproducts Rule requires routine monitoring of which regulated contaminants/disinfectants?
    1. Bromate
    2. Chlorine/chloramines
    3. Chlorite
    4. DBP precursors
    5. TTHM/HAA5
43. Which items are possible public relations objectives?
    1. Better customer relations
    2. Enhanced organizational credibility
    3. Greater water conservation
    4. Increased O&M budgets
    5. More certified operators

44. What kinds of information should be in a brochure that supplements a plant tour?
    1. Accomplishments
    2. Goals
    3. Processes
    4. Regulatory fines
    5. Violations or noncompliance issues

45. Which items are part of the total operating expenses of a water utility?
    1. Administrative costs
    2. Chemicals
    3. Energy
    4. Salaries
    5. Supplies

46. Which items are the benefits of an effective operation and maintenance program?
    1. Collecting accurate information and data on which to base the operations and maintenance of the system and to justify requests for financial resources
    2. Ensuring the availability of facilities and equipment as intended
    3. Maintaining the reliability of the equipment and facilities as designed
    4. Maintaining the value of the investment
    5. Obtaining full use of the system throughout its design life

47. Responsibilities in the implementation of cross-connection control programs are shared by which groups?
    1. Environmental advocates
    2. Health agencies
    3. Plumbing officials
    4. Water suppliers
    5. Water users (businesses and industries)

48. Which actions should a utility take to increase security and reduce threats from terrorism?
    1. Coordinate actions for effective emergency response
    2. Guard against unplanned physical intrusion
    3. Inform media of results of vulnerability assessments
    4. Invest in security and infrastructure improvements
    5. Make security a priority for employees

49. When determining the emergency limits of a contaminant (the amount of contaminant necessary to cause injury), which factors must be considered?
    1. Attraction of the contaminant to the organism being contaminated
    2. Duration of exposure to the contaminant
    3. Quantity or concentration of the contaminant
    4. Sensitivity of the organism to the contaminant
    5. Solubility of the contaminant in the organism

50. Which items are the basic elements of a hazard communication program?
    1. Define hazardous conditions
    2. Identify hazardous materials
    3. Obtain chemical information
    4. Properly label hazards
    5. Train operators

Short Answer

1. Why should iron and manganese be controlled in drinking water?
2. What happens when an ion exchange resin becomes fouled with iron rust or insoluble manganese dioxide?
3. How does an operator decide when to backwash a manganese greensand filter?
4. Why are drinking waters fluoridated?
5. How can water be softened prior to use with fluoridation equipment?
6. How would you protect yourself from the dust of dry fluoride compounds?
7. Why should water be softened?
8. How can operators protect themselves from lime?
9. Why is sludge sometimes recirculated back into the primary mix area of conventional plants?
10. How would you ensure that large amounts of resin are not being lost during the backwash stage?
11. What happens if an ion exchange softener removes iron in the ferrous (soluble) or ferric (solid) form?
12. How would you determine if iron has fouled the resin of an ion exchange softener?
13. How are trihalomethanes formed in drinking water?
14. What are the advantages of removing precursors before free chlorine is added?
15. What items must be considered before an alternative disinfectant is applied to any system?
16. What should a water utility do before considering treatment of a water source to reduce or remove arsenic?
17. What are the required types of arsenic sampling/monitoring?
18. How can membrane fouling be described?
19. What usually happens to water flux with time? Explain.
20. What is the most common and serious problem resulting from concentration polarization?
21. Why does water to be demineralized require pretreatment?
22. What will happen in a reverse osmosis plant if the brine flow valves are accidentally closed during operation?

23. An excessive concentration of any specific ion in the feedwater to an electrodialysis unit can cause what problem?
24. The amount of sludge accumulation at a typical water treatment plant depends on what factors?
25. What precaution must be taken before draining any in-ground tank?
26. Why should the back of the sludge truck face downhill when releasing wet sludge at one point?
27. Why should operators thoroughly read and understand manufacturers' literature before attempting to maintain plant equipment?
28. Why should your plant have an emergency team to repair chlorine leaks?
29. Why must motor nameplate data be recorded and filed?
30. Why should a pump never be allowed to run dry?
31. When two or more pumps of the same size are installed, why should they be operated alternately?
32. How can you determine if a new pump will turn in the direction intended?
33. What factors can cause wear on gate valve seats?
34. Why is rust a problem in water-cooled systems?
35. Why should solid feeders and hoppers be kept clean and dry?
36. Why should operators understand instrumentation and control systems?
37. What is the difference between accuracy and precision?
38. Why is it poor practice to ignore the lamps that are lit up (alarm conditions) on an annunciator panel?
39. Why should plant measuring systems be periodically calibrated?
40. Who is responsible for safety?
41. What special precautions should be taken when handling and storing caustic soda?
42. What safety precautions should an operator take when handling alum?
43. How can an operator prevent fires in a water treatment plant?
44. How would you remove a manhole cover?
45. How should gas cylinders be stored in laboratories?
46. What is the most common reason safety equipment falls into disrepair?
47. What is the purpose of spectrophotometer calibration curves?
48. Why are algae counts in raw water important to operators?
49. Why is iron undesirable in a domestic water supply?
50. How would you interpret changes (away from normal) in conductivity in water?
51. How would you attempt to reduce unpleasant flavors and odors in drinking water?
52. What criteria must be met before EPA selects a contaminant for regulation?
53. Why are coliform bacteria found in drinking water and what does their presence indicate?
54. What do secondary drinking water regulations control?
55. Why is the absence of flavors and odors in drinking water important?
56. How is a sampling route selected?
57. What can happen without adequate utility planning?
58. When has a supervisor successfully delegated?
59. When hiring new employees, the selection method and examination process used to evaluate applicants must be based on what criteria?
60. How should discipline problems be solved?
61. With whom do managers need to communicate?
62. What attitude should management try to develop among its employees regarding the customer?
63. What does a SCADA system do?
64. How would you assess the vulnerability of a water supply system?
65. What topics should be included in a safety officer's talk to new operators?
66. Why do water utilities need an effective water and energy conservation program?

Problems

1. What is the setting on a potassium permanganate chemical feeder in pounds per day if the required chemical dose is 2.4 mg/L and the flow is 0.18 MGD?
2. A flow of 200 GPM is to be treated with a 2.4 percent (0.2 pound per gallon) solution of sodium fluoride (NaF). The water to be treated contains 0.5 mg/L of fluoride ion and the desired fluoride ion concentration is 1.4 mg/L. What is the sodium feed rate in gallons per day? Assume the sodium fluoride has a fluoride purity of 43.4 percent.
3. How many gallons of water with a hardness of 12 grains per gallon may be treated with an ion exchange softener with an exchange capacity of 8,000 kilograins?
4. Calculating the mineral rejection as a percent, what is an estimate of the ability of a reverse osmosis plant to reject minerals? The feedwater contains 1,600 mg/L TDS and the product water TDS is 130 mg/L.
5. What is the feed rate of a dry chemical feeder in pounds per day if 3.0 pounds of chemical are caught in a weighing tin during 15 minutes?

SUGGESTED ANSWERS TO COMPREHENSIVE REVIEW QUESTIONS

VOLUME II

True-False

1. True Both iron and manganese are essential to the growth of many plants and animals, including humans.
2. True Only one cell of iron bacteria is needed to start an infestation of iron bacteria in a well or distribution system.
3. False pH is increased by the removal of carbon dioxide.
4. True Fluoride saturators are a special application of a solution feeder.
5. True Fluoride solutions are extremely corrosive.
6. True Excessively hard water can cause scaling on the inside of pipes and thereby restrict flow.
7. True The stability of treated water is determined by measuring the pH and calculating the Langelier Index.
8. False The addition of acid to prevent calcium carbonate scale will not accomplish the same things as recarbonation and the addition of polyphosphate.
9. False The feed rates determined by jar tests do not always produce the exact same results in the actual plant.
10. True The effluent from an ion exchange unit usually will have zero hardness until the unit needs regeneration.
11. False The overwhelming bulk of THM precursors in water are from natural organic compounds, not from industrial organic chemicals.
12. False When collecting water samples for THM analysis, be sure to allow time for the water temperature to become constant.
13. False The higher the pH of the water, the faster the production of THMs.
14. True Low levels of arsenic consumed over a long period of time can cause cancer and noncancerous toxic effects on humans.
15. False Electrical conductivity (EC) cannot usually be correlated directly with arsenic reduction.
16. True All waters contain various amounts of total dissolved solids (TDS), including fresh water.
17. False The rate of hydrolysis is accelerated by increased temperature.
18. False The output of a positive-displacement pump may not be throttled.
19. True At a minimum, the rate of membrane hydrolysis is at a pH of about 4.7, and it increases with both increasing and decreasing pH.
20. True Trash and grit may be disposed of in landfills.
21. True Before filling a drained tank after removing sludge, thoroughly disinfect the tank.
22. False The application of nitrogen fertilizers causes a reduction in soil pH.
23. False One person should never be permitted to attempt an emergency repair alone.
24. True Operators must be familiar with equipment and understand what it is intended to do before developing a preventive maintenance program and maintaining equipment.
25. False Check the temperature of a centrifugal pump bearing with a thermometer.
26. True Always be sure a centrifugal pump is primed before starting the pump.
27. True Diesel engines can be air or water cooled.
28. True Never pour water on a chlorine leak because this will only create a bigger problem by enlarging the leak.
29. False Digital instruments provide a direct numerical readout.
30. False Long detection lags can result in unstable processes and poor control.
31. True Entry into confined spaces is never permitted until the space has been properly ventilated using specially designed forced-air ventilators.
32. True Pneumatic instrumentation depends upon a constant source of clean, dry, pressurized air for reliable operation.
33. True The fumes from hydrofluosilicic acid can corrode metal and etch glass on protective equipment.

34. True In reactions with most metals, hydrochloric acid will produce hydrogen gas.
35. False Anhydrous ammonia (gas) may not be stored in the same room with chlorine (gas).
36. True When working with chlorine, be equipped to control chlorine leaks, which are most often found in the control valve.
37. True Ferric chloride is a very corrosive compound and should be treated as you would treat any acid.
38. False The most common form of iron in solution in groundwater and in water under anaerobic conditions (bottom of a lake or reservoir) is ferrous ion.
39. False Nitrate represents the most completely oxidized form of nitrogen found in water.
40. True Chloroform is the most common THM found in drinking water and is usually present in the highest concentration.
41. True EPA regulations change frequently as new information becomes available.
42. True MCLs apply whether the contaminant is from naturally occurring sources or from man-made pollution.
43. False The health risk from mercury is greater from mercury in fish than simply from waterborne mercury.
44. True The presence of coliform bacteria indicates that other harmful organisms may be present in the water.
45. True Tier 1 violations are any violations or situations in which even a brief violation or short-term exposure to a contaminant has the potential to cause a health problem.
46. True No discipline problem ever solves itself and the sooner you deal with the problem, the better will be the outcome.
47. False Written communication is more demanding than oral communication and requires more careful preparation.
48. True Customer calls are frequently your first indication that something may be wrong.
49. True Corrective maintenance, including emergency maintenance, is considered reactive maintenance.
50. True The SCADA system allows operators to enhance the efficiency of their water facility by keeping them fully informed and fully in control.
51. True The operator should always call to the supervisor's attention unsafe conditions, environment, and equipment or other concerns about the work they are performing.
52. False On-the-Job Training (OJT) is not a good way of preventing accidents with an inexperienced operator.

Best Answer

1. 2 The growth of iron bacteria can be controlled by chlorination.
2. 4 The best way to determine if there is an iron and manganese problem in a water supply is to look for stained plumbing fixtures in a couple of houses.
3. 4 When operating an ion exchange unit and iron and manganese start to appear in treated water, the operator should regenerate with a brine solution.
4. 3 If children drink the recommended dose of fluoride, the children will have fewer dental caries (decay or cavities).
5. 3 When working with fluoridation systems using a sodium fluoride solution, the hardness of water is very important because scaling can be caused by hardness.
6. 4 Hardness is a measure of the soap- or detergent-consuming power of water.
7. 2 Bagged quicklime should never be stored close to combustible materials because considerable heat will be generated if the lime accidentally gets wet.
8. 1 Scale formation on the filter sand, distribution mains, and household plumbing can be prevented by converting the excess caustic and unprecipitated carbonate ions to soluble forms.
9. 2 Restrictions are placed on the location where THM samples are collected to avoid the possibility of contaminating samples or losing THMs.
10. 2 Some THM sample bottles do not contain any reducing agent to determine the maximum concentration of TTHMs that can form over an extended period of time.
11. 1 A major advantage of removing precursors before chlorine is added is that it allows the continued use of free chlorine as a disinfectant.
12. 2 Arsenic split-sample analyses are performed by a certified laboratory to confirm results from arsenic test kit analyses.
13. 3 Filtration is the most common pretreatment process to remove turbidity and suspended solids.
14. 2 The growth of bacterial slimes and molds on membrane surfaces can be inhibited or retarded by the application of chlorine.
15. 1 Acid addition is a part of the pretreatment of the feedwater supply for electrodialysis demineralization to prevent carbonate scaling.
16. 4 The key to speedy sludge drying in a sludge drying bed is the regular removal of the water on top of the sludge.
17. 1 The principal advantage of using centrifuges for sludge dewatering is the density of the sludge cake can be varied.
18. 4 An ampere is the unit of electric current or an indication of how hard the electricity is working.

19. 1 The circuit breaker has been tripped when the handle is at the mid position between on and off.

20. 2 Bearing grease should be purchased in cartridge form to minimize the chance of getting dirt into the lubricant.

21. 3 A pump is primed when it is completely filled with water.

22. 4 A compressor unloader is a device that allows the compressor to start under a no-load condition.

23. 3 The purpose of cathodic protection is to prevent the rusting of base metal surfaces.

24. 2 Accuracy is how closely an instrument measures the true or actual value of a process variable.

25. 1 A very precise method of measuring fluctuating water levels is the bubbler tube.

26. 1 An annunciator panel "test" button allows an operator to confirm that no alarm lamps have burned out.

27. 4 A multimeter is the most useful piece of general electrical test equipment.

28. 3 Bases can be neutralized with dilute acids.

29. 4 Carbon dioxide gas is considered hazardous because it can cause suffocation due to lack of oxygen.

30. 3 High levels of oxygen are dangerous because they can increase the possibility of an explosion.

31. 1 A ground-fault circuit interrupter (GFCI) must be used when operating power tools outdoors for protection against electric shock.

32. 4 The purpose of a calibration curve is to determine the concentration of the water quality indicator contained in a sample.

33. 4 The apparent color in water is caused by suspended materials (including colloidal substances).

34. 1 Specific conductance or conductivity is a numerical expression of the ability of water to conduct an electric current.

35. 4 The Disinfectants/Disinfection Byproducts (D/DBP) Rule is designed to protect the public from harmful concentrations of disinfectants.

36. 3 The purpose of the Total Coliform Rule (TCR) is to improve public health protection by reducing fecal pathogens to minimal levels through control of total coliform bacteria.

37. 2 The purpose of the Filter Backwash Recycling Rule (FBRR) is to improve public health protection by assessing and changing, where needed, recycling practices for improved contaminant control.

38. 2 The important difference between MCL and MCLG is that the MCLG is nonenforceable and MCL is an enforceable standard.

39. 2 Backflow situations in distribution systems are caused by either back pressure or backsiphonage.

40. 3 The National Terrorism Advisory System is a system to issue formal, detailed alerts about specific or credible terrorist threats.

41. 2 An acute toxic agent is an agent that causes injury quickly.

42. 3 If contamination of the water supply is discovered, the immediate concern is the safety of the public.

43. 1 The purpose of the SDS is to have a readily available reference document that includes complete information on the hazardous substance.

Multiple Choice

1. 1, 2, 3, 5 The presence of iron and manganese in drinking water is objectionable because they may stain clothes laundered in the water; promote the growth of iron bacteria; stain plumbing fixtures, bathtubs, and sinks; and create slimes that cause foul flavors and odors.

2. 2, 4 The pH of water being treated with a continuous regeneration (CR) manganese greensand plant can be increased by adding sodium carbonate and sodium hydroxide.

3. 1, 2, 4, 5 To minimize red water or dirty water complaints deliver stable (noncorrosive) water to the distribution system, ensure that water pumped to the distribution system contains no iron and manganese, maintain a free chlorine residual in all water throughout the distribution system, and provide adequate treatment to control iron and manganese.

4. 1, 2, 3, 4, 5 Fluoridation plant warning system alarms indicate downtime measured by a time meter, high or low fluoride levels, low injection water pressure, low water flow in the main pipeline, and power outages.

5. 1, 2, 3, 4, 5 Before starting up a chemical feeder system, check that safety equipment, such as eyewash, drench showers, dust masks, face shields, gloves, and vent fans, are in place and functional; that alarms and safety shutoffs operate properly; that safety guards are in place; that auxiliary equipment, including dust collectors, fans, cooling water, mixing water, and safety equipment are operational; and that plant files include all important nameplate data.

6. 1, 2, 3, 5 An operator can come in contact with fluoride by eye contact, ingestion, inhalation, and skin contact.

7. 1, 2, 3, 4, 5 Excessive hardness in water is undesirable because it causes damage in some industrial processes, deposition of scale in boilers, formation of soap curds, increased use of soap, and objectionable flavors.

8. 2, 4, 5 The limitations of the ion exchange softening process include the increase in the sodium content of softened water, ultimate disposal of rinse water, and the ultimate disposal of spent brine.

9. 1, 2, 3, 4, 5 When troubleshooting the operation of the backwash stage, the operator should check for adequate flow rates, excessive flow rates to prevent washing resin from the unit, and higher than normal head loss, as well as the condition of the resin by visually inspecting the top layer of the bed, and the timer to make sure the unit is washed for the required length of time.

10. 1, 2, 4 Treatment processes that are available to remove THMs after they have been formed include adsorption, aeration, and oxidation.

11. 1, 2, 3, 4, 5 Oxidation processes that may remove or modify THM precursors include chlorine dioxide, hydrogen peroxide, ozone, ozone/ultraviolet light, and permanganate.

12. 1, 2, 3, 4, 5 The Best Available Technology (BAT) processes available for treating arsenic include activated alumina, coagulation/filtration, ion exchange, oxidation filtration, and reverse osmosis.

13. 1, 4, 5 Technologies used by POU and POE devices to remove arsenic from drinking water include activated alumina, iron-based sorbents, and reverse osmosis.

14. 1, 2, 3, 4 Distillation, electrodialysis, ion exchange, and reverse osmosis can be used to remove minerals from water.

15. 3, 4, 5 Nanofiltration membranes are used for demineralization to reject hardness, TDS, and THM formation potential.

16. 1, 2, 3, 4, 5 Constituents in feedwater to a reverse osmosis unit that require pretreatment to prevent scaling include carbonates, ferric iron, hydrogen sulfide, silica, and sulfate.

17. 1, 2, 3, 4, 5 Commonly used chemicals that require special handling in a reverse osmosis plant operation include acids, chlorine, citric acid, formaldehyde, and sodium hexametaphosphate.

18. 1, 2, 5 Sludges and brines produced at water treatment plants are disposed of by depositing in landfills, spreading on land, and discharging to wastewater collection systems (sewers).

19. 2, 3, 4 Extreme caution should be exercised whenever drainage waters may be recycled to avoid flavor and odor problems, health hazards, and operational problems.

20. 1, 2, 3, 4, 5 Voltage imbalance could be caused by changing energy demands, circuit breakers, high-resistance contacts, loose connections, and motor starters.

21. 1, 2, 3, 4, 5 Items that could be the cause of an electric motor control problem include moisture, overload, poor design, voltage, and wear.

22. 1, 3, 4 Possible causes of pump misalignment include bolts loosening, foundations settling unevenly, and piping changing positions.

23. 2, 3, 4, 5 Pump operating troubles include high power requirements, noisy pump, pump will not start, and reduced rate of discharge.

24. 1, 2, 3, 4, 5 Factors that could cause a reduced rate of discharge from a pump include air in the water, stuck or clogged check valves, discharge head too high, clogged impeller, and greater than anticipated suction lift.

25. 1, 2, 3, 4, 5 Maintenance procedures that should be performed on compressors include cleaning the cylinder or casing fins weekly, draining the condensate from the air receiver, inspecting the suction filter of the compressor regularly, lubricating the compressor regularly, and testing the safety valves.

26. 1, 2, 3, 4, 5 Examples of modulating control systems for water treatment process variables include chlorine residual analyzer/controller, continuous level control process basins, flow-paced chemical feeders, pressure- or flow-regulating valves, and variable-speed pumping systems for flow/level control.

27. 1, 2, 3, 4, 5 When combined with air, gasoline vapors, hydrogen, manufactured fuel gas, methane, and natural gas may develop into explosive gas mixtures.

28. 1, 2, 3, 4, 5 Excessive heat, static electricity, temporary power failures, tripped panel circuit breakers, and voltage surges may cause electronic instrument problems.

29. 1, 2, 3, 4, 5 Critical preventive maintenance measures for instrumentation systems include bleeding pneumatic systems regularly as instructed; not lubricating, "tweaking," fixing, calibrating, freeing up, or modifying any component of a system arbitrarily; keeping instrument components clean on the outside and closed/sealed against inside contamination; keeping recorder pens and charts functioning as designed by frequent checking and service; and protecting all instrumentation from moisture, vibration, and mechanical shock.

30. 1, 2, 3, 4, 5 To safely handle chemicals, an operator must be aware of the tendency of chemicals to form an arch in a storage bin, know how to feed the chemicals, know how to make up solutions, know how to store chemicals, and understand the fire problem associated with each chemical.

31. 1, 2, 3 Flammable or explosive bases used in water treatment include ammonia, caustic soda, and hypochlorite compounds.

32. 1, 3, 5 Safety precautions that should be taken when handling most salts used in water treatment plants include eye protection, respiratory protection, and ventilation.

33. 1, 2, 3, 4, 5 Before entry into a confined space, empty, flush, or purge the space of any harmful substance(s) to the greatest extent possible; identify and close off or reroute any lines that may convey harmful substance(s) to, or through, the work area; if an atmospheric hazard is noted, use portable blowers to further ventilate the area; monitor the atmosphere at the work site and within the space to determine if dangerous air contamination and/or oxygen deficiency exists; and record the atmospheric test results and keep them at the site throughout the work period.

34. 2, 4, 5 Steps that must be completed in order to prepare a spectrophotometer calibration graph include preparing a series of standards, treating a standard solution and a blank with the color-developing reagent, and using the spectrophotometer to determine the absorbance or transmittance at the specified wave length.

35. 1, 4, 5 Color can be measured in water by using color comparator kits, Nessler tubes, and spectrophotometers.

36. 1, 3 High levels of nitrate in raw water samples indicate biological wastes in the final stage of stabilization and runoff from fertilized areas.

37. 1, 2, 3, 4, 5 Topics included in the 1996 amendments to the SDWA are backwash water recycling, bottled water, capacity development, consumer awareness, and disinfectants and disinfection byproducts.

38. 1, 4, 5 Bacteria, protozoa, and viruses are considered microbial contaminants.

39. 1, 2, 5 Before selecting a contaminant for regulation, the EPA must consult with the scientific community, demonstrate that the contaminant actually occurs in public water systems, and solicit public comments.

40. 1, 2, 3, 4, 5 Synthetic organic chemicals (SOCs) include food additives, man-made compounds, paints, pesticides, and plastics.

41. 1, 2, 3, 5 Disinfection methods in use today include application of ozone, chloramination, free chlorination, and use of chlorine dioxide.

42. 1, 2, 3, 4, 5 The Stage 1 Disinfectants and Disinfection Byproducts Rule requires routine monitoring of bromate, chlorine/chloramines, chlorite, DBP precursors, and TTHM/HAA5.

43. 1, 2, 3 Possible public relations objectives include better customer relations, enhanced organizational credibility, and greater water conservation.

44. 1, 2, 3 Information that should be in a brochure that supplements a plant tour includes accomplishments, goals, and processes.

45. 1, 2, 3, 4, 5 Items that are part of the total operating expenses of a water utility include administrative costs, chemicals, energy, salaries, and supplies.

46. 1, 2, 3, 4, 5 The benefits of an effective operation and maintenance program include collecting accurate information and data on which to base the operations and maintenance of the system and to justify requests for financial resources, ensuring the availability of facilities and equipment as intended, maintaining the reliability of the equipment and facilities as designed, maintaining the value of the investment, and obtaining full use of the system throughout its design life.

47. 2, 3, 4, 5 Responsibilities in the implementation of cross-connection control programs are shared by health agencies, plumbing officials, water suppliers, and water users.

48. 1, 2, 4, 5 Actions that utilities should take to increase security and reduce threats from terrorism include coordinating actions for effective emergency response, guarding against unplanned physical intrusion, investing in security and infrastructure improvements, and making security a priority for employees.

49. 1, 2, 3, 4, 5 When determining the emergency limits of a contaminant, factors considered include attraction of the contaminant to the organism being contaminated, duration of exposure to the contaminant, quantity or concentration of the contaminant, sensitivity of the organism to the contaminant, and solubility of the contaminant in the organism.

50. 1, 2, 3, 4, 5 The basic elements of a hazard communication program include defining hazardous conditions, identifying hazardous materials, obtaining chemical information, properly labeling hazards, and training operators.

Short Answer

1. Iron and manganese should be controlled in drinking water because when clothes are laundered in water containing iron and manganese, they come out stained. Also, waters containing iron and manganese promote the growth of iron bacteria. These bacteria form thick slimes on the walls of the distribution system mains. The slimes are rust colored from iron and black from manganese. Variations in flow cause these slimes to slough, which results in dirty water. Furthermore, these slimes will cause foul flavors and odors in the water.
2. If an ion exchange resin becomes fouled with iron rust or insoluble manganese dioxide, the resin can be cleaned but this is expensive and the resin capacity is reduced.
3. Typically, a manganese greensand filter should be backwashed when head loss reaches 10 psi or after treating a predetermined number of gallons of water.
4. Drinking waters are fluoridated to reduce the incidence (number) of dental caries (tooth decay) in children.
5. Water can be softened prior to use with fluoridation equipment by the use of zeolite ion exchangers.
6. To protect yourself from the dust of dry fluoride compounds, be sure the dust collector system works properly. Even with the use of dust collector systems, dust will circulate in the air. Always use approved respirators equipped with cartridges for organic dusts and vapors, protective coveralls, and gloves when emptying sacks or cleaning up equipment and plant surfaces.
7. Water should be softened because hard waters cause difficulties in doing the laundry and in dishwashing in the household. There is also a coating that forms inside the hot water heater. Many industrial processes are adversely affected by hard water.
8. Operators can protect themselves from lime by wearing protective clothing. Protective clothing includes long-sleeved shirt with sleeves and collar buttoned, trousers with legs down over tops of shoes or boots, head protection, and gloves. A protective cream should be applied to exposed parts of the body. Operators should also wear a lightweight filter mask and tight-fitting safety glasses with side shields for protection from lime dust.
9. Sludge may be recirculated back into the primary mix area of conventional plants to help seed the process. The advantages are: (1) recirculation speeds up the precipitation process, and (2) some reduction of chemical requirements may result.
10. To ensure that large amounts of resin are not being lost during the backwash stage, a glass beaker can be used to catch a sample of the effluent while the unit is backwashing. A trace amount of resin should cause no alarm, but a steady loss of resin could indicate a problem in the unit and the cause should be located and corrected as soon as possible.
11. If iron is removed in the ferrous (soluble) form, the media can become iron coated and the efficiency of the softener will be greatly reduced. If iron is removed in the ferric (solid) form, the bed could become plugged and force the water to channel or short-circuit through the bed. The result is incomplete contact between the water and media, thus creating hardness leakage and loss of softening efficiency.
12. To determine if iron is fouling the resin of an ion exchange softener, look at the color of the resin. If iron has fouled the resin, the bed will be an orange, rusty color, while the backwash effluent will appear a light orange at the end of the backwash stage. Also, the head loss on the unit will run higher than normal as the bed becomes plugged with iron.
13. Trihalomethanes are formed in drinking water when free chlorine comes in contact with naturally occurring organic compounds (THM precursors). Trihalomethanes are a class of organic compounds in which there has been a replacement of the three hydrogen atoms in the methane molecule with three halogen atoms (chlorine or bromine). The production of THMs can generally be shown as:

   Free Chlorine + Natural Organics (precursor) + Bromide
   $\rightarrow$ THMs + Other Products
14. Removing precursors before free chlorine is added allows the continued use of free chlorine as a disinfectant and the formation of fewer "other products" whose health significance is not known.
15. Before an alternative disinfectant is applied to any system, the source of the water supply, water quality, and treatment effectiveness for bacteriological control must be evaluated.
16. Before considering treatment of a water source to reduce or remove arsenic, a water utility should investigate a new source as an alternative water supply.
17. Types of required arsenic sampling/monitoring include compliance, process control, calibration/verification, and special sampling for troubleshooting.
18. Membrane fouling can be described by whether the cause of fouling can be removed (reversible or irreversible), by the material causing the fouling (biological, organic, particulate, or dissolved), and by the means of fouling (cake formation or membrane pore blockage).
19. Even under ideal conditions (pure feedwater and no fouling of the membrane surface), there is a decline in water flux with time. This decrease in flux is due to membrane compaction. Flux decline also results from foulants and bacterial growth.
20. The most common and serious problem resulting from concentration polarization is the increasing tendency for precipitation of sparingly soluble salts and the deposition of particulate matter on the membrane surface.
21. Water to be demineralized always contains impurities that should be removed by pretreatment to protect the membrane and to ensure maximum efficiency of the reverse osmosis process.
22. If the brine flow valves are accidentally closed during the operation of a reverse osmosis plant, 100 percent recovery will result in almost certain damage to the membranes due to the precipitation of inorganic salts ($CaSO_4$).

23. An excessive concentration of any specific ion in the feedwater to an electrodialysis unit could lead to chemical fouling due to scaling.

24. The amount of sludge accumulation at a typical water treatment plant depends on the type and amount of suspended matter in the source water being treated as well as on the level of dosage and the type of coagulant used. Also, by using a source water stabilizing reservoir, a plant will accumulate less sludge because the turbidity level of the water to be treated will be reduced, providing a more constant quality of water that requires less alum.

25. Before draining any in-ground tank, always determine the level of the water table. If the water table is high, an empty tank could float like a cork on the water surface and cause considerable damage to the tank and piping.

26. When releasing wet sludge at one point, the back of the sludge truck should face downhill so:
    1. The tank will drain faster
    2. The tank will empty completely
    3. The truck will not become stuck in the sludge

27. The operator should thoroughly read and understand manufacturers' literature before attempting to maintain plant equipment so that the job required to keep the equipment operating can be done properly.

28. An emergency team to repair chlorine leaks is important because chlorine is a hazardous chemical. An emergency team is specially trained and qualified to control and repair emergency chlorine leaks. Unqualified and untrained personnel may injure themselves and create hazards for others.

29. Motor nameplate data must be recorded and filed so the information is available when needed to repair the motor or to obtain replacement parts.

30. A pump should never be allowed to run dry because water acts as a lubricant between the rings and the impeller. Running the pump without this lubrication could cause severe pump damage.

31. When two or more pumps of the same size are installed, they should be operated alternately to equalize wear, keep motor windings dry, and distribute lubricant in the bearings.

32. To determine if a new pump will turn in the direction intended, momentarily start the motor by a quick electrical contact and check to be sure the motor will turn the pump in the direction indicated by the rotational arrows on the pump.

33. Wear on gate valve seats can be caused by: (1) minute particles transported in the water, and (2) operation of the gate valve partially opened.

34. Rust is a problem in water-cooled systems because: (1) rust does not allow good heat transfer, and (2) the sloughing of rust scale can block narrow passages.

35. Solid feeders and hoppers should be kept clean and dry in order to prevent "bridging" of the chemical in the hopper and clogging in the feeder.

36. Operators must have a good working familiarity with instrumentation and control systems to properly monitor and control the treatment processes.

37. Accuracy refers to how closely an instrument measures the true or actual value of a process variable and precision refers to how closely the device (instrument) can reproduce a given reading (or measurement) time after time.

38. It is poor practice to ignore the lamps that are lit up (alarm conditions) on an annunciator panel because a true alarm condition requiring immediate operator attention may be lost in the resulting general indifference to the alarm system.

39. Plant measuring systems must be periodically calibrated to ensure accurate measurements.

40. Safety is the responsibility of everyone, from top management to all employees.

41. Special precautions to be taken when handling and storing caustic soda include preventing eye and skin contact, not breathing dusts or mists, and avoiding storing this chemical next to strong acids.

42. For safety precautions for the dry type of alum, the operator should avoid breathing the powder and skin contact. For the liquid type of alum, the operator should avoid exposures to the skin and eyes, as well as ingestion.

43. Fires can be prevented by good housekeeping. This means keeping a well-kept, neat, and orderly plant.

44. A manhole cover should be lifted with a special tool such as a pick with a bent point or a manhole hook. Use your legs, not your back, for lifting. Locate the cover outside the working area to provide adequate working area around the manhole opening.

45. Gas cylinders must be kept in place by safety retaining devices, such as chains around the cylinders, to keep them from falling. The valve and cylinder regulator should be protected from being struck by stools, ladders, and other objects.

46. Safety equipment may fall into disrepair because it is only used occasionally and may deteriorate due to heat, time, and other environmental factors.

47. Spectrophotometer calibration curves are used to determine the concentrations of water quality indicators by the use of a spectrophotometer.

48. Algae counts in raw waters are important to operators to help explain the cause of color and turbidity and the presence of flavors and odors in the water, explain the clogging of screens or filters, and document variability in the water quality.

49. Iron is undesirable in a domestic water supply because it can stain laundry, concrete, and porcelain. A bitter, astringent flavor can be detected by some people at levels above 0.3 mg/L.

50. Changes in conductivity from normal levels may indicate changes in mineral composition of the water, seasonal variations in lakes and reservoirs, or intrusion of pollutants.

51. Corrective measures designed to reduce unpleasant flavors and odors include aeration, or the addition of chlorine, chlorine dioxide, potassium permanganate, or activated carbon. (For additional details, see Volume I, Chapter 9, "Taste and Odor Control.")

52. To regulate a substance, EPA must demonstrate that the contaminant meets three criteria:
    1. The contaminant has an adverse effect on human health
    2. It occurs, or is likely to occur, in public water systems at a frequency and concentration of significance to public health
    3. Regulation of the contaminant offers a meaningful opportunity to reduce health risks for people served by public water systems

53. Many types of coliform bacteria from human and animal wastes may be found in drinking water if the water is not properly treated. Often the bacteria themselves do not cause diseases transmitted by water, although certain coliforms have been identified as the cause of "traveler's" diarrhea. In general, however, the presence of coliform bacteria indicates that other harmful organisms may be present in the water.

54. Secondary drinking water regulations control contaminants that primarily affect the aesthetic qualities relating to the public acceptance of drinking water. At considerably higher concentrations of these contaminants, health implications may also exist as well as aesthetic degradation.

55. The absence of flavors and odors helps to maintain the consumers' confidence in the quality of their water, even though it does not guarantee that the water is safe.

56. The sampling route should be arranged so that samples that must be analyzed immediately are not delayed while other sampling is done.

57. Without adequate planning, your utility will be facing system failures, inability to meet compliance regulations, and inadequate service capacity to meet community needs.

58. A supervisor has successfully delegated when proper job assignments, authority, and responsibility are all present.

59. When hiring new employees, the selection method and examination process used to evaluate applicants must be based on the applicant's knowledge, skills, and abilities to perform relevant job-related activities.

60. Discipline problems should be solved quickly. The sooner you deal with the problem, the better the outcome will be.

61. Managers need to communicate with employees, the governing body, and the public.

62. Management should try to develop among its employees the attitude that even though the customer is not always right, every customer is always entitled to courteous treatment and a proper explanation of anything the customer does not understand.

63. A SCADA system collects, stores, and analyzes information about all aspects of operation and maintenance; transmits alarm signals, when necessary; and allows fingertip control of alarms, equipment, and processes.

64. The following steps should be taken in assessing the vulnerability of a system:
    1. Identify and describe the system components.
    2. Assign assumed disaster characteristics.
    3. Estimate disaster effects on system components.
    4. Estimate customer demand for service following a potential disaster.
    5. Identify key system components that would be primarily responsible for system failure.

65. The safety officer should tell new employees about utility policy, safety reports, past accidents, and orient the new operator toward the importance of safety to operators and to the organization.

66. An effective water and energy conservation program is an excellent means of saving money. Water conservation can reduce energy costs. Reducing water and energy consumption results in conservation of natural resources.

Problems

1. What is the setting on a potassium permanganate chemical feeder in pounds per day if the required chemical dose is 2.4 mg/L and the flow is 0.18 MGD?

Known		Unknown
Flow, MGD	= 0.18 MGD	Chemical Feeder, lb/day
Dose, mg/L	= 2.4 mg/L	

Determine the chemical feeder setting in pounds per day.

Chemical Feeder, lb/day = (Flow, MGD)(Dose, mg/L)(8.34 lb/gal)

= (0.18 MGD)(2.4 mg/L)(8.34 lb/gal)

= 3.6 lb/day

2. A flow of 200 GPM is to be treated with a 2.4 percent (0.2 pound per gallon) solution of sodium fluoride (NaF). The water to be treated contains 0.5 mg/L of fluoride ion and the desired fluoride ion concentration is 1.4 mg/L. What is the sodium feed rate in gallons per day? Assume the sodium fluoride has a fluoride purity of 43.4 percent.

Known		Unknown
Flow, GPM	= 200 GPM	Feed Rate, gal/day
NaF Solution, %	= 2.4%	
NaF Solution, lbs/gal	= 0.2 lb/gal	
Desired F, mg/L	= 1.4 mg/L	
Actual F, mg/L	= 0.5 mg/L	
Purity, %	= 43.4%	

1. Convert the flow from gallons per minute to million gallons per day.

$$\text{Flow, MGD} = \frac{(\text{Flow, gal/min})(60\ \text{min/hr})(24\ \text{hr/day})(1\ \text{million})}{1{,}000{,}000}$$

$$= \frac{(200\ \text{gal/min})(60\ \text{min/hr})(24\ \text{hr/day})(1\ \text{million})}{1{,}000{,}000}$$

$$= 0.29\ \text{MGD}$$

2. Determine the fluoride feed dose in milligrams per liter.

$$\text{Feed Dose, mg/L} = \text{Desired Dose, mg/L} - \text{Actual F, mg/L}$$

$$= 1.4\ \text{mg/L} - 0.5\ \text{mg/L}$$

$$= 0.9\ \text{mg/L}$$

3. Calculate the feed rate in pounds of fluoride per day.

$$\text{Feed Rate, lb/day} = (\text{Flow, MGD})(\text{Feed Dose, mg/L})(8.34\ \text{lb/gal})$$

$$= (0.29\ \text{MGD})(0.9\ \text{mg/L})(8.34\ \text{lb/gal})$$

$$= 2.2\ \text{lb F/day}$$

4. Convert the feed rate from pounds of fluoride per day to gallons of sodium fluoride solution per day.

$$\text{Feed Rate, gal/day} = \frac{(\text{Feed Rate, lbs F/day})(100\%)}{(\text{NaF Solution, lbs F/gal})(\text{Purity, \%})}$$

$$= \frac{(2.2\ \text{lbs F/day})(100\%)}{(0.2\ \text{lb/gal})(43.4\%)}$$

$$= 25.3\ \text{gal/day or } 25\ \text{gal/day}$$

3. How many gallons of water with a hardness of 12 grains per gallon (gpg) may be treated with an ion exchange softener with an exchange capacity of 8,000 kilograins?

Known		Unknown
Hardness, gpg	= 10 gpg	Water Treated, gallons
Exchange Capacity, grains	= 8,000,000 grains	

Calculate the gallons of water that may be treated.

$$\text{Water Treated, gal} = \frac{\text{Exchange Capacity, grains}}{\text{Hardness, gpg}}$$

$$= \frac{8{,}000{,}000\ \text{grains}}{10\ \text{gpg}}$$

$$= 800{,}000\ \text{gal}$$

$$\text{or} = 0.80\ \text{M gal}$$

4. Calculating the mineral rejection as a percent, what is an estimate of the ability of a reverse osmosis plant to reject minerals? The feedwater contains 1,600 mg/L TDS and the product water TDS is 130 mg/L.

Known		Unknown
Feedwater TDS, mg/L	= 1,600 mg/L	Mineral Rejection, %
Product Water TDS, mg/L	= 130 mg/L	

Calculate the mineral rejection as a percent.

$$\text{Mineral Rejection, \%} = \left(1 - \frac{\text{Product TDS, mg/L}}{\text{Feed TDS, mg/L}}\right)(100\%)$$

$$= \left(1 - \frac{130\ \text{mg/L}}{1{,}600\ \text{mg/L}}\right)(100\%)$$

$$= (1 - 0.08)(100\%)$$

$$= 92\%$$

5. What is the feed rate of a dry chemical feeder in pounds per day if 3.0 pounds of chemical are caught in a weighing tin during 15 minutes?

Known		Unknown
Chemical, lb	= 3.0 lb	Chemical Feed, lb/day
Time, min	= 15 min	

Calculate the chemical feed rate in pounds of chemical per day.

$$\text{Chemical Feed, lb/day} = \frac{(\text{Chemical, lb})(60\ \text{min/hr})(24\ \text{hr/day})}{\text{Time, min}}$$

$$= \frac{(3.0\ \text{lb})(60\ \text{min/hr})(24\ \text{hr/day})}{15\ \text{min}}$$

$$= 288\ \text{lb/day}$$

APPENDIX

HOW TO SOLVE WATER TREATMENT PLANT ARITHMETIC PROBLEMS

(VOLUME II)

by

Ken Kerri

TABLE OF CONTENTS

HOW TO SOLVE WATER TREATMENT PLANT ARITHMETIC PROBLEMS

HOW TO SOLVE WATER TREATMENT PLANT ARITHMETIC PROBLEMS (VOLUME II)

A.1 BASIC CONVERSION FACTORS (ENGLISH SYSTEM)

UNITS

1,000,000	= 1 Million	1,000,000/1 Million

LENGTH

12 in	= 1 ft	12 in/ft
3 ft	= 1 yd	3 ft/yd
5,280 ft	= 1 mi	5,280 ft/mi

AREA

144 in^2	= 1 ft^2	144 in^2/ft^2
43,560 ft^2	= 1 acre	43,560 ft^2/ac

VOLUME

7.48 gal	= 1 ft^3	7.48 gal/ft^3
1,000 mL	= 1 liter	1,000 mL/L
3.785 L	= 1 gal	3.785 L/gal
231 in^3	= 1 gal	231 in^3/gal

WEIGHT

1,000 mg	= 1 g	1,000 mg/g
1,000 g	= 1 kg	1,000 g/kg
454 g	= 1 lb	454 g/lb
2.2 lb	= 1 kg	2.2 lb/kg

POWER

0.746 kW	= 1 HP	0.746 kW/HP

DENSITY

8.34 lb	= 1 gal	8.34 lb/gal
62.4 lb	= 1 ft^3	62.4 lb/ft^3

DOSAGE

17.1 mg/L	= 1 grain/gal	17.1 mg/L/gpg
64.7 mg	= 1 grain	64.7 mg/grain

PRESSURE

2.31 ft water	= 1 psi	2.31 ft water/psi
0.433 psi	= 1 ft water	0.433 psi/ft water
1.133 ft water	= 1 in mercury	1.133 ft water/in mercury

FLOW

694 GPM	= 1 MGD	694 GPM/MGD
1.55 CFS	= 1 MGD	1.55 CFS/MGD

TIME

60 sec	= 1 min	60 sec/min
60 min	= 1 hr	60 min/hr
24 hr	= 1 day	24 hr/day*

*This may be written either as 24 hr/day or 1 day/24 hours depending on which units we wish to convert to obtain our desired results.

A.2 BASIC FORMULAS

IRON AND MANGANESE CONTROL

1a. $$\text{Stock Solution, mg/mL} = \frac{(\text{Polyphosphate, grams})(1{,}000\text{ mg/g})}{(\text{Solution, liter})(1{,}000\text{ mL/L})}$$

1b. $$\text{Dose, mg/L} = \frac{(\text{Stock Solution, mg/mL})(\text{Volume Added, mL})}{\text{Sample Volume, L}}$$

1c. $$\text{Dose, lb/MG} = \frac{(\text{Dose, mg/L})(3.785\text{ L/gal})(1{,}000{,}000/\text{Million})}{(1{,}000\text{ mg/g})(454\text{ g/lb})}$$

2. Chemical Feeder, lb/day $= (\text{Flow, MGD})(\text{Dose, mg/L})(8.34\ \text{lb/gal})$

3. Detention Time, min $= \dfrac{(\text{Basin Vol, gal})(24\ \text{hr/day})(60\ \text{min/hr})}{\text{Flow, gal/day}}$

4. $KMnO_4$ Dose, mg/L $= 0.2(\text{Iron, mg/L}) + 2.0(\text{Manganese, mg/L})$

FLUORIDATION

5. Feed Rate, gal/day $= \dfrac{(\text{Feed Rate, lb F/day})(100\%)}{(\text{NaF Solution, lb F/gal})(\text{Purity, \%})}$

6. Feed Solution, gal/day $= \dfrac{(\text{Flow, gal/day})(\text{Feed Dose, mg/L})}{\text{Feed Solution, mg/L}}$

7. Fluoride Ion Purity, % $= \dfrac{(\text{Molecular Weight of Fluoride})(100\%)}{\text{Molecular Weight of Chemical}}$

8a. Feed Dose, mg/L $= \text{Desired Dose, mg/L} - \text{Actual Conc, mg/L}$

8b. Feed Rate, lb/day $= \dfrac{\text{Feed Rate, lb F/day}}{\text{lb F/lb Commercial } Na_2SiF_6}$

9. Feed Solution, gal $= \dfrac{(\text{Flow Vol, gal})(\text{Feed Dose, mg/L})}{\text{Feed Solution, mg/L}}$

10. Mixture Strength, % $= \dfrac{(\text{Tank, gal})(\text{Tank, \%}) + (\text{Vendor, gal})(\text{Vendor, \%})}{\text{Tank, gal} + \text{Vendor, gal}}$

SOFTENING

11. Total Hardness, mg/L as $CaCO_3$ = Calcium Hardness, mg/L as $CaCO_3$ + Magnesium Hardness, mg/L as $CaCO_3$

12. If alkalinity is greater than total hardness,

    Carbonate Hardness, mg/L as $CaCO_3$ = Total Hardness, mg/L as $CaCO_3$

    and

    Noncarbonate Hardness, mg/L as $CaCO_3$ = 0

13. If alkalinity is less than total hardness,

    Carbonate Hardness, mg/L as $CaCO_3$ = Alkalinity, mg/L as $CaCO_3$

    and

    Noncarbonate Hardness, mg/L as $CaCO_3$ = Total Hardness, mg/L as $CaCO_3$ − Alkalinity, mg/L as $CaCO_3$

14a. Phenolphthalein Alkalinity, mg/L as $CaCO_3$ $= \dfrac{A \times N \times 50{,}000}{\text{mL of sample}}$

14b. Total Alkalinity, mg/L as $CaCO_3$ $= \dfrac{B \times N \times 50{,}000}{\text{mL of sample}}$

15a. Hydrated Lime ($Ca(OH)_2$) Feed, mg/L $= \dfrac{(A + B + C + D)1.15}{\text{Purity of Lime as a decimal}}$

15b. Soda Ash (Na_2CO_3) Feed, mg/L = (Noncarbonate Hardness, mg/L as $CaCO_3$)(106/100)

15c. Total CO_2 Feed, mg/L = ($Ca(OH)_2$ excess, mg/L)(44/74) + (Mg^{2+} residual, mg/L)(44/24.3)

16. Feeder Setting, lb/day = (Flow, MGD)(Conc, mg/L)(8.34 lb/gal)

17. Feed Rate, lb/min $= \dfrac{\text{Feeder Setting, lb/day}}{(60\text{ min/hr})(24\text{ hr/day})}$

18. Hardness, mg/L $= \dfrac{(\text{Hardness, gpg})(17.1\text{ mg/L})}{1\text{ gpg}}$

19. Exchange Capacity, grains $= (\text{Resin Vol, ft}^3)(\text{Removal Capacity, grains/ft}^3)$

20. Water Treated, gal $= \dfrac{\text{Exchange Capacity, grains}}{\text{Hardness, grains/gal}}$

21. Operating Time, hr $= \dfrac{\text{Water Treated, gal}}{(\text{Avg Daily Flow, gal/min})(60\text{ min/hr})}$

22. Salt Needed, lb $= (\text{Salt Required, lb/1,000 gr})(\text{Hardness Removed, gr})$

23. Bypass Flow, GPD $= \dfrac{(\text{Total Flow, GPD})(\text{Plant Effl Hardness, gpg})}{\text{Raw Water Hardness, gpg}}$

SPECIALIZED TREATMENT PROCESSES

24. Avg TTHM, µg/L $= \dfrac{\text{Sum of Measurements, µg/L}}{\text{Number of Measurements}}$

25. Annual Running TTHM Average, µg/L $= \dfrac{\text{Sum of Average TTHM for Four Quarters}}{\text{Number of Quarters}}$

26. Blended Concentration, mg/L $= \dfrac{(\text{Flow 1, GPM})(\text{Conc 1, mg/L}) + (\text{Flow 2, GPM})(\text{Conc 2, mg/L})}{\text{Flow 1, GPM} + \text{Flow 2, GPM}}$

MEMBRANE TREATMENT PROCESSES

27. Flow, GPD/ft² $= \dfrac{(\text{Flux, g/cm}^2\text{-sec})(2.54\text{ cm/in})^2(12\text{ in/ft})^2(60\text{ sec/min})(60\text{ min/hr})(24\text{ hr/day})}{(1{,}000\text{ g/L})(3.785\text{ L/gal})}$

28. Mineral Rejection, % $= \left(1 - \dfrac{\text{Product TDS, mg/L}}{\text{Feed TDS, mg/L}}\right)(100\%)$

29. Recovery, % $= \dfrac{(\text{Product Flow, MGD})(100\%)}{\text{Feed Flow, MGD}}$

MAINTENANCE

30. Pump Capacity, GPM $= \dfrac{\text{Tank Volume, gal}}{\text{Pumping Time, min}}$

31. Flow, GPD $= \dfrac{(\text{Volume Pumped, gal})(24\text{ hr/day})}{\text{Time, hr}}$

32. Polymer Feed, lb/day $= \dfrac{(\text{Poly Conc, mg/L})(\text{Vol Pumped, mL})(60\text{ min/hr})(24\text{ hr/day})}{(\text{Time Pumped, min})(1{,}000\text{ mL/L})(1{,}000\text{ mg/g})(454\text{ g/lb})}$

33. Chemical Feed, lb/day $= \dfrac{(\text{Chemical, g})(60\text{ min/hr})(24\text{ hr/day})}{(454\text{ g/lb})(\text{Time, min})}$

ADVANCED LABORATORY PROCEDURES

34. Threshold Odor Number (TON) $= \dfrac{\text{Sample Size, mL} + \text{Odor-Free Water, mL}}{\text{Sample Size, mL}}$

35. Geometric Mean $= (X_1 \times X_2 \times X_3 \times \ldots X_n)^{1/n}$

36. Flavor Threshold Number (FTN) $= \dfrac{\text{Sample Size, mL} + \text{Flavor-Free Water, mL}}{\text{Sample Size, mL}}$

37a. Arithmetic Mean, $\bar{X}$, Flavor Rating $= \dfrac{X_1 + X_2 + X_3 + \ldots X_n}{n}$

37b. Standard Deviation, S, Flavor Rating $= \left[\dfrac{(X_1 - \bar{X})^2 + (X_2 - \bar{X})^2 + \ldots (X_n - \bar{X})^2}{n - 1}\right]^{0.5}$

or $= \left[\dfrac{(X_1{}^2 + X_2{}^2 + \ldots X_n{}^2) - (X_1 + X_2 + \ldots X_n)^2/n}{n - 1}\right]^{0.5}$

REGULATIONS

38. Portions Positive, %/mo $= \dfrac{\text{(Number Positive/mo)(100\%)}}{\text{Total Portions Tested}}$

SAFETY

39. Injury Frequency Rate $= \dfrac{\text{(Number of Injuries/yr)(1,000,000)}}{\text{Number of Hours Worked/yr}}$

40. Injury Severity Rate $= \dfrac{\text{(Number of Hours Lost/yr)(1,000,000)}}{\text{Number of Hours Worked/yr}}$

LOG REMOVALS

46a. Log Removal, particles/mL = Log Influent, particles/mL – Log Effluent, particles/mL

46b. Log Removal, particles/mL $= \text{Log}\left(\dfrac{\text{Influent, particles/mL}}{\text{Effluent, particles/mL}}\right)$

A.3 TYPICAL WATER TREATMENT PLANT PROBLEMS (ENGLISH SYSTEM)

A.30 Iron and Manganese Control

Example 1

A standard polyphosphate solution is prepared by mixing and dissolving 1.0 gram of polyphosphate in a container and adding distilled water to the 1-liter mark. Determine the concentration of the stock solution in milligrams per milliliter. If 6.0 milliliters of the stock solution are added to a 1-liter sample, what is the polyphosphate dose in milligrams per liter and pounds per million gallons?

Known		Unknown
Polyphosphate, g	= 1.0 g	1. Stock Solution, mg/mL
Solution, L	= 1.0 L	2. Dose, mg/L
Stock Solution, mL	= 6 mL	3. Dose, lb/MG
Sample, L	= 1 L	

1. Calculate the concentration of the stock solution in milligrams per milliliter.

$$\text{Stock Solution, mg/mL} = \frac{(\text{Polyphosphate, g})(1{,}000\text{ mg/g})}{(\text{Solution, L})(1{,}000\text{ mL/L})}$$

$$= \frac{(1.0\text{ g})(1{,}000\text{ mg/g})}{(1\text{ L})(1{,}000\text{ mL/L})}$$

$$= 1.0\text{ mg/mL}$$

2. Determine the polyphosphate dose in the sample in milligrams per liter.

$$\text{Dose, mg/L} = \frac{(\text{Stock Solution, mg/mL})(\text{Vol Added, mL})}{\text{Sample Volume, L}}$$

$$= \frac{(1.0\text{ mg/mL})(6\text{ mL})}{1\text{ L}}$$

$$= 6.0\text{ mg/L}$$

3. Determine the polyphosphate dose in the sample in pounds of phosphate per million gallons of water.

$$\text{Dose, lb/MG} = \frac{(\text{Dose, mg/L})(3.785\text{ L/gal})(1{,}000{,}000/\text{Mil})}{(1{,}000\text{ mg/g})(454\text{ g/lb})}$$

$$= \frac{(6.0\text{ mg/L})(3.785\text{ L/gal})(1{,}000{,}000/\text{Mil})}{(1{,}000\text{ mg/g})(454\text{ g/lb})}$$

$$= 50\text{ lb/MG}$$

Example 2

Determine the chemical feeder setting in pounds of polyphosphate per day if 0.62 MGD is treated with a dose of 6 mg/L.

Known		Unknown
Flow, MGD	= 0.62 MGD	Chemical Feeder, lb/day
Dose, mg/L	= 6 mg/L	

Determine the chemical feeder setting in pounds per day.

$$\text{Chemical Feeder, lb/day} = (\text{Flow, MGD})(\text{Dose, mg/L})(8.34\text{ lb/gal})$$

$$= (0.62\text{ MGD})(6\text{ mg/L})(8.34\text{ lb/gal})$$

$$= 31\text{ lb/day}$$

Example 3

A reaction basin 14 feet in diameter and 4 feet deep treats a flow of 240,000 gallons per day. What is the average detention time in minutes?

Known		Unknown
Diameter, ft	= 14 ft	Detention Time, min
Depth, ft	= 4 ft	
Flow, GPD	= 240,000 GPD	

1. Calculate the basin volume in cubic feet.

$$\text{Basin Vol, ft}^3 = (0.785)(\text{Diameter, ft})^2(\text{Depth, ft})$$

$$= (0.785)(14\text{ ft})^2(4\text{ ft})$$

$$= 615\text{ ft}^3$$

2. Convert the basin volume from cubic feet to gallons.

$$\text{Basin Vol, gal} = (\text{Basin Vol, ft}^3)(7.48\text{ gal/ft}^3)$$

$$= (615\text{ ft}^3)(7.48\text{ gal/ft}^3)$$

$$= 4{,}600\text{ gal}$$

3. Determine the average detention time in minutes for the reaction basin.

$$\text{Detention Time, min} = \frac{(\text{Basin Vol, gal})(24\text{ hr/day})(60\text{ min/hr})}{\text{Flow, gal/day}}$$

$$= \frac{(4{,}600\text{ gal})(24\text{ hr/day})(60\text{ min/hr})}{240{,}000\text{ gal/day}}$$

$$= 28\text{ minutes}$$

Example 4

Calculate the potassium permanganate dose in milligrams per liter for a well water with 2.4 mg/L iron before aeration and 0.3 mg/L after aeration. The manganese concentration is 0.8 mg/L both before and after aeration.

Known		Unknown
Iron, mg/L	= 0.3 mg/L	$KMnO_4$ Dose, mg/L
Manganese, mg/L	= 0.8 mg/L	

Calculate the potassium permanganate dose in milligrams per liter.

$$\text{KMnO}_4 \text{ Dose, mg/L} = 0.2(\text{Iron, mg/L}) + 2.0(\text{Manganese, mg/L})$$

$$= 0.2(0.3 \text{ mg/L}) + 2.0(0.8 \text{ mg/L})$$

$$= 1.66 \text{ mg/L}$$

NOTE: If there are any oxidizable compounds (organic color, bacteria, or hydrogen sulfide) in the water, the dose will have to be increased.

A.31 Fluoridation

Example 5

Determine the setting for a chemical feed pump in gallons per day when the desired fluoride dose is 1.8 pounds of fluoride per day. The sodium fluoride solution contains 0.2 pound of fluoride per gallon and the fluoride purity is 43.4 percent.

Known		Unknown
Feed Rate, lb F/day	= 1.8 lb F/day	Feed Rate, gal/day
NaF Solution, lb F/gal	= 0.2 lb F/gal	
Purity, %	= 43.4%	

Determine the setting on the chemical feed pump in gallons per day.

$$\text{Feed Rate, gal/day} = \frac{(\text{Feed Rate, lb F/day})(100\%)}{(\text{NaF Solution, lb F/gal})(\text{Purity, \%})}$$

$$= \frac{(1.8 \text{ lb F/day})(100\%)}{(0.2 \text{ lb F/gal})(43.4\%)}$$

$$= 20.7 \text{ gal/day}$$

$$\text{or} = 21 \text{ gal/day}$$

Example 6

Determine the setting on a chemical feed pump in gallons per day if 500,000 gallons per day of water must be treated with 0.9 mg/L of fluoride. The fluoride feed solution contains 18,000 mg/L of fluoride.

Known		Unknown
Flow, gal/day	= 500,000 gal/day	Feed Pump, gal/day
Fluoride, mg/L	= 0.9 mg/L	
Feed Solution, mg/L	= 18,000 mg/L	

Determine the setting on the chemical feed pump in gallons per day.

$$\text{Feed Pump, gal/day} = \frac{(\text{Flow, gal/day})(\text{Feed Dose, mg/L})}{\text{Feed Solution, mg/L}}$$

$$= \frac{(500{,}000 \text{ gal/day})(0.9 \text{ mg/L})}{18{,}000 \text{ mg/L}}$$

$$= 25 \text{ gal/day}$$

Example 7

Determine the fluoride ion purity of Na_2SiF_6 as a percent.

Known	Unknown
Fluoride Chemical, Na_2SiF_6	Fluoride Ion Purity, %

Determine the molecular weight of fluoride and Na_2SiF_6.

Symbol	(No. Atoms)	(Atomic Wt)	=	Molecular Wt
Na_2	(2)	(22.99)	=	45.98
Si	(1)	(28.09)	=	28.09
F_6	(6)	(19.00)	=	114.00
Molecular Weight of Chemical			=	188.07

Calculate the fluoride ion as a percent.

$$\text{Fluoride Ion Purity, \%} = \frac{(\text{Molecular Weight of Fluoride})(100\%)}{\text{Molecular Weight of Chemical}}$$

$$= \frac{(114.00)(100\%)}{188.07}$$

$$= 60.62\%$$

Example 8

A flow of 1.7 MGD is treated with sodium silicofluoride. The raw water contains 0.2 mg/L of fluoride ion and the desired fluoride concentration is 1.1 mg/L. What should be the chemical feed rate in pounds per day? Assume each pound of commercial sodium silicofluoride (Na_2SiF_6) contains 0.6 pound of fluoride ion.

Known		Unknown
Flow, MGD	= 1.7 MGD	Feed Rate, lb/day
Raw Water F, mg/L	= 0.2 mg/L	
Desired F, mg/L	= 1.1 mg/L	
Chemical, lb F/lb	= 0.6 lb F/lb	

1. Determine the fluoride feed dose in milligrams per liter.

$$\text{Feed Dose, mg/L} = \text{Desired Dose, mg/L} - \text{Actual Conc, mg/L}$$

$$= 1.1 \text{ mg/L} - 0.2 \text{ mg/L}$$

$$= 0.9 \text{ mg/L}$$

2. Calculate the fluoride feed rate in pounds per day.

$$\text{Feed Rate, lb F/day} = (\text{Flow, MGD})(\text{Feed Dose, mg/L})(8.34 \text{ lb/gal})$$

$$= (1.7 \text{ MGD})(0.9 \text{ mg/L})(8.34 \text{ lb/gal})$$

$$= 12.8 \text{ lb F/day}$$

3. Determine the chemical feed rate in pounds of commercial sodium silicofluoride per day.

$$\text{Feed Rate, lb/day} = \frac{\text{Feed Rate, lb F/day}}{\text{lb F/lb Commercial } Na_2SiF_6}$$

$$= \frac{12.8 \text{ lb F/day}}{0.6 \text{ lb F/lb Commercial } Na_2SiF_6}$$

$$= 21.3 \text{ lb/day Commercial } Na_2SiF_6$$

Example 9

The feed solution from a saturator containing 1.8 percent fluoride ion is used to treat a total flow of 250,000 gallons of water. The raw water has a fluoride ion content of 0.2 mg/L and the desired fluoride level in the treated water is 0.9 mg/L. How many gallons of feed solution are needed?

Known		Unknown
Flow Vol, gal	= 250,000 gal	Feed Solution, gal
Raw Water F, mg/L	= 0.2 mg/L	
Desired F, mg/L	= 0.9 mg/L	
Feed Solution, % F	= 1.8% F	

1. Convert the feed solution from a percentage fluoride ion to milligrams fluoride ion per liter of water.

$$1.0\% \text{ F} = 10{,}000 \text{ mg F/L}$$

$$\text{Feed Solution, mg/L} = \frac{(\text{Feed Solution, \%})(10{,}000 \text{ mg/L})}{1.0\%}$$

$$= \frac{(1.8\% \text{ F})(10{,}000 \text{ mg/L})}{1.0\%}$$

$$= 18{,}000 \text{ mg/L}$$

2. Determine the fluoride feed dose in milligrams per liter.

$$\text{Feed Dose, mg/L} = \text{Desired Dose, mg/L} - \text{Raw Water F, mg/L}$$

$$= 0.9 \text{ mg/L} - 0.2 \text{ mg/L}$$

$$= 0.7 \text{ mg/L}$$

3. Calculate the gallons of feed solution needed.

$$\text{Feed Solution, gal} = \frac{(\text{Flow Vol, gal})(\text{Feed Dose, mg/L})}{\text{Feed Solution, mg/L}}$$

$$= \frac{(250{,}000 \text{ gal})(0.7 \text{ mg/L})}{18{,}000 \text{ mg/L}}$$

$$= 9.7 \text{ gallons}$$

Example 10

A hydrofluosilicic acid (H_2SiF_6) tank contains 350 gallons of acid with a strength of 19.3 percent. A commercial vendor delivers 2,500 gallons of acid with a strength of 18.1 percent to the tank. What is the resulting strength of the mixture as a percentage?

Known		Unknown
Tank Contents, gal	= 350 gal	Mixture Strength, %
Tank Strength, %	= 19.3%	
Vendor, gal	= 2,500 gal	
Vendor Strength, %	= 18.1%	

Calculate the strength of the mixture as a percentage.

$$\text{Mixture Strength, \%} = \frac{(\text{Tank, gal})(\text{Tank, \%}) + (\text{Vendor, gal})(\text{Vendor, \%})}{\text{Tank, gal} + \text{Vendor, gal}}$$

$$= \frac{(350 \text{ gal})(19.3\%) + (2{,}500 \text{ gal})(18.1\%)}{350 \text{ gal} + 2{,}500 \text{ gal}}$$

$$= \frac{6{,}755 + 45{,}250}{2{,}850}$$

$$= 18.2\%$$

A.32 Softening

Example 11

Determine the total hardness as $CaCO_3$ for a sample of water with a calcium content of 33 mg/L and a magnesium content of 6 mg/L.

Known		Unknown
Calcium, mg/L	= 33 mg/L	Total Hardness, mg/L as $CaCO_3$
Magnesium, mg/L	= 6 mg/L	

Calculate the total hardness as milligrams per liter of calcium carbonate equivalent.

$$\text{Total Hardness, mg/L as } CaCO_3 = \text{Calcium Hardness, mg/L as } CaCO_3 + \text{Magnesium Hardness, mg/L as } CaCO_3$$

$$= 2.5(\text{Ca, mg/L}) + 4.12(\text{Mg, mg/L})$$

$$= 2.5(33 \text{ mg/L}) + 4.12(6 \text{ mg/L})$$

$$= 82 \text{ mg/L} + 25 \text{ mg/L}$$

$$= 107 \text{ mg/L as } CaCO_3$$

Example 12

The alkalinity of a water is 120 mg/L as $CaCO_3$ and the total hardness is 105 mg/L as $CaCO_3$. What is the carbonate and noncarbonate hardness in mg/L as $CaCO_3$?

Known		Unknown
Alkalinity, mg/L	= 120 mg/L as $CaCO_3$	1. Carbonate Hardness, mg/L as $CaCO_3$
Total Hardness, mg/L	= 105 mg/L as $CaCO_3$	2. Noncarbonate Hardness, mg/L as $CaCO_3$

1. Determine the carbonate hardness in mg/L as $CaCO_3$.

   Since the alkalinity is greater than the total hardness (120 mg/L > 105 mg/L),

$$\text{Carbonate Hardness, mg/L as } CaCO_3 = \text{Total Hardness, mg/L as } CaCO_3$$

$$= 105 \text{ mg/L as } CaCO_3$$

2. Determine the noncarbonate hardness in mg/L as $CaCO_3$.

   Since the alkalinity is greater than the total hardness,

$$\text{Noncarbonate Hardness, mg/L as } CaCO_3 = 0$$

   In other words, all of the hardness is in the carbonate form.

Example 13

The alkalinity of a water is 92 mg/L as $CaCO_3$ and the total hardness is 105 mg/L. What is the carbonate and noncarbonate hardness in mg/L as $CaCO_3$?

Known		Unknown
Alkalinity, mg/L	= 92 mg/L as $CaCO_3$	1. Carbonate Hardness, mg/L as $CaCO_3$
Total Hardness, mg/L	= 105 mg/L as $CaCO_3$	2. Noncarbonate Hardness, mg/L as $CaCO_3$

1. Determine the carbonate hardness in mg/L as $CaCO_3$.

   Since the alkalinity is less than the total hardness (92 mg/L < 105 mg/L),

$$\text{Carbonate Hardness, mg/L as } CaCO_3 = \text{Alkalinity, mg/L as } CaCO_3$$

$$= 92 \text{ mg/L as } CaCO_3$$

2. Determine the noncarbonate hardness in mg/L as $CaCO_3$.

   Since the alkalinity is less than the total hardness (92 mg/L < 105 mg/L),

$$\text{Noncarbonate Hardness, mg/L as } CaCO_3 = \text{Total Hardness, mg/L as } CaCO_3 - \text{Alkalinity, mg/L as } CaCO_3$$

$$= 105 \text{ mg/L} - 92 \text{ mg/L}$$

$$= 13 \text{ mg/L as } CaCO_3$$

Example 14

Results from alkalinity titrations on a water sample were as follows:

Known

Sample Size, mL	= 100 mL
mL Titrant Used to pH 8.3, A	= 1.1 mL
Total mL of Titrant Used, B	= 12.4 mL
Acid Normality, *N*	= 0.02 $N\ H_2SO_4$

Unknown

1. Total Alkalinity, mg/L as $CaCO_3$
2. Bicarbonate Alkalinity, mg/L as $CaCO_3$
3. Carbonate Alkalinity, mg/L as $CaCO_3$
4. Hydroxide Alkalinity, mg/L as $CaCO_3$

See Table 14.4, page 81, for alkalinity relationships among constituents.

1. Calculate the phenolphthalein alkalinity in mg/L as $CaCO_3$.

$$\text{Phenolphthalein Alkalinity, mg/L as } CaCO_3 = \frac{A \times N \times 50{,}000}{\text{mL of Sample}}$$

$$= \frac{(1.1 \text{ mL})(0.02\ N)(50{,}000)}{100 \text{ mL}}$$

$$= 11 \text{ mg/L as } CaCO_3$$

2. Calculate the total alkalinity in mg/L as $CaCO_3$.

$$\text{Total Alkalinity, mg/L as } CaCO_3 = \frac{B \times N \times 50{,}000}{\text{mL of Sample}}$$

$$= \frac{(12.4 \text{ mL})(0.02\ N)(50{,}000)}{100 \text{ mL}}$$

$$= 124 \text{ mg/L as } CaCO_3$$

3. Refer to Table 14.4 for alkalinity constituents. The second row indicates that since P is less than ½T (11 mg/L < ½(124 mg/L)), bicarbonate alkalinity is T – 2P and carbonate alkalinity is 2P.

$$\text{Bicarbonate Alkalinity, mg/L as } CaCO_3 = T - 2P$$
$$= 124 \text{ mg/L} - 2(11 \text{ mg/L})$$
$$= 102 \text{ mg/L as } CaCO_3$$

$$\text{Carbonate Alkalinity, mg/L as } CaCO_3 = 2P$$
$$= 2(11 \text{ mg/L})$$
$$= 22 \text{ mg/L as } CaCO_3$$

$$\text{Hydroxide Alkalinity, mg/L as } CaCO_3 = 0 \text{ mg/L as } CaCO_3$$

Example 15

Calculate the hydrated lime ($Ca(OH)_2$) with 90 percent purity, soda ash, and carbon dioxide requirements in milligrams per liter for the water shown below.

Known

Constituents	Source Water	Softened Water After Recarbonation and Filtration
CO_2, mg/L	= 7 mg/L	= 0 mg/L
Total Alkalinity, mg/L	= 125 mg/L as $CaCO_3$	= 22 mg/L as $CaCO_3$
Total Hardness, mg/L	= 240 mg/L as $CaCO_3$	= 35 mg/L as $CaCO_3$
Mg^{2+}, mg/L	= 38 mg/L	= 8 mg/L
pH	= 7.6	= 8.8
Lime Purity, %	= 90%	

Unknown

1. Hydrated Lime, mg/L
2. Soda Ash, mg/L
3. Carbon Dioxide, mg/L

1. Calculate the hydrated lime ($Ca(OH)_2$) required in milligrams per liter.

$$A = (CO_2, \text{ mg/L})(74/44)$$
$$= (7 \text{ mg/L})(74/44)$$
$$= 12 \text{ mg/L}$$

$$B = (\text{Alkalinity, mg/L})(74/100)$$
$$= (125 \text{ mg/L} - 22 \text{ mg/L})(74/100)$$
$$= 76 \text{ mg/L}$$

$C = 0$ Hydroxide Alkalinity = 0

$$D = (Mg^{2+}, \text{ mg/L})(74/24.3)$$
$$= (38 \text{ mg/L} - 8 \text{ mg/L})(74/24.3)$$
$$= 91 \text{ mg/L}$$

$$\text{Hydrated Lime } (Ca(OH)_2) \text{ Feed, mg/L} = \frac{(A + B + C + D)1.15}{\text{Purity of Lime, as a decimal}}$$
$$= \frac{(12 \text{ mg/L} + 76 \text{ mg/L} + 0 + 91 \text{ mg/L})1.15}{0.90}$$
$$= \frac{(179 \text{ mg/L})(1.15)}{0.90}$$
$$= 229 \text{ mg/L}$$

2. Calculate the soda ash required in milligrams per liter.

$$\text{Total Hardness Removed, mg/L as } CaCO_3 = \text{Total Hardness, mg/L as } CaCO_3 - \text{Total Hardness Remaining, mg/L as } CaCO_3$$
$$= 240 \text{ mg/L} - 35 \text{ mg/L}$$
$$= 205 \text{ mg/L as } CaCO_3$$

$$\text{Noncarbonate Hardness, mg/L as } CaCO_3 = \text{Total Hardness Removed, mg/L as } CaCO_3 - \left(\text{Carbonate Hardness, mg/L as } CaCO_3 - \text{Carbonate Hardness Remaining, mg/L as } CaCO_3\right)$$
$$= 205 \text{ mg/L} - (125 \text{ mg/L} - 22 \text{ mg/L})$$
$$= 102 \text{ mg/L as } CaCO_3$$

$$\text{Soda Ash } (Na_2CO_3) \text{ Feed, mg/L} = \left(\text{Noncarbonate Hardness, mg/L as } CaCO_3\right)(106/100)$$
$$= (102 \text{ mg/L})(106/100)$$
$$= 108 \text{ mg/L}$$

3. Calculate the dosage of carbon dioxide required for recarbonation.

$$\text{Excess Lime, mg/L} = (A + B + C + D)(0.15)$$
$$= (12 \text{ mg/L} + 93 \text{ mg/L} + 0 + 116 \text{ mg/L})(0.15)$$
$$= (221 \text{ mg/L})(0.15)$$
$$= 33 \text{ mg/L}$$

$$\text{Total } CO_2 \text{ Feed, mg/L} = (Ca(OH)_2 \text{ Excess, mg/L})(44/74) + (Mg^{2+} \text{ Residual, mg/L})(44/24.3)$$
$$= (33 \text{ mg/L})(44/74) + (8 \text{ mg/L})(44/24.3)$$
$$= 20 \text{ mg/L} + 14 \text{ mg/L}$$
$$= 34 \text{ mg/L}$$

Example 16

The optimum lime dosage from the jar tests is 180 mg/L. If the flow to be treated is 1.7 MGD, what is the feeder setting in pounds per day and the feed rate in pounds per minute?

Known		Unknown
Lime Dose, mg/L	= 180 mg/L	1. Feeder Setting, lb/day
Flow, MGD	= 1.7 MGD	2. Feed Rate, lb/min

1. Calculate the feeder setting in pounds per day.

$$\text{Feeder Setting, lb/day} = (\text{Flow, MGD})(\text{Lime, mg/L})(8.34\text{ lb/gal})$$

$$= (1.7\text{ MGD})(180\text{ mg/L})(8.34\text{ lb/gal})$$

$$= 2{,}550\text{ lb/day}$$

2. Calculate the feed rate in pounds per minute.

$$\text{Feed Rate, lb/min} = \frac{\text{Feeder Setting, lb/day}}{(60\text{ min/hr})(24\text{ hr/day})}$$

$$= \frac{2{,}550\text{ lb/day}}{(60\text{ min/hr})(24\text{ hr/day})}$$

$$= 1.8\text{ lb/min}$$

Example 17

How much soda ash is required (pounds per day and pounds per minute) to remove 40 mg/L noncarbonate hardness as $CaCO_3$ from a flow of 1.7 MGD?

Known		Unknown
Noncarbonate Hardness Removed, mg/L as $CaCO_3$	= 40 mg/L	1. Feeder Setting, lb/day
Flow, MGD	= 1.7 MGD	2. Feed Rate, lb/min

1. Calculate the soda ash dose in milligrams per liter. See Section 14.316, "Calculation of Chemical Dosages," page 85, for the following formula.

$$\text{Soda Ash, mg/L} = \left(\text{Noncarbonate Hardness, mg/L as } CaCO_3\right)(106/100)$$

$$= (40\text{ mg/L})(106/100)$$

$$= 42.4\text{ mg/L}$$

2. Determine the feeder setting in pounds per day.

$$\text{Feeder Setting, lb/day} = (\text{Flow, MGD})(\text{Soda Ash, mg/L})(8.34\text{ lb/gal})$$

$$= (1.7\text{ MGD})(42.4\text{ mg/L})(8.34\text{ lb/gal})$$

$$= 601\text{ lb/day}$$

3. Calculate the soda ash feed rate in pounds per minute.

$$\text{Feed Rate, lb/min} = \frac{\text{Feeder Setting, lb/day}}{(60\text{ min/hr})(24\text{ hr/day})}$$

$$= \frac{601\text{ lb/day}}{(60\text{ min/hr})(24\text{ hr/day})}$$

$$= 0.42\text{ lb/min}$$

Example 18

What is the hardness in milligrams per liter for a water with a hardness of 12 grains per gallon (gpg)?

Known		Unknown
Hardness, gpg	= 12 gpg	Hardness, mg/L

Calculate the hardness in milligrams per liter.

$$\text{Hardness, mg/L} = \frac{(\text{Hardness, gpg})(17.1\text{ mg/L})}{1\text{ gpg}}$$

$$= \frac{(12\text{ gpg})(17.1\text{ mg/L})}{1\text{ gpg}}$$

$$= 205\text{ mg/L}$$

Example 19

Estimate the exchange capacity in grains of hardness for an ion exchange unit that contains 600 cubic feet of resin with a removal capacity of 25,000 grains per cubic foot.

Known		Unknown
Resin Volume, ft^3	= 600 ft^3	Exchange Capacity, grains
Removal Capacity, grains/ft^3	= 25,000 grains/ft^3	

Estimate the exchange capacity in grains of hardness.

$$\text{Exchange Capacity, grains} = (\text{Resin Vol, ft}^3)(\text{Removal Capacity, grains/ft}^3)$$

$$= (600\text{ ft}^3)(25{,}000\text{ grains/ft}^3)$$

$$= 15{,}000{,}000\text{ grains of hardness}$$

Example 20

How many gallons of water with a hardness of 12 grains per gallon may be treated by an ion exchange softener with an exchange capacity of 15,000,000 grains?

Known		Unknown
Hardness, gpg	= 12 gpg	Water Treated, gallons
Exchange Capacity, grains	= 15,000,000 grains	

Calculate the gallons of water that may be treated.

$$\text{Water Treated, gal} = \frac{\text{Exchange Capacity, grains}}{\text{Hardness, gpg}}$$

$$= \frac{15{,}000{,}000 \text{ grains}}{12 \text{ gpg}}$$

$$= 1{,}250{,}000 \text{ gal}$$

Example 21

How many hours will an ion exchange softening unit operate when treating an average daily flow of 750 GPM? The unit is capable of softening 1,250,000 gallons of water before requiring regeneration.

Known		Unknown
Avg Daily Flow, GPM	= 750 GPM	Operating Time, hr
Water Treated, gal	= 1,250,000 gal	

Estimate how many hours the softening unit can operate before requiring regeneration.

$$\text{Operating Time, hr} = \frac{\text{Water Treated, gal}}{(\text{Avg Daily Flow, gal/min})(60 \text{ min/hr})}$$

$$= \frac{1{,}250{,}000 \text{ gal}}{(750 \text{ gal/min})(60 \text{ min/hr})}$$

$$= 27.8 \text{ hours}$$

Example 22

Determine the pounds of salt needed to regenerate an ion exchange softening unit capable of removing 15,000,000 grains of hardness if 0.25 pound of salt is required for every 1,000 grains of hardness removed.

Known		Unknown
Hardness Removed, gr	= 15,000,000 gr	Salt Needed, lb
Salt Required, lb/1,000 gr	= 0.25 lb Salt/1,000 gr	

Calculate the pounds of salt needed to regenerate the ion exchange softening unit.

$$\text{Salt Needed, lb} = (\text{Salt Required, lb/1,000 gr})(\text{Hardness Removed, gr})$$

$$= \left(\frac{0.25 \text{ lb Salt}}{1{,}000 \text{ grains}}\right)(15{,}000{,}000 \text{ grains})$$

$$= 3{,}750 \text{ lb of Salt}$$

Example 23

Estimate the bypass flow in gallons per day around an ion exchange softener if the plant treats 250,000 gallons per day with a source water hardness of 20 grains per gallon if the desired product water hardness is 5 grains per gallon.

Known		Unknown
Total Flow, GPD	= 250,000 GPD	Bypass Flow, GPD
Source Water Hardness, gpg	= 20 gpg	
Plant Effl Hardness, gpg	= 5 gpg	

Estimate the bypass flow in gallons per day.

$$\text{Bypass Flow, GPD} = \frac{(\text{Total Flow, GPD})(\text{Plant Effl Hardness, gpg})}{\text{Source Water Hardness, gpg}}$$

$$= \frac{(250{,}000 \text{ GPD})(5 \text{ gpg})}{20 \text{ gpg}}$$

$$= 62{,}500 \text{ GPD}$$

A.33 Specialized Treatment Processes

Example 24

A water utility collected and analyzed eight samples from a water distribution system on the same day for TTHMs. The results are shown below.

Sample No.	1	2	3	4	5	6	7	8
TTHM, µg/L	80	90	100	90	110	100	100	90

What was the average TTHM for the day?

Known	Unknown
Results from analyses of 8 TTHM samples	Average TTHM level for the day

Calculate the average TTHM level in micrograms per liter.

$$\text{Avg TTHM, µg/L} = \frac{\text{Sum of Measurements, µg/L}}{\text{Number of Measurements}}$$

$$= \frac{80 \text{ µg/L} + 90 \text{ µg/L} + 100 \text{ µg/L} + 90 \text{ µg/L} + 110 \text{ µg/L} + 100 \text{ µg/L} + 100 \text{ µg/L} + 90 \text{ µg/L}}{8}$$

$$= \frac{760 \text{ µg/L}}{8}$$

$$= 95 \text{ µg/L}$$

Example 25

The results of the quarterly average TTHM measurements for two years are given below. Calculate the running annual average of the four quarterly measurements in micrograms per liter.

Quarter	1	2	3	4	1	2	3	4
Avg Quarterly TTHM, μg/L	77	88	112	95	83	87	109	89

Known	Unknown
Results from analyses of two years of TTHM sampling	Running Annual Average of quarterly TTHM measurements

Calculate the running annual average of the quarterly TTHM measurements.

$$\text{Annual Running TTHM Average, μg/L} = \frac{\text{Sum of Average TTHM for Four Quarters}}{\text{Number of Quarters}}$$

QUARTERS 1, 2, 3, AND 4

$$\text{Annual Running TTHM Average, μg/L} = \frac{77\ \text{μg/L} + 88\ \text{μg/L} + 112\ \text{μg/L} + 95\ \text{μg/L}}{4}$$

$$= \frac{372\ \text{μg/L}}{4}$$

$$= 93\ \text{μg/L}$$

QUARTERS 2, 3, 4, AND 1

$$\text{Annual Running TTHM Average, μg/L} = \frac{88\ \text{μg/L} + 112\ \text{μg/L} + 95\ \text{μg/L} + 83\ \text{μg/L}}{4}$$

$$= \frac{378\ \text{μg/L}}{4}$$

$$= 95\ \text{μg/L}$$

QUARTERS 3, 4, 1, AND 2

$$\text{Annual Running TTHM Average, μg/L} = \frac{112\ \text{μg/L} + 95\ \text{μg/L} + 83\ \text{μg/L} + 87\ \text{μg/L}}{4}$$

$$= \frac{377\ \text{μg/L}}{4}$$

$$= 94\ \text{μg/L}$$

QUARTERS 4, 1, 2, AND 3

$$\text{Annual Running TTHM Average, μg/L} = \frac{95\ \text{μg/L} + 83\ \text{μg/L} + 87\ \text{μg/L} + 109\ \text{μg/L}}{4}$$

$$= \frac{374\ \text{μg/L}}{4}$$

$$= 94\ \text{μg/L}$$

QUARTERS 1, 2, 3, AND 4

$$\text{Annual Running TTHM Average, μg/L} = \frac{83\ \text{μg/L} + 87\ \text{μg/L} + 109\ \text{μg/L} + 89\ \text{μg/L}}{4}$$

$$= \frac{368\ \text{μg/L}}{4}$$

$$= 92\ \text{μg/L}$$

SUMMARY OF RESULTS

Quarter	1	2	3	4	1	2	3	4
Avg Quarterly TTHM, μg/L	77	88	112	95	83	87	109	89
Annual Running TTHM Avg, μg/L				93	95	94	94	92

Example 26

Well number 1 delivers a flow of 50 GPM with an arsenic concentration of 0.024 mg/L. Well number 2 delivers a flow of 500 GPM with an arsenic concentration of 0.002 mg/L. What would be the blended arsenic concentration in milligrams per liter if the flows from these two wells were blended together?

Known		Unknown
Well 1 Q (Q_1), GPM	= 50 GPM	Blended Conc (C_b), mg/L
Well 2 Q (Q_2), GPM	= 500 GPM	
Well 1 Conc (C_1), mg/L	= 0.024 mg/L	
Well 1 Conc (C_2), mg/L	= 0.002 mg/L	

Calculate the blended arsenic concentration in milligrams per liter.

$$Q_b = Q_1 + Q_2 = 50\ \text{GPM} + 500\ \text{GPM} = 550\ \text{GPM}$$

$$C_b = \frac{Q_1C_1 + Q_2C_2}{Q_b}$$

$$= \frac{50\ \text{GPM} \times 0.024\ \text{mg/L} + 500\ \text{GPM} \times 0.002\ \text{mg/L}}{550\ \text{GPM}}$$

$$= \frac{2.2}{550}$$

$$= 0.004\ \text{mg/L}$$

A.34 Membrane Treatment Processes

Example 27

Convert a water flux of 12×10^{-4} g/cm²-sec to gallon per day per square foot.

Known		Unknown
Water Flux, g/cm²-sec	$= 12 \times 10^{-4}$ g/cm²-sec	Flow, GPD/ft²

Convert the water flux from g/cm²-sec to flow in GPD/ft².

$$\text{Flow, GPD/ft}^2 = \frac{\left(\text{Flux}, \frac{\text{g}}{\text{cm}^2\text{-sec}}\right)\left(\frac{2.54\ \text{cm}}{\text{in}}\right)^2\left(\frac{12\ \text{in}}{\text{ft}}\right)^2\left(\frac{60\ \text{sec}}{\text{min}}\right)\left(\frac{60\ \text{min}}{\text{hr}}\right)\left(\frac{24\ \text{hr}}{\text{day}}\right)}{\left(\frac{1{,}000\ \text{g}}{\text{L}}\right)\left(\frac{3.785\ \text{L}}{\text{gal}}\right)}$$

$$= \frac{\left(\frac{0.0012\ \text{g}}{\text{cm}^2\text{-sec}}\right)\left(\frac{2.54\ \text{cm}}{\text{in}}\right)^2\left(\frac{12\ \text{in}}{\text{ft}}\right)^2\left(\frac{60\ \text{sec}}{\text{min}}\right)\left(\frac{60\ \text{min}}{\text{hr}}\right)\left(\frac{24\ \text{hr}}{\text{day}}\right)}{\left(\frac{1{,}000\ \text{g}}{\text{L}}\right)\left(\frac{3.785\ \text{L}}{\text{gal}}\right)}$$

$$= 25.5\ \text{GPD/ft}^2$$

Example 28

Estimate the ability of a reverse osmosis plant to reject minerals by calculating the mineral rejection as a percent. The feedwater contains 1,800 mg/L TDS and the product water TDS is 120 mg/L.

Known		Unknown
Feedwater TDS, mg/L	= 1,800 mg/L	Mineral Reaction, %
Product Water TDS, mg/L	= 120 mg/L	

Calculate the mineral rejection as a percent.

$$\text{Mineral Rejection, \%} = \left(1 - \frac{\text{Product TDS, mg/L}}{\text{Feed TDS, mg/L}}\right)(100\%)$$

$$= \left(1 - \frac{120 \text{ mg/L}}{1{,}800 \text{ mg/L}}\right)(100\%)$$

$$= (1 - 0.067)(100\%)$$

$$= 93.3\%$$

Example 29

Estimate the percent recovery of a reverse osmosis unit with a 4-2-1 arrangement if the feed flow is 2.0 MGD and the product flow is 1.75 MGD.

Known		Unknown
Product Flow, MGD	= 1.75 MGD	Recovery, %
Feed Flow, MGD	= 2.0 MGD	

Calculate the recovery as a percent.

$$\text{Recovery, \%} = \frac{(\text{Product Flow, MGD})(100\%)}{\text{Feed Flow, MGD}}$$

$$= \frac{(1.75 \text{ MGD})(100\%)}{2.0 \text{ MGD}}$$

$$= 87.5\%$$

A.35 Maintenance

Example 30

Calculate the pumping capacity of a pump in gallons per minute when 12 minutes are required for the water to rise 3 feet in an 8-foot by 6-foot rectangular tank.

Known		Unknown
Length, ft	= 8 ft	Pump Capacity, GPM
Width, ft	= 6 ft	
Depth, ft	= 3 ft	
Time, min	= 12 min	

1. Calculate the volume pumped in cubic feet.

$$\text{Volume Pumped, ft}^3 = (\text{Length, ft})(\text{Width, ft})(\text{Depth, ft})$$

$$= (8 \text{ ft})(6 \text{ ft})(3 \text{ ft})$$

$$= 144 \text{ ft}^3$$

2. Convert the volume pumped from cubic feet to gallons.

$$\text{Volume Pumped, gal} = (\text{Volume Pumped, ft}^3)(7.48 \text{ gal/ft}^3)$$

$$= (144 \text{ ft}^3)(7.48 \text{ gal/ft}^3)$$

$$= 1{,}077 \text{ gal}$$

3. Calculate the pump capacity in gallons per minute.

$$\text{Pump Capacity, GPM} = \frac{\text{Volume Pumped, gal}}{\text{Pumping Time, min}}$$

$$= \frac{1{,}077 \text{ gal}}{12 \text{ min}}$$

$$= 90 \text{ GPM}$$

Example 31

A small chemical feed pump lowered the chemical solution in a 2.5-foot diameter tank 2.25 feet during 7 hours. Estimate the flow delivered by the pump in gallons per minute and gallons per day.

Known		Unknown
Tank Diameter, ft	= 2.5 ft	Flow, GPM
Chemical Drop, ft	= 2.25 ft	Flow, GPD
Time, hr	= 7.0 hr	

1. Determine the gallons of chemical solution pumped.

$$\text{Volume, gal} = (0.785)(\text{Diameter, ft})^2(\text{Drop, ft})(7.48 \text{ gal/ft}^3)$$

$$= (0.785)(2.5 \text{ ft})^2(2.25 \text{ ft})(7.48 \text{ gal/ft}^3)$$

$$= 83 \text{ gallons}$$

2. Estimate the flow delivered by the pump in gallons per minute and gallons per day.

$$\text{Flow, GPM} = \frac{\text{Volume Pumped, gal}}{(\text{Time, hr})(60 \text{ min/hr})}$$

$$= \frac{83 \text{ gallons}}{(7 \text{ hr})(60 \text{ min/hr})}$$

$$= 0.2 \text{ GPM}$$

or

$$\text{Flow, GPD} = \frac{(\text{Volume Pumped, gal})(24 \text{ hr/day})}{\text{Time, hr}}$$

$$= \frac{(83 \text{ gallons})(24 \text{ hr/day})}{7 \text{ hr}}$$

$$= 285 \text{ GPD}$$

Example 32

Determine the chemical feed in pounds of polymer per day from a chemical feed pump. The polymer solution is 1.8 percent or 18,000 mg polymer per liter. Assume a specific gravity of the polymer solution of 1.0. During a test run, the chemical feed pump delivered 650 mL of polymer solution in 4.5 minutes.

Known		Unknown
Polymer Solution, %	= 1.8%	Polymer Feed, lb/day
Polymer Conc, mg/L	= 18,000 mg/L	
Polymer Sp Gr	= 1.0	
Volume Pumped, mL	= 650 mL	
Time Pumped, min	= 4.5 min	

Calculate the polymer fed by the chemical feed pump in pounds of polymer per day.

$$\text{Polymer Feed, lb/day} = \frac{(\text{Poly Conc, mg/L})(\text{Vol Pumped, mL})(60\ \text{min/hr})(24\ \text{hr/day})}{(\text{Time Pumped, min})(1{,}000\ \text{mL/L})(1{,}000\ \text{mg/g})(454\ \text{g/lb})}$$

$$= \frac{(18{,}000\ \text{mg/L})(650\ \text{mL})(60\ \text{min/hr})(24\ \text{hr/day})}{(4.5\ \text{min})(1{,}000\ \text{mL/L})(1{,}000\ \text{mg/g})(454\ \text{g/lb})}$$

$$= 8.2\ \text{lb/day}$$

Example 33

Determine the actual chemical feed in pounds per day from a dry chemical feeder. A pie tin placed under the chemical feeder caught 824 grams of chemical in 5 minutes.

Known		Unknown
Chemical, g	= 824 g	Chemical Feed, lb/day
Time, min	= 5 min	

Determine the chemical feed in pounds per day.

$$\text{Chemical Feed, lb/day} = \frac{(\text{Chemical, g})(60\ \text{min/hr})(24\ \text{hr/day})}{(454\ \text{g/lb})(\text{Time, min})}$$

$$= \frac{(824\ \text{g})(60\ \text{min/hr})(24\ \text{hr/day})}{(454\ \text{g/lb})(5\ \text{min})}$$

$$= 523\ \text{lb/day}$$

A.36 Advanced Laboratory Procedures

Example 34

Calculate the threshold odor number (TON) for a sample when the first detectable odor occurred when the 70 mL sample was diluted to 200 mL (130 mL of odor-free water was added to the 70 mL sample).

Known		Unknown
A or Sample Size, mL	= 70 mL	TON
B or Odor-Free Water, mL	= 130 mL	

Calculate the threshold odor number (TON).

$$\text{TON} = \frac{A + B}{A}$$

$$= \frac{70\ \text{mL} + 130\ \text{mL}}{70\ \text{mL}}$$

$$= 3$$

Example 35

Determine the geometric mean threshold odor number for a panel of six testers given the results shown below.

Known		Unknown
Tester 1, X_1	= 2	Geometric Mean Threshold Odor Number
Tester 2, X_2	= 4	
Tester 3, X_3	= 3	
Tester 4, X_4	= 8	
Tester 5, X_5	= 6	
Tester 6, X_6	= 2	

Calculate the geometric mean.

$$\text{Geometric Mean TON} = (X_1 \times X_2 \times X_3 \times X_4 \times X_5 \times X_6)^{1/n}$$

$$= (2 \times 4 \times 3 \times 8 \times 6 \times 2)^{1/6}$$

$$= (2{,}304)^{0.167}$$

$$= 3.6$$

Example 36

Calculate the flavor threshold number (FTN) for a sample when the first detectable flavor occurred when the 8.3 mL sample was diluted to 200 mL (191.7 mL of flavor-free water was added to the 8.3 mL sample).

Known		Unknown
A or Sample Size, mL	= 8.3 mL	FTN
B or Flavor-Free Water, mL	= 191.7 mL	

Calculate the flavor threshold number (FTN).

$$\text{FTN} = \frac{A + B}{A}$$

$$= \frac{8.3\ \text{mL} + 191.7\ \text{mL}}{8.3\ \text{mL}}$$

$$= 24$$

Example 37

Determine the flavor rating for a water by calculating the arithmetic mean and standard deviation for the panel ratings given below.

Known	Unknown
Tester 1, $X_1 = 2$	1. Arithmetic Mean, $\overline{X}$
Tester 2, $X_2 = 5$	2. Standard Deviation, S
Tester 3, $X_3 = 3$	
Tester 4, $X_4 = 6$	
Tester 5, $X_5 = 2$	
Tester 6, $X_6 = 6$	

1. Calculate the arithmetic mean, $\overline{X}$, flavor rating.

$$\text{Arithmetic Mean, } \overline{X}, \text{ Flavor Rating} = \frac{X_1 + X_2 + X_3 + X_4 + X_5 + X_6}{n}$$

$$= \frac{2 + 5 + 3 + 6 + 2 + 6}{6}$$

$$= \frac{24}{6}$$

$$= 4$$

2. Calculate the standard deviation, S, of the flavor rating.

$$\text{Standard Deviation, S} = \left[\frac{(X_1 - \overline{X})^2 + (X_2 - \overline{X})^2 + (X_3 - \overline{X})^2 + (X_4 - \overline{X})^2 + (X_5 - \overline{X})^2 + (X_6 - \overline{X})^2}{n - 1}\right]^{0.5}$$

$$= \left[\frac{(2-4)^2 + (5-4)^2 + (3-4)^2 + (6-4)^2 + (2-4)^2 + (6-4)^2}{6-1}\right]^{0.5}$$

$$= \left[\frac{(-2)^2 + (1)^2 + (-1)^2 + (2)^2 + (-2)^2 + (2)^2}{5}\right]^{0.5}$$

$$= \left[\frac{4 + 1 + 1 + 4 + 4 + 4}{5}\right]^{0.5}$$

$$= \left[\frac{18}{5}\right]^{0.5}$$

$$= (3.6)^{0.5}$$

$$= 1.9$$

or

$$\text{Standard Deviation, S} = \left[\frac{(X_1^2 + X_2^2 + X_3^2 + X_4^2 + X_5^2 + X_6^2) - (X_1 + X_2 + X_3 + X_4 + X_5 + X_6)^2/n}{n - 1}\right]^{0.5}$$

$$= \left[\frac{(2^2 + 5^2 + 3^2 + 6^2 + 2^2 + 6^2) - (2 + 5 + 3 + 6 + 2 + 6)^2/6}{6 - 1}\right]^{0.5}$$

$$= \left[\frac{(4 + 25 + 9 + 36 + 4 + 36) - (24)^2/6}{5}\right]^{0.5}$$

$$= \left[\frac{114 - 96}{5}\right]^{0.5}$$

$$= \left[\frac{18}{5}\right]^{0.5}$$

$$= (3.6)^{0.5}$$

$$= 1.9$$

A.37 Regulations

Example 38

A small water system collected 14 samples during one month. After each sample was collected, 10 mL of each sample was placed in each of 5 fermentation tubes. At the end of the month, the results indicated that 2 out of a total of 70 fermentation tubes were positive. What percent of the portions tested during the month were positive?

Known		Unknown
Number Positive/mo	= 2 Positive/mo	Portions Positive, %/mo
Total Portions Tested	= 70 Portions	

Calculate the percent of the portions tested during the month that were positive.

$$\text{Portions Positive, \%/mo} = \frac{(\text{Number Positive/mo})(100\%)}{\text{Total Portions Tested}}$$

$$= \frac{(2\ \text{Positive/mo})(100\%)}{70\ \text{Portions}}$$

$$= 3\%/\text{mo}$$

A.38 Safety

Example 39

Calculate the injury frequency rate for a water utility where there were four injuries in one year and the operators worked 97,120 hours.

Known		Unknown
Number of Injuries/yr	= 4 Injuries/yr	Injury Frequency Rate
Number of Hours Worked/yr	= 97,120 hr/yr	

Calculate the injury frequency rate.

$$\text{Injury Frequency Rate} = \frac{(\text{Number of Injuries/yr})(1{,}000{,}000)}{\text{Number of Hours Worked/yr}}$$

$$= \frac{(4\ \text{Injuries/yr})(1{,}000{,}000)}{97{,}120\ \text{hr/yr}}$$

$$= 41.2$$

Example 40

Calculate the injury severity rate for a water company that experienced 57 operator-hours lost due to injuries while the operators worked 97,120 hours during the year.

Known		Unknown
Number of Hours Lost/yr	= 57 hr/yr	Injury Severity Rate
Number of Hours Worked/yr	= 97,120 hr/yr	

Calculate the injury severity rate.

$$\text{Injury Severity Rate} = \frac{(\text{Number of Hours Lost/yr})(1{,}000{,}000)}{\text{Number of Hours Worked/yr}}$$

$$= \frac{(57\ \text{hr/yr})(1{,}000{,}000)}{97{,}120\ \text{hr/yr}}$$

$$= 587$$

A.4 BASIC CONVERSION FACTORS (METRIC SYSTEM)

LENGTH		
100 cm	= 1 m	100 cm/m
3.281 ft	= 1 m	3.281 ft/m
AREA		
2.4711 ac	= 1 ha*	2.4711 ac/ha
10,000 m^2	= 1 ha	10,000 m^2/ha
VOLUME		
1,000 mL	= 1 liter	1,000 mL/L
1,000 L	= 1 m^3	1,000 L/m^3
3.785 L	= 1 gal	3.785 L/gal
WEIGHT		
1,000 mg	= 1 g	1,000 mg/g
1,000 g	= 1 kg	1,000 g/kg
DENSITY		
1 kg	= 1 liter	1 kg/L
PRESSURE		
10.015 m	= 1 kg/cm^2	10.015 m/kg/cm^2
1 Pascal	= 1 N/m^2	1 Pa/N/m^2
1 psi	= 6,895 Pa	1 psi/6,895 Pa
FLOW		
3,785 m^3/day	= 1 MGD	3,785 m^3/day/MGD
3.785 ML/day	= 1 MGD	3.785 ML/day/MGD

*hectare

A.5 TYPICAL WATER TREATMENT PLANT PROBLEMS (METRIC SYSTEM)

A.50 Iron and Manganese Control

Example 1

A standard polyphosphate solution is prepared by mixing and dissolving 1.0 gram of polyphosphate in a container and adding distilled water to the 1-liter mark. Determine the concentration of the stock solution in milligrams per milliliter. If 6.0 milliliters of the stock solution are added to a 1-liter sample, what is the polyphosphate dose in milligrams per liter and milligrams per kilogram?

Known		Unknown
Polyphosphate, g	= 1.0 g	1. Stock Solution, mg/mL
Solution, L	= 1.0 L	2. Dose, mg/L
Stock Solution, mL	= 6 mL	3. Dose, lb/MG
Sample, L	= 1 L	

1. Calculate the concentration of the stock solution in milligrams per milliliter.

$$\text{Stock Solution, mg/mL} = \frac{(\text{Polyphosphate, g})(1{,}000 \text{ mg/g})}{(\text{Solution, L})(1{,}000 \text{ mL/L})}$$

$$= \frac{(1.0 \text{ g})(1{,}000 \text{ mg/g})}{(1 \text{ L})(1{,}000 \text{ mL/L})}$$

$$= 1.0 \text{ mg/mL}$$

2. Determine the polyphosphate dose in the sample in milligrams per liter.

$$\text{Dose, mg/L} = \frac{(\text{Stock Solution, mg/mL})(\text{Vol Added, mL})}{\text{Sample Volume, L}}$$

$$= \frac{(1.0 \text{ mg/mL})(6 \text{ mL})}{1 \text{ L}}$$

$$= 6.0 \text{ mg/L}$$

3. Determine the polyphosphate dose in the sample in milligrams of phosphate per kilogram of water.

$$\text{Dose, mg/kg} = \frac{(\text{Stock Solution, mg/L})(\text{Vol Added, mL})}{(\text{Sample Volume, L})(1 \text{ kg/L})}$$

$$= \frac{(1.0 \text{ mg/mL})(6 \text{ mL})}{(1 \text{ L})(1 \text{ kg/L})}$$

$$= 6.0 \text{ mg/kg}$$

Example 2

Determine the chemical feeder setting in grams per second and kilograms per day if 2.4 MLD (mega or million liters per day) are treated with a dose of 5 mg/L.

Known		Unknown
Flow, MLD	= 2.4 MLD	1. Chemical Feeder, g/sec
Dose, mg/L	= 5 mg/L	2. Chemical Feeder, kg/day

1. Determine the chemical feeder setting in grams per second.

$$\text{Chemical Feeder, g/sec} = \frac{(\text{Flow, MLD})(\text{Dose, mg/L})(1{,}000{,}000/\text{M})}{(24 \text{ hr/day})(60 \text{ min/hr})(60 \text{ sec/min})(1{,}000 \text{ mg/g})}$$

$$= \frac{(2.4 \text{ MLD})(5 \text{ mg/L})(1{,}000{,}000/\text{M})}{(24 \text{ hr/day})(60 \text{ min/hr})(60 \text{ sec/min})(1{,}000 \text{ mg/g})}$$

$$= 0.139 \text{ g/sec}$$

$$\text{or} = 139 \text{ mg/sec}$$

2. Determine the chemical feeder setting in kilograms per day.

$$\text{Chemical Feeder, kg/day} = \frac{(\text{Flow, MLD})(\text{Dose, mg/L})(1{,}000{,}000/\text{M})}{(1{,}000 \text{ mg/g})(1{,}000 \text{ g/kg})}$$

$$= \frac{(2.4 \text{ MLD})(5 \text{ mg/L})(1{,}000{,}000/\text{M})}{(1{,}000 \text{ mg/g})(1{,}000 \text{ g/kg})}$$

$$= 12 \text{ kg/day}$$

Example 3

A reaction basin 4 meters in diameter and 1.2 meters deep treats a flow of 0.9 MLD. What is the average detention time in minutes?

Known		Unknown
Diameter, m	= 4 m	Detention Time, min
Depth, m	= 1.2 m	
Flow, MLD	= 0.9 MLD	

1. Calculate the basin volume in cubic meters.

$$\text{Basin Vol, m}^3 = (0.785)(\text{Diameter, m})^2(\text{Depth, m})$$

$$= (0.785)(4 \text{ m})^2(1.2 \text{ m})$$

$$= 15.1 \text{ m}^3$$

2. Determine the average detention time in minutes for the reaction basin.

$$\text{Detention Time, min} = \frac{(\text{Basin Vol, m}^3)(24 \text{ hr/day})(60 \text{ min/hr})(1{,}000 \text{ L/m}^3)}{(\text{Flow, MLD})(1{,}000{,}000/\text{M})}$$

$$= \frac{(15.1 \text{ m}^3)(24 \text{ hr/day})(60 \text{ min/hr})(1{,}000 \text{ L/m}^3)}{(0.9 \text{ MLD})(1{,}000{,}000/\text{M})}$$

$$= 24 \text{ minutes}$$

Example 4

Calculate the potassium permanganate dose in milligrams per liter for a well water with 2.4 mg/L iron before aeration and 0.3 mg/L after aeration. The manganese concentration is 0.8 mg/L both before and after aeration.

Known		Unknown
Iron, mg/L	= 0.3 mg/L	$KMnO_4$ Dose, mg/L
Manganese, mg/L	= 0.8 mg/L	

Calculate the potassium permanganate dose in milligrams per liter.

$$\text{KMnO}_4 \text{ Dose, mg/L} = 0.2(\text{Iron, mg/L}) + 2.0(\text{Manganese, mg/L})$$

$$= 0.2(0.3 \text{ mg/L}) + 2.0(0.8 \text{ mg/L})$$

$$= 1.66 \text{ mg/L}$$

NOTE: If there are any oxidizable compounds (organic color, bacteria, or hydrogen sulfide) in the water, the dose will have to be increased.

A.51 Fluoridation

Example 5

Determine the setting for a chemical feed pump in liters per day and milliliters per second when the desired fluoride dose is 0.9 kilogram of fluoride per day. The sodium fluoride solution contains 0.025 kilogram of fluoride per liter and fluoride purity is 43.4 percent.

Known		Unknown
Feed Rate, kg F/day	= 0.9 kg F/day	1. Feed Rate, liters/day
NaF Solution, kg F/L	= 0.025 kg F/L	2. Feed Rate, mL/sec
Purity, %	= 43.4%	

1. Determine the setting on the chemical feed pump in liters per day.

$$\text{Feed Rate, L/day} = \frac{(\text{Feed Rate, kg F/day})(100\%)}{(\text{NaF Solution, kg F/L})(\text{Purity, \%})}$$

$$= \frac{(0.9 \text{ kg F/day})(100\%)}{(0.025 \text{ kg F/L})(43.4\%)}$$

$$= 83 \text{ liters/day}$$

2. Convert the feed rate from kilograms per day to grams per second.

$$\text{Feed Rate, g/sec} = \frac{(\text{Feed Rate, kg F/day})(1{,}000 \text{ g/kg})}{(24 \text{ hr/day})(60 \text{ min/hr})(60 \text{ sec/min})}$$

$$= \frac{(0.9 \text{ kg F/day})(1{,}000 \text{ g/kg})}{(24 \text{ hr/day})(60 \text{ min/hr})(60 \text{ sec/min})}$$

$$= 0.010 \text{ g/sec}$$

$$\text{or} = 10 \text{ mg/sec}$$

3. Determine the setting on the chemical feed pump in milliliters per second.

$$\text{Feed Rate, mL/sec} = \frac{(\text{Feed Rate, g/sec})(100\%)(1{,}000 \text{ mL/L})}{(\text{NaF Solution, kg F/L})(\text{Purity, \%})(1{,}000 \text{ g/kg})}$$

$$= \frac{(0.010 \text{ g/sec})(100\%)(1{,}000 \text{ mL/L})}{(0.025 \text{ kg F/L})(43.4\%)(1{,}000 \text{ g/kg})}$$

$$= 0.92 \text{ mL/sec}$$

Example 6

Determine the setting on a chemical feed pump in liters per day and milliliters per second if 2 megaliters per day of water must be treated with 0.9 mg/L of fluoride. The fluoride feed solution contains 18,000 mg/L of fluoride.

Known		Unknown
Flow, MLD	= 2 MLD	1. Feed Pump, liters/day
Fluoride, mg/L	= 0.9 mg/L	
Feed Solution, mg/L	= 18,000 mg/L	2. Feed Pump, mL/sec

1. Determine the setting on the chemical feed pump in liters per day.

$$\text{Feed Pump, liters/day} = \frac{(\text{Flow, MLD})(\text{Feed Dose, mg/L})(1{,}000{,}000/\text{M})}{\text{Feed Solution, mg/L}}$$

$$= \frac{(2.0 \text{ MLD})(0.9 \text{ mg/L})(1{,}000{,}000/\text{M})}{18{,}000 \text{ mg/L}}$$

$$= 100 \text{ liters/day}$$

2. Determine the setting on the chemical feed pump in milliliters per second.

$$\text{Feed Pump, mL/sec} = \frac{(\text{Flow, MLD})(\text{Feed Dose, mg/L})(1{,}000{,}000/\text{M})}{(\text{Feed Solution, mg/L})(24 \text{ hr/day})(60 \text{ min/hr})(60 \text{ sec/min})}$$

$$= \frac{(2.0 \text{ MLD})(0.9 \text{ mg/L})(1{,}000{,}000/\text{M})}{(18{,}000 \text{ mg/L})(24 \text{ hr/day})(60 \text{ min/hr})(60 \text{ sec/min})}$$

$$= 1.16 \text{ mL/sec}$$

Example 7

Determine the fluoride ion purity of Na_2SiF_6 as a percent.

Known	Unknown
Fluoride Chemical, Na_2SiF_6	Fluoride Ion Purity, %

Determine the molecular weight of fluoride and Na_2SiF_6.

Symbol	(No. Atoms)	(Atomic Wt)	=	Molecular Wt
Na_2	(2)	(22.99)	=	45.98
Si	(1)	(28.09)	=	28.09
F_6	(6)	(19.00)	=	114.00
Molecular Weight of Chemical			=	188.07

Calculate the fluoride ion as a percent.

$$\text{Fluoride Ion Purity, \%} = \frac{(\text{Molecular Weight of Fluoride})(100\%)}{\text{Molecular Weight of Chemical}}$$

$$= \frac{(114.00)(100\%)}{188.07}$$

$$= 60.62\%$$

Example 8

A flow of 6.5 MLD is treated with sodium silicofluoride. The raw water contains 0.2 mg/L of fluoride ion and the desired fluoride concentration is 1.1 mg/L. What should be the chemical feed rate in kilograms per day and milligrams per second? Assume each gram of commercial sodium silicofluoride (Na_2SiF_6) contains 0.6 gram of fluoride ion.

Known		Unknown
Flow, MLD	= 6.5 MLD	1. Feed Rate, kg/day
Raw Water F, mg/L	= 0.2 mg/L	2. Feed Rate, mg/sec
Desired F, mg/L	= 1.1 mg/L	
Chemical, g F/g	= 0.6 g F/g	

1. Determine the fluoride feed dose in milligrams per liter.

$$\text{Feed Dose, mg/L} = \text{Desired Dose, mg/L} - \text{Actual Conc, mg/L}$$

$$= 1.1 \text{ mg/L} - 0.2 \text{ mg/L}$$

$$= 0.9 \text{ mg/L}$$

2. Calculate the chemical feed rate in kilograms per day.

$$\text{Feed Rate, kg/day} = \frac{(\text{Flow, MLD})(\text{Feed Dose, mg/L})(1{,}000{,}000/\text{M})}{(\text{Purity, g F/g Chemical})(1{,}000 \text{ mg/g})(1{,}000 \text{ g/kg})}$$

$$= \frac{(6.5 \text{ MLD})(0.9 \text{ mg/L})(1{,}000{,}000/\text{M})}{(0.6 \text{ g F/g Chemical})(1{,}000 \text{ mg/g})(1{,}000 \text{ g/kg})}$$

$$= 9.75 \text{ kg/day}$$

3. Calculate the chemical feed rate in milligrams per second.

$$\text{Feed Rate, mg/sec} = \frac{(\text{Flow, MLD})(\text{Feed Dose, mg/L})(1{,}000{,}000/\text{M})}{(\text{Purity, g F/g Chem})(24 \text{ hr/day})(60 \text{ min/hr})(60 \text{ sec/min})}$$

$$= \frac{(6.5 \text{ MLD})(0.9 \text{ mg/L})(1{,}000{,}000/\text{M})}{(0.6 \text{ g F/g Chem})(24 \text{ hr/day})(60 \text{ min/hr})(60 \text{ sec/min})}$$

$$= 113 \text{ mg/sec}$$

Example 9

The feed solution from a saturator containing 1.8 percent fluoride ion is used to treat a total flow of 0.95 megaliter (ML) of water. The raw water has a fluoride ion content of 0.2 mg/L and the desired fluoride level in the treated water is 0.9 mg/L. How many liters of feed solution are needed?

Known		Unknown
Flow Vol, ML	= 0.95 ML	Feed Solution, liters
Raw Water F, mg/L	= 0.2 mg/L	
Desired F, mg/L	= 0.9 mg/L	
Feed Solution, % F	= 1.8% F	

1. Convert the feed solution from a percentage fluoride ion to milligrams fluoride ion per liter of water.

$$1.0\% \text{ F} = 10{,}000 \text{ mg F/L}$$

$$\text{Feed Solution, mg/L} = \frac{(\text{Feed Solution, \%})(10{,}000 \text{ mg/L})}{1.0\%}$$

$$= \frac{(1.8\% \text{ F})(10{,}000 \text{ mg/L})}{1.0\%}$$

$$= 18{,}000 \text{ mg/L}$$

2. Determine the fluoride feed dose in milligrams per liter.

$$\text{Feed Dose, mg/L} = \text{Desired Dose, mg/L} - \text{Raw Water F, mg/L}$$

$$= 0.9 \text{ mg/L} - 0.2 \text{ mg/L}$$

$$= 0.7 \text{ mg/L}$$

3. Calculate the liters of feed solution needed.

$$\text{Feed Solution, L} = \frac{(\text{Flow Vol, ML})(\text{Feed Dose, mg/L})(1{,}000{,}000/\text{M})}{\text{Feed Solution, mg/L}}$$

$$= \frac{(0.95 \text{ ML})(0.7 \text{ mg/L})(1{,}000{,}000/\text{M})}{18{,}000 \text{ mg/L}}$$

$$= 37 \text{ liters}$$

Example 10

A hydrofluosilicic acid (H_2SiF_6) tank contains 1,300 liters of acid with a strength of 19.3 percent. A commercial vendor delivers 10,000 liters of acid with a strength of 18.1 percent to the tank. What is the resulting strength of the mixture as a percentage?

Known		Unknown
Tank Contents, liters	= 1,300 L	Mixture Strength, %
Tank Strength, %	= 19.3%	
Vendor, L	= 10,000 L	
Vendor Strength, %	= 18.1%	

Calculate the strength of the mixture as a percentage.

$$\text{Mixture Strength, \%} = \frac{(\text{Tank, L})(\text{Tank, \%}) + (\text{Vendor, L})(\text{Vendor, \%})}{(\text{Tank, L} + \text{Vendor, L})}$$

$$= \frac{(1{,}300 \text{ L})(19.3\%) + (10{,}000 \text{ L})(18.1\%)}{(1{,}300 \text{ L} + 10{,}000 \text{ L})}$$

$$= \frac{25{,}090 + 181{,}000}{11{,}300}$$

$$= 18.2\%$$

A.52 Softening

Example 11

Determine the total hardness as $CaCO_3$ for a sample of water with a calcium content of 33 mg/L and a magnesium content of 6 mg/L.

Known		Unknown
Calcium, mg/L	= 33 mg/L	Total Hardness, mg/L as $CaCO_3$
Magnesium, mg/L	= 6 mg/L	

Calculate the total hardness as milligrams per liter of calcium carbonate equivalent.

$$\text{Total Hardness, mg/L as } CaCO_3 = \text{Calcium Hardness, mg/L as } CaCO_3 + \text{Magnesium Hardness, mg/L as } CaCO_3$$

$$= 2.5(\text{Ca, mg/L}) + 4.12(\text{Mg, mg/L})$$

$$= 2.5(33 \text{ mg/L}) + 4.12(6 \text{ mg/L})$$

$$= 82 \text{ mg/L} + 25 \text{ mg/L}$$

$$= 107 \text{ mg/L as } CaCO_3$$

Example 12

The alkalinity of a water is 120 mg/L as $CaCO_3$ and the total hardness is 105 mg/L as $CaCO_3$. What is the carbonate and noncarbonate hardness in mg/L as $CaCO_3$?

Known		Unknown
Alkalinity, mg/L	= 120 mg/L as $CaCO_3$	1. Carbonate Hardness, mg/L as $CaCO_3$
Total Hardness, mg/L	= 105 mg/L as $CaCO_3$	2. Noncarbonate Hardness, mg/L as $CaCO_3$

1. Determine the carbonate hardness in mg/L as $CaCO_3$.

   Since the alkalinity is greater than the total hardness (120 mg/L >105 mg/L),

$$\text{Carbonate Hardness, mg/L as } CaCO_3 = \text{Total Hardness, mg/L as } CaCO_3$$

$$= 105 \text{ mg/L as } CaCO_3$$

2. Determine the noncarbonate hardness in mg/L as $CaCO_3$.

   Since the alkalinity is greater than the total hardness,

$$\text{Noncarbonate Hardness, mg/L as } CaCO_3 = 0$$

   In other words, all of the hardness is in the carbonate form.

Example 13

The alkalinity of a water is 92 mg/L as $CaCO_3$ and the total hardness is 105 mg/L. What is the carbonate and noncarbonate hardness in mg/L as $CaCO_3$?

Known		Unknown
Alkalinity, mg/L	= 92 mg/L as $CaCO_3$	1. Carbonate Hardness, mg/L as $CaCO_3$
Total Hardness, mg/L	= 105 mg/L as $CaCO_3$	2. Noncarbonate Hardness, mg/L as $CaCO_3$

1. Determine the carbonate hardness in mg/L as $CaCO_3$.

   Since the alkalinity is less than the total hardness (92 mg/L <105 mg/L),

$$\text{Carbonate Hardness, mg/L as } CaCO_3 = \text{Alkalinity, mg/L as } CaCO_3$$

$$= 92 \text{ mg/L as } CaCO_3$$

2. Determine the noncarbonate hardness in mg/L as $CaCO_3$.

   Since the alkalinity is less than the total hardness (92 mg/L <105 mg/L),

$$\text{Noncarbonate Hardness, mg/L as } CaCO_3 = \text{Total Hardness, mg/L as } CaCO_3 - \text{Alkalinity, mg/L as } CaCO_3$$

$$= 105 \text{ mg/L} - 92 \text{ mg/L}$$

$$= 13 \text{ mg/L as } CaCO_3$$

Example 14

Results from alkalinity titrations on a water sample were as follows:

Known

Sample Size, mL	= 100 mL
mL Titrant Used to pH 8.3, A	= 1.1 mL
Total mL of Titrant Used, B	= 12.4 mL
Acid Normality, *N*	= 0.02 $N\ H_2SO_4$

Unknown

1. Total Alkalinity, mg/L as $CaCO_3$
2. Bicarbonate Alkalinity, mg/L as $CaCO_3$
3. Carbonate Alkalinity, mg/L as $CaCO_3$
4. Hydroxide Alkalinity, mg/L as $CaCO_3$

See Table 14.4, page 81, for alkalinity relationships among constituents.

1. Calculate the phenolphthalein alkalinity in mg/L as $CaCO_3$.

$$\text{Phenolphthalein Alkalinity, mg/L as } CaCO_3 = \frac{A \times N \times 50{,}000}{\text{mL of Sample}}$$

$$= \frac{(1.1 \text{ mL})(0.02\ N)(50{,}000)}{100 \text{ mL}}$$

$$= 11 \text{ mg/L as } CaCO_3$$

2. Calculate the total alkalinity in mg/L as $CaCO_3$.

$$\text{Total Alkalinity, mg/L as } CaCO_3 = \frac{B \times N \times 50{,}000}{\text{mL of Sample}}$$

$$= \frac{(12.4\ \text{mL})(0.02\ N)(50{,}000)}{100\ \text{mL}}$$

$$= 124\ \text{mg/L as } CaCO_3$$

3. Refer to Table 14.4 for alkalinity constituents. The second row indicates that since P is less than ½T (11 mg/L < ½(124 mg/L)), bicarbonate alkalinity is T – 2P and carbonate alkalinity is 2P.

$$\text{Bicarbonate Alkalinity, mg/L as } CaCO_3 = T - 2P$$

$$= 124\ \text{mg/L} - 2(11\ \text{mg/L})$$

$$= 102\ \text{mg/L as } CaCO_3$$

$$\text{Carbonate Alkalinity, mg/L as } CaCO_3 = 2P$$

$$= 2(11\ \text{mg/L})$$

$$= 22\ \text{mg/L as } CaCO_3$$

$$\text{Hydroxide Alkalinity, mg/L as } CaCO_3 = 0\ \text{mg/L as } CaCO_3$$

Example 15

Calculate the hydrated lime ($Ca(OH)_2$) with 90 percent purity, soda ash, and carbon dioxide requirements in milligrams per liter for the water shown below.

Known

Constituents	Source Water	Softened Water After Recarbonation and Filtration
CO_2, mg/L	= 7 mg/L	= 0 mg/L
Total Alkalinity, mg/L	= 125 mg/L as $CaCO_3$	= 22 mg/L as $CaCO_3$
Total Hardness, mg/L	= 240 mg/L as $CaCO_3$	= 35 mg/L as $CaCO_3$
Mg^{2+}, mg/L	= 38 mg/L	= 8 mg/L
pH	= 7.6	= 8.8
Lime Purity, %	= 90%	

Unknown

1. Hydrated Lime, mg/L
2. Soda Ash, mg/L
3. Carbon Dioxide, mg/L

1. Calculate the hydrated lime ($Ca(OH)_2$) required in milligrams per liter.

$$A = (CO_2,\ \text{mg/L})(74/44)$$

$$= (7\ \text{mg/L})(74/44)$$

$$= 12\ \text{mg/L}$$

$$B = (\text{Alkalinity, mg/L})(74/100)$$

$$= (125\ \text{mg/L} - 22\ \text{mg/L})(74/100)$$

$$= 76\ \text{mg/L}$$

$$C = 0 \qquad \text{Hydroxide Alkalinity} = 0$$

$$D = (Mg^{2+},\ \text{mg/L})(74/24.3)$$

$$= (38\ \text{mg/L} - 8\ \text{mg/L})(74/24.3)$$

$$= 91\ \text{mg/L}$$

$$\text{Hydrated Lime } (Ca(OH)_2) \text{ Feed, mg/L} = \frac{(A + B + C + D)1.15}{\text{Purity of Lime, as a decimal}}$$

$$= \frac{(12\ \text{mg/L} + 76\ \text{mg/L} + 0 + 91\ \text{mg/L})1.15}{0.90}$$

$$= \frac{(179\ \text{mg/L})(1.15)}{0.90}$$

$$= 229\ \text{mg/L}$$

2. Calculate the soda ash required in milligrams per liter.

$$\text{Total Hardness Removed, mg/L as } CaCO_3 = \text{Total Hardness, mg/L as } CaCO_3 - \text{Total Hardness Remaining, mg/L as } CaCO_3$$

$$= 240\ \text{mg/L} - 35\ \text{mg/L}$$

$$= 205\ \text{mg/}L\ \text{as } CaCO_3$$

$$\text{Noncarbonate Hardness, mg/L as } CaCO_3 = \text{Total Hardness Removed, mg/L as } CaCO_3 - \left(\text{Carbonate Hardness, mg/L as } CaCO_3 - \text{Carbonate Hardness Remaining, mg/L as } CaCO_3\right)$$

$$= 205\ \text{mg/L} - (125\ \text{mg/L} - 22\ \text{mg/L})$$

$$= 102\ \text{mg/L as } CaCO_3$$

$$\text{Soda Ash } (Na_2CO_3) \text{ Feed, mg/L} = \left(\text{Noncarbonate Hardness, mg/L as } CaCO_3\right)(106/100)$$

$$= (102\ \text{mg/L})(106/100)$$

$$= 108\ \text{mg/L}$$

3. Calculate the dosage of carbon dioxide required for recarbonation.

$$\text{Excess Lime, mg/L} = (A + B + C + D)(0.15)$$

$$= (12\ \text{mg/L} + 93\ \text{mg/L} + 0 + 116\ \text{mg/L})(0.15)$$

$$= (221\ \text{mg/L})(0.15)$$

$$= 33\ \text{mg/L}$$

$$\text{Total } CO_2 \text{ Feed, mg/L} = (Ca(OH)_2\ \text{Excess, mg/L})(44/74) + (Mg^{2+}\ \text{Residual, mg/L})(44/24.3)$$

$$= (33\ \text{mg/L})(44/74) + (8\ \text{mg/L})(44/24.3)$$

$$= 20\ \text{mg/L} + 14\ \text{mg/L}$$

$$= 34\ \text{mg/L}$$

Example 16

The optimum lime dosage from the jar tests is 180 mg/L. If the flow to be treated is 6.5 MLD, what is the feeder setting in kilograms per day and the feed rate in grams per second?

Known		Unknown
Lime Dose, mg/L	= 180 mg/L	1. Feeder Setting, kg/day
Flow, MLD	= 6.5 MLD	2. Feed Rate, g/sec

1. Calculate the feeder setting in kilograms per day.

$$\text{Feeder Setting, kg/day} = \frac{(\text{Flow, MLD})(\text{Lime, mg/L})(1{,}000{,}000/\text{M})}{(1{,}000\text{ mg/g})(1{,}000\text{ g/kg})}$$

$$= \frac{(6.5\text{ MLD})(180\text{ mg/L})(1{,}000{,}000/\text{M})}{(1{,}000\text{ mg/g})(1{,}000\text{ g/kg})}$$

$$= 1{,}170\text{ kg/day}$$

2. Calculate the feed rate in grams per second.

$$\text{Feed Rate, g/sec} = \frac{(\text{Flow, MLD})(\text{Lime, mg/L})(1{,}000{,}000/\text{M})}{(1{,}000\text{ mg/g})(24\text{ hr/day})(60\text{ min/hr})(60\text{ sec/min})}$$

$$= \frac{(6.5\text{ MLD})(180\text{ mg/L})(1{,}000{,}000/\text{M})}{(1{,}000\text{ mg/g})(24\text{ hr/day})(60\text{ min/hr})(60\text{ sec/min})}$$

$$= 13.5\text{ g/sec}$$

Example 17

How much soda ash is required (kilograms per day and grams per second) to remove 40 mg/L noncarbonate hardness as $CaCO_3$ from a flow of 6.5 MLD?

Known		Unknown
Noncarbonate Hardness Removed, mg/L as $CaCO_3$	= 40 mg/L	1. Feeder Setting, kg/day
Flow, MLD	= 6.5 MLD	2. Feed Rate, g/sec

1. Calculate the soda ash dose in milligrams per liter. See Section 14.316, "Calculation of Chemical Dosages," page 85, for the following formula.

$$\text{Soda Ash, mg/L} = \left(\begin{array}{c}\text{Noncarbonate Hardness,}\\ \text{mg/L as } CaCO_3\end{array}\right)(106/100)$$

$$= (40\text{ mg/L})(106/100)$$

$$= 42.4\text{ mg/L}$$

2. Determine the feeder setting in kilograms per day.

$$\text{Feeder Setting, kg/day} = \frac{(\text{Flow, MLD})(\text{Soda Ash, mg/L})(1{,}000{,}000/\text{M})}{(1{,}000\text{ mg/g})(1{,}000\text{ g/kg})}$$

$$= \frac{(6.5\text{ MLD})(42.4\text{ mg/L})(1{,}000{,}000/\text{M})}{(1{,}000\text{ mg/g})(1{,}000\text{ g/kg})}$$

$$= 276\text{ kg/day}$$

3. Calculate the soda ash feed rate in grams per second.

$$\text{Feed Rate, g/sec} = \frac{(\text{Flow, MLD})(\text{Soda Ash, mg/L})(1{,}000{,}000/\text{M})}{(1{,}000\text{ mg/g})(24\text{ hr/day})(60\text{ min/hr})(60\text{ sec/min})}$$

$$= \frac{(6.5\text{ MLD})(42.4\text{ mg/L})(1{,}000{,}000/\text{M})}{(1{,}000\text{ mg/g})(24\text{ hr/day})(60\text{ min/hr})(60\text{ sec/min})}$$

$$= 3.2\text{ g/sec}$$

Example 18

What is the hardness in grains per gallon (gpg) for a water with a hardness of 200 mg/L?

Known		Unknown
Hardness, mg/L	= 200 mg/L	Hardness, gpg

Calculate the hardness in grains per gallon.

$$\text{Hardness, gpg} = \frac{(\text{Hardness, mg/L})(1\text{ gpg})}{17.1\text{ mg/L}}$$

$$= \frac{(200\text{ mg/L})(1\text{ gpg})}{17.1\text{ mg/L}}$$

$$= 11.7\text{ gpg}$$

Example 19

Estimate the exchange capacity in milligrams of hardness for an ion exchange unit that contains 20 cubic meters of resin with a removal capacity of 14,000 milligrams per cubic meter.

Known		Unknown
Resin Volume, m^3	= 600 m^3	Exchange Capacity, milligrams
Removal Capacity, mg/m^3	= 14,000 mg/m^3	

Estimate the exchange capacity in grains of hardness.

$$\text{Exchange Capacity, mg} = (\text{Resin Vol, m}^3)(\text{Removal Capacity, mg/m}^3)$$

$$= (20\text{ m}^3)(14{,}000\text{ mg/m}^3)$$

$$= 280{,}000\text{ mg of hardness}$$

Example 20

How many liters of water with a hardness of 200 mg/L may be treated by an ion exchange softener with an exchange capacity of 280,000 milligrams?

Known		Unknown
Hardness, mg/L	= 200 mg/L	Water Treated, liters
Exchange Capacity, mg	= 280,000 mg	

Calculate the liters of water that may be treated.

$$\text{Water Treated, liters} = \frac{\text{Exchange Capacity, mg}}{\text{Hardness, mg/L}}$$

$$= \frac{280{,}000 \text{ mg}}{200 \text{ mg/L}}$$

$$= 1{,}400 \text{ liters}$$

Example 21

How many hours will an ion exchange softening unit operate when treating an average daily flow of 50 liters per second? The unit is capable of softening 4,500,000 liters of water before requiring regeneration.

Known		Unknown
Avg Daily Flow, L/sec	= 50 L/sec	Operating Time, hr
Water Treated, L	= 4,500,000 L	

Estimate how many hours the softening unit can operate before requiring regeneration.

$$\text{Operating Time, hr} = \frac{\text{Water Treated, L}}{(\text{Avg Daily Flow, L/sec})(60 \text{ sec/min})(60 \text{ min/hr})}$$

$$= \frac{4{,}500{,}000 \text{ L}}{(50 \text{ L/sec})(60 \text{ sec/min})(60 \text{ min/hr})}$$

$$= 25 \text{ hours}$$

Example 22

Determine the kilograms of salt needed to regenerate an ion exchange softening unit capable of removing 225,000 milligrams of hardness if 7 kilograms of salt are required for every 1,000 milligrams of hardness removed.

Known		Unknown
Hardness Removed, mg	= 225,000 mg	Salt Needed, kg
Salt Required, kg/1,000 mg	= 7 kg Salt/1,000 mg	

Calculate the kilograms of salt needed to regenerate the ion exchange softening unit.

$$\text{Salt Needed, kg} = (\text{Salt Required, kg/1,000 mg})(\text{Hardness Removed, mg})$$

$$= \frac{(7 \text{ kg Salt})(225{,}000 \text{ mg})}{1{,}000 \text{ mg}}$$

$$= 1{,}575 \text{ kilograms of Salt}$$

Example 23

Estimate the bypass flow in cubic meters per day and megaliters per day around an ion exchange softener in a plant that treats 1,000 cubic meters per day with a source water hardness of 350 mg/L if the desired product water hardness is 80 mg/L.

Known		Unknown
Total Flow, m^3/day	= 1,000 m^3/day	1. Bypass Flow, m^3/day
Source Water Hardness, mg/L	= 350 mg/L	2. Bypass Flow, MLD
Plant Effluent Hardness, mg/L	= 80 mg/L	

1. Estimate the bypass flow in cubic meters per day.

$$\text{Bypass Flow, m}^3\text{/day} = \frac{(\text{Total Flow, m}^3\text{/day})(\text{Plant Effl Hardness, mg/L})}{\text{Source Water Hardness, mg/L}}$$

$$= \frac{(1{,}000 \text{ m}^3\text{/day})(80 \text{ mg/L})}{350 \text{ mg/L}}$$

$$= 229 \text{ m}^3\text{/day}$$

2. Estimate the bypass flow in megaliters per day.

$$\text{Bypass Flow, MLD} = \frac{(\text{Total Flow, m}^3\text{/day})(\text{Plant Effl Hardness, mg/L})(1{,}000 \text{ L/m}^3)}{(\text{Source Water Hardness, mg/L})(1{,}000{,}000\text{/M})}$$

$$= \frac{(1{,}000 \text{ m}^3\text{/day})(80 \text{ mg/L})(1{,}000 \text{ L/m}^3)}{(350 \text{ mg/L})(1{,}000{,}000\text{/M})}$$

$$= 0.229 \text{ MLD}$$

A.53 Specialized Treatment Processes

Example 24

A water utility collected and analyzed eight samples from a water distribution system on the same day for TTHMs. The results are shown below.

Sample No.	1	2	3	4	5	6	7	8
TTHM, µg/L	80	90	100	90	110	100	100	90

What was the average TTHM for the day?

Known	Unknown
Results from analyses of 8 TTHM samples	Average TTHM level for the day

Calculate the average TTHM level in micrograms per liter.

$$\text{Avg TTHM, µg/L} = \frac{\text{Sum of Measurements, µg/L}}{\text{Number of Measurements}}$$

$$= \frac{80 \text{ µg/L} + 90 \text{ µg/L} + 100 \text{ µg/L} + 90 \text{ µg/L} + 110 \text{ µg/L} + 100 \text{ µg/L} + 100 \text{ µg/L} + 90 \text{ µg/L}}{8}$$

$$= \frac{760 \text{ µg/L}}{8}$$

$$= 95 \text{ µg/L}$$

Example 25

The results of the quarterly average TTHM measurements for two years are given below. Calculate the running annual average of the four quarterly measurements in micrograms per liter.

Quarter	1	2	3	4	1	2	3	4
Avg Quarterly TTHM, μg/L	77	88	112	95	83	87	109	89

Known	Unknown
Results from analyses of two years of TTHM sampling	Running Annual Average of quarterly TTHM measurements

Calculate the running annual average of the quarterly TTHM measurements.

$$\text{Annual Running TTHM Average, μg/L} = \frac{\text{Sum of Average TTHM for Four Quarters}}{\text{Number of Quarters}}$$

QUARTERS 1, 2, 3, AND 4

$$\text{Annual Running TTHM Average, μg/L} = \frac{77\ \mu g/L + 88\ \mu g/L + 112\ \mu g/L + 95\ \mu g/L}{4}$$

$$= \frac{372\ \mu g/L}{4}$$

$$= 93\ \mu g/L$$

QUARTERS 2, 3, 4, AND 1

$$\text{Annual Running TTHM Average, μg/L} = \frac{88\ \mu g/L + 112\ \mu g/L + 95\ \mu g/L + 83\ \mu g/L}{4}$$

$$= \frac{378\ \mu g/L}{4}$$

$$= 95\ \mu g/L$$

QUARTERS 3, 4, 1, AND 2

$$\text{Annual Running TTHM Average, μg/L} = \frac{112\ \mu g/L + 95\ \mu g/L + 83\ \mu g/L + 87\ \mu g/L}{4}$$

$$= \frac{377\ \mu g/L}{4}$$

$$= 94\ \mu g/L$$

QUARTERS 4, 1, 2, AND 3

$$\text{Annual Running TTHM Average, μg/L} = \frac{95\ \mu g/L + 83\ \mu g/L + 87\ \mu g/L + 109\ \mu g/L}{4}$$

$$= \frac{374\ \mu g/L}{4}$$

$$= 94\ \mu g/L$$

QUARTERS 1, 2, 3, AND 4

$$\text{Annual Running TTHM Average, μg/L} = \frac{83\ \mu g/L + 87\ \mu g/L + 109\ \mu g/L + 89\ \mu g/L}{4}$$

$$= \frac{368\ \mu g/L}{4}$$

$$= 92\ \mu g/L$$

SUMMARY OF RESULTS

Quarter	1	2	3	4	1	2	3	4
Avg Quarterly TTHM, μg/L	77	88	112	95	83	87	109	89
Annual Running TTHM Avg, μg/L				93	95	94	94	92

Example 26

Well number 1 delivers a flow of 0.25 MLD with an arsenic concentration of 0.024 mg/L. Well number 2 delivers a flow of 2.50 MLD with an arsenic concentration of 0.002 mg/L. What would be the blended arsenic concentration in milligrams per liter if the flows from these two wells were blended together?

Known		Unknown
Well 1 Q (Q_1), MLD	= 0.25 MLD	Blended Conc (C_b), mg/L
Well 2 Q (Q_2), MLD	= 2.50 MLD	
Well 1 Conc (C_1), mg/L	= 0.024 mg/L	
Well 1 Conc (C_2), mg/L	= 0.002 mg/L	

Calculate the blended arsenic concentration in milligrams per liter.

$$Q_b = Q_1 + Q_2 = .25\text{ MLD} + 2.50\text{ MLD} = 2.75\text{ MLD}$$

$$C_b = \frac{Q_1C_1 + Q_2C_2}{Q_b}$$

$$= \frac{0.25\text{ MLD} \times 0.024\text{ mg/L} + 2.50\text{ MLD} \times 0.002\text{ mg/L}}{2.75\text{ MLD}}$$

$$= \frac{0.011}{2.75}$$

$$= 0.004\text{ mg/L}$$

A.54 Membrane Treatment Processes

Example 27

Convert a water flux of 12×10^{-4} g/cm²-sec to liters per second per square centimeter and liters per day per square centimeter.

Known	Unknown
Water Flux, g/cm²-sec = 12×10^{-4} g/cm²-sec	1. Flow, liters per cm²-sec 2. Flow, liters per cm²-day

1. Convert the water flux from g/cm²-sec to flow in liters per second per square centimeter.

$$\text{Flow, L/cm}^2\text{-sec} = \frac{\text{Flux, g/cm}^2\text{-sec}}{1{,}000\text{ g/L}}$$

$$= \frac{12 \times 10^{-4}\text{ g/cm}^2\text{-sec}}{1{,}000\text{ mL}}$$

$$= 12 \times 10^{-7}\text{ L/cm}^2\text{-sec}$$

2. Convert the water flux from g/cm²-sec to flow in liters per day per square centimeter.

$$\text{Flow, L/cm}^2\text{-day} = \frac{(\text{Flux, g/cm}^2\text{-sec})(60 \text{ sec/min})(60 \text{ min/hr})(24 \text{ hr/day})}{1{,}000 \text{ g/L}}$$

$$= \frac{(0.0012 \text{ g/cm}^2\text{-sec})(60 \text{ sec/min})(60 \text{ min/hr})(24 \text{ hr/day})}{1{,}000 \text{ g/L}}$$

$$= 0.10 \text{ L/cm}^2\text{-day}$$

Example 28

Estimate the ability of a reverse osmosis plant to reject minerals by calculating the mineral rejection as a percent. The feedwater contains 1,800 mg/L TDS and the product water TDS is 120 mg/L.

Known		Unknown
Feedwater TDS, mg/L	= 1,800 mg/L	Mineral Reaction, %
Product Water TDS, mg/L	= 120 mg/L	

Calculate the mineral rejection as a percent.

$$\text{Mineral Rejection, \%} = \left(1 - \frac{\text{Product TDS, mg/L}}{\text{Feed TDS, mg/L}}\right)(100\%)$$

$$= \left(1 - \frac{120 \text{ mg/L}}{1{,}800 \text{ mg/L}}\right)(100\%)$$

$$= (1 - 0.067)(100\%)$$

$$= 93.3\%$$

Example 29

Estimate the percent recovery of a reverse osmosis unit with a 4-2-1 arrangement if the feed flow is 8.0 MLD and the product flow is 7.0 MLD.

Known		Unknown
Product Flow, MLD	= 7.0 MLD	Recovery, %
Feed Flow, MLD	= 8.0 MLD	

Calculate the recovery as a percent.

$$\text{Recovery, \%} = \frac{(\text{Product Flow, MLD})(100\%)}{\text{Feed Flow, MLD}}$$

$$= \frac{(7.0 \text{ MLD})(100\%)}{8.0 \text{ MLD}}$$

$$= 87.5\%$$

A.55 Maintenance

Example 30

Calculate the pumping capacity of a pump in liters per second when 12 minutes are required for the water to rise 1.0 meter in a 2.5-meter by 2.0-meter rectangular tank.

Known		Unknown
Length, m	= 2.5 m	Pump Capacity, L/sec
Width, m	= 2.0 m	
Depth, m	= 1.0 m	
Time, min	= 12 min	

1. Calculate the volume pumped in cubic meters.

$$\text{Volume Pumped, m}^3 = (\text{Length, m})(\text{Width, m})(\text{Depth, m})$$

$$= (2.5 \text{ m})(2.0 \text{ m})(1.0 \text{ m})$$

$$= 5.0 \text{ m}^3$$

2. Calculate the pump capacity in liters per second.

$$\text{Pump Capacity, liters/sec} = \frac{(\text{Volume Pumped, m}^3)(1{,}000 \text{ L/m}^3)}{(\text{Pumping Time, min})(60 \text{ sec/min})}$$

$$= \frac{(5.0 \text{ m}^3)(1{,}000 \text{ L/m}^3)}{(12 \text{ min})(60 \text{ sec/min})}$$

$$= 6.9 \text{ L/sec}$$

Example 31

A small chemical feed pump lowered the chemical solution in a 0.8-meter diameter tank 0.7 meter during 7.0 hours. Estimate the flow delivered by the pump in liters per second and milliliters per second.

Known		Unknown
Tank Diameter, m	= 0.8 m	1. Flow, L/sec
Chemical Drop, m	= 0.7 m	2. Flow, mL/sec
Time, hr	= 7.0 hr	

1. Determine the liters of chemical solution pumped.

$$\text{Volume, liters} = (0.785)(\text{Diameter, m})^2(\text{Drop, m})(1{,}000 \text{ L/m}^3)$$

$$= (0.785)(0.8 \text{ m})^2(0.7 \text{ m})(1{,}000 \text{ L/m}^3)$$

$$= 352 \text{ liters}$$

2. Estimate the flow delivered by the pump in liters per second.

$$\text{Flow, L/sec} = \frac{\text{Volume Pumped, L}}{(\text{Pumping Time, hr})(60 \text{ min/hr})(60 \text{ sec/min})}$$

$$= \frac{352 \text{ L}}{(7.0 \text{ hr})(60 \text{ min/hr})(60 \text{ sec/min})}$$

$$= 0.014 \text{ L/sec}$$

3. Estimate the flow delivered by the pump in milliliters per second.

$$\text{Flow, mL/sec} = \frac{(\text{Volume Pumped, L})(1{,}000 \text{ mL/L})}{(\text{Pumping Time, hr})(60 \text{ sec/min})}$$

$$= \frac{(352 \text{ L})(1{,}000 \text{ mL/L})}{(7.0 \text{ hr})(60 \text{ min/hr})(60 \text{ sec/min})}$$

$$= 14 \text{ mL/sec}$$

Example 32

Determine the chemical feed in kilograms of polymer per day and grams per second from a chemical feed pump. The polymer solution is 1.8 percent or 18,000 mg polymer per liter. Assume a specific gravity of the polymer solution of 1.0. During a test run, the chemical feed pump delivered 650 mL of polymer solution in 4.5 minutes.

Known		Unknown
Polymer Solution, %	= 1.8%	1. Polymer Feed, kg/day
Polymer Conc, mg/L	= 18,000 mg/L	
Polymer Sp Gr	= 1.0	2. Polymer Feed, g/sec
Volume Pumped, mL	= 650 mL	
Time Pumped, min	= 4.5 min	

1. Calculate the polymer fed by the chemical feed pump in kilograms of polymer per day.

$$\text{Polymer Feed, kg/day} = \frac{(\text{Vol Pumped, mL})(\text{Poly Conc, mg/L})(60 \text{ min/hr})(24 \text{ hr/day})}{(\text{Time Pumped, min})(1{,}000 \text{ mL/L})(1{,}000 \text{ mg/g})(1{,}000 \text{ g/kg})}$$

$$= \frac{(650 \text{ mL})(18{,}000 \text{ mg/L})(60 \text{ min/hr})(24 \text{ hr/day})}{(4.5 \text{ min})(1{,}000 \text{ mL/L})(1{,}000 \text{ mg/g})(1{,}000 \text{ g/kg})}$$

$$= 3.7 \text{ kg/day}$$

2. Calculate the polymer fed by the chemical feed pump in grams of polymer per second.

$$\text{Polymer Feed, g/sec} = \frac{(\text{Vol Pumped, mL})(\text{Poly Conc, mg/L})}{(\text{Time Pumped, min})(1{,}000 \text{ mL/L})(60 \text{ sec/min})(1{,}000 \text{ mg/g})}$$

$$= \frac{(650 \text{ mL})(18{,}000 \text{ mg/L})}{(4.5 \text{ min})(1{,}000 \text{ mL/L})(60 \text{ sec/min})(1{,}000 \text{ mg/g})}$$

$$= 0.043 \text{ g/sec or } 43 \text{ mg/sec}$$

Example 33

Determine the actual chemical feed in kilograms per day and grams per second from a dry chemical feeder. A pie tin placed under the chemical feeder caught 824 grams of chemical in 5 minutes.

Known		Unknown
Chemical, g	= 824 g	1. Chemical Feed, kg/day
Time, min	= 5 min	2. Chemical Feed, g/sec

1. Determine the chemical feed in kilograms per day.

$$\text{Chemical Feed, kg/day} = \frac{(\text{Chemical, g})(60 \text{ min/hr})(24 \text{ hr/day})}{(\text{Time, min})(1{,}000 \text{ g/kg})}$$

$$= \frac{(824 \text{ g})(60 \text{ min/hr})(24 \text{ hr/day})}{(5 \text{ min})(1{,}000 \text{ g/kg})}$$

$$= 237 \text{ kg/day}$$

2. Determine the chemical feed in grams per second.

$$\text{Chemical Feed, g/sec} = \frac{\text{Chemical, g}}{(\text{Time, min})(60 \text{ sec/min})}$$

$$= \frac{824 \text{ g}}{(5 \text{ min})(60 \text{ sec/min})}$$

$$= 2.75 \text{ g/sec}$$

A.56 Advanced Laboratory Procedures

Example 34

Calculate the threshold odor number (TON) for a sample when the first detectable odor occurred when the 70 mL sample was diluted to 200 mL (130 mL of odor-free water was added to the 70 mL sample).

Known		Unknown
A or Sample Size, mL	= 70 mL	TON
B or Odor-Free Water, mL	= 130 mL	

Calculate the threshold odor number (TON).

$$\text{TON} = \frac{A + B}{A}$$

$$= \frac{70 \text{ mL} + 130 \text{ mL}}{70 \text{ mL}}$$

$$= 3$$

Example 35

Determine the geometric mean threshold odor number for a panel of six testers given the results shown below.

Known		Unknown
Tester 1, X_1	= 2	Geometric Mean Threshold Odor Number
Tester 2, X_2	= 4	
Tester 3, X_3	= 3	
Tester 4, X_4	= 8	
Tester 5, X_5	= 6	
Tester 6, X_6	= 2	

Calculate the geometric mean.

$$\text{Geometric Mean TON} = (X_1 \times X_2 \times X_3 \times X_4 \times X_5 \times X_6)^{1/n}$$

$$= (2 \times 4 \times 3 \times 8 \times 6 \times 2)^{1/6}$$

$$= (2{,}304)^{0.167}$$

$$= 3.6$$

Example 36

Calculate the flavor threshold number (FTN) for a sample when the first detectable flavor occurred when the 8.3 mL sample was diluted to 200 mL (191.7 mL of flavor-free water was added to the 8.3 mL sample).

Known		Unknown
A or Sample Size, mL	= 8.3 mL	FTN
B or Flavor-Free Water, mL	= 191.7 mL	

Calculate the flavor threshold number (FTN).

$$\text{FTN} = \frac{A + B}{A}$$

$$= \frac{8.3 \text{ mL} + 191.7 \text{ mL}}{8.3 \text{ mL}}$$

$$= 24$$

Example 37

Determine the flavor rating for a water by calculating the arithmetic mean and standard deviation for the panel ratings given below.

Known	Unknown
Tester 1, $X_1 = 2$	1. Arithmetic Mean, $\overline{X}$
Tester 2, $X_2 = 5$	2. Standard Deviation, S
Tester 3, $X_3 = 3$	
Tester 4, $X_4 = 6$	
Tester 5, $X_5 = 2$	
Tester 6, $X_6 = 6$	

1. Calculate the arithmetic mean, $\overline{X}$, flavor rating.

$$\text{Arithmetic Mean, } \overline{X}, \text{ Flavor Rating} = \frac{X_1 + X_2 + X_3 + X_4 + X_5 + X_6}{n}$$

$$= \frac{2 + 5 + 3 + 6 + 2 + 6}{6}$$

$$= \frac{24}{6}$$

$$= 4$$

2. Calculate the standard deviation, S, of the flavor rating.

$$\text{Standard Deviation, S} = \left[\frac{(X_1 - \overline{X})^2 + (X_2 - \overline{X})^2 + (X_3 - \overline{X})^2 + (X_4 - \overline{X})^2 + (X_5 - \overline{X})^2 + (X_6 - \overline{X})^2}{n - 1}\right]^{0.5}$$

$$= \left[\frac{(2 - 4)^2 + (5 - 4)^2 + (3 - 4)^2 + (6 - 4)^2 + (2 - 4)^2 + (6 - 4)^2}{6 - 1}\right]^{0.5}$$

$$= \left[\frac{(-2)^2 + (1)^2 + (-1)^2 + (2)^2 + (-2)^2 + (2)^2}{5}\right]^{0.5}$$

$$= \left[\frac{4 + 1 + 1 + 4 + 4 + 4}{5}\right]^{0.5}$$

$$= \left[\frac{18}{5}\right]^{0.5}$$

$$= (3.6)^{0.5}$$

$$= 1.9$$

or

$$\text{Standard Deviation, S} = \left[\frac{(X_1^2 + X_2^2 + X_3^2 + X_4^2 + X_5^2 + X_6^2) - (X_1 + X_2 + X_3 + X_4 + X_5 + X_6)^2/n}{n - 1}\right]^{0.5}$$

$$= \left[\frac{(2^2 + 5^2 + 3^2 + 6^2 + 2^2 + 6^2) - (2 + 5 + 3 + 6 + 2 + 6)^2/6}{6 - 1}\right]^{0.5}$$

$$= \left[\frac{(4 + 25 + 9 + 36 + 4 + 36) - (24)^2/6}{5}\right]^{0.5}$$

$$= \left[\frac{114 - 96}{5}\right]^{0.5}$$

$$= \left[\frac{18}{5}\right]^{0.5}$$

$$= (3.6)^{0.5}$$

$$= 1.9$$

A.57 Regulations

Example 38

A small water system collected 14 samples during one month. After each sample was collected, 10 mL of each sample was placed in each of 5 fermentation tubes. At the end of the month, the results indicated that 2 out of a total of 70 fermentation tubes were positive. What percent of the portions tested during the month were positive?

Known		Unknown
Number Positive/mo	= 2 Positive/mo	Portions Positive, %/mo
Total Portions Tested	= 70 Portions	

Calculate the percent of the portions tested during the month that were positive.

$$\text{Portions Positive, \%/mo} = \frac{(\text{Number Positive/mo})(100\%)}{\text{Total Portions Tested}}$$

$$= \frac{(2\ \text{Positive/mo})(100\%)}{70\ \text{Portions}}$$

$$= 3\%/\text{mo}$$

A.58 Safety

Example 39

Calculate the injury frequency rate for a water utility where there were four injuries in one year and the operators worked 97,120 hours.

Known		Unknown
Number of Injuries/yr	= 4 Injuries/yr	Injury Frequency Rate
Number of Hours Worked/yr	= 97,120 hr/yr	

Calculate the injury frequency rate.

$$\text{Injury Frequency Rate} = \frac{(\text{Number of Injuries/yr})(1{,}000{,}000)}{\text{Number of Hours Worked/yr}}$$

$$= \frac{(4\ \text{Injuries/yr})(1{,}000{,}000)}{97{,}120\ \text{hr/yr}}$$

$$= 41.2$$

Example 40

Calculate the injury severity rate for a water company that experienced 57 operator-hours lost due to injuries while the operators worked 97,120 hours during the year.

Known		Unknown
Number of Hours Lost/yr	= 57 hr/yr	Injury Severity Rate
Number of Hours Worked/yr	= 97,120 hr/yr	

Calculate the injury severity rate.

$$\text{Injury Severity Rate} = \frac{(\text{Number of Hours Lost/yr})(1{,}000{,}000)}{\text{Number of Hours Worked/yr}}$$

$$= \frac{(57\ \text{hr/yr})(1{,}000{,}000)}{97{,}120\ \text{hr/yr}}$$

$$= 587$$

A.6 CALCULATION OF CT VALUES

The Surface Water Treatment Rule requires all surface water systems and all systems using ground water under the influence of surface water to achieve 3-log removal of *Giardia* and 4-log removal or inactivation of viruses. This level of treatment can be accomplished by disinfection alone if the source water is relatively free of turbidity and the system meets other specific conditions. Or, the treatment requirements can be met by a combination of filtration and disinfection.

When both filtration and disinfection are used, it is first necessary to know the effectiveness of filtration in removing *Giardia* and viruses before it will be possible to determine the level of disinfection needed to reach the 3-log and 4-log treatment levels. Research studies of conventional, direct, slow sand, and diatomaceous earth filtration have measured the ability of each of these systems to remove *Giardia* cysts. For example, a well-operated conventional filtration system should be able to achieve 2.5-log removal of *Giardia* and 2.0-log removal/inactivation of viruses. Table A.1 lists the treatment removal efficiencies of each of these four types of filtration systems. Using the example above, if filtration achieves 2.5-log removal, disinfection processes will need to achieve at least 0.5-log *Giardia* removal to meet the SWTR requirement of 3-log *Giardia* removal.[1]

1. Please refer to Chapter 22, "Drinking Water Regulations," for additional information. Also, drinking water supply and distribution utilities and operators are strongly urged to stay in close contact with their state and federal regulatory agencies to keep informed of regulations such as the Surface Water Treatment Rules, the Disinfection Byproducts Rules (DBPR), and the Filter Backwash Rule.

TABLE A.1 LOG REMOVAL EFFICIENCY

Type of Filtration	Treatment Removal Credit (Log Removals)		Required Disinfection (Log Inactivations[a])	
	Giardia	Viruses	*Giardia*	Viruses
Conventional	2.5	2.0	0.5	2.0
Direct	2.0	1.0	1.0	3.0
Slow Sand	2.0	2.0	1.0	2.0
Diatomaceous Earth	2.0	1.0	1.0	3.0

a Assumed 3-log *Giardia* and 4-log virus removal or inactivation (can be increased based on sanitary hazards in watershed).

Not all disinfectants are equally effective at inactivating *Giardia* and viruses. Research has shown that a free chlorination system that achieves 3-log *Giardia* inactivation also achieves 4-log virus inactivation. On the other hand, a system using chloramination must disinfect to the 4-log level for virus inactivation before the water will be considered safe and free of harmful organisms.

The effectiveness of disinfection by free chlorination also depends on pH and temperature. The required CT value for any given free chlorine residual increases with increasing pH and decreasing temperature. Tables A.2 through A.7 (pages 762–767) provide the CT values for inactivation of *Giardia* cysts by free chlorine, and the remainder of this section provides examples of how to calculate disinfection CT (concentration × contact time) values using free chlorination.

Example 41

A 10 MGD direct filtration plant applies free chlorine as a disinfectant. The disinfectant has a contact time of 26 minutes under peak flow conditions. The pH of the water is 7.5 and the temperature is 10°C. The free chlorine residual is 2.0 mg/L. Table A.1 above indicates that direct filtration is capable of achieving 2-log (99 percent) removal of *Giardia* cysts. Therefore, 1-log (90 percent or CT-90) *Giardia* inactivation must be achieved by disinfection.

SOLUTION

1. Using Table A.4 (page 764) with a pH of 7.5, a temperature of 10°C, 1.0-log inactivation, and a free chlorine residual of 2.0 mg/L, find the required CT value of 50 mg/L-min.
2. Calculate the CT provided.

$$\begin{aligned}\text{CT Provided, mg/L-min} &= (\text{C, mg/L})(\text{T, min})\\ &= (2\text{ mg/L})(26\text{ min})\\ &= 52\text{ mg/L-min}\end{aligned}$$

Since 52 mg/L-min is greater than the required 50 mg/L-min, the free chlorine residual is adequate for the peak flow conditions.

3. Using the above procedure, assume appropriate free chlorine residuals for lower flows and determine if the CT values provided are greater than the required CT-90 (1-log) values.

Flow, MGD	Contact Time, min	Free Chlorine Residual (assumed), mg/L	CT Provided, mg/L-min	CT-90 Required,* mg/L-min
10	26	2	52	50
7.5	39	1.4	55	47
5	52	1	52	45
2.5	104	0.5	52	43

*From Table A.4 (CT-90 = 1-log inactivation).

Changing the disinfectant dose to meet the required CT values as the flow changes is not a practical operational procedure. The suggested operational procedure is to adjust the disinfectant dose to meet the highest flow requirement with various flow ranges.

Flow Range, MGD	Free Chlorine Residual, mg/L
7.5–10	2.0
5.0–7.5	1.4
2.5–5.0	1.0

Example 42

The 10 MGD plant in Example 41 is a conventional filtration plant and only needs to achieve 0.5-log *Giardia* inactivation (CT-70). Rework Example 41 for 0.5-log *Giardia* inactivation assuming a free chlorine residual of 1.0 mg/L for peak flow conditions.

Flow, MGD	Contact Time, min	Free Chlorine Residual (assumed), mg/L	CT Provided, mg/L-min	CT-70 Required, mg/L-min
10	26	1.0	26	22
7.5	39	0.7	27	22
5	52	0.5	26	21
2.5	104	0.4	42	21

Example 43

A conventional water treatment plant consists of flocculation, sedimentation, filtration, and a clear well. Prechlorination is applied before flocculation and postchlorination after the clear well. A dye tracer study has been done to determine the chlorine contact times ($T_{10,}$ time for 10 percent of the tracer to pass through a tank). Assume a water temperature of 15°C and a pH of 7.5. The plant needs to achieve 0.5-log inactivation of *Giardia* cysts. Determine the fraction of the CT value (CTF) the plant provides the first customer in the distribution system given the following information:

Process	Cl_2 Free Residual, mg/L	T_{10}, min
Flocculation	0.8	15
Sedimentation	0.4	75
Filtration	0.2	15
Clear Well	0.6	120
Pipe to Customer	0.4	20

SOLUTION

Calculate the actual CT value for each process, determine the CT required from Table A.5 (page 765), calculate the CT actual/CT required ratio, and sum the ratios. The sum of the ratios is the CTF value and should be greater than 1.0.

Process	Cl Conc, mg/L	T_{10}, min	CT Actual	CT Required*	CT Ratio
Floc.	0.8 mg/L	15 min	12	15	0.80
Sed.	0.4 mg/L	75 min	30	14	2.14
Filt.	0.2 mg/L	15 min	3	14	0.21
CW	0.6 mg/L	120 min	72	14	5.14
Pipe	0.4 mg/L	20 min	8	14	0.57
					8.86

*From Table A.5.

Example 44

If the sum of the CT actual/CT required ratios for the flocculation, sedimentation, and filtration processes for Example 43 was 0.72, determine the free chlorine residual from the clear well necessary to achieve a total CTF (CT fraction) value greater than 1.0 for the plant only (neglect pipe CT). Assume the T_{10} value (actual detention time for the clear well) is 30 minutes.

SOLUTION

Determine the minimum CT ratio for the clear well that will produce a total CTF value greater than 1.0 for the plant.

$$\text{Min Clear Well CT Ratio Needed} = \text{Min CTF} - \text{CTF Provided}$$

$$= 1.0 - 0.72$$

$$= 0.28$$

Determine the CT ratios (CT actual/CT required) for various chlorine residuals in the clear well. Start with the minimum free chlorine residual for the clear well.

Cl_2 Free Residual, mg/L	$T_{10,}$ min	CT Actual	CT Required	CT Ratio
0.2 mg/L	30 min	6	14	0.43
0.4 mg/L	30 min	12	14	0.86
0.6 mg/L	30 min	18	14	1.29
0.8 mg/L	30 min	24	15	1.60

Since the free chlorine residual of 0.2 mg/L produces a CT ratio of 0.43, which is greater than 0.28, the free chlorine residual of 0.2 mg/L will be adequate according to these calculations.

Example 45

If prechlorination were not practiced in Example 44, what free chlorine residual would be necessary for the clear well alone to provide adequate disinfection?

SOLUTION

The required CT value is 14 mg/L-min.

$$\text{Free Chlorine Residual, mg/L} = \frac{\text{CT, mg/L-min}}{T_{10}\text{, min}}$$

$$= \frac{14\text{ mg/L-min}}{30\text{ min}}$$

$$= 0.5\text{ mg/L or greater}$$

A.7 CALCULATION OF LOG REMOVALS

Regulations may require the calculation of log removals for inactivation of *Giardia* cysts, viruses, or particle counts. How are log removals calculated? The following example illustrates two methods of calculating log removals.

Example 46

Calculate the log removal of 5- to 15-micron particles per milliliter if the influent particle count to a water treatment filter reported 2,100 particles in the 5- to 15-microns range per milliliter of water and the filter effluent reported 30 particles in the 5- to 15-microns range per milliliter of filtered water.

Known		Unknown
Influent, particles, mL	= 2,100 particles/mL	Log Removal, particles/mL
Effluent, particles, mL	= 30 particles/mL	

Calculate the log removal of particles in the 5- to 15- microns range per mL by the filter.

PROCEDURE 1

$$\text{Log Removal, particles/mL} = \text{Log Influent, particles/mL} - \text{Log Effluent, particles/mL}$$

$$= \text{Log } 2{,}100 \text{ particles/mL} - \text{Log } 30 \text{ particles/mL}$$

$$= 3.3 - 1.5$$

$$= 1.8$$

PROCEDURE 2

$$\text{Log Removal, particles/mL} = \text{Log}\left(\frac{\text{Influent, particles/mL}}{\text{Effluent, particles/mL}}\right)$$

$$= \text{Log}\left(\frac{2{,}100 \text{ particles/mL}}{30 \text{ particles/mL}}\right)$$

$$= \text{Log } 70$$

$$= 1.8$$

TABLE A.2 CT VALUES FOR INACTIVATION OF *GIARDIA* CYSTS BY FREE CHLORINE AT 0.5°C OR LOWER (1)

CHLORINE CONCENTRATION (mg/L)	pH<=6 Log Inactivations						pH=6.5 Log Inactivations						pH=7.0 Log Inactivations						pH=7.5 Log Inactivations					
	0.5	1.0	1.5	2.0	2.5	3.0	0.5	1.0	1.5	2.0	2.5	3.0	0.5	1.0	1.5	2.0	2.5	3.0	0.5	1.0	1.5	2.0	2.5	3.0
<=0.4	23	46	69	91	114	137	27	54	82	109	136	163	33	65	98	130	163	195	40	79	119	158	198	237
0.6	24	47	71	94	118	141	28	56	84	112	140	168	33	67	100	133	167	200	40	80	120	159	199	239
0.8	24	48	73	97	121	145	29	57	86	115	143	172	34	68	103	137	171	205	41	82	123	164	205	246
1	25	49	74	99	123	148	29	59	88	117	147	176	35	70	105	140	175	210	42	84	127	169	211	253
1.2	25	51	76	101	127	152	30	60	90	120	150	180	36	72	108	143	179	215	43	86	130	173	216	259
1.4	26	52	78	103	129	155	31	61	92	123	153	184	37	74	111	147	184	221	44	89	133	177	222	266
1.6	26	52	79	105	131	157	32	63	95	126	158	189	38	75	113	151	188	226	46	91	137	182	228	273
1.8	27	54	81	108	135	162	32	64	97	129	161	193	39	77	116	154	193	231	47	93	140	186	233	279
2	28	55	83	110	138	165	33	66	99	131	164	197	39	79	118	157	197	236	48	95	143	191	238	286
2.2	28	56	85	113	141	169	34	67	101	134	168	201	40	81	121	161	202	242	50	99	149	198	248	297
2.4	29	57	86	115	143	172	34	68	103	137	171	205	41	82	124	165	206	247	50	99	149	199	248	298
2.6	29	58	88	117	146	175	35	70	105	139	174	209	42	84	126	168	210	252	51	101	152	203	253	304
2.8	30	59	89	119	148	178	36	71	107	142	178	213	43	86	129	171	214	257	52	103	155	207	258	310
3	30	60	91	121	151	181	36	72	109	145	181	217	44	87	131	174	218	261	53	105	158	211	263	316

CHLORINE CONCENTRATION (mg/L)	pH=8.0 Log Inactivations						pH=8.5 Log Inactivations						pH<=9.0 Log Inactivations					
	0.5	1.0	1.5	2.0	2.5	3.0	0.5	1.0	1.5	2.0	2.5	3.0	0.5	1.0	1.5	2.0	2.5	3.0
<=0.4	46	92	139	185	231	277	55	110	165	219	274	329	65	130	195	260	325	390
0.6	48	95	143	191	238	286	57	114	171	228	285	342	68	136	204	271	339	407
0.8	49	98	148	197	246	295	59	118	177	236	295	354	70	141	211	281	352	422
1	51	101	152	203	253	304	61	122	183	243	304	365	73	146	219	291	364	437
1.2	52	104	157	209	261	313	63	125	188	251	313	376	75	150	226	301	376	451
1.4	54	107	161	214	268	321	65	129	194	258	323	387	77	155	232	309	387	464
1.6	55	110	165	219	274	329	66	132	199	265	331	397	80	159	239	318	398	477
1.8	56	113	169	225	282	338	68	136	204	271	339	407	82	163	245	326	408	489
2	58	115	173	231	288	346	70	139	209	278	348	417	83	167	250	333	417	500
2.2	59	118	177	235	294	353	71	142	213	284	355	426	85	170	256	341	426	511
2.4	60	120	181	241	301	361	73	145	218	290	363	435	87	174	261	348	435	522
2.6	61	123	184	245	307	368	74	148	222	296	370	444	89	178	267	355	444	533
2.8	63	125	188	250	313	375	75	151	226	301	377	452	91	181	272	362	453	543
3	64	127	191	255	318	382	77	153	230	307	383	460	92	184	276	368	460	552

Notes:

(1) $CT_{99.9}$ = CT for 3-log inactivation

Reprinted from *Guidance Manual for Compliance with the Filtration and Disinfection Requirements for Public Water Systems Using Surface Water Sources*, a publication of the American Water Works Association. Reproduced with permission.

TABLE A.3 CT VALUES FOR INACTIVATION OF *GIARDIA* CYSTS BY FREE CHLORINE AT 5°C (1)

CHLORINE CONCENTRATION (mg/L)	pH<=6 Log Inactivations						pH=6.5 Log Inactivations						pH=7.0 Log Inactivations						pH=7.5 Log Inactivations					
	0.5	1.0	1.5	2.0	2.5	3.0	0.5	1.0	1.5	2.0	2.5	3.0	0.5	1.0	1.5	2.0	2.5	3.0	0.5	1.0	1.5	2.0	2.5	3.0
<=0.4	16	32	49	65	81	97	20	39	59	78	98	117	23	46	70	93	116	139	28	55	83	111	138	166
0.6	17	33	50	67	83	100	20	40	60	80	100	120	24	48	72	95	119	143	29	57	86	114	143	171
0.8	17	34	52	69	86	103	20	41	61	81	102	122	24	49	73	97	122	146	29	58	88	117	146	175
1	18	35	53	70	88	105	21	42	63	83	104	125	25	50	75	99	124	149	30	60	90	119	149	179
1.2	18	36	54	71	89	107	21	42	64	85	106	127	25	51	76	101	127	152	31	61	92	122	153	183
1.4	18	36	55	73	91	109	22	43	65	87	108	130	26	52	78	103	129	155	31	62	94	125	156	187
1.6	19	37	56	74	93	111	22	44	66	88	110	132	26	53	79	105	132	158	32	64	96	128	160	192
1.8	19	38	57	76	95	114	23	45	68	90	113	135	27	54	81	108	135	162	33	65	98	131	163	196
2	19	39	58	77	97	116	23	46	69	92	115	138	28	55	83	110	138	165	33	67	100	133	167	200
2.2	20	39	59	79	98	118	23	47	70	93	117	140	28	56	85	113	141	169	34	68	102	136	170	204
2.4	20	40	60	80	100	120	24	48	72	95	119	143	29	57	86	115	143	172	35	70	105	139	174	209
2.6	20	41	61	81	102	122	24	49	73	97	122	146	29	58	88	117	146	175	36	71	107	142	178	213
2.8	21	41	62	83	103	124	25	49	74	99	123	148	30	59	89	119	148	178	36	72	109	145	181	217
3	21	42	63	84	105	126	25	50	76	101	126	151	30	61	91	121	152	182	37	74	111	147	184	221

CHLORINE CONCENTRATION (mg/L)	pH=8.0 Log Inactivations						pH=8.5 Log Inactivations						pH<=9.0 Log Inactivations					
	0.5	1.0	1.5	2.0	2.5	3.0	0.5	1.0	1.5	2.0	2.5	3.0	0.5	1.0	1.5	2.0	2.5	3.0
<=0.4	33	66	99	132	165	198	39	79	118	157	197	236	47	93	140	186	233	279
0.6	34	68	102	136	170	204	41	81	122	163	203	244	49	97	146	194	243	291
0.8	35	70	105	140	175	210	42	84	126	168	210	252	50	100	151	201	251	301
1	36	72	108	144	180	216	43	87	130	173	217	260	52	104	156	208	260	312
1.2	37	74	111	147	184	221	45	89	134	178	223	267	53	107	160	213	267	320
1.4	38	76	114	151	189	227	46	91	137	183	228	274	55	110	165	219	274	329
1.6	39	77	116	155	193	232	47	94	141	187	234	281	56	112	169	225	281	337
1.8	40	79	119	159	198	238	48	96	144	191	239	287	58	115	173	230	288	345
2	41	81	122	162	203	243	49	98	147	196	245	294	59	118	177	235	294	353
2.2	41	83	124	165	207	248	50	100	150	200	250	300	60	120	181	241	301	361
2.4	42	84	127	169	211	253	51	102	153	204	255	306	61	123	184	245	307	368
2.6	43	86	129	172	215	258	52	104	156	208	260	312	63	125	188	250	313	375
2.8	44	88	132	175	219	263	53	106	159	212	265	318	64	127	191	255	318	382
3	45	89	134	179	223	268	54	108	162	216	270	324	65	130	195	259	324	389

Notes:

(1) $CT_{99.9}$ = CT for 3-log inactivation

TABLE A.4 CT VALUES FOR INACTIVATION OF *GIARDIA* CYSTS BY FREE CHLORINE AT 10°C (1)

CHLORINE CONCENTRATION (mg/L)	pH<=6 Log Inactivations						pH=6.5 Log Inactivations						pH=7.0 Log Inactivations						pH=7.5 Log Inactivations					
	0.5	1.0	1.5	2.0	2.5	3.0	0.5	1.0	1.5	2.0	2.5	3.0	0.5	1.0	1.5	2.0	2.5	3.0	0.5	1.0	1.5	2.0	2.5	3.0
<=0.4	12	24	37	49	61	73	15	29	44	59	73	88	17	35	52	69	87	104	21	42	63	83	104	125
0.6	13	25	38	50	63	75	15	30	45	60	75	90	18	36	54	71	89	107	21	43	64	85	107	128
0.8	13	26	39	52	65	78	15	31	46	61	77	92	18	37	55	73	92	110	22	44	66	87	109	131
1	13	26	40	53	66	79	16	31	47	63	78	94	19	37	56	75	93	112	22	45	67	89	112	134
1.2	13	27	40	53	67	80	16	32	48	63	79	95	19	38	57	76	95	114	23	46	69	91	114	137
1.4	14	27	41	55	68	82	16	33	49	65	82	98	19	39	58	77	97	116	23	47	70	93	117	140
1.6	14	28	42	55	69	83	17	33	50	66	83	99	20	40	60	79	99	119	24	48	72	96	120	144
1.8	14	29	43	57	72	86	17	34	51	67	84	101	20	41	61	81	102	122	25	49	74	98	123	147
2	15	29	44	58	73	87	17	35	52	69	87	104	21	41	62	83	103	124	25	50	75	100	125	150
2.2	15	30	45	59	74	89	18	35	53	70	88	105	21	42	64	85	106	127	26	51	77	102	128	153
2.4	15	30	45	60	75	90	18	36	54	71	89	107	22	43	65	86	108	129	26	52	79	105	131	157
2.6	15	31	46	61	77	92	18	37	55	73	92	110	22	44	66	87	109	131	27	53	80	107	133	160
2.8	16	31	47	62	78	93	19	37	56	74	93	111	22	45	67	89	112	134	27	54	82	109	136	163
3	16	32	48	63	79	95	19	38	57	75	94	113	23	46	69	91	114	137	28	55	83	111	138	166

CHLORINE CONCENTRATION (mg/L)	pH=8.0 Log Inactivations						pH=8.5 Log Inactivations						pH<=9.0 Log Inactivations					
	0.5	1.0	1.5	2.0	2.5	3.0	0.5	1.0	1.5	2.0	2.5	3.0	0.5	1.0	1.5	2.0	2.5	3.0
<=0.4	25	50	75	99	124	149	30	59	89	118	148	177	35	70	105	139	174	209
0.6	26	51	77	102	128	153	31	61	92	122	153	183	36	73	109	145	182	218
0.8	26	53	79	105	132	158	32	63	95	126	158	189	38	75	113	151	188	226
1	27	54	81	108	135	162	33	65	98	130	163	195	39	78	117	156	195	234
1.2	28	55	83	111	138	166	33	67	100	133	167	200	40	80	120	160	200	240
1.4	28	57	85	113	142	170	34	69	103	137	172	206	41	82	124	165	206	247
1.6	29	58	87	116	145	174	35	70	106	141	176	211	42	84	127	169	211	253
1.8	30	60	90	119	149	179	36	72	108	143	179	215	43	86	130	173	216	259
2	30	61	91	121	152	182	37	74	111	147	184	221	44	88	133	177	221	265
2.2	31	62	93	124	155	186	38	75	113	150	188	225	45	90	136	181	226	271
2.4	32	63	95	127	158	190	38	77	115	153	192	230	46	92	138	184	230	276
2.6	32	65	97	129	162	194	39	78	117	156	195	234	47	94	141	187	234	281
2.8	33	66	99	131	164	197	40	80	120	159	199	239	48	96	144	191	239	287
3	34	67	101	134	168	201	41	81	122	162	203	243	49	97	146	195	243	292

Notes:

(1) $CT_{99.9}$ = CT for 3-log inactivation

Reprinted from *Guidance Manual for Compliance with the Filtration and Disinfection Requirements for Public Water Systems Using Surface Water Sources*, a publication of the American Water Works Association. Reproduced with permission.

TABLE A.5 CT VALUES FOR INACTIVATION OF *GIARDIA* CYSTS BY FREE CHLORINE AT 15°C (1)

CHLORINE CONCENTRATION (mg/L)	pH<=6 Log Inactivations						pH=6.5 Log Inactivations						pH=7.0 Log Inactivations						pH=7.5 Log Inactivations					
	0.5	1.0	1.5	2.0	2.5	3.0	0.5	1.0	1.5	2.0	2.5	3.0	0.5	1.0	1.5	2.0	2.5	3.0	0.5	1.0	1.5	2.0	2.5	3.0
<=0.4	8	16	25	33	41	49	10	20	30	39	49	59	12	23	35	47	58	70	14	28	42	55	69	83
0.6	8	17	25	33	42	50	10	20	30	40	50	60	12	24	36	48	60	72	14	29	43	57	72	86
0.8	9	17	26	35	43	52	10	20	31	41	51	61	12	24	37	49	61	73	15	29	44	59	73	88
1	9	18	27	35	44	53	11	21	32	42	53	63	13	25	38	50	63	75	15	30	45	60	75	90
1.2	9	18	27	36	45	54	11	21	32	43	53	64	13	25	38	51	63	76	15	31	46	61	77	92
1.4	9	18	28	37	46	55	11	22	33	43	54	65	13	26	39	52	65	78	16	31	47	63	78	94
1.6	9	19	28	37	47	56	11	22	33	44	55	66	13	26	40	53	66	79	16	32	48	64	80	96
1.8	10	19	29	38	48	57	11	23	34	45	57	68	14	27	41	54	68	81	16	33	49	65	82	98
2	10	19	29	39	48	58	12	23	35	46	58	69	14	28	42	55	69	83	17	33	50	67	83	100
2.2	10	20	30	39	49	59	12	23	35	47	58	70	14	28	43	57	71	85	17	34	51	68	85	102
2.4	10	20	30	40	50	60	12	24	36	48	60	72	14	29	43	57	72	86	18	35	53	70	88	105
2.6	10	20	31	41	51	61	12	24	37	49	61	73	15	29	44	59	73	88	18	36	54	71	89	107
2.8	10	21	31	41	52	62	12	25	37	49	62	74	15	30	45	59	74	89	18	36	55	73	91	109
3	11	21	32	42	53	63	13	25	38	51	63	76	15	30	46	61	76	91	19	37	56	74	93	111

CHLORINE CONCENTRATION (mg/L)	pH=8.0 Log Inactivations						pH=8.5 Log Inactivations						pH<=9.0 Log Inactivations					
	0.5	1.0	1.5	2.0	2.5	3.0	0.5	1.0	1.5	2.0	2.5	3.0	0.5	1.0	1.5	2.0	2.5	3.0
<=0.4	17	33	50	66	83	99	20	39	59	79	98	118	23	47	70	93	117	140
0.6	17	34	51	68	85	102	20	41	61	81	102	122	24	49	73	97	122	146
0.8	18	35	53	70	88	105	21	42	63	84	105	126	25	50	76	101	126	151
1	18	36	54	72	90	108	22	43	65	87	108	130	26	52	78	104	130	156
1.2	19	37	56	74	93	111	22	45	67	89	112	134	27	53	80	107	133	160
1.4	19	38	57	76	95	114	23	46	69	91	114	137	28	55	83	110	138	165
1.6	19	39	58	77	97	116	24	47	71	94	118	141	28	56	85	113	141	169
1.8	20	40	60	79	99	119	24	48	72	96	120	144	29	58	87	115	144	173
2	20	41	61	81	102	122	25	49	74	98	123	147	30	59	89	118	148	177
2.2	21	41	62	83	103	124	25	50	75	100	125	150	30	60	91	121	151	181
2.4	21	42	64	85	106	127	26	51	77	102	128	153	31	61	92	123	153	184
2.6	22	43	65	86	108	129		52	78	104	130	156	31	63	94	125	157	188
2.8	22	44	66	88	110	132	27	53	80	106	133	159	32	64	96	127	159	191
3	22	45	67	89	112	134	27	54	81	108	135	162	33	65	98	130	163	195

Notes:

(1) $CT_{99.9}$ = CT for 3-log inactivation

TABLE A.6 CT VALUES FOR INACTIVATION OF *GIARDIA* CYSTS BY FREE CHLORINE AT 20°C (1)

CHLORINE CONCENTRATION	pH<=6 Log Inactivations						pH=6.5 Log Inactivations						pH=7.0 Log Inactivations						pH=7.5 Log Inactivations					
(mg/L)	0.5	1.0	1.5	2.0	2.5	3.0	0.5	1.0	1.5	2.0	2.5	3.0	0.5	1.0	1.5	2.0	2.5	3.0	0.5	1.0	1.5	2.0	2.5	3.0
<=0.4	6	12	18	24	30	36	7	15	22	29	37	44	9	17	26	35	43	52	10	21	31	41	52	62
0.6	6	13	19	25	32	38	8	15	23	30	38	45	9	18	27	36	45	54	11	21	32	43	53	64
0.8	7	13	20	26	33	39	8	15	23	31	38	46	9	18	28	37	46	55	11	22	33	44	55	66
1	7	13	20	26	33	39	8	16	24	31	39	47	9	19	28	37	47	56	11	22	34	45	56	67
1.2	7	13	20	27	33	40	8	16	24	32	40	48	10	19	29	38	48	57	12	23	35	46	58	69
1.4	7	14	21	27	34	41	8	16	25	33	41	49	10	19	29	39	48	58	12	23	35	47	58	70
1.6	7	14	21	28	35	42	8	17	25	33	42	50	10	20	30	39	49	59	12	24	36	48	60	72
1.8	7	14	22	29	36	43	9	17	26	34	43	51	10	20	31	41	51	61	12	25	37	49	62	74
2	7	15	22	29	37	44	9	17	26	35	43	52	10	21	31	41	52	62	13	25	38	50	63	75
2.2	7	15	22	29	37	44	9	18	27	35	44	53	11	21	32	42	53	63	13	26	39	51	64	77
2.4	8	15	23	30	38	45	9	18	27	36	45	54	11	22	33	43	54	65	13	26	39	52	65	78
2.6	8	15	23	31	38	46	9	18	28	37	46	55	11	22	33	44	55	66	13	27	40	53	67	80
2.8	8	16	24	31	39	47	9	19	28	37	47	56	11	22	34	45	56	67	14	27	41	54	68	81
3	8	16	24	31	39	47	10	19	29	38	48	57	11	23	34	45	57	68	14	28	42	55	69	83

CHLORINE CONCENTRATION	pH=8.0 Log Inactivations						pH=8.5 Log Inactivations						pH<=9.0 Log Inactivations					
(mg/L)	0.5	1.0	1.5	2.0	2.5	3.0	0.5	1.0	1.5	2.0	2.5	3.0	0.5	1.0	1.5	2.0	2.5	3.0
<=0.4	12	25	37	49	62	74	15	30	45	59	74	89	18	35	53	70	88	105
0.6	13	26	39	51	64	77	15	31	46	61	77	92	18	36	55	73	91	109
0.8	13	26	40	53	66	79	16	32	48	63	79	95	19	38	57	75	94	113
1	14	27	41	54	68	81	16	33	49	65	82	98	20	39	59	78	98	117
1.2	14	28	42	55	69	83	17	33	50	67	83	100	20	40	60	80	100	120
1.4	14	28	43	57	71	85	17	34	52	69	86	103	21	41	62	82	103	123
1.6	15	29	44	58	73	87	18	35	53	70	88	105	21	42	63	84	105	126
1.8	15	30	45	59	74	89	18	36	54	72	90	108	22	43	65	86	108	129
2	15	30	46	61	76	91	18	37	55	73	92	110	22	44	66	88	110	132
2.2	16	31	47	62	78	93	19	38	57	75	94	113	23	45	68	90	113	135
2.4	16	32	48	63	79	95	19	38	58	77	96	115	23	46	69	92	115	138
2.6	16	32	49	65	81	97	20	39	59	78	98	117	24	47	71	94	118	141
2.8	17	33	50	66	83	99	20	40	60	79	99	119	24	48	72	95	119	143
3	17	34	51	67	84	101	20	41	61	81	102	122	24	49	73	97	122	146

Notes:

(1) $CT_{99.9}$ = CT for 3-log inactivation

Reprinted from *Guidance Manual for Compliance with the Filtration and Disinfection Requirements for Public Water Systems Using Surface Water Sources*, a publication of the American Water Works Association. Reproduced with permission. Copyright © 1991 by American Water Works Association.

TABLE A.7 CT VALUES FOR INACTIVATION OF *GIARDIA* CYSTS BY FREE CHLORINE AT 25°C (1)

CHLORINE CONCENTRATION (mg/L)	pH<=6 Log Inactivations						pH=6.5 Log Inactivations						pH=7.0 Log Inactivations						pH=7.5 Log Inactivations					
	0.5	1.0	1.5	2.0	2.5	3.0	0.5	1.0	1.5	2.0	2.5	3.0	0.5	1.0	1.5	2.0	2.5	3.0	0.5	1.0	1.5	2.0	2.5	3.0
<=0.4	4	8	12	16	20	24	5	10	15	19	24	29	6	12	18	23	29	35	7	14	21	28	35	42
0.6	4	8	13	17	21	25	5	10	15	20	25	30	6	12	18	24	30	36	7	14	22	29	36	43
0.8	4	9	13	17	22	26	5	10	16	21	26	31	6	12	19	25	31	37	7	15	22	29	37	44
1	4	9	13	17	22	26	5	10	16	21	26	31	6	12	19	25	31	37	8	15	23	30	38	45
1.2	5	9	14	18	23	27	5	11	16	21	27	32	6	13	19	25	32	38	8	15	23	31	38	46
1.4	5	9	14	18	23	27	6	11	17	22	28	33	7	13	20	26	33	39	8	16	24	31	39	47
1.6	5	9	14	19	23	28	6	11	17	22	28	33	7	13	20	27	33	40	8	16	24	32	40	48
1.8	5	10	15	19	24	29	6	11	17	23	28	34	7	14	21	27	34	41	8	16	25	33	41	49
2	5	10	15	19	24	29	6	12	18	23	29	35	7	14	21	27	34	41	8	17	25	33	42	50
2.2	5	10	15	20	25	30	6	12	18	23	29	35	7	14	21	28	35	42	9	17	26	34	43	51
2.4	5	10	15	20	25	30	6	12	18	24	30	36	7	14	22	29	36	43	9	17	26	35	43	52
2.6	5	10	16	21	26	31	6	12	19	25	31	37	7	15	22	29	37	44	9	18	27	35	44	53
2.8	5	10	16	21	26	31	6	12	19	25	31	37	8	15	23	30	38	45	9	18	27	36	45	54
3	5	11	16	21	27	32	6	13	19	25	32	38	8	15	23	31	38	46	9	18	28	37	46	55

CHLORINE CONCENTRATION (mg/L)	pH=8.0 Log Inactivations						pH=8.5 Log Inactivations						pH<=9.0 Log Inactivations					
	0.5	1.0	1.5	2.0	2.5	3.0	0.5	1.0	1.5	2.0	2.5	3.0	0.5	1.0	1.5	2.0	2.5	3.0
<=0.4	8	17	25	33	42	50	10	20	30	39	49	59	12	23	35	47	58	70
0.6	9	17	26	34	43	51	10	20	31	41	51	61	12	24	37	49	61	73
0.8	9	18	27	35	44	53	11	21	32	42	53	63	13	25	38	50	63	75
1	9	18	27	36	45	54	11	22	33	43	54	65	13	26	39	52	65	78
1.2	9	18	28	37	46	55	11	22	34	45	56	67	13	27	40	53	67	80
1.4	10	19	29	38	48	57	12	23	35	46	58	69	14	27	41	55	68	82
1.6	10	19	29	39	48	58	12	23	35	47	58	70	14	28	42	56	70	84
1.8	10	20	30	40	50	60	12	24	36	48	60	72	14	29	43	57	72	86
2	10	20	31	41	51	61	12	25	37	49	62	74	15	29	44	59	73	88
2.2	10	21	31	41	52	62	13	25	38	50	63	75	15	30	45	60	75	90
2.4	11	21	32	42	53	63	13	26	39	51	64	77	15	31	46	61	77	92
2.6	11	22	33	43	54	65	13	26	39	52	65	78	16	31	47	63	78	94
2.8	11	22	33	44	55	66	13	27	40	53	67	80	16	32	48	64	80	96
3	11	22	34	45	56	67	14	27	41	54	68	81	16	32	49	65	81	97

Notes:

(1) $CT_{99.9}$ = CT for 3-log inactivation

ABBREVIATIONS

°C	degrees Celsius
°F	degrees Fahrenheit
μ	micron
μg	microgram
μm	micrometer
ac	acres
ac-ft	acre-feet
amp	amperes
atm	atmosphere
CFM	cubic feet per minute
CFS	cubic feet per second
Ci	Curie
cm	centimeters
cm^2	square centimeters
cm^3	cubic centimeters
D	Dalton
dB	decibel
ft	feet or foot
ft^2	square feet
ft^3	cubic feet
ft-lb/min	foot-pounds per minute
g	gravity
gal	gallons
gal/day	gallons per day
GFD	gallons of flux per square foot per day
g	grams
GPCD	gallons per capita per day
GPD	gallons per day
gpg	grains per gallon
GPM	gallons per minute
GPY	gallons per year
gr	grains
ha	hectares
HP	horsepower
hr	hours
Hz	hertz
in	inches
in^2	square inches
in^3	cubic inches
J	joules
k	kilos
kg	kilograms
km	kilometers
kN	kilonewtons
kPa	kiloPascals
kW	kilowatts
kWh	kilowatt-hours
L	liters
lb	pounds
lb/in^2	pounds per square inch
M	mega
M	million
M	molar (or molarity)
m	meters
m^2	square meters
m^3	cubic meters
mA	milliampere
meq	milliequivalent
mg	milligrams
MGD	million gallons per day
mg/L	milligrams per liter
min	minutes
mL	milliliters
mm	millimeters
N	Newton
N	normal (or normality)
nm	nanometer
ohm	ohm
Pa	Pascal
pCi	picoCurie
pCi/L	picoCuries per liter
ppb	parts per billion
ppm	parts per million
psf	pounds per square foot
psi	pounds per square inch
psig	pounds per square inch gauge
RPM	revolutions per minute
SCFM	standard cubic feet per minute
sec	seconds
SI	Le Système International d'Unités
W	watt
yd	yards
yd^2	square yards
yd^3	cubic yards

WATER WORDS

A Summary of the Words Defined

in

WATER TREATMENT PLANT OPERATION,

WATER DISTRIBUTION SYSTEM OPERATION AND MAINTENANCE,

and

SMALL WATER SYSTEM OPERATION AND MAINTENANCE

PROJECT PRONUNCIATION KEY

by Warren L. Prentice

The Project Pronunciation Key is designed to aid you in the pronunciation of new words. While this key is based primarily on familiar sounds, it does not attempt to follow any particular pronunciation guide. This key is designed solely to aid operators in this program.

You may find it helpful to refer to other available sources for pronunciation help. Each current standard dictionary contains a guide to its own pronunciation key. Each key will be different from each other and from this key. Examples of the difference between the key used in this program and the *Webster's New World College Dictionary* [1] key are shown below.

In using this key, you should accent (say louder) the syllable that appears in capital letters. The following chart presents examples of how to pronounce words using the project key.

	SYLLABLE				
WORD	1st	2nd	3rd	4th	5th
aerobic	air	O	bick		
bacteria	back	TEER	e	uh	
contamination	kun	TAM	uh	NAY	shun

The first word, *AEROBIC*, has its second syllable accented. The second word, *BACTERIA*, has its second syllable accented. The third word, *CONTAMINATION*, has its second and fourth syllables accented.

1. The *Webster's New World College Dictionary*, Fourth Edition, was chosen rather than an unabridged dictionary because of its availability to the operator. Other editions may be slightly different.

WATER WORDS

SYMBOLS

> GREATER THAN — > GREATER THAN

DO >5 mg/L would be read as DO GREATER THAN 5 mg/L.

< LESS THAN — <LESS THAN

DO <5 mg/L would be read as DO LESS THAN 5 mg/L.

A

ABC — ABC

See ASSOCIATION OF BOARDS OF CERTIFICATION (ABC).

ABSORPTION (ab-SORP-shun) — ABSORPTION

The taking in or soaking up of one substance into the body of another by molecular or chemical action (as tree roots absorb dissolved nutrients in the soil).

ACCOUNTABILITY — ACCOUNTABILITY

When a manager gives power/responsibility to an employee, the employee ensures that the manager is informed of results or events.

ACCURACY — ACCURACY

How closely an instrument measures the true or actual value of the process variable being measured or sensed.

AceOps — AceOps

See ALLIANCE OF CERTIFIED OPERATORS, LABORATORY ANALYSTS, INSPECTORS, AND SPECIALISTS (AceOps).

ACIDIC (uh-SID-ick) — ACIDIC

The condition of water or soil that contains a sufficient amount of acid substances to lower the pH below 7.0.

ACIDIFICATION (uh-SID-uh-fuh-KAY-shun) — ACIDIFICATION

The addition of an acid (usually nitric or sulfuric) to a sample to lower the pH below 2.0. The purpose of acidification is to fix a sample so it will not change until it is analyzed.

ACID RAIN — ACID RAIN

Precipitation that has been rendered (made) acidic by airborne pollutants.

ACRE-FOOT — ACRE-FOOT

A volume of water that covers 1 acre to a depth of 1 foot, or 43,560 cubic feet (1,233.5 cubic meters).

ACTIVATED ALUMINA — ACTIVATED ALUMINA

A charged form of aluminum, used with a synthetic, porous media in an ion exchange adsorption process to remove charged contaminants.

ACTIVATED CARBON ACTIVATED CARBON

Adsorptive particles or granules of carbon usually obtained by heating carbon (such as wood). These particles or granules have a high capacity to selectively remove certain trace and soluble materials from water.

ACUTE HEALTH EFFECT ACUTE HEALTH EFFECT

An adverse effect on a human or animal body, with symptoms developing rapidly.

ADSORBATE (add-SORE-bait) ADSORBATE

The material being removed by the adsorption process.

ADSORBENT (add-SORE-bent) ADSORBENT

The material (activated carbon) that is responsible for removing the undesirable substance in the adsorption process.

ADSORPTION (add-SORP-shun) ADSORPTION

The gathering of a gas, liquid, or dissolved substance on the surface or interface zone of another material.

ADVANCED METERING INFRASTRUCTURE (AMI) ADVANCED METERING INFRASTRUCTURE (AMI)

Refers to a system that measures, collects, and analyzes energy usage while interacting with advanced devices, such as water meters, through various communication media, either on demand or on predefined schedules. This infrastructure includes hardware, software, communications, consumer energy displays and controllers, customer associated systems, meter data management software, and supplier and network distribution systems.

AERATION (air-A-shun) AERATION

The process of adding air to water. Air can be added to water by either passing air through water or passing water through air.

AEROBIC (air-O-bick) AEROBIC

A condition in which atmospheric or dissolved oxygen is present in the aquatic (water) environment.

AESTHETIC (es-THET-ick) AESTHETIC

Attractive or appealing.

AGE TANK AGE TANK

A tank used to store a chemical solution of known concentration for feed to a chemical feeder. An age tank usually stores sufficient chemical solution to properly treat the water being treated for at least 1 day. Also called a day tank.

AIR BINDING AIR BINDING

The clogging of a filter, pipe, or pump due to the presence of air released from water. Air entering the filter media is harmful to both the filtration and backwash processes. Air can prevent the passage of water during the filtration process and can cause the loss of filter media during the backwash process.

AIR GAP AIR GAP

An open, vertical drop, or vertical empty space, between a drinking (potable) water supply and the nonpotable point of use. This gap prevents the contamination of drinking water by backsiphonage because there is no way wastewater can reach the drinking water supply. Air gap devices are also used to provide adequate space above the top of a manhole and the end of the hose from the fire hydrant.

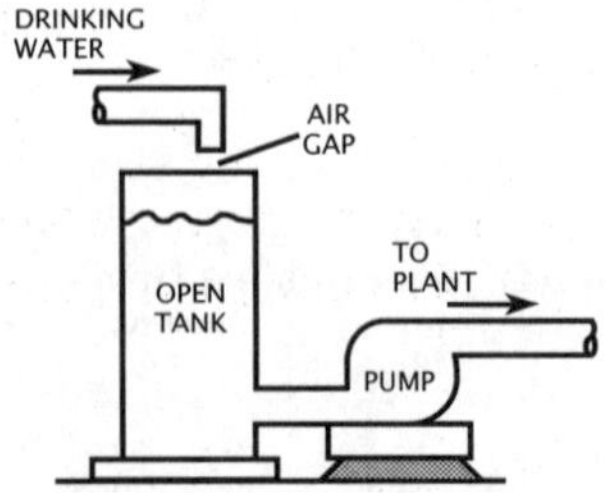

AIR PADDING AIR PADDING

Pumping dry air (dew point –40°F [–40°C]) into a container to assist with the withdrawal of a liquid or to force a liquefied gas, such as chlorine or sulfur dioxide, out of a container.

AIR STRIPPING — AIR STRIPPING

A physical treatment process used to remove volatile substances from water or wastestreams. The process uses large volumes of air to transfer volatile pollutants from a high concentration in the water or wastestream into a lower concentration in an air stream.

ALARM CONTACT — ALARM CONTACT

A switch that operates when some preset low, high, or abnormal condition exists.

ALGAE (AL-jee) — ALGAE

Microscopic plants containing chlorophyll that live floating or suspended in water. They also may be attached to structures, rocks, or other submerged surfaces. Excess algal growths can impart tastes and odors to potable water. Algae produce oxygen during sunlight hours and use oxygen during the night hours. Their biological activities appreciably affect the pH, alkalinity, and dissolved oxygen of the water.

ALGAL (AL-gull) BLOOM — ALGAL BLOOM

Sudden, massive growths of microscopic and macroscopic plant life, such as green or blue-green algae, which can, under the proper conditions, develop in lakes, reservoirs, and ponds.

ALGICIDE (AL-juh-side) — ALGICIDE

Any substance or chemical specifically formulated to kill or control algae.

ALIPHATIC (al-uh-FAT-ick) HYDROXY ACIDS — ALIPHATIC HYDROXY ACIDS

Organic acids with carbon atoms arranged in branched or unbranched open chains rather than in rings.

ALIQUOT (AL-uh-kwot) — ALIQUOT

Representative portion of a sample. Often, an equally divided portion of a sample.

ALKALI (AL-kuh-lie) — ALKALI

Any of certain soluble salts, principally of sodium, potassium, magnesium, and calcium, that have the property of combining with acids to form neutral salts and may be used in chemical water treatment processes.

ALKALINE (AL-kuh-line) — ALKALINE

The condition of water or soil that contains a sufficient amount of alkali substances to raise the pH above 7.0.

ALKALINITY (AL-kuh-LIN-it-tee) — ALKALINITY

The capacity of water or wastewater to neutralize acids. This capacity is caused by the water's content of carbonate, bicarbonate, hydroxide, and occasionally borate, silicate, and phosphate. Alkalinity is expressed in milligrams per liter of equivalent calcium carbonate. Alkalinity is not the same as pH because water does not have to be strongly basic (high pH) to have a high alkalinity. Alkalinity is a measure of how much acid must be added to a liquid to lower the pH to 4.5.

ALLIANCE OF CERTIFIED OPERATORS, LABORATORY ANALYSTS, INSPECTORS, AND SPECIALISTS (AceOps) — ALLIANCE OF CERTIFIED OPERATORS, LABORATORY ANALYSTS, INSPECTORS, AND SPECIALISTS (AceOps)

A professional organization for operators, laboratory analysts, inspectors, and specialists dedicated to improving professionalism; expanding training, certification, and job opportunities; increasing information exchange; and advocating the importance of certified operators, lab analysts, inspectors, and specialists. For information on membership, visit AceOps at aceops.org or phone (515) 266-0248.

ALLUVIAL (uh-LOO-vee-ul) — ALLUVIAL

Relating to mud or sand deposited by flowing water. Alluvial deposits may occur after a heavy rainstorm.

ALTERNATING CURRENT (AC) — ALTERNATING CURRENT (AC)

An electric current that reverses its direction (positive/negative values) at regular intervals.

ALTITUDE VALVE — ALTITUDE VALVE

A valve that automatically shuts off the flow into an elevated tank when the water level in the tank reaches a predetermined level. The valve automatically opens when the pressure in the distribution system drops below the pressure in the tank.

AMBIENT (AM-bee-ent) TEMPERATURE — AMBIENT TEMPERATURE

Temperature of the surrounding air (or other medium). For example, temperature of the room where a gas chlorinator is installed.

AMERICAN WATER WORKS ASSOCIATION (AWWA) — AMERICAN WATER WORKS ASSOCIATION (AWWA)

A professional organization for all persons working in the water utility field. This organization develops and recommends goals, procedures, and standards for water utility agencies to help them improve their performance and effectiveness. For information on AWWA membership and publications, visit AWWA at awwa.org or phone (800) 926-7337.

AMPERAGE (AM-purr-age) — AMPERAGE

The strength of an electric current measured in amperes. The amount of electric current flow, similar to the flow of water in gallons per minute.

AMPERE (AM-peer) — AMPERE

The unit used to measure current strength. The current produced by an electromotive force of 1 volt acting through a resistance of 1 ohm.

AMPEROMETRIC (am-purr-o-MET-rick) — AMPEROMETRIC

A method of measurement that records electric current flowing or generated, rather than recording voltage. Amperometric titration is a means of measuring concentrations of certain substances in water.

AMPEROMETRIC (am-purr-o-MET-rick) TITRATION — AMPEROMETRIC TITRATION

A means of measuring concentrations of certain substances in water (such as strong oxidizers) based on the electric current that flows during a chemical reaction. Also see TITRATE.

AMPLITUDE — AMPLITUDE

The maximum strength of an alternating current during its cycle, as distinguished from the mean or effective strength.

ANAEROBIC (AN-air-O-bick) — ANAEROBIC

A condition in which atmospheric or dissolved oxygen (DO) is *NOT* present in the aquatic (water) environment.

ANALOG — ANALOG

The continuously variable signal type sent to an analog instrument (for example, 4–20 mA).

ANALOG READOUT — ANALOG READOUT

The readout of an instrument by a pointer (or other indicating means) against a dial or scale. Also see DIGITAL READOUT.

ANALYZER — ANALYZER

A device that conducts a periodic or continuous measurement of turbidity or some factor such as chlorine or fluoride concentration. Analyzers operate by any of several methods, including photocells, conductivity, or complex instrumentation.

ANGSTROM (ANG-strem) — ANGSTROM

A unit of length equal to one-tenth of a nanometer or one ten-billionth of a meter (1 Angstrom = 0.000 000 000 1 meter). One Angstrom is the approximate diameter of an atom.

ANION (AN-EYE-en) — ANION

A negatively charged ion in an electrolyte solution, attracted to the anode under the influence of a difference in electrical potential. Chloride ion (Cl^-) is an anion.

ANIONIC (AN-eye-ON-ick) POLYMER — ANIONIC POLYMER

A polymer having negatively charged groups of ions; often used as a filter aid and for dewatering sludges.

ANNULAR (AN-yoo-ler) SPACE — ANNULAR SPACE

A ring-shaped space located between two circular objects. For example, the space between the outside of a pipe liner and the inside of a pipe.

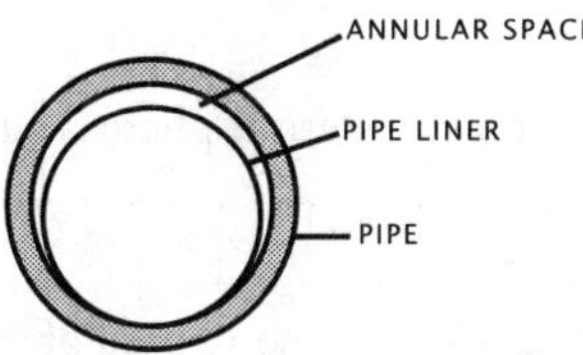

ANODE (AN-ode) — ANODE

The positive pole or electrode of an electrolytic system such as a battery. The anode attracts negatively charged particles or ions (anions).

APPARENT COLOR — APPARENT COLOR

Color of the water that includes not only the color due to substances in the water but suspended matter as well.

APPROPRIATIVE RIGHTS — APPROPRIATIVE RIGHTS

Water rights to or ownership of a water supply that is acquired for the beneficial use of water by following a specific legal procedure.

APPURTENANCE (uh-PURR-ten-nans) — APPURTENANCE

Machinery, appliances, structures, and other parts of the main structure necessary to allow it to operate as intended, but not considered part of the main structure.

AQUEOUS (ACK-wee-us) — AQUEOUS

Something made up of, similar to, or containing water; watery.

AQUIFER (ACK-wi-fer) — AQUIFER

A natural, underground layer of porous, water-bearing materials (sand, gravel) usually capable of yielding a large amount or supply of water.

ARCH — ARCH

(1) The curved top of a sewer pipe or conduit.

(2) A bridge or arch of hardened or caked chemical that will prevent the flow of the chemical.

ARTESIAN (are-TEE-zhun) — ARTESIAN

Pertaining to groundwater, a well, or underground basin where the water is under a pressure greater than atmospheric and will rise above the level of its upper confining surface if given an opportunity to do so.

ASEPTIC (a-SEP-tick) — ASEPTIC

Free from the living germs of disease, fermentation, or putrefaction. Sterile.

ASSET MANAGEMENT — ASSET MANAGEMENT

The process of maintaining the functionality and value of a utility's assets through repair, rehabilitation, and replacement. Examples of utility assets include buildings, tools, equipment, pipes, and machinery used to operate a water or wastewater system. The primary goal of asset management is to provide safe, reliable, and cost-effective service to a community over the useful life of a utility's assets. Another benefit of an effective asset management program is the extension of the life of the asset.

ASSOCIATION OF BOARDS OF CERTIFICATION (ABC) — ASSOCIATION OF BOARDS OF CERTIFICATION (ABC)

An international organization representing over 150 boards that certify the operators of waterworks and wastewater facilities. For information on ABC publications regarding the preparation of and how to study for operator certification examinations, visit ABC at www.abccert.org or phone (515) 232-3623.

ASYMMETRIC (A-sim-MET-rick) ASYMMETRIC

Not similar in size, shape, form, or arrangement of parts on opposite sides of a line, point, or plane.

atm atm

The abbreviation for atmosphere. One atmosphere is equal to 14.7 psi or 100 kPa.

ATOM ATOM

The smallest unit of a chemical element; composed of protons, neutrons, and electrons.

AUDIT, WATER AUDIT, WATER

A thorough examination of the accuracy of water agency records or accounts (volumes of water) and system control equipment. Water managers can use audits to determine their water distribution system efficiency. The overall goal is to identify and verify water and revenue losses in a water system.

AUTHORITY AUTHORITY

The power and resources to do a specific job or to get that job done.

AUTOMATIC METER READING (AMR) AUTOMATIC METER READING (AMR)

Technology that automatically collects usage, diagnostic, and status data from water metering devices and transfers that data to a central database for billing, troubleshooting, and analyzing.

AVAILABLE CHLORINE AVAILABLE CHLORINE

A measure of the amount of chlorine available in chlorinated lime, hypochlorite compounds, and other materials that are used as a source of chlorine when compared with that of elemental (liquid or gaseous) chlorine.

AVAILABLE EXPANSION AVAILABLE EXPANSION

The vertical distance from the sand surface to the underside of a trough in a sand filter. This distance is also called freeboard.

AVERAGE AVERAGE

A number obtained by adding quantities or measurements and dividing the sum or total by the number of quantities or measurements. Also called the arithmetic mean.

$$\text{Average} = \frac{\text{Sum of Measurements}}{\text{Number of Measurements}}$$

AVERAGE DEMAND AVERAGE DEMAND

The total demand for water during a period of time divided by the number of days in that time period. This is also called the average daily demand.

AWWA AWWA

See AMERICAN WATER WORKS ASSOCIATION (AWWA).

AXIAL TO IMPELLER AXIAL TO IMPELLER

The direction in which material being pumped flows around the impeller or flows parallel to the impeller shaft.

AXIS OF IMPELLER AXIS OF IMPELLER

An imaginary line running along the center of a shaft (such as an impeller shaft).

B

BACKFILL BACKFILL

The material placed over a pipe up to the ground surface. This consists of initial and final backfill. Also see FINAL BACKFILL and INITIAL BACKFILL.

BACKFLOW BACKFLOW

A reverse flow condition, created by a difference in water pressures, that causes water to flow back into the distribution pipes of a potable water supply from any source or sources other than an intended source. Also see BACKSIPHONAGE.

BACK PRESSURE BACK PRESSURE

A pressure that can cause water to backflow into the water supply when a user's water system is at a higher pressure than the public water system.

BACKSIPHONAGE BACKSIPHONAGE

A form of backflow caused by a negative or below atmospheric pressure within a water system. Also see BACKFLOW.

BACKWASHING BACKWASHING

The process of reversing the flow of water back through the filter media to remove the entrapped solids.

BACKWATER GATE BACKWATER GATE

A gate installed at the end of a drain or outlet pipe to prevent the backward flow of water or wastewater. Generally used on storm sewer outlets into streams to prevent backward flow during times of flood or high tide. Also called a tide gate. Also see CHECK VALVE and FLAP GATE.

BACTERIA (back-TEER-e-uh) BACTERIA

Bacteria are living organisms, microscopic in size, that usually consist of a single cell. Most bacteria use organic matter for their food and produce waste products as a result of their life processes.

BAFFLE BAFFLE

A flat board or plate, deflector, guide, or similar device constructed or placed in flowing water, wastewater, or slurry systems to cause more uniform flow velocities, to absorb energy, and to divert, guide, or agitate liquids (water, chemical solutions, slurry).

BAILER (BAY-ler) BAILER

A length of pipe equipped with a valve at the lower end used to remove slurry from the bottom or the side of a well as it is being drilled.

BALANCED SCORECARD BALANCED SCORECARD

A strategic planning and management system used extensively by water and wastewater facility managers to align business activities to the vision and strategy of the organization, improve communications, and monitor organizational performance against strategic goals. More information is available at balancedscorecard.org.

BALLAST BALLAST

A type of transformer that is used to limit the current to an ultraviolet (UV) lamp.

BASE-EXTRA CAPACITY METHOD BASE-EXTRA CAPACITY METHOD

A cost allocation method used by water utilities to determine water rates for various water user groups. This method considers base costs (O&M expenses and capital costs), extra capacity costs (additional costs for maximum day and maximum hour demands), customer costs (meter maintenance and reading, billing, collection, accounting), and fire protection costs.

BASE METAL BASE METAL

A metal (such as iron) that reacts with dilute hydrochloric acid to form hydrogen. Also see NOBLE METAL.

BATCH PROCESS BATCH PROCESS

A treatment process in which a tank or reactor is filled, the water is treated or a chemical solution is prepared, and the tank is emptied. The tank may then be filled and the process repeated.

BEDDING BEDDING

The soil placed in the bottom of a trench on top of the foundation soil to provide uniform support for a pipe. In addition to bringing the trench bottom to required grade, the bedding levels out any irregularities and ensures uniform support along the length of the pipe.

BENCHMARK | BENCHMARK

A standard or point of reference used to judge or measure quality or value.

BENCHMARKING | BENCHMARKING

A process an agency uses to gather and compare information about the productivity and performance of other similar agencies with its own information. The purpose of benchmarking is to identify best practices, set improvement targets, and measure progress.

BENCH-SCALE ANALYSIS (TEST) | BENCH-SCALE ANALYSIS (TEST)

A method of studying different ways or chemical doses for treating water or wastewater and solids on a small scale in a laboratory. Also see JAR TEST.

BIOCHEMICAL OXYGEN DEMAND (BOD) | BIOCHEMICAL OXYGEN DEMAND (BOD)

See BOD.

BIOFILTRATION | BIOFILTRATION

The process of filtering water through a filter medium that has been allowed to develop a microbial biofilm that assists in the removal of fine particulate matter and dissolved organic materials. Also see BIOLOGICAL FILTRATION.

BIOLOGICAL FILTRATION | BIOLOGICAL FILTRATION

The process of filtering water through a filter medium that has been allowed to develop a microbial biofilm that assists in the removal of fine particulate matter and dissolved organic materials. Also see BIOFILTRATION.

BIOLOGICAL GROWTH | BIOLOGICAL GROWTH

The activity and growth of any and all living organisms.

BLANK | BLANK

A bottle containing only dilution water or distilled water; the sample being tested is not added. Tests are frequently run on a sample and a blank and the differences are compared. The procedure helps to eliminate or reduce test result errors that could be caused when the dilution water or distilled water used is contaminated.

BLOCKOUT | BLOCKOUT

The physical prevention of the unexpected movement of machinery, working parts, or other sources of energy within a machine.

BLOCKOUT DEVICE | BLOCKOUT DEVICE

A device that prevents the unexpected movement of machinery, working parts, or other sources of energy within a machine.

BOD (pronounce as separate letters) | BOD

Biochemical Oxygen Demand. A measure of the amount of dissolved oxygen consumed (exertion) by aerobic organisms while stabilizing decomposable organic matter under aerobic conditions. The BOD value is commonly expressed as BOD_5 and refers to the milligrams of oxygen consumed per liter of water during a 5-day incubation period at 68°F (20°C). In decomposition, organic matter serves as food for the organisms and energy results from its oxidation. BOD measurements are used as a measure of the organic strength of wastes in water.

BOND | BOND

(1) A written promise to pay a specified sum of money (called the face value) at a fixed time in the future (called the date of maturity). A bond also carries interest at a fixed rate, payable periodically. The difference between a note and a bond is that a bond usually runs for a longer period of time and requires greater formality. Utility agencies use bonds as a means of obtaining large amounts of money for capital improvements.

(2) A warranty by an underwriting organization, such as an insurance company, guaranteeing honesty, performance, or payment by a contractor.

BONNET (BON-it) | BONNET

The cover on a gate valve.

BOWL, PUMP BOWL, PUMP

The submerged pumping unit in a well, including the shaft, impellers, and housing.

BRAKE HORSEPOWER (BHP) BRAKE HORSEPOWER (BHP)

(1) The horsepower required at the top or end of a pump shaft (input to a pump).

(2) The energy provided by a motor or other power source.

BREAKPOINT CHLORINATION BREAKPOINT CHLORINATION

Addition of chlorine to water or wastewater until the chlorine demand has been satisfied. At this point, further additions of chlorine will result in a free chlorine residual that is directly proportional to the amount of chlorine added beyond the breakpoint.

BREAKTHROUGH BREAKTHROUGH

A crack or break in a filter bed allowing the passage of floc or particulate matter through a filter. This will cause an increase in filter effluent turbidity. A breakthrough can occur (1) when a filter is first placed in service, (2) when the effluent valve suddenly opens or closes, and (3) during periods of excessive head loss through the filter (including when the filter is exposed to negative heads).

BRINELLING (bruh-NEL-ing) BRINELLING

Tiny indentations (dents) high on the shoulder of the bearing race or bearing. A type of bearing failure.

BUFFER BUFFER

A solution or liquid whose chemical makeup neutralizes acids or bases without a great change in pH.

BUFFER CAPACITY BUFFER CAPACITY

A measure of the capacity of a solution or liquid to neutralize acids or bases. This is a measure of the capacity of water or wastewater for offering a resistance to changes in pH.

C

CAISSON (KAY-sawn) CAISSON

A structure or chamber that is usually sunk or lowered by digging from the inside. Used to gain access to the bottom of a stream or other body of water.

CALCIUM CARBONATE ($CaCO_3$) EQUILIBRIUM CALCIUM CARBONATE ($CaCO_3$) EQUILIBRIUM

A water is considered stable when it is just saturated with calcium carbonate. In this condition, the water will neither dissolve nor deposit calcium carbonate. Thus, in this water the calcium carbonate is in equilibrium with the hydrogen ion concentration.

CALCIUM CARBONATE ($CaCO_3$) EQUIVALENT CALCIUM CARBONATE ($CaCO_3$) EQUIVALENT

An expression of the concentration of specified constituents in water in terms of their equivalent value to calcium carbonate. For example, the hardness in water that is caused by calcium, magnesium, and other ions is usually described as calcium carbonate equivalent. Alkalinity test results are usually reported as mg/L $CaCO_3$ equivalents. To convert chloride to $CaCO_3$ equivalents, multiply the concentration of chloride ions in mg/L by 1.41, and for sulfate, multiply by 1.04.

CALIBRATION CALIBRATION

A procedure that checks or adjusts an instrument's accuracy by comparison with a standard or reference.

CALL DATE CALL DATE

First date a bond can be paid off.

CAPILLARY (KAP-uh-larry) ACTION CAPILLARY ACTION

The movement of water through very small spaces due to molecular forces.

CAPILLARY (KAP-uh-larry) FORCES — CAPILLARY FORCES

The molecular forces that cause the movement of water through very small spaces.

CAPILLARY (KAP-uh-larry) FRINGE — CAPILLARY FRINGE

The porous material just above the water table that may hold water by capillarity (a property of surface tension that draws water upward) in the smaller void spaces.

CAPITAL IMPROVEMENT PLAN (CIP) — CAPITAL IMPROVEMENT PLAN (CIP)

A detailed plan that identifies requirements for the repair, replacement, and rehabilitation of facility infrastructure over an extended period, often 20 years or more. A utility usually updates or prepares this plan annually. The plan is often a part of a water system master plan that combines water demand projections with supply alternatives and facility requirements.

CARCINOGEN (kar-SIN-o-jen) — CARCINOGEN

Any substance that tends to produce cancer in an organism.

CATALYST (KAT-uh-list) — CATALYST

A substance that changes the speed or yield of a chemical reaction without being consumed or chemically changed by the chemical reaction.

CATALYZE (KAT-uh-lize) — CATALYZE

To act as a catalyst. Or, to speed up a chemical reaction.

CATALYZED (KAT-uh-lized) — CATALYZED

To be acted upon by a catalyst.

CATHODE (KATH-ode) — CATHODE

The negative pole or electrode of an electrolytic cell or system. The cathode attracts positively charged particles or ions (cations).

CATHODIC (kath-ODD-ick) PROTECTION — CATHODIC PROTECTION

An electrical system for prevention of rust, corrosion, and pitting of metal surfaces that are in contact with water, wastewater, or soil. A low-voltage current is made to flow through a liquid (water) or a soil in contact with the metal in such a manner that the external electromotive force renders the metal structure cathodic. This concentrates corrosion on auxiliary anodic parts, which are deliberately allowed to corrode instead of letting the structure corrode.

CATION (KAT-EYE-en) — CATION

A positively charged ion in an electrolyte solution, attracted to the cathode under the influence of a difference in electrical potential. Sodium ion (Na^+) is a cation.

CATIONIC (KAT-eye-ON-ic) POLYMER — CATIONIC POLYMER

A polymer having positively charged groups of ions; often used as a coagulant aid.

CAUTION — CAUTION

This word warns against potential hazards or cautions against unsafe practices. Also see DANGER, NOTICE, and WARNING.

CAVITATION (kav-uh-TAY-shun) — CAVITATION

The formation and collapse of a gas pocket or bubble on the blade of an impeller or the gate of a valve. The collapse of this gas pocket or bubble drives water into the impeller or gate with a terrific force that can knock metal particles off and cause pitting on the impeller or gate surface. Cavitation is accompanied by loud noises that sound like someone is pounding on the impeller or gate with a hammer.

CENTRATE — CENTRATE

The water leaving a centrifuge after most of the solids have been removed.

CENTRIFUGAL (sen-TRIF-uh-gull) PUMP — CENTRIFUGAL PUMP

A pump consisting of an impeller fixed on a rotating shaft that is enclosed in a casing, and having an inlet and discharge connection. As the rotating impeller whirls the liquid around, centrifugal force builds up enough pressure to force the water through the discharge outlet.

CENTRIFUGE — CENTRIFUGE

A mechanical device that uses centrifugal or rotational forces to separate solids from liquids.

CERTIFICATION EXAMINATION — CERTIFICATION EXAMINATION

An examination administered by a state agency or professional association that operators take to indicate a level of professional competence. In the United States, certification of operators of water treatment plants, wastewater treatment plants, water distribution systems, and small water supply systems is mandatory. In many states, certification of wastewater collection system operators, industrial wastewater treatment plant operators, pretreatment facility inspectors, and small wastewater system operators is voluntary; however, current trends indicate that more states, provinces, and employers will require these operators to be certified in the future. Operator certification is mandatory in the United States for the Chief Operators of water treatment plants, water distribution systems, and wastewater treatment plants.

CERTIFIED OPERATOR — CERTIFIED OPERATOR

A person who has the education and experience required to operate a specific class of treatment facility as indicated by possessing a certificate of professional competence given by a state agency or professional association.

C FACTOR — C FACTOR

A factor or value used to indicate the smoothness of the interior of a pipe. The higher the C Factor, the smoother the pipe, the greater the carrying capacity, and the smaller the friction or energy losses from water flowing in the pipe. To calculate the C Factor, measure the flow, pipe diameter, distance between two pressure gauges, and the friction or energy loss of the water between the gauges.

$$\text{C Factor} = \frac{\text{Flow, GPM}}{193.75(\text{Diameter, ft})^{2.63}(\text{Slope})^{0.54}}$$

or

$$\text{C Factor} = \frac{\text{Flow, m}^3\text{/sec}}{0.278(\text{Diameter, m})^{2.63}(\text{Slope})^{0.54}}$$

CHARGE CHEMISTRY — CHARGE CHEMISTRY

A branch of chemistry in which the destabilization and neutralization reactions occur between stable negatively charged and stable positively charged particles.

CHECK SAMPLING — CHECK SAMPLING

Whenever an initial or routine sample analysis indicates that a Maximum Contaminant Level (MCL) has been exceeded, check sampling is required to confirm the routine sampling results. Check sampling is in addition to the routine sampling program.

CHECK VALVE — CHECK VALVE

A special valve with a hinged disk or flap that opens in the direction of normal flow and is forced shut when flows go in the reverse or opposite direction of normal flows. Also see FLAP GATE and TIDE GATE.

CHELATING (KEY-LAY-ting) AGENT — CHELATING AGENT

A chemical used to prevent the precipitation of metals (such as copper).

CHELATION (key-LAY-shun) — CHELATION

A chemical complexing (forming or joining together) of metallic cations (such as copper) with certain organic compounds, such as EDTA (ethylene diamine tetracetic acid). Chelation is used to prevent the precipitation of metals (copper). Also see SEQUESTRATION.

CHEMTREC (KEM-trek) — CHEMTREC

Chemical Transportation Emergency Center. A public service of the American Chemistry Council dedicated to assisting emergency responders deal with incidents involving hazardous materials. Their toll-free 24-hour emergency phone number is (800) 424-9300.

CHLORAMINATION (KLOR-uh-min-NAY-shun) CHLORAMINATION

The application of chlorine and ammonia to water to form chloramines for the purpose of disinfection.

CHLORAMINES (KLOR-uh-means) CHLORAMINES

Compounds formed by the reaction of hypochlorous acid (or aqueous chlorine) with ammonia.

CHLORINATION (klor-uh-NAY-shun) CHLORINATION

The application of chlorine to water, generally for the purpose of disinfection, but frequently for accomplishing other biological or chemical results (aiding coagulation and controlling tastes and odors).

CHLORINATOR (KLOR-uh-nay-ter) CHLORINATOR

A metering device that is used to add chlorine to water.

CHLORINE DEMAND CHLORINE DEMAND

Chlorine demand is the difference between the amount of chlorine added to water or wastewater and the amount of chlorine residual remaining after a given contact time. Chlorine demand may change with dosage, time, temperature, pH, and nature and amount of the impurities in the water.

Chlorine Demand, mg/L = Chlorine Applied, mg/L – Chlorine Residual, mg/L

CHLORINE REQUIREMENT CHLORINE REQUIREMENT

The amount of chlorine that is needed for a particular purpose. Some reasons for adding chlorine are reducing the MPN (Most Probable Number) of coliform bacteria, obtaining a particular chlorine residual, or oxidizing some substance in the water. In each case, a definite dosage of chlorine will be necessary. This dosage is the chlorine requirement.

CHLORINE RESIDUAL CHLORINE RESIDUAL

The concentration of chlorine present in water after the chlorine demand has been satisfied. The concentration is expressed in terms of the total chlorine residual, which includes both the free and combined or chemically bound chlorine residuals. Also called residual chlorine.

CHLOROPHENOLIC (KLOR-o-fee-NO-lick) CHLOROPHENOLIC

Chlorophenolic compounds are phenolic compounds (carbolic acid) combined with chlorine.

CHLOROPHENOXY (KLOR-o-fuh-NOX-ee) CHLOROPHENOXY

A class of herbicides that may be found in domestic water supplies and cause adverse health effects. Two widely used chlorophenoxy herbicides are 2,4-D (2,4-dichlorophenoxy acetic acid) and 2,4,5-TP (2,4,5-trichlorophenoxy propionic acid [silvex]).

CHLORORGANIC (klor-or-GAN-ick) CHLORORGANIC

Organic compounds combined with chlorine. These compounds generally originate from, or are associated with, living or dead organic materials, including algae in water.

CHRONIC HEALTH EFFECT CHRONIC HEALTH EFFECT

An adverse effect on a human or animal body with symptoms that develop slowly over a long period of time or that recur frequently.

CIRCLE OF INFLUENCE CIRCLE OF INFLUENCE

The circular outer edge of a depression produced in the water table by the pumping of water from a well. Also see CONE OF INFLUENCE and CONE OF DEPRESSION.

[SEE DRAWING ON PAGE 783]

CIRCUIT CIRCUIT

The complete path of an electric current, including the generating apparatus or other source; or, a specific segment or section of the complete path.

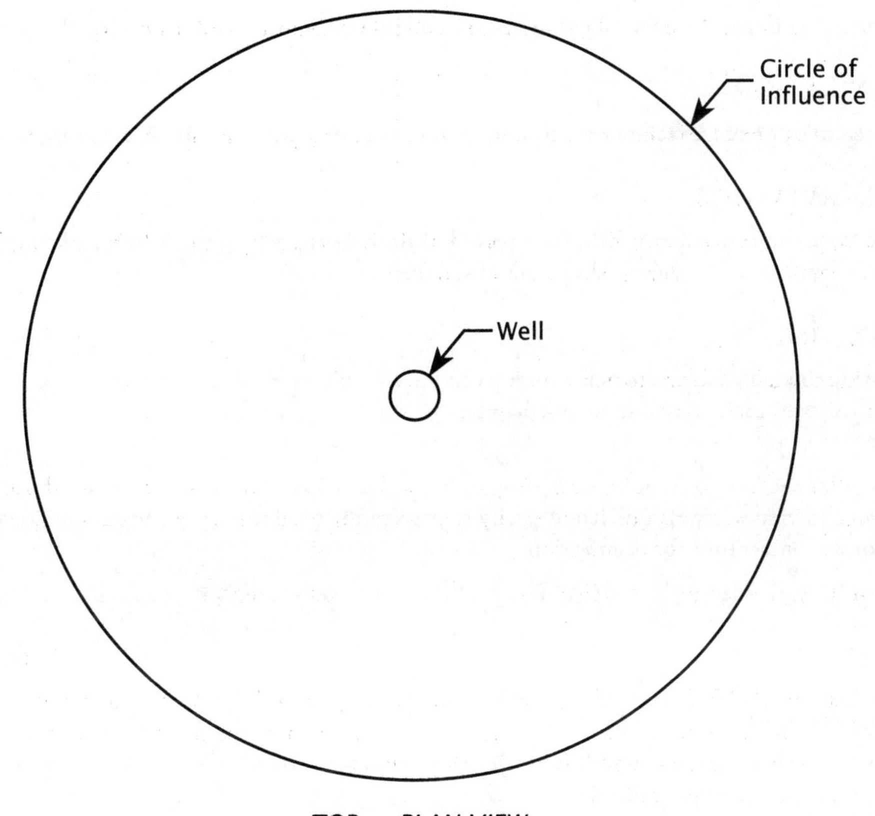

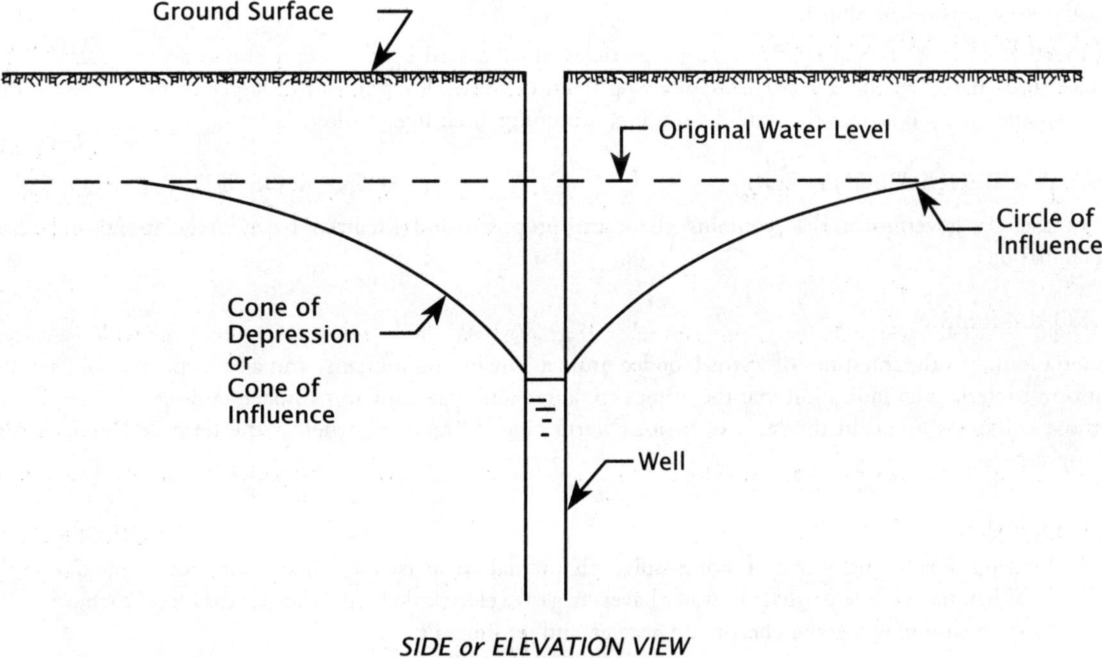

Circle of influence and cone of depression/cone of influence

CIRCUIT BREAKER — CIRCUIT BREAKER

A safety device in an electric circuit that automatically shuts off the circuit when it becomes overloaded. The device can be manually reset.

CISTERN (SIS-turn) — CISTERN

A small tank (usually covered) or a storage facility used to store water for a home or farm. Often used to store rainwater.

CLARIFIER (KLAIR-uh-fire) — CLARIFIER

A tank or basin in which water or wastewater is held for a period of time during which the heavier solids settle to the bottom and the lighter materials float to the surface. Also called settling tank or sedimentation basin.

CLASS, PIPE AND FITTINGS — CLASS, PIPE AND FITTINGS

The working pressure rating, including allowances for surges, of a specific pipe for use in water distribution systems. The term is used for cast iron, ductile iron, asbestos cement, and some plastic pipe.

CLEAR WELL — CLEAR WELL

A reservoir for the storage of filtered water of sufficient capacity to prevent the need to vary the filtration rate with variations in demand. Also used to provide chlorine contact time for disinfection.

CLEAR ZONE — CLEAR ZONE

See SUPERNATANT.

COAGULANT (ko-AGG-yoo-lent) — COAGULANT

A chemical that causes very fine particles to clump (floc) together into larger particles. This makes it easier to separate the solids from the liquids by settling, skimming, draining, or filtering.

COAGULANT (ko-AGG-yoo-lent) AID — COAGULANT AID

Any chemical or substance used to assist or modify coagulation.

COAGULATION (ko-agg-yoo-LAY-shun) — COAGULATION

The clumping together of very fine particles into larger particles (floc) caused by the use of chemicals (coagulants). The chemicals neutralize the electrical charges of the fine particles, allowing them to come closer and form larger clumps. This clumping together makes it easier to separate the solids from the water by settling, skimming, draining, or filtering.

CODE OF FEDERAL REGULATIONS (CFR) — CODE OF FEDERAL REGULATIONS (CFR)

A publication of the US government that contains all of the proposed and finalized federal regulations, including safety and environmental regulations.

COLIFORM (KOAL-i-form) — COLIFORM

A group of bacteria found in the intestines of warm-blooded animals (including humans) and also in plants, soil, air, and water. The presence of coliform bacteria is an indication that the water is polluted and may contain pathogenic (disease-causing) organisms. Fecal coliforms are those coliforms found in the feces of various warm-blooded animals, whereas the term coliform also includes other environmental sources.

COLLOIDS (KALL-loids) — COLLOIDS

Very small, finely divided solids (particles that do not dissolve) that remain dispersed in a liquid for a long time due to their small size and electrical charge. When most of the particles in water have a negative electrical charge, they tend to repel each other. This repulsion prevents the particles from clumping together, becoming heavier, and settling out.

COLOR — COLOR

The substances in water that impart a yellowish-brown color to the water. These substances are the result of iron and manganese ions, humus and peat materials, plankton, aquatic weeds, and industrial waste present in the water. Also see TRUE COLOR.

COLORIMETRIC MEASUREMENT COLORIMETRIC MEASUREMENT

A means of measuring unknown chemical concentrations in water by measuring a sample's color intensity. The specific color of the sample, developed by addition of chemical reagents, is measured with a photoelectric colorimeter or is compared with color standards using, or corresponding with, known concentrations of the chemical.

COMBINED AVAILABLE CHLORINE COMBINED AVAILABLE CHLORINE

The total chlorine, present as chloramine or other derivatives, that is present in a water and is still available for disinfection and for oxidation of organic matter. The combined chlorine compounds are more stable than free chlorine forms, but they are somewhat slower in disinfection action.

COMBINED AVAILABLE CHLORINE RESIDUAL COMBINED AVAILABLE CHLORINE RESIDUAL

The concentration of chlorine residual that is combined with ammonia, organic nitrogen, or both in water as a chloramine (or other chloro derivative) and yet is still available to oxidize organic matter and help kill bacteria.

COMBINED CHLORINE COMBINED CHLORINE

The sum of the chlorine species composed of free chlorine and ammonia, including monochloramine, dichloramine, and trichloramine (nitrogen trichloride). Dichloramine is the strongest disinfectant of these chlorine species, but it has less oxidative capacity than free chlorine.

COMBINED RESIDUAL CHLORINATION COMBINED RESIDUAL CHLORINATION

The application of chlorine to water or wastewater to produce a combined available chlorine residual. The residual may consist of chlorine compounds formed by the reaction of chlorine with natural or added ammonia (NH_3) or with certain organic nitrogen compounds.

COMMODITY-DEMAND METHOD COMMODITY-DEMAND METHOD

A cost allocation method used by water utilities to determine water rates for the various water user groups. This method considers the commodity costs (water, chemicals, power, amount of water use), demand costs (treatment, storage, distribution), customer costs (meter maintenance and reading, billing, collection, accounting), and fire protection costs.

COMPETENT PERSON COMPETENT PERSON

A competent person is defined by OSHA as a person capable of identifying existing and predictable hazards in the surroundings, or working conditions that are unsanitary, hazardous, or dangerous to employees, and who has authorization to take prompt corrective measures to eliminate the hazards.

COMPLETE TREATMENT COMPLETE TREATMENT

A method of treating water that consists of the addition of coagulant chemicals, flash mixing, coagulation-flocculation, sedimentation, and filtration. Also called conventional filtration. Also see DIRECT FILTRATION and IN-LINE FILTRATION.

COMPOSITE (PROPORTIONAL) SAMPLE COMPOSITE (PROPORTIONAL) SAMPLE

A composite sample is a collection of individual samples obtained at regular intervals, usually every 1 or 2 hours during a 24-hour time span. Each individual sample is combined with the others in proportion to the rate of flow when the sample was collected. Equal volume individual samples also may be collected at intervals after a specific volume of flow passes the sampling point or after equal time intervals and still be referred to as a composite sample. The resulting mixture (composite sample) forms a representative sample and is analyzed to determine the average conditions during the sampling period.

COMPOUND COMPOUND

A pure substance composed of two or more elements whose composition is constant. For example, table salt (sodium chloride, NaCl) is a compound.

CONCENTRATION POLARIZATION CONCENTRATION POLARIZATION

(1) A buildup of retained particles on the membrane surface due to dewatering of the feed closest to the membrane. The thickness of the concentration polarization layer is controlled by the flow velocity across the membrane.

(2) Used in corrosion studies to indicate a depletion of ions near an electrode.

(3) The basis for chemical analysis by a polarograph.

CONDITIONING CONDITIONING

Pretreatment of sludge to facilitate removal of water in subsequent treatment processes.

CONDUCTANCE CONDUCTANCE

A rapid method of estimating the total dissolved solids content of a water supply. The measurement indicates the capacity of a sample of water to carry an electric current, which is related to the concentration of ionized substances in the water. Also called specific conductance.

CONDUCTIVITY CONDUCTIVITY

A measure of the ability of a solution (water) to carry an electric current.

CONDUCTOR CONDUCTOR

(1) A pipe that carries a liquid load from one point to another point. In a wastewater collection system, a conductor is often a large pipe with no service connections. Also called a conduit. Also see INTERCEPTOR (INTERCEPTING) SEWER or INTERCONNECTOR.

(2) In plumbing, a pipe installed to drain water from the roof gutters or roof catchment to the storm drain or other means of disposal. Also called a downspout, roof leader, or roof drain.

(3) In electricity, a substance, body, device, or wire that readily conducts or carries electric current. Also called a conduit.

CONDUCTOR CASING CONDUCTOR CASING

The outer casing of a well. The purpose of this casing is to prevent contaminants from surface waters or shallow groundwaters from entering a well.

CONDUIT CONDUIT

Any artificial or natural duct, either open or closed, for carrying fluids from one point to another. An electrical conduit carries electricity.

CONE OF DEPRESSION CONE OF DEPRESSION

The depression, roughly conical in shape, produced in the water table by the pumping of water from a well. Also called the cone of influence. Also see CIRCLE OF INFLUENCE.

[SEE DRAWING ON PAGE 783]

CONE OF INFLUENCE CONE OF INFLUENCE

See CONE OF DEPRESSION.

CONFINED SPACE CONFINED SPACE

A space that has the following characteristics:

(1) Is large enough and so configured that an employee can bodily enter and perform assigned work

(2) Has limited or restricted means for entry or exit (for example, manholes, tanks, vessels, silos, storage bins, hoppers, vaults, and pits are spaces that may have limited means of entry)

(3) Is not designed for continuous employee occupancy

Also see DANGEROUS AIR CONTAMINATION and OXYGEN DEFICIENCY.

CONFINED SPACE, CLASS A CONFINED SPACE, CLASS A

A confined space that presents a situation that is immediately dangerous to life or health (IDLH). These include but are not limited to oxygen deficiency, explosive or flammable atmospheres, and concentrations of toxic substances.

CONFINED SPACE, CLASS B CONFINED SPACE, CLASS B

A confined space that has the potential for causing injury and illness, if preventive measures are not used, but is not immediately dangerous to life and health.

CONFINED SPACE, CLASS C CONFINED SPACE, CLASS C

A confined space in which the potential hazard would not require any special modification of the work procedure.

CONFINED SPACE, NON-PERMIT CONFINED SPACE, NON-PERMIT

A non-permit confined space is a confined space that does not contain or, with respect to atmospheric hazards, have the potential to contain any hazard capable of causing death or serious physical harm.

CONFINED SPACE, PERMIT-REQUIRED (PERMIT SPACE) CONFINED SPACE, PERMIT-REQUIRED (PERMIT SPACE)

A confined space that has one or more of the following characteristics:

(1) Contains or has a potential to contain a hazardous atmosphere

(2) Contains a material that has the potential for engulfing an entrant

(3) Has an internal configuration such that an entrant could be trapped or asphyxiated by inwardly converging walls or by a floor that slopes downward and tapers to a smaller cross section

(4) Contains any other recognized serious safety or health hazard

CONFINING UNIT CONFINING UNIT

A layer of rock or soil of very low hydraulic conductivity that hampers the movement of groundwater in and out of an aquifer.

CONSOLIDATED FORMATION CONSOLIDATED FORMATION

A geologic material whose particles are stratified (layered), cemented, or firmly packed together (hard rock); usually occurring at a depth below the ground surface. Also see UNCONSOLIDATED FORMATION.

CONSUMER CONFIDENCE REPORT (CCR) CONSUMER CONFIDENCE REPORT (CCR)

An annual report prepared by a water utility to communicate with its consumers. The report provides consumers with information on the source and quality of their drinking water and is an opportunity for positive communication with consumers.

CONTACTOR CONTACTOR

An electric switch, usually magnetically operated.

CONTAMINATION (kun-TAM-uh-NAY-shun) CONTAMINATION

The introduction into water of microorganisms, chemicals, toxic substances, wastes, or wastewater in a concentration that makes the water unfit for its next intended use.

CONTINUOUS SAMPLE CONTINUOUS SAMPLE

A flow of water from a particular place in a plant to the location where samples are collected for testing. This continuous stream may be used to obtain grab or composite samples. Frequently, several taps (faucets) will flow continuously in the laboratory to provide test samples from various places in a water treatment plant.

CONTROLLER CONTROLLER

A device that controls the starting, stopping, or operation of a device or piece of equipment.

CONTROL LOOP CONTROL LOOP

The combination of one or more interconnected instrumentation devices that are arranged to measure, display, and control a process variable. Also called a loop.

CONTROL SYSTEM CONTROL SYSTEM

An instrumentation system that senses and controls its own operation on a close, continuous basis in what is called proportional (or modulating) control.

CONVENTIONAL FILTRATION CONVENTIONAL FILTRATION

A method of treating water that consists of the addition of coagulant chemicals, flash mixing, coagulation-flocculation, sedimentation, and filtration. Also called complete treatment. Also see DIRECT FILTRATION and IN-LINE FILTRATION.

CONVENTIONAL TREATMENT CONVENTIONAL TREATMENT

(1) The common wastewater treatment processes, such as preliminary treatment, sedimentation, flotation, trickling filter, rotating biological contactor, activated sludge, and chlorination wastewater treatment processes, used by POTWs.

(2) The hydroxide precipitation of metals processes used by pretreatment facilities.

COPRECIPITATION COPRECIPITATION

A treatment process that occurs when ferrous iron is added to water or metallic wastestreams and subsequently oxidized in an aerator. The oxidized iron, which is insoluble, precipitates along with other metallic contaminants present in the water or wastestream, thereby enhancing metals removal.

CORPORATION STOP CORPORATION STOP

A water service shutoff valve located at a street water main. This valve cannot be operated from the ground surface because it is buried and there is no valve box. Also called a corporation cock.

CORROSION CORROSION

The gradual decomposition or destruction of a material by chemical action, often due to an electrochemical reaction. Corrosion may be caused by (1) stray current electrolysis, (2) galvanic corrosion caused by dissimilar metals, or (3) differential-concentration cells. Corrosion starts at the surface of a material and moves inward.

CORROSION INHIBITORS CORROSION INHIBITORS

Substances that slow the rate of corrosion.

CORROSIVE GASES CORROSIVE GASES

Gases that, when dissolved in water, can oxidize materials of construction such as steel and concrete.

CORROSIVITY CORROSIVITY

An indication of the corrosiveness of a water. The corrosiveness of a water is described by the water's pH, alkalinity, hardness, temperature, total dissolved solids, dissolved oxygen concentration, and the Langelier Index.

COULOMB (KOO-lahm) COULOMB

A measurement of the amount of electrical charge carried by an electric current of 1 ampere in 1 second. One coulomb equals about 6.25×10^{18} electrons (6,250,000,000,000,000,000 electrons).

COUPON COUPON

A steel specimen inserted into water or wastewater to measure corrosiveness. The rate of corrosion is measured as the loss of weight of the coupon or change in its physical characteristics. Measure the weight loss (in milligrams) per surface area (in square decimeters) exposed to the water or wastewater per day. 1 meter = 10 decimeters = 100 centimeters.

COVERAGE RATIO COVERAGE RATIO

The coverage ratio is a measure of the ability of the utility to pay the principal and interest on loans and bonds (this is known as debt service) in addition to any unexpected expenses.

CROSS-CONNECTION CROSS-CONNECTION

(1) A connection between a drinking (potable) water system and an unapproved water supply.

(2) A connection between a storm drain system and a sanitary collection system.

(3) Less frequently used to mean a connection between two sections of a collection system to handle anticipated overloads of one system.

CROSS-FLOW FILTRATION CROSS-FLOW FILTRATION

A type of membrane filtration where the water being filtered flows across the surface of the membrane to keep the particle buildup and fouling to a minimum. The flow that is not filtered becomes concentrated and flows out the end of the membrane fiber as a wastestream.

CRYPTOSPORIDIUM (KRIP-toe-spo-RID-ee-um) CRYPTOSPORIDIUM

A waterborne intestinal parasite that causes a disease called cryptosporidiosis (KRIP-toe-spo-rid-ee-O-sis) in infected humans. Symptoms of the disease include diarrhea, cramps, and weight loss. *Cryptosporidium* contamination is found in most surface waters and some groundwaters. Commonly referred to as crypto.

CT VALUE CT VALUE

Residual concentration of a given disinfectant in mg/L times the disinfectant's contact time in minutes.

CURB STOP CURB STOP

A water service shutoff valve located in a water service pipe near the curb and between the water main and the building. This valve is usually operated by a wrench or valve key and is used to start or stop flows in the water service line to a building. Also called a curb cock.

CURIE (KYOOR-ee) CURIE

A measure of radioactivity. One Curie of radioactivity is equivalent to 3.7×10^{10} or 37,000,000,000 nuclear disintegrations per second.

CURRENT CURRENT

A movement or flow of electricity. Electric current is measured by the number of coulombs per second flowing past a certain point in a conductor. A coulomb is equal to about 6.25×10^{18} electrons (6,250,000,000,000,000,000 electrons). A flow of 1 coulomb per second is called 1 ampere, the unit of the rate of flow of current.

CYCLE CYCLE

A complete alternation of voltage or current in an alternating current (AC) circuit.

D

DANGER DANGER

This word is used where an immediate hazard presents a threat of death or serious injury to employees. Also see CAUTION, NOTICE, and WARNING.

DANGEROUS AIR CONTAMINATION DANGEROUS AIR CONTAMINATION

An atmosphere presenting a threat of causing death, injury, acute illness, or disablement due to the presence of flammable or explosive, toxic, or otherwise injurious or incapacitating substances.

(1) Dangerous air contamination due to the flammability of a gas, vapor, or mist is defined as an atmosphere containing the gas, vapor, or mist at a concentration greater than 10 percent of its lower explosive (lower flammable) limit (LEL).

(2) Dangerous air contamination due to a combustible particulate is defined as a concentration that meets or exceeds the particulate's lower explosive limit (LEL).

(3) Dangerous air contamination due to the toxicity of a substance is defined as the atmospheric concentration that could result in employee exposure in excess of the substance's permissible exposure limit (PEL).

NOTE: A dangerous situation also occurs when the oxygen level is less than 19.5 percent by volume (OXYGEN DEFICIENCY) or more than 23.5 percent by volume (OXYGEN ENRICHMENT).

DATEOMETER (day-TOM-uh-ter) DATEOMETER

A small calendar disk attached to motors and equipment to indicate the year in which the last maintenance service was performed.

DATUM LINE DATUM LINE

A line from which heights and depths are calculated or measured. Also called a datum plane or a datum level.

DAY TANK DAY TANK

A tank used to store a chemical solution of known concentration for feed to a chemical feeder. A day tank usually stores sufficient chemical solution to properly treat the water being treated for at least one day. Also called an age tank.

DBP DBP

See DISINFECTION BYPRODUCT (DBP).

DEAD END DEAD END

The end of a water main that is not connected to other parts of the distribution system by means of a connecting loop of pipe.

DEAD-END FILTRATION DEAD-END FILTRATION

A type of membrane filtration where the water being filtered flows through the membrane, but there is no wastestream from the system. All solids accumulate on the membrane during filtration and are removed during backwash.

DEBT SERVICE DEBT SERVICE

The amount of money required annually to pay the (1) interest on outstanding debts or (2) funds due on a maturing bonded debt or the redemption of bonds.

DECANT (de-KANT) DECANT

To draw off the upper layer of liquid (water) after the heavier material (a solid or another liquid) has settled.

DECANT (de-KANT) WATER DECANT WATER

Water that has separated from sludge and is removed from the layer of water above the sludge.

DECHLORINATION (DEE-klor-uh-NAY-shun) DECHLORINATION

The deliberate removal of chlorine from water. The partial or complete reduction of chlorine residual by any chemical or physical process.

DECIBEL (DES-uh-bull) DECIBEL

A unit for expressing the relative intensity of sounds on a scale from zero for the average least perceptible sound to about 130 for the average level at which sound causes pain to humans. Abbreviated dB.

DECOMPOSITION or DECAY DECOMPOSITION or DECAY

The conversion of chemically unstable materials to more stable forms by chemical or biological action.

DEFLUORIDATION (DEE-floor-uh-DAY-shun) DEFLUORIDATION

The removal of excess fluoride in drinking water to prevent the mottling (brown stains) of teeth.

DEGASIFICATION (DEE-gas-if-uh-KAY-shun) DEGASIFICATION

A water treatment process that removes dissolved gases from water. The gases may be removed by either mechanical or chemical treatment methods or a combination of both.

DELEGATION DELEGATION

The act in which power is given to another person in the organization to accomplish a specific job.

DEMINERALIZATION (DEE-min-er-al-uh-ZAY-shun) DEMINERALIZATION

A treatment process that removes dissolved minerals (salts) from water.

DENSITY DENSITY

A measure of how heavy a substance (solid, liquid, or gas) is for its size. Density is expressed in terms of weight per unit volume, that is, grams per cubic centimeter or pounds per cubic foot. The density of water at 39°F (4°C) is about 62.4 pounds per cubic foot or 1.0 gram per cubic centimeter.

DEPOLARIZATION DEPOLARIZATION

The removal or depletion of ions in the thin boundary layer adjacent to a membrane or pipe wall.

DEPRECIATION DEPRECIATION

The gradual loss in service value of a facility or piece of equipment due to all the factors causing the ultimate retirement of the facility or equipment. This loss can be caused by sudden physical damage, wearing out due to age, obsolescence, inadequacy, or availability of a newer, more efficient facility or equipment. The value cannot be restored by maintenance.

DESALINIZATION (DEE-SAY-lin-uh-ZAY-shun) DESALINIZATION

The removal of dissolved salts (such as sodium chloride, NaCl) from water by natural means (leaching) or by specific water treatment processes.

DESICCANT (DESS-uh-kant) DESICCANT

A drying agent that is capable of removing or absorbing moisture from the atmosphere in a small enclosure.

DESICCATION (dess-uh-KAY-shun) DESICCATION

A process used to thoroughly dry air; to remove virtually all moisture from air.

DESICCATOR (DESS-uh-kay-tor) DESICCATOR

A closed container into which heated weighing or drying dishes are placed to cool in a dry environment in preparation for weighing. The dishes may be empty or they may contain a sample. Desiccators contain a substance (DESICCANT), such as anhydrous calcium chloride, that absorbs moisture and keeps the relative humidity near zero so that the dish or sample will not gain weight from absorbed moisture.

DESTRATIFICATION (DEE-strat-uh-fuh-KAY-shun) DESTRATIFICATION

The development of vertical mixing within a lake or reservoir to eliminate (either totally or partially) separate layers of temperature, plant life, or animal life. This vertical mixing can be caused by mechanical means (pumps) or through the use of forced air diffusers that release air into the lower layers of the reservoir.

DETENTION TIME DETENTION TIME

(1) The theoretical (calculated) time required for a small amount of water to pass through a tank at a given rate of flow.

(2) The actual time in hours, minutes, or seconds that a small amount of water is in a settling basin, flocculating basin, or rapid-mix chamber. In storage reservoirs, detention time is the length of time entering water will be held before being drafted for use (several weeks to years, several months being typical).

$$\text{Detention Time, hr} = \frac{(\text{Basin Volume, gal})(24\ \text{hr/day})}{\text{Flow, gal/day}}$$

or

$$\text{Detention Time, hr} = \frac{(\text{Basin Volume, m}^3)(24\ \text{hr/day})}{\text{Flow, m}^3\text{/day}}$$

DEWATER DEWATER

(1) To remove or separate a portion of the water present in a sludge or slurry. To dry sludge so it can be handled and disposed of.

(2) To remove or drain the water from a tank or a trench. A structure may be dewatered so that it can be inspected or repaired.

DEW POINT DEW POINT

The temperature to which air with a given quantity of water vapor must be cooled to cause condensation of the vapor in the air.

DIATOMACEOUS (DYE-uh-toe-MAY-shus) EARTH DIATOMACEOUS EARTH

A fine, siliceous (made of silica) earth composed mainly of the skeletal remains of diatoms.

DIATOMS (DYE-uh-toms) DIATOMS

Unicellular (single cell), microscopic algae with a rigid, box-like internal structure consisting mainly of silica.

DIELECTRIC (DYE-ee-LECK-trick) DIELECTRIC

Does not conduct an electric current. An insulator or nonconducting substance.

DIGITAL DIGITAL

The encoding of information that uses binary numbers (ones and zeros) for input, processing, transmission, storage, or display, rather than a continuous spectrum of values (an analog system) or non-numeric symbols such as letters or icons.

DIGITAL READOUT DIGITAL READOUT

The readout of an instrument by a direct, numerical reading of the measured value or variable.

DILUTE SOLUTION DILUTE SOLUTION

A solution that has been made weaker, usually by the addition of water.

DIMICTIC (dye-MICK-tick) DIMICTIC

Lakes and reservoirs that freeze over and normally go through two stratification and two mixing cycles within a year.

DIRECT CURRENT (DC) DIRECT CURRENT (DC)

Electric current flowing in one direction only and essentially free from pulsation. Also see ALTERNATING CURRENT.

DIRECT FILTRATION DIRECT FILTRATION

A method of treating water that consists of the addition of coagulant chemicals, flash mixing, coagulation, minimal flocculation, and filtration. The flocculation facilities may be omitted, but the physical-chemical reactions will occur to some extent. The sedimentation process is omitted. Also see CONVENTIONAL FILTRATION and IN-LINE FILTRATION.

DIRECT RUNOFF DIRECT RUNOFF

Water that flows over the ground surface directly into streams, rivers, or lakes. Also called storm runoff.

DISCHARGE HEAD DISCHARGE HEAD

The pressure (in pounds per square inch [psi] or kilopascals [kPa]) measured at the centerline of a pump discharge and very close to the discharge flange, converted into feet or meters. The pressure is measured from the centerline of the pump to the hydraulic grade line of the water in the discharge pipe.

$$\text{Discharge Head, ft} = (\text{Discharge Pressure, psi})(2.31\ \text{ft/psi})$$

or

$$\text{Discharge Head, m} = (\text{Discharge Pressure, kPa})(1\ \text{m}/9.8\ \text{kPa})$$

DISCRETE CONTROL DISCRETE CONTROL

ON/OFF control; one of the two output values is equal to zero.

DISCRETE I/O (INPUT/OUTPUT) DISCRETE I/O (INPUT/OUTPUT)

A digital signal that senses or sends either ON or OFF signals. For example, a discrete input would sense the position of a switch; a discrete output would turn on a pump or light.

DISINFECTION (dis-in-FECT-shun) DISINFECTION

The process designed to kill or inactivate most microorganisms in water or wastewater, including essentially all pathogenic (disease-causing) bacteria. There are several ways to disinfect, with chlorination being the most frequently used in water and wastewater treatment plants. Compare with STERILIZATION.

DISINFECTION BYPRODUCT (DBP) DISINFECTION BYPRODUCT (DBP)

A contaminant formed by the reaction of disinfection chemicals (such as chlorine) with other substances in the water being disinfected.

DISSOLVED ORGANIC CARBON (DOC) — DISSOLVED ORGANIC CARBON (DOC)

That portion of the organic carbon in water that passes through a 0.45 μm pore-diameter filter.

DISSOLVED ORGANIC MATTER (DOM) — DISSOLVED ORGANIC MATTER (DOM)

That portion of the organic matter in water that passes through a 0.45 μm pore-diameter filter.

DISTILLATE (DIS-tuh-late) — DISTILLATE

In the distillation of a sample, a portion is collected by evaporation and recondensation; the part that is recondensed is the distillate.

DISTRIBUTED CONTROL SYSTEM (DCS) — DISTRIBUTED CONTROL SYSTEM (DCS)

A computer control system having multiple microprocessors to distribute the functions performing process control, thereby distributing the risk from component failure. The distributed components (input/output devices, control devices, and operator interface devices) are all connected by communications links and permit the transmission of control, measurement, and operating information to and from many locations.

DIVALENT (dye-VAY-lent) — DIVALENT

Having a valence of two, such as the ferrous ion, Fe^{2+}. Also called bivalent.

DIVERSION — DIVERSION

Use of part of a stream flow as a water supply.

DOC — DOC

See DISSOLVED ORGANIC CARBON (DOC).

DOM — DOM

See DISSOLVED ORGANIC MATTER (DOM).

DOWNSPOUT — DOWNSPOUT

In plumbing, a pipe installed to drain water from the roof gutters or roof catchment to the storm drain or other means of disposal. Also called a conductor, roof leader, or roof drain.

DPD METHOD — DPD METHOD

A method of measuring the chlorine residual in water. The residual may be determined by either titrating or comparing a developed color with color standards. DPD stands for N,N-diethyl-p-phenylenediamine.

DRAFT — DRAFT

(1) The act of drawing or removing water from a tank or reservoir.

(2) The water that is drawn or removed from a tank or reservoir.

DRAWDOWN — DRAWDOWN

(1) The drop in the water table or level of water in the ground when water is being pumped from a well.

(2) The amount of water used from a tank or reservoir.

(3) The drop in the water level of a tank or reservoir.

DRIFT — DRIFT

The difference between the actual value and the desired value (or set point); characteristic of proportional controllers that do not incorporate reset action. Also called offset.

DYNAMIC HEAD DYNAMIC HEAD

When a pump is operating, the vertical distance (in feet or meters) from a point to the energy grade line. Also see ENERGY GRADE LINE (EGL), STATIC HEAD, and TOTAL DYNAMIC HEAD (TDH).

DYNAMIC PRESSURE DYNAMIC PRESSURE

When a pump is operating, pressure resulting from the dynamic head.

$$\text{Dynamic Pressure, psi} = (\text{Dynamic Head, ft})(0.433\ \text{psi/ft})$$

or

$$\text{Dynamic Pressure, kPa} = (\text{Dynamic Head, m})(9.8\ \text{kPa/m})$$

E

EDUCTOR (e-DUCK-ter) EDUCTOR

A hydraulic device used to create a negative pressure (suction) by forcing a liquid through a restriction, such as a Venturi. An eductor or aspirator (the hydraulic device) may be used in the laboratory in place of a vacuum pump. As an injector, it is used to produce vacuum for chlorinators. Sometimes used instead of a suction pump.

EFFECTIVE RANGE EFFECTIVE RANGE

That portion of the design range (usually from 10 to 90+ percent) in which an instrument has acceptable accuracy. Also see RANGE and SPAN.

EFFECTIVE SIZE (ES) EFFECTIVE SIZE (ES)

The diameter of the particles in a granular sample (filter media) for which 10 percent of the total grains are smaller and 90 percent larger on a weight basis. Effective size is obtained by passing granular material through sieves with varying dimensions of mesh and weighing the material retained by each sieve. The effective size is also approximately the average size of the grains.

EFFLUENT (EF-loo-ent) EFFLUENT

Water or other liquid—raw (untreated), partially treated, or completely treated—flowing from a reservoir, basin, treatment process, or treatment plant.

EGL EGL

See ENERGY GRADE LINE (EGL).

EJECTOR EJECTOR

A device used to disperse a chemical solution into water being treated.

ELECTRICAL LOGGING ELECTRICAL LOGGING

A procedure used to search for water-bearing formations (aquifers) by determining the porosity (spaces or voids) of geologic materials. Electrical probes are lowered into wells, an electric current is induced at various depths, and the resistance measured indicates the porosity of the formation.

ELECTROCHEMICAL REACTION ELECTROCHEMICAL REACTION

Chemical changes produced by electricity (electrolysis) or the production of electricity by chemical changes (galvanic action). In corrosion, a chemical reaction is accompanied by the flow of electrons through a metallic path. The electron flow may come from an external source and cause the reaction, such as electrolysis caused by a DC (direct current) electric railway, or the electron flow may be caused by a chemical reaction, as in the galvanic action of a flashlight dry cell.

ELECTROCHEMICAL SERIES ELECTROCHEMICAL SERIES

A list of metals with the standard electrode potentials given in volts. The size and sign of the electrode potential indicates how easily these elements will take on or give up electrons, or corrode. Hydrogen is conventionally assigned a value of zero.

ELECTROLYSIS (ee-leck-TRAWL-uh-sis) ELECTROLYSIS

The decomposition of material by an outside electric current.

ELECTROLYTE (ee-LECK-tro-lite) ELECTROLYTE

A substance that dissociates (separates) into two or more ions when it is dissolved in water.

ELECTROLYTIC (ee-LECK-tro-LIT-ick) CELL ELECTROLYTIC CELL

A device in which the chemical decomposition of material causes an electric current to flow. Also, a device in which a chemical reaction occurs as a result of the flow of electric current. Chlorine and caustic (NaOH) are made from salt (NaCl) in electrolytic cells.

ELECTROMOTIVE FORCE (EMF) ELECTROMOTIVE FORCE (EMF)

The electrical pressure available to cause a flow of current (amperage) when an electric circuit is closed. Also called voltage.

ELECTROMOTIVE SERIES ELECTROMOTIVE SERIES

A list of metals and alloys presented in the order of their tendency to corrode (or go into solution). Also called the galvanic series. This is a practical application of the theoretical ELECTROCHEMICAL SERIES.

ELECTRON ELECTRON

(1) A very small, negatively charged particle that is practically weightless. According to the electron theory, all electrical and electronic effects are caused either by the movement of electrons from place to place or because there is an excess or lack of electrons at a particular place.

(2) The part of an atom that determines its chemical properties.

ELEMENT ELEMENT

A substance that cannot be separated into its constituent parts and still retain its chemical identity. For example, sodium (Na) is an element.

EMPTY BED CONTACT TIME (EBCT) EMPTY BED CONTACT TIME (EBCT)

A measure of the time during which a water to be treated is in contact with the treatment medium in a contact vessel, assuming that all liquid passes through the vessel at the same velocity. EBCT is equal to the volume of the empty bed divided by the flow rate.

ENCLOSED SPACE ENCLOSED SPACE

See CONFINED SPACE.

END BELLS END BELLS

Devices used to hold the rotor and stator of a motor in position.

ENDEMIC (en-DEM-ick) ENDEMIC

Something peculiar to a particular people or locality, such as a disease that is always present in the population.

ENDOCRINE (EN-doe-krin) EFFECTS ENDOCRINE EFFECTS

An altering of the organs in the human body responsible for secreting hormones into the bloodstream. Endocrine glands include the thyroid gland, the pancreas, and the adrenal glands.

END POINT END POINT

The completion of a desired chemical reaction. Samples of water or wastewater are titrated to the end point. This means that a chemical is added, drop by drop, to a sample until a certain color change (blue to clear, for example) occurs. This is called the end point of the titration. In addition to a color change, an end point may be reached by the formation of a precipitate or the reaching of a specified pH. An end point may be detected by the use of an electronic device, such as a pH meter.

ENDRIN (EN-drin) ENDRIN

A pesticide toxic to freshwater and marine aquatic life that produces adverse health effects in domestic water supplies.

ENERGY GRADE LINE (EGL) ENERGY GRADE LINE (EGL)

A line that represents the elevation of energy head (in feet or meters) of water flowing in a pipe, conduit, or channel. The line is drawn above the hydraulic grade line (gradient) a distance equal to the velocity head ($V^2/2g$) of the water flowing at each section or point along the pipe or channel. Also see HYDRAULIC GRADE LINE (HGL).

[SEE DRAWING ON PAGE 797]

ENERGY-ISOLATING DEVICE ENERGY-ISOLATING DEVICE

A mechanical device that physically prevents the transmission or release of energy, including but not limited to the following: A manually operated electric circuit breaker; a disconnect switch; a manually operated switch by which the conductors of a circuit can be disconnected from all ungrounded supply conductors, and, in addition, no pole can be operated independently; a line valve; a block; and any similar device used to block or isolate energy. Push buttons, selector switches, and other control circuit type devices are not energy-isolating devices.

ENTERIC ENTERIC

Of intestinal origin, especially applied to wastes or bacteria.

ENTRAIN ENTRAIN

To trap bubbles in water either mechanically through turbulence or chemically through a reaction; or to trap one substance or material by another substance or material.

ENVIRONMENTAL PROTECTION AGENCY ENVIRONMENTAL PROTECTION AGENCY

See EPA.

ENZYMES (EN-zimes) ENZYMES

Organic substances (produced by living organisms) that cause or speed up chemical reactions. Organic catalysts or biochemical catalysts.

EPA EPA

United States Environmental Protection Agency. A regulatory agency established by the US Congress to administer the nation's environmental laws. Also called the US EPA.

EPIDEMIC (EP-uh-DEM-ick) EPIDEMIC

A disease that occurs in a large number of people in a locality at the same time and spreads from person to person.

EPIDEMIOLOGY (EP-uh-DE-me-ALL-o-jee) EPIDEMIOLOGY

A branch of medicine that studies epidemics (diseases that affect significant numbers of people during the same time period in the same locality). The objective of epidemiology is to determine the factors that cause epidemic diseases and how to prevent them.

EPILIMNION (EP-uh-LIM-nee-on) EPILIMNION

The upper layer of water in a thermally stratified lake or reservoir. This layer consists of the warmest water and has a fairly uniform (constant) temperature. The layer is readily mixed by wind action.

EQUILIBRIUM, CALCIUM CARBONATE EQUILIBRIUM, CALCIUM CARBONATE

A water is considered stable when it is just saturated with calcium carbonate. In this condition, the water will neither dissolve nor deposit calcium carbonate. Thus, in this water the calcium carbonate is in equilibrium with the hydrogen ion concentration.

EQUITY EQUITY

The value of an investment in a facility.

EQUIVALENT WEIGHT EQUIVALENT WEIGHT

That weight that will react with, displace, or is equivalent to 1 gram atom of hydrogen.

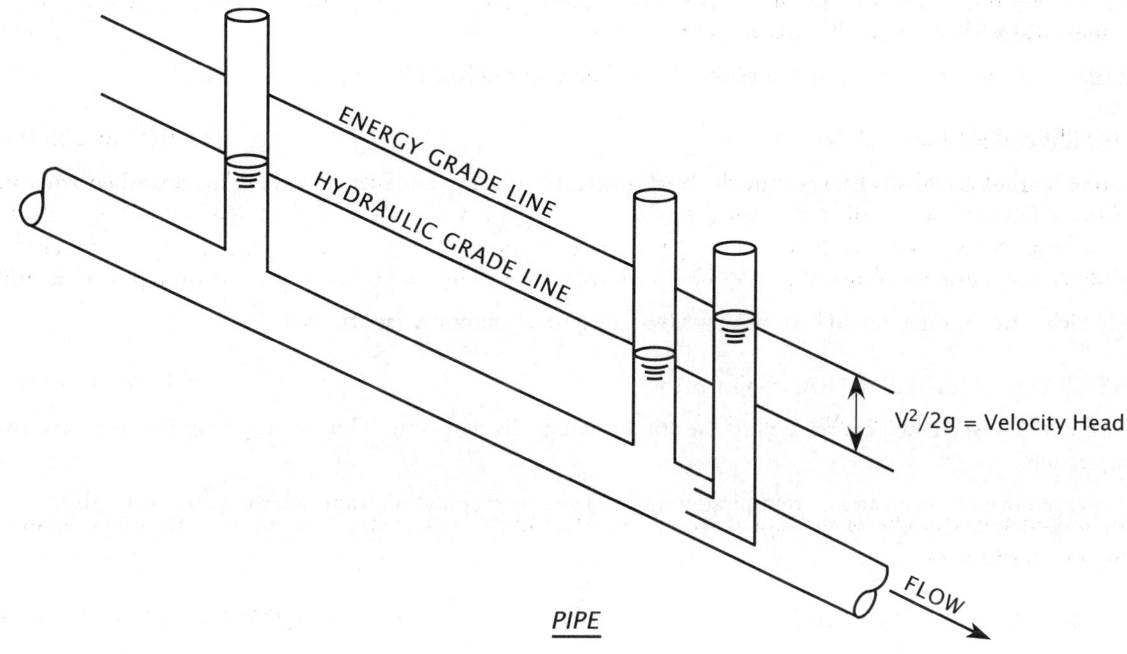

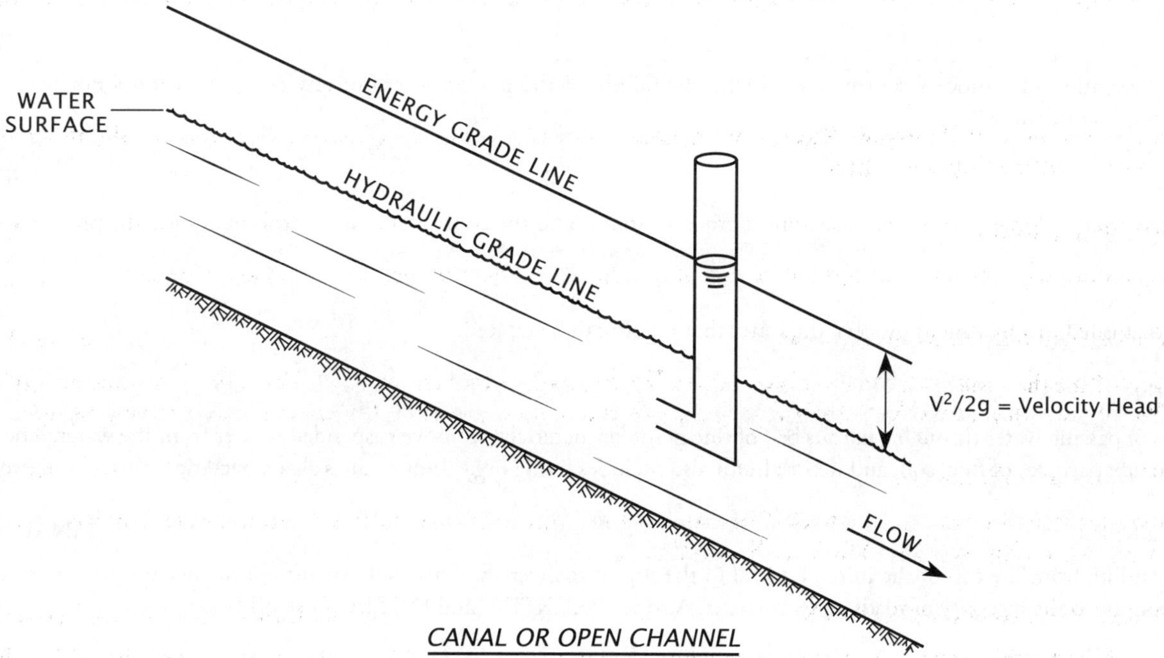

Energy grade line and hydraulic grade line

ESTER ESTER

A compound formed by the reaction between an acid and an alcohol with the elimination of a molecule of water.

EUTROPHIC (yoo-TRO-fick) EUTROPHIC

Reservoirs and lakes that are rich in nutrients and very productive in terms of aquatic animal and plant life.

EUTROPHICATION (YOO-tro-fi-KAY-shun) EUTROPHICATION

The increase in the nutrient levels of a lake or other body of water; this usually causes an increase in the growth of aquatic animal and plant life.

EVAPORATION EVAPORATION

The process by which water or other liquid becomes a gas (water vapor or ammonia vapor).

EVAPOTRANSPIRATION (ee-VAP-o-TRANS-purr-A-shun) EVAPOTRANSPIRATION

(1) The process by which water vapor is released to the atmosphere by living plants. This process is similar to people sweating. Also called transpiration.

(2) The total water removed from an area by transpiration (plants) and by evaporation from soil, snow, and water surfaces.

F

FACULTATIVE (FACK-ul-tay-tive) BACTERIA FACULTATIVE BACTERIA

Facultative bacteria can use either dissolved oxygen or oxygen obtained from food materials such as sulfate or nitrate ions. In other words, facultative bacteria can live under aerobic, anoxic, or anaerobic conditions.

FAIL-SAFE FAIL-SAFE

Design and operation of a process control system whereby failure of the power system or any component does not result in process failure or equipment damage.

FEEDBACK FEEDBACK

The circulating action between a sensor measuring a process variable and the controller that controls or adjusts the process variable.

FEEDWATER FEEDWATER

The water that is fed to a treatment process; the water that is going to be treated.

FILTRATION FILTRATION

The process of passing water through a porous bed of fine granular material to remove suspended matter from the water. The suspended matter is mainly particles of floc, soil, and debris; but it also includes living organisms such as algae, bacteria, viruses, and protozoa.

FINAL BACKFILL FINAL BACKFILL

Backfill extending from the top of the initial backfill to the top of the trench. This zone has little influence on pipe performance, but can be important to the integrity of roads and structures. Also see BACKFILL and INITIAL BACKFILL.

FINISHED WATER FINISHED WATER

Water that has passed through a water treatment plant. All the treatment processes are completed or finished. This water is the product from the water treatment plant and is ready to be delivered to consumers. Also called product water.

FIXED COSTS FIXED COSTS

Costs (rent, insurance, interest) that a utility must cover or pay even if there is no demand for water or no water to sell to customers. Also see VARIABLE COSTS.

FIXED SAMPLE FIXED SAMPLE

A sample is fixed in the field by adding chemicals that prevent the water quality indicators of interest in the sample from changing before final measurements are performed later in the laboratory.

FLAGELLATES (FLAJ-el-lates) FLAGELLATES

Microorganisms that move by the action of tail-like projections.

FLAME POLISHED FLAME POLISHED

Melted by a flame to smooth out irregularities. Sharp or broken edges of glass (such as the end of a glass tube) are rotated in a flame until the edge melts slightly and becomes smooth.

FLAP GATE FLAP GATE

A hinged gate that is mounted at the top of a pipe or channel to allow flow in only one direction. Flow in the wrong direction closes the gate. Also see CHECK VALVE and TIDE GATE.

FLOAT ON THE SYSTEM FLOAT ON THE SYSTEM

A method of operating a water storage facility. Daily flow into the facility is approximately equal to the average daily demand for water. When consumer demands for water are low, the storage facility will be filling. During periods of high demand, the facility will be emptying.

FLOC FLOC

Clumps of bacteria and particles or coagulants and impurities that have come together and formed a cluster. Found in flocculation tanks and settling or sedimentation basins.

FLOCCULATION (flock-yoo-LAY-shun) FLOCCULATION

The gathering together of fine particles after coagulation to form larger particles by a process of gentle mixing. This clumping together makes it easier to separate the solids from the water by settling, skimming, draining, or filtering.

FLOW LINE FLOW LINE

(1) The top of the wetted line, the water surface, or the hydraulic grade line of water flowing in an open channel or partially full conduit.

(2) The lowest point of the channel inside a pipe, conduit, canal, or manhole. This term is used by some contractors, however, the preferred term for this usage is invert.

FLUIDIZED (FLOO-id-i-zd) FLUIDIZED

A mass of solid particles that is made to flow like a liquid by injection of water or gas is said to have been fluidized. In water and wastewater treatment, a bed of filter media is fluidized by backwashing water through the filter.

FLUORIDATION (floor-uh-DAY-shun) FLUORIDATION

The addition of a chemical to increase the concentration of fluoride ions in drinking water to a predetermined optimum limit to reduce the incidence (number) of dental caries (tooth decay) in children. Defluoridation is the removal of excess fluoride in drinking water to prevent the mottling (brown stains) of teeth.

FLUSHING FLUSHING

The removal of deposits of material that have lodged in water distribution lines or sewers because of inadequate velocity of flows. Water is discharged into the lines at such rates that the larger flow and higher velocities are sufficient to remove the material.

FLUX FLUX

A flowing or flow.

FOOT VALVE FOOT VALVE

A special type of check valve located at the bottom end of the suction pipe on a pump. This valve opens when the pump operates to allow water to enter the suction pipe but closes when the pump shuts off to prevent water from flowing out of the suction pipe.

FOUNDATION FOUNDATION

The material in the bottom of a trench. It may or may not have a layer of bedding soil placed over it. The foundation soil may be (1) undisturbed and remain in place; (2) unsuitable and must be removed and replaced with another material; (3) so wet and soft that it must be displaced by dumping rock into the trench; or (4) removed from the trench, placed back in the trench, and then compacted. Over-excavation and backfill of the foundation is required only when the native trench bottom does not provide a firm working platform for placement of the pipe bedding material.

FREE AVAILABLE CHLORINE RESIDUAL FREE AVAILABLE CHLORINE RESIDUAL

That portion of the total available chlorine residual composed of dissolved chlorine gas (Cl_2), hypochlorous acid (HOCl), or hypochlorite ion (OCl^-) remaining in water after chlorination at the end of a specified contact period. This does not include chlorine that has combined with ammonia, nitrogen, or other compounds.

FREEBOARD FREEBOARD

(1) The vertical distance from the normal water surface to the top of the confining wall.

(2) The vertical distance from the sand surface to the underside of a trough in a sand filter. This distance is also called available expansion.

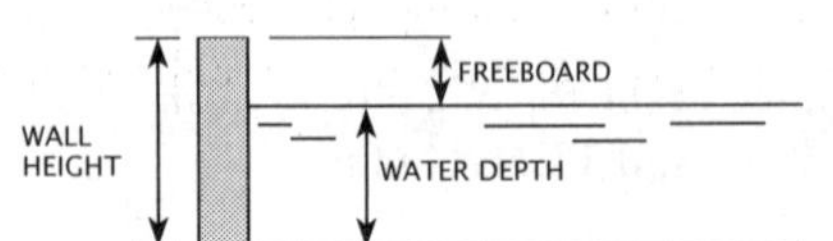

FREE RESIDUAL CHLORINATION FREE RESIDUAL CHLORINATION

The application of chlorine to water to produce a free available chlorine residual equal to at least 80 percent of the total chlorine residual (sum of free and combined available chlorine residual).

FRICTION LOSS FRICTION LOSS

The head, pressure, or energy (they are the same) lost by water flowing in a pipe or channel as a result of turbulence caused by the velocity of the flowing water and the roughness of the pipe, channel walls, or restrictions caused by fittings. Water flowing in a pipe loses head, pressure, or energy as a result of friction. Also called head loss.

FULVIC ACID FULVIC ACID

A complex organic compound that can be derived either from soil or water. Aquatic fulvic acids are major precursors of disinfection byproducts.

FUNGI (FUN-ji) FUNGI

Mushrooms, molds, mildews, rusts, and smuts that are small non-chlorophyll-bearing plants lacking roots, stems, and leaves. They occur in natural waters and grow best in the absence of light. Their decomposition may cause objectionable tastes and odors in water.

FUSE FUSE

A protective device having a strip or wire of fusible metal that, when placed in a circuit, will melt and break the electric circuit if heated too much. High temperatures will develop in the fuse when a current flows through the fuse in excess of that which the circuit will carry safely.

G

GALVANIC CELL GALVANIC CELL

An electrolytic cell capable of producing electric energy by electrochemical action. The decomposition of materials in the cell causes an electric (electron) current to flow from cathode to anode.

GALVANIC SERIES GALVANIC SERIES

A list of metals and alloys presented in the order of their tendency to corrode (or go into solution). Also called the electromotive series. This is a practical application of the theoretical ELECTROCHEMICAL SERIES.

GALVANIZE GALVANIZE

To coat a metal (especially iron or steel) with zinc. Galvanization is the process of coating a metal with zinc.

GARNET GARNET

A group of hard, reddish, glassy, mineral sands made up of silicates of base metals (calcium, magnesium, iron, and manganese). Garnet has a higher density than sand.

GAUGE PRESSURE GAUGE PRESSURE

The pressure within a closed container or pipe as measured with a gauge. In contrast, absolute pressure is the sum of atmospheric pressure (14.7 lb/in^2 [1.0 atm.]) *PLUS* pressure within a vessel (as measured with a gauge). Most pressure gauges read in gauge pressure or pounds per square inch gauge (psig) pressure or kilopascals.

GEOGRAPHIC INFORMATION SYSTEM (GIS) GEOGRAPHIC INFORMATION SYSTEM (GIS)

An integrated system of computer hardware, software, and trained personnel linking topographic, demographic, utility, facility, images, and other resource data that are geographically referenced. A GIS is used to help operators and maintenance personnel locate utility system features or structures, such as pipes, valves, and manholes, and to assist with the scheduling and performance of maintenance activities.

GEOLOGICAL LOG GEOLOGICAL LOG

A detailed description of all underground features discovered during the drilling of a well (depth, thickness, and type of formations).

GEOPHYSICAL LOG GEOPHYSICAL LOG

A record of the structure and composition of the earth encountered when drilling a well or similar type of test hole or boring.

GERMICIDE (JERM-uh-side) GERMICIDE

A substance formulated to kill germs or microorganisms. The germicidal properties of chlorine make it an effective disinfectant.

GHS GHS

See GLOBALLY HARMONIZED SYSTEM OF CLASSIFICATION AND LABELING OF CHEMICALS (GHS).

GIARDIA (jee-ARE-dee-ah) GIARDIA

A waterborne intestinal parasite that causes a disease called giardiasis in infected humans. Symptoms of the disease include diarrhea, cramps, and weight loss. *Giardia* contamination is found in most surface waters and some groundwaters.

GIARDIASIS (jee-are-DYE-uh-sis) GIARDIASIS

Intestinal disease caused by an infestation of *Giardia* flagellates.

GIS GIS

See GEOGRAPHIC INFORMATION SYSTEM (GIS).

GLOBALLY HARMONIZED SYSTEM OF CLASSIFICATION AND LABELING OF CHEMICALS (GHS) GLOBALLY HARMONIZED SYSTEM OF CLASSIFICATION AND LABELING OF CHEMICALS (GHS)

A worldwide initiative to promote standard criteria for classifying chemicals according to their health, physical, and environmental hazards. It uses harmonized pictograms, hazard statements, precautionary statements, and the signal words "Danger" and "Warning" to communicate hazard information on product labels and safety data sheets (SDSs) in a logical and comprehensive way. The primary goals of the GHS are the following:

(1) To enhance the protection of human health and the environment by providing an internationally comprehensible system for hazard communication

(2) To provide a recognized framework for those countries without an existing system

(3) To reduce the need for testing and evaluation of chemicals

(4) To facilitate international trade in chemicals whose hazards have been properly assessed and identified on an international basis

GRAB SAMPLE GRAB SAMPLE

A single sample of water collected at a particular time and place that represents the composition of the water only at that time and place.

GRADE GRADE

(1) The elevation of the invert (or bottom) of a pipeline, canal, culvert, sewer, or similar conduit.

(2) The inclination or slope of a pipeline, conduit, stream channel, or natural ground surface; usually expressed in terms of the ratio or percentage of number of units of vertical rise or fall per unit of horizontal distance. A 0.5 percent grade would be a drop of 0.5 feet per 100 feet (0.5 meters per 100 meters) of pipe.

GRAVIMETRIC GRAVIMETRIC

A means of measuring unknown concentrations of water quality indicators in a sample by weighing a precipitate or residue of the sample.

GRAVIMETRIC FEEDER GRAVIMETRIC FEEDER

A dry chemical feeder that delivers a measured weight of chemical during a specific time period.

GREENSAND GREENSAND

A mineral (glauconite) material that looks like ordinary filter sand except that it is green in color. Greensand is a natural ion exchange material that is capable of softening water. Greensand that has been treated with potassium permanganate ($KMnO_4$) is called manganese greensand; this product is used to remove iron, manganese, and hydrogen sulfide from groundwaters.

GROUND GROUND

An expression representing an electrical connection to earth or a large conductor that is at the earth's potential or neutral voltage.

H

HALOACETIC (HAL-o-uh-SEE-tick) ACID (HAA) HALOACETIC ACID (HAA)

A class of disinfection byproducts, formed mainly during the chlorination of water, containing natural organic matter. HAA5 is the sum of the concentrations, in milligrams per liter, of five haloacetic acid compounds.

HARDNESS, WATER HARDNESS, WATER

A characteristic of water caused mainly by the salts of calcium and magnesium, such as bicarbonate, carbonate, sulfate, chloride, and nitrate. Excessive hardness in water is undesirable because it causes the formation of soap curds, increased use of soap, deposition of scale in boilers, damage in some industrial processes, and sometimes causes objectionable tastes in drinking water.

HARD WATER HARD WATER

Water having a high concentration of calcium and magnesium ions. A water may be considered hard if it has a hardness greater than the typical hardness of water from the region. Some textbooks define hard water as a water with a hardness of more than 100 mg/L as calcium carbonate.

HARMFUL PHYSICAL AGENT or TOXIC SUBSTANCE HARMFUL PHYSICAL AGENT or TOXIC SUBSTANCE

Any chemical substance, biological agent (bacteria, virus, or fungus), or physical stress (noise, heat, cold, vibration, repetitive motion, ionizing and non-ionizing radiation, hypo- or hyperbaric pressure) that meets any of the following criteria:

(1) Is regulated by any state or federal law or rule due to a hazard to health

(2) Is listed in the latest printed edition of the National Institute of Occupational Safety and Health (NIOSH) Registry of Toxic Effects of Chemical Substances (RTECS)

(3) Has yielded positive evidence of an acute or chronic health hazard in human, animal, or other biological testing conducted by, or known to, the employer

4) Is described by a Material Safety Data Sheet (MSDS) available to the employer that indicates that the material may pose a hazard to human health

Also see ACUTE HEALTH EFFECT and CHRONIC HEALTH EFFECT.

HAZARD COMMUNICATION STANDARD HAZARD COMMUNICATION STANDARD

The Occupational Safety and Health Administration (OSHA) mandate, 29 CFR 1910.1200, states that companies producing and using hazardous materials must provide employees with information and training on the proper handling and use of these materials. OSHA's Hazard Communication Standard (HCS) requires the development and dissemination of the following information:

(1) Chemical manufacturers and importers are required to evaluate the hazards of the chemicals they produce or import, and prepare labels and safety data sheets to convey the hazard information to their downstream customers.

(2) All employers with hazardous chemicals in their workplaces must have labels and safety data sheets for their exposed workers, and train them to handle the chemicals appropriately.

The requirements of this standard are intended to be consistent with the provisions of the United Nations Globally Harmonized System of Classification and Labeling of Chemicals (GHS), Revision 3.

HAZARD STATEMENT HAZARD STATEMENT

A statement assigned to a hazard class and category that describes the nature of the hazard(s) of a chemical, including, where appropriate, the degree of hazard.

HEAD HEAD

The vertical distance, height, or energy of water above a reference point. A head of water may be measured in either height (feet or meters) or pressure (pounds per square inch or kilograms per square centimeter). Also see DISCHARGE HEAD, DYNAMIC HEAD, STATIC HEAD, SUCTION HEAD, SUCTION LIFT, and VELOCITY HEAD.

HEADER HEADER

A large pipe to which the ends of a series of smaller pipes are connected. Also called a manifold.

HEAD LOSS HEAD LOSS

The head, pressure, or energy (they are the same) lost by water flowing in a pipe or channel as a result of turbulence caused by the velocity of the flowing water and the roughness of the pipe, channel walls, or restrictions caused by fittings. Water flowing in a pipe loses head, pressure, or energy as a result of friction. The head loss through a filter is due to friction caused by material building up on the surface or by the water flowing through the filter media. Also called friction loss.

[SEE DRAWING ON PAGE 804]

HEAT SENSOR HEAT SENSOR

A device that opens and closes a switch in response to changes in the temperature. This device might be a metal contact, or a thermocouple that generates a minute electric current proportional to the difference in heat, or a variable resistor whose value changes in response to changes in temperature. Also called a temperature sensor.

HECTARE (HECK-ter) HECTARE

A measure of area in the metric system similar to an acre. One hectare is equal to 10,000 square meters and 2.4711 acres.

HEPATITIS (HEP-uh-TIE-tis) HEPATITIS

Hepatitis is an inflammation of the liver caused by an acute viral infection. Yellow jaundice is one symptom of hepatitis.

HERBICIDE (HERB-uh-side) HERBICIDE

A compound, usually a manmade organic chemical, used to kill or control plant growth.

HERTZ (Hz) HERTZ (Hz)

The number of complete electromagnetic cycles or waves in one second of an electric or electronic circuit. Also called the frequency of the current.

HETEROTROPHIC (HET-er-o-TROF-ick) HETEROTROPHIC

Describes organisms that use organic matter for energy and growth. Animals, fungi, and most bacteria are heterotrophs.

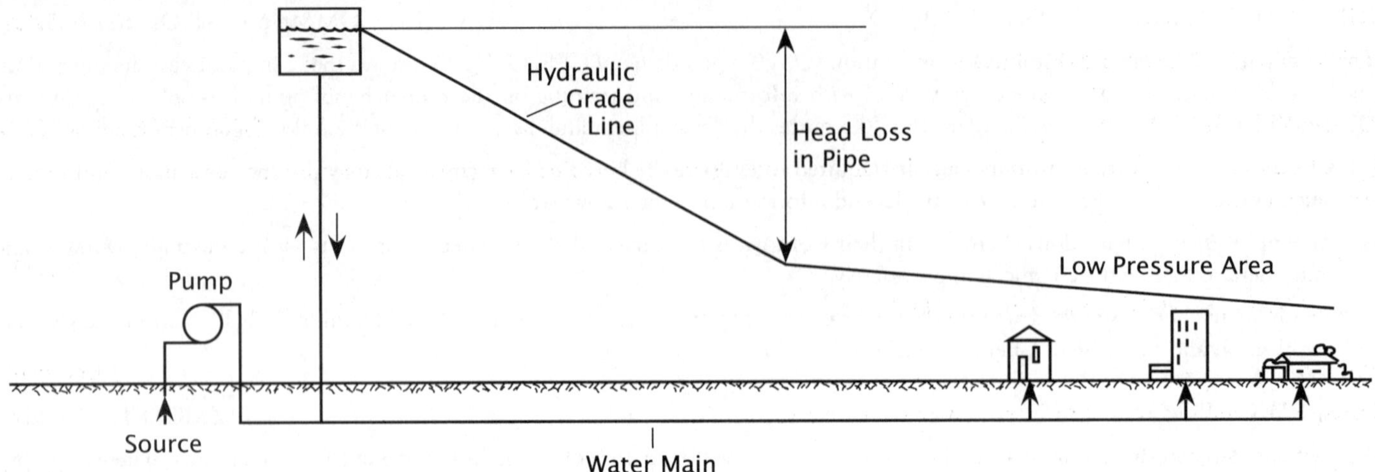

HEAD LOSS IN PIPE

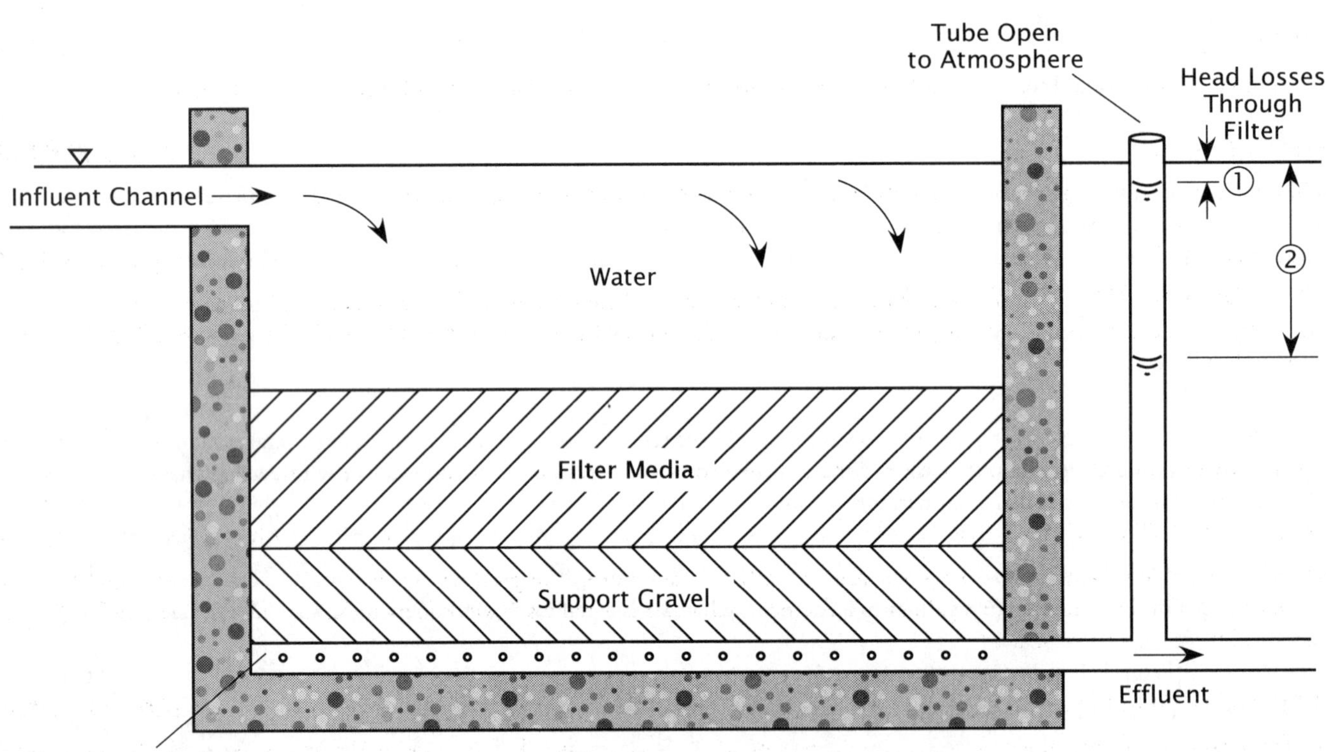

HEAD LOSS THROUGH FILTER

Head loss

HGL HGL

See HYDRAULIC GRADE LINE (HGL).

HIGH-LINE JUMPERS HIGH-LINE JUMPERS

Pipes or hoses connected to fire hydrants and laid on top of the ground to provide emergency water service for an isolated portion of a distribution system.

HMI HMI

See HUMAN MACHINE INTERFACE (HMI).

HOSE BIB HOSE BIB

Faucet. A location in a water line where a hose is connected.

HTH HTH

High Test Hypochlorite. Calcium hypochlorite or $Ca(OCl)_2$.

HUMAN MACHINE INTERFACE (HMI) HUMAN MACHINE INTERFACE (HMI)

The device at which the operator interacts with the control system. This may be an individual instrumentation and control device or the graphic screen of a computer control system. Also see MAN MACHINE INTERFACE (MMI) and OPERATOR INTERFACE.

HUMIC ACID HUMIC ACID

A polymeric constituent of soils, lignite, and peat. Aquatic humic acids are major precursors of disinfection byproducts.

HUMIC SUBSTANCES HUMIC SUBSTANCES

Natural organic matter resulting from partial decomposition of plant or animal matter and forming the organic portion of soil.

HYDRATED LIME HYDRATED LIME

Limestone that has been burned and treated with water under controlled conditions until the calcium oxide portion has been converted to calcium hydroxide ($Ca(OH)_2$). Hydrated lime is quicklime combined with water. $CaO + H_2O \rightarrow Ca(OH)_2$. Also called slaked lime. Also see QUICKLIME.

HYDRAULIC HYDRAULIC

Referring to water flowing through manmade structures such as pipes or channels or natural environments such as rivers.

HYDRAULIC CONDUCTIVITY (K) HYDRAULIC CONDUCTIVITY (K)

A coefficient describing the relative ease with which groundwater can move through a permeable layer of rock or soil. Typical units of hydraulic conductivity are feet per day, gallons per day per square foot, or meters per day (depending on the unit chosen for the total discharge and the cross-sectional area).

HYDRAULIC GRADE LINE (HGL) HYDRAULIC GRADE LINE (HGL)

The surface or profile of water flowing in an open channel or a pipe flowing partially full. If a pipe is under pressure, the hydraulic grade line is that level water would rise to in a small, vertical tube connected to the pipe. Also see ENERGY GRADE LINE (EGL).

[SEE DRAWING ON PAGE 797]

HYDRAULIC GRADIENT HYDRAULIC GRADIENT

The slope of the hydraulic grade line. This is the slope of the water surface in an open channel, the slope of the water surface of the groundwater table, or the slope of the water pressure for pipes under pressure.

HYDROGEOLOGIST (HI-dro-jee-ALL-uh-jist) HYDROGEOLOGIST

A person who studies and works with groundwater.

HYDROLOGIC (HI-dro-LOJ-ick) CYCLE — HYDROLOGIC CYCLE

The process of evaporation of water into the air and its return to earth by precipitation (rain or snow). This process also includes transpiration from plants, groundwater movement, and runoff into rivers, streams, and the ocean. Also called the water cycle.

HYDROLYSIS (hi-DROLL-uh-sis) — HYDROLYSIS

(1) A chemical reaction in which a compound is converted into another compound by taking up water.

(2) Usually a chemical degradation of organic matter.

HYDROPHILIC (hi-dro-FILL-ick) — HYDROPHILIC

Having a strong affinity (liking) for water. The opposite of HYDROPHOBIC.

HYDROPHOBIC (hi-dro-FOE-bick) — HYDROPHOBIC

Having a strong aversion (dislike) for water. The opposite of HYDROPHILIC.

HYDROPNEUMATIC (hi-dro-new-MAT-ick) — HYDROPNEUMATIC

A water system, usually small, in which a water pump is automatically controlled (started and stopped) by the air pressure in a compressed-air tank.

HYDROSTATIC (hi-dro-STAT-ick) PRESSURE — HYDROSTATIC PRESSURE

(1) The pressure at a specific elevation exerted by a body of water at rest.

(2) In the case of groundwater, the pressure at a specific elevation due to the weight of water at higher levels in the same zone of saturation.

HYGROSCOPIC (hi-grow-SKAWP-ick) — HYGROSCOPIC

Absorbing or attracting moisture from the air.

HYPOCHLORINATION (HI-poe-klor-uh-NAY-shun) — HYPOCHLORINATION

The application of hypochlorite compounds to water or wastewater for the purpose of disinfection.

HYPOCHLORINATORS (HI-poe-KLOR-uh-nay-tors) — HYPOCHLORINATORS

Chlorine pumps, chemical feed pumps, or devices used to dispense chlorine solutions made from hypochlorites, such as bleach (sodium hypochlorite) or calcium hypochlorite into the water being treated.

HYPOCHLORITE (HI-poe-KLOR-ite) — HYPOCHLORITE

Chemical compounds containing available chlorine; used for disinfection. They are available as liquids (bleach) or solids (powder, granules, and pellets) in barrels, drums, and cans. Salts of hypochlorous acid.

HYPOLIMNION (HI-poe-LIM-nee-on) — HYPOLIMNION

The lowest layer in a thermally stratified lake or reservoir. This layer consists of colder, denser water, has a constant temperature, and no mixing occurs.

I

ICR — ICR

See INFORMATION COLLECTION RULE (ICR).

IDLH — IDLH

Immediately Dangerous to Life or Health. The atmospheric concentration of any toxic, corrosive, or asphyxiant substance that poses an immediate threat to life or would cause irreversible or delayed adverse health effects or would interfere with an individual's ability to escape from a dangerous atmosphere.

IMHOFF CONE　　IMHOFF CONE

A clear, cone-shaped container marked with graduations. The cone is used to measure the volume of settleable solids in a specific volume (usually 1 liter) of water or wastewater.

IMPELLER　　IMPELLER

A rotating set of vanes in a pump or compressor designed to pump or move water or air.

IMPERMEABLE (im-PURR-me-uh-bull)　　IMPERMEABLE

Not easily penetrated. The property of a material or soil that does not allow, or allows only with great difficulty, the movement or passage of water.

INDICATOR　　INDICATOR

(1) (Chemical indicator) A substance that gives a visible change, usually of color, at a desired point in a chemical reaction, generally at a specified end point.

(2) (Instrument indicator) A device that indicates the result of a measurement, usually using either a fixed scale and movable indicator (pointer), such as a pressure gauge, or a moving chart with a movable pen like those used on a circular flow-recording chart. Also called a receiver.

INFILTRATION　　INFILTRATION

The seepage of groundwater into a pipe through defects in a pipe wall system, including service connections. Seepage frequently occurs through defective or cracked pipes, pipe joints and connections, interceptor access risers and covers, or manhole walls. Similar to, but the opposite of, EXFILTRATION.

INFLUENT　　INFLUENT

Water or other liquid—raw (untreated) or partially treated—flowing into a reservoir, basin, treatment process, or treatment plant.

INFORMATION COLLECTION RULE (ICR)　　INFORMATION COLLECTION RULE (ICR)

The Information Collection Rule (ICR) was promulgated on May 14, 1996 and approved by the Director of the Federal Register on June 18, 1996. It was to remain effective until December 31, 2000. The rule specified requirements for monitoring microbial contaminants and disinfection byproducts (DBPs) by large public water systems (PWSs). It required large PWSs to conduct either bench- or pilot-scale testing of advanced treatment techniques. The data reported under the ICR were used by EPA to learn more about the occurrence of microbial contamination and disinfection byproducts, the health risks posed, appropriate analytical methods, and effective forms of treatment. The ICR data form the scientific basis for EPA's development of the Enhanced Surface Water Treatment Rule and the Disinfectants and Disinfection Byproducts (D/DBP) Rule.

INITIAL BACKFILL　　INITIAL BACKFILL

Backfill extending from the springline or horizontal centerline of a pipe to a point above the top of the pipe. This zone provides some pipe support and helps to prevent damage to the pipe during placement of the final backfill. Also see BACKFILL and FINAL BACKFILL.

INITIAL SAMPLING　　INITIAL SAMPLING

The very first sampling conducted under the Safe Drinking Water Act (SDWA) for each of the applicable contaminant categories.

INJECTOR WATER　　INJECTOR WATER

Service water in which chlorine is added (injected) to form a chlorine solution.

IN-LINE FILTRATION　　IN-LINE FILTRATION

The addition of chemical coagulants directly to the filter inlet pipe. The chemicals are mixed by the flowing water. Flocculation and sedimentation facilities are eliminated. This pretreatment method is commonly used in pressure filter installations. Also see CONVENTIONAL FILTRATION and DIRECT FILTRATION.

INORGANIC | INORGANIC

Used to describe material such as sand, salt, iron, calcium salts, and other mineral materials. Inorganic materials are chemical substances of mineral origin, whereas organic substances are usually of animal or plant origin. Also see ORGANIC.

INORGANIC WASTE | INORGANIC WASTE

Waste material such as sand, salt, iron, calcium, and other mineral materials that are only slightly affected by the action of organisms. Inorganic wastes are chemical substances of mineral origin; whereas organic wastes are chemical substances usually of animal or plant origin. Also see NONVOLATILE MATTER, ORGANIC WASTE, and VOLATILE SOLIDS.

INPUT HORSEPOWER | INPUT HORSEPOWER

The total power used in operating a pump and motor.

$$\text{Input Horsepower, HP} = \frac{(\text{Brake Horsepower, HP})(100\%)}{\text{Motor Efficiency, \%}}$$

INSECTICIDE | INSECTICIDE

Any substance or chemical formulated to kill or control insects.

INSOLUBLE (in-SAWL-yoo-bull) | INSOLUBLE

Something that cannot be dissolved.

INTEGRATOR | INTEGRATOR

A device or meter that continuously measures and sums a process rate variable in cumulative fashion over a given time period. For example, total flows displayed in gallons per minute, million gallons per day, cubic feet per second, or some other unit of volume per time period. Also see TOTALIZER.

INTERCEPTOR (INTERCEPTING) SEWER | INTERCEPTOR (INTERCEPTING) SEWER

A large sewer that receives flow from a number of sewers and conducts the wastewater to a treatment plant. Often called an interceptor. The term interceptor is sometimes used in small communities to describe a septic tank or other holding tank that serves as a temporary wastewater storage reservoir for a septic tank effluent pump (STEP) system.

INTERCONNECTOR | INTERCONNECTOR

A sewer installed to connect two separate sewers. If one sewer becomes blocked, wastewater can back up and flow through the interconnector to the other sewer.

INTERFACE | INTERFACE

The common boundary layer between two substances, such as water and a solid (metal); or between two fluids, such as water and a gas (air); or between a liquid (water) and another liquid (oil).

INTERLOCK | INTERLOCK

A physical device, equipment, or software routine that prevents an operation from beginning or changing function until some condition or set of conditions is fulfilled. An example would be a switch that prevents a piece of equipment from operating when a hazard exists.

INTERNAL FRICTION | INTERNAL FRICTION

Friction within a fluid (water) due to cohesive forces.

INTERSTICE (in-TUR-stuhz) | INTERSTICE

A very small open space in a rock or granular material. Also called a pore, void, or void space. Also see VOID.

INVERT (IN-vert) | INVERT

The lowest point of the channel inside a pipe, conduit, canal, or manhole. Also called flow line by some contractors, however, the preferred term is invert.

ION ION

An electrically charged atom, radical (such as SO_4^{2-}), or molecule formed by the loss or gain of one or more electrons.

ION EXCHANGE ION EXCHANGE

A water or wastewater treatment process involving the reversible interchange (switching) of ions between the water being treated and the solid resin contained within an ion exchange unit. Undesirable ions are exchanged with acceptable ions on the resin or recoverable ions in the water being treated are exchanged with other acceptable ions on the resin.

ION EXCHANGE RESINS ION EXCHANGE RESINS

Insoluble polymers, used in water or wastewater treatment, that are capable of exchanging (switching or giving) acceptable cations or anions to the water being treated for less desirable ions or for ions to be recovered.

IONIC CONCENTRATION IONIC CONCENTRATION

The concentration of any ion in solution, usually expressed in moles per liter.

IONIZATION (EYE-on-uh-ZAY-shun) IONIZATION

(1) The splitting or dissociation (separation) of molecules into negatively and positively charged ions.

(2) The process of adding electrons to, or removing electrons from, atoms or molecules, thereby creating ions. High temperatures, electrical discharges, and nuclear radiation can cause ionization.

J

JAR TEST JAR TEST

A laboratory procedure that simulates coagulation/flocculation with differing chemical doses. The purpose of the procedure is to estimate the minimum coagulant dose required to achieve certain water quality goals. Samples of water to be treated are placed in six jars. Various amounts of chemicals are added to each jar, stirred, and the settling of solids is observed. The lowest dose of chemicals that provides satisfactory settling is the dose used to treat the water.

JOGGING JOGGING

The frequent starting and stopping of an electric motor.

JOULE (JOOL) JOULE

A measure of energy, work, or quantity of heat. One joule is the work done when the point of application of a force of one newton is displaced a distance of one meter in the direction of the force. Approximately equal to 0.7375 ft-lb (0.1022 m-kg).

K

KELLY KELLY

The square section of a rod that causes the rotation of the drill bit. Torque from a drive table is applied to the square rod to cause the rotary motion. The drive table is chain- or gear-driven by an engine.

KILO KILO

(1) Kilogram.

(2) Kilometer.

(3) A prefix meaning thousand used in the metric system and other scientific systems of measurement.

KINETIC ENERGY KINETIC ENERGY

Energy possessed by a moving body of matter, such as water, as a result of its motion.

KJELDAHL (KELL-doll) NITROGEN KJELDAHL NITROGEN

Nitrogen in the form of organic proteins or their decomposition product ammonia, as measured by the Kjeldahl Method.

L

LAG TIME LAG TIME

The time period between the moment a process change is made and the moment such a change is finally sensed by the associated measuring instrument.

LANGELIER INDEX (LI) LANGELIER INDEX (LI)

An index reflecting the equilibrium pH of a water with respect to calcium and alkalinity. This index is used in stabilizing water to control both corrosion and the deposition of scale.

$$\text{Langelier Index} = pH - pH_s$$

where pH = actual pH of the water

pH_s = pH at which water having the same alkalinity and calcium content is just saturated with calcium carbonate

LAUNDERING WEIR LAUNDERING WEIR

Sedimentation basin overflow weir. A plate with V-notches along the top to ensure a uniform flow rate and avoid short-circuiting.

LAUNDERS LAUNDERS

Sedimentation basin and filter discharge channels consisting of overflow weir plates (in sedimentation basins) and conveying troughs.

LEAD (LEED) LEAD

A wire or conductor that can carry electric current.

LEATHERS LEATHERS

O-rings or gaskets used with piston pumps to provide a seal between the piston and the side wall.

LEL LEL

See LOWER EXPLOSIVE LIMIT (LEL).

LESS THAN (<) LESS THAN (<)

DO <5 mg/L would be read as DO LESS THAN 5 mg/L.

LEVEL CONTROL LEVEL CONTROL

A float device (or pressure switch) that senses changes in a measured variable and opens or closes a switch in response to that change. In its simplest form, this control might be a floating ball connected mechanically to a switch or valve, such as is used to stop water flow into a toilet when the tank is full.

LINDANE (LYNN-dane) LINDANE

A pesticide that causes adverse health effects in humans and is toxic to freshwater and marine aquatic life.

LINEARITY (lin-ee-AIR-it-ee) LINEARITY

How closely an instrument measures actual values of a variable through its effective range.

LITTORAL (LIT-or-ul) ZONE LITTORAL ZONE

(1) That portion of a body of fresh water extending from the shoreline lakeward to the limit of occupancy of rooted plants.

(2) The strip of land along the shoreline between the high and low water levels.

LOCKOUT LOCKOUT

The placement of a lockout device on an energy-isolating device, in accordance with an established procedure, ensuring that the energy-isolating device and the equipment being controlled cannot be operated until the lockout device is removed.

LOCKOUT DEVICE LOCKOUT DEVICE

A device that uses a positive means such as a lock, either key or combination type, to hold an energy-isolating device in a safe position and prevent the energizing of a machine or equipment. Included are blank flanges and bolted slip blinds.

LOCKOUT/TAGOUT LOCKOUT/TAGOUT

The practices and procedures necessary to disable machinery or equipment, thereby preventing the release of hazardous energy while employees perform servicing and maintenance activities. OSHA's standard, The Control of Hazardous Energy (Lockout/Tagout) (Title 29 CFR Part 1910.147) outlines measures for controlling hazardous energies—electrical, mechanical, hydraulic, pneumatic, chemical, thermal, and other energy sources.

LOGARITHM (LOG-uh-rith-um) (LOG) LOGARITHM (LOG)

The exponent that indicates the power to which a number must be raised to produce a given number. For example, if $B^2 = N$, the 2 is the logarithm of N (to the base B), or $10^2 = 100$ and $\log_{10} 100 = 2$.

LOWER EXPLOSIVE LIMIT (LEL) LOWER EXPLOSIVE LIMIT (LEL)

The lowest concentration of a gas or vapor (percentage by volume in air) below which a flame will not spread in the presence of an ignition source (arc, flame, or heat). Concentrations lower than LEL are "too lean" to burn. Also called lower flammable limit (LFL). Also see UPPER EXPLOSIVE LIMIT (UEL).

LOWER FLAMMABLE LIMIT (LFL) LOWER FLAMMABLE LIMIT (LFL)

See LOWER EXPLOSIVE LIMIT (LEL).

M

M (MOLAR) *M* (MOLAR)

A molar solution consists of 1 gram molecular weight of a compound dissolved in enough water to make 1 liter of solution. A gram molecular weight is the molecular weight of a compound in grams. For example, the molecular weight of sulfuric acid (H_2SO_4) is 98. A 1 *M* solution of sulfuric acid would consist of 98 grams of H_2SO_4 dissolved in enough distilled water to make 1 liter of solution.

MACROSCOPIC (MACK-row-SKAWP-ick) ORGANISMS MACROSCOPIC ORGANISMS

Organisms big enough to be seen by the eye without the aid of a microscope.

MANDREL (MAN-drill) MANDREL

(1) A special tool used to push bearings in or to pull sleeves out.

(2) A testing device used to measure for excessive deflection in a flexible conduit.

MANIFOLD MANIFOLD

A large pipe to which the ends of a series of smaller pipes are connected. Also called a header.

MAN MACHINE INTERFACE (MMI) MAN MACHINE INTERFACE (MMI)

The device at which the operator interacts with the control system. This may be an individual instrumentation and control device or the graphic screen of a computer control system. Also see HUMAN MACHINE INTERFACE (HMI) and OPERATOR INTERFACE.

MANOMETER (man-NAH-mut-ter) MANOMETER

An instrument for measuring pressure. Usually, a manometer is a glass tube filled with a liquid that is used to measure the difference in pressure across a flow measuring device, such as an orifice or a Venturi meter. The instrument used to measure blood pressure is a type of manometer.

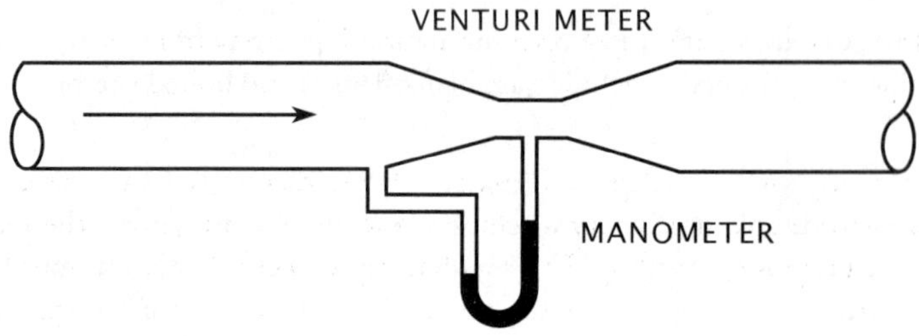

MATERIAL SAFETY DATA SHEET (MSDS) MATERIAL SAFETY DATA SHEET (MSDS)

A document that provides pertinent information and a profile of a particular hazardous substance or mixture. An MSDS is normally developed by the manufacturer or formulator of the hazardous substance or mixture. The MSDS is required to be made available to employees and operators or inspectors whenever there is the likelihood of the hazardous substance or mixture being introduced into the workplace. Some manufacturers are preparing MSDSs for products that are not considered to be hazardous to show that the product or substance is not hazardous. Also see SAFETY DATA SHEET (SDS).

MAXIMUM CONTAMINANT LEVEL (MCL) MAXIMUM CONTAMINANT LEVEL (MCL)

The largest allowable amount. MCLs for various water quality indicators are specified in the National Primary Drinking Water Regulations (NPDWR).

MAXIMUM RESIDUAL DISINFECTANT LEVEL (MRDL) MAXIMUM RESIDUAL DISINFECTANT LEVEL (MRDL)

The highest level of a disinfectant allowed in drinking water without causing an unacceptable possibility of adverse health effects.

MBAS MBAS

Methylene Blue Active Substance. Another name for surfactants or surface active agents. The determination of surfactants is accomplished by measuring the color change in a standard solution of methylene blue dye.

MCL MCL

Maximum Contaminant Level. The largest allowable amount. MCLs for various water quality indicators are specified in the National Primary Drinking Water Regulations (NPDWR).

MCLG MCLG

Maximum Contaminant Level Goal. MCLGs are health goals based entirely on health effects. They are a preliminary standard set but not enforced by EPA. MCLs consider health effects, but also take into consideration the feasibility and cost of analysis and treatment of the regulated MCL. Although often less stringent than the corresponding MCLG, the MCL is set to protect health.

MEASURED VARIABLE MEASURED VARIABLE

A factor (flow, temperature) that is sensed and quantified (reduced to a reading of some kind) by a primary element or sensor.

MECHANICAL JOINT MECHANICAL JOINT

A flexible device that joins pipes or fittings together by the use of lugs and bolts.

MEG MEG

(1) Abbreviation of MEGOHM.

(2) A procedure used for checking the insulation resistance on motors, feeders, bus bar systems, grounds, and branch circuit wiring. Also see MEGGER.

MEGGER (from megohm) MEGGER

An instrument used for checking the insulation resistance on motors, feeders, bus bar systems, grounds, and branch circuit wiring. A megger reads in millions of ohms. Also see MEG.

MEGOHM (MEG-ome) MEGOHM

Millions of ohms. Mega- is a prefix meaning 1 million, so 5 megohms means 5 million ohms.

MEMBRANE FOULING MEMBRANE FOULING

The cause of a loss of flow through the membrane as a result of material being retained on the surface of the membrane or within the membrane pores. Membrane fouling may be reversible or irreversible. Loss of flow through the membrane by reversible fouling can be recovered by regular backwashing of the membrane surface. Flow loss by irreversible fouling cannot be recovered.

MENISCUS (meh-NIS-cuss) MENISCUS

The curved surface of a column of liquid (water, oil, mercury) in a small tube. When the liquid wets the sides of the container (as with water), the curve forms a valley. When the confining sides are not wetted (as with mercury), the curve forms a hill or upward bulge.

MESH MESH

One of the openings or spaces in a screen or woven fabric. The value of the mesh is usually given as the number of openings per inch. This value does not consider the diameter of the wire or fabric; therefore, the mesh number does not always have a definite relationship to the size of the hole.

MESOTROPHIC (MESS-o-TRO-fick) MESOTROPHIC

Reservoirs and lakes that contain moderate quantities of nutrients and are moderately productive in terms of aquatic animal and plant life.

METABOLISM METABOLISM

All of the processes or chemical changes in an organism or a single cell by which food is built up (anabolism) into living protoplasm and by which protoplasm is broken down (catabolism) into simpler compounds with the exchange of energy.

METALIMNION (met-uh-LIM-nee-on) METALIMNION

The middle layer in a thermally stratified lake or reservoir. In this layer there is a rapid decrease in temperature with depth. Also called thermocline.

METHOXYCHLOR (meth-OX-e-klor) METHOXYCHLOR

A pesticide that causes adverse health effects in domestic water supplies and is toxic to freshwater and marine aquatic life. The chemical name for methoxychlor is 2,2-bis(p-methoxyphenol)-1,1,1-trichloroethane.

METHYL ORANGE ALKALINITY METHYL ORANGE ALKALINITY

A measure of the total alkalinity in a water sample. The alkalinity is measured by the amount of standard sulfuric acid required to lower the pH of the water to a pH level of 4.5, as indicated by the change in color of methyl orange from orange to pink. Methyl orange alkalinity is expressed as milligrams per liter equivalent calcium carbonate.

MF MF

See MICROFILTRATION (MF).

mg/L mg/L

See MILLIGRAMS PER LITER, mg/L.

MICROBIAL (my-KRO-bee-ul) GROWTH MICROBIAL GROWTH

The activity and growth of microorganisms such as bacteria, algae, diatoms, plankton, and fungi.

MICROFILTRATION (MF) MICROFILTRATION (MF)

A pressure-driven membrane filtration process that separates particles down to approximately 0.1 μm diameter from influent water using a sieving process.

MICRON (MY-kron) MICRON

μm, Micrometer or Micron. A unit of length. One millionth of a meter or 1 thousandth of a millimeter. One micron equals 0.00004 of an inch.

MICROORGANISMS (MY-crow-OR-gan-is-ums) MICROORGANISMS

Living organisms that can be seen individually only with the aid of a microscope.

MIL MIL

A unit of length equal to 0.001 of an inch. The diameter of wires and tubing is measured in mils, as is the thickness of plastic sheeting.

MILLIGRAMS PER LITER, mg/L MILLIGRAMS PER LITER, mg/L

A measure of the concentration by weight of a substance per unit volume. For practical purposes, 1 mg/L of a substance in fresh water is equal to 1 part per million parts (ppm). Thus, a liter of water with a specific gravity of 1.0 weighs 1 million milligrams. If 1 liter of water contains 10 milligrams of calcium, the concentration is 10 milligrams per million milligrams, or 10 milligrams per liter (10 mg/L), or 10 parts of calcium per million parts of water, or 10 parts per million (10 ppm).

MILLIMICRON (MILL-uh-MY-kron) MILLIMICRON

A unit of length equal to $10^{-3}\mu$ (one thousandth of a micron), 10^{-6} millimeters, or 10^{-9} meters; correctly called a nanometer, nm.

MOLAR MOLAR

See *M* (MOLAR).

MOLARITY MOLARITY

A measure of concentration defined as the number of moles of solute per liter of solution. Also see *M* or MOLAR.

MOLE MOLE

The name for a quantity of any chemical substance whose mass in grams is numerically equal to its atomic weight. One mole equals 6.02×10^{23} molecules or atoms. Also see MOLECULAR WEIGHT.

MOLECULAR WEIGHT MOLECULAR WEIGHT

The molecular weight of a compound in grams per mole is the sum of the atomic weights of the elements in the compound. The molecular weight of sulfuric acid (H_2SO_4) in grams is 98.

Element	Atomic Weight	Number of Atoms	Molecular Weight
H	1	2	2
S	32	1	32
O	16	4	64
			98

MOLECULE MOLECULE

The smallest division of a compound that still retains or exhibits all the properties of the substance.

MONOMER (MON-o-mer) MONOMER

A molecule of low molecular weight capable of reacting with identical or different monomers to form polymers.

MONOMICTIC (mah-no-MICK-tick) MONOMICTIC

Lakes and reservoirs that are relatively deep, do not freeze over during the winter months, and undergo a single stratification and mixing cycle during the year. These lakes and reservoirs usually become destratified during the mixing cycle, usually in the fall of the year.

MONOVALENT MONOVALENT

Having a valence of one, such as the cuprous (copper) ion, Cu^{+}.

MOST PROBABLE NUMBER (MPN) MOST PROBABLE NUMBER (MPN)

See MPN.

MOTILE (MO-till) MOTILE

Capable of self-propelled movement. A term that is sometimes used to distinguish between certain types of organisms found in water.

MOTOR EFFICIENCY MOTOR EFFICIENCY

The ratio of energy delivered by a motor to the energy supplied to it during a fixed period or cycle. Motor efficiency ratings will vary depending on motor manufacturer and usually will be near 90.0 percent.

MPN MPN

MPN is the Most Probable Number of coliform-group organisms per unit volume of sample water. Expressed as a density or population of organisms per 100 mL of sample water.

MRDL MRDL

Maximum Residual Disinfectant Level. The highest level of a disinfectant allowed in drinking water without causing an unacceptable possibility of adverse health effects.

MSDS MSDS

See MATERIAL SAFETY DATA SHEET (MSDS).

MUDBALLS MUDBALLS

Material, approximately round in shape, that forms in filters and gradually increases in size when not removed by the backwashing process. Mudballs vary from pea-sized up to golf-ball-sized or larger.

MULTISTAGE PUMP MULTISTAGE PUMP

A pump that has more than one impeller. A single-stage pump has one impeller.

N

N (NORMAL) *N* (NORMAL)

A normal solution contains 1 gram equivalent weight of reactant (compound) per liter of solution. The equivalent weight of an acid is that weight that contains 1 gram atom of ionizable hydrogen or its chemical equivalent. For example, the equivalent weight of sulfuric acid (H_2SO_4) is 49 (98 divided by 2 because there are two replaceable hydrogen ions). A 1 *N* solution of sulfuric acid would consist of 49 grams of H_2SO_4 dissolved in enough water to make 1 liter of solution.

NAMEPLATE NAMEPLATE

A durable, metal plate found on equipment that lists critical installation and operating conditions for the equipment.

NANOFILTRATION (NF) NANOFILTRATION (NF)

A pressure-driven membrane filtration process that separates particles down to approximately 0.002 to 0.005 μm diameter from influent water using a sieving process.

NATIONAL ENVIRONMENTAL, SAFETY & HEALTH TRAINING ASSOCIATION (NESHTA) (formerly NATIONAL ENVIRONMENTAL TRAINING ASSOCIATION (NETA)) NATIONAL ENVIRONMENTAL, SAFETY & HEALTH TRAINING ASSOCIATION (NESHTA)

A professional organization devoted to serving the environmental trainer and promoting better operation of waterworks and pollution control facilities. For information on NESHTA membership and publications, visit NESHTA at neshta.org or phone (602) 956-6099.

NATIONAL ENVIRONMENTAL TRAINING ASSOCIATION (NETA) NATIONAL ENVIRONMENTAL TRAINING ASSOCIATION (NETA)

See NATIONAL ENVIRONMENTAL, SAFETY & HEALTH TRAINING ASSOCIATION.

NATIONAL INSTITUTE OF OCCUPATIONAL SAFETY AND HEALTH (NIOSH) NATIONAL INSTITUTE OF OCCUPATIONAL SAFETY AND HEALTH (NIOSH)

See NIOSH.

NATURAL ORGANIC MATTER (NOM) NATURAL ORGANIC MATTER (NOM)

Humic substances composed of humic acid and fulvic acid that come from decayed vegetation. Also see HUMIC SUBSTANCES, HUMIC ACID, and FULVIC ACID.

NEPHELOMETRIC (neff-el-o-MET-rick) NEPHELOMETRIC

A means of measuring turbidity in a sample by using an instrument called a nephelometer. A nephelometer passes light through a sample and the amount of light deflected (usually at a 90-degree angle) is then measured.

NESHTA (formerly NETA) NESHTA

See NATIONAL ENVIRONMENTAL, SAFETY & HEALTH TRAINING ASSOCIATION (NESHTA).

NETA NETA

See NATIONAL ENVIRONMENTAL, SAFETY & HEALTH TRAINING ASSOCIATION (NESHTA).

NEWTON NEWTON

A force that, when applied to a body having a mass of 1 kilogram, gives it an acceleration of 1 meter per second per second.

NF NF

See NANOFILTRATION (NF).

NIOSH (NYE-osh) NIOSH

The National Institute of Occupational Safety and Health is an organization that tests and approves safety equipment for particular applications. NIOSH is the primary federal agency engaged in research in the national effort to eliminate on-the-job hazards to the health and safety of working people. A complete list of all publications issued by NIOSH is available at the Centers for Disease Control and Prevention (CDC) website at cdc.gov/niosh/.

NITRIFICATION (NYE-truh-fuh-KAY-shun) NITRIFICATION

An aerobic process in which bacteria oxidize the ammonia and organic nitrogen in water into nitrite and then nitrate.

NITROGENOUS (nye-TRAH-jen-us) NITROGENOUS

A term used to describe chemical compounds (usually organic) containing nitrogen in combined forms. Proteins and nitrate are nitrogenous compounds.

NOBLE METAL NOBLE METAL

A chemically inactive metal (such as gold). A metal that does not corrode easily and is much scarcer (and more valuable) than the so-called useful or base metals. Also see BASE METAL.

NOM NOM

See NATURAL ORGANIC MATTER (NOM).

NOMINAL DIAMETER NOMINAL DIAMETER

An approximate measurement of the diameter of a pipe. Although the nominal diameter is used to describe the size or diameter of a pipe, it is usually not the exact inside diameter of the pipe.

NONIONIC (NON-eye-ON-ick) POLYMER NONIONIC POLYMER

A polymer that has no net electrical charge.

NON-PERMIT CONFINED SPACE NON-PERMIT CONFINED SPACE

See CONFINED SPACE, NON-PERMIT.

NONPOINT SOURCE NONPOINT SOURCE

A runoff or discharge from a field or similar source, in contrast to a point source, which refers to a discharge that comes out the end of a pipe or other clearly identifiable conveyance. Also see POINT SOURCE.

NONPOTABLE (non-POE-tuh-bull) NONPOTABLE

Water that may contain objectionable pollution, contamination, minerals, or infective agents and is considered unsafe or unpalatable for drinking.

NONVOLATILE MATTER NONVOLATILE MATTER

Material such as sand, salt, iron, calcium, and other mineral materials that are only slightly affected by the actions of organisms and are not lost on ignition of the dry solids at 1,022°F (550°C). Volatile materials are chemical substances usually of animal or plant origin. Also see INORGANIC WASTE and VOLATILE SOLIDS.

NORMAL NORMAL

See *N* (NORMAL).

NORMALITY NORMALITY

The number of gram-equivalent weights of solute in 1 liter of solution. The equivalent weight of any material is the weight that would react with or be produced by the reaction of 8.0 grams of oxygen or 1.0 gram of hydrogen. Normality is used for certain calculations of quantitative analysis. Also see *N* or NORMAL.

NOTICE NOTICE

This word calls attention to information that is especially significant in understanding and operating equipment or processes safely. Also see CAUTION, DANGER, and WARNING.

NPDES PERMIT NPDES PERMIT

National Pollutant Discharge Elimination System permit is the regulatory agency document issued by either a federal or state agency that is designed to control all discharges of potential pollutants from point sources and stormwater runoff into US waterways. NPDES permits regulate discharges into US waterways from all point sources of pollution, including industries, municipal wastewater treatment plants, sanitary landfills, large animal feedlots, and return irrigation flows.

NPDWR NPDWR

National Primary Drinking Water Regulations.

NSDWR NSDWR

National Secondary Drinking Water Regulations.

NSF NSF

NSF International is a noncommercial, not-for-profit organization concerned with public health safety and environmental protection. NSF Standard 60 lists certified drinking water chemicals and NSF Standard 61 lists certified drinking water system components.

NTU NTU

Nephelometric Turbidity Units. See TURBIDITY UNITS (TU).

NUTRIENT NUTRIENT

Any substance that is assimilated (taken in) by organisms and promotes growth. Nitrogen and phosphorus are nutrients that promote the growth of algae. There are other essential and trace elements that are also considered nutrients. Also see NUTRIENT CYCLE.

NUTRIENT CYCLE NUTRIENT CYCLE

The transformation or change of a nutrient from one form to another until the nutrient has returned to the original form, thus completing the cycle. The cycle may take place under either aerobic or anaerobic conditions.

O

OCCUPATIONAL SAFETY AND HEALTH ACT OF 1970 (OSHA) OCCUPATIONAL SAFETY AND HEALTH ACT OF 1970 (OSHA)

See OSHA.

ODOR THRESHOLD ODOR THRESHOLD

The minimum odor of a gas or water sample that can just be detected after successive dilutions with odorless gas or water. Also called threshold odor.

OFFSET OFFSET

(1) The difference between the actual value and the desired value (or set point); characteristic of proportional controllers that do not incorporate reset action. Also called drift.

(2) A pipe fitting in the approximate form of a reverse curve or other combination of elbows or bends that brings one section of a line of pipe out of line with, but into a line parallel with, another section.

(3) A pipe joint that has lost its bedding support, causing one of the pipe sections to drop or slip, thus creating a condition where the pipes no longer line up properly.

OHM OHM

The unit of electrical resistance. The resistance of a conductor in which 1 volt produces a current of 1 ampere.

OLFACTORY (all-FAK-tore-ee) FATIGUE OLFACTORY FATIGUE

A condition in which a person's nose, after exposure to certain odors, is no longer able to detect the odor.

OLIGOTROPHIC (ah-lig-o-TRO-fick) OLIGOTROPHIC

Reservoirs and lakes that are nutrient poor and contain little aquatic plant or animal life.

OPERATING PRESSURE DIFFERENTIAL OPERATING PRESSURE DIFFERENTIAL

The operating pressure range for a hydropneumatic system. For example, when the pressure drops below 40 psi in a system designed to operate between 40 psi and 60 psi, the pump will come on and stay on until the pressure builds up to 60 psi. When the pressure reaches 60 psi the pump will shut off. The operating pressure differential in this example is 20 psi.

OPERATING RATIO OPERATING RATIO

The operating ratio is a measure of the total revenues divided by the total operating expenses.

OPERATOR INTERFACE OPERATOR INTERFACE

The device at which the operator interacts with the control system. This may be an individual instrumentation and control device or the graphic screen of a computer control system. Also see HUMAN MACHINE INTERFACE (HMI) and MAN MACHINE INTERFACE (MMI).

ORGANIC ORGANIC

Used to describe chemical substances that come from animal or plant sources. Organic substances always contain carbon. (Inorganic materials are chemical substances of mineral origin.) Also see INORGANIC.

ORGANICS ORGANICS

(1) A term used to refer to chemical compounds made from carbon molecules. These compounds may be natural materials (such as animal or plant sources) or manmade materials (such as synthetic organics). Also see ORGANIC.

(2) Any form of animal or plant life. Also see BACTERIA.

ORGANIC WASTE ORGANIC WASTE

Waste material that may come from animal or plant sources. Natural organic wastes generally can be consumed by bacteria and other small organisms. Manufactured or synthetic organic wastes from metal finishing, chemical manufacturing, and petroleum industries may not normally be consumed by bacteria and other organisms. Also see INORGANIC WASTE and VOLATILE SOLIDS.

ORGANISM ORGANISM

Any form of animal or plant life. Also see BACTERIA.

ORGANIZING ORGANIZING

Deciding who does what work and delegating authority to the appropriate persons.

ORIFICE (OR-uh-fiss) ORIFICE

An opening (hole) in a plate, wall, or partition. An orifice flange or plate placed in a pipe consists of a slot or a calibrated circular hole smaller than the pipe diameter. The difference in pressure in the pipe above and at the orifice may be used to determine the flow in the pipe. In a trickling filter distributor, the wastewater passes through an orifice to the surface of the filter media.

ORP (pronounce as separate letters) ORP

See OXIDATION-REDUCTION POTENTIAL.

ORTHOTOLIDINE (or-tho-TOL-uh-dine) ORTHOTOLIDINE

Orthotolidine is a colorimetric indicator of chlorine residual. If chlorine is present, a yellow-colored compound is produced. This reagent is no longer approved for chemical analysis to determine chlorine residual.

OSHA (O-shuh) OSHA

The Williams-Steiger Occupational Safety and Health Act of 1970 (OSHA) is a federal law designed to protect the health and safety of industrial workers and the operators of water supply systems and treatment plants. The act regulates the design, construction, operation, and maintenance of water supply systems and water treatment plants. OSHA also refers to the federal and state agencies that administer the OSHA regulations.

OSMOSIS (oz-MOE-sis) OSMOSIS

The passage of a liquid from a weak solution to a more concentrated solution across a semipermeable membrane. The membrane allows the passage of the water (solvent) but not the dissolved solids (solutes). This process tends to equalize the conditions on either side of the membrane.

OUCH PRINCIPLE OUCH PRINCIPLE

This principle says that as a manager when you delegate job tasks you must be **O**bjective, **U**niform in your treatment of employees, and the tasks must be **C**onsistent with utility policies, and **H**ave job relatedness.

OVERALL EFFICIENCY, PUMP OVERALL EFFICIENCY, PUMP

The combined efficiency of a pump and motor together. Also called the wire-to-water efficiency.

OVERDRAFT OVERDRAFT

The pumping of water from a groundwater basin or aquifer in excess of the supply flowing into the basin. This pumping results in a depletion or mining of the groundwater in the basin.

OVERFLOW RATE — OVERFLOW RATE

One of the guidelines for the design of settling tanks and clarifiers in treatment plants. Used by operators to determine if tanks and clarifiers are hydraulically (flow) over- or underloaded. Also called surface loading.

$$\text{Overflow Rate, GPD/ft}^2 = \frac{\text{Flow, gal/day}}{\text{Surface Area, ft}^2}$$

or

$$\text{Overflow Rate, } \frac{\text{m}^3\text{/day}}{\text{m}^2} = \frac{\text{Flow, m}^3\text{/day}}{\text{Surface Area, m}^2}$$

OVERHEAD — OVERHEAD

Indirect costs necessary for a water utility to function properly. These costs are not related to the actual treatment and delivery of water to consumers, but include the costs of rent, lights, office supplies, management, and administration.

OVERTURN — OVERTURN

The almost spontaneous mixing of all layers of water in a reservoir or lake when the water temperature becomes similar from top to bottom. This may occur in the fall/winter when the surface waters cool to the same temperature as the bottom waters and in the spring when the surface waters warm after the ice melts. This is also called turnover.

OXIDATION — OXIDATION

Oxidation is the addition of oxygen, removal of hydrogen, or the removal of electrons from an element or compound; in the environment and in wastewater treatment processes, organic matter is oxidized to more stable substances. The opposite of REDUCTION.

OXIDATION-REDUCTION POTENTIAL (ORP) — OXIDATION-REDUCTION POTENTIAL (ORP)

The electrical potential required to transfer electrons from one compound or element (the oxidant) to another compound or element (the reductant); used as a qualitative measure of the state of oxidation in water and wastewater treatment systems. ORP is measured in millivolts, with negative values indicating a tendency to reduce compounds or elements and positive values indicating a tendency to oxidize compounds or elements.

OXIDIZING AGENT — OXIDIZING AGENT

Any substance, such as oxygen (O_2) or chlorine (Cl_2), that will readily add (take on) electrons. When oxygen or chlorine is added to water or wastewater, organic substances are oxidized. These oxidized organic substances are more stable and less likely to give off odors or to contain disease-causing bacteria. The opposite is a REDUCING AGENT.

OXYGEN DEFICIENCY — OXYGEN DEFICIENCY

An atmosphere containing oxygen at a concentration of less than 19.5 percent by volume.

OXYGEN ENRICHMENT — OXYGEN ENRICHMENT

An atmosphere containing oxygen at a concentration of more than 23.5 percent by volume.

OZONATION (O-zoe-NAY-shun) — OZONATION

The application of ozone to water for disinfection or for taste and odor control.

P

PACKER ASSEMBLY — PACKER ASSEMBLY

An inflatable device used to seal the tremie pipe inside the well casing to prevent the grout from entering the inside of the conductor casing.

PALATABLE (PAL-uh-tuh-bull) — PALATABLE

Water at a desirable temperature that is free from objectionable tastes, odors, colors, and turbidity. Pleasing to the senses.

PARSHALL FLUME (PAR-shul FLOOM) PARSHALL FLUME

A device used to measure the flow in an open channel. The flume narrows to a throat of fixed dimensions and then expands again. The rate of flow can be calculated by measuring the difference in head (pressure) before and at the throat of the flume.

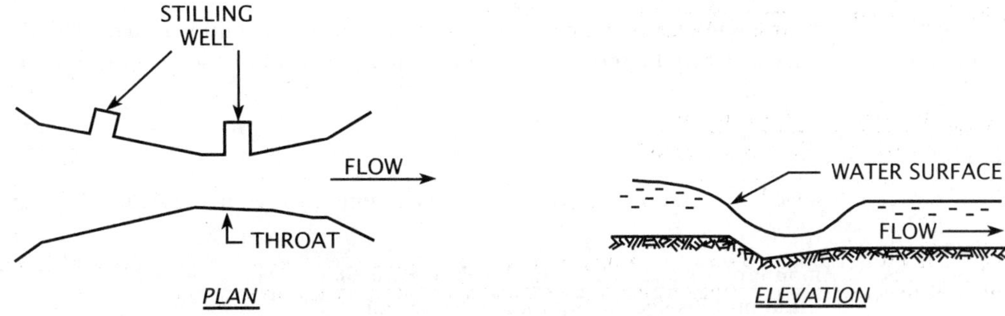

PARTICLE COUNT PARTICLE COUNT

The results of a microscopic examination of treated water with a special particle counter, which classifies suspended particles by number and size.

PARTICLE COUNTER PARTICLE COUNTER

A device that counts and measures the size of individual particles in water.

PARTICLE COUNTING PARTICLE COUNTING

A procedure for counting and measuring the size of individual particles in water. Particles are divided into size ranges and the number of particles is counted in each of these ranges. The results are reported in terms of the number of particles in different particle diameter size ranges per milliliter of water sampled.

PARTICULATE (par-TICK-yoo-let) PARTICULATE

A very small solid suspended in water that can vary widely in size, shape, density, and electrical charge. Colloidal and dispersed particulates are artificially gathered together by the processes of coagulation and flocculation.

PARTS PER MILLION (PPM) PARTS PER MILLION (PPM)

Parts per million parts, a measurement of concentration on a weight or volume basis. This term is equivalent to milligrams per liter (mg/L), which is the preferred term.

PASCAL (Pa) PASCAL (Pa)

The pressure or stress of one newton per square meter.

$$1 \text{ psi} = 6{,}895 \text{ Pa} = 6.895 \text{ kN/m}^2 = 0.0703 \text{ kg/cm}^2$$

PATHOGENIC (path-o-JEN-ick) ORGANISMS PATHOGENIC ORGANISMS

Organisms, including bacteria, viruses, protozoa, or internal parasites, capable of causing diseases (such as giardiasis, cryptosporidiosis, typhoid fever, cholera, or infectious hepatitis) in a host (such as a person). There are many types of organisms that do not cause disease. These organisms are called nonpathogenic.

PATHOGENS (PATH-o-jens) PATHOGENS

See PATHOGENIC ORGANISMS.

PCBs PCBs

Polychlorinated biphenyls. A class of organic compounds that cause adverse health effects in domestic water supplies.

pCi/L pCi/L

picoCurie per liter. A picoCurie is a measure of radioactivity. One picoCurie of radioactivity is equivalent to 0.037 nuclear disintegrations per second.

pcu pcu

Platinum cobalt units. A measure of color using platinum cobalt standards by visual comparison.

PEAK DEMAND PEAK DEMAND

The maximum momentary load placed on a water treatment plant, pumping station, or distribution system. This demand is usually the maximum average load in 1 hour or less, but may be specified as the instantaneous load or the load during some other short time period.

PERCENT SATURATION PERCENT SATURATION

The amount of a substance that is dissolved in a solution compared with the amount dissolved in the solution at saturation, expressed as a percent.

$$\text{Percent Saturation, \%} = \frac{\text{Amount of Substance That Is Dissolved} \times 100\%}{\text{Amount Dissolved in Solution at Saturation}}$$

PERCOLATING (PURR-ko-lay-ting) WATER PERCOLATING WATER

Water that passes through soil or rocks under the force of gravity.

PERCOLATION (purr-ko-LAY-shun) PERCOLATION

The slow passage of water through a filter medium; or, the gradual penetration of soil and rocks by water.

PERFORMANCE INDICATOR PERFORMANCE INDICATOR

A measurable goal used to determine system performance and level of service provided. Examples of performance indicators include the number of stoppages per 100 miles of sewer per year and the number of lost time accidents per year—measurements of how *well* a utility is doing rather than how *much* a utility is doing. Also see PRODUCTION INDICATOR.

PERIPHYTON (pair-e-FI-tawn) PERIPHYTON

Microscopic plants and animals that are firmly attached to solid surfaces under water such as rocks, logs, pilings, and other structures.

PERMEABILITY (PURR-me-uh-BILL-uh-tee) PERMEABILITY

The property of a material or soil that permits considerable movement of water through it when it is saturated.

PERMEATE (PURR-me-ate) PERMEATE

(1) To penetrate and pass through, as water penetrates and passes through soil and other porous materials.

(2) The liquid (demineralized water) produced from the reverse osmosis process that contains a low concentration of dissolved solids.

PERMIT-REQUIRED CONFINED SPACE (PERMIT SPACE) PERMIT-REQUIRED CONFINED SPACE (PERMIT SPACE)

See CONFINED SPACE, PERMIT-REQUIRED (PERMIT SPACE).

PESTICIDE PESTICIDE

Any substance or chemical designed or formulated to kill or control animal pests. Also see INSECTICIDE and RODENTICIDE.

PET COCK PET COCK

A small valve or faucet used to drain a cylinder or fitting.

pH (pronounce as separate letters) pH

pH is an expression of the intensity of the basic or acidic condition of a liquid. Mathematically, pH is the logarithm (base 10) of the reciprocal of the hydrogen ion activity.

$$\text{pH} = \text{Log}\,\frac{1}{\{H^+\}}$$

If $\{H^+\} = 10^{-6.5}$, then pH = 6.5. The pH may range from 0 to 14, where 0 is most acidic, 14 most basic, and 7 neutral.

PHENOLIC (fee-NO-lick) COMPOUNDS — PHENOLIC COMPOUNDS

Organic compounds that are derivatives of benzene. Also called phenols (FEE-nolls).

PHENOLPHTHALEIN (FEE-nol-THAY-leen) ALKALINITY — PHENOLPHTHALEIN ALKALINITY

The alkalinity in a water sample measured by the amount of standard acid required to lower the pH to a level of 8.3, as indicated by the change in color of phenolphthalein from pink to clear. Phenolphthalein alkalinity is expressed as milligrams per liter of equivalent calcium carbonate.

PHOTOSYNTHESIS (foe-toe-SIN-thuh-sis) — PHOTOSYNTHESIS

A process in which organisms, with the aid of chlorophyll, convert carbon dioxide and inorganic substances into oxygen and additional plant material, using sunlight for energy. All green plants grow by this process.

PHYTOPLANKTON (FIE-tow-plank-ton) — PHYTOPLANKTON

Small, usually microscopic plants (such as algae), found in lakes, reservoirs, and other bodies of water.

PICO- (PEE-ko) — PICO-

A prefix used in the metric system and other scientific systems of measurement which means 10^{-12} or 0.000 000 000 001.

PICOCURIE (PEE-ko-KYOOR-ee) — PICOCURIE

A measure of radioactivity. One picoCurie (pCi) of radioactivity is equivalent to 0.037 nuclear disintegrations per second.

PICTOGRAM — PICTOGRAM

A graphical composition that may include a symbol plus other graphic elements, such as a border, background pattern, or color, that is intended to convey specific information about the hazards of a chemical. There are nine pictograms under the Globally Harmonized System (GHS) to convey the health, physical, and environmental hazards.

PILOT-SCALE STUDY — PILOT-SCALE STUDY

A method of studying different ways of treating water or wastewater and solids or to obtain design criteria on a small scale in the field.

PINPOINT FLOC — PINPOINT FLOC

Very small floc (the size of a pin point) that does not settle out of the water in a sedimentation basin or clarifier. Also see FLOC.

PIPE EMBEDMENT — PIPE EMBEDMENT

The material placed around a pipe that supports the pipe.

PIPE GAUGE — PIPE GAUGE

A number that defines the thickness of the sheet used to make steel pipe. The larger the number, the thinner the pipe wall.

PIPE SCHEDULE — PIPE SCHEDULE

A sizing system of numbers that specifies the inside diameter (ID) and outside diameter (OD) for each diameter pipe. The schedule number is the ratio of internal pressure in psi divided by the allowable fiber stress multiplied by 1,000. Typical schedules of iron and steel pipe are Schedules 40, 80, and 160. Other forms of piping are divided into various classes with their own schedule schemes.

PITLESS ADAPTER — PITLESS ADAPTER

A fitting that allows the well casing to be extended above ground while having a discharge connection located below the frost line. Advantages of using a pitless adapter include the elimination of the need for a pit or pump house and it is a watertight design, which helps maintain a sanitary water supply.

PITOT (PEA-toe) TUBE — PITOT TUBE

An instrument used to measure fluid (liquid or air) velocity by means of the differential pressure between the tip (dynamic) and side (static) openings.

PLAN PLAN

A drawing or photo showing the top view of sewers, manholes, streets, or structures.

PLANKTON PLANKTON

(1) Small, usually microscopic, plants (phytoplankton) and animals (zooplankton) in aquatic systems.

(2) All of the smaller floating, suspended, or self-propelled organisms in a body of water.

PLANNING PLANNING

Management of utilities to build the resources and financial capability to provide for future needs.

PLC PLC

Programmable logic controller. A microcomputer-based control device containing programmable software; used to control process variables.

PLUG FLOW PLUG FLOW

A type of flow that occurs in tanks, basins, or reactors when a slug of water or wastewater moves through a tank without ever dispersing or mixing with the rest of the water or wastewater flowing through the tank.

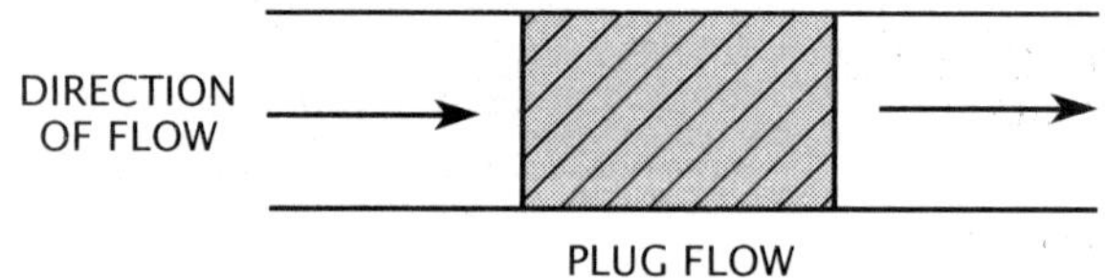

PMCL PMCL

Primary Maximum Contaminant Level. Primary MCLs for various water quality indicators are established to protect public health.

POINT-OF-USE/POINT-OF-ENTRY (POU/POE) POINT-OF-USE/POINT-OF-ENTRY (POU/POE)

Point-of-use applications refer to a water treatment device that treats water at the location of the end user. Point-of-entry application refers to a water treatment device that is located at the inlet to an entire building or facility.

POINT SOURCE POINT SOURCE

A discharge that comes out the end of a pipe or other clearly identifiable conveyance. Examples of point source conveyances from which pollutants may be discharged include ditches, channels, tunnels, conduits, wells, containers, rolling stock, concentrated animal feeding operations, landfill leachate collection systems, vessels, or other floating craft. A NONPOINT SOURCE refers to runoff or a discharge from a field or similar source.

POLARIZATION POLARIZATION

The concentration of ions in the thin, boundary layer adjacent to a membrane or pipe wall.

POLE SHADER POLE SHADER

A copper bar circling the laminated iron core inside the coil of a magnetic starter.

POLLUTION POLLUTION

The impairment (reduction) of water quality by agricultural, domestic, or industrial wastes (including thermal and radioactive wastes) to a degree that the natural water quality is changed to hinder any beneficial use of the water or render it offensive to the senses of sight, taste, or smell or when sufficient amounts of wastes create or pose a potential threat to human health or the environment.

POLYANIONIC (poly-AN-eye-ON-ick) POLYANIONIC

Characterized by many active negative charges especially active on the surface of particles.

POLYCHLORINATED BIPHENYLS (PCBs) (POLY-KLOR-uh-nate-ed BI-FEEN-alls) POLYCHLORINATED BIPHENYLS (PCBs)

A class of organic compounds that cause adverse health effects in domestic water supplies.

POLYELECTROLYTE (POLY-ee-LECK-tro-lite) POLYELECTROLYTE

A high-molecular-weight (relatively heavy) substance, having points of positive or negative electrical charges, that is formed by either natural or synthetic (manmade) processes. Natural polyelectrolytes may be of biological origin or obtained from starch products or cellulose derivatives. Synthetic polyelectrolytes consist of simple substances that have been made into complex, high-molecular-weight substances. Used with other chemical coagulants to aid in binding small suspended particles to larger chemical flocs for their removal from water. Often called a polymer.

POLYMER (POLY-mer) POLYMER

A long-chain molecule formed by the union of many monomers (molecules of lower molecular weight). Polymers are used with other chemical coagulants to aid in binding small suspended particles to larger chemical flocs for their removal from water. Also see POLYELECTROLYTE.

PORE PORE

A very small open space in a rock or granular material. Also called an interstice, void, or void space. Also see VOID.

POROSITY POROSITY

(1) A measure of the spaces or voids in a material or aquifer.

(2) The ratio of the volume of spaces in a rock or soil to the total volume. This ratio is usually expressed as a percentage.

$$\text{Porosity, \%} = \frac{(\text{Volume of Spaces})(100\%)}{\text{Total Volume}}$$

POSITIVE BACTERIOLOGICAL SAMPLE POSITIVE BACTERIOLOGICAL SAMPLE

A water sample in which gas is produced by coliform organisms during incubation in the multiple tube fermentation test.

POSITIVE DISPLACEMENT PUMP POSITIVE DISPLACEMENT PUMP

A type of piston, diaphragm, gear, or screw pump that delivers a constant volume with each stroke. Positive displacement pumps are used as chemical solution feeders.

POSTCHLORINATION POSTCHLORINATION

The addition of chlorine to the plant discharge or effluent, following plant treatment, for disinfection purposes.

POTABLE (POE-tuh-bull) WATER POTABLE WATER

Water that does not contain objectionable pollution, contamination, minerals, or infective agents and is considered satisfactory for drinking.

POU/POE POU/POE

See POINT-OF-USE/POINT-OF-ENTRY (POU/POE).

POWER FACTOR POWER FACTOR

The ratio of the true power passing through an electric circuit to the product of the voltage and amperage in the circuit. This is a measure of the lag or lead of the current with respect to the voltage. In alternating current, the voltage and amperes are not always in phase; therefore, the true power may be slightly less than that determined by the direct product.

PPM PPM

See PARTS PER MILLION (PPM).

PRECAUTIONARY STATEMENT PRECAUTIONARY STATEMENT

A phrase that describes recommended measures to be taken to minimize or prevent adverse effects resulting from exposure to a hazardous chemical, or improper storage or handling of a hazardous chemical.

PRECHLORINATION PRECHLORINATION

The addition of chlorine at the headworks of the plant prior to other treatment processes mainly for disinfection and control of tastes, odors, and aquatic growths. Also applied to aid in coagulation and settling.

PRECIPITATE (pre-SIP-uh-TATE) PRECIPITATE

(1) An insoluble, finely divided substance that is a product of a chemical reaction within a liquid.

(2) The separation from solution of an insoluble substance.

PRECIPITATION PRECIPITATION

(1) The total measurable supply of water received directly from clouds as rain, snow, hail, or sleet; usually expressed as depth in a day, month, or year, and designated as daily, monthly, or annual precipitation.

(2) The process by which atmospheric moisture is discharged onto a land or water surface.

(3) The chemical transformation of a substance in solution into an insoluble form (precipitate).

PRECISION PRECISION

The ability of an instrument to measure a process variable and repeatedly obtain the same result. The ability of an instrument to reproduce the same results.

PRECURSOR, THM (PRE-curse-or) PRECURSOR, THM

Natural, organic compounds found in all surface and groundwaters, which may react with halogens (such as chlorine) to form trihalomethanes (THMs); they must be present in order for THMs to form.

PRESCRIPTIVE (pre-SKRIP-tive) RIGHTS PRESCRIPTIVE RIGHTS

Water rights that are acquired by diverting water and putting it to use in accordance with specified procedures. These procedures include filing a request (with a state agency) to use unused water in a stream, river, or lake.

PRESENT WORTH PRESENT WORTH

The value of a long-term project expressed in today's dollars. Present worth is calculated by converting (discounting) all future benefits and costs over the life of the project to a single economic value at the start of the project. Calculating the present worth of alternative projects makes it possible to compare them and select the one with the largest positive (beneficial) present worth or minimum present cost.

PRESSURE CONTROL PRESSURE CONTROL

A switch that operates on changes in pressure. Usually this is a diaphragm pressing against a spring. When the force on the diaphragm overcomes the spring pressure, the switch is activated.

PRESSURE HEAD PRESSURE HEAD

The vertical distance (in feet or meters) equal to the pressure (in psi or kPa) at a specific point. The pressure head is equal to the pressure in psi (or kPa) times 2.31 ft/psi (or 1.0 m/9.81 kPa).

PRESTRESSED PRESTRESSED

A prestressed pipe has been reinforced with wire strands (which are under tension) to give the pipe an active resistance to loads or pressures on it.

PREVENTIVE MAINTENANCE PREVENTIVE MAINTENANCE

Regularly scheduled servicing of machinery or other equipment using appropriate tools, tests, and lubricants. This type of maintenance can prolong the useful life of equipment and machinery and increase its efficiency by detecting and correcting problems before they cause a breakdown of the equipment.

PREVENTIVE MAINTENANCE UNITS PREVENTIVE MAINTENANCE UNITS

Crews assigned the task of cleaning sewers (for example, high-velocity cleaning crews) to prevent stoppages and odor complaints. Also see PREVENTIVE MAINTENANCE.

PRIMACY PRIMACY

Under the Safe Drinking Water Act (SDWA), primacy is the responsibility for ensuring that a law is implemented, and the authority to enforce a law and related regulations (40 CFR 142.2). A primacy agency has the primary responsibility for administrating and enforcing regulations.

PRIMARY ELEMENT PRIMARY ELEMENT

(1) A device that measures (senses) a physical condition or variable of interest. Floats and thermocouples are examples of primary elements. Also called a sensor.

(2) The hydraulic structure used to measure flows. In open channels, weirs and flumes are primary elements or devices. Venturi meters and orifice plates are the primary elements in pipes or pressure conduits.

PRIME PRIME

The action of filling a pump casing with water to remove the air. Most pumps must be primed before startup or they will not pump any water.

PROCESS VARIABLE PROCESS VARIABLE

A physical or chemical quantity that is usually measured and controlled in the operation of a water, wastewater, or industrial treatment plant. Common process variables are flow, level, pressure, temperature, turbidity, chlorine, and oxygen levels.

PRODUCT WATER PRODUCT WATER

Water that has passed through a water treatment plant. All the treatment processes are completed or finished. This water is the product from the water treatment plant and is ready to be delivered to consumers. Also called finished water.

PROFILE PROFILE

A drawing showing elevation plotted against distance such as the vertical section or side view of sewers, manholes, or a pipeline.

PROGRAMMABLE LOGIC CONTROLLER (PLC) PROGRAMMABLE LOGIC CONTROLLER (PLC)

A microcomputer-based control device containing programmable software; used to control process variables.

PROPORTIONAL WEIR (WEER) PROPORTIONAL WEIR

A specially shaped weir in which the flow through the weir is directly proportional to the head.

PRUSSIAN BLUE PRUSSIAN BLUE

A blue paste or liquid (often on a paper like carbon paper) used to show a contact area. Used to determine if gate valve seats fit properly.

PSIG PSIG

Pounds per square inch gauge pressure. The pressure within a closed container or pipe measured with a gauge in pounds per square inch. Also see GAUGE PRESSURE.

PUMP BOWL PUMP BOWL

The submerged pumping unit in a well, including the shaft, impellers, and housing.

PUMPING WATER LEVEL PUMPING WATER LEVEL

The vertical distance from the centerline of the pump discharge to the level of the free pool while water is being drawn from the pool.

PURVEYOR (purr-VAY-or), WATER PURVEYOR, WATER

An agency or person that supplies water (usually potable water).

PUTREFACTION (PYOO-truh-FACK-shun) PUTREFACTION

Biological decomposition of organic matter, with the production of foul-smelling and -tasting products, associated with anaerobic (no oxygen present) conditions.

Q

QUICKLIME QUICKLIME

A material that is mostly calcium oxide (CaO) or calcium oxide in natural association with a lesser amount of magnesium oxide. Quicklime is capable of combining with water, that is, becoming slaked. Also see HYDRATED LIME.

R

RADIAL TO IMPELLER RADIAL TO IMPELLER

Perpendicular to the impeller shaft. Material being pumped flows at a right angle to the impeller.

RADICAL RADICAL

A group of atoms that is capable of remaining unchanged during a series of chemical reactions. Such combinations (radicals) exist in the molecules of many organic compounds; sulfate (SO_4^{2-}) is an inorganic radical.

RANGE RANGE

The spread from minimum to maximum values that an instrument is designed to measure. Also see EFFECTIVE RANGE and SPAN.

RANNEY COLLECTOR RANNEY COLLECTOR

This water collector is constructed as a dug well from 12 to 16 feet (3.5 to 5 m) in diameter that has been sunk as a caisson near the bank of a river or lake. Screens are driven radially and approximately horizontally from this well into the sand and gravel deposits underlying the river.

[SEE DRAWING ON PAGE 829]

RATE OF RETURN RATE OF RETURN

A value that indicates the return of funds received on the basis of the total equity capital used to finance physical facilities. Similar to the interest rate on savings accounts or loans.

RAW WATER RAW WATER

(1) Water in its natural state, prior to any treatment.

(2) Water entering the first treatment process of a water treatment plant.

REACTIVE MAINTENANCE REACTIVE MAINTENANCE

Maintenance activities that are performed in response to problems and emergencies after they occur.

READOUT READOUT

The reading of the value of a process variable from an indicator or recorder or on a computer screen.

REAERATION (RE-air-A-shun) REAERATION

The introduction of air through forced air diffusers into the lower layers of the reservoir. As the air bubbles form and rise through the water, oxygen from the air dissolves into the water and replenishes the dissolved oxygen. The rising bubbles also cause the lower waters to rise to the surface where oxygen from the atmosphere is transferred to the water. This is sometimes called surface reaeration.

REAGENT (re-A-gent) REAGENT

A pure, chemical substance that is used to make new products or is used in chemical tests to measure, detect, or examine other substances.

RECARBONATION (re-kar-bun-NAY-shun) RECARBONATION

A process in which carbon dioxide is bubbled into the water being treated to lower the pH. The pH may also be lowered by the addition of acid. Recarbonation is the final stage in the lime–soda ash softening process. This process converts carbonate ions to bicarbonate ions and stabilizes the solution against the precipitation of carbonate compounds.

RECEIVER RECEIVER

A device that indicates the result of a measurement, usually using either a fixed scale and movable indicator (pointer), such as a pressure gauge, or a moving chart with a movable pen like those used on a circular flow-recording chart. Also called an indicator.

RECORDER RECORDER

A device that creates a permanent record, on a paper chart, magnetic tape, or in a computer, of the changes in a measured variable.

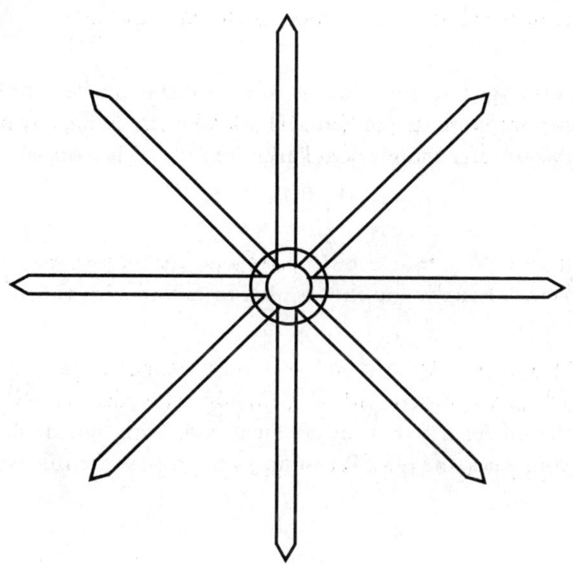

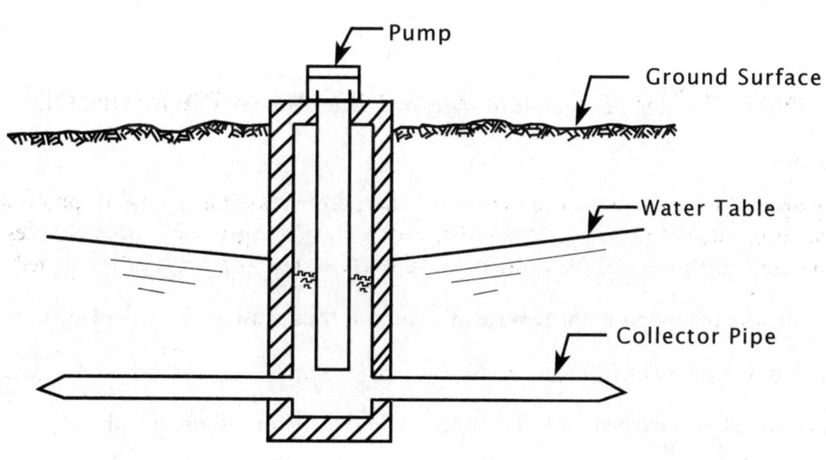

Ranney collector

REDUCING AGENT | REDUCING AGENT

Any substance, such as base metal (iron) or the sulfide ion (S^{2-}), that will readily donate (give up) electrons. The opposite is an OXIDIZING AGENT.

REDUCTION (re-DUCK-shun) | REDUCTION

Reduction is the addition of hydrogen, removal of oxygen, or the addition of electrons to an element or compound. Under anaerobic conditions (no dissolved oxygen present), sulfur compounds are reduced to odor-producing hydrogen sulfide (H_2S) and other compounds. In the treatment of metal finishing wastewaters, hexavalent chromium (Cr^{6+}) is reduced to the trivalent form (Cr^{3+}). The opposite of OXIDATION.

REFERENCE | REFERENCE

A physical or chemical quantity whose value is known exactly, and thus is used to calibrate instruments or standardize measurements. Also called a standard.

REGULATORY NEGOTIATION | REGULATORY NEGOTIATION

A process whereby the US Environmental Protection Agency acts on an equal basis with outside parties to reach consensus on the content of a proposed rule. If the group reaches consensus, the US EPA commits to propose the rule with the agreed upon content.

RELIQUEFACTION (re-lick-we-FACK-shun) | RELIQUEFACTION

The return of a gas to the liquid state; for example, a condensation of chlorine gas to return it to its liquid form by cooling.

REPRESENTATIVE SAMPLE | REPRESENTATIVE SAMPLE

A sample portion of material, water, or wastestream that is as nearly identical in content and consistency as possible to that in the larger body being sampled.

RESIDUAL CHLORINE | RESIDUAL CHLORINE

The concentration of chlorine present in water after the chlorine demand has been satisfied. The concentration is expressed in terms of the total chlorine residual, which includes both the free and combined or chemically bound chlorine residuals. Also called chlorine residual.

RESIDUE | RESIDUE

The dry solids remaining after the evaporation of a sample of water or sludge. Also see TOTAL DISSOLVED SOLIDS.

RESINS | RESINS

See ION EXCHANGE RESINS.

RESISTANCE | RESISTANCE

That property of a conductor or wire that opposes the passage of a current, thus causing electric energy to be transformed into heat.

RESPIRATION | RESPIRATION

The process in which an organism takes in oxygen for its life processes and gives off carbon dioxide.

RESPONSIBILITY | RESPONSIBILITY

Answering to those above in the chain of command to explain how and why you have used your authority.

REVERSE OSMOSIS (oz-MOE-sis) (RO) | REVERSE OSMOSIS (RO)

The application of pressure to a concentrated solution, which causes the passage of a liquid from the concentrated solution to a weaker solution across a semipermeable membrane. The membrane allows the passage of the water (solvent) but not the dissolved solids (solutes). In the reverse osmosis process, two liquids are produced: (1) the reject (containing high concentrations of dissolved solids) and (2) the permeate (containing low concentrations). The clean water (permeate) is not always considered to be demineralized. Also see OSMOSIS.

RIPARIAN (ri-PAIR-ee-an) RIGHTS RIPARIAN RIGHTS

Water rights that are acquired together with title to the land bordering a source of surface water. The right to put to beneficial use surface water adjacent to your land.

RO RO

See REVERSE OSMOSIS (RO).

RODENTICIDE (row-DENT-uh-side) RODENTICIDE

Any substance or chemical formulated to kill or control rodents.

ROOF LEADER ROOF LEADER

In plumbing, a pipe installed to drain water from the roof gutters or roof catchment to the storm drain or other means of disposal. Also called a conductor, downspout, or roof drain.

ROTAMETER (ROTE-uh-ME-ter) ROTAMETER

A device used to measure the flow rate of gases and liquids. The gas or liquid being measured flows vertically up a tapered, calibrated tube. Inside the tube is a small ball or bullet-shaped float (it may rotate) that rises or falls depending on the flow rate. The flow rate may be read on a scale behind or on the tube by looking at the middle of the ball or at the widest part or top of the float.

ROTOR ROTOR

The rotating part of a machine. The rotor is surrounded by the stationary (nonmoving) parts (stator) of the machine.

ROUTINE SAMPLING ROUTINE SAMPLING

Sampling repeated on a regular basis.

S

SACRIFICIAL ANODE SACRIFICIAL ANODE

An easily corroded material deliberately installed in a pipe or tank. The intent of such an installation is to give up (sacrifice) this anode to corrosion while the water supply facilities remain relatively corrosion free.

SAFE DRINKING WATER ACT (SDWA) SAFE DRINKING WATER ACT (SDWA)

An act passed by the US Congress in 1974. The act establishes a cooperative program among local, state, and federal agencies to ensure safe drinking water for consumers. The act has been amended several times, including the 1980, 1986, and 1996 amendments.

SAFETY DATA SHEET (SDS) SAFETY DATA SHEET (SDS)

Safety data sheets (SDSs) are an essential component of the Globally Harmonized System of Classification and Labeling of Chemicals (GHS) and are intended to provide comprehensive information about a substance or mixture for use in workplace chemical management. They are used as a source of information about hazards, including environmental hazards, and to obtain advice on safety precautions. In the GHS, they serve the same function that the material safety data sheet (MSDS) does in OSHA's Hazard Communication Standard. The SDS is normally product related and not specific to the workplace; nevertheless, the information on an SDS enables the employer to develop an active program of worker protection measures, including training, which is specific to the workplace, and to consider measures necessary to protect the environment.

SAFE WATER SAFE WATER

Water that does not contain harmful bacteria, or toxic materials or chemicals. Water may have taste and odor problems, color, and certain mineral problems and still be considered safe for drinking.

SAFE YIELD SAFE YIELD

The annual quantity of water that can be taken from a source of supply over a period of years without depleting the source permanently (beyond its ability to be replenished naturally in wet years).

SALINITY — SALINITY

(1) The relative concentration of dissolved salts, usually sodium chloride, in a given water.

(2) A measure of the concentration of dissolved mineral substances in water.

SANITARY SURVEY — SANITARY SURVEY

A detailed evaluation or inspection of a source of water supply and all conveyances, storage, treatment, and distribution facilities to ensure protection of the water supply from all pollution sources.

SAPROPHYTES (SAP-row-fights) — SAPROPHYTES

Organisms living on dead or decaying organic matter. They help natural decomposition of organic matter in water or wastewater.

SATURATION — SATURATION

The condition of a liquid (water) when it has taken into solution the maximum possible quantity of a given substance at a given temperature and pressure.

SATURATOR (SAT-yoo-ray-tor) — SATURATOR

A device that produces a fluoride solution for the fluoridation process. The device is usually a cylindrical container with granular sodium fluoride on the bottom. Water flows either upward or downward through the sodium fluoride to produce the fluoride solution.

SCADA (SKAY-dah) SYSTEM — SCADA SYSTEM

Supervisory Control And Data Acquisition system. A computer-monitored alarm, response, control, and data acquisition system used to monitor and adjust treatment processes and facilities.

SCALE — SCALE

(1) A combination of mineral salts and bacterial accumulation that sticks to the inside of a collection pipe under certain conditions. Scale, in extreme growth circumstances, creates additional friction loss to the flow of water. Scale may also accumulate on surfaces other than pipes.

(2) The marked plate against which an indicator or recorder reads, usually the same as the range of the measuring system. Also see RANGE.

SCFM — SCFM

Standard Cubic Feet per Minute. Cubic feet of air per minute at standard conditions of temperature, pressure, and humidity (0°C, 14.7 psia, and 50 percent relative humidity).

SCHEDULE, PIPE — SCHEDULE, PIPE

See PIPE SCHEDULE.

SCHMUTZDECKE (shmoots-DECK-ee) — SCHMUTZDECKE

A layer of trapped matter at the surface of a slow sand filter in which a dense population of microorganisms develops. These microorganisms within the film or mat feed on and break down incoming organic material trapped in the mat. In doing so, the microorganisms both remove organic matter and add mass to the mat, further developing the mat and increasing the physical straining action of the mat.

SDS — SDS

See SAFETY DATA SHEET (SDS).

SDWA — SDWA

See SAFE DRINKING WATER ACT (SDWA).

SECCHI (SECK-key) DISK — SECCHI DISK

A flat, white disk lowered into the water by a rope until it is just barely visible. At this point, the depth of the disk from the water surface is the recorded Secchi disk transparency.

SEDIMENTATION (SED-uh-men-TAY-shun) SEDIMENTATION

The process of settling and depositing of suspended matter carried by water or wastewater. Sedimentation usually occurs by gravity when the velocity of the liquid is reduced below the point at which it can transport the suspended material.

SEDIMENTATION (SED-uh-men-TAY-shun) BASIN SEDIMENTATION BASIN

A tank or basin in which water or wastewater is held for a period of time during which the heavier solids settle to the bottom and the lighter materials float to the surface. Also called settling tank or clarifier.

SEIZING or SEIZE UP SEIZING or SEIZE UP

Seizing occurs when an engine overheats and a part expands to the point where the engine will not run. Also called freezing.

SENSITIVITY SENSITIVITY

The smallest change in a process variable that an instrument can sense.

SENSITIVITY (PARTICLE COUNTERS) SENSITIVITY (PARTICLE COUNTERS)

The smallest particle that a particle counter will measure and count.

SENSOR SENSOR

A device that measures (senses) a physical condition or variable of interest. Floats and thermocouples are examples of sensors. Also called a primary element.

SEPTIC (SEP-tick) SEPTIC

A condition produced by bacteria when all oxygen supplies are depleted. If severe, the bottom deposits produce hydrogen sulfide, the deposits and water turn black, give off foul odors, and the water has a greatly increased chlorine demand.

SEQUESTRATION (SEE-kwes-TRAY-shun) SEQUESTRATION

A chemical complexing (forming or joining together) of metallic cations (such as iron) with certain inorganic compounds, such as phosphate. Sequestration prevents the precipitation of the metals (iron). Also see CHELATION.

SERVICE PIPE SERVICE PIPE

The pipeline extending from the water main to the building served or to the consumer's system.

SET POINT SET POINT

The position at which the control or controller is set. This is the same as the desired value of the process variable. For example, a thermostat is set to maintain a desired temperature.

SEWAGE SEWAGE

The used household water and water-carried solids that flow in sewers to a wastewater treatment plant. The preferred term is WASTEWATER.

SHEAVE SHEAVE

V-belt drive pulley, which is commonly made of cast iron or steel.

SHIM SHIM

Thin metal sheets that are inserted between two surfaces to align or space the surfaces correctly. Shims can be used anywhere a spacer is needed. Usually shims are 0.001 to 0.020 inch (0.025 to 0.50 mm) thick.

SHOCK LOAD SHOCK LOAD

The arrival at a water treatment plant of raw water containing unusual amounts of algae, colloidal matter, color, suspended solids, turbidity, or other pollutants.

SHORT-CIRCUITING | SHORT-CIRCUITING

A condition that occurs in tanks or basins when some of the flowing water entering a tank or basin flows along a nearly direct pathway from the inlet to the outlet. This is usually undesirable because it may result in shorter contact, reaction, or settling times in comparison with the theoretical (calculated) or presumed detention times.

SIGNAL WORD | SIGNAL WORD

A single word used to indicate the relative level of severity of a chemical hazard and alert the reader to a potential hazard on the label. The signal words used are "Danger" and "Warning." "Danger" is used for the more severe hazards, while "Warning" is used for less severe hazards.

SIMULATE | SIMULATE

To reproduce the action of some process, usually on a smaller scale.

SINGLE-STAGE PUMP | SINGLE-STAGE PUMP

A pump that has only one impeller. A multistage pump has more than one impeller.

SLAKE | SLAKE

To mix with water so that a true chemical combination (hydration) takes place, such as in the slaking of lime.

SLAKED LIME | SLAKED LIME

See HYDRATED LIME.

SLOPE | SLOPE

The slope or inclination of a trench bottom or a trench side wall is the ratio of the vertical distance to the horizontal distance or rise over run. Also see GRADE (2).

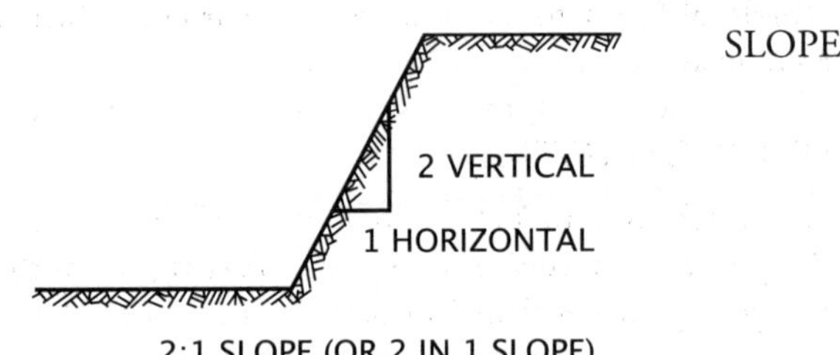

SLUDGE (SLUJ) | SLUDGE

(1) The settleable solids separated from liquids during processing.

(2) The deposits of foreign materials on the bottoms of streams or other bodies of water or on the bottoms and edges of wastewater collection lines and appurtenances.

SLURRY | SLURRY

A watery mixture or suspension of insoluble (not dissolved) matter; a thin, watery mud or any substance resembling it (such as a grit slurry or a lime slurry).

SMCL | SMCL

Secondary Maximum Contaminant Level. Secondary MCLs for various water quality indicators are established to protect public welfare.

SNARL | SNARL

Suggested No Adverse Response Level. The concentration of a chemical in water that is expected not to cause an adverse health effect.

SOFTWARE PROGRAM | SOFTWARE PROGRAM

Computer program; the list of instructions that tell a computer how to perform a given task or tasks. Some software programs are designed and written to monitor and control treatment processes.

SOFT WATER | SOFT WATER

Water having a low concentration of calcium and magnesium ions. According to US Geological Survey guidelines, soft water is water having a hardness of 60 milligrams per liter or less.

SOLENOID (SO-luh-noid) SOLENOID

A magnetically operated mechanical device (electric coil). Solenoids can operate small valves or electric switches.

SOLUTION SOLUTION

A liquid mixture of dissolved substances. In a solution it is impossible to see all the separate parts.

SOUNDING TUBE SOUNDING TUBE

A pipe or tube used for measuring the depths of water.

SPAN SPAN

The scale or range of values an instrument is designed to measure. Also see RANGE.

SPECIFIC CAPACITY SPECIFIC CAPACITY

A measurement of well yield per unit depth of drawdown after a specific time has passed, usually 24 hours. Typically expressed as gallons per minute per foot or cubic meters per day per meter (GPM/ft or m^3/day/m).

SPECIFIC CAPACITY TEST SPECIFIC CAPACITY TEST

A testing method used to determine the adequacy of an aquifer or well by measuring the specific capacity.

SPECIFIC CONDUCTANCE SPECIFIC CONDUCTANCE

A rapid method of estimating the total dissolved solids content of a water supply. The measurement indicates the capacity of a sample of water to carry an electric current, which is related to the concentration of ionized substances in the water. Also called conductance.

SPECIFIC GRAVITY SPECIFIC GRAVITY

(1) Weight of a particle, substance, or chemical solution in relation to the weight of an equal volume of water. Water has a specific gravity of 1.000 at 39°F (4°C). Particulates with specific gravity less than 1.0 float to the surface and particulates with specific gravity greater than 1.0 sink.

(2) Weight of a particular gas in relation to the weight of an equal volume of air at the same temperature and pressure (air has a specific gravity of 1.0). Chlorine gas has a specific gravity of 2.5.

SPECIFIC YIELD SPECIFIC YIELD

The quantity of water that a unit volume of saturated permeable rock or soil will yield when drained by gravity. Specific yield may be expressed as a ratio or as a percentage by volume.

SPLIT SAMPLE SPLIT SAMPLE

A single grab sample that is separated into at least two parts such that each part is representative of the original sample. Often used to compare test results between field kits and laboratories or between two laboratories.

SPOIL SPOIL

Excavated material, such as soil, from the trench of a water main or sewer.

SPORE SPORE

The reproductive body of certain organisms, which is capable of giving rise to a new organism either directly or indirectly. A viable (able to live and grow) body regarded as the resting stage of an organism. A spore is usually more resistant to disinfectants and heat than most organisms. Gangrene and tetanus bacteria are common spore-forming organisms.

SPRINGLINE SPRINGLINE

The horizontal centerline of a pipe.

STALE WATER STALE WATER

Water that has not flowed recently and may have picked up tastes and odors from distribution lines or storage facilities.

STANDARD | STANDARD

A physical or chemical quantity whose value is known exactly, and thus is used to calibrate instruments or standardize measurements. Also called a reference.

STANDARD DEVIATION | STANDARD DEVIATION

A measure of the spread or dispersion of data.

STANDARDIZE | STANDARDIZE

To compare with a standard.

(1) In wet chemistry, to find out the exact strength of a solution by comparing it with a standard of known strength. This information is used to adjust the strength by adding more water or more of the substance dissolved.

(2) To set up an instrument or device to read a standard. This allows you to adjust the instrument so that it reads accurately, or enables you to apply a correction factor to the readings.

STANDARD METHODS | STANDARD METHODS

Standard Methods for the Examination of Water and Wastewater, 22nd Edition. A joint publication of the American Public Health Association (APHA), American Water Works Association (AWWA), and the Water Environment Federation (WEF) that outlines the accepted laboratory procedures used to analyze the impurities in water and wastewater. Available from American Water Works Association at awwa.org or phone (800) 926-7337. Order No. 10085. Also available from Water Environment Federation at wef.org or phone (800) 666-0206. Order No. S82210.

STANDARD SOLUTION | STANDARD SOLUTION

A solution in which the exact concentration of a chemical or compound is known.

STARTERS (MOTOR) | STARTERS (MOTOR)

Devices used to start up large motors gradually to avoid severe mechanical shock to a driven machine and to prevent disturbance to the electrical lines (causing dimming and flickering of lights).

STATIC HEAD | STATIC HEAD

When water is not moving, the vertical distance (in feet or meters) from a reference point to the water surface is the static head. Also see DYNAMIC HEAD, DYNAMIC PRESSURE, and STATIC PRESSURE.

STATIC PRESSURE | STATIC PRESSURE

The static pressure in psi (or kPa) is the STATIC HEAD in feet times 0.433 psi/ft (or meters × 9.81 kPa/m). Also see DYNAMIC HEAD, DYNAMIC PRESSURE, and STATIC HEAD.

STATIC WATER DEPTH | STATIC WATER DEPTH

The vertical distance in feet (or meters) from the centerline of the pump discharge down to the surface level of the free pool while no water is being drawn from the pool or water table.

STATIC WATER LEVEL | STATIC WATER LEVEL

(1) The elevation or level of the water table in a well when the pump is not operating.

(2) The level or elevation to which water would rise in a tube connected to an artesian aquifer, basin, or conduit under pressure.

STATOR | STATOR

That portion of a machine that contains the stationary (nonmoving) parts that surround the moving parts (rotor).

STERILIZATION (STAIR-uh-luh-ZAY-shun) | STERILIZATION

The removal or destruction of all microorganisms, including pathogens and other bacteria, vegetative forms, and spores. Compare with DISINFECTION.

STETHOSCOPE | STETHOSCOPE

An instrument used to magnify sounds and carry them to the ear.

STORATIVITY (S) | STORATIVITY (S)

The volume of groundwater an aquifer releases from or takes into storage per unit surface area of the aquifer per unit change in head. Also called the storage coefficient.

STORM RUNOFF | STORM RUNOFF

Water that flows over the ground surface directly into streams, rivers, or lakes. Also called direct runoff.

STORMWATER | STORMWATER

The excess water running off from the surface of a drainage area during and immediately after a period of rain. Also see STORM RUNOFF.

STRATIFICATION (STRAT-uh-fuh-KAY-shun) | STRATIFICATION

The formation of separate layers (of temperature, plant life, or animal life) in a lake or reservoir. Characteristics within each layer are similar; for instance, all water in the same layer has the same temperature. Also see THERMAL STRATIFICATION.

STRAY CURRENT CORROSION | STRAY CURRENT CORROSION

A corrosion activity resulting from stray electric current originating from some source outside the plumbing system such as DC grounding on phone systems.

SUBMERGENCE | SUBMERGENCE

The distance between the water surface and the media surface in a filter.

SUBSIDENCE (sub-SIDE-ence) | SUBSIDENCE

The dropping or lowering of the ground surface as a result of removing excess water (overdraft or overpumping) from an aquifer. After excess water has been removed, the soil will settle, become compacted, and the ground surface will drop, which can cause the settling of underground utilities.

SUCTION HEAD | SUCTION HEAD

The positive pressure (in feet [meters] of water or pounds per square inch [kilograms per square centimeter] of mercury vacuum) on the suction side of a pump. The pressure can be measured from the centerline of the pump up to the elevation of the hydraulic grade line on the suction side of the pump.

SUCTION LIFT | SUCTION LIFT

The negative pressure (in feet [meters] of water or inches [centimeters] of mercury vacuum) on the suction side of a pump. The pressure can be measured from the centerline of the pump down to (lift) the elevation of the hydraulic grade line on the suction side of the pump.

SUPERCHLORINATION (SOO-per-KLOR-uh-NAY-shun) | SUPERCHLORINATION

Chlorination with doses that are deliberately selected to produce free or combined residuals so large as to require dechlorination.

SUPERNATANT (soo-per-NAY-tent) | SUPERNATANT

The relatively clear water layer between the sludge on the bottom and the scum on the surface of a basin or container. Also called clear zone.

SUPERSATURATED | SUPERSATURATED

An unstable condition of a solution (water) in which the solution contains a substance at a concentration greater than the saturation concentration for the substance.

SURFACE LOADING — SURFACE LOADING

One of the guidelines for the design of settling tanks and clarifiers in treatment plants. Used by operators to determine if tanks and clarifiers are hydraulically (flow) over- or underloaded. Also called overflow rate.

$$\text{Surface Loading, GPD/ft}^2 = \frac{\text{Flow, gal/day}}{\text{Surface Area, ft}^2}$$

or

$$\text{Surface Loading, } \frac{\text{m}^3\text{/day}}{\text{m}^2} = \frac{\text{Flow, m}^3\text{/day}}{\text{Surface Area, m}^2}$$

SURFACTANT (sir-FAC-tent) — SURFACTANT

Abbreviation for surface-active agent. The active agent in detergents that possesses a high cleaning ability.

SURGE CHAMBER — SURGE CHAMBER

A chamber or tank connected to a pipe and located at or near a valve that may quickly open or close or a pump that may suddenly start or stop. When the flow of water in a pipe starts or stops quickly, the surge chamber allows water to flow into or out of the pipe and minimize any sudden positive or negative pressure waves or surges in the pipe.

[SEE DRAWING ON PAGE 839]

SUSPENDED SOLIDS — SUSPENDED SOLIDS

(1) Solids that either float on the surface or are suspended in water, wastewater, or other liquids, and that are largely removable by laboratory filtering.

(2) The quantity of material removed from water or wastewater in a laboratory test, as prescribed in *Standard Methods for the Examination of Water and Wastewater*, and referred to as Total Suspended Solids Dried at 103–105°C.

T

TAGOUT — TAGOUT

The placement of a tagout device on an energy-isolating device, in accordance with an established procedure, to indicate that the energy-isolating device and the equipment being controlled may not be operated until the tagout device is removed.

TAGOUT DEVICE — TAGOUT DEVICE

A prominent warning device, such as a tag and a means of attachment, that can be securely fastened to an energy-isolating device in accordance with an established procedure to indicate that the energy-isolating device and the equipment being controlled may not be operated until the tagout device is removed.

TAILGATE SAFETY MEETING — TAILGATE SAFETY MEETING

Brief (10 to 20 minutes) safety meetings held every 7 to 10 working days. The term comes from the safety meetings regularly held by the construction industry around the tailgate of a truck.

TCE — TCE

See TRICHLOROETHANE (TCE).

TDS — TDS

See TOTAL DISSOLVED SOLIDS (TDS).

TELEMETRY (tel-LEM-uh-tree) — TELEMETRY

The electrical link between a field transmitter and the receiver. Telephone lines are commonly used to serve as the electrical line.

TEMPERATURE SENSOR — TEMPERATURE SENSOR

A device that opens and closes a switch in response to changes in the temperature. This device might be a metal contact, or a thermocouple that generates a minute electric current proportional to the difference in heat, or a variable resistor whose value changes in response to changes in temperature. Also called a heat sensor.

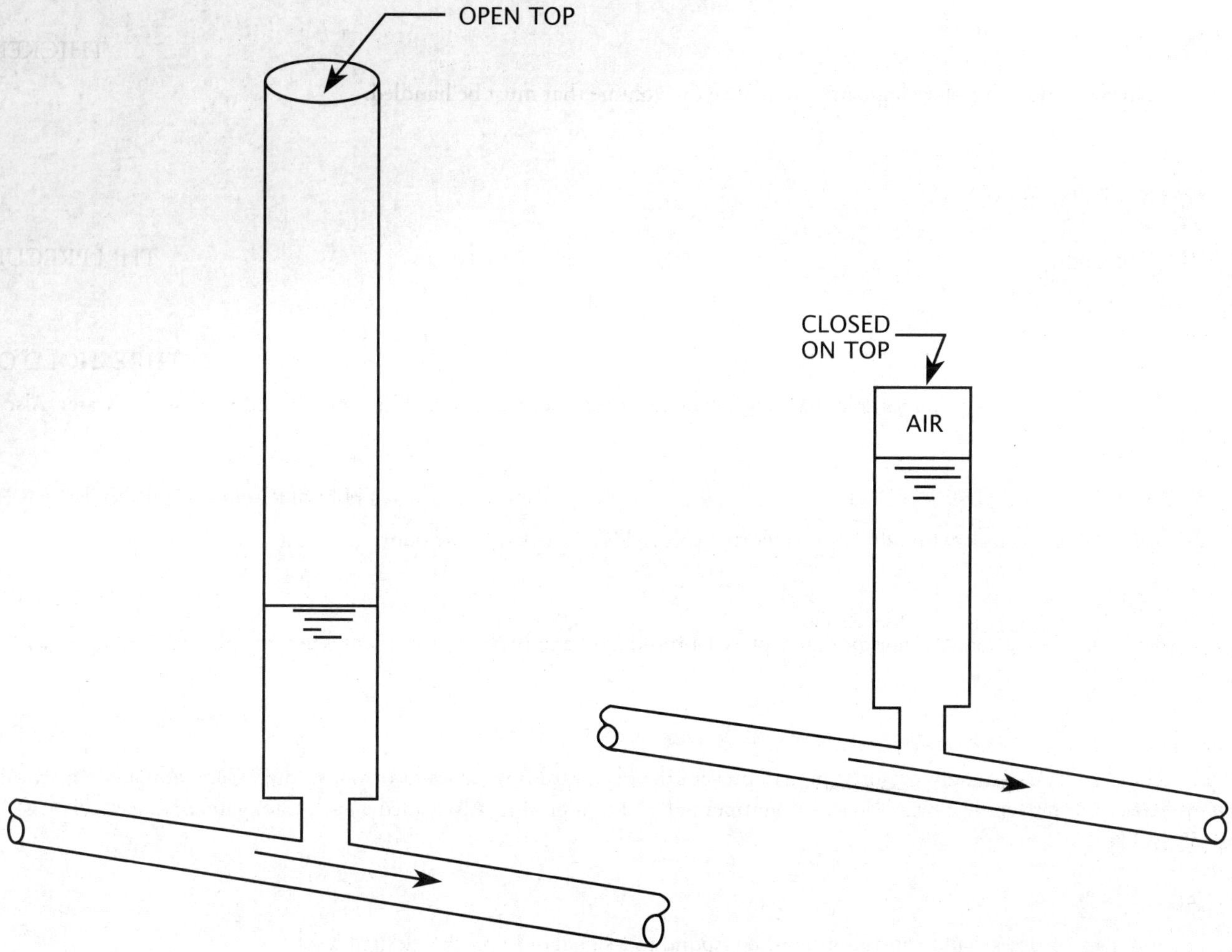

Types of surge chambers

THERMAL STRATIFICATION (strat-uh-fuh-KAY-shun) THERMAL STRATIFICATION

The formation of layers of different temperatures in a lake or reservoir. Also see STRATIFICATION.

THERMOCLINE (THUR-moe-kline) THERMOCLINE

The middle layer in a thermally stratified lake or reservoir. In this layer there is a rapid decrease in temperature with depth. Also called the metalimnion.

THERMOCOUPLE THERMOCOUPLE

A heat-sensing device made of two conductors of different metals joined together. An electric current is produced when there is a difference in temperature between the ends.

THICKENING THICKENING

Treatment to remove water from the sludge mass to reduce the volume that must be handled.

THM THM

See TRIHALOMETHANES (THMs).

THM PRECURSOR THM PRECURSOR

See PRECURSOR, THM.

THRESHOLD ODOR THRESHOLD ODOR

The minimum odor of a gas or water sample that can just be detected after successive dilutions with odorless gas or water. Also called odor threshold.

THRESHOLD ODOR NUMBER (TON) THRESHOLD ODOR NUMBER (TON)

The greatest dilution of a sample with odor-free water that still yields a just-detectable odor.

THRUST BLOCK THRUST BLOCK

A mass of concrete or similar material appropriately placed around a pipe to prevent movement when the pipe is carrying water. Usually placed at bends and valve structures.

TIDE GATE TIDE GATE

A gate installed at the end of a drain or outlet pipe to prevent the backward flow of water or wastewater. Generally used on storm sewer outlets into streams to prevent backward flow during times of flood or high tide. Also called a backwater gate. Also see CHECK VALVE and FLAP GATE.

TIME LAG TIME LAG

The time required for processes and control systems to respond to a signal or to reach a desired level.

TIMER TIMER

A device for automatically starting or stopping a machine or other device at a given time.

TITRATE (TIE-trate) TITRATE

To titrate a sample, a chemical solution of known strength is added drop by drop until a certain color change, precipitate, or pH change in the sample is observed (end point). Titration is the process of adding the chemical reagent in small increments (0.1–1.0 milliliter) until completion of the reaction, as signaled by the end point.

TOPOGRAPHY (toe-PAH-gruh-fee) TOPOGRAPHY

The arrangement of hills and valleys in a geographic area.

TOTAL CHLORINE TOTAL CHLORINE

The total concentration of chlorine in water, including the combined chlorine (such as inorganic and organic chloramines) and the free available chlorine.

TOTAL CHLORINE RESIDUAL — TOTAL CHLORINE RESIDUAL

The total amount of chlorine residual (including both free chlorine and chemically bound chlorine) present in a water sample after a given contact time.

TOTAL DISSOLVED SOLIDS (TDS) — TOTAL DISSOLVED SOLIDS (TDS)

All of the dissolved solids in a water. TDS is measured on a sample of water that has passed through a very fine mesh filter to remove suspended solids. The water passing through the filter is evaporated and the residue represents the total dissolved solids. Also see SPECIFIC CONDUCTANCE.

TOTAL DYNAMIC HEAD (TDH) — TOTAL DYNAMIC HEAD (TDH)

When a pump is lifting or pumping water, the vertical distance (in feet or meters) from the elevation of the energy grade line on the suction side of the pump to the elevation of the energy grade line on the discharge side of the pump. The total dynamic head is the static head plus pipe friction losses.

TOTALIZER — TOTALIZER

A device or meter that continuously measures and sums a process rate variable in cumulative fashion over a given time period. For example, total flows displayed in gallons per minute, million gallons per day, cubic feet per second, or some other unit of volume per time period. Also see INTEGRATOR.

TOTAL ORGANIC CARBON (TOC) — TOTAL ORGANIC CARBON (TOC)

TOC is a measure of the amount of organic carbon in water.

TOXAPHENE (TOX-uh-feen) — TOXAPHENE

A chemical that causes adverse health effects in domestic water supplies and is toxic to freshwater and marine aquatic life.

TOXIC — TOXIC

A substance that is poisonous to a living organism. Toxic substances may be classified in terms of their physiological action, such as irritants, asphyxiants, systemic poisons, and anesthetics and narcotics. Irritants are corrosive substances that attack the mucous membrane surfaces of the body. Asphyxiants interfere with breathing. Systemic poisons are hazardous substances that injure or destroy internal organs of the body. Anesthetics and narcotics are hazardous substances that depress the central nervous system and lead to unconsciousness.

TOXIC SUBSTANCE — TOXIC SUBSTANCE

See HARMFUL PHYSICAL AGENT and TOXIC.

TRANSDUCER (trans-DUE-sir) — TRANSDUCER

A device that senses some varying condition measured by a primary sensor and converts it to an electrical or other signal for transmission to some other device (a receiver) for processing or decision making.

TRANSMISSION LINES — TRANSMISSION LINES

Pipelines that transport raw water from its source to a water treatment plant. After treatment, water is usually pumped into pipelines (transmission lines) that are connected to a distribution grid system.

TRANSMISSIVITY (TRANS-miss-SIV-it-tee) — TRANSMISSIVITY

A measure of the ability to transmit (as in the ability of an aquifer to transmit water).

TRANSPIRATION (TRAN-spur-RAY-shun) — TRANSPIRATION

The process by which water vapor is released to the atmosphere by living plants. This process is similar to people sweating. Also called evapotranspiration.

TREMIE (TREH-me) — TREMIE

A device used to place concrete or grout under water.

TRICHLOROETHANE (TCE) (try-KLOR-o-ETH-hane) TRICHLOROETHANE (TCE)

An organic chemical used as a cleaning solvent that causes adverse health effects in domestic water supplies.

TRIHALOMETHANES (THMs) (tri-HAL-o-METH-hanes) TRIHALOMETHANES (THMs)

Derivatives of methane, CH_4, in which three halogen atoms (chlorine or bromine) are substituted for three of the hydrogen atoms. Often formed during chlorination by reactions with natural organic materials in the water. The resulting compounds (THMs) are suspected of causing cancer.

TRUE COLOR TRUE COLOR

Color of the water from which turbidity has been removed. The turbidity may be removed by double filtering the sample through a Whatman No. 40 filter when using the visual comparison method.

TU TU

See TURBIDITY UNITS (TU).

TUBERCLE (TOO-burr-kull) TUBERCLE

A crust of corrosion products (rust) that builds up over a pit caused by the loss of metal due to corrosion.

TUBERCULATION (too-BURR-kyoo-LAY-shun) TUBERCULATION

The development or formation of small mounds of corrosion products (rust) on the inside of iron pipe. These mounds (tubercles) increase the roughness of the inside of the pipe thus increasing resistance to water flow (decreases the C Factor).

TUBE SETTLER TUBE SETTLER

A device that uses bundles of small-bore (2 to 3 inches or 50 to 75 mm) tubes installed on an incline as an aid to sedimentation. The tubes may come in a variety of shapes including circular and rectangular. As water rises within the tubes, settling solids fall to the tube surface, and as the resulting sludge gains weight, it moves down the tubes and settles to the bottom of the basin for removal by conventional sludge collection means. Also called high-rate settlers.

TURBID TURBID

Having a cloudy or muddy appearance.

TURBIDIMETER TURBIDIMETER

See TURBIDITY METER.

TURBIDITY (ter-BID-it-tee) TURBIDITY

The cloudy appearance of water caused by the presence of suspended and colloidal matter. In the waterworks field, a turbidity measurement is used to indicate the clarity of water. Technically, turbidity is an optical property of the water based on the amount of light reflected by suspended particles. Turbidity cannot be directly equated to suspended solids because white particles reflect more light than dark-colored particles and many small particles will reflect more light than an equivalent large particle.

TURBIDITY (ter-BID-it-tee) METER TURBIDITY METER

An instrument for measuring and comparing the turbidity of liquids by passing light through them and determining how much light is reflected by the particles in the liquid. The normal measuring range is 0 to 100 and is expressed as nephelometric turbidity units (NTUs). Also called a turbidimeter.

TURBIDITY (ter-BID-it-tee) UNITS (TU) TURBIDITY UNITS (TU)

Turbidity units are a measure of the cloudiness of water. If measured by a nephelometric (deflected light) instrumental procedure, turbidity units are expressed in nephelometric turbidity units (NTU) or simply TU. Those turbidity units obtained by visual methods are expressed in Jackson turbidity units (JTU), which are a measure of the cloudiness of water; they are used to indicate the clarity of water. There is no real connection between NTUs and JTUs. The Jackson turbidimeter is a visual method and the nephelometer is an instrumental method based on deflected light.

TURN-DOWN RATIO TURN-DOWN RATIO

The ratio of the design range to the range of acceptable accuracy or precision of an instrument. Also see EFFECTIVE RANGE.

U

UEL UEL

See UPPER EXPLOSIVE LIMIT (UEL).

UF UF

See ULTRAFILTRATION (UF).

ULTRAFILTRATION (UF) ULTRAFILTRATION (UF)

A pressure-driven membrane filtration process that separates particles down to approximately 0.01 μm diameter from influent water using a sieving process.

ULTRAVIOLET (UV) ULTRAVIOLET (UV)

Pertaining to a band of electromagnetic radiation just beyond the visible light spectrum. Ultraviolet radiation is used in water treatment to disinfect the water. When ultraviolet radiation is absorbed by the cells of microorganisms, it damages the genetic material in such a way that the organisms are no longer able to grow or reproduce, thus ultimately killing them.

UNCONSOLIDATED FORMATION UNCONSOLIDATED FORMATION

A sediment that is loosely arranged or unstratified (not in layers) or whose particles are not cemented together (soft rock); occurring either at the ground surface or at a depth below the surface. Also see CONSOLIDATED FORMATION.

UNIFORMITY COEFFICIENT (UC) UNIFORMITY COEFFICIENT (UC)

The ratio of (1) the diameter of a grain (particle) of a size that is barely too large to pass through a sieve that allows 60 percent of the material (by weight) to pass through, to (2) the diameter of a grain (particle) of a size that is barely too large to pass through a sieve that allows 10 percent of the material (by weight) to pass through. The resulting ratio is a measure of the degree of uniformity in a granular material, such as filter media.

$$\text{Uniformity Coefficient} = \frac{\text{Particle Diameter}_{60\%}}{\text{Particle Diameter}_{10\%}}$$

UPPER EXPLOSIVE LIMIT (UEL) UPPER EXPLOSIVE LIMIT (UEL)

The highest concentration of a gas or vapor (percentage by volume in air) above which a flame will not spread in the presence of an ignition source (arc, flame, or heat). Concentrations higher than UEL are “too rich” to burn. Also called upper flammable limit (UFL). Also see LOWER EXPLOSIVE LIMIT (LEL).

UPPER FLAMMABLE LIMIT (UFL) UPPER FLAMMABLE LIMIT (UFL)

See UPPER EXPLOSIVE LIMIT (UEL).

US EPA US EPA

United States Environmental Protection Agency. A regulatory agency established by the US Congress to administer the nation’s environmental laws.

UV UV

See ULTRAVIOLET (UV).

V

VARIABLE, MEASURED VARIABLE, MEASURED

A factor (flow, temperature) that is sensed and quantified (reduced to a reading of some kind) by a primary element or sensor.

VARIABLE, PROCESS VARIABLE, PROCESS

A physical or chemical quantity that is usually measured and controlled in the operation of a water, wastewater, or industrial treatment plant.

VARIABLE COSTS | VARIABLE COSTS

Costs that a utility must cover or pay that are associated with the actual treatment and delivery of water. These costs vary or fluctuate on the basis of the volume of water treated and delivered to customers (water production). Also see FIXED COSTS.

VARIABLE FREQUENCY DRIVE | VARIABLE FREQUENCY DRIVE

A control system that allows the frequency of the current applied to a motor to be varied. The motor is connected to a low-frequency source while standing still; the frequency is then increased gradually until the motor and pump (or other driven machine) are operating at the desired speed.

VELOCITY HEAD | VELOCITY HEAD

The energy in flowing water as determined by a vertical height (in feet or meters) equal to the square of the velocity of flowing water divided by twice the acceleration due to gravity ($V^2/2g$).

VENTURI (ven-TOOR-ee) METER | VENTURI METER

A flow-measuring device placed in a pipe. The device consists of a tube whose diameter gradually decreases to a throat and then gradually expands to the diameter of the pipe. The flow is determined on the basis of the difference in pressure (caused by different velocity heads) between the entrance and throat of the Venturi meter.

NOTE: Most Venturi meters have pressure sensing taps rather than a manometer to measure the pressure difference. The upstream tap is the high-pressure tap or side of the manometer.

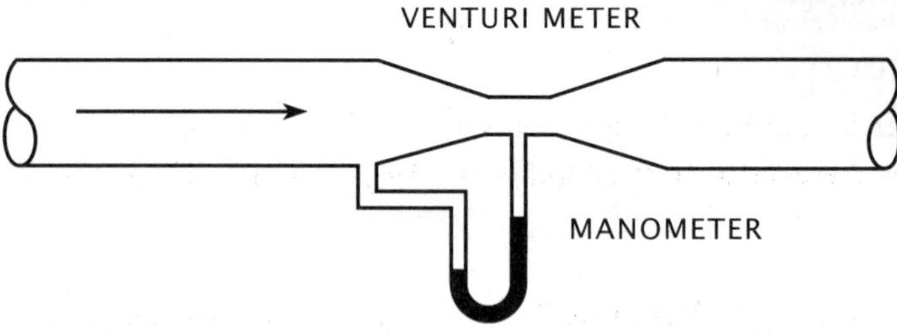

VISCOSITY (vis-KOSS-uh-tee) | VISCOSITY

A property of water, or any other fluid, that resists efforts to change its shape or flow. Syrup is more viscous (has a higher viscosity) than water. The viscosity of water increases significantly as temperatures decrease. Motor oil is rated by how thick (viscous) it is; 20-weight oil is considered relatively thin while 50-weight oil is relatively thick or viscous.

VOID | VOID

A pore or open space in rock, soil, or other granular material, not occupied by solid matter. The pore or open space may be occupied by air, water, or other gaseous or liquid material. Also called an interstice, pore, or void space.

VOLATILE (VOL-uh-tull) | VOLATILE

(1) A volatile substance is one that is capable of being evaporated or changed to a vapor at relatively low temperatures. Volatile substances can be partially removed from water or wastewater by the air stripping process.

(2) In terms of solids analysis, volatile refers to materials lost (including most organic matter) upon ignition in a muffle furnace for 60 minutes at 1,022°F (550°C). Natural volatile materials are chemical substances usually of animal or plant origin. Manufactured or synthetic volatile materials, such as plastics, ether, acetone, and carbon tetrachloride, are highly volatile and not of plant or animal origin. Also see NONVOLATILE MATTER.

VOLATILE ACIDS | VOLATILE ACIDS

Fatty acids produced during digestion that are soluble in water and can be steam-distilled at atmospheric pressure. Also called organic acids. Volatile acids are commonly reported as equivalent to acetic acid.

VOLATILE LIQUIDS | VOLATILE LIQUIDS

Liquids that easily vaporize or evaporate at room temperature.

VOLATILE SOLIDS VOLATILE SOLIDS

Those solids in water, wastewater, or other liquids that are lost on ignition of the dry solids at 1,022°F (550°C). Also called organic solids and volatile matter.

VOLTAGE VOLTAGE

The electrical pressure available to cause a flow of current (amperage) when an electric circuit is closed. Also called electromotive force (EMF).

VOLUMETRIC VOLUMETRIC

A measurement based on the volume of some factor. Volumetric titration is a means of measuring unknown concentrations of water quality indicators in a sample by determining the volume of titrant or liquid reagent needed to complete particular reactions.

VOLUMETRIC FEEDER VOLUMETRIC FEEDER

A dry chemical feeder that delivers a measured volume of chemical during a specific time period.

VORTEX VORTEX

A revolving mass of water that forms a whirlpool. This whirlpool is caused by water flowing out of a small opening in the bottom of a basin or reservoir. A funnel-shaped opening is created downward from the water surface.

W

WARNING WARNING

This word is used to indicate a hazard level between "Caution" and "Danger." Also see CAUTION, DANGER, and NOTICE.

WASTEWATER WASTEWATER

A community's used water and water-carried solids (including used water from industrial processes) that flow to a treatment plant. Stormwater, surface water, and groundwater infiltration also may be included in the wastewater that enters a wastewater treatment plant. The term sewage usually refers to household wastes, but this word is being replaced by the term wastewater.

WATER AUDIT WATER AUDIT

A thorough examination of the accuracy of water agency records or accounts (volumes of water) and system control equipment. Water managers can use audits to determine their water distribution system efficiency. The overall goal is to identify and verify water and revenue losses in a water system.

WATER CYCLE WATER CYCLE

The process of evaporation of water into the air and its return to earth by precipitation (rain or snow). This process also includes transpiration from plants, groundwater movement, and runoff into rivers, streams, and the ocean. Also called the hydrologic cycle.

WATER HAMMER WATER HAMMER

The sound like someone hammering on a pipe that occurs when a valve is opened or closed very rapidly. When a valve position is changed quickly, the water pressure in a pipe will increase and decrease back and forth very quickly. This rise and fall in pressures can cause serious damage to the system.

WATER PURVEYOR (purr-VAY-or) WATER PURVEYOR

An agency or person that supplies water (usually potable water).

WATERSHED WATERSHED

The region or land area that contributes to the drainage or catchment area above a specific point on a stream or river.

WATER TABLE WATER TABLE

The upper surface of the zone of saturation of groundwater in an unconfined aquifer.

WATER TABLE HEAD — WATER TABLE HEAD

See PRESSURE HEAD.

WATT — WATT

A unit of power equal to 1 joule per second. The power of a current of 1 ampere flowing across a potential difference of 1 volt.

WEIR (WEER) — WEIR

(1) A wall or plate placed in an open channel and used to measure the flow of water. The depth of the flow over the weir can be used to calculate the flow rate, or a chart or conversion table may be used to convert depth to flow. Also see PROPORTIONAL WEIR.

(2) A wall or obstruction used to control flow (from settling tanks and clarifiers) to ensure a uniform flow rate and avoid short-circuiting.

WEIR (WEER) DIAMETER — WEIR DIAMETER

Many circular clarifiers have a circular weir within the outside edge of the clarifier. All the water leaving the clarifier flows over this weir. The diameter of the weir is the length of a line from one edge of a weir to the opposite edge and passing through the center of the circle formed by the weir.

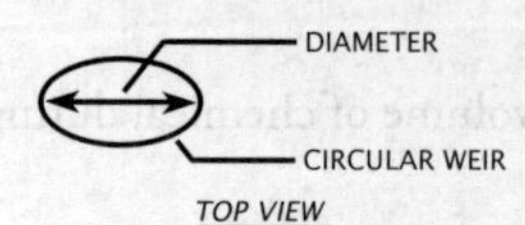

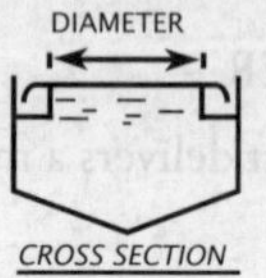

WEIR (WEER) LOADING — WEIR LOADING

A guideline used to determine the length of weir needed on settling tanks and clarifiers in treatment plants. Used by operators to determine if weirs are hydraulically (flow) overloaded.

$$\text{Weir Loading, GPM/ft} = \frac{\text{Flow, GPM}}{\text{Length of Weir, ft}}$$

or

$$\text{Weir Loading, m}^3\text{/day/m} = \frac{\text{Flow, m}^3\text{/day}}{\text{Length of Weir, m}}$$

WELLHEAD PROTECTION AREA (WHPA) — WELLHEAD PROTECTION AREA (WHPA)

The surface and subsurface area surrounding a public water system water well or well field, through which contaminants are reasonably likely to move toward and reach that water well or well field. Also see WELL ISOLATION ZONE.

WELL ISOLATION ZONE — WELL ISOLATION ZONE

A surface or zone with restricted land uses surrounding a public water system water well or well field. The zone is established to prevent contaminants from a nonpermitted land use to move toward and reach the water well or well field. Also see WELLHEAD PROTECTION AREA.

WELL LOG — WELL LOG

A record of the thickness and characteristics of the soil, rock, and water-bearing formations encountered during the drilling (sinking) of a well.

WET CHEMISTRY — WET CHEMISTRY

Laboratory procedures used to analyze a sample of water using liquid chemical solutions (wet) instead of, or in addition to, laboratory instruments.

WHOLESOME WATER — WHOLESOME WATER

Water that is safe and palatable for human consumption.

WIRE-TO-WATER EFFICIENCY — WIRE-TO-WATER EFFICIENCY

The combined efficiency of a pump and motor together. Also called the overall efficiency.

WYE STRAINER — WYE STRAINER

A screen shaped like the letter Y. Water flows through the upper parts of the Y and the debris is trapped by the screen at the fork.

X

(NO LISTINGS)

Y

YIELD — YIELD

The quantity of water (expressed as a rate of flow—GPM, GPH, GPD, m^3/day, ML/day, or total quantity per year) that can be collected for a given use from surface or groundwater sources. The yield may vary with the use proposed, with the plan of development, and with economic considerations. Also see SAFE YIELD.

Z

ZEOLITE — ZEOLITE

A type of ion exchange material used to soften water. Natural zeolites are siliceous compounds (made of silica) that remove calcium and magnesium from hard water and replace them with sodium. Synthetic or organic zeolites are ion exchange materials that remove calcium or magnesium and replace them with either sodium or hydrogen. Manganese zeolites are used to remove iron and manganese from water.

ZETA POTENTIAL — ZETA POTENTIAL

In coagulation and flocculation procedures, the difference in the electrical charge between the dense layer of ions surrounding the particle and the charge of the bulk of the suspended fluid surrounding this particle. The zeta potential is usually measured in millivolts.

ZONE OF AERATION — ZONE OF AERATION

The comparatively dry soil or rock located between the ground surface and the top of the water table.

ZONE OF SATURATION — ZONE OF SATURATION

(1) The soil or rock located below the top of the groundwater table. By definition, the zone of saturation is saturated with water. Also see WATER TABLE.

(2) Where raw wastewater is exfiltrating from a sewer pipe, the area of soil that is moistened around the leak point is often called the zone of saturation.

ZOOPLANKTON (ZOE-uh-PLANK-ton) — ZOOPLANKTON

Small, usually microscopic animals (such as protozoans), found in lakes and reservoirs.

INDEX

A

E

F

S

W

X

Y

Z

NOTES

International
Association
of Fire Chiefs

Fundamentals of

Firefighter Skills and Hazardous Materials Response

FIFTH EDITION

Fundamentals of

Firefighter Skills and Hazardous Materials Response

FIFTH EDITION

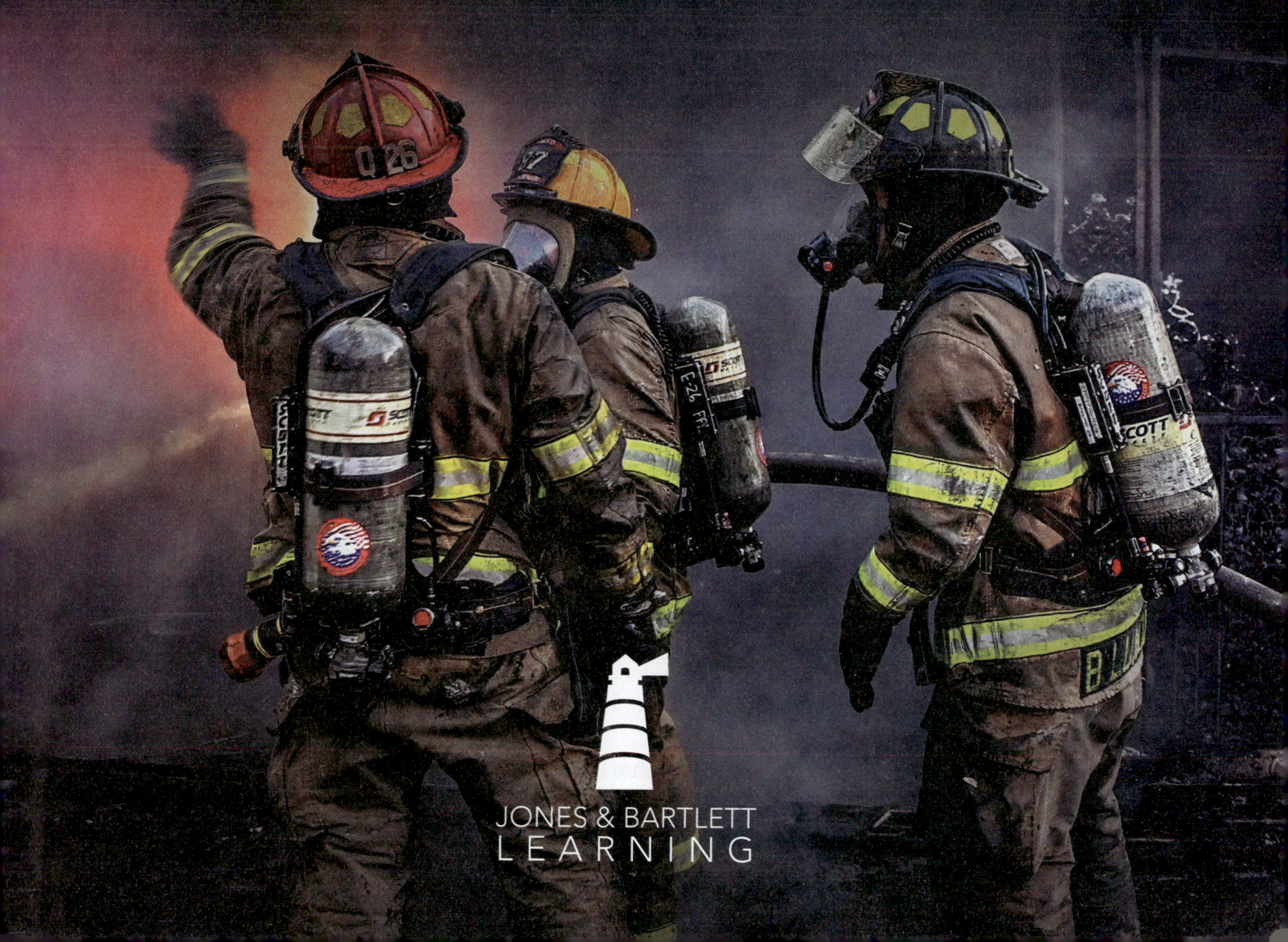

JONES & BARTLETT
LEARNING

World Headquarters
Jones & Bartlett Learning
25 Mall Road
Burlington, MA 01803
978-443-5000
info@jblearning.com
www.psglearning.com

National Fire Protection Association
1 Batterymarch Park
Quincy, MA 02169-7471
www.NFPA.org

International Association of Fire Chiefs
8251 Greensboro Drive, Suite 650
McLean, VA 22102
www.IAFC.org

Jones & Bartlett Learning books and products are available through most bookstores and online booksellers. To contact the Jones & Bartlett Learning Public Safety Group directly, call 800-832-0034, fax 978-443-8000, or visit our website, www.psglearning.com.

Substantial discounts on bulk quantities of Jones & Bartlett Learning publications are available to corporations, professional associations, and other qualified organizations. For details and specific discount information, contact the special sales department at Jones & Bartlett Learning via the above contact information or send an email to specialsales@jblearning.com.

29827-7

Production Credits
General Manager, Jones & Bartlett Learning: Josh Braunstein
Vice President, Content Strategy and Implementation: Christine Emerton
Director, Product Management: Cathy Esperti
Product Manager: Janet Maker
Director, Content Management: Donna Gridley
Manager, Content Strategy: Kim Crowley
Manager, Content Strategy: Tiffany Sliter
Content Strategist: Jennifer Deforge-Kling
Content Strategist (Digital): Elena Sorrentino
Content Strategist (Digital): Summer Ibrahim
Content Strategist (Digital): Alexander Belloli
Developmental Editor: Katherine Pinard
Director, Project Management and Content Services: Karen Scott
Manager, Project Management: Jackie Reynen
Project Manager: John Fuller
Senior Digital Project Specialist: Angela Dooley
Director, Marketing: Andrea DeFronzo
Director, Sales: Brian Hendrickson
Vice President, Enterprise Solution Development and Support: Matthew Maniscalco
Content Services Manager: Colleen Lamy
Product Fulfillment Manager: Wendy Kilborn
Composition: S4Carlisle Publishing Services
Cover and Text Design: Scott Moden
Media Development Editor: Faith Brosnan
Rights & Permissions Manager: John Rusk
Cover Image (Title Page, Part Opener, Chapter Opener): © Glen E. Ellman
Printing and Binding: Lakeside Book Company
Cover Printing: Lakeside Book Company

Library of Congress Cataloging-in-Publication Data
Names: National Fire Protection Association, author. | International Association of Fire Chiefs, author.
Title: Fundamentals of firefighter skills and hazardous materials response / International Association of Fire Chiefs, National Fire Protection Association.
Other titles: Fundamentals of firefighter skills
Description: Fifth edition. | Burlington, MA : Jones & Bartlett Learning, [2025] | Includes bibliographical references and index.
Identifiers: LCCN 2023040536 | ISBN 9781284298260 (paperback)
Subjects: LCSH: Fire prevention. | Fire extinction.
Classification: LCC TH9146 .F855 2025b | DDC 628.9/25--dc23/eng/20240109
LC record available at https://lccn.loc.gov/2023040536

6048

Printed in the United States of America
29 28 27 26 25 10 9 8 7 6 5 4 3

Brief Contents

SECTION 5 Hazardous Materials Operations Level: Mission Specific

Contents

CHAPTER **33**

Estimating Potential Harm and Planning a Response 1319

CHAPTER **34**

Implementing the Planned Response 1343

SECTION **5**

Hazardous Materials Operations Level: Mission Specific

CHAPTER **35**

Hazardous Materials Responder Personal Protective Equipment 1365

Skill Drills

CHAPTER **11**

CHAPTER **12**

CHAPTER **13**

CHAPTER 16

CHAPTER 17

CHAPTER 18

CHAPTER 19

Preface

A Note on the Relationship between NFPA 1010 and NFPA 1001

The fifth edition of *Fundamentals of Firefighter Skills and Hazardous Materials Response* covers the job performance requirements that candidates must master in order to become certified as a Firefighter I or Firefighter II. As part of the National Fire Protection Association's (NFPA) standard consolidation initiative, multiple standards that cover firefighter, fire apparatus driver/operator, airport firefighter, and marine firefighting training have been combined in a single document.

NFPA 1010, *Standard on Professional Qualifications for Firefighters,* 2024 Edition was published in December 2023 with an effective date of January 2024. NFPA 1010, 2024 Edition represents a consolidation and revision of the following stand-alone standards:

- **NFPA 1001**, *Standard for Fire Fighter Professional Qualifications,* 2019 Edition
- **NFPA 1002**, *Standard for Fire Apparatus Driver/Operator Professional Qualifications,* 2017 Edition
- **NFPA 1003**, *Standard for Airport Fire Fighter Professional Qualifications,* 2019 Edition
- **NFPA 1005**, *Standard for Professional Qualifications for Marine Fire Fighting for Land-Based Fire Fighters,* 2019 Edition

Appendix A and the Learning Objectives at the end of each chapter are correlated to relevant professional qualifications in the following chapters in NFPA 1010, 2024 Edition:

- Chapter 6, "Firefighter I" (NFPA 1001)
- Chapter 7, "Firefighter II" (NFPA 1001)

A Note on the Relationship between NFPA 1010 and 470

The fifth edition of *Fundamentals of Firefighter Skills and Hazardous Materials Response* covers the job performance requirements (JPRs) that candidates must master in order to become certified as a Firefighter I and/or Firefighter II. This includes the requirements listed in 6.1 in Chapter 6: Firefighter I from NFPA 1010, 2024 Edition, which requires candidates to meet the JPRs as defined in Chapters 5: Awareness (NFPA 1072), Chapter 7: Operations (NFPA 1072) and 9.2 and 9.6 in Chapter 9: Operations/Mission Specifics Sections (NFPA 1072) from NFPA 470, 2022 Edition.

As part of the NFPA's standard consolidation initiative, multiple standards that cover hazardous materials have been combined in a single document. **NFPA 470**, *Hazardous Materials/Weapons of Mass Destruction (WMD) Standard for Responders,* 2022 Edition was published in August 2021 with an effective date of September 15, 2021. NFPA 470, 2022 Edition represents a consolidation and revision of the following stand-alone standards:

- **NFPA 472**, *Standard for Competence of Responders to Hazardous Materials/Weapons of Mass Destruction Incidents,* 2018 Edition
- **NFPA 473**, *Standard for Competencies for EMS Personnel Responding to Hazardous Materials/Weapons of Mass Destruction Incidents,* 2018 Edition
- **NFPA 1072**, *Standard for Hazardous Materials/Weapons of Mass Destruction Emergency Response Personnel Professional Qualifications,* 2017 Edition

Appendix A and the Learning Objectives at the end of Chapters 29–36 are correlated to relevant professional qualifications in the following chapters in NFPA 470, 2022 Edition:

- Chapter 5, "Professional Qualifications for Hazardous Materials/WMD Awareness Level Responders" (NFPA 1072)
- Chapter 7, "Professional Qualifications for Hazardous Materials/WMD Operations Level Responders" (NFPA 1072)
- 9.2: Personal Protective Equipment and 9.6: Product Control in Chapter 9, "Professional Qualifications for Hazardous Materials/WMD Operations Level Responders Assigned Mission-Specific Responsibilities" (NFPA 1072)

Acknowledgments

The Jones & Bartlett Public Safety Group, the National Fire Protection Association, and the International Association of Fire Chiefs would like to thank all of the authors, contributors, advisors, and reviewers of *Fundamentals of Firefighter Skills and Hazardous Materials Response, Fifth Edition.*

Contributors

Toby Ballard
Captain
Missoula Rural Fire District
Missoula, Montana

Bret A. Davidson
Deputy Fire Chief
City of Vista
Vista, California

Marc Davidson
Captain
Fairfax County Fire & Rescue Department
Fairfax, Virginia

Christopher DeSantis
Yonkers, New York Fire Department
Yonkers, New York

Rommie L. Duckworth, MPA, LP, EFO, FO
Ridgefield Fire Department
Ridgefield, Connecticut

Todd Eddy, BS
Alabama Fire College
Tuscaloosa, Alabama

Brandon Hausbeck, BASc, CFI-I
Fire Chief
Saginaw Fire Department
Saginaw, Michigan

Michael Heffner, MS, EMT-P
Eastern Oregon University
La Grande, Oregon

Dustin Housewright, CFO, CEMSO, NRP
Eastman Emergency Services
Kingsport, Tennessee

James P. Kenney
Assistant Chief, Warwick, Rhode Island Fire Department (retired)
Rope Coordinator for the Connecticut State Fire Academy
Warwick, Rhode Island

Chad Landis
Training Officer
Rapides Parish Fire District 3/Alpine FD
Pineville, Louisiana

Kevin Lewis
Battalion Chief
Cobb County Fire Department
Marietta, Georgia

Aaron Miranda
Poway Fire Department
Poway, California

Guy Peifer, BS, NRP
Captain (retired)
Yonkers Fire Department
Yonkers, New York

Becki Rowan-White
Battalion Chief
Chanhassen Fire Department
Chanhassen, Minnesota

Rob Schnepp
Alameda County (CA) Fire Department (retired)
Alameda, California

Michael A. Smith, MPA
Lieutenant, Lynn Fire Department, Lynn, Massachusetts
Adjunct Professor, Anna Maria College, Paxton, Massachusetts
Adjunct Professor, North Shore Community College, Danvers, Massachusetts
Adjunct Professor, Bunker Hill Community College, Boston, Massachusetts

Sean Wilson
Captain, Royal Oak, Michigan Fire Department
Owner, Rise Above Fire Training
Royal Oak, Michigan

Ancillary Contributors

Travis Adkison, MEd
Public Safety Academy Director
Chapel Hill - Carrboro City Schools
Chapel Hill, North Carolina

Tim Cassidy, CFO, TPM, FD Safety Officer, MPPA
Battalion Chief of Training and Safety
Mount Prospect Fire Department
Mount Prospect, Illinois

Chad R. Chambers, BS, FF, FI-2, EMT-P, IC
Training Officer
Benton Township (Potterville), Michigan

Matthew Claflin, MSOL, OFO
University of Akron
Akron Fire Department
Akron, Ohio

Kimberly C. Cox, Captain (Retired)
Eden Prairie Fire Department
Eden Prairie, Minnesota

Timothy W. Guffey BS, EFO, CFO, CFI
Fire Chief (Retired), Adjunct Instructor
Johnston Community College
Smithfield, North Carolina

Jonathan Hinson
Battalion Chief, Chesapeake Fire Department
Chesapeake, Virginia

Alan E. Joos, MS, EFO, CFO, FIFireE
Chief Training Division
Nebraska State Fire Marshal Agency
Grand Island, Nebraska

Jeff Laskowske
Fire Chief, Albert Lea Fire Rescue
Albert Lea, Minnesota

Jake Martin
Lieutenant, Battle Creek Fire Department
Battle Creek, Michigan

Reviewers

Travis Adkison
Chapel Hill–Carrboro City Schools
Raleigh, North Carolina

Craig Anders
Gatlinburg FD/Cosby VFD Training Officer
Newport, Tennessee

David Aten
Blair Volunteer Fire and Rescue
Blair, Nebraska

Gary Bacon
Montrose Fire Academy
Morris, Michigan

Gary Baum
Fox Valley Career Center
Maple Park, Illinois

Holly Bilbo
Hancock County Board of Supervisors
Kiln, Mississippi

Chester W. Bombriant IV
Palm Bay Magnet High School
Palm Bay, Florida

Justin Bouler
Jackson County Office of Emergency Services
Pascagoula, Mississippi

David Briggs
City of Wausau Fire Department
Mid-State Technical College
Northcentral Technical College
Firefighter Training, Development, and Coaching LLC
Wisconsin Rapids, Wisconsin

Donald Burns
City of Wilmington Fire Department
Wilmington, North Carolina

Brendon Caid
Battle Creek Fire Department
Lansing Community College
Battle Creek, Michigan

Tim Cassidy
Mount Prospect Fire Department (Training Officer)
Mount Prospect, Illinois

Jason Caughey
Laramie County Fire District #2
Cheyenne, Wyoming

Chad R. Chambers
Benton Township Fire Department
Mason, Michigan

Matthew Claflin
Akron Fire Department
Akron, Ohio

Stephen Cook
McLennan Community College
Waco, Texas

Jon Culberson
Joint Base Myer-Henderson Hall Fire and Emergency Services. Assistant Chief (Retired)
Virginia Department of Fire Programs Adjunct Instructor
Spotsylvania County Fire Rescue and Emergency Management Instructor Cadre
Spotsylvania Volunteer Fire Department Training Officer
Spotsylvania, Virginia

Shawn Dillingham
Mid-State Technical College School of Public Safety
Wisconsin Rapids, Wisconsin

Ed Dolan
Catskill Fire Department
Catskill, New York

Scott E. Exo
Northeastern Illinois Public Safety Training Academy
Woodridge, Illinois

Tom Fentress
Keener Township Volunteer Fire Department
Demotte, Indiana

Mark J. Finucane
Johnson City Fire Department
Johnson City, Tennessee

Dustin T. Gladwell
Rockingham County Fire Rescue (Virginia)
Grottoes, Virginia

Tim Hartz
MU Fire and Rescue Training
Harrisburg, Missouri

Jonathan Hinson
Chesapeake Fire Department
Newsoms, Virginia

Brian P. Kazmierzak
Benton Harbor Department of Public Safety
Benton Harbor, Michigan

Jeffery Laskowske
Albert Lea Fire Academy
Albert Lea, Minnesota

Michael Martin
Lansing Community College Regional Fire
Lansing, Michigan

Pat McAuliff
Collin College
McKinney, Texas

Ben McMinn
Desoto County Emergency Services
Nesbit, Mississippi

Alima Mims
Georgia Emergency Management Agency/Homeland Security
Savannah, Georgia

Nicholas Morgan
St Louis Fire Department
St. Louis, Missouri

Kevin Neely
Lansing Community College
Lansing, Michigan

Chris Pabst
Columbus State Community College
Columbus, Ohio

Robert L. Rayburn
Jefferson City Fire
Jefferson City, Tennessee

Dan Reed
Canton City Fire Department/Stark State College
Canton, Ohio

Kevin S. Reed
Hillsborough Community College
Tampa, Florida

Lester Remington
Morris, Michigan

Leigh H. Shapiro
Hartford Fire Department
Hartford, Connecticut

Timothy Stainer
Delaware Area Career Center
Delaware, Ohio

Michael Swiman
Rocky Mount Fire Department
Rocky Mount, North Carolina

Jeff Tanner
Amherst Fire Department
Madison Heights, Virgina

Theodore Tautges
City of Wausau Fire Department
Northcentral Technical College
Fire Innovations, LLC
Wausau, Wisconsin

Roland Tenley
Peoria Fire Department
Peoria, Iowa

Eric Valles
Gulfport Fire Department
Gulfport, Mississippi

Scott Woodward
Gallatin Fire Department
Gallatin, Tennessee

David Yakowenko
Blackhawk Technical College
Kenosha, Wisconsin

Thank You

Thank you to the Aurora, Colorado Fire Department and AIMS Community College Public Safety Institute for hosting photoshoots for the *Fifth Edition* and to the following individuals:

Juan Calderoa
Brandon Fryman
Austin Hackbarth
Commander Mark Hays
Captain Rob Hulse
John McDougall
Jamie Meyers
Travis Miller
Susan Moreland
Jason Natzke
Ryan Nelson
Steve Reynoldson
Ronnie Riedel
Jared Scott
Jonathan Timms
Lee Vidal
Andrew Welsh

SECTION

1

Firefighter I

CHAPTER

1

Firefighter I

The Fire Service

KNOWLEDGE OBJECTIVES

After studying this chapter, you will be able to:

- Explain the mission of the fire service.
- Understand the history of the fire service and how it affected the modern fire service and the evolution of building and fire codes.
- Describe the modern fire service.
- Understand the culture of the fire service and the recent shift to include community risk reduction.
- Describe how department policies and standard operating procedures or guidelines apply to firefighters.
- Describe types of apparatus, companies, roles, and the command structure within a fire department.
- Describe the qualifications needed to become a firefighter and the roles and responsibilities of support persons and firefighters.

SKILLS OBJECTIVES

After studying this chapter, you will be able to perform the following skills:

- Locate specific information that identifies current best practices.

ADDITIONAL NFPA STANDARDS AND CODES

- **NFPA 1**, *Fire Code, 2021 Edition*
- **NFPA 13**, *Standard for the Installation of Sprinkler Systems, 2022 Edition*
- **NFPA 101**, *Life Safety Code, 2021 Edition*
- **NFPA 470**, *Hazardous Materials/Weapons of Mass Destruction (WMD) Standard for Responders, 2022 Edition*
- **NFPA 1006**, *Standard for Technical Rescue Personnel Professional Qualifications, 2021 Edition*
- **NFPA 1026**, *Standard for Incident Management Personnel Professional Qualifications, 2024 Edition*
- **NFPA 1030**, *Standard for Professional Qualifications for Fire Prevention Program Positions, 2024 Edition*
- **NFPA 1091**, *Standard for Traffic Incident Management Personnel Professional Qualifications, 2024 Edition*
- **NFPA 1225**, *Standard for Emergency Services Communications, 2022 Edition*
- **NFPA 1401**, *Recommended Practice for Fire Service Training Reports and Records, 2017 Edition*
- **NFPA 1410**, *Standard on Training for Emergency Scene Operations, 2020 Edition*
- **NFPA 1452**, *Guide for Training Fire Service Personnel to Conduct Community Risk Reduction for Residential Occupancies, 2020 Edition*
- **NFPA 1550**, *Standard for Emergency Responder Health and Safety, 2024 Edition*
- **NFPA 1582**, *Standard on Comprehensive Occupational Medical Program for Fire Departments, 2022 Edition*
- **NFPA 1700**, *Guide for Structural Fire Fighting, 2021 Edition*
- **NFPA 1710**, *Standard for the Organization and Deployment of Fire Suppression Operations, Emergency Medical Operations, and Special Operations to the Public by Career Fire Departments, 2020 Edition*

- **NFPA 1900**, *Standard for Aircraft Rescue and Firefighting Vehicles, Automotive Fire Apparatus, Wildland Fire Apparatus, and Automotive Ambulances, 2024 Edition*
- **NFPA 1910**, *Standard for the Inspection, Maintenance, Refurbishment, Testing, and Retirement of In-Service Emergency Vehicles and Marine Firefighting Vessels, 2024 Edition*
- **NFPA 1970**, *Standard on Protective Ensembles for Structural and Proximity Firefighting, Work Apparel and Open-Circuit Self-Contained Breathing Apparatus (SCBA) for Emergency Services, and Personal Alert Safety Systems (PASS), 2024 Edition*

CASE STUDY

You Are the Firefighter

It is your first day in the station after having completed 4 months of recruit academy. Your new crew welcomes you. As you are shown around the station, the doorbell rings, and you are told your first official duty is to answer it. At the door, you find a woman and her young child. Excitedly, she tells you her child has been begging to meet the firefighters because, "Riley wants to grow up to be a firefighter."

You agree to give a quick tour of the station. You first show them the living quarters and then move to the apparatus floor. The child is full of questions and asks, "Do you go to fires all day long? Do you know how to use all those tools? Do you report to the fire chief? Do you work with police officers?" The mother tells her child to give you a chance to explain before asking any more questions.

1. What are the duties of a firefighter?
2. What tools and equipment do you need specialized training to operate?
3. What are the roles and responsibilities of a firefighter versus a company officer?
4. With what other agencies do firefighters typically interact?

Introduction

Training to become a firefighter is challenging. **Fire suppression**—that is, the various activities involved in controlling and extinguishing fires—and rescue operations are much more complex than most people imagine. You will be challenged in many ways during this course. You must keep your body in excellent condition so you can complete your assignments. You must also remain mentally alert to cope with the situations and conditions you will encounter.

When you join the fire service, you join a profession with a long and noble history of protecting and serving the community (**FIGURE 1-1**). After you complete this course, you will be well equipped to continue a centuries-old tradition of preserving lives and property threatened by fire.

Mission of the Fire Service

Individual departments may have different needs and capabilities, but the core mission of the fire service remains the same. The primary mission of the fire service is to save lives, but the fire service also works to protect property and the environment. The missions of the fire service are accomplished through prevention, education, suppression, rescue, and salvage activities. This is typically presented in a department's mission statement. For example, *"We protect the lives and property of residents and visitors through fire prevention, education, and delivery of fire, rescue, and emergency medical services."* A mission statement communicates the purpose and goals of the department both to its firefighters and to the public. This example mission statement makes it clear that the fire department protects lives and property by providing fire prevention and education, as well as providing fire suppression, rescue services, and emergency medical services (EMS), and disaster services. The core mission of the fire service remains the same for every firefighter in every department.

FIGURE 1-1 The National Fallen Firefighters Memorial is located at the National Fire Academy in Emmitsburg, Maryland. The memorial honors those who have died in the line of duty. The eternal flame at the base of the memorial symbolizes the spirit of all past, present, and future firefighters.

© Bill Ryan/AP Photo

Brief History of the Fire Service

Communities have needed to fight fires and prevent them from happening since the first humans walked the earth. As human societies grew into more dense communities, this need became more acute because one person's fire could quickly become an entire community's decimation. Some of the most important lessons have been learned in the aftermath of tragic fires that resulted in a devastating loss of life. Governments became involved in firefighting and fire prevention efforts early in human history.

TIP

Many resources describing the history of firefighting are available on websites and in books. Learn about how things have changed—and not changed—over the years. For example, buckets made of leather did not hold water for any length of time, so sand was sometimes substituted for water in those buckets. After the sand was thrown on the fire, the bucket would be refilled with water and used in a bucket brigade.

Voice of Experience

In order to appreciate today's fire service, the aspiring firefighter should know the evolutionary progression of the fire service. Although the fire service in the United States dates back to Benjamin Franklin, the global fire service can trace its origins back to the bucket brigades of the ancient Egyptians. The fire service has developed from a small group of neighbors throwing water on a fire to an organization that is a cultural force and a family that works tirelessly for the betterment of society.

Change often occurs due to incidents such as the Great Chicago Fire of 1871, the Hotel Vendome Fire in Boston in 1972, the World Trade Center Attacks on September 11, 2001, The Station Nightclub Fire in Rhode Island in 2003, and the Oakland, California, Warehouse Fire in 2016. The fire service has grown and progressed due to these past events and sacrifices. In order to embrace the future, the fire service must fully understand the importance of these incidents and the bravery shown at these sites. Never forget the past, because it informs us about how we became who and what we are today. To honor those sacrifices, always strive for a better tomorrow working within this great service.

Firefighting is not just an occupation, it is a way of life, rich with great pride and tradition. The fire service has developed and changed over the years and will continue to do so long after you have retired. It is your job to help write the pages of the history of the fire service, with the pride and integrity, as so many others have done before you.

Dana Peloso
Oxford, MA

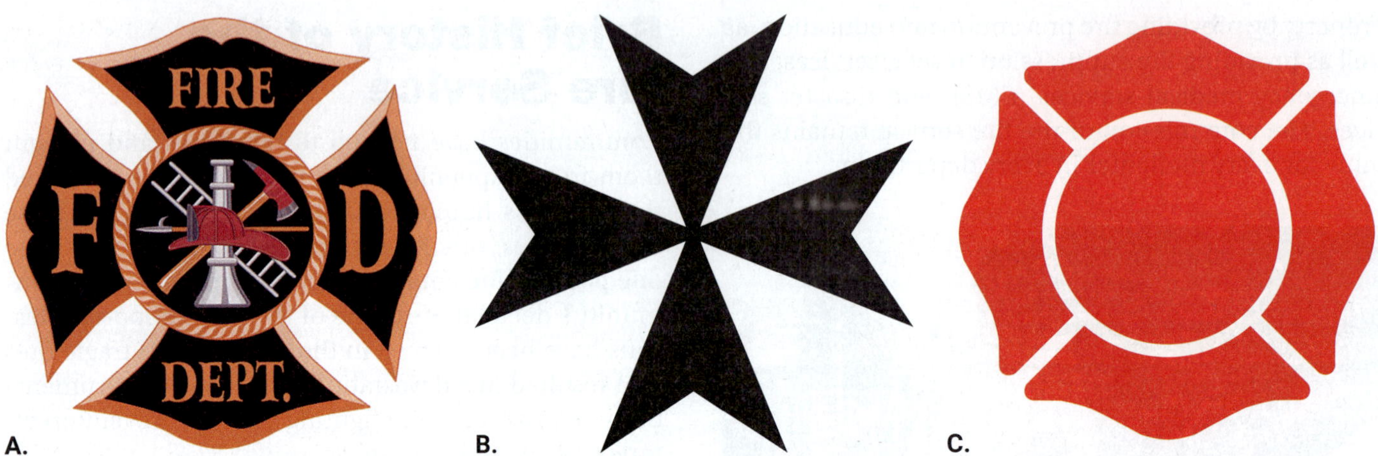

FIGURE 1-2 **A.** The Firefighter's Cross is widely recognized as a symbol of the fire service and is displayed on most fire vehicles, on firefighter uniforms, and on firefighter badges. **B.** The Maltese cross has been a symbol of the fire service since about the 12th century. **C.** The Florian cross has represented the fire service since the 4th century and is named after a man charged with organizing firefighting brigades in Rome.

The Firefighter's Cross

The Firefighter's Cross is a symbol of the fire service used across the United States (**FIGURE 1-2A**). You will see this symbol on most fire vehicles, on firefighter uniforms, and on firefighter badges. The origins of the Firefighter's Cross are not clear. It is often referred to as the Maltese cross, which was originally a symbol of the knights of Malta (**FIGURE 1-2B**). The Maltese Cross dates back to the 16th century and may have been used to represent the fire service as early as the 12th and 13th century. It is an international symbol of the fire service. Each of the eight points on the Maltese cross serves as a reminder of eight key attributes of a good first responder: observant, tactful, resourceful, persevering, dexterous, clear, sympathetic, and always using good judgment. Another cross symbol known as the Florian cross (**FIGURE 1-2C**) has also been used to represent the fire service. It is named after a Roman officer in the 4th century who was assigned to organize firefighting brigades in Rome (Martinez-Granata 2017).

The important thing to remember is that the Firefighter's Cross continues to represent the firefighter's life of service, dedication, and sacrifice. After you become a firefighter and you wear this cross on your uniform, you must wear it proudly and strive to uphold the best traditions of the fire service.

Fire Protection in England

As early as 1066, William the Conqueror decreed that all home fires in England were to be extinguished and covered every evening with a metal lid called a "couvre feu," likely the source of the modern concept of a curfew. In 1500, English cities passed ordinances regulating hazardous trades such as baking and kettle making as well as governing fire hazards such as wooden chimneys and thatched roofs. Despite early successes in fire prevention, the Great Fire of London struck in 1666, destroying more than 13,000 homes. This disaster led to improvements in fire protection in England for more than 100 years (Coleman 1988, 133).

Early America

The first documented structure fire in North America occurred in Jamestown, Virginia, in 1608. This fire quickly spread in the fort in which the settlers had built their houses and almost burned down the entire settlement. At that time, most structures were built entirely of combustible materials such as straw and wood. In 1630, the city of Boston, Massachusetts, established the first fire regulations in North America when it banned wood chimneys and thatched roofs. In 1648, in the Dutch colony of New Amsterdam (which became New York), Governor Peter Stuyvesant enacted the same ban and required that chimneys be swept out regularly. Stuyvesant also appointed fire wardens to impose fines on homeowners who did not obey these regulations. The money collected was used to pay for firefighting equipment (Merrimack Fire and Rescue n.d.).

The first fire department with paid firefighters in the United States was established in 1678 in Boston. Boston also had the first fire stations and fire engines (Boston Fire Historical Society n.d.). The first volunteer fire department was created in Philadelphia in 1736 under the leadership of Benjamin Franklin (Benjamin Franklin Historical Society n.d.). Franklin recognized the many dangers of fire and continually sought ways to prevent it. For example, citizens in Philadelphia were required to keep buckets filled with water outside their doors, readily available to fight fires. Franklin also developed the lightning rod to help draw lightning strikes (a common cause of fires) away from homes. Fire prevention and fire suppression efforts have gone hand in hand since the very inception of the American fire service.

Early Firefighting Equipment

Colonial-era firefighters had only buckets, ladders, and fire hooks (long poles with metal hooks at the end used to pull down burning structures) at their disposal. Homeowners were required to keep buckets filled with sand or water and to bring them to the scene of a fire. Some towns also required that ladders be available so that firefighters could access the roof to extinguish small fires. If all else failed, the **fire hook** was used to pull down a burning building and prevent the fire from spreading to nearby structures. (The hook-and-ladder truck evolved from this early equipment.)

Buckets gave way to hand-powered pumpers around 1721, when Richard Newsham developed the first one in London, England. The pump made it possible to propel a steady stream of water from a safe distance (Noonan 2017). In the early 1800s, more powerful steam-powered pumpers replaced the hand-powered pumpers. Steam-powered pumpers were heavy machines that were pulled to the fire by a trained team of horses. They required constant attention, which limited their use to larger cities that could bear the costs of maintaining the horses and the pumpers.

The advent of the internal combustion engine in the early 1900s greatly changed the fire service and enabled even small towns to have motorized pumpers. Although pumpers powered by an internal combustion engine required regular maintenance, those engines did not require the constant attention that horses or steam pumpers did.

The progress in equipment in the fire service extends beyond trucks. Firefighters without an adequate water supply are severely hampered in their ability to extinguish fires. Romans developed the first municipal water systems, just as they had developed the first fire companies. But it was not until the 1800s that water distribution systems were used for fire suppression by fire departments. Frederick Graff Sr., a firefighter in New York City, developed the first modern fire hydrant in 1801. He used a valve to control access to the water in city pipes (FireHydrant.org 2004). This enabled firefighters to tap into the water system whenever a fire occurred. This type of valve, called a **fireplug**, was used with both aboveground and below-ground piping systems.

Early Communications

Good communication is vital for effective firefighting. When a fire is discovered, firefighters must be summoned and citizens must be alerted to the danger. While they are fighting a fire, officers must be able to communicate with firefighters and summon additional resources. Not surprisingly, improvements in communication systems are tied to improvements in the fire service.

During the colonial period, a **fire warden** or night watchman patrolled neighborhoods and sounded the alarm if a fire was discovered. Some towns, including Charleston, South Carolina, built a series of fire towers from which wardens could watch for fires. In many towns, ringing the community fire bell or church bells alerted citizens to a fire.

In the late 1800s, telegraph fire alarm systems were installed in large cities. These systems allowed fires to be reported to the fire departments more quickly. They also made it possible for officers to request additional resources or to let others know when the fire was extinguished. In small towns, community sirens mounted on poles or tall rooftops were used signal that assistance was needed to fight a fire.

Communications while fighting a fire is also important. One of the earliest methods chief officers used was to shout commands through a speaking trumpet. The **chief's trumpet**, or **chief's bugle**, eventually became a symbol of authority (**FIGURE 1-3**). Although chief officers no longer use trumpets for communicating, the use of trumpets to symbolize ranks in the fire service signifies the officer's need to communicate as well as to lead (**FIGURE 1-4**). The more trumpets, the higher the rank.

Paying for Fire Service

Fire insurance companies were established in England soon after the Great Fire of London in 1666 to help victims cope with the financial loss from fires. Benjamin Franklin, who coined the phrase "an ounce of prevention is worth a pound of cure," organized one of the first insurance companies in the United States, the Philadelphia Contributionship.

Because insurance companies saved money if a fire was put out before much damage was done, they agreed to pay fire companies to extinguish fires at the homes

FIGURE 1-3 The chief's trumpet was once used to amplify the commander's voice. Today it serves as a symbol of authority in the fire service.

they insured. To identify these homes, early insurance companies marked them with a **fire mark**, which is a plaque that displayed the name or logo of the insurance company (**FIGURE 1-5**).

Sometimes more than one fire company showed up to fight a fire. When that happened, a dispute arose over which company would fight the fire and collect the money. Consequently, municipalities began assuming the role of providing fire protection. Today almost all fire protection in the United States is funded directly or indirectly through tax dollars.

FIGURE 1-4 The historic symbol of the officer's trumpet is still used on an officer's badge. This image of crossed trumpets is one of the cherished traditions of the fire service.

Courtesy of Rom Duckworth.

Evolution of Building and Fire Codes

Throughout history, the threat of fire, not to mention the experience of devastating fires, have driven communities to establish fire and building codes. A **code** is a set of legally adopted rules that specify the minimum requirements and standard practices in an industry or profession. A **fire code** does the following:

- Specifies practices and procedures to prevent fires from occurring
- Requires structures to be constructed in a way that will help prevent fires from spreading by both suppressing fire (for example, by requiring automatic sprinkler systems) and preventing it from spreading (for example, requiring doors that unlock when alarms sound or doors that swing shut to contain fire to a particular area when alarms sound)
- Protects lives in the event of a fire by specifying how occupants will be evacuated

FIGURE 1-5 Fire marks were originally symbols affixed to the front of a building designating the insurance company responsible for covering that fire.

FIGURE 1-6 The Great Chicago Fire of 1871 caused the deaths of 300 people.

A **building code** specifies how structures are designed, constructed, and remodeled so that buildings are safe for people who live and work in them. Building codes reference fire codes and require that buildings meet the established fire code for the municipality.

In 1871, two major fires significantly affected the development of both the fire service and fire codes. At the time, the city of Chicago was a "boom town" with 60,000 buildings. Of these buildings, 40,000 were constructed of wood and had roofs of tar and felt or wooden shingles. There were few regulations for construction, and the rules that existed were not enforced. On October 8, 1871, there had been no rain for 3 weeks, which primed the city for a major fire. On that day, a fire started in a barn on the west side of the city. Initial communication and coordination failures resulted in a delayed response time. In addition, the fire department was exhausted from fighting a four-block fire earlier that day. The Great Chicago Fire burned the city for 3 days (**FIGURE 1-6**). When it was over, more than 2000 acres and more than 17,000 homes had been destroyed, the city had suffered more than $200 million in damage, 300 people were dead, and more than 100,000 were homeless (Smith 2020, 22).

At the same time, another major fire was raging north of Chicago in Peshtigo, Wisconsin. This fire was not as highly publicized as the Great Chicago Fire even though it was more deadly. That summer, the north woods of Wisconsin had experienced drought-like conditions. In addition, logging operations had left pine branches carpeting the forest floor. A flash forest fire created a "tornado of fire." Approximately 2500 people lost their lives and 1.2 million acres of forest land burned.

As a result of the Great Chicago Fire and the Peshtigo Fire, communities in the United States began to enact stricter building and fire codes. The development of water pumping systems, advances in firefighting equipment, and improvements in communications and alarm systems helped ensure that such tragedies did not reoccur.

Several other significant fires in the United States contributed to the current building and fire codes. Five weeks after the doors opened, on December 30, 1903, the Iroquois Theater in Chicago, Illinois, became a fire deathtrap for 602 people, mostly women and children (NFPA 2008). The fire ranks as America's worst single building fire. Lack of preplanning, sprinklers, fire alarms, and outward opening exits as well as hidden exits contributed to the problem. This fire resulted in important changes for other theaters, including mandatory upgrades in theaters to include exit lighting; automatic sprinklers; standpipes; fire alarms; and flame-resistant scenery, props, and curtains.

The Triangle Shirtwaist Factory Fire in Manhattan, New York, in March 1911 took the lives of 146 individuals (NFPA 2008). This landmark fire prompted the NFPA to create NFPA 101, *Life Safety Code*, which is still in use today. Other code changes brought about by this fire included the following requirements:

- Stairways in buildings must be enclosed.
- Elevators must have fireproof shafts.
- In buildings where more than 25 people work above ground level, sprinkler systems must be installed.
- Posted maps showing escape routes in case of fire.
- Doors must be outward opening and remain unlocked during business hours.

In 1958, the Our Lady of the Angels School Fire in Chicago, Illinois, claimed the lives of 95 people, 92 of whom were children. The school did not have any fire detection equipment, so when the fire broke out, it went unnoticed for several minutes. Once it was noticed, there was a delay in notifying the fire department, and the fire alarms were not connected to the fire department. The building had only one fire escape, and the second-floor windows were 25 feet from the ground, making it difficult for children to escape from the second story. Fire extinguishers were out of the reach of both the children and most adults (OLA Memorial n.d.). After this fire, the changes to the codes and practices surrounding school safety were substantial and included the following requirements:

- Schools must be equipped with automatic building alarm systems that are connected to the fire department.
- Schoolroom occupancy must be limited to the number allowed by the fire code based on the size of the room.
- School personnel must be trained in fire safety.

By 1977, an adequate number of exits were required at clubs, flammable materials were prohibited, and fire doors on stairwells were required. But time had passed since the last catastrophic fire, and fire prevention took a back seat to unsafe nightclub conditions. Too many people, lack of an evacuation plan, code violations, and poor inspection efforts led to The Beverly Hills Supper Club Fire in Southgate, Kentucky, in which 165 people lost their lives (NFPA 2008). This fire triggered requirements for automatic sprinklers and fire alarm systems in new and existing buildings when the number of people allowed in the building is greater than 300.

In 2003, The Station nightclub fire in Warwick, Rhode Island, took the lives of 100 people (NFPA 2008). The band performing on stage included pyrotechnics as part of the show. The walls of the nightclub were lined with foam intended to soundproof the building, but the foam was not approved for that use and was highly flammable. When the pyrotechnics ignited the walls, overcrowded conditions, an inadequate number of and poorly marked or obscured exits, and an untrained staff set the stage for the disaster. In the aftermath of this fire, fire codes were changed to include the following:

- Prohibit unreserved seating in buildings that accommodated more than 250 people without a special evaluation by the fire marshal
- Provide crowd managers when there are more than 250 people in a building
- Require that automatic sprinklers be installed in bars, dance halls, discotheques, nightclubs, and other similar buildings that can accommodate more than 100 people
- Require that automatic sprinklers be installed in all new bars, dance halls, discotheques, nightclubs, and assembly occupancies with unreserved seating
- Conduct an inspection of the means of egress (exit) before opening buildings to the public, and maintain records of inspections

As the country grew, each community established its own rules for building safe buildings with an eye toward fire prevention. As larger government jurisdictions were established, however, a more uniform code was adopted across the United States, reflecting a minimum standard. Local communities were permitted to make this code stricter as needed.

Today, U.S. building and fire codes are developed by the **National Fire Protection Association (NFPA)**—a nonprofit association that develops and maintains nationally recognized fire codes and standards for fire safety and handling hazardous materials—and by the **International Code Council (ICC)**—an international association that creates standards and codes for buildings. A **standard** describes how tasks should be done, what equipment should be used, and how that equipment should be used to accomplish the tasks. For example, NFPA 1550, *Standard for Emergency Responder Health and Safety*, is a standard that "specifies the minimum requirements for an occupational safety and health program for fire departments or organizations that provide rescue, fire suppression, emergency medical services, hazardous materials mitigation, special operations, and other emergency services" (NFPA n.d.). The ICC, also a nonprofit association, develops and maintains building and other public safety codes and standards. Both associations develop their codes and standards by consensus among their members. This means the members of the organization contribute to writing the standards and vote on whether to accept the final standard.

Standards are not laws. They are developed by nongovernmental agencies and are voluntary. Basically, this means they are recommendations that can be adopted into law in a jurisdiction. A **regulation**, on the other hand, is developed by government or government-authorized organizations to implement a law. However, regulations often refer to the standards. For example, a regulation might include something like the following: "It is required that new buildings include sprinkler systems as specified in the NFPA 13, *Standard for the Installation of Sprinkler Systems*." The **authority having jurisdiction (AHJ)** is the person or organization who is responsible for ensuring that the requirements of a standard adopted into law by a jurisdiction are upheld. The AHJ is who decides whether to adopt a standard as a regulation.

It is important to understand that standards are designed to help you become a more efficient, competent, and safe firefighter. You will see many NFPA standards referenced throughout this course. The NFPA provides free online access to these standards at www.nfpa.org.

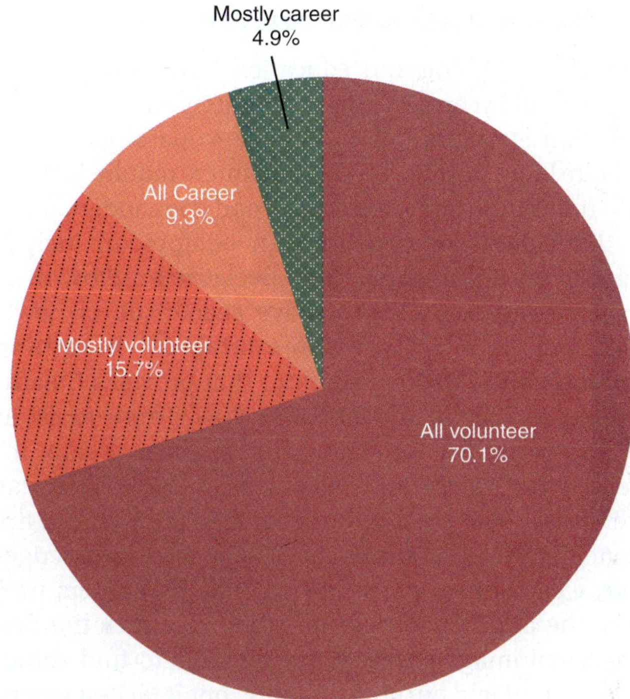

FIGURE 1-7 The majority of fire departments consist of firefighters who are volunteers.

Data from U.S. Fire Administration. 2023. "National Fire Department Registry Quick Facts." Accessed April 15, 2023. https://apps.usfa.fema.gov/registry/summary.

Fire Service in the United States Today

Today, the fire service in the United States is the product of evolution occurring over the past 400 years. As a beginning firefighter, it is helpful for you to learn from the past as well as to study the modern-day U.S. fire service.

According to the NFPA, there were 1,041,200 firefighters in the United States in 2020. Of this number, approximately 35 percent are full-time, career firefighters, and 65 percent are volunteers. Most career firefighters (71 percent) work in communities with populations of 25,000 or more. Most volunteer firefighters (95 percent) work in fire departments that protect fewer than 25,000 people (Fahy, Evarts, and Stein 2021, 1–2).

The firefighters served in approximately 30,000 fire departments throughout the United States. These fire departments responded to 36,416,000 calls for help (Fahy, Evarts, and Stein 2021, 4–5). Departments are categorized based on whether the firefighters are career, volunteer, or a combination of the two (**FIGURE 1-7**):

- All career fire departments: A career fire department is one in which all members are paid, full-time firefighters.
- All volunteer fire departments: A volunteer fire department is one in which all members are volunteer firefighters.
- Combination fire departments: These departments include both paid, full-time firefighters and either on-call firefighters or volunteers:
  - Mostly career fire departments: Between 51 percent and 99 percent of the members are paid, full-time firefighters.
  - Mostly volunteer fire departments: Between 51 percent and 99 percent of the members are volunteer firefighters.

In 2021, fire departments in the United States responded to approximately 1.35 million fires. Those fires caused 3800 civilian fire fatalities, 14,700 civilian fire injuries, and $15.9 billion in property damage. Although only 25 percent of the fires were in residences, that 25 percent caused 75 percent of the civilian fire fatalities and 76 percent of the civilian fire injuries (Hall and Evarts 2021, 1).

TIP

Although career and mostly career fire departments make up only 18 percent of the total number of fire departments in the United States, they serve 69 percent of the U.S. population.

Training and Education

Firefighter training and education have come a long way over the years. The first firefighters simply needed sufficient strength and endurance to pass buckets or operate hand pumps. Today, firefighters require formalized training, attention to safety, and good judgment.

Firefighters operate modern equipment, including million-dollar apparatus, specialized radios, thermal imaging devices, protective gear made of newly developed fabrics, and **self-contained breathing apparatus (SCBA)**, a respirator that provides breathable air to the wearer (**FIGURE 1-8**). These tools, as well as better fire detection devices, have greatly increased the safety and effectiveness of modern-day firefighters. The most important "machines" at the fire scene, however, remain the intelligent, knowledgeable, well-trained, physically capable firefighters who have the ability and determination to attack the fire. A thermal imaging device may be able to find someone trapped in a burning building, but it takes a smart, able-bodied firefighter to remove the victim safely.

The world and the science of firefighting is constantly changing. This means that firefighters must continually sharpen their skills and increase their knowledge of potential hazards. The need for ongoing education is why training courses such as this one are just as important as good physical fitness to today's firefighters.

At the end of your training, you will likely have to pass a test or a series of tests to successfully complete your course of study. Certification is available at all levels of the fire and emergency services, from entry-level firefighters through chief-level officers and everything in between. There are currently two national organizations that accredit or recognize emergency service certification systems: the **National Board on Fire Service Professional Qualifications** (**Pro Board**) and the **International Fire Service Accreditation Congress (IFSAC)**. These organizations establish the criteria used to ensure that testing does the following:

- Aligns with NFPA professional qualifications standards
- Uses valid and reliable testing methodologies
- Provides a mechanism by which to certify those that successfully complete the testing

FIGURE 1-8 A thermal imaging device is one of many tools available to modern-day firefighters.

SAFETY TIP

Firefighting is a dangerous activity. Firefighters work in an environment that is subject to unscheduled and unexpected emergency events. Always remember that your safety and mental well-being are directly related to your personal health and physical fitness. For more information about safety and physical and behavioral health, see Chapter 2, *Firefighter Health and Safety*.

Communications

Many of the early communications systems have been replaced by hard-wired and cell phones that enable citizens to report an emergency from almost anywhere. Use of telephones has greatly reduced the time and difficulty of reporting a fire. The introduction of computer-aided dispatch facilities has likewise improved response times because the closest available fire units can quickly be sent to the site of the emergency. Fire departments that do not have a dispatch center rely on pagers or two-way radios to summon part-time or volunteer firefighters to emergencies.

Communication at the fire scene also has improved over the years, from simply shouting loud enough to be heard over the chaos to using two-way radios. Today's two-way radios enable fire units and individual firefighters to remain in contact with one another at all times. In addition, voice amplification systems, SCBA alarms, and **personal alert safety system (PASS)** devices—devices worn by firefighters that sound an alarm if the firefighter activates it or if the device does not detect motion for 30 seconds—communicate potential safety issues to the firefighter and to the crew.

Source of Authority

Governments—whether municipal, state, provincial, or national—are charged with protecting the welfare of the public against common threats. Citizens accept certain restrictions on their behavior and pay taxes to protect themselves and the common good. Fire is one such peril. An uncontrolled fire threatens everyone in the community.

The fire service draws its authority from the governing entity responsible for protecting the public from fire, whether it is a town, city, county, township, or special fire district. Federal and state governments also grant authority to fire departments. For example, firefighters can legally enter a locked home without permission to extinguish a fire and protect the public. In some states, private corporations have contracts to provide fire protection to municipalities or government agencies. The head of the fire department (the fire chief) is accountable to the elected leaders of the governing body, such as the city council, the county commission, the mayor, or the city manager. Because of the relationship between fire departments and local government, firefighters are often (but not always) civil servants, working for the citizens of their community.

Culture of the Fire Service and Community Risk Reduction

The culture of any organization influences the behavior of its members. Culture is the collection of expectations, values, and practices that guide the actions of the members of an organization. In short, it is the personality of the organization.

Every fire department has its own culture. Often, each group within a department may even have its own sub-culture, a way they behave that is somewhat different than other groups in the department. The culture of the fire service is known for the following characteristics:

- Duty: Firefighters accept the obligation to make performing the requirements of their mission the top priority in all that they do.
- Honor: Firefighters hold a special, trustworthy position in their communities. Firefighters should adhere to the highest standard of moral and ethical conduct to maintain that trust.
- Service: An inherent part of being a firefighter is to put the needs of others above your own needs.
- Safety: To be a firefighter is to accept an element of danger in your work. However, taking unnecessary risks endangers the firefighting operation, yourself, your firefighting family, and the public.
- Excellence: Firefighters must strive to be more than "good enough." Firefighters demand excellence of themselves and those around them. They are prepared to put in the work to achieve that excellence.

Recently, there has been a significant shift in the culture of the fire service. In the past, the fire service considered fire suppression to be their primary responsibility. Over the past several years, members of the fire service have realized that fire prevention is just as important. If fires can be prevented, fewer lives will be lost, fewer people and firefighters will be injured, and less property will be destroyed. **Community risk reduction (CRR)** is a comprehensive approach to reduce the overall incidence and impact of emergencies within a community by identifying the potential risks and creating plans and procedures to mitigate those risks. This differs from traditional ideas of fire prevention because it is not a "one size fits all" approach. Instead, each department takes a systematic approach to identify and prioritize risks in their community, and then plan methods to mitigate those risks.

The first CRR program was introduced in Merseyside, England. As a result of this program, fire fatalities were significantly reduced. Using that program as a model, the fire service in North America has made CRR an integral part of their mission. This is largely due to the work of the Vision 20/20 project, an initiative that brought together experts to develop a national plan to coordinate fire prevention activities and efforts. Departments use CRR to identify the most dangerous risks and the risks that are most likely to occur in their communities. After identifying these risks, departments create a plan to remove or mitigate them. The basic strategies of fire prevention are summarized by the 5 Es: education, enforcement, engineering, economic incentives, and emergency response. These originally were known as the **5 Es of fire prevention**, but as CRR has gained more prominence, they are often called the **5 Es of community risk reduction (CRR)**.

- Education: Change behavior by teaching people about fire and emergency prevention and response.
- Engineering: Use technology, such as smoke alarms, sprinkler systems, child-resistant medication caps, and car seats to make buildings, cars, and products safer.
- Enforcement: Enforce fire and building codes to create and maintain the safety of, and reduce risk in, the buildings in the community.

- Economic incentives: Provide financial motivation to encourage beneficial behaviors and choices by developers; for example, a community could provide tax credits to a developer who installs sprinkler systems or fine a developer who does not install or removes smoke alarms.
- Emergency response: Develop a well-trained and prepared emergency response force that is responsive to the unique aspects of the community.

Fire Department Governance

Governance is the framework and procedures for managing and operating an organization. The governance of a fire department depends on its policies and standard operating procedures (SOPs). Fire department **policies** outline expectations for firefighter performance and procedures in different circumstances. These policies often require personnel to make judgments to determine the best course of action within the stated policy. Policies governing parts of a fire department's operations may be developed by other government agencies, such as personnel policies that cover all employees of a city or county.

Standard operating procedures (SOPs) provide specific information describing the actions that should be taken to accomplish tasks within the department or at an emergency scene (**FIGURE 1-9**). An example would be your policy and procedure manuals. SOPs are developed within the fire department, are approved by the chief of the department, and ensure that all members of the department perform a given task in the same manner. They provide a uniform way of dealing with emergency situations, enabling firefighters from different stations or companies to work together smoothly, even if they have never encountered one another before. These procedures are vital because they enable everyone in the department to work together and know what is expected for each task. Firefighters must learn and frequently review departmental SOPs.

In some fire departments, **standard operating guidelines (SOGs)** are utilized. SOGs are similar to SOPs; however, SOGs may vary due to circumstances surrounding a particular incident. SOGs are not as strict as SOPs, because conditions may dictate that the firefighter or officer uses their personal judgment in completing the procedure. This flexibility allows the responder to deviate from a set procedure yet still be held accountable for that action.

Each department member should have access to up-to-date department policies and a physical or digital up-to-date SOP or SOG manual. The SOP or SOG manual should be organized in sections, such as administration, safety, scene operations, apparatus and equipment, station duties, uniforms, and miscellaneous. Many fire departments maintain their SOPs on a computer network, which simplifies the process of providing all employees with up-to-date SOPs. One of the first things you should do as a new recruit in a fire department is ask where the policies and SOP or SOG manual is located.

Fire Apparatus

In the fire service, the term *equipment* generally refers to the tools a firefighter uses to fight fires and rescue victims, protective clothing, and devices firefighters use to help them breathe. The term **fire apparatus**, often shortened to just **apparatus**, are the vehicles firefighters drive to an emergency scene or other incident and then use to help fight the fire. (The **emergency scene**, sometimes referred to as simply **on-scene**, is the location of an emergency incident and the surrounding area where responders set up and where apparatus are located. When the incident is a fire, this is also called the **fireground** or **fire scene**.) Modern apparatus carry water; a pumping mechanism; hose; equipment such as the tools a firefighter uses to fight fires and rescue victims, protective clothing, and devices that help them breathe in hazardous environments; as well as personnel. In this sense, a single apparatus has replaced several single-function vehicles from the past. Apparatus continue to evolve as new inventions are adapted to the needs of the fire service.

The two main types of apparatus are pumpers and aerial fire apparatus. A **pumper**, also referred to as an **engine**, is the apparatus that has a pump, carries hose, and maintains a booster tank of water (**FIGURE 1-10**). Pumpers also carry a limited quantity of ladders and hand tools.

An aerial fire apparatus, also referred to as a **truck**, **ladder truck**, **aerial apparatus** or just **aerial**, or **tower ladder truck**, is the apparatus that carries several ground ladders, ranging from 8 feet to 50 feet (2.4 to 50 m) in length, as well as an extensive quantity of tools. Trucks are also equipped with an aerial device, such as an aerial ladder, a tower ladder, or an elevating platform (**FIGURE 1-11**). These aerial devices can be raised and positioned above a roof to provide firefighters with a stable, safe work zone.

In addition to pumpers and trucks, most fire departments have additional specialized apparatus, including the following:

- **Quint apparatus**: Often shortened to simply **quint**. "Quint" is short for "quintuple," meaning

Anytown Fire Department

Standard Operating Procedure
Date: 1/1/24
Section: Administration SOP 01-01
Maintaining Station Logbooks

Page 1 of 1

Purpose
This guideline is provided to ensure that information documented in station logbooks is maintained in a consistent manner, station to station, shift to shift, throughout the department. The station commander shall have discretion in formatting the information.

Scope
This guideline shall be followed by all authorized department personnel entering information into the station logbooks. Any deviations from this guideline will be the responsibility of the individual making the deviation.

Policy
The station log is an important component of the fire department's record-keeping system. All entries must be legible, written in black ink, and the writer identified by name and computer identification number. All members should treat the logbook as a legal document and record all of the station's activities as well as other information deemed important by the station commander.

Procedure
The following guidelines should be used when making logbook notations:

1. The day, date, and shift noted on the top line of the page.
2. A list of each person assigned to that shift to include their computer identification numbers, and the apparatus to which they are assigned for the shift. Personnel on leave should be identified with the type of leave (ie, annual, sick, leave without pay, funeral, military, training) noted. A notation should be made by the name of any member temporarily assigned to another station during that shift.
3. Daily safety talks should be recorded prior to the section dedicated to emergency responses. For more information on safety talks, refer to SOP 31.3
4. The section of the log dedicated to recording chronological activity should have two (2) columns on the side of the book: the far left should note time and the second column the incident number. A column to the far right should be used to note the identification number of the person making the entry.
5. Incident numbers for medic unit responses should be noted in blue ink.
6. Incident numbers for fire suppression responses should be noted in red ink.
7. Units responding, the address to which the response is being made, and the type of situation found should be noted behind the message number (ie, E-5 responded to 304 Albemarle Drive/Gas Leak).
8. The term fill-in should be used in the logbook to denote one station standing by another station to cover a company that has responded to an incident.
9. Entries for medic units should include the hospital to which the patient was transported.
10. Station maintenance/repairs and apparatus maintenance should be documented in the section of the logbook dedicated to that purpose. This documentation may also be maintained in a separate logbook dedicated to those types of entries.

Although the information requirements provided here must be maintained in the station's logbook, the format used for ensuring that documentation is left to the station commander's discretion.

Approved: ______________________
Fire Chief

Date: 1/1/24

FIGURE 1-9 A sample standard operating procedure.

five. The quint apparatus has five components: a pump, a water tank, a fire hose storage area, an aerial device, and ground ladders.

- **Initial attack apparatus**: The primary purpose of the initial attack apparatus is to initiate a fire suppression attack on structural, vehicular, or vegetation fires. It is sometimes referred to as a **quick attack apparatus**.
- **Mobile water supply apparatus**: The mobile water supply apparatus is designed to transport water to emergency scenes. It may or may not have a pump. This apparatus is also referred to as a **tanker** or **water tender**, or just **tender**.
- **Rescue apparatus**: Rescue apparatus carry an extensive array of regular and specialized tools and equipment that are used to rescue victims. Most of this equipment is too heavy to carry on pumpers.
- **Ambulance** or first response vehicle: An EMS vehicle carries medical supplies such as medications, defibrillators, and other equipment that is used to stabilize a critical patient prior to and during transport to a hospital (**FIGURE 1-12**).

FIGURE 1-10 A pumper brings water, a pump, and hose to a fire scene.

Courtesy of Rom Duckworth.

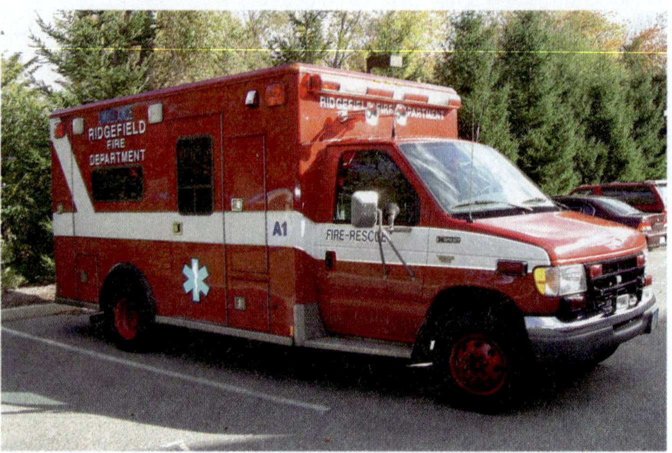

FIGURE 1-12 An EMS company delivers medical services.

Courtesy of Rom Duckworth.

FIGURE 1-11 An aerial fire apparatus carries ground ladders and an aerial device.

- Community health vehicle: Some departments use a community health vehicle to provide medical care to citizens without bringing them to a hospital.
- **Wildland apparatus**: A four-wheel drive vehicle that is used to transport firefighters closer to wildfires over rough, uneven terrain. It often carries a tank of water and a pump that enables them to pump water while the truck is moving. It also carries special firefighting equipment such as portable pumps, rakes, shovels, and chainsaws.

Fire Department Organization

A **fire department** is the organization that fights fires and promotes fire prevention in a municipality. A **fire station** or **fire house** is a building that houses fire apparatus and equipment for a geographic area within a fire department. Departments in larger jurisdictions have multiple stations.

Departments in very large cities are further subdivided. The names of the subdivision vary from one department to the next. For example, in Boston, Massachusetts, the department is divided into two divisions, and these divisions are divided into districts, each of which includes several fire stations. In New York City, a much larger municipality, the department is divided into boroughs, which are divided into divisions, which are divided into battalions consisting of several fire stations each.

Companies

Fire stations are often made up of smaller units called companies, which are groups of people assigned to a specific apparatus who have specific responsibilities at an emergency scene under the command of an officer. A **company** may be composed of various combinations of people and equipment; in smaller departments, one company may fill many roles. The most common types of companies are the following:

- **Engine company**: An engine company rides on the pumper and is responsible for securing a water source, unloading hoses that are held by hand, conducting search and rescue operations, and putting water on the fire.
- **Truck company**: A truck company, also referred to as a **ladder company** or a **tower ladder company**, rides on the aerial fire apparatus and specializes in forcible entry, ventilation, roof operations, search and rescue, and deployment of ground ladders.
- **Rescue company**: A rescue company usually is responsible for rescuing victims from fires, confined spaces, trenches, and high-angle situations.

- **Wildland company**: A wildland company, also called a **brush company**, is dispatched to vegetation fires.
- **Hazardous materials company**: A hazardous materials company responds to and controls scenes involving spilled or leaking hazardous chemicals. These companies have special equipment, personal protective equipment (PPE), and training to handle most emergencies involving chemicals.
- **Emergency medical services (EMS) company**: Sometimes referred to as an **emergency medical services (EMS) squad**, or just **squad**, an EMS company rides in an ambulance or first-response vehicle. These companies treat medical and trauma victims so they are stable enough to transport to medical facilities for further treatment. Engine or truck companies may be staffed with EMS providers who provide medical care until an ambulance arrives.

The terms **crew** and **unit** are often used in casual conversation as synonyms for company. They can also indicate any small collection of firefighters doing a similar job.

Roles in the Fire Department

There are many roles in fire departments. Not every department has someone in every position. No matter what your position is, every member of the fire service will interact with the public. In addition to interacting with citizens at the scene of incidents, firefighters encounter people who visit the fire station requesting a tour or asking questions about specific fire safety issues (**FIGURE 1-13**). Firefighters should be prepared to assist these visitors and use this opportunity to provide them with additional fire safety information. Keep in mind that interaction with the public occurs both on and off duty. Use every instance to deliver positive public relations and send educational messages.

FIGURE 1-13 Any contact with the public should be used as an educational opportunity.

Courtesy of Rom Duckworth.

Roles in the fire department can be sorted into two categories: general and specialized response. Although every department is different, generally firefighters seeking specialty roles in a department do not need to do this in a specific order. In other words, a firefighter does not necessarily need to be a driver/operator of an engine before they become a driver/operator of an aerial apparatus or a fire inspector. In fact, a driver/operator of an engine does not necessarily need to be a firefighter.

General Roles

General roles do not require advanced, specialized training. Common positions include the following:

- **Firefighter**: A firefighter may be assigned any task from routine cleaning and maintenance duties, placing hose lines to extinguishing fires, to assisting with a public fire prevention program. Generally, firefighters are not responsible for command functions and do not supervise other personnel, except on a temporary basis as a senior firefighter or an acting officer. A firefighter may fulfill many roles during their career through additional training, testing, and promotion.
- **Support person**: A support persons is a fire department member who is not trained to the Firefighter I level, but assists members of a fire department by performing duties in environments that are not hazardous. These duties can include assisting with communications, identifying hazardous environments, connecting a pumper to a water supply, opening and closing fire hydrants, operating emergency scene lighting, refilling SCBA air cylinders, and cleaning and checking equipment and tools.
- **Driver/operator**: Often called an **engineer** or a **technician**, the driver/operator is responsible for driving the fire apparatus to the scene safely, and then setting up and running the pump on the engine or operating the aerial device on the scene. In some departments, this function is a full-time role; in other departments, it is rotated among members.
- **Company officer**: The company officer is usually a lieutenant or captain who oversees the crew of an apparatus. This person leads the company both on-scene and at the station. The company officer is responsible for initial firefighting strategy, personnel safety, and the overall activities of the firefighters on their apparatus. Once command is established, the company officer focuses on tactics.

- **Incident safety officer (ISO)**: The ISO observes the overall operation for unsafe practices. The ISO has the authority to halt any firefighting activity, allowing the activity to resume only when it can be done safely and correctly. The senior ranking officer may act as the ISO until the appointed ISO arrives or until another officer is delegated those duties.
- **Training officer**: The training officer is responsible for updating the training of current firefighters and for training new firefighters. They must be aware of the most current techniques of firefighting, EMS, and other specialized services provided by the fire department. The training officer coordinates various aspects of the fire department training program and maintain documentation of all training activities.
- **Fire marshal**: The fire marshal delivers, manages, and/or administers fire protection and life safety-related codes and standards, investigations, education, and/or prevention services.
- **Fire inspector**: The fire inspector inspects businesses and enforces public safety laws and fire codes.
- **Fire investigator**: The fire investigator responds to fire scenes to help investigate the cause of a fire. The fire investigator may have full police powers to investigate and arrest suspected arsonists and people causing false alarms.
- **Fire and life safety educator (FLSE)**: The FLSE educates the public about fire safety and injury prevention and presents juvenile fire safety programs. The FLSE and the public information officer (PIO) may be part of a larger community risk reduction program within the department.
- **Telecommunicator**: Also called a **dispatcher**, the telecommunicator takes calls from the public, sends appropriate units to the scene, assists callers with emergency medical information (treatment that can be performed until the medical unit arrives), and assists the incident commander (IC) with obtaining needed resources from the communications center.
- **Emergency vehicle technician (EVT)**: An EVT repairs and services fire and EMS vehicles, keeping them ready to respond to emergencies. These individuals are usually trained by equipment manufacturers to repair vehicle engines, lights, and all parts of the fire pump and aerial ladders.
- **Fire police officer**: Fire police officers are usually firefighters who control traffic and secure the scene from public access. Many fire police officers are also sworn peace officers.
- **Public information officer (PIO)**: The PIO serves as a liaison between the IC and the news media.
- **Fire protection engineer**: The fire protection engineer reviews building plans and works with building owners to ensure that their fire suppression and detection systems meet the applicable codes and function as needed. Some fire protection engineers design these systems, and most have a degree in fire engineering.

Specialized Response Roles

Many assignments require specialized training. Most large departments have teams of firefighters who respond to specific types of calls. Members of these teams are usually required to be firefighters before they begin additional training. Specialist positions include the following:

- **Emergency medical services (EMS) personnel**: EMS personnel administer care to people who are sick or injured at the scene and during transport to a medical facility. Calls for this type of service account for the majority of responses in many departments today, so EMS personnel are often firefighters with additional training. EMS training levels are normally divided into the following four categories:
  - **Emergency Medical Responder (EMR)**: EMRs have basic training for providing initial medical assistance. They often perform in an assistant role in the ambulance and have training in bleeding control and cardiopulmonary resuscitation (CPR). Police officers, firefighters, lifeguards, or other rescuers who are often the first to arrive at the scene of an emergency are often trained as EMRs.
  - **Emergency Medical Technician (EMT)**: Most EMS providers are EMTs. They have training in basic emergency care skills, including bleeding control, CPR, using automated external defibrillators, oxygen therapy, using basic airway devices, and assisting patients with certain medications.
  - **Advanced Emergency Medical Technician (AEMT)**: AEMTs have advanced training and can often perform more procedures than EMTs. They have training in specific aspects of advanced life support, such as intravenous (IV) therapy, interpretation of cardiac rhythms, and advanced airway management.
  - **Paramedic**: A paramedic has completed the highest level of training in EMS. They have extensive training in advanced life support, administering drugs, cardiac monitoring, inserting

advanced airways, manual defibrillation, and other advanced assessment and treatment skills.

- **Technical rescuer**: Also called **rescue technician**, these firefighters are trained in special rescue techniques for incidents involving structural collapse, trench rescue, swift water rescue, confined-space rescue, high-angle rescue, and other unusual situations. The units they work in are sometimes called urban search and rescue teams.
- **Hazardous materials technician**: Hazardous materials technicians have training and certification in chemical identification, leak control, decontamination, and clean-up procedures.
- **Airport firefighter**: Airport firefighters are based at airports. They receive specialized training in extinguishing aircraft fires and extricating victims from an aircraft. They wear special PPE and respond in specialized fire apparatus that protects them from high-temperature fires fueled by substances such as jet fuel.

TIP

The NFPA reported that in 2020, 63 percent of the fire departments in the United States provided emergency medical services (Fahy, Evarts, and Stein 2020, 1).

Chain of Command

The fire department uses a paramilitary style of leadership. Firefighters operate under a rank system, which establishes the chain of command. The **chain of command** is a rank-based, hierarchical structure that creates an orderly line of authority (**FIGURE 1-14**). Although the precise ranks may vary from department to department, the basic concept remains the same across the fire service. The chain of command creates a structure for managing the department and the fireground operations. Larger departments have more levels in their chain of command than smaller departments.

Firefighters usually report to a **lieutenant**, who is responsible for a single fire company (such as an engine company) on a shift. The next level in the chain of command is the captain. A **captain** is responsible not only for managing a fire company on their shift but also for coordinating the company's activities with other shifts. A captain may also be in charge of the activities of a fire station. Both captains and lieutenants are company officers.

Captains report to chief officers. A **battalion chief** is the first level of chief officer. They are typically

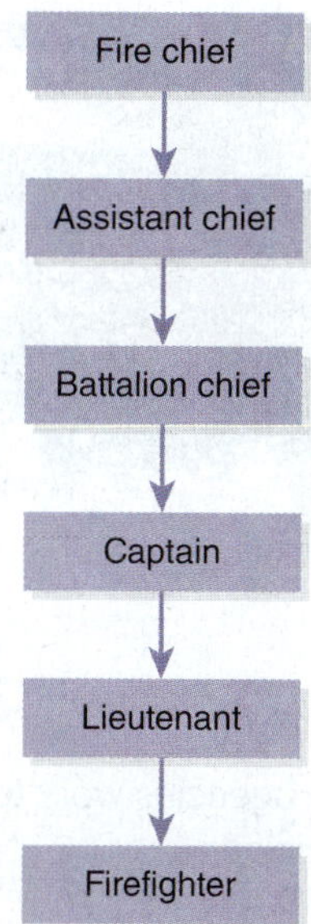

FIGURE 1-14 The chain of command ensures that the department's mission is carried out effectively and efficiently.

responsible for coordinating the activities of several fire companies in a defined geographic area such as a district or a division. A battalion chief is usually the officer in charge of a single-alarm working fire.

Above the battalion chief is the **assistant chief**, **deputy chief**, or **division chief**. Chiefs at this level are usually in charge of a functional area, such as training, within the department or in command of a group of battalions or districts. These officers report directly to the chief of the department.

The **fire chief** is the highest-ranking chief officer. The fire chief is in charge of the entire department and has overall responsibility for the administration and operations of the fire department. The chief can delegate responsibilities to other members of the department but is still responsible for ensuring that these activities are carried out properly. The chief usually answers directly to the mayor, town manager, or other designated public official.

Although firefighters often do not need to advance in a specific order to be assigned to a different company or specialty role in a department, to advance in rank, promotion is typically needed. For example, a firefighter must first be promoted to lieutenant before becoming eligible for promotion to captain or battalion chief.

Working with Other Organizations

Modern fire departments do not respond only to fires. They also respond to motor vehicle collisions, elevator

FIGURE 1-15 Many agencies work together at the scene of a motor vehicle collision.

© Patrick Kane/The Progress-Index/AP Photo

emergencies, victims trapped in a collapsed building, accidents in manufacturing plants, and more. Consider the response to a motor vehicle collision (**FIGURE 1-15**). In an area with a centralized 911 call center, the fire department, EMS, law enforcement officials, and tow-truck operators are all notified of this event. If EMS is not part of the fire department, a separate EMS provider is contacted. If the crash affects community infrastructure, the fire department may interact with the local environmental protection agency, utility company, highway department, building department, or other related agencies. If the emergency is large-scale, the fire department may work alongside local, regional, or state emergency management personnel. If the fire department responds to a school, firefighters will interact with school staff. Firefighters also frequently interact with hospital personnel who assume care for the ill or injured victims who are transported to emergency departments.

When only one agency is involved in an incident, a single **incident commander (IC)** is responsible for all aspects of managing the incident. All personnel, apparatus, and resources work under the authority and direction of the IC. When multiple agencies work together at an incident, a unified command must be established as part of the Incident Command System (ICS). A **unified command** creates a single set of incident goals and objectives and fosters mutual communication and cooperation among agencies. When a unified command is established, each responding agency assigns one or more representatives to the unified command and these representatives have input on how the incident will be handled instead of having a single IC manage the incident. Even when a formal unified command is not established, emergency and non-emergency personnel from different agencies must work together in a coordinated manner.

Nongovernmental organizations may also be involved in incidents. For example, the Salvation Army and the Red Cross may be part of the team responding to a structure fire to assist with care for non-injured victims or to provide fluids, food, or medical care for firefighters. Wildland fires may involve representatives from various government agencies and jurisdictions.

Firefighter Qualifications

Someone who applies to become a firefighter is called a **candidate**. If their application is accepted, they become a **recruit**. Not every candidate will become a recruit. To achieve this accomplishment, a candidate must be in good physical condition and well-trained. Candidates must also have a positive attitude towards the community they will serve as well as the teammates with whom they will work. A successful firefighter is a person who has the desire to learn, the will to practice, and the ability to apply the proper knowledge, skills, and attitudes (**FIGURE 1-16**). A firefighter must also constantly learn to stay educated about the expanding body of knowledge about fires and the technology used to fight them.

Each department establishes their own age, background, education, medical, physical fitness, and emergency medical care requirements.

Age and Background Requirements

Most career fire departments require that candidates be at least 18 years old, although some require a minimum age of 21. Candidates should have a valid driver's license, have a clean driving record, have no criminal record, and be drug free.

Volunteer fire departments often have different age requirements for each position within the department. Some volunteer fire departments allow people younger than 18 to apply to join as junior members but restrict their activities until they reach age 18.

Education Requirements

Most career fire departments require that candidates have at least a high school diploma or equivalent. Some require candidates to have completed college-level courses in an emergency services field. Virtually all departments require firefighters to take additional training and certification courses for promotion or assignment to specialty teams.

FIGURE 1-16 Firefighters understand the mission of the fire department: to save lives and protect property and the environment through prevention, education, suppression, rescue, and salvage.

Courtesy of Rom Duckworth.

Medical Requirements

Firefighting is both stressful and physically demanding. Before a candidate is accepted as a recruit, they must undergo a medical evaluation before their training begins. This medical evaluation identifies most medical conditions or physical limitations that could increase the risk of injury or illness to the candidate or other firefighters. Part of the medical requirement is to successfully pass a drug screening test. Medical requirements for firefighters are specified in NFPA 1582, *Standard on Comprehensive Occupational Medical Program for Fire Departments.*

Physical Fitness Requirements

Physical fitness requirements ensure that firefighters have the strength, stamina, and flexibility needed to perform the tasks associated with emergency operations. It is important that firefighters maintain a high level of fitness throughout their careers to meet the physical rigors required of firefighters and to help prevent injuries and illness. NFPA 1010, *Standard for Firefighter, Fire Apparatus Driver/Operator, Airport Firefighter, and Marine Firefighting for Land-Based Firefighters Professional Qualifications,* allows individual fire departments to choose the fitness-testing method that will be used for firefighter candidates. Many departments require candidates to complete a standardized physical ability test. These tests are designed to measure the strength and endurance required by active firefighters. One of the most widely used tests is the Candidate Physical Ability Test (CPAT). This testing process was developed by the International Association of Firefighters (IAFF) and the International Association of Fire Chiefs (IAFC). The test consists of the following eight scenarios, which are completed while wearing a 50-pound vest to simulate the weight of equipment worn during a fire (IAFC n.d.):

- Stair Climb: Climbing stairs while carrying an additional 25-pound (11 kg) simulated hose pack
- Ladder Raise and Extension: Placing a ground ladder at the fire scene and extending the ladder to the roof or a window
- Hose Drag: Stretching uncharged hose lines and advancing lines
- Equipment Carry: Removing and carrying equipment from fire apparatus to the fireground
- Forcible Entry: Penetrating a locked door and breaching a wall
- Search: Crawling through dark, unpredictable areas to search for victims
- Rescue Drag: Removing a victim or partner from a fire building
- Ceiling Pull: Locating fire and checking for fire extension

Firefighting candidates have 10 minutes and 20 seconds to complete all eight events. Many fire departments require firefighters to complete this series of events every year to ensure that they maintain the level of physical fitness required for active firefighting.

Emergency Medical Care Requirements

Delivering EMS is an important function of most fire departments. NFPA 1010 allows individual fire departments to specify the level of emergency medical care training required for candidates accepted as recruits. At a minimum, you will need to understand infection control procedures and be able to perform CPR, control bleeding, and manage shock. Many departments require firefighters to become certified as EMS personnel at least at the EMR level.

Roles and Responsibilities of Support Persons, Firefighter I, and Firefighter II

The training and performance qualifications for firefighters are specified by NFPA 1010. This standard defines the parameters of the support persons, Firefighter I, and Firefighter II courses. As stated earlier, support persons are people who are not firefighters but who assist members of a fire department performing duties in environments that are not hazardous. A **Firefighter I** is a person at the first level of progression in becoming a firefighter as defined in Chapter 6 of NFPA 1010. A Firefighter I has the knowledge and skills to function as a member of a firefighting team under direct supervision.

A **Firefighter II** is a person at the second level of progression in becoming a firefighter as defined in Chapter 7 of NFPA 1010. A Firefighter II has a higher level of training that allows them to more fully assist in mitigating emergency situations. For example, the Firefighter II can coordinate and direct other firefighters and assume a greater level of responsibility for incident management. While the Firefighter I must work under direct supervision, the Firefighter II works under general supervision.

The first step in understanding the organization of the fire service is to learn your roles and responsibilities as a Firefighter I or Firefighter II. The roles and responsibilities of Firefighter II are built upon the roles and responsibilities of Firefighter I. You should also understand the role of support persons. The roles and responsibilities of support persons are a subset of the roles and responsibilities of Firefighter I. As you progress through this text, you will learn what to do and how to do it so that you can take your place confidently among the ranks. In some cases, the same or similar roles and responsibilities may appear under Firefighter I and Firefighter II. This enables firefighters to learn the fundamentals of those concepts as a Firefighter I and advance their knowledge, skills, and attitudes as a Firefighter II.

Roles and Responsibilities of Support Persons

- Put on and take off PPE properly.
- Decontaminate PPE in the field and prepare it to be used again.
- Locate information in departmental documents and standard operating procedures.
- Understand and correctly apply appropriate communication protocols.
- Identify hazardous environments that require respiratory protection and avoid those areas.
- Respond on apparatus to an emergency scene.
- Establish and operate safely in emergency work areas.
- Connect a fire department engine to a water supply as a team.
- Extinguish fires in their initial stages using portable fire extinguishers.
- Illuminate an emergency scene.
- Turn off utilities.
- Hoist hand tools using appropriate ropes and knots.
- Refill SCBA cylinders.
- Clean and maintain equipment.
- Operate as part of a team.

Roles and Responsibilities of Firefighter I

- Put on and take off PPE properly.
- Decontaminate PPE in the field and prepare it to be used again.
- Locate information in departmental documents and standard operating procedures.
- Understand and correctly apply appropriate communication protocols.
- Use SCBA.
- Respond on apparatus to an emergency scene.
- Establish and operate safely in emergency work areas.
- Force entry into a structure.
- Exit a hazardous area safely as a team.
- Set up and use ground ladders safely and correctly.
- Attack a passenger vehicle fire, an exterior Class A fire, and an interior structure fire.
- Conduct search and rescue in a structure as a team.
- Ventilate an involved structure as a team.
- Overhaul a fire scene by ensuring a fire is completely extinguished.
- Conserve property with salvage tools and equipment as a team.

- Connect a fire department engine to a water supply as a team.
- Extinguish fires in their initial stages using portable fire extinguishers.
- Illuminate an emergency scene.
- Turn off utilities.
- Hoist hand tools using appropriate ropes and knots.
- Combat ground cover fires as a team.
- Use an air monitor properly.
- Clean and maintain equipment.
- Operate as part of a team.

NFPA 1010 also requires the Firefighter I to meet some of the requirements in NFPA 470, *Hazardous Materials/Weapons of Mass Destruction (WMD) Standard for Responders*. In NFPA 470, the Firefighter I must meet the requirements in Chapter 7, "Professional Qualifications for Hazardous Materials/WMD Operations Level Responders (NFPA 1072)." In addition, in NFPA 470, Chapter 9, "Professional Qualifications for Hazardous Materials/WMD Operations Level Responders Assigned Mission-Specific Responsibilities (NFPA 1072)," the Firefighter I must meet the requirements in in Section 9.2, Personal Protective Equipment, and in Section 9.6, Product Control.

Roles and Responsibilities of Firefighter II

- Perform an analysis of a scene.
- Determine the level of the ICS needed to safely and efficiently mitigate the emergency.
- Arrange and coordinate the ICS until command is transferred.
- Prepare reports.
- Communicate the need for assistance.
- Coordinate an interior attack line team.
- Extinguish an ignitable liquid fire.
- Control a flammable gas cylinder fire.
- Locate the origin area of a fire.
- Protect evidence of fire cause and origin.
- Assess and disentangle victims from motor vehicle accidents.
- Assist special rescue team operations.
- Perform a fire safety survey.
- Present fire safety information.
- Prepare descriptions and drawings of areas in the community to be used by firefighters during an incident.
- Maintain fire power equipment.
- Perform annual service tests on fire hose.

CASE STUDY

You Are the Firefighter CONCLUSION

It is your first day in the station after having completed 4 months of recruit academy. Your new crew welcomes you. As you are shown around the station, the doorbell rings, and you are told your first official duty is to answer it. At the door, you find a woman and her young child. Excitedly, she tells you her child has been begging to meet the firefighters because, "Riley wants to grow up to be a firefighter."

You agree to give a quick tour of the station. You first show them the living quarters and then move to the apparatus floor. The child is full of questions and asks, "Do you go to fires all day long? Do you know how to use all those tools? Do you report to the fire chief? Do you work with police officers?" The mother tells her child to give you a chance to explain before asking any more questions.

1. What are the duties of a firefighter?

Answer: The specific duties of a firefighter will depend on the firefighter's department and their assignment within that department. However, the overall mission of the fire service, and of every firefighter, is to save lives and protect property and the environment through prevention, education, suppression, rescue, and salvage activities. This is typically reflected in the mission statement of the department.

2. **What tools and equipment do you need specialized training to operate?**

   Answer: Beyond the qualifications listed in NFPA 1010 for Firefighter I and Firefighter II, some of the equipment that you need additional training to operate includes: driving/operating pumping, aerial, and airport rescue apparatus; equipment required for special rescues such as rescues from confined spaces, a collapsed structure, or swift water; equipment required for handling hazardous materials; and equipment required to provide medical assistance to people.

3. **What are the roles and responsibilities of a firefighter versus a company officer?**

   Answer: The basic responsibilities of a firefighter are outlined in the Firefighter I and Firefighter II job performance requirements of NFPA 1010. Company officers lead the company both on-scene and at the station. At the emergency scene, the company officer is responsible for establishing the initial firefighting strategy for their company, monitoring the safety of the personnel, and supervising the overall activities of the firefighters on their apparatus.

4. **With what other agencies do firefighters typically interact?**

   Answer: Firefighters typically interact with other public safety agencies including law enforcement, EMS, or emergency management. If an emergency involves parts of the community infrastructure, firefighters may interact with highway departments, environmental protection services, highway department, building department, utility companies, and other related agencies.

WRAP-UP

SUMMARY

KNOWLEDGE OBJECTIVES

- Explain the mission of the fire service. (**NFPA 1010: 6.1.1**, pp. 4–5)
- Understand the history of the fire service and how it affected the modern fire service and the evolution of building and fire codes.
  - Explain the evolution of the methods and tools of firefighting from the time of the Roman Empire to the present. (pp. 5–8, 11–13)
  - Describe the evolution of fire equipment for fire department services. (pp. 7, 14–16)
  - Describe the evolution of communications for fire department services. (pp. 7, 12)
  - Describe the evolution of funding for fire department services. (pp. 7–8, 14)
- Describe the modern fire service.
  - Explain how building codes prevent the loss of life and property. (pp. 8–11)
  - Explain the difference between career, volunteer, and combination fire departments. (**NFPA 1010: 6.1.1**, p. 11)
  - Describe the evolution of training and education for fire department services. (p. 12)
  - Describe methods of communication used in the fire service. (p. 12)
  - Explain where the fire service draws its authority from. (p. 13)
- Understand the culture of the fire service and the recent shift to include community risk reduction.
  - Describe the culture of the fire service. (pp. 13–14)
  - Describe the characteristics of a CRR program. (pp. 13–14)
  - List the five Es of CRR. (pp. 13–14)
- Describe how department policies and standard operating procedures or guidelines apply to firefighters.
  - Explain the concept of governance and describe how policies and standard operating procedures affect it. (**NFPA 1010: 6.1.1**, p. 14)

- Describe how to locate information in departmental documents and standard operating procedures. (**NFPA 1010: 6.1.2**, p. 14)
- Describe the fire department's standard operating procedures (SOPs) and rules and regulations as they apply to the Firefighter I. (**NFPA 1010: 6.1.1**, pp. 22–23)

- Describe types of apparatus, companies, roles, and the command structure within a fire department.
  - Describe the different types of fire apparatus. (pp. 14–16)
  - Describe the organization of the fire service. (**NFPA 1010: 6.1.1**, pp. 16–20)
  - Describe how to organize a fire department in terms of staffing, function, and geography. (**NFPA 1010: 6.1.1**, pp. 16–20)
  - List the different types of fire department companies and describe their functions. (**NFPA 1010: 6.1.1**, pp. 16–17)
  - Describe the common roles of firefighters within the organization of fire department. (**NFPA 1010: 6.1.1**, pp. 17–18)
  - Describe the specialized response roles within the fire department. (**NFPA 1010: 6.1.1**, pp. 18–19)
  - Explain the basic structure of the chain of command within the fire department. (**NFPA 1010: 6.1.1**, p. 19)
  - Describe a situation in which you will interact with other organizations within your community. (**NFPA 1010: 6.1.1**, pp. 19–20)
- Describe the qualifications needed to become a firefighter and the roles and responsibilities of support persons and firefighters.
  - Describe the general qualifications for becoming a firefighter. (pp. 20–21)
  - Outline the roles and responsibilities of support persons. (pp. 22–23)
  - Outline the roles and responsibilities of a Firefighter I. (**NFPA 1010: 6.1.1**, pp. 22–23)
  - Outline the roles and responsibilities of Firefighter II. (p. 23)

SKILLS OBJECTIVES

- Locate information that identifies current best practices.
  - Know how to access specific information in fire service code and standards documents. (**NFPA 1010: 6.1.2**, pp. 10–11)
  - Locate information in departmental documents and standard operating procedures. (**NFPA 1010: 6.1.2**, pp. 14–15)

KEY TERMS

5 Es of community risk reduction (CRR) Education, enforcement, engineering, economic incentives, and emergency response. Also called *5 Es of prevention.*

5 Es of prevention See *5 Es of community risk reduction (CRR).*

Advanced Emergency Medical Technician (AEMT) Emergency medical services (EMS) personnel who can do everything an EMT can do, and who have advanced training in specific areas of advanced life support, including IV therapy, interpretation of cardiac rhythms, and advanced airway management.

aerial See *aerial fire apparatus.*

aerial apparatus See *aerial fire apparatus.*

aerial fire apparatus A vehicle equipped with an aerial ladder, elevating platform, or water tower that is designed and equipped to support firefighters and rescue operations by positioning personnel, handling materials, providing continuous egress, or discharging water at positions elevated from ground. (NFPA 1900)

airport firefighter The Firefighter II who has demonstrated the skills and knowledge necessary to function as an integral member of an aircraft rescue and firefighting (ARFF) team. (NFPA 1010)

ambulance A vehicle used for out-of-hospital medical care and patient transport that provides a driver's compartment; a patient compartment to accommodate an emergency medical services provider (EMSP) and at

KEY TERMS CONTINUED

least one patient located on the primary cot positioned so that the primary patient can be given emergency care during transit; equipment and supplies at the scene as well as during transport; safety, comfort, and avoidance of aggravation of the patient's injury or illness; two-way radio communication; and audible and visual warning devices. (NFPA 1900)

apparatus See *fire apparatus.*

assistant chief A midlevel chief who often has a functional area of responsibility, such as training, or who is responsible for a group of battalions or districts and who answers directly to the fire chief. Also called *deputy chief* or *division chief.*

authority having jurisdiction (AHJ) An organization, office, or individual responsible for enforcing the requirements of a code or standard or for approving equipment, materials, an installation, or a procedure. (NFPA 1)

battalion chief Usually the first level of chief, the person in charge of running calls and supervising multiple stations or districts within a city.

brush company See *wildland company.*

building code A regulation that specifies how structures are designed, constructed, and remodeled so that buildings are safe for people who live and work in them.

candidate A person who applies to become a firefighter.

captain The second rank of promotion in the fire service, between the lieutenant and the battalion chief. Captains are responsible for managing a fire company and for coordinating the activities of that company among the other shifts.

chain of command A rank-based, hierarchical structure that creates an orderly line of authority.

chief's bugle See *chief's trumpet.*

chief's trumpet An obsolete amplification device that was a precursor to a bullhorn and that enabled a chief officer to give orders to firefighters during an emergency. Also called *chief's bugle.*

code A standard that is an extensive compilation of provisions covering broad subject matter or that is suitable for adoption into law independent of other codes and standards. (NFPA 1)

community risk reduction (CRR) A process to identify and prioritize local risks, followed by the integrated and strategic investment of resources to reduce their occurrence and impact. (NFPA 1452)

company The basic firefighting organizational unit staffed by various grades of firefighters under the supervision of an officer and assigned to one or more specific pieces of apparatus. (NFPA 1410)

company officer The individual responsible for command of a company, a designation not specific to any particular fire department rank (can be a firefighter, lieutenant, captain, or chief officer, if responsible for command of a single company). (NFPA 1026)

crew A collective term used casually to refer to a group of firefighters in a department with similar duties or responsibilities. See also *company* and *unit.*

deputy chief See *assistant chief.*

dispatcher See *telecommunicator.* (NFPA 1225)

division chief See *assistant chief.*

driver/operator A person qualified to operate a fire apparatus. Also called *engineer* or *technician.* (NFPA 1910)

Emergency Medical Responder (EMR) Emergency medical services (EMS) personnel who have basic training for providing initial medical assistance, have training in bleeding control and CPR, and often perform in an assistant role within the ambulance.

emergency medical services (EMS) company A company that may include medical units and first-response vehicles and that responds to and assists in the transport of medical and trauma victims to medical facilities for further treatment. Also called *emergency medical services (EMS) squad* or *squad.*

emergency medical services (EMS) personnel Personnel responsible for administering care to people who are sick and injured.

emergency medical services (EMS) squad See *emergency medical services (EMS) company.*

Emergency Medical Technician (EMT) Emergency medical services (EMS) personnel who can do everything an emergency medical responder (EMR) can do, and who have training in basic emergency care skills, including oxygen therapy, bleeding control, CPR, automated external defibrillation, use of basic airway devices, and assisting patients with certain medications.

emergency scene The area encompassed by the incident and the surrounding area needed by the emergency forces to stage apparatus and mitigate the incident. Also called *on-scene.* See also *fireground* and *fire scene.* (NFPA 901)

emergency vehicle technician (EVT) The individual who repairs and performs service on emergency vehicles.

engine See *pumper.*

engine company A piece of fire apparatus along with firefighters that have the primary responsibility to deliver a fire stream or streams to extinguish the fire in coordination with ventilation (truck company) and rescue operations. (NFPA 1700)

engineer See *driver/operator.*

fire and life safety educator (FLSE) An individual who has demonstrated the ability to coordinate, create, administer, prepare, deliver, and evaluate educational programs and information.

fire apparatus A vehicle designed to be used under emergency conditions to transport personnel and equipment or to support the suppression of fires and mitigation of other hazardous situations. Also called *apparatus.* (NFPA 1010)

fire chief The highest-ranking officer in charge of a fire department. (NFPA 1550)

fire code The code that specifies practices and procedures to prevent fires, prevent fires that start from spreading by suppressing them and blocking them, and protect lives in the event of a fire by specifying how occupants will be evacuated.

fire department An organization providing rescue, fire suppression, and related activities, including any public, governmental, private, industrial, or military organization engaging in this type of activity. (NFPA 1010)

firefighter A member of a fire department who is assigned to do routine cleaning and maintenance, place hose line to extinguish fires, and assist with a public fire prevention program.

Firefighter I A person, at the first level of progression as defined in Chapter 6 of NFPA 1010, who has demonstrated the knowledge and skills to function as an integral member of a firefighting team under direct supervision in hazardous conditions. (NFPA 1010)

Firefighter II A person, at the second level of progression as defined in Chapter 7 of NFPA 1010, who has demonstrated the skills and depth of knowledge to function under general supervision. (NFPA 1010)

fireground Another name for *emergency scene* when the incident is a fire. Also called *fire scene.*

fire hook A tool used to pull down burning structures.

fire house See *fire station.*

fire inspector An individual who conducts fire code inspections and applies codes and standards. (NFPA 1030)

fire investigator An individual who has demonstrated the skills and knowledge necessary to conduct, coordinate, and complete an investigation. (NFPA 1030)

fire mark A plaque displayed on a building with the name or logo of a fire insurance company informing firefighters that the building was insured by that insurance company, which means that insurance company would pay the firefighters for extinguishing the fire.

fire marshal A person designated to provide delivery, management, and/or administration of fire protection– and life safety–related codes and standards, investigations, education, and/or prevention services for local, county, state, provincial, federal, tribal, or private sector jurisdictions as adopted or determined by that entity. (NFPA 1030)

fireplug Historically speaking, a plug installed to control water accessed from wooden pipes, but today is slang for fire hydrant.

fire police officer An individual officially deployed who provides scene security, directs traffic, and conducts other duties as determined by the authority having jurisdiction (AHJ). (NFPA 1091)

fire protection engineer A member of the fire department or an employee of an architectural firm who is responsible for reviewing building plans and working with building owners to ensure that the design of and systems for fire detection and suppression meet applicable codes and function as needed.

fire scene Another name for *emergency scene* when the incident is a fire. Also called *fireground.*

fire station A building that houses fire apparatus and equipment for a geographic area within a fire department. Also called *fire house.*

fire suppression The activities involved in controlling and extinguishing fires. (NFPA 1500)

fire warden An individual charged with enforcing fire regulations in colonial America.

FLSE See *fire and life safety educator.*

governance The framework and procedures for managing and operating an organization.

hazardous materials company A company that responds to and controls scenes where hazardous

KEY TERMS CONTINUED

materials have spilled or leaked and whose members wear special suits and are trained to deal with most chemicals.

hazardous materials technician A person who responds to hazardous materials/weapons of mass destruction (WMD) incidents using a risk-based response process to analyze a problem involving hazardous materials/WMD, plan a response to the problem, implement the planned response, evaluate progress of the planned response and adjust accordingly, and assist in terminating the incident. (NFPA 470)

incident commander (IC) The individual responsible for all incident activities, including the development of strategies and tactics and the ordering and release of resources. (NFPA 1410)

incident safety officer (ISO) A member of the command staff responsible for monitoring and assessing safety hazards and unsafe situations and for developing measures for ensuring personnel safety. (NFPA 1700)

initial attack apparatus Fire apparatus with a fire pump of at least 250 gpm (946 L/min) capacity, water tank, and hose body, whose primary purpose is to initiate a fire suppression attack on structural, vehicular, or vegetation fires and to support associated fire department operations. Also called *quick attack apparatus.* (NFPA 1900)

International Code Council (ICC) An international association that creates standards and codes for buildings.

International Fire Service Accreditation Congress (IFSAC) A national organization that accredits or recognizes emergency service certification systems.

ladder company See *truck company.*

ladder truck See *aerial fire apparatus.*

lieutenant The first level of officer and the person who is usually responsible for a single fire company on a shift.

mobile water supply apparatus A vehicle designed primarily for transporting (pickup, transporting, and delivering) water to fire emergency scenes to be applied by other vehicles or pumping equipment. Also called *tanker, tender,* or *water tender.* (NFPA 1900)

National Board on Fire Service Professional Qualifications A national organization that accredits or recognizes emergency service certification systems. Also called *Pro Board.*

National Fire Protection Association (NFPA) A nonprofit association that develops and maintains nationally recognized minimum consensus standards and fire codes for fire safety and handling hazardous materials.

on-scene See *emergency scene.*

Paramedic Emergency medical services (EMS) personnel who can do everything an advanced emergency medical technician (AEMT) can do, and who have extensive training in advanced life support, including administering drugs, cardiac monitoring, inserting advanced airways, manual defibrillation, and other advanced assessment and treatment skills.

personal alert safety system (PASS) A device worn by firefighters that sound an alarm if the firefighter activates it or if the device does not detect motion for 30 seconds.

policies Formal statements that outline expectations for performance and procedures in different circumstances, but usually require personnel to make judgments to determine the best course of action within the stated study.

Pro Board See *National Board on Fire Service Professional Qualifications.*

public information officer (PIO) A member of the command staff responsible for interfacing with the public and media or with other agencies with incident-related information requirements. (NFPA 1550)

pumper Fire apparatus with a permanently mounted fire pump of at least 750 gpm (1300 L/min) capacity water tank, and a hose body whose primary purpose is to control structural and associated fires. (NFPA 1900)

quick attack apparatus See *initial attack apparatus.*

quint See *quint apparatus.*

quint apparatus Fire apparatus with a permanently mounted fire pump, a water tank, a hose storage area, an aerial device or elevating platform with a permanently mounted waterway, and a complement of ground ladders. Also called *quint.* (NFPA 1710)

recruit A candidate whose application to become a firefighter is accepted.

regulation A mandate issued and enforced by governmental bodies such as the U.S. Occupational Safety and Health Administration (OSHA) and the U.S. Environmental Protection Agency (EPA).

rescue apparatus Apparatus that carry an extensive array of regular and specialized tools and equipment that are used to rescue victims.

rescue company A piece of fire apparatus along with firefighters that are generally utilized for search and rescue at fire incidents. (NFPA 1700)

rescue technician See *technical rescuer.*

self-contained breathing apparatus (SCBA) An atmosphere-supplying respirator that supplies a respirable air atmosphere to the user from a breathing air source that is independent of the ambient environment and designed to be carried by the user. (NFPA 1970)

squad See *emergency medical services (EMS) company.*

standard Documents, the main text of which contain only mandatory provisions using the word "shall" to indicate requirements and that is in a form generally suitable for mandatory reference by another standard or code or for adoption into law. Nonmandatory provisions are not to be considered a part of the requirements of a standard and shall be located in an appendix or annex, footnote, informational note, or other means as permitted in the NFPA Manuals of Style. (NFPA 1)

standard operating guidelines (SOGs) Written organizational directives that establish or prescribe specific operational or administrative methods to be followed routinely, which can be varied due to operational need in the performance of designated operations or actions. (NFPA 1550)

standard operating procedures (SOPs) Written organizational directives that establish or prescribe specific operational or administrative methods to be followed routinely for the performance of designated operations or actions. (NFPA 1550)

support person A fire department member who is not a firefighter but who assists members of a fire department performing duties in environments that are not hazardous.

tanker See *mobile water supply apparatus.*

technical rescuer A person who is trained to perform or direct a technical rescue. Also called *rescue technician.* (NFPA 1006)

technician See *driver/operator.*

telecommunicator An individual whose primary responsibility is to receive, process, or disseminate information of a public safety nature via telecommunication devices. Also called *dispatcher.* (NFPA 1225)

tender See *mobile water supply apparatus.*

tower ladder company See *truck company.*

tower ladder truck See *aerial fire apparatus.*

training officer The person designated by the fire chief with authority for overall management and control of the organization's training program. (NFPA 1401)

truck See *aerial fire apparatus*

truck company A company of firefighters who are equipped with one or more pieces of aerial fire apparatus. Also called *ladder company* or *tower ladder company.* (NFPA 1700)

unified command A team effort that allows all agencies with jurisdictional responsibility for an incident or planned event, either geographical or functional, to manage the incident or planned event by establishing a common set of incident objectives and strategies. (NFPA 1026)

unit A collective term used casually to refer to a group of firefighters in a department with similar duties or responsibilities. See also *company* and *crew.*

water tender See *mobile water supply apparatus.*

wildland apparatus A four-wheel drive vehicle used to transport firefighters closer to wildfires over rough, uneven terrain and that carries a tank of water and a pump that enables them to pump water while the truck is moving and special firefighting equipment such as portable pumps, rakes, shovels, and chainsaws.

wildland company A company who fight vegetation fires where larger pumpers cannot gain access and who are equipped with four-wheel drive vehicles and special firefighting equipment. Also called *brush company.*

REVIEW QUESTIONS

1. What is the primary mission of the fire service?
2. Where were the first fire regulations in the United States enacted?
3. What precipitated most of the major advances in fire safety codes?
4. What is a combination department?
5. What is a unified command?
6. Which aspect of fire service culture means putting the needs of others before your own needs?
7. What are the Five Es of fire prevention?

REVIEW QUESTIONS CONTINUED

8. What are SOPs?
9. Which company in a fire department specializes in forcible entry, ventilation, roof operations, search and rescue, and deployment of ground ladders?
10. Which fire department member is responsible for setting up and running the fire pump or operating the aerial device?
11. What is the minimum education requirement that most career departments require a candidate has?
12. What is the difference between support persons, Firefighter I, and Firefighter II?

DISCUSSION QUESTIONS

1. Does a culture of safety mean that firefighters should never do anything that is dangerous?
2. How is community risk reduction (CRR) different from traditional fire prevention activities?
3. How should firefighters interact with representatives of other agencies at emergency scenes?
4. Why does a firefighter need to know and understand departmental SOPs?
5. Why should a firefighter today care about fires that happened decades or even hundreds of years ago?

APPLYING THE CONCEPTS

You are walking to meet friends from your unit at a festival downtown. On your way you strike up a conversation with a woman at one of the vendor booths. She notices you're wearing a T-shirt from your department. "Ah, you're a firefighter! My grandfather was a firefighter in Chicago in the 40s and 50s. I loved hearing stories about his adventures. I grew up thinking he was a superhero. He was on the scene of that big school fire in the late 50s where almost 100 children died. After that he spent the rest of his career working with the city to change fire codes so a tragedy like that would never happen again. You would've loved talking to him."

1. What fire service cultural values do you and your peers share with firefighters from past generations?
2. Up until recently, the fire service focused primarily on suppressing fires. How has this focus shifted? Why?
3. What is this fire prevention approach called and how does it work?

You share how your station is using CRR in town to target the most dangerous fire risks and how you've seen how these efforts have actually helped reduce fire calls. "That's interesting," she says. "Tell me, how else has firefighting changed since the days of my grandfather?"

4. In your reply, how would you summarize key improvements to modern firefighting equipment?
5. How would you describe your key responsibilities as a firefighter?

"You know, you've rekindled my fascination with firefighting...no pun intended," she laughed. "I'd love to learn more about what it looks like today. Plus...I need to get ready. My son seems to have inherited his great-grandfather's passion and says he wants to be a firefighter when he grows up!" You invite her and her son to the open house your fire station is having next month. "I'll be happy to give you a personalized tour."

6. Why is it important for firefighters to engage with the community?
7. What do you do if you're asked a question you don't know the answer to?

REFERENCES

Ahrens, Marty, and Ben Evarts. 2021. *Fire Loss in the United States during 2020.* National Fire Protection Association (NFPA), September 2021. Accessed September 6, 2022. https://www.nfpa.org/-/media/Files/News-and-Research/Fire-statistics-and-reports/US-Fire-Problem/osFireLoss.pdf.

Benjamin Franklin Historical Society. n.d. "Union Fire Company." Accessed March 28, 2023. http://www.benjamin-franklin-history.org/union-fire-company/.

Boston Fire Historical Society. n.d. "Boston History before 1859." Accessed March 28, 2023. https://bostonfirehistory.org/boston-history-before-1859.

Coleman, Ronny J. 1988. *Managing Fire Services.* Edited by John A. Granito. Washington, DC: International City/County Management Association.

Fahy, Rita, Ben Evarts, and Gary P. Stein. *U.S. Fire Department Profile 2019.* National Fire Protection Association (NFPA), December 2021. Accessed September 12, 2022. https://web.archive.org/web/20220324020917id_/https://www.nfpa.org/-/media/Files/News-and-Research/Fire-statistics-and-reports/Emergency-responders/osfdprofile.pdf.

Federal Emergency Management Agency (FEMA). 2018. *ICS Organizational Structure and Elements.* March 2018. Accessed February 23, 2023. https://training.fema.gov/emiweb/is/icsresource/assets/ics%20organizational%20structure%20and%20elements.pdf.

Federal Emergency Management Agency (FEMA). n.d. *ICS 100 – Incident Command System.* Accessed March 24, 2023. https://www.usda.gov/sites/default/files/documents/ICS100.pdf.

FireHydrant.org. 2004. "A Brief History of the Hydrant." Revised January 28, 2003. Accessed March 24, 2023. http://www.firehydrant.org/pictures/hydrant_history.html.

Giesler, Marsha. 2018. *Fire and Life Safety Educator: Principles and Practice,* 2nd ed. Burlington, MA: Jones & Bartlett Learning.

Hall, Shelby, and Ben Evarts. 2022. *Fire Loss in the United States during 2021.* National Fire Protection Association (NFPA), September 2022. Accessed September 21, 2022. https://www.nfpa.org.

International Association of Fire Chiefs (IAFC). n.d. "Candidate Physical Ability Test." Accessed September 14, 2022. https://www.iafc.org/topics-and-tools/safety-health/wellness-fitness-task-force/candidate-physical-ability-test.

International Association of Fire Fighters (IAFF). 2007. *Candidate Physical Ability Test,* 2nd ed. Accessed September 14, 2022. https://www.iaff.org/wp-content/uploads/2019/04/CPAT-2nd-Edition.pdf.

Merrimack Fire and Rescue. n.d. "The History of Firefighting." Accessed March 26, 2023. https://www.merrimacknh.gov/about-fire-rescue/pages/the-history-of-firefighting.

National Fire Protection Association (NFPA). 2008. "Deadliest Single Building or Complex Fires and Explosions in the U.S." Updated March 2008. Accessed February 20, 2023. https://www.nfpa.org.

National Fire Protection Association (NFPA). 2016. *NFPA 1401: Recommended Practice for Fire Service Training Reports and Records.* 2017 Edition. Quincy, MA: NFPA.

National Fire Protection Association (NFPA). 2019. *NFPA 1410: Standard on Training for Emergency Scene Operations.* 2020 Edition. Quincy, MA: NFPA.

National Fire Protection Association (NFPA). 2019. *NFPA 1452: Guide for Training Fire Service Personnel to Conduct Community Risk Reduction for Residential Occupancies.* 2020 Edition. Quincy, MA: NFPA.

National Fire Protection Association (NFPA). 2019. *NFPA 1710: Standard for the Organization and Deployment of Fire Suppression Operations, Emergency Medical Operations, and Special Operations to the Public by Career Fire Departments.* 2020 Edition. Quincy, MA: NFPA.

National Fire Protection Association (NFPA). 2020. *NFPA 1: Fire Code.* 2021 Edition. Quincy, MA: NFPA.

National Fire Protection Association (NFPA). 2020. *NFPA 101, Life Safety Code.* 2021 Edition. Quincy, MA: NFPA.

National Fire Protection Association (NFPA). 2020. NFPA 901: *Standard Classifications for Fire and Emergency Services Incident Reporting.* 2020 Edition. Quincy, MA: NFPA.

National Fire Protection Association (NFPA). 2020. *NFPA 1006: Standard for Technical Rescue Personnel Professional Qualifications.* 2021 Edition. Quincy, MA: NFPA.

National Fire Protection Association (NFPA). 2020. *NFPA 1500: Standard on Fire Department Occupational Safety, Health, and Wellness Program.* 2021 Edition. Quincy, MA: NFPA.

National Fire Protection Association (NFPA). 2020. *NFPA 1700: Guide for Structural Fire Fighting.* 2021 Edition. Quincy, MA: NFPA.

National Fire Protection Association (NFPA). 2021. *NFPA 13, Standard for the Installation of Sprinkler Systems.* 2022 Edition. Quincy, MA: NFPA.

National Fire Protection Association (NFPA). 2021. *NFPA 470, Hazardous Materials/Weapons of Mass Destruction (WMD) Standard for Responders.* 2022 Edition. Quincy, MA: NFPA.

National Fire Protection Association (NFPA). 2021. *NFPA 1225: Standard for Emergency Services Communications.* 2022 Edition. Quincy, MA: NFPA.

National Fire Protection Association (NFPA). 2021. *NFPA 1582: Standard on Comprehensive Occupational Medical Program for Fire Departments.* 2022 Edition. Quincy, MA: NFPA.

National Fire Protection Association (NFPA). 2024. NFPA 1010: *Standard on Professional Qualifications for Firefighters.* 2024 Edition. Quincy, MA: NFPA.

National Fire Protection Association (NFPA). 2024. *NFPA 1026: Standard for Incident Management Personnel Professional Qualifications.* 2024 Edition. Quincy, MA: NFPA.

National Fire Protection Association (NFPA). 2023. *NFPA 1030: Standard for Professional Qualifications for Fire Prevention Program Positions.* 2024 Edition. Quicny, MA: NFPA.

REFERENCES CONTINUED

National Fire Protection Association (NFPA). 2023. *NFPA 1091: Standard for Traffic Incident Management Personnel Professional Qualifications.* 2024 Edition. Quincy MA, NFPA.

National Fire Protection Association (NFPA). 2023. *NFPA 1550, Standard for Emergency Responder Health and Safety.* 2024 Edition. Quincy, MA: NFPA.

National Fire Protection Association (NFPA). 2023. *NFPA 1900: Standard for Aircraft Rescue and Firefighting Vehicles, Automotive Fire Apparatus, Wildland Fire Apparatus, and Automotive Ambulances.* 2024 Edition. Quincy, MA: NFPA.

National Fire Protection Association (NFPA). 2023. *NFPA 1910: Standard for the Inspection, Maintenance, Refurbishment, Testing, and Retirement of In-Service Emergency Vehicles and Marine Firefighting Vessels.* 2024 Edition. Quincy, MA: NFPA.

National Fire Protection Association (NFPA). 2023. *NFPA 1970: Standard on Protective Ensembles for Structural and Proximity Firefighting, Work Apparel and Open-Circuit Self-Contained Breathing Apparatus (SCBA) for Emergency Services, and Personal Alert Safety Systems (PASS).* 2024 Edition. Quincy, MA: NFPA.

National Fire Protection Association (NFPA). 2023. *NFPA 101: Life Safety Code.* 2024 Edition. Quincy, MA: NFPA.

Noonan, Travis. 2017. "History of the Modern Fire Truck." DriveZing, November 29, 2017. Accessed March 24, 2023. https://drivezing.com/history-modern-fire-truck.

Our Lady of Angels (OLA) Fire Memorial. n.d. Accessed March 28, 2023. https://olafire.com/FireSummary.asp.

Rosenfeld, Everett. 2011. "Top 10 Devastating Wildfires." Time, June 8, 2011. Accessed March 24, 2023. https://content.time.com/time/specials/packages/article/0,28804,2076476_2076484_2076503,00.html.

Smith, Carl. 2020. *Chicago's Great Fire: The Destruction and Resurrection of an Iconic American City.* New York: Atlantic Monthly Press.

U.S. Fire Administration (USFA). 2023. "National Fire Department Registry Quick Facts." Last updated March 26, 2023. Accessed March 28, 2023. https://apps.usfa.fema.gov/registry/summary.

Vision 2020 Project. n.d. "Model Performance Criteria (MPC) Template & Guidance." Accessed September 24, 2022. https://strategicfire.org/model-performance/template-and-guidance.

CHAPTER

2

Firefighter I

Firefighter Health and Safety

KNOWLEDGE OBJECTIVES

After studying this chapter, you will be able to:

- Describe the value of fire and life safety initiatives in support of the fire department mission to reduce firefighter line-of-duty injuries and fatalities.
- Describe the importance of physical fitness and healthy lifestyle to the ability of firefighters to perform their duties.
- Describe the signs and symptoms of behavioral and emotional distress, including the role of the fire department's member assistance program.
- Identify proper procedures to increase safety and operations at emergency scenes.
- Identify potential hazards involved in operating on emergency scenes and how to mitigate those hazards.
- Describe how to promote a culture of safety inside and outside the workplace.

SKILLS OBJECTIVES

After studying this chapter, you will be able to perform the following skills:

- Mount and dismount a fire apparatus safely.

ADDITIONAL NFPA STANDARDS

- **NFPA 1250**, *Recommended Practice in Fire and Emergency Services Organization Risk Management, 2020 Edition*
- **NFPA 1451**, *Standard for a Fire and Emergency Service Vehicle Operations Training Program, 2018 Edition*
- **NFPA 1500**, *Standard on Fire Department Occupational Safety, Health, and Wellness Program, 2021 Edition*
- **NFPA 1581**, *Standard on Fire Department Infection Control Program, 2022 Edition*
- **NFPA 1582**, *Standard on Comprehensive Occupational Medical Program for Fire Departments, 2022 Edition*

CASE STUDY

You Are the Firefighter

You and your crew are at a large fire in an apartment building that started on the third floor, extended into the attic, and is now venting through the roof. Your captain gave the incident commander the accountability tags, and you were assigned to confine the fire on the third floor. You gather your gear and, as you approach the apartment building, you consider the ways you and your crew can accomplish your mission while avoiding preventable injury or death.

1. What is the leading cause of firefighter injury and death?
2. What are the potential hazards during this incident?
3. What safety measures should be taken during this incident?

Introduction

This chapter covers the topics of injury and illness prevention, means of reducing firefighter deaths, and safety and health measures needed during all activities performed by firefighters, including training, response, fireground operations, and fire station duties. Firefighting, by its nature, is dangerous. Every firefighter must be aware of the risks inherent in all job responsibilities and activities and must learn safe methods of confronting all the risks.

For firefighters, the goal is never to eliminate all risks. Rather, it is to reduce preventable deaths and injuries by reducing unnecessary risks. For example, potential carcinogens in smoke put firefighters at a high risk of cancer. However, proper decontamination of protective gear, tools, equipment, and firefighters after a fire can greatly reduce that risk without compromising the ability of the firefighter to do their job. A commitment to this kind of safety is integral to a firefighter's integrity and professionalism.

Every fire department must do what it can to reduce the hazards and dangers of the job and help prevent firefighter injuries, illness, and deaths. Each fire department must have a strong commitment to firefighter health and safety, with firefighters taking the lead for their own health and safety. When firefighters understand the various risks associated with their job responsibilities and activities, and the actual threat that they present, safety measures can become routine, consistent, and fully integrated into every activity, procedure, and job description.

Advances in standards, technology, and equipment require fire departments to review and revise their health and safety policies and procedures regularly. Safety officers are responsible for evaluating the hazards of various situations and recommending appropriate safety measures to the incident commander (IC). Each accident, injury, or near miss must be thoroughly investigated to learn why it happened and how it can be avoided in the future. After-incident reviews and research by designated health and safety officers can identify new hazards as well as appropriate safety measures. In addition, reports of accidents, fatalities, and near misses from other fire departments can help identify common problems and lead to the development of effective preventive actions.

Because firefighters must be ready to react immediately to an alarm, preparations for response begin long before the alarm is sounded. These preparations include physical and mental readiness, checking personal equipment, ensuring that the fire apparatus is ready, and making sure that all equipment carried on the apparatus is ready for use. Firefighters also should be familiar with their response district, know the buildings under their protection, and understand their department's standard operating procedures (SOPs).

Response actions for the apparatus driver also include considering road and traffic conditions, determining the best route to the incident, identifying nearby hydrant locations or water sources, and selecting the best position for the apparatus at the incident scene.

Causes of Firefighter Deaths and Injuries

The NFPA reports that 135 firefighters were killed in the line of duty in 2020. These deaths resulted from emergency operations such as emergency medical services (EMS), fire suppression, rescue, and hazardous materials calls, and non-emergency situations, such as public service calls, training, and responding to or returning from emergency situations (**FIGURE 2-1**).

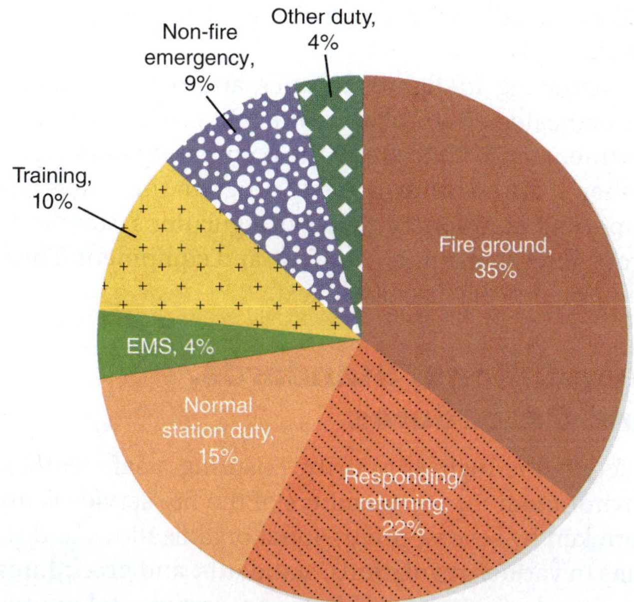

FIGURE 2-1 Firefighter deaths in the United States by type of duty. In 2020, 135 firefighters died in the line of duty.

Reproduced from Campbell, Richard and Jay T. Petrillo. 2023. "Fatal Firefighter Injuries in the United States Overview of Fatal Firefighter Injuries in 2022." National Fire Protection Association. https://www.nfpa.org/education-and-research/research /nfpa-research/fire-statistical-reports/fatal-firefighter-injuries.

TABLE 2-1 Firefighter Deaths by Cause of Injury, 2020

Cause	Percent
Overexertion, stress, medical	57%
Vehicle accidents	14%
Struck by vehicles	9%
Structural collapse	6%
Rapid fire progress/explosions	4%
Struck by objects	3%
Falls	3%
Exposure to electricity	1%
Fatal assault	1%
Lost inside	1%

National Fire Protection Association, *Firefighter Injuries in the United States*, 2021.

The largest share of deaths occurred from stress, overexertion, and medical issues (**TABLE 2-1**). Cardiac events accounted for 38 percent of these deaths. The second leading cause of death was vehicle crashes. In 2020, 15 firefighters died in vehicle accidents, a majority of whom died responding to or returning from incidents.

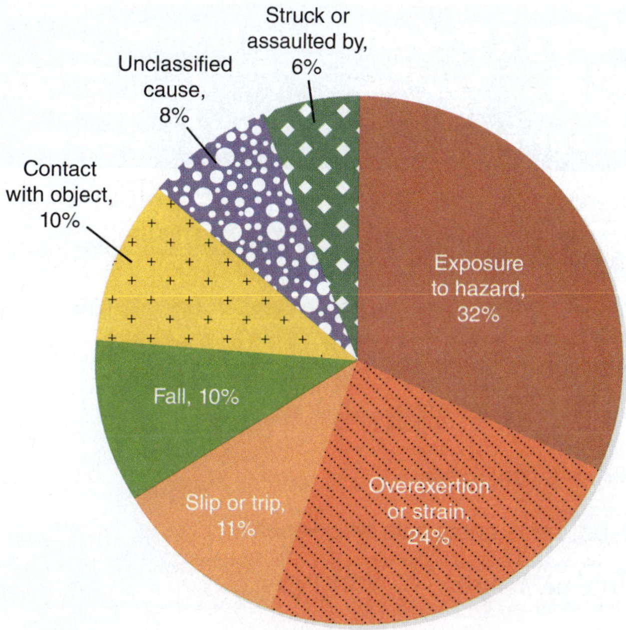

FIGURE 2-2 Firefighter injuries in the United States by type of duty. In 2020, 64,875 firefighters were injured in the line of duty.

Data from Campbell, Richard and Jay T. Petrillo. 2023. "Overview of Fatal Firefighter Injuries in 2022." National Fire Protection Association. https://www.nfpa.org /education-and-research/research/nfpa-research/fire-statistical-reports/fatal -firefighter-injuries.

The NFPA estimates that 64,875 firefighters were injured in the line of duty in 2020, a 7 percent increase from the previous year. **FIGURE 2-2** shows the breakdown of injuries by type of duty.

Fewer than half of the injuries occurred on the fireground. The leading cause of fireground injuries was overexertion or strain (**TABLE 2-2**).

SAFETY TIP

Prevention of illness and injury is a responsibility shared by each member of the firefighting team. Firefighters must always consider three groups when ensuring safety at the scene:

- Their personal safety
- The safety of their team members
- The safety of everyone at an emergency scene

Reducing Firefighter Deaths and Injuries

In 1992, Congress created the National Fallen Firefighters Foundation to lead a nationwide effort to remember the U.S. fallen firefighters. In the years since then, this

TABLE 2-2 Fireground Injuries by Cause

Types of Injury	Percentage of Total Injuries
Overexertion/strain	31%
Fall, jump, slip	21%
Other	16%
Exposure to fire products	15%
Contact with an object	10%
Struck by an object	5%
Exposure to chemicals or radiation	3%
Extreme weather	3%

National Fire Protection Association, *Firefighter Injuries in the United States*, 2021.

foundation has expanded its programs to sponsor the annual National Fallen Firefighters Memorial Weekend, offer support programs for family members and other survivors, award scholarships to fire service members' survivors, and work to prevent line-of-duty injuries and deaths.

Although firefighting is an inherently dangerous activity, most firefighter injuries and deaths are the result of preventable situations. Recognizing this fact, organizations such as the International Association of Fire Chiefs (IAFC) and the National Fallen Firefighters Foundation have developed programs with the goal of reducing line-of-duty deaths. For example, developed in 2005, the Near Miss Reporting System provides a method for reporting situations that could have resulted in injuries or deaths. This system, which is accessible online, provides a means for all firefighters to learn from situations that occur both rarely and frequently.

In an effort to do more to prevent line-of-duty deaths and injuries, the first National Firefighter Life Safety Summit convened in 2004, uniting fire service leaders and organizations across the United States. The result was the creation of the Everyone Goes Home program and a set of key initiatives, the 16 Firefighter Life Safety Initiatives. The goal of the Everyone Goes Home program is to raise awareness of life safety issues, improve safety practices, and allow everyone to return home at the end of their shift. The 16 Firefighter Life Safety Initiatives describe steps that need to be taken to change the current culture of the fire service to help make it a safe work environment. They are listed in **TABLE 2-3**.

Reducing firefighter injuries and deaths requires the dedicated efforts of every firefighter, of every fire department, and of the entire fire community working together. It also requires a safety program that integrates important components such as regulations, standards, procedures, personnel, training, and equipment. These components are discussed next.

Regulations, Standards, and Procedures

Safety is the highest priority. Ensuring a safe working environment for the members of the fire service is undertaken by several professional organizations and results in various regulations, standards, and procedures.

Standards are issued by nongovernmental entities and are generally consensus based. Standards are discussed in greater detail in Chapter 1, *The Fire Service*. The National Fire Protection Association (NFPA) develops standards that help to standardize training courses, apparatus, equipment, and operations. Their mission is focused on safety: to save lives and reduce loss with information, knowledge, and passion. It is important to understand that these standards are designed to help you become a safe, efficient, and competent firefighter. You will see many NFPA standards referenced throughout this course. Most fire departments have access to these standards.

To reduce the risks of accidents, injuries, occupational illnesses, and fatalities, a successful health and safety program that complies with NFPA 1500, *Standard on Fire Department Occupational Safety, Health, and Wellness Program*, must be established. NFPA 1500 includes guidance on several key aspects of health and safety, including policies, training and education, apparatus operation, personal protective equipment (PPE), emergency operations, station safety, medical and physical requirements, and health and safety programs. This chapter addresses each of these areas briefly.

NFPA 1500 provides a template for implementing a comprehensive health and safety program. Additional NFPA standards focus on specific subjects directly related to health and safety—for example, NFPA 1581, *Standard on Fire Department Infection Control Program*, and NFPA 1582, *Standard on Comprehensive Occupational Medical Program for Fire Departments.*

Regulations are issued and enforced by governmental bodies. The federal **Occupational Safety and Health Administration (OSHA)**, along with a variety of state and provincial health and safety agencies,

TABLE 2-3 16 Firefighter Life Safety Initiatives

1. **Cultural change:** Define and advocate the need for a cultural change within the fire service relating to safety and incorporating leadership, management, supervision, accountability, and personal responsibility.
2. **Accountability:** Enhance the personal and organizational accountability for health and safety throughout the fire service.
3. **Risk management:** Focus greater attention on the integration of risk management with incident management at all levels, including strategic, tactical, and planning responsibilities.
4. **Empowerment:** All firefighters must be empowered to stop unsafe practices.
5. **Training and certification:** Develop and implement national standards for training, qualifications, and certification (including regular recertification) that are equally applicable to all firefighters based on the duties they are expected to perform.
6. **Medical and physical fitness:** Develop and implement national medical and physical fitness standards that are equally applicable to all firefighters based on the duties they are expected to perform.
7. **Research agenda:** Create a national research agenda and a data collection system that relates to the 16 Firefighter Life Safety Initiatives.
8. **Technology:** Utilize available technology whenever it can produce higher levels of health and safety.
9. **Fatality, near-miss investigation:** Thoroughly investigate all firefighter fatalities, injuries, and near misses.
10. **Grant support:** Grant programs should support the implementation of safe practices and procedures and/or mandate safe practices as an eligibility requirement.
11. **Response policies:** National standards for emergency response policies and procedures should be developed and championed.
12. **Violent incident response:** National protocols for response to violent incidents should be developed and championed.
13. **Psychological support:** Firefighters and their families must have access to counseling and psychological support.
14. **Public education:** Public education must receive more resources and be championed as a critical fire and life safety program.
15. **Code enforcement and sprinklers:** Advocacy must be strengthened for the enforcement of codes and the installation of home fire sprinklers.
16. **Apparatus design and safety:** Safety must be a primary consideration in the design of apparatus and equipment.

National Fallen Firefighters Association. *Everyone Goes Home. 16 Firefighter Life Safety Initiatives.*

develops and enforces government regulations on workplace safety and, in some cases, responder safety. NFPA standards often are incorporated by reference in government regulations.

Each state in the United States has the right to *adopt and/or supersede* workplace health and safety regulations put forth by OSHA. States that have adopted the OSHA regulations are called state-plan states. California, for example, is a state-plan state; its regulatory body is called Cal-OSHA. About half of the states in the United States are state-plan states. States that have not adopted the OSHA regulations are non-plan states. Non-plan states are EPA states because they follow Title 40 of the **Code of Federal Regulations (CFR)**, *Protection of the Environment, Part 311, Worker Protection.* The CFR is a collection of permanent rules published by the federal government. It includes 50 titles that represent broad areas of interest that are governed by federal regulation.

TIP

It is important to understand the relationship between OSHA regulations and NFPA standards. OSHA regulations are the law; NFPA standards are generally utilized as guidelines that under some circumstances have the force of law.

Every fire department should have a set of SOPs, or standard operating guidelines (SOGs) that provide specific information on the actions that should be taken to accomplish a certain task. These procedures are vital because they enable everyone in the department to function properly and know what is expected for each task. Each firefighter is responsible for understanding and following these procedures. This enables firefighters from different stations or companies to work together safely and smoothly.

The fire department chain of command also enforces safety goals and procedures. In particular, the command structure keeps everyone working toward common goals in a safe manner. An Incident Command System (ICS) is a nationally recognized plan to establish command and control of emergency incidents. The ICS is flexible enough to meet the needs of any emergency situation, so it should be implemented at every emergency scene—from a routine auto accident to a major disaster involving responders from numerous agencies. More information on the ICS is presented in Chapter 22, *Establishing and Transferring Command.*

Many fire departments have a health and safety committee that is responsible for establishing policies and monitoring firefighter health and safety. Members of this committee should include representatives from every area, component, and level within the department, from firefighters to chief officers. The health and safety officer and the fire department physician also should be members of the committee.

Personnel

A health and safety program is only as effective as the individuals who implement it. Safety officers are members of the fire department whose primary responsibility is safety (discussed further in Chapter 1, *The Fire Service*). At the emergency scene, the designated safety officer reports directly to the IC and has the authority to correct or stop any action that is judged to be unsafe. Safety officers observe operations and conditions, evaluate risks, and work with the IC to identify hazards and ensure the safety of all personnel. They also determine when firefighters can work without self-contained breathing apparatus (SCBA) after a fire is extinguished. Safety officers can enhance safety in the workplace, at emergency incidents, and at training exercises. Even so, it is important to remember that each and every member of the fire department shares the responsibility for promoting safety, both as an individual and as a member of the team.

Teamwork is an essential element of safe emergency operations. On the fireground and during any hazardous activity, firefighters must work together to get the job done. Freelancing has no place on the fireground; it poses a danger to the firefighter who acts independently and every other firefighter on the emergency scene. **Freelancing** is acting independently of a superior's orders or the fire department's SOPs. Freelancing is discussed in more detail later in the chapter.

Training

Adequate training is essential for firefighter safety. The initial firefighter training covers the potential hazards of each skill and outlines the steps necessary to avoid injury. Firefighters must avoid sloppy practices or shortcuts that might potentially contribute to injuries. They also must learn how to identify hazards and unsafe conditions.

The knowledge and skills developed during training classes are essential to maintain safety at actual emergency scenes. The initial training course is just the beginning—firefighters must continually seek out additional courses to keep their skills current.

Equipment

A firefighter's equipment ranges from power and hand tools to PPE and electronic instruments. Firefighters must know how to use equipment in the correct manner and then operate it safely at all times. Equipment also must be properly maintained. Poorly maintained equipment can create additional hazards to the user or fail to operate when needed.

Manufacturers usually supply operating instructions and safety procedures for their equipment. These instructions cover proper use of the equipment, its limitations, and warnings about potential hazards. Firefighters must read and heed these warnings and instructions. In addition, new equipment must meet applicable standards to ensure that it can perform under the difficult and dangerous conditions often encountered on the fireground.

Personal Health and Well-Being

Safety and well-being are directly related to personal health and physical fitness. Although fire departments regularly monitor and evaluate the health of firefighters, each department member is responsible for their own personal health, conditioning, and nutrition. To be an effective firefighter, you must exercise regularly, eat a healthy diet, get an adequate amount of sleep, and take preventive measures to avoid illnesses such as heart disease and cancer.

Physical Fitness

All firefighters—whether career or volunteer—should spend at least an hour each day in physical fitness activities of some kind (**FIGURE 2-3**). Firefighters should be examined by either a departmental or personal

FIGURE 2-3 Regular exercise will help you to stay healthy and enable you to perform your job effectively.

physician before beginning any new workout routine. An exercise routine that includes weight training, cardiovascular workouts, and stretching with a concentration on job-related exercises is ideal. For example, many firefighters use a stair-climbing machine to focus on the muscle groups used for climbing. This type of exercise also builds cardiovascular endurance for the fireground; however, other muscle groups should not be neglected. Physical fitness must be a career-long activity, not something left at the fire academy when you graduate. Firefighting is a stressful activity that demands you to maintain a good fitness level throughout your career.

Nutrition

The adage, "you are what you eat," is critical for firefighters. If you expect to perform the duties of a firefighter, you have to give your body the right nutrition, or it will fail when you need it. A healthy diet includes good proportions of fruits, vegetables, healthy fats, whole grains, and lean protein. Pay attention to portion sizes. Unfortunately, most people eat larger portions than their bodies need. Meals can still be satisfying without excess calories.

Hydration

Hydration is an important part of staying healthy (**FIGURE 2-4**). Water is generally the best fluid available because the body absorbs it faster than any other fluid. Avoid fluids that contain high levels of sugar. Sugary fluids can actually slow the rate of fluid absorption by the body and cause abdominal discomfort. A good guideline is to consume 8 to 10 ounces (oz; 0.2 to 0.3 liter [L]) of water for every 5 to 10 minutes of physical exertion. Do not wait until you feel thirsty to start

FIGURE 2-4 Consume 8 to 10 oz (0.2 to 0.3 L) of water for every 5 to 10 minutes of physical exertion.

Courtesy of Sean Wilson.

hydrating. In fact, firefighters should drink up to 1 gallon (gal; 3.8 L) of water each day to keep properly hydrated. The amount of water needed to maintain adequate hydration will depend on the type of work you are doing and the ambient temperature.

One indication of adequate hydration is frequent urination. Infrequent urination or urine that has a deep yellow color indicates dehydration. Remember, anytime you are working in full PPE, your internal environment will rapidly become hot, and you cannot dissipate body heat normally to the outside environment. Proper hydration enables muscles to work longer and reduces the risk of illness and injuries at the emergency scene. Some recent studies have indicated that maintaining a good level of hydration while engaged in firefighting activities also may reduce your chance of having a heart attack. More information on hydration is presented in Chapter 20, *Firefighter Rehabilitation*.

SAFETY TIP

Maintaining proper hydration is essential to performing at your peak physical level. A working firefighter must be proactive when it comes to hydration and not wait until they feel thirsty or too tired to begin drinking water.

Sleep

Good health requires that you get an adequate amount of uninterrupted sleep to maintain alertness, prevent stress, and avoid illnesses and injuries. The nature of shift work and the intensity of firefighting means that many firefighters find themselves chronically fatigued and experiencing sleep deprivation issues. The National Sleep Foundation and the American Academy of Sleep Medicine recommend that adults obtain a minimum of 7 to 9 hours of sleep per night. In the long term, sleep deprivation has been shown to lead to hypertension, sleep apnea, respiratory issues, diabetes, depression, and other medical conditions. Sleep deprivation and fatigue issues can increase stress and, in turn, increased stress can contribute to sleep deprivation and fatigue issues. Establish a consistent sleep schedule and sleep routine, such as turning off all electronic devices one-half hour before bedtime to allow your mind to wind down and prepare for sleep.

Heart Disease

Heart disease is the leading cause of death in the United States as a whole and a leading cause of death among firefighters in particular. A healthy lifestyle that includes a balanced diet, effective hydration, physical conditioning, and proper sleep can help reduce many risk factors for heart disease and enable firefighters to meet the physical demands of the job. It is also important that you have regular physical examinations to identify heart disease at an early stage.

Cancer

An increase in the use of synthetic products has led to an increase in the toxicity of today's modern fires. Cancer, now considered to be the leading cause of death among firefighters, can be caused by a wide variety of cancer-causing substances (carcinogens) entering the body (International Association of Firefighters, 2017). These include exhaust from diesel engines, poisonous gases in smoke, and a wide variety of chemical particles. The dirt and soot that attaches itself to a firefighter's turnout gear and uniform contains large quantities of substances known to cause cancer. Firefighter's hoods and gloves are thought to contain especially high concentrations of carcinogens. Carcinogens can be ingested through the mouth, injected into the body, absorbed through the respiratory system, or absorbed through the skin. Firefighters are most likely to absorb carcinogens through their skin and through their respiratory systems.

The Firefighter Cancer Support Network estimates that firefighters have a 9 percent higher risk of being diagnosed with cancer than the general U.S. population (Firefighter Cancer Support Network, 2017). Along the same lines, a study conducted by the National Institute for Occupational Safety and Health (NIOSH) concluded that the nearly 30,000 participants had a greater number of cancer diagnoses and cancer-related deaths than the general U.S. population. Most were digestive, oral, respiratory, and urinary cancers. When comparing firefighters in this study to each other, they found that the chance of lung cancer increased with the amount of time spent at fires, and the chance of death from leukemia increased with the number of fire incidents (Centers for Disease Control and Prevention, 2016).

Remember, contaminated objects that are placed in the cab of a fire engine or in the trunk of a firefighter's personal vehicle continue to release cancer-causing substances into the area around them. The longer contaminated objects are present, the more these objects continue to release toxic substances. Remove your **structural firefighting protective equipment**—personal protective clothing that provides full-body coverage that covers every inch of the body and provides protection from heat and fire, keeps water away from the body, and helps reduce injuries from cuts or falls—and all other contaminated clothing as soon as possible. All structural firefighting protective equipment should be transported away from the riding compartment in a fire apparatus and be thoroughly washed immediately according to the manufacturer's instructions (**FIGURE 2-5**). Additional information on the care and maintenance of PPE can be found in departmental SOPs as well as NFPA 1581, *Standard on Selection, Care, and Maintenance of Protective Ensembles for Structural Firefighting and Proximity Fire Fighting.*

The Firefighter Cancer Support Network (FCSN)[1] suggests actions you can take to protect yourself.

1. Use SCBA from initial attack to finish of overhaul. (Not wearing SCBA in both active and post-fire

1. Firefighter Cancer Support Network. 2013. "Taking Action Against Cancer in the Fire Service." Accessed November 15, 2023. https://www.firefightercancersupport.org/resources/library.

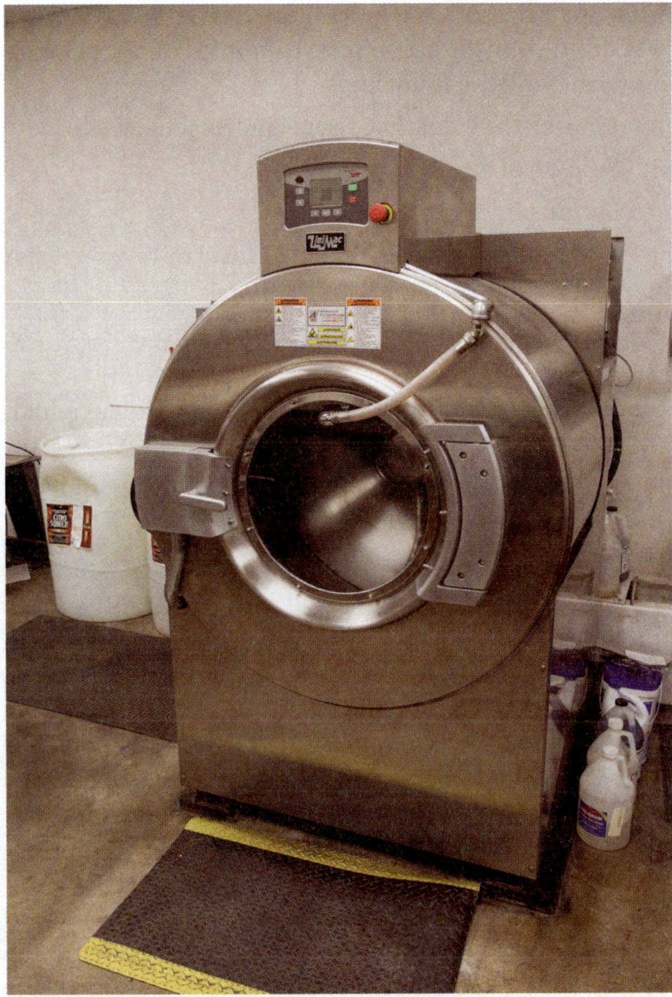

FIGURE 2-5 Special washing machines are available to launder personal protective clothing.

environments is the most dangerous voluntary activity in the fire service today.)

2. Do gross field decontamination of PPE to remove as much soot and particulates as possible.
3. Use cleansing wipes to remove as much soot as possible from the head, neck, jaw, throat, underarms, and hands immediately and while still on the scene.
4. Change your clothes and wash them immediately after a fire.
5. Shower thoroughly after a fire.
6. Clean your PPE, gloves, hood, and helmet immediately after a fire.
7. Do not take contaminated clothes or PPE home or store them in your vehicle.
8. Perform decontamination of fire apparatus interior after fires.
9. Keep structural firefighting protective equipment out of living and sleeping quarters.
10. Stop using tobacco products.
11. Use sunscreen or sun block.

The importance of annual medical examinations cannot be overstated—early detection and early treatment are essential to increasing survival. These 11 lifesaving actions are excerpted from FCSN's "Taking Action Against Cancer" white paper; it is available free of charge at firefightercancersupport.org.

Some cancers do not present for 20 years or more after exposure to a carcinogen. Therefore, it is important to reduce your exposure to cancer-causing substances starting on your first day of service as a firefighter. More information on performing field reduction of contaminants can be found in Chapter 3, *Personal Protective Equipment*.

Smoking and Tobacco Products, Alcohol, and Illicit Drugs

Many fire departments have adopted policies that prohibit the use of smoking products (including tobacco, marijuana, and vaping devices) by firefighters, both on duty and off duty. Smoking, especially of tobacco products, is a major risk factor for cardiovascular disease, it reduces the efficiency of the body's respiratory system, and it increases the risk of infections as well as lung and other types of cancer. Firefighters should avoid tobacco products entirely for both health and insurance reasons.

Firefighters who have consumed alcohol within the previous 8 hours must not be permitted to engage in training or emergency operations. Excessive alcohol use can damage the body and affect performance. In addition, alcohol use increases the risk of mouth, throat, larynx, esophagus, liver, colon, and breast cancers (American Cancer Society, 2017).

Both prescription medications and illegal drugs may be abused or misused. Many fire departments have drug-testing programs to ensure that firefighters do not use or abuse legal or illegal drugs. Substance abuse endangers lives.

SAFETY TIP

Everyone is subject to an occasional illness or injury. You should not try to work when ill or injured. Operating safely as a member of a team requires both fitness and concentration. Do not compromise the safety of the team or your personal health by trying to work while you are ill or injured.

Counseling and Critical Incident Stress Management

Many firefighters see more traumatic situations in a short time than most citizens see in their entire lifetime. Firefighting involves not only the stresses directly connected to fighting the fire and to rendering emergency medical care but also the added burdens of disrupted sleep patterns, rotating work schedules, unscheduled overtime, and interrupted meals. High levels of stress can produce a variety of symptoms. Some people are not able to sleep well, others tend to gain or lose weight. Many people become irritable when stressed. Overeating, increased consumption of alcoholic beverages, and the use of nonprescribed drugs may be the result of stress. Stress may produce depression or suicidal thoughts in some people.

TIP

Find time for yourself, your family, and friends as a mental buffer from the stressors of the job.

To diminish the effect of these stressors, it is important to get adequate sleep; maintain a healthy, balanced diet; and get adequate exercise. It is also important for you to balance your work schedule with other activities and monitor your behavioral health. Take vacations to lower your stress levels and improve your physical health so you will be ready to respond the next time you are needed.

If at any point you feel the stress of work seems overwhelming, say something. Talking with family, friends, and co-workers may be a starting point and peer counselors or mental health professionals can provide helpful strategies. Remember, seeking help does not make you weak in the eyes of others—it shows that you are in control of your life.

It is essential to identify and utilize the resources that are available to assist in maintaining behavioral health. Some of these resources are provided by employers, such as the employee assistance programs, which will be discussed later. Other resources and assistance are provided by professional organizations. For example, the International Association of Firefighters (IAFF) offers information, available through their website, as well as the inpatient IAFF Center of Excellence for Behavioral Health Treatment and Recovery.

Critical Incident Stress Management

Critical incidents challenge the capacity of most individuals to deal with stress. It is important to understand what the stressors in this job are and to learn how to work to diminish their effect. Examples of critical incidents include the following:

- Line-of-duty deaths (police, fire/rescue, EMS)
- Suicide of a colleague
- Serious injury to a colleague
- Situations that involve a high level of personal risk to firefighters
- Events in which the victim is known to the firefighters
- Multiple-casualty/disaster/scenes of violence incidents
- Events involving death or life-threatening injury or illness to a victim, especially a child

This list is not complete, nor is it necessarily a fact that any of these situations will seriously trouble every individual. Normal coping mechanisms help many firefighters to handle many situations. Some individuals deal with stressful situations through exercise, talking to friends and family, or turning to their religious beliefs. These are healthy, nondestructive ways to manage the pressures of being exposed to critical incidents.

Post-traumatic stress disorder (PTSD) may develop after a person has experienced a critical incident. It is characterized by reexperiencing the event and overresponding to stimuli that recalls the event. Some of the symptoms of PTSD include depression, startle reactions, flashback phenomena, and dissociative episodes (e.g., amnesia of the event).

Critical incident stress management (CISM) was developed to address acute stress situations and potentially decrease the likelihood that PTSD will develop after such an incident. CISM is used to confront the responses to critical incidents and defuse them, directing the firefighters toward physical and emotional equilibrium. CISM can occur formally, as a debriefing for those who were on scene. In such situations, trained CISM teams of peers and mental health professionals may facilitate this.

These symptoms can occur in anyone, even individuals who normally have healthy coping skills. Reactions vary from one individual to the next, both in type and severity. Many times firefighters do not realize they are affected in a deeply negative manner. A somewhat routine incident, however, may trigger negative reactions from a critical incident that occurred in the past.

A **critical incident stress debriefing (CISD)** is held as soon as possible after a traumatic call. It provides a forum for firefighting and EMS personnel to discuss the anxieties, stress, and emotions triggered by a difficult

SAFETY TIP

Co-workers often notice a change in behavior or attitude before an officer or chief does. This is because firefighters develop close relationships between people who work together and share rooms, meals, and social interactions. Being a firefighter means helping a friend. Talk to your partner about changes you may notice in their behavior. If you are the firefighter having trouble dealing with a crisis, remember, you are not alone. Talk with a trusted officer, chief, or member of your crew.

call. Follow-up sessions can be arranged for individuals who continue to experience stressful or emotional responses after a challenging incident. With peer support teams, specially trained department members help those struggling with day-to-day issues by communicating with them regularly and recommending resources to assist them. Firefighters should understand the resources available to them and know how they can access them. Maintaining good emotional health is a simple but important part of firefighter survival. The aim of counseling, peer support teams, employee assistance programs, and CISM programs is to prevent these emotional reactions from having a negative impact on the firefighter's work and life, over both the short term and the long term (**FIGURE 2-6**). The major difference between CISM and peer support teams is that CISM teams usually respond immediately after a crisis incident, and peer support teams provide continuous, ongoing support.

Defusing sessions are the first to occur. These sessions are held during the event or immediately afterward. A group informally discusses events that they experienced together. Defusing sessions are designed to educate the participants about the expectations over the next few days and give guidance on proper techniques to manage the feelings they may be experiencing. One example is to discourage drinking alcohol during this stressful time.

Debriefing sessions are held within 24 to 72 hours of a major incident. These meetings are held by a CISM team consisting of peers and mental health professionals. At the debriefing session, pent up emotions can be properly expressed. It is more likely you will be ready to express your emotions more freely a few days following the event.

One of the important rules associated with the debriefing session is to not turn it into an operational critique. No one is right. No one is wrong. No one is to blame. Only emotions about the specific event are to be relayed. These debriefing sessions may also need to be repeated at a later time.

FIGURE 2-6 Group stress defusing sessions are sometimes used to alleviate stress reactions generated by high-stress emergency situations.

TIP

CISM programs are located throughout the United States. You can locate a CISM program in your area via the Internet, or it can be requested through your employer.

TIP

CISM can be a helpful strategy, but it may not be effective for everyone. Some people are not receptive to openly discussing psychologically traumatic memories. When an individual's behavior is noticeably different after a traumatic event and CISM is not an option, private counseling by a mental health professional may be invaluable.

Burnout

Critical incident stress also can be cumulative, building up over time. This condition, which is called burnout, cannot be traced to any one incident. Like other negative aspects of stress, burnout affects not only the personal well-being of the firefighter, it also affects their colleagues and the victims they care for due to increased errors and decreased performance. Firefighters suffering from burnout contribute to decreases in work morale, overall work effort, and effective teamwork, as well as an increase in job turnover.

Compassion Fatigue

Sometimes referred to as "the cost of caring", **compassion fatigue** is common among those who work in health care and disaster and emergency services. Compassion fatigue, also known as secondary stress disorder, is a disorder characterized by gradual lessening of compassion over time. It differs from PTSD in that PTSD is caused by direct exposure to a traumatic incident or series of traumatic incidents, whereas compassion fatigue is a reaction to caring for others who have experienced trauma themselves. Symptoms of compassion fatigue include the following:

- High absenteeism
- Difficult relationships with colleagues and co-workers
- Inability to work in teams
- Aggressive behavior toward victims
- Strong negative attitudes toward work
- Lack of empathy for patients
- Judgmental attitude toward victims
- Preoccupation with non-work issues while on duty
- Other symptoms of increased stress

Supporting victims in emergency situations is difficult. It is stressful for them and for you. Firefighters are vulnerable to all the stresses that go with the profession. It is critical that firefighters recognize the signs of cumulative stress so it does not interfere with their work or life away from work, including their family life. The signs and symptoms of cumulative stress may not be obvious at first. Rather, they may be subtle and not present all the time, such as the following:

- Irritability toward co-workers, family, and friends
- Inability to concentrate
- Difficulty sleeping, increased sleeping, or nightmares
- Feelings of sadness, anxiety, or guilt
- Indecisiveness
- Loss of appetite (gastrointestinal disturbances)
- Loss of interest in sexual activities
- Isolation
- Loss of interest in work
- Increased use of alcohol
- Recreational drug use
- Physical symptoms such as chronic pain (headache, backache)
- Feelings of hopelessness

Firefighting is a job that requires close teamwork. It is the responsibility of all firefighters to monitor themselves and other members of their team for signs of mental stress. We are all interdependent on the members of our team in order to accomplish our goal. The team needs to function well, not only at emergency scenes but also when performing routine tasks or relaxing between runs. It is important that you respect each member of your team. Bullying or discrimination should never be tolerated. It is everyone's job to be on the lookout for signs of unhealthy behavior or stress. If you think a fellow firefighter is exhibiting signs of stress or experiencing depression, there are several ways in which you may be able to help. Although the ways in which this outcome is accomplished vary, the purpose remains the same.

Sometimes it may be helpful to sit down with a co-worker and ask them how they are doing or to let them know that you are concerned about them. You may be able to encourage them to seek help. If you see signs of stress, talk with your officer or someone who can help that person receive assistance.

Suicide Awareness and Prevention

While suicide is the 11th leading cause of death in the United States, the rate among emergency responders is much higher. According to the Ruderman White Paper on Mental Health and Suicide of First Responders, firefighters and law enforcement personnel are more likely to die of suicide than from life-of-duty death. In 2016, there were 99 reported firefighter deaths as a result of suicide (Firefighter Behavioral Health Alliance, 2017). Estimates reveal that a fire department is three times more likely to experience a suicide in any given year than a line-of-duty death (The National Fallen Firefighters Association, 2017). This fact reminds us that stressful occupations may experience a higher rate of suicides than the general population. Many people are concerned about the incidence of suicide in firefighters and other emergency care providers.

While firefighters are trained to be the best they can be, many are not prepared for the ill effects or aftermath of stress or a traumatic situation. It is important to understand that by recognizing signs of stress or depression in a team member you may be able to help them get assistance for their problem before it becomes worse.

Numerous national services are available to firefighters. These include the following:

- Fire/EMS Helpline at 1-888-731-FIRE (3473)
- National Suicide Prevention Lifeline at 988

TIP

Just as 911 is called for help in an emergency, 988 is the number to call for help during in a behavioral health emergency. The Suicide and Crisis Lifeline, available by calling 988, is free, confidential, and staffed with trained professionals who are ready to listen.

TIP

According to the Firefighter Behavioral Health Alliance (FBHA), the top five warning signs of job-related stress include the following:

- Recklessness/impulsiveness
- Anger
- Isolation
- Loss of confidence in abilities or skills
- Sleep deprivation

If you believe you are suffering from these issues please seek help from your employee assistance program, chaplain, peer support team, or a qualified counselor in your community.

Employee Assistance Programs

There are many formal programs to support firefighter behavioral health. These programs, also known as member assistance programs, are maintained by fire departments, unions, local governments, and even charitable organizations. Support is provided through special events, projects, peer support, chaplain programs, and education. A traditional and common form of support for firefighters is the employee assistance program. **Employee assistance programs (EAPs)** provide confidential help with a wide range of problems that might affect performance. Many fire departments have established EAPs so that firefighters can get counseling, support, or other assistance in dealing with physical, financial, emotional, or substance use issues. EAPs include a variety of helpful resources, including access to qualified counselors and chaplains who have a working knowledge of the fire service. Some fire departments have qualified counselors available 24 hours a day. The initial counseling may consist of a group session for all firefighters and rescuers; alternatively, it can also be done on a one-on-one basis or in smaller groups. A fire officer may refer a firefighter to an EAP if a problem starts to affect the individual's job performance. Firefighters who take advantage of an EAP can do so with complete confidentiality and without fear of retribution.

According to the FBHA, many do not seek help, mainly due to confidentiality, job promotion concerns, or simply because counselors may not know the culture of the fire service. The FBHA recommends inviting counselors to the station, including them on ride time, in training, and during meals. Another suggestion is to have the EAP counselors create a video biography so department members and their families can see who they are and get to know them.

Safety during Training

During training, firefighters learn and practice the skills that they will use later under emergency conditions. Typically, the patterns that develop during training continue during actual emergency incidents. Thus, developing the proper working habits during training courses helps ensure safety later. Use of proper protective gear and good teamwork are as important during training as they are on the fireground.

Instructors and veteran firefighters are virtually always willing to share their experiences and advice. They can explain and demonstrate every skill and point out the safety hazards involved because they have performed these skills hundreds of times and know what to do. But here, too, safety is a shared responsibility. Do not attempt anything you feel is beyond your ability or knowledge. If you see something that you believe is an unsafe practice, bring it to the attention of your instructors or a designated safety officer.

Avoid freelancing. Wait for instructions or orders before beginning any task. Do not assume that something is safe and act independently. Follow instructions and learn to work according to the proper procedures.

Teamwork is important during training exercises. Assignments are given to firefighting teams during most live fire exercises. Teams must stay together. If any member of the team becomes fatigued, is in pain or discomfort, or needs to leave the training area for any reason, notify the instructor or safety officer. EMS personnel should be available to perform an examination and to transport ill or injured personnel to further treatment if necessary. A firefighter who is injured during training should not return until medically cleared for duty.

Safety during the Emergency Response

Safety during an emergency response must begin before there is an actual response. Safety begins with

Voice of Experience

My department is a small one, and we are familiar with our residents because most of our members grew up here. On March 5, 2016, we encountered a tragedy that was inconceivable. At approximately 4:30 in the afternoon we received a call that changed our members and community forever. It was a reported structure fire in a single-family residence with possible people trapped. Three family members had escaped, but unfortunately, two small children were trapped on the second floor of the residence. An aggressive rescue attempt was performed, but it was too late. We lost two innocent children from our community. Although our performance was never questioned, many members blamed themselves, and the event changed the mentality and drive of our department.

The day started off fairly normally. We were sitting on the bumper of the engine talking shop and enjoying the beautiful weather. Within seconds, our lives changed forever. The call came in and the crew scrambled to get ready. As we mounted the apparatus, dispatch announced that there were two children trapped on the second floor. My position on the truck that day was the officer's seat. As we rolled out of the station, I remember looking over my right shoulder toward the address and seeing a column of heavy black smoke. The scene was approximately one-third of a mile from the firehouse. As we approached the scene there was heavy fire coming from the second floor of the front of the home. As I exited the engine I was met by three individuals who had been burned. The female, who turned out to be the mother, grabbed the leg of my bunker pants and stated "My babies are up there." No training in the world can prepare you to hear cries from a mother. An aggressive rescue attempt was made and the children were located, but it was too late.

As the scene progressed, our chief removed us from the scene. At the time, not being able to continue what I had started made me extremely angry; however, as the days went by, I realized that my chief had done the right thing. As we made our way back to the firehouse to grab our belongings and head home, there were a number of questions that kept entering my head. Why? What did these innocent children do to anyone? What did we do to deserve this? Unfortunately, there were no answers to these questions.

In the days following the fire, I tried to pretend that everything was okay, but it wasn't. I was unwilling to admit that I needed help. I would lie in bed at night and have nightmares that I was back in the house standing over the two children. I had suicidal thoughts, and my family was torn apart because I had refused to accept the fact that I needed help. All of this was a part of me trying to deal with this tragedy. Finally, after about 2 weeks, my chief, who is also a close friend, came to me and suggested I see someone to get the help I needed. I took his advice. I was put in contact with a grief counselor and spent an entire week at a treatment program. The treatment program gave me the courage I needed to realize that I would be okay. While I was told to never forget the incident, I was shown how to deal with it properly.

Part of the way I chose to properly deal with this tragedy is through training. I have always been active in fire training but this event showed me the reasoning behind it. Now, I train for those two children. Training keeps my mind focused on the positives. If we are the best at what we do, then we will never blame ourselves or question our actions.

In closing, don't ever be afraid to ask for help. Your family, friends, and co-workers are there for you. If you try to deal with it on your own, you have the potential to lose it all. If it wasn't for my son, my family, and a department that truly cares about our mental health, the outcome may have been much different.

James Hopkins
Orange Fire Department
Orange, MA

firefighters being prepared for emergencies. Firefighters must be ready to respond to an emergency at any time during their tour of duty. This process begins by ensuring that your PPE is complete, ready for use, and in good condition. At the beginning of each tour of duty, place your PPE in its designated location, which depends on your assigned riding position on the apparatus (**FIGURE 2-7**).

You should also conduct a daily inspection of the SCBA for your riding position at the beginning of each

FIGURE 2-7 Protective clothing should be properly positioned so that you can quickly don it.

FIGURE 2-8 Conduct the daily check of your SCBA at the beginning of each tour of duty.

tour of duty (**FIGURE 2-8**). The air cylinder should be full, the face piece clean, and the personal alert safety system (PASS) operable. Also check the availability and operation of your hand light and any hand tools you

FIGURE 2-9 Check the tools assigned to your riding position.

might require, based on your assigned position on the apparatus (**FIGURE 2-9**). Recheck your PPE and tools thoroughly when you return from each emergency response and clean them whenever they are used. Personal protective gear should be properly positioned so that you can don it quickly. This routine will help ensure that your gear and equipment are fully functional when the next alarm comes.

Every response to an incident has a high potential for accidents, injuries, and death. **Response** actions include receiving the alarm, donning protective clothing and equipment, mounting and dismounting the apparatus, and transporting equipment and personnel to and from the emergency incident quickly and safely.

Alarm Receipt

When an alarm is received, the response should be prompt and efficient. Responding firefighters should walk briskly to the apparatus. There is no need to run; the objective is to respond quickly, without injuring anyone or causing any damage. Follow established

SAFETY TIP

Volunteer firefighters who are not assigned to specific tours of duty or riding positions should check their PPE, SCBA, and associated tools and equipment on a regular basis to ensure that these items are ready for use whenever an emergency response becomes necessary. After each use, all PPE should be carefully cleaned and checked before it is put away.

procedures to ensure that stoves, faucets, and other appliances at the station are shut off. Wait until the apparatus bay doors are fully open before leaving the station.

Riding the Apparatus

Expectations for donning PPE are frequently established by the company officer. A common practice is for firefighters to don PPE prior to mounting the apparatus. While a fire apparatus is in motion, all crew members should be wearing seat belts properly. Do not attempt to don PPE while the apparatus is on the road. Wait until you dismount at the incident scene to don any protective clothing that was not donned prior to mounting the apparatus. Don an SCBA only after the apparatus stops at the scene, unless the SCBA is seat mounted.

All equipment should be properly mounted, stowed, or secured on the fire apparatus. Unsecured equipment in the crew compartment can prove dangerous if the apparatus must stop or turn quickly, because a flying tool, map book, or PPE can seriously injure a firefighter.

Be careful when mounting apparatus, because the steps on fire apparatus are often high and can be slippery. Use handrails when mounting or dismounting the apparatus. Follow the steps in **SKILL DRILL 2-1** to mount an apparatus properly.

All firefighters must be seated in their assigned riding positions with seat belts and/or harnesses fastened before the apparatus begins to move. NFPA 1500, *Standard on Fire Department Occupational Safety, Health, and Wellness Programs*, and NFPA 1010, *Standard for Firefighter Professional Qualifications*, require all firefighters to be in their seats, with seat belts secured, whenever the vehicle is in motion. Do not unbuckle your seat belt to don any clothing or equipment while the apparatus is en route to an incident. Vehicle occupants who are not wearing seat belts are much more likely to suffer serious injuries or death than are occupants who have their seat belts properly fastened. Air bags are most effective when seat belts are properly fastened; they are not effective without properly applied

SKILL DRILL 2-1

Mounting Apparatus Firefighter I, NFPA 1010: 6.3.2

1. When mounting (climbing aboard) fire apparatus, always have at least one hand firmly grasping a handhold and at least one foot firmly placed on a foot surface. Maintain the one hand and one foot placement until you are seated.

2. Fasten your seat belt, and leave it fastened until the apparatus is stopped at its destination. Don any other required safety equipment for the response, such as hearing protection and intercom systems.

seat belts. Seat belts also greatly reduce the possibility that vehicle occupants will be ejected from the vehicle.

For safety's sake, SOPs often prohibit specific actions during response. As noted previously, firefighters must remain seated with their seat belts securely fastened while the emergency vehicle is in motion. Never unfasten your seat belt to retrieve or don equipment. Do not dismount the apparatus until the vehicle comes to a complete stop. In addition, you should never stand up while riding on apparatus. Do not hold on to the side of a moving vehicle or stand on the rear step. When a vehicle is in motion, everyone aboard must be seated and belted in an approved riding position.

The noise produced by sirens and air horns can have long-term, damaging effects on a firefighter's hearing. For this reason, your department should provide hearing protection for personnel riding on fire apparatus. Some of these devices include radio and intercom capabilities so that firefighters can talk to one another and hear information from the dispatcher or IC.

During response, limit conversation to the exchange of pertinent information. Listen for instructions from the IC, for instructions from your company officer, and for additional information about the incident over the radio. The vehicle operator's attention should be focused on driving the apparatus safely to the scene of the incident.

The ride to the incident is a good time to consider any relevant factors that could affect the situation. These factors could include the time of day or night, the temperature, the presence of precipitation or wind, the type of occupancy, the type of construction, and the location and type of incident. Using this time to think ahead will help you mentally prepare for the various possibilities that you might encounter at the scene.

When the apparatus arrives at the incident scene, the driver/operator will park it in a location that is both safe and functional. Wait until the vehicle comes to a complete stop before dismounting. Always check for traffic before opening the doors or stepping out of the apparatus. During the dismount, watch for other hazards—for example, ice and snow, downed power lines, uneven terrain, or hazardous materials—that could be present.

Be careful when dismounting apparatus. The increased weight of PPE and adverse conditions can contribute to slips, strains, and sprains. Follow the steps in **SKILL DRILL 2-2** to dismount an apparatus safely.

Traffic Safety on the Scene

An emergency incident scene presents several risks to firefighters in addition to the hazards of fighting fires and performing other duties. One of these dangers is traffic, particularly when the incident scene is on a

SKILL DRILL 2-2

Dismounting a Stopped Apparatus Firefighter I, NFPA 1010: 6.3.2

1. Become familiar with your riding position and the safest way to dismount.

2. Maintain the one hand and one foot placement when leaving the apparatus, especially on wet or potentially icy roadway surfaces.

street or highway. Traffic safety should be a major concern for the first-arriving units because approaching drivers might not see emergency workers or realize how much room firefighters need to work safely.

The first unit or units to arrive at the incident scene have a dual responsibility. Not only must the firefighters focus on the emergency situation facing them, they also must consider approaching traffic, including other emergency vehicles, and other, less obvious hazards. Always check for traffic before opening doors and dismounting the apparatus and watch out for traffic when working in the street.

One of the most dangerous work areas for firefighters is on a roadway, where traffic can be approaching at high speeds. Follow departmental SOPs to close streets quickly and to block access to areas where operations are being conducted. Place traffic cones, flares, emergency scene signage, and other warning devices far enough away from the incident to allow inbound traffic to slow down and be directed away from the work area (**FIGURE 2-10**). Police officers should assist by diverting traffic at a safe distance outside the hazard area. It is not uncommon to set large control zones at the onset of an incident, only to discover that the zones may have been established too liberally. At the same time, control zones should not be defined too narrowly. As the IC gets more information about the incident, the control zones may be expanded or reduced. Wind shifts are a common reason why control zones are modified during the incident. If there is a prevailing wind pattern in your area, then factor that information into your decision-making process when it comes to control zones.

Keep in mind that even a work area that is properly marked and blocked by apparatus, the work area remains a dangerous place. Firefighters must maintain situational awareness for those who may present hazards to an emergency scene, especially drivers who are distracted, drowsy, or intoxicated.

FIGURE 2-10 The scene of a crash should be marked properly, and traffic should be diverted so that responders have enough room to work.

Placement of emergency vehicles on the scene is also critical. With proper placement, such vehicles can act as a barrier between oncoming traffic and the scene. All firefighters working at highway incidents should wear high-visibility safety vests in addition to their normal PPE. These vests should meet the ANSI 207 standard for public safety vests. Many fire departments have specific SOPs covering required safety procedures for these incidents.

SAFETY TIP

- Do not attempt to mount or dismount a moving vehicle.
- Do not remove your seat belt until the apparatus comes to a complete stop.
- Do not stand directly behind an apparatus that is backing up. Instead, stand off to one side, where the driver can see you in the rear view mirror. All fire apparatus should have working, audible alarms when in reverse gear.

Safe Driving Practices

In 2020, the NFPA estimated there were 15,675 collisions involving fire department emergency vehicles responding to or returning from incidents (NFPA, U.S. Firefighter Injuries, 2020). As discussed earlier, in 2020, 15 firefighters died in vehicle-related incidents. In addition, 4975 firefighters were injured responding to or returning from incidents (NFPA, Firefighter Fatalities in the United States, 2020).

Drivers of fire apparatus have a great responsibility to get the fire apparatus and the crew members to the emergency scene without having or causing a traffic accident en route. Drivers must always exercise caution when driving to an incident. Fire apparatus can be large, heavy, and difficult to maneuver, and they must know the streets in their first-due area and any target hazards. They must be able to operate the vehicle skillfully and keep it under control at all times. In addition, they must anticipate all responses from other drivers who might not see or hear an approaching emergency vehicle or know what to do when confronted by one. Operating an emergency vehicle without the proper regard for safety can endanger the lives of both the firefighters on the vehicle and any civilian drivers and pedestrians encountered along the way.

Prompt response is a goal, but safe response is a much higher priority. A major factor in crashes is the attitude and ability of the vehicle operator. A competent emergency vehicle operator needs to have a confident, but not arrogant, attitude. Aggressive driving has no place in the fire service. The emergency vehicle operator should have good judgment and reactions, mental fitness, maturity, physical fitness, alertness, and good driving habits. It is important to maintain a clean driving record, because a person who has been cited for multiple moving violations in their personal vehicle will usually be an increased risk as the driver of an emergency vehicle. To be a good driver, it is important to know the state and local laws relating to motor vehicle operations. In addition, an emergency vehicle operator needs to understand the reaction time, braking distance, and stopping distance of the vehicle.

Emergency driving, even when properly performed, increases risks. Impaired driving dramatically increases those risks. Impairment can result from many different sources—for example, using some prescribed medicines, using some over-the-counter medicines, and being overly fatigued. Anyone who has been drinking alcoholic beverages should not drive. Driving while eating, texting, or talking on a communications device are all forms of distracted driving. Even the use of devices with navigation technology should not be used in such a way that the driver is distracted. Distracted driving during routine driving or when responding to an emergency is to be avoided—it is the cause of many collisions and deaths. In some cases, this has led to criminal prosecution and conviction.

Firefighters who drive emergency vehicles must have special driver training and know the laws and regulations that apply to emergency responses. Many jurisdictions require a special driver's license to operate fire apparatus. The rules that apply to emergency vehicles are specific, and the driver/operator is legally responsible for the safe operation of the vehicle and the safety of the vehicle occupants at all times. These driving skills are not required for the Firefighter I and Firefighter II courses and are beyond the scope of this resource.

Although most states and provinces permit drivers of emergency vehicles to take exception to specific traffic regulations when responding to emergency incidents, driver/operators must always consider the potential actions of other drivers before making such a decision. For example, traffic laws require other drivers to yield the right of way to an emergency vehicle. There is no assurance, however, that other drivers will do so when an emergency vehicle approaches. The driver/operator also must anticipate which routes other units responding to the same incident will take. All passengers and drivers of emergency vehicles should wear seat belts on routine and emergency responses.

Many collisions occur when the motor vehicle operator loses control of the vehicle. These crashes may be caused by driving too fast for the prevailing conditions, braking inappropriately, changing directions too abruptly, or tracking around a curve too fast. In addition, many collisions involving emergency vehicles occur on open roads, where excessive speed is often cited as a primary cause. Intersections are common sites of collisions involving emergency vehicles; most of these collisions occur when the emergency vehicle operator fails to stop at an intersection to ensure that other traffic has stopped before proceeding through the intersection.

Finally, a motor vehicle collision itself consists of a series of separate collision events. The first collision occurs when the vehicle collides with a second vehicle or with a stationary object; the second collision occurs when the occupants of the vehicle collide with the interior of the vehicle.

SAFETY TIP

Always fasten your seat belt each and every time you get into a motor vehicle. It is vital that you follow this rule each and every time you climb into a fire apparatus.

Laws and Regulations Governing Emergency Vehicle Operation

Four general principles govern emergency vehicle operation:

- Emergency vehicle operators are subject to all traffic regulations unless a specific exemption is made. A specific exemption is a statement that appears in a statute, such as "The driver of an authorized vehicle may exceed the maximum speed limits so long as he or she does not endanger life or property."
- Exemptions are legal only when the vehicle is operating in emergency mode.
- Even with an exemption, the emergency vehicle operator can be found criminally or civilly liable if involved in a crash.
- An exemption does not relieve the operator of an authorized emergency vehicle from the duty to drive with reasonable care for all persons using the highway.

Laws governing emergency vehicle operation vary from one state to another. You must follow the laws and regulations of your state regarding emergency vehicle operations. Some states permit private vehicles to operate as emergency vehicles when responding to a fire station or to the scene of an emergency; other states outline a limited use of emergency equipment and dictate certain restrictions. You must understand and follow the specific laws and regulations of your state as well as your department's SOPs.

In many cases, one of the best predictors of future performance is past behavior. Your driving record is an important consideration in your career in a fire department. A person who has past moving violations for speeding, reckless operation, aggressive driving, chargeable crashes, or driving under the influence of alcohol or drugs may be excluded from driving any emergency vehicles. Your driving record, both on duty and off duty, is important to your career as a firefighter.

SAFETY TIP

Developing situational awareness on the emergency scene is essential to your safety and survival. This concept can be summed up by saying that you must maintain a "big picture" view of the emergency scene and avoid "tunnel vision." Situational awareness requires that firefighters observe, not just see. Observation is deliberate. Seeing is casual. This ability develops through experience and with well-developed competency in basic firefighter skills. The less you have to concentrate on the components of your task, the more you are able to observe your surroundings. This quality emerges through hours and hours of practice and skill repetition. Do not stop practicing until your basic skills become procedural memories. If you can do your job with your eyes closed, you will be able to observe much more when they are open!

Standard Operating Procedures for Personal Vehicles

In some departments where firefighters are not on duty at the fire station, it may be necessary for members to respond to the emergency scene in their private vehicles. For you to use your personal vehicle in a situation requiring an emergency response, this use must be permitted by state laws and regulations, and this operation must also be permitted by your local fire department. If your local fire department does not permit the use of personal vehicles for emergency responses, then you are not permitted to engage in this behavior regardless of what is permitted by your state's laws.

Firefighters who respond to emergency incidents in their personal vehicles must follow the specific laws and regulations of their state or province and follow departmental SOPs related to this issue. Some jurisdictions require firefighters to equip their privately owned vehicles with warning devices and to operate them as emergency vehicles; others grant no special status to privately owned vehicles driven during an emergency response. In some areas, volunteer firefighters responding to an emergency incident use colored lights to request the right of way from other drivers.

Your fire department should have SOPs that spell out whether private vehicles can be used for emergency response. These procedures typically address which kind of training you must complete before you are permitted to use your personal vehicle for emergency response. This training usually includes a course in defensive driving and functioning as an emergency vehicle operator.

Vehicle Collision Prevention

Safe driving practices will prevent most vehicle collisions. If you are using your personal vehicle to respond to an emergency, it is important to take the characteristics of your vehicle into account while driving. For example, four-wheel drive pickup trucks and large sport utility vehicles have a higher center of gravity than sleek sports cars, and sudden changes in direction are more likely to result in roll-over crashes when a vehicle has a high center of gravity. For this reason, it is critical to learn the characteristics and limitations of your vehicle.

Anticipate the road and road conditions. If you regularly travel the same roads to report to your fire station, learn the characteristics of that strip of roadway, and be able to anticipate when you need to slow down or stop. Always drive at a safe speed for that road: Traveling on a limited-access highway is much different from traveling through a school zone, for example. Observe the traffic conditions, and slow down when traffic congestion is present. Expect that motorists around you may do anything at any time. Upon the approach of an emergency vehicle, for example, some motorists will slow down, some will stop, some will speed up, some will pull to the right, and others will turn left in front of you. Expect the unexpected.

Make allowances for weather conditions. On a clear day, it might seem that you can see forever, but, really, you cannot. In conditions of rain, snow, fog, dust, or darkness, the distance over which you can see is greatly reduced; reduce your speed to compensate for the limited visibility. There also may be vehicle collisions, downed trees or power lines, or debris in the road. At night, your vision is limited by the distance your headlights reach, so reduce your speed accordingly. Recognize that you cannot see objects outside the projection of the headlight beams. Understand that the goal of an emergency response is not to drive as fast as you can, but rather to arrive on the scene as quickly as you can while maintaining safety.

Adjust your speed of response to accommodate any storm conditions. Rainstorms reduce visibility and produce slippery road surfaces. When water collects on a roadway, your vehicle can hydroplane on the thin film of water that separates your tires and the road surface. You have no control over your vehicle when hydroplaning occurs. Be aware of high winds from tornadoes, hurricanes, or other windstorms that can topple trees onto the roadway. During heavy rainstorms, watch for water flowing over the road. You cannot see how deep the water is, and you cannot tell whether the floodwater has washed out part of the roadway. Slow down when driving following an earthquake or tornado because you will not know where the roadway, bridges, and overpasses have been blocked, damaged, or destroyed.

When operating an emergency vehicle, you are not exempt from the laws of physics. The laws of physics tell us that when the speed of a vehicle doubles, the force exerted by that vehicle increases by a factor of four. Thus, when the speed of a vehicle increases from 20 miles per hour to 40 miles per hour (32 kilometers per hour to 64 kilometers per hour), the force exerted by that vehicle is increased by a factor of four. Higher speeds require more braking power and a longer distance to bring the vehicle to a stop.

To drive safely, you need to understand where the greatest risks are located. As mentioned earlier, studies have shown that many crashes involving emergency vehicles occur at intersections (**FIGURE 2-11**). Most fire departments require all emergency vehicles to come to a full stop when the emergency vehicle operator encounters a stop sign or a red traffic light. If you enter an intersection without stopping, you may not be able to see approaching traffic, and many motorists do not drive defensively. They assume that if there is no stop sign or red light, they can proceed through the intersection without looking for oncoming vehicles.

In addition, many emergency vehicle crashes occur on open stretches of straight roads during daylight hours. The primary cause of these crashes is excessive speed. Most of us have limited experience with high-speed driving, so it is important to keep within the limits of your driving skills and within the limits of the vehicle you are operating. Many departments limit the speed during emergency responses to 5 miles per hour (8 kilometers per hour) over the posted speed limit. Multiple studies have shown that little time is saved by driving at high rates of speed.

During an emergency response, be alert for the presence of other emergency vehicles. These vehicles may carry other firefighters responding from their homes or fire, EMS, or law enforcement personnel responding to the same incident. It is difficult to hear the sounds of other apparatus when you are on an emergency response.

Drive with a cushion of safety around you—that is, in such a manner that even if a motorist does the unexpected, you still have enough time and distance to avoid a collision. Remember that the sound of a siren carries a limited distance. Sometimes in an urban environment motorists may not be able to determine the location of an approaching emergency vehicle because of surrounding buildings, which distort the source of the siren. Similarly, emergency lights have limitations and cannot be depended on to clear a path of travel for you.

TABLE 2-4 summarizes the guidelines to follow when driving to respond to an emergency call.

The Importance of Vehicle Maintenance

Do not underestimate the importance of proper vehicle maintenance for both fire department vehicles and your private vehicle. It is important to perform regular maintenance of the engine, transmission, and other equipment to ensure dependable starting and running. Keep the brakes and tires in proper condition to ensure dependable stopping. In addition, note that tire tread depth for emergency vehicles should exceed the minimum required for non-emergency vehicles. The suspension system and steering system need to be properly maintained to ensure safe handling of the vehicle. Likewise, the windshield wipers and windshield washers must be kept in good condition to ensure clear vision. Regularly check all headlights, taillights, and turn signals for proper operation. Private vehicles that are used for emergency response should be maintained at a higher level than vehicles that are not subject to the stresses of emergency operation.

Safety at the Incident

At the emergency scene, firefighters should never charge blindly into action. From the moment that firefighters arrive at an emergency incident scene, their department's SOPs and the structured ICS must guide all

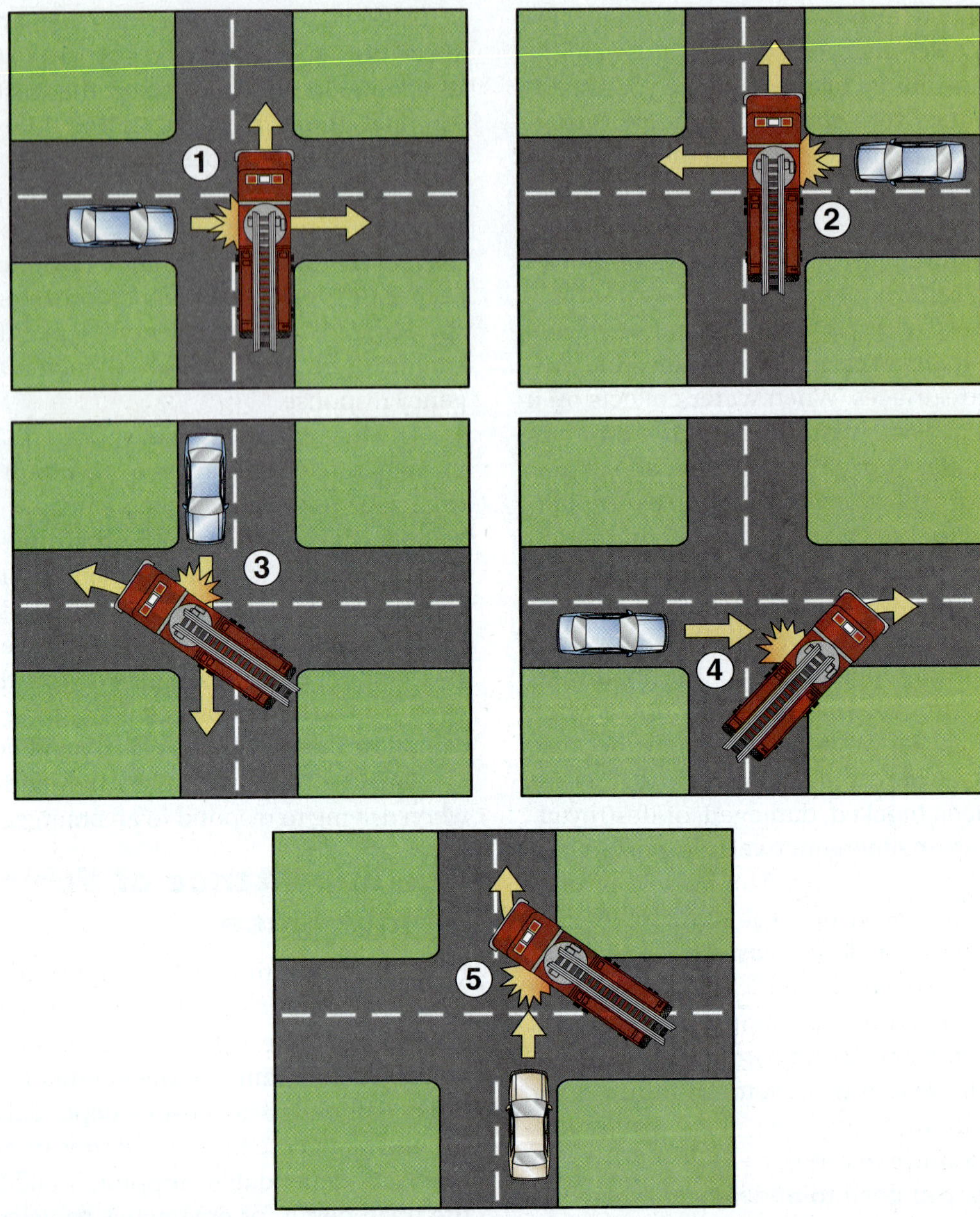

FIGURE 2-11 Every two-lane intersection has five potential crash points, making intersections very dangerous.

TABLE 2-4 Guidelines for Safe Emergency Vehicle Response

1. Drive defensively.
2. Follow agency policies in regard to posted speed limits.
3. Always maintain a safe distance. Use the "4-second rule." Stay at least 4 seconds behind another vehicle in the same lane.
4. Maintain an open space or cushion in the lane next to you as an escape route in case the vehicle in front of you stops suddenly.
5. Always assume that other drivers will not hear your siren or see your emergency lights.
6. Select the shortest and least congested route to the scene at the time of dispatch.
7. Visually clear all directions of an intersection before proceeding.
8. Go with the flow of traffic.
9. Watch carefully for bystanders and pedestrians. They may not move out of your way or could move the wrong way.

of their actions. The officer in command will "size up" the situation, carefully evaluating the conditions to determine whether the burning building is safe to enter or what needs to be done to make the scene safe for action to take place. Your job is to pay attention to your surroundings and to report things that do not match your expectations. For example, observe the time of day, the weather conditions, the people on the scene, and fire and smoke conditions. Wait for instructions and follow directions for the specific tasks to be performed. Firefighters should always work in assigned teams (companies or crews) and be guided by a strategic plan for the incident. Teamwork and disciplined action are essential to provide for the safety of all firefighters and the effective, efficient conduct of operations.

Freelancing, whether done by individual firefighters, groups of firefighters, or full companies, is unacceptable. Freelancing cannot be tolerated at any emergency incident. The safety of every person on the scene can be compromised by firefighters who do not work within the system.

For example, a firefighter who enters a burning structure without informing a superior may be trapped by rapidly changing conditions. By the time the firefighter is missed, it may be too late to perform a rescue. Searching for a missing firefighter exposes others to unnecessary risk.

Personnel must not respond to an emergency incident scene unless they have been dispatched or have an assigned duty to respond. Unassigned units and individual personnel arriving on the scene can overload the IC's ability to manage the incident effectively. Individuals who simply show up and find something to do are likely to compromise their own safety and create more problems for the command staff. All personnel must operate within the established system, reporting to a designated supervisor under the direction of the IC.

SAFETY TIP

Teamwork is a hallmark of effective fireground operations. Training and teamwork produce effective, coordinated operations. Unassigned individual efforts and freelancing can disrupt the operations and endanger the lives of both firefighters and civilians.

Teamwork

On the fireground or emergency scene, a firefighting team should always consist of at least two firefighters who work together and are in constant communication with each other. Some departments call this scheme the buddy system (**FIGURE 2-12**). In some cases, firefighters work directly with the company officer, and all crew members function as a team. In other situations, two individual firefighters may form a team that is assigned to carry out a specific task. In either case, the company officer must always know where teams are and what they are doing.

FIGURE 2-12 A firefighting team should consist of at least two members who work together.

Partners or assigned team members should enter together, work together, and leave together. If one member of a team must leave the fire building for any reason, the entire team must leave together, regardless of whether it is a two-person team or an entire crew working as a team.

Before entering a burning building to perform interior search and rescue or fire suppression operations, firefighters must be properly equipped with approved and appropriate PPE. Partners should check each other's PPE to ensure it has been donned and is working correctly before they enter a hazardous area. Team members should maintain visual, vocal, or physical contact with one another at all times. At least one member of each team should have a portable radio to maintain contact with the IC or a designated individual in the chain of command who remains outside the hazardous area. Two firefighters must remain outside the hazardous area, properly equipped to respond immediately if the entry team needs to be rescued. This rapid intervention crew/company (also known as a rapid intervention team) must be able to communicate with the entry team, either by sight or by radio, and should be ready to provide assistance. For more information about the importance of teamwork and a rapid intervention crew/company, see Chapter 19, *Firefighter Survival.*

Personnel Accountability

Every fire department must have a **personnel accountability system** to track personnel and assignments on the emergency scene. This system should record the individuals assigned to each company, crew, or entry team; the assignments for each team; and the team's current activities. If any firefighters are reported missing, lost, or injured, the accountability system can identify who is missing and what their last assignment was. Several kinds of accountability systems are acceptable, ranging from paper assignments or display boards to laptop computers and electronic tracking devices.

Accountability systems provide an up-to-date accounting of everyone who is working at the incident and how they are organized. At set intervals, an accountability check, or personnel accountability report, is completed to account for everyone. A **personnel accountability report (PAR)**—an accountability check or roll call taken by a supervisor at an emergency incident that may be conducted as part of the personnel accountability system—is also completed when the operational strategy changes or when a situation occurs that could endanger firefighters. Usually, a company officer reports on the status of each crew. The company officer should always know exactly where each crew is and what it is doing. If a crew splits into two or more teams, the company officer should be in contact with at least one member of each team. A list of personnel and assignments should be readily available at the command post.

Firefighters must learn which kind of accountability system their department uses, how to work within it, and how this system works within the ICS. They are responsible for complying with this system and staying in contact with a company officer or assigned supervisor at all times. Most importantly, firefighters must always remember that teams must stay together.

Chapter 19, *Firefighter Survival,* covers many of these accountability systems, self-survival skills, and the steps for calling for help ("mayday").

Scene Hazards

Firefighters must be aware of their surroundings when performing their assigned tasks at an emergency scene. At an incident, make a safe exit from the fire apparatus and look at the building or situation for safety hazards such as traffic, downed utility wires, and adverse environmental conditions. The introduction of new technologies requires firefighters to be familiar with their run areas and observant of new hazards. This includes the increasingly common photovoltaic power systems (solar power systems) and battery energy storage systems within electrical power grids. An incident on a street or highway must first be secured with proper traffic- and scene-control devices. A variety of traffic incident management techniques may be appropriate. Flares, traffic cones, and barrier tape are all measures that can help keep the public at a safe distance from the scene and to divert traffic around the area. Emergency vehicles can be placed to block traffic and protect the incident scene. Always operate within established boundaries and protected work areas.

Changing fire conditions also will affect safety. Through situational awareness, firefighters monitor the changing conditions and maintain safety. This should continue even during the overhaul phase and while picking up equipment—watch out for falling debris, smoldering areas of fire, and sharp objects. If a safety officer is not on the scene, another qualified person should be assigned to monitor the atmosphere to ensure it is safe to remove SCBA or enter the area without SCBA. Because the chance for injury increases when you are tired, do not let down your guard even though the main part of the firefighting operation is over. For more information about hazard recognition, see Chapter 19, *Firefighter Survival.*

Utilities

Controlling utilities is one of the first tasks that must be accomplished at many working structure fires. Once utilities have been controlled, this information should be communicated to the IC. If additional utility hazards are found while conducting operations, the IC should be notified immediately. Everyone working on the fireground should know the status of the utilities. If control of the utilities cannot be completed, the IC should determine the safest course of action. Most departments have written SOPs that define when the utilities are to be shut off. Although this responsibility is often assigned to a particular company or crew, all firefighters should know how to shut off a building's electrical, gas, and water service. Chapter 14, *Fire Suppression* covers how to shut off utilities safely in detail.

Controlling utilities is particularly important if firefighters need to open walls or ceilings to look for fire in the void spaces, to cut ventilation holes in roofs, or to penetrate through floors. Because these spaces often contain both electrical lines and gas pipes, the danger of electrocution or explosion exists unless these utilities are disconnected.

Electrical, gas, and water utilities at a burning building may be disconnected for any of several reasons. For example, if faulty electrical equipment or a gas appliance caused the fire, shutting off the supply will help alleviate the problem. It also will reduce the risk of injury to firefighters.

Electrical Service

Electricity is present in most structure fires, many outdoor emergencies, and in many traffic collisions. Disconnecting electrical service reduces the risk of electrical injury to firefighters. For example, a firefighter using a saw to open a void space could be electrocuted if the tool contacts energized electrical wires. Having the electrical service disconnected can prevent problems such as short circuits and electrical arcing that could result from fire or water damage. In addition, disconnecting electrical service eliminates potential ignition sources that might cause an explosion if leaking gas has accumulated inside the building. The electrical supply must be disconnected at a location outside the area where gas might be present. Interrupting the power from a remote location alleviates the risk of an electrical arc that could cause a gas explosion.

The type of electrical service delivery depends on the utility company providing that service and the age of the system. The most common installation in older areas is a service drop from above-ground utility wires to the electric meter, which is typically mounted on the outside of the building (**FIGURE 2-13**). Some newer developments include underground distribution systems that include underground cables and transformers. Electrical meters may be located inside or outside of a building. There may be separate meters for each tenant or one meter for multiple occupancies.

Electrical wires may be energized yet nearly invisible when lying on the ground or when dangling from a pole or a building. This is especially true in conditions of limited visibility such as darkness, fog, and smoke. Always check carefully for overhead power lines

FIGURE 2-13 The electrical service often has an exterior meter connected to above-ground utility wires.

before raising ladders. During any major fire, the electrical power supply to the building should be turned off as part of the fireground task called "controlling the utilities."

In addition to structure fires, fire departments are often called to electrical emergencies, such as those involving downed power lines, fires, arcing wires, and transformer fires. Whenever there is a possibility that there are live power lines, have the power disconnected or turn off the power if you are trained and permitted by department policy.

Fire apparatus should be parked outside the area and away from power lines when responding to a call for an electrical emergency. A downed power line should always be considered energized until the power company confirms that it is dead. Secure the area around the power line and keep the public at a safe distance. Never drive fire apparatus over a downed line or attempt to move it using tools. If a sparking power line causes a brush fire, and it is safe to do so, attempt to contain the fire from a safe distance, but do not use water near the power line.

Fire departments must work with their local electrical utility companies to identify the different types of electrical services and arrangements used in the area and to determine the proper procedures for dealing with each type. Often, a main disconnect switch can be operated to interrupt the power. In general, to turn off an electrical service to a structure, you must first identify the circuit breaker box that supplies the building or part of the building in which you need to disconnect the power. Then you can turn off the main circuit breaker that supplies the whole breaker box. It is important to leave a firefighter at the circuit breaker box, or to use a lockout system that tags the box, to ensure that the power does not get turned on accidentally. Most utility companies will train firefighters to identify and operate shut-off devices on typical installations. Other utility companies request that the fire department call a company representative to shut off the power. Shutting off large systems that involve high-voltage equipment should be done by a company technician or a trained individual from the premises.

Sometimes a utility company representative needs to be called to interrupt power from a remote location, such as a utility pole. This step may be necessary if the outside wires have been damaged by fire, if firefighters are working with ladders or aerial apparatus, or if an explosion is possible. In urban areas, the electrical utility company can usually dispatch a qualified technician or crew to respond quickly to a fire or emergency incident. In many cases, qualified technicians are dispatched automatically to all working fires. Some utility companies can interrupt the service to an area from a remote location in response to a fire department request.

Gas Service

Shutting down gas service eliminates the potential for explosion due to damaged, leaking, or ruptured gas piping. For example, a power saw slicing through a gas pipe can cause a rapid release of gas and create the potential for an explosion. A structural collapse can rupture gas lines or create leaks in the stressed piping.

Both natural gas and liquefied petroleum gas (LPG) are used for heating and cooking. Generally, natural gas is delivered through a network of underground pipes. By contrast, LPG is usually delivered by a tank truck and stored in a container on the premises. In some areas, LPG is distributed from one large storage tank through a local network of underground pipes to several customers.

A single valve usually controls the natural gas supply to a building. This valve is generally located outside the building at the entry point of the gas piping; natural gas service has a distinctive piping arrangement. In older buildings, the shut-off valve for a natural gas system may be located in the basement. The shut-off valve for a natural gas system is usually a quarter-turn valve with a locking device so it can be secured in the off position (**FIGURE 2-14**). When the handle is in line with the pipe, the valve is open; when the handle is at a right angle to the pipe, the valve is closed. A special key, an adjustable wrench, or a spanner wrench can be used to turn the handle.

FIGURE 2-14 The natural gas shut-off valve is generally located outside the building at the entry point of the gas piping. In older buildings, the shut-off valve for a natural gas system may be located in the basement.

The shut-off valve for an LPG (propane) system is usually located at the storage tank (**FIGURE 2-15**). The more common type of LPG valve, however, has a distinctive handle that indicates the proper rotation direction to open or close the valve. To shut off the flow of gas, rotate the handle to the fully closed position.

A gas valve that has been shut off must not be reopened until the system piping has been inspected by a qualified person. Air must be purged from the system, and any pilot lights must be reignited to prevent gas leaks.

Water Service

If a serious water leak has occurred inside the building, shutting off the water supply may help to minimize additional water damage to the structure and contents.

FIGURE 2-15 The shut-off valve for an LPG system is usually located at the storage tank.

Water service to a building can usually be shut off by closing one valve at the entry point. Many communities permit the water service to be turned off at the connection between the utility pipes and the building's system. This underground valve, which is often accessed through a curb box, is located outside the building and can be operated with a special wrench or key. In most cases, another valve is found inside the building, usually in the basement (if there is one), where the water line enters. In warmer climates, water supply valves are sometimes located above ground.

Lifting and Moving

Lifting and moving objects are part of a firefighter's daily duties. Do not try to move something that is too heavy alone—ask for help. Never bend at the waist to lift an object; instead, bend at the knees, and use your legs to lift the weight. If you must move objects over a long distance, use equipment such as handcarts, hand trucks, and wheelbarrows.

In addition to moving inanimate objects, firefighters must often move sick or injured victims. Do not move an injured person, unless their life is in danger, until the appropriate medical personnel are on scene to stabilize the victim. Discuss and evaluate the options before moving a victim, and then proceed carefully. If necessary, request help. Never be afraid to call for additional resources, such as another engine or truck company, to assist in lifting and moving a heavy victim.

Adverse Weather Conditions

In adverse weather conditions, firefighters must dress appropriately. A structural firefighting protective coat, pants, boots, and helmet can keep you warm and dry in rain, snow, or ice. Firefighting gloves and knit caps also help retain body heat and keep you warm. If conditions are icy, make smaller movements, watch your step, and keep your balance. During the hot summer months, it is difficult to stay cool in PPE. Many departments have SOPs covering appropriate PPE to use in summer months. Check your department's SOPs for guidance on acceptable PPE use in adverse weather.

Rehabilitation

Never be afraid or embarrassed to admit you need a break when on the emergency scene. Typically, breaks are in the form of recycling and rehabilitation. Recycling is brief and consists of taking a few minutes to rest, hydrate, and change the SCBA bottle before getting back to work.

Rehabilitation is a systematic process that provides longer periods of rest and recovery for emergency workers during an incident. It is usually conducted in a designated area away from the hazards of the emergency scene. The rehabilitation area, or "rehab," is usually staffed by fire and EMS personnel.

While in rehabilitation, firefighters should take advantage of the opportunity to rest, rehydrate, have their vital signs checked, and receive treatment for minor injuries (**FIGURE 2-16**). Rehabilitation gives firefighters the chance to cool off in hot weather and to warm up in cold weather. Rehabilitation time can also be used to replace SCBA cylinders, obtain new batteries for portable radios, and make repairs or adjustments to tools or equipment. Firefighting teams can discuss recently completed assignments and plan their next work cycle. When a crew is released from rehabilitation, its members should be rested, refreshed, and ready for another work cycle. If members of the crew are too exhausted or unable to return to work or have not been medically cleared to return to work, they should be replaced and released from the incident. More information on firefighter rehabilitation is presented in Chapter 20, *Firefighter Rehabilitation.*

FIGURE 2-16 In the rehabilitation area, firefighters can rest and rehydrate.

Violence

Firefighter safety is jeopardized by violence. Firefighters are sometimes dispatched to calls involving domestic disputes, active shooters, injuries from an assault, or other violent scenes. In such incidents, the staging area and the fire apparatus should be located at some distance from the scene until the police arrive, investigate, and declare the scene safe. Only then should firefighters proceed to enter the emergency scene. There have been violent incidents where firefighters have been targeted, and in some cases, they have been mistaken for intruders. This increases the need for situational awareness. A firefighter's personal safety should always be paramount. If there is any threat to your personal safety, retreat as quickly as possible from the emergency scene to a safe distance and request the police to secure the scene. Do not become a victim.

If you are confronted with a potentially violent situation, do not respond violently. Remain calm, speak quietly, and attempt to gain the person's trust and call for police assistance. You might consider taking additional classes to increase your understanding and develop appropriate skills for these situations.

Safety at the Fire Station

The fire station is just as much a workplace as the fireground. Indeed, firefighters spend much of their time during a shift at their fire station. The length of each shift varies, from 12, to 24, to 48, to 72 hours and, in some cases, to a 96-hour shift. Some may be longer. The need for safety around the fire station is as important as it is on the fireground or at other emergency scenes. While in the fire station, be careful when working with power tools and equipment, ladders, electrical appliances, pressurized cylinders, and hot surfaces. Practice using tools and equipment properly and safely before attempting to use them at an actual emergency incident. Follow the proper procedures and safety precautions both in training and at an incident scene.

Equipment should always be in excellent condition and ready for use. Proper maintenance of tools includes sharpening, lubricating, and cleaning each tool. All firefighters should be able to do basic repairs such as changing a saw blade or a hand light battery. Practice using and maintaining tools and equipment at the fire station until you can perform these tasks quickly and safely. Remember, injuries that occur at the firehouse can be just as devastating as those that occur at an emergency incident scene.

SAFETY TIP

Remember these guidelines to stay safe—both on and off the job:

- You are personally responsible for safety. Keep yourself safe. Keep your teammates safe. Keep citizens—your customers—safe.
- Work as a team. The safety of the entire firefighting unit depends on the efforts of each member. Become a dependable member of the team.
- Follow orders. Freelancing can endanger other firefighters as well as yourself.
- Think! Before you act, think about what you are doing. Many people are depending on you.

Safety Outside Your Workplace

Continue to follow safe practices when you are off duty. An accident or injury, regardless of where it happens, can end your career as a firefighter. For example, if you are using a ladder while off duty, follow the same safety practices that you would use while on duty. Keep your seat belt fastened in your personal vehicle, just as you are required to do when you are on duty.

CASE STUDY

You Are the Firefighter CONCLUSION

You and your crew are at a large fire in an apartment building that started on the third floor, extended into the attic, and is now venting through the roof. Your captain gave the IC the accountability tags, and you were assigned to confine the fire on the third floor. You gather your gear and as you approach the apartment building, you consider the ways you and your crew can accomplish your mission while avoiding preventable injury or death.

1. **What is the leading cause of firefighter injury and death?**

   Answer: The largest share of firefighter deaths occurs from stress, overexertion, and medical issues. Before the incident, proper nutrition, hydration, physical conditioning, sleep, and stress management will reduce these risks. During the incident, rehabilitation and medical monitoring can further reduce the risk of firefighter injury and death. In addition, rehabilitation allows a fire crew to ensure that they are working together and helps reduce freelancing. After the incident, decontamination and, where possible, a return to proper nutrition, hydration, physical conditioning, sleep, and stress management can further reduce these risks.

2. **What are the potential hazards during this incident?**

   Answer: Potential hazards found on the fireground include the changing conditions of the fire itself as well as the effect of the fire on the building. In addition, utilities, poorly maintained equipment, poorly trained and/or freelancing firefighters, extreme environmental conditions, uneven terrain, hazardous materials, and possibly roadway traffic near the operational area all pose significant fireground hazards.

3. **What safety measures should be taken during this incident?**

   Answer: Hazards that can be safely mitigated should be immediately addressed, for example, removing a ladder with a broken rung from service and replacing it with a functional ladder. Hazards that cannot be mitigated should be marked off and brought to the attention of the next officer in your chain of command or to the incident safety officer, for example, using traffic cones and banner tape to mark off an area that has live broken electrical lines.

WRAP-UP

SUMMARY

KNOWLEDGE OBJECTIVES

- Describe the value of fire and life safety initiatives in support of the fire department mission to reduce firefighter line-of-duty injuries and fatalities.
  - List the major causes of death and injury in firefighters. (**NFPA 1010: 6.3.10**, pp. 34–35)
  - Describe some of the organizations that set the regulations, standards, and procedures intended to ensure a safe working environment for the fire service. (**NFPA 1010: 6.1.1**, pp. 36–38)
  - Describe the 16 firefighter life safety initiatives. (**NFPA 1010: 6.1.1**, p. 37)
- Describe the importance of physical fitness and healthy lifestyle to the ability of firefighters to perform their duties.
  - Explain the practices firefighters should take to promote optimal physical and mental health and well-being. (**NFPA 1010: 6.1.1**, pp. 38–45)
  - Describe the connection between physical fitness and firefighter safety. (**NFPA 1010: 6.1.1**, pp. 38–39)
  - Describe the components of a well–rounded physical fitness program. (**NFPA 1010: 6.1.1**, pp. 38–39)

KNOWLEDGE OBJECTIVES CONTINUED

- Describe the signs and symptoms of behavioral and emotional distress; including the role of the fire department's member assistance program.
  - Explain the role of a critical incident stress debriefing in preserving the mental well-being of firefighters. (**NFPA 1010: 6.1.1**, pp. 42–43)
  - List signs and symptoms of burnout and compassion fatigue. (**NFPA 1010: 6.1.1**, pp. 43–45)
  - Describe the purpose of an employee assistance program. (**NFPA 1010: 6.1.1**, p. 45)
- Identify proper procedures to increase safety and operations at emergency scenes.
  - Explain how firefighter candidates, instructors, and veteran firefighters work together to ensure safety during training. (p. 45)
  - Describe how to safely mount an apparatus. (**NFPA 1010: 6.3.2**, p. 48)
  - Describe how to safely ride a fire apparatus to an emergency scene. (**NFPA 1010: 6.3.2**, pp. 48–49)
  - Describe how to safely dismount an apparatus. (**NFPA 1010: 6.3.2**, p. 49)
  - Describe hazards and safety measures associated with riding apparatus. (**NFPA 1010: 6.3.2**, pp. 48–49)
  - List the NFPA standards that require firefighters to wear safety belts while riding in a fire apparatus. (**NFPA 1010: 6.3.2**, p. 48).
  - List the prohibited practices when riding in a fire apparatus to an emergency scene. (**NFPA 1010: 6.3.2**, pp. 48–49)
  - Describe how to manage traffic safely at an emergency scene. (**NFPA 1010: 6.3.3**, pp. 49–50)
  - List the four general principles that govern emergency vehicle operation. (pp. 50–52)
- Identify potential hazards involved in operating on emergency scenes and how to mitigate those hazards.
  - Explain how the teamwork concept is applied during every stage of an emergency incident to ensure the safety of all firefighters. (pp. 55–56)
  - Describe how the personnel accountability system is implemented during an emergency incident. (p. 56)
  - Explain considerations for hazard and scene control. (**NFPA 1010: 6.3.3**, p. 56)
  - List the common hazards at an emergency incident. (**NFPA 1010: 6.3.3**, p. 56)
  - Describe the measures firefighters follow to ensure safety around utilities at an emergency incident. (**NFPA 1010: 6.3.3**, pp. 56–59)
  - Describe how to lift and move objects safely. (p. 59)
  - Explain how rehabilitation is used to protect the safety of firefighters during an emergency incident. (pp. 59–60)
- Describe how to promote a culture of safety inside and outside the workplace.
  - Describe how to ensure safety at the fire station. (p. 60)
  - Describe how to ensure safety outside of the workplace. (p. 60)

SKILLS OBJECTIVES

- Mount and dismount a fire apparatus safely.
  - Mount an apparatus safely. (**NFPA 1010: 6.3.2**, p. 48)
  - Dismount from an apparatus safely. (**NFPA 1010: 6.3.2**, p. 49)

KEY TERMS

Code of Federal Regulations (CFR) A collection of permanent rules published in the Federal Register by the executive departments and agencies of the U.S. federal government. Its 50 titles represent broad areas of interest that are governed by federal regulation. Each volume of the CFR is updated annually.

critical incident stress debriefing (CISD) A post-incident meeting designed to assist rescue personnel in dealing with psychological trauma as the result of an emergency. (NFPA 1006)

critical incident stress management (CISM) A program designed to reduce acute and chronic effects of stress related to job functions. (NFPA 450)

employee assistance programs (EAPs) An employer-sponsored service designed for personal or family problems, including mental health, substance abuse, various addictions, marital problems, parenting problems, emotional problems, or financial or legal concerns. (NFPA 450)

freelancing The dangerous practice of acting independently of command instructions.

Occupational Safety and Health Administration (OSHA) The U.S. federal agency that regulates worker safety and, in some cases, responder safety. OSHA is part of the U.S. Department of Labor.

personnel accountability report (PAR) Periodic reports verifying the status of responders assigned to an incident or planned event. (NFPA 1026)

personnel accountability system A system that readily identifies both the location and function of all members operating at an incident scene. (NFPA 1550)

post-traumatic stress disorder (PTSD) A behavioral disorder that develops after a person has experienced a critical incident; characterized by reexperiencing the event and overresponding to stimuli that recalls the event; symptoms include depression, startle reactions, flashback phenomena, and dissociative episodes (e.g., amnesia of the event).

response Immediate and ongoing activities, tasks, programs, and systems to manage the effects of an incident that threaten life, property, operations, or the environment. (NFPA 1600).

structural firefighting protective clothing All of the clothing elements of the structural firefighting protective ensemble.

REVIEW QUESTIONS

1. What should a firefighter's approach to safety be?
2. What is the top cause of firefighter deaths by cause of injury?
3. What is the cause of most firefighter injuries and deaths?
4. What should a firefighter incorporate into their daily routine?
5. A firefighter injured during training should not return to duty until ________.
6. What actions should a firefighter perform during a daily inspection of their equipment to be ready for assignment?
7. What is the highest priority when responding in an emergency apparatus?
8. What should firefighters be able to do to promote safety at the fire station?
9. What should firefighters do to stay safe while off duty?

DISCUSSION QUESTIONS

1. How can a firefighter be "safe" in an inherently dangerous profession?
2. How can firefighters reduce their risk of dying or being injured due to stress and overexertion?
3. What steps can be taken to increase safety and effectiveness overall in the fire service?
4. How does working in teams increase firefighter safety and effectiveness?
5. How does freelancing impact safety and effectiveness of emergency operations?

APPLYING THE CONCEPTS

You and your engine unit are on the scene of fire in a three-story commercial building in the downtown area of your town. Thick, dark smoke is billowing from the window of the third floor and flames are visible. Fire attack teams from another department are inside the building conducting an offensive attack to knock down the fire. The building is connected to a large multistory apartment building. Rescue teams are evacuating residents from the apartment building. After donning PPE and SCBA your officer tells you the IC has ordered your crew to assist with fire attack on the third floor.

1. It is important you protect yourself with PPE and SCBA while on scene, but what can you do before you arrive to keep yourself safe?

APPLYING THE CONCEPTS CONTINUED

2. Let's say you eat well, exercise every day, hydrate, get plenty of sleep, and find healthy ways to de-stress. What responsibility do the incident commander and safety officer have to keep you safe at an emergency response? Or, are you responsible for your own safety?

You and your four-person crew advance hose lines through the entrance on side Bravo and proceed up the stairwell. There is minimal smoke on the first floor, but it gets heavier as you climb the stairs. As you pass the second floor you see a frightened worker in the hallway, crying and seemingly in shock. You alert your crew, "Hey! I see someone over there. I'm going to help get them out and I'll catch back up to you in a couple of minutes."

3. Is this the right thing to do? Why or why not?
4. So, what *should* you do in this situation?

The IC radios your crew, "The rescue team is 60 seconds out. Maintain your position and hand-off when they arrive. Then proceed to the third floor to assist with firefighting operations. Remember, maintain situational awareness." The rescue team arrives and you continue to the third floor. You advance hose lines as instructed by the fire attack crew officer. Working with the other teams, you make progress to knock down the fire.

5. What is situational awareness and why is it essential to firefighter safety?
6. When you rotate or the incident ends, what do you do to handle your gear to keep yourself safe?

REFERENCES

Firefighter Behavioral Health Alliance. 2016. Accessed October 30, 2023. *Firefighter Behavioral Health Alliance.* https://www.ffbha.org/.

Firefighter Cancer Support Network. 2022. Accessed October 30, 2023. *Firefighter Cancer Support Network: Together we can...* https://firefightercancersupport.org/

International Association of Fire Fighters. March 27, 2019. *Taking Action Against Occupational Cancer.* Accessed October 30, 2023. https://www.iaff.org.

Laroche, Elena, and Sylvain L'Espérance, S. 2021. "Cancer Incidence and Mortality among Firefighters: An Overview of Epidemiologic Systematic Reviews." *International Journal of Environmental Research and Public Health, 18*(5), 2519. https://doi.org/10.3390/ijerph18052519.

National Cancer Institute. 2021, July 14. *Alcohol and Cancer Risk.* Accessed October 30, 2023. https://www.cancer.gov/about-cancer/causes-prevention/risk/alcohol/alcohol-fact-sheet.

National Fallen Firefighter's Foundation. 2022. *Everyone Goes Home—Firefighter Life Safety Initiatives.* Accessed October 30, 2023. https://www.everyonegoeshome.com/.

National Fire Protection Association (NFPA). 2018. *NFPA 1451: Standard for a Fire and Emergency Service Vehicle Operations Training Program.* 2018 Edition. Quincy, MA: NFPA.

National Fire Protection Association (NFPA). 2018. NFPA 1600, *Standard on Continuity, Emergency, and Crisis Management.* 2019 Edition. Quincy, MA: NFPA.

National Fire Protection Association (NFPA). 2019. NFPA 1951, *Standard on Protective Ensembles for Technical Rescue Incidents.* 2020 Edition. Quincy, MA: NFPA.

National Fire Protection Association (NFPA). 2020. NFPA 450, *Emergency Medical Services and Systems.* 2021 Edition. Quincy, MA: NFPA.

National Fire Protection Association (NFPA). 2020. NFPA 1006, *Standard for Technical Rescue Personnel Professional Qualifications.* 2021 Edition. Quincy, MA: NFPA.

National Fire Protection Association (NFPA). 2020. *NFPA 1250: Recommended Practice in Fire and Emergency Service Organization Risk Management.* 2020 Edition. Quincy, MA: NFPA.

National Fire Protection Association (NFPA). 2021. *Firefighter injuries in the United States.* Accessed October 30, 2023. https://www.nfpa.org/News-and-Research/Data-research-and-tools/Emergency-Responders/Firefighter-injuries-in-the-United-States.

National Fire Protection Association (NFPA). 2021. *NFPA 1500: Standard on Fire Department Occupational Safety, Health, and Wellness Program.* 2021 Edition. Quincy, MA: NFPA.

National Fire Protection Association (NFPA). 2022. *NFPA 1582: Standard on Comprehensive Occupational Medical Program for Fire Departments.* 2022 Edition. Quincy, MA: NFPA.

National Volunteer Fire Council. 2022. *Share the Load.* Accessed October 30, 2023https://www.nvfc.org/programs/share-the-load-program/.

Suicide Prevention Resource Center. 2022. *Suicide Prevention Resource Center.* Accessed October 30, 2023. https://www.sprc.org/.

United State Fire Administration. 2022, April 7. *Firefighter Fatalities in the United States.* U.S. Fire Administration. Accessed June 7, 2024. https://www.usfa.fema.gov/statistics/reports/firefighters-departments/firefighter-fatalities.html.

CHAPTER

3

Firefighter I

Personal Protective Equipment

KNOWLEDGE OBJECTIVES

After studying this chapter, you will be able to:

- Explain how the structural firefighting ensemble works to protect the body.
- Describe the specialized protective equipment required for vehicle extrication and wildland fires.
- Describe the procedure for donning and doffing personal protective clothing.
- Describe the procedure for inspecting, maintaining, and cleaning personal protective equipment (PPE).
- Describe the different types of respiratory protection and why they are required.
- Describe the limitations of respiratory protection.
- Describe the major components of the self-contained breathing apparatus (SCBA).
- Describe the importance of inspecting and maintaining SCBA and explain how to inspect them.

SKILLS OBJECTIVES

After studying this chapter, you will be able to perform the following skills:

- Don and doff approved personal protective clothing.
- Don an SCBA using various methods.
- Doff an SCBA.
- Perform a visible and operational inspection of an SCBA.
- Replace an SCBA air cylinder.
- Refill an SCBA air cylinder from a cascade or compressor system.
- Clean an SCBA.

ADDITIONAL NFPA STANDARDS

- **NFPA 853**, *Standard for the Installation of Stationary Fuel Cell Power Systems, 2020 Edition*
- **NFPA 1404**, *Standard for Fire Service Respiratory Protection Training, 2018 Edition*
- **NFPA 1550**, *Standard for Emergency Responder Health and Safety, 2024 Edition*
- **NFPA 1700**, *Guide for Structural Fire Fighting, 2021 Edition*
- **NFPA 1851**, *Standard on Selection, Care, and Maintenance of Protective Ensembles for Structural Fire Fighting and Proximity Fire Fighting, 2020 Edition*
- **NFPA 1852**, *Standard on Selection, Care, and Maintenance of Open-Circuit Self-Contained Breathing Apparatus (SCBA), 2021 Edition*
- **NFPA 1900**, *Standard for Aircraft Rescue and Firefighting Vehicles, Automotive Fire Apparatus, Wildland Fire Apparatus, and Automotive Ambulances, 2024 Edition*
- **NFPA 1970**, *Standard on Protective Ensembles for Structural and Proximity Firefighting, Work Apparel and Open-Circuit Self-Contained Breathing Apparatus (SCBA) for Emergency Services, and Personal Alert Safety Systems (PASS), 2024 Edition*
- **NFPA 1977**, *Standard on Protective Clothing and Equipment for Wildland Fire Fighting and Urban Interface Fire Fighting, 2022 Edition*
- **NFPA 1984**, *Standard on Respirators for Wildland Fire Fighting Operations and Wildland Urban Interface Operations, 2022 Edition*
- **NFPA 1990**, *Standard for Protective Ensembles for Hazardous Materials and CBRN Operations, 2022 Edition*

CASE STUDY

You Are the Firefighter

One cool fall evening, you are dispatched for a report of a smoke odor in the area. These calls are normal for this time of year as residents start to use their fireplaces for the first time. As the engine drives slowly through the neighborhood, you notice the distinctive smell of a working house fire rather than that of firewood. When you turn the corner, you see smoke drifting from the basement window of a house. Your captain completes a quick size-up and requests additional resources from dispatch. The captain then tells you and another firefighter to pull a cross-lay and attack the fire.

1. Which personal protective equipment would you need for this situation?
2. What is the most efficient way to don a self-contained breathing apparatus? Why?
3. Once the fire is over, what is the proper way to care for your protective equipment?

Introduction

Two safety components used by firefighters require special consideration: personal protective equipment and self-contained breathing apparatus. **Personal protective equipment (PPE)** is the broad term that includes clothing and equipment worn to protect the wearer from the hazards they encounter as they do their job. **Self-contained breathing apparatus (SCBA)** is a component of PPE and is a personal respirator worn by a firefighter that provides an air supply to the firefighter. For you to remain safe and do your job effectively, you need to understand the purposes for which this equipment is made, what it can do, and what it cannot do. If you exceed the protection offered by PPE, severe injuries and death can occur.

Personal Protective Equipment

PPE is an essential component of a firefighter's safety system. It enables a person to survive under conditions that might otherwise result in death or serious injury. Different PPE ensembles are designed for specific hazardous conditions, such as structural firefighting, wildland firefighting, airport rescue and firefighting, hazardous materials operations, and emergency medical operations. The various PPE ensembles provide specific protections, so an understanding of their designs, applications, and limitations is critical. It is important for a firefighter to understand what protection PPE can provide, so you can adequately choose which is best for each situation.

The complete set of clothing and equipment that a firefighter wears when attacking a structure fire (**structural firefighting**) is referred to as the **structural firefighting protective ensemble**. This ensemble consists of **structural firefighting protective clothing**—personal protective clothing that provides full-body coverage that covers every inch of the body and provides protection from heat and fire, keeps water away from the body, and helps reduce injuries from cuts or falls—SCBA for respiratory protection, and a **personal alert safety system (PASS)** device, which is an electronic device that emits a loud, audible signal when a firefighter becomes trapped or injured (**FIGURE 3-1**). Some departments have their firefighters carry additional equipment such as hand lights, radios, etc., to provide more complete protection.

To be effective, the entire ensemble must be worn whenever potential exposure to those hazards exists. Structural firefighting protective ensembles allow

FIGURE 3-1 The structural firefighting protective ensemble provides protection from multiple hazards.

Courtesy of Central County Fire & Rescue.

FIGURE 3-2 The structural firefighting ensemble consists of a helmet, protective hood, protective coat, protective pants, boots, gloves, SCBA, and a PASS device.

firefighters to enter burning buildings and work in areas with elevated temperatures and concentrations of toxic gases. Without PPE, firefighters would be unable to conduct search and rescue operations or perform fire suppression activities. The structural firefighting ensemble consists of the following (**FIGURE 3-2**):

- Personal protective clothing
  - Helmet
  - Hood
  - Coat
  - Trousers (pants or coveralls)
  - Boots
  - Gloves
- SCBA
- PASS device

All of these elements must be correctly worn together to provide the necessary level of protection.

Personal Protective Clothing

Personal protective clothing is the first layer of PPE for firefighters. Without personal protective clothing, firefighters would not be able to enter a burning building to conduct an interior attack or rescue people from toxic environments.

Work Uniforms

The lowest level of personal protective clothing is normal street clothing or work uniforms. Firefighters who are not responding to a call, police officers, and emergency medical services (EMS) providers typically wear this as a part of their daily uniform. In fact, at this level, the personal protective clothing is the PPE system, since personnel wearing these uniforms do not wear devices to protect their airways or alert others if they are injured, disabled, or killed.

Certain synthetic fabrics, such as nylon and polyester, melt at relatively low temperatures, even when worn under a complete personal protective ensemble. This can cause severe burns because as the material melts, it sticks to the skin, and when that material is removed, it removes layers of skin. Therefore, clothing containing nylon or polyester, even if these materials are blended with natural fibers, should not be worn in a firefighting environment. Clothing made of natural fibers, such as cotton or wool, may char or burn if exposed to flames, but it does not melt. When worn under a protective ensemble, these materials will not be exposed to flame and are generally safer to wear. NFPA 1970, *Standard on Protective Ensembles for Structural and Proximity Firefighting, Work Apparel and Open-Circuit Self-Contained Breathing Apparatus (SCBA) for Emergency Services, and Personal Alert Safety Systems (PASS), 2024 Edition,* defines criteria for selecting appropriate fabrics for work uniforms for firefighters.

SAFETY TIP

Do not mix and match different brands and styles of personal protective clothing. Some styles are not compatible with others and may leave gaps that will expose the wearer.

Protective Clothing for Structural Firefighting

The next level of protection is provided by structural firefighting protective clothing, which is personal protective clothing that provides full-body coverage that covers every inch of the body and provides protection from heat and fire, keeps water away from the body, and helps reduce injuries from cuts or falls. Structural firefighting protective clothing is often referred to as **turnout gear**, or **bunker gear**.

TIP

Special synthetic fibers such as Nomex and PBI, which are used in structural firefighting protective clothing, have excellent resistance to high temperatures.

SAFETY TIP

The structural firefighting protective ensemble protects the wearer from the wide range of interior and exterior fires to which a fire department responds. It is not designed to protect firefighters against the hazards encountered when responding to hazardous materials, water rescue, high-angle rescue, or wildland fire incidents. Instead, each of these specialized functions requires a specific set of PPE. If your department performs specialized operations, you need the appropriate PPE for these activities.

A structural firefighting protective ensemble provides full-body coverage and offers the following:

- Provides thermal protection
- Repels water
- Provides impact protection
- Protects against cuts and abrasions
- Furnishes padding against injury
- Allows firefighters to be more easily seen by others in smoky conditions

For example, a moisture barrier between different layers of PPE keeps liquids and vapors, such as hot water or steam, from reaching the skin. The knees in the pants may be reinforced with pads for greater protection when crawling. Reflective trim adds visibility in dark or smoky environments.

Each item of protective clothing is designed to overlap from one item to the next. For example, boots overlap the pants, coats overlap the pants, and coats overlap the gloves. This ensures that the entire body is protected at all times. It is vitally important that the PPE a firefighter wears is sized appropriately and does not fit too tight or too loose.

There are a number of different manufacturers and styles of structural firefighting protective ensembles. All protective clothing for firefighters must be manufactured according to exacting standards. The requirements for protective clothing for firefighters are outlined in two standards:

- NFPA 1970, *Standard on Protective Ensembles for Structural and Proximity Firefighting, Work Apparel and Open-Circuit Self-Contained Breathing Apparatus (SCBA) for Emergency Services, and Personal Alert Safety Systems (PASS), 2024 Edition*
- NFPA 1977, *Standard on Protective Clothing and Equipment for Wildland Fire Fighting and Urban Interface Fire Fighting, 2022 Edition*

According to NFPA 1970, each item must have a permanent label verifying that the particular item meets the requirements of a certification organization (**FIGURE 3-3**). Labels also include the appropriate term for the piece of PPE and manufacturer information, including date of manufacture, model name or number, size or size range, and principal material of construction. Usage limitations as well as cleaning and maintenance instructions should also be provided.

TIP

Most manufacturers measure firefighters for the appropriate-sized coats and pants. It is important to be sure that PPE is adequately sized for the firefighter to whom it is assigned.

Structural firefighting PPE will protect you under limited conditions. The longer you remain in an environment with elevated temperatures, the more heat energy the clothing absorbs. After a certain period of time in a high temperature environment, your gear will start to transfer more heat through the coat, and you will start to feel this stored heat energy transmitted to your body. This can happen quickly in a fire situation. Once the firefighter starts to feel this heat transfer take place and it becomes uncomfortable, they should start planning to exit the environment.

Helmet

The **structural firefighting protective helmet** provides protection to the firefighter's head from falling debris, and the attached face shield provides eye protection (**FIGURE 3-4**). Helmets are manufactured in several designs and shapes using a variety of materials. Every design must meet the requirements specified in NFPA 1970.

The components of a fire helmet are as follows:

- Hard outer shell: This must be lined with energy-absorbing material and have a suspension system to provide impact protection against falling objects. The outer shell also repels water, protects against steam, and creates a thermal barrier against heat and cold. Its shape helps to deflect water away from the head and neck.
- Face and eye protection: The face shield is attached to the helmet and provides face and eye protection. Some helmets also have goggles attached to provide additional eye protection. The face shield (and goggles) is used when an SCBA is not needed or when the SCBA face piece is not in place.

Garment Safety Label

⚠ DANGER

Please read to confirm your understanding of these warnings and instructions.
Serious injury or death can result from a failure to heed these warnings. **1234**

This garment should be worn for FIREFIGHTING DUTIES ONLY.
THIS GARMENT CANNOT AND DOES NOT PROVIDE PROTECTION AGAINST CBRN TERRORISM AGENTS.

Before wearing this garment, please read to confirm your understanding of the *Instruction, Safety and Training Guide for Users* accompanying this garment.

The Guide for Users provides:
- Serious safety information and protective clothing restrictions.
- Proper sizing.
- Measures for dressing and removing protective clothing.
- Cleaning, decontamination, inspection and storing directives.
- NFPA 1500-compatible usage.
- Useful information limitations and procedures for retiring garments.

Only those who are properly trained in firefighting techniques should wear this garment, and the wearer should have knowledge of the proper selection, fit, use, care and limitations of protective clothing and equipment.
- PLEASE BE AWARE that this garment provides only limited protection against heat and flames.
- The wearer should minimize exposure to heat. Burns without warning may occur, even without receiving damage to garment. Avoid contact with hot objects.
- Burns to skin happen when skin reaches a temperature of 118°F. And fires can burn at temperatures up to 2000°F.
- Protection can be diminished by even a small amount of dampness or moisture and/or compression in your garment.
- Exertion in hot conditions may result in heat fatigue or poor decision-making. If you feel dizzy, dehydrated, unfocused, or short of breath, PLEASE find a safe area, remove this garment, and find medical attention.
- If this clothing is soiled or damaged in any way, DO NOT USE. Soiled or damaged garments will NOT offer the desired protection. ALWAYS follow manufacturer's cleaning instructions.
- This garment has limited useful life. Inspect this garment regularly and discard when appropriate, paying full attention to the *Guide for Users*. See also NFPA 1700.

DO NOT REMOVE OR WRITE ON THIS LABEL! REV. 4.0 12/31

Garment Cleaning Label

Call immediately with any questions. **1234**
J&B
1-800-555-1212

CLEANING AND STORAGE INSTRUCTIONS
- Use the *Guide for Users* for cleaning, inspection, storing, and altering instructions.
- Never use chlorine bleach or the protection used in the formation of this garment will be severely compromised.
- For coats, remove DRD and launder using mild detergent and warm water by hand only.
- Fasten all hooks and D-rings, turn garment inside out, and launder in a laundry bag, machine washing in warm water, with mild detergent and only when necessary, liquid chlorine bleach. Rinse twice in cool water but DO NOT USE FABRIC SOFTENERS.
- DO NOT DRY CLEAN.
- Hang garment to dry, away from sunlight and/or fluorescent light, and store similarly.
- THIS PROTECTIVE FIRE FIGHTING GARMENT MEETS THE REQUIREMENTS OF NFPA 1971, 2018 EDITION. PROTECTIVE GARMENT FOR STRUCTURAL FIRE FIGHTING IN ACCORDANCE WITH NFPA 1971-2018.
- When worn with the inner liner and outer shell assembled together, this garment meets the PPE criteria of US Dept. of Labor OSHA Bloodborne Pathogens Standard, Title 29 CFR, Part 1910, 1030, and CAL-OSHA Standard Title 3 Section 3406.

Rev. 4.0 12/31 **DO NOT REMOVE OR WRITE ON THIS LABEL**

Garment Information Label

TECHSQUARE MOISTURE BARRIER (TEPR)
GLOSS 4G GUDERE F-92 (G) THERM.LINER
NOMEX F-92 QUILT
REQ:091772
MFG DATE:10/15/2018
CUT:01202012
MODEL:SRDM
LINER:0120SRDM
SIZE:051711T

Garment Liner Attachment Safety Label

⚠ WARNING

The inner liner on its own WILL NOT protect against heat, flame, chemical or biological hazards. DO NOT wear the inner liner without the SAME SIZE AND MODEL outer shell, as identified on labels located on each detachable component. Please assemble and wear together ALL of the following items in order to avoid the risk of injury or death: 1. protective coat and pant with outer shell, attached inner liner and DRD installed in coat 2. gloves 3. boots 4. helmet with eye protection 5. protective hood 6. SCBA 7. PASS device

ALWAYS confirm that all ensemble layers have the suitable overlap and all layers fit with satisfactory slackness. A tight fit will lower insulation protection and will limit mobility.

I

MADE IN THE U.S.A.
DO NOT REMOVE OR WRITE ON THIS LABEL!

FI# 0120
REV. 4.0 12/31

FIGURE 3-3 Mandatory labels provide important information about each item of protective clothing.

- Chin strap: This strap maintains the helmet in the proper position and helps keep the helmet on the firefighter's head during an impact.
- Inner liner: The inner layer provides added thermal protection and protects the wearer's neck. These liners have ear tabs that the firefighter pulls down before entering a burning building to provide maximum protection.
- Adjustable inner suspension system: The suspension system holds the shell away from the head and cushions the head against impacts. It must be adjusted to fit each firefighter. To ensure proper fit, the firefighter should put the helmet on and adjust it with the protective hood in place. Then the firefighter should make further adjustments as needed.

In many departments, helmet shells are color coded according to the firefighter's rank and function. Some fire departments include a shield or badge on the front of the helmet that identifies the firefighter's rank and company. Each department is unique, but it is

FIGURE 3-4 A helmet is constructed with multiple layers and components.

FIGURE 3-5 A protective hood.

fairly standard for chief officers to wear a white helmet. Bright, fluorescent reflective materials or decals are applied to helmets to make the firefighter more visible in all types of lighting conditions. Some decals are luminescent (glow in the dark).

Hood

Although the helmet's ear tabs cover the ears and neck, this area is still at risk for burns when the head is turned or the neck is flexed. A **structural firefighting protective hood** is constructed of a flame-resistant material such as Nomex, PBI, or carbon fiber and it covers the whole head and neck (except for that part of the face that would be protected by the SCBA face piece) to provide additional thermal protection for these areas (**FIGURE 3-5**). The lower part of the protective hood, which is called the **bib**, drapes down inside the protective coat. Because of the danger from the contaminants in smoke itself, some protective hoods provide particulate protection to reduce the amount of contaminants that reach the skin. Protective hoods are available in various styles, lengths, materials, and colors. One type is a particulate barrier hood, which filters out many of the contaminants firefighters encounter to reduce their exposure to carcinogens. Per NFPA 1971, a particulate hood must block a minimum of 90 percent of particulates ranging in size from 0.1 to 1.0 microns.

FIGURE 3-6 A structural firefighting protective coat.

Coat

The **structural firefighting protective coat**, sometimes called a **turnout coat** or **bunker coat**, protects the firefighter's arms and body (**FIGURE 3-6**). Structural firefighting protective coats consist of three layers: the outer shell, the moisture barrier, and the thermal barrier.

The outer layer or shell is constructed of a sturdy, flame-resistant material such as Nomex, Kevlar, or PBI. Fluorescent reflective material applied to the outer shell makes the firefighter more visible in smoky conditions and at night. When light-colored fabrics are used for protective coats, it makes it easier to identify contaminants such as hydrocarbons, blood, and body fluids on the coat.

The second layer of the protective coat is the moisture barrier, which usually consists of a flexible membrane attached to the thermal barrier material. The moisture barrier helps prevent the transfer of water, steam, and other fluids to the skin. It is critical because water applied to a fire generates large amounts of superheated steam, which can engulf firefighters and burn unprotected skin.

The third layer of the protective coat creates a thermal barrier, which is made of a multilayered or quilted material that insulates the body from external temperatures. It enables firefighters to operate in the elevated temperatures generated by a fire. It also keeps the body warm during cold weather.

The front of the protective coat has an overlapping flap to provide a secure seal. The inner closure is secured first, and then the outer flap is secured, creating a double seal. Several different combinations of zippers, Velcro, snaps, and D-rings can be used to secure the inner and outer closures.

The collar of the protective coat works with the protective hood to protect the neck. The collar has snaps or a Velcro closure system in front to keep it in a raised position.

The coat's sleeves include thumb wristlets (straps or reinforced openings). Wristlets prevent the sleeves from riding up on the wrists so that no skin is exposed between the gloves and the sleeves, which could result in wrist burns. It is important that the wearer use the thumb wristlets.

Protective coats have an additional integrated safety component. The **drag rescue device (DRD)** is webbing integrated into the back of the protective coat with a handle that is accessible just below the collar. A rescuer can grab this handle and use it to drag the incapacitated firefighter to safety.

Pockets in the coat can be used for carrying small tools or extra gloves. Additional pockets or loops can be installed to hold radios, microphones, flashlights, or other accessories.

Trousers

Structural firefighting protective trousers, also called **bunker pants** or **turnout pants**, are constructed in a waist-length design (pants) or a bib-overall configuration (**FIGURE 3-7**). Like protective coats, protective pants must satisfy the conditions specified in NFPA 1970.

Protective pants are constructed with the same three layers as a protective coat. The outer shell resists abrasions and repels water. The second layer is a moisture barrier that protects the skin from liquids and steam burns. The third, inner layer is a quilted, thermal barrier that protects the body from elevated temperatures.

FIGURE 3-7 Structural firefighting protective pants in a waist-length design.

Protective pants are reinforced around the ankles and knees with leather or extra padding to protect those areas from injury, especially if the firefighter needs to crawl to stay below hot gases and smoke.

Protective pants are manufactured with a double-fastener system at the waist, similar to the front flap of the protective coat. Suspenders hold the pants up. Florescent or reflective stripes around the ankles provide added visibility. Pants should be large enough that you can put them on quickly. They also need to be roomy enough to allow you to bend your knees and crawl easily, but they should not be bigger than necessary.

Footwear

Structural firefighting protective footwear (boots) protects the feet and ankles from heat sources, keeps them dry, prevents puncture injuries, and protects the toes from crushing injuries. Structural firefighting boots can be constructed of rubber or leather, and they are available in a variety of styles. Taller boots provide shin protection. Rubber firefighting boots come in a step-in style without laces (**FIGURE 3-8**). Leather firefighting boots are available in pull-on style or in a shorter version with laces (**FIGURE 3-9**). Many firefighters install a zipper on the laced boots to make it faster for the firefighter to put on and take off this component of PPE.

Both rubber and leather styles must meet the requirements specified in NFPA 1970, and some boots are dual certified to meet the requirements of NFPA 1970 and the more stringent requirements of NFPA 1990, *Standard for Protective Ensembles for Hazardous Materials and CBRN Operations, 2022 Edition*, which requires that boots provide liquid splash protection to guard against hazardous materials. The outer layer of boots repels water and is flame and cut resistant. Boots also must have a heavy sole with a slip-resistant design, a puncture-resistant sole, and a reinforced toe to

FIGURE 3-8 Rubber firefighting boots.

FIGURE 3-9 Leather firefighting boots.

FIGURE 3-10 Firefighting gloves.

prevent injury from falling objects. An inner liner constructed of a material such as Nomex or Kevlar adds thermal protection.

Like all protective clothing, boots must be the correct size for a firefighter's feet. The foot should be secure within the boot to prevent ankle injuries and provide secure footing on ladders or uneven surfaces. Improperly sized boots cause blisters and other problems.

Gloves

A **structural firefighting protective glove** protects a firefighter's hand from heat, cuts, and abrasions (**FIGURE 3-10**). Gloves are an important part of the firefighting ensemble because most fire suppression tasks require the use of the hands. Gloves must provide adequate protection while permitting the manual dexterity needed to accomplish tasks at an emergency scene. NFPA 1970 specifies that gloves must be resistant to heat, liquid absorption, vapors, cuts, and penetration. A liner adds thermal protection and serves as a moisture barrier.

Many firefighters carry a second set of gloves in their gear or on the apparatus so they can change them when the first pair becomes dirty, wet, or damaged. Do not wring or twist wet gloves, as this motion can tear or damage the inner liners.

Although gloves furnish needed protection, they inevitably reduce manual dexterity. For this reason, firefighters need to practice manual skills while wearing gloves to become accustomed to working with them on and to learn to adjust their movements accordingly.

SAFETY TIP

Plain leather work gloves or plastic-coated gloves should never be used for structural firefighting. Gloves used for firefighting must be labeled as such and meet the specifications given by NFPA 1970. Latex or nitrile gloves are required for rendering emergency medical care.

Donning and Doffing Personal Protective Clothing

To **don**—that is, put on—the structural firefighting protective ensemble, the process must be performed properly and quickly. Protective clothing must be donned in a specific order to obtain maximum protection. Following a set pattern of donning PPE reduces the time it takes to dress. This exercise does not include donning an SCBA, which will be discussed later. First become proficient in donning personal protective clothing, and then add the SCBA. Protective clothing is donned either in the firehouse before mounting the apparatus or on scene after you arrive. *Do not* don protective clothing inside the apparatus while en route to an emergency incident. Instead, stay in your assigned seat, properly secured by a seat belt or safety harness, while the vehicle is in motion.

To **doff**—that is, remove—the structural firefighting protective ensemble, reverse the procedure you used when donning it. Doffing should be done at the scene so that the interior of the apparatus is not contaminated with anything from the emergency scene.

Current NFPA standards require decontaminated gear to be stored in a separate room of the fire station that has good ventilation and limited sunlight. The separation and ventilation aid in keeping cancer-causing agents away from firefighters, and reduced sunlight helps improve the life of PPE by preventing unnecessary exposure to ultraviolet (UV) rays which can break down materials. Even after it is thoroughly cleaned, PPE should not be kept in living areas, sleeping areas, or vehicles because even when cleaned, it may still contain some carcinogens. Wherever it is stored, PPE must be properly maintained, organized, and ready for the next emergency response. It should be laid out in a logical order for donning quickly, and the boots should be inside the legs of the protective pants so that you can step into the boot and pants at the same time.

SAFETY TIP

Do not ride back to the station wearing or sitting next to contaminated PPE.

Follow the steps in **SKILL DRILL 3-1** to don the structural firefighting protective ensemble. Follow the steps in **SKILL DRILL 3-2** to doff the structural firefighting protective ensemble.

SKILL DRILL 3-1

Donning Personal Protective Clothing Firefighter I, NFPA 1010: 6.1.2

1. Place your protective hood over your head and down around your neck.

2. Step into your boots and protective pants, and then pull up the pants. Place the suspenders over your shoulders, and then secure the front of the pants.

Continues.

SKILL DRILL 3-1 CONTINUED

Donning Personal Protective Clothing Firefighter I, NFPA 1010: 6.1.2

3. Put on your protective coat and close the front of it.

4. Place your helmet on your head with the ear tabs extended and adjust the chin strap securely. Turn up your coat collar and secure it in front.

5. Put on your gloves and secure the wristlet.

6. Have your partner check your clothing.

SKILL DRILL 3-2

Doffing Personal Protective Clothing Firefighter I, NFPA 1010: 6.1.2

1. Remove your gloves.

2. Open the collar of your protective coat.

3. Release the helmet chin strap, and then remove your helmet.

Continues.

SKILL DRILL 3-2 CONTINUED

Doffing Personal Protective Clothing Firefighter I, NFPA 1010: 6.1.2

4. Remove your protective coat.

5. Remove your protective pants and boots.

6. Remove your protective hood.

Specialized Protective Clothing

Firefighters who rescue and extricate victims from a vehicle and those who fight wildfires face different hazards than when fighting structure fires. This means they require some specialized items in their protective clothing ensemble.

Vehicle Extrication Protective Clothing

Due to the risk of fire at the scene of a vehicle extrication incident, members of the emergency team wear full protective gear. The officer in charge also may designate one or more members to put on their SCBA and stand by with a charged hose line. A structural firefighting protective ensemble protects against many of the hazards present at a vehicle extrication incident, such as broken glass and sharp metal objects. Some protective clothing, such as special gloves and coveralls or jumpsuits, has been specifically designed for vehicle extrication. These items are generally lighter in weight and more flexible than structural firefighting protective clothing, although they may be constructed of the same basic materials. If firefighters are operating on a roadway, they should also wear a high-visibility safety vest.

Firefighters performing vehicle extrication must always be aware of the possibility of contact with blood or other body fluids. Latex or nitrile gloves should be worn when providing patient treatment. Eye protection should always be worn, given the possibilities of breaking glass, contact with body fluids, metal debris, and accidents with tools. Helmet shields or other helmet mounted face shields do not take the place of proper eye protection.

Proper cleaning of extrication gear is especially important. Many of the fluids encountered during vehicle extrication are petroleum-based—for example, gasoline, diesel fuel, transmission fluid, and brake fluid—and oil-based fluids that can soak into the fabric of your protective clothing. Because these fluids are flammable, they pose a hazard for firefighters who enter a burning building while wearing gear contaminated with petroleum-based fluids.

Wildland Fire Protective Clothing

Structural firefighting protective clothing is not designed for extended wildland firefighting because it may cause high body temperatures, resulting in heat exhaustion or heat stroke. Protective clothing designed specifically for fighting wildland or brush fires must meet the requirements outlined in NFPA 1977. In this type of firefighting ensemble, the jacket and pants are made of fire-resistant materials, such as Nomex or specially treated cotton, that are designed for comfort and maneuverability while working in the wilderness. Wildland firefighters wear a helmet made of a thermoresistant plastic, which is lighter than the material used in a structural firefighting protective helmet; eye protection, which is typically goggles; and pigskin or leather gloves, which allow more dexterity and provide resistance against blisters, cuts, and so on (more like a work glove than a structural firefighting protective glove). The boots are designed to provide comfort and sure footing while hiking to a remote fire scene and to provide protection against crush and puncture injuries.

TIP

Sometimes referred to as a proximity suit, high temperature–protective equipment allows a properly trained firefighter to work in extreme fire conditions, such as those posed by aircraft fires. This type of PPE shields the wearer during short-term exposures to high temperatures.

Additional Personal Equipment

A firefighter should always carry a **hand light**, given that most interior firefighting takes place in near-dark, zero-visibility conditions. A good working hand light can illuminate your surroundings, mark your location, and help you find your way under difficult conditions. In addition to carrying a hand light, hands-free options such as helmet-mounted lights or other lights that clip to clothing or SCBA may be useful.

Two-way radios link the members of a firefighting team. It is ideal for all personnel to have a radio when working inside a burning building or in a hazardous atmosphere. If radios are not available for everyone, at least one team member should have a radio. Some fire departments provide a radio for every on-duty firefighter. Follow your department's standard operating procedure (SOP) on radio use. A radio should be considered part of PPE and always carried with you.

Any time you are operating in a roadway or are close to traffic, you are required to wear a brightly colored reflective safety vest that meets the latest iteration of the American National Standards Institute (ANSI) and the International Safety Equipment Association (ISEA) *High Visibility Safety Apparel and Accessories* standard (ANSI/ISEA 107-2020).

Some eye protection is provided by the face shield mounted on the fire helmet. When additional eye protection is needed, such as when using power saws or hydraulic rescue tools, and if goggles are not part of the helmet, firefighters can use approved goggles or safety glasses. Goggles can be carried easily in the pocket of the protective coat.

Firefighters are often exposed to loud noises such as sirens and engines. Because hearing loss is cumulative, it is important to limit exposure to loud sounds. Headsets with a boom microphone and an intercom system on the apparatus provide hearing protection by reducing engine and siren noise, permit communications between crew members, and enable everyone to hear radio communications (**FIGURE 3-11**). A small speaker is incorporated into large earmuffs located at each riding and operating position. Flexible ear plugs are useful in other situations involving loud sounds. Firefighters should use the hearing protection supplied by their departments to prevent hearing loss when using power equipment or other equipment that produces noise.

FIGURE 3-11 Some fire apparatus are equipped with a headset and intercom systems. These systems can provide hearing protection, permit communications between crew members, and enable everyone to hear radio communications.

Firefighters have personal preferences about the additional tools and equipment they carry in their pockets. Observe which items seasoned members of your department carry in their pockets. Their choices will help you decide whether you need to carry similar equipment. The following are some common items:

- Spare flashlight
- Multi-tool
- Lineman's cutters
- Wedges
- Webbing
- Screwdriver with an interchangeable head
- Shove knife
- Folding knife

Inspection and Maintenance of Personal Protective Clothing

Approved personal protective clothing is built to rigorous standards. It requires proper care if it is to continue to afford maximum protection. A complete set of approved clothing (excluding SCBA) costs more than $3000. The gear is expensive, so it should be properly maintained to ensure adequate protection for those wearing it. Avoid unnecessary cuts or abrasions on the outer material. This material already meets NFPA standards—do not look for new opportunities to test its effectiveness.

Inspecting PPE

NFPA 1851, *Standard on Selection, Care, and Maintenance of Protective Ensembles for Structural Fire Fighting and Proximity Fire Fighting, 2020 Edition*, recommends a routine inspection before each shift and after each time structural firefighting gear is worn. In addition to the routine inspection, an advanced inspection should take place annually or when damage is noticed during the routine inspection. The advanced inspection can be performed by a third party or by trained personnel within the department. During the routine inspection, look for excessive wear, contamination, dirt, and damage. If the fabric is damaged, it must be taken out of service and properly repaired to retain its protective qualities. Clothing that is worn or damaged beyond repair must be replaced immediately because it will not be able to protect you. Follow the manufacturer's instructions for repairing or replacing the protective clothing.

The chin strap and suspension system must be properly adjusted and all parts of the helmet kept in good repair. Never modify the inside of the helmet basket because it can reduce the protection level the helmet offers.

> **SAFETY TIP**
>
> It is recommended that all personnel have access to an extra set of personal protective clothing for times when their gear is dirty or has become damaged and needs repair.

Cleaning PPE

PPE should be cleaned any time it becomes dirty—for example, by being exposed to smoke, blood or body fluids, or chemicals—to ensure your safety and health. Smoke contains carcinogens, body fluids can transmit diseases, and chemicals can cause burns. It is equally important to remove all of your clothing and thoroughly shower as soon as possible to reduce your exposure to these hazardous materials. In the field, you can use wet wipes to remove dirt from your face and neck so you are not breathing it in on your way back to the firehouse.

Protective clothing must be cleaned in accordance with the manufacturer's instructions as soon as possible after it has been exposed to a fire or other situation involving any hazardous materials. Dirt builds up in the clothing fibers during routine use and exposure to fire environments. Smoke particles also become embedded in the outer shell and continue to off-gas if not properly cleaned. The interior layers will frequently be soaked with perspiration which is another reason to clean it regularly. Allowing fire by-products to stay embedded in protective clothing will break down the protective components of the gear, causing the potential for the material to fail.

Other contaminants are formed from the by-products of burnt plastics, melted tar, petroleum products, other synthetic products, or other contaminants. These residues, which are flammable, can become trapped between the fibers or build up on the outside of protective clothing, damaging the materials and reducing their protective qualities. A firefighter who is wearing protective clothing contaminated with flammable residues is, in essence, bringing additional fuel into the fire on the clothing.

Keeping personal protective clothing clean reduces your exposure to carcinogens and maintains the protective properties of the clothing. Regular and routine cleaning should remove most of these contaminants.

FIGURE 3-12 Perform a field reduction of contaminants to remove dirt and debris before returning to the station.

Items that have been exposed to other chemicals or hazardous materials may need to be cleaned by a professional cleaning company or be impounded for decontamination or disposal.

The first step in cleaning protective clothing that has been exposed to hazardous conditions is at the incident scene. While the firefighter is still wearing the protective clothing, another firefighter should spray them down using a small diameter hose to remove surface contaminants (**FIGURE 3-12**). Next, doff the PPE at the scene, and then use a brush or water to remove as much dirt and debris as possible. When you return to the station, thoroughly clean the PPE following the manufacturer's instructions.

Fire departments should have special washing machines that are approved for cleaning protective clothing. Do not launder protective clothing in personal or public washing machines. This could cause cross contamination with home laundry. If a department does not have approved cleaning devices, they can contract with an outside firm to clean and repair protective clothing. Follow your fire department's policies in regard to cleaning contaminated PPE. Your fire department may require that the PPE be bagged in a biohazard bag and professionally cleaned. In all cases, the manufacturer's instructions for cleaning and maintaining the garment must be followed. These instructions can be found on a label affixed to the inside of the garments by the manufacturer. Always follow the manufacturer's cleaning instructions and use only an approved detergent. If you use other methods or cleaning solutions, you may reduce the effectiveness of the garment and create an unsafe situation for the wearer. According to NFPA 1851, advanced cleaning should be done every 6 months.

Advanced cleaning is performed by someone who has received additional training to do so; sometimes this is an outside agency. If you wash your protective clothing in house, follow the manufacturer's instructions for drying these garments properly.

SAFETY TIP

Make sure your PPE is dry before using it on the fireground. If wet protective clothing is exposed to the high temperatures of a structural fire, the water trapped in the liner materials will turn into steam and be trapped inside the moisture barrier, which could cause painful steam burns.

Helmets also require regular cleaning and maintenance. The outer shell of your helmet should be cleaned with a mild soap as recommended by the manufacturer. The inner parts of helmets should be removed and cleaned according to the manufacturer's instructions.

Protective hoods and gloves get dirty quickly and should be cleaned according to the manufacturer's instructions. Dirty protective hoods may contain large quantities of carcinogens following a fire. Most protective hoods can be washed with appropriate soaps or detergents. Do not use gloves or protective hoods that have holes in them; repair or discard them. Even a small cut or opening in these items can result in a burn injury.

Boots should be maintained according to the manufacturer's instructions. Do not store these items or any other PPE in direct sunlight. Leather boots must be properly maintained to keep them supple and in good repair. All types of boots need to be repaired or replaced if the outer shell is damaged. Exposure of the metal safety toe or boot shank could allow the body to conduct electricity if used in an electrified environment.

As discussed, correctly storing your PPE is another important step in the proper care and maintenance of your PPE. PPE should be stored only in the proper PPE storage areas.

SAFETY TIP

When doffing PPE, *do not* leave a contaminated protective hood or SCBA mask around your neck and below your nose and mouth. You will absorb or inhale additional toxins.

Respiratory Protection

Respiratory protection equipment is an essential component of the structural firefighting PPE. Respiratory protection is both expensive and complicated; using this equipment confidently requires practice. You must be proficient in using it before you engage in fire suppression activities.

The atmosphere of a burning building is considered to be an **immediately dangerous to life and health (IDLH)** environment due to the presence of airborne contaminants. An IDLH environment is one in which conditions pose an immediate threat to life or could cause irreversible or delayed adverse health effects. Attempting to work in such an environment without proper respiratory protection can cause serious injury or death. Never enter or operate in a hazardous atmosphere without first ensuring that you have appropriate respiratory protection.

More fire deaths are caused by smoke inhalation than burns. Clearly, adequate respiratory protection is essential to your safety as a firefighter. In fact, the products of combustion from house fires and commercial fires may be so toxic that just a few breaths can result in death. This is why you must always take respiratory protection precautions before entering the fire.

When you first arrive at the scene of a fire, you do not have any way to measure the immediate danger to your life and your health posed by that fire. You must don full PPE and approved breathing apparatus before you enter and operate within this atmosphere. This includes response to outside fires, such as vehicle and dumpster fires. Respiratory protection and PPE also must be used in any situation where there is a possibility of toxic gases or oxygen deficiency, such as in a confined space. Always assume that the atmosphere is hazardous until it has been tested and proven to be safe.

Continue to wear full PPE and approved breathing apparatus during **overhaul**—the process of finding, exposing, and suppressing any smoldering or hidden pockets of fire in an area that has been burned—until the air has been tested and proven to be safe. Do not remove your respiratory protection even if it appears the fire has been suppressed. Wait for the signal from command that overhaul has been completed and it is safe to do so.

Respiratory Hazards of Fires

Firefighters need respiratory protection because a fire involves a complex series of chemical reactions that can rapidly affect the atmosphere in unpredictable ways.

Overview of Smoke and Products of Combustion

The most readily evident by-product of a fire is **smoke**, which is the airborne products of combustion consisting of solids (particles), liquids (vapors), and gases. Think of smoke as unburned fuel. Even though smoke is the constant companion of the firefighter, there is not always a thorough understanding of the link between the components of smoke and the reason people get sick or die when they breathe smoke, or why firefighters may contract heart disease or experience neurological dysfunction, cancer, or other illnesses related to a single or repeated exposure to smoke.

Essentially, smoke is a collection of gaseous products from burning materials—especially organic materials such as plastic, nylon, rubber products, wood, and paper—made visible by the presence of small particles of incomplete combustion (carbon). Some of the harmful substances found in smoke include soot, carbon monoxide, cyanide compounds, formaldehyde, and the oxides of nitrogen. These substances are dangerous if inhaled and repeated exposures to these materials may have negative health effects. Cancer is prevalent in the fire service, more so than in most any other profession, and this high rate should come as no surprise once the by-products of combustion are understood. Many known and suspected human carcinogens are liberated as fire gases and particulates.

Most people identify carbon monoxide as the main harmful component of fire smoke but struggle to list more than a handful of substances beyond that. Less often acknowledged are compounds such as ammonia, hydrogen chloride, sulfur dioxide, hydrogen sulfide, hydrogen cyanide, carbon dioxide, the oxides of nitrogen, formaldehyde, acrolein, polycyclic aromatic hydrocarbons, and soot (**FIGURE 3-13**).

FIGURE 3-13 A different way to look at fire smoke: If this cargo trailer were carrying smoke, this is how it might be placarded. If these placards were present at the next house fire you responded to, would you think differently about the potential health hazards?

Smoke production depends on several factors, including the chemical make-up of the burning material, the temperature of the combustion process, and the influence of ventilation (oxygenation). Make no mistake: Fire (combustion) is a complex process, and the smoke produced during a fire event is an intricate collection of particulates, superheated air, and gaseous chemical compounds.

The extensive use of synthetic manufacturing and construction materials (for example, plastics, nylon, rubber products, fire retardants, laminates, and foams such as Styrofoam and polystyrene) has significant effects on fire behavior and smoke production. Synthetic substances ignite and burn quickly, causing rapidly developing fires and toxic smoke and making structural firefighting today more dangerous than ever before.

As an example, consider polyurethane foam, the most predominant substance in a typical mattress. It consists of many different chemicals, including polyol (an organic alcohol molecule and the majority of the polyurethane compound), toluene diisocyanate (TDI), methylene chloride, and ammonia-based catalysts. When polyurethane foam is exposed to heat, the parent substances break down and bond with each other, creating many new compounds. Some of those compounds are irritants, such as hydrogen chloride and ammonia, which may cause eye irritation or airway problems in smoke exposures. Other compounds, such as carbon monoxide and cyanide compounds, are acutely toxic when inhaled.

The thermal decomposition of polyurethane foam in the mattress fire scenario is broadly representative of the way gases are liberated by a working fire. (Keep in mind this is a limited representation of the process; many more substances are also liberated.) Certainly, each substance is present in varying levels depending on the material(s) involved, the heat of combustion, and the available oxygen (ventilation of the fire). Aside from the toxic gases produced, it is equally important to recognize the visible part of combustion (aside from the flames), which consists of carbon particles of varying sizes, ranging from large embers to particles you cannot see with the naked eye (**FIGURE 3-14**).

This collection of particulates—what we traditionally called smoke—is actually **soot**. Soot is a known human carcinogen, just like benzene and formaldehyde

FIGURE 3-14 Notice how the beam of light is visible due to the particulates in the air.

FIGURE 3-15 The thermal decomposition of muscle meats such as chicken, beef, and pork generates polycyclic aromatic hydrocarbons (PAHs) and heterocyclic amines (HCAs) when they are cooked over an open flame.

TABLE 3-1 Smoke Matrix

	Harm You Now	Harm You Later
What you see	Air	Soot, particulates
What you don't see	Carbon monoxide (CO), hydrogen cyanide (HCN), oxides of nitrogen (NOx), sulfur dioxide (SO_2), hydrogen chloride (HCl), hydrogen sulfide (H_2S)	Aldehydes, benzene

(also by-products of combustion). With that point in mind, take a moment and think about your perception of smoke and revise it to this: The combustion process liberates things you can see (particulates) and things you cannot see (gases). **TABLE 3-1** breaks smoke down into these two main categories to make it easier to understand and separate the hazards. This matrix is not inclusive of all potentially harmful substances but it does illustrate the point that many properties of smoke can cause acute and chronic health effects. It also shows that smoke is not just *one thing*, but rather a dynamic, multifaceted mixture of gases and particulates that changes from minute to minute at any fire.

Another group of compounds commonly found in fire smoke comprises the polycyclic aromatic hydrocarbons (PAHs), which are classified as probable or possible human carcinogens. More than 100 PAHs have been identified and categorized by various regulatory agencies. This group of substances occurs naturally in materials such as coal and crude oil. These substances also are generated during the combustion of organic materials and can be found in vehicle exhaust, tobacco smoke, and the smoke generated from structure fires, vehicle fires, wildland fires, or any other type of fire. PAHs can exist as particles or gases. Chances are that you have been exposed to PAHs throughout your life from a common and perhaps surprising activity—grilling food (**FIGURE 3-15**).

When PAHs are generated during a structure fire, they may bind with the soot, resulting in dermal and inhalation exposures. PAHs are believed to be immunosuppressants, a property perhaps contributing to the mechanism by which they are suspected to cause cancer. Examples of PAHs include the following chemicals:

- Anthracene
- Benzopyrene
- Methylchrysene
- Phenanthrene
- Pyrene

Responders not only have to be concerned about acute exposure to PAHs and other materials, but also need to consider this group of chemicals (and many others) as potential contaminants of their structural firefighting protective clothing. NFPA 1851 offers detailed guidance on many aspects of cleaning and maintaining structural firefighting protective clothing.

Products of Combustion

Smoke particles consist of unburned, partially burned, and completely burned substances. Many smoke particles are so small that they can pass through the natural protective mechanisms of the respiratory system and enter the lungs or pass through the skin, move into the bloodstream, and be transported throughout the body. Some are toxic to the body and can result in severe injuries or death if they are inhaled. These particles can also be extremely irritating to the eyes and digestive system.

Smoke vapors are finely suspended liquids—that is, aerosols. This is similar to fog, which consists of small water droplets suspended in the air. When oil-based compounds burn, they produce small hydrocarbon droplets that become part of the smoke. If the droplets are inhaled or ingested, the hydrocarbons can affect the respiratory and circulatory systems. In addition, some toxic droplets in smoke vapor can cause poisoning if they are absorbed through the skin.

A fire also produces several types of gases. The amount of oxygen available to the fire and the type of fuel being burned determine which gases are produced. Many of the gases produced by residential or commercial fires are highly toxic. Carbon monoxide (CO), hydrogen cyanide, and phosgene are three gases commonly present in smoke. These gases are discussed in more detail in Chapter 5, *Fire Behavior*.

The harmful products in smoke require firefighters to use respiratory protection in all fire environments, regardless of whether the environment is known to be contaminated, suspected of being contaminated, or could possibly become contaminated without warning. The use of respiratory protection allows firefighters to enter and work in a hazardous atmosphere with a safe, independent air supply.

SAFETY TIP

Research conducted by Underwriters Laboratories (UL) demonstrated that residual smoke on gear retains many heavy metals, including arsenic, cobalt, chromium, lead, mercury, phosphorus, and phthalates. Exposure to these chemicals has been linked to increased risk of cancer. Dirty PPE is a danger to your health!

Carbon Monoxide and Hydrogen Cyanide: Silent Killers

Smoke is one of the first observable signs of a working fire. Firefighters note the volume, color, and force of smoke as it exits a fire building—all good indicators of what the fire is doing inside—and use that information to implement appropriate fireground tactics. Firefighters may aggressively enter smoky buildings to search for victims but rarely perform a conscious evaluation of the toxic substances lurking in the smoke. This could be a significant oversight in terms of treating smoke inhalation victims, according to studies performed in Paris, France, and Dallas County, Texas. These studies, though conducted many years ago, focused on CO and hydrogen cyanide (HCN) specifically because they are acutely toxic, present to some degree in nearly all fires, and have clinical interventions available to reverse the adverse health effects of the exposure. Clearly, many toxic substances are generated during a typical structure fire, and smoke inhalation is a complicated illness. It is also clear that successful medical treatment cannot be accomplished by a single drug or single action and that smoke inhalation victims are quite sick.

The Paris and Dallas County studies were done more than 20 years ago, which underscores the fact that the presence of HCN and CO in smoke is not a new revelation; however, the fire service and other disciplines have recently been connecting the dots to understand that smoke is a bigger health and safety issue than previously imagined. In short, these studies identify and evaluate the impact of HCN and CO on victims of smoke inhalation. The discussion here does not focus on the fact that those two substances make smoke toxic; rather, the important point is that both are toxic substances that are common by-products of combustion (present at some level at most fires) and their effects are treatable with clinical interventions. Again, many other toxins can be found in smoke—the goal here is to highlight two fire gases found at nearly any kind of fire.

The Paris study was designed to prospectively assess the role of cyanide in smoke-related morbidity and mortality. Morbidity is the incidence of injury or disease within a population; mortality is the incidence of death in a population. In the Paris study, blood samples were drawn from fire survivors as well as fatalities (at the time of exposure), and the cyanide and carbon monoxide levels were measured. In several deaths, cyanide levels registered in the lethal range, whereas carbon monoxide levels were in nontoxic concentrations. This suggests cyanide toxicity as the primary cause of death. The study also revealed another bit of interesting information: Deaths also occurred in victims with both cyanide *and* carbon monoxide levels in the nontoxic range, perhaps revealing a relationship between carbon monoxide and cyanide in victims of smoke inhalation. The results of the Paris study can be summarized as follows:

- Cyanide and carbon monoxide are both important determinants of smoke inhalation–associated morbidity and mortality.
- Cyanide concentrations are directly related to the probability of death.
- Cyanide poisoning may be more predominant than carbon monoxide poisoning as a cause of death in certain fire victims.
- Cyanide and carbon monoxide may potentiate each other's harmful effects.

The Dallas County study measured blood cyanide levels in victims after exposure to fire smoke and in many respects echoed the findings of the Paris study (Silverman et al 1988). Over a 2-year period, researchers collected blood samples from a total of 187 smoke inhalation patients, within 8 hours of exposure. There were 144 viable patients at the University of Texas Health Sciences emergency department; 43 victims were dead on arrival at the Dallas County Medical Examiner's office.

Of the 144 living patients who reached the emergency room, 12 had blood cyanide concentrations exceeding 1 mg/L. Of these 12 patients, 8 eventually died. None had blood carboxyhemoglobin (COHb) concentrations suggesting carbon monoxide as the cause of death (i.e., ≥50 percent). Although some of the patients had extensive burns that may have contributed to their death, three of them (patients 2, 3, and 9) had total body surface area burns of 4 percent or more. According to the study, blood cyanide levels greater than 1 mg/L had a significant impact on patient outcome. More importantly, the study found that elevated cyanide levels were pervasive in victims of smoke inhalation, and cyanide concentrations were directly related to the probability of death.

Lastly, both studies dispelled a long-held belief in the fire service—that carbon monoxide is the sole (or at least predominant) killer in fire smoke. In fact, it appears that cyanide plays a role in smoke-related deaths and injuries, perhaps more often than we think. Therefore, any victim(s) exposed to significant amounts of smoke or rescued from a closed-space structure fire may be suffering from cyanide toxicity. However, remember that smoke has many properties that are capable of causing acute illness and leaving an indelible impression on the body.

Depending on the dose, hydrogen cyanide has the ability to incapacitate a victim, preventing escape from the fire environment and thereby increasing the exposure to more cyanide, carbon monoxide, and other toxic by-products of combustion. Although this theory is currently unsupported with human data related to smoke exposures, there is information to substantiate the "knock down" potential of cyanide. In the mid-1980s, studies were conducted on monkeys exposed to the fumes of heated polyacrylonitrile. When polyacrylonitrile is broken down by pyrolysis, cyanide is liberated. Cyanide-exposed monkeys first hyperventilated and then rapidly lost consciousness at a dose-dependent concentration. A concentration of 200 parts per million (ppm) was associated with rapid incapacitation but not with elevated blood cyanide concentrations measured hours after exposure. The direct correlation to human data is unknown at present but could be interpreted in the following way: Hydrogen cyanide could be partly responsible for rendering firefighters and civilians incapable of self-rescue when exposed to smoke.

SAFETY TIP

Cyanide exposures can be fatal in the presence of normal oxygen levels.

To appreciate cyanide's mechanism of action, it is first necessary to understand the process of oxygen transportation and use in the body and the basic idea of aerobic metabolism. Aerobic metabolism is the process of creating energy through the breakdown of nutrients in the presence of oxygen. Its by-products are carbon dioxide and water, which the body disposes of by breathing and sweating. To simplify the concept, imagine the circulatory system as an efficient public transit system, full of "buses" (red blood cells) carrying passengers (oxygen) to and from a multitude of bus stops (the cells). The circulatory system, similar to a network of streets, is loaded with red blood cells—hemoglobin buses—each carrying four oxygen passengers (**FIGURE 3-16**).

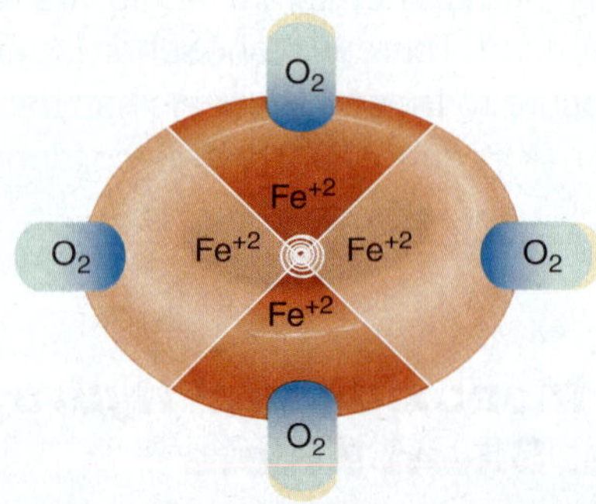

FIGURE 3-16 The circulatory system is loaded with red blood cells, each of which carries four oxygen "passengers."

During normal cellular respiration, the bus system transports oxygen passengers to the bus stops (cells). At the appropriate stop, four oxygen molecules get off and move through an electron chain, ultimately combining with the final electron acceptor—cytochrome oxidase (an enzyme that acts as an agent of electron transfer from certain cytochrome molecules to oxygen molecules to aid in cell respiration)—before entering the mitochondria of each cell. Mitochondria are responsible for converting nutrients into energy-yielding molecules of adenosine triphosphate (ATP) to fuel the cell's activities (**FIGURE 3-17**). ATP production is highly dependent on oxygen (and glucose); without it, normal aerobic metabolism is impossible. If this process becomes seriously compromised, death is imminent.

Cyanide compounds, once absorbed in the body, "poison" the cytochrome oxidase, blocking oxygen from entering the mitochondria and effectively shutting down the process of aerobic metabolism. In short, the buses may be transporting some amount of oxygen passengers, but when they get off the bus, they find their ultimate destination locked. Without oxygen, the cells switch to anaerobic metabolism, the creation of energy through the breakdown of glucose without oxygen. This process produces toxic by-products such as lactic acid, which ultimately destroys the cell. Therefore, cyanide toxicity is not about the amount of oxygen available to the body, but rather hinges on the body's inability to *use* oxygen for aerobic (life-sustaining) metabolism. Consequently, an elevated lactic acid level is a key indicator of cyanide toxicity.

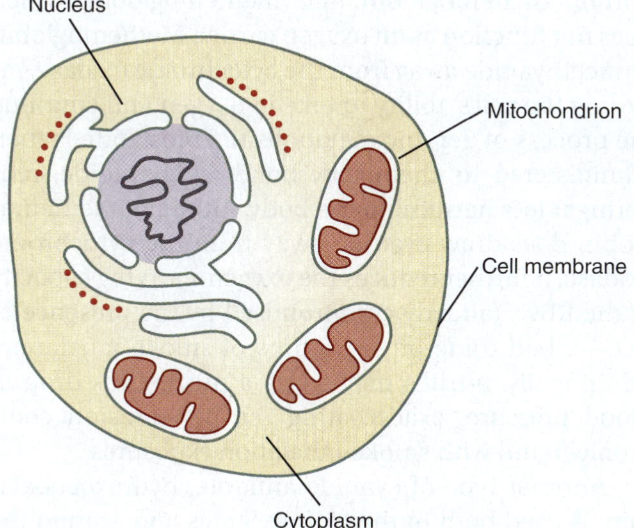

FIGURE 3-17 Mitochondria are responsible for converting nutrients into energy-yielding molecules of adenosine triphosphate (ATP) to fuel the cell's activities.

The signs and symptoms of acute cyanide toxicity are similar to those of CO poisoning and mimic the nonspecific signs and symptoms of oxygen deprivation, including headache, dizziness, stupor, anxiety, rapid breathing, and increased heart rate. In extreme cases of cyanide poisoning, patients may present with seizures, a significantly altered level of consciousness (including coma), severe respiratory depression or respiratory arrest, and complete cardiovascular collapse.

According to the studies done in Paris and Dallas County, a deleterious relationship between CO and cyanide may potentially occur in victims of smoke inhalation. This relationship could be attributed to the cells' inability to use oxygen (due to cyanide's blocking effects), coupled with the adverse impact of CO on the RBCs. Carbon monoxide binds in place of oxygen on the RBCs, thereby reducing the ability of oxygen to ride on the hemoglobin bus. (The oxygen-carrying capacity of the RBCs becomes limited or nonexistent, which reduces the amount of oxygen transported to the cells.)

Carbon monoxide is one of the most commonly encountered industrial hazards. This colorless, odorless gas is produced during incomplete combustion. As mentioned earlier, it affects the RBCs' oxygen-carrying capacity, causing hypoxia—that is, "inadequate oxygenation of the blood and tissue sufficient to cause impairment of function" (NFPA 99). Consequently, the signs and symptoms of CO exposure will be similar to those associated with cyanide poisoning and consistent with hypoxia—headache, nausea, vomiting, disorientation, and, if the exposure is extreme, seizures, coma, and death. **FIGURE 3-18** depicts the signs and symptoms of a CO exposure. As the arrow indicates, the severity of health effects is related to the severity of the exposure.

In summary, you can look at CO and HCN exposures in this way: Carbon monoxide reduces the amount of oxygen carried to the cells; cyanide renders the cells incapable of using whatever oxygen is present.

Smoke Inhalation Treatment

Smoke inhalation is one of the most complex and challenging patient presentations faced by medical care providers. Patient outcomes vary greatly and are influenced by such factors as the extent and duration of the smoke exposure, amounts and nature of the toxicants in the smoke, degree of thermal burns to the skin and lungs, quantity and size of inhaled particulates (soot), patient's age, and underlying medical conditions. In the United States, there is no nationwide standard protocol for treating smoke inhalation, leaving paramedics and other prehospital care providers with limited

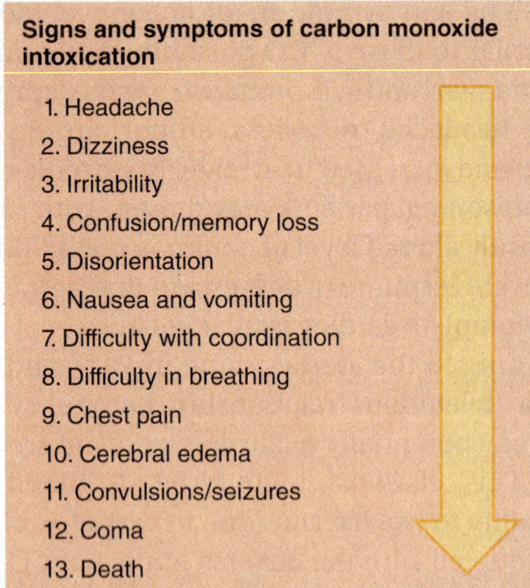

FIGURE 3-18 As the level of carboxyhemoglobin (HbCO) rises, the severity of the signs and symptoms increases. The downward arrow indicates the progression from the most common symptom of carbon monoxide (CO) poisoning—headache—to the most extreme symptoms—seizures, coma, and death.

guidance and training to properly care for victims of such exposures. In many EMS systems, treatment of smoke inhalation outside the hospital mostly focuses on supportive care: monitoring vital signs, providing high-flow oxygen, establishing intravenous (IV) lines, performing advanced airway management techniques such as endotracheal intubation, cardiac monitoring, and rapid transport.

The studies done in Paris and Dallas County, however, make it clear that supportive care alone will not correct the underlying cause of death in victims of smoke inhalation—namely, the adverse effects of HCN and CO on the body's oxygen transportation and utilization system, which cause asphyxia at the cellular level. The bottom line is this: Until the underlying cause of asphyxia is reversed at the cellular level, normal oxygenation is not possible. This requires a clinical intervention—administering an antidote—to restore the body's ability to use oxygen.

SAFETY TIP

Hydrogen cyanide poisoning should be suspected in patients with smoke inhalation who have significant hypotension, soot in the nose or mouth, or an altered level of consciousness.

Oxygen is the natural antidote for CO poisoning, and high-flow oxygen should be administered in all cases of smoke inhalation. In patients with CO poisoning, which primarily targets the heart and brain, the administration of high-flow oxygen reduces the half-life of CO in the body to around 1 hour. If hyperbaric therapy is indicated, the amount of CO in the body can be reduced by one-half in approximately 30 minutes. Note that these recommendations are merely approximations based on dose, and patient responses may vary from case to case. The point is that oxygen is good for victims who have inhaled CO, whether from fire smoke or some other source.

When it comes to treating the HCN portion of smoke inhalation, a different kind of antidote is indicated along with oxygen administration. Currently, in the United States there are a number of Food and Drug Administration (FDA)–approved interventions for treating cyanide poisoning. Numerous fire departments and EMS agencies carry and administer cyanide antidotes for smoke inhalation with success. Nithiodote is a cyanide antidote that contains sodium nitrite and sodium thiosulfate. Alternatively, sodium nitrite and sodium thiosulfate can be administered as separate interventions, rather than as part of the Nithiodote product.

The nitrite component of these treatments is administered to convert hemoglobin in the RBCs to methemoglobin, a substance formed by a change in the iron atom in hemoglobin from the ferrous (+2) to the ferric state (+3). It is normal to have a small amount of methemoglobin in the blood, but substances such as nitrites and nitrile-containing substances convert a larger proportion of hemoglobin into methemoglobin, which does not function as an oxygen carrier. Methemoglobin attracts cyanide away from the cytochrome oxidase, restoring the cell's ability to take in oxygen and continue the process of aerobic metabolism. Thiosulfate is then administered to chemically bond with cyanide, rendering it less harmful to the body. Although methemoglobin does draw cyanide away from the cytochrome oxidase, it further reduces the oxygen-carrying capacity of the RBCs (already compromised by the presence of CO)—a bad trade-off in victims of smoke inhalation. Additionally, nitrites may cause a precipitous drop in blood pressure, exacerbating the hypotension commonly found with smoke inhalation exposures.

Another type of cyanide antidote, hydroxocobalamin, is used both in the United States and around the globe. This drug (a precursor to vitamin B_{12}) is a relatively benign substance, with minimal side effects, making it well suited for use in the prehospital setting. Hydroxocobalamin has no adverse effect on RBCs' oxygen-carrying capacity and no negative effect on the patient's blood pressure—significant benefits when treating victims of smoke inhalation. Surprisingly, its mechanism of action is simple: Hydroxocobalamin binds to cyanide, forming vitamin B_{12} (cyanocobalamin), a nontoxic compound that is ultimately excreted in the urine. Hydroxocobalamin is the antidote found in the Cyanokit, another cyanide antidote kit often used by fire departments and EMS agencies. The Cyanokit can be administered to a victim of smoke inhalation without first verifying the presence of cyanide in the body, with little fear of making the patient worse.

It is important to understand how individuals who experience smoke inhalation are treated in your jurisdiction and whether a cyanide antidote is carried on advanced life support transport units or stocked in the local hospitals. It is also important to understand that smoke inhalation is an illness just like heart failure, asthma, or any other medical condition you may encounter in the field, and that aggressive clinical intervention is critical to achieving a good outcome. As a reference source, consult NFPA 470, chapter 48, *Competencies for Hazardous Materials/WMD Advanced Life Support (ALS) Responders Assigned Mission-Specific Responsibilities.* This chapter provides background information and guidance on treating patients with smoke inhalation.

Oxygen Deficiency

Normal outside or room air contains approximately 21 percent oxygen. A decrease in the amount of oxygen in the air drastically affects an individual's ability to function (**TABLE 3-2**). An atmosphere with an oxygen concentration of less than 19.5 percent is considered oxygen deficient (OSHA 1970). If the oxygen level drops below 19.5 percent, people can experience disorientation, an inability to control their muscles, and irrational thinking, which can make escaping a fire much more difficult.

TABLE 3-2 Physiological Effects of Reduced Oxygen Concentration

Oxygen Concentration	Effect
21%	Normal breathing air
17%	Judgment and coordination impaired; lack of muscle control
12%	Headache, dizziness, nausea, fatigue
9%	Unconsciousness
6%	Respiratory arrest, cardiac arrest, death

SAFETY TIP

Smoky environments and oxygen-deficient atmospheres are deadly! Always use an SCBA when operating in a smoky environment, and because smoke is not just a respiratory danger—toxins can enter your body through the skin—always wear the required protective clothing.

When a fire is burning within an enclosed area, oxygen depletion occurs in two ways. First, the fire consumes large quantities of the available oxygen, decreasing the concentration of oxygen in the atmosphere. Second, the fire produces large quantities of other gases, which displace the oxygen that would otherwise be present inside the enclosed area.

Increased Temperature

Heat is also a respiratory hazard of fire. The temperature of the smoke varies, depending on the fire conditions and the distance the smoke has traveled. Inhaling superheated smoke can cause severe burns of the respiratory tract immediately. If the smoke is hot enough, a single inhalation can cause fatal respiratory burns. More information about fire behavior and products of combustion appears in Chapter 5, *Fire Behavior.*

Respiratory Hazards in Other Toxic Environments

Not all hazardous atmospheric conditions are caused by fires. Indeed, firefighters may encounter oxygen-deficient atmospheres in numerous types of emergency situations. For example, toxic gases may be released at hazardous materials incidents from leaking storage containers or industrial equipment, from chemical reactions, or from the normal decay of organic materials. Internal combustion engines or improperly operating heating appliances can produce carbon monoxide. Respiratory protection is just as important in these situations as it is in a fire suppression operation.

Types of Breathing Apparatus

There are two basic categories of breathing apparatus: atmosphere-supplying respirators and air purifying respirators.

An **atmosphere-supplying respirator (ASR)** uses air that is supplied from a source independent of the ambient (room) air. There are two types of ASRs. One type uses air carried in a compressed gas cylinder, such as the fire service SCBA. The construction and operation of SCBA will be more fully described in the following sections.

The second type of ASR is a supplied-air respirator. A **supplied-air respirator (SAR)**, also known as an **air-line respirator**, uses an external source for the breathing air (**FIGURE 3-19**). In this type of device, a hose line is connected to a breathing-air compressor or to compressed air cylinders located outside the hazardous area. The user breathes air through the line and exhales through a one-way valve. Supplied-air respirators are more frequently used in industrial settings and in some hazardous materials situations. Their use will be described in more detail as part of your hazardous materials training.

The second major type of breathing apparatus is an air-purifying respirator. An **air-purifying respirator (APR)** has an air-purifying filter, cartridge, or canister that removes specific air contaminants by passing ambient (room) air through the air-purifying element (**FIGURE 3-20**). APRs remove particulate matter or specific gases and vapors. They are used in some industrial settings where the air is monitored carefully to determine the precise quantity of contaminants that are present. APRs are not suitable for firefighting operations because you need to know the type and concentration of contaminants present so that you can use the correct filter. This is not possible at a fire scene. Furthermore, APRs do not supply oxygen in an oxygen-deficient atmosphere.

A **powered air-purifying respirator (PAPR)** is similar to the standard APR, but it includes a small fan to help circulate air into the mask (**FIGURE 3-21**). The fan draws outside air through the filters and into the mask via a low-pressure hose. PAPRs make it easier for the wearer to breathe, help reduce fogging in the mask, and provide a constant flow of cool air across the face. PAPRs are not considered to be true positive-pressure units—that is, a unit that creates higher pressure inside the mask than the atmospheric pressure outside the mask. This is because it is possible for the wearer to "outbreathe" the flow of supplied air—that is, inhale more air than is being supplied via the low-pressure hose. This creates negative pressure—pressure less than the atmospheric pressure outside the mask—inside the

FIGURE 3-19 A supplied-air respirator (SAR) may be needed for special rescue operations.

Courtesy of 3M.

FIGURE 3-20 Air-purifying respirators (APRs) offer specific degrees of protection if the hazard present is known and the appropriate filter canister is used.

Courtesy of Sperian Respiratory Protection.

FIGURE 3-21 Powered air-purifying respirators (PAPRs) have a small fan to help circulate air into the mask.

Courtesy of Chris Hawley.

mask. When this happens, the seal with the face is compromised, which will possibly allow contaminants to enter the face mask.

APRs and PAPRs are typically worn to respond to a hazardous materials situation where the type and quantity of contaminants are known. These respirators should not be used in structural firefighting situations.

Standards and Regulations

In the United States, the **National Institute for Occupational Safety and Health (NIOSH)** is the federal agency that specifies the design, testing, and certification requirements for respiratory protection. NFPA 1970 requires that all SCBA equipment used in the emergency service meet the design, testing, and certification requirements established by NIOSH. The NFPA has developed three standards directly related to SCBA:

- NFPA 1550, *Standard for Emergency Responder Health and Safety, 2024 Edition,* includes the basic requirements for SCBA use and program management.
- NFPA 1404, *Standard for Fire Service Respiratory Protection Training, 2018 Edition,* sets the requirements for an SCBA training program within a fire department.
- NFPA 1970, *Standard Protective Ensembles for Structural and Proximity Firefighting, Work Apparel and Open-Circuit Self-Contained Breathing Apparatus (SCBA) for Emergency Services, and Personal Alert Safety Systems (PASS), 2024 Edition,* includes requirements for the design, performance, testing, and certification of open-circuit SCBA for the fire service.

These standards specify approved SCBA equipment, initial training for each firefighter, and annual retraining.

The **Occupational Safety and Health Administration (OSHA)** is the U.S. federal agency responsible for research and development on occupational safety and health issues. OSHA and state agencies are responsible for establishing and enforcing regulations for respiratory protection programs. Fire departments in states that are governed by OSHA regulations are required to comply with the OSHA respiratory protection regulations, such as OSHA 1910.134, which covers the fit testing, medical screening, and training of all firefighters in the use of SCBA. In some states and in Canadian provinces, individual occupational safety and health agencies establish and enforce these regulations. Each fire department must follow applicable standards and regulations to ensure safe working conditions for all personnel.

Self-Contained Breathing Apparatus and Personal Alert Safety System

The SCBA is an essential component of the PPE used for structural firefighting. An SCBA provides clean breathing air (the same as environmental air) to firefighters who are working in the hostile environment of a fire. Without adequate respiratory protection, firefighters would not be able to enter toxic environments. The SCBA must meet rigid manufacturing specifications so that it can function in the increased temperatures and

FIGURE 3-22 Open-circuit SCBA.

Courtesy of 3M.

FIGURE 3-23 A closed-circuit SCBA is commonly referred to as a "rebreather."

© Jones & Bartlett Learning. Photographed by Glen E. Ellman.

smoke-filled environments that firefighters encounter. When properly maintained, this equipment will provide enough air to enable firefighters to perform rigorous tasks.

TIP

Do not confuse SCBA and SCUBA. An SCBA is used by firefighters during fire suppression activities. A self-contained underwater breathing apparatus (SCUBA) is used by divers while swimming underwater.

The two main types of SCBA are open-circuit and closed-circuit devices. An **open-circuit self-contained breathing apparatus (open-circuit SCBA)** uses a cylinder of compressed air that provides a limited air supply to the user, and exhaled air is released into the atmosphere through a one-way valve (**FIGURE 3-22**). Open-circuit SCBAs are the type most commonly used in the fire service

With a **closed-circuit self-contained breathing apparatus (closed-circuit SCBA)**, commonly called a **rebreather**, exhaled air is not released to the outside environment. Instead the user's exhaled air passes through a mechanism that removes carbon dioxide and chemically generates oxygen. The air is then "rebreathed" by the wearer (**FIGURE 3-23**). Many closed-circuit SCBA units also include a small oxygen cylinder. A closed-circuit SCBA is often used for extended operations, such as mine rescue work and operations in long tunnels where breathing apparatus must be worn for a long time.

Using an SCBA requires that firefighters develop unique skills, including different breathing techniques. Proficiency in the use of SCBA and other PPE requires ongoing training and practice.

Limitations of SCBA

Like any type of equipment, SCBA has its limitations. Some of these limitations apply to the equipment; others apply to the user's physical and psychological abilities. Because an open-circuit SCBA carries its own air supply in a pressurized air cylinder, its use is limited by the amount of air in the cylinder. SCBA air cylinders for structural firefighting must carry enough air for a minimum of 30 minutes. Air cylinders rated for 45 minutes, 60 minutes, and 75 minutes are also available. These duration ratings are based on ideal laboratory conditions. The realistic useful life of an SCBA air cylinder for firefighting operations is usually much less than the rated duration, and actual use time depends on the size of the user, their physical fitness and conditioning, their physical exertion, and how calm they are. An SCBA air cylinder generally has a realistic useful life of no more than 50 percent of the rated time. For example, an SCBA air cylinder rated for 30 minutes can be expected to last for a maximum of 15 minutes during strenuous firefighting.

Firefighters must manage their working time while using SCBA so that they have enough time to exit the hazardous area before exhausting the air supply. To properly manage the air supply, a firefighter must consider the following factors:

- The time and the effort it will take to reach the task destination. Climbing stairs takes more energy and air than walking across a flat floor, for example.
- The amount of air that will be available upon reaching the task destination.
- The amount of time necessary to complete the task and the air that will be used during that period. Some tasks take more energy and air.
- The amount of time it will take to reach a safe area. At the end of this time, the firefighter must have a reserve of air for unexpected emergencies.

An SCBA provides a limited window of time for firefighting and a safe exit from the hazardous conditions of the fire. It is essential that you have a margin of safety built into your air supply for the unexpected. In some cases, you may need to begin exiting from the fire scene before half of your air supply is exhausted.

The weight of an SCBA varies, based on the manufacturer and the type and size of the air cylinder. Generally, an SCBA weighs at least 25 lb (11 kg). The added weight and bulk of the SCBA decrease the user's flexibility and mobility and shift the user's center of gravity.

The size and shape of an SCBA may make it difficult for you to fit through tight openings when wearing this equipment (**FIGURE 3-24**). Several techniques may help you navigate these spaces:

- Change your body position. Rotate your body by 45 degrees and try again.
- Loosen one shoulder strap and change the location of the SCBA on your back.
- If you have no other choice, you may have to remove your SCBA. In this case, do not let go of the harness for any reason. Keep the unit in front of you as you navigate through the tight space. Reattach the SCBA harness as soon as you are through the restricted space. *Note*: This is a last resort procedure!

FIGURE 3-24 An SCBA expands a user's profile, making it more difficult to pass through tight spaces.

FIGURE 3-25 This SCBA face piece shows damage from heat.

Always remember that this equipment limits normal sensory awareness—the senses of smell, hearing, and sight are all affected by the apparatus. As previously noted, the design of the SCBA face piece limits the firefighter's vision—particularly their peripheral vision. The face piece lens may fog up under some conditions, further limiting visibility. Face pieces also can fail if the temperature and radiant heat are too high (**FIGURE 3-25**).

NIST studies show that the face piece lens can soften at approximately 302°F (150°C) with the lens starting to "craze" or bubble at higher temperatures and then actually fail at temperatures above 419°F (215°C) (Mensch, Braga, and Bryner 2011, 1). SCBA may affect the user's ability to communicate, depending on the type of face piece and any additional hardware provided, such as voice amplification and radio microphones. The equipment can be noisy during inhalation and exhalation, which may limit the user's hearing as well, especially when coupled with a protective hood and helmet.

Physical Limitations of the SCBA User

Conditioning is important for SCBA users. A person in good physical condition will be able to perform more work per cylinder of air than a person who is overweight or out of shape. An out-of-shape or obese firefighter will consume the air supply from this equipment more quickly and will have to exit the fire building long before a well-conditioned firefighter must do so. Overweight or poorly conditioned firefighters also are at greater risk for heart attack due to physical stress.

Altogether, the protective clothing and SCBA that must be worn when fighting fires can weigh more than 50 lb (23 kg). Moving with this extra weight requires additional energy, which, in turn, increases air consumption and body temperature. Taken collectively, this activity places additional stress on a firefighter's body.

The weight and bulk of the complete PPE ensemble limit a firefighter's ability to walk, climb ladders, lift objects, and crawl through restricted spaces. Firefighters must become accustomed to these limitations and learn to alter their movements accordingly. Practice and conditioning are key to becoming proficient in wearing and using PPE while fighting fires.

Psychological Limitations of the SCBA User

In addition to the physical limitations, the user must make mental adjustments when wearing an SCBA. Breathing through an SCBA is different from normal breathing, and it can be stressful. Covering your face with a face piece, the noises of the air rushing in and valves opening and closing, and exhaling against positive pressure are all foreign sensations. The surrounding environment, which is often dark and filled with smoke, is foreign as well.

Firefighters must adjust so that they can operate effectively under these stressful conditions. Practice in donning PPE, breathing through SCBA, and performing firefighting tasks in darkness helps build confidence, not only in the equipment but also in the firefighter's personal skills. Practice also helps the firefighter improve their air consumption rate. Training generally introduces one skill at a time. Practice each skill as it is introduced and try to become proficient in that skill. As your skills improve, you will be able to tackle tasks characterized by increasing levels of difficulty.

TIP

The goal of training is to bring you to a level of comfort and proficiency in using your equipment. Start with a friendly environment until you are used to your equipment. Then add stressors such as darkness, smoke, and heat, one at a time.

Components of SCBA

Like protective clothing, the NFPA specifies the minimum requirements for SCBA and PASS devices as outlined in the following standards:

- NFPA 1852, *Standard on Selection, Care, and Maintenance of Open-Circuit Self-Contained Breathing Apparatus (SCBA), 2019 Edition*
- NFPA 1970, *Standard on Protective Ensembles for Structural and Proximity Firefighting, Work Apparel and Open-Circuit Self-Contained Breathing Apparatus (SCBA) for Emergency Services, and Personal Alert Safety Systems (PASS), 2024 Edition*
- NFPA 1984, *Standard on Respirators for Wildland Fire Fighting Operations and Wildland Urban Interface Operations, 2022 Edition*

An SCBA consists of four main parts: the harness, the air cylinder, the regulator assembly, and the face piece assembly (**FIGURE 3-26**). Although the basic features and operations of all models are similar, you need to become familiar with the specific model of SCBA used by your department.

FIGURE 3-26 The components of self-contained breathing apparatus.

Courtesy of 3M.

FIGURE 3-27 SCBA harnesses come in a variety of models.

© Jones & Bartlett Learning. Photographed by Glen E. Ellman.

Harness

The **SCBA harness** consists of the backpack or frame for mounting the working parts of the SCBA and the straps and fasteners used to attach the SCBA to the firefighter (**FIGURE 3-27**). It is usually constructed of a lightweight metal or composite material. Most SCBA harnesses have two adjustable shoulder straps and a waist belt. These must be constructed of a material, such as Kevlar, that is strong and able to withstand elevated temperatures. Depending on the specific model of SCBA, the waist belt and shoulder straps carry different proportions of the pack's weight. The procedures for tightening and adjusting the straps also vary based on the model. The SCBA harness must be secure enough to keep the SCBA firmly fastened to the user but not so tight that it interferes with breathing or movements. The waist belt must be tight enough to keep the SCBA from moving from side to side or getting caught on obstructions. Some SCBA models are equipped with a reinforced harness or hand loop, similar to the DRD on a protective coat, that can be used to help drag a fallen firefighter out of danger.

Breathing Air Cylinder

The **breathing air cylinder** (or **air cylinder**) on open-circuit SCBA holds the compressed breathing air for an SCBA. This removable air cylinder is attached to the SCBA harness and can be swapped with another cylinder quickly in the field. An experienced firefighter should be able to remove and replace the air cylinder in complete darkness.

Firefighters should be familiar with the type of air cylinders used in their departments. Air cylinders are marked with the materials used in their construction, the working pressure, and the rated duration.

The air pressure in filled SCBA air cylinders ranges from 2200 to 5500 pounds per square inch (psi; 15,168 to 37,920 kilopascals [kPa]). The greater the air pressure, the more air that can be stored in the cylinder. Low-pressure air cylinders, which are pressurized at 2200 psi (15,168 kPa), can be constructed of steel or aluminum and are usually rated for 30 minutes of use. Composite air cylinders are generally constructed of an aluminum shell wrapped with carbon, Kevlar, or glass fibers. They are significantly lighter in weight; can be pressurized up to 5500 psi (37,920 kPa); and are rated for 30, 45, 60, or 75 minutes of use.

As previously noted, the rated duration times are established under laboratory conditions. A working firefighter can quickly use up the air because of exertion, so the ratings should be viewed with caution. Generally, the working time available for a particular air cylinder is half the rated duration.

FIGURE 3-28 SCBA regulator.

Courtesy of 3M.

The neck of an air cylinder is equipped with a hand-operated shut-off valve that controls the flow of air leaving the air cylinder. Be careful not to damage the threads or let any dirt get into the outlet of the air cylinder. The **air-cylinder pressure gauge** located near the air cylinder valve shows the amount of air currently in the cylinder. This gauge cannot be viewed by the user while wearing the SCBA, so a second pressure gauge called the **remote pressure gauge** is located on the shoulder strap or in another location where it can be seen while the SCBA is being used. If the two pressure gauges are working correctly, the readings should be within 10 percent of each other.

Regulator

An **SCBA regulator** controls the flow of air to the user (**FIGURE 3-28**). The regulator may be mounted on the waist belt or shoulder strap of the SCBA harness or attached directly to the face piece. Inhaling decreases the air pressure in the face piece. This change in pressure opens the regulator, which in turn releases air from the cylinder. When inhalation stops, the regulator shuts off the air supply. Exhaling opens a second valve (the exhalation valve), thereby expelling the exhaled air into the atmosphere. Most SCBA units are equipped with a **dual-path pressure reducer**, a feature that automatically provides a backup method for air to be supplied to the regulator if the primary passage malfunctions.

A proper face piece-to-face seal must always be maintained. In addition to making sure your face piece fits well and is correctly donned, SCBA regulators maintain a slightly higher air pressure to the face piece in relation to the ambient air pressure outside the face piece. This is called positive pressure. This helps prevent the hazardous atmosphere outside the face piece from leaking into the face piece during inhalation. If any leakage occurs in the area where the face piece and the face make the seal, the positive-pressure breathing air inside the face piece keeps the hazardous atmosphere from entering the device. Breathing with this slight positive pressure may require some practice. New firefighters often report that it takes more energy to breathe when first using positive-pressure SCBA, but this sensation gradually decreases.

An SCBA also has a **regulator purge/bypass valve**, which allows the air from the air cylinder to enter the face piece without passing through the regulator. This will create a constant flow of air into the face piece; however, this action will rapidly deplete the remaining air supply in the air cylinder. You would use this valve if the air supply is partially or completely cut off during use. In the event that it is necessary to open the regulator purge/bypass valve, you should immediately exit from the IDLH area. During normal use, the regulator purge/bypass valve can be momentarily opened to remove condensation or residual air from the respirator after the air cylinder valve is turned off.

The regulator may have an air saver/donning switch that prevents the rapid loss of the air supply if the cylinder valve is open and the face piece is removed from the face or the regulator is removed from the face piece. In other words, if you open the air cylinder valve, air will not flow unless you depress the air saver/donning switch.

Many models of SCBA regulators exist. Firefighters must learn how to operate the particular model that is used in their department. They should be able to operate the regulator in the dark and while wearing firefighting gloves.

Face Piece

The **face piece** delivers breathing air to the firefighter and protects the face from high temperatures and smoke (**FIGURE 3-29**). It consists of a face mask with a clear lens and an exhalation valve. On SCBA

FIGURE 3-29 SCBA face pieces come in several sizes.

models with a harness-mounted regulator, the face piece also include a flexible low-pressure supply hose. In some models, the regulator is attached directly to the face piece.

Full face pieces cover the nose, mouth, and eyes. Half face pieces cover just the nose and mouth. The part that comes in contact with the skin is made of special rubber or silicone, because these materials provide for a tight seal. Exhaled air is expelled from the face piece through the one-way exhalation valve, which has a spring mechanism to maintain positive pressure inside the face piece. Because it is difficult to communicate through a face piece, a voice amplification device or mechanical diaphragm is used to facilitate communication. A mechanical diaphragm is a vibrating airtight membrane that transmit the firefighter's voice without the use of electricity.

Face pieces are equipped with **nose cups**, an insert inside the face piece that fits over the user's mouth and nose. The nose cup has two uses. It prevents the build-up of carbon dioxide (CO_2) by directing exhaled air toward the exhalation valve. It also helps prevent fogging of the clear lens. Fogging occurs because the compressed air you breathe is dry, but the air you exhale is moist. If fogging occurs, the regulator purge/bypass valve can be opened slightly for a second or two to clear the condensation.

TIP

Fogging is a greater problem in colder climates.

A leak in the face piece seal may result from an improperly sized face piece, an improper donning procedure, or facial hair around the edge of the face piece. In particular, the following factors can affect the seal on an SCBA face piece:

- Facial hair, sideburns, or beard
- A low hairline that interferes with the sealing surface
- Ponytails or buns that interfere with the smooth and close fit on the head harness
- A skull cap that projects under the face piece or temple pieces
- The absence of teeth
- Improper size face piece

An improperly fitted face piece may lead to exposure to a hazardous environment. Such leaks are dangerous for two reasons. First, a large leak could overcome the positive pressure in the face piece and allow contaminated air to enter the face piece. Second, a leak of any size will deplete the breathing air and reduce the amount of time available for firefighting.

When wearing SCBA, the protective hood is worn over the face piece and under the helmet. After securing the face piece straps, the firefighter should carefully fit the protective hood around the face piece so that no areas of bare skin are left exposed. The protective hood must fit snugly around the clear area of the face piece so that vision is not compromised and hot gases cannot leak between the face piece and the protective hood.

Face pieces are manufactured in several sizes. NFPA 1550 requires that all firefighters have their face pieces fit-tested annually to ensure that they are wearing the proper size. Some departments issue individual face pieces to each firefighter; others provide a selection of sizes on each apparatus. NFPA 1550 also requires that the sealing surface of the face piece be in direct contact with the user's skin; that is, hair or a beard cannot be in the seal area.

Additional Features

An SCBA is also required to have a **heads-up display (HUD)**, which must be visible to the user while wearing the face piece, enabling the user to constantly monitor the amount of air in the air cylinder. Most SCBA face pieces contain **light-emitting diodes (LEDs)** that indicate the amount of air remaining in the cylinder. These indicate whether the air cylinder is 100%, 75%, 50%, or 35% full. Other LEDs on the display may

FIGURE 3-30 A heads-up display enables the user to constantly monitor the amount of air in the air cylinder.

FIGURE 3-31 The rapid intervention crew/company (RIC) uses the RIC UAC to refill the SCBA cylinder of a trapped firefighter who is running out of air. This connection is used only for emergency refilling of an air cylinder.

provide additional information—for example, low batteries or other problems (**FIGURE 3-30**).

NFPA standards require that SCBA include two **end-of-service-time-indicator (EOSTI)** or low-air alarms, which operate independently of each other and activate different senses. For example, one alarm might ring a bell, whereas the second alarm might vibrate, whistle, or flash an LED.

This warning device tells the user that the end of the breathing air supply is approaching. Currently, the EOSTI or low-air warning alarm is constructed to sound when the pressure in the air cylinder is down to 35 percent of its capacity. At this pressure level you must begin to exit from the IDLH environment (NFPA 1970 2023). Most fire departments require firefighters to exit from the IDLH area before the EOSTI alarm sounds because the low-air alarm does not sound until two-thirds of the air supply has been exhausted. This reserve of one-third of the air supply is not always adequate for a safe escape (NFPA 1970 2023).

Communications between firefighters wearing SCBA are difficult. To facilitate communication, an SCBA is required to be equipped with a voice communication system. This functionality may be as simple as a mechanical voice diaphragm, or it may be as sophisticated as an electronic system. Some SCBA are also equipped with voice communications systems that utilize wireless communication interoperability to interface with mobile radios. Recent changes have been made in the NFPA 1970 standards to improve the face piece voice communication intelligibility and volume.

NFPA 1970 allows for open-circuit SCBA to have an **emergency breathing safety system (EBSS)** integrated. The EBSS allows for SCBA users to "buddy breathe" or share their available air supply when one is low or out of air. An SCBA is required to be equipped with a **rapid intervention crew/company universal air connection (RIC UAC)** (**FIGURE 3-31**). A **rapid intervention crew/company (RIC)**, sometimes called a **rapid intervention team (RIT)**, is a team of firefighters who stand by, fully dressed in PPE and SCBA, during emergency operations so that they can be quickly deployed to rescue a trapped firefighter. The RIC uses this connection to refill the SCBA cylinder of a trapped firefighter who is running out of air. A universal air connection (UAC) is attached to a hose from a full air cylinder and brought to the downed firefighter by a RIC, who refills the downed firefighter's cylinder. This connection should not be used for the routine refilling of a cylinder. The RIC UAC is designed only to be used for emergency refilling of an air cylinder. It is important that the dust cap is in place over the UAC to prevent damage. This universal connection can be used between SCBA packs of any manufacturer.

Fire department SCBA are required to be certified to provide protection against certain chemical, biological, radiological, and nuclear agents—that is, agents that could be released as a result of a terrorist attack. An SCBA provides protection against these agents by preventing chemical fumes, disease-causing biological organisms, and radioactive particles from being inhaled and entering the firefighter's respiratory system. However, an SCBA does not protect from contamination by other means of transmission.

Many accessories are available for SCBA, including data logging, unit IDs, tracking devices, corrective eye lenses for face pieces, thermal imaging capabilities, and electronic communications devices (NFPA 1970 2023). Data logging and unit IDs aid in keeping track of the use of each SCBA and checking the proper functioning of each unit. Some of these tracking devices work by sending out sounds of varying pitch. Corrective lenses for face pieces enable SCBA users to use their face pieces as corrective eyewear. Electronic communications systems integrate radio communications between SCBA users. Some SCBA offer a thermal imaging device that can be added into the unit to provide each firefighter with the ability to detect heat sources. If your SCBA is equipped with any of these accessories, you must become competent in the operation of these extra devices. To ensure proper functioning of your SCBA, do not add any devices or accessories that are not approved by the manufacturer of the equipment.

Pathway of Air through an SCBA

In an open-circuit SCBA, the breathing air is stored under pressure in the air cylinder. This air passes through the cylinder valve into the high-pressure air line (hose), which then carries it to the regulator. The regulator opens when the user inhales, reducing the air pressure on the downstream side. In an SCBA unit with a face piece–mounted regulator, this low pressure air goes directly into the face piece. In units with a harness-mounted regulator, the low-pressure air travels from the regulator through a low-pressure hose into the face piece. From the face piece, the air is inhaled through the user's air passages and into the lungs. When the user exhales, the used air is returned to the face piece. The exhaled air is exhausted from the face piece through the exhalation valve. As the pressure in the face piece drops, the exhalation valve closes, and the regulator opens. This cycle repeats with every inhalation.

Breathing Techniques to Conserve Air

Because the air supply is limited in an open-circuit SCBA, firefighters can use breathing techniques to conserve air. These techniques include skip breathing, Reilly (humming) breathing, and controlled breathing.

To use the skip breathing technique, the firefighter takes a short breath, holds it, takes a second short breath (without exhaling in between breaths), and then relaxes with a long exhale. Each breath should take 5 seconds. You can confirm the benefits of skip breathing by conducting a simple drill. One firefighter dons PPE and an SCBA with a full air cylinder and walks in a circle around a set of traffic cones, the track at the local school, or, if safety permits, the parking lot at the fire station. A second firefighter times how long it takes for the firefighter to completely deplete the air in the SCBA. After the first firefighter is completely rested, the air cylinder is replaced, and the same drill is repeated by the first firefighter using the skip breathing technique. A comparison of the times after completion of both cycles will confirm that skip breathing conserves the air in the cylinder.

Another breathing technique for conserving air is the Reilly (humming) technique. To use this technique, slowly inhale, and then exhale while making a humming sound. Humming allows for a longer exhalation, and this slows the breathing rate.

The controlled breathing technique also helps extend the SCBA air supply. To use this technique, make a conscious effort to inhale naturally through the nose and to force exhalation from the mouth by pursing your lips and blowing out slowly.

Practicing controlled breathing during training will help you to maximize the efficient use of air. When you are able to decrease the rate and depth of your breathing, you increase the amount of time you can breathe from a single cylinder of air. Keep calm, perform your assignments efficiently, and do not start breathing from your air supply until you need to.

Personal Alert Safety System

As described earlier, a personal alert safety system (PASS) is an electronic device that emits a loud audible signal when a firefighter becomes trapped or injured to help colleagues locate the downed firefighter. Newer PASS devices may include a radio transmitter that sends a signal to the command post when the alarm sounds. Many SCBA models are manufactured with an integrated PASS device (**FIGURE 3-32**).

The PASS device combines an electronic motion sensor with an alarm system. If the user remains motionless for 30 seconds, it produces a low warning tone before sounding a full alarm. The user can reset the device by moving during this warning period. A firefighter in distress can manually activate this device.

Turning on the air supply in an SCBA with an integrated PASS device automatically activates the PASS

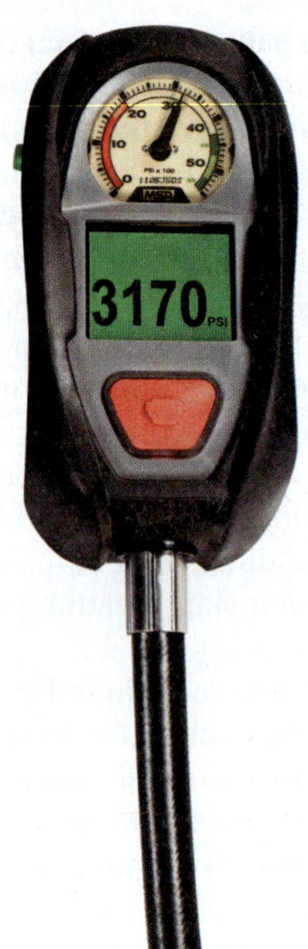

FIGURE 3-32 An integrated PASS device can save a firefighter's life.

Courtesy of MSA.

device. This ensures that a firefighter does not forget to turn the PASS device on when entering a hazardous area. Nonintegrated PASS devices, found with older SCBA, must be turned on manually. Some PASS devices may be used by firefighters not wearing SCBA. All firefighters must confirm that their PASS devices are on and working properly before they enter a burning building or hazardous area. It is important to learn how to activate and deactivate the PASS device on your SCBA.

Donning an SCBA

Donning an SCBA is an important skill. Firefighters should be able to don and activate an SCBA in 1 minute. Both the firefighter's personal safety and the effectiveness of the firefighting operation depend on mastery of this skill. Firefighters must already be wearing personal protective clothing before they don an SCBA. However, the protective hood should be around the wearer's neck, not over their head, because the hood must be pulled over the SCBA.

Before beginning the actual donning process, firefighters must carefully check the SCBA to ensure it is ready for operation:

- Check whether the air-cylinder pressure gauge is registering at least 90 percent of the tank's rated pressure.
- If the SCBA has an air saving/donning switch, confirm that it is activated to prevent air from flowing.
- Open the air cylinder valve two or three turns, listen for the low-air alarm to sound, and then open the valve fully.
- Test the PASS device.
- Check the remote pressure gauge and the air-cylinder pressure gauge. The reading for these gauges should be within 10 percent of each other.
- Check all harness and face piece straps to ensure that they are fully extended.
- Check all valves to ensure that they are in the correct position. (An open regulator purge/bypass valve will waste air.)

To activate the SCBA, it is necessary to open the air cylinder valve; don the SCBA harness; don the face piece; and, depending on the type of regulator you are using, attach the regulator to the face piece or connect the low-pressure air supply hose to the regulator and breathe.

There are several ways to don SCBA. Three of the most common methods are to don it from an apparatus seat mount, from an apparatus compartment mount, or away from the apparatus from the ground or a storage case. Donning SCBA from an apparatus seat mount is the fastest and most efficient method because firefighters can dismount the apparatus ready to attack the fire without delay. However, not all apparatus are equipped with seat-mounted brackets for SCBA. Donning SCBA from a compartment mount is also fairly fast because firefighters do not need to bend over to reach the SCBA.

The skill drills in the following sections will need to be modified for different SCBA models. In particular, the sequence for adjusting the shoulder straps and the waist belt varies with different models. Modifications also must be made for SCBA models with waist-mounted regulators. Refer to the specific manufacturer's instructions supplied with each unit and follow the SOPs for your department.

SAFETY TIP

Make sure that your face piece is the correct size for your face and is properly fitted.

Donning an SCBA from an Apparatus Seat Mount

An SCBA should be located so that firefighters can don it quickly when they arrive at the scene of a fire. Seat-mounted brackets enable firefighters to don SCBA en route to an emergency scene, without unfastening their seat belts or otherwise endangering themselves (**FIGURE 3-33**). This enables firefighters to begin work as soon as they arrive on scene.

Several types of apparatus seat-mounting brackets are available. Some hold the SCBA with the friction of a clip. Others are equipped with a mechanical hold-down device that must be released to remove the SCBA. Regardless of which mounting system is employed, it must hold the SCBA securely in the bracket. A collision or sudden stop should not dislodge the SCBA from the brackets, because a loose SCBA can be a dangerous projectile. The firefighter who dons an SCBA from a seat-mounted bracket should not tighten the shoulder straps while seated, so as not to dislodge the SCBA in a sudden-stop situation. Firefighters must remain securely restrained by a seat belt or combination seat belt/shoulder harness any time the apparatus is moving.

FIGURE 3-33 Seat-mounted brackets hold the SCBA securely in place and enable firefighters to don an SCBA en route to an emergency scene.

TIP

To help firefighters differentiate the SCBA waist belt from a seat belt, NFPA 1900, *Standard for Aircraft Rescue and Firefighting Vehicles, Automotive Fire Apparatus, Wildland Fire Apparatus, and Automotive Ambulances, 2024 Edition*, requires fire apparatus seat belts to be either bright orange or bright red.

Donning an SCBA while en route to an emergency can save valuable time because firefighters can dismount the engine ready to immediately attack the fire. This maneuver requires that you place your protective hood around your neck and don your protective boots, pants, and coat before mounting the apparatus. Keep your helmet and gloves close by. You should not don your helmet in the apparatus because if the apparatus is involved in a motor vehicle accident, the helmet will interfere with the protection provided by the head rests (**FIGURE 3-34**).

When you exit the apparatus, be sure to take a face piece with you. Face pieces should be kept either in a storage bag close to each seat-mounted SCBA or attached to the SCBA harness. The lens of the face piece should be protected from contaminants and damage while stored.

Follow the steps in **SKILL DRILL 3-3** to don an SCBA from a seat-mounted bracket. Before beginning this skill drill, confirm that the SCBA has been inspected and is ready for service.

FIGURE 3-34 Do not don your helmet until you arrive at the scene.

SKILL DRILL 3-3

Donning an SCBA from an Apparatus Seat Mount Firefighter I, NFPA 1010: 6.3.1

1. Put your protective hood over your head and leave it around your neck. Don your protective boots, pants, and coat. Safely mount the apparatus and sit in the seat. Place your arms through the SCBA shoulder straps. Partially tighten the shoulder straps; do not fully tighten them.

2. Fasten your SCBA waist belt.

3. Fasten your seat belt. When the apparatus stops at the emergency scene, release the seat belt, and release the SCBA from its brackets. If the apparatus has an SCBA locking device, detach the SCBA from the locking device.

4. Carefully exit the apparatus. Maintain three points of contact with the vehicle while exiting.

SKILL DRILL 3-3 CONTINUED

Donning an SCBA from an Apparatus Seat Mount Firefighter I, NFPA 1010: 6.3.1

5. Tighten the SCBA waist belt.

6. Adjust shoulder straps by pulling them straight down until they are snug.

7. Open the air cylinder valve and check the air pressure. Activate the air saver/donning switch to prevent the flow of air, if needed. Prepare to don the face piece (Skill Drill 3-7).

Donning an SCBA from an Apparatus Compartment Mount

Compartment-mounted SCBA units may be mounted on apparatus and can be donned quickly. These units are used by firefighters who arrive in vehicles without seat-mounted SCBA or whose seats were not equipped with them. Driver/operators and firefighters who arrive in their private vehicles often use these types of SCBA. The mounting brackets should be positioned high enough to allow for easy donning of the equipment. Some mounting brackets allow the firefighter to lower the SCBA without removing it from the mounting bracket. Older apparatus may have the brackets mounted on the exterior of the vehicle. Any exterior-mounted SCBA should be protected from weather and dirt by a secure cover. The SCBA should be stored on the apparatus in ready-for-use condition, with the air cylinder valve closed.

Follow the steps in **SKILL DRILL 3-4** to don an SCBA from a side-mounted compartment or bracket. Before beginning this skill drill, confirm that the SCBA has been inspected and is ready for service. Examine the equipment to see how it is mounted in the compartment. Check the release mechanism and determine how it operates. If the SCBA is mounted on an exterior bracket, remove the protective cover before beginning the donning sequence.

SKILL DRILL 3-4

Donning an SCBA from an Apparatus Compartment Mount Firefighter I, NFPA 1010: 6.3.1

1. Put your protective hood over your head and leave it around your neck. Don your protective boots, pants, and coat. Place your helmet and gloves nearby. Stand in front of the SCBA bracket, and fully open the air cylinder valve. Check the air pressure to make sure there is at least 90 percent of the rated pressure available.
2. Turn your back toward the SCBA and slide your arms through the shoulder straps. Partially tighten the shoulder straps; do not fully tighten them.
3. Release the SCBA from the bracket, and then step away from the apparatus.
4. Attach the waist belt and tighten it.
5. Adjust the shoulder straps by pulling straight down until they are snug. Fully open the main air cylinder valve. Activate the air saver/donning switch to prevent the flow of air, if needed. Prepare to don the face piece (Skill Drill 3-7).

Donning an SCBA from the Ground or a Storage Case

Firefighters sometimes must don an SCBA that is on the ground, or the SCBA may be kept in a storage case. Storage cases are usually used for transporting extra SCBA units. It should not be used to transport SCBA that will be used during the initial phase of operations at a fire scene as it can add considerable time to remove the unit from the case and prepare it for use. Two methods of donning can be used in these situations: the over-the-head method and the coat method.

Over-the-Head Method. Follow the steps in **SKILL DRILL 3-5** to don an SCBA using the over-the-head method. Before starting, confirm that the SCBA has been inspected and is ready for service. Don your protective pants, boots, and coat. Place your protective hood around your neck. Place your helmet and gloves close by.

SKILL DRILL 3-5

Donning an SCBA Using the Over-the-Head Method Firefighter I, NFPA 1010: 6.3.1

1. Put your protective hood over your head and leave it around your neck. Don your protective boots, pants, and coat. Place your helmet and gloves nearby. Lay out the SCBA so that the cylinder is resting on the floor or ground, the backplate is facing up, and the air cylinder valve is facing away from you. Move the shoulder straps to the sides.

2. Fully open the main air cylinder valve and check the air pressure to make sure there is at least 90 percent of the rated pressure available. Activate the air saver/donning switch to prevent the flow of air, if needed.

Continues.

SKILL DRILL 3-5 CONTINUED

Donning an SCBA Using the Over-the-Head Method Firefighter I, NFPA 1010: 6.3.1

3. Bend down and grasp the SCBA backplate with both hands. Using your knees to support and lift the extra weight, lift the SCBA up and over your head. Once the SCBA clears your head, rotate it 180 degrees so that the waist belt straps are pointed toward the ground.

4. Slowly slide the pack down your back. Make sure that your arms slide into the shoulder straps. Once the SCBA is in place, tighten the shoulder straps by pulling straight down until they are snug, and then secure and tighten the waist belt. Prepare to don the face piece (Skill Drill 3-7).

Coat Method. Follow the steps in **SKILL DRILL 3-6** to don an SCBA using the coat method. Before beginning this skill drill, confirm that the SCBA has been inspected and is ready for service.

Donning the Face Piece

To perform properly, the face piece must be the correct size and must be adjusted to fit your face. Fit testing needs to be performed before firefighters are permitted to use SCBA. To ensure a proper fit, testing should be repeated every year. Make sure you have been tested to determine the proper size for you. The requirements for face piece fit testing are described in NFPA 1550.

No facial hair can be present in the seal area. Eyeglasses that pass through the seal area cannot be worn with a face piece because they can cause leakage between the face piece and your skin. If you wear glasses, custom spectacle kits for SCBA are available.

Face pieces for various brands and models of SCBA may differ slightly. Some have the regulator mounted on the face piece; others have it mounted on the harness straps. Your face piece must match your SCBA—you cannot interchange a face piece from a different SCBA model. Firefighters must learn about the specific face pieces used by their departments.

The face piece is held in place with a weblike series of straps or a net and straps. Face pieces should be stored with the straps in the longest position to make them easier to don. To tighten the straps and ensure a snug fit, pull the end of the straps toward the back of the head (not out to the sides).

SKILL DRILL 3-6

Donning an SCBA Using the Coat Method Firefighter I, NFPA 1010: 6.3.1

1. Put your protective hood over your head and leave it around your neck. Don your protective boots, pants, and coat. Place your helmet and gloves nearby. Lay out the SCBA so that the cylinder is resting on the floor or ground, the backplate is facing up, and the air cylinder valve is facing toward you. Move the shoulder straps to the sides. Fully open the air cylinder valve and check the air pressure to make sure there is at least 90 percent of the rated pressure available. Activate the air saver/donning switch to prevent the flow of air, if needed. Place your dominant hand on the opposite shoulder strap. For safety reasons, be sure to grasp the strap as close to the backplate as possible.

2. Lift the SCBA, and swing it over your dominant shoulder, being mindful of people or objects around you.

3. Slide your other hand between the SCBA cylinder and the corresponding shoulder strap.

Continues.

SKILL DRILL 3-6 CONTINUED

Donning an SCBA Using the Coat Method Firefighter I, NFPA 1010: 6.3.1

4. Tighten the shoulder straps by pulling straight down.

5. Attach the waist belt and tighten it. Prepare to don the face piece (Skill Drill 3-7).

The following method is one way to check the seal of a face mask:

1. Don the SCBA, and then begin breathing cylinder air.
2. Completely close the air cylinder valve.
3. Breathe through the face piece until all air stops flowing from the breathing regulator.
4. Inhale slowly, and then hold your breath. Your face piece should be slightly drawn to your face.
5. Listen and feel for air leakage around the face piece seal. Check that the negative pressure in the face piece does not change.
6. If no change occurs, no leaks are present.

Follow the steps in **SKILL DRILL 3-7** to don the face piece. Before beginning this procedure, make sure the protective hood is around your neck and you have donned your protective boots, pants, and coat, and that your helmet and gloves are nearby.

Safety Precautions for SCBA

As you practice using your SCBA, remember that this equipment is your protection against serious injury or death in hazardous conditions. Practice safe procedures from the beginning.

Before you enter a hazardous environment, make sure that you have an adequate supply of air in your air cylinder and that your PASS device is activated. Be sure that you are properly entered into your personnel accountability system. Always work in teams of two in IDLH environments. In addition, always have an RIC outside at the ready whenever two firefighters are working in an IDLH environment.

SKILL DRILL 3-7

Donning a Face Piece Firefighter I, NFPA 1010: 6.3.1

1. With the protective hood around your neck, fully extend the straps on the face piece.

2. Rest your chin in the chin pocket at the bottom of the face mask. Fit the face piece to your face, bringing the straps or webbing over your head.

3. Tighten the lowest two straps by pulling the straps straight back, rather than out to the sides. Check the head harness net to make sure it is lying flat against the back of your head. If there is a pair of straps at your temple, tighten these straps. If your model has additional top straps, tighten these strap(s) last.

4. Confirm that your nose fits in the nose cup, and then check for a proper seal. This process depends on the model and type of face piece you use.

Continues.

SKILL DRILL 3-7 CONTINUED

Donning a Face Piece Firefighter I, NFPA 1010: 6.3.1

5. Pull the protective hood into position on your head so that the hood covers the sides of the face piece. Make sure it does not get under your face piece or obscure your vision.

6. Don your helmet and secure the chin strap.

7. If needed, attach the regulator to your face piece or attach the low-pressure air supply hose to the regulator. Inhale sharply to start the flow of air. If the PASS device is not integrated in the SCBA and activated automatically, activate your PASS device.

Many fire departments need to take precautions to ensure proper operation of SCBA in cold temperatures. SCBA should be stored in a location that is kept at a temperature above freezing (32°F [0°C]). After use, SCBA equipment should be cleaned and dried thoroughly to avoid retained moisture that might freeze moving parts. When SCBA units are doffed in below-freezing temperatures, the face piece and the regulator should be placed inside the protective coat to keep it warm and prevent any moisture from freezing.

SCBA Use during Emergency Operations

As a firefighter, your job is to practice using SCBA until you are confident that you can carry out a variety of tasks in hazardous conditions while depending on your equipment to supply you with safe air to breathe. Because harsh environments are often unpredictable, firefighters must be prepared to react if an emergency occurs while they are using an SCBA. Keep calm, stop, and think. Panic increases air consumption. Try to control your breathing by maintaining a steady rate of respirations. A calm person has a greater chance of surviving an emergency.

If you experience a problem with your SCBA, try to exit the IDLH area to a safe environment. If your cylinder contains air but no air comes out of your regulator, open the regulator purge/bypass valve to release a constant supply of air. Remember, however, that this measure will rapidly empty your cylinder. In this circumstance, you must immediately exit from the hazardous environment. If you are in danger, follow the steps for self-survival and call a mayday. This topic is discussed in Chapter 19, *Firefighter Survival.*

You will be using your SCBA in a variety of conditions that most people would consider to be emergencies. Because you are a firefighter, these activities are predictable for you. Therefore, you need to master a wide variety of firefighting skills while wearing SCBA. You need to practice most of the skills you will perform as a firefighter—for example, advancing hose lines, climbing ladders, crawling through windows, performing rescues, providing medical assistance, and crawling through confined spaces—while wearing SCBA. Practice these tasks first in conditions of good visibility, and then progress to doing the same tasks in conditions of limited visibility.

One skill you should practice while wearing SCBA is performing on a search and rescue team. The Conducting a Primary Search Using the Standard Search Method skill drill in Chapter 17, *Search and Rescue,* describes how to perform search and rescue while wearing SCBA. Another skill you should practice is passing through a restricted space while wearing an SCBA. This activity is illustrated in the Opening a Wall to Escape skill drill in Chapter 19, *Firefighter Survival.* A third activity to familiarize you with your SCBA simulates the activities you would perform when ventilating a roof. These activities are described in Chapter 16, *Ventilation.*

Doffing an SCBA

The procedure for doffing an SCBA depends on which model you use and whether it has a face piece–mounted regulator or a harness-mounted regulator. Follow the procedures recommended by the manufacturer and your department's SOPs. Your department may teach some variation of these steps. Follow the procedure taught by your department.

Follow the steps in **SKILL DRILL 3-8** to doff your SCBA. After you have taken your PPE off, do not leave a contaminated hood or face piece around your neck, below your nose and mouth. You will inhale the toxins.

SKILL DRILL 3-8

Doffing an SCBA Firefighter I, NFPA 1010: 6.3.1

1. Remove the regulator from your face piece or disconnect the low-pressure air supply hose from the regulator.

2. Close the air cylinder valve or fully depress the air saver/donning switch to stop the flow of air.

Continues.

SKILL DRILL 3-8 CONTINUED

Doffing an SCBA Firefighter I, NFPA 1010: 6.3.1

3. Release your waist belt.

4. Loosen the shoulder straps, and then remove the SCBA harness. If you have not already done so, close the air cylinder valve.

5. Open the regulator purge/bypass valve to bleed the air pressure from the regulator.

6. Ensure that the PASS device is turned off. Place the SCBA in a safe location.

SKILL DRILL 3-8 CONTINUED

Doffing an SCBA Firefighter I, NFPA 1010: 6.3.1

7. Remove your gloves, then remove your helmet. Pull your protective hood down around your neck. Loosen the straps on your face piece.

8. Remove your face piece. Remove your protective hood. Clean your SCBA as soon as possible, following the manufacturer's instructions.

Putting It All Together: Donning the Entire PPE Ensemble

The complete PPE ensemble consists of both personal protective clothing and respiratory protection. Although donning personal protective clothing and donning respiratory protection can be learned and practiced separately, you must be able to integrate these skills to have a complete PPE ensemble. Each part of the complete ensemble must be in the proper place to provide whole-body protection.

The steps for donning complete PPE (personal protective clothing and SCBA) are summarized as follows:

1. Place the protective hood over your head and bring it down around your neck.
2. Put on your protective boots and pants. Adjust the suspenders and secure the front flap of the pants.
3. Put on your protective coat and secure the front.
4. Open the air cylinder valve on your SCBA and check the air pressure to make sure there is at least 90 percent of the rated pressure available. Press the air saver/donning switch to prevent air flow, if needed.
5. Put on your SCBA harness.
6. Tighten both shoulder straps of the SCBA harness.
7. Attach the waist belt of the SCBA harness and tighten it. Tighten the chest straps, if present.
8. Fit the face piece to your face.
9. Tighten the face piece straps, beginning with the lowest straps.
10. Check the face piece for a proper seal. (Follow the manufacturer's instructions.)
11. Pull the protective hood into position over the face piece straps so that it covers all bare skin but does not obscure your vision.
12. Place your helmet on your head with the ear tabs extended, and then secure the chin strap.
13. Turn up your coat collar and secure it in front.

14. Put on your gloves.
15. Check your clothing to be sure it is properly secured.
16. If your PASS device is not integrated with your SCBA, turn your PASS device on.
17. Attach your regulator or open the air cylinder valve to start the flow of breathing air.
18. Work safely!

Limitations of the Structural Firefighting Protective Ensemble

Unfortunately, even today's advanced PPE has drawbacks and limitations. Understanding those limitations will help you avoid situations that could result in serious injury or death.

First, the structural firefighting protective ensemble is not easy to don. It includes several components, all of which must be put on in the proper order and correctly secured. You must be able to don your equipment quickly and correctly, either at the fire station before you respond to an emergency or after you arrive at the scene. Practice donning your protective clothing ensemble until you can do so quickly and smoothly (**FIGURE 3-35**).

The ensemble is heavy, weighing nearly 50 lb (23 kg). This increased weight means that everything you do—even walking—requires more energy and strength. It also limits mobility. Full bunker gear not only limits the range of motion, it also makes movements awkward and difficult. Tasks such as advancing an attack line up a stairway or using an axe to ventilate a roof can be difficult, even for a firefighter in excellent physical condition.

FIGURE 3-35 It takes practice to don the full structural firefighting protective ensemble.

Because PPE retains body heat and perspiration, it is difficult for the body to cool itself when wearing this equipment. Perspiration is retained inside the protective clothing rather than evaporating to cool you. As a consequence, firefighters in full protective gear can rapidly develop elevated body temperatures, even when the ambient temperature is cool. The problem of overheating is more acute when surrounding temperatures are high, which is one reason why firefighters must undergo regular rehabilitation and adequate fluid replacement. Removing your ensemble will help you to cool down quickly.

Wearing PPE also decreases normal sensory abilities. Wearing heavy gloves, for example, reduces the sense of touch. Structural firefighting protective coats and pants protect skin but reduce its ability to determine the temperature of hot air. Sight is restricted when you wear an SCBA. The plastic face piece reduces peripheral vision, and the helmet, protective hood, and coat make turning the head difficult. Both the helmet ear tabs and the protective hood over the ears limit hearing. Speaking becomes muffled and distorted by the SCBA face piece, even if it is equipped with a special voice amplification system.

PPE absorbs smoke particles that continue to **off-gas**—emit harmful chemicals in the form of a gas—toxins from the hazardous conditions from which they just came. Remove as much of your PPE as possible when you arrive at rehabilitation, and then place the PPE in an area away from direct contact with yourself and others. Much of the structural firefighting gear worn by firefighters contains a chemical known as PFAS. PFAS is a known carcinogen and has been linked to several health problems. Research is being conducted to find an alternative. This is just another reason to always decontaminate your gear and keep it stored separately from living quarters and passenger vehicles.

For these reasons, firefighters must become accustomed to wearing and using PPE. Practicing skills while wearing PPE will help you become comfortable with its operation and limitations.

SCBA Inspection and Maintenance

An SCBA must be properly cleaned, inspected, and prepared for the next use each time it is used, whether in an actual emergency incident or as part of a training exercise. The air cylinder must be changed or refilled, the face piece and regulator must be sanitized according to the manufacturer's instructions, and the unit must be cleaned, inspected, and checked for proper operation.

After operating in a fire environment, your SCBA will be coated with dirt and soot, which contains many dangerous and carcinogenic substances. It is important to thoroughly clean all parts of your SCBA as soon as possible. Cleaning SCBAs is just as important as cleaning your firefighting gear to rid it of any carcinogens or other harmful substances. Follow the instructions of the SCBA manufacturer and of your department. It is the user's responsibility to ensure that the SCBA is in good working order and in ready-to-use condition before it is returned to the fire apparatus.

The U.S. Department of Transportation (DOT) limits the number of years that a cylinder can be used. In addition, the DOT requires that SCBA cylinders are tested on a periodic basis to identify any defects or damage that might render the pressurized cylinder unsafe. This test is called a **hydrostatic test**. The date and results of hydrostatic tests are listed on a label on the cylinder. Hydrostatic testing is usually handled by certified third parties. NFPA 1852 acknowledges the requirements from the DOT within the standard.

Cylinders constructed of different materials have different testing requirements. Aluminum, steel, and carbon-fiber cylinders must be hydrostatically tested every 5 years (CFR 49 180.205). Cylinders constructed of composite materials such as Kevlar or fiberglass fibers must be tested every 3 years (CFR 49 180.205). Firefighters must know which types of cylinders are used by their departments and must check each cylinder for a current hydrostatic test date before filling it.

Inspection

In addition, each SCBA must be inspected and tested on a regular basis to ensure that it will function properly at an emergency scene. In career departments, inspection and testing are done at the beginning of each shift. In volunteer departments, this step is commonly performed on a weekly schedule. A complete annual inspection and maintenance procedure must also be performed on each SCBA unit. The annual inspection must be performed by a certified manufacturer's representative or a person who has been trained and certified to perform this work.

If an SCBA inspection reveals any problems that cannot be remedied by routine maintenance, the SCBA must be removed from service for repair. Only properly trained and certified personnel are authorized to repair SCBA.

The purpose of a visual SCBA inspection is to identify any parts of the SCBA that are visibly damaged and need to be repaired or replaced to ensure continued safe operation. This visual inspection can be done in conjunction with the operational testing sequence discussed next. Follow the steps in **SKILL DRILL 3-9** to conduct a visual inspection of an SCBA. A more detailed inspection is required if a cylinder has been exposed to excessive heat, has come into contact with flame, has been exposed to chemicals, or has been dropped.

SKILL DRILL 3-9

Visual SCBA Inspection Firefighter I, NFPA 1010: 6.5.1

1. Visually inspect the air cylinder and valve assembly for dents and gouges. Look for black or discolored areas that indicate exposure to flame.

2. Check the cylinder for the current hydrostatic test date and date of manufacture. Check the air-cylinder pressure gauge to be sure it is full.

Continues.

SKILL DRILL 3-9 CONTINUED

Visual SCBA Inspection Firefighter I, NFPA 1010: 6.5.1

3. Inspect hose and rubber parts for damage or deterioration.

4. Inspect the SCBA harness, webbing, buckles, fasteners, and cylinder retention system for damage.

5. Verify that the SCBA has been cleaned according to the manufacturer's and department's recommendations. Inspect the regulator for intact gaskets and visible damage.

6. Inspect the face piece for damage and worn components. Look for damage to the lenses, and check for the presence of a nose cup.

7. Inspect the head harness to confirm that all parts are present and working properly.

8. Check the quick disconnects and the RIC UAC to make sure they are not damaged, they are operating properly, and that the dust cap is in place.

Operational Testing

A pressurized SCBA cylinder contains a tremendous amount of potential energy. Not only does the air within the cylinder exert considerable pressure on its walls, but the cylinder itself is used under extreme conditions on the fireground. If the cylinder ruptures and suddenly releases this energy, it can cause serious injury or death. For this reason, cylinders must be regularly inspected and tested to ensure they are safe. The operational testing sequence is designed to check the function of the many parts of the SCBA to ensure safe use of the device. It concentrates on the working parts of the SCBA. Follow the steps in **SKILL DRILL 3-10** for operational testing of an SCBA.

SKILL DRILL 3-10

SCBA Operational Inspection Firefighter I, NFPA 1010: 6.5.1

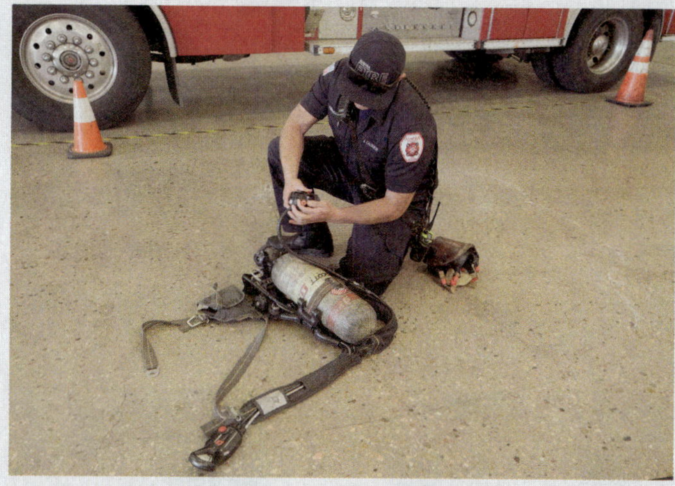

1. Check the regulator purge/bypass valve to be sure it is closed.

2. Depress the air saver/donning switch, if present, to start the flow of air.

3. Slowly open the air cylinder valve. Check for proper operation of the heads-up display and of the low-battery indicator. Confirm that the low-air alarm and the PASS device are working.

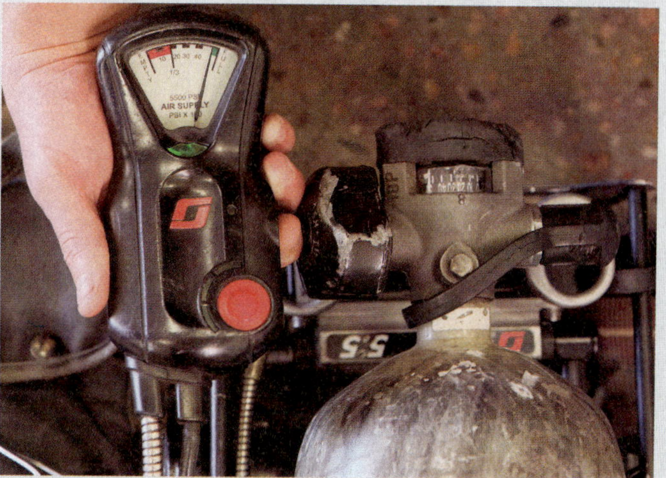

4. Check the remote pressure gauge for proper operation.

Continues.

SKILL DRILL 3-10 CONTINUED

SCBA Operational Inspection Firefighter I, NFPA 1010: 6.5.1

5. Don the face piece. Adjust it to obtain a good seal. Inhale sharply to start the flow of air. Breathe normally to check for proper operation.

6. Remove the regulator or face piece; air should flow freely.

7. Depress the air saver/donning switch to stop the flow of air.

8. Open the regulator purge/bypass valve to check for air flow.

SKILL DRILL 3-10 CONTINUED

SCBA Operational Inspection Firefighter I, NFPA 1010: 6.5.1

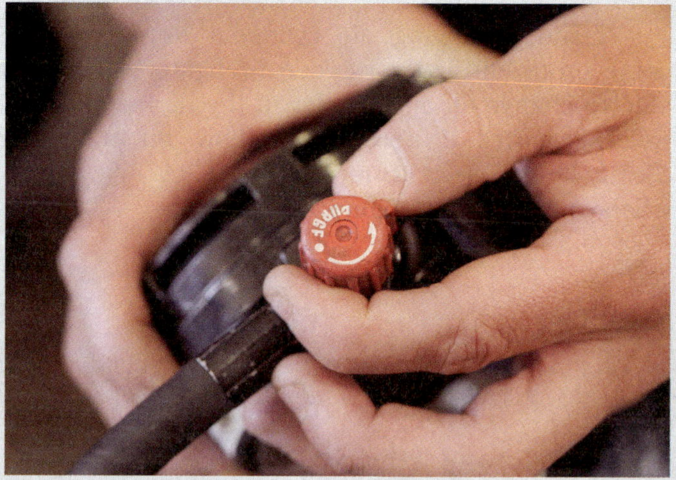

9. Close the regulator purge/bypass valve to stop the flow of air.

10. Rotate the air cylinder valve to close it.

11. Open the regulator purge/bypass valve slightly to vent residual air pressure from the system. Watch the heads-up display to verify its proper operation as the air pressure is exhausted.

12. Once the air flow stops, close the regulator purge/bypass valve. Complete any reporting that is required.

Replacing SCBA Cylinders

A used air cylinder can be quickly replaced with a full cylinder in the field to enable you to continue firefighting activities. Be sure that you are physically able to fight a second round with the fire. It is better to allow yourself a few minutes in rehab than to get back into the fire immediately and require rescue from other crew members. Do not overtax yourself by replacing the air cylinder and going back to work without adequate rest when you need it.

A firefighter who is working alone must doff their SCBA harness to replace the air cylinder. Two firefighters who are working together can change each other's cylinders without removing their SCBA harness. This procedure may vary slightly depending on the model of SCBA being used. Follow the procedure recommended by the SCBA manufacturer and by your department's SOPs.

Practice changing air cylinders until you become proficient at this task. A firefighter should be able to change an air cylinder in the dark and while wearing gloves if necessary. Follow the steps in **SKILL DRILL 3-11** to replace an SCBA air cylinder when you are working alone.

SKILL DRILL 3-11

Replacing an SCBA Cylinder Firefighter I, NFPA 1010: 6.3.1

1. Place the SCBA on the floor or a bench.

2. Close the air cylinder valve.

3. Open the regulator purge/bypass valve to bleed off the pressure.

4. Disconnect the high-pressure supply hose. Keep the ends clean.

SKILL DRILL 3-11 CONTINUED

Replacing an SCBA Cylinder Firefighter I, NFPA 1010: 6.3.1

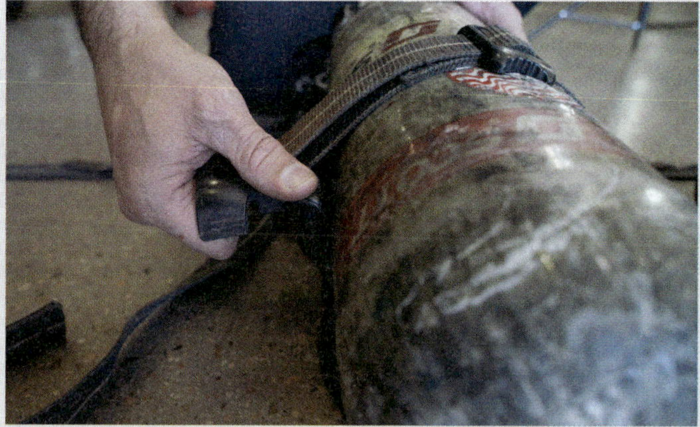

5. Release the air cylinder from the SCBA harness, and then remove the depleted air cylinder.

6. Slide a full air cylinder into the SCBA harness. Align the outlet to the supply hose. Lock the air cylinder in place.

7. Check that the "O" ring is present and in good shape.

8. Connect the high-pressure hose to the air cylinder. Hand-tighten only.

9. Open the air cylinder valve. Check the air-cylinder pressure gauge and the remote pressure gauge.

If you need to quickly reenter a fire scene, a second firefighter can replace your air cylinder while you continue to wear your SCBA harness. Follow the steps in **SKILL DRILL 3-12** to replace an SCBA air cylinder on another firefighter.

Refilling SCBA Cylinders

SCBA air cylinders can be refilled by equipment permanently located at a maintenance facility or at a firehouse or mounted on a truck or a trailer for mobile use.

SKILL DRILL 3-12

Replacing an SCBA Cylinder on Another Firefighter Firefighter I, NFPA 1010: 6.3.1

1. Remove the regulator from the face piece or remove the face piece so that the firefighter can breathe ambient air.

2. Close the air cylinder valve on the used air cylinder.

3. Open the regulator purge/bypass valve to bleed off pressure from the high-pressure supply line.

4. Disconnect the high-pressure supply line.

SKILL DRILL 3-12 CONTINUED

Replacing an SCBA Cylinder on Another Firefighter Firefighter I, NFPA 1010: 6.3.1

5. Release the SCBA air cylinder from the SCBA harness and set it aside.

6. Slide the full SCBA air cylinder into the SCBA harness.

7. Check for the presence and satisfactory condition of the "O" ring.

8. Lock the SCBA air cylinder into place.

9. Connect the high-pressure line to the SCBA cylinder.

10. Open the air cylinder valve, and then notify the firefighter that the SCBA cylinder change is complete. State the pressure reading of the SCBA air cylinder to the firefighter.

FIGURE 3-36 Compressors filter and compress atmospheric air before transferring it to an SCBA air cylinder.

FIGURE 3-37 Cascade systems store filtered and compressed breathing air in storage cylinders and transfer it to SCBA air cylinders.

Mobile filling units often are brought to the scene of a large fire. Before an air cylinder is refilled, the hydrostatic test date must be checked to ensure that its certification has not expired.

Two types of systems are used to refill air cylinders. A **compressor** filters atmospheric air, compresses it to a high pressure, and transfers it to the SCBA air cylinders (**FIGURE 3-36**). A **cascade system** has several large storage cylinders of compressed breathing air with pressure ranging from lower pressure to higher pressure connected by a high-pressure manifold system (**FIGURE 3-37**). The empty SCBA air cylinder is connected to the cascade system, and compressed air is transferred from the storage cylinders to the SCBA air cylinder. The specific steps for filling SCBA air cylinders vary with different systems.

Proper training is required to fill SCBA cylinders. Refilling SCBA cylinders requires special precautions because of the high pressures involved in this procedure. While it is being refilled, the SCBA cylinder must be placed in a shielded container designed to prevent injury if the cylinder ruptures. In addition, special procedures must be followed to ensure that the air used to fill the SCBA cylinder is not contaminated. Only those firefighters who have been trained on the safe use of your department's equipment should refill air cylinders. Follow the manufacturer's recommendations. Follow the steps in **SKILL DRILL 3-13** to safely fill SCBA air cylinders.

A **firefighter breathing air replenishment system (FBAR)** is required in some high-rise buildings based on the local jurisdiction's code. FBARS offer firefighters the option to refill SCBA air cylinders from inside the building on any floor. A fire department air connection is installed outside the building, and plumbing within the structure carries the air throughout the building. Fill stations are located in designated areas throughout the building.

Air cylinders can also be refilled using the RIC UAC. Refilling air cylinders using these connections allows the SCBA to be trans-filled which equalizes the air inside the fill bottle and the firefighters bottle. When you use the RIC/UAC to refill the SCBA air cylinder, the downed firefighter is never exposed to the atmosphere. Before doing this, however, you need to ensure their SCBA is functioning as designed and their face mask has not been compromised.

SKILL DRILL 3-13

Refilling SCBA Cylinders Firefighter I, NFPA 1010: 6.5.1

1. Complete the fill record form, including the date, hydrostatic test date, and air cylinder serial number.

2. Ensure the air cylinder is safe to fill by checking its date of manufacture and the hydrostatic test date.

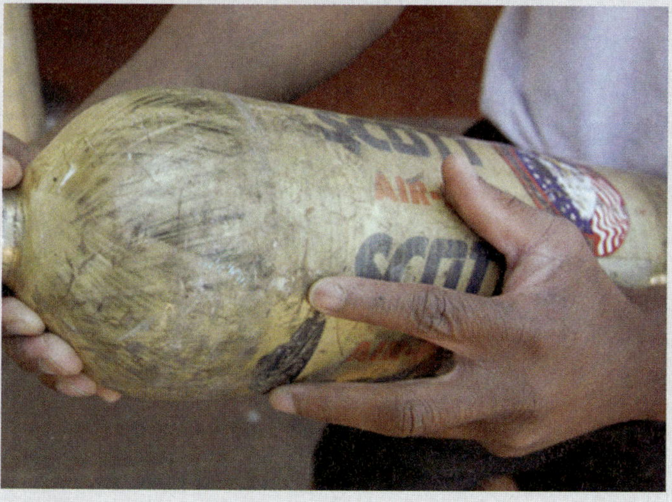

3. Check the air cylinder for visible damage. Follow the compressor or cascade system filling procedures. Secure the system after use according to system procedures.

However, using the RIC UAC to refill an air cylinder in below-freezing conditions presents two concerns. The first concern relates to the protective cap on the RIC UAC coupling. Water on the cover of the coupling may freeze, preventing a connection between the filling hose and the universal air connection. The second concern is that if the UAC is used to fill an SCBA cylinder in below-freezing temperatures, and then the SCBA is taken into a warmer temperature, the difference in temperature could cause the pressure in the cylinder to expand to unsafe levels. If there is excess pressure, open the air cylinder valve and the regulator purge/bypass valve to release the excess pressure.

Cleaning and Sanitizing SCBA

An SCBA must be cleaned after each use. The first step in cleaning the SCBA is to rinse the entire unit with clean water using a hose. It is important to clean the outside of the SCBA *before* taking it apart. This prevents foreign substances getting into SCBA connectors and internal parts. The SCBA harness assembly and air cylinder can be cleaned with a mild detergent or soap-and-water solution. If additional cleaning is needed, the unit can be scrubbed with a stiff brush. After scrubbing, the SCBA harness and air cylinder should be rinsed with clean water. Most SCBA manufacturers provide specific instructions for the care and cleaning of their models. Follow the manufacturer's cleaning instructions and the protocols of your department.

After a fire, face pieces and regulators can be cleaned with a mild detergent or soap-and-water solution or with a disinfectant cleaning solution. The face piece should be fully submerged in the cleaning solution. If additional cleaning is needed, a soft brush can be used to scrub the face piece. During the cleaning process, avoid scratching the lens or damaging the exhalation valve. The regulator can be cleaned with the same solution, but it should not be submerged. The face piece and regulator should then be rinsed with clean water. Follow the manufacturer's recommendations, and avoid the use of aerosol cleaners or any alcohol-containing cleaner, as these materials degrade the rubber material in the face piece. Face pieces should be air dried or wiped with a soft, nonabrasive cloth to avoid scratching the lens.

Allow the SCBA time to dry completely before returning it to service. Also check for any damage before returning the equipment to service. Follow the steps in **SKILL DRILL 3-14** to clean and sanitize an SCBA.

SKILL DRILL 3-14

Cleaning an SCBA Firefighter I, NFPA 1010: 6.5.1

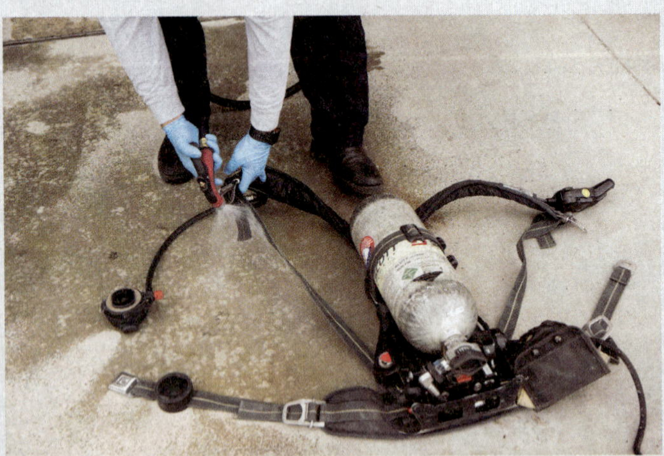

1. Rinse the entire unit using a hose with clean water. Inspect the SCBA for any damage that might have occurred before cleaning.

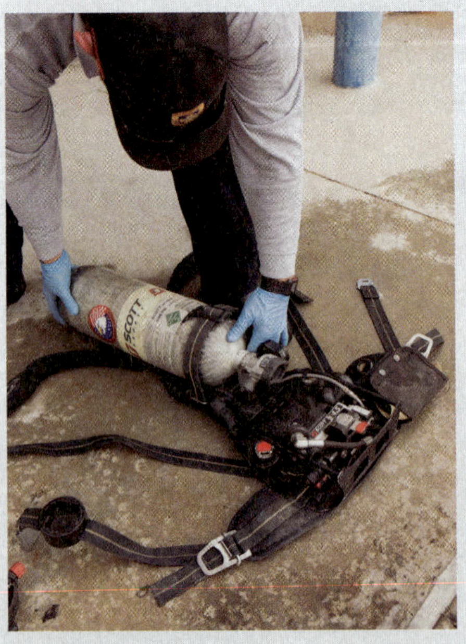

2. Detach the SCBA air cylinder from the harness. On some models, the regulator can be removed from the SCBA harness.

SKILL DRILL 3-14 CONTINUED

Cleaning an SCBA Firefighter I, NFPA 1010: 6.5.1

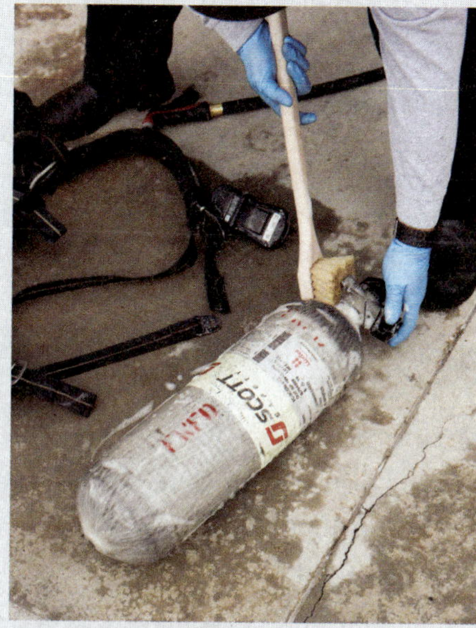

3. Using a stiff brush, along with a mild detergent or soap-and-water solution, scrub the SCBA air cylinder and harness. Rinse and set these pieces aside to dry.

4. In a 5-gallon (19-liter) bucket, make a mild detergent-and-water or soap-and-water solution or prepare the manufacturer's recommended cleaning and disinfecting solution and water. Submerge the SCBA face piece in the cleaning solution. For heavier cleaning, allow the face piece to soak.

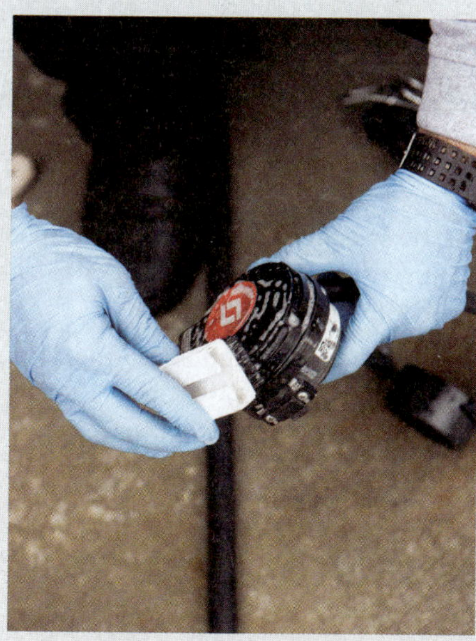

5. Clean the regulator with the soapy water or cleaning solution, following the manufacturer's instructions. Use a soft brush, if necessary, to scrub contaminants from the face piece and regulator.

6. Completely rinse the face piece and the regulator with clean water. Do *not* submerge the regulator. Set them aside and allow them to dry. Reassemble and inspect the entire SCBA before placing it back in service.

CASE STUDY

You Are the Firefighter CONCLUSION

One cool fall evening, you are dispatched for a report of a smoke odor in the area. These calls are normal for this time of year as residents start to use their fireplaces for the first time. As the engine drives slowly through the neighborhood, you notice the distinctive smell of a working house fire rather than that of firewood. When you turn the corner, you see smoke drifting from the basement window of a house. Your captain completes a quick size-up and requests additional resources from dispatch. The captain then tells you and another firefighter to pull a cross-lay and attack the fire.

1. **Which personal protective equipment would you need for this situation?**

   Answer: Knowing that you will be participating in fire attack, you need to put on the entire structural firefighting ensemble and wear an SCBA.

2. **What is the most efficient way to don a self-contained breathing apparatus? Why?**

   Answer: The most efficient way to don an SCBA is from the apparatus seat mount while riding in the engine. With this method, firefighters can dismount the engine ready to immediately attack the fire.

3. **Once the fire is over, what is the proper way to care for your protective equipment?**

   Answer: First, have another firefighter spray you down with a small diameter hose to remove surface contaminants. Then doff your PPE and use a brush and water to remove as much dirt and debris as possible. Transport the PPE to the station according to your department's SOPs and then thoroughly clean it following the manufacturer's instructions.

WRAP-UP

SUMMARY

KNOWLEDGE OBJECTIVES

- Explain how the structural firefighting ensemble works to protect the body.
  - List the components of personal protective equipment (PPE). (pp. 66–67)
  - Explain the role of the firefighter's work clothing as part of the PPE ensemble. (pp. 67–68)
  - Describe the type of protection provided by the structural firefighting protective ensemble. (pp. 67–68)
  - Explain how each design element of a fire helmet works to protect the head, face, and eyes. (pp. 68–70)
  - Explain why protective hoods are a part of the structural firefighting ensemble. (p. 70)
  - Explain how each design element of a structural firefighting protective coat works to protect the upper body. (pp. 70–71)
  - Explain how each design element of structural firefighting protective pants works to protect the lower body. (p. 71)
  - Describe how each design element of boots works to protect the feet. (pp. 71–72)
  - Describe how each design element of gloves works to protect the hands and wrists. (p. 72)
  - Explain how a personal alert safety system (PASS) helps to ensure firefighter safety. (**NFPA 1010: 6.3.1**, pp. 97–98)
  - List the limitations of PPE. (p. 112)
- Describe the specialized protective equipment required for vehicle extrication and wildland fires. (p. 77)
- Describe the procedure for donning and doffing personal protective clothing.

- Describe the procedure for donning personal protective clothing. (**NFPA 1010: 6.1.2**, pp. 72–74)
- Describe the procedure for doffing personal protective clothing. (**NFPA 1010: 6.1.2**, pp. 72, 75–76)
- List the complete sequence of donning PPE. (**NFPA 1010: 6.1.2, 6.3.1**, pp. 72–74)

- Describe the procedure for inspecting, maintaining, and cleaning PPE.
  - Describe how to inspect the condition of PPE. (**NFPA 1010: 6.1.2**, pp. 78–79)
  - Describe how to properly maintain PPE. (**NFPA 1010: 6.1.2, 6.5.1**, pp. 78–80)
  - Describe why thoroughly cleaning PPE immediately after it has been exposed to smoke or fire conditions is an important step in reducing your chance of developing cancer. (**NFPA 1010: 6.1.2, 6.5.1**, pp. 79–80)
- Describe the different types of respiratory protection and why they are required.
  - List the respiratory hazards posed by smoke and fire. (**NFPA 1010, 6.3.1, 6.3.10,** pp. 80–87)
  - List the conditions that require respiratory protection or self-contained breathing apparatus (SCBA). (**NFPA 1010, 6.3.1**, pp. 80–90)
  - Describe the types of breathing apparatus. (pp. 87–89)
  - Describe the differences between open-circuit self-contained breathing apparatus and closed-circuit self-contained breathing apparatus. (pp. 89–90)
- Describe the limitations of respiratory protection.
  - Describe the limitations of SCBA. (**NFPA 1010, 6.3.1**, pp. 90–92)
  - Describe the physical and psychological limitations of an SCBA user. (**NFPA 1010, 6.3.1**, p. 92)
- Describe the major components of the SCBA.
  - List and describe the major components of SCBA. (**NFPA 1010, 6.3.1**, pp. 92–97)
  - Describe the devices on an SCBA that can assist the user in air management. (**NFPA 1010, 6.3.1**, pp. 92–97)
  - Describe the pathway that air travels through an SCBA. (p. 97)
  - Explain the breathing techniques used to conserve air supply. (**NFPA 1010, 6.3.1**, p. 97)
- Describe the importance of inspecting and maintaining SCBA and explain how to inspect them.
  - Describe the importance of SCBA inspections and SCBA operational testing. (**NFPA 1010: 6.5.1**, pp. 112–117)
  - Explain how to inspect an SCBA to ensure that it is operation ready. (**NFPA 1010: 6.5.1**, pp. 115–117)
  - Explain the procedures for refilling SCBA air cylinders. (pp. 120–124)
  - Describe how to clean SCBA. (**NFPA 6.5.1**, pp. 124–125)

SKILLS OBJECTIVES

- Don and doff approved personal protective clothing.
  - Don approved personal protective clothing. (**NFPA 1010: 6.1.2**, pp. 73–74)
  - Doff approved personal protective clothing. (**NFPA 1010: 6.1.2**, pp. 75–76)
- Don an SCBA using various methods.
  - Don an SCBA from an apparatus seat mount. (**NFPA 1010: 6.3.1**, pp. 100–101)
  - Don an SCBA from an apparatus compartment mount. (**NFPA 1010: 6.3.1**, p. 102)
  - Don an SCBA from a storage case using the over-the-head method. (**NFPA 1010: 6.3.1**, pp. 103–104)
  - Don an SCBA from a storage case using the coat method. (**NFPA 1010: 6.3.1**, pp. 105–106)
  - Don a face piece. (**NFPA 1010: 6.3.1**, pp. 107–108)
- Doff an SCBA.
  - Doff an SCBA. (**NFPA 1010: 6.3.1**, pp. 109–111)
- Perform a visible and operational inspection of an SCBA.
  - Perform a visible inspection of an SCBA. (**NFPA 1010: 6.5.1**, pp. 113–114)
  - Perform an operational inspection of an SCBA. (**NFPA 1010: 6.5.1**, pp. 115–117)
- Replace an SCBA air cylinder.
  - Replace an SCBA air cylinder. (**NFPA 1010: 6.3.1**, pp. 118–119)
  - Replace an SCBA air cylinder on another firefighter. (**NFPA 1010: 6.3.1**, pp. 120–121)
- Refill an SCBA air cylinder from a cascade or compressor system.
  - Refill an SCBA air cylinder from a compressor or a cascade system. (**NFPA 1010: 6.5.1**, p. 123)
- Clean an SCBA. (**NFPA 1010: 6.5.1**, pp. 124–125)

KEY TERMS

air cylinder See *breathing air cylinder.*

air-cylinder pressure gauge The device on an SCBA that measures and displays pressure readings to indicate the quantity of breathing air available.

air-line respirator See *supplied-air respirator.*

air-purifying respirator (APR) A respirator that removes specific air contaminants by passing ambient air through one or more air purification components. (NFPA 1984)

atmosphere-supplying respirator (ASR) A respirator that supplies the respirator user with breathing air from a source independent of the ambient atmosphere and includes self-contained breathing apparatus (SCBA) and supplied-air respirators (SARs). (NFPA 1970)

bib The lower part of the protective hood that is part of the structural firefighting ensemble.

breathing air cylinder The pressure vessel or vessels that are an integral part of the SCBA and that contain the breathing gas supply; can be configured as a single cylinder or other pressure vessel or as multiple cylinders or pressure vessels. (NFPA 1970)

bunker coat See *structural firefighting protective coat.*

bunker gear See *structural firefighting protective clothing.*

bunker pants See *structural firefighting protective trousers.*

cascade system A method of piping air tanks together to allow air to be supplied to the SCBA fill station using a progressive selection of tanks, each with a higher pressure level. (NFPA 1900)

closed-circuit self-contained breathing apparatus (closed-circuit SCBA) A recirculation-type SCBA in which the exhaled gas is rebreathed by the wearer after the carbon dioxide has been removed from the exhalation gas and the oxygen content within the system has been restored from sources such as compressed breathing air, chemical oxygen, liquid oxygen, or compressed gaseous oxygen. Also called *rebreather.* (NFPA 1970)

compressor A device used for increasing the pressure and density of a gas. (NFPA 853)

doff The process of properly removing a member's PPE and respiratory protection to limit additional contamination and exposure. (NFPA 1700)

don The process of properly dressing in full PPE, ensuring all exposed skin and airway are protected. (NFPA 1700)

drag rescue device (DRD) A fabric handle integrated just below the collar at the back of the protective coat that a rescuer can grab to drag an incapacitated firefighter to safety.

dual-path pressure reducer A feature that automatically provides a backup method for air to be supplied to the regulator of an SCBA if the primary passage malfunctions.

emergency breathing safety system (EBSS) A device on an SCBA that allows users to share their available air supply in an emergency situation.

end-of-service-time-indicator (EOSTI) A warning device on an SCBA that alerts the user that the reserved air supply is being utilized. (NFPA 1970)

face piece The part of the SCBA consisting of the face mask and exhalation valve that delivers breathing air to the firefighter and protects the face from high temperatures and smoke.

firefighter breathing air replenishment system (FBAR) A system that allows firefighters to refill their SCBA cylinders on any floor in a high-rise building.

hand light A small, portable light carried by firefighters to improve visibility at emergency scenes; it is often powered by rechargeable batteries.

heads-up display (HUD) Visual display of information and system conditions status that is visible to the wearer. (NFPA 1970)

hydrostatic test A test performed by filling pressure-containing components completely with water or other incompressible fluid while expelling all contained air, closing or capping all open ports of the pressure-containing components, and then raising and maintaining the contained pressure to pressurize the pressure-containing components to a prescribed value through an externally supplied pressure-generating device. (NFPA 1900)

immediately dangerous to life and health (IDLH) Any condition that would pose an immediate or delayed threat to life, cause irreversible adverse health effects, or interfere with an individual's ability to escape unaided from a hazardous environment. (NFPA 1700)

light-emitting diodes (LEDs) Electronic semiconductors that emit a single-color light when activated. LEDs are used for operational displays in SCBA.

National Institute for Occupational Safety and Health (NIOSH) The U.S. federal agency responsible for research and development on occupational safety and health issues.

nose cups An insert inside the face piece of an SCBA that fits over the user's mouth and nose.

Occupational Safety and Health Administration (OSHA) The U.S. federal agency that regulates worker safety and, in some cases, responder safety. It is part of the U.S. Department of Labor.

off-gas To emit harmful chemicals in the form of a gas.

open-circuit self-contained breathing apparatus (open-circuit SCBA) An SCBA in which the exhaled air is released into the atmosphere and is not reused.

overhaul A firefighting term involving the process of final extinguishment after the main body of the fire has been knocked down. All traces of fire must be extinguished at this time. (NFPA 1700)

personal alert safety system (PASS) A device that continually monitors for lack of movement of the wearer and automatically activates an alarm signal, indicating the wearer is in need of assistance; can also be manually activated to trigger the alarm signal. (NFPA 1970)

personal protective equipment (PPE) The full complement of garments fire fighters are required to wear while on an emergency scene, including turnout coat, protective trousers, firefighting boots, firefighting gloves, a protective hood, self-contained breathing apparatus (SCBA), a personal alert safety system (PASS) device, and a helmet with eye protection. (NFPA 1010)

powered air-purifying respirator (PAPR) An air-purifying respirator that uses a powered blower to force the ambient air through one or more air-purifying components to the respiratory inlet covering. (NFPA 1984)

rapid intervention crew/company (RIC) A dedicated crew of at least one officer and three members, positioned outside the IDLH, trained and equipped as specified in NFPA 1407, who are assigned for rapid deployment to rescue lost or trapped members. Also called *rapid intervention team (RIT)*. (NFPA 1550)

rapid intervention crew/company universal air connection (RIC UAC) A system that allows emergency replenishment of breathing air to the SCBA of disabled or entrapped fire or emergency services personnel. (NFPA 1970)

rapid intervention team (RIT) See *rapid intervention crew/company (RIC).*

rebreather See *closed-circuit self-contained breathing apparatus (closed-circuit SCBA)*

regulator purge/bypass valve A device or devices designed to bypass a regulator.

remote pressure gauge The device on an SCBA that measures and displays pressure readings to indicate the quantity of breathing air available and that is located on the shoulder strap or in another location where it can be seen by the user while the SCBA is in use.

SCBA harness The backpack or frame for mounting the working parts of the SCBA and the straps and fasteners used to attach the SCBA to the firefighter.

SCBA regulator The part of the SCBA that reduces the high pressure in the cylinder to a usable lower pressure and controls the flow of air to the user.

self-contained breathing apparatus (SCBA) An atmosphere-supplying respirator that supplies a respirable air atmosphere to the user from a breathing air source that is independent of the ambient environment and designed to be carried by the user. (NFPA 1970)

smoke The airborne solid and liquid particulates and gases evolved when a material undergoes pyrolysis or combustion, together with the quantity of air that is entrained or otherwise mixed into the mass. (NFPA 1700)

soot Black particles of carbon produced in a flame. (NFPA 1700)

structural firefighting The activities of rescue, fire suppression, and property conservation in buildings or other structures, vehicles, railcars, marine vessels, aircraft, or like properties. (NFPA 1010)

structural firefighting protective clothing All of the clothing elements of the structural firefighting protective ensemble.

structural firefighting protective coat The element of the protective ensemble that provides protection to the upper torso and arms, excluding the hands and head. Also called *bunker coat* or *turnout coat.* (NFPA 1970)

structural firefighting protective ensemble Multiple elements of compliant protective clothing and equipment that when worn together provide protection from some risks, but not all risks, of emergency incident operations. (NFPA 1970)

structural firefighting protective footwear The element of the protective ensemble that provides protection to the foot, ankle, and lower leg. (NFPA 1970)

structural firefighting protective glove The element of the protective ensemble that provides protection to the hand and wrist. (NFPA 1970)

KEY TERMS CONTINUED

structural firefighting protective helmet The element of the protective ensemble that provides protection to the head. (NFPA 1970)

structural firefighting protective hood The interface element of the protective ensemble that provides limited protection to the coat/helmet/SCBA facepiece interface area. (NFPA 1970)

structural firefighting protective trousers The element of the protective ensemble that provides protection to the lower torso and legs, excluding the ankles and feet. Also called *bunker pants* or *turnout pants.* (NFPA 1970)

supplied-air respirator (SAR) An atmosphere-supplying respirator for which the source of breathing air is not designed to be carried by the user. Also called *air-line respirator.* (NFPA 1852)

turnout coat See *structural firefighting protective coat.*

turnout gear See *structural firefighting protective clothing.*

turnout pants See *structural firefighting protective trousers.*

two-way radios Portable communication devices used by firefighters. Every firefighting team should carry at least one radio to communicate distress, progress, changes in fire conditions, and other pertinent information.

REVIEW QUESTIONS

1. What are the six clothing items that make up the set of protective clothing for structural firefighting?
2. Which NFPA standard do structural firefighting protective coats need to meet?
3. How many layers are structural firefighting protective coats and pants constructed with and what do they consist of?
4. How often should a routine inspection be performed on structural firefighting gear?
5. What percentage of oxygen does outside or room air contain? At what percentage of oxygen do people start experiencing disorientation and other side effects?
6. What are the two categories of breathing apparatus?
7. Which type of SCBA is typically used for structural firefighting?
8. What is the realistic useful life of an SCBA air cylinder?
9. What are the four main parts of an SCBA?
10. What is the range of air pressure in filled SCBA air cylinders?
11. What does the EOSTI alarm indicate?
12. What components make up the complete PPE ensemble?
13. Which agency requires periodic hydrostatic testing of SCBA air cylinders?
14. How often should an SCBA be cleaned?

DISCUSSION QUESTIONS

1. What are some of the reasons you should practice donning all the components of the structural firefighting protective ensemble quickly and correctly?
2. Discuss some of the psychological limitations of the SCBA user and explain why they might be stressful.
3. Explain why it is important to clean and store your firefighting gear appropriately.

APPLYING THE CONCEPTS

Your engine unit is the first responding unit to an alarm at an industrial facility on a Saturday afternoon. Dispatch reported a worker was exposed to a toxic chemical and is having severe difficulty breathing. The chemical was not identified, and the worker is still alone inside the plant. It was reported that a vapor cloud is visible. Because it is a Saturday, only a few other workers are in the facility. The incident commander does an initial size-up and determines that safely extracting the exposed worker and evacuating the building are the primary objectives.

1. Before entering the building, what information do you need to know?
2. If the chemical cannot be quickly identified, do you risk entering the facility with your standard structural firefighting PPE?

A worker exits the building and runs up to your crew. "I'm one of the managers here. We identified the substance as an overheated lithium-ion battery. Here's the safety data sheet." The IC reviews the SDS and determines that with proper PPE the firefighting team on the scene can enter the facility to extract the exposed worker, evacuate the building, and stabilize the situation.

3. What is the proper PPE for this situation?
4. What are the pros and cons of using an air-purifying respirator (APR)?
5. What are the pros and cons of using an SCBA?

The IC orders, "Team A, we need to act quickly to extract the victim. Use structural firefighting gear and SCBA with full face piece and make sure everything is donned and secured properly."

6. What is your checklist to make sure the SCBA is ready for operation?

REFERENCES

American National Standards Institute and the International Safety Equipment Association (ANSI)/(ISEA). 2020. *American National Standard for High-Visibility Safety Apparel* (ANSI/ISEA 107-2020), September 2020. Arlington, VA: ISEA.

Centers for Disease Control and Prevention (CDC). 2021. "Respirators." Last reviewed August 22, 2021. Accessed June 23, 2023. https://www.cdc.gov/niosh/topics/respirators/default.html.

Mensch, Amy, George Braga, and Nelson Bryner. 2011. *Fire Exposures of Fire Fighter Self-Contained Breathing Apparatus Facepiece Lenses.* NIST Technical Note 1724. National Institute of Standards and Technology (NIST), November 2011. Accessed June 22, 2023. http://ws680.nist.gov/publication/get_pdf.cfm?pub_id=909917.

National Fire Protection Association (NFPA). 2017. *NFPA 1404, Standard for Fire Service Respiratory Protection Training.* 2018 Edition. Quincy, MA: NFPA.

National Fire Protection Association (NFPA). 2018. *NFPA 1852, Standard on Selection, Care, and Maintenance of Open-Circuit Self-Contained Breathing Apparatus (SCBA).* 2019 Edition. Quincy, MA: NFPA.

National Fire Protection Association (NFPA). 2019. *NFPA 853, Standard for the Installation of Stationary Fuel Cell Power Systems.* 2020 Edition. Quincy, MA: NFPA.

National Fire Protection Association (NFPA). 2019. *NFPA 1851, Standard on Selection, Care, and Maintenance of Protective Ensembles for Structural Fire Fighting and Proximity Fire Fighting.* 2020 Edition. Quincy, MA: NFPA.

National Fire Protection Association (NFPA). 2019. *NFPA 1852, Standard on Selection, Care, and Maintenance of Open-Circuit Self-Contained Breathing Apparatus (SCBA).* 2019 Edition. Quincy, MA: NFPA.

National Fire Protection Association (NFPA). 2020. *NFPA 99, Health Care Facilities Code.* 2021 Edition. Quincy, MA: NFPA.

National Fire Protection Association (NFPA). 2023. *NFPA 1010, Standard on Professional Qualifications for Firefighters.* 2024 Edition. Quincy, MA: NFPA.

National Fire Protection Association (NFPA). 2020. *NFPA 1700, Guide for Structural Fire Fighting.* 2021 Edition. Quincy, MA: NFPA.

National Fire Protection Association (NFPA). 2021. *NFPA 1977, Standard on Protective Clothing and Equipment for Wildland Fire Fighting.* 2022 Edition. Quincy, MA: NFPA.

National Fire Protection Association (NFPA). 2021. *NFPA 1984, Standard on Respirators for Wildland Fire Fighting Operations and Wildland Urban Interface Operations.* 2022 Edition. Quincy, MA: NFPA.

National Fire Protection Association (NFPA). 2021. *NFPA 1990, Standard for Protective Ensembles for Hazardous Materials and CBRN Operations.* 2022 Edition. Quincy, MA: NFPA.

National Fire Protection Association (NFPA). 2023. *NFPA 1550, Standard for Emergency Responder Health and Safety.* 2024 Edition. Quincy, MA: NFPA.

National Fire Protection Association (NFPA). 2023. *NFPA 1900, Standard for Aircraft Rescue and Firefighting Vehicles, Automotive Fire Apparatus, Wildland Fire Apparatus, and Automotive Ambulances.* 2024 Edition. Quincy, MA: NFPA.

National Fire Protection Association (NFPA). 2023. *NFPA 1970, Standard on Protective Ensembles for Structural and Proximity Firefighting, Work Apparel and Open-Circuit Self-Contained Breathing Apparatus (SCBA) for Emergency Services, and Personal Alert Safety Systems (PASS).* 2024 Edition. Quincy, MA: NFPA.

Occupational Safety and Health Administration (OSHA). 1970. "Occupational Safety and Health Standards: Personal Protective Equipment, Respiratory Protection." CFR 1910.124, updated September 26, 2019. Accessed July 26, 2023. https://www.osha.gov/laws-regs/regulations/standardnumber/1910/1910.134.

REFERENCES CONTINUED

Occupational Safety and Health Administration (OSHA). n.d. "Oxygen Concentration Efforts." Accessed July 27, 2023. https://www.osha.gov/sites/default/files/2018-12/fy15_sh-27664-sh5_Confined_Space_Handout_Effects_of_Oxygen.pdf.

Silverman, Steven H., Gary F. Purdue, John L. Gary F., Hunt, and Robert O. John L., Bost., Robert O. 1988. "Cyanide Toxicity in Burned Patients." *The Journal of Trauma: Injury, Infection, and Critical Care* 28 (2): 171–176.

U.S. Department of Transportation. 2017. 49 *Code of Federal Regulations* 180.205 https://www.ecfr.gov/cgi-bin/text-idx?SID=475de645c98062d4043077d9794a583f&mc=true&node=se49.3.180_1205&rgn=div8. Accessed June 18, 2018.

CHAPTER

4

Firefighter I

Fire Service Communications

KNOWLEDGE OBJECTIVES

After studying this chapter, you will be able to:

- Define the role of a public safety communications center, the responsibilities of the dispatcher, and the equipment used in a communications center.
- Describe the steps the dispatcher takes to process a reported emergency incident and how computer-aided dispatch assists in dispatching the correct resources to an emergency incident.
- Explain the procedures for handling emergency and non-emergency calls at a fire station and the procedures for transmitting the information to fire service responders.
- Identify modes of fire service communication and the types of radios used in the fire service.
- Describe how radios work and how a repeater and trunked systems enhance communications at an incident scene.
- Describe how to use a radio at an incident scene and how to convey location and emergency information.

SKILLS OBJECTIVES

After studying this chapter, you will be able to perform the following skills:

- Receive a call at a fire station and initiate a response to an emergency.
- Use a radio and communicate information properly and according to your department's standard operating procedures.

ADDITIONAL NFPA STANDARDS

- **NFPA 72**, *National Fire Alarm and Signaling Code, 2022 Edition*
- **NFPA 450**, *Standard for Emergency Services Communications, 2021 Edition*
- **NFPA 1225**, *Standard for Emergency Services Communications, 2022 Edition*
- **NFPA 1410**, *Standard on Training for Emergency Scene Operations, 2020 Edition*
- **NFPA 1900**, *Standard for Aircraft Rescue and Firefighting Vehicles, Automotive Fire Apparatus, Wildland Fire Apparatus, and Automotive Ambulances, 2024 Edition*

CASE STUDY

You Are the Firefighter

At 0304 hours, you are dispatched to a report of a fire at 3256 West Madison. While en route, the dispatcher advises that the caller reported that smoke was coming from the eaves of the building and then hung up. With this new information, you are preparing yourself for a possible house fire. The dispatcher radios back and says that this may be a false report because the caller used a cell phone that was identified as being on the east side of town. Thinking you had been woken up for another false call, the dispatcher comes back on the radio with a frantic sound in her voice and says that a home security company just reported a fire alarm at 3256 *East* Madison Street.

1. What is the process for receiving 911 calls in an emergency communications center?
2. How does the dispatcher know where the cell phone call originated?
3. How are alarm systems monitored and then reported to the communications center?

Introduction

Fighting a fire requires the coordination of numerous resources and people. Different crews and equipment fill various functions on the fireground, many of which occur simultaneously or must not start until something else has been accomplished. Ensuring a smooth interaction between all of the firefighting tasks being performed requires good communications among fire suppression crews, officers, and members of the Incident Command System (ICS). In addition, fire departments need to have good communications protocols, techniques, and equipment to manage incoming calls for emergency assistance. Finally, fire service communications also encompass post-incident reporting because sharing data and information about the cause of the fire and activities undertaken to suppress a fire is important for preventing fires and being able to suppress them more safely and efficiently.

Every fire department depends on a functional **public safety communications center**, or simply **communications center**, which is the location where 911 calls for that community or jurisdiction are directed. When someone requests assistance or an alarm sounds, the communications center dispatches the appropriate units to the incident. **Dispatch** means to select units to respond to a reported incident, alert those units, and then communicate the request for service by quickly and accurately transmitting the information to them. The communications center maintains communication with those units during the incident. It also tracks the location and status of every other fire department unit. The communications center must always know which units are available to be dispatched to an incident, and it must be able to contact those units promptly. The communications center is responsible for redeploying units to maintain adequate coverage for all areas. It is the link between firefighters on the scene, the rest of the organization, and others. The communications center monitors communications from the incident scene and processes all requests for assistance or special resources.

At the scene, firefighters need to communicate with one another so that the incident commander (IC) can manage the operation efficiently based on progress reports or requests for assistance from firefighters. The ICS depends on the presence of a functional onsite communications system. During incidents, firefighters must be able to communicate not only with one another, but also with other emergency response agencies.

In addition to these special communications requirements, a fire department must have a communications infrastructure that allows it to function as an effective organization. Basic administration and day-to-day management require an efficient communications network, including telephone and data links with every fire station and work site. Rapidly developing technology and advanced communications systems are improving fire department communications. As a firefighter, you must be familiar with the communications systems, equipment, and procedures used in your department.

Communications Centers

A large percentage of the United States has access to some type of 911 system to report an emergency. These calls are directed to a **public safety answering point**

(PSAP), also known as a 911 call center. PSAPs are often co-located within the jurisdiction's public safety communications center, or simply communications center. There are several types of communications centers. A communications center may be a **stand-alone communications center**, serving a single specific fire department, or a regional center, serving many fire departments. In addition, it may be co-located with other communications centers of various public safety agencies, or it may be integrated, bringing police, fire, and EMS into one location. The primary responsibilities of the communications center are to process 911 calls, as well as other non-emergency public safety calls.

SAFETY TIP

The 911 emergency number should always be pronounced as "nine-one-one." It should not be pronounced as "nine-eleven," because there is no "11" on a telephone.

When fire, EMS, and law enforcement communications are in the same facility, calls can be answered, processed, and dispatched immediately. A communications center may have independent personnel and systems for each emergency responding agency, or all employees may be trained to receive calls and dispatch responders to any type of emergency incident. If each of these agencies has a stand-alone communications center that operates in separate facilities, each call must be transferred to the appropriate communications center. Some agencies have a mobile communications center that allows dispatchers to be onsite and run communications for a larger incident.

Regardless of how or where the call is transferred, the public safety communications center is the hub of the emergency response system. It serves as the central processing point for all information relating to an emergency incident and all of the information relating to the location, status, and activities of responding units.

The size and complexity of the communications center varies depending on the needs of the department. For example, the communications center for a small, rural department might be a small room in a fire station staffed by one person at a time. The communications center for a large department with multiple stations in an urban area might be a specially designed, highly sophisticated facility with advanced technological equipment that requires special training and multiple personnel to be onsite. Regardless of their size, location, and configuration, all communications centers perform the same basic functions (**FIGURE 4-1**).

FIGURE 4-1 All communications centers perform the same basic functions.

Dispatchers

The employee who staffs a communications center is called a **telecommunicator**, more commonly called a **dispatcher**. Dispatchers receive, process, and disseminate information; understand and follow complicated procedures; perform multiple tasks effectively; memorize information; and make decisions quickly. Dispatchers should be trained to work in a public safety communications environment and have completed advanced training and professional certification programs to ensure that skilled, competent individuals can fulfill these critical roles in the public safety system.

The job of a dispatcher can be complicated, demanding, and extremely stressful. Just as special qualities can make an individual a good firefighter, a dispatcher must possess certain qualities to meet the modern challenges to be successful. For example, one of the dispatcher's most important skills is the ability to communicate effectively to obtain critical information, even when the caller is highly stressed or in extreme personal danger. If an emotional caller criticizes or insults the dispatcher, the dispatcher must still respond professionally and focus on obtaining the essential information. Voice control and the ability to maintain composure under pressure are important qualities for dispatchers. They must always be clear, calm, and in control.

Dispatchers must be skilled in operating the systems and equipment in the communications center. They must understand and follow the fire department's operational procedures, particularly those relating to dispatch policies and protocols, radio communications, and incident management. The dispatcher must always keep track of the status and location of each unit and must monitor the overall deployment and availability of resources throughout the system. NFPA 1225, *Standard for Emergency Services Communications*, contains a complete list of qualifications for dispatcher candidates.

Communications Facility Center Requirements

The communications center should be designed and operated to ensure that its critical mission can be performed with a high degree of reliability. The performance requirements in NFPA 1225 govern the design and construction of public safety communications centers. These requirements apply whether the communications center serves a small community with only one or two fire stations or a metropolitan area with dozens of stations.

The communications center must be well protected against natural and human-made threats, and it must be able to withstand predictable damaging forces such as floods, earthquakes, snowstorms, tornadoes, hurricanes, and other severe storms. It must be located so that it can function during times of civil unrest. The communications center must be able to operate at maximum capacity, without interruption, even when other community services are severely affected. The building must be equipped with emergency generators and other systems so that it can continue to operate for several days in even the most challenging conditions.

NFPA 1225 also requires backup systems for all of the critical equipment in a communications center, so that the failure of a single component or system will not disable the entire operation. For example, there must be more than one way of transmitting a message from the communications center to each agency. A backup radio transmitter should be available, and the telephone system must be able to receive calls even if part of the system is damaged. The design of the facility must minimize its vulnerability to a fire originating inside the building as well as to nearby fires. The center also must be secured to prevent unauthorized entry.

A backup communications center at a different location must be established. If some unanticipated situation makes it impossible to operate from the primary location, this backup location can be activated to ensure ongoing operation of the communications function.

Plans need to be in place describing the procedure for reporting emergencies if the emergency communications center fails. Some communities locate fire department and law enforcement vehicles at strategic locations in the community so citizens can report emergencies. Agencies may plan to staff a watch desk to take reports of emergencies from citizens who walk into the building.

Communications Center Equipment

Depending on the size of the communications center and the specific role of each, most communications centers have the following equipment:

- Phone lines dedicated to receiving 911 calls
- Phone lines that connect directly to other agencies
- Non-emergency phone lines
- Equipment to receive alarms from public or private fire alarm systems
- TTY/TDD equipment
- Computers that use a **geographic information system (GIS)**, which is an application that layers a wide variety of data about a community on top of a map of the community. For example, users can see the location of gas lines and fire hydrants; demographic data; topographical data, including which areas are prone to flooding; and much more.
- Equipment for alerting and dispatching units to emergency calls
- Two-way radio system(s)
- Recording systems to record phone calls and radio communications
- System for maintaining and managing records
- Backup generators to supply electricity in case of an electrical outage

Computer-Aided Dispatch

Many communications centers have **computer-aided dispatch (CAD)**, which is a combination of hardware and software that assists dispatchers by performing specific functions more quickly and efficiently than doing them manually (**FIGURE 4-2**). For example, once the address of the incident is determined and the incident description is in the CAD system, the system can make a recommendation on the appropriate units to dispatch very quickly. Data links that provide the location of the caller and automatically enter the address can also save time. If duplicate addresses exist or if the address entered is not a valid location, the CAD system prompts the dispatcher to ask the caller for more information.

FIGURE 4-2 A CAD system enables a dispatcher to work more quickly and efficiently.

FIGURE 4-3 Some CAD systems transmit dispatch information directly to terminals in fire stations and mobile data terminals in the apparatus.

The most advanced type of CAD system utilizes **global positioning system (GPS)** data devices to confirm the location of the fire department units. The CAD system determines which units are available to respond to a call and the closest fire stations in order of response for any location, and it keeps track of units assigned to incidents and units that are temporarily assigned to cover different areas. The CAD system then indicates the units that can respond quickly to an alarm, even if some of the units that would normally respond to the call are currently unavailable.

A major advantage of a CAD system is that it automatically captures and stores every event as it occurs. Before such systems were developed, someone had to write down and timestamp every response, including every radio transmission, and all of these hard-copy logs had to be retained for future reference.

Some CAD systems use the radio system or cellular signals to transmit dispatch information directly to **mobile data terminals (MDTs)** on an apparatus or to computers in the fire station (**FIGURE 4-3**). In addition, CAD systems can also provide immediate access to information such as preincident plans, hazardous materials lists, lockbox locations, and information such as whether persons with limited mobility reside at the address of the emergency call. By linking the CAD system to other data files, firefighters gain access to even more useful information such as travel route instructions, maps, and reference materials.

Voice Recorders and Activity Logs

Almost everything that happens in a communications center is recorded, 24 hours a day, by a voice recording system or an activity logging system. A **voice recording system** is a system that records communications over the phones and the radio. Digital recording systems, many of which are cloud-based, allow a recording to be recalled at any time. Older recording systems have an instant playback unit that allows the dispatcher to replay conversations only from the previous 10 to 15 minutes. Replaying calls is valuable if the caller speaks quickly, has challenging to understand manner of speaking, hangs up, or is disconnected.

An **activity logging system** is a computer system that keeps a detailed record of every incident and activity that occurs. These records include every call received or made; every unit dispatched; and every significant event related to emergency incidents, such as reports when a unit is en route and when it has arrived at the scene, when the incident is under control, and when the last unit leaves the scene.

Voice recorders and activity logs are maintained for several reasons. First, they serve as legal records of the official delivery of a government service by a public agency. These records may be required for legal proceedings, sometimes years after the incident occurred. They may be needed to defend a department's actions when questions are raised about an unfortunate outcome. These records document the events and often demonstrate that the organization and its employees performed ethically, responsibly, and professionally. They also make it difficult to hide an error if a mistake was made.

Second, records are valuable for reviewing and analyzing information about department operations. Good record-keeping enables a fire department to examine what happened on a particular call as well as to measure workloads, system performance, activity trends, and other factors as part of its planning and budget preparation. For example, analysts and planners can use data from CAD systems to study deployment strategies and to make the most efficient use of fire department resources.

Communications Center Operations

Basic functions that a dispatcher performs in a communications center include the following:

- Receiving calls reporting emergency incidents and dispatching the appropriate units
- Supporting the operations of fire department and other response units delivering emergency services
- Coordinating operations among responding agencies
- Always keeping track of the status of each response unit
- Monitoring the level of coverage and dispatching additional units as needed
- Notifying designated individuals, such as on-call chiefs, and agencies, such as utility companies or critical incident stress debriefing teams, of particular events and situations
- Maintaining information required for dispatch purposes

All of these activities must be performed accurately and efficiently, even in the most challenging circumstances. For example, if a disastrous event that involves many members of the community occurs, the activity level in a communications center will increase rapidly as calls come in. Dispatchers must not let the chaos outside affect the operations inside. If the communications center fails to perform its mission, the emergency responders will not be able to deliver much-needed emergency services.

Most emergency responses begin when a call comes into the communications center which then dispatches appropriate units. When a unit goes to an incident, it is called going on a **run**. When the dispatcher dispatches units, it is called generating or creating a run. If a citizen reports an emergency directly to a fire station or if a crew discovers a situation, the information and the request for help must immediately be relayed to the communications center. The communications center can then initiate the emergency response process.

The dispatcher's first responsibility is to obtain the information that is required to dispatch the appropriate

FIGURE 4-4 Dispatchers must obtain the needed information and relay it accurately to the appropriate responders.

units to the correct location (**FIGURE 4-4**). Each jurisdiction will have slightly different standard operating procedures (SOPs), but usually the dispatcher will use the CAD system information to confirm which agencies or units should respond and transmit the necessary information to them. As stated in NFPA 1225, the generally accepted performance objective is to dispatch units to the highest priority calls within 60 seconds, 90 percent of the time, from the time a call reaches the communications center (NFPA 1225). Responding to a reported emergency without using the proper dispatch procedures can result in confusion and delay in the necessary resources reaching the emergency.

The communications center must have both a primary and a backup method of transmitting alarms to stations. Although radio, telephone, and public address systems often are used to transmit information to fire stations, some communications centers use computers to transmit dispatch messages to MDTs in apparatus or to printers at fire stations. Some fire departments still use a system of bells or tones to transmit alarms. Volunteers or rural departments may use outdoor sirens or horns to summon firefighters to an emergency. Dispatch messages can also be transmitted over pagers or cell phones.

The major steps in processing an emergency incident include the following:

1. Call arrival
2. Location validation
3. Classification and prioritization
4. Unit selection
5. Dispatch

Call Arrival

Call arrival is the process of receiving a call for service and obtaining the information necessary to initiate a response. Calls come from the following sources:

- Telephone calls from the general public or other municipal agencies
- Municipal fire alarm systems
- Private automatic fire alarm systems
- Calls reported directly to fire stations
- Calls initiated via radio from fire or police units that are already "on the road"

TIP

Dispatchers must be careful about information they disclose to the public. Information such as the availability of individual fire departments, apparatus placement, and available equipment should only be disclosed with permission from a chief officer. Each dispatcher must know their department's SOPs describing which types of information are allowed to be broadcast and which types are not.

Telephone Communication

The 911 calls received by a PSAP or communications center are generated mostly from a person calling on a phone, such as a cell phone, home or business land line, or a roadside emergency telephone. According to the National Emergency Number Association (NENA), 80 percent of 911 calls come from wireless devices (2021). Some calls are made from a phone using **Voice over Internet Protocol (VoIP)**. VoIP converts a person's voice into a digital signal that is sent via the Internet instead of over traditional phone lines. In most communities, phone calls to 911 connect the caller with a PSAP. The PSAP telecommunicator either takes the information directly from the caller or transfers the call to the appropriate agency, based on the nature of the emergency. In some systems, calling 911 connects the caller directly with a dispatcher, who obtains the required information and assigns the appropriate units.

Each communications center also has a normal 10-digit, non-emergency phone number. Some people use that number to report an emergency because they are not sure that their situation is serious enough to be considered a true emergency. Many communities have implemented the 311 system to handle non-emergency calls. The purpose of the 311 system is to reduce the number of non-emergency calls to 911. This system often links callers with community resources that can assist with a problem when it the problem is not life threatening. Any number, whether it is listed as an emergency or non-emergency number, that is published as a fire department phone number should be always answered, including nights and weekends.

The dispatcher who takes the call must conduct a telephone interview, asking the caller questions to obtain the required information. The first pieces of information the dispatcher needs to know are the location of the emergency and the nature of the problem. The dispatcher needs to remember that the caller thinks the situation is an emergency, and they must treat every call as an emergency until it is determined that no emergency exists. The caller may be distressed, scared, angry, or excited, and unable to organize their thoughts to communicate the problem clearly. If English is not the caller's first language, the caller might not know the right words to explain the situation and might not understand any questions the dispatcher asks. Dispatchers must not allow a caller's strong emotions affect their ability to do their job. They must speak calmly, remain professional, and use active listening techniques, following the agency's SOPs, to interpret the information. Many communications centers provide dispatchers with a structured set of questions designed to obtain accurate information on the nature of the situation.

Dispatchers should not allow gaps of silence to occur while questioning the caller. If the caller is suddenly silent, something may have happened to them or the caller may be in personal danger. Conversely, if the dispatcher is silent, the caller might think that the dispatcher is no longer on the line or no longer listening.

Disconnects are another problem dispatchers need to be prepared to handle. Callers to 911 might hang up accidentally or be disconnected after reaching the communications center and before describing the reason for the call. The dispatcher should first attempt to call the person back using the automatic number identification from the initial call. If the dispatcher cannot reach the caller and the caller's location is already known, the dispatcher can dispatch a police officer to the location to determine if more help is needed.

With just two critical pieces of information—the location and nature of the problem—the dispatcher can initiate a response. Local guidelines or SOPs, however, may require the dispatcher to obtain additional information. Getting the caller's name and confirming their phone number is useful in case it is necessary to call them back to obtain additional information.

If the caller is in danger or distress, the dispatcher should try to keep the line open and remain in contact with the caller until help arrives or in some cases, the dispatcher can advise the caller what to do next. For example, if a building fire is being reported, the dispatcher might advise the occupants to evacuate and wait outside. If the 911 call is for a medical incident, it will be routed to a dispatcher trained to provide medical self-help instructions to callers based on the caller's description of the patient's symptoms. These specially trained dispatchers are called **emergency medical dispatchers (EMDs)**.

On occasion, a communications center may receive calls about issues that cannot be handled by the fire department or other agencies participating in the communications system. If this happens, the dispatcher should make every effort to accommodate the caller, and if possible, transfer the caller to the proper agency and provide the caller with the proper contact information in case the transfer fails. The caller will appreciate the assistance and retain a positive image of the department's service.

In addition to receiving calls from different types of phones, communications centers receive calls from the following:

- TTY/TDD systems
- Text-to-911
- Direct-line phones
- Multi-line telephone systems

TTY/TDD Systems. People with a speech or hearing impairment can communicate by phone using a **TTY/TDD system**, which has a keyboard and a screen for displaying text rather than transmitting audio, physically connected to a phone or a computer (**FIGURE 4-5**). (TTY stands for teletype; TDD stands for telecommunications device for the deaf.) The Americans with Disabilities Act (ADA) requires that every call-taking position within a PSAP must have its own TTY or TTY-compatible equipment and be able to receive calls via TTY/TDD systems (ADA.gov 2016, 35.161 and 35.162). Dispatchers must know how to operate this equipment. A communications center might have a dedicated telephone line with its own 10-digit number, but the FCC states where 911 is not available and a PSAP provides emergency services via a 7- or 10-digit number, it still must provide direct, equal access to TTY callers.

FIGURE 4-5 TTY/TDD systems enable people with speech or hearing impairments to communicate via text using a phone.

Text-to-911. Text-to-911 is currently available only in certain locations. The **Federal Communications Commission (FCC)**, which regulates interstate and international communications between the United States and other countries by radio, television, wire, satellite, and cable, encourages emergency call centers to begin accepting texts, but it is up to each call center to decide the particular method in which to implement and deploy text-to-911 technology (FCC 2015.)

Direct-Line Phones. It is important to have a second means to communicate with emergency agencies and responding units if the cellular network or radio system becomes inoperable. A **direct-line phone** (or **ring-down phone**) connects two predetermined points. When there is a phone at either end of the direct line, picking up the phone at one end causes the phone at the other end to ring. Direct lines may connect the communications center to each fire station in its jurisdiction. They also often link police and fire communications centers or two fire communications centers that serve adjacent areas. Direct lines also may connect hospitals, private alarm companies, utility companies, airports, and similar facilities with the communications center and each other.

Multi-line Telephone Systems. A multi-line telephone system (MLTS) is a phone system that allows multiple users throughout an organization, such as a business, a hotel, or a university, make and receive calls at the same time. In most cases, when someone calls 911 from a phone connected to an MLTS, the caller's location within the organization, such as a floor or room number, is transmitted along with the street address.

Municipal Fire Alarm Systems

Some communities still have a **municipal fire alarm system**, which is a network of fire alarm boxes and

FIGURE 4-6 Fire alarm boxes are still in use in some parts of the country.

© Beschi Mauro/Shutterstock

emergency telephones on street corners or in public places. Although these have largely been replaced by landline and cell phones, they still are used in areas where there are no landline and cell phone service is unreliable. A **fire alarm box** transmits a coded signal to the communications center and, in some cases, directly to individual fire departments. A manual fire alarm box, often called a "pull box," is activated when the lever is pulled (**FIGURE 4-6**). The coded signal identifies the location of the box, but it does not indicate the type of emergency that is occurring. Fire alarm boxes in municipal fire alarm systems may also be connected to automatic fire alarm systems in private commercial, industrial, and residential buildings. When the building's fire alarm system is activated, the connected municipal fire alarm box transmits the alarm to the communications center. Municipal fire alarm boxes are usually connected by networks of dedicated cables or through a dedicated radio frequency.

If a community chooses to keep its municipal fire alarm system, in many cases, they replace the boxes connected to the system by underground cables with radio-operated fire alarm boxes, which use radio waves to transmit the signal (**FIGURE 4-7**).

FIGURE 4-7 A radio-operated fire alarm box.

Courtesy of Signal Communications Corporation (Sigcom).

Some communities have converted their municipal fire alarm systems into call box systems. A **call box** is a direct-line phone or a radio that connects a caller directly to a dispatcher. The caller can request a full range of emergency assistance—from police, fire, or emergency medical assistance to a tow truck. Emergency call boxes are often located along major highways in remote areas and near bridges, tunnels, subways, large complexes such as universities, and in other locations where nearby phones are not available and where cell phones do not work. Call boxes may be solar powered so they can be installed in locations without electric or telephone service (**FIGURE 4-8**).

Private Automatic Fire Alarm Systems

Many private, commercial, industrial, school, and residential buildings have automatic fire alarm systems. These systems use heat detectors, smoke detectors, water flow detectors, or other devices to initiate an alarm. Chapter 24, *Systems of Fire Detection, Suppression, and Smoke Control,* discusses these systems in detail.

The connection used to transmit an alarm from a private fire alarm system to a communications center depends on many factors. These systems may be monitored by a privately operated alarm monitoring service. When the monitoring service detects that an alarm has been initiated at a location, staff at the monitoring service alerts the communications center, sometimes via a direct line. Some communications centers provide monitoring services. No matter how the alarm reaches the communications center, the result is the creation of an incident report and the dispatch of fire department resources.

FIGURE 4-8 Wireless call boxes are sometimes found in places without other kinds of telephone service.

Walk-Ins

Most emergencies are reported by telephone, but sometimes people come to a fire station seeking assistance. When a **walk-in** arrives, someone at the station should immediately contact the communications center and describe the situation, even if the units at the station can handle the situation without assistance. If the walk-in did not clearly provide information about the incident, the firefighter interacting with the walk-in should consider putting the walk-in on the phone to convey information directly to the dispatcher. The dispatcher needs to generate a run and create an incident report, as well as dispatch any needed additional units.

People should be able to come to a fire station and report an emergency at any time, even when the station is unoccupied. Many departments install a direct-line phone to the communications center just outside each fire station. These phones should be labeled with a sign instructing the public that if the station is vacant, they can pick up the phone in the red box to report an emergency.

TIP

While technology may be of great help in emergency situations, it is important to keep in mind that sharing private and protected information can cause great grief and harm to a person and violates the privacy rights of that person. Become familiar with your department's policy regarding the use of private devices, such as cell phones, in emergency situations. Never post pictures from an incident on the Internet. You and your department may be subject to serious legal action.

Radio Calls from Fire or Police Units

A fire or police unit responding to a call or on patrol may come upon an emergency situation or is notified by a bystander, referred to as a "flag down." The fire or police unit then contacts the communications center via radio to report the emergency and the need to assign units to the call. At other times, a change in status at an emergency incident may require additional or specialty units. For example, if a unit arrives at a fire scene and find that the fire is much larger than the original report, they would request additional units to be dispatched to the scene. Likewise, if additional units are en route but the units on scene have the situation under control, the IC on scene will notify the communications center so that the additional units can be recalled.

Location Validation

To process the request for service and before dispatching units, the dispatcher must confirm that the location information received is adequate and there is no confusion about the address. For example, two streets in the same town or city might have similar names, such as 123 East Main Street and 123 West Main Street or 456 Spring Street and 456 Spring Avenue. Some larger cities even have two streets with the same name. This process of elimination is done through the GIS system in CAD. The information provided must point to a valid location on a map or in a street index system, and that location must be within the geographic jurisdiction of the potential dispatch units.

Enhanced 911

To help identify the location of an incident, Enhanced 911 was developed. If the call is made with a land line,

Voice of Experience

On an early fall evening our dispatch center received a 911 emergency call from a frantic woman. The woman was overly excited, and the telecommunicator could tell that this was a real emergency by the sound of her voice. The woman was reporting that her daughter's house was on fire. Instead of the daughter calling the dispatch center directly, she called her mother on instinct. More than likely she had more confidence in her mother being able to make the call than if she made it herself. This shows how some people can react when they are under extreme pressure or, in this case, in an emergency situation.

Once our dispatch center received the call, our SOP was to activate a general alarm for the department bringing two engines, one ladder tower, and one ladder truck to the scene. As the first engine and department chief arrived at the reported address, there were no visible signs of a fire or smoke in the area. They continued to investigate the address and surrounding area but came up with nothing. The chief notified dispatch of the situation and requested that they verify the information with the initial caller. The telecommunicator restated that the call seemed legitimate due to the nature of the caller's message and the urgency in her voice.

At this point the chief decided to have all responding apparatus proceed with caution to the scene for further investigation and until verification with the caller could be completed. When the telecommunicator called the woman back for verification of the correct address, he was advised by the caller that the house that was on fire was her other daughter's home. She had provided the wrong address. The home with the fire was located two towns over and 15 miles (24 kilometers) away.

In this case, the telecommunicator asked all of the right questions and the caller's voice was legitimate, as was the nature of the incident and the homeowner's relationship to the caller; however, the dispatcher was given an incorrect address from a third party. This is a perfect example of how misinformation can be transmitted when someone is overly excited during an emergency situation.

Richard J. Kosmoski, BA, MS
President
New Jersey Volunteer Fire Chief's Association

Enhanced 911 (E911) provides the phone number of the phone making the call and the location where the call originated. Most 911 systems currently in operation are E911 systems.

In E911 systems, **automatic number identification (ANI)** displays the phone number of the phone making the call. **Automatic location identification (ALI)** shows the location of the telephone making the call, the subscriber's name, and other details (**FIGURE 4-9**). If the call comes from a cell phone, the signal is picked up by the closest cell tower, which may not be in the same town or city. If the cell user has location services activated, the caller can be located using GPS. Otherwise, cell carriers can locate a phone through triangulation by calculating the distance from multiple cell phone towers to which the cell phone connected.

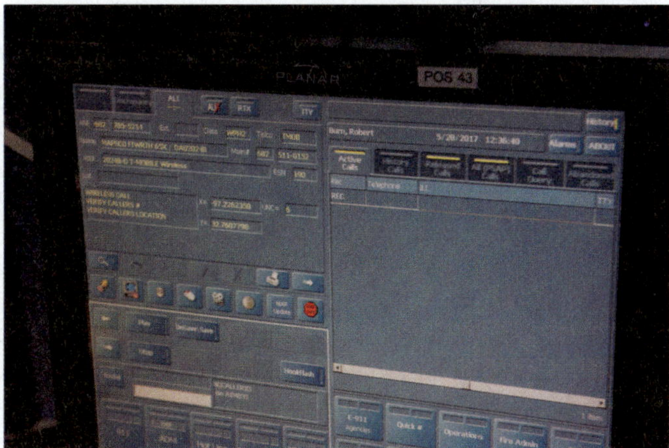

FIGURE 4-9 An automatic location identification (ALI) system provides information about the origin of a 911 call.

If the call is made using VoIP, the user must have provided the VoIP provider with their address in order for E911 to be able to retrieve that location. However, because VoIP calls can be made from anywhere with a connection to the Internet, that address might not be the location where the call originated. Efforts to provide location validation for calls received from cell phones and VoIP are advancing; however, in some areas the dispatcher cannot determine the location of 911 calls made from these devices (911.gov 2023).

If the ANI and ALI are provided, the dispatcher should do their best to confirm that the information that is generated by E911 is correct and identifies the actual location of the emergency. The provided information might be inaccurate for a variety of reasons, such as the following:

- If the caller recently moved, the database may contain the old address.
- Sometimes a billing address for the phone service is provided rather than the physical location of the telephone.
- If the call was made from a location other than the site of the emergency incident, the location provided by the E911 system would not be the correct location to send responders.
- If the call was made from a cell phone or VoIP, the ANI and the ALI may not display.
- Many people calling 911 may not know their exact location.
- Numerous calls from cell phones may come in for the same incident, making it hard to determine the exact location of the emergency and overloading the communications center.

Next Generation 911

Next Generation 911 (NG911) systems allow all types of digital information, including voice, text messages, photos, and video, to be transmitted to a PSAP or stand-alone communications center. In addition to being able to receive digital data, PSAPs with NG911 are also able to receive data and alerts from safety devices, medical devices, and sensors, and they will be able to issue emergency alerts to wireless devices and highway alert systems. NG911 relies on a GIS to determine the location from which a call is made. Next Generation communications centers also provide CAD to CAD services; permit dispatch centers to receive data, texts, video clips, and other communications from the general public; incorporate CAD and GPS; and automate 911 hang-ups.

Call Classification and Prioritization

Call classification and prioritization is the process of assigning a response category based on the nature of the reported problem. Response categories in a department might be as follows:

- Critical/severe
- Major/high
- Medium
- Minor/low

Some calls qualify for emergency response (lights and siren); others are classified as non-emergency. Fire departments respond to many types of situations, ranging from a fire in a dumpster, to a fire in a high-rise building, to someone who collapsed, to an incident with multiple casualties. The nature of the call dictates its urgency and priority. High-priority calls are dispatched immediately. Sometimes it is necessary to delay dispatch of a lower-priority call if a more urgent incident is reported or if several calls come in at the same time.

Unit Selection

The call classification and prioritization help the dispatcher select the appropriate units to respond to the incident. **Unit selection** is the process of determining exactly which unit or units should be dispatched, based on the location and classification of the incident. The department's SOPs and dispatch protocols dictate the type and number of units to dispatch to each category of incident. The response can range from assigning a single unit to assigning multiple units and dozens of firefighters. Generally, the standard assignment to each type of incident in each geographic response area is stored on **run cards**. Run cards list units in the order of response based on response distance or estimated response time. They often specify the units that should be dispatched if there are multiple alarms. Run cards are prepared in advance. At communications centers with a CAD system, the information is entered into the CAD, but the cards are retained so they can be used if the CAD fails.

Most CAD systems are programmed to recommend which units to dispatch to an incident based on the location of the incident, call type and classification, and the status of all units. The dispatcher can accept the recommendation or adjust based on the circumstances or additional information. When additional units are needed at an incident, the same process is used.

Selecting units requires quick decision-making skills, even when the CAD system is programmed to follow set policies for various situations. Usually, the policy is to dispatch the closest available unit that can provide the needed assistance. When all units are available and in their fire stations, the dispatcher can make this determination quickly based on information already programmed into the CAD. Unit selection becomes more complicated when units are not in their fire stations, are on an assignment, or are unavailable. Some communications centers have vehicle locator systems that track apparatus using GPS devices. These systems allow dispatchers to quickly identify the closest units in service. Departments sometimes enter into automatic mutual aid agreements with other departments to dispatch the closest available units even if those units are not from the same jurisdiction as the incident.

Dispatch

After the dispatcher selects the unit or units to respond to the incident, the next step is dispatch. Fire departments use a variety of dispatch systems. The communications center must have at least two ways to send a dispatch message from the communications center to each fire station.

Fire station alerting systems receive a tone or series of tones to get the attention of firefighters in the fire station. This is followed by a verbal or automated message providing the call type and units to respond. The communications center and the fire station alerting system are connected for dispatch via one of the following:

- Hard-wired circuit that directly connects the communications center to the alerting system, and usually used when the communications center and the firehouse are located in the same building
- Phone line, usually a fiber optic line
- Computer-based data link
- Microwave transmission system that transmits call information to a printer located in the firehouse
- Radio system that sends a signal to the station alerting system when firehouses are located great distances from the communications center

Regardless of how the dispatch message is relayed, it is broadcast over speakers in the fire station so that everyone in the station knows the location and nature of the incident and which units are being dispatched. If a unit is out of the station and available, the dispatcher contacts them via the radio system. The fire station or each unit must confirm that the message was received and the unit is responding. If the communications center does not receive confirmation within a set time, it must dispatch substitute units.

CAD systems can be programmed to alert the appropriate fire station automatically. It can send dispatch information to computer terminals or printers, sound distinctive tones, turn on lights and public address speakers, turn off the stove, and open the garage bay doors. The dispatch message can be sent directly to each individual vehicle that has an MDT. The CAD notification is often accompanied by a verbal announcement over the fire station speakers and the radio.

Fire departments with volunteer responders must be able to reach members with a dispatch message. Some departments issue pagers, use cell phones, or have online or mobile apps to reach volunteers. Some CAD systems send text messages to cell phones. Some volunteer fire departments rely on outdoor sirens, horns, or whistles to notify their members of an emergency. These audible devices usually can be activated by remote control from the communications center. Volunteers then contact the communications center by phone or radio to receive specific instructions.

Operational Support and Coordination

After the communications center dispatches the units, it begins providing operational support and coordination. A dispatcher in the communications center must remain in contact with the responding units throughout the entire incident. They need to confirm that the dispatched units received the alarm, record the en route and on-scene arrival times of the units, receive updates from the units on scene, and provide additional and updated information to the units.

Generally, the IC uses the radio to communicate with a dispatcher operating a radio in the communications center. The communications center closely monitors the radio and provides needed support for the incident. Communications from the IC to the communications center include the following:

- Progress and incident status reports
- Requests for additional units or the release of extra units when they are no longer needed
- Requests for outside resources, such as a utility company's vacuum truck or equipment from a local construction company
- Notifications to municipal officials and on-call personnel
- Requests for information such as building plans

Each part of the public safety network—fire, EMS, and police—needs to know what other agencies are doing at an incident. The communications center coordinates the fire department's activities and requirements with other agencies and resources. For example, the dispatcher might need to notify a non-emergency resource such as a utility company about the incident and request that they take certain actions such as shutting off a gas line.

TIP

The communications center should have accurate, current telephone numbers and contact information for every relevant agency.

Status Tracking and Deployment Management

The communications center must always know the location and status of every fire department unit. Units should never get "lost" in the system. As conditions change at an incident, units may need to be reassigned or additional units may need to be requested from outside the normal response area or from other districts.

Tracking the status of units can be difficult if the dispatcher must rely on radio reports and colored magnets or tags on a map or status board. CAD systems that use GPS data make this job much easier because as status changes are entered, the map and status descriptions are automatically updated.

Communications centers continually monitor the availability of units in each geographic area and redeploy units to balance coverage, when necessary, even when no major incidents are in progress. This might mean requesting coverage from surrounding jurisdictions. Many fire departments list both unit relocations and multiple alarm units—that is, units to be dispatched if an incident requires additional responders—on the run cards. Usually, the supervisor in a communications center is responsible for determining when and where to redeploy units as well as requesting coverage from surrounding jurisdictions. Sometimes companies from the same department are relocated to another station; other times, units from a different jurisdiction may be redeployed to respond for coverage if a large-scale incident depletes local resources. This redeployment plan is often described in a formal memorandum of understanding (MOU) under regional or statewide mutual aid plans. These plans must include a system for tracking every unit and a designated communications center for maintaining contact with all units.

Taking Calls at the Fire Station

One of the first things you should learn when you are assigned to a fire station is how to use the telephone, intercom, and radio systems. You must be able to use them to answer a call or to initiate a response. When answering a call at the fire station, similar to what a dispatcher does, firefighters should solicit needed information from the caller, and then, if needed, initiate a response to an emergency by contacting the communications center or transferring the call to the communications center.

A firefighter who answers the telephone in a fire station, fire department facility, or communications center is a representative of the fire department. Use your department's standard greeting when you answer the phone: "Good afternoon, Pleasant Town Fire Department, Firefighter Smith speaking. How may I help you?" Be prompt, polite, professional, and concise. Based on the caller's response, decide whether it is an emergency or a non-emergency call. Keep non-emergency calls as short as possible so incoming phone lines are available to receive emergency calls.

An emergency call can come in on any fire department phone line, for example, if someone calls a fire station directly instead of dialing 911. If this happens, the firefighter who answers the call is responsible for ensuring that the caller receives the appropriate emergency assistance. Detailed SOPs for obtaining information and processing calls should be provided to the firefighter assigned to answer incoming emergency telephone lines. The department's SOPs should outline exactly the steps to take, such as whether the firefighter who answered the call should take the information from the caller or connect the caller directly to the communications center. Even though firefighters are not specifically trained in **telephone interrogation** (the phase in a 911 call during which the telecommunicator asks questions to obtain vital information such as the location of the emergency), the SOPs and understanding the nature of the information needed to dispatch the appropriate response should guide the firefighter as they communicate with the caller.

Follow the steps in **SKILL DRILL 4-1** to receive a call, obtain the essential information from a caller, and initiate a response to an emergency.

New firefighters should tour the communications center for their department. Observing the operation of the center provides a much better understanding of the role of the dispatcher.

SKILL DRILL 4-1

Receiving a Call at a Fire Station and Initiating a Response to an Emergency Firefighter I, NFPA 1010: 6.2.1, and 6.2.2

1. Answer the phone promptly and professionally. Identify yourself, your agency, and your company. Determine immediately whether the caller is reporting an emergency, and if they are, follow your department's SOPs. Obtain the following information from the caller:
   - Incident location (including cross streets and identifying landmarks)
   - Type of incident or situation
   - Safety concerns at the scene
   - When the incident occurred
   - Caller's name
   - Location of the caller if it is different from the incident location
   - Caller's callback number

   Always terminate the call in a courteous manner, and let the caller hang up first.

2. Record the date and time of the call and the responses to your questions. Initiate a response following the department SOPs. The SOPs in your department may vary from the steps listed here. Follow the SOPs of the authority having jurisdiction (AHJ) for your department's communications.

Radio Systems

Fire department communications systems rely on two-way radios. As discussed earlier in the chapter, radios are one of the ways a communications center transmits dispatch information to fire stations. Radios are also used in many other situations including:

- Communication between the communications center and individual units
- Communication between units at an incident scene
- To page firefighters
- To link mobile data terminals

A radio system is an integral component of the ICS because it links the IC or unified command with all the units at an incident, up and down the chain of command. A radio may be a firefighter's only means to call for help in a dangerous situation.

Fire departments use many types of radios and radio systems, and technological advances are rapidly adding new features and system configurations.

Although it would be impossible to describe all of the possible features and principles of operation of every radio system here, this section describes some common systems and operating features. Be sure you learn how to operate the radio assigned to you and to learn your department's radio SOPs.

Radio Equipment

The fire service primarily uses three types of radios: portable radios, mobile radios, and a base station.

A **portable radio** is a hand-held, two-way radio that is small enough for a firefighter to always carry (**FIGURE 4-10**). The body of a portable radio contains a speaker and microphone, on/off switch or knob, volume control, channel select switch, and a push-to-talk (PTT) button. At least one—if not every—member of a team carries a portable radio during an emergency incident.

Portable radios have an antenna to receive and transmit signals. A popular optional attachment is an extension microphone/speaker unit that can be clipped to a collar or shoulder strap while the radio remains in a pocket or pouch.

Portable radios are usually powered by a rechargeable battery. Battery-operated portable radios also have limited transmitting power, usually 1 to 5 watts (w). This means the signal can be transmitted only within a certain range and is easily blocked or overpowered by a stronger signal.

A **mobile radio** is permanently mounted in the vehicle and powered by the electrical system in the vehicle. They are more powerful than portable radios, usually 25 to 50 w, so they have a greater range than portable radios (**FIGURE 4-11**). In most departments, every fire department vehicle has a mobile radio.

Mobile radios usually have a fixed speaker and an attached, hand-held microphone on a coiled cord. The PTT button is on the microphone, and the antenna is usually mounted on the exterior of the vehicle. Fire apparatus often include protective headsets with a combined intercom/radio system that enable crew members to talk to each other in the apparatus and hear any announcements over the radio.

A **base station** is a radio permanently mounted in a building, such as a fire station, communications center, or a remote transmitter site. Base station radios are more powerful than either portable or mobile radios, usually 50 to 100 w. The antenna of the base station is often mounted on a radio tower, so communications

FIGURE 4-10 A portable radio should be carried by each individual firefighter.

FIGURE 4-11 A mobile radio is permanently mounted in a vehicle.

FIGURE 4-12 The antenna for a base station is often mounted on a radio tower to provide maximum coverage.

FIGURE 4-13 SCBA with a voice amplifier.

Courtesy of 3M.

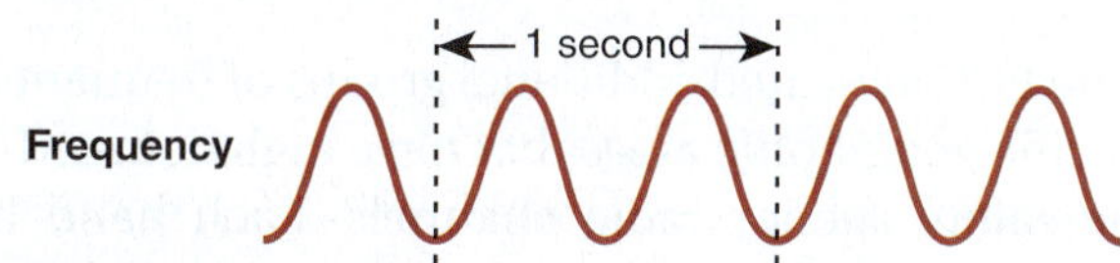

FIGURE 4-14 The frequency of a radio wave is the number of times the wave oscillates per second.

Courtesy of NASA.

are transmitted over the widest possible coverage area (**FIGURE 4-12**). Public safety radio systems for a large geographic area often have additional transmitters installed at different locations to extend the range of the transmission.

MDTs transmit data using both the radio system and cellular signals. Firefighters can press buttons on the MDT to communicate *en route* or *on-scene* to dispatch. MDTs also have GPS technology so satellites can track the location of the apparatus responding to and arriving at the incident scene. MDTs also allow firefighters to verify the location of an incident. For example, instead of having to radio the dispatcher to ask whether they said 11345 Main Street or 11354 Main Street, firefighters can look at the MDT where the address is displayed. Some departments use MDTs to track unit status, transmit dispatch messages, and exchange many other types of information.

To enable firefighters to communicate more effectively while wearing self-contained breathing apparatus (SCBA), some manufacturers provide accessories for SCBA face pieces that amplify a firefighter's voice for face-to-face communication. Another accessory connects the SCBA face mask to a firefighter's portable radio via Bluetooth (**FIGURE 4-13**). These devices enable firefighters to hear messages more clearly while wearing SCBA and to transmit information over their portable radios. If your department uses this type of SCBA, learn how to use it effectively.

Radio Operation

In the United States, the FCC has established strict limitations governing the assignment of radio channels (frequencies) to ensure that all users have adequate access. Every radio system must be licensed and operated within these guidelines. To do this, the FCC licenses an agency to operate on one or more specific frequencies called bands. The most commonly used bands for public safety communications are the **very high-frequency (VHF) band** and the **ultrahigh-frequency (UHF) band**. The frequencies available to the U.S. fire service are in several different ranges, including 33 to 46 megahertz (MHz; VHF low band), 150 to 174 MHz (VHF high band), 450 to 460 MHz (UHF band), 700 MHz, and 800 to 900 MHz (**FIGURE 4-14**).

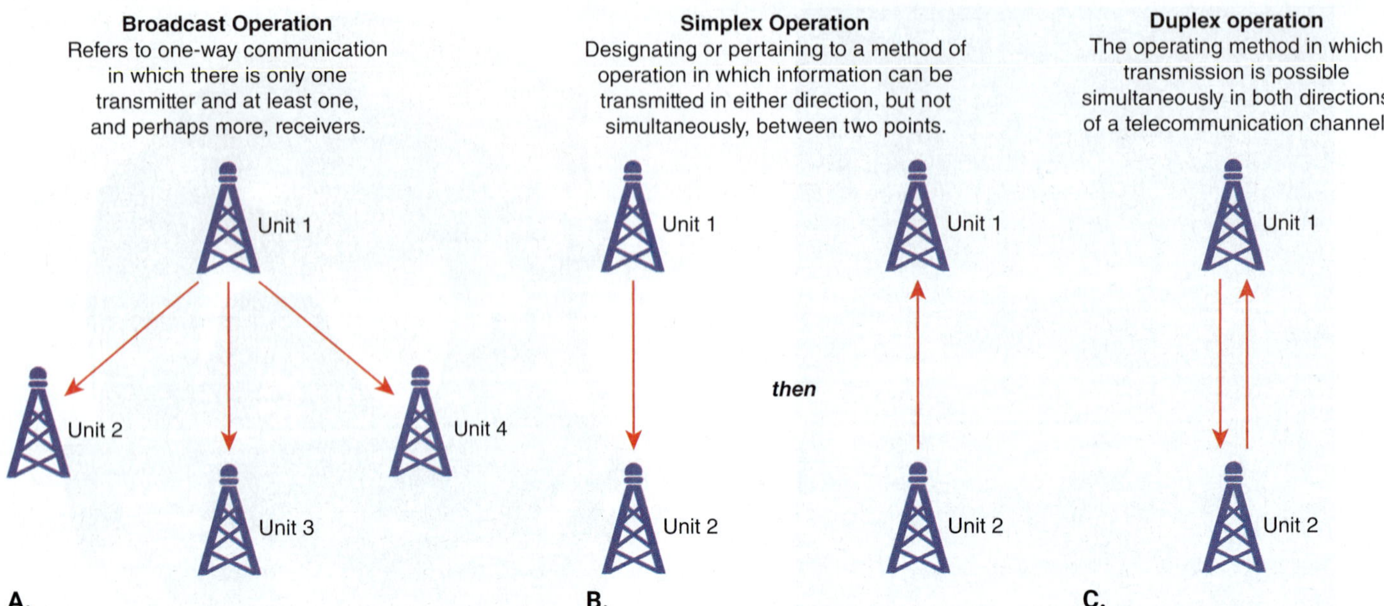

FIGURE 4-15 **A.** On a broadcast channel, transmission goes in only one direction. **B.** On a simplex channel, transmission goes in both directions, but only one party can transmit at a time. **C.** On a duplex channel, transmission goes in both directions and both parties can transmit at the same time.

The FCC allocated additional groups of frequencies in specific geographic areas that have high demand for public safety agency radio channels. Each band has certain advantages and disadvantages relating to geographic coverage, topography (hills and valleys), and penetration into structures. Some bands have many different users, which can cause interference problems, particularly in densely populated metropolitan areas.

Generally, a radio can be programmed by a qualified technician to operate on several frequencies in a particular band. Communication may be affected if neighboring police, fire, and EMS systems within the same jurisdiction operate on different bands. When different agencies working at the same incident communicate via different bands, they must make complicated arrangements so that they can communicate with one another. Being able to communicate across different radio bands and between different agencies is called **interoperability**. Public safety agencies rely on three main types of radio systems to overcome interoperability barriers:

- Simplex or line-of-sight systems
- Conventional radio repeater systems
- Trunked radio repeater systems

A **simplex communication system**, sometimes referred to as a **line-of-sight system**, allows communication to flow in only one direction at a time (**FIGURE 4-15**). A drawback of a simplex or line-of-sight system is that it has limited range, providing radio coverage to a small area, such as one city block. This limited coverage is not adequate for dispatching units or communicating between the communications center and apparatus assigned to a run. It is used by firefighters when they need their signals to penetrate a specific building or when operational security is critical. Some fire departments use a simplex communication system for on-scene communications. This configuration is sometimes called a **talk-around channel** because the radios transmit and receive on the same frequency and the signal is not sent over a **repeater channel**, which is a channel that transmits to a repeater. A **repeater** is a combination of a radio receiver and a transmitter that receives a signal and retransmits—repeats—it to a wider geographic location. A talk-around channel often works well for short-distance communications, such as from the location of the IC to crews inside a house fire or from one unit to another inside a building. Conversations on a talk-around channel are not routinely monitored by the communications center. If the IC wants to maintain contact with the communications center, they must use a radio channel designated to transmit to a repeater.

Radio messages can only be transmitted over a limited distance for two reasons. First, the signal weakens as it travels farther from the source and eventually

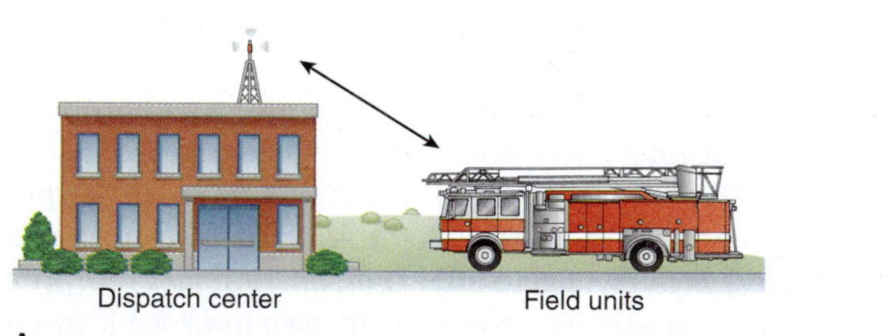

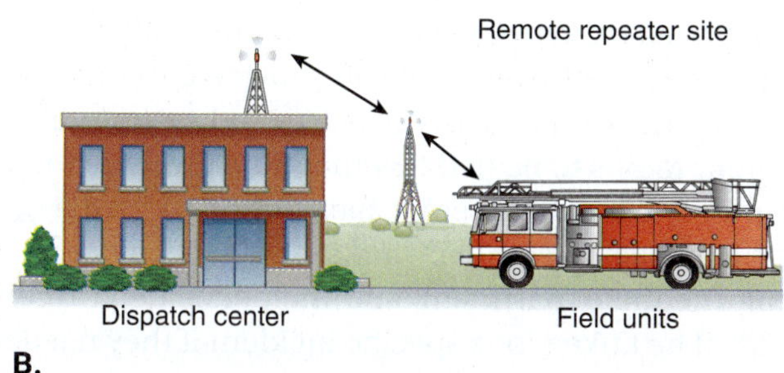

FIGURE 4-16 **A.** A transmission from a portable or mobile radio may not be strong enough to travel to everyone it needs to reach. **B.** A repeater intercepts the transmissions and retransmits them with more power, so the message travels to a wider geographic area.

becomes undetectable. Second, buildings, tunnels, and topography create interference and block the signal from traveling. These problems are more significant for hand-held portable radios, which have limited transmitting power. Mobile radios that operate in systems that cover large geographic areas face similar problems. To compensate for these shortcomings, fire departments use radio repeater systems, which use a repeater installed in a location that has line of sight to a large geographic area, greatly extending the operating range of radios. This location could be the top of a building or a hill. A radio repeater system can be digital or analog, and it typically can receive and send communications over multiple channels. A drawback to repeater systems is that they are inefficient when used by large numbers of public safety personnel.

A repeater has a large antenna that can receive lower-power signals, such as those from a portable radio. That signal is then retransmitted to all the radios set on the designated channel with all the power of a base station. Communications systems that use repeaters usually have outstanding systemwide communications and can get the best signal from portable and mobile radios. This approach enables the transmission to reach a wider coverage area (**FIGURE 4-16**).

Many public safety repeater systems have multiple receivers called voting receivers, which are geographically distributed over the service area to capture weak signals. A **voting receiver** receives radio signals and forwards the strongest signal to the repeater. If the original signal is strong enough to reach one of the voting receivers, the system is effective. If the signal does not reach a repeater, no one will receive it.

A **mobile repeater** boosts signals at an incident scene by capturing the weak signal from portable radios and rebroadcasting them. It can be permanently mounted in a vehicle or set up near the IC. Some fire departments use a **cross-band repeater**, which boosts the power of a weak signal and then retransmits it over a different radio band. For example, a cross-band repeater allows a firefighter using a UHF portable radio inside a building to communicate with an IC who is outside on a VHF radio. Similar systems are often used in large buildings or underground structures so crews working inside can communicate with units on the outside or with the communications center.

A **trunked radio system** is a repeater system that operates on the same principles as a conventional radio repeater system, except that a computer connected to a control channel sets the operating frequencies. When users press the PTT button, the radio transmits a "key" that signals to the control channel that a message to the trunk group is being transmitted. The system automatically directs that signal to the channel assigned to the trunk group. Trunked radio systems support many more users than conventional repeater systems.

A **digital radio** allows the transmission of digital signals to transmit data or analog signals to transmit voices that have been digitized and compressed by a computer. Using digital technology is more efficient than using analog technology, and it also enables more users to communicate at the same time in a limited portion of the radio spectrum. When digital radios are used in a trunked radio system, instead of the trunk group being assigned to only one or two frequencies, many frequencies are assigned to a group. These can be thought of as virtual channels that appear as conversations occur and disappear when they are completed. As a radio conversation in a trunk group begins, the computer scans for the next open frequency and directs all the radios in the trunk group to receive the message on that frequency. As the conversation continues, the frequency may change automatically because the computer constantly monitors the frequency load and reassigns the transmissions to unused frequencies.

Trunked systems make the use of available frequencies more efficient and can provide coverage over multiple trunk groups. Because trunked systems allow different radios to be tied together for a given incident, it is possible to link different agencies on the same system. For instance, public works, the highway department, the fire department, and the police department could all be linked for a specific incident if they needed to share radio communications.

A **multiplex channel** combines analog and digital signals so they can simultaneously transmit two or more different types of information, such as voice and data, in both directions over the same frequency.

FirstNet

Cell phones are used by emergency responders to assist them in daily operations. The First Responder Network Authority (FirstNet Authority), part of the U.S. Department of Commerce, is a nationwide broadband network for communication among first responders and communications centers to provide a national coordinated response to emergencies. Officially launched in 2018, the mission of the FirstNet Authority is "to ensure the building, deployment, and operation of the nationwide broadband network that equips first responders to save lives and protect U.S. communities" (FirstNet Authority n.d.). That network is called FirstNet. FirstNet was designed to work with NG911 so that responders can get the same information that is sent to the communications center. The FirstNet dedicated network allows first responders to have priority cell service.

Using a Radio

As a firefighter, you must learn how to operate a radio assigned to you as well as how to use the radio systems used by your department. You must know when to identify the channel you are using, such as "Channel 3," "Tac 5," or "Charlie 2"; which buttons to press; and which knobs to turn. Study the training materials and SOPs provided by your department to learn how to properly use and maintain the radio and don't be afraid to ask questions if there is something you don't understand.

When a radio is assigned to you, make sure you receive an explanation on how to operate and maintain it. Learn the protocol for recharging the battery or replacing the batteries if the device is not rechargeable. Whether the radio assigned to you has a rechargeable battery or one that needs to be replaced, check it at the beginning of each shift to make sure you have full power. Because the battery has a limited capacity, it must be recharged or replaced after extended operations, so check it again when you return to the station after a run. On the fireground, your radio is your link to the outside world. It must work properly—your life and the lives of your fellow firefighters depend on it.

If you hold a portable radio right-side-up, perpendicular to the ground with the antenna pointing toward the sky, you will get better transmission and reception. Range and transmission quality also improve if you remove the radio from the radio pocket or belt clip before you use it.

All radio transmissions should be pertinent to the situation. Avoid unnecessary radio traffic. Keep in mind that most radio communications are automatically recorded. These recordings provide a complete record of everything that is said and can be admissible as evidence in a legal case.

TIP

When using a radio, always maintain a professional demeanor. In addition to the fact that radio communications are recorded, remember that anyone in your department and in the communications center, as well as anyone with a radio scanner, including the news media and the public, can hear everything you say. Don't say anything of a sensitive nature or anything you might regret later.

A common way to communicate by radio is the "Hey you, it's me" method. This method helps ensure that radio traffic reaches the appropriate person and is properly understood. In the following example of the "Hey you, it's me" method, E1 stands for the crew leader of Engine 1:

IC: Engine 1 (Hey you), Elm Street Command (it's me).

E1: Go ahead Elm Street Command (you), this is Engine 1 (me).

IC: Engine 1, I want you to exit the building and report upon exiting.

E1: Engine 1 copies. We will exit the building and report upon exiting.

In this example, the IC was clear about which unit they were trying to reach. That crew leader then acknowledged that they heard the communication. Once the IC gave their orders, the unit crew leader repeated the orders to confirm that they accurately understood the command.

The National Incident Management System (NIMS) and the National Fire Protection Agency (NFPA) recommend that fire departments use plain English for radio communications as in the previous example. Using this method helps reduce miscommunications. However, some departments use codes for standard messages. A few departments have developed intricate systems of numeric codes to fit any situation. **Ten-codes**, a system of coded messages that begin with the number 10, were once widely used, and are still commonly used by law enforcement. Ten-codes, however, are not standardized across the country, with a few exceptions—for example, almost everyone knows that "10-4" means acknowledgment, or OK. To be effective, codes must be understood by all parties and mean the same thing to all users. Ten-code systems are problematic when units from different jurisdictions need to communicate with one another or when multiple agencies need to communicate at a large emergency incident.

If your department uses radio codes, you must learn all the codes and understand how and when they are used. If your department uses plain English, be aware that even plain English may include coded terminology, such as "code blue" for a cardiac arrest or "MVC" for motor vehicle collision. The objective is to be clearly and easily understood, without having to explain every message in detail. If your department has a few codes or special words for special situations, they should be specified in the SOPs.

To use a radio, follow the steps in **SKILL DRILL 4-2**.

Incident Command System and Fire Department Communications

According to the National Institute for Occupational Safety and Health (NIOSH), firefighter injuries and deaths are often caused by poor communications (Ludwig 2019). Effective communication is essential to reduce complications at incident scenes and training drills and to be able to successfully complete an incident operation. The ICS, which should be utilized at every incident, helps ICs, fire officers, and firefighters communicate more effectively. **Unity of command**—the concept that each person reports to only one direct supervisor, each supervisor answers to only one boss, and so on—minimizes confusion by ensuring that each individual at an incident reports to only one supervisor so that important messages are communicated to the correct person. Unity of command also reduces occurrences of freelancing. Yonkers Fire Department Chief Chris Kiernan use to say that a good command structure created an environment of "controlled chaos." One of the benefits of the ICS is that it allows ICs to create sections to manage different aspects of the incident, and within those sections, smaller groups can be created. As early as possible, the IC should assign a communications unit leader.

During an incident, the IC or the communications unit leader should identify and assign specific radio channels to each company, crew, or unit as the incident determines. When multiple agencies, especially multiple agencies from different jurisdictions, respond to the same incident, it is essential that a means of interoperable communication is established as soon as possible. Some agencies have radio systems that allow for interoperability. Other agencies have adopted VoIP communication systems that allow transmission between radios and cell phones by adding an accessory device to portable radios (**FIGURE 4-17**).

In disaster situations, the larger cellular network providers can deploy 5G-capable, portable cell signal towers called cell on wheels (COWS). One major cellular network provider has flying COWS (cell on wings), which are 4G- and 5G-capable drones.

Size-Up and Progress Reports

A **size-up** is the rapid evaluation and analysis of an incident by an officer or the IC to determine which resources need to be deployed and which actions can be undertaken safely. The first-arriving unit at an incident—whether a firefighter, a company officer, or a chief—needs to give a brief initial size-up report and establish command. This initial report should convey a preliminary assessment of the situation and give other units a sense of what is happening so that they can anticipate what their assignments may be when they arrive at the scene. The communications center should repeat the initial size-up information. For example, on arrival at the scene, an officer, or an engine company firefighter if there is no officer currently on scene, might make the following report:

> Communications, this is Engine 410. Engine 410 is on location at the intersection of High Street and Chambersburg Road. We have an electric vehicle fully involved in fire. Engine 410 is assuming High Street Command. This will be an offensive operation.

The communications center would respond by announcing:

> Engine 410, I have you on the scene assuming High Street Command. Electric vehicle fire fully involved. Offensive operation.

SKILL DRILL 4-2

Using a Radio Firefighter I, NFPA 1010: 6.2.2

1. If you are holding the radio, position it right-side-up, perpendicular to the ground with the antenna pointing toward the sky. If the radio is in your pocket or a pouch, try to position it in the same manner.

2. Listen to determine that the channel is clear of any other traffic. Press and hold the PTT button, then wait at least 2 seconds before speaking to allow the system to capture the channel without cutting off the first part of the message. Some systems sound a distinctive tone when the channel is ready after the PTT button is pressed.

3. Hold the microphone 1 to 2 inches (2.5 to 5 centimeters) from your mouth, then speak across the microphone at a 45-degree angle. Do not have something in your mouth when speaking on the radio. Know what you are going to say before you start talking. Speak clearly and keep the message brief and to the point.

4. Release the PTT button only after you have finished speaking so the message does not get cut off.

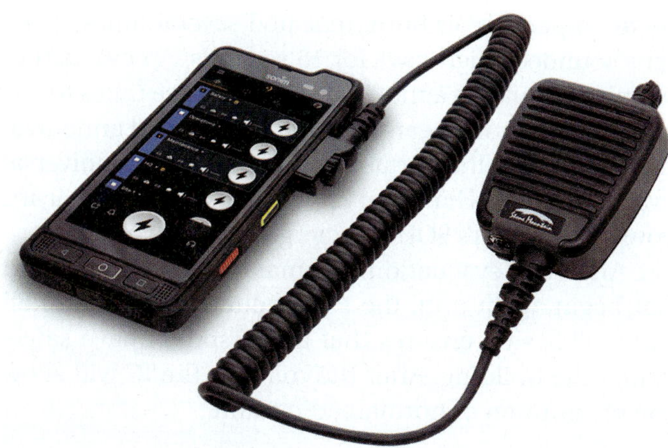

FIGURE 4-17 PTT accessories for cell phones allow transmission between cell phones and radios.
Courtesy of Instant Connect Software LLC.

SAFETY TIP

Verifying the information received over the radio is vital. Always restate an important message or instruction to confirm that it has been received and understood.

As the incident progresses, and for the duration of an incident, the IC should provide regular progress reports or updates to the communications center. The communications center should prompt the IC to report at set intervals, known as **time marks**. Time marking allows the IC to assess the progress of the incident and decide whether the strategy or tactics should be changed. All progress reports are noted in the activity log to document the incident.

If additional resources, such as apparatus, equipment, or personnel, are needed, the IC transmits these requests to the communications center. The communications center will advise the IC of additional resources that are dispatched.

Size-up and how the IC should handle progress reports are described in Chapter 22, *Establishing and Transferring Command.*

Emergency Messages

Sometimes the dispatcher must interrupt normal radio transmissions for emergency traffic. **Emergency traffic** is an urgent message that takes priority over all other communications. When a unit needs to transmit emergency traffic, the dispatcher generates a distinctive alert tone to notify everyone on the frequency to stand by so that the channel is available for the emergency communication. Once the emergency message is complete, the dispatcher notifies all units to resume normal radio traffic.

To alert the communications center that emergency traffic needs to be transmitted, some radio systems have emergency buttons located on portable and mobile units that a firefighter in trouble can push to transmit an emergency signal to the dispatcher. This button alerts the communications center that the ID assigned to that radio needs to transmit emergency traffic. Some radio systems can also identify the location of the unit in trouble. If your radios are equipped with this feature, learn the procedure for using it.

Mayday

The most important emergency traffic is a firefighter's call for help. Most departments use the word **mayday** to indicate that a firefighter is in imminent danger and requires immediate assistance, is lost, or is missing. If a mayday call is transmitted, all other radio traffic should stop immediately. The firefighter making the mayday call must describe the situation, identify their location, and describe the help needed.

Some agencies use the acronym LUNAR to report a mayday, which stands for the following:

- **Location**, your location in the building/incident
- **Unit**, the unit to which you are assigned
- **Name**, who you are
- **Air**, the amount of air you have in your cylinder, or **Assignment**, where you were last assigned
- **Resources**, what you need to get you out of the mayday situation.

In other agencies, the person reporting the mayday reports **who** (they are), **what** (the situation is), and **where** (they are located) instead of providing a LUNAR report. An example of a mayday call is:

Firefighter: MAYDAY, MAYDAY, MAYDAY.

[All radio traffic stops.]

IC: Unit calling MAYDAY, go ahead.

Firefighter: This is Engine 403. We are on the second floor and running out of air. Fire has cut off our escape route. We request a ladder to the window on the Charlie side (rear of building) of the building so we can evacuate.

IC: Command copied. Engine 403, your escape route is cut off by fire. I am sending the rapid intervention crew to Charlie side (rear of building) with a ladder.

The procedure for initiating and responding to a mayday call should be studied and practiced frequently. The steps for initiating a mayday are explained more fully in Chapter 19, *Firefighter Survival*.

Evacuation Message

Another emergency traffic message is "evacuate the building," which directs all units inside a structure to abandon the building immediately. Most fire departments have specified a standard **evacuation signal** to warn all personnel to pull back to a safe location. The evacuation signal that is commonly used is a sequence of three blasts on an apparatus air horn, repeated several times, or sirens sounded "high-low" for 15 seconds. An evacuation warning should be announced at least three times to ensure that everyone hears it; the IC also should announce the warning on the portable radio. Because no universal evacuation signal has been established, you need to learn your department's SOP for emergency evacuation.

After the evacuation, normal radio traffic should not begin again until the IC conducts a roll call of all units. This step ensures that all personnel have safely exited the building. After the roll call, the IC will allow the resumption of normal radio traffic.

CASE STUDY

You Are the Firefighter CONCLUSION

At 0304 hours, you are dispatched to a report of a fire at 3256 West Madison. While en route, the dispatcher advises that the caller reported that smoke was coming from the eaves of the building and then hung up. With this new information you are preparing yourself for a possible house fire. The dispatcher radios back and says that this may be a false report because the caller used a cell phone that was identified as being on the east side of town. Thinking you had been woken up for another false call, the dispatcher comes back on the radio with a frantic sound in her voice and says that a home security company just reported a fire alarm at 3256 *East* Madison Street.

1. **What is the process for receiving 911 calls in an emergency communications center?**

   Answer: Upon receiving a call, the dispatcher determines the type of emergency that exists; the location of the emergency; and the type of resources required, such as police, fire, and ambulance. The dispatcher then assigns the call to all needed resources using the jurisdiction's SOPs. Once the units are dispatched, the dispatcher continues to interview the caller for additional information. If the dispatcher is trained as an EMD, they provide the caller with instructions for emergency medical care over the phone.

2. **How does the dispatcher know where the cell phone call originated?**

   Answer: The best way for a dispatcher to know the location of a cell phone caller is to have the caller tell them the address directly. Current technology does not provide an exact location of a caller from a cell phone, although because many cell phones have GPS, cell phone carriers may be able to provide the latitude and longitude to the communications center. Otherwise, the approximate location of the caller is provided by the cellular carrier by using cell tower triangulation.

3. **How are alarm systems monitored and then reported to the communications center?**

   Answer: When heat, fire, or smoke is detected by an alarm system, the communications center needs to be notified. Notification may occur via a direct call from someone at the location; a transmitter at the location dialing the communications center and playing a prerecorded message; or a central monitoring station receiving notification of the alarm causing monitoring personnel to contact the communications center.

WRAP-UP

SUMMARY

KNOWLEDGE OBJECTIVES

- Define the role of a public safety communications center, the responsibilities of the dispatcher, and the equipment used in a communications center.
  - Describe the role of the communications center in the fire service. (pp. 134–135)
  - Describe the role and responsibilities of a dispatcher. (pp. 135–136)
  - List the requirements of a communications center. (p. 136)
  - Describe the equipment used in a communications center. (pp. 136–138)
  - Describe how computer-aided dispatch assists in dispatching the correct resources to an emergency incident. (pp. 136–137)
- Describe the steps the dispatcher takes to process a reported emergency incident and how computer-aided dispatch assists in dispatching the correct resources to an emergency incident. (pp. 138–152)
- Explain the procedures for handling emergency and non-emergency calls at a fire station and the procedures for transmitting the information to fire service responders.
  - Explain the importance of following department SOPs for receiving and processing communications. (**NFPA 1010: 6.2.1**, p. 138)
  - Describe the procedures for handling emergency calls. (**NFPA 1010: 6.2.1**, pp. 138–146)
  - Identify the information to be obtained when taking a report of an emergency to enable necessary assistance to be dispatched. (**NFPA 1010: 6.2.1**, pp. 139–140)
  - Explain the methods of receiving emergency and non-emergency fire department communications. (**NFPA 1010: 6.2.1**, pp. 139–142)
  - Describe how dispatchers conduct telephone interviews. (pp. 139–140)
  - Describe how location validation systems operate. (pp. 142–144)
  - Determine if a communication is emergent or non-emergent. (**NFPA 1010: 6.2.1**, p. 144)
  - Explain the procedures for transmitting the emergency information to a dispatch center. (**NFPA 1010: 6.2.1**, p. 145)
- Identify the modes of fire service communication and the types of radios used in the fire service radios.
  - Identify other modes of fire service communication. (**NFPA 1010: 6.2.1**, pp. 140–141)
  - Describe the three types of fire service radios. (pp. 148–149)
- Describe how radios work and how a repeater and trunked systems enhance communications at an incident scene.
  - Describe how two-way radio systems operate. (pp. 148–152)
  - Explain how a repeater system works to enhance fire service communications. (pp. 148–152)
  - Explain how digital radios and trunked systems work to enhance fire service communications. (pp. 151–152)
- Describe how to use a radio at an incident scene and how to convey location and emergency information.
  - Describe the basic principles of effective radio communication. (**NFPA 1010: 6.2.3**, pp. 149–152)
  - Identify radio departmental procedures and codes for using fire department radios. (**NFPA 1010: 6.2.1** and **6.2.3**, pp. 152–154)
  - Describe when to use plain language and how ten-codes are implemented in fire service communications. (**NFPA 1010: 6.2.3**, pp. 152–153)
  - Outline the information provided in size-up and progress reports. (pp. 153–155)
  - Recognize routine traffic, emergency traffic, and emergency evacuation signals. (**NFPA 1010: 6.2.3**, pp. 155–156)

SKILLS OBJECTIVES

After studying this chapter, you will be able to perform the following skills:

- Receive a call at a fire station and initiate a response to an emergency.
  - Receive a phone call, and obtain, route, and document information according to department procedures. (**NFPA 1010: 6.2.2**, p. 147)
- Use a radio and communicate information properly and according to your department's standard operating procedures.
  - Send and receive messages over the fire department radio. (**NFPA 1010: 6.2.2, and 6.2.3**, p. 154)
  - Determine if a radio communication is routine or emergency traffic. (**NFPA 1010: 6.2.3**, p. 155)

KEY TERMS

activity logging system A device that keeps a detailed record of every incident and activity that occurs.

automatic location identification (ALI) The automatic display at the PSAP of the caller's telephone number, the address/location of the telephone, and supplementary emergency services information about the location from which a call originates.

automatic number identification (ANI) A series of alphanumeric characters that informs the recipient of the source of an event or power supply module. (NFPA 1225)

base station A stationary radio transceiver with an AC or DC power supply. (NFPA 1225)

call box A system of telephones connected by phone lines, radio equipment, or cellular technology to a communications center or fire department.

call classification and prioritization The process of assigning a response category based on the nature of the reported problem.

call arrival The process of receiving a call for service and obtaining the information necessary to initiate a response.

communications center See *public safety communications center.*

computer-aided dispatch (CAD) A combination of hardware and software that provides data entry, makes resource recommendations, and notifies and tracks those resources before, during, and after fire service alarms, preserving records of those alarms and status changes for later analysis. (NFPA 1225)

cross-band repeater A repeater that boosts a weak signal and then retransmits it over a different radio band.

digital radio A radio that transmits information via radio waves using digital data or analog (voice) signals that have been converted to a digital signal and compressed.

direct-line phone A phone that connects two predetermined points and does not require the user on either end to dial to cause the phone at the other end to ring. Also called *ring-down phone.*

dispatch To send resources to an address or incident location for a specific purpose. (NFPA 450)

dispatcher See *telecommunicator.*

emergency medical dispatcher (EMD) Personnel specifically trained and certified in interviewing techniques, pre-arrival instructions, and call prioritization. (NFPA 450)

emergency traffic An urgent message, such as a call for help or evacuation, transmitted over a radio that takes precedence over all normal radio traffic.

evacuation signal A distinctive signal intended to be recognized by the occupants as requiring evacuation of the building.

Federal Communications Commission (FCC) The federal regulatory authority that oversees radio communications in the United States.

fire alarm box A device connected via an underground cable to a municipal fire alarm system.

geographic information systems (GIS) A system of computer software, hardware, data, and personnel to describe information tied to a spatial location. (NFPA 450)

global positioning systems (GPS) A satellite-based radio navigation system composed of three segments: space, control, and user. (NFPA 1900)

interoperability The ability to communicate across different radio bands and between different agencies.

line-of-sight system See *simplex communication system.*

mayday A verbal declaration indicating that a firefighter is lost, missing, or trapped, and requires immediate assistance.

mobile data terminals (MDTs) Technology that allows firefighters to receive data while in the fire apparatus or at the station.

mobile radio A two-way radio that is permanently mounted in a fire apparatus.

mobile repeater A repeater used at an incident scene to boost signals at the scene by capturing the weak signals from portable radios and rebroadcasting them.

multiplex channel Simultaneous transmission of multiple data streams, most often voice signals, in either or both directions over the same frequency.

municipal fire alarm system A network of fire alarm boxes and emergency telephones on street corners or in public places

Next Generation 911 (NG911) A system designed to replicate traditional Enhanced 911 (E911) features and functions and provide additional capabilities to provide access to emergency services from all connected communications sources and provide multimedia data capabilities for public safety answering points (PSAPs) and other emergency service organizations. (NFPA 1225)

portable radio A battery-operated, hand-held transceiver. (NFPA 1225)

public safety answering point (PSAP) A facility equipped and staffed to receive emergency and non-emergency calls requesting public safety services via telephone and other communication devices. (NFPA 1225)

public safety communications center A building or portion of a building that is specifically configured for the primary purpose of providing emergency communications services or public safety answering point (PSAP) services to one or more public safety agencies under the authority or authorities having jurisdiction. Also called *communications center.* (NFPA 1225)

repeater A combination of a radio receiver and transmitter that receives radio signals, usually over multiple channels, and retransmits—repeats—them to a wider geographic location.

repeater channel A radio channel that transmits to a repeater.

ring-down phone See *direct-line phone.*

run The act of a unit travelling to an incident.

run cards Information prepared in advance and stored that describes a predetermined response to an emergency.

simplex communication system A radio system that allows communication to flow in only one direction at a time. Also called *line-of-sight system* or *talk-around channel.*

size-up The process of gathering and analyzing information to help fire officers make decisions regarding the deployment of resources and the implementation of tactics. (NFPA 1410)

stand-alone communications center A public safety communications center that serves and dispatches a single emergency response agency.

talk-around channel See *simplex communication system.*

telecommunicator An individual whose primary responsibility is to receive, process, or disseminate information of a public safety nature via telecommunication devices. Also called *dispatcher.* (NFPA 1225)

telephone interrogation The phase in a 911 call during which the telecommunicator asks questions to obtain vital information such as the location of the emergency.

ten-codes A system of predetermined coded messages, such as "What is your 10-20?" used by responders over the radio.

time marks Status updates provided to the communications center every 10 to 20 minutes. Such an update should include the type of operation, the progress of the incident, the anticipated actions, and the need for additional resources.

trunked radio system A repeater system that uses a computerized shared bank of frequencies to make the most efficient use of radio resources.

TTY/TDD systems User devices that allow speech- and/or hearing-impaired citizens to communicate over a telephone system. TTY stands for teletype, and TDD stands for telecommunications device for the deaf; the displayed text is the equivalent of a verbal conversation between two hearing persons.

ultrahigh-frequency (UHF) band Radio frequencies between 300 and 3000 MHz.

unit selection The process of determining exactly which unit or units should be dispatched after a call to 911 is received by a communications center, based on the location and classification of the incident.

unity of command The concept that each person reports to only one direct supervisor.

very high-frequency (VHF) band Radio frequencies between 30 and 300 MHz; the VHF spectrum is further divided into high and low bands.

Voice over Internet Protocol (VoIP) Technology that converts a person's voice into a digital signal that can be sent via the Internet to another device.

KEY TERMS CONTINUED

voice recording system A system that records communications over the phones and the radio.

voting receiver A device in a repeater system that receives signals from radio transmissions and then sends the strongest signal to the repeater.

walk-in A person who comes to a fire station seeking assistance rather than calling 911. Also called *door banger*.

REVIEW QUESTIONS

1. What does *dispatch* mean?
2. List 11 types of equipment that are in most communications centers.
3. Describe Enhanced 911.
4. What is VoIP?
5. What is an emergency medical dispatcher (EMD)?
6. What happens if the dispatcher does not receive confirmation that a unit is responding within a set time?
7. What are the three types of radios that fire service personnel primarily use?
8. In what type of system are radio signals transmitted between the sender and the receiver on the same frequency?
9. What is a base station that uses two separate frequencies—one to receive messages and one to transmit messages—to retransmit the message to a wider coverage area?

DISCUSSION QUESTIONS

1. Describe reasons that explain why the job of dispatcher is so stressful.
2. A walk-in arrives at your fire station and tells you that scaffolding collapsed two blocks from the station and there are many injuries. What actions should you take?
3. If you answer the phone at the fire station and the caller is reporting an incident, what information do you need to obtain from the caller?

APPLYING THE CONCEPTS

You are a new firefighter who has been working at your fire department for 3 months. Your company just returned from a call for a fire in a dumpster outside at a grocery store. You and your crew are doing preventive maintenance on the portable radios, headsets, and other communications equipment used during the call. As you are walking over to one of the engines to gather some of the equipment, the firehouse doorbell rings. You answer the door, and a man breathlessly says, "There's a fire down the street! People are trapped inside!" You look outside and can see heavy smoke in the distance.

1. Do you tell him to hold on while you go find a higher-ranking colleague?
2. Do you invite the passerby inside to sit down so you can gather as much information as possible, like his name, contact information, location of the incident, safety concerns at the scene, and a description of the structure?
3. The quickest way to contact the communications center is by using your hand-held radio. Is this your best choice?

Using the base station, you call the communications center, giving them the location of the incident and basic details. You then use the firehouse intercom to notify the officer in charge and announce the call over the firehouse's public address system. You thank the passerby and ask him to contact 911 to provide additional information. All available personnel arrive on the apparatus floor and prepare to leave. While en route, the officer contacts the communications center via the mobile radio and requests additional resources.

4. How does a mobile radio differ from a portable radio? Why is it a better choice in this situation?

5. What is a drawback of radio communication, and how does a repeater address this problem?

Your engine unit arrives on scene to find the house fully involved. As you are donning PPE, a woman runs over to you, frantically yelling, "My grandmother was in the window right up there on the second floor. But I don't see her anymore! You have to go in and get her . . . hurry!"

6. The safety of the grandmother is obviously in jeopardy. Do you rush into the building to try and find her?

7. So, what should you do?

REFERENCES

911.gov. 2023. "Frequently Asked Questions." Updated March 8, 2023. Accessed May 30, 2023. https://www.911.gov/calling-911/frequently-asked-questions.

ADA.gov. 2016. "Americans with Disabilities Act Title II Regulations." 2016. ADA.gov, October 11, 2016. Accessed March 12, 2023. https://www.ada.gov/law-and-regs/title-ii-2010-regulations/#-35161-telecommunications.

Dolcourt, Jessica. 2014. "Text to 9-1-1: What You Need to Know (FAQ)." CNET, May 5, 2014. Accessed May 27, 2023. https://www.cnet.com/news/text-to-911-what-you-need-to-know-faq.

Federal Communications Commission (FCC). 2015. "Best Practices for Implementing Text-to-911." Updated December 22, 2015. Accessed May 30, 2023. https://www.fcc.gov/general/best-practices-implementing-text-911.

Federal Communications Commission (FCC). 2024. "Enhanced 911 - Wireless Services." Accessed June 7, 2024. https://www.fcc.gov/general/enhanced-9-1-1-wireless-services.

Federal Communications Commission (FCC). 2020. "Text to 911: What You Need to Know." Updated January 6, 2020. Accessed June 29, 2023. https://www.fcc.gov/consumers/guides/what-you-need-know-about-text-911.

FirstNet Authority. n.d. Home page. Accessed March 12, 2023. https://www.firstnet.gov.

Hawkins, Dan. 2007. *Communications in the Incident Command System.* United States Department of Justice. March 2007. Accessed June 7, 2024. https://www.ojp.gov/ncjrs/virtual-library/abstracts/issue-brief-2-communications-incident-command-system.

Ludwig, Gary. 2019. "Examining the Role of Culture in Firefighter Deaths." FireRescue1, June 10, 2019. Accessed May 30, 2023. https://www.firerescue1.com/fire-chief/articles/examining-the-role-of-culture-in-firefighter-deaths-8ndTaFl986zrxm81.

National Emergency Number Association (NENA). 2021. "9-1-1 Statistics." February 2021. Accessed March 9, 2023. https://www.nena.org/page/911Statistics.

National Fire Protection Association (NFPA). 2019. *NFPA 1410: Standard on Training for Emergency Scene Operations.* 2020 Edition. Quincy, MA: NFPA.

National Fire Protection Association (NFPA). 2020. *NFPA 450: Emergency Medical Services and Systems.* 2021 Edition. Quincy, MA: NFPA.

National Fire Protection Association (NFPA). 2021. *NFPA 72®: National Fire Alarm and Signaling Code.* 2022 Edition. Quincy, MA: NFPA.

National Fire Protection Association (NFPA). 2021. NFPA 1225: *Emergency Services Communications.* 2022 Edition. Quincy, MA: NFPA.

National Fire Protection Association (NFPA). 2023. *NFPA 1900. Standard for Aircraft Rescue and Firefighting Vehicles, Automotive Fire Apparatus, Wildland Fire Apparatus, and Automotive Ambulances.* 2024 Edition. Quincy, MA: NFPA.

U. S. Department of Transportation (DOT). *NG911 & FirstNet.* Accessed May 28, 2023. https://www.npstc.org/download.jsp?tableId=37&column=217&id=3974&file=NG911_FirstNet_Guide_170820.pdf.

CHAPTER

5

Firefighter I

Fire Behavior

KNOWLEDGE OBJECTIVES

After studying this chapter, you will be able to:

- Describe the basic chemistry of fire, including the types of fuels, conditions, and reactions necessary for combustion, as well as how fire spreads, transfers heat, and releases by-products.
- Identify the five classes of fire and how each type can be extinguished.
- Describe fire behavior, including the stages of fire; how to recognize changing fire conditions; and how factors such as building construction, building contents, flow paths, and wind impact fire growth.
- Describe the characteristics and basic chemical properties of liquid- and gas-fuel fires.
- Describe the risks and safety considerations associated with fires involving batteries and stored energy systems.
- Describe how to read smoke.

SKILLS OBJECTIVES

There are no skills objectives for Firefighter I candidates. NFPA 1010 contains no Firefighter I Job Performance Requirements for this chapter.

ADDITIONAL NFPA STANDARDS

- **NFPA 1**, *Fire Code, 2024 Edition*
- **NFPA 10**, *Standard for Portable Fire Extinguishers, 2022 Edition*
- **NFPA 24**, *Standard for the Installation of Private Fire Service Mains and Their Appurtenances, 2022 Edition*
- **NFPA 53**, *Recommended Practice on Materials, Equipment, and Systems Used in Oxygen-Enriched Atmospheres, 2021 Edition*
- **NFPA 67**, *Guide on Explosion Protection for Gaseous Mixtures in Pipe Systems, 2019 Edition*
- **NFPA 99**, *Health Care Facilities Code, 2024 Edition*
- **NFPA 115**, *Standard for Laser Fire Protection, 2020 Edition*
- **NFPA 268**, *Standard Test Method for Determining Ignitibility of Exterior Wall Assemblies Using a Radiant Heat Energy Source, 2022 Edition*
- **NFPA 291**, *Recommended Practice for Water Flow Testing and Marking of Hydrants, 2022 Edition*
- **NFPA 402**, *Guide for Aircraft Rescue and Fire-Fighting Operations, 2019 Edition*
- **NFPA 1403**, *Standard for Live Fire Training, 2018 Edition*
- **NFPA 1410**, *Standard on Training for Emergency Scene Operations, 2020 Edition*
- **NFPA 1700**, *Guide for Structural Fire Fighting, 2021 Edition*

CASE STUDY

You Are the Firefighter

It is 0346 and you arrive at a modern-looking two-story house for the report of a garage fire called in by a neighbor. Through the smoke and flames, you observe what appears to be a popular model of an electric vehicle parked inside the garage and involved in the fire. As you begin to apply water to the body of fire, you witness fire conditions growing rapidly and extending into the house. The hair on the back of your neck rises as you get an unsettled feeling.

1. Why would this situation make you uneasy?
2. How might the structure's construction and contents contribute to fire behavior?
3. What elements of this scene may make extinguishment of the fire challenging?

Introduction

Since prehistoric times, when humans first discovered how to use fire for cooking food and keeping warm, fire has fueled our lives. Although we have progressed from using open fires for cooking and for warmth, we continue to depend on fires to provide the energy for modern society. Burning fossil fuels creates most of our electric power. Our gasoline- and diesel-powered vehicles depend on small explosions (internal combustion) to propel them. Open flames in well-designed furnaces heat most of our homes.

Unfortunately, destruction of lives and property by uncontrolled fires also has been occurring since ancient times. Despite our advanced technology for detecting and combating fires, we continue to wage an ongoing battle with fire. In the United States, the fire death rate steadily increased between 2012 and 2021 (USFA 2023). According to the National Fire Protection Association (NFPA), in the United States in 2021, there were 3800 civilian fire deaths and 14,700 civilian fire injuries. Fire departments responded to 1,353,500 fires. These fires resulted in a property loss of $15.9 billion. Seventy-nine percent of the civilian fire deaths and 86 percent of the civilian fire injuries were caused by home structure fires (Hall and Evarts 2021, 1–2). Understanding how fires start, the factors that cause them to grow, the characteristics of different types of fires, and the information provided by smoke is a good start to learning how to prevent unintended fires and how to extinguish them safely when they occur.

What Is Fire?

A well-trained and experienced firefighter can put out more fire with less water if they understand the chemistry of fire. **Fire** is the visible result of combustion. **Combustion** is a rapid chemical chain reaction between a material or a substance and oxygen that produces heat and usually light. For firefighters' purposes, the terms *combustion* and *fire* can be used interchangeably. Materials and substances that are capable of catching fire are **combustible**. Fire is characterized by the production of a flame, which can be many different colors, depending on what is burning and how much heat is produced. The location where a fire starts is the **area of origin**. This is sometimes called the fire compartment when the fire is in a structure, the **fire seat**, or the **seat of the fire** in casual conversation.

Matter

Matter is anything that has mass and volume—in simple terms, it weighs something and it takes up space. Matter is made up of elements; an **element** is a substance that cannot be chemically broken down into simpler substances. Elements are often referred to by their symbol. For example, oxygen is an element and its symbol is O. Hydrogen is another element, and its symbol is H. Elements are made up of atoms; an **atom** is the smallest unit of an element that retains the properties of the element. Some atoms, such as oxygen, exist in pairs because they are more stable that way. Therefore the symbol for stable oxygen is O_2, which essentially means two oxygen atoms bonded together. A **molecule** is made up of atoms that are chemically bonded together. This type of bond is called a **molecular bond**. For example, oxygen and hydrogen are elements. When two atoms of hydrogen (H) bond with one atom of oxygen (O), one molecule of water (H_2O) is formed.

The physical form of a material is referred to as its **state of matter** (**FIGURE 5-1**). There are three states of matter: solid, liquid, and gas.

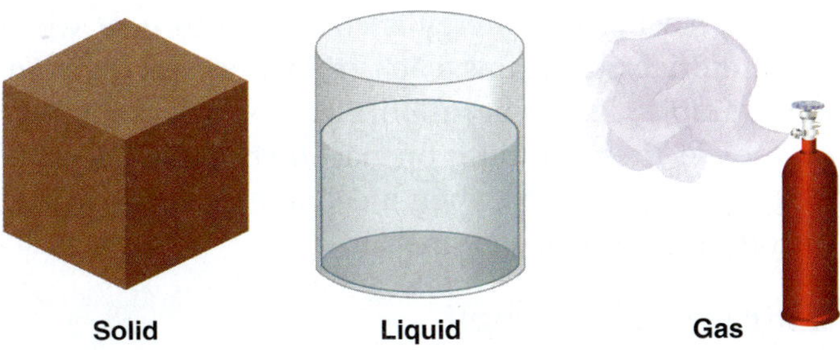

FIGURE 5-1 The three states of matter are solid, liquid, and gas.

A **solid** has a specific size and shape. Cold makes most solids more brittle, whereas heat makes them more flexible. Most of the molecules that make up a solid are inside of it—that is, they are cushioned or insulated by the outer surface of the solid. Only a limited number of the molecules that make up the solid are present on its surface.

A **liquid** has a specific volume but does not have a specific size or shape. A liquid assumes the shape of the container in which it is placed. Most liquids expand when heated, and if heated enough, turn into gases. Liquids cannot be compressed. This means that when you apply pressure to one end of a container, the liquid will move. This characteristic allows firefighters to pump water for long distances through pipelines or hose.

A **gas**, sometimes called a **vapor**, does not have a specific size, shape, or volume. Instead, a gas expands to fill the volume of the container into which it is released. The gas we most commonly encounter is air, the mixture of invisible, odorless, tasteless gases that surrounds the earth. Air is composed of 21 percent oxygen, 78 percent nitrogen, 1 percent argon, and trace amounts of other gases such as carbon dioxide, neon, methane, helium, krypton, hydrogen, and xenon.

Matter can change from one physical state to another under specific conditions. For example, water is a liquid at room temperature. When it reaches 32°F (0°C), it freezes—that is, it changes to a solid. When it reaches 212°F (100°C), it vaporizes—that is, it changes to a gas. The water molecules do not change—the molecules are all still H_2O—only the physical state of the water changes.

Energy

Energy is the ability to do work. The law of conservation of energy states that energy cannot be created or destroyed. Instead, energy is converted from one form to another. Think of an automobile. Chemical energy in the gasoline is converted to mechanical energy when the car moves down the road. When you apply the brakes to stop the car, the mechanical energy used to do this is converted to heat energy by the friction between the wheel rotors and the brake pads. Electrical energy can be converted to heat or to light. Mechanical energy can be converted to electrical energy through a generator. In a house fire, the stored chemical energy in the wood structure and plastic-based contents of a house is converted into heat and light energy during the fire. Energy exists in many forms, including chemical, mechanical, electrical, light, nuclear, and heat.

Energy is kinetic or potential. **Kinetic energy** is the energy possessed by an object because of its motion. **Potential energy** is energy stored by an object as a result of its position or condition. For example, if you hold a ball, it has potential energy. If you drop the ball, the potential energy changes to kinetic energy and the ball falls to the ground.

Heat Energy

Heat energy (also referred to as **thermal energy**) is the potential energy of a combustible material and the kinetic energy released when heat is applied to the material. Heat can be measured using different units, including the following:

- **British thermal unit (BTU)**: A BTU is the amount of heat energy required to raise 1 pound of water at sea level by 1 degree Fahrenheit. BTU is a common measure of heat used in the United States. For example, architects designing a new fire station might select a commercial heating system based on its BTU rating to ensure it has the capacity to adequately heat the building.
- **Calorie**: A calorie is the amount of heat energy required to raise 1 gram of water at sea level by 1 degree Celsius.
- **Joule**: A joule is a measure of heat energy equal to 0.4 calorie. One calorie is equal to about 4.2 joules and 1 BTU is equal to approximately 1055 joules. Firefighters may recognize joules as the unit used to describe the amount of energy delivered by a defibrillator for treating cardiac arrest.

Most people use the terms *heat* and *temperature* interchangeably; however, they have different meanings. Heat is a form of energy. As heat is applied to a material, the material's molecules increasingly move and vibrate, causing its temperature to rise. **Temperature** is the measurement of this movement and tells us how hot or cold something is. The two most common units of measurement for temperature are degrees Fahrenheit (°F) and degrees Celsius (°C). Residential structure fires can reach well over 1000°F (538°C).

Changing the temperature of a material can change its physical state. For example, when water is heated, the molecules move and vibrate until the water is boiling, and when it reaches 212°F (100°C), the physical state of the water molecules change to gas—steam. When the temperature of water is lowered—as heat leaves or is removed from the water—the movement of the molecules slows. When the temperature reaches 32°F (0°C), it changes to a solid—ice.

Chemical Energy

Chemical energy is the stored potential energy in molecular bonds and the kinetic energy created by a chemical reaction. Some chemical reactions are **exothermic**, which means they produce, or give off, heat. Others are **endothermic**, which means they absorb heat. Combustion is an example of an exothermic reaction. The reaction between water and ammonium salts in an instant ice pack is an example of an endothermic reaction—when combined, the two rapidly absorb heat. Most chemical reactions occur because the molecular bonds established between elements are broken when energy is applied.

Oxidation is the process during which oxygen combines chemically with another substance to create a new compound. For example, steel that is exposed to oxygen results in rust. The process of oxidation can be extremely slow; indeed, it can take years for oxidation to become evident. Slow oxidation does not produce easily measurable heat. Combustion is a type of fast oxidation.

Heat is produced whenever oxygen combines with a combustible material. If the reaction occurs slowly in a well-ventilated area, the heat is released harmlessly into the air. If the reaction occurs rapidly or within an enclosed space, the material can be heated to its ignition temperature. **Ignition temperature** is the minimum temperature at which a material ignites in the presence of oxygen. This process is called **autoignition**. The fire that results is energy released as a result of the chemical reaction between oxygen and the fuel. An example of this occurs when a bundle of rags soaked with linseed oil releases enough heat through oxidization, causing the rags to ignite spontaneously.

Mechanical Energy

Mechanical energy is the potential energy stored because of the position of an object (for example, a ball you hold) or the kinetic energy of an object in motion. Water falling over a dam is an example of mechanical energy. Mechanical energy is converted to heat when two materials rub against each other and cause friction. For example, a fan belt rubbing against a seized pulley or vehicle tires spinning on pavement produce heat. Heat also is produced when mechanical energy is used to compress air in a compressor.

Electrical Energy

Electrical energy is energy from an electrical charge. It is carried through the electrical wires inside homes and can be stored in batteries that convert chemical energy to electrical energy. Electrical energy is converted to heat energy in several different ways. For example, electricity produces heat when it flows through a wire or any other conductive material. The greater the flow of electricity and the greater the resistance of the material, the greater the amount of heat produced. Examples of electrical energy that can produce enough heat to start a fire include electric heating elements, overloaded wires, electrical arcs, batteries and other energy storage systems, and lightning.

Light Energy

Light energy is produced by electromagnetic waves packaged in discrete bundles called photons. This energy travels as thermal radiation, a form of heat. When light energy is hot enough, it can sometimes be seen in the form of visible light. One example of light energy is the radiant energy we receive from the sun. We think of candles, fires, light bulbs, and lasers as forms of light energy. We should recognize that, while these objects do produce light energy, they also produce heat. If light energy is of a frequency that we cannot see, the energy may be felt as heat but not seen as visible light.

Nuclear Energy

Nuclear energy is the potential energy stored within the nucleus of an atom or the kinetic energy released by splitting the nucleus of an atom into two smaller nuclei (fission) or by combining two small nuclei into one large nucleus (fusion). Nuclear energy is stored in radioactive materials. Nuclear reactions release large amounts of energy in the form of heat. These reactions can be controlled, as in a nuclear power plant, or uncontrolled, as in an atomic bomb explosion. In a nuclear power plant, the nuclear reaction releases carefully controlled amounts of heat. This is used to heat water to produce steam, which powers a steam turbine generator. The mechanical energy from the generator is converted to electrical energy. Both uncontrolled explosions and controlled reactions release radioactive material, which can cause injury or death.

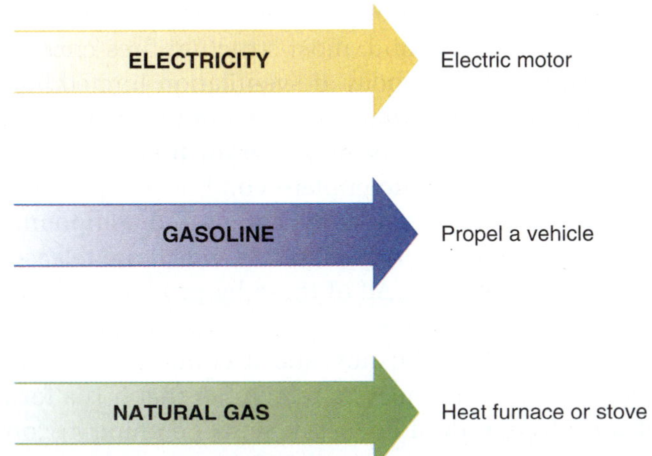

FIGURE 5-2 Energy being converted to work.

FIGURE 5-3 A fire needs all three components of the fire triangle: fuel, oxygen, and heat.

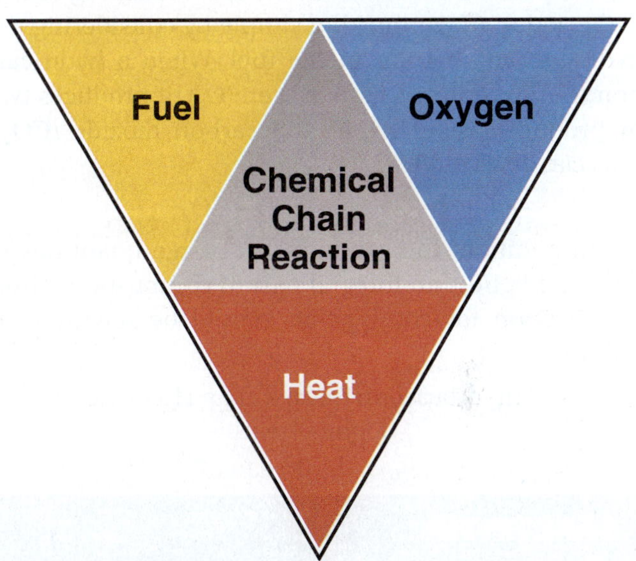

FIGURE 5-4 A chemical chain reaction unites fuel, oxygen, and heat in the fire tetrahedron.

Fuels

Fuel is matter in any state—solid, liquid, or gas—that stores energy and is combustible. When the energy in fuel is released, for example, when it is burning, it is converted into another type of energy. Think of the vast amount of kinetic energy in the form of heat and light that is released during a large fire. This energy released in the form of heat and light was stored in the fuel as potential energy before it is burned. The release of the potential energy in a gallon of gasoline, for example, can move a car many miles down the road (**FIGURE 5-2**).

Fire requires fuel that is in the form of combustible gas in order for the chemical process of combustion to occur. When solid and liquid fuels are heated, they produce a gas (they **off-gas**) and that is what ultimately ignites. Materials that remain in a solid or liquid state will not burn while in that state. Solid fuels such as wood, liquid fuels such as gasoline, and gaseous fuels such as propane all burn after they have changed into a gas.

Conditions Needed for Fire

To understand the behavior of fire, you need to consider the three basic elements needed for combustion to occur: fuel, oxygen, and heat. First, a combustible fuel must be present. Second, oxygen must be available in sufficient quantities. Third, a source of ignition (heat) must be present. These three components together form the **fire triangle** (**FIGURE 5-3**).

The fourth factor is combustion, the chemical chain reaction that occurs when the fuel, oxygen, and heat interact. One way of visualizing this process is to show the chemical chain reaction joining the elements of the fuel, oxygen, and heat. This depiction reflects the central role that the chemical reaction plays in maintaining the process of flaming combustion. This relationship is sometimes characterized as the **fire tetrahedron**. A tetrahedron is a four-sided, three-dimensional figure—a four-sided pyramid. Each side of the fire tetrahedron represents one of the four elements needed for a fire to occur (**FIGURE 5-4**).

The key point to remember is that fuel, oxygen, and heat must all be present for a fire to start and to continue burning. If you remove any of these elements, the fire will go out. A **fuel-limited fire** is a fire that has sufficient oxygen for fire growth but has a limited amount of fuel available for burning. When all of the fuel is consumed, the fire stops burning. One example of a fuel-limited fire occurs in cases where an outdoor fire consumes the fuel supply. It also can occur in a room or compartment fire when the fuel supplied by the compartment contents is being consumed and the fire has not extended beyond the compartment because of fire-resistant construction. A **ventilation-limited fire** is a fire that has sufficient fuel for fire growth but has a limited amount of oxygen available, so it doesn't burn as rapidly as it would with an unlimited supply of oxygen. A ventilation-limited fire can grow quickly if more oxygen is added.

Chemistry of Combustion

Almost all fuels consist of hydrogen (H) and carbon (C) atoms; hence, they are called hydrocarbons. Fuels may contain a wide variety of other elements, but for our purpose we can describe the self-sustaining chemical reaction of combustion simply by considering the hydrogen and carbon in the fuel. When a hydrocarbon (H–C) combines with oxygen (O_2), it produces two by-products, water (H_2O) and carbon dioxide (CO_2). This can be written as:

$$H\text{–}C + O_2 = H_2O + CO_2$$

In addition, the production of large quantities of heat and light is an integral part of combustion. Thus, the reaction for combustion should be rewritten as follows:

$$H\text{–}C \text{ (hydrocarbon fuels)} + O_2 = H_2O + CO_2 + \text{Light} + \text{Heat}$$

SAFETY TIP

Smoke consists mainly of unburned forms of hydrocarbon fuels. As a consequence, when smoke is combined with adequate oxygen and heat, it will burn. Smoke contains large amounts of potential energy that can be released violently under the right conditions.

Products of Combustion

Of course, the reactions that occur in most actual fires are not as simple as this basic reaction that results in only water and carbon dioxide suggests. First, hydrocarbon fuels are complex molecules that contain complex chains of atoms in addition to hydrogen and carbon. As a consequence, the combustion process produces numerous toxic by-products in addition to water and carbon dioxide. Second, most structure fires encountered by firefighters today are ventilation-limited fires. This results in **incomplete combustion**, which is a combustion process during which the fuel is not completely consumed. Incomplete combustion produces significant quantities of deadly gases and compounds and a variety of other by-products, which are released into the atmosphere. All of these by-products are collectively called **smoke**. Because smoke is the product of incomplete combustion and it contains unburned hydrocarbons, you need to remember that it is a form of fuel. A column of hot black smoke coming in contact with an adequate supply of oxygen and an ignition source can ignite suddenly and violently. Smoke is composed of three components: solids in the form of **smoke particles**, liquids in the form of an **aerosol**—tiny water droplets suspended in the air similar to fog—and gases.

The particles in smoke include unburned, partially burned, and completely burned materials. The unburned particles are usually readily visible. Partially burned particles become part of the smoke because inadequate oxygen is available to allow for their complete combustion. Completely burned particles are primarily ash. The unburned and partially burned particles include a variety of substances. Some smoke particles are small enough that they get past the protective mechanisms of the respiratory system and enter the lungs. Most smoke particles are toxic. The concentration of the three types of particles depends on the amount of oxygen that was available to fuel the fire.

Smoke aerosol is formed when water is applied to a fire and tiny water droplets become suspended in the smoke or haze that forms. The vaporization of these water droplets may produce steam contained within the smoke. That steam may not be visible and thus firefighters risk steam burns if not adequately protected. If the fuel contains oil-based compounds, small oil-based droplets become part of the smoke. Oil-based or lipid compounds can cause great harm when their smoke is inhaled. Some types of toxic aerosols cause poisoning if absorbed through the skin.

Smoke contains a wide variety of gases. The composition of gases in smoke varies depending on the amount of oxygen available to the fire at any instant. The composition of the gases produced by a burning substance influences the composition of the gases in the smoke. In other words, a fire fueled by wood produces a different composition of gases than a fire fueled by a petroleum-based fuel such as plastic.

Almost all gases produced by a fire are toxic, including the following:

- **Carbon monoxide (CO)**, produced by incomplete combustion, is deadly in fairly small quantities. Hemoglobin molecules in the blood normally carry oxygen (O_2) to the body's cells. When inhaled, CO quickly replaces the oxygen in the bloodstream because it binds with the hemoglobin molecules 200 times more readily than oxygen does. Even a small concentration of CO can quickly disable and kill a firefighter.
- **Hydrogen cyanide (HCN)** is formed from the incomplete combustion of plastic and foam products, such as furniture and polyvinyl chloride (PVC) pipe used in residential construction. HCN is quickly absorbed by the blood and interferes with cellular respiration. Low-level exposure can cause cyanosis, headache, dizziness, unsteady gait, and nausea. Just a small amount of HCN can render a person unconscious. Recent studies have indicated that cyanide poisoning in firefighters may be far more common than previously recognized.
- **Phosgene** is a gas formed from incomplete combustion of many common household products, including vinyl materials. This gas affects the body in several ways. At low levels, it causes itchy eyes, a sore throat, and a burning cough. At higher levels, phosgene gas can cause pulmonary edema (fluid retention in the lungs) and death.
- Fires produce many other gases that may affect the human body in harmful ways. Among these toxic gases are hydrogen chloride (HCl) and several compounds containing different combinations of nitrogen (N) and oxygen (O_2). Many of the compounds cause cancer, years after exposure. Higher rates of exposure to smoke increase the possibility of firefighters contracting cancer.

SAFETY TIP

Smoky environments are deadly! An Underwriters Laboratories (UL) smoke study determined that 97 percent of a fire's particulates are too small to be visible to the human eye (Fabian et al. 2010, iv). It is essential to use a self-contained breathing apparatus (SCBA) whenever you are operating in a smoky environment. This rule applies whether you are dealing with a structure fire, a dumpster fire, a car fire, or any other type of fire that generates smoke. It also applies during overhaul, after the fire has been extinguished. Toxic gases are still present during the overhaul stage of a fire, even when little smoke is visible.

Carbon dioxide (CO_2), which is produced when sufficient oxygen is available for complete combustion, is not toxic, but it displaces oxygen in the atmosphere and can cause hypoxia and asphyxiation. Small amounts of CO_2 do exist naturally in our atmosphere, so an increase may seem entirely normal, but in fact, CO_2 in amounts above normal is immediately dangerous to life and health (IDLH) because of the lack of sufficient oxygen. Because CO_2 is heavier than air, it may pool in low-lying areas such as a basement or confined space. You must use an SCBA in such a setting.

A discussion of the by-products of combustion would be incomplete without considering heat. Because smoke is the result of fire, it is hot. The temperature of smoke varies depending on the conditions of the fire and the distance the smoke travels from the fire. Injuries from smoke occur when the particles, droplets, and gases that make up smoke are inhaled. In addition, inhaling the hot gases in smoke can cause severe burns to the respiratory tract. In addition, all of the components in smoke can cause severe burns to the skin.

SAFETY TIP

Gases can quickly fill confined spaces, and gases that are heavier than air, such as CO_2 and phosgene, can quickly fill below-grade structures. Any confined space or below-grade area must be treated as a hazardous atmosphere until it has been tested to ensure that the concentration of oxygen is adequate and that no hazardous or dangerous gases are present.

Fire Spread and Heat Transfer

The exchange of heat energy between materials is known as **heat transfer**. When there is a difference in temperature between two objects, heat transfers from a hotter object to a cooler object until they reach equal temperatures. When two objects have the same temperature, heat transfer does not occur.

The **heat release rate (HRR)** is the rate at which heat energy is generated and it is measured in watts (1 watt is equivalent to 1 joule per second). **Heat flux** is the measure of the heat transfer to or from one surface to another. For example, if a sofa is actively burning and heat is moving to the ceiling, the heat flux would indicate how much heat was being transferred to the ceiling. Heat flux is expressed as kilowatts per square meter (kW/m^2), watts per square centimeter (W/cm^2), or BTUs per square foot (BTU/ft^2). Heat flux depends

Voice of Experience

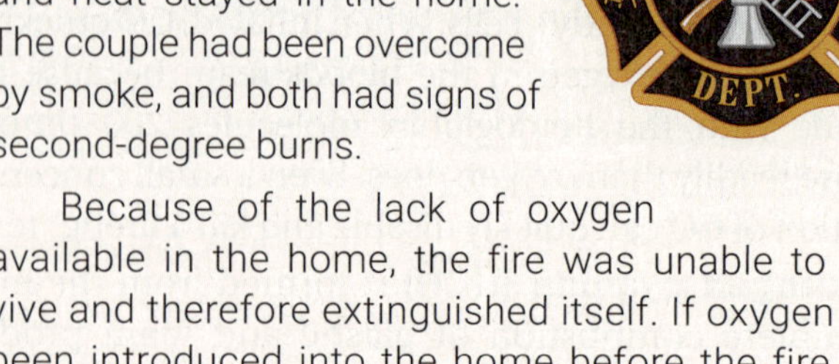

We responded to a call involving the recovery of an elderly couple in a house that had had a fire inside. When the female occupant of the house did not report to work, a fellow employee went to the house to check on her. The employee discovered the deceased couple and called 911.

It was determined that a large fire started in the kitchen and produced enough fire, smoke, and heat to overcome the entire residence. It was evident that the couple had tried to extinguish the fire on their own as well as collect valuables before leaving. The doors to the bedrooms and bathroom were wide open, and this had allowed the smoke to fill the entire home. The ceilings and the tops of the walls had peeled and cracked drywall and layers of dark, heavy smoke stains. At about eye level the smoke stains were lighter, and they disappeared at about waist level. This was visible throughout the entire house. In addition, we could see that the heat had started to transfer to the contents of the home. The tops of the couches and chairs in the room had char marks and other signs of combustion.

The house had been renovated with newer windows and doors. The windows and doors created a seal to the outside that did not allow air in or out. It was so well sealed that the smoke and heat stayed in the home. The couple had been overcome by smoke, and both had signs of second-degree burns.

Because of the lack of oxygen available in the home, the fire was unable to survive and therefore extinguished itself. If oxygen had been introduced into the home before the fire had cooled off, the superheated gases could have reignited and most likely would have caused a backdraft situation.

If the occupants had kept the doors to the bathroom and bedrooms closed, they could have prevented the spread of the smoke, heat, and fire throughout the house. Possibly, they could have gotten out through a window or called for help. Instead, the couple tried to extinguish the fire themselves and collect valuables prior to leaving. This course of action ultimately got them killed.

Matt Rector
South Sioux City Fire Department
South Sioux City, Nebraska

on two factors: the difference in temperature between the two substances and the ability of each material to conduct heat. The higher the difference in temperature between the two substances, the faster the rate of transfer.

Heat is transferred in three ways: conduction, convection, and radiation.

Conduction

Conduction is the process of transferring heat between solids that are touching each other (**FIGURE 5-5**). This heat transfer occurs because of the kinetic energy of the molecules within the solid materials. Conduction transfers energy directly from one molecule to another, much the same way as a billiard ball transfers energy from one billiard ball to the next.

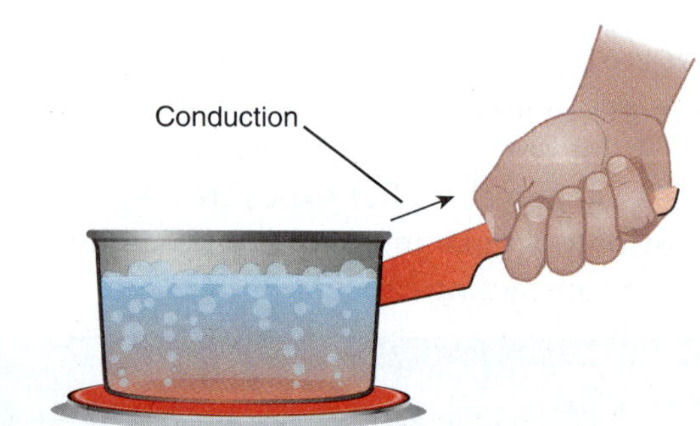

FIGURE 5-5 Conduction through a solid object.

The ability of a material to conduct heat is known as **thermal conductivity**. Objects vary in their ability to conduct energy. The thermal conductivity of an object depends on its composition, its moisture content, and how porous it is; that is, objects that have more tightly packed molecules are more efficient in conducting heat than objects that are less densely constructed. For example, inorganic materials such as metals are good conductors of heat; organic materials such as plastic, rubber, and wood are weak conductors of heat. If one end of a copper pipe is heated, the heat will be readily conducted along the pipe.

Heat transfer by conduction is dependent on three factors. The first factor is the thermal conductivity of the material being heated—better conductors transfer more heat. The second factor is the size of the area being heated. If all other factors remain the same, heating a small area results in less heat transfer than heating a larger area. The third factor is the difference in temperature between the heated object and the object that is being heated. As discussed previously, the greater the temperature difference or energy difference between the two materials, the greater the rate of heat transfer.

Insulating materials limit the transfer of heat by conduction. Air or other gas is trapped in small pockets within the insulating material, and gases are not efficient conductors of heat from one solid to another because the gas molecules are farther apart than the molecules in denser materials. It is important to note, however, that although some insulating materials, such as polyurethane foam, are poor conductors, they can be highly combustible.

Convection

Convection is the transfer of heat in a gas or liquid by circulating the material from hotter areas to cooler areas (**FIGURE 5-6**). During a fire, the smoke and hot gases generated by the fire move by convection. The heat of the fire warms the gases and particles in the smoke. The hotter and less dense column of gases rises and displaces cooler, denser gases downward. A large fire burning in the open can generate a **plume**, also called a **thermal column**, which is an elongated column of heated gases and smoke that rises high in the air by convection. This convection stream can carry smoke and large bands of burning fuel for several blocks before the gases cool and fall back to the earth.

When a fire occurs in a building, the convection currents generated by the fire rise in the room and travel along the ceiling. This is known as a **ceiling jet**. The hot gases carried by these currents may ultimately heat combustible room surfaces and contents to their ignition temperatures and those materials will ignite.

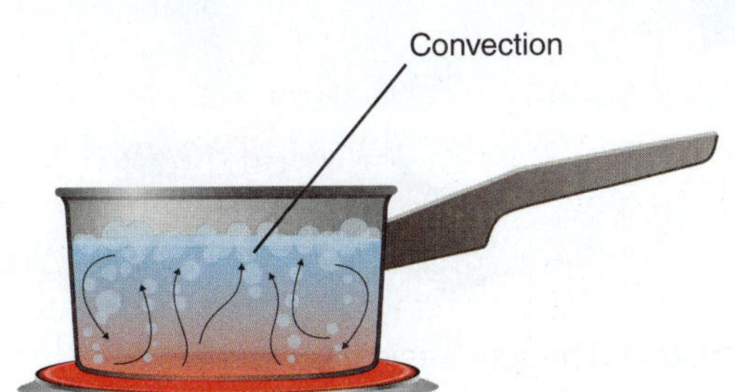

FIGURE 5-6 Convection transfers heat by the flow of gases or fluids from hotter areas to cooler areas.

Firefighters can take advantage of convection currents by using the cooler, lower layer to advance a hose line or conduct a search for occupants.

As the fire grows, the volume of hot gases and smoke increases. If the pressure in the area of origin, or **fire compartment**—the room in which the fire started—is sufficient, the hot gases push laterally outside the fire compartment. Because the pressure in the lower parts of the fire compartment is lower, cooler air is drawn into the lower levels of the fire compartment. This route along which the heat, smoke, and hot gases move from the higher pressure area inside the fire compartment to the lower pressure area outside the fire compartment is the **flow path**. A flow path needs at least one intake vent, such as the front opening of a fireplace, and at least one exhaust port, such as the chimney above the fireplace. In other words, the flow path transfers heat, hot gases, and smoke from the higher pressure within the fire area toward the lower pressure areas through doorways, window openings, and roof openings. Fire growth progresses along the flow path towards the exhaust vent. Convection currents, and therefore the flow path, are influenced by the layout of the building. For example, air flows quickly through a structure with an open floor plan and less quickly through a structure made up of small, closed rooms. Chapter 16, *Ventilation*, contains more information on flow paths.

Thermal Radiation

Thermal radiation is the transfer of heat through the emission of energy in the form of invisible electromagnetic waves. If you hover your hand over a burner on a stove, you feel the radiant heat from that burner (**FIGURE 5-7**). The sun radiates energy to the earth. When this energy is absorbed—for instance, as the sun's energy touches your body on a warm day—it is converted to heat. The direction in which the thermal radiation travels can be changed or redirected, as when a mirror reflects the sun's rays and bounces the energy in another direction. For example, heat from a fire is transferred in a direct line away from the source object

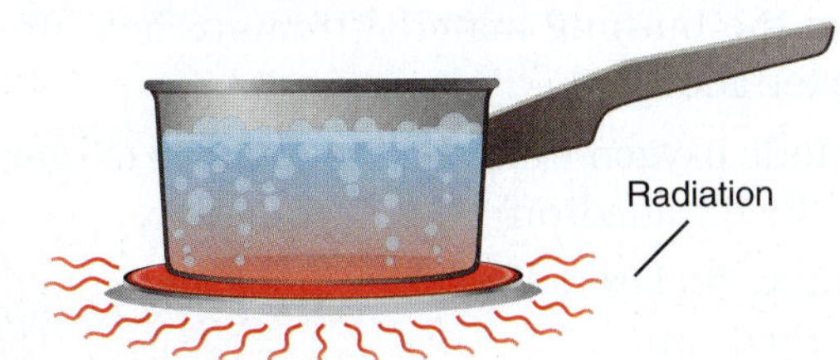

FIGURE 5-7 Thermal radiation.

and absorbed by cooler objects, including liquids and gases. If not properly protected, vehicles and structures next to a house on fire may become exposed to thermal radiant heat and ignite into additional fire problems for firefighters to address. Thermal radiation also occurs in a fire compartment when firefighters encounter heat from the fire and heat that is radiated back from the surfaces in the area of origin.

SAFETY TIP

Radiant heat can travel through the air gaps in a water spray.

Thermal radiation from a heat source travels in all directions. The effect of thermal radiation, however, is not seen or felt until the radiation strikes an object and heats the surface of the object. Thermal radiation is a significant factor in the growth of a campfire from a small flicker of flame to a fire hot enough to ignite large logs. The growth of a small fire in a wastebasket to a full-blown room-and-contents fire is due in part to the effect of thermal radiation. A building that is fully involved in fire radiates a tremendous amount of energy in all directions. Indeed, the radiant heat from a large building fire can travel several hundred feet to ignite an unattached building.

The solid fuel in modern structures contains large amounts of petroleum-based materials, and this burning material produces a large amount of heat energy. This increased HRR creates conditions in which firefighters are exposed to higher levels of energy in a shorter amount of time, potentially saturating their personal protective equipment (PPE). If this happens, that heat energy is then transferred to the firefighter's body, potentially leading to dangerous conditions.

Methods of Extinguishment

There are four methods of extinguishing fires (**FIGURE 5-8**):

- Cool the burning material (remove heat from the fire tetrahedron)
- Exclude oxygen from the fire (remove oxygen from the fire tetrahedron)
- Remove fuel from the fire (remove fuel from the fire tetrahedron)
- Interrupt the chemical reaction with a flame inhibitor (remove heat from the fire tetrahedron).

FIGURE 5-8 The four basic methods of fire extinguishment. **A.** Cool the burning material. **B.** Exclude oxygen from the fire. **C.** Remove fuel from the fire. **D.** Interrupt the chemical reaction with a flame inhibitor.

There are many variations on the way that these methods can be implemented. Sometimes a combination of these methods is used to achieve suppression of fires.

The most common method of extinguishing a fire is to cool the burning material. The actions firefighters take to set up a water supply, lay hose lines to the fire, and apply water to the fire are all steps in implementing this method of fire extinguishment.

A second method of extinguishing a fire is to exclude oxygen from the fire. A simple way to do this is to place the lid on an unvented charcoal grill or to close the door to a fire compartment. Reducing the amount of air reaching a fire will retard fire growth or extinguish the fire. To extinguish petroleum fires, firefighters use foam instead of water. They spray the foam in such a way that it creates a blanket that smothers the fire by separating the flammable gas from the oxygen in the air.

A third method of extinguishing a fire is to remove the fuel. For example, if a fire is being fed by a supply of natural gas, shutting off the supply extinguishes that fire. Remember that smoke is a product of incomplete combustion and is fuel, so reducing or cooling hot gases and smoke also diminishes the supply of fuel available to a fire. In wildland fires, a **firebreak**—a swath where the fuel (trees and brush) are removed—cut around a fire puts further fuel out of reach of the fire.

The fourth method of extinguishing a fire is to interrupt the chemical reaction with a **flame inhibitor**, a chemical extinguishing agent that reacts with the fuel to chemically disrupt the combustion process. These fire-extinguishing agents are applied with portable extinguishers or through a fixed suppression system designed to flood an enclosed space. They leave no residue and are often used to protect electronics and computer systems.

FIGURE 5-9 A Class A fire involves wood, paper, or other ordinary combustibles.

FIGURE 5-10 A Class B fire involves flammable liquids such as gasoline.

Classes of Fire

Fires are categorized into one of five classes based on the type of fuel: Class A, Class B, Class C, Class D, and Class K.

Class A Fires

A **Class A fire** involves ordinary solid combustible materials such as wood, paper, plastics, and cloth (**FIGURE 5-9**). Natural vegetation such as the grass that burns in ground cover fires also is considered to be part of this group of materials. The methods most commonly used to extinguish Class A fires are cooling the fuel with water to a temperature that is below the ignition temperature or using a combination of limiting ventilation and applying water.

Class B Fires

A **Class B fire** involves flammable or combustible liquids such as gasoline, kerosene, diesel fuel, grease, tar, lacquer, oil-based paints, and motor oil, or gases such as propane and natural gas (**FIGURE 5-10**). These fires can be extinguished by shutting off the supply of fuel or by using foam to separate the oxygen in the air from the fuel.

Class C Fires

A **Class C fire** involves energized electrical equipment (**FIGURE 5-11**). A Class C fire could involve building wiring and outlets, fuse boxes, circuit breakers, transformers, generators, or electric motors. Power tools, lighting fixtures, household appliances, and electronic devices such as televisions, radios, and computers could be involved in Class C fires as well. Class C fires are extinguished with an extinguishing agent that does

FIGURE 5-11 A Class C fire involves energized electrical equipment.

not conduct electricity. Incorrectly attacking a Class C fire with an extinguishing agent that conducts electricity can result in injury or death to the firefighter. Once the power is cut to a Class C fire, the fire is treated as a Class A or Class B fire depending on the type of material that is burning.

FIGURE 5-12 A Class D fire involves metals such as magnesium, sodium, or titanium.

Class D Fires

A **Class D fire** involves combustible metals such as sodium, magnesium, zirconium, lithium, potassium, and titanium (**FIGURE 5-12**). Class D fires cannot be extinguished with water. In fact, applying water to fires involving these metals will result in violent explosions. This is because when these metals are heated and water is applied to them, the water molecules (H_2O) react with the heated metal and produce hydrogen gas (H_2), which burns with an explosive force. Instead, these fires must be attacked with extinguishing agents that prevent explosions and smother the fire and snuff out the supply of oxygen. Many automobile manufacturers use magnesium components to reduce the weight and increase the strength of vehicles, so firefighters must take care when extinguishing a vehicle fire.

Class K Fires

A **Class K fire** involves combustible cooking oils and fats and often occurs in kitchens (**FIGURE 5-13**). Heating vegetable fats, animal fats, or oils in appliances such as deep-fat fryers can result in serious fires that are difficult to fight with ordinary fire extinguishers. Class K extinguishing agents contain a wet agent that combines with cooking oils to produce a soap-like substance. The resulting soapy foam absorbs heat from the fire.

FIGURE 5-13 A Class K fire involves combustible cooking oils and fats.

Fires Involving Mixed Materials

Some fires fit into more than one class. For example, a fire involving a wood building—Class A—also could involve petroleum-based contents—Class B. Likewise, a fire involving energized electrical circuits—a Class C fire—also might involve Class A or Class B materials. A vehicle fire may involve Class A, B, C, and even D materials, depending on the components used to construct the vehicle. If a fire involves live electrical sources, it should be treated as a Class C fire until the source of electricity has been disconnected to isolate the ignition source and protect firefighters. Once isolated, firefighters may treat the fire based on the classes of materials involved.

Characteristics of Solid-Fuel Fires

Most building fires that firefighters encounter are fed by solid fuels. In structure fires, the building and most of the contents exist as solids. A variety of solid fuels are found in most buildings. These include wood and wood-based products, fabrics, paper, and carpeting, as well as petroleum-based fuels, such as plastics and petroleum-based foam cushions. Wood is a commonly encountered building material. Older furniture was commonly constructed of solid-wood frames, and natural materials such as cotton were used for padding and upholstery. Modern furniture is more frequently constructed from petroleum-based foams and fabrics and is made with plastic frames and engineered wood that includes glue or adhesive that make the wood more flammable. Wall coverings in older houses are often paper. Today, wall coverings are more likely to be latex-based paint or other petroleum-based materials.

Solid fuels burn when they are heated sufficiently to change them into flammable vapors. This process of liberating gaseous fuel vapors due to the heating of a solid fuel is called **pyrolysis**. Pyrolysis is evident when wood, heated sufficiently, breaks down into vapors and char (**FIGURE 5-14**). Some fuels change from a solid form directly into a vapor. Other solid fuels first change into a liquid before becoming a vapor. Plastics and petroleum-based materials pyrolyze faster than wood-based products. This means that it requires less heat to decompose and ignite these fuels than it does to ignite wood.

Wood contains varying amounts of moisture. When wood is heated, the first change that occurs is that the water is vaporized and escapes. This causes the wood to begin to char. Charring results in a number of protective mechanisms, including burning off the softer exterior cellulose and exposing a tougher, inner layer that requires higher heat to ignite. Also, the carbon in the charring creates a protective layer. As the wood chars, it begins to lose mass, which can compromise its load-carrying capabilities. When the wood is heated to about 425°F (218°C), pyrolysis begins. As the wood is heated, the energy required to vaporize the water delays the beginning of pyrolysis.

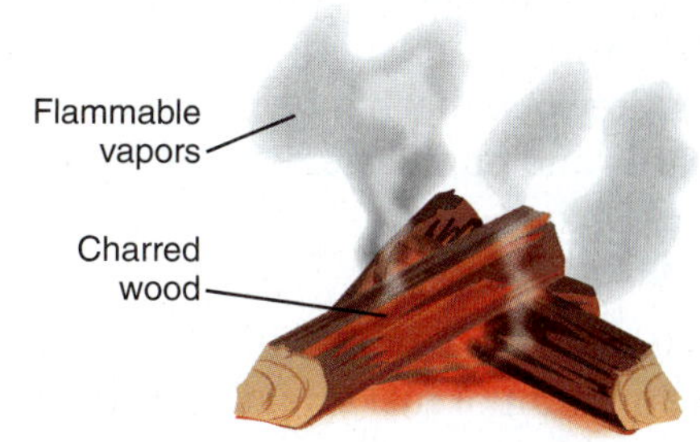

FIGURE 5-14 Pyrolysis is evident when wood, heated sufficiently, breaks down into vapors and char.

The characteristics of solid fuels that influence their combustion can be broken into three major categories: the composition of the fuel, the amount of the fuel, and the configuration of the fuel.

Composition of Solid Fuel

The chemical composition of the fuel has a significant impact on how the fuel burns. For example, wood is composed primarily of a natural, combustible fiber called cellulose. Most of the contents of modern structures, including synthetic foams, upholstery coverings, wall coverings, and plastic furniture, are manufactured from petroleum products. Petroleum-based products generally contain more potential heat energy than products made from natural products. This means that fires fueled by modern petroleum-based products have a much higher HRR than fires fueled by natural products such as wood and cotton.

> **TIP**
>
> Remember that smoke is one of the most common forms of fuel that firefighters encounter. Hot smoke is a powerful and potentially deadly fuel and is present at almost every fire you will encounter.

The amount of moisture content in the fuel is another factor that influences how the fire burns and the amount of heat that is released. Fuel with more moisture takes longer to ignite than fuel with less moisture. The water contained in a fuel serves as a cooling agent. It requires heat to vaporize the water from the fuel before the fuel can be pyrolyzed to create a flammable vapor. For example, green firewood from a recently cut tree contains a lot of moisture and is difficult to burn in a fireplace, whereas firewood that has been allowed to dry will ignite more easily and burn faster with a higher HRR.

Amount of Fuel

The second characteristic of fuel that determines how it burns is the amount of fuel available to the fire. If all other factors are the same, when more fuel is available, the HRR is higher than when less fuel is available. For example, under identical conditions, a fire fueled with four wooden pallets will have a much lower HRR than a fire fueled with 40 wood pallets.

Configuration of Fuel

The third characteristic of fuel that determines how it burns is the configuration of the fuel. Three factors contribute to this: the surface-to-mass ratio, the orientation of the fuel, and the continuity of the fuel.

Surface-to-Mass Ratio

The size and shape of a solid fuel greatly impact the ability of the fuel to ignite, the time it takes the fuel to be consumed, and the HRR of the burning fuel. For example, consider a large wood log that is 10 feet (3 meters) long. This log has a large mass or weight, but the surface area of the log is relatively small compared to its mass. If we cut the log into 4 by 4 in. (10 by 10 centimeter [cm]) beams, the total mass remains the same, but the surface area is much greater than it was when in the form of a single log. If the original log were cut into thin shingles, it would again have the same original mass as the log but would have a far greater total surface area than the original log or the 4 by 4 in. (10 by 10 cm) beams. The energy required to ignite the low surface-to-mass ratio log is much greater than the energy required to ignite the higher surface-to-mass ratio shingles. The higher the surface-to-mass ratio, the less energy that is required to ignite the fuel. This is because less energy is required to heat the material to a temperature sufficient to burn. For example, holding a match to a large log usually will not result in ignition of the log, but holding a match to a thin piece of shingle from the same source usually will result in ignition of the shingle.

Orientation of the Fuel

The second configuration factor that affects how a fuel burns is the orientation of the fuel. For example, a board in a horizontal position burns more slowly than the same size board in a vertical position. A board in a horizontal position that is ignited on one corner will produce hot fire gases that rise away from the surface of the board. As convection currents carry the heat from the fire upward, that heat is carried away from the surface of the board. If the same board is in a vertical position and it is ignited on one of the lower corners, the convection created by the hot fire gases will transfer some of the heat along the vertical surface of the board. The heat absorbed by convective heat transfer on the vertical board will be much greater than the heat that is absorbed by the board in a horizontal position.

Continuity of the Fuel

The third configuration factor that impacts how a fuel burns is the continuity of the fuel, or how close one piece of fuel is to the next piece of fuel. Fuel that is in contact with other fuel will ignite faster and reach a peak HRR more quickly. The closer the fuel is, the easier and more quickly it will ignite. This rate of growth is partly due to spread by radiation, convection, and in some cases, close proximity conduction. Continuity can occur in a horizontal direction along floors, ceilings, or horizontal surfaces of building contents. It also can occur in a vertical direction along walls or the vertical surfaces of building contents.

Solid-Fuel Fire Development

A ventilated, fuel-limited, solid-fuel fire progresses through four classic stages of growth: the incipient stage, the growth stage, the fully developed stage, and the decay stage (**FIGURE 5-15**).

1. Incipient Stage

The first stage of fire development is the **incipient stage**. The incipient stage occurs when there is an adequate supply of fuel, oxygen, and heat or ignition (**FIGURE 5-16**). At the incipient stage, a fire is small and confined to the initial fuel that was ignited. Because the fire is small at this stage, the initial growth is largely dependent on the type of fuel and how much of it can be pyrolyzed into a gas. Radiant heat from the fire begins to pyrolyze increasing amounts of fuel into flammable gases. A fire at the incipient stage consumes relatively small amounts of oxygen, so at this stage the fire is fuel dependent.

As the fire grows and produces more heat, a plume of hot gases and smoke begins to rise from the fire. As this plume rises, it is cooled by the surrounding cooler air. As increasing amounts of heat are generated, the plume carries more heat and hot gases upward. If it

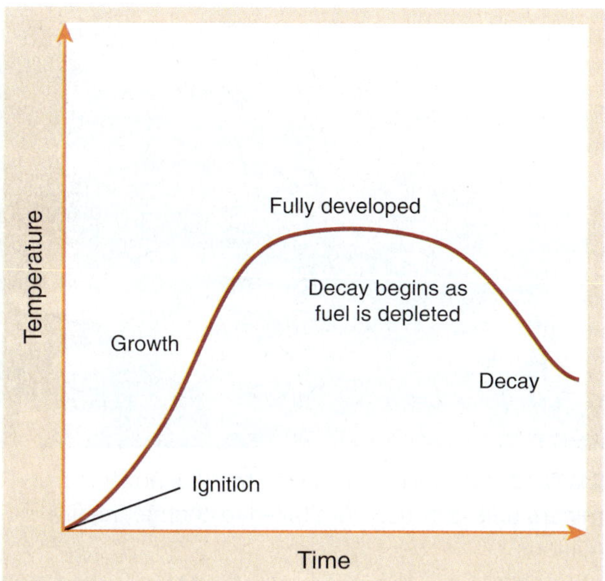

FIGURE 5-15 Illustration of a traditional fire growth curve with a fuel-limited fire.

Modified from National Institute of Standards and Technology. "Fire dynamics." Updated June 2, 2021. https://www.nist.gov/el/fire-research-division-73300/firegov-fire-service/fire-dynamics.

FIGURE 5-16 The incipient stage of a fire.

Courtesy of the National Institute of Standards and Technology.

reaches the ceiling, it will spread out in a circle that increases in size. These hot gases begin to heat the surfaces with which they make contact.

In the incipient stage, because the amount of heat is limited, there is little increase in the temperature of the room. The amount of smoke and combustion gases in the compartment usually is not high enough to pose a serious threat to occupants if they exit quickly. Many fires in the incipient stage can be extinguished safely with a portable fire extinguisher.

It is important to understand, however, that the stages of fire growth do not have a clear dividing line between them. A fire can progress rapidly from the incipient stage to the growth stage in a short period of time. Because fire can grow rapidly, building occupants and firefighters who are not equipped with a complete ensemble of PPE, including SCBA, should exit from even a small fire.

FIGURE 5-17 The growth stage of a fire.

Courtesy of the National Institute of Standards and Technology.

2. Growth Stage

The second stage of fire development is the **growth stage**. A fire in the growth stage produces more interaction and is more dependent on the environment in the compartment around it than a fire in the incipient stage (**FIGURE 5-17**). Especially important are the composition of the compartment surfaces and contents and the placement and configuration of the compartment contents. The amount of ventilation is also an important factor.

As a fire begins to grow, it produces larger amounts of heat, producing a higher HRR. The hot gases and smoke increase in density and turbulence as additional fuel is consumed. The location of the fuel in a compartment influences the fire development. If the fuel is in the center of the compartment, it may **entrain**, that is, encircle, draw along, and transport air from four sides, and the hot gases and smoke in the plume mix with the cooler air. This cooling reduces the rate of growth, the length of the flame, and the length of the plume, thereby reducing the vertical extension of the fire. These factors help to slow the rate of fire growth. If the burning fuel is located close to a corner of the compartment, the plume can entrain air from only two sides. As a result, the rate of growth will be faster and the length of the flame and the plume will be longer than if the burning fuel were in the middle of the room. If the fuel is located close to a wall, it entrains air on three sides, resulting in a growth rate slower than a fire located in a corner but faster than a fire with fuel in the middle of the room. A higher fire plume increases the hot gases at the ceiling of the compartment. More hot gases at the ceiling radiates more heat back to the room surfaces, the room contents, and to the burning fuel, further increasing the rate of fire growth.

Specific types of fire conditions are described next. These include thermal layering, rollover, flashover, backdraft, the behavior of ventilation-limited fires, and rapid fire growth. These conditions are not limited to the growth stage of fire development but are introduced here because they are likely to occur in the growth stage.

Thermal Layering

As a fire continues to grow, more hot gases are generated. Hot gases are lighter and buoyant and therefore tend to seek a higher level in the compartment. Because cooler gases are heavier, they settle at lower levels in the compartment. The phenomenon of gases forming into layers according to temperature is called **thermal layering**, also known as **heat stratification**. Any time the ventilation of the fire changes, it changes the thermal layering.

As the hot fire gases rise, they increase the rate of heat transfer back to the contents and walls of the fire compartment. The layer of hot gases in the upper level of the compartment creates an increase in pressure. This layer of hot gases tends to push out of the upper level through any available openings such as doors or windows. Because the cooler gases at the bottom of the room exert less pressure than the hotter gases, cooler air is drawn into the fire compartment, usually at lower levels.

As the hot gases exit the fire compartment at the upper levels and the cooler air enters the fire compartment at a lower level, there is a level in the compartment where the pressure exerted by the lighter hot gases flowing out and the pressure of the cooler air flowing in is equal. This is called the **neutral plane**. In order for a neutral plane to exist, there must be a flow of cooler air entering the compartment and a flow of hot gases exiting the compartment.

SAFETY TIP

Firefighters should avoid working in the layer of hot gases. Stay low whenever possible to remain under the layer of hot gases.

FIGURE 5-18 Rollover is a sign that flashover is imminent unless actions are taken to change the fire conditions.

As the fire continues to grow and the amount of heat generated increases, small flames "dance" in the hot gas layer. These isolated flames are an indication that the gases in the hot layer are near their ignition temperature. Often, these flames appear around the edges of the hot plume because this area has sufficient oxygen for combustion. These isolated flames may indicate that the fire is a ventilation-limited fire. This also may indicate that conditions are approaching a point where the room and contents will autoignite.

Rollover

Rollover, also called **flameover**, is the spontaneous ignition of hot gases in the upper levels of a room or compartment (**FIGURE 5-18**). During the growth stage of the fire, the hottest gases rise to the top of the room. If any of these gases reach their ignition temperature from the radiation of heat coming from the flaming fire or from direct contact with open flames, they will ignite. In other words, the upper layer of flammable gases catches fire. These flames can flicker across the ceiling and then go out. Alternatively, when they get a few degrees hotter, they can extend throughout the room at ceiling level. Rollover is a sign that the temperature is rising, and if it continues to rise, the temperature throughout the room will reach the point of flashover, when the room and contents spontaneously and rapidly ignite.

Flashover

Flashover is the rapid transition from a fire that is growing by igniting one type of fuel, to a fire where all of the exposed surfaces have ignited. It is a rapid change or transition from the growth stage to the fully developed

FIGURE 5-19 Flashover is the rapid change or transition from the growth stage to the fully developed stage.

stage. In a fire, it may be hard to determine which stage a fire is in. Fires do not always follow the four classic stages of development. A flashover can occur any time heat and sufficient oxygen are available to support combustion (**FIGURE 5-19**). Flashovers are less likely to occur if the fire is in a large room or a compartment that supplies more air during fire development. Larger areas entrain more air in the fire plume, which cools the plume and reduces the radiation and convection of heat to the room surfaces and contents.

SAFETY TIP

Flashover can be prevented by cooling the fuel, removing the fuel, or reducing the amount of oxygen present by controlling ventilation.

The critical temperature for a flashover to occur is approximately 1000°F (540°C). Once this temperature is reached, all fuels in the room are involved in the fire, including the floor coverings. This means that the temperature at the floor level may be as high as the temperature of the ceiling just before flashover. Firefighters, even with full PPE, cannot survive for more than a few seconds in a flashover, so it is important to be able to recognize the signs that may indicate a flashover is imminent. These signs include off-gassing or smoke from furniture and carpeting, a sudden increase in heat, and zero visibility. Even when you see that these conditions are present, you often will not have time to escape from danger. Work to avoid placing yourself in these conditions by either cooling the environment as you go or avoiding locations where flashover indicators are present. Firefighters must consider the potential risk level of entering a structure showing signs of an impending flashover and consider the possible benefits. If flashover has occurred, there is extreme risk for firefighters and no benefit, as any victims exposed to the flashover will be deceased. There is no sense in firefighters risking their lives to try to save someone who is already dead.

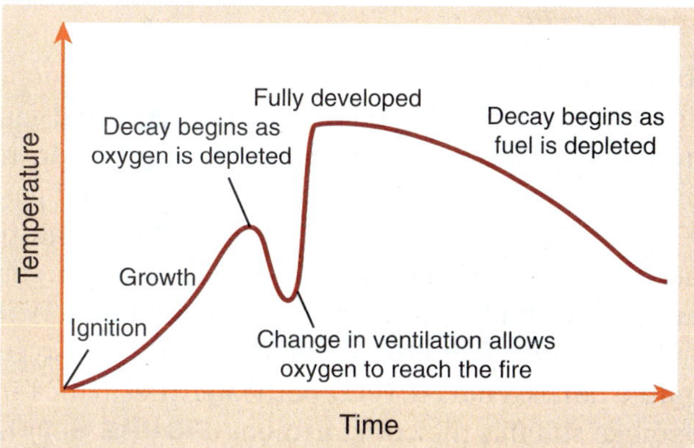

FIGURE 5-20 Illustration of a ventilation-limited fire, which often occurs in houses with lightweight construction and petroleum-based contents.

Modified from National Institute of Standards and Technology. "Fire dynamics." Updated June 2, 2021. https://www.nist.gov/el/fire-research-division-73300/firegov-fire-service/fire-dynamics.

Flashover may not occur if the fire remains ventilation-limited. A ventilation-limited fire produces a limited amount of heat energy. If the supply of oxygen is not increased, the fire may enter a decay stage. During this decay stage, heat continues to pyrolyze fuels in the compartment, producing additional vaporized fuel in the form of smoke. A fire in the decay stage will flash over quickly, however, if temperatures are maintained and sufficient oxygen is added (**FIGURE 5-20**). Keeping doors and windows to the fire compartment closed helps to reduce the amount of oxygen available to the fire and may help prevent a flashover. It is important to note that even with increased ventilation, many compartment fires remain ventilation-limited rather than advancing to a fully developed fire because there is insufficient oxygen to permit the fire to freely burn.

SAFETY TIP

Firefighters must learn to recognize the warning signs of flashover and use the reach of their hose stream to cool as close to the fire as possible, thereby reducing the chance of a flashover. A firefighter—even one who is wearing full PPE—has little chance of avoiding severe injury or death if they are caught in a flashover.

Backdraft

A **backdraft** is an explosion caused by the introduction of oxygen into a compartment where superheated gases and contents are hot enough for ignition but do not have sufficient oxygen to combust. Superheated gases are gases heated above their ignition temperatures. Backdrafts require a "closed box"—that is, a room, compartment, or building that is ventilation limited. When the fire compartment has a limited amount of oxygen, combustion is reduced, yet superheated fuel in the form of smoke still fills the compartment (**FIGURE 5-21**). If a supply of oxygen is introduced into the room by ventilation, such as opening the front door or an accidental event such as a window breaking because of excess heat, sudden and explosive combustion can occur owing to the presence of the superheated flammable gases. This explosive combustion may exert enough force to cause severe injury or death to firefighters.

FIGURE 5-21 These photos show the rapid transition to a backdraft. The photos were taken only seconds after the compartment door was opened.

Signs and symptoms of an impending backdraft include the following:

- A confined fire with a large heat build-up
- Little or no flame visible from the exterior of the building
- A "living fire," where the building appears to be breathing due to smoke puffing out of and then being sucked back into the building
- Smoke that seems to be ventilating out under pressure
- Smoke-stained windows, which indicates a significant fire
- No visible smoke
- Turbulent smoke
- Thick yellowish smoke, which indicates that it contains sulfur compounds

It is important for firefighters to stay alert for the conditions that signal a possible backdraft and to reduce the possibility of a backdraft developing. Chapter 14, *Fire Suppression*, and Chapter 16, *Ventilation*, describe ways to limit the supply of oxygen to the fire and to cool a ventilation-limited fire.

Behavior of Ventilation-Limited Fires

Experiments conducted by UL, the National Institute of Standards and Technology (NIST), and the New York City Fire Department (FDNY) have consistently demonstrated that many building fires become ventilation-limited because of a limited supply of oxygen. Newer houses are tightly sealed, with added insulation and caulking, double-paned windows, and storm windows. A fire in a building during the growth stage can consume enough oxygen to reduce the concentration of oxygen below the level needed to maintain a growing fire. When this occurs, the fire becomes ventilation-limited. Assessing a ventilation-limited fire presents a challenge to firefighters. A fire in this condition may be misleading and incorrectly appear to be a small fire in the incipient stage even though it contains a large amount of energy in the form of hot gases and smoke. Research indicates that fires in modern residential occupancies are likely to enter a ventilation-limited decay stage prior to the arrival of the first-due engine company (Kerber 2009).

Introducing air into a ventilation-limited fire can result in explosive, rapid fire growth. Research has demonstrated that firefighters making entry through the front door can introduce enough air into the fire area to produce rapid fire growth and flashover. This change can occur so fast that it is not possible to escape this deadly environment. Firefighters need to consider carefully what constitutes ventilation. Opening any door, window, skylight, or roof introduces oxygen into a burning building. Repeated experiments have been conducted that produced rapid fire growth after the front door was opened. Rapid fire growth can be prevented by cooling the fuel, removing the fuel, or controlling the amount of oxygen present. During rapid growth, a fire can progress to its maximum HRR unless actions are taken to limit the supply of oxygen or apply water to the fire.

From the outside, a ventilation-limited fire may be deceiving—it may look like a small fire. Often flames are not visible. All that is needed is more oxygen to produce explosive fire growth. Be aware that a fire that appears to be "small and relatively harmless" may in fact be a ventilation-limited fire containing enormous amounts of energy that just needs a healthy dose of oxygen to grow rapidly.

3. Fully Developed Stage

The third stage of fire development is the **fully developed stage**. During this stage, the fire is consuming the maximum amount of fuel possible, and it is achieving the maximum HRR possible for the fuel supply and oxygen present (**FIGURE 5-22**). During the fully developed stage, the fire may be ventilation limited or fuel limited. When a fire has an unlimited supply of oxygen, the fire is fuel limited. Consider a building under construction that has been framed with wood and combustible sheathing but lacks windows and doors. If this building catches fire, the fully developed fire may be fuel limited because it has an abundant supply of oxygen. In contrast, most structure fires that firefighters encounter have limited openings from the fire compartment to the outside. Because there are limited openings to admit oxygen and to exhaust the hot fire gases and smoke, these fires are ventilation limited.

FIGURE 5-22 The fully developed stage of a fire.

Courtesy of the National Institute of Standards and Technology.

During the fully developed stage, as the fire releases large quantities of heat, the energy pyrolyzes large amounts of fuel, generating large quantities of smoke and fire gases. The smoke and fire gases increase pressure in the fire compartment and force hot fire gases from the fire compartment. The HRR will be dependent on the size of the compartment openings, which provide a supply of oxygen and openings for hot fire gases to exit.

When the hot fire gases are exhausted from the fire compartment, if they are above the ignition temperature of the gases, they may ignite upon mixing with a fresh supply of oxygen. This is what produces flames from compartment openings during this active burning state. If these hot fire gases travel to another room or compartment adjacent to the fire compartment, they can transfer enough heat to spread the fire to those rooms or compartments. As the hot vaporized fuel travels to adjacent areas, it can ignite when it reaches a new supply of oxygen. The large quantities of heat transferred to adjacent areas will pyrolyze fuel in these areas, increasing the supply of vaporized fuel available to the fire. This provides an additional source of fuel for the fire.

Not all fires reach the fully developed stage. If the fire compartment has limited openings that result in the fire being ventilation limited, the fire probably will not reach the conditions needed to achieve the maximum rate of heat production of a fully developed fire that has an unlimited supply of oxygen.

4. Decay Stage

The fourth and final stage in fire development is the **decay stage**. This condition occurs because of a decreasing fuel supply or because of a limited oxygen supply (**FIGURE 5-23**).

During the decay stage of a fuel-limited fire, active flaming combustion decreases or stops. The heat continues to pyrolyze the fuel and create flammable gases and vapors. These flammable fuel products will continue to burn, but the rate of combustion will continue to decrease and eventually stop when the fuel supply is exhausted.

During the decay stage of a ventilation-limited fire, the rate of combustion slows, and visible flames decrease or disappear. Because there is still significant heat in and around the fire and the fire compartment, fuels continue to pyrolyze and create additional

FIGURE 5-23 The decay stage of a fire.

Courtesy of the National Institute of Standards and Technology.

flammable vapors and gases. The rate of pyrolysis slows as the rate of combustion slows, but large quantities of flammable fuel may still be present. If additional oxygen is introduced into the fire compartment, rapid or violent fire growth can develop quickly and the fire will return to the fully developed stage.

During the decay stage, smoke may travel inside the structure some distance from the fire. When it comes in contact with a source of ignition, the flammable mixture will ignite—often in a violent manner. This is known as a **smoke explosion**. Smoke explosions can occur during the decay stage of a fire, but they also sometimes occur during final extinguishment or overhaul. A smoke explosion does not occur because of a change to the ventilation profile such as an open door or window. Instead, it occurs when smoke travels within the structure to an ignition source.

Fire Behavior in Modern Structures

Modern construction techniques and synthetic contents have altered fire behavior in today's fires. Houses built in the last 40 years are constructed to be energy and cost efficient, so they contain insulation to prevent heat loss in the winter and loss of cool air in the summer. They are also sealed to prevent air exchange through the walls and around the doors and windows. As a consequence, when a fire occurs in the house, only a limited quantity of air enters the structure when the doors and windows are closed. In many parts of the country, doors and windows are kept closed most of the year because buildings are being heated or cooled by heating and cooling systems.

Many newer houses are constructed with lightweight manufactured building components. These manufactured components contain many highly combustible products, such as glues and coatings, most of which are made from petroleum-based products. They contain less material (less mass) than houses built with dimensional lumber. Petroleum-based products contain more potential heat energy. These building products are not only combustible, but they also burn hotter and faster than a house that is constructed primarily of wood. This means that fires in newer houses have a higher HRR than fires in older houses constructed primarily of wood. The lightweight and new-growth frames of newer houses also fail much faster than the heavier wood framing of older homes. The lightweight construction often provides greater surface area of structural members and the potential for greater void spaces between floors. Even the floor plans of modern homes—incorporating large rooms, vaulted ceilings, and insulated windows—contribute to more destructive and hurried fire growth.

Modern houses are furnished with furniture and accessories made of plastics and petroleum-based products. These products catch fire easily, reach high temperatures quickly, and have a high HRR. When they burn, they release huge quantities of thick, dark smoke. This rapid combustion requires huge amounts of oxygen to maintain the growth stage (that is, to remain burning) (**FIGURE 5-24**). Compare these newer homes to homes built a few decades ago that were constructed and furnished with natural materials, such as wood, cotton fabrics, ceramics, and metal. Natural materials have more fire-resistive properties; ignite more slowly; and ultimately, burn cleaner, producing less smoke.

These changes in building construction materials and techniques and changes in the building contents have, in turn, changed the dynamics of fire behavior inside a structure. Research scientists studying fire dynamics constructed mock-ups of two separate living rooms, one using legacy materials and one with modern materials. In carefully monitored experiments, they ignited the rooms to analyze how fire behaves with these two different fire loads. The data consistently demonstrated that modern fires build up superheated gases, reach flashover, and experience structural failure much more quickly than legacy fires.

Fires in modern structures progress to the fully developed stage quickly. In most cases, these fires enter a ventilation-limited stage because there is a limited supply of oxygen available. During this ventilation-limited stage, the heat from the fire continues to pyrolyze available fuels, thereby increasing the supply of vaporized

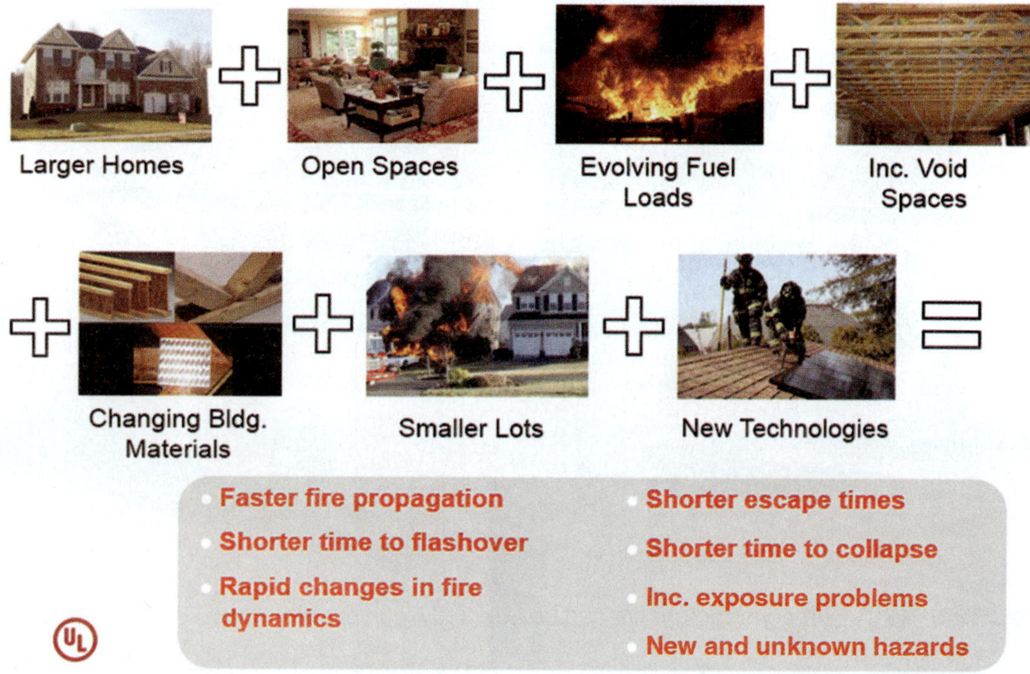

FIGURE 5-24 Modern fire environment.

Modified from Kerber, Stephen. 2011. "Analysis of Changing Residential Fire Dynamics and Its Implications on Firefighter Operational Timeframes." Fire Technology 48, no. 4: 865–891. https://doi.org/10.1007/s10694-011-0249-2. Creative Commons Attribution 4.0 International License (http://creativecommons.org/licenses/by/4.0/).

fuel. All that is needed is an increased supply of oxygen to change the fire into a rapid growth stage, often in the form of a flashover.

In these cases, the fire department often arrives on the scene while the fire is in a ventilation-limited decay stage. One of the first actions taken by some fire departments is to open the front door to gain access to the building for an interior attack. Unfortunately, opening the door has the unintentional effect of introducing a fresh supply of oxygen to the fire. As discussed previously, when a fresh supply of oxygen comes in contact with superheated flammable fuel, the result is rapid growth of the fire, often in the form of a flashover.

The intensity and speed with which a fire burns depend on the concentration of oxygen available to the fire. During the ignition stage, there is usually sufficient oxygen available for the fire to grow. In the growth, fully developed, and decay stages, a fire in an enclosed compartment may be regulated by the amount of oxygen available to it. If the atmosphere contains 16 to 21 percent oxygen, the fire typically will burn with a visible, open flame. If the concentration of oxygen drops to less than 16 percent, flame size will decrease. If the concentration of oxygen drops to less than 10 percent of the atmosphere, most substances will stop burning (Gann and Friedman 2015). If oxygen is introduced into a dying fire, the conditions can change quickly and result in a dangerous situation for firefighters. To prevent this scenario from occurring, firefighters must limit the amount of airflow to the fire. This requires coordinated timing in implementing building entry, ventilation, and application of water to the fire. Properly coordinated ventilation and fire attack procedures are designed to prevent these conditions from occurring. More information on these topics is presented in Chapter 14, *Fire Suppression*; Chapter 15, *Forcible Entry*; and Chapter 16, *Ventilation*.

Being aware of whether a home or structure is built and furnished with modern versus legacy materials is critical for firefighters to safely select appropriate and effective victim rescue and fire suppression tactics.

Wind Effect

One factor that greatly influences the behavior of a fire is the wind. Wind changes the behavior of fire drastically. For example, when the air is relatively still, the best approach to a fire may be through the "A" side of a building. Conversely, if the wind is blowing at 20 to 25 miles per hour (mph; 32 to 40 kilometers per hour

FIGURE 5-25 An interior structure fire can be affected by an exterior wind.

[km/h]) from the "C" side of the same building, entering on the "A" side may be a deadly miscalculation because the wind is pushing the fire to the "A" side of the building (**FIGURE 5-25**). The impact of the wind is similar to placing several large ventilation fans on the "C" side of the building and pushing huge quantities of air and fire toward the "A" side of the building.

The effect of wind may not be evident during the initial size-up when the structure is tightly closed. However, because wind can have a huge effect on fire behavior, the initial size-up of a structure fire must always include an evaluation of the impact of the wind. This evaluation will sometimes change the plan for attacking the fire. For example, if the fire's area of origin is located on the "C" side of the building and a strong wind is blowing from that direction, the incident commander (IC) might choose not to begin ventilation operations on the "C" side windows. Opening up a window on the "C" side could cause the fire to spread rapidly toward the "A" side. NIST research has shown that wind speeds as low as 10 mph (16 km/h) can create wind-driven fire conditions inside the structure (NIST 2016).

Characteristics of Liquid-Fuel Fires

Fires involving liquid fuels have some different characteristics from fires that involve solid fuels. Recall that solid fuels do not burn in the solid state but instead must be converted to a gas and mixed with oxygen before they will burn. Liquids share the same characteristic. They must be converted to a gas and mix with oxygen before they will burn. Like solid fuel, heat is what causes this state change. Three conditions must be present for a vapor and air mixture to ignite:

- The fuel and air must be present at a concentration within a flammable range.
- There must be an ignition source with enough energy to start ignition.
- The ignition source and the fuel mixture must make contact for long enough to transfer the energy to the air–fuel mixture.

As liquids are heated, the molecules in the liquid become more active, and the speed of vaporization of the molecules increases. Most liquids eventually reach their boiling point during a fire. **Boiling point** is the temperature at which a liquid continually vaporizes in sustained amounts and, if held at that temperature long enough, turns completely into a gas. As the boiling point is reached, the amount of flammable gas generated increases significantly (**FIGURE 5-26**). Because most liquid fuels are a mixture of compounds (for example, gasoline contains approximately 100 compounds), the fuel does not have a single boiling point. The single compound with the lowest ignition temperature determines the flammability of the mixture—the temperature at which that fuel will spontaneously ignite.

FIGURE 5-26 As liquids are heated, the molecules become more active, and the speed of vaporization increases.

Liquids that have a lower molecular weight tend to vaporize more readily than liquids with a higher molecular weight. Liquids that vaporize more readily are more volatile than liquids that vaporize more slowly. In addition, the higher the temperature, the quicker the liquid will evaporate. As more of the liquid vaporizes, it may reach a point where enough gas is present in the air to create a flammable vapor–air mixture.

Two additional terms are used to describe the flammability of liquids: flash point and fire point. The **flash point** is the lowest temperature at which a liquid or solid produces a flammable gas. It is measured by determining the lowest temperature at which a liquid produces enough vapor to support a small flame or flash fire for a short period of time until the fuel is consumed (the flame may go out quickly) (**TABLE 5-1**). The **fire point** (also known as the **flame point**) is the lowest temperature at which a liquid produces enough vapor to sustain a continuous fire. For most materials, the fire point is only slightly higher than the flash point.

TABLE 5-1 Flash Point

Liquid	Flash Point
Water	N/A
Gasoline	−45°F/−43°C
Acetone	−4°F/−20°C
#2 grade diesel	125°F/52°C

Modified from Friedman, Raymond. 1998. *Principles of Fire Protection Chemistry and Physics*. 3rd edition, p. 115. National Fire Protection Association.

Specific Gravity

A significant challenge faced when extinguishing flammable liquid-fuel fires is that many flammable liquids float on water. The term **specific gravity** refers to the density of a liquid compared to water, which has a specific density of 1.0. Gasoline, for example, has a specific gravity ranging from 0.72 to 0.76 and thus floats on top of water. **TABLE 5-2** lists the specific gravity of several liquids.

TABLE 5-2 Specific Gravity Examples

Liquid	Specific Gravity
Water	1.0
Gasoline	0.72–0.76
Acetone	0.79
#2 grade diesel	0.84

Data from Barsan, Michael E., ed. September 2007. *NIOSH Pocket Guide to Chemical Hazards*. Accessed August 23, 2023. https://stacks.cdc.gov/view/cdc/21265.

When flammable liquids have a specific gravity that is lighter than water, they float on water. If those liquids catch fire, the flames will continue to burn on the water's surface, and attempts to extinguish the flames with more water will simply spread the pool of burning flammable liquids across the water's surface. Alternative extinguishing agents, such as firefighting foam, may need to be applied to the floating pool of flammable liquids. In some cases, it may actually be the most practical to allow the pool of flammable liquids to burn while firefighters protect nearby structures and objects.

Heat-Induced Tears in Containers

Some liquid fuels are stored and transported in nonpressurized containers. A **heat-induced tear (HIT)** occurs when a nonpressurized vessel is exposed to direct **flame impingement**—flames in direct contact with the surface of a material transferring radiant heat—for a prolonged period, causing the structural integrity of the container to fail. When the container reaches a critical temperature, it ruptures and suddenly releases the flammable liquid inside. Once the escaping flammable liquid reaches its flash point and the gas finds an ignition source, the fuel rapidly ignites into a large fireball. A fireball is different from an explosion. A **fireball** is a burst of flames that rapidly ignites available flammable vapors but is not under pressure. An **explosion** is a violent and pressurized release of energy.

For example, when crude oil is transported in tank cars along the railroad, it is normally transported under very little, if any, pressure. If a tank car carrying crude oil derails, it might appear to catch fire and explode, but it is more probable that it experienced a HIT. A key distinction between HITs and explosions is that a HIT does not typically result in fragments of the container being launched outward.

Characteristics of Gas-Fuel Fires

Gas-fuel fires present unique challenges to firefighters. By learning about the characteristics of flammable gas fuels, you can help prevent injuries or deaths in emergency situations and work to mitigate the conditions causing the problem.

Vapor Density

Vapor density is the weight of a gas compared to an equal volume of dry air (**TABLE 5-3**). The weight of air is assigned the value of 1. A gas with a vapor density less than 1 rises to the top of a confined space or rises in the atmosphere. For example, hydrogen gas, which has a vapor density of 0.07, is a very light gas. Conversely, a gas with a vapor density greater than 1 is heavier than air and settles close to the ground. For example, propane gas, which has a vapor density of 1.55, settles to the ground when it is released from a container. Carbon monoxide has a vapor density of 0.97—almost the same as that of air—so it mixes readily with all layers of the air. In situations where a flammable gas is present, firefighters need to know the vapor density of the escaping fuel so that they can take actions to prevent the ignition of the fuel and allow the gaseous fuel to safely escape into the atmosphere.

TABLE 5-3 Vapor Density of Common Gases

Gaseous Substance	Vapor Density
Carbon monoxide	0.97
Hydrogen	0.07
Methane	0.55
Propane	1.55

Data from Barsan, Michael E., ed. September 2007. *NIOSH Pocket Guide to Chemical Hazards*. Accessed August 23, 2023. https://stacks.cdc.gov/view/cdc/21265.

Flammable Range

Flammable gases need oxygen to burn. Mixtures of flammable gas and air will burn only when they are mixed in specific proportions. If too much fuel is present in the mixture, there will not be enough oxygen to support the combustion process; if too little fuel is present in the mixture, there will not be enough fuel to support the combustion process. The range of air–fuel mixtures that will burn varies from one fuel to another. For example, carbon monoxide will burn when mixed with air in concentrations between 12.5 percent and 74 percent. Natural gas, however, will burn only when it is mixed with air in concentrations between 4.5 percent and 15 percent.

The lower percentage is the **lower explosive limit (LEL)**, also referred to as the **lower flammable limit (LFL)**. The LEL is the minimum percentage of gaseous fuel that must be present in an air–fuel mixture for the mixture to be flammable. For example, the LEL of carbon monoxide is 12.5 percent. The higher percentage is the **upper explosive limit (UEL)**, also referred to as the **upper flammable limit (UFL)**. The UEL is the maximum percentage of gaseous fuel that can be present in an air–fuel mixture for the mixture to be flammable. The UEL of carbon monoxide is 74 percent. The **flammable range**, also called the **explosive limits**, is the range between the lower and upper flammable limits. The terms *flammable range* and *explosive limits* are used interchangeably because under most conditions, if the flammable air–fuel mixture will not explode, it will not ignite. Test instruments are available to measure the percentage of fuels in air–fuel mixtures to determine when an emergency scene is safe.

Boiling Liquid/Expanding Vapor Explosions

One potentially deadly set of circumstances involving fuels that are gases in normal atmospheric conditions but are stored as a liquid under pressure in a container is a **boiling liquid/expanding vapor explosion (BLEVE)**. A BLEVE can occur when a liquid fuel is stored in a closed vessel under pressure. If the vessel is filled with propane, for example, the bottom part of the vessel would contain liquid propane, and the upper part of the vessel would contain gaseous propane (vapor) (**FIGURE 5-27**). If the sealed container is subjected to a source of high heat, the heat causes the liquid fuel to convert to its gaseous form. As the heat causes the liquid fuel to convert to its gaseous form, the vapor pressure inside the tank increases

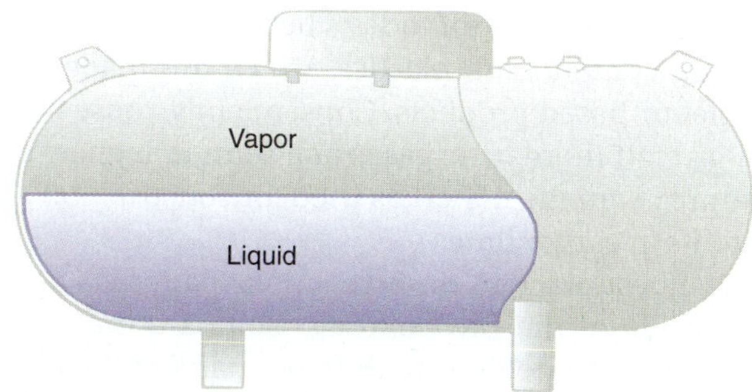

FIGURE 5-27 A propane tank contains both liquid and vapor.

© Jones & Bartlett Learning

FIGURE 5-28 This photo shows a fireball formed from an ignited BLEVE.

© Ivan Cholakov/Shutterstock

until gas escapes from the pressure relief valve. If the fire is not extinguished quickly, more and more gas escapes, the level of the liquid fuel inside the tank drops, and the surface of the tank begins to weaken due to extreme temperatures. If the internal pressure exceeds the strength of the container, the container can catastrophically rupture, and any propane still in its liquid form will vaporize and ignite in an expanding fireball (**FIGURE 5-28**).

The key to preventing a BLEVE is to cool the top of the tank, which contains the vapor. This action will prevent the fuel from building up enough pressure to cause a catastrophic rupture of the container. Prevention of BLEVEs is covered in more detail in Chapter 23, *Advanced Fire Suppression*.

FIGURE 5-29 The power source for electric vehicles is often a collection of numerous battery cells connected together and built into the floor of the car.

© leonello/iStock/Getty Images Plus/Getty Images

Characteristics of Battery Fires

Batteries and stored energy systems are increasingly used to power and propel everything from electronic devices to scooters, bicycles, cars, and even as backup power to buildings (**FIGURE 5-29**). Lithium-ion technology is often used as a rechargeable battery solution because it can deliver sustained amounts of power while charged, versus older alkaline battery technology where the power will fade as the amount of charge declines with use.

Lithium-ion batteries are small, lightweight, and have a high energy density. They are fabricated with a positive electrode, called a cathode; a negative electrode, called an anode; and an electrolyte (a liquid, gel, or paste-like substance that conducts electrical current) filling the space between the two electrodes. When lithium-ion batteries are charging, lithium ions (positively charged atoms) move from the positive cathode through the electrolyte to the negative anode. When these batteries discharge power, the lithium ions move from the anode to the cathode.

Although designs vary and the technology continues to evolve, electric vehicles are often constructed with a power source that comprises a collection of smaller battery cells connected together. It may be helpful to think of a skateboard when visualizing where the battery cells are positioned in an electric vehicle. In many modern electric vehicles, battery cells form the base of the car, much like a skateboard's deck. If an electric vehicle is involved in a motor vehicle crash and flipped onto its side or top, firefighters may be more directly exposed to the battery cells forming the floor of the car. Batteries in electric vehicles should always be treated as charged and having the ability to energize the car.

One of the dangers that electric vehicles pose are fires. When battery cells are compromised because of damage, extreme heat, or corrosion, the individual cells may self-discharge, overheat, catch fire, and even explode. Because these battery cells are generally constructed within a sealed metal case, overheating may cause the battery to swell from within, potentially causing the battery to explode due to overpressurization. When lithium-ion batteries catch fire, they can release immense amounts of heat, toxic gases, and smoke. Identifying electric vehicles and suppressing vehicle fires are discussed in more detail in Chapter 23, *Advanced Fire Suppression*, and in Chapter 26, *Vehicle Rescue and Extraction*.

TIP

Flames indicate what is happening now, whereas smoke gives a more complete picture of the characteristics of the fire and where it is going.

Reading Smoke

Learning the principles underlying fire behavior helps firefighters understand the rules that govern the way fires burn. One practical application of these principles is "reading" the smoke at a fire. Reading smoke helps you identify where the fire is, how big it is, and the direction it is moving. At a fire, most untrained people look at flames. Flames indicate where the fire is now; however, they do not tell you how big the fire is and where it is moving. At a structure fire, often there are no visible flames, because the fire is occurring inside the building. The ability to read smoke gives firefighters information that they need to mount a more effective attack on a fire, and this may help to save either their lives or the lives of the building's occupants.

Remember that smoke is a fuel. Fuels behave in predictable ways when they are combined with the right mixture of oxygen and heat. Most of the fuel in a flashover and in a backdraft is in the form of smoke. As you study the art of smoke reading, think of it as studying the fuel that is all around you at a fire.

Hot smoke is extremely flammable and will ultimately dictate the fire's behavior. Recall that smoke is composed of three major components: solids in the form of particles, liquids in the form of an aerosol, and gases. The solids consist of carbon, soot, dust, and fibers; the aerosol contains suspended hydrocarbons and water; and the gases consist of carbon monoxide and a wide variety of other gases. These three components of smoke represent fuels that will burn when heated sufficiently in the presence of oxygen. Also remember that many of today's building materials and building contents are plastics, which are made from petroleum-based products. Consequently, these materials give off more toxic gases and burn at higher temperatures than the materials used for construction and contents in earlier times.

The best place to observe smoke patterns is from the outside of the fire building. Compare the smoke coming out of openings in different parts of the building. It will tell you more than looking at the flames.

Step 1: Determine the Key Attributes of Smoke

The first step in reading smoke is to consider four key attributes of the smoke:

- Volume (amount)
- Velocity (speed)
- Density (thickness)
- Color

You also need to factor in the size of the building, the layout of the building, and any other characteristics of the building that might change the appearance of these key attributes. Finally, wind affects smoke and must be taken into considerations as well.

Volume

The volume of smoke coming from the fire provides some idea of how much fuel is being heated to the point that it off-gasses. As you assess the smoke volume, also consider the size of the burning building (the box). It takes relatively little smoke to fill up a small building and a lot of smoke to fill up a large building. Thus, you must consider the size of the box when considering the amount of smoke coming from it. Assessing the volume of smoke alone does not provide a complete picture of the fire, but it sets the stage for better understanding the fire.

Velocity

The velocity—speed—at which smoke leaves the building suggests how much pressure is accumulating inside the building. Smoke is pushed both by heat and by volume. In either case, smoke first rises fairly quickly. When smoke is pushed by heat, it then slows down gradually. When smoke is pushed by volume, it slows down immediately. Assess the smoke velocity to determine whether it is being pushed by heat or by volume.

While you are assessing the smoke velocity, consider whether the smoke has a turbulent or a laminar

FIGURE 5-30 Turbulent smoke flow.

Courtesy of Keith Muratori.

FIGURE 5-31 Laminar smoke flow.

© Robert Lessmann/Shutterstock

flow. A **turbulent smoke flow** is agitated, boiling, and angry. A turbulent flow is caused by rapid molecular expansion of the gases within the smoke and the restrictions created by the building (the box) (**FIGURE 5-30**). This rapid expansion occurs when the box cannot absorb any more heat. When this happens, the heat that the box has absorbed is radiated back into the smoke. Turbulent smoke contains an immense amount of heat energy. When this energy reaches a point where the smoke is heated to its ignition temperature in the presence of sufficient oxygen, flashover occurs. As the smoke travels away from the fire and radiates some of its energy, the flow smooths out. This smooth or streamlined flow is called a **laminar smoke flow**. It indicates that the smoke has less heat energy, that the box and its contents are absorbing heat, and that the pressure in the box is not too high, either because the fire is small or because the smoke has traveled some distance and has lost the high energy of turbulent smoke (**FIGURE 5-31**).

SAFETY TIP

Whenever you see turbulent smoke, be aware that flashover is likely to occur very soon.

By comparing the speed and type of flow streaming from similar-sized openings in the building, firefighters can get a good idea of where the fire is located. Look for the fastest smoke coming from the most restrictive opening. A similar-sized opening with a higher velocity of smoke is closer to the location of the fire. Studying the velocity of smoke provides valuable information as firefighters prepare to attack the fire.

Density

Evaluate the density—thickness—of the smoke. Smoke density suggests how much fuel is contained in the smoke. The denser the smoke, the more fuel it contains. Dense smoke can fuel a powerful flashover when conditions are right. In fact, dense smoke can flash over even without turbulent flow because the supply of fuel reaches from the heart of the fire to the location where the smoke is issuing from the building.

When you think of smoke as fuel, you quickly realize that firefighters who are surrounded by dense smoke are working in a pool of flammable fuel. Although no one would send a firefighter into a pool of petroleum, we are doing just that when we send a firefighter into dense smoke. Moreover, dense smoke contains many poisonous and cancer-causing substances. When firefighters are in an environment of dense smoke that has banked down to the floor, they are in an environment that will not support life for a building occupant. Just a few breaths of this toxic brew will result in death.

Color

The color of smoke may give you some indication of what substances are burning, which stage the fire is in, and the location of the fire's area of origin in the building. If a single substance is burning, the color of the smoke may provide clues to help you determine the identity of that substance. Of course, most fires involve multiple substances burning, so using smoke color alone to determine what is burning does not always yield a clear answer.

Smoke color can also help you determine the stage of heating. Some solid fuels such as wood emit white smoke when they are first heated. This white color is primarily a result of moisture being released from the material. When firefighters apply water to a fire, the same effect happens and the smoke changes from black to white. As fuel continues to burn and dries out, the color of the smoke changes. Smoke from burning wood changes to tan or brown. Plastics and painted or stained surfaces emit gray smoke, which is a combination of black smoke from the hydrocarbons and white smoke from the escaping moisture. Synthetic materials such as polyester emit white smoke, which is actually uncombusted fuel.

The color of smoke can also help firefighters determine the location of the fire. Smoke is darkest at the fire's area of origin. As smoke travels, its heat causes moisture to evaporate from some of the materials it passes through. This added moisture changes black smoke to lighter smoke the farther it travels from the fire. In addition, the carbon suspended in black smoke settles out and is filtered by the materials it passes through and this also causes the color to become lighter as it travels farther from the fire's area of origin.

How do you differentiate between white smoke caused by early heating and white smoke from a hot fire that has traveled a distance? White smoke that is lazy or slow indicates early heating. White smoke that has its own pressure indicates smoke from a hot fire that has traveled for some distance.

Black Fire

Black fire is a high-volume, high-velocity, turbulent, ultra-dense, black smoke. It is referred to as black *fire* even though it is smoke because in many ways it is a form of fire. Black fire is so hot, at temperatures up to 1000°F (540°C), that it creates a deadly environment. Black fire produces charring as well as heat damage to steel and concrete. Its presence indicates impending autoignition and flashover. Black fire indicates that the environment is just seconds away from becoming deadly—even for firefighters wearing full PPE, and therefore obviously not tenable for unprotected occupants of a building—due to flashover or autoignition. When black fire occurs, there are usually no lives to be saved in that location.

Step 2: Determine What Is Influencing the Key Attributes

The second step in reading smoke is to evaluate the influences on the four key attributes of the smoke. Start by considering the size of the structure that is on fire. How big is the box in which the fire is contained? A small amount of smoke rapidly fills up a small building, so it does not take a big fire to fill a small building with smoke. Conversely, in a large building (such as a "big box" store or a warehouse), it takes a large fire to produce enough smoke to fill the whole building and pressurize the smoke coming out of the building. A firefighter who pulls up to a large building that has signs of smoke from the outside should realize that only a significant fire would produce enough smoke to vent to the outside. Always consider the size of the box when examining smoke for clues about the fire.

Next, evaluate the following factors:

- Is there wind? Wind can change the direction in which the smoke travels.
- What is the normal thermal balance in a room or building? That is, is the temperature of the room or building affected by the heating and air-conditioning systems? This is especially relevant in a high-rise building.
- What fire streams are being applied? Fog nozzle or straight tip? Is the water being applied to cool superheated gases overhead or applied to the source of the fire?
- Has a ventilation team created any ventilation openings? Ventilation openings are made with the intention of changing the normal flow of smoke, so the characteristics of the smoke would naturally be expected to change as the building is ventilated.
- Has an operating fire sprinkler system been activated? An operating sprinkler system causes smoke to be cooled, and this cooled smoke will hang close to the bottom of the room, making it more difficult to see.

Step 3: Determine the Rate of Change

The third step in reading smoke is to determine the rate of change. Remember that flames indicate what is happening now, whereas smoke gives a more complete picture of the characteristics of the fire and where it is going. Are the volume, velocity, density, and color of the smoke changing? If they are, how are they changing and how rapidly are these changes occurring? Do the changes indicate improving or worsening conditions? What do these changes suggest about the progression of the fire?

Step 4: Predict the Event

The final step in reading smoke is to put the pieces of the puzzle together to determine the location of the fire's area of origin, the size of the fire, and the potential for a hostile fire event such as a backdraft or a flashover. Communicate the key parts of these observations to your company officer. With practice, this assessment process will help you systematically evaluate the information the smoke signals are giving you about the fire.

One way to become more proficient in smoke reading is to review videos of fires. Assess the smoke at the beginning of the tape and try to identify the location of the fire's area of origin and the stage of burning. You will be surprised at how much information you can obtain from a short video clip. Smoke is deadly. If you want to stay alive, you need to understand it.

Reading Smoke at Doors and Windows

When you see indications of a hot fire, such as darkened windows but little visible smoke moving around closed doors and windows, you may be dealing with a fire that is in a decay stage because it is ventilation-limited. This is a sign of great danger. As soon as this type of fire receives a fresh supply of oxygen by opening a door or window or other means of ventilation, it will likely produce a rapid fire growth event and possibly a backdraft.

When you open a door, watch what the smoke does and identify the neutral plane to help you identify the location of the fire. If smoke exits through the top half of the door and clean air enters through the bottom half of the opening, then the fire is probably on the same level. This phenomenon is sometimes called "smoke that has found balance." If smoke rises and the opening clears out, fresh air is being pulled into the building. This indicates the fire is probably above the level of the opening. If the smoke thins when the door is opened but smoke still fills the doorway, the fire may be below the level of the opening.

CASE STUDY

You Are the Firefighter CONCLUSION

It is 0346 and you arrive at a modern-looking two-story house for the report of a garage fire called in by a neighbor. Through the smoke and flames, you observe what appears to be a popular model of an electric vehicle parked inside the garage and involved in the fire. As you begin to apply water to the body of fire, you witness fire conditions growing rapidly and extending into the house. The hair on the back of your neck rises as you get an unsettled feeling.

1. **Why would this situation make you uneasy?**

   Answer: A car in the garage suggests that there are occupants inside the burning structure, necessitating efforts to coordinate search and rescue operations while also conducting fire attack operations.

2. **How might the structure's construction and contents contribute to fire behavior?**

   Answer: Modern structures are constructed with lightweight, manufactured building components that contain many highly combustible products, such as glues and coatings, most of which are made from petroleum-based products that contain significant potential heat energy. They also contain less material (less mass) than houses built with lumber, provide greater surface area of structural members, and fail much faster than the heavier wood framing of older homes.

3. **What elements of this scene may make extinguishment of the fire challenging?**

   Answer: Electric vehicles have batteries that pose significant risks when involved in fire, and extinguishing these fires can prove to be very difficult, often requiring thousands of gallons of water or a strategy of isolating the vehicle away from other combustibles to allow it to burn out on its own.

WRAP-UP

SUMMARY

KNOWLEDGE OBJECTIVES

- Describe the basic chemistry of fire, including the types of fuels, conditions, and reactions necessary for combustion, as well as how fire spreads, transfers heat, and releases by-products.
  - Describe the chemistry of fire. (pp. 164–166)
  - List the three states of matter. (**NFPA 1010: 6.3.10**, pp. 164–165)
  - List the forms of energy. (pp. 165–166)
  - List different ways heat energy is measured. (pp. 165–166)
  - Explain the concept of the fire triangle. (**NFPA 1010: 6.11**, p. 167)
  - Explain the concept of the fire tetrahedron. (**NFPA 1010: 6.3.11**, p. 167)
  - Describe the chemistry of combustion. (**NFPA 1010: 6.3.11**, p. 168)
  - Describe the by-products of combustion. (**NFPA 1010: 6.3.11**, pp. 168–169)
  - Explain how fires are spread by conduction, convection, and thermal radiation. (**NFPA 1010: 6.3.12**, pp. 169–172)
  - Define flow path and describe how it influences the growth of a building fire. (**NFPA 1010: 6.3.11**, pp. 169–172)
- Identify the five classes of fire and how each type can be extinguished.
  - Describe the four methods of extinguishing fires. (pp. 172–173)
  - Define Class A, B, C, D, and K fires. (pp. 173–174)
  - Describe fires involving mixed materials. (pp. 174–175)
- Describe fire behavior, including the stages of fire; how to recognize changing fire conditions; and how factors such as building construction, building contents, flow paths, and wind impact fire growth.
  - Describe the importance of the following characteristics in solid-fuel fires: composition of fuel, amount of fuel, and configuration of fuel. (**NFPA 1010: 6.3.11**, pp. 175–176)
  - Describe the four stages of fire development: incipient stage, growth stage, fully developed stage, and decay stage. (**NFPA 1010: 6.3.11**, pp. 176–182)
  - Define the following terms: thermal layering, neutral plane, rollover, flashover, backdraft, fuel-limited fires, ventilation-limited fires, and smoke explosion. (pp. 177–182)
  - Describe the conditions that cause thermal layering. (**NFPA 1010: 6.3.12**, p. 178)
  - Describe the conditions that lead to rollover. (**NFPA 1010: 6.3.12**, p. 178)
  - Describe the conditions that lead to flashover. (**NFPA 1010: 6.3.12**, pp. 178–179)
  - Describe the conditions that lead to a backdraft. (**NFPA 1010: 6.3.11**, p. 180)
  - Describe the conditions that lead to rapid fire growth. (pp. 180–181)
  - Describe the conditions that lead to a fuel-limited fire and a ventilation-limited fire. (pp. 180–181)
  - Describe the conditions that lead to a smoke explosion. (p. 182)
  - Describe how fire behaves in modern structures. (pp. 182–183)
  - Describe how the wind effect impacts fire behavior. (pp. 183–184)
- Describe the characteristics and basic chemical properties of liquid- and gas-fuel fires.
  - Describe the characteristics of liquid-fuel fires. (pp. 184–186)
  - Explain the concept of specific gravity. (p. 185)
  - Define the following term: heat-induced tear. (pp. 185–186)
  - Define the following terms: boiling point, flash point, and fire point. (pp. 184–186)
  - Define the characteristics of gas-fuel fires. (p. 186)
  - Explain the concept of vapor density. (p. 186)
  - Explain the concept of flammable range. (p. 186)
  - Define the following terms: lower explosive limit (LEL) and upper explosive limit (UEL). (p. 186)
  - Describe the causes and effects of a boiling liquid/expanding vapor explosion (BLEVE). (pp. 186–187)
- Describe the risks and safety considerations associated with fires involving batteries and stored energy systems.

- Explain the concept of a stored energy system. (pp. 187–188)
- Describe how to read smoke.
  - List the steps of reading smoke. (pp. 188–191)
  - List the attributes of smoke that are assessed. (pp. 188–190)
  - Describe how smoke is assessed at doors and windows. (p. 191)

SKILLS OBJECTIVES

There are no skills objectives for Firefighter I candidates. NFPA 1001 contains no Firefighter I Job Performance Requirements for this chapter.

KEY TERMS

aerosol An intimate mixture of a liquid or a solid in a gas; the liquid or solid, called the dispersed phase, is uniformly distributed in a finely divided state throughout the gas, which is the continuous phase or dispersing medium. (MED) (NFPA 99)

area of origin The room or general area where a fire started. Also called *fire compartment, fire seat, or seat of the fire.*

atom The smallest particle of an element that retains the properties of that element.

autoignition Initiation of combustion by heat but without a spark or flame. (NFPA 921)

backdraft A deflagration (explosion) resulting from the sudden introduction of air into a confined space containing oxygen-deficient products of incomplete combustion. (NFPA 1403)

black fire A hot, high-volume, high-velocity, turbulent, ultra-dense black smoke that can produce charring and cause heat damage to steel and concrete and indicates an impending flashover or autoignition.

boiling liquid/expanding vapor explosion (BLEVE) An explosion that occurs when pressurized liquefied materials (e.g., propane or butane) in a closed container are exposed to a source of high heat, releasing the fuel which instantly vaporizes and ignites.

boiling point The temperature at which the vapor pressure of a liquid equals the surrounding atmospheric pressure. (NFPA 1)

British thermal unit (BTU) The amount of heat energy required to raise 1 pound of water at sea level by 1 degree Fahrenheit.

calorie The amount of heat energy required to raise 1 gram of water (at sea level) by 1 degree Celsius.

carbon dioxide (CO_2) A nontoxic gas produced when sufficient oxygen is available for complete combustion that can displace oxygen in the atmosphere. Also, a colorless, odorless, electrically nonconductive inert gas that is a suitable medium for extinguishing Class B and Class C fires. (NFPA 10)

carbon monoxide (CO) A toxic gas produced through incomplete combustion.

ceiling jet A strong, turbulent convection current that rose to the ceiling and traveled along it.

chemical energy Potential energy in molecular bonds and the kinetic energy created by a chemical reaction.

Class A fire A fire in ordinary combustible materials, such as wood, cloth, paper, rubber, and many plastics. (NFPA 1)

Class B fire A fire in flammable liquids, combustible liquids, petroleum greases, tars, oils, oil-based paints, solvents, lacquers, alcohols, and flammable gases. (NFPA 1)

Class C fire A fire that involves energized electrical equipment. (NFPA 1)

Class D fire A fire in combustible metals, such as magnesium, titanium, zirconium, sodium, lithium, and potassium. (NFPA 1)

Class K fire A fire in a cooking appliance that involves combustible cooking media (vegetable or animal oils and fats). (NFPA 1)

combustible Capable of undergoing combustion. (NFPA 1700)

combustion A chemical process of oxidation that occurs at a rate fast enough to produce heat and usually light in the form of either a glow or a flame. (NFPA 1)

KEY TERMS CONTINUED

conduction Heat transfer to another body or within a body by direct contact. (NFPA 921)

convection Heat transfer by circulation within a medium such as a gas or a liquid. (NFPA 921)

convection column See *plume.*

decay stage The stage of fire development within a structure characterized by either a decrease in the fuel load or available oxygen to support combustion, resulting in lower temperatures and lower pressure in the fire area. (NFPA 1410)

electrical energy Energy is produced by an electrical charge.

element A substance that cannot be chemically broken down into a simpler substance.

endothermic A chemical reaction that absorbs heat.

energy The ability to do work.

entrain To encircle, draw along, and transport.

exothermic A chemical reaction that produces heat.

explosion A violent and pressurized release of energy.

explosive limits See *flammable range.*

fire The visible result of combustion.

firebreak A swath where the fuel (trees and brush) are removed.

fire compartment The area of origin when the fire is in a structure.

fire point The lowest temperature at which a liquid will ignite and achieve sustained burning when exposed to a test flame in accordance with ASTM 92, Standard Test Method for Flash and Fire Points by Cleveland Open Cup Tester. Also called *flame point.* (NFPA 1)

fire seat See *area of origin.*

fire tetrahedron A geometric shape used to depict the four components—fuel, oxygen, heat, and chemical chain reactions—required for a fire to occur.

fire triangle The three components—fuel, oxygen, and heat—required for combustion.

fireball A burst of flames that rapidly ignites available flammable vapors but is not under pressure.

flame impingement Flames in direct contact with the surface of a material transferring radiant heat.

flame inhibitor A chemical extinguishing agent that reacts with the fuel to chemically disrupt the combustion process.

flameover See *rollover.*

flame point See *fire point.*

flammable range The range in concentration between the lower and upper flammable limits. Also called *explosive limits.* (NFPA 67)

flashover A transition phase in the development of a compartment fire in which surfaces exposed to thermal radiation reach ignition temperature more or less simultaneously, and fire spreads rapidly throughout the space, resulting in full room involvement or total involvement of the compartment or enclosed space. (NFPA 921)

flash point The minimum temperature at which a liquid or a solid emits vapor sufficient to form an ignitable mixture with air near the surface of the liquid or the solid. (NFPA 115)

flow path The movement of heat and smoke from the higher pressure within the fire area toward the lower pressure areas accessible via doors, window openings, and roof structures. (NFPA 1410)

fuel A material that will maintain combustion under specified environmental conditions. (NFPA 53)

fuel-limited fire A fire in which the heat release rate and fire growth are controlled by the characteristics of the fuel because there is adequate oxygen available for combustion. (NFPA 1410)

fully developed stage The stage of fire development where heat release rate has reached its peak within a compartment. (NFPA 1410)

gas The physical state of a substance that has no shape or volume of its own and will expand to take the shape and volume of the container or enclosure it occupies. (NFPA 921)

growth stage The stage of fire development where the heat release rate from an incipient fire has increased to the point where heat transferred from the fire and the combustion products are pyrolyzing adjacent fuel sources and the fire begins to spread across the ceiling of the fire compartment (rollover). (NFPA 1410)

heat energy The potential energy of a combustible material and the kinetic energy released when heat is applied to the material. Also called *thermal energy.*

heat flux The measure of the rate of heat transfer to a surface, typically expressed in kilowatts per meter squared (kW/m^2) or BTU/ft^2. (NFPA 268)

heat-induced tear (HIT) A tear in a nonpressurized container that occurs when the container is exposed to direct or indirect flame impingement for a prolonged period, causing the structural integrity of the container to fail.

heat release rate (HRR) The rate at which heat energy is generated by burning. (NFPA 921)

heat stratification See *thermal layering*.

heat transfer The movement of heat energy from a hotter medium to a cooler medium by conduction, convection, or radiation.

hydrogen cyanide (HCN) An extremely toxic gas produced by the incomplete combustion of many common plastic-based materials.

ignition temperature Minimum temperature a substance should attain in order to ignite under specific test conditions. (NFPA 402)

incipient stage The early stage of fire development where the fire's progression is limited to a fuel source and the thermal hazard is localized to the area of the burning material. (NFPA 1410)

incomplete combustion A combustion process during which the fuel is not completely consumed, usually due to a limited supply of oxygen.

joule A measure of heat energy equal to 0.4 calorie.

kinetic energy Energy possessed by an object because of its motion.

laminar smoke flow The smooth or streamlined movement of smoke.

light energy Energy produced by electromagnetic waves packaged in discrete bundles called photons.

liquid Matter that has a specific volume but does not have a specific size or shape.

lower explosive limit (LEL) The minimum concentration of a combustible vapor or combustible gas in a mixture of the vapor or gas and gaseous oxidant, above which propagation of flame will occur on contact with an ignition source. (NFPA 115)

lower flammable limit (LFL) See *lower explosive limit*.

matter Anything that has mass and volume.

mechanical energy The potential energy stored because of the position of an object or the kinetic energy of an object in motion.

molecular bond The connection between atoms in a molecule.

molecule Atoms chemically bonded together.

neutral plane The interface at a vent, such as a doorway or a window opening, between the hot gas flowing out of a fire compartment and the cool air flowing into the compartment where the pressure difference between the interior and exterior is equal.

nuclear energy Potential energy stored in the nucleus of an atom or the kinetic energy released by splitting the nucleus of an atom into two smaller nuclei (fission) or by combining two small nuclei into one large nucleus (fusion).

off-gas To emit a gas, often harmful, organic vapors.

oxidation Reaction with oxygen either in the form of the element or in the form of one of its compounds. (NFPA 53)

phosgene A gas formed from incomplete combustion of many common household products.

plume The column of hot gases, flames, and smoke rising above a fire. Also called *convection column, thermal updraft*, or *thermal column*. (NFPA 921)

potential energy Energy stored by an object as a result of its position or condition.

pyrolysis A process in which material is decomposed, or broken down, into simpler molecular compounds by the effects of heat alone; pyrolysis often precedes combustion. (NFPA 921)

rollover The condition in which unburned fuel (pyrolysate) from the originating fire has accumulated in the ceiling layer to a sufficient concentration (i.e., at or above the lower flammable limit) that it ignites and burns. This can occur without ignition of, or prior to the ignition of, other fuels separate from the origin. Also called *flameover*. (NFPA 921)

seat of the fire See *area of origin*.

smoke The airborne solid and liquid particulates and gases evolved when a material undergoes pyrolysis or combustion, together with the quantity of air that is entrained or otherwise mixed into the mass. (NFPA 1404)

smoke explosion A violent release of energy that occurs when smoke travels away from its source to a void area or other area separate from the fire compartment and comes in contact with a source of ignition without any change to the ventilation profile.

smoke particles The unburned, partially burned, and completely burned substances found in smoke.

solid Matter that has a specific size and shape; one of the three states of matter.

specific gravity The density of a liquid compared to water (which is 1.0).

state of matter The physical state of a material (solid, liquid, or gas).

KEY TERMS CONTINUED

temperature The measurement of the movement of molecules used to describe how hot or cold something is.

thermal column See *plume.*

thermal conductivity The ability of a material to conduct heat.

thermal energy See *heat energy.*

thermal layering The phenomenon of gases forming into layers according to their temperatures. Also called *heat stratification.*

thermal radiation The means by which heat is transferred to other objects.

thermal updraft See *plume.*

turbulent smoke flow The agitated, boiling, and angry movement of smoke caused by rapid molecular expansion of the gases within the smoke and the restrictions of the box containing the smoke.

upper explosive limit (UEL) The maximum amount of gaseous fuel that can be present in the air if the air–fuel mixture is to be flammable or explosive.

upper flammable limit (LFL) See *upper explosive limit.*

vapor See *gas.*

vapor density The weight of a gas compared to an equal volume of dry air.

ventilation-limited fire A fire in which the heat release rate and fire growth are regulated by the available oxygen within the space. (NFPA 1410)

REVIEW QUESTIONS

1. What is the fire triangle? What is the fire tetrahedron?
2. What are the three methods of heat transfer?
3. What are the four methods of extinguishing fire?
4. Describe the five classes of fire.
5. How do solid materials burn?
6. What are the four classic stages of fire development?
7. Describe rollover, flashover, and backdraft.
8. How do modern homes differ from those constructed several decades ago?
9. What effect does wind have on fire behavior?
10. What three conditions that must be present for liquid fuels to ignite?
11. What is vapor density?
12. What type of incident can occur when a liquid fuel is stored in a closed vessel under pressure and the container is exposed to heat?
13. What happens when lithium-ion batteries catch fire?
14. What are the four key attributes of smoke that you need to assess in the first step of reading smoke?

DISCUSSION QUESTIONS

1. Why it is important to understand the five classes of fire?
2. How do factors such as building construction, building contents, flow paths, and wind impact fire growth?
3. Discuss the risks, safety considerations, and tactical priorities associated with fires involving batteries and stored energy systems.

APPLYING THE CONCEPTS

Your engine unit is en route to a fire in a two-story house in a residential neighborhood. Dispatch reported a teenage resident made the call and stated the fire started on an enclosed back patio when he and his friends were making a fire. No victims are reported to be inside. You learned at your morning shift briefing that today's weather is overcast with 15 mph (24 km/h) winds, gusting up to 25 mph (40 km/h). Your crew arrives on the scene, and you see an active fire with the wind blowing toward side Charlie of the newer-construction house. During the 360-degree size-up it is confirmed that the fire started in the rear of the house on the patio. It is already progressed into the house with smoke and flames visible on the first floor, and smoke visible on the second floor. Additional resources are on their way.

1. What environmental factors must be considered when determining an initial plan of attack?
2. What structural factors must be considered when doing the size-up?

The IC radios detailed findings from the size-up including fire behavior, wind conditions, and house construction, and calls for additional resources. The IC commands, "Unit A, we need to obstruct flow paths. Confirm all doors and windows are closed. Unit B, advance a 1¾-in. (44-mm) hose line to side Charlie."

3. Why is it a priority to close doors and windows?
4. Although a ventilation-limited fire burns more slowly as oxygen is reduced, what dangers does it present?

You and your crew are conducting a transitional attack on side Charlie toward side Alpha to cool the environment and begin an offensive attack. Neighbors have gathered out front to watch and to console family members. One of the children cries, "Boots?! Where's Boots? I think he's still inside!" Hoping to create a way for the pet to escape, a well-meaning bystander runs to the front door of the house and opens it.

5. Why is attacking the fire from side Alpha the wrong decision?
6. Besides potentially causing a backdraft, what effect does opening the front door have on the fire?
7. You notice the house starts to look like it's breathing as smoke puffs out of it and is sucked back inside. What does this mean?

REFERENCES

Barsan, Michael E., ed. 2007. NIOSH Pocket Guide to Chemical Hazards. Accessed August 23, 2023. https://www.cdc.gov/niosh/npg/default.html.

Durham, Patrick. 2022. *Electric Vehicle Response: Fire Attack and Extrication Basics.* FireRescue1, March 1, 2022. Accessed December 1, 2022. https://www.firerescue1.com/electric-fire/articles/electric-vehicle-response-fire-attack-and-extrication-basics-PwPBmx8uuMuMOR2G.

Fabian, Thomas, Jacob L. Borgerson, Stephen I. Kerber, Pravinray D. Gandhi, C. Stuart Baxter, Clara Sue Ross, James E. Lockey, and James M. Dalton. 2010. *Firefighter Exposure to Smoke Particulates.* Underwriters Laboratories (UL), April 1, 2010. Accessed July 22, 2023. https://fsri.org/sites/default/files/2021-07/EMW-2007-FP-02093-Executive-Summary.pdf.

Gann, Richard, and Raymond Friedman. 2015. *Principles of Fire Behavior and Combustion,* 4th edition. Burlington, MA: Jones & Bartlett Learning.

Hall, Shelby, and Ben Evarts. 2022. *Fire Loss in the United States during 2021.* National Fire Protection Association (NFPA), September 2022. Accessed July 19, 2023. http://www.nfpa.org/news-and-research/fire-statistics-and-reports/fire-statistics/fires-in-the-us.

International Association of Fire Chiefs (IAFC). 2021. *IAFC Bulletin: Fire Department Response to Electric Vehicle Fires.* October 15, 2021. Accessed December 1, 2022. https://www.iafc.org/docs/default-source/1haz/respondingtoelectricalvehiclefires.pdf?sfvrsn=9421650c_6.

Kerber, Stephen. 2009. *Impact of Ventilation on Fire Behavior in Legacy and Contemporary Residential Construction.* Underwriters Laboratories (UL). Accessed June 18, 2024. https://fsri.org/research/impact-ventilation-fire-behavior-legacy-and-contemporary-residential-construction.

Minos, Scott. 2023. *How Lithium-Ion Batteries Work.* U.S. Department of Energy (DOE), February 28, 2023. Accessed December 1, 2022. https://www.energy.gov/eere/articles/how-does-lithium-ion-battery-work.

National Fire Protection Association (NFPA). 2023. *NFPA 1, Fire Code.* 2024 Edition. Quincy, MA: NFPA.

National Fire Protection Association (NFPA). 2021. *NFPA 10, Standard for Portable Fire Extinguishers.* 2022 Edition. Quincy, MA: NFPA.

National Fire Protection Association (NFPA). 2023. *NFPA 24, Standard for the Installation of Private Fire Service Mains and Their Appurtenances.* 2022 Edition. Quincy, MA: NFPA.

National Fire Protection Association (NFPA). 2020. *NFPA 53, Recommended Practice on Materials, Equipment, and Systems Used in Oxygen-Enriched Atmospheres.* 2021 Edition. Quincy, MA: NFPA.

National Fire Protection Association (NFPA). 2018. *NFPA 67, Guide on Explosion Protection for Gaseous Mixtures in Pipe Systems.* 2019 Edition.

National Fire Protection Association (NFPA). 2023. *NFPA 99, Health Care Facilities Code.* 2024 Edition. Quincy, MA: NFPA.

National Fire Protection Association (NFPA). 2019. *NFPA 115, Standard for Laser Fire Protection.* 2020 Edition. Quincy, MA: NFPA.

National Fire Protection Association (NFPA). 2021. *NFPA 268, Standard Test Method for Determining Ignitability of Exterior Wall Assemblies Using a Radiant Heat Energy Source.* 2022 Edition. Quincy, MA: NFPA.

National Fire Protection Association (NFPA). 2021. *NFPA 291, Recommended Practice for Water Flow Testing and Marking of Hydrants.* 2022 Edition. Quincy, MA: NFPA.

National Fire Protection Association (NFPA). 2021. *NFPA 550, Guide to the Fire Safety Concepts Tree.* 2022 Edition. Quincy, MA: NFPA.

REFERENCES CONTINUED

National Fire Protection Association (NFPA). 2018. *NFPA 402, Guide for Aircraft Rescue and Fire-Fighting Operations.* 2019 Edition. Quincy, MA: NFPA.

National Fire Protection Association (NFPA). 2017. *NFPA 1403, Standard on Live Fire Training Evolutions.* 2018 Edition. Accessed October 10, 2023. Quincy, MA: NFPA.

National Fire Protection Association (NFPA). 2017. *NFPA 1404, Standard for Fire Service Respiratory Protection Training.* 2018 Edition. Quincy, MA: NFPA.

National Fire Protection Association (NFPA). 2019. *NFPA 1410, Standard on Training for Emergency Scene Operations.* 2020 Edition. Quincy, MA: NFPA.

National Fire Protection Association (NFPA). 2020. *NFPA 1, Fire Code.* 2021 Edition. Quincy, MA: NFPA.

National Fire Protection Association (NFPA). 2020. *NFPA 921, Guide for Fire and Explosion Investigations.* 2021 Edition. Quincy, MA: NFPA.

National Fire Protection Association (NFPA). 2020. *NFPA 1700, Guide for Structural Fire Fighting.* 2021 Edition. Quincy, MA: NFPA.

National Institute of Standards and Technology (NIST). 2016. Videos for Wind-Driven Fires: Governors Island and Laboratory Experiments. Updated September 21, 2016. Accessed July 24, 2023. https://www.nist.gov/el/fire-research-division-73300/firegov-fire-service/videos-wind-driven-fires-governors-island.

United States Fire Administration (USFA). 2023. Fire Death and Injury Risk. Last reviewed August 2, 2023. Accessed August 6, 2023. https://www.usfa.fema.gov/statistics/deaths-injuries/.

CHAPTER

6

Firefighter I

Building Construction

KNOWLEDGE OBJECTIVES

After studying this chapter, you will be able to:

- Describe the basic principles of building construction.
- Describe how occupancy classifications and the contents of a structure affect fire suppression operations.
- Identify the various materials and components utilized in building construction.
- Describe the types of construction and effects of fire on their structural integrity.
- Describe single-family dwelling types and their characteristics.
- Describe the hazards created by fire conditions and the potential for building collapse.
- Identify the information to collect to determine a structure's type of building construction.

SKILLS OBJECTIVES

- Evaluate and forecast a fire's growth and development and determine developing hazardous building or fire conditions.
- Recognize hazards.

ADDITIONAL NFPA STANDARDS

- **NFPA 80**, *Standard for Fire Doors and Other Opening Protectives, 2022 Edition*
- **NFPA 101**, *Life Safety Code, 2024 Edition*
- **NFPA 220**, *Standard on Types of Building Construction, 2024 Edition*
- **NFPA 5000**, *Building Construction and Safety Code, 2024 Edition*

CASE STUDY

You Are the Firefighter

Your engine company has been dispatched to a house fire. You remember from your area familiarization training that the subdivision is mostly 1950s ranch houses. You pull up on the Bravo side of the structure and your first impression is that the house is not a ranch-style; it shows two stories and looks much older than the surrounding houses. You note that there is a large volume of fire from the roof. You also see that the windows on the first and second floor align and you see that the front door and the window above line up over each other.

1. What building construction characteristics would you expect to see on a ranch built in the 1950s or 1960s?
2. Instead of a ranch, what style of construction does the building appear to be?
3. How will the tactical considerations change now that you know the type of building construction?

Introduction

Knowledge of all aspects of building construction is critical for every firefighter. A firefighter who does not have a sound understanding of building construction and materials is like a baseball player who doesn't know the distance between the bases, or, more to the point, like a soldier who doesn't understand the field of battle on which they will fight. A firefighter must have a thorough understanding of the basics of building construction. A building under attack by fire is literally being deconstructed every minute the fire is not suppressed. The firefighter armed with foundational knowledge about basic principles of building construction, construction materials, building components, types of construction, and, most importantly, how all of these interact under emergency conditions to contribute to fire growth and potential collapse of the building has the best chance of operating effectively during an emergency incident. This, in turn, gives the occupants, and firefighters, the best chance for a successful outcome.

Basic Principles of Building Construction

Understanding the basics underlying all aspects of building construction—force and load—allows firefighters to more ably predict how a building will react under emergency conditions. A force is a push or a pull on an object. **Loads** are the forces to which a structure is subjected due to superimposed weight or pressures. Forces and loads work in tandem on the building materials and components that form any structure and are calculated to ensure the integrity of a structure before it is subjected to fire conditions. These fundamental principles largely hold true regardless of the type of building, its components, or the construction materials used.

Forces

Forces are a fundamental physical presence; for example, gravity and wind are forces. Gravity is a constant force, and the design and construction of any building continually fights to withstand gravity. Buildings must also be able to withstand forces that are not constant, such as those generated by wind.

Understanding not only gravity, but the contribution of additional forces that can degrade the structural stability of any building, is vital. During fire or other emergency conditions, a building's ability to withstand gravity while it is being *de*constructed affects every decision during an incident. For example, if structural elements in an attic catch fire, it is only a matter of time before gravity exerts its influence and the roof collapses. Forces and loads work in tandem on the building materials and components that form any structure and are calculated to ensure the integrity of any structure before it is subjected to fire conditions.

Three main forces interact in structures, and these interactions keep buildings standing despite the pull of the force of gravity. Understanding these forces forms the basis for building design. The forces are as follows (**FIGURE 6-1**):

- **Compressive force** pushes a material together. For example, the walls in a house are in compression, which means that everything above them is pressing down with compressive force.
- **Tensile force** pulls a material apart.

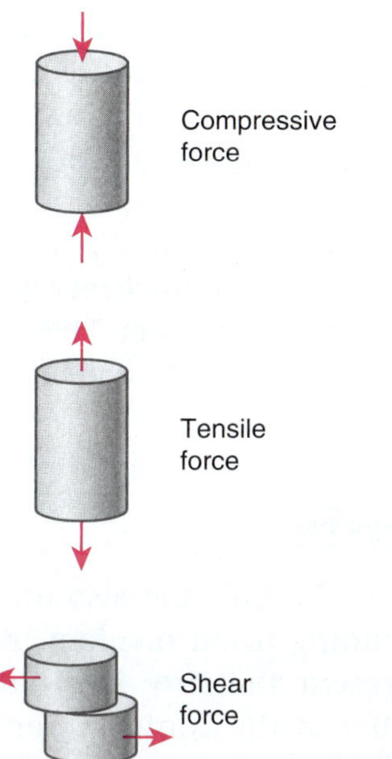

FIGURE 6-1 Compressive forces push a material together, tensile forces pull a material apart, and shear forces occur when materials slide past each other.

- **Shear force** occurs when materials slide past each other. For example, when the force from an earthquake causes a multi-story structure to shift and move floors in opposite directions.

These forces deform the materials or objects with which they interact. A stack of tissues, a cardboard tube from a paper towel roll standing on its end, and an iron post are technically all examples of columns. If you add weight to the top of a column, it compresses or shortens that column. The amount the column is compressed depends on the strength of the material. That compression will be more obvious in the stack of tissues, but it is present—if less visible—in any material, including in the iron post. Materials in tension become longer as they deform. An example of this are the links in a chain holding a large overhead light fixture. If the light is heavier than the links are rated for, they will begin to stretch out. Many materials deform when extreme heat is applied. For example, if those links in the light chain are heated, they will soften and stretch out. Another key example is the compression or tension exhibited on a truss, where the top chord is in compression, and the bottom cord is under tension. (A **truss** is a structural component that is composed of smaller pieces called members joined to form a triangle or a system of triangles. A chord is the straight part of the truss to which the triangles are connected.) The deformities that occur under normal conditions are magnified as greater force, heat, or loads from emergency or fire conditions are introduced.

Loads

Loads can be broadly categorized as dead or live. A **dead load** is a constant load in a structure, such as the roof, the flooring or anything else permanently attached to the building. Firefighters must recognize that over time, additional dead loads not accounted for in the original design may have been added, such as new heating or air conditioning units or cell towers installed on the roof. A **live load** is essentially all other loads in a building, including people, furniture, or equipment, as well as natural forces, such as snow, wind, rain, or water introduced during firefighting operations. Dead loads can be measured and are known. Live loads are constantly changing, particularly during a dynamic emergency event. If live loads are not taken into account, catastrophic failure can result.

Two types of live loads are impact loads and concentrated loads. An **impact load** is a load imposed quickly on a structure. The classic example of an impact load is a firefighter who jumps onto a roof from a ladder. The weight and load exerted by that firefighter is magnified greatly compared to stepping slowly onto the roof. A **concentrated load** is a load in one specific area of a building. For example, a large safe brought into a building and placed in the middle of a floor is a concentrated load; its weight is not necessarily being evenly distributed across the structural components.

Understanding how loads are placed on structural components means that firefighters will be able to better predict its ability to sustain damage. **Axial loads** go directly through the center, **eccentric loads** are off center, and **torsional loads** create twisting (**FIGURE 6-2**). Each of these loads affects a structure differently. They can affect a structure at different times, or they can be cumulative and pile on top of each other, depending on the situation. Firefighters need to be able to assess the structure and the current conditions, decide whether the structure is already degraded or is in the process of degrading, and decide whether fire or other conditions such as an explosion, a weather-related event, or a structural failure due to human error is causing the degradation. In those moments, they also need to take into account whether any additional forces and loads will contribute to the incident. These decisions will allow them to choose the best way to mitigate the incident effectively.

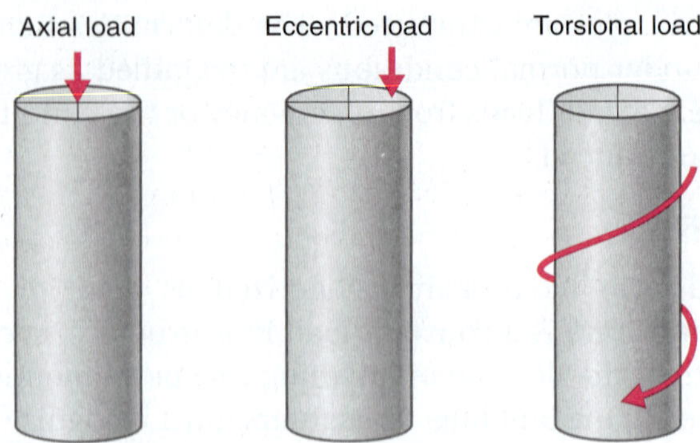

FIGURE 6-2 Axial, eccentric, and torsional loads.

Occupancy

The term **occupancy** describes how a building is used. The building's occupancy classification will help a firefighter understand how many people are potentially in a building and what they are likely to be doing. Occupancy also suggests the types of hazards and situations that might be encountered in the building.

Occupancy classifications are used in conjunction with building and safety codes, specifically the International Building Code (IBC); NFPA 101, *Life Safety Code, 2021 Edition*; and NFPA 5000, *Building Construction and Safety Code, 2024 Edition*, to establish regulatory requirements, including the types of construction that can be used to construct buildings of a particular size, use, or location. Regulations classify building occupancies into major categories, such as residential, health care, business, and industrial, based on common characteristics.

Occupancy classifications can be used to predict the number of occupants who are likely to be at risk in a fire. For example, hospitals and nursing homes are occupied 24 hours a day by persons who will probably need assistance to evacuate. An office building likely has a large population during the day, but most of the workers should be able to evacuate the site without assistance. A daycare center might be filled with young children during the day but is likely to be empty at night. A nightclub is typically empty during the day and crowded with adults at night. Firefighters must consider these factors when responding to a particular building.

Building codes require that certain types of building construction be used for specific types of occupancies. In many communities, however, buildings that were originally built for one purpose often have been repurposed for another use. In addition, a building may be used for a variety of occupancies. For example, a single commercial building may contain many different types of businesses with vastly different contents and hazards. A mixed-use building may include several levels of underground parking, multiple commercial occupancies on the lower floors, and residential units on the upper floors. These types of building present a wide variety of challenges for responding firefighters.

Contents

The contents of a building also must be considered when responding to an incident. Building contents largely represent the **fire load**—the quantity and combustibility of the contents and structural components in a space, compartment, or structure—in a building. While the structural components may eventually become part of the fire load, typically they do so only in the latter stages of a fire event. Although the modern construction environment seeks to limit the fire load from structural components, understanding what these components are and how much heat and toxic gases they can generate is critical. Building contents vary widely but are usually closely related to the occupancy of a building. The contents of a sporting goods store are different from those in an auto tire or repair shop. Some buildings have noncombustible contents that would not feed a fire, whereas the contents of other buildings could be so dangerous that firefighters could not safely attack any fire at the site. For example, the fire load in a factory that manufactures cast-iron pipe fittings will be vastly different from the fire load in a factory that produces lithium-ion batteries.

Even similar occupancies can pose different levels of risk. For example, three warehouses that are identical on the exterior could have very different contents that would either increase or reduce the risks to firefighters. One might contain ceramic floor tiles, the second might be filled with wooden furniture, and the third might serve as a storehouse for swimming pool chemicals. The tiles will not burn, the furniture will burn readily, and the chemicals will create toxic products of combustion and contaminated runoff.

When evaluating the risks at any response, firefighters need to consider the occupancy, the contents, and the building construction. Together, they represent an interwoven package of risks that face firefighters responding to the emergency scene.

Types of Construction Materials

An understanding of building construction begins with the materials used. The properties of these materials is part of what determines the fire characteristics of the building. Architects often place a priority on functionality and aesthetics when selecting building materials and construction methods. Builders are often concerned about price and ease of construction. Building owners usually prioritize durability and maintenance expenses, although they also care about the initial cost of the materials and aesthetics. The chief concern of firefighters, however, is the behavior of the building materials under fire conditions. In particular, they want to know if a building material or construction method has the potential to create serious problems or deadly hazards for firefighters.

The most commonly used building materials are wood and engineered wood products, steel, masonry, concrete, aluminum, glass, gypsum board, and plastics. Within these basic categories, hundreds of variations are possible. Different materials react differently under fire conditions due to the following characteristics:

- **Combustibility**: This is the property describing whether a material will burn and how quickly it will burn. Materials such as wood burn when they are ignited, and they release heat, light, and smoke until they are completely consumed by the fire. Concrete, brick, and steel are noncombustible materials that cannot be ignited and consumed by a fire.
- **Thermal conductivity**: This is the property that describes how quickly a material will conduct heat. Metals such as steel and aluminum are good heat conductors. Brick, concrete, and gypsum board are poor conductors of heat.
- Strength at elevated temperatures: Many materials lose strength at elevated temperatures. For example, steel loses strength and bends or buckles when exposed to the high temperatures associated with fires. Aluminum melts in a fire. Some engineered building products are constructed of wood components held together by glue or plastics, and the glue in these products begins to fail long before the wood is hot enough to burn. In contrast, bricks and concrete can generally withstand high temperatures for extended periods of time before losing strength.
- Degree of thermal expansion: Some materials—steel, in particular—expand significantly when they are heated. A steel beam exposed to a fire will elongate (stretch). If steel is restrained so that it cannot elongate, it will sag, warp, or twist. As a general rule, a steel beam will elongate at a rate of 1 inch (in.; 2.5 centimeters [cm]) per 10 feet (ft; 3 meters [m]) of length at a temperature of 1000°F (540°C). Because steel beams are used to create the structure of buildings, this expansion can push out the walls and cause the building to collapse.

Wood

Wood is one of the most commonly used building materials. It is the material most often used to construct single-family dwellings (in which a majority of civilian deaths due to fire occur in the United States (NFPA 2022, 1, 5). It is economical to produce, is easy to use, and can be shaped into many different forms, ranging from heavy structural supports to thin strips of exterior siding.

For firefighters, the most important characteristic of wood is its high combustibility. Wood acts as fuel in a structural fire and provides a path for the fire to spread. It ignites at fairly low temperatures and gradually becomes consumed by the fire. In addition, burning wood creates great quantities of heat and hot gases. A burning wood structure becomes weaker every minute and eventually collapses. Eventually, all that remains after the fuel is consumed is residual ash. Wood is a solid fuel (as discussed in Chapter 5, *Fire Behavior*), and the rate at which wood combusts depends on several factors, including the following:

- Ignition source: A small ignition source contains less energy and takes longer to ignite a fire. A large ignition source ignites more quickly, and if an accelerant is used, this process is even faster.
- Moisture content: Damp or moist wood takes longer to ignite and burn.
- Density: Heavy, dense wood is harder to ignite than lighter, less dense wood.
- Whether the wood is preheated: The more the wood is preheated, the faster it ignites.
- Surface-to-mass ratio: Wood with a higher surface-to-mass ratio ignites and burns more quickly than wood with a lower surface-to-mass ratio (wood shavings compared to a log).
- Orientation: The way in which the wood is oriented, for example, horizontally versus vertically, affects the ability of heat and flame to spread over its surface.

Wood products are broadly grouped into three categories: solid lumber, engineered wood, and mass lumber.

Solid Lumber

Older buildings were built with **nominal lumber**, which is lumber that is true to its stated dimensions. In other words, a two-by-four stud actually measured 2 in. by 4 in. (5 cm by 10 cm). Starting in the early 1920s, large lumber became more expensive and wood was kiln dried. This led to the use of **dimensional lumber**, which is lumber that is cut to nominal sizes, then dried and cut down to new standard sizes that are smaller than nominal sizes. For example, a two-by-four stud of dimensional lumber measures 1½ in. by 3½ in. (4 cm by 8 cm) or less. And, more critically, even smaller wood components are becoming more common. Dimensional lumber is often referred to as **solid lumber**.

Solid lumber is used in a wide variety of applications in a building, from heavy timbers used in mills and barns, to smaller two-by-four studs in walls, to wood siding, trim, floors, and shingles. Solid lumber is found in multiple sizes in both **legacy construction**—older buildings constructed before 1970—and **contemporary construction**—newer buildings constructed since about 1970 that incorporate lightweight construction techniques and engineered wood components.

Engineered Wood

Engineered wood, also referred to as **manufactured board**, **manmade wood**, or **composite wood**, is manufactured from smaller pieces of wood held together with glue or adhesive. The adhesives include urea-formaldehyde resins, phenol-formaldehyde resins, melamine-formaldehyde resins, and polyurethane resins. Examples of engineered wood products include plywood, fiberboard, oriented strand board (OSB), and particle board. These materials are used in roofs, walls, floors, countertops, doors, and cabinets and other locations.

Engineered wood products are preferred for certain uses over natural wood for a variety of reasons, including the following:

- They are less expensive than large, solid planks of natural wood.
- They can be fabricated in longer lengths than natural wood.
- They can maximize the natural characteristics of wood, and through engineering, have greater strength and stiffness than natural wood.
- They are easy to work with using ordinary tools and can be readily shaped into curved surfaces.

There are some drawbacks to engineered wood products. They are generally lightweight and smaller, which results in a much lower surface-to-mass ratio. This in turn means that time to failure is much more rapid as the integrity of the wood component will be reduced much more quickly. Additionally, the glues and resins in these products include chemicals such as formaldehyde, that, when subjected to heat, produce toxic gases including hydrogen cyanide, carbon monoxide, and polycyclic aromatic hydrocarbons (PAHs). The biggest disadvantage from the firefighters' perspective is that these products contribute to faster fire spread when compared to natural lumber equivalents by shortening the time to flashover and structural failure.

SAFETY TIP

The glues that are used to create engineered wood products produce toxic gases when exposed to heat, degrade when exposed to moisture over time, and are highly flammable and fail quickly under fire conditions.

Mass Lumber Products

Mass lumber (or **mass timber**) are large-scale engineered wood products designed to replace structural materials such as concrete and steel. These heavier engineered wood products allow high-rise and other large structures to be built from wood. They rely on increased mass to make them more resistive to fire than their lightweight counterparts, and they are structurally equivalent to metal or masonry.

Mass lumber products come in three broad categories. Each of the following has different specific manufacturing processes so they can be used in different applications in construction based on their dimensions and predicted loads (**FIGURE 6-3**):

- Cross-laminated timber (CLT): Layers of lumber glued together in which each layer is perpendicular to the previous layer
- Glue laminated timber (Glulam): Layers of **laminated wood** glued together
- Structural composite lumber (SCL): Layers of wood veneers, strands, or flakes oriented in the same direction and bonded together. There are four types of SCL:
    - Laminated veneer lumber (LVL)
    - Parallel strand lumber (PSL)
    - Laminated strand lumber (LSL)
    - Oriented strand lumber (OSL)

FIGURE 6-3 An example of a mass lumber product.

It is not important for the firefighter to remember each type of engineered or mass wood and its definition or common uses. What is important is that firefighters recognize that these products are becoming more common and to understand their characteristics. It is anticipated they will eventually be found in every community in the United States.

SAFETY TIP

Fire retardant treated (FRT) wood, particularly in the 1980s, was found to be structurally deficient under normal conditions. FRT plywood that was used as roof sheathing deteriorated within 1 to 3 years of installation under the typical heat conditions to which they were subjected. Thousands of roofs during that era had to be replaced. There were reports of firefighters falling through these roofs, even under nonfire conditions. Modern FRT chemicals and products have largely removed those structural issues; however, the toxic chemicals generated when these products are subjected to fire conditions are still a cause for concern.

Masonry Units

Masonry is blocks of stone, concrete, or brick stacked on top of one another. The individual blocks are bonded together into a solid mass, usually with **mortar**, which is produced by mixing sand, lime, water, and cement. Blocks made of brick or concrete are also known as concrete–masonry units (CMUs). Similar to concrete, CMUs can be assembled into walls. A single layer of masonry may be placed over a wood-frame building to make it appear more substantial (**FIGURES 6-4A** and **6-4B**). CMUs come in multiple shapes and sizes.

A.

B.

FIGURE 6-4 Masonry materials are inherently fire resistive. **A.** Solid masonry wall. **B.** Veneer wall.

Masonry materials are inherently fire resistive; that is, they do not burn or deteriorate at the temperatures normally encountered in building fires. Masonry is also a poor conductor of heat. For these reasons, masonry is often used to construct fire walls to protect vulnerable materials. A **fire wall** is a wall made of fire-resistive material and is designed to remain standing under fire conditions to help prevent the spread of a fire from one side of the wall to the other (**FIGURE 6-5**). Although masonry materials do not conduct heat well, they can act as a heat sink or reservoir, causing excessive heat to be held in the building after the fire is extinguished.

Not all masonry walls are necessarily fire walls. If a masonry wall contains unprotected openings, a fire can spread through them. If the mortar has deteriorated or the wall has been exposed to fire for a prolonged time, a masonry wall can collapse during a fire.

A masonry structure can collapse under fire conditions if the roof or floor assembly collapses. In such a case, the masonry fails because of the mechanical action of the collapse. Likewise, aged, weakened mortar can contribute to a collapse. Regardless of the cause, a collapsing masonry wall can be a deadly hazard to firefighters.

FIGURE 6-5 Fire walls help prevent the spread of fire from one side of the wall to the other.

Courtesy of Achim Hering.

Concrete

Concrete is a mixture of cement, aggregates such as sand and gravel, and water. Concrete is inexpensive and easy to shape and form. It is often used for foundations, columns, floors, walls, roofs, and pavement.

Under compression, concrete is strong and can support a great deal of weight. Under tension or shearing forces, however, it is very weak. When concrete is used in building construction, steel is often embedded into the concrete to strengthen it. The result is referred to as **reinforced concrete**. The steel can be either solid reinforcing bars, commonly called **rebar**, or steel cables under tension, commonly called **tendons**. Under typical conditions, the concrete insulates the reinforcing steel from heat (**FIGURE 6-6**).

Like masonry materials, concrete is naturally fire-resistive. In addition, concrete does not have a high degree of thermal expansion—that is, it does not expand greatly when exposed to heat and fire—nor will it lose strength when exposed to high temperatures. Because it does not burn or conduct heat well, it is often used to insulate other building materials from fire.

Although it is fire resistive, concrete can be damaged by exposure to a fire. For example, a fire can convert moisture trapped in the concrete to steam. As the steam expands, it creates internal pressure that can cause sections of the concrete surface to break off. This process is called **spalling** (**FIGURE 6-7**). Severe spalling in reinforced concrete can expose the steel reinforcing rods to the heat of the fire. If the fire is hot enough, the steel could weaken, resulting in a structural collapse, although this is a rare event.

FIGURE 6-6 An example of tilt-up construction.

© Piyapong Tangteerasunant/Shutterstock

FIGURE 6-7 Spalled concrete that has broken to expose the rebar reinforcement.

SAFETY TIP

Heat transfer can occur through conduction to exposed steel reinforcing rods. This can accelerate the potential for spalling or complete structural failure depending on the orientation and loading of the concrete component.

Steel

Steel is an alloy (a combination of two or more metals) of iron and carbon. Other metals may be added to the mix to produce steel with special properties, such as stainless steel or galvanized steel, which are types of steel that resist corrosion. Steel can be produced in a wide variety of shapes and sizes, ranging from heavy beams and columns to thin sheets. It is resistant to aging and does not rot, although most types of steel will rust unless they are protected from exposure to air and moisture.

Steel is the strongest building material in widespread use in terms of both compression and tension. It is often used in the structural framework of commercial buildings to support floors and the roof. Many newer occupancies use steel studs instead of wood studs.

Steel is a strong conductor of heat, and when heated, it will quickly reach critical temperatures. Steel begins to lose strength at temperatures as low as 400°F (204.4°C), and it loses half its strength at 650°F (343.3°C). The amount of heat absorbed by steel depends on the mass of the object and the amount of protection surrounding it. Smaller, lighter pieces of steel absorb heat more easily than larger and heavier pieces. For this reason, other materials, such as concrete, masonry, or layers of gypsum board, are often used to protect steel from the heat of a fire. Steel can also be sprayed with an **intumescent coating**, a paint-like coating of mineral materials or cement-like materials to insulate steel from heat in case of fire.

Failure of a steel structure depends on three factors: the mass of the steel components, the loads placed upon them, and the methods used to connect the steel pieces. Horizontal steel beams in tension that are failing sag and twist, whereas vertical steel columns in compression tend to buckle as they lose strength. While immediate application of a water stream cools and potentially stops this process, it should be viewed as a sign of imminent collapse.

SAFETY TIP

Bending, sagging, stretching, or buckling of steel structural members should be considered a warning of an immediate risk of failure.

Other Metals

A variety of other metals, including copper, zinc, iron, and aluminum, are also used in building construction. Copper is used primarily for electrical wiring and piping, and it is sometimes used for decorative roofs, gutters, and downspouts. Zinc is used primarily as a coating to protect metal parts from rust and corrosion.

Two types of iron are used in building construction: wrought iron and cast iron. Wrought iron, which is iron that has been shaped, is used for water pipes, rivets, and other decorative work. Cast iron, which is iron that has been poured into a mold, is used primarily for columns and support beams.

Aluminum is occasionally used as a structural material in building construction. It is more expensive and not as strong as steel, so its use is generally limited to light-duty applications such as awnings, siding, window frames, door frames, roof panels, and sunshades. Aluminum expands more extensively than steel when heated and loses strength quickly when exposed to a fire. It has a lower melting point than steel, so it often melts and drips in a fire.

Membrane Materials

Membrane materials are used to insulate and provide a finished appearance in the interior compartments of a building. They play a key fire protection role by protecting the structural components from direct flame impingement. This also means that they can hide fire and make accessing other compartments and void spaces more difficult. Membrane materials are not strong structural materials. They are attached to the building's structural components.

Gypsum Board

Gypsum is a naturally occurring mineral composed of calcium sulfate and water molecules. It is used to make plaster, which is the material that was commonly used prior to the early 20th century to create smooth walls over laths or rough brick or stonework. Gypsum is a good insulator and is noncombustible; it will not burn, even in atmospheres of pure oxygen. **Gypsum board** (also called drywall, sheetrock, or plasterboard) is the most commonly used membrane material and is used to cover the interior walls and ceilings of residential living areas and commercial spaces. It is manufactured in large sheets (typically 4 ft by 8 ft [122 cm by 244 cm]) consisting of a layer of compacted gypsum sandwiched between two layers of specially produced paper (**FIGURE 6-8**). It is manufactured in a variety of widths. Wider sheets are heavier, stronger, and more fire resistive. In a building, gypsum board is nailed or screwed in place on a framework of wood or metal studs, and the edges are secured with a special tape. The nail or screw heads and tape are then typically covered with a thin layer of joint compound (drywall mud).

Gypsum board can be reinforced with other materials when additional strength is needed. The resulting product is referred to as impact- or abuse-resistant drywall. Impact-resistant drywall can be difficult to breach in a fire situation. Common methods for reinforcing gypsum board include adding heavy paper or fiberglass backing or mixing fiber in with the gypsum. Gypsum may also be mixed with cement to produce a cement board that can be used for exterior sheathing or as an underlayment for ceramic floors or wall tiles. Another example of impact-resistant drywall is gypsum board with a thin sheet of Lexan backing (a plastic) laminated to it. When exposed to fire, this plastic smolders and melts and may burn as a combustible liquid. Use of impact-resistant drywall is usually limited to high security environments and typically would not be associated with residential structures.

FIGURE 6-8 Gypsum board is used to cover the interior walls and ceilings of residential living areas and commercial spaces.

Gypsum board has limited combustibility. Although gypsum does not burn, the paper covering burns slowly when exposed to a fire. Because gypsum does not conduct or release heat to an extent that would contribute to fire spread, it is often used in a **fire barrier wall**, which is an interior wall that extends from a floor to the underside of the floor above to prevent fire from spreading from one area to another or to protect building components from fire. Gypsum board is required in some applications, such as residential garages, to create a rated fire-resistant barrier. When mounted on wood framing, gypsum board protects the wood from fire for a period of time based on its dimensions and installation.

When gypsum board is heated, some of the water present in the calcium sulfate evaporates, and this causes the board to deteriorate. Despite its limited combustibility, if gypsum board is exposed to fire for a long time, it will fail. If gypsum board retains its integrity after a fire, sections that were directly exposed to the fire cannot be trusted to serve as effective fire barriers and should be replaced after the incident. Water also weakens and permanently damages gypsum board.

Plaster and Lath

Plaster and lath historically was used to create walls and ceilings. Plaster is made by combining gypsum, lime, or cement; water; and sand to form a paste that hardens as it dries. It is not common in modern construction, but it can be prevalent in communities with structures built through the 1950s (**FIGURE 6-9**). The plaster (or stucco) was added over wood laths, which are horizontal wood strips attached to the studs. Wire mesh was sometimes used in place of wood laths to secure the plaster or stucco. which can make it significantly more difficult to break through the walls or ceiling for which it was used.

Other Membrane Materials

Materials such as tiles made from mineral fibers are commonly used for suspended ceilings. These tiles

FIGURE 6-9 Plaster and lath is often used to build ceilings.

© Jordan Feeg/Shutterstock

FIGURE 6-10 Tempered glass will shatter into small pieces that do not have the sharp edges of ordinary glass.

© Sylvie Bouchard/Shutterstock, Inc.

FIGURE 6-11 Laminated glass is much stronger than ordinary glass and will usually deform instead of breaking.

© Salazar Benjamin/Shutterstock

are fire resistive and are easily displaced by either fire hooks or water streams. This also means they can be easily removed or damaged by occupants, which can lead to unprotected structural components in the event of a fire. Wall coverings and other ceiling finishings made from light metals, cork, and wood products provide varying degrees of fire protection. Some of these products do not provide protection from fire and are actually combustible.

Glass

Almost all buildings contain glass. It is used in windows, doors, skylights, and sometimes walls. Ordinary glass fails easily when impacted or when exposed to fire. However, glass can be manufactured to resist breakage and to withstand impacts or high temperatures. For example, specially formulated glass can be used as a fire barrier in certain situations.

Many types of glass are available for use in building construction:

- Ordinary (or annealed) window glass is the glass used most often for windows. When ordinary window glass breaks, it breaks into large shards that usually have sharp edges.
- **Tempered glass** is treated so that it is much stronger than ordinary glass and more difficult to break (**FIGURE 6-10**). When it does break, it shatters into small pieces that do not have the sharp edges of ordinary glass. Tempered or laminated glass usually is required in sliding glass doors in residences and in glass doors and windows in commercial buildings to prevent injuries from shards of broken glass.
- **Laminated glass** is made by placing a thin sheet of plastic between two sheets of glass (**FIGURE 6-11**). Like tempered glass, the resulting product is much stronger than ordinary glass and it is difficult to break. When it does break, it usually deforms instead of shattering into pieces. When exposed to a fire, laminated glass windows are likely to crack, but they usually remain in place. Laminated glass is sometimes used in buildings to help soundproof areas, and it has evolved as the glass of choice for hurricane- or theft-proof windows.
- **Wired glass** is made by molding tempered glass with a reinforcing wire mesh (**FIGURE 6-12**). When wired glass is subjected to heat or an impact, the wire holds the glass together and prevents it from breaking. Wired glass is often used in fire doors and in windows designed to prevent fire spread.
- **Glass blocks** are thick pieces of glass similar to bricks or tiles. They are designed to be built into a wall with mortar so that light can be transmitted through the wall. Glass blocks have limited

FIGURE 6-12 Wired glass in a door.

© oradige59/Shutterstock

FIGURE 6-13 Plastic building materials can be used for siding, windows, window frames, and interior finishes.

© AbleStock

strength and are not intended to be used as part of a load-bearing wall, but they can usually withstand a fire. Some glass blocks are approved for use with masonry walls that are intended to with stand fire and can be considered more fire resistive than typical glass sheet products. They are not commonly used in modern construction.

Plastics

Plastics are synthetic materials that are used in a wide variety of products (**FIGURE 6-13**). These materials may be transparent or opaque, stiff or flexible, and tough or brittle. Plastics are rarely used for structural support; however, they may be found throughout a building. When used to produce building products, they are often combined with other materials.

There are many types of plastics. For example, building exteriors may include vinyl siding and window frames. Plastic can also be used in place of glass. Skylights are made of plastic panels, and some windows are made of clear plastic panes. Two plastics commonly used in place of glass are polycarbonates and acrylics. These plastics may be used for added strength and resistance against breakage. Foam plastic materials may be used as insulation. Plastic pipe and fittings, plastic tub and shower enclosures, and plastic lighting fixtures are also commonly used. Plastics are used in most of the contents of modern homes and buildings, including carpeting, floor and window coverings, and furniture of every kind.

The combustibility of plastics varies greatly. Some plastics ignite easily and burn quickly, but others burn only when an external source of heat is present. Some plastics can withstand high temperatures and exposure to fire without igniting, but they will still off-gas at some point. All plastics produce toxic gases as well as heavy, dense, dark smoke when they burn. This smoke resembles smoke from a petroleum fire, which should not be surprising given that plastics are produced from petroleum products.

Thermoplastic material is plastic material that is softened by heating and hardened by cooling. It melts and drips when exposed to high temperatures, even those as low as 500°F (260°C). The dripping, burning plastic can rapidly cause a fire to spread. In contrast, **thermoset material** is plastic material fused by heat. Once thermoset materials are formed into a shape, they will not melt or soften when exposed to heat. While thermoset material can withstand higher temperatures than thermoplastic material, it still produces toxic gases when exposed to heat and will eventually combust.

SAFETY TIP

Any modern furnished structure subjected to fire—even small fires—will contain plastics that will burn and produce toxic gases. Always wear full personal protective equipment (PPE) and self-contained breathing apparatus (SCBA).

Building Components

Now that you have an understanding of the typical materials used in building construction, you need to understand how the materials are brought together to create the larger components that create a structure. A critical part of this knowledge is knowing whether a material or component can withstand fire. **Fire resistance** is the ability of a material or component to withstand fire. **Fire resistance ratings** are assigned to materials and building components based on the length of time that the material or component can maintain its integrity when subjected to fire. Firefighters who understand how building components function, how they are built, and whether they are fire resistant will have a better understanding of the risks involved when entering a burning building.

Building components are made from building materials, assemblies, and systems. An **assembly** is various construction materials joined together to form a component of a structure. For example, one type of a roof assembly consists of the framing—the beams that support the roof—and the plywood that covers the beams. A **system** is a collective term for the assemblies and materials that together make up a building component. For example, a simple roof system consists of the roof assembly and the waterproofing on top of the plywood.

Understanding basic typical building construction from the ground up allows firefighters to visualize how a structure is built and understand how they interrelate to create a building.

The major components of a building are as follows:

- Connections
- Foundations
- Basements
- Floors and ceilings
- Walls
- Attics
- Roofs
- Doors and windows
- Exterior finishes
- Interior finishes

Connections

Materials are brought together in assemblies. The components of an assembly are joined together and the different assemblies are connected to one another at connection points. The types of connections and the connection points vary depending on their purpose and location within the assembly and the assembly's purpose and location in the structure. The connections range from gravity, such as resting one material on another (beam resting in a pocket joint), to a simple nail (holding a joist hanger to a ledger board), to a nut and bolt (connecting a column to a beam). In just about every case, connection points are the weakest point of an assembly or component, especially when a component is attacked by fire.

FIGURE 6-14 An example of a connection.

A connection's most important function is to ensure that a load is transferred from one component to another. Aside from gravity, this is accomplished in several ways (**FIGURE 6-14**):

- **Pinning**, which means making a connection using bolts or rivets.
- Bonding with concrete means using wet concrete to join concrete elements.
- **Welding** joins pieces of metal together in the same way wet connections with concrete join concrete elements.
- Wood materials are connected with glue, fasteners such as nails or screws, or joints such as a finger joint in which the end of one piece of wood is cut to fit precisely in a notch in the end of another piece of wood.

The strength of a connection relies on the quality of the work during construction and how the connections are maintained. For example, a bolted connection that is properly applied and protected from rust will maintain its strength over time. However, a column designed for an axial load that suddenly needs to support an eccentric load during a renovation is subject to failure. Even if a connection was constructed with a high degree of quality, was maintained correctly, and is loaded correctly, fire will degrade the connecting point or the materials around the connection and failure becomes less a matter of if, but when.

Foundations

The primary purpose of a building foundation is to transfer the weight of the building and its contents to the ground (**FIGURE 6-15**). The foundation ensures that the base of the building is planted firmly in a fixed location, which helps keep all other components connected.

Modern foundations are usually constructed of concrete or masonry, although wood pilings may be used in some areas. Foundations can be either shallow or deep, depending on the type of building and the soil composition. Some buildings are built on concrete footings or piers; others are supported using steel piles (hollow steel tubes in the ground that are filled with concrete) or wooden posts driven into the ground. In a residential setting, a foundation may also be a concrete pad that is usually 4 to 6 in. (10 to 15 cm) thick.

As long as the foundation remains stable and intact, it usually is not a major concern for firefighters. Most foundation problems are caused by circumstances other than a fire, such as poor construction, shifting soil conditions, or earthquakes. Although fire can damage timber post and other wood foundations, most foundations remain intact even after a severe fire has damaged the rest of the building. If the foundation shifts or is in poor condition before the fire starts, it can be a critical concern. A burning building with a shifted or weak foundation could be in imminent danger of collapse.

When examining a building, take a close look at the foundation. Look for cracks that indicate movement of the foundation. If the building has been modified or remodeled, look for areas where the support could be compromised.

FIGURE 6-15 The foundation supports the entire weight of the building.

Basements

The walls of a foundation form a basement or cellar. While there are some specific technical details that differentiate the two, from a firefighting perspective, the important difference is that cellars typically are not a livable space, but basements can be. Basements have some sort of egress other than the stairs to the first floor of the building. In a basement, this is usually windows, slider doors, or a doorway (leading out or to a set of exterior stairs). A cellar may not have any means of egress other than the stairs to the first floor of the building. In either case, this space or compartment beneath the first floor of the building is the lowest level of the structure. It is often used as storage, which means this space also may or may not have a ceiling that is fire resistant. If the ceiling is not fire resistant, fire may cause structural failure much sooner than it would if the ceiling of the basement or cellar was fire resistant. It is critical for firefighters to note these spaces during their size-up to determine the lowest level of the structure as they may not always be clearly apparent from the street upon arrival.

Ceilings

From a structural standpoint, ceilings are part of floor assemblies. The primary function of ceilings is to hold light fixtures and diffuse light. Ceilings also improve the appearance of a building because they conceal heating, ventilation, and air-conditioning (HVAC) systems, as well as wiring and fire sprinkler systems.

Ceilings can be part of the fire-resistive package. For example, they can be covered with plaster or gypsum board, or they can consist of drop ceilings covered with fire-resistant tiles. Some ceilings are fire resistant because they are part of a fire-resistant floor assembly.

Firefighters should be aware that often there is a hollow, void space between a ceiling and the floor above it. This can contribute to the horizontal spread of fire because the fire can easily spread through that space. Similarly, holes in the ceiling and the floor above it can contribute to the vertical spread of fire.

Floors

Floor construction is important to firefighters for three major reasons. First, firefighters who are working inside a building must rely on the floor to support their weight. A floor that fails could drop the firefighters into a fire on a lower level. Second, in a multistory structure, firefighters may be working below a floor that would fall on them if it collapsed. Third, although some floor systems are designed to resist fires, others have no fire-resistant capabilities. A floor system with fire-resistant capabilities contains a fire on a single level. A floor system without fire-resistant capabilities can allow a fire to spread vertically from one floor to another.

In multistory buildings, floors and ceilings are generally considered to be a combined structural system. This means that the structure that supports the floor also supports the ceiling of the story below. In a fire-resistive building, this prevents a fire from spreading vertically and is intended to prevent a collapse when a fire occurs in the space below the floor–ceiling assembly. Fire-resistive floor–ceiling systems are rated in terms of hours of fire resistance based on a standard test fire.

Fire-Resistive Floors

Floors made of concrete are fire-resistive. The concrete can be cast in place or assembled from panels or beams of precast concrete, which are made at a factory and then transported to the construction site for placement. The thickness of the concrete floor depends on the load that the floor needs to support. Concrete floors can be either self-supporting or supported by a system of steel beams or trusses. If steel beams are used, the steel is usually protected by a sprayed-on intumescent coating or by concrete, masonry, or gypsum board.

If the floor is part of a floor–ceiling system and the system has a fire rating, the ceiling of the floor below is included as part of that rating. A fire-resistant ceiling can be constructed from plaster or gypsum board, or it can be a system of tiles suspended from the floor structure. It provides a thermal barrier to protect the steel members from a fire in the area below the ceiling. In many cases, the void space between the ceiling and the floor above it contains building systems and equipment such as electrical or telephone wiring, heating and air-conditioning ducts, and plumbing and fire sprinkler system pipes (**FIGURE 6-16**). If the space above the ceiling is not subdivided by fire-resistant partitions or protected by automatic sprinklers, a fire can quickly extend horizontally across a large area.

FIGURE 6-16 The space between the ceiling and the underside of the floor above it often holds electrical and communications wiring and other building systems.

Wood-Supported Floors

Wood floors are common in buildings that are not fire resistant. Wood floor systems range from heavy timber construction, which is often found in old mill buildings, to modern, lightweight truss or engineered wooden I-beam construction.

Although heavy timber construction provides a huge fuel load for a fire, it also can contain and withstand a fire for a considerable length of time without collapsing. Heavy timber construction uses posts and beams that are at least 8 in. (20 cm) on the smallest side and often as large as 14 in. (35 cm) in depth. Floor decking—the part of the floor system that consists of the part you walk on—is often assembled from solid wood boards, 2 or 3 in. (5 to 8 cm) thick, which are covered with an additional 1 in. (2.5 cm) of finished wood flooring. In this case, the depth of the wood used for the decking often will contain a fire for an hour or more before the floor fails or burns through.

Conventional wood flooring systems, which were widely used for many decades, are much lighter than heavy timber flooring. They use solid lumber as beams, floor joists, decking, and finished flooring. Conventional wood flooring systems burn readily when exposed to a fire but, depending on the materials used, can burn through or reach structural failure, but at a rate slower than lightweight construction.

Modern lightweight flooring systems use structural elements that are much less substantial than conventional lumber. For example, lightweight wooden trusses or engineered plywood **I-beams**—beams that look like an uppercase letter *I*—are often used as supporting structures. Plywood or other wood sheet products such as OSB are used as decking, with finishing materials on top (i.e. carpet, tile, wood flooring). This type of floor system provides little resistance to fire. The lightweight structural elements can fail quickly when exposed to fire, and the fire can burn through the decking quickly. A study conducted by the National Institute for Occupational Safety and Health (NIOSH) in 2009 found that while it took 19 minutes for an unprotected traditional residential floor assembly (dimensional lumber) to fail, an unprotected lightweight engineered I-joist failed in only 6 minutes.

It is impossible to tell how a floor is constructed by looking at it from above. The important information about a floor can be observed only from below if it is visible at all. The building's age and local construction methods can provide significant clues to a floor's construction, but you should be aware that many older buildings have been renovated using modern lightweight systems and materials. For example, in some reconstructions of office or computer rooms, the original floor systems/assemblies are replaced with removable floor panels to hide computer cables under the floor.

Trusses and Plywood I-Beams

A truss is a structural component that is composed of smaller pieces called members joined to form a triangle or a system of triangles. Trusses are used extensively in support systems for floors because triangles create a strong, rigid structure that can support a load much greater than its own mass. For example, both a solid beam and a simple truss with the same overall dimensions can support the same load. The strength of a truss depends on both its members and the connections between them. A properly assembled and installed truss is strong. However, if any member or connector begins to fail, the strength of the entire truss will be compromised.

Trusses require much less material than the girder, beams, or joists they are replacing., Trusses are much lighter, and can span long distances without supports. Trusses often are prefabricated and transported to the building location. They are widely used in new construction, and they are often used to replace heavier solid beams and **joists**—beams that hold up a floor—in renovated or modified older buildings. They are frequently used in residential construction, apartment buildings, small office buildings, commercial buildings, warehouses, fast-food restaurants, airplane hangars, churches, and even firehouses. They may be clearly visible, or they may be concealed within the construction.

Trusses used for ceiling and floor assemblies typically have parallel chords. A chord is the straight part of the truss to which the triangles are connected. A **parallel chord truss** has two parallel chords connected by a system of diagonal and sometimes vertical members called **web members**. The void spaces between the parallel chords allow builders to run wiring, plumbing, and HVAC components through them.

In lightweight construction, an engineered **wood truss** is often assembled with wood chords and either wood or lightweight steel web members. A steel bar joist is another example of a parallel chord truss.

FIGURE 6-17 A parallel chord truss assembled with wood chords and wood web members.

Plywood I-beams (ply Is) have become even more lightweight and cheaper to produce. A plywood I-beam is constructed of either dimensional or engineered lumber as the top and bottom chord, with thin piece of plywood or OSB glued into a slot between the two. Structurally, this creates a component as strong as truss or dimensional lumber, but with even less mass. This also means that instead of a truss that allows a builder to pass wiring, plumbing, and HVAC components through an open triangle, they must cut the web. This significantly weakens the structural strength even before fire conditions are present.

In some buildings, trusses or plywood I-beams are exposed and readily visible as part of a ceiling (**FIGURE 6-17**). This means that they are not protected by any fire-resistant membrane and are directly exposed to heat and flame in a fire. In other buildings, the floor or roof trusses are not visible because they are protected by a membrane of some kind.

Walls

Walls are the most visible parts of a building because they shape the exterior and define the interior. Walls are either load-bearing or non-load-bearing. A **load-bearing wall** provides structural support by supporting both a portion of the building's dead load and live load (**FIGURE 6-18**). Damaging or removing a load-bearing wall can result in a partial or total collapse of the building.

A **non-load-bearing wall**, also called a **nonbearing wall** or **partition wall**, by contrast, does not support any of the dead or live load of the building; it supports only its own weight (**FIGURE 6-19**). For this reason, most nonbearing walls can be breached or

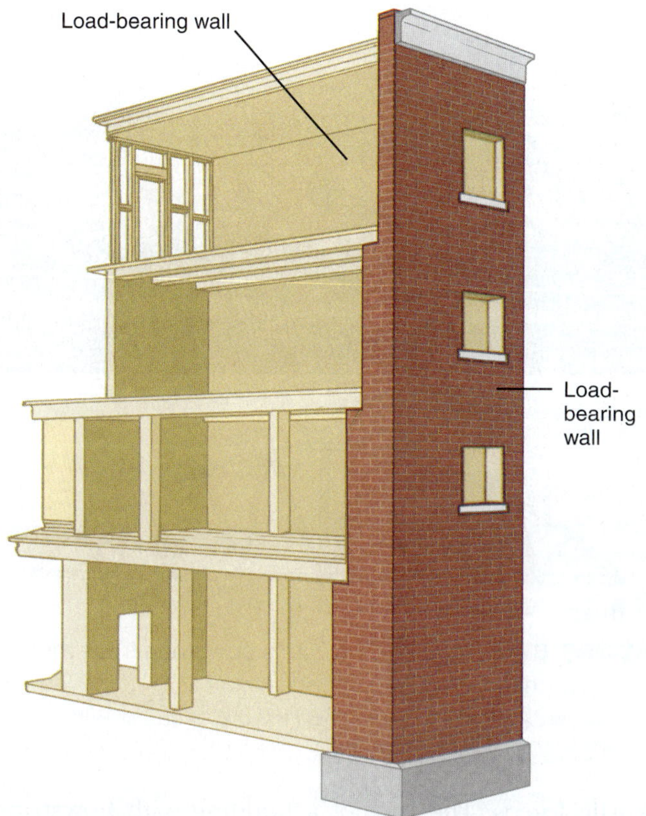

FIGURE 6-18 A load-bearing wall provides structural support.

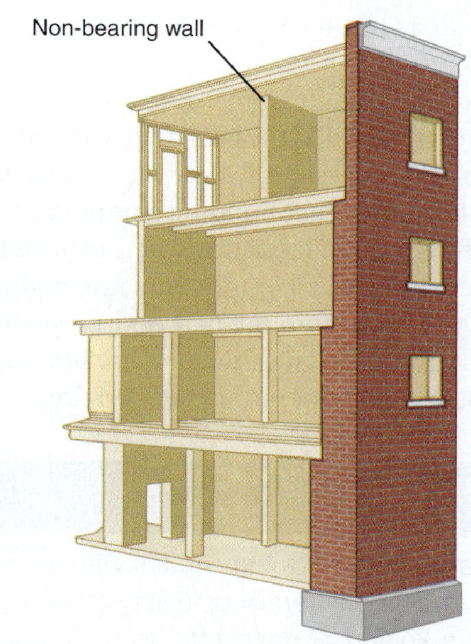

FIGURE 6-19 A nonbearing wall supports only its own weight.

removed without compromising the structural integrity of the building. Nonbearing walls are usually interior walls that divide the building into rooms and spaces. The exterior walls of a building can be nonbearing when a system of columns supports the building. Differentiating between bearing and non-load-bearing walls will typically not be possible under emergency conditions.

Several specialized wall types also may be used in construction:

- A fire wall is a wall made of fire-resistive material that is designed to remain standing under fire conditions to help prevent the spread of a fire from one side of the wall to the other. A fire wall might divide a large building into sections or separate two attached buildings. Fire walls usually extend from the foundation up to and through the roof of a building. They may be fire rated.
- A fire barrier wall is an interior wall that extends from a floor to the underside of the floor above. They often enclose fire-rated interior corridors or divide a floor area into separate fire compartments.
- A **fire separation** encloses interior vertical openings, such as stairwells, elevator shafts, and chases for building utilities, to prevent fire and smoke from spreading from floor to floor via the opening. Fire separations also protect the occupants when they are using the exit stairways.
- A **curtain wall** is a nonbearing exterior wall attached to the outside of a building. These walls often serve as the exterior skin on a steel-frame, high-rise building.
- A **party wall** is constructed on the line between two properties and is shared by a building on either side of the line. They are almost always load-bearing walls. A party wall is often—but not always—constructed as a fire wall between the two properties.

The walls in most houses and small commercial buildings are constructed with wood framing. Each vertical member of the frame is called a **stud**, and the studs support the walls. If fire resistance is critical, steel studs are used to frame walls.

The outside of the wood framing used for exterior walls can be covered with a variety of materials, including wooden siding, vinyl siding, aluminum siding, stucco, and masonry veneer. Moisture barriers, wind barriers, and other types of insulation are usually applied to the outside of the vertical studs of the exterior walls before the outer covering is applied.

The inside of the wood framing for exterior walls, as well as the inside of wood framing of interior walls, is usually covered with gypsum board and a variety of interior finish materials. The space between each side of the finished wall may be empty, contain thermal and sound insulating materials, or contain electrical wiring, telephone wires, and plumbing. These spaces often provide pathways through which fire can spread.

Walls may also be constructed of masonry. Load-bearing masonry walls, at least 6 to 8 in. (15 to 20 cm) thick, can be used for buildings up to six stories high. Nonbearing masonry walls can be almost any height. Older buildings often have masonry load-bearing walls that are several feet thick at the bottom and decrease in thickness as the building's height increases. Modern masonry walls are typically reinforced with steel rods or concrete to provide a stronger structural system.

Masonry walls provide a durable, fire-resistant outer covering for a building. They are also often used as fire walls. A well-designed masonry fire wall is often completely independent of the structures on either side. Even if the building on one side burns completely and collapses, the fire wall prevents the fire from spreading to the building on the other side. When properly constructed and maintained, masonry walls are strong and can withstand a vigorous assault by fire. If the interior structure begins to collapse and exerts unanticipated forces on exterior solid masonry walls, however, those walls may fail during a fire.

Attics and Roofs

The shape and design of attics and roofs are based on the arrangement of the structural components that create them. Knowing the how those components are laid out allows the firefighter to understand their strengths or weaknesses, particularly under fire conditions.

Sloping roofs are typically supported by pitched chord trusses. A **pitched chord truss** is a truss created by three chords that form a triangle, with additional triangle shapes in the interior (**FIGURE 6-20**). Most modern residential construction uses a series of prefabricated, wooden pitched chord trusses to support the roof. The roof decking is supported by the top chords, and the ceilings of the occupied rooms are attached to the bottom chords.

A **bowstring truss** has the same shape as an archery bow, where the top chord is curved and the bottom chord is straight (**FIGURE 6-21A**). The top chord resists compressive forces, and the bottom chord resists tensile forces. The roof of a building with bowstring trusses has a distinctive curved shape (**FIGURE 6-21B**). Bowstring trusses are usually quite large and widely spaced in a structure. They were once popular in warehouses, supermarkets, and similar buildings with large, open floor areas but have since faded from popularity (Jakubowski 2013).

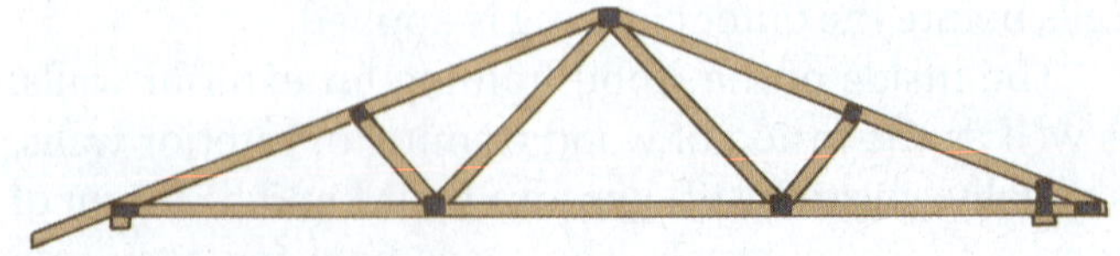

FIGURE 6-20 An illustration of a pitched chord truss.

© Jones & Bartlett Learning

A.

B.

FIGURE 6-21 **A.** A building with a bowstring truss roof. **B.** A bowstring truss.

A: Courtesy of Captain David Jackson, Saginaw Township Fire Department; **B:** © Jones & Bartlett Learning

Under normal circumstances, a properly engineered and installed truss will do a satisfactory job of supporting the roof or floor assembly for which it was designed. However, like many other building components, trusses display different characteristics when exposed to heat and fire. Because these components are engineered to work with the least amount of wood or steel needed to support a given weight, they provide little margin of safety for firefighters operating during a fire.

SAFETY TIP

Trusses and roof sheathing can fail without warning early in a fire. Firefighters should stay off of roofs any time fire is known or suspected to be in an unprotected void space below where they are working.

Trusses consist of a small cross-section of wood with a high surface-to-mass ratio. As a consequence, a fire burning for a short period of time can destroy enough wood to weaken a truss. All it takes is a failure at one point in a truss to produce a failure in the entire truss. Wooden trusses that are constructed using a metal **gusset plate**—a connecting plate made of a thin sheet of steel that connects the components of a truss—are especially prone to failure. Because the gusset plates are embedded in the wood at a depth of only ⅜ in. (1 cm), heating the metal gusset plate decomposes the wood fibers, quickly leading to failure of the truss. Likewise, wooden trusses held together with steel pins can fail in a fire. When heated, the steel pins conduct heat into the wood, which in turn causes decomposition of the wood. When the wood weakens, failure initially occurs at that point, followed by failure of the entire truss. Fire also can weaken the glue in glued trusses, causing the truss to weaken.

Steel trusses are also prone to failure during fires. Automatic sprinklers can protect the steel from the heat; however, without such protection, the steel will eventually elongate and twist. A 100-ft (30-m) beam or truss can elongate by as much as 9 in. (23 cm) when heated to 1000°F (540°C) (NFPA 2015). Elongated steel beams can rotate and drop wooden trusses, with deadly consequences. Elongation also can push down the exterior walls of the building. Cracks in the outside walls at the roof level are a warning indicator of this condition.

Truss spaces can contribute to the spread of fire. Spaces under floors occupied by trusses can serve as conduits for the spread of fires. These fires often are difficult to detect and may spread rapidly, much as fires spread quickly through any open spaces. Fires that occur in the space above a truss-supported ceiling are challenging to detect. Whenever a truss-supported ceiling has fire above it, be prepared for the possibility of a rapid collapse of the trusses supporting the roof.

Because trusses exist in many types of buildings and give firefighters few signs of impending failure, some jurisdictions have passed laws requiring that buildings constructed with trusses be marked with a specified sign. Identifying the presence of trusses enables firefighters to better predict the behavior of a particular building under fire conditions and to evaluate the risk and benefit of different approaches to fire suppression in that building.

Roofs

The primary purpose of a roof is to protect the inside of a building from the weather. In some cases, the roof is also vital to the stability of the building. Roof assemblies consist of three major components: the supporting structure, the roof decking, and the roof covering (**FIGURE 6-22**).

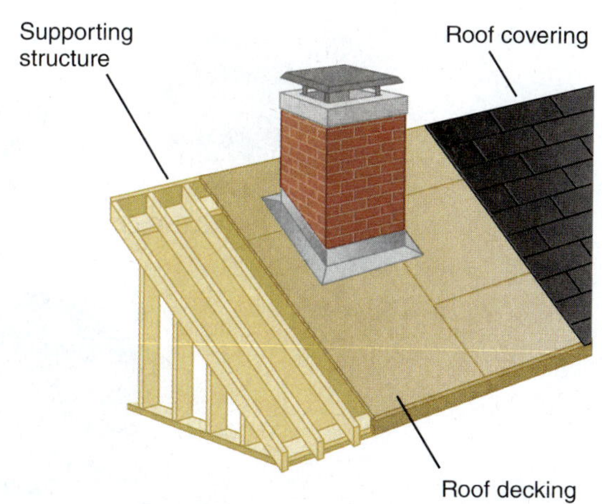

FIGURE 6-22 Roof assemblies consist of a supporting structure, decking, and covering.

The most common types of roof supports are solid beams of wood, steel, or concrete or a system of trusses. Solid-beam construction uses solid components, such as girders, beams, and **rafters** (solid wood joist mounted in an inclined position), to support the roof.

The roof decking is the portion of roof between the roof supports and the roof covering. It is composed of a rigid layer of wooden boards, plywood sheets, or metal panels anchored to the supporting structure and then treated or covered to prevent water penetration. The roof covering is applied on top of the deck; it is the part of the roof exposed to the weather.

Several materials and methods are used for roof construction. Roof materials may be combustible, noncombustible, or both. Roofs are constructed in three primary designs: pitched, curved, and flat. A building may include a combination of these designs. Each type of roof is different and presents different challenges.

Pitched Roofs

A **pitched roof** has sloping or inclined surfaces. Pitched roofs are used on many houses and some commercial buildings. The pitch or angle of the roof can vary depending on local architectural styles and climate. Variations of pitched roofs include gable, hip, mansard, gambrel, and lean-to roofs (**FIGURE 6-23**).

The slope of a pitched roof can present either a minor inconvenience or a complete lack of secure footing, depending on its angle. Firefighters working on a pitched roof are always in danger of falling, but the risk is particularly high when the roof is wet or icy, when the roof covering is not secure, or when smoke or darkness obscures vision. Roof ladders and aerial apparatus are used to provide a secure working platform for firefighters working on a pitched roof.

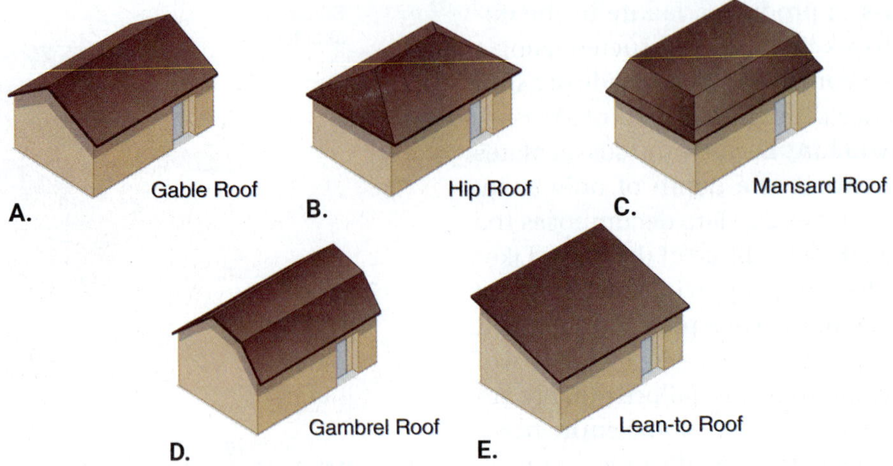

FIGURE 6-23 Examples of pitched roofs: **A.** A gable roof. **B.** A hip roof. **C.** A mansard roof. **D.** A gambrel roof. **E.** A lean-to roof.

Pitched roofs are usually supported by either rafters or trusses. Modern lightweight construction uses manufactured wood trusses to construct most pitched roofs. Many lightweight trusses are manufactured using metal gusset plates that penetrate the wood no more than ⅜ in. (1 cm) and tie the chords and members of the truss together. The decking usually consists of thin plywood or a sheeting material such as wood particleboard. When trusses and decking are exposed to heat and fire, they often fail after a short period of time with no prior warning. Firefighters cannot work safely on this type of roof when the supporting structures become involved in a fire.

Curved Roofs

A **curved roof** is often used in buildings that require large, open interiors, such as supermarkets, warehouses, industrial buildings, arenas, auditoriums, bowling alleys, churches, airplane hangars, and others. Steel or wood bowstring trusses or arches usually are used to support these kinds of roofs.

The decking on curved roof buildings can range from solid wooden boards or plywood to corrugated steel sheets. The type of material in use must be identified before ventilation openings can be made. Often, the roof covering consists of layers that include felt, mineral fibers, and asphalt, although some curved roofs are covered with foam plastics or plastic panels.

Flat Roofs

A **flat roof** can be found on houses, apartment buildings, shopping centers, warehouses, factories, schools, and hospitals (**FIGURE 6-24**). Most flat roofs have a slight slope so that water can drain off the structure. If the roof does not have the proper slope or if the drains are not maintained, water may pool on the roof, overloading the structure and causing a collapse.

FIGURE 6-24 An example of a flat roof.

The support systems for most flat roofs are constructed of either wood or steel. A wood support structure uses solid wooden beams (a **wooden beam** is a load-bearing member assembled from individual wood components) and joists; laminated wood beams may be included to provide extra strength or to span long distances. Lightweight construction uses wood trusses or wooden I-beams as the supporting members. Such lightweight assemblies are much less fire resistant and collapse more quickly than do solid-wood systems.

Steel also may be used to support flat roofs. Open-web steel trusses (sometimes called bar joists) can span long distances and remain stable during a fire if they are not subjected to excessive heat. The fire resistance of steel trusses is directly related to how well the steel is protected from the heat of the fire.

Flat roof decks can be constructed of wood planking, plywood, corrugated steel, gypsum, or concrete. The material used depends on the building's age and size, the climate, the cost of materials, and the type of roof covering.

SAFETY TIP

When working on a flat roof, stay away from the edges and avoid walking on skylights and other openings that will not support your weight. Fiberglass panels are sometimes used as skylights and can be hard to distinguish from metal roofs, especially at night.

Roof Covering Materials

Several roof-covering materials are used on pitched roofs, usually in the form of shingles or tiles (**FIGURE 6-25**). Shingles usually are made from felt or mineral fibers impregnated with asphalt, although metal and fiberglass shingles also are used. Wood shingles, often made from cedar and called cedar shakes, are popular in some areas. In older construction, wood shingles often were mounted on individual wood slats instead of continuous decking; this configuration of shingles burns rapidly in dry weather conditions.

Shingles are generally durable, economical, and easily repaired. A shingle roof that has aged and deteriorated should be removed completely and replaced. Some older buildings may have newer layers of shingles on top of older layers, which can make it difficult to cut an opening for ventilation.

Roofing tiles are usually made from clay or concrete products. Clay tiles, which have been used for roofing since ancient times, can be either flat or rounded. Rounded clay tiles are sometimes called mission tiles. Clay tiles are both durable and fire resistant.

Slate tiles are produced from thin sheets of rock. Slate is an expensive, long-lasting roofing material, but it is brittle and becomes slippery when wet. Slate tiles also pose a threat to firefighters below when a pitched roof is being ventilated, because falling tiles can become airborne missiles.

Metal panels of galvanized steel, copper, and aluminum also may be used on pitched roofs. Expensive metals such as copper often are used for their decorative appearance, whereas lower-cost galvanized steel is

A.

B.

C.

FIGURE 6-25 Several roof-covering materials are used on pitched roofs. **A.** A tile roof. **B.** A wood shingle roof. **C.** A slate roof.

A: © Paul Springett/Up the Resolution/Alamy Stock Photo; **B:** © Ron Chapple/Thinkstock/Alamy Stock Photo; **C:** © TH Photo/Alamy Stock Photo.

used on barns and industrial buildings. Corrugated, galvanized metal panels are strong enough to be mounted on a roof without roof decking. Metal roof coverings will not burn; however, they can conduct heat to the roof decking.

Most flat roofs are covered with multiple-layer, built-up roofing systems that help prevent leaks. This is due to the fact that flat roofs do not shed water as well as pitched roofs, and this can lead to leaks if the flat roof is not sealed well. A typical built-up roof covering contains five layers: a vapor barrier, thermal insulation, a waterproof membrane, a drainage layer, and a top layer called a **wear course**, which is a layer that can be walked on. The outside of the flat roof reveals only the top layer, which is often gravel; it serves as the wear course and protects the underlying layers from wear and tear. Cutting through all of the layers to open a ventilation hole can be quite a challenge.

The coverings used on flat roofs typically contain highly combustible materials, including asphalt, roofing felt, tar paper, rubber or plastic membranes, and plastic insulation layers (**FIGURE 6-26**). These materials are difficult to ignite. Once lit, however, they burn readily and release great quantities of heat and black smoke.

Flat roofs can present unique problems during vertical ventilation operations. A roof with a history of leaks could have several patches where additional layers of roofing material make for an extra-thick covering. Sometimes a whole new roof is constructed on top of the old one. After opening a hole in the new roof, firefighters might discover the old roof and have to cut into that as well. Firefighters also might discover fire burning in the void space between the two roofs.

FIGURE 6-26 Flat roofs typically contain highly combustible materials, including asphalt, roofing felt, tar paper, rubber or plastic membranes, and plastic insulation layers.

A similar problem is often encountered in remodeled buildings or buildings with additions. An old flat roof may be found under a new, pitched section, resulting in two separate void spaces. The old roof may provide a large supply of fuel for the fire. Such a situation is difficult to predict unless the incident commander (IC) has the building plan in advance.

Doors and Windows

Doors and windows are important components of any building. Although they generally have different functions—doors provide entry and exit, whereas windows provide light and ventilation—in an emergency, doors and windows are almost interchangeable. A window can serve as an entry or an exit, for example, whereas a door can provide light and ventilation.

Hundreds of door and window designs exist, with many different applications. Fire doors and fire windows are designed to prevent the spread of fire, but even ordinary wood doors provide some barrier to fire spread. Closing any door during a fire may give you a few additional minutes to search a room and minimize a potential flowpath.

Door Assemblies

Most doors are constructed of either wood or metal. Hollow-core wooden doors are often used inside buildings. A typical hollow-core door has an internal framework whose outer surfaces are covered by thin sheets of wood. Although a fire will burn through a hollow-core door, this type of door can act as a fire barrier for a short period of time. Additionally, these doors can be easily opened with simple hand tools, which means they are generally not used in buildings where security is a concern.

Solid-core wooden doors are used where a more substantial door is required. These types of doors are manufactured from solid panels or blocks of wood and are more difficult to force open. A solid-core door provides some fire resistance and often can keep a fire contained within a room for 20 minutes or more, giving firefighters a chance to arrive on the scene and mount an effective attack.

Metal doors are more durable and fire resistant than wooden doors. Some metal doors have a solid wood core that is covered on both sides by a thin sheet of metal; others are constructed entirely of metal and reinforced for added security. The interior of a metal door can be either hollow or filled with wood, sound-deadening material, or thermal insulation. Although most approved fire doors are metal, you should not assume that all metal doors are fire doors.

Window Assemblies

Firefighters sometimes need to use windows as entry points to attack a fire and as emergency exits. Like opening a door, opening a window can greatly change the flow path of a fire by providing ventilation, allowing smoke and heat to escape the building and cooler, fresh air to enter it.

Windows come in many shapes, sizes, and designs for different buildings and occupancies (**FIGURE 6-27**). Firefighters must become familiar with the specific types of windows found in local occupancies, learn to recognize them, and understand how they operate. It is especially important to know if a particular type of window is difficult to open or cannot be opened. Modern windows are built with energy-conserving features. As part of this trend, double- and triple-pane glass windows are common in some parts of the country. Window coatings and stronger types of glass, including hurricane-proof or burglar-resistant glass, may make forcible entry more difficult during firefighting operations. Chapter 15, *Forcible Entry,* contains more information on window construction.

Fire Doors and Fire Windows

A **fire door assembly** and a **fire window** are constructed to prevent the passage of flames and heat through an opening during a fire. They must be tested and meet the standards specified in NFPA 80, *Standard for Fire Doors and Other Opening Protectives, 2022 Edition.* Fire doors and fire windows come in many different shapes and sizes, and they provide different levels of fire resistance. For example, fire doors can swing on hinges, slide down or across an opening, or roll down to cover an opening.

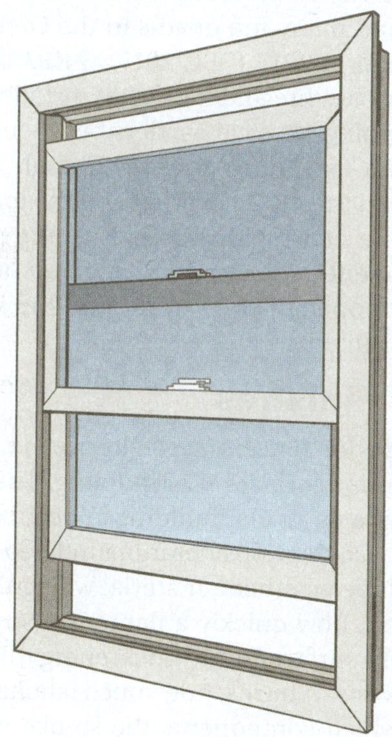

FIGURE 6-27 Windows can open in many different ways. The double-hung window shown here is just one example.

The fire rating on a door or window covers the actual door or window, the frame, the hinges or closing mechanism, the latching hardware, and any other equipment that is required to operate the door or window. All fire doors must have a mechanism that keeps the door closed or automatically closes the door when a fire occurs. For example, doors that are normally open can be closed by the release of a fusible link—a substance that melts when exposed to heat to release a connection—or an electromagnet. The closing of these doors is triggered by the activation of a smoke detector, heat detector, or fire alarm system.

Fire doors and fire windows are rated for a particular duration of fire resistance to a standard test fire. Fire resistance is the length of time that a building or a building component can withstand a fire before igniting. Note that a 1-hour rating does not guarantee that the door will resist any fire for 60 minutes; rather, it simply establishes that the door resisted the standard test fire for 60 minutes. In general, a door rated at 1 hour will last about twice as long as a door rated 30 minutes.

Fire windows are used when a window is needed in a required fire-resistant wall. These windows often are made of wired glass, which is designed to withstand exposures to high temperatures without breaking. Fire-resistant glass without wires is available for some applications; in such cases, special steel window frames are required to keep the glass firmly in place.

Wired glass is used to provide vision panels in or next to fire doors. Vision panels allow a person to view conditions on the opposite side of a door without opening the door or to make sure no one is standing in front of the door before opening it. In these configurations, the components making up the entire assembly—including the door, the window, and the frame—are tested and approved together.

When a window is required only for light passage, glass blocks are sometimes used instead of wired glass. Glass blocks resist high temperatures and remain in place during a fire (assuming they are properly installed). The size of the opening is limited and depends on the fire-resistance rating of the wall.

TIP

Remember, after occupants have been removed from the building, no building is worth a firefighter's life.

Exterior Finishes and Siding

The outsides of buildings are covered with a wide variety of materials, including stucco, brick, stone, metal, wood, and plastics. Some may not ignite or burn under ordinary circumstances, and some siding materials actually help prevent the spread of fire.

Stucco is a thin concrete surface that can be used over any material, such as brick, tile, or wood. Compared to wood siding, stucco offers better protection against fire exposure; however, it may be more difficult to make openings in walls covered with this type of siding.

Brick veneer siding is popular for wood-frame residences, garden apartments, and smaller commercial buildings in areas where brick is economical. This siding consists of a single layer of brick or stone applied to the exterior walls of the building to give the appearance of a durable and architecturally desirable outer covering. Brick veneer is not structural; it is attached to a load-bearing wall. Natural or artificial stone can be used in the same manner. This means that even though a building has an outer layer of brick or stone, firefighters should never assume that the exterior walls are masonry. Many buildings that look like masonry are actually constructed using wood-frame techniques and materials. If the wood structure is damaged during a fire, this veneer is likely to collapse.

SAFETY TIP

To help determine whether a brick wall is a solid masonry wall or a brick veneer, look at the pattern of the bricks and the age of the building. If every seventh row of bricks is turned sideways so that the end of the brick is visible (kings row pattern), it may be a solid masonry wall. If all of the rows of brick are the same, it may be brick veneer, particularly if it is modern construction.

Metal siding, when used on residences, is usually made up to resemble some other material. For example, aluminum siding is made to look like clapboards, and embossed sheet metal may be made to look like stone. Metal siding can present severe electrical hazards, both from stray electrical currents and from lightning. Although bonding and grounding of the siding may prevent such hazards, this is rarely done, and firefighters need to be aware of the possibility that the siding is energized.

Wood siding may consist of thick or thin boards or cedar shake shingles. Some forms of wood siding are more flammable than others. For example, thick boards on a barn require more heat to ignite; however, they contain large amounts of energy that, once ignited, result in increased fire growth and sustainability. Thin boards, on the other hand, require less heat to ignite and burn quickly. Wooden shake shingles can ignite easily and spread flames quickly.

Plastic or vinyl siding is often made to look like wood siding and may be used to cover up old wood siding. This type of siding melts at relatively low temperatures and will deform, burn, and drip when on fire.

Some types of siding are bonded with insulation. Referred to as aluminum composite materials (ACM), metal composite materials (MCM), or high pressure laminates (HPL), these cladding components come in different variations but are generally considered to be flammable due to not only the hydrocarbon-based insulation, but the lightweight aluminum that begins to burn as well. The use of ACM contributed to the rapid fire spread and many fire deaths in the Grenfell Tower Fire in London in 2017 (BBC 2017) (**FIGURE 6-28**). If the siding catches fire and it was not installed correctly with fire stopping physical barriers between floors that stop fire from extending upward through the hollow space between the ACM covering and the load-bearing walls—the fire can rapidly spread up the side of the building, and enter the interior of the building through soffit vents or open doors or windows, and into the interior of the building.

Interior Finishes

The term **interior finish** is typically used to refer to the exposed interior surfaces of a building. These materials affect how a particular building or occupancy reacts when a fire occurs. Considerations related to interior finishes include whether a material will ignite easily or resist ignition, how quickly a flame will spread across the material's surface, how much energy the material will release when it burns, how much smoke it will produce, and which components the smoke will contain (e.g., toxic particles or gases).

A room with a bare concrete floor, concrete block walls, and a concrete ceiling has no interior finishes

FIGURE 6-28 The use of ACM contributed to the rapid fire spread and many fire deaths in the Grenfell Tower Fire in London in 2017.

that will increase the fire load. However, when the same room has an acrylic carpet with rubber padding on the floor, wooden baseboards, varnished wood paneling on the walls, and foam plastic acoustic insulation panels on the ceiling, the situation is vastly different. These interior finishes ignite quickly, spread flames quickly across other surfaces, and release significant quantities of heat, flames, and toxic smoke.

Different interior finish materials contribute in various ways to a building fire. Each individual material has certain characteristics, and firefighters must evaluate the particular combination of materials in a room or space. Typical wall coverings include painted plaster or gypsum board, wallpaper or vinyl wall coverings, wood paneling, and many other surface finishes. Floor coverings might include different types of carpet, vinyl floor tiles, or finished wood flooring. All of these products will burn to some extent, and each one involves a different set of fire risk factors. Firefighters must understand the hazards posed by different interior finish materials if they are to operate safely at the scene of a fire. Keep in mind that most of the furnishings and interior finishes in a modern house consist of some type of plastic that is derived from petroleum products.

Types of Construction

Firefighters need to understand and recognize the five types of building construction, designated as Type I through Type V. These classifications are directly related to fire protection and fire behavior in the building.

Buildings are classified based on the combustibility of the structure and the fire resistance of its components. Buildings using Type I or Type II construction are assembled with primarily noncombustible materials and limited amounts of wood and other materials that will burn. This does not mean, however, that Type I and Type II buildings cannot be damaged by a fire. In fact, if the contents of the building burn, the fire can seriously damage or destroy the structure.

In buildings using Type III, Type IV, or Type V construction, both the structural components and the building contents will burn. Wood and wood products are used to varying degrees in these buildings. If the wood ignites, the fire weakens and consumes both the structure and its contents. Structural elements of these buildings can be damaged or destroyed in a very short time.

As mentioned earlier, a building is classified based on the fire resistance of its components. Fire-resistance ratings are stated in hours, based on the time that a test assembly withstands or contains a standard test fire. For example, walls are rated based on whether they stop the progress of the test fire for periods ranging from 20 minutes to 4 hours. A floor assembly is rated based on whether it supports a load for 1 hour or 2 hours. As explained earlier, because ratings are based on a standard test fire, an actual fire could be more or less severe than the test fire. Fire ratings are considered merely guidelines; that is, an assembly with a 2-hour rating will not necessarily withstand an actual fire for 2 hours, but it will withstand a fire for about twice as long as another component in the same fire that is rated for 1 hour.

Building codes specify the type of construction to be used, based on the height, area, occupancy classification, and location of the building. Other factors, such as the presence of automatic sprinklers, also affect the required construction classification. NFPA 220, *Standard on Types of Building Construction, 2024 Edition,* provides detailed requirements for each type of building construction.

Type I Construction: Fire-Resistive

Type I construction (fire-resistive) is the most fire-resistive category of building construction. It is used for buildings designed for large numbers of

FIGURE 6-29 Type I construction is commonly found in schools, hospitals, and high-rise buildings.

people, buildings with a high life safety hazard, tall buildings, large-area buildings, and buildings containing special hazards. Type I construction is commonly found in schools, hospitals, and high-rise buildings (**FIGURE 6-29**).

Buildings with Type I construction can withstand and contain a fire for a specified period of time. The fire resistance and combustibility of all construction materials are carefully evaluated, and each building component must be engineered to contribute to fire resistance of the entire building to warrant this classification.

All of the structural members and components used in Type I construction must be made of noncombustible materials, such as steel, concrete, or gypsum board. In addition, the structure must be constructed or protected so that it has at least 2 hours of fire resistance. Building codes specify the fire-resistance requirements for different components. For example, columns and load-bearing walls in multistory buildings could have a fire-resistance requirement of 3 or 4 hours, whereas a floor might be required to have a fire-resistance rating of only 2 hours. Some jurisdictions allow Type I buildings to use a limited amount of combustible materials as part of the interior finish.

If a Type I building exceeds specific height and area limitations, codes generally require the use of fire-resistive walls and/or floors to subdivide it into fire compartments. A **fire compartment** is an area in a building that is constructed so that if a fire is burning within the compartment, the fire will be confined to that compartment for a specified length of time. It might consist of a single floor in a high-rise building or a part of a floor in a large-area building. In any event, a fire in one compartment should not spread to any other parts of the building. To ensure that fire is contained, stairways, elevator shafts, and utility shafts should be enclosed in construction that prevents fire from spreading from floor to floor or from compartment to compartment.

Type I buildings typically use reinforced concrete and protected steel-frame construction. As mentioned earlier, concrete is noncombustible and provides thermal protection around the steel reinforcing rods. Reinforced concrete can fail, however, if it is subjected to a fire for a long period of time or if the building contents create an extreme fire load. Given its tendency to lose strength in the face of high temperatures, structural steel framing must be protected from the heat of a fire. In Type I construction, the structural steel members are generally encased in concrete, shielded by a fire-resistive ceiling, covered with multiple layers of gypsum board, or protected by a sprayed-on insulating material. An unprotected steel beam exposed to a fire can fail and cause the entire building to collapse.

By themselves, Type I building materials should not provide enough fuel to create a serious fire. Thus it is the contents of the building that determine the severity of a Type I building fire. In theory, a fire could consume all of the combustible contents of a Type I building yet leave the structure basically intact.

Although Type I construction is designed to give firefighters time to conduct an interior attack, it can be very difficult to extinguish a fire in a Type I building. A fire that is fully contained within fire-resistive construction can be very hot and difficult to ventilate because the burning contents may produce copious quantities of heat and smoke. For this reason, fire-resistive construction in modern high-rise buildings is typically combined with an automatic fire detection system and automatic sprinklers.

Sometimes, the burning contents of a building provide sufficient fuel and generate enough heat to overwhelm the fire-resistant properties of the construction. In these cases, the fire can escape from its compartment and damage or destroy the structure. Similarly, inadequate or poorly maintained construction can undermine the fire-resistive properties of the building materials. Under extreme fire conditions, a Type I building can collapse. Even so, these buildings have a much lower potential for structural failure than do other buildings.

Type II Construction: Noncombustible

In **Type II construction (noncombustible)** (**FIGURE 6-30**), all of the structural components

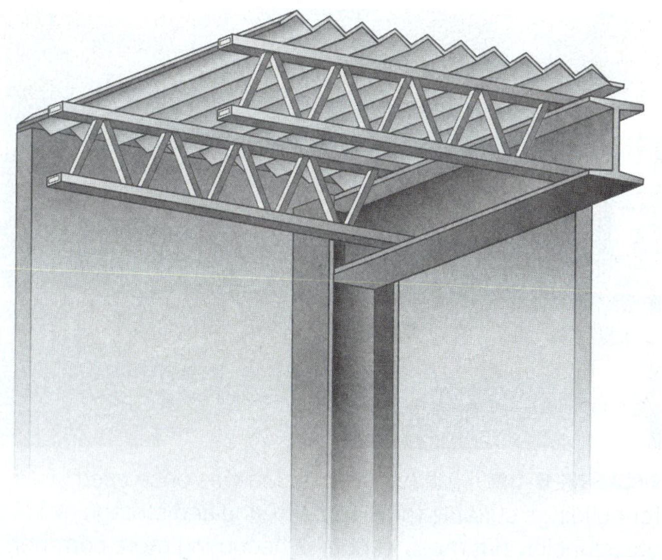

FIGURE 6-30 Type II construction is also referred to as noncombustible construction.

© Jones & Bartlett Learning

must be made of noncombustible materials. The fire-resistive requirements, however, are less stringent for Type II construction than for Type I construction. In some cases, no fire-resistance requirements apply to the building. In other cases, known as protected noncombustible construction, a fire-resistance rating of 1 or 2 hours may be required for certain elements.

Type II construction is most common in single-story story commercial (strip shopping centers or "big box" stores) buildings, warehouses, or factory buildings, where vertical fire spread is not an issue. Unprotected construction of this type is generally limited to a maximum of two stories, although some multistory buildings are constructed using protected noncombustible construction.

Steel is the most common structural material in Type II buildings. Insulating materials can be applied to the steel when fire resistance is required. A typical example of Type II construction is a large-area, single-story building with a steel frame, metal or concrete block walls, and a metal roof deck. Fire walls are sometimes used to subdivide these large-area buildings and prevent catastrophic losses. Undivided floor areas must be protected with automatic sprinklers to limit the fire risk.

Fire severity in a Type II building is determined by the contents of the building, because the structural components contribute little or no fuel, and interior finish materials are limited. If the building contents provide a high fuel load, a fire can collapse and destroy the structure. In this kind of construction, automatic sprinklers should be used to protect combustible and valuable contents. Fires in buildings equipped with monitored fire detection systems are more likely to be reported earlier than fires in buildings without monitored fire detection systems.

FIGURE 6-31 Type III construction is also referred to as ordinary construction because it is used in a wide variety of buildings.

© Ken Hammon/USDA

Type III Construction: Ordinary

Type III construction (ordinary) is used in a wide variety of buildings, ranging from commercial strip malls to small apartment buildings (**FIGURE 6-31**). Ordinary construction is usually limited to buildings of no more than four stories, but it can sometimes be found in buildings as tall as six or seven stories.

Type III buildings have masonry exterior walls that support the floors and the roof structure. The interior structural and nonstructural members—including the walls, floor, and roof—are constructed of wood. In most ordinary construction, gypsum board or plaster is used as an interior finish material, covering the wood framework and providing minimal fire protection.

Fire-resistance requirements for interior construction of Type III buildings are limited. The gypsum or plaster coverings over the interior wood components provide some fire resistance. Key interior structural components may be required to have fire-resistance ratings of 1 or 2 hours, or there may be no requirements at all. Some Type III buildings use interior masonry load-bearing walls to meet the requirements for fire-resistant structural support.

A building of Type III construction has two separate fire loads: the contents and the combustible building materials used. Even a vacant Type III construction building contains a sufficient quantity of wood and other combustible components to produce a large fire. Given the combustibility of these materials, a fire involving both the contents and the structural components can quickly destroy the building.

Fortunately, most fires originate with the building contents and are extinguished before the flames ignite the Type III structure. Once a fire extends into the structure and begins to consume the fuel within the walls and above the ceilings, it becomes much more difficult to control and prone to roof or floor collapse.

Type III construction presents several problems for firefighters. For example, an electrical fire can begin inside the void spaces within the walls, floors, and roof assemblies and extend to the contents. The void spaces allow a fire to extend vertically and horizontally, spreading from room to room and from floor to floor. In such a case, Firefighters will need to open the void spaces to fight the fire. An uncontrolled fire within the void spaces is likely to destroy the building. Firefighters need to be especially aware that fires in void spaces under floors or under the roof can quickly destroy the strength of these components. Do not be surprised by these "hidden" fires; instead, anticipate them.

The fire resistance of interior structural components often depends on the age of the building and on local building codes. Older Type III buildings were built from solid dimensional lumber, which can contain or withstand a fire for a limited time. Newer or renovated buildings might contain lightweight assemblies that can be damaged much more quickly and are prone to early failure. Floors and roofs in these buildings collapse suddenly and without warning. Because firefighters cannot always determine the type of structure during a fire, you should assume that lightweight components may be present in any Type III building.

Exterior walls can collapse if a fire causes significant damage to the interior structure. Because the exterior walls, the floors, and the roof are connected in a stable building, the collapse of the interior structure could make the free-standing masonry walls unstable and prone to collapse.

Type IV Construction: Heavy Timber

In a **Type IV construction (heavy timber)** (**FIGURE 6-32**) building, the exterior walls consist of masonry construction and the interior walls, columns, beams, floor assemblies, and roof structure are made of wood. The exterior walls are usually brick and are extra thick to support the weight of the building and its contents.

FIGURE 6-32 Type IV construction was once used for buildings suitable for manufacturing and storage occupancies, but mass lumber is becoming more common.

The wood used in Type IV construction is much heavier than that used in Type III construction. It is more difficult to ignite and will withstand a fire for a much longer time before the building collapses. A typical heavy timber building might have 8-in. (20-cm)-square wood posts supporting 14-in. (35-cm)-deep wood beams and 8-in. (20-cm) floor joists. The floors could be constructed of solid wood planks, 2 or 3 in. (5 or 8 cm) thick, with a top layer of wood serving as the finished floor surface.

Heavy timber construction, if it meets building codes, has no concealed spaces or voids. This structure helps reduce the horizontal and vertical fire spread that often occurs in ordinary construction buildings. Unfortunately, many heavy timber buildings do contain vertical openings for elevators, stairs, or machinery, which can provide a path for a fire to travel from one floor to another.

In the past, heavy timber construction was used to construct buildings as tall as six to eight stories with open spaces suitable for manufacturing and storage occupancies. The heavy, solid wood columns, support beams, floor assemblies, and roof assemblies used in heavy timber construction will withstand a fire much longer than the smaller wood members used in ordinary and lightweight combustible construction. Nevertheless, once involved in a fire, the structure of a heavy timber building can burn for many hours. A fire that ignites the combustible portions of heavy timber construction is likely to burn until it runs out of fuel and the building is reduced to a pile of rubble. As the fire consumes the heavy timber support members, the masonry walls will become unstable and collapse.

Type IV construction is also referred to as **mill construction** because these buildings were commonly constructed during the 1800s in the United States, especially in northeastern states, as large mill buildings to be used as factories and warehouses. Mill construction was state of the art for its time. Automatic sprinkler systems were developed to protect mill buildings. As long as the sprinklers are properly maintained, these buildings have a good safety record. Without sprinklers, however, a heavy timber mill building becomes a major fire hazard.

As a new subset of Type IV construction, mass lumber, is becoming more common. Mass lumber differs from heavy timber in that it relies on large engineered wood products to make up the structural components. With several variations designed for different applications and load configurations, mass lumber can be used to build high-rise structures that would not have been possible with heavy timber due to the limitations in wood size required in today's forestry environment. Some of the same concepts apply with respect to surface-to-mass ratio and the ability of mass lumber to withstand direct flame contact for longer periods of time than smaller typical lumber components.

Type V Construction: Wood-Frame

In **Type V construction (wood-frame)**, all of the major components are constructed of wood or other combustible materials (**FIGURE 6-33**). Wood-frame construction is used not only in one- and two-family dwellings and small commercial buildings, but also in larger structures such as apartment and condominium complexes and office buildings up to four stories in height.

FIGURE 6-33 Type V (wood-frame) construction is the most common type of construction used today.

Many wood-frame buildings do not have any fire-resistive components. Some codes require a 1-hour fire-resistance rating for limited parts of wood-frame buildings, particularly those of more than two stories. Plaster or gypsum board barriers often are used to achieve this rating, but firefighters should not assume that these barriers are present or effective in all Type V buildings.

Because all of the structural components are combustible, wood-frame buildings can rapidly become totally involved in a fire. In addition, Type V construction usually creates voids and channels that allow a fire within the structure to spread quickly. A fire that originates in a Type V building can easily extend to other nearby buildings.

Wood-frame buildings often collapse and suffer major destruction from fires. Smoke alarms are essential to warn building residents early if a fire occurs. Although compartmentalization can help limit the spread of a fire, fire detection devices and automatic sprinklers are the most effective way of protecting lives and property in Type V buildings.

Wood-frame buildings are constructed in a variety of ways. Older wood-frame construction was assembled from solid lumber, which relied on its mass to provide strength. To reduce costs and create the largest building with the least amount of material, lightweight construction makes extensive use of engineered wood products such as plywood I-beams and wooden trusses. With these construction techniques, structural assemblies are engineered to be just strong enough to carry the required load. As a result, there is little built-in safety margin, and these buildings can collapse early, suddenly, and completely during a fire.

The presence of flammable building contents may greatly increase the severity of a fire and accelerate the speed with which the fire destroys a wood-frame building. Indeed, the structural elements of such a building can be damaged or destroyed in a very short time. Whenever a fire has been burning under a floor or roof, you should assume that the structure is unsafe and may collapse. Fires that start to consume structural parts of a Type V building present a high risk of building collapse.

Wood-frame buildings might be covered with wood siding, vinyl siding, aluminum siding, brick veneer, or stucco. Remember, however, that the covering on a building does not reflect the type of construction of that building. Just because you see a brick covering does not mean that the building is constructed of solid brick walls. During a structural fire, the brick veneer is likely to collapse or peel away as the wall behind it burns or collapses. In addition, a single thickness of bricks is rarely strong enough to stand independently. Firefighters should be aware of buildings constructed in this manner and maintain a safe distance from potentially falling bricks. For this reason, it is imperative that a collapse zone be set up around a burning building.

Two systems are used to assemble wood-frame buildings: balloon-frame construction and platform-frame construction. Because these systems developed in different eras, firefighters often can anticipate the type of construction based on the age of a building.

SAFETY TIP

Many buildings have a brick veneer on the exterior walls of wood-frame construction. A thin layer of brick or stone might be applied to enhance the appearance of the building or to reduce the risk of fire spread. Unfortunately, this practice can give the impression that a building is Type III (ordinary) construction when it is really Type V (wood-frame) construction.

Balloon-Frame Construction

Balloon-frame construction was popular from the late 1800s until the 1950s. In a balloon-frame building, the exterior walls are assembled with wood studs that run continuously from the basement to the roof (**FIGURE 6-34**). In a two-story building, the floor joists that support the first and second floors are nailed to these continuous studs. As a result, there is an open channel between each pair of studs that extends from the foundation to the attic. Each of these channels provides a path that enables a fire to spread from the basement to the attic without being visible on the first- or second-floor levels. When attacking a fire in a balloon-frame building, firefighters must anticipate that a fire that originated on a lower level will quickly extend through these voids, and they should open the void spaces or check the lowest level of the structure to check for hidden fires and to prevent rapid vertical extension and working above the seat of the fire.

Because these stud channels have a consistent width, builders typically took advantage of this and installed doors and windows over these spaces. This usually meant that the windows on the second floor were directly aligned over windows or doors on the first floor. This can be used as a potential indicator for balloon-frame construction versus platform-frame construction.

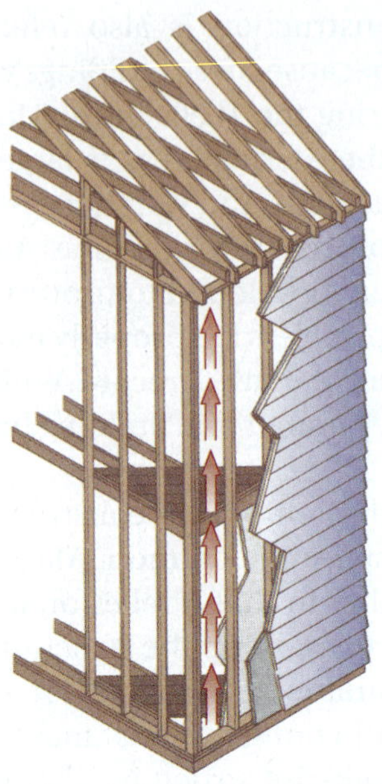

FIGURE 6-34 In a balloon-frame building, the exterior walls create channels that enable a fire to spread from the basement to the attic.

Platform-Frame Construction

Platform-frame construction became the method of choice for single family dwelling construction starting the 1950s and is used for almost all modern wood-frame construction. Platform framing prevents fire from spreading from one floor to another through continuous stud spaces. In a building with platform construction, the exterior wall studs are not continuous. Instead, each floor is constructed as a platform, and the studs for the exterior walls of that floor are erected on top of the platform. Each set of studs extends only to the underside of the platform of the floor above it. The platform for each successive floor blocks vertical void spaces. This means that at each level, the floor platform blocks the path of any fire rising within the void spaces (**FIGURE 6-35**). Although a fire can eventually burn through the wood platform, the platform will slow the fire spread.

TIP

Buildings may be designed specifically for the local climate and uses of a community. Additionally, building materials may vary from one region to another depending on cost and what is readily available. Learning about the unique types of buildings that exist in a community will help you predict the fire behavior of those buildings.

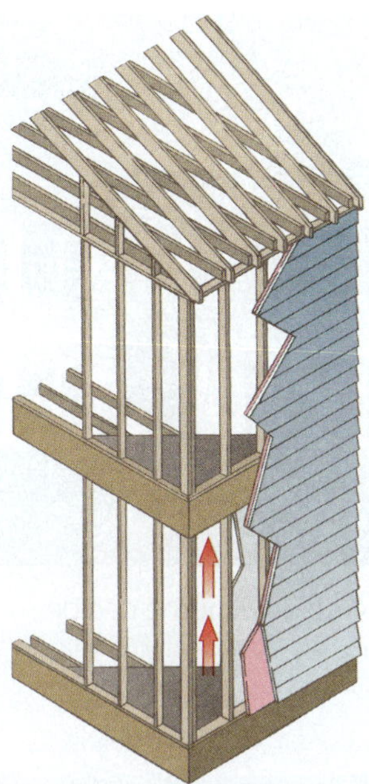

FIGURE 6-35 In a platform-frame building, the floor platform blocks the path of fire, preventing it from spreading from one floor to another.

FIGURE 6-36 This residence was initially constructed with Type III ordinary legacy construction in the 1920s and was renovated with contemporary Type V lightweight construction in 2016. This resulted in a house consisting of Type III ordinary construction on the left half and Type V lightweight wood-frame construction on the right half. If a fire occurs, fire spread and collapse will occur much faster in the right half (Type V) than in the left half (Type III).

Hybrid Building Construction

Learning about the different types of construction is important in understanding the stability of each type of construction and how each type of building will react under fire conditions. It is equally important to understand that a single building may contain more than one type of construction. A **hybrid building** is one that does not fit entirely into any of the five construction types because it incorporates building materials of more than one type, such as wooden beams and steel columns.

A hybrid building may result from initial construction or from renovations made to the structure. Some buildings are initially constructed with one part of the structure adhering to one type of construction and other parts of the structure adhering to a different type of construction. For example, a building may be constructed with a parking garage on its first level and residential apartments or condo units on its upper levels. In this case, local building codes may require that the first level be constructed with fire-resistive construction (Type I) but permit the upper levels to be constructed with contemporary lightweight wood framing (Type V), provided that the building is protected with a fire sprinkler system. Until the fire detection and sprinkler systems are installed and fully operational, while under construction, this type of building is at high risk of being completely destroyed by a fire in a short period of time. Because there is such a huge fire load with no fire stops, a fire can rapidly engulf and destroy the building in spite of aggressive firefighting efforts.

Extensive renovations to a structure also may result in a hybrid building. For example, a structure initially constructed of legacy construction (Type III) may be renovated to include an addition constructed using contemporary lightweight construction (Type V). Renovated buildings can be dangerous because they can give the impression that they are one type of construction when they are really a mixture of construction types. The original part of a renovated building may consist of a stronger or more fire-resistive construction, while the newer part may consist of lightweight construction that will quickly fail under fire conditions (**FIGURE 6-36**). The result is a building in which fire spread and collapse will occur much faster in one section.

Single-Family Dwelling Types

The majority of fires and fire deaths in the United States occur in residential structures, and of those, an overwhelming majority are in single-family dwellings. Single-family dwelling fires are considered "bread and butter" operations due to their frequency, as well as the fact that they represent the largest percentage of the building stock in most districts covered by the fire service. The six most common types of single-family

FIGURE 6-37 A Cape Cod home design.

© Susan Law Cain/Shutterstock

FIGURE 6-38 A colonial home design.

© PT Hamilton/Shutterstock

homes are in the United States are Cape Cod, colonial, ranch, split level, split foyer, and contemporary/modern. Each of these types has distinct features that have tactical implications for the firefighters arriving on the scene of a fire.

Cape Cod

A Cape Cod house, sometimes referred to simply as a cape, dates back to the colonial era. Key characteristics of this design are a high-pitched roof and a half story on the second floor (**FIGURE 6-37**). The high-pitched roofs were built to prevent snow accumulation. Typically these are smaller footprint structures compared to the average colonial-style house. They feature a front door centered at the front and flanked by windows with living, dining room and kitchen on the first floor and bedrooms typically on the second (half) floor. This half story presents a key tactical challenge because it includes knee walls used to finish the space, creating void spaces that can hide fire extension.

Colonial

The colonial house dates back to the 1600s and comes in multiple styles (**FIGURE 6-38**). They usually consist of at least two stories with the staircase located in the center of the hallway aligned near the front door. This means firefighters will typically be able to gain quick access to the second floor where the bedrooms are normally located as long as the stairway is not compromised. The first floor usually has the common areas, such as the living room or family room, kitchen, and dining room. These houses normally have a symmetrical structure (square or rectangular) on all floors and often have high-pitched roofs. Windows are typically balanced on either side of the front door.

FIGURE 6-39 A ranch home.

© Ursula Page/Shutterstock

Ranch/Rambler

The typical ranch house (also known as a rambler or simply ranch) is a single-story house that features an open layout for the living, dining, and kitchen areas (**FIGURE 6-39**). They often have long, low-pitch rooflines and large picture windows along the front of the house. Most are rectangular, but they can also be built in a *U* or *L* shape. Other common features of ranch homes include sliding glass doors that open onto the patio or backyard, wide roof eaves, and an attached garage. While most ranches don't have a second floor, they do have full basements that are often partially or fully finished. Many ranches were built after World War II because of their relatively low cost and the ability of developers to build entire communities simply. Construction of this style of house peaked in the 1950s to the 1960s.

Split Level

A split-level home features multiple floor levels that are staggered. There are usually two sets of stairs off the home's main story, leading upstairs to bedrooms and downstairs to the basement. They have been described as the marriage of a ranch and a split foyer. They were produced en masse during the late 1960s and early 1970s because they were cheap, easy to build, and offered more space for growing families than a ranch. Split levels typically have three distinct levels joined by short flights of stairs. The main living space connects to two other stories, often a basement and an upstairs area used for bedrooms. Split-level homes share one common feature with ranches; a large picture window is usually found in the living room on the main floor. Split levels always have a floor located below the point of entry. Lower-level basements commonly have a direct entrance into the garage. It is critical to understand that split-level homes can have as many as four or five levels, which will not present themselves on size-up from the front. Labeling can be confusing, so it is important that a common system is adopted by jurisdictions to avoid issues during fireground operations.

Split Foyer

The split-foyer house is often described as a variation of the split level (**FIGURE 6-40**). A key difference is that split foyers are almost always only two levels compared to the three to five levels in a split level. The front door is at ground level and is located between the two floors. Upon entering, there is a landing with a set of stairs going up and a set going down. The top floor usually contains the kitchen, living room, dining room, bedrooms, and bathrooms. The lower floor typically is partially below ground level and has storage and laundry areas, family rooms, and access to the garage. Roofs are typically low-pitched. The split foyer was popular in the 1970s.

FIGURE 6-40 A split-foyer home.

Contemporary/Modern

Modern (contemporary) homes typically have open floor plans and share common design elements such as tall, irregularly shaped windows and asymmetrical façades and floor plans (**FIGURE 6-41**). Stairs joining multiple levels can be in a variety of locations. Open floor plans mean a lack of compartmentation and an inability to predict what "typical" layouts may be, possibly inhibiting a faster search since orientation may be more difficult. Building materials include concrete, steel, and glass. Roofs are typically flat.

Building Collapse

Partial or complete building collapse has the potential to kill or injure building occupants, firefighters, and rescuers (**FIGURE 6-42**). Each department should

FIGURE 6-41 A modern or contemporary home.

FIGURE 6-42 A building collapse can kill or injure building occupants, firefighters, and rescuers.

have information and knowledge about the local types of building construction and hazards. Often, this information is obtained through preincident planning, and it should be available to firefighters who are dispatched to an emergency scene. It should include characteristics of the building that might lead to building collapse. Preincident planning is discussed later in this chapter. (See also Chapter 25, *Fire and Life Safety Initiatives*.)

To fully understand the potential for building collapse, you must first understand the forces that act on every building. Gravity exerts downward force on a building, which, in turn, exerts stress on the sides of a building. Buildings are constructed to resist the force of gravity. Structural elements and the connections that tie the building together help to transfer the load and contents of the building to another part of the structure. For example, the roof is supported by the walls of a building. When fire or lateral forces, such as wind or an explosion, damage or degrade the walls, a collapse is more likely to occur. The failure of any structural building component, such as a vertical column or a horizontal beam or truss, whether due to fire or poor building renovation, can begin a chain of events that places more load onto adjoining components which in turn may lead to their failure and then a localized or general collapse of the floor it is supporting.

Indicators of Potential Collapse

Once a collapse begins, the results are unpredictable. It is important that firefighters do everything possible to anticipate collapse. The following list is a review of the indicators of possible collapse:

- Cracks in walls, especially cracks that develop or grow during a fire
- Leaning walls
- Pitched or sagging doors
- Doors stuck in shifted frames
- Moaning, groaning, or cracking sounds from the structure
- Any type of movement or vibration
- Movement or shifting of water on the floor
- Smoke pushing through cracks in the wall
- Lack of water runoff from firefighting operations (because the added water weight adds to the live load in the building)

In addition to recognizing the signs, it is important to be able to identify the factors that increase the likelihood of building collapse so you can take the steps needed to reduce the risk of injury or death. These factors include underlying structural instability, fire and explosion damage, and the environment.

Underlying Structural Instability

The stability of a building is dependent on the interaction of all its structural components. The strength of each structural component relies on the engineering and construction methods that were used to develop and build them individually, all the way down to the smallest connection point. Although underlying structural instability often results from the state of the building *before* an emergency event occurs, the firefighter should also take into account the structural events that occur while the incident is ongoing. It is critical that the firefighter understand the importance of seizing opportunities to identify these issues prior to a collapse. Given the reality of the scope of building stock in any district, it becomes incumbent on the firefighter to always operate with high levels of situational awareness and be prepared to adapt to the quickly changing dynamics of any incident with respect to the building's condition.

Lightweight Construction

With the trend toward even lighter and smaller structural components, firefighters should assume that any newer construction will contain these construction elements. Keep in mind that they are generally hidden from view. This type of construction collapses more frequently and approximately three times faster than other types of construction (FSRI 2009). This fact, coupled with the sheer number of structures that use these elements, means that lightweight construction potentially poses the greatest collapse danger to firefighters today. In general, trusses, both wood and steel, are engineered and constructed with a margin of safety under normal conditions. During a fire, the protective membranes or other materials that normally shield these structural components burn away. Without those protective membranes or materials shielding them, the structural components cannot withstand fire conditions and the fire will quickly degrade the integrity of the component. Failure of any one element, such as a chord, a web, or an attachment point, leads to a weakening of the entire truss system. Failure of one truss increases the load on adjoining trusses. If the adjoining trusses are also degrading, then the cascading chain of events can lead to localized or generalized collapse in very short time.

In some cases, trusses may fail because of overload. As buildings are modified, or when the use of buildings changes, more weight may be placed on floors or roofs than they were originally intended to carry. In the event of a fire, the added weight contributes to the rapid failure of a truss-supported floor or roof.

In the past, firefighters were taught that sounding the floor or roof with a hand tool to see if it was "spongy" would help determine if that part of the structure was safe. While this method may indicate which parts of a floor to avoid, just because a floor does *not* feel spongy does not mean it is safe to step on. A better indicator of potential collapse when dealing with structural components hidden from view is to conduct an effective size-up and ensure that the area the firefighter is going to be working in or on is not directly above the fire. This helps define the difference between a contents fire and a structural (not structure) fire. Once structural components have started degrading, it is only a matter of time before collapse becomes likely.

SAFETY TIP

Merely sounding the floor or roof to see if it is spongy is not on its own a valid test to determine the integrity of the component.

Buildings under Construction or Demolition

Buildings under construction or renovation or in the process of demolition present a variety of problems and additional hazards for firefighters. In many cases, the fire protection features found in finished buildings are missing from these structures. For example, sprinklers and standpipe systems might not have been installed yet or might have been disconnected. Smoke alarms might not have been installed or might be inoperable. There may be no coverings on the walls or membrane on the ceiling leaving the structural components fully exposed. Missing doors and windows can provide an unlimited supply of fresh air to feed the fire and allow the flames to spread rapidly. In addition, construction and demolition sites often have large quantities of building materials stored close to buildings. These materials can add a huge additional fire load that contributes to the rapid spread of fire, often from one building to another. All of these factors can lead to rapid fire growth and quick structural failure.

Many fires at construction and demolition sites are inadvertently caused by workers using torches to weld, cut, or take apart pieces of the structure. Tanks of flammable gases and piles of highly combustible construction materials might be left in locations where they could add even more fuel to a fire. Buildings under construction or demolition often are left unoccupied for many hours, resulting in the delayed discovery and reporting of fires. In some cases, it might prove difficult for fire apparatus to approach the structure or for firefighters to access working hydrants. All of these problems must be anticipated when considering the fire risks associated with a construction or demolition site.

During construction of a building, structural components may be unstable. For example, buildings constructed with poured concrete floors are initially supported with shoring—temporary vertical supports. If this shoring is removed before the concrete is properly cured, the top floor can collapse. This can result in the progressive collapse of the floors below, even if they were properly cured. Precast concrete panels are constructed in one location and then hoisted into place on-site. This means, by definition, they are properly cured. However, they are unstable until all connections have been completed.

Construction workers sometimes decide to cut holes in supporting members, such as joists, beams, or trusses, to feed wires and piping through. If too many holes are cut into these members, they may fail during construction. A failure of one member can result in a localized or general collapse.

Buildings that are being renovated present concerns similar to those for buildings that are under construction. Load-bearing members may be structurally compromised during renovation. Damage to any structural component can cause a building to collapse. If a building is renovated to change from one occupancy type to another, the new contents of the building can exceed the load that the building was initially designed to support.

SAFETY TIP

Buildings under construction or demolition are at high risk for a major fire. Such was the case at the Deutsche Bank building in New York City, where two firefighters lost their lives in 2007 (NIOSH 2010). This building was damaged in the terrorist attacks of September 11, 2001, and was set for demolition. The New York City Fire Department responded to a working fire in the building and attempted to use the standpipe system to fight the fire. A part of the standpipe system had been removed, however, and the system was inoperable.

Building Load

The live load of the building may contribute to building collapse if the floors and/or roofs are overloaded. Roofs are usually not designed to be as strong as floors, especially in warmer climates that do not experience heavy snowfall. If the space under the roof is used for storage, or if extra HVAC equipment is mounted on the roof, the load can exceed the designer's expectations. During a fire, adding the weight of several firefighters to an overloaded roof can be disastrous.

Waterlogged building contents can contribute to building collapse. During a fire, the contents of a building may become soaked and retain the water that was used to extinguish the fire due to firefighting operations. Water weighs 8.34 pounds per gallon (1 kilograms per liter), so a nozzle that flows 100 gallons per minute (379 liters per minute) adds about 830 pounds (380 kilograms) of weight to the building each minute. It is important that firefighters calculate the total water load on the building. Waterlogged contents may cause a floor to collapse, which, in turn, may cause a wall to collapse.

Fire and Explosion Damage

Sustained moderate to heavy fire conditions can result in building collapse. Firefighters performing ventilation activities on the roof to release heated gases should be supported on an aerial apparatus if possible, rather than standing on the roof. Opening a ceiling, floor, or attic void space could produce a backdraft, flashover, or explosion, any of which can result in a sudden roof collapse. Additionally, the collapse itself can result in a sudden blast of fire. If the fire attack is a defensive strategy, firefighters should stay away from doors and windows in the event that a collapse occurs.

Building collapse can also occur during overhaul activities after the fire is extinguished. Speed is not paramount during overhaul activities, so use caution when switching from active firefighting to overhaul activities. A fire also may decrease the mass of wooden structural components. As the contents and structural components cool after the fire is extinguished, the contraction of the structural components and building contents can cause building movements that lead to collapse. If there is any doubt about the structural integrity of the building, it may be necessary to overhaul the building from outside and have it inspected by an engineer.

Environmental Factors

Building codes require structures be constructed to withstand heavy snow, rain, winds, and earthquakes in a distributed system; however, under fire conditions, these environmental factors can quickly lead to building collapse. As fire begins to alter the structural stability of the building, the additional stress from the snow, rain, or wind can cause the already weakened structural members to collapse prematurely.

Types of Floor Collapse

It is important for the firefighter to know the types of floor collapse that can occur in a building that they may encounter, particularly if a firefighter has initiated a mayday (call for help). This will directly affect the decision-making process with regard to identifying survivable spaces and search patterns and techniques needed to execute immediate rescues. (NOTE: This discussion must not be confused with the need for specialized technical rescue skills needed for larger scale events.) Floor collapses in a building happen in the following ways:

- Lean-to collapse: A partial collapse that results when the middle section of a floor fails and is still attached or leaning against one load-bearing wall.
- V-shaped collapse: A partial collapse that results when the middle section of a floor fails still attached or leaning against two load-bearing walls.
- A-frame collapse: A partial collapse that results when the sections of floor attached to the load-bearing wall fail and the middle remains intact.
- Pancake collapse: A complete collapse that results when the entire floor fails and drops onto the floor below.

Collecting the Needed Information

When a building is on fire, how do you determine which type of construction was used to build the building? Even experienced firefighters need to assess a building carefully to make this determination.

The best way to gather this information is to conduct surveys of buildings in your jurisdiction and collect this information in a preincident plan that describes the types of building construction as well as specific characteristics of various buildings (**FIGURE 6-43**). Consider the following questions:

- Which type of building construction is present?
- Which type of occupancy is this building?
- What type of contents does it contain?
- Which type of support do the floors have?

FIGURE 6-43 Preincident inspections help you determine the type of building construction before a fire occurs.

- Is it safe to climb onto the roof?
- Does the roof include photovoltaic solar panels or grow areas?
- Does the building contain roof trusses, floor trusses, or manufactured wood I-joists that might collapse suddenly?
- Which types of fire detection and fire suppression systems are installed in the building?

The importance of preplanning cannot be overemphasized. An excellent way to identify building construction is to inspect the building multiple times while it is being built. Your preplanning will be greatly advanced by this exercise. More detailed information about preplanning is presented in Chapter 25, *Fire and Life Safety Initiatives*.

In addition to formal preincident planning, firefighters should always pay attention to their surroundings while in the course of non-emergency functions (physical training, public education events, meal shopping, and so on) or after emergency events are completed, and conduct informal planning. Noting one or two unique items about a structure or identifying specific characteristics of a building in the course of clearing a non-suppression event can pay dividends later. It is vital that this information is shared with your crew as well as other members in your organization in the same way formal preincident plans are shared.

Of course, it is not possible to do preincident planning for every structure in your district or response area. In cases where you have not conducted a preincident inspection, it will be necessary to rely on an incident size-up to determine the type of construction present in the fire building. By learning the general characteristics of the building types in your jurisdiction, you will get some idea of the type of construction in a given neighborhood. In a neighborhood where the houses were built in the 1920s, for example, you are likely to find balloon-frame construction. In a neighborhood populated by recently completed houses, you are likely to encounter lightweight construction with ply-Is and wooden trusses.

Use the information you have learned about building construction to make firefighting safer both for you and for building residents. Your knowledge of building construction, preincident inspections, and incident size-up will help you gather the information you need to remain a safe and effective firefighter.

Over the course of your career as a firefighter, it is important to continue to keep up with the changes in building construction. Spend some time at construction sites. Review the latest fire studies from places like the Firefighter Safety Research Institute (FSRI) and the National Institute of Standards and Technology (NIST), which are available online. The information you gain may be valuable at your next fire scene.

TIP

It is important to know which types of construction are found in your response area. When a building is under construction, contact the builder or site foreperson and arrange for a tour. Ask questions about the structure and its construction. Inquire about any special features and its fire protection system.

CASE STUDY

You Are the Firefighter CONCLUSION

Your engine company has been dispatched to a house fire. You remember from your area familiarization training that the subdivision is mostly 1950s ranches. You pull up on the Bravo side of the structure and your first impression is that the house is not a ranch-style; it shows two stories and looks much older than the surrounding houses. You note that there is a large volume of fire from the roof. You also see that the windows on the first and second floor align and you see that the front door and the window above line up over each other.

1. **What building construction characteristics would you expect to see on a ranch built in the 1950s or 1960s?**

   Answer: In a typical ranch house, you can anticipate that dimensional lumber is used for all the structural components and the attic is made with rafters, the ceilings are shorter (8 ft; 2.4 m), and there is a single floor over a basement.

2. **Instead of a ranch, what style of construction does the building appear to be?**

   Answer: It appears to be an older structure than would have been in the original subdivision. In addition, the windows on the first and second floor align, as do the front door and a second floor window. Older houses (pre-1950s) with windows and doors that align are possible indicators for balloon-frame construction.

3. **How will the tactical considerations change now that you know the type of building construction?**

   Answer: You can assume that the fire has extended internally through the stud channels since the fire is displaying from the attic. However, since the fire may have started in the basement, early confirmation of no fire on the lower floors is critical.

WRAP-UP

SUMMARY

KNOWLEDGE OBJECTIVES

- Describe the basics principles of building construction.
  - List the types of forces and how structures must be constructed to withstand them. (pp. 200–201)
  - Identify the types of loads and describe how loads impact the integrity of structural components. (p. 201)
- Describe how occupancy classifications and the contents of a structure affect fire suppression operations. (p. 201)
- Identify the various materials and components utilized in building construction.
  - Describe the characteristics of wood building materials, including solid lumber, engineered wood, and mass lumber products. (pp. 203–206)
  - Describe the characteristics of masonry building materials. (pp. 205–206)
  - Describe the characteristics of concrete building materials. (pp. 206–207)
  - Describe the characteristics of steel and other mental building materials. (p. 207)
  - Describe the characteristics of membrane materials, including gypsum board and plaster and lath. (pp. 207–209)
  - Describe the characteristics of glass building materials. (pp. 209–210)
  - Describe the characteristics of plastic building materials. (p. 210)
  - Describe the types of connections used in building construction. (p. 211)

- Describe the purpose of a foundation in a structure. (p. 212)
- Describe the characteristics of basements. (p. 212)
- Describe the characteristics of ceilings. (p. 212)
- Explain how floor construction affects fire suppression operations. (pp. 212–214)
- Describe the characteristics of fire-resistive floors. (p. 213)
- Describe the characteristics of wood-supported floors. (p. 213)
- Describe the characteristics of trusses and plywood I-beam floors. (p. 214)
- Describe the characteristics of trusses. (p. 214)
- List the types of trusses. (p. 214)
- Describe the effects of fires on trusses. (p. 214)
- Describe the characteristics of walls. (pp. 214–216)
- List the common types of walls in structures. (p. 215)
- List the three primary components of roof assemblies. (pp. 216–217)
- List the three primary types of roofs. (pp. 217–219)
- Describe the characteristics of doors and windows. (pp. 220–221)
- Describe the characteristics of door assemblies. (p. 220)
- Describe the characteristics of window assemblies. (p. 221)
- Describe the characteristics of fire doors and windows. (p. 221)
- Explain the effect that exterior finishes and siding have on fire suppression operations. (p. 222)
- Explain the effect that interior finishes have on fire suppression operations. (pp. 222–223)

- Describe the types of construction and effects of fire on their structural integrity.
  - List the five types of building construction. (p. 223)
  - Describe the characteristics of Type I construction. (pp. 223–224)
  - Describe the effects of fire on Type I construction. (pp. 223–224)
  - Describe the characteristics of Type II construction. (pp. 224–225)
  - Describe the effects of fire on Type II construction. (pp. 224–225)
  - Describe the characteristics of Type III construction. (pp. 225–226)
  - Describe the effects of fire on Type III construction. (pp. 225–226)
  - Describe the characteristics of Type IV construction. (pp. 226–227)
  - Describe the effects of fire on Type IV construction. (pp. 226–227)
  - Describe the characteristics of Type V construction. (pp. 227–228)
  - Describe the effects of fire on Type V construction. (pp. 227–228)
  - Describe the characteristics of balloon-frame construction. (p. 228)
  - Describe the effects of fire on balloon-frame construction. (p. 228)
  - Describe the characteristics of platform-frame construction. (p. 228)
  - Describe the effects of fire on platform-frame construction. (p. 228)
  - Describe the challenges associated with fighting a fire in a hybrid building. (p. 229)
- Describe single-family dwelling types and their characteristics.
  - Describe the common types of single-family dwellings. (pp. 229–231)
- Describe the hazards created by fire conditions and the potential for building collapse.
  - Identify the signs of potential building collapse. (pp. 232)
  - Describe the structural factors that increase the chance of building collapse. (pp. 232–233)
  - Describe how building load can increase the chance of building collapse. (p. 234)
  - Describe how fire and explosion damage can increase the chance of building collapse. (p. 234)
  - Describe how environmental factors can increase the chance of building collapse. (p. 234)
  - Describe the types of building collapse. (p. 234)
- Identify the information to collect to determine a structure's type of building construction. (pp. 234–235)

SKILLS OBJECTIVES

- Evaluate and forecast a fire's growth and development and determine developing hazardous building or fire conditions.
- Recognize hazards. (pp. 231–234)

KEY TERMS

assembly Various construction materials joined together to form a component of a structure.

axial loads Loads that bear directly through the center of the structure.

balloon-frame construction An older type of wood frame construction in which the wall studs extend vertically from the basement of a structure to the roof without any fire stops.

bowstring truss A truss that is curved on the top and straight on the bottom.

combustibility The property describing whether a material will burn and how quickly it will burn.

compressive force The force that pushes a material together.

concentrated load A load focused in one specific area of a building.

concrete A mixture of cement, aggregates such as sand and gravel, and water.

contemporary construction Buildings constructed since about 1970 that incorporate lightweight construction techniques and engineered wood components. These buildings exhibit less resistance to fire than older buildings.

curtain wall Nonbearing walls that separate the inside and outside of the building but are not part of the support structure for the building.

curved roof A roof with a curved shape.

dead load Dead loads consist of the weight of all materials of construction incorporated into the building including but not limited to walls, floors, roofs, ceilings, stairways, built-in partitions, finishes, cladding and other similarly incorporated architectural and structural items, and fixed service equipment including the weight of cranes. (NFPA 5000)

dimensional lumber Lumber cut to nominal sizes, then dried and cut down to new standard sizes that are smaller than the nominal sizes. See also *solid lumber.*

eccentric loads Loads that are off center.

engineered wood Manufactured building material made of smaller pieces of wood held together with glue or adhesive; examples include plywood, fiberboard, oriented strand board (OSB), and particle board. See also *manufactured board, manmade wood,* or *composite wood.*

fire barrier wall A wall, other than a fire wall, having a fire resistance rating. (NFPA 5000)

fire compartment An area in a structure constructed so that if a fire is burning within the compartment, the fire will be confined to that compartment for a specified length of time.

fire door assembly Any combination of a fire door, a frame, hardware, and other accessories that together provide a specific degree of fire protection to the opening. (NFPA 80)

fire load The overall quantity and combustibility of the contents and structural components in a space, compartment, or structure

fire resistance The measure of the ability of a material, product, or assembly to withstand fire or give protection from it. (NFPA 251)

fire separation A horizontal or vertical fire resistance–rated assembly of materials that have protected openings and are designed to restrict the spread of fire. (NFPA 45)

fire wall A wall separating buildings or subdividing a building to prevent the spread of fire and having a fire-resistance rating and structural stability. (NFPA 5000)

fire window A window assembly rated in accordance with NFPA 257 and installed in accordance with NFPA 80. (NFPA 5000)

fire resistance ratings Ratings that are assigned to materials and building components based on the length of time that the material or component can maintain its integrity when subjected to fire.

flat roof A horizontal roof; often found on commercial or industrial occupancies.

forces A fundamental physical presence (i.e., gravity)

glass blocks Thick pieces of glass that are similar to bricks or tiles.

gusset plate Connecting plate made of a thin sheet of steel used to connect the components of the truss.

gypsum A naturally occurring material consisting of calcium sulfate and water molecules.

gypsum board The generic name for a family of sheet products consisting of a noncombustible core primarily of gypsum with paper surfacing. (NFPA 5000)

hybrid building A building that does not fit entirely into any of the five construction types because it incorporates building materials of more than one type.

I-beams Beams that look like an uppercase letter *I*.

impact load A load imposed quickly on a structure.

interior finish The exposed surfaces of walls, ceilings, and floors within buildings. (NFPA 5000)

intumescent coating Paint-like coatings of mineral materials or cement-like materials to insulate steel from heat in case of fire.

joists Beams that hold up a floor.

laminated glass Safety glass; the lamination process places a thin layer of plastic between two layers of glass so that the glass does not shatter and fall apart when broken.

laminated wood Pieces of wood that are glued together.

legacy construction An older type of construction that used dimensional lumber and was built before about 1970.

live load The load produced by the use and occupancy of the building or other structure, which does not include construction or environmental loads such as wind load, snow load, rain load, earthquake load, flood load, or dead load. Live loads on a roof are those produced (1) during maintenance by workers, equipment, and materials and (2) during the life of the structure by movable objects such as planters and by people. (NFPA 5000)

load-bearing wall A wall that supports any vertical load in addition to its own weight or any lateral load.

loads The forces to which a structure is subjected due to superimposed weight or pressures

masonry Built-up unit of construction or combination of materials such as clay, shale, concrete, glass, gypsum, tile, or stone set in mortar. (NFPA 5000)

mass lumber Large-scale engineered wood products designed to replace structural materials such as concrete and steel in smaller structures such as single-family dwellings. See also *mass timber*.

mass timber See *mass lumber*.

mill construction See *Type IV construction*.

mortar An adhesive produced by mixing sand, lime, water, and cement.

nominal lumber Lumber that is true to its stated dimensions.

nonbearing wall Any wall that is not a bearing wall. (NFPA 5000)

non-load-bearing wall Walls that only support their own weight and do not support any of the dead or live load of the building. See also *partition wall* or *nonbearing wall*.

occupancy The purpose for which a building or other structure, or part thereof, is used or intended to be used. (NFPA 5000)

partition wall See *non-load-bearing wall*.

parallel chord truss A truss in which the top and bottom chords are parallel.

party wall A wall constructed on the line between two properties.

pinning To make a connection using bolts or rivets between building materials.

pitched chord truss A type of truss typically used to support a sloping roof.

pitched roof A roof with sloping or inclined surfaces.

platform-frame construction Construction technique for building the frame of the structure one floor at a time. Each floor has a top and bottom plate that acts as a firestop.

rafters Joists that are mounted in an inclined position to support a roof.

rebar Solid reinforcing bars.

reinforced concrete Concrete embedded with steel for additional strength.

shear force A force that occurs when materials slide past each other.

solid lumber See *dimensional lumber*.

spalling Chipping or pitting of concrete or masonry surfaces. (NFPA 921)

steel An alloy or a combination of two or more metals of iron and carbon.

KEY TERMS CONTINUED

stud The vertical member of the frame of a structure.

system A collective term for the assemblies and materials that together make up a building component.

tempered glass A type of safety glass that is heat treated so that, under stress or fire, it will break into small pieces that are not as dangerous.

tendons Steel cables under tension.

tensile force The force that pulls a material apart.

thermal conductivity A property that describes how quickly a material will conduct heat.

thermoplastic material Plastic material capable of being repeatedly softened by heating and hardened by cooling and, that in the softened state, can be repeatedly shaped by molding or forming. (NFPA 5000)

thermoset material Plastic material that, after having been cured by heat or other means, is substantially infusible and cannot be softened and formed. (NFPA 5000)

torsional loads Loads that create twisting.

truss A collection of lightweight structural components joined in a triangular configuration that can be used to support either floors or roofs.

Type I construction (fire resistive) The type of construction in which the fire walls, structural elements, walls, arches, floors, and roofs are of approved noncombustible or limited-combustible materials that have a specified fire resistance.

Type II construction (noncombustible) The type of construction in which the fire walls, structural elements, walls, arches, floors, and roofs are of approved noncombustible or limited-combustible materials without fire resistance.

Type III construction (ordinary) The type of construction in which exterior walls and structural elements that are portions of exterior walls are of approved noncombustible or limited-combustible materials and in which fire walls, interior structural elements, walls, arches, floors, and roofs are entirely or partially of wood of smaller dimensions than required for Type IV construction or are of approved noncombustible, limited-combustible, or other approved combustible materials. (NFPA 14)

Type IV construction (heavy timber) The type of construction in which fire walls, exterior walls, and interior bearing walls and structural elements that are portions of such walls are of approved noncombustible or limited-combustible materials. Other interior structural elements, arches, floors, and roofs are constructed of solid or laminated wood or cross-laminated timber without concealed spaces within allowable dimensions of the building code. (NFPA 14)

Type V construction (wood frame) The type of construction in which structural elements, walls, arches, floors, and roofs are entirely or partially of wood or other approved material. (NFPA 14)

wear course The layer of the roof that is sound enough to walk on.

web members A system of diagonal and sometimes vertical members between two parallel chords.

wired glass A glazing material with embedded wire mesh.

wooden beam Load-bearing member assembled from individual wood components.

wood truss An assembly of small pieces of wood or wood and metal.

REVIEW QUESTIONS

1. Which type of load is focused in one specific area of a building?
2. Masonry components offer what benefit with regard to firefighting activities?
3. At what temperature will steel lose approximately half of its strength?
4. What type of glass will usually shatter into small pieces that do not have sharp edges?
5. What characteristic typically denotes the difference between a cellar and a basement?
6. What is the term used to describe a wall that is shared by two buildings?
7. Pitched roofs typically have what kind of roof covering material?
8. Name a defining characteristic of Type I construction.
9. What is another name for Type III construction?
10. In which style of single-family dwelling are half-stories more common?
11. What is the term used to describe a complete collapse?

DISCUSSION QUESTIONS

1. What are the key differences between Type I and II construction?
2. What are the characteristics of split-foyer house compared to a split-level house?
3. If there was a V-shaped collapse in a structure, where could you anticipated finding a survivable space?
4. What action can a firefighter take to stop the sagging or bending of a structural steel component?
5. Explain specifically how truss construction differs from dimensional lumber.
6. What are the differences between flat and pitched roof covering materials?
7. Describe how a roof and the load imposed on it by accumulated snow is transmitted down through a building. Also describe why the removal of a column supporting a floor may eventually lead to a collapse.

APPLYING THE CONCEPTS

You and your crew are responding to the report of a fire in a residence in the northwest quadrant of town. Dispatch reports that neighbors walking their dog saw heavy smoke coming from the top floor of the three-story house. It is unknown how the fire started but the structure is being renovated and is unoccupied. You have been studying the buildings in your community, and you know that this neighborhood is made up of a mixture of colonial and undefined houses from the early 1900s, split levels and ranches from the 1960s and 1970s, and new construction single-family houses. As your engine arrives on the scene, you see a significant amount of smoke pushing from the attic of an old colonial-style house. One of your crew members comments as you pull up that the windows and doors line up over each other.

1. Why is it important to know what style a house is and when it was built?
2. How does platform construction differ from balloon-frame construction? And how does it prevent fire from spreading?
3. Based on your observations, what type of construction do you think the house is?
4. What should your immediate action be?

The IC thinks the fire started in a lower level of the house and spread through the stud channels into the attic. He orders an offensive fire attack through side Alpha door to locate the seat of the fire. You and your crew advance an 1¾-in. (44-millimeter [mm]) hose line. As you enter, you encounter significant smoke and see evidence of remodeling in progress. You progress down the hallway and see heavy smoke streaming from under a door you assume is the basement. The temperature increases as you move farther into the structure.

5. If the house is being remodeled using modern materials, how might this affect fire behavior?
6. How do plaster and lath found in old homes challenge fire response?

Your crew is able to attack the fire and achieve knockdown from a well window.

7. What risks and challenges does Type V construction present firefighters?
8. If the IC had incorrectly identified the construction of the house and made the initial fire attack in the attic, what danger would that present to firefighters?

REFERENCES

American Wood Council. 2023. "Understanding the Tall Mass Code Changes." Accessed November 14, 2023. https://awc.org/wp-content/uploads/2022/01/tmt_toolkit.pdf.

BBC. 2017, July 19. "London Fire: What Happened at Grenfell Tower?" Accessed November 20, 2023. http://www.bbc.com/news/uk-england-london-40272168.

National Fire Protection Association. *NFPA 80, Standard for Fire Doors and Other Opening Protectives.* Quincy, MA: NFPA.

National Fire Protection Association (NFPA). 2019. *Brannigan's Building Construction for the Fire Service, Sixth Edition.* Burlington, MA: Jones & Bartlett Learning.

National Fire Protection Association (NFPA). 2020. *NFPA 921, Guide for Fire and Explosion Investigations.* 2021 Edition. Quincy, MA: NFPA.

National Fire Protection Association (NFPA). 2021. NFPA 220, *Standard on Types of Building Construction.* 2021 Edition. Quincy, MA: NFPA.

REFERENCES CONTINUED

National Fire Protection Association (NFPA). *NFPA 251, Standard Methods of Tests of Fire Resistance of Building Construction and Materials.* 2006 edition. Quincy, MA: NFPA.

National Fire Protection Association (NFPA). *NFPA 45, Standard on Fire Protection for Laboratories Using Chemicals Book.* 2024 edition. https://link.nfpa.org/free-access/publications/45/2024.

National Fire Protection Association (NFPA). 2022. *NFPA 80, Standard for Fire Doors and Other Opening Protectives.* 2022 Edition. Quincy, MA: NFPA.

National Fire Protection Association. *NFPA 5000. Building Construction and Safety Code.* 2024 edition. Quincy, MA: NFPA.

National Fire Protection Association (NFPA). 2024. *NFPA 14, Standard for the Installation of Standpipe and Hose Systems.* 2024 Edition. Quincy, MA: NFPA.

National Institute for Occupational Safety and Health (NIOSH). 2009, February. "Preventing Deaths and Injuries of Firefighters Working Above Fire-Damaged Floors." DHHS (NIOSH) Publication No. 2009–114. Accessed November 20, 2023. https://www.cdc.gov/niosh/docs/wp-solutions/2009-114/default.html.

National Institute for Occupational Safety and Health (NIOSH). 2010, August 5. "Two Career Firefighters Die Following a Seven-Alarm Fire in a High-Rise Building Undergoing Simultaneous Deconstruction and Asbestos Abatement—New York." Accessed November 20, 2023. https://www.cdc.gov/niosh/fire/reports/face200737.html.

U.S. Department of Energy. n.d. "Energy-Efficient Windows." Accessed November 20, 2023. https://energy.gov/energysaver/energy-efficient-windows.

Weyerhauser Company. "Oriented Strand Board." 2023. Accessed November 14, 2023. https://www.weyerhaeuser.com/woodproducts/osb-panels/osb/.

Chapter Opener: © Glen E. Ellman.

CHAPTER

7

Firefighter I

Firefighter Tools and Equipment

KNOWLEDGE OBJECTIVES

After studying this chapter, you will be able to:

- Describe the general considerations of tools and equipment, including safety considerations and effective usage.
- Describe how the correct tool or piece of equipment is utilized for firefighting operations.
- Explain the procedures for cleaning and inspecting tools and equipment, and for properly documenting and reporting any tool or equipment that is taken out of service.

SKILL OBJECTIVES

After studying this chapter, you will be able to perform the following skills:

- Select the appropriate cleaning supplies, and demonstrate the proper technique for cleaning and inspecting tools and equipment.
- Demonstrate the proper technique for cleaning and inspecting power tools.

ADDITIONAL NFPA STANDARDS

- **NFPA 440**, *Guide for Aircraft Rescue and Firefighting Operations and Airport/Community Emergency Planning, 2024 Edition*
- **NFPA 1407**, *Standard for Training Fire Service Rapid Intervention Crews, 2020 Edition*
- **NFPA 1410**, *Standard on Training for Emergency Scene Operations, 2020 Edition*
- **NFPA 1550**, *Standard for Emergency Responder Health and Safety, 2024 Edition*
- **NFPA 1660**, *Standard for Emergency, Continuity, and Crisis Management: Preparedness, Response, and Recovery, 2024 Edition*
- **NFPA 1700**, *Guide for Structural Fire Fighting, 2021 Edition*
- **NFPA 1960**, *Standard for Fire Hose Connections, Spray Nozzles, Manufacturer's Design of Fire Department Ground Ladders, Fire Hose, and Powered Rescue Tools, 2024 Edition*

CASE STUDY

You Are the Firefighter

Upon reporting for duty, you are notified that you have been detailed to another station to work with the engine company for the tour. This will be your first time working with an engine company, and you are excited about the opportunity. It isn't long before an alarm sounds. "Person trapped, motor vehicle collision, Fourth and Elm," directs the dispatcher. You slip into your gear, take your seat, and put on your seat belt. As the engine pulls out from the station, you think how this changes your plan for what you thought you would be doing today.

1. How will your role be different for this call?
2. What kinds of tools will you need for this job?
3. What other kind of knowledge will you need?

Introduction

Firefighters use tools and equipment to perform a wide range of activities. The same tools may be used in different ways for fire suppression, rescue operations, and overhaul. Firefighters must know how to use these tools effectively, efficiently, and safely, even when it is dark or when visibility is limited. They also need to know how to maintain the tools and equipment so that they will always be ready for use.

General Considerations

Hand tools are used to extend or multiply the actions of your body and to increase your effectiveness in performing specific functions. Most of these tools operate using simple mechanical principles. For example, a pike pole extends your reach, allowing you to penetrate through a ceiling, and enables you to apply force to pull down ceiling material; an axe multiplies the cutting force you can exert on a given area. In contrast, power tools and equipment have a mechanically driven source of power, such as an electric motor or an internal combustion engine. In certain cases, they are faster and more efficient than hand tools. A **hydraulic tool** uses pressurized fluid to exert force. They can be manually or mechanically powered. If they are mechanically powered, the same hydraulic power source can be used to power multiple hydraulic tools.

General Tool Safety

Safety is a high priority at any incident scene. Before using any kind of tool or equipment, you must be fully aware of how it works and know both the tool's and your individual limitations. You need to operate the equipment so that you, your fellow firefighters, victims, and bystanders are not unintentionally injured. Using any tool without the proper training, or using the wrong tool for the application, could result in serious injury or death. Personal safety also requires the use of proper personal protective equipment (PPE), including self-contained breathing apparatus (SCBA) and a personal alert safety system (PASS).

The best way to learn how to use tools and equipment properly is under optimal conditions of visibility and safety and with adequate supervision. Your first exposure and initial use of any tool should be done in an environment where you are able to see what you are doing and practice without endangering yourself and others. As you become more proficient, you should practice using tools and equipment under more difficult working conditions. Eventually, you should be able to use tools and equipment safely and effectively even when darkness or smoke decreases visibility. You must be able to work safely in any environment. For example, you might be working on a ladder or on a pitched roof, or you might be surrounded by noise and other emergency scene activities. You also need to be able to use tools while wearing your protective clothing and using your SCBA. Some departments require firefighters to train and practice certain skills and evolutions in total darkness or with their face masks covered to simulate the darkness of actual fires.

Effective, efficient use of tools and equipment requires that you understand how the tool or device works and the best way to use it to its full potential while using the least amount of energy to accomplish the task. Being effective means that you achieve the desired goal; being efficient means that you produce the desired effect without wasting time or energy. When you are assigned a task on the fireground, your objective is to complete that task safely and quickly. If you waste energy by working inefficiently, you will not be able to perform additional tasks. Knowing which tool is the best one for the task at hand will help you achieve the desired objective. Knowing how to use the tool effectively, following the manufacturer's instructions, and using good technique will ensure you have the energy needed to complete other tasks. Let the tool do the work, don't force it, and pace yourself; exceeding your personal limits will prevent you from completing the required tasks. Many fire departments have standard operating procedures (SOPs) or standard operating guidelines (SOGs) that suggest which tools and equipment could be used in various situations. However, even if these SOPs or SOGs exist, a firefighter should understand which tool will allow them to accomplish the task in the most expeditious and efficient manner possible.

New firefighters are often surprised by the strength and energy required to perform many tasks. An aggressive, continuous program of physical fitness and training in full gear will enable you to maintain your body in the optimal state of readiness. It does little good to practice using a tool for hours if you are not in good physical condition.

As a firefighter, you must know where every tool and piece of equipment is carried on your apparatus. Knowing how to use a piece of equipment does you no good if you cannot find it quickly. Your riding position, the seat you are sitting in, will often dictate your responsibilities and what tools you are expected to carry. Some firefighters carry a selection of small tools and equipment in the pockets of their coats or pants. Check whether your department requires you to always carry certain tools and equipment, and ask senior firefighters for recommendations about which tools and equipment to carry.

General Carrying Tips

Tools can cause injuries even when they are not in use. Carrying a tool improperly can result in muscle strains, abrasions, or lacerations. Keep the following general carrying tips in mind when you need to carry tools:

- Do not try to carry a tool or equipment that is too heavy or designed to be used by more than one person. Request assistance from another firefighter to help you carry it.
- Use your legs, not your back, when lifting heavy tools or equipment.
- Keep sharp edges and points away from your body at all times. Cover or shield these edges with a gloved hand to protect those around you.
- Carry long tools with the head of the tool down toward the ground. Be aware of overhead obstructions and wires, especially when using pike poles and ladders.
- When on scene, do not leave tools lying on the ground or floor. Tools that are left lying around are tripping hazards. Return unused tools to the tool staging area or to their proper place on the apparatus.

Functions

Some tools are carried on all fire apparatus, while other tools are carried only by specific types of companies. A good way to remember the function of each tool is to categorize them by their function. Most of the tools carried by fire departments fit into the functional categories described in **TABLE 7-1**.

TABLE 7-1 Common Firefighter Tools Categorized by Function

Function	Tool
Rotate	Box-end wrench Gripping pliers Hydrant wrench Open-end wrench Pipe wrench Screwdriver Socket wrench Spanner wrench
Push/pull	Ceiling hook Clemens hook Drywall hook K tool Multipurpose hook Pike pole Plaster hook Roofman's hook Shove knife San Francisco hook
Pry/spread	Claw bar Crowbar Flat bar Halligan tool Hux bar Hydraulic spreader Kelly tool Pry bar Rabbit tool

Continues.

TABLE 7-1 Common Firefighter Tools Categorized by Function CONTINUED

Function	Tool
Strike	Battering ram Chisel Flat-head axe Hammer Mallet Maul Pick-head axe Sledgehammer Spring-loaded center punch
Cut	Axe Bolt cutter Chainsaw Circular saw Cutting torch Hacksaw Handsaw Hydraulic shears Reciprocating saw Rotary saw Seat belt cutter

Courtesy of Guy Peifer.

Rotating Tools

Tools that rotate things apply a rotational force to make something turn. Many rotating tools are basic hand tools. The rotating tools used most often by firefighters are screwdrivers, wrenches, and pliers, which are used to assemble or disassemble parts that are connected with threaded fasteners. Rotating tools are described in **TABLE 7-2** and shown in **FIGURE 7-1**.

Assembling and disassembling are basic mechanical skills that are routinely employed by firefighters to solve problems. Most fire apparatus carry a tool kit with a selection of open-end wrenches, box-end wrenches, socket wrenches, adjustable wrenches, pipe wrenches, and pliers in a variety of shapes and sizes for different applications. These tool kits also usually include a variety of sizes and types of screwdrivers, including slotted-head, Phillips-head, Roberts-head, Torx, and others. A screwdriver with interchangeable heads is sometimes more useful than a selection of different screwdrivers. This type of screwdriver is carried by many firefighters in a pocket of their turnout coat.

Cutting Tools

Cutting tools have sharp edges that sever objects. They come in several forms and are used to cut a wide variety of materials. Firefighters use a wide variety of cutting tools, many of which are described in **TABLE 7-3**.

TABLE 7-2 Tools That Rotate Fasteners for Assembly and Disassembly

Tool	Description
Gripping pliers	A hand tool with a pincer-like working end. In addition to rotating things, gripping pliers can also be used to bend wire or hold smaller objects.
Screwdriver	A tool that is used to turn screws. Two common types of screwdrivers are flat-head, which insert into a screw that has a slot across its head, and Phillips-head, which insert into a screw that has a cross-shaped indentation on its head.
Wrench	Hand tool used to tighten or loosen bolts or turn pipes. Most wrenches come in several sizes.
Box-end wrench	A wrench with a closed end. Some box-end wrenches have a ratchet—a mechanism that rotates in only one direction—on the inside of the closed box-end that makes it easier to use.
Open-end wrench	A wrench with an open end.
Combination wrench (**FIGURE 7-2**)	A wrench with a box-end on one side and an open-end on the other.
Adjustable wrench	An open-ended wrench with one fixed grip and one movable grip so that the opening can be adjusted to accommodate bolts of different sizes.

Tool	Description
Pipe wrench	An adjustable wrench that can be adjusted to fit securely around pipes and other tubular objects.
Hydrant wrench (**FIGURE 7-3**)	A wrench used to open or close a hydrant by rotating the **stem nut** on the top of the hydrant to start or stop the flow of water into the hydrant and to remove the caps from the hydrant outlets. Some hydrant wrenches have a ratchet.
Socket wrench	A wrench that has a socket that fits over a nut or bolt and has a ratcheting handle that you use to tighten or loosen the nut or bolt. You can swap out the socket with other sockets of different sizes.
Speed socket wrench	A long, curved handle with a socket wrench that fits over a nut or bolt at the end. To tighten or loosen the nut or bolt, the handle is rotated.
Spanner wrench	A wrench used to tighten or loosen a fire hose **coupling**—a connection device that you use to connect—or couple—individual lengths of fire hose together and to connect a hose to a fire hydrant or a pump or to nozzles and hose appliances.

Courtesy of Guy Peifer.

FIGURE 7-1 Hydrant wrench, two spanner wrenches, pipe wrench, combination wrench, gripping pliers, Phillips-head screwdriver, and speed socket wrench.

© Jones & Bartlett Learning

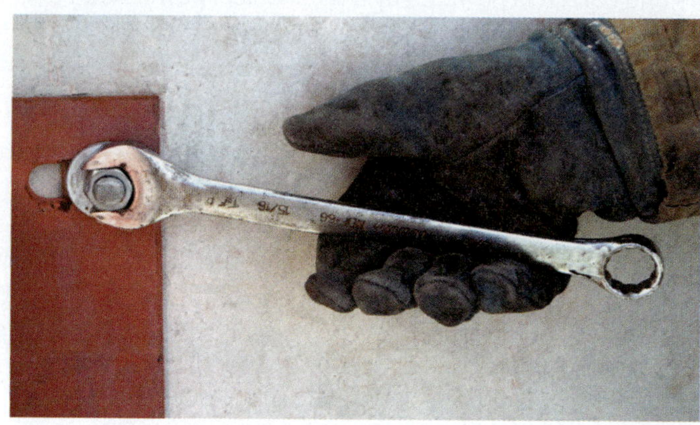

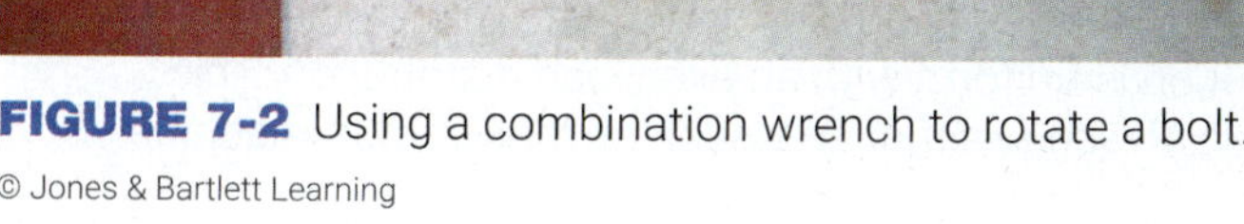

FIGURE 7-2 Using a combination wrench to rotate a bolt.

© Jones & Bartlett Learning

FIGURE 7-3 Using a hydrant wrench to rotate the stem nut on a hydrant to open or close the hydrant.

© Jones & Bartlett Learning. Photographed by Glen E. Ellman.

TABLE 7-3 Cutting Tools

Tool	Description
Axes	Axes are one of the most basic tools in the fire service. The axe head has a wide cutting blade that can be used to chop into a wall, roof, or door. Axe heads vary in weight from 4 to 8 pounds (lb; 1.8 to 3.6 kilograms [kg]), and handles range from 28 to 36 inches (in.; 71 to 94 centimeters [cm]). The weight, size, and type of axe you use depends on the job to be accomplished and your individual ability. Axes are often used to cut through chains or padlocks to open doors or gates. By concentrating the cutting force on a small area, it is possible to break through many chains in just a few seconds.
Flat-head axe (**FIGURE 7-4**)	This axe has a flat head opposite the blade that can be used for striking objects. It can also be used as a striking tool for forcible entry, usually in combination with a prying tool, such as a Halligan.
Pick-head axe (**FIGURE 7-5**)	This axe has a pick or a point opposite the blade that can be used for puncturing, pulling, and prying. It can also be used to establish a foothold while working on a sloped roof; for this reason, they are often carried by firefighters assigned to ladder trucks.
Bolt cutter (**FIGURE 7-6**)	This scissors-like hand tool is used to cut through items such as chains or padlocks or to cut through obstacles such as metal fences.
Cutting torch (**FIGURE 7-7**)	A tool that produces an extremely high-temperature flame capable of heating steel until it melts, effectively cutting it. Cutting torches are sometimes used for rescue situations and for cutting through heavy steel objects. These torches produce flames at extremely high temperatures—5700°F (3148°C)—so operators must be specially trained before using them. One drawback of this equipment is that it cannot be used in situations where flammable fuels are present.
Handsaw	A saw that is manually powered
Carpenter's handsaw (**FIGURE 7-8**)	This saw is designed to cut wood. Saws with large teeth are effective in cutting large timbers or tree branches. Saws with finer teeth are designed for cutting finished lumber.
Coping saw	This saw is used to cut curves in wood. It consists of a handsaw with a narrow blade set between the ends of a U-shaped frame.
Hacksaw (**FIGURE 7-9**)	This saw is designed to cut metal. Different blades can be used, depending on the type of metal being cut. Hacksaws are useful when metal needs to be cut under closely controlled conditions.
Keyhole saw	This specialty saw is narrow and slender and is used to cut keyholes in wood and drywall.
Hydraulic shears (**hydraulic cutters**)	This tool cuts quickly through metal posts and bars. It is powered by a mechanical power source and is used along with hydraulic spreaders and rams to extricate victims from motor vehicles. The same hydraulic power source can be used with all three types of tools.
Power saws	Power saws are mechanically powered by electric motors or gasoline-powered engines, although battery-powered, cordless models of some types of power saws are available. Power saws can accomplish more work than a handsaw in a shorter period of time, allowing firefighters to conserve energy, resulting in less fatigue. Because power saws are dangerous, only trained operators should use them. Some disadvantages to power saws are that they are heavy to carry, can be difficult to start, and they may require an electrical connection.
Band saw (**FIGURE 7-10**)	This electrically powered saw has a toothed metal blade stretched over two pulleys in a loop. Some people call it an endless blade. The blade cuts in only one direction. The benefit of using a band saw, unlike a reciprocating saw, is that there is no pulling or pushing of the object being cut and very little vibration. The cutting action is uniform with an even load distribution. This makes it ideal for victim extrication in special rescue situations, such as man-versus-machine or impalement injuries.

Tool	Description
Chainsaw	This saw has a gasoline-powered engine or electrically powered motor and is commonly used to cut wood, particularly trees. Firefighters often use special chainsaws called **ventilation saws** to cut ventilation openings in roofs constructed of wood, metal, tar, gravel, or insulating materials (**FIGURE 7-11**). Ventilation saws are specifically designed for roof ventilation and have different cutting chains than chainsaws used to cut wood. They also may have a depth gauge on the bar.
Reciprocating saw (**FIGURE 7-12**)	This saw is powered by either an electric or battery motor that rapidly pulls a saw blade back and forth. Different blades are used to cut different materials. Reciprocating saws can be used to cut metal during extrication of a victim from a motor vehicle.
Rotary saw (**FIGURE 7-13**)	This saw may be powered by either gasoline-powered engines or electric motors. Some rotary saws have a round metal blade with teeth. Different blades are used depending on the type of material being cut. Other rotary saws have a flat, abrasive disk made of composite materials designed to wear down as they are used. It is important to match the appropriate saw blade or disk to the material being cut.
Rescue knife	A spring-assisted folding knife that can be used with one hand. Often has a seat belt cutter and a window breaker as part of the knife.
Seat belt cutter (**FIGURE 7-14**)	A specialized tool that quickly cuts through a seat belt in a motor vehicle.
Wire cutter or **diagonal cutter**	A hand tool used to cut wire and small-diameter cable.

Courtesy of Guy Peifer.

FIGURE 7-4 Flat-head axe.

Courtesy of Sean Wilson.

Each of these tools is designed to work on certain types of materials. Reminder, firefighters can be injured and tools can be damaged if the tools are used incorrectly.

> **SAFETY TIP**
>
> Any saw—or any tool—that could generate a spark should not be used in a flammable environment.

Pushing/Pulling Tools

Tools used for pushing and pulling are metal hooks on a head at the end of a pole (**FIGURE 7-15**). These tools extend the reach of the firefighter as well as increase the firefighter's mechanical advantage on an object.

The poles can be made of wood, fiberglass, or sometimes aircraft steel and come in various lengths from 2 to 12 feet (ft; 0.6 to 3.7 meters [m]). A 4- to 6-ft (1.2- to 1.8-m) pole enables a firefighter to stand on a floor and pull down a 8- to 10-ft (2.4- to 3-m) high ceiling. Poles

A.

B.

FIGURE 7-5 **A.** Pick-head axe. **B.** A pick-head axe can be used to pry up boards.

A: Courtesy of Sean Wilson; **B:** © Jones & Bartlett Learning. Photographed by Glen E. Ellman.

FIGURE 7-6 Bolt cutters.

Courtesy of Sean Wilson.

FIGURE 7-7 A cutting torch can be used to cut through a metal door.

© Jones & Bartlett Learning

12- to 14-ft (1.2- to 4.3-m) long are used in rooms with very high ceilings.

The metal hooks attached to the end of the poles are available in many styles that are suited for different applications. The different hook designs are intended for different types of ceilings and come in a variety of configurations. Many fire departments use one type of hook for plaster ceilings and another type for drywall ceilings. The end of the pole handles opposite the metal hook might have a "D"-shaped handle for better pulling power. **TABLE 7-4** describes several types of hooks used in fire department operations.

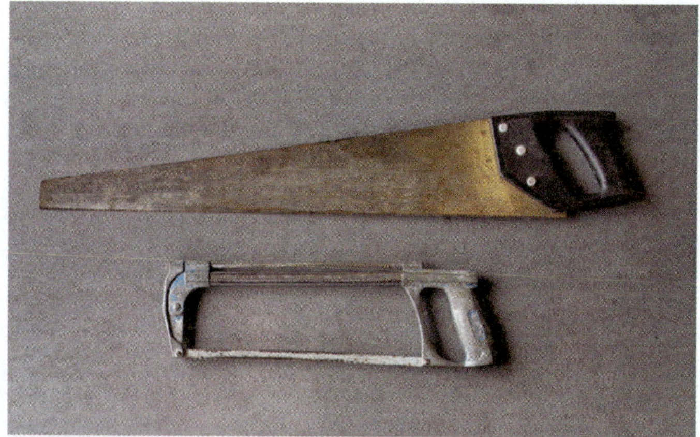

FIGURE 7-8 **A.** Carpenter's handsaw. **B.** Hacksaw.

FIGURE 7-9 A firefighter using a hacksaw to cut through metal.

FIGURE 7-10 A band saw cutting a metal pipe.

FIGURE 7-11 Firefighters using a ventilation saw to cut through a roof.

A.

B.

FIGURE 7-12 **A.** The components of a reciprocating saw. **B.** A reciprocating saw being used during the extrication of a victim from a vehicle.

FIGURE 7-13 A rotary saw can cut through metal or wood, depending upon the type of blade used. **A.** A rotary saw with a metal cutting blade. **B.** A diamond rescue disk. **C.** A composite abrasive disk. **D.** A carbide-tip disk.

Courtesy of Sean Wilson.

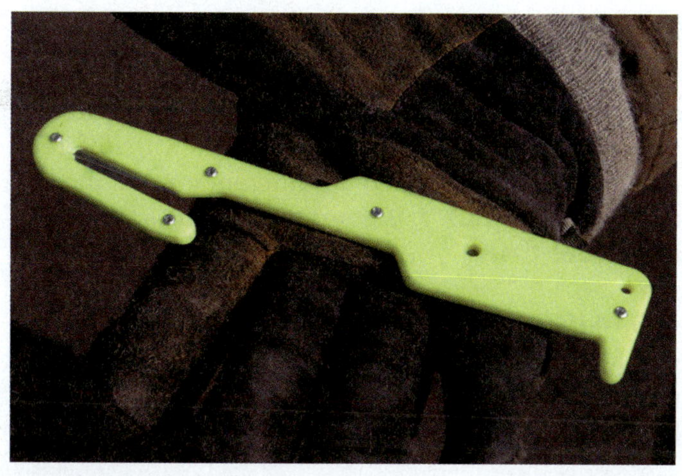

FIGURE 7-14 Seat belt cutter.

SAFETY TIP

Hooks should never be used for prying because they are not levers and will break if used inappropriately.

FIGURE 7-15 Hooks come in different sizes and with metal heads that are different shapes.

TABLE 7-4 Tools for Pushing and Pulling (Pike Poles and Hooks)

Tool	Description
Pike pole (**FIGURE 7-16**)	This tool is an all-purpose hook and is one of the most versatile and most common tools in a firefighter's "tool box." The pole is made of wood or fiberglass, and the metal head has a sharpened point that you can punch through a ceiling and a hook that can grab and pull the ceiling down to get to the seat of a fire burning above or to locate an extension of the fire. It can also be used to break windows on a higher floor (**FIGURE 7-17**). Pike poles typically come in lengths of 4 to 12 ft (1.2 to 3.7 m). It is important to size up the building upon arrival and bring the right size pole inside. If the pole is too short, you will not be able to reach the ceiling. If it is too long, you will not be able to use it in a room with a low ceiling.
Ceiling hook	The pole on this tool is wood or fiberglass, and the metal head has a spur at right angles that can be used to probe ceilings and pull the material down to expose the wood behind it.
Clemens hook	A multipurpose tool that can be used for forcible entry and ventilation applications because of its unique head design.
Closet hook	This tool is usually 2 to 4 ft (0.6 to 1 m) long and is used in smaller, tight spaces such as closets and crawl spaces. It is also often used to overhaul upholstery.
Drywall hook	A specialized hook designed to remove drywall, but is also used on other materials. It has a honed head to allow for better penetration and a large, toothed head to rip away large areas of material. It is sometimes referred to as the *universal hook*.
Multipurpose hook	A long pole with a wooden or fiberglass handle and a metal hook.
Plaster hook	A long pole with a pointed head and two retractable cutting blades on the side.
Roofman's hook or **New York hook**	A long pole with a solid metal hook on one end a chisel on the other end that is often used to force open roof scuttles and hatches as well as pull ceilings and skylights.
San Francisco hook	A multipurpose tool used for forcible entry and ventilation applications. It includes a built-in gas shut-off, allowing firefighters to turn off the gas at the street.

Note: A pick-head axe can also be used to push and pull. Hydraulic spreaders can be used in combination with chains to pull a steering column or a dashboard, or even to move a vehicle.
Courtesy of Guy Peifer.

FIGURE 7-16 Using a pike pole to push into a ceiling.

Prying/Spreading Tools

Tools used for prying or spreading may be as simple as a **pry bar** or as mechanically complex as a hydraulic spreader. They come in several sizes and with different features that are designed for different applications as described in **TABLE 7-5**. Some of these tools are shown in **FIGURE 7-18**.

Striking Tools

Striking tools are used to apply an impact force to an object. They often are employed to gain entrance to a building or a vehicle or to make an opening in a wall or

FIGURE 7-17 Firefighter using a pike pole to break a window.

TABLE 7-5 Tools for Prying and Spreading

Tool	Description
Claw bar	A tool with a pointed claw-hook on one end and a forked- or flat-chisel pry on the other end that can be used for forcible entry.
Crowbar	A straight bar made of steel or iron with a forked chisel on the working end.
Flat bar	A specialized prying tool made of flat steel with prying ends suitable for performing forced entry.
Halligan tool, **Halligan bar**, or simply **Halligan** (**FIGURE 7-19**)	A tool that incorporates three working ends, making its versatility in prying unmatched. One end of the tool is a bifurcated fork or claw that is rounded on one side; this rounded side is often referred to as the bevel. The other end of the tool has two working sides—a pick and an adze. The pick is round and tapers to a sharp point. The **adze** is a curved or straight wedge. Both the pick and the adze are set 90 degrees to the shaft of the tool. Though shorter and longer models are available, a standard Halligan is 30 in. (0.8 m) long. It is used to pry open doors or windows. It was designed in 1948 by New York City firefighter Deputy Chief Hugh Halligan (Firefighter Nation 2007). Many variations of this tool exist, and many names are used to describe it and its parts.

Tool	Description
Hux bar	A multipurpose tool that can be used for forcible entry and ventilation applications because of its unique design. A Hux bar also can be used as a hydrant wrench.
Hydraulic spreader (**FIGURE 7-20**)	A mechanically driven—that is, powered by gasoline, electricity, or batteries—rescue tool that enables you to apply several tons of force on a small area. You must have special training to operate these machines safely. Fire and rescue departments most commonly use this tool for extrication of victims from motor vehicles and machinery. Hydraulic spreaders are available as both full-size corded units used with a gasoline engine intended for vehicle extrication, or as small battery-operated units that have changeable tips for extrication or forcible entry. Battery-powered hydraulic tools do not have hoses or generators, so rescue crews can quickly reach the vehicle involved, especially if it is far away from a roadway. Battery-powered tools also do not produce exhaust fumes, making the scene safer.
Kelly tool	A steel bar with two main features—a large pick and a large chisel or fork.
Rabbet tool (also called the **bunny tool**) and **Hydra-Ram** (**FIGURE 7-21**)	Manually powered hydraulic spreading tools used to pry open doors that swing inward by prying the door away from the door jamb at the point where the lock is located.

Courtesy of Guy Peifer.

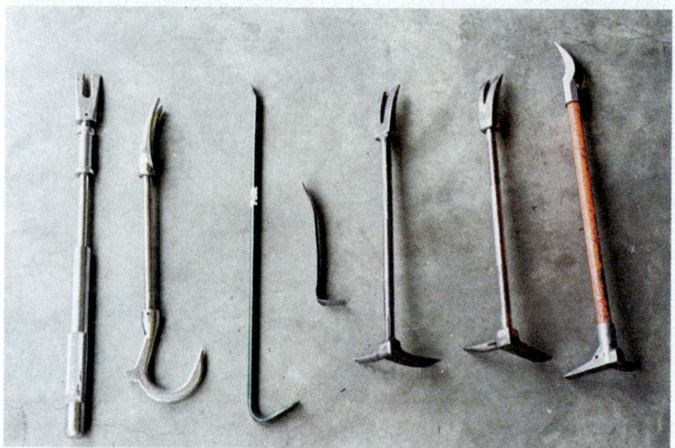

FIGURE 7-18 Tools used for prying and spreading.

© 2003, Berta A. Daniels

roof. This equipment also can be used to force the end of a prying tool into a small opening. Striking tools are described in (**TABLE 7-6**) and some of them are shown in (**FIGURE 7-22**).

SAFETY TIP

Some tools require a considerable amount of movement and room to operate. Always confirm that no one is in danger of being injured before you use these tools. Look around and make sure you can operate an axe or sledgehammer safely and effectively.

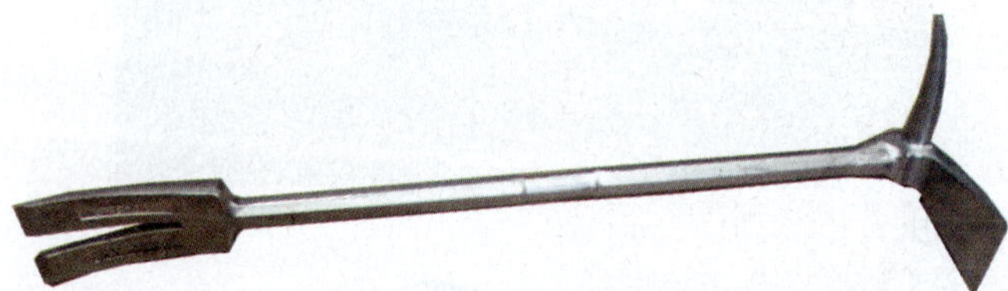

FIGURE 7-19 Using a Halligan tool to pry open a door.

© Jones & Bartlett Learning. Photographed by Glen E. Ellman.

Multiple-Function Tools

Certain tools are designed to perform multiple functions, thereby reducing the number of tools needed to achieve a goal. For example, a flat-head axe can be used as either a cutting tool or a striking tool (**FIGURE 7-24**). Some

A.

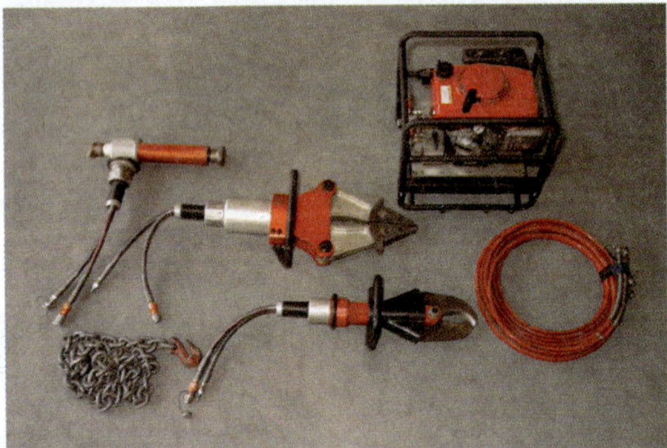

B.

C.

FIGURE 7-20 **A.** Battery-operated hydraulic spreader. **B.** Components of a gasoline-powered hydraulic rescue tool. **C.** A powered hydraulic spreader prying open a door on a motor vehicle.

A: Courtesy of Sean Wilson; **B:** © Jones & Bartlett Learning; **C:** © Jones & Bartlett Learning. Photographed by Glen E. Ellman.

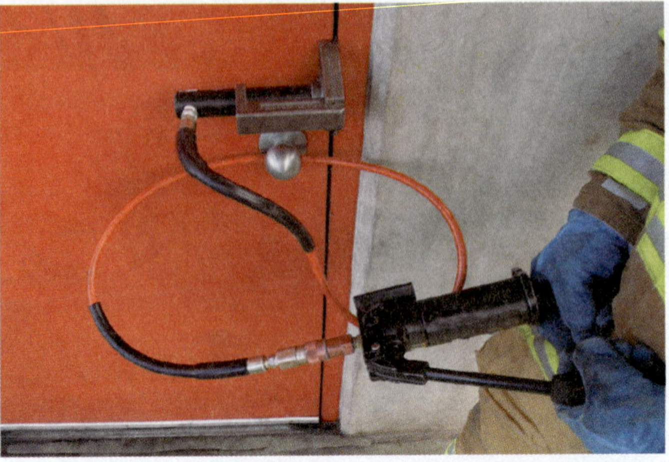

A.

B.

FIGURE 7-21 **A.** Rabbet tool. **B.** Hydra-Ram.

A: © Jones & Bartlett Learning; **B:** Courtesy of Sean Wilson.

FIGURE 7-22 Striking tools (from top): hammer, maul, mallet, sledgehammer.

© Jones & Bartlett Learning. Photographed by Glen E. Ellman.

TABLE 7-6 Striking Tools

Tool	Description
Battering ram (**FIGURE 7-23**)	A heavy metal bar used to break down doors and breach walls. Battering rams come in different sizes and weights. Single-person battering rams are available, but most battering rams are made to be used by two to four people. Battering rams are more commonly used by law enforcement agencies than by fire departments.
Chisel	A metal tool with one sharpened end that can be used to break apart material when used in conjunction with a hammer, mallet, or sledgehammer.
Hammer	A hand tool constructed of solid material with a long handle and a head affixed to the top of the handle, with one side of the head used for striking and the other side used for prying.
Mallet	A short-handled hammer with a round head.
Maul	A specialized striking tool that weighs 6 lb (3 kg) or more. It has an axe on one side of the head and a sledgehammer on the other side.
Sledgehammer	A hammer that can be one of a variety of weights and sizes. The head of the hammer can weigh from 2 to 20 lb (1 to 9 kg) and the handle may be short like a carpenter's hammer or long like an axe handle.

Note: THE PIG and the flat-head axe also can be used to strike objects or as a striking tool for forcible entry.
Courtesy of Guy Peifer.

FIGURE 7-23 A battering ram might be used to break down doors but is more often used by firefighters to breach concrete walls.

FIGURE 7-24 Breeching a wall with a multiple-purpose tool.

combination tools can be used to cut, to pry, to strike, and to turn off utilities. The **multi-tool** (**FIGURE 7-25**) is a compact, pocket-size, multiple-function tool that combines several different tools, such as a knife, scissors, wire cutters, pliers, and screwdriver.

Another example of a multiple-function tool is THE PIG. Designed by a firefighter, it is a combination flat-head and pick-head axe (**FIGURE 7-26**). The pick-head can be used for venting a roof (cutting) and to pull roof materials and for overhaul (pushing/pulling), and the flat-head axe blade is used for forcible entry (cutting and striking).

FIGURE 7-25 A multi-tool.

FIGURE 7-26 THE PIG combination pick-head and flat-head axe.

Courtesy of Sean Wilson.

Special-Use Tools

Some fire situations require special-use tools that perform other functions. For example, fire departments located in areas where wildland fires occur frequently may need to carry fire rakes, firefighting brooms, shovels, and combination tools that can be used for raking, chopping, cutting, and leaf blowing. These tools are described in Chapter 21, *Wildland and Ground Cover Fires.*

Most apparatus also carry tools for forcible entry. One example is a **spring-loaded center punch**, which is used to break tempered glass in automobile windows and windows in some buildings (**FIGURE 7-27**). It exerts a large amount of force on a pinpoint-size portion of tempered glass. This disrupts the integrity of the tempered glass and causes the window to shatter into small, uniform-sized pieces. Another forcible entry tool is a **shove knife**, which is used to release the latch on an outward-swinging door (**FIGURE 7-28**). Pulling the tool down and outward releases the locking mechanism of the door.

There are several types of forcible entry tools that pull lock cylinders so that the lock can be released. This "through the lock" method of forcing a door open minimizes damage to the door because it leaves the door and most of the locking mechanism undamaged. The building owner can have the lock cylinder replaced at a relatively low cost. For example, a **K tool** is a lock pulling tool that is used with a Halligan and a mallet to pull out a lock cylinder mounted in a wood or heavy metal door (**FIGURE 7-29**). Other tools used for pulling locks are discussed in Chapter 15, *Forcible Entry.*

FIGURE 7-27 A spring-loaded center punch can be used to break a car window safely.

FIGURE 7-28 Shove knife.

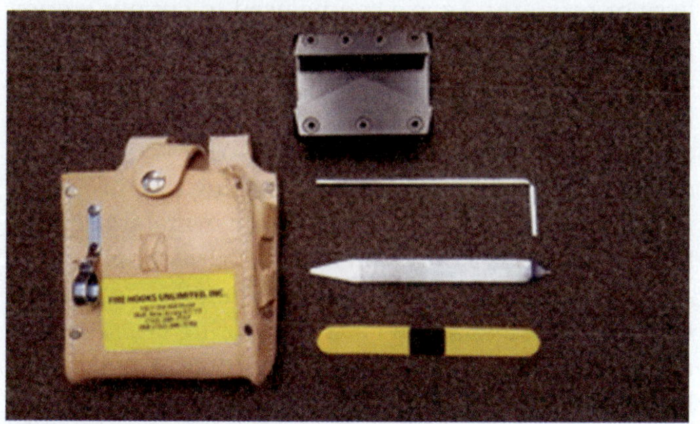

FIGURE 7-29 Components of a K tool.

Fire companies may also carry specialized equipment such as jacks and air bags for lifting heavy objects (**FIGURE 7-30**), a **come along** (a small, hand-operated winch) for lifting or moving heavy objects or bending objects, and tripods. You can learn more about the proper use of this special equipment by taking special rescue courses or during in-service training.

Operations

Efficient and effective incident scene operations usually involves a sequence of steps or stages. Each phase of a fireground operations may require the use of certain types of tools and equipment. The phases of incident scene operations are summarized here:

- **Response**—the deployment of fire companies to an emergency incident—and **size-up**—the rapid evaluation and analysis of an incident by an officer or the incident commander (IC) to determine which resources need to be deployed and which actions can be undertaken safely. This phase begins with response, when the emergency call is received, and continues as the units travel to the incident scene and with the initial size-up when the first unit arrive on scene.

FIGURE 7-30 Heavy-duty air bags can be used to lift vehicles in rescue situations.

- **Forcible entry**: This phase begins when entry to buildings, vehicles, aircraft, or other confined areas is locked or blocked, requiring firefighters to use force to break windows or break down a door to gain access.
- **Fire suppression, which may include an attack**: During this phase, a team of firefighters is assigned to enter the fire structure, locate the fire, and extinguish the fire.
- **Search and Rescue**: This phase involves searching for any victims trapped by the fire and extricating them from the building.
  - Rapid intervention: During this phase, a **rapid intervention crew/company (RIC)**, also known as a **rapid intervention team (RIT)** or **firefighter assist and search team (FAST)**, provides immediate assistance to injured or trapped firefighters. (RICs are discussed in Chapter 14, *Fire Suppression*, and in Chapter 19, *Firefighter Survival.*)
- **Ventilation**: During this phase, the air within a compartment is changed by the controlled and coordinated removal of heat and smoke from a structure by natural or powered means.
- **Salvage** and **overhaul**: The final phase is the process of finding, exposing, and suppressing any smoldering or hidden pockets of fire in an area that has been burned after a fire has been suppressed.

You may need to use different tools in each of these phases, or you may need to use the same tools to complete different tasks.

Response and Size-Up

During the response and size-up phase, consider what your initial actions will be and which tools you will need for the job even with the limited amount of information available at this point. Consider the information from the dispatcher along with preincident response plan information about the location. For example, an automobile fire on the highway presents different problems and requires different tools than a call for smoke coming from a single-family house. You might make different decisions when you are dispatched at midnight to a house fire that may have trapped a family inside than when you respond to a report of a kitchen stove fire at suppertime. Upon arrival at the incident, the company officer will perform a size-up and develop the incident action plan (IAP). This action plan determines the tools that will be needed for each assigned task.

Fire departments have SOPs or SOGs that suggest the tools and equipment required for different types of incidents. Some fire companies use the seat assignment as a guide to determine which specific tools and equipment each crew member is expected to bring from the apparatus. These requirements consider the type of incident and roles that different units have within the fire department.

Forcible Entry

Gaining entrance to a locked building or vehicle can present a challenge to even the most seasoned firefighter. Most are equipped with security devices designed to keep unwanted people out, but these same devices make it difficult for firefighters to gain access to them. The skills required during this phase are discussed in depth in Chapter 15, *Forcible Entry*.

Several types of tools can be used for forcible entry. Prying tools, including the Halligan, pry bar, flat bar, crowbar, Hux bar, hydraulic-powered rabbet tool, and hydraulic-powered Hydra-Ram, can be used for prying open windows or to force open doors. A shove knife is used to open outward-swinging doors by inserting it between the door and door frame and then pulling down and outward. This leaves the door and locking mechanism undamaged. Hydraulic spreaders can be used to force open a door and breach walls, fences, or gates. Lock-pulling tools can pull the lock cylinder to release the lock with comparatively little damage to the door or door frame.

Sometimes the easiest—or only—way to gain access is to use cutting tools. Although cutting is a destructive process, it is justified when required to save lives or property. An axe can be used to cut out a door panel or break a window. A power saw can be used to cut openings through obstacles including doors, walls, fences, gates, security bars, and other barriers. Bolt cutters can be used to remove a padlock or to cut through obstacles such as a metal fence. Cutting torches or rotary saws can be used to cut through metal security bars, hinges, or padlocks on metal security doors.

FIGURE 7-31 A flat-head axe can be used with a Halligan tool to force open a door.

A variety of striking tools, including flat-head axes, hammers, sledgehammers, and battering rams, can be used for forcible entry when brute force is needed to break into a building. A sledgehammer can be used to breach walls or break a window. It also can be used in conjunction with other tools, such as a Halligan tool, to force open a door or break off a padlock. A hammer or mallet may be used in conjunction with other tools such as a chisel to force entry through windows or doors. A spring-loaded center punch (special-use tool) is used to break tempered glass.

A flat-head axe (striking) and a Halligan (prying), collectively called the **irons**, are used in combination to pry open a door (**FIGURE 7-31**). This combination is highly effective in most forcible-entry situations, although they may permanently damage both the door and the frame. In addition, company officers often carry a tool that has many functions to assist forcible entry.

SAFETY TIP

When choosing the means and tool to use for forcible entry, always remember that life safety is your top priority. Although the use of some tools will cause considerable damage to property, being able to quickly access a building or a vehicle may make all the difference in saving a life.

Voice of Experience

One of the best sets of tools that a firefighter can grab is the irons—the flat-head axe and Halligan tool strapped together. These tools are married together using a heavy-duty Velcro or rubber strap. They are connected by placing the fork end of the Halligan onto the head of the axe between the tip and the handle. This enables the firefighter to carry the two tools with one hand. Working together, the irons can make quick work of forcing open conventional doors.

Using the irons inside a structure is beneficial as well. For example, when you make entry into the structure for fire suppression or search and rescue activities, remember to bring the irons with you. When conducting search and rescue operations, the flat-head axe is a good tool to assist in searching for victims.

Great emphasis is put on forcible entry, but forcible exit also must be considered. Always ask the question, "If things go bad, what can I carry in that will allow me to breach a wall and get out?" The irons could be the tool that saves your life by allowing you to breach a wall and exit the structure.

Joseph Ramsey
Winston-Salem Fire Department
Winston-Salem, North Carolina

Many techniques and types of tools can be used to gain entry into secured structures. The exact tool needed will depend on the method of entry and the type of obstacle. Because having greater experience usually suggests the best way to gain entry in each situation, rely on the direction of your officer and suggestions from fellow firefighters.

Interior Attack

Every crew working inside a burning building should carry some basic tools and equipment. These basic tools enable them to solve problems they may encounter while performing interior operations. For example, crew members may encounter obstacles such as locked doors, or they may need to open an emergency escape route. They may need to establish horizontal ventilation by forcing, opening, or breaking a window. They may need to gain access to the space above the ceiling by using a pike pole or making a hole in a wall or floor with an axe. A powerful light is also important, because smoke can quickly reduce interior visibility to just a few inches.

The basic set of tools for interior firefighting includes the following items:

- A prying tool, such as a Halligan
- A striking tool, such as a flat-head axe or a sledgehammer
- A cutting tool, such as an axe
- A pushing/pulling tool, such as a pike pole

In addition, firefighters should carry a portable radio if available, a **hand light**, and a length of webbing or rope. Webbing is an important and versatile piece of equipment. It can be used to control a door during forcible entry, as a harness to rescue a downed firefighter or civilian, or to lash a patient to a backboard or basket stretcher. Crews also may carry specialized tools and equipment needed for their individual assignment, such as attic ladders, a portable fire extinguisher, or **thermal imager device**.

The specific tools that must be carried by each crew are usually defined in the fire department's training manuals and SOPs. These requirements are based on local conditions and preferences and may differ depending on the type of company and the assignment. See Chapter 14, *Fire Suppression*, for more detailed information on interior attack operations.

Search and Rescue

Search and rescue should be carried out quickly, shortly after arrival on the fireground. (See Chapter 17, *Search and Rescue*, for more detailed information on search and rescue operations.) A search team should carry the same basic hand tools as the interior attack team. These tools can be used both to gain access to an area for a search and, if necessary, to create an emergency exit path. For example, a search team may use a tool to sweep under beds for unconscious victims, such as the handle of an axe or the handle end of a short pike pole, which is relatively light and may reduce the time needed to search an area by extending the firefighter's reach.

Members of a RIC should carry the standard set of tools for interior firefighting as well as extra tools and equipment that are particularly important for search and rescue tasks. The extra tools and equipment should help the RIC find and gain access to a firefighter who is in trouble, extricate a firefighter who is trapped under debris, provide breathing air for a firefighter who has experienced an SCBA failure or run out of air, and remove an injured or unconscious firefighter from the building. This equipment should be gathered and staged with the RIC, so that it will be immediately available if needed. See Chapter 14, *Fire Suppression*, and Chapter 19, *Firefighter Survival*, for more information about RICs.

Ventilation

The objective of ventilation is to provide openings so that fresh air can enter a burning structure and hot gases and products of combustion can escape from the building. (See Chapter 16, *Ventilation*, for more detailed information on ventilation operations.) Many of the same tools used for forcible entry can also be used to provide ventilation. For example, power saws and hand tools are commonly used to create vent openings. Axes, Halligan tools, pry bars, tin cutters, pike poles, and other hooks can all be used to remove coverings from existing openings, cut through roof decking, remove sections of the roof, and punch holes in the interior ceiling.

To provide adequate ventilation, unlocked or easily released windows and doors should be opened normally. Locked or jammed windows and doors may have to be broken or forced open using a hand tool (Halligan tool, axe, or pike pole) or ladder. It also may be necessary to create interior openings within the building so that contaminated air can reach the exterior openings.

In addition to making or controlling openings to influence ventilation, firefighters can use hydraulic or mechanical ventilation to direct the flow of fresh air and combustion gases. These fans can be powered either by electric or gasoline motors or by water pressure. A series of electrically powered fans may be placed throughout a large structure to help move smoke in the desired direction. A gasoline-powered fan may not be suitable in some situations, because it can introduce carbon monoxide into a structure if an exhaust hose is not available. If the building atmosphere contains potentially dangerous levels of flammable gases, a water-powered fan may be the best choice. The use of fans is discussed more thoroughly in Chapter 16, *Ventilation*.

Salvage and Overhaul

The purpose of salvage and overhaul is to protect property and belongings from damage and ensure that all hidden fire is located and extinguished. During the overhaul phase, potential hot spots in enclosed spaces behind walls, above ceilings, and under floors need to be exposed. To do this, crew use pushing/pulling tools such as pike poles and ceiling hooks of various lengths. Prying tools, such as crowbars and Halligans, can be used to open closed spaces and remove baseboards and window or door casings. Cutting tools, such as axes and power saws, can be used to open walls, floors, and ceilings. Hand tools such as pitchforks, shovels, brooms, rakes, buckets, wheelbarrows, and carryalls help remove burned debris. For more information about salvage and overhaul and additional tools used during these operations, see Chapter 18, *Salvage and Overhaul*.

Maintenance

All tools and equipment must be properly maintained so that they will be ready for use when they are needed. This means that a tool or other equipment should be in proper working order, in its proper storage location, and ready for immediate use. Every crew member should be able to locate the right tool immediately and be confident that it is ready for use. You must keep equipment clean and free from rust, keep cutting blades sharpened, and keep fuel tanks filled and batteries charged. A tool that is improperly maintained can be dangerous to both the operator and other nearby persons. Every tool and piece of equipment must be ready for use before you respond to an emergency incident. Research and follow the manufacturer's instructions for cleaning, inspecting, and maintaining each piece of equipment. After cleaning and inspecting tools, place them back in their proper location, ready for use. If a tool is damaged or inoperative, immediately tag it, take it out of service, and document the damage following department procedures, and report it to your supervisor. Generally, tools should be inspected after every use. In addition, thorough, conscientious daily or weekly checks of these items should be performed, particularly with infrequently used tools.

Cleaning and Inspecting Hand Tools

All hand tools should be cleaned completely and inspected, and their conditions documented after each use. Remove all dirt, debris, and rust from the tools. If appropriate, use water streams to remove the debris and mild soap to clean the equipment thoroughly. Learn how to safely use the cleaning solutions that your department and the manufacturer specify for cleaning tools and equipment.

Before any tool is placed back into service, it should be inspected for damage. Avoid painting tools, as paint

SKILL DRILL 7-1

Cleaning and Inspecting Hand Tools Firefighter I, NFPA 1010: 6.5.1

1. Clean and dry all metal parts. Metal tools must be dried completely, either by hand or by air, before being returned to the apparatus. Remove rust with steel wool. Coat unpainted metal surfaces with a light film of lubricant to help prevent rusting. Do not oil the striking surface of metal tools, as this treatment may cause them to slip.
2. Inspect wood handles for damage such as cracks and splinters. Repair or replace any damaged handles. Sand the handle if necessary. Do not paint or varnish a wood handle; instead, apply a coat of boiled linseed oil. Check that the tool head is tightly fixed to the handle.
3. Clean fiberglass handles with soap and water. Inspect for damage. Repair or replace any damaged handles. Check that the tool head is tightly fixed to the handle.
4. Inspect cutting edges for nicks or other damage. Cutting tools should be sharpened after each use. File and sharpen as needed. Power grinding may weaken some tools, so hand sharpening may be required.

may hide defects or visible damage. If painting of a tool is required for identification, do *not* paint axe heads. Painting an axe head hides defects and stress fractures in the metal and may cause the axe head to stick and bind when using it. Keep the number of markings on a tool to a minimum. If an axe needs to be sharpened, it is best to sharpen it by hand with a file. If the axe head edge is sharpened so the edge is thin, the axe will be prone to chips and cracks.

To clean and inspect hand tools, follow the steps in **SKILL DRILL 7-1**.

Cleaning and Inspecting Power Tools

Just like hand tools, all power tools should be cleaned completely and inspected and their conditions documented after each use. Remove all dirt and debris from the tools. You will likely have to disassemble the tool somewhat to clean it correctly. This is especially true of power saws. If appropriate, use an approved solvent and parts washer to remove the debris and to clean the equipment thoroughly. Learn how to safely use the cleaning solutions that your department and the manufacturer specify for power tools and equipment. While cleaning cutting tools, inspect the blade and make sure it is ready for use.

When tools powered by a rechargeable battery are used, make sure the battery is replaced and the used battery put on a charger. Ensure that tools powered by gasoline engines are refueled with the correct fuel. Some use straight gasoline, while others use a 50:1 or 100:1 gasoline and oil mixture. If a gasoline-powered tool is not likely to be used in the next 12 to 15 days, add a fuel stabilizer. If it is expected to be out of service for more than 15 days, its fuel should be drained and stored in a separate container. To avoid these problems that can occur with ordinary gasoline, many fire agencies use commercial fuel specifically designed for power and emergency equipment. These fuels do not contain ethanol and evaporate cleanly, leaving no residue, which allows a fire agency to keep their tools permanently fueled.

It is poor practice to place a power tool back on the apparatus without cleaning it, ensuring that it is fueled or charged and that any parts that need replacement are replaced, such as a chain or blade from a rescue saw.

TIP

Avoid storing composite rotary saw blades in compartments where gasoline vapors might accumulate. There is a possibility that the hydrocarbons in the vapors could break down the glues in the blade, causing it to wear quickly or fail during use.

SAFETY TIP

Before any tool is placed back into service, it should be inspected for damage.

CASE STUDY

You Are the Firefighter CONCLUSION

Upon reporting for duty, you are notified that you have been detailed to another station to work with the engine company for the tour. This will be your first time working with an engine company, and you are excited about the opportunity. It isn't long before an alarm sounds. "Person trapped, motor vehicle collision, Fourth and Elm," directs the dispatcher. You slip into your gear, take your seat, and put on your seat belt. As the engine pulls out from the station, you think how this changes your plan for what you thought you would be doing today.

1. How will your role be different for this call?

   Answer: Even though you are not responding to a fire, safety should still be a high priority. Traffic is a major safety concern, and you need to ensure that you do not exit the fire engine into oncoming traffic. Your role at this incident might be that of a medical first responder, or you might be tasked with stabilizing the vehicle or assisting with forcible entry. If extrication of the victim is required, your role might be that of the nozzle firefighter to provide a safety hose line in case of spark.

2. What kinds of tools will you need for this job?

   Answer: The tools required depend on what your assignment is. If you are assigned the role of a medical first responder, you will need medical equipment. If you are assigned to assist with forcible entry, you will need tools to pry, spread, and possibly cut. You might also need a glass center punch to break a tempered glass window or maybe a tool to cut through the laminated windshield. Whatever your role, ensure you are wearing the proper PPE, including eye protection.

3. What other kind of knowledge will you need?

   Answer: You will need to know which tools are carried on the apparatus, where the tools and any accessories for them are located on the apparatus, and how each tool works. For example, if you choose a tool to pry with, you should be able to use it without the tool breaking. You might need to decide if it would be faster to cut through the side pillar of a vehicle with a battery-powered saw than to wait for the hydraulic cutters to be set up.

WRAP-UP

SUMMARY

KNOWLEDGE OBJECTIVES

- Describe the general considerations of tools and equipment, including safety considerations and effective usage.
    - Describe the general purposes of tools and equipment. (p. 244)
    - Describe the safety considerations for the use of tools and equipment. (pp. 244–245)
    - Describe the guidelines for carrying a tool safely. (p. 245)
- Describe how the correct tool or piece of equipment is utilized for firefighting operations.
    - List and describe tools and equipment that are used for rotating. (p. 246)
    - List and describe tools and equipment that are used for cutting. (pp. 246, 248–249)
    - List and describe tools and equipment that are used for pushing or pulling. (pp. 249–253)
    - List and describe tools and equipment that are used for prying or spreading. (pp. 254–255)
    - List and describe tools and equipment that are used for striking. (pp. 254–258)
    - List and describe multiple-function tools and special-use tools. (pp. 255, 258–259)
    - Describe the tools used in response and size-up activities. (p. 260)
    - Describe the tools used in a forcible entry. (pp. 260–261)
    - Describe the tools used during an interior attack. (p. 261)
    - Describe the tools used during search and rescue operations. (pp. 261–262)
    - Describe the tools used during ventilation operations. (p. 262)
    - Describe the tools used during salvage and overhaul operations. (p. 262)
- Explain the procedures for cleaning and inspecting tools and equipment, and properly documenting and reporting any tool or equipment that is taken out of service.
    - Describe the importance of properly maintaining tools and equipment. (**NFPA 1010: 6.5.1**, pp. 262–264)
    - Describe the supplies needed to clean and inspect hand tools. (**NFPA 1010: 6.5.1**, pp. 262–263)
    - Describe the supplies needed to clean and inspect power tools. (pp. 263–264)
    - Explain the importance of replacing tools in their assigned locations. (**NFPA 1010: 6.5.1**, p. 262)
    - Identify procedures, including reporting requirements, for removing a damaged tool from service. (**NFPA 1010: 6.5.1**, pp. 262)
    - Describe why it is important for you to know where tools are stored. (**NFPA 1010: 6.5.1**, p. 262)

SKILL OBJECTIVES

- Select the appropriate cleaning supplies, and demonstrate the proper technique for cleaning and inspecting tools and equipment.
    - Clean and inspect hand tools. (**NFPA 1010: 6.5.1**, p. 263)
    - Demonstrate the proper technique for cleaning and inspecting power tools. (pp. 263–264)

KEY TERMS

adjustable wrench An open-ended wrench whose opening can be adjusted to accommodate bolts of different sizes.

adze The curved or straight wedge part of a Halligan tool.

KEY TERMS CONTINUED

axes Cutting tools that have a wide cutting blade that can be used to chop into a wall, roof, or door.

band saw An electrically powered saw that has a toothed metal blade stretched over two pulleys in a loop.

battering ram A tool made of hardened steel with handles on the sides used to force doors and to breach walls. Larger versions may be used by as many as four people; smaller versions are made for one or two people.

bolt cutter A cutting tool used to cut through thick metal objects such as bolts, locks, and wire fences.

box-end wrench A hand tool used to tighten or loosen bolts. The end is enclosed, as opposed to an open-end wrench. Each wrench is a specific size, and most have ratchets for easier use.

bunny tool See *rabbet tool.*

carpenter's handsaw A saw designed for cutting wood.

ceiling hook A tool with a long wooden or fiberglass pole that has a metal point with a spur at right angles at one end. It can be used to probe ceilings and pull down plaster lath material.

chainsaw A power saw that uses the rotating movement of a chain equipped with sharpened cutting edges. It is typically used to cut through wood.

chisel A metal tool with one sharpened end that is used to break apart material in conjunction with a hammer, mallet, or sledgehammer.

claw bar A tool with a pointed claw-hook on one end and a forked- or flat-chisel pry on the other end. It is often used for forcible entry.

Clemens hook A multipurpose tool that can be used for several forcible entry and ventilation applications because of its unique head design.

closet hook A type of pike pole intended for use in tight spaces, commonly 2 to 4 ft (0.6 to 1.2 m) in length.

combination wrench A hand tool with an open-end wrench on one end and a box-end wrench on the other.

come along A hand-operated tool used for dragging or lifting heavy objects that uses pulleys and cables or chains to multiply a pulling or lifting force.

coping saw A saw designed to cut curves in wood.

coupling One set or pair of connection devices attached to a fire hose that allow the hose to be interconnected to additional lengths of hose or adapters and other firefighting appliances. (NFPA 1960)

crowbar A straight bar made of steel or iron with a forked chisel on the working end that is suitable for performing forcible entry.

cutting torch A torch that produces a high-temperature flame capable of heating metal to its melting point, thereby cutting through an object. Because of the high temperatures (5700°F [3148°C]) that these torches produce, the operator must be specially trained before using this tool.

diagonal cutter See *wire cutter.*

drywall hook A specialized version of a pike pole that can remove drywall more effectively because of its hook design.

firefighter assist and search team (FAST) See *rapid intervention crew/company (RIC).*

flat bar A specialized type of prying tool made of flat steel with prying ends suitable for performing forcible entry.

flat-head axe A tool that has a head with an axe on one side and a flat head on the opposite side.

forcible entry Techniques used by fire personnel to gain entry into buildings, vehicles, aircraft, or other areas of confinement when normal means of entry are locked or blocked. (NFPA 440)

gripping pliers A hand tool with a pincer-like working end that can be used to bend wire or hold smaller objects.

hacksaw A cutting tool designed for use on metal. Different blades can be used for cutting different types of metal.

Halligan See also *Halligan tool.*

Halligan bar See also *Halligan tool.*

Halligan tool A prying tool that incorporates a sharp tapered pick, a blade (either an adze or wedge), and a fork; it is specifically designed for use in the fire service. See also *Halligan* and *Halligan bar.*

hammer A striking tool.

hand light A small, portable light carried by firefighters to improve visibility at emergency scenes. It is often powered by rechargeable batteries.

handsaw A manually powered saw designed to cut different types of materials. Examples include hacksaws, carpenter's handsaws, keyhole saws, and coping saws.

Hux bar A multipurpose tool that can be used for several forcible entry and ventilation applications because of its unique design. It also may be used as a hydrant wrench.

hydrant wrench A hand tool that is used to operate the valves on a hydrant; it also may be used as a spanner wrench. Some models are plain wrenches, whereas others have a ratchet feature.

Hydra-Ram A one-piece integrated hydraulic forcible entry tool.

hydraulic cutters See *hydraulic shears.*

hydraulic shears A lightweight, hand-operated tool that can produce up to 10,000 lb (4500 kg) of cutting force. See also *hydraulic cutters.*

hydraulic spreader A lightweight, hand-operated tool that can produce up to 10,000 lb (4500 kg) of prying and spreading force.

hydraulic tool A power tool that uses pressurized fluid to exert force.

interior attack The assignment of a team of firefighters to enter a structure and attempt fire suppression.

irons A combination of tools, usually consisting of a Halligan tool and a flat-head axe, that are commonly used for forcible entry.

Kelly tool A steel bar with two main features: a large pick and a large chisel or fork.

keyhole saw A saw designed to cut keyhole circles in wood and drywall.

K tool A tool that is used to remove lock cylinders from structural doors so that the locking mechanism can be unlocked.

mallet A short-handled hammer.

maul A specialized striking tool, weighing 6 lb (3 kg) or more, with an axe on one side of the head and a sledgehammer on the other side.

multipurpose hook A long pole with a wooden or fiberglass handle and a metal hook on one end used for pulling.

multi-tool A compact, pocket-size, multiple-function tool that combines several different tools, such as a knife, scissors, wire cutters, pliers, and screwdriver.

New York hook See *roofman's hook.*

open-end wrench A hand tool that is used to tighten or loosen bolts. The end is open, as opposed to a box-end wrench. Each wrench is a specific size.

overhaul A firefighting term involving the process of final extinguishment after the main body of the fire has been knocked down. All traces of fire must be extinguished at this time. (NFPA 1700)

pick-head axe A tool that has a head with an axe on one side and a pointed end ("pick") on the opposite side.

pike pole A pole with a sharp point ("pike") on one end coupled with a hook. It is used to make openings in ceilings and walls. Pike poles are manufactured in different lengths for use in rooms of different heights.

pipe wrench A wrench having one fixed grip and one movable grip that can be adjusted to fit securely around pipes and other tubular objects.

plaster hook A long pole with a pointed head and two retractable cutting blades on the side.

power saw A saw that usually is powered by an electric motor or a gasoline engine. The three primary types of mechanical saws are chainsaws, rotary saws, and reciprocating saws.

pry bar A specialized prying tool made of a hardened steel rod with a tapered end that can be inserted into a small area.

rabbet tool A hydraulic spreading tool designed to pry open doors that swing inward. See also *bunny tool.*

rapid intervention crew/company (RIC) A minimum of two fully equipped personnel onsite, in a ready state, for immediate rescue of disoriented, injured, lost, or trapped rescue personnel. (NFPA 1550)

rapid intervention team (RIT) See *rapid intervention crew/company (RIC).*

reciprocating saw A saw that is powered by an electric motor or a battery motor and whose blade moves back and forth.

rescue knife A spring-assisted folding knife that can be used with one hand; often includes a seat belt cutter and a window breaker.

response Immediate and ongoing activities, tasks, programs, and systems to manage the effects of an incident that threatens life, property, operations, or the environment. (NFPA 1660)

roofman's hook A long pole with a solid metal hook used for pulling. See also *New York hook.*

rotary saw A saw that is powered by an electric motor or a gasoline engine and that uses a large rotating blade to cut through material. The blades can be changed depending on the material being cut.

San Francisco hook A multipurpose tool that can be used for several forcible entry and ventilation applications because of its unique design, which includes a built-in gas shut-off and directional slot.

KEY TERMS CONTINUED

salvage The process of protecting property and a structure from additional damage from fire suppression operations.

screwdriver A tool used for turning screws.

search and rescue The process of searching a building for a victim and extricating the victim from the building.

seat belt cutter A specialized cutting device that cuts through seat belts.

shove knife A forcible entry tool used to trip the latch on outward-swinging doors.

size-up The process of gathering and analyzing information to help fire officers make decisions regarding the deployment of resources and the implementation of tactics. (NFPA 1410)

sledgehammer A hammer that can be one of a variety of weights and sizes.

socket wrench A wrench that fits over a nut or bolt and uses the ratchet action of an attached handle to tighten or loosen the nut or bolt.

spanner wrench A type of tool used to couple or uncouple hose by turning the rocker lugs or pin lugs on the connections.

speed socket wrench A long, curved handle with a socket wrench that fits over a nut or bolt at the end. To tighten or loosen the nut or bolt, the handle is rotated.

spring-loaded center punch A spring-loaded punch used to break automobile glass.

stem nut The large nut at the top of the operating stem in a dry barrel hydrant that is turned to open the hydrant valve.

thermal imager device An electronic device that detects differences in temperature based on infrared energy and then generates images based on those data. It is commonly used in smoke-filled environments to locate victims as well as to search for hidden fire during size-up and overhaul.

ventilation The controlled and coordinated removal of heat and smoke from a structure, replacing the escaping gases with fresh air. (NFPA 1410)

ventilation saws Cutting tools designed for roof ventilation that have a different cutting chain than chainsaws used to cut wood; also may have a depth gauge on its bar.

wire cutter A hand tool used to cut wire and small diameter cable. See also *diagonal cutter*.

wrench A hand tool that comes in several sizes and is used to tighten or loosen bolts.

REVIEW QUESTIONS

1. What is considered a high priority at any incident scene? How does this relate to using tools and equipment.
2. If you are assigned the task of looking for pockets of fire in the ceiling, what functional category would the tool you use fall under? List at least three tools from this category that you could use.
3. List the three working ends on a Halligan.
4. What are the seven phases of incident scene operations?
5. Where can firefighters find instructions for cleaning and maintaining tools and equipment?
6. What needs to happen after a hand tool is used?

DISCUSSION QUESTIONS

1. If the hydrant wrench is missing from the hydrant bag, what can you use to remove the hydrant caps and turn the stem nut?
2. If your company is assigned to go to the roof at a structure fire to assist with ventilation, which cutting hand tool should you bring?
3. As a new firefighter, what could you do if you are unsure of which tool to choose for the task at hand?

APPLYING THE CONCEPTS

Your engine company is enroute to a call for a fire at a professional office in a three-story, Type V construction building in the historic area of your jurisdiction. You're riding in the "hydrant" seat and are assigned to establish the water supply at the hydrant, then assist where needed. You've helped connect to a hydrant before, but this is the first time you're doing it solo. The engine driver stops near the hydrant. Before getting out of the engine, you secure your gloves and helmet, and grab your tools. You set the tools by the hydrant and pull a length of supply hose from the hose bed. You tie it around the hydrant and signal the driver who does a forward lay to the fire.

1. Every second counts, so it's important to arrive at the hydrant equipped with the right tools. How do you know what tools you'll need?
2. If this is a typical hydrant with standard stem nut and national standard thread (NST), what tools do you need to connect the supply hose and open the hydrant?

You connect the supply hose and charge the line at the driver's signal. You follow the hose back to the engine to ensure it doesn't have any kinks. Then you radio your company officer, "Water supply established. Ready for next assignment." He instructs you to get tools off the engine to help the nozzle team make forcible entry into the alpha door of the structure.

3. What do you need to know before deciding what forcible opening tools to gather?
4. The company officer radios that the door is inward opening, wood construction with a keyed deadbolt. What door opening tools do you select? Why?

The nozzle crew is conducting an interior attack, but the source of the fire hasn't been located. Because the house is balloon frame construction, the IC suspects there are hidden fire pockets and extensions in the structure. You're assigned to assist the overhaul crew to locate the hidden fire. The officer instructs you to gather pike poles, Halligans, and a PIG.

5. Why is a pike pole an especially useful tool for overhaul?
6. How do you select the right size pike pole?

REFERENCES

Firefighter Nation. 2014. "Effective and Safe Use of Axes on the Fireground." May 1, 2014. Accessed October 5, 2023. https://www.firefighternation.com/firerescue/effective-and-safe-use-of-axes-on-the-fireground/.

Long, Merritt. 2007. "History of the Halligan Tool." Firefighter Nation, December 24, 2007. Accessed October 4, 2023. http://my.firefighternation.com/group/firefightinghistorymyths/forum/topics/889755:Topic:233830.

National Fire Protection Association (NFPA). 2019. *NFPA 1407, Standard for Training Fire Service Rapid Intervention Crews.* 2020 Edition. Quincy, MA: NFPA.

National Fire Protection Association (NFPA). 2019. *NFPA 1410, Standard on Training for Emergency Scene Operations.* 2020 Edition. Accessed April 1, 2023. Quincy, MA: NFPA.

National Fire Protection Association (NFPA). 2020. *NFPA 1700, Guide for Structural Fire Fighting.* 2021 Edition. Accessed April 1, 2023. Quincy, MA: NFPA.

National Fire Protection Association (NFPA). 2023. *NFPA 440, Guide for Aircraft Rescue and Firefighting Operations and Airport/Community Emergency Planning.* 2024 Edition. Quincy, MA: NFPA.

National Fire Protection Association (NFPA). 2023. *NFPA 1550, Standard for Emergency Responder Health and Safety.* 2024 Edition. Quincy, MA: NFPA.

National Fire Protection Association (NFPA). 2023. *NFPA 1660, Standard for Emergency, Continuity, and Crisis Management: Preparedness, Response, and Recovery.* 2024 Edition.

National Fire Protection Association (NFPA). 2023. *NFPA 1960, Standard for Fire Hose Connections, Spray Nozzles, Manufacturer's Design of Fire Department Ground Ladders, Fire Hose, and Powered Rescue Tools.* 2024 Edition. Quincy, MA: NFPA.

National Fire Protection Association (NFPA). 2023. *NFPA 1960, Standard for Aircraft Rescue and Firefighting Vehicles, Automotive Fire Apparatus, Wildland Fire Apparatus, and Automotive Ambulances.* 2024 Edition. Quincy, MA: NFPA.

CHAPTER

8

Firefighter I

Ropes and Knots

KNOWLEDGE OBJECTIVES

After studying this chapter, you will be able to:

- List the four primary types of fire service rope and describe the characteristics of each type.
- List the types of materials rope can be made from and describe the characteristics of each material.
- List three types of rope construction and describe the method in which they are constructed.
- List the four parts of rope maintenance and the key components of each part.
- Understand rope terminology and describe the 10 common knots and hitches used in the fire service.
- Describe how to use rope to safely hoist tools and equipment.

SKILLS OBJECTIVES

After studying this chapter, you will be able to perform the following skills:

- Demonstrate how to care for, clean, inspect, and place life safety rope in a bag.
- Tie the 10 common knots used in the fire service.
- Tie the appropriate knots onto common firefighting equipment and hoist them safely.

ADDITIONAL NFPA STANDARD

- **NFPA 2500**, *Standard for Operations and Training for Technical Search and Rescue Incidents and Life Safety Rope and Equipment for Emergency Services*

CASE STUDY

You Are the Firefighter

You and your partner are on the fourth floor of a multistory apartment building and find yourself trapped by a rapidly progressing fire. You get to a bedroom at the front of the building and know your only escape route is out the window. You don't have time to wait for the aerial ladder to get to you, so you prepare for self-rescue. You pull out your rope and begin the most important task you have ever done in your life.

1. What type of rope should you use for this type of activity?
2. How many people is this type of rope designed to support?
3. After you have used your fire escape rope in a fire environment, what should you do next?

Introduction

The fire service often utilizes different types of rope to support or achieve tactical objectives. Rope is often used in emergency situations to directly or indirectly save or protect the lives of civilians and firefighters. This chapter will provide the basic information for firefighters to begin developing the skills needed to accomplish those essential objectives.

You will learn about the different types of ropes used, the materials they are made of, their construction and characteristics, and how to maintain them. Firefighters must be able to choose the appropriate rope for a variety of situations. Firefighters must also know how to tie essential knots quickly, accurately, and securely. They should be able to deploy and use ropes quickly to move equipment or people in a controlled, safe manner.

Understanding Ropes and Webbing

Before we dive into the world of ropes and knots, let's consider some key concepts. A **rope** is an object constructed of fibers that is designed to sustain a weight (**FIGURE 8-1**). In the fire service, rope is used in tasks ranging from lifting and lowering tools to raising victims from a confined space. **Webbing** is a woven material of flat or tubular weave in a long strip that is often used in conjunction with rope. All ropes are able to stretch and elongate, which is one reason why ropes are able to absorb the forces created when objects pull their fibers. There is a limit to how far a rope can stretch, which can be measured in the rope's minimum breaking strength. The **minimum breaking strength (MBS)** is the weight limit that a rope can safely support.

Rope Types and Webbing

Four types of rope are primarily used in the fire service. Each of these rope types has a specific function.

- Life safety rope
- Escape rope
- Throwline for water rescues
- Utility rope

FIGURE 8-1 Rope is used in tasks ranging from lifting and lowering tools to raising victims.

SAFETY TIP

Many fire departments use color coding or other visible markings to identify different types of rope. This allows a firefighter to determine quickly if a rope is a life safety rope or a utility rope. The length of each rope should be clearly marked by a tag or a label on the rope bag.

In addition, webbing is also used in conjunction with rope in certain conditions. A firefighter must be able to recognize the type of rope instantly from its appearance and markings.

NFPA 2500, *Standard for Operations and Training for Technical Search and Rescue Incidents and Life Safety Rope and Equipment for Emergency Services, 2022 Edition,* specifies the criteria for the design, construction, and performance of ropes, webbing, and related equipment. NFPA 2500 also requires rope manufacturers to provide detailed instructions for the proper use, maintenance, and inspection of rope and webbing, including the conditions for removing the rope or webbing from service. In addition, NFPA 2500 requires rope manufacturers to supply a list of things to look for when a rope is inspected after it has been used. If the rope does not meet all of the inspection criteria, it must be retired from service.

Life Safety Rope

Only **life safety rope** is used for suspending people. It must be used whenever a rope is needed to support the weight of one or more persons, whether during training, or during firefighting, rescue, or other emergency operations. Life safety rope has the most stringent performance requirements for rope in the fire service because when rope is supporting a person, rope failure could result in serious injury or death. Life safety rope is a critical tool and should be used *only* for life-saving purposes.

Life safety rope is identified by the paperwork provided by the manufacturer when it is purchased. There is also a label called a **trailer** along the entire length of the rope inside the outer sheath that indicates that it is life safety rope. The trailer also includes the name of the manufacturer, model number, make number, serial number, and date of manufacture. The paperwork that accompanies the rope upon purchase usually provides a comprehensive description of the classification of use, strength rating, elongation, core and sheath materials, length, diameter, weight, durability, among other items specific to the type of rope desired.

There are two basic types of life safety rope: general use life safety rope and technical use life safety rope. **General use life safety rope** is the most common life safety rope carried by the fire service in the United States. NFPA 2500 lists the requirements for the performance of life safety rope. These requirements are gauged by MBS, elongation, and minimum and maximum diameter of the ropes. The diameter of a general use life safety rope must be at least 7/16 inch (in.; 11 millimeter [mm]) but not larger than 5/8 in. (16 mm). In addition, general use life safety rope must have an MBS strength of 8992 pound-force (lbf; 40 kilonewtons [kN]). **Technical use life safety rope**, per NFPA 2500, must have a diameter that is at least 3/8 in. (9.5 mm) but not larger than 1/2 in. (12.5 mm). The MBS of technical use safety rope is 4496 lbf (20 kN), which is half of the required breaking strength of general use life safety rope. NFPA 2500 also specifies the limits of elongation of life safety ropes and the testing requirements for life safety ropes with minimum and maximum elongation.

All life safety rope must be resistant to abrasions so that they do not fray. Life safety rope must also be rope made using **block creel construction**, which means the rope must be made from fibers that run the entire length of the rope without knots or splices. The continuous filaments produce a rope that is stronger than one constructed of shorter fibers that are twisted or braided together. Chapter 29 in NFPA 2500 provides information on how to select a life safety rope.

Although technical use life safety rope can be significantly lighter than general use life safety rope, it is not as strong. Highly trained rescue teams who deploy to technical environments, such as mountainous or wilderness terrain, carry this smaller, lighter rope so they can carry more ropes and longer lengths over rough terrain. Although general use life safety rope allows a much greater margin of safety, the highly trained technical rescue technicians know how to use technical use life safety rope within the limits of its specifications. Rope with a diameter of 7/16 in. (11 mm) falls within the specified range for both general use and technical use life safety rope. But because general use life safety rope has a breaking strength that is double that of technical use life safety rope, technical rescue teams often choose 7/16 in. (11 mm) general use life safety rope instead of the same diameter technical rescue safety rope.

Escape Rope

When you are responding to an emergency, you should always have a safe way to get out of a situation and reach a safe location. You might be able to go back through the door that you entered, or you might have another exit route, such as through a different door, through a window, or down a ladder. If conditions suddenly change for the worse, having an escape route can save your life. Sometimes, however, you might find yourself in a situation during which conditions deteriorate so rapidly that you cannot use your planned exit route. For example, the stairway you used might collapse behind you, or the room you are in might suddenly flash over, blocking your planned exit route. In such a situation, you might need to take extreme measures to get out of the building.

Two types of ropes were developed for this type of situation. The **escape rope** is intended for emergency, self-rescue situations. The **fire escape rope** is intended specifically for emergency self-rescue from an immediately hazardous environment in which fire or fire products are involved. The performance requirements for fire escape rope state that this type of rope must be able to hold 300 pounds (lb; 136 kilograms [kg]) for 45 seconds at 1112°F (600°C) and for 5 minutes at 750°F (400°C).

Both types of escape rope are designed to carry the weight of only one person and to be used only one time (**FIGURE 8-2**). Both types of escape rope must have a diameter of at least 19/64 in. (7.5 mm) but not larger than 3/8 in. (9.5 mm), and an MBS of 3034 lbf (13.5 kN). Because they are thinner and lighter than life safety ropes, they fit easily into a small packet or pouch and are easy to carry. An escape rope should be replaced with a new rope if it is exposed to an immediately dangerous to life or health (IDLH) environment, even if it was not used. When replacing any rope or equipment, follow the proper procedure for your department, notifying your supervisor or appropriate personnel and using the proper forms to record all of the information.

FIGURE 8-2 An escape rope is designed to be used by only one person.

TIP

Escape ropes are not classified as life safety ropes.

Throwline for Water Rescue

A **throwline** for water rescue is a type of rope that is used during water rescue operations and that floats on top of the water. They are 50 to 75 feet (ft.; 15 to 30 meter [m]) in length and are stored in a **throw bag**, which is attached to one end of the rope and has a layer of sponge foam so that it floats (**FIGURE 8-3A**). The throw bag is thrown to a victim in the water while the rescuer holds on to the other end of the rope. As the bag travels through the air, the rope unravels (**FIGURE 8-3B**). The victim can grab the bag or the rope, and then the rescuer can pull the victim toward them. Throwlines also can be used as a tether for rescuers entering the water.

Throwlines must be at least 19/64 in. (7 mm) in diameter but not larger than 3/8 in. (9.5 mm) in diameter, and they must have an MBS of 2923 lbf (13 kN). They are made of material that floats, such as polypropylene, so the rope does not snag on underwater hazards or get entangled in motorboat propellers. Throwlines do not have the strength or abrasion resistance required to be used as a life safety rope.

Utility Rope

Utility rope is used in most cases when it is not necessary to support the weight of a person, such as when hoisting or lowering tools or equipment (**FIGURE 8-4**). Utility ropes can also be used to mark off areas or stabilize objects. NFPA 2500 does not specify performance requirements for utility ropes and they must never be used in situations where life safety rope is required.

Webbing

As stated in the beginning of the chapter, webbing is a woven material of flat or tubular weave in a long strip (**FIGURE 8-5**). Although it is not actually

A.

B.

FIGURE 8-3 **A.** Water rescue throwlines are kept in throw bags. **B.** To use, the rescuer holds onto the end of the rope sticking out of the top of the bag and throws the bag, which has the other end of the rope attached to it, to the victim in the water.

FIGURE 8-4 Utility ropes are used for hoisting and lowering tools.

FIGURE 8-5 Webbing is available in tubular (left) or flat (right) construction.

considered a type of rope, webbing is often used in conjunction with rope. Because of its special characteristics, webbing may even be preferable to rope in certain situations. For example, webbing is better suited to rig anchor points around objects with abrupt bends or corners on them, such as steel I-beams or square columns. In addition, because of its wide, flat surface, webbing may be more abrasion resistant in some rigging applications. **Rigging** is a general term used when building a lifting and lowering system with ropes and rope equipment.

There are hundreds of types and sizes of webbing. Most webbing is made of nylon or polyester and ranges in width. Flat webbing is constructed of a single layer of material; tubular webbing consists of a flattened tube of material. Although flat webbing is less expensive, tubular webbing is more supple and easier to work with. Escape webbing must have an MBS of 3034 lbf (13.5 kN), as well as meeting all of the other performance requirements as escape rope in NFPA 2500.

Rope Materials

Ropes are made from many different materials. The earliest ropes were made from naturally occurring vines or natural fibers, such as manila or cotton, that were woven together. The natural fibers were twisted together to form strands. Each strand could contain hundreds of individual fibers of different lengths.

Natural fiber ropes have the following drawbacks:

- Lose their load-carrying ability over time
- Are subject to mildew, which weakens the fibers
- Absorb water when wet, making them susceptible to deterioration
- Once wet, are difficult to dry
- Degrade quickly, even when properly stored

Today, ropes made from natural fibers are no longer acceptable as life safety ropes and are rarely even used as utility ropes. Life safety ropes are always made from synthetic fibers.

Ever since the invention of nylon in 1938, synthetic fibers have been used to make ropes. In addition to nylon, several newer synthetic materials, such as polyester, polypropylene, and polyethylene, are now also used to make ropes. Synthetic fibers have several advantages over natural fibers:

- Synthetic fibers are generally stronger than natural fibers, so it is possible to use a smaller diameter rope without sacrificing strength.
- Synthetic materials can produce very long fibers that run the full length of a rope to provide greater strength and added safety.
- Synthetic fibers have a longer life span than natural fibers
- Synthetic ropes are more resistant to rotting and mildew than natural fiber ropes and do not degrade as rapidly.
- Depending on the material, synthetic ropes might also be more resistant to melting and burning than natural fiber ropes.
- Synthetic ropes absorb much less water when they get wet than natural fiber ropes, and they can be washed and dried.
- Some types of synthetic rope can float on water, which is a major advantage in water rescue situations.

As with all ropes, prolonged exposure to ultraviolet light and exposure to strong acids or alkalis can damage synthetic ropes and decrease their life expectancy. Additionally, all ropes need to be protected from abrasion and sharp objects that can damage or cut the rope fibers.

Nylon is commonly used for ropes because it has a high melting temperature compared to other synthetic materials, it has a high resistance to abrasion, and it is strong but lightweight. Nylon ropes are also resistant to most acids and alkalis. Polyester is another commonly used synthetic fiber for ropes. Some ropes are made of a combination of nylon and polyester or other synthetic fibers. Polypropylene is the lightest of the synthetic fibers used for ropes. Because it does not absorb water and it floats, throwlines are often made of polypropylene (**FIGURE 8-6**). Polypropylene is not, however, suitable material for life safety or escape rope because it is not as strong as other synthetic materials, it is hard to knot, and it has a low melting point.

FIGURE 8-6 Polypropylene rope is often used in water rescues.

Rope Construction

There are several different types of rope construction. The three most common types of rope construction that a firefighter will encounter are twisted rope, braided rope, and kernmantle rope.

Twisted Rope

Twisted rope, which is also called **laid rope**, is made of individual fibers twisted into strands. The strands are then twisted together to make the rope (**FIGURE 8-7**). This method of rope construction has been used for hundreds of years. Both natural and synthetic fibers can be used to make twisted rope.

There are a few disadvantages to twisted ropes. In a twisted rope, all of the fibers are exposed and subject to abrasion, which can damage the rope fibers and reduce rope strength. Twisted ropes also are more prone to stretching and unraveling when a load is applied. Some fire departments have used twisted ropes for their utility ropes because they are easily distinguishable from life safety ropes. Twisted ropes are suitable for many activities on the fireground, including hoisting tools. They need to be maintained properly per the manufacturers' directions, have sufficient strength for what they will lift or stabilize, have minimal stretch (or elongation), and be replaced when damaged or fail inspection. Finally, twisted ropes are not life safety ropes.

Braided Rope

Braided rope is constructed by weaving or intertwining strands—typically synthetic fibers—together in the same way that hair is braided (**FIGURE 8-8**). Similar to twisted rope, all of the strands in a braided rope are exposed and subject to abrasion. A double-braided rope is a braided rope covered by a protective braided sleeve. This means that only the fibers in the outer sleeve are exposed. The inner core is protected from abrasion (**FIGURE 8-9**).

Like twisted rope, braided rope stretches under a load, but it is not prone to unraveling or twisting. Braided ropes are also often used as utility rope. They serve well as utility rope for moving and hauling equipment, and holding items in place. The main disadvantage to using braided rope as utility rope is that it looks very similar to life safety rope, which may cause confusion.

Braided and double-braided ropes are often used in the marine and arborist industry. They are used to lift, lower, and stabilize equipment, vessels, and objects. The fire service typically uses braided rope as utility rope or in the fire/marine divisions.

FIGURE 8-7 Twisted rope.

FIGURE 8-8 Braided rope.

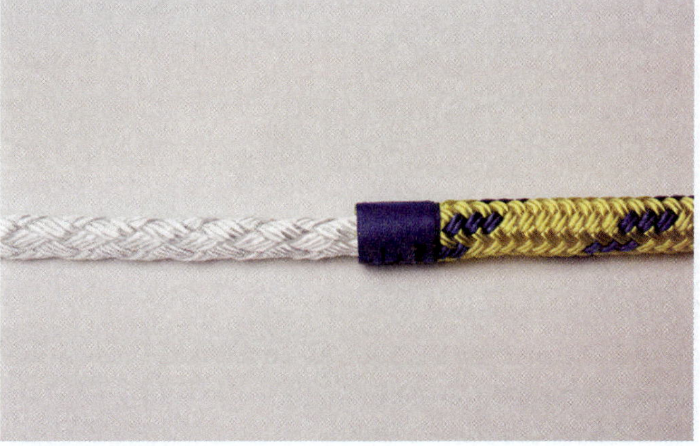

FIGURE 8-9 Double-braided rope.

Courtesy of Steve Hudson.

Kernmantle Rope

Kernmantle rope consists of two distinct parts: the kern and the mantle. The **kern** is the center or core of the rope. The **mantle**, or **sheath**, is a braided covering that protects the kern from dirt and abrasion. Although both parts of a kernmantle rope are made with synthetic fibers, each part may be made from different synthetic fibers. Kernmantle ropes are made with block creel construction, where each individual fiber continues from end to end without splices

Kernmantle rope is strong and flexible yet relatively thin and lightweight (**FIGURE 8-10**). This construction is well suited for rescue work and is very popular for life safety rope.

Static and Dynamic Kernmantle Rope

All ropes stretch or elongate. A kernmantle rope can be either dynamic or static, depending on how it reacts to an applied load. The differences between dynamic and static ropes result from both the fibers used and the construction method. A **dynamic rope** is designed to stretch more than a static rope when it is loaded. Some dynamic rope can elongate up to 40 percent of the length of the entire rope. Dynamic rope is typically used in safety lines in climbing. The safety rope stretches and cushions the shock when a sudden load is applied—for example, if a climber falls.

A **static rope** has a lower range of elasticity than dynamic rope. A static rope is designed to stretch less when a load is applied and is, therefore, more suitable for most fire rescue situations in which sudden shock loads are less likely and when limited elongation is desired to have more control over a system as when lowering, raising or rappelling for a rescue. Decreased elongation is also desirable where rope haul systems are used. NFPA 2500 outlines how much life safety ropes can elongate.

FIGURE 8-10 Kernmantle rope with the kern exposed.
Courtesy of Steve Hudson.

A dynamic kernmantle rope is constructed with overlapping or woven fibers in the core (**FIGURE 8-11A**). When the rope is loaded, the core fibers are pulled tighter, which gives the rope its additional elasticity. In the core of a static kernmantle rope, all of the fibers are parallel, which decreases its elongation. (**FIGURE 8-11B**).

Most fire departments use life safety ropes that are made with static kernmantle construction. This type of rope is well suited for lowering a person and can be used with a mechanical advantage pulley system for lifting individuals. It can also be used to create a **highline system**, which is a system of ropes or cables that allow rescuers to travel or send an object or a victim in a litter through the air from one building or object to another. Decreased elongation is desirable in these circumstances.

Rope Maintenance

All ropes—especially life safety ropes—need proper care. Proper care ensures a long life for your rope and reduces the chance of equipment failure and accident. If a rope is not properly maintained, it will degrade much more quickly and have a shorter life span than it should.

Your life and the lives of others depend on the proper maintenance of your ropes. All ropes should be maintained and cared for with the same concern as life safety rope because all ropes perform critical tasks. For example, a piece of equipment could be needed on the roof of a building to prevent the spread of fire toward firefighters and a utility rope could be hoisting that tool. For more information on the care and maintenance of life safety rope, see Chapter 29 in NFPA 2500. There are four parts to the maintenance formula:

- Care
- Clean
- Inspect
- Store

Care for the Rope

You must follow certain principles to preserve the strength and integrity of rope:

- Protect the rope from sharp and abrasive surfaces. Use edge protectors when the rope must pass over a sharp or unpadded surface.
- Protect the rope from rubbing against another rope or webbing. Friction generates heat, which can damage or destroy the rope.

A.

B.

FIGURE 8-11 **A.** Dynamic kernmantle rope core with overlapping fibers in the core. Note the trailer label that travels the length of the core under the mantle. **B.** Static kernmantle rope core with parallel fiber in the core. Note the trailer label that travels the length of the core under the mantle.

A. Courtesy of Pigeon Mountain Industries; **B.** © Jones & Bartlett Learning. Courtesy of Steve Hudson.

- Protect the rope from heat, chemicals, and flames.
- Protect the rope from prolonged exposure to sunlight. Ultraviolet radiation can damage rope.
- Do not step on the rope! Your footstep could force shards of glass, splinters, or abrasive particles into the core of the rope, damaging the rope fibers.
- Follow the manufacturer's recommendations for rope care.
- Protect the rope from shock loads.

A **shock load** is a load on a rope that suddenly drops or stops. A piece of equipment being raised or lowered that drops suddenly would shock load a utility rope. A rescuer who steps over an edge or off a ledge with slack in the rope would shock load a life safety rope. A shock load places a rope under extra tension. If a rope undergoes shock load, it should be carefully inspected for tears, flat spots, bulges, or any deformations. Although there might not be any visible damage, a severe shock load should be reason to consider retiring the rope from service. Repeated shock loads can severely weaken a rope so that it can no longer be used safely. Keeping accurate rope records helps identify potentially damaged rope.

To care for ropes properly, including life safety ropes, follow the steps in **SKILL DRILL 8-1**.

Clean the Rope

Dirt can work its way into rope fibers, weakening the rope, so part of maintaining ropes is to wash them after use. There are several ways to clean ropes. Many synthetic ropes can be washed by hand with a scrub brush and mild soap and water. If the manufacturer recommends it, you can attach a rope washer to a garden hose that you can run a rope through. Some manufacturers recommend placing the rope in a mesh bag and washing it in a front-loading washing machine. Do not use washing machines with agitators to wash ropes because they can bind or wind the rope and cause unpredictable or undetectable stresses on the rope. Follow the manufacturer's recommendations for specific care of your rope. When using a washing machine, use the gentle cycle and use a mild detergent that is safe for nylon such as Woolite or similar type of product sold by rope manufacturers. Do not use bleach because it can damage rope fibers.

Ropes should be dried before they are put away. Do not pack or store wet or damp rope. Most manufacturers recommend air drying ropes by suspending them out of direct sunlight. Do not air dry a rope by laying it on the floor, floors often have had oils or other chemicals

SKILL DRILL 8-1

Caring for Life Safety Ropes Firefighter I, NFPA 1010: 6.5.1

1. Protect the rope from sharp and abrasive edges. Use edge protectors if necessary.
2. Protect the rope from rubbing against other ropes.
3. Protect the rope from heat, chemicals, flames, and sunlight.
4. Avoid stepping on the rope.

Courtesy of Pigeon Mountain Industries.

on them at one time or another. The use of mechanical drying devices is not recommended because the heat could damage the fibers. Also, rope should never be dried or stored in direct sunlight because the ultraviolet rays will damage the fibers.

Follow the steps in **SKILL DRILL 8-2** to clean fire department ropes.

Inspect the Rope

Life safety ropes must be inspected after each use, whether the rope was used for an emergency incident or in a training exercise. Unused life safety rope should be inspected on a regular schedule. Some departments inspect all rope, including escape, throwlines, and utility ropes, every 3 months. Each department should obtain the recommended inspection criteria from the manufacturer of each of their ropes.

When you inspect a rope, ask the following questions:

- Has the rope been exposed to heat or flame?
- Has the rope been exposed to abrasion?
- Has the rope been exposed to chemicals?
- Has the rope been exposed to shock loads?
- Are there any depressions, discolorations, or lumps in the rope?

Inspect the rope visually, looking for cuts, frays, discoloration, shiny marks from heat or friction, or other damage as you run it through your fingers. Feel for inconsistencies in the thickness of the rope. Make sure that your grasping hand is gloved to protect yourself from sharp objects that may be embedded in the rope. Because you cannot see the inner core of a kernmantle rope, feel for any flat spots or lumps on the inside.

SKILL DRILL 8-2

Cleaning Fire Department Ropes Firefighter I, NFPA 1010: 6.5.1

1. Wash the rope by hand with mild soap and water and use a soft brush to clean dirtier areas if needed.

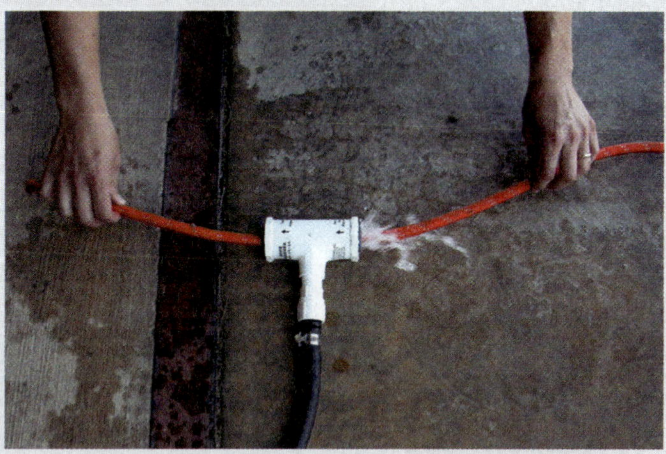

2. Use a rope washer or put the rope into a mesh bag and then wash it in a washing machine without an agitator if recommended by the rope's manufacturer.

3. Hang the rope out of direct sunlight to air dry it. Inspect the rope and replace it in the rope bag so that it is ready for use.

Examine the mantle for discolorations, abrasions, or flat spots, and check for fibers from the kern poking through the mantle. If you have any doubt about whether the rope has been damaged, consult your company officer.

A life safety rope that is no longer usable must be pulled from service and either destroyed or designated as a utility rope. A downgraded rope must be clearly marked so that it cannot be confused with a life safety rope.

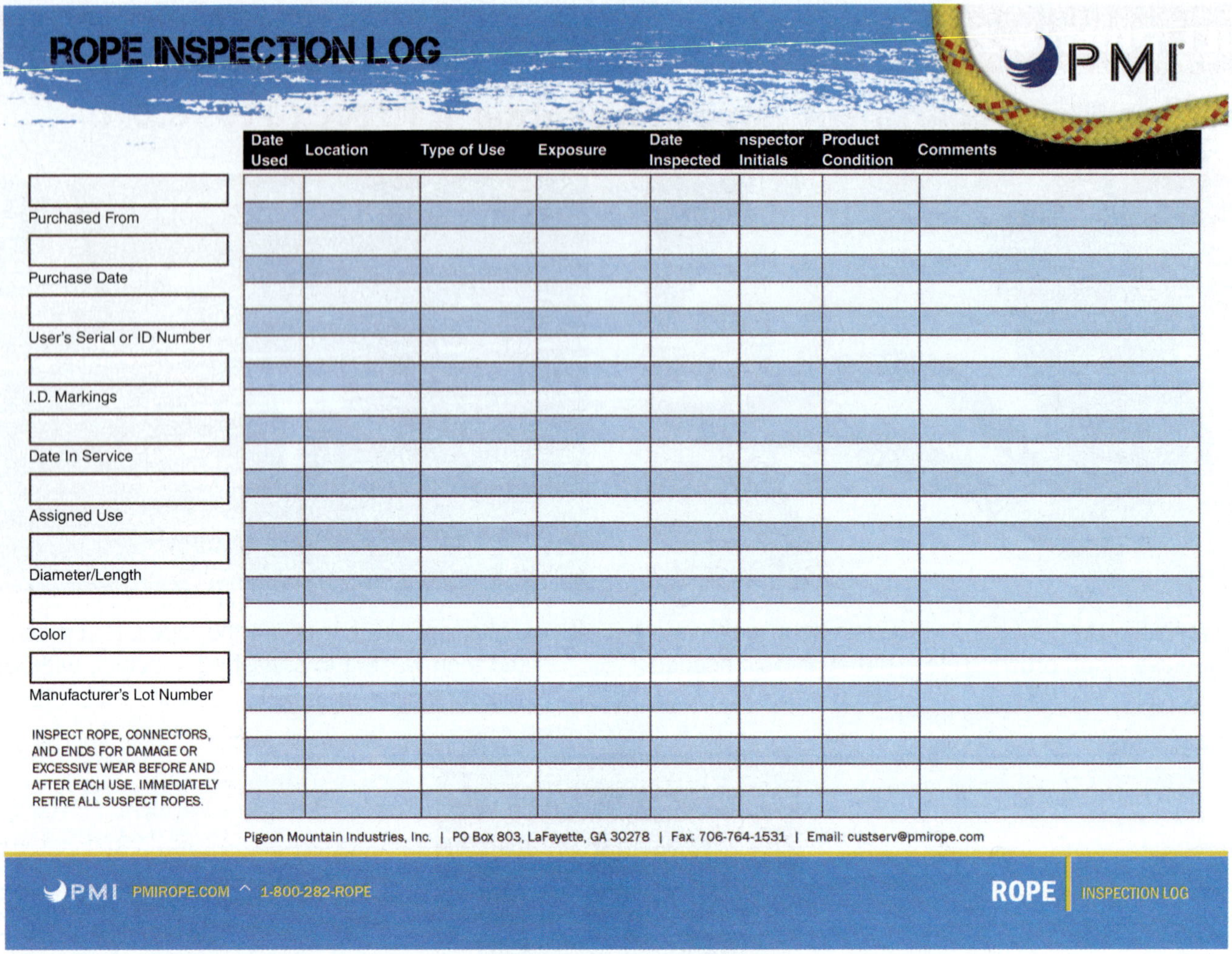

ROPE INSPECTION LOG

PMI

Purchased From

Purchase Date

User's Serial or ID Number

I.D. Markings

Date In Service

Assigned Use

Diameter/Length

Color

Manufacturer's Lot Number

INSPECT ROPE, CONNECTORS, AND ENDS FOR DAMAGE OR EXCESSIVE WEAR BEFORE AND AFTER EACH USE. IMMEDIATELY RETIRE ALL SUSPECT ROPES.

Date Used	Location	Type of Use	Exposure	Date Inspected	nspector Initials	Product Condition	Comments

Pigeon Mountain Industries, Inc. | PO Box 803, LaFayette, GA 30278 | Fax: 706-764-1531 | Email: custserv@pmirope.com

PMI PMIROPE.COM ^ 1-800-282-ROPE ROPE INSPECTION LOG

FIGURE 8-12 Rope inspection should be recorded in a rope record.

Courtesy of Pigeon Mountain Industries.

Each inspection of a life safety rope or utility rope should be recorded in a **rope record** (**FIGURE 8-12**). NFPA 2500 indicates that at least the following record shall be kept for each life safety rope:

- Individual identification of the rope
- The date the rope was purchased
- Date the rope was placed in service
- The manufacturer's contact information
- The item or model number
- The month and year of manufacture
- The date of use, including how it was used, the weather conditions, any damage, and other conditions related to use
- The date of cleaning and inspection
- Removal from service and date of return
- Other notes deemed relevant by the authority having jurisdiction

Follow the steps in **SKILL DRILL 8-3** to inspect fire department ropes.

Store the Rope

Ropes need to be stored in areas where there is some air circulation, away from temperature extremes, and out of sunlight. Ropes should also not be placed where they are exposed to fumes from gasoline, oils, hydraulic fluids, or other petroleum products because those fumes might damage the ropes. This means that apparatus compartments used to store ropes should be separated from compartments used to store any oil-based products or machinery powered by gasoline or diesel fuel. Finally, do not place any heavy objects on top of the rope to avoid damaging the rope and to ensure that the rope is readily accessible (**FIGURE 8-13**).

SKILL DRILL 8-3

Inspecting Fire Department Ropes Firefighter I, NFPA 1010: 6.5.1

1. Inspect the rope after each use and at regular intervals.
   - Examine the core by looking and feeling for depressions, flat spots, and lumps.
   - Examine the sheath by looking for discolorations, abrasions, flat spots, and embedded objects.

2. Record the inspection in the rope record. Remove the rope from service if it is damaged.

Courtesy of Steve Hudson.

FIGURE 8-13 Ensure that ropes are stored safely.

Some types of rope are stored in a **rope bag** to protect them and to make it easy to deploy them. If a rope bag is used, the entire rope should be stored in the bag. If you leave any section of the rope outside of the bag, that section is exposed to the environment outside the bag and could be damaged by coming into contact with objects. Some rope bags have a hole in the bottom that rope can fit through. Some agencies may advise leaving a small length of rope outside through the hole, but this is not advisable as it exposes that section of rope to the elements.

Ropes stored in rope bags should be placed in the bag so that the rope is easy to deploy. That is, a firefighter should be able to reach into the rope bag, quickly find the end of the rope, and then pull to get the rope out of the bag without any snags. Avoid storing

SKILL DRILL 8-4

Placing a Life Safety Rope in a Rope Bag Firefighter I, NFPA 1010: 6.5.1

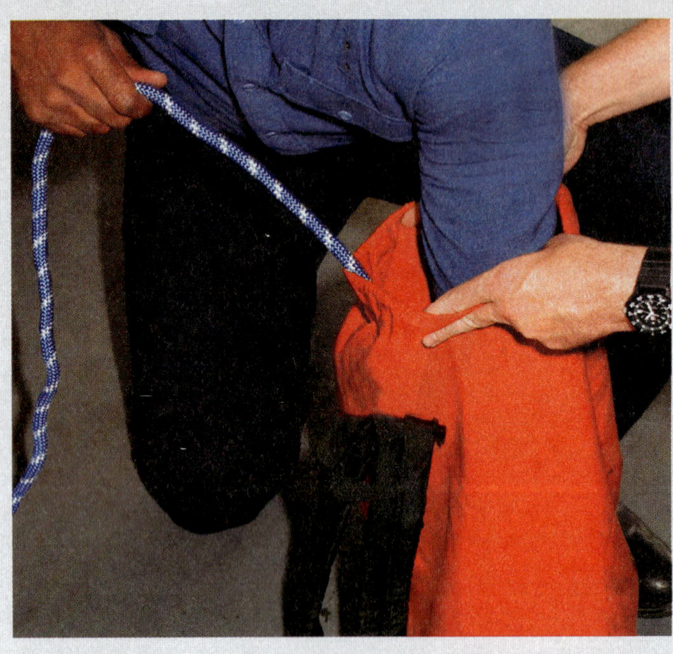

1. Inspect the rope to be certain it is fit for service. Stuff the life safety rope into the rope bag in a random pattern so it deploys properly. Do not try to coil the rope in the bag because this action will cause the rope to kink and become tangled when it is pulled out.
2. Return the rope and the rope bag to its proper location for storage or service. The ends of the rope should not protrude out of the bag.

ropes with pre-made knots in bags unless they are of the immediate pre-rigged deployable types such as throwlines, escape ropes, and fire escape ropes, which need to be ready for use immediately and are stored and rigged to do so.

Rope bags can be used to aid in identifying the type of rope stored in the bag. For example, some agencies store all of their utility ropes in black bags; life safety rope in orange, red, or yellow bags; and water rescue rope in blue bags. The color of the bag used can be any variation. The color of the bag does not matter as much as the fact that the agency has a policy so that all members know which rope is in which color bag. If an agency uses colored rope bags to help identify the type of rope, the rope should still have an identifying label on each end to identify it in.

Follow the steps in **SKILL DRILL 8-4** to place a life safety rope into a rope bag.

Manufacturers typically package rope on spools or rope is coiled and wrapped in plastic or in a sealed plastic bag to protect it during transport. Ropes are not typically coiled or spooled for storage when they are in service and stored on apparatus. Ropes stored coiled or on spools are typically ropes that an agency has recently purchased, is keeping in storage, or has not been issued for department use yet. Coiled and spooled ropes are exposed to the elements around them, so if these storage methods are used, extra consideration should be given to where they are stored so that they are not exposed to extreme temperatures or stored adjacent to objects that can damage them.

Knots

A **knot** is a prescribed way of fastening lengths of rope or webbing to objects or to each other. As a firefighter, you must know how to tie certain knots correctly and when to use each type of knot. Knot tying has often been described as a perishable skill. It cannot be emphasized enough that firefighters will need to practice tying and applying these knots frequently throughout their fire service. Practice, practice, practice!

Firefighters should be aware that knots decrease the load-carrying capacity of rope by a certain percentage. The reduction varies based on the type of knot, the type of rope, the method by which the rope is loaded, and the test itself. Therefore, providing a percentage reduction would be somewhat arbitrary, because it can vary from rope to rope and test to test. Many technical experts consider that the knot will reduce the strength of the rope by up to 50 percent, which still keeps their safety margin within a working range for most situations. With the exception of high-tensioned systems such as horizontal highlines utilized by technical rope rescuers, most firefighters do not need to be concerned with knots reducing the load-carrying capacity. Experienced rope technicians select the knot that best fits the situation and understand the safety margin calculated with respect to the load applied and the strength of the rope used. Knowing which knot to use for particular situation is more important than knowing the percentage of load-bearing reduction the knot will cause.

Instructions for tying knots refer to three parts of the rope (**FIGURE 8-14**):

- The **working end** is the part of the rope you hold as you form the knot.
- The **running end**, or standing end, is the other end of the rope, the part that is not used for forming the knot.
- The **standing part** is the part of the rope between the working end and the running end.

When a knot is tied, a bend is formed in the rope. Three types of bends are:

- A **bight**, which is formed by reversing the direction of the rope to form a U bend without crossing the rope so it has two parallel ends (**FIGURE 8-15**).
- A **loop**, which is formed by forming a circle in the rope by crossing the rope (**FIGURE 8-16**).
- A **round turn**, which is formed by making a loop and then bringing the two ends of the rope parallel to each other (**FIGURE 8-17**).

A firefighter should know how to tie and properly use the following common knots:

- Safety knot
- Figure eight
- Half hitch
- Clove hitch
- Figure eight on a bight
- Figure eight follow-through
- Bowline

FIGURE 8-14 The sections of a rope used in tying knots.

FIGURE 8-15 A bight.

FIGURE 8-16 A loop.

- Sheet bend (Becket bend)
- Figure eight bend
- Water knot (ring bend)

There are many ways to correctly tie each of these knots. Find the method that works for you and use it all the time. Your department might require that you learn how to tie other knots in addition to those listed here.

After you tie a knot, you should properly **dress** the knot by tightening and removing twists, kinks, and slack from the knot. The configuration of a properly dressed knot should be evident so that it can be easily inspected.

It is important to become proficient in tying knots. Knot-tying skills can be quickly lost without adequate practice (**FIGURE 8-18**). You never know when you will need to use these skills in an emergency situation. For added practice, try tying these knots with your gloves on or in darkness to simulate the conditions you will often encounter as a firefighter.

Safety Knot

The **safety knot** is used to back up or finish knots, such as the clove hitch, bowline, or sheet bend, that have a tendency to become loose when they are not continuously loaded or that can slip when loaded.

FIGURE 8-17 A round turn.

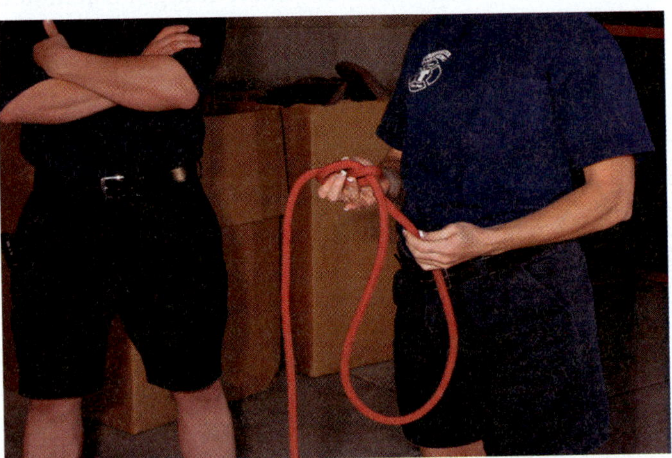

FIGURE 8-18 To maintain your knot-tying skills, practice tying different knots frequently.

Voice of Experience

Firefighters need to frequently practice skills that are used infrequently in order to excel during emergencies. Maintaining dexterity and improving your speed in accurately tying knots should be part of your daily readiness check, just like inspecting your personal protective equipment and self-contained breathing apparatus at the beginning of your shift. Practice—or the lack of practice—will quickly become evident during an emergency. Practice will enable you to swiftly and correctly tie the knot that prevents a car from going over an embankment or enables you to safely escape a burning room through the window. During emergencies, the time to get the job done is limited and practice will shave precious seconds off of your task time. If you see a length of rope hanging on the wall of your station, take advantage of the opportunity to practice.

James P. Kenney
Assistant Chief of Warwick, Rhode Island Fire Department (retired)
Rope Coordinator for the Connecticut State Fire Academy
Rhode Island

Safety knots are not needed on many of the knots commonly used in the fire service and by rope rescue operations- and technician-level personnel. When properly tied and dressed, these knots are secure on their own as long as you leave a 4- to 6-in. (10- to 15-centimeter [cm]) tail on the leftover working end of the rope. When a safety knot is used to finish or back up another knot, it should be tied close to the primary knot and be properly dressed. Using a safety knot is commonly termed to as "backing up" or "finishing the knot." The overhand knot is often used as a safety knot in the fire service for the knots that need them. Follow the steps in **SKILL DRILL 8-5** to tie an overhand knot as a safety knot.

SKILL DRILL 8-5

Tying an Overhand Knot as a Safety Knot Firefighter I, NFPA 1010: 6.1.2, 6.3.20

1. Take the end of the rope that exits the main knot, usually the shorter end of the rope near the accompanying main knot, and form a loop around the standing part of the rope as close as possible to the accompanying knot.

2. Pass the end of the rope through the loop.

3. Tighten the safety knot by pulling on both ends at the same time.

4. When complete, the safety knot should sit directly next to the accompanying knot.

Figure Eight Knot

A figure eight knot is used to prevent the end of a rope from passing through a device such as a pulley. For this reason, the figure eight knot is sometimes known as a stopper knot. When you pull the figure eight knot tight, it has the shape of a figure eight. This knot is the basis of other figure eight knots, including the figure eight on a bight, the figure eight follow-through, and the figure eight bend (all three of these are discussed later in this chapter). When a rope is stored in a rope bag, the figure eight knot is often loosely tied at the end of a rope to make it easy to identify the end of the rope inside the bag.

Follow the steps in **SKILL DRILL 8-6** to tie a figure eight knot.

SKILL DRILL 8-6

Tying a Figure Eight Knot Firefighter I, NFPA 1010: 6.1.2, 6.3.20

1. Form a loop in the rope.

2. Pass the end of the rope around the front of the rope.

3. Thread the end or the rope through the loop from the back.

4. Pull both ends of the rope in opposite directions to tighten the knot.

Hitches

A **hitch** is a knot that can be slipped over or tied around an object to secure the rope to the object and is easily untied when the object is removed. They are typically used on round or cylindrically shaped objects. Hitches are ideal for hoisting many fire service tools such as axes, pike poles, and hose lines. Other types of hitches can be used to adjust the tension in a rope that is used to secure or stabilize an object. The advantage of using hitches is that they can be tied and removed quickly. However, hitches can move or turn around the object they are tied to, so you should tie a safety knot to prevent the hitch from moving if that is a concern. There are many types of hitches, but half hitches and clove hitches are the types that firefighters use the most often on the fireground.

Half Hitch

The half hitch is a loop of rope passed around or over an object or around the rope itself. It is not a secure knot, so it is used only in conjunction with other knots. For example, when hoisting an axe or a pike pole, you use the half hitch to keep the rope aligned with the handle so the object won't swing on the rope and the entire object can be controlled throughout the hoist. On long objects, you might need to use several half hitches.

Follow the steps in **SKILL DRILL 8-7** to tie a half hitch around an object.

SKILL DRILL 8-7

Tying a Half Hitch Around an Object Firefighter I, NFPA 1010: 6.1.2, 6.3.20

1. Grab the rope with your palms facing away from you.

2. Rotate one hand so that the palm is facing you to form a loop in the rope.

Continues.

SKILL DRILL 8-7 CONTINUED

Tying a Half Hitch Around an Object Firefighter I, NFPA 1010: 6.1.2, 6.3.20

3. Pass the loop over the end of the object.

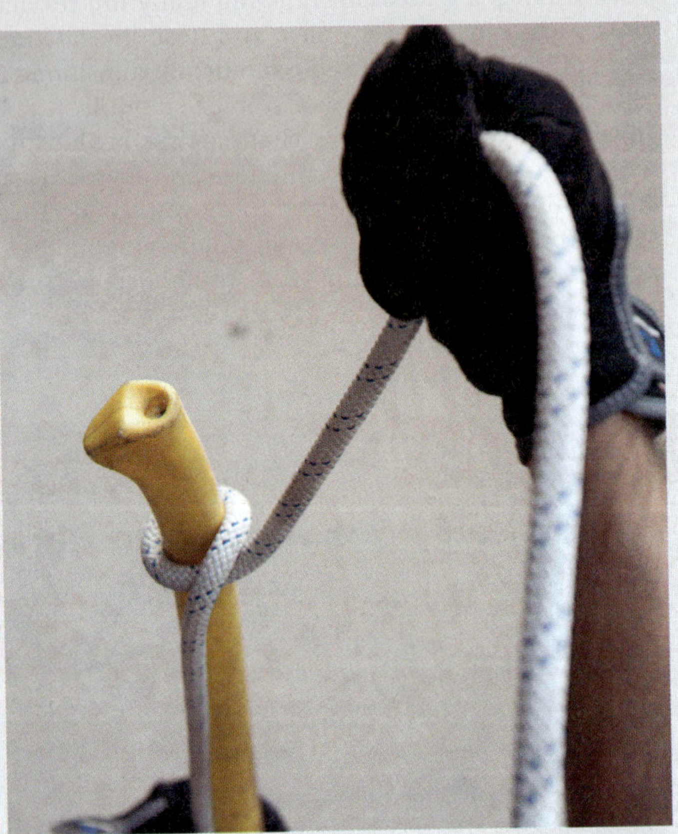

4. Finish the half-hitch knot by positioning it and pulling tight, maintaining tension on the rope.

Clove Hitch

A clove hitch is essentially two half hitches oriented in opposite directions. The clove hitch can be used to tie a hoisting rope around an axe or pike pole. A clove hitch can be tied anywhere in a rope and will hold equally well if tension is applied to either end of the rope or to both ends simultaneously.

There are two methods of tying this knot. A clove hitch tied in the open is used when the knot can be formed and then slipped over the end of an object, such as an axe or a pike pole. It is tied by making two consecutive loops in the rope. Follow the steps in **SKILL DRILL 8-8** to tie a clove hitch in the open.

SKILL DRILL 8-8

Tying a Clove Hitch in the Open Firefighter I, NFPA 1010: 6.1.2, 6.3.20

1. Thread the end of the rope through the loop from the back.

2. Holding on to the rope, uncross your hands to create a loop in each hand with the end of the rope in one hand hanging behind the rope between the two loops and the end of the rope in the other hand hanging in front of the rope between the two loops.

3. Place the loop with the end of the rope hanging in front of the rope between the two loops behind the other loop so the two loops are on top of each other.

4. Slide the loops over the end of the object. Pull the ends of the rope in opposite directions perpendicular to the object to tighten the knot. If needed, tie a safety knot in the working end of the rope.

A clove hitch may or may not require a safety knot to finish it. The type of object and where on the rope the knot is tied will dictate the need for a safety knot. For a round object such as a large tree or fence post, a clove hitch is tied around the object by passing the end of the rope around the object twice to form the clove hitch. A safety knot is used to finish the knot because the clove hitch will only have tension on one side of the rope and needs the additional stability of a safety knot. For objects with handles such as pike poles and axes, a clove hitch can be formed separately or in the open near the end of the rope and then slipped over the object. A safety knot is then used to finish the knot and a tag line is attached to guide the object during hoisting. When a clove hitch can be tied on an object using the middle of the rope, then a safety knot typically is not used. One end of the rope is used for hoisting and the other end is used to guide the object.

If the object is configured so the clove hitch cannot be slipped over one end, the same knot can be tied around the object by tying two half hitches. For example, if you need to tie a clove hitch around a tall tree, utility pole, or rung of a ladder, or to the rail of a litter basket, you can't create the loops and then slide them over the end of those objects. Follow the steps in **SKILL DRILL 8-9** to tie a clove hitch around an object.

SKILL DRILL 8-9

Tying a Clove Hitch Around an Object Firefighter I, NFPA 1010: 6.1.2, 6.3.20

1. Wrap the working end of the rope around the object.

2. Pass the working end of the rope in front of the standing part of the rope to create a loop, and then pass it over the object again, a short distance above or next to the first loop.

3. Pass the working end of the rope under the second loop so that it goes in front of the object between the two loops.

SKILL DRILL 8-9 CONTINUED

Tying a Clove Hitch Around an Object Firefighter I, NFPA 1010: 6.1.2, 6.3.20

4. Pull the ends of the rope in opposite directions perpendicular to the object to tighten the knot.

5. If needed, pass the working end of the rope over the object and then tie a safety knot close to the clove hitch around the standing part of the rope.

Loop Knots

A **loop knot** is used to form a loop in the end of a rope. Loops can be used to hoist tools, to secure a person during a rescue, or to secure a rope to a fixed object. When tied properly, these loop knots will not slip yet are easy to untie. Like hitches, there are many types of loop knots. Three common loop knots that firefighters need to know how to tie are the figure eight on a bight, the figure eight follow-through, and the bowline.

Figure Eight on a Bight

The figure eight on a bight knot creates a secure loop at the working end of a rope. This loop can be used to attach the end of the rope to a fixed object or a piece of equipment. Follow the steps in **SKILL DRILL 8-10** to tie a figure eight on a bight.

Figure Eight Follow-Through

A figure eight follow-through knot is used to create a secure loop at the end of the rope when the rope must be wrapped around an object or passed through an opening before the loop is formed. It is very useful for attaching a rope to a fixed ring or object that is closed at both ends like the handle to a power saw. Follow the steps in **SKILL DRILL 8-11** to tie a figure eight follow-through.

SKILL DRILL 8-10

Tying a Figure Eight on a Bight Firefighter I, NFPA 1010: 6.1.2, 6.3.20

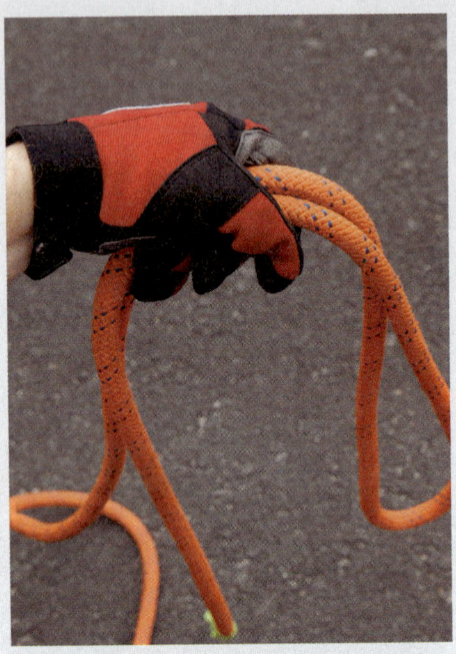

1. Double over a section of the rope to form a bight. Hold the two sides of the bight together as if they were one rope. The closed end of the bight becomes the working end of the rope.

2. Form a loop in the doubled section of the rope by passing the working end of the rope behind the standing part of the rope.

3. Pass the working end of the rope around the standing part of the rope and feed it through the bight from the back.

4. While holding onto the bight part of the working end, pull the ends of the rope that exit the knot, making sure you pull both pieces of rope evenly and maintaining the figure eight shape to form a secure loop.

SKILL DRILL 8-11

Tying a Figure Eight Follow-Through Firefighter I, NFPA 1010: 6.1.2, 6.3.20

1. Tie a simple figure eight knot in the rope, leaving enough rope between the knot and the working end to wrap around the object or pass through the opening and still have enough left to tie the follow through. Do not tighten the knot; leave it loose. Then thread the working end around the object or through the opening.

2. Pass the working end back through the original figure eight knot, following the path of the rope in the original knot in the opposite direction.

3. Once the end of the rope has been threaded through the entire knot, pull the knot tight and evenly in the same manner as to dress the figure eight on a bight.

Bowline

A bowline is also used to form a loop in the end of a rope. The bowline can be used as an alternate knot for a figure eight on a bight or figure eight. This type of knot is frequently used to secure the end of a rope to an object or an anchor point. It also can be used to hoist equipment. A bowline can be tied in the open or after you wrap the rope around something. One advantage to using a bowline is that it can be tied and untied quickly, which is obviously very useful at an emergency scene. The bowline can become loose when it is not under load. For this reason, a bowline needs to be finished or backed up with a safety knot. Follow the steps in **SKILL DRILL 8-12** to tie a bowline.

SKILL DRILL 8-12

Tying a Bowline Firefighter I, NFPA 1010: 6.1.2, 6.3.20

1. If you are tying a bowline in the open, form a small loop near one end of the rope with the part of the rope leading to the working end on the top of the loop. The point at which you make this small loop determines how big the bowline loop will be. If you are tying the bowline around an object, such as a tree, wrap the rope around the object. You should have a part of the rope in each hand, one from the working end and the other going to the standing end. Form a loop of the rope in the rope going toward the standing end with the part of the rope leading to the working end on top of the loop.

2. Thread the working end up through the loop from the bottom.

3. Pass the working end under the standing part.

SKILL DRILL 8-12 CONTINUED

Tying a Bowline Firefighter I, NFPA 1010: 6.1.2, 6.3.20

4. Pass the working end over the standing part and then back down through the same small loop.

5. Tighten the knot by pulling the working end and the standing or running end in opposite directions.

6. Tie a safety knot in the working end of the rope. The overhand safety knot gets tied around the rope that forms the loop.

Bend Knots

A **bend knot** is used to join two ropes or pieces of webbing together. These knots also can be used to join a rope to a chain or a cable. The three types of bend knots a firefighter will most often use are the sheet bend, the figure eight bend, and the water knot.

Sheet Bend

The sheet bend (or Becket bend) is used to join two ropes of unequal diameters. A sheet bend is also used to join a rope to a chain or to hoist a rope of a different diameter. It should not be relied upon to support a human load. Follow the steps in **SKILL DRILL 8-13** to tie a sheet bend.

SKILL DRILL 8-13

Tying a Sheet Bend Firefighter I, NFPA 1010: 6.1.2, 6.3.20

1. Form a bight at the working end of one rope (the blue rope) and position your finger along the side of the bight. If the ropes are of unequal diameters, form the bight in the larger rope.

2. Thread the working end of the second rope (the red rope) up through the bight from the back side.

SKILL DRILL 8-13 CONTINUED

Tying a Sheet Bend Firefighter I, NFPA 1010: 6.1.2, 6.3.20

3. Wrap the second rope (the red rope) completely around your finger and the top side of the bight (the blue rope) and continue wrapping all the way around the bight. Thread the working end of the second rope (the red rope) between the original bight and under the second rope.

4. Hold the bight on the first rope (the blue rope) in one hand and the running end of the second rope (the red rope) in the other hand, then pull in opposite directions to tighten the knot.

5. Tie a safety knot in the working end of each rope.

Figure Eight Bend

The figure eight bend (or tracer eight) is used to join two ropes that have a similar diameter. When used with life safety rope, this knot can support a human load. Follow the steps in **SKILL DRILL 8-14** to tie a figure eight bend.

Water Knot

The water knot (or ring bend) is used to join webbing of the same or different widths together. When a single piece of webbing is used and the opposite ends are tied to each other, a loop or sling is created. The water knot could also be used to make to several shorter lengths into a single longer length. Follow the steps in **SKILL DRILL 8-15** to tie a water knot.

SKILL DRILL 8-14

Tying a Figure Eight Bend Firefighter I, NFPA 1010: 6.1.2, 6.3.20

1. Tie a figure eight near the end of one rope.

2. Thread the working end of the second rope through the knot from the opposite end and follow the path of the rope in the original knot in the opposite direction.

SKILL DRILL 8-14 CONTINUED

Tying a Figure Eight Bend Firefighter I, NFPA 1010: 6.1.2, 6.3.20

3. Pull the knot tight.

SKILL DRILL 8-15

Tying a Water Knot Firefighter I, NFPA 1010: 6.1.2, 6.3.20

1. At one end of the webbing, approximately 6 in. (152 mm) from the end, tie an overhand knot. Do not tighten this knot, keep it loose.

Continues.

SKILL DRILL 8-15 CONTINUED

Tying a Water Knot Firefighter I, NFPA 1010: 6.1.2, 6.3.20

2. Thread the other end of the webbing through the knot and follow the path of the webbing in the original knot in the opposite direction. Thread it through the knot until approximately 6 in. (152 mm) is poking out from the other side of the knot.

Hoisting

Tying knots and using rope to hoist equipment in emergency situations is a skill at which you need to become proficient. In emergency situations, you may need to raise or lower a tool or equipment to other firefighters. When you gain more knowledge and experience, you may need to hoist a vehicle or people using ropes or cables with other devices such as pulleys and winches. In all of these situations, it is essential that the correct knots are used, tied correctly, and secured, and that the movement of the objects is controlled throughout the hoist.

It is important for you to learn how to raise and lower an axe, a pike pole, a ladder, a charged hose line, an uncharged hose line, an exhaust fan, power tools, and any other item your department uses or carries. When you are hoisting or lowering a tool or any object, make sure no one is standing under the object in case there is a failure and it falls. Also keep the scene clear of people to avoid any chance of an accident.

When hoisting a large object, such as an exhaust fan, attach a **tag line**—a separate rope that personnel on the ground use to guide an object that is being hoisted or lowered—to the object at the opposite end or corner from where the hoisting rope is secured to the object. This is used to help control the path of the object and prevent it from spinning or getting caught on an obstruction, such as an electrical line, a utility pole, or a structural feature. The tag line should be attached with a knot that is easily untied, such as a bowline, so that it can be removed easily to allow the equipment to be placed into service quickly.

To hoist an object, one firefighter carries a utility rope up to the roof, an upper floor, or higher elevation; secures one end of the rope; and then tosses the other end of the rope down to a firefighter on the ground. If possible, the firefighter at the higher elevation should secure the end they are holding to an anchor point, but if there is nothing stable to use as an anchor point, that firefighter will need to hold tightly to their end of the rope. The firefighter on the ground then ties the tool or piece of equipment to the utility rope using the appropriate knot. If needed, the firefighter on the ground also ties a tag line to the piece of equipment. Then the firefighter on the ground signals the firefighter at the higher elevation, and the firefighter at the higher elevation begins pulling on the rope to hoist the tool or piece of equipment. If there is a tag line attached, the firefighter on the ground holds the end of the tag line to help control the object on its way up.

To lower a tool or piece of equipment, essentially the reverse happens. The firefighter at the higher elevation ties the utility rope to the object and secures that end of the rope. If a tag line is needed, the firefighter at the higher elevation ties a tag line to the object. The firefighter at the higher elevation tosses the other ends of the utility rope and the tag line down to the firefighter on the ground. The firefighter at the higher elevation feeds the rope out to lower the object, ensuring that a reserve of rope remains on the roof. For example, if the utility rope is 150 ft (46 m) and the distance to lower the tool is 40 ft (12 m), then the firefighter would ensure that at least 100 ft (30 m) of rope remains on the roof. If a tag line is used, the firefighter on the ground helps control the descent.

Hoisting an Axe

An axe should be hoisted in a vertical position with the head of the axe lower than the handle. After receiving the end of the utility rope from the firefighter above, follow the steps in **SKILL DRILL 8-16** to hoist an axe.

SKILL DRILL 8-16

Hoisting an Axe Firefighter I, NFPA 1010: 6.1.2, 6.3.20

1. Tie the end of the hoisting rope around the handle of the axe near the head using a figure eight on a bight. Slip the knot down the handle from the end to the head.

2. Pass the loop under the head.

3. Place the standing part of the rope parallel to the axe handle.

4. Tie a half hitch along the axe handle to keep the handle parallel to the rope.

Continues.

SKILL DRILL 8-16 CONTINUED

Hoisting an Axe Firefighter I, NFPA 1010: 6.1.2, 6.3.20

5. Add a second half hitch along the axe handle if necessary.

6. Communicate with the firefighter above that the axe is ready to raise. To release the axe, hold the middle of the handle, release the half hitches, and slip the knot up and off.

Hoisting a Pike Pole

A pike pole is hoisted in a vertical position with the head at the top. After receiving the end of the utility rope from the firefighter above, follow the steps in **SKILL DRILL 8-17** to hoist a pike pole.

Hoisting a Ladder

A ladder is hoisted in a vertical position. A tag line attached to the bottom helps keep the ladder under control as it is hoisted. If it is a roof ladder, the hooks on the ladder should be in the retracted position. After receiving the end of the utility rope from the firefighter above, follow the steps in **SKILL DRILL 8-18** to hoist a ladder.

SKILL DRILL 8-17

Hoisting a Pike Pole Firefighter I, NFPA 1010: 6.1.2, 6.3.20

1. Place a clove hitch over the bottom of the handle and secure it close to the bottom of the handle, finishing it with a safety knot or leaving enough length of rope below the clove hitch to be used as a tag line while raising the pike pole.

2. Place a half hitch around the middle of handle about halfway up the handle to keep the rope parallel to the handle.

3. Place a second half hitch over the handle, and then secure it under the head of the pike pole.

4. Communicate with the firefighter above that the pike pole is ready to raise.

SKILL DRILL 8-18

Hoisting a Ladder Firefighter I, NFPA 1010: 6.1.2, 6.3.20

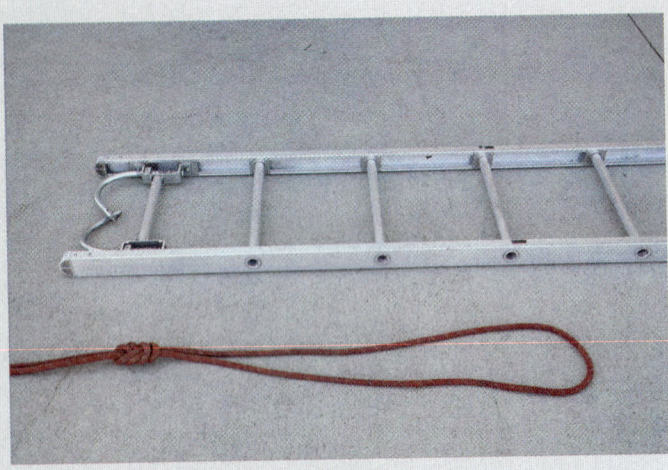

1. Tie a figure eight on a bight to create a loop approximately 3 to 4 ft (1 to 1.2 m) in diameter and large enough to fit around both ladder beams (the long vertical sides of the ladder).

2. Starting at the top of the ladder, pass the loop under the rungs, between the beams, until the loop is three or four rungs from the top of the ladder. Pull the end of the loop from under the rungs and toward the top of the ladder.

3. Place the loop over the top tip of the ladder, over the outside of both beams.

SKILL DRILL 8-18 CONTINUED

Hoisting a Ladder Firefighter I, NFPA 1010: 6.1.2, 6.3.20

4. Pull the running end of the rope to remove the slack from the rope.

5. Attach a tag line to the bottom rung of the ladder to stabilize it as it is being hoisted.

6. Communicate with the firefighter above that the ladder is ready to be raised. Hold on to the tag line to stabilize the bottom of the ladder as it is being hoisted.

Hoisting a Charged Hose Line

A **charged hose line** is a hose line filled with water and under pressure from the pump. It is almost always preferable to hoist a dry or uncharged hose line because water adds considerable weight to a hose line. Water weighs 8.33 lb per gallon (3.79 kg per liter), which makes hoisting a charged line much more difficult. For example, a 50-ft (15-m) dry hose that is 2½-in. (64-mm) in diameter weighs about 30 lb (14 kg). When that hose is charged and filled with water, it weighs about 140 lb (64 kg). However, there will be occasions when it is necessary to hoist a charged hose line. After receiving the end of the utility rope from the firefighter above, follow the steps in **SKILL DRILL 8-19** to hoist a charged hose line.

SKILL DRILL 8-19

Hoisting a Charged Hose Line Firefighter I, NFPA 1010: 6.1.2, 6.3.20

1. Make sure that the nozzle shut-off handle is in the forward position so that it is completely closed.

2. Tie a clove hitch, 1 to 2 ft (30 to 60 cm) behind the nozzle. Add a safety knot to secure the loose end of the rope on the side of the clove hitch away from the nozzle.

3. At the end of the rope closest to the nozzle, make a bight in the rope even with the nozzle shut-off handle. Insert the bight through the handle opening.

4. Slip the bight loop over the end of the nozzle.

SKILL DRILL 8-19 CONTINUED

Hoisting a Charged Hose Line Firefighter I, NFPA 1010: 6.1.2, 6.3.20

5. Pull the working end of the rope to pull the bight tight. This creates a half hitch and secures the handle in the off position while the charged hose line is hoisted.

6. Communicate with the firefighter above that the hose line is ready to hoist. The knot can be released after the line is hoisted by removing the tension from the rope and slipping the bight back over the end of the nozzle.

Hoisting an Uncharged Hose Line

An **uncharged hose line** is a hose line that is not filled with water and not under pressure from a pump. Before hoisting an uncharged hose line, fold the hose back on itself and place the nozzle on top of the hose about 3 ft (1 m) from the bend in the hose that you created by folding the hose. This action eliminates unnecessary stress on the couplings by ensuring that the rope pulls on the hose and not directly on the nozzle. After receiving the end of the utility rope from the firefighter above, follow the steps in **SKILL DRILL 8-20** to hoist an uncharged hose line.

Hoisting an Exhaust Fan or Power Tool

Several types of tools and equipment, such as exhaust fans, chainsaws, circular saws, and any other object that has a strong, closed handle, are hoisted using the same technique. The hoisting rope is secured to the object by passing the rope through the opening in the handle. A figure eight follow-through knot or bowline with a safety knot is then used to close the loop.

You should practice hoisting the actual tools and equipment used in your department. You should be able to perform this task automatically and in adverse conditions.

Remember that you always use utility rope for hoisting tools. You do not want to get oil or grease on designated life safety ropes. If a life safety rope becomes oily or greasy, it should be taken out of service and destroyed so that it will not be used again mistakenly as a life safety rope. The damaged rope can be cut into short lengths and designated and downgraded to be used as utility rope.

After receiving the end of the utility rope from the firefighter above, follow the steps in **SKILL DRILL 8-21** to hoist an exhaust fan.

SKILL DRILL 8-20

Hoisting an Uncharged Hose Line Firefighter I, NFPA 1010: 6.1.2, 6.3.20

1. Fold about 3 ft (1 m) of hose back on itself and place the nozzle on top of the hose.

2. Tie a clove hitch around both the nozzle and the hose it is laying on top of, and then tighten the knot securely.

3. Tie a safety knot around the nozzle and the hose below the clove hitch.

4. Tie two half hitches around the folded-over hose, evenly spaced between the nozzle and the bend in the hose created when you folded the hose.

SKILL DRILL 8-20 CONTINUED

Hoisting an Uncharged Hose Line Firefighter I, NFPA 1010: 6.1.2, 6.3.20

5. Communicate with the firefighter above that the hose line is ready to hoist. Hoist the hose with the fold at the top and the nozzle pointing down. Before releasing the rope, the firefighters at the top must pull up enough hose so that the weight of the hanging hose does not drag down the hose.

SKILL DRILL 8-21

Hoisting an Exhaust Fan or Power Tool Firefighter I, NFPA 1010: 6.1.2, 6.3.20

1. Tie a figure eight knot in the rope about 3 ft (1 m) from the working end of the rope. Loop the working end of the rope around the fan handle and back to the figure eight knot.

2. Secure the rope by tying a figure eight follow-through.

Continues.

SKILL DRILL 8-21 CONTINUED

Hoisting an Exhaust Fan or Power Tool Firefighter I, NFPA 1010: 6.1.2, 6.3.20

3. Attach a tag line to the opposite corner on the same side of the fan for better control.

4. Communicate with the firefighter above that the exhaust fan is ready to hoist.

Choosing the Appropriate Knot

Now that we have discussed the basic knots and hitches used in the fire service, you may be asking, with so many options, how do I choose which knot to use? Remember, task objectives drive tactics. For example, the task objective is to get a particular tool to a location on the fireground as fast and safely as possible so it can be deployed as quickly as possible. With the length of rope available and the type of tool, you must decide which knot or hitch will accomplish that task.

How do you choose? You draw upon your knowledge to match the right solution to the issue at hand. For example, you need to hoist a pike pole to the second floor of a structure so the salvage and overhaul team can carry out their operations. You know that clove hitches are ideal for tools that are cylindrical such as an axe or a pike pole and you apply that solution. If we changed the tool to a chainsaw, then you would use a bowline or figure eight follow-through because there is a handle to tie the knot around. Keep the basics in mind and use that knowledge to make swift and accurate decisions.

TIP

The fire and rescue business is dynamic; this is why firefighters need to constantly train, read, study, and stay current with new techniques and standards. Learning about ropes and knots will not end with this chapter.

CASE STUDY

You Are the Firefighter CONCLUSION

You and your partner are on the fourth floor of a multistory apartment building and find yourself trapped by a rapidly progressing fire. You get to a bedroom at the front of the building and know your only escape route is out the window. You don't have time to wait for the aerial ladder to get to you, so you prepare for self-rescue. You pull out your rope and begin the most important task you have ever done in your life.

1. **What type of rope should you use for this type of activity?**

   Answer: You should use a fire escape rope because you are in a hazardous environment that is involved with fire or fire products.

2. **How many people is this type of rope designed to support?**

   Answer: Fire escape rope, like all escape rope, is designed to carry the weight of only one person.

3. **Now that you used your fire escape rope in a fire environment, what should you do next?**

   Answer: Immediately report the situation to the appropriate personnel, make sure the rope is replaced, and update the rope record to accurately reflect the incident.

WRAP-UP

SUMMARY

KNOWLEDGE OBJECTIVES

- List the four primary types of fire service rope and describe the characteristics of each type.
  - Describe the four primary types of fire service rope. (**NFPA 1010: 6.3.20**, pp. 272–273)
  - List the two types of life safety rope and their minimum breaking strength. (**NFPA 1010: 6.3.20**, p. 273)
  - Describe the characteristics of escape rope and fire escape rope. (**NFPA 1010: 6.3.20**, p. 274)
  - Describe the characteristics of water rescue throwlines. (**NFPA 1010: 6.3.20**, p. 274)
  - Describe the characteristics of utility ropes. (**NFPA 1010: 6.3.20**, p. 274)
  - Describe the characteristics of webbing. (**NFPA 1010: 6.3.20**, pp. 274–276)
- List the types of materials rope can be made from and describe the characteristics of each material.
  - List the disadvantages of natural fiber ropes. (**NFPA 1010: 6.3.20**, p. 276)
  - List the advantages of synthetic fiber ropes. (**NFPA 1010: 6.3.20**, p. 276)
  - List the disadvantages of synthetic fiber ropes. (**NFPA 1010: 6.3.20**, p. 276)
  - List common types of synthetic fibers that are used in fire service rope. (**NFPA 1010: 6.3.20**, p. 276)
- List three types of rope construction and describe the method in which they are constructed.
  - Describe how twisted ropes are constructed. (**NFPA 1010: 6.3.20**, p. 277)
  - Describe how braided ropes are constructed. (**NFPA 1010: 6.3.20**, p. 277)
  - Describe how kernmantle ropes are constructed. (**NFPA 1010: 6.3.20**, p. 278)
  - Explain the differences between dynamic kernmantle rope and static kernmantle rope. (**NFPA 1010: 6.3.20**, p. 278)

KNOWLEDGE OBJECTIVES CONTINUED

- List the four parts of rope maintenance and the key components of each part.
  - List the four components of the rope maintenance formula. (**NFPA 1010: 6.3.20**, p. 278)
  - Describe how to preserve rope strength and integrity. (**NFPA 1010: 6.3.20**, pp. 278–284)
  - Describe how to clean rope. (**NFPA 1010: 6.5.1**, pp. 279–280)
  - Describe how to inspect rope. (**NFPA 1010: 6.5.1**, pp. 280–282)
  - Describe how to store rope properly. (**NFPA 1010: 6.5.1**, pp. 282–284)
  - Describe how to keep an accurate rope record. (**NFPA 1010: 6.5.1**, pp. 280–282)
- Understand rope terminology and describe the 10 common knots and hitches used in the fire service.
  - List the terminology used to describe the parts of a rope when tying knots. (**NFPA 1010: 6.3.20**, pp. 284–286)
  - List the terminology used to describe the bends in rope that are formed when a knot is tied. (**NFPA 1010: 6.3.20**, pp. 284–286)
  - List the 10 common types of knots that are used in the fire service. (**NFPA 1010: 6.3.20**, pp. 285–286)
  - Describe the characteristics of a safety knot. (**NFPA 1010: 6.3.20**, pp. 286–287)
  - Describe the characteristics of a figure eight knot. (**NFPA 1010: 6.3.20**, p. 288)
  - Describe the characteristics of a half hitch. (**NFPA 1010: 6.3.20**, p. 289)
  - Describe the characteristics of a clove hitch. (**NFPA 1010: 6.3.20**, p. 290)
  - Describe the characteristics of a figure eight on a bight. (**NFPA 1010: 6.3.20**, pp. 293–294)
  - Describe the characteristics of a figure eight follow-through. (**NFPA 1010: 6.3.20**, p. 293)
  - Describe the characteristics of a bowline. (**NFPA 1010: 6.3.20**, p. 295)
  - Describe the characteristics of a sheet bend. (**NFPA 1010: 6.3.20**, p. 298)
  - Describe the characteristics of a figure eight bend. (**NFPA 1010: 6.3.20**, p. 300)
  - Describe the characteristics of a water knot. (**NFPA 1010: 6.3.20**, p. 300)
- Describe how to use rope to safely hoist tools and equipment.
  - Describe the methods used to hoist tools or equipment. (**NFPA 1010: 6.3.20**, pp. 302–312)

SKILLS OBJECTIVES

- Demonstrate how to care for, clean, inspect, and place life safety rope in a bag.
  - Care for fire department ropes. (**NFPA 1010: 6.5.1**, p. 280)
  - Clean fire department ropes. (**NFPA 1010: 6.5.1**, p. 281)
  - Inspect fire department ropes. (**NFPA 1010: 6.5.1**, p. 283)
  - Place a life safety rope in a rope bag. (**NFPA 1010: 6.5.1**, p. 284)
- Tie the 10 common knots used in the fire service.
  - Tie a safety knot. (**NFPA 1010: 6.1.2, 6.3.20**, p. 287)
  - Tie a figure eight knot. (**NFPA 1010: 6.1.2, 6.3.20**, p. 288)
  - Tie a half hitch. (**NFPA 1010: 6.1.2, 6.3.20**, pp. 289–290)
  - Tie a clove hitch in the open. (**NFPA 1010: 6.1.2, 6.3.20**, p. 291)
  - Tie a clove hitch around an object. (**NFPA 1010: 6.1.2, 5.3.20**, pp. 292–293)
  - Tie a figure eight on a bight. (**NFPA 1010: 6.1.2, 6.3.20**, p. 294)
  - Tie a figure eight follow-through. (**NFPA 1010: 6.1.2, 6.3.20**, p. 295)
  - Tie a bowline. (**NFPA 1010: 6.1.2, 6.3.20**, pp. 296–297)
  - Tie a sheet bend. (**NFPA 1010: 6.1.2, 6.3.20**, pp. 298–299)
  - Tie a figure eight bend. (**NFPA 1010: 6.1.2, 6.3.20**, pp. 300–301)
  - Tie a water knot. (**NFPA 1010: 6.1.2, 6.3.20**, pp. 301–302)

- Tie appropriate knots onto firefighting equipment and hoist them safely.
  - Hoist an axe. (**NFPA 1010: 6.1.2, 6.3.20**, pp. 303–304)
  - Hoist a pike pole. (**NFPA 1010: 6.1.2, 6.3.20**, p. 305)
  - Hoist a ladder. (**NFPA 1010: 6.1.2, 6.3.20**, pp. 306–307)
  - Hoist a charged hose line. (**NFPA 1010: 6.1.2, 6.3.20**, pp. 308–309)
  - Hoist an uncharged hose line. (**NFPA 1010: 6.1.2, 6.3.20**, pp. 310–311)
  - Hoist an exhaust fan or power tool. (**NFPA 1010: 6.1.2, 6.3.20**, pp. 311–312)

KEY TERMS

bend A knot that joins two ropes or webbing pieces together. (NFPA 2500)

bight The open loop in a rope or piece of webbing formed when it is doubled back on itself. (NFPA 1006)

block creel construction Rope constructed without knots or splices in the yarns, ply yarns, strands or braids, or rope. (NFPA 2500)

braided rope Rope constructed by intertwining strands in the same way that hair is braided.

charged hose line A hose line filled with water and under pressure from the pump.

dress To tighten and remove twists, kinks, and slack from the rope after tying a knot.

dynamic rope A rope typically used for climbing that is designed to be elastic and stretch when loaded.

escape rope Rope dedicated solely for the purpose of supporting people during emergency self-escape (self-rescue); not intended for use in a hazardous environment involving fire or fire products; not classified as a life safety rope. (NFPA 2500)

fire escape rope Rope dedicated solely for the purpose of supporting people during emergency self-escape (self-rescue) from an immediately hazardous environment involving fire or fire products; not classified as a life safety rope. (NFPA 2500)

general use life safety rope A life safety rope with a diameter that is at least 7/8 in. (11 mm) but not larger than 5/8 in. (16 mm), with a minimum breaking strength of 8992 lbf (40 kN).

highline system A system of using rope or cable suspended between two points for movement of persons or equipment over an area that is a barrier to the rescue operation, including systems capable of movement between points of equal or unequal height. (NFPA 1006)

hitch A knot that attaches to or wraps around an object so that when the object is removed, the knot will fall apart. (NFPA 2500)

kern In a kernmantle rope, the center or core of the rope.

kernmantle rope Rope made of two parts, the kern and the mantle.

knot A fastening made by tying rope or webbing in a prescribed way. (NFPA 2500)

laid rope See *twisted rope.*

life safety rope Rope dedicated solely for the purpose of supporting people during rescue, firefighting, other emergency operations, or during training evolutions. (NFPA 2500)

loop A piece of rope formed into a circle by crossing the rope.

loop knot A knot that forms a secure loop in the end of a rope.

mantle In a kernmantle rope, the braided covering that protects the kern from dirt and abrasion. Also called *sheath.*

minimum breaking strength (MBS) The result of subtracting three standard deviations from the mean result of the lot being tested using the formula in 28.2.5.2. (NFPA 2500)

rigging The process of building a system to move or stabilize a load. (NFPA 1006)

rope A compact but flexible, torsionally balanced, continuous structure of fibers produced from strands that are twisted, plaited, or braided together, and that serve primarily to support a load or transmit a force from the point of origin to the point of application. (See also 3.3.153.2, Life Safety Rope.) (NFPA 1006)

rope bag A bag used to protect and store rope so that the rope can be easily and rapidly deployed without kinking.

rope record A record for each piece of rope that includes a history of when the rope was placed in service, when it was inspected, when and how it was used, and which types of loads were placed on it.

KEY TERMS CONTINUED

round turn A piece of rope looped to form a complete circle with the two ends parallel.

running end The part of a rope that is not used to form a knot.

safety knot A knot used to back up another knot that has a tendency to become loose when not continuously loaded by or that can slip when loaded by securing the leftover working end of the rope.

sheath See *mantle.*

shock load An instantaneous load that places a rope under extreme tension, such as when a falling load is suddenly stopped as the rope becomes taut.

standing part The part of a rope between the working end and the running end.

static rope A rope that stretches very little under load.

tag line A rope that personnel on the ground can use to guide an object that is being hoisted or lowered.

technical use life safety rope A life safety rope with a diameter that is at least 3/8 in. (9.5 mm) but not larger than 1/2 in. (12.5 mm), with a minimum breaking strength of 4496 lbf (20 kN) and that is used by highly trained rescue teams who deploy to technical environments such as mountainous or wilderness terrain.

throw bag A water rescue system that includes 50 ft to 75 ft (15.24 m to 22.86 m) of water rescue rope, an appropriately sized bag, and a closed-cell foam float. (NFPA 1006)

throwline A floating rope that is intended to be thrown to a person during water rescues or as a tether for rescuers entering the water. (NFPA 2500)

trailer A label that travels the entire length of a life safety rope under the outer sheath that identifies the rope as a life safety rope.

twisted rope Rope constructed of fibers twisted into strands, which are then twisted together. Also called *laid rope.*

uncharged hose line A hose line that is not filled with water and not under pressure from a pump.

utility rope Rope used for securing objects, hoisting equipment, or securing a scene to prevent bystanders from being injured; utility rope must never be used in life safety operations.

webbing Woven material of flat or tubular weave in the form of a long strip. (NFPA 2500)

working end The part of the rope used for forming a knot.

REVIEW QUESTIONS

1. What are the four primary types of ropes used in the fire service?
2. What type of fiber is commonly used for fire service rope?
3. What are the three main rope construction?
4. What type of construction is required for all life safety rope?
5. What are the four parts of the rope maintenance formula?
6. What does a knot do to the load-carrying capacity of the rope?
7. When hoisting equipment, what should you use to guide it and prevent it from striking obstructions?

DISCUSSION QUESTIONS

1. Describe the characteristics of each of the four types of rope.
2. Describe the four parts of rope maintenance and explain why each part is necessary.
3. Describe some key factors to consider when using rope to hoist tools or equipment.

APPLYING THE CONCEPTS

You are on the scene of an active fire in a five-story residential building. Several units have already been dispatched inside the structure for firefighting operations. An engine company calls for a hose line to be hoisted to the third floor. To facilitate this, a firefighter lowers a rope out of a third-floor window that has heavy smoke venting from the top of it.

1. What type of rope did the firefighter lower and why?
2. What are the pros and cons of using an uncharged hose line?
3. What are the pros and cons of using a charged hose line?

Given the urgency to get water to the firefighter as quickly as possible, you and your partner retrieve a charged hose line. You open the nozzle and bleed off the air, then close the nozzle completely. During your spare time at the station house you have been practicing knot tying and hoisting. This is your first opportunity to apply these skills in an actual fire situation.

4. What knot should you use to hoist a hose line and where should it be tied?
5. What other knots should be used to securely attach the hose line?
6. How do you make sure the handle stays in the off position while the charged line is being hoisted?

There is a large oak tree outside the third-floor window. Your partner surveys the area and says, "We gotta watch out for the tree branches . . . and that light post. The hoseline or rope could get caught up in them! Do you think we should attach a tag line?"

7. Is it necessary to attach a tag line to the charged hoseline?

You determine a tag line won't be needed. Your partner clears bystanders away from the hoisting area so no one is struck as the hoseline is being lifted or injured if it falls. The hoseline is ready to be hoisted.

8. Do you wait for the firefighter above to start hoisting the hoseline?

REFERENCES

Reproduced from National Fire Protection Association (NFPA). 2020. *NFPA 1006, Standard for Technical Rescue Personnel Professional Qualifications.* 2021 Edition. Quincy, MA: NFPA.

National Fire Protection Association (NFPA). 2021. *NFPA 2500, Standard for Operations and Training for Technical Search and Rescue Incidents and Life Safety Rope and Equipment for Emergency Services.* 2022 Edition. Quincy, MA: NFPA.

CHAPTER

9

Firefighter I

Portable Fire Extinguishers

KNOWLEDGE OBJECTIVES

After studying this chapter, you will be able to:

- Explain the purpose, classification, and rating system of fire extinguishers and how portable fire extinguishers suppress fires.
- Define the classes of fire and select the type of fire extinguisher used for each.
- State the classification of fire extinguishers, the method used to rate each class, and the labeling system used to identify each.
- Describe how the placement of fire extinguishers is determined, including how to estimate fire load while being able to discuss the three area hazard classifications.
- Describe how fire extinguishers are designed and their basic components.
- Describe the types of agents and operating systems used in fire extinguishers and fire extinguishing systems and give examples of the steps to correctly operate a fire extinguisher.
- Explain the basic steps of inspecting, maintaining, recharging, and hydrostatic testing of fire extinguishers.

SKILLS OBJECTIVES

After studying this chapter, you will be able to perform the following skills:

- Inspect a fire extinguisher using an inspection checklist.
- Locate and transport the fire extinguisher to the location of the fire.
- Identify the class of fire and extinguish the fire with the correct type of extinguisher and extinguishing agent.
- Correctly operate each type of fire extinguisher.
- Fill and pressurize a stored pressure water extinguisher.

ADDITIONAL NFPA STANDARDS

- **NFPA 1**, *Fire Code, 2021 Edition*
- **NFPA 10**, *Standard for Portable Fire Extinguishers, 2022 Edition*
- **NFPA 11**, *Standard for Low-, Medium-, and High-Expansion Foam, 2021 Edition*
- **NFPA 13**, *Standard for the Installation of Sprinkler Systems, 2022 Edition*
- **NFPA 408**, *Standard for Aircraft Hand Portable Fire Extinguishers, 2022 Edition*
- **NFPA 440**, *Guide for Aircraft Rescue and Firefighting Operations and Airport/Community Emergency Planning, 2024 Edition*
- **NFPA 557**, *Standard for Determination of Fire Loads for Use in Structural Fire Protection Design, 2023 Edition*
- **NFPA 1660**, *Standard for Emergency, Continuity, and Crisis Management: Preparedness, Response, and Recovery, 2024 Edition*
- **NFPA 1700**, *Guide for Structural Fire Fighting, 2021 Edition*
- **NFPA 1900**, *Standard for Aircraft Rescue and Firefighting Vehicles, Automotive Fire Apparatus, Wildland Fire Apparatus, and Automotive Ambulances, 2024 Edition*

CASE STUDY

You Are the Firefighter

You are the designated on-duty cook for the day, and the kitchen is bare. You go into the grocery store while the rest of your crew waits in the apparatus. As you make your way through the produce section, an employee runs up to you and says that there is a fire in the back room. You quickly radio your officer to inform them of the situation, and then immediately proceed to the back room. You locate a small fire in a trash can that is just starting to spread. You quickly scan the room looking for a fire extinguisher. Not seeing one, you ask the employee where the closest extinguisher is located.

1. What type of fire extinguisher would be most effective for this fire? Why?
2. As the fire grows, at what point will using a fire extinguisher become ineffective?
3. How do fire inspectors ensure the fire extinguisher will work when needed?

Introduction

Portable fire extinguishers are required in many types of occupancies as well as in commercial vehicles, boats, aircraft, and various other locations. Fire prevention efforts encourage citizens to learn how to use and keep fire extinguishers in their homes, particularly in their kitchens. Fire extinguishers have been used successfully to put out small fires, preventing millions of dollars in property damage as well as saving lives. Most fire extinguishers are easy to operate and can be used effectively by an individual with only basic training. Firefighters often provide fire extinguisher training for the public, so it is essential that you understand the characteristics and operations of each type of fire extinguisher. You need to be able to select the most appropriate extinguisher to use for different types of fires; you also need to know which extinguishers must not be used for certain fires. (The selection of the proper fire extinguisher builds on the information presented in Chapter 5, *Fire Behavior.*) You must be able to operate the most common types of portable fire extinguishers correctly and effectively to reduce the risk of personal injury and property damage, and know how to inspect and maintain them.

Purpose of a Fire Extinguisher

Fire extinguishers range in size from models that can be operated with one hand to large, wheeled models that contain several hundred pounds of **extinguishing agent** (material used to stop the combustion process) (**FIGURE 9-1**). Extinguishing agents include water, water with additives, dry chemicals, wet chemicals, dry powders, and gaseous agents. Each agent is suitable for a specific type of fire.

Portable fire extinguishers are used to extinguish incipient stage fires and to control fires where traditional methods of fire suppression are not recommended.

Extinguishing Incipient Stage Fires

Fire extinguishers should be placed according to NFPA 1, *Fire Code, 2021 Edition,* so that they will be

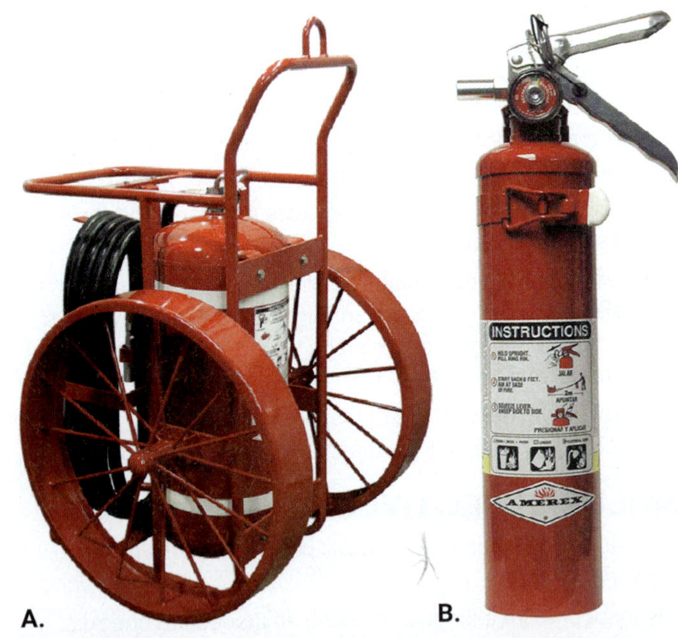

A. B.

FIGURE 9-1 Portable fire extinguishers can be large or small. **A.** A wheeled fire extinguisher. **B.** A hand-held fire extinguisher.

Courtesy of Amerex Corporation.

available for immediate use on small, incipient stage fires, such as a fire in a wastebasket. A person familiar with fire extinguishers, with access to a suitable fire extinguisher, can usually control this type of fire (**FIGURE 9-2**). If flames spread beyond the area of origin to other contents of the room, the fire may become increasingly difficult and dangerous to control with only a portable fire extinguisher.

One advantage of fire extinguishers is their portability and ease of deployment. A firefighter assigned to a ladder company might have the responsibility of being the "can" person, meaning they carry an extinguisher into a building on each assignment, usually a 2-A water can, which carries 2.5 gallons (gal; 9.5 liters [L]) of water, immediately after arriving at the scene of a suspected structure fire. If a fire is located and it is still in its incipient stage, the "can" person controls the growth of the fire—or possibly extinguishes it—before the engine crew is able to stretch a hose line to the fire. At times, a firefighter may use a fire extinguisher from the fire-site premises to control an incipient stage fire. NFPA 1900, *Standard for Aircraft Rescue and Firefighting Vehicles, Automotive Fire Apparatus, Wildland Fire Apparatus, and Automotive Ambulances, 2024 Edition,* recommends that fire apparatus be equipped with a minimum of one dry-chemical, portable fire extinguisher, and one pressurized water extinguisher with a minimum capacity of 2.5 gal (9.5 L). Fire department vehicles that are not equipped with water or fire hose usually carry at least one multipurpose fire extinguisher.

FIGURE 9-2 A trained individual with a suitable fire extinguisher can usually control an incipient stage fire.

A disadvantage of fire extinguishers is that they are "one-shot" devices. In other words, once the contents of a fire extinguisher have been discharged, the device cannot be used to fight fires until it is recharged or replaced. If the fire extinguisher does not control the fire before it is completely discharged, some other device or method must be employed. This is a serious limitation when compared to a fire hose with a continuous water supply. Therefore, when you use a portable fire extinguisher to control an incipient stage fire, make sure you do not place yourself in a dangerous situation.

SAFETY TIP

Do not place yourself in a dangerous situation by trying to fight a large fire with a small fire extinguisher. You cannot fight a fire or protect yourself with an empty extinguisher, an undersized extinguisher, or the inappropriate type of extinguisher.

Extinguishing Fires with the Appropriate Extinguishing Agent

Remember that fires are categorized into five classes (**TABLE 9-1**). The right fire extinguisher with the right extinguishing agent needs to be utilized. For example, using water on Class C fires that involve energized electrical equipment increases the risk of electrocution to firefighters and can cause extensive damage to the electrical equipment. Water should also not be used on fires that involve flammable liquids or gases such as propane (Class B), cooking oils (Class K), and combustible metals (Class D).

As a firefighter, you must know which fires require which extinguishing agents, which type of fire extinguisher should be used, and how to operate the

TABLE 9-1 Types of Fires

Class A	Ordinary combustibles
Class B	Flammable or combustible liquids or gases
Class C	Energized electrical equipment
Class D	Combustible metals
Class K	Kitchen fires involving oils and fats

different types of fire extinguishers. Using an improper type of fire extinguisher or extinguishing agent can spread burning material, cause unnecessary damage, and pose a danger to the fire extinguisher operator.

Methods of Fire Extinguishment

Understanding the nature of fire is key to understanding how fire extinguishers work and how they differ from one another. All fires require four elements: fuel, heat, oxygen, and a sustained chemical reaction (fire tetrahedron). The fire tetrahedron is discussed in greater detail Chapter 5, *Fire Behavior*. If you remove any of these four elements, a fire will not ignite or the fire will be extinguished. The combustion process begins when the fuel is heated to its ignition temperature, which is the temperature at which it begins to burn. The energy that initiates this process can come from a spark or flame, friction, electrical energy, or a chemical reaction. Once a substance begins to burn, it will generally continue burning as long as an adequate supply of oxygen and fuel to sustain the chemical reaction are present. Removing the heat, oxygen, or fuel interrupts the process.

The extinguishing agents in portable fire extinguishers disrupt the fire tetrahedron by a variety of methods. Some fire extinguishers stop the combustion process by cooling the fuel to below its ignition temperature, some work by blocking the supply of oxygen, and some work by performing both actions. Others release extinguishing agents that interrupt the complex chemical reaction that occurs between the heated fuel and the oxygen.

Water extinguishes fires by cooling the fuel. If the temperature of the fuel falls below its ignition temperature, the combustion process stops.

Creating a barrier that interrupts the flow of oxygen to the flames also extinguishes a fire. Putting a lid on a pan of burning food is an example of this technique (**FIGURE 9-3A**). Applying a blanket of foam to the surface of a burning liquid or gas is another example (**FIGURE 9-3B**). Similarly, surrounding the fuel with a layer of **carbon dioxide (CO_2)**—a colorless, odorless, nontoxic gas that is 1.5 times heavier than air—can cut off the supply of oxygen necessary to sustain the burning process. When the combustion process is stopped by only interrupting the flow of oxygen, be aware that the fuel has not been cooled. This means that if oxygen is reintroduced, the fuel is likely to ignite again.

Some extinguishing agents work by interrupting the molecular chain reactions required to sustain combustion. In some cases, a very small quantity of the agent accomplishes this objective.

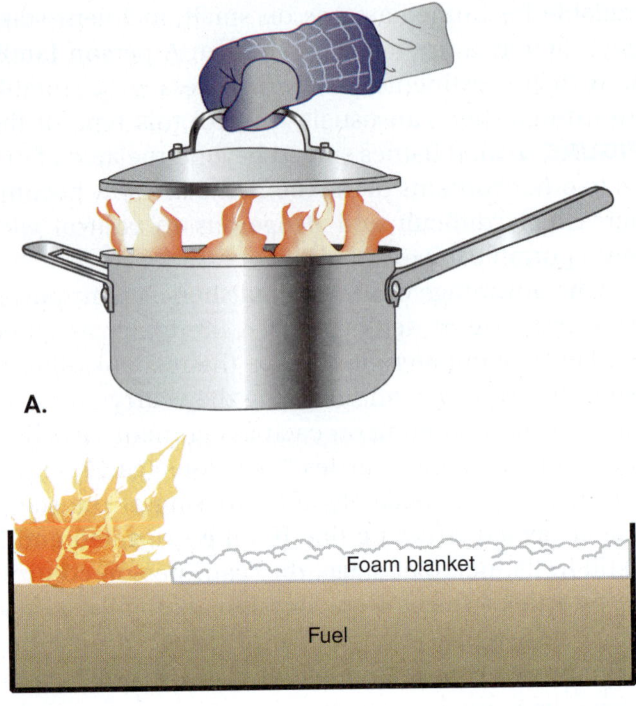

FIGURE 9-3 **A.** Covering a pan of burning food with a lid will extinguish a fire by cutting off the supply of oxygen. **B.** One of the ways a blanket of foam extinguishes fire is by separating the fuel from oxygen.

Classification and Rating of Fire Extinguishers

Portable fire extinguishers are classified and rated based on their extinguishing properties and capabilities. In the United States, Underwriters Laboratories Standards & Engagement is the organization that developed the standards, classification, and rating system for portable fire extinguishers. Each fire extinguisher has a specific rating that identifies the classes of fires for which it is both safe and effective. This information is important for selecting the proper fire extinguisher to fight a particular fire. It is also used to determine which types of fire extinguishers should be placed in each location so that incipient stage fires can be controlled quickly. For example, a commercial kitchen would be an appropriate location for a Class K portable extinguisher.

All portable fire extinguishers are rated with letters that indicate the classes of fire for which the fire extinguisher can be used. Fire extinguishers that are safe and effective for more than one class will be rated with multiple letters. For example, a fire extinguisher that is safe and effective for Class A fires is rated with an *A*; an extinguisher that is safe and effective for Class B fires is rated with a *B*; and one that is safe and effective for both Class A and Class B fires is rated with both an *A* and a *B*.

The rating on Class A and Class B fire extinguishers also includes a number. On Class A fire extinguishers, this number indicates the equivalent amount of water it contains after multiplying by 1.25. So a Class A fire extinguisher that is rated 1-A contains the equivalent of 1.25 gal (4.7 L) of water, and a Class A fire extinguisher that is rated 5-A contains the equivalent of 6.25 gal (23.7 L) of water. The higher the number, the greater the extinguishing capability of the fire extinguisher. A typical Class A fire extinguisher carried on a fire apparatus contains 2.5 gal (9.5 L) of water and has a 2-A rating.

On Class B fire extinguishers, the number in the rating indicates the approximate area in square feet (ft^2) of burning fuel that these devices are capable of extinguishing. Certification testing of these fire extinguishers is performed by trained experts who can control a larger fire than a nonexpert user; therefore, the numerical rating is about 40 percent of the area of burning fuel that an expert can consistently extinguish (Underwriters Laboratories 2017). For example, a 10-B rating indicates that a nonexpert user should be able to extinguish a fire in a pan of flammable liquid that is 10 ft^2 (0.9 square meters [m^2]) in surface area, while an expert user should be able to extinguish a fire that is 25 ft^2 (2.3 m^2) in surface area. A nonexpert user should be able to use a fire extinguisher rated 40-B to control a flammable liquid pan fire with a surface area of 40 ft^2 (4.7 m^2), while an expert user should be able to extinguish a fire with a surface area of 100 ft^2 (9 m^2).

Numbers are used to rate a fire extinguisher's effectiveness only for Class A and Class B fires; Class C, D, and K fire extinguishers are rated only with the letter indicating the class. For example, if the fire extinguisher can also be used for Class C fires, it contains an agent proven to be nonconductive to electricity and safe for use on energized electrical equipment. So a fire extinguisher that carries a 2-A:10-B:C rating can be used on Class A, Class B, and Class C fires. It has the extinguishing capabilities of a 2-A fire extinguisher when applied to Class A fires, has the capabilities of a 10-B fire extinguisher when applied to Class B fires, and can be used safely on energized electrical equipment.

Use fire extinguishers labeled as appropriate for Class B and Class C fires with caution on Class A fires. They are less effective in extinguishing a common combustible fire than a comparable Class A fire extinguisher would be.

Independent laboratories use standard test fires to rate the effectiveness of fire extinguishers. This testing may involve different agents, amounts, application rates, and application methods. Fire extinguishers are rated not only for their ability to control a specific type of fire, but also for the extinguishing agent's ability to prevent rekindling. Some agents successfully suppress a fire by breaking the chemical chain reaction but are unable to prevent the fuel from reigniting because they do not cool the fuel. A rating is given only if the fire extinguisher completely extinguishes the standard test fire and prevents rekindling.

Fire extinguishers that have been tested and approved by an independent laboratory are labeled to clearly designate the classes of fire the unit is capable of extinguishing safely. The traditional lettering system has been used for many years and is still found on many fire extinguishers. The labels consisted of the letters corresponding to the class the extinguisher is rated for on colored backgrounds (**FIGURE 9-4**). The shape of the colored backgrounds for Class A, B, and C mirrors the shape of the letter—*A* is in a green triangle, *B* is in a red square, and *C* is in a blue circle. The letter *D* is in a yellow star, and the letter *K* is in a black hexagon. If a

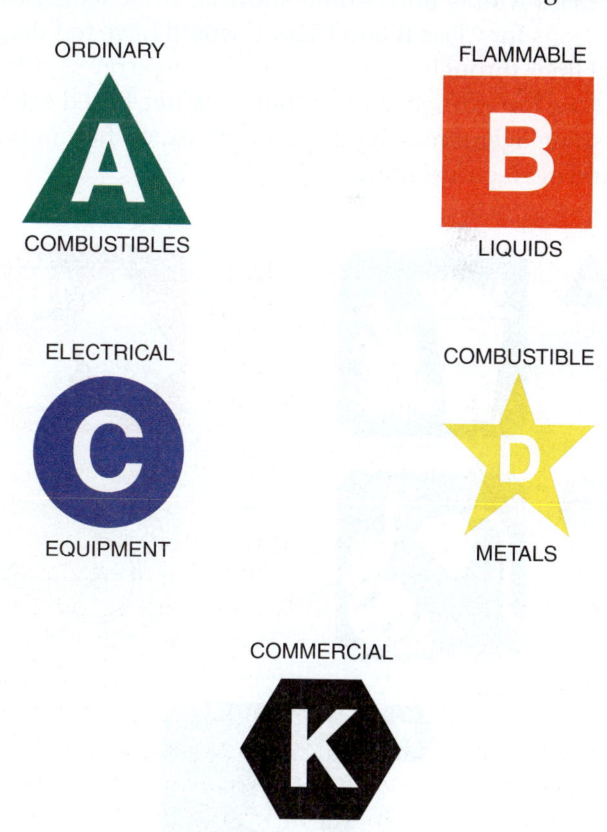

FIGURE 9-4 Traditional letter labels on fire extinguishers often incorporate a shape as well as a letter.

letter appears on an extinguisher, it can safely be used on that class of fire. If an extinguisher has more than one letter label, it can safely be used on all the classes of fire identified by those labels.

A universal pictograph system was developed that does not require the user to be familiar with the alphabetic codes for the different classes of fires. This system indicates whether a fire extinguisher is appropriate for use on a particular class of fire. The pictographs are all square icons, each of which is designed to represent a certain class of fire (**FIGURE 9-5**):

- Class A: Burning trash can beside a wood fire
- Class B: A flame and a gasoline can
- Class C: A flame and an electrical plug and socket
- Class D A flame and a metal gear
- Class K: Flames in a frying pan

Like the lettering system, the presence of an icon indicates that the fire extinguisher has been rated for that class of fire. In addition, if a pictograph label has a red slash across the icon, it means that the fire extinguisher must *not* be used on that type of fire because doing so would create additional risk. A fire extinguisher rated for Class A fires only would show all three icons, but the icons for Class B and Class C would have red diagonal lines through them. This three-icon array signifies that the fire extinguisher contains a water-based extinguishing agent, making it unsafe to use on flammable liquid or electrical fires.

FIGURE 9-5 Icons used to label fire extinguishers.

SAFETY TIP

The safest and surest way to extinguish a Class C fire is to turn off the power and treat it like a Class A or B fire. If you are unable to turn off the power, you should be prepared for reignition, because the electricity could reignite the fire after it has been extinguished.

Fire Extinguisher Placement

Fire and building codes and regulations require the installation of fire extinguishers in multifamily, public, and commercial buildings so that they will be available to fight incipient stage fires. NFPA 10, *Standard for Portable Fire Extinguishers, 2022 Edition*, lists the requirements for placing and mounting portable fire extinguishers as well as the appropriate mounting heights in different types of occupancies. The regulations for each type of occupancy specify the maximum floor area that can be protected by each fire extinguisher, the maximum travel distance from the closest fire extinguisher to a potential fire, and the types of fire extinguishers that should be provided. Two key factors must be considered when determining which type of fire extinguisher should be placed in each area: the class of fire that is likely to occur and the potential size of an incipient stage fire.

Fire extinguishers should be mounted so they are readily visible and easily accessible (**FIGURE 9-6**). Heavy fire extinguishers should not be mounted high on a wall. If the fire extinguisher is mounted too high, the user might be unable to lift it off its hook or could be injured in the attempt. NFPA 10 requires the following mounting heights for fire extinguishers:

- Fire extinguishers weighing up to 40 pounds (lb; 18 kilograms [kg]) must be mounted so that the top of the extinguisher is not more than 5 feet (ft; 2 meters [m]) above the floor.
- Fire extinguishers weighing more than 40 lb (18 kg) must be mounted so that the top of the extinguisher is not more than 3.5 ft (1 m) above the floor.
- The bottom of an extinguisher must be at least 4 inches (in.; 10 centimeters [cm]) above the floor.

Classifying Area Hazards

Several factors must be considered when determining the number and types of fire extinguishers that should be placed in each area of an occupancy. Among these

FIGURE 9-6 Fire extinguishers should be mounted in locations with unobstructed access and visibility.

factors are the types of fuels found in the area and the quantities of those materials. Areas where portable fire extinguishers are required are categorized into three risk classifications—light or low hazard, ordinary or moderate hazard, and extra or high hazard—based on the amount and type of combustibles that are present, including building materials, decorations, furniture, and all other contents, including materials in storage. The total quantity of combustible materials present is the building's **fire load** and is measured as the average weight of combustible materials per square foot (or per square meter). The larger the fire load, the larger the potential fire.

Note that the hazard classification for each area is based on the actual amount and type of combustibles that are present, not the occupancy type. Although there are recommended hazard classifications for different types of occupancies, these are simply guidelines based on typical situations.

Light (Low) Hazard

A **light hazard area**, or **low hazard area**, is an area where the quantity, combustibility, and heat release rate of the materials are low, and most materials are arranged so that a fire is not likely to spread. Light hazard areas usually contain limited amounts of Class A combustibles. The area might also contain some Class B combustibles such as copy machine chemicals or modest quantities of paints and solvents, but all Class B materials in a light hazard area must be kept in closed containers and stored safely. Common light hazard areas are offices, classrooms, churches, assembly halls, and hotel guest rooms (**FIGURE 9-7**).

FIGURE 9-7 Light hazard areas include offices, churches, and classrooms.

Courtesy of Bill Larkin.

Ordinary (Moderate) Hazard

An **ordinary hazard area**, or **moderate hazard area**, contains more Class A and Class B materials than light hazard locations, and the combustibility and heat release rate of the materials are moderate. Typical examples of ordinary hazard areas are retail stores with onsite storage areas, light manufacturing facilities, auto showrooms, parking garages, research facilities, hotel laundry rooms, and restaurant kitchens (**FIGURE 9-8**). Ordinary hazard areas also include warehouses that contain Class I and Class II commodities. Class I commodities are noncombustible products stored on wooden pallets or in corrugated cartons that are shrink-wrapped or wrapped in paper. Class II commodities are noncombustible products stored in wooden crates or multilayered corrugated cartons.

Extra (High) Hazard

An **extra hazard area**, or **high hazard area**, contains more Class A and Class B materials than ordinary hazard locations, and the combustibility and heat release rate of the materials are high. Examples of extra hazard areas are woodworking shops; service and repair facilities for cars, aircraft, or boats; and kitchens and other

FIGURE 9-8 Auto showrooms, hotel laundry rooms, and parking garages are classified as ordinary hazard areas.

cooking areas that have deep fryers, flammable liquids, or gases under pressure (**FIGURE 9-9**). In addition, areas used for manufacturing processes such as painting, dipping, or coating, as well as facilities used for storing or handling flammable liquids, are classified as extra hazard environments. Warehouses containing products that meet the definitions of Class III, Class IV, and Class V materials also are considered extra hazard locations. These commodities are made of natural fibers, paper products, and products containing plastic.

Determining the Most Appropriate Placement

Most buildings require fire extinguishers that are suitable for fighting Class A fires because ordinary combustible materials, such as furniture, partitions, interior finish materials, paper, and packaging products, are so common. Even where other classes of products are used or stored, there is still a need to defend the facility from a fire involving common combustibles.

In some facilities, a variety of conditions are present. In these occupancies, each area must be individually evaluated, and the extinguisher installation tailored to its circumstances. A restaurant is a good example of this situation. The dining areas contain common combustibles, such as furniture, tablecloths, and paper products, that would require a fire extinguisher rated for Class A fires. In the restaurant's kitchen, where the risk of fire involves cooking oils, a Class K fire extinguisher would provide the best defense.

Similarly, within a hospital, fire extinguishers for Class A fires would be appropriate in hallways, offices, lobbies, and patient rooms. Class B fire extinguishers should be mounted in laboratories and areas where

FIGURE 9-9 Kitchens, woodworking shops, and auto repair shops are considered possible extra hazard locations.

flammable anesthetics are stored or handled. Electrical rooms should have fire extinguishers that are approved for use on Class C fires, whereas kitchens need Class K fire extinguishers.

Some areas may need more than one type of fire extinguisher or fire extinguishers with more than one rating. For example, areas that contain both Class A and Class B combustibles require either a fire extinguisher that is rated for both types of fires or a separate fire extinguisher for each class of fire. A single multipurpose fire extinguisher is generally less expensive than two individual fire extinguishers and eliminates the problem of selecting the proper fire extinguisher for a particular fire. However, it is sometimes more appropriate to install Class A fire extinguishers in general-use areas and to place fire extinguishers that are especially effective in fighting Class B or Class C fires near those specific hazards.

Fire Extinguisher Design

Portable fire extinguishers contain a variety of extinguishing agents—the substances that put out a fire. These extinguishing agents are expelled from the fire extinguisher using pressure, and many fire extinguishers rely on pressurized gas, such as nitrogen, to eject the extinguishing agent. This gas is stored either with the extinguishing agent in the body of the fire extinguisher or externally in a separate cartridge or cylinder. When the gas is stored externally, the extinguishing agent is put under pressure only when it is used. A **stored-pressure fire extinguisher** holds both the extinguishing agent, in wet or dry form, and the expeller gas under pressure in the body of the extinguisher. A **cartridge/cylinder-operated fire extinguisher** relies on an external cartridge of pressurized gas, which is released only when the fire extinguisher is to be used by pushing down on a lever that punctures the cartridge and pressurizes the cylinder.

Some extinguishing agents, such as CO_2, are a **self-expelling agent**, which is an extinguishing agent that has sufficient vapor pressure at normal operating temperatures to expel itself from a fire extinguisher. Most self-expelling agents are normally gases but are stored as liquids under pressure in the fire extinguisher. When the confining pressure is released, the agent rapidly expands, causing it to self-discharge. Hand-operated pumps are used to expel the agent when water or water with additives is the extinguishing agent.

Portable Fire Extinguisher Components

Most hand-held portable fire extinguishers have six basic parts (**FIGURE 9-10**):

- A cylinder that holds the extinguishing agent
- A carrying handle
- A nozzle or horn
- A trigger and discharge valve assembly
- A locking mechanism to prevent accidental discharge
- A pressure indicator

Cylinder

The body of the fire extinguisher, known as the **cylinder** or **container**, holds the extinguishing agent. Nitrogen, compressed air, or CO_2 can be used to pressurize the cylinder to expel the agent.

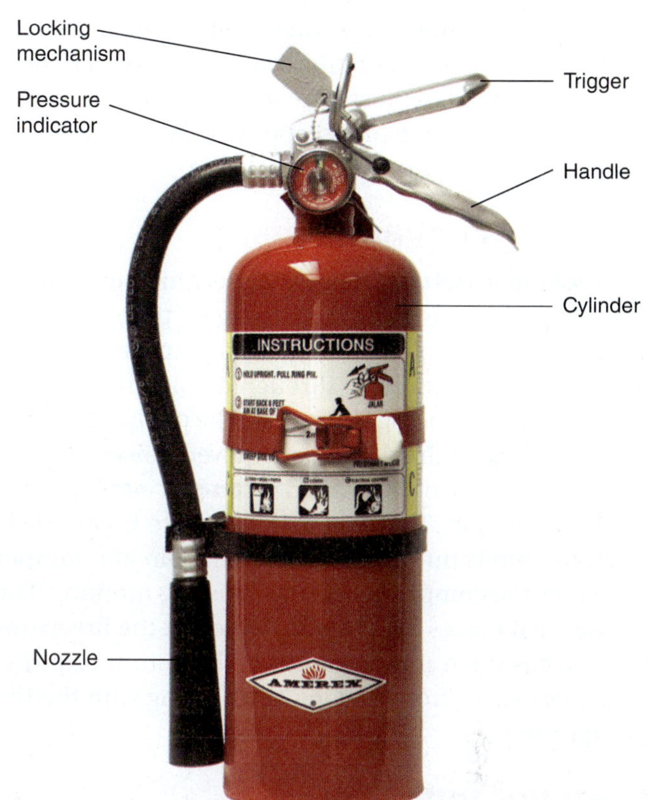

FIGURE 9-10 Portable fire extinguishers have six basic parts.

Courtesy of Amerex Corporation.

Handle

The **handle** is used to carry the portable fire extinguisher and, in many cases, to hold it during use. The actual design of the handle varies from model to model, but all fire extinguishers that weigh more than 3 lb (1.4 kg) have handles. In many cases, the handle is located just below the trigger mechanism.

Nozzle or Horn

The extinguishing agent is expelled through a **nozzle** or **horn**. In some fire extinguishers, the nozzle is attached directly to the valve assembly at the top of the extinguisher. In other models, the nozzle is at the end of a short hose.

Trigger

The **trigger** is the mechanism that is squeezed or depressed to discharge the extinguishing agent. On most portable fire extinguishers, the trigger is a lever located above the handle. On some models, the trigger is a button positioned just above the handle. To release the extinguishing agent, the operator lifts the fire extinguisher by the handle and simultaneously squeezes down on the trigger.

Cartridge/cylinder-operated fire extinguishers usually have a two-step operating sequence. First, a handle or lever is pushed to pressurize the stored agent; then, a trigger-type mechanism incorporated in the nozzle assembly is used to control the discharge.

Locking Mechanism

The **locking mechanism** is a simple, quick-release device that prevents accidental discharge of the extinguishing agent. The simplest form of locking mechanism is a stiff pin, which is inserted through a hole in the trigger to prevent it from being depressed. The pin usually has a ring at the end so that it can be removed quickly.

A special plastic tie, called a **tamper seal**, is used to secure the pin. This seal is designed to break easily when the pin is pulled. (Removing the pin and tamper seal is best accomplished with a twisting motion.) The tamper seal makes it easy to see whether the fire extinguisher has been used and not recharged. It also discourages people from playing or tinkering with the fire extinguisher.

Pressure Indicator

The **pressure indicator** is a gauge that shows whether a fire extinguisher has sufficient pressure to operate properly. Over time, the pressure in a fire extinguisher may dissipate. Checking the gauge first will tell you whether the fire extinguisher is ready for use. Not all fire extinguishers have pressure indicators, so other measures, such as weighing the extinguisher, may be required to determine if the extinguisher is ready to operate. For example, CO_2 extinguishers do not have gauges and must be weighed to determine if they are fully charged.

Pressure indicators vary in terms of both design and sophistication. Most fire extinguishers have a needle gauge. Pressure may be shown in pounds per square inch (psi) or on a three-step scale that indicates if the pressure is too low, in the proper range, or too high. Pressure gauges are usually color-coded, with a green area indicating the proper pressure zone.

Some disposable fire extinguishers intended for home use have an even simpler pressure indicator—a plastic pin built into the cylinder. Pressing on the pin tests the pressure within the fire extinguisher. If the pin pops back up, the fire extinguisher has enough pressure to operate; if it remains depressed, the pressure has dropped below the acceptable level.

Wheeled Fire Extinguishers

A **wheeled fire extinguisher** is a large unit mounted on a wheeled carriage that typically contains between 150 and 350 lb (68 and 158 kg) of extinguishing agent. The wheeled design allows one person to transport the fire extinguisher to the fire. If a wheeled fire extinguisher is intended for indoor use, doorways and aisles must be wide enough to allow for its passage to every area where it could be needed.

Wheeled fire extinguishers usually have a long delivery hose, so the unit can stay in one spot as the operator moves around to attack the fire from more than one side. Usually, a separate cylinder containing nitrogen or some other compressed gas provides the pressure necessary to operate the fire extinguisher.

Wheeled fire extinguishers are most often installed in special hazard areas, such as at military bases, airports, and industrial settings. Models with rubber tires or wide-rimmed wheels are available for outdoor installations.

Types of Fire Extinguishers

Portable fire extinguishers vary according to their extinguishing agent, capacity, effective range, and the time it takes to completely discharge the extinguishing agent. They also have different mechanical designs. This section describes the basic types of extinguishing agents used in portable fire extinguishers, as well as the basic characteristics of the different types of extinguishers. The seven types of fire extinguishers discussed in this section are organized by type of extinguishing agent:

- Water-type fire extinguishers
- Dry-chemical fire extinguishers
- CO_2 fire extinguishers
- Class B foam fire extinguishers
- Wet-chemical fire extinguishers
- Halogenated-agent fire extinguishers
- Dry-powder fire extinguishers

Water-Type Fire Extinguishers

Water is an efficient, plentiful, and inexpensive extinguishing agent that extinguishes fire by cooling the fuel. When it is applied to a fire, it is quickly converted from a liquid into steam, expanding 1600 times, absorbing great quantities of heat in the process. As the heat is removed from the combustion process, the fuel cools below its ignition temperature, and the fire stops burning.

Water-type fire extinguishers are intended for use primarily on Class A fires. Many Class A fuels absorb water, which further lowers the temperature of the fuel and prevents rekindling. Water is a much less effective extinguishing agent for other classes of fires. For example, using a water-type fire extinguisher on hot cooking oil, for example, can cause explosive splattering, which can spread the fire and endanger the operator of the fire extinguisher. Many burning flammable liquids simply float on top of water. Because streams of water conduct electricity, it is dangerous to apply a stream of water to any fire that involves energized electrical equipment. If a water-type fire extinguisher is used on a burning combustible metal, a violent reaction can occur. Because of these limitations, plain water is used only in Class A fire extinguishers. Class B foam fire extinguishers, which are a specific type of water extinguisher, are intended for fires involving flammable liquids.

Wetting-Agent and Class A Foam Fire Extinguishers

Plain water beads up and rolls off the surface of the material it is applied to. In 1984, a **Class A foam concentrate**, sometimes called **wet water**, was developed. This concentrate adds surfactants to water. A **surfactant** is a substance that reduces the surface tension of water (the physical property that causes water to bead or form a puddle on a flat surface). This allows the water to better penetrate and soak into Class A materials. The National Institute of Standards and Technology (NIST) estimates that treated water can "wet" Class A materials 20 times more effectively than plain water. A **Class A foam fire extinguisher** contains a solution of water and Class A foam concentrate. This extinguishing agent has foaming properties that allow it to cling to the material with minimal run-off, as well as the ability to reduce surface tension. Class A foam has greater cooling abilities and requires less water to extinguish a fire. A **wetting-agent fire extinguisher** expels water that contains a solution intended to reduce surface tension. Reducing the surface tension allows water to spread over the fire and penetrate more efficiently into Class A fuels.

Both wetting-agent and Class A foam fire extinguishers are available in the same configurations as other water-type fire extinguishers, including hand-held stored-pressure models and wheeled units. These fire extinguishers should not be exposed to temperatures below 40°F (4°C).

Stored-Pressure Water-Type Fire Extinguishers

A **stored-pressure water-type fire extinguisher** contains water or a water-based extinguishing agent stored under pressure. The most popular kind of stored-pressure water-type fire extinguisher is the 2.5-gal (9.5-L) model with a 2-A rating (**FIGURE 9-11**). Many fire department vehicles carry this type of fire extinguisher for use on incipient stage Class A fires. This type of fire extinguisher expels water in a solid stream with a range of 35 to 40 ft (9 to 12 m) through a nozzle at the end of a short hose. The discharge time is approximately 55 seconds if the fire extinguisher is used continuously. A full fire extinguisher weighs about 30 lb (14 kg).

FIGURE 9-11 Most fire apparatus carry stored-pressure water-type fire extinguishers.

The recommended procedure for operating a stored-pressure water-type fire extinguisher is to set it on the ground, grasp the handle with one hand, and pull out the ring pin or release the locking latch with the other hand. At this point, the fire extinguisher can be lifted and used to douse the fire. Use one hand to aim the stream at the fire and squeeze the trigger with the other hand. The stream of water can be turned into a spray by putting a thumb at the end of the nozzle. This technique is often used after the flames have been extinguished as part of the effort to thoroughly soak the fuel.

Stored-pressure water-type fire extinguishers can be recharged at any location that provides water and a source of compressed air. It is recommended that recharging be performed by a trained individual or a company that specializes in fire extinguishers. If that is not possible, however, be sure to follow the manufacturer's instructions to ensure proper and safe recharging.

Water Mist Fire Extinguishers

A **water mist fire extinguisher** is constructed in a manner similar to the stored-pressure water-type extinguisher. They are typically white in color (**FIGURE 9-12**).

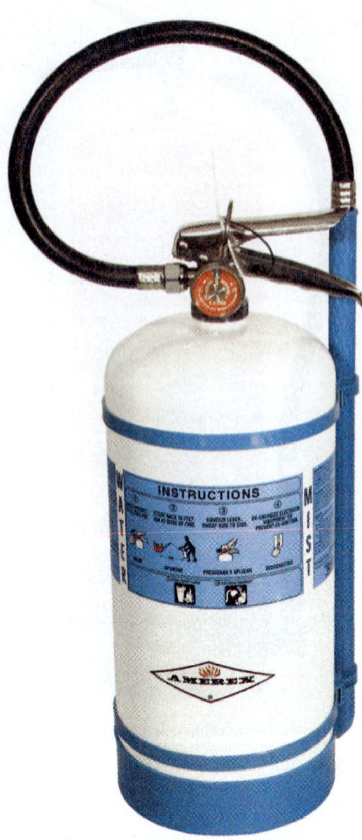

FIGURE 9-12 Water mist fire extinguishers are easily identifiable because they are typically white in color.
Courtesy of Amerex Corporation.

Instead of the discharge hose and nozzle assembly of a stored-pressure water-type extinguisher, they have a discharge hose that is connected to an applicator wand and a misting nozzle. They are commonly available in 1.75 gal (6.6 L) and 2.5 gal (9.5 L) sizes.

Water mist fire extinguishers contain distilled or de-ionized water, which conducts less electricity than tap water. The water is discharged from the misting nozzle as a fine spray that provides safety from electrical shock. For this reason, water mist fire extinguishers are safe to use on Class A and Class C fires and are rated at 2-A:C. The operator must be within 5 to 12 ft (2 to 4 m) of the fire in order for this type of fire extinguisher to be effective.

Water mist fire extinguishers are used where reduced water application is required and regular fire extinguishers might cause excessive damage. Typical uses include museums, rare book collections, hospital environments, telecommunication facilities, and "clean room" manufacturing facilities.

Loaded-Stream Fire Extinguishers

One notable disadvantage of water as an extinguishing agent is that it freezes at 32°F (0°C). In areas that are subject to below-freezing temperatures, a **loaded-stream fire extinguisher** can be used to counteract this limitation. Loaded-stream fire extinguishers discharge a solution of water containing an alkali metal salt that prevents freezing at temperatures as low as −40°F (−40°C). Pressure for these fire extinguishers is supplied by a separate cylinder of CO_2.

The most common model is the 2.5-gal (9-L) unit, which is identical to a typical stored-pressure water-type extinguisher. Hand-held models are available with capacities of 1 to 2.5 gal (4 to 9.5 L) of water and are rated from 1-A to 3-A. Larger units, including a 17-gal (64-L) unit rated 10-A and a 33-gal (124-L) unit rated 20-A, are also available.

Pump Tank Fire Extinguishers

A **pump tank fire extinguisher** uses water as an extinguishing agent, but the water in these devices is not stored under pressure. Instead, the pressure needed to expel the water is provided by a hand-operated, double-acting, vertical piston pump, which moves water out through a short hose on both the up and the down strokes. The manually operated pump may be mounted directly on the cylinder of the fire extinguisher, or it may be part of the nozzle assembly. This type of extinguisher is available in sizes ranging from 1-A-rated, 1.5-gal (6-L) units to 4-A-rated, 5-gal (19-L) units. This type of fire extinguisher sits upright on the ground during use. A small bracket at the bottom allows the operator to steady the fire extinguisher with one foot while pumping.

Pump tank fire extinguishers can be used with antifreeze agents. The manufacturer should be consulted for details because some antifreeze agents (such as common salt) can corrode the fire extinguisher or damage the pump. Fire extinguishers with steel cylinders corrode more easily than those with copper or nonmetallic cylinders.

Backpack Fire Extinguishers

A **backpack fire extinguisher** is used primarily outdoors for fighting brush and grass fires. Most of these units have a tank capacity of 5 gal (19 L) and weigh approximately 50 lb (23 kg) when full. Backpack fire extinguishers are rated for Class A fires but do not carry numeric ratings.

The water tank in a backpack fire extinguisher is made of fiberglass, stainless steel, galvanized steel, nylon, canvas, or brass. Backpack fire extinguishers are designed to be refilled easily in the field, such as from a lake or a stream, through a wide-mouth opening at the top. A filter keeps dirt, stones, and other contaminants from entering the tank. Antifreeze agents, wetting agents, or other special water-based extinguishing agents can be used with backpack fire extinguishers.

Most backpack fire extinguishers are operated via hand pumps. The most common design has a trombone-type, double-acting piston pump located at the nozzle, which is attached to the tank by a short rubber hose. To discharge the fire extinguisher, the operator holds the pump in both hands and moves the piston back and forth.

Some models have a compression pump built into the side of the tank. On these devices, it takes about 10 strokes of the pump handle to build up the initial pressure, which is maintained through continuous slow strokes. The operator uses the other hand to control the discharge. A lever-operated shut-off nozzle is provided at the end of a short hose.

Dry-Chemical Fire Extinguishers

A **dry chemical** is an extinguishing agent composed of fine particles that extinguish fire three ways. First, the finer particles of the chemical vaporize when they reach the high temperature of the flame, and this vapor interrupts the chemistry of the flame. Second, the particles of the dry chemical shield the surface of the fuel from the heat radiation from the flame, thereby reducing the rate at which the burning fuel is being vaporized. And third, when a dry chemical is extensively applied and reaches the surface of the fuel, it smothers the fire by forming an insulating blanket. A **dry-chemical fire extinguisher** delivers a stream of dry chemicals onto a fire. This type of extinguisher is available as both stored-pressure and cartridge/cylinder-operated fire extinguishers.

The following five compounds are used as dry-chemical extinguishing agents:

- Sodium bicarbonate (rated for Class B and C fires only)
- Potassium bicarbonate (rated for Class B and C fires only)
- Urea-based potassium bicarbonate (rated for Class B and C fires only)
- Potassium chloride (rated for Class B and C fires only)
- Monoammonium phosphate (rated for Class A, B, and C fires)

Sodium bicarbonate (baking soda), the original dry-chemical agent, is often used in small household fire extinguishers. Potassium bicarbonate, potassium chloride, and urea-based potassium bicarbonate all have greater fire-extinguishing capabilities per unit volume for Class B fires than does sodium bicarbonate. Potassium chloride is more corrosive than the other dry-chemical extinguishing agents. **Monoammonium phosphate**, also known as **ammonium phosphate**, is a finely ground substance that looks like yellow talcum powder. It is the only dry-chemical extinguishing agent that is rated as suitable for use on Class A fires. Although the other types of dry-chemical fire extinguishers can be used to extinguish Class A fires, the other chemicals do not cool the fuel so a water dousing would be needed to completely extinguish any smoldering embers and prevent rekindling.

Dry-chemical fire extinguishers offer several advantages over water-type fire extinguishers:

- They are effective on Class B fires.
- They can be used on Class C fires because the chemicals are nonconductive.
- They can be stored and used in areas where temperatures fall below the freezing point.

Dry-chemical fire extinguishers can be used on Class C fires that involve energized electrical equipment, but the chemicals can damage computers, electronic devices, and electrical equipment. The fine particles are carried in the air and settle like a fine dust inside the equipment. Over a period of months, this residue can corrode metal parts, causing considerable damage. Another disadvantage of dry-chemical fire extinguishers is that although the chemicals are not toxic, the cloud of fine dust created when they are discharged in an enclosed environment can make breathing more difficult and impair vision. Firefighters should use self-contained breathing apparatus (SCBA) to protect them from both the toxic gases from the fire and the dry-chemical dust discharged from the fire extinguisher.

The trigger on a dry-chemical fire extinguisher allows it to be discharged intermittently; releasing the trigger stops the flow of the agent. This does not mean, however, that the fire extinguisher can be put aside and used again later. Dry-chemical fire extinguishers usually continue to lose pressure after a partial discharge. Pressure loss occurs even when only a small amount of agent has been discharged and the pressure gauge still indicates that the fire extinguisher is properly charged. This loss of pressure occurs because the agent leaves residue in the valve assembly that allows the stored pressure to leak out slowly.

According to NFPA 10, depending on the fire extinguisher's size, the horizontal range of the discharge stream can be from 5 ft to 30 ft (1.5 m to 9.2 m). Some models have special nozzles that allow for a longer range. The long-range nozzles are useful when the fire involves burning gas or a flammable liquid under pressure or when the operator is working in a strong wind. It is best to have the wind at your back when using a handheld dry-chemical fire extinguisher.

Most small, hand-held, dry-chemical fire extinguishers are available with capacities ranging from 1 to 30 lb (0.45 to 14 kg) of agent and are designed to discharge their contents completely in as little as 8 to 20 seconds. Larger units may discharge for as long as 30 seconds. Wheeled fire extinguishers are available with capacities up to 350 lb (159 kg) of agent. Large dry-chemical fire extinguishers also may be mounted on fire apparatus to deal with special risks.

The selection of which dry-chemical fire extinguisher to use depends on the compatibility of different agents with one another and with any products that they might contact. Some dry-chemical extinguishing agents cannot be used in combination with particular types of foam.

SAFETY TIP

The different types of dry-chemical extinguishers look similar. Make sure you check the label before using one to ensure you have the desired extinguishing agent.

Ordinary Dry-Chemical Fire Extinguishers

The first dry-chemical fire extinguishers were introduced in the 1950s and were rated for Class B and C fires only. The industry term for this type of B:C-rated unit that uses sodium bicarbonate as the extinguishing agent is **ordinary dry-chemical extinguisher**. Ordinary dry-chemical fire extinguishers are available in hand-held models with ratings up to 160-B:C. Larger, wheeled units carry ratings up to 640-B:C.

In 1959, the U. S. Navy Research Laboratory developed a dry chemical that is twice as effective in extinguishing Class B fires as sodium bicarbonate and five times more effective than CO_2. This new extinguishing agent was named Purple K due to the purple color of the substance and the characteristic lavender tint it gives to flames. The main component is potassium bicarbonate, and it also contains sodium bicarbonate, mica, Fuller's earth, and amorphous silica. It is made **hydrophobic** (repels or does not mix with water) by adding methyl hydrogen polysiloxane. Purple K extinguishers are rated for Class B fires and Class C fires that involve a flammable liquid as they are nonconductive.

SAFETY TIP

Purple K should not be confused with Class K extinguishers, as they are not effective in class K fires.

Multipurpose Dry-Chemical Fire Extinguishers

During the 1960s, the **multipurpose dry-chemical fire extinguisher** was introduced. These fire extinguishers contain monoammonium phosphate and are rated for Class A, B, and C fires. The chemicals in these fire extinguishers take the form of fine particles that are treated with other chemicals to prevent the particles from absorbing moisture, which could cause packing or caking and interfere with the extinguisher's discharge and flow. When discharged, these chemicals form a crust over Class A combustible fuels, thereby preventing rekindling (**FIGURE 9-13**).

Multipurpose dry-chemical fire extinguishers should never be used on Class K fires, such as those involving deep-fat fryers located in commercial kitchens. The monoammonium phosphate–based extinguishing agent is acidic and does not react with cooking oils to produce the smothering foam needed to extinguish this type of fire. Even worse, the acid will counteract the foam-forming properties of any alkaline extinguishing agent that is applied to the same fire.

Multipurpose dry-chemical fire extinguishers are available as stored-pressure, cartridge, or nitrogen cylinder–type hand-held models with ratings ranging from 1-A to 20-A and from 10-B:C to 120-B:C. Larger wheeled models have ratings ranging from 20-A to 40-A and from 60-B:C to 320-B:C.

Carbon Dioxide Fire Extinguishers

When CO_2 is discharged on a fire, it forms a dense cloud that displaces the air surrounding the fuel because it is heavier than air. This effect interrupts the combustion process by reducing the amount of oxygen that can reach the fuel. The placement of a blanket of CO_2 over the surface of a liquid fuel also disrupts the fuel's ability to vaporize.

CO_2 is both a self-expelling agent and an extinguishing agent. In a portable **carbon dioxide (CO_2) fire extinguisher**, CO_2 is stored under a pressure of 800 psi

FIGURE 9-13 Multipurpose dry-chemical fire extinguishers can be used for Class A, B, and C fires.

(5500 kPa), which keeps the CO_2 in liquid form at room temperature. When the pressure is released, the liquid CO_2 rapidly converts to a gas, and the expansion of the gas forces the agent out of the container.

The CO_2 is discharged through a siphon tube that reaches to the bottom part of the storage cylinder; it is then forced through a hose and expelled through a horn or cone-shaped applicator that directs the flow of the agent on the fire. The rapidly moving product through the applicator sometimes creates static electricity and can release a static charge. This is not dangerous and will not ignite any unburnt fuel.

When discharged, the CO_2 is very cold and contains a mixture of CO_2 gas and dry ice, which is quickly converted to a gas. The gas forms a visible cloud of dry ice that freezes the moisture in the air when the moisture comes in contact with the CO_2. This helps cool the burning materials and surrounding areas. After the fire is extinguished, the CO_2 should continue to be applied to promote cooling and reduce the chance of reignition.

CO_2 fire extinguishers are rated for Class B and C fires only. This extinguishing agent does not conduct electricity and has two significant advantages over dry-chemical agents: It is not corrosive, and it does not leave any residue because it dissipates into the air.

Gaseous extinguishing agents are a type of **clean agent** that extinguish fires using extinguishing agents that do not leave a residue. Because of this, gaseous agents such as CO_2 can be effective in suppressing fires in areas that contain computers and other sensitive electronic equipment because they can extinguish a fire without causing significant damage to the room or contents. CO_2 fire extinguishers are also used in food preparation areas and in laboratories.

CO_2 fire extinguishers have several limitations and disadvantages:

- CO_2 fire extinguishers are heavier than similarly rated extinguishers that use other extinguishing agents (**FIGURE 9-14**).
- CO_2 fire extinguishers have a short discharge range (3 to 8 ft [1 to 2.5 m]), requiring the operator to be close to the fire, which increases the risk of personal injury. Using a CO_2 extinguisher from an upwind (wind at your back) location will give you greater reach.
- CO_2 fire extinguishers do not perform well at temperatures below 0°F (–18°C) or in windy or drafty conditions because the extinguishing agent dissipates before it reaches the fire.
- When CO_2 fire extinguishers are used in confined spaces, the extinguishing agent dilutes the oxygen in the air. If the air is diluted enough, people in the area begin to suffocate.
- CO_2 fire extinguishers are not suitable for use on fires involving pressurized fuel or on Class K fires.

SAFETY TIP

CO_2 discharged into a confined space will reduce the oxygen level in that space. For this reason, anyone entering the confined area must wear SCBA.

Smaller CO_2 fire extinguishers contain from 2.5 to 5 lb (1 to 2 kg) of agent and are designed to be operated with one hand. The horn is attached directly to the discharge valve on the top of the fire extinguisher by a hinged metal tube. In larger models, the horn is attached at the end of a short hose and requires two-handed operation. Depending on their size, CO_2 fire extinguishers can discharge completely in 8 to 30 seconds.

FIGURE 9-14 Carbon dioxide fire extinguishers are heavy due to the weight of the container and the large quantity of agent needed to extinguish a fire. They also have a large discharge nozzle, making them easily identifiable.

CO_2 extinguishers are not single use; they can be recharged after they are used. The trigger mechanism on CO_2 fire extinguishers can be operated intermittently to preserve any remaining agent. The pressurized CO_2 remains in the fire extinguisher, but the extinguisher must still be recharged after use. The fire extinguisher can be weighed to determine how much agent is left in the storage cylinder.

SAFETY TIP

Do not aim the CO_2 fire extinguisher discharge at anyone or allow it to come in contact with exposed skin; frostbite could result.

Class B Foam Fire Extinguishers

Class B foam fire extinguishers are similar in appearance and operation to water-type, wetting agent, and Class A foam fire extinguishers. They discharge a solution of water and concentrates of either **aqueous film-forming foam (AFFF)**—a foam designed to form a blanket over spilled flammable liquids to suppress vapors or on actively burning pools of flammable liquids—or **film-forming fluoroprotein (FFFP) foam**—a foam that contains surfactants that produce a fluid film for suppressing hydrocarbon fuel vapors (**FIGURE 9-15**). The agent is discharged through an **air-aspirating nozzle**, which draws air into the water stream. The result is a foam solution that floats over the surface of a burning liquid, creating a blanket that separates the fuel from oxygen. This blanket prevents the fuel from vaporizing and forms a barrier between the fuel and the oxygen, extinguishing the flames and preventing reignition.

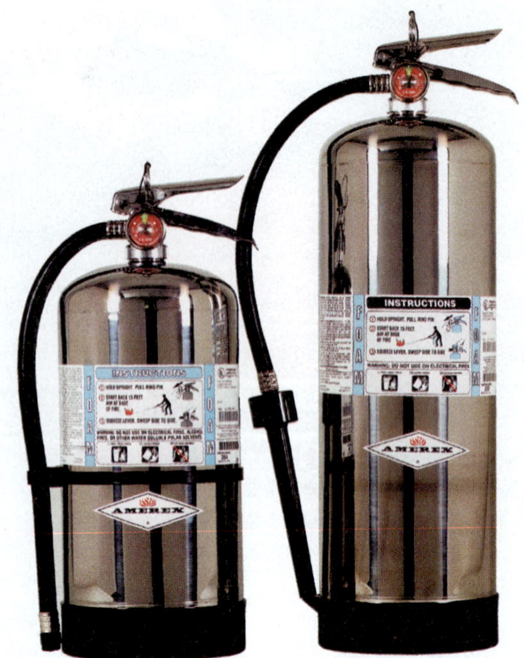

FIGURE 9-15 An AFFF fire extinguisher produces an effective foam for use on Class B fires.

Class B foam fire extinguishers are effective fighting Class A fires, but they are not suitable for Class C fires or for fires involving flammable liquids or gases under pressure. They are also not intended for use on Class K fires. In addition, although fires involving a **polar solvent**—a water-soluble, flammable liquid, such as alcohol, acetone, ester, and ketone—are considered to be Class B fires, only specifically labeled foam fire extinguishers can be used on these fires. Class B foam fire extinguishers also are not effective at freezing temperatures. Consult the fire extinguisher manufacturer for information on using foam agents effectively at low temperatures.

AFFF and FFFP hand-held stored-pressure fire extinguishers are available in two sizes: 1.6 gal (6 L), rated 2-A:10-B, and 2.5 gal (9 L), rated 3-A:20-B. A wheeled model with a 33-gal (125-L) capacity and a rating of 20-A:160-B is also available.

Both AFFF and FFFP concentrates produce highly effective foams. The decision of which one should be used depends on the foam's compatibility with the flammable liquid involved and other extinguishing agents that could be used on the same fire. Detailed information on the use of AFFF and FFFP is available in NFPA 11, *Standard for Low-, Medium-, and High-Expansion Foam, 2021 Edition.*

Wet-Chemical Fire Extinguishers

A **wet-chemical extinguishing agent** converts the fatty acids in cooking oils or fats to a soap or foam, a process known as **saponification**. They include aqueous solutions of potassium acetate, potassium carbonate, and potassium citrate, either singly or in various combinations. Wet-chemical extinguishing agents are specifically formulated for use in commercial kitchens and food-product manufacturing facilities, especially where food is cooked in a deep fryer. A **wet-chemical fire extinguisher** uses wet-chemical agents to extinguish Class K fires.

Portable Class K wet-chemical fire extinguishers are available in three sizes: 0.8 gal (3 L), 1.5 gal (6 L), and 2.5 gal (9 L). There are no numerical ratings for these fire extinguishers. Built-in, automatic extinguishing systems in commercial kitchens, which are equipped with deep-fat fryers, cooking oils, and grills, use wet-chemical extinguishing agents. The fixed, automatic fire-extinguishing systems discharge the agent directly over the cooking surfaces. Wet-chemical extinguishing agents are discharged as a fine spray, which reduces the risk of splattering. Clean-up afterward is much easier than after other types of extinguishing agents are used, allowing a business to reopen sooner.

Halogenated-Agent Fire Extinguishers

Almost all fuels consist of hydrogen (H) and carbon (C) atoms, so they are called hydrocarbons. Hydrocarbons are discussed further in Chapter 5, *Fire Behavior.* **Halogen** is the family of elements (chemicals listed in the periodic table) that includes fluorine (F), bromine (Br), iodine (I), and chlorine (Cl), which are chemically related. A **halogenated hydrocarbon**, also known as a **halocarbon**, is a hydrocarbon in which at least one hydrogen atom of the hydrocarbon is replaced by a halogen. A common type of halocarbon used in fire suppression is **hydroflurocarbon (HFC)**. A **halogenated extinguishing agent** is an extinguishing agent made from halogenated hydrocarbons (halocarbons), and a **halogenated-agent fire extinguisher**—also called **clean-agent fire extinguisher**—uses halogenated agents.

There are two broad categories of halogenated agents: Halons and newer, less ozone-depleting Halon alternatives. Halons and Halon alternatives are used primarily for Class B and Class C fires because they are clean agents that extinguish fire by interrupting the chemical reaction of the fire tetrahedron. Remember that clean agents are nonconductive, noncorrosive, and leave no residue, so they will not harm sensitive equipment or people. Per pound, they are approximately twice as effective at extinguishing fires as CO_2, which means less extinguishing agent is required to extinguish a fire.

Commonly used Halon for fire suppression exists in two forms: as Halon 1211, which is Bromochlorodifluoromethane ($CBrClF_2$), also known as BCF, and as Halon 1301, Bromotrifluoromethane ($CBrF_3$). Halon 1211 is used only in portable extinguishers and is a streaming agent, whereas Halon 1301 is used only in fixed extinguisher installations, typically in cargo holds or engines, and it is a total flooding agent. Both types of Halon are rated for Class B and C fires. Larger-capacity extinguishers are also rated for use on Class A fires.

Halon 1211 is available in hand-held, stored-pressure fire extinguishers with capacities that range from 2 lb (0.9 kg) rated 2-B:C to 22 lb (10 kg) rated 4-A:80-B:C. These fire extinguishers are not rated for use on Class B fires involving pressurized fuels or on Class K fires. Wheeled Halon 1211 models are available with capacities up to 150 lb (68 kg) with a rating of 30-A:160-B:C. The wheeled fire extinguishers use a nitrogen booster charge from an auxiliary cylinder to expel the agent. Portable fire extinguishers that use Halon 1211 as the extinguishing agent store the agent as a liquid and discharge it under relatively high pressure.

A 1987 international agreement known as the Montreal Protocol limited Halon production and importation because these agents severely damage the earth's ozone layer. As a result of this agreement, Halons can no longer be manufactured or imported. Since then, Halons have been replaced by more environmentally friendly halocarbons that are not subject to the same environmental restrictions. However, Halons are the extinguishing agent in many existing systems and portable fire extinguishers still in use today. Halons can be reclaimed—that is, removed from existing systems or extinguishers that are being taken out of service and processed to remove contaminants—and reclaimed Halons can be used to recharge a system or a portable fire extinguisher that has discharged. Eventually, this supply will be depleted. If no reclaimed Halon is available, that Halon system or extinguisher must be replaced with a new, different type of system or extinguisher. Halon systems that are no longer used must be decommissioned by a certified company that reclaims the existing Halon agent according to strict environmental policies.

A fire extinguisher containing a halogenated extinguishing agent releases a mist of vapor and liquid droplets that disrupts the molecular chain reactions within the combustion process, thereby extinguishing the fire. Halogenated-agent fire extinguishers have a horizontal discharge stream range of 6 to 35 ft (2 to 11 m). These agents dissipate rapidly in windy conditions, as does CO_2, so their effectiveness is limited outdoors. Because halogenated agents also displace oxygen, they should be used with care in confined areas.

Two common halogenated agents, 3M Novec 1230 Fire Protection Fluid and FM-200, have an ozone depletion potential of zero, which means they are safe for the earth's ozone layer. Both of these halogenated agents extinguish fire by rapidly removing heat. 3M Novec 1230 Fire Protection Fluid evaporates 50 times faster than water. Although these particular Halon alternatives do not contain Halon gas, they still use HFC to suppress fires, which is not as eco-friendly.

Dry-Powder Fire Extinguishers and Extinguishing Agents

Not to be confused with dry-chemical extinguishers, which are rated for Class A, B, and C fires, a **dry-powder fire extinguisher** uses **dry powder**, a chemical compound that is stored in fine granular or powdered form. Dry powder is used to extinguish Class D fires involving combustible metals. Flammable metal fires are not easily extinguished with water, and some metals react violently when water is applied, so a dry-powder extinguishing agent is applied to smother the fire. The dry-powder agent forms a solid crust over the burning metal, which blocks out oxygen and absorbs heat. The extinguishing agents and application methods required to put out Class D fires vary greatly, depending on the specific metal, the quantity of metal involved, and the physical form of the fuel (whether the metal is a solid object or is in the form of grindings or shavings.

SAFETY TIP

Do not confuse dry-chemical and dry-powder agents. Dry-chemical fire extinguishers are rated for Class B and C fires or for Class A, B, and C fires. Dry-powder fire extinguishers are designed for use in suppressing Class D fires.

The most commonly used dry-powder extinguishing agent is formulated from finely ground sodium chloride (table salt) plus additives to help it flow freely over a fire. A thermoplastic material mixed with the agent binds the sodium chloride particles into a solid mass when they come into contact with a burning metal. Dry-powder fire extinguishers using sodium chloride–based agents are available with a 30-lb (14-kg) capacity in either stored-pressure or cylinder/cartridge models. Wheeled models are available with 150- and 350-lb (68- to 160-kg) capacities.

Dry-powder fire extinguishers are often carried on specialty apparatus such as hazardous materials units, but they might be carried on frontline apparatus if there is a known hazard within the primary response area. They also are used in businesses or manufacturing plants where specific hazards are present. It is a good idea to identify the location of these extinguishers during building inspections and in the **preincident plan**—data about a location in your community that describe the layout and hazards at the location and identify response essentials.

Dry-powder extinguishers are easily identified by their yellow cylinder (**FIGURE 9-16**). They have adjustable nozzles that allow the operator to vary the flow of the extinguishing agent. When the nozzle is fully

FIGURE 9-16 Dry-powder fire extinguishers are easily identified by their yellow cylinder.

Voice of Experience

I once was assigned to conduct a search at a reported structure fire. My crew and I made entry with a portable fire extinguisher to find a mattress in a bedroom on fire. We extinguished the fire with the portable fire extinguisher while the engine company was stretching a hose line. Because the fire was in its incipient stage, it was easily contained with the portable fire extinguisher and the hose line was not needed. We were able to limit the damage to the room of origin. We overhauled the bedroom, packed up, and went home.

That one act of using the portable fire extinguisher saved the property owner the added expense of having to repair any water damage from the hose lines. The insurance agent gave a donation to the fire department out of his own pocket to thank us for using the portable fire extinguisher instead of automatically dousing the whole place with water from our hose line.

Rob Gaylor
City of Westfield Fire Department
Westfield, Indiana

opened, the hand-held models have a range of 6 to 8 ft (2 to 2.5 m). Extension wand applicators are available to direct the discharge from a more distant position.

Class D agents must be applied carefully so that the molten metal does not splatter. Care must be taken to ensure that no water comes in contact with the burning metal because even a trace quantity of moisture can cause a violent reaction. Flammable metals can burn at more than 5000 degrees, hot enough to separate water into its component parts of hydrogen and oxygen, and hydrogen is flammable and has explosive properties.

The same sodium chloride–based dry-powder agent that is used in portable fire extinguishers can be stored in bulk form and applied by hand. Another dry-powder extinguishing agent for Class D fires, graded granular graphite mixed with phosphorus-containing compounds, cannot be expelled from a portable fire extinguisher. Instead, this agent is produced in bulk and must be applied manually from a pail or other container using a shovel or scoop. When applied to a metal fire, the phosphorus compounds release gases that blanket the fire and cut off its supply of oxygen, and the graphite absorbs heat from the fire, allowing the metal to cool to below its ignition temperature. Bulk dry-powder agents are available in 40- and 50-lb (18- and 23-kg) pails and 350-lb (159-kg) drums.

More information about Class D extinguishing agents and fire extinguishers can be found by in the NFPA's *Fire Protection Handbook*.

Use of Fire Extinguishers

Fire extinguishers are designed to be simple to operate. Every portable fire extinguisher is labeled with printed operating instructions. An individual with only basic training should be able to use most fire extinguishers safely and effectively.

There are six basic steps in extinguishing a fire with a portable fire extinguisher:

1. Locate the fire extinguisher.
2. Select the proper classification of fire extinguisher.
3. Ensure your personal safety by having an exit route.
4. Transport the fire extinguisher to the location of the fire.
5. Activate the fire extinguisher to release the extinguishing agent.
6. Apply the extinguishing agent to the fire for maximum effect.

SAFETY TIP

Proper technique in using a portable fire extinguisher is important for both safety and effectiveness. Practice using a portable fire extinguisher under the careful supervision of a trained instructor.

Although these steps are not complicated, practice and training are essential for effective fire suppression with a fire extinguisher. Research has shown that the effective use of Class B portable fire extinguishers depends heavily on user training and expertise. A trained expert can extinguish a larger fire than a nonexpert can, using the same extinguisher. As a firefighter, you should be able to operate any fire extinguisher that you might be required to use, whether it is carried on your fire apparatus, hanging on the wall of your firehouse, or placed in some other location in your community.

SAFETY TIP

Always test a fire extinguisher before entering the area involved. Activate the trigger long enough to ensure that the agent discharges properly.

Locating Fire Extinguishers

As a firefighter, you should know which types of fire extinguishers are carried on your department's apparatus and where each one is located. You should also know where fire extinguishers are located in and around the fire station and other workplaces in your department. You should have at least one fire extinguisher in your home and another in your personal vehicle, and you should know exactly where they are located. Knowing the exact locations of fire extinguishers can save valuable time in an emergency.

Selecting the Proper Fire Extinguisher

Selecting the proper fire extinguisher requires an understanding of the classification and rating system for fire extinguishers. Knowing which types of agents are available, how they work, which ratings the fire extinguishers carried on your fire apparatus have, and which extinguisher is appropriate for a particular fire situation is also important. Understanding the fire extinguisher rating system and the different types of agents will enable a firefighter to determine whether it is suitable for a particular fire situation. A quick look at the label should be all that is needed.

Firefighters should be able to assess a fire quickly, determine whether the fire can be controlled by a fire extinguisher, and identify the appropriate extinguisher to use. Using a fire extinguisher with an insufficient rating may not completely extinguish the fire, which can place the operator in danger of being burned or otherwise injured. If the fire is too large for the fire extinguisher, consider other options, such as obtaining additional extinguishers or making sure that a charged hose line is ready to provide backup.

SAFETY TIP

Before deciding to use a fire extinguisher, size up the fire to ensure that the extinguisher has an adequate capacity and holds the proper extinguishing agent.

Firefighters should be able to determine the most appropriate type of fire extinguisher to place in a given area based on the hazards that are present and the types of fires that could occur. In some cases, one type of fire extinguisher might be preferred over another. For example, an extinguishing agent such as CO_2 or a halogenated agent is a better choice than other agents for a fire involving sensitive electronic equipment because it leaves no residue. A dry-chemical fire extinguisher is generally more appropriate than a CO_2 fire extinguisher to fight an outdoor fire because wind will quickly dissipate CO_2. A dry-chemical fire extinguisher also would be the best choice for a fire involving a flammable liquid leaking under pressure from a pipe, whereas foam would be a better choice for a fire involving a liquid spill on the ground.

Ensuring Your Personal Safety

Before using a fire extinguisher, be sure to approach the fire with an exit plan. If the fire suddenly expands or the fire extinguisher fails to control it, you must have a planned escape route. Never let the fire get between you and a safe exit. After the fire has been suppressed, do not turn your back on it. Always watch it and be prepared for it to rekindle until the fire has undergone complete **overhaul**—the process of finding, exposing, and suppressing any smoldering or hidden pockets of fire in an area that has been burned.

When ordinary civilians use fire extinguishers on incipient stage fires, they are probably wearing their normal clothing. As a firefighter, however, you should wear your personal protective clothing and use appropriate personal protective equipment (PPE). Take advantage of the protection they provide.

If you must enter an enclosed area where a fire extinguisher has been discharged, wear full PPE and use SCBA. The atmosphere within the enclosed area will contain a mixture of combustion products and extinguishing agents, and the oxygen content within the space may be dangerously low.

Transporting a Fire Extinguisher

The best method of transporting a hand-held portable fire extinguisher depends on the extinguisher's size, weight, and design. Hand-held portable models can weigh as little as 1 lb (0.45 kg) or as much as 50 lb (23 kg). The ability to handle the heavier fire extinguishers depends on an individual operator's personal strength.

Fire extinguishers with a fixed nozzle should be carried in the favored or stronger hand. This approach enables the operator to depress the trigger and direct the discharge easily. Fire extinguishers that have a hose between the trigger and the nozzle should be carried in the weaker or less-favored hand so that the favored hand can grip and aim the nozzle.

Heavier fire extinguishers may have to be carried as close as possible to the fire and placed upright on the ground. The operator can then depress the trigger with one hand, while holding the nozzle and directing the stream with the other hand.

To transport a fire extinguisher, follow the steps in **SKILL DRILL 9-1**.

SKILL DRILL 9-1

Transporting a Fire Extinguisher Firefighter I, NFPA 1010: 6.3.16

1. Locate the closest fire extinguisher.

2. Assess that the fire extinguisher is safe and effective for the type of fire being attacked. Release the mounting bracket straps.

Continues.

SKILL DRILL 9-1 CONTINUED

Transporting a Fire Extinguisher Firefighter I, NFPA 1010: 6.3.16

3. Lift the fire extinguisher using good body mechanics. Lift small fire extinguishers with one hand and large extinguishers with two hands.

4. Walk briskly—do not run—toward the fire. If the fire extinguisher has a hose and nozzle, carry the extinguisher with one hand, and grasp the nozzle with the other hand.

Operating a Fire Extinguisher

Practice discharging different types of fire extinguishers in training situations to build confidence in your ability to use them properly and effectively. Most fire extinguishers have simple operation systems. Activating a fire extinguisher to apply the extinguishing agent is a simple operation that involves four steps. The **PASS** acronym is a helpful way to remember these steps:

1. Pull the safety pin.
2. Aim the nozzle at the base of the flames.
3. Squeeze the trigger to discharge the agent.
4. Sweep the nozzle across the base of the flames.

To extinguish a Class A fire with a stored-pressure water-type fire extinguisher, follow the steps in **SKILL DRILL 9-2**. *Note:* Stored-pressure water-type fire extinguishers are safe and effective to use on Class A fires. They are not effective on Class B fires, and they are not safe to use on Class C fires.

SKILL DRILL 9-2

Extinguishing a Class A Fire with a Stored-Pressure Water-Type Fire Extinguisher Firefighter I, NFPA 1010: 6.3.16

1. Size up the fire to determine whether a stored-pressure water-type fire extinguisher is safe and effective for the fire. Ensure the fire extinguisher is large enough to be safe and effective. Ensure your safety. Make sure you have an exit route from the fire. Do not turn your back on the fire.

2. Remove the hose and nozzle from its holder if present. Quickly check the pressure gauge to verify that the fire extinguisher is adequately charged.

3. Pull the pin to release the fire extinguisher control valve.

4. Aim the nozzle at the base of the flames, squeeze the trigger, and then sweep the water stream at the base of the flames. You must be within 35 to 40 ft (11 to 12 m) of the fire to be effective.

Continues.

SKILL DRILL 9-2 CONTINUED

Extinguishing a Class A Fire with a Stored-Pressure Water-Type Fire Extinguisher Firefighter I, NFPA 1010: 6.3.16

5. Overhaul the fire by breaking apart tightly packed fuel to prevent rekindling. Summon additional help if needed.

SKILL DRILL 9-3

Extinguishing a Class A Fire with a Multipurpose Dry-Chemical Fire Extinguisher Firefighter I, NFPA 1010: 6.3.16

1. Size up the fire to determine whether a multipurpose dry-chemical fire extinguisher is safe and effective for this fire. Ensure the fire extinguisher is large enough to be safe and effective. Ensure your safety. Make sure you have an exit route from the fire. Do not turn your back on the fire. Remove the hose and nozzle.
2. Quickly check the pressure gauge to verify that the fire extinguisher is adequately charged.
3. Remove the hose and nozzle. Pull the pin to release the fire extinguisher control valve.
4. Aim the nozzle at the base of the flames, squeeze the trigger, and then sweep the dry-chemical discharge at the base of the flames. Depending on the size of the fire and fire extinguisher, you must be within 5 to 45 ft (2 to 14 m) of the fire to be effective. Coat the burning fuel with dry chemical.
5. Overhaul the fire by breaking apart tightly packed fuel to prevent rekindling. Summon additional help if needed.

To extinguish a Class A fire with a multipurpose dry-chemical fire extinguisher, follow the steps in **SKILL DRILL 9-3**. *Note:* Multipurpose dry-chemical fire extinguishers are safe and effective on Class A, B, and C fires.

To extinguish a Class B flammable liquid spill or pool fire with a dry-chemical extinguisher, follow the steps in **SKILL DRILL 9-4**. *Note:* Dry-chemical fire extinguishers are safe to use on Class B and C fires. They are effective on Class B fires, but they are not rated for Class A fires.

To extinguish a Class B flammable liquid spill or pool fire with a stored-pressure fire extinguisher containing an AFFF or FFFP extinguishing agent, follow the steps in **SKILL DRILL 9-5**. *Note:* Stored-pressure foam fire extinguishers are safe to use on Class A, B,

SKILL DRILL 9-4

Extinguishing a Class B Flammable Liquid Fire with a Dry-Chemical Fire Extinguisher Firefighter I, NFPA 1010: 6.3.16

1. Size up the fire to determine whether a dry-chemical fire extinguisher is safe and effective for the fire. Quickly check the pressure gauge to verify that the fire extinguisher is adequately charged. Ensure the fire extinguisher is large enough to be safe and effective. Ensure your safety. Make sure you have an exit route from the fire. Do not turn your back on the fire.

2. Remove the hose and nozzle. Pull the pin to release the fire extinguisher valve.

3. Aim the nozzle at the near edge of the surface of the burning liquid, squeeze the trigger, and then sweep the dry-chemical discharge across the surface of the burning liquid. Depending on the size of the fire and fire extinguisher, you must be within 5 to 45 ft (2 to 14 m) of the fire to be effective. Start at the near edge of the fire, and work toward the back.

4. Overhaul the fire by keeping a blanket of dry chemical over the fuel. Summon additional help if needed.

SKILL DRILL 9-5

Extinguishing a Class B Flammable Liquid Fire with a Stored-Pressure Foam Fire Extinguisher (AFFF or FFFP) Firefighter I, NFPA 1010: 6.3.16

1. Size up the fire to determine whether a stored-pressure foam fire extinguisher is safe and effective for the fire. Ensure the fire extinguisher is large enough to be safe and effective. Ensure your safety. Make sure you have an exit route from the fire. Do not turn your back on the fire. Remove the hose and nozzle. Quickly check the pressure gauge to verify that the fire extinguisher is adequately charged. Pull the pin to release the fire extinguisher control valve.

2. Aim the nozzle at the edge of the burning liquid, squeeze the trigger, and then discharge the stream of foam so the foam drops gently onto the surface of the burning liquid at the front or the back of the container. You must be within 20 to 25 ft (6 to 8 m) of the fire to be effective. Let the foam blanket flow across the surface of the burning liquid. Avoid splashing foam on the burning liquid because it can cause the burning fuel to splatter. Continue to apply the agent until the foam blanket has extinguished all of the flames.

3. Overhaul the fire by keeping a thick blanket of foam intact over the hot liquid and reapplying foam over any hot spots. Summon additional help if needed.

and C fires. They are effective on Class B fires by forming a foam blanket. They are effective on Class A fires by cooling and penetrating the fuel.

To operate a CO_2 fire extinguisher, follow the steps in **SKILL DRILL 9-6**. *Note:* CO_2 fire extinguishers are most effective when used in areas without wind. They are designed for Class B and C fires only.

To extinguish a fire in an electrical equipment room with a halogenated stored-pressure fire extinguisher, follow the steps in **SKILL DRILL 9-7**. *Note:*

SKILL DRILL 9-6

Operating a Carbon Dioxide Fire Extinguisher Firefighter I, NFPA 1010: 6.3.16

1. Size up the fire to determine whether CO_2 is a safe and effective agent for this fire. Ensure the fire extinguisher is large enough to be safe and effective. Ensure your safety. Make sure you have an exit route from the fire. Do not turn your back on a fire. Take hold of the hose and the horn or nozzle. Pull the pin to release the fire extinguisher control valve.

2. Quickly squeeze the trigger to verify that the fire extinguisher is charged; CO_2 fire extinguishers do not have pressure gauges.

3. Aim the horn or nozzle at the base of the flames, squeeze the trigger, and then sweep at the base of the flames. You must be within 3 to 8 ft (1 to 2.5 m) of the fire to be effective.

4. Overhaul the fire by taking steps to prevent rekindling depending on what type of fuel was involved. You may have to re-apply to keep material cooled. Summon additional help if needed.

SKILL DRILL 9-7

Operating a Halogenated Stored-Pressure Fire Extinguisher Firefighter I, NFPA 1010: 6.3.16

1. Size up the fire to determine whether a halogenated stored-pressure fire extinguisher is safe and effective for this fire. Ensure the fire extinguisher is large enough to be safe and effective. Ensure your safety. Turn off electricity if possible. Make sure you have an exit route from the fire. Do not turn your back on the fire.
2. Remove the hose and nozzle. Quickly check the pressure gauge to verify that the fire extinguisher is adequately charged. Pull the pin to release the fire extinguisher control valve.
3. Aim the nozzle at the base of the flames, squeeze the trigger, and then sweep the flames off the surface starting at the near edge of the flames. Depending on the size of the fire and fire extinguisher, you must be within 3 to 35 ft (1 to 11 m) of the fire to be effective.
4. Overhaul the fire by continuing to apply the extinguishing agent to cool the fuel. Summon additional help if needed.

SKILL DRILL 9-8

Extinguishing an Incipient Stage Fire Involving Combustible Metal Filings or Shavings with a Dry-Powder Fire Extinguisher Firefighter I, NFPA 1010: 6.3.16

1. Size up the fire to determine whether a dry-powder fire extinguisher is safe and effective for this fire. Ensure the fire extinguisher is large enough to be safe and effective. Ensure your safety. Make sure you have an exit route from the fire. Do not turn your back on the fire. Avoid water or other extinguishing agents that might react with combustible metals.
2. Remove the hose and nozzle. Quickly check the pressure gauge to ensure the fire extinguisher is charged. Pull the pin to release the control valve.
3. Aim the nozzle at the base of the flames, squeeze the trigger to fully open the valve to provide a maximum range, and then reduce the valve to produce a soft, heavy flow while sweeping the dry-powder discharge at the base of the flames to completely cover the burning metal. You must be within 6 to 8 ft (2 to 2.5 m) of the fire to be effective. The method of application may vary depending on the type of metal burning and the extinguishing agent being used.
4. Overhaul the fire by continuing to place a thick layer of the extinguishing agent over the hot metal to form an airtight blanket and allow the hot metal to cool. Summon additional help if needed.

Halogenated stored-pressure fire extinguishers are safe to use on Class A, B, and C fires. They are effective on Class B fires but require large quantities to be effective on Class A fires.

To extinguish an incipient stage fire involving combustible metal filings or shavings (Class D fire) with a dry-powder fire extinguisher, follow the steps in **SKILL DRILL 9-8**. *Note:* Dry-powder fire extinguishers are safe to use only on Class D fires. They are effective only on the metals for which they are intended.

To extinguish a fire in a deep-fat fryer with a wet-chemical fire extinguisher, follow the steps in **SKILL DRILL 9-9**. *Note:* Wet-chemical fire extinguishers are designed to be used on Class K fires—that is, fires in deep-fat fryers and on grills. They are not rated for other classes of fire.

SKILL DRILL 9-9

Extinguishing a Fire in a Deep-Fat Fryer with a Wet-Chemical Fire Extinguisher Firefighter I, NFPA 1010: 6.3.16

1. Size up the fire to determine whether a wet-chemical fire extinguisher is safe and effective for this fire. Ensure the fire extinguisher is large enough to be safe and effective. Ensure your safety. Make sure you have an exit route from the fire. Do not turn your back on the fire.
2. Remove the hose and nozzle. Quickly check the pressure gauge to verify that the fire extinguisher is adequately charged. Pull the pin to release the fire extinguisher control valve.
3. Aim the nozzle at the surface of the burning oil, squeeze the trigger, and then discharge the wet-chemical stream so it drops gently into the surface of the burning liquid at the front or the back of the deep-fat fryer. You must be within 8 to 12 ft (2.5 to 4 m) of the fire to be effective. Let the deep foam blanket flow across the surface of the burning liquid. Avoid splashing foam on the burning liquid.
4. Do not disturb the foam blanket even after all of the flames have been suppressed. If reignition occurs, repeat these steps.

The Care of Fire Extinguishers

Fire extinguishers must be regularly inspected and properly maintained to ensure that they will be available for use in an emergency. Records must be kept confirming that the required inspections and maintenance have been performed on schedule. The individuals assigned to perform these functions must be properly trained and must always follow the manufacturer's recommendations for inspecting, maintaining, recharging, and testing the equipment. Persons performing maintenance or recharging of portable fire extinguishers should be qualified according to the requirements in NFPA 10.

Inspection

According to NFPA 10, an inspection is a quick check to verify that a fire extinguisher is available and ready for immediate use. Fire extinguishers on fire apparatus should be inspected as part of the regular equipment checks mandated by your department once per month. The firefighter charged with inspecting the fire extinguishers should perform the following tasks:

- Ensure that all tamper seals are intact.
- Examine all parts for signs of physical damage, corrosion, or leakage.
- Check the pressure gauge to confirm that it is in the operable range. Some stored-pressure fire extinguishers use compressed air, whereas others use compressed nitrogen. The weight of the fire extinguisher and the presence of an intact tamper seal should indicate that the unit is full of extinguishing agent.
- If the extinguisher is a cartridge-type fire extinguisher, check that the cartridge has not been punctured.
- If the extinguisher is a CO_2 fire extinguisher, determine fullness by weighing the fire extinguisher. The proper weight is labeled on the outside of the fire extinguisher.
- Ensure that the fire extinguisher is properly identified by type and rating.
- Check the hose and nozzle for damage or obstruction by foreign objects.
- Check the hydrostatic test date of the fire extinguisher.

If an inspection reveals any problems, the fire extinguisher should be removed from service until the required maintenance procedures are performed. Spare fire extinguishers should be used until the problem is corrected.

Maintenance

The maintenance requirements and intervals for various types of fire extinguishers are outlined in NFPA 10. Maintenance includes an internal inspection as well as any repairs that may be required. These procedures must be performed periodically, depending on the type of fire extinguisher.

Maintenance procedures must always be performed by a qualified person. Some procedures can be performed only at a properly licensed facility. The specific qualifications and training requirements are determined by the manufacturer and the jurisdictional authority. Untrained personnel should never be allowed to perform fire extinguisher maintenance.

Common indications that a fire extinguisher needs maintenance include the following:

- The reading on the pressure gauge is outside the normal range.
- The inspection tag is out-of-date.
- The tamper seal is broken, especially in fire extinguishers with no pressure gauge.
- The fire extinguisher does not appear to be full of extinguishing agent.
- The hose or nozzle assembly is obstructed.
- There are signs of physical damage, corrosion, or rust.
- There are signs of leakage around the discharge valve or nozzle assembly.

Recharging

All performance standards assume that a fire extinguisher will be fully charged when it has to be used. A rechargeable fire extinguisher must be taken out of service immediately and recharged after each and every use, even if it was not completely discharged. This guideline also applies to any fire extinguisher that leaks or has a pressure gauge reading above or below the proper operating range. Most rechargeable fire extinguishers must be recharged by qualified personnel. Non-rechargeable fire extinguishers should be replaced after any use.

When a fire extinguisher is recharged, the extinguishing agent is refilled, and the system that expels the agent is properly pressurized. Both the quantity of the agent and the pressurization must be verified. After recharging is complete, a tamper seal is installed to provide assurance that the fire extinguisher has not been fully or even partially discharged since the last time it was recharged.

Fire extinguishers should be recharged by a trained individual with the proper equipment or a fire extinguisher company that maintains fire extinguishers. If that is not possible, however, typical 2.5-gal (9-L) stored-pressure water-type fire extinguishers can be recharged by firefighters using water and a source of compressed air. Before the fire extinguisher is refilled, all remaining stored pressure must be discharged so that the valve assembly can be safely removed. Water is then added up to the water-level indicator, and the valve assembly is replaced. Finally, compressed air is introduced to raise the pressure to the level indicated on the gauge.

Hydrostatic Testing

Most fire extinguishers are pressurized vessels, meaning that they are designed to hold a steady internal pressure. The ability of a fire extinguisher to withstand this internal pressure is measured by periodic **hydrostatic testing**. The hydrostatic testing requirements for fire extinguishers are established by the U.S. Department of Transportation. These tests are conducted in a special test facility and involve filling the fire extinguisher with water and applying above-normal pressure.

Each fire extinguisher has an assigned maximum interval between hydrostatic tests, usually 5 or 12 years, depending on the construction material and vessel type (**TABLE 9-2**). The date of the most recent hydrostatic test must be indicated on the outside of the fire extinguisher. A fire extinguisher may not be refilled if the date of the most recent hydrostatic test is not within the prescribed limit. Any fire extinguisher that is out-of-date should be removed from service and sent to the appropriate maintenance facility for hydrostatic testing.

TABLE 9-2 Hydrostatic Test Interval for Fire Extinguishers

Fire Extinguisher Type	Test Interval (Years)
Stored-pressure water-loaded stream and/or antifreeze agent	5
Wetting agent	5
AFFF (aqueous film-forming foam)	5
FFFP (film-forming fluoroprotein foam)	5
Dry chemical with stainless-steel shells	5
Carbon dioxide	5
Wet chemical	5
Dry chemical, stored pressure, with mild steel shells, brazed brass shells, or aluminum shells	12
Dry chemical, cartridge or cylinder operated, with mild steel shells	12
Halogenated agents	12
Dry powder, stored pressure, cartridge or cylinder operated, with mild steel shells	12

Source: Reproduced from National Fire Protection Association (NFPA). 2021. NFPA 10, *Standard for Portable Fire Extinguishers*. 2022 Edition. Quincy, MA: NFPA. https://www.nfpa.org/codes-and-standards/1/0/10.

CASE STUDY

You Are the Firefighter CONCLUSION

You are the designated on-duty cook for the day, and the kitchen is bare. You go into the grocery store while the rest of your crew waits in the apparatus. As you make your way through the produce section, an employee runs up to you and says that there is a fire in the back room. You quickly radio your officer to inform them of the situation, and then immediately proceed to the back room. You locate a small fire in a trash can that is just starting to spread. You quickly scan the room looking for a fire extinguisher. Not seeing one, you ask the employee where the closest extinguisher is located.

1. **What type of fire extinguisher would be most effective for this fire? Why?**

   Answer: An extinguisher rated for Class A fires would be most effective. Class A fires involve ordinary solid combustible materials such as wood, paper, cloth, rubber, and household rubbish. A water-type fire extinguisher would be ideal as the water would cool the fuel, removing heat from the fire tetrahedron. If a water-type extinguisher is not available, than a multipurpose dry-chemical extinguisher would be the next best choice.

2. **As the fire grows, at what point will using a fire extinguisher become ineffective?**

   Answer: Fire extinguishers work best when the fire is in its incipient stage. If the fire were to spread beyond this stage, attempting to control it with only a fire extinguisher would most likely not be successful and would prove to be dangerous to the user.

3. **How do fire inspectors ensure the fire extinguisher will work when needed?**

   Answer: Fire extinguishers should be inspected monthly to ensure that all tamper seals are intact; the fire extinguisher is fully charged; there are no signs of physical damage, corrosion, or leakage; the hose and nozzle are not damaged or obstructed by foreign objects; and the hydrostatic test date of the fire extinguisher is current.

WRAP-UP

SUMMARY

KNOWLEDGE OBJECTIVES

- Explain the purpose, classification, and rating system of fire extinguishers and how portable fire extinguishers suppress fires.
  - State the primary purposes of fire extinguishers. (**NFPA 1010: 6.3.16**, pp. 320–321)
- Define the classes of fire and select the type of fire extinguisher used for each.
  - Define Class A fires. (**NFPA 1010: 6.3.16**, pp. 321–324)
  - Define Class B fires. (**NFPA 1010: 6.3.16**, pp. 321–324)
  - Define Class C fires. (**NFPA 1010: 6.3.16**, pp. 321–324)
  - Define Class D fires. (**NFPA 1010: 6.3.16**, pp. 321–324)
  - Define Class K fires. (**NFPA 1010: 6.3.16**, pp. 321–324)
- State the classification of fire extinguishers, method used to rate each class, and the labeling system used to identify each.
  - Describe the classification and rating system for fire extinguishers. (**NFPA 1010: 6.3.16**, pp. 322–324)
  - Describe the labeling system for fire extinguishers. (**NFPA 1010: 6.3.16**, pp. 322–324)
- Describe how the placement of fire extinguishers is determined, including how to estimate fire load while being able to discuss the three area hazard classifications.
  - Describe the three risk classifications for area hazards. (**NFPA 1010: 6.3.16**, pp. 324–326)

KNOWLEDGE OBJECTIVES CONTINUED

- Describe how fire extinguishers are designed and their basic components.
  - Identify the purpose of the components of a portable fire extinguisher, including cylinders, handles, nozzles, triggers, locking mechanisms, and pressure indicators. (pp. 327–328)
- Describe the types of agents and operating systems used in fire extinguishers and fire extinguishing systems, and give examples of the steps to correctly operate a fire extinguisher.
  - Describe the types of agents and operating systems used in fire extinguishers. (**NFPA 1010: 6.3.16**, pp. 328–337)
  - Describe the basic steps of fire extinguisher operation. (**NFPA 1010: 6.3.16**, p. 337)
  - Select the proper class of fire extinguisher. (**NFPA 1010: 6.3.16**, p. 338)
- Explain the basic steps of inspecting, maintaining, recharging, and hydrostatic testing of fire extinguishers.
  - Explain the basic steps of inspecting and maintaining fire extinguishers. (**NFPA 1010: 6.3.16**, pp. 347–348)
  - Explain the basic steps of recharging and hydrostatic testing of fire extinguishers. (**NFPA 1010: 6.3.16**, p. 348)

SKILLS OBJECTIVES

- Locate and transport the fire extinguisher to the location of the fire.
  - Transport the fire extinguisher to the location of the fire. (**NFPA 1010: 6.3.16**, pp. 339–340)
- Select and extinguish a Class A, B, C, D, and K fire with a fire extinguisher.
  - Extinguish a Class A fire with a stored-pressure water-type fire extinguisher. (**NFPA 1010: 6.3.16**, pp. 341–342)
  - Extinguish a Class A fire with a multipurpose dry-chemical fire extinguisher. (**NFPA 1010: 6.3.16**, p. 342)
  - Extinguish a Class B flammable liquid fire with a dry-chemical fire extinguisher. (**NFPA 1010: 6.3.16**, p. 343)
  - Extinguish a Class B flammable liquid fire with a stored-pressure foam fire extinguisher. (**NFPA 1010: 6.3.16**, p. 344)
  - Operate a carbon dioxide fire extinguisher. (**NFPA 1010: 6.3.16**, p. 345)
  - Operate a halogenated agent–type fire extinguisher. (**NFPA 1010: 6.3.16**, p. 346)
  - Operate a dry-powder fire extinguisher. (**NFPA 1010: 6.3.16**, p. 346)
  - Operate a wet-chemical fire extinguisher. (**NFPA 1010: 6.3.16**, p. 347)
- Inspect a fire extinguisher using an inspection checklist. (p. 347)
- Fill and pressurize a stored pressure water extinguisher. (p. 348)

KEY TERMS

air-aspirating nozzle A nozzle that draws air into the water stream, creating an aerated or foamy spray that increases the surface area of water droplets, allowing for better heat absorption and faster cooling of a fire, and when used with firefighting foam solutions, aerates the foam mixture.

ammonium phosphate See *monoammonium phosphate*.

aqueous film-forming foam (AFFF) A concentrate based on fluorinated surfactants plus foam stabilizers to produce a fluid aqueous film for suppressing hydrocarbon fuel vapors and usually diluted with water to a 1 percent, 3 percent, or 6 percent solution. (NFPA 11)

backpack fire extinguisher A portable fire extinguisher usually consisting of a 5-gal (19-L) water tank that is worn on the user's back and features a hand-powered piston pump for discharging the water and that is primarily used to fight brush and grass fires.

carbon dioxide (CO_2) A colorless, odorless, electrically nonconductive inert gas that is a suitable medium for extinguishing Class B and Class C fires. (NFPA 10)

carbon dioxide (CO_2) fire extinguisher A fire extinguisher that uses carbon dioxide gas as the extinguishing agent. It is rated for use on Class B and C fires.

cartridge/cylinder-operated fire extinguisher A fire extinguisher in which the expellant gas is in a separate container from the agent storage container. (NFPA 10)

Class A foam concentrate A concentrate that when combined with water reduces the surface tension of the water and creates a foam. Also called *wet water.*

Class A foam fire extinguisher A fire extinguisher that contains a solution of water and Class A foam concentrate.

clean agent Electrically nonconducting, volatile, or gaseous fire extinguishant that does not leave a residue upon evaporation. (NFPA 10)

clean-agent fire extinguisher A fire extinguisher that uses a halogenated extinguishing agent. Also called *halogenated-agent fire extinguisher.*

container See *cylinder.*

cylinder The body of the fire extinguisher where the extinguishing agent is stored. Also called *container.*

dry chemical A powder composed of very small particles, usually sodium bicarbonate, potassium bicarbonate, or ammonium phosphate based with added particulate material supplemented by special treatment to provide resistance to packing, resistance to moisture absorption (caking), and the proper flow capabilities. (NFPA 10)

dry-chemical fire extinguisher A fire extinguisher that uses a dry-chemical extinguishing agent and is usually rated for use on Class B and C fires, and sometimes on Class A fires.

dry powder Solid materials in powder or granular form designed to extinguish Class D combustible metal fires by crusting, smothering, or heat-transferring means. (NFPA 10)

dry-powder fire extinguisher A fire extinguisher that uses solid materials in powder or granular form to extinguish Class D combustible metal fires by crusting, smothering, or heat-transferring means.

extinguishing agent A material used to stop the combustion process. Extinguishing agents may include liquids, gases, dry-chemical compounds, and dry-powder compounds.

extra hazard area An occupancy where the total amount of Class A combustibles and Class B flammables is greater than expected in occupancies classed as ordinary (moderate) hazards and the combustibility and heat release rate of the materials are high. Also called *high hazard area.*

film-forming fluoroprotein (FFFP) foam A protein-foam solution that uses fluorinated surfactants to produce a fluid aqueous film for suppressing liquid fuel vapors. (NFPA 10)

fire load The total energy content of combustible materials in a building, space, or area including furnishing and contents and combustible building elements expressed in MJ. (NFPA 557)

halocarbon See *halogenated hydrocarbon.*

halogen A family of elements (chemicals listed in the periodic table), including fluorine (F), bromine (Br), iodine (I), and chlorine (Cl), that are chemically related.

halogenated-agent fire extinguisher A fire extinguisher that uses a halogenated extinguishing agent. Also called *clean-agent fire extinguisher.*

halogenated extinguishing agent A liquefied gas extinguishing agent that extinguishes fire by chemically interrupting the combustion reaction between fuel and oxygen. Halogenated agents leave no residue. (NFPA 402)

halogenated hydrocarbon A hydrocarbon in which at least one hydrogen atom of the hydrocarbon is replaced by a halogen. Also called *halocarbon.*

handle The grip used for holding and carrying a portable fire extinguisher.

high hazard area See *extra hazard area.*

horn The tapered discharge nozzle of a carbon dioxide fire extinguisher.

hydroflurocarbon (HFC) A common type of halocarbon used in fire suppression.

hydrophobic The quality of repelling or being unable to mix with water.

hydrostatic testing Pressure testing of a fire extinguisher to verify its strength against unwanted rupture. (NFPA 10)

light hazard area An occupancy where the quantity, combustibility, and heat release of the materials is low, and the majority of materials are arranged so that a fire is not likely to spread. Also called *low hazard area.*

loaded-stream fire extinguisher A stored-pressure water-type fire extinguisher that uses an alkali metal salt as a freezing-point depressant.

KEY TERMS CONTINUED

locking mechanism A device that locks a fire extinguisher's trigger to prevent its accidental discharge.

low hazard area See *light hazard area.*

moderate hazard area See *ordinary hazard area.*

monoammonium phosphate A finely ground substance that looks like yellow talcum powder that is used as an extinguishing agent in dry-chemical fire extinguishers that are rated for Class A, B, and C fires. Also called *ammonium phosphate.*

multipurpose dry-chemical fire extinguisher A fire extinguisher that uses a monoammonium phosphate–based extinguishing agent that is effective on fires involving ordinary combustibles, such as wood or paper, and fires involving flammable liquids and that is rated to fight Class A, B, and C fires.

nozzle A device for use in applications requiring special water discharge patterns, directional spray, or other unusual discharge characteristics. (NFPA 13)

ordinary dry-chemical fire extinguisher A dry-chemical fire extinguisher rated for only Class B and Class C fires.

ordinary hazard area An area that contains more Class A and Class B materials than a light hazard area and the combustibility and heat release rate of the materials is moderate. Also called *moderate hazard area.*

overhaul A firefighting term involving the process of final extinguishment after the main body of the fire has been knocked down. All traces of fire must be extinguished at this time. (NFPA 1700)

PASS Acronym for the steps involved in operating a portable fire extinguisher: Pull pin, Aim nozzle, Squeeze trigger, Sweep across burning fuel.

polar solvent A water-soluble flammable liquid such as alcohol, acetone, ester, and ketone.

preincident plan A document developed by gathering general and detailed data that is used by responding personnel in effectively managing emergencies for the protection of occupants, responding personnel, property, and the environment. (NFPA 1660)

pressure indicator A gauge on a pressurized portable fire extinguisher that indicates the internal pressure of the expellant.

pump tank fire extinguisher A nonpressurized, manually operated water-type fire extinguisher that is rated for use on Class A fires. Discharge pressure is provided by a hand-operated, double-acting piston pump.

saponification The process of converting the fatty acids in cooking oils or fats to soap or foam; the action caused by a Class K fire extinguisher.

self-expelling agent An agent that has sufficient vapor pressure at normal operating temperatures to expel itself from a fire extinguisher.

stored-pressure fire extinguisher A fire extinguisher in which both the extinguishing agent and expellant gas are kept in a single container and that includes a pressure indicator or gauge. (NFPA 10)

stored-pressure water-type fire extinguisher A fire extinguisher in which water or a water-based extinguishing agent is stored under pressure.

surfactant A compound that lowers the surface tension (or interfacial tension) between two liquids, between a gas and a liquid, or between a liquid and a solid, and which can act as a detergent, wetting agent, emulsifier, foaming agent, and dispersant. (NFPA 1700)

tamper seal A retaining device that breaks when the locking mechanism is released.

trigger The button or lever used to discharge the agent from a portable fire extinguisher.

water mist fire extinguisher A fire extinguisher containing distilled or de-ionized water and employing a nozzle that discharges the agent in a fine spray. (NFPA 10)

wet-chemical extinguishing agent Normally, an aqueous solution of organic or inorganic salts or a combination thereof that forms an extinguishing agent. (NFPA 10)

wet-chemical fire extinguisher A fire extinguisher containing a wet-chemical extinguishing agent for use on Class K fires.

wetting-agent fire extinguisher A fire extinguisher that expels water combined with a concentrate to reduce the surface tension and increase its ability to penetrate and spread.

wet water See *Class A foam concentrate.*

wheeled fire extinguisher A portable fire extinguisher equipped with a carriage and wheels intended to be transported to the fire by one person. (NFPA 10)

REVIEW QUESTIONS

1. What is an extinguishing agent?
2. Which part of the fire tetrahedron should you remove to stop the combustion process?
3. What does a fire extinguisher rating of 2-A:10-B:C tell you?
4. What is the minimum distance above the floor that the bottom of a fire extinguisher should be in multifamily, public, and commercial buildings?
5. What is the term for the retaining device that breaks when the locking mechanism is released?
6. Which type of fire extinguisher can typically be used to extinguish Class A, Class B, or Class C fires?
7. What is a polar solvent?
8. Which extinguishing agent should be used on Class K fires?
9. What does the acronym PASS stand for?
10. How do you determine if a carbon dioxide fire extinguisher is completely filled?
11. What does hydrostatic testing confirm?

DISCUSSION QUESTIONS

1. What type of extinguishing agent would you use on a Class C fire? Why? Describe which is best for areas that contain sensitive electronic equipment and explain why.
2. Why is it essential for firefighters to understand the characteristics and operation of the various types of fire extinguishers?
3. While doing an apparatus check at the start of your shift, you notice that the tamper seal is missing from the locking pin on the CO_2 extinguisher. What do you do?
4. You are assigned to a ladder company and are dispatched to a fire on a stove in a small local restaurant. The officer yells back that the engine is delayed, and your company will arrive first. The officer tells your crew to grab a fire extinguisher to bring into the restaurant. What type of extinguisher do you choose?

APPLYING THE CONCEPTS

Your fire company receives a call for an outdoor fire at a salvage yard that's located two blocks from your fire station. Dispatch reports car parts are on fire and the flames are "dazzling white." Your four-person crew gears up and heads to the fire in a fire engine. A ladder truck from another unit was called, as was a hazardous materials unit. The mobile data terminal shows that the closest fire hydrant is 300 yards (274 m) from the salvage yard entrance. You'll be the first arriving unit on scene.

1. Based on this information, what do you consider as you mentally size-up the situation while enroute?

When you arrive, the owner tells your crew the fire contains magnesium car parts. This explains the bright white flames. During size-up, you determine the fire is still in the incipient stage. From preplanning this business, your crew knows it has ABC extinguishers and 30-pound (14 kg) Class D extinguishers. It also has a fire suppression system; however, since the fire is outside, this system won't help. You and your crew consider extinguishment strategies.

2. Should you run hose lines from the hydrant or use water from the fire engine? Why or why not?
3. So, what type of extinguishing method should you use? Why?

The owner tells your crew where the extinguishers in the yard area are stored. You're assigned to retrieve and use the 30-lb (14 kg) class D extinguisher, which is larger than the one stored on your fire engine. When you get to the storage area, you see several different types of extinguishers.

4. What clues tell you the correct extinguisher to choose?
5. How does the dry powder in the fire extinguisher work to put out a fire?
6. What's the proper way to use a dry-powder fire extinguisher?

REFERENCES

CHEMGUARD. n.d. "Use and Benefit of Class 'A' Foam Concentrate in Water." Accessed November 12, 2022. https://www.chemguard.com/about-us/documents-library/foam-info/class-a-foam-concentrate.htm.

Conroy, Mark. 2003. *NFPA Guide to Portable Fire Extinguishers.* Burlington, MA: Jones & Bartlett Learning.

Finnerty, Anthony E., and Lawrence J. Vande Kieft. 1997. "Fire-Extinguishing Powders." Paper presented at Halon Options Technical Working Conference, May 6–8, 1997. Accessed September 16, 2023. https://www.nist.gov/system/files/documents/el/fire_research/R0301118.pdf.

Firetrace International. 2019. "Exploring the Benefits of Clean Agents." April 22, 2019. Accessed November 12, 2022. https://www.firetrace.com/fire-protection-blog/exploring-the-benefits-of-clean-agents?hsLang=en.

Firetrace International. n.d. "FM-200™ Fire Suppression Systems." Accessed November 12, 2022. https://www.firetrace.com/en/fm-200.

Griffin, Bill, and Mike Morgan. *60 Years of Commercial Kitchen Fire Suppression. ASHRAE Journal* 56, no. 7 (June 2014): 48–58.

National Fire Protection Association (NFPA). 2017. *Home Structure Fires.* Quincy, MA.

National Fire Protection Association (NFPA). 2018. *NFPA 402, Guide for Aircraft Rescue and Fire-Fighting Operations.* 2019 Edition. Quincy, MA: NFPA.

National Fire Protection Association (NFPA). 2020. *NFPA 11, Standard for Low-, Medium-, and High-Expansion Foam.* 2021 Edition. Quincy, MA: NFPA.

National Fire Protection Association (NFPA). 2020. *NFPA 1700, Guide for Structural Fire Fighting.* 2021 Edition. Quincy, MA: NFPA.

National Fire Protection Association (NFPA). 2021. *NFPA 1, Fire Code.* 2021 Edition. Quincy, MA: NFPA.

National Fire Protection Association (NFPA). 2021. *NFPA 10, Standard for Portable Fire Extinguishers.* 2022 Edition. Quincy, MA: NFPA.

National Fire Protection Association (NFPA). 2021. *NFPA 13, Standard for the Installation of Sprinkler Systems.* 2022 Edition. Quincy, MA: NFPA.

National Fire Protection Association (NFPA). 2021. *NFPA 408, Standard for Aircraft Hand Portable Fire Extinguishers.* 2022 Edition. Quincy, MA: NFPA.

National Fire Protection Association (NFPA). 2022. *NFPA 557, Standard for Determination of Fire Loads for Use in Structural Fire Protection Design.* 2023 Edition. Quincy, MA: NFPA.

National Fire Protection Association (NFPA). 2023. *NFPA 440, Guide for Aircraft Rescue and Firefighting Operations and Airport/Community Emergency Planning.* 2024 Edition. Quincy, MA: NFPA.

National Fire Protection Association (NFPA). 2023. *NFPA 1660, Standard for Emergency, Continuity, and Crisis Management: Preparedness, Response, and Recovery.* 2024 Edition. Quincy, MA: NFPA.

National Fire Protection Association (NFPA). 2023. *NFPA 1900, Standard for Aircraft Rescue and Firefighting Vehicles, Automotive Fire Apparatus, Wildland Fire Apparatus, and Automotive Ambulances.* 2024 Edition. Quincy, MA: NFPA.

Reliable Fire & Security. n.d. "MET-L-X™ and LITH-X™ Extinguishers." Accessed November 14, 2022. https://reliablefire.com/fire-protection/fire-extinguishers-2/met-l-x-and-lith-x-extinguishers/.

Underwriters Laboratories. 2017. "Extinguishers and Extinguishing System Units, FWFZ." Accessed June 18, 2018. http://productspec.ul.com/document.php?id=FWFZ.GuideInfo.

CHAPTER

10

Firefighter I

Water Supply Systems

KNOWLEDGE OBJECTIVES

After studying this chapter, you will be able to:

- Describe the types of water sources and water distribution systems.
- Describe the types of fire hydrants and how they operate.
- Describe the principles of hydraulics and explain how these principles impact field operations.
- Describe how to maintain fire hydrants, including inspecting and testing fire hydrants.
- Describe the apparatus and procedures that are utilized to safeguard the water supply for fire suppression in rural areas.

SKILLS OBJECTIVES

After studying this chapter, you should be able to perform the following skills:

- Operate dry barrel and wet barrel fire hydrants.
- Test fire hydrants.
- Secure rural water supplies.

ADDITIONAL NFPA STANDARDS

- **NFPA 24**, *Standard for the Installation of Private Fire Service Mains and Their Appurtenances, 2022 Edition*
- **NFPA 291**, *Recommended Practice for Water Flow Testing and Marking of Hydrants, 2022 Edition*
- **NFPA 1140**, *Standard for Wildland Fire Protection, 2022 Edition*
- **NFPA 1142**, *Standard on Water Supplies for Suburban and Rural Fire Fighting, 2022 Edition*
- **NFPA 1900**, *Standard for Aircraft Rescue and Firefighting Vehicles, Automotive Fire Apparatus, Wildland Fire Apparatus, and Automotive Ambulances, 2024 Edition*
- **NFPA 1910**, *Standard for the Inspection, Maintenance, Refurbishment, Testing, and Retirement of In-Service Emergency Vehicles and Marine Firefighting Vessels, 2024 Edition*
- **NFPA 1960**, *Standard for Fire Hose Connections, Spray Nozzles, Manufacturer's Design of Fire Department Ground Ladders, Fire Hose, and Powered Rescue Tools, 2024 Edition*
- **NFPA 1962**, *Standard for the Care, Use, Inspection, Service Testing, and Replacement of Fire Hose, Couplings, Nozzles, and Fire Hose Appliances, 2018 Edition*

CASE STUDY

You Are the Firefighter

Your department has been dispatched to a reported building fire. Upon arrival you find a barn that is fully involved. The fire attack is clearly going to be a defensive operation, but aggressive actions will be required to protect a house and other buildings that are located approximately 100 yards (91 meters) away from the barn, which is located on a narrow, tree-lined lane. The incident commander says he needs a 500 gallons per minute (gpm; 1893 liters per minute [L/min]) flow to be established.

1. What water sources might be available?
2. Which type of water supply would be most dependable?
3. Which type of water source would require the most resources to implement?

Introduction

After life safety, one of the primary objectives at a fire is to secure an adequate water supply to get water on the fire and protect exposures. Water cools the fuel, allowing the fire to be extinguished. **Water supply**, the source of water utilized by firefighters; hydraulics, the behavior of flowing water; rural and municipal water supplies; and types of fire hydrants and their operation and maintenance are all critical concepts that a firefighter needs to understand. This chapter covers the water flow during the first half of its journey for the fire service—from the water source to the fire pumper or fire engine.

When determining a water supply, you need to make sure it is reliable and has sufficient volume and pressure to meet the needs of the incident. The importance of a dependable and adequate water supply for fire-suppression operations is self-evident. The charged hose line, that is, one or more lengths of hose coupled together, is not only the primary weapon for fighting fire, but also the firefighter's primary defense against being burned or driven out of a burning building. The basic plan for fighting most fires depends on having an adequate supply of water to safely confine, control, and extinguish the fire.

Water supply cannot be interrupted while crews are working inside a building; firefighters can be injured or killed without an adequate water supply. Firefighters entering a burning building need to be confident that their water supply is both reliable and adequate—for their protection and to extinguish the fire.

The source of water varies in different communities. Some communities have a **municipal water system**, a public water supply network owned and maintained by a government or municipality. Other communities have a **private water system**, a system owned and maintained by a private business entity. These water systems provide water under pressure through fire hydrants (**FIGURE 10-1**). Fire hydrants are usually a reliable and practical source of water, but fire hydrants have limited capacity and will not provide an endless supply of water. Automatic sprinkler systems and standpipe systems may be connected directly to a municipal water source, affecting the available water

FIGURE 10-1 The water that comes from a hydrant is provided by a municipal or private water system.

FIGURE 10-2 Mobile water supply apparatus can deliver limited quantities of water to the scene of a fire.

FIGURE 10-3 Impurities are removed at the water treatment facility.

from the hydrants. Rural or remote areas may depend on a **static water source**, a water source that is not under pressure such as a pond, lake, or stream, or even a swimming tank and private cistern or other body of water. Some areas depend on large water tenders (often called tankers) that carry up to 5000 gallons (gal; 18,927 liters [L]) of water as their water supply. These sources serve as drafting sites for fire department apparatus to obtain and deliver water to the fire scene. To **draft** is the process of drawing water up through a hose from a static water source and into the pump on an apparatus.

Often, the first available water source is the water carried in a tank on the first-arriving fire apparatus. Fire engines carry a minimum of 300 gal (1100 L) of water and may carry up to 1000 gal (3785 L) of water. This tank water can be used in the initial fire attack while the establishment of an adequate, continuous water supply is secured to support continued fire attack and to ensure the safety of the firefighters on the scene (**FIGURE 10-2**). Departments that do not rely on a public water supply and fire hydrants often have a mobile water supply apparatus that can carry 1000 to 5000 gal (3785 to 18,927 L) of water. The operational plan must ensure that an adequate and reliable water supply is available before the tank becomes empty.

Municipal Water Systems

As the name suggests, most municipal water systems are owned and operated by a local government agency, such as a city, county, or special water district, although some municipal water systems are privately owned. Municipal water is supplied to homes, commercial establishments, and industries, as well as to hydrants and fire protection systems in buildings.

A municipal water system has three major components: a water source, a water treatment facility, and a distribution system.

Water Sources

The water source for a municipal water system can be wells, rivers, streams, lakes, or human-made water storage facilities. The source depends on the geographic and hydrologic features of the area. Underground pipelines or open canals supply some cities with water from sources that are many miles away. The water source needs to be large enough to meet the total demands of the service area. Many municipal water systems draw water from multiple sources to ensure a sufficient supply. In addition, most municipal water systems include large storage facilities to ensure that they will be able to meet the community's water supply demands if access to the primary source or courses is interrupted. The backup supply for some systems can provide water for several months or years. In other systems, however, the supply may last only a few days.

Water Treatment Facilities

Municipal water systems include a water treatment facility, where impurities are removed from the water (**FIGURE 10-3**). The nature of the treatment system depends on the quality of the untreated source water. Source water that is clean and clear requires little treatment, but some systems must use extensive filtration to remove impurities and foreign substances. Some treatment facilities use chemicals to remove impurities, kill bacteria and harmful organisms and improve the water's taste. Ultraviolet (UV) light sources are also used to kill bacteria and harmful organisms and to keep the water pure as it moves through the distribution system to individual homes or businesses. All the water in the system must be suitable for drinking.

Water Distribution System

After the water has been treated, it enters the distribution system. The distribution system delivers water from the treatment facility to the end users and fire hydrants through a complex network of underground pipes. This type of pipe is called a **water main**. In most cases, the distribution system also includes pumps, storage tanks, control valves, reservoirs, and other necessary components to ensure that the required volume of water can be delivered where and when it is needed at the required pressure.

Water pressure requirements differ, depending on how the water will be used. Generally, water pressure ranges from 20 to 80 pounds per square inch (psi; 138 to 551 kilopascals [kPa]) at the delivery point. The recommended minimum pressure for water coming from a fire hydrant is 20 psi (138 kPa), but it is possible to operate with lower hydrant pressures under some circumstances. In some locations with high fire risk, higher pressure systems may be present.

Most water distribution systems rely on an arrangement of pumps to provide the required water pressure. Some systems use pumps to provide pressure. If the pumps stop operating, the pressure is lost, and the system is unable to deliver adequate water pressure or volume to the end users or to hydrants. Most systems that use pumps have multiple pumps and backup power supplies to reduce the risk of a service interruption due to a pump failure. The extra pumps can be used to boost the flow when the pressures in the system drop below a predetermined setting, such as for a major fire or during other high demand periods.

SAFETY TIP

During operations at a major fire, the water department should be contacted to boost the water supply and pressure to the area of the alarm.

In a **gravity-feed system**, the water source, water treatment facility, and water storage facilities are located on high ground while the end users live in lower-lying areas, such as a community in a valley (**FIGURE 10-4**). This type of system may not require any pumps, because gravity provides the necessary pressure to deliver the water downhill. In some systems, the difference between the elevation of the water and the end users is so great that the water pressure created by gravity is so high that pressure-control devices are needed to keep the system from being subjected to excessive pressure.

FIGURE 10-4 A gravity-feed system can deliver water to a low-lying community without the need for pumps.

FIGURE 10-5 Water stored in an elevated water storage tower can be delivered to end users under pressure.

Many municipal water supply systems use a combination pump and gravity-feed system to deliver water. Pumps can be used to deliver water from the water treatment facility to an **elevated water storage tower** or to storage tanks located on hills or high ground. The elevated water storage facilities maintain the desired water pressure in the distribution system, ensuring that water can be delivered under pressure even if the pumps are not operating (**FIGURE 10-5**). When the elevated water storage facilities need refilling, large supply pumps are used. Additional pumps may be installed to increase the pressure in particular areas, such as for a neighborhood on a hilltop.

Combination pump and gravity-feed systems must maintain enough water in elevated storage tanks and in reservoirs to meet anticipated demands. When more water is used than the pumps can supply or if the pumps are out of service, some systems can operate for several days using their elevated storage reserves, but others may be able to function for only a few hours before the system will become empty.

The underground distribution system that delivers water to end users is made up of three sizes of water main (**FIGURE 10-6**). A large main is called a **primary feeder** or **trunk line** and carries large quantities of water to a section of the municipality. A smaller main is called a **secondary feeder** or **branch line**, and is connected to the primary feeder and distributes water to a smaller area such as a neighborhood. The smallest type of main, called a **distributor pipe**, or simply **distributor**, carries water to the users and to hydrants along individual streets.

The size of water mains varies depending on the amount of water needed both for normal consumption and for fire protection in each location. Most jurisdictions specify the minimum size main that can be installed in to ensure an adequate flow. Older systems, however, may have undersized water mains. In addition, the volume of water delivered through a water main may decrease if the pipe becomes corroded or partly filled with sediment.

Water mains in a well-designed distribution system are laid out in a grid pattern. A grid arrangement provides water flow to a fire hydrant from two or more directions and establishes multiple paths from the water source to each area. This helps ensure an adequate flow of water for firefighting. The grid design also helps minimize downtime for the other portions of the system if a water main breaks or needs maintenance work because the water flow can be diverted around the affected section.

Older water distribution systems may have **dead-end water main**, which is a main that is supplied from only one direction. Fire hydrants on a dead-end main have a limited water supply. If two or more hydrants on the same dead-end main are used to fight a fire, the hydrant closest to the water source may have more water and greater water pressure than the hydrants farther from the water supply.

Control valves installed at intervals throughout a water distribution system allow different sections of the water supply to be turned off or isolated. These valves

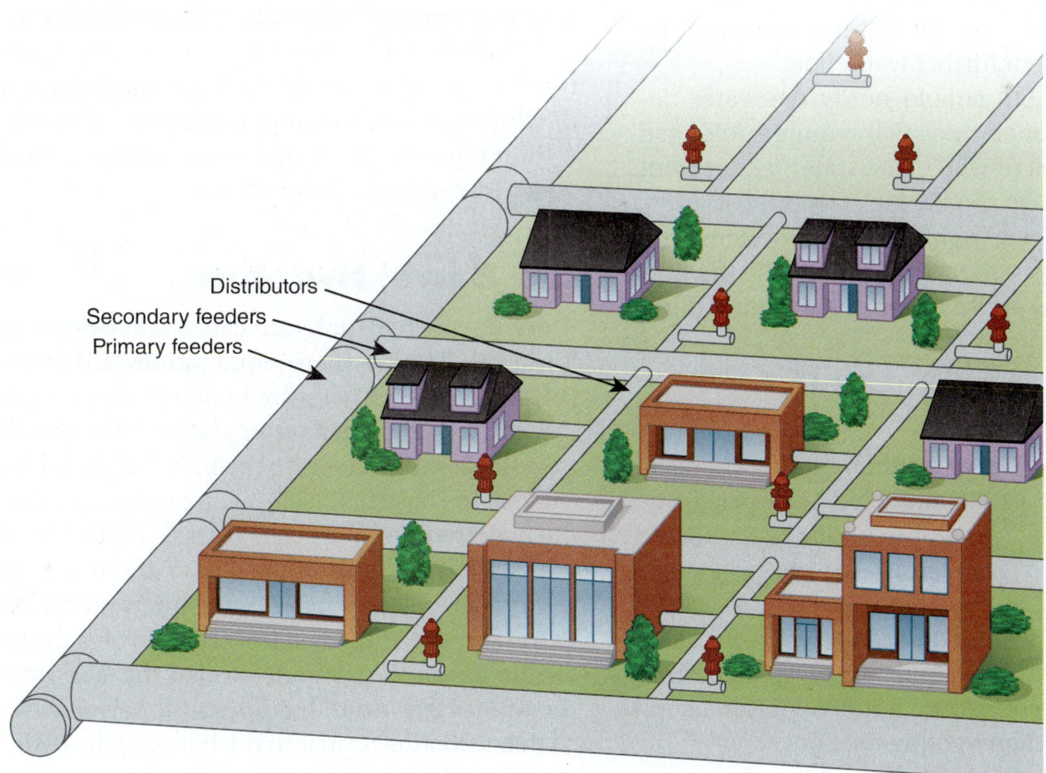

FIGURE 10-6 The underground distribution system includes three different sizes of water main: primary feeders, secondary feeders, and distributors.

FIGURE 10-7 A shut-off valve controls the water supply to an individual user or fire hydrant. The shut-off valve is located under the cover.

are used when a water main breaks or when work must be performed on a section of the system.

A **shut-off valve** is located at each connection point where the underground mains meet the distributor pipes. These valves control the flow of water to individual customers or to individual fire hydrants (**FIGURE 10-7**). If the water system in a building or to a fire hydrant is damaged, firefighters can close the shut-off valve to prevent further water flow.

The fire department should notify the water department when fire operations will require prolonged use of large quantities of water. The water department may be able to increase the normal volume and pressure by starting additional pumps. In some systems, the water department can open valves to increase the flow to a certain area in response to fire department operations at major fires.

Fire Hydrants

Fire hydrants provide water for firefighting purposes. Fire hydrants are part of the municipal water distribution system and connect directly to public water mains. Fire hydrants are also installed on private water systems supplied by the municipal water system or by a separate source. The water source, as well as the adequacy and reliability of the supply to private hydrants, must be identified to ensure that a sufficient water supply will be available when fighting fires.

Most fire hydrants consist of an upright steel casing—the **barrel**—attached to the underground water distribution system. An **outlet** or outlets are provided to connect fire department hose to the hydrant. The outlets may be of various sizes depending on the local jurisdiction, although 2½-inch (in.; 64-millimeter [mm]) connections and one larger connection (4½, 5, or 6 in. [114, 127, or 154 mm])—called a **steamer port**—are common. Fire hydrants are equipped with one or more valves to control the flow of water through the hydrant.

Fire hose needs to be connected to the fire hydrant. The connection on the end of the hose that you use to connect hoses to outlets on fire hydrants and fire pumps is called a **coupling**. Couplings are either threaded or nonthreaded. A **threaded hose coupling** has a male hose coupling with threads on the outside on one end of the hose and a female hose coupling with threads on the inside on the other end of the hose. The female hose coupling also has a **swivel**, which is a ring around the coupling that you turn to secure the female coupling around a male coupling without twisting the hose. Nonthreaded couplings are the same on both ends of the hose and interlock with another nonthreaded coupling of the same type. Most hydrants and fire pumps have male outlets and therefore require a female coupling. (Couplings are discussed in more detail in Chapter 11, *Fire Hose, Fire Appliances and Tools, and Nozzles.*)

Threads on fire hydrant outlets are usually the national standard type, although some jurisdictions have their own thread type. When not in use, outlets are covered by a **hydrant cap** (also called a **discharge cap**).

The two most common types of fire hydrants are the dry barrel hydrant and the wet barrel hydrant, both of which are connected to a pressurized water source. A third type of hydrant is the dry hydrant, which drafts water from a static water source.

Dry Barrel Hydrants

A **dry barrel hydrant**, also called **frost-proof hydrant**, is used in climates where temperatures fall below freezing (**FIGURE 10-8**). These fire hydrants are connected to a pressurized municipal water system. The valve that controls the flow of water into the barrel of the hydrant is located at the base of the barrel below the frost line to keep the hydrant from freezing (**FIGURE 10-9**). The length of the barrel depends on the climate and the depth of the water main below ground. Water enters the barrel of a dry barrel hydrant only when it is used. Turning the **stem nut** on the top of the hydrant—called the **bonnet**—rotates the **operating stem** that opens the valve below ground so that water flows up into the barrel of the hydrant. Taking this action will **charge** the hydrant.

Whenever this type of hydrant is not in use, the barrel must remain dry. If the barrel contains standing water,

FIGURE 10-8 A dry barrel hydrant.

Courtesy of American AVK Company.

FIGURE 10-9 A dry barrel hydrant is controlled by an underground valve.

© Jones & Bartlett Learning

FIGURE 10-10 Dry barrel hydrant with a gate valve attached.

© Jerry Bergquist/Shutterstock

it will freeze in cold weather and render the hydrant inoperable. An opening at the bottom of the barrel allows the water to drain out after each use. When the hydrant valve is fully closed, the drain is fully open. When the hydrant valve is opened, the drain closes, which prevents water from being forced out of the drain when the hydrant is under pressure. If the drain becomes clogged, the hydrant may not drain. In this case, it may be necessary to pump water out of the hydrant before it freezes.

Hydrant valves should always be fully opened or fully closed. If the valve is partially opened, the drain is also partially open, so pressurized water can flow out. This leakage can erode the soil around the base of the hydrant and may damage the hydrant or cause the water main to break. A fully opened hydrant valve also allows the maximum flow of water available to fight a fire.

Most dry barrel hydrants contain only one large valve that controls the flow of water to all outlets. Before that valve is opened, each outlet must be connected to a hose or a gate valve, or it must have a hydrant cap firmly in place before the valve is turned on. A **gate valve** is a valve that firefighters attach to a hydrant outlet that can be used to turn off the flow of water at that outlet. This allows firefighters to connect hose to that outlet after the hydrant is charged (**FIGURE 10-10**). While it is not required to use a gate valve, it is the best practice to use one when connecting to a dry barrel hydrant. If a gate valve is not attached before the hydrant is charged, the hydrant would need to be shut off before additional lines could be added.

Wet Barrel Hydrants

A **wet barrel hydrant** is used in locations where temperatures do not drop below freezing. Like dry barrel hydrants, wet barrel hydrants are connected to a

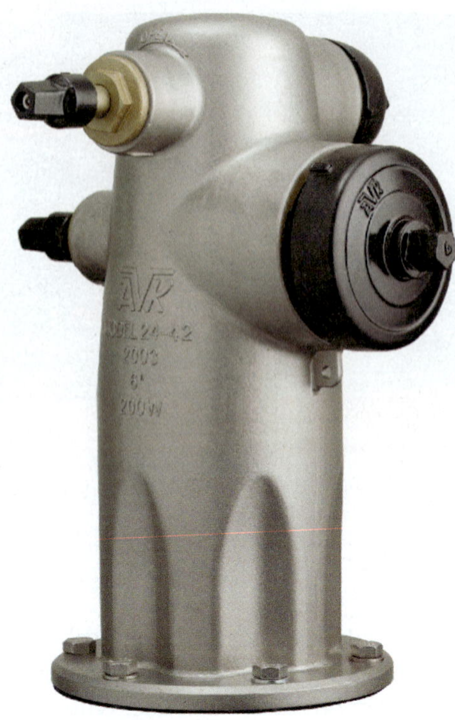

FIGURE 10-11 Each outlet on a wet barrel hydrant has its own valve that controls the flow of water to that outlet.
Courtesy of American AVK Company.

FIGURE 10-12 A wet barrel fire hydrant after being struck by a car.
© Don Bartell/Alamy Stock Photo

pressurized water source. These fire hydrants always have water in the barrel and do not have to be drained after each use. Each outlet on a wet barrel hydrant has its own valve that controls the flow of water to that outlet (**FIGURE 10-11**). The operating nut for each individual valve is on the opposite side of each outlet. A firefighter can hook up one hose line and begin flowing water, and then later attach a second hose line and open the valve for that outlet without first shutting down the hydrant.

Because wet barrel hydrants are always full of water and the control valve is above ground, there is a chance that water will flow uncontrolled if the hydrant is damaged (**FIGURE 10-12**). Newer wet barrel hydrants have a flapper valve to prevent water flow if the hydrant is damaged or struck by a vehicle.

Voice of Experience

One of my most memorable experiences was the very first time I was responsible for obtaining a water supply at a fire. We arrived on scene at a residential working fire and located a wet barrel hydrant approximately 50 feet (ft; 15 meters [m]) in front of the apparatus. I checked the operation of the hydrant and flushed out some debris. Next, I grabbed a 50-ft (15-m) section of 5-in. (127-mm) supply hose to connect to the steamer port on the hydrant and realized I was about 10 ft (3 m) short. I had already charged one attack line from the engine's booster tank, and that was being utilized for fire attack, so I quickly ran back to the truck to grab another, short section of 5-in. (127-mm) supply hose. I raced back to connect this section to the 50-ft (15-m) section of hose and realized that the couplings would not connect. Luckily for me, others came to check and ensure a water supply was established. With their help and the use of a 5-in. (127-mm) spanner wrench, we were finally able to get the hose to connect and complete the establishment of a water supply.

The lesson I learned that day as a driver was that no matter what order you arrive on scene, before considering fire attack, you must ensure a sufficient water supply is established. Also, always check on the person establishing that water supply; they may need your help!

Dena M. Ali

Raleigh Fire Department
Raleigh, North Carolina

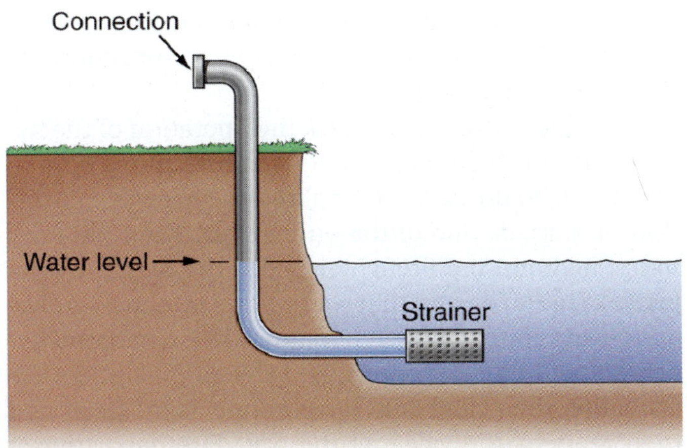

FIGURE 10-13 A dry hydrant, or drafting hydrant, can be placed at an accessible location near a static water source.

© Jones & Bartlett Learning

FIGURE 10-14 The height of the dry hydrant or drafting hydrant connection is convenient for hook-up to an engine or a portable pump.

Courtesy of Ryan Van Buskirk.

Dry Hydrants

A **dry hydrant**, also called a **drafting hydrant**, is a fixed piping system with one end in a static water source and the other end accessible to attach a hose. Do not confuse the terms *dry barrel hydrant* and *dry hydrant*—they are very different. Dry barrel hydrants are connected to a pressurized municipal water system in areas where the temperature drops below freezing. A dry hydrant is a permanent piping system connected to a static water source such as a stream, pond, or lake. The end of the pipe in the water has a strainer on it. The strainer end should be located below the water's surface and away from any silt or potential obstructions. The other end has a connection for a hard suction hose (**FIGURE 10-13**). The connection should be accessible to fire apparatus, and the connection should be at a convenient height for hook-up (**FIGURE 10-14**). To get water from a dry or drafting hydrant, you must draft the water using a fire pump.

Dry hydrants are often installed in lakes and rivers and close to clusters of buildings where there is a recognized need for fire protection. They may also be installed in farm cisterns or connected to swimming tanks on private property to make water available for the local fire department. In some areas, dry hydrants provide access to static water sources in areas that are inaccessible to fire apparatus. They are also used to enable firefighters to reach water under the frozen surface of a lake or river. NFPA 1142, *Standard on Water Supplies for Suburban and Rural Fire Fighting, 2022 Edition*, has more information about dry hydrants.

New Developments

Until recently, fire hydrant design had not changed in over 100 years. Today, we are seeing new hydrant designs and an increased use of mechanisms to prevent tampering and to prevent the public from opening hydrants. For example, the Sigelock Spartan fire hydrant is a new dry barrel hydrant design that offers improved water flow and prevents a dangerous high-pressure surge caused by shutting off water flow in a pressurized system. Because it requires a special wrench that is only available to the fire service, it also reduces tampering (**FIGURE 10-15**). Another anti-tampering device is the Hydra Shield security cap. This hydrant cap can be attached to most existing hydrants and can only be opened with the manufacturer's mating wrench.

Fire Hydrant Locations

Fire hydrants are located according to local standards and nationally recommended practices. They may be placed a certain distance apart, perhaps every 500 ft (152 m) in residential areas and every 300 ft (91 m) in high-value commercial and industrial areas. In many communities, hydrants are located at every street intersection, with mid-block hydrants being installed if the distance between intersections exceeds a specified limit.

In some cases, the requirements for locating hydrants are based on the occupancy, construction type, and size of a building. A builder may be required to install additional hydrants when a new building is

FIGURE 10-15 The Sigelock Spartan fire hydrant.
Courtesy of Sigelock Systems.

constructed so that no part of the building will be more than a specified distance from the closest hydrant.

Knowing the location of fire hydrants in your jurisdiction makes them easier to find in emergency situations. The locations and flow rates of fire hydrants should be identified during preincident surveys and included in preincident plans. Many departments access their community's geographic information system (GIS)—an online mapping system that identifies many types of data that can be located on a map, such as the locations and flow rates of hydrants—using the mobile data terminal (MDT) in the front seat of the apparatus.

Fire Hydrant Operation

Firefighters must be proficient in operating fire hydrants. First, always ensure that the hydrant is operational before use. Debris in the water distribution system may be forced into the barrel of a hydrant, a leaking ground valve may allow water to enter the hydrant barrel and freeze during inclement weather, or vandals may remove caps and place trash or foreign objects into the hydrant. These materials can obstruct the water flow or damage a fire department pumper if they are drawn into the pump.

It is good practice to check the operation of the hydrant and flush out debris before connecting a hose to a hydrant. To do so, the firefighter making the connection first opens one of the outlet caps. Then on a dry barrel hydrant, the firefighter opens the hydrant valve just enough to ensure that water flows into the hydrant and flushes out any foreign matter. On a wet barrel hydrant, the firefighter opens the valve to the outlet. The firefighter then closes the valve, connects the hose, and then reopens the valve all the way. Fire departments should ensure that inspections and tests to keep hydrants operating smoothly are performed regularly.

Operating a Dry Barrel Fire Hydrant

To open a dry barrel hydrant, you remove the cap from the outlet you will be using, and then slowly open the hydrant valve to flush out any debris. After closing the hydrant valve, you attach a hose or gate valve to the outlet, and then slowly and completely reopen the hydrant valve.

Individual fire departments may have their own variations on this procedure. For example, some departments specify that the wrench used to open the hydrant valve be left on the hydrant. Other departments require that the wrench be removed and returned to the apparatus so that an unauthorized person cannot interfere with the operation. Always follow the standard operating procedures (SOPs) for your department.

SKILL DRILL 10-1 outlines the steps for getting water from a dry barrel fire hydrant efficiently and safely.

Shutting a hydrant down properly is just as important as opening a hydrant properly. If the hydrant is damaged during shutdown, it cannot be used until it has been repaired. Dry barrel fire hydrants are located in areas where the temperatures dip below freezing. After the fire is out and before you leave the fire scene, you need to make sure that dry barrel hydrants are completely drained, even if the weather is warm. During winter months, any water left in the hydrant can freeze. Firefighters may lose valuable time connecting a hose to a frozen hydrant, only to discover that it will not operate. If this happens, firefighters will be without water until they can locate a working hydrant and can reposition and reconnect the hose lines. A properly draining hydrant will create suction against a hand placed over the outlet opening. After the hydrant is fully drained, replace the cap. To shut down a dry barrel hydrant efficiently and safely, follow the steps in **SKILL DRILL 10-2**.

SKILL DRILL 10-1

Opening a Dry Barrel Fire Hydrant Firefighter I, NFPA 1010: 6.3.15

1. Remove the cap from the outlet you will be using.

2. Look inside the hydrant opening for debris.

3. Check that the remaining caps are tight.

4. Attach the hydrant wrench to the stem nut located on top of the hydrant. Check the top of the hydrant for an arrow indicating the direction to turn to open.

Continues.

SKILL DRILL 10-1 CONTINUED

Opening a Dry Barrel Fire Hydrant Firefighter I, NFPA 1010: 6.3.15

5. Open the hydrant valve enough to verify flow of water and to flush out any debris that may be in the hydrant.

6. Close the hydrant valve to stop the flow of water. Attach the hose or gate valve to the hydrant outlet.

7. When instructed to do so by your officer or the pump driver/operator, start the flow of water by turning the hydrant wrench slowly to avoid a pressure surge. Continue turning the wrench to fully open the valve. This may take 12 or more turns depending on the type of hydrant.

8. Once the flow of water has begun, you can open the hydrant valve more quickly. Make sure that you open the hydrant valve completely. If the valve is not opened fully, the drain hole will remain open.

SKILL DRILL 10-2

Shutting Down a Dry Barrel Fire Hydrant Firefighter I, NFPA 1010: 6.3.15

1. Turn the hydrant wrench until the stem valve is closed.

2. Allow the hose to drain by opening a drain valve or disconnecting a hose connection downstream. Slowly disconnect the hose from the hydrant outlet, allowing any remaining pressure to escape. If a gate valve is attached, slowly open the gate valve to release pressure in the system, disconnect the hose from the gate valve, and then disconnect the gate valve from the hydrant outlet.

3. Leave one hydrant outlet open until the hydrant is fully drained. If you feel suction on your hand when you place it over the opening, the hydrant is still draining. In very cold weather, you may have to use a hydrant pump to remove all of the water and prevent freezing.

4. Replace the hydrant cap. Do not leave or replace the caps on a dry barrel hydrant until you are sure that the water has completely drained from the barrel.

Operating a Wet Barrel Fire Hydrant

Wet barrel fire hydrants are located in areas where the temperatures do not normally dip below freezing. These hydrants always have water in their barrels and do not require draining. The steps for operating them are different from the steps required to operate a dry barrel hydrant. **SKILL DRILL 10-3** lists the steps for opening a wet barrel fire hydrant efficiently and safely. Follow the steps in **SKILL DRILL 10-4** to shut down a wet barrel fire hydrant.

Fire Hydraulics

Fire-suppression companies are often assigned to test the flow from hydrants in their districts. The procedures for testing hydrants are relatively simple, but a basic understanding of the concepts of hydraulics and careful attention to detail are required. **Fire hydraulics** deals with the properties of energy, pressure, and water flow as related to fire suppression. When operating hose lines at a fire, it is important to understand some basic principles of hydraulics—namely, friction loss in different sizes of hose lines, elevation-related changes in pressure, and the development of water hammers. Firefighters who advance to the position of pump operator will learn more about fire service hydraulics.

Water Flow and Pressure

Water flow and water pressure are two different, but mathematically related, measurements. **Water flow** is the volume of water moving through a pipe, hose, or nozzle over a period of time. It is measured in terms of

SKILL DRILL 10-3

Operating a Wet Barrel Fire Hydrant Firefighter I, NFPA 1010: 6.3.15

1. Remove the cap from the outlet you will be using.
2. Look inside the hydrant opening for debris.
3. Check that the remaining caps are tight.
4. Attach the hydrant wrench to the stem nut located behind the outlet you will be using. Check the hydrant for an arrow indicating the direction to turn to open.
5. Open the hydrant valve enough to verify flow of water and to flush out any debris in the hydrant.
6. Close the hydrant valve to stop the flow of water.
7. Attach the hose or valve to the hydrant outlet.
8. When instructed to do so by your officer or the pump driver/operator, start the flow of water by turning the hydrant wrench slowly to avoid a pressure surge. Continue turning the wrench to fully open the valve. This may take 12 or more turns, depending on the type of hydrant.
9. Once the flow of water has begun, you can open the hydrant valve more quickly. Make sure that you open the hydrant valve completely.

SKILL DRILL 10-4

Shutting Down a Wet Barrel Fire Hydrant Firefighter I, NFPA 1010: 6.3.15

1. Turn the hydrant wrench until the valve opposite the outlet you are using is closed.
2. Allow the hose to drain by opening a drain valve or disconnecting a hose connection downstream. Slowly disconnect the hose from the hydrant outlet, allowing any remaining pressure to escape.
3. Replace the hydrant cap.

its **flow rate**, usually specified in units of gallons per minute (gpm). (Canadian fire departments use the metric system and measure flow in liters per minute [L/min]. The conversion factor is 1 gal = 3.785 L.) **Water pressure** is the force, or amount of energy, per unit area. It is measured in pounds per square inch (psi). (In the metric system, pressures are measured in kilopascals [kPa]. The conversion factor is 1 psi = 6.894 kPa.) Water is a liquid, which means that it cannot be compressed. Water pressure pushes water through a hose, expels water through a nozzle, or lifts water up to a higher level. A pump adds energy to a water stream, causing an increase in pressure.

Water that is not moving has **potential energy**, or energy that it has stored up as a result of its position or condition. When a stream of water flows out through an opening, the energy is converted to kinetic energy. **Kinetic energy** is the energy possessed by an object because of its motion. When the water is moving, it has a combination of potential (stored) energy and kinetic (in motion) energy.

Static means unchanging or at rest. **Static pressure** is the amount of pressure in a system when the water is not moving. Static pressure is potential energy, because it would cause the water to move if there were some place the water could go. Static pressure is what causes the water to flow out of an opened fire hydrant. If no static pressure were present, nothing would happen when the hydrant was opened and the fire hydrant would be inoperable. To measure static pressure in a water distribution system, place a pressure gauge cap on a hydrant outlet, and then open the hydrant valve. No water can be flowing out of the hydrant when a static pressure measurement is being taken.

The static pressure reading assumes that there is no flow in the system. Because municipal water systems deliver water to hundreds or thousands of users, there is almost always some water flowing within the system. Thus, in most cases, a static pressure reading actually measures the normal operating pressure of the system. **Normal operating pressure** is the amount of pressure in a water distribution system during a period of normal consumption. In a residential neighborhood, for example, people are constantly using water to care for lawns, wash clothes, bathe, cook, and conduct other household activities. In an industrial or commercial area, normal consumption occurs during a normal business day as water is used for various purposes. The water distribution system uses some of the static pressure to deliver this water to residents and businesses. A pressure gauge connected to a hydrant during a period of normal consumption will indicate the normal operating pressure of this system, which in this case is recorded as the static pressure. Because the regular users of the system will be drawing off a normal amount of water even during firefighting operations, the normal operating pressure is sufficient for measuring available water. (*Note*: The normal operating pressure may change according to the time of day in some areas owing to high demand during certain hours.)

Residual pressure is the amount of pressure that remains in the system when water is flowing out of it. For example, when firefighters open a hydrant and start to draw large quantities of water out of the system, some of the potential energy of still water has converted into the kinetic energy of moving water. However, not all the potential energy turns into kinetic energy—some of it is used to overcome friction in the pipes. The pressure remaining after this loss is the residual pressure.

Residual pressure is important because it provides the best indication of how much more water is available in the system. As more water flows, there is less residual pressure in the system. In theory, when the maximum amount of water is flowing, the residual pressure is zero, and there is no more potential energy to push more water through and out of the system. The minimum usable residual pressure necessary to reduce the risk of damage to underground water mains or pumps by collapse and prevent damage to the fire engine pump is 20 psi (138 kPa).

Knowing the static pressure, the water flow in gallons (liters) per minute, and the residual pressure enables firefighters to calculate the amount of water that can be obtained from a hydrant or a group of hydrants on the same water main. At the scene of a fire, the pump operator uses the difference between static pressure and residual pressure to determine how many more attack lines or appliances can be operated from the available water supply. To do this, the pump operator refers to a set of tables that are based on the static and residual pressure readings taken during hydrant testing.

Friction Loss

Remember that when water is flowing, not all the potential energy turns into kinetic energy because some of it is used to overcome friction in the pipes. This decrease in pressure caused by the resistance as water moves through a pipe or hose is called **friction loss**. This loss of pressure represents the energy required to push the water through the hose. Friction loss is influenced by the following:

- Flow rate of the water traveling through the hose
- Diameter of the hose
- Distance the water travels
- Valves or hose appliances applied to the hose

A higher flow rate produces more friction loss. In a 2½-in. (64-mm) hose, for example, a 300-gpm (1136-L/min) flow causes much more friction loss than a 200-gpm (757-L/min) flow. At any given flow rate, the smaller the diameter of the hose, the greater the friction loss. At a flow rate of 500 gpm (1893 L/min), the friction loss in a 2½-in. (64-mm) hose is much greater than the friction loss in a 4-in. (101-mm) hose. With any combination of flow and diameter, the friction loss is directly proportional to the distance the water travels. At a flow of 250 gpm (946 L/min) in a 2½-in. (64-mm) hose, the friction loss in 200 ft (61 m) of hose is double the friction loss in 100 ft (30 m) of hose. The pump operator must take friction loss into account when setting the discharge pressure on the pump.

Elevation Pressure

Static pressure is generally created by elevation pressure, pump pressure, or both. **Elevation pressure**, sometimes referred to as **head pressure**, is the pressure created by gravity as the water flows from a hilltop reservoir to the water mains in the valley below. An elevated water tank, for example, creates elevation pressure in the water mains due to the difference in height between the water in the water tank and the underground delivery pipes. Similarly, if a fire hose is laid down a hill, the water at the bottom will have additional pressure because of the change in elevation. Conversely, if a fire hose is advanced upstairs to the third floor of a building, it will lose water pressure because of the energy required to lift the water. Pumps create pressure by applying energy from an external source into the system.

The pump operator must take elevation changes into account when setting the discharge pressure on the pump. For every 10 ft (3 m) of elevation, the pump pressure must be increased 4.34 psi (29.92 kPa). Conversely, for every 10 ft (3 m) of elevation loss, there is an additional downward pressure of 4.34 psi (29.92 kPa) and the pump pressure must be adjusted to account for this. Many pump operators round up to 5 psi (35 kPa) per story, but don't count the first floor. So, if an engine company is working on the fifth floor of a multistory building, the pump operator would add 20 psi (138 kPa) to account for elevation loss.

Water Hammer

Water hammer is a surge in pressure caused by suddenly stopping the flow of a stream of water. A fast-moving stream of water has a large amount of kinetic energy. If the water suddenly stops moving when a valve is closed abruptly, all the kinetic energy is converted to an instantaneous increase in pressure. Because water cannot be compressed, the additional pressure is transmitted along the hose or pipe as a shock wave. This water hammer can rupture a hose, cause a coupling to separate, or damage the plumbing on a piece of fire apparatus. Severe water hammer can damage an underground piping system. Firefighters have been injured by equipment that was damaged by a water hammer.

A similar situation can occur if a valve is opened too quickly, and a surge of pressurized water suddenly fills a hose. The surge in pressure can damage the hose or cause the firefighter at the nozzle to lose control of the stream.

To prevent water hammer, always open and close fire hydrant valves slowly. Pump operators need to open and close the valves on fire engines slowly. When operating the nozzle on an attack line, the firefighter must open the nozzle slowly. Just as important, when closing the shut-off valve on an attack line, the firefighter must do it slowly.

Inspecting, Maintaining, and Testing Fire Hydrants

Because fire hydrants are essential to fire-suppression efforts, firefighters must understand how to inspect, maintain, and test them. Fire hydrants should be checked on a regular schedule—no less than once per year—to ensure that they are in proper operating condition. During inspections, firefighters may encounter some common problems, and they should know how to correct them.

TIP

Depending on the jurisdiction, inspections of fire hydrants may fall to the fire department that utilizes the system or the water department that maintains it. Regardless, firefighters should understand the inspection process and what is required in their jurisdiction.

FIGURE 10-16 Fire hydrants should not be hidden or obstructed.

Courtesy of Captain David Jackson, Saginaw Township Fire Department.

Inspecting Fire Hydrants

When inspecting a fire hydrant, the first thing to check is its visibility and accessibility. Fire hydrants should be visible from every direction so they can be easily spotted. They should not be hidden by tall grass, brush, fences, debris, dumpsters, or any other obstructions (**FIGURE 10-16**). In many communities, fire hydrants are painted in bright reflective colors for increased visibility. Colored reflectors are sometimes mounted next to fire hydrants or placed in the pavement in front of them to make them more visible at night. In winter, fire hydrants must be clear of snow. In areas that experience large amounts of snow, fire hydrants are sometimes marked with a small flag mounted on top of a pole that is attached to the hydrant. In addition, all jurisdictions have laws prohibiting parking within a specified distance from a fire hydrant but sometimes the public takes a chance. During an inspection, look for evidence of someone habitually parking in front of the hydrant.

Fire hydrants are classified based on their flow rate. The bonnet and hydrant caps may be color coded to indicate the flow rate of the fire hydrant. While some jurisdictions use their own color-coding system, the National Fire Protection Association (NFPA) specifies colors to indicate the water flow available from each hydrant at 20 psi (140 kPa) residual pressure in NFPA 291, *Recommended Practice for Fire Flow Testing and Marking of Hydrants, 2022 Edition* (**TABLE 10-1**).

Fire hydrants should be installed at an appropriate height above the ground. Their outlets should not be so high or so low that firefighters have difficulty removing

TABLE 10-1 Fire Hydrant Colors Indicating Available Flow Rates

Class	Flow Available at 20 psi (138 kPa)	Color
Class AA	1500 gpm (5700 L/min) and higher	Light blue
Class A	1000–1499 gpm (3800–5699 L/min)	Green
Class B	500–999 gpm (1900–3799 L/min)	Orange
Class C	Less than 500 gpm (1900 L/min)	Red

Modified from National Fire Protection Association (NFPA). 2021. NFPA 291, *Recommended Practice for Fire Flow Testing and Marking of Hydrants.* 2022 Edition. Quincy, MA: NFPA.

the hydrant caps with a wrench and connecting hose lines to them. NFPA 24, *Standard for the Installation of Private Fire Service Mains and Their Appurtenances, 2022 Edition*, requires a minimum of 18 in. (450 mm) from the center of a hose outlet to the finished grade. Fire hydrants that experience snow accumulations may require hydrants to be placed farther above the finish grade. Fire hydrants should be positioned so that the connections—especially the steamer port—face the street.

During a fire hydrant inspection, check the exterior of the hydrant for signs of damage. Open the steamer port of dry barrel hydrants to ensure that the barrel is dry and free of debris (**FIGURE 10-17**). Make sure that all caps are present and that the outlet threads are in good working order and lightly greased. If the threads on the discharge ports need cleaning, use a steel wire brush and a small triangular file to remove any burrs in the threads. Check the gaskets in the caps to make sure they are not cracked, broken, or missing, and replace worn gaskets with new ones. Follow the manufacturer's recommendations for maintaining any parts that require lubrication.

The second part of the inspection ensures that the hydrant works properly. Open the hydrant valve just enough to confirm that water flows out and flushes any debris out of the barrel. After flushing, shut down the hydrant. A properly draining hydrant will create suction against a hand placed over the outlet opening. When the hydrant is fully drained, replace the cap.

FIGURE 10-17 All hydrants should be checked at least annually.

Testing Fire Hydrants

Firefighters need to know if hydrants can deliver enough water at the needed pressure to enable them to control a fire at that location. If a hydrant cannot do this, the firefighters need to know what can be done to improve the water supply and how additional water will be obtained if needed. To determine this, the water flow at hydrants must be tested. NFPA 291 recommends that hydrants should be flow tested at least every 5 years.

The procedure for testing hydrant flows requires two adjacent hydrants on the same water main and elevation, a flow meter or a Pitot gauge, and a **cap gauge**, which is an outlet cap with a built-in pressure gauge (**FIGURE 10-18**). As part of testing, firefighters first measure the static and residual pressures

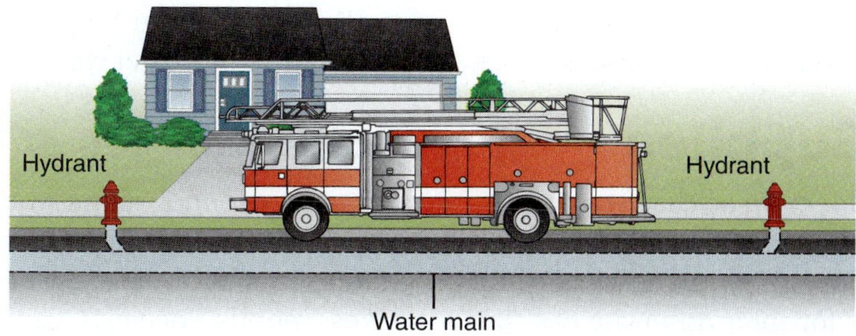

FIGURE 10-18 Testing hydrant flow requires two hydrants on the same water main.

© Jones & Bartlett Learning

at one hydrant. This hydrant is referred to as the residual hydrant. Then the other hydrant is opened to let water flow out. This second hydrant is the flow hydrant.

The cap gauge is placed on one of the outlets on the residual hydrant. If the hydrant is a dry barrel hydrant, that hydrant valve is opened to allow water to fill the hydrant barrel. If the hydrant is a wet barrel hydrant, the valve for that outlet is opened. The initial pressure reading on this gauge is recorded as the static pressure.

To find the water flow at the flow hydrant, firefighters use either a flow meter or a Pitot gauge. A **flow meter** measures water flow in gallons (or liters) per minute from an orifice. An **orifice** is simply an opening, but in fire hydraulics, it is an opening whose diameter is used in calculations. A **Pitot gauge** is a type of pressure gauge that measures the water pressure at the center of the water stream as it passes through an orifice (**FIGURE 10-19**). This measurement, along with the size and flow characteristics of the orifice, is then used to calculate the water flow. Although using a flow meter is easier, they are not always available, so you also should be familiar with using a Pitot gauge.

At the flow hydrant, firefighters remove one of the outlet caps. If they are using a flow meter, they attach it to the outlet. If they are using a Pitot gauge, they need to measure the internal diameter of the outlet. Then they open the hydrant to flow water from that outlet. If the firefighters at the flow hydrant are using a flow meter, they record this reading as the flow rate in gallons (or liters) per minute. If they are using a Pitot gauge, they place the tip in the middle of the stream. This reading is recorded as the Pitot pressure or the flow pressure. At the same time, firefighters at the residual hydrant record the residual pressure reading at that hydrant.

FIGURE 10-19 The Pitot gauge measures water pressure at the center of the water stream as it passes through the orifice on a hydrant.

Courtesy of Eric Scruggs.

TIP

Outlet attachments have smooth tips and brackets that hold the Pitot gauge in the exact required position.

If the firefighters at the flow hydrant used a Pitot gauge, there is an extra step to determine the water flow in gallons (or liters) per minute. To do this, they can either plug the internal diameter of the outlet opening (usually 2½ in. [64 mm]) and the Pitot pressure into a formula or using these two measurements look up the water flow in a table.

After determining the flow rate, firefighters can use special graph paper or software to plot the static pressure and the residual pressure at the test flow rate. The line defined by these two points shows the number of gallons (or liters) per minute that is available at the residual pressure, which should be at least 20 psi (138 kPa).

To test the operability and flow of a fire hydrant using a Pitot gauge, follow the steps in **SKILL DRILL 10-5**.

Rural Water Supplies

Many fire departments protect areas that are not serviced by municipal or private water systems. In these areas, residents usually depend on individual wells or cisterns to supply water for their domestic use. Because there are few or no fire hydrants in these areas, firefighters need to know where water sources are located and how to get water from those sources.

Static Sources of Water

Several potential static water sources can be used for fighting fires in rural areas. Both natural and human-made bodies of water such as rivers, streams, lakes, ponds, oceans, canals, reservoirs, swimming tanks, and cisterns can be used to supply water for fire

SKILL DRILL 10-5

Conducting a Fire Hydrant Flow Test Using a Pitot Gauge Firefighter I

1. At the residual hydrant, remove the cap from the hydrant outlet, open the hydrant, and allow water to flow until it runs clear. Close the hydrant valve. Place a cap gauge on one of the outlets.

2. At the residual hydrant, open the hydrant valve to fill the hydrant barrel. No water should be flowing. Record the initial pressure reading on the gauge as the static pressure.

3. At the flow hydrant, remove one of the discharge caps, measure the inside diameter of the orifice, and then open this hydrant.

4. Place the Pitot gauge tip in the water stream one-half the diameter of the orifice away from the opening. Record this pressure as the Pitot or flow pressure. At the same time, firefighters at the residual hydrant record a second pressure reading and record this as the residual pressure. Use the recorded pressure readings to calculate or look up the flow rates at 20 psi (138 kPa) residual pressure. Document your findings.

FIGURE 10-20 Any accessible body of water can be used as a static source.

© Jones & Bartlett Learning. Photographed by Glen E. Ellman.

FIGURE 10-21 A portable pump can be used if the water source is inaccessible to a fire department engine.

Courtesy of the National Interagency Fire Center.

FIGURE 10-22 Drafting with a floating strainer.

Courtesy of Fol-Da-Tank.

suppression (**FIGURE 10-20**). Some areas have many different static sources, whereas others have few or none.

Water from a static source can be used to fight a fire directly if it is close enough to the fire scene. Otherwise, it must be transported to the fire using long hose lines, engine relays, or mobile water supply apparatus.

Static water sources must be accessible to a fire engine or a portable fire pump. If a road or hard surface is located within 20 ft (6 m) of the static water source, a fire engine can drive close enough to draft water directly from the water source into the pump through a hard suction hose. Some fire departments construct special access points so engines can approach the water source. The portable pump can assist firefighters in accessing water in areas that are inaccessible to fire apparatus (**FIGURE 10-21**). The portable pump can be hand-carried or transported using an off-road vehicle to the water source. Portable pumps can deliver as much as 500 gpm (1893 L/min).

When selecting a drafting site, take into consideration the vertical lift that will be required. Remember for every 10 ft (3 m) in elevation, there is a loss of 4.34 psi (29.92 kPa). Vertical lifts greater than 10 ft (3 m) reduce pumping capacity—the maximum water flow that the pump discharges. A vertical lift of 20 ft (6 m) can reduce an engine's pumping capacity by 40 percent. Fire departments should identify and preplan static water sources in the area and practice establishing drafting operations at all of these locations.

Once the drafting location is identified, firefighters should inspect the swivel gaskets on the coupling. The **swivel gasket** is an O-shaped piece of rubber that sits inside the swivel section of the female hose coupling. When a male hose coupling is tightened against it, the swivel gasket forms a seal that stops water from leaking. If the swivel gasket is damaged or missing, the hose coupling will leak.

Next, the appropriate length of hose is selected and connected, and that a strainer is on the end of the hose. The strainer prevents large debris such as trash, rocks, weeds, small twigs, and animals/fish from entering the pump. Some drafting situations require a floating strainer to prevent suctioning debris and bottom silt (**FIGURE 10-22**). Once the strainer is in place, drafting can begin.

Follow the steps in **SKILL DRILL 10-6** to assist the pump operator with assembling the equipment needed to draft water from a static water supply.

Portable Tanks

Portable tanks carried on fire apparatus can be quickly set up at a fire scene. These tanks typically hold between 600 and 5000 gal (2271 and 18,927 L) of water and should be placed so that they can be accessed from multiple directions. There are two types of portable tanks: metal-frame and self-expanding. Metal-frame portable tanks need to be set up by firefighters before water is added to them. Self-expanding portable tanks are laid flat, and then expand vertically as water is added. Sometimes a firefighter will need to hold the collar of the tank up until the water level is high enough so that the tank can support itself.

With this supply method, a mobile water supply apparatus is used to fill the portable tank while an engine drafts from the portable tank a (**FIGURE 10-23**). As soon as the mobile water supply apparatus discharges all of its water into the portable tank, the driver/operator of that apparatus leaves to get another load of water.

Speed is a primary advantage when using a portable tank system because tankers do not have to hook up to the pumper to transfer the water into the portable tank. Instead, a **dump valve** on the mobile

SKILL DRILL 10-6

Assisting the Pump Operator with Drafting Firefighter I, NFPA 1010: 6.3.15

1. After the pump operator has positioned the engine at the draft site, inspect the coupling for damage or debris.

2. Connect each section of suction hose together and connect the strainer to the end of the hose that will be placed in the water.

SKILL DRILL 10-6 CONTINUED

Assisting the Pump Operator with Drafting Firefighter I, NFPA 1010: 6.3.15

3. The pump operator will connect the other end of the suction hose to the fire pump.

4. Advance the suction hose assembly into position with the strainer in the water.

5. Ensure that the strainer assembly has at least 24 in. (0.6 m) of water in all directions around the strainer.

FIGURE 10-23 An engine may be set up to draft water from a portable tank.

Courtesy of Captain David Jackson, Saginaw Township Fire Department.

FIGURE 10-24 A dump valve allows a mobile water supply apparatus to discharge water into the portable tank quickly.

© Jones & Bartlett Learning. Photographed by Glen E. Ellman.

water supply apparatus enables the tankers to offload as much as 3000 gal (11,356 L) of water in 1 minute (**FIGURE 10-24**). Some apparatus use a pump system to offload the water even faster. The faster the mobile water supply apparatus can offload its water, the quicker it can return to the fill site for another load.

Another advantage of the portable tank system is its ability to expand rapidly. Additional portable tanks can be set up and linked together to increase water storage capacity, additional engines can be used to draft water from the portable tanks, and additional tankers can be used to deliver more water at a faster rate. A series of mobile water supply apparatus may be used as shuttles, dumping water either simultaneously or in sequence.

To set up a portable tank, follow the steps in **SKILL DRILL 10-7**.

SAFETY TIP

Never get between two apparatus or tankers!

Water Shuttle Operations

Water shuttle operations are conducted when large volumes of water from a static water source are needed at a fire scene. They are conducted using mobile water supply apparatus. The number of mobile water supply apparatus needed depends on the distance between the fill site and the fire scene, the time it takes the apparatus to dump and refill, and the flow rate required at the fire scene. In rural areas, several mobile water supply apparatus may be dispatched for a structural fire.

TIP

Mobile water supply apparatus are commonly referred to as tankers or water tenders. In the western part of the United States where aerial water tankers (helicopters and fixed-wing aircraft) are used commonly in fighting wildland fires, the term ***water tender*** means a truck-mounted mobile water supply apparatus. In parts of the country where aerial water tankers are not used commonly, the term ***tanker*** means truck-mounted mobile water supply apparatus.

Fire department engines usually carry at least 500 gal (1893 L) of water in the booster tank, whereas fire department mobile water supply apparatus generally carry 1000 to 3500 gal of water (3785 to 13,249 L). Some tankers can transport as much as 5000 gal (18,927 L) of water.

All the components of a water shuttle operation must be set up so that water moves efficiently from the fill site to the fire scene. If a mobile water supply apparatus is the only source of water for fighting a structural fire, the attack must be carefully planned. Enough water must be available on the scene to supply the required hose lines. Some fire departments begin the attack using water from the booster tank of the first-arriving

SKILL DRILL 10-7

Setting Up a Portable Tank Firefighter I, NFPA 1010: 6.3.15

1. Firefighters lift the portable tank off the apparatus. This tank may be mounted on a side rack or on a hydraulic rack that lowers it to the ground. Place the portable tank on as level ground as possible beside the engine. The pump operator will indicate the best location.

2. Expand the tank (metal-frame type) or lay it flat (self-expanding type).

3. One firefighter helps the pump operator place the strainer on the end of the suction hose, put the suction hose into the tank, and connect it to the engine.

4. A second firefighter helps the mobile water supply apparatus driver/operator discharge water into the portable tank. If the tank is self-expanding, the firefighters may need to hold the collar until the water level is high enough for the tank to support itself.

unit. The mobile water supply apparatus then pump additional water directly into the attack pumper to keep it full. They can also pump into a portable tank, sometimes called a pond.

At the fill site, the tankers must be refilled without delay. The routes in both directions must be planned so that tankers make efficient round trips, without having to back up or make U-turns. At the fire scene, the tankers should be able to drive up, dump their water into the portable tanks, and immediately return to the fill site. The portable tanks must be large enough to receive the full load of each tanker as it arrives. An effective water shuttle operation can deliver several hundred gallons of water per minute without interruption. If your department uses mobile water supply apparatus, you need to learn the specific system used by your area.

Departments must practice water shuttle operations to ensure proper coordination and effective water delivery. An effective operation requires preplanning and excellent leadership. If the water supply is exhausted before the fire is extinguished, the attack team will be in serious danger, and the building will probably be lost. Conversely, if the use of water is too conservative, fire-suppression efforts will probably be unsuccessful.

CASE STUDY

You Are the Firefighter CONCLUSION

Your department has been dispatched to a reported building fire. Upon arrival you find a barn that is fully involved. The fire attack is clearly going to be a defensive operation, but aggressive actions will be required to protect a house and other buildings that are located approximately 100 yards (91 meters) away from the barn, which is located on a narrow, tree-lined lane. The incident commander says he needs a 500 gallons per minute (gpm; 1893 liters per minute [L/min]) flow to be established.

1. **What water sources might be available?**

   Answer: **A working fire hydrant would be the ideal water source, but since this fire involves a barn, it is most likely quite a distance from a fire hydrant. In this case you need to think outside of the box. Is there a pump that feeds the irrigation system for the fields? These pumps usually have a decent water flow and their own well or water source. A pond, lake, or stream can be a good source for drafting. A nearby swimming pool will provide you with a limited amount of water, maybe enough to extinguish the fire. If not, it will provide enough water to help keep the fire controlled until another water source is located.**

2. **Which type of water supply would be most dependable?**

   Answer: **The most dependable water supply would be the one that is readily available and provides a constant amount of available flow. Fire hydrants supplied by a municipal water source would seem like the most dependable, but you need to consider the size of the water main and the age of the system. For example, a fire hydrant that flows 125 gpm (473 L/min) is not sufficient for a fire that requires 250 gpm (946 L/min). A fire engine connected to a dry hydrant at a good water source, such as a large pond or lake, that is able to supply the attack fire engine using a short stretch of large diameter hose would be the most dependable in this scenario.**

3. **Which type of water source would require the most resources to implement?**

   Answer: **Drafting from a water source that is a distance from the fire and requiring the use of mobile water supply apparatus (tenders or tankers) to deliver the water to the incident would require the most resources. A fire engine at the source would be required to draft the water and fill the tenders. Depending on the terrain and location of the water source, and therefore the mobile water supply apparatus, a fire engine pumping from the mobile water supply apparatus to the attack engines would be needed. Finally, a number of water supply apparatus would be required to shuttle water from the drafting site to the portable ponds.**

WRAP-UP

SUMMARY

KNOWLEDGE OBJECTIVES

- Describe the types of water sources and water distribution systems.
  - Describe how municipal water systems supply water to communities. (pp. 357–360)
- Describe the types of fire hydrants and how they operate.
  - List the types of fire hydrants. (pp. 360–363)
  - Describe the characteristics of dry barrel hydrants. (**NFPA 1010: 6.3.15**, p. 363)
  - Describe the characteristics of wet barrel hydrants. (**NFPA 1010: 6.3.15**, pp. 361–363)
  - Describe the common guidelines that govern the location of fire hydrants. (pp. 363–364)
- Describe the principles of hydraulics and explain how these principles impact field operations.
  - Explain the principles of fire hydraulics. (pp. 368–371)
  - Describe how water flow is measured. (pp. 368–371)
  - Describe how water pressure is measured. (pp. 368–371)
  - Compare potential and kinetic energy. (p. 369)
  - Describe the similarities between static pressure and normal operating pressure of a system. (pp. 369–370)
  - Explain how friction loss affects water pressure. (p. 370)
  - Explain how elevation pressure affects water pressure. (p. 370)
  - Describe how to prevent water hammer. (p. 371)
- Describe how to maintain fire hydrants, including inspection and testing fire hydrants.
  - Describe how to inspect a fire hydrant. (pp. 371–372)
  - Describe how to test a fire hydrant. (pp. 372–374)
- Describe the apparatus and procedures that are utilized to safeguard the water supply for fire suppression in rural areas.
  - Describe the equipment and procedures that are used to access static sources of water. (**NFPA 1010: 6.3.15**, pp. 374–380)
  - Describe the advantages of a portable tank system. (**NFPA 1010: 6.3.15**, pp. 376–378)
  - Describe the characteristics of a mobile water supply apparatus. (**NFPA 1010: 6.3.15**, pp. 378–380)

SKILLS OBJECTIVES

- Operate dry barrel and wet barrel fire hydrants.
  - Operate a dry barrel fire hydrant. (**NFPA 1010: 6.3.15**, pp. 365–366)
  - Shut down a dry barrel fire hydrant. (**NFPA 1010: 6.3.15**, p. 367)
  - Operate a wet barrel fire hydrant. (**NFPA 1010: 6.3.15**, pp. 368–369)
  - Shut down a wet barrel fire hydrant. (**NFPA 1010: 6.3.15**, p. 369)
- Test fire hydrants.
  - Conduct a hydrant flow test. (p. 374)
- Secure rural water supplies.
  - Assist the pump driver/operator with drafting. (**NFPA 1010: 6.3.15**, pp. 376–377)
  - Set up a portable tank. (**NFPA 1010: 6.3.15**, p. 379)

KEY TERMS

barrel The upright steel casing that is the main part of a fire hydrant.

bonnet The top of a hydrant.

branch line See *secondary feeder.*

cap gauge An outlet cap with a built-in pressure gauge.

charge To fill with water under pressure.

coupling A connection device that connects (couples) individual lengths of fire hose together connects a hose to a fire hydrant or a pump or to nozzles and hose appliances.

KEY TERMS CONTINUED

dead-end water main A water main that is supplied from only one direction.

discharge cap See *hydrant cap.*

distributor See *distributor pipe.*

distributor pipe The smallest-diameter underground water main pipes in a water distribution system that deliver water to local users within a neighborhood. Also called *distributor.*

draft The use of suction to move a liquid (such as water) from a vessel or source that is below the intake of a pump. (NFPA 1910)

drafting hydrant See *dry hydrant.*

dry barrel hydrant A type of hydrant with the main control valve below the frost line between the footpiece and the barrel. Also called *frost-proof hydrant.* (NFPA 24)

dry hydrant An arrangement of pipe permanently connected to a water source other than a piped, pressurized water supply system that provides a ready means of water supply for firefighting purposes and that utilizes the drafting (suction) capability of a fire department pump. Also called *drafting hydrant.* (NFPA 1142)

dump valve A large opening from the water tank of a mobile water supply apparatus for unloading purposes. (NFPA 1900)

elevated water storage tower An aboveground water storage tank that is designed to maintain pressure on a water distribution system.

elevation pressure The amount of pressure created by gravity. Also called *head pressure.*

fire hydraulics The physical science of how water flows through a pipe or hose.

flow meter A device that measures water flow in gallons (or liters) per minute from an orifice.

flow rate The quantity of water flowing, usually measured in gallons (or liters) per minute.

friction loss The reduction in pressure resulting from the water being in contact with the side of the hose. This contact requires force to overcome the drag that the wall of the hose creates.

frost-proof hydrant See *dry barrel hydrant.*

gate valve A valve firefighters attach to an outlet on a dry hydrant that can turn off the flow of water at that outlet.

gravity-feed system A water distribution system that depends on gravity to provide the required pressure. The system storage is usually located at a higher elevation than the end users of the water.

head pressure See *elevation pressure.*

hydrant cap The cover on a fire hydrant outlet that is in place when the hydrant is not in use. Also called *discharge cap.*

kinetic energy The energy possessed by an object as a result of its motion.

municipal water system A system having water pipes servicing fire hydrants and designed to furnish, over and above domestic consumption, a minimum of 250 gpm (946 L/min) at 20 psi (138 kPa) residual pressure for a 2-hour duration. (NFPA 1140)

normal operating pressure The observed static pressure in a water distribution system during a period of normal demand.

operating stem The steel rod extending from the top of a dry barrel hydrant to the hydrant valve that firefighters can turn using the stem nut at the top to open and close the main valve.

orifice In fire hydraulics, an opening whose diameter is used in calculations.

outlet An opening on a fire hydrant through which water is discharged.

Pitot gauge A type of gauge that is used to measure the velocity pressure of water that is being discharged from an opening. It is used to determine the flow of water from a hydrant or nozzle.

portable tanks Folding or collapsible tanks that are used at the fire scene to hold water for drafting.

potential energy The energy that an object has stored up as a result of its position or condition. A raised weight and a coiled spring have potential energy.

primary feeder The largest-diameter water main pipe in a water distribution system that carries the greatest amounts of water. Also called *trunk line.*

private water system A privately owned water system that operates separately from the municipal water system.

residual pressure The pressure that exists in the distribution system, measured at the residual hydrant at the time the flow readings are taken at the flow hydrants. (NFPA 24)

secondary feeder The smaller-diameter water main pipe in a water distribution system that connects a primary feeder to a distributor. Also called *branch line.*

shut-off valve A valve whose primary function is to operate in either a fully shut off or a fully open condition. (NFPA 1960)

static pressure The pressure that exists at a given point under normal distribution system conditions measured at the residual hydrant with no hydrants flowing. (NFPA 24)

static water source A water source such as a pond, river, stream, or other body of water that is not under pressure.

steamer port The large-diameter port on a fire hydrant.

stem nut The large nut at the top of the operating stem in a dry barrel hydrant that is turned to open the hydrant valve.

swivel A ring around female couplings that you turn to secure the female coupling around the male coupling without twisting the hose.

swivel gasket An O-shaped piece of rubber inside the swivel section of a female hose coupling that forms a seal that stops water from leaking when a male hose coupling is tightened against it.

threaded hose coupling A type of coupling that requires a male fitting and a female fitting to be screwed together.

trunk line See *primary feeder.*

water flow The volume of water moving through a pipe, hose, or nozzle over a period of time, usually expressed in gallons (liters) per minute (gpm or L/min).

water hammer The surge of pressure that occurs when a high-velocity flow of water is abruptly shut off. The pressure exerted by the flowing water against the closed system can be seven or more times that of the static pressure. (NFPA 1962)

water main A generic term for any underground water pipe.

water pressure The application of force by one object against another. When water is forced through the distribution system, it creates water pressure.

water shuttle operations A method of transporting water from a source to a fire scene using a number of mobile water supply apparatus.

water supply A source of water for firefighting activities. (NFPA 1140)

wet barrel hydrant A type of hydrant that is intended for use where there is no danger of freezing weather and where each outlet is provided with a valve and an outlet. (NFPA 24)

REVIEW QUESTIONS

1. What are the three major components of a municipal water system?
2. What is the recommended minimum pressure for water coming from a fire hydrant?
3. What type of fire hydrant is used in areas subjected to below-freezing temperatures?
4. Why is it important when operating from a dry barrel fire hydrant to ensure the hydrant valve is fully opened?
5. How are the locations of fire hydrants in residential areas identified and documented?
6. When hooking to a fire hydrant, either wet barrel or dry barrel, what is the first thing you should do before hooking the hose line to it?
7. In a water system, what is static pressure and what is residual pressure?
8. What is water hammer? Why is this action not recommended?
9. What class and color is the fire hydrant with the greatest amount of flow?
10. When selecting a drafting site at a static water source, why is a vertical lift greater than 10 ft (3 m) a concern?

DISCUSSION QUESTIONS

1. Describe the differences between a dry barrel hydrant and a wet barrel hydrant and explain why one is installed in a jurisdiction over the other.
2. Why is it important for firefighters to understand friction loss?
3. What is a dead-end main and why is it important for you to know where they are and what fire hydrants are connected to them?

APPLYING THE CONCEPTS

You are a volunteer firefighter in a rural fire department. At 1530, you and your engine unit are responding to report of a structure on fire several miles out of town. Dispatch said a passerby saw smoke and made the call. You are familiar with the property and know that it has been vacant for over a year. The area does not have a municipal water source and is not supplied by fire hydrants. Your engine is carrying 1000 gallons (3785 L) of water. As you arrive, you see a growing fire in the two-story clapboard house. There is a large, detached aluminum-clad garage/warehouse behind the house on side Delta, and several other wooden storage structures about 60 feet (18 m) from the house. You exit the fire engine in full personal protective equipment and do a 360-degree inspection.

1. As you size up the situation, what is your first concern?
2. What is the greatest challenge to extinguishing this fire?

During size-up it was noted the wind speed is 8 to 12 miles per hour (13 to 19 kilometers per hour), blowing toward side Delta—and the wooden storage structures. You also saw a large pond about 50 yards (46 m) behind the house. Because your department frequently responds to calls where water source is an issue, your crew has practiced static water operations.

3. How does water supply influence fire attack strategies?
4. What are the options for getting water to the scene?
5. What factors must be considered to determine if using the pond is a viable option?

Based on the size of the fire, limited water supply, and the threat to exposures, the incident commander (IC) opts for a defensive strategy to prevent the spread of the fire. There is a paved road leading to the pond, and the ground has a 4-ft (1.2 m) elevation, so it's possible to use the pond as a water source. Your department's 5000-gallon (18,927 L) water tender will draft from the pond and shuttle the water to a portable tank. Then, the fire engine will draft from the portable tank to supply the attack lines. The IC assigns you and your partner to set up and assist with the portable tank.

6. The pump operator selects the location for the portable tank. After setting it up, how do you and your partner assist the drafting and water supply operations?
7. What challenges do water shuttle operations present and specifically how can they be overcome?

REFERENCES

National Fire Protection Association (NFPA). 2017. *NFPA 1962, Standard for the Care, Use, Inspection, Service Testing, and Replacement of Fire Hose, Couplings, Nozzles, and Fire Hose Appliances.* 2018 Edition. Quincy, MA: NFPA.

National Fire Protection Association (NFPA). 2021. *NFPA 24, Standard for the Installation of Private Fire Service Mains and Their Appurtenances.* 2022 Edition. Quincy, MA: NFPA.

National Fire Protection Association (NFPA). 2021. *NFPA 291, Recommended Practice for Water Flow Testing and Marking of Hydrants.* 2022 Edition. Quincy, MA: NFPA.

National Fire Protection Association (NFPA). 2021. *NFPA 1142, Standard on Water Supplies for Suburban and Rural Fire Firefighting.* 2022 Edition. Quincy, MA: NFPA.

National Fire Protection Association (NFPA). 2023. *NFPA 1140, Standard for Wildland Fire Protection.* 2024 Edition. Quincy, MA: NFPA.

National Fire Protection Association (NFPA). 2023. *NFPA 1900, Standard for Aircraft Rescue and Firefighting Vehicles, Automotive Fire Apparatus, Wildland Fire Apparatus, and Automotive Ambulances.* 2024 Edition. Quincy, MA: NFPA.

National Fire Protection Association (NFPA). 2023. *NFPA 1910, Standard for the Inspection, Maintenance, Refurbishment, Testing, and Retirement of In-Service Emergency Vehicles and Marine Firefighting Vessels.* 2024 Edition. Quincy, MA: NFPA.

National Fire Protection Association (NFPA). 2023. *NFPA 1960, Standard for Fire Hose Connections, Spray Nozzles, Manufacturer's Design of Fire Department Ground Ladders, Fire Hose, and Powered Rescue Tools.* 2024 Edition. Quincy, MA: NFPA.

CHAPTER

11

Firefighter I

Fire Hose, Fire Appliances and Tools, and Nozzles

KNOWLEDGE OBJECTIVES

After studying this chapter, you will be able to:

- List the various types and sizes of fire hose and how they are used.
- Describe the characteristics of the different types of fire hose.
- Describe the characteristics of common types of couplings.
- Identify key usage applications of various types of fire hose, including supply, suction, and attack hose and how hoses are rolled for transport.
- Describe the different types of hose nozzles.
- List the common hose appliances used in conjunction with fire hose.
- Describe how to organize, clean, and maintain hose and the importance of a hose inspection.

SKILLS OBJECTIVES

After studying this chapter, you will be able to perform the following skills:

- Replace the swivel gasket on a fire hose.
- Perform various methods of coupling and uncoupling a fire hose.
- Perform various hose rolls.
- Operate a fog-stream nozzle and smooth-bore nozzle.
- Clean and dry hose.
- Inspect and mark a defective hose.

ADDITIONAL NFPA STANDARDS

- **NFPA 1142**, *Standard on Water Supplies for Suburban and Rural Firefighting, 2022 Edition*
- **NFPA 1410**, *Standard on Training for Emergency Scene Operations, 2020 Edition*
- **NFPA 1900**, *Standard for Aircraft Rescue and Firefighting Vehicles, Automotive Fire Apparatus, Wildland Fire Apparatus, and Automotive Ambulances, 2024 Edition*
- **NFPA 1960**, *Standard for Fire Hose Connections, Spray Nozzles, Manufacture's Design of Fire Department Ground Ladders, Fire Hose, and Power Rescue Tools, 2024 Edition*
- **NFPA 1962**, *Standard for the Care, Use, Inspection, Service Testing, and Replacement of Fire Hose, Couplings, Nozzles, and Fire Hose Appliances, 2018 Edition*

CASE STUDY

You Are the Firefighter

Shortly after beginning recruit fire training, you are assigned to ride along with an engine company for a shift. After a relatively uneventful day, your company is dispatched for a working fire at a large warehouse. When you arrive, there are flames coming from the front of the building. You hear the incident commander's instructions to lay a 4-in. (101-mm) supply line and to prepare for an attack with the deck gun.

1. Why is it important to understand the function and proper uses of each type of hose carried on your apparatus?
2. What are the advantages of using a master stream appliance on this fire?

Introduction

One of the primary functions of firefighting is to quickly and effectively apply sufficient quantities of water to extinguish it. In order to achieve this objective, it is necessary to understand the tools that enable you to complete this task, namely fire hose, fire hose appliances, hose tools, and nozzles.

Fire Hose

Fire hose is a flexible tube used to convey water or other extinguishing agents. It is used for two main purposes: as supply hose and as attack hose. **Supply hose** (or **supply line**) is used to deliver water to an **attack engine**—an engine used to pump water through attack lines at the fireground—from a pressurized source such as a fire hydrant or from a **supply engine**—an engine used to pump water to an attack engine using supply lines. These topics are discussed further in Chapter 10, *Water Supply Systems*. The supply engine may be connected to a pressurized water source such as a fire hydrant or to a static (unpressurized) water source. An **attack hose** (or **attack line**) is used for fire suppression by discharging water under pressure onto a fire. Most attack hose carry water directly from the attack engine to a nozzle that is used to direct the water onto the fire. Attack hose that is held by firefighters to attack fire is called a **handline**. NFPA 1960, *Standard for Fire Hose Connections, Spray Nozzles, Manufacture's Design of Fire Department Ground Ladders, Fire Hose, and Power Rescue Tools, 2024 Edition*, defines the specifications and the inspection and testing requirements for the design and construction of new fire hose.

Hose Sizes

Hose size is the inside diameter of the hose when it is filled with water and under pressure. Fire hose ranges in size from ¾ to 6 inches (in.; 19 to 152 mm) (**FIGURE 11-1**).

Small-Diameter Hose

Small-diameter hose (SDH) is either 1½-in. (38-mm) or 1¾-in. (44-mm) hose and is most often used as the primary attack hose for most fires. Both sizes use the same 1½-in. (38-mm) couplings. These two sizes of attack hose are the sizes used most often during basic fire training. SDH can usually be operated by one firefighter, although having a second firefighter on the line makes it much easier to advance and control the hose. SDH usually comes in 50-foot (ft; 15-meter [m]) lengths.

The primary difference between 1½-in. (38-mm) and 1¾-in. (44-mm) hose is the amount of water that can flow through the hose. Depending on the pressure in the hose and the type of nozzle used, a 1½-in. (38-mm) hose can generally flow between 100 and 200 gallons per minute (gpm; 379 and 757 liters per minute

FIGURE 11-1 Fire hose comes in a wide range of sizes for different uses and situations.

[L/min]). An equivalent 1¾-in. (44-mm) hose can flow between 150 and 250 gpm (568 and 946 L/min). This difference is important because the amount of fire that can be extinguished is directly related to the amount of water that is applied to it. A 1¾-in. (44-mm) hose can deliver much more water and is only slightly heavier and more difficult to advance than a 1½-in. (38-mm) hose.

Medium-Diameter Hose

Medium-diameter hose (MDH) is 2½-in. (64-mm) and 3-in. (76-mm). The 2½-in. (64-mm) hose can be used as either attack hose or supply hose, but it is most often used as attack hose. A 2½-in. (64-mm) hose is used as an attack hose for fires that are too large to be controlled by a 1½-in. (38-mm) or 1¾-in. (44-mm) hose. A 2½-in. (64-mm) hose delivers a flow of approximately 250 gpm (946 L/min). It takes at least two firefighters to safely control a 2½-in. (64-mm) hose due to the weight of the hose, the water, and the nozzle reaction force. A 50-ft (15-m) length of dry 2½-in. (64-mm) hose weighs about 30 pounds (lb; 14 kilograms [kg]). When the hose is filled with water, however, it can weigh as much as 140 lb (64 kg). A 2½-in. (64-mm) hose is most often used for interior attacks in large buildings and for exterior attacks.

The 3-in. (76-mm) hose is also used as both attack and supply hose. When used as an attack hose, the 3-in. (76-mm) hose is used to deliver water to master stream appliances, which are devices that discharge between 350 gpm (1325 L/min) and 1500 gpm (5678 L/min) or more. Like 1½-in. (38-mm) or 1¾-in. (44-mm) hose, MDH usually comes in 50-ft (15-m) lengths.

Large-Diameter Hose

Large-diameter hose (LDH) has a diameter of 3½ in. (89 mm) or more. Standard LDH sizes include 4-in. (101-mm) and 5-in. (127-mm) diameters. Most LDH is constructed as supply hose, but some fire departments use special LDH that can withstand higher pressures. The largest LDH size is 6 in. (152 mm) in diameter. LDH is commonly available in standard lengths of 100 ft (30 m) but can range from 25 ft to 350 ft (7.6 m to 106.7 m) depending on a department's needs. These lengths can produce a water flow between 350 gpm (1325 L/min) and 1500 gpm (5678 L/min). Like 3-in. (76-mm) hose, LDH is also used to deliver water to master stream appliances.

Hose Construction

Most fire hose is constructed with an inner waterproof liner surrounded by one or multiple outer layers or reinforcements. The outer layer or layers is commonly referred to as the jacket. A **single-jacket hose** is constructed with one layer of woven fiber. A **multiple-jacket** hose is constructed with two or more layers of woven fibers. In a multiple-jacket hose with two layers—**double-jacket hose**—the outer layer is bonded to the inner woven layer. The jacket serves as a protective covering, while the inner layer provides most of the strength needed to keep the hose from rupturing under pressure. The woven fibers of the jacket are treated to resist water and provide added protection from many common hazards that are likely to be encountered at the scene of a fire. The jacket also provides the strength needed to withstand the high pressures exerted by the water inside the hose. This strength is provided by a woven mesh made from high-strength synthetic fibers such as nylon that are resistant to high temperatures, mildew, and many chemicals. These fibers can also withstand some mechanical abrasion.

SAFETY TIP

Gloves should always be worn when handling hose. Metal shavings, glass shards, or other sharp objects may potentially become embedded in the fibers of the hose. These objects could easily sever a muscle or tendon in a bare hand—and end a career.

Rubber-covered hose is a type of double-jacket fire hose that features a durable rubber-like compound on the outer layer or jacket. This construction provides additional protection and durability to the hose, making it suitable for various firefighting applications. In rubber-covered hose, the inner and outer layers of the jacket are bonded together with woven fibers contained between these layers (**FIGURE 11-2**). This bonding process ensures that the layers remain securely attached and prevents separation or delamination during use. The woven fibers provide reinforcement and strength to the hose, allowing it to withstand high pressure and resist abrasion and damage.

Collapsible fire hose is designed to be compact and lay flat, and it is easily stored when not in use. It is typically made of synthetic materials, such as polyester, nylon, or a blend of synthetic fibers, which means the hose is flexible and foldable. The collapsible nature of this hose makes it convenient for transportation and storage, as it takes up less space compared to non-collapsible, rigid hose. When pressurized with water, the hose expands and becomes rigid, allowing for efficient water flow and firefighting operations.

FIGURE 11-2 Rubber-covered hose.

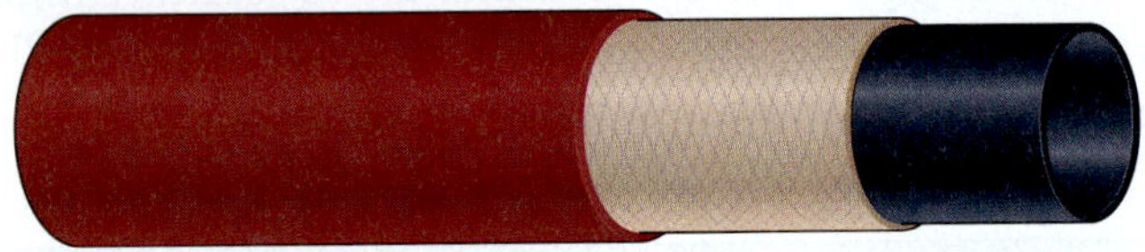

FIGURE 11-3 The hose liner inside a fire hose can be made from synthetic rubber or a variety of other membrane materials.

Non-collapsible fire hose, also known as **rigid fire hose**, is a type of fire hose designed to maintain its circular shape and cross-sectional integrity even when not filled with water and pressurized. Unlike regular collapsible hose, non-collapsible hose incorporates an inner liner or element that provides rigidity and prevents the hose from flattening out when not charged. It is typically constructed from durable materials such as PVC (polyvinyl chloride) or synthetic rubber. The reinforcement prevents the hose from collapsing or kinking when pressurized, ensuring a consistent water flow.

It is important to note that both collapsible and non-collapsible fire hose serve specific purposes and have advantages and disadvantages depending on the firefighting scenario. Each fire department and agency needs to select the appropriate type of hose based on their operational needs, available resources, and the specific requirements of the firefighting situation.

The inner waterproof liner of the hose is the **hose liner**. The hose liner is usually made of a synthetic rubber compound or a thin, flexible membrane material that can be flexed and folded without developing leaks (**FIGURE 11-3**). The hose liner prevents water from leaking out of the hose and provides a smooth inside surface for water to move against. Without this smooth surface, excessive friction would arise between the moving water and the inside of the hose, causing friction loss and reducing the water pressure and flow rate that reaches the nozzle.

Hose Couplings

Hose usually comes in lengths of 50 to 100 ft (15 to 30 m), although some hose is shorter. At a fire scene, the hose often needs to be much longer to reach the fire. To achieve this, both ends of a length of hose have a coupling permanently attached. A **coupling** is a connection device that you use to connect—or couple—individual lengths of fire hose together. Couplings are also used to connect a hose to a fire hydrant or a pump or to nozzles and appliances. NFPA 1960 defines the performance requirements for hose couplings, as well as the specifications for the connections of the couplings. The two most common types of supply hose and attack hose couplings are threaded hose couplings and Storz hose couplings.

Threaded Hose Couplings

A **threaded hose coupling** is used on most hose that is 3 in. (76 mm) or less in diameter. A length of fire hose with threaded couplings has a male hose coupling with threads on the outside on one end of the hose and a female hose coupling with threads on the inside on the other end of the hose. The female hose coupling also has a **swivel**, which is a ring around the coupling that you turn to secure the female coupling around a male coupling without twisting the hose (**FIGURE 11-4**). Most hydrants and fire pumps have male outlets and therefore require a female coupling.

When connecting fire hose with threaded couplings, the threads must be properly aligned so that the male and female hose couplings fully engage with minimal resistance. When properly coupled, threaded hose couplings provide a secure connection between two sections of hose and are unlikely to become disconnected during proper use. The swivel on the female hose coupling should be turned until the connection is snug, but only hand tight, so that the hose couplings can be easily disconnected.

A disadvantage of threaded hose couplings is that they are prone to cross-threading, meaning the threads of each coupling are not properly aligned with each other. This can result in leakage and possible separation. If leakage does occur after the hose is filled with water, further tightening may be needed, in which case you can use a spanner wrench to gently tighten the couplings until the leakage is stopped (**FIGURE 11-5**). Note that even if you only hand-tighten the connection, you may need to use a spanner wrench to uncouple the

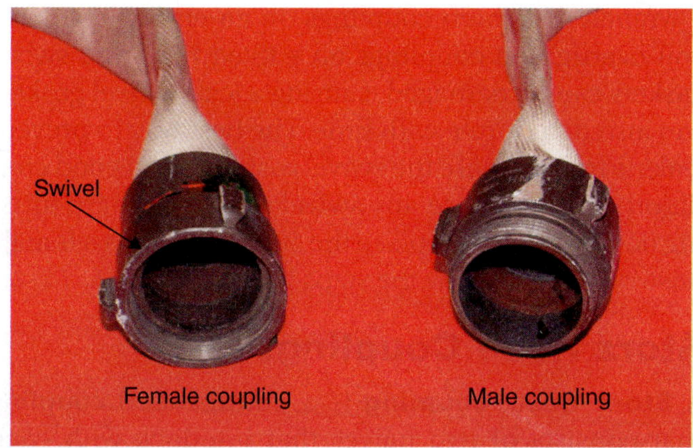

FIGURE 11-4 A set of threaded hose couplings includes one male coupling and one female coupling. The female coupling has threads on the inside of the coupling and a swivel. The male coupling has exposed threads on the outside of the coupling.

FIGURE 11-6 Lugs are extensions or indentations on couplings that provide leverage to aid in the connection and disconnection of hose couplings. **A.** Pin lugs are cylinder-like extensions. **B.** Recessed lugs are circular indentations. **C.** Rocker lugs are rectangular extensions.

FIGURE 11-5 Spanner wrenches are used to couple or uncouple hose couplings.

hose after it has been pressurized with water. If you do need to use a spanner wrench to couple or uncouple the hose couplings, normally, two spanner wrenches are used together to rotate the two hose couplings in opposing directions.

Threaded hose couplings have a **lug**, which is a protrusion or indentation that provides leverage as you screw couplings together or unscrew them to separate them. On male couplings, the lugs are on the outside of the coupling. On female couplings, the lugs are on the outside of the swivel. Three types of lugs are used: rocker lugs, pin lugs, and recessed lugs (**FIGURE 11-6**). A **rocker lug** (or **rocker pin**) is a rectangular-shaped extension and is the type of lug most commonly used on threaded fire hose couplings. The edges of rocker lugs are beveled to prevent them from catching on objects as the hose is dragged across a surface. A **pin lug** looks like a small cylinder that extends outward from the hose coupling. They are rarely found on fire hose today because the pins tend to snag as the hose is being pulled over rough surfaces. They are, however, found on a **fire department connection (FDC)**—the fire hose connection that fire departments use to pump water into a sprinkler system or a standpipe (a **standpipe** is a vertical pipe that connects a water supply to hose outlets in a multilevel building or sprinkler system). Instead, pin lugs have largely been replaced by rocker lugs. A **recessed lug** is a circular indentation and requires a specially designed spanner wrench called a booster hose wrench to engage them. Recessed lugs are usually found on the couplings for ¾-in. (19-mm) or 1-in. (25-mm) booster hose.

A **Higbee indicator** is a notch or cut on one of the rocker lugs that indicates the position of the first thread on a coupling. If you align the rocker lugs with the Higbee indicators on a male coupling and a female coupling, the threads will align and you will be able to screw the couplings together quickly with a low risk of cross-threading (**FIGURE 11-7**).

To create an airtight seal between couplings, it is important to have a flexible gasket—an O-shaped piece of rubber—to facilitate the seal. On threaded couplings, the gasket sits inside the swivel on the female hose coupling and it is called a **swivel gasket**. When a male hose coupling is tightened against it, the swivel gasket forms a seal that stops water from leaking. If the swivel gasket is damaged or missing, the hose coupling will leak. Although a leaking hose coupling is not a critical problem during most firefighting operations, it can result in unnecessary water damage. During cold weather, a leaking hose coupling can cause ice to form and create a significant safety hazard. If you use a wrench to tighten couplings on an empty hose or if you overtighten couplings on a filled hose, the swivel gaskets can be damaged, causing a leak. Swivel gaskets can also deteriorate with time. The best way to prevent leaks is to make sure the swivel gaskets are in good condition and to replace any swivel gaskets that are missing or damaged. To replace the swivel gasket in a female hose coupling, follow the steps in **SKILL DRILL 11-1**.

FIGURE 11-7 Higbee indicators are notches or cuts in rocker lugs that show the position where the threads on a pair of couplings properly align with each other. Higbee indicators are especially helpful when the threads are not visible, such as at night.

Storz Hose Couplings

A **Storz hose coupling** is a nonthreaded coupling that consists of two halves that interlock with each other to form a tight seal (**FIGURE 11-8**). Storz couplings are often referred to as sexless couplings because they do not have distinct male or female ends. Instead, they have identical symmetrical lugs and slots that allow them to connect with other Storz couplings of the same size. There are other types of nonthreaded couplings, but Storz couplings are the most common. Storz hose couplings are made for all hose sizes; however, in North America, they are most often used on LDH. Many fire departments use LDH with Storz hose couplings as a supply line between a fire hydrant and an engine.

SKILL DRILL 11-1

Replacing the Swivel Gasket Firefighter I, NFPA 1010: 6.5.2

1. Feel the swivel gasket with your fingers. If the gasket is dry, brittle, or missing, it must be replaced. If the old gasket is still in place, pull it out with your fingers. Then fold the new swivel gasket in half by bringing the thumb and the forefinger together to create two loops.

2. Place either of the two loops inside the hose coupling, and position it against the gasket seat.

SKILL DRILL 11-1 CONTINUED

Replacing the Swivel Gasket Firefighter I, NFPA 1010: 6.5.2

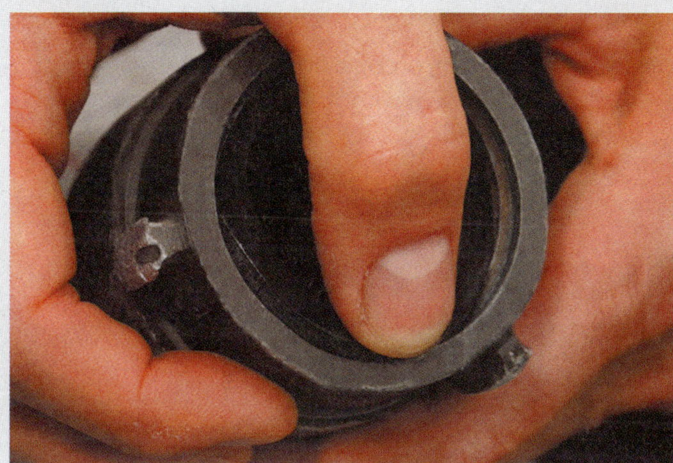

3. Using the thumb, push the remaining unseated portions of the swivel gasket into the hose coupling until the entire swivel gasket is properly positioned against the gasket seat inside the coupling.

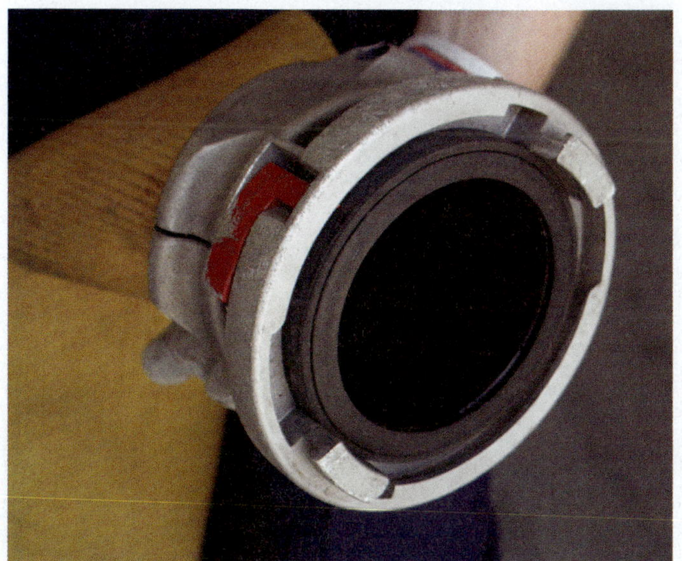

FIGURE 11-8 Storz hose couplings on both ends of a length of hose are the same and can be attached to any other Storz coupling of the same diameter.

Because there are no threads to make a connection, Storz hose couplings are connected by mating the two couplings face-to-face and then turning them clockwise into a locking position. To disconnect a set of Storz-type hose couplings, the two parts are rotated counterclockwise until they release. They may require spanner wrenches for complete coupling and uncoupling.

Storz couplings are more prone to accidental disconnect than threaded couplings if they are not completely coupled. Some Storz hose couplings have a small pin or lever that must be released before uncoupling the hose to prevent an accidental uncoupling of the hose. Storz couplings have gaskets, but they are much more durable and long-lasting compared to gaskets on threaded couplings and rarely require maintenance or replacement.

Storz couplings are also found on some FDCs.

Coupling and Uncoupling Hose

Several techniques are used for connecting and disconnecting hose. Depending on the circumstances, one technique may be more effective than another. A firefighter should learn how to perform each skill. The following two skill drills describe how to couple hose if you are alone and how to couple hose working with a second firefighter. To perform the one-firefighter foot-tilt method of coupling fire hose, follow the steps in **SKILL DRILL 11-2**. To perform the two-firefighter method for coupling a fire hose, follow the steps in **SKILL DRILL 11-3**.

SAFETY TIP

Always wear your protective gloves when coupling and uncoupling hose connections.

SKILL DRILL 11-2

Performing the One-Firefighter Foot-Tilt Method of Coupling a Fire Hose with Threaded Couplings Firefighter I, NFPA 1010: 6.3.10

1. Place one foot on one hose behind the male coupling. Push down with your foot to tilt the male coupling upward.

2. Place one hand behind the female coupling on the other hose and grasp the hose.

3. Place the other hand on the swivel of the female coupling. Bring the two couplings together and align the Higbee indicators. Turn the female coupling counterclockwise until it clicks, which indicates that the threads are aligned. Turn the swivel in a clockwise direction to connect the hose.

SKILL DRILL 11-3

Performing the Two-Firefighter Method of Coupling a Fire Hose with Threaded Couplings Firefighter I, NFPA 1010: 6.3.10

1. One firefighter picks up one hose, grasping it directly behind the male coupling, and holds it tightly against their body.

2. The second firefighter holds the female coupling on the other hose firmly with both hands.

3. The second firefighter brings the female coupling to the male coupling and aligns the Higbee indicators on the couplings.

4. The second firefighter turns the female coupling counterclockwise until it clicks, which indicates that the threads are aligned, and then turns the swivel on the female coupling clockwise to couple the hose.

Hose line sections should never be uncoupled when the hose line is charged—to **charge** a hose line is to fill it with water under pressure. Not only does the water pressure in a charged hose line make it difficult to uncouple the charged line, but the loosened couplings can also become uncoupled and flail around wildly and cause serious injury to firefighters or bystanders in the vicinity. Always shut off the water supply and bleed the pressure from the hose before uncoupling a hose line. If the coupling resists an attempt to uncouple it, check to make sure the pressure is relieved before using spanner wrenches to loosen the coupling. Like coupling hose, you can uncouple hose by yourself or working with another firefighter. To perform the one-firefighter knee-press method of uncoupling a fire hose, follow the steps in **SKILL DRILL 11-4**. To perform the two-firefighter stiff-arm method of uncoupling a hose, follow the steps in **SKILL DRILL 11-5**. To uncouple a hose using spanner wrenches, follow the steps in **SKILL DRILL 11-6**.

SAFETY TIP

Never attempt to uncouple a charged hose line.

SKILL DRILL 11-4

Performing the One-Firefighter Knee-Press Method of Uncoupling a Fire Hose with Threaded Couplings Firefighter I, NFPA 1010: 6.3.10

1. Pick up the female coupling.

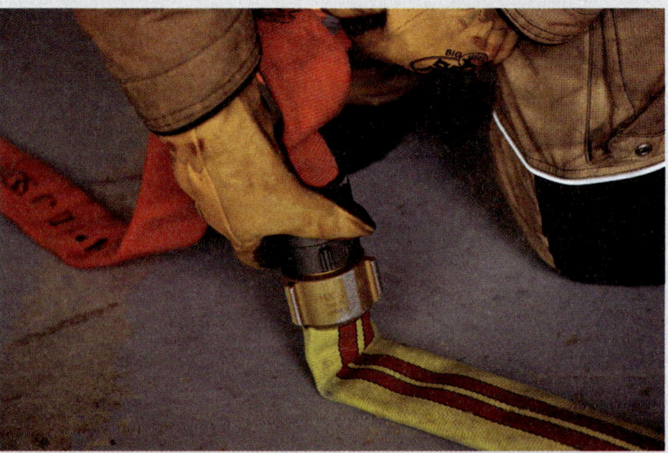

2. Turn the connection upright, resting the male coupling on a firm surface.

3. Place a knee on the female coupling and press down on it with your body weight to compress the swivel gasket. Turn the swivel on the female coupling counterclockwise to loosen the coupling.

SKILL DRILL 11-5

Performing the Two-Firefighter Stiff-Arm Method of Uncoupling a Fire Hose with Threaded Couplings Firefighter I, NFPA 1010: 6.3.10

1. Two firefighters face each other and firmly grasp their respective coupling.
2. With elbows locked straight, they push toward each other to compress the swivel gasket.
3. While continuing to push toward each other, one firefighter turns the swivel on the female coupling counterclockwise, loosening the coupling.

SKILL DRILL 11-6

Uncoupling a Hose with Threaded Couplings Using a Spanner Wrench Firefighter I, NFPA 1010: 6.3.10

1. With the connection on the ground, straddle the hose above the female coupling.

2. Place a spanner wrench on the swivel of the female coupling, with the handle of the wrench to the left.

Continues.

SKILL DRILL 11-6 CONTINUED

Uncoupling a Hose with Threaded Couplings Using a Spanner Wrench
Firefighter I, NFPA 1010: 6.3.10

3. Place a second spanner wrench on the male coupling, with the handle of the wrench to the right.

4. Push both spanner wrench handles down toward the ground, loosening the connection.

Types of Hose

As mentioned earlier, there are two main types of hose: supply hose and attack hose.

Supply Hose

The **hose bed**—the main storage area on an apparatus for carrying hose—on fire department engines is normally loaded with at least one bed of hose that can be used as supply hose. Sometimes engines are loaded with two beds of supply hose so they can easily drop a supply hose in two directions.

Supply hose is used to do the following:

- Connect the intake (suction) side of the pump on an attack or supply engine to a fire hydrant
- Connect the discharge side of a pump on a supply engine to the intake side of the pump on an attack engine
- To **draft**—draw water up through a hose—from a **static water source**—a source that is not under pressure, such as a pond, a lake, a stream, or a swimming pool—to the intake side of a pump on an engine or a mobile water supply apparatus or to a portable pump

The size of supply hose varies. It can be 2½ in. (64 mm), 3 in. (76 mm), 4 in. (101 mm), 5 in. (127 mm), or 6 in. (152 mm) in diameter. The choice of diameter is based on the preferences and operating requirements of each fire department. It also depends on the amount of water needed to supply the attack engine, the distance from the source to the attack engine, and the pressure that is available at the source.

A 2½-in. (64-mm) hose may be used as either a supply hose or an attack hose. This size hose has a limited flow capacity as supply hose, but it can be effective at a low to moderate **flow rate**—the number of gallons or liters per minute—and over short distances. When this size hose is used as supply hose, sometimes two parallel lines of 2½-in. (64-mm) hose are used to provide a more effective water supply.

LDH is much more efficient than 2½-in. (64-mm) or 3-in (76-mm) hose for moving large volumes of water over longer distances. Many fire departments use 4-in. (101-mm) or 5-in. (127-mm) hose as their standard supply hose. A single 5-in. (127-mm) supply hose can deliver flow rates exceeding 1500 gpm (5678 L/min) under some conditions. LDH is heavy and difficult to move after it is filled with water, however. NFPA 1900, *Standard for Aircraft Rescue and Firefighting Vehicles, Automotive Fire Apparatus, Wildland Fire Apparatus, and Automotive Ambulances, 2024 Edition*, does not specify the amount of hose that an engine must carry, but a typical engine may carry anywhere from 800 ft (243 m) to 1250 ft (381 m) of supply hose.

Supply hose must be tested annually at a pressure of at least 200 pounds per square inch (psi; 1379 kilopascals [kPa]) or at a pressure not to exceed the service test pressure marked on the hose. It is intended to be used at pressures up to 185 psi (1275 kPa).

FIGURE 11-9 Soft sleeve hose.

© Jones & Bartlett Learning. Photographed by Glen E. Ellman.

Suction Hose

The hose used to connect the intake side of a fire pump to a hydrant or to a static water source is a special type of supply hose called **suction hose**. There are two types of suction hose: soft sleeve hose and hard suction hose.

Soft sleeve hose (known historically as a **soft suction hose**) is a short section (10 to 25 ft [3 to 7.6 m]) of LDH used to transport water from a pressurized source, such as a fire hydrant, to the intake side of the fire pump (**FIGURE 11-9**). The soft sleeve hose allows as much water as possible to flow from the pressurized water source to the intake side of the fire pump through a single line. A soft sleeve hose has threaded female couplings or a Storz or other nonthreaded coupling on both ends. The couplings have rocker lugs to allow for quick tightening by hand. Soft sleeve hose ranges from 2½ in. (64 mm) to 6 in. (152 mm) in diameter and is usually between 10 ft (3 m) and 25 ft (7.6 m) in length.

FIGURE 11-10 Hard suction hose.

© Jones & Bartlett Learning. Photographed by Glen E. Ellman.

A **hard suction hose** is a short section of rigid fire hose that is used to draft water from a static water source (**FIGURE 11-10**). It can also be used to carry water from a low-pressure hydrant or draft hydrant to the pumper for drafting operations. The diameter is based on the capacity of the pump but can be as large as 6 in. (152 mm). It normally comes in 10-ft (3-m) or 20-ft (6-m) sections. Hard suction hose can be made from either rubber or plastic. The plastic versions are much lighter and more flexible.

Like soft sleeve hose, hard suction hose has either threaded female couplings or a Storz or other nonthreaded coupling on both ends. Long rocker lugs on the female couplings of hard suction hose assist in tightening the connection. To draft water, it is essential to have an airtight connection at each coupling. Sometimes it may be necessary to gently tap the lugs on the female couplings with a rubber mallet to tighten the hose or to loosen the connection before disconnecting it. Never tap the lugs with anything metal, however, or you could cause damage to the lugs or to the coupling itself.

Attack Hose

Attack hose is for fire suppression. This hose carries the water from the attack engine to the fire or to an FDC, or from a standpipe system to the fire. The common diameters for attack hose are 1½-in. (38-mm) or 1¾-in. (44-mm), and some fire departments also use 2½-in. (64-mm) attack hose (**FIGURE 11-11**). Each section of attack hose is usually 50 ft (15 m) long. Attack hose is often stored on an engine as a **preconnected attack line**, sometimes called a **preconnect**, in lengths

Voice of Experience

After 20 years on the job, I've seen my fair share of fires and have learned to appreciate the importance of maintaining your equipment. There are just some fires that you will never forget, and I'll never forget the four-alarm warehouse fire back in '08. That cold February day is seared into my memory. I only had four years on then, but I've never forgotten the lessons from that fire.

We were dispatched on the second alarm. As we raced across the city, sirens blaring, we could see the pillar of dark smoke from over a mile away. When we arrived on the scene, flames were already through the roof. The incident commander started barking orders to hook up to a hydrant and get water on the fire fast. I was on the back step, so I grabbed the hydrant assist valve and struggled to drag the supply line to hook up to the hydrant. Sure, supply line is heavy, but this was almost unmovable. It turns out the line wasn't fully drained from the last fire, and the remaining water had frozen in the line. The line was partially frozen into an accordion shape. A few guys saw me struggling and gave me a hand hooking up the supply line.

Once the supply line was hooked up and water was feeding the engine, we were ordered to take a 2 1/2" (64 cm) line and make entry into the loading dock entrance of the warehouse. My officer and I were masked up and ready to make entry to attack this fire. I opened the nozzle only to have water just barely sputter out the fog nozzle with no pressure. The nozzle was either frozen or damaged; we weren't sure which. I scrambled to swap out the nozzle as the fire consumed more of the warehouse by the minute. Finally, we got water flowing into the fire. We were at that fire for over 14 hours. We ended up losing the whole building.

Losing that building was not the fault of a frozen line or a broken nozzle, but I'll never forget how those equipment issues delayed our attack and the frustration I felt at that moment. Since that day, I've drilled into all the recruits I've trained the critical importance of meticulous hose, nozzle, and equipment maintenance. Inspect your inventory religiously after every run. Ensure hoses get cleaned and dried, couplings get brushed out, jackets are checked for wear, and nozzles are inspected and cleaned. Our gear has to work flawlessly when lives are on the line. Those smoky, chaotic minutes outside that burning warehouse taught me that preventive maintenance is one of the most vital parts of this job. I'll never let us get caught in that position again.

Michael A. Smith
Lieutenant
Lynn Fire Department
Lynn, Massachusetts

FIGURE 11-11 Attack hose used during fire suppression is most commonly 1½-, 1¾-, or 2½-inch (38-, 44-, or 64-mm) lines.

Courtesy of Steve Redick.

ranging from 150 to 350 ft (45 to 106 m). These travel on the engine already equipped with a nozzle and connected to a pump discharge outlet so they are ready for immediate use.

Attack hose can be either multiple-jacket or rubber-covered construction. It must withstand high pressure. Because it is used during fire suppression, it will be subjected to high temperatures, sharp surfaces, abrasion, and other potentially damaging conditions. For this reason, attack hose must be tough. Because firefighters need to be able to maneuver with it, it also needs to be flexible and lightweight.

Attack hose must be tested annually at a pressure of at least 300 psi (2068 kPa) and is intended to be used at pressures up to 275 psi (1896 kPa).

Attack hose can be either multiple-jacket or rubber-covered construction.

Booster Hose and Forestry Fire Hose

Booster hose and forestry fire hose are specialized attack hose. **Booster hose** (or **booster line**) is 1-in. (25-mm) diameter, non-collapsible, rubber-covered hose usually carried on a hose reel that holds 150 or 200 ft (45 or 61 m) of hose. Booster hose contains a **braided reinforcement**—layers of braided strands of yarn or wire with a layer of rubber between each braid—that gives it a rigid shape. This rigid shape allows the hose to flow water without pulling all of the hose off the reel. Booster hose is lightweight and can be advanced quickly by one person.

Booster hose should not be used for structural or vehicle firefighting because it has limited flow. The normal flow from a 1-in. (38-mm) booster hose is in the range of 40 to 50 gpm (151 to 189 L/min), which is not adequate for extinguishing structure fires. Booster hose use is typically limited to small outdoor fires and dumpster fires.

SAFETY TIP

Although booster lines are classified as attack hose, they should never be used for structural or vehicle fires.

Forestry fire hose is lightweight, collapsible, ¾-in. (19-mm), 1-in. (25-mm), or 1½-in. (38-mm) diameter hose often used to fight wildland and ground cover fires. Large amounts of water are usually not needed for these fires, and the small diameter provides much better maneuverability through brush and trees. This type of hose is sometimes extended for hundreds of feet.

Hose Rolls

An efficient way to transport a single section of fire hose is in the form of a roll. Rolled hose is both compact and easy to manage. A fire hose can be rolled in many different ways, depending on how it will be used. Follow the standard operating procedures (SOPs) of your department when rolling hose.

Straight or Storage Hose Roll

The **straight hose roll**, or **storage hose roll**, is a simple and frequently used hose roll. It is used for general handling and transportation of hose as well as to prepare hose to be stored on a rack (**FIGURE 11-12**). With this roll, the male coupling is at the center of the roll, and the female coupling is on the outside of the roll. To perform a straight hose roll, follow the steps in **SKILL DRILL 11-7**.

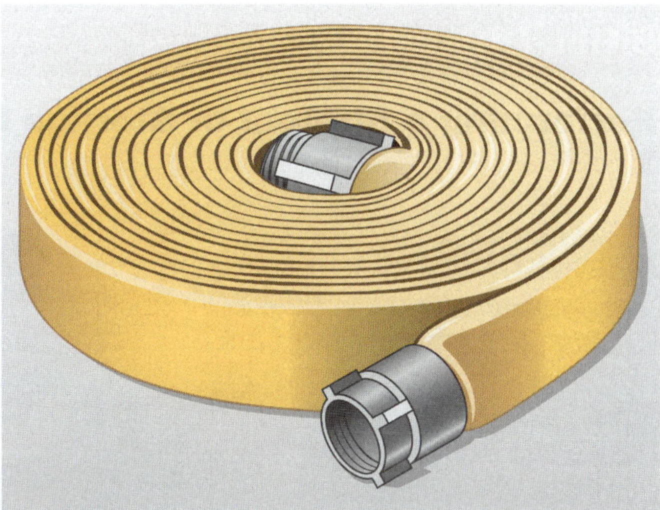

FIGURE 11-12 A straight or storage hose roll is used for transporting and storing hose. The exposed threads of the male coupling stay protected at the center of the roll.

Single-Doughnut Hose Roll

The **single-doughnut hose roll** is used when the hose will be put into use directly from its rolled state. With this arrangement, both couplings are on the outside of the roll, and the rolled hose can be connected and extended by one firefighter (**FIGURE 11-13**). With this roll, the threads of the male coupling are protected by the hose rolled on top of it. To perform a single-doughnut roll, follow the steps in **SKILL DRILL 11-8**.

Twin-Doughnut Hose Roll

The twin-doughnut hose roll is used primarily to make a small compact roll that can be carried easily (**FIGURE 11-14**). This roll can be carried by hand, by a rope, or by webbing. To perform a twin-doughnut hose roll, follow the steps in **SKILL DRILL 11-9**.

Self-Locking Twin-Doughnut Hose Roll

The self-locking twin-doughnut hose roll is similar to the twin-doughnut roll, except that it forms its own carry loop (**FIGURE 11-15**). To perform a self-locking twin-doughnut roll, follow the steps in **SKILL DRILL 11-10**.

SKILL DRILL 11-7

Performing a Straight or Storage Hose Roll Firefighter I, NFPA 1010: 6.5.2

1. Lay the hose flat and in a straight line. Stand at the end with the male coupling. Fold the male coupling over on top of the hose.

2. Fold the male coupling over on top of the hose again so that the opening is face down.

3. Continue rolling the hose until you reach the female coupling at the other end.

4. Set the hose roll on its side, and then tap any protruding hose flat with your foot.

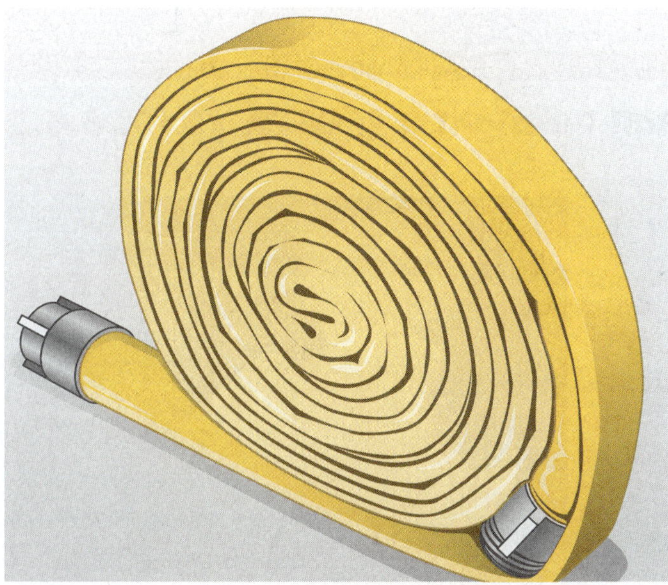

FIGURE 11-13 A single-doughnut hose roll leaves both couplings easily accessible, so it is used when the hose will be put into use directly from its rolled state.

Nozzles

A **nozzle** is attached to the discharge end of attack hose to give shape and direction to the **fire stream**, which is the stream of water or other extinguishing agent such as foam. The nozzle also increases the flow rate at which water is expelled from the hose. Without a nozzle, the water discharged from the end of a hose would extend only a short distance. Nozzles are used on all sizes of hose and on master stream appliances.

Nozzles are classified into three groups based on the size of the fire stream:

- A **low-volume nozzle** flows 40 gpm (151 L/min) or less. In structural firefighting, low-volume nozzles are primarily used for booster hose; in other words, their use is limited to small outdoor fires. In wildland firefighting, some nozzles flow as little as 15 gpm (57 L/min).
- A **handline nozzle** flows between 60 and 350 gpm (227 and 1325 L/min) and is used on

SKILL DRILL 11-8

Performing a Single-Doughnut Hose Roll Firefighter I, NFPA 1010: 6.5.2

1. Lay the hose flat and in a straight line.

2. Locate the midpoint of the hose. From the midpoint, move 5 ft (1.5 m) toward the male coupling end. Pinch the hose at this point.

Continues.

SKILL DRILL 11-8 CONTINUED

Performing a Single-Doughnut Hose Roll Firefighter I, NFPA 1010: 6.5.2

3. Start rolling the pinched bit of hose toward the female coupling end.

4. At the end of the roll, wrap the rest of the hose over the male coupling to protect the threads on the male coupling. Set the hose roll on its side, and then tap any protruding hose flat with your foot.

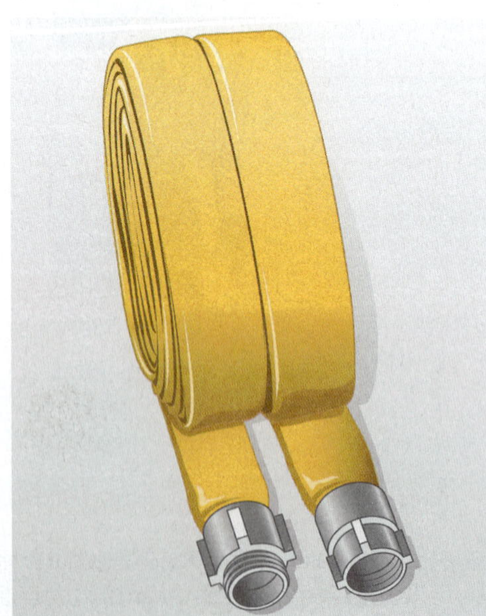

FIGURE 11-14 A twin-doughnut hose roll is used to make a small compact roll that can be carried easily.

handlines—hose ranging from 1½ in. (38 mm) to 2½ in. (64 mm) in diameter.

- A **master stream nozzle** flows more than 350 gpm (1325 L/min) and is used on master stream appliances.
- An **air-aspirating nozzle** draws air into the water stream, creating an aerated or foamy spray. By combining air with water or foam, these nozzles enhance fire suppression capabilities. The aerated spray increases the surface area of water droplets, which allows for better heat absorption and faster cooling of the fire. When used with firefighting foam solutions, air-aspirating nozzles also aerate the foam mixture. The turbulent water flow creates small bubbles that produce a thick blanket of foam. This provides faster knockdown and suppression of flammable liquid fires. Air-aspirating nozzles offer increased reach and coverage, reduced water usage, and versatility in different firefighting applications. Firefighters use these nozzles to optimize fire

SKILL DRILL 11-9

Performing a Twin-Doughnut Hose Roll Firefighter I, NFPA 1010: 6.5.2

1. Lay the hose flat and in a straight line.

2. Bring the male coupling alongside the female coupling, basically laying each half of the hose next to the other half.

3. With the two halves of the hose laid next to each other, fold over the bend in the hose, and then roll both sections of hose toward the couplings, creating a double roll.

4. Continue rolling until you reach the two ends of the hose. Keep the roll tight and compact as you progress. This will help create a small, easily carried roll.

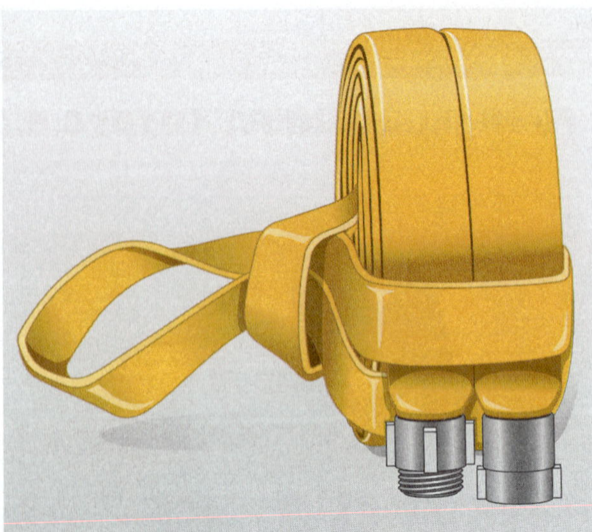

FIGURE 11-15 A self-locking twin-doughnut hose roll forms its own carry loop.

suppression, making them valuable tools in structural, wildland, aircraft, and industrial firefighting scenarios.

Nozzles must have an adequate volume of water and adequate pressure to produce a good fire stream. The volume and pressure requirements vary according to the type and size of the nozzle. Some nozzles incorporate a mechanism that can automatically adjust the flow based on the water's volume and pressure.

Nozzles have three basic parts: a handle, a tip, and a shut-off valve with a way to operate it. The handle is the part of the nozzle that you hold when streaming water. The tip is what determines the type of stream that will flow. The **shut-off valve** is used to control the flow of water. Low-volume and handline nozzles have a nozzle shut-off valve. The handle that controls this valve is called a **bale**. Some nozzles have a **rotary control valve** that is operated by rotating the nozzle in one direction to open (turn on) and in the opposite direction

SKILL DRILL 11-10

Performing a Self-Locking Twin-Doughnut Hose Roll Firefighter I, NFPA 1010: 6.5.2

1. Lay the hose flat and straighten it out, ensuring there are no knots or tangles. Fold the hose in half, bringing the male coupling and female coupling close to each other. The two couplings should be aligned and facing the same direction.

2. With the hose folded in half, create a small loop or doughnut shape near the folded end of the hose, passing the folded end through the loop. This creates the first doughnut shape.

SKILL DRILL 11-10 CONTINUED

Performing a Self-Locking Twin-Doughnut Hose Roll Firefighter I, NFPA 1010: 6.5.2

3. Bring the top of the loop back toward the couplings to the point where the hose crosses, creating two loops on either side. One loop should be slightly larger than the other.

4. Grasp the inner part of each of the two loops, and then begin rolling the two sides of the hose toward the couplings.

5. When you reach the couplings, pass the larger loop over the rolled hose and through the smaller loop.

6. Hold the roll by the smaller loop, which you can place over your shoulder to carry the hose roll.

to close (shut off) the flow of water. The shut-off valve for a master stream appliance is usually separate from the nozzle itself. Some nozzles are made so that the tip of the nozzle can be separated from the shut-off valve. A **breakaway fire nozzle** allows firefighters to shut off the flow, unscrew the nozzle tip, and then add more lengths of hose to extend the hose without shutting off the valve at the engine.

Types of Nozzles

Nozzles direct the water stream into one of the following shapes or patterns (**FIGURE 11-16**):

- **Solid stream**: A solid, unbroken column of water that provides both reach and penetration. The path of a solid stream is fully visible from the nozzle to the target.

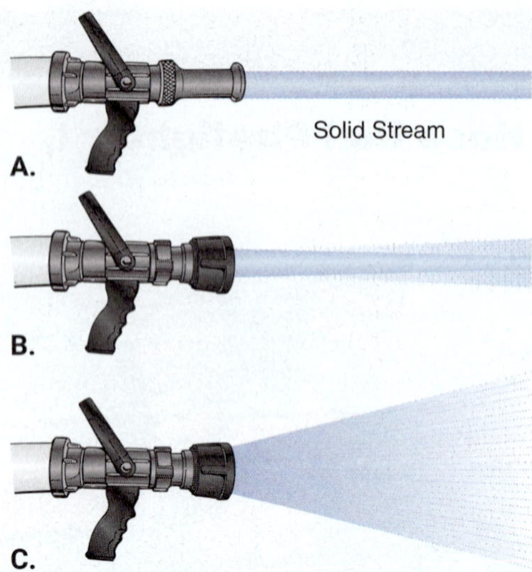

FIGURE 11-16 **A.** A solid stream is a solid, unbroken column of water. **B.** A straight stream is composed of droplets that form a tight cone pattern with an opening in the center and disperse sooner than a solid stream. **C.** A fog stream is also composed of droplets and is dispersed as it leaves the nozzle.

- **Straight stream**: A hollow cone pattern of water composed of tiny droplets. The path of a straight stream is initially easy to see as the stream leaves the nozzle. However, the droplets in the stream break up and start to scatter fairly quickly. After that point, the visibility of the stream path decreases rapidly. Because a straight stream is more dispersed and breaks up faster than a solid stream, it does not have the reach of a solid stream.
- **Fog stream**: A wide spray pattern composed of tiny droplets—basically a cloud (fog) of fine water droplets. This stream is very dispersed and provides the shortest reach. The path of a fog stream is obscured. It is hard to see exactly where droplets are landing.

Solid streams and straight streams seem similar, but they are not the same. Solid streams are a solid, unbroken column of water. Straight streams have openings in the center, like a hollow tube, and the stream is composed of droplets. Straight streams are also affected more dramatically by wind than a solid stream is.

Different types of nozzles produce different types of streams. The two types used most often in the fire service are the smooth-bore nozzle, which produces a solid stream, and the adjustable fog-stream nozzle, which produces a straight stream or a fog stream.

FIGURE 11-17 A smooth-bore nozzle.

Smooth-Bore Nozzles

A **smooth-bore nozzle** produces a solid stream. It discharges water in a compact, unbroken flow with limited turbulence or friction loss. The simplest smooth-bore nozzle consists of a shut-off valve and a smooth-bore tip that gradually decreases the diameter of the solid stream to a size smaller than the hose diameter (**FIGURE 11-17**). Smooth-bore nozzles are manufactured to fit both handlines and master stream appliances. Those that are used for master stream appliances often consist of a set of stacked tips, where each successive tip in the stack has a smaller-diameter opening. Tips can be quickly added or removed to provide the desired stream size. This approach allows different sizes of streams to be produced under different conditions.

As a general rule of thumb, smooth-bore nozzle orifices should not exceed half the diameter of the attaching hose. For example, on a 2½-in (64 mm) diameter fire hose, the maximum smooth-bore nozzle orifice should be 1¼ in. (32 mm) or less. On a 1¾-in. (45 mm) hose, the nozzle orifice should be 7/8 in. (22 mm) or less. Exceeding the half the diameter guideline can result in excessive friction loss and inefficient flow, which makes the stream difficult to control and reduces the reach of the stream. Using a properly sized smooth-bore nozzle helps concentrate the available water flow into an effective firefighting stream. This allows the full capabilities of the hose line to be utilized on the fireground.

The smooth-bore nozzle offers several advantages over a fog-stream nozzle:

- The straight stream it produces has a longer reach than a fog-stream nozzle operating at a straight stream setting.

- It operates at lower pressures than fog-stream nozzles do. Most smooth-bore nozzles operate at 50 psi (345 kPa), whereas fog-stream nozzles generally require pressures of 75 to 100 psi (517 to 689 kPa). Lower nozzle pressure makes it easier for a firefighter to handle the nozzle.
- A solid stream extinguishes a fire with less air movement and less disturbance of the thermal layering than does a fog stream, which in turn makes the heat conditions less intense for firefighters during an interior attack.
- It is easier for the operator to see the pathway of a solid stream than a fog stream.
- They have a basic, uncomplicated design and there are fewer internal components to inspect, clean, or repair. This makes them easier to use and maintain.
- The operation of a smooth-bore nozzle is straightforward; the flow rate is controlled by bale position.
- The unobstructed, uniform internal passage resists debris buildup, and high flow rates help flush away accumulations.
- They are more durable than a fog-stream nozzle because they do not have fragile fog teeth that can wear down over time.

Smooth-bore nozzles remain popular choices for busy fire departments despite modern advancements in fog nozzle designs. Their simplicity and reliability make them well-suited for challenging fireground conditions.

There are some disadvantages associated with smooth-bore nozzles. They are not as effective for hydraulic ventilation as fog-stream nozzles. (Hydraulic ventilation is discussed in Chapter 16, *Ventilation.*) Also, a firefighter cannot change the setting of a smooth-bore nozzle to produce a fog stream. In contrast, a fog-stream nozzle can be set to produce a straight stream.

To operate a smooth-bore nozzle, follow the steps in **SKILL DRILL 11-11**.

SKILL DRILL 11-11

Operating a Smooth-Bore Nozzle Firefighter I, NFPA 1010: 6.3.10

1. If the nozzle is a breakaway fire nozzle, select the desired tip size, and then attach it to the nozzle shutoff valve. Attain a stable stance by standing with feet shoulder-width apart, knees bent, and body turned about 45 degrees toward the nozzle or kneeling on one knee, with the other leg out to the side to form a triangle base and body turned about 45 degrees toward the nozzle, leaning into the nozzle reaction. Place one hand on the nozzle handle and tuck the hose line under that arm and hold it tight to the body. Place the other hand on the bale or rotary control valve.

2. Slowly open the valve by pulling the bale toward you or by turning the rotary control valve, allowing water to flow.

Continues.

SKILL DRILL 11-11 CONTINUED

Operating a Smooth-Bore Nozzle Firefighter I, NFPA 1010: 6.3.10

3. Open the valve completely to achieve maximum water flow.

4. Direct the stream to the desired location.

Fog-Stream Nozzles

A **fog-stream nozzle**, sometimes called a **spray nozzle** or an **adjustable fog-stream nozzle**, produces fine droplets of water and can be adjusted to produce a variety of stream patterns ranging from a straight stream to a narrow fog cone of less than 45 degrees to a wide-angle fog pattern that is close to 90 degrees (**FIGURE 11-18**). The size of the water droplets and the discharge pattern are varied by adjusting the nozzle setting.

These droplets of water absorb heat much more quickly and efficiently than a solid column of water. That is, the smaller droplets absorb more heat per gallon than water from a straight stream. Using a fog-stream nozzle to discharge 1 gallon of water in 100 cubic feet (3 liters in 2 cubic meters) of involved interior space can extinguish a fire in 30 seconds. A fog-stream nozzle is a good choice for reducing the room temperature to prevent a flashover.

Fog-stream nozzles move large volumes of air along with the water. This can be an advantage or a disadvantage, depending on the situation. A fog stream can be used to exhaust smoke and gases through hydraulic ventilation. Unfortunately, this type of air movement can result in sudden heat inversion in a room, which then pushes hot steam and gases down onto the firefighters. Research has shown that if used incorrectly, some applications of a fog pattern can push the fire into unaffected areas of a building by the addition of airflow.

FIGURE 11-18 A fog-stream nozzle.

To produce an effective stream using a fog-stream nozzle, the nozzle must be operated at the pressure recommended by the manufacturer. For many years, the standard operating pressure for fog-stream nozzles was 100 psi (689 kPa). Some manufacturers produce low-pressure fog-stream nozzles that are designed to operate at 50 or 75 psi (345 or 517 kPa), which makes it easier to control and advance the hose line. Lower nozzle pressure also decreases the risk that the nozzle will get out of control.

There are three types of fog-stream nozzles: fixed-gallonage, adjustable-gallonage, and automatic-adjusting.

A **fixed-gallonage fog nozzle** delivers a specific flow rate when it is operated at its rated discharge pressure. Fixed-gallonage fog nozzles are available in a range of flow rates from around 30 gpm (114 L/min) up to 300 gpm (1136 L/min) or more. The rated flow is based on the hose line size and desired firefighting tactics. Handlines that are 1½ in. (38 mm) and 1¾ in. (44 mm) in diameter usually use fixed-gallonage fog nozzles that flow 30, 60, and 95 gpm (114, 227, and 360 L/min). This combination produces a wide fog pattern for exposure protection. Handlines that are 2½ in. (64 mm) in diameter are often paired with fixed-gallonage nozzles that flow 150 and 200 gpm (568 and 757 L/min). The nozzle that flows 150 gpm (568 L/min) allows good flow rate with handline maneuverability, and the nozzle that flows 200 gpm (757 L/min) provides a high flow for offensive fire attack. Hose lines that are 2½ in. (64 mm) in diameter and that supply master stream devices are often paired with a fixed-gallonage nozzle that flows 300 gpm (1136 L/min) to provide high-volume application.

An **adjustable-gallonage fog nozzle**, also called a **variable-flow fog nozzle**, allows the operator to vary the flow rate on demand. The operator can select the flow rate by rotating a selector bezel to adjust the size of the opening. For example, a nozzle could have the options of flowing 95, 125, or 150 gpm (360, 473, or 568 L/min). This flexibility allows you to increase or decrease the flow based on fire conditions, attack needs, the available water supply, and the operating pressure at the nozzle.

An **automatic-adjusting fog nozzle** automatically regulates the flow rate based on the incoming water pressure while maintaining 100 psi (689 kPa) discharge pressure without requiring manual adjustment by the firefighter. As the pressure changes, an internal, spring-loaded piston moves to adjust the size of the opening and therefore adjusts the flow rate to maintain the rated pressure and produce a good stream. The automatic adjustment helps maintain a consistent reach and spray pattern despite pressure changes. Because they require minimal monitoring or adjustment by the nozzle operator during use, they are well-suited for master stream applications. A typical automatic nozzle could have an operating range of 95 to 225 gpm (360 to 852 L/min). Automatically adjusting nozzles provide consistency and reduce guesswork for the nozzle operator. This makes them well-suited for critical master stream applications.

To operate a fog-stream nozzle, follow the steps in **SKILL DRILL 11-12**.

SKILL DRILL 11-12

Operating a Fog-Stream Nozzle Firefighter I, NFPA 1010: 6.3.10

1. Select the desired nozzle. Attain a stable stance by standing with feet shoulder-width apart, knees bent, and body turned about 45 degrees toward the nozzle or kneeling on one knee, with the other leg out to the side to form a triangle base and body turned about 45 degrees toward the nozzle, leaning into the nozzle reaction. Place one hand on the nozzle handle and tuck the hose line under that arm and hold it tight to the body. Place the other hand on the bale or rotary control valve.

Continues.

SKILL DRILL 11-12 CONTINUED

Operating a Fog-Stream Nozzle Firefighter I, NFPA 1010: 6.3.10

2. Slowly open the valve by pulling the bale toward you or by turning the rotary control valve, and allow water to flow.

3. Open the valve completely to achieve maximum water flow.

4. Select the desired water pattern by rotating the bezel of the nozzle. Direct the stream to the desired location.

Other Types of Nozzles

Several other types of nozzles are used for special purposes. If your fire department has other types of specialty nozzles, you need to become proficient in their use and operation.

A **piercing nozzle** is used to make a hole in automobile sheet metal, aircraft, or a building's walls, floors, or roof to extinguish fires behind these surfaces. Piercing nozzles can be attached to hand lines or attached to an aerial water tower (**FIGURE 11-19**).

A **cellar nozzle** and **Bresnan distributor nozzle** are used to fight fires in cellars or basements and other inaccessible places such as attics and cocklofts (**FIGURE 11-20**). These nozzles discharge water in a wide circular pattern as the nozzle is lowered vertically through a hole into the space. They work like a large sprinkler head.

A **chimney nozzle** is used by firefighters for extinguishing fires inside chimneys. They often have a 45-degree elbow attached to a long discharge pipe that sprays water upward to extinguish a chimney fire. Some

A.

B.

FIGURE 11-19 **A.** Handheld piercing nozzle. **B.** Mounted piercing nozzle.

A. Courtesy of Task Force Tips; **B.** Courtesy of Pierce Manufacturing.

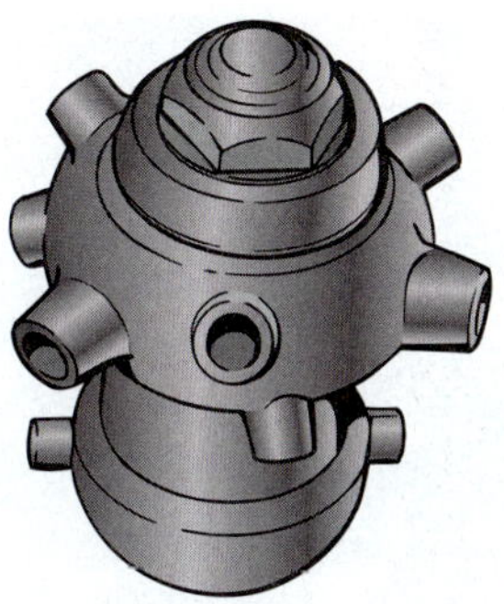

FIGURE 11-20 A Bresnan distributor nozzle is used to fight fires in inaccessible places such as cellars or basements, and attics.

Courtesy of Akron Brass Company.

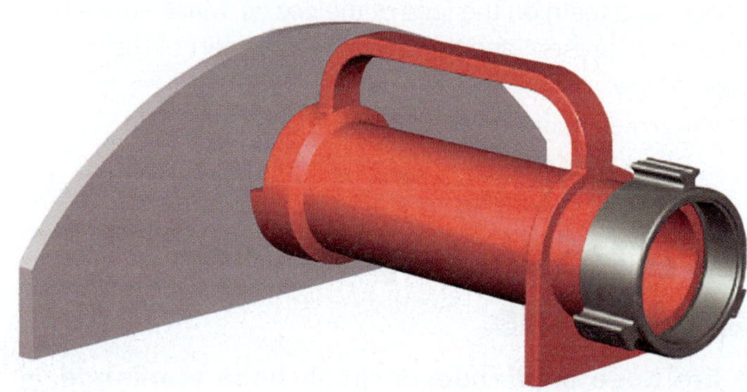

FIGURE 11-21 A water curtain nozzle delivers a flat screen of water that then forms a protective sheet (curtain) of water.

Courtesy of POK of North America, Inc.

chimney nozzles have a weighted nozzle head that is designed to be lowered into a chimney from above. Both types can quickly suppress a chimney fire by discharging an ultrafine water mist spray. This greatly reduces potential water damage to the property over traditional extinguishing methods.

A **water curtain nozzle** delivers a flat screen of water that then forms a protective sheet (curtain) of water on the surface of an exposed building (**FIGURE 11-21**). The water curtains must be directed onto the exposed building because radiant heat can pass through the air gaps in the water curtain.

Nozzle Maintenance and Inspection

Nozzles need to be inspected on a regular basis and after each use before they are placed back in service on the apparatus. They must be kept clean. Dirt and grit inside the nozzle will affect the performance of the nozzle, possibly reducing flow. Dirt and grit can also interfere with the valve operation and prevent the nozzle from opening and closing fully. A light grease on the surface of the ball on the shut-off valve keeps it operating smoothly.

On fog-stream nozzles, inspect the **fingers**—the individual, adjustable vanes or protrusions on the face of the fog-stream nozzle that are responsible for shaping the water stream into a specific pattern—and the **teeth**—the serrations or grooves around the fog-stream nozzle's orifice that help to break up the water stream and create a fog pattern—on the face of the nozzle (**FIGURE 11-22**). Make sure all fingers are present and that the moveable part of the nozzle spins freely. Any missing fingers or failure of the ring to spin will drastically affect the fog pattern. Make sure all the teeth are intact and free from damage or distortion. Any problems noted should be referred to a competent technician for repair.

FIGURE 11-22 On fog-stream nozzles, inspect the fingers and teeth on the face of the nozzle. Make sure all fingers are present and that the moveable part of the nozzle spins freely.

FIGURE 11-23 Double-female and double-male adapters are used to join two couplings of the same sex.

FIGURE 11-24 A reducer is used to connect a smaller-diameter hose to a larger-diameter hose.

Fire Hose Appliances and Tools

A **fire hose appliance**, or simply **hose appliance**, is any device used in conjunction with a fire hose for the purpose of delivering water. A **fire hose tool**, or just **hose tool**, is a device that assists firefighters with handling, manipulating, connecting, and using a fire hose. It is important to learn how to use the hose appliances and tools required by your fire department. In particular, you should understand the purpose of each device and be able to use each appliance correctly. Some hose appliances and tools are used primarily with supply hose, and others are most often used with attack hose. Many appliances and tools have applications with both supply hose and attack hose.

Adapters and Reducers

An **adapter** is a hose appliance that allows you to connect couplings with different threads or mating surfacing, or to connect hose couplings to other appliances. Although most fire departments use standardized hose threads, the use of standardized hose threads is not universal among all jurisdictions. Dissimilar threads could also be encountered in industrial settings where the hose threads of the building's equipment do not match the threads of the fire department. Some private fire hydrants have different threads from the municipal system. Adapters are also used to connect threaded couplings to Storz or other unthreaded couplings.

Adapters are also used when it is necessary to connect two female couplings or two male couplings. A **double-female adapter** is used to join two male hose couplings. A **double-male adapter** is used to join two female hose couplings (**FIGURE 11-23**).

A **reducer** is an appliance used to attach a smaller-diameter hose to a larger-diameter hose (**FIGURE 11-24**). Usually, the larger end has a female coupling, and the smaller end has a male coupling. For example, a reducer is used to attach a 2½-in. (64-mm) hose to a 1½-in. (38-mm) hose so that the 1½-in. (38-mm) hose can be attached for overhaul, during which lower water pressure is preferable because it allows for better maneuverability and conserves water. In fact, many 2½-in. (64-mm) nozzles are constructed with a built-in reducer to make this easier. Reducers are also used to attach a 2½-in. (64-mm) supply hose to a larger intake inlet on a fire pump.

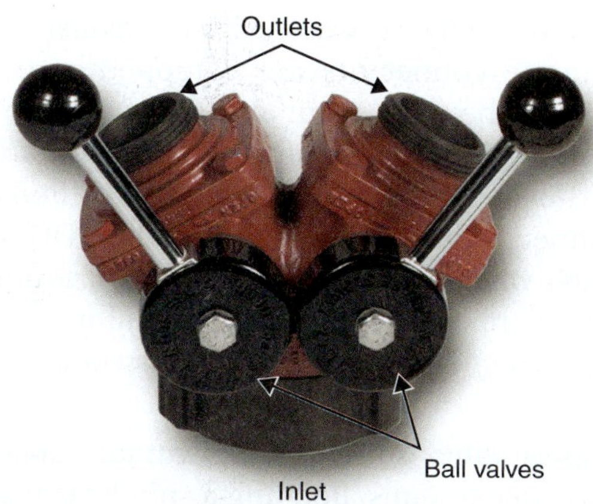

FIGURE 11-25 On the gated wye, two valves allow the flow of water to each outlet to be controlled independently.
Courtesy of Akron Brass Company.

When firefighters need to attach a smaller hose to a larger hose, or when they need to connect two hose lines that possess different threads, a reducer or adapter can be added to both the male and female hose couplings. This arrangement is called a five-piece coupling set (the five pieces are the male and female couplings on the first hose, the male and female couplings on the second hose, and the adapter).

Appliances That Split Water Streams

A **wye** is an appliance with one inlet and two outlets that splits one hose into two or more separate lines and the flow rate through both lines is the same. The word wye refers to a Y-shaped part or object. When threaded couplings are used, a wye has one female connection and two or more male connections. The wye that is most commonly used in the fire service splits one 2½-in. (64-mm) hose into two 1½-in. (38-mm) hose lines. This appliance is used primarily on attack hose.

A **gated wye** is a wye equipped with two valves so that the flow of water to each line can be controlled independently (**FIGURE 11-25**). This allows firefighters to attach and operate one hose and then add a second hose later, if necessary, without shutting down the water supply. In addition to gated wyes that split one 2½-in. (64-mm) hose into two 1½-in. (38-mm) hose lines, some gated wyes are used on LDH. These wyes have an inlet for LDH (4 in. [101 mm] or 5 in. [127 mm]) and several 2½-in. (64-mm) outlets.

A **water thief** is similar to a gated wye, but it has three outlets instead of two (**FIGURE 11-26**). It is used to divert water from a supply hose to another hose line or appliance. The outlets on water thieves come in various sizes to match different supply hose diameters, from a 2½-in. (64-mm) to 6-inch (152 mm). The 2½-in. (64-mm) water thief is the size most commonly used on the fireground. This is the standard size used with 1½-in. (38-mm) and 2½-in. (64-mm) handlines and as outlets on gated wyes. The larger 3- to 6-in. (76- to 152-mm) sizes are used for handling very high water volumes. They allow diversion from the largest supply hose during major firefighting operations. Many fire departments carry a selection of water thief sizes to handle the various supply line and water demands encountered. Having multiple options allows greater flexibility in how water can be diverted and distributed at a fire scene. Under most conditions, it is not possible to supply all three outlets at the same time because the capacity of the supply hose is limited. A water thief is used primarily on attack hose, and larger ones are used with LDH.

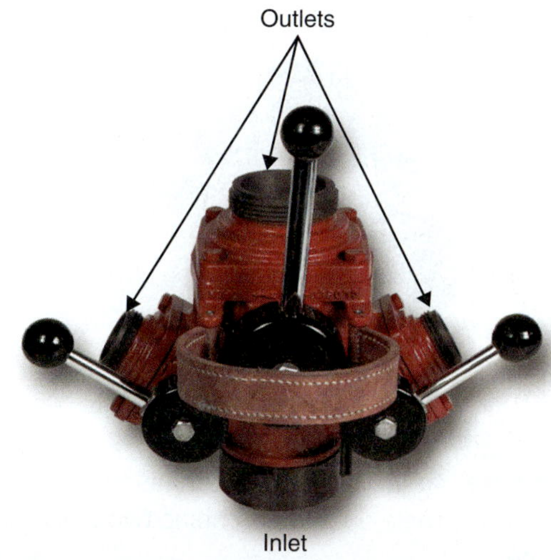

FIGURE 11-26 A water thief is similar to a gated wye, but it has three outlets instead of two.
Courtesy of Akron Brass Company.

Siamese Connector

A **Siamese connector** is a hose appliance that combines two or more hose lines into one. It has two inlets and one outlet (**FIGURE 11-27**). This increases the flow of water on the outlet side of the Siamese connector. The most commonly used type of Siamese connector combines two 2½-in. (64-mm) hose lines into a single 2½-in. (64-mm) line. A Siamese connector that is used with threaded couplings has two female inlets and one male outlet. Like gated wyes, Siamese connectors have two valves so that firefighters can attach a second hose without shutting down the water supply from the first hose.

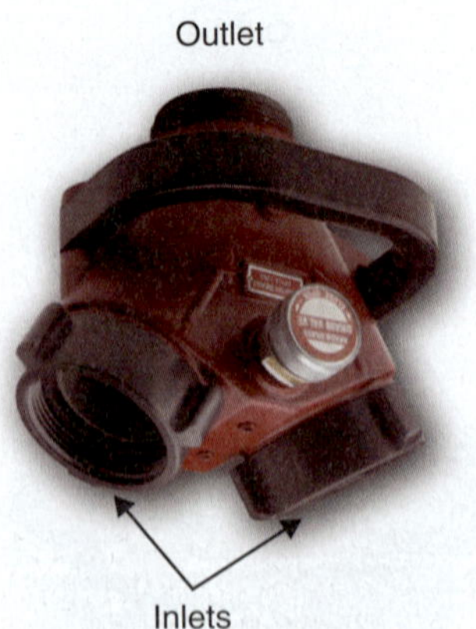

FIGURE 11-27 A Siamese connector has two female inlets and a single male outlet to combine two or more hose lines into one.

Courtesy of Akron Brass Company.

Siamese connectors can be used with supply hose and with some attack hose. It is sometimes used on the inlet side of a fire pump to allow water to be received from two different supply lines. They are also used to supply master stream appliances. Siamese connectors are commonly installed on FDCs. Firefighters should review their department's policies on the correct procedures for supplying fire department Siamese connectors.

Valves

Valves are hose appliances that control the flow of water in a hose. Valves are also used on pipes in sprinkler systems. It is important to remember when opening and closing any valve or nozzle to do it s-l-o-w-l-y. This prevents **water hammer**, a surge in pressure caused by suddenly stopping the flow of a stream of water. It also prevents damage to the hose or causing the firefighter at the nozzle to lose control of the stream if a valve is opened too quickly.

The following types of valves are commonly used in hose and appliances:

- **Ball valves** consist of a ball with a hole in the middle (**FIGURE 11-28A**). When the hole is lined up with the inlet and the outlet, water flows through it. As the ball is rotated, the flow of water is gradually reduced until it is shut off completely. Ball valves are the most common type of nozzle shut-off valve. They are also used in gated wyes and Siamese connectors.
- A **clapper valve**, also called a **clapper mechanism**, prevents water from flowing backward by closing automatically if water flows against it. Some Siamese connectors are equipped with clapper valves. The most common use of a clapper valve is in FDCs for standpipe systems and automatic sprinkler systems. They allow the firefighter to connect a hose to one side of the FDC and flow water while the other side of the FDC is uncapped. These are discussed in more detail in Chapter 24, *Systems of Fire Detection, Suppression, and Smoke Control*.
- A **gate valve** is used in fire hydrant outlets and in sprinkler systems. Rotating a spindle causes a gate to move slowly across the opening. The spindle is rotated by turning it with a wrench or a wheel-type handle (**FIGURE 11-28B**).
- **Butterfly valves** are often used in pump intake connections where a suction hose is connected. They are opened or closed by rotating a handle one-quarter turn (**FIGURE 11-28C**).
- A **four-way hydrant valve** is attached to a fire hydrant outlet so that the attack engine can connect to the water supply, and then a second engine can connect to provide additional pressure without first disconnecting the attack engine's supply. By placing a second fire pumper at the fire hydrant, it is possible to boost the pressure in the supply hose line by changing the position of the four-way hydrant valve to enable the second-arriving fire pumper to pump from the hydrant into the supply line. It can boost pressure in a supply hose without interrupting the flow of water. The use of a four-way hydrant valve is covered in detail in Chapter 13, *Supply Line and Attack Line Evolutions*.
- A **remote-controlled hydrant valve** is attached to a fire hydrant. This device allows the operator to turn the hydrant on without flowing water into the hose line. These valves are operated by the pump operator using a radio control. This frees up the personnel assigned to hydrant duty for other tasks and gives the pump driver/operator control over the hydrant from a distance.

Master Stream Appliances

A **master stream appliance**, also called **master stream device**, is used to produce high-volume water streams for large fires. Most master stream appliances discharge between 350 gpm (1325 L/min) and 1500 gpm (5678 L/min), though much larger capacities are available. Master stream appliances can discharge

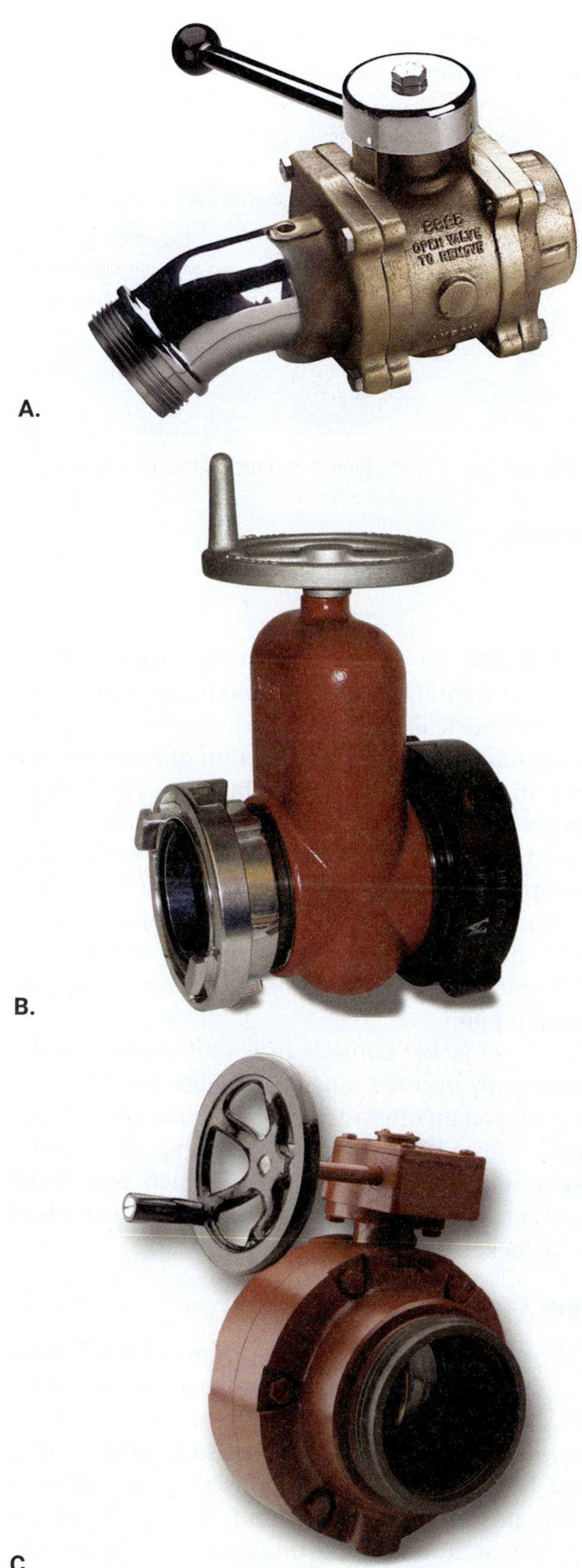

FIGURE 11-28 Valves are used to control the flow of water in a hose or a pipe. **A.** Ball valve. **B.** Gate valve. **C.** Butterfly valve.

FIGURE 11-29 A deck gun is permanently mounted on an apparatus and equipped with a piping system that delivers water to the device.

a stream of water farther than a hand line. A master stream appliance can be either manually or remotely operated from a fixed position. There are three main types of master stream appliances.

A **deck gun** is permanently mounted on and operated from a vehicle, and it is equipped with a piping system that delivers water to the gun (**FIGURE 11-29**). These devices are also called turret pipes or wagon pipes. Sometimes a hose must be connected to the deck gun in order to place it in operation.

A **portable monitor** is a master stream appliance that can be carried on apparatus and removed from the apparatus when needed (**FIGURE 11-30**). It is placed on the ground, and hose lines are connected to the portable monitor to supply the water. Most of these devices come equipped with either one, two, or three inlets. Portable monitors are built with 2½-in. (64-mm) inlets or with an LDH inlet. Smaller portable monitors are sometimes set up attached to preconnected hose. This allows the monitor to be quickly placed in service by one firefighter.

An **elevated master stream appliance** is mounted on aerial apparatus such as aerial ladders, tower ladders, elevated platforms, or special hydraulically operated master stream booms (**FIGURE 11-31**). A **ladder pipe** is a removable elevated master stream device that is mounted close to the tip of an aerial ladder or tower ladder and supplied by a hose. Aerial apparatus are usually equipped with permanently mounted waterways that supply the master stream appliances that are mounted close to the top of the aerial device.

FIGURE 11-30 A portable monitor is placed on the ground and supplied with water from one or more hose lines.

FIGURE 11-31 Elevated master stream devices can be mounted on aerial apparatus.

Additional information on the uses and operation of these devices is presented in Chapter 14, *Fire Suppression*.

Hose Jacket

A **hose jacket** is a hose tool designed to temporarily stop leaks in sections of hose, providing a quick fix until the hose can be replaced properly. They are useful in emergency situations where immediate replacement is not possible, and they can be used on both supply and attack hose. Metal hose jackets are typically made from durable metals like aluminum or brass (**FIGURE 11-32**). They usually have screws or clamps that tighten around the leaking section of the hose to seal the leak. Leather hose jackets are wrapped around a damaged or leaking section of hose and are secured in place with straps or buckles. The leather conforms to the shape of the hose and creates a tight seal, helping to minimize water leakage. A hose jacket is a temporary fix and should only be used as a last resort until the hose can be properly replaced or repaired. It does not provide the same level of durability and integrity as a complete hose replacement or professional repair.

FIGURE 11-32 A hose jacket is used to repair a leaking hose.

The hose jacket consists of a split metal cylinder that fits tightly over the outside of a hose line. This cylinder is hinged on one side to allow it to be placed over the leak. A fastener is then used to clamp the cylinder tightly around the hose. Gaskets on each end of the hose jacket prevent water from leaking out the ends of the hose jacket.

Hose Roller

A **hose roller** is a hose tool made of metal that is used to prevent chafing or kinking at a sharp edge when the hose is being hoisted over the edge of a roof or over a windowsill (**FIGURE 11-33**). A hose roller is also called a **hose hoist** because it can be used while raising or lowering hose to and from a roof, through windows, over fence, or around many other obstacles. Hose rollers are also used to protect ropes when hoisting an object over the edge of a building and during rope rescue operations. They are typically used with attack hose.

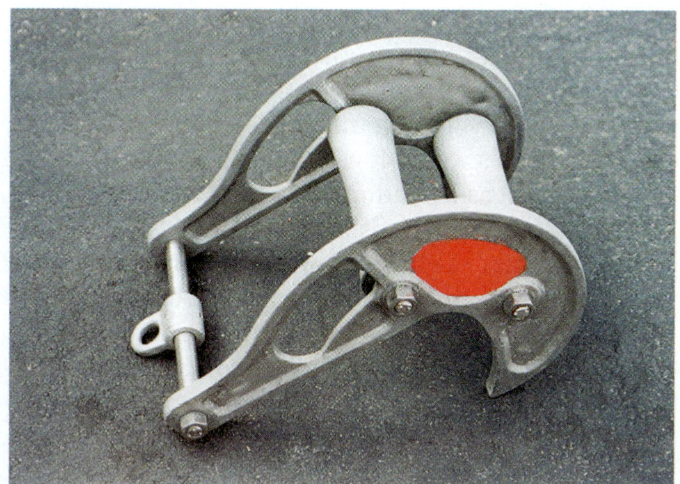

FIGURE 11-33 A hose roller can be used to protect a hose when it is hoisted over a sharp edge of a roof or a windowsill.

Chafing Blocks

A **chafing block** is a sturdy rubber, plastic, or wooden block that is placed under a fire hose where it lays on the ground or rests against hard surfaces to raise the hose off the ground and provide a smooth contact surface. This protects the hose from abrasion, friction, and wear against rough surfaces like asphalt, concrete, or debris. Some departments just use sections of old hose also to accomplish this. Chafing blocks are especially critical when pulling large-diameter supply hose because the high weight and water pressure inside these hose lines can quickly cause damage if the hose rubs against the ground.

Hose Bridge

A **hose bridge**, also called a **hose ramp**, is a hose tool that protects a hose when it is necessary to drive a vehicle over a hose line. They spread the weight of the vehicle over a wider area to reduce pressure on the hose. Hose bridges are made of rubber or metal. Metal hose bridges are placed over the hose, and the hose is placed in an opening in the center of the hose bridge. Some rubber hose bridges go over the hose, but with others, the bridge is placed under the hose, and the hose is placed into a trough in the center of the hose bridge. Hose bridges come in various sizes to accommodate different hose diameters. The most common sizes used by fire departments are 2½-in. and 3-in. (64-mm and 76-mm) hose bridges, because these match typical attack line and water supply hose diameters. There are

FIGURE 11-34 Hose clamps are used to temporarily interrupt the flow of water in a hose.

also adjustable hose ramps and bridges that can accommodate a wide range of hose sizes, usually from 1 to 5 in. (25 to 125 mm) in diameter. Of course, whenever possible, it is best to avoid driving over hose at all, even with a hose bridge, because this practice may still damage hose.

Hose Clamp

A **hose clamp** is an adjustable metal device used to temporarily cut off water flow in a fire hose when needed rather than having to shut down the entire hose system. For example, it can be used to temporarily shut off a charged hose line in order to make adjustments, replace damaged sections, or replace appliances. The two most popular styles of hose clamps are lever-operated and screw-down. Lever-operated hose clamps have a ratcheting lever mechanism that allows quick, one-handed operation. This type of hose clamp is ideal for small- and medium-diameter hose like 1½-in., 1¾-in., and 2½-in. (38-mm, 44-mm, and 64-mm) attack and supply lines. Screw-down hose clamps operate by tightening a metal band around the hose using a mechanical screw that is turned manually or with tools (**FIGURE 11-34**). They provide tight clamping and fine incremental adjustment, making them well-suited for LDH of 3 to 5 in. (76 to 127 mm) in diameter, but they require two hands to operate.

Hose Care, Maintenance, and Inspection

Fire hose should be regularly inspected and tested following the procedures in NFPA 1962, *Standard for the Care, Use, Inspection, Service Testing, and Replacement of Fire Hose, Couplings, Nozzles, and Fire Hose Appliances, 2018 Edition.* Hose that is not properly maintained can deteriorate over time and eventually burst. In addition, the swivel gaskets in female couplings need to be checked regularly and replaced when they are worn or damaged.

Causes and Prevention of Hose Damage

Fire hose is a lifeline for firefighters. Every time firefighters respond to a fire, they rely on fire hose to deliver the water needed to attack the fire and protect themselves from it. Fire hose is a highly engineered product designed to perform well under adverse conditions. Firefighters must be careful to prevent damage to the hose that could result in premature or unexpected failure. The factors that most commonly damage fire hose include mechanical causes, heat, cold, UV radiation, chemicals, and mildew.

Mechanical Damage

Mechanical damage can occur from many sources. For example, hose that is dragged over rough objects or along a roadway can be damaged by abrasion. Broken glass and sharp objects can cut through the hose. Particles of grit caught in the fibers can damage the jacket or puncture holes in the liner. Reloading dirty hose can cause damage to the fibers in the jacket of the hose. Be especially careful if you need to place a hose through a broken window; if you need to do this, remove any protruding sharp edges of glass first. Fire hose is also likely to be damaged if it is run over by a vehicle; if traffic must drive over a hose in a roadway, use a hose bridge.

Hose couplings can be damaged by dropping them on the ground. In particular, the exposed threads on male couplings are easily damaged if they are dropped. Avoid dragging hose couplings, as this practice can cause damage to the threads and to the swivels.

Heat, Cold, and UV Radiation

Hose can be damaged by heat and cold as well as by prolonged exposure to UV radiation (sunlight). Heat is an obvious concern when fighting a fire. Hose is not fireproof. A hose that is directly exposed to a fire can burn through and burst quickly. Burning embers and hot coals can cause small leaks or weaken the hose enough that it will burst under pressure. Avoid storing a hose in places where it will come in contact with hot surfaces, such as a heating unit or the exhaust pipe on a vehicle. If the apparatus is parked outside, use a hose cover to protect the hose from sunlight. It is important to keep the hose from becoming brittle. Also, to prevent UV radiation damage, do not dry hose in sunlight.

Cold temperatures can also damage hose. If a hose freezes, the hose liner can rupture and fibers in the jacket can break. When firefighters are working in below-freezing temperatures, water should be kept flowing through the hose to prevent it from freezing. If a line must be shut down temporarily, the nozzle should be left partly open to keep the water moving. If this is necessary, direct the stream to a safe location away from the fire scene where it will not cause additional water damage or create a slipping hazard when frozen. When a line is no longer needed, the hose should be drained and rolled before it freezes. Taking steps to prevent ice buildup helps maintain hose integrity and firefighter safety in cold weather operations.

If a hose does freeze, do not attempt to bend a section of frozen hose. Hose that is frozen or encased in ice can be thawed out with a generator that produces steam. You can also carefully use the blunt side of a flat-head axe to lightly pound on the ice encasing the hose, being careful not to damage the hose itself. The frozen hose is then transported back to the fire station to thaw. In situations where a hose is frozen solid, it may be necessary to transport the hose back to the fire station on a flatbed truck.

Chemicals

Many chemicals can damage fire hose. You may encounter chemicals at incidents in facilities where chemicals are manufactured, stored, or used, as well as in locations where their presence is not anticipated. Most vehicles contain a wide variety of chemicals that can damage fire hose, including battery acid, gasoline, diesel fuel, antifreeze, motor oil, and transmission fluid. Hose may come in contact with these chemicals at vehicle fires or at the scene of a collision where the chemicals are spilled on the roadway. Supply hose in particular often comes in contact with residues from these chemicals when hose lines are laid in the roadway. Remove chemicals from the hose as soon as possible and then wash the hose with an approved detergent, thoroughly rinse it, and let it dry completely.

Mildew

Mildew is a fungus that can grow on fabrics and materials in warm, moist conditions. A fire hose that has been packed away while it is still wet and dirty is a natural breeding ground for mildew. This fungus feeds on nutrients found in many natural fibers and this can cause the fibers to rot and deteriorate.

In the days when cotton fibers were used in hose jackets, mildew was a major problem. Hose had to be washed and completely dried after every use before it could be placed back on the apparatus. Modern fire hose is made from synthetic fibers that are resistant to mildew, and most types can be repacked without drying first. The fibers in rubber-covered hose are protected from mildew. Nevertheless, mildew may still grow on exposed fibers if they are soiled with contaminants that will provide mildew with the necessary nutrients.

SAFETY TIP

Whenever a fire hose has suffered possible damage, it should be thoroughly inspected and tested according to chapter 4 in NFPA 1962 before it is returned to service.

Cleaning and Maintaining Hose

It is important to properly clean fire hose. Because hose can be made from different materials, the exact steps needed to clean and maintain it varies. When hose becomes dirty, it is important to clean it as soon as possible. For a mild cleaning, cool water and a soft brush may be adequate. For dirty hose, it may be necessary to use a mild detergent, especially if the hose has come in contact with petroleum-based products.

TIP

In general, try to prevent hose from coming in contact with petroleum and abrasive substances whenever possible.

Some fire departments use specially manufactured hose washers for washing and drying fire hose. A simple hose washer is a cylindrical device with water jets inside to which you attach a supply line. To use it, first lay the hose out straight and scrub it with a mild detergent to remove debris and dirt. Then insert one end of the scrubbed hose into the cylinder of the hose washer. When the water is turned on to the hose washer, it sprays out of the jets inside the cylinder onto the hose. Slowly pull the hose through the washer, rinsing off the detergent and debris (**FIGURE 11-35**).

More complex hose washers consist of a large cabinet-style mechanical device. These may contain an automatic feed that moves hose through a power washing cycle and then squeegees some of the water off the hose before it leaves the washing machine (**FIGURE 11-36**). This type of machine enables one person to wash a large quantity of hose in a fairly short period of time.

Although most fire hose is rubber-jacketed hose, which is more resistant to water damage than other fabrics, fire hose—even rubber-jacketed hose—should not routinely be stored wet for the following reasons:

FIGURE 11-35 A simple hose washer.

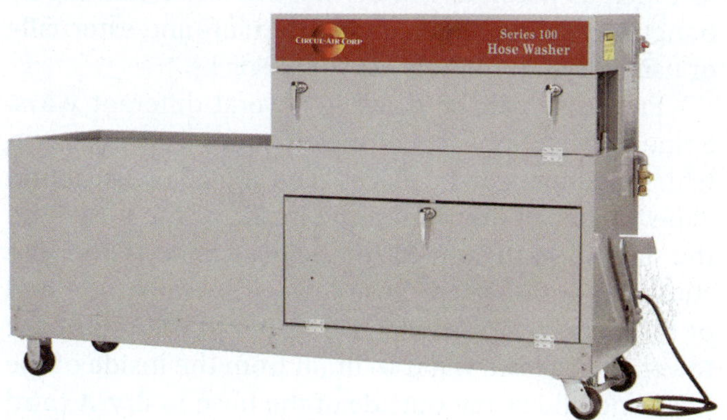

FIGURE 11-36 A mechanical hose washer.

FIGURE 11-37 A hose drying cabinet.

Courtesy of Circul-Air Corp.

- Storing wet hose promotes mold and mildew growth internally, which can weaken the hose liner.
- Water trapped inside hose can corrode metal couplings over time.
- The additional weight of water makes hose heavier to handle and deploy.
- Hard water deposits can build up on the exterior jacket.
- Ice that forms in freezing temperatures can cause internal damage.

The best practice is to limit wet storage duration as much as possible and dry hose before reloading by hanging it to drain and air dry internally and externally or using drying cabinets, towers, or racks.

Fire hose can be dried in several different ways. Some fire stations have angled racks with slats on which wet hose can be placed. The angled construction allows water to drain from the inside of the hose, and the slats allow the outside of the hose to dry. Other fire stations are built with hose towers in which one end of the clean and wet hose is hoisted to the top of the tower. This allows water to drain from the inside of the hose and allows the outside of the hose to dry. A third method of drying hose is to place wet hose that has been loosely folded or coiled in a heated hose drying cabinet (**FIGURE 11-37**). Heated air is then circulated through the cabinet. Some heated hose drying cabinets can also be used for drying personal protective equipment.

Always follow the hose manufacturer's guidelines and procedures for washing and drying fire hose. If your department has hose washing and drying equipment, follow the instructions from those manufacturers to be sure the hose is properly cleaned and dried. Some hose washing equipment does not completely clean the hose couplings. Hose couplings may require special attention to be sure they are clean and will operate easily and effectively under emergency conditions.

To manually clean hose that is dirty or contaminated, follow the steps in **SKILL DRILL 11-13**.

Hose Inspections

Visual hose inspections should be performed at least quarterly. A visual inspection should also be performed after each use, either while the hose is being cleaned and dried or when it is reloaded onto the apparatus. Always visually inspect any hose that has come in direct contact with a fire. In addition, hose that has not been used in 30 days should be unpacked, inspected, cleaned, and reloaded. If any defects are found, that length of hose should be immediately removed from service and tagged with a description of the problem. The appropriate notifications must be made to have the hose repaired.

To inspect hose for defects and mark a defective hose, follow the steps in **SKILL DRILL 11-14**.

Hose Testing and Records

Each length of hose should be tested at least annually according to the procedures listed in NFPA 1962. The procedures are complicated and require special equipment. This equipment must be operated according to the manufacturer's instructions.

A **hose record** is a written history of each individual length of fire hose. Each length of hose is labeled with a unique number stenciled or painted on it. The hose record contains detailed information about the hose such as the hose size, type, manufacturer, date of manufacture, date of purchase, and testing dates. Hose testing and recording hose records are Firefighter II tasks. To perform an annual service test on a fire hose, record the results of the service test, and maintain records, see Chapter 23, *Advanced Fire Suppression*.

SKILL DRILL 11-13

Cleaning and Drying Hose Firefighter I, NFPA 1010: 6.5.2

1. Lay the hose out flat. Rinse the hose with water.

2. Gently scrub the hose with mild detergent, paying attention to soiled areas.

3. Turn over the hose and repeat steps 1 and 2. Give a final rinse to the hose with water.

SKILL DRILL 11-14

Inspecting a Hose for Defects and Marking a Defective Hose
Firefighter I, NFPA 1010: 6.5.2

1. Thoroughly inspect the fire hose and couplings to identify any defects or damage. Look for common defects to the hose such as holes, cuts, bulges, cracks, or other signs of wear and tear that compromise the hose's integrity. Also inspect the couplings, looking carefully for damage to the threads, lugs, swivels, and gaskets.
2. If a defect is identified, assess the severity and extent of the defect. It is essential to determine whether the hose can still be used or if it needs to be removed from service.
3. Mark the damaged area so that it is easily visible and identifiable to all personnel who handle the hose. Use marking tools approved by your fire department or agency, such as colored tape or paint, or other marking devices that are durable and visible.
4. Document the hose defect, including its location, type, and extent by recording this information in the hose records.
5. If the defect is severe and compromises the hose's safety and performance, remove the hose from service immediately, and place an out-of-service tag (or other method dictated by your department's SOPs) on it, and notify your superiors. Replace the hose with a new one or have it repaired by qualified professional following manufacturer's guidelines.

CASE STUDY

You Are the Firefighter CONCLUSION

Shortly after beginning recruit fire training, you are assigned to ride along with an engine company for a shift. After a relatively uneventful day, your company is dispatched for a working fire at a large warehouse. When you arrive, there are flames coming from the front of the building. You hear the incident commander's instructions to lay a 4-in. (101-mm) supply line and to prepare for an attack with the deck gun.

1. **Why is it important to understand the function and proper uses of each type of hose carried on your apparatus?**

   Answer: One of the primary functions of firefighting is to quickly and effectively apply sufficient quantities of water to the fire to extinguish it. In order to achieve this objective, it is necessary to understand the tools that enable you to complete this task.

2. **What are the advantages of using a master stream appliance on this fire?**

   Answer: A master stream appliance is able to produce high-volume water streams for this large warehouse fire. Most master stream appliances discharge between 350 gpm (1325 L/min) and 1500 gpm (5678 L/min). A master stream appliance can also discharge a stream of water farther than a hand line, allowing the firefighters to keep a distance away until it is safe to advance.

WRAP-UP

SUMMARY

KNOWLEDGE OBJECTIVES

- List the various types and sizes of fire hose and how they are used.
  - List the two types of fire hose. (**NFPA 1010: 6.3.15**, pp. 386–387)
  - Identify the sizes of fire hose and their usage. (pp. 386–387)
- Describe the characteristics of the different types of fire hose.
  - Explain how fire hose is constructed. (**NFPA 1010: 6.3.15**, pp. 387–388)
  - Describe the characteristics of single-jacket hose. (**NFPA 1010: 6.3.15**, p. 387)
  - Describe the characteristics of multiple-jacket hose. (**NFPA 1010: 6.3.15**, p. 387)
  - Describe the characteristics of rubber-covered hose. (**NFPA 1010: 6.3.15**, p. 387)
- Describe the characteristics of common types of couplings.
  - Describe the characteristics of couplings. (**NFPA 1010: 6.3.10**, pp. 388–396)
  - List the common types of couplings. (**NFPA 1010: 6.3.10**, pp. 388–396)
- Identify key usage applications of various types of fire hose, including supply, suction, and attack hose and how hoses are rolled for transport.
  - Describe how supply hose is utilized in the field. (**NFPA 1010: 6.3.15**, pp. 396–397)
  - Describe the two types of suction hose. (**NFPA 1010: 6.3.15**, p. 397)
  - Describe the characteristics of attack hose. (**NFPA 1010: 6.3.10**, pp. 397–399)
  - Describe the types of rolls used to safely prepare hose for transport. (NFPA 1010: 6.5.2, pp. 399–401)
- Describe the different types of hose nozzles.
  - Discuss the differences between smooth-bore nozzles and fog-stream nozzles. (**NFPA 1010: 6.3.10**, pp. 401–410)
  - Describe specialized nozzles and how they are utilized. (pp. 410–411)
  - Describe how to inspect and maintain nozzles. (pp. 411–412)
- List the common hose appliances used in conjunction with fire hose.
  - List the common hose appliances used in conjunction with fire hose. (**NFPA 1010: 6.3.15**, pp. 412–417)
  - Describe the characteristics of adapters and reducers. (**NFPA 1010: 6.3.15**, pp. 412–413)
  - Describe the characteristics of wyes. (**NFPA 1010: 6.3.10**, p. 413)
  - Describe the characteristics of water thieves. (**NFPA 1010: 6.3.15**, p. 413)
  - Describe the characteristics of Siamese connectors. (**NFPA 1010: 6.3.15**, pp. 413–414)
  - Describe the types of valves used to control water in pipes or hose lines. (**NFPA 1010: 6.3.15**, p. 414)
  - Describe the characteristics of master stream appliances. (**NFPA 1010: 6.3.15**, pp. 414–416)
  - Describe the characteristics of hose jackets. (**NFPA 1010: 6.3.10**, p. 416)
  - Describe the characteristics of hose rollers. (**NFPA 1010: 6.3.10**, p. 416)
  - Describe the characteristics of chafing blocks. (p. 417)
  - Describe the characteristics of hose bridges. (**NFPA 1010: 6.3.15**, p. 417)
  - Describe the characteristics of hose clamps. (**NFPA 1010: 6.3.10**, p. 417)
- Describe how to organize, clean, and maintain hose and the importance of a hose inspection.
  - List the common types of hose damage and how to prevent them. (**NFPA 1010: 6.5.2**, pp. 418–419)
  - Describe how to clean and maintain hose. (**NFPA 1010: 6.5.2**, pp. 419–420)
  - Describe the importance of a hose inspection. (**NFPA 1010: 6.5.2**, pp. 420, 422)

SKILLS OBJECTIVES

- Replace the swivel gasket on a fire hose. (**NFPA 1010: 6.5.2**, pp. 390–391)
- Perform various methods of coupling and uncoupling a fire hose.
  - Perform the one-firefighter foot-tilt method of coupling a fire hose. (**NFPA 1010: 6.3.10**, p. 392)
  - Perform the two-firefighter method of coupling a fire hose. (**NFPA 1010: 6.3.10**, p. 393)
  - Perform the one-firefighter knee-press method of uncoupling a fire hose. (**NFPA 1010: 6.3.10**, p. 394)
  - Perform the two-firefighter stiff-arm method of uncoupling a fire hose. (**NFPA 1010: 6.3.10**, p. 395)
  - Uncouple a hose with a spanner wrench. (**NFPA 1010: 6.3.10**, pp. 395–396)
- Perform various hose rolls.
  - Perform a straight or storage hose roll. (**NFPA 1010: 6.5.2**, p. 400)
  - Perform a single-doughnut hose roll. (**NFPA 1010: 6.5.2**, pp. 401–402)
  - Perform a twin-doughnut hose roll. (**NFPA 1010: 6.5.2**, p. 403)
  - Perform a self-locking twin-doughnut hose roll. (**NFPA 1010: 6.5.2**, pp. 404–405)
- Operate a smooth-bore nozzle and a fog-stream nozzle.
  - Operate a smooth-bore nozzle. (**NFPA 1010: 6.3.10**, pp. 407–408)
  - Operate a fog-stream nozzle. (**NFPA 1010: 6.3.10**, pp. 409–410)
- Clean and dry hose. (**NFPA 1010: 6.5.2**, p. 421)
- Inspect and mark a defective hose. (**NFPA 1010: 6.5.2**, p. 422)

KEY TERMS

adapter Any device that allows fire hose couplings to be safely interconnected with couplings of different sizes, threads, or mating surfaces, or that allows fire hose couplings to be safely connected to other appliances. (NFPA 1960)

adjustable fog-stream nozzle See *fog-stream nozzle.*

adjustable-gallonage fog nozzle A nozzle that allows the operator to select a desired flow from several settings. Also called *variable-flow fog nozzle.*

air-aspirating nozzle A nozzle that draws air into the water stream, creating an aerated or foamy spray that increases the surface area of water droplets, allowing for better heat absorption and faster cooling of a fire, and when used with firefighting foam solutions, aerate the foam mixture.

attack engine An engine used to pump water through attack lines at the fireground.

attack hose Hose designed to be used by trained firefighters and fire brigade members to combat fires beyond the incipient stage. Also called *attack line.* (NFPA 1962)

attack line See *attack hose.*

automatic-adjusting fog nozzle A nozzle that can deliver a wide range of water stream flows. As the pressure at the nozzle increases or decreases, an internal spring-loaded piston moves in or out to adjust the size of the opening.

bale A shut-off valve on a nozzle that is a handle that you pull to control the flow of water from the nozzle.

ball valves Valves used on nozzles, gated wyes, and engine discharge gates. They consist of a ball with a hole in the middle of the ball.

booster hose A non-collapsible hose used under positive pressure having an elastomeric or thermoplastic tube, a braided or spiraled braided reinforcement, and an outer protective cover. Also called *booster line.* (NFPA 1962)

booster line See *booster hose.*

braided reinforcement A hose reinforcement consisting of one or more layers of interlaced spiraled strands of yarn or wire, with a layer of rubber between each braid. (NFPA 1962)

breakaway fire nozzle A nozzle with a tip that can be separated from the shut-off valve.

Bresnan distributor nozzle A nozzle that can be placed in confined spaces such as cellars or basements. The nozzle spins, spreading water over a large area.

butterfly valves Valves that are found on the large pump intake connections where the suction hose connects to the suction side of the fire pump.

cellar nozzle A nozzle used to fight fires in cellars or basements and other inaccessible places that spreads water in a wide pattern.

chafing block A sturdy rubber, plastic, or wooden block placed under a fire hose where it lays on the ground or rests against hard surfaces to raise the hose off the ground and provide a smooth contact surface and protect the hose from abrasion, friction, and wear against rough surfaces like asphalt, concrete, or debris.

charge To fill with water under pressure.

chimney nozzle A nozzle that has a 45-degree elbow attached to a long discharge pipe that sprays water upward.

clapper mechanism See *clapper valve.*

clapper valve A mechanical device installed within a piping system that allows water to flow in only one direction. Also called *clapper mechanism.*

collapsible fire hose Fire hose typically made from synthetic materials that make the hose flexible and foldable.

coupling A connection device that connects (couples) individual lengths of fire hose together, or connects a hose to a fire hydrant or a pump or to nozzles and hose appliances.

deck gun A device that is permanently mounted on and operated from a vehicle and equipped with a piping system that delivers water to the gun.

double-female adapter A hose adapter that is used to join two male hose couplings.

double-jacket hose A hose constructed with two layers of woven fibers.

double-male adapter A hose adapter that is used to join two female hose couplings.

draft The use of suction to move a liquid (such as water) from a vessel or source that is below the intake of a pump. (NFPA 1910)

elevated master stream appliance A nozzle mounted on the end of an aerial device that is capable of delivering large amounts of water into a fire or exposed building from an elevated position.

fingers Individual, adjustable vanes or protrusions on the face of a fog-stream nozzle that are responsible for shaping the water stream into a specific pattern.

fire department connection (FDC) A connection through which the fire department can pump supplemental water into the sprinkler system, standpipe, or other system furnishing water for fire extinguishment to supplement existing water supplies. (NFPA 13)

fire hose A flexible conduit used to convey water or other extinguishing agents (NFPA 1960).

fire hose appliance A piece of hardware (excluding nozzles) generally intended for connection to fire hose to control or convey water. Also called *hose appliance.* (NFPA 1962)

fire hose tool A device that assists firefighters with handling, manipulating, connecting, and using a fire hose.

fire stream Stream of water or extinguishing agents.

fixed-gallonage fog nozzle A nozzle that delivers a set number of gallons per minute (liters per minute) as per the nozzle's design, no matter what pressure is applied to the nozzle.

flow rate The number of gallons or liters of fluid streamed per unit of time.

fog stream A stream of water that is flowed in the form of small water droplets. (NFPA 1700)

fog-stream nozzle A nozzle that is placed at the end of a fire hose and can be adjusted to produce a straight stream or to separate the water into droplets to produce a variety of fog streams. Also called *spray nozzle* or *adjustable-fog-stream nozzle.*

forestry fire hose A hose designed to meet specialized requirements for fighting wildland fires. (NFPA 1960)

four-way hydrant valve A specialized type of valve that can be placed on a hydrant and that allows another engine to increase the supply pressure without interrupting flow.

gate valve A valve found on hydrants and sprinkler systems with a rotating a spindle that causes the gate to move slowly across the opening.

gated wye A valved device that splits a single hose into two separate hose lines, allowing each hose to be turned on and off independently.

handline A hose and nozzle that can be held and directed by hand. (NFPA 11)

KEY TERMS CONTINUED

handline nozzle A nozzle with a rated discharge of less than 350 gpm (1325 L/min).

hard suction hose A short section of supply hose that is used to draft water from a static source such as a river, lake, or portable drafting basin to the suction side of the fire pump on a fire department engine or into a portable pump.

Higbee indicator An indicator on both the male and female threaded couplings that indicates where the threads start. These indicators should be aligned before firefighters start to thread the couplings together.

hose appliance See *fire hose appliance.*

hose bed The main storage area on an apparatus for carrying hose.

hose bridge A device that protects a hose when it is necessary for a vehicle to drive over a hose. Also called *hose ramp.*

hose clamp A device used to compress a fire hose to stop water flow.

hose hoist See *hose roller.*

hose jacket A device used to stop a leak in a fire hose or to join hose lines that have damaged couplings.

hose liner The inside portion of a hose that is in contact with the flowing water; also called *hose inner jacket.*

hose ramp See *hose bridge.*

hose record A written history of each individual length of fire hose.

hose roller A device that is placed on the edge of a roof and is used to protect hose as it is hoisted up and over the roof edge. Also called *hose hoist.*

hose size An expression of the internal diameter of the hose. (NFPA 1962)

hose tool See *fire hose tool.*

ladder pipe A monitor that attaches to the rungs of a vehicle-mounted aerial ladder. (NFPA 1960)

large-diameter hose (LDH) A hose 3.5 in. (89 mm) or larger that is designed to move large volumes of water to supply master stream appliances, portable hydrants, manifolds, standpipe and sprinkler systems, and fire department pumpers from hydrants and in relay. (NFPA 1410)

low-volume nozzle A nozzle that flows 40 gallons per minute (151 L/min) or less.

lug A protrusion or indentation on a hose coupling that aids in securing and tightening the connection between two couplings.

master stream appliance A device that discharges high-volume water streams, usually between 350 gpm (1325 L/min) and 1500 gpm (5678 L/min), though much larger capacities are available. Also called *master stream device.*

master stream device See *master stream appliance.*

master stream nozzle A nozzle with a rated discharge of 350 gpm (1325 L/min) or greater. (NFPA 1960)

medium-diameter hose (MDH) A hose 2½-in. (64-mm) or 3-in. (76-mm) in diameter most often used as attack hose, but can be used as supply hose.

multiple-jacket A construction consisting of a combination of two separately woven reinforcements (double jacket) or two or more reinforcements interwoven. (NFPA 1962)

non-collapsible fire hose Fire hose typically made from durable materials such as PVC or synthetic rubber so that it maintains its shape and structure. Also called *rigid fire hose.*

nozzle A constricting appliance attached to the end of a fire hose or monitor to increase the water velocity and form a stream. (NFPA 1960)

piercing nozzle A nozzle that can be driven through sheet metal or other material to deliver a water stream to that area.

pin lug A lug that looks like a small cylinder that extends outward from a hose coupling.

portable monitor A monitor that can be lifted from a vehicle-mounted bracket and moved to an operating position on the ground by not more than two people. (NFPA 1960)

preconnect See *preconnected attack line.*

preconnected attack line Attack hose that travels on the engine already equipped with a nozzle and connected to a pump discharge outlet so it is ready for immediate use. Also called *preconnect.*

recessed lug A lug that is circular indentation and requires a specially designed spanner wrench called a booster hose wrench to engage.

reducer A fitting used to connect a small hose line or pipe to a larger hose line or pipe. (NFPA 1142)

remote-controlled hydrant valve A valve that is attached to a fire hydrant to allow the operator to turn the hydrant on without flowing water into the hose line.

rigid fire hose See *non-collapsible fire hose.*

rocker lug A rectangular-shaped, bevel-edged lug that extends from a coupling. Also called *rocker pin.*

rocker pin See *rocker lug.*

rotary control valve A shut-off valve on a nozzle that is a dial that you rotate to control the flow of water from the nozzle.

rubber-covered hose Hose whose outside covering is made of rubber, which is said to be more resistant to damage.

shut-off valve On a nozzle, a device that enables the firefighter at the nozzle to start or stop the flow of water.

Siamese connector A hose appliance that allows two hose to be connected together and flow into a single hose.

single-doughnut hose roll A hose roll that has both female couplings on the outside of the roll and the male coupling is protected by the hose rolled on top of it and that is used when the hose will be put into use directly from its rolled state.

single-jacket hose A construction consisting of one woven jacket. (NFPA 1962)

small-diameter hose (SDH) A hose 1½-in. (38-mm) or 1¾-in. (44-mm) hose that is most often used as the primary attack hose for most fires.

smooth-bore nozzle A nozzle that produces a solid stream, or a solid column of water.

soft sleeve hose A short section of large-diameter supply hose that is used to provide water from the large steamer outlet (the large-diameter port) on a fire hydrant or other pressurized water source to the suction side of the fire pump. Also called *soft suction hose.*

soft suction hose See *soft sleeve hose.*

solid stream A solid column of water.

spray nozzle See *fog-stream nozzle.*

standpipe A pipe and attached hose valves and hose (if provided) used for conveying water to various parts of a building for fire-fighting purposes. (NFPA 1140)

static water source A water source that is not under pressure, such as a pond, a lake, a stream, or a swimming pool.

storage hose roll See *straight hose roll.*

Storz hose coupling A hose coupling that has the property of being both the male and the female coupling. It is connected by engaging the lugs and turning the coupling a one-third turn.

straight hose roll A hose roll that has the male coupling at the center of the roll and the female coupling on the outside of the roll and that is used for general handling and transportation of hose as well as to prepare hose to be stored on a rack. Also called *storage hose roll.*

straight stream A semi-compact stream of water composed of tiny droplets that are all moving relatively close to one another, much like a solid stream but not quite as compact.

suction hose A hose that is designed to prevent collapse under vacuum conditions so that it can be used for drafting water from below the pump (lakes, river, wells, etc.) (NFPA 1960)

supply engine An engine used to pump water to an attack engine using supply lines that may be connected to a pressurized water source such as a fire hydrant or to a static (unpressurized) water source.

supply hose Hose designed for the purpose of moving water between a pressurized water source and a pump that is supplying attack lines. Also called *supply line.* (NFPA 1960)

supply line See *supply hose.*

swivel A ring around female couplings that you turn to secure the female coupling around the male coupling without twisting the hose.

swivel gasket An O-shaped piece of rubber inside the swivel section of a female hose coupling that forms a seal that stops water from leaking when a male hose coupling is tightened against it.

teeth Serrations or grooves around the fog-stream nozzle's orifice that help to break up the water stream and create a fog pattern.

threaded hose coupling A type of coupling that requires a male fitting and a female fitting to be screwed together.

variable-flow fog nozzle See *adjustable-gallonage fog nozzle.*

water curtain nozzle A nozzle used to deliver a flat screen of water that forms a protective sheet of water to protect exposures from fire.

water hammer The surge of pressure caused when a high-velocity flow of water is abruptly shut off. The pressure exerted by the flowing water against the closed system can be seven or more times that of the static pressure. (NFPA 1962)

KEY TERMS CONTINUED

water thief A device that has a 2½-in. (64-mm) inlet and a 2½-in. (64-mm) outlet in addition to two 1½-in. (38-mm) outlets. It is used to supply more than one hose from one source.

wye A device used to split a single hose into two or more separate lines.

REVIEW QUESTIONS

1. Fire hose is used for what two main purposes?
2. What size hose is usually used as the primary attack hose?
3. What is the 2½-in. (64-mm) hose most often used as?
4. What sizes can large-diameter hose (LDH) be?
5. What is the difference between single-jacket and multiple-jacket hose?
6. What is the difference between threaded couplings and Storz couplings?
7. What is supply hose used for?
8. List four common hose rolls.
9. List the two main types of nozzles and their associated streams.
10. What advantages does a smooth-bore nozzle have over a fog-stream nozzle?
11. What advantages does a fog-stream nozzle have over a smooth-bore nozzle?
12. Define what a hose appliance is and list several examples.
13. How often should fire hose be visually inspected for serviceability?

DISCUSSION QUESTIONS

1. Why is proper selection and deployment of the right type of fire hose for the specific conditions so critical for effective firefighting operations?
2. In what types of situations might a firefighter need to evaluate the tradeoffs between using a fog pattern nozzle versus a smooth-bore nozzle when attacking a fire?
3. How can improper storage and cleaning of fire hose lead to potential damage over time?
4. Why is routine inspection and testing important for maintaining fire hose in good operating condition?

APPLYING THE CONCEPTS

It's 0200 and your fire department is en route to a house fire in a densely populated residential neighborhood. As your engine arrives, you see heavy smoke and flames on the second floor, side alpha of the structure. Neighbors have gathered on the sidewalk.

1. As you size up the situation, what information do you gather?

Size-up revealed there are no cars in the driveway and a neighbor said the family is on vacation. The house is Type V construction with vinyl siding. The wind is calm but the fire is growing, and nearby exposures are threatened by the flames. A fire hydrant is located 50 feet (15 m) away. Your engine carries 1¾-inch, 2½-inch, and 4-inch (44-, 64-, 101-mm) hose lines. The IC calls for an offensive strategy to attack the fire.

2. What hoses can be used for fire attack?
3. Which hose should be used to establish a water supply? Why?

The IC directs you and your crew to advance the hose line to initiate a transitional attack to cool the environment before attacking the fire. You and your crew work together to deploy 150 feet (46 m) 1¾-inch (44-mm) hose. Then you are ready to perform a transitional attack with a fog-stream nozzle.

4. Why is the 1¾-inch (44-mm) hose a better choice for the initial fire attack than the 2½-inch (64-mm) hose?
5. Why is using a fog-stream nozzle a better choice to knock down the fire than a smooth-bore nozzle?

6. When would a smooth-bore nozzle be a better choice?

You and your crew advance hose lines to the Alpha side of the structure and direct a fog pattern around the entrance to cool the environment and improve visibility, reporting your progress and status to the IC. You enter the structure and proceed toward the stairs. The heat is intense and the smoke extremely dense. The IC radios, "Interior team, the fire has significantly intensified, and conditions are deteriorating. Evacuate immediately." Just after you and your crew safely exit, a portion of the roof collapses. The IC calls for master streams to be used for defensive operations on the structure and exposures.

7. For a master stream what size hose would you use for the hose lines? Why?
8. If an aerial ladder is used to initiate a defensive operation on the roof of the structure, what type of master stream appliance would be used?

REFERENCES

National Fire Protection Association (NFPA). 2017. *NFPA 1962, Standard for the Care, Use, Inspection, Service Testing, and Replacement of Fire Hose, Couplings, Nozzles, and Fire Hose Appliances.* 2018 Edition. Quincy, MA: NFPA.

National Fire Protection Association (NFPA). 2019. *NFPA 1410, Standard on Training for Emergency Scene Operations.* 2020 Edition. Quincy, MA: NFPA.

National Fire Protection Association (NFPA). 2020. *NFPA 11, Standard for Low-, Medium-, and High-Expansion Foam.* 2021 Edition. Quincy, MA: NFPA.

National Fire Protection Association (NFPA). 2020. *NFPA 1700, Guide for Structural Fire Fighting.* 2021 Edition. Quincy, MA: NFPA.

National Fire Protection Association (NFPA). 2021. *NFPA 13, Standard for the Installation of Sprinkler Systems.* 2022 Edition. Quincy, MA: NFPA.

National Fire Protection Association (NFPA). 2021. *NFPA 24, Standard for the Installation of Private Fire Service Mains and Their Appurtenances.* 2022 Edition. Quincy, MA: NFPA.

National Fire Protection Association (NFPA). 2021. *NFPA 1140, Standard for Wildland Fire Protection.* 2022 Edition. Quincy, MA: NFPA.

National Fire Protection Association (NFPA). 2021. *NFPA 1142, Standard on Water Supplies for Suburban and Rural Firefighting* 2022 Edition. Quincy, MA: NFPA.

National Fire Protection Association (NFPA). 2023. *NFPA 1900, Standard for Aircraft Rescue and Firefighting Vehicles, Automotive Fire Apparatus, Wildland Fire Apparatus, and Automotive Ambulances.* 2024 Edition. Quincy, MA: NFPA.

National Fire Protection Association (NFPA). 2023. *NFPA 1910, Standard for the Inspection, Maintenance, Refurbishment, Testing, and Retirement of In-Service Emergency Vehicles and Marine Firefighting Vessels.* 2024 Edition. Quincy, MA: NFPA.

National Fire Protection Association (NFPA). 2023. *NFPA 1960, Standard for Fire Hose Connections, Spray Nozzles, Manufacturer's Design of Fire Department Ground Ladders, Fire Hose, and Powered Rescue Tools.* 2024 Edition. Quincy, MA: NFPA.

CHAPTER

12

Firefighter I

Ladders

KNOWLEDGE OBJECTIVES

After studying this chapter, you will be able to:

- List and describe the parts of a ladder and identify the different types of ladders.
- Describe the methods of inspection, maintenance, cleaning, and the service testing process for ladders.
- Specify the hazards associated with ground ladders and describe the measures firefighters should take to maximize safety when working on and with ladders.
- Describe proper ladder selection and subsequent proper ladder placement based on common fireground tasks.

SKILLS OBJECTIVES

After studying this chapter, you will be able to perform the following skills:

- Clean, inspect, and maintain a ladder.
- Demonstrate all identified ladder carries.
- Demonstrate the various raises used with ladders and properly tie a halyard.
- Climb a ladder and then utilize a leg lock to work from a ladder.

ADDITIONAL NFPA STANDARDS

- **NFPA 1900**, *Standard for Aircraft Rescue and Firefighting Vehicles, Automotive Fire Apparatus, Wildland Fire Apparatus, and Automotive Ambulances, 2024 Edition*
- **NFPA 1932**, *Standard on Use, Maintenance, and Service Testing of In-Service Fire Department Ground Ladders, 2020 Edition*
- **NFPA 1960**, *Standard for Fire Hose Connections, Spray Nozzles, Manufacturer's Design of Fire Department Ground Ladders, Fire Hose, and Powered Rescue Tools, 2024 Edition*
- **NFPA 2500**, *Standard for Operations and Training for Technical Search and Rescue Incidents and Life Safety Rope and Equipment for Emergency Services, 2022 Edition*

CASE STUDY

You Are the Firefighter

At a fire incident at a large residential home, a mayday is announced over the radio: "MAYDAY, MAYDAY, MAYDAY, members trapped on the third floor in the rear!" There is no aerial access at the rear of the building, so you and your crew are assigned to deploy ground ladder(s) for immediate egress for the trapped members. Time is critical and the stress level is high as these members need immediate rescue.

1. Having multiple ground ladders to choose from, which will you take to the rear?
2. What is the best way to rapidly carry and raise the ladder(s) to the trapped members?
3. What safety concerns could you foresee during this scenario?

Introduction

Ladders are undoubtedly one of the most versatile and valuable tools available to firefighters. Firefighters use ladders for a variety of tasks at an incident scene, including rescuing fellow firefighters, rescuing victims, gaining access to upper floors, and delivering a master stream. Ladders have many additional uses at an incident scene. Every firefighter must be knowledgeable about the many uses of ladders and be proficient in using them.

Uses of Ladders

At a fire scene, ground ladders are used to gain access to upper floors to search for possible victims, to remove victims, or to give firefighters emergency egress. They can provide access to and egress from a window, a cockloft, or an attic. Ladders can also provide a safe pathway between floors, enabling firefighters to avoid a damaged or unsafe stairway. Ladders may be used to gain access to a second-floor window or balcony or to provide stable footing and distribute the weight of firefighters during roof operations. They also can be deployed across a small space to create a bridge or placed to enable a firefighter to climb over a fence or obstruction. Sometimes the best escape route from a fire may be to travel horizontally across a ladder used as a bridge to a nearby building. During salvage operations, a ladder covered with a waterproof tarp can be used as a chute to funnel water out of a building.

Firefighters also use ground ladders in a wide variety of non-firefighting situations. A ground ladder can be used to reach an injured person in a ravine or down a steep highway embankment. When rescuing victims from a collapsed trench, firefighters use ground ladders to access a victim or to aid in bringing them to the surface. Ground ladders are also used to provide access to or escape from other below-grade locations such as a maintenance hole (the opening to access utility equipment underground). Specialized rescue teams use ladders to provide access to and exit from buses, railcars, or airplanes. Ladders are also used for a variety of functions during a hazardous materials incident. Finally, ladders can be used as ramps to assist in moving equipment during emergencies.

Ladder Construction and Components

Ladders are constructed of metal, fiberglass, or wood. Most fire departments use ladders constructed of metal, usually aluminum, and a few departments use wood ladders. Fiberglass is less frequently used in the fire service. Each material has advantages and disadvantages (**TABLE 12-1**). Following the manufacturer's instructions and cautions for using your ladders is essential.

Ladders must meet the performance requirements described in NFPA 1960, *Standard for Fire Hose Connections, Spray Nozzles, Manufacturer's Design of Fire Department Ground Ladders, Fire Hose, and Powered Rescue Tools.* This standard requires manufacturers to adhere to basic requirements such as specifications for the diameter and spacing of rungs; the inclusion of specific labels and markings, including ladder positioning stickers, heat sensor labels, and electrocution hazard warning labels; and length and width requirements.

All ladders have two beams connected by a series of parallel rungs. The **beam** is the main structural component that runs the length of a ladder. A **rung** is the crosspiece between the two beams, which serves as a step. Per NFPA 1960, the surface area of rungs is required to be skid resistant (NFPA 1960 2023).

TABLE 12-1 Advantages and Disadvantages of Ladder Materials

Material	Advantages	Disadvantages
Metal	▪ Relatively light in weight ▪ Inexpensive ▪ Easy to repair	▪ Conducts electricity ▪ Subject to failure when exposed to high heat
Wood	▪ Less likely to conduct electricity ▪ Retains more strength when exposed to heat ▪ More flexible than metal ladders	▪ Higher cost ▪ Requires more maintenance ▪ Heavier weight ▪ Combustible
Fiberglass	▪ Less likely to conduct electricity ▪ Increased strength	▪ Increased chance of failure if overloaded ▪ May be combustible

There are three main types of ladders, characterized by the type of beam (**FIGURE 12-1**):

- **Trussed beam ladder**: The beams on a trussed beam ladder consist of a top and bottom **rail** connected by a smaller piece called a **truss block**. The rungs are attached to the truss blocks. Trussed beam ladders are usually constructed of aluminum or wood. Longer extension ladders are usually trussed beam ladders because this type of construction results in a lighter-weight ladder.
- **I-beam ladder**: The beams on an I-beam ladder are continuous pieces shaped like an uppercase letter "I". The top and bottom of the I-beam are a **flange** and the part between the flanges is the **web**. The rungs are attached to the web. In the fire service, this type of beam is usually made from fiberglass but may also be constructed of metal.
- **Solid beam ladder**: The beams on a solid beam ladder are rectangular-shaped, continuous pieces. Many wood ladders have solid beams. Fiberglass and metal ladders can also have solid beams. Hollow or C-shaped rectangular aluminum beams are also classified as solid beams.

To use and maintain ladders properly, firefighters must be familiar with all the parts of a ladder and the terms used to describe those parts (**FIGURE 12-2**):

- The **tip** is the very top of the ladder.
- The **butt** is the end of the ladder that is placed against the ground when the ladder is raised. It is sometimes called the **heel** or **base**.
- **Butt spurs** are metal spikes attached to the butt of a ladder to prevent the butt from slipping out of position (**FIGURE 12-3**). A **butt plate**, an alternative to butt spurs, is a flat plate attached to the butt of the ladder that swivels to maintain contact with the ground and has spurs on the edges of the plate and cleats on the bottom of the plate.
- The **heat sensor label** changes color when the ladder is exposed to the amount of heat that could damage the structural integrity of the ladder. They are applied to metal and fiberglass ladders. Heat sensors activate when exposed to a temperature of 300°F (149°C) plus or minus 5 percent. They have an expiration date, at which point the sensor needs to be replaced (**FIGURE 12-4**).
- **Protection plates** are reinforcing pieces on metal and fiberglass ladders that are placed on a ladder at chafing and contact points to prevent damage to the ladder from friction or contact with other surfaces.
- A **tie rod** is a metal bar under each rung of some wood ladders. Tie rods run from one beam to the other and help keep the beams from separating from the rungs.

Types of Ladders

Ladders used in the fire service can be classified into two broad categories: ground and aerial. A **fire department ground ladder**, or simply a **ground ladder**, is a portable ladder carried on engines and removed to be used as needed. An **aerial ladder** is a ladder that is a permanently mounted power-operated ladder on an apparatus with a working length of between 50 feet (ft; 15 meters [m]) and 137 ft (41 m) and operated from that apparatus.

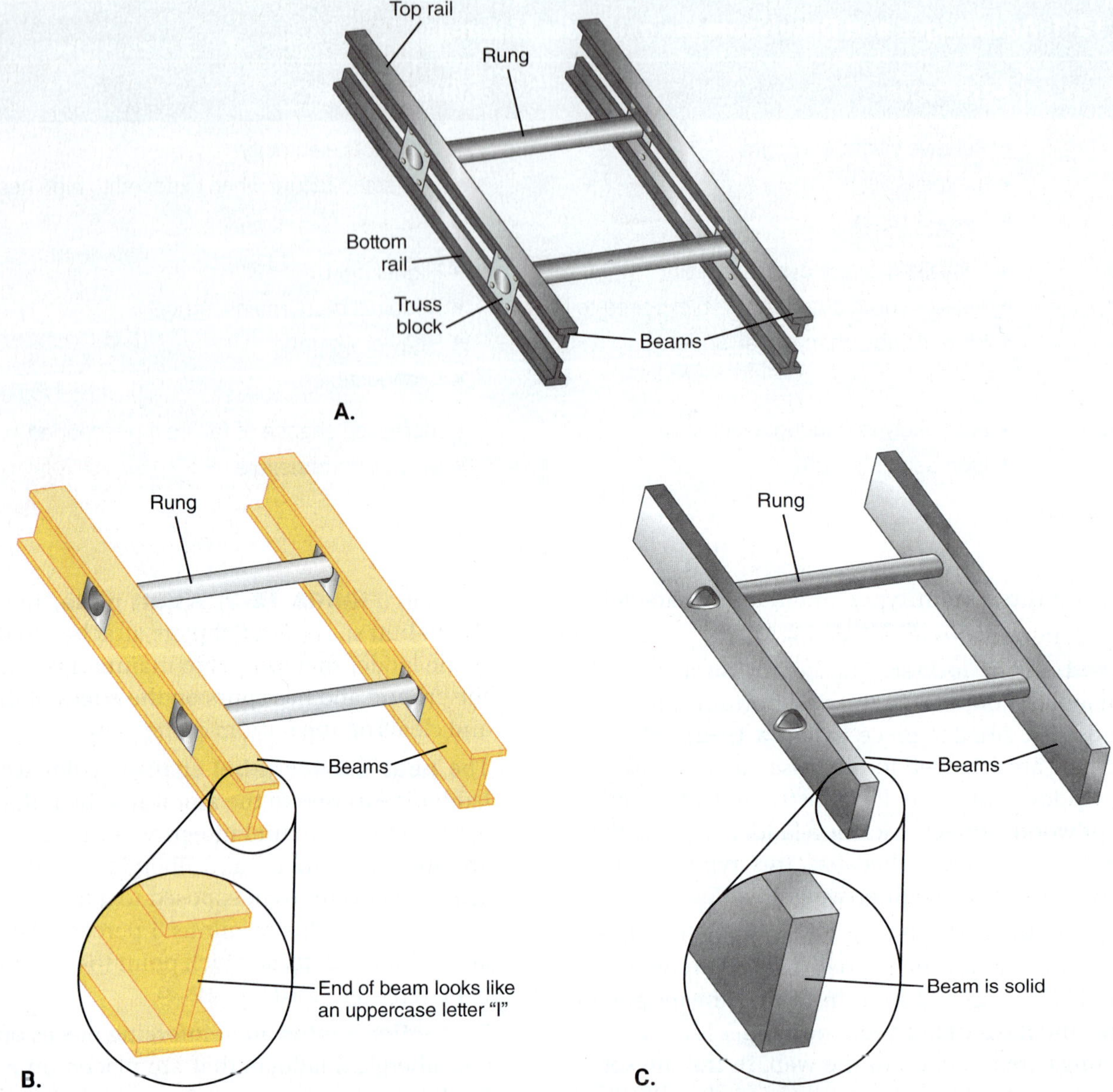

FIGURE 12-1 Three types of beam construction are found in ladders: **A.** Trussed beam. **B.** I-beam. **C.** Solid beam.

Ground Ladders

The first broad category of ladders used in the fire service is ground ladders. Most fire apparatus carry an assortment of ground ladders. Types of ground ladders include the following:

- Straight ladder
- Roof ladder
- Extension ladder
- A-frame ladder
- Combination ladders
- Folding ladder

Straight Ladder

A **straight ladder**, also called a **single ladder** or a **wall ladder**, is a single-section, fixed-length ground ladder. These lightweight ladders can be raised quickly, and they can reach the windows and roofs of one- and two-story structures. Straight ladders are commonly 12 to 14 ft (4 to 4.2 m) long but can be as long as 20 ft (6 m).

Roof Ladder

A **roof ladder**, sometimes called a **hook ladder**, is a straight ground ladder that is equipped with roof hooks at one or both ends. **Roof hooks** are spring-loaded,

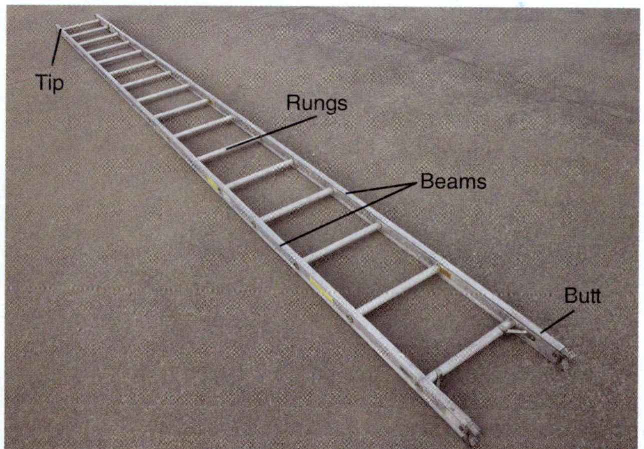

FIGURE 12-2 The basic components of a ladder.
© Jones & Bartlett Learning. Photographed by Glen E. Ellman.

FIGURE 12-3 Butt spurs dig into the ground to prevent the ladder from slipping.
Courtesy of Kevin Lewis.

FIGURE 12-4 A heat sensor label indicates whether the ladder has been exposed to excessive heat.
Courtesy of Kevin Lewis.

retractable, curved, metal pieces attached to the beams at the tip or at both ends of a roof ladder to secure the tip of the ladder to the peak of a pitched roof when the ladder lies flat on the roof. If a straight ladder has roof hooks on both ends, butt spurs are also on both ends.

Roof ladders provide stable footing and distribute the weight of firefighters and their equipment, thereby helping to reduce the risk of structural failure in the roof assembly. They can also be used on flat roofs to distribute weight. In addition to being used on roofs, roof ladders can be used in situations where a straight ladder is needed. Roof ladders can also be used for access to the second floor or a balcony.

Roof ladders in the fire service are usually 12 to 20 ft (4 to 6.6 m) long (**FIGURE 12-5**). Roof ladders are not designed as free-hanging ladders, meaning that they are not specifically designed to support weight when hanging exclusively from the hooks. Nevertheless, they are tested to make sure they can support 1000 pounds (lb; 454 kilograms [kg]) for 1 minute when hanging only by their hooks (this is called hang testing) per NFPA 1932, *Standard on Use, Maintenance, and Service Testing of In-Service Fire Department Ground Ladders*.

Extension Ladder

An **extension ladder** is an adjustable-length ground ladder with two or more connected sections that can be extended or retracted to adjust the length of the ladder (**FIGURE 12-6**). An extension ladder is usually heavier than a straight ladder of the same length and normally

A.

B.

FIGURE 12-5 **A.** The main use of roof ladders is to distribute weight on a roof. Roof hooks on a straight ground ladder means that the ladder can be used as a roof ladder. **B.** Many roof ladders have a set of roof hooks at both ends, which makes it easier to use on both sides of a roof because you do not need to spin it around.

requires more than one person to set it up. Because the length is adjustable, an extension ladder can replace several straight ladders, and it can be stored in places where a longer straight ladder would not fit.

FIGURE 12-6 An extension ladder can be used at any length, from fully retracted to fully extended.

Extension ladders have additional parts that are not found on straight ladders (**FIGURE 12-7**):

- The **bed section** is the widest section of an extension ladder. It is also called the **base section** because it serves as the base of the ladder—all other sections are raised from the bed section. The butt of an extension ladder is at the bottom of the bed section.
- The bed section supports one or more fly sections. A **fly section** is raised or extended from the bed section or from the previous section if the extension ladder has more than one fly section. When the fly section is not extended—that is, it is all the way down—it is called fully bedded. An extension ladder with a bed and one fly section is a two-section ladder, one with a bed and two fly sections is a three-section ladder, and one with three fly sections is a four-section ladder. (Straight ladders are single section ladders.)
- The **halyard** is a rope or cable attached to the fly sections and that runs through a pulley (or multiple

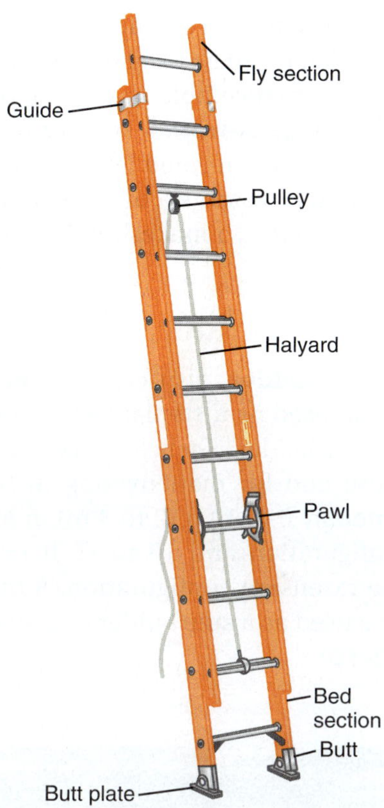

FIGURE 12-7 Extension ladders have additional parts that are not found in straight ladders.

© Jones & Bartlett Learning

pulleys) near the top of the bed section and down to the ground so that it can be pulled to hoist the fly sections. On extension ladders with three and four sections, wire-rope halyard extensions are attached to the halyard so that the last fly section can be extended fully.

- A **guide** is a strip of metal, wood, or fiberglass that wraps around a fly section and the section below it or slots in the fly section or the section below it to guide the fly section as it is being extended. Some manufacturers recommend that wax be applied between the guides and the beams to ensure smooth operation when extending and collapsing the sections.
- A **pawl** is a mechanical locking device used in pairs to secure and hold extended fly sections in place. They are sometimes called a **ladder lock**, a **dog**, or a **rung lock**.
- A **stop** is a piece of metal, wood, or fiberglass that prevents the fly sections of a ladder from overextending and collapsing the ladder. This is also referred to as a **stop block**.

Fresno Ladder. A **Fresno ladder** is a narrow, two-section extension ladder specifically designed to provide attic access and to be used in any tight space. For example, a Fresno ladder is used as a bridge over a damaged section of an interior stairway. A Fresno ladder is generally short—just 10 to 14 ft (3 to 4 m)—so it has no halyard and is extended manually (**FIGURE 12-8**).

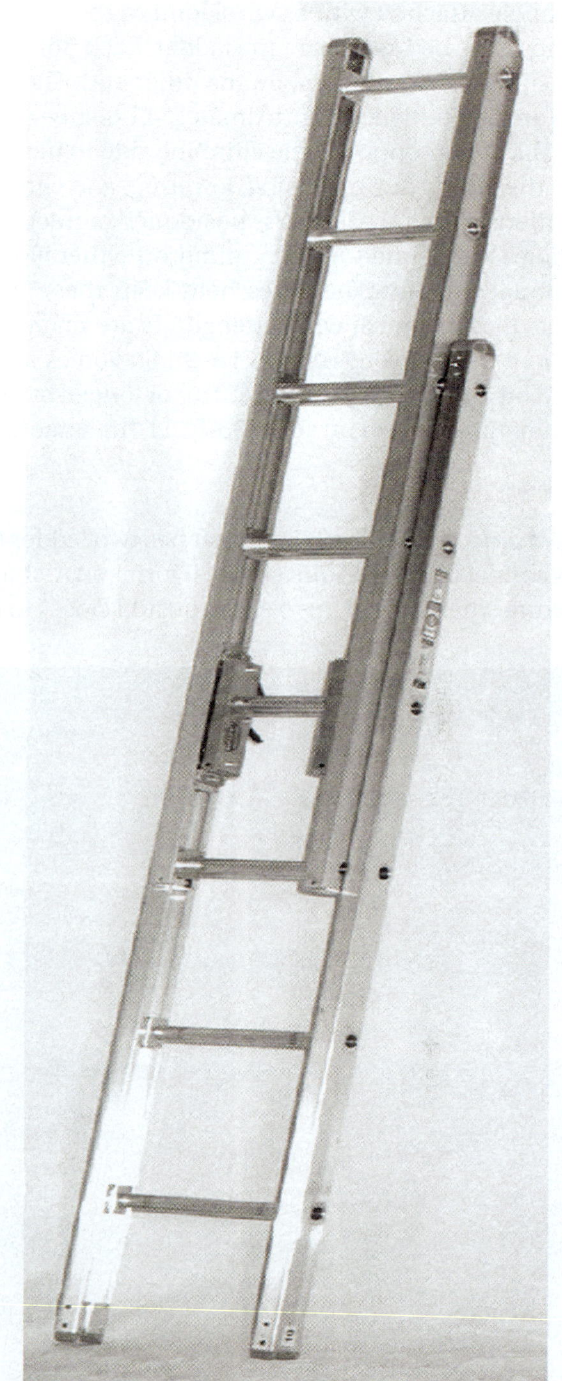

FIGURE 12-8 Fresno ladders are usually short and are extended manually rather than with a halyard.

Courtesy of Duo-Safety Ladder Corporation.

Bangor Ladder. A **Bangor ladder**, also called a **tormentor ladder**, is an extension ladders with staypoles. A **staypole**, also called a **tormentor**, is a long metal pole attached with a swivel joint to the beams at the top of the bed section of the ladder. Each pole has a spur, similar to a butt spur, on the other end. The staypoles are positioned at approximately 45-degree angles from the ladder opposite the climbing side to help stabilize the ladder during raising, lowering, and climbing operations. When the ladder is positioned correctly, the staypoles are planted in the ground on either side for additional stability. The poles help keep these heavy ladders under control while firefighters are maneuvering them into place (**FIGURE 12-9**). Staypoles are required on ladders that are 40 ft (12 m) or longer, but they are sometimes found on some 35-ft (11-m) ladders.

A-Frame Ladder

An **A-frame ladder** is a ladder that has two ladder sections connected with a joint so that it forms an A-shaped structure when it is set up to climb on and closes to fold flat for storage. A common example of an A-frame ladder is a stepladder. It can be used to lower rescuers into a trench or maintenance hole or to raise or lower victims safely. A **ladder A-frame**, sometimes referred to as an **A-frame hoist**, is formed by using rope to attach two straight ladders at the tip in an A shape. A ladder A-frame can be used as a makeshift lift when raising a trapped person.

Combination Ladder

A **combination ladder**, also called a **multipurpose ladder**, can be used as a stepladder (A-frame ladder) or an extension ladder. Such ladders are convenient for indoor use and for maneuvering in tight spaces. They are generally 6 to 10 ft (2 to 3 m) in length in the A-frame configuration and 10 to 15 ft (3 to 5 m) in length in the extension configuration. A multipurpose ladder can be used as a step ladder or a straight ladder (**FIGURE 12-10**).

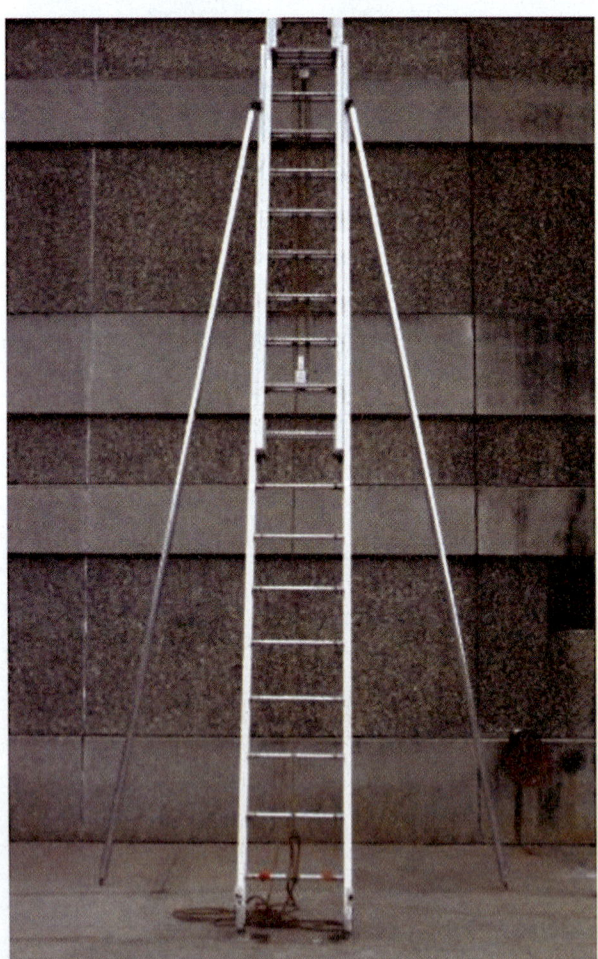

FIGURE 12-9 Staypoles stabilize a Bangor ladder while it is being raised or lowered.

Reproduced from The Connecticut Fire Academy. 2018. "Practical Skill Training: Skill Sheet 13.4.3: Ground Ladders." Updated March 5, 2018. https://portal.ct.gov/-/media/CFPC/files/NEW-ITEMS-2019/Uploaded-Files/Instructor-Lesson-Plans/Uploaded-Files/Unit-13/SS-1343-Instructor-Reference-Material.pdf

FIGURE 12-10 Combination ladders can be used in several different configurations.

Folding Ladder

A **folding ladder**, also called an **attic ladder**, is a narrow, collapsible ladder that is designed to allow access to attic scuttle holes and confined areas (**FIGURE 12-11**). The two beams fold in to enhance the ladder's portability. Folding ladders are commonly available in 8- to 14-ft (2- to 2.4-m) lengths.

TIP

A pompier ladder is a single beam with rungs placed across the beam or a narrow two-beam ladder and with a large hook at one end. They are rarely used in the United States anymore.

Ground Ladder Requirements for Apparatus

The NFPA does not specify minimum lengths for ground ladders. Most engines carry 24- or 28-ft (7- or 8-m) ladders, which can reach the second-floor windows and the roof of a typical two-story building. Generally, fire service ground ladders are limited to a maximum length of 50 ft (15 m). If a building's height exceeds the capabilities of these ladders, aerial apparatus must be used. Truck companies usually carry 35- or 40-ft (11- or 12-m) ladders, which can reach the windows and roofs of most three-story buildings.

A.

B.

FIGURE 12-11 Folding ladders are designed to be used in narrow spaces or restrictive passages. **A.** A folding ladder in the portable position. **B.** A folding ladder unfolded and ready to use.

A. © Jones & Bartlett Learning. Photographed by Glen E. Ellman; **B.** Courtesy of Kevin Lewis.

Voice of Experience

While some may undervalue the importance of ladder training and drills, when the time comes and ground ladders need to be placed rapidly, firefighters will fall back on the training and evolutions they practiced until they became second nature. My fire department once received a call for a structure fire in a building in an older section of the downtown area. Upon arrival, several occupants were leaning out of the second story windows. The fire had trapped them in their apartments.

Using an aerial ladder was not an option due to the power lines in front of the building. Ground ladders were the safest option. We positioned the ground ladders and accomplished four rescues in a matter of minutes from when we arrived at the scene. Relying on our training enabled us to "ladder the building" in a safe and efficient manner and quickly perform the rescues. This was a clear example of how training had reinforced the necessary skills needed in the entire team and led to a successful conclusion to an incident.

Chris Harner
Deputy Chief—Operations
Pueblo Fire Department
Pueblo, Colorado

A.

B.

C.

FIGURE 12-12 There are three types of aerial devices. **A.** Aerial ladder. **B.** Elevated platform. **C.** Water tower.

Aerial Devices

The second broad category of ladders in the fire service are aerial devices. Aerial devices vary greatly in terms of their design and function. (Detailed requirements for trucks are documented in NFPA 1900, *Standard for Aircraft Rescue and Firefighting Vehicles, Automotive Fire Apparatus, Wildland Fire Apparatus, and Automotive Ambulances.*) Aerial devices are classified as aerial ladders, elevated platforms, and water towers (**FIGURE 12-12**).

A.

B.

FIGURE 12-13 **A.** A steel aerial ladder. **B.** The permanently mounted waterway running up the ladder is visible.

Courtesy of Kevin Lewis.

Aerial ladders are made of metal and contain two or more sections that telescope into one another (**FIGURE 12-13**). Most aerial ladders have a permanently mounted waterway and a monitor nozzle.

FIGURE 12-14 Tiller truck, also known as a tractor-drawn aerial.

Courtesy of Kevin Lewis.

Some departments use a tiller truck to carry an aerial device. A **tiller truck**, also called a **tractor-drawn aerial**, is a tractor-trailer apparatus that requires a second operator positioned at the back of the trailer to steer the back wheels (**FIGURE 12-14**). Tractor-drawn aerials can be maneuvered into extremely tight places.

The **elevated platform** aerial device has a passenger-carrying platform (often referred to as the bucket) attached to the tip of an aerial ladder or a boom. A boom is a series of telescoping tubes that extend and retract. An elevated platform attached to an aerial ladder allows for continuous access to the platform. An elevated platform that is 110 ft (33 m) or less in length must have a permanently mounted waterway and a monitor nozzle.

Water towers are aerial devices that deliver streams of water and contain high-capacity pre-piped waterways. The water streams are directed from the truck. They do not have ladders or platforms. Some aerial ladders and elevated platforms also have permanently installed waterways.

Cleaning, Inspection, Maintenance, Service Testing, and Storing Ladders

Ground ladders used by the fire service must be able to withstand extreme conditions. They might be dropped, overloaded, or exposed to temperature extremes during use. Ladders must meet the performance requirements described in NFPA 1960. These requirements are based on probable conditions during emergency operations. NFPA 1932 provides general guidance for the use of ground ladders. Ladders must be regularly inspected, maintained, and service tested, following the NFPA standard. In addition, ladders should always be inspected and maintained in accordance with the manufacturer's recommendations.

Cleaning

Ladders must be cleaned regularly to remove any road grime and dirt that have built up on the apparatus during storage. In addition, ladders should be cleaned after each use and before each inspection to ensure that any hidden faults can be observed.

Use a soft-bristle brush and water to clean ladders. A mild, diluted detergent may be used, if allowed by the manufacturer's recommendations. Remove any tar, oil, or grease deposits with a safety solvent as recommended by the manufacturer. Do not get any solvent on the halyard of an extension ladder because contact with solvents can damage halyard ropes. Rinse and dry the cleaned ladder before placing it back on the apparatus.

SAFETY TIP

Do not get any solvent on the halyard of an extension ladder. Contact with solvents can damage halyard ropes.

Inspection and Maintenance

Ladders must be visually inspected at least once per month. Ladders should also be inspected after each use. Ladders are easily damaged while in use. For example, a ladder might be overloaded or used at a low angle during a rescue, or an unexpected shift in fire conditions could bring the ladder in direct contact with flames. Whenever a ladder is used outside of its recommended limits, it should be taken out of service for inspection and testing, even if there is no visible damage. If an inspection reveals any deficiencies, the ladder must be removed from service until repairs are made. All inspections should be recorded according to your department's SOPs.

Maintenance is simply the regular process of keeping the ladder in proper operating condition. All firefighters should be able to perform routine ladder maintenance, such as replacing a halyard or an expired heat sensor label. But only qualified personnel at a properly equipped repair facility should perform repairs involving the structural or mechanical components of a ladder. Like inspections, all maintenance should be recorded according to your department's SOPs.

TABLE 12-2 describes things to look for during a ladder inspection and the actions you can take to correct deficiencies and maintain the ladders.

TABLE 12-2 Ladder Inspection and Maintenance

Component	What to Look For	Action to Take If a Deficiency Is Found
All ladders		
Beams	Check the beams for cracks, splintering, breaks, gouges, checks, wavy conditions, or deformations.	Immediately place out of service
Rungs	Check all rungs for snugness, tightness, punctures, wavy conditions, worn serrations, splintering, breaks, gouges, checks, or deformations.	Immediately place out of service
Butt spurs and butt plates	Check the butt spurs and butt plates for excessive wear or other defects.	Immediately place out of service
Heat sensor labels	▪ Check to see if the sensor has expired. ▪ Check to see if the sensors indicate that the ladder has been exposed to excessive heat.	▪ If the heat sensor labels have reached their expiration date, replace them. ▪ If the heat sensor label indicates that a ladder has been exposed to excessive heat, remove the ladder from service and conduct a service test.

Component	What to Look For	Action to Take If a Deficiency Is Found
Bolts and rivets	Check all bolts and rivets for tightness. Bolts on wood ladders should be snug and tight without crushing the wood.	Immediately place out of service
Welded joints	Check all welded joints on metal ladders for cracks or defects.	Immediately place out of service
Metal surfaces	Check metal surfaces for signs of surface corrosion.	Immediately place out of service
Fiberglass surfaces	Check fiberglass ladders for loss of gloss on the beams.	Maintain the finish in accordance with the manufacturer's recommendations. This should be done on a regular basis according to the manufacturer's recommendations in addition to when an inspection reveals the need.
Wood surfaces	▪ Check for damage to the varnish finish on wood ground ladders. ▪ Check for signs of wood rot—that is, dark soft spots.	▪ Maintain the finish in accordance with the manufacturer's recommendations. This should be done on a regular basis according to the manufacturer's recommendations in addition to when an inspection reveals the need. ▪ If there are signs of wood rot, take the ladder out of service
Painted surfaces	Because paint can hide structural defects in a ladder, ensure that ladders are *not* painted except for the top and bottom 18 inches (in.; 0.47 m) of each section, which are painted for purposes of identification and visibility.	▪ Remove the paint.
Roof ladders (**FIGURE 12-15**)	▪ Check the roof hooks for sharpness. ▪ Check that the roof hooks operate properly. ▪ Check that the roof hooks do not have rust and other contaminants. ▪ Check for sufficient lubrication.	▪ Take out of service and replace ▪ Take out of service and replace ▪ Remove rust and other contaminants. ▪ Lubricate roof hooks.
Extension ladders		
Halyards (**FIGURE 12-16**)	Check the halyard for kinking and fraying and ensure that it moves smoothly through the pulleys.	Replace the halyard if it is kinked, frayed, or otherwise worn or damaged.
Wire-rope halyard extensions	Check the wire-rope halyard extensions on three- and four-section ladders for snugness. This check should be performed when the fly sections are fully bedded to ensure that the upper sections will align properly during operation.	
Guides	▪ Check the guides for chafing. ▪ If the manufacturer requires wax, check for adequate wax between the guides and the beams.	▪ Lubricate the guides in accordance with the manufacturer's recommendations.
Pawls (**FIGURE 12-17**)	Check the pawls for proper operation.	Lubricate the pawls following the manufacturer's instructions.

Courtesy of Kevin Lewis.

SAFETY TIP

Repairs involving structural or mechanical components of a ladder should be performed only by trained repair personnel.

FIGURE 12-15 Roof hooks must operate smoothly.

FIGURE 12-16 Replace the halyard if it is worn or damaged.

FIGURE 12-17 Extension ladder pawls must operate smoothly.

Service Testing

Service tests measure the structural integrity of a ladder, ensuring that ladders are safe to use for their intended purposes. Service tests must be performed both on new ladders and periodically on ladders already in use to evaluate the continued usefulness of a ladder. According to NFPA 1932, service tests should be conducted before a new ladder is used and at least once per year while the ladder remains in service. A ladder that has been exposed to extreme heat, has been overloaded, has been impact or shock loaded, is visibly damaged, or is suspected of being unsafe for any other reason must be removed from service until it has passed a service test. A ladder that has been repaired—unless a halyard replacement was the only repair—must also undergo a service test before it can be returned to service.

Service testing of ladders requires special training and equipment and must be conducted only by qualified personnel. Many fire departments use outside contractors to perform these tests. The results of all service tests must be recorded and kept for future reference.

Storage

When ladders are not in use, they should be placed on racks or in brackets and protected from the weather. Fiberglass ladders can be damaged by prolonged exposure to direct sunlight (**FIGURE 12-18**).

To clean, inspect, maintain, and store ladders, follow the steps in **SKILL DRILL 12-1**.

FIGURE 12-18 Ladders should be stored on racks or in brackets, out of the weather or direct sunlight.

SKILL DRILL 12-1

Clean, Inspect, Maintain, and Store a Ladder Firefighter I, NFPA 1010: 6.5.1

1. Clean all components following national standards and the manufacturer's recommendations. Visually inspect each component of the ladder for wear and damage.

2. On extension ladders, lubricate the ladder pawls, guides, and pulleys using the recommended material and according to the manufacturer's recommendations.

3. Perform a functional check of all components.

Continues.

SKILL DRILL 12-1 CONTINUED

Clean, Inspect, Maintain, and Store a Ladder Firefighter I, NFPA 1010: 6.5.1

4. If deficiencies are found and cannot be addressed at the fire station, tag and remove the ladder from service, and if needed, schedule the ladder to go to a repair facility. If the ladder passes inspection or is repaired at the fire station, return the ladder to the apparatus or to the storage area. Complete the maintenance record for the ladder.

Using Ground Ladders

Ground ladders are often urgently needed during emergency incidents. However, ladders must be used with caution. Several potential hazards are associated with ladder use, and unfortunately these risks are easily overlooked during emergency operations. Consequently, many firefighters have been seriously injured or killed in ladder accidents, both on the fireground and during training sessions. Because an accident or error in handling or using a ladder can result in death or serious injury, all firefighters must know how to work with ladders. Firefighters need to learn standard operating procedures (SOPs) for handling ladders and regularly reinforce basic skills through training.

Using a ladder requires that firefighters complete a series of tasks:

- Select the best ladder for the job from those available.
- Remove the ladder from the apparatus.
- Lift and carry the ladder to the location where it will be used.
- Place the ladder in the safest location to conduct the task at hand.
- Raise the ladder with it positioned at the appropriate angle for the task at hand.
- Secure the ladder.
- Climb the ladder.
- Work on the ladder.
- Dismount the ladder.
- Conduct rescue operations using a ladder.
- Descend a ladder.
- Lower the ladder and return it to the apparatus.

In any firefighting operation, personal protective equipment (PPE) is essential for working with ladders. PPE provides protection from mechanical injuries as well as from fire. Helmets, coats, pants, gloves, and protective footwear provide protection from falling debris, impact injuries, and pinch injuries. Firefighters must also be able to work with and on ladders while wearing self-contained breathing apparatus (SCBA). You should practice working with ladders while wearing all of your PPE.

Selecting the Ladder

The first step in using a ladder is to select the appropriate ladder from those available. Firefighters must be familiar with all of the ladders carried on their apparatus. Engine and truck company apparatus usually carry several ground ladders of various lengths. Many other types of apparatus, such as mobile water supply apparatus and rescue units, also carry ladders.

Selecting an appropriate ladder requires that firefighters estimate the heights of windows and rooflines. General guidelines have been developed for estimating the heights for residential and commercial buildings (**TABLE 12-3**).

TABLE 12-3 Approximate Distances for Residential and Commercial Construction

Type of Construction	Approximate Height (Bottom of Sill)
Residential floor-to-floor height	8–10 ft (2–3 m)
Residential floor-to-windowsill height	3 ft (1 m)
Commercial floor-to-floor height	10–12 ft (3–4 m)
Commercial floor-to-windowsill height	4 ft (1 m)

In addition to estimating the height where the ladder needs to be used, firefighters need to consider what they need to accomplish when they climb the ladder. When the ladder is used to access a roof, the tip of the ladder should extend several feet above the roofline to provide a convenient handhold and secure footing for firefighters as they mount and dismount. A general rule is to ensure that at least five ladder rungs show above the roofline. This also makes the tip of the ladder visible to firefighters who are working on the roof (**FIGURE 12-19**).

FIGURE 12-19 The tip of the ladder should extend above the roofline during roof operations.

Whether you need to break a window for venting, enter a window for rescue operations, or help a victim climb out of a window, it is best to place the tip of the ladder immediately below the windowsill (**FIGURE 12-20**). This position prevents the ladder from obstructing the window opening while a trapped occupant is removed. This position is also safer for venting because the firefighter does not need to consider the direction of the wind when placing the ladder. For venting, the ladder can be placed to the left or right of a window, but you need to consider the wind direction (**FIGURE 12-21**). When the window is vented, the wind could push the smoke and flames toward you if you pick the wrong side.

Vertical reach is the height a ladder can reach. If a ladder is perpendicular to the ground, the vertical reach is the same as the length of the ladder. When a ladder is positioned at an angle for climbing, the vertical reach is less than the length of the ladder. If you are going to climb the ladder to access the roof or ventilate a window, an angle of approximately 75 degrees between the base of the ladder and the ground provides the best combination of strength, stability, and vertical reach. A 75-degree angle is sometimes referred to as the **climbing angle**. The **base distance**—the distance from the base of the building to the butt of the ladder—of a ladder placed at the climbing angle is equal to one-fourth of the ladder's height. For example, if a 20-ft (6-m) ladder is raised to a roof 18 ft

FIGURE 12-20 For firefighter ingress or egress or for victim removal down the ladder, the ladder tip should be positioned just below the windowsill.

FIGURE 12-21 If the ladder is positioned to the side of a window for venting, the tip of the ladder should be on the upwind side of the window and even with the top of the window opening.

(5.5 m) above the ground and placed at the climbing angle, the base distance should be 5 ft (1.5 m) (**FIGURE 12-22A**).

Although comfortable for ascending and descending ladders, a 75-degree angle is too steep for a firefighter to safely climb out of a window and onto the ladder or to remove a victim. Moving the butt of the ladder farther away from the building to increase the base distance and create a 60-degree angle—often called the **rescue angle**—provides a more gradual and controllable slope for both exiting firefighters and victims. The height of the sill must be considered. If the sill is 13 ft (4 m) above the ground, the butt of a 20-ft (6-m) ladder placed 10 ft (3 m) from the building will create a 60-degree angle (**FIGURE 12-22B**).

At a 75-degree angle, the vertical reach is approximately 1 ft (0.3 m) shorter than the length of the ladder for every 15 ft (5 m) of vertical rise (that is, the height of the building). At a 60-degree angle, the vertical reach of a ladder is 1 ft (0.3 m) shorter than the length of the ladder for every 10 ft (5 m) of vertical height. For example, suppose there is a window with the top of the opening 30 ft (9 m) above the ground. To reach this window for access operations with a ladder placed at an angle of approximately 75 degrees, the ladder would need to be at least 32 ft (9.8 m) long to be even with the top of the window opening. If the bottom of the window opening was 30 ft (9 m) above the ground, to reach that height for rescue operations, the ladder would need to be at least 33 ft (10 m) long positioned at a 60-degree angle. Because ladders used in roof operations need to extend above the roofline by five rungs, accessing a roof that is 30 ft (9 m) above the ground positioned at a 75-degree angle would require a ladder at least 35 ft (11 m) long.

Removing the Ladder from the Apparatus

Ladders are mounted on apparatus in various ways, and firefighters must be familiar with how ladders are mounted on their apparatus and practice removing them safely and quickly. Ladders are often nested one inside another and mounted on brackets on the side of the engine. Firefighters should note the nesting order and location of the ladders relative to the brackets when removing or replacing them. Ladders that are not needed or put aside should not be placed on the ground in front of an exhaust pipe where hot exhaust could damage them.

Ladders also can be stored on overhead hydraulic lifts (**FIGURE 12-23**). The hydraulic mechanism keeps them out of the way until they are needed. If your department's apparatus has a hydraulic lift, it is important to review its operation so that you can operate it efficiently and safely during an emergency.

On some vehicles, ground ladders are stored in compartments under the hose bed or aerial device. The ladders may lie flat or vertically on one beam (**FIGURE 12-24**). To remove or replace them,

FIGURE 12-22 Achieving the proper ladder angle is critical to your safety and the safety of people being rescued. **A.** The climbing angle of 75 degrees is achieved by placing the base of a ladder out from the building at a distance equal to one-fourth of the ladder's height. **B.** The rescue angle of 60 degrees is safer for firefighters climbing out of a window onto a ladder and for rescuing victims.

firefighters slide the ladders out of the rear of the vehicle. It is important to leave adequate room behind apparatus that store ladders in this manner so that the ladders can be removed at a fire scene.

Lifting Ladders

Many ladders are heavy and awkward to carry and maneuver. To prevent lifting injuries while handling ladders, firefighters must work together to lift and carry long or heavy ladders. Teamwork is essential when working with ladders. Never attempt to lift or move a ladder that weighs more than you are capable of lifting safely. It is better to ask for help in moving a ladder than to risk injury or delay the placement of the ladder.

When firefighters work as a team to lift or carry a ladder, one firefighter acts as the leader. Departments should establish a method for determining the lead firefighter for ladder lifts. In some departments, the firefighter on the right side at the butt end of the ladder is the leader. In other departments, the firefighter at the tip of the ladder on the right side may be the leader. The departmental policy should be consistent for all ladder operations.

The lead firefighter calls out the intended movements using two-step commands. For example, when

FIGURE 12-23 Some engines have a hydraulic mechanism to lower the ladders to a convenient height when they are needed and raise them for storage when not in use. This photo shows the ladder being raised for storage.

Courtesy of Kevin Lewis.

FIGURE 12-24 Ladders in rear compartments can be stored either flat or vertically. This configuration requires adequate space behind the truck to remove ladders at an emergency scene.

Courtesy of Kevin Lewis.

lowering a ladder, the leader would say, "Prepare to lower," followed by the command, "Lower." All team members should use the same terminology for commands. Because terminology may differ among jurisdictions, firefighters who are newly or temporarily assigned to an apparatus should verify the commands used by that company and what they mean so there is no confusion about the specific meaning of each command.

Always use proper lifting techniques when handling ladders. When bending to pick up a ladder, bend at the knees and keep your back straight (**FIGURE 12-25**). Use your legs when you lift and lower the load, rather than your back. Take care to avoid twisting motions while lifting and lowering, as these motions often lead to back strains.

TIP

Never attempt to lift or move a ladder that weighs more than you are capable of lifting safely. Remember, teamwork is essential when working with ladders!

Most ladders are carried with the butt end forward and pointed slightly downward so that the ladder can be positioned quickly. Roof ladders, however, are carried tip first with the hooks toward the front. Roof ladders usually must be carried up another ladder to reach the roof. The roof ladder is then pushed up the pitched roof slope from the butt end until the hooks grab on to the peak.

Carrying Ladders

Once a ladder has been removed from the apparatus or lifted from the ground, it must be carried to the

FIGURE 12-25 Bend your knees and keep your back straight when lifting or lowering a ladder.

placement site. Ladders can be carried at shoulder height or at arm's length and either on edge or flat. The most common carries are described in the following sections.

One-Firefighter Shoulder Carry

Most straight and roof ladders less than 20 ft (6.6 m) long can safely be carried by a single firefighter. To perform the steps for the one-firefighter shoulder carry, follow the steps in **SKILL DRILL 12-2**.

Two- and Three-Firefighter Shoulder Carry

The two-firefighter shoulder carry can be used with extension ladders up to 35 ft (11 m) long. It also can be used to carry straight ladders or roof ladders that are too long for one person to handle. Three firefighters may be needed to carry a heavier ladder. This carry is similar to the two-firefighter shoulder carry, with the additional firefighter positioned at the middle of the ladder. To perform the two- or three-firefighter shoulder carry, follow the steps in **SKILL DRILL 12-3**.

SKILL DRILL 12-2

One-Firefighter Shoulder Carry Firefighter I, NFPA 1010: 6.3.6

1. Start with the ladder mounted in a bracket or standing on one beam. Locate the center of the ladder. Facing the butt end of the ladder, grasp the two rungs on either side of the middle rung.

2. Lift the ladder and rest it on your shoulder.

Continues.

SKILL DRILL 12-2 CONTINUED

One-Firefighter Shoulder Carry Firefighter I, NFPA 1010: 6.3.6

3. Walk carefully with the butt end first and pointed slightly downward.

SKILL DRILL 12-3

Two- and Three-Firefighter Shoulder Carry Firefighter I, NFPA 1010: 6.3.6

1. Start with the ladder mounted in a bracket, standing on one beam, or resting on the ground. All firefighters are on the same side of the ladder facing the butt end of the ladder. One firefighter is positioned at the butt end of the ladder and one firefighter is positioned near the tip of the ladder.

If three firefighters are carrying the ladder, the third firefighter is positioned at the middle of the ladder.

SKILL DRILL 12-3 CONTINUED

Two- and Three-Firefighter Shoulder Carry Firefighter I, NFPA 1010: 6.3.6

2. If the ladder is resting on the ground, the firefighters, on the leader's command, rotate it so it is positioned on a beam.

Each firefighter places an arm between two rungs, and, on the leader's command, all firefighters lift the ladder onto their shoulders.

3. The firefighter closest to the butt end covers the sharp butt spur with a gloved hand to prevent injury to other firefighters.

Two- and Three-Firefighter Suitcase Carry

The two-firefighter suitcase carry is commonly used with straight and extension ladders. With this technique, the ladder is carried on its edge by a beam at arm's length. A three-firefighter suitcase carry can be used for heavier ladders. The additional firefighter is positioned at the center of the ladder. To perform the two- or three-firefighter suitcase carry, follow the steps in **SKILL DRILL 12-4**.

SKILL DRILL 12-4

Two- and Three-Firefighter Suitcase Carry Firefighter I, NFPA 1010: 6.3.6

1. Start with the ladder standing on one beam. All firefighters are on the same side of the ladder facing the butt end of the ladder. One firefighter is positioned at the butt end of the ladder and one firefighter is positioned near the tip of the ladder.

If three firefighters are carrying the ladder, the third firefighter is positioned at the middle of the ladder.

2. Each firefighter reaches down and grasps the upper beam of the ladder.

SKILL DRILL 12-4 CONTINUED

Two- and Three-Firefighter Suitcase Carry Firefighter I, NFPA 1010: 6.3.6

3. On the leader's command, the firefighters pick the ladder up from the ground and carry it at arm's length.

4. The firefighter closest to the butt end covers the sharp butt spur with a gloved hand to prevent injury to other firefighters.

Three- and Four-Firefighter Flat Carry

The three-firefighter flat carry is typically used with extension ladders up to 35 ft (11 m) long. The four-firefighter flat carry is similar to the three-firefighter flat carry, except that two firefighters are positioned on each side of the ladder at the tip and the butt end of the ladder. To perform the three- or four-firefighter flat carry, follow the steps in **SKILL DRILL 12-5**.

SKILL DRILL 12-5

Three- and Four-Firefighter Flat Carry Firefighter I, NFPA 1010: 6.3.6

1. Start with the bed section of the ladder flat on the ground. All firefighters face the butt end of the ladder.

If three firefighters are carrying the ladder, two firefighters stand on one side of the ladder, one at the butt and one at the tip. The third firefighter stands on the opposite side of the ladder at the middle of the ladder.

If four firefighters are carrying the ladder, one firefighter is positioned at each corner of the ladder, with two firefighters at the butt end and two firefighters at the tip end of the ladder.

2. All firefighters kneel and grasp the beam on their side at arm's length.

SKILL DRILL 12-5 CONTINUED

Three- and Four-Firefighter Flat Carry Firefighter I, NFPA 1010: 6.3.6

3. On the leader's command, the firefighters pick the ladder up from the ground and carry it at arm's length.

4. The firefighters closest to the butt end cover the sharp butt spurs with gloved hands to prevent injury to other firefighters.

Three- and Four-Firefighter Flat Shoulder Carry

The three-firefighter flat shoulder carry is similar to the three-firefighter flat carry, except that the ladder is carried on the shoulders instead of at arm's length. This carry is useful when firefighters must carry the ladder over short obstacles. Raising the ladder to shoulder height increases the potential for back strain, however, so firefighters must follow proper lifting techniques. A minimum of three firefighters should perform the flat shoulder carry. The four-firefighter flat shoulder carry is similar to the three-firefighter flat shoulder carry except that two firefighters are positioned on each side of the ladder. On each side, one firefighter is at the tip and the other at the butt end of the ladder. To perform the three- or four-firefighter flat shoulder carry, follow the steps in **SKILL DRILL 12-6**.

SKILL DRILL 12-6

Three- and Four-Firefighter Flat Shoulder Carry Firefighter I, NFPA 1010: 6.3.6

1. Start with the bed section of the ladder flat on the ground. All firefighters face the tip of the ladder.

   If three firefighters are carrying the ladder, two firefighters stand on one side of the ladder, one at the butt and one at the tip. The third firefighter stands on the opposite side of the ladder at the middle of the ladder.

   If four firefighters are carrying the ladder, one firefighter is positioned at each corner of the ladder, with two firefighters at the butt end and two firefighters at the tip end of the ladder.

2. All firefighters kneel and grasp the beam on their side at arm's length.

SKILL DRILL 12-6 CONTINUED

Three- and Four-Firefighter Flat Shoulder Carry Firefighter I, NFPA 1010: 6.3.6

3. On the leader's command, the firefighters stand, raising the ladder. As the ladder approaches chest height, the leader instructs the firefighters to pivot into the ladder. As they pivot, they place the hand not holding the ladder on the underside of the beam. The firefighters continue turning until they are facing the butt end of the ladder, and they place the beam on their shoulders.

4. The firefighters closest to the butt end cover the sharp butt spurs with gloved hands to prevent injury to other firefighters.

SAFETY TIP

When firefighters of different heights are working together and carrying ladders, you can make these operations safer and smoother by considering the placement of firefighters. When performing three-person carries, arrange the firefighters in order from shortest to tallest, with the shortest firefighter at the butt end of the ladder and the tallest firefighter at the tip end of the ladder. This strategy prevents a shorter firefighter from being placed between two taller firefighters and having to carry the ladder at an awkward and unsafe height. When performing four-person carries, place shorter firefighters at the butt end of the ladder to prevent the ladder from being tipped to one side while it is being carried.

Placing a Ladder

The first step in placing a ladder is selecting the proper location for the ladder. Generally, an officer or senior firefighter selects the general area for ladder placement, and the firefighter at the butt of the ladder determines the exact site at which to position it. Both the officer ordering the ladder and the firefighter placing the ladder need to consider several factors in their decisions. Before placing a ground ladder, the officer should survey the area where a ladder will be used. If possible, inspect the area before retrieving the ladder from the apparatus. If you identify any hazards, try to find a different location. Sometimes, despite best efforts, a ladder must be placed in a potentially hazardous location. If you are aware of the hazard, you can take corrective actions and implement special precautions to avoid accidents.

One of the most important things to assess is the location of overhead utility lines. If a ladder comes in contact with or is positioned too close to an electrical wire, the firefighters who are handling it can be electrocuted. If the line falls, it can electrocute other firefighters in the area as well. Do not assume that it is safe to use wood or fiberglass ladders around power lines. Metal ladders will certainly conduct electricity, but a wet or dirty wood or fiberglass ladder can also conduct electricity.

If possible, do not raise ladders anywhere near overhead wires, and avoid placing a ladder against a surface that has been energized by a damaged or fallen power line. At a minimum, keep ladders at least 10 ft (3 m) away from power lines while using or raising them. If a power line is nearby, make sure that enough firefighters are present to keep the ladder under control as it is raised. One firefighter should watch to make sure that the ladder does not come too close to the line. If the ladder falls into a power line, the results can be deadly.

SAFETY TIP

Always maintain at least 10 ft (3 m) of clearance between a ladder and utility lines to prevent electrocution.

Other types of overhead obstructions may also be hazards when raising a ladder. If the ladder hits something as it is being raised, the weight and momentum of the ladder will shift, which can cause firefighters to lose control and drop the ladder.

Do not place the ladder where it will be exposed to direct flames or extreme temperatures. Excessive heat can cause permanent damage to or catastrophic failure of the ladder. Check the heat sensor labels immediately after a ladder has been exposed to high temperatures.

Ladders must be placed on stable and relatively level surfaces (**FIGURE 12-26**). As a firefighter climbs a ladder, the moving weight causes the load to shift back and forth between the two beams. The shifting

A.

B.

FIGURE 12-26 **A.** Ladders must be placed on stable and relatively level surfaces. **B.** A ladder placed on an uneven surface is unsafe because it can easily tip over.

weight of a climber can easily tip a ladder placed on a slope or unstable surface. Unless both butt ends are firmly placed on solid ground, this uneven load can start a rocking motion that will tip the ladder over. Always evaluate the stability of the surface before placing the ladder and check it again during the course of the firefighting operations. A ladder placed on snow or mud, for example, may become unstable as the snow melts or as the base of the ladder sinks into the mud. In addition, water from the firefighting operations can make soft ground even more unstable. Ladders should also not be placed on unstable structures such as maintenance holes or trap doors. The combined weight of the ladder, firefighters, and equipment could cause the cover or door to fail, injuring the firefighters.

It is also important to ensure that the tip of the ladder is placed against a stable surface such as the solid wall of a structure. Placing the tip of a ladder against an unstable surface can result in the ladder falling forward into a fire or being pushed backward.

SAFETY TIP

Avoid injury or death—do not place the tip of a ladder against any unstable surface.

As mentioned earlier, ladders should be placed in areas with no overhead obstructions. If a ladder comes into contact with overhead utility lines—particularly electric power lines—both the firefighters working on the ladder and those stabilizing it could be killed or injured. This is true for all ladders, regardless of their composition (fiberglass, wood, or metal).

Be aware that a ladder can become energized even if it does not actually touch an electric line. For example, a ladder that enters the electromagnetic field surrounding a power line can become energized. For this reason, ladders should be placed at least 10 ft (3 m) away from energized power lines.

Finally, ground ladders should not be placed in high-traffic areas unless no other alternative is available. For example, because the main entrance to a structure is heavily used, a ground ladder should not be placed where it would obstruct the door.

SAFETY TIP

Ladders can become energized without making contact with overhead electric lines. An electromagnetic field surrounds electric lines, and the size of this field increases as the voltage increases. The conductivity of the air is also increased as the relative humidity increases. Always keep ladders at least 10 ft (3 m) away from electric lines.

Raising a Ladder

Once the position has been selected, the ladder must be raised for climbing. Before climbing a ladder, make sure it is set at the proper angle for the task to be performed. Remember, if you need to access the roof or a window for access operations, position the ladder at the climbing angle of 75 degrees, and if a ladder is going to be used for rescue operations, it should be placed at the rescue angle of 60 degrees. The shallower slope of a ladder placed at a 60-degree angle is safer to bring a victim down than a ladder placed at a 75-degree angle.

If you need the ladder placed at a 75-degree angle, a way to check to make sure the angle is correct is to stand on a rung of the ladder and stretch your hands in front of you. At a 75-degree angle, the rung at shoulder height is approximately an arm's length away (**FIGURE 12-27**). Also, most new ladders have a sticker

FIGURE 12-27 To confirm a 75-degree angle, stand on the bottom rung and hold the arms straight out.

on their beams positioned so that the bottom line of the indicator on the sticker is parallel to the ground when the ladder is positioned properly.

There may be a time when it is impossible to place a ladder at an angle of at least 60 degrees. If this happens, it is critical that at least one firefighter stabilizes the ladder while it is being used.

SAFETY TIP

Placing a ladder at the wrong angle will make the ladder unstable. If the butt of the ladder is placed too close to the building—in other words, at an angle greater than 75 degrees—the climbing angle will be too steep and the tip of the ladder could pull away from the building as the firefighter climbs. Conversely, if the butt is too far from the structure, the climbing angle will be too shallow, reducing the load capacity of the ladder and increasing the risk that the butt could slip out from under the ladder.

Two commonly used techniques for raising ground ladders are the beam raise and the flat raise. A **flat raise** (also called a **rung raise**) is a method of raising a ladder by positioning the ladder on the ground with the butt near the building and then lifting the tip to lean it against the building. A flat raise is used when the ladder can safely be raised perpendicular to the target surface. A **beam raise** is a method of raising a ladder by laying the ladder on one beam and them raising it by lifting the tip of the ladder while resting the butt of the ladder on the single beam. A beam raise is used when the ladder must be raised parallel to the target surface. (Note that these ladder raises are referred to by many different terms. Make sure you understand the terminology used in your department.)

SAFETY TIP

Always check for overhead hazards before raising a ladder.

The number of firefighters required to raise a ladder depends on the length and weight of the ladder, as well as on the available clearance from obstructions. A single firefighter can safely raise many straight ladders and lighter extension ladders. Two or more firefighters are required to raise longer and heavier ladders.

Raising Extension Ladders

When raising extension ladders, firefighters must know whether the fly section should be facing toward the building (fly in) or away from the building (fly out). The ladder manufacturer and department SOPs will specify whether the fly should be facing in or out. Manufacturers of metal and fiberglass ladders generally recommend that the fly sections are facing away (fly out) from the structure. In contrast, wood fire-service ladders often are designed to be used with the fly facing in. With some ladders, it is not possible to raise the fly section of the ladder from the fly side of the ladder.

SAFETY TIP

When raising the fly section of an extension ladder, never wrap the halyard around your hand. If the ladder falls or if the fly section retracts unexpectedly, your hand could be caught in the halyard, and you could be seriously injured. Also, never place your foot under the fly sections. If the rope slips, your foot will be crushed.

The halyard of an extension ladder should always be tied after the ladder has been extended and lowered into place. A tied halyard stays out of the way and provides a safety backup to the pawls for securing the fly section. One method of tying a halyard is shown in **SKILL DRILL 12-7**.

One-Firefighter Flat Raise

The one-firefighter flat raise can be used to raise a small, straight ladder, typically 14 ft (4 m) or less in length or with straight ladders longer than 14 ft (4 m), including extension ladders that can be safely handled by one firefighter. Every firefighter has a different physical limit, depending on their strength and the weight of the ladder.

With practice, it is possible to move the ladder from the carrying position to placing the butt of the ladder directly into the position where it should be when the ladder is raised. However, in the beginning it is generally easier and more efficient to lay the ladder down flat with the butt against the building before moving the ladder into the vertical position.

To perform the one-firefighter flat raise, follow the steps in **SKILL DRILL 12-8**.

SKILL DRILL 12-7

Tying the Halyard Firefighter I, NFPA 1010: 6.3.6

1. Wrap the excess halyard rope around two rungs of the ladder and pull the rope tight across the upper of the two rungs.

2. Tie a clove hitch around the upper rung and the vertical section of the halyard. Refer to Chapter 8, *Ropes and Knots*, to review how to tie a clove hitch.

3. Pull the clove hitch tight.

4. Place an overhand safety knot as close to the clove hitch as possible to prevent slipping.

SKILL DRILL 12-8

One-Firefighter Flat Raise for Ladders Less Than 14 Feet (4 m) Long
Firefighter I, NFPA 1010: 6.3.6

1. Place the butt of the ladder on the ground directly against the structure so that both spurs contact the ground and the structure, then lay the ladder on the ground. If the ladder is an extension ladder, place the fly section on the ground so that the ladder will be placed fly out.

2. Stand at the tip of the ladder and check for overhead hazards. Take hold of a rung near the tip, bring that end of the ladder to chest height, and then step beneath the ladder.

3. Raise the ladder using a hand-over-hand motion as you walk toward the structure until the ladder is vertical and against the structure.

SKILL DRILL 12-8 CONTINUED

One-Firefighter Flat Raise for Ladders Less Than 14 Feet (4 m) Long
Firefighter I, NFPA 1010: 6.3.6

4. If the ladder is an extension ladder, hold the ladder vertical against the structure, and extend the fly section by pulling the halyard smoothly, with a hand-over-hand motion, until the desired height is reached and the pawls are locked.

5. Grip a lower rung, then lift slightly while pulling outward and, at the same time, apply pressure to an upper rung to keep the tip of the ladder against the structure. Pull the butt of the ladder out from the structure to create a 75-degree climbing angle or a 60-degree rescue angle, depending on the situation.

6. If the ladder is an extension ladder, tie the halyard. Check that the ladder's tip and butt are positioned securely.

Two-Firefighter Flat Raise

The two-firefighter flat raise is commonly used with midsized extension ladders up to 35 ft (11 m) long. This raise has one firefighter positioned at each end of the ladder as it starts to get raised. To prevent the butt of the ladder from slipping as the tip is raised, the firefighter at the butt of the ladder needs to **heel** (or **foot**) it to secure it. Heeling uses the firefighter's weight to keep the base of the ladder from slipping. There are several methods of heeling a ladder (**FIGURE 12-28**):

- A firefighter on the outside of the ladder facing the structure places the toes or instep of one foot

A.

B.

C.

FIGURE 12-28 To prevent a ladder from slipping, firefighters should secure the base by heeling the ladder. **A.** One foot or instep on one of the beams and the heel of that foot on the ground. **B.** One foot or both feet on the bottom rung and the heel(s) on the ground. **C.** Kneeling in front of the ladder with a thigh against the lowest rung.

on one of the beams and the heel of that foot on the ground.

- A firefighter on the outside of the ladder facing the structure places the toes of one foot or both feet on the bottom rung and the heel(s) on the ground.
- A firefighter on the outside of the ladder facing the structure kneels in front of the ladder.

To perform the two-firefighter flat raise, follow the steps in **SKILL DRILL 12-9**.

SKILL DRILL 12-9

Two-Firefighter Flat Raise Firefighter I, NFPA 1010: 6.3.6

1. Start with a two-firefighter shoulder or suitcase carry. The firefighter at the butt of the ladder places the butt of the lower beam on the ground, while the firefighter at the tip holds the tip up. The firefighter at the tip rotates the ladder so that the bed section is on the bottom and the butts of both beams are in contact with the ground.

2. The firefighter at the butt of the ladder stands on the bottom rung, crouches down, grasps a higher rung with both hands, and leans backward to heel the ladder with both feet.

3. The firefighter at the tip of the ladder checks for overhead hazards, takes hold of a rung near the tip, brings that end of the ladder to chest height, steps beneath the ladder, and then walks toward the butt while raising the ladder using a hand-over-hand motion until the ladder is vertical. This positions the firefighter who raised the tip on the bed side of the ladder and the firefighter who stabilized the ladder at the butt on the side of the ladder with the fly section.

Continues.

SKILL DRILL 12-9 CONTINUED

Two-Firefighter Flat Raise Firefighter I, NFPA 1010: 6.3.6

4. The two firefighters grasp the beams and pivot the ladder into position so that the fly side is out and the firefighter on the fly side of the ladder is facing the structure. Each firefighter heels the ladder by placing the toe or instep of one boot against opposing beams.

5. The firefighter on the outside of the ladder facing the structure extends the fly section by pulling the halyard smoothly with a hand-over-hand motion until the desired height is reached and the pawls are locked. The firefighter on the inside of the ladder stabilizes the ladder by holding the outside of the beams on the bed section so that if the fly comes down suddenly, it will not strike that firefighter's hands.

6. The firefighter on the outside of the ladder facing the structure continues to heel the ladder while the firefighter on the inside of the ladder gradually steps back as both firefighters lean the ladder into place. The firefighter on the inside of the ladder ties the halyard. The firefighters check the ladder for a 75-degree climbing angle or a 60-degree rescue angle, depending on the situation, and then check that the ladder's tip and butt are positioned securely.

Two-Firefighter Beam Raise

The two-firefighter beam raise can also be used with midsized extension ladders up to 35 ft (11 m) long. In most cases the beam raise should be performed with two or more firefighters. If the ladder is small, this raise can be performed by one firefighter. To perform the two-firefighter beam raise, follow the steps in **SKILL DRILL 12-10**.

SKILL DRILL 12-10

Two-Firefighter Beam Raise Firefighter I, NFPA 1010: 6.3.6

1. Start with a two-firefighter shoulder or suitcase carry. The firefighter at the butt of the ladder places the butt of the lower beam on the ground, while the firefighter at the tip holds the tip up.

2. The firefighter at the butt of the ladder places a foot on the butt of the beam that is in contact with the ground and grasps the upper beam.

3. The firefighter at the tip of the ladder checks for overhead hazards, brings that end of the ladder to chest height, steps beneath the ladder, and then walks toward the butt while raising the beam, using a hand-over-hand motion until the ladder is vertical. The firefighters position themselves so that one is on the bed side of the ladder and one is on the side with the fly section.

Continues.

SKILL DRILL 12-10 CONTINUED

Two-Firefighter Beam Raise Firefighter I, NFPA 1010: 6.3.6

4. The two firefighters, one positioned on the fly side of the ladder and the other positioned on the bed side of the ladder, grasp the beams and pivot the ladder into position with the fly side out. Each firefighter heels the ladder by placing the toe or instep of one boot against opposing beams.

5. The firefighter on the outside of the ladder facing the structure extends the fly section by pulling the halyard smoothly with a hand-over-hand motion until the desired height is reached and the pawls are locked. The firefighter on the inside of the ladder stabilizes the ladder by holding the outside of the beams on the bed section so that if the fly comes down suddenly it will not strike that firefighter's hands.

6. The firefighter on the outside of the ladder facing the structure continues to heel the ladder while the firefighter on the inside of the ladder gradually steps back as both firefighters lean the ladder into place. The firefighter on the inside of the ladder ties the halyard. The firefighters check the ladder for a 75-degree climbing angle or a 60-degree rescue angle, depending on the situation, and then check that the ladder's tip and butt are positioned securely.

Three- and Four-Firefighter Flat Raises

The three- and four-firefighter flat raises are used for heavier ladders. The steps for these raises are similar to the two-firefighter flat raise. In a three-firefighter flat raise, the third firefighter faces the building and assists with the hand-over-hand raising of the ladder. When four firefighters are raising the ladder, two firefighters raise the ladder and two firefighters anchor the butt by placing their inside foot on the bottom rung of the ladder and bending over and grasping a rung.

To perform the three-firefighter flat raise, follow the steps in **SKILL DRILL 12-11**. To perform the four-firefighter flat raise, follow the steps in **SKILL DRILL 12-12**.

SKILL DRILL 12-11

Three-Firefighter Flat Raise Firefighter I, NFPA 1010: 6.3.6

1. Start with a three-firefighter shoulder or suitcase carry. The firefighter at the butt of the ladder places the butt of the lower beam on the ground, while the firefighter at the tip holds the tip up. The firefighter in the middle moves to the tip. The firefighters at the tip rotate the ladder so that the fly section is on the bottom and the butts of both beams are in contact with the ground.

2. The firefighter at the butt of the ladder stands on the bottom rung, crouches down, grasps a higher rung with both hands, and leans backward to heel the ladder with both feet.

3. The firefighters at the tip of the ladder check for overhead hazards, take hold of the same rung near the tip, bring that end of the ladder to chest height, step beneath the ladder, and then walk towards the butt while raising the ladder using a hand-over-hand motion until the ladder is vertical. This positions the firefighters who raised the tip on the fly side of the ladder and the firefighter who stabilized the ladder at the butt on the side of the ladder with the bed section.

Continues.

SKILL DRILL 12-11 CONTINUED

Three-Firefighter Flat Raise Firefighter I, NFPA 1010: 6.3.6

4. The three firefighters grasp the beams and pivot the ladder into position so that the fly side is out and the firefighters on the fly side of the ladder are facing the structure. The two firefighters on the outside of the ladder facing the structure heel the ladder by placing the toe or instep of their inside leg against the closest beam, and the firefighter on the inside of the ladder heels one of the beams in the same manner.

5. One of the firefighters on the outside of the ladder facing the structure extends the fly section by pulling the halyard smoothly with a hand-over-hand motion until the desired height is reached and the pawls are locked. The other firefighters stabilize the ladder by holding the outside of the beams on the bed section so that if the fly comes down suddenly, it will not strike the hands of those firefighters.

6. The two firefighters on the outside of the ladder facing the structure continue to heel the ladder while the firefighter on the inside of the ladder gradually steps back as all the firefighters lean the ladder into place. The firefighter on the inside of the ladder ties the halyard. The firefighters check the ladder for a 75-degree climbing angle or a 60-degree rescue angle, depending on the situation, and then check that the ladder's tip and butt are positioned securely.

SKILL DRILL 12-12

Four-Firefighter Flat Raise Firefighter I, NFPA 1010: 6.3.6

1. Start with a four-firefighter flat carry. The firefighters at the butt of the ladder place the butt end of the ladder on the ground, while the firefighters at the tip hold the tip up.

2. The two firefighters at the butt of the ladder stand side by side, facing the ladder. Each firefighter places their inside foot on the bottom rung and the other foot on the ground outside the beam, then they crouch down, grab the same higher rung with their inside hand, grasp the beam with their outside hand, and lean backward to heel the ladder with their inside foot.

3. The firefighters at the tip of the ladder check for overhead hazards, take hold of the same rung near the tip, bring that end of the ladder to chest height, step beneath the ladder, and then walk towards the butt while raising the ladder using a hand-over-hand motion until the ladder is vertical. This positions two firefighters on the bed side of the ladder and two firefighters on the side of the ladder with the fly section.

4. The four firefighters grasp the beams and pivot the ladder into position so that the fly side is out and the firefighters on the fly side of the ladder are facing the structure. The four firefighters heel the ladder by placing the toe or instep of their inside leg against the closest beam.

Continues.

SKILL DRILL 12-12 CONTINUED

Four-Firefighter Flat Raise Firefighter I, NFPA 1010: 6.3.6

5. One of the firefighters on the outside of the ladder facing the structure extends the fly section by pulling the halyard smoothly with a hand-over-hand motion until the desired height is reached and the pawls are locked. The other firefighters stabilize the ladder by holding the outside of the beams on the bed section so that if the fly comes down suddenly, it will not strike the hands of those firefighters.

6. The two firefighters on the outside of the ladder facing the structure continue to heel the ladder while the two firefighters on the inside of the ladder gradually step back as all four firefighters lean the ladder into place. One of the firefighters on the inside of the ladder ties the halyard. The firefighters check the ladder for a 75-degree climbing angle or a 60-degree rescue angle, depending on the situation, and then check that the ladder's tip and butt are positioned securely.

Securing the Ladder

Before climbing a ground ladder, check the angle to make sure it is positioned properly for the task at hand. Before climbing an extension ladder, verify that the pawls are locked and the halyard is secured. In addition, ground ladders should be secured. One way of securing a ladder is to have a firefighter heel it to prevent it from moving once it is in place.

SAFETY TIP

Take special care to make sure the heel firefighter is in position prior to climbing a ladder positioned at the rescue angle of 60 degrees.

Another method of securing a ladder is to tie the lower part of the ladder to any solid object using a rope, a rope–hose tool, or webbing to keep the base from kicking out. The tip of the ladder can also be tied to a secure object near the top to keep it from pulling away from the building. If you tie the lower part of the ladder or the tip, it is best to also secure the other end.

SAFETY TIP

The base should always be secured if the ladder is used at an angle of less than 60 degrees.

Climbing the Ladder

Once a ladder is secured, it can be climbed. Climbing a ladder at a fire or emergency scene should be done in a deliberate and controlled manner.

Be aware of the weight limits of any ladder you climb. No more than two firefighters should be on a ladder simultaneously. If more than one firefighter is on a ladder at the same time, the weight should be distributed along the length of the ladder. Only one firefighter should be on each section of an extension ladder at any time. Most fire department ground ladders—straight ladders, roof ladders, extension ladders, and combination ladders—are required to support a weight of 750 lb (340 kg) (NFPA 1932 2019). In a rescue situation, this weight limit is equivalent to one victim and two firefighters wearing full PPE and equipment. In contrast, ground ladders designed for construction and home use may be rated for as little as 225 lb (102 kg). Given the potential difference in ratings, it is important to avoid using unapproved ladders for fire department operations. Construction ladders are not rated for fire department use and should not be used for fire department operations.

While climbing a ladder, keep any bouncing and shifting to a minimum. Keep your eyes focused forward, with only occasional glances upward. This approach prevents debris, such as falling glass, from injuring your face or eyes. Lower the protective face shield for additional protection.

As you climb, use a hand-over-hand motion on the rungs of the ladder or slide both hands along the underside of the beams. Sliding the hands along the beams is more secure because it maintains three contact points with the ladder at all times (both hands and one foot).

While climbing a ladder, you need to be mindful that misguided hose streams, falling debris, or even falling people may come in contact with you. Stay alert and hold on to the ladder. Your bunker gear will help protect you.

If tools must be moved up or down, it is better to hoist them than to carry them up a ladder. Carrying tools on a ladder reduces the firefighter's grip. If the tool slips or is dropped, both the firefighter on the ladder and firefighters on the ground could potentially be injured. However, most times you will need to carry a tool up a ladder, so you should practice doing it safely. To climb a ladder while holding a tool, hold the tool against one beam with one hand, and maintain contact with the opposite beam with the other hand. Tools such as pike poles can be hooked onto the ladder and moved up every few rungs.

One method of climbing a ladder while carrying a tool is described in **SKILL DRILL 12-13**.

Working from a Ladder

Firefighters must often work from a ladder. A firefighter working on a ladder is in a less stable position than one who is working on the ground. There is always the danger of falling as well as a risk to people below if something falls or drops. While working on a ladder, you must constantly adjust your balance, especially when swinging a hand tool or reaching for a trapped occupant. You must also always be prepared to climb down quickly if fire conditions change. For example, if the fire suddenly flashes over or if flames break through a window located near the ladder, you need to move out of danger quickly. Your gear will not protect you from direct exposure to flames for more than a few seconds.

SKILL DRILL 12-13

Climbing a Ladder While Carrying a Tool Firefighter I, NFPA 1010: 6.3.12

1. Check the ladder to ensure that it is at a safe angle to climb and is secure. Grasp the tool comfortably and securely in one hand and hold that hand and the tool against the beam of the ladder.

2. Wrap the other hand around the underside of the other beam, then begin climbing the ladder.

3. Climb smoothly and safely by maintaining contact between your free hand and the beam and by sliding the tool along the opposite beam.

To minimize the risk of falling, the firefighter should be secured to the ladder. Firefighters use different methods to secure themselves to a ladder. One method is to use a **ladder belt**, a piece of equipment worn as a belt, which has a carabiner clip attached to it that the firefighter can attach to a rung or a solid object when working on an elevated surface (**FIGURE 12-29**). Firefighters should use only ladder belts designed and certified under the requirements specified by NFPA 2500, *Standard for Operations and Training for Technical Search and Rescue Incidents and Life Safety Rope and Equipment for Emergency Services*. Utility belts designed to carry tools should never be used as ladder belts.

There are a few methods of securing a firefighter to a ladder that do not require special equipment. These methods keep both hands free for a variety of tasks, including placing a roof ladder or controlling a hose stream. One method is to secure yourself by spreading your legs outward, pressing your knees against the beams. Another method is the leg lock, which is a way of wrapping one of your legs around a rung and then hooking your foot on a beam. To apply a leg lock, follow the steps in **SKILL DRILL 12-14**.

FIGURE 12-29 A ladder belt has a large clip that is designed to secure a firefighter to a ladder.

SKILL DRILL 12-14

Using a Leg Lock to Work from a Ladder Firefighter I, NFPA 1010: 6.3.6

1. Climb to the desired work height. On the side of the ladder where the work will be performed, step up to one more rung with the leg on the other side.

2. Extend the leg on the higher rung between the rungs.

Continues.

SKILL DRILL 12-14 CONTINUED

Using a Leg Lock to Work from a Ladder Firefighter I, NFPA 1010: 6.3.6

3. Bend the knee of the leg between the rungs, and then push your foot under the rung and back through the two rungs below the rung your knee is hooked around so your foot is on the climbing side of the ladder.

4. Secure the foot you just pushed back to the climbing side of the ladder against the rung below the rung your knee is hooked around or against the beam of the ladder. Use the thigh of that leg for support by pressing it against the beam, and then step down one rung with the opposite foot (the foot that is not hooked around the rung).

5. Work over the side of the ladder opposite the side with the locked leg.

Dismounting the Ladder

When a ladder is used to reach a roof or a window entry, the firefighter needs to dismount the ladder. Firefighters can minimize their risk of slipping and falling by ensuring that the surface is stable by testing its stability with a tool before dismounting. Firefighters who dismount from a ladder onto a roof typically **sound** the roof—strike it with a tool such as an axe to check for stability (or soundness)—before stepping onto the surface (**FIGURE 12-30**). Remember, sounding alone does not provide a complete picture of a structure's stability. Always look for signs of potential structural collapse. For more information about this, see Chapter 14, *Fire Suppression*.

Try to maintain contact with the ladder at three points when dismounting. For example, a firefighter who steps onto a roof should keep two hands and one foot on the ladder while checking the footing. This consideration is critical if the surface slopes or is covered with rain, snow, or ice. Do not shift your weight onto the roof until you have tested the footing.

It is also essential to check the footing when entering a window. Before climbing through the window, confirm that the interior surface is structurally sound and offers secure footing by sounding it with a tool. The three points of contact rule also applies in this situation.

Under heavy fire and smoke conditions, firefighters sometimes dismount from a ladder at a window by climbing over the tip and sliding over the windowsill to the floor inside the building. Sound the floor inside the window before entering to confirm that it is solid and stable, keeping in mind that sounding alone does not provide a complete picture of a structure's stability and you should always look for signs of potential structural collapse. (For more information about this, see Chapter 14, *Fire Suppression.*)To remount a ladder under these conditions, back out the window feet first, and rest your abdomen on the sill until you can feel the ladder under your feet. When the ladder is positioned to the side of a window, sit on the windowsill with your legs out and roll sideways onto the ladder.

FIGURE 12-30 Check the stability of the roof before dismounting from the ladder.

Placing a Roof Ladder

Several methods can be used to properly position a roof ladder on a sloping roof. Carrying a roof ladder up a ladder requires strength and practice. One commonly used method is described in **SKILL DRILL 12-15**.

SAFETY TIP

When deploying a roof ladder, try to keep the roof hooks pointed away from you as you raise and lower the ladder on the roof. Should the roof ladder slip or fall from the roof, the hooks could strike you if they are positioned toward you.

SKILL DRILL 12-15

Deploying and Removing a Roof Ladder Firefighter I, NFPA 1010: 6.3.12

1. Carry the roof ladder to the base of the climbing ladder that is already in place to provide access to the roofline.

2. Place the butt end of the roof ladder on the ground, then rotate the hooks of the roof ladder to the open position.

Continues.

SKILL DRILL 12-15 CONTINUED

Deploying and Removing a Roof Ladder Firefighter I, NFPA 1010: 6.3.12

3. Raise and lean the roof ladder against one beam of the climbing ladder with the hooks oriented outward, away from you. Climb the climbing ladder until you reach the midpoint of the roof ladder that is positioned next to you, slip one shoulder between two rungs of the roof ladder, and then shoulder the roof ladder.

4. Climb the ladder until your midsection is at the roofline of the structure, carrying the roof ladder on one shoulder. Secure yourself to the climbing ladder.

5. Place the roof ladder on the roof surface with the tip towards the peak and the roof hooks down. Push the ladder up toward the peak of the roof with a hand-over-hand motion.

6. Once the hooks have passed the peak, pull back on the roof ladder to set the hooks, then check that they are secure.

To remove a roof ladder from the roof, reverse the process just described. Push the ladder up to release the hooks from the peak, and then pull the ladder toward you. If the hooks catch on the peak again or on the roofing material, turn the ladder on one of its beams or flip the ladder so the hooks are pointing up, then slide the ladder down the roof without catching the hooks on the roofing material.

Conducting Rescue Operations Using a Ladder

Ladders can be an essential tool for rescue. For example, ground ladders are often used for search operations and for reaching and removing trapped occupants from the upper stories of a building. As described earlier, when positioning a ladder for rescue at a window opening, place the tip of the ladder at or slightly below the windowsill. This provides a direct pathway for the firefighter entering the window and a direct route for exiting the structure. The ladder should not block any part of the window opening.

A trapped person might try to jump onto the tip of an approaching ladder or to reach out for anything or anyone nearby. As a firefighter, you might be pulled or pushed off the ladder by the person you are trying to rescue. If several people are trapped, they might all try to climb down the ladder at the same time. A person in extreme danger may not wait to be rescued and jump out of a window before a ladder can be raised. This action not only risks their own lives, it endangers the firefighters trying to rescue them. Firefighters have been seriously injured by persons who jumped before a rescue could be completed. It is important to make verbal contact with any person you are trying to rescue. You must remain in charge of the situation and not allow the individual to panic. Tell the person to remain calm and wait to be rescued. If enough firefighters are present, one could maintain contact with the person while others raise the ladder.

After you reach the person, you still have to get them down to the ground safely. Try to make the person realize that any erratic movements could cause the ladder to tip over. It is often helpful to have one firefighter assist the person, while a second firefighter acts as a guide and backup to the rescuer. Although modern fire ladders are designed to support the load of two firefighters and a victim, a ladder that is overloaded during a rescue operation must be removed from service for inspection and service testing. Refer to Chapter 17, *Search and Rescue*, for specific ladder rescue techniques.

Descending a Ladder

As you prepare to descend a ladder, apply the same safety practices as when climbing a ladder. Until you reach the ground, you are in an elevated position and at risk of slipping and falling.

Before starting to descend, ensure that the ladder is still at the correct angle and that someone is heeling the ladder or that it is still safely tied off. Communicate to the person heeling the ladder that you are about to descend. You cannot see the rungs during descent, so first check that they are not slippery from water or, in the case of cold weather, from ice. If you are holding a tool, establish a secure grip on it that will allow you to maintain good contact with the ladder as you descend. Consider lowering tools using a utility rope.

As you descend, face the ladder, keeping your back perpendicular to the ground and your arms almost fully extended. Use a hand-over-hand motion on the rungs of the ladder or slide both hands along the underside of the beams to maintain three-point contact as you descend. As you step on each rung, position your foot so that the arch of your foot is on each rung. Descend slowly, avoid sudden movements, and stay in the middle of the ladder to increase stability. As when you climbed the ladder, minimize bouncing and swaying back and forth. Do not overload the ladder. Unless it is an emergency, only one person should descend at a time.

Ladder Safety

As you work on learning the skills needed to perform ladder operations, do not lose sight of the dangers inherent in working with ladders. As you practice the skills described in this chapter, follow these safety rules:

- Wear appropriate PPE. This equipment should include SCBA when you are operating on the fireground.
- Choose the proper ladder for the job.
- Choose the proper number of people for the job.
- Move smoothly, safely, and efficiently, always keeping the ladder under your control.
- Lift with your legs, keeping your back straight.
- Use proper hand and foot placement during carries, raises, climbing, and lowering of ladders.
- Look overhead for wires and obstructions before raising or lowering a ladder. Communicate the results of this check by stating "clear overhead" to your partner.
- Allow the heel person to take charge of the processes of raising and lowering the ladder.
- After extending an extension ladder, check the pawls to confirm that they are set, and make sure the halyard is tied off.
- Check the ladder for the proper angle.
- Ensure that the wall or roof will support the ladder before climbing it.
- Remain aware of hazards to yourself and to others.
- Maintain hand contact with the ladder. Watch your hands when climbing up, and watch your legs when climbing down.
- Use an appropriate method to secure yourself to the ladder.

CASE STUDY

You Are the Firefighter CONCLUSION

At a fire incident at a large residential home, a mayday is announced over the radio: "MAYDAY, MAYDAY, MAYDAY, members trapped on the third floor in the rear!" There is no aerial access at the rear of the building, so you and your crew are assigned to deploy ground ladder(s) for immediate egress for the trapped members. Time is critical and the stress level is high as these members need immediate rescue.

1. **Having multiple ground ladders to choose from, which will you take to the rear?**

   Answer: It would be best to take at least one 28-ft (8-m) extension ladder or one 35-ft (11-m) extension ladder to assist the members out of the window.

2. **What is the best way to rapidly carry and raise the ladder(s) to the trapped members?**

   Answer: Depending on staffing, two or three firefighters could use a suitcase carry for both the 28-ft (8-m) and the 35-ft (11-m) ladders. At least two firefighters would be needed to raise the 28-ft (8-m) ladder and at least three firefighters would be needed to raise the 35-ft (11-m) ladder.

3. **What safety concerns could you foresee during this scenario?**

   Answer: There may be overhead obstructions, such as wires and trees that could get in the way of safely raising the ladder(s). The ground surface at the back of the residence may be sloped or too soft for safe ladder placement, and it might be difficult to place the ladder(s) at the safest angle for rescue.

WRAP-UP

SUMMARY

KNOWLEDGE OBJECTIVES

After studying this chapter, you will be able to:

- List and describe the parts of a ladder and identify the different types of ladders.
  - List and describe the parts of a ladder. (**NFPA 1010: 6.3.6**, pp. 432–433)
  - Categorize the different types of ladders. (**NFPA 1010: 6.3.6**, pp. 433–439)
- Describe the methods of inspection, maintenance, cleaning, and the service testing process for ladders.
  - Inspect ladders. (**NFPA 1010: 6.5.1**, pp. 441–443)
  - Maintain ladders. (**NFPA 1010: 6.5.1**, pp. 441–443)
  - Clean ladders. (**NFPA 1010: 6.5.1**, p. 442)
  - Describe when, where, and who performs service testing on ladders. (**NFPA 1010: 6.5.1**, p. 444)
- Specify the hazards associated with ground ladders and describe the measures firefighters should take to maximize safety when working on and with ladders.
  - Specify the hazards associated with ladders. (**NFPA 1010: 6.3.6**, pp. 446–481)
  - Itemize the measures firefighters should take to ensure safety when working with and on ladders. (**NFPA 1010: 6.3.6**, p. 481)
- Describe proper ladder selection and subsequent proper ladder placement based on common fireground tasks.
  - Cite the factors and guidelines used to select the appropriate ladder from the fire apparatus. (**NFPA 1010: 6.3.6**, pp. 446–448)
  - Describe how to remove a ladder from the apparatus. (pp. 448–449)
  - Describe how to lift ladders. (pp. 449–450)
  - Determine appropriate ladder placement for common fireground tasks. (**NFPA 1010: 6.3.6**, pp. 460–461)

SKILLS OBJECTIVES

After studying this chapter, you will be able to perform the following skills:

- Clean, inspect, and maintain a ladder. (**NFPA 1010: 6.5.1**, pp. 445–446)
- Demonstrate ladder carries.
  - Carry a ladder using the one-firefighter shoulder carry. (**NFPA 1010: 6.3.6**, pp. 451–452)
  - Carry a ladder using the two-firefighter shoulder carry. (**NFPA 1010: 6.3.6**, pp. 452–453)
  - Carry a ladder using the three-firefighter shoulder carry. (**NFPA 1010: 6.3.6**, pp. 452–453)
  - Carry a ladder using the two-firefighter suitcase carry. (**NFPA 1010: 6.3.6**, pp. 454–455)
  - Carry a ladder using the three-firefighter suitcase carry. (**NFPA 1010: 6.3.6**, pp. 454–455)
  - Carry a ladder using the three-firefighter flat carry. (**NFPA 1010: 6.3.6**, pp. 456–457)
  - Carry a ladder using the four-firefighter flat carry. (**NFPA 1010: 6.3.6**, pp. 456–457)
  - Carry a ladder using the three-firefighter flat-shoulder carry. (**NFPA 1010: 6.3.6**, pp. 458–459)
  - Carry a ladder using the four-firefighter flat-shoulder carry. (**NFPA 1010: 6.3.6**, pp. 458–459)
- Demonstrate the various raises used with ladders and properly tie a halyard.
  - Tie the halyard. (**NFPA 1010: 6.3.6**, p. 463)
  - Raise a ladder using the one-firefighter flat raise. (**NFPA 1010: 6.3.6**, pp. 464–465)
  - Raise a ladder using the two-firefighter flat raise. (**NFPA 1010: 6.3.6**, pp. 467–468)
  - Raise a ladder using the two-firefighter beam raise. (**NFPA 1010: 6.3.6**, pp. 469–470)
  - Raise a ladder using the three-firefighter flat raise. (**NFPA 1010: 6.3.6**, pp. 471–472)
  - Raise a ladder using the four-firefighter flat raise. (**NFPA 1010: 6.3.6**, pp. 473–474)
- Climb a ladder and then utilize a leg lock to work from a ladder.
  - Climb a ladder while carrying a tool. (**NFPA 1010: 6.3.6**, p. 476)
  - Use a leg lock to work from a ladder. (**NFPA 1010: 6.3.6**, pp. 477–478)
  - Dismount a ladder. (**NFPA 1010: 6.3.6**, pp. 479–480)
  - Place a roof ladder. (**NFPA 1010: 6.3.6, 6.3.12**, pp. 479–480)
  - Descend a ladder. (p. 481)

KEY TERMS

A-frame hoist See *ladder A-frame.*

A-frame ladder A ladder with two ladder sections connected with a joint so that it forms an A-shaped structure when it is set up to climb on and that closes to fold flat for storage.

aerial ladder A self-supporting, turntable-mounted, power-operated ladder of two or more sections permanently attached to a self-propelled automotive fire apparatus and designed to provide a continuous egress route from an elevated position to the ground. (NFPA 1900)

attic ladder See *folding ladder.*

Bangor ladder A ladder equipped with tormentor poles, or staypoles, that stabilize the ladder during raising and lowering operations.

base See *butt.*

base distance The distance from the base of the building to the butt of the ladder when the ladder is raised.

base section See *bed section.*

beam The main structural side of a ground ladder. (NFPA 1960)

beam raise A method of raising a ladder by laying the ladder on one beam and them raising it by lifting the tip of the ladder while resting the butt of the ladder on the single beam.

bed section The lowest or widest section of an extension ladder. Also called *base section.* (NFPA 1960)

butt The end of the beam that is placed on the ground, or other lower support surface, when ground ladders are in the raised position. Also called *heel* or *base.* (NFPA 1960)

butt plate An alternative to a simple butt spur; a swiveling plate with both a spur and a cleat or pad that is attached to the butt of the ladder.

butt spurs That component of ground ladder support that remains in contact with the lower support surface to reduce slippage. (NFPA 1960)

KEY TERMS CONTINUED

climbing angle The angle that provides the best combination of strength, stability, and vertical reach for accessing the roof or for venting operations.

combination ladder A ground ladder that is capable of being used both as a stepladder and as a single or extension ladder. Also called *multipurpose ladder.* (NFPA 1960)

dog See *pawl.*

elevated platform An apparatus that includes a passenger-carrying platform (bucket) attached to the tip of a boom or ladder.

extension ladder A non-self-supporting ground ladder that consists of two or more sections traveling in guides, brackets, or the equivalent arranged so as to allow length adjustment. (NFPA 1960)

fire department ground ladder Any portable ladder specifically designed for fire department use in rescue, firefighting operations, or training. Also called *ground ladder.* (NFPA 1960)

flange On an I-beam ladder, the top and bottom of a beam.

flat raise A method of raising a ladder by positioning the ladder on the ground with the butt near the building and then lifting the tip to lean it against the building. Also called *rung raise.*

fly section The upper section of an extension ladder or any section of an aerial telescoping device beyond the base section. (NFPA 1960 and 1900)

folding ladder A single-section ladder with rungs that can be folded or moved to allow the beams to be brought into a position touching or nearly touching each other. Also called *attic ladder.* (NFPA 1960)

foot See *heel (v.).*

Fresno ladder A narrow, two-section extension ladder that has no halyard. Because of its limited length, it can be extended manually.

ground ladder See *fire department ground ladder.*

guide A strip of metal or wood that serves to guide a fly section during extension.

halyard Rope used on extension ladders for the purpose of raising a fly section(s). (NFPA 1960)

heat sensor label A label that changes color at a preset temperature to indicate a specific heat exposure. (NFPA 1960)

heel (v.) To use your weight to keep the base of a ladder from slipping by placing the toes or instep of one foot on one of the beams and the heel of that foot on the ground, placing the toes of one foot or both feet on the bottom rung and the heel(s) on the ground, or kneeling in front of the ladder. Also called *foot.*

heel (n.) See *butt.*

hook ladder See *roof ladder.*

I-beam ladder A ladder beam constructed of one continuous piece of I-shaped metal or fiberglass to which the rungs are attached.

ladder A-frame An A-shaped structure formed by using rope to attach two straight ladders at the tip in an A-shape and that can be used as a makeshift lift when raising a trapped person. Also called *A-frame hoist.*

ladder belt A compliant equipment item that is intended for use as a positioning device for a person on a ladder. (NFPA 2500)

ladder lock See *pawl.*

multipurpose ladder See *combination ladder.*

pawl A device attached to a fly section(s) to engage ladder rungs near the beams of the section below for the purpose of anchoring the fly section(s). Also called *ladder lock, dog,* or *rung lock.* (NFPA 1960)

protection plates Reinforcing material placed on a ladder at chafing and contact points to prevent damage from friction and contact with other surfaces.

rail The top or bottom piece of a trussed beam assembly used in the construction of a trussed ladder. Also, the top and bottom surfaces of an I-beam ladder. Each beam has two rails.

rescue angle The angle that provides a more gradual and controllable slope than the climbing angle for both firefighters to safely climb out of a window and onto the ladder or to remove a victim.

roof hooks The spring-loaded, retractable, curved metal pieces that allow the tip of a roof ladder to be secured to the peak of a pitched roof.

roof ladder A single ladder equipped with hooks at the top end and/or the bottom of the ladder. Also called *hook ladder.* (NFPA 1960)

rung The ladder crosspieces, on which a person steps while ascending or descending. (NFPA 1960)

rung lock See *pawl.*

rung raise See *flat raise.*

single ladder See *straight ladder.*

solid beam ladder A ladder beam constructed of a solid rectangular piece of material (typically wood), to which the ladder rungs are attached.

sound To strike a roof or floor system for stability with a tool.

staypole A pole attached to each beam of the base section of extension ladders, which assist in raising the ladder and help provide stability of the raised ladder. Also called *tormentor.* (NFPA 1960)

stop A piece of material that prevents the fly sections of a ladder from becoming overextended, leading to collapse of the ladder. Also called *stop block.*

stop block See *stop.*

straight ladder A single-section, fixed-length ground ladder. Also called *single ladder* or *wall ladder.*

tie rod Typically found on wood ladders, a metal rod that runs from one beam of the ladder to the other to keep the beams from separating.

tiller truck A tractor-trailer apparatus that requires a second operator positioned at the back of the trailer to steer the back wheels. Also called a *tractor-drawn aerial.*

tip The very top of the ladder.

tormentor See *staypole.*

tormentor ladder See *Bangor ladder.*

tractor-drawn aerial See *tiller truck.*

truss block A piece of wood or metal that ties the two rails of a trussed beam ladder together and serves as the attachment point for the rungs.

trussed beam ladder A ladder beam constructed of top and bottom rails joined by truss blocks that tie the rails together and support the rungs.

vertical reach The height a ladder can reach.

wall ladder See *straight ladder.*

water tower Aerial devices that deliver streams of water and contain high-capacity pre-piped waterways.

web On an I-beam ladder, the part the connects the top and bottom flanges on a beam.

REVIEW QUESTIONS

1. Which NFPA standard specifies the performance requirements ladders?
2. Describe some of the uses of ground ladders.
3. What are the three types of beam construction in ladders?
4. List the ladder components on fire service ground ladders.
5. List the six types of ground ladders used in the fire service in the United States.
6. How long are the roof ladders used in the fire service?
7. List the parts of extension ladders.
8. How often must ladders be visually inspected?
9. What is the closest a ladder should be positioned near power lines while raising or using them?
10. Where should the ladder tip be positioned for firefighter ingress or egress or for victim removal down the ladder?

DISCUSSION QUESTIONS

1. At what angle should a ground ladder be placed for access operations? For rescue operations? Why?
2. Describe the two types of fire service ladders and their characteristics.
3. What are some of the dangers a firefighter might face when climbing, working on, and dismounting a ladder?

APPLYING THE CONCEPTS

It is late on a snowy night and you are at the scene of a fire on the first floor of a three-story row house in an urban area. Dispatch reported the call was placed by the babysitter and a Christmas tree is suspected to be the source of the fire. The babysitter and two children exited the structure safely. Thick smoke and intense flames engulf the front of the first floor. Firefighting operations are underway, and it is discovered that

APPLYING THE CONCEPTS CONTINUED

the 3-year-old boy ran back into the house to a get a toy he had just gotten for Christmas. He is stuck in his bedroom located on the second floor at the rear of the structure. The incident commander orders you to get a ladder from the engine for rescue operations.

1. The engine has one 12-ft (3.7-m) folding ladder, one 16-ft (4.9-m) roof ladder, and one 24-ft (7.3-m) extension ladder stored on the side of the engine. What factors do you need to consider when selecting a ladder?

As you are considering ladder selection, incident command reports there are overhead utility lines in the alley behind the structure. You also note that the snow is accumulating. Given this information, you select the 24-ft (7.3-m) extension ladder.

2. Why is this a better choice than the roof ladder?
3. Given the urgency to deliver the ladder to the scene, do you carry it on your own? Why or why not?

You radio for assistance. As the lead firefighter you direct instructions to lift and carry the ladder. Working as a team, you and your partner use a shoulder carry to transport the ladder with the butt end forward. When you arrive behind the structure, an area has been cleared of snow and the officer in charge indicates where the ladder should be placed. An additional firefighter is assigned to make sure the ladder stays clear of the power lines.

4. How far should the ladder be from utility lines?
5. To safely maneuver around the utility lines, you use a beam raise to place the ladder under the windowsill. If you are not instructed by an officer, how do you determine where to place the butt of the ladder?

Because you have selected the appropriate ladder for the operation and followed safety procedures, the boy is quickly and safely rescued.

REFERENCES

National Fire Protection Association (NFPA). 2019. NFPA 1932, *Standard on Use, Maintenance, and Service Testing of In-Service Fire Department Ground Ladders*. 2020 Edition. Quincy, MA: NFPA.

National Fire Protection Association (NFPA). 2021. *NFPA 2500, Standard for Operations and Training for Technical Search and Rescue Incidents and Life Safety Rope and Equipment for Emergency Services*. 2022 Edition. Quincy, MA: NFPA.

National Fire Protection Association (NFPA). 2023. *NFPA 1900, Standard for Aircraft Rescue and Firefighting Vehicles, Automotive Fire Apparatus, Wildland Fire Apparatus, and Automotive Ambulances*. 2024 Edition. Quincy, MA: NFPA.

National Fire Protection Association (NFPA). 2023. *NFPA 1960, Standard for Fire Hose Connections, Spray Nozzles, Manufacturer's Design of Fire Department Ground Ladders, Fire Hose, and Powered Rescue Tools*. 2024 Edition. Quincy, MA: NFPA.

CHAPTER

13

Firefighter I

Supply Line and Attack Line Evolutions

KNOWLEDGE OBJECTIVES

After studying this chapter, you will be able to:

- Describe how to connect, lay, load, and advance supply hose.
- Describe how to load, deploy, and advance attack lines.
- Describe how to attach supply hose to a fire department connection for a standpipe and/or sprinkler system.
- List the steps involved in replacing a section or sections of an attack or supply line.
- Describe the importance and best practices of regularly unloading and reloading fire hose.

SKILLS OBJECTIVES

After studying this chapter, you will be able to perform the following skills:

- Attach a supply hose to a hydrant.
- Deploy the appropriate supply hose load for a given apparatus and fire attack.
- Perform the appropriate supply hose load for a given apparatus and fire attack.
- Extend a supply line.
- Perform the appropriate attack hose load for a given apparatus and fire attack.
- Advance and operate the appropriate attack hose load.
- Attach fire hose to a fire department connection for a standpipe and/or sprinkler system.
- Replace a section or sections of an attack or supply line.

ADDITIONAL NFPA STANDARDS

- **NFPA 11**, *Standard for Low-, Medium-, and High-Expansion Foam, 2021 Edition*
- **NFPA 13**, *Standard for the Installation of Sprinkler Systems, 2022 Edition*
- **NFPA 25**, *Standard for the Inspection, Testing, and Maintenance of Water-Based Fire Protection Systems, 2021 Edition*
- **NFPA 1140**, *Standard for Wildland Fire Protection, 2022 Edition*
- **NFPA 1410**, *Standard on Training for Emergency Scene Operations, 2020 Edition*
- **NFPA 1962**, *Standard for the Care, Use, Inspection, Service Testing, and Replacement of Fire Hose, Couplings, Nozzles, and Fire Hose Appliances, 2018 Edition*

CASE STUDY

You Are the Firefighter

In the early morning hours before sunrise, your engine company is dispatched to a report of a residential structure fire. The driver/operator sees another company approaching an intersection just ahead and close to the dispatched address. The officer on your engine radios the other company and advises them to help perform a split hose lay. The officer then advises dispatch that your company will deploy the attack line once the supply line is in place.

1. Why is it important to understand supply line evolutions and attack line evolutions?
2. What is the primary purpose of supply line evolutions?
3. What is the primary purpose of establishing an attack line?

Introduction

Putting water on the fire is an essential part of what we do. Ensuring a dependable water supply is a critical fireground operation that must be accomplished as soon as possible. This topic is discussed further in Chapter 10, *Water Supply Systems.* Ideally, a water supply should be established at the same time as size-up. The methods of moving water through fire hose must be efficient during an emergency when time is of the essence. Fire hose is used for two main purposes: to supply water from a water source to the pumper (supply hose) and to carry water from the pumper to a nozzle for fire attack (attack hose). These methods are discussed further in Chapter 11, *Fire Hose, Fire Appliances and Tools, and Nozzles.* When lengths of hose are connected to each other, the result is referred to as a hose line. A hose line is a single appliance in a water delivery system. A supply line or an attack line is the system with all of the included appliances.

To become proficient at moving and placing fire hose, firefighters practice fire hose evolutions. An **evolution** is a standard set of steps for working on the fireground and during training. Each fire department must set up equipment and conduct regular training so that firefighters are prepared to deploy and use hose lines and at a fire scene. Every firefighter should know how to perform all of the standard evolutions quickly and efficiently. When an officer calls for a particular evolution to be performed, each crew member should know exactly what to do.

Supply Line Evolutions

Supply line evolutions, also called **supply line operations**, are the steps involved with deploying and laying supply hose between a water source and an attack or supply engine, attaching hose to the connections on a pump and a hydrant, carrying and advancing supply hose, and loading the hose onto the apparatus so that it can be deployed quickly and easily. To **deploy hose**—or to **pull hose**—means to remove the hose from the hose bed or other storage location. To **lay** or **flake out** hose means arranging hose so it can easily be pulled toward the fire. An **attack engine** is an engine used to pump water through attack lines at the fireground. A **supply engine** is an engine used to pump water to an attack engine using supply lines. Most hose is carried in the **hose bed** on the apparatus—the main storage area for carrying hose.

Drafting from a static water source is one way to deliver water, but the most common way to deliver water to the fire pump on an engine is from a pressurized water source such as a fire hydrant. Drafting is described in Chapter 10, *Water Supply Systems.* The descriptions in this chapter involve connections between a pump and a hydrant.

Connecting an Engine to a Fire Hydrant

Supply hose is used to deliver the water needed to fight the fire over a short distance. In most cases, a short length of soft sleeve hose is used to transport water from a pressurized source, such as a fire hydrant, to the suction side of the fire pump. A soft sleeve hose is a large diameter supply hose, usually 10- to 25-feet (ft; 3- to 7.6-meter [m]) long, used to connect the steamer port on the hydrant to the suction side of the fire pump. See Chapter 11, *Fire Hose, Fire Appliances and Tools, and Nozzles*, for a detailed description. Although it is uncommon, the connection can also be made with a hard suction hose. The short length of soft sleeve hose is usually not carried on the hose bed. Instead, it is carried in a compartment next to the pump panel or sometimes in a compartment in the front bumper. On some engines, it is preconnected to an intake on the pump. To attach a soft sleeve hose to a fire hydrant, follow the steps in **SKILL DRILL 13-1**.

SKILL DRILL 13-1

Attaching a Soft Sleeve Hose to a Fire Hydrant Firefighter I, NFPA 1010: 6.3.15

1. After the engine driver/operator positions the engine so that the suction side of the pump is approximately 10 ft (3 m) from the fire hydrant, step off of the apparatus, and get the hydrant wrench and all necessary tools. Remove the hose from the hose bed along with any needed adaptors.

2. If the soft sleeve hose is not preconnected, attach it to the suction side of the pump on the engine. In some cases, it may be necessary to use an adaptor.

3. Unroll the hose toward the hydrant.

4. Remove the cap on the steamer port on the hydrant. Visually inspect the inside of the outlet for debris, and if needed, carefully insert a gloved hand and remove the debris. Attach the hydrant wrench to the stem nut on the fire hydrant, and then check the hydrant for an arrow indicating the direction to turn the stem nut to open the hydrant. Open the hydrant valve enough to verify flow of water and to flush out any debris that may be in the hydrant. Close the valve to stop the flow of water.

Continues.

SKILL DRILL 13-1 CONTINUED

Attaching a Soft Sleeve Hose to a Fire Hydrant Firefighter I, NFPA 1010: 6.3.15

5. Attach the soft sleeve hose to the fire hydrant.

6. Ensure that there are no kinks or sharp bends in the hose that might restrict the flow of water.

7. When the engine driver/operator signals to charge the hose, open the hydrant valve slowly to avoid water hammer. Check all connections for leaks and tighten the couplings if necessary.

8. Where required, place chafing blocks under the hose where it contacts the ground to prevent mechanical abrasion.

Laying Supply Hose

The objective of laying supply hose is to remove supply hose from the hose bed and place it between a water supply source and an engine. This can be done using a forward hose lay, a reverse hose lay, or a split hose lay (**FIGURE 13-1**). After the hose is laid, it is connected to the water supply and to the pump. Each fire department will determine its own preferred methods and procedures for supply line operations based on available apparatus, water supply, and regional considerations.

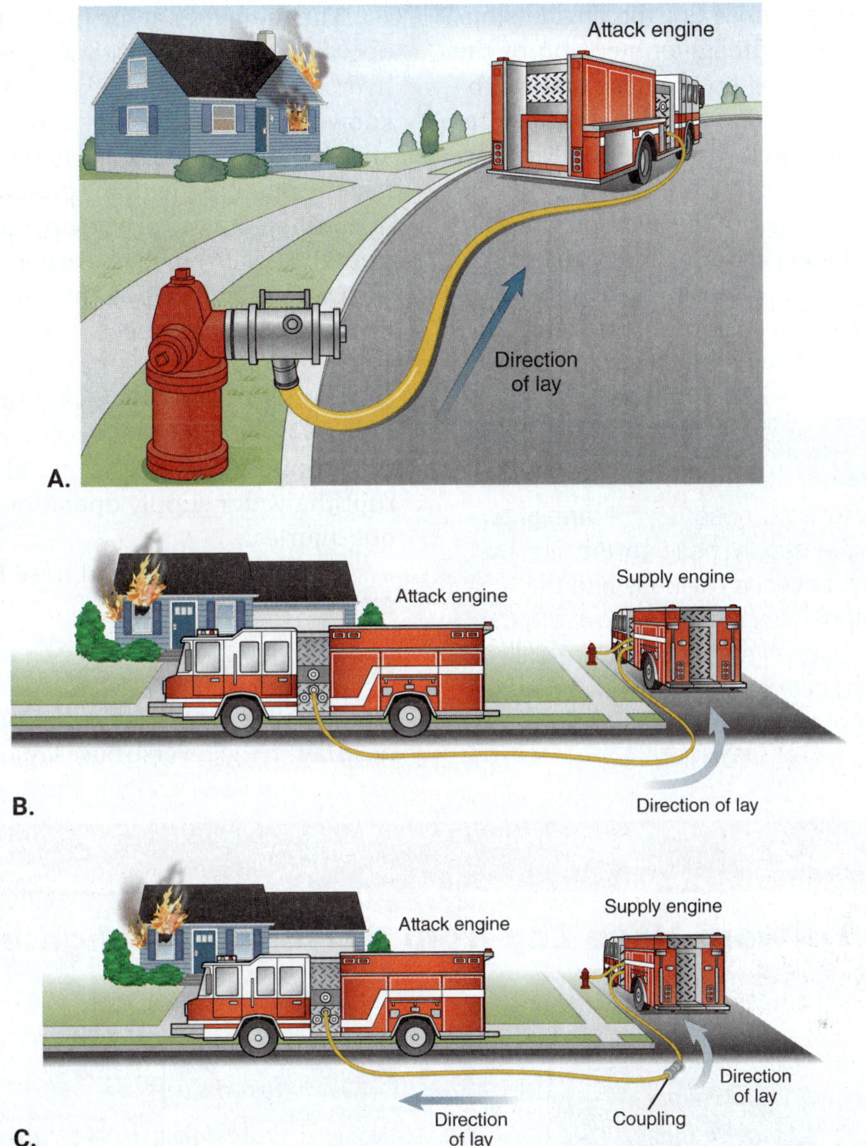

FIGURE 13-1 Laying supply hose can be done using a forward hose lay, a reverse hose lay, or a split hose lay. **A.** A forward hose lay is made from the water source to the attack engine, which is then driven forward toward the fire. **B.** A reverse hose lay is usually done with two engines and is made from the attack engine at the fireground to a water source. **C.** A split hose lay is done when the water source is too far from the fire to use a single supply hose. The attack engine performs a forward hose lay from a specific point, usually an intersection. The supply engine stops at the location where the attack engine's hose was dropped and connects a supply hose to the attack engine's hose. The supply engine then performs a reverse hose lay to the water source.

Forward Hose Lay

The most common way of laying hose is the forward hose lay. To execute the **forward hose lay**, also called a **straight hose lay**, the supply hose is pulled from the hose bed on the attack engine and anchored near the water source, usually the hydrant. The engine is slowly driven forward toward the fire, unfolding the hose from the hose bed as it drives—that is, laying out the hose. This method allows the engine company to establish a water supply without assistance from another company and places the attack engine close to the fire. It is most often used by the first-arriving engine company at the scene of a fire. A benefit of the forward hose lay is that the pumper is on scene, allowing firefighters to access additional hose and the tools and equipment that are carried on the apparatus. A forward hose lay can be performed using 2½-inch (in.; 64-millimeter [mm]) hose or larger.

To perform the forward hose lay, the engine stops near the fire hydrant and a firefighter steps off of the apparatus, grasps the coupling on the hose that will be used, and walks toward the hydrant, pulling the hose behind them. When they have walked about 6 ft (1.8 m) past the hydrant, they return to the hydrant and loop the end of supply hose around the base of the hydrant. Once the hose is secured, the firefighter at the hydrant signals to the engine driver/operator that the apparatus can proceed to the fire. As the apparatus slowly moves forward, the hose unfolds from the apparatus and onto the ground.

SAFETY TIP

When performing a forward hose lay, the firefighter who is connecting the supply hose to the fire hydrant must not stand between the hose and the fire hydrant. When the apparatus starts to move off, the hose could become tangled and suddenly be pulled taut. Anyone standing between the hose and the fire hydrant could be seriously injured.

The firefighter at the hydrant waits until the driver/operator signals that they are ready for you to open the hydrant and charge the hose line. Make sure that you know your department's signal to charge a hose line, and do not become so excited or rushed that you mistakenly open the hydrant prematurely. If you open the hydrant before the driver/operator uncouples the hose from the hose bed, the hose bed could become charged with water. If you open the hydrant before the driver/operator either connects the other end to the suction side of the pump or clamps the hose, depending on the situation and department's standard operating procedure (SOP), the loose hose line will discharge water on the ground at the fire scene. Either situation will disrupt the water supply operation and could cause serious injuries.

To perform a forward hose lay, follow the steps in **SKILL DRILL 13-2**.

Reverse Hose Lay

The **reverse hose lay** is the opposite of the forward hose lay. In the reverse hose lay, the hose is laid out from

SKILL DRILL 13-2

Performing a Forward Hose Lay from a Hydrant Firefighter I, NFPA 1010: 6.3.15

1. The pump driver/operator stops the fire apparatus approximately 10 ft (3 m) past the fire hydrant.

2. Step off of the apparatus and get the hydrant wrench and all necessary tools. Grasp the coupling on the hose in the hose bed, and then walk toward the hydrant pulling the hose behind you. When you have walked about 6 ft (1.8 m) past the hydrant, walk back to the hydrant, and then wrap the hose around the base of the hydrant to form one full loop or secure the hose as specified by your department's SOP. Do not stand between the hose and the fire hydrant.

SKILL DRILL 13-2 CONTINUED

Performing a Forward Hose Lay from a Hydrant Firefighter I, NFPA 1010: 6.3.15

3. Signal the engine driver/operator to proceed to the fire.

4. Once the apparatus has moved off and a length of supply hose has been removed from the apparatus and is lying on the ground, remove the appropriate-size fire hydrant cap from the outlet nearest to the fire, and then check the operating condition of the hydrant.

5. Unwrap the hose from the base of the hydrant, then attach the hose to the open outlet on the hydrant. An adaptor may be needed if a large-diameter hose (LDH) with a Storz-type coupling is used.

6. Attach the hydrant wrench to the stem nut on the fire hydrant and check the hydrant for an arrow indicating the direction to turn the stem nut to open the hydrant. Wait for the attack engine driver/operator to signal that they are ready for the line to be charged.

Continues.

SKILL DRILL 13-2 CONTINUED

Performing a Forward Hose Lay from a Hydrant Firefighter I, NFPA 1010: 6.3.15

7. When the attack engine driver/operator signals to charge the hose, open the hydrant valve slowly to avoid water hammer. Make sure you open it fully. Follow the supply hose to the engine and remove any kinks.

the fire to the water source, such as a fire hydrant, in the direction opposite to the flow of the water. This can be a useful tactic when multiple pumpers arrive on a fire scene close to the same time. The attack engine begins fire attack using the booster tank on the engine while the supply engine does a reverse lay to the water source.

TIP

The supply pumper can drop crew members at the scene to help with fire attack before starting the reverse lay.

To perform the reverse hose lay with two engines, the supply engine stops with its hose bed close to an intake on the attack engine. The supply hose is pulled from the hose bed of the supply engine and, if possible, is anchored to a stationary object. The supply engine is then driven to the fire hydrant (or alternative water source), laying out the hose as it goes. At the fire area, the supply hose is connected to an intake on the attack engine. At the water source, the other end is uncoupled from the hose bed on the supply engine and connected to a discharge outlet on the supply engine. The supply engine driver/operator then uses additional sections of supply hose to connect the intake side of the pump on the supply engine to the water source, and then pumps water to the attack engine. Like the forward lay, the reverse lay can be performed using a 2½-in. (64-mm) hose or larger.

When performing a reverse lay with threaded couplings, you may find yourself with the wrong end of the hose to make the required connection. Make sure a double-male adaptor and a double-female adaptor are easily accessible so that you can join two couplings of the same sex if needed.

To perform a reverse hose lay, follow the steps in **SKILL DRILL 13-3**.

Split Hose Lay

A **split hose lay** is performed by two engine companies in situations where hose can or must be laid in two different directions to establish a water supply. This evolution could be used when the attack engine must approach a fire either along a dead-end street with no hydrant or down a long driveway. To perform a split hose lay, the attack engine drops the end of its supply hose where the hose from the two engines will meet, such as at a corner or an intersection. The attack engine then performs a forward lay toward the fire. The supply engine stops at the same location, pulls off enough hose to connect to the end of the supply line that is already there, and then performs a reverse hose lay to

SKILL DRILL 13-3

Performing a Reverse Hose Lay with Two Engines Firefighter I, NFPA 1010: 6.3.15

1. The attack engine driver/operator positions the attack engine at the fireground. The supply engine driver/operator positions the supply engine so its hose bed is close to an intake on the attack engine. Pull sufficient hose from the hose bed on the supply engine to reach from the supply engine to an intake on the pump on the attack engine. Anchor the hose to a stationary object if possible. Do not stand between the hose and the stationary object.

2. The supply engine driver/operator slowly drives to the fire hydrant, laying out the supply hose.

3. The attack engine driver/operator connects their end of the supply hose to an intake on the pump on the attack engine.

4. At the supply engine, the driver/operator uncouples the supply hose that was laid out from the hose bed and attaches it to a discharge outlet on the supply engine's pump.

Continues.

SKILL DRILL 13-3 CONTINUED

Performing a Reverse Hose Lay with Two Engines Firefighter I, NFPA 1010: 6.3.15

5. The supply engine driver/operator connects another line to the intake side of the pump on the supply engine. Connect the other end of that line to the hydrant after checking the operating condition of the hydrant (see Skill Drill 13-1). Upon the signal from the attack engine driver/operator, the supply engine driver/operator charges the supply line. If the attack engine driver/operator was using the booster tank while waiting for the supply engine to charge the line, they switch over to the supply hose.

the hydrant. With the two lines connected together, the supply engine can pump water to the attack engine. A split hose lay often requires coordination by two-way radio because the attack engine must advise the supply engine of the plan and indicate where the end of the supply line will be dropped. In many cases, the attack engine is out of sight when the supply engine arrives at the location where the attack engine's line was dropped. And as with a reverse hose lay performed with two engines, when performing a split lay with threaded couplings, make sure a double-male adaptor and a double-female adaptor are easily accessible so that you can join two couplings of the same sex if needed.

To perform a split hose lay, follow the steps in **SKILL DRILL 13-4**.

Four-Way Hydrant Valve

A **four-way hydrant valve** allows the attack engine to connect to the water supply, and then allows a second engine to connect to provide additional pressure without first disconnecting the attack engine's supply (**FIGURE 13-2**). This operation can be accomplished without disconnecting any lines or interrupting the flow of water. It is used where long supply lines are needed or when the flow pressure from the hydrant is not adequate.

When the four-way hydrant valve is placed on a fire hydrant, the attack engine's supply hose is connected to the main discharge outlet on the valve. When the hydrant is opened, the water flows from the hydrant through the valve to the connected supply hose, delivering water to the attack engine. When the supply engine arrives, a supply line can be connected between the suction side of the pump on the supply engine and the additional discharge outlet on the four-way hydrant valve without disconnecting the attack engine's line. Then a second supply hose is connected between the discharge outlet on the pump on the supply engine and the inlet on the four-way valve. Next, the position of the four-way hydrant valve is changed to redirect the water flow from the hydrant to the supply engine.

Basically, two separate water flows have been created. The first is from the hydrant through the four-way valve to the intake side of the pump on the supply engine. The pump on the supply engine increases the water pressure and discharges the water into the supply hose connected to the discharge outlet on the supply engine. This second water flow moves through the four-way valve to the supply hose that is connected to the attack engine, boosting the flow of pressurized water to the attack engine. To use a four-way hydrant valve, follow the steps in **SKILL DRILL 13-5**.

SKILL DRILL 13-4

Performing a Split Hose Lay Firefighter I, NFPA 1010: 6.3.15

1. The attack engine driver/operator stops at the location where the supply line will be dropped. A firefighter from the attack engine removes the end of the supply hose from the hose bed and anchors it to a stationary object if possible. This firefighter stays at the drop location until either the attack engine has driven off or the supply engine arrives, according to the department's SOPs, and then they walk to the fire.
2. The attack engine driver/operator slowly drives toward the fire, laying out the hose line. When the attack engine arrives at the fireground, the driver/operator can start pumping from the booster tank on the engine.
3. When the supply engine arrives at the location where the supply hose was dropped, it stops. A firefighter from the supply engine pulls off enough supply hose to connect to the hose end laid in the street by the attack engine, and then connects the two lines (see Figure 13-1C). If threaded couplings are used, a double-male or double-female adaptor may be required. Remount the apparatus, then the supply engine driver/operator proceeds slowly to the hydrant.
4. When the supply pumper arrives at the water source, uncouple the supply line from the hose bed. The supply engine driver/operator connects this line to a discharge outlet on the supply engine's pump.
5. The supply engine driver/operator connects another section of supply hose to the intake side of the pump on the supply engine. Connect the other end of this line to the hydrant after checking the operating condition of the hydrant (see Skill Drill 13-1).
6. Upon the signal from the attack engine driver/operator, the supply engine driver/operator charges the supply line. If the attack engine driver/operator was using the booster tank while waiting for the supply engine to charge the line, they switch over to the supply hose.

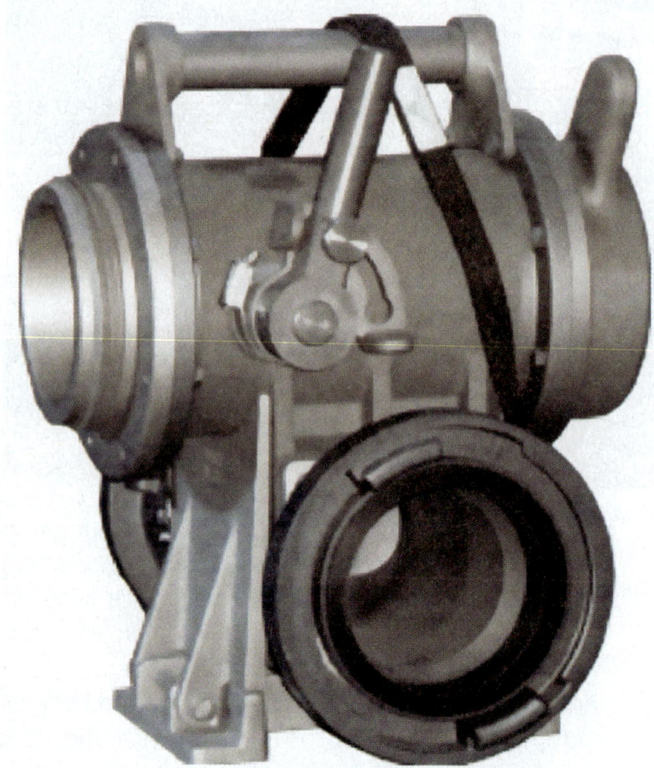

FIGURE 13-2 A four-way hydrant valve.

Courtesy of Humat Incorporated.

Loading Supply Hose

Hose can be loaded into the hose bed in several ways, depending on how it will be laid out at a fire scene. Hose must be easily removable from the hose bed, without kinks or twists, and without the possibility of becoming caught or tangled. The ideal hose load is easy to load, avoids wear and tear on the hose, has minimum sharp bends, and allows the hose to lay out of the hose bed smoothly and easily.

SAFETY TIP

Many modern hose beds are several feet above the ground and a fall could result in significant injuries, so when loading hose on an apparatus, always use caution in climbing up and down on the apparatus. If you are loading hose at a fire scene, watch out for wet, slippery surfaces; ice; or other hazards. Also, wet hose can be heavy, and you may need to reach, stretch, or lift the hose to get it into the hose bed. Use caution! Also, wear appropriate personal protective equipment (PPE) when loading hose. At a minimum, wear your helmet, gloves, and boots.

SKILL DRILL 13-5

Using a Four-Way Hydrant Valve Firefighter I, NFPA 1010: 6.3.15

1. The attack engine driver/operator stops the attack engine approximately 10 ft (3 m) past the fire hydrant in preparation for a forward hose lay.

2. Grasp the four-way hydrant valve and the attached hose, and carry the four-way hydrant valve and supply hose, along with the hydrant wrench and any other needed tools, to the hydrant.

3. Secure the hose for a forward lay as specified by your department's SOP. Do not stand between the fire hydrant and hose. Signal the attack engine driver/operator to proceed to the fire.

4. Remove the steamer port and check the operating condition of the hydrant. Once enough hose has been removed from the apparatus and is lying on the ground, attach the four-way hydrant valve to the hydrant outlet. (An adaptor may be needed.) Attach the hydrant wrench to the stem nut on the hydrant and check the hydrant for an arrow indicating the direction to turn the stem nut to open the hydrant.

SKILL DRILL 13-5 CONTINUED

Using a Four-Way Hydrant Valve Firefighter I, NFPA 1010: 6.3.15

5. At the same time, a firefighter on the attack engine, which is now at the fireground, uncouples the hose from the hose bed and the attack engine driver/operator attaches the end of that supply line to the suction side of the pump on the attack engine. The attack engine driver/operator signals to charge the line. Open the hydrant to supply the attack engine with water.

6. When the supply engine arrives at the fire scene, the supply engine driver/operator stops at the fire hydrant that has the four-way valve. The supply engine driver/operator attaches a hose to the suction side of the pump on the supply engine. The other end of that line is connected to the second discharge outlet on the four-way hydrant valve outlet. The supply engine driver/operator attaches a second supply hose to the discharge side of the pump on the supply engine. The other end of that line is connected to the inlet on the four-way hydrant valve.

7. The position of the four-way hydrant valve is changed to redirect the flow of water from the hydrant so it discharges to the supply engine instead of to the attack engine.

Steps 1 to 5: Courtesy of Todd Eddy; **Steps 6 to 7:** © Jones & Bartlett Learning. Photographed by Glen E. Ellman.

When loading hose, it is best to observe the following practices:

- Drain all of the water out of the hose before loading it.
- Rolling the hose first will result in a flatter hose load because there will be no air in the hose. (Rolling hose is described in Chapter 11, *Fire Hose, Fire Appliances and Tools, and Nozzles.*)

FIGURE 13-3 The Dutchman fold prevents the coupling from turning and becoming stuck as the hose is laid out.

Courtesy of the Westbury Fire Department.

- Do not load hose too tightly. Leave enough room so that you can slide a hand between the folds of hose. If hose is loaded too tightly, it may not lay out properly.
- Load hose so that couplings do not have to turn around as the hose is pulled out of the hose bed. Make a short fold in the hose close to the coupling to keep the hose properly oriented. This short fold is called a Dutchman. The Dutchman fold allows the hose to deploy or unload smoothly, and it prevents the coupling from turning and becoming stuck as the hose is laid out (**FIGURE 13-3**).
- Couple sections of hose with the flat sides of the hose oriented in the same direction.
- Check that swivel gaskets are in place before coupling hose.
- Tighten couplings so that they are hand tight only. With a good gasket, the hose should not leak.
- If your department's SOPs require it, load the appliances for hydrant connections.

Three hose loads are commonly used for loading supply hose: the flat hose load, the horseshoe hose load, and the accordion hose load (**FIGURE 13-4**). Any one

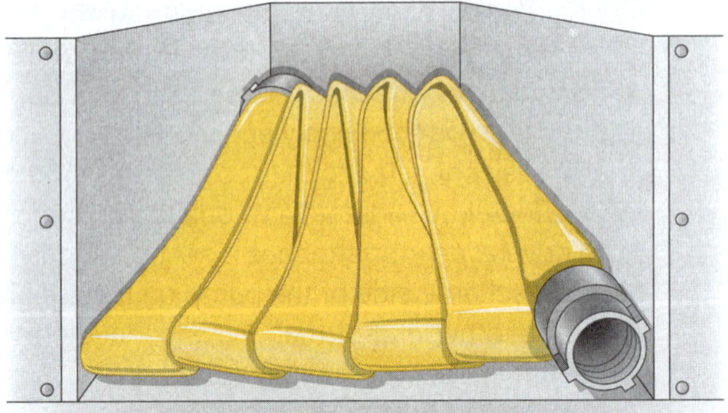

A.

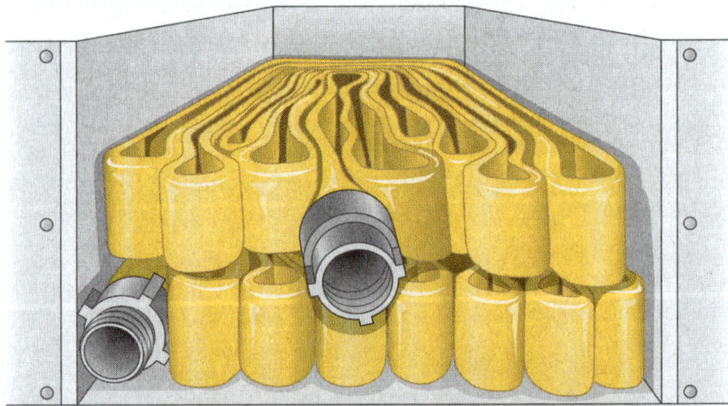

B.

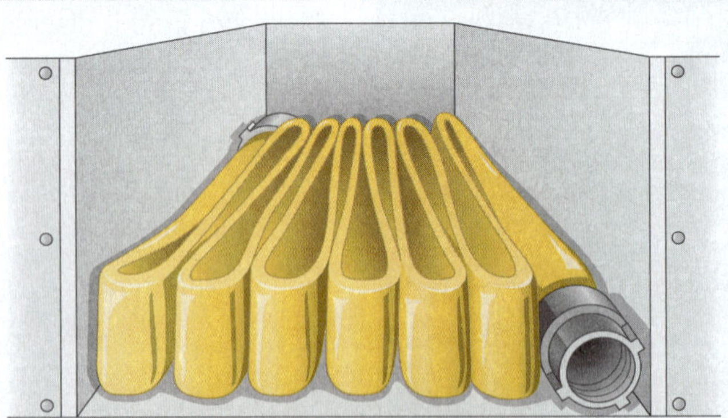

C.

FIGURE 13-4 Three hose loads are commonly used to load supply hose onto the apparatus. **A.** The flat hose load. **B.** The horseshoe hose load. **C.** The accordion hose load.

TIP

When loading hose, remember that the time and attention that go into loading the hose properly will be valuable when it you need to unload the hose at a fire.

of these methods can be used to load hose for either a forward hose lay or a reverse hose lay. You need to learn the specific hose loads used by your fire department.

When referring to the hose bed, the end closest to the cab (at the front of the vehicle) is the front of the hose bed. The end closest to the tailboard (at the back of the vehicle) is the rear of the hose bed.

If you are loading supply hose with threaded couplings, first determine whether the hose will be used for a forward hose lay or a reverse hose lay. Generally, hydrants require a female coupling. Therefore, to load hose for a forward hose lay, place the male coupling in the hose bed first, and to load hose for a reverse hose lay, place the female coupling in the hose bed first.

Before loading any hose, the amount of hose that will be loaded needs to be determined.

Flat Hose Load

With the **flat hose load**, the hose is laid flat and stacked on top of the previous section. This method is the easiest loading technique to implement, and it can be used for any size hose, including LDH. In fact, the flat hose load is the only load that should be used with LDH. Because the hose is placed flat in the hose bed, it should lay out flat without twists or kinks. The flat hose load minimizes wear and tear on the edges of the hose from the movement and vibration of the vehicle during travel. Many variations of the flat hose load exist, so follow your department's SOPs. To perform a flat hose load, follow the steps in **SKILL DRILL 13-6**.

Horseshoe Hose Load

To load hose with the **horseshoe hose load**, you place the hose on its edge and position it around the perimeter of the hose bed in a U-shape. At the completion of the first U-shape, the hose is folded inward to form another U-shape in the opposite direction. This continues until a complete layer is filled; then another layer is started above the first. When the hose load is complete,

SKILL DRILL 13-6

Performing a Flat Hose Load Firefighter I, NFPA 1010: 6.5.2

1. Start the hose load by placing the coupling at the front of the hose bed and run the hose to the rear of the hose bed aligned up against one side of the hose bed. If you are loading supply hose with threaded couplings, place the male coupling in the hose bed first to set up for a forward hose lay, and place the female coupling in the hose bed first to set up for a reverse hose lay.

2. When you reach the rear of the hose bed, fold the hose back on itself at the rear of the hose bed.

Continues.

SKILL DRILL 13-6 CONTINUED

Performing a Flat Hose Load Firefighter I, NFPA 1010: 6.5.2

3. Run the hose back to the front of the hose bed on top of the previous length of hose. While laying the hose to the front of the hose bed, angle the hose to the side of the previous fold.

4. Continue laying the hose in neat folds until the required amount of hose is loaded. To avoid the ends getting too high because the folds are on top of one another, make every other layer of hose slightly shorter, or alternate the folds. Position the last coupling as close as possible to the rear of the bed so it is within reach from the rear of the apparatus.

the hose in each layer is in the shape of a horseshoe. A major advantage of the horseshoe hose load is that it contains fewer sharp bends than the other hose loads.

There are a few disadvantages to the horseshoe load. A horseshoe hose load cannot be used for LDH because the hose tends to fall over when it stands on edge. The load also causes more wear on the hose than the flat hose load because the weight of the hose is supported only by the edges, and these edges are subject to the vibration of the hose bed while the motor is running. When laying out a horseshoe hose load, the hose tends to lay out in a wave-like manner from one side of the street to the other.

Many variations of the horseshoe hose load exist, so follow your department's SOPs. To perform a horseshoe hose load, follow the steps in **SKILL DRILL 13-7**.

Accordion Hose Load

The **accordion hose load**, like the horseshoe hose load, requires that the hose is placed on its edge. But instead of forming a U-shape along the edges of the hose bed, the hose is laid side to side in the hose bed. This load is easy to implement and makes the hose easy to deploy and carry. Each firefighter can carry multiple folds from the hose bed. When the hose load is complete, the hose in each layer is in the shape of accordion pleats.

Because the hose is stacked on its side in an accordion hose load, this load has the same disadvantages as the horseshoe hose load. That is, this load cannot be used for LDH, and it causes more wear on the hose than the flat hose load.

Many variations of the accordion hose load exist, so follow your department's SOPs. To perform the accordion hose load, follow the steps in **SKILL DRILL 13-8**.

Combination Hose Load in a Split Hose Bed

A **split hose bed** is a hose bed that is divided into two or more sections. Split hose beds have several advantages:

- One compartment in a split hose bed can be loaded for a forward hose lay with the female coupling at the rear of the hose bed, and the other side can be loaded for a reverse hose lay with the male coupling at the rear of the hose bed. This allows a line to be laid in either direction without adaptors.

SKILL DRILL 13-7

Performing a Horseshoe Hose Load Firefighter I, NFPA 1010: 6.5.2

1. Start the hose load by placing the coupling at the rear of the hose bed. If you are loading supply hose with threaded couplings, place the male coupling in the rear of the hose bed to set up for a forward hose lay, and place the female coupling in the rear of the hose bed to set up for a reverse hose lay. Lay the first length of hose on its edge against one of the walls of the hose bed and run the hose to the front of the hose bed along the wall.

2. When you reach the front of the hose bed, lay the hose across the width of the bed and continue down the opposite side toward the rear of the hose bed.

3. When the hose reaches the rear of the hose bed, fold the hose back on itself, then lay it back toward the front of the hose bed along the second side. Keep the hose tight to the previous row of hose around the hose bed until it is back to the rear of the hose bed on the starting side. Fold the hose back on itself again, then run the hose back to the rear of the hose bed, packing the hose tight to the previous row.

4. Continue to pack the hose in the same manner. Each fold of hose will decrease the amount of space available inside the horseshoe. Once the center of the horseshoe is filled in, begin a second layer by bringing the hose from the rear of the hose bed and laying it around the perimeter of the hose bed on top of the first layer. Complete additional layers using the same pattern you used for the first layer. Position the last coupling as close as possible to the rear of the bed so it is within reach from the rear of the apparatus.

SKILL DRILL 13-8

Performing an Accordion Hose Load Firefighter I, NFPA 1010: 6.5.2

1. Start the hose load by placing the coupling at the rear of the hose bed. If you are loading supply hose with threaded couplings, place the male coupling in the rear of the hose bed first to set up for a forward hose lay, and place the female coupling in the rear of the hose bed first to set up for a reverse hose lay. Lay the first length of hose on its edge against one of the walls of the hose bed and run the hose to the front of the hose bed along the wall.

2. When you reach the front of the hose bed, fold the hose back on itself. Run the hose back to the rear of the hose bed, then fold the hose back on itself so that the bend is even with the edge of the hose bed.

SKILL DRILL 13-8 CONTINUED

Performing an Accordion Hose Load Firefighter I, NFPA 1010: 6.5.2

3. Continue to lay folds of hose along the hose bed until the required amount of hose is loaded. Alternate the length of the hose folds at each end to allow more room for the folded ends.

4. Continue to pack the hose in the same manner. Once the bottom layer is completed, begin a second layer by angling the hose upward to begin the second tier. Complete additional layers using the same pattern you used for the first layer.

- When 2½-in. (64-mm) or 3-in. (76-mm) hose is used, two parallel hose lines can be laid at the same time. This is referred to as **dual hose lines**. Dual hose lines are beneficial if the situation requires more water than one hose line can supply.
- The split hose beds can be used to store hose of different size. For example, one side of the hose bed could be loaded with 2½-in. (64-mm) hose that can be used as supply hose or as attack hose. The other side of the hose bed could be loaded with 5-in. (127-mm) hose for use as a supply hose.
- Hose from all sections of the hose bed can be laid out as a single hose line by coupling the end of the last length of hose in one bed to the beginning of the first length of hose in the next bed. This is called a **combination hose load**. When laying a combination hose load, all of the hose is laid out of one hose bed first, and then the hose continues to lay

out from the second hose bed. When the two sides of a split bed are loaded with the hose in opposite directions, either a double-female or double-male adaptor is used to make the connection between the two hose beds.

Supply Hose Carries and Advances

Several different techniques are used to carry and advance supply hose. The best technique for a particular situation will depend on the size of the hose, the distance over which it must be moved, and the number of firefighters available to perform the task. Although the same techniques can be used for both supply lines and preconnected attack lines, the working hose drag and the shoulder carry are almost always used with supply lines.

Whenever possible, a hose line should be laid out and positioned as close as possible to where it will be operated before it is charged with water. A charged hose line is much heavier and more difficult to maneuver than a dry hose line. A suitable amount of extra hose should be available to allow for maneuvering after the line is charged.

Working Hose Drag

The working hose drag is used to deploy hose from a hose bed and advance the line over a relatively short distance to the desired location. Depending on the size and length of the hose, several firefighters may be required to perform this task. To perform a working hose drag, follow the steps in **SKILL DRILL 13-9**.

Shoulder Carry

The shoulder carry is used to transport multiple sections of connected hose over a longer distance than is practical to drag the hose. It is also useful when a hose line must be advanced around obstructions. For example, this technique can be used to stretch an attack line from the front of a building around to the rear entrance and up to the second floor. The shoulder carry can also be used to stretch a supply line from a water source to an attack engine when the hose cannot be laid out by another engine. This technique requires practice and good teamwork to be successful. To perform a shoulder carry, follow the steps in **SKILL DRILL 13-10**.

TIP

By working together to complete tasks efficiently, firefighters can achieve their goal of extinguishing the fire in the shortest period of time.

SKILL DRILL 13-9

Performing a Working Hose Drag Firefighter I, NFPA 1010: 6.3.10

1. Grasp the coupling on the hose in the hose bed and place the end of the hose over one shoulder so that the coupling is at chest height. Hold on to the coupling with the other hand.

2. Walk in the direction you want to advance the hose.

SKILL DRILL 13-9 CONTINUED

Performing a Working Hose Drag Firefighter I, NFPA 1010: 6.3.10

3. As the next hose coupling is ready to come off the hose bed, a second firefighter grasps that coupling and places the hose over their shoulder.

4. Continue this process until enough hose has been pulled out of the hose bed, and then another firefighter uncouples it from the hose bed.

SKILL DRILL 13-10

Performing a Shoulder Carry from a Flat or Horseshoe Load Firefighter I, NFPA 1010: 6.3.10

1. Stand on the ground at the tailboard of the apparatus. Grasp the end of the hose and place it over one shoulder so that the coupling is at chest height. Hold on to the coupling with the other hand. Have another firefighter place additional hose on your shoulder so that the ends of the folds reach about knee level, front and back. Have the other firefighter continue to place folds on your shoulder, but only as much as you can safely carry.

2. Hold the hose to prevent it from falling off your shoulder and move forward about 15 ft (4.6 m).

Continues.

SKILL DRILL 13-10 CONTINUED

Performing a Shoulder Carry from a Flat or Horseshoe Load
Firefighter I, NFPA 1010: 6.3.10

3. A second firefighter stands on the ground at the tailboard, another firefighter places additional hose on their shoulder in the same manner, and then both firefighters holding hose move forward about 15 ft (4.6 m). When enough hose has been unloaded onto firefighters, the hose is uncoupled from the hose bed and carried to the desired location.

4. All of the firefighters start walking in the direction the hose is needed. As they walk, the rear firefighter starts offloading hose from their shoulder. Once the rear firefighter has laid out all of their hose, the next firefighter starts laying out hose from their shoulder. Each firefighter lays out their supply of hose in turn until the entire length is laid out.

If hose is loaded with the accordion load, one firefighter should be able to load hose onto their own shoulder. To advance an accordion load using a shoulder carry, follow the steps in **SKILL DRILL 13-11**.

Attack Line Evolutions

Attack hose is used to discharge water from an engine onto a fire. This process is explained further in Chapter 11, *Fire Hose, Fire Appliances and Tools, and Nozzles*. **Attack line evolutions**, also called **attack line operations**, are standard methods of working with attack lines to deliver water from an attack engine to a handline, which discharges the water onto the fire.

The attack engine is usually positioned close to the fire, and attack lines are stretched from the attack engine to the fire manually by firefighters. In some situations, an attack engine drops an attack line at the fire and drives from the fire to a fire hydrant or other water source. This procedure is similar to the reverse hose lay for supply lines.

Most engines are equipped with preconnected attack lines, which are lengths of attack hose that travel on the engine equipped with a nozzle and connected to a pump discharge outlet. The connection to the discharge outlet may be in the hose bed or in another location, depending on the configuration of the pumper. Preconnected hose lines are intended for immediate use as attack lines. Additional attack hose that is not preconnected is also carried on the apparatus. To create an attack line with this additional hose, the desired length of hose is removed from the bed, disconnected from the next coupling in the hose bed, and attached to the pump discharge outlet. This hose can also be used to extend the length of a preconnected attack line or to attach to a wye or a water thief.

Loading Preconnected Attack Lines

Attack hose is loaded on the apparatus in such a manner that it can be quickly and easily deployed. There are

SKILL DRILL 13-11

Advancing an Accordion Load Firefighter I, NFPA 1010: 6.3.10

1. Find the end of the accordion load. Using two hands, grasp the end of the load and the number of folds it will take to make an adequate shoulder load.

2. Pull the folds about one-third of the way off the apparatus.

3. Rotate the folds so that they will lie flat on your shoulder with the end of the accordion load on the bottom of the now flat shoulder load.

4. Transfer the hose to your shoulder while turning so that you face the direction you will walk.

Continues.

SKILL DRILL 13-11 CONTINUED

Advancing an Accordion Load Firefighter I, NFPA 1010: 6.3.10

5. Place the shoulder load over your shoulder and grasp it tightly with both hands.

6. Walk away from the apparatus, pulling the shoulder load out of the hose bed. If additional hose is needed, additional firefighters should follow the same steps to load the hose on their shoulders. When sufficient hose has been removed from the hose bed, another firefighter uncouples it from the hose bed.

many ways to load attack lines into a hose bed, and this section presents only a few of the most common hose loads. Your department might use a variation of one of these techniques. It is important for you to master the hose loads used by your department.

The most common preconnected attack line is 200 ft (61 m) of 1¾-in. (44-mm) hose. Many engines are also equipped with preconnected 2½-in. (64-mm) hose lines to enable them to make a quick attack on larger fires. Some departments use different lengths, for example, 150 ft (46 m) or 250 ft (76 m), depending on the size of the structures in their area. Firefighters should be able to pull an attack line that is long enough to reach the fire but not so long that an excess of hose might slow down the operation and become tangled.

Preconnected attack lines can be placed in several locations on the apparatus. For example, a section of a hose bed at the rear of the apparatus can be loaded with a preconnected attack line. Many engines have a transverse hose bed installed above the pump and loaded so that the hose can be pulled off from either side of the apparatus, sometimes referred to as "cross-lays." Preconnected attack lines can also be loaded into trays mounted on the side of the apparatus. Short preconnected attack lines can be stored in a compartment in the front bumper on many engines. This attack line is often used for vehicle or dumpster fires because the apparatus can drive up close to the incident and a longer hose line is not needed. (This hose line is sometimes referred to as the "trash line.") Booster hose, 1-in. (25-mm) hose stored on reels, is another type of preconnected attack line. Booster hose reels can be mounted in a variety of locations on fire apparatus. Remember that booster hoses should not be used for structural or vehicle firefighting because they cannot flow the amount of water required.

Attack lines are loaded in a manner that ensures that one or two firefighters can remove the hose quickly from the hose bed and advance the hose to the fire. Whichever hose load technique is used, the hose must be able to be deployed efficiently and completely—efficiently so it does not become tangled as it is being removed from the hose bed and advanced, and completely so it can be charged with water. Laying out attack hose should not require multiple trips between the engine and the point of fire attack.

The hose should also be able to be deployed efficiently around obstacles and corners. If additional firefighters are available, they can help move hose line around obstacles. It should also be possible to reload the hose quickly and with minimal personnel.

There is no perfect hose load that works well for every situation. The three most common hose loads for preconnected attack lines are the flat hose load, the minuteman hose load, and the triple-layer hose load. Variations in circumstances may make one type of hose load preferable for your department.

Preconnected Flat Hose Load

The **preconnected flat hose load** is loaded in a similar manner to the flat hose load used for supply lines (**FIGURE 13-5**). The flat hose load is easy to load and easy to deploy. It can be deployed on the shoulder or by grasping the loop of the hose line with your arm. It can be used in jurisdictions that have varying types of structures. To perform a preconnected flat hose load, follow the steps in **SKILL DRILL 13-12**.

Minuteman Hose Load

The **minuteman hose load** can be deployed and carried over one shoulder by a single firefighter (**FIGURE 13-6**). The firefighter drops one fold or loop

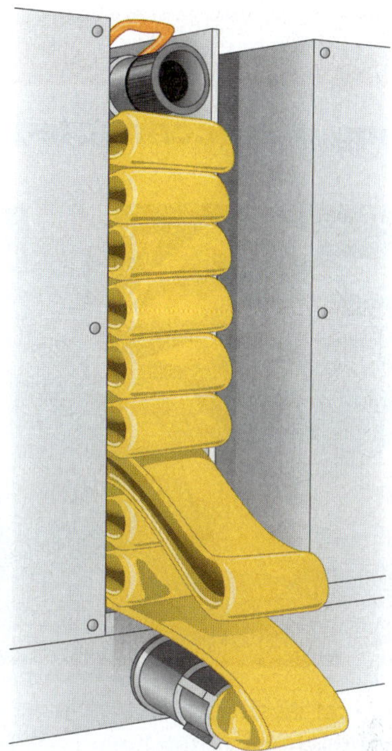

FIGURE 13-5 The preconnected flat hose load is loaded in a similar manner to the flat hose load used for supply lines, except the female end of the hose line is preconnected to a pump discharge outlet, and a nozzle is attached to the male end of the hose line.

SKILL DRILL 13-12

Performing a Preconnected Flat Hose Load Firefighter I, NFPA 1010: 6.5.2

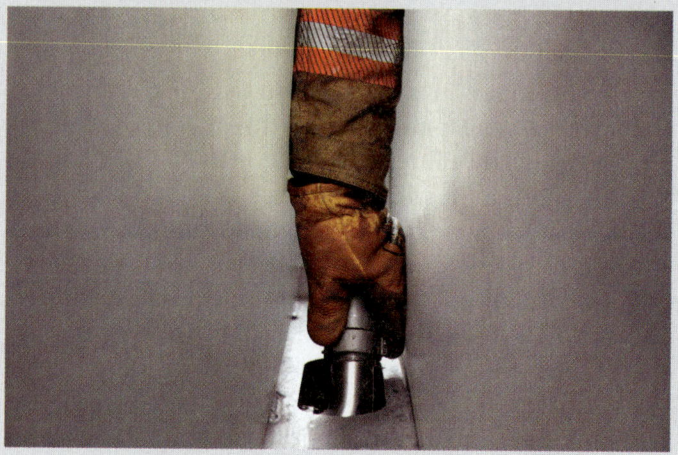

1. Attach the female end of the hose to the preconnect discharge outlet.

2. Begin flat loading the hose in the hose bed.

Continues.

SKILL DRILL 13-12 CONTINUED

Performing a Preconnected Flat Hose Load Firefighter I, NFPA 1010: 6.5.2

3. Load the first layer of hose and make the first fold even with the edge of the hose bed. Then, make a loop, or ear, on the second or higher fold. This loop will be used as a pulling handle.

4. Flat load the remainder of the hose in the hose bed. A second loop can be made toward the top of the load and used as an arm loop. Attach the nozzle to the male coupling at the end of the hose.

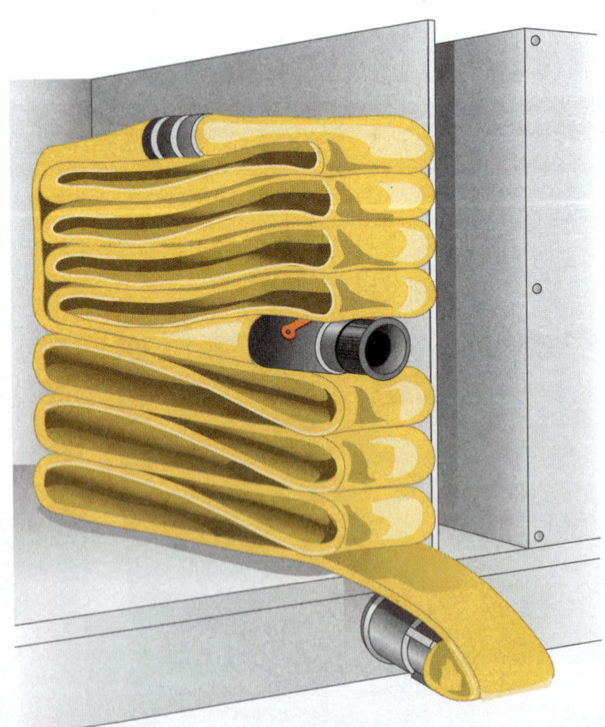

FIGURE 13-6 The minuteman hose load allows a single firefighter to drop one fold or loop of hose at a time from the shoulder as they advance toward the fire.

of hose at a time from the shoulder as they advance toward the fire. This load avoids needing to maneuver the hose line around obstacles, helps to prevent sharp kinks, and avoids dragging the hose on the ground. Because you carry this load on your shoulder, you should have sufficient hose that you can position to pull into the structure if needed. To perform a minuteman hose load, follow the steps in **SKILL DRILL 13-13**.

Triple-Layer Hose Load

The **triple-layer hose load** is suited for departments that generally respond to fires in one- or two-story single-family dwellings (**FIGURE 13-7**). It is a hose loading method in which the hose is folded back onto itself to reduce the overall length to one-third before loading the hose in the bed. This method is used by engine companies with minimal personnel. One person can clear the hose bed and deploy this type of hose load. It can be deployed and charged before entering the building. The triple-layer hose load requires at least two firefighters and must be practiced often. It also requires space in which to lay out the hose behind the engine. To perform a triple-layer hose load, follow the steps in **SKILL DRILL 13-14**.

SKILL DRILL 13-13

Performing a Minuteman Hose Load Firefighter I, NFPA 1010: 6.5.2

1. Connect the female end of the first length of hose to the preconnect discharge outlet on the engine.

2. Lay the first section of hose in the hose bed using a flat hose load. At the second fold, make a loop or ear. This loop will be a handle to grab when deploying the hose.

3. Flat load the second section of hose. Place the male coupling end of the second section of hose at the side of the hose bed.

4. Couple the remaining two sections of hose together, and then attach the nozzle to the male coupling on the last section of hose. Place the connected nozzle on top of the second section previously loaded.

Continues.

SKILL DRILL 13-13 CONTINUED

Performing a Minuteman Hose Load Firefighter I, NFPA 1010: 6.5.2

5. Continue flat loading the remaining sections of hose into the hose bed.

6. When you reach the female coupling of the last section of hose, connect the female coupling from the last section of hose to the male coupling from the second section of hose. When you are finished, the nozzle and the loops should be sticking out of the hose bed. This will make it easy to quickly grab the nozzle and the folds when you need to deploy the hose.

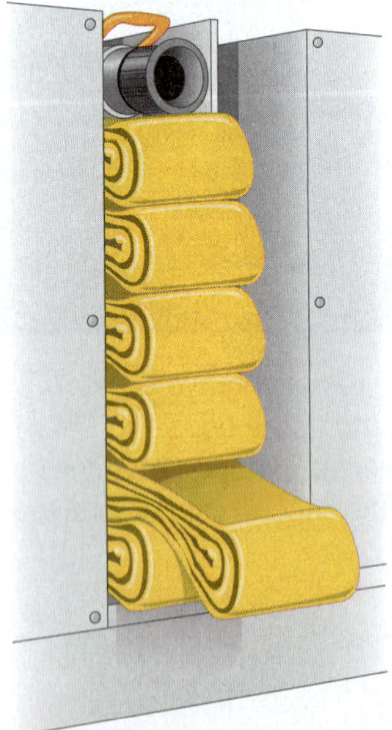

FIGURE 13-7 The triple-layer hose load method is a hose loading method in which the hose is folded back onto itself to reduce the overall length to one-third before loading the hose in the bed.

Deploying and Advancing Attack Lines from the Attack Engine to the Fire

When a building is on fire, fire attack can be conducted from outside the building or from inside the building. When a fire attack is conducted from outside a structure, it is called an **exterior attack**. When firefighters enter a building to fight the fire, it is called an **interior attack**. For both types of attack, first the hose needs to be advanced to the fire. If an interior attack is then required, there are techniques for maneuvering inside a building with a charged hose line.

There are many techniques for advancing attack lines from the attack engine to the fire. Regardless of the technique used, the objective is to lay out the hose in a single pass using limited personnel so that the line can be charged quickly.

If the incident is a structure fire and it is necessary for firefighters to conduct an interior attack, the hose must be laid out in a manner that permits the attack crew to advance the hose line from the entry point to the seat of the fire. The technique chosen depends on the distance from the apparatus to the entry point, the

SKILL DRILL 13-14

Performing a Triple-Layer Hose Load Firefighter I, NFPA 1010: 6.5.2

1. Attach the female end of the hose to the preconnect discharge outlet on the engine. Connect the sections of hose together.

2. Extend the hose from the hose bed. Pick up the hose two-thirds of the distance from the preconnect discharge outlet to the hose nozzle.

3. Carry this fold back to the apparatus, and then place it on the ground at the rear of the apparatus, forming three layers of hose.

4. With a second firefighter—and possibly additional firefighters—pick up the entire length of folded hose.

Continues.

SKILL DRILL 13-14 CONTINUED

Performing a Triple-Layer Hose Load Firefighter I, NFPA 1010: 6.5.2

5. Lay the triple-folded hose in the hose bed just like you would a flat load with the nozzle on top and as close as possible to the rear of the bed so it is within reach from the rear of the apparatus.

number of obstacles between the apparatus and the entry point, the size of the structure, and the distance from the entry point to the seat of the fire. All of these factors impact the type of hose load firefighters choose and the deployment technique they use to advance the hose to the entry point. When the attack line is laid out to the entry point, it should be laid in a serpentine pattern with lengths of hose running parallel to the front of the fire building so that it can be easily advanced into the building, and it will not become tangled when it is charged (**FIGURE 13-8**). It should also be set back from the doorway so that it does not obstruct the entry and exit path.

Preconnected attack hose loaded with the minuteman hose load, preconnected flat hose load, and the triple-layer hose load require different techniques to deploy and advance them. To deploy and advance a minuteman hose load, follow the steps in **SKILL DRILL 13-15**. To deploy and advance a preconnected flat hose load, follow the steps in **SKILL DRILL 13-16**. To deploy and advance a triple-layer hose load, follow the steps in **SKILL DRILL 13-17**.

Once the line is charged, the hose becomes much more difficult to maneuver and advance. It is also important to verify that the hose length will reach the location where the line is needed inside the building. To charge the hose line and prepare to attack the fire, follow the steps in **SKILL DRILL 13-18**.

TIP

It is preferable to have a thermal imager (TI) and flashlight if you need to enter a burning building.

Deploying and Advancing Wyed Lines

To reach a fire that is some distance from the engine, it can be necessary to advance a supply line with a larger diameter, such as a 2½-in. (64-mm) hose line, and then split it into two 1¾-in. (44-mm) attack lines using a gated wye or a water thief. One method of unloading wyed lines is shown in **SKILL DRILL 13-19**.

Advancing Attack Lines from the Entry Point to the Fire

Advancing a hose line from the entry point of the building to the seat of the fire requires a team effort. A hose team may consist of as few as two members. Whenever possible, placing more people on this team will result in a more efficient and smoother operation. Ideally, a hose line team consists of at least three members at the nozzle and a fourth member at the door. As you move inside the building, stay low to avoid the greatest amount of heat and smoke.

If it is necessary to conduct an interior attack, keep in mind that visibility in a smoke-filled structure is limited. You need to rely on your senses of feel and hearing as you advance. Communication with the other members of the nozzle team is critical as you advance.

TIP

If someone on your crew is trained to use a thermal imager (TI), that tool can help you operate in limited visibility situations.

FIGURE 13-8 Before advancing an attack line into a building, the hose should be laid in a serpentine pattern outside the building entrance. This ensures that the hose will not become tangled once it is charged.

SKILL DRILL 13-15

Deploying and Advancing a Minuteman Hose Load Firefighter I, NFPA 1010: 6.3.10

1. Grasp the nozzle and the folds on top of it. Pull this top part of the load approximately one-third out of the hose bed.

2. Turn away from the hose bed, place this top part of the hose load on your shoulder, and hold the load against your body.

Continues.

SKILL DRILL 13-15 CONTINUED

Deploying and Advancing a Minuteman Hose Load Firefighter I, NFPA 1010: 6.3.10

3. Turn around and grasp the lower loop or ear from the first section of hose with your other hand. Walk away from the apparatus until all hose is clear from the hose bed. Continue walking and allow the rest of the hose to deploy from the top of the load on your shoulder to the ground.

SKILL DRILL 13-16

Deploying and Advancing a Preconnected Flat Hose Load Firefighter I, NFPA 1010: 6.3.10

1. Place your arm through the top loop and grasp the bottom loop.

2. Grasp the nozzle with your free hand.

SKILL DRILL 13-16 CONTINUED

Deploying and Advancing a Preconnected Flat Hose Load Firefighter I, NFPA 1010: 6.3.10

3. Turn and walk away from the truck to stretch the hose behind you as you walk.

SKILL DRILL 13-17

Deploying and Advancing a Triple-Layer Hose Load Firefighter I, NFPA 1010: 6.3.10

1. Grasp the nozzle and the top fold.

2. Turn away from the hose bed and place the hose on the shoulder. Walk away from the apparatus until the entire load is out of the bed.

Continues.

SKILL DRILL 13-17 CONTINUED

Deploying and Advancing a Triple-Layer Hose Load Firefighter I, NFPA 1010: 6.3.10

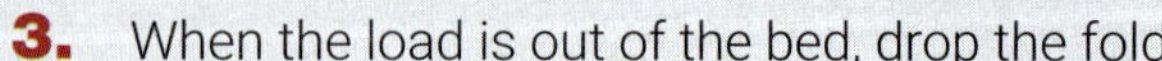

3. When the load is out of the bed, drop the fold.

4. Extend the nozzle the remaining distance.

SKILL DRILL 13-18

Charging the Hose Line and Preparing to Attack the Fire Firefighter I, NFPA 1010: 6.5.2

1. Once the hose has been advanced and laid out, signal the driver/operator to charge the line.
2. Open the nozzle slowly to bleed out any trapped air and to make sure the hose is operating properly. If you are using an adjustable nozzle, make sure the nozzle is set to deliver the appropriate stream. Once this is done, slowly close the nozzle.
3. Quickly recheck all parts of your PPE. Make sure your coat is fastened and your collar is turned up and fastened in front. Check your gloves, hood, mask, and helmet. Check your partner's equipment, and have your partner check your equipment. If you have time, catch your breath while you are breathing ambient air before entering the area that is immediately dangerous to life and health (IDLH). Be ready to start breathing air from your self-contained breathing apparatus (SCBA) and to advance the charged hose line as soon as your officer directs you to do so.
4. When you are given the command to advance the charged line, keep safety as your first priority. Make sure the other members of the nozzle team are ready. If the entry point needs to be opened, do not stand in front of it because fire and superheated gases can exit violently when it is opened.

SKILL DRILL 13-19

Deploying and Advancing Wyed Lines Firefighter I, NFPA 1010: 6.3.10

1. Grasp the wye that is attached to the end of a 2½-in. (64-mm) attack line and pull it from the bed.

2. Advance the 2½-in. (64-mm) attack line toward the fire.

3. Attach the female end of a 1¾-in. (44-mm) attack line to one outlet on the gated wye.

4. Attach the female end of a second 1¾-in. (44-mm) attack line to the second outlet of the gated wye. The individual 1¾-in. (44-mm) attack lines can now be extended to the desired positions.

As you advance the hose line, you need to have enough hose to enable you to move forward. It is not easy to advance charged hose lines through a building involved in a fire. Successful hose advancement relies on teamwork. As resistance is encountered in advancing the line, the firefighter at the nozzle can help to advance more hose, while the firefighter at the door helps feed more hose into the building. Move the hose using the large muscles of your legs rather than tiring the smaller, weaker muscles of your arms. Use your arms for small movements.

The goal should be to have surplus hose to reach the target in one pass, rather than having to go back to get more hose. If everyone on the line initially moves an appropriate amount of hose, you should have enough to reach the target with a little surplus. The objective of moving hose is to move the amount that is needed, not to move as much hose as possible. Too much hose placed close to the nozzle can cause as much of a problem as having too little hose at the nozzle.

The amount of hose that each member of the team takes is dependent on several factors, including the number of people on the team, the length of the hose, and the distance between friction points. A **friction point** is a place where you can predict that the hose will become caught and will slow or stop the advance of the hose line, such as where the line changes directions, turns a corner, goes through a doorway, or goes up or down stairs. All members of the hose line team should work to avoid problems at friction points and to provide the best angle from which to apply water to the fire. For example, when approaching inward-opening doors, approach from the hinge side to allow the nozzle person to direct the stream without obstruction. When approaching outward-opening doors, the unhinged side provides clear access to the room and allows the nozzle team to make a bight (a U-shaped bend in the hose with two parallel ends) with the line outside the door and advance the hose line on the unhinged side of the door to avoid the friction point on the hinged side of the door.

Create surplus hose by pulling extra hose in front of a friction point. This hose can be pulled into a bight or a loop. The bight or loop can be flat on the ground, or it can be placed against the wall. When moving hose forward between two friction points, it is usually best to work in the middle between the two friction points. When you are located at a corner, position yourself to the outside of the line and pull the hose to the outside of the corner to create surplus hose and to avoid contact with the inside corner friction point. This surplus hose allows the nozzle person enough hose that they can choose the best angle of attack for the stream. The nozzle person must communicate continuously with other team members so they will know how much more hose will be needed to reach the fire.

Members must space out on the line. Try to maximize the amount of hose carried by each person. This will help to maintain a fast, efficient stretch. Each member on the line takes control over the span of hose between themself and the next member. The length of the hose span is based on the distance between the friction points. It is a good idea to have someone tending each corner and to be ready to move in an organized fashion.

The use of bights or loops will help to keep surplus hose available and ready for any needed advancement. An option when advancing attack hose is to form an upright loop about 4 ft (1.2 m) in diameter outside the door to the fire building. If more hose is needed, one person can roll this loop toward the nozzle. This will yield several additional feet of hose at the nozzle.

Helpful tips for successful hose movement include the following:

- Communicate hose needs, distance, and direction.
- Hose advancement is seldom successful when you rely on brute strength alone.
- Move hose line with your legs, and feed with your arms from the corners.
- Create bights or loops for surplus spans.
- Position team members at corners to feed hose around the corner.
- Overshoot corners and off load hose in the direction of travel.

Advancing an Attack Line Up a Stairway

When advancing a hose line up stairs, you need enough hose at the bottom of the staircase to make it to the top. When advancing up more than one staircase, you need a firefighter at each turn to feed hose around the friction point. Make sure all members of the team are ready to move together. It is difficult to move a charged hose line up a set of stairs while flowing water through the nozzle. Shutting down the hose line while you are moving up the stairs allows you to get to the top of the stairs more quickly and safely. If you are advancing into the fire floor from a protected stairway, lay excess hose up the stairs toward the next floor. This will make advancement into the fire area more efficient because gravity will help as you advance the hose. If the stairs are not threatened by fire, as in a protected stairway, you can advance an uncharged line. In this case, a shoulder carry such as the minuteman deployment allows for efficient advancement. To advance an uncharged attack line up a stairway, follow the steps in **SKILL DRILL 13-20**.

SKILL DRILL 13-20

Advancing an Uncharged Attack Line Up a Stairway Firefighter I, NFPA 1010: 6.3.10

1. Use a shoulder carry to advance the uncharged line up the stairs when possible. When ascending the stairway, lay the hose against the outside of the stairs to avoid sharp bends and kinks and to reduce tripping hazards.

2. Arrange excess hose so that it is available to firefighters entering the fire floor.

Advancing an Attack Line Down a Stairway

Advancing a charged hose line down a stairway is difficult and dangerous. The smoke and flames from the fire below you tend to travel up the stairway. This means firefighters will be operating in the exhaust portion of the flow path. Due to the risks of operating in a potential flow path, the incident commander (IC) should consider cooling the fire environment from the safest location possible—the exterior—prior to sending in hose teams to conduct an interior attack. Applying a straight or solid stream into an exterior window prior to entry cools the fire environment, limits fire damage to the floor system, and reduces the thermal assault to the operating members.

If the application of an exterior attack line is not possible and an interior attack must be made, get down the stairway and position yourself below the heat and smoke as quickly as you can. Keep as low as possible to avoid the worst of the heat and smoke. Make sure the hose does not get caught or kink on a friction point, as this can restrict the flow of water and potentially compromise safety. When you advance a hose line down a stairway, you have a major advantage on your side: Gravity is working with you to bring the hose line down the stairs.

Wearing PPE and SCBA changes your center of gravity. If you try to crawl down a stairway headfirst, you are likely to find yourself tumbling head over heels. Instead, move down the stairway feet first, using your feet to feel for the next step and to check for stability of the staircase. You may find it necessary to sit on the stairs and lower yourself one step at a time.

To advance a charged attack line down a stairway, follow the steps in **SKILL DRILL 13-21**.

Advancing an Attack Line Up a Ladder

If a hose line needs to be advanced up a ladder, it should be done before the line is charged if not threatened by fire. The hose line should be placed across your chest, with the nozzle draped over your shoulder. With

SKILL DRILL 13-21

Advancing a Charged Attack Line Down a Stairway Firefighter I, NFPA 1010: 6.3.10

1. Advance forward with the charged hose line.

2. Descend stairs. If there is smoke, position yourself underneath it and crawl down the stairs, feet first and facing forward.

3. Position firefighters at friction points where hose lines could snag.

the hose arranged in this way, the line will not push you away from the ladder if it is mistakenly charged while you are climbing. Additional firefighters should pick up the hose about every 25 ft (7.6 m) and help to advance it up the ladder. Additional hose should be fed up the ladder until sufficient hose is inside the building to reach the fire. To advance an attack line up a ladder, follow the steps in **SKILL DRILL 13-22**.

A hose stream can be operated from a ladder and directed into a building through a window or other opening. To operate an attack line from a ladder, follow the steps in **SKILL DRILL 13-23**.

Extending an Attack Line

It is better to have too much hose than not enough. With a hose that is longer than necessary, you can lay out the excess hose. If a hose is too short, you cannot advance it to the seat of the fire without shutting down the line and taking the time to extend it by adding

SKILL DRILL 13-22

Advancing an Uncharged Attack Line up a Ladder Firefighter I, NFPA 1010: 6.3.10

1. Pick up the nozzle and place the uncharged hose across your chest with the nozzle draped over the shoulder. Climb up the ladder with the uncharged hose line. When you reach the first fly section of the ladder, a second firefighter shoulders the hose and starts climbing up the ladder. To avoid overloading the ladder, enforce a limit of one firefighter per fly section.

2. At the top of the ladder, place the nozzle over the top rung of the ladder and advance into the fire area.

3. Additional hose can be fed up the ladder until sufficient hose is in position. The hose can be secured to the ladder with a hose strap to support its weight and keep it from becoming dislodged.

SKILL DRILL 13-23

Operating an Attack Line from a Ladder Firefighter I, NFPA 1010: 6.3.10

1. Climb the ladder with a hose line to the height at which the line will be operated. Secure yourself by applying a leg lock or using a ladder belt.

2. Secure the hose to the ladder near the nozzle with rope or webbing.

3. Carefully operate the hose stream from the ladder. Be careful when opening and closing the nozzle and when redirecting the stream because the nozzle backpressure could destabilize the ladder.

additional lengths of hose. Sometimes, however, it will be necessary to extend a hose line—for example, if the fire is farther from the apparatus than initially estimated.

There are two basic ways to extend a hose line after the valve has been closed by the operator. First, you can disconnect the hose from the discharge outlet on the attack engine and add the extra hose at that location. A disadvantage of this method is you will have to advance the entire attack line to take advantage of the extra hose, and this will take extra time and effort.

Alternatively, you can add extra hose to the discharge end of the charged attack line. This is especially efficient if the nozzle is a breakaway fire nozzle that can be separated from the shut-off valve. With this type of nozzle, you close the valve to stop the flow of water, remove the nozzle tip, then attach the extra hose to the shut-off valve. The nozzle tip can then be installed on another nozzle at the male end of the added hose. This method allows firefighters to lengthen the attack hose without shutting off the valve at the pump on the engine. If it is not a breakaway nozzle, the valve must be closed at the pump by the operator. This evolution should occur in an area not threatened by fire.

Voice of Experience

"More line!" Jim called from the bottom of the basement stairs.

I was supplying hose as fast as I could. Don't let anybody tell you different: it's pretty hard to bend charged 1¾-in. (44-mm) hose lengths into loops, especially once you've already done three of them. My partner was waiting for me to finish and join him at the bottom of the basement stairs.

"How deep are you, Jim?" I called down, my voice muffled by my mask. There was no answer. Had he made the bottom step yet? He had plenty of line to do so. Maybe he had advanced a little beyond the base of the stairs and run out of slack. I called out again. No reply. I kept making my loops. "Joe! Feed me more hose!" I yelled to our man outside. Shouting uses precious air, and the smoke made it difficult to see the gauge on my shoulder strap in order to see just how much air, and time, I had left.

As another section of hose came through, more light broke through the door behind me and a cool breeze followed. A bit of the smoke dissipated, and I realized that there were only two loops done, lying there in front of me against the wall. Not three. Had I miscounted? Had I made two loops, or had Jim dragged one of the three into the basement with him, leaving two behind?

Suddenly I heard Jim's muffled voice from the darkness below. "More line!" cried Jim. "I need more line!" If he was at the bottom step, he should be able to assume a defensive position from there, if necessary, or hightail it back up to me if conditions at basement level were completely unbearable. I called out, yet again, in an effort to figure out what was going on down there. There was no response.

My heart was pounding, my breathing was labored, and despite my efforts to create a surplus of hose that would be easy to maneuver, the hose was heavy and twisted. There was some intermittent light, but the sweat in my eyes made it difficult to see. I stopped making loops amidst his renewed shouting and shoveled what hose I had down into the darkness. "Here it comes, Jim!" I kept pushing hose down the stairs, as it came to me through the door.

His shouting continued, and then there came a tug on the line. He couldn't have used up all the slack so quickly! Was there a kink in the hose? Had it turned around on itself? Was he pulling on the line as a reference point, trying to find his way out of the building? I needed to get down into that basement and figure out what was going on.

As I turned around and reached for a length of hose to guide myself down the steps feet-first, a sudden jerk on the line pulled the hose and me down several steps into the blackness. Luckily, I was able to stabilize myself and continue to the bottom of the stairs. My partner, as it turns out, had enough hose and was several feet beyond the bottom step. As we proceeded about 50 ft (15 m) through the compartmentalized basement to the fire area, the radiating heat got more intense. We cooled the fire room before making entry to extinguish the fire.

Mark W. Ingoglio

Kingston Fire Department
Kingston, New York

TIP

A standpipe kit can be used to supply the added section of hose. Standpipe kits include the appropriate hose and nozzle, a spanner wrench, and any required adaptors.

Connecting Lines to Standpipes

Another water supply evolution is furnishing water to standpipe and sprinkler systems. A **standpipe** is a vertical pipe that connects a water supply to hose outlets in a multilevel building. Standpipes have outlets equipped with valves to which firefighters can connect hose. A **standpipe system** consists of a water supply and a network of horizontal and vertical piping built into a structure to carry the water to the vertical standpipes. The firefighters inside the building depend on firefighters outside the building to supply the water to the **fire department connection (FDC)** on the building (**FIGURE 13-9**). FDCs allow the fire department to pump water into standpipe and sprinkler systems. There are two types of standpipe systems. A dry standpipe system depends on the fire department to provide the primary water supply—that is, all of the water. A wet standpipe system has a built-in water supply, but the

FIGURE 13-9 A standpipe connection.

FDC allows a fire department to deliver a higher flow or to boost the pressure. The pressure requirements for standpipe systems depend on the height where the water will be used inside the building.

Connecting Supply Hose Lines to Standpipes

The FDC for a sprinkler system allows the fire department to supplement the normal water supply. The hose connected to the FDC is considered to be a supply line because it supplies water to the standpipe and sprinkler systems. However, this supply line is connected to the discharge side of the pump on the engine.

The required pressures and flows for different types of sprinkler systems can vary significantly. Large-diameter supply hose is rated for a safe working pressure of 185 psi (1276 kPa). Generally, sprinkler systems require at least 150 psi (1034 kPa) to be provided at the FDC unless more specific information is available. A water hammer can cause a spike of much higher pressure. The excess pressure could burst the hose and place firefighters who are working inside the building in danger. Because of this, many fire department SOPs require two hose lines to be connected to an FDC. If one line breaks or if the flow is interrupted, water will still be delivered to the system through the other line.

To connect a hose line to supply an FDC, follow the steps in **SKILL DRILL 13-24**.

Advancing an Attack Line from a Standpipe Outlet

Standpipe outlets inside a building are provided for firefighters to connect attack hose. Their availability eliminates the need to advance hose lines from the attack engine to an upper floor or to a fire deep inside a large building. In tall buildings, it would be impossible to advance hose lines up the stairways in a reasonable time and to supply sufficient pressure to fight a fire on an upper floor. Instead, firefighters carry standpipe kits, which contain hose, a nozzle, a spanner wrench, and adaptors, into the building and connect to the standpipe outlet.

Standpipe outlets are often located in stairways. Standard procedure is to connect attack hose to an outlet one floor below the fire. Before charging the hose line and opening the door to the fire floor, it is important to lay the hose line so it will be ready to advance into the fire floor. The hose line should be laid up the stairs from the standpipe to the fire floor. Then continue laying the hose up one side of the next set of stairs to the next landing, form a U-shape, and then lay it back down the stairs along the other side to the fire floor. When the hose line is charged and you advance into the fire floor, gravity helps move the line forward. This technique is much easier than trying to pull the charged hose line up the stairs. It is also better to have too much hose than not enough hose. To connect and advance an attack line from a standpipe outlet, follow the steps in **SKILL DRILL 13-25**.

Replacing a Defective Section of Hose

Each length of hose should be tested at least annually, according to the procedures listed in NFPA 1962, *Standard for the Care, Use, Inspection, Service Testing, and Replacement of Fire Hose, Couplings, Nozzles, and Fire Hose Appliances*. Because of proper maintenance and testing, the risk of fire hose failure should be low, but it is always a possibility. Firefighters need to know what to do if a section of supply or attack hose bursts or develops a major leak while flowing water and be prepared to quickly replace a length of defective hose and restore the flow.

A burst hose line should be shut down as soon as possible. The best way to stop flow from a burst hose line is to shut the flow off with the control valve at the engine or fire hydrant. If this is not possible, then a hose clamp can be placed on an undamaged section of the hose upstream from the damaged section on the supply side of the next coupling. After the water flow has been shut off, quickly remove the damaged section of hose and replace it with two sections of hose. Using two sections of hose ensures that the replacement hose is long enough to replace the damaged section. When you

SKILL DRILL 13-24

Connecting a Supply Line to a Fire Department Connection
Firefighter I, NFPA 1010: 6.3.15

1. Locate the FDC to the standpipe or sprinkler system. Extend a hose line from the discharge side of the pump on the engine to the FDC using the size of hose required by your department's SOPs. Some departments use a single hose line, whereas others call for two or more lines to be connected.

2. Remove the caps on the standpipe inlet. Some caps are threaded into the connections and must be unscrewed. Other caps are designed to break away when struck with a tool such as a hydrant wrench or spanner wrench.

3. Visually inspect the interior of the connection on the FDC to ensure that it does not contain any debris that might obstruct the water flow. Never stick your ungloved hand or fingers inside the connections; firefighters have been injured from sharp debris left inside these connections. Attach the hose line to the FDC. Notify the pump driver/operator when the connection has been completed.

SKILL DRILL 13-25

Connecting and Advancing an Attack Line from a Standpipe Outlet
Firefighter I, NFPA 1010: 6.3.10

1. Carry a standpipe kit to the standpipe outlet located one floor below the fire.

2. Remove the cap from the standpipe outlet. Open the standpipe valve to flush the standpipe. Close the valve, and then attach the proper adaptor or an appliance such as a gated wye to the standpipe outlet.

3. Prevent kinks by laying the hose up along the outside wall of the stairs to the landing above the fire floor or along a hallway outside the fire compartment.

4. Form a U-shape with the hose and extend the hose back down to the fire floor. Prepare for the fire attack.

return to the station, mark the damaged hose as defective and report the damage to your officer or supervisor. To replace a damaged section of hose, follow the steps in **SKILL DRILL 13-26**.

TIP

Firefighters operating a hose line that bursts should leave the area if the situation is hazardous.

Draining and Picking Up Hose

After hose is used, it must be drained of water before it is put back into service. To do this, lay a section of hose straight on a flat surface, and then lift one end of the hose to shoulder level. Gravity causes the water to flow to the lower portion of the hose. Proceed down the length of hose, letting it feed over the shoulder and back to the ground behind the firefighter to be rolled or folding the hose back and forth over your shoulder to be carried to another location. As you do this, all the water will eventually flow out of the hose. When you reach the end of the section, the hose is ready to be rolled or carried. To drain a hose, follow the steps in **SKILL DRILL 13-27**.

Unloading the Hose Bed

Hose should be unloaded from the hose bed and then reloaded on a regular basis to place the bends in different portions of the hose. Leaving bends in the same locations for long periods of time weakens the hose at the bends. Hose might also need to be unloaded to transfer it to another apparatus. It could also be necessary to unload the hose for annual testing to be conducted. Before the hose is unloaded, you should take advantage of the fact that the hose bed is empty to clean the hose bed. To unload and clean the hose bed, follow the steps in **SKILL DRILL 13-28**.

SKILL DRILL 13-26

Replacing a Damaged Section of Hose Firefighter I, NFPA 1010: 6.3.10, 6.5.2

1. Shut down or clamp off the damaged hose line.
2. Remove the damaged section of hose.
3. Replace the damaged section of hose with two new sections to ensure that the length of the hose will be adequate.
4. Restore the water flow.
5. When you return to the station, mark the damaged hose as defective and report to your officer or supervisor.

SKILL DRILL 13-27

Draining a Hose Firefighter I, NFPA 1010: 6.5.2

1. Lay the section of hose straight on a flat surface.

Continues.

SKILL DRILL 13-27 CONTINUED

Draining a Hose Firefighter I, NFPA 1010: 6.5.2

2. Starting at one end of the hose, lift the hose to shoulder level.

3. Move down the length of hose, laying it on the ground or folding it back and forth over the shoulder.

4. Continue down the length until the entire hose is drained on the ground or on your shoulder.

SKILL DRILL 13-28

Unload and Clean a Hose Bed Firefighter I, NFPA 1010: 6.5.2

1. Set up in a large area such as a parking lot. Be sure the area is clean.
2. Disconnect gate valves and nozzles from the hose before you begin.
3. Grasp the end of a hose and pull it off the hose bed in a straight line.
4. When a coupling comes off the engine, uncouple the hose, and place it on the ground. Continue pulling off sections of hose until all of the hose sections are on the ground and the hose bed is empty.
5. Use a broom to brush off any dirt or debris on both sides of the hose.
6. Sweep out any debris and dirt from the hose bed.
7. Store hose rolls off the floor on a rack in a cool dry area until they are reloaded onto the apparatus.

CASE STUDY

You Are the Firefighter CONCLUSION

In the early morning hours before sunrise, your engine company is dispatched to a report of a residential structure fire. The driver/operator sees another company approaching an intersection just ahead and close to the dispatched address. The officer on your engine radios the other company and advises them to help perform a split hose lay. The officer then advises dispatch that your company will deploy the attack line once the supply line is in place.

1. **Why is it important to understand supply line evolutions and attack line evolutions?**

   Answer: Firefighters must be efficient at deploying and placing hose into the needed positions during a fire response. Hose evolutions prepare us to do that.
2. **What is the primary purpose of supply line evolutions?**

   Answer: To become more efficient at deploying and handling fire hose during emergencies.
3. **What is the primary purpose of establishing an attack line?**

   Answer: To put water on the fire in the most effective and efficient method.

WRAP-UP

SUMMARY

KNOWLEDGE OBJECTIVES

- Describe how to connect, lay, load, and advance supply hose.
  - Describe the procedures used to connect supply lines to a fire hydrant. (**NFPA 1010: 6.3.15**, pp. 488–497)
- Describe the common types of supply line evolutions. (**NFPA 1010: 6.3.15**, pp. 488–508)
- Describe the common techniques used to load supply hose. (**NFPA 1010: 6.3.15, 6.5.2** pp. 497–506)

KNOWLEDGE OBJECTIVES CONTINUED

- Describe the common techniques to offload a supply hose. (**NFPA 1010: 6.3.15**, pp. 506–508)
- List describe how to load, deploy, and advance attack lines.
  - Describe the types of loads used to organize attack hose. (**NFPA 1010: 6.5.2**, pp. 508–516)
  - Describe the common techniques used to deploy attack hose. (**NFPA 1010: 6.3.10**, pp. 509–522)
  - Describe the general procedures that are followed during attack line evolutions. (**NFPA 1010: 6.3.10**, pp. 509–522)
  - Describe the procedures to follow when advancing attack hose. (**NFPA 1010: 6.3.10**, p. 522)
  - Describe how to extend an attack line. (**NFPA 1010: 6.3.10**, pp. 522–526)
- Describe how to attach supply hose to a fire department connection for a standpipe and/or sprinkler system.
  - Describe the two types of standpipe systems. (pp. 527–528)
  - Describe how to connect and advance an attack line from a standpipe outlet. (**NFPA 1010: 6.3.10**, p. 528)
- List the steps involved in replacing a section or sections of an attack or supply line.
  - Describe how to replace a damaged section of attack hose. (**NFPA 1010: 6.3.10**, pp. 528–531)
- Describe the importance and best practices of regularly unloading and reloading fire hose.
  - Describe why hose should be unloaded and reloaded on a regular basis. (**NFPA 1010: 6.5.2** p. 531)
  - Describe how to unload and reload fire hose. (**NFPA 1010: 6.5.2** pp. 531–533)

SKILLS OBJECTIVES

- Attach a supply hose to a hydrant.
  - Attach a soft sleeve hose to a fire hydrant. (**NFPA 1010: 6.3.15**, pp. 489–490)
- Deploy the appropriate supply hose load for a given apparatus and fire attack.
  - Perform a forward hose lay. (**NFPA 1010: 6.3.15**, pp. 492–494)
  - Perform a reverse hose lay. (**NFPA 1010: 6.3.15**, pp. 495–496)
  - Perform a split hose lay. (**NFPA 1010: 6.3.15**, p. 497)
  - Attach a fire hose to a four-way hydrant valve. (**NFPA 1010: 6.3.15**, pp. 498–499)
- Perform the appropriate supply hose load for a given apparatus and fire attack.
  - Perform a flat hose load. (**NFPA 1010: 6.5.2** pp. 501–502)
  - Perform a horseshoe hose load. (**NFPA 1010: 6.5.2** p. 503)
  - Perform an accordion hose load. (**NFPA 1010: 6.5.2** pp. 504–505)
- Extend a supply line.
  - Perform a working hose drag. (**NFPA 1010: 6.3.10**, pp. 506–507)
  - Perform a shoulder carry. (**NFPA 1010: 6.3.10**, pp. 507–508)
  - Advance an accordion load. (**NFPA 1010: 6.3.10**, pp. 509–510)
- Perform the appropriate attack hose load for a given apparatus and fire attack.
  - Perform a preconnected flat hose load. (**NFPA 1010: 6.5.2**, pp. 511–512)
  - Perform a minuteman hose load. (**NFPA 1010: 6.5.2**, pp. 513–514)
  - Perform a triple-layer hose load. (**NFPA 1010: 6.5.2**, pp. 515–516)
  - Deploy and advance a minuteman hose load. (**NFPA 1010: 6.3.10**, pp. 517–518)
  - Deploy and advance a preconnected flat hose load. (**NFPA 1010: 6.3.10**, pp. 518–519)
  - Deploy and advance a triple-layer hose load. (**NFPA 1010: 6.3.10**, pp. 519–520)
  - Charge a hose line and prepare to attack a fire. (**NFPA 1010: 6.5.2**, p. 520)
  - Deploy and advance wyed lines. (**NFPA 1010: 6.3.10**, p. 521)
- Advance and operate the appropriate attack hose load.

  - Advance an uncharged attack line up the stairway. (**NFPA 1010: 6.3.10**, p. 523)
  - Advance a charged attack line down a stairway. (**NFPA 1010: 6.3.10** p. 524)
  - Advance an uncharged attack line up a ladder. (**NFPA 1010: 6.3.10**, p. 525)
  - Operate an attack line from a ladder. (**NFPA 1010: 6.3.10**, p. 526)
- Attach fire hose to a fire department connection for a standpipe and/or sprinkler system.
  - Connect a hose line to supply a fire department connection. (**NFPA 1010: 6.3.15**, p. 529)
  - Connect and advance an attack line from a standpipe outlet. (**NFPA 1010: 6.3.10**, p. 530)
- Replace a section or sections of an attack or supply line.
  - Replace a damaged section of hose. (**NFPA 1010: 6.5.2**, p. 531)
  - Drain a fire hose. (**NFPA 1010: 6.5.2**, pp. 531–532)
  - Unload and clean a hose bed. (**NFPA 1010: 6.5.2**, p. 533)

KEY TERMS

accordion hose load A method of loading hose on a vehicle that results in a hose appearance that resembles accordion sections. This is achieved by standing the hose on its edge and laying it side to side in the hose bed.

attack engine An engine used to pump water through attack lines at the fireground.

attack line evolutions The delivery of water from an attack engine to a handline, which discharges the water onto the fire. Also called *attack line operations*.

attack line operations See *attack line evolutions*.

combination hose load A hose loading method used when one long hose line is needed; the end of the last length of a hose in one bed of a split hose bed is coupled to the beginning of the first length of hose in the opposite bed.

deploy hose To remove hose from the hose bed or other storage location. Also called *pull hose*.

dual hose lines Two parallel hose lines laid at the same time.

evolution A set of prescribed actions that result in an effective fireground activity. (NFPA 1410)

exterior attack A fire attack conducted from outside a structure.

fire department connection (FDC) A connection through which the fire department can pump supplemental water into the sprinkler system, standpipe, or other system furnishing water for fire extinguishment to supplement existing water supplies. (NFPA 13)

flake out See *lay*.

flat hose load A hose loading method in which the hose is laid flat and stacked on top of the previous section.

forward hose lay A method of laying a supply line where the line starts at the water source and ends at the attack engine. Also called *straight hose lay*.

four-way hydrant valve A specialized type of valve that can be placed on a hydrant and that allows another engine to increase the supply pressure without interrupting flow.

friction point A place where a hose line encounters resistance, such as a corner, a doorway, and stairs.

horseshoe hose load A hose loading method in which the hose is laid on its edge around the perimeter of the hose bed so that it resembles a horseshoe.

hose bed The main storage area on an apparatus for carrying hose.

interior attack A fire attack conducted inside a structure.

lay To arrange hose on the ground. Also called *flake out*.

minuteman hose load A hose loading method that allows a single fire fighter to deploy and advance the necessary amount of hose from the shoulder and avoids having to maneuver the hose lines around obstacles and helps to prevent sharp kinks.

preconnected flat hose load A hose loading method in which the hose is laid flat and stacked on top of the previous section, similar to the flat hose load but with one or two loops sticking out of the load to make it easier to deploy.

pull hose See *deploy hose*.

reverse hose lay A method of laying a supply line where the supply line starts at the attack engine and ends at the water source.

split hose bed A hose bed arranged to enable the engine to lay out either a single supply line or two supply lines simultaneously.

KEY TERMS CONTINUED

split hose lay A scenario in which the attack engine forward lays a supply line from an intersection to the fire, and the supply engine reverse lays a supply line from the hose left by the attack engine to the water source.

standpipe A pipe and attached hose valves and hose (if provided) used for conveying water to various parts of a building for firefighting purposes. (NFPA 1140)

standpipe system An arrangement of piping, valves, hose connections, and associated equipment installed in a building or structure, with the hose connections located in such a manner that water can be discharged in streams or spray patterns through attached hose and nozzles, for the purpose of extinguishing a fire, thereby protecting a building or structure and its contents in addition to protecting the occupants. (NFPA 25)

straight hose lay See *forward hose lay.*

supply engine An engine used to pump water to an attack engine using supply lines.

supply line evolutions The process of laying supply hose to deliver water from a water supply source to an attack engine. Also called *supply line operations.*

supply line operations See *supply line evolutions.*

triple-layer hose load A hose loading method in which the hose is folded back onto itself to reduce the overall length to one-third before loading the hose into the hose bed. This load method reduces deployment distances.

REVIEW QUESTIONS

1. Which supply hose load must be used with LDH?
2. Which attack hose load can be quickly loaded onto one firefighter's shoulder?
3. Why is the hose connected between the FDC and the discharge side of the pump on the attack engine considered to be a supply line?
4. Where is a hose clamp placed to stop the flow of water in a damaged fire hose?
5. What causes the water to flow out of the fire hose when you are picking it up?
6. What happens to fire hose if bends are left in the same place for long periods of time?

DISCUSSION QUESTIONS

1. What are the advantages and disadvantages of the forward lay?
2. Why is it important to stretch the attack hose completely when deploying attack lines?
3. What are the advantages of placing firefighters at corners when advancing attack lines?
4. What are the advantages of picking up the fire hose as you are draining the hose?

APPLYING THE CONCEPTS

You have just arrived on the scene of a car fire located in the lower level of a three-story, 8500 square foot (790 square meter) residence. The structure is located at the end of a 300-ft (91-m) curved driveway. The garage entry is on side-Delta. Fire hydrants are approximately 500 ft (152 m) in either direction. The pumper is carrying 1000 ft (305 m) of 2½-in. (64-mm) hose flat loaded in the rear hose bed. It also has two 200-ft (61 m) 1¾-in. (44-mm) preconnected minuteman loads and 400 ft (122 m) of rolled 1¾ in. (44-mm) hose. Size-up revealed that there is a culvert at the entrance to the driveway that cannot handle the weight of your department's 20-ton (18,143 kilogram) pumper.

1. What are the main challenges to laying attack hose?
2. In the attack line evolution, do you start with the preconnected 1¾-in. (44-mm) hose? Why or why not?
3. In this scenario, what is the best attack hose evolution?

Your four-person crew dons their PPE and remains in the stand-by position. Smoke coming from the structure has intensified and a few flames are visible. IC commands, "2½" attack lines, 200 feet. Gated wye with 200 feet "1¾" lines. Attack side-Delta."

4. Should you form a firefighter chain to quickly pull the hose from the pumper?

5. What technique should you use to stretch the attack hoses?

6. When the attack line has been laid out to the structure, how do you handle the extra length of hose?

Backup personnel on a supply engine arrive. They stop at the fire hydrant where the attack engine crew attached a four-way hydrant valve. The supply engine driver attaches a hose to the outlet side of the four-way valve and connects it to the suction side of the engine pump. Then he attaches a hose to the inlet side of the four-way hydrant valve and connects it to the discharge side of the engine's pump. He moves the lever on the valve to redirect water flow through the supply engine.

7. What is a four-way hydrant valve? When is it used? What benefits does it provide?

REFERENCES

National Fire Protection Association (NFPA). 2017. *NFPA 1962, Standard for the Care, Use, Inspection, Service Testing, and Replacement of Fire Hose, Couplings, Nozzles, and Fire Hose Appliances.* 2017 Edition. Quincy, MA: NFPA.

National Fire Protection Association (NFPA). 2019. *NFPA 1410, Standard on Training for Emergency Scene Operations.* 2020 Edition. Quincy, MA: NFPA.

National Fire Protection Association (NFPA). 2021. *NFPA 13, Standard for the Installation of Sprinkler Systems.* 2022 Edition. Quincy, MA: NFPA.

National Fire Protection Association (NFPA). 2021. *NFPA 1140, Standard for Wildland Fire Protection.* 2022 Edition. Quincy, MA: NFPA.

National Fire Protection Association (NFPA). 2022. *NFPA 25, Standard for the Inspection, Testing, and Maintenance of Water-Based Fire Protection Systems.* 2023 Edition. Quincy, MA: NFPA.

National Fire Protection Association (NFPA). 2023. *NFPA 1960, Standard for Fire Hose Connections, Spray Nozzles, Manufacturer's Design of Fire Department Ground Ladders, Fire Hose, and Powered Rescue Tools.* 2024 Edition. Quincy, MA: NFPA.

CHAPTER

14

Firefighter I

Fire Suppression

KNOWLEDGE OBJECTIVES

After studying this chapter, you will be able to:

- Describe the objectives of defensive and offensive operations and methods of direct, indirect, combination, and transitional fire attack strategies.
- Describe the characteristics of handlines and master stream appliances.
- Explain when you should shut off building utilities and how to do this.
- Describe specific fireground operations for different circumstances.
- Explain how to identify the origin and cause of fires.
- Describe the types of motor vehicles, characteristics of vehicle fires, the tactics used to suppress vehicle fires, and how to overhaul a vehicle fire.

SKILLS OBJECTIVES

After studying this chapter, you will be able to perform the following skills:

- Perform an offensive attack.
- Perform the one- and two-firefighter method for operating a large handline.
- Operate a portable monitor and a deck gun.
- Shut off gas and electric utilities.
- Locate and suppress concealed-space fires and extinguish an outside Class A fire.
- Extinguish a vehicle fire and immobilize and disable a vehicle following a vehicle fire or motor vehicle accident.

ADDITIONAL NFPA STANDARDS

- **NFPA 440**, *Guide for Aircraft Rescue and Firefighting Operations and Airport/Community Emergency Planning, 2024 Edition*
- **NFPA 780**, *Standard for the Installation of Lightning Protection Systems, 2023 Edition*
- **NFPA 921**, *Guide for Fire and Explosion Investigations, 2024 Edition*
- **NFPA 1145**, *Guide for the Use of Class A Foams in Fire Fighting, 2022 Edition*
- **NFPA 1550**, *Standard for Emergency Responder Health and Safety, 2024 Edition*
- **NFPA 1700**, *Guide for Structural Fire Fighting, 2021 Edition*

CASE STUDY

You Are the Firefighter

This is it—your first shift at the fire house after graduating from the fire academy. You are preparing lunch for your new crew as the fire house tradition goes for your first shift. Just as you are preparing the plates for your crew members, the alarm sounds for a house fire in your fire house's first-due district. You turn off the stove and cover the food, then quickly slip on your gear before climbing into the cab of the fire engine. Your heart is racing as the other crew members mount the engine like it is just another day at the office. As you get close to the scene, you can see dark black smoke off in the distance: The incident is declared a working fire by on-scene command.

1. What is the difference between an offensive operation and a defensive operation?
2. When would you use an indirect attack rather than a direct attack?
3. Which type of nozzle would you use for an indirect attack?

Introduction

Fire suppression is the tactics and tasks performed at a fire scene to stop the combustion process and achieve the final goal of extinguishing the fire. **Fire attack** is the process of cooling the fire gases and suppressing the fire. To stop the combustion process, one of the three components of the fire triangle—fuel, oxygen, or heat—needs to be removed to stop the chemical chain reaction that causes combustion. The apparatus and equipment employed by most fire departments are designed to apply large volumes of water to a fire in an attempt to remove the heat by cooling the fuel to below its ignition temperature. Although fire departments deploy a variety of extinguishing agents for different situations, water is used more often than any other agent.

Fire Attack Strategies

When the first-arriving officer at a fire scene conducts the size-up, one of the decisions made is how to attack the fire. As a new firefighter, you are not responsible for determining the type of fire attack that will be used. Nevertheless, you should understand the various factors that go into making these decisions and recognize why different types of fire attacks are used for different types of fires.

When a building is on fire, fire attack can be conducted from outside the building or from inside the building. When a fire attack is conducted from outside a structure, it is called an **exterior attack**. When the firefighters enter a building to fight the fire, it is called an **interior attack**. A firefighter team conducting an interior attack should always be alert for indicators of possible structural collapse.

When conducting an interior attack, NFPA 1550, *Standard for Emergency Responder Health and Safety, 2024 Edition,* requires that a team of at least two firefighters must enter together, and that at least two other

SAFETY TIP

When you flow water into a building, you add weight to it. For example, a nozzle flowing 1000 gallons per minute (gpm; 3785 liters per minute [L/min]) adds 4.2 tons (8400 pounds or 3810 kilograms) to the building every minute. As you continue to flow water, the building is at increasing risk of structural collapse. Indicators of possible structural collapse include the following:

- Cracks in walls, especially cracks that develop or grow during a fire
- Leaning walls
- Pitched or sagging doors, roofs, or floors
- Doors stuck in shifted frames
- Any type of movement or vibrations
- Smoke emitting from cracks in the wall
- Moaning, groaning, creaking, or cracking sounds
- Lack of runoff from firefighting operations

In addition to recognizing the signs, it is important to be able to identify the factors that increase the likelihood of structural collapse so you can take the steps needed to reduce the risk of injury or death. These factors include the environment, building occupancy, existing structural instability, fire and explosion damage, and lightweight construction.

firefighters be positioned outside the danger area, dressed in full personal protective equipment (PPE) and self-contained breathing apparatus (SCBA), ready to rescue firefighters who are inside the building. The two firefighters standing ready outside of the danger area are called the **rapid intervention crew (RIC)** (or sometimes **rapid intervention company [RIC]** or **rapid intervention team [RIT]**). This policy is sometimes called the **two-in/two-out rule**. For more information about RICs, see Chapter 19, *Firefighter Survival*.

The initial goal of fire suppression is to **knock down** the fire, which means to attack the fire to the point where it is almost extinguished and it is unlikely to extend further. Once knockdown is achieved, **overhaul**—the process of finding, exposing, and suppressing any smoldering or hidden pockets of fire in an area that has been burned—can begin. After overhaul is complete, the fire can be considered completely extinguished.

Firefighters must learn how to operate both small and large handlines. Small handlines, which are operated by a single firefighter, can be as large as 2 inches (in.; 51 millimeters [mm]) in diameter. The most frequently used handlines for interior fire attack are 1½-in. or 1¾-in. (38- or 44-mm) handlines. Large handlines are 2½ in. (64 mm) or more in diameter. Water flows through 2½-in. (64-mm) hose lines at 250 gallons per minute (gpm) (946 liters per minute [L/min]), and at a faster flow rate through hoses that have a larger diameter. In an interior, offensive attack, a 2½-in. (64-mm) handline can be advanced into a building to apply a heavy stream of water onto a large volume of fire. That size handline can overwhelm a substantial interior fire if it can be discharged directly into the involved area. The extra reach of the stream from a hose this size can also prove valuable when making an interior attack in a large building.

Defensive Operations

A **defensive operation**, also called a **defensive attack** or **defensive strategy**, is usually an exterior attack and is conducted by directing water streams toward the fire from a safe distance from the exterior of the building or the room involved in the fire or while moving away from the building or fire. A defensive operation is typically an exterior attack. Defensive attacks are implemented when the structure is heavily involved in fire or when the level of risk to firefighters makes an interior, offensive attack unacceptable.

The primary objective in a defensive operation is to prevent the fire from spreading. During a defensive operation, firefighters often use high-volume handlines or master streams from a safe position to control the fire. Water is directed into the building through doorways, windows, and openings in the roof or onto exposures to keep the fire from spreading. Exposure protection should be a high priority during a defensive operation. A defensive operation can also be conducted inside a large structure if the fire is contained to one area. Fire streams can be directed toward the involved area from a safe location to keep the fire from spreading to other parts of the building. The distance that firefighters should stand away from the structure is determined by the **collapse zone**, which is the area where debris will fall if the structure collapses. A general rule of thumb is to establish this zone in a perimeter with a radius of one-and-a-half times the height of the building.

Offensive Operations

An **offensive operation**, also called an **offensive attack** or **offensive strategy**, is usually an interior attack and is conducted by advancing hose lines toward a building or the fire seat to allow firefighters to get close enough to the fire to apply extinguishing agents directly onto the fire so that the extinguishing agents overpower and extinguish it. Because an offensive operation typically exposes firefighters to the heat and smoke of the fire inside the building as well as to several other risk factors, such as the possibility of being trapped by a structural collapse, it is used in situations where the fire is not too large or too dangerous to be extinguished by applying water using handlines. An offensive operation requires well-planned coordination among crews performing different tasks, such as ventilating, operating hose lines, and conducting aggressive search and rescue. The primary objective of an offensive operation is to get to the seat of the fire for extinguishment. When an offensive attack is successful, the fire can be controlled with the least amount of property damage.

SAFETY TIP

During an offensive operation, it is imperative that you extinguish all fire as you proceed to the seat of the fire. If you fail to do this, the fire may grow behind you, trapping you by cutting off your escape route.

Once the type of fire attack is determined during the size-up—before operations begin—it is clearly communicated to and understood by everyone involved in the operation. The incident commander (IC) continuously evaluates the decision throughout the incident. If the IC decides to switch from one strategy to another at any point during an operation, this change must be clearly communicated and understood by all firefighters. A change from one type of attack to another could be warranted if the first attack is unsuccessful or if the risk factors change. Specific strategies and tactics are usually driven by consideration of firefighter safety and victim survivability.

There is no room for confusion or error. It could be a dangerous situation if one group of firefighters initiates one type of attack on the exterior of the building while another group of firefighters conducts a different type of attack inside the building.

Before beginning any type of fire attack, you should always don full PPE and SCBA. Then you need to select the proper attack line and nozzle based on the fire's size, location, and the type of fire attack planned. See Chapter 11, *Fire Hose, Fire Appliances and tools, and Nozzles,* for more information about hose lines and nozzles and Chapter 13, *Supply Line and Attack Line Evolutions,* for more information about working with attack lines. Advance the hose line from the apparatus to the appropriate position depending on the type of attack planned. To ensure that the hose will not kink as you advance the attack line, flake out the excess hose. When you are ready, let the pump driver/operator know that you are ready for water.

Before entering a building through a door for an interior attack or for rescue, check for high heat behind the door. To do this, aim a straight stream at the top of the door. If the water converts to steam, the door is extremely hot (**FIGURE 14-1**). Many fires today are likely to be ventilation-limited when you arrive. Remember that even the act of opening the front door can introduce additional oxygen to the fire, contributing to a violent flashover in a short period of time!

Before entering a building to attack fire, there are several steps you should always take. You should always don full PPE and SCBA. You need to select an attack line following the instructions of the IC, your officer, or your department's standard operating procedures (SOPs), and then advance it to the entry point of the building. If the nozzle you are using is an adjustable nozzle, make sure it is set to straight stream. When you are ready, you notify the pump operator to charge the line, and then you purge the air from the line. To prepare to make an interior, offensive fire attack, follow the steps in **SKILL DRILL 14-1**.

There are four types of fire attack the IC can choose from to conduct an offensive operation: direct, indirect, combination, and transitional.

FIGURE 14-1 Steam from a quick burst of water from a nozzle onto an entry door can indicate high heat.

Direct Attack

In a **direct attack**, the firefighter delivers water directly onto the base of the fire using a solid or a straight stream to cool the burning surface until its temperature is below its ignition temperature. Water should be applied until flame is no longer visible. To perform a direct attack, follow the steps in **SKILL DRILL 14-2**.

Indirect Attack

In an **indirect attack**, firefighters apply water to the hot gases created by combustion to cool them by using a straight or solid stream and sweeping the ceiling with water. The stream should flow between 120 to 180 gpm (454 and 681 L/min). This will cool everything quickly without generating significant amounts of steam. The intention of this indirect attack is to absorb the heat radiating from the walls and ceiling and to cool the hot gases in the fire compartment. In an indirect attack, the water stream may also hit and cool some of the solid fuel and burning surfaces. This cooling only lasts for a short period of time unless the stream is sustained. The application of water in an indirect attack can be accomplished from a safer location, such as from a doorway, a hallway, or the exterior of the building, to reduce the thermal exposure to operating firefighters. To perform an indirect attack, follow the steps in **SKILL DRILL 14-3**.

SKILL DRILL 14-1

Preparing to Make an Interior, Offensive Fire Attack Firefighter I, NFPA 1010: 6.3.10

1. Acknowledge the assignment, and then check into the personnel accountability system. Don full personal protective equipment (PPE) and self-contained breathing apparatus (SCBA). Select the proper attack line to be used to attack the fire based on the fire's size, location, and type. Advance the hose line from the apparatus to the entry point of the structure. Flake out excess hose in front of the entry point. Check the hose for any bends or kinks.

2. Don the face piece and activate the SCBA and personal alert safety system (PASS) device prior to entering the building.

3. Notify the pump driver/operator that you are ready for water. This is usually done with hand signals or a radio if not within the line of sight.

4. Open the nozzle to purge air from the system and make sure water is flowing. If using an adjustable nozzle, ensure that it is set to a straight stream. Shut down the nozzle until you are in a position to apply water.

SKILL DRILL 14-2

Performing a Direct Attack Firefighter I, NFPA 1010: 6.3.10

1. Prepare to make an interior, offensive attack (see Skill Drill 14-1). Open the nozzle in a controlled fashion, and then spray the top of the door with water to check the entry door for heat before making entry.

2. Control the flow path as crews advance toward the fire.

3. If it is safe to do so, advance into the room involved in fire, and apply water to the base of the fire.

4. Shut the nozzle off and reassess the fire conditions. Confirm that ventilation has been completed. Locate and extinguish hot spots until the fire is completely extinguished.

SKILL DRILL 14-3

Performing an Indirect Attack Firefighter I, NFPA 1010: 6.3.10

1. Prepare to make an interior, offensive attack (see Skill Drill 14-1). Open the nozzle in a controlled fashion, and then spray the top of the door with water to check the entry door for heat before making entry.

2. Control the flow path as crews advance toward the fire.

3. From the safest location, such as a hallway, adjoining room, or doorway, apply water to the superheated gases at the ceiling, and move the stream back and forth. Flow water until the room begins to darken.

4. Shut the nozzle off and reassess the fire conditions. Confirm that ventilation has been completed. Locate and extinguish hot spots until the fire is completely extinguished.

Combination Attack

A **combination attack** is an interior, indirect attack followed by a direct attack. This strategy is used when a room's interior has been heated to the point that it is nearing a flashover condition. Firefighters first use an indirect attack method from the safest location, such as a hallway, an adjoining room, or a doorway, to cool the fire gases down to prevent flashover from occurring using a solid or a straight stream. As soon as the water reduces the temperature in the fire compartment, the firefighters can transition to a direct attack. To perform a combination attack, follow the steps in **SKILL DRILL 14-4**.

SKILL DRILL 14-4

Performing a Combination Attack Firefighter I, NFPA 1010: 6.3.10

1. Prepare to make an interior, offensive attack (see Skill Drill 14-1). Open the nozzle in a controlled fashion, and then spray the top of the door with water to check the entry door for heat before making entry.

2. Control the flow path as crews advance toward the fire.

SKILL DRILL 14-4 CONTINUED

Performing a Combination Attack Firefighter I, NFPA 1010: 6.3.10

3. From the safest location, such as a hallway, adjoining room, or doorway, apply water to the superheated gases at the ceiling, and move the stream back and forth. Flow water until the room begins to darken.

4. If it is safe to do so, advance into the fire compartment, and apply water to the base of the fire.

5. Shut the nozzle off and reassess the fire conditions. Confirm that ventilation has been completed. Locate and extinguish hot spots until the fire is completely extinguished.

Transitional Attack

A **transitional attack**, sometimes referred to as a **blitz attack**, is similar to a combination attack, except it begins with a brief exterior, indirect attack to start the cooling process in the fire compartment and stop the progress of the fire. After a brief application of water to the compartment, firefighters proceed to the interior of the building, in coordination with ventilation operations, to suppress the fire. Because the first part of a transitional attack is conducted outside the building, a transitional attack can be started using a preconnected hose line or a master stream appliance using water from the tank on the engine before a permanent water supply has been established. A transitional attack "softens the target"—reduces the fire before the offensive operation begins—or "resets the fire"—reduces the fire to its beginning stages—buying time for more personnel to arrive on scene or to get a permanent water supply established. In fact, two other terms for a transitional attack are **softening the target** and **resetting the fire**.

The transitional attack places water on the seat of the fire sooner, thus making conditions more tenable for any trapped occupants because it cools the fire gases, reduces the risk of flashover, improves visibility, and allows firefighters to enter the fire compartment quickly (**FIGURE 14-2**). With the fire knocked down, in addition to beginning an offensive operation, more aggressive ventilation can follow without the risk of inducing flashover. As visibility continues to improve, firefighters can move through the building more effectively, allowing them to conduct rescue operations, final suppression, and overhaul more safely and efficiently. A transitional attack must be coordinated with ventilation. A ventilation team should not be sent until an attack stream is ready.

FIGURE 14-2 Firefighters can use a transitional attack to reduce the burning and quickly cool the interior before entering a burning building.

Courtesy of UL.

The transitional attack has gained in popularity over the past decade. Traditionally, it was thought that attacking the fire from the outside pushes the fire to other parts of the building. That led to the assumption that offensive operations must be conducted exclusively from the interior of the structure and from the unburned side of the structure to "push" the fire away from the uninvolved portion of the building and any occupants who are behind the advancing hose team. This required firefighters to enter the burning building, often with low visibility and little idea of where the seat of the fire is located. Research conducted by the National Institute of Standards and Technology (NIST), the Underwriters Laboratories (UL), and the New York City Fire Department (FDNY) demonstrated that these assumptions are not always accurate.

In 2013, UL released a report, titled *Innovating Fire Attack Tactics*. This report outlined the results of a series of experiments conducted in burning structures by UL with NIST and FDNY. The researchers found that applying water onto the fire through an open window, door, or roof opening before entering the structure can improve interior conditions both for possible victims and for firefighters entering the structure to rescue victims or conduct an offensive operation. (**FIGURE 14-3**).

FIGURE 14-3 Prior to making entry, the nozzle firefighter directs a straight stream through the top of the door or window and deflects the stream off of the ceiling. Evaluate fire conditions before transitioning to an interior fire attack.

NIST and UL determined that a properly applied straight or solid stream does not "push" fire. Instead, they determined that the movement of the fire is dependent on the flow path. The **flow path** is the route within a structure along which hot gases and smoke travel from areas of higher pressure within the fire to areas of lower pressure accessible through doorways, window openings, and roof openings (discussed further in Chapter 5, *Fire Behavior*). Water application is most effective if a straight or solid stream is aimed through the smoke into the ceiling of the fire compartment. This technique allows heated gases to continue to vent from the fire compartment while cooling the hot fuel inside. However, if the stream or streams are directed through all of the ventilation points, the water blocks the flow path of the gases that are trying to move from the higher pressure area (where the fire is) to a lower pressure area (outside). If this happens, the gases will seek another way out, creating an alternate flow path.

Fog patterns should not be used in this application. The fog pattern entrains large volumes of air and pushes air into the building. A fog stream can also block a ventilation opening, changing the fire flow path as described previously.

Firefighters should consider using a transitional attack any time fire has already self-vented through a door or window. That way, the inside of the structure can be cooled down before making entry to the seat of the fire. The UL NIST experiments showed that when a transitional attack is applied properly, it does not push fire or hot gases, nor does it create dangerous steam conditions. However, it is not always possible to use a transitional attack. For example, if the fire has not self-ventilated or if the fire is deep inside a building with no opening to achieve self-ventilation, a transitional attack is not an option. Fire crews will have to go inside the structure to cool the fire and find the seat of the fire for extinguishment.

The researchers also discovered that if water was introduced into the fire from outside, it cooled not just the fire compartment but also other areas in the building (**FIGURE 14-4**). Temporary cooling was achieved even when the water entered the structure at a distance from the fire seat.

These experiments showed that the effect of this brief offensive, exterior attack is significant. Introducing water from the outside to an interior fire does not completely extinguish a fire at its seat, but it does slow the growth of the fire by cooling quantities of hot gaseous and solid fuel below their ignition temperature, and it reduces temperatures in other parts of a building some distance from the seat of the fire. However, although the effect of this offensive, exterior attack is significant, it is short-lived. The gains are only temporary unless additional water is applied to the seat of the fire. To prevent the fire from regaining its former state and to search for and remove any occupants, it is necessary to quickly enter the building and complete extinguishment of the remaining fire immediately following the initial knockdown from the exterior.

The results of the experiments conducted by NIST and UL require us to reconsider our long-standing approaches to ventilation and the application of water into the fire building. Here are the conclusions the researchers drew from their experiments:

- A properly applied offensive, exterior fire attack through a window or door, even when it is the only exterior vent, will not push fire.
- Water application is most effective if a straight or solid stream is aimed through the smoke into the ceiling of the room involved with fire to allow heated gases to continue to vent from the fire compartment while cooling the hot fuel inside. Fog

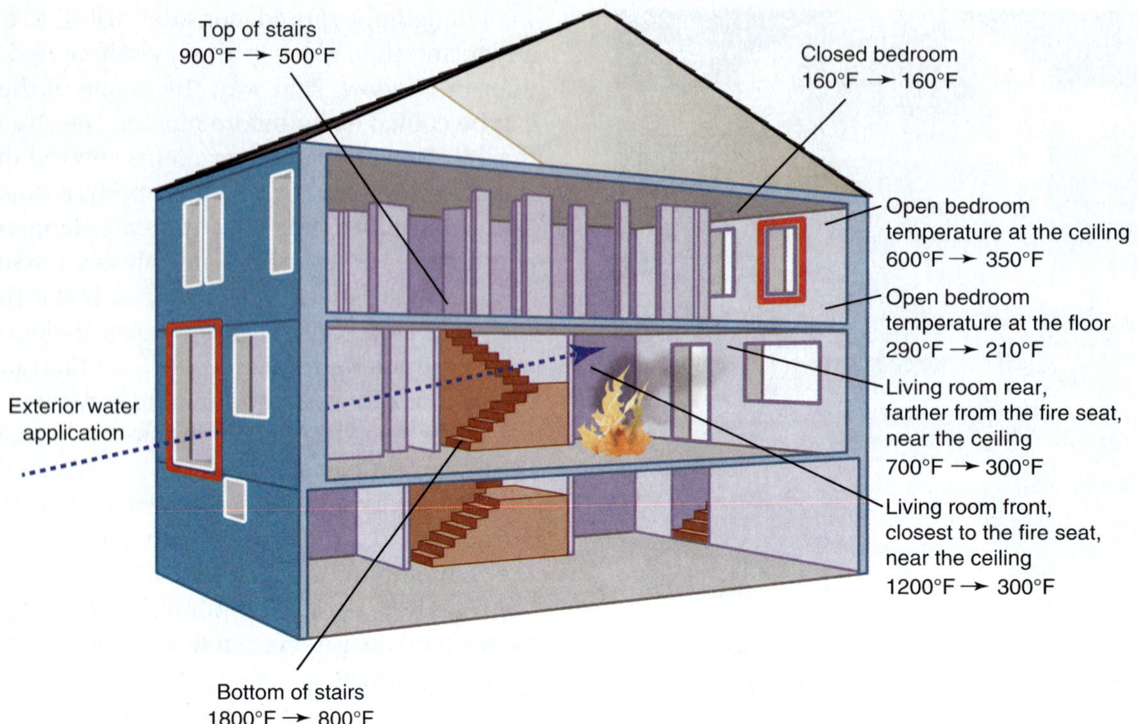

FIGURE 14-4 A transitional attack begins with an exterior, indirect attack conducted some distance from the fire; this reduces temperatures throughout the house and partially suppresses the fire. The temperatures listed are from before and after the exterior application of water.

Courtesy of the National Institute of Standards and Technology.

patterns should not be used because the stream includes large volumes of air and pushes this air into the building, and it can block the ventilation opening and change the flow path.

- Applying a straight or solid stream through a ventilation opening into a room involved in a fire resulted in improved conditions throughout the structure.
- Even in cases where the front and rear doors were open and windows had been vented, application of water through one of the ventilation openings improved conditions throughout the structure.
- Applying water directly into the fire compartment as soon as possible resulted in the most effective means of suppressing the fire.
- Coordinate the fire attack with ventilation—do not ventilate before an attack stream is ready.

Data from Underwriters Laboratories (2013).

To perform a transitional attack, follow the steps in **SKILL DRILL 14-5**.

SAFETY TIP

The acronym S.L.I.C.E.-R.S. has been developed by fire service instructors to help first-arriving company officers remember the steps to be taken for initial engine company operations upon arrival at a structure fire scene. The steps are as follows:

Sequential Actions

S Size-up

L Locate

I Identify and control the fire flow path

C Cool the space from the safest location

E Extinguish the fire

Actions of Opportunity

R Rescue

S Salvage

When using any acronym, keep in mind the objectives of your actions. It may be necessary to modify your actions to fit a specific situation.

SKILL DRILL 14-5

Performing a Transitional Attack Firefighter I, NFPA 1010: 6.3.10

1. Prepare to make an interior, offensive attack (see Skill Drill 14-1). If fire has vented from a door or window, open the nozzle in a controlled fashion, and then apply a solid or straight stream through the top of the opening from a safe, exterior location so it deflects off the ceiling. Evaluate the effectiveness of the hose stream before transitioning to an interior fire attack.

2. Spray the top of the door with water to check the entry door for heat before making entry.

Continues.

SKILL DRILL 14-5 CONTINUED

Performing a Transitional Attack Firefighter I, NFPA 1010: 6.3.10

3. Control the flow path as crews advance toward the fire.

4. If it is safe to do so, advance into the fire compartment and apply water to the base of the fire. If it is not safe to advance into the fire compartment, apply water from a safe location such as a hallway, adjoining room, or doorway, until the room begins to darken.

5. Shut the nozzle off and reassess the fire conditions. Confirm that ventilation has been completed.

6. Locate and extinguish hot spots until the fire is completely extinguished.

Voice of Experience

Several years ago our department was dispatched to a possible structure fire inside of a large big-box furniture supply store. The initial report from the 911 call was that the smoke detectors had activated and there was light smoke in the building. This situation escalated quickly. When the first-arriving fire company reached the location, heavy smoke and fire were showing from the roof of the building. The first-arriving engine company performed a complete size-up of the structure and decided that the fire conditions would still allow firefighters to safely perform an interior, offensive attack on the fire.

Upon making entry, one of the most difficult challenges we faced was trying to find the seat of the fire. With any large building of this type, the floor plan is unpredictable. Because the building was a furniture warehouse, there was no direct route through the front entrance to the location of the fire. We found ourselves navigating in low visibility on the first floor, which was filled with a variety of household furniture. Firefighters attempted for several minutes to find the seat of the fire but were unsuccessful. Inspection holes were cut into the ceiling in an attempt to find the location of the fire. Several times firefighters believed that they had extinguished the fire only to be told by the IC that he could see the fire growing in intensity and the roof of the building was going to fail.

At this point, the IC made a decision that likely saved the lives of everyone inside of the building. The officer in charge of this incident had recently completed an inspection and preplan of the building. During this inspection he realized that the building was constructed using a bowstring truss roof. In his notes he made mention that if this building were to catch fire, firefighters would have only a few minutes to make an interior attack before the roof would collapse.

Based on this knowledge, the IC ordered an immediate evacuation of the structure. The conditions inside of the building consisted of limited visibility with no visible smoke or flames. This was due to the fact that the main body of fire was in the roof line and was consuming the bowstring trusses. Within 5 minutes of all fire companies evacuating, the entire roof collapsed, revealing a large working fire inside the building. Had firefighters been allowed to stay inside the structure, it is possible that multiple firefighters could have been injured or killed.

The firefighters transitioned into a defensive attack and ultimately extinguished the fire. It was the decision made by the IC that day, based on his training and experience, that allowed all of us to return home safely after our shift. The ultimate lesson here is to make sure you study the building construction in your fire district. Firefighters need to know what potential hazards exist in these large, high fire load buildings.

Terry Blackmer
Mayville Fire Department
Mayville, Michigan
Port Huron Fire Department
Port Huron, Michigan

Handlines and Master Stream Appliances

Every firefighter must be able to advance and operate a handline and a master stream appliance effectively to extinguish a fire. The proper operation of a handline is essential to protect yourself, your crew, and trapped victims from the fire.

Before operating a handline, you need to decide which type of nozzle and stream to use. The nozzle defines the shape and direction of the water or extinguishing agent that is discharged onto the fire (**TABLE 14-1**). Chapter 11, *Fire Hose, Fire Appliances and Tools, and Nozzles,* describes these factors in greater detail. Fire department policies and SOPs usually dictate the types of nozzles that are used with different types of hose lines. The nozzle operator must know which type of nozzle should be used in the specific situation at hand.

SAFETY TIP

Care should be taken not to have two crews operating opposing handlines. If this happens so that the two crews are working "against" each other, firefighters could potentially be injured from the water streams and debris.

TABLE 14-1 Types of Fire Streams

Type of Stream	Characteristics	Nozzle Type
Solid stream	Water is discharged as a continuous column of water, which means it can be operated at a lower pressure and gives it greater reach and penetrating power than a straight stream of water.	Smooth-bore
Straight stream	Water is discharged as a highly concentrated pattern of droplets all discharged in the same direction, which keeps the water concentrated in a small area so it can penetrate through a hot atmosphere to reach and cool the burning surfaces.	Adjustable-gallonage fog-nozzle or an automatic-adjusting fog nozzle
Fog stream	Water is divided into droplets that have a large surface area and absorb heat efficiently, which is useful when the temperature in a building needs to be lowered quickly or firefighters need to be protected from the radiant heat of a large fire.	Fixed-gallon fog nozzle, adjustable-gallonage fog-nozzle, or an automatic-adjusting fog nozzle

Courtesy of Todd Eddy.

Handling a charged fire hose can be strenuous and demanding. A charged handline pushes back with significant force and makes the hose hard to control. Fatigue can set in quickly if firefighters do not maintain correct body posture to control the hose and the forces applied to it. One reliable stance firefighters use to maintain correct body posture is a low kneeling position with the underside of the hose grasped with both hands about an arm's length from the nozzle and the hose under an armpit and pinned against a hip. In this position, the arm that is pinning the fire hose needs to have its elbow creating a contact point on the leg closest to the line. This technique grounds energy in three ways: through the kneeling outside leg, through the elbow down to the interior leg, and back out through the hose line. During an interior fire attack, you can use the space around you to ground the nozzle reaction. For example, you can pin the hose against door thresholds, interior walls, and even furniture. This allows the operator to transfer the energy to solid structures rather than through the body. If you are standing, maintain a good stance with the feet about shoulder-width apart and leaning forward a little to offset the force of the hose when the nozzle is opened. This keeps the weight on your hips and off your back, using your legs for support while keeping a good body posture.

To be able to correctly manage the nozzle, position the nozzle about an arm's length away from the firefighter holding the nozzle. When the firefighter's arm is fully extended, the firefighter should be able to grip the bale of the nozzle in the closed position. With the nozzle in this position, the firefighter will be able to direct the stream of water in any direction by simply moving the nozzle as opposed to moving the firefighter's whole body. Once the nozzle is open, the other hand can hold the hose just behind the coupling to aid in the control of the nozzle.

Operating Large Handlines

It is difficult for firefighters to advance and maneuver a large handline inside a building, particularly in tight quarters or around corners. Although one firefighter can control a large handline if it is firmly anchored, at least two firefighters are needed to advance and maneuver a 2½-in. (64-mm) handline inside a building. In addition, firefighters using large handlines must contend with the nozzle reaction force as well as the combined weight of the hose and the water. In situations where the hose line must be advanced over a considerable distance into a building, additional firefighters are required to move the line. The extra effort required to deploy such a line, however, is balanced by the powerful fire suppression capabilities of a large handline.

Large handlines are often used in defensive, exterior attacks to direct a heavy stream of water onto a large exterior fire or to protect exposures. In that case, the nozzle is usually positioned so that it can be operated from a fixed location by one or two firefighters. It can also be used in an indirect attack by directing the stream into a building through a doorway or window opening to knock down a large volume of fire inside. If the exterior attack is successful in reducing the volume of fire, the IC might decide to switch to an interior attack to complete extinguishment of the fire.

One-Firefighter Method for Operating a Large Handline

One firefighter can operate a large handline if it is firmly anchored. To do this, first form a large loop of hose about 2 feet (ft; 0.6 meters [m]) behind the nozzle. Make sure the nozzle end is underneath the end attached to the apparatus so that the weight of the hose stabilizes the nozzle and reduces the nozzle reaction. To anchor it, sit on top of the crossed hose line or use rope or webbing to lash the hose loop to the hose behind the nozzle where they cross. By anchoring the handline, you reduce the energy needed to control the line, which is helpful if it is necessary to maintain the water stream for a long period of time. Although this method does not allow the hose to be moved while water is flowing, it is a good choice for protecting exposures during a defensive operation. To perform the one-firefighter method for operating a large handline, follow the steps in **SKILL DRILL 14-6**.

SKILL DRILL 14-6

Performing the One-Firefighter Method for Operating a Large Handline
Firefighter I, NFPA 1010: 6.3.8

1. Prepare to make an interior, offensive fire attack (see Skill Drill 14-1). Make a loop with the hose. Ensure that the nozzle is under the hose line that is coming from the fire apparatus. Using rope or a strap, secure the hose sections together where they cross, or use your body weight to kneel or sit on the hose line at the point where the hose crosses itself.

2. Allow enough hose to extend past the section where the line crosses itself for maneuverability.

Continues.

SKILL DRILL 14-6 CONTINUED

Performing the One-Firefighter Method for Operating a Large Handline
Firefighter I, NFPA 1010: 6.3.8

3. Open the nozzle in a controlled fashion and direct water onto the designated area.

Two-Firefighter Method for Operating a Large Handline

When two firefighters operate a large handline, one is the nozzle operator, and the other serves as a backup to help stabilize the hose. The backup firefighter can use a hose strap to maintain a better hand grip on a large handline. When the handline is operated from a fixed position, the second firefighter might kneel because it can quickly get tiring to stabilize the charged large handline. That firefighter might also kneel on the hose with one knee to stabilize it against the ground. To perform the two-firefighter method for operating a large handline, follow the steps in **SKILL DRILL 14-7**.

If it is necessary to advance a flowing 2½-in. (64-mm) handline over a short distance and only two firefighters are available, be aware of the large reaction force exerted by the flowing water. It is much easier to shut down the nozzle momentarily and move it to the new position than to relocate a flowing line. If the line must be moved while water is flowing, both firefighters must brace the hose against their bodies to keep it under control. Three firefighters can stabilize and advance a large handline more comfortably and safely than can two firefighters.

Operating Master Stream Appliances

Master stream appliances are used when large quantities of water are needed to control a large fire. Chapter 11, *Fire Hose, Fire Appliances and Tools, and Nozzles,* discusses these in greater detail. A master stream delivers water at a flow rate of at least 350 gpm (1325 L/min). The most commonly used master stream appliances deliver flows between 350 and 1500 gpm (1325 and 5678 L/min), although some master stream appliances can flow more than 2000 gpm (7571 L/min). The stream discharged from a master stream appliance has a greater range than the stream from a handline, so it can be effective from a greater distance.

SKILL DRILL 14-7

Performing the Two-Firefighter Method for Operating a Large Handline
Firefighter I, NFPA 1010: 6.3.8

1. Prepare to make an interior, offensive fire attack (see Skill Drill 14-1). Before attacking the fire, the firefighter on the nozzle should cradle the hose on their hip while grasping the nozzle with one hand and supporting the hose with the other hand. The second firefighter should stay 3 ft (0.9 m) behind them and grasp the hose, securing it with two hands.

2. The firefighter on the nozzle opens the nozzle in a controlled fashion and directs water onto the fire or designated exposure.

Master stream appliances are either manually operated or directed by remote control. Many of these devices can be set up and then left to operate unattended. This capability is extremely valuable in high-risk situations because it eliminates the need to leave a firefighter in an unsafe location or a hazardous environment to operate the device.

Master streams should never be directed into a building while firefighters are operating inside the structure because they can push heat, smoke, and fire onto the firefighters. The force and impact of the stream can also dislodge loose materials or cause a structural collapse.

There are three main types of master stream appliances: deck guns, portable monitors, and elevated stream devices. They are operated from a fixed position—either on top of a piece of fire apparatus (deck gun), on the ground (portable monitor), or on an aerial device. They are typically used for defensive operations, although master streams can be used to blitz a fire during the first part of a transitional attack so that crews can then stretch handlines into the site and extinguish the remaining fire.

Portable Monitors

A portable monitor is a master stream appliance that is carried on the apparatus and placed on the ground wherever a master stream is needed. Hose lines are connected to the portable monitor to supply the water. Most portable monitors have either one, two, or three inlets. Portable monitors can be built with multiple 2½-in. (64-mm) inlets or with one large-diameter hose inlet. Smaller portable monitors are sometimes set up attached to a preconnected hose. This allows one firefighter to quickly place the portable monitor in operation.

If the portable monitor is not adequately secured, the nozzle reaction force can cause it to move from the position where it was originally placed. A moving portable monitor poses a danger to anyone in its path. Many of these master stream appliances are equipped with a strap or chain that may be secured to a fixed object to prevent the monitor from moving. Pointed feet on the base also help to keep a portable monitor from moving. If the stream is operated at a low angle, the reaction force will tend to make the monitor unstable. For this reason, a safety lock is usually provided to prevent the monitor from being lowered beyond a safe angle of 35 degrees. When setting up any portable monitor, always follow the manufacturer's instructions and your department's SOPs to ensure its safe and effective operation. To set up and operate a portable monitor, follow the steps in **SKILL DRILL 14-8**.

Deck Guns

A deck gun is a master stream appliance that is permanently mounted on a vehicle and equipped with a piping system to deliver water to the appliance. Some deck guns can be taken off the fire apparatus and used as portable monitors. If the vehicle is equipped with a pump, the pump driver/operator can usually open a valve to start the flow of water. Sometimes, however, a hose must be connected to an inlet on the deck gun to deliver water to it. If your fire apparatus is equipped with a deck gun, you need to learn your role when placing it in operation. To operate a deck gun, follow the steps in **SKILL DRILL 14-9**.

Elevated Master Stream Appliances

Elevated master stream appliances are mounted on aerial devices—aerial ladders, elevated platforms, or water towers. Most elevated master stream appliances are equipped with a fixed piping system called a waterway that delivers water to a permanently mounted master stream appliance at the top. This arrangement saves valuable setup time at a fire scene. Some elevated master stream appliances have a ladder pipe mounted at the tip of an aerial ladder or elevated platform only when needed, and a hose is run up the ladder to deliver water to the device. In this case, the hose must be secured to the ladder with ropes or straps before flowing water. If your apparatus is equipped with a ladder pipe, you need to learn how to assist in its setup.

SKILL DRILL 14-8

Operating a Portable Monitor Firefighter I, NFPA 1010: 6.3.8

1. Remove the portable monitor from the fire apparatus and move it into the desired position.
2. Attach the necessary hose lines to the monitor as per your department's SOPs or the manufacturer's instructions.
3. Loop the hose lines in front of the monitor to counteract the force created by water flowing out of the nozzle.
4. Signal the pump driver/operator that you are ready for water.
5. Aim the water stream at the fire or onto the designated exposure and adjust the stream as necessary.

SKILL DRILL 14-9

Operating a Deck Gun Firefighter I, NFPA 1010: 6.3.8

1. Make sure that all firefighting personnel are out of the structure. Place the deck gun in the correct position. Aim the deck gun at the fire or at the target exposure. Signal the pump driver/operator that you are ready for water.

2. Once water is flowing, adjust the angle, aim, or flow water as necessary.

Shutting Off Building Utilities

Because most structures use electricity or gaseous fuel to operate lighting, heating, cooling, cooking, and manufacturing equipment, it is important to shut these utilities off as early as possible during fire suppression activities. The IC should identify which utilities are present during size-up and request that the appropriate utility companies respond to disconnect service to the building. If it is safe to do so, firefighters can often shut off these services to the structure before the utility company arrives.

Shutting Off Gas Service

Many structures use either natural gas or propane for heating and cooking. These two energy sources also have many industrial applications. If a gas line inside a structure becomes compromised during a fire, the escaping gas can add fuel to the fire. The means by which the gas is supplied to the structure must be located to stop the flow.

Most residential gas supplies are delivered through a gas meter connected to an underground utility network or from a storage tank located outside the building. If the gas is supplied by an underground distribution system, the flow can be stopped by closing a quarter-turn valve on the gas meter. This meter is usually located outside of the structure. If the gas is supplied from an outside LPG storage cylinder, closing the cylinder valve will stop the flow. After the gas service has been shut off, attach a lockout or shut-off tag or lock to ensure that it is not turned back on (**FIGURE 14-5**). Only a professional from the utility company can reestablish the flow of gas to a structure.

To shut off gas utilities, follow the steps in **SKILL DRILL 14-10**.

Shutting Off Electrical Service

When a fire occurs in a building, the electrical service should be turned off quickly to reduce the risk of injury or death to firefighters, even if there is no involvement of electrical equipment in the fire. If possible, this shut-off should take place at the main circuit breaker box. As with shutting off gas service, attach a lockout tag to prevent someone from accidentally turning the electrical current back on. Some buildings have an exterior main disconnect switch that shuts off electrical power to the building. These should be identified during preincident planning.

SAFETY TIP

Firefighters should never attempt to remove an electric meter to disconnect power.

A.

B.

FIGURE 14-5 **A.** An example of a residential gas meter. **B.** Lockout tag on a gas meter.

SKILL DRILL 14-10

Shutting Off Gas Utilities Firefighter I, NFPA 1010: 6.3.18

1. Don PPE, including SCBA, and enter the personnel accountability system.
2. Acknowledge the assignment.
3. Locate the exterior gas shut-off valve and close it.
4. Attach a shut-off or lockout tag and lock if required.
5. Notify command that the gas is shut off.

SKILL DRILL 14-11

Shutting Off Electrical Utilities Firefighter I, NFPA 1010: 6.3.18

1. Don PPE, including SCBA; enter the personnel accountability system.
2. Acknowledge the assignment.
3. Locate the appropriate disconnect switch or main breaker, and then shut the power off.
4. Attach a lockout tag and lock if required.
5. Notify command that the electricity is shut off.

If it is not possible to turn the electricity off at the breaker box or the main disconnect switch, firefighters must notify the electrical utility to send a representative to the fire scene to disconnect the service from a location outside the building. In many cases, the local utility company is automatically notified of any working fire. All electrical equipment should be considered as potentially energized until the power company or a qualified electrical professional confirms that the power is off.

To shut off the electrical utilities, follow the steps in **SKILL DRILL 14-11**.

Shutting Off Water Service

If a serious water leak has occurred inside a building, shutting off the water supply may help minimize additional water damage to the structure and contents. Water service to a building can usually be shut off by closing one valve at the entry point. Many communities permit the water service to be turned off at the connection between the utility pipes and the building's system. This underground valve is located outside the building and can be operated with a special wrench or key. In most cases, another valve is found inside the building, usually in the basement (if there is one), where the water line enters. In warmer climates, water supply valves are sometimes located above ground.

Specific Operations on the Fireground

The way a fire is attacked depends on where the fire is located. For example, a fire in a basement or attic of a house is approached differently than a fire on the first floor in a kitchen or bedroom. Alternatively, a fire in a building under construction or in a trash container presents firefighters with hazards they would not encounter in an ordinary residence. Firefighters must be aware of the potential hazards at the scene of fires in a variety of locations.

Protecting Exposures

Although protecting exposures is a consideration at every fire, it is even more important with a large fire. If a fire is relatively small and contained within a limited area, usually the best way to protect exposures is to extinguish the fire. In cases where the fire is too large to be controlled, exposure protection becomes a priority. In some cases, the best outcome that firefighters can hope to obtain is to stop the fire from spreading.

The IC must consider the size of the fire and the risk to exposures in relation to how much firefighting capability is available and how quickly those resources can be assembled. In some cases, the IC will direct the first-arriving companies to protect exposures while a

FIGURE 14-6 Protecting an exposure from radiant heat.

second group of companies prepares to attack the fire. At other times, the IC must identify a point where the progress of the fire can be stopped and direct all firefighting efforts toward that objective.

Protecting exposures is part of a defensive operation. At a large, free-burning fire, the first priority is to protect exposed buildings and property from the three primary mechanisms by which heat is transferred: conduction, convection, and radiation (**FIGURE 14-6**). Modern fires produce heat flux levels that may affect exposures earlier in the event. (**Heat flux** is the measure of the rate of heat transfer to or from one surface to another; see Chapter 5, *Fire Behavior*, for further details.) The best option in these circumstances is usually to direct the first hose streams at the exposures rather than at the fire itself. Wetting the exposures will keep the fuel from reaching its ignition temperature. Master stream appliances are excellent tools for protecting exposures. They also ensure that large volumes of water are directed onto the exposures from a safe distance without putting firefighters in the path of excess heat or in danger from structural collapse. Fog streams can sometimes be used to absorb some of the heat coming from the main body of fire, by spreading water more efficiently.

Concealed-Space Fires

Fire in a structure can extend into concealed spaces. For example, fires in ordinary and wood-frame construction can extend to combustible void spaces behind walls and under subfloors and ceilings. Fire in concealed spaces is not always immediately visible. To prevent the fire from spreading, firefighters need to look for **fire extension**—the spreading of the fire to adjacent areas from the seat of the fire—in these spaces and suppress it. To locate and suppress fires behind walls and under subfloors, follow the steps in **SKILL DRILL 14-12**.

Basement Fires

Fires in basements or below grade level present several different challenges. First and foremost, it is often hard to recognize a basement fire because although they start in the basement, they can spread to the first floor or above. Thermal imaging devices may help indicate whether there is a basement fire, but they cannot be used to assess structural integrity or fire severity from above. Many vertical void spaces in a building originate or terminate in the basement, providing ample opportunities for fire gases to spread throughout the building. By the time firefighters arrive on the scene, they may assume that the fire is in one of the main rooms of the house, because that is where the fire is most visible.

In many parts of the country, especially in subdivisions with lightweight construction built prior to the adoption of the 2012 International Residential Code, most basement ceilings are left unfinished when the house is constructed. Fires originating in the basement of those houses quickly involve the floor and support

SKILL DRILL 14-12

Locating and Suppressing Concealed-Space Fires Firefighter I, NFPA 1010: 6.3.8

1. Locate the area of the building where a hidden fire is believed to exist. Look for signs of fire such as smoke coming from cracks or openings in walls, charred areas with no outward evidence of fire, and peeling or bubbled paint or wallpaper. Listen for cracks and pops or hissing steam. Use a thermal imaging device to look for areas of heat that may indicate a hidden fire. Use the back of your hand to feel for heat coming from a wall or floor.

2. If a hidden fire is suspected, use a tool such as an axe or Halligan tool to remove the building material over the area. If fire is found, expose the area as much as possible without causing unnecessary damage, and extinguish the fire using conventional firefighting methods.

system, resulting in a failure and early collapse of the floor over the fire. If firefighters do not identify the fire in the basement prior to entering the building above that basement fire, they are at an increased risk for falling through the damaged floor and ending up in the burning basement below.

In houses with balloon-frame construction, a basement fire can spread to upper floors much more quickly than in a house constructed with modern platform-framing. In balloon-frame construction, the exterior walls are assembled with wood studs that run continuously from the basement to the roof. As a result, an open channel between each pair of studs extends from the foundation to the attic. Each channel provides a path that enables a fire to spread from the basement to the attic without being visible on the first- or second-floor levels.

Basements are difficult and dangerous spaces to access because they have limited routes for entering and exiting. Many older buildings have only one interior stairwell to the basement, so they may be difficult to ventilate because of the limited openings to the outside. Basements are often used for storage, so firefighters may find narrow, disorganized, or cluttered spaces. Basements may contain a high volume of flammable materials, which results in a high fuel load and the production of large quantities of heat. All of these risks highlight the importance of completing a 360-degree size-up prior to entering a structure below grade.

Traditionally, firefighters were taught to advance down to the basement using the interior staircase to put themselves in position to attack the seat of the fire. The UL Firefighter Safety Research Institute (FSRI) conducted a study in 2018 to better understand and fight basement fires. One of the findings of this research is that water applied via the interior stairs has a limited effect on cooling the basement or extinguishing the fire. Researchers also discovered that the temperatures at the bottom of the basement stairs are not sufficiently cool to enable firefighters to operate safely in that environment. In addition, attacking a basement fire from a stairway places firefighters in a high-risk location if the floor above the fire collapses and creates a new flow path up the stairs. Therefore, it is now recommended that if the only point of entry into a basement is an interior stairway, firefighters must look for alternative openings to apply an exterior stream to the basement to reduce temperatures and the fire's impact on the floor system (Madrzykowski and Weinschenk 2018, 89–93).

The research from UL FSRI that investigated the characteristics of basement fires has provided valuable information about how firefighters can use a safer and more effective approach to attack basement fires. If the basement has exterior windows or an external door, these should be considered as points for ventilation and fire attack. Once an external window or basement door is opened, it becomes a fuel-limited fire instead of a ventilation-limited fire and the flow path changes. The growth of basement fires can be slowed and the basement temperatures can be temporarily reduced through the use of a transitional attack. Coordinating ventilation is extremely important. If hose lines are not in place and fire crews are not ready for fire attack, ventilating the basement may create a flow path up the stairs and out through the front door of the structure, almost doubling the speed of the hot gases and increasing the temperature of the gases to levels that could cause injury or death, even to a fully protected firefighter.

According to the results of the UL FSRI study, to conduct a safe and effective attack on a basement fire, identify any external openings from the basement to the outside and advance a charged hose line with an adjustable fog-steam nozzle adjusted to produce a straight stream to those openings. If the basement does not have external windows or doors, the fire can be attacked by cutting a small opening in the floor above the fire and inserting a charged handline with a fog-stream nozzle, cellar nozzle, or a Bresnan distributor nozzle to flow water, temporarily cooling the basement compartment. At the same time, make sure a crew is prepared to ventilate through one or more openings. As soon as the ventilation crew is ready to ventilate, apply a straight stream through the opening. Following this, firefighters can advance into the basement to complete the knockdown, ventilation, extinguishment, and overhaul of the fire.

Fires above Ground Level

Advancing charged handlines up stairs and along narrow hallways requires much more physical effort than advancing a charged handline on a level surface. It is much more labor intensive. A hose crew consisting of two or more firefighters have to work together to advance the hose in a timely fashion for fire attack.

If you need to move to an upper floor in a multilevel structure when fighting a fire, it is important to protect stairways and other vertical openings between floors. Specifically, handlines must be placed on the floor where the fire is and on the floor above the fire to keep the fire from extending vertically and to ensure that exit paths remain available. You might want to also consider placing a handline on the top floor if resources allow so that if the fire extends vertically all the way up to that top floor, you can knock it down. In addition, interior fire crews who move to an upper floor must always look for a secondary egress in case their entry route becomes blocked by the fire or by a structural collapse. This could be a second interior stairway, an outside fire escape, a ground ladder placed to a window, or an aerial device.

In high-rise buildings, the standpipe system is typically used to supply water for hose lines. A standpipe is a large vertical pipe connected to a water supply in a multilevel building that has outlets to which firefighters can connect hose. A standpipe system is the standpipes, a water supply, and the network of inlets, pipes, and outlets that provide water to the standpipes. Firefighters must practice connecting hose lines to standpipe outlets and extending lines from stairways into remote floor areas. Additional hose lines, tools, air cylinders, and emergency medical services (EMS) equipment should be staged one or two floors below the fire.

SAFETY TIP

Be alert to the risk of structural instability and collapse. Remember that modern construction does not always provide clear signs of weakness before collapse. Studies have shown firefighters cannot rely on sounding the floor for accurate stability indicators.

Attic Fires

Fires that involve attics can start in several different locations. Some start in the attic area. An example of this is a fire that starts from an electrical short circuit in the attic. Some attic fires start in a location apart from the attic. For example, a fire that starts in the basement of an older house built using balloon-frame construction can quickly spread from the basement to the attic through the open channels between each pair of studs that extend from the basement to the attic. A third location where attic fires can start is outside. These fires often start as a **ground cover fire**—a fire that burns loose debris on the surface of the ground—or a grilling adventure gone awry. The fire then impinges on the siding. If the siding is combustible, it ignites and spreads up the side of the house. If the siding is vinyl, the fire melts the siding, causing it to fall away from the exterior wall and, in many cases, ignite the combustible foam insulation. The fire spreads up the side of the house and rapidly transitions into the attic space through the soffit.

Attics typically contain large quantities of wood, possibly other combustible materials, open air spaces, and limited or absent fire stops, which all contribute to the spread of a fire. Convection currents in the attic and radiant heat also contribute to the growth of a fire.

The attic spaces in most houses are designed to have a limited yet constant exchange of air to prevent excess moisture from building up in the attic and to remove excess heat during hot weather. During normal conditions, convection currents exhaust warmer air from the attic through vents (**FIGURE 14-7**). In modern construction, fresh air enters the attic space through vents mounted in the soffit that usually run the length of the house on both sides and exhaust air and heat through vents along the peak or ridge of the roof. Older houses may be constructed with gable vents at each end of the house just below the roof instead of soffit and ridge vents. Commercial buildings have multiple options for roof vents. The three most common types are static vents, dormer vents, and whirlybird vents. Static vents and dormer vents are essentially holes in a roof that are placed in an even line to prevent and regulate air leakage. They are often used on flat and metal commercial roofs. Whirlybird vents provide mechanical ventilation with a single-point or multi-point fans.

The air supply available through vents and the open channels between studs is limited. The exhaust flow path is limited by the fact that the ridge vent is often made of plastic, and the heat from the fire often melts the ridge vent to seal this opening. This means that although attic fires contain large quantities of fuel, they usually become ventilation-limited fires. Because of this, use a transitional fire attack on attic fires, if possible, in coordination with **vertical ventilation**—openings made in roofs or floors so that heat, smoke, and toxic gases can escape from the structure in a vertical direction. It is important to understand how flow paths operate in attic fires. Remember that these fire flow paths will change if there is wind blowing that either enhances or retards the air flow.

You need to be aware of the different types of roof vents in your district and understand how the vents affect the flow path of fire. When there is a fire in the attic of a house, the flow path is from the soffit vents to the ridge vent or from one gable vent to the other. In older houses constructed using balloon-frame construction, part of the flow path inlet is through the open channels

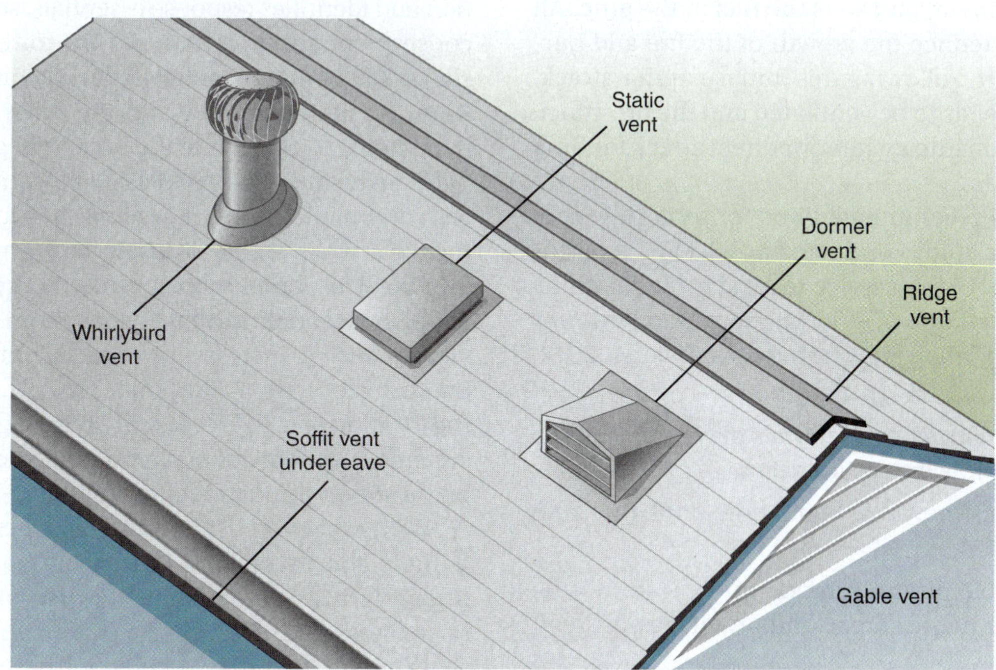

FIGURE 14-7 Roof vents affect how fires grow and spread.

between the wall studs that run from the basement to the attic. A trend in the building industry is to create temperature-controlled attic spaces with the use of spray polyurethane foam to seal the attic and soffit. When you use spray foam in an attic, no vents are used. The spray foam insulates and seals the attic, and no ventilation is needed. The foam works great as a thermal insulator. However, it is highly flammable in fire conditions and burns fast. It also produces highly toxic gases during a fire.

Because air enters the fire compartment through the soffit vents, that is where you should direct a stream of water. To do this on modern-construction houses, quickly remove the soffit vents from the underside of the soffit. In most modern-constructed houses, these are made of plastic and, in the case of a single-story house, can be easily removed with a pike pole. Once the vents are removed, use a solid or a straight stream to spray a stream of water into the soffit vent opening, directing water all the way along one side of the house. In the case of a two-story or taller house, this initial attack may be accomplished using aerial devices to apply a stream of water through the soffit vent openings. At older houses built with gable vents, apply the water through these openings. This initial stream of water wets the underside of the sheathing on that side. The water also deflects off the ridge line and runs down the other side of the interior roof sheathing.

This application of water serves several purposes. It removes large quantities of the fire fuel gases, cools the fire compartment, and wets the fuel in the attic. All of these actions reduce the growth of the fire and partially extinguish it. Following this initial exterior attack, the attic space needs to be ventilated and the fire attack needs to transition into an interior direct attack for final suppression.

Many of the recommendations for attacking attic fires come from a study conducted by UL FSRI. According to this study, 12 fire service tactical considerations for use in the mitigation of attic, knee-wall, and exterior fires include the following (Kerber and Zevotek 2014, 246–72):

- Synthetic construction materials may cause the fire to spread rapidly into attic spaces.
- Delay or limit vertical ventilation until the attic space is cooled.
- If fire is visible on the exterior, apply water to the exterior to limit fire spread and protect exposures.
- Apply water under the exterior of the eaves; remove soffit vents with a pike pole, and apply water through the openings to cool the attic space.
- If fire conditions in the attic are severe, consider flowing master streams up through soffit vents or into gable vents instead of down through the vertical vent opening.
- Watch for rapid fire growth and spread behind knee walls that may encircle firefighters inside the structure; cool by flowing water through the soffit vents.

Fires in Large Buildings

Large buildings have many different uses and layouts. For example, big-box stores contain one large, open space surrounded by smaller rooms and storage areas. These stores are usually one story tall. Buildings constructed for multiple offices may be many stories tall. The fire protection for most large buildings is dependent on fire-resistive construction, fire detection, and automatic sprinkler systems. The fire load in large buildings varies greatly depending on the building contents. For example, large stores and home improvement centers contain an extensive amount of flammable contents ranging from lumber, to flammable fuels, to large quantities of home furnishings and clothing. Many large buildings have floor plans that can cause firefighters to become lost or disoriented while working inside, particularly in low-visibility or zero-visibility conditions.

Well-organized preincident plans for large buildings can be essential when fighting fires in these buildings. A **preincident plan** is data about a location in your community that describes the layout and hazards at the location and identifies response essentials, such as the type of construction, occupancy type, fire control and suppression systems, and potential issues with access and egress. Knowing the occupancy and the other hazards beforehand helps to determine the best strategy and tactics for fighting the fire. For example, in large buildings equipped with automatic sprinkler systems, it is important to augment the water supply by using a fire apparatus to pump into the **fire department connection (FDC)**—the hose coupling or couplings that connect to an automatic sprinkler system—to increase the water volume and pressure in the sprinkler system. The preincident plan should identify the location of the FDC. In large buildings with standpipe systems, the preincident plan should identify the location of the outlets throughout the building.

Whether there is a preincident plan in place, firefighters should remember that big buildings can produce big fires and require the use of large-diameter hose lines with high volume water supplies. Recognize that the fire load may be high in these buildings. Also, in large buildings with complicated floor plans, firefighters should be prepared to use a **search rope**, also called a **search line**, which is a rope with an anchor point to keep them in contact with each other and with the egress route from a burning building.

Fires in Large Buildings during Construction, Renovation, or Demolition

Buildings under construction, renovation, or demolition are at increased risk for destruction by fire because of the following reasons:

- They often have large quantities of combustible materials exposed.
- They lack the fire-resistant features of a finished building.
- They are often unoccupied and are therefore easy targets for arsonists.
- If the building does not have intact windows and doors, an almost unlimited supply of oxygen is available to fuel a fire.
- Fire detection, fire alarm, and automatic fire suppression systems often are not installed or are inoperable.
- Construction workers using torches and other flame-producing devices pose a notable fire risk.

If no life-safety hazards are involved, firefighters may need to use a defensive operation for fires in these buildings. In such a case, firefighters should not enter the building, and a collapse zone should be established. Inspections of construction and demolition sites can give firefighters indications of whether construction is following code.

SAFETY TIP

It is especially important to maintain safety precautions when fighting a fire in a large building during construction, renovation, or demolition. Unsupported or fire-damaged walls may collapse hours after the main body of fire has been knocked down.

Fires in Confined Spaces

Fires and other types of emergencies can occur in confined spaces. The Occupational Safety and Health Administration (OSHA) requires specially trained entry teams to manage emergencies in confined spaces. Never attempt to enter a confined space for fire suppression or rescue operations unless you have been trained as a technical rescuer, and refer to your department's SOPs for guidance on how to respond to confined space emergencies. Fires in confined spaces that house cables and other electrical equipment, utilities, or transformers—called underground vaults, utility rooms, or transformer vaults—are too dangerous to enter. In these cases, firefighters should summon the utility company and keep the area around manhole covers and other openings clear while awaiting the arrival of utility company personnel.

Gas Fire Suppression

Fires involving gaseous fuels such as natural gas or propane are extremely dangerous. **Fugitive gas**—gaseous fuel that escapes from the system—can travel between floors, inside walls or other void spaces, and even underground. If the correct mixture of fuel and air comes into contact with an ignition source, an explosion can occur. This is why it is important to shut off the gas valve (see Skill Drill 14-10). Once the fuel source is shut off, completely extinguish and overhaul any remaining Class A fire. If the gas source is a propane tank, refer to Chapter 23, *Advanced Fire Suppression*, for information on compressed gas cylinder fires.

Electrical Fire Suppression

Fire suppression methods for fires involving electrical equipment vary according to the type of equipment and the power supply. The greatest danger with most fires involving electrical equipment is electrocution. In many cases, the best approach is to wait until the power is disconnected and then use the appropriate extinguishing agents to control the fire. All electrical equipment should be considered as potentially energized until the power company or a qualified electrical professional confirms that the power is off. For this reason, only Class C extinguishing agents should be used when energized equipment is involved in a fire. If the power cannot be disconnected or the situation requires immediate action, only Class C extinguishing agents—such as halogenated extinguishing agents, CO_2, or dry chemicals, or water mist—should be used. Once the electrical service has been disconnected, most fires in electrical equipment can be controlled using the same tactics and procedures that are used for a Class A fire.

When delicate electronic equipment is involved in a fire, halogenated extinguishing agents, carbon dioxide, or water mist should be used to limit the damage as much as possible. These agents cause less damage to computers and sensitive equipment than do water or dry-chemical agents.

When power distribution lines or transformers are involved in a fire, special care must be taken to ensure the safety of both emergency personnel and the public. No attempt should be made to attack these fires until the power has been disconnected. In some cases, firefighters may need to protect exposures or extinguish a fire that has spread to other combustible materials, if this can be accomplished without coming in contact with the electrically energized equipment. If a hose stream comes in contact with the energized equipment, the current can flow back through the water to the nozzle and electrocute firefighters who are in contact with the hose line.

Do not apply water to a burning transformer. Many electrical transformers contain a cooling liquid that includes polychlorinated biphenyls (PCBs), a cancer-causing material. Some large transformers contain large quantities of cooling oil. Water can cause the transformer's cooling liquid to spill or splash, contaminating both firefighters and the environment. If the transformer is located on a pole, it should be allowed to burn until electrical utility professionals arrive and disconnect the power. Dry-chemical extinguishers can then be used to control the fire. Fires in ground-mounted transformers can also be extinguished with dry-chemical agents after the power has been disconnected. Until the power is off, firefighters can still protect exposures while taking care to avoid contamination. Firefighters should stay out of the smoke and away from any liquids that are discharged from a transformer, and they must wear full PPE and SCBA to attack the fire.

Underground power lines and transformers are often located in underground vaults, and explosive gases can build up within these vaults. If a spark ignites these gases, the resulting explosion can lift a manhole cover from a vault and hurl it for a considerable distance. Products of combustion can also leak into buildings through underground conduits. Firefighters should never enter an underground electrical vault while the equipment is energized. Even after the power has been disconnected, these vaults should be considered to be confined spaces containing potentially toxic gases, explosive atmospheres, or oxygen-deficient atmospheres.

Large commercial and residential structures often have high-voltage electrical service connections and interior rooms containing transformers and distribution equipment. These areas should be clearly marked with electrical hazard signs, and firefighters should not enter them unless there is a rescue to be made. Until the power has been disconnected, fire suppression efforts should be limited to protecting exposures. Firefighters must wear full PPE and SCBA owing to the inhalation hazards presented by the burning of the plastics and cooling liquids that are often used with equipment of this size.

Fires in Stacked or Piled Materials

Fires in stacked or piled materials, such as plastics, tires, and packaging materials, present a variety of hazards. The greatest danger to firefighters is the possibility that a stack of heavy material, such as rolled paper or baled rags, collapses without warning. For example, a fire could damage stacked materials, making the stack unstable. Water sprayed onto a stack of materials to suppress a fire can soak into the materials. Absorbed water increases the weight of many materials and weakens cardboard and paper products. A tall stack of material that falls on top of a team of firefighters can cause injury or death. Therefore, fires in stacked materials should be approached cautiously. Firefighters must remain outside potential collapse zones, and mechanical equipment should be used to move material that has been partially burned or water soaked.

Conventional methods of fire attack can often be used to gain control of a fire in stacked or piled materials. However, firefighters need to ensure that enough water is used to penetrate into the stacked material to fully extinguish residual combustion. Class A foam and wetting agents can be applied to extinguish smoldering fires in tightly packed combustible materials. (See Chapter 23, *Advanced Fire Suppression*, for more information about firefighting foam application.) During the overhaul phase, firefighters need to separate the materials to expose any remaining **deep-seated fire**, which is a fire burning below the surface. This can be a labor-intensive operation unless mechanical equipment can be used to separate the damaged material.

Fires in Lumberyards

Lumberyards contain large quantities of highly combustible materials—that is, piles of stacked lumber—stored in the open or in sheds where plenty of air is available to support combustion. Given this rich fuel supply, a lumberyard fire can produce tremendous quantities of radiant heat. It also releases burning embers that cause the fire to spread quickly from stack to stack. Fires in lumber stacks can quickly extend to nearby buildings and other structures. Large buildings in lumberyards are often constructed with trusses and other lightweight building techniques, so be aware of the potential for rapid structural collapse at these incidents.

Lumberyard fires are prime candidates for a defensive firefighting operation. When there is little life-safety risk, protecting exposures is often the primary objective at a lumberyard fire. Collapse zones must be established around stacks of burning material and buildings.

Trash Container and Rubbish Fires

Trash container (dumpster) fires usually occur outside of a structure and generally present fewer challenges than fires inside buildings. Some trash containers may contain hazardous materials or materials that are highly flammable or explosive. Residential dumpsters are usually constructed from polyethylene, which burns rapidly. In fact, polyethylene and other hazardous materials could be in any dumpster because people illegally throw these materials away all the time, so firefighters should be prepared to treat a dumpster fire as if it contained hazardous materials. Preincident plans should include the locations of dumpsters where known hazardous waste is disposed of, such as hospitals, dialysis centers, and other medical facilities. Because there is no way of knowing what might be included in a collection of trash, firefighters must be vigilant and wear full PPE and use a SCBA when fighting trash container or trash pile fires.

If the fire is deep-seated, firefighters need to overhaul the trash to make sure that the fire is completely extinguished. To do this, they need to pull the contents of a trash container apart with pike poles and other hand tools so that water can reach the burning material. This process can be labor intensive and involves considerable risk to firefighters because they are exposed to contaminants in the container as well as to the risks of injury from burns, smoke, or other causes. Considering the low value of the contents of a trash container, it is difficult to justify any risk to firefighters' safety. Instead, Class A foam is often used because it allows water to soak into the materials, eliminating the need for manual overhaul. Some fire departments use the deck gun on the top of an engine or an aerial ladder truck to extinguish large trash container fires, and then complete extinguishment by filling the container with water. This is done by pointing the deck gun at the dumpster and slowly opening the discharge gate to lob a water stream into the container.

Trash containers are often placed behind large buildings and businesses. If the container is close to the structure, be sure to check for fire extension. Also look around the container for the presence of telephone, cable, and power lines that might have been damaged by the fire.

To extinguish an outside trash fire or other outside class A fire, follow the steps in **SKILL DRILL 14-13**.

SKILL DRILL 14-13

Extinguishing an Outside Class A Fire Firefighter I, NFPA 1010: 6.3.8

1. Don full PPE, including SCBA; enter the personnel accountability system; and work as a team. Perform size-up, and then give an arrival report. Call for additional resources if needed. Ensure that apparatus is positioned uphill and upwind of the fire and that it protects the scene from traffic.

2. Deploy an appropriate attack line (at least 1½ in. [38 mm] in diameter).

Continues.

SKILL DRILL 14-13 CONTINUED

Extinguishing an Outside Class A Fire Firefighter I, NFPA 1010: 6.3.8

3. Open the nozzle to purge air from the system, and make sure water is flowing. If using an adjustable nozzle, ensure that it is set to the proper nozzle pattern. Shut down the nozzle until you are in a position to apply water.

4. Direct the crew to attack the fire in a safe manner—uphill and upwind from the fire.

5. Break up compact materials with hand tools or hose streams.

6. Notify command when the fire is under control, and then overhaul the fire. Identify obvious signs of the origin and cause of the fire. Preserve any evidence of arson. Return the equipment and crew to service.

Chimney Fires

Chimneys pass from floor to floor, through the attic, and above the roof line. Newer chimneys are usually made from double-wall or triple-wall metal and have sufficient clearance between the hot metal and the combustible structural members of the building. Many chimney fires are caused by lack of maintenance and cleaning. Older chimneys were made from brick and mortar or clay tiles that deteriorate over time. The mortar in old chimneys may crumble, leaving cracks and openings between the bricks where flames and embers can escape into void spaces in the wall or attic. Creosote, a by-product of burning wood, is highly combustible and builds up over time on the inside of chimneys. Chimney fires involving creosote are extremely hot and can cause flames to shoot out of the top of the chimney.

When responding to a reported chimney fire, look for smoke or flames coming from the top of the chimney. If you see smoke coming from the eaves or attic vents, the fire has most likely extended into void spaces or into the attic itself. Dry-chemical extinguishers are commonly used to extinguish chimney fires. Unless it is necessary because a dry-chemical extinguisher is not available, avoid using water to extinguish the fire—the water may crack or damage the hot flue. If the fire is in a chimney connected to a fireplace and it is not severe, cover the floor and contents near the fireplace with salvage tarps to reduce damage from the extinguisher discharge. While wearing full PPE and SCBA, discharge the extinguisher into the firebox (the part of the fireplace where wood is placed) and up into the chimney. It may be necessary to attack the fire from both the top and the bottom of the chimney. Some departments carry small plastic bags filled with a dry-chemical extinguishing agent. When these bags are dropped from the top of the chimney, they melt and release the contents of the bag. Some departments also carry Chimfex fire suppressant, a product designed specifically to suppress chimney fires. You light the Chimfex stick, similar to a flare, and place it next to the flames in the firebox. The fumes it emits displace the oxygen, extinguishing the fire.

SAFETY TIP

When operating on a roof, be sure to use safe roof operation techniques, including using a roof ladder. Because chimney fires usually occur in cold weather, watch for snow and ice on the roof.

It is important to thoroughly overhaul chimney fires. Many chimney fires have rekindled with disastrous results several hours after the fire department has left the scene. Use a thermal imaging device to check for hot spots behind walls and in the attic. Visually check the attic for fire extension and overhaul any areas where fire has breached the chimney. If smoke is still being produced after the fire is extinguished in the firebox and chimney, remove drywall or siding the entire length of the chimney to find areas where the fire is still burning. Remove all debris and ensure that all burning embers are extinguished.

Solar Photovoltaic Systems

A **solar photovoltaic (PV) system** generates electrical power from sunlight by converting particles of sunlight (photons) from light energy into electrical energy. These systems consist of individual solar cells mounted together to form thin **solar panels**. Multiple solar panels are usually mounted close together on the roof of a building, although installations on the ground are gaining in popularity. The solar panels are connected to one another by electrical cables.

A PV system generates electrical current any time light strikes the solar panels. The current generated by solar panels is direct current (DC). The direct current is transmitted to an inverter that changes the direct current into alternating current (AC). The energy created by the solar panels can be used in that building, sold back to the power company, or stored in a battery bank on the premises. It is hard to completely mitigate the electrocution hazard from PV systems because most PV systems cannot be turned off with a single switch that disables the entire system, and all of the electrical shutoffs may not be adequately marked.

Almost all parts of the PV system, including the solar panels and battery storage systems, pose potential electrocution hazards. Firefighters should assume that all components of these systems are energized. Be alert for metal roofs where solar panels are present. Metal roofs can conduct electrical hazards to other parts of the building. Also, exercise care during ventilation and overhaul operations. Fire or tools may damage the insulation around electrical components, which poses a shock or a fire hazard.

Electrical shock hazards increase when water is applied to electrical systems. Firefighters should be aware that electrical enclosures are not resistant to water penetration due to the force of a fire hose stream. Firefighter gloves and boots offer limited protection against electrical shock, and the amount of protection from electric shock is reduced if the gloves or boots are wet. The degree of the hazard depends on the following:

- Voltage of the system
- Distance from which the water is applied
- Spray pattern
- Conductivity of the water

Tests conducted by the UL FSRI demonstrated that operating a hose stream at a distance of at least 20 ft (6 m) from solar panels using a fog pattern of at least 10 degrees is safe (Backstrom and Dini 2011, 3). Because salt water conducts electrical current more readily than fresh water, it should not be used around PV systems.

Fall and collapse hazards are another concern with PV systems installed on a roof. Because solar panels are heavy, there is an increased chance for roof failure during a fire. During fire conditions, roof panels may become dislodged from the roof and slide off, so it is important to stay away from the roof line underneath solar roof panels.

Lithium-Ion Batteries

The batteries on e-bikes can present challenges to firefighters, just like any lithium-ion operated device. Lithium-ion batteries store a large amount of energy and can pose a threat if not treated properly. They can overheat, catch fire, or explode.

Origin and Cause of Fires

Although a firefighter's primary concern is saving lives and property, you will inevitably make observations and gather information as you perform your duties. Note the time of day, the weather conditions, and any route obstructions. As the incident progresses, make a mental note of the building, fire and smoke conditions, bystanders, and any other information you observe. This information could help the fire investigator determine the area and point of origin and the cause of the fire. The **area of origin** is the room or general area where the fire started. The **point of origin** is the specific spot where the fire first ignited. For example, if a fire started on a stove, the area of origin is the kitchen, and the point of origin is the specific burner on the stove where the fire first ignited.

In addition to determining the point of origin, the fire investigator will be looking for indications of arson. Some common signs of arson include the following:

- Significant damage
- Unusual burn patterns
- Valuable items moved
- Multiple points of origin
- Prescence of accelerants
- Damage to automatic sprinkler systems
- Evidence of forced entry

Firefighters should make every attempt to protect and preserve the entire fire scene, intact and undisturbed, and keep the contents, fixtures, and furnishings in their pre-fire locations. If an object must be disturbed or moved, take pictures and document where the object was originally, who moved it, why it was moved, and where it was moved to. If you must move objects out of a room, make sure you keep all the items that were in a room together. Don't put items from multiple rooms together; keep them separated for the fire investigator. Disturb as little as possible in or near the area of origin until a fire investigator determines the cause of the fire and gives permission to clear the area. For more information on preserving evidence of the fire origin and cause, see Chapter 18, *Salvage and Overhaul.*

Vehicle Fires

Although vehicle fires are one of the most common types of fires handled by fire departments, they are an often-overlooked part of the fire problem. Fire departments respond to an average of one highway vehicle fire every 3 minutes and 2 seconds. In 2021, an estimated 208,500 vehicle fires (15 percent of the total reported fires that year) caused 680 civilian fire deaths, 1500 civilian fire injuries, and $2.1 billion in direct property damage (Hall and Evarts 2022, 5, 9). These fires have a variety of causes. For example, discarded smoking materials can cause fires in upholstery, electrical short circuits can cause fires in many parts of a vehicle, dragging brakes or defective wheel bearings create friction that can cause a fire, and ruptured fuel lines as a result of a motor vehicle accident (MVA) spill fuel that result in fires. In addition to common passenger vehicles, fires occur in trucks, buses, and recreational vehicles. Fires in these types of vehicles present additional hazards and operational challenges. For example, these larger vehicles carry more fuel, or a truck could be transporting hazardous materials.

Modern vehicles contain a variety of gas-filled, pressurized cylinders and containers that may explode when exposed to fire. One example is hydraulic pistons, which support hatch backs, trunks, tailgates, and automobile hoods; are part of energy-absorbing bumper systems; and are a component in a MacPherson strut suspension system. When these pressurized cylinders are quickly heated to high temperatures in a fire, they can release pressure explosively, sending metal parts hurtling away from the vehicle. In addition, the supplemental restraint systems (SRS) found in most vehicles include air bags and air curtains that contain chemicals that can ignite explosively if exposed to heat from a fire.

Modern automobiles are constructed from hundreds of pounds of plastics, which produce large quantities of heat and toxic smoke when they burn. They also contain a variety of petroleum products, including gasoline or diesel fuel, motor oil, brake fluid, and automatic transmission fluid. These products ignite easily, burn with high intensity, and produce large quantities of toxic gases.

Types of Vehicles

At a vehicle fire, firefighters need to be able to quickly determine how the vehicle is powered. Is it a conventional vehicle or an alternative-fuel vehicle? If it is an alternative-fuel vehicle, what type is it?

Conventional Vehicles

Most vehicles on the road today are **conventional vehicles** that use internal combustion engines for power. Internal combustion engines burn gasoline or diesel fuel to produce power. This fuel creates a hazard if it leaks after an MVA. Other hazards associated with conventional vehicles include electrical short circuits and battery acid leaks, as well as a wide variety of combustibles in the passenger compartment and trunk. The 12-volt electrical systems in these vehicles pose a minimal threat to rescuers.

Alternative-Fuel Vehicles

Alternative-fuel vehicles use anything other than petroleum-based motor fuel (gasoline or diesel fuel) for power. Alternative-fuel vehicles are usually identified by markings on the vehicle (**FIGURE 14-8**). Alternative-fuel vehicles may be powered by electricity via a battery, electricity and liquid fuel (hybrid), blended liquid fuels such as ethanol, compressed gases such as propane or natural gas, fuel cells powered by hydrogen, and rarely methanol. Alternative-fuel vehicles are growing in popularity. Some states have even passed laws that prohibit sales of new conventional vehicles after 2035. Following an MVA or fire, these vehicles present hazards that are not encountered in incidents involving conventional vehicles, including the following:

- Electrical shock from high-voltage batteries in electric-drive vehicles
- Exposure to toxic, flammable gases that can be carried in smoke or steam from electric-drive vehicles
- If the vehicle is still on and has not been placed in park, it can roll unexpectedly, but because alternative-fuel vehicles have a silent powered drive, no engine noise is produced to alert those nearby that the vehicle is on.

FIGURE 14-8 Alternative-fuel vehicles are usually identified by markings on the vehicle.

It is important for rescuers to recognize the hazards these vehicles pose both to rescuers and to victims and to be familiar with the additional steps needed to mitigate these hazards. As with conventional vehicles, never approach alternative-fuel vehicles without proper PPE and SCBA.

The technology behind alternative-fuel vehicles is still evolving, which means the best practices for mitigating hazards when these vehicles are part of an emergency scene will continue to change as well. Firefighters need to stay current on all of the new vehicle technology as it evolves. Follow the car manufacturers for updates and changes to new vehicle technology as it is released and look at the vehicles in your district to see the new technology first hand.

Electric-Drive Vehicles. An **electric-drive vehicle** is a vehicle that is powered solely by a high-voltage battery or by a combination of a high-voltage battery and a liquid fuel. The high-voltage batteries that power electric-drive vehicles are made up of multiple cells, each of which produces an electrical current.

A **battery electric vehicle (BEV)** is propelled solely by an electric motor that is powered by a high-voltage battery. Usually, these batteries must be recharged from an external source. These vehicles have the 12-volt system found in conventional vehicles to power accessories in addition to the high-voltage battery that powers the vehicle. Hazards posed by BEVs include the potential for a firefighter to receive an electrical shock or from fire caused by arcing. Additional hazards include off-gassing and delayed-reaction electrical fire from damaged batteries.

A **hybrid electric vehicle (HEV)** uses both a battery-powered electric motor and a liquid-fueled engine (**FIGURE 14-9**). A **plug-in hybrid electric vehicle (PHEV)** is a vehicle that uses both a battery-powered electric motor and a liquid-fueled engine to propel the vehicle and whose battery can be recharged by connecting it to an external power source. The batteries power the electric motor that drives the wheels, much like a train locomotive. The gasoline-powered engine is used for auxiliary power and to recharge the batteries. HEVs contain all of the same hazards associated with conventional gasoline-powered vehicles, plus the additional hazards associated with BEVs.

FIGURE 14-9 A hybrid sticker indicates that the vehicle is powered by both a liquid-fuel engine and an electric motor. Like a battery electric vehicle, it contains a high-voltage battery.

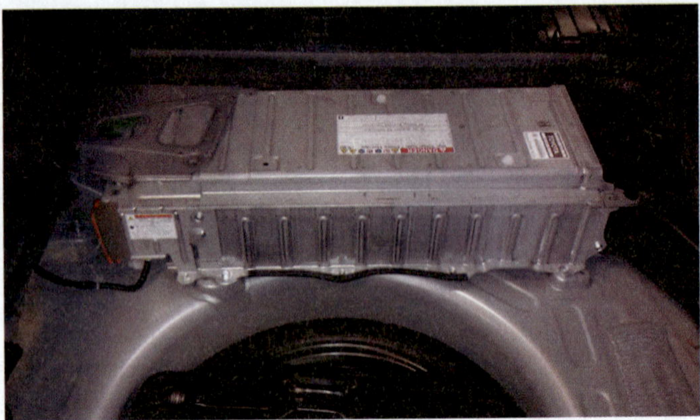

FIGURE 14-10 A high-voltage battery installed in the trunk of a Toyota Prius.

FIGURE 14-11 High-voltage cable in the engine compartment of an electric-drive vehicle.

In the event of an MVA, the high-voltage system in electric-drive vehicles is automatically disabled. In most electric-drive vehicles, the high-voltage system is isolated once the 12-volt battery is disconnected. However, because the battery that powers these vehicles has a high voltage, it may take several minutes for the high-voltage system to discharge after the 12-volt battery is turned off. The batteries in electric-drive vehicles are usually not located in the engine compartment but in other areas, such as the trunk or under the seats or floorboards (**FIGURE 14-10**).

Brightly colored—typically orange for high-voltage cables and blue or yellow for medium/intermediate voltage cables—connect the high-voltage battery to the inverter/converter, which is connected to the electric motor. These cables are usually located beneath the passenger compartment toward the center of the vehicle, so they do not interfere with most extrication procedures. Those high-voltage cables can be located nearly anywhere in the vehicle (**FIGURE 14-11**). Because these cables carry high voltage, they pose a risk to rescuers. Look for placards and high-voltage labels that identify a potential shock hazard. Avoid contact and remain vigilant and aware of the dangers.

Electric-drive vehicles that use lithium-ion traction battery packs are of particular concern during an incident. When the cells in these types of batteries are damaged, they short circuit and start to overheat. This condition is known as **thermal runaway**. As the temperature in the cell continues to rise, gas is produced and starts to off-gas into the surrounding area, eventually igniting. When the heat from one cell damages surrounding cells within the high-voltage battery, it causes other cells go into thermal runaway and ignite. The gases produced during thermal runaway are toxic and flammable. It is recommended that when firefighters approach electric-drive vehicles involved in an MVA that they use a multi-gas meter to detect the presence of toxic or combustible gases. Additionally, a thermal imaging camera (TIC) or a surface thermometer can be used to determine if the high-voltage battery is producing excessive heat above manufacturer's recommended operating temperatures. Battery pack temperature above 200°F (93.3°C) or a rapid increase in temperature is of concern and should be suspected as a sign of thermal runaway until proven otherwise. Other signs of thermal runaway include sparks, smoke (commonly white in color), and gurgling or popping sounds coming from the high-voltage battery pack.

SAFETY TIP

The current best practice is to treat every high-voltage battery in an electric-drive vehicle as a thermal runaway.

When attacking a fire in an electric or hybrid vehicle, it is important to establish a water supply because fires in these vehicles require more water and a longer period of time to extinguish than fires in conventional vehicles. According to the current manufacturers' recommendations, water remains the extinguishing agent of choice in case of a fire involving an electric-drive vehicle. Controlled fires in hybrid electric vehicles and battery electric vehicles have not shown any electrical hazards to firefighters while they are applying water to fires in these vehicles (NFPA 2018, 21). Apply water even after the flames are no longer visible to cool the high-voltage battery. These batteries can reheat and reignite for a long period of time after the flames are extinguished. Listen for popping sounds, which may indicate that the battery is heating up. In cases where you can see the battery, you can use a thermal imaging device to see if the battery is cooling or heating up. After a fire, electric-drive vehicles should be stored at least 50 ft (15.2 m) from any exposures because of the possibility of a late rekindle.

The components of these vehicles continue to change, and the design and location of each component vary from one manufacturer to another. It is important to keep up with changes in vehicle architecture, as they affect the response to incidents involving the vehicles. The best practices a few years from now might be different from the recommended best practices now. Most manufacturers offer a wealth of information that can be viewed or downloaded from the manufacturer's website. It is a good idea for firefighters to receive additional training on alternative-fuel fires.

Blended Liquid Fuel–Powered Vehicles. **Blended liquid fuel–powered vehicles** are powered by a blend of liquid fuels. This type of fuel is often called flex fuel, and most gas stations label it as E85 (**FIGURE 14-12**). The most common combination is gasoline and another flammable liquid, such as methanol or ethanol. The mixture of 10 percent alcohol and 90 percent gasoline is sold in many cities in the United States as a measure to reduce pollution. Diesel fuel also may be blended with other combustibles to produce a blended diesel fuel, and gasoline may be blended with biofuels, derived from agriculturally produced products such as grains, grasses, or recycled animal oils.

FIGURE 14-12 A flex fuel identification badge.

Biofuels are derived from agriculturally produced products such as grains, grasses, or recycled animal oils. They are processed to produce a type of alcohol that is added to gasoline to produce a blended fuel. Vehicles operating on these blended fuels are usually identified by special designations. They are singled out for attention primarily because of their environmental benefits.

MVAs involving vehicles that rely on blended fuels are handled in the same manner as MVAs involving vehicles powered by conventional fuels.

Compressed Gas–Powered Vehicles. **Compressed gas–powered vehicles** use one of three forms of compressed gas to power vehicles: compressed natural gas, liquefied natural gas, and liquefied petroleum gas.

Compressed natural gas (CNG) is commonly stored in steel, aluminum, or carbon fiber–wound tanks, under pressures ranging from 3000 to 3600 pounds per square inch (psi; 20,684 to 24,821 kilopascals [kPa]). In particular, buses, delivery vans, and fleet sedans may be powered by CNG (**FIGURE 14-13**). Many cities power their municipal bus fleets with CNG. Automobiles powered by CNG contain storage cylinders that are similar to SCBA cylinders. These cylinders are usually located in the trunk of smaller vehicles or on the roof of larger vehicles and contain CNG at high pressures (**FIGURE 14-14**). These vehicles can be identified by the "pregnant shape" of the rooftop. When a fire occurs, they must be cooled and protected just like any gas cylinder. CNG is a nontoxic, lighter-than-air gas that will rise and dissipate if it is released into the atmosphere. Any sign of fire in a vehicle powered by CNG should be treated with extreme caution. At no time should water be applied directly to composite cylinders as it can prevent the temperature pressure relief

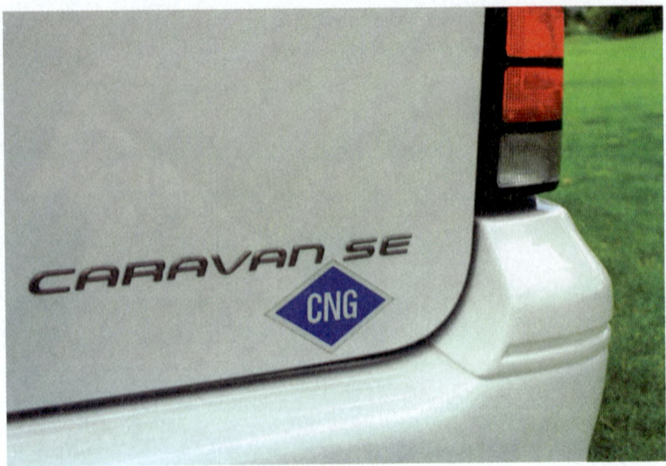

FIGURE 14-13 The CNG sticker on this van indicates the that it is powered by compressed natural gas.

FIGURE 14-14 These CNG tanks are mounted to the roof of a bus.

device (TPRDs) from functioning properly. This could result in catastrophic container failure. If the cylinders are already involved in fire, do not approach the vehicle. Establish a safe perimeter of at least 80 to 100 ft (24 to 30 m) away and allow it to burn while protecting any exposures (NFPA 2018, 24)

Liquefied natural gas (LNG) is stored at very low temperatures. Storage tanks for LNG are double walled and must be insulated to keep the liquefied gas cold. The use of LNG in compressed gas–powered vehicles is not as common as the use of CNG.

Liquefied petroleum gas (LPG), also called propane, is popular for light-duty vehicles, medium-duty trucks, and shuttle vans used in high-mileage fleet applications, for example, taxis and delivery vehicles (**FIGURE 14-15**). Propane is stored in a liquid state at a pressure of 50 psi (345 kPa).

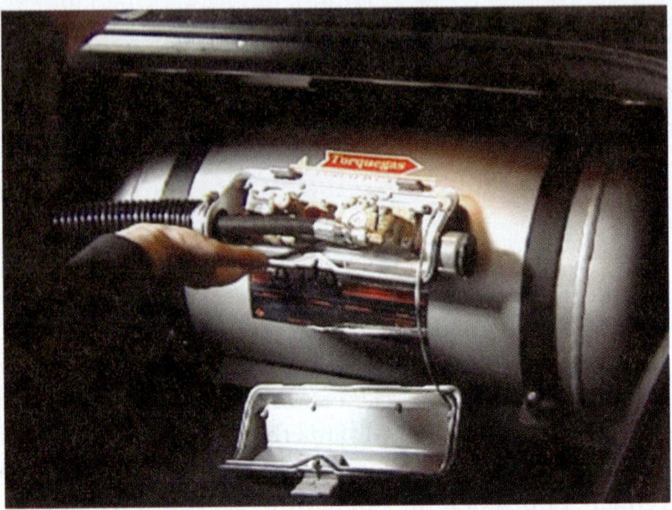

FIGURE 14-15 The LPG tanks used in automobiles such as taxis are usually found in the trunk of the vehicle.

Look for signage or placards on the vehicle identifying it as being powered by a compressed gas, as shown in Figure 14-13. If a compressed gas is present, check for the presence of leaking cylinders or fuel lines. Damage to the gas cylinders or fuel lines could result in the escape of flammable natural gas. In addition, a fire in a gas-powered vehicle poses the threat of a boiling liquid/expanding vapor explosion (BLEVE). In the event of a fire, keep the gas cylinders cool to avoid a BLEVE. If you suspect that a leak is present, keep all sources of ignition away from the vehicle and take measures to stop the leak if possible. Fully involved fires in vehicles powered by compressed gas should be fought with a handline from a protected location or with an unmanned master stream to prevent injuries from exploding gas cylinders. If LNG is leaking, avoid spraying a stream of water on the escaping liquid, as this action heats the cold liquid; it may then convert to a gaseous state with explosive force. It is acceptable to spray a fog stream on an escaping cloud of vapor, however. Ensure that a firefighter in full protective equipment is standing by with a charged hose line that is at least 1½ in. (38-mm) in diameter.

SAFETY TIP

Vehicles that run on compressed gas can produce flammable vapors, just like conventional gasoline- and diesel-powered vehicles. It is important to stretch a charged hose line at the scene of all motorized vehicle/equipment extrications.

FIGURE 14-16 The Honda Clarity was a hybrid car powered by a fuel cell and gasoline.

© VDWI Automotive/Alamy Stock Photo

Fuel Cell–Powered Vehicles. A **fuel cell–powered vehicle** is powered by **fuel cells**, which generate electricity through a chemical reaction between hydrogen gas and oxygen gas to produce water, and in the process, produce electricity and heat. The purpose of fuel cells is not to produce water, however, but rather to produce electricity that can be used to drive an electric motor. Fuel cell–powered vehicles are essentially electric-drive vehicles with a different device for generating the electricity. In a vehicle powered by a fuel cell, the electric motor is powered by electricity generated by the fuel cell. The major components needed for a fuel cell–powered vehicle are a hydrogen gas storage tank, a high-output battery, a group of fuel cells, and an electric motor. The use of fuel cells is being advocated because they produce less pollution than internal combustion engines (**FIGURE 14-16**). Although only a few fuel cell–powered vehicles are being produced today, new technology is continually being developed, so you should expect to see more of them on the road in the future.

Arrival and Size-Up of the Scene

Many hazards are associated with vehicle fires, including traffic hazards, electrical hazards, fuel, pressurized cylinders, fire, and toxic smoke. Upon arrival at the scene of a vehicle fire, it is important to assess the hazards present and determine the types of vehicles involved. Scan the entire area for hazards. For example, did the vehicle hit an electrical source or rupture a gas line (**FIGURE 14-17**)? Hazards are everywhere, and a quick scan might save everyone on the scene from further injury. Be sure to wear full PPE and SCBA and follow your department's SOPs to guard against hazards when responding to a vehicle fire.

FIGURE 14-17 Assume electrical lines are live until it is confirmed that the power has been shut off by the utility company.

Courtesy of Aaron Miranda.

The IC will usually perform a size-up of the scene by conducting a 360-degree walk-around of the scene. During this size-up, the IC will evaluate the hazards present and determine the number of victims. Using the information obtained from the scene size-up, the IC can create an action plan and call for additional resources, if needed.

While the IC conducts the size-up, conduct your own so that you are better prepared to attack the fire. Perform a risk–benefit analysis—that is, look at the big picture and weigh the options. Do not risk injury when fighting a vehicle fire. Only when a viable victim is trapped in a burning vehicle does the scenario become a life-or-death situation. If a victim is visible in the vehicle, immediate rescue is the priority if it is safe to approach the vehicle. One or more firefighters should attempt to rapidly remove the victim from the vehicle while another firefighter provides protection with a fog stream. (Rescuing victims from a vehicle is a Firefighter II skill. See Chapter 26, *Vehicle Rescue and Extrication*, for more information about this.) If the driver is outside of the burning vehicle, ask about any specific hazards that may be present in the vehicle, such as portable propane cylinders, propane torches, medical oxygen equipment, cans of spray paint, and other hazardous materials.

To reduce confusion and minimize the potential for mistakes at these scenes, it is important to use standardized terminology when referring to specific parts of vehicles. In North America, the driver's seat is on the left side of the vehicle, so that side of the vehicle is the driver's side. The right side is called the passenger's side. Vehicles generally consist of three main compartments: the engine compartment, the passenger compartment, and the trunk (cargo area).

Because vehicle fires usually occur on streets and highways, one of the biggest hazards faced by firefighters is the danger posed by traffic. Drivers are easily distracted by the sight of a burning vehicle. The hazards posed by traffic need to be handled quickly before they lead to additional MVAs or injuries. Sometimes the most important action to take at the scene of an MVA is to slow, stop, or divert the flow of traffic before proceeding with additional actions. If possible, position large emergency vehicles so that they provide a barrier against motorists who fail to recognize or heed emergency warning lights without disrupting traffic any more than necessary. Fire apparatus operators also often place their vehicles 100 ft (30 m) behind the burning vehicle to stop traffic and to position the apparatus a safe distance from the burning vehicle. This position helps to push the apparatus to the side of the accident in the event that the emergency apparatus is struck from behind. Avoid positioning emergency vehicles in a manner that could confuse oncoming motorists. Make sure that the apparatus headlights do not shine in the direction of approaching drivers to avoid blinding them.

Firefighters need to work together with law enforcement officials to control traffic in a manner that is safe for emergency workers, victims of the MVA, and motorists. If law enforcement officials are not on the scene when you arrive, verify that they are aware of the incident and that they have been dispatched. Do not hesitate to request that law enforcement close the road or divert traffic flow if necessary to ensure safety for firefighters. Traffic cones or flares can be placed to direct motorists away from the accident (**FIGURE 14-18**). Be sure to look for leaking fuels before using flares. Always refer to your department's SOPs for guarding against traffic hazards.

Before exiting fire apparatus at an emergency scene, be alert for vehicles that might cause injury to firefighters. Do not assume that motorists will heed the warning lights. Let law enforcement personnel coordinate traffic control.

Attacking Vehicle Fires

Conduct a quick size-up and look for any people that may be in the vehicle. Take immediate action to protect and remove them if it is safe to do so. (Rescuing victims from a vehicle is a Firefighter II skill. See Chapter 26, *Vehicle Rescue and Extrication*, for more information about this.) As you prepare to approach a vehicle fire, make sure that you have created a safe perimeter around the vehicle, and remove all bystanders from the area.

It is often difficult to differentiate conventional vehicles and alternative-fuel vehicles from a distance. Do not assume that because a vehicle doesn't look modern that it isn't an electric-drive vehicle or some other type of alternative-fuel vehicle. Consumers can now modify and retrofit their vehicles to convert them from conventional to alternative-fuel vehicles (**FIGURE 14-19**). For this reason, it is recommended that firefighters and rescuers approach all vehicles involved in a fire or an MVA as if they are alternative-fuel vehicles until proven otherwise.

FIGURE 14-18 Traffic cones or flares can be placed to direct motorists away from an incident involving a vehicle.

Courtesy of Captain David Jackson, Saginaw Township Fire Department.

FIGURE 14-19 A conventional vehicle can be converted to a battery electric vehicle. Do not assume that a vehicle is not an alternative-fuel vehicle because it doesn't look modern.

Courtesy of Aaron Miranda.

Use a hose line at least 1½ in. (38-mm) in diameter. This size hose will provide sufficient cooling power to suppress the fire and provide protection from a sudden flare-up. Use an adjustable fog-stream nozzle so that you can start with a straight stream from a distance, then as you approach the car, you can switch to a narrow fog stream.

Charge the hose line from a safe location, and slowly open the nozzle to bleed the air from the hose. With the nozzle adjusted to deliver a straight stream, start applying water from as far away as possible, approaching the vehicle from an uphill and upwind position, and moving in at a 45-degree angle (**FIGURE 14-20**). Approaching at a 45-degree angle helps prevent being hit or run over by inadvertent vehicle movement. Also, some vehicles have compressed gas pistons in the front and rear bumpers that may explode in a fire and approaching at an angle moves you out of the direct line of fire.

Open the nozzle and sweep the ground and the bottom part of the vehicle using a horizontal motion. Extinguish all visible fire while advancing toward the vehicle. Sweeping along the undercarriage helps to cool the bumper pistons, shock absorbers, and hydraulic struts; cool the tires before they explode; and cool the fuel tank before it fails. Remember when fighting an electric vehicle fire to spray water under the vehicle on the high-voltage battery. By applying a sufficient quantity of water to the lower part of the vehicle, you reduce the chance of an explosive event.

As you approach the vehicle, observe the area under the car during the approach for any sign of leaking flammable liquids. Adjust the nozzle to a narrow fog pattern as you get closer to the vehicle, and then continue to widen the fog pattern as you approach the vehicle. MVAs that pose large fire hazards or actual fires may require additional fire suppression resources, which should be requested as soon as possible. This is especially true with lithium-ion battery fires, which will require large amounts of water to suppress the fire and keep the temperature within the battery down to stop the propagation process. Some departments use compressed air foam or Class B foam hose lines to attack vehicle fires or if a large quantity of fuel is burning. See Chapter 23, *Advanced Fire Suppression*, for more information about using Class B extinguishing agents.

SAFETY TIP

Some vehicle engines or bodies may be constructed from magnesium or other flammable or explosive metals that can react violently when water is applied during fire suppression. If possible, a Class D extinguishing agent should be used in these cases instead of water.

SAFETY TIP

As soon as it is safe to do so, chock the wheels to prevent the vehicle from moving. You can chock a vehicle that is on fire without touching the vehicle at all. This is important because vehicles can move while a vehicle fire is attacked, and the rolling vehicle that is on fire can hit other vehicles or hit and pin firefighters.

Once this sweep of the exterior is complete, firefighters can begin to attack fire in the passenger area, the engine compartment, and the cargo compartment.

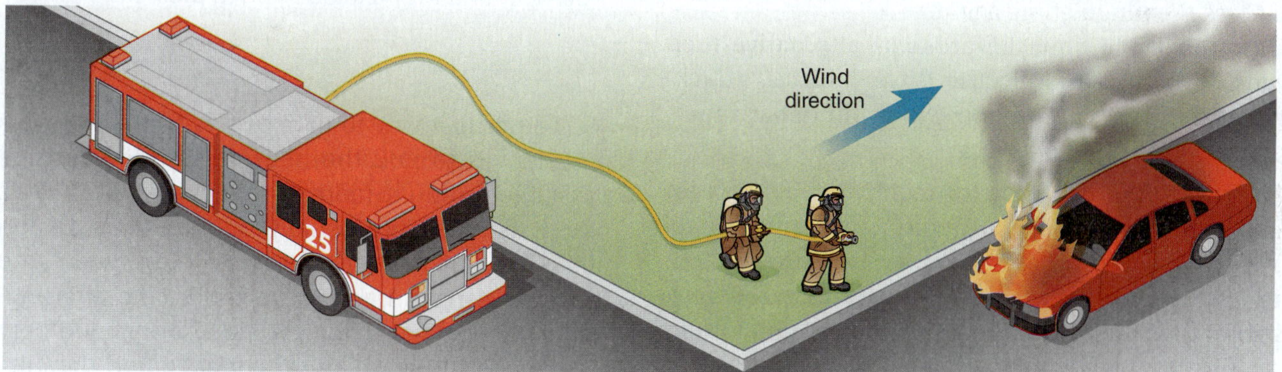

FIGURE 14-20 When possible, approach vehicle fires in the hood or engine area at a 45-degree angle from the uphill and upwind side.

Fire in the Passenger Area

Fires in the passenger area of a vehicle are more visible and accessible than fires in the engine compartment or the cargo area. Usually, it is logical to extinguish the fire in the passenger compartment before moving on to extinguish any fire in the engine and cargo compartments. Often the windows are broken or a door is ajar, presenting an opening through which firefighters can direct a stream of water. If the doors will not open, stand upwind from the window, and use a striking tool to break out one or more windows. Be cautious because a backdraft may occur when the car is vented.

As you get closer to the vehicle, change the nozzle to produce a wider pattern that will cool a wider area and give you some protection from the heat of the fire. Pay special attention to cooling areas such as the steering column and the dashboard on the passenger side. Also cool areas that contain side-curtain air bags. Cooling these areas will greatly reduce the chance for accidental deployment of the SRS. Electric vehicles require special attention. You need to spray water under the vehicle on the high-voltage battery to keep it cool. In addition, you might need to secure a water supply to sustain continuous operations for a long period of time.

Fire in the Engine Compartment

The engine compartment of a vehicle is filled with a variety of devices that use petroleum products to power or lubricate them. It also contains components made from plastics and rubber. As a consequence, these devices produce a large amount of smoke when they burn. Other hazards present in the engine compartment include suspension struts and hydraulic lift cylinders for the hood. The 12-volt battery in vehicles contains sulfuric acid, which can cause serious burns. Some vehicles have magnesium parts within the engine compartment; when magnesium is present, you must use a Class D extinguishing agent.

Be aware of additional hazards with alternative-fuel vehicles. In hybrid and electric drive vehicles, attack the fire as normal in accordance with your department SOPs. Be aware of the following:

- Hydrogen fuel cell vehicles require firefighters to use a thermal imager to determine the presence of fire. Hydrogen fires should not be extinguished unless the flow of gas can be stopped. Hydrogen is lighter than air. In fires fed by gas that is lighter than air, directing water at the gas-fed flame and extinguishing the gas-fed fire before the flow is shut off creates an even more hazardous situation. Because the escaping gas is no longer burning, it will collect in the atmosphere until an ignition source reignites the gas, creating an explosion with potentially devastating consequences.
- CNG vehicles require special care when they are involved in fire. They must be approached cautiously depending on the size of the fire and if it is close to the cylinder location.
- LNG vehicles can be attacked using normal tactics unless the fire is being fueled by an active leak. If the fire is being fueled by an active leak, let it burn and protect exposures.
- LPG vehicles can be attacked using normal tactics unless the fire is impinging on the tanks. If that is the case, apply copious amounts of water for cooling the tanks to prevent a BLEVE.

One of the challenges of extinguishing a fire in the engine compartment is gaining access to the fire. An initial attack can be made through the wheel well. The plastic liner between the engine compartment and the wheel well is often consumed during such a fire, so it may be possible to spray water into the engine compartment through this opening. An alternative initial approach is to spray water through the grille after thoroughly cooling the area around the bumper. Although neither of these two methods is effective at totally extinguishing the fire, they will help to diminish the volume of fire while firefighters are gaining access to the engine compartment.

SAFETY TIP

When spraying water through the grille, stand at a 45-degree angle. Avoid standing directly in front of the bumper until this area is thoroughly cooled in case the vehicle has compressed gas pistons in the bumper and they explode due to the heat.

The fastest way to gain access to the engine compartment is to pull the hood-release lever inside the passenger compartment to open the hood. Unfortunately, because fire usually damages the cable connected to this lever, pulling it will not work, so other methods usually need to be tried. One alternative is to insert a pry bar along the side of the hood between the edge of the hood and the fender, and then pry the side of the hood away from the fender to produce an opening big enough to apply a stream of water into the engine compartment. Another alternative is to break out the plastic grille, locate the hood-release cable, and then do one of the following actions:

- With a gloved hand, pull on the cable to release the hood.
- Use the forked end of a Halligan tool to twist the cable until the hood releases.
- Use pliers to grasp and pull the cable.
- Use a hood release tool and wrap the cable around the tool till the hood releases.

Once the hood is raised, firefighters should have good access to any remaining fire and hot spots. A firefighter should be ready in full PPE and SCBA with a charged hose line before the hood is raised. Use plenty of water or dry chemical extinguishing agents to cool this area.

Fire in the Cargo Area (Trunk)

A fire in the trunk of a vehicle presents unknown hazards because it is not possible to know what the trunk contains. Also, if the vehicle is an electric-drive vehicle, the high-voltage battery may be located in the trunk. In some vehicles, the 12-volt battery is located in the trunk. If the vehicle is a compressed gas–powered vehicle, the CNG or propane cylinders might be located in the trunk.

Fires in this area are challenging to access. Initial access can sometimes be made by knocking out the taillight assembly on one side, enabling firefighters to direct a hose stream into the trunk to cool down and partially extinguish fire in this area. A fire in the trunk area of an automobile can also be accessed by using the pick of a Halligan tool to force the lock into the trunk, and then using a screwdriver or key tool (K-tool lock puller) to turn the lock cylinder in a clockwise direction. A charged hose line must be ready when the trunk lid is raised.

Fires in the rear of light trucks and vans must always be approached cautiously. Vans are often used by couriers and could contain medical waste, laboratory specimens, or radioactive material.

To extinguish a vehicle fire, follow the steps in **SKILL DRILL 14-14**.

Overhauling Vehicle Fires

Overhaul is the process of finding, exposing, and suppressing any smoldering or hidden pockets of fire in an area that has been burned. Overhaul of vehicle fires is just as important as overhaul of structure fires. If the type of vehicle has not already been identified, determine whether the vehicle is a conventional vehicle or an alternative-fuel vehicle by checking the vehicle for markings that identify it as an alternative-fuel vehicle. As soon as it is safe to approach the vehicle, chock the wheels to prevent the vehicle from moving. If a vehicle fire erupted quickly, the driver may not have had time to set the brake and parking gear before exiting the vehicle. Also, fire can damage cables and wires that control the operation of parking and braking mechanisms. After all visible fire has been knocked down and the vehicle type has been identified and the vehicle is immobilized and disabled, allow a few minutes for the steam and smoke to dissipate before starting overhaul. This delay will allow visibility to improve so that overhaul can be completed safely.

SKILL DRILL 14-14

Extinguishing a Vehicle Fire Firefighter I, NFPA 1010: 6.3.7

1. Don full PPE, including SCBA; enter the personnel accountability system; and work as a team. Perform size-up, and give an arrival report. Call for additional resources if needed. Ensure that apparatus is positioned uphill and upwind of the fire and that it protects the scene from traffic.

Continues.

SKILL DRILL 14-14 CONTINUED

Extinguishing a Vehicle Fire Firefighter I, NFPA 1010: 6.3.7

2. Deploy an appropriate attack line (at least 1½ in. [38-mm] in diameter).

3. Open the nozzle to purge air from the system, and make sure water is flowing. If using an adjustable nozzle, ensure that it is set to the proper nozzle pattern. Shut down the nozzle until you are in a position to apply water.

4. Direct the crew to attack the fire in a safe manner. Attack from uphill and upwind of the fire and at a 45-degree angle and extinguish any fire under the vehicle. Observe the area under the car during the approach for any sign of leaking flammable liquids.

SKILL DRILL 14-14 CONTINUED

Extinguishing a Vehicle Fire Firefighter I, NFPA 1010: 6.3.7

5. Carefully approach the vehicle, and completely suppress the fire. Adjust the nozzle to a narrow fog pattern as you get closer to the vehicle and continue to widen the fog pattern as you approach the vehicle.

6. Overhaul all areas of the vehicle, including the passenger compartment, engine compartment, and cargo area (trunk).

7. Notify command when the fire is under control. Identify obvious signs of the origin and cause of the fire. Preserve any evidence of arson. Return the equipment and crew to service.

SAFETY TIP

During overhaul of interior fires, remember that air bags can deploy without warning in a burned automobile. Never place any part of your body in the path of a front or side air bag.

As you overhaul the vehicle, be systematic and thorough. All areas of the vehicle must be checked for fire extension and for potential victims—there have been numerous cases of vehicle fires that were used to cover a crime. Do not miss areas that may contain lingering sparks. Direct the hose stream under the dashboard, and soak and remove smoldering upholstery. Apply water over and under all parts of the engine compartment. Confirm that no fluids are leaking from the vehicle. If contents in the car might potentially be salvaged, treat them with respect. Continue to use your SCBA as long as smoke or fumes are present.

The scene needs to be stabilized by reducing, removing, or mitigating the hazards at the scene. This is done to make it safe for firefighters and rescuers to operate prior to working on a vehicle involved in an MVA. The order in which these hazards are addressed depends on the conditions at a scene and the amount of risk that each hazard poses. Generally, you should do the following:

1. Identify the type of vehicle.
2. Immobilize the vehicle.
3. Disable the vehicle.

The National Fire Protection Association (NFPA) created a quick reference guide describing best practices for managing the hazards associated with electric-drive vehicles. The information on the first page of this guide can be applied to conventional vehicles as well. If victims need to be extricated or if the vehicle is in danger of tipping, it also needs to be stabilized. For information on stabilizing a vehicle and extricating victims, see Chapter 26, *Vehicle Rescue and Extrication.*

Identify

First conduct an initial scene assessment to identify the type of drive system—that is, conventional or alternative-fuel—so that the potential hazards may be identified. To positively identify alternative-fuel vehicles, search for formal marking such as identification badges, emblems, and labels, and look for manufacturer design features used solely for alternative-fuel vehicles or models that are available only as an alternative-fuel vehicle such as Tesla and Toyota Prius, among others.

FIGURE 14-21 While labels can identify alternative-fuel vehicles, do not solely rely upon them when assessing the potential hazards of a scene.

However, do not rely only on badges, emblems, and labels to identify alternative-fuel vehicles (**FIGURE 14-21**). The identifiers might not be visible, or they might have fallen or gotten destroyed during the incident. Look for informal clues to identify alternative-fuel vehicles such as the following:

- High-voltage (orange) and medium/intermediate voltage (blue or yellow) electrical cables
- Battery vents: High-voltage battery packs produce a lot of heat and require adequate ventilation. Vents near the battery prevent overheating under normal conditions by keeping airflow around the case.
- Charging ports: BEVs and PHEVs have a charging port to connect to an external power source. It has a cover similar to a gas cap and is usually located at the front of the vehicle or in the front or rear fenders (**FIGURE 14-22**).
- Information on the instrument panel

Immobilize

After identifying the vehicle type, the next step is to immobilize the vehicle by chocking the wheels, setting the emergency brake, and putting it in park.

As stated before, to prevent being hit or run over by inadvertent vehicle movement, approach every vehicle at a 45-degree angle or from the side whenever possible until the vehicle has been immobilized or positively shut down. As soon as it is deemed safe to touch the vehicle, deploy wheel chocks in front and behind one or more wheels. Wheel chocks should be

FIGURE 14-22 Charging port on a BEV.

Courtesy of Aaron Miranda.

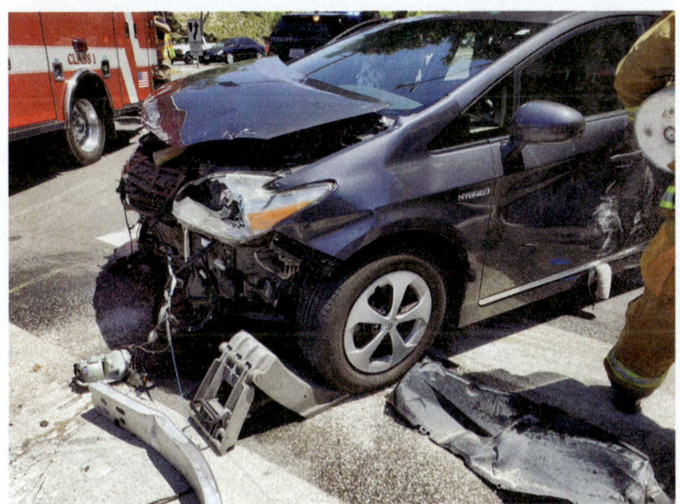

FIGURE 14-23 Place wheel chocks in front and behind one or more wheels to prevent forward and rearward vehicle movement.

Courtesy of Aaron Miranda.

strategically placed to prevent forward and rearward movement but not interfere with stabilization efforts (**FIGURE 14-23**).

The next step in the immobilization efforts is to place the vehicle in "park" and to engage the parking brake. If the driver is still inside the vehicle and is able to follow commands, ask them to perform these tasks. Some vehicles have joystick shifters that return to the same position regardless of the gear selected or a push-button electronic shifter. Also, some vehicles have electronically operated parking brakes. This is why it is important to shift into park and engage the parking brake at this stage prior to disabling the low-voltage electrical system (**FIGURE 14-24**).

FIGURE 14-24 Some vehicles have push-button shifters and electric parking brakes. Make sure you engage the parking brake before disabling the low-voltage electric system.

Courtesy of Aaron Miranda.

Disable

After immobilizing the vehicle, the next step is to disable it by turning the power off, and then disconnecting the 12-volt battery. Before disabling the vehicle, and if needed, roll down windows, move seats, unlock the doors, and open the trunk, as these systems will not work once the low-voltage power has been removed.

In some vehicles, a "READY" or "AUTO STOP" light, or a green car symbol on the instrument panel can indicate that the vehicle is on, and that, therefore, when placed in gear, the vehicle can move (**FIGURE 14-25**). Again, if the driver is still inside the vehicle and is able to follow commands, ask them to turn off the motor and hand you the key. If you need to perform these tasks, turn the key and remove it if it is a traditional key ignition, or push the start button and remove the proximity key from the vehicle if it can be located.

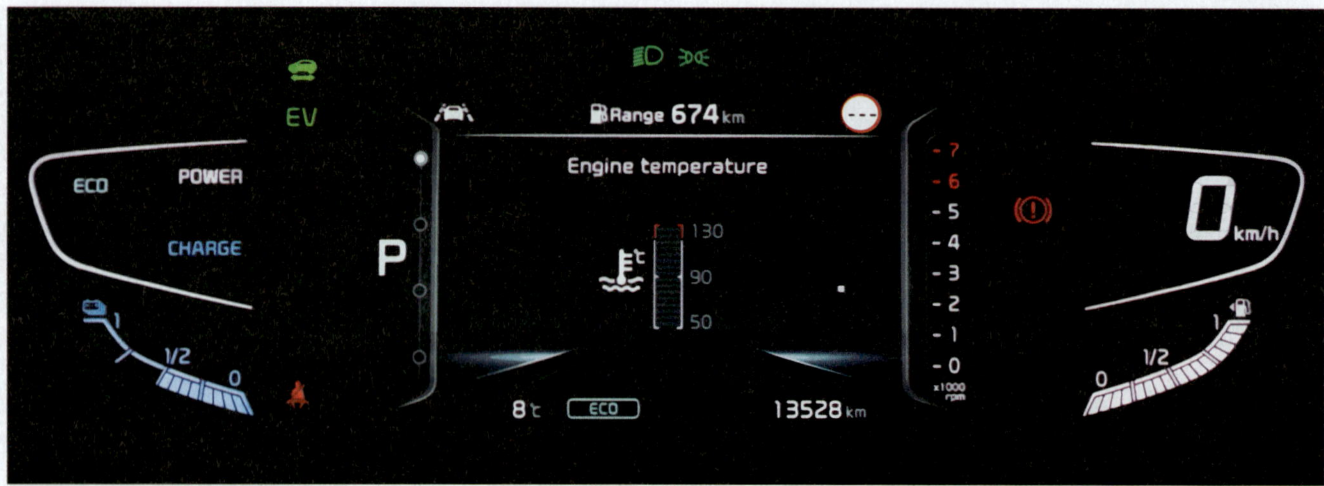

FIGURE 14-25 Instrument panel indicating that the vehicle is not shut down.

The next step is to disconnect the 12-volt battery by cutting its cables. This ensures that nothing is powered accidentally, and in electric-drive vehicles, isolates the high-voltage battery system. Although shutting off the ignition in electric-drive vehicles disengages the high-voltage systems, including the high-voltage battery, disconnecting the low-voltage battery prevents the high-voltage system from accidentally reenergizing. Disconnecting the 12-volt battery also disables the occupant SRS—that is, the air bags. Some vehicles have a disconnect switch or fuses that can be removed to disconnect the motors from the power source.

Usually, the 12-volt battery is located in the engine compartment. Open the hood if it has not already been opened, and then verify that any fire in the engine compartment has been extinguished. If the battery is not under the hood, check the wheel wells, under the back seat, under the floor board, or in the trunk. To disconnect the power to the 12-volt battery, remove a section at least 6 in. (15 centimeter [cm]) long of the cable attached to the negative terminal by making two cuts with wire cutters. This is usually the black cable. The cable to the negative terminal is cut first to avoid producing any sparks. Removing a section instead of just making one cut prevents the two ends from accidentally touching. Then, remove a similar section of the cable attached to the positive terminal—this is usually the red cable. If the car is an electric-drive vehicle, do not cut the high- or medium/intermediate voltage cables that connect the high-voltage battery to the electric motor.

SAFETY TIP

Make sure to perform this step for all 12-volt batteries in vehicles with multiple batteries.

SAFETY TIP

Regardless of how the vehicle was shut down and disabled, remember that the high-voltage battery will still retain its charge. Always treat high-voltage components as if they are still energized.

To immobilize and disable the electrical system of conventional and electric-drive vehicles, follow the steps in **SKILL DRILL 14-15**.

Terminating a Vehicle Fire Incident

Terminating a vehicle fire incident includes securing the scene by removing the damaged vehicle and equipment and ensuring the scene is left in a safe condition. Law enforcement will secure the scene by conducting their own on-scene investigation. If the incident is a potential crime scene, law enforcement will manage the scene, preserve and secure any evidence, and close off the scene or roadway. Once a scene has been released by law enforcement, the removal of vehicles can be coordinated with a tow agency.

The high-voltage batteries in electric-drive vehicles may cause a delayed fire, especially if the high-voltage battery sustained direct physical damage. Firefighters should be proactive and stand by with a charged hose line. Be aware of smoke, sparks, irritating fumes, or a popping or gurgling sound coming from the battery.

SKILL DRILL 14-15

Immobilize and Disable a Vehicle Following a Vehicle Fire or a Motor Vehicle Accident NPFA 1010: 6.3.7

1. Immobilize the vehicle by chocking the wheels. Chock both sides of one tire to prevent the vehicle from rolling by placing one wheel chock in front of a wheel and a second wheel chock in back of the wheel. If possible, chock more than one wheel. If you can access the controls, set the parking brake and place the vehicle in park.
2. If possible, lower all automatic windows, move seats, unlock the doors, and open the trunk before shutting the motor off and disabling the electrical system. Shut the motor off by turning the key and removing it if it is a traditional key ignition or by pushing the start button and removing the proximity key (if possible). If you locate the proximity key, move it as far away from the vehicle as possible—at least 16 ft (5 m)—to prevent the possibility of an unintentional restart.
3. Determine the location of the 12-volt, low-voltage battery. Whenever possible, remove a section at least 6 in. (15 cm) long of the cable attached to the negative terminal (the black cable) by making two cuts with wire cutters, and then repeat this with the cable attached to the positive terminal (the red cable).

Courtesy of Aaron Miranda.

These are warning signs of a damaged high-voltage battery that is undergoing a thermal event and is at risk for fire, hours or even days later. In addition, there is a risk of toxic off-gassing. If personnel detect any of these warning signs, the vehicle must be ventilated immediately by opening all doors, rolling down or breaking the windows, and opening the cargo compartment to prevent gas build-up and a battery fire.

Any fluid hazards that have spilled from the vehicle, such as gasoline, motor oil, transmission fluid, or radiator fluid, will be removed by the towing agency. Towing agencies must be licensed to transport and dispose of any hazardous substances; fire rescue agencies are typically not licensed to do so. Ask the tow agency representative if you may assist the towing agency by placing any vehicle parts, such as doors, roofs, and fenders, back into a heavily damaged vehicle. The towing agency may have their own SOPs for stowing and transporting loose objects. Being considerate of their preferences will help to maintain a good working relationship with the towing agency for future endeavors such as acquiring vehicles for use in an extrication class. It is advisable to carry a contact list of various private and public organizations or businesses that can offer a particular resource to be utilized on the incident, such as public works/utilities, the department of transportation (DOT), or a heavy equipment company.

Electric-drive vehicles should be towed using a flatbed truck. Towing them with the drive wheels on the ground could result in an electrical fire. If law enforcement officers take command of the scene and take possession of the vehicle to conduct their investigation, brief them on the hazards associated with electric-drive vehicles.

After you have secured the scene and packed your equipment, return to the station and fully inventory, clean, service, and maintain all of the equipment per the manufacturer's instructions and your department's SOPs to prepare it for the next call. Some items will need repair, but most will need simple maintenance before they are placed back on the apparatus and considered in service.

TIP

Organizations such as the Energy Security Agency (ESA) offer on-scene electric-drive vehicle support free of charge to first and second responders. The ESA can perform risk analysis services and provide guidance to make the vehicle safe for release. They will also recommend isolation procedures if the vehicle is at high risk for thermal runaway. Visit their website at energysecurityagency.com for more information.

CASE STUDY

You Are the Firefighter CONCLUSION

This is it—your first shift at the fire house after graduating from the fire academy. You are preparing lunch for your new crew as the fire house tradition goes for your first shift. Just as you are preparing the plates for your crew members, the alarm sounds for a house fire in your fire house's first-due district. You turn off the stove and cover the food then quickly slip on your gear before climbing into the cab of the fire engine. Your heart is racing as the other crew members mount the engine like it is just another day at the office. As you get close to the scene, you can see dark black smoke off in the distance: The incident is declared a working fire by on-scene command.

1. **What is the difference between an offensive operation and a defensive operation?**

   Answer: The main difference is that offensive fire operations are typically done from inside a building or close to the fire to perform a direct attack and defensive fire operations are typically conducted from the exterior of the building or from a safe position away from the fire. An offensive operation or an offensive attack typically exposes firefighters to the heat and smoke of the fire inside the building as well as to other risk factors, such as the possibility of being trapped by a structural collapse. A defensive operation or defensive attack is implemented in situations where the fire is heavily involved and in situations where the level of risk to firefighters makes an interior, offensive attack unacceptable.

2. **When would you use an indirect attack rather than a direct attack?**

   Answer: Indirect application of water is used in situations where the interior temperature is increasing and it appears that the room or space is ready to flash over, potentially causing injury or death to firefighters. An indirect attack can be accomplished from a safer location, such as an adjoining room, doorway, or hallway, to reduce the thermal exposure to operating firefighters.

3. **Which type of nozzle would you use for an indirect attack?**

   Answer: A straight or solid stream flowing between 120 to 180 gpm (454 and 681 L/min), which will cool everything quickly without generating significant amounts of steam.

WRAP-UP

SUMMARY

KNOWLEDGE OBJECTIVES

- Describe the objectives of defensive and offensive operations and methods of direct, indirect, combination, and transitional fire attack strategies.
  - Describe the objectives of a defensive operation. (**NFPA 1010: 6.3.10**, p. 541)
  - Describe the operations performed during a defensive operation. (**NFPA 1010: 6.3.10**, p. 541)
  - Describe the objectives of an offensive operation. (**NFPA 1010: 6.3.10**, pp. 541–542)
  - Describe the operations performed during an offensive operation. (**NFPA 1010: 6.3.10**, pp. 541–542)
  - Describe the objectives of a direct attack. (**NFPA 1010: 6.3.10**, p. 542)
  - Describe the objectives of an indirect attack. (**NFPA 1010: 6.3.10**, p. 542)
  - Describe the objectives of a combination attack. (**NFPA 1010: 6.3.10**, p. 546)
  - Describe the objectives of a transitional attack. (**NFPA 1010: 6.3.10**, pp. 548–550)
  - Describe the operations performed during a transitional attack. (**NFPA 1010: 6.3.10**, pp. 548–553)
- Describe the characteristics of handlines and master stream appliances.
  - Describe the techniques used to operate large handlines. (**NFPA 1010: 6.3.8**, pp. 553–556)
  - Describe the characteristics of a master stream appliance. (**NFPA 1010: 6.3.8**, pp. 556–559)
  - Describe the characteristics of a portable monitor. (**NFPA 1010: 6.3.8**, pp. 557–558)
  - Describe the characteristics of a deck gun. (**NFPA 1010: 6.3.8**, p. 558)
  - Describe the characteristics of elevated master stream appliances. (**NFPA 1010: 6.3.8**, p. 558)
- Explain when you should shut off building utilities and how to do this.
  - Describe when gas service should be shut off. (**NFPA 1010: 6.3.18**, pp. 559–560)
  - Describe when the electrical service should be shut off. (**NFPA 1010: 6.3.18**, pp. 560–561)
  - Describe when water service should be shut off. (**NFPA 1010: 6.3.18**, p. 561)
- Describe specific fireground operations for different circumstances.
  - Describe the tactics used to protect exposures. (**NFPA 1010: 6.3.8**, pp. 561–562)
  - Describe the characteristics of concealed-space fires. (**NFPA 1010: 6.3.10**, p. 562)
  - Describe the tactics used to suppress concealed-space fires. (**NFPA 1010: 6.3.10**, p. 562)
  - Describe the characteristics of basement fires. (**NFPA 1010: 6.3.10**, pp. 562–564)
  - Describe the tactics used to suppress basement fires. (**NFPA 1010: 6.3.10**, pp. 562–564)
  - Describe the characteristics of fires above ground level. (**NFPA 1010: 6.3.10**, p. 564)
  - Describe the tactics used to suppress fires above ground level. (**NFPA 1010: 6.3.10**, p. 564)
  - Describe the characteristics of attic fires. **NFPA 1010: 6.3.10**, pp. 564–566)
  - Describe the tactics used to suppress attic fires. (**NFPA 1010: 6.3.10**, pp. 564–566)
  - Describe the characteristics of fires in large buildings. (**NFPA 1010: 6.3.10**, pp. 566–567)
  - Describe the tactics used to suppress fires in large buildings. (**NFPA 1010: 6.3.10**, pp. 566–567)
  - Describe the characteristics of fires in buildings under construction, renovation, or demolition. (**NFPA 1010: 6.3.10**, p. 567)
  - Describe the tactics used to suppress fires in buildings under construction, renovation, or demolition. (**NFPA 1010: 6.3.10**, p. 567)
  - Describe the characteristics of fires in confined spaces. (**NFPA 1010: 6.3.8**, p. 567)
  - Describe the tactics of gas fire suppression. (p. 567)
  - Describe the hazards posed by electrical fires. (**NFPA 1010: 6.3.18**, pp. 567–568)
  - Describe the tactics used to suppress an electrical fire. (pp. 567–568)
  - Describe the characteristics of fires in stacked or piled materials. (**NFPA 1010: 6.3.8**, p. 568)
  - Describe the tactics used to suppress fires in stacked or piled materials. (**NFPA 1010: 6.3.8**, p. 568)

KNOWLEDGE OBJECTIVES CONTINUED

- Describe the characteristics of fires in lumberyards. (**NFPA 1010: 6.3.8**, p. 568)
- Describe the tactics used to suppress fires in lumberyards. (**NFPA 1010: 6.3.8**, p. 568)
- Describe the characteristics of fires in trash containers. (**NFPA 1010: 6.3.8**, pp. 569–570)
- Describe the tactics used to suppress fires in trash containers. (**NFPA 1010: 6.3.8**, pp. 569–570)
- Describe the characteristics of chimney fires. (**NFPA 1010: 6.3.10**, pp. 570–571)
- Describe the tactics used to suppress fires in chimneys. (**NFPA 1010: 6.3.10**, pp. 570–571)
- Describe the characteristics of fires on buildings with solar photovoltaic systems. (pp. 571–572)
- Describe the tactics used to suppress fires on buildings with solar photovoltaic systems. (pp. 571–572)
- Describe the hazards of lithium batteries. (p. 572)

- Explain how to identify the origin and cause of fires.
  - Describe the importance of preserving evidence. (**NFPA 1010: 6.3.8**, p. 572)
- Describe the types of motor vehicles, characteristics of vehicle fires, the tactics used to suppress vehicle fires and how to overhaul a vehicle fire.
  - Describe the characteristics of vehicle fires. (**NFPA 1010: 6.3.7**, p. 572)
  - Describe the types of motor vehicles. (**NFPA 1010: 6.3.7**, pp. 573–577)
  - Describe the different types of alternative fuels that power motor vehicles. (**NFPA 1010: 6.3.7**, pp. 573–577)
  - Describe the tactics used to suppress vehicle fires. (**NFPA 1010: 6.3.7**, pp. 578–579)
  - Describe the tactics used to suppress fires in the passenger area of a vehicle. (**NFPA 1010: 6.3.7**, p. 580)
  - Describe the tactics used to suppress fires in the engine compartment of a vehicle. (**NFPA 1010: 6.3.7**, pp. 580–581)
  - Describe the tactics used to suppress fires in the trunk of a vehicle. (**NFPA 1010: 6.3.7**, p. 581)
  - Describe how to overhaul a vehicle fire. (**NFPA 1010: 6.3.7**, pp. 581–587)

SKILLS OBJECTIVES

- Perform an offensive attack.
  - Prepare to perform an interior, offensive attack. (**NFPA 1010: 6.3.10**, p. 543).
  - Perform a direct attack. (**NFPA 1010: 6.3.10**, p. 544)
  - Perform an indirect attack. (**NFPA 1010: 6.3.10**, p. 545)
  - Perform a combination attack. (**NFPA 1010: 6.3.10**, pp. 546–547)
  - Perform a transitional attack. (**NFPA 1010: 6.3.10**, pp. 551–552)
- Perform the one- and two-firefighter method for operating a large handline.
  - Perform the one-firefighter method for operating a large handline. (**NFPA 1010: 6.3.8**, pp. 555–556)
  - Perform the two-firefighter method for operating a large handline. **NFPA 1010: 6.3.8**, p. 557)
- Operate a portable monitor and a deck gun.
  - Operate a deck gun. (**NFPA 1010: 6.3.8**, p. 559)
  - Deploy and operate a portable monitor. (**NFPA 1010: 6.3.8**, p. 558)
- Shut off gas and electric utilities.
  - Shut off gas utilities. (**NFPA 1010: 6.3.18**, p. 560)
  - Shut off electric utilities. (**NFPA 1010: 6.3.18**, p. 561)
- Locate and suppress concealed-space fires and extinguish an outside Class A fire.
  - Locate and suppress concealed-space fires. (**NFPA 1010: 6.3.10**, p. 563)
  - Extinguish an outside trash fire or other outside Class A fire. (**NFPA 1010: 6.3.10**, pp. 569–570)
- Extinguish a vehicle fire and immobilize and disable a vehicle following a vehicle fire or motor vehicle accident.
  - Extinguish a vehicle fire. (**NFPA 1010: 6.3.7**, pp. 581–583)
  - Immobilize a vehicle. (**NFPA 1010: 6.3.7**, p. 587)

KEY TERMS

alternative-fuel vehicles A vehicle that uses anything other than a petroleum-based motor fuel (gasoline or diesel fuel) to propel a motorized vehicle.

area of origin A structure, part of a structure, or general geographic location within a fire scene, in which the "point of origin" of a fire or explosion is reasonably believed to be located. (NFPA 921)

battery electric vehicle (BEV) A vehicle propelled solely by an electric motor that is powered by a high-voltage battery.

blended liquid fuel–powered vehicle A vehicle powered by a blend of liquid fuels.

blitz attack See *transitional attack.*

collapse zone The area that is exposed to trauma, debris, and/or thrust should a building or part of a building collapse. (NFPA 1550)

combination attack An indirect attack followed by a direct attack.

compressed gas–powered vehicle A vehicle powered by one of three forms of compressed gas (compressed natural gas (CNG), liquefied natural gas (LNG), and liquefied petroleum gas (LPG or propane).

conventional vehicles Vehicles that use internal combustion engines for power.

deep-seated fire A fire burning below the surface.

defensive attack See *defensive operation.*

defensive operation Actions that are intended to control a fire by limiting its spread to a defined area, avoiding the commitment of personnel and equipment to dangerous areas. Also called *defensive attack* or *defensive strategy.* (NFPA 1550)

defensive strategy See *defensive operation.*

direct attack Firefighting operations involving the application of extinguishing agents directly onto the burning fuel. (NFPA 1145)

electric-drive vehicle A vehicle that is powered solely by a high-voltage battery or by a combination of a high-voltage battery and a liquid fuel.

exterior attack A fire attack conducted from outside a structure.

fire attack The process of cooling the fire gases and suppressing a fire.

fire department connection (FDC) The hose coupling or couplings that connect to an automatic sprinkler system.

fire extension The spreading of the fire to adjacent areas from the seat of the fire.

fire suppression The tactics and tasks performed at a fire scene to stop the combustion process and achieve the final goal of extinguishing the fire.

flow path The route within a structure along which hot gases and smoke travel from areas of higher pressure within the fire to areas of lower pressure accessible through doorways, window openings, and roof.

fuel cell–powered vehicle A vehicle powered by fuel cells.

fuel cells Cells that generate electricity through a chemical reaction between hydrogen gas and oxygen gas to produce water and in the process produce electricity to propel a vehicle.

fugitive gas Gaseous fuel that escapes from the system.

ground cover fire A fire that burns loose debris on the surface of the ground.

heat flux The measure of the rate of heat transfer to or from one surface to another.

hybrid electric vehicle (HEV) A vehicles that use both a battery-powered electric motor and a liquid-fueled engine to propel a vehicle.

indirect attack Firefighting operations involving the application of extinguishing agents to reduce the build-up of heat released from a fire without applying the agent directly onto the burning fuel. (NFPA 1145)

interior attack A fire attack conducted inside a structure.

knock down To attack the fire to the point where it is almost extinguished and it is unlikely to extend further.

offensive attack See *offensive operation.*

offensive operation Actions generally performed in the interior of involved structures that involve a direct attack on a fire to directly control and extinguish the fire. Also called *offensive attack* or *offensive strategy.* (NFPA 1550)

offensive strategy See *offensive operation.*

KEY TERMS CONTINUED

overhaul The process of final extinguishment after the main body of a fire has been knocked down. All traces of fire must be extinguished at this time. (NFPA 440).

plug-in hybrid electric vehicle (PHEV) A vehicle that uses both a battery-powered electric motor and a liquid-fueled engine to propel the vehicle and whose battery can be recharged by connecting it to an external power source.

point of origin The physical location within the area of origin where a heat source, a fuel, and an oxidizing agent first interact, resulting in a fire or explosion. (NFPA 921)

preincident plan Data about a location in your community that describes the layout and hazards at the location and identifies response essentials, such as the type of construction, occupancy type, fire control and suppression systems, and potential issues with access and egress.

rapid intervention company (RIC) See *rapid intervention crew (RIC).*

rapid intervention crew (RIC) A dedicated crew of at least one officer and three members, positioned outside the immediately dangerous to life or health area, trained and equipped as specified in NFPA 1407, who are assigned for rapid deployment to rescue lost or trapped member. Also called *rapid intervention crew (RIC)* or *rapid intervention team (RIT).* (NFPA 1700)

rapid intervention team (RIT) See *rapid intervention crew (RIC).*

resetting the fire See *transitional attack.*

search line See *search rope.*

search rope A rope with an anchor point to keep firefighters in contact with each other and with the egress route from a burning building during an interior attack or a search and rescue operation. Also called *search line.*

softening the target See *transitional attack.*

solar panels A general term for thermal collectors or photovoltaic (PV) modules. (NFPA 780)

solar photovoltaic (PV) system A power system designed to convert solar energy into electrical energy; may also be referred to as a PV system or a solar power system.

thermal runaway In a high-voltage battery, the condition when the temperature in the cell continues to rise, producing gas, and eventually igniting.

transitional attack An offensive fire attack initiated by an exterior, indirect handline operation into the fire compartment to initiate cooling while transitioning into interior direct fire attack in coordination with ventilation operations. Also called *blitz attack, softening the target,* and *resetting the fire.*

two-in/two-out rule A safety procedure that requires a minimum of two personnel to enter a hazardous area and a minimum of two backup personnel to remain outside the hazardous area during the initial stages of an incident.

vertical ventilation Openings made in roofs or floors so that heat, smoke, and toxic gases can escape from the structure in a vertical direction.

REVIEW QUESTIONS

1. What is the primary objective in a defensive operation?
2. What is a direct attack?
3. What type of fire attack allows you to buy time for more manpower to arrive on scene by resetting the fire?
4. What size is the most frequently used handline for interior fire attack?
5. What is the lowest safe angle for operating a portable monitor?
6. Which class of extinguishing agents should be used when energized equipment is involved in a fire?
7. Why are basements dangerous spaces to access?
8. What affects how attic fires grow and spread?
9. What are the four categories of alternate-fuel vehicles on the road today?
10. From what direction should you approach a vehicle fire?

DISCUSSION QUESTIONS

1. What is your department's go-to method for operating large handlines on a fire scene?
2. Describe your department's fire suppression tactics for attic and basement fires, and explain why they are effective.
3. At a vehicle fire, how critical is a 360-degree scene size-up and identification of the type of vehicle for your safety at a vehicle fire?
4. What is the greatest danger with most fires involving electrical equipment, and what can you do to decrease this danger at an emergency scene?

APPLYING THE CONCEPTS

Your engine company is the first unit dispatched to a fire at a single-story family dwelling in a rural area. Dispatch reported everyone exited the house safely. The cause of the fire is unknown. An initial size-up by the incident commander (IC) reveals there is heavy smoke and fire visible from the main entrance (side alpha door) of the house. No other structures are threatened by the fire and there is a fire hydrant nearby. A 10-mile per hour (16 kilometer per hour) wind is blowing from the east, hitting the left side of the structure.

1. If you were the IC, what fire suppression operations do you consider? What are the objectives of each strategy?
2. In this situation, what are the disadvantages of a defensive strategy?

The IC calls for a transitional attack and orders, "Use a 1¾ preconnected handline to attack exterior side alpha front door. We need to create a path for the firefighters to get in there and get to the fire seat. Move it!" You make sure the handline is properly flaked out with no kinks or twists in the hose. Then you advance the handline and get into position near the front door. "Ready for water," you tell the pump operator. You open the nozzle to purge air from the system and make sure the water is flowing. If using an adjustable nozzle, ensure that it is set to the proper nozzle pattern for entry. You then direct the nozzle on a straight stream through the smoke-filled doorway, deflecting off the ceiling. After 60 seconds, the IC orders your crew to enter the structure to begin a direct attack.

3. Why did you aim the water stream toward the ceiling?
4. Why is it important for this first phase of the transitional attack to happen quickly?

You and your crew proceed into the house. As you enter, the smoke is thick, but you have moderate visibility. Your crew determines the seat of the fire is in the living room, near the fireplace. You position and advance the handline to begin a direct attack on the base of the fire. Suddenly a wind shift sends a gust of wind through the open door. The fire immediately intensifies.

5. Do you continue with the direct attack on the fire to try to knock it down?
6. So, what *do* you do?

The IC gives the order, "Unit A, fall back to side alpha front door and proceed with an indirect attack."

7. What is the objective of the indirect attack? And what is the danger if it's not successful?

REFERENCES

Backstrom, Robert, and David A. Dini. 2011. *Firefighter Safety and Photovoltaic Installations Research Project.* Underwriters Laboratories (UL), November 29, 2011. Accessed June 18, 2024. https://fsri.org/research-update/firefighter-safety-and-photovoltaic-installations-research-project-released.

Energy Security Agency (ESA). n.d. *Risk Analysis and Guidance for First Responders.* Accessed May 25, 2023. https://energysecurityagency.com/hybrid-electric-vehicle-risk-analysis-for-fire-service.

Hall, Shelby, and Ben Evarts. 2022. *Fire Loss in the United States during 2021.* National Fire Protection Association (NFPA), September 2022. Accessed April 27, 2023. https://www.nfpa.org.

Kerber, Stephen, and Robin Zevotek. 2014. *Study of Residential Attic Fire Mitigation Tactics and Exterior Fire Spread Hazards on Firefighter Safety.* UL Fire Safety Research Institute (FSRI), November 26, 2014. Accessed May 8, 2023. https://d1gi3fvbl0xj2a.cloudfront.net/public/2021-07/Attic-Final-Report-Online.pdf.

Madrzykowski, Daniel, and Craig Weinschenk. 2018. *Understanding and Fighting Basement Fires.* UL Fire Safety Research Institute (FSRI), August 3, 2018. Accessed May 6, 2023. https://d1gi3fvbl0xj2a.cloudfront.net/public/2021-07/Understanding_and_Fighting_Basement_Fires.pdf.

Kerber, Stephen, and Robin Zevotek. 2014. *Study of Residential Attic Fire Mitigation Tactics and Exterior Fire Spread Hazards*

REFERENCES CONTINUED

on Firefighter Safety. UL Fire Safety Research Institute (FSRI), November 26, 2014. Accessed January 3, 2024. https://www.fstaresearch.org/resource/?FstarId=11436.

National Fire Protection Association (NFPA). 2018. *Emergency Field Guide: Hybrid, Electric, Fuel Cell, and Gaseous Fuel Vehicles.* 2018 Edition. Quincy, MA: NFPA.

National Fire Protection Association (NFPA). 2020. NFPA 1700, *Guide for Structural Fire Fighting.* 2021 Edition. Quincy, MA: NFPA.

National Fire Protection Association (NFPA). *2021. NFPA 1145, Guide for the Use of Class A Foams in Fire Fighting.* 2022 Edition. Quincy, MA: NFPA.

National Fire Protection Association (NFPA). 2022. NFPA 780, *Standard for the Installation of Lightning Protection Systems.* 2023 Edition. Quincy, MA: NFPA.

National Fire Protection Association (NFPA). 2021. NFPA 1145, *Guide for the Use of Class A Foams in Fire Fighting.* 2022 Edition. Quincy, MA: NFPA.

National Fire Protection Association (NFPA). 2023. NFPA 440, *Guide for Aircraft Rescue and Firefighting Operations and Airport/Community Emergency Planning.* 2024 Edition. Quincy, MA: NFPA.

National Fire Protection Association (NFPA). 2023. NFPA 921, *Guide for Fire and Explosion Investigations.* 2024 Edition.

National Fire Protection Association (NFPA). 2023. NFPA 1550, *Standard for Emergency Responder Health and Safety.* 2024 Edition. Quincy, MA: NFPA.

Underwriters Laboratories (UL). 2013. *Innovating Fire Attack Tactics,* Summer 2013.

CHAPTER

15

Firefighter I

Forcible Entry

KNOWLEDGE OBJECTIVES

After studying this chapter, you will be able to:

- Identify common situations where forcible entry is required in the fire service.
- Describe the building components that are utilized to force entry into a structure.
- List the rules of forcible entry.
- Describe the tools utilized to force entry into a structure.
- Describe the types of forcible entry techniques utilized to gain entry into a structure.

SKILLS OBJECTIVES

After studying this chapter, you will be able to perform the following skills:

- Force entry through a door.
- Pull a lock cylinder.
- Force entry by cutting with a rotary saw.
- Force entry through a window.
- Breach a wall.

ADDITIONAL NFPA STANDARDS

- **NFPA 80**, *Standard for Fire Doors and Other Opening Protectives, 2022 Edition*

CASE STUDY

You Are the Firefighter

There is a working fire on the second floor of a local hotel. The hotel is a three-story, wood-frame building with a hallway down the center. It was built in 1981 and renovated last year. Your crew is assigned to search the third floor for occupants who may still be in the building. As you dismount the fire apparatus, your captain tells you to grab the tools for forcible entry as the manager cannot find his master key.

1. What kind of access issues would you expect in this structure?
2. Which forcible entry tools would you bring with you?
3. How would the forcible entry challenges for this structure differ from a single-family home?

Introduction

In order to access a fire or a trapped victim, firefighters are expected to gain entry into houses and commercial buildings that are on fire but have locked or blocked doors. This often means they need to use brute force to break windows or break down doors because they need to gain immediate entry. As the modern fire service has taken on many new roles, firefighters are encountering many more situations that require them to gain access to a structure whose entrances are locked or blocked. Today's firefighters respond to a good many medical alarms, emergency medical service (EMS) calls, lift-assists for people who have fallen but are uninjured, odor investigations, as well as fire alarms. These calls might be true emergencies or they might be false alarms caused by malfunctioning equipment. Many of these situations are not as immediately life-threatening as working fires. Firefighters, therefore, are expected to tailor their efforts to ensure that the level of damage caused to the building matches the severity of the incident as reported and assessed once on scene. **Forcible entry** is the process of gaining access to a structure when the normal means of entry are locked, secured, obstructed, blocked, or unable to be used for some other reason. Techniques for forcible entry include everything from breaking down doors, potentially ruining the door frame, to pulling out the cylinder of a lockset so that only the lockset needs to be replaced.

Size-Up and Priority of Entry

Although it is typically the company officer who decides when and where to force entry and which tactics will be used, it is the responsibility of every crew member to assist in the size-up process and to carry out the order to force entry when the time comes. Once the order is given, each crew member needs to be confident that the tools they select to bring to the structure are the correct tools given the circumstances. When selecting the proper tools and tactics to force entry into a building, the key factors to keep in mind are as follows:

- Nature of the reported emergency
- Conditions at the scene upon arrival
- Type of building involved (residential or commercial)
- Type of doors, locks, and windows present
- Tools available on the responding apparatus
- How the building will be secured after the emergency is over

Proper analysis of these key factors helps to determine whether the priority for entering the structure is speed or cost. For instance, at a working fire, the priority for entering the structure is speed, with little consideration given to the damage done to the building. Incidents that require speed of entry as the priority include the following:

- Working fires
- Fire alarms with visible smoke, a heavy odor of smoke, or where automatic sprinklers have been activated
- EMS responses where the victim needs immediate medical attention, such as an incident where the victim is unconscious, has chest pain, has difficulty breathing, and other symptoms that indicate a potentially life-threatening concern (**FIGURE 15-1**).

However, if the incident is a fire alarm activation but no smoke or fire is showing on arrival, the tactics

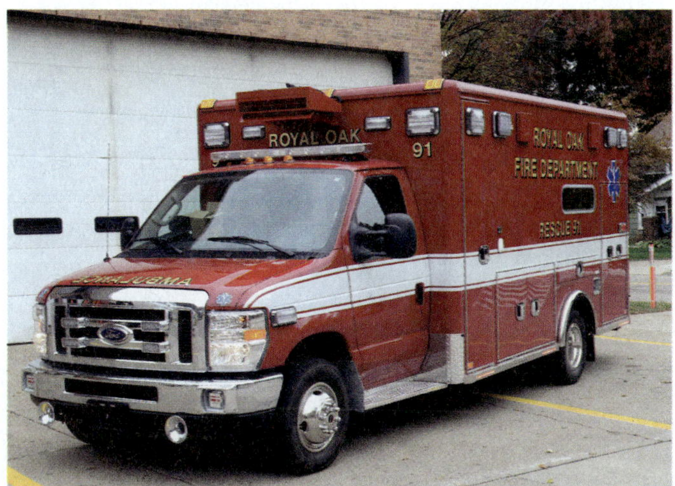

FIGURE 15-1 The fire service responds to many types of emergency incidents, including medical emergencies.

Courtesy of Dennis Walus.

used to force entry can probably be less aggressive and less damaging than at a working fire. In this case, the priority is cost—that is, the cost that will be incurred by the property owner to repair damage done while gaining entry. Incidents where the priority is determined to be cost can be approached more slowly with a greater focus on minimizing damage and resecuring the building at the termination of the call. Incidents where cost is determined to be an important factor when deciding on how to forcibly enter a structure include the following:

- Fire alarms with no smoke or odors of smoke on arrival
- Odor investigations with no odor upon arrival
- EMS calls with a minor complaint or a need for only a lift-assist

Building Components Relevant to Forcible Entry

In order to negotiate the physical obstacles barring them from entry, firefighters must have at least a rudimentary understanding of the following building components that protect the building's interior from the elements and from unauthorized entry:

- Door systems
- Locksets
- Windows
- Walls
- Security gates and bars

Door Systems

Three main elements make up a **door system**. (**FIGURE 15-2**):

- Door: A moveable panel in a wall opening that allows entry into a building or room.
- **Door frame**: The outer housing for a door that is attached to the wall into which the door is set. It includes **hardware**, which is the hinges that attach the door to the frame or the tracks that a sliding door moves in, and the lockset.
- **Lockset**: The hardware, including the doorknob, handle, or push bar used to latch and unlatch a door, that physically holds the door shut within its frame. Although locksets are part of door systems, they will be discussed in the next section.

Swinging Doors

Swinging doors are the most common type of pedestrian door. They have hinges on one side and swing

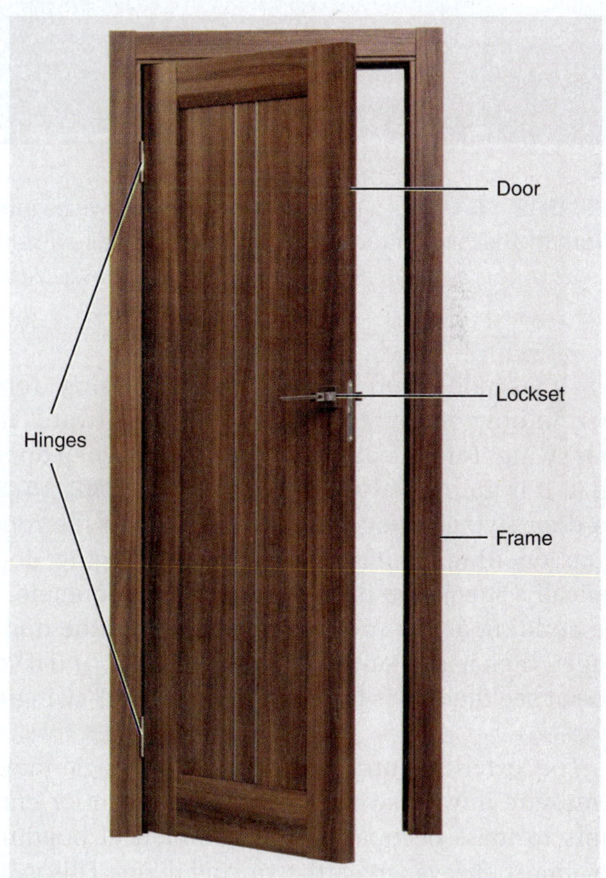

FIGURE 15-2 The parts of a common door system. Door systems are categorized by how they open and close. The most common door systems in North America are swinging, sliding, revolving, and overhead.

© Ivan Nakonechnyy/Shutterstock

A.

B.

FIGURE 15-3 **A.** An inward-swinging door swings into the building or room. **B.** An outward-swinging door swings away from the building or room. The hinges are usually visible on outward-swinging doors.

from that side when opened. Most swinging doors open in only one direction—inward or outward. If a door swings into the building or the room when opening it, it is an inward-swinging door (**FIGURE 15-3**). If a door swings away from the building or the room when opening it, it is an outward-swinging door. Typically speaking, if you are standing outside of the building or the room and you can see the door's hinges, then it is an outward-swinging door, and if you cannot see the door's hinges, it is an inward-swinging door.

The exterior entry doors to most single-family homes are inward-swinging doors. The exterior entry doors to mass occupancy and commercial buildings are almost always outward-swinging doors. This is because an outward-swinging door allows for quick egress in the event of a fire or other incident and most modern building codes require outward-swinging doors in these types of buildings. The exterior entry doors to common areas and hallways in buildings such as apartment and condominium buildings and hotels are often outward-swinging doors, but entry doors into individual apartment units are usually inward-swinging doors.

Types of Swinging Doors. Most swinging doors are solid-core, hollow-core, or glass storefront doors found in commercial buildings.

A **solid-core door** is a door made of solid wood panels that are fit together to look and feel almost like a single, solid piece of wood (**FIGURE 15-4**). Metal-clad doors are solid-core doors covered in sheet metal to provide increased fire resistance. Metal-clad doors are rare in recent construction, but they are still commonly found in all regions in older commercial buildings (**FIGURE 15-5**).

Despite the name, a **hollow-core door** is usually not hollow. Instead, this type of door has an outer covering—veneer—over a framework that gives the door its shape. Most types of hollow-core doors also have a filling material. The veneer is wood, fiberglass, or metal. The

FIGURE 15-4 A solid-core wood door.

Courtesy of Sean Wilson.

FIGURE 15-5 A metal-clad door with a solid wood core.

Courtesy of Sean Wilson.

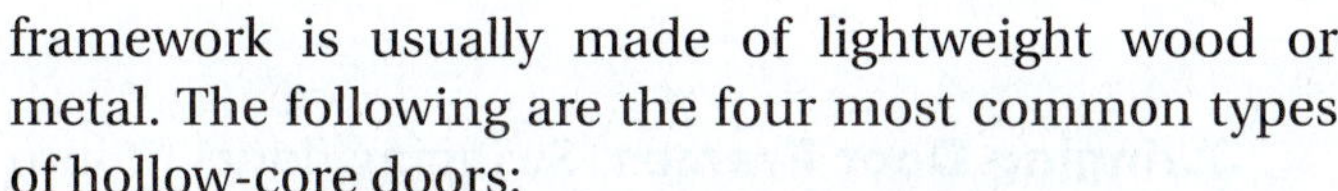

framework is usually made of lightweight wood or metal. The following are the four most common types of hollow-core doors:

- Air-core door: The lightest type of hollow-core door, an air-core door is typically composed of thin wood paneling over a lightweight wood framework with no additional filling material. These doors are used as interior doors in residential homes and apartments (**FIGURE 15-6**).
- Foam-core door: Foam-core doors are made of a metal, wood, or fiberglass covering placed over a framework with foam filling material (**FIGURE 15-7A**). The foam provides either thermal insulation or fire resistance, depending on the type of foam used. Foam-core doors are used as exterior residential doors as well as interior and exterior commercial doors.
- Cardboard-core door: Cardboard-core doors are made of a metal, wood, or fiberglass covering placed over a framework filled with cardboard baffles that resemble a honeycomb (**FIGURE 15-7B**).

FIGURE 15-6 Interior of an air-core door.

Courtesy of Sean Wilson.

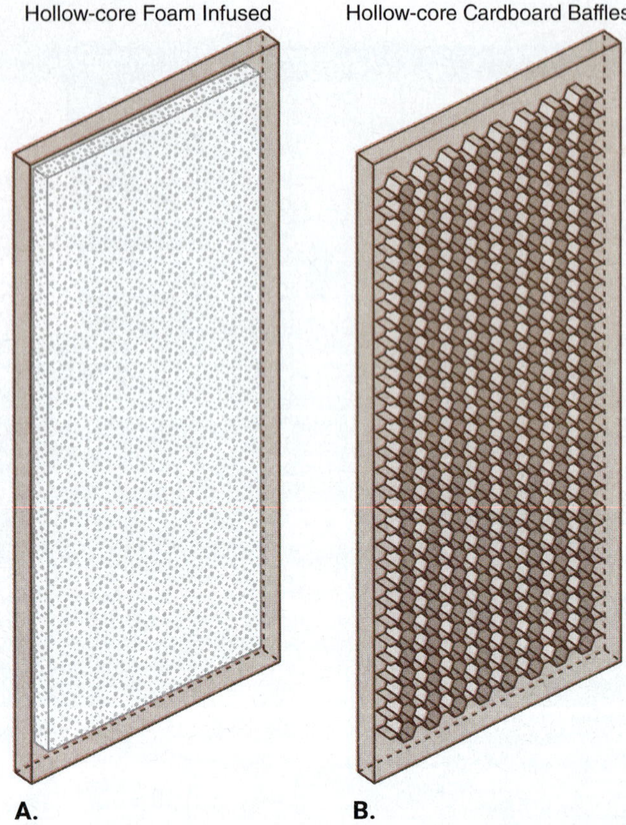

FIGURE 15-7 **A.** Interior of a foam-core door. **B.** Interior of a cardboard-core door.

FIGURE 15-8 A glass storefront door.

Courtesy of Sean Wilson.

The cardboard is treated with a fire-resistant resin that greatly increases the fire resistance of the door. These doors are usually found as exterior or interior commercial doors.

- Mineral-core door: Mineral-core doors are made of a metal, wood, or fiberglass covering placed over a framework filled with mineral material, usually gypsum. The mineral core makes this type of door more fire resistant than other types of hollow-core doors. Mineral-core doors are heavy compared to other types of hollow-core doors and are sometimes mistaken for a solid-core door because of this. The mineral core, however, does not provide much additional structural integrity for the door, which means that when this type of door is forced by firefighters, it will breach more easily than a solid-core wood door.

Glass storefront doors consist of a piece of commercial glass usually framed in hollow aluminum tubing (**FIGURE 15-8**). These doors are typically found on the street side of stores, restaurants, office buildings, schools, apartment buildings, and many other types of commercial buildings.

Swinging Door Frames. Swinging doors fit into a door frame that is attached to the wall at the door opening. The two most common materials used in the construction of door frames are wood and metal. In addition to being the attachment point of the door to the building, the door frame for swinging doors needs to have a way to prevent the door from traveling in the opposite direction from the door's intended swing. There are two types of wood door frames. All wood door frames are made of three pieces of flat lumber cut to fit into the opening in the wall where the door system will be installed. Two of the pieces fit vertically in the opening and the third piece fits horizontally at the top of the opening. The hinges are attached to one of the vertical pieces of the door frame. Wood-stopped door frames have a slender piece of lumber attached with nails to the middle portion of the door frame to act as a door stop—that is, to prevent the door from swinging in the opposite direction (**FIGURE 15-9A**). Wood-rabbeted frames are made

A.

B.

FIGURE 15-9 **A.** A wood-stopped door frame. **B.** Often, wood-rabbeted door frames are covered in aluminum siding.

Courtesy of Sean Wilson.

from a thicker pieces of lumber and have a notch cut into the frame to act as the door stop (**FIGURE 15-9B**). Wood-rabbeted frames are often covered in aluminum siding. The door stop of a wood-stopped frame can be removed, which can make forcible entry easier. The door stop of a wood-rabbeted frame cannot be removed.

Metal door frames are usually made of steel or aluminum (**FIGURE 15-10**). Glass storefront doors are set into a frame made of the same hollow aluminum square tubing that the door is made with and contain an aluminum door stop that is bolted or snapped into place. Steel door frames are made of folded steel, with a door stop that is an integral part of the frame and cannot be removed (**FIGURE 15-11**). These frames are commonly found in commercial buildings and many apartment buildings. When set into hollow partition walls, the metal frame is usually left hollow. When set into exterior or load-bearing walls, the metal frame is typically filled with concrete. Concrete-filled, metal door frames increase the overall strength of the door system, which makes forcible entry much more difficult.

FIGURE 15-10 Outward-swinging door set into a metal frame.

Courtesy of Sean Wilson.

Hinges for Swinging Doors. Hinges attach a swinging door to the frame. Most hinges include a frame-side hinge plate that connects to the frame, a door-side hinge plate that connects to the door, and a pin that goes through the openings to connect the two hinge plates. It is sometimes possible to remove the hinge pin from interior doors to disconnect the hinge-plates and open the door. Most exterior doors, however, have hinges in which the pin is sealed, welded, or

A.

B.

FIGURE 15-11 **A.** Metal frames have a door stop that is integral to the design and cannot be removed. **B.** Metal frames are made of folded steel. They are often filled with concrete on load-bearing walls and left hollow on non-load-bearing walls.

Courtesy of Sean Wilson.

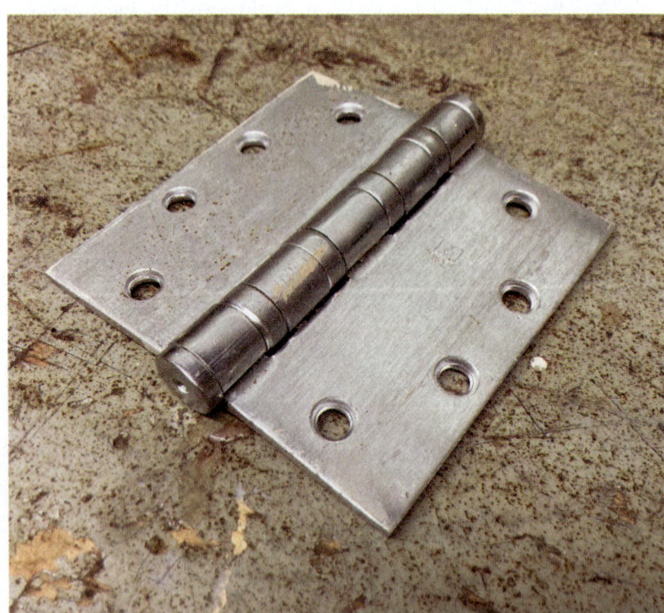

FIGURE 15-12 Standard barrel hinge.

Courtesy of Sean Wilson.

otherwise not removable without destroying the hinge. There are several types of door hinges for swinging doors:

- Barrel hinge: Two hinge plates connected by a single pin run through small tubes or "barrels" welded to each plate (**FIGURE 15-12**).
- Studded hinge: A barrel hinge with a stud or bolt on each hinge plate that connects the two plates when the door is closed, preventing the door from being opened from the hinge side should the hinge pin be removed or cut off (**FIGURE 15-13**).
- Gear hinge: A type of hinge composed of gears that run the full length of the door and turn when the door is opened (**FIGURE 15-14**). Gear hinges are usually found on commercial and school buildings where it is desirable that the doors open flat to the wall.

Sliding Glass Doors

A sliding glass door, sometimes called a doorwall or arcadia door, is a large sheet of tempered glass framed in wood, vinyl, or metal that moves within a track in the floor and ceiling of a home or building. These doors are most often found in residential homes and balconies of apartments and hotels. Automatic-opening sliding glass doors are often found on glass storefronts, hotels, and apartment and condominium buildings.

The weak point in this door system is the simple locking mechanism that can sometimes be pried with a Halligan or removed with a lock-pulling tool. However, it is common for people in residential homes to place a metal rod or wooden dowel into the track to block the door shut from unauthorized entry

A.

B.

FIGURE 15-13 **A.** Studded security hinge. **B.** Studded hinges secure the plates together when closed, making it impractical to cut the hinges to open the door.

Courtesy of Sean Wilson.

(**FIGURE 15-15**). If this is the case, another entry point should be explored because otherwise there will be little choice but to break the large sheet of glass. Whether in a residential home or in a commercial

A.

B.

FIGURE 15-14 **A.** A door with gear hinges. **B.** A close-up view of gear hinges.

Courtesy of Sean Wilson.

FIGURE 15-15 A sliding glass door that has a metal rod or wooden dowel in the track may be difficult to open without breaking the glass.

Courtesy of Dennis Walus.

FIGURE 15-16 Building codes may require outward-opening doors next to a revolving door, through which forcible entry is more easily made.

Courtesy of Sean Wilson.

building, these door systems are expensive and should rarely be the first choice for entry except in obviously high-priority emergencies.

SAFETY TIP

Consideration should be given prior to breaking a glass door or window at a working fire. Once broken, that opening can no longer be easily closed should it negatively impact the flow path of the fire.

Revolving Doors

Revolving doors are found in large commercial buildings or mass occupancies. They are usually made of several tempered glass panels with metal frames. The panels are designed to collapse outward onto one another with a certain amount of pressure to allow for rapid escape during an emergency.

Forcible entry through revolving doors should be avoided whenever possible. If it is absolutely necessary for the evacuation of a large number of building occupants, force entry through a revolving door by attacking the locking mechanism directly, or by breaking the glass. Because of building code requirements, standard outward-opening swinging doors are usually present adjacent to the revolving doors (**FIGURE 15-16**). It may be easier to attempt to force entry through these doors instead of going through the revolving doors. In addition, repairing any damage to these swinging doors will be less expensive.

Overhead Doors

Overhead doors are doors that swing or roll upward from the ground level to above the door frame to open. Many residential garages have overhead doors made of wood, fiberglass, or light-gauge sheet metal. Commercial overhead doors are almost always made of metal.

Forcible entry through overhead doors is usually accomplished by cutting through them. Some residential garage doors and some commercial overhead doors are insulated with foam panels that increase the overall thickness of the door and may slow the cutting process. Metal overhead doors are most easily cut with a rotary saw fitted with a diamond rescue blade or a composite metal-cutting blade. Older wood and fiberglass garage doors can be cut with a rotary saw fitted with a wood-cutting blade. Knowing the differences among

FIGURE 15-17 Tilt-up overhead doors are found on garages in many older neighborhoods.

Courtesy of Sean Wilson.

the types of overhead doors can help firefighters perform a quick size-up at a glance and select an appropriate method for forcible entry at the scene.

Modern building codes may require the placement of an adjacent swinging door near an overhead door in commercial buildings. It is usually easier and quicker to force this swinging door and then open the overhead door in the normal fashion than it is to force the overhead door.

There are three types of overhead doors:

- Tilt-up overhead doors: These doors are one large panel made of wood or sheet metal that remain flat and tilt upward when opened. This type of door is usually found on residential garages in older neighborhoods. They are rarely installed today on newly built or newly renovated homes (**FIGURE 15-17**).
- Sectional overhead doors: This type of overhead door consists of five or more short, wide, hinged panels that allow the door to move along a track that stores the door flat overhead when opened. These doors are commonly found in both residential garages and commercial buildings and are either insulated or uninsulated (**FIGURE 15-18**).
- Rolling steel overhead doors: This type of overhead door is made of short, wide, interlocking steel slats that hinge and roll up into a cylindrical barrel when opened. These doors pivot similarly to sectional overhead doors, but the steel slats are much shorter than the panels used in sectional overhead doors, and the slats fold to fit together without any hinge hardware (**FIGURE 15-19**). This type of overhead

A.

B.

FIGURE 15-18 **A.** Sectional overhead doors are commonly found on commercial buildings and residential garages. **B.** Sectional overhead doors have hinged panels that pivot when opened. They open on a track and are stowed parallel to the ceiling.

Courtesy of Sean Wilson.

A.

B.

FIGURE 15-19 **A.** A rolling steel door is made of short, wide steel slats. **B.** The slats from rolling steel overhead doors are folded and are slid over one another to lock together.

Courtesy of Sean Wilson.

door is strong, is almost never insulated, and is found almost exclusively on commercial buildings. The barrel the door sits in when opened can be on the interior or exterior of the building (**FIGURE 15-20**).

A.

B.

FIGURE 15-20 The barrel that rolling steel doors roll up into can be housed either on the inside or outside of the building. **A.** Inside. **B.** Outside.

Courtesy of Sean Wilson.

All three types of overhead doors can be opened or closed manually, by a chain-hoist, or by an electric motor operated by a wall switch or a remote control. All three types operate on a track that can be locked. The locking mechanism is either a padlock on the chain-hoist or on the track itself or a sliding bar lock built into the track. If an overhead door is operated by an electric motor, the door will not move in the track unless the motor is activated.

> **SAFETY TIP**
>
> When operating at a fire where an overhead door has been opened, be sure to block the track so the door cannot accidentally close. This can be done by clamping locking pliers onto the track or by blocking the track with a pike pole or other tool. Firefighters have been trapped when gravity has brought down a manually or chain-operated door, or a short in the electrical system has closed a motor-operated door.

Locksets

As mentioned in the "Door Systems" section in this chapter, a lockset is the hardware in a door system that physically holds the door shut within its frame. The purpose of a lockset is to secure a door to its frame and, if necessary, prevent unauthorized entry. Locksets can be fairly simple or quite elaborate. They keep out unauthorized persons (entry lockset), give a person temporary privacy while using a bathroom or restroom (privacy lockset), or simply keep a door closed when a strong or air current blows (passage lockset). Entry locksets usually contain keyed cylinders that require a key to open from the outside when locked. Rather than a keyed cylinder, privacy locksets have a simple mechanism that is often operated by depressing a button or other simple action from the inside, temporarily offering privacy to the user. Passage locksets have no mechanism to prevent the door from being opened.

Elements Common to All Locksets

Although there are different types of locksets, elements common to all types are the strike plate, latch, operating spindle, knob or lever, and cylinder (**FIGURE 15-21**).

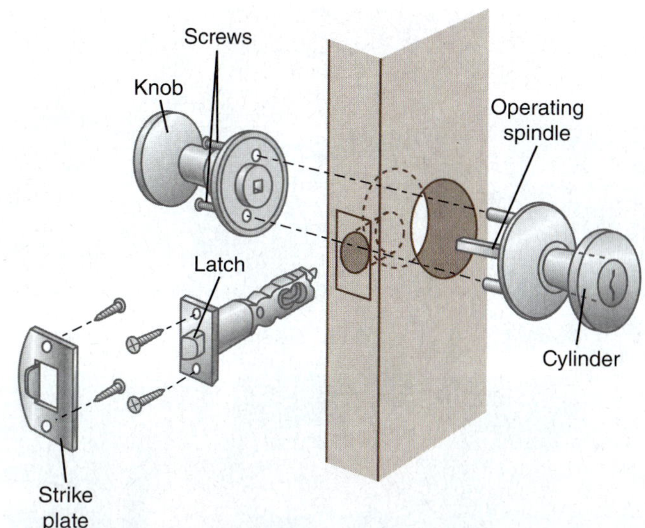

FIGURE 15-21 The elements of a lockset.

The strike plate and latch are the prime components of the lockset. The **strike plate** is a metal plate with a hole in it that is mounted on the door frame. The **latch** is a moveable bolt that protrudes from the edge of the door and slides into the strike plate when the door is closed. The latch meeting the strike plate is what physically holds the door shut.

The two most common types of latches are the slam latch and the dead latch. A **slam latch** is a short, angled piece of metal that operates on a spring (**FIGURE 15-22**). When the door is closed—or slammed—the latch is pushed into its housing until the corresponding hole in the strike plate and indentation in the frame allows the latch to pop back out to secure the door. A **dead latch** is longer than a slam latch and flat at the end. When a door with a dead latch is closed, the dead latch must be retracted or the door will not close (**FIGURE 15-23**). After the door is closed, the mechanism that extends the dead latch into place and into the corresponding hole in the strike plate and indentation in the frame secures the door.

Most modern slam latches have a small pin called an anti-loiding pin integrated into the latch (or above or below the latch). When this pin is depressed, it prevents the latch from being pushed in manually. When the door is closed, the anti-loiding pin is depressed by the strike plate. This prevents someone from using a shove knife or a piece of hard plastic such as a credit card from sticking it between the door and the frame and manipulating the tool to depress the latch.

The **operating spindle** is a shaft that connects the hardware that a user turns, pulls, pushes, or otherwise manipulates to the mechanism that retracts the latch from the strike plate. The most common types of hardware used to rotate the operating spindle and retract the latch are ball-shaped doorknobs, levers, and push bars. Because of the Americans with Disabilities Act of 1990, lever-style knobs are slowly replacing ball-shaped doorknobs in newly built and renovated homes and in commercial buildings. Push bars, which are installed on outward-opening doors in commercial buildings, are often referred to as panic hardware because they are designed to allow a panicked crowd to exit a building efficiently and quickly.

The cylinder is a mechanism within an entry lockset that prevents unauthorized entry. When the correct key is inserted, the notches and ridges on the

A.

B.

FIGURE 15-22 A. Slam latch on a bored lockset. A slam latch is angled and operates on a spring so the door can be slammed and closed without turning the knob or lever. **B.** Slam latch on a mortise lockset.

Courtesy of Sean Wilson.

A.

B.

FIGURE 15-23 A. Dead latches on bored locksets. A dead latch has a blunt end and operates only when its lever is turned. The door cannot be slammed and closed. **B.** A dead latch on a mortise lockset.

Courtesy of Sean Wilson.

key align with the corresponding pin pattern inside the cylinder mechanism, which allow the key to turn the cylinder to lock or unlock it. When the cylinder is turned and unlocked, it allows the knob or lever to rotate, thus retracting the latch and opening the door. Most exterior locksets and some interior locksets include a cylinder that can be locked or unlocked with the proper key.

Types of Locksets

Most locksets are categorized as one of the following five types:

- Bored lockset
- Mortise lockset
- Rim lockset
- Smart lockset
- Padlock

Firefighters should familiarize themselves with the common locksets that they see every day in their homes, apartments, schools, offices, and stores so that when it comes time to size up a building for forcible entry, they will be able to do so quickly. Being able to quickly identify the type of lockset on a door will help firefighters to rapidly devise the correct strategy for forcible entry through that door.

Bored Locksets. **Bored locksets** take their name from the fact that there are two holes bored into the door to install this type of lockset: a large, 2-inch (in.; 51-millimeter [mm]) hole bored into the door face where the operating spindle is inserted, and a smaller, 1-in. (25-mm) hole bored into the door edge where the latch mechanism is inserted (**FIGURE 15-24**).

The two most common types of bored locksets are key-in-knob locksets and tubular deadbolts. Key-in-knob locksets are so called because the cylinder sits within in the knob or lever of the lockset, and the key is inserted there to lock or unlock the cylinder (**FIGURE 15-25**). When the device that you turn to open the door is a lever, this type of lockset is called

FIGURE 15-24 A bored lockset gets its name because it is installed on a door that has two holes bored into the door.
Courtesy of Sean Wilson.

A.

B.

FIGURE 15-25 **A.** Key-in-knob bored lockset. **B.** Many bored locksets are held together by two screws run through the lockset from the interior side, as shown in this key-in-knob lockset.
Courtesy of Sean Wilson.

FIGURE 15-26 Key-in-lever versions of bored locksets are required in many buildings to comply with accessibility laws.

Courtesy of Sean Wilson.

a key-in-lever lockset (**FIGURE 15-26**). Both locksets have slam latches.

A **tubular deadbolt** has a cylinder that is protected by a thick, angled collar protruding approximately 2 in. (51 mm) from the door (**FIGURE 15-27**). When the door is locked, a dead latch extends 1 in. (25 mm) or more into the strike plate. On the outside of the door, a key turned in the cylinder moves the dead latch into the strike plate (locked) or retracts it into the door (unlocked). Most tubular deadbolts have a thumb-turn mounted to the back plate on the inside of the door that is rotated to extend or retract the latch, but some use a key on the inside of the door. These types of bored locksets are found on most residential doors and doors in many commercial buildings. Both key-in-knob locksets and tubular deadbolt locksets are usually mounted to the door with two screws that are inserted through a back plate on the interior side of the door.

Mortise Locksets. A **mortise lockset** consists of a rectangular assembly that fits into a similarly shaped cavity within the door. In woodworking, this type of cavity is called a *mortise*, which is where the name of the lockset comes from (**FIGURE 15-28**). Unlike a key-in-knob lock, the key is not inserted into the knob or lever, but into a cylinder attached to the housing through a hole in the door face above the knob or lever. Mortise locksets may have a slam latch, a dead latch, or both

A.

B.

FIGURE 15-27 **A.** A tubular deadbolt. **B.** Interior view of a tubular deadbolt.

Courtesy of Sean Wilson.

a slam latch and a dead latch that are both operated with the same key inserted into the same cylinder. Cylinders for mortise locks are threaded and screwed into the housing and held in place with a small **set screw**, which is a screw used to secure an object by pressure or friction, usually within or against another object.

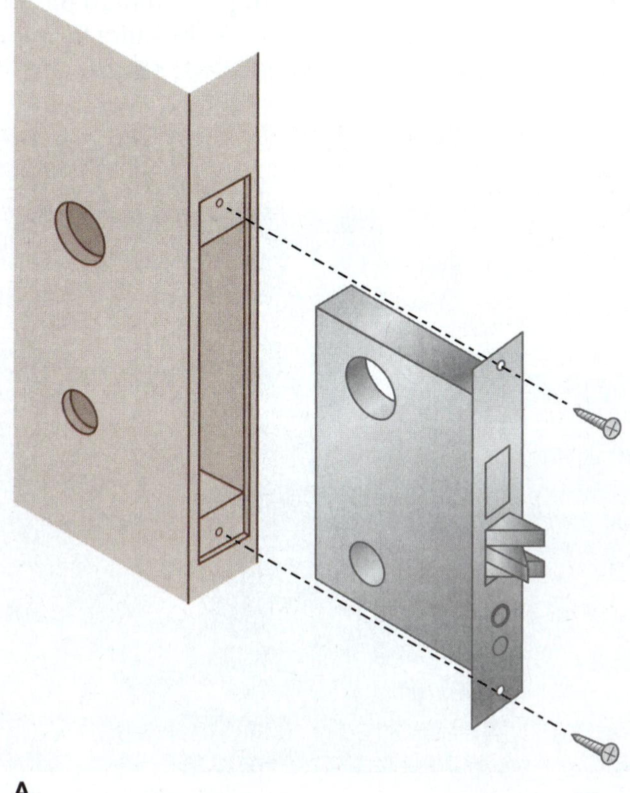

A.

B.

FIGURE 15-28 **A.** A mortise lockset is a rectangular assembly that fits into a rectangular cavity that is "mortised" into the door. **B.** It has a cylinder that is threaded and screws into the lockset. The keyway is in the cylinder above or below the knob or lever, not in the knob or lever itself.

A.

B.

FIGURE 15-29 **A.** Knob-style commercial mortise lockset. Note that the cylinder is above the knob in a mortise lockset, not in the knob or lever. **B.** Lever-style commercial mortise lockset. This style is more common in newer buildings for compliance with accessibility laws.

Mortise locksets are primarily found on commercial buildings, but they can also be found in older residential construction.

The two primary types of mortise locksets that firefighters need to familiarize themselves with are standard mortise locksets and Adams Rite mortise locksets. Standard mortise locksets are large locks that are installed on solid-core wood doors and hollow-core metal commercial doors (**FIGURE 15-29**). These doors are pre-cut at the factory to accept a mortise lockset. Adams Rite mortise locksets are smaller than standard mortise locksets and are installed on glass storefront doors. They have the same type of threaded cylinder as standard mortise locksets, but the latches on them can be quite formidable when prying against them (**FIGURE 15-30**).

A.

B.

C.

D.

FIGURE 15-30 **A.** Adams Rite mortise lockset with a hooked dead latch. **B.** Adams Rite mortise cylinder and set screw. **C.** Threaded mortise cylinders and collar. **D.** The Adams Rite mortise lockset is often found on glass storefront doors.

Courtesy of Sean Wilson.

Rim Locksets. A **rim lockset** is mounted on the interior surface (inside rim) of the door, usually with three to eight screws depending on the size of the lockset. The cylinder is inserted into a 1-in. (25-mm) hole drilled through the door and mounted to the door with two screws attached to a back plate. The spindle of the cylinder is inserted into the lock housing, and the lock housing is then attached to the interior surface of the door (**FIGURE 15-31**). Rim locksets may or may not include a cylinder that can be opened with a key from the exterior.

Rim locksets are found on doors with panic hardware in many commercial buildings where the building code requires that a door swing out (**FIGURE 15-32**). Residential rim locksets in are more common in older occupancies than in new construction (**FIGURE 15-33**). They were often installed as add-on locks for additional security. They were popular because they were inexpensive and easy to install for the average homeowner.

One newer type of commercial rim lockset is the Securitech Trident lockset, commonly found on national chain pharmacies around North America. The

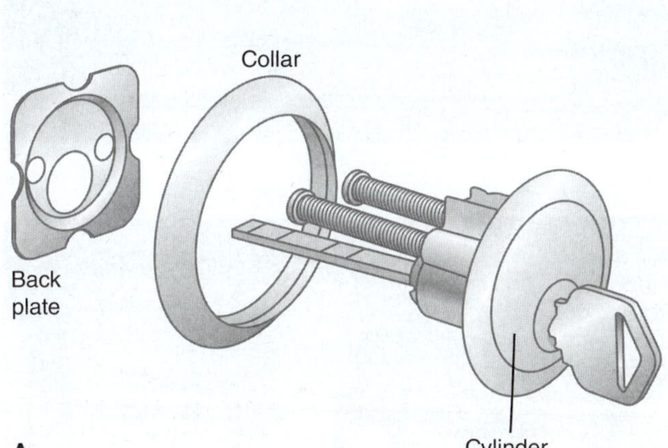

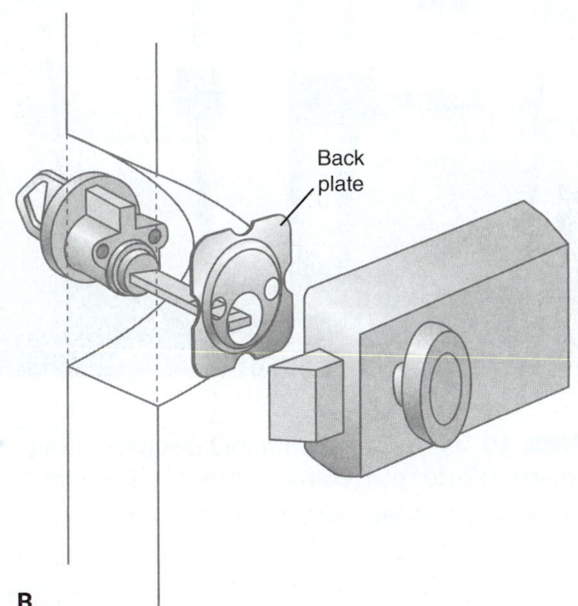

FIGURE 15-31 **A.** The cylinder of a rim lockset looks similar to a mortise lock cylinder when installed, but the rim cylinder is held in place by two screws attached to a back plate and cannot be unscrewed. **B.** The lock body of a rim lockset is mounted to the inside "rim" of the door, hence the name.

FIGURE 15-32 Most types of commercial panic hardware are a form of rim lockset. **A.** Newer style push-bar panic hardware. **B.** Older style push-bar panic hardware.

A.

B.

FIGURE 15-33 **A.** Rim lockset cylinder. **B.** Residential rim locksets: the Segal lock (sometimes called a vertical deadbolt) and the night latch.

Courtesy of Sean Wilson.

A.

B.

FIGURE 15-34 **A.** Outside view of a Securitech Trident commercial rim lockset. **B.** Inside view of a Securitech Trident commercial rim lockset.

Courtesy of Sean Wilson.

Securitech Trident consists of a series of three latches operated by a single push bar and installed on any outward-swinging, commercial door. Some versions of the Securitech Trident have a keyed cylinder that can be pulled, but most have no cylinder and must be forced by conventional means discussed later in the chapter (**FIGURE 15-34**). An older type of rim lockset no longer

A.

B.

FIGURE 15-35 **A.** Electrically activated mortise lockset with fob/card reader. **B.** This electrically activated lockset is a standard mortise lock with an electric strike plate mounted into the frame that allows entry when the correct fob or card is swiped.

Courtesy of Sean Wilson.

in production is the Fox Police Lock. The Fox Police Lock consists of a rim cylinder installed into the center of a door and shielded with a plate installed onto the door to prevent the cylinder from being pulled. The Fox Police Lock Company went out of business in the early 1980s, but many of their rim locksets are still found in larger cities in the northeast United States. If either the Securitech Trident or Fox Police Locks are encountered in the field, it is usually best to find another door to force unless absolutely necessary.

Electrically Activated Locksets. Electrically activated locksets (sometimes called "smart locks") are used in buildings where many people require access but security is a concern, such as apartment buildings and college dorms, or in private residences where a using a key is not desirable to the resident. Many firefighters find the addition of electricity to a lockset to be confusing and intimidating. In most cases, however, these locksets fall into one of the previously discussed categories—bored, mortise, or rim—and can be removed in a manner consistent with their type. The fact that they may not require a physical key to open the door often does not change the forcible entry strategy or techniques.

In place of a standard cut metal key, these locksets allow entry by either entering a numerical code, swiping a fob or key card, placing a finger or thumb on a pad, or sending the door system a signal through cell phone or hardwired technology (**FIGURE 15-35**). In commercial buildings, these locks often have a standard cylinder that can be unlocked with a cut metal key as a backup. If this is the case, then that cylinder can be removed by standard lock-pulling techniques.

Some keyless locksets are electric and have an alkaline or lithium battery that powers a small motor that turns the operating spindle and extends or retracts the latch (**FIGURE 15-36**). Many electrically activated locksets, especially in commercial buildings, are standard, mechanical, commercial locksets that are paired with an electrically activated strike plate that is mounted into the door frame. When the correct code is entered or the correct fob or key card is swiped on a pad mounted to the door frame, the electric strike plate disengages and opens, allowing the latch to clear the strike plate and the door to swing open.

The only keyless lockset that is truly unique is the electromagnetic lock found in some commercial buildings. This type of lockset has a thick metal plate attached to the door on the interior side with a corresponding plate mounted on the door frame. The plate on the door frame is an electromagnet, which means

A.

B.

FIGURE 15-36 **A.** Electrically activated tubular deadbolt (a bored lockset) with a numeric keypad. **B.** This electrically activated tubular deadbolt is simply a standard bored lock whose mechanism is operated with an alkaline or lithium battery. This lockset also accepts a key.

Courtesy of Sean Wilson.

the magnetic field is produced by an electric current (**FIGURE 15-37**). This type of door is unlocked when the current is disrupted, which temporarily disables the magnetic property of the plate. Like other keyless locksets, these doors can be unlocked when the correct code is entered or the correct fob or key card is swiped on the pad. They can also be unlocked when an authorized user "buzzes" the door open from a remote location. These locks can also be locked or unlocked on a timer.

If power is lost in a building with electromagnetic locks, the locksets are programmed to fail in either the locked position (fail-secure) or in the unlocked position (fail-safe), depending on the type of occupancy. For example, magnetic locksets used in a jail or prison would be programmed to be fail-secure, whereas in a hospital these locks would be programmed to be fail-safe.

Padlocks

Padlocks are free-standing locks designed to be used with a chain or a hasp and staple to secure a door, window, or gate. Although there are many types of padlocks, three of the most common are standard padlocks, high-security padlocks, and hockey-puck padlocks.

Standard padlocks and high-security padlocks typically have a rectangular body with a rounded, *U*-shaped **shackle** that is permanently attached on one side of the padlock. The other side of the shackle opens when the padlock is unlocked. The cylinder where a key is inserted or a combination is entered to unlock the padlock is usually on the bottom or side of the padlock.

Standard padlocks have shackles that are ¼ in. (6 mm) in diameter or less and are made of steel or other metal. The shackle on standard padlocks can often be cut with a medium to large (18- to 36-in. [46- to 91-centimeter (cm)]) bolt cutters. Standard padlocks lock on only one side of the shackle, the side that opens when unlocked (**FIGURE 15-38**).

High-security padlocks have shackles thicker than ¼ in. (6 mm) in diameter and are made of case-hardened steel, a type of steel that is created using a process that hardens the outer portion of a steel component. High-security padlocks lock on both sides of the shackle, which is called "heel and toe locking" (**FIGURE 15-39**). Case-hardened steel can be cut only with specialized tools. The shackle of high-security padlocks may be too hard or too thick to cut with bolt cutters and may actually damage the bolt cutters. If the padlock is secured using a chain, cutting the last link in the chain may be the better option; this option also allows the lock and chain to be reused.

A.

B.

C.

D.

FIGURE 15-37 **A.** Door system with a magnetic lockset. **B.** Magnetic locksets can often be identified from the outside by a bolt head that holds the metal plate to the door. **C.** Magnetic lockset frame plate on the door. **D.** Electromagnet that holds the frame plate attached to the door.

Courtesy of Sean Wilson.

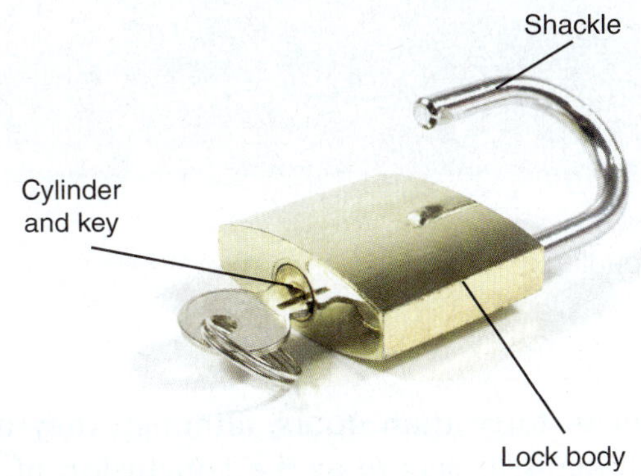

FIGURE 15-38 Parts of a padlock.

© Lik Studio/Shutterstock

A **hockey puck padlock** is easy to spot because of its unique shape (**FIGURE 15-40**). They have a hidden shackle on the back of the lock and are used most often to secure roll-up steel doors, security gates, and work vans and trucks. In most cases, hockey-puck locks are attached to the specific hardware they were designed to fit into, often with a shield surrounding the outer edge of the lock (**FIGURE 15-41**). If the outer edges of the lock are left unshielded, it is possible to remove the lock using a 36-in. (91-cm) pipe wrench to spin the lock and break the shackle or the mounting hardware. If shielded, the lock would need to be cut using a rotary saw.

A.

B.

FIGURE 15-39 **A.** Standard and high-security padlocks. **B.** High-security padlocks often lock on both sides of the shackle. This is called "heel and toe locking" and requires that both sides of the shackle be cut.

Courtesy of Sean Wilson.

A.

B.

FIGURE 15-40 **A.** Locked hockey puck padlock. **B.** Unlocked hockey puck padlock.

Courtesy of Sean Wilson.

Window Systems

Windows provide both airflow and light to the interior of buildings. They can also provide occupants or first responders emergency entry or exit. Windows are often easier to force than doors, although they may be more difficult to secure at the conclusion of the emergency or should a fire's flow path be negatively affected. Understanding how and when to force entry

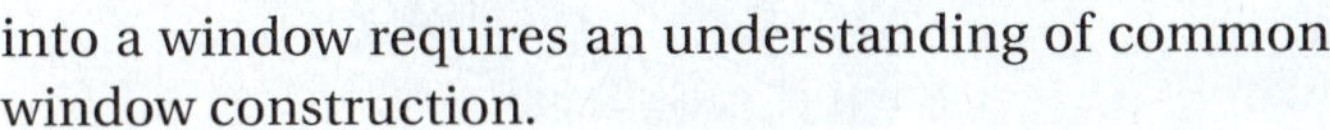

FIGURE 15-41 Hockey puck padlock hardware.

Courtesy of Sean Wilson.

into a window requires an understanding of common window construction.

The goal of forcible entry through windows is to gain entry with minimal or no damage to the glass or frame. It is also important to bear in mind that some glass is easy to break and inexpensive to replace, whereas some glass is strong or decorative and may be expensive to match or replace.

Safety

Although breaking the glass in a window may be the easiest way to gain entry into a building, it can also be dangerous. There are three chief concerns when it comes to the safety of forcing entry through windows: injuries from broken glass, changes to the flow path, and surprised occupants.

Serious injuries can occur from broken pieces of wood, jagged metal, and sharp glass. Firefighters who break a window should wear full personal protective equipment (PPE), including eye and hand protection, and stand on the upwind side of the window with their hands higher than the breaking point, if possible. This positioning ensures that the broken glass will fall away from your hands and body. Placing the tip of the tool in a corner of the window will give you more control when breaking the glass. After the glass is broken, remove any loose, moveable framing and screens, and clear the glass from the entire frame so that no glass shards are left to cause injury to the firefighters who enter or exit through the window or to victims during rescue.

Remember that opening or forcing entry through a window during a working fire will change the flow path and can cause a ventilation-limited fire to rapidly grow or change direction. Do not attempt to open or force a window unless a hose line is in place attacking the fire and you have been ordered to do so by your company officer or the incident commander (IC), as ventilation performed without a simultaneous fire attack is likely to increase the volume of fire.

At non-fire incidents where entry to a building must be made to access a patient or investigate an odor, first responders should not open or force entry through a window unless voice contact with a resident or occupant inside the home or building can be made. Firefighters have been mauled by dogs and shot by residents mistaking them for burglars. Forcing entry through a door can bring these same dangers, but it is much harder to quickly exit a window should danger present itself.

Window System Components

Similar to a door system, a window system includes several elements:

- Glazing, which is a glass or thermoplastic sheet (such as Plexiglas or Lexan) through which light enters the building but rain and the extremes of temperature are kept out. The term *glazing* includes both the glass or thermoplastic sheet and the method used to secure it to the sash frame.

- **Sash frames**, or simply **sashes**, are the part of the system that moves to open or close the window by sliding or swinging out. (Windows that are fixed and do not open generally do not contain sash frames, although some do for appearances.) These frames can be made of wood, metal, or vinyl.
- **Window frames** attach the window system to the building.
- **Sash locks** prevent the window from being opened from the outside.
- Screens allow natural ventilation of a room or building but keep insects and animals out.

Types of Window Glass

Glass can be easy to break, or it can be surprisingly strong and difficult to break. What determines a pane of glass's resistance to our efforts to break and remove it from its frame is how the glass was manufactured.

Single-Pane Annealed Glass. **Annealed** glass, often just called window glass, is standard glass that is often used in residential windows and small commercial windows because it is relatively inexpensive. Annealed glass is easily broken, and firefighters frequently use a pike pole or a roof hook to do this. When broken, annealed glass creates large shards that can cut or penetrate boots, gloves, and other PPE and cause serious lacerations. **Single-pane glass windows** have only one pane of annealed glass.

Double-Pane and Triple-Pane Annealed Glass. **Double-pane glass** windows and **triple-pane glass** windows are used in many homes because they improve the thermal insulation of a building by sandwiching air or an inert gas such as argon between two or three layers of annealed glass. These windows are sealed units, which makes them more expensive to replace than single-pane glass windows. Forcing entry through these insulated windows is similar to forcing entry through a single-pane window except that the second and third pane of glass may need to be broken separately. Because these windows are made of annealed glass, breaking these windows can also produce large and dangerous shards.

Plate Glass. **Plate glass** is a stronger, thicker type of annealed glass that was commonly used in large commercial windows and residential sliding doors from the 1950s to the 1980s. It is not commonly installed today, but many older buildings still contain plate glass windows and storefront doors. It can be broken easily with a sharp object such as a Halligan tool, pick-head axe, or roof hook. As with annealed glass, breaking plate glass can create large and dangerous shards that can cut through a firefighter's PPE, therefore great care should be taken when breaking plate glass (**FIGURE 15-42**). If you need to break a plate glass window, consider having two firefighters hold up a salvage cover over the window while a third firefighter strikes the window with a tool. Breaking the window in this manner reduces the chance of injury from falling shards.

FIGURE 15-42 Plate glass shatters relatively easily and breaks into large shards that can be dangerous.

Courtesy of Sean Wilson.

Laminated Glass. **Laminated safety glass**, or simply **laminated glass**, consists of two layers of annealed glass bonded together with a clear polymer film. This type of glass does not shatter when broken. Instead, it cracks in a star shape when it is broken, or with stronger forces, a pattern resembling a spider web. This type of glass is most commonly found in automotive windshields and is starting to be used in the driver- and passenger-side windows. Other types are sometimes found in windows in residential homes and commercial buildings.

Two newer types of residential windows made with laminated glass are hurricane windows and impact windows. The polymer used in these windows is 10 to 20 times the strength of windshield laminate and can be used in any style of window (single-hung, double-hung, casement, projected, etc.). Although the glass can be broken, the polymer is intended to provide impact resistance from wind or objects striking the window. Hurricane windows are primarily used in areas prone to heavy storms. Impact windows are used when security is a concern and are constructed with stronger reinforced frames. These types of glass are expensive and difficult to force. Forcing hurricane and impact windows should be avoided unless absolutely necessary. Should it become necessary, force these windows in the same manner you would use to force wired glass.

Tempered Glass. **Tempered safety glass**, or simply **tempered glass**, is created by exposing annealed glass to extreme heat and then rapidly cooling it. As a result, the outer surface of the glass goes into compression (pushing inward) and the center goes into tension (pushing outward). This gives tempered glass its added strength. When tempered glass breaks, the opposing forces holding the glass together suddenly release, and the glass breaks into small, mostly dull-edged pieces. This type of glass is commonly found in modern storefront windows and doors, residential sliding doors, and side and rear car windows. Tempered glass is best broken using a sharp pointed tool to strike one of the corners of the glass. During vehicle extrication, a center punch is often used to break tempered glass.

Wired Glass. Wired glass is plate glass or tempered glass that has been reinforced with wire (**FIGURE 15-43**). This kind of glass is often used in fire-rated doors with a window. Wired glass windows are difficult to force because once the glass is broken, each individual wire must be cut or broken in order to clear the window.

FIGURE 15-43 Wired glass.

Courtesy of Sean Wilson.

Should forcing a wired glass window become necessary, create a small hole with a Halligan or pick-head axe and use a reciprocating saw to cut the glass and wire around the inner margin of the frame. If a reciprocating saw is not available, the same can be done with an axe, although it will be more time consuming and labor intensive.

FIGURE 15-44 Glass block window.

Courtesy of Sean Wilson.

Glass Block. Glass blocks, sometimes called glass bricks, are hollow bricks constructed out of glass filled with air or argon for insulation. These blocks can be used as windows, walls, or even floors, and are stacked and held together using a mortar similar to the mortar used in masonry wall construction. They are often used for basement windows as they allow light into an area but are more secure against break-in than standard window glass. Glass blocks are strong, but they can be broken one at a time using an axe, sledge, or other hand tool (**FIGURE 15-44**).

Types of Windows

A huge variety of window types have been installed on homes and buildings in the last century. Some window types are easy to force with little to no damage; however, the only option for other types may be to break the glass or find another entry point. Due to differences in function, materials, and decorative features, it can be difficult to keep current with the ever-changing styles of windows and how they operate. Most, however, fall into one of the following six categories:

- Storefront
- Single- and double-hung
- Slider
- Casement
- Projected
- Awning and jalousie

Storefront Windows. **Storefront windows** are large windows that do not move, slide, or open (**FIGURE 15-45**). They are typically made of tempered glass or plate glass and are found in many commercial occupancies to showcase merchandise and let in natural light.

Single- and Double-Hung Windows. Single- and **double-hung windows** have two sashes made of wood, metal, or vinyl. With single-hung windows, only the lower sash slides up and down in a track mounted into the window frame. With double-hung windows, both sashes slide up and down (**FIGURE 15-46**). Most single- and double-hung windows are secured by one or two sash locks that hold the bottom of the top sash to the top of the bottom sash, which prevents either sash from sliding in its track.

The sash locks on wood-frame single- and double-hung windows can often be defeated using a hammer to drive a strong slender tool such as a flat-head screwdriver or a scraper bar up between the two sashes where the sash lock is mounted. The tool contacts the sash lock and drives the screws holding the sash lock out of the wood sash frame.

The sash locks on vinyl-frame single- and double-hung often have a small lip that covers the area between the two sashes. The best way to force this type of window is to place the adze of a Halligan or a similar tool between the two sashes where the lock sits and slowly rock the tool from side to side. Vinyl is a rubbery plastic, so doing this will often bend and distort the vinyl sash until the two halves of the sash lock separate.

Slider Windows. **Slider windows** are similar to single- and double-hung windows, but they slide horizontally rather than vertically (**FIGURE 15-47**). The frames are made out of the same materials as single- and double-hung windows, and the same types of sash locks are used to secure them. For this reason, the same techniques used to force single- and double-hung windows work well against slider windows. As with sliding

FIGURE 15-45 Non-opening storefront windows next to a storefront door.

Courtesy of Sean Wilson.

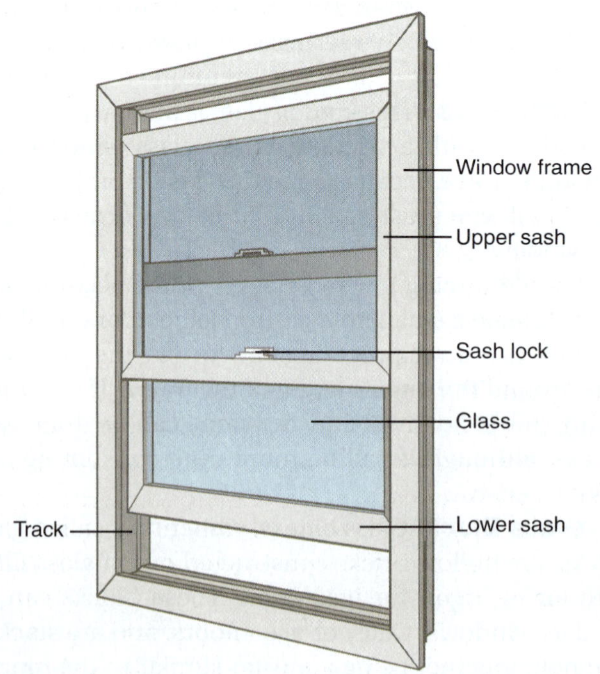

FIGURE 15-46 Components of a double-hung window.

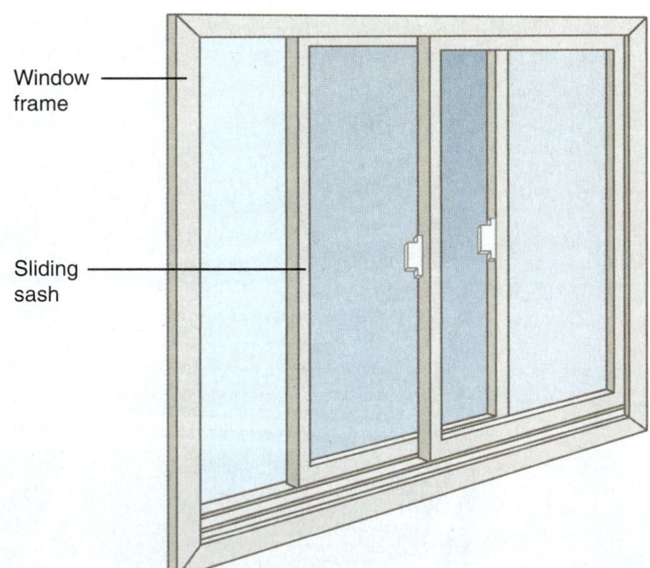

FIGURE 15-47 Slider windows are similar in design to single- and double-hung windows except they are oriented horizontally instead of vertically.

glass doors, residents sometimes place a bar or dowel into the track to prevent outside entry. If this is the case, you will be forced to either break the glass or find another point of entry.

Casement Windows. **Casement windows** have a metal, vinyl, or wood frame and open out from the building on a hinge operated by a crank (**FIGURE 15-48**). Most have a sash lock to keep the window shut tight against drafts when closed. These windows are expensive and should be avoided as an entry point unless absolutely necessary. It is not possible to bypass the crank mechanism from the outside. In order to force this type of window, break the glass, reach in and unlock the sash lock, and then crank out the sash frame.

Projected Windows. **Projected windows** are windows that can project inward or outward from a hinge located at the top or bottom of the window, or pivot at the middle of the window (**FIGURE 15-49**). These windows may be one pane of glass or multiple panes of glass framed in metal. Older style projected windows are found in older warehouse and factory buildings. They were installed for the purpose of ventilation and rarely have screens. They often are installed adjacent to or within fixed windows that do not open. The sash locks are usually simple quarter-turn locks or hand cranks.

Forcing projected windows should be avoided if possible because these windows are often not large enough for a person to enter through, and the sash locks usually

FIGURE 15-48 Casement windows open to the side with a crank mechanism.

FIGURE 15-49 Projected windows open inward or outward on a top or bottom hinge.

A.

B.

FIGURE 15-50 **A.** A jalousie window. **B.** An awning window. Even when open, these windows are usually too small for a person to make entry through without breaking out and removing each glass section. This is time-consuming and costly to replace. When possible, choose another entry point for forcible entry.

cannot be opened from the outside without breaking the glass. If it is necessary to force a projected window, break the window or a panel of the window, reach in and unlock the sash lock, and then open the window.

Awning and Jalousie. **Awning windows** and **jalousie windows** are both made of short, wide sections of annealed or tempered glass set into a metal or vinyl frame (**FIGURE 15-50**). These sections may or may not overlap when closed. They are operated by a small wheel or crank in the corner or center of the window or a slider bar on the side or center of the window. The only appreciable difference between the function of these two types of windows is that jalousie windows have shorter glass sections and more of them, and the sections on an awning window are larger than those on a jalousie window. These types of windows are found in commercial buildings, older homes, and mobile homes.

Walls

The construction of a wall may dictate the amount of flex and movement between a door and its frame when forced. Understanding how walls are constructed can help firefighters size up a door for forcible entry. Walls can be classified as either load-bearing or non-load-bearing. A load-bearing wall is one that helps to bear the structural weight of the building components resting on it. A non-load-bearing wall, often called an **interior wall** or a **partition wall**, does not bear significant weight of the structure. The **exterior wall** of a building is almost always load-bearing. Depending on the design of the building, some interior walls may be load-bearing, but most are non-load-bearing partition walls.

Masonry Walls

Masonry walls are usually load-bearing, but they can be non-load-bearing. The three most common types of masonry walls are masonry block, brick, and poured concrete.

Masonry block walls are constructed using large blocks called mason blocks, cinder blocks, or concrete blocks (**FIGURE 15-51**). These blocks are available in

A.

B.

FIGURE 15-51 **A.** Masonry block. **B.** Masonry block wall.

Courtesy of Sean Wilson.

different shapes, sizes, and exterior finishes, but most are hollow and made of concrete. Their ease of use in construction and low cost make them ubiquitous in North American construction.

Brick walls that are exterior walls may be load-bearing or they may be a façade that is attached to a load-bearing wall behind the brick. If a brick wall is load-bearing, it will be constructed with at least two layers of brick and will have a row of bricks laid on its end every six to eight rows to tie the layers together. This is called a header row, although it may be referred to by different names in different regions or varying style (**FIGURE 15-52**). This second layer of brick

A.

B.

C.

FIGURE 15-52 **A.** Load-bearing brick wall with header rows. **B.** Brick façade wall that is non-load-bearing. **C.** Decorative, half-brick veneer found on many modern buildings.

Courtesy of Sean Wilson.

FIGURE 15-53 Poured concrete wall.
Courtesy of Sean Wilson.

FIGURE 15-54 Wood stud walls can be load-bearing or non-load-bearing.
Courtesy of Tony Cattini.

behind the first helps to stabilize the wall and distribute the weight that the wall bears. A non-load-bearing brick wall is made of a single layer of bricks attached to a load-bearing wall behind it. In modern construction, it is also common to find exterior wall coverings made of a half-brick veneer that is decorative, but not load-bearing in any way.

As the name implies, poured concrete walls are made by pouring a wet concrete slurry into pre-built forms and letting them harden in place (**FIGURE 15-53**). These walls can be reinforced with metal rebar to give them tensile strength. These types of walls are most often found in newer high-rise buildings and new construction residential basement walls.

Stud Walls

Stud walls are constructed of vertical framing members called studs that run from floor to ceiling on each floor or from foundation to attic in older balloon-frame construction and can be load-bearing or non-load-bearing (**FIGURE 15-54**). Studs are made of dimensional lumber or hollow, C-shaped metal channels. Stud walls are finished by covering the studs with wood paneling, wood lath and plaster, or drywall. The studs are placed anywhere between 16 and 24 in. (41 to 61 cm) on center—that is, from the center of one stud to the center of the next one. When used for load-bearing walls, the studs are typically 16 in. (41 cm) on center. Stud walls are usually hollow between the studs. Plumbing and electrical wiring are often run through these voids. Exterior stud walls are often insulated with rolled fiberglass or blown-in insulation.

Security Bars and Gates

Security bars and gates can be found in just about any city or town. Some are fixed to buildings and immovable, and some roll down or flex into place and are padlocked. Some are mounted to the exterior and are easily visible; others sit inside doors or windows and are less perceptible from the outside.

Bars and gates are usually attached to the building in one of three ways. For masonry construction, bars can be directly set into the masonry of the building when it is built (**FIGURE 15-55**), or they can be installed later using expandable masonry anchors (**FIGURE 15-56**). For buildings with wood stud walls covered by siding or half-brick veneer, bars and gates are typically installed with large, threaded screws called lag bolts that are attached directly into the studs (**FIGURE 15-57**). Folding and hinged security gates are also attached to the building on one side of a door or window and extend or swing into place where they are secured with a padlock from the inside or outside (**FIGURE 15-58** and **FIGURE 15-59**).

The easiest and quickest method of removing bars and gates from a building is to use power tools. If they are secured with a padlock, cut the shackle of the lock if it is exposed. If the bars or gates are fixed and immobile, cut the attachment points with a rotary saw or hydraulic cutters on one side of the assembly, and then swing the rest outward. It may be necessary to cut all attachment points if the assembly cannot be swung out.

If no power tools are immediately available, use hand tools to demolish the anchor points where the

FIGURE 15-55 Security bars set directly into masonry.

Courtesy of Sean Wilson.

FIGURE 15-56 Security bars attached to masonry with expandable anchors.

Courtesy of Sean Wilson.

fixtures attach to the building. If bars are set directly into masonry, use a sledgehammer to destroy the masonry around the bars. If bars are attached using expandable anchors, attempt to pry the bars from the wall with a Halligan, or use a sledgehammer to destroy the masonry in which the expandable anchors are set.

If bars are set into wood stud walls with lag bolts, use a flat-head axe or a sledgehammer to drive the pick of the Halligan into the wood where the lag bolts hold the assembly to the building. This will split the wood, which will loosen the lag bolt enough that the assembly can be pried away from the building.

FIGURE 15-57 Security bars attached to a wood-frame building with lag screws.

Courtesy of Sean Wilson.

FIGURE 15-58 Folding security gates can be located on the inside or outside of doors or windows and are secured with a padlock.

Courtesy of Sean Wilson.

Water Damage Mitigation Devices

Water damage mitigation devices are increasingly found in coastal and other areas that are prone to flooding and heavy storms. There are two types of devices. The first type protects doors and windows from wind, rain, and flying debris. The most common version of this first type is a storm shutter, which is a wood, fiberglass, or metal covering attached to the area over doors

FIGURE 15-59 Hinged security gate secured with a padlock.

Courtesy of Sean Wilson.

and windows on the exterior of the structure. There are many types of storm shutters. Some are removeable and are affixed onto the building before a storm. Others are permanently attached and operated like a roll-up security door or are hinged like decorative window shutters. Some storm shutters are lockable and present an obstacle to firefighters who need to gain quick access to the structure. These lockable shutters should be treated like security gates and security bars on doors and windows and be forced in a manner consistent with those obstacles.

The second type of water damage mitigation device acts as a flood barrier to protect the structure and its contents from water damage due to flooding. These flood barriers are metal or heavy plastic gates that are dropped or fitted into hardware that is permanently attached to the door frame or window frame, and then secured by nuts screwed into existing bolts or simply held in place by the force of floodwaters pressing the barrier against the building. Often, they do not cover the entire door or window. Flood barriers are rarely locked, as they are not security gates, and they can usually can be removed by firefighters in the same manner they were put in place by the homeowner or building owner.

Rules of Forcible Entry

To ensure that the quickest possible entry is made and that unnecessary building damage is not incurred, firefighters follow four rules at the scene of every emergency incident:

1. Try before you pry. The number one rule of forcible entry has always been, and will always be: *Try before you pry*. Opening an unlocked door or window in the normal fashion is far easier and quicker than forcing one you assumed was locked. Therefore, any door or window you attempt entry through should be checked first prior to any forcible entry efforts.
2. Look for a lock box. A **lock box** is a small safe, usually located near the main entrance of a commercial building, that contains the keys to the building's exterior doors and possibly other interior doors (**FIGURE 15-60**). Many municipalities have a lock box program in which building owners can participate. Participation in the lock box program may be voluntary or it may be required by local ordinance. With this type of program, the fire department needs only one key, which is kept on the apparatus, to access all the lock boxes in district or municipality. Having quick access to the building's keys is far faster and less damaging than forcible entry for every type of situation, including working fires. Lock boxes also allow firefighters to easily re-secure the building at the termination of the call.

A.

B.

FIGURE 15-60 **A.** A commercial lock box. **B.** A residential lock box.

Courtesy of Sean Wilson.

Commercial lock boxes are so successful that some fire departments offer a residential lock box program to citizens for their homes or apartments. These residential lock boxes are popular with senior citizens and other residents with mobility issues that might prevent them from being able to open the door in the event of an emergency.

Bear in mind, however, that having a lock box program in your district does not relieve you of the responsibility of learning and understanding forcible entry theory and techniques. It is all too common for a lock box to be frozen or rusted shut, or to open a lock box only to discover that it is empty or contains keys that were not updated when a new building tenant moved in.

3. Consider the ventilation flow path. Before forcing entry at the scene of a working fire, you need to consider the flow path and how it will be affected by ventilation. Opening any door or window changes the flow path and feeds the fire with oxygen it did not have a moment before the opening was made. For example, if you enter through a door, that door should be closed after entry is made so that the fire is not supplied with additional oxygen. Or, if you break the glass out of a large picture window or glass storefront door, keep in mind that there is no easy way to close that opening if fire conditions worsen after the glass is gone.
4. Re-secure the building. At the conclusion of the incident, the building will need to be either be secured or turned over to the owner, a building representative, or another responsible individual or agency. Firefighters may be able to secure the building themselves, but, depending on the techniques used and the damage done, a locksmith or fire restoration company (a company that boards up buildings after fire) may need to be summoned to the scene before the fire personnel can leave (**FIGURE 15-61**).

Forcible Entry Tools

Selecting the right forcible entry tools for a particular situation is an important decision. Fire departments use many different types of tools, ranging from basic hand tools to sophisticated mechanical and hydraulic equipment. Firefighters everywhere must know the following:

- Which forcible entry tools their department uses
- Where forcible entry tools are carried on the apparatus (usually in a designated compartment)

FIGURE 15-61 After a fire or other emergency, the building must be secured or turned over to the owner or other responsible party.

Courtesy of Sean Wilson.

- What the uses and limitations of each tool are
- How to select the proper tool for the situation
- How to safely operate each tool
- How to safely carry each tool
- How to inspect and maintain each tool

Many of the tools you will use for forcible entry were discussed in Chapter 7, *Firefighter Tools and Equipment*. They are summarized in **TABLE 15-1**.

In addition to the basic tools listed in Table 15-1, other specialized tools used for forcible entry include the following:

- Lock-pulling tools: These tools are used to pull lock cylinders or manipulate latch mechanisms in combination with hammers, mallets, or other striking tools, which set the lock-pulling tools in place. They are often carried together in a tool bag for use in situations when the priority of entry is keeping damage to a minimum.
  - **A tool**: The A tool is used to pull bored locks. It resembles a small Halligan tool, but it has an A-shaped notch cut into the adze (**FIGURE 15-62**). They are usually 16 to 20 in. (41 to 51 cm) in length. The A tool was the first lock-pulling tool to be used in the fire service. The earliest version was nothing more than a hardware store nail-puller tool with the notched widened so it could pull lock cylinders. Modern versions resemble the type shown in Figure 15-62.
  - **K tool**: So-called because it resembles the letter *K*, this tool is used to pull mortise and rim lockset

TABLE 15-1 Tools Used for Forcible Entry

Category	Tools
Cutting	▪ Axes ▪ Flat-head axe (6 or 8 pounds [lb]; 3.6 to 4.5 kilograms [kg]) ▪ Pick-head axe ▪ Bolt cutter ▪ Power saws ▪ Rotary saw (with wood, metal, and masonry blades) ▪ Reciprocating saw (with wood and metal blades)
Pushing/pulling	Pike poles and other hooks have limited use in forcible entry beyond breaking glass. Using these tools exclusively to attempt forcible entry through a door or other entry point can damage the tool. *Note: A pick-head axe can be used to push and pull.*
Prying/spreading	▪ Halligan tool ▪ Hydra-Ram ▪ Hydraulic spreader ▪ Kelly tool ▪ Pinch-point pry bar ▪ Rabbit tool
Rotating	▪ Pliers ▪ Screwdrivers ▪ Wrenches
Striking	▪ Battering ram ▪ Hammer ▪ Mallet ▪ Splitting maul ▪ Sledgehammer *Note: The flat-head axe can be used as a striking tool.*
Multiple-function	▪ THE PIG ▪ Fire Maul (10 lb [4.5 kg])

Courtesy of Sean Wilson.

FIGURE 15-62 The A tool, which is used to pull bored locksets.

Courtesy of Sean Wilson.

FIGURE 15-64 The Rex tool, which is used on mortise, rim, and key-in-knob locksets.

Courtesy of Sean Wilson.

FIGURE 15-63 The K tool, which is used to pull mortise and rim locksets.

Courtesy of Sean Wilson.

FIGURE 15-65 An S&D tool, which is used to pull any type of door lock using one of the two heads.

Courtesy of Sean Wilson.

cylinders when used in combination with a Halligan (**FIGURE 15-63**). It was designed in the early 1970s by a firefighter who was also a licensed locksmith.

- **Rex tool**: This 24-in. (61-cm) steel shaft with a large lock-pulling wedge welded onto one end is used to pull mortise and rim cylinders and key-in-knob locks. The wedge has a notch milled into it in the shape of a *U* (**FIGURE 15-64**). It was designed as an update to the K tool in the late 1990s.
- **S&D tool**: This double-headed, lock-pulling tool combines the head of an A tool and the head of a Rex tool onto an 18-in. (46-cm) shaft. It is used to pull any type of common commercial or residential door lock (**FIGURE 15-65**).

- **Key tools**: These are three simple tools—a ¼-in. (6-mm) slotted screwdriver, needle-nose pliers, and a bent-end key tool—often carried together as a set used to manipulate most locks after the cylinder is pulled or removed (**FIGURE 15-66**).
- Locking pliers and slip-joint pliers: Pliers are used to grab the face of mortise cylinders and spin them out to expose the latch mechanism so that it can be manipulated using a key tool (**FIGURE 15-67**).
- **Shove knife**: A thin, flexible piece of flat steel with a handle and a notch cut into the blade, it is used to manually push the slam latch of a key-in-knob or mortise lock out of the strike plate to open the door (**FIGURE 15-68**). These tools can occasionally be

A.

B.

FIGURE 15-66 **A.** Key tools: small flat-head screwdriver, bent-end key tool, and small needle-nose pliers. **B.** Close-up of the bent-end key tool.

Courtesy of Sean Wilson.

FIGURE 15-67 Locking pliers and slip-joint pliers.

Courtesy of Sean Wilson.

FIGURE 15-68 Various shove knives.

Courtesy of Sean Wilson.

FIGURE 15-69 A wooden door wedge and various aluminum force wedges.

Courtesy of Sean Wilson.

FIGURE 15-70 The J tool.

Courtesy of Sean Wilson.

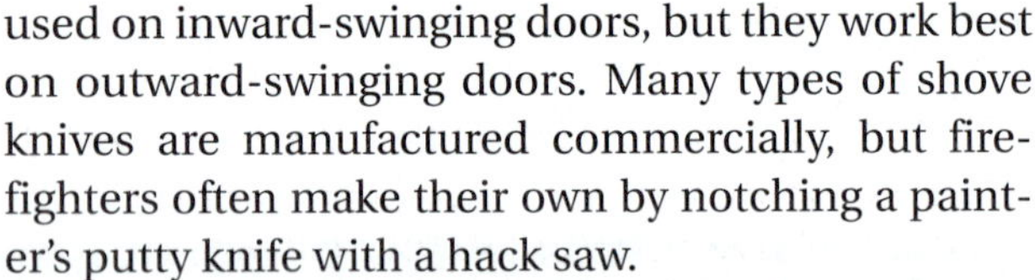

used on inward-swinging doors, but they work best on outward-swinging doors. Many types of shove knives are manufactured commercially, but firefighters often make their own by notching a painter's putty knife with a hack saw.

- Door wedge and force wedge: Wedges are used to keep doors open or widen the gap between a door and its frame when using a Halligan or a shove knife (**FIGURE 15-69**). Wood door wedges, usually made from dimensional lumber cut by individual firefighters, are often carried by firefighters in a gear pocket. Aluminum force wedges are used for similar purposes but are commercially made and more durable.
- **J tool**: A small diameter metal rod bent into a large *J* or *C* shape, usually between 14 and 24 in. (36 and 61 cm) in overall length, it is used to open panic hardware on double, outward-swinging doors by fitting into the seam between the two doors and engaging the panic hardware (**FIGURE 15-70**). It is sold commercially but also is sometimes made by firefighters out of the handle of a paint roller or the metal framing from an election yard sign.
- **Scraper bar**: This small, flat bar about 10 in. (25 cm) long is designed to scrape paint but is used by firefighters to drive sash locks off of certain types of windows (**FIGURE 15-71**).

FIGURE 15-71 Scraper bar.

Courtesy of Sean Wilson.

- **Duckbill padlock breaker**: As the name implies, this tool comprises a large, steel wedge resembling

FIGURE 15-72 Duckbill padlock breaker.

a duck's bill welded onto a metal shaft that serves as a handle that is used for breaking padlocks (**FIGURE 15-72**). One firefighter fits the narrow end of the wedge into the shackle of a standard padlock and holds the handle of the tool. A second firefighter strikes the back of the wedge with a sledgehammer or the flat head of a flat-head axe. With each blow of the striking tool, the increasing height of the wedge forces the shackle out of the lock-body until the locking mechanism is overcome.

Forcible Entry Categories

There are four categories of forcible entry that firefighters typically use: conventional, through-the-lock, rotary saw, and specialty. Firefighters should understand the differences between the strategies used in each category and why a technique in one category might be employed over a technique in another category based on a given scenario. Selection of one technique over another may hinge on the nature of the emergency, the tools available on scene, and the experience and training levels of the firefighters tasked with forcible entry.

Conventional Forcible Entry

Conventional forcible entry means using standard fire service hand tools to push or pull a swinging door through its normal range of motion. This typically means using a prying tool like a Halligan with a striking tool like a sledgehammer or a heavy axe to pry an outward-swinging door out or push an inward-swinging door in.

Damage caused to the door system by conventional forcible entry is usually significant. If the damage is catastrophic, it can require the replacement of the door, the lockset, and the frame. For this reason, conventional forcible entry is typically only used during high-priority emergencies such as working fires, fire alarm responses with visible smoke, and serious medical emergencies. Conventional forcible entry may also be used as an eventual "Plan B" technique for when other, less-damaging techniques have failed at the scene of a low-priority emergency.

TIP

Remember, always try before you pry!

Conventional Forcible Entry with Wood Frame Doors

Forcing entry into a house on fire is usually simple—though not necessarily easy—for two reasons: the door usually swings inward and it is usually set in a wood frame. In most instances, this type of door system can be defeated quickly by striking the door using one of two methods: the sledgehammer method and the stutter-step method. Both methods apply force to the door, and this force is transmitted to the latch inserted into the strike plate, which in turn cracks the wood door frame, thus allowing entry into the home. Both methods usually require striking the door repeatedly until the frame cracks and the door opens.

Kicking or shouldering a door are not safe or appropriate techniques. Although weak doors may sometimes be openable using these techniques, they carry a great potential for injury to the firefighter. In addition, these techniques reflect poorly on the professionalism of the crew and fire department who use them. A properly equipped crew who are knowledgeable about their role on the fireground and who are expecting to fight a fire have no need to bring their feet to an irons fight.

Sledgehammer Method. The sledgehammer method is performed on inward-swinging doors with a tubular deadbolt. To do this, select a sledgehammer, THE PIG, a heavy axe, or another striking tool of similar weight (at least 8 lb [3.6 kg]). Then, swing the tool to strike the door on or above the tubular deadbolt. If six to eight strikes do not open the door, you will need to switch to prying techniques, which are discussed later in this chapter. To force an inward-swinging door with a tubular deadbolt using the sledgehammer method, follow the steps in **SKILL DRILL 15-1**.

SKILL DRILL 15-1

Forcing an Inward-Swinging Door Set in a Wood Frame with a Tubular Deadbolt Using the Sledgehammer Method Firefighter I, NFPA 1010: 6.3.4

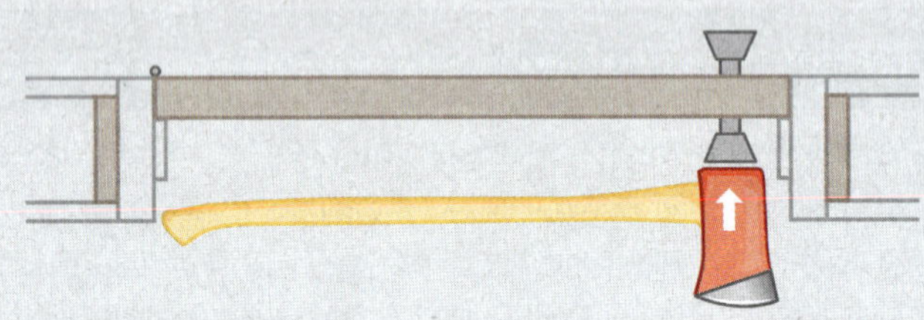

1. Try the door to ensure it is locked; size up the number of locksets; and verify that the locksets are tubular deadbolts, that the door is set in a wood door frame, and that the door swings inward. Ensure that the area is free of obstructions and that other firefighters are standing clear. Select a striking tool with an 8- to 12-lb (3.6- to 5.4-kg) head. Stand in front of the door and place the head of the tool in contact with the door on the deadbolt or just above it. This area is the striking target.

2 – Draw back

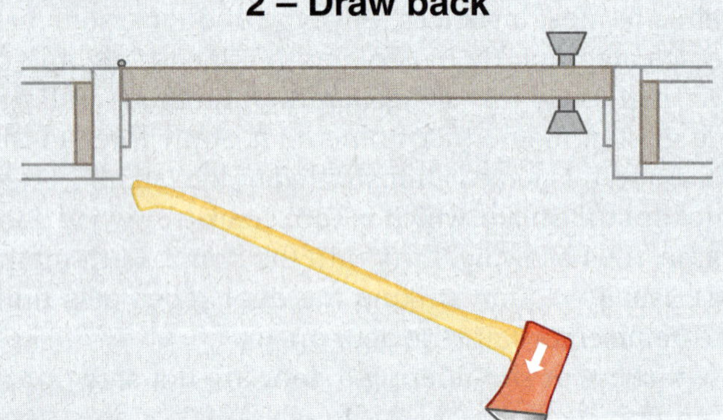

2. Draw the tool back in a ready stance.

3 – Swing forward

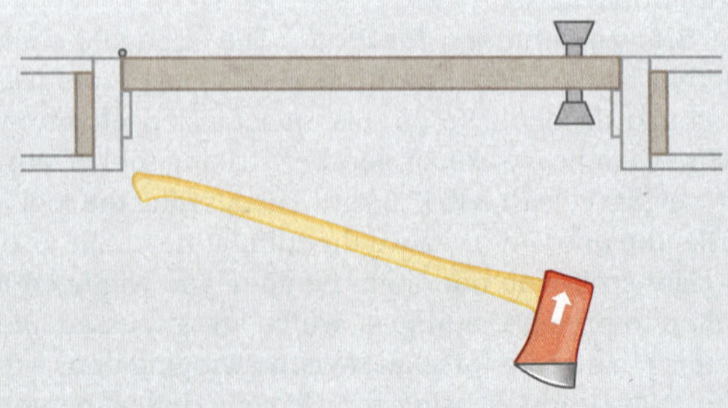

3. Swing the tool and strike the face of the deadbolt or just above it.

SKILL DRILL 15-1 CONTINUED

Forcing an Inward-Swinging Door Set in a Wood Frame with a Tubular Deadbolt Using the Sledgehammer Method Firefighter I, NFPA 1010: 6.3.4

4 – Strike the door

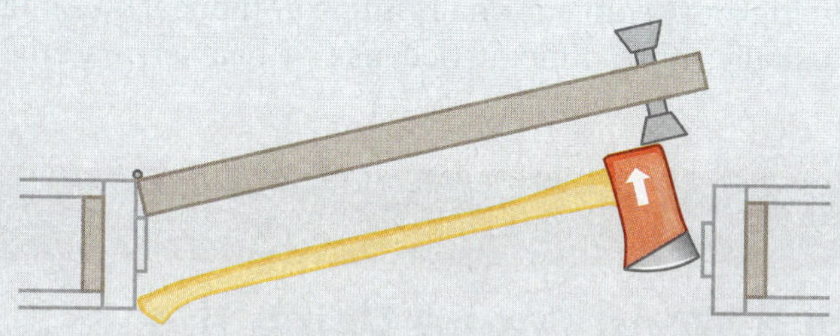

4. Strike the door repeatedly in this spot until the door opens. After six to eight strikes, if no progress has been made opening the door or breaking the frame, consider switching to a prying technique.

5 – Grab the door

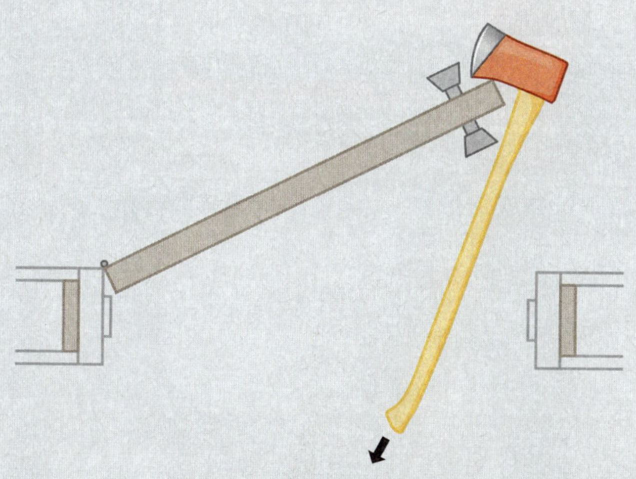

5. If the structure is on fire and if the door is successfully forced open, reach in with the head of the tool and pull the door shut on the tool shaft so that the door closes off the flow path without re-locking.

6 – Control the door

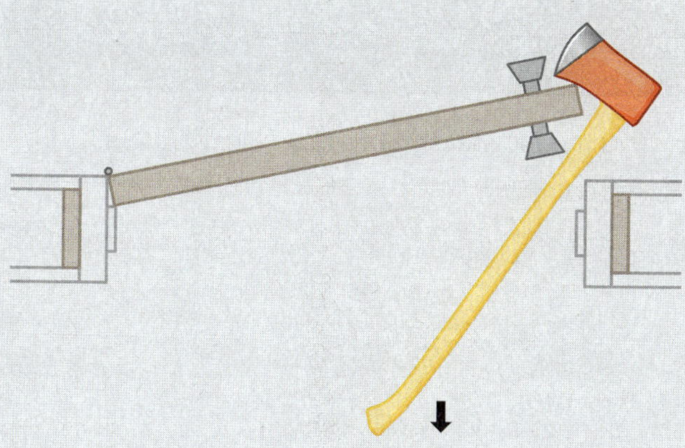

6. Keep the door closed until the crew is ready to enter the building.

Stutter-Step Method. The stutter-step method is a forcible entry method that forces an inward-swinging door in a wood frame by using a Halligan or almost any striking tool as a battering ram. It is often used when there is not enough space to swing the tool using the sledgehammer method, but there is linear space leading to the door. To perform the stutter-step method, stand sideways to the door with feet shoulder-width apart. Grasping the tool with two hands, line up the tool on the striking point—usually below the knob or lever—step back with both feet while drawing the tool backward, and then step forward with both feet while swinging the tool forward forcefully to strike the door. The momentum generated by striding forward and swinging the tool forward at the same time is the force that cracks the frame and opens the door. To force open a door using the stutter-step method, follow the steps in **SKILL DRILL 15-2**.

Conventional Forcible Entry with Metal Frame Doors

Forcing entry conventionally into commercial buildings is usually more difficult because of heavier materials

SKILL DRILL 15-2

Forcing an Inward-Swinging Door Set in a Wood Frame Using the Stutter-Step Method Firefighter I, NFPA 1010: 6.3.4

1. Try the door to ensure it is locked, size up the type and number of locksets, and verify that the door is an inward-swinging door set in a wood door frame. Ensure that the area is free of obstructions and that other firefighters are standing clear. Stand sideways to the door with feet shoulder-width apart and grasp an axe, sledgehammer, or Halligan with both hands. Place the head of the tool in contact with the door just below the knob or lever. This area is the striking target.

2. Pull the tool backward away from the door.

SKILL DRILL 15-2 CONTINUED

Forcing an Inward-Swinging Door Set in a Wood Frame Using the Stutter-Step Method Firefighter I, NFPA 1010: 6.3.4

3. Step back from the door with both feet.

4. Step forward forcefully with both feet while swinging the tool forward and strike the target.

5. Repeat this move until the door opens. After six to eight strikes, if no progress has been made opening the door or breaking the frame, consider switching to a prying technique. If the structure is on fire and if the door has been successfully forced open, reach in with the head of the tool and pull the door shut on the tool shaft so that the door closes off the flow path without re-locking. Keep the door closed until the crew is ready for entry.

Courtesy of Sean Wilson.

used in commercial doors and locksets, and because the door frame is typically made of metal and often filled with concrete. Also, doors on commercial buildings often swing outward. For these reasons, it is unlikely that the sledgehammer or stutter-step method will be effective to force entry through these doors. You also may encounter a door on a residential home that cannot be opened using the sledgehammer or stutter-step method because it has three, four, five, or even more locksets.

When faced with door with a metal frame in a commercial building—or any formidable door, whether residential or commercial, inward- or outward-swinging—the best practice is to use a more measured approach called the GAP-SET-FORCE technique using a Halligan and an axe or a sledgehammer (an irons set) to pry the door open. The GSF technique is a prying attack in which one of the working ends of the Halligan is pushed into the tight seam between the door and door frame (or between the door and door stop) and used to pry the elements apart. The force delivered usually bends or breaks the latch of the lockset, as it is the latch in the strike plate on the door frame that physically holds the door shut.

As the name implies, GSF is delivered as a three-step process:

- Gap: One firefighter holds the Halligan adze or fork up to the seam between the door and door frame, about 1 foot (ft; 0.3 meter [m]) above or below the lockset. A second firefighter strikes the Halligan with an axe or sledgehammer until the adze or fork is inserted into the seam. If it is an outward-swinging door, the adze or fork will be resting on the door stop. The seam is now gapped.
- Set: The firefighter holding the Halligan pulls or pushes the Halligan to change the angle of the adze or fork so it can slip past the door stop on an outward-swinging door or past the door frame on an inward-swinging door, and the firefighter holding the axe or sledgehammer strikes the Halligan to drive it past the door stop or the inner portion of the door frame. The Halligan is now set into place.
- Force: Once the tool is set, the firefighter holding the Halligan adjusts their stance and grip on the tool and attempts to force the door by applying force to the bar to break or bend the weakest part of the door system, which is usually the lock.

When performing conventional forcible entry using an irons set, the team concept is used. The irons team comprises two firefighters: the Halligan firefighter and the axe firefighter. The Halligan firefighter's job is to try the door to see if it is unlocked, size up the door for forcible entry, quickly develop a plan, and then communicate commands to the axe firefighter. The axe firefighter's job is to listen for commands and to safely deliver blows to the Halligan at the direction of the Halligan firefighter. The Halligan firefighter must communicate their needs with a series of simple commands. These commands are used because they are simple words that are easy to say, easy to understand, and cannot be confused for a different command. The commands are as follows:

- "HIT" means hit the Halligan with the axe one time, and then wait for further instructions.
- "DRIVE" means keep hitting the Halligan until I tell you to stop.
- "STOP" means no more hits until further instructions are given.
- "WEDGE" means insert the head of the axe into the door to maintain the progress that you have made (called "holding the purchase") while the Halligan is being moved or adjusted.
- "HELP" means help is needed pushing or pulling the Halligan.

Forcing an Outward-Swinging Door with a Halligan Fork. The simplest way method of using the GSF technique is to use the fork end of the Halligan against an outward-swinging door. First, the Halligan firefighter brings the tool to the door at a 90-degree angle and places the fork of the Halligan in the seam between the door and door frame approximately 1 ft (0.3 m) above or below the lockset, with the bevel or rounded side of the fork facing the frame. Next the axe firefighter strikes the Halligan, driving the fork into the seam until the fork rests on the door stop. The Halligan firefighter then pushes or pulls the bar horizontally to change the angle of the fork while the axe firefighter drives the Halligan into place until the fork is set past the door stop. The tool is now in position where the Halligan firefighter can push or pull the Halligan to force the door out of its frame and away from the building. To force an outward-swinging door using GSF and a Halligan fork, follow the steps in **SKILL DRILL 15-3**.

Forcing an Outward-Swinging Door with a Halligan Adze. There are times when the range of motion necessary to use the fork of the Halligan may not exist, or the seam between the door and frame is too tight to insert the fork. In these instances, using the adze of the Halligan is preferred because the adze is slimmer than the fork and it is oriented 90 degrees to the tool shaft, which allows a different range of motion. To force an outward-swinging door using GSF and a Halligan adze, follow the steps in **SKILL DRILL 15-4**.

SKILL DRILL 15-3

Using the GSF Technique and a Halligan Fork on an Outward-Swinging Door Firefighter I, NFPA 1010: 6.3.4

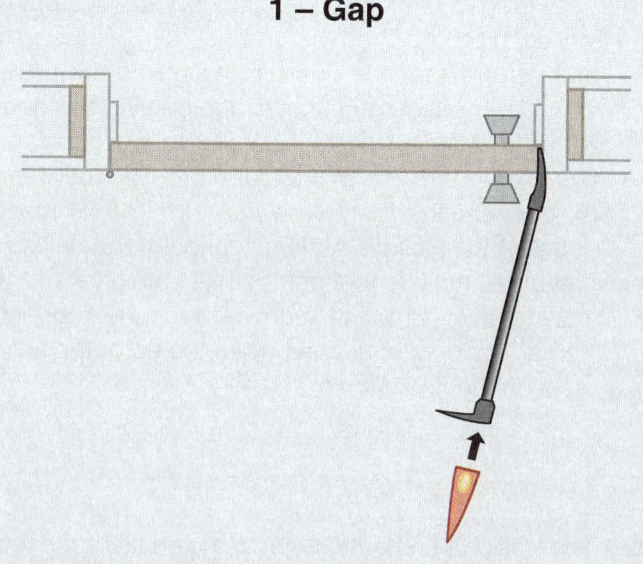

1. Try the door to ensure it is locked. Size up the type and number of locksets and verify that the door is an outward-swinging door. The Halligan firefighter approaches the door with the shaft of the tool at a 90-degree angle to the door, and then places the fork end into the seam between the door and the frame about 1 ft (0.3 m) above or below the lockset with the bevel or rounded side of the fork facing the frame. At the direction of the Halligan firefighter, the axe firefighter strikes the adze of the Halligan to drive the fork into the seam between the door and frame until the fork tips are resting on the door stop to set the gap.

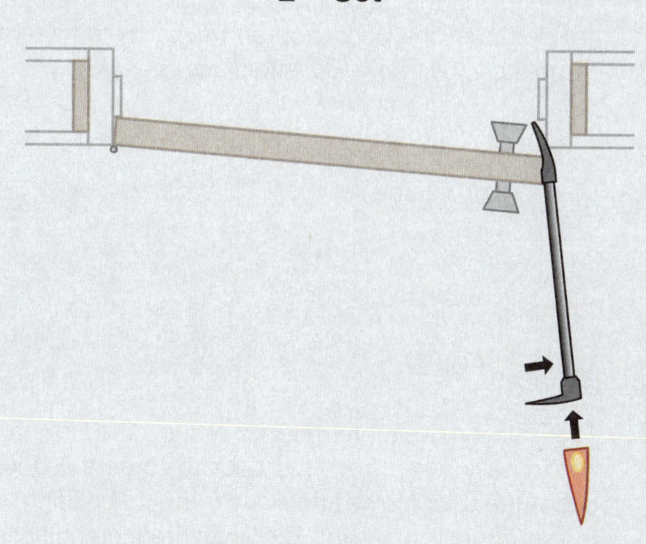

2. The Halligan firefighter adjusts the angle of the tool slightly, and then directs the axe firefighter to continue striking the adze of the tool until the fork tips have traveled past the door stop and through the seam completely to set the tool into place.

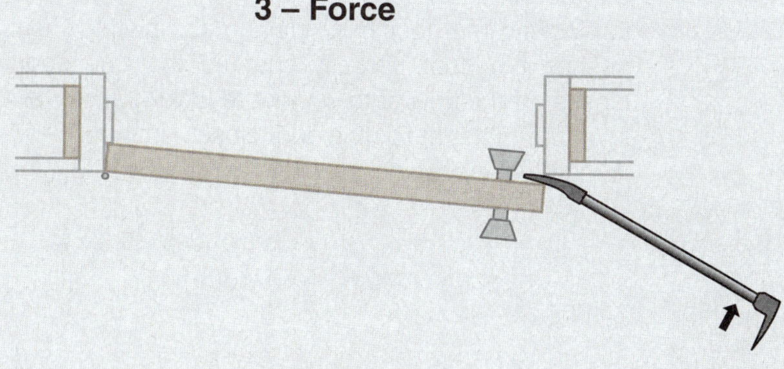

3. The Halligan firefighter pushes or pulls the Halligan horizontally away from the door to force the door out of the door frame and away from the building. If the structure is on fire, the Halligan firefighter places the tool on the ground so that a portion of the tool is blocking the door from closing completely, then pushes the door partially shut so that the door closes off the flow path without re-locking. The Halligan firefighter keeps the door closed until the crew is ready for entry.

SKILL DRILL 15-4

Using the GSF Technique and a Halligan Adze on an Outward-Swinging Door Firefighter I, NFPA 1010: 6.3.4

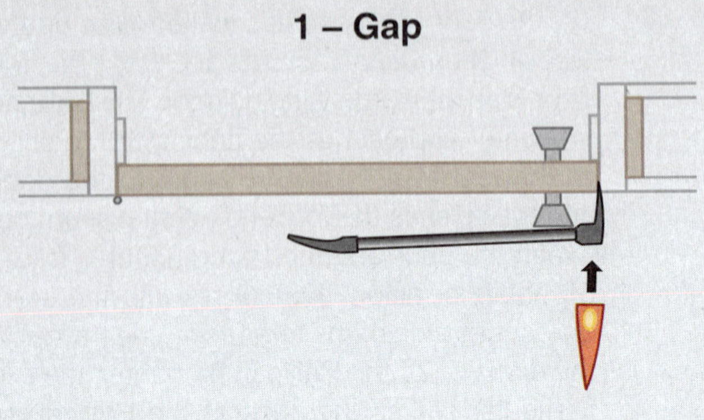

1. The Halligan firefighter tries the door to ensure it is locked, sizes up the type and number of locksets, and verifies that the door is an outward-swinging door. The Halligan firefighter approaches the door with the shaft of the tool parallel to the door and then places the adze of the tool in the seam between the door and the door frame about 1 ft (0.3 m) above or below the lockset. At the direction of the Halligan firefighter, the axe firefighter strikes the back of the adze to drive it into the seam between the door and frame. The door is gapped when the tip of the adze rests on the door stop.

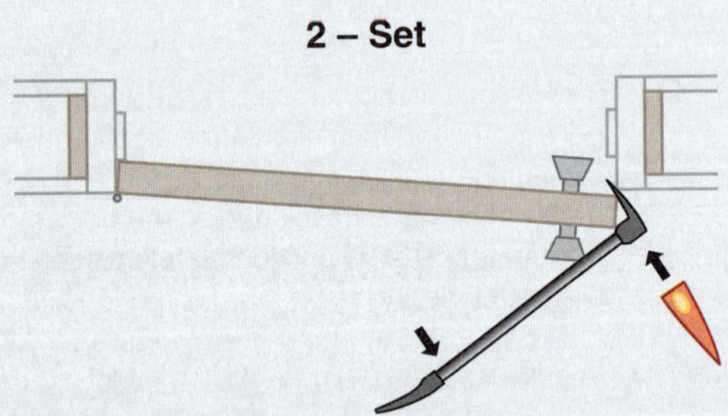

2. The Halligan firefighter adjusts the angle of the tool slightly, and then directs the axe firefighter to continue striking the back of the adze until the tool has traveled past the door stop and through the seam completely to set the tool into place.

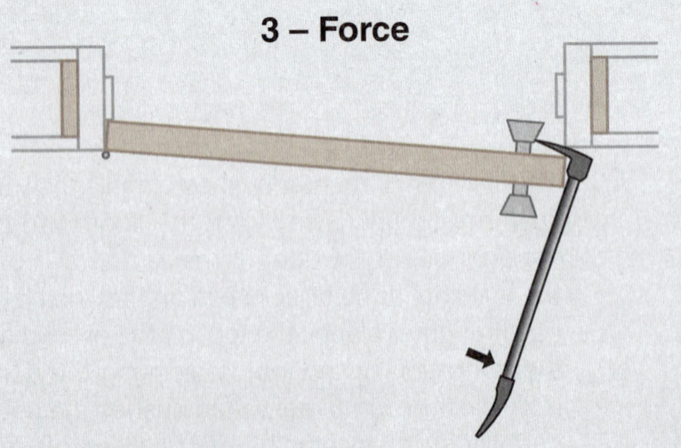

3. The Halligan firefighter pushes or pulls the Halligan horizontally away from the door to force the door out of the door frame and away from the building. If the structure is on fire, the Halligan firefighter places the tool on the ground so that a portion of the tool is blocking the door from closing completely, then pushes the door partially shut so that the door closes off the flow path without re-locking. The Halligan firefighter keeps the door closed until the crew is ready for entry.

Forcing an Inward-Swinging Door with a Halligan Fork. Forcing inward-swinging doors using the GSF technique can be challenging because in an inward-swinging door system, the door stop covers the seam. This can make it difficult to create the gap between door and the door frame. However, if the crew is disciplined and has a clear understanding of the door system and the technique, the odds of success are high.

To force an inward-swinging door with a Halligan fork, you use the same GSF technique you use when you force an outward-swinging door with a Halligan fork with the following modifications:

- The Halligan firefighter brings the tool to the door at a 45-degree angle instead of the 90-degree angle used for forcing an outward-swinging door.
- The Halligan firefighter places the fork of the Halligan into the seam between the door and the door stop so bevel (rounded side) of the fork faces the door instead of the door stop or door frame.
- To gap the door, the axe firefighter drives the fork into the seam until the fork tips rest on the door frame between the door and the door stop.
- To set the tool into place, the Halligan firefighter slowly pulls the bar horizontally to change the angle of the fork, while the axe firefighter drives the Halligan so that the fork tips travel past the interior frame casing.
- To force the door, the Halligan firefighter pushes the Halligan forcefully toward the door horizontally.

To force an inward-swinging door using GSF and a Halligan fork, follow the steps in **SKILL DRILL 15-5**.

Forcing an Inward-Swinging Door with a Halligan Adze. Using a Halligan adze to force an inward-swinging door is almost the same as using a Halligan fork. This method, however, uses the adze of the Halligan to create a gap between the door stop and the door, and then uses the head of the axe or an aluminum force wedge to hold the gap and widen it in order to set the Halligan into place to force the door. This method is useful on tight door systems where inserting the fork into the seam is difficult.

First, the Halligan firefighter brings the tool to the door with the shaft at a 90-degree angle to the door and places the tip of the adze into the seam between the door and door stop about 1 ft (0.3 m) above or below

SKILL DRILL 15-5

Using the GSF Technique and a Halligan Fork on an Inward-Swinging Door Firefighter I, NFPA 1010: 6.3.4

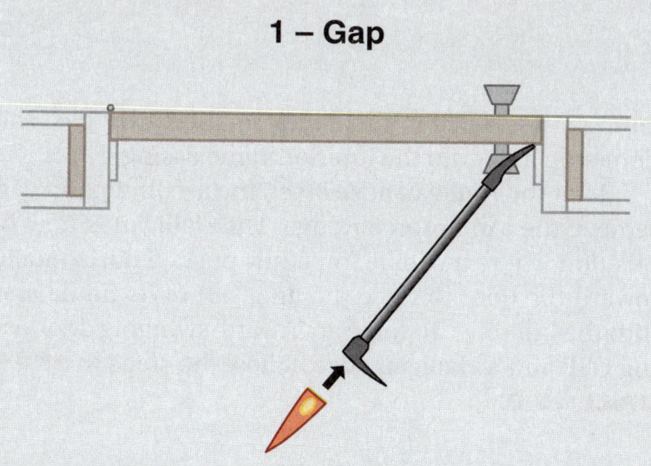

1. The Halligan firefighter tries the door to ensure it is locked, sizes up the type and number of locksets, and verifies that the door is an inward-swinging door. The Halligan firefighter approaches the door with the shaft of the tool at a 45-degree angle to the door, and then places the fork end into the seam between the door and the door stop about 1 ft (0.3 m) above or below the lockset with the bevel (the rounded side) of the fork facing the door. At the direction of the Halligan firefighter, the axe firefighter strikes the adze of the Halligan to drive the fork into the seam between the door and door stop until the fork tips are resting on the door frame to create the gap.

Continues.

SKILL DRILL 15-5 CONTINUED

Using the GSF Technique and a Halligan Fork on an Inward-Swinging Door Firefighter I, NFPA 1010: 6.3.4

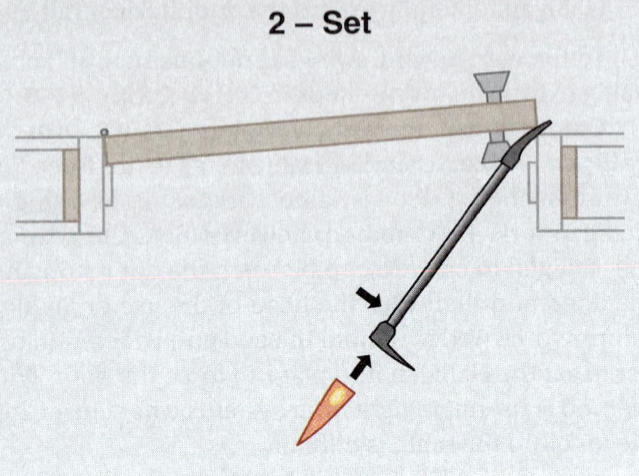

2. The Halligan firefighter adjusts the angle of the tool slightly, and then directs the axe firefighter to continue striking the adze of the tool until the fork tips have traveled through the seam and past the inner frame casing completely to set the tool into place.

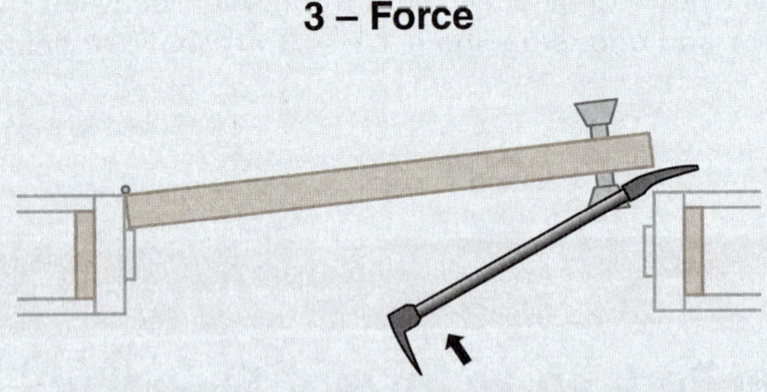

3. The Halligan firefighter forcefully pushes the tool horizontally toward the door to force the door into the building and out of the door frame. If the structure is on fire, the Halligan firefighter reaches in with the head of the tool and pulls the door shut on the tool shaft so that the door closes off the flow path without re-locking. The Halligan firefighter keeps the door closed until the crew is ready for entry.

the lockset. If necessary, the axe firefighter strikes the Halligan to set the adze behind the door stop. Next, the Halligan firefighter slowly rolls the bar vertically (up or down) to create a gap between the door and door frame. The axe firefighter then brings the head of the axe or an aluminum force wedge in to capture this space temporarily. Once captured, the back of the axe or force wedge can be struck with the Halligan to widen the gap between door and door frame. Once that the gap is wide enough, the Halligan firefighter sets the adze past the door stop and onto the interior frame casing.

After the Halligan adze is set, the axe firefighter can remove the axe or force wedge. The Halligan is now in position where it can be forcefully pushed horizontally toward the door to force the door out of its frame and into the building. To force an inward-swinging door using GSF and a Halligan adze, follow the steps in **SKILL DRILL 15-6**.

SKILL DRILL 15-6

Using the GSF Technique with a Halligan Adze on an Inward-Swinging Door Firefighter I, NFPA 1010: 6.3.4

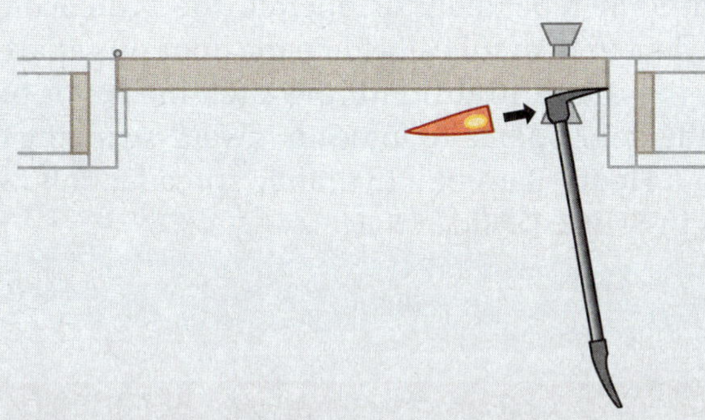

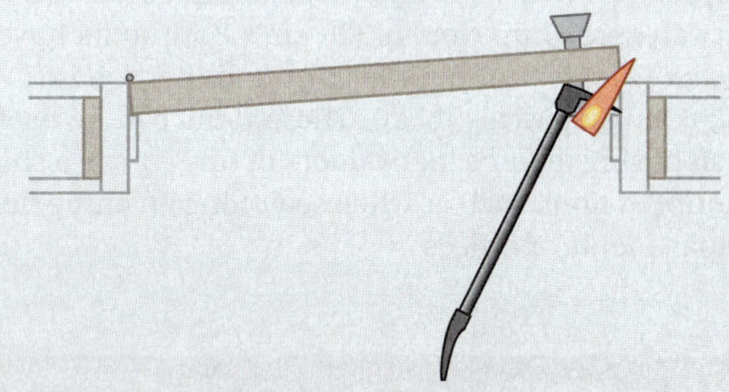

1. The Halligan firefighter tries the door to ensure it is locked, sizes up the type and number of locksets, and verifies that the door is an inward-swinging door. The Halligan firefighter approaches the door with the shaft of the tool at a 90-degree angle to the door, places the adze of the tool between the door and the door stop about 1 ft (0.3 m) above or below the lockset with the top of the adze facing the door, and then rolls the tool vertically (up or down) to expose a triangular-shaped gap between the door and the door frame. (If the door is tight in its frame, the axe firefighter will strike the Halligan to set the adze behind the door stop).

2. The axe firefighter inserts the blade of the axe or a force wedge into the gap.

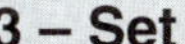

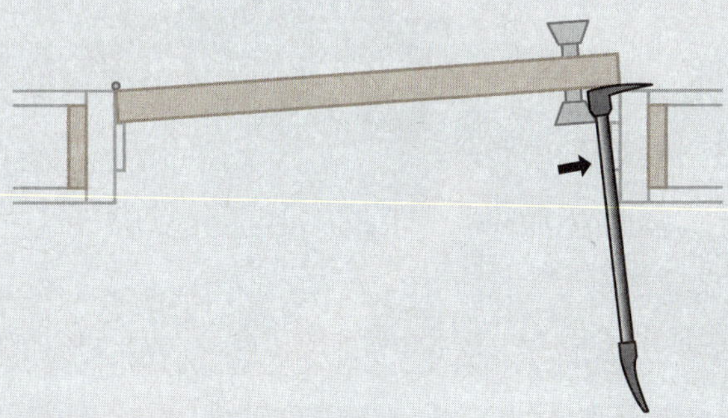

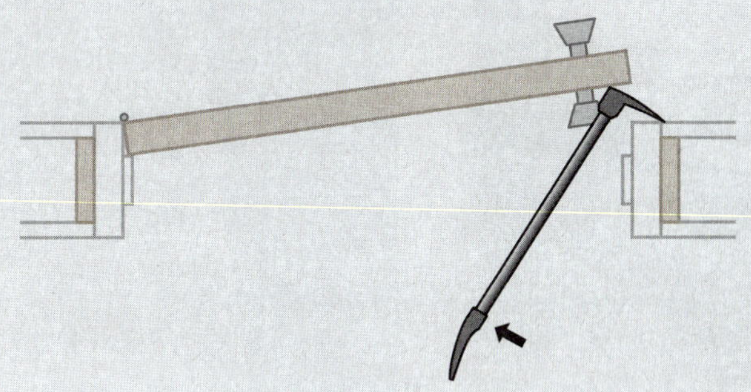

3. The Halligan firefighter removes the Halligan from the gap, and then strikes the back of the axe head or force wedge with the Halligan to widen the gap, and then inserts the adze (or fork) into the widened gap to set the tool onto the inner frame casing.

4. The Halligan firefighter brings the tool to a 90-degree angle to the door, and then forcefully pushes the tool either vertically (up or down) or horizontally (toward the door) to force the door into the building and out of the door frame. If the structure is on fire, the Halligan firefighter reaches in with the head of the tool and pulls the door shut on the tool shaft so that the door closes off the flow path without re-locking. The Halligan firefighter keeps the door closed until the crew is ready for entry.

Forcing an Inward-Swinging Door with a Hydra-Ram or Hydraulic Spreaders. Forcing an inward-swinging door with a Hydra-Ram or hydraulic spreaders with door tips is relatively easy. These units were designed specifically to force inward-swinging doors set in metal frames; they have limited use on other door systems. The Hydra-Ram weighs 12 lb (5.4 kg) and can spread up to 4 in. (10 cm). The hydraulic spreaders with door tips weigh about 25 lb (11.3 kg) and can spread up to 8 in. (20 cm). Both units have a spreading force of about 10,000 pounds per square inch (psi; 6.90 kilopascals [kPa]). These units can be useful when having to force many doors of this type in a short amount of time, such as when searching an entire floor of apartments or offices.

To use one of these hydraulic tools to force an inward-swinging door, first set the door tips of the tool into place between the door and the door stop just above or just below the lockset. In tight door systems, you might need to tap the unit with a rubber mallet to set the door tips into place. (Do not use an axe or hammer for this purpose as it can damage the tool.) Once the tool is set into place, pump the hand pump of the Hydra-Ram repeatedly or operate the switch on the spreaders to hydraulically force the door tips apart and push the door inward until the latch breaks or bends and the door opens. To force an inward-swinging door using a Hydra-Ram or a hydraulic spreader, follow the steps in **SKILL DRILL 15-7**.

SKILL DRILL 15-7

Using the GSF Technique on an Inward-Swinging Door with a Hydra-Ram or Hydraulic Spreaders with Door-Opening Tips
Firefighter I, NFPA 1010: 6.3.4

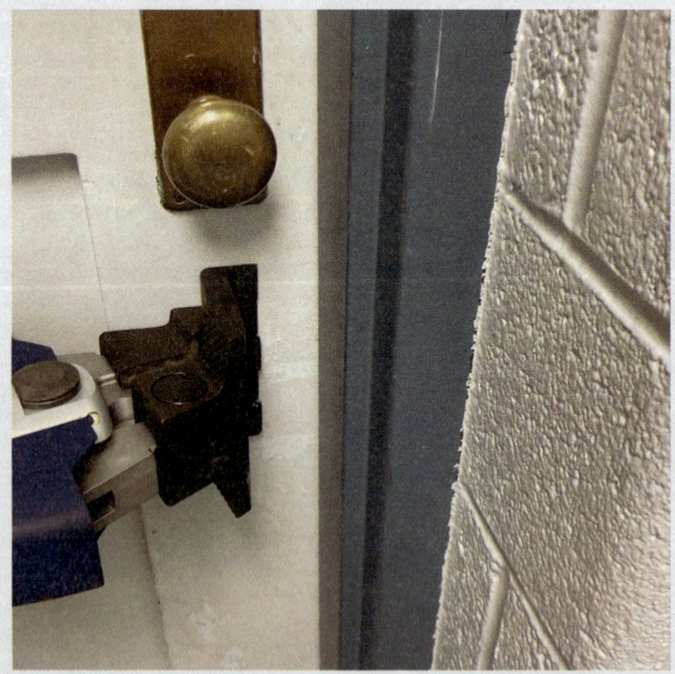

1. Try the door to ensure it is locked, size up the type and number of locksets, and verify that the door is an inward-swinging door. Place the tips of the tool into the seam between the door and the door stop just above or just below the lockset.

2. If necessary, tap the tool into place with a rubber mallet. Do not strike the tool with an axe or hammer because it may damage the unit.

SKILL DRILL 15-7 CONTINUED

Using the GSF Technique on an Inward-Swinging Door with a Hydra-Ram or Hydraulic Spreaders with Door-Opening Tips
Firefighter I, NFPA 1010: 6.3.4

3. Open the spreading tips on the tool by pumping the hand-pump of the Hydra-Ram repeatedly or by operating the switch on the spreaders.

4. Continue applying force until the door clears the door frame and is opened. If the structure is on fire, reach in with a hand tool and pull the door shut on the tool shaft so that the door closes off the flow path without re-locking. Keep the door closed until the crew is ready for entry.

Courtesy of Sean Wilson.

Through-the-Lock Forcible Entry

Through-the-lock forcible entry is the removal or bypass of the cylinder of a lockset so that the latch can be manipulated. It is often performed in two separate stages: the removal of the cylinder or lock face and the direct manipulation of the latch mechanism. If one of these two stages is performed incorrectly, then the lockset may not be able to be opened and more destructive techniques may need to be used. By itself, through-the-lock forcible entry is often damaging only to the lockset, which is the cheapest portion of the door system to replace.

This type of forcible entry often requires specialized tools and some basic knowledge of common locksets. It is often performed at low-priority emergencies, such as routine fire alarms and medical alarms, but is also sometimes appropriate at working fires depending on the type of door and lockset encountered and whether lock-pulling tools are immediately available.

Through-the-lock forcible entry tools are often carried in a small tool bag along with sprinkler tongs, elevator keys, door wedges, screwdrivers, and other small tools needed at routine emergencies. The irons should accompany this tool bag because conventional forcible entry is "Plan B" should less-destructive methods fail.

Selection of the proper tool and technique when pulling a lock is largely dictated by the type of lock you are faced with and the tools available on scene. Some fire departments are well equipped with lock-pulling tools on all apparatus, and some have no tools of this kind at all. Learning to tailor your actions based on the conditions you find in the field and the tools in your arsenal is essential.

Forcing a Bored Lockset

Bored locksets are relatively easy to defeat because the cylinder is housed in the knob or lever. The easiest method is to use the large jaws on the S&D or Rex tool to grab the knob or lever and rock it back and forth until the crimping that holds the knob to the lockset becomes loose and can be removed. This exposes the operating spindle that connects to the latch mechanism. An operating spindle can easily be rotated with a flat-head screwdriver or pair of needle-nose pliers, which will unlock the door. If an S&D or Rex tool is unavailable, the entire face of the exterior lockset can be removed by using the adze of an A tool or a Halligan to pry the lock face and expose the latch mechanism so that it can be manipulated with a screwdriver. This technique will work against many tubular deadbolts as well. To force a bored lockset with an S&D or Rex tool, follow the steps in **SKILL DRILL 15-8**. To force a bored lockset with an A tool or a Halligan, follow the steps in **SKILL DRILL 15-9**.

Forcing a Mortise Lockset

The cylinder is often protected by a large metal collar or by a pull-handle on the door that can make it difficult to apply lock-pullers to the door. However, given a little time and practice, these obstacles can often be overcome.

During a high-priority emergency such as a working fire, mortise cylinders can quickly be pulled using a K tool, an S&D tool, or a Rex tool. These tools can be placed over the top of a mortise cylinder and driven downward by striking them with a flat-head axe or hammer. The narrowing blades of the lock-puller tool bite into the body of the cylinder as it is driven down over the cylinder. Once set into place on the cylinder, the lock-puller is pried horizontally or vertically until the cylinder is forcibly removed from the lock-body. There is no "right" direction to pry in. Prying can be horizontal or vertical as long as the lock-puller blades are set firmly into the cylinder. Because of wrap-around pull-handles or other obstructions, the lock-puller may need to be set into place from the side or even diagonally. To force

SKILL DRILL 15-8

Forcing a Bored Lockset with an S&D or Rex Tool Firefighter I, NFPA 1010: 6.3.4

1. Try the door to ensure it is locked and size up the type and number of locksets and the direction the door swings. Drop the large head of the S&D tool or Rex tool behind the knob or lever.

SKILL DRILL 15-8 CONTINUED

Forcing a Bored Lockset with an S&D or Rex Tool Firefighter I, NFPA 1010: 6.3.4

2. Rock the tool slowly from side to side to loosen the knob or lever from the rest of the lockset.

3. Pull up sharply to remove the knob or lever and expose the latch mechanism.

Continues.

SKILL DRILL 15-8 CONTINUED

Forcing a Bored Lockset with an S&D or Rex Tool Firefighter I, NFPA 1010: 6.3.4

4. Use a flat-head screwdriver or needle-nose pliers to turn the latch mechanism and retract the latch.

Courtesy of Sean Wilson.

SKILL DRILL 15-9

Forcing a Bored Lock with an A Tool or a Halligan Adze Firefighter I, NFPA 1010: 6.3.4

1. Try the door to ensure it is locked and size up the type and number of locksets and the direction the door swings. Place the adze of the A tool or the Halligan on top of the lockset at approximately a 25-degree angle from the door.

SKILL DRILL 15-9 CONTINUED

Forcing a Bored Lock with an A Tool or a Halligan Adze Firefighter I, NFPA 1010: 6.3.4

2. Strike the back of the tool with a hammer or flat-head axe until it is set behind the collar of the lockset.

3. Lift up on the tool to put tension on the lockset, and then rock the tool sharply from side to side to break or strip the two screws holding the lockset together.

Continues.

SKILL DRILL 15-9 CONTINUED

Forcing a Bored Lock with an A Tool or a Halligan Adze Firefighter I, NFPA 1010: 6.3.4

4. Remove the collar to expose the latch mechanism and use a flat-head screwdriver or needle-nose pliers to turn the latch mechanism and retract the latch.

Courtesy of Sean Wilson.

a mortise lockset with an S&D or Rex tool or a K tool, follow the steps in **SKILL DRILL 15-10**.

For lower-priority calls such as fire alarms where there is no smoke or fire upon arrival, you can use locking pliers or angled slip-joint pliers to grip the face of the mortise cylinder and spin the cylinder out. This will damage the cylinder and set screw, but usually the rest of the lock will not be damaged. This technique may take several minutes because the threads of the cylinder are fine and 15 to 30 full rotations of the cylinder may be required to fully unscrew it. If there is a security collar on the cylinder preventing your pliers from gripping it, tap a flat-head screwdriver firmly into the keyway of the cylinder, and then grip the screwdriver with the pliers and turn it like an axle, unscrewing the cylinder to which it is attached. To force a mortise lock with locking pliers or angled slip-joint pliers, follow the steps in **SKILL DRILL 15-11**.

Adams Rite mortise locksets, which are often found on glass storefront doors, often have a dead latch operated by a small metal wheel in the cylinder hole in either the 5 o'clock or the 7 o'clock position. All that is required to retract this latch is to use a bent-end key tool to push the wheel down and rotate it over to the opposite position. Standard mortise locksets, which are found mostly on hollow-core, steel commercial doors, have mechanisms that are less standardized. Most, however, can be manipulated using a bent-end key tool to move the latch mechanisms in the cylinder hole between the 3 o'clock and 9 o'clock positions.

Forcing a Rim Lockset

From the exterior of a building, rim cylinders and mortise cylinders look similar. Both sit flush to the door and usually have a security collar or ring around the outside of the cylinder to protect it. The collars typically used with rim cylinders are flatter and smoother than those used on mortise cylinders. Fortunately, if armed with a good lock-puller, it won't matter which type of cylinder it is because both types are removed using the same technique. However, because rim cylinders are held in place with two screws attached to a back plate, it is not possible to grip the cylinder with pliers and unscrew

SKILL DRILL 15-10

Forcing a Mortise Lock with a K Tool, an S&D Tool, or a Rex Tool
Firefighter I, NFPA 1010: 6.3.4

1. Try the door to ensure it is locked and size up the type and number of locksets and the direction the door swings. Place the K tool or the large head of the S&D or Rex tool above, below, or to the side of the cylinder. If there is collar on the cylinder, aim the tool behind the collar.

2. Strike the lock-puller several times with a hammer or the back of a flat-head axe to set the blades of the lock-puller into the cylinder.

Continues.

SKILL DRILL 15-10 CONTINUED

Forcing a Mortise Lock with a K Tool, an S&D Tool, or a Rex Tool
Firefighter I, NFPA 1010: 6.3.4

3. If you are using a K tool, insert the adze of a Halligan into the strap on the tool, and then strike the Halligan several times to set the K tool firmly into the cylinder.

4. Rock the tool slowly from side to side to "walk" the threaded cylinder out of its threads and expose the latch mechanism.

SKILL DRILL 15-10 CONTINUED

Forcing a Mortise Lock with a K Tool, an S&D Tool, or a Rex Tool
Firefighter I, NFPA 1010: 6.3.4

5. Use a bent-end key tool to manipulate the latch mechanism and retract the latch.

Courtesy of Sean Wilson.

SKILL DRILL 15-11

Forcing a Mortise Lock with Locking Pliers or Angled Slip-Joint Pliers
Firefighter I, NFPA 1010: 6.3.4

1. Try the door to ensure it is locked and size up the type and number of locksets and the direction the door swings. Grab the face of the cylinder with locking pliers or angled slip-joint pliers.

Continues.

SKILL DRILL 15-11 CONTINUED

Forcing a Mortise Lock with Locking Pliers or Angled Slip-Joint Pliers
Firefighter I, NFPA 1010: 6.3.4

2. If the cylinder has a security collar on it, drive a ¼- or 5/16-in. (6- or 8-mm) flat-head screwdriver into the keyway with a hammer, and then grab the handle of the screwdriver with the pliers.

3. Spin the cylinder counter-clockwise to unscrew it from the lock housing and expose the latch mechanism. This may require 15 to 30 full rotations of the cylinder depending on the length of the cylinder, the type of lockset, and the thickness of door. Use a bent-end key tool to manipulate the latch mechanism and retract the latch.

Courtesy of Sean Wilson.

it from the lockset in the same manner you use to remove the cylinder from a mortise lockset. Instead, a lock-puller must be used. To force a rim lockset with an S&D tool, Rex tool, or a K tool, follow the steps in **SKILL DRILL 15-12**.

Residential-grade rim lockssets often have additional security features to prevent manipulation of the latch should the cylinder be removed. If the latch mechanism will not turn or is not accessible, it may not be possible to retract the latch. Rim locksets, however, are only surface-mounted to the interior side of the door. This means you can drive the entire lock assembly off the door by having one firefighter insert the pick of a Halligan into the cylinder hole and having a second firefighter strike the back of the Halligan with an axe or sledgehammer. Rim locksets are the only type of lock that can be driven from the door in this manner. Attempting this technique with bored or mortise locksets will merely jam the lockset in the locked position.

SKILL DRILL 15-12

Forcing a Rim Lockset with an S&D Tool, a Rex Tool, or a K Tool
Firefighter I, NFPA 1010: 6.3.4

1. Try the door to ensure it is locked and size up the type and number of locksets and the direction the door swings. Place the large head of an S&D tool or a Rex tool above, below, or to the side of the cylinder. If there is collar on the cylinder, aim the tool behind the collar.

2. Strike the lock-puller several times with a hammer or the back of a flat-head axe to set the blades into the cylinder. If you are using a K tool, insert the adze of a Halligan into the strap on the tool, and then strike the Halligan several times to set the K tool into the cylinder.

Continues.

SKILL DRILL 15-12 CONTINUED

Forcing a Rim Lockset with an S&D Tool, a Rex Tool, or a K Tool
Firefighter I, NFPA 1010: 6.3.4

3. Pull the tool sharply from side to side to break the two screws or pull the screws through the back plate to expose the latch mechanism.

4. The latch mechanism in a rim lockset is often a small cross-shaped receptable into which the tail of the cylinder fits. Insert a slotted screwdriver into the hub and rotate the latch mechanism until the latch is retracted from the strike plate.

SKILL DRILL 15-12 CONTINUED

Forcing a Rim Lockset with an S&D Tool, a Rex Tool, or a K Tool
Firefighter I, NFPA 1010: 6.3.4

5. If the mechanism is damaged or if additional security measures prevent manipulation of the latch, insert the pick of a Halligan into the cylinder hole and have a second firefighter strike the back of the tool to drive the latch housing off of the interior of the door.

Courtesy of Sean Wilson.

Forcing an Outward-Swinging Door with a Shove Knife

Shove knives can be effective at opening some doors at low-priority calls. The technique requires patience, concentration, and finesse, but because it is also completely non-destructive, it is worth dedicating time to learn it. To be successful with a shove knife, the door must be outward-swinging, the latch must be a slam latch on a bored lockset or on a mortise lockset, and the anti-loiding pin must be disengaged.

If the door swings outward and the lockset is either a bored or mortise type, the latch should be visible in the seam between the door and the door frame. Insert the shove knife into the seam above or below the lock with the notch facing the latch, and then place the notch onto the angle of the slam latch and pull back slightly toward your body. Keep the tension constant and work the shove knife sideways in a scooping motion. Doing these two things at once will push the latch back into its housing in the door and maintain any progress that is made. Once the latch is removed from the strike plate, use your free hand to pull the knob or lever and open the door. To use a shove knife on a bored or mortise lockset in an outward-swinging door, follow the steps in **SKILL DRILL 15-13**.

If the latch will not be pushed back into the door, it is likely the anti-loiding pin is engaged and the latch cannot be pushed in manually. (If the anti-loiding pin is depressed, it is engaged.) You may be able to disengage the anti-loiding pin by pushing hard on the door so that the pin pops out into the latch hole in the strike plate. You can also try tapping a wooden door wedge or aluminum force wedge into the seam just below the lock to spread the seam wider so that the anti-loiding pin slips out enough to disengage. If neither of these methods works, move to another door or switch to a more aggressive technique.

SKILL DRILL 15-13

Using a Shove Knife on a Bored or Mortise Lockset in an Outward-Swinging Door Firefighter I, NFPA 1010: 6.3.4

1. Try the door to ensure it is locked and size up the type and number of locksets and the direction the door swings. Insert the shove knife into the seam between the door and the door frame above or below the visible slam latch.

2. Place the hook of the shove knife behind the angled portion of the slam latch and draw the shove knife outward toward your body with slight tension.

SKILL DRILL 15-13 CONTINUED

Using a Shove Knife on a Bored or Mortise Lockset in an Outward-Swinging Door Firefighter I, NFPA 1010: 6.3.4

3. While maintaining this tension, make a side-to-side scooping motion with the tool to push the latch back into the edge of the door.

4. Once the latch is pushed in, hold the latch in place with the shove knife and pull on the knob or lever with your free hand to open the door (or have a second firefighter do so).

Courtesy of Sean Wilson.

Forcing an Outward-Swinging Double-Door Set with a J Tool

The J tool is single-use tool used exclusively for opening outward-swinging doors affixed with panic hardware. The J tool is most effective in forcing double-door sets, but it can be used on single doors as well. To use the J tool, first widen the gap between the two doors using a door wedge or force wedge. Next, insert the J tool and work it down near the area of the panic hardware. Lastly, rotate the J tool and pull back on it to engage the panic hardware and open the door. To use a J tool on panic hardware on an outward-swinging door, follow the steps in **SKILL DRILL 15-14**.

SKILL DRILL 15-14

Using a J Tool on Panic Hardware on an Outward-Swinging Door
Firefighter I, NFPA 1010: 6.3.4

1. Try the door to ensure it is locked and size up the type and number of locksets and the direction the door swings. Gap the door by pulling on it or using a Halligan or other tool. Insert the front of the J tool into the gap just above the panic hardware.

2. On the other side of the door, if the J tool was inserted above the lock, the J tool is vertical and facing down as shown here. If it was inserted below the lock, the J tool needs to be facing up.

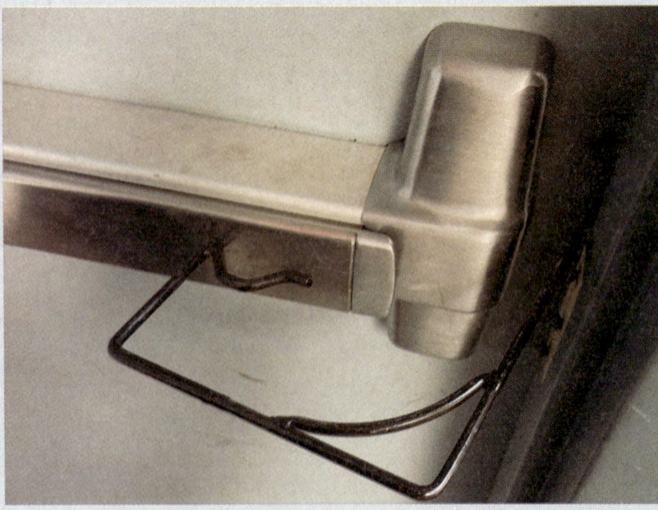

3. Rotate the J tool 45 degrees and pull the tool outward toward your body, engaging the panic hardware and opening the door.

Courtesy of Sean Wilson.

Rotary Saw Forcible Entry

A rotary saw fitted with a diamond rescue blade or composite metal cutting blade can be quite effective at forcing entry through certain types of doors, locksets, and security bars and gates. There are two types of modern rotary saws available: 14-in. (35.6-cm) gasoline-powered or battery-powered saws, and 9-in. (23-cm) battery-powered saws. The 14-in. (36-cm) saw offers a cutting depth of 4 to 5 in. (10 to 13 cm). The 9-in. (23-cm) saw only has a cutting depth of 3 in. (8 cm), which is deep enough for some forcible entry problems, but not all. The technology is rapidly improving, so it is likely that battery-powered fire service saws will become more widely used in the near future.

As discussed, conventional forcible entry is typically used at high-priority situations, and through-the-lock forcible entry is most often used during more routine emergencies. Rotary saw forcible entry is unique in that it is used across the full spectrum of situations where forcible entry may be performed. Rotary saws can be appropriate for situations ranging from working commercial fires to routine fire alarms. This is because these saws offer the ability, when used properly, to be surgically precise in cutting the specific area necessary to open a door or lockset. In the right hands, the use of a rotary saw can lead to a quick entry while keeping damage at a minimum.

Using Saws Safely

As with all power tools, care must be taken while operating rotary saws to prevent serious injury. The following safety guidelines should be adhered to any time rotary saws are used, whether in the field or during training:

- Never use a power tool that you have not been trained to use.
- Wear full PPE when using a saw, including helmet and eye protection, turnout coat and pants, boots, and fire gloves.
- Wear hearing protection when it is practical to do so. This may not be possible during some emergencies.
- Use only manufacturer-approved blades and fuel for the saw.
- Match the proper blade to the material being cut.
- Ensure the blade and fuel cap are tight before cutting.
- Maintain situational awareness of personnel and objects around you when cutting.
- When another firefighter is cutting, give them plenty of room to work.
- Never hand a spinning rotary saw to another firefighter. Wait for it to stop spinning or place it on the ground for them to pick up.
- Do not carry a running saw more than 10 ft (3 m). A running saw may still be spinning and can be a hazard to anyone nearby who may walk too near or trip and fall into your saw. If a running saw must be walked a longer distance to where it will be used, shut off the saw, walk to your destination, and restart it.
- Do not hoist a running saw with a rope or webbing.
- Do not climb a ladder with a running saw.
- If a saw must be hoisted or carried up a ladder, start the saw on the ground to warm it up, then shut the saw down before hoisting it or carrying it up a ladder.
- Most cuts should be made by first revving the saw to full revolutions per minute (RPMs) before cutting.

Cutting Latches on Outward-Swinging Doors

Outward-swinging doors with bored or mortise locksets are particularly vulnerable to a rotary saw. A general rule of thumb is, if you can see a latch in the seam between door and the door frame, you can cut that latch with a rotary saw. If the seam is too tight to fit the width of the saw blade or if you cannot see the latch, you can spread it slightly by tapping a door wedge, force wedge, or axe head into the seam with an axe or a Halligan above or below the cutting area. If you still cannot see the latch even after spreading the seam with a wedge, then the door is likely secured with a rim lockset with the latch mounted on the other side of the door, out of reach of the saw. In this case, another forcible entry strategy must be devised.

To cut the latch, start the saw, let it warm up for a few seconds, then bring the saw to full RPMs and lay it into the area of the seam just above the lockset. Slowly move the saw into the seam between the door and the door frame above the latch until the spinning blade makes contact with the door stop, and then draw the saw downward to cut the latch. You will not be able to see the latch as it is being cut, so you will need to be patient and thorough as you cover the entire area. If even ¼ in. (6 mm) of the latch is not cut through, the door will remain closed. To cut a latch with a rotary saw, follow the steps in **SKILL DRILL 15-15**.

SKILL DRILL 15-15

Cutting a Latch with a Rotary Saw Firefighter I, NFPA 1010: 6.3.4

1. Try the door to ensure it is locked, size up the type and number of locksets, and verify that the door is an outward-swinging door. Insert a door wedge, force wedge, or axe head into the seam between the door and the door frame 1 ft (0.3 m) or more below the lock to allow for room for the saw, and then tap it into place with a Halligan or other striking tool. Verify that you can see the latch in the seam.

2. Start the saw, bring it to full RPMs, and then slowly lay the blade into the seam above the latch.

SKILL DRILL 15-15 CONTINUED

Cutting a Latch with a Rotary Saw Firefighter I, NFPA 1010: 6.3.4

3. Slowly draw the saw down to the latch, allowing the weight of the saw to be the only pressure applied to the latch while it is being cut.

4. After you have cut all the way through the latch, remove the saw blade from the seam, shut the saw off and set it down, remove the wedge or axe, and open the door.

Courtesy of Sean Wilson.

Cutting Hinges on Outward-Swinging Doors

Much is made of attacking the hinges of an outward-swinging door, either by cutting the hinges or removing the hinge-pins, but these techniques are rarely selected as the first-choice strategy for forcible entry because of the time they take and the potential problems they can create.

The first issue is that exterior hinges are sealed in some way, preventing you from simply popping the pins out. Therefore, to attack the hinges, they need to be cut with a rotary saw. Heavy, commercial hinges typically take about 1 full minute to cut each hinge, with an average of three hinges on most doors. Once all the hinges are cut, the door must still be pried out of the frame using the irons because there is no knob or lever to grab onto on the hinge side of the door. In other words, a light version of conventional forcible entry must still be employed after using the rotary saw to cut the hinges.

The next issue stems from the possibility that the hinges are studded hinges. As discussed earlier, studded hinges have bolts or flaps that hold the hinge-plates together when the door is shut. This means that if you cannot pry the door out from the frame on the hinge side without additional cutting. It is often not possible to identify studded hinges from the outside.

Finally, once the door is finally removed from its frame, it cannot quickly be closed again if that opening negatively affects the flow path. Once opened, a door whose hinges are cut or removed stays open.

If, for some reason, it is decided that cutting the hinges on an outward-swinging door is the best option for gaining entry, bring the rotary saw up to full RPMs, and then bring the blade in at a diagonal angle to the top hinge and cut the hinge plates attached to the barrel of the hinges rather than the barrel itself (**FIGURE 15-73**). Start with the top hinge and work your way down.

Cutting Window Bars and Security Gates

If a set of window bars or security gates are held in place by a padlock that is accessible from the exterior, it is usually faster to cut the padlock than to attack the bars. However, if there is no padlock present or the padlock is inaccessible, quickly size up and cut the attachment points where the bars or gates attach to the building. Attacking the attachment points usually removes the obstacle faster and with fewer cuts than attempting to cut individual bars out of the way.

FIGURE 15-73 A hinge on an outward-swinging door being cut with a rotary saw.

Courtesy of Sean Wilson.

Cutting Standard and High-Security Padlocks

So long as access to the shackle is possible, both standard and high-security padlocks are easily cut with a rotary saw. High-security padlocks may be too thick or too strong for bolt cutters, but when using a rotary saw, the only real difference between the two grades is that high-security padlocks lock on both sides of the shackle. It is always good practice to cut each side of the shackle, but in this case, both sides *must* be cut.

When a padlock secures a gate or other obstacle with a chain, it is usually easier to cut the last link of the chain and to cut the padlock shackle. This allows the owner to reuse the lock and possibly the chain as well.

Cutting a padlock with a rotary saw is a two-firefighter operation. One firefighter secures the padlock in place using either a pair of locking pliers on a lanyard clipped onto the **lock body** (the structural housing of a padlock that holds the main locking mechanisms and secures the shackle) or a short length of rope or webbing looped through the shackle that they hold in tension. This prevents the lock from jumping around when the spinning blade makes contact with the shackle. Both techniques also keep the firefighter securing the padlock safely away from the blade. The second firefighter brings the saw up to full RPMs, and then leans the saw into the shackle. The second firefighter should cut both sides of the shackle so that the shackle is completely separated from the lock body. To cut a padlock with a rotary saw, follow the steps in **SKILL DRILL 15-16**.

SKILL DRILL 15-16

Cutting a Padlock with a Rotary Saw Firefighter I, NFPA 1010: 6.3.4

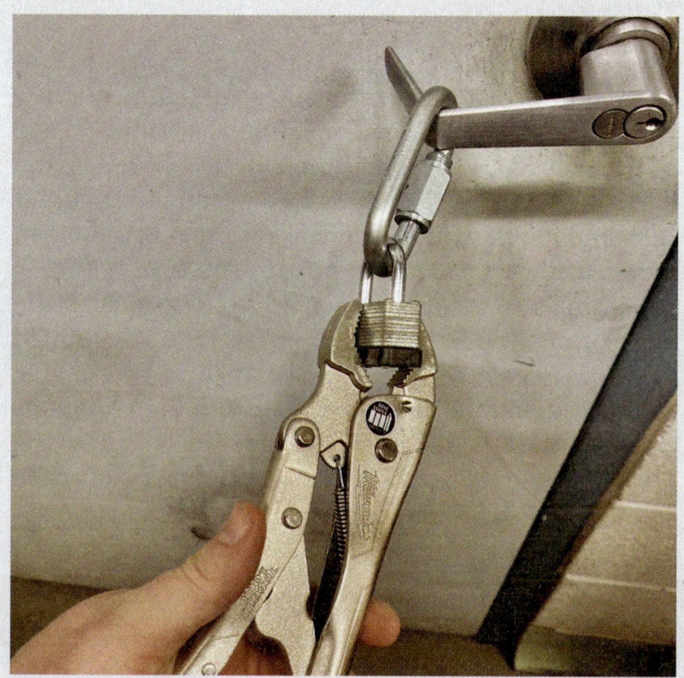

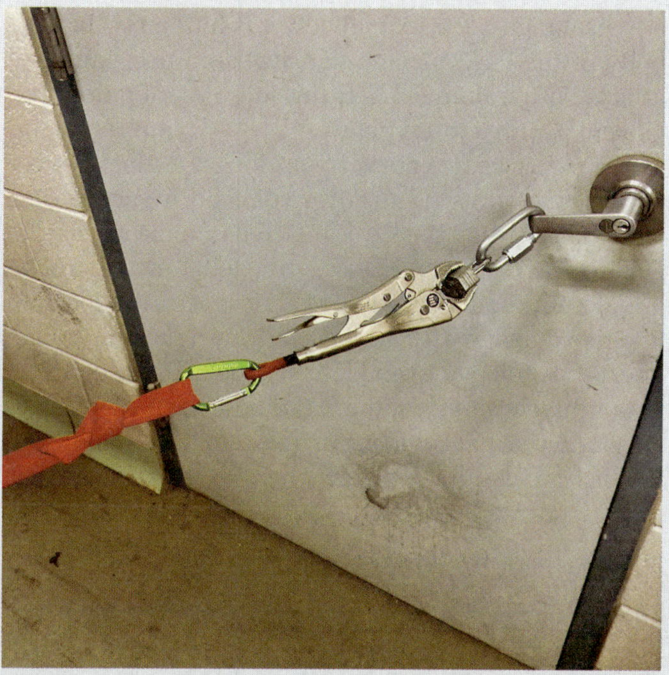

1. Try the padlock to ensure that it is locked and size up the attachment point and associated hardware. Direct another firefighter to secure the padlock using a pair of locking pliers on a lanyard or a short length of rope or webbing, and pull on the pliers, rope, or webbing to hold the padlock under tension.

2. Start the saw, bring it to full RPMs, and lean the blade onto the shackle on one side of the padlock until it is cut through.

3. Repeat on the other end of the shackle so that the padlock drops from the lock body.

Courtesy of Sean Wilson.

Cutting a Hockey-Puck Lock

Hockey-puck locks are used on roll-down security gates and work vehicles or trailers that store expensive tools or equipment. Although the shackle of a hockey puck lock is not any thicker or stronger than the shackle on a typical high-security padlock, it is hidden within the body of the lock, making the shackle much harder to attack. With training in how this lock is configured, you can cut it with a rotary saw.

The shackle in a hockey puck lock is a straight piece of steel that moves in and out in a direction directly in line with the key when the key is inserted into the keyway. To cut the shackle, find the keyway and follow it in a straight line across the lock body. This line traces the location and movement of the shackle. Imagine a second straight line, perpendicular to and intersecting with the first line, about three-quarters of the way across the lock from the keyway. Make the cut along this second line until the entire lock is cut into two pieces. This cuts through both the shackle and the lock body, and it separates the lock from the mounting hardware. To cut a hockey puck lock with a rotary saw, follow the steps in **SKILL DRILL 15-17**.

Cutting Overhead Doors with the Triangle Cut

Firefighters have successfully used the triangle cut for decades against all kinds of overhead doors. Its primary advantage is that it is a simple cut that can be used against any type of overhead door. Its disadvantages are that it creates a small opening when compared to other cuts, and if a hose line is deployed through the hole in the door, it will be harder to raise the door.

To make a triangle cut, stand in the center of the door, bring the saw to full RPMs, raise the saw as high as you can reach, and then make a cut at a 45-degree angle, bringing the saw down as low to the ground as possible. Start the second cut at the same height as the first cut but leave several inches (centimeters) of space

SKILL DRILL 15-17

Cutting a Hockey Puck Lock with a Rotary Saw Firefighter I, NFPA 1010: 6.3.4

1. Rattle the hockey puck lock to ensure it is locked and size up the orientation of the lock body, the attachment points, and any associated hardware. Find the keyway, and then draw an imaginary line from it straight across the lock body.

SKILL DRILL 15-17 CONTINUED

Cutting a Hockey Puck Lock with a Rotary Saw Firefighter I, NFPA 1010: 6.3.4

2. Draw a second imaginary line perpendicular to and intersecting with the first line about three-fourths of the distance across the lock body from the keyway to determine where you will make the cut.

3. Start the saw, bring it to full RPMs, and lean the blade into the lock at the location of the second imaginary line until both the lock body and the shackle fall from the attachment hardware.

Courtesy of Sean Wilson.

between the two cuts. This prevents massive amounts of smoke from inside the building pouring out and obscuring your vision as the second cut is made. Once the second cut is completed, go back to the top and make a third cut that connects the first two cuts. Kick the portion of the door into the opening that has been created. To make a triangle cut in an overhead door with a rotary saw, follow the steps in **SKILL DRILL 15-18**.

SKILL DRILL 15-18

Making a Triangle Cut in an Overhead Door with a Rotary Saw
Firefighter I, NFPA 1010: 6.3.4

1. Stand at the center of the door, bring the saw to full RPMs, and then raise the saw as high as possible. Lay the saw slowly into the door at a 45-degree angle, and cut down to as close to the ground as possible.

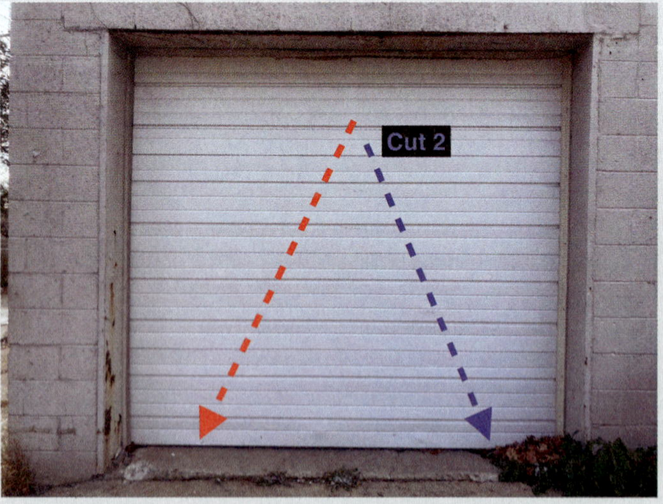

2. Return to the center of the door, again raise the saw as high as possible, positioned several inches (centimeters) from the first cut, and then make the second cut at a 45-degree angle as a mirror image to the first; that is, cut down toward the ground at a 45-degree angle but in the opposite direction of the first cut.

SKILL DRILL 15-18 CONTINUED

Making a Triangle Cut in an Overhead Door with a Rotary Saw
Firefighter I, NFPA 1010: 6.3.4

3. Return to the center of the door, and then connect the two cuts at the top. Push the cut portion inward at the top to drop the cut portion out of the door. Kick the cut portion inward into the building. Shut the saw off and set it down.

Courtesy of Sean Wilson.

Cutting Overhead Doors with Rectangle Box Cut

The rectangle box cut is an effective technique for cutting a large opening in a commercial or residential overhead door. The advantages of this technique are its simplicity and the fact that any type of overhead door can be opened for almost the entire width of the door to shoulder or head height. The disadvantage of the rectangle box cut is that it requires a lot of cutting time.

Start the rectangle box cut by making a long horizontal cut at shoulder-height, beginning 1 ft (0.3 m) from one side of the door and terminating 1 ft (0.3 m) from the opposite side. The second cut is a vertical cut starting just below the horizontal cut on one side of the door, but not overlapping the first cut. The third cut is the same as the second cut but on the opposite side. Finally, connect the vertical cuts to the horizontal cut and let the door fall. Waiting to connect these cuts keeps the door in place and minimizes smoke that could impede visibility while cutting. To make an overhead door rectangle box cut with a rotary saw, follow the steps in **SKILL DRILL 15-19**.

Cutting Overhead Doors with Rectangle Hinged Cut

The rectangle hinged cut is an effective method of creating a large opening in the door for most residential garage doors and some light commercial overhead doors. It consists of two long cuts and two short cuts. For this cut to be successful, the door must be a light-gauge steel door without thick insulation panels and have a light-gauge steel bottom edge that can be bent by hand at the hinge point. This cut is not the best choice for insulated doors because their thickness might prevent you from making the long cuts quickly. This cut is also not suitable for heavy commercial doors that have doubled angle iron at the bottom edge of the door, which is far too strong to bend by hand at the hinge point.

The primary advantage of this cut over the triangle cut is that the opening created is much larger than the opening made with the triangle cut. This is safer for crews working inside the building should conditions change and an emergency evacuation be ordered. Another advantage is that the hinged portion of the door can be put back in place to lessen the negative impact a hole this size may have on the flow path. The disadvantages of this cut are that it requires more cutting to complete than the triangle cut and it is not practical or effective against heavier commercial doors.

The first cut is a long vertical cut near one side of the door starting at the head height of the firefighter holding the saw and terminating as low to the ground as possible. The second cut is a diagonal cut that overlaps the first from about knee height of the firefighter holding the saw and terminating as low to the ground

SKILL DRILL 15-19

Making an Overhead Door Rectangle Box Cut with a Rotary Saw
Firefighter I, NFPA 1010: 6.3.4

1. Bring the saw to full RPMs. Raise the saw to shoulder height about 1 ft (0.3 m) from one side of the door, and then make the first cut as a long horizontal cut extending to about 1 ft (0.3 m) from the other side of the door.

2. Start the second cut just below one end of the horizontal cut close to one side of the door without overlapping the horizontal cut, and then cut vertically down as low as you can reach.

3. Make a third cut identical to the second cut but on the other end of the door.

SKILL DRILL 15-19 CONTINUED

Making an Overhead Door Rectangle Box Cut with a Rotary Saw
Firefighter I, NFPA 1010: 6.3.4

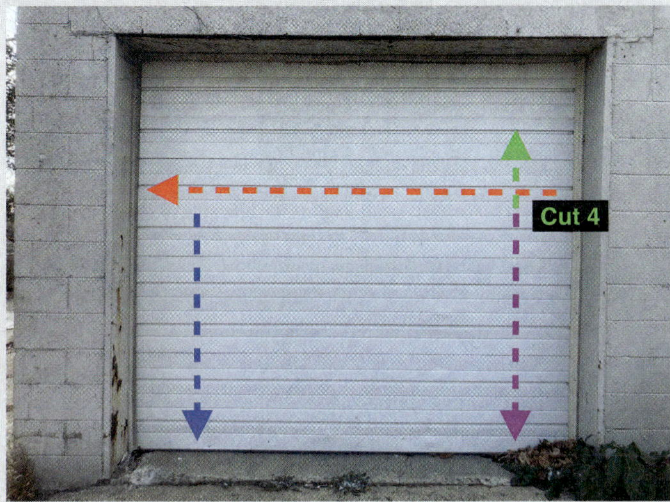

4. Connect the third cut to the horizontal cut.

5. Connect the second cut to the horizontal cut. Shut the saw off and set it down, and then push the cut portion of the door into the building.

Courtesy of Sean Wilson.

as possible. After kicking the small triangle of door material that was created with the first two cuts into the building, the firefighter makes a short, third cut all the way through along the bottom edge of the door. The fourth cut is a long horizontal cut on the opposite side of the door from where the first cut was made. The firefighter makes this cut by placing the saw on their shoulder and cutting until they overlap the long vertical cut.

Once all four cuts are made, a second firefighter grabs the edge of the vertical cut with a roof hook or a Halligan and pries the rectangle portion out until it swings out from the door and bends back on itself like a hinge. Some doors may need to be battered in a straight vertical line to crease the hinge area and allow it to bend more easily. If, after battering the door, the area that is intended to hinge will not bend, consider completing an additional long vertical cut in the hinge area to make this a rectangle box cut.

To make a rectangle hinged cut in an overhead door with a rotary saw, follow the steps in **SKILL DRILL 15-20**.

SKILL DRILL 15-20

Making a Rectangle Hinged Cut in an Overhead Door with a Rotary Saw Firefighter I, NFPA 1010: 6.3.4

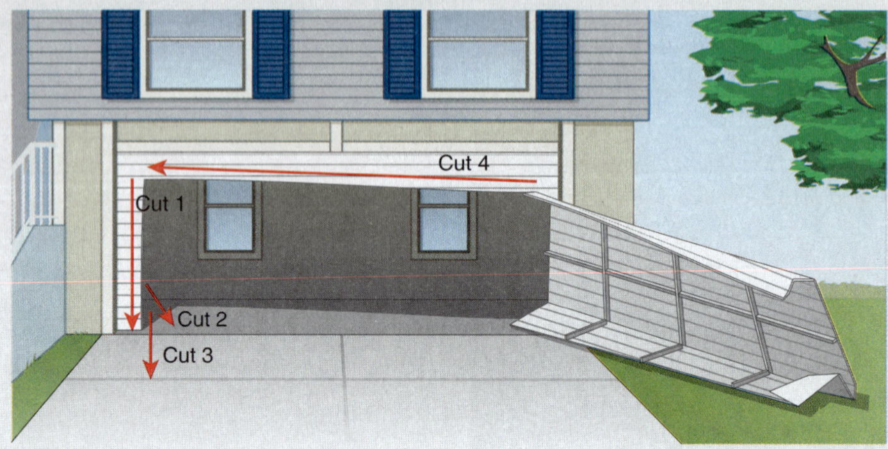

1. Bring the saw to full RPMs, and then make the first cut as a long vertical cut near one side of the door, starting at head height and terminating as low to the ground as you can reach.
2. Position the saw at about knee height, and then make a second cut overlapping the first cut and angling down at a 45-degree angle, terminating as low to the ground as can be reached.
3. Kick in the small triangle of cut material, and then make a small third cut through the bottom rail of the door.
4. Make the fourth and final cut by beginning on the opposite side of the door from the first cut at shoulder height and cutting horizontally back to shoulder height at the first cut, terminating after this cut overlaps the first cut. Shut the saw off and set it down.
5. Have a second firefighter use a Halligan or roof hook to grab the edge of the door along the vertical cut and pull the rectangle portion out away from the building until it swings out and bends back on itself.
6. If the door will not bend outward, batter the door in a straight line along the side without the vertical cut—the side that will form the "hinge"—to weaken it.
7. If battering the door does not weaken the hinge area, or if the bottom rail is too substantial to be bent, change the technique to use the rectangle box cut. Make a fifth cut vertically down the hinge area—the side of the door that does not have a vertical cut. This cut should mirror the first cut and overlap the horizontal cut at shoulder height. Shut the saw off and set it down, and then push the cut portion of the door into or out of the building.

Cutting Rolling Steel Doors with the Slice and Pull Method

Because rolling steel overhead doors are made of short steel slats that are folded so that they lock onto one another without hinge hardware, it is possible to make a single vertical cut in the center of the door and pull the slats on either side of the cut, dropping the rest of the door below the pulled slat. To use this method, make a single vertical cut in the center of the door from as high as you can reach to as low as the saw will go. Attach locking pliers onto the cut end of one of the slats at about shoulder-height, and then pull the slat until it is completely free of the door. Repeat this with the same slat on the opposite side of the cut, dropping the door to the ground. To cut a rolling steel door with a rotary saw using the slice and pull method, follow the steps in **SKILL DRILL 15-21**.

SKILL DRILL 15-21

Cutting a Rolling Steel Door with a Rotary Saw Using the Slice and Pull Method Firefighter I, NFPA 1010: 6.3.4

1. Stand at the center of the door, bring the saw to full RPMs, raise the saw as high as possible, and then make a long vertical cut terminating as low to the ground as possible. Shut the saw off and set it down.

2. Use your shoulder to push the door slightly into the building on one side of the cut. Using locking pliers, grab one of the cut slats at about chest or head height and pull the slat horizontally out from the door until the slat is separated from the door. All the slats below the slat you pulled will drop from the door on that side of the cut.

Continues.

SKILL DRILL 15-21 CONTINUED

Cutting a Rolling Steel Door with a Rotary Saw Using the Slice and Pull Method Firefighter I, NFPA 1010: 6.3.4

3. Using locking pliers, pull a slat on the other side of the cut.

4. All the slats below the slat you pulled will fall from the door on that side of the cut.

Courtesy of Sean Wilson.

Putting It All Together: Forcing Open Inward-Swinging and Outward-Swinging Doors

Inward-swinging doors set in wood door frames are found on residential homes and some apartments and condominiums. Outward- and inward-swinging doors set in metal frames that are sometimes filled with concrete are often found on commercial buildings and in some apartment and condominium buildings. There are many techniques for forcing each type of door system in different situations, and firefighters facing a working fire must be able to size up the priority level of the situation and identify the number of locksets, the type of door frame, and the direction of the door swing. Then they need to be able to quickly put their training into practice and decide whether to use conventional, through-the-lock, rotary saw, or specialty forcible entry techniques, and, if it is a metal door frame, whether to apply the GSF method to successfully force the locked door in front of them.

The steps for conventionally forcing a door at a structure fire are as follows:

1. Wear full PPE, including turnout gear, helmet, gloves, eye protection, and SCBA.
2. Bring the irons and lock box keys (if available) to the door.
3. Try before you pry!
4. Look for a lock box. If there is one, use it!
5. Consider the type of structure, whether it is commercial or residential.
6. Determine whether the door is an inward-swinging or an outward-swinging door.
7. Determine if the door frame is wood or metal.
8. Size up the number and types of locksets visible on the door.
9. Attempt to determine if all the locksets are locked.
10. Evaluate whether this is the best door to force or if another door may be a better choice.
11. If the incident is a residential house fire and the door is an inward-swinging door set in a wood frame, split the irons and use the axe or sledge to forcefully strike the door near the cylinder of the lockset until the door opens.
12. If the incident is a commercial building fire or another building in which the door is set in a metal frame—whether inward- or outward-swinging—split the irons, give the axe or sledgehammer to your partner, and keep the Halligan.
13. Use the Halligan and axe or sledgehammer to create a gap between the door and the door stop or the door frame.
14. Use the axe or force wedge to capture the gap created with the Halligan and widen it if necessary.
15. Set the tool into place past the door stop and onto the inner portion of the door or interior casing of the frame.
16. Force the door by pushing or pulling the Halligan forcefully until the door opens.

Specialty Forcible Entry

The specialized techniques for forcible entry discussed in this section are not normally performed at routine fires and incidents. However, they may be necessary or useful in some situations.

Forcing Windows

Remember, making entry through windows can be hazardous in non-fire situations due to the difficulty in getting back out should a first responder encounter a dog or a surprised homeowner with a firearm. Avoid making entry through windows in non-fire situations unless you are in voice contact with a resident inside.

Forcing Wood Frame Single- and Double-Hung Windows. Sash locks on wood single- and double-hung windows are made of two pieces, one on the lower sash and one on the upper sash. The pieces are installed with wood screws. Most wood frame windows of this type have a slender seam between the two sashes that can be exploited, and this can easily be fixed by the homeowner later.

To force these sash locks, place a small, steel scraper bar or flat-head screwdriver between the two sashes where the lock is located and drive the tool upward with a hammer. The tool will contact the sash lock and drive the screws out of the wood, unlocking the window. To force a wood frame single- or double-hung window, follow the steps in **SKILL DRILL 15-22**.

Forcing Vinyl-Frame Single-Hung, Double-Hung, and Slider Windows. Single- and double-hung windows made with vinyl frames are similar in design to wood single- and double-hung windows with one important difference. On vinyl windows, there is often a small lip that covers the seam between the two sashes. This prevents you from sliding a tool between the sashes as in the technique used for wood frame single- and double-hung windows. But because vinyl is a plastic product, it has some elasticity when stressed. This means you can apply gentle, steady force to the sashes in an attempt to separate the two halves of the sash lock.

SKILL DRILL 15-22

Forcing Wood Frame Single- and Double-Hung Windows Firefighter I, NFPA 1010: 6.3.4

1. Try the window to ensure it is locked. Insert a flat scraper bar or a slender but sturdy flat-head screwdriver into the seam between the two sashes in line with the sash lock. Drive this tool upward with a hammer or a mallet, taking care to not strike the glass with the hammer. The tool should slip through the sashes and make contact with one or both halves of the sash lock.

2. Confirm that contact with the sash lock has been made, and then continue to drive the tool upward with the hammer or mallet until the sash lock screws are driven out of the sash or sashes or the two halves of the lock separate and the window is unlocked.

Courtesy of Sean Wilson.

First, partially insert the adze of a Halligan or A tool upward into the seam between the two sashes at the point where the lock is located. Then gently rock the bar from left to right in a progressively larger arc until the sash-lock separates and unlocks. Take care to move the tool slowly and to avoid tool contact with the window glass. Windows of this type may have two sash locks. If there is more than one lock, perform this technique on one lock at a time.

This technique can also be used on vinyl-frame slider windows because they are similar in design to single- and double-hung windows except they are oriented horizontally rather than vertically. Like single- and double-hung windows, slider windows may have one or two locks. To force vinyl-frame single- and double-hung windows, follow the steps in **SKILL DRILL 15-23**.

Techniques for Breaching Walls

Breaching a wall for forcible entry is a massively destructive and labor-intensive undertaking, and generally not considered unless a genuine need exists. There are two situations where breaching a wall is most likely to be employed. One is during normal forcible entry operations into a building or room, when the door system is so heavily fortified that making a hole in an adjacent wall will actually be quicker and easier than forcing the

SKILL DRILL 15-23

Forcing Vinyl-Frame Single- and Double-Hung Windows Firefighter I, NFPA 1010: 6.3.4

1. Try the window to ensure it is locked. Insert the small head of an S&D tool, the adze of a Halligan, or an A tool into the seam between the two sashes in line with the sash lock. Usually, the tool will not need to be struck or driven into place.

2. Taking care not to contact the window glass, gently rock the tool back and forth from left to right in a progressively larger arc until the two halves of the sash lock separate and unlock.

Courtesy of Sean Wilson.

door. The other is when there is a trapped firefighter at a location where there is no door or other opening, or when a trapped or disoriented firefighter needs to make an immediate, forcible exit. It should go without saying that some walls are easier to breach than others based on their design and materials used.

Breaching a Stud Wall. Stud walls can be breached by first finding the **stud space**—the hollow spot between two studs—piercing the plaster, drywall, sheetrock, or paneling in the stud space with an axe or a Halligan, and then clearing out the entire stud space on both sides of the wall so that none of the wall covering remains. The hole should be 3 to 4 ft (0.9 to 1.2 m) tall.

It is best to breach the wall at a stud space that is not adjacent to an electrical outlet or other obvious utility that might run through that part of the wall. If a wider hole is needed, break out two (or more) adjacent stud spaces, knock the middle stud loose from where it is secured at the floor by striking it with a tool, and then remove the middle stud or push it aside. To breach a stud wall, follow the steps in **SKILL DRILL 15-24**.

Breaching a Masonry Wall. Making a hole in a masonry wall is a big job. In most cases, you will use a sledgehammer, THE PIG, an 8-lb (3.6-kg) flat-head axe, or another similar heavy striking tool. A rotary saw equipped with a masonry blade or a diamond rescue blade is also helpful for outlining the hole and weakening one side of the area to breached.

When breaching a masonry wall, the use of a team circuit system is beneficial. In this system, all crew members line up and take turns swinging the sledgehammer at the wall. Each member takes a predetermined

SKILL DRILL 15-24

Breaching a Stud Wall Firefighter I, NFPA 1010: 6.3.4

1. Sound the wall with a tool to locate the stud space between two studs. If possible, avoid walls likely to contain electrical wiring, plumbing, ductwork, or other utilities.

2. Pierce the plaster, drywall, sheetrock, or paneling with an axe or a Halligan. Create an inspection hole to verify conditions and that a full breach can be accomplished.

SKILL DRILL 15-24 CONTINUED

Breaching a Stud Wall Firefighter I, NFPA 1010: 6.3.4

3. Clear out the entire stud space on both sides of the wall up to a height of 3 to 4 ft (0.9 to 1.2 m). If a wider hole is needed, the firefighter will break out an additional stud space (or more) and knock the middle stud loose from the floor by striking it with a tool to push it aside or remove it.

number of swings, usually 5 to 10, hands the tool off to the next firefighter, and then goes to the back of the line. This allows each firefighter to give a small number of high-quality swings without becoming overly fatigued, and allows them to have a short rest before needing to swing again. So long as the tool is handed off quickly, this is an incredibly efficient method to overcome a difficult obstacle in the shortest amount of time with the physical exertion minimized by sharing it equally among members.

To breach a masonry block wall, use a sledgehammer or THE PIG to destroy the hollow section on the exterior of three to four blocks. Next, break out the interior side of the same blocks until they are demolished or pushed into the building. Finally, attack the mortar joints where the blocks meet one another. Once three to four blocks are broken out, the blocks above the hole often loosen up, and you should be able to push intact blocks into the building by striking at the mortar joints between blocks.

Brick walls can be breached in a similar fashion, but it may be more difficult to make the initial hole in the wall because the cavities within standard bricks are much smaller and often filled with mortar.

Breaching poured concrete walls is challenging and should be avoided if at all possible. If no other option exists, cut the outline of the required hole with a masonry blade or diamond rescue blade on a rotary saw. Although this saw will be able to penetrate only 3 to 5 in. (8 to 13 cm) into the wall, it will weaken the area substantially. Next, attack the wall using a sledgehammer and a team circuit of as many firefighters as are available. Keep the rotary saw handy to cut any rebar discovered in the concrete.

CASE STUDY

You Are the Firefighter CONCLUSION

There is a working fire on the second floor of a local hotel. The hotel is a three-story, wood-frame building with a hallway down the center. It was built in 1981 and renovated last year. Your crew is assigned to search the third floor for occupants who may still be in the building. As you dismount the fire apparatus, your captain tells you to grab the tools for forcible entry as the manager cannot find his master key.

1. **What kind of access issues would you expect in this structure?**

   Answer: Access to upper floors may be difficult if stairwells from the lobby floor are locked or if there is no elevator key to take control of the elevator. If an entire floor of hotel rooms has to be searched, forcible entry into all or most rooms may make this an extended operation.

2. **Which forcible entry tools would you bring with you?**

   Answer: You would bring Irons, Hydra-Ram or hydraulic spreaders with door tips (if available), lock-puller, key tools, and any keys available from the building's lock box.

3. **How would the forcible entry challenges for this structure differ from a single-family home?**

   Answer: Although many of the doors on hotel rooms will likely swing inward, they will likely be set in a metal frame with heavier commercial locks, making them dlfficult and time-consuming to force.

WRAP-UP

SUMMARY

KNOWLEDGE OBJECTIVES

- Identify common situations where forcible entry is required in the fire service.
  - Describe the situations and circumstances that require forcible entry into a structure. (**NFPA 1010: 6.3.4**, pp. 596–597)
- Describe the building components that are utilized to force entry into a structure.
  - Describe the basic components of a door. (**NFPA 1010: 6.3.4**, p. 597)
  - Describe the basic classifications of doors by opening type and how the opening type affects forcible entry operations. (**NFPA 1010: 6.3.4**, pp. 597–607)
  - Explain the differences between a solid-core and a hollow-core door. (**NFPA 1010: 6.3.4**, pp. 598–600)
  - Describe the major components of a lockset. (**NFPA 1010: 6.3.4**, pp. 607–608)
  - Describe the five major types of locksets and how the lockset type affects forcible entry operations. (**NFPA 1010: 6.3.4**, pp. 609–618)
  - Describe the basic configurations of window construction. (**NFPA 1010: 6.3.4**, pp. 618–620)
  - Describe how the window type and glass affects forcible entry operations. (**NFPA 1010: 6.3.4**, pp. 620–624)
  - Describe the common types of window frames and glass. (**NFPA 1010: 6.3.4**, pp. 622–624)
  - Describe the types of wall construction. (**NFPA 1010: 6.3.4**, pp. 624–626)
  - Explain the differences between load-bearing and non-load-bearing walls. (**NFPA 1010: 6.3.4**, p. 624)

- Describe the materials used in exterior and interior walls. (**NFPA 1010: 6.3.4**, pp. 624–626)
- Describe how security bars and gates affect forcible entry operations. (**NFPA 1010: 6.3.4**, pp. 626–628)
- List the rules of forcible entry. (pp. 629–630)
- Describe the tools utilized force entry into a structure.
  - List the types of tools used in forcible entry. (**NFPA 1010: 6.3.4**, pp. 630–635)
  - List the special-use and lock tools used in forcible entry. (**NFPA 1010: 6.3.4**, pp. 631–635)
- Describe the types of forcible entry techniques utilized to gain entry into a structure.
  - Describe the types of forcible entry techniques utilized to gain entry via a door. (**NFPA 1010: 6.3.4**, pp. 635–677)
  - Describe the types of forcible entry techniques utilized to gain entry via window. (**NFPA 1010: 6.3.4**, (pp. 677–678)
  - Describe how to force entry through security gates and windows. (**NFPA 1010: 6.3.4**, (pp. 677–678)

SKILLS OBJECTIVES

- Force entry through a door.
  - Force an inward-swinging door set in a wood frame with a tubular deadbolt using the sledgehammer method. (**NFPA 1010: 6.3.4**, pp. 636–637)
  - Force an inward-swinging door set in a wood frame using the stutter-step method. (**NFPA 1010: 6.3.4**, pp. 638–639)
  - Use the GSF technique with a Halligan fork on an outward-swinging door. (**NFPA 1010: 6.3.4**, p. 641)
  - Use the GSF technique with a Halligan adze on an outward-swinging door. (**NFPA 1010: 6.3.4**, p. 642)
  - Use the GSF technique with a Halligan fork on an inward-swinging door. (**NFPA 1010: 6.3.4**, pp. 643–644)
  - Use the GSF technique with a Halligan adze on an inward-swinging door. (**NFPA 1010: 6.3.4**, p. 645)
  - Use the GSF technique on an inward-swinging door with a Hydra-Ram or hydraulic spreaders with door-opening tips. (**NFPA 1010: 6.3.4**, pp. 646–647)
- Pull a lock cylinder.
  - Force a bored lockset with an S&D or Rex tool. (**NFPA 1010: 6.3.4**, pp. 648–650)
  - Force a bored lockset with an A tool or Halligan adze tool. (**NFPA 1010: 6.3.4**, pp. 650–652)
  - Force a mortise lock with a K tool, an S&D tool, or a Rex tool. (**NFPA 1010: 6.3.4**, pp. 653–655)
  - Force a mortise lock with locking pliers or angled slip-joint pliers. (**NFPA 1010: 6.3.4**, pp. 655–656)
  - Force a rim lockset with an S&D tool, a Rex tool, or a K tool. (**NFPA 1010: 6.3.4**, pp. 657–659)
  - Use a shove knife on a bored or mortise lockset in an outward-swinging door. (**NFPA 1010: 6.3.4**, pp. 660–661)
  - Use a J tool on panic hardware on an outward-swinging door. (**NFPA 1010: 6.3.4**, p. 662)
- Force entry by cutting with a rotary saw.
  - Cut a latch with a rotary saw. (**NFPA 1010: 6.3.4**, pp. 664–665)
  - Cut a padlock with a rotary saw. (**NFPA 1010: 6.3.4**, p. 667)
  - Cut a hockey puck lock with a rotary saw. (**NFPA 1010: 6.3.4**, pp. 668–669)
  - Make a triangle cut in an overhead door with a rotary saw. (**NFPA 1010: 6.3.4**, pp. 670–671)
  - Make an overhead door rectangle box cut with a rotary saw. (**NFPA 1010: 6.3.4**, pp. 672–673)
  - Make a rectangle hinged cut in an overhead door with a rotary saw. (**NFPA 1010: 6.3.4**, p. 674)
  - Cut a rolling steel door with a rotary saw using the slice and pull method. (**NFPA 1010: 6.3.4**, pp. 675–676)
- Force entry through a window.
  - Force a wood frame single- and double-hung windows. (**NFPA 1010: 6.3.4**, p. 678)
  - Force a vinyl-frame single- and double-hung windows. (**NFPA 1010: 6.3.4**, p. 679)
- Breach a wall
  - Breach a stud wall. (**NFPA 1010: 6.3.4**, pp. 680–681)

KEY TERMS

annealed The process of forming standard window glass.

A tool A special-use tool used for pulling bored locksets.

awning windows Windows that have one large or multiple medium-size panels that do not overlap when they are closed. The window is operated by a hand crank from the corner of the window. The hinge is on the top.

bored locksets The most common fixed locks in use today. The locks and handles are placed into a predrilled hole in the door. The outside of the doorknob will usually have a key-in-knob lock; the inside will usually have a keyway, a button, or another type of locking/unlocking mechanism.

case-hardened steel Steel created in a process that hardens the outer portion of a steel component while the inner core remains soft.

casement windows Windows in a steel or wood frame that open away from the building via a crank mechanism. These windows have a side hinge.

dead latch A manually operated latch bolt that is squared off that does not allow a door to be closed when slammed. (NFPA 80)

door frame The upright or vertical parts of a door system onto which a door is secured.

door system The door, door frame, and lockset.

double-hung windows Windows that have two movable panels, or sashes, that can move up and down.

double-pane glass A window design that traps air or inert gas in the space between two pieces of glass to help insulate a house.

duckbill padlock breaker A tool with a point that can be inserted into the shackles of a padlock. As the point is driven farther into the lock, it gets larger and forces the shackle out of the lock body.

exterior wall A wall—often made of wood, brick, metal, or masonry—that makes up the outer perimeter of a building. Exterior walls are often load-bearing.

forcible entry Techniques used by fire personnel to gain entry into buildings, vehicles, aircraft, or other areas of confinement when normal means of entry are locked or blocked. (NFPA 402)

hardware The parts of a door or window that enable it to be locked or opened.

hockey puck padlock A type of padlock with hidden shackles that cannot be forced open through conventional methods.

hollow-core door A door with a sheet metal covering or wood or fiberglass veneer over a light framework. These doors are sometimes left completely hollow, but are usually filled with cardboard baffles, gypsum, or foam.

interior wall A wall inside a building that divides a large space into smaller areas.

jalousie windows Windows made of small slats of tempered glass, which overlap each other when the window is closed. Often found in trailers and mobile homes, jalousie windows are held together by a metal frame and operated by a small hand wheel or crank found in the corner of the window.

J tool A tool that is designed to fit between double doors that are equipped with push bars or panic bars.

K tool A tool that is used to remove mortise and rim lockset cylinders from structural doors so the locking mechanism can be unlocked.

key tools Three simple tools—a ¼-in. (6-mm) slotted screwdriver, needle-nose pliers, and a bent-end key tool—often carried together as a set; used to manipulate most locks after the cylinder is pulled or removed.

laminated glass Also known as safety glass. The lamination process places a thin layer of plastic between two layers of glass so that the glass does not shatter and fall apart when broken. See also *laminated safety glass.*

laminated safety glass See *laminated glass.*

latch A component of a lockset; a moveable bolt that protrudes from the edge of the door and slides into the strike plate when the door is closed.

lock body The structural housing of a padlock that holds the main locking mechanisms and secures the shackle.

lock box A small safe containing the keys to a building's exterior doors and possibly interior doors.

lockset A standard doorknob lock, deadbolt lock, sliding latch, or padlock.

mortise lockset Rectangular-shaped lockset that fits into a rectangular cavity within a door. Often found in commercial buildings or older homes.

operating spindle The handle, doorknob, or keyway of a door that turns the latch to open it.

padlocks The most common types of locks on the market today, portable locks built to provide regular-duty or heavy-duty service. Several types of locking mechanisms are available, including keyways, combination wheels, and combination dials.

partition wall A non-load-bearing interior wall that spans horizontally or vertically from support to support. The supports may be the basic building frame, subsidiary structural members, or other portions of the partition system.

plate glass A thicker type of annealed glass that is commonly used for large windows in commercial buildings.

projected windows Windows that project outward on a top hinge or that pivot at the middle of the window. They are usually found in older warehouses or factories.

rex tool A 24-in. (61-cm) steel shaft with a large lock-pulling wedge welded onto one end; used to pull mortise and rim cylinders and key-in-knob locks.

rim lockset A lock mounted to the interior surface of a door.

sash frames Part of the window system that moves to open or close the window by sliding or swinging out; frames can be made of wood, metal, or vinyl. See also *sashes.*

sash locks Components that prevent the window from being opened from the outside.

sashes See *sash frames.*

S&D tool A double-headed lock-puller tool; large head pulls rim cylinders, mortise cylinders, and key-in-knob/lever locksets; small head pulls tubular deadbolts.

scraper bar A small tool, usually 9 to 10 in. (23 to 25 cm) in length designed for scraping paint and pulling wood trim and molding.

set screw A screw that is used to secure an object by pressure or friction, usually within or against another object.

shackle The U-shaped part of a padlock that runs through a hasp and then is secured back into the lock body.

shove knife A forcible entry tool used to trip the latch of outward-swinging doors.

single-pane glass windows Windows that have only one pane of annealed glass.

slam latch An angled latch bolt that is spring-loaded. After release by physical action, it returns to its operating position and automatically engages the strike plate when the door is slammed or is returned to the closed position. (NFPA 80)

slider windows Windows that slide open horizontally.

solid-core door A door design that consists of solid wood panels or wood pieces inside the door. This construction creates a stronger door that may be fire rated.

storefront windows Large windows that do not move, slide, or open found in many commercial occupancies; typically made of tempered glass or plate glass.

strike plate A component of a lockset; a metal plate with a hole in it that is mounted on the door frame.

stud space The hollow space between two vertical members in a stud wall.

stud walls Walls constructed of vertical framing members called studs that run from floor to ceiling on each floor or from foundation to attic in older balloon-frame construction; can be load-bearing or non-load-bearing.

tempered glass A type of safety glass that is heat treated so that it will break into small pieces that are not as dangerous. See also *tempered safety glass.*

tempered safety glass See *tempered glass.*

triple-pane glass A window design that traps air or inert gas in the two spaces between three pieces of glass to help insulate a house.

tubular deadbolt Surface- or interior-mounted lock on or in a door with a bolt that provides additional security, a type of bored lockset.

window frames Components that attach the window system to the building.

REVIEW QUESTIONS

1. What is forcible entry?
2. At a working fire, what is the priority for entering the structure?
3. What is a lock box?
4. What does a door system consist of?
5. What are the three types of overhead doors?
6. How did bored locksets get their name?
7. What type of padlock has a hidden shackle that cannot be reached with bolt cutters?
8. What type of glass would you expect to find in a glass storefront door in new construction?
9. Are stud walls load-bearing or non-load-bearing?
10. What is the name of the lock-pulling tool that combines the head of an A tool and the head of a Rex tool on an 18-in. (46-cm) shaft?
11. What are the four categories of forcible entry?
12. The three-step technique used in conventional forcible entry against a door in a metal frame is called GSF. What do those letters stand for?
13. What are the two stages of through-the-lock forcible entry?
14. Which type of lockset is threaded into the lock housing and held in place with a set screw?
15. Which tool can be used to cut standard and high-security padlocks and hockey puck locks?
16. What type of overhead door consists of five or more short, wide-hinged panels that allow the door to move along a track that stores the door flat when opened?

DISCUSSION QUESTIONS

1. What are the dangers of making entry through a window of a home at the scene of a medical emergency where voice contact with persons inside cannot be made?
2. How can removing a door from its hinges to gain entry to building on fire affect the flow path?
3. Why is damage from through-the-lock forcible entry typically cheaper to fix than damage done by using conventional forcible entry techniques?

APPLYING THE CONCEPTS

At 2230 your engine company is responding to a fire alarm at a furniture store downtown. Dispatch reported the business is closed and there is no keyholder information on file. The alarm company reported the alarm is coming from a duct sensor in zone 11. When you arrive, you do not see smoke or flames through the glass windows of the store, but you hear an alarm and see strobes inside. A size-up reveals there are glass double doors with a mortise lock on the front of the store, a large hollow-core metal door secured with two sliding bolts and panic hardware on side Charlie, and a single glass door with a mortise lock on side Delta, obscured by a dumpster. Passersby from local bars and restaurants are starting to gather.

1. Given this information, what is the priority of entry at the scene? Why?
2. Before a forced entry is made, what three things should firefighters do?

The IC commands, "Alright, it looks like side Delta is our best point of entry. Crew A, initiate the operation to gain access. Bring the correct tools and make sure to minimize damage to the door and surrounding area as much as possible. Report your findings. Crew B, keep the crowd at a safe distance and maintain situational awareness."

3. Why is entering through the door on side Delta the best choice?
4. Why is entering through side Charlie door the worst choice?

You and your crew gather tools and head to side Delta door. You note the door swings inward and use your tools to open the lock. You gain entry quickly. As you move into the structure to locate the alarm panel, the beam of your flashlight shines on a puddle of water. Following the water, you and your crew discover an active sprinkler. You don't see smoke or fire, but you smell residual smoke.

5. What are the different ways you could force entry through the door?
6. What is the best choice and why?
7. What tools did you select to force the mortise lock?

REFERENCES

National Fire Protection Association (NFPA). 2019. *NFPA 402, Guide for Aircraft Rescue and Fire-Fighting Operations.* 2019 Edition. Quincy, MA: NFPA.

National Fire Protection Association (NFPA). 2021. *NFPA 80, Standard for Fire Doors and Other Opening Protectives.* 2022 Edition. Quincy, MA: NFPA.

New York City Fire Department (FDNY). 2010. *Forcible Entry Reference Guide: Techniques and Procedures.* New York: FDNY.

Phillips, Bill. 2005. *The Complete Guide to Locks & Locksmithing, Seventh Edition.* New York: McGraw-Hill.

CHAPTER

16

Firefighter I

Ventilation

KNOWLEDGE OBJECTIVES

After studying this chapter, you will be able to:

- Describe how fire behavior impacts ventilation.
- Describe how size-up findings affect ventilation operations.
- Describe the importance of the timing and coordination of ventilation and suppression.
- Identify the steps that can be taken to minimize backdrafts and flashovers.
- Describe the types of ventilation, how each type removes contaminated atmospheres, and the techniques used to provide ventilation to a structure.
- Describe the hazards of vertical ventilation and how those hazards are mitigated.
- Describe the common techniques of vertical ventilation.
- Describe ventilation techniques for special conditions.
- Describe the tasks involved in maintaining ventilation equipment.

SKILLS OBJECTIVES

After studying this chapter, you will be able to perform the following skills:

- Begin the ventilation process with either a hand tool or a ladder.
- Deliver negative-pressure, positive-pressure, or hydraulic ventilation.
- Perform vertical ventilation by making a roof cut.
- Maintain equipment so it is in a ready state.

ADDITIONAL NFPA STANDARDS

- **NFPA 92**, *Standard for Smoke Control Systems, 2021 Edition*
- **NFPA 853**, *Standard for the Installation of Stationary Fuel Cell Power Systems, 2020 Edition*
- **NFPA 1410**, *Standard on Training for Emergency Scene Operations, 2020 Edition*
- **NFPA 5000**, *Building Construction and Safety Code, 2024 Edition*

CASE STUDY

You Are the Firefighter

It is a routine day at the hardware store where you work full time. You are busy doing inventory when your pager goes off, alerting you to a structure fire only a few blocks away. On your arrival, you find a one-story, wood-frame house with dirty brown smoke pushing out through the eaves, siding, and other cracks. The windows are stained black. The neighbors say they do not think anyone is home because the occupants work during the day.

1. What is the best way to ventilate this structure?
2. What concerns do you have about ventilating this structure?
3. When would you want to ventilate this structure?

Introduction

Ventilation is the controlled and coordinated removal of heat and smoke from a structure. Effective ventilation not only removes heat and smoke, it also replaces the escaping gases with cooler, cleaner, oxygen-rich air. This firefighting operation must be planned and systematic. Understanding how and when to perform ventilation is vital to conducting safe and effective fire suppression. When properly performed and coordinated, ventilation helps remove hot gases from the fire compartment; makes it easier to locate the seat of the fire; improves visibility; and contributes to faster and safer knockdown, more effective fire suppression, and improved efficiency of searching. Poorly performed and poorly timed ventilation can quickly add oxygen to a fire and, if not coordinated with fire attack and other fireground operations, can result in increased hazards and rapid fire growth.

Ventilation and Fire Behavior

Fire service ventilation can be looked at as a two-step process of assessment and execution. The first step, which can be done during size-up, is a determination of the need for ventilation, an assessment of the location and amount of ventilation needed, and the coordination of ventilation operations with other fire-suppression operations. The second step of ventilating a fire is the mechanical operations of actually creating an opening by opening or closing doors and windows, opening skylights, and cutting openings in the roof.

Before firefighters open a door, take out a window, or cut a hole in the roof, they should be aware that all of these actions provide ventilation and should ask themselves the following questions:

- Why am I ventilating? Is it to make entry or gain access to the fire or occupants? How will those actions impact the environment for possible occupants and firefighters?
- Where do I want to accomplish the ventilation? Do I want to perform **horizontal ventilation**—open or remove windows or doors so that air flows horizontally through the structure—or **vertical ventilation**—open bulkhead doors and skylights and cut openings in the roof so the airflow is vertical? Where is the fire in relation to my anticipated ventilation opening?
- When do I want to perform the ventilation? Have I coordinated my efforts with the actions of the suppression team and the search and rescue teams?

Before ventilating, it is important to understand how a given action will affect the behavior of the fire. First, firefighters must determine if they are dealing with a ventilation-limited fire or a fuel-limited fire. These type of fires are described in Chapter 5, *Fire Behavior*. A ventilation-limited fire is a fire that has sufficient fuel for fire growth but has a limited amount of oxygen available so it does not burn as rapidly as it would with an unlimited supply of oxygen. This type of fire can grow very quickly if more oxygen is added. In other words, all a ventilation-limited fire needs is a supply of oxygen to rapidly convert it to an active-burning state.

Buildings and homes constructed with modern techniques and materials, and the modern contents made of synthetic materials found inside these structures, cause fires to grow much faster and burn hotter than fires involving older, natural-based contents. Any action that results in the introduction of oxygen to the fire is likely to change the fire from a ventilation-limited fire in the decay stage to a fuel-limited fire—that is, a fire that has sufficient oxygen for fire growth but has a limited amount of fuel available for burning. In a fuel-limited fire, when all of the fuel is consumed, the fire stops burning. The action of introducing oxygen to the fire can be as simple as opening a door in the building to make entry.

SAFETY TIP

Many structure fires today are ventilation limited and are discovered during the ventilation-limited stage.

Any opening establishes a potential flow path within the structure (described in Chapter 5, *Fire Behavior*.) The flow path is the route along which the heat, smoke, and hot gases move from the higher pressure area inside the fire compartment to the lower pressure area outside the fire compartment. The heat, smoke, and hot gases are transferred from the higher pressure area within the fire compartment toward the lower pressure areas accessible through doorways, window openings, and roof openings. As the hot gases exit the fire compartment at the upper levels and the cooler air enters the fire compartment at a lower level, there is a level in the compartment where the pressure exerted by the lighter hot gases flowing out and the pressure of the cooler air flowing in is equal. This is called the **neutral plane**. In order for a neutral plane to exist, there must be a flow of cooler air entering the compartment and a flow of hot gases exiting the compartment.

The flow path can be unidirectional—flow in one direction between the fire and an opening to the outside—or bidirectional—flow in two directions, both from the fire to an opening to the outside and from an opening to the outside toward the fire. When openings are at two different levels, such as a window on the first floor and a window on the second floor at a multilevel structure, unidirectional flow may occur. With unidirectional flow, air flows in through the opening below or at the same level as the fire compartment (the inlet), then along the flow path and out of the other opening above the inlet (the exhaust). The inlet opening is positioned below the neutral plane, and the exhaust opening is positioned above the neutral plane. A bidirectional flow often occurs when there is one opening at the same level as the fire. The fire gases flow out and air flows in at the same opening, so the opening is acting both as an inlet for fresh air and an exhaust for the fire gases. In a bidirectional flow, a firefighter may identify the neutral plane that defines the inlet and outlet.

The flow path is determined by the building design and which doors and windows are open to the outside. Every new ventilation opening may provide a new flow path between the fire and the opening. Any operations conducted in the flow path place firefighters at significant risk for injury or death because of the large amount of heat and smoke that travel along the flow path. A firefighter should never be between where the fire is and where it wants to go without access to water or a door to close.

Several factors influence the effectiveness of ventilation and the speed with which the ventilation changes the flow path. These factors include door control, the impact of the ventilation location, ventilation hole size, wind, and exterior suppression.

Impact of Door Control

Following a thorough **size-up**—a rapid evaluation and analysis of an incident, usually conducted by the first-arriving company officer, to determine which resources need to be deployed and which actions can be undertaken safely—one of the first actions taken by most fire departments at a structure fire is to gain access to the building by opening a door. Although this action may be necessary to fight the fire, it also changes the ventilation profile of the fire by supplying additional oxygen to the fire. Once the door is opened, the clock is ticking before either the fire gets extinguished or it grows until conditions are untenable, potentially jeopardizing the safety of everyone in the structure and at the immediate fireground. When you open a door, pay attention to the air flow through the door. A rapid rush of air out or a tunneling effect—where hot smoke flows out through the upper part of the doorway and fresh air pushes in the bottom, creating a tunnel of clear space inside the doorway—could indicate a ventilation-limited fire to which you just provided oxygen.

Sometimes the most effective act in ventilation is to limit ventilation until a coordinated fire attack can be made. Limiting the air inlet limits the fire's ability to grow. The simple act of closing the door after entering will limit the air supply to the fire and slow the fire growth until a crew is ready to make access to the building and perform a coordinated fire attack. Leaving the door closed as long as possible aids significantly in controlling a fire by limiting the amount of oxygen supplied to it. When making an interior fire attack or rescue effort, try to maintain control of the door. Keep it closed as much as possible and open it just enough to allow the

FIGURE 16-1 When making an interior fire attack or rescue effort, try to maintain control of the exterior door, keeping it closed as much as possible to limit the flow of air.

hose line to be advanced to the fire. It may be necessary to station a firefighter at the door to maintain this partial door closure (**FIGURE 16-1**). This tactic is known as "manning the door." This firefighter can also assist in advancing the hose line as they control the door.

Fire growth also may be limited by closing interior doors. Closing the door after searching a room potentially limits the supply of air available to the fire. It also helps to reduce the risk of the fire spreading into that room or other areas connected to that room. When closing the door to the room of origin or the involved room, it can help to contain the fire to that room until water can be applied.

Impact of the Ventilation Location

The location of the ventilation opening influences how quickly a fire reacts to changes in ventilation and determines how effectively the hot gases and smoke are removed from the fire location. Ventilation should occur as close to the fire as possible. The most desirable options are to provide a ventilation opening directly over the seat of the fire or to open a door or window in the room where the fire is burning to release heat and smoke from the fire directly to the exterior. Ventilating directly over the fire produces the fastest impact on the behavior of the fire and exhausts the greatest amount of combustion products. However, if this ventilation is done without coordinated water application, the result will most likely be added fire growth because of the added supply of oxygen. In other words, ventilating over the fire is helpful only if ventilation is coordinated with fire attack. The application of water absorbs some of the heat produced by the fire and helps remove more energy than the fire is creating. Once the fire growth has been slowed, the ventilation opening continues to exhaust hot gases and smoke, improving the conditions in the fire compartment by reducing the air temperature and improving visibility.

TIP

Opening a door is an act of ventilation that should be conducted in coordination with suppression efforts, the same as any other ventilation tactic. If you limit air inlets to the fire, you can limit the ability of the fire to grow. Sometimes the best ventilation is less ventilation.

If it is not possible to create a ventilation opening in the immediate area of the fire, firefighters must be able to predict how ventilation from another location will affect the fire. Fire travels toward a ventilation opening, unless influenced by outside forces such as strong winds along the flow path. Additionally, if the ventilation opening is farther from the fire, it will take longer for any changes in ventilation to impact the fire. Although these distant openings are less effective in removing smoke and hot gases from the fire, they may still contribute significantly to its growth.

Impact of the Ventilation Hole Size

Fire studies conducted by UL's Fire Safety Research Institute (FSRI), a part of the UL Research Institute (formerly known as Underwriters Laboratories [UL]), demonstrated that larger-sized vertical ventilation openings do not localize the growth of a fire. The studies compared the impact of cutting a 4-foot by 4-foot (ft; 1.2-meter by 1.2-meter [m]) roof ventilation hole with the effect of cutting a 4-ft by 8-ft (1.2-m by 2.4-m) roof ventilation hole. Neither the small hole nor the larger hole lowered the temperature in the fire area when only ventilation was performed. However, they found that vertical ventilation in coordination with an exterior or interior application of water as close to the fire as possible improves visibility, reduces the temperature in the fire compartment, and temporarily limits the growth of the fire. When vertical ventilation is conducted in conjunction with water application, the larger ventilation hole was found to be more effective than the smaller ventilation hole in lowering the temperature of the fire compartment and in other parts of the building (Kerber 2013, 289–290).

Impact of Wind

Wind is a powerful force that can change the direction and speed of a fire and its flow path rapidly. Any time a window or door is opened on the side of a building that faces the wind—the upwind or the windward side—an unlimited amount of oxygen under high pressure is introduced into the fire. This action may be unintentional, such as a glass window or patio slider door breaking because of heat from the fire, or it can be an intentional act of ventilation. A sudden increase in oxygen, combined with hot flammable smoke (fuel) from the fire, will result in rapid fire growth and can produce a sudden change in the direction of the flow path due to the pressure imposed by the wind. Think of the wind as a giant positive-pressure fan forcing huge quantities of oxygen into a ventilation-limited fire (**FIGURE 16-2**). Firefighters should remember to keep the wind at their backs during a fire attack and avoid ventilating on the upwind or downwind side of a fire unless it is part of a well-organized suppression effort. The National Institute of Standards and Technology (NIST) has conducted experiments on wind-driven fires. These tests were conducted using high-rise structures, but wind is also a very real threat in one- and two-family structure fires. (More information about these studies is available at the NIST website at www.nist.gov/fire.)

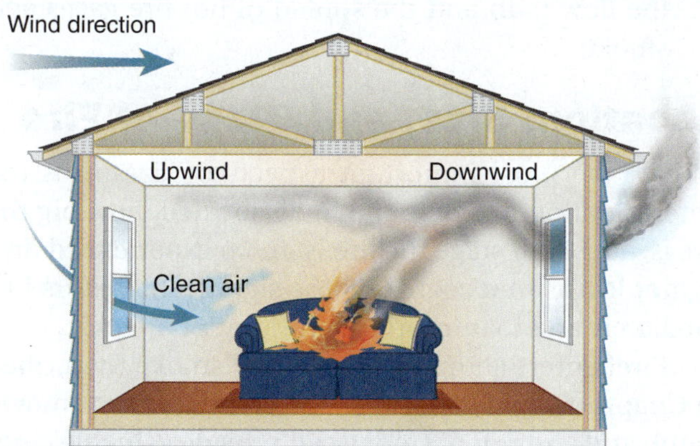

FIGURE 16-2 On a windy day, proper ventilation helps to remove smoke and heat from a structure.

Impact of Exterior Suppression

The FSRI and NIST have conducted fire research studies that demonstrated that our traditional assumptions about the effects of ventilation need to be modified. Ventilation used to be defined as the systematic removal of heat, smoke, and fire gases from a building and replacement of those elements with cool, fresh air. Although this definition is technically true, it can lead to a misapplication of the principle. Because of the extremely fuel-rich environment found on today's fireground ventilation that is not coordinated with effective suppression will introduce enough oxygen to rapidly bring the fire area to flashover. The following describes the conclusions for residential fire behavior that were drawn from the fire experiments conducted by the FSRI, NIST, and the New York City Fire Department (FDNY) in 2012:

- Increasing the air flow to a ventilation-limited structure fire by opening doors, windows, or roof openings increases the fire hazard from the fire for both building occupants and firefighters.
- Increasing the air flow to a ventilation-limited structure fire may lead to a rapid transition to flashover.
- There is a need for systematic fire attack to coordinate the activities of size-up, entry, ventilation, search and rescue, and application of water.*

For the greatest reduction in the production of hot gases and smoke, use a coordinated fire attack by performing ventilation and applying water to the fire from a safe location as close to the fire as possible. Applying even a small amount water to the fire as quickly as possible from either the exterior or interior can help to improve conditions in the entire fire structure. Well-planned ventilation, when coordinated with the proper application of water, has several benefits. It removes large amounts of highly flammable fuels in the form of superheated gases and smoke and cools the fire building, which greatly reduces the chance for flashovers or backdrafts. Ventilation also helps to improve the visibility within the fire building, thereby helping firefighters maintain their orientation within the building and making it easier to search for fire victims. Reduced temperatures improve firefighter safety and increase the potential for the survival of building occupants.

* List content courtesy of the NIST.

A transitional attack is a fire attack that starts with a quick, indirect, exterior attack into the fire compartment before transitioning to an interior, offensive attack. The transitional attack is described in greater detail in Chapter 14, *Fire Suppression*. The initial exterior attack initiates cooling and darkens the fire, allowing firefighters to quickly transition into an interior attack for final suppression. When used, this technique should be performed prior to entry, search, and suppression. FSRI/NIST research demonstrated the application of water from the outside, when properly applied, did *not* push the fire from one place to another (Kerber 2013, 6). The water introduced into the fire from the outside produced cooling, not just in the fire compartment but also in other areas of the fire building. Significant cooling was achieved even when the water was applied to a body of fire in the interior of the structure at a distance from the seat of the fire. While this application of water did not totally extinguish the fire, it cooled the fire gases throughout most parts of the fire building, making it safer for firefighters to enter and complete the suppression of the fire. This greatly improved the survival profile for building occupants.

It is important to stress that the transitional fire attack does not reduce the importance of proper ventilation and that ventilation remains a crucial part of fire suppression. Close coordination between the fire attack team and the ventilation crew is always required at a fire. Providing ventilation without this coordination can contribute to the growth of the fire and may lead to dangerous conditions (**FIGURE 16-3**).

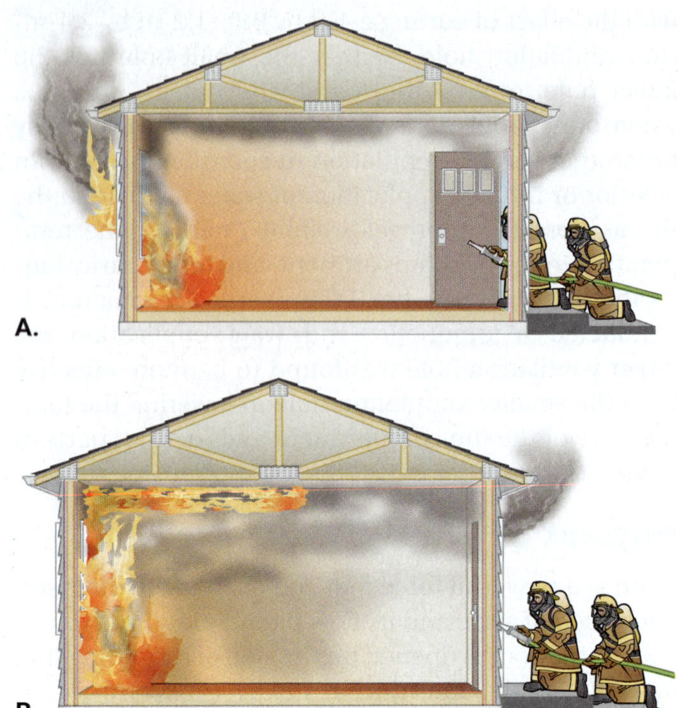

FIGURE 16-3 Proper ventilation enables firefighters to control a fire rapidly. **A.** Vented structure. **B.** Unvented structure.

Size-Up and Ventilation

A thorough size-up of each fire is important to develop a plan not only for rescue and suppression, but also for the tactics needed to ventilate the fire structure. Ventilation is not a one-size-fits-all type of operation, where firefighters perform exactly the same actions at every fire. The size-up helps determine which actions need to be taken and in which order these actions need to be initiated. The size-up should include consideration of the following factors:

- Location, size, and stage of the fire
- Fire department arrival time
- Size of the building
- Type of building construction
- Geometry or shape of the building
- Potential for rescue
- Potential for building collapse
- Fuel load, including the amount and type of fuel in the building contents—that is, does the building contain modern, synthetic materials or traditional, natural materials, and are there any potentially hazardous materials such as chemicals or pressurized gas containers
- Best initial type of fire attack
- Possible locations to ventilate while keeping in mind how these ventilation options will impact the flow path and the spread of hot fire gases and smoke

Location, Size, and Stage of Fire

One of the most important aspects of size-up is the determination of where the fire is located, how big the fire is, and what stage the fire is in. An experienced firefighter learns to recognize these significant factors immediately and to interpret each situation quickly.

Firefighters should learn to read smoke (described in Chapter 5, *Fire Behavior*). The color, location, movement, and amount of smoke can provide valuable clues about the fire's size, intensity, and fuel (**FIGURE 16-4**). A turbulent smoke flow is produced by a very hot fire. If the smoke flow grows more turbulent, it indicates that

FIGURE 16-4 The color, location, and amount of smoke can provide valuable clues for firefighters.

Courtesy of District Chief Chris E. Mickal/New Orleans Fire Department, Photo Unit.

the fire is growing hotter and the fire gases are expanding, raising the pressure in the structure. A cooler fire produces a slower, laminar smoke flow.

> **SAFETY TIP**
>
> The seemingly benign appearance of light-colored smoke should not fool firefighters. Such smoke can be just as dangerous as thick black smoke. Always remember that these unburned products of combustion produce toxic gases and vapors, no matter what color the smoke is.

A different phenomenon occurs in buildings with automatic sprinkler systems. The water discharged from the sprinklers cools the smoke and produces a cold smoke that may hardly move within the building. This type of smoke can fill a large warehouse space from floor to ceiling with an opaque mixture that behaves much like fog on a damp day. Because this cold smoke has a lower temperature than the surrounding atmosphere, it sits at the lower levels of the room, making it hard to see and hard to locate the seat of a fire. Mechanical ventilation is often needed to clear this type of smoke from the building. A similar situation occurs when smoke becomes trapped within any building long enough for it to cool to the ambient temperature.

A fire where there is little or no visible smoke may indicate that it is a fire that has exhausted most of its fuel supply or it is a small fire in the incipient stage that can be easily extinguished with a small amount of water and little fire damage. However, a fire that initially has no visible flames and very little or no smoke also may be a ventilation-limited fire.

> **TIP**
>
> Visible flames do not mean the fire is vented—it means it is venting and additional ventilation points will grow the fire if water is not applied.

Type of Building Construction

The way a building is constructed affects ventilation operations. The different types of construction are discussed in Chapter 6, *Building Construction.* Each construction type presents its own set of problems in regard to ventilation operations. For example, modern construction built with new-growth lumber and lightweight, manufactured building components that contain resins and glues, and a fuel load of petroleum-based furnishings not only provide more fuel, but these fuels reach flashover faster, and generate greater amounts of energy, which cause building components to fail much sooner than older, traditional construction and furnishings. Additionally, most residences built since the 1990s are sealed from the outside with tight windows, vapor barriers, and insulation. The wide use of heating and air conditioning systems means that in many parts of the country, houses are tightly closed for most of the year. The result is that during a fire event, these buildings develop higher interior temperatures much more quickly. If the doors and windows remain closed, the fire will reach a point where it has such a limited supply of oxygen that it will not sustain open combustion. When this happens, the fire, which is still hot and has an abundance of fuel, becomes a ventilation-limited fire.

TIP

Concerns about saving energy have led to increased use of insulated thermal glass, both in new construction and as a replacement for conventional glass in older windows. Insulated thermal glass panes have multiple layers of glass sealed together with a small air space between the layers. These windows are much more difficult to break than conventional glass windows, so a firefighter might have to use multiple strikes and additional force to break this type of glass. This is also true of hurricane glass, which is found in some windows in hurricane-prone areas. This type of glass can be difficult to penetrate. In these cases, firefighters should look for alternative means to effect horizontal ventilation, such as by opening the window or removing the window frame.

FIGURE 16-5 Vertical fire extension can occur when the fire jumps from a lower floor to a higher floor through exterior windows.

Type I Construction (Fire-Resistive)

Type I construction (fire-resistive) describes a building in which all of the structural components are made of noncombustible materials, such as steel or concrete. This type of construction generally has spaces divided into compartments, which limits potential fire spread. A fire in a Type I building is most likely to involve the combustible contents of an interior space in the building.

Although Type I construction is intended to confine a fire, it does contain potential avenues for fire spread. Openings for mechanical systems such as heating, ventilation, and air conditioning (HVAC); plumbing and electrical chases; elevator shafts; and stairwells all provide paths for fire and smoke to spread. A fire can also spread from a lower floor to a higher floor through exterior windows, a phenomenon called **vertical fire extension** or **auto-exposure** (**FIGURE 16-5**).

Because the windows in these buildings are fixed glass—that is, they do not open—and are not broken easily, there are limited opportunities for creating ventilation openings. In some cases, one stairway can be used as an exhaust shaft, while other stairways are cleared for rescue and access. To accomplish this, ventilation fans are often required to direct the flow of smoke. Some of these buildings contain a number of glass panels with a symbol (sometimes a Maltese cross) in a lower corner, indicating that the panel is made of tempered safety glass and can be broken easily with a sharp, pointed object.

In many of these buildings, engineered designs are installed to assist firefighters with ventilation or controlling the movement of hot gases and smoke throughout the building including the following:

- Smoke and heat vents are installed in the ventilation systems.
- Stairwells and elevator shafts are designed with positive pressure—that is, the air pressure inside the stairwells and elevator shafts is higher than the air pressure in the rest of the building. This limits smoke migration.
- HVAC systems include features that arriving firefighters can use to pressurize specific areas to limit fire spread.

In some cases, the most valuable person on scene may be the building engineer, who can assist in the operations of many of these systems from the fire command center.

The roof on a Type I building usually is supported by concrete or steel roof decking. It can be difficult or impossible to make vertical ventilation holes in these types of roofs. These roofs might, however, have openings for skylights or for HVAC ducts.

Type II Construction (Noncombustible)

Type II construction (noncombustible) describes buildings in which all of the structural components are made of noncombustible materials. A typical example of Type II construction is a large-area, single-story building with a steel frame, metal or concrete block walls, and a metal roof deck.

Type II construction is most common in single-story warehouse or factory buildings, where vertical fire spread is not an issue, although fire walls are sometimes used to subdivide these large-area buildings and limit the potential fire spread. Similar to Type I construction, a fire in this type of building is most likely to involve the combustible contents of an interior space within the building, as the structural components contribute little or no fuel, and interior finish materials are limited.

Many Type II buildings have few window openings, so horizontal ventilation is limited to existing doors. Vertical ventilation of this type of building should be attempted only with aerial devices. If vertical ventilation is attempted, keep in mind that the building contents may provide a high fuel load, and a fire in this type of building can result in failure of the building walls and collapse of the roof at any time. Additionally, the roof on a Type II building usually is supported by metal roof decking. It can be more challenging to make vertical ventilation holes in these types of roofs.

Type III Construction (Ordinary)

Type III construction (ordinary) describes building with exterior walls made of noncombustible or limited-combustible materials that support the roof and floor assemblies. The interior walls and floors are usually wood construction, and the roof usually has wood decking and a wood structural support system. Type III construction buildings usually have windows and doors that can be used for horizontal ventilation. The roof on this type of building can be cut with power saws or axes to open vertical ventilation holes.

The walls and floors in Type III buildings often include numerous openings for plumbing and electrical chases. If a fire breaks through the interior finish materials (plaster or drywall), it can spread undetected within these void spaces to other portions of the structure. Interior stairwells generally allow a fire to extend vertically within the building. Often, Type III construction buildings have been remodeled and have multiple roofs, such as when a newer roof is constructed on top of an older roof, making ventilation more difficult. Although ventilation can be performed either horizontally or vertically in such a case, it can be hard to ventilate in the area close to where the fire is burning.

The vertical openings in Type III construction frequently provide a path for fire to extend into an attic or cockloft. Heat and smoke can accumulate within these spaces and then spread laterally between sections of a large building or into attached buildings.

Type IV Construction (Heavy Timber)

Type IV construction (heavy timber) describes buildings with exterior walls that consist of masonry construction and interior walls, columns, beams, floor assemblies, and a roof structure that are made of wood. The exterior walls are usually brick and are extra thick to support the weight of the building and its contents.

The heavy, solid wood used in Type IV construction is difficult to ignite, but once involved in a fire, the structure of this type of building can burn for many hours. As the fire consumes the heavy timber support members, the masonry walls become unstable and collapse.

Heavy timber construction should have no concealed spaces or voids, reducing the risk of horizontal and vertical fire spread that often occurs in ordinary construction buildings. However, many Type IV buildings have been converted into small shops, galleries, office buildings, and residential occupancies. These conversions tend to divide the open spaces into smaller compartments and create void spaces within the structure. Additionally, many heavy timber buildings contain vertical openings for elevators, stairs, or machinery, all of which provide a path for a fire and smoke to travel from one floor to another.

Type IV buildings usually contain a large number of windows. A large fire in this type of construction often self-ventilates as the windows break from the heat of the fire. Vertical ventilation, however, may be difficult because of the thick layers of wood sheathing and the build-up of multiple layers of roofing materials.

Type V Construction (Wood Frame)

Type V construction (wood frame) describes buildings that have many of the same features as Type III buildings, except that the exterior walls in a Type V building are not required to be constructed of masonry or noncombustible materials.

Type V buildings often contain many void spaces where fire can spread, including attics or cocklofts. Additionally, they are commonly built with modern, fast-growth lumber; lightweight wood-truss roofs; and manufactured I-beam floors, which can fail quickly under fire conditions.

Older Type V buildings were often assembled with balloon-frame construction. This type of construction includes direct vertical channels within the exterior walls, allowing a fire to spread quickly from a basement or lower level to the attic or cockloft. Heated smoke and gases can accumulate under the roof of these buildings, which requires rapid vertical roof ventilation.

Modern Type V construction typically uses platform-frame construction. In these buildings, the structural frame is built one floor at a time. Between each floor, a plate at the floor and the ceiling acts as a fire stop. Its presence limits the fire from spreading upward and helps contain and limit the fire on a single floor.

Modern Type V construction typically uses lightweight components to save money and provide longer and more open areas, which presents problems when a fire occurs. Engineered wooden structural components used in residential construction, such as trusses and I-beams, contain much less wood than the solid beams used in older construction. These products burn more rapidly and could fail quickly. The fire service cannot become complacent with modern engineered lumber.

TIP

The changes in residential building construction materials and techniques and the changes in the fuel load of building contents due to modern furnishings have caused changes in fire behavior in residential fires. This has required modification of some of the ventilation tactics that were used in the past. The recommended ventilation tactics are based on the observations of firefighters on the fireground and controlled tests conducted by NIST and the FSRI. These studies are available at www.nist.gov/fire or at www.fsri.org.

Timing and Coordination of Ventilation and Suppression

Limiting the amount of air entering the fire compartment limits the fire's ability to grow. Keeping the front door and other ventilation openings closed until firefighters are prepared to apply water to the fire reduces the speed of the fire growth. This is a temporary action until water can be applied to the fire. When considering ventilation options and timing, remember that opening the building does not guarantee positive results. Firefighters need to make sure the fire does not get larger by applying water to change it from a ventilation-limited fire to a fuel-limited fire. Once the water cools the fire so that more thermal energy is absorbed by the water than is being generated by the fire, ventilation will begin to work as intended.

Because it takes time to get a charged hose line in place, opening a vent should sometimes be delayed until the charged hose line is in place and the crew is ready to apply water to the fire from a close and safe location. Effective ventilation requires the incident commander (IC) to coordinate timing and fire attack. When thinking about ventilation, think about the Three Ws—when, where, and why—the "when" describes timing.

Minimizing Backdrafts and Flashovers

Ventilation is a major factor in two significant fireground phenomena: backdraft and flashover. (See Chapter 5, *Fire Behavior*, for more information about these conditions.) Both can be deadly, and firefighters should exercise great caution when conditions indicate that either is possible. An indirect or transitional fire attack reduces the chance of a backdraft or flashover. Apply water to the fire from a safe location as close to the fire as possible. This will reduce heat and fuel in the form of hot gases and smoke from the fire compartment. It will also reduce the temperature of the space, ideally converting the ventilation-limited fire to a fuel-limited fire. At that point, ventilation will be more effective in removing hot gases and smoke from the fire space, which will allow entry into the interior space for final fire suppression.

Backdrafts and Ventilation

A backdraft can occur when a building is charged with superheated gases and most of the available oxygen has been consumed. Few flames might be apparent, but the hot gases can contain large amounts of unburned or partially burned fuel (**FIGURE 16-6A**). If oxygen is introduced to the mixture, the fuel can ignite and explode. The hot vaporized fuel bursts into flames and the gases explode through the opening, resulting in a backdraft. To reduce the chance of a backdraft, firefighters try to release as much heat and unburned products of combustion as possible.

Placing a ventilation opening as high as possible within the building or area can help to eliminate potential backdraft conditions. A roof opening will draw the hot gases up and relieve the interior pressure. Because the mixture rising into the open atmosphere is exposed to oxygen, it might ignite, so the ventilation crew should not ventilate until the attack crew is ready to advance the hose line.

SAFETY TIP

Remember that opening an exterior door of a structure is an act of ventilation. The door should remain closed until the attack team is ready to advance the hose line into the building.

A.

B.

FIGURE 16-6 **A.** Building structure and fire conditions should be evaluated for signs of potential backdraft before any type of ventilation opening is made. **B.** Improperly venting a fire in an oxygen-deficient atmosphere can cause a backdraft.

> **SAFETY TIP**
>
> The flow of hot gases in a flow path has been measured at speeds as high as 15 miles per hour (24 kilometers per hour) with no wind present (Norwood and Ricci 2014). This is faster than a firefighter can run to try to escape the path of the flashover.

Flashovers and Ventilation

Flashover is the rapid transition from a fire that is growing by igniting one type of fuel to a fire where all of the exposed surfaces have ignited. (Flashover is described further in Chapter 5, *Fire Behavior*.) The critical temperature for gases to begin to flash is approximately 400°F (204°C), with the room undergoing total flashover at 1000°F (538°C) (NIST 2021).

Because the potential harm from a flashover is so great, firefighters need to be able to recognize any conditions that might indicate the possibility of a flashover and take steps to prevent it whenever possible. These conditions include:

- Off-gassing or smoke from furniture and carpeting
- A sudden increase in heat
- Zero visibility
- Fire conditions with such high heat that when a fog stream is directed at the ceiling, the water droplets vaporize rather than fall back to the floor
- Smoky conditions in the room but visible flames dancing along the ceiling
- Flames that appear intermittently
- Lowering of the neutral plane
- Window glass breaking due to rapid heat increase
- Smoke rolling out of a window changing so it seems like it's being pushed out faster because the temperature inside the compartment increased dramatically
- Smoke that changes color from lighter to black
- Light colored smoke suddenly appearing from furnishings or other décor

> **SAFETY TIP**
>
> Do not enter an environment if signs of an impending flashover are present.

If you recognize the potential signs of pending flashover, either exit from the nearest egress or apply water to the upper areas of the compartment immediately and monitor conditions closely. Use a transitional attack to cool the fire compartment, vent, and then enter.

Principles of Ventilation

To remove the products of combustion and other airborne contaminants from a structure, firefighters use horizontal and vertical ventilation. The type of ventilation used depends on several factors, including the ease and effectiveness of executing (performing) that type of ventilation. The effectiveness of each type of ventilation depends on the size of the opening and its location. A larger opening creates a larger air flow, so opening a door usually creates more ventilation than opening a window, and a ventilation opening closer to the fire provides more air flow than an opening farther from the fire.

During ventilation, the **contaminated atmosphere**—the products of combustion that must be removed from a building—is exhausted through the ventilation openings. **Clean air** is the outside air that replaces the contaminated atmosphere. The contaminated atmosphere can include any combination of heat, smoke, and gases produced by combustion, including toxic gases. A contaminated atmosphere can also be defined as any dangerous or undesirable atmosphere even if it was not caused by a fire. For example, natural gas, carbon monoxide, ammonia, and many other products can contaminate the atmosphere as a result of a leak, spill, or similar event. The same basic principles of ventilation can be applied to many of these situations.

Ventilation can be intentional or unintentional. **Intentional ventilation** occurs when the ventilation is planned and done on purpose. **Unintentional ventilation** occurs when a window or door fails because of the fire or a window or door is left open by mistake. Any ventilation may cause a rapid change in the direction or intensity of a fire, and unexpected ventilation can jeopardize firefighters operating inside a structure due to sudden changes in the fire's behavior.

Finally, ventilation can be either natural or mechanical. **Natural ventilation** depends on convection currents and other natural forces, such as the wind, to move heat and smoke out of a building and to allow clean air to enter. Natural ventilation can be used only when the natural convection air currents or wind is adequate to move the contaminated atmosphere out of the building and replace it with clean air. **Mechanical ventilation** requires fans or other powered equipment to introduce clean air or to exhaust heat and smoke. Mechanical devices can be used if natural forces do not provide adequate ventilation.

Horizontal Ventilation

Horizontal ventilation takes advantage of the doors, windows, and other openings at the same level as the fire. It is usually easiest to use doors for ventilation. They often need to be opened anyway to provide access to or egress from the building. It is possible to open some doors without tools. Additionally, the door may have been left open by the building residents. If a door is locked, it is usually not too difficult to perform forcible entry with simple tools. Windows are the next easiest to open (**FIGURE 16-7**). Remember, try before you pry—that is, try manually opening a window before removing the glass and frame. Second-floor windows are usually readily accessible using ground ladders. In some cases, firefighters might make additional openings in a wall to provide horizontal ventilation. Horizontal ventilation can be employed in many situations, particularly in small fires. It is commonly used in residential fires, room-and-contents fires, and fires that can be controlled quickly by the attack team.

FIGURE 16-7 Windows are frequently used in horizontal ventilation.

Courtesy of District Chief Chris E. Mickal/New Orleans Fire Department, Photo Unit.

Horizontal ventilation can be a rapid, generally easy way to clear a contaminated atmosphere. Search and attack teams operating inside a structure, as well as firefighters operating outside the building, may use horizontal ventilation to reduce heat and smoke conditions. Horizontal ventilation can be used in less urgent situations if the building can be ventilated without additional structural damage. Low-urgency situations might include residual smoke caused by a stovetop fire or a small natural gas leak.

Horizontal ventilation tactics include both natural and mechanical methods. It is most effective when an opening goes directly into the room or space where the fire is located. Opening an exterior door or window, for example, allows heat and smoke to flow directly outside and eliminates the problems that can occur when these products travel through the interior spaces of a building. Horizontal ventilation is more difficult if there are no direct openings to the outside or if the openings are inaccessible. In these situations, it might be necessary to direct the air flow through other interior spaces, create openings, or to use vertical ventilation.

SAFETY TIP

Closing a door provides a temporary barrier to the spread of fire. This can protect parts of a building from fire and smoke.

Natural Ventilation

Horizontal natural ventilation is often used when quick ventilation is needed, such as when attacking a room-and-contents fire or a first-floor residential fire with people trapped on the second floor. For search and attack teams to enter these buildings and act quickly, smoke and heat must be ventilated immediately. Natural ventilation uses existing or created openings in a building.

Wind speed and direction play an important role in natural ventilation. If possible, windows on the leeward side of a building—the side of the building that faces away from the wind—should be opened first so that the contaminated atmosphere flows out. Openings on the windward side can then be used to bring in clean air. Conversely, opening a window on the windward side first could push the fire into uninvolved areas of the structure, particularly on a windy day.

Breaking Glass

The methods used to create horizontal openings for natural ventilation require a variety of tools and techniques. For example, some windows can simply be opened by hand. If the window cannot be opened and the need for ventilation is urgent, firefighters should not hesitate to radio their officer and ask permission to break the glass. Although breaking a window is a fast and simple way to create a ventilation opening, it is important to communicate with the officer before employing this technique because the firefighter may not be able to see how the building is reacting to the fire or where other crews are working.

When breaking glass, the firefighter should always use a hand tool (Halligan tool, axe, or pike pole) wearing full PPE with eye protection and keep their hands above or to the side of the falling glass. This tactic prevents pieces of glass from sliding down the tool and potentially injuring the firefighter. The tool should then be used to clear the entire opening of all remaining pieces of glass. This creates the largest opening possible and provides a way for firefighters to enter or exit through the window in the event of an emergency.

Clearing the broken glass also reduces the risk of injury to anyone who might be in the area and eliminates the danger of sharp pieces falling out later. Before breaking glass—particularly if the window is above the ground floor—firefighters must look out to ensure that no one will be struck by the falling glass.

To break glass on the ground floors with a hand tool, follow the steps in **SKILL DRILL 16-1**.

Breaking a Window from a Ladder. A similar technique can be used to break a window on an upper floor, with the firefighter working from a ladder. The ladder should be positioned to the upwind side of the window so that smoke is carried away from the firefighter. The tip of the ladder should be even with the top of the window (**FIGURE 16-8**). The ladder should not be positioned below the window, where glass could slide down the handle of the tool onto the firefighter. The firefighter in full PPE climbs to a position level with the window and uses a leg lock or a ladder belt to secure themselves to the ladder. Using a hand tool, the firefighter strikes and breaks the window, and then clears the remaining glass from the frame completely.

Breaking a Window with a Ladder. Firefighters on the ground can use the tip of a ladder to break and clear a window when immediate ventilation is needed on an upper floor. This technique requires proper ladder selection. Usually, second-floor windows can be reached by 16- or 20-ft (5- or 6-m) roof ladders, as well as by 24- or 28-ft (7- or 8.5-m) extension ladders. Extension ladders are needed to break windows on the third or higher floors.

To perform this maneuver, the ladder is either raised directly into the top half of the window or raised next to the window to determine the proper height for the tip, and then rotated so the tip is on the top third of the window. Pull the tip of the ladder back so the tip is not on the glass, then forcibly drop the tip into the top third of the window. The objective is to push the broken glass into the window opening. There is always a risk that some of the glass will fall outward, so firefighters working below the window must wear full PPE to shield themselves from falling glass. If the opening will be used for access, the remaining glass needs to be cleared from the window frame with a tool before anyone enters through it.

SKILL DRILL 16-1

Breaking Glass with a Hand Tool Firefighter I, NFPA 1010: 6.3.11

1. Wearing full personal protective equipment (PPE), including eye protection and self-contained breathing apparatus (SCBA), size up the window by trying to open the window and looking at the window construction. Select a hand tool, and position yourself to the side of the window.

2. Standing off to the side of the window, break the window glass. If the window has multiple panes of glass, strike the lowest pane of glass; if the window has a single pane of glass, strike the top of the pane. Always use controlled, accurate swings to maintain control of your tool.

3. Clear the remaining glass from the window frame with the hand tool.

FIGURE 16-8 When breaking an upper-story window, the firefighter should place the ladder to the side of the window.

SAFETY TIP

The higher the window, the greater the danger of falling glass.

To break a window with a ladder, follow the steps in **SKILL DRILL 16-2**.

Opening Doors

Like windows, door openings can be used for horizontal natural ventilation operations. Treat opening a door as a major act of ventilation. It should be opened only when the hose line is charged and the attack team is ready to advance. Doorways have certain advantages when used for ventilation. Specifically, they are large openings—often twice the size of an average window. In fact, some doorways can be large enough to drive a truck through.

The problem with using doorways as ventilation openings is that their use for this purpose compromises entry and exit for interior fire attack teams and search teams. If heat and smoke are pouring through a doorway, firefighters cannot enter or exit through that opening. Because protecting exit paths is a priority,

SKILL DRILL 16-2

Breaking Windows with a Ladder Firefighter I, NFPA 1010: 6.3.11

1. Wearing full PPE, including eye protection and SCBA, select the proper size ladder for the job. Check for overhead lines.

2. Raise the ladder next to the window using standard procedures to perform a ladder raise. Extend the tip so that it is even with the top third of the window. If a roof ladder is used, extend the hooks toward the window.

Continues.

SKILL DRILL 16-2 CONTINUED

Breaking Windows with a Ladder Firefighter I, NFPA 1010: 6.3.11

3. Position the ladder in front of the window. Coordinate operations with the incident's tactical objectives.

4. Forcibly drop the ladder into the window. Exercise caution—falling glass can cause serious injury.

5. Raise the ladder from the window, and then move it to the next window to be ventilated. Either carry the ladder vertically or pivot the ladder on its feet.

doorways are more suitable as entries for clean air, which means the contaminated atmosphere should be discharged through a different opening.

Firefighters often approach a fire through the interior of a building to attack from the unburned side. Opening an exterior door that leads directly to the fire can create an opening for oxygen-rich air to enter the building that can quickly produce a backdraft or a vent that pushes the heat and smoke out of the building.

Before opening an exterior door, examine the area around the door. A small amount of smoke can indicate that the fire might be small or at the incipient stage, or it could indicate a ventilation-limited fire. Proceed with extreme caution. Once the exterior door is opened, watch how the air moves through the exterior door. A rapid movement of air into the structure can indicate a ventilation-limited fire. If the fire is actively burning, clean air may be seen entering through the bottom half of the exterior door and smoke escaping through the top half (bidirectional flow).

Mechanical Ventilation

In addition to making or controlling openings to influence natural ventilation, firefighters can use mechanical ventilation to direct the flow of combustion gases. There are three different methods of mechanical ventilation: negative-pressure ventilation, positive-pressure ventilation, and hydraulic ventilation.

Doorways are good places to put mechanical ventilation devices when natural ventilation needs to be augmented. Fans can be used to create positive-pressure ventilation by pushing clean air in or negative-pressure ventilation by pulling contaminated air out. In some buildings, HVAC systems can be used to provide mechanical ventilation. The HVAC system must be configured so that the zones can be individually controlled. Some HVAC systems can be used to exhaust smoke or supply clean air if a fire occurs, whereas others should be shut down immediately. The building engineer or maintenance staff might have the specific knowledge required to assist in this effort. Information about HVAC systems should be obtained during preincident planning surveys.

Negative-Pressure Ventilation

Negative-pressure ventilation pulls or draws smoke, heat, and air out through openings in a structure using a type of fan called a **smoke ejector**. To create negative-pressure ventilation, firefighters need to locate the seat of the fire, and then use a smoke ejector to exhaust the products of combustion out through a window or door. The smoke ejector creates a **negative pressure**—lower pressure— inside the building. This draws the heat, smoke, and fire gases out of the building toward the higher atmospheric pressure outside. Smoke ejectors are usually 16 to 24 inches (in.; 40 to 60 centimeters [cm]) in diameter, although larger models are available. They can be powered by electricity, gasoline, or water pressure (**FIGURE 16-9**).

FIGURE 16-9 A smoke ejector pulls products of combustion out of a structure. Gasoline-powered smoke ejectors might have problems running because of reduced oxygen or carbon particles in the smoke.

Courtesy of Super Vacuum Mfg. Co., Inc.

Negative-pressure ventilation can be used to clear smoke out of a structure after a fire, particularly if natural ventilation would be too slow or if no natural cross-ventilation exists. In large buildings, negative-pressure ventilation can be used to pull heat and smoke out while firefighters are working, although it is not generally used before or during a fire attack. This type of ventilation is usually associated with horizontal ventilation, but a smoke ejector can also be used in a roof-vent operation to pull heated gases out of an attic. Smoke ejectors usually have explosion-proof motors, which makes them excellent choices for venting flammable or combustible gases and other hazardous environments.

Negative-pressure ventilation has several limitations, including the following:

- Often, firefighters must enter the heated and smoke-filled environment to set up the smoke ejector, and they might need to use braces and hangers to position it properly.

- Because most smoke ejectors run on electricity, a power source and a cord are required for them to function.
- Smoke ejectors get very dirty while they are running, so they must be taken apart and cleaned after each use.
- Air flow control with a negative-pressure system is difficult. The smoke ejector must be completely sealed so that the exhausted air is not immediately drawn back into the building. This phenomenon, which is called **recirculation**, reduces the effectiveness of the smoke ejector. Recirculation can be eliminated by completely blocking the opening around the ejector.

To perform negative-pressure ventilation, follow the steps in **SKILL DRILL 16-3**.

Positive-Pressure Ventilation

Positive-pressure ventilation introduces clean air into a structure using a type of fan called a **positive-pressure fan** placed in an opening. These fans create **positive pressure**—higher

SKILL DRILL 16-3

Delivering Negative-Pressure Ventilation Firefighter I, NFPA 1010: 6.3.11

1. Determine the area to be ventilated and the direction of the outside wind. Place the smoke ejector so that it is not on the upwind side of the building.

2. Hang the smoke ejector in the upper part of the ventilation opening.

3. Use a salvage cover to prevent recirculation if needed, for example, if the structure's windows are not intact. Provide an opening on the upwind side of the structure to provide cross ventilation if needed.

pressure—inside the structure. The clean air displaces the contaminated atmosphere and pushes heat and other products of combustion out through another opening. Positive-pressure fans are used to reduce interior temperatures and smoke conditions in coordination with an initial fire attack or to clear a contaminated atmosphere after a fire has been extinguished. When positive-pressure fans are used to control the flow path of the products of combustion prior to fire control, this is known as **positive-pressure attack**. A positive-pressure attack can provide increased visibility and tenability for firefighters and potential occupants while fire suppression efforts are under way.

Positive-pressure fans are usually set up outside the structure at exterior doorways, often at the same opening used by the attack team (**FIGURE 16-10**). This creates a cone of fresh air that is pushed in the same direction as the advancing hose line so that firefighters are enveloped in a stream of clean, cool air while they move toward the fire.

FIGURE 16-10 Positive-pressure fans are usually set up at exterior doorways, often at the same opening used by the attack team.

When using positive-pressure ventilation, there must also be an exhaust opening to release the positive pressure created by the fan. This opening must be located in the fire room to effectively allow the heat and smoke to escape (**FIGURE 16-11**). The size of the exhaust opening should be a 2:1 ratio (or, at minimum, a 1:1 ratio) to the entry opening to create the desired positive pressure within the room. It is also important that the interior door of the fire room—not the exterior door of the structure—dictate the size of the exhaust opening. A higher pressure must be maintained at this opening in order to retain an efficient flow. If there is no exhaust opening or if the opening is too small or located in other areas of the structure, the heat and smoke can flow into the other areas of the building or back toward the attack team. In addition, this kind of system will not work effectively if the building is not intact.

To increase the efficiency of positive-pressure attack or ventilation and to prevent smoke and fire from reaching other parts of the building, firefighters should close doors to unaffected areas of the structure. This is an effective action to maintain a tenable environment for people trapped in other parts of the building.

TIP

It is important to assess and monitor both the inlet and outlet to assess the effectiveness of your ventilation. Positive-pressure attack may not be effective in houses with open floor plans due to the difficulty of establishing and maintaining the minimum 1:1 ratio of exhaust opening to entry opening.

FIGURE 16-11 When using positive-pressure ventilation, there must be an opening in the fire room to allow heat and products of combustion to escape.

When planning positive-pressure attack or ventilation, the force and direction of the wind need to be taken into consideration. Positive-pressure fans should never be placed in a position where they are blowing against the wind. Check the direction and speed of the wind before placing fans and monitor the wind for changes in direction or intensity.

Natural ventilation should be used after the structure is cleared to prevent carbon monoxide build-up. It is recommended to open all windows of the structure immediately after fire suppression operations. This is the most effective technique for fully ventilating a single-family or two-family residence. Note that this recommendation applies only to positive-pressure ventilation after the fire has been extinguished. During positive-pressure attack, on the other hand, only openings in the fire room are effective. Openings other than those in the fire room permit the fire to spread to other areas of the structure.

Like negative-pressure ventilation, positive-pressure ventilation can be used in vertical ventilation as well. For example, in a high-rise building, positive pressure fans can be positioned to blow fresh air up through a stair shaft. By opening the doors, firefighters can clear one floor at a time. Very large structures can be ventilated by placing multiple fans side by side.

Positive-pressure ventilation has several advantages over negative-pressure ventilation. For instance, a single firefighter can set up a fan very quickly. Because the fan is positioned outside the structure, the firefighter does not have to enter a hazardous environment, and the fan does not interfere with interior operations. Positive pressure is both quick and efficient because the entire space within the structure is under pressure.

When used with a well-placed exhaust opening, positive-pressure ventilation can help keep a fire confined to a smaller area and increase safety for interior operating crews, as well as any building occupants. In addition, positive-pressure fans do not require as much cleaning and maintenance as negative-pressure fans because the products of combustion never move through the fan.

Positive-pressure ventilation does have some disadvantages, however. The importance of close coordination among members of the ventilation crew and the attack team cannot be overstressed. Good communication and proper coordination of operations can prevent rapid fire spread. Ineffective positive-pressure ventilation occurs if positive pressure is used and there are no openings for exhaust gases to release the positive pressure from the building. For example, if the fire is burning in a structural space, such as in a pipe chase or above a ceiling, positive pressure can push it into unaffected areas of the structure if there is no exhaust opening. If the fire is located in structural void spaces, positive-pressure ventilation should not be used until access to these spaces is available and attack crews are in place.

Positive-pressure fans operate at high velocity and can be noisy. Most are powered by internal combustion engines, so they can increase carbon monoxide levels if they are run for significant periods of time after the fire is extinguished. On some fans, if the fan is tipped down while removing it from the apparatus, the oil sensor will prevent the fan from being started. Firefighters need to know the procedure to use to get the fan to start. If the ventilation causes conditions to worsen, the fan should be turned off.

SKILL DRILL 16-4 outlines the steps for performing positive-pressure ventilation.

Hydraulic Ventilation

Hydraulic ventilation, sometimes called **fog ventilation**, moves hot gases and smoke from a room by directing a fog stream placed 2 to 4 ft (0.6 to 1.2 m) from an open window from inside the structure. This creates a pressure differential behind and in front of the nozzle by lowering the pressure behind of the nozzle. To use this ventilation technique, the firefighter working the hose line directs a narrow fog stream out of the building through an opening, and then widens the stream until it covers about 85 to 90 percent of the window opening. The contaminated atmosphere is drawn into the low-pressure area created behind the nozzle. An induced draft created by the high-pressure stream of water then pulls the smoke and gas out through the opening (**FIGURE 16-12**). A well-placed fog stream can move a tremendous volume of air through an opening.

Hydraulic ventilation can move several thousand cubic feet of air per minute and is effective in clearing both heat and smoke from a fire room. It is most useful after the fire is under control. Because hydraulic ventilation does not require any specialized equipment, it can be performed by the attack team with the same hose line that was used to control the fire.

Hydraulic ventilation has some disadvantages. Firefighters must enter the heated, toxic environment and remain in the path of the products of combustion as they are being exhausted. If the water supply is limited, using water for ventilation must be balanced with using water for fire attack. This technique can

SKILL DRILL 16-4

Delivering Positive-Pressure Ventilation Firefighter I, NFPA 1010: 6.3.12

1. Place the fan outside of the structure in front of the opening to be used for the fire attack. The exact position depends on the size of the opening, the size of the fan, and the direction of the wind, preferably on the downwind side of the structure.

2. Provide an exhaust opening in the fire room. This opening can be made either immediately before the fan is started or when the fan is started.

3. Start the fan.

4. Check the cone of air produced; it should completely cover the opening. This can be checked by running a hand around the door frame to feel the direction of air currents. Monitor the exhaust opening to ensure there is a unidirectional flow. Allow the smoke to clear.

FIGURE 16-12 Hydraulic ventilation can be used to draw smoke out of a building after the fire is controlled.

cause excessive water damage if it is used improperly or for long periods of time. Also, in cold climates, ice build-up can occur on the ground, creating a safety hazard.

Ideally, the ventilation opening should be created before the hose line advances into the fire area. If needed, the hose stream itself can be used to break a window before it is directed on the fire. The steam created when the water hits the fire will push the heat and smoke out through the vent. Hydraulic ventilation can then be used to clear the remaining heat and smoke from the room.

To perform hydraulic ventilation, follow the steps in **SKILL DRILL 16-5**.

SKILL DRILL 16-5

Delivering Hydraulic Ventilation Firefighter I, NFPA 1010: 6.3.12

1. Enter the room, and position yourself close to the ventilation opening. Place the nozzle through the opening, and then open the nozzle to a narrow fog or broken-pattern spray.

2. Keep directing the stream outside and back into the room until the stream almost fills the opening. The nozzle should be 2 to 4 ft (0.6 to 1.2 m) inside the opening.
3. Stay low, out of the heat and smoke, or to one side to keep from partially obstructing the opening.

Vertical Ventilation

Pathways for vertical ventilation include ceilings, stairwells, exhaust vents, and roof openings such as skylights, scuttles, or monitors. Ventilating skylights and roofs usually requires gaining access using ground ladders or aerial devices. Additional openings can be created by cutting holes in the roof or the floor and making sure that the opening extends through every layer of the roof or floor. Vertical ventilation typically also incorporates horizontal openings, such as doors and windows, to provide an intake for clean air to replace the smoke that is being exhausted through the vertical vent.

Vertical ventilation occurs naturally. If an opening is available above a fire, convection causes the products of combustion to rise and flow through the opening. In some situations, vertical ventilation can be assisted by mechanical means such as fans or hose streams.

Although vertical ventilation may involve any opening that allows the products of combustion to travel up and out, this term is most often applied to describe operations on the roof of a structure. The ventilation openings can be existing features, such as skylights or bulkheads, or holes created by firefighters who cut through the roof covering. The choice of ventilation openings depends primarily on the building's roof construction.

A vertical ventilation opening should be made as close as possible to the seat of the fire. Smoke issuing from the roof area, melted asphalt shingles, or steam coming from the roof surface are all signs that firefighters can use to identify the hottest point. A thermal imaging device also may be helpful in locating the part of the roof closest to the fire.

In order for a vertical exhaust vent to be effective, there must be a horizontal intake vent to admit air. In addition, if horizontal vents are made on a floor above the fire, they can produce the effect of vertically venting the fire. Intake is arguably the most important part of an attic ventilation system. When there is inadequate intake or none at all, an exhaust vent may not function properly.

TIP

Vertical ventilation is the most efficient type of natural ventilation. It allows for most hot gases to exit the structure; however, it also allows the most air to be entrained into the structure. Like horizontal ventilation, vertical ventilation must be coordinated with fire attack.

Safety Considerations in Vertical Ventilation

Vertical ventilation should be performed only when it is necessary and can be done safely. Rooftop operations involve significant inherent risks. If horizontal ventilation can sufficiently control the situation, vertical ventilation might be unnecessary. Before performing vertical ventilation, firefighters must evaluate all pertinent safety issues and seek to avoid unnecessary risks. In particular, they should assess the roof for skylights, roof scuttles, heat vents, plumbing vents, louver ventilation, translucent roof panels, cell phone towers, photovoltaic panels (solar panels), and fan shafts.

Falling from a roof—either off of the building or through a ventilation opening, open shaft, or skylight into the building—is a risk firefighters encounter when operating on roofs. Smoke reduces visibility and can cause disorientation, increasing the risk of falling off the edge of a roof. Factors such as darkness, snow, ice, or rain can create hazardous situations. Not surprisingly, a pitched roof presents even greater risks than a flat roof. Safe operations on pitched roofs require firefighters to operate from an aerial device or to use a roof ladder.

The most obvious risk in a rooftop operation is the possibility that the roof will collapse if the fire compromises the structural integrity of the roof support system. If the structural system supporting the roof fails, both the roof and the firefighters standing on it will fall into the fire. Firefighters also can fall through an area of the roof decking that has been weakened by the fire.

Roof operations should be conducted from a well-secured roof ladder or an aerial device, especially on pitched roofs (**FIGURE 16-13**). Roofs—pitched or

FIGURE 16-13 Firefighters operating from the safety of an aerial device.

Courtesy of Captain David Jackson, Saginaw Township Fire Department.

flat—can fail suddenly and without warning due to either fire damage or modern lightweight construction. Modern lightweight construction frequently uses thin plywood for roof decking, with a variety of waterproof and insulating coverings applied to the plywood. A relatively minor fire under this type of roof can cause the plywood to delaminate, with no visible indication of damage or weakness from above. The weight of a firefighter on a damaged area can be enough to open a hole and cause the firefighter to fall through the roof. Determine the type of roof construction before beginning roof ventilation operations. If the roof is constructed with lightweight trusses, ventilating the roof from an aerial device is recommended for added safety. Remember, roofs that are built using lightweight manufactured trusses can fail quickly without any signs or warning.

Effective vertical ventilation may assist with prompt fire control and reduces the risks to firefighters operating inside the building. In some situations, prompt, proper, and coordinated vertical ventilation can save the lives of building occupants by relieving interior conditions, opening exit paths, and allowing firefighters to enter and perform search and rescue operations.

Vertical ventilation should always be performed as quickly, efficiently, and safely as possible. If they provide an adequate opening, skylights, ventilators, bulkheads, and other existing openings should be opened first. If firefighters must cut through the roof covering to create a hole, the location should be identified and the operation performed promptly and efficiently. It must be done in coordination with the other firefighters operating on the fire scene because vertical ventilation can create dangerous flow paths for firefighters located between an uncontrolled fire and the ventilation opening.

A firefighter's path to a proposed vent-hole site should follow the areas of greatest support and strength. These locations include the roof edges, which are supported by bearing walls, and the hips and valleys of the roof, where structural materials are doubled for strength. These areas should be stronger than interior sections of the roof.

Firefighters should always be aware of their surroundings when working on a roof. Knowing where the roof's edge is located, for example, can help avoid accidentally walking or falling off the roof. Maintaining this knowledge is especially important—although difficult—in darkness and in adverse weather or heavy smoke conditions.

Firefighters who are working on the roof should always have two safe exit routes. A second ground ladder or aerial device should be positioned to provide a quick alternative exit in case conditions on the roof deteriorate rapidly. The two exit routes should be separate from each other, preferably in opposite directions from the operation site. The team working on the roof should always know where these exit routes are located, and they should plan ventilation operations to work toward these locations.

FIGURE 16-14 Be prepared for smoke and flames that are released when the roof is opened.

The ventilation opening should never be located between the firefighters and their exit routes. A ventilation opening releases smoke and hot gases from the fire (**FIGURE 16-14**). If this exhaust flow path of hot gases and smoke separates the firefighters from their exit from the roof, they will be in serious danger.

SAFETY TIP

Firefighters operating on roofs must be alert for the presence of skylights, translucent panels, and photovoltaic panels (solar panels). Firefighters have been injured or killed due to stepping on and falling through these types of panels. These panels are not required to support the weight of a firefighter or to be marked so firefighters can identify them during limited visibility operations. Additionally, photovoltaic panels are energized and can result in electrocution. Preincident planning and careful operations are required.

Once a ventilation opening has been made, firefighters should withdraw to a safe location. There is no good reason to stand around the vent hole and look at it. This behavior exposes the firefighters to an unnecessary risk.

Venting a roof should not be standard practice. Firefighters should operate on a roof only if absolutely necessary. Any rooftop operation should be conducted only after a complete size-up and assessment have been conducted and only when it is specifically needed to achieve a positive outcome.

The order of the cuts in making a ventilation opening should be planned carefully. Firefighters should be upwind (winds blowing away from them), have clear exit routes, and be standing on a firm section of the roof or using a roof ladder. If firefighters are upwind from the open hole, the wind will push the heat and smoke away from them. This means the escaping heat and smoke should not block their exit route. To prevent an accidental fall into the building, firefighters should stand on a portion of the roof that is firmly supported or work from a roof ladder or aerial device, particularly when making the final cuts.

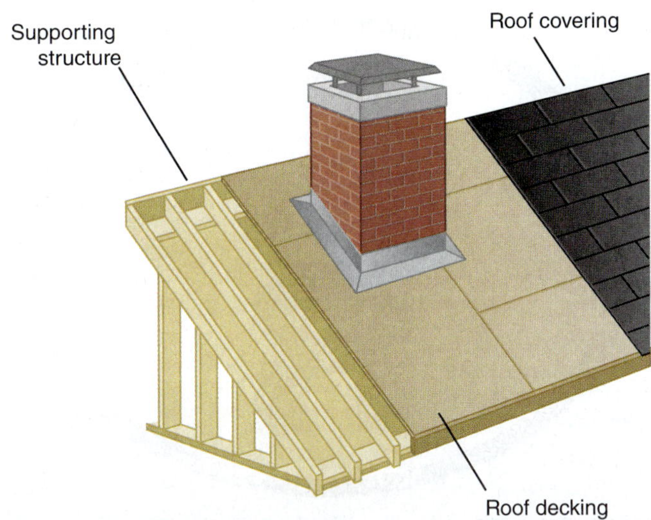

FIGURE 16-15 Roof assemblies consist of a supporting structure, decking, and covering.

Basic Indicators of Roof Collapse

Roof collapse poses the greatest risk to firefighters performing vertical ventilation. Some roofs—particularly roofs supported by trusses—give little or no warning that they are about to collapse.

Firefighters who are assigned to vertical ventilation tasks should always be aware of the condition of the roof. They should always be sounding the roof with a tool as they work and immediately retreat from the roof if they notice any of the following signs:

- Visible indication of sagging roof supports
- Any indication that the roof assembly is separating from the walls, such as the appearance of fire or smoke near the roof edges
- Structural failure of any portion of the building, even if it is some distance from the ventilation operation
- A sudden increase in the intensity of the fire from the roof opening (for example, fire showing around roof vents, melting snow, and evaporating or steaming water)
- High heat indicators on a thermal imaging device

Roof Construction

Roofs can be constructed of many different types of materials in several configurations. Nevertheless, all roofs have three major components: the supporting structure, the **roof decking**—the rigid portion of roof between the roof supports and the roof covering—and the **roof covering**—the membrane that resists fire and provides weather protection to the building against water infiltration, wind, and impact (**FIGURE 16-15**). Roof construction and design are discussed in Chapter 6, *Building Construction.*

Vertical ventilation operations often involve cutting a hole through the roof covering. The selection of tools, the technique used, the time required to make an opening, and the personnel requirements all depend on the roofing materials and the layering configurations.

Solid-Beam versus Lightweight Construction

The two major structural support systems for residential or wood-frame roofs use either solid-beam or lightweight construction. It can be impossible to determine which system is used simply by looking at the roof from the exterior of the building, particularly when the building is on fire. Given this fact, information on the roof support system should be obtained during preincident planning surveys. Some states and jurisdictions require building owners to place a sign near the building entrance to indicate that the roof is constructed using lightweight construction (**FIGURE 16-16**).

The basic difference between solid-beam and lightweight construction is in the way individual load-bearing components are made. Solid-beam construction uses solid components, such as girders, beams, and rafters, to support the roof. Lightweight construction is assembled from smaller, individual components and includes trusses or engineered systems such as I-joists. Trusses are constructed by assembling relatively small and lightweight components of wood, steel, or a combination of wood and steel in a series of triangles (**FIGURE 16-17**). Engineered systems are constructed from composite materials that are

FIGURE 16-16 Some jurisdictions post a sign near building entrances to indicate that the roof is constructed using lightweight construction. The symbol shown here indicates truss roof construction in New York state.

Courtesy of New York State Department of State.

FIGURE 16-17 Truss construction uses smaller, individual components to support the roof.

© Jones & Bartlett Learning. Photographed by Glen E. Ellman.

usually glued together. In most cases, it makes no difference whether solid-beam or lightweight construction is used. For firefighters, however, there is an important difference: Lightweight construction can collapse quickly and without warning when it is exposed to a fire even for a short period of time.

The disadvantage of lightweight construction is that some systems fail completely when only one of the smaller components is weakened or when only one of the connections between components fails. If only one component fails, other components in the system might be able to absorb the additional load, but the overall strength of the system is still compromised. A fire that causes one component to fail probably weakens additional components in the same area. These components can also fail, sometimes as soon as the additional load is transferred. In most cases, a series of trusses or engineered components fail in rapid succession, resulting in a total collapse of the roof. Sometimes just the weight of a firefighter walking on the roof is enough to trigger a collapse.

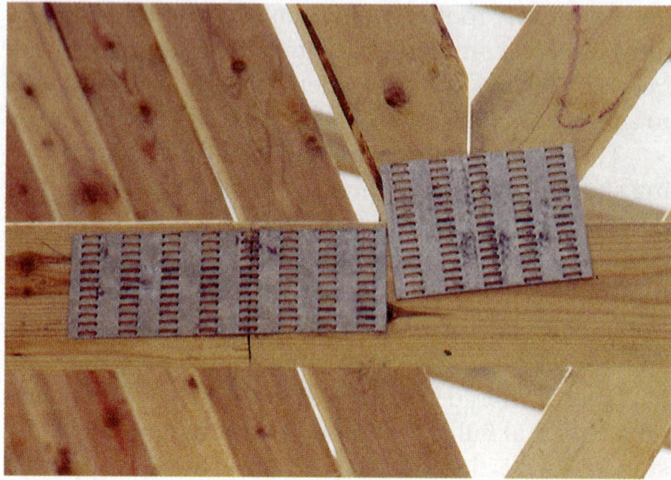

FIGURE 16-18 Gusset plates can fail in a fire, destroying the truss.

Courtesy of Captain David Jackson, Saginaw Township Fire Department.

Lightweight construction is not necessarily bad or inherently weak construction. Indeed, some lightweight construction is very strong and is assembled with sufficient protection from fire retardant materials, so that it can resist fire just as well as solid beams. Many of these systems, however, are made of lightweight materials and are designed to carry as much load as possible with as little mass as possible. The individual components of lightweight construction used for roof support are often wood 2- by 4-in. (50.8- by 101.6-millimeter [mm]) sections or a combination of wood and lightweight steel bars or tubes. Because these components are small, they can be weakened quickly if not protected from a fire.

Even more critical, however, are the points of the lightweight construction where the individual components join together. The components of a truss constructed with 2- by 4-in. (50.8- by 101.6-mm) wood pieces can be connected with heavy-duty staples or with gusset plates (**FIGURE 16-18**). These connection points are often the weakest portion of the lightweight construction and the most common point of failure in a fire. When one connection fails, the system loses its ability to support a load.

Trusses also may be made of individual steel bars or angle sections that are welded together. Lightweight steel trusses, known as bar joists, often support flat roofs on commercial or industrial buildings. When exposed to the heat of a fire, these metal trusses expand and lose strength. A roof supported by bar joists will probably sag before it collapses, a warning that firefighters should heed. The expansion can elongate the trusses, which may cause the supporting walls to collapse suddenly. Horizontal cracks in the upper part of a wall may indicate that the steel roof supports are pushing outward.

Lightweight construction can be found in almost any type of roof or floor, including flat roofs, pitched roofs, or curved roofs. Firefighters should assume that any modern construction uses lightweight construction for the roof support system until proven otherwise.

Effects of Roof Construction on Fire Resistance

Fire can quickly burn through some roofs, whereas others retain their integrity for long periods. The difference arises because each type of material used in roof construction is affected differently by fire. These reactions either increase or decrease the time available to perform roof operations. For example, solid beams are generally more fire resistant than truss systems because solid beams are larger and heavier (more mass). Wooden beams are affected by fire more quickly than concrete beams. Steel, unlike wood, does not burn, but it does elongate and lose strength when heated.

Most roofs eventually fail as a result of fire exposure—some quickly and others more gradually. In some situations, the fire weakens or burns through the roof supporting system, which collapses even if the roof covering remains intact. Failure of the roof supporting system usually results in sudden and total collapse of the roof. A structural failure involving other building components could also cause the roof to collapse.

In other cases, the fire burns through the roof covering, but the supporting structure is still sound. This type of roof failure usually begins with a "burn through" close to the seat of the fire or at a high point above the fire. As combustible products in the roof covering become involved in the fire, the opening spreads, eventually causing the roof to collapse.

The inherent strength and fire resistance of a roof are often determined by local climate conditions. In areas that receive winter snow loads, roof supporting structures can usually support a substantial weight and might include multiple layers of insulating material. A serious fire could burn under this type of roof with little visible evidence from above. Heavily constructed roofs, supported by solid structural elements, can burn for hours without collapsing.

In warmer climates, however, roof construction is often very light. Roofs in these geographic areas might simply function as umbrellas. The supporting structure can be just strong enough to support a thin roof decking with a waterproof, weather-resistant membrane. This type of roof can burn through quickly, and the supporting structure can collapse after a short fire exposure.

Roof Design

Firefighters must be able to identify the various types of roof designs and the materials used in their construction. Creating ventilation openings requires the use of many different tools and techniques, depending on the type of design and roof decking.

Flat Roofs. Flat roofs can be constructed with many different support systems, roof decking systems, and materials. Although they are classified as flat, most roofs have a slight slope so that water can drain off the structure.

Flat roofs often have vents, skylights, scuttles, or other features that penetrate the roof decking. Removing the covers from these openings provides vertical ventilation without the need to make cuts through the roof decking. Flat roofs also can have a **parapet**, a free-standing wall that extends above the normal roofline (**FIGURE 16-19**). A parapet can be an extension of a firewall, a division wall between two buildings, or a decorative addition to the exterior wall.

Pitched Roofs. Pitched roofs have a visible slope that provides for rain, ice, and snow runoff. The pitch or angle of the roof can vary depending on local climate and architectural styles. A pitched roof is usually supported by either trusses or rafters. Most pitched roofs have a layer of thin plywood or a sheeting material such

FIGURE 16-19 Flat roofs can have parapet walls that extend above the normal roofline.

as wood particleboard as decking that is covered by a weather-resistant membrane and an outer covering such as shingles or tiles. Some pitched roofs have a system of laths instead of solid sheeting to support the roof covering. A **lath** is a thin strip of wood used to make the supporting structure for roof tiles.

As with a flat roof, the type of roof construction material dictates how to ventilate a pitched roof. For example, a tile pitched roof can be opened by breaking the tiles and pushing them through the supporting laths. A tin roof can be cut and peeled back like the lid on a can. Opening a wooden roof usually requires cutting, chopping, or sawing.

Roof ladders should be used to provide a stable support while working on a pitched roof. A ground ladder or aerial device can be used to access the lower part of the roof, with the roof ladder placed on the sloping surface and hooked over the center ridge of the roof.

Curved Roofs. Curved roofs are generally found in commercial structures because they create large open spans without requiring the use of columns. This type of roof is common in supermarkets, warehouses, industrial buildings, arenas, auditoriums, bowling alleys, churches, airplane hangars, and other similar buildings.

Steel or wood bowstring trusses or arches are usually used to support curved roofs. The decking on curved roof buildings can range from solid wooden boards or plywood to corrugated steel sheets. Often, the roof covering consists of layers that include felt, mineral fibers, and asphalt, although some curved roofs are covered with foam plastics or plastic panels.

Although curved roofs are quite distinctive when seen from above or outside, their structure might not be evident from inside the building because a flat ceiling is attached to the bottom chords of the trusses. This creates a huge attic space that is often used for storage. A hidden fire within this space can quickly and severely weaken the bowstring trusses. Firefighters should be wary of any fire involving the truss space in these buildings. The collapse of a bowstring truss roof is usually sudden. Firefighters have been killed when a bowstring truss roof collapsed.

Buildings with this type of roof construction should be identified and documented during preincident planning surveys. Parapet walls, smoke, and other obstructions make it difficult to recognize the distinctive curved shape of a bowstring truss roof during an incident.

Vertical Ventilation Techniques

The objective of any vertical ventilation operation is simple: to provide the largest opening in the appropriate location, using the least amount of time and the safest technique. Among the different roof openings that can provide vertical ventilation are built-in roof openings, examination openings to locate the optimal place to vent, primary expandable openings located directly over the fire, and defensive secondary openings intended to prevent fire spread.

Before starting any vertical ventilation operation, firefighters must make an initial assessment with direct coordination with incident command. They should note construction features and indications of possible fire damage, establish safety zones and exit paths, and identify built-in roof openings that can be used immediately.

Ventilation operations must not be conducted in unsafe locations. A less efficient opening in a safe location is better than an optimal opening in a location that jeopardizes the lives of the ventilation team. In some cases, it might not be safe to conduct any rooftop operations unless they can be performed from the safety of an aerial device. If vertical ventilation cannot be performed, firefighters must rely on horizontal ventilation techniques.

Information on the fire's location as well as visible clues from the roof can be used to pinpoint the best location to vent. The IC or attack crews might be able to provide some direction. The use of a drone with a thermal imager can be a great asset to find the heat signature in a building from an aerial view (**FIGURE 16-20**).

An **examination opening** is used to determine how large an area is involved, whether a fire is spreading, and in which direction it is moving. These openings

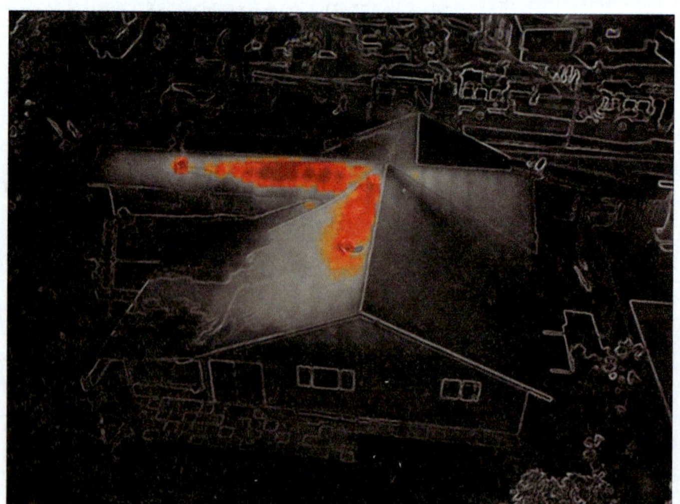

FIGURE 16-20 The fire department deployed its drone at a recent fire to help find hot spots in the roof. This allowed the ventilation team to accurately vent the roof at the hottest area of the structure.

Courtesy of Lauderhill Fire Rescue.

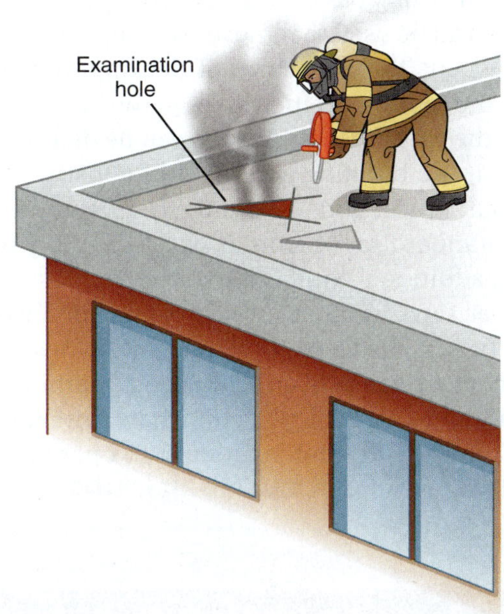

FIGURE 16-21 A triangular cut can be used as an examination opening.

© Jones & Bartlett Learning

FIGURE 16-22 Built-in rooftop openings such as the skylights shown here often can provide vertical ventilation quickly.

© Jones & Bartlett Learning. Photographed by Glen E. Ellman.

allow the ventilation team to evaluate conditions under the roof and to verify the proper location for a ventilation opening. Sometimes a single slit cut (a **kerf cut**) into the roof with a power saw can give firefighters some indication of the location of the fire. If a single cut is not effective, a triangular examination opening can be created quickly by making three small cuts in the roof (**FIGURE 16-21**).

Once the optimal place to vent is located, the ventilation team should determine the most appropriate type of opening to make. Built-in rooftop openings provide readily available ventilation openings (**FIGURE 16-22**). This approach also results in less property damage to the building. Many residential roofs have a ridge vent installed along the ridge line of the roof to keep the attic space ventilated. Often, these can be removed quickly because they are attached with small nails.

If the roof must be cut to provide a ventilation opening, cutting one large hole is better than making several small ones. Some departments recommend starting with a square opening about 4 ft long by 4 ft wide (1.2 m long by 1.2 m wide). If necessary, this hole can be expanded by continuing the cuts to make a larger opening (**FIGURE 16-23**). If the crews inside the building report that smoke conditions are lifting and temperature levels are dropping, the vent is effective and probably large enough. If interior crews do not see a difference, the opening might be obstructed or need to be expanded.

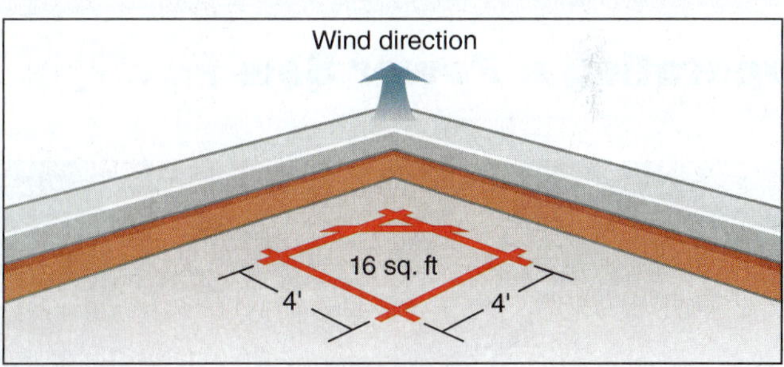

Total 16 sq. ft

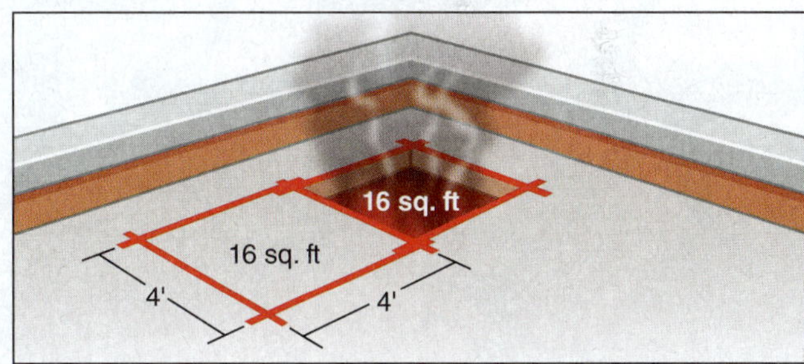

Total 32 sq. ft

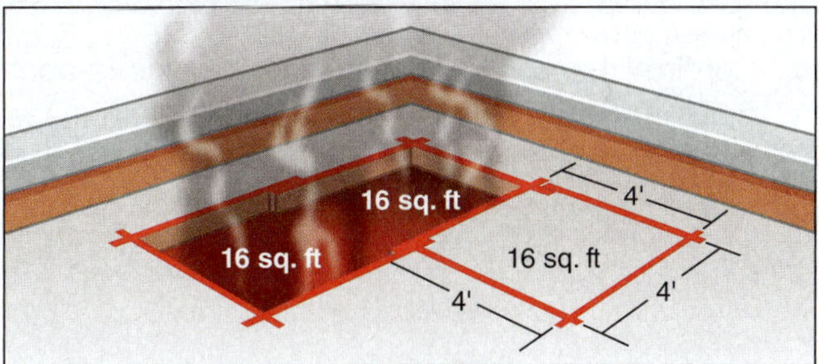

Total 48 sq. ft

FIGURE 16-23 A ventilation hole can be quickly expanded by extending the cuts.

© Jones & Bartlett Learning

Once the roof opening is made, a hole of the same size should be made in the ceiling material below to allow heat and smoke to escape from the interior of the building. If the ceiling is not opened, only the heat and smoke from the space between the ceiling and the roof will escape through the roof opening. The blunt end of a pike pole or hook can be used to push down as much of the ceiling material as possible. A roof opening is much less effective if there is no ceiling hole or if the ceiling hole is too small.

Tools Used in Vertical Ventilation

Several tools can be used in roof ventilation. Firefighters commonly use power saws to cut vent openings, but they also use many hand tools. For example, axes, Halligan tools, pry bars, tin cutters, pike poles, and other hooks can all be used to remove coverings from existing openings, cut through the roof decking, remove sections of the roof, and punch holes in the interior ceiling. Usually the specific tools needed can be determined by looking at the building and the roof construction features, but ventilation team members should always carry a standard set of tools as well as a utility rope for hoisting additional equipment if needed.

Power saws, such as a rotary saw with a wood- or metal-cutting blade or a chain saw, effectively cut through most roof coverings. Special carbide-tipped saw blades are especially useful when cutting through typical roof construction materials. To operate a power saw, follow the steps in **SKILL DRILL 16-6**.

SKILL DRILL 16-6

Operating a Power Saw Firefighter I, NFPA 1010: 6.3.12

1. Confirm that the cutting device or blade is appropriate for the material anticipated. Briefly inspect the blade or chain for obvious damage. Ensure that proper protective gear is in place.

2. Before going to the roof, start the saw to ensure that it runs properly. Set the choke to the halfway position or as recommended by the manufacturer.

SKILL DRILL 16-6 CONTINUED

Operating a Power Saw Firefighter I, NFPA 1010: 6.3.12

3. Stay clear of any moving parts of the saw. Use a foot or knee to anchor the saw to the ground and pull the starter cord as recommended to start the saw.

4. Run the saw briefly at full throttle to verify its proper operation.

5. Shut down the saw, wait for the blade or chain to stop completely, and then carry the saw to the roof.

6. Whenever possible, work off of a roof ladder or an aerial device for added safety.

Continues.

SKILL DRILL 16-6 CONTINUED

Operating a Power Saw Firefighter I, NFPA 1010: 6.3.12

7. Start the saw in an area slightly away from where you intend to cut.

8. Always run the saw at maximum throttle when cutting. The saw should be running at full speed before the blade or chain touches the roof decking. Keep the throttle fully open while cutting and removing the blade or chain from the cut to reduce the tendency for the blade or chain to bind. For the saw to cut efficiently, adjust the speed of your cut so that the saw remains at maximum speed throughout the process.

Types of Roof Cuts

Roof construction is a major consideration in determining which type of cut to use. Some roofs are thin and easy to cut with an axe or a power saw; others have multiple layers that are difficult to cut. Preplanning, being familiar with the types of roof construction in your jurisdiction, and understanding the best methods for opening them increases the efficiency and effectiveness of ventilation operations.

Rectangular or Square Cut. A 4-ft by 4-ft (1.2-m by 1.2-m) **rectangular cut** or **square cut** is the most common vertical ventilation opening. It requires the firefighter to make four cuts completely through the roof decking, using an axe or power saw. When using a power saw, the firefighter must carefully avoid cutting through the structural supports. To do this, the firefighter should sound the roof to locate the structural supports before starting to cut. Once the structural supports are located, the firefighter should cut beside the supports rather than through them. After the four cuts are completed, the section of the roof decking can be removed.

The firefighter should stand upwind of the opening, with two unobstructed exit routes. The first and last cuts should be made parallel to and just inside the roof supports. The firefighter making the cuts must always stand on a solid portion of the roof or on a roof ladder. A triangular cut in one corner of the planned opening can be used as a starting point for prying the decking up with a hand tool. Depending on the roof construction, it might be possible to lift out the entire section at once. If several layers of roofing material are present, firefighters might have to peel them off in layers. The decking itself could consist of plywood sheets or individual boards that have to be removed one at a time. To perform the rectangular or square cut, follow the steps in **SKILL DRILL 16-7**.

Seven, Nine, Eight (7, 9, 8) Rectangular Cut. Large commercial buildings require larger ventilation openings to exhaust the heat and smoke from a fire.

SKILL DRILL 16-7

Making a Rectangular or Square Cut Firefighter I, NFPA 1010: 6.3.12

1. Locate the roof supports by sounding. Make the first cut parallel to the roof support. The support should be outside the area that will be opened.

2. Make a triangular cut at the first corner of the opening. Then make two cuts perpendicular to the roof supports.

3. Make the final cut parallel to and slightly inside another roof support. Be sure to stand on the solid portion of the roof when making this cut.

4. Use a hand tool to pull out or push in the triangle cut to create a small starter hole. If possible, pull the entire roof section free, and flip it over onto the solid roof. It might be necessary to pull the decking out one board at a time.

Continues.

SKILL DRILL 16-7 CONTINUED

Making a Rectangular or Square Cut Firefighter I, NFPA 1010: 6.3.12

5. Use a pike pole to punch out the ceiling below. The hole should be the same size as the opening in the roof decking. Be wary of a sudden updraft of hot gases or flames.

When these buildings have flat roofs, the **seven, nine, eight (7, 9, 8) rectangular cut** is an effective ventilation technique. The steps to create this cut are similar to the steps for creating a square or rectangular cut. Seven cuts are produced, resulting in a 4-ft by 8-ft (1.2-m by 2.4-m) ventilation hole. To perform the seven, nine, eight (7, 9, 8) cut, follow the steps in **SKILL DRILL 16-8**.

Louver Cut. Another common cut is the **louver cut**, which is particularly suitable for flat or sloping roofs with plywood decking. To make a louver cut, firefighters use a power saw or axe to make two parallel cuts, approximately 4 ft (1.2 m) apart, perpendicular to the roof supports. The firefighters then make cuts parallel to the roof supports, approximately halfway between each pair of supports. Using each roof support as a fulcrum, the cut sections are tilted to create a series of louvered openings. Louver cuts can quickly create a large opening. Continuing the same cutting pattern in any direction creates additional louver sections. To make a louver cut, follow the steps in **SKILL DRILL 16-9**.

Triangular Cut. The **triangular cut** works well on metal roof decking because it prevents the decking from rolling away as it is cut. Using saws or axes, the firefighter removes a triangle-shaped section of decking. Smaller triangular cuts can be made between supports (so that the decking falls into the opening) or over supports (to create a louver effect). Because triangular cuts are generally smaller than other types of roof ventilation openings, several might be needed to create an adequately sized vent. To make a triangular cut, follow the steps in **SKILL DRILL 16-10**.

Peak Cut. A **peak cut** is used only on pitched roofs sheeted with plywood. In these structures, the 4-ft by 8-ft (1.2-m by 2.4-m) sheets of plywood are applied starting at the bottom edge of the roof, working toward the top. A partial sheet is usually placed at the top, along the roof peak. A small gap is left between the two sides of the roof to allow the plywood sheeting to expand and contract with temperature changes.

To make a peak cut, the firefighter first uses a hand tool to clear the outer covering along the peak (roof cap), which reveals the roof supports through the gap in the sheeting. Using a power saw or axe, the firefighter makes a series of vertical cuts between the supports from the top to the bottom of the plywood sheet. The individual panels are then struck with an axe and louvered or pried up with a hook.

Repeating this technique on both sides of the roof essentially removes the entire peak. Unless multiple layers of roof decking are present, there is no need to make horizontal cuts. The opening can be expanded either vertically or horizontally with a minimum of cutting.

To make a peak cut, follow the steps in **SKILL DRILL 16-11**.

SKILL DRILL 16-8

Making a Seven, Nine, Eight (7, 9, 8) Rectangular Cut Firefighter I, NFPA 1010: 6.3.12

1. Locate the roof supports by sounding. Make the first cut parallel to the roof supports. The cut should be 3.5 to 4 ft (1 to 1.2 m) long.

2. Make the second cut at a 45-degree angle to the first cut. The length of this cut should be 1 to 2 ft (0.3 to 0.6 m) in length.

3. Make the third cut perpendicular to the first cut from the corner where the knock-out cut was made. This cut should be about 8 ft (2.4 m) long. The three cuts should produce the shape of the number 7.

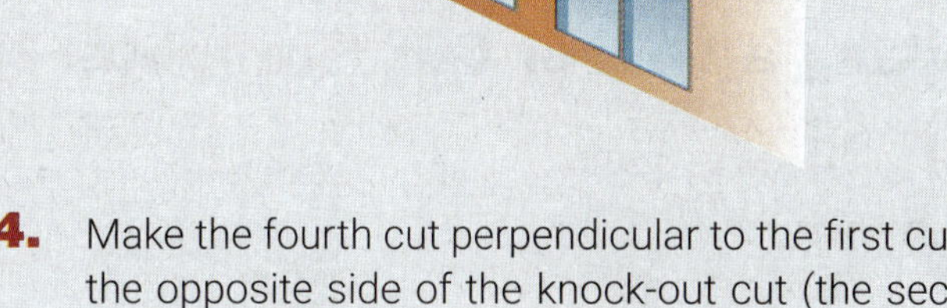

4. Make the fourth cut perpendicular to the first cut on the opposite side of the knock-out cut (the second cut). The cut should be about 4 ft (1.2 m) long.

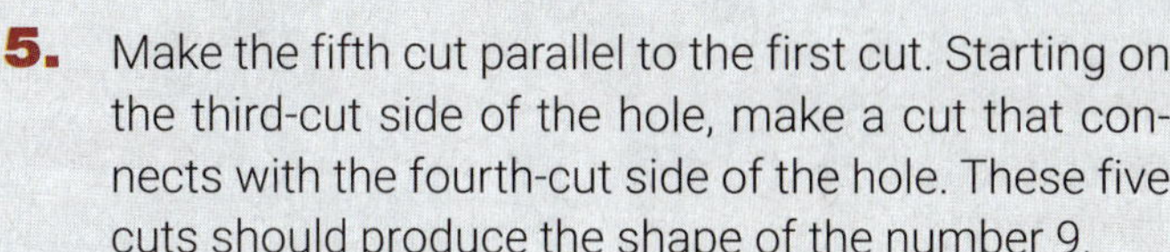

5. Make the fifth cut parallel to the first cut. Starting on the third-cut side of the hole, make a cut that connects with the fourth-cut side of the hole. These five cuts should produce the shape of the number 9.

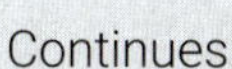

Continues.

SKILL DRILL 16-8 CONTINUED

Making a Seven, Nine, Eight (7, 9, 8) Rectangular Cut Firefighter I, NFPA 1010: 6.3.12

6. Make the sixth cut perpendicular to the first cut. Extend the fourth cut approximately 4 ft (1.2 m) until it is even with the third cut.

7. Make the seventh cut parallel to the first cut and the fifth cut. Start on the side of the sixth cut, and cut toward the side of the third cut, until the cuts are connected. These seven cuts should produce the shape of the number 8.

SKILL DRILL 16-9

Making a Louver Cut Firefighter I, NFPA 1010: 6.3.12

1. Locate the roof supports by sounding.

2. Make two parallel cuts perpendicular to the roof supports.

SKILL DRILL 16-9 CONTINUED

Making a Louver Cut Firefighter I, NFPA 1010: 6.3.12

3. Cut parallel to the supports and between pairs of supports in a rectangular pattern.

4. Tilt the panel to a vertical position. Open the interior ceiling area below the opening by using the butt end of a pike pole. This hole should be the same size as the opening made in the roof decking.

SKILL DRILL 16-10

Making a Triangular Cut Firefighter I, NFPA 1010: 6.3.12

1. Locate the roof supports.

2. Make the first cut from just inside a support member in a diagonal direction toward the next support member.

Continues.

SKILL DRILL 16-10 CONTINUED

Making a Triangular Cut Firefighter I, NFPA 1010: 6.3.12

3. Begin the second cut at the same location as the first, and make it in the opposite diagonal direction, forming a V shape.

4. Make the final cut along the support member to connect the first two cuts. Cutting from this location allows firefighters the full support of the member directly below them while performing ventilation.

SKILL DRILL 16-11

Making a Peak Cut Firefighter I, NFPA 1010: 6.3.12

1. Locate the roof supports.

2. Clear the roofing materials away from the roof peak.

SKILL DRILL 16-11 CONTINUED

Making a Peak Cut Firefighter I, NFPA 1010: 6.3.12

3. Make the first cut vertically, at the farthest point away. Start at the roof peak in the area between the support members, and then cut down to the bottom of the first plywood panel.

4. Make parallel downward cuts between supports, moving horizontally along the roofline to make additional ventilation openings.

5. Strike the nearest side of the roofing material with an axe or maul, pushing it in, using the support located at the center as a fulcrum. This causes one end of the roofing material to go downward into the opening and the other to rise up. If necessary, repeat this process on both sides of the peak, horizontally across the peak, or vertically toward the roof edge.

6. Open the interior ceiling area below the opening by using the butt end of a pike pole. This hole should be the same size as the vent opening made in the roof decking.

Trench Cut. A **trench cut** (or **strip cut**) is used to stop fire spread in long narrow buildings, such as strip malls and small storage complexes. A trench cut creates a large opening ahead of the fire, removing a section of fuel and letting heated smoke and gases flow out of the building. Essentially, it is a firebreak in the roof. Water is sometimes applied through the trench cut.

Trench cuts are a defensive ventilation tactic intended to stop the progress of a large fire, particularly one that is advancing through an attic or cockloft. The IC who chooses this tactic is writing off part of the building and identifying a point where crews will be able to stop the fire.

A trench cut is made from one exterior wall across to the other. It begins with two parallel cuts, spaced 2 to 4 ft (0.6 to 1.2 m) apart. Approximately every 4 ft (1.2 m), firefighters make short perpendicular cuts between the two parallel cuts. They can then lift the roof covering out in sections, completely opening a section of the roof. On a pitched roof, the trench should run from the peak down.

A trench cut must be made far in advance of the fire. The ventilation crew must be able to complete the cut and the interior crew must be able to get into position before the fire passes the trench. Examination openings should be made to ensure that the fire has not already passed the chosen site before the trench cut is completed. The crew members' lives will be in danger and the tactic is useless if the fire advances beyond the trench before it is opened.

Although trench cuts are effective, they require both time and personnel to complete. As with all types of vertical ventilation, careful coordination between the ventilation crew and the interior attack crew is essential. While the trench cut is being made, attack teams should be deployed inside the building to defend the area in front of the cut.

To make a trench cut, follow the steps in **SKILL DRILL 16-12**.

Special Considerations

Many obstacles can be encountered during ventilation operations. Firefighters must be creative and remember the basic objectives of ventilation—to create high openings as rapidly as possible so that hose lines can be advanced into the building.

Poor access or obstructions such as trees, fences, electric lines, or tight exposures can prevent firefighters from getting close enough to place ladders. Steel bars, shutters, and other security features can also hinder ventilation efforts. Window openings in abandoned buildings might be boarded or sealed. A building that requires forcible entry is also likely to present ventilation challenges. To enhance their security, some commercial buildings have roofs made of steel plating. In these buildings, firefighters should not attempt vertical ventilation through the roof, but rather should use alternative ventilation techniques.

Many residential and commercial roofs have multiple layers; some even have a new roof built on top of an older one. Buildings that have two roofs present a challenge for firefighters. A well-executed roof ventilation

SKILL DRILL 16-12

Making a Trench Cut Firefighter I, NFPA 1010: 6.3.12

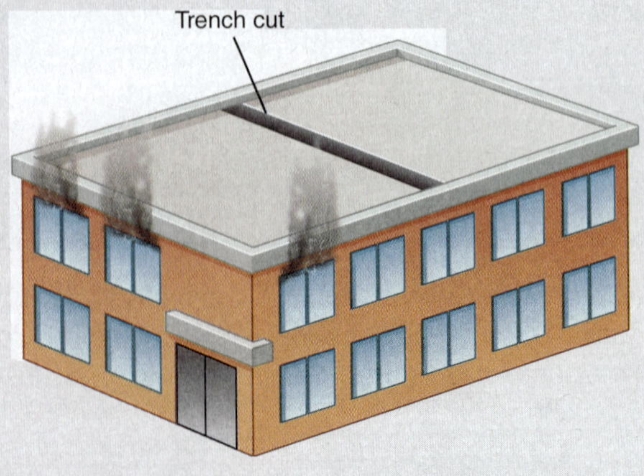

1. After a primary cut has been made over the seat of the fire, cut a number of small inspection holes to identify a point sufficiently far ahead of the fire travel.
2. Make two parallel cuts, 2 to 4 ft (0.6 to 1.2 m) apart, across the entire roof, starting at the ridge pole (for pitched roofs) or a load-bearing wall (for flat roofs).
3. Cut between the two long cuts to make a row of rectangular sections.
4. Remove the rectangular panels to open the trench.

evolution may not result in effective ventilation for these buildings. In this case, it is necessary to determine the reason the ventilation is ineffective to overcome the issue. Additional holes may need to be cut in the second roof, or another type of ventilation may need to be performed. Whenever possible, buildings with two roofs should be identified during preincident surveys.

Ventilating Basements

Research involving basement fires has provided valuable information about safe and effective ways to ventilate and extinguish basement fires. Many vertical voids within a building, such as those found in balloon construction frames, originate or terminate in the basement, providing ample opportunities for fire gases to spread throughout the building. In many parts of the country, especially in newer subdivisions with lightweight construction built prior to the adoption of the 2012 International Residential Code, basement ceilings are left unfinished when the home is sold. Under these conditions, fires originating in the basement will quickly involve the floor system, resulting in a failure of the floor over the fire and early collapse, with the potential for causing injury or death to firefighters entering the building with conditions of limited visibility.

SAFETY TIP

Remember that basement fires often do not look like basement fires. Because of the many voids between the basement and other parts of the building, a fire that started in the basement may at first appear to be a kitchen fire or a fire on the first or second floor of the building. As a safety measure, operate as though the floor, your working platform, has been weakened by a fire underneath until you can prove that there is no fire beneath you.

When attacking basement fires, the use of interior stairs for ventilation should not be considered a safe option because of the potential for failure of the floor system and the danger of firefighters being in the exhaust flow path of hot gases from the fire. In some basement fires, the temperature reaches untenable levels even for firefighters in full PPE and the floor level temperature is not lower than other parts of the basement. The high temperatures may be significant from floor to ceiling. This seems to indicate that it may not always be a safe idea to send firefighters into basement fires before the compartment has been cooled.

Whenever possible, basement fires should be ventilated through exterior windows or doors. Ventilation might also be challenging since basement windows are almost always small. You may need to open up every access point to the basement that you can find to get any sort of relief from the heavy heat build-up that will likely be present. This ventilation results in oxygen being added to a ventilation-limited fire and may cause rapid growth of the fire unless a hose stream is quickly directed into the volume of fire. The application of a straight stream as close to the fire as possible until the fire is darkened is effective in cooling the fire compartment and the rest of the building. Once this offensive exterior attack is completed, it will be safer and easier to enter the basement to complete extinguishment of the remaining fire.

Ventilating Concrete Roofs

Some commercial or industrial structures have concrete roofs. Concrete roofs are generally flat and difficult to breach. The roof decking is usually very stable, but fire conditions underneath could weaken the supporting structural components or load-bearing walls, leading to failure and collapse. There are few options for ventilating concrete roofs. Even special concrete-cutting saws are generally ineffective against these structures. In these cases, firefighters should use alternative ventilation openings such as vents, skylights, and other roof penetrations or horizontal ventilation.

Ventilating Metal Roofs

Metal roofs and metal roof decking present many challenges for the ventilation crew. Ventilation on these types of roofs should only be deployed when necessary and when staffing and time permit. Because metal conducts heat more quickly than other roofing materials, discoloration and warping of this material can indicate the seat of the fire. Tin-cutter hand tools can be used to slice through thin metal coverings, whereas special saw blades might be needed to cut through metal roof decking. In many cases, the metal is on the bottom and supports a built-up or composite roof covering.

Metal roof decking is often supported by lightweight steel bar joists, which can sag or collapse when exposed to a fire. Because the metal decking is lightweight, the supporting structure can be relatively weak, with widely spaced bar joists. The resulting assembly can fail quickly with only limited fire exposure.

As the fire heats the metal decking, the tar roof covering can melt and leak through the joints into the building, where it releases flammable vapors. When this sequence of events occurs, it can quickly spread the fire over a wide area under the roof decking. Firefighters should look for indications of dripping or melting tar and begin rapid ventilation to dissipate the flammable vapors before they can ignite. Hose streams should be used to cool the roof decking from below to stop the tar from melting and producing vapors.

When metal roof decking is cut, the metal can roll down and create a dangerous slide directly into the opening. The triangular cut prevents the decking from rolling away as easily, so it is the preferred option, even though several cuts can be needed to create an adequately sized vent.

Ventilating High-Rise Buildings

Ventilating a high-rise building can be challenging. A high-rise building resembles a stack of individual floor compartments connected by stairways, elevator shafts, and other vertical passages. Most high-rise buildings have sealed windows that are difficult to break. In addition, high-rise buildings havc unique patterns of smoke movement. Smoke might be trapped on individual floors, or it may move up or down within the vertical shafts.

Many newer, high-rise buildings have smoke management capabilities built into their HVAC systems. Use of this type of system enables different areas to be pressurized with fresh air and contaminated air to be exhausted directly to the outside. If the HVAC system does not have this capacity, it can complicate problems by circulating smoke to different areas of the building. High-rise buildings also often have design features known as smoke containment systems. These features, such as fire walls and automatic doors, help keep the smoke in the fire compartment and prevent it from traveling to other areas of the building. (For more information about smoke management and smoke control systems, see Chapter 24, Systems of *Fire Detection, Suppression, and Smoke Control*)

A **stack effect** is the vertical air flow within a building caused by temperature differences between the interior and the exterior of the building. If it is cold outdoors, a heated interior causes smoke to rise quickly through stairways, elevator shafts, and other vertical openings, filling the upper levels of the building (**FIGURE 16-24**). The opposite situation can occur on a hot summer day, when the interior temperature is much cooler than the outside atmosphere. In this scenario, the heavy cooler air pushes the smoke down the vertical openings, toward a lower-level exit (**FIGURE 16-25**).

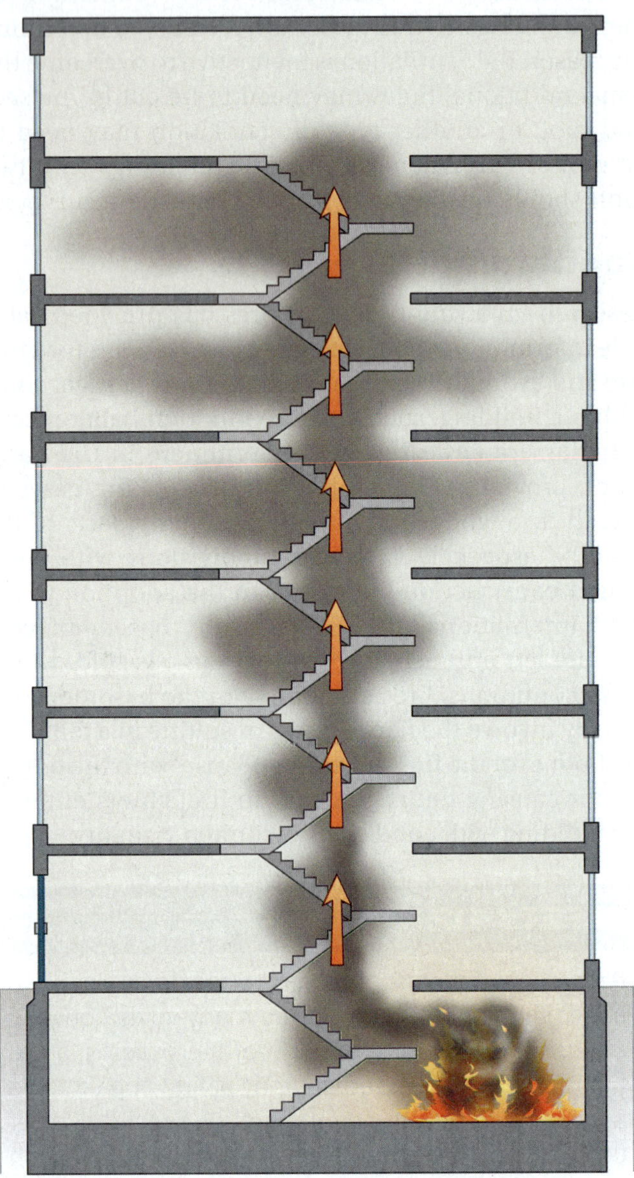

FIGURE 16-24 A winter stack effect occurs when the outside air is much cooler than the interior temperature.

The situation can change if the fire produces enough heat to alter the temperature profile within the building. The air currents within a tall building might, for example, be stronger than the convection currents that normally govern the flow of smoke and other products of combustion. A strong wind through an open or broken window can change the direction of smoke. Contaminated air might suddenly move toward the opening or be pushed back as a strong draft of fresh air enters the building.

After smoke mixes with fresh air or is hit by water from sprinklers or hose streams, it cools and then sits in at lower levels in one location. This cold smoke can

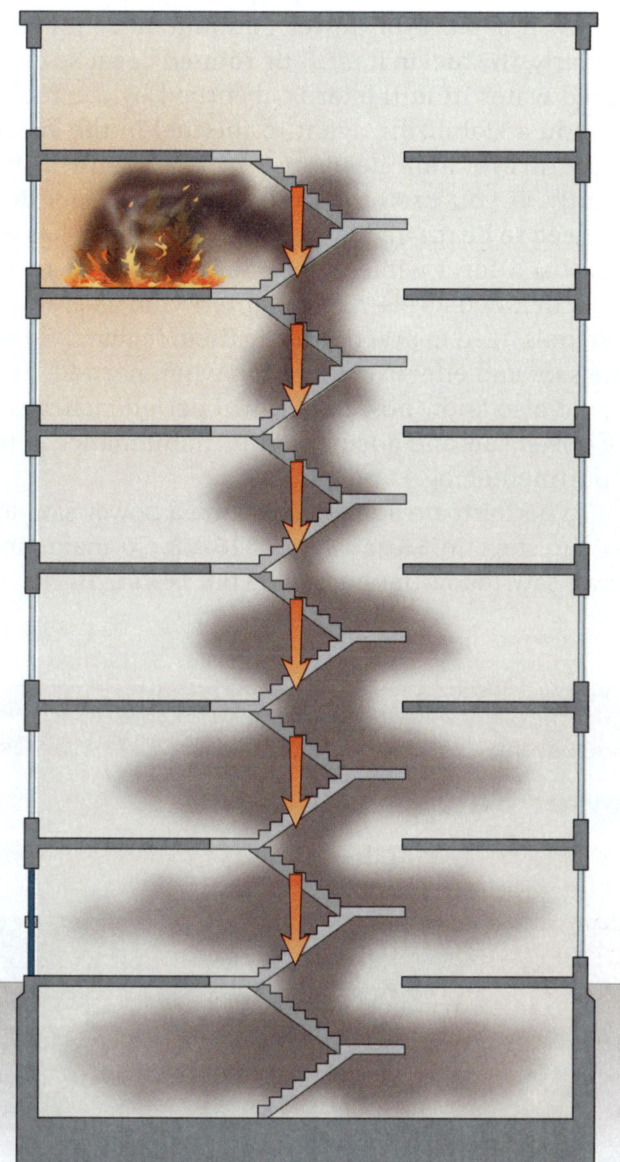

FIGURE 16-25 A summer stack effect occurs when the interior temperature is much cooler than the outside air.

© Jones & Bartlett Learning

fill several floors and usually needs to be cleared by mechanical means.

A key objective in ventilating a high-rise building is to manage the air movement in stairways and elevator shafts. Positive-pressure ventilation can help keep stairways clear of smoke and ventilate individual floors in high-rise structures.

At least one stairwell should be designated as the evacuation stairwell and announced to the operating companies. This stairwell should be kept well-ventilated and designated as an occupant and rescue route. These stairs must be used by rescuers and escaping victims only. Some newer buildings have smoke-proof stair towers or pressurized stair shafts that are designed to keep smoke out of the stairway. Otherwise, positive-pressure fans can be used in a stairway to keep smoke out. Place the positive-pressure fans at a ground-floor doorway, positioning them so that fresh air blows into the stairway. In emergencies, positive-pressure fans can also be used to introduce clean, oxygen-rich air into the potentially toxic atmosphere of the evacuation stairwell. Make certain that the engines of the positive-pressure fans are well ventilated because they can increase carbon monoxide levels.

A pressurized stairway also can be used to clear smoke from a floor. Opening a door from the stairway allows the higher pressure, fresh air to enter the floor. The contaminated atmosphere can then be vented out through a window or another stairway.

FIGURE 16-26 Structures without windows pose significant problems in ventilation.

© olaf schlueter/Shutterstock.

Ventilating Windowless Buildings

Many structures do not have window openings (**FIGURE 16-26**). Some buildings are designed without windows; others have bricked-up or covered windows. These buildings pose two significant risks to firefighters: Heat and products of combustion are trapped, and firefighters have no secondary exit route.

Windowless buildings are similar to basements in terms of the ventilation approach to be used. Any ventilation needs to be as high as possible and probably requires mechanical assistance. Using existing rooftop openings, cutting openings in the roof, reopening boarded-up windows or doors, and making new openings in exterior walls are all possible ways to ventilate windowless structures.

Ventilating Large Buildings

Providing adequate ventilation is more difficult in large buildings than in smaller ones. In a large building, a ventilation hole placed in the wrong location can draw the fire toward the opening, spreading the fire to an area that was not previously involved. This underscores the importance of coordinating ventilation operations with the overall fire attack strategy.

If possible, firefighters should use interior walls and doors to create several smaller compartments in a large building, thereby limiting the spread of heat and smoke. The smaller areas can be cleared one at a time with positive-pressure fans. Several fans can be used in a series or in parallel lines to clear smoke from a large area.

Equipment Maintenance

It is important that all equipment used for ventilation is kept in good repair and ready to operate at peak efficiency. It is important to read and follow the manufacturer's instructions for maintaining power equipment at regular and properly documented intervals. If a piece of power equipment is not used regularly, the fuel in it must be rotated because evaporated water from the air is absorbed by the fuel. If you add a stabilizing agent to the fuel in the equipment, this rotation needs to be performed less frequently. If you use ethanol-free fuel, the fuel does not need to be rotated, even if the equipment has not been used for a while. Keep fuel tanks filled to the recommended levels. All firefighters who use ventilation tools need to practice using them regularly to ensure safe and effective operation when needed. They should also know how to perform periodic checks of the power saw and document the maintenance in the departmental log.

To perform a readiness check on a power saw, follow the steps in **SKILL DRILL 16-13**. To maintain a power saw, follow the steps in **SKILL DRILL 16-14**.

SKILL DRILL 16-13

Performing a Readiness Check on a Power Saw Firefighter I, NFPA 1010: 6.5.1

1. Ensure that the fuel tank is full. Make certain that the bar and chain oil reservoir is full.

2. Check the throttle trigger for smooth operation.

SKILL DRILL 16-13 CONTINUED

Performing a Readiness Check on a Power Saw Firefighter I, NFPA 1010: 6.5.1

3. Ensure that the saw, bar, chain, air filter, and chain brake are clean and working. Inspect the blade for even wear, and then lubricate the sprocket tip if needed.

4. Check the chain for wear, missing teeth, or other damage. Check the chain for proper tension. Check the chain catcher.

5. Check for loose bar nuts and screws, and tighten them if needed. Check the starter and starter cord for wear.

6. Inspect the spark plugs.

Continues.

SKILL DRILL 16-13 CONTINUED

Performing a Readiness Check on a Power Saw Firefighter I, NFPA 1010: 6.5.1

7. Start the saw. Make certain that both the bar and the chain are being lubricated while the saw is running. Check the chain brake.

8. Make certain that the stop switch functions. Record the results of the inspection.

SKILL DRILL 16-14

Maintaining a Power Saw Firefighter I, NFPA 1010: 6.5.1

1. Remove, clean, and inspect the clutch cover, bar, and chain for damage and wear. Replace if necessary.

2. Inspect the air filter and clean/replace as needed.

SKILL DRILL 16-14 CONTINUED

Maintaining a Power Saw Firefighter I, NFPA 1010: 6.5.1

3. Lubricate components as recommended by the manufacturer. Reinstall the bar and chain, flipping the bar over each time to help wear the bar evenly. Replace the clutch cover.

4. Adjust the chain tension (make sure the bar and chain cool before adjusting).

5. Fill the power saw with fuel. Fill the bar and chain oil reservoir.

CASE STUDY

You Are the Firefighter CONCLUSION

It is a routine day at the hardware store where you work full time. You are busy doing inventory when your pager goes off, alerting you to a structure fire only a few blocks away. On your arrival, you find a one-story, wood-frame house with dirty brown smoke pushing out through the eaves, siding, and other cracks. The windows are stained black. The neighbors say they do not think anyone is home because the occupants work during the day.

1. **What is the best way to ventilate this structure?**

   Answer: Place a ventilation opening as high as possible within the building or fire area to help eliminate the backdraft conditions. Ventilation must be coordinated with the attack crew, and not started until the attack crew is ready to advance on the fire.

2. **What concerns do you have about ventilating this structure?**

   Answer: A backdraft could occur when oxygen is introduced into the structure that is concealing hot gases containing large amounts of unburned or partially burned fuel. The mixture of oxygen and superheated gases can ignite and explode in as little as 10 seconds after ventilation occurs.

3. **When would you want to ventilate this structure?**

   Answer: After arriving on scene and doing a thorough scene size-up to develop a plan not only for rescue and suppression but also for the tactics needed to ventilate the fire structure. Ventilation must be coordinated with the attack crew and not started until the attack crew is ready to advance on the fire. Once all crews are in place and everyone is ready, ventilation can start by placing a ventilation opening as high as possible within the structure.

WRAP-UP

SUMMARY

KNOWLEDGE OBJECTIVES

- Describe how fire behavior impacts ventilation.
  - Describe the characteristics of a ventilation-limited fire. (pp. 690–694)
  - Describe the impact of door control on ventilation. (pp. 691–692)
  - Describe how location impacts ventilation operations. (p. 692)
  - Describe the impact of the size of the ventilation hole. (pp. 692–693)
  - Describe the impact of wind on fire behavior. (p. 693)
  - Describe the impact of exterior suppression on fire behavior. (pp. 693–694)
- Describe how size-up findings affect ventilation operations.
  - Describe the importance of including ventilation considerations in a size-up. (**NFPA 1010: 6.3.11**, p. 694)
  - Describe how the location, size, and stage of fire affect ventilation operations. (**NFPA 1010: 6.3.11**, pp. 694–695)
  - Describe how the characteristics of different construction types affect ventilation operations. (**NFPA 1010: 6.3.11**, pp. 695–698)

- Describe the importance of the timing and coordination of ventilation and suppression.
  - List the three Ws in timing ventilation with suppression. (**NFPA 1010: 6.3.11**, p. 698)
- Identify how to minimize backdrafts and flashovers.
  - Describe steps that can be taken to minimize the risk of backdrafts. (**NFPA 1010: 6.3.11**, p. 698)
  - Describe steps that can be taken to minimize the risk of flashovers. (**NFPA 1010: 6.3.11**, p. 699)
- Describe the types of ventilation, how each type removes contaminated atmospheres, and the techniques used to provide ventilation to a structure.
  - List the two basic types of ventilation. (**NFPA 1010: 6.3.11**, pp. 699–700)
  - Explain how horizontal ventilation removes contaminated atmosphere from a structure. (**NFPA 1010: 6.3.11**, pp. 700–701)
  - List the two methods of horizontal ventilation. (**NFPA 1010: 6.3.11**, pp. 700–701)
  - Explain how natural ventilation removes contaminated atmosphere from a structure. (**NFPA 1010: 6.3.11**, pp. 701–705)
  - Describe the techniques used to provide natural ventilation to a structure. (**NFPA 1010: 6.3.11**, pp. 701–705)
  - Explain how mechanical ventilation removes contaminated atmosphere from a structure. (**NFPA 1010: 6.3.11**, pp. 705–708)
  - Describe the techniques used to provide mechanical ventilation to a structure. (**NFPA 1010: 6.3.11**, pp. 705–710)
  - Describe how negative-pressure ventilation removes contaminated atmosphere from a structure. (**NFPA 1010: 6.3.11**, pp. 705–706)
  - Describe the techniques used to provide negative-pressure ventilation to a structure. (**NFPA 1010: 6.3.11**, pp. 705–706)
  - Describe how positive-pressure ventilation removes contaminated atmosphere from a structure. (**NFPA 1010: 6.3.11**, pp. 706–708)
  - Describe the techniques used to provide positive-pressure ventilation to a structure. (**NFPA 1010: 6.3.11**, pp. 706–708)
  - Describe how hydraulic ventilation removes contaminated atmosphere from a structure. (**NFPA 1010: 6.3.11**, pp. 708–710)
  - Describe the techniques used to provide hydraulic ventilation to a structure. (**NFPA 1010: 6.3.11**, pp. 708–710)
  - Describe how vertical ventilation removes contaminated atmosphere from a structure. (**NFPA 1010: 6.3.12**, pp. 711–718)
- Describe the hazards of vertical ventilation and how those hazards are mitigated.
  - Describe how to ensure firefighter safety during vertical ventilation operations. (**NFPA 1010: 6.3.12**, pp. 711–713)
  - Identify the warning signs of roof collapse. (**NFPA 1010: 6.3.12**, p. 713)
  - Describe the components and characteristics of roof assemblies. (**NFPA 1010: 6.3.12**, pp. 713–716)
  - List the differences in solid-beam construction and lightweight construction in roofs. (**NFPA 1010: 6.3.12**, pp. 713–715)
  - Explain how roof construction affects fire resistance. (**NFPA 1010: 6.3.12**, p. 715)
  - List the basic types of roof design. (**NFPA 1010: 6.3.12**, pp. 715–716)
  - Describe the characteristics of flat roofs. (**NFPA 1010: 6.3.12**, pp. 715–716)
  - Describe the characteristics of pitched roofs. (**NFPA 1010: 6.3.12**, pp. 715–716)
  - Describe the characteristics of curved roofs. (**NFPA 1010: 6.3.12**, p. 716)
- Describe the common techniques of vertical ventilation.
  - Describe the techniques of vertical ventilation. (**NFPA 1010: 6.3.12**, pp. 716–728)
  - List the tools utilized in vertical ventilation. (**NFPA 1010: 6.3.12**, pp. 718–720)
  - List the types of roof cuts utilized in vertical ventilation operations. (**NFPA 1010: 6.3.12**, pp. 720–728)
  - Describe the characteristics of a rectangular or square cut. (**NFPA 1010: 6.3.12**, pp. 720–721)
  - Describe the characteristics of a seven, nine, eight (7, 9, 8) rectangular cut. (**NFPA 1010: 6.3.12**, pp. 721–722)
  - Describe the characteristics of a louver cut. (**NFPA 1010: 6.3.12**, pp. 722, 724–725)
  - Describe the characteristics of a triangular cut. (**NFPA 1010: 6.3.12**, pp. 722, 725–726)

KNOWLEDGE OBJECTIVES CONTINUED

- Describe the characteristics of a peak cut. (**NFPA 1010: 6.3.12**, pp. 722, 726–727)
- Describe the characteristics of a trench cut. (**NFPA 1010: 6.3.12**, p. 728)
- Describe ventilation techniques for special conditions.
  - Describe the special considerations in ventilating basements. (**NFPA 1010: 6.3.12**, p. 729)
  - Describe the special considerations in ventilating concrete roofs. (**NFPA 1010: 6.3.12**, p. 729)
  - Describe the special considerations in ventilating metal roofs. (**NFPA 1010: 6.3.12**, pp. 729–730)
  - Describe the special considerations in ventilating high-rise buildings. (**NFPA 1010: 6.3.12**, pp. 730–731)
  - Describe the special considerations in ventilating windowless buildings. (**NFPA 1010: 6.3.12**, p. 731)
  - Describe the special considerations in ventilating large buildings. (**NFPA 1010: 6.3.12**, pp. 731–732)
- Describe the tasks involved in maintaining ventilation equipment.
  - Explain how to ensure that ventilation equipment is in a state of readiness. (**NFPA 1010: 6.5.1**, pp. 732–735)

SKILLS OBJECTIVES

- Begin the ventilation process with either a hand tool or a ladder.
  - Break glass with a hand tool. (**NFPA 1010: 6.3.11**, p. 702)
  - Break glass with a ladder. (**NFPA 1010: 6.3.11**, pp. 703–704)
- Deliver negative-pressure, positive-pressure, or hydraulic ventilation.
  - Deliver negative-pressure ventilation. (**NFPA 1010: 6.3.11**, p. 706)
  - Deliver positive-pressure ventilation. (**NFPA 1010: 6.3.11**, p. 709)
  - Perform hydraulic ventilation. (**NFPA 1010: 6.3.11**, p. 710)
- Perform vertical ventilation by making a roof cut.
  - Operate a power saw. (**NFPA 1010: 6.3.12**, pp. 718–720)
  - Make a rectangular or square cut to deliver vertical ventilation. (**NFPA 1010: 6.3.12**, pp. 721–722)
  - Make a seven, nine, eight (7, 9, 8) rectangular cut to deliver vertical ventilation. (**NFPA 1010: 6.3.12**, pp. 723–724)
  - Make a louver cut to deliver vertical ventilation. (**NFPA 1010: 6.3.12**, pp. 724–725)
  - Make a triangular cut to deliver vertical ventilation. (**NFPA 1010: 6.3.12**, pp. 725–726)
  - Make a peak cut to deliver vertical ventilation. (**NFPA 1010: 6.3.12**, pp. 726–727)
  - Make a trench cut to deliver vertical ventilation. (**NFPA 1010: 6.3.12**, p. 728)
- Maintain equipment so it is in a ready state.
  - Perform a readiness check on a power saw. (**NFPA 1010: 6.5.1**, pp. 732–734)
  - Maintain a power saw. (**NFPA 1010: 6.5.1**, pp. 734–735)

KEY TERMS

auto-exposure See *vertical fire extension.*

clean air The outside air that replaces a contaminated atmosphere.

contaminated atmosphere The products of combustion that must be removed from a building.

examination opening A small opening in a roof used to determine how large an area is involved, whether a fire is spreading, and in which direction it is moving, and to evaluate conditions under the roof and verify the proper location for a ventilation opening.

fog ventilation See *hydraulic ventilation.*

horizontal ventilation The opening or removal of windows or doors on any floor of a fire building to create flow paths for fire conditions. (NFPA 1410)

hydraulic ventilation Ventilation that relies on the movement of air caused by a fog stream that is placed 2 to 4 ft (0.6 to 1.2 m) in front of an open window. Also referred to as *fog ventilation.*

intentional ventilation Ventilation that is planned and done on purpose.

kerf cut A cut that is the width and depth of the saw blade. It is used to inspect cockloft spaces from the roof.

lath Thin strips of wood used to make the supporting structure for roof tiles.

louver cut A cut that is made using power saws and axes to cut along and between roof supports so that the sections created can be tilted into the opening.

mechanical ventilation A process of removing heat, smoke, and gases from a fire area by using exhaust fans, blowers, air-conditioning systems, or smoke ejectors. (NFPA 440)

natural ventilation The flow of air or gases created by the difference in the pressures or gas densities between the outside and inside of a vent, room, or space. (NFPA 853)

negative pressure The condition that exists when the air pressure in one area is lower than the atmospheric pressure in another area.

negative-pressure ventilation Ventilation that relies on electric fans to pull or draw the smoke, heat, and air by creating negative pressure in a structure or an area.

parapet The part of a wall entirely above the roofline. (NFPA 5000)

peak cut A ventilation opening that runs along the top of a pitched roof.

positive pressure The condition that exists when the air pressure in one area is higher than the atmospheric pressure in another area.

positive-pressure attack Ventilation that controls the flow of products of combustion before the fire is controlled by introducing clean air into a structure, which pushes the contaminated atmosphere out.

positive-pressure fan A large, powerful fan that is used to create positive-pressure ventilation by forcing clean air into a structure.

positive-pressure ventilation Ventilation that relies on fans to push or force clean air into a structure after a structure fire has been controlled.

recirculation The phenomenon that occurs when a smoke ejector is not completely sealed so that exhausted air is drawn back into the building.

rectangular cut The most common vertical ventilation opening that is usually 4-ft by 4-ft (1.2-m by 1.2-m) in size. Also called *square cut.*

roof covering The membrane, which may also be the roof assembly, that resists fire and provides weather protection to the building against water infiltration, wind, and impact. (NFPA 5000)

roof decking The rigid portion of roof between the roof supports and the roof covering.

seven, nine, eight (7, 9, 8) rectangular cut A ventilation opening that is usually about 8 ft by 4 ft (1.2 m by 2.4 m) in size; it is primarily used for large commercial buildings with flat roofs.

size-up A rapid evaluation and analysis of an incident, usually conducted by the first-arriving company officer, to determine which resources need to be deployed and which actions can be undertaken safely.

smoke ejector A mechanical device, similar to a large fan, that can be used to force heat, smoke, and gases from a post-fire environment and draw in fresh air. (NFPA 440)

square cut See *rectangular cut.*

stack effect The vertical air flow within buildings caused by the temperature-created density differences between the building interior and exterior or between two interior spaces. (NFPA 92)

strip cut See *trench cut.*

trench cut A roof cut that is made from one load-bearing wall to another load-bearing wall and that is intended to prevent horizontal fire spread in a building. Also called *strip cut.*

triangular cut A triangle-shaped ventilation cut in the roof decking that is made using a saw or an axe.

unintentional ventilation Ventilation that occurs when a window or door fails because of the fire or when a window or door is left open by mistake.

ventilation The controlled and coordinated removal of heat and smoke from a structure, replacing the escaping gases with fresh air. (NFPA 1410)

vertical fire extension Fire spread from a lower floor to a higher floor through an exterior window. Also called *auto-exposure.*

vertical ventilation The vertical venting of structures involving the opening of bulkhead doors, skylights, scuttles, and roof cutting operations to release smoke and heat from inside the fire building. (NFPA 1410)

REVIEW QUESTIONS

1. What does effective ventilation do?
2. What influences how quickly a fire reacts to changes in ventilation?
3. What is Type I construction (fire-resistive)?
4. Why should you try to limit the amount of air entering the fire compartment before fire attack and ventilation begin?
5. When a backdraft situation exists in a building, which actions will most likely reduce the probability of an explosion?
6. What is the effect of creating a larger ventilation opening compared to a smaller one, and what is the effect of creating a ventilation opening closer to the seat of the fire compared to farther than from the seat of the fire?
7. When a firefighter is to perform ventilation through a window by climbing a ladder and breaking the glass, where should the ladder be placed?
8. When executed properly, how does a trench cut prevent fire spread?
9. What is a key objective in ventilating a high-rise building?
10. When doing power saw maintenance, what do you want to make sure functions correctly?

DISCUSSION QUESTIONS

1. What is your go-to ventilation tactic after the size-up of a fire scene, and what type of ventilation are you most comfortable with doing on a fire scene?
2. How does the type of building construction affect how you conduct ventilation operations?
3. How does the type of roof construction affect your ventilation choices on a fire scene?
4. What obstacles might you encounter at a fire scene during ventilation operations?

APPLYING THE CONCEPTS

Your fire department is on the scene of a multi-alarm fire in an older three-story apartment building. It houses about 50 students from the local university. It is summer and not many of the students are living in the building; however, it is unknown how many residents are still inside the structure. The apartment is a brick-clad, Type III structure. Size-up determined that the fire seat is on the second floor, in an apartment on the side Charlie–side Delta corner of the building. The fire is in the growth stage and ventilation limited. An interior crew is on the second floor conducting a direct attack on the fire. A rescue crew is looking for potential victims trapped inside. The IC is determining ventilation strategies.

1. Why would the IC decide to ventilate the fire?
2. What are the risks of ventilating? How are these risks minimized?
3. When considering ventilation, what are the three Ws?

You and your four-firefighter crew have been assigned to ventilation operations. The IC radios, "Ventilation crew, don full PPE and SCBA. Initiate vertical ventilation on side Charlie–side Delta corner of the roof. Bring your tools and establish a secure position for the opening. Stay in radio contact so we can coordinate the interior fire attack. And remember, safety first."

4. What tools will you take to make an opening in the roof?
5. How will you decide whether to use a ground ladder or a ladder truck to access the roof?

Because obstructions block access by a ladder truck, your crew positions a ground ladder on side Charlie of the structure. You sling your tools and power saw across your back and climb up the ladder, radioing the interior crew to alert them of your progress. The heat intensifies as you ascend, and you see a unit on the third floor is filled with heavy smoke.

6. Before stepping onto the roof, what do you need to consider?

You determine the roof is safe and you proceed to the area specified by incident command. Coordinating with the interior crew, you use the power saw to quickly make a small triangular cut to examine the opening. You report your findings and get the go ahead to make a 4-ft by 4-ft (1.2-m by 1.2-m) rectangular cut in the roof.

7. Four-by-four cuts are the most common vertical ventilation opening. Before making a cut, what should you do to avoid cutting through structural roof supports?
8. What happens if you do not cut the ventilation hole directly over the seat of the fire?
9. What are the most important safety measures to remember when doing vertical ventilation on a roof?

REFERENCES

Jakubowski, Greg. 2013. "Bowstring Truss Roof Construction Hazards." *Firefighter Nation*, January 10, 2013. Accessed September 8, 2023. https://www.firefighternation.com/firerescue/bowstring-truss-roof-construction-hazards/.

Kerber, Stephen. 2012. "Analysis of Changing Residential Fire Dynamics and Its Implications on Firefighter Operational Timeframes." *Fire Technology*, 48, no. 3 (November 29, 2011): 865–891. Accessed September 4, 2023. https://doi.org/10.1007/s10694-011-0249-2.

Kerber, Stephen. 2013. *Study of the Effectiveness of Fire Service Vertical Ventilation and Suppression Tactics in Single Family Homes.* UL's Firefighter Safety Research Institute (FSRI), July 1, 2013. Accessed September 6, 2023. https://d1gi3fvbl0xj2a.cloudfront.net/2021-10/UL-FSRI-2010-DHS-Report_Comp_0.pdf.

National Fire Chiefs Council (NFCC). n.d. "Control Measure: Understand Signs and Symptoms of Flashover." Accessed September 8, 2023. https://www.ukfrs.com/guidance/search/understand-signs-and-symptoms-flashover.

National Fire Protection Association (NFPA). 2019. *NFPA 853, Standard for the Installation of Stationary Fuel Cell Power Systems.* 2020 Edition. Quincy, MA: NFPA.

National Fire Protection Association (NFPA). 2019. *NFPA 1410, Standard on Training for Emergency Scene Operations.* 2020 Edition. Quincy, MA: NFPA.

National Fire Protection Association (NFPA). 2020. *NFPA 92, Standard for Smoke Control Systems.* 2021 Edition. Quincy, MA: NFPA.

National Fire Protection Association (NFPA). 2023. *NFPA 5000, Building Construction and Safety Code.* 2024 Edition. Quincy, MA: NFPA.

National Fire Protection Association (NFPA). 2023. *NFPA 440, Guide for Aircraft Rescue and Firefighting Operations and Airport/Community Emergency Planning.* 2024 Edition. Quincy, MA: NFPA.

National Institute of Standards and Technology (NIST). 2021. "Fire Dynamics." Updated June 2, 2021. Accessed September 20. 2023. https://www.nist.gov/el/fire-research-division-73300/firegov-fire-service/fire-dynamics.

Norwood, P. J., and Frank Ricci. 2021. "Ventilation Limited Fire: Keeping It Rich and Other Tactics Based Off Science." *Fire Engineering*, January 2, 2021. Accessed September 6, 2023. https://www.fireengineering.com/firefighting/ventilation-limited-fire-keeping-it-rich-and-other-tactics-based-off-science/.

Regan, John, Julie Bryant, and Craig Weinschenk. 2020. *Analysis of the Coordination of Suppression and Ventilation in Single-Family Homes.* UL's Fire Safety Research Institute, March 19, 2020. Accessed September 8, 2023. https://d1gi3fvbl0xj2a.cloudfront.net/public/2021-07/Coord_Tactics_Single_Family_Homes.pdf.

UL's Firefighter Research and Safety Institute (FSRI). 2006. "Structural Stability of Engineered Lumber in Fire Conditions." Accessed September 6, 2023. https://fsri.org/sites/default/files/2021-07/NC9140-20090512-Report-Independent.pdf.

UL's Firefighter Research and Safety Institute (FSRI). 2011. "Effectiveness of Fire Service Vertical Ventilation and Suppression Tactics in Single Family Homes." August 23, 2013. Accessed September 8, 2023. https://fsri.org/research-update/fsri-launches-vertical-ventilation-and-suppression-online-training.

UL's Firefighter Research and Safety Institute (FSRI). 2013. "Governors Island Experiments." May 21, 2013. Accessed September 8, 2023. https://fsri.org/research/governors-island-experiments.

CHAPTER

17

Firefighter I

Search and Rescue

KNOWLEDGE OBJECTIVES

After studying this chapter, you will be able to:

- Describe the mission of search and rescue operations.
- Explain how search and rescue operations are coordinated with other fire suppression operations.
- Identify the factors to evaluate during a search and rescue size-up.
- List the priorities of search operations.
- Describe the objectives of a primary and secondary search.
- List the common types of simple victim carries performed during rescue operations.
- List the six emergency drags performed during rescue operations.
- Describe the conditions that may require a ground or aerial ladder rescue.

SKILLS OBJECTIVES

After studying this chapter, you will be able to perform the following skills:

- Conduct a primary search using the standard, oriented, oriented-vent-enter-isolate-search (O-VEIS), and team search methods.
- Conduct a secondary search.
- Perform one- and two-person carries and drags.
- Perform rescues by ladder to conscious and unconscious victims.

ADDITIONAL NFPA STANDARDS

- **NFPA 92**, *Standard for Smoke Control Systems, 2021 Edition*
- **NFPA 921**, *Guide for Fire and Explosion Investigations, 2024 Edition*
- **NFPA 1550**, *Standard for Emergency Responder Health and Safety, 2024 Edition*
- **NFPA 1660**, *Standard for Emergency, Continuity, and Crisis Management: Preparedness, Response, and Recovery, 2024 Edition*
- **NFPA 1700**, *Guide for Structural Fire Fighting, 2021 Edition*
- **NFPA 1801**, *Standard on Thermal Imagers for the Fire Service, 2021 Edition*
- **NFPA 1950**, *Standard on Protective Ensembles for Technical Rescue Incidents, 2020 Edition*
- **NFPA 2500**, *Standard for Operations and Training for Technical Search and Rescue Incidents and Life Safety Rope and Equipment for Emergency Services, 2022 Edition*

CASE STUDY

You Are the Firefighter

You arrive on the scene of a fast-moving apartment fire. The structure is a three-story, center-hallway, Type III (ordinary construction) apartment building with fire showing on the second floor. The second- and third-floor hallways are smoke charged. Dispatch is reporting possible occupants trapped. You are assigned to conduct a primary search on the third floor while another crew begins the initial fire attack. This is a call that will tax the resources of your department.

1. How will you conduct this search?
2. What will you do if you find a victim in the first unit you search?
3. Will you take a hose line with you on the search?

Introduction

The mission of the fire department is to save lives and protect property. Saving lives is always the highest priority at a fire scene. The first firefighters to arrive must always consider the possibility that lives could be in danger and act accordingly.

Saving lives remains the highest priority until it is determined that every occupant who was in danger has been found and moved to a safe location or until it is no longer possible to rescue anyone successfully. The first situation occurs when a thorough search is performed and confirms that there are no occupants remaining to be rescued. The second situation occurs when fire conditions make it unlikely or impossible that any occupants could still be alive to be rescued.

Search and Rescue Operations

Search and rescue are almost always performed in tandem, yet they are actually separate actions. The purpose of **search** operations is to locate victims. The purpose of the **primary search** is to locate individuals who may be trapped, injured, or in immediate danger. The purpose of **rescue** operations is to physically remove an occupant or victim from a hazardous environment. For example, rescue occurs when a firefighter leads an occupant to an exit, carries an unconscious victim out of a burning building, or rescues a victim using a ladder. In all three cases, the occupant or the victim was physically removed from imminent danger through the actions of a firefighter.

Any fire department company or unit can be assigned to search and rescue operations. Many fire departments routinely assign this responsibility to truck companies, rescue companies, or designated engine companies. However, all firefighters must be trained and prepared to perform search and rescue functions and should practice the specific search and rescue procedures used by their departments.

Search and rescue operations must be conducted quickly and efficiently. A systematic approach ensures that every occupant who can possibly be saved is successfully located and removed from danger. Often, the first-arriving firefighters will not know whether any occupants are inside a burning building or how many occupants might be inside. Firefighters should never assume that a building is unoccupied. The only way to confirm that every occupant has safely evacuated a building is to conduct a thorough search. The decision whether to enter a burning building to search for living victims is made by the incident commander (IC) and is based on a risk–benefit analysis. A room that has flashed over is not survivable, even for a firefighter in full personal protective equipment (PPE), for more than a few seconds. The IC will not risk the lives of firefighters when the conditions are not survivable for occupants or even fully protected firefighters.

Coordination with Fire Suppression Operations

Although search and rescue is a top priority, it is rarely the only action taken by the first-arriving companies. The IC and firefighters must plan and coordinate all fire suppression operations—from ventilation to fire attack—to support the search and rescue priority.

One of the first considerations all search and rescue efforts must take into account is the flow path of the fire. Any operation conducted in the flow path increases the risks posed to the firefighter. Getting caught in a flow path that is carrying hot gases away from the fire can result in a deadly situation in a matter of seconds. It is critical that ventilation operations are coordinated with search and rescue operations. Any change in the ventilation of the fire building, whether intentional, such as the opening of a door, window, or roof, or unintentional by glass failure or a change in the wind direction, can be deadly to both rescuers and persons trapped in the fire.

In some cases, the best way to save lives is to first cool down the interior with water applied from the exterior using a transitional attack. This process is described in Chapter 14, *Fire Suppression*. For example, it might be necessary to keep the fire away from potential victims or to protect the entry and exit paths by controlling, limiting, or redirecting the flow path so that the victims can be found and safely removed. In a transitional attack, firefighters first conduct a quick, indirect, exterior attack into the fire compartment to improve conditions before entering the building. Because smoke is fuel, an initial cooling of hot gases, particles, and aerosols in the room of origin can result in a dramatic slowing of the fire growth and a dramatic reduction in the temperature of not only the room of origin but also of other parts of the building. Firefighters then quickly transition to an interior attack for final suppression. It is at this point that the search and rescue crew enters.

Other fire scene activities must also be coordinated with search and rescue. For example, forcible entry is sometimes needed to provide entry for search and rescue teams. Sometimes, well-placed ventilation will be needed, but at other times, it may be best to limit ventilation and protect trapped occupants by closing doors and not opening windows. Ladders may be raised to second-story windows to provide an escape route if needed. Portable lighting can also provide valuable assistance to interior search crews.

Often, the search for potential victims provides valuable information about the location and extent of the fire within a building. In essence, the searchers act as a reconnaissance team to determine which areas in the building are involved and where a fire might spread. They report this information to the IC, who develops the overall incident action plan (IAP).

Search and Rescue Size-Up

The size-up process at every fire should include a specific evaluation of the critical factors for search and rescue—namely, occupant information, including the following:

- The number of occupants in the building
- Their location(s)
- The degree of risk to their lives
- The probability of finding them
- Their ability to evacuate by themselves

Usually, this information is not immediately available, so firefighters' actions must be based on a combination of observations and expectations that might indicate whether and how many people may need to be rescued, including the following:

- Type of occupancy
- Size, construction, and arrangement of the building
- Time of day and the day of the week

In addition, the IC must determine the level of risk faced by the possible occupants of a burning building, including the following:

- The location of the fire within the building
- The direction of the fire spread
- The speed and direction of the wind
- The volume and intensity of the fire
- Visible smoke and fire conditions in different areas of the structure

A search and rescue plan can then be developed based on this information. This plan identifies the areas to be searched, the priorities for searching different areas, the number of search teams required, and any additional actions needed to support the search and rescue activities. One search team might be sufficient for a small building, whereas multiple teams might be needed to search a large building.

Occupant Information

The initial size-up of an incident can provide valuable information about possible building occupants. But even if there is no definitive information provided, firefighters should never automatically assume that a building is unoccupied. An observant firefighter will notice clues that indicate whether a building is occupied and how many occupants are likely to be present (**FIGURE 17-1**). For example, the first-responding unit at a residential fire might see two cars in the driveway and numerous toys on the front lawn. These signs indicate that the house is probably occupied by a family with children. In such a situation, the IC would likely assign several search teams and emphasize the importance of quickly searching every room where victims might be located. Conversely, an absence of cars, locked doors, and an overflowing mailbox might suggest that the house is vacant or unoccupied.

Similar observations can be made for commercial structures. Although an empty parking lot does not guarantee that a building is empty, the presence of a car could indicate that at least one maintenance worker or security guard is inside. However, it could also mean that a car is simply parked in the lot. Boarded-up

A.

B.

FIGURE 17-1 Exterior observations can often provide a good indication of whether a building is occupied. **A.** Parked cars indicate the residence may be occupied. **B.** Boarded-up windows indicate the residence is probably unoccupied.

A: © Imagenet/Shutterstock; **B:** Courtesy of Captain David Jackson, Saginaw Township Fire Department.

windows and doors on a building surrounded by a chain-link fence are indications that the building is probably unoccupied.

If there is a chance that savable victims might be present, the IC will order a search in case someone is inside. Search teams should be assigned on the basis of these priorities. Firefighters should first rescue the occupants who are in the most immediate danger, followed by those who are in less danger. In establishing search priorities, firefighters should also consider where occupants are most likely to be located. Occupants who are close to the fire, above the fire, or in the fire flow path are usually at greater risk than occupants who are located farther away from the fire. An occupant in the apartment where the fire started is probably in immediate danger, whereas the occupants of a lower or upper floor may be relatively safe. This decision will be somewhat dependent on the type of construction used in the fire building.

Building occupants who are at windows or on balconies calling for help obviously realize that they are in danger and want to be rescued. There may be other occupants in the building who cannot be seen from the exterior. For example, occupants could be asleep, unconscious, incapacitated, or trapped. Firefighters must make sure that everyone is accounted for. The dispatcher may relay information on the status of the occupants. Neighbors are another source of information about the status of the occupants. An important question to ask is, "Are you sure that no one else was home?" However, it is often difficult to obtain accurate information from occupants who have just escaped from a burning building, especially if they believe that family members, close friends, or a beloved pet might still be inside. Occupants may be too emotional to speak or even to think clearly.

If someone says that they think occupants are still inside, firefighters should obtain as much information as possible. Ask specific questions, such as "Who is still inside?"; "Where is their room located?"; and "Where were they when you last saw them?" It is always easier to find someone if you know who you are looking for and where to look. Many firefighters can tell stories about searching for a missing baby, only to discover that "Baby" is the family dog.

SAFETY TIP

Do not consider all of the information gathered from neighbors and bystanders to be accurate. In the stress of the moment, well-intentioned people may make incorrect statements. Do not rush into an empty burning house because of a bad tip from a bystander.

Occupancy Type

Young children and elderly people, who are more likely to need assistance to evacuate, are at greater risk than adolescents and adults. Individuals who are confined to a bed or wheelchair are at greater risk than occupants who can move freely. Consequently, more firefighters and resources will be needed for search and rescue operations in certain occupancies.

A fire in a single-family dwelling generally places the members of one family at risk, whereas a fire in a large apartment building may endanger multiple families. A warehouse fire probably presents a direct risk to relatively few occupants.

Building Size, Construction, and Arrangement

The size, type of construction, and arrangement of the building are other important factors to consider when planning and conducting a search. A small, one-story house is simpler to search than a larger building with multiple units or a large commercial building. Even though the complexity of the search increases as a building gets larger, searchers need to be aware of the fire flow path in the building and avoid actions that could place them in a dangerous position.

The risk posed to victims and firefighters may also increase depending on the building construction: Occupants of a Type V, unprotected, wood-frame building are in greater danger than those in a Type II building with fire-resistive construction and automatic sprinklers.

The arrangement of the interior of the building affects the risk to firefighters when conducting a search. To ensure that each area is searched promptly and that no areas are omitted, specific areas must be assigned to different search teams in an organized manner. Assignments are often based on the stairway locations, corridor arrangements, and the apartment or room numbering system. Teams must practice good door control at the point of entry into the building and report when they have completed searching each area.

Access to an interior layout or floor plan is often helpful when planning and assigning teams to search a building. Because it is difficult to determine a building's layout or floor plan in the midst of an emergency, firefighters often conduct preincident surveys of buildings such as apartment complexes and offices. The information assembled during the preincident survey is used to prepare a **preincident plan**, so firefighters will be prepared when an emergency occurs. In some departments, this information is available to responding companies electronically or through paper forms (**FIGURE 17-2**).

Preincident plans can include a variety of valuable information about a building, including the following:

- Hydrant locations
- Fire department connection (FDC) location
- Knox box location
- Corridor layouts
- Exit locations
- Stairway locations
- Apartment layouts
- Number of bedrooms in apartments
- Locations of handicapped residents' apartments
- Special function rooms or areas

FIGURE 17-2 Preincident plans may be available electronically or through paper forms.

Firefighters should note how the floors of a building are numbered. Buildings constructed on a slope may appear to have a different number of levels when viewed from the "A" or front side than when viewed from the "C" or back side. For example, a fire may appear to be on the third floor to a firefighter standing in front of the building, while a firefighter at the rear of the building might report that it is on the fifth floor. Without knowing how floors are numbered, firefighters can be sent to work above a fire instead of below it. This situation resulted in multiple firefighter deaths in Pittsburgh, Pennsylvania, in 1995 (USAF 1995).

Firefighters should also conduct an exterior search for any missing occupants. An occupant who escaped from the fire may be lying unconscious on the ground nearby or in the care of neighbors, for example. An occupant who jumped from a window could be injured and unable to move.

Visible Smoke and Fire Conditions

The visible smoke and fire conditions can provide clues to the location and intensity of the fire (**FIGURE 17-3**).

FIGURE 17-3 The visible smoke and fire conditions at the scene can help firefighters determine the location and intensity of the fire.

Observe the volume and density of the smoke and the velocity of the smoke movement. More volume, darker color, and greater speed may indicate more severe fire conditions.

Assess the direction of the wind and determine where it may push the fire. Look for the amount of visible flames. Determine the number of openings where flames are visible. All of these factors will help you to determine the location and size of the fire, the potential for growth, and the direction of travel. Careful assessment of these factors will help you to develop a search and rescue plan.

Time of Day and Day of Week

Generally, the life risk in residential occupancies is higher at night and on weekends when more people are at home. Risk is greatest late at night when the occupants are probably asleep. A fire that occurs in a residential occupancy on a weekday afternoon would present a significantly different rescue problem than a fire in the same location at 2:00 AM. An office building that is fully occupied on a weekday afternoon is typically unoccupied or has a low occupancy at midnight. Conversely, a nightclub fire on a Friday or Saturday night could endanger hundreds of lives.

Search Coordination

All incidents handled by the fire service have three priorities. These priorities are, in order:

1. Life safety
2. Incident stabilization
3. Property conservation

The IAP for the incident must focus on the life safety priority as long as search and rescue operations are still underway. The IC makes search assignments and designates a fire officer or team leader to be in charge of the search effort (search officer). As the searchers complete a search of each area, the search officer notifies the IC of the status and results of the search effort. An "all clear" report indicates that an area has been searched and all victims have been removed.

If firefighters who have been assigned to perform some other task, discover a victim, or come upon a critical rescue situation, they must notify the IC immediately. Because life safety is always the highest priority, the IC may send a rescue team to meet the search team so that the search team does not lose the continuity of its search.

Another critical aspect of search coordination is keeping track of everyone who was rescued or who escaped without assistance. This information should be tracked at the incident command post so that reports of missing occupants can be matched to reports of rescued victims.

Search Priorities

A search begins in the areas where victims are at the greatest risk. Search teams must work together closely and coordinate their searches to ensure that all areas are covered. One or two search teams can usually go through all of the rooms in a single-family dwelling in less than 15 minutes. Multiple search teams and a systematic division of the building are needed to cover larger structures such as apartment buildings and high-rise buildings.

Area search assignments should be based on a system of priorities such as the following:

- The first area search priority is to search the areas immediately around the fire where live victims may be located, followed by searching the rest of the fire floor.
- The second area search priority is to search the area directly above the fire and the rest of that floor.
- The third area search priority is higher-level floors, typically working from the top floor down, because smoke and heat are likely to accumulate in these areas.
- Generally, areas below the fire floor are the lowest priority unless there is a danger of collapse.

These priorities are guidelines and need to be modified based on occupancy, construction type, and the circumstances of the incident.

When addressing the third priority of area search, firefighters typically work from the top floor down. High-rise buildings, however, are an exception to this rule due to the impact of the stack effect. **Stack effect** is the vertical airflow within a building caused by the differences in temperature inside and outside the building. Stack effect is discussed more thoroughly in Chapter 16, *Ventilation*. When it is colder outside and the interior is heated, smoke rises quickly through stairways, elevator shafts, and other vertical openings, filling the upper levels of the building. This phenomenon may result in large quantities of products of combustion being present in the upper levels of the building. In the case of high-rise buildings, officers must take stack effect into account and initiate search operations from the highest floor impacted by the fire's by-products. Unaffected floors can be searched after these floors have been swept. Note that the opposite situation can occur on a hot summer day when the interior temperature is much cooler than the outside atmosphere. In this scenario, the heavy cooler air pushes the smoke down the vertical openings, toward a lower-level exit. This could impact search and rescue operations by creating potentially hazardous environments in areas well below the fire in areas that would normally be considered low priority.

Search and Rescue Safety

Search and rescue operations present a high risk to firefighters. During searches, firefighters are exposed to the same hazards that endanger the lives of potential victims. Even though firefighters have the advantages of protective clothing and equipment, training, teamwork, and standard operating procedures (SOPs), they can still be seriously or fatally injured during these operations. Safety is an essential consideration in all search and rescue operations.

Risk Management

Search and rescue situations require a special type of risk management. Although every emergency operation involves a degree of unavoidable, inherent risk, firefighters can encounter situations during search and rescue operations that involve a significantly higher degree of personal risk. The risk involved in conducting search and rescue operations must always be weighed against the probability of finding someone who is savable—that is, who is still alive to be rescued.

The IC is responsible for managing the level of risk during emergency operations. They must perform a risk–benefit analysis to determine which actions will be taken in each situation. Actions that present a high level of risk to the safety of firefighters are justified only if victims are known or believed to be in immediate danger, and if there is a reasonable probability that lives can be saved. In contrast, only a limited risk level is acceptable to save property. The risk–benefit analysis should consider the stage of the fire, the condition of the building, and the presence of any other hazards.

When there is no possibility of saving either lives or property, no risk is acceptable, and search and rescue operations cannot be performed at all. For example, if a building is fully involved in flames, a room has flashed over, or a building demonstrates potential backdraft conditions, the risk to firefighters might be too high, and the possibility of saving any occupants inside the structure might be too low to justify sending a search and rescue team into the building (**FIGURE 17-4**). A similar decision might be made if a fire occurs in an abandoned building or a lightweight construction building in danger of structural collapse. The IC may be able to identify these conditions from the exterior, or they may learn of them from a team assigned to conduct a search. A search team that encounters conditions that make entry impossible should report their findings back to the IC immediately. In these situations, it might not be possible to conduct an interior search until the fire has been extinguished.

A risk–benefit analysis of the scene is often guided by established policies. For example, many fire departments have policies that prohibit firefighters from entering abandoned structures known to be in poor structural condition. This kind of policy reflects the fact that such a building has a high risk of collapse and a low probability that anyone would be inside.

FIGURE 17-4 The IC must weigh the risk to the firefighters against the possibility of saving any occupant before authorizing an interior search.

Courtesy of Joseph Sperber.

To summarize, the IC must always balance the risks involved in an emergency operation with the potential benefits:

- Actions that present a high level of risk to the safety of firefighters are justified only if there is a potential to save lives.
- Only a limited level of risk is acceptable to save valuable property.
- It is not acceptable to risk the safety of firefighters when there is no chance of saving lives or property.

In order to assess the degree of risk and minimize the chance of injuries, it is important that the IC do the following:

- Conduct an **initial size-up**—the rapid evaluation and analysis of an incident to determine which resources need to be deployed and which actions can be undertaken safely, usually conducted by the first-arriving company officer—also called a **windshield size-up**. This action is followed by a thorough **360-degree walk-around**, or simply **360**, to complete the size-up.
- Use a thermal imager (discussed later in this chapter).
- Evaluate the possibility of a ventilation-limited fire, which could lead to a backdraft.
- Practice good door control at the point of entry into the building to limit the supply of oxygen to the fire.
- Consider the use of a transitional attack to cool the fire before starting the search process.

Search and Rescue Tools and Equipment

To perform search and rescue properly, firefighters must have the appropriate tools and equipment, such as the following:

- PPE
- Hand light or flashlight
- Portable radio
- Thermal imager
- Forcible entry (exit) tools
- Hose lines
- Ladders
- Search rope(s)
- A 16- to 24-foot (ft; 5- to 7-meter [m]) piece of tubular webbing or short rope
- Chalk, crayons, felt tip markers, spray paint, or masking tape

As always, full structural firefighting PPE is essential. Each firefighter must use self-contained breathing apparatus (SCBA) and carry a flashlight or hand light and portable radio. Before entering the building, firefighters must verify that their personal alert safety system (PASS) devices are activated.

A **thermal imager** is helpful during search and rescue operations. This tool, which is similar to a video camera, can be used in smoke-filled buildings to locate victims and search for hidden fires during size-up. The use of thermal imaging devices is discussed later in this chapter.

In addition, each firefighter assigned to a search and rescue team should carry the same basic hand tools as the interior attack team, including at least one forcible entry tool, such as an axe, sledgehammer, Halligan tool, or short pike pole. These tools can be used both to open an area for a search and, if necessary, to open an emergency exit path. A hand tool can also be used to extend the firefighter's reach to **sweep**—using your arm or a tool to swipe across areas to feel for victims or objects—for unconscious victims. Some departments advocate that each searcher carry a tool to extend their reach. Others believe that the tool reduces the searcher's ability to feel the texture and composition of an object.

A search team that is working close to the fire should carry a hose line or be accompanied by another team with a hose line. A hose line can protect firefighters and enable them to search a structure more efficiently. This measure is essential when a search team is working close to or directly above the fire. The hose line can be used to knock down the fire, protect a means of egress (stairway or corridor), and protect the victims as they are escaping.

A search team may need to use ground ladders to gain access to a search site that is located on a second or higher floor. Ladders may also be used to rescue victims from dangerous areas if regular exits cannot be used. Using a ground ladder to rescue a trapped occupant is a stressful and demanding task that carries a considerable risk of injury to both firefighters and victims. Proper ladder rescue techniques are discussed later in this chapter.

A **search rope**, also referred to as a **search line**, may be used during search operations to provide searchers with an anchor point that will help them maintain their orientation while searching for victims. Search ropes are especially useful when a search is being conducted in large, open areas. Webbing can be used when it is necessary to drag or carry a victim from a hazardous environment.

Once a room or apartment is searched, searchers should use chalk, felt tip markers, spray paint, or masking tape to mark the door. This marking lets other personnel know that the room has already been searched.

Methods to Determine Whether an Area Is Tenable

Firefighters must make rapid, accurate, and ongoing assessments about the safety of the building while working at a structure fire. When they arrive, firefighters must quickly determine the type of structure involved, the stability of the structure and the possibility of collapse, and the life safety risk involved. Try to determine how long the structure has been on fire. Are there reports of fire showing before your arrival? Consider the possibility that if no fire is showing, this could indicate a ventilation-limited fire condition. A room that contains a ventilation-limited fire may, at first glance, appear to be a small incipient fire, but in reality, it may be a larger fire in a ventilation-limited state. Dark black and turbulent smoke, blackened windows, the appearance of a "breathing" building, soot around windows or doors indicating smoke travel, puffing or little to no smoke showing, and intense heat are signs of possible flashover or backdraft. Cracks in walls, leaning walls, pitched or sagging doors, and any partial collapse are significant indicators of an impending collapse. Additional signs of collapse are discussed in Chapter 6, *Building Construction*.

Even after the decision has been made to enter a burning structure, firefighters must continually reevaluate the safety of the operation. Those working on the search team rely on fire officers, team members, and the IC to notify them of changing fire conditions. Fire officers must be alert for changing conditions inside the building and for any information about changing conditions that are from other parts of the fire scene. The IC may know or see things that firefighters working inside the building might not be aware of, for example, and may call for an evacuation as the fire situation changes. For their part, firefighters in the interior of the building should check and recheck the surfaces they are working on. Floors that have burned through, that are sagging, or that are warm; furniture or carpeting off-gassing; or flame "rollover" should be avoided. A rapid rise in the amount of heat or flame "rollover" may indicate the potential for flashover.

Evaluate the floor before entering the building. Entering a burning building when there is a fire burning beneath you is an act of great risk. Firefighters have been taught to sound the floor in front of them with a tool, such as an axe, Halligan, or pike pole. While this is good practice to make sure that the floor in front of you is intact, sounding the floor is not a valid measurement of structural stability. Relying on this method alone is dangerous and has led numerous firefighters to injuries and deaths. Floors that are supported by lightweight wooden trusses and other manufactured components fail quickly when exposed to fire. The collapse times of unprotected wood floor systems exposed to fire can vary depending on several factors, including the intensity and duration of the fire, the condition of the wood, and the ventilation available. As construction materials change over time, so too must the fire service's approach to residential fires. Research from UL focused on examining the hazards of basement fires involving various residential flooring systems. Floor collapse times ranged from 3:28 minutes to 12:45 minutes, with an average of 11:57 minutes (Kerber et al. 2012, 25). Extreme caution should be used when operating on any floor above a fire.

Firefighters should take actions at a higher level of risk only in imminent, life-threatening situations when immediate action can prevent the loss of life or serious injury. The IC must evaluate the situation and determine whether this level of risk is justified. The IC also must be prepared to explain their decision.

SAFETY TIP

During search and rescue operations, firefighters should follow these guidelines:

- Work from a single plan.
- Maintain radio contact with the IC, both through the chain of command and via portable radios.
- Monitor fire conditions during the search.
- Coordinate ventilation operations with search and rescue activities.
- Adhere to the personnel accountability system.
- Stay with a partner.

Primary Search

A primary search is a quick search of a structure to locate any potential victims who are in danger. This search should be as thorough as time permits and should cover any places where victims are likely to be found. Fire conditions might make it impossible to conduct a primary search in some areas, or they may limit the time available for an exhaustive search. A primary search should be completed in 15 minutes or less.

During the primary search, firefighters rapidly search the accessible areas of a burning structure where conditions may still permit human life to exist. The objective of the primary search is to find any potential victims as quickly as possible and to remove them from danger. When firefighters complete the primary search, they have gone as far as they can and have removed anyone who was rescued. The phrase "primary search is all clear" is used to report that the primary search has been completed.

By necessity, the primary search is conducted quickly and gives priority to those areas where victims are most likely to be located. Time is a critical concern because firefighters must reach potential victims before they are burned, overcome by smoke and toxic gases, or trapped by a structural collapse. Search teams may have only a few minutes to conduct a primary search. In that limited amount of time, firefighters must try to find anyone who could be in danger and remove those individuals to a safe area. Active fire conditions may limit the areas that can be searched quickly as well as the time that can be spent in each area.

Firefighters should try to check all of the areas where victims might be, such as beds, cribs, and sofas. Adults who tried to escape on their own are often found near doors or windows. Some people—particularly children—may try to hide in a closet, under a bed, in a bathtub or shower, or behind a piece of furniture (**FIGURE 17-5**). Be aware that young children may be frightened by the appearance of a firefighter wearing full PPE.

The primary search is frequently conducted in conditions that expose both firefighters and victims to the risks presented by heavy smoke, heat, structural collapse, and entrapment by the fire. Search teams must often work in conditions of zero visibility and may have to crawl along the floor to stay below layers of hot gases. When conducting a primary search in a structure fire, speed is generally prioritized over crawling. Although crawling can be beneficial in certain situations to avoid heat and smoke, it may not be the most efficient approach during a primary search. Although firefighters should crawl during search operations in some circumstances to protect themselves from heat, smoke, and toxic gases, the primary search is a critical operation aimed at quickly locating and rescuing any occupants who may be trapped inside.

Because the beam from a powerful hand light might be visible only a few inches inside a smoke-filled building, search teams must practice searching for victims in total darkness. They must know how to keep track of their location and how to get back to their entry point. Practicing these skills in a controlled environment will enable firefighters to perform confidently under similar conditions at a real emergency incident.

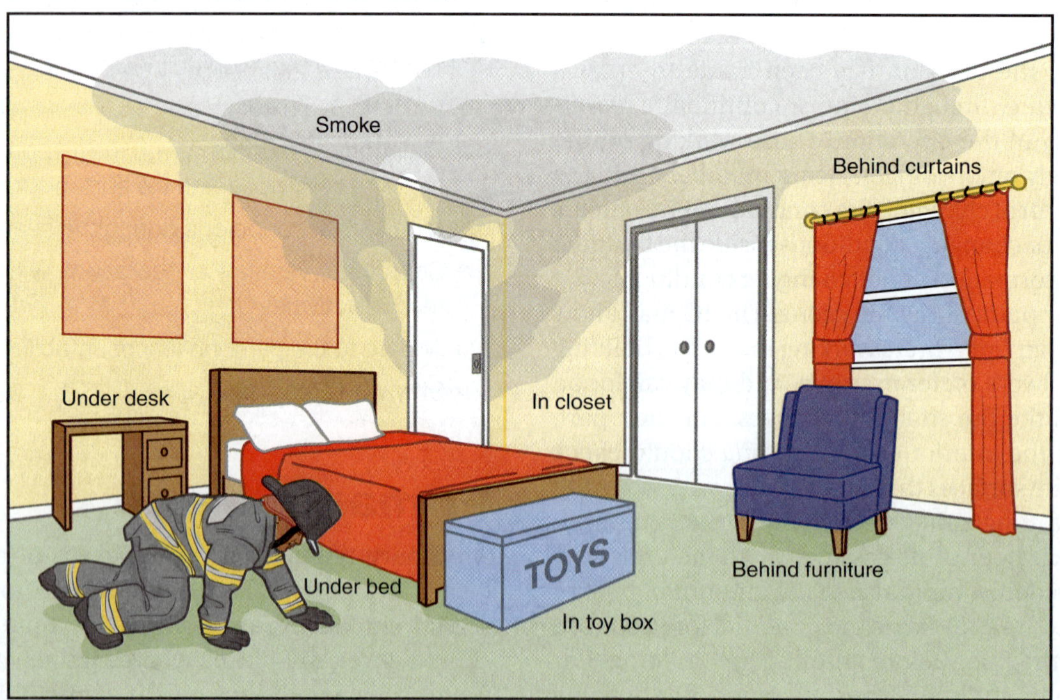

FIGURE 17-5 During a fire, children can become scared and confused. They tend to hide somewhere they feel safe.

Courtesy of Michael Smith.

Firefighters must rely on their senses when they search a building:

- Sight: Can you see anything?
- Sound: Can you hear someone calling for help, moaning, or groaning?
- Touch: Do you feel a victim's body?

If smoke and fire restrict visibility, firefighters must use sound and touch to find victims. Every few seconds a member of the search team should yell out, "Is anyone in here? Can anyone hear me?" Then listen. Hold your breath to quiet your breathing regulator and stop moving. People who have suffered burns or are semiconscious may not be able to speak intelligibly, so you will need to listen for guttural sounds, cries, faint voices, groans, and moans. Focus on the direction of any sounds you hear.

As you search, feel around for the hands, legs, arms, and torsos of potential victims. Use side crawls to extend your reach sideways; sweep ahead and to the sides as you crawl. Practice until you can tell the difference between a piece of furniture and a human body. The type of furniture you encounter may also provide clues for finding victims. Spindle-type legs in a bedroom may indicate a crib that needs to be searched. An unusually low bed may be the bottom berth of a children's bunk bed, indicating that both the bottom bed and top bed need to be searched.

Zero-visibility conditions can be disorienting. In such circumstances, firefighters must follow walls and note turns and doorways to avoid becoming lost. The use of a thermal imager can be beneficial in improving the speed and efficiency of a search. After locating a victim or completing a search of an area, the search team must be able to retrace its path and return to its entry point.

Search teams must also identify a secondary escape route in case fire conditions change and block their planned exit route. Thus, during the search, firefighters should note the locations of stairways, doors, and windows. Always be aware of the nearest exit and an alternate exit. It is also important to know the location of interior doors in the case that firefighters need to isolate themselves from rapid fire growth. If the situation deteriorates rapidly, a window can serve as an emergency exit from a room. Firefighters should keep track of the exterior walls and window locations, even reaching up to feel for the windows if conditions force them to crawl.

The search team must remain in voice contact with one another and in radio contact with the IC or someone outside the building. The search team members must be able to give their location, including the building section and floor, to the IC if they need assistance. If the search team needs a ladder for the rescue, the IC will need to know where the ladder should be placed (which side of the building) and how long of a ladder will be needed (which floor). Firefighters should use standard Incident Command System (ICS) terminology to describe their location over the radio.

SAFETY TIP

While searching for victims, you also need to focus on your safety. Continuously monitor your air supply and the conditions of the fire and the structure. You should also be aware of the status of other crews, such as the fire attack crew and the ventilation crew.

General Search Techniques

Firefighters should employ standard techniques to search assigned areas quickly, efficiently, and safely. They should operate in teams of two or more and should always stay close together (**FIGURE 17-6**). Partners must remain in direct visual, voice, or physical contact with each other.

As discussed previously, at least one member of each search team should also have a radio to maintain contact with the command post or someone outside the building. The search team can use this radio to call for help if they become disoriented, trapped by fire, or need assistance. If the search team finds a victim, they can notify the IC so that help will be available to remove the victim from the building and begin medical treatment. The search team must notify the IC when the search of each area has been completed so the IC can make informed decisions about which steps to take next.

Remember, the primary search is a quick attempt to locate any potential victims who are in danger. The primary search should be efficient and as thorough as possible. It should cover any places that can be safely searched where a victim might be located. Fire conditions may make it impossible or unsafe to search in some parts of a building. If it is safe to do so, the search team usually starts a search walking. However, if conditions require it, they crawl instead. Changing fire conditions will also limit the time available for a primary search.

Team members must keep themselves safe. They need to monitor changes in the fire conditions to make sure their position is not in danger of an immediate flashover and to monitor changes in the smoke level and characteristics, the heat level, and the location of the flames. Possible indications of structural changes, such as unusual sounds or sagging, must be noted and

FIGURE 17-6 Search teams always include at least two members.

reported to the IC. They also need to carefully monitor their air supplies to ensure they know when to exit in order to be out of the IDLH zone before their end-of-service-time indicator (EOSTI) sounds.

Changes in fire suppression activities are important, too. The search team needs to know if the fire attack team is making progress or if they are having difficulty in establishing a water supply or advancing the attack line. The search team members must have a plan for conducting a search that will maximize their efforts. Perhaps most importantly, they need to keep track of the progress they are making in completing the search.

Using Thermal Imagers

A thermal imager is a valuable tool for conducting a search in a smoke-filled building. This piece of equipment is similar to a video camera, except that it displays images of heat emitted from surfaces instead of visible light images. The images appear on a display screen and show the relative amounts of heat being radiated by different objects. A major benefit of using a thermal imager is that it can "see" the heat signature of a person in conditions of total darkness or through smoke that totally obscures normal vision (**FIGURE 17-7**). Firefighters can identify the shape of a human body with a thermal image scan because the body will be either warmer or cooler than its surroundings. It is important that you learn how to properly set any controls on your thermal imager and practice operating it in order to be proficient in its use during an emergency situation.

FIGURE 17-7 A thermal imager device can capture an image of a person through thick smoke.

A thermal imager can show furniture, walls, doorways, and windows. This information may enable a firefighter to navigate through the interior of a smoke-filled building. A thermal imager can also sometimes be used to locate a fire in a smoke-filled building or behind doors, walls, or ceilings.

One of the limitations of thermal imagers is that the user of the device is the only person who can see the images. The person operating the thermal imaging device must communicate the information seen on the camera screen to the search team members. Additionally, firefighters can find themselves in a difficult situation if the thermal imager device fails and leaves the team members without this special set of eyes. It is important to use good search methods and to encourage all team members to remain aware of their surroundings despite the use of a thermal imager.

Remember that the thermal imager does not "see" fire. It measures the **heat flux**—the heat that is given off by an object—and shows differences in the amount of heat given off by different objects or from different parts of the room. By itself, it does not show you the integrity of a structure. Be alert for areas that show small heat

FIGURE 17-8 Some SCBA include an integrated thermal imager.

Courtesy of MSA.

signatures (areas of higher temperature), which may indicate that a floor is not safe to walk on or an area of a roof that is unsafe to be under. If the thermal imager gives any indication of an unsafe condition, make sure that a thorough exterior size-up has been conducted.

Several types of thermal imagers are available. Most are small, hand-held devices. Some manufacturers integrate thermal imagers directly into the SCBA (**FIGURE 17-8**). There are also helmet-mounted devices that produce an image in front of the facepiece of the user's SCBA. Each type of thermal imager has different controls and functions, and the images may be displayed in a variety of ways. It is important to carefully read and understand the instructions for the thermal imagers used by your department. Firefighters need training and practice to become proficient in using these devices, interpreting the images, and maneuvering confidently through a building while using a thermal imager. Regular equipment checks are required to ensure adequate battery life and proper operation.

Primary Search Methods

There are several methods of conducting a primary search, including the following:

- Standard search method
- Oriented search method
- Oriented-vent-enter-isolate-search (O-VEIS) method
- Team search method

It is important to understand how each method is conducted and under what circumstances each method is most appropriately used. All four methods require a thorough training program that includes sufficient practice in order for you to perform efficient and effective searches. You will need regular practice in simulated fire conditions to maintain proficiency.

Standard Search Method

The **standard search method** is conducted with a search team of two or more members. All members of the search team cover the same area at the same time. The standard search method is the most commonly used primary search method. The main use of the standard search method is in residential fires.

The search pattern is dictated by the size of the room, scene conditions, the potential number of victims, and the number of available team members. Searchers conduct a quick and systematic search by following an outside wall of the room using a left-handed or a right-handed search pattern. The firefighter closest to the wall keeps in contact with the wall at all times (**FIGURE 17-9**). Firefighters must follow walls and note turns and doorways to avoid becoming lost. Searchers should use the most efficient movement based on the hazard encountered and should maintain team integrity during the search by using visual, voice, or direct contact. Depending on the department's procedures, searchers can mark the door as searched or verbally report their progress to the IC after completing the search.

Not all rooms have just four walls; some have five or more. To determine how many walls a room has, check for walls on each side of the door as you enter the room. If there is a wall parallel to the door, on the right of the door, and to the left of the door, then you will need to search five wall surfaces. If there is a parallel wall only to the right or to the left of the door, then the room has four walls. Noting the number of walls in a room about to be searched is a good way to ensure that the entire room has been searched and to maintain orientation.

If the room is large, it may be necessary for searchers to perform side crawls to check the center parts of the room. To do this, one searcher stays close to the wall at all times, and a partner may move closer to the center of the room with two or three side crawls.

When search and rescue efforts are coordinated with fire suppression and ventilation operations, the operational plan may include closing the door to the room as it is being searched. When properly coordinated, this action

FIGURE 17-9 When conducting a search using the standard search method, follow an outside wall using either a left-handed or right-handed search pattern.

Courtesy of Chris Rimm.

can reduce smoke and heat from entering the room and allows the firefighters to open a window to clear smoke from the room if needed. After completing the search, the searchers should close any open windows before opening the door of the room, reentering the hallway, closing the door behind them, and proceeding to the next room. The use of this technique will depend on department procedures and the overarching operational plan.

A primary search using the standard search method keeps searchers together and in voice contact with each other. A search officer or team leader may not be available to help monitor the searchers' air supply and safety, the stability of the structure, and other fireground operations. The searchers must be careful to pay attention to their surroundings and watch out for deteriorating conditions. If a victim is located, the searchers must exit with that victim. For this reason, the continuity of the search is more likely to be lost than if using the oriented search method.

Specific safety requirements for search and rescue operations are defined in NFPA 1550, *Standard for Emergency Responder Health and Safety, 2024 Edition*, and in regulations enforced by the Occupational Safety and Health Administration (OSHA). The NFPA requirements state that if only one search crew is operating in the hazardous area at the initial stages of a structural fire, a minimum of four firefighters are required. Two of those firefighters work as a crew in the hazardous area, and two firefighters in full PPE and SCBA stand by outside the hazardous area ready to assist or rescue the two firefighters in the hazardous area. This policy is sometimes called the **two-in/two-out rule**.

Follow the steps in **SKILL DRILL 17-1** to perform a primary search of a residential structure and rescue victims using the standard search method.

Oriented Search Method

The **oriented search method** is conducted with a search team consisting of an officer or team leader and one to three searchers. The officer or team leader remains outside the individual rooms that are being searched, and a single searcher or multiple searchers systematically search one room at a time using the right-handed or left-handed search method. The officer or team leader should have a thermal imager and is responsible for the following:

- Maintaining the searchers' safety
- Monitoring their air supplies
- Monitoring the progression of the fire and the fire suppression efforts
- Developing a systematic search plan
- Keeping track of which rooms have been searched and which are yet to be searched
- Assessing the progress of the search effort
- Monitoring the activities on the rest of the fire scene
- Coordinating activities with the IC

This method of search should only be conducted if approved by your department.

By staying outside the rooms being searched, the officer is in a better position to note changes than the search team who is searching inside the rooms. This provides a greater degree of safety for the search team and allows the searchers to focus on the search itself. The officer must maintain voice contact with each member of the search team. The officer controls the door and monitors the means of exit for the search team. If possible, there should be a primary exit and a secondary exit that can be used if the first exit becomes untenable.

Depending on department procedures, searched rooms or apartments can be marked so other personnel will know that they have been searched. Chalk, crayons, felt tip markers, spray paint, or masking tape can be used to mark the door for this purpose. Some fire departments use a two-part marking system to indicate when a search

SKILL DRILL 17-1

Conducting a Primary Search Using the Standard Search Method
Firefighter I, NFPA 1010: 6.3.9

1. Don your PPE, including SCBA, and enter the personnel accountability system. Bring forcible entry tools, a flashlight, a radio, a thermal imager, and search ropes if indicated. Notify the IC that the search is starting and indicate the area to be searched and the direction of the search. Use hand tools or ground ladders if needed to gain access to the site. Conduct a quick and systematic search by staying on an outside wall and searching from room to room. Maintain contact with an outside wall.

2. Maintain crew integrity using visual, voice, or direct contact. Use the most efficient movement based on the hazard encountered: duck walk, crawl, stand only when you can see your feet and it is not hot. Use tools to extend your reach if recommended by your department. Clear each room visually or by touch, and then close the door. Search the area, including stairs up to the landing on the next floor.

3. Periodically listen for victims and sounds of fire. Communicate the locations of doors, windows, and inside corners to other team members. Observe fire, smoke, and heat conditions and update the IC. If victims are located, remove them and notify the IC. When the search is complete, conduct a personnel accountability report. Report the results of the search to your officer.

is in progress and when it has been completed. In this system, a slash (/) indicates that a search is in progress, and an *X* indicates that the search has been completed (**FIGURE 17-10**). Other fire departments place an object in the doorway or attach a tag or latch strap to the doorknob to indicate that the room has been searched.

The continuity of the search can be easily lost when a victim is found and removed from the hazardous environment by a single search team. By the time a second search team has been assembled and arrives at the location where the victim was found to continue the search, it is hard to pick up the search without duplication of efforts and wasting valuable time. In addition, when switching from one search team to another, it is easy to miss areas that need to be searched. With the oriented search method, the most efficient way to maintain the continuity of the search is for the officer or team leader to notify the IC that a victim has been found. The search team then moves toward the exit with the victim, where they are met by a second team who removes the victim the rest of the way out of the building. This approach enables the search team to return to the location they were searching when they found the victim and continue the search for additional victims without losing much time and without losing the continuity of the search. This operation requires planning, a lot of practice, and a well-coordinated fire suppression operation.

Follow the steps in **SKILL DRILL 17-2** to conduct a primary search using the oriented search method. The oriented search method takes less time than using the standard search method.

FIGURE 17-10 In large buildings, doors should be marked after the rooms are searched.

SKILL DRILL 17-2

Conducting a Primary Search Using the Oriented Search Method
Firefighter I, NFPA 1010: 6.3.9

1. A search team consisting of one officer or team leader and one to three searchers is assembled. Searchers don PPE, including SCBA and forcible entry tools, and enter the personnel accountability system. The officer or team leader should be equipped with a thermal imager and a way to mark the doors of rooms that have been searched.

2. The officer or team leader notifies the IC that the search is starting and directs the searchers to the area to be searched.

SKILL DRILL 17-2 CONTINUED

Conducting a Primary Search Using the Oriented Search Method
Firefighter I, NFPA 1010: 6.3.9

3. The officer or team leader remains outside the rooms to be searched to monitor safety conditions, air supplies, and the status of the fire. The officer or team leader maintains a systematic search pattern and coordinates activities with the IC.

4. Searchers use a left-handed or a right-handed search pattern and perform two to three side crawls as necessary to extend the search toward the center of the room. Upon completion of the search of each room, the officer directs searchers to the next room to be searched, closes the door of the room that was just searched, and then marks the door to indicate that it has been searched according to department procedures. If a victim is found, the officer notifies the IC and requests a second team to help remove the victim. The search team moves the victim toward the exit and turns over the care and removal of the victim to the second team.

5. The search team then returns to the last location searched and continues the systematic search of the building. When the search is complete, the officer or team leader conducts a personnel accountability report and reports the results of the search and the personnel accountability report to the IC.

Oriented-Vent-Enter-Isolate-Search (O-VEIS) Method

The original vent-enter-search (VES) method of conducting a primary search was developed for situations in which a porch is located in front of a bedroom window, and a person in the bedroom needs to be rescued. VES is a dangerous method and should not be used. This method of primary search was intended to be used in extreme situations in which there is a shortage of personnel. One firefighter can quickly place a ladder to the porch roof, quickly open or break the bedroom window, enter the bedroom, perform a quick search of the room, and exit back onto the porch roof. If it is possible to access a second bedroom from the same porch, this method allows for a search to be conducted in two bedrooms in a short period of time. This evolution is dangerous for several reasons. Opening a bedroom window is a type of ventilation that adds oxygen to the fire, contributing to a flashover or directing the hot fire gases toward the open window. Additionally, the VES method violates the two-in/two-out rule, and because it is often done before any hose line is in place, the fire may expand rapidly.

The **oriented-vent-enter-isolate-search (O-VEIS) method** is conducted using a search team consisting of an officer or team leader and one searcher. The team places a ladder in front of the window leading to the room or to the porch in front of the room to be searched. Both team members climb to the window or porch roof and assess the situation. If they determine that the room seems tenable for a victim and for the firefighters:

- They vent—that is, they open or remove the window.
- The searcher enters the room. The officer or team leader remains outside—on the porch roof or, if no porch is present, on the ladder—to continue to assess the situation in the room, maintain radio contact, and monitor activities in other parts of the fire ground (**FIGURE 17-11**).
- The searcher isolates the room from the flow path by locating the door and closing it as quickly as possible. Even a thin hollow core door will provide some protection from the flow path for a short period of time. This isolation also reduces the heat and smoke in that room for a few minutes and decreases the chance that the room will quickly become untenable for the rescuer and any victims present. In addition, it prevents unplanned ventilation, which could rapidly accelerate the fire growth, it improves visibility in the bedroom, and it reduces the chance of flashover.
- The searcher quickly searches the room.

FIGURE 17-11 In the oriented-vent-enter-isolate-search (O-VEIS) method of primary search, the search officer remains outside to monitor the fire and safety conditions and to assist the inside searcher.

If a victim is found, the officer or team leader can assist in removing the victim to the porch roof or ladder. If a victim is found and there is another bedroom to be searched from the same porch, it may be more efficient for the officer or team leader to request a second team to remove the first victim from the porch roof or ladder so the original search team can begin to search the second bedroom.

A primary search using the O-VEIS method should be considered only in dire emergencies in which a sizeable risk has a large potential benefit. It should not be attempted if the room is fully charged with smoke and in danger of flashing over when additional oxygen is introduced.

Team Search Method

The **team search method**, also known as the **rope search method**, of conducting a primary search is conducted by a team consisting of an officer or team leader who acts as the anchor person and two or more searchers. The anchor person is responsible for securely holding and managing the rope that connects to the search team or teams performing the search. The anchor person ties off a search rope to a stable, stationary object.

If you anchor the search rope to a stationary object, you can do this either outside or inside the building.

Anchoring the search rope outside of the building provides a secure point of attachment without impeding the movement of the search team inside the building. Anchoring the search rope inside the building may limit the movement and range of the search team because they would be tethered to a fixed location within the structure. If you anchor the rope outside of the building, locate a stationary object about 10 ft (3 m) outside the exit door. An anchor point at this distance is less likely to be affected by the potential hazards inside the structure, such as collapsing walls or falling debris, and won't obstruct the search team as they exit the building. If you need to anchor the rope inside the structure, tie it to a stable, stationary object within the structure, such as a column, beam, or a piece of heavy furniture.

Tying the rope to a stationary object provides a secure point of attachment and helps maintain tension in the rope. If there are no suitable stationary objects, a member of the search team may use specialized anchoring devices carried on themselves or on the apparatus, such as mechanical anchors or anchor straps. In some cases, the anchor person may use their own body as the anchor point by tying the end of the search rope around their waist or holding the end securely. Whether it is better to tie the rope to an object or be the anchor yourself depends on various factors, including the specific circumstances of the search and the resources available. Tying the rope to a stable, stationary object outside the building can provide a more secure anchor point, especially if there are no suitable objects inside the structure or if the hazards inside make it impractical to use them as anchor points. This method allows the anchor person to focus on managing the rope and supporting the search team's efforts. However, if there are no suitable external anchor points or if the search team is equipped with specialized anchoring devices, the anchor person can be the anchor themselves. This allows for greater flexibility and adaptability during the search, as the anchor person can move along with the search team and adjust their position as needed.

As the searchers progress through the building, the anchor person feeds the line out while keeping slight tension on it at a comfortable height about 18 to 20 inches (in.; 0.5 m) above the floor. This method gives the searchers an identifiable egress route out of the building if the direction of travel inside the building needs to be changed. Additionally, the anchor person can let the searchers know how far they have progressed into the building.

The team search method of conducting a primary search is more complicated than the other three methods discussed and requires a lot of training and coordination. It is also time-consuming. It is, however, well-suited for locating and removing lost occupants or injured firefighters. It is especially valuable for a rapid intervention crew/company (RIC) working to locate a missed, downed, or trapped firefighter in a large space. Its use also enables other firefighters to locate the search team quickly. The team search method is not often used in a residential fire unless the house is large or has large open spaces. It can be a valuable method to locate an occupant in a large warehouse or other large open structure when visibility is severely limited.

The team search method of conducting a primary search also helps firefighters keep in contact with the exit in a building that contains cold smoke. **Cold smoke** is generated when a fire sprinkler system is activated during a fire. The sprinkler cools the smoke and causes it to bank down to the floor. Cold smoke and condensed water droplets severely limit visibility.

With the team search method, the officer or team leader is responsible for developing the search plan, keeping the team safe, and communicating with the IC. The officer or team leader stays in contact with the search rope and maintains voice contact with the searchers to help keep the searchers oriented. The officer or team leader also ensures that each member is monitoring their air supply. In addition, the officer or team leader is usually responsible for operating the thermal imager and directing the team based on the images seen on the screen. They must also assess any changing conditions around the search team and communicate these changes to the IC.

The members of the search team are responsible for following the search rope into the building and conducting the primary search off the rope. This is done by placing one searcher on each side of the search rope. When searching large areas, each of the searchers performs two or three side crawls at a right angle to the rope. The searcher on the left side of the rope side crawls to the left of the rope, and the searcher on the right side of the rope side crawls to the right of the rope. Some departments use a short lateral rope that clips to the main rope called a **tether line** or **tag line**. This enables them to move a limited distance from the main search rope without losing contact with the rope. When the search team is ready to exit, the officer or team leader assumes the anchor position on the search line and maintains tension on the line as the team works its way along the line to the exit.

The team search method keeps searchers together, preventing individuals from getting lost or separated. It also keeps searchers oriented to the exit and provides a safe, quick, and direct route out of the building. This method is useful when searching large areas, especially where there are limited landmarks and limited visibility, to provide directional guidance. It also allows a rescue team to find the searchers if they need to enter to remove a victim who has been located by the search team.

However, because the team search method relies solely on the search rope, the search team spends more time holding on to the rope than engaging in active searching. It is also hard to maintain the continuity of the search if a victim is found and removed. If the search team locates a victim and leaves to remove the victim, it is hard for a new crew to return to the same place to continue a systematic search.

When a rope is used as a search tool, knots are sometimes placed every 20 ft (6 m) to enable searchers to know how far they have progressed into the building. The number of knots indicates the number of feet from the end of the rope. For example, four knots equals 80 ft (24 m) of travel into the building.

Some fire departments prefer to use a fire hose in place of a search rope because it is larger, always available, and easier to locate when visibility is limited. In cases of emergency, a firefighter can determine which direction leads to the exit by identifying the male and female couplings on the hose—the male coupling leads back to the fire engine and safety. This is commonly referred to as "long lug out" because the lug on the male coupling is longer them the lug on the female coupling.

Secondary Search

The purpose of the secondary search is to locate any occupants or victims who are unaccounted for and were not found and removed during the primary search. A secondary search is conducted as soon as the fire is under control or fully extinguished and there are sufficient resources available (**FIGURE 17-12**). At this point, any remaining occupants may have been exposed to conditions that cannot support life, and it is likely a recovery effort. If possible, it should be conducted by a different search team so that each area of the building is examined with a fresh set of eyes to lessen the possibility that a victim will be missed a second time. The secondary search is conducted slowly and methodically to ensure that no areas are overlooked. Areas that could not be covered during the primary search are covered during the secondary search.

FIGURE 17-12 A secondary search is conducted after the fire is under control.

The time needed to complete a secondary search depends on the building size, the type of occupancy, the type of construction, the fire severity, the amount of contents or clutter in the building, and the probability of collapse.

There are fewer risks to firefighters during the secondary search than during the primary search. Once the fire is under control, conditions during the secondary search should be much different. Ventilation should have cleared out most of the smoke, and visibility should be greatly improved. Lights can be set up to make it easier to see the area being searched. It is important to remain alert for conditions indicating an unstable structure that may result in a collapse. In addition, firefighters conducting a secondary search should watch for signs of a fire **rekindle**—a return to flaming combustion after apparent but incomplete extinguishment.

The secondary search team needs to wear proper PPE, including SCBA, helmets and gloves, and structural firefighting protective gear. They should continue to use SCBA until the air quality has been thoroughly tested. The air inside a structure where a fire occurred can look clear but can contain high concentrations of poisonous gases and small suspended particles of toxic by-products of the fire. Approximately 97 percent of these particulates are invisible to the human eye. In addition, the by-products of combustion (soot, smoke, and ash) contain cancer-causing substances that can be absorbed through the skin. Air monitoring is discussed in Chapter 18, *Salvage and Overhaul.*

Like the primary search, the secondary search should be conducted in an orderly, systematic manner to ensure that all areas are searched. It may be most efficient to start the secondary search in the areas less affected by the fire. This is especially true when overhaul is still occurring in the burned parts of the building. The search team can then systematically work their way toward the area affected by the fire. In other instances, it is important to begin the secondary search in the area affected by the fire—for example, when the fire is small or when you have a strong reason to believe a victim is located close to the fire.

Pay special attention to any witnesses, especially if they have knowledge of the building occupants. Be methodical and coordinated. Move furniture away from walls, and carefully look under beds, furniture, boxes, and clothes that might hide a child or an adult. Be especially thorough in checking all places where a child could hide, including closets, bathtubs, shower enclosures, behind doors, under windows, under beds, and inside toy chests. If debris is present from the fire, remove it to thoroughly search all areas of the room.

When searching a large commercial or residential building, the secondary search should include the floor above the fire, the floor below the fire, stairwells, elevator shafts, the roof, the outside perimeter of the building, under trees and shrubs around the building, and even the rooftops of exposed buildings. As you are searching, be alert for fire rekindle and any signs of the origin of the fire. These need to be reported to the IC. When the secondary search has been completed, the search officer or team leader should notify the IC by reporting "secondary search is clear."

SAFETY TIP

Even though the fire is under control, safety remains a top consideration during a secondary search. SCBA should be used until the air is tested and deemed safe to breathe because levels of carbon monoxide and other poisonous gases often remain high for a long time after a fire is extinguished.

Follow the steps in **SKILL DRILL 17-3** to perform a secondary search.

SKILL DRILL 17-3

Performing a Secondary Search Firefighter I, NFPA 1010: 6.3.9

1. Don PPE, including SCBA, and enter the personnel accountability system. Bring hand tools, a flashlight, a radio, a thermal imaging device, and search ropes if indicated. Notify command that the search is starting and indicate the area to be searched and the direction of the search.

Continues.

SKILL DRILL 17-3 CONTINUED

Performing a Secondary Search Firefighter I, NFPA 1010: 6.3.9

2. Conduct a systematic, thorough search for any possible victims. In addition to checking for victims, be alert for hidden fire; salvage needs; and indicators of the fire's cause, its origin, or arson. Protect any evidence. Maintain crew integrity using visual, voice, or direct contact. Notify the IC of the results of the search and personnel accountability report.

Rescue Techniques

Rescue is the second component of search and rescue. When you locate a victim during a search, you must direct, assist, or carry that person to a safe area. Rescue can be as simple as verbally directing an occupant toward an exit, or it can be as demanding as extricating a trapped, unconscious victim and physically carrying that person out of the building. The term *rescue* is generally applied to situations where the rescuer physically assists or removes the victim from the dangerous area.

Most people who realize that they are in a dangerous situation will attempt to escape on their own. Children and elderly persons, physically or developmentally challenged persons, impaired people, and ill or injured persons, however, may be unable to escape and will need to be rescued. People who are sleeping, are under the influence of alcohol or drugs, or have diminished mental capacity may not become aware of the danger in time to escape from the structure on their own. Toxic gases may incapacitate even healthy individuals before they can reach an exit. Victims also may become trapped when a rapidly spreading fire, an explosion, or a structural collapse cuts off potential escape routes.

Because fires threaten the lives of both victims and rescuers, the first priority is to remove the victim from the fire building or dangerous area as quickly as possible. It is usually better to move the victim to a safe area first and then provide any necessary medical treatment. Always use the safest and most practical means of egress when removing a victim from a dangerous area. A building's normal exit system, such as interior corridors and stairways, should be used if it is open and safe. If the regular exits cannot be used, an outside fire escape, a ladder, or some other method of egress must be found. Ladder rescues can be both difficult and dangerous, whether the victim is conscious and physically fit or is unconscious and injured (**FIGURE 17-13**).

As a firefighter, you must learn and practice the various methods of rescuing people from fires. After mastering these techniques, you will be able to rescue victims from life-threatening situations. Firefighters rescue people not only from fires but also from a wide variety of accidents and mishaps. Although the techniques explained in this section refer primarily to rescuing occupants from burning buildings, many of the techniques discussed here can be used in other situations. Note that the assists, lifts, and carries described in this section should not be used if you suspect that the victim has a spinal injury unless there is no other way to remove the person from the life-threatening situation.

FIGURE 17-13 Bringing an unconscious victim or a person who needs assistance down a ladder can be difficult and dangerous.

Exit Assist

An **assist** is a rescue technique in which the victim is responsive and able to walk but needs assistance. The simplest rescue is the exit assist, in which the victim is responsive and able to walk without assistance or with little assistance. Even if the victim can walk without assistance, the firefighter should take the person's arm or use the one-person walking assist to make sure that the victim does not fall or become separated from the rescuer. The firefighter simply guides the person to safety or provides a minimal level of physical support. There are two types of exit assists: the one-person walking assist and the two-person walking assist.

One-Person Walking Assist

The one-person walking assist can be used if the person is capable of walking. To perform a one-person walking assist, follow the steps in **SKILL DRILL 17-4**.

Two-Person Walking Assist

The two-person walking assist is useful if the victim cannot stand and bear weight without assistance. To perform this assist technique, the two rescuers completely support the victim's weight. It may be difficult to walk through doorways or narrow passages using this type of assist. To perform a two-person walking assist, follow the steps in **SKILL DRILL 17-5**.

Simple Victim Carries

A **carry** is a technique that moves a victim who is conscious and responsive but incapable of standing or walking. There are four simple carry techniques:

- Two-person extremity carry
- Two-person seat carry
- Two-person chair carry
- Cradle-in-arms carry

SKILL DRILL 17-4

Performing a One-Person Walking Assist Firefighter I, NFPA 1010: 6.3.9

1. Help the victim stand next to you, facing the same direction.

2. Have the victim place their arm around your neck, and hold on to the victim's wrist, which should be draped over your shoulder. Put your free arm around the victim's waist, and then help the victim to walk.

SKILL DRILL 17-5

Performing a Two-Person Walking Assist Firefighter I, NFPA 1010: 6.3.9

1. Two firefighters stand facing the victim, one on each side of the victim. Both firefighters assist the victim to a standing position.

2. Once the victim is fully upright, place the victim's right arm around the neck of the firefighter on the right side. Place the victim's left arm around the neck of the firefighter on the left side. The victim's arms should drape over the firefighters' shoulders. The firefighters hold the victim's wrist in one hand.

3. Both firefighters put their free arms around the victim's waist for added support, and then slowly assist the victim to walk. Firefighters must coordinate their movements and move slowly.

Two-Person Extremity Carry

The two-person extremity carry requires no equipment and can be performed in tight or narrow spaces, such as the corridors of mobile homes, small hallways, and narrow spaces between buildings. The focus of this carry is on the victim's extremities. To perform a two-person extremity carry, follow the steps in **SKILL DRILL 17-6**.

SKILL DRILL 17-6

Performing a Two-Person Extremity Carry Firefighter I, NFPA 1010: 6.3.9

1. Two firefighters help the victim to sit up.

2. The first firefighter kneels behind the victim, reaches under the victim's arms, and grasps the victim's wrists.

3. The second firefighter backs in between the victim's legs, reaches around, and grasps the victim behind the knees.

4. The first firefighter gives the command to stand and carry the victim away, walking straight ahead. The firefighters must coordinate their movements.

Two-Person Seat Carry

The two-person seat carry is used with victims who are disabled or paralyzed. This type of carry requires the assistance of two firefighters, and moving through doors and down stairs may be difficult. To perform a two-person seat carry, follow the steps in **SKILL DRILL 17-7**.

SKILL DRILL 17-7

Performing a Two-Person Seat Carry Firefighter I, NFPA 1010: 6.3.9

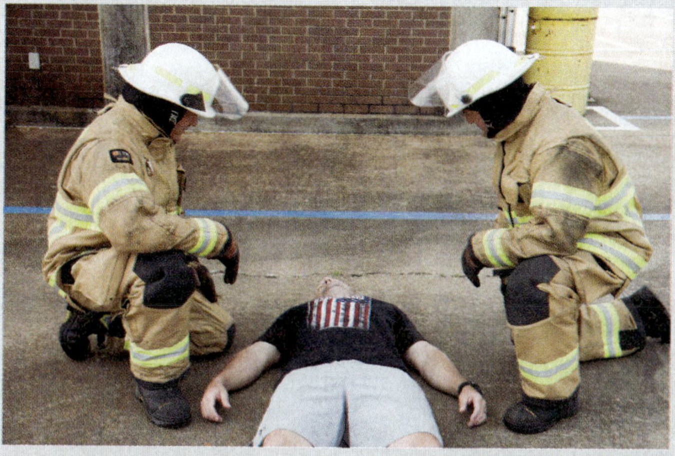

1. Two firefighters kneel near the victim's hips, one on each side of the victim.

2. Both firefighters raise the victim to a sitting position and link arms behind the victim's back.

3. The firefighters place their free arms under the victim's knees. If possible, the victim puts their arms around the firefighters' necks for additional support. The rescuers stand and move toward safety.

Two-Person Chair Carry

The two-person chair carry is particularly suitable when a victim must be carried through doorways, along narrow corridors, or up or down stairs. In this technique, two rescuers use a chair to transport the victim. The chair must be strong enough to support the weight of the victim while they are being carried. A folding chair should not be used for this purpose because it could collapse while carrying a victim. The victim should feel much more secure with this carry than with the two-person seat carry, and they should be encouraged to hold on to the chair. To perform a two-person chair carry, follow the steps in **SKILL DRILL 17-8**.

SKILL DRILL 17-8

Performing a Two-Person Chair Carry Firefighter I, NFPA 1010: 6.3.9

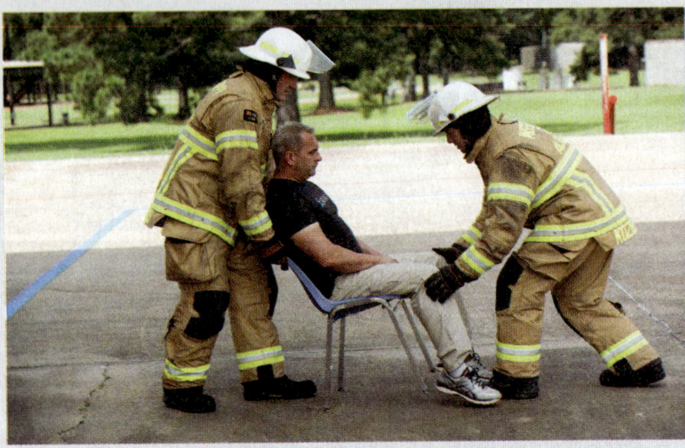

1. Tie the victim's hands together, or have the victim clasp their hands together or hold onto the chair. This prevents the victim from reaching for a stationary object while you are moving them. One firefighter stands behind the seated victim, reaches down, and grasps the back of the chair.

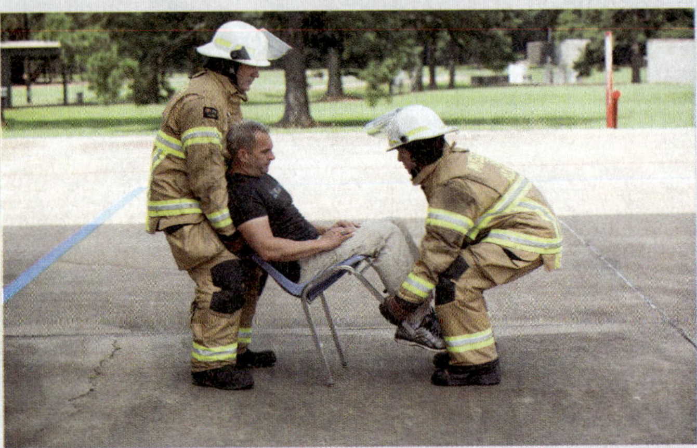

2. The firefighter behind the victim tilts the chair slightly backward on its rear legs so that the second firefighter can step back between the legs of the chair and grasp the tips of the chair's front legs. The victim's legs should be between the legs of the chair.

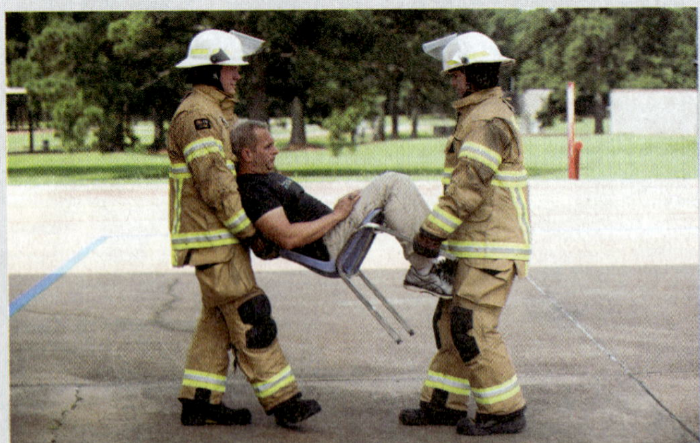

3. When both firefighters are correctly positioned, the firefighter behind the chair gives the command to lift and walk away. Because the chair carry may force the victim's head forward, watch the victim for airway problems.

Cradle-in-Arms Carry

The cradle-in-arms carry can be used by one firefighter to carry a child or a small adult. With this technique, the firefighter should be careful of the victim's head when moving through doorways or down stairs. To perform the cradle-in-arms carry, follow the steps in **SKILL DRILL 17-9**.

SKILL DRILL 17-9

Performing a Cradle-in-Arms Carry Firefighter I, NFPA 1010: 6.3.9

1. Kneel beside the child, and place one arm around the child's back and the other arm under the thighs.

2. Lift slightly and roll the child into the hollow formed by your arms and chest.

3. Be sure to use your leg muscles to stand.

Emergency Drags

A drag is a rescue technique that moves an unconscious victim from a dangerous location. The most efficient method of removing an unconscious or unresponsive victim from a dangerous location is a drag. There are six types of emergency drags that can be used to remove unresponsive victims from a hazardous situation:

- Clothes drag
- Blanket drag
- Standing drag
- Webbing sling drag
- Firefighter drag
- Emergency drag from a vehicle

When using an emergency drag, the rescuer should make every effort to pull the victim in line with the long axis of the body to provide as much spinal protection as possible. The victim should be moved head first to protect the head.

Clothes Drag

The clothes drag is used to move a victim who is on the floor or the ground and is too heavy for one rescuer to lift and carry alone. In this technique, the rescuer drags the person by pulling on the clothing in the neck and shoulder area. The rescuer grasps the clothes just behind the collar, using their arms to support the victim's head, and drags the victim away from danger. To perform the clothes drag, follow the steps in **SKILL DRILL 17-10**.

SKILL DRILL 17-10

Performing a Clothes Drag Firefighter I, NFPA 1010: 6.3.9

1. Crouch behind the victim's head and grab the shirt or jacket around the collar and shoulder area. Support the head with your arms.

2. Lift with your legs until you are fully upright. Walk backward, dragging the victim to safety.

Blanket Drag

The blanket drag is used to move a victim who is not dressed or who is dressed in clothing that is too flimsy for the clothes drag (for example, a nightgown). This procedure requires the use of a large sheet, blanket, curtain, or rug. Place the item on the floor, roll the victim onto it, and then pull the victim to safety by dragging the sheet or blanket. To perform a blanket drag, follow the steps in **SKILL DRILL 17-11**.

Standing Drag

The standing drag can be used to move a victim who is unconscious or disoriented. To perform the standing drag, follow the steps in **SKILL DRILL 17-12**.

SKILL DRILL 17-11

Performing a Blanket Drag Firefighter I, NFPA 1010: 6.3.9

1. Lay the victim supine (face up) on the ground. Stretch out the material you are using next to the victim.

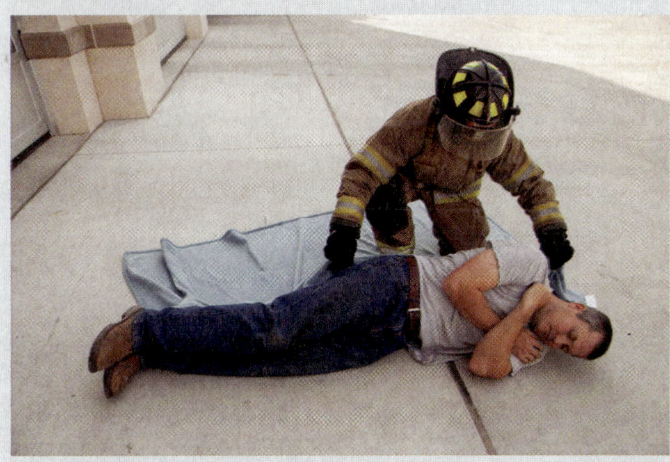

2. Roll the victim onto one side. Neatly bunch one-third of the material against the victim's body so the victim will lie approximately in the middle of the material.

3. Lay the victim back down (supine). Pull the bunched material out from underneath the victim and wrap it around the victim.

4. Grab the material at the head and drag the victim backward to safety.

SKILL DRILL 17-12

Performing a Standing Drag Firefighter I, NFPA 1010: 6.3.9

1. Kneel at the head of the supine victim.

2. Raise the victim's head and torso by 90 degrees so that the victim is leaning against you.

SKILL DRILL 17-12 CONTINUED

Performing a Standing Drag Firefighter I, NFPA 1010: 6.3.9

3. Reach under the victim's arms, wrap your arms around the victim's chest, and lock your arms.

4. Stand straight up using your legs. Drag the victim backward to safety.

Webbing Sling Drag

The webbing sling drag provides a secure grip around the upper part of a victim's body, allowing for a faster removal from the dangerous area. In this drag, a sling made out of webbing is placed around the victim's chest and under the armpits. The rescuer then uses the webbing to drag the victim. The webbing sling helps support the victim's head and neck. The sling is prepared ahead of time and can be rolled and kept in a bunker coat pocket. A carabiner can be attached to such a sling to secure the straps under the victim's arms and provide additional protection for the victim's head and neck. To perform the webbing sling drag, follow the steps in **SKILL DRILL 17-13**.

SKILL DRILL 17-13

Performing a Webbing Sling Drag Firefighter I, NFPA 1010: 6.3.9

1. Using a prepared webbing sling, place the victim in the center of the loop so the webbing is behind the victim's back in the area just below the armpits.

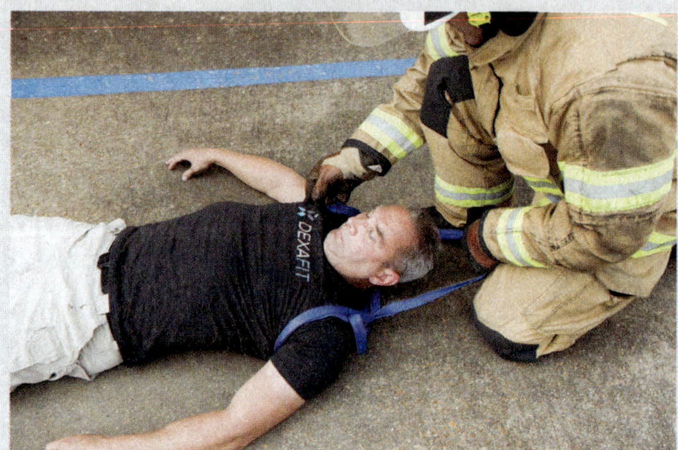

2. Take the large loop over the victim and place it above the victim's head. Reach through, grab the webbing behind the victim's back, and pull through all the excess webbing. This creates a loop at the top of the victim's head and two loops around the victim's arms.

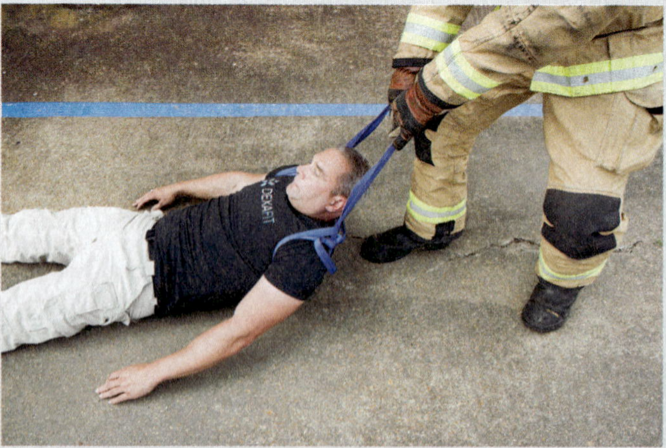

3. If necessary, adjust your hand placement to protect the victim's head while dragging.

Firefighter Drag

The firefighter drag can be used if the victim is heavier than the rescuer because it does not require lifting or carrying the victim. To perform the firefighter drag, follow the steps in **SKILL DRILL 17-14**.

Emergency Drag from a Vehicle

An emergency drag from a vehicle is performed when the victim must be quickly removed from a vehicle to save their life. The drag described in this section might be used, for example, if the vehicle is on fire or if the victim requires cardiopulmonary resuscitation (CPR). There is no effective way for one person to remove a victim from a vehicle without some movement of the neck and spine. It is important, however, to prevent excess movement of the victim's neck. To perform an emergency drag from a vehicle, follow the steps in **SKILL DRILL 17-15**.

Assisting a Person Down a Ground Ladder

Using a ground ladder to rescue a trapped occupant is one of the most critical, stressful, and demanding tasks

SKILL DRILL 17-14

Performing a Firefighter Drag Firefighter I, NFPA 1010: 6.3.9

1. Tie the victim's wrists together with anything that is handy—a cravat (a folded triangular bandage), gauze, a belt, or a necktie.

2. Get down on your hands and knees and straddle the victim.

3. Pass the victim's tied hands around your neck, straighten your arms, and drag the victim across the floor by crawling on your hands and knees.

SKILL DRILL 17-15

Performing an Emergency Drag from a Vehicle Firefighter I, NFPA 1010: 6.3.9

1. Grasp the victim under the arms and cradle their head between your arms.

2. Gently pull the victim out of the vehicle. Try to stabilize the victim's head and neck against your chest.

3. Lower the victim down into a horizontal position in a safe place.

performed by firefighters. Assisting someone down a ladder carries a considerable risk of injury to both firefighters and victims. Whenever it is possible to use a stairway, fire escape, or aerial tower for rescue, these options should be considered before using ground ladders. Firefighters must use proper techniques to safely accomplish a ladder rescue. In addition, they must have the physical strength and stamina needed to accomplish the rescue without causing injury to anyone involved. Although circumstances could require an individual firefighter to work alone, at least two firefighters should work as a team to rescue a victim whenever possible.

Time is a critical factor in many rescue situations—someone waiting to be rescued is often in immediate danger and may be preparing to jump. As a consequence, firefighters may have only a limited time to work. They must quickly and efficiently raise a ladder and assist the victim to safety.

Ladder rescue begins with proper placement and securement of the ladder. This is described in

Chapter 12, *Ladders*. A ladder used to rescue a person from a window should have its tip placed just below the windowsill. Placing it at the 60-degree rescue angle can make it easier for a conscious victim to mount the ladder; however, always follow your department's SOPs. All ladder rescues should be performed with at least two firefighters whenever possible. If possible, one or more firefighters in the interior of the building should help the victim onto the ladder, and one firefighter should stay on the ladder to assist the individual down. Any ladder used for rescue should be stabilized by heeling it or tying it in. The weight of an occupant and one or two firefighters, all moving on the ladder at the same time, can easily destabilize a ladder that is not adequately secured.

It is not often that firefighters need to perform a rescue using a ground ladder. This type of rescue is physically taxing and requires carefully executed skills. It is important to practice the skills to be prepared for this event, as those skills may be needed at any time. Using a live person to practice this skill can be dangerous. For this reason, many departments use a mannequin when performing this drill.

Rescuing a Conscious Person from a Window

A ladder rescue is often frightening to conscious victims. When a rescue involves a conscious person, firefighters should establish verbal contact as quickly as possible to reassure the victim that help is on the way. People have jumped to their deaths just seconds before a ladder could be raised to a window. To rescue a conscious person from a window, follow the steps in **SKILL DRILL 17-16**.

SAFETY TIP

It is possible for a single firefighter to rescue a conscious person from a window, but this should be avoided if possible.

SKILL DRILL 17-16

Rescuing a Conscious Victim from a Window Firefighter I, NFPA 1010: 6.3.9

1. The rescue team sets up and secures the ladder in the rescue position, with the tip of the ladder just below the windowsill and placed at the 60-degree rescue angle.

Continues.

SKILL DRILL 17-16 CONTINUED

Rescuing a Conscious Victim from a Window Firefighter I, NFPA 1010: 6.3.9

2. The first firefighter climbs the ladder, makes contact with the victim, and climbs inside the window to assist the victim. The firefighter should make contact as soon as possible to calm the victim and encourage the victim to stay at the window until the rescue can be performed.

3. The second firefighter climbs up to the window, leaving at least one rung available for the victim. When ready, the firefighter advises the victim to slowly come out onto the ladder, feet first and facing the ladder.

SKILL DRILL 17-16 CONTINUED

Rescuing a Conscious Victim from a Window Firefighter I, NFPA 1010: 6.3.9

4. The second firefighter forms a semi-circle around the victim, with both hands on the beams of the ladder.

5. The second firefighter and victim proceed slowly down the ladder, one rung at a time, with the firefighter always staying one rung below the victim. If the victim slips or loses their footing, the firefighter's legs should keep the victim from falling. The firefighter can take control of the victim at any time by leaning in toward the ladder and squeezing the victim against the ladder. The firefighter should verbalize each step and talk to the person being rescued to help reassure and calm them and encourage the person to keep their gaze forward.

Rescuing an Unconscious Person from a Window

Rescuing an unconscious victim by a ladder is dangerous and difficult, but it may be the only way to save the victim's life. To get an unconscious person onto a ladder, one or more firefighters need to climb inside the building and pass the person out of the window to a firefighter on a ladder. Caution should be used when lowering an unconscious victim down a ladder because it is easy for the victim's arms or legs to get caught in the ladder. To rescue an unconscious person via a ladder, follow the steps in **SKILL DRILL 17-17**.

SKILL DRILL 17-17

Rescuing an Unconscious Victim from a Window Firefighter I, NFPA 1010: 6.3.9

1. The rescue team sets up and secures the ladder in rescue position, with the tip of the ladder just below the windowsill and placed at the 60-degree rescue angle.

2. One firefighter climbs up the ladder and enters the window to rescue the victim. The second firefighter climbs up to the window opening and waits for the victim.

SKILL DRILL 17-17 CONTINUED

Rescuing an Unconscious Victim from a Window Firefighter I, NFPA 1010: 6.3.9

3. The second firefighter places both hands on the rungs of the ladder, with one leg straight and the other horizontal to the ground with the knee at an angle of 90 degrees. The foot of the straight leg should be one rung below the foot of the bent leg. When both firefighters are ready, the first firefighter passes the victim out through the window and onto the ladder, keeping the victim's back toward the ladder.

4. The victim is lowered so that they straddle the second firefighter's leg. The firefighter's arms should be positioned under the victim's arms, holding on to the rungs. The firefighter keeps the balls of both feet on the rungs of the ladder to make it easier to move their feet. The firefighter climbs down the ladder slowly, one rung at a time, transferring the victim's weight from one leg to the other. The victim's arms can also be secured around the firefighter's neck.

Rescuing an Unconscious Child or a Small Adult from a Window

Small adults and children can be cradled across a firefighter's arms during a rescue. To use this rescue technique, the child must be light enough that the firefighter can descend safely while using only arm strength to support the victim. To carry an unconscious child down a ladder, follow the steps in **SKILL DRILL 17-18**.

SKILL DRILL 17-18

Rescuing an Unconscious Child or a Small Adult from a Window
Firefighter I, NFPA 1010: 6.3.9

1. The rescue team sets up and secures the ladder in rescue position, with the tip below the windowsill and placed at the 60-degree rescue angle.

2. The first firefighter climbs the ladder and enters the window to assist the victim. The second firefighter climbs the ladder to the window opening and waits to receive the victim. Both of the second firefighter's arms should be level with their hands on the beams.

SKILL DRILL 17-18 CONTINUED

Rescuing an Unconscious Child or a Small Adult from a Window
Firefighter I, NFPA 1010: 6.3.9

3. When ready, the first firefighter passes the victim to the second firefighter, so the victim is cradled across the second firefighter's arms.

4. The second firefighter climbs down the ladder slowly, with the victim being held in the firefighter's arms. The firefighter's arms should stay level, and their hands should slide down the beams.

Rescuing a Large Adult from a Window

Three or four firefighters using two ladders may be needed to rescue very tall or heavy adults. To rescue a large adult using a ladder, follow the steps in **SKILL DRILL 17-19**.

Removing a Victim by Aerial Ladder or Platform

An aerial ladder or platform can be used for rescue operations. The same basic rescue techniques used with ground ladders can be used with aerial ladders, but aerial ladders offer several advantages over ground

SKILL DRILL 17-19

Rescuing a Large Adult from a Window Firefighter I, NFPA 1010: 6.3.9

1. The rescue team places and secures two ladders, side by side, in the rescue position with the tips of the two ladders just below the windowsill and placed at the 60-degree rescue angle.

2. Multiple firefighters may be required to enter the window to assist from the inside.

SKILL DRILL 17-19 CONTINUED

Rescuing a Large Adult from a Window Firefighter I, NFPA 1010: 6.3.9

3. Two firefighters, one on each ladder, climb up to the window opening and wait to receive the victim.

4. When ready, the victim is lowered down across the arms of the firefighters, with one firefighter supporting the victim's legs and the other supporting the victim's arms. Once in place, the firefighters can slowly descend the ladder, using both hands to hold on to the ladder rungs.

ladders. Aerial ladders are much stronger and have a longer reach. In addition, they are wider and more stable than ground ladders, with side rails that provide greater security for both rescuers and victims (**FIGURE 17-14**).

Aerial platforms are even more suitable than aerial ladders for rescue operations. These devices reduce the risk of slipping and falling because the victim is lowered to the ground mechanically. An aerial platform is usually preferred for rescue work if one is available.

Shelter-in-Place

In some situations, the best option is to shelter the occupants in place instead of trying to remove them from the building. This option should be considered when the occupants are conscious and are found in a part of the building that is adequately protected from the fire by fire-resistive construction or fire suppression systems. If smoke and fire conditions block the exits, victims might be safer staying in the sheltered location than attempting to evacuate through a hazardous environment.

One example of when shelter-in-place is probably the safest option is in a high-rise apartment building when a fire is confined to one apartment or one floor. In this scenario, the stairways and corridors may be filled with smoke, but the occupants who are remote from the fire would be safe on their balconies or in their apartments with the windows open to provide fresh air. They would be exposed to greater levels of risk if they attempted to exit than if they remain in their apartments until the fire is extinguished.

When occupants remain in place, it is important for them to isolate the room where they are from the fire. This can be done by practicing good door control or by determining whether they are safer with the windows open or with the windows closed. Open windows can serve as an entry point for smoke from the outside or actually draw the fire to the occupant. The decision to follow a shelter-in-place strategy must be made by the IC or the officer in charge of the rescue team(s). Potential locations to shelter-in-place should be identified on the preincident plan.

FIGURE 17-14 An aerial ladder is stronger and more stable than a ground ladder.

CASE STUDY

You Are the Firefighter CONCLUSION

You arrive on the scene of a fast-moving apartment fire. The structure is a three-story, center-hallway, Type III (ordinary construction) apartment building with fire showing on the second floor. The second- and third-floor hallways are smoke charged. Dispatch is reporting possible occupants trapped. You are assigned to conduct a primary search on the third floor while another crew begins the initial fire attack. This is a call that will tax the resources of your department.

1. **How will you conduct this search?**

   Answer: In a search team of two or more members, a primary search would be conducted quickly, efficiently, and systematically, using a search method determined by the fire conditions as prescribed by the IC or department SOP.

2. **What will you do if you find a victim in the first unit you search?**

   Answer: Immediately notify the IC. If resources allow, the IC will send a rescue team to meet the search team so that the search team does not lose the continuity of its search. Otherwise, the first search team needs to remove the victim.

3. **Will you take a hose line with you on the search?**

   Answer: A search team that is working close to the fire should carry a hose line or be accompanied by another team with a hose line. The hose line can protect firefighters and enable them to search a structure more efficiently.

WRAP-UP

SUMMARY

KNOWLEDGE OBJECTIVES

- Describe the mission of search and rescue operations.
  - Describe the mission of search operations. (NFPA 1010: 6.3.9, p. 744)
  - Describe the mission of rescue operations. (NFPA 1010: 6.3.9, p. 744)
- Explain how search and rescue operations are coordinated with other fire suppression operations. (pp. 744–745)
- Identify the factors to evaluate during a search and rescue size-up.
  - Describe the information that impacts a search and rescue size-up. (NFPA 1010: 6.3.9, pp. 745–748)
  - Explain how search operations are coordinated. (NFPA 1010: 6.3.9, p. 748)
  - Describe the role of a fire officer during search operations. (NFPA 1010: 6.3.9, p. 748)
- List the priorities of search operations.
  - Describe the priorities of search operations. (NFPA 1010: 6.3.9, pp. 748–749)
  - Explain how firefighters maintain safety through risk management. (NFPA 1010: 6.3.9, pp. 749–750)
  - List the tools and equipment used in search and rescue operations. (NFPA 1010: 6.3.9, pp. 750–751)
  - Describe the methods firefighters use to determine whether an area is tenable. (NFPA 1010: 6.3.9, p. 751)
- Describe the objectives of a primary and secondary search.
  - Describe the objectives of a primary search. (NFPA 1010: 6.3.9, pp. 751–753)
  - Describe the search patterns commonly used in primary search operations. (NFPA 1010: 6.3.9, pp. 753–755)
  - Explain how thermal imaging devices are used during search operations. (NFPA 1010: 6.3.9, pp. 754–755)
  - Describe a primary search using the standard search method. (NFPA 1010: 6.3.9, pp. 755–756)
  - Describe a primary search using the oriented search method. (NFPA 1010: 6.3.9, pp. 756–758)
  - Describe a primary search using the oriented-vent-enter-isolate-search (O-VEIS) method. (NFPA 1010: 6.3.9, p. 760)
  - Describe a primary search using the team search method. (NFPA 1010: 6.3.9, pp. 760–762))
  - Describe the objectives of a secondary search. (NFPA 1010: 6.3.9, pp. 762–763)
  - List the major types of rescue methods. (NFPA 1010: 6.3.9, pp. 764–787)
  - Describe the concept of sheltering-in-place. (NFPA 1010: 6.3.9, p. 788)
- List the common types of simple victim carries performed during rescue operations.
  - Describe how to assist a victim to an exit. (NFPA 1010: 6.3.9, p. 765)
  - List the common types of simple victim carries performed during rescue operations. (NFPA 1010: 6.3.9, pp. 765–772)
- List the six emergency drags performed during rescue operations. (pp. 772–777)
- Describe the conditions that may require a ground or aerial ladder rescue.
  - Describe why a ground ladder rescue may be required. (NFPA 1010: 6.3.9, pp. 777–786)
  - Describe the advantages of using aerial ladders and platforms during rescue operations. (NFPA 1010: 6.3.9, pp. 786–787)

SKILLS OBJECTIVES

- Conduct a primary search using the standard, oriented, oriented-vent-enter-isolate-search (O-VEIS), and team search methods.
  - Conduct a primary search using the standard search method. (NFPA 1010: 6.3.9, p. 757)
  - Conduct a primary search using the oriented search method. (NFPA 1010: 6.3.9, pp. 758–759)
  - Conduct a primary search using the oriented -vent-enter-isolate-search (O-VEIS) method. (NFPA 1010: 6.3.9, p. 760)

SKILLS OBJECTIVES CONTINUED

- Conduct a primary search using the team search method. (**NFPA 1010: 6.3.9**, pp. 760–762)
- Conduct a secondary search. (**NFPA 1010: 6.3.9**, pp. 763–764)
- Perform one and two-person carries and drags.
  - Perform a one-person walking assist. (**NFPA 1010: 6.3.9**, p. 766)
  - Perform a two-person walking assist. (**NFPA 1010: 6.3.9**, p. 767)
  - Perform a two-person extremity carry. (**NFPA 1010: 6.3.9**, p. 768)
  - Perform a two-person seat carry. (**NFPA 1010: 6.3.9**, p. 769)
  - Perform a two-person chair carry. (**NFPA 1010: 6.3.9**, p. 770)
  - Perform a cradle-in-arms carry. (**NFPA 1010: 6.3.9**, p. 771)
  - Perform a clothes drag. (**NFPA 1010: 6.3.9**, p. 772)
  - Perform a blanket drag. (**NFPA 1010: 6.3.9**, p. 773)
  - Perform a standing drag. (**NFPA 1010: 6.3.9**, pp. 774–775)
  - Perform a webbing sling drag. (**NFPA 1010: 6.3.9**, p. 776)
  - Perform a firefighter drag. (**NFPA 1010: 6.3.9**, p. 777)
  - Perform a one-person emergency drag from a vehicle. (**NFPA 1010: 6.3.9**, p. 778)
- Perform rescues by ladder to conscious and unconscious victims.
  - Rescue a conscious victim from a window. (**NFPA 1010: 6.3.9**, pp. 779–781)
  - Rescue an unconscious victim from a window. (**NFPA 1010: 6.3.9**, pp. 782–783)
  - Rescue an unconscious child or small adult from a window. (**NFPA 1010: 6.3.9**, pp. 784–785)
  - Rescue a large adult from a window. (**NFPA 1010: 6.3.9**, pp. 786–787)

KEY TERMS

360 See *360-degree walk-around.*

360-degree walk-around An evaluation and analysis of an incident conducted by the incident commander (IC) or someone designated by the IC after the initial size-up is completed. Also called *360.*

assist Rescue technique in which the victim is responsive and able to walk but needs support.

carry Technique that physically moves a victim who is conscious and responsive but incapable of standing or walking.

cold smoke Smoke in a room where a fire sprinkler system was activated so that the smoke is cooled and banked down to the floor.

heat flux The measure of the rate of heat transfer to a surface or an area. (NFPA 921)

initial size-up Usually conducted by the first-arriving company officer, a rapid evaluation and analysis of an incident upon arrival at the incident and before the 360-degree walk-around is conducted. Also called *windshield size-up* because it is generally given by the engine officer before they get out of the engine.

oriented search method A primary search method in which a search team consisting of an officer or team leader and one to three searchers conduct a systematic search where the searchers search the rooms using a left-handed or right-handed search pattern along the outside wall of rooms and the officer or team leader stays at the door to direct the search.

oriented-vent-enter-Isolate-search method (O-VEIS) A primary search method in which a search team consisting of an officer or team leader and one searcher approach a window, determine that the room is tenable, open the window so that the searcher can quickly enter the room, close the door, and search the room.

preincident plan A document developed by gathering general and detailed data that is used by responding personnel in effectively managing emergencies for the protection of occupants, participants, responding personnel, property, and the environment. (NFPA 1660)

primary search A quick search of the structures likely to contain survivors. (NFPA 2500)

rekindle A return to flaming combustion after apparent but incomplete extinguishment. (NFPA 921)

rescue Those operations directed at locating and removing endangered persons and treating the injured at an emergency incident. (NFPA 1951)

rope search method See *team search method.*

search Land-based efforts to find victims or recover bodies. (NFPA 1951)

search line See *search rope.*

search rope A rope with an anchor point to keep firefighters in contact with each other and with the egress route from a burning building during an interior attack or a search and rescue operation. Also called *search line.*

secondary search A detailed, systematic search of an area. (NFPA 2500)

stack effect The vertical air flow within buildings caused by the temperature-created density differences between the building interior and exterior or between two interior spaces. (NFPA 92)

standard search method A primary search method in which a search team consisting of at least two firefighters conducts a systematic search using a left-handed or right-handed search pattern along the outside wall of rooms.

sweep To use your arm or a tool to swipe across areas to feel for victims or objects.

tag line See *tether line.*

team search method A primary search method in which a search team consisting of an officer or team leader who acts as the anchor person and two or more searchers conduct a systematic search where the anchor person ties the rope to a stationary object and the searchers search the rooms while holding on to the rope so that they can quickly exit by following the rope out if needed.

tether line A smaller line, typically thinner and more lightweight than the main search rope, attached to the main search rope during large area searches to serve as a secondary line to aid in navigation and orientation within the search area. Also called *tag line.*

thermal imager Special electronic equipment that creates a picture based on the heat produced by a person or object. (NFPA 1801)

two-in/two-out rule A safety procedure that requires a minimum of two personnel to enter a hazardous area and a minimum of two backup personnel to remain outside the hazardous area during the initial stages of an incident.

windshield size-up See *initial size-up.*

REVIEW QUESTIONS

1. What is the purpose of search operations? What is the purpose of rescue operations?
2. Why must search and rescue operations be coordinated with other fire suppression operations?
3. What critical factors of search and rescue should be evaluated during size-up?
4. List and explain the priorities of search operations.
5. What must an IC do before sending firefighters into a structure for search and rescue operations?
6. What is the objective of a primary search? What is the objective of a secondary search?
7. List the four methods of conducting a primary search.
8. What is the first priority when finding a victim during rescue operations?
9. List the four common types of simple victim carries performed during rescue operations.
10. List the six emergency drags performed during rescue operations.

DISCUSSION QUESTIONS

1. From the first day of training, the importance of the two-in/two-out rule is stressed to all firefighters. The importance of this rule is further stressed in NFPA 1550, *Standard for Emergency Responder Health and Safety,* and OSHA standard 1910.134(g)(4), *Two-In/Two-Out Rule for Interior Structural Firefighting.* Why is the two-in/two-out rule important? Is there ever a reason to violate that rule? Provide examples and insights from the chapter, your own research, or first-person experiences to support your response.
2. In search and rescue operations, balancing the risks involved with the potential benefits is crucial for firefighter safety and the successful outcome of the operation. Based on the information provided in this chapter, discuss the importance of risk management in search and rescue operations and the factors that should be considered when assessing the degree of risk. How can incident commanders effectively manage risks to ensure the safety of firefighters while maximizing the chances of

DISCUSSION QUESTIONS CONTINUED

saving lives? Provide examples and insights from the chapter and your own research to support your response.

3. Discuss the different methods of conducting a primary search and the circumstances under which each method is most appropriately used. How can firefighters ensure efficient and effective searches while maintaining their safety and the safety of potential victims?

APPLYING THE CONCEPTS

At 0130, your fire company is on-scene at an active fire at a four-story apartment building near the university that houses students. Flames and heavy smoke are visible on side alpha of the third floor. The building is Type III with a mix of concrete block and wood construction. It is summer session, so the building is partially occupied. Some residents evacuated safely, but dispatch reported multiple occupants are trapped on the third and fourth floors. Your department preplanned this building, so you know the layout, entry points, suppression system, and FDC location.

1. Because there are victims needing rescue, is the first action to send in a rescue team? Why or why not?
2. What information is important to assess when determining a search and rescue plan?

Flames break through a window on the third floor, sending glass flying. The IC orders a fire suppression team to initiate a transitional attack to cool the environment and knock down the fire. You and your crew are assigned to search and rescue.

3. Why is it crucial for suppression, ventilation, and rescue teams to communicate and coordinate their actions?
4. Where will the search begin? The area closest to the fire? Top-down starting at the highest floors where victims were reported? Or bottom-up?

The IC radios, "Rescue teams A and B, don your SCBA and gather your equipment. Acknowledge when ready. You will enter through side delta and conduct an oriented search on the fourth floor. Mark doors as you go. Maintain radio comm throughout."

5. What are the advantages of using the oriented search method?

Working closely with your team and reporting your progress, you work your way up the stairwell. Your team enters the fourth floor. You see the hallway is filled with medium gray smoke and has limited visibility. You crouch down to stay below the layout of hot gases and shout, "We're here to help! If you can hear me, make a noise." Your team leader is using thermal imaging and yells, "I see a heat signature. Over here!" Just then you notice the smoke in the hallway has changed. It's quickly getting darker and thicker.

6. Do you continue with search operations or evacuate?
7. The rescue teams finish the primary search and victims are rescued. Are search and rescue operations finished?

REFERENCES

Coleman, John "Skip". 2011. *Searching Smarter.* Tulsa, OK: PennWell Corporation.

Kerber, Stephen, Daniel Madrzykowski, James Dalton, and Bob Backstrom. 2012. *Improving Fire Safety by Understanding the Fire Performance of Engineered Floor Systems and Providing the Fire Service with Information for Tactical Decision Making.* Underwriters Laboratories Firefighter Safety Research Institute (FSRI), March 1, 2012. Accessed June 25, 2023. https://d1gi3fvbl0xj2a.cloudfront.net/files/2021-07/2009_NIST_ARRA_Compilation_Report.pdf.

National Fire Protection Association (NFPA). 2019. *NFPA 1950, Standard on Protective Ensembles for Technical Rescue Incidents.* 2020 Edition. Quincy, MA: NFPA.

National Fire Protection Association (NFPA). 2019. *NFPA 1951, Standard on Protective Ensembles for Technical Rescue Incidents.* 2020 Edition. Quincy, MA: NFPA.

National Fire Protection Association (NFPA). 2020. *NFPA 92, Standard for Smoke Control Systems.* 2021 Edition. Quincy, MA: NFPA.

National Fire Protection Association (NFPA). 2020. *NFPA 921, Guide for Fire and Explosion Investigations.* 2021 Edition. Quincy, MA: NFPA.

National Fire Protection Association (NFPA). 2020. *NFPA 1700, Guide for Structural Fire Fighting.* 2021 Edition. Quincy, MA: NFPA.

National Fire Protection Association (NFPA). 2020. *NFPA 1801, Standard on Thermal Imagers for the Fire Service.* 2021 Edition. Quincy, MA: NFPA.

National Fire Protection Association (NFPA). 2023. *NFPA 921, Guide for Fire and Explosion Investigations.* 2024 Edition. Quincy, MA: NFPA.

National Fire Protection Association (NFPA). 2023. *NFPA 1550, Standard for Emergency Responder Health and Safety.* 2024 Edition. Quincy, MA: NFPA.

National Fire Protection Association (NFPA). 2023. *NFPA 1660, Standard for Emergency, Continuity, and Crisis Management: Preparedness, Response, and Recovery.* 2024 Edition. Accessed June 16, 2023. Quincy, MA: NFPA.

National Fire Protection Association (NFPA). 2023. *NFPA 2500, Standard for Operations and Training for Technical Search and Rescue Incidents and Life Safety Rope and Equipment for Emergency Services.* 2022 Edition. Quincy, MA: NFPA.

Occupational Safety and Health Administration (OSHA). 1998. *Two-In/Two-Out Rule for Interion Structural Firefighting.* Standard Number: 1910.124(g)(4). November 13, 1998. Accessed June 16, 2023. https://www.osha.gov.

Underwriters Laboratories (UL). 2010. *Firefighter Exposure to Smoke Particulates.* April 1, 2010. Accessed June 25, 2023. https://ul.org/library/1187.

Underwriters Laboratories Firefighter Safety Research Institute (FSRI). 2013. *Governors Island Experiments.* May 21, 2013. Accessed June 16, 2023. https://ulfirefightersafety.org/research-projects/governors-island-experiments.html.

U.S. Fire Administration (USFA). *Three Firefighters Die in Pittsburgh House Fire.* Investigated by J. Gordon Routley. 1995. USFA-TR-078 February, 1995. Accessed June 16, 2023. https://www.hsdl.org/?view&did=484126.

CHAPTER

18

Firefighter I

Salvage and Overhaul

KNOWLEDGE OBJECTIVES

After studying this chapter, you will be able to:

- Describe the various lighting equipment and options to illuminate a work area.
- Describe the various salvage tools and their use in salvage techniques.
- Explain how to prevent excess water damage, including how to shut off a fire sprinkler system or activated sprinkler head.
- Describe the salvage techniques to limit smoke and heat damage.
- Describe the purpose of and how safety is maintained during the overhaul phase.
- Describe the techniques utilized during the overhaul phase.

SKILLS OBJECTIVES

After studying this chapter, you will be able to perform the following skills:

- Illuminate and provide a safe working area for firefighters completing salvage and overhaul operations.
- Shut down an activated fire sprinkler system.
- Deploy, fold, and store salvage covers.
- Perform overhaul techniques utilizing firefighting tools and equipment.

ADDITIONAL STANDARDS

- **NFPA 1**, *Fire Code, 2024 Edition*
- **NFPA 901**, *Standard Classifications for Fire and Emergency Services Incident Reporting, 2021 Edition*
- **NFPA 1700**, *Guide for Structural Fire Fighting, 2021 Edition*

CASE STUDY

You Are the Firefighter

Your shift has just started with Engine Company 3. You are doing your morning truck check when your company gets dispatched for a residential structure fire. You arrive and find a large, two-story residential structure with a fire on the first floor. The neighborhood you are in was developed in the early 1900s, and the structure appears to be a Type V structure with a basement and is balloon-frame construction. Your company is the fourth engine company on scene. Your crew approaches the incident commander and is told that interior crews are reporting the fire on the first floor is knocked down but they have not yet opened up any walls or ceilings. Command assigns your four-person crew to salvage operations.

1. On which areas of the building would you focus your salvage operations?
2. What equipment will you need to prevent further damage to this house and its contents?
3. Given the building's construction and fire location, where is the overhaul crew most likely to find hidden fire that has spread from the area of origin?

Introduction

Remember that the three priorities at every fire incident are first to save lives (life safety), second to control the fire (incident stabilization), and third to protect property (property conservation). **Salvage** efforts protect property and belongings from damage during and after an incident, particularly from the effects of smoke and water, although some salvage techniques are useful for incidents not involving fire, such as a call involving a burst pipe. **Overhaul** ensures that a fire is completely extinguished by finding, exposing, and suppressing any smoldering or hidden pockets of fire in an area that has been burned after visible fire has been suppressed. The overhaul process ensures that a fire is completely extinguished so that it does not reignite.

Salvage and overhaul are usually conducted in close coordination with each other. If a department has enough personnel on scene to address more than one incident priority, salvage can be performed concurrently with fire suppression. Overhaul, however, can begin only when the fire is believed to be under control.

Lighting

Lighting is an important concern at many fires and other emergency incidents. Firefighters performing salvage and overhaul must be able to see where they are going, what they are doing, and which potential hazards, if any, are present. Lights can be portable or mounted on an apparatus. Many emergency incidents occur at night, so exterior lighting is required to illuminate the scene and enable safe, efficient operations. Inside fire buildings, heavy smoke can completely block out natural light. Even if the smoke has cleared, it can still be too dark to safely perform salvage and overhaul operations without additional light. Electrical service often is interrupted or disconnected for safety purposes, so lights mounted on an apparatus and portable lights should be set up to fully illuminate areas during salvage and overhaul. Most fire departments use different types of lighting equipment depending on the particular situation. A **spotlight**, for example, projects a narrow, concentrated beam of light. A **floodlight** projects a more diffuse light over a wide area. Lights can be portable or permanently mounted on fire apparatus.

Although the primary responsibility for fire-scene lighting usually belongs to truck companies or support units, any company can be assigned this duty. Lighting equipment can be mounted on any apparatus, and some fire departments equip special vehicles with high-powered lights to illuminate incident scenes. Departmental standard operating procedures (SOPs) or standard operating guidelines (SOGs), or the incident commander (IC) at the incident scene will dictate the responsibility for setting up lighting.

Exterior lighting should be provided at all incident scenes during hours of darkness so that firefighters can see what they are doing, recognize hazards, and find victims who need rescuing. Exterior lighting also makes firefighters more visible to drivers who are approaching the scene or maneuvering emergency vehicles. Powerful exterior lights can often be seen through the windows and doors of a dark, smoke-filled building and can therefore guide disoriented firefighters or victims to safety.

Apparatus operators should turn on their apparatus-mounted floodlights and position them for maximum effectiveness so that as much of the area as

FIGURE 18-1 The particles in smoke make it impossible for light to cut through the smoke. Using hand lights during fire attack can be counterproductive as the light is diffused by smoke, creating a "washout" effect. The denser the smoke, the more pronounced the light diffusion will be.
Courtesy of Brandon Hausbeck.

possible is illuminated. The entire area around a fire building should be illuminated. If some areas cannot be covered with apparatus-mounted lights, use portable lights.

Portable lights can provide interior lighting at a fire scene. Position portable lights inside the building to illuminate interior areas as needed and as time permits. It is important to remember that smoke, in part, consists of particulates. Lighting cannot penetrate or "cut through" smoke. In many cases, smoke will diffuse light, causing a beam of light to be directed in many directions and angles as the beam bounces off the particulates. This can make it difficult to see the glow of a fire and actually decrease visibility as the room becomes "washed" in light (**FIGURE 18-1**).

When operations reach the salvage and overhaul phase, adequate interior lighting must be provided in all interior areas so that crews can work safely. Proper lighting enables firefighters to see what they are doing and to observe any dangerous conditions that need to be addressed. Although exterior lighting is generally needed only at night, interior lighting may be needed even when it is a bright day outside.

Lighting Equipment

Portable lights can be taken into buildings to illuminate the interior. They can also be set up outside the structure to illuminate the fire or emergency incident scene (**FIGURE 18-2**). Portable lights usually range from 300 to 1500 watts and can use several types of bulbs, including quartz bulbs, halogen bulbs, and light-emitting diodes (LEDs).

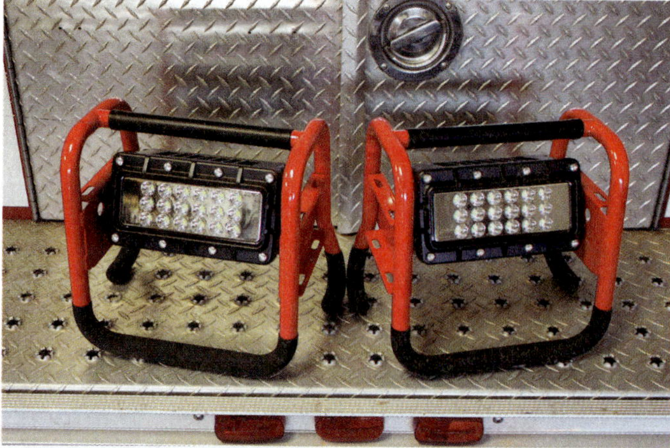

FIGURE 18-2 Portable emergency lights come in various sizes and levels of brightness.
Courtesy of Eric Scruggs.

The electricity for portable lights is supplied by a generator, an inverter, a building's electrical system, or a battery. (Chapter 27, *Assisting Special Rescue Teams*, provides further detail about these sources of electricity.) Portable light fixtures are connected to generators and inverters with electrical cords. These electrical cords should be stored neatly coiled or on permanently mounted reels attached to the fire apparatus. The cord is then pulled from the reel to the place where the power is needed. Portable reels can be taken from the apparatus into the fire scene.

A **junction box** is a device that attaches to electrical cords to provide mobile power outlets. They are placed in convenient locations so that cords for individual lights and electrical equipment can be attached to them. Junction boxes used by fire departments are protected by waterproof covers and are often equipped with small lights so that they can be easily located in the dark.

The connectors and plugs used for fire department lighting have special connectors that attach with a slight clockwise twist. This setup keeps the power cords from becoming unplugged during fire department operations. You should become familiar with the types of electrical connectors in use in your department, and practice making connections until you can connect the pieces of equipment in the dark.

Lights can be permanently mounted on fire apparatus to illuminate incident scenes (**FIGURE 18-3**). In this case, the vehicle operator can immediately illuminate the emergency scene simply by pressing the generator starter button. Some vehicle-mounted lights

FIGURE 18-3 A side apparatus-mounted light.

FIGURE 18-4 An apparatus-mounted light tower can provide near-daylight conditions at a nighttime incident scene.

can be manually raised to illuminate a larger area. Mechanically operated light towers, which can be raised and rotated by remote control, can create near-daylight conditions at a night incident scene (**FIGURE 18-4**).

Battery-powered lights are lightweight, easily transported, do not require power cords, and can be used immediately (**FIGURE 18-5**). Battery-powered lights can provide either personal lighting or scene lighting. Personal lights are carried by the firefighter by hand or attached to their personal protective equipment (PPE). Scene lights are placed to illuminate a specific area. Battery-powered lights are powered by either disposable or rechargeable batteries, so they have a limited operating time before the batteries need to be recharged or replaced.

Large hand lights that project a powerful beam of light are often preferred for interior operations over smaller, personal flashlights. Hand lights offer increased candlepower over smaller, personal flashlights and can be equipped with a shoulder strap or hooked to a belt for easy transport. Every crew member entering a fire building should be equipped with a high-powered hand light.

A personal flashlight is another type of battery-operated light used by firefighters. Always carry a flashlight as part of your PPE. It should be rugged and project a strong light beam. It could be a lifesaver

FIGURE 18-5 Hand lights manufactured for use by firefighters have light outputs ranging from 3500 to 75,000 candlepower.

SKILL DRILL 18-1

Illuminating an Emergency Scene Firefighter I, NFPA 1010: 6.3.17

1. Wear PPE. Depending on scene conditions, you may or may not wear self-contained breathing apparatus (SCBA).
2. Inspect all equipment while setting up.
3. Start the portable generator, engage the inverter, or check that there is electrical power in the building.
4. Connect cords, plug adaptors, and lighting equipment.
5. Ensure proper grounding and GFI use.
6. Position lights so that the scene is adequately and safely illuminated.

Courtesy of Brandon Hausbeck.

if your primary light source fails. Flashlights are not as powerful as larger hand lights, and they will not operate for as long on one set of batteries.

Safety Principles and Practices

The lighting and power equipment used at a fire scene that is not battery powered generally operates on 110-volt AC (alternating current), which is the same as standard household current. Some systems require higher voltage. All electrical cords, junction boxes, lights, and power tools must be maintained properly and handled carefully to avoid electrical shocks. In addition, all electrical equipment must be properly grounded. Electrical cords must be well insulated, without cuts or defects, and properly sized to handle the required amperage.

Generators should have a **ground-fault interrupter (GFI)** to prevent a firefighter from receiving a potentially fatal electric shock. A GFI senses when there is a problem with an electrical ground and interrupts the current, shutting down both the power source and the equipment it is feeding.

Some portable generators are equipped with a grounding rod that must be inserted into the ground. Always use a grounding rod if one is provided. Avoid areas of standing or flowing water when placing power cords and junction boxes at a fire scene, and place electrical equipment on higher ground whenever possible.

Follow the steps in **SKILL DRILL 18-1** to illuminate an emergency scene.

SAFETY TIP

A good rule to remember: Light early, light often, and light safely.

Salvage Overview

The damage caused by smoke and water at a fire scene can be more extensive and costly to repair or replace than the property that is burned. Salvage operations are also conducted in non-fire situations that endanger property, such as floods or water leaking from burst pipes. At the scene of the emergency incident, the victims' property is entrusted to the care of firefighters, who are expected to protect that property from avoidable loss or damage to the best of their ability. Salvage efforts usually are aimed at preventing or limiting **secondary loss** that results from smoke and water damage, fire suppression efforts, and other causes. These efforts must be performed promptly to preserve the building and protect the contents from avoidable damage. Salvage operations include removing smoke, removing heat, controlling water runoff, removing water from the building, securing a building after a fire, covering broken windows and doors, and providing temporary patches for ventilation openings in the roof to protect the structure and contents.

Throughout the salvage and overhaul process, firefighters must attempt to preserve evidence related to the cause of the fire, particularly when working near the suspected area of fire origin and when the fire cause is not obvious. In general, disturb as little as possible in or near the **area of origin** until a fire investigator determines the cause of the fire and gives permission to clear the area. Knowing what caused a fire can help prevent future fires and enable victims to recoup losses under their insurance coverage. This subject is covered in greater depth in Chapter 28, *Understanding Fire Investigation.*

Salvage efforts at residential fires often focus on protecting personal property. These items can have personal implications for the victims' lives. Firefighters may be able to identify and protect critical personal belongings, such as wallets, purses, car keys, or medications. The value of some items, such as photographs and family heirlooms, cannot be measured in dollars, but these items are certainly treasured by victims and are often irreplaceable. In some communities it is common for occupants to not have homeowner's or renter's insurance. In these situations, the salvaged items may be the only possessions the occupant has.

TIP

One way many fire departments help homeowners after a fire is by having a list of board-up and fire restoration companies available that they can give to the homeowner. Board-up companies are prepared to respond quickly and secure a property to protect it from rain, snow, or damage from vandals. Fire restoration companies are skilled in removing smoke and water from a building and its contents before more damage occurs. Because many building owners are not aware of these companies, recommendations are greatly appreciated by the owner or occupant of a damaged building at this highly stressful time.

At commercial or industrial fires, the salvage priorities are different from those at residential fires. The preincident plan should identify critical items to enable firefighters to focus their salvage efforts appropriately. For example, a law firm might identify its files as the most valuable property, and a library might place more importance on books and computer files than on furniture. In a factory, machinery may be a higher priority than any finished products stored on the premises. A preincident plan should note whether the contents of a warehouse are valuable electronic components or industrial steel pipe. After the operation, the building must be secured to prevent break-ins or theft.

If forcible entry was required during the fire attack, the type of forcible entry used has a direct impact on salvage operations. If through-the-lock methods were used, a hasp and padlock may be all that are needed to secure a site. A resourceful crew might screw a forced door shut from the inside and leave from a door that has not been forced. Broken windows or doors can be boarded up with plywood. Do everything possible to reduce the amount of damage caused by the fire, fire suppression activities, and your own actions. Homeowners will be especially appreciative of support from the fire department after the fire is extinguished. (Chapter 15, *Forcible Entry,* discusses forcible entry techniques in more detail.)

Safety Considerations during Salvage Operations

Safety is a primary concern during salvage operations. Salvage operations often begin while the fire is still burning and may continue for several hours after it has been extinguished. During salvage operations, as with firefighting operations, firefighters should wear a full set of protective clothing and equipment, including SCBA. The PPE required depends on the site of the salvage operations, the stage of the fire, any hazards present, and department policy.

If salvage is conducted in a different part of the building than where the fire occurred, several floors below the fire, or in a neighboring building, firefighters may have other protective clothing requirements because hazards in these areas might be different from those in the fire area. Firefighters conducting salvage operations in remote areas can usually wear partial PPE—boots, turnout pants, helmet, and gloves—but only with the approval of the incident safety officer (ISO) and in accordance with department policy.

Structural collapse is always a possibility during salvage operations because of fire damage to the building's structural components. For example, lightweight trusses can collapse quickly when damaged by fire. Heavy objects, such as air-conditioning units and appliances, can crash through roofs or ceilings weakened by fire damage. Even the water used to douse the fire can contribute to structural collapse. Recall that each gallon of water weighs 8.34 pounds (lb; 4 kilograms [kg]). A ladder pipe discharging 800 gallons (gal; 2903 liters) of water per minute into a building adds 6400 lb (3.2 tons [2.9 tonnes]) of weight to the building every minute. Extinguishing a major fire can load the structure with many tons of extra water weight. If the structure is already weakened by the fire, the extra water weight can cause a catastrophic structural collapse, with the potential to injure or kill firefighters.

Sometimes this danger may be masked by absorbent materials such as piles of cloth or bales of rags stored in the building. Uneven absorption of water by these materials also presents a danger. For example, stacked rolls of newsprint in a printing factory can collapse if the bottom layer becomes waterlogged and, therefore, unable to support the stack.

Water creates potential hazards for firefighters conducting salvage operations on floors below the fire. Water used to extinguish a fire inevitably leaks down through the building and can cause major damage. If it becomes trapped in the void space above a ceiling, it may cause sections of the ceiling to collapse. Firefighters should watch for warning signs such as water leaking through ceiling joints or light fixtures and sagging areas of ceilings. Water that accumulates in a basement can hide a stairway and present additional hazards if utility connections are located in the basement.

Damage to the building's utility systems also presents significant risks to firefighters who are conducting salvage operations. Gas and electrical services should always be shut off to eliminate the potential hazards of electrocution or explosion stemming from gas leaks.

Salvage Tools

Firefighters conducting salvage operations use a variety of special tools and equipment to shield and cover building contents from water runoff, falling debris, soot, and particulate matter in smoke residue, and to remove smoke, water, and other products that can damage property, including the following:

- **Salvage cover** (large sheet of heavy canvas or plastic material) or rolls of polyethylene film
- Box cutter (for cutting plastic)
- **Floor runner** (piece of canvas or plastic material, usually 3 to 4 feet (ft; 91 to 122 centimeters [cm]) wide in various lengths that are used to protect flooring from dropped debris and dirt from shoes and boots)
- Wet/dry vacuums
- Squeegees
- Drainage pumps and hose
- Sprinkler shut-off kit
- Ventilation fans, power blowers
- Small tool kit
- Pike poles (to construct water chutes to catch dripping water)

Using Salvage Techniques to Prevent Water Damage

The best way to prevent water damage at a fire scene is to limit the amount of water used to fight the fire. Try to adhere to the following practices:

- Use only enough water to cool surfaces and knock down the fire quickly. Do not use more water than is needed. Do not continue to douse a fire that is already out, because this action merely creates unnecessary water damage.
- For materials that require extensive overhaul, such as mattresses, consider moving the material to the exterior of the structure.
- Turn off hose nozzles when they are not in use. If a nozzle must be left partly open to prevent freezing during cold weather, direct the flow outward through a window.
- Take the time to tighten any leaking hose couplings after the fire has been knocked down to prevent additional water damage during overhaul.
- Look for leaking water pipes in the building. If you find any, shut off the building's water supply.

Deactivating Sprinklers

Buildings equipped with sprinkler systems require special steps to limit water damage. Sprinklers can control a fire efficiently, but they will continue flowing until the activated sprinklers are shut off or the entire system is shut down. This ongoing flow can cause significant water damage. Responding fire companies should know where and how to shut down any automatic sprinkler system, if needed. If the system is accidentally activated, it should be shut down as soon as possible to avoid excessive water damage.

In an actual fire, only the IC should give the order to shut off the sprinkler system. Generally, the sprinkler system should not be shut down until the fire is completely extinguished. If sprinklers are shut down prematurely, the fire may rekindle, causing major damage. Between the time when sprinklers are shut down and when overhaul is completed, a firefighter with a portable radio should stand by and be ready to reactivate the sprinklers if necessary.

FIGURE 18-6 Most fires are controlled by the activation of only one or two sprinkler heads.

Courtesy of Tyco Fire and Building Products.

FIGURE 18-7 A sprinkler stop can be used to stop the flow of water from a sprinkler head that has been activated.

SAFETY TIP

Many structures have been destroyed after shutting off the sprinkler system because the fire was thought to be under control.

Sprinkler systems are usually designed so that only those sprinkler heads directly over or close to the fire will be activated. Indeed, most fires can be contained or fully extinguished by the release of water from only one or two sprinkler heads (**FIGURE 18-6**). Although movies may show all sprinkler heads activating at once, this does not occur in real life except in specially designed deluge sprinkler systems. More information on sprinkler systems is presented in Chapter 24, *Systems of Fire Detection, Suppression, and Smoke Control.*

If only one or two sprinkler heads have been activated, you can insert a sprinkler wedge or a sprinkler stop to quickly stop the flow. A **sprinkler wedge** is a small triangular piece of wood. Inserting two wedges on opposite sides, between the orifice (the opening through which water flows) and the deflector (the part of the sprinkler head that directs the spray pattern of the water) of the sprinkler, and pushing them together plugs the opening effectively, stopping the flow from standard upright and pendant sprinkler heads. A **sprinkler stop** is a more sophisticated mechanical device with a rubber stopper that can be inserted into a sprinkler head (**FIGURE 18-7**). Several types of sprinkler stops are available, including some that work only with specific types of sprinkler heads. Some types of sprinkler stops have a fusible link (a solder with a low melting point linking two pieces of metal and when it melts, the

FIGURE 18-8 This sprinkler stop shuts off the sprinkler and provides a temporary fusible link to keep the sprinkler head in service.

Courtesy of Scott Dornan, ConocoPhillips, Alaska.

link breaks and the water flows). Using this type of stop allows the sprinkler head to remain functional until it can be permanently replaced (**FIGURE 18-8**). Using either sprinkler wedges or stops to stop the flow at one sprinkler head allows the rest of the system to remain operational during overhaul.

If a sprinkler system is inoperable but the structure is otherwise capable of occupancy, the occupant will need to provide a **fire watch**—a temporary measure intended to ensure continuous and systematic surveillance of a building, or portion of the building, by one or more qualified individuals for the purpose of identifying and controlling fire hazards, detecting early signs of fire, activating an alarm, and notifying the fire department—until the sprinkler system is returned to normal operation.

To stop the flow of water from a sprinkler using a pair of sprinkler wedges, follow the steps in **SKILL DRILL 18-2**. To stop the flow of water from a sprinkler using a sprinkler stop, follow the steps in **SKILL DRILL 18-3**.

Not all sprinklers can be shut off with a sprinkler wedge or a sprinkler stop. Recessed sprinkler heads, which are often installed in buildings with finished ceilings, are usually difficult to shut off. If the individual heads cannot be shut off or if several sprinklers have been activated, you can stop the flow by closing the main sprinkler control valve. After the valve is closed, the water trapped in the piping system will drain out through the activated sprinklers for several minutes. If there is a drain valve near the main sprinkler control valve, open it to drain the system quickly. This step directs the water flow to a location where it will not cause additional

SKILL DRILL 18-2

Using Sprinkler Wedges Firefighter I, NFPA 1010: 6.3.14

1. Hold one wedge in each hand.

2. Insert the two wedges, one from each side, between the discharge orifice and the sprinkler deflector.

3. Bump the wedges securely into place to stop the water flow.

SKILL DRILL 18-3

Using a Sprinkler Stop Firefighter I, NFPA 1010: 6.3.14

1. Hold the sprinkler stop in one hand.

2. Place the flat-coated part of the sprinkler stop over the sprinkler orifice and between the frame of the sprinkler.

3. Push the lever to expand the sprinkler stop until it snaps into position.

damage and ensures the entirety of the system is drained via accessing the lowest point in the system.

The following are three of the most common main sprinkler control valves used as main water supply valves:

- Outside screw and yoke (OS&Y) valve
- Butterfly valve
- Post indicator valve (PIV)

An **outside screw and yoke (OS&Y) valve**—also called an **outside stem and yoke valve**—has a threaded stem that moves in and out as the valve is opened or closed. If the stem is out, the valve is open; if the stem is in, the valve is closed (**FIGURE 18-9**). OS&Y valves are typically found in a mechanical room in the basement or on the ground-floor level of a building, where the water supply for the sprinkler system enters the building. In warmer climates, they may be found outside.

Butterfly valves have an internal disk the same diameter as the pipe that either blocks the pipe and prevents water from flowing into the system or is rotated to allow the water to flow past it. The valve can be rotated to any position to control the water flow. When it is rotated a full 90 degrees, the valve is completely open for maximum water flow. An external indicator on the

FIGURE 18-9 The screw or stem on an OS&Y valve indicates whether the valve is open or closed. The screw on this OS&Y valve is out, indicating that this valve is open.

Courtesy of Chris DeSantis.

FIGURE 18-10 The yellow butterfly valve indicator is parallel to the pipe, indicating that the valve is open.

stem shows if the valve is open or closed. If the indicator is perpendicular to the pipe, the valve is closed. If the indicator is parallel (in line) with the pipe, the valve is open (**FIGURE 18-10**).

A.

B.

FIGURE 18-11 **A.** A ground PIV is mounted in the ground to open or close an underground valve. **B.** A wall PIV is mounted on the outside wall of a building.

A. © Jones & Bartlett Learning; **B.** © Jones & Bartlett Learning. Courtesy of MIEMSS.

A **post indicator valve (PIV)** is, as its name implies, a post with an indicator that reads either "OPEN" or "SHUT" (**FIGURE 18-11**). A PIV is usually located outside the building or on an exterior wall. Opening or closing a PIV requires a wrench, which is usually attached to the side of the valve.

In larger facilities, the sprinkler system may be divided into zones, with zone valves that control the flow of water to sprinklers in different areas (zones) of the building. These valves allow a zone to be shut off without turning the entire system off. High-rise buildings, for example, usually have a zone valve for each floor. Closing a zone valve stops the flow to sprinklers in that zone, but the sprinklers in the rest of the building remain operational. The locations of the main control valves and zone valves should be identified during preincident planning visits to a building.

Sprinkler control valves should always be locked in the open position, usually with a chain and padlock, and should be equipped with a tamper alarm that is tied into the fire alarm system. Firefighters who are sent to shut off a sprinkler control valve should take a pair of bolt cutters to remove the lock if the key is not available.

To close and reopen an OS&Y valve, follow the steps in **SKILL DRILL 18-4**. To close and reopen a butterfly valve, follow the steps in **SKILL DRILL 18-5**. To close and reopen a PIV, follow the steps in **SKILL DRILL 18-6**.

Before a sprinkler system can be restored to normal operation, the sprinkler heads that were activated must be replaced. Every sprinkler system should have spare heads and head wrenches sized for the sprinkler heads used in the system stored in a head box, usually near the main control valves. The spare heads must be the same type and size and have the same temperature

SKILL DRILL 18-4

Closing and Reopening a Main OS&Y Valve Firefighter I, NFPA 1010: 6.3.14

1. Locate the OS&Y valve as indicated on the preincident plan. If there is more than one OS&Y valve, identify the valve that controls sprinklers in the fire area. If the valve is locked in the open position with a chain and padlock and the key is readily available, unlock the padlock and remove the chain. If no key is available, cut the lock or the chain close to the padlock with a pair of bolt cutters. (Cut a link close to the padlock so that the chain can be reused.)

2. Turn the valve handle clockwise to close the valve. Keep turning until resistance is strong and little of the valve stem is visible.

SKILL DRILL 18-4 CONTINUED

Closing and Reopening a Main OS&Y Valve Firefighter I, NFPA 1010: 6.3.14

3. To reopen the OS&Y valve, turn the handle counter-clockwise until resistance is strong and the valve stem is visible again. Lock the valve in the open position.

SKILL DRILL 18-5

Closing and Reopening a Main Butterfly Valve Firefighter I, NFPA 1010: 6.3.14

1. Locate the butterfly valve as indicated on the preincident plan. If the valve is locked in the open position with a chain and padlock and the key is readily available, unlock the padlock and remove the chain. If no key is available, cut the lock or the chain close to the padlock with a pair of bolt cutters. (Cut a link close to the padlock so that the chain can be reused.)
2. Check the top of the valve for an arrow indicating the direction to turn to close, and then turn the valve by hand 90 degrees in that direction to close the valve.
3. To reopen the butterfly valve, turn the valve 90 degrees in the opposite direction to fully open the valve. Lock the valve in the open position.

rating as the heads used in the system. Before replacing any sprinkler heads that had been activated, the main sprinkler control valve or the appropriate zone valve must be closed, and the system must be drained. After the heads that had been activated are replaced, the sprinkler system can be restored to service. Both replacing sprinkler heads and restoring a sprinkler system to service takes special training and should be performed only by qualified individuals.

SAFETY TIP

Do not shut down the water to a sprinkler system until ordered to do so by the IC. The fire must be completely out, and hose lines must be available should the fire reignite. A firefighter should be stationed at the main sprinkler control valve with a portable radio, ready to reopen the valve if necessary.

SKILL DRILL 18-6

Closing and Reopening a Main Post Indicator Valve Firefighter I, NFPA 1010: 6.3.14

1. Locate the PIV as indicated on the preincident plan. If the valve is locked in the open position with a chain and padlock and the key is readily available, unlock the padlock and remove the chain. If no key is available, cut the lock or the chain with a pair of bolt cutters. Cut a link close to the padlock so that the chain can be reused.
2. Remove the handle from its storage position on the PIV, and place it on top of the valve, similar to the use of a hydrant wrench. Turn the valve stem in the direction indicated on top of the valve to close the valve. Keep turning until resistance is strong and the visual indicator changes from "OPEN" to "SHUT."
3. To reopen the PIV, turn the valve stem in the opposite direction until resistance is strong and the indicator changes back to "OPEN." Lock the valve in the open position.

Removing Water

Water that accumulates within a building or drips down from higher levels should be channeled to a drain or to the outside of the building to prevent or limit water damage to the structure. A salvage pump may be needed to help remove the water in some cases. Some buildings have floor drains that funnel water into a below-ground sewer system. These floor drains should be kept free from debris so that the water can drain freely. Firefighters can use squeegees to direct the water into the drain.

Water on a floor at ground level can often be channeled to flow outside of the structure through a doorway or other opening. Water on a floor above ground level can sometimes be drained to the outside by making an opening at floor level in an exterior wall of the building.

Water chutes or water catch-alls are often used to collect water leaking down from firefighting operations on higher floor levels and to help protect property on calls involving burst pipes or leaking roofs. A **water chute** catches dripping water and directs it toward a drain or to the outside through a window or doorway. A **water catch-all** is a temporary pond that holds dripping water in one location. The accumulated water must then be drained to the outside of the building. Water chutes and catch-alls can be constructed quickly using salvage covers or a roll of polyethylene film. To construct a water chute using a salvage cover, follow the steps in **SKILL DRILL 18-7**. To construct a water catch-all, follow the steps in **SKILL DRILL 18-8**.

A **wet vacuum** (also known as a **water vacuum**) is a special vacuum cleaner that sucks up liquids and is often used during salvage operations. Two types of wet vacuums are available: a small-capacity, backpack type and a larger wheeled unit. The backpack wet vacuum cannot be used by someone who is wearing SCBA. Wet/dry shop vacuums can be used as a low-cost alternative to wet vacuums.

Drainage pumps remove water that has accumulated in basements or below ground level. Portable electric submersible pumps can be lowered directly into the water to pump it out of a building. Gasoline-powered portable pumps must be placed outside a building because they exhaust poisonous carbon monoxide gas. These pumps draft water out of the building by lowering a hard suction hose into the basement through a window or other opening.

SKILL DRILL 18-7

Constructing a Water Chute Firefighter I, NFPA 1010: 6.3.14

1. Fully open a large salvage cover flat on the ground.

2. If using pike poles, lay one pole on one edge of the cover, and roll the cover around the handle. Roll the cover tightly toward the middle.

3. Repeat the actions in Step 2 on the opposite edge of the cover, rolling the opposite edge tightly toward the middle. If using pike poles, make sure the metal head is on the same side of the cover as the head on the other pike pole. Continue rolling both edges tightly toward the middle until the two rolls are 1 to 3 ft (30 to 91 cm) apart.

4. Turn the cover upside down.

Continues.

SKILL DRILL 18-7 CONTINUED

Constructing a Water Chute Firefighter I, NFPA 1010: 6.3.14

5. Position the chute with one end on the floor and the other end propped up by a chair, stepladder, or other tall object so that it collects dripping water and channels it toward a drain or outside opening.

SKILL DRILL 18-8

Constructing a Water Catch-All Firefighter I, NFPA 1010: 6.3.14

1. Fully open a large salvage cover flat on the ground. Roll one of the edges of the cover toward the middle about 2 ft (61 cm). Repeat with the opposite edge.

2. Fold each corner over at a 90-degree angle creating triangles, starting each fold approximately 3 ft (91 cm) in from the edge.

SKILL DRILL 18-8 CONTINUED

Constructing a Water Catch-All Firefighter I, NFPA 1010: 6.3.14

3. Continue rolling one of the edges that is already rolled toward the middle until the roll covers the two triangles at that end.

4. Lift the rolled edge over the corner flaps of the triangles, and then tuck it in under the flaps of the triangles to lock the corners in place. Repeat Steps 3 and 4 with the opposite rolled edge.

SAFETY TIP

You cannot operate a gasoline-powered pump safely inside a structure because the operation of the pump generates deadly carbon monoxide gas.

Using Salvage Techniques to Limit Smoke and Heat Damage

One way to reduce property loss is to keep heat and smoke out of areas that are not involved in the fire. A closed door can effectively keep smoke and heat out of a room or area. Firefighters performing search and rescue operations should remember to close doors after searching a room.

Properly timed and effective ventilation practices using exhaust fans and natural ventilation can also limit smoke and heat damage. Rapid ventilation often reduces smoke damage in an area already filled with smoke. The visible components of smoke are mostly soot particles and other products of combustion, including corrosive chemicals that settle on horizontal surfaces as the smoke cools. Blowing the smoke out of the building is a better option than allowing these contaminants to settle on the contents. Fans should not be used to clear smoke out of an area to help prevent damage until it is confirmed that the fire has been extinguished and there are no more hot spots. All members of the fire suppression team—not just the salvage crews—should recognize opportunities to prevent smoke and heat damage.

Salvage Covers

The most common method of protecting building contents is to cover them with salvage covers. As previously mentioned, salvage covers are large square or rectangular sheets of heavy canvas or plastic material that are used to protect furniture and other items from water runoff, falling debris, soot, and particulate matter in smoke residue. Salvage crews usually begin their work on the floor immediately below the fire, with the goal of preventing water damage to furniture and other contents on lower floors. If the fire is in an attic, firefighters may have enough time to spread covers over the furniture in the rooms directly below the fire before pulling the ceilings to attack the flames.

The most efficient way to protect a room's contents is to move all the furniture to the center of the room, away from the walls, where water could damage the

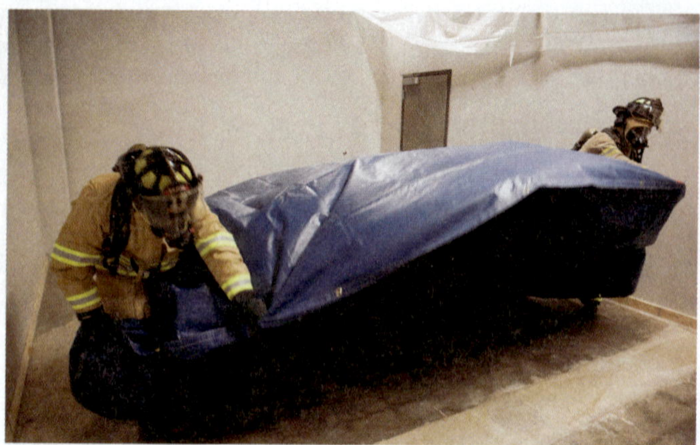

FIGURE 18-12 Move furniture to the center of the room so that a single salvage cover can protect all of the contents.

backs of the furniture. This approach reduces the total area that must be covered, enabling one or two firefighters to cover the pile with a salvage cover quickly and move on to the next room (**FIGURE 18-12**).

Remove any pictures from the walls and place them with the furniture. Put smaller pictures and valuable objects in drawers or wherever they will be protected from breakage. If you have enough time, roll up any rugs, and place them on the pile.

Some departments use rolls of construction-grade polyethylene film for protecting a structures' contents instead of salvage covers. This material comes in rolls as long as 120 ft (36.6 meters), so it can be unrolled over the room's contents and cut to the correct length with a box cutter. Polyethylene film is particularly useful for covering long surfaces, such as retail display counters. This material is disposable and can be left behind after the fire; in contrast, traditional salvage covers must be picked up, washed, dried, and properly folded for storage in fire apparatus compartments after each use. Occasionally, salvage covers may be left behind as protection for the building contents, but they should be picked up when they are no longer needed.

Storing and Deploying Salvage Covers

Special folding and rolling techniques are used to store salvage covers so that they can be deployed quickly by one or two firefighters. Be certain to follow your fire department's SOPs or SOGs when folding or rolling a salvage cover.

To fold a salvage cover to prepare it for one firefighter to deploy, follow the steps in **SKILL DRILL 18-9**. To fold a salvage cover to prepare it for two firefighters to deploy, follow the steps in **SKILL DRILL 18-10**.

SKILL DRILL 18-9

Performing a Salvage Cover Fold for One-Firefighter Deployment
Firefighter I, NFPA 1010: 6.3.14

1. Spread the salvage cover flat on the ground with a partner facing you.

2. At one end of the salvage cover, make a fold at the quarter point of the cover. Next, make a second fold that ends in the middle of the cover.

SKILL DRILL 18-9 CONTINUED

Performing a Salvage Cover Fold for One-Firefighter Deployment
Firefighter I, NFPA 1010: 6.3.14

3. At the other end of the salvage cover, make a fold at the quarter point of the cover. Next, make a second fold that ends in the middle of the cover.

4. Fold the two halves together, and flatten the salvage cover to remove any trapped air.

5. Make 1-ft (30-cm) folds from each end of the cover until you reach the center of the salvage cover.

6. Fold the two halves together.

SKILL DRILL 18-10

Performing a Salvage Cover Fold for Two-Firefighter Deployment
Firefighter I, NFPA 1010: 6.3.14

1. Spread the salvage cover flat on the ground with a partner facing you. Together, fold the cover in half.

2. Together, grasp the unfolded edge, and fold the cover in half again. Flatten the salvage cover to remove any trapped air.

3. Move to the newly created narrow ends of the salvage cover, and then fold the salvage cover in half lengthwise.

4. Fold the salvage cover in half lengthwise again. Make certain that the open end is on top.

5. Fold the cover in half a third time.

You can also store a salvage cover in a roll rather than folded flat. Although it often takes two people to fold and roll a salvage cover for storage, when it is stored in a roll, one person can unroll and deploy the cover easily and quickly. To fold and roll a salvage cover, follow the steps in **SKILL DRILL 18-11**.

A single firefighter can deploy a rolled salvage cover or a salvage cover folded for one-firefighter deployment. To perform the one-person salvage cover roll with a rolled salvage cover, follow the steps in **SKILL DRILL 18-12**. To perform a salvage cover shoulder toss, follow the steps in **SKILL DRILL 18-13**

Two firefighters can use a balloon toss to cover a pile of building contents quickly. To perform a balloon toss, follow the steps shown in **SKILL DRILL 18-14**.

Salvage Cover Maintenance

Salvage covers must be adequately maintained to preserve their useful life. Proper maintenance depends on the salvage cover material. A canvas salvage cover can usually be cleaned with a scrub brush and clean water. If the cover becomes particularly dirty, it may be necessary to clean it with a mild detergent. Covers should

SKILL DRILL 18-11

Folding and Rolling a Salvage Cover Firefighter I, NFPA 1010: 6.3.14

1. Spread the salvage cover flat on the ground with a partner facing you.

2. Together, fold the outside edge in to the middle of the cover, creating a fold at the quarter point.

3. Fold the outside fold in to the middle of the cover, creating a second fold.

4. Repeat Step 2 from the opposite side of the cover.

Continues.

SKILL DRILL 18-11 CONTINUED

Folding and Rolling a Salvage Cover Firefighter I, NFPA 1010: 6.3.14

5. Repeat Step 3 from the opposite side of the cover so the folded edges meet at the middle of the cover, with the folds touching but not overlapping.

6. Tightly roll up the folded salvage cover from the end.

SKILL DRILL 18-12

Deploying a Salvage Cover with the One-Person Salvage Cover Roll Firefighter I, NFPA 1010: 6.3.14

1. Stand in front of the end of the object that you are going to cover.

2. Start to unroll the cover over one end of the object.

SKILL DRILL 18-12 CONTINUED

Deploying a Salvage Cover with the One-Person Salvage Cover Roll
Firefighter I, NFPA 1010: 6.3.14

3. Continue unrolling the cover until you reach the top of the object. Allow the remainder of the cover to unroll and settle at the end of the object.

4. Spread the cover, unfolding each side outward over the object to the first fold.

5. Unfold the second fold on each side, and then drape the cover completely over the object.

6. Tuck in all loose edges of the cover around the object so that no one trips over the loose ends.

SKILL DRILL 18-13

Performing a One-Firefighter Salvage Cover Shoulder Toss Firefighter I, NFPA 1010: 6.3.14

1. Place the folded salvage cover over one arm.

2. Toss the cover over the salvaged object with a straight-arm movement.

3. Unfold the cover until it completely drapes the object.

be adequately rinsed when a detergent is used. Canvas covers must be properly hung and dried before being returned to service to reduce mildewing. Vinyl salvage covers are easily maintained by rinsing. Although they do not mildew as easy as canvas covers, they should also be hung to dry before storing them.

Once dried, salvage covers should be inspected for tears and holes. Any damage found can be patched by following the manufacturer's guidelines.

Floor Runners

As previously mentioned, floor runners protect carpets or hardwood floors from water, debris, firefighters' boots, and firefighting equipment. Firefighters entering an area for salvage operations should unroll this protective material ahead of themselves and stay on the floor runner while working in the area.

Other Salvage Operations

The best way to protect the contents of a building may be to move them to a safe location. The IC makes this determination. Any items removed from the building should be placed in a dry, secure area—preferably a single location. In some cases, salvaged items can be moved to a suitable location within the same building. If items are moved outside, they should be protected from further

SKILL DRILL 18-14

Performing a Two-Firefighter Salvage Cover Balloon Toss Firefighter I, NFPA 1010: 6.3.14

1. Place the salvage cover on the ground beside the object.

2. Unfold the cover so that it runs along the entire base of the object. Each firefighter grabs one edge of the cover and brings it up to waist height.

3. Together, lift the cover quickly so that it fills with air like a balloon.

4. Move quickly to the other side of the item, using the air to help support the cover, and spread the entire cover over the object.

damage caused by firefighting operations or the weather. Valuable items should be placed in the care of a law enforcement officer if the property owner is not present.

Sometimes the building contents can provide clues about the cause or spread of the fire. In these situations, fire investigators should be consulted, and they should supervise the removal process. If possible, it is advantageous to get pictures of the building contents in their original location prior to moving.

Salvage operations can extend outside the building to include valuable property such as vehicles or machinery. Use a salvage cover to protect outside property or move vehicles if this action can be done without compromising the fire suppression efforts.

Overhaul Overview

After a fire has been knocked down, a new fire remains a possibility even if 99 percent of a fire is out and just 1 percent is smoldering. Indeed, a single pocket of embers can rekindle after firefighters leave the scene and cause more damage and destruction than the original fire. A fire cannot be considered fully extinguished until the overhaul process is complete.

The process of overhaul begins after the fire is brought under control. Overhaul can be a time-consuming, physically demanding process. The greatest challenge during this phase of fire operations is to identify and open void spaces in a building where the fire might be burning undetected. If the fire extends into any void spaces, firefighters must open the walls and ceilings to expose the burned area. Any materials that are still burning must be soaked with water or physically removed from the building. This process continues until all of the burned material is located and unburned areas are exposed. Overhaul is also required for non-structure fires such as those occurring in automobiles, bulk piles (tires or woodchips), vegetation (wildland), and even garbage, as well as any other object where fire is prone to hide and be difficult to reach.

Health Considerations during Overhaul

During overhaul, you need to be aware of the threats to your health. Overhaul, even after ventilation, is conducted in an environment that has unknown amounts of smoke, soot, and other products of incomplete combustion. Smoke and soot contain particles from burned materials. The air contains poisonous gases such as carbon monoxide and hydrogen cyanide in unknown quantities. If overhaul is being conducted in an enclosed compartment, the atmosphere may contain elevated levels of carbon dioxide or decreased levels of oxygen.

In addition to carbon monoxide and hydrogen cyanide, smoke and soot contain toxic compounds such as ammonia and formaldehyde. These airborne compounds present two types of health hazards to firefighters. Some of these gases, such as carbon monoxide and hydrogen cyanide, can result in immediate and deadly illness. Others are carcinogens that present long-term health hazards by increasing the chance of contracting a variety of cancers later in life.

The poisonous substances present during overhaul can enter the body in three ways: absorption (through the skin), inhalation (through the lungs), and accidental ingestion. To prevent the poisonous particles in smoke and soot from being absorbed through your skin, wear full PPE, including SCBA. In addition to protecting your face from absorption, SCBA prevents poisonous gases and particles from being inhaled into the lungs. When overhaul is complete, carefully remove and clean contaminated PPE as soon as possible. This as described further in Chapter 3, *Personal Protective Equipment*. Be sure to thoroughly shower to remove any toxic substances from your hair and skin and clean your hands and body before eating to prevent inadvertently ingesting toxic products. These steps will go a long way toward preventing immediate and long-term health problems.

SAFETY TIP

SCBA must be worn during overhaul to protect the wearer from inhaling toxins and poisons that are present in smoke and off-gassing materials.

An additional health consideration is adequate rehabilitation. Working for extended periods in full PPE creates heat stress and dehydration. Overhaul crews should work for short periods and take frequent breaks. Fresh crews or crews that have been properly rehabilitated should replace fatigued crews. Chapter 20, *Firefighter Rehabilitation*, discusses this topic in more detail.

Because there is such a wide variety of toxic substances present at the fire scene, there is no single best practice when it comes to detection and monitoring air quality in the fire environment, especially during overhaul. Although most fire gases and particulates can be detected and measured by one type of technology or another, there is no single device that can detect all of them.

The most common method of gas detection involves the use of single-gas detection devices; however, detection of a single gas does not determine whether the entire building is safe or unsafe to operate in without SCBA. For example, testing the air for carbon monoxide levels does nothing to measure the amount of complex carcinogens present. Additional methods of detection and monitoring include using multi-gas meters configured with a variety of sensors (e.g., oxygen, carbon monoxide, hydrogen sulfide, flammable gas) and using more than one technology to look for multiple gases. To use a multi-gas meter to provide atmospheric monitoring, follow the steps in **SKILL DRILL 18-15**. The authority having jurisdiction will determine the exact monitoring device to be used for the particular situation, the alarm levels and configuration of that device, and the SOPs to follow if the device's high- or low-level alarm activates.

SKILL DRILL 18-15

Using a Multi-Gas Meter Firefighter I, NFPA 1010: 6.3.21

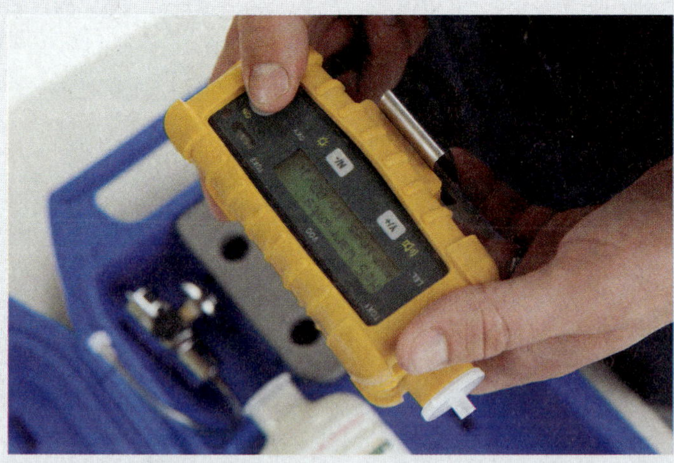

1. Turn the unit on and let it warm up (usually 5 minutes is sufficient) in a well-ventilated area, away from the fire scene and any vehicles. Ensure the battery has sufficient life for the operational period. Identify the installed sensors (e.g., oxygen, flammability, hydrogen sulfide, carbon monoxide), and verify that the sensors are not expired. Review alarm limits and the audio and visual alarm notifications associated with those limits. Review and understand the types of gases and vapors that could harm or destroy the sensors. Use other methods to check for those substances (e.g., pH paper) to ensure they are not present in the atmosphere to be sampled. Care must be taken to avoid pulling liquids into the device—it is designed to sample air, not liquid!

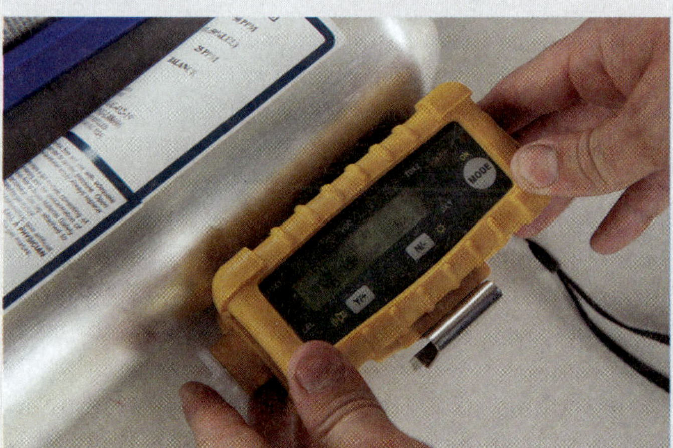

2. Perform a test on the pump by occluding the inlet and ensuring the appropriate alarm sounds.

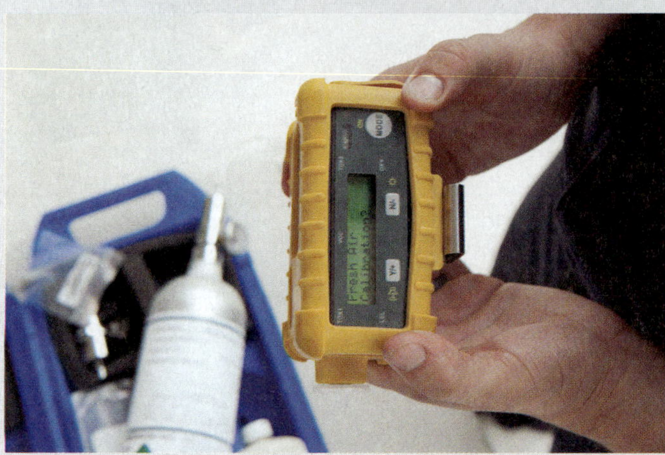

3. Perform a fresh air calibration, and "zero" the unit.

Continues.

SKILL DRILL 18-15 CONTINUED

Using a Multi-Gas Meter Firefighter I, NFPA 1010: 6.3.21

4. Ensure the meter is operating correctly by exposing the unit to a substance or substances that the unit should detect and react to accordingly. In essence, you are making sure the unit will "see" what it is supposed to see before it is called upon to see it in a real situation.

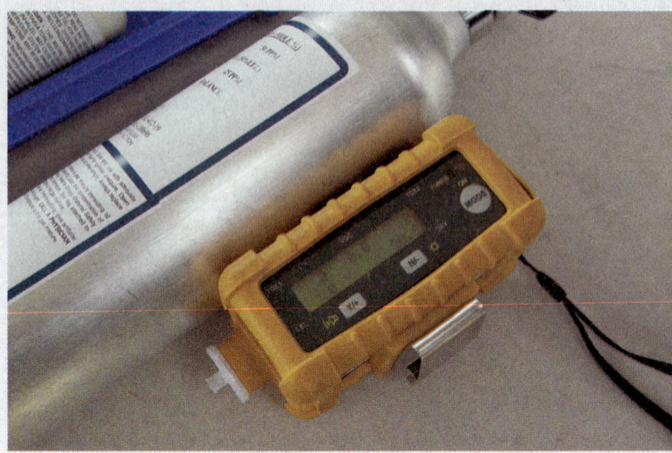

5. Allow the device to reset, or return to fresh air calibration state, then review the alarm levels and resetting procedures for addressing sensors that become saturated, or exposed to too much gas or vapor.

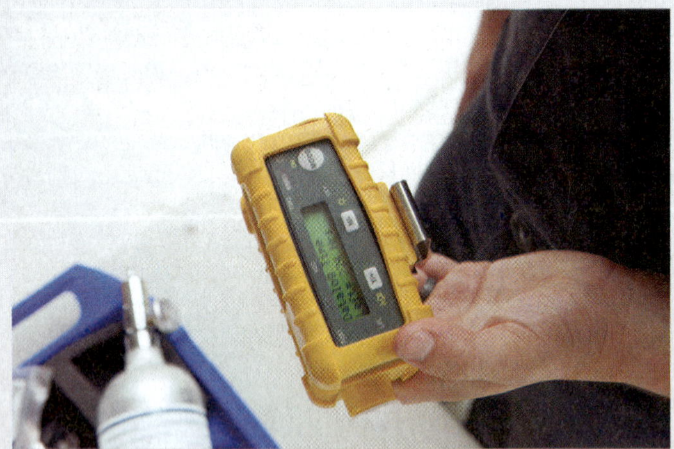

6. Review other device functions such as screen illumination, data logging (if available), and low-battery alarm.

7. Review decontamination procedures. Carry out monitoring and detection per the device manufacturer's instructions. If the high- or low-level alarm activates, follow the SOPs established by the authority having jurisdiction.

FIGURE 18-13 Electrochemical sensors ready to be installed and used.

Courtesy of Rob Schnepp.

Electrochemical sensors are one of the most common technologies used in post-fire detection and monitoring devices. These sensors are filled with a chemical reagent—a substance that is designed to react when it encounters another specific substance. Typical electrochemical sensors used in post-fire detection and monitoring test for oxygen, hydrogen cyanide, carbon monoxide, hydrogen sulfide, ammonia, and chlorine. When the reagent reacts with the target gas, a meter reading results (**FIGURE 18-13**). If the amount of gas present is above a preset level, the low-level or high-level alarm will sound. For example, a detection device with an electrochemical sensor used to detect carbon monoxide may have a low-level alarm setting of 35 ppm. A low-level alarm sounding during post-fire detection would indicate that the level of carbon monoxide in the area of detection has reached 35 ppm, and personnel should take protective action, such as donning an SCBA mask or performing additional ventilation. Electrochemical sensors are commonly found in single-gas devices.

FIGURE 18-14 A photoionization detector (PID) in service during a session of fire investigation research.

Courtesy of Rob Schnepp.

Another type of sensor is a photoionization detector (PID). A PID is available as a stand-alone unit, or it may be incorporated into a multi-gas meter (**FIGURE 18-14**). In either case, the PID uses ultraviolet light to ionize the gases that move through the sensor. When the gases are ionized, sensors are able to determine the ionization of the molecules. The PID detects organic materials (materials that contain carbon as part of their molecular structure), such as benzene, acetone, toluene, ethanol, butane, and many others. It will not detect the presence of carbon monoxide, hydrogen cyanide, or compounds such as natural gas. Additionally, the PID does not identify which toxic gas is in the air—it only alerts the user that a toxic gas is present.

Another option for the detection of toxic gases is colorimetric tubes, which can be used to detect the presence and/or levels of specific chemical vapors and/or to confirm the readings of the electrochemical sensors or other technologies. The glass tubes are filled with reagents that react to a unique substance at a particular concentration when a certain amount of sample air moves through the tube. The tubes are set up to detect single substances and/or chemical families (groups).

If there is concern that a department does not have the appropriate metering device to measure an airborne contaminant, firefighters can use a monitor that measures oxygen to determine if something abnormal is present. Because the typical oxygen level in air is 20.9 percent, any deviation from this percentage, usually shown as a reduction, will indicate that something abnormal is present and accounting for a portion of the air in the area. Keep in mind that a 0.1 percent reduction in oxygen, from 20.9 to 20.8 percent, indicates a 1000 ppm reduction. Some chemicals, such as hydrogen cyanide, are lethal at concentrations much lower than 1000 ppm, and may not show a change in the oxygen levels before reaching lethal concentrations.

Regardless of the technology used, air monitoring does not address airborne contamination in total. It is up to the user to understand the alarm levels of each type of device used by their department and to know the established emergency actions to take if the low- or high-level alarm on the device sounds. Ultimately, the safest way for a firefighter to reduce the possibility of inhalation exposures is to keep their SCBA on!

Safety Considerations during Overhaul

Many injuries can occur during overhaul. As mentioned earlier, overhaul is strenuous work conducted in an area already damaged by fire. Firefighters who were involved in the fire suppression efforts may be physically fatigued and overlook hazards during overhaul, and as previously mentioned, firefighters conducting overhaul need adequate breaks for rehabilitation.

Overhaul should not be rushed and should occur at a steady pace. When a structure is being overhauled, the IC must keep in mind the unique characteristics of the fire building. Quickly beginning overhaul in a balloon-frame structure is much more important than timeliness of overhaul in a single-story structure built on a slab because fire can quickly travel in the void spaces of balloon-frame homes. For example, a basement fire in a balloon-frame structure that has been knocked down but not yet overhauled can quickly advance to the attic and burn the roof off if firefighters do not prioritize accessing void spaces and overhauling the fire area. Remember that smoke is unburned fuel. If large quantities of smoke reach ignition temperature, it may ignite and create a backdraft or provide a fire with a fuel supply.

Additional hazards may be present when overhauling a structure fire. Notably, the structural safety of the building is often compromised. Catastrophic building collapses have occurred during overhaul. Heavy objects could lead to roof or ceiling collapse, debris could litter the area, and there could be holes in the floor. Visibility is often limited, so firefighters may have to depend on portable lighting. The presence of wet or icy surfaces makes falls more likely. Utilities, if not secured, can be dangerous to a firefighter opening a void space (**FIGURE 18-15**). In addition, during overhaul operations, dangerous equipment, including axes, pike poles, and power tools, is used in close quarters.

FIGURE 18-15 Electrical wires can pose a shock hazard to a firefighter opening an area to expose hidden fire.

Courtesy of Brandon Hausbeck.

Firefighters must be aware of all of these hazards and proceed with caution. When necessary, take extra time to evaluate the hazards and determine the safest way to proceed.

For example, a smoldering fire in a void space could reignite suddenly when firefighters open the space and allow fresh air to enter. This fire could quickly become intense and dangerous. In addition, fine-grained combustible materials such as sawdust can smolder for a long time and then ignite explosively when they are disturbed and oxygenated. A charged hose line must always be ready for use during overhaul operations because flare-ups may occur (**FIGURE 18-16**).

A safety officer should always be present during overhaul operations to note any hazards and ensure that operations are conducted safely. Company officers should supervise operations, look for hazards, and make sure that all crew members work carefully. The work should proceed at a steady pace, with an appropriate number of crew members. Too many people working in a small area creates limited work space and can also overload fire-weakened structural members.

FIGURE 18-16 A charged hose line should be in place and ready to extinguish fire that can flare up once a void space is opened.

Courtesy of Brandon Hausbeck.

Always evaluate the structural condition of a building before beginning overhaul. Chapter 6, *Building Construction*, discusses building collapse. Signs of collapse may include the following:

- Cracked or leaning walls
- Pitched or sagging doors or doors stuck in shifted frames
- Moaning/groaning or cracking sounds
- Any type of movement or vibrations
- Movement or shifting of water on the floor
- Significant fire damage to structural members
- Excessive standing water
- Pre-existing decay of structural components

During the overhaul phase, make sure you do not compromise the structural integrity of the building. When opening walls and ceilings, remove only the outer coverings, and leave the structural members in place. If more invasive overhaul is required, be careful not to compromise the structure's load-bearing members. Avoid cutting lightweight wood trusses, load-bearing wall studs, and floor or ceiling joists, especially when using power tools.

If fire damages the structural integrity of a building, the IC may call for a "hydraulic overhaul" rather than a standard overhaul. In this situation, large-caliber hose streams are used to completely extinguish a fire from the exterior. This strategy is appropriate if the site poses excessive risks to firefighters and damage to the property is so extensive that it has no salvage value. Heavy mechanical equipment may be used to demolish unsafe buildings and expose any remaining hot spots. A condemned or abandoned building that is going to be torn down is not a place to risk the life or health of a firefighter.

If a complete overhaul cannot be conducted, the IC can establish a fire watch. The fire watch team remains at the fire scene with a charged hose line and watches for signs of rekindling. This team can request additional help if the fire reignites.

Preserving Evidence during Overhaul

Overhaul crews must work with fire investigators to ensure that important evidence that could indicate the cause or the area of origin of a fire is not lost or destroyed as a result of their efforts to extinguish all remnants of the fire. Ideally, a fire investigator should examine the area before overhaul operations begin, identifying evidence and photographing the scene before it is disturbed.

If a fire investigator is not immediately available, the overhaul crews should make careful observations and report to the investigator later. To help the investigator piece together the location of building contents, pictures of the fire area should be taken prior to moving items and overhauling. When performing overhaul in or near the suspected area of origin, note the fire patterns and smoke residue on the walls or ceilings that could indicate the exact site of origin or the path the fire traveled. Often, the point of origin is directly below the most damaged area on the ceiling, where the heat of the fire was most intensely concentrated. Fire patterns and damage often spread outward from the room or area where the damage is most severe. When moving appliances or other electrical items, note whether they were plugged in or turned on. Always look for evidence that the fire investigator can use to determine the cause of the fire.

If you discover anything suspicious—particularly indications of arson—you should leave everything the way it was found, make sure that no one interferes with the item or the surrounding area, and then notify a fire officer or fire investigator immediately. Continue carefully overhauling the area to ensure that the fire does not rekindle, but do not do anything to unnecessarily compromise the scene once the fire is under control. Never discard debris until the fire investigator gives their approval to do so. Chapter 28, *Understanding Fire Investigation*, provides additional information on preservation of evidence in case of fire.

Where to Overhaul

Overhaul must be thorough and extensive enough to ensure that the fire is completely out. At the same time, firefighters should try not to destroy more property than is necessary. Determining when, where, and how much property needs to be overhauled requires good judgment and understanding of the relationship between fire behavior and building construction. Generally, it is better to make sure that the fire is definitely out than it is to be too careful about damaging property. If the fire rekindles, more property damage will occur, and the fire department could be held responsible for the additional losses.

The area that must be overhauled depends on the building's construction, its contents, and the size of the fire. All areas directly involved in the fire must be overhauled. If the fire was confined to a single room, all of the furniture in that room must be checked for smoldering fire. The overhaul process must ensure that the fire did not extend into any void spaces in the walls, above the ceiling, or into the floor. If signs indicate that the fire spread into the structure itself or into the void spaces, all suspect areas must be opened to expose any hidden fire.

TIP

A piercing nozzle can be used to extinguish hidden fire in void areas around the suspected area of origin.

If the fire involved more than one room, overhaul must include all possible paths of fire extension. The paths available for a fire to spread within a building are directly related to the type of construction. As a firefighter, you must learn to anticipate where and how a fire is likely to spread in different types of buildings to find any hidden pockets of fire. Understanding the basics of the type of construction and the age of the building and its contents will allow you to more quickly identify the most likely paths of fire extension and the potential areas of concern, such as void spaces between floors in balloon-frame construction.

Areas that might need to be overhauled are areas that show **thermal degradation**—that is, heat damage. Firefighters must understand the difference between heat damage and smoke staining. Smoke can easily travel to remote parts of the building and discolor and stain walls and ceilings (**FIGURE 18-17**). Smoke stains do not indicate heat transfer or fire extension; smoke stains indicate only that smoke traveled to those locations. Therefore, smoke stains do not mean that those areas need to be overhauled. In order for fire to spread, heat must be transferred. (Methods of heat transfer are discussed in Chapter 5, *Fire Behavior*.) Areas displaying evidence of thermal degradation should be checked for hidden fire (**FIGURE 18-18**). Examples of thermal

FIGURE 18-17 Smoke staining does not indicate heat transfer or fire extension and does not indicate an area that requires overhaul.

FIGURE 18-18 Areas that show evidence of thermal degradation need to be thoroughly overhauled.

degradation include blistered paint/finishings, discoloration from exposure to high heat, and charring.

Fire-resistive construction can help contain a fire and keep it from spreading within a building, although this outcome is not guaranteed. Look for any openings that would allow the fire to spread, including utility shafts, pipe chases, and fire doors or dampers that failed to close tightly.

Wood-frame and ordinary construction buildings may contain several areas where a hidden fire could be burning. In particular, these structures often have void spaces under the wall-covering materials. When a serious fire strikes a building of ordinary or wood-frame construction, firefighters may need to open many walls, ceilings, and void spaces to check for fire extension. Neighboring buildings may need to be overhauled if the fire has spread into them.

In balloon-frame construction, a fire can extend directly from the basement to the attic, without obvious signs of fire on any other floor (**FIGURE 18-19**). For this reason, these buildings require a thorough floor-by-floor overhaul. The problems could be compounded if the attic insulation consists of blown-in cellulose materials—these materials can smolder for a long time.

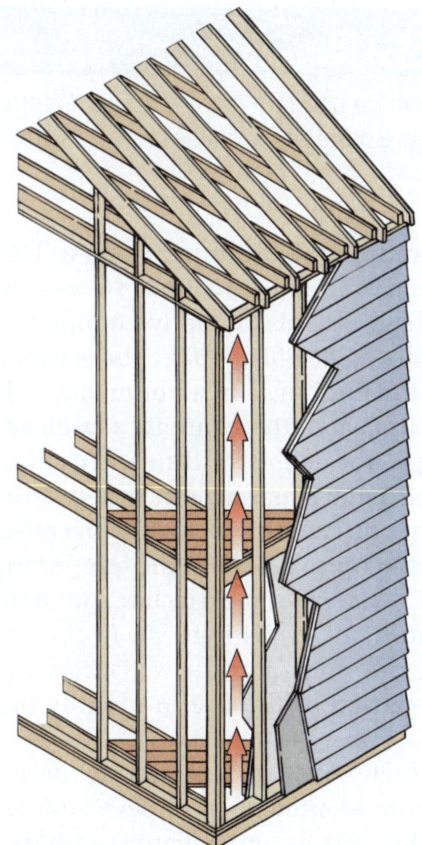

FIGURE 18-19 A fire in a balloon-frame construction building can spread directly from the basement to the attic.

TIP

To help determine where overhaul is needed, think of a fire compartment as having seven sides: the four walls of the compartment, the ceiling, the floor, and any void spaces. Thoroughly check each of these areas to be certain that the fire is fully extinguished.

An extensively remodeled building presents challenges for overhaul similar to the challenges faced by the ventilation team. Fire can hide in the space between a drop ceiling and the original ceiling, or it may extend into a different section of the building through doors and windows covered by new construction. Some buildings have two roofs—one original and one added later—with a void space between them. A fire in this void space presents especially difficult overhaul problems. Expect the unexpected; in many cases, the proper renovation permits were not obtained and renovations may not be compliant with code.

The cause of the fire can indicate the extent of necessary overhaul. A kitchen stove fire may extend into the kitchen exhaust duct or cabinetry above and could ignite combustible materials in the immediate vicinity. Follow the path of the duct and open the areas around it to locate any residual fire (**FIGURE 18-20**). A lightning

FIGURE 18-20 An improperly insulated stove pipe can cause fires in void spaces. The fire will travel in the void space, following the joists.

strike releases enormous energy. This energy can travel through the wiring and piping systems in a building, which can start multiple fires in different parts of the building. Overhaul after this type of incident must be extensive.

Using Your Senses

Efficient and effective overhaul requires the use of all of your senses. Look, listen, and feel to detect signs of potential burning.

Look for the following signs:

- Smoke seeping from cracks or from around doors and windows
- Fresh or new smoke
- Red, glowing embers in dark areas
- Burned areas
- Discolored material
- Water that quickly evaporates after being applied to a material
- Peeling paint or cracked plaster
- Hot spots shown on a thermal imager

Listen for the following sounds:

- Crackling sounds that indicate active fire
- Hissing sounds that indicate water has touched hot objects

Feel for this condition:

- Heat shown on a thermal imager or felt on the back of your hand (only if it is safe to remove your glove)

TIP

Learning the difference between smoke and steam can help prevent unnecessary water damage.

Experience is a valuable trait during overhaul situations. By using their senses in a systematic manner, an experienced firefighter can often determine which parts of the structure need to be opened and which areas were untouched by the fire.

Using Thermal Imagers

Thermal imagers are a valuable high-tech tool that is used during overhaul. Thermal imagers can be used to locate hidden hot spots or residual pockets of fire. It can quickly differentiate between unaffected areas and areas that need to be opened. Using a thermal imager device (TID) can decrease the amount of time needed to overhaul a fire scene and reduce the amount of physical damage to the building.

FIGURE 18-21 Thermal imagers can help locate hot spots.

SAFETY TIP

Drones can be used to visually inspect a structure to assess the potential for collapse.

Interpreting the readings from a TID requires practice and training (**FIGURE 18-21**). Remember that this device displays relative temperature differences, so an object will appear to be warmer or cooler than its surroundings. If a room has been superheated by fire, all of the contents as well as the walls, the floor, and the ceiling, will appear as hot. This phenomenon is known as washout, where a thermal imager screen goes completely white because all of the objects in the room are the same temperature. Under these circumstances, an extra-hot spot behind a wall could be difficult to distinguish. In this situation, it might be better to look at the walls from the side that was not exposed to the fire to identify hot spots or cool the objects.

Do not rely solely on the information from a thermal imager to identify areas of persistent fire because anything that acts as an insulator can hide hot spots. Hot spots behind heavy padding and carpeting, in insulated walls, or behind double-layer drywall, for example, might not show up as hot spots on the thermal imager screen. Firefighters must treat the thermal imager as just one tool in the complement of tools available to assist with the overhaul operation. While thermal imagers can help identify hot spots, they cannot rule out the possibility of concealed fire. The only way to ensure that no hidden fire exists is to create an inspection hole to verify TID readings.

Overhaul Techniques

The objective of overhaul is to find and extinguish any fires that could still be burning after a fire is brought under control. Any fire uncovered during overhaul must be thoroughly extinguished. During overhaul operations, a charged hose line should be available to douse any sudden flare-ups or exposed pockets of fire. When wetting down wall studs, soffits, and other areas of the structure that were involved in fire, use a solid or straight stream to ensure penetration of the water to effectively cool the material. Remember that the surfactants in Class A foam reduce the surface tension of water, which increases the ability for water to saturate and cool a material. To avoid unnecessary water damage, extinguish small pockets of fire or smoldering materials with the least possible amount of water by using a short burst from the hose line or simply drizzling water from the nozzle directly onto the fire. Because overhaul situations usually do not require high pressures or large volumes of water, a 1½- or 1¾-in. (38- or 44-mm) hose line is usually sufficient to extinguish hot spots. Follow your department's SOPs when choosing a hose line during overhaul.

SAFETY TIP

Before overhaul operations begin, the safety officer should confirm that electric power and gas service in the overhaul area are shut off. Be careful: after the gas supply is turned off, gas under pressure might still be present in the lines and will need to be bled off.

Extinguish smoldering objects that can be safely picked up by dropping them into a bathtub or bucket filled with water. Remove materials prone to smoldering, such as mattresses and cushioned furniture, from the building and thoroughly soak them outside. If possible, roll mattresses and secure them with rope or webbing to move them out of the building and decrease the possibility of rekindling while being moved out of the building. Once outside, the mattress can be unrolled and fully overhauled.

Place debris that is moved outside far enough away from the building without blocking entrances or exits to prevent any additional damage if it reignites. In some cases, a window opening can be enlarged so that debris can be removed more easily. Heavy machinery, such as a front-end loader, may be used to remove large quantities of debris from commercial buildings.

Tools used during overhaul can cut, pry, and pull so that firefighters can access spaces that might contain hidden fires. Many of these tools are also used for ventilation and forcible entry. The following tools are frequently required in overhaul operations:

- Pike poles and ceiling hooks are used to pull ceilings and remove gypsum wallboard.
- Crowbars and Halligan-type tools are used to remove baseboards and window or door casings.
- Axes are used to chop through wood, such as floor boards and roofing materials.
- Power tools such as battery-powered saws are used to open up walls and ceilings.
- Pitchforks and shovels are used to remove debris.
- Rubbish hooks and rakes are used to pull things apart.
- Thermal imagers are used to identify hot spots.
- Buckets, tubs, wheelbarrows, and carryalls are used to remove debris from a building. A **carryall** is a piece of heavy canvas with handles (**FIGURE 18-22**).

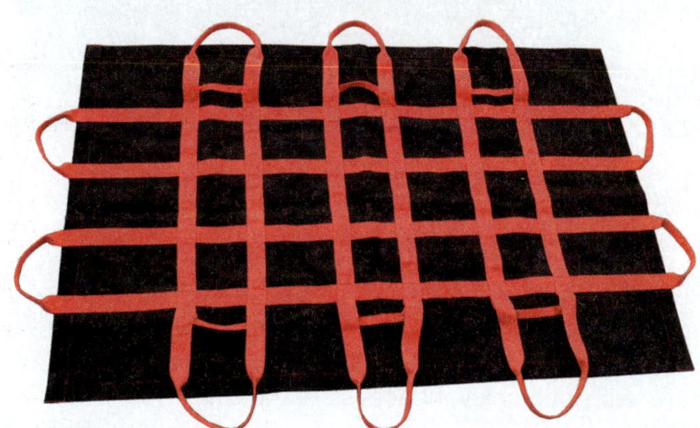

FIGURE 18-22 A carryall is used to remove debris during overhaul.

Courtesy of Cascade Fire Equipment Company.

TIP

Overhaul is physically demanding work. Pace yourself, use proper technique to increase your efficiency, and remember to take breaks for rehabilitation. Ask experienced members of your team for tips on improving your technique.

To pull down ceilings made with gypsum board, use the pike pole to break through the ceiling and pull down large sections. Pulling down laths and breaking through plaster ceilings require more force. Power saws may be required to cut through ceilings made with plywood or solid boards.

To open a hole in a wall, use pike poles, axes, power saws, and handsaws. Use the same technique to open a wall with a pike pole as you do to open a ceiling. When using an axe, make vertical cuts with the blade of the axe, and then pull the wallboard away from the studs using gloved hands or the pick end of the axe. You could also use a power saw to make vertical cuts, and then pull the wall section away with another tool or by hand.

To pull down a ceiling with a pike pole, follow the steps in **SKILL DRILL 18-16**. To open an interior wall with a pick-head axe, follow the steps in **SKILL DRILL 18-17**.

TIP

Overhaul operations should continue until the IC is satisfied that no smoldering fires are left.

SKILL DRILL 18-16

Pulling a Ceiling Using a Pike Pole Firefighter I, NFPA 1010: 6.3.13

1. Select the appropriate length of pike pole based on the height of the ceiling. Determine which area of the ceiling will be opened. Typically, the most heavily damaged areas are opened first, followed by the surrounding areas. Position yourself to begin work with your back toward a door so the debris you pull down will not block your access to the exit.

2. Hold the pike pole so that the hook side of the tip is away from you, and then use a strong, upward-thrusting motion to penetrate the ceiling with the tip of the pike pole.

SKILL DRILL 18-16 CONTINUED

Pulling a Ceiling Using a Pike Pole Firefighter I, NFPA 1010: 6.3.13

3. Pull down and away from your body, so that the ceiling material falls away from you.

4. Continue pulling down sections of the ceiling until the desired area is opened. Pull down any insulation, such as rolled fiberglass, found in the ceiling.

SKILL DRILL 18-17

Opening an Interior Wall Firefighter I, NFPA 1010: 6.3.13

1. Determine which area of the wall will be opened. Open those areas most heavily damaged by the fire first, followed by the surrounding areas.

2. Use the axe blade to begin cutting near the top of the wall, and then cut downward between wall studs. Be alert for electrical switches or receptacles, as they indicate the presence of electrical wires.

Continues.

SKILL DRILL 18-17 CONTINUED

Opening an Interior Wall Firefighter I, NFPA 1010: 6.3.13

3. Make a second vertical cut. Working from top to bottom, use the pick end of the axe to pull the wall material away from the studs and open the wall. Remove items such as baseboards or window and door trim with a Halligan tool or axe.

4. Continue opening additional sections of the wall until the desired area is open. Pull out any insulation, such as fiberglass, found behind the wall.

CASE STUDY

You Are the Firefighter CONCLUSION

Your shift has just started with Engine Company 3. You are doing your morning truck check when your company gets dispatched for a residential structure fire. You arrive and find a large, two-story residential structure with a fire on the first floor. The neighborhood you are in was developed in the early 1900s, and the structure appears to be a Type V structure with a basement and is balloon-frame construction. Your company is the fourth engine company on scene. Your crew approaches the incident commander and is told that interior crews are reporting the fire on the first floor is knocked down but they have not yet opened up any walls or ceilings. Command assigns your four-person crew to salvage operations.

1. On which areas of the building would you focus your salvage operation?

Answer: To protect the property most likely to be damaged, you would focus salvage operations on the areas closest to the fire location and the basement. If overhaul crews find fire extension to the second floor, you would extend the salvage operations to that floor as well.

2. What equipment will you use to prevent further damage to this house and its contents?

   Answer: A salvage crew would typically use the following to prevent further damage to the house and its contents:
   - Salvage covers or rolls of construction-grade polyethylene film to protect contents from water damage
   - Floor runners to protect carpet or hardwood floors from water, debris, and scratches or other damage
   - Ventilation fans to remove smoke once the fire is confirmed to be extinguished
3. Given the building's construction and fire location, where is the overhaul crew most likely to find hidden fire that has spread from the area of origin?

   Answer: Balloon-frame construction allows fire to spread upward in the walls of the structure, eventually reaching the attic space. This means the overhaul crew should check void spaces above the area of origin as well as the void spaces at the top of the house (the attic or the cockloft).

WRAP-UP

SUMMARY

KNOWLEDGE OBJECTIVES

- Describe the various lighting equipment and options to illuminate a work area.
  - Describe the types of lights used to illuminate exterior and interior scenes. (**NFPA 1010: 6.3.17**, pp. 796–797)
  - Describe the equipment used to illuminate an emergency scene. (**NFPA 1010: 6.3.17**, pp. 797–799)
  - Describe the safety precautions to take when working with lighting equipment. (**NFPA 1010: 6.3.17**, p. 799)
  - Describe how to operate lighting equipment to light exterior and interior scenes. (**NFPA 1010: 6.3.17**, p. 799)
- Describe the various salvage tools and their use in salvage techniques.
  - Explain the purpose of salvage operations. (**NFPA 1010: 6.3.14**, pp. 799–800)
  - List the tasks involved in a salvage operation. (**NFPA 1010: 6.3.14**, pp. 799–800)
  - Describe how forcible entry operations affect salvage operations. (**NFPA 1010: 6.3.14**, p. 800)
  - Describe the safety precautions that need to be considered when performing salvage. (**NFPA 1010: 6.3.14**, pp. 800–801)
  - List the tools used to perform salvage operations. (**NFPA 1010: 6.3.14**, p. 801)
- Explain how to prevent excess water damage, including how to shut off a fire sprinkler system or activated sprinkler head.
  - Describe the salvage techniques commonly used to prevent water damage. (**NFPA 1010: 6.3.14**, p. 801)
  - Describe the general procedures for preventing excess water damage from fire sprinklers. (**NFPA 1010: 6.3.14**, pp. 801–811)
  - List the equipment used to shut down fire sprinklers. (**NFPA 1010: 6.3.14**, pp. 801–803)
  - Describe the identifying characteristics of a main control valve of a fire sprinkler system. (**NFPA 1010: 6.3.14**, pp. 803–804)
  - Describe the general procedures and equipment used to remove excess water from a structure. (**NFPA 1010: 6.3.14**, pp. 808–811)
- Describe the salvage techniques to limit smoke and heat damage.
  - Describe the general procedures and equipment used to limit smoke and heat damage. (**NFPA 1010: 6.3.14**, pp. 811–819)
  - Describe how to maintain salvage covers. (**NFPA 1010: 6.5.1**, pp. 812–819)
  - Explain when fire investigators should become involved in salvage operations. (**NFPA 1010: 6.3.14**, pp. 825–826)

KNOWLEDGE OBJECTIVES CONTINUED

- Describe the purpose of and how safety is maintained during the overhaul phase.
  - Describe the purpose of overhaul operations. (**NFPA 1010: 6.3.13**, p. 820)
  - List the concerns that must be addressed to ensure the health of firefighters who are performing overhaul. (**NFPA 1010: 6.3.13**, p. 820)
  - Describe the common methods of air monitoring at the fire scene. (**NFPA 1010: 6.3.21**, pp. 821–824)
  - List the indicators of possible structural collapse. (**NFPA 1010: 6.3.13**, p. 825)
  - Explain how to preserve structural integrity during overhaul. (**NFPA 1010: 6.3.13**, p. 825)
  - Describe how to preserve evidence during overhaul operations. (**NFPA 1010: 6.3.13**, pp. 825–826)
  - Explain how firefighters determine overhaul locations. (**NFPA 1010: 6.3.13**, pp. 826–829)
- Describe the techniques utilized during the overhaul phase.
  - List the tools that are used for overhaul operations. (**NFPA 1010: 6.3.13**, pp. 829–830)
  - Describe the general techniques used in overhaul operations. (**NFPA 1010: 6.3.13**, pp. 829–832)

SKILLS OBJECTIVES

- Illuminate and provide a safe working area for firefighters completing salvage and overhaul operations.
  - Illuminate an emergency scene. (**NFPA 1010: 6.3.17**, p. 799)
- Shut down an activated fire sprinkler system.
  - Use a sprinkler wedge to shut down a sprinkler head. (**NFPA 1010: 6.3.14**, p. 803)
  - Use a sprinkler stop to shut down a sprinkler head. (**NFPA 1010: 6.3.14**, p. 804)
  - Close and reopen a main OS&Y valve. (**NFPA 1010: 6.3.14**, pp. 806–807)
  - Close and open a main butterfly valve. (**NFPA 1010: 6.3.14**, p. 807)
  - Close and open a main post indicator valve. (**NFPA 1010: 6.3.14**, p. 808)
- Deploy, fold, and store salvage covers.
  - Construct a water chute. (**NFPA 1010: 6.3.14**, pp. 809–810)
  - Construct a water catch-all. (**NFPA 1010: 6.3.14**, pp. 810–811)
  - Fold a salvage cover for one-firefighter deployment. (**NFPA 1010: 6.3.14**, pp. 812–813)
  - Fold a salvage cover for two-firefighter deployment. (**NFPA 1010: 6.3.14**, p. 814)
  - Fold and roll a salvage cover. (**NFPA 1010: 6.3.14**, pp. 815–816)
  - Perform a one-firefighter salvage cover roll. (**NFPA 1010: 6.3.14**, pp. 816–817)
  - Perform a salvage cover shoulder toss. (**NFPA 1010: 6.3.14**, p. 818)
  - Perform a salvage cover balloon toss. (**NFPA 1010: 6.3.14**, p. 819)
- Perform overhaul techniques utilizing firefighting tools and equipment.
  - Use a multi-gas air monitoring device. (**NFPA 1010: 6.3.21**, pp. 821–822)
  - Open a ceiling to check for fire using a pike pole. (**NFPA 1010: 6.3.13**, pp. 830–831)
  - Open an interior wall to check for fire. (**NFPA 1010: 6.3.13**, pp. 831–832)

KEY TERMS

area of origin A structure, part of a structure, or general geographic location within a fire scene, in which the "point of origin" of a fire or explosion is reasonably believed to be located. (NFPA 901)

butterfly valve A control valve in an automatic sprinkler system or on large pump intake connections where the suction hose connects to the suction side of the fire pump that either blocks the water and prevents it from flowing into the system or is rotated 90 degrees to allow the water to flow past it and that has an external indicator on the stem that is parallel or in-line with the piping to signify the valve is open or perpendicular to the piping to indicate the valve is closed.

carryall A piece of heavy canvas with handles, which can be used to tote debris, ash, embers, and burning materials out of a structure.

fire watch The assignment of a person or persons to an area for the express purpose of notifying the fire department, the building occupants, or both of an emergency; preventing a fire from occurring; extinguishing small fires; protecting the public from fire and life safety dangers. (NFPA 1)

floodlight A light that illuminates a broad area.

floor runner A piece of canvas or plastic material, usually 3 to 4 ft (91 to 122 cm) wide in various lengths, that is used to protect flooring from dropped debris and dirt from shoes and boots.

ground fault interrupter (GFI) A device that senses when there is a problem with an electrical ground and interrupts the current, shutting down both the power source and the equipment it is feeding.

junction box A device that attaches to an electrical cord to provide additional outlets.

outside screw and yoke (OS&Y) valve A sprinkler control valve with a valve stem that moves in and out as the valve is opened or closed. Also called *outside stem and yoke (OS&Y) valve.*

outside stem and yoke (OS&Y) valve See *outside screw and yoke (OS&Y) valve.*

overhaul A firefighting term involving the process of final extinguishment after the main body of the fire has been knocked down. All traces of fire must be extinguished at this time. (NFPA 1700)

post indicator valve (PIV) A sprinkler control valve with an indicator that reads either "open" or "shut" depending on its position.

salvage The process of protecting the contents within a building during and following the fire incident. (NFPA 1700)

salvage cover A large square or rectangular sheet made of heavy canvas or plastic material that is spread over furniture and other items to protect them from water runoff and falling debris.

secondary loss Property damage that occurs due to smoke, water, or other measures taken to extinguish the fire or mitigate the emergency incident.

spotlight A light designed to project a narrow, concentrated beam of light.

sprinkler stop A mechanical device inserted between the deflector and the orifice of a sprinkler head to stop the flow of water.

sprinkler wedge A piece of wedge-shaped wood placed between the deflector and the orifice of a sprinkler head to stop the flow of water.

thermal degradation Damage caused by heat transfer.

water catch-all A salvage cover that has been folded to form a container to hold water until it can be removed.

water chute A salvage cover that has been folded to direct water flow out of a building or away from sensitive items or areas.

water vacuum See *wet vacuum.*

wet vacuum A device similar to a wet/dry shop vacuum cleaner that can pick up liquids. It is used to remove water from buildings. See also *water vacuum.*

REVIEW QUESTIONS

1. What is the difference between a hand light and a personal flashlight?
2. What do salvage operations in residential structures often focus on? How do salvage operations in commercial structures differ?
3. Describe the difference between a water chute and water catch-all.
4. Explain why it is generally more desirable to shut off the water flow to an individual activated sprinkler head than shutting down the entire system at the main sprinkler valve.
5. What are the two pieces of equipment that can be used to stop the flow of water from a flowing sprinkler head without shutting down the entire sprinkler system?
6. Why is it important to ensure salvage covers are cleaned and dried after use?
7. What are signs of a compromised structure and possible building collapse?
8. What tools are frequently used in overhaul operations?
9. What piece of equipment is useful when breaching ceilings constructed with plywood or solid boards?

DISCUSSION QUESTIONS

1. Explain the importance of ensuring electrical equipment is properly grounded.
2. Describe the secondary losses that are encountered at a fire scene.
3. What are some signs of potential burning in a concealed void space?

APPLYING THE CONCEPTS

Your crew has just arrived at an active fire on the second floor of a three-story apartment building. The building is wood-frame construction with a brick-clad exterior, and it is equipped with a sprinkler system. All residents have been evacuated from the building. Crews onsite are conducting an interior attack to suppress the fire. There is no preincident plan for this building. Incident dispatch reports that the crews inside have knocked down the fire in the back of the building and are working to contain the fire in the front of the building. Sprinklers have activated. The IC orders your crew to begin salvage operations.

1. Based on this information, what salvage options do you consider?
2. What are the salvage operation priorities?

The salvage officer divides your crew into two teams. "Team A, go to the second floor. Shut off the sprinklers and start removing water. Close doors to isolate unaffected areas and set up fans to remove smoke."

3. Since the fire is almost under control, should Team A shut down the sprinkler system to limit water damage? Why or why not?
4. How should firefighters shut off the sprinklers?

"OK, Team A, let's go to the first floor. Our primary objectives are to limit water damage and save as much of the personal property as we can." You don your protective gear and get going. Upon entering, you lay out floor runners to protect the floor. The salvage officer directs your team to the area under the fire room where water is draining through the ceiling. He directs you and your partner to construct a water catch-all.

5. What is a water catch-all and what tools will you need to make one?

Incident command reports the fire on the second floor is extinguished and a unit has begun overhaul in the fire room to locate and extinguish hot spots. While doing this, they discover the structure is balloon-frame construction. The salvage officer receives orders and tells your crew, "We're ceasing salvage operations and moving to the third floor for overhaul."

6. What is balloon-frame construction and why is it important to conduct overhaul on the floors above the fire?
7. When you arrive on the third floor, you do not observe any obvious fire infiltration. Where should overhaul efforts be focused primarily and what tools should be used?

REFERENCES

National Fire Protection Association (NFPA). 2022. *NFPA 901, Standard Classifications for Fire and Emergency Services Incident Reporting, 2021 Edition*. Quincy, MA: NFPA.

National Fire Protection Association (NFPA). 2022. *NFPA 1700, Guide for Structural Fire Fighting, 2021 Edition*. Quincy, MA: NFPA.

National Fire Protection Association (NFPA). 2024. *NFPA 1, Fire Code, 2024 Edition*. Quincy, MA: NFPA.

CHAPTER

19

Firefighter I

Firefighter Survival

KNOWLEDGE OBJECTIVES

After studying this chapter, you will be able to:

- Describe how to determine potential personal hazards during an emergency incident.
- Describe standard safe operating procedures for firefighters, including the personnel accountability system, emergency communications, and rapid intervention crews.
- Describe firefighter survival techniques.
- Describe how to rescue a downed firefighter and maintain their air supply.

SKILLS OBJECTIVES

After studying this chapter, you will be able to perform the following skills:

- Demonstrate emergency communication skills and self-rescue techniques.
- Demonstrate rescuing a downed firefighter.

ADDITIONAL NFPA STANDARDS

- **NFPA 101**, *Life Safety Code, 2021 Edition*
- **NFPA 704**, *Standard System for the Identification of the Hazards of Materials for Emergency Response, 2022 Edition*
- **NFPA 1006**, *Standard for Technical Rescue Personnel Professional Qualifications, 2021 Edition*
- **NFPA 1026**, *Standard for Incident Management Personnel Professional Qualifications, 2024 Edition*
- **NFPA 1407**, *Standard for Training Fire Service Rapid Intervention Crews, 2020 Edition*
- **NFPA 1550**, *Standard on Emergency Responder Health and Safety, 2024 Edition*

CASE STUDY

You Are the Firefighter

You are dispatched as the first-due engine company to a two-story single-family residence. All the houses in the district to which you have been dispatched were built over the last 10 years using lightweight construction techniques and materials. While en route, the dispatcher states that they have received multiple calls reporting heavy smoke showing from the outside of the residence. You arrive on scene, and your officer reports a working fire.

1. Why is a personnel accountability system so important for firefighter survival when working at a fire such as this?
2. What information about hazardous conditions can firefighters identify before they arrive on scene?
3. How can firefighters stay safe and effective as a crew in low-visibility conditions such as heavy smoke?

Introduction

Firefighting is an inherently dangerous activity. Firefighters often encounter situations where their survival depends on making the right decisions and taking the appropriate actions. Firefighters work hard to achieve their tactical objectives. Without firefighter survival, there is no achieving tactical objectives. They must work just as hard to learn and practice the specialized techniques for surviving the life-threatening environments in which they work as they do to learn the techniques for fire suppression and other firefighting activities. Firefighter survival training focuses on the practices that will enable them to successfully get themselves or another firefighter out of a life-threatening situation. "Everybody goes home" should be the goal of every member of the team. This chapter introduces the actions, attitudes, and systems necessary to achieve that goal.

Risk–Benefit Analysis

A **risk–benefit analysis** is an assessment of the potential risks and the potential benefits of a specific course of action. At an emergency scene, the person conducting the risk–benefit analysis considers the actions firefighters might take and weighs the potential positive results that can be achieved against the probability and severity of the harm that might occur if things go wrong. A risk–benefit analysis is a key decision-making tool that all firefighters, officers, and chiefs use to ensure effective emergency operations and reduce firefighter deaths and injuries. Risk–benefit analyses should be part of the standard operating procedures (SOPs) for all fire departments. NFPA 1550, *Standard on Emergency Responder Health and Safety, 2024 Edition,* section 6.3, provides an excellent reference guide.

A risk–benefit analysis should be practiced by everyone on the fireground. The incident commander (IC) is responsible for analyzing the overall scene. At the same time, company officers and firefighters also need to process the risks and benefits of the strategies implemented by command and be cognizant of changes on the fireground. For example, if a building is known to be occupied, firefighters accept a higher level of risk to accomplish a rescue. If a building is known to be unoccupied, firefighters still perform suppression but tolerate far less risk because no property has a value greater than a firefighter's life. If a fire has reached the stage where no occupants can survive and no property of value can be saved, there is no justification for risking the lives of firefighters in an interior attack (**FIGURE 19-1**).

FIGURE 19-1 If a fire has already reached the stage where no occupants can survive and no property of value can be saved, there is no justification for risking the lives of firefighters in an interior attack.

Courtesy of Rom Duckworth.

The IC must continually assess the risks and benefits before committing crews to the interior of a burning structure. In addition, each firefighter involved in the operation should conduct a risk–benefit analysis from their own perspective. A report from an observant firefighter to their company officer can be crucial to the safety of an operation. The IC must reevaluate the risk–benefit balance whenever conditions change and decide whether the strategy must be changed. Chief Alan Brunacini, who was a well-known chief of the Phoenix, Arizona, Fire Department, summed this up concisely when he said, "Risk a lot to save a lot, risk a little to save a little, and risk nothing for what is already lost."

Hazard Indicators

Hazardous conditions are any conditions that may directly or indirectly harm firefighters in the course of their duties. Direct hazards are hazards that cause immediate harm. Some examples of direct hazards are heat, smoke, unstable structures, unstable surfaces, exposed electrical elements, sharp objects, and hazardous chemicals. Indirect hazards are conditions that could potentially cause harm and include stress, fatigue, and exhaustion. Hazardous conditions may not always be evident by simple observation. For example, smoke and flames may be hidden in a wall and detectable only if searched for. Firefighters train to identify, understand, and properly respond to both visible and hidden hazards. Three observable factors a firefighter should evaluate are the type of construction, type of occupancy, and weather conditions.

Knowledge of building construction helps firefighters anticipate fire behavior and recognize the potential for structural collapse. When approaching a building, firefighters must consider if it is designed to withstand a fire for some time or is likely to weaken quickly. Has the building been modified from its original use? Does the structure incorporate void spaces that allow a fire to spread quickly? Is it constructed using lightweight construction techniques? Does the building have truss roofs or floors or lightweight construction that could fail quickly? Chapter 6, *Building Construction*, has more information on hazard indicators related to building construction.

The occupancy and use of a building can give clues to potential hazards. Common commercial buildings such as hardware stores and home improvement centers contain huge amounts of highly flammable and hazardous materials. A name such as "Central Plating Company" on a building should cause firefighters to anticipate that hazardous materials could be present. A warning placard on the outside of a building is a more specific hazardous materials indicator. For example, NFPA 704, *Standard System for the Identification of the Hazards of Materials for Emergency Response, 2022 Edition*, describes a system of identifying specific hazards in a building or a container by using a combination of colors and numbers (**FIGURE 19-2**).

FIGURE 19-2 The NFPA 704 diamond indicates that hazardous materials are present. This diamond indicates that under emergency conditions, the hazardous materials in the building or container are potentially hazardous to health (blue) and could cause significant irritation (1), are flammable (red) under almost all ambient conditions (3), and are normally stable (yellow and 0). The absence of a symbol on the white diamond indicates that there are no special hazards—materials that react violently with water or can oxidize, or gases that can asphyxiate a person—present.

Weather is also a factor when evaluating hazards. While inclement weather can be easy to spot, its effects are not always obvious. For example, it is obvious that there is snow on the ground, but it is easy to forget that snow may add enough weight to a roof to cause it to collapse. Similarly, it is obvious when it is windy, but not as obvious that wind can rapidly change a fire's flow path or rapidly increase the intensity of a fire, often with deadly results.

These are just a few examples of observable factors that might indicate a hazard. At every incident, firefighters should look for critical indications of a hazard.

Safe Operating Procedures

SOPs describe the way a fire department conducts operations at an emergency incident. Many of these procedures were developed to protect firefighters' health and safety. SOPs must be learned and practiced so they can be performed correctly and under stressful conditions. Training and experience are the only way for individuals to become proficient at performing any set of skills. To be effective and efficient, a firefighter must practice the same skills routinely and consistently at every training session, and then apply this knowledge at every emergency incident. In the moments when conditions are most difficult, firefighters fall back on performing skills in the manner that they practiced them. When survival is at stake, well-practiced habits can be a lifesaver.

Rules of Engagement for Firefighter Survival

To help define safe practices of structural firefighting, the International Association of Fire Chiefs (IAFC) and the National Volunteer Fire Council (NVFC) developed following 11 Rules of Engagement for Firefighter Survival:

1. **Size up your tactical area of operation.** At an emergency incident, firefighters and company officers need to pause for a moment and look over their area of responsibility. During this size-up, evaluate your personal risk exposure and determine a safe approach to completing your assigned objectives.
2. **Determine the occupant survival profile.** Firefighters and company officers need to consider the fire conditions to determine if an occupant could be alive in the conditions that present themselves. They must maintain an ongoing assessment of these conditions.
3. ***DO NOT* risk your life for lives or property that cannot be saved.** Do not engage in high-risk search and rescue and firefighting activities when the fire conditions prevent occupant survival or if property destruction is inevitable.
4. **Extend *LIMITED* risk to protect *SAVABLE* property.** Limit your risk to a reasonable, cautious, and conservative level when trying to save property.
5. **Extend *VIGILANT* and *MEASURED* risk to protect and rescue *SAVABLE* lives.** Engage in search and rescue and firefighting operations in a calculated, controlled, and safe manner, while remaining alert to changing conditions during high-risk search and rescue operations.
6. **Go in together, stay together, and come out together.** Always enter burning buildings as a team of two or more. Ensure that no firefighter is allowed to be alone at any time while entering, operating in, or exiting a hazardous building.
7. **Maintain continuous awareness of your air supply, situation, location, and fire conditions.** Maintain awareness of what is happening in your area of operation and elsewhere on the fireground so that you are alert to conditions that may affect your risk and safety. Monitor your air supply and make sure that you exit the hazard zone before your low-air alarm activates.
8. **Constantly monitor fireground communications for critical radio reports.** Maintain awareness of fireground radio communications on your assigned channels for progress reports, critical messages, and other information that may affect your risk and safety.
9. **You are required to report unsafe practices or conditions that can harm you. Stop, evaluate, and decide.** Report unsafe practices and exposure to unsafe conditions to your supervisor. There should be no penalty for this reporting.
10. **You are required to abandon your position and retreat before deteriorating conditions can harm you.** Exit immediately to a safe area when you are exposed to deteriorating conditions, unacceptable risk, or a life-threatening condition.
11. **Declare a mayday as soon as you *THINK* you are in danger.** Declare a mayday as soon as you think you are in trouble. Failure to declare a mayday as soon as possible puts your life and the lives of fellow firefighters in even more danger.

Team Integrity

Teamwork is an essential requirement in firefighting (**FIGURE 19-3**). A team is two or more individuals working together to achieve a common goal. Given the intense physical labor necessary to pull hose or raise ladders, team efforts are imperative in firefighting. The team leader should also always maintain **team integrity**, also referred to as **crew integrity**. That is, they must ensure that the team stays together and works as a team, and that no single firefighter goes off on their own. The most important reason for maintaining team integrity in an emergency operation is for safety. Team members keep track of one another and

FIGURE 19-3 A team works together to achieve a common goal.

Courtesy of Rom Duckworth.

FIGURE 19-4 Personnel accountability system.

Courtesy of Mike Legeros.

provide immediate assistance to other team members if they get hurt or require assistance.

Team integrity means that the members of a team should always be oriented to one another's location, activities, and condition, even in vision-obscured conditions. In a hazardous atmosphere, such as inside a burning structure, a firefighter should always be able to contact their team members by voice, sight, or touch. When air cylinders need to be replaced, all team members should leave the building together, change cylinders together, and return to work together. If any team members require rehabilitation, the whole team should go to rehabilitation together and return to action together.

The standard operational team in firefighting is a company, usually consisting of two to five firefighters working under the direct supervision of a company officer. Chapter 1, *The Fire Service*, discusses the various types of companies. Career firefighters may arrive as a company on the same apparatus, whereas on-call or volunteer firefighters should assemble in companies upon arrival. Individual companies are assigned as a group to perform tasks at the scene of an emergency incident. Sometimes, a company is divided into smaller teams. For example, a truck company might be split into roof and search teams. Or the initial interior attack on a fire in room in a single-family dwelling might be made by two firefighters from a company while two other firefighters from that company remain outside, following the two-in/two-out rule.

For safety and survival reasons, firefighters should always use a buddy system, working in teams of at least two. The two team members must remain in direct contact with each other. If one team member should become incapacitated, that is an appropriate time to use the universal distress signal word, *mayday*. Their partner can provide help immediately and call for assistance. At least one member of every team must have a portable radio to maintain contact with the IC. In many fire departments, every member carries a portable radio.

The IC keeps track of the companies involved in the response and communicates with company officers. Every company officer must have a portable radio to maintain contact with the IC. The company officer is responsible for knowing the location of every company member; they should know exactly what each individual firefighter is doing at all times.

Personnel Accountability System

Every fire department must have a **personnel accountability system**, which provides a systematic way to keep track of every company, crew, or team of firefighters operating at the scene of an incident from the time they arrive until the time they are released (**FIGURE 19-4**). In addition, it must be able to identify each individual firefighter who is assigned to a particular company, crew, or team; the location of each team; the assignments for each team; and the team's current activities. The system assumes that all company members will be together working as a unit under the supervision of the company officer. If this is not the case because a company has been divided into smaller teams, the IC must be advised of the change. At any point in the operation, the IC should be able to identify the location and function of each team and, within seconds, determine which individual firefighters are assigned to that team. If firefighters are reported missing, lost, or injured, the IC uses the accountability system to identify who is missing and what their last assignment was.

SAFETY TIP

It is essential that fire departments, whether career, volunteer, or combination, have a method to keep track of all members responding to an incident and the teams to which they are assigned. Career departments typically have established shifts and riding positions, but many volunteer departments do not know who is able and available to respond to the incident until they arrive at the emergency scene. This makes an accountability system harder to implement but no less important.

The personnel accountability systems used by fire departments take many different forms. Some departments use electronic tracking devices. Others use simpler systems consisting of Velcro or magnetic nametags affixed to a display board. If a company travels together in an apparatus, this board is carried in the cab. In the case of a volunteer company, the tags can be affixed to the board upon arrival at the incident. This board, sometimes called a passport, must be kept up-to-date throughout the tour of duty. At the scene of an incident, these boards are left with a designated person at the command post or the entry point to a hazardous area to indicate that the company and those individual firefighters have entered the hazardous area. Later, after the entire company leaves the hazardous area or is released from the scene, the company officer retrieves the board.

Firefighters must learn which kind of accountability system their department uses, how to work within it, and how this system works within the Incident Command System (ICS). Some fire departments start with a written roster of the team members assigned to each company. Others use a computer database connected to a daily staffing program. Lists, roll calls, or electronic tracking devices should be carried on each apparatus. These are used to account for all personnel at the beginning of every tour of duty and when changes are made to assignments throughout the tour. Whatever type of system is used, the only way for it to be truly effective is if every firefighter takes responsibility for its use at all incidents, even the smallest ones.

A **personnel accountability report (PAR)** is an accountability check or roll call taken by a supervisor at an emergency incident that may be conducted as part of the personnel accountability system. Many departments conduct PARs at regular intervals. The IC should also request a PAR at tactical benchmarks, such as when the operational strategy changes or when a situation occurs that could endanger firefighters. When the IC requests a PAR, each company officer physically verifies that all assigned members are present and confirms this information to the IC. To do this, the company officer must be in verbal, visual, or physical contact with all crew members to verify their status. If a crew splits into two or more teams, the company officer must be in contact with at least one member of each team. Once the company officer verifies the status of the members, they report the PAR to the IC, usually by radio. If a firefighter cannot be accounted for, they are considered missing until proven otherwise. A report of a missing firefighter always becomes the highest priority at the incident scene.

A PAR should always be performed if unusual or unplanned events occur at an incident. For example, if there is a report of an explosion, a structural collapse, or a firefighter missing or in need of assistance, a PAR should immediately take place so that command can identify any missing individuals and establish their last known locations and assignments. These locations would then serve as the starting points for search and rescue crews.

Rapid Intervention Crews/ Companies

A **rapid intervention crew/company (RIC)**, sometimes known as a **rapid intervention team (RIT)**, a **firefighter assist and search team (FAST)**, or a **rescue assist team (RAT)**, is a crew consisting of three firefighters and an officer that is established for the sole purpose of rescuing firefighters who are operating at emergency incidents. The firefighters in a RIC must be fully dressed in personal protective equipment (PPE) and SCBA and equipped for action, ready to deploy immediately to assist firefighters in danger when assigned to do so by the IC.

A RIC is an extension of the two-in/two-out rule. During the early stages of an interior fire attack, a minimum of two firefighters are required to establish an entry team, and a minimum of two additional firefighters are required to remain outside the hazardous area. The two firefighters who remain outside the hazardous area can perform other functions, but they must have donned full PPE and SCBA to immediately enter and assist the entry team if the interior firefighters need to be rescued. The two firefighters who remain outside the hazardous area fulfill the first stage of a RIC procedure.

The second stage involves a dedicated RIC assigned to stand by for firefighter rescue assignments.

NFPA 1407, *Standard for Training Fire Service Rapid Intervention Crews, 2020 Edition*, includes training procedures for RIC operations, and NFPA 1550 contain a complete description of RIC procedures and provides examples of their use at typical scenes. Several options are available for establishing and organizing RICs, including dispatching one or more additional companies to incidents with this specific assignment. Some small fire departments rely on an automatic response from an adjoining community to provide a dedicated RIC at working incidents.

SOPs should dictate when a RIC is required, how it is assigned, and where it should be positioned at an incident scene. A RIC should be in place at any incident where firefighters are operating in conditions that are immediately dangerous to life and health (IDLH), which includes any structure fire where firefighters are working inside a building and using SCBA. Many fire departments specify a unique set of rescue equipment, including extra SCBA air cylinders, that is brought to the standby location. All RIC members must wear full PPE and SCBA and be ready for action in an instant.

The IC should quickly deploy the RIC to any situation where a firefighter needs immediate assistance. Situations that call for using the RIC include a lost or missing firefighter, an injured firefighter who has to be removed from a hazardous location, and a firefighter trapped by a structural collapse. The U.S. federal regulation that establishes requirements for using respiratory protection (OSHA 29 CFR 1910.134) mandates the use of backup teams at incidents where operations are conducted in an unsafe atmosphere (potentially IDLH).

Firefighter Survival Procedures

Some of the most important procedures you need to practice as a firefighter are those related to your safety and the safety of the firefighters working with you, including the following:

- Maintaining your orientation in smoke-filled buildings
- Rescuing yourself
- Identifying safe locations in a burning building when you are awaiting rescue
- Managing the supply of air in your SCBA so that it lasts as long as possible

These procedures will keep you from entering dangerous situations and guide you in situations where you could be in immediate danger. Your personal safety depends on learning, practicing, and consistently following these procedures.

SAFETY TIP

Never lose sight of the fact that preventing life-threatening situations is preferable to utilizing survival procedures.

Maintaining Orientation

It is easy to become disoriented in a dark, smoke-filled building, and although SCBA is what allows you to be in the smoky environment, it can make your disorientation worse. The visibility inside a burning building is often only a few feet, and sometimes it is reduced to zero. In these conditions, it is vital to stay oriented. If you become lost, you could run out of air before you find your way out. The atmosphere outside of your facepiece could be fatal in one or two breaths. Even if you are able to call for assistance, rescuers might be unable to find you.

To help you stay oriented inside a smoke-filled building, do the following:

- Before entering a building, look at it from the outside to get an idea of the size, shape, arrangement, and number of stories.
- After entering the structure, paint a picture in your mind of your surroundings.
- Use your sense of touch. Since most of your senses will be disrupted by dense smoke, touch may become your only means of moving through a structure. By feeling your surroundings, you can tell which type of room you have entered. Feel for changes in floor coverings.
- Always look for alternative means of egress as you maneuver through the building.

Preincident plans help with orientation. Reviewing the preincident plan while en route to the the structure will give you a general idea of the structure's floor plan. See Chapter 25, *Fire and Life Safety Initiatives*, for more information.

One of the most basic methods of remaining oriented is to always stay in contact with a hose line by always keeping a hand or foot on the hose line. To find your way out, feel for the couplings, and place one hand on the male coupling and the other on the female coupling (know the hose configuration of your fire department). The male coupling leads back to the fire engine and safety. To exit the building, travel in the direction of the hand holding the male coupling. This is referred to as "long lug out" because the lug on the male coupling is longer than the lug on the female coupling. This method is further discussed in Chapter 17, *Search and Rescue.* Some firefighters use the phrase, "Smooth, bump, bump—back to the pump" to help them remember which direction to travel.

SAFETY TIP

To exit a building by following a hose line, travel in the direction of the male hose coupling.

Team integrity is an essential factor in maintaining orientation. When the team members cannot see one another, they must remain in direct physical contact or close enough that they can hear each other speak. If you are part of a search team, follow your department's SOPs. Additional recommendations are described in Chapter 17, *Search and Rescue.*

Training sessions in conditions with no visibility build confidence. This type of training can be conducted inside any building while using an SCBA face piece with the lens covered. More realistic conditions can be simulated by using smoke machines inside training facilities. Firefighters should practice navigating around furniture, following charged hose lines, climbing and descending stairs, and fitting through narrow spaces. Distracting noises, such as operating fans, can add a realistic feel. Team integrity must always be maintained in these practice sessions.

Safe Locations

A **safe location**, sometimes called a **safe haven**, is a temporary location at an emergency scene that provides refuge for firefighters awaiting rescue or trying to find a method of self-rescue from an extremely hazardous situation. In this case, the term *safe* is relative. The location may not be exceptionally safe, but it is less dangerous than the alternatives. Safe locations are important when situations become critical and few alternatives are available. When a critical situation occurs, firefighters should know where to look for a safe location and recognize one when it presents itself.

TIP

Wildland firefighters carry lightweight reflective emergency shelters to act as a form of safe location. When overrun by a fire, taking refuge in the shelter might be the firefighter's only chance for survival.

If firefighters become trapped inside a burning building, a room with a door and a window could be a safe location. Closing the door could keep the fire out of the room for at least a few minutes, and opening the window may provide fresh air and an escape route. In this case, the safe location provides time for a rescue team to reach the firefighters, a ladder to be raised to the window, or another group of firefighters to control the fire. A roof or floor collapse often leaves a void adjacent to an exterior wall. If trapped firefighters can reach this void, it could be the best place to go. Maintaining team integrity is important when escaping to a safe location. In some cases, firefighters have survived in safe locations long enough to work together to create an opening.

Like all the emergency procedures described in this chapter, these activities require good instruction and practice. Always follow your department's SOPs when it comes to self-rescue techniques.

Air Management

Air management is vital to all firefighters when using SCBA. Air equals time. How long a firefighter can survive in a dangerous atmosphere depends on how much air is in the SCBA and how quickly the firefighter uses it. Chapter 3, *Personal Protective Equipment,* discusses breathing techniques that can be used to conserve air. Firefighters must manage their air supply and time in a hazardous atmosphere. The person in the hazardous atmosphere must consider the time it takes to enter and get to the location where the firefighter plans to operate and the time it takes to exit safely. The work time is limited by the amount of air in the SCBA cylinder after allowing for entry and exit time. If it takes 6 minutes to reach a safe atmosphere and the firefighter has only 3 minutes of air left, the results could be disastrous.

Recall that the time rating for an SCBA is based on a standard rate of consumption for a typical adult under low-exertion conditions in a test laboratory. Firefighters typically consume air much more rapidly than occurs in the standard test. Firefighters often consume a 45-minute-rated air supply in just 20 to 25 minutes. The actual rate of air consumption varies significantly among individual firefighters and depends on the activities that they are performing. For example, chopping holes with an axe or pulling ceilings with a pike pole will cause a firefighter to consume air more rapidly than conducting a survey with a thermal imager. Even given the same workload, no two individuals will use exactly the same amount of air from their SCBA cylinders. Differences in body size, fitness levels, and even hereditary factors cause air consumption rates to be faster or slower than the average.

Air management is a team effort as well as an individual responsibility. Firefighters always work as part of a team when using SCBA, and team members inevitably use air at different rates. The member who uses their air supply most rapidly determines the working time for the team.

TIP

Keeping physically fit results in a more efficient use of oxygen and allows you to work for longer periods with a limited air supply.

A good way to determine your personal air usage rate is to participate in an SCBA consumption exercise. This drill puts firefighters through a series of exercises and obstacles to learn how quickly each firefighter uses air when performing different activities. Repeated practice can help you learn to use air more efficiently by building your confidence and learning to control your breathing rate. Practice sessions should take place in a nonhazardous atmosphere. For example, while wearing full PPE and breathing air from SCBA, perform typical activities such as carrying hose packs up stairs, hitting an object with an axe or sledgehammer, or crawling along a hallway and record the following information:

- The amount of air in the cylinder at the start
- The length of time you work
- A description of the physical labor and whether it is light or heavy
- The number of minutes from the start of the exercise to when the low-air alarm sounds
- The number of minutes from the start of the exercise to when the cylinder is completely empty

This exercise will allow you to determine your personal consumption rate and work on improving your times. With practice, you will be able to predict how long your air supply should last under a given set of conditions.

Knowing your team members' physical conditions and workloads can help you look out for their safety. Because many tasks are divided up in any given operation, one member may be working to their maximum level while another team member is acting only in a supportive role. This difference in exertion levels could cause the first team member to use up their air supply much faster without realizing it.

When using SCBA at an emergency incident, be aware of the limitations of the device. Do not enter a hazardous area unless your air cylinder is full. Keep track of your air supply by using the heads-up display to determine how much air you have left. Always follow the rule of air management: Know how much air you have in your SCBA before you go in and manage that amount so that you leave the hazardous environment *before* your SCBA low-air alarm activates. A good rule of thumb is to check your air supply level whenever you change assignments or locations. Do not wait until the low-air alarm sounds to start thinking about leaving the hazardous area. You should be out of the hazardous area before this alarm sounds.

SAFETY TIP

Follow the rule of air management: Know how much air you have in your SCBA before you go in and manage that amount so that you leave the hazardous environment *before* your SCBA low-air alarm activates.

Emergency situations can occur even when firefighters employ the best possible air management practices. An SCBA can malfunction, or firefighters may be trapped inside a building by a structural collapse or an unexpected situation. In these circumstances, remaining calm and knowing how to use all of the emergency features of the SCBA can be critically important. Remaining calm, controlling your breathing rate, and employing air management breathing techniques can slow air consumption. All of these techniques require practice to build confidence.

SAFETY TIP

Before entering a hazard zone:

- Ensure that all crew members have portable radios and appropriate hand tools.
- Make sure that all members of a crew stay together.
- Work under the Incident Command System.
- Follow guidelines for managing your air supply.

Emergency Communications Procedures

Communication is critical to any emergency operation, but it is especially important to firefighter safety and survival (**FIGURE 19-5**). Breakdowns in communication are a significant contributor to firefighter injuries

FIGURE 19-5 Communication is a critical part of any emergency operation.
Courtesy of Rom Duckworth.

and deaths. Conversely, good communication can improve operations and save the lives of firefighters in dangerous situations. Standard communications procedures ensure that the message is clearly stated and then repeated as confirmation that it has been received and understood. The following is a model of good radio communication:

"Locust Lane Command to Ladder 4."

"This is Ladder 4. Go ahead, Locust Lane Command."

"Ladder 4 ventilate the roof."

"Ladder 4 copies. Beginning roof ventilation."

Note that some fire departments might open radio communications differently. Chapter 4, *Fire Service Communications*, describes proper radio communications procedures. Follow the SOPs outlined by your department.

When a firefighter is missing or in trouble, or when an imminent safety hazard is detected, it is even more important to communicate clearly and to ensure that the message is understood. For this reason, some critical communications should be standardized. Specific, dedicated phrases, sounds, and signals for emergency messages should be a part of every fire department's SOPs. Standard phrases should be used for conveying information about emergencies and hazardous conditions, and these phrases should be known and practiced by everyone in the department. The same procedures and terminology should be used by all the fire departments that can be expected to work together at emergency incidents. In many areas, these procedures are coordinated throughout an entire region.

Emergency traffic, also referred to as an **emergency message**, is an urgent message that takes priority over all other communications, except a mayday call. Many departments use it to indicate an imminent fireground hazard, such as a potential explosion or structural collapse. This kind of message would also be used to order firefighters to immediately withdraw from interior offensive attack positions and switch to a defensive strategy.

All imminent hazards and emergency instructions should be announced to capture the attention of everyone at the incident scene and ensure that the message is clearly understood. In many departments, the communications center uses a special tone to alert all members that an emergency situation is in progress. It then repeats the information to ensure everyone at the incident scene heard the message correctly. Some departments sound a horn in such circumstances.

The word **mayday** indicates that a firefighter is in trouble and requires immediate assistance. It could be used if a firefighter is lost, trapped, or injured; has an SCBA failure; or runs out of air. A firefighter can transmit a mayday message to request assistance or report that a firefighter is missing or in trouble. Any use of the term *mayday* should bring an immediate response and take precedence over all other radio communications.

Analysis of firefighter fatalities and serious injuries has shown that firefighters often wait too long before initiating a mayday call. Instead of doing this when they first get into trouble, firefighters often wait until the situation is critical before requesting help. This delay could be because of a fear of embarrassment if the situation turns out to be less severe than anticipated, combined with the hope that the firefighter will be able to resolve the problem without assistance. This failure to act promptly can be fatal in many situations. Systems have been designed to do everything possible to protect firefighters and to rescue firefighters from dangerous situations, but those systems are effective only when firefighters act appropriately when they find themselves in trouble. Do not hesitate to call for help when you think you need it.

When the decision is made to initiate a mayday call, follow your fire department's SOPs and *do not move from your location.* In most cases, the firefighter in trouble transmits the message "Mayday! Mayday! Mayday!" followed by *who* (they are), *what* (the situation is), and *where* (they are located) over the radio to initiate the process. The IC immediately interrupts all other communications and requests additional information so that the right rescue actions can be taken. The IC may follow a format such as the LUNAR report. LUNAR stands for the following:

Location: the firefighter's location in the building/incident

Unit: the unit the firefighter is assigned to

Name: the name of the firefighter

Air: the amount of air the firefighter has in their cylinder, or **A**ssignment, where the firefighter was last assigned

Resources: what the firefighter needs to get out of the mayday situation

Some departments use other formats such as LIP, which stands for:

Location: the firefighter's location in the building/incident

Identification: the firefighter's name and assignment

Problem: the specific problem(s) such as lost, trapped, low on air, injured, etc.

The next actions could include committing the RIC to the rescue effort, calling for additional resources, and redirecting other teams to support a search and rescue operation. An example of a mayday call is as follows:

Firefighter: Mayday! Mayday! Mayday!

[All radio traffic stops.]

IC: Unit calling mayday, go ahead.

Firefighter: This is Engine 403. We are on the second floor and running out of air. Fire has cut off our escape route. We request a ladder to the window on the Charlie side of the building so we can evacuate.

IC: Engine 403, your escape route cut off by fire. I am sending the rapid intervention crew to Charlie side with a ladder.

Firefighters should make their mayday call and then manually activate their personal alert safety system (PASS) device so that other firefighters can hear the signal and come to the firefighter's location. If they have a portable radio, they should also activate the radio's emergency alert button if it has this feature. Both of these audible alert signals make it difficult for the firefighter to be heard over the radio, so they must call their mayday first. If the PASS device has already activated, the firefighter should deactivate it before talking into the radio, and then reactivate it after they have transmitted their message. To create a visual signal, the firefighter should shine their flashlight at the ceiling. These actions increase the probability that the RIC or other firefighters will be able to locate the firefighter in trouble. It is also crucial that firefighters continue to work to free themselves from the dangerous situation.

To initiate a mayday call, follow the steps in **SKILL DRILL 19-1**.

SKILL DRILL 19-1

Initiating a Mayday Call for Emergency Assistance Firefighter I, NFPA 1010: 6.2.3

1. If your PASS device has activated, deactivate it. Use your radio to call "Mayday! Mayday! Mayday!" If possible, wait to hear that your mayday call was received. Report who you are, what is wrong, and where you can be found. If your department uses a different format of mayday report, follow the requirements of your SOPs.

Continues.

SKILL DRILL 19-1 CONTINUED

Initiating a Mayday Call for Emergency Assistance Firefighter I, NFPA 1010: 6.2.3

2. Manually activate your PASS device and your radio's emergency alert button.

3. Slow your breathing as much as possible to conserve your air supply. Attempt self-rescue. If you can move, identify a safe location to await rescue. If unable to self-rescue, lie on your side in a fetal position with your PASS device pointing out so that it can be heard and with your flashlight pointed toward the ceiling.

Self-Rescue

If you become separated from your team, or if you become lost, disoriented, or trapped inside a structure, you can use several techniques for self-rescue. The first step is always to call for assistance by issuing a mayday. Do not wait until it is too late for anyone to find you before making it known that you are in trouble. As previously discussed, time is critical when your safety is at risk. You need to initiate the process as soon as you *think* you are in trouble. Do not wait until you are absolutely sure you are in grave danger because it may be too late by then.

The safest option is to self-rescue using normal egress routes. If this is not possible, firefighters can use more involved techniques to escape dangerous situations. such as breaching a wall. Like all procedures, self-rescue methods must be practiced regularly so that you can perform them without hesitation.

If you are unable to self-rescue using normal egress routes, determine if you are in a safe location or can easily get to one. If not, or if you must evacuate your location because it is too dangerous, your focus will be on using the techniques that will increase your survivability while waiting for help, such as properly calling for a mayday, using effective air-management techniques, and assisting rescuers in finding and assisting you.

Using Ordinary Egress Routes

If you are separated from your team, the best option for self-rescue may be to take a normal route of egress to the exterior. Follow a hose line back to an open doorway, descend a ladder, or locate a window or door. Each of these options can provide a direct exit and require no special tools or techniques. Once successful, you must immediately communicate with your company officer or the IC to inform them that you have exited safely. To rescue yourself by locating and following a hose line to an exit, follow the steps in **SKILL DRILL 19-2**. To locate a door or window that can be used for an exit, follow the steps in **SKILL DRILL 19-3**.

SKILL DRILL 19-2

Performing Self-Rescue Using a Hose Line Firefighter I, NFPA 1010: 6.3.5

1. Initiate a mayday. Stay calm and control your breathing. Systematically search the room to locate a hose line using an oriented search pattern (described in Chapter 17, ***Search and Rescue***). The hose most likely will be on the ground, so stay low or crawl while you are searching.

2. Once you find the hose, follow the hose line to a hose coupling, and then identify the male and female ends of the coupling.

3. Move from the female coupling to the male coupling and follow the hose out. Exit the hazard area. Notify command of your location.

Breaching a Wall

Breaching a wall means knocking your way through it. Most walls are constructed with studs. To escape through a wall, you need to expose two studs and then reduce the size of your profile to get your body and your SCBA through the narrow opening between two studs, typically 14.5 inches (in.; 37 centimeters [cm]. (Remember that the distance between the studs will vary depending on the date of construction, the type of construction materials used, and the size of the building.) The specific technique for accomplishing this that works for you depends on your size. A small firefighter may be able to squeeze through a limited-size opening without removing their SCBA. A larger person who tries

SKILL DRILL 19-3

Locating a Door or Window for Emergency Exit Firefighter I, NFPA 1010: 6.3.5

1. Initiate a mayday. Stay calm, control your breathing, and stay low. Systematically locate a wall.

2. Use a sweeping motion on the wall to locate an exit. Identify the opening as a window, an interior door, or an exterior door. Beware of closets, bathrooms, and other openings without egress. If the first opening is not an exit, continue to search. Maintain your orientation and stay low. When you find an opening, exit the room safely if possible.

3. If unable to exit, find a safe location and assume the downed firefighter position. Keep command informed of your situation.

to escape through the same small opening may have to remove their SCBA backpack to fit through it no matter what technique they use.

One method of getting through a wall after you have exposed two studs is to loosen your waist strap and one strap of your SCBA, and then reposition your SCBA as you twist through the opening. This method is called the backhanded swim technique. Opening a wall and using the backhanded swim technique to escape through the wall opening is shown in **SKILL DRILL 19-4**.

SKILL DRILL 19-4

Using the Backhanded Swim Technique to Escape through a Wall
Firefighter I, NFPA 1010: 6.3.1, 6.3.5, 6.3.9

1. Identify deteriorating conditions that require exiting through a wall. Initiate a mayday. Use a hand tool or your feet to open a hole in the wall between two studs. If using a hand tool, drive the tool completely through the wall to check for obstacles and fire conditions on the other side.

2. If conditions are safer on the other side, enlarge the hole so it is tall enough to fit a seated firefighter. Check for wires and other obstructions in the wall. Place your tools through the breached opening in the wall.

3. Sit with your back to the hole so your SCBA bottle fits into the hole.

4. Raise your right arm up and over in a backwards swim motion to fit between the studs. Larger firefighters may need to first loosen their right SCBA shoulder strap. Continue the motion so you turn onto your right side through the opening in the wall.

Continues.

SKILL DRILL 19-4 CONTINUED

Using the Backhanded Swim Technique to Escape through a Wall
Firefighter I, NFPA 1010: 6.3.1, 6.3.5, 6.3.9

5. Continue to maneuver through the breach opening until you are on the other side. Readjust your SCBA straps and retrieve your tools.

Another method of escaping through a wall opening is the forward swim technique. This technique may be preferred if the breach opening is smaller or if the wall has entanglement hazards such as wiring that must be avoided. In this technique, after exposing the studs, you lie face down and stretch your arms through the opening. Then you rotate your shoulders and waist sideways without lifting yourself up on your elbows because that will raise your profile. To open a wall and use the forward swim technique to escape through the wall opening, follow the steps shown in **SKILL DRILL 19-5**.

If these methods fail, you may need to remove your SCBA backpack while keeping your face piece on. You will then have to basically perform the forward swim technique except you will need to push your SCBA pack in front of you. Hold on to the backpack to prevent it from becoming separated from your face piece. This technique should not be used unless simpler techniques will not work.

Disentangling Yourself

When crawling through a smoke-filled building, a firefighter can easily become entangled in fallen wires, cables, and debris. Many commercial buildings have a huge and varied mass of electrical wires and conduit, computer, and TV cables; plumbing and sewage pipes; telephone cables; and heating and air-conditioning tubing and ducts located above dropped ceilings. Many dropped ceilings are held up only with wires, so it does not take much fire or water damage to bring this massive tangle of utilities crashing down. If this tangled debris falls on your back while wearing SCBA, you could easily be trapped, so all firefighters should practice skills for disentangling themselves. Many firefighters carry small tools in the pockets of their PPE that they can use to cut through wires or small cables. However, this procedure can be difficult if poor-visibility conditions prevent the firefighter from seeing and identifying the entangling material. Also, small tools might be completely ineffective for dealing with thicker cables or pipes. To escape from an entanglement, follow the steps in **SKILL DRILL 19-6**.

Using Tools and Equipment in Unconventional Ways

Some additional self-rescue methods involve using tools and equipment in ways in which they were not designed. These "last resort" methods should be taught only by experienced and qualified instructors and must be practiced with strict safety measures in place. Some of these methods are controversial, and some fire departments consider them unacceptable because of the high risk of injury when they are used or even practiced. Other departments provide training in them because these measures could save the life of a firefighter in an extreme situation. This policy decision must be made by each fire department and most likely reflect the circumstances of their community discovered during preincident surveys.

SKILL DRILL 19-5

Using the Forward Swim Technique to Escape through a Wall
Firefighter I, NFPA 1010: 6.3.1, 6.3.5, 6.3.9

1. Identify deteriorating conditions that require exiting through a wall. Initiate a mayday. Use a hand tool or your feet to open a hole in the wall between two studs. If using a hand tool, drive the tool completely through the wall to check for obstacles on the other side. Enlarge the hole. Look first to check the floor and fire conditions on the other side of the wall. Loosen the SCBA shoulder straps if necessary, but do not remove them.

2. Lie prone—that is, with your stomach flat on the ground—and your head pointed toward the wall opening and about 6 in. (15 cm) from the opening. Stretch your arms toward your head and attempt to touch your ears with your upper arms to reduce your profile.

3. Extend your arms, and head into the wall opening first.

4. Rotate your shoulders, SCBA, and waist as necessary to escape through the opening. Do not lift yourself up on your elbows, as this will raise your profile. Report your status to command.

SKILL DRILL 19-6

Escaping from an Entanglement Firefighter I, NFPA 1010: 6.3.1, 6.3.5, 6.3.9

1. Initiate a mayday. Stay calm and control your breathing. Change your position—back up and turn on your side and attempt to free yourself.

2. If you are not successful, swing your arms up like a swimmer to catch any wires and cables that you might be tangled in and then lift the wires and cables off of yourself.

3. If you are still trapped, loosen the SCBA straps, remove one arm, and slide the air pack to the front of your body to free the SCBA.

4. If all else fails, cut the wires or cables causing the entanglement. Be aware of any possible electrocution risk. If you are able to exit, notify command that you are out of danger. If you are unable to disentangle yourself, notify command of your situation.

Rescuing a Downed Firefighter

One of the most critical and demanding situations in the fire service arises when firefighters must rescue another firefighter who is in trouble. Air management is critical in a firefighter rescue situation for both the rescuers and the firefighters who are in trouble.

When you reach a downed firefighter, the first step is to assess the firefighter's condition. Is the firefighter conscious? Are they breathing? Is the firefighter trapped or injured? Is the firefighter's SCBA working and how much air do they have? The firefighter needs enough air to keep them breathing for the duration of the rescue operation. Make a rapid assessment and notify the IC of your situation and location. Have the RIC deployed to your location with a spare SCBA cylinder or **rapid intervention pack (RIC pack)**—a portable air supply that provides an emergency source of breathing air for a single firefighter who has run out of air or whose air supply is insufficient to safely exit from an IDLH atmosphere. Quickly determine and request the additional resources needed.

Removing a Downed Firefighter

When removing a downed firefighter quickly, you can use the firefighter's SCBA straps to create a rescue harness. To rescue a downed firefighter by making a rescue harness from the firefighter's SCBA straps, follow the steps in **SKILL DRILL 19-7**.

SKILL DRILL 19-7

Rescuing a Downed Firefighter Using the Waist Straps of the SCBA Harness as a Rescue Harness Firefighter I, NFPA 1010: 6.3.9

1. After locating the downed firefighter, activate the mayday procedure if that step has not already been taken. Shut off the PASS device as needed to aid in communication. Assess the situation and the condition of the downed firefighter, and make sure their SCBA is working and they have sufficient air supply.

2. Position yourself at the legs of the downed firefighter. Lift one leg of the downed firefighter up onto your shoulder.

3. Locate and loosen the waist straps and shoulder straps (if needed) of the downed firefighter's SCBA harness.

4. Unbuckle the downed firefighter's waist strap and buckle the waist strap under the lifted leg. Tighten the straps. Remove the downed firefighter from the hazard area to a safe area.

You can also use a downed firefighter's drag rescue device (DRD) that is integrated in the structural firefighting protective coat to remove them from the hazardous area. To rescue a downed firefighter who is wearing PPE and SCBA from the immediate hazard using their DRD, follow the steps in **SKILL DRILL 19-8**.

As a member of a two-person team, you can also remove a downed firefighter with PPE and SCBA from a hazardous area by having one firefighter pull on the downed firefighter's SCBA straps and the second firefighter support the legs of the downed firefighter. To do so, follow the steps in **SKILL DRILL 19-9**.

SKILL DRILL 19-8

Rescuing a Downed Firefighter Using a Drag Rescue Device Firefighter I, NFPA 1010: 6.3.9

1. After locating the downed firefighter, activate the mayday procedure if that step has not already been taken. Shut off the PASS device to aid in communication. Assess the situation and the condition of the downed firefighter, and make sure their SCBA is working and they have sufficient air supply.

2. Access the firefighter's drag rescue device (DRD) just below the collar at the back of their coat.

3. Secure the DRD by grabbing it with your hand, attaching webbing to the DRD, or inserting a tool through the DRD. Remove the downed firefighter from the hazard area to a safe area.

SKILL DRILL 19-9

Rescuing a Downed Firefighter as a Two-Person Team Firefighter I, NFPA 1010: 6.3.9

1. After locating the downed firefighter, activate the mayday procedure if that step has not already been taken. Shut off the PASS device to aid in communication. Assess the situation and the condition of the downed firefighter, and make sure their SCBA is working and they have sufficient air supply.

2. The second firefighter loosens both the waist straps and the shoulder straps of the downed firefighter's SCBA harness, and then converts the waist straps into a rescue harness (as described in Skill Drill 19-7).

3. The first firefighter grabs the shoulder straps or uses the webbing to create a handle to pull the downed firefighter.

4. The second firefighter stays at the downed firefighter's legs and supports the legs of the downed firefighter. The first firefighter pulls the downed firefighter using the straps, and the second firefighter pushes the downed firefighter by pushing their legs forward.

Continues.

SKILL DRILL 19-9 CONTINUED

Rescuing a Downed Firefighter as a Two-Person Team Firefighter I, NFPA 1010: 6.3.9

5. Remove the downed firefighter from the hazard area to a safe area.

Supplying Air to a Downed Firefighter

If the downed firefighter has a working SCBA, determine how much air remains in their cylinder. A firefighter who is breathing and has an adequate air supply is not in immediate, life-threatening danger, and you can continue your rescue efforts. However, if their SCBA contains little or no air, you need to provide an additional air supply. A spare SCBA air cylinder or a rapid intervention pack should be brought to the location and connected to the downed firefighter's SCBA through the rapid intervention crew/company universal air connection (RIC UAC). This provides rapid emergency refilling of breathing air to an SCBA for the downed or trapped firefighter.

If the firefighter's SCBA is not working, check whether a simple fix can make it operational while you are waiting for the RIC. Did a hose or regulator come off? Was the cylinder valve accidentally closed? If regulator failure is a possibility, open the bypass valve. If everything fails, use the emergency breathing safety system (EBSS) to share your air supply by buddy breathing.

Using a Rapid Intervention Pack or a Spare SCBA Cylinder

If a firefighter needs an additional air supply and their SCBA is working properly, the SCBA cylinder can be replaced with another, full cylinder. If their SCBA is not working properly, the RIC team carries a RIC pack. RIC packs include the same basic components as an SCBA, including the following (**FIGURE 19-6**):

FIGURE 19-6 A rapid intervention pack is a portable air supply system used by rapid intervention crews/companies.

- A full air cylinder
- A valve at the neck of the air cylinder that controls the flow of air leaving the air cylinder
- A pressure gauge located near the air cylinder valve
- A regulator assembly to control airflow to the user

- A low-pressure hose through which air flows through the RIC pack regulator
- A high-pressure hose through which the emergency air supply passes from the air cylinder at full pressure
- An emergency face piece that differs from the facepiece on SCBA because it is not equipped with a nose piece assembly and may not have voice transmitting capability (although it will have a heads-up display visible to the user while wearing the face piece indicating the amount of air in the air cylinder).

Most rapid intervention packs are also supplied with a heavy-duty container to store all the components needed. Some have other compartments to carry additional supplies needed by the RIC.

There are many situations that may call for the use of a rapid intervention pack. These include a downed firefighter who cannot be quickly moved because they are injured or trapped and a firefighter who has suffered some type of SCBA failure. Like SCBA, a rapid intervention pack uses an air cylinder for breathing air. Different packs may contain cylinders of different sizes and pressures. The cylinder must be compatible with the downed firefighter's SCBA equipment. Follow the manufacturer's recommendations for use.

Rapid intervention packs can be connected to a trapped or downed firefighter using either the low-pressure or the high-pressure hose on the pack. To decide whether to use the high- or low-pressure hose, and, if you use the low-pressure hose, which connection method you should use, consider the following conditions:

- If the downed firefighter's SCBA is operating properly, if their face piece and regulator are attached, and if their air supply is adequate and will likely last the duration of the rescue effort, start efforts to rescue the firefighter. The RIC should bring the rapid intervention pack anyway, in case the downed firefighter's air supply gets too low.
- If the downed firefighter's SCBA is operating properly and if their face piece and regulator are attached, but their air supply is low, connect the high-pressure hose from the rapid intervention pack to the RIC UAC on the downed firefighter's SCBA. Leave it in place until the pressure is equalized between the two air cylinders. You could also replace the downed firefighter's cylinder with a new cylinder.
- If no air is being supplied to the downed firefighter's low-pressure SCBA hose but the rest of the firefighter's SCBA, including the face piece and the regulator, are in place and functioning properly, disconnect the downed firefighter's low-pressure SCBA hose from the firefighter's SCBA regulator, and connect the low-pressure hose from the rapid intervention pack to the downed firefighter's SCBA regulator.
- If the firefighter's SCBA regulator is missing or not operating properly, first attach the low-pressure hose of the rapid intervention pack to the pack's regulator. Then detach the downed firefighter's regulator from their SCBA and attach the pack's regulator and low-pressure hose to the firefighter's SCBA.
- If any part of the firefighter's face piece or regulator is damaged or missing, remove the downed firefighter's helmet, hood, and facepiece. Then place the emergency face piece from the rapid intervention pack on the firefighter and attach the rapid intervention regulator and low-pressure hose to the emergency face piece. This method requires time and additional movement of the firefighter because you will need to remove the downed firefighter's helmet, hood, face piece, and regulator, leaving the firefighter without air. This method should only be used if the other methods will be ineffective.

When the high-pressure hose is connected to the firefighter's SCBA through the RIC UAC, air is transferred from the higher pressure in the rapid intervention pack air cylinder to the lower pressure in the firefighter's air cylinder. Once the pressure in the two cylinders is equalized, the high-pressure hose should be disconnected from the firefighter's RIC UAC. This may take up to 60 seconds to complete.

If necessary, a second air transfer can be done with another RIC pack; however, there will be significantly less air pressure available the second time than there was initially. During this operation, one person must monitor the downed firefighter's SCBA for proper operation and remaining air.

When you need to supply air to a downed firefighter, follow the steps in **SKILL DRILL 19-10**.

Using the Emergency Breathing Safety System

Since 2018, all NFPA-certified open-circuit SCBA must include a universally compatible **emergency breathing safety system (EBSS)**, sometimes known as a **buddy breather**. The purpose of the EBSS is to allow two firefighters to share air if needed in an emergency situation. Unlike the high-pressure UAC, which equalizes the air pressure between the two connected bottles, the EBSS shares the air, maximizing the available air in

SKILL DRILL 19-10

Supplying Air to a Downed Firefighter Using the Low-Pressure Hose from a Rapid Intervention Pack Firefighter I, NFPA 1010: 6.3.9

1. Inspect the rapid intervention pack for proper operation, and make sure that the air cylinder is full. Turn on the air cylinder valve.

2. Check the downed firefighter's SCBA to determine if it is operating properly. If it is and if their face piece and regulator are attached and their air supply is adequate and will likely last the duration of the rescue effort, start efforts to rescue the firefighter.

3. If the downed firefighter's SCBA is operating properly and their face piece and regulator are attached, but their air supply is low, inspect the firefighter's RIC UAC to ensure it is clean and undamaged, then connect the high-pressure hose from the rapid intervention pack to the RIC UAC on the downed firefighter's SCBA. Leave it in place until the pressure is equalized between the two air cylinders, then disconnect the high pressure hose from the RIC UAC. Or, if you have an extra, filled SCBA air cylinder, replace the downed firefighter's cylinder with the new one.

4. If no air is being supplied to the downed firefighter's low-pressure SCBA hose but the rest of the firefighter's SCBA, including the face piece and the regulator, are in place and functioning properly, disconnect the downed firefighter's low-pressure SCBA hose from the firefighter's SCBA regulator, then connect the low-pressure hose from the rapid intervention pack to the downed firefighter's SCBA regulator.

SKILL DRILL 19-10 CONTINUED

Supplying Air to a Downed Firefighter Using the Low-Pressure Hose from a Rapid Intervention Pack Firefighter I, NFPA 1010: 6.3.9

5. If the firefighter's SCBA regulator is missing or not operating properly, attach the low-pressure hose of the rapid intervention pack to the pack's regulator, then detach the downed firefighter's regulator from their SCBA and attach the pack's regulator and low-pressure hose to the firefighter's SCBA.

6. If any part of the firefighter's face piece or regulator is damaged or missing, remove the downed firefighter's helmet, hood, and facepiece. Then place the emergency face piece from the rapid intervention pack on the firefighter and attached the rapid intervention regulator and low-pressure hose to the emergency face piece.

7. Continue to monitor the air pressure and operation of the downed fire fighter's SCBA and the pressure in the rapid intervention pack air cylinder.

both systems. While the exact configuration will vary slightly from manufacturer to manufacturer, most EBSS consist of a pouch on the waist belt of the SCBA that contains a hose that has one end connected to the firefighter's air supply. The other end of the hose has both a male and a female connector. You connect the male connector on one firefighter's EBSS hose to the female connector on the other firefighter's EBSS hose. As soon as they are connected, the air from the full air cylinder is shared between the two firefighters. Follow the steps in **SKILL DRILL 19-11** to use an EBSS to share air between two SCBAs in an emergency situation.

SKILL DRILL 19-11

Supplying Air to a Downed Firefighter Using the Emergency Breathing Safety System Firefighter I, NFPA 1010: 6.3.9

1. Check your air supply, and consider the effort that will be needed to successfully remove the firefighter in distress. Once the EBSS is connected, both firefighters will be using the remaining air and their movement will be restricted.

2. Remove the end of the EBSS hose with the two connectors from the pouch on your SCBA waist belt. Remove any protective caps from your connectors.

3. If possible, have the firefighter in distress remove their EBSS hose from the pouch on their SCBA waist belt and remove any protective caps from their connectors. If the firefighter in distress cannot assist, remove their EBSS hose from the pouch on their SCBA waist belt and remove any protective caps from the connectors.

4. Connect the male connector on one of the EBSS to the female connector on the other EBSS hose. It does not matter which male connector attaches to which female connector.

SKILL DRILL 19-11 CONTINUED

Supplying Air to a Downed Firefighter Using the Emergency Breathing Safety System Firefighter I, NFPA 1010: 6.3.9

5. As soon as the EBSSs are connected, move with the downed firefighter in a coordinated manner to immediately exit the IDLH area.

6. If it becomes necessary to disconnect the EBSS, push the couplings together and slide the collar of the female connector away from the male connector. The couplings will disconnect and the EBSS will self-seal. The firefighters are then able to move independently.

SAFETY TIP

If it has not already been done, a mayday should be called in any situation where the use of EBSS is being considered.

Firefighter Down CPR

For a firefighter in cardiac arrest, the turnout gear and SCBA they are wearing can present significant challenges to providing immediate high-quality cardiopulmonary resuscitation (CPR). **Firefighter down CPR (FD-CPR)** is a technique for doing CPR that allows firefighters to begin providing immediate life-saving care to a firefighter in cardiac arrest while their SCBA and gear are removed in the most efficient way. This 10-step process for FD-CPR presumes the downed firefighter is in full turnout gear and wearing SCBA and is still using the SCBA air supply. These steps assume at least three rescuers are available to assist with the procedure.

1. The first rescuer drags the firefighter in cardiac arrest to a safe location to begin care. The rescuer should then drop to a seated position behind the firefighter with the downed firefighter's SCBA bottle between the rescuer's legs so that the rescuer can use their legs to stabilize the SCBA pack.
2. If possible, a second "fresh" firefighter (one who has not just been involved in strenuous activity) should be available to be a second rescuer The second rescuer positions themselves to the side of the downed firefighter and unclips the chest clip of the downed firefighter's SCBA if it has one. Rescuer 2 then begins high-quality chest compressions through the turnout gear without removing the downed firefighter's helmet, SCBA mask, or any other equipment.
3. (OPTIONAL) The first rescuer (the firefighter stabilizing the SCBA) may open the downed firefighter's SCBA bypass valve to free-flow air to the downed firefighter. The first rescuer may also shut off the downed firefighter's PASS alarm if it is sounding.

Steps 4–8 should be performed together, coordinated between all rescuers. The second rescuer continues to perform chest compressions while the other rescuers perform the actions described in Steps 4–8.

4. While the second rescuer continues to perform chest compressions, the first rescuer removes the downed firefighter's, helmet, SCBA mask, and hood.
5. The third rescuer positions themselves at the downed firefighter's waist and unbuckles/unzips the downed firefighter's turnout coat.
6. The first rescuer loosens the shoulder straps of the downed firefighter's SCBA, and then raises the downed firefighter's arms above their head. Rescuer 1 removes the downed firefighter's gloves and unhooks their thumbs from the thumb loops in the coat sleeves.
7. The first and third rescuers together to open the buckles, zippers, Velcro, or other fasteners on the downed firefighter's turnout coat.
8. The first rescuer grabs the downed firefighter's turnout coat and SCBA straps and prepares to give the order of "Pull down!" The third rescuer (and, if available, a fourth rescuer) prepares to pull the downed firefighter by the legs down and away from the first rescuer, who will remain in place holding the coat and SCBA.
9. On the "pull down" command, the third rescuer (and, if available, the fourth rescuer) pull the downed firefighter by the legs to slide the downed firefighter away from the first and second rescuers.
10. A new resuscitation crew assumes resuscitation of the downed firefighter.

SAFETY TIP

Rehabilitation helps firefighters retain the ability to perform at the current incident and restores their capacity to work at later incidents. It is integral to firefighter safety and survival. See Chapter 20, *Firefighter Rehabilitation,* for more information about this topic.

CASE STUDY

You Are the Firefighter CONCLUSION

You are dispatched as the first-due engine company to a two-story single-family residence. All the houses in the district to which you have been dispatched were built over the last 10 years using lightweight construction techniques and materials. While en route, the dispatcher states that they have received multiple calls reporting heavy smoke showing from the outside of the residence. You arrive on scene, and your officer reports a working fire.

1. **Why is a personnel accountability system so important for firefighter survival when working at a fire such as this?**

   Answer: Accountability is essential for firefighters to be both an effective part of a coordinated fire suppression or rescue effort and to help ensure that firefighters who cannot call for help do not become "lost" on the fireground. Accountability helps reduce freelancing on the fireground and keeps firefighters in communication with their teammates. This keeps firefighters safer and makes the command structure more effective.

2. **What information about hazardous conditions can firefighters identify before they arrive on scene?**

   Answer: Firefighters may be aware of general hazards in their response district, such as lightweight building construction. Firefighters must also understand the hazards of the reported heavy smoke conditions. Most importantly, firefighters must consider how reports and observation such as heavy smoke and a working fire present a critical hazard when combined with other known issues such as the lightweight construction.

3. **How can firefighters stay safe and effective as a crew in low visibility conditions such as heavy smoke?**

   Answer: Firefighters must work together in crews to improve their effectiveness in accomplishing their assigned tasks, maintaining group situational awareness, and increasing their ability to help each other if any individual firefighter has a problem. Firefighters on a crew should stay in visual, verbal, or touch communication, even in low-visibility situations.

WRAP-UP

SUMMARY

KNOWLEDGE OBJECTIVES

- Describe how to determine potential personal hazards during an emergency incident.
  - Describe how to apply a risk–benefit analysis to an emergency incident. (pp. 838–839)
  - List the common hazard indicators that should alert firefighters to a potentially life-threatening situation. (**NFPA 1010: 6.3.5**, p. 839)
- Describe standard safe operating procedures for firefighters, including the personnel accountability system, emergency communications, and rapid intervention crews.
  - List the 11 Rules of Engagement for Firefighter Survival. (p. 840)
  - Explain how to maintain team integrity during emergency operations. (**NFPA 1010: 6.3.5**, pp. 840–841)
  - Define personnel accountability system. (**NFPA 1010: 6.3.5**, pp. 841–842)
  - Describe the types of personnel accountability systems and how they function. (**NFPA 1010: 6.3.5**, pp. 841–842)
  - Explain how a personnel accountability report is taken. (**NFPA 1010: 6.3.5**, p. 842)
  - Define rapid intervention crews/companies. (**NFPA 1010: 6.3.10**, pp. 842–843)
- Describe firefighter survival techniques.
  - Describe the methods used for maintaining orientation. (pp. 843–844)
  - Describe how to find a safe location while awaiting rescue. (**NFPA 1010: 6.3.5**, p. 844)
  - Describe air management procedures. (**NFPA 1010: 6.3.1, 6.3.5**, pp. 844–845)
  - Describe how to initiate emergency communications procedures. (**NFPA 1010: 6.2.3**, pp. 845–848)
  - Describe the information that should be included in a mayday call for emergency assistance. (**NFPA 1010: 6.2.3**, pp. 845–848)
  - Describe common self-rescue techniques. (**NFPA 1010: 6.3.5**, pp. 848–854)
- Describe how to rescue a downed firefighter and maintain their air supply.
  - Describe common techniques for rescuing a downed firefighter. (**NFPA 1010: 6.3.9**, pp. 854–863)
  - Describe how a rapid intervention pack or emergency breathing safety system can provide an emergency air supply to a downed or trapped firefighter. (**NFPA 1010: 6.3.9**, pp. 858–863)

SKILLS OBJECTIVES

- Demonstrate emergency communication skills and self-rescue techniques.
  - Initiate a mayday call for emergency assistance. (**NFPA 1010: 6.2.3**, pp. 847–848)
  - Perform a self-rescue using a hose line. (**NFPA 1010: 6.3.5**, p. 849)
  - Locate a door or window for an emergency exit. (**NFPA 1010: 6.3.5**, p. 850)
  - Use the backhanded swim technique to escape through a wall. (**NFPA 1010: 6.3.1, 6.3.5, 6.3.9**, pp. 851–852)
  - Use the forward swim technique to escape through a wall. (**NFPA 1010: 6.3.1, 6.3.5, 6.3.9**, p. 853)
  - Escape from an entanglement. (**NFPA 1010: 6.3.5**, p. 854)
- Demonstrate rescuing a downed firefighter.
  - Rescue a downed firefighter using the firefighter's SCBA straps. (**NFPA 1010: 6.3.9**, p. 855)
  - Rescue a downed firefighter using a drag rescue device. (**NFPA 1010: 6.3.9**, p. 856)
  - Rescue a downed firefighter as a two-person team. (**NFPA 1010: 6.3.9**, pp. 857–858)
  - Supply air to a downed firefighter using the low-pressure hose from a rapid intervention pack. (**NFPA 1010: 6.3.9**, pp. 860–861)
  - Supply air to a downed firefighter using the emergency breathing safety system (**NFPA 1010: 6.3.9**, pp. 862–863)

KEY TERMS

buddy breather See *emergency breathing safety system (EBSS).*

crew integrity See *team integrity.*

emergency breathing safety system (EBSS) A device on an SCBA that allows users to share their available air supply in an emergency situation. Also called *buddy breather.* (NFPA 1970)

emergency message See *emergency traffic.*

emergency traffic An urgent message that takes priority over all other communications except a mayday call. Also called *emergency message.*

firefighter assist and search team (FAST) See *rapid intervention crew/company (RIC).*

firefighter down CPR (FD-CPR) A technique for doing CPR that allows firefighters to begin providing immediate life-saving care to a firefighter in cardiac arrest while their SCBA and gear are removed in the most efficient way.

mayday A verbal declaration indicating that a firefighter is lost, missing, or trapped, and requires immediate assistance.

personnel accountability report (PAR) Periodic reports verifying the status of responders assigned to an incident or planned event. (NFPA 1026)

personnel accountability system A system that readily identifies both the location and function of all members operating at an incident scene. (NFPA 1550)

rapid intervention crew/company (RIC) A minimum of two fully equipped personnel on site, in a ready state, for immediate rescue of disoriented, injured, lost, or trapped rescue personnel. Also called *rapid intervention team (RIT), firefighter assist and search team (FAST),* and *rescue assist team (RAT).* (NFPA 1550)

rapid intervention pack (RIC pack) A portable air supply that provides an emergency source of breathing air for a single firefighter who has run out of air or whose air supply is insufficient to safely exit from an IDLH atmosphere.

rapid intervention team (RIT) See *rapid intervention crew/company (RIC).*

rescue assist team (RAT) See *rapid intervention crew/company (RIC).*

risk–benefit analysis An assessment of the risk to rescuers versus the benefits that can be derived from their intended actions. (NFPA 1006)

safe haven See *safe location.*

safe location A location remote or separated from the effects of a fire so that such effects no longer pose a threat. Also called *safe haven.* (NFPA 101)

team integrity The concept of keeping a crew together and working as a team so that no single firefighter goes off on their own. Also called *crew integrity.*

REVIEW QUESTIONS

1. On the fireground, who is responsible for making a risk–benefit analysis?
2. Describe the difference between direct and indirect hazards and give examples of each.
3. As a firefighter operating on an emergency scene, which three Rules of Engagement for Firefighter Survival deal with firefighter risk–benefit analysis?
4. How can members of a firefighting team maintain team integrity in conditions where they cannot see each other?
5. When firefighters are not accounted for, they are considered what until they are located?
6. How can a firefighter make sure that their survival skills are ready when they need them?
7. What is the time rating for an SCBA based on?
8. If you become lost during your search, what should your first action be?
9. What is the first step you should take if you are trapped in an entanglement during an emergency situation?
10. When you encounter a downed firefighter, what about their condition should you check immediately?

DISCUSSION QUESTIONS

1. Why do you use the high-pressure hose to connect to the RIC UAC when rescuing a downed firefighter with a rapid intervention pack?
2. Why is it important to call a mayday immediately upon realizing that you or another firefighter is lost, missing, or in trouble?
3. What is the purpose of a personnel accountability system?

APPLYING THE CONCEPTS

You and your four-person crew have just arrived at an active fire in a two-story, single-family residence in a suburban neighborhood. The structure is a Type V wood frame construction. All residents have escaped safely. The glow of flames is visible through the second-floor windows and smoke is showing through the gable vent on side alpha of the attic. Your crew is experienced, but you've been a volunteer firefighter for 4 months and this is your first active house fire. Units are on the ground conducting interior fire attack, evacuating neighbors, and undertaking other tactical operations. You and your crew report to the command post and the IC orders you to vertical ventilation operations in the attic. Eager to get going, you rush off to gather your tools and head to the attic.

1. By heading off by yourself to your assignment, what rule are you breaking?
2. Why is teamwork so important for firefighters' safety?

As you rush to the structure on your own, you pause and remember your training...and rule six. You turn around and head back to your crew, who are gearing up to enter the structure. Your company officer reviews the hazards, assigns buddies, and lays out objectives. She ends by saying, "Remember your buddy...communicate...and stay situationally aware. Everyone goes home today."

3. When determining what strategies and tactics to take at an emergency response, how is risk–benefit analysis used and who uses it? How does it apply to you?
4. In what situations do the **benefits** of taking action outweigh the risks, and in what situations do the **risks** of taking action outweigh the benefits?

Armed with tools and a ladder, you and your crew enter the structure through side Charlie and head to the stairs. The smoke gets thicker and the heat increases as you ascend to the second floor. Your crew maintains visual and verbal contact as you proceed, sharing observations. Your buddy locates the attic access and you set the ladder. After observing the attic's small size, your company officer decides on a two-in/two-out approach and radios the IC your crew's status and this tactical adjustment. She then instructs, "Team A, let's climb up to start vertical ventilation. Team B, remain here. Observe the fire situation, remain in radio contact, and communicate any changes. Be prepared to rotate in."

5. What is a personnel accountability system and how is the above scene an example of this in action?

While Team A is in the attic, the IC radios for a PAR report. Your company officer confirms the presence of each member of your team and reports this to the IC. However, one firefighter from the interior fire attack team is not accounted for.

6. How does this information change the incident strategy?

REFERENCES

International Association of Fire Chiefs (IAFC). n.d. *Rules of Engagement for Firefighter and Incident Commander Survival.* Accessed July 13, 2023. https://www.iafc.org/topics-and-tools/resources/resource/rules-of-engagement-training-poster.

National Fire Protection Association (NFPA). 2019. *NFPA 1407, Standard for Training Fire Service Rapid Intervention Crews.* 2020 Edition. Quincy, MA: NFPA.

National Fire Protection Association (NFPA). 2020. *NFPA 101, Life Safety Code.* 2021 Edition. Quincy, MA: NFPA.

National Fire Protection Association (NFPA). 2020. *NFPA 704, Standard System for the Identification of the Hazards of Materials for Emergency Response.* 2022 Edition. Quincy, MA: NFPA.

REFERENCES CONTINUED

National Fire Protection Association (NFPA). 2020. *NFPA 1006, Standard for Technical Rescue Personnel Professional Qualifications.* 2021 Edition. Quincy, MA: NFPA.

National Fire Protection Association (NFPA). 2023. *NFPA 1026, Standard for Incident Management Personnel Professional Qualifications.* 2024 Edition. Quincy, MA: NFPA.

National Fire Protection Association (NFPA). 2023. NFPA *1550, Standard for Emergency Responder Health and Safety.* 2024 Edition. Quincy, MA: NFPA.

National Fire Protection Association (NFPA). 2023. *NFPA 1970, Standard on Protective Ensembles for Structural and Proximity Firefighting, Work Apparel and Open-Circuit Self-Contained Breathing Apparatus (SCBA) for Emergency Services, and Personal Alert Safety Systems (PASS).* 2024 Edition. Quincy, MA: NFPA.

Occupational Safety and Health Administration (OSHA). *Respiratory Protection, 29 CFR 1910.134.* Accessed November 11, 2022. https://www.osha.gov/laws-regs/regulations/standardnumber/1910/1910.134.

CHAPTER

20

Firefighter I

Firefighter Rehabilitation

KNOWLEDGE OBJECTIVES

After studying this chapter, you will be able to:

- Describe the need for and function of firefighter rehabilitation.
- Describe the factors that necessitate the setting up of a rehabilitation station.
- Describe the process of rehabilitation and each individual's role.

SKILLS OBJECTIVES

There are no skills objectives for Firefighter I candidates. NFPA 1010 contains no Firefighter I Job Performance Requirements for this chapter.

ADDITIONAL NFPA STANDARDS

- **NFPA 704**, *Standard System for the Identification of the Hazards of Materials for Emergency Response, 2022 Edition*
- **NFPA 1550**, *Standard for Emergency Responder Health and Safety, 2024 Edition*
- **NFPA 1584**, *Standard on the Rehabilitation Process for Members During Emergency Operations and Training Exercises, 2022 Edition*

CASE STUDY

You Are the Firefighter

You are working out early to get a jump on the day. You have just completed a run on the treadmill when you are dispatched to a house fire in an older section of town that dates back to the 1890s. Upon arrival, you are assigned to fire attack and are told that the fire is in the basement. Your crew has trouble locating the seat of the fire in the dirty brown smoke. When you finally reach it, you realize it is a large fire that is beginning to spread to the upper floors. As you knock down the fire in the basement, you notice you are getting low on air. You exit with your crew, get your air cylinder changed out, and are reassigned to the first floor to begin opening up walls for overhaul. As you are pulling walls, you start to feel that you are breathing harder than usual. You begin to get mild cramps in your arms as you tire from swinging the axe. You strive to keep up with your partner. When you notice you are getting low on air again, you let your partner know that you need to exit and you are going to head to rehabilitation.

1. What is the purpose of firefighter rehabilitation?
2. When should a firefighter seek rehabilitation on the fireground?
3. How does firefighter rehabilitation help with firefighter safety and performance?

Introduction

Firefighters have a deep commitment to help others. If you were to put your personal comfort first, you probably would not want to get up in the middle of a cold winter night to answer a call. Firefighters must take care of themselves, both physically and mentally, so that they can continue to help others. Firefighters must prioritize the health and well-being of their teammates for the good of the fire department and the community they protect.

Firefighting operations can require the kind of short-term physical demands that put rescuers on par with professional athletes. However, unlike professional athletes, firefighting operations require more than just practicing on a set schedule, showing up on game day, doing the work, and going home. The demand on firefighters is more intense; not only do they not know when and where they will have to perform, the quality of their performance and that of their teammates can be a matter of life or death. Firefighters must be ready to respond anytime, anywhere. Only firefighters who are physically fit and mentally alert will be able to perform the tasks necessary to save civilian and their teammates' lives and protect property.

The Purpose of Rehabilitation

Fatigue is frequently cited as a critical factor in workplace-related injuries. It impairs situational awareness, decision making, reaction time, and the physical ability to perform tasks. Firefighters cannot work indefinitely. During prolonged physically and mentally draining operations, firefighters require relief from the extreme environmental conditions. They need rest and recovery, cooling or rewarming, rehydration, calorie and electrolyte replacement, medical monitoring, and possibly treatment by emergency medical services (EMS). All of this must happen while each firefighter remains accounted for and evaluated for fitness to return to duty. These are the concepts behind **emergency incident rehabilitation**.

To **rehabilitate** (usually called simply **rehab**) means to restore someone or something to a condition of health or a state of useful and constructive activity. Fighting fires is a job that requires excellent physical conditioning to combat the rigors of heat, cold, smoke, flames, physical exertion, and emotional stress that team members routinely encounter. Even a seasoned, well-conditioned firefighter can quickly become fatigued when battling a tough fire (**FIGURE 20-1**).

At most fire department incidents, only one rehabilitation area is established. At large-scale incidents, it may be necessary to set up multiple rehabilitation areas. In some communities, rehabilitation is initiated entirely by fire department personnel. In other communities, establishing the rehabilitation facility is a combined effort with the Salvation Army, the American Red Cross, the fire department auxiliary unit, or a separate EMS agency.

Because firefighters must be ready to respond at any time, they must do what they can to ensure that, like any other equipment they will use, their bodies

FIGURE 20-1 Firefighting often involves extreme physical exertion and results in rapid fatigue.

Courtesy of Rom Duckworth.

FIGURE 20-2 Rehabilitation allows firefighters to take a break for rest, cooling or rewarming, rehydration, calorie replacement, medical monitoring, and ensuring accountability.

Courtesy of Rom Duckworth.

are always prepared for the next call. This is the concept of **pre-habilitation**—maintaining physical readiness by ensuring adequate hydration, nutrition, rest, clothing appropriate for anticipated weather conditions, and more.

New firefighters are frequently amazed at the energy expended during fire suppression activities. The exertion takes its toll on your body, leaving you hot, dehydrated, hungry, and tired. The goal of rehabilitation is to take time to recharge your body's batteries so you can continue to be a productive member of the team. Rehabilitation is a critical factor in maintaining your health and well-being. It is so vital that it is the focus of several NFPA standards, including NFPA 1550, *Standard for Emergency Responder Health and Safety, 2024 Edition* and NFPA 1584, *Standard on the Rehabilitation Process for Members During Emergency Operations and Training Exercises, 2022 Edition.*

Emergency incident rehabilitation is part of the overall emergency effort. This aspect of the emergency response ensures that firefighters and other emergency workers who are exhausted, dehydrated, hungry, ill, injured, or emotionally upset can take a break for rest, fluids, food, medical monitoring, and treatment of illnesses or injuries (**FIGURE 20-2**). Fighting fire takes a toll on the body, and rehabilitation gives firefighters the opportunity to evaluate and address these issues.

SAFETY TIP

A tired or dehydrated firefighter is more likely to be injured. Rehabilitation is essential to correct imbalances in the body that, if left untreated, could reduce your effectiveness, endangering you, your crew, and others at the scene.

Rehabilitation enables firefighters to perform more safely and effectively at an emergency scene. A tired or dehydrated firefighter cannot accomplish as much work as and is more likely to become ill or injured than a rested and hydrated firefighter. The effort required to rescue an ill or injured firefighter takes time and resources away from fire suppression activities. Rehabilitation is necessary for firefighter safety and is crucial to maintaining effective firefighter operations.

Many conditions come together during a firefighting operation to create a stressful environment. Consider the stresses involved in a typical, middle-of-the-night call for a working fire. The loud, jarring sound of the alarm jolts your sleeping body awake. Without hesitation, you must immediately get up, get dressed, and put on your personal protective equipment (PPE). You do not have time to eat or get something to drink. You may have to drive an emergency vehicle, perform forcible entry, haul hose, position a ladder, and climb to the roof to cut a ventilation hole. All these tasks require a significant amount of energy and concentration, and you must be able to move into action quickly with no time to warm up your muscles as athletes do before an event.

SAFETY TIP

Stress management skills are crucial for high-quality performance, regardless of the type of stressor involved and the role that the person plays in emergency assistance. A firefighter needs to implement basic stress management skills to deliver high-quality performance. Exercise, proper nutrition, and emotional support are all helpful in managing stress.

You may be called to a fire on the hottest day of the year, the coldest day of the year, or under all types of other adverse circumstances. Because you know that lives and property are at risk, you may feel an added emotional stress, which also affects your body. Finally, firefighters often work in unfamiliar, smoke-filled environments. All of these factors make firefighting difficult and stressful.

Rehabilitation provides firefighters with rest and time to recover from the fatigue and stresses of fighting fires and participating in emergency operations. Studies have shown that proper pre-habilitation and rehabilitation will help prevent firefighters from collapsing or suffering injuries during fire suppression and emergency operations. Taking short breaks, replacing fluids, ingesting healthy food, and cooling or rewarming are all measures that reduce the risks of injury and illness (**FIGURE 20-3**). Rehabilitation also helps to improve the quality of decision making, because people who are tired tend to make poor decisions.

FIGURE 20-3 Taking short breaks to rehabilitate lowers your risk of injury and illness.

Courtesy of Rom Duckworth.

TIP

The SAID principle—Specific Adaptation to Imposed Demands—identifies a need for training that mimics the type of work to be performed. In other words, the best way to prepare for physical work is to work up to performing activities that match the type, intensity, and duration of the work. The optimal training program should include both high- and low-intensity cardiovascular training and muscular work that involves strength, endurance, and power.

Personal Protective Equipment

Wearing PPE contributes to **heat stress**, which is the condition that occurs when the body cannot get rid of excess heat; in other words, when the body's response to heat begins to fail (**FIGURE 20-4**). PPE creates a protective envelope around a firefighter that protects their body from the smoke, flames, heat, and steam of a fire. At the same time, however, it traps almost all body heat inside the protective envelope. Also, PPE can weigh 40 pounds (lb; 18 kilograms [kg]) or more, and the extra weight increases the energy needed to move around.

When a firefighter is wearing PPE, perspiration soaks the inner clothing. Normally, evaporating perspiration helps cool your body so that it does not overheat—this is called evaporative cooling. However when you wear PPE, the same vapor barrier that keeps hot liquids and steam from getting to the body also prevents most of your perspiration from evaporating. Because evaporative cooling cannot occur, a firefighter's core temperature quickly rises to dangerous levels.

Regardless of the temperature and relative humidity in the ambient air, the temperature and relative humidity inside your PPE will quickly rise to high levels. When the body cannot get rid of excess heat, it is subject to heat stress. Heat stress reduces your ability

FIGURE 20-4 The personal protective equipment worn to protect a firefighter contributes to heat stress.

Courtesy of Rom Duckworth.

to do work, and excessive amounts can lead to collapse if not treated. Because your body continues to sweat in an attempt to cool your body, you can quickly become dehydrated in addition to being overheated under conditions of heat stress. The heat stress index combines temperature and relative humidity to measure the degree of heat stress. For example, if the relative humidity is 70 percent and the temperature is 88°F (31°C), the heat stress index is 100°F (38°C)—in other words, it feels like the temperature is 100°F (38°C) (**TABLE 20-1**).

Dehydration

Dehydration is when the body's fluid losses are greater than its fluid intake. If left untreated, this imbalance can lead to electrolyte imbalance, shock, and death. Fighting fires is a strenuous activity, and the large amounts of muscular energy required during firefighting activities produce a significant amount of heat. During this exertion, the body loses a substantial amount of water through perspiration. Firefighters in action can lose as much as 1 quart (qt; 1 liter [L]) of fluid in less than 1 hour.

Virtually everyone gets somewhat dehydrated overnight. Firefighters who begin their day by drinking caffeinated beverages or sodas and/or who drink sports or energy drinks throughout the day should be aware that those drinks cause the kidneys to eliminate water, causing further dehydration. Often, this dehydration goes unnoticed until the firefighter is subjected to physical exertion or heat stress. It is at that point when the body begins to weaken and fail.

Dehydration reduces strength, endurance, and mental judgment, as evidenced by a variety of signs and symptoms (**TABLE 20-2**). Extreme dehydration can lead to confusion and total collapse. It is crucial to prevent dehydration or correct it as quickly as possible. Drink at least a quart (liter) of water per hour during heavy physical exertion. It is simple to plan for and accomplish this fluid intake when you have a scheduled physical conditioning session or in-service drill. Because fires are unscheduled events, you must try to keep your body pre-hydrated when you are on duty. In addition, replenishing fluids during rehabilitation is essential to correct any fluid imbalance in your body. It is better to prevent dehydration by drinking adequate amounts of water before an incident than trying to correct severe dehydration during or after an incident.

SAFETY TIP

Always wash or sanitize your hands before you eat.

Energy Consumption

Food provides the fuel your muscles need to work correctly. During strenuous activity, the body burns carbohydrates and fats for energy, and these energy sources need to be replenished. Without a sufficient supply of the right types of food for energy, your body cannot continue to perform at peak levels for extended periods. The number of calories you need depends on the duration of the activity, the time since your last meal, and your general physical condition. For example, a 6-foot (ft) 3-inch (in.; 191-centimeter [cm]) muscular athlete who has not eaten in 4 hours will need more calories to fuel their body than a 5-ft 2-in. (157-cm) person who maintains a sedentary lifestyle. The type of calories that you consume is also important. It is important to refuel the body with a balanced diet consisting of nutritious food such as lean proteins, complex carbohydrates, and fruits and vegetables. If your diet consists mainly of processed foods that are mainly simple carbohydrates and added sugar, you will not have the energy stores you need to work safely at emergency incidents.

TABLE 20-1 Heat Stress Index

Relative humidity (%)	Temperature (°F)															
	80	82	84	86	88	90	92	94	96	98	100	102	104	106	108	110
40	80	81	83	85	88	91	94	97	101	105	109	114	119	124	130	136
45	80	82	84	87	89	93	96	100	104	109	114	119	124	130	137	
50	81	83	85	88	91	95	99	103	108	113	118	124	131	137		
55	81	84	86	89	93	97	101	106	112	117	124	130	137			
60	82	84	88	91	95	100	105	110	116	123	129	137				
65	82	85	89	93	98	103	108	114	121	128	136					
70	83	86	90	95	100	105	112	119	126	134						
75	84	88	92	97	103	109	116	124	132							
80	84	89	94	100	106	113	121	129								
85	85	90	96	102	110	117	126	135								
90	86	91	98	105	113	122	131									
95	86	93	100	108	117	127										
100	87	95	103	112	121	132										

Likelihood of Heat Disorders with Prolonged Exposure or Strenuous Activity

Caution | Extreme caution | Danger | Extreme danger

Modified from National Weather Service (NWS), https://www.weather.gov/safety/heat-index

Relative humidity (%)	Temperature (°C)															
	26	27	28	30	31	32	33	34	35	36	37	38	40	41	42	43
40	26	27	27	29	31	32	34	36	38	40	42	45	48	51	130	57
45	26	27	28	30	31	33	35	37	40	42	45	48	51	54	58	
50	27	27	29	31	32	35	37	39	42	45	47	51	131	58		
55	27	28	30	31	33	36	38	41	44	47	51	54	58			
60	27	28	31	32	35	37	40	43	46	50	53	58				
65	27	29	31	33	36	39	42	45	49	53	57					
70	28	30	32	35	37	40	44	48	52	56						
75	28	31	33	36	39	42	46	51	55							
80	28	31	34	37	41	45	49	53								
85	29	32	35	38	43	47	53	57								
90	30	32	36	40	45	50	55									
95	30	33	37	42	47	52										
100	30	35	39	44	49	55										

Likelihood of Heat Disorders with Prolonged Exposure or Strenuous Activity

Caution | Extreme caution | Danger | Extreme danger

TABLE 20-2 Signs and Symptoms of Dehydration

Percentage of Body Weight Lost	Signs and Symptoms
1%	Increased thirst
2%	Loss of appetite Dry skin Dark urine Fatigue Dry mouth
3%	Increased heart rate
4–5%	Decreased work capacity by up to 30%
5%	Increased respiration Nausea Decreased sweating Decreased urine output Increased body temperature Markedly increased fatigue Muscle cramps Headache
10%	Muscle spasms Markedly elevated pulse rate Vomiting Dim vision Confusion Altered mental status

Reproduced from International Association of Fire Chiefs. *A Guide for Best Practices: An Introduction to NFPA 1584 (2008 Standards).*

Tolerance for Stress

Everyone has a different tolerance level for the stresses that occur when fighting fires. For example, younger people tend to have greater endurance and can tolerate higher levels of stress than older people. Likewise, a well-rested person in good physical and mental condition will have a greater level of endurance than someone who is tired and in poor physical and mental condition. A person who carries extra weight and performs strenuous tasks strains their cardiovascular system and significantly increases their risk of a heart attack. Finally, certain medications and substances, such as energy drinks and workout beverages and supplements, may cause a person to lose more fluids than they would if they were not taking those substances. Conditioning plays a significant role in a firefighter's level of endurance. A well-conditioned person with good cardiovascular capacity, good flexibility, and well-developed muscles is better able to tolerate the stresses of fighting fires than someone who is out of shape (**FIGURE 20-5**). Nevertheless, even the most impressive conditioning will not keep a firefighter from becoming exhausted under physically stressful situations.

FIGURE 20-5 A well-conditioned firefighter will have a greater tolerance for the stresses encountered when fighting fires.

When Rehabilitation Is Needed

It is essential to know when rehabilitation is needed during an emergency incident. A fire or emergency crew should perform self-rehabilitation—resting with fluid replacement—for a minimum of 20 minutes following the depletion of one self-contained breathing apparatus (SCBA) cylinder or after 40 minutes of intense work without SCBA. They should also rest with fluid replacement for a minimum of 20 minutes whenever a full encapsulating suit is worn. It is important to understand that these guidelines and the "one-bottle rule" are used as a matter of convenience because SCBA bottle usage is easy to track. This does not reliably identify firefighters who may need rehabilitation before the SCBA bottle is depleted.

Follow the rehabilitation guidelines established by your department. After a period of rehabilitation has been completed, company officers or other crew members should ensure that members are fit to return to duty. During larger incidents, it may be necessary to designate a person to serve as a time recorder to ensure that all crew members rotate out to rehabilitation as often as needed.

Although the concept of rehabilitation needs to be addressed at all incidents, it is not necessary to implement all components of a rehabilitation area for every incident. For example, firefighters who put out a small fire in a single room might require only water for rehydration. In contrast, those involved in extinguishing a major wildland fire may need a full-fledged rehabilitation station.

SAFETY TIP

During firefighting operations, firefighters may be reluctant to admit that they need a rest. Asking an obviously exhausted firefighter if they need to go to rehabilitation is likely to bring a response such as, "No. I'm okay." All company members need to watch out for one another, and company officers must monitor their crews for indications of fatigue. It is better to go to rehabilitation a few minutes early to stay ready to work than to wait too long and risk the consequences of injury or exhaustion.

The introduction of structured rehabilitation procedures has reduced the number of injuries caused by heat stress and exhaustion. The old philosophy was to "fight hard until you drop." It was not unusual to see exhausted firefighters being carried out of burning buildings to waiting ambulances. Of course, this procedure disrupted effective firefighting operations because the firefighters who could still function were busy rescuing their injured and exhausted teammates instead of attacking the fire. Structured rehabilitation procedures address issues before they cause a problem on the fireground by not allowing exhausted and dehydrated firefighters to continue working without periodic rests.

SAFETY TIP

The amount of rest needed to recover from physical exertion is directly related to the intensity of the work performed. Firefighters who have expended a tremendous amount of energy will require a more extended recovery period than those who have performed moderate work.

Extended Fire Incidents

Significant emergency incidents require full-scale rehabilitation efforts. Firefighters cannot always predict when a fire incident will turn into a large-scale incident that requires many hours on scene. Therefore, they should pre-hydrate whenever possible.

SAFETY TIP

Reducing the amount of clothing worn at a safe distance from the fire can help avoid heat stress before engaging in firefighting activities.

Structure Fires

The extended on-scene time at major structure fires is hard on crews (**FIGURE 20-6**). It can be easy for firefighters to work beyond their limits. While the initial attack and rescue operations may occur quickly at a structure fire, the overall incident is much more like a marathon than a sprint. Part of the purpose of rehab is to rest and refuel for these extended operations.

High-Rise Fires

High-rise fires place increased stress on firefighters while they drain crew members' energy resources. It takes considerable energy to walk up many flights of stairs and work in PPE and SCBA. Hose packs, hand tools, and extra SCBA cylinders must be carried up staircases that may be crowded with people exiting the building. Getting everyone and everything up to the location of the fire may require a marathon effort, even before the

FIGURE 20-6 Major fires often require a strong effort that lasts for an extended period of time. Rehabilitation and crew rotation are essential measures that limit the risk of exhaustion and injuries to firefighters.

FIGURE 20-7 Wildland fires often involve hundreds of firefighters who work for days or weeks at the site.
Courtesy of Michael Rieger/FEMA News Photo.

FIGURE 20-8 The same type of rehabilitation procedures should be implemented for training activities as are used for actual emergency incidents.
© Jones & Bartlett Learning

attack on the fire begins. It is no wonder that fatigue and dehydration can occur rapidly in these circumstances.

During a high-rise fire incident, the incident commander (IC) may assign three companies to work normally done by one company. This strategy enables the companies to rotate between attacking the fire; replenishing their air supplies; and meeting their needs for rest, fluids, and nutrition. At high-rise fires, the rehabilitation area is often located two or three floors below the level of the fire to reduce the time and energy expended walking up and down the stairs.

Wildland Fires

Emergency incident rehabilitation was first used on a widespread basis during firefighting operations carried out in wildland areas such as grasslands and forests. Given the size and intensity of major wildland fires, crews need to work in shifts so that their bodies can recover from the heat, smoke, and stresses of the fire. Minimal rehabilitation efforts may be all that is needed for a small ground fire that is easily extinguished. In contrast, a large forest fire that requires hundreds of firefighters and takes several weeks to control and extinguish will require a significant and ongoing rehabilitation effort for all crews involved (**FIGURE 20-7**).

Proper nutrition and hydration are essential for keeping the crews healthy and ready to resume duty during wildland fires. Some studies have shown that wildland firefighters may require as much as 5000 to 6000 calories per day (Sharkey, Ruby, and Cox 2002).

Training Exercises

Training exercises, including live-burn exercises in training towers or actual structures, simulate many conditions of actual firefighting. Firefighters may be wearing full PPE, including SCBA. The activities may be conducted on days when the ambient temperature is hot or cold and may last for several hours or an entire day.

Just as with actual fire operations, pre-hydration and rehabilitation should be incorporated into the planning for training activities (**FIGURE 20-8**). Exercises conducted over a full day should be set up to allow the participants to rehabilitate between strenuous activities. Rehabilitation is no less critical during training activities than it is during emergency incidents.

Other Types of Incidents Requiring Rehabilitation

Other types of incidents may require extensive rehabilitation efforts to maintain the health and safety of firefighters. For example, hazardous materials incidents that require responders to wear a **fully encapsulated suit** (a protective suit that fully covers the responder, including the breathing apparatus) are incredibly stressful (**FIGURE 20-9**). These kinds of incidents can expose responders to strenuous conditions for extended periods of time. In addition, they may require that the staging area be located at a significant distance from the hazardous materials, requiring responders to walk for long distances while wearing heavy PPE. For these reasons, hazardous materials incidents require adequate rehabilitation for responders.

On other occasions, the fire department may be involved in long-duration search and rescue activities or other incidents that require the presence of public safety agencies for extended periods. These situations

FIGURE 20-9 Responders wearing fully encapsulated suits must be carefully monitored for symptoms of heat stress.

can be both mentally and physically taxing. Establishing a rehabilitation area where firefighters can recuperate during these stressful incidents is essential.

The need for rehabilitation is not limited to emergencies. Athletic events and stand-by assignments also may require rehabilitation. Whenever firefighters need to be ready for action for an extended period, some provision should be made for replenishing fluids, nourishing foods, and warming or cooling for crew members.

How Does Rehabilitation Work?

One way to understand how an emergency incident rehabilitation area works is to look at the functions it is designed to perform. A widely used model of rehabilitation includes seven parts:

- Relief from climatic conditions
- Rest and recovery
- Passive or active cooling or warming
- Rehydration and calorie replacement
- Medical monitoring
- Member accountability
- Release and reassignment

Although these seven activities do not necessarily need to be completed in the order listed, it is important that each step be addressed.

Field Reduction of Contaminants

Before entering a rehabilitation area, it is important to minimize exposure to carcinogens and other contaminants on your PPE and skin by performing a field reduction of contaminants. This process will vary depending on the firefighter's proximity to the fire scene, the temperature of the rehabilitation area, and the amount of time needed at the rehabilitation area. All PPE exposed to hazardous conditions should be removed and cleaned. This process is further described in Chapter 3, *Personal Protective Clothing*. Field reduction of contaminants begins while the firefighter is still wearing the protective clothing, when another firefighter should spray them down using a small diameter hose to remove surface contaminants, using a brush if needed. Next, clean any exposed skin as much as possible before starting the steps of rehabilitation. If you doff your PPE, doff it before you enter the rehabilitation area, and then use a brush and water to remove as much dirt and debris as possible. Store only in the proper PPE storage area to avoid continued exposure to potential carcinogens. Always follow your department's protocols.

Relief from Climatic Conditions

Few days provide the perfect climatic conditions for rehabilitation. When the temperature is between approximately 68°F and 72°F (20°C and 22°C), the humidity is low, and plenty of shade is available, an indoor rehabilitation area may not be needed. Conversely, on days when the temperature is warmer or colder than 68°F to 72°F (20°C and 22°C) or when there is high humidity or wind, rehabilitation needs to take place in a climate-controlled location. In that case, a rehabilitation area may be set up in a nearby building, such as a church, school, business, or shopping center, or rehabilitation services may be provided in a transit bus that is brought to the emergency scene. Whether conditions are hot or cold or wet or dry, returning the body's temperature back to normal is one of the primary goals of rehabilitation.

FIGURE 20-10 In the rehabilitation area, firefighters should be able to remove personal protective clothing and get some rest before returning to action.

When entering the rehabilitation area, it is important to take off your PPE to help your body warm up or cool down as needed (**FIGURE 20-10**). If your clothing is wet, it may be necessary to change. Many firefighters carry an extra pair of socks and a second pair of gloves to handle such emergencies. Use a towel to dry off, if needed.

When the temperature is very hot, the need to rotate crews and provide for extensive rehabilitation at an emergency scene is increased. Crews working on the interior fire attack may not notice much difference because their environment during the attack is hot regardless of the external temperature. Hot weather, however, will affect crews working on the outside. These firefighters tend to become dehydrated and fatigued much faster than they would if the ambient temperature was in a more comfortable range.

Another factor that must be considered is the humidity. Remember that high humidity reduces evaporative cooling, making it more difficult for the body to regulate its internal temperature. If the body cannot disperse heat because it is covered by PPE, the firefighter experiences heat stress, which can lead to serious short- and long-term medical situations, including the following:

- **Heat rash**: An itchy rash that develops on skin that is wet from prolonged sweating
- **Heat cramp**: Painful muscle spasm, usually in muscles in the leg or abdomen, that occurs suddenly during or after physical exertion
- **Heat exhaustion**: A mild form of shock that occurs when the circulatory system begins to fail because of the body's inadequate effort to give off excessive heat
- **Heat stroke**: A severe, sometimes fatal condition that occurs when the body's temperature-regulating capacity fails and that is indicated by a reduction or cessation of sweating so that the skin is hot and dry, body temperature of 105°F (40.5°C) or higher, rapid pulse, hot and dry skin, headache, confusion, unconsciousness, and convulsions

Because heat emergencies are so common, all firefighters must be able to identify when heat stress is taking its toll. If a firefighter is sweating heavily and complaining of fatigue, muscle cramps, headache, and dizziness or nausea, they should be moved to the rehabilitation area for evaluation if possible, or at least a safe cooling area. Their PPE should be removed, and they should be cooled with cool, wet towels; water mist; cold air; or other active cooling methods described later in this chapter. Firefighters in this condition should also consume small sips of water or rehydrating fluids. If a firefighter has severe fatigue and exhibits signs of heat stroke, they must be moved to a safe area, have their PPE removed, and be immediately cooled with a cold water or ice bath. Heat stroke is an emergency in which the body has become so overheated, it has lost its ability to cool itself, and the brain is beginning to "cook." Emergency treatment such as an ice bath may seem extreme, but it is necessary, life-saving care. Pre-hydration and early identification of the need for rehabilitation can prevent heat stroke and help firefighters avoid injury and continue to work effectively.

SAFETY TIP

Firefighters who are overheated should remove their PPE as soon as possible to permit evaporative cooling. PPE must be removed only in a safe environment outside the hazard area.

Cold weather also increases the need for rehabilitation and crew rotation. Indeed, low temperatures can be just as dangerous as high temperatures. Water is used for fire suppression because of its tremendous capacity for cooling. If firefighters are wet from firefighting, weather conditions, or sweating under their PPE, the cooling capacity of water increases a firefighter's heat loss drastically. Even in cold weather, the weight of PPE and the physical exertion of fighting a fire cause the body to sweat inside the protective clothing. The combination of damp clothing and cold temperatures can quickly lead to hypothermia. **Hypothermia**, a condition in which the internal body temperature falls below 95°F (35°C), can lead to loss of coordination,

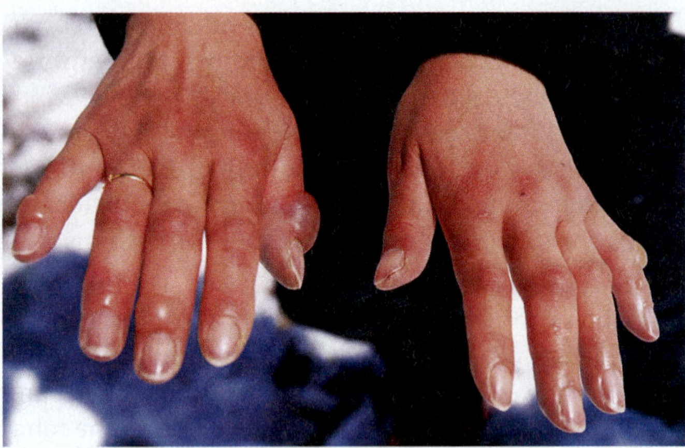

FIGURE 20-11 Prolonged exposures to freezing weather can result in severe frostbite.

Courtesy of Neil Malcom Winkelmann.

muscle stiffness, coma, and death. In cold weather, exposed skin is also susceptible to **frostbite** (tissue damage resulting from prolonged exposure to cold) (**FIGURE 20-11**). The U.S. Fire Administration (USFA) recommends that rehabilitation operations be initiated whenever the wind chill factor drops to 10°F (–12°C) or lower (USFA 2008, 83).

Rest and Recovery

Rest should begin as soon as the firefighter arrives for rehabilitation. The rehabilitation area should be located away from the central activity of the emergency, so firefighters can disengage from all other stressful activities and remove personal protective clothing. Rest continues as firefighters go through each aspect of revitalization. Rehabilitation areas can be equipped with chairs or cots so firefighters can sit or lie down and relax in a comfortable position.

TIP

The rehabilitation area should provide some type of restroom facility. For example, the existing restrooms in the building may be used for the rehabilitation area, or portable restrooms may be brought to the emergency scene.

Passive or Active Cooling or Warming

The third step of rehabilitation is stabilizing the body's internal temperature. For the overheated firefighter, PPE should be removed as soon as possible to allow the body to cool. Two types of cooling are possible: passive and active.

Passive cooling is performed by removing PPE and other clothing and moving to a cooler environment. This allows the body to cool through sweating. Passive cooling works best when both the air temperature and relative humidity are low. Stabilizing a firefighter's body

Voice of Experience

Firefighter rehabilitation is often overlooked, even by many experienced firefighters, because it is a relatively new component of the fire service. At the Connecticut Fire Academy, we stress to our fire recruits the importance of rehabilitation and provide a formal rehabilitation during their live fire training. Each recruit is evaluated after every live fire training session. In addition, the instructors follow a two-in, two-out rotation.

When conversing with a colleague, he told me about a fire that he had worked on with three recent graduates of our recruit firefighter program. He stated that they came out of the fire after consuming their second cylinder of oxygen and said "Cap, it's time to go to rehab." The captain looked at them and said, "Yeah, okay." He was impressed that these new firefighters knew and understood the need for rehabilitation, and I was impressed that the new firefighters were able to take their seasoned captain into rehab with them.

The importance of rehabilitation is being spread by the new firefighters graduating from fire academies. They understand that rehabilitation is not a means to pull them out of the fire but instead a means to keep them in the fight. Through our training program, our recruits become comfortable with the rehab process, and my hope is that they continue to educate seasoned firefighters on the importance of rehabilitation.

Debra L. Burch

Connecticut Fire Academy
Windsor Locks, Connecticut

FIGURE 20-12 Winter weather presents a different set of problems for firefighters than summer weather. Rehabilitation procedures must be conducted in a warm space, and firefighters should be given hot liquids, food, and dry clothing.

temperature through passive cooling may not be possible during periods of hot weather or high humidity. When the weather is hot and humid, rehabilitation areas must be climate-controlled so firefighters can achieve a normal body temperature before resuming active duty. Some fire departments have air-conditioned vehicles available for rehabilitation. Others use buses or establish rehabilitation areas in an already-cooled area, such as a shopping center.

Active cooling may be required if passive cooling is not adequate to reduce body temperature. Active cooling is performed by engaging a misting fan; placing cool, wet towels around the head and neck; or dipping both forearms in cool water. If a firefighter exhibits signs of heat stroke, they must be immediately actively cooled with a cold water or ice bath.

For firefighters to avoid hypothermia and frostbite in cold weather, rehabilitation areas need to be heated so that firefighters can rewarm before returning to the cold environment (**FIGURE 20-12**). Passive warming occurs when the firefighter gets out of the cold, and the body restores itself to an equilibrium temperature.

Active heating may be needed when a firefighter's body is unable to restore itself to a normal temperature balance. Medical professionals achieve active warming by using warming blankets, heated and humidified oxygen, and warmed intravenous fluids. Wet or severely chilled firefighters should be wrapped in warming blankets and moved into a well-heated area before they remove their PPE. All damp clothing should be replaced with warm, dry clothing as soon as possible.

Rehydration and Calorie Replacement

The third step of rehabilitation is rehydration and calorie replacement. Both are important steps in keeping the body fueled and ready to work after the rehabilitation period.

Rehydration

When the body becomes overheated, it sweats, so evaporative cooling can reduce body temperature. Perspiration is composed of water and other dissolved substances, such as salt. During fire suppression activities, a firefighter can perspire enough to lose 2 qt (2 L) of water in the time it takes to go through two cylinders of air (Selkirk, McLellan, and Wong 2006, 425). Because 1 pint of water weighs 1 lb (1 L weighs 1 kg), losing 2 qt (2 L) of water is approximately equivalent to a 2 percent loss in body weight for a 200-lb (91-kg) firefighter. The loss of that much body fluid can result in impaired body temperature regulation. Further dehydration can result in reduced muscular endurance, reduced strength, and heat cramps. In addition, severe cases of dehydration can contribute to heat stroke and death. Because a firefighter can lose such a large quantity of water during fire suppression activities, it is essential to replace fluids early—before severe dehydration occurs (**FIGURE 20-13**).

When the body perspires, it loses **electrolytes** (salts and other chemicals that are dissolved in body fluids and cells) as well as water. If a firefighter loses large amounts of water through sweat, their electrolytes must be replenished in addition to the water. Sports drinks are often used in rehabilitation because they contain water, balanced electrolytes, and some sugars.

Rehydrating the body is not as simple as drinking lots of fluids quickly. Drinks such as colas, coffee, and tea should be avoided because they contain caffeine, and caffeine acts as a diuretic that causes the body to excrete more water. Sugar-rich carbonated beverages are not tolerated or absorbed as well as plain water or sports drinks because large amounts of sugar are difficult to digest and cause swings in the body's energy level. In addition, drinks that are too cold or hot may be difficult to consume and ultimately may prevent the firefighter from ingesting enough liquids.

FIGURE 20-13 The rehabilitation area should have plenty of fluids available to rehydrate firefighters. Plain water or diluted sports drinks are preferred.

An additional concern is the rate at which fluids are absorbed from the stomach. Drinking too much too quickly can cause bloating, a condition in which air fills the stomach, and this causes a feeling of fullness and can lead to discomfort, nausea, and even vomiting. Studies have demonstrated that the stomach can absorb approximately 1 qt (1 L) of fluid per hour. However, as previously noted, the body can also lose up to 1 qt (1 L) of fluid per hour during intense activity. Thus the body may lose fluids as rapidly as they can be replaced. Once the initial 1 qt (1 L) of water is lost, the body will require 1 to 2 hours to recover.

One key to remaining hydrated while fighting fires is ensuring you are hydrated before reaching the fireground. To do so, drink enough fluids to keep your body properly hydrated whenever you are on duty, particularly during hot weather. This habit will decrease your risk of dehydration and the need to play catch-up with fluids when an emergency incident occurs. Adequate hydration also ensures peak physical and mental performance. More information on hydration is presented in Chapter 2, *Firefighter Health and Safety*.

Calorie Replacement

A firefighter performing intense physical work burns many calories, and it is necessary to replace those calories during rehabilitation. In much the same way that an engine runs on diesel fuel, the human body runs on glucose. **Glucose** (also known as **blood sugar**) is carried throughout the body by the bloodstream and is needed to burn fat efficiently and release energy. For the body to work properly, its glucose levels need to be in balance. If blood sugar levels drop too low, the body becomes weak and shaky. If blood sugar levels are too high, the body becomes sluggish. Blood glucose levels can be balanced by eating a proper diet of carbohydrates, proteins, and fats.

SAFETY TIP

The sensation of thirst is not a reliable indicator of the amount of water the body has lost. Thirst develops only after the body is already dehydrated. Do not wait until you feel thirsty to focus on your hydration. Instead, start to drink before you get thirsty, and remember to drink early and often. Firefighters should strive to drink a quart (liter) of water per hour during periods of heavy physical exertion.

Carbohydrates—a major source of fuel for the body—are found in grains, vegetables, and fruits. The body converts carbohydrates into glucose, so "carbs" are an excellent energy source. A common dietary myth is that carbohydrates are fattening and should be avoided. In fact, carbohydrates are needed for a balanced diet. Carbohydrates have the same number of calories per gram as proteins and fewer calories than fat. Additionally, carbohydrates are the only fuel that the body can readily use during high-intensity physical activities such as fighting fires.

Proteins perform many vital functions within the body. Most protein comes from meats and dairy products, but smaller amounts are found in grains, nuts, legumes, and vegetables. The body uses proteins (amino acids) to grow and repair tissues; proteins are used as a primary fuel source only in extreme conditions, such as during starvation. Like other nutrients, excess proteins are converted and stored in the body as fat.

Fats are also essential for life. These nutrients are used as a source of energy, for insulating and protecting organs, and for breaking down certain vitamins. Some fats, such as those found in fish and nuts, are healthier and more beneficial to the body than others, such as the fats found in margarine. However, excess fat consumption—particularly of saturated fats (which come mostly from animal products)—is linked to high cholesterol, high blood pressure, and cardiovascular disease.

Both candy and soft drinks contain sugar (unless they are specifically labeled as "sugar-free"), so the body can quickly absorb and convert these foods to fuel. However, simple sugars also stimulate the production

FIGURE 20-14 Plans for rehabilitation at extended-duration incidents should include complete, balanced meals.

of insulin, which reduces blood glucose levels. As a consequence, eating a lot of sugar can actually result in lower energy levels.

For firefighters to sustain peak performance levels, it is necessary for them to "refuel" during rehabilitation. During short-duration incidents, low-sugar, high-protein sports bars can be used to keep the glucose balance steady. During extended-duration incidents, however, firefighters should eat a more complete meal (**FIGURE 20-14**). The proper balance of carbohydrates, proteins, and fats will maintain energy levels throughout the emergency. To ensure peak performance, the meal should include complex carbohydrates such as whole-grain breads, whole-grain pasta, rice, and vegetables. It is also better to eat a series of smaller meals over time, rather than one or two large meals, because larger meals can increase glucose levels and slow down the body.

Medical Monitoring

Medical monitoring during rehabilitation is important to identify potentially dangerous health problems. Checking vital signs is one way to identify potential problems. Many rehabilitation protocols specify the following types of medical monitoring:

- Check the firefighter's heart rate, respiratory rate, blood pressure, and temperature.
- Determine how much oxygen is in the firefighter's bloodstream by taking a pulse oximetry reading.
- Perform a carbon monoxide assessment.

If the firefighter's vital signs or test results are abnormal upon entering the rehabilitation area, the firefighter should be reexamined before being permitted to return to firefighting.

In addition, signs of illness or injury should be assessed in the rehabilitation area before the firefighter resumes active duty. As a firefighter, you should be alert for any of the following signs and symptoms:

- Chest pain
- Weakness
- Dizziness
- Shortness of breath
- Headache
- Cramps
- Mental status changes
- Behavioral changes
- Changes in speech
- Changes in gait
- Abnormal body temperature

If you notice any of these signs in yourself or in one of your crew members, report them to medical personnel at the rehabilitation area.

If any firefighters are ill or injured, the rehabilitation area arranges for transportation to a hospital. At most emergency incidents, an ambulance should be available to ensure that ill or injured firefighters receive prompt transport.

SAFETY TIP

Many fire departments have found that assigning a fire or EMS officer to be in charge of the rehabilitation area ensures that firefighters stay in rehabilitation as long as needed and that firefighters who need to be transported to a hospital for further evaluation are transported rapidly.

Member Accountability

The seventh and last part of rehabilitation is maintaining crew integrity. Crew integrity should be maintained during rehabilitation—crew members should be released to and released from rehabilitation together. This practice makes accountability on the fireground much easier for the IC to control. It also ensures that all members of the crew report to rehabilitation to receive the rest, cooling or warming, hydration, food, and medical monitoring that they require. In the event that one member of a crew is not able to return to the active fire scene, the IC must determine how to handle the situation.

Release and Reassignment

Once they are rested, rehydrated, refueled, and rechecked to make certain that they are fit for duty, firefighters can be released from rehabilitation and reassigned to active duty. Release and reassignment is the last step in the rehabilitation process. Reassignments may be to the same job performed before the trip to rehabilitation or to a different task, depending on the decision of the IC.

Personal Responsibility in Rehabilitation

The goal of every firefighting team is to save lives first and property second. To achieve this goal, you need to take care of yourself first, take care of the rest of your crew second, and then take care of the people involved in the incident. Another way of looking at this issue is to remember that safety begins with you. The other members of your crew depend on you to bear your share of the load. To meet their expectations, you need to maintain your body in peak condition.

Part of your responsibility is to know your own limits. No one else can know what you ate, how much water you have had throughout the day versus coffee, soda, or energy drinks, whether you are lightheaded or dehydrated, whether you are feeling ill, or whether you need a breather. You are the only person who knows these things. Therefore, you may be the only person who knows when you need to request rehabilitation. It can be difficult to say, "I need a break," while the rest of the crew is still hard at work. It is better to take a break when you need it than to push yourself too far and have to be rescued by other members of your crew.

Regular rehabilitation enables firefighters to accomplish more work during a major incident and decreases the risk of stress-related injuries, illness, and death. Remember—safety begins and ends with you.

CASE STUDY

You Are the Firefighter CONCLUSION

You are working out early to get a jump on the day. You have just completed a run on the treadmill when you are dispatched to a house fire in an older section of town that dates back to the 1890s. Upon arrival, you are assigned to fire attack and are told that the fire is in the basement. Your crew has trouble locating the seat of the fire in the dirty brown smoke. When you finally reach it, you realize it is a large fire that is beginning to spread to the upper floors. As you knock down the fire in the basement, you notice you are getting low on air. You exit with your crew, get your air cylinder changed out, and are reassigned to the first floor to begin opening up walls for overhaul. As you are pulling walls, you start to feel that you are breathing harder than usual. You begin to get mild cramps in your arms as you tire from swinging the axe. You strive to keep up with your partner. When you notice you are getting low on air again, you let your partner know that you need to exit and you are going to head to rehabilitation.

1. **What is the purpose of firefighter rehabilitation?**

   Answer: The purpose of firefighter rehabilitation (and pre-habilitation) is to keep firefighters safe and healthy, reduce the chance of illness or injury, and maintain an effective firefighting force for fireground operations.

2. **When should a firefighter seek rehabilitation on the fireground?**

   Answer: Firefighters should seek rehabilitation according to their local policies and procedures, as well as any time they feel the signs and symptoms of exhaustion or heat or cold stress, or other indicators that they are approaching their physical limits.

3. How does firefighter rehabilitation help with firefighter safety and performance?

Answer: In the heat of battle, firefighters typically work up to and beyond their physical capability. While the drive to perform beyond physical limits is admirable, the reality is that a firefighter working near or past their physical limits is, at best, not contributing much to the firefighting effort and, at worst, they are a liability, as an exhausted, sick, or injured firefighter draws efforts away from firefighting operations. Firefighter rehabilitation (and pre-habilitation) helps to ensure that firefighters remain ready for action and within their effective operational capacity. In other words, proactive pre-habilitation, along with taking breaks for rehabilitation, helps maintain firefighter effectiveness in fireground operations of all kinds.

WRAP-UP

SUMMARY

KNOWLEDGE OBJECTIVES

- Describe the need for and function of firefighter rehabilitation.
  - Define rehabilitation. (p. 870)
  - Describe the factors and causes that require rehabilitation for firefighters. (pp. 870–873)
  - Explain how heat stress and personal protective equipment tax the firefighter's body. (pp. 872–874)
  - Describe the hazards of dehydration, and explain how dehydration can be prevented. (p. 873)
  - List the signs of dehydration. (p. 873)
- Describe the factors that necessitate the setting up of a rehabilitation station.
  - Describe the types of extended fire incidents during which firefighters need rehabilitation. (pp. 875–876)
  - Describe the other types of incidents during which firefighters need rehabilitation. (pp. 877–878)
  - Explain why the body needs rehabilitation during severe weather conditions. (pp. 878–881)
- Describe the process of rehabilitation and each individual's role.
  - List the steps in rehabilitation. (p. 878)
  - Describe the types of fluids that are ideal for firefighters to drink during rehabilitation. (pp. 881–882)
  - Describe the types of food that are ideal for firefighters to eat during rehabilitation. (pp. 882–883)
  - Explain what the individual firefighter's personal responsibilities are in rehabilitation. (p. 884)

SKILLS OBJECTIVES

There are no skills objectives for Firefighter I candidates. NFPA 1001 contains no Firefighter I Job Performance Requirements for this chapter.

KEY TERMS

blood sugar See *glucose.*

dehydration A state in which the body's fluid losses are greater than fluid intake. If left untreated, dehydration may lead to shock and even death.

electrolytes Certain salts and other chemicals that are dissolved in body fluids and cells. Proper levels of electrolytes need to be maintained for good health and strength.

emergency incident rehabilitation A function on the emergency scene that cares for the well-being of the firefighters. It includes relief from climatic conditions, rest, cooling or warming, rehydration, calorie replacement, medical monitoring, member accountability, and release.

frostbite A localized condition that occurs when the layers of the skin and deeper tissue freeze. (NFPA 704)

KEY TERMS CONTINUED

fully encapsulated suit A protective suit that completely covers the firefighter, including the breathing apparatus, and does not let any vapor or fluids enter the suit. It is commonly used in hazardous materials emergencies.

glucose The source of energy for the body. One of the basic sugars, it is the body's primary fuel, along with oxygen. Also called *blood sugar.*

heat cramp A painful muscle spasm, usually in muscles in the leg or abdomen, that occurs suddenly during or after physical exertion.

heat exhaustion A mild form of shock that occurs when the circulatory system begins to fail because of the body's inadequate effort to give off excessive heat.

heat rash Itchy rash on skin that is wet from sweating; seen after prolonged sweating.

heat stress The condition that occurs when the body cannot get rid of excess heat, possibly leading to heat rash, heat cramps, heat exhaustion, and heat stroke.

heat stroke A severe, sometimes fatal condition that occurs when the body's temperature-regulating capacity fails and that is indicated by a reduction or cessation of sweating, body temperature of 105°F (40.5°C) or higher, rapid pulse, hot and dry skin, headache, confusion, unconsciousness, and convulsions.

hypothermia A condition in which the internal body temperature falls below 95°F (35°C), usually a result of prolonged exposure to cold or freezing temperatures.

pre-habilitation The concept of ensuring that a firefighter sustains physical readiness at all times through adequate hydration, nutrition, rest, and awareness of the state of their body and the surrounding environment.

rehab See *rehabilitate.*

rehabilitate To restore someone or something to a condition of health or to a state of useful and constructive activity. Also referred to as *rehab.*

REVIEW QUESTIONS

1. What term describes the concept that firefighters must do what they can to ensure that, like any other equipment they use, their bodies are always ready for the next call?
2. What pre-habilitation measures can firefighters take to reduce the risk of fatigue, injury, and illness?
3. Is firefighter rehabilitation necessary during training exercises? Why or why not?
4. What are the indicators of heat stroke?
5. What is the immediate care for a firefighter who is suffering heat stroke?
6. What are the seven parts typically used in most firefighter rehabilitation processes?
7. Who is ultimately responsible for knowing when you, as an individual firefighter, are in need of rehabilitation?

DISCUSSION QUESTIONS

1. How do pre-habilitation and rehabilitation affect a firefighter's ability to deal with the mental stresses of the job?
2. Why was the use of a SCBA cylinder chosen as a starting point for firefighter rehabilitation?
3. In addition to the specific types of incidents described in the chapter that require rehabilitation, describe some emergency incidents where firefighters might need rehabilitation for physical and mental reasons.
4. Describe some specific environmental conditions that would affect the need for firefighter rehabilitation.

APPLYING THE CONCEPTS

You and your engine unit are on-scene of a residential fire in a rural area. The fire started in the basement of a two-story house with Type V construction. The owner has piles of clutter, boxes, magazines, and debris filling the basement and lining the stairs, making it difficult to maneuver and giving the fire lots of fuel. You and your crew are assigned to conduct fire suppression in the basement. The heat is intense, and the smoke is heavy. After working for 45 minutes, your partner's SCBA gives a "low air" alarm. You check your own air and realize it's getting low but you have time left. The fire has not been knocked down and there is still a lot of work to do.

1. How do you handle this situation? Do you exit with your partner or continue working until you're out of air?
2. Do you and your partner hurry to the engine to replace your SCBA bottles so you can return to work as quickly as possible?
3. What should you do instead?

You and your partner exit the structure and first go to the decontamination area to remove and clean your PPE. Then you use wipes to clean your hands and exposed skin. After doffing your PPE, you head to the rehab area that's in a popup tent located a safe distance from the structure. It's set up with chairs, cots, beverages, food, towels, and spare clothes. Other firefighters are there eating and drinking, and some are resting on the cots. A Salvation Army volunteer greets you and offers you beverages with electrolytes and high-protein bars.

4. What are the objectives of rehabilitation and why is it so important for firefighters?
5. How are heat and dehydration intertwined?

After 30 minutes in rehabilitation, you ask your partner if he's ready to go back in. He says, "You bet! Let do it." But when you look at him you feel in your gut something's off and he's not okay. You ask the EMT to check him out. While being examined, your partner admits he feels nauseated and has some leg cramps. His heart rate is rapid, despite resting, but he insists he's fine and wants to finish the job.

6. Should you encourage your partner to return to work? Why or why not?
7. Do you wait with your partner, or do you return to work?

REFERENCES

Bledsoe, Bryan E., James J. Augustine, and the International Association of Fire Chiefs (IAFC) 2009. *Rehabilitation and Medical Monitoring: An Introduction to NFPA 1584 (abstract).* Accessed September 4, 2023. https://www.iafc.org/topics-and-tools/resources/resource/rehabilitation-and-medical-monitoring-an-introduction-to-nfpa-1584-(abstract).

Campbell, Richard, and Shelby Hall. 2022. *United States Firefighter Injuries in 2021.* National Fire Protection Association (NFPA), December 2022. Accessed September 4, 2023. https://www.nfpa.org/-/media/Files/News-and-Research/Fire-statistics-and-reports/Emergency-responders/osffinjuries.pdf.

Lerman, Steven E., Evamaria Eskin, David J. Flower, Eugenia C. George, Benjamin Gerson, Natalie Hartenbaum, Steven R. Hursh, Martin Moore-Ede; American College of Occupational and Environmental Medicine (ACOEM) Presidential Task Force on Fatigue Risk Management. 2012. "Fatigue Risk Management in the Workplace." *Journal of Occupational and Environmental Medicine,* 54, no. 2 (February 2012): 231–258. https://doi.org/10.1097/JOM.0b013e318247a3b0.

National Fire Protection Association (NFPA). 2021. *NFPA 704, Standard System for the Identification of the Hazards of Materials for Emergency Response.* 2022 Edition. Accessed July 12, 2023. https://www.nfpa.org/codes-and-standards/7/0/4/704.

National Fire Protection Association (NFPA). 2021. *NFPA 1582: Standard on Comprehensive Occupational Medical Program for Fire Departments.* 2022 Edition. Accessed April 1, 2023. https://www.nfpa.org/codes-and-standards/1/5/8/1582.

National Fire Protection Association (NFPA). 2021. *NFPA 1584, Standard on the Rehabilitation Process for Members during Emergency Operations and Training Exercises.* 2022 Edition. Accessed April 1, 2023. https://www.nfpa.org/codes-and-standards/1/5/8/1584.

National Fire Protection Association (NFPA). 2023. *NFPA 1550: Standard for Emergency Responder Health and Safety.* 2024 Edition. Accessed April 1, 2023. https://www.nfpa.org/codes-and-standards/1/5/5/1550.

National Weather Service. n.d. "Heat Cramps, Exhaustion, Stroke." Accessed September 4, 2023. https://www.weather.gov/safety/heat-illness.

Occupational Safety and Health Administration (OSHA). n.d. "Heat-Related Illnesses and First Aid". Accessed September 4, 2023. https://www.osha.gov/heat-exposure/illness-first-aid.

Selkirk, Glen A., Tom M. McLellan, and Jonathan Wong. 2006. "The Impact of Various Rehydration Volumes for Firefighters Wearing Protective Clothing in Warm Environments." *Ergonomics,* 49, no. 4: 418–433 (published online February 20, 2007). Accessed September 12, 2023. https://doi.org/10.1080/00140130600558007.

Sharkey, Brian, Brent Ruby, and Carla Cox. 2002. "Feeding the Wildland Firefighter." U.S. Forest Service, 0251-2323-MTDC, July 2022. Accessed September 4, 2023. https://www.fs.usda.gov/t-d/pubs/htmlpubs/htm02512323/index.htm.

U.S. Fire Administration (USFA). 2008. Emergency Incident Rehabilitation, February 2008. Accessed September 2008. https://www.usfa.fema.gov/downloads/pdf/publications/fa_314.pdf.

Walker, Adam, Rodney Pope, and Robin Marc Orr. 2016. "The Impact of Fire Suppression Tasks on Firefighter Hydration: A Critical Review with Consideration of the Utility of Reported Hydration Measures." *Annals of Occupational and Environmental Medicine,* 28, no. 63 (November 15, 2016). Accessed September 4, 2023. https://doi.org/10.1186/s40557-016-0152-x.

Woodbury, Blair, and Stefan A. Merrill. 2023. "EMS Emergency Incident Rehabilitation." *StatPearls.* StatPearls Publishing. National Center for Biotechnology Information (NCBI), August 14, 2023. Accessed September 4, 2023. http://www.ncbi.nlm.nih.gov/books/NBK537174/.

CHAPTER

21

Firefighter I

Wildland and Ground Cover Fires

KNOWLEDGE OBJECTIVES

After studying this chapter, you will be able to:

- Discuss the history of wildland fire management.
- Identify the anatomy of a wildland fire and explain the effects of fuel, weather, and topography on the behavior of a wildland fire.
- Describe wildland terminology, parts of a wildland fire, and the strategy and tactics in suppressing wildland fires.
- Describe the role of fire apparatus to suppress wildland fires.
- Describe wildland safety considerations.
- Describe the triage categories, considerations, strategies, and tactics to defend structures during a wildland/urban interface fire.

SKILLS OBJECTIVES

After studying this chapter, you will be able to perform the following skills:

- Demonstrate how to suppress wildland fires given hand tools and water delivery devices.
- Demonstrate how to deploy a fire shelter for firefighter survival.

ADDITIONAL NFPA STANDARDS

- **NFPA 901**, *Standard Classifications for Fire and Emergency Services Incident Reporting, 2021 Edition*
- **NFPA 1140**, *Standard for Wildland Fire Protection, 2022 Edition*
- **NFPA 1550**, *Standard for Emergency Responder Health and Safety, 2024 Edition*
- **NFPA 1900**, *Standard for Aircraft Rescue and Firefighting Vehicles, Automotive Fire Apparatus, Wildland Fire Apparatus, and Automotive Ambulance, 2024 Edition*
- **NFPA 1977**, *Standard on Protective Clothing and Equipment for Wildland Fire Fighting and Urban Interface Fire Fighting, 2022 Edition*
- **NFPA 5000**, *Building Construction and Safety Code, 2021 Edition*

CASE STUDY

You Are the Firefighter

You have been dispatched to a wildland fire that is approaching your community. Normally, the state forestry service handles these calls, but they call local resources for assistance when they are out of resources, and that is what happened this time. Winds are high and the fire is moving rapidly. Your engine is assigned to a position on the edge of town, and your company has been assigned to evacuate people and protect as many homes as possible.

1. How do you determine if you are in a safety zone?
2. What does it mean to determine the fire behavior?
3. Based on the expected fire behavior, what strategy or tactics are effective and safe under these conditions?

Introduction

Throughout the history of the United States, we have had conflagrations, or large fires that burned significant portions of cities. The Chicago Fire of 1871 and the Great Boston Fire in 1872 are examples of fires in which people died and that burned thousands of homes. These fires occurred at the same time as weather events that caused high winds.

When a fire occurs on **wildland**—open land and forested area—it is called a **wildland fire**. Wildland fires represent fire in the ground, on the surface, and in aerial fuels in trees. A wildland fire that burns fuels such as grass, leaves, and branches on the ground is sometimes called a **ground cover fire**.

History of Wildland Fire Management

As with conflagrations in cities, devastating wildland fires throughout history also occur with high winds. In 1910, the United States experienced large, wind-driven fires across Idaho, Montana, Washington, and Oregon. This was called the "Big Blowup." The U.S. Forest Service (USFS), which operates under the Department of Agriculture, realized that wildland fires posed a significant risk to communities and to natural and man-made resources. They decided foresters and firefighters at the federal level were needed to suppress these fires, and they established policies to suppress fires that pose a risk to resources and infrastructure in wildland areas. From the beginning, the USFS recognized that fighting these fires would be a dangerous occupation. They established a network of fire science laboratories around the country to study wildland fires. These laboratories are collectively called the fire labs. Over time, specialized tools, equipment, apparatus, and protective clothing were developed to keep personnel safe while combating wildland fires.

In addition to the USFS, several other federal land management organizations have responsibility to suppress wildland fires, including the Bureau of Land Management (BLM), the Bureau of Indian Affairs (BIA), the National Park Service (NPS), and the U.S. Fish and Wildlife Service (FWS). Each of these agencies manages fire suppression crews and equipment to combat fires. In addition, each state has their own version of a forestry agency that is responsible primarily for wildland fires.

In many local communities, structural firefighters and wildland firefighters are separate specialties and work independently of one another. Local and state agencies with primary wildland responsibilities need to ensure all personnel are well educated and equipped to handle this risk. However, in many areas of the country, wildland fires are an immediate threat to the community and by necessity firefighters in those communities are equally trained in structural and wildland firefighting. However, even structural firefighters who do not have wildland as a primary responsibility should have a basic understanding of wildland fire behavior, tools, and equipment, as well as a basic understanding of the best strategy and tactics to use to combat wildland fires effectively and safely.

Anatomy of a Wildland Fire

Firefighters need to understand the terminology used on the wildland fireground so that they can both understand directions from command and report information back to fire officers.

The location where a fire begins is the area of origin (explained in Chapter 5, *Fire Behavior*). As a wildland fire grows and moves into new fuel, the most rapidly moving area—the traveling edge of the fire—is called the **head of the fire** (**FIGURE 21-1**). As the fire gets bigger, the area close to the area of origin is often referred to as the **heel of the fire** or the **rear of the fire**. The **flank of the fire** is the edge between the head and heel of the fire that runs parallel to the direction of the fire spread. The **fire line** is the crew of firefighters conducting fire attack.

As the fire grows, a change in weather, topography, or fuel may cause it to move in such a way that it projects out into a long, narrow extension called a **finger**. A finger can produce a secondary direction of travel for the fire. The unburned area between a finger and the main body of the fire is called a **pocket**. A pocket is a dangerous place for firefighters because this area of unburned fuel is surrounded on three sides by fire. An area of land that is left untouched by the fire, but is surrounded by burned land, is called an **island**.

A **spot fire** is a new fire that starts outside the perimeter of the main fire. Spot fires usually begin when flaming vegetation, in the form of an ember or spark, is picked up by the convection currents generated by the fire and dropped some distance away from the original fire.

Green and black are terms often used by wildland firefighters to describe areas in wildland fires. The **black area** is an area that has already been burned. Because black areas contain few unburned fuels, they are often relatively safe for firefighters. A **green area** is an area containing unburned fuels. Green areas are more likely to become involved in the fire than black areas.

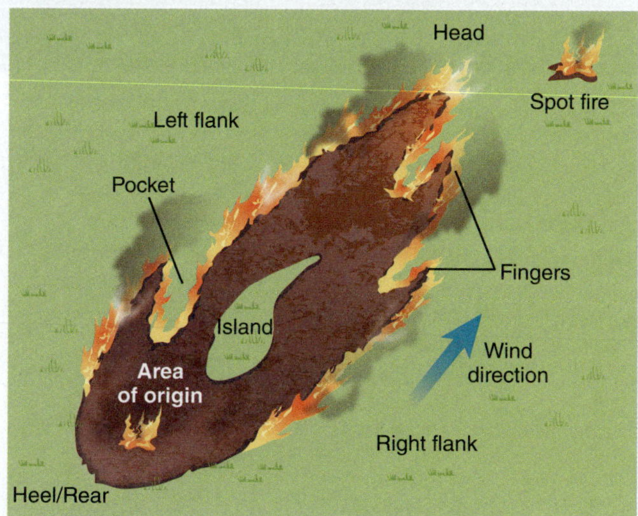

FIGURE 21-1 The anatomy of a wildland fire.

Fire Behavior

Wildland fire behavior is influenced by fuel, weather, and topography. This is often referred to as the wildland fire triangle. While fire behavior in structures is complex because of compartmentalization and the flow path of air and products of combustion, fire behavior in a wildland setting is simple. **Wildland fire behavior** is the flame length and the rate of spread of a fire, which are all affected by the fuel, weather, and topography.

Flame length is the flame height measured from the burning fuel to the tip of the flame. This is different from **flame height**, which is measured from the ground to the tip of the flame. Flame length is important because the USFS fire labs have conducted studies that determined that different flame lengths require different suppression techniques.

Flame length is a factor in determining the **safety zone**, which is an area big enough to accommodate all of the personnel working in an area to be safe as the fire moves around them. Generally, this area is four times the flame length to the nearest firefighter (**FIGURE 21-2**).

Rate of spread (ROS) is the speed at which the flaming front, or the head of the fire, is moving. A slow ROS is slower than a person normally walks; a moderate ROS is as fast as a person walking with purpose; a rapid ROS is faster than a person walking with purpose. A fourth ROS—critical or extreme—applies to wind-driven fires. The ROS is defined as critical or extreme when fuel ahead of the fire has been superheated from convection or conduction and is burning rapidly enough for area ignition *(NWCG 2022, 50)*. Many states describe ROS in feet per hour; the federal

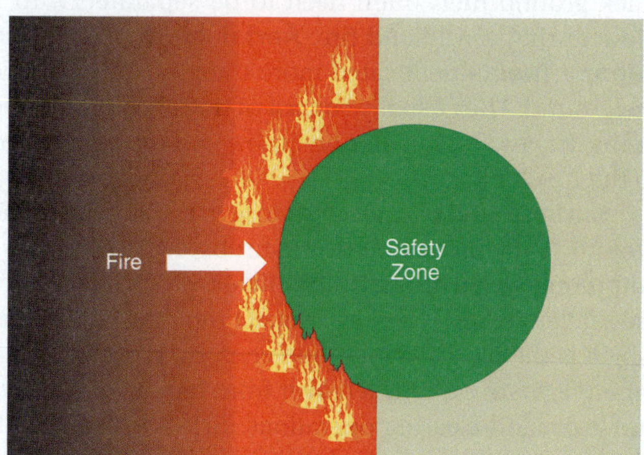

FIGURE 21-2 Safety zones must have enough space to allow the fire to burn all the way around your location and should be calculated based on flame length.

agencies use chains per hour. A chain is a surveying term that was adopted by the fire service; one **chain** equals 66 feet (ft; 20 meters [m]). Firefighters generally convert chains per hour to feet per minute—if a grass fire is burning at 60 chains per hour, firefighters should visualize the fire moving a little faster than 60 ft (18 m) per minute.

It is important to be aware of a fire's ROS. If you are working ahead of the fire, knowing the ROS can give you an idea of how much time you have before you need to move to the safety zone. ROS is also used to determine what flank of the fire is the priority. Generally, the priority flank is the one with the fastest ROS. Incident commanders (ICs) use the ROS to determine which communities may be affected by the fire and how much time is needed to evacuate those communities.

SAFETY TIP

To be safe, and in case you underestimate the ROS, you should give yourself double the time anticipated to get into your safety zone.

Fuel, weather, and topography all play an important role in influencing the flame lengths and rate of spread of a wildland fire.

Fuel

Fuels in wildland fires are categorized by their location. **Ground fuel** refers to fuels on the ground such as vegetation and tree roots. **Subsurface fuels** are partially decomposed matter beneath the ground, such as roots, moss, and decomposed stumps. In Florida, thick ground fuels often need to be separated with a plow during a wildland fire. **Aerial fuels** (also called **canopy fuels**) are located more than 6 ft (2 m) above the ground. They usually consist of trees, including tree limbs, leaves and needles on limbs, and moss attached to the tree limbs.

Surface fuels are located close to the surface of the ground. They include grasses; shrubs, which are more commonly called brush; **timber litter**, which is pine needles, fallen leaves, twigs, and small branches; and **slash**, which is the pieces of logs, branches, bark, stumps, and other vegetative debris left over from high winds, heavy snow, or land-clearing operations. Brush less than 6 ft (2 m) above the ground is also classified as a surface fuel. Surface fuels are involved in ground cover fires.

The amount of moisture in the fuel, or **fuel moisture**, affects how easily it will ignite and how intensely it will burn. The moisture content of fuel is classified in two ways. One is dead fuel moisture, which is the amount of moisture in a fuel that is not living. Dead fuel moisture changes only through absorbing the moisture content in the air. The second is live fuel moisture or the amount of moisture in living, growing fuels. Live fuels gain moisture primarily from the plant root system, but is also affected by the moisture content in the air. Live fuel moisture depends on the vegetation's growing cycle and varies greatly between species and seasons. Both dead fuel moisture and live fuel moisture are measured as a percentage of the weight of the fuel.

Time lag is the amount of time it takes for the moisture content of a live or dead fuel to increase or decrease based on the moisture content in the surrounding air. For example, with a light afternoon rain, light grass will absorb moisture and its moisture content would increase. However, if the air was warm and dry, these fuels could dry out in an hour and be more susceptible for ignition.

Grass is considered a 1-hour time lag fuel. It takes about 1 hour of air exposure to bring the fuel moisture level of grass up or down. This means that grass can go from 100% fuel moisture to a combustible fuel in 1 hour. Small to medium-sized brush is classified as 10-hour fuel, and logs or downed trees are classified as 100-hour fuel. The large timber and large down trees are 1000-hour fuels. This time relates to how long firefighters have until the fuel is dry enough to ignite.

Firefighters need to be aware of the fire behavior changes that occur between the fuel groups (**TABLE 21-1**). For example, grass fires have the fastest

TABLE 21-1 ROS and Flame Length of Major Live Fuel Categories

Type of Fuel	Flame Length	Rate of Spread
Grass	2 to 11 feet	Slow to critical/ extreme depending on wind conditions
Shrubs/brush	3 to 20, but as high as 80 to 100 feet when combined with high winds and/or slope	Slow to critical/ extreme depending on wind conditions and slope
Timber litter	1 to 4 feet	Slowest rate of spread of all fuel groups
Slash	2 to 6 feet	Slow and long burn time, slow rate of spread

Courtesy of Bret Davidson.

ROS and the most variable flame length. In a grass fire, flame lengths can go from 2 feet to 12 feet very quickly. More firefighters have died in grass fires than in fires fueled by any other type of fuel. Shrubs (brush) burn with the longest flame lengths of all surface fuels and have a variable ROS when combined with wind or steep slopes.

As a firefighter, you should be able to predict the change in fire behavior based on the available fuel. For example, if 6-ft (2-m) flames are moving rapidly in grass and the head of the fire is moving into the brush. Although grass burns faster, so the ROS will slow down, brush produces much longer flame lengths. This should be taken into consideration in the strategy, tactics, and safety zone ahead of the fire. The practical application is to estimate the change of flame length from 6 ft (1.8 m) in grass to 12 ft (3.7 m) in brush. If this is what you anticipate, then the safety zone in front of the fire needs to be 48 ft (14.6 m) all the way around. (Remember that you should not confront a fire with flame lengths over 8 ft [2.4 m] at the head without being in a safety zone.)

Weather

Weather is the most powerful and dangerous of three major influences of wildland fire behavior, because it can change so quickly compared to fuels and topography. Two components of weather that affect wildland fire are moisture and air movement.

SAFETY TIP

The National Weather Service issues "Red Flag Warnings" and "Fire Weather Watches" for particularly dangerous weather conditions. Even small fires can rapidly spin out of control during these times, so extra vigilance must be used. Do *not* expect normal fire behavior during periods when these warnings are issued!

Moisture

Moisture is present in the air in the form of relative humidity or precipitation. **Relative humidity** is the ratio of the amount of water vapor in the air compared to the maximum amount of water vapor that the air can hold at a given temperature. Warm air has a higher capacity for moisture content than cool air. Relative humidity is expressed as a percentage. If the air contains the maximum amount of water vapor possible, the relative humidity is 100 percent—in other words, extremely muggy or raining. In very dry climates, the average relative humidity may be 20 percent or less.

Relative humidity is a major factor in the behavior of wildland fires. When the relative humidity is low, fire behavior increases because vegetative fuels dry out, making them more susceptible to ignition. Conversely, when the relative humidity is high, fire behavior decreases because the moisture from the air is absorbed by the vegetative fuels, making them less susceptible to ignition. Changes in relative humidity affect **fine fuel**—fuel that ignites easily, such as pine needles and leaves—more dramatically than they affect **heavy fuel**—fuel that ignites slowly because it has a large diameter, such as large brush or branches. If the relative humidity drops and stays at 10 percent, for example, eventually all fuels will equalize at that level, but fine 1-hour dead fuels will reach that level much more quickly than heavy fuels.

Relative humidity typically varies with the time of day. As the temperature increases throughout the day, the relative humidity generally drops. Conversely, as the temperature decreases during the evening and nighttime hours, the relative humidity rises. A rise in the relative humidity can help slow the spread of a wildland fire and aid the work of fire crews who are trying to control the fire.

SAFETY TIP

Heavy fuels such as trees and stumps require close attention. Spraying water on the outside of these objects will not extinguish the fire because heavy fuels have a tendency to continue to burn on the inside. Heavy fuels need to be cut out to confirm that firefighters have extinguished all of the fire and to prevent rekindling.

SAFETY TIP

All fires should be treated with the same caution and safety, whether the fuels are heavy or light. Many firefighters have died because they underestimated fire behavior. Treat all wildland fires with respect.

In some parts of the United States, the average relative humidity also varies with the season of the year. These changes contribute to variations in the moisture content of vegetative fuels. This is one of the reasons why many regions of the United States demonstrate seasonal patterns of wildland fires. The seasons with the most wildland fires are often the seasons with the lowest relative humidity.

The second type of moisture that influences the behavior of wildland fires is precipitation. Moisture falling from the sky is absorbed by plants, which increases their moisture content and makes them less susceptible to combustion. In years with adequate precipitation, the incidence of wildland fires is typically much lower than in years with below-average precipitation. However, in desert regions, years with high precipitation can cause abnormally high grass growth. When those grasses dry out, fires can occur where they would not normally occur.

Air Movement

Two types of air movement are important to understand when fighting wildland fires: air stability and wind.

Air Stability. **Air stability** is the vertical movement of air into the atmosphere (**FIGURE 21-3**). Unstable environments allow the free flow of air from the ground into the upper atmosphere. Stable air resists vertical air movement. The best way to understand stability is to visualize a fireplace where air enters the fire from the front and goes up the chimney in the convection column. This is an unstable atmosphere because the vertical air flow is uninhibited. With the proper air flow into the fire and up the chimney, flame length and intensity will be high. If you close the damper in the chimney, preventing air from freely flowing up the chimney, the air pressure above the fire increases, which slows the air coming into the fire and decreases the flame length and intensity of the fire. When the damper is closed, the environment is considered to be stable.

Stable environments limit the vertical movement of air and keep flame lengths lower and fire spread slower. At wildland fires, it is common to have an inversion layer over the fire in the morning. An **inversion layer** is a layer of warmer air above the cooler air layer on the ground. An inversion layer slows fire growth by limiting air rising into the atmosphere. This is like the damper being shut on the fire. It is important to understand what time the inversion layer is expected to weaken and disappear because this is when the damper will open and air flow and fire behavior will increase. Inversion layers disappear when the sun hits the ground and heats up the air at ground level and the temperature of that air increases to become the same temperature or hotter than the air mass above. Firefighters need to expect this change in fire behavior and build it into the fire attack plan.

Wind

The most powerful component of weather is wind. **Wind** is horizontal movement of air over the surface of the Earth. Wind dramatically increases both flame length and ROS. At a grass fire, a fire attack with hand tools may seem safe and not a big deal. If wind hits that fire, it can cause it to change direction, make a direct attack dangerous, and potentially cause injuries to firefighters. It is important to be able to anticipate wind changes and include wind changes in fire behavior predictions.

There are two types of winds firefighters should be familiar with: local winds caused by the heating and cooling of the topography and general winds produced by cold and warm fronts. Local winds normally change during the day as temperatures heat up and cool down. The greater the difference between the low

FIGURE 21-3 A well-defined smoke column is an unstable air mass indicator.

and high temperatures during the day, the stronger the local wind will blow up from a valley or a canyon or on shore (from water onto the land) (**FIGURE 21-4**). At night, this changes and the stronger local wind will blow down into a valley or a canyon or off shore (from the land to the water).

General winds are classified as either a low-pressure weather system or a high-pressure weather system. Weather systems consist of air moving across the United States, changing our weather every 5 to 7 days. Low-pressure systems are usually associated with moving cold air—cold fronts—and high-pressure systems are normally associated with moving warm air—warm fronts. In the western United States, cold fronts come down from Alaska and cross over the continental United States. Many warm fronts travel from southern, tropical regions up north into the United States. These general winds are stronger and dominate the local winds.

Fire fatalities have occurred on many fires because of the effects of thunderstorms or the beginning or end of Foehn winds. **Foehn winds** are called **east winds**, **chinook winds**, or **Santa Ana winds**, depending on the part of the country in which you are located. These winds are caused by a high-pressure system of warm air replacing cold air and bringing with it unusually high winds in the opposite direction of the local winds. They usually occur for 2 to 3 days at a time. These winds are responsible for the largest fatality and structural loss fires in United States history, including the devastating fires in Hawaii in 2023.

It is important to understand and anticipate wind behavior because of the significant change in fire behavior it can cause. Every firefighter must pay attention to wind direction and, whenever possible, anticipate changes. If a wind change is anticipated, it should be considered when implementing tactics and safety considerations at the fire.

Topography

Topography—the way land is configured—is the most constant of the three major influences on wildland fire behavior. If there is an area in your jurisdiction that is identified as an area that has a high threat of wildland fires, it is likely due to topography. Topography does not change unless the area undergoes major development. Often, the only way to make areas that are prone to wildland fire less dangerous is to remove the combustible fuels, for example, when natural vegetation is removed and subdivisions of homes are built or replaced with irrigated crops.

Topography, such as narrow canyons, saddles (the low point between two peaks), ridgelines, and box canyons, can cause many challenges to firefighters approaching fires. Many of these features limit access or increase the dangers because of the wind's effect on the topography in these areas. The decision to approach fires in dangerous topography should be made with caution.

In areas where there are mountains, two topographical features directly affect flame length and ROS: aspect and slope. **Aspect** is the direction the side of the mountain is facing. Aspect affects fire behavior primarily because of sunlight's effect on fuel and moisture content in the fuels. In the Northern Hemisphere, southern aspects get the most direct sun exposure. Because of this, the weather on that side tends to be hotter and dryer and the fuels are drier and lighter. Because of the hotter temperatures and lighter fuels, wildland fires on southern aspects occur more often and become bigger than on the other aspects. Conversely, northern aspects get the least direct sun exposure, so temperatures are lower, the moisture content of fuels is higher, and the fuels are heaviest. Eastern aspects get early morning sun and western aspects get sunlight later in the day.

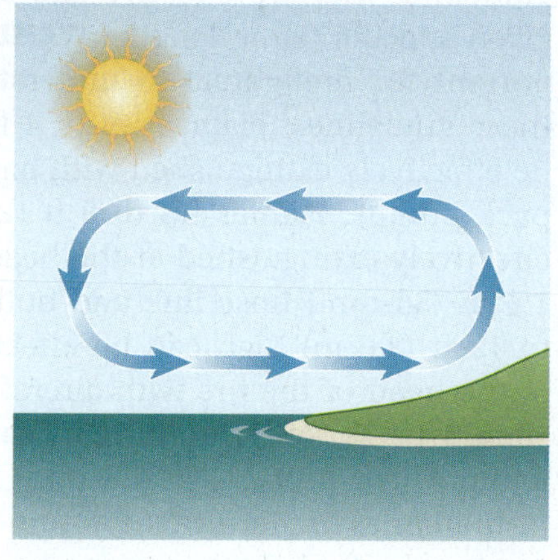
A.

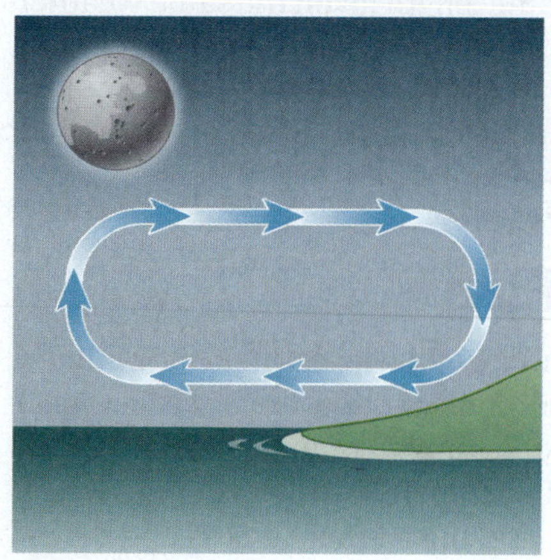
B.

FIGURE 21-4 A. Sea breeze development **B.** Land breeze development.

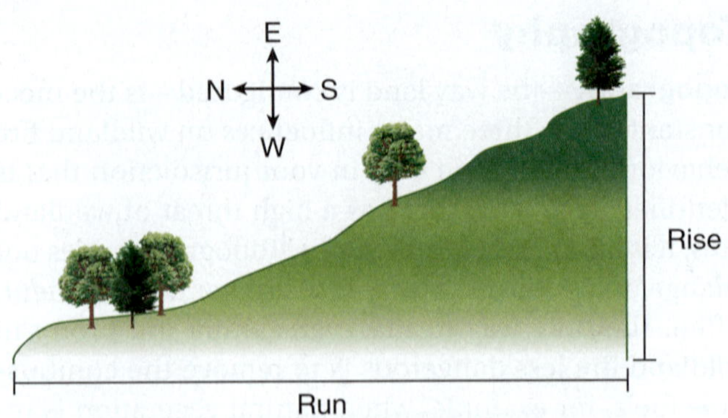

FIGURE 21-5 Slope increases flame lengths due to the increased influence of flame impingement, convection, and conduction.

Slope is basically how steep the mountainside is (**FIGURE 21-5**). Slope is an immediate concern for firefighters when fighting fire. The steeper the slope, the more influence direct flame impingement, conduction, and convection affect fuels ahead of the fire. In turn, this increases flame length and ROS. Slope is described in percentages or degrees, and is measured on a topographical map or in the field. To calculate the slope percentage, first divide the rise (the vertical distance between two points on the slope) by the run (the distance between two horizontal points on the slope)—this is often called "rise over run." Then multiply that result by 100. For example, if the slope is 45 degrees, that means for every 1-ft (30.5-cm) change in elevation, you move 1 ft (30.5 cm) forward. In this case, rise over run is 1/1, which equals 1. Multiply that by 100 to calculate the slope percentage of 100.

A fire that starts at the top of a hill and moves down the hill as it burns in brush may have flame lengths only a few feet tall. When the fire reaches the bottom and starts burning uphill, firefighters should anticipate a major change in flame length given the same weather and fuel conditions. This is called a **slope reversal**. These should be easy to identify when on the fire line and built into any strategic and safety plans.

Strategy for Wildland Fires

To safely attack and effectively extinguish wildland fires, firefighters must consider how the three factors of fuel, weather, and topography come together to influence the behavior of the fire. Before determining which method of attack to use, the IC must assess and evaluate the fire behavior and priorities for preserving lives and property. The top priority, of course, is to ensure the safety of both firefighters and citizens. Wildland fires can advance and change directions quickly. Given their unpredictability, they present a hazardous environment for firefighters. For these reasons, it is important for the IC to match the firefighting attack strategy to the conditions present at the fire.

There are generally two strategies to contain and extinguish wildland fires: offensive, direct attack strategies and defensive, indirect strategies. A **direct attack** is a method of wildland fire attack in which firefighters focus on containing and extinguishing the fire at its **burning edge**—the active perimeter of the fire. An **indirect attack** is a method of wildland fire attack in which a **fire control line**—a constructed or natural barrier used to control the fire—is located along natural fuel breaks, such as a lake; at favorable breaks in the topography, such as along the ridgeline at the top of the slope; or at considerable distance from the fire. The fuel between the fire control line and the fire is removed by hand tools or bulldozers, or it is burned off with a controlled fire. At a wildland fire, many of these strategies can be used at the same time, but they need to be coordinated to be effective and safe. The safest strategy is a direct attack because firefighters are located directly on the fire line, which means they are in a black area. However, if the fire behavior becomes too intense, direct attacks become less effective, and then the majority of the strategies become defensive.

Direct fire attacks should normally start from an **anchor point**—an area close to the area of origin and from which the vegetation has already burned. The anchor point provides an area of safety for firefighters. Ideally, it should be located close to a highway or road that can be used as an escape route. This helps prevent a situation in which the firefighters could get trapped if the fire changes direction and circles back on the firefighters.

The federal fire labs have established tactical guidelines that are effective and safe for fire attacks given specific flame lengths (**TABLE 21-2**). It is important for firefighters to understand how to apply these guidelines. Flames up to 4 ft (1.2 m) high can be effectively extinguished with hand tools and backpack pumps. Flames up to 8 ft (2.4 m) high can be effectively extinguished at the head of the fire with a 1½-in (38-mm) hose line and bulldozers. Flames up to 12 ft (3.6 m) high can be effectively extinguished at the head of the fire with aircraft. These guidelines should be seriously applied to each fire. Many firefighters have used 1½-in (38-mm) hose to attack 20-ft (6-m) flames or more, but not from the head of the fire. Any time a firefighter is at the head of a fire and the flames are over 8 ft (2.4 m) high, the fire should not be

TABLE 21-2 Fire Suppression Limitations Based on Flame Length

Flame Length	Method of Attack
0 to 4 ft (0 to 1.2 m)	Direct attack hand tools
4 to 8 ft (1.2 to 2.4 m)	Hose lays, retardant, bulldozers
8 to 11 ft (2.4 to 3.4 m)	Aircraft may be effective
More than 11 ft (3.4 m)	Indirect attack, control efforts at head ineffective, spotting, major runs

Data from Andrews, Patricia L. and Richard C. Rothermel. 1982. Charts for Interpreting Wildland Fire Behavior Characteristics. U.S. Department of Agriculture. https://gacc.nifc.gov/nwcc/content/products/fwx/publications/Charts_for_Interpreting_Wildland_Fire_Behavior_Characteristics_int_gtr131.pdf

FIGURE 21-6 This photo of the Gavilan Fire, a wildland fire that occurred in California in 2002, shows an example of dangerous flame lengths and ROS in brush.

Courtesy of Bret Davidson.

engaged unless the firefighters are located in a safety zone (**FIGURE 21-6**).

Direct Attack Strategies

The types of offensive attack strategies for wildland fires are the anchor, flank, tandem, and pincer attack; the flanking attack; and tandem attack.

The **anchor, flank, tandem, and pincer attack** (also known as the **pincer attack**) is the safest, most reliable, and most widely accepted method of attack. A successful attack using this method requires the availability of sufficient personnel to be able to mount a two-pronged attack. This method requires two or more teams of firefighters to establish one or more solid anchor points. One team establishes an anchor point on

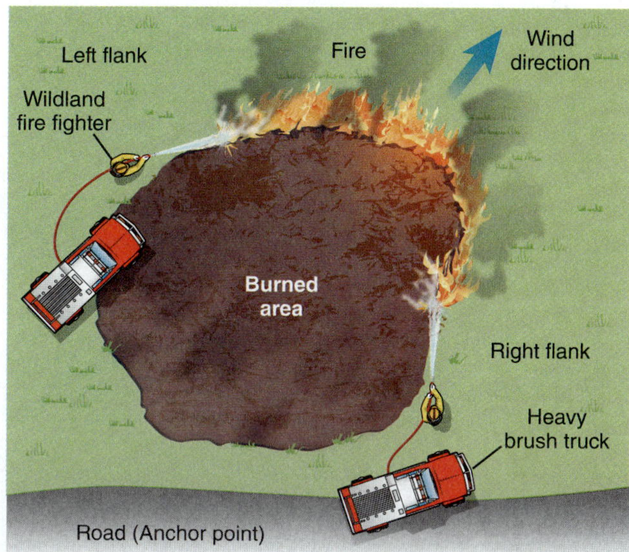

FIGURE 21-7 The anchor, flank, tandem, and pincer attack involves engagement on both flanks of a wildland fire.

the left side of the fire near the area of origin, and then mounts a direct attack along the left flank of the fire, working toward the head of the fire. A second team of firefighters establishes an anchor point on the right side of the fire near the area of origin, and then mounts a direct attack along the right flank of the fire, also working toward the head of the fire. As the teams advance, the fire gets "pinched" between them, which reduces its growth. By starting at a safe anchor point, the risk to firefighters is minimized. (**FIGURE 21-7**).

The **flanking attack** is a method of direct attack that requires only one team of firefighters. This team starts at an anchor point and attacks along one flank of the fire. This type of attack is used when sufficient resources are not available to mount two simultaneous attacks along both flanks or when the risks posed by the fire are high on one flank and lower on the other flank. The decision regarding which flank of the fire to attack is based on the determination of which side of the fire poses the greatest risk (**FIGURE 21-8**).

Indirect Attack Strategies

An indirect attack is most often used for large wildland fires that are too dangerous to approach through a direct attack because of the flame lengths, the ROS, or both. Three types of indirect attack are parallel attack, burning out, and backfiring. Backfiring is discussed later in the chapter.

An indirect attack is mounted by building a fire control line along a predetermined route, based on natural fuel breaks or favorable breaks in the topography,

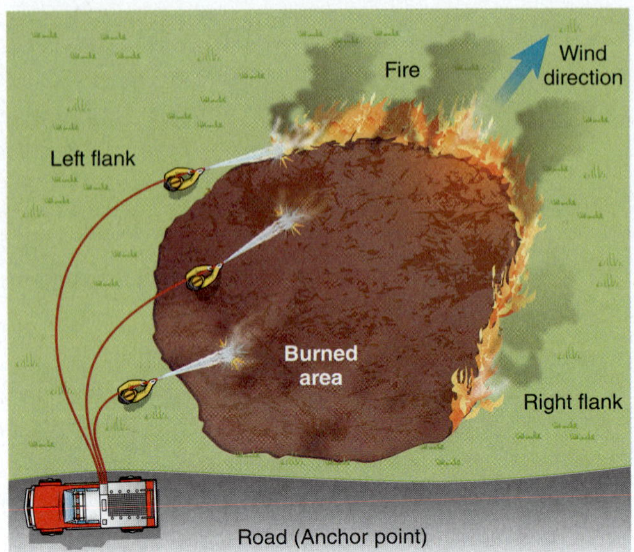

FIGURE 21-8 A flanking attack is made on only one flank of a wildland fire.

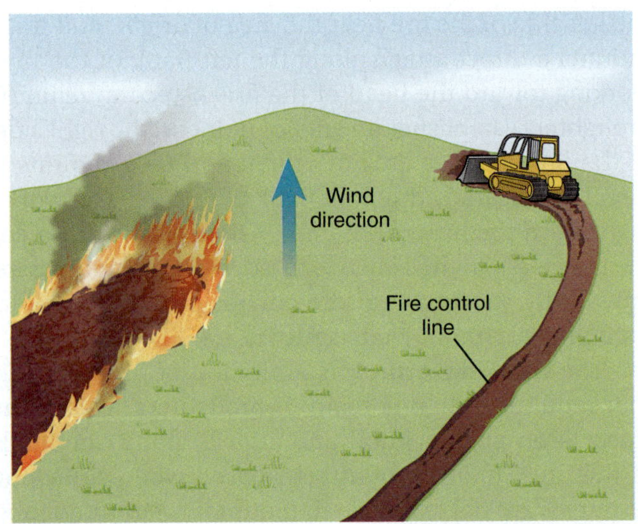

FIGURE 21-9 An indirect attack is made along natural fuel breaks, at favorable breaks in the topography, or at a considerable distance from the fire.

and then burning the fuel between the fire control line and the fire (**FIGURE 21-9**). The fire control line may be constructed fairly close to the fire, or it may be constructed several miles away. An indirect attack requires only one team of firefighters and can be mounted using either hand tools or mechanized equipment.

A **parallel attack** is a type of indirect method of attack used when the fire edge is too hot to approach through a direct attack. In a parallel attack, a fire control line is constructed parallel to the edge of the fire about 5 to 50 ft (1.5 to 15 m) from the edge of the fire—far

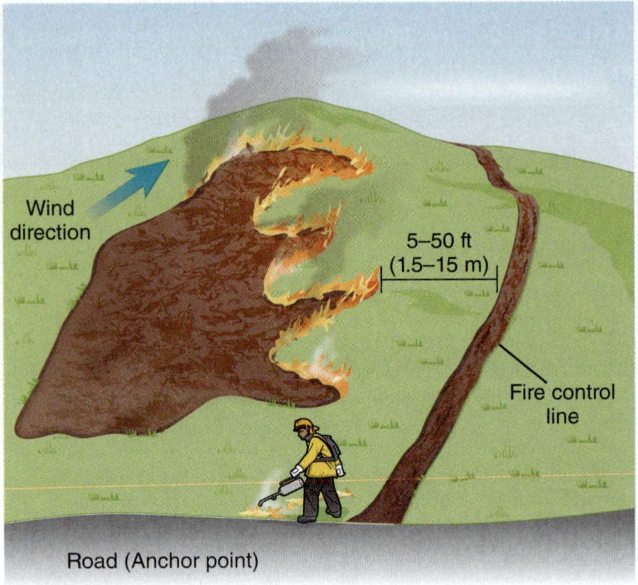

FIGURE 21-10 A parallel attack is made parallel to the fire's edge.

enough from the fire that workers and equipment can continue to work effectively if the intensity of the fire grows (**FIGURE 21-10**). In some cases, the fire control line can be shortened by constructing it across the fire's fingers—that is, from a burned edge, across an unburned area, to another burned edge. The unburned fuel between the edge of the fire and the fire control line is typically burned out before the main fire moves toward the control line, but it can also burn out unassisted with the main fire as long as the control line is wide enough to prevent the main fire from crossing the fire line.

Wildland Strategy and the PACE Model

Many times, both direct and indirect methods need to be considered and used at the same time on a wildland fire. When you are considering attack strategies, determining priorities, and allocating resources, keep the acronym PACE in mind—Primary plan, Alternative plan, Contingency plan, and Emergency plan. PACE is a military acronym adapted to the wildland environment used both strategically and tactically. The tactical PACE will be discussed later in this chapter. Strategically, the primary plan is the first priority and at a wildland fire, and it should be the most aggressive plan possible given the fire behavior. Normally, this is anchoring and flanking the fire using a variety of direct methods of attack. The secondary or alternative plan is the next priority. At a wildland fire, the alternative plan usually includes evacuating structures and defending infrastructure such as homes. The third priority, the contingency plan, is to build measurable objectives around the fire if it is not extinguished in the initial attack perimeter. This is sometimes referred to as "determining the fire box." For example, the objectives might be to keep the fire south of a ridgeline, north of a highway, east of a housing subdivision, and west of a creek. The emergency plan is a bigger fire box if the fire is not contained with the contingency plan. The emergency plan is used to coordinate evacuations and road closures with law enforcement. It is also used as a secondary set of objectives if the first contingency plan fails.

Methods of Extinguishment

Both direct and indirect attacks focus on disrupting the conditions represented by the fire triangle to extinguish the fire. Like any other fire, wildland fires can be controlled and extinguished by cooling the fuel, removing fuel, or by smothering the fire.

Direct Method Tactics

In direct attacks, water is the primary method used to cool the fuels that drive wildland fires. Although the principles involved in cooling wildland fires are the same as those used for structural fires, the methods used to apply the water may be different. Direct attack on wildland fires can be undertaken with hand tools if flames are 4 ft (1.2 m) or less and with mechanized equipment if flames are 8 ft (2.4 m) or less. When flames are higher than 8 ft (2.4 m), the guidelines are to use the indirect tactic of removing the fuel in the area ahead of the fire.

For small fires with a light fuel load, backpack fire extinguishers can be effective (**FIGURE 21-11**). These devices can be transported to the fire and quickly put into operation before the fire has an opportunity to grow. Backpack fire extinguishers are discussed in detail in Chapter 9, *Portable Fire Extinguishers*. Most portable backpack fire extinguishers hold 5 gallons (gal; 19 liters [L]) of water and are designed to be refilled easily in the field, such as from a lake or stream. Firefighters expel the water by using a hand pump.

FIGURE 21-11 Backpack fire extinguishers can be effective for small fires with a light fuel load.

Water from booster tanks carried on structural fire apparatus or on special wildland fire trucks is another means of cooling a wildland fire's fuel. Unfortunately, it is not always possible to get trucks into a position where they can attack a wildland fire. Structural fire apparatus is not designed to be taken off the road, and wildland fire apparatus can reach only limited parts of wildland. Forestry fire hose is sometimes extended for hundreds of feet. This type of hose is typically single-jacket, lightweight, collapsible, and smaller in diameter than the hose required for structural firefighting. The water supply may be limited to the amount of water carried by an apparatus, unless static water sources can be tapped by fire pumpers or portable pumps. Backpack fire extinguishers and forestry fire hose are used to extinguish most ground cover fires.

A third means of applying water to a wildland fire is from aircraft. Airplanes can take on a load of water from a lake or river and deliver it to firefighters at the scene. Fire retardant or other extinguishing agents can also be delivered by airplanes (**FIGURE 21-12**). These fire suppression agents must be loaded at a special airport and may not be used within a minimum of 300 ft (91.4 m) of a waterway (USFS 2019). Helicopters can either carry water in an internal tank and drop it directly on the fire, or they can use a long cable and bucket system to transport water to the firefighters working at the scene (**FIGURE 21-13**). Helicopters can also carry fire retardant or foam extinguishing agents. The use of aircraft can be an effective means to control wildland fires, but it requires close communication and coordination between ground and air resources.

FIGURE 21-12 Airplanes can be used for delivering water or fire retardant to wildland fires.

Courtesy of Jeff Pricher.

A.

B.

FIGURE 21-13 Helicopters can be used to apply water to a wildland fire by either **(A)** carrying the water in an internal tank and dropping it directly on the fire, or **(B)** by using a long cable and bucket system to transport the water to the scene.

Courtesy of Jeff Pricher.

Indirect Method Tactics

The second method of controlling and extinguishing wildland and ground cover fires is to remove the fuel from the fire. This can be accomplished using hand tools or mechanized equipment. Wildland fire guidelines state that the indirect attack tactics that remove the fuel from the area ahead of the fire should be used when flames are higher than 8 ft (2.4 m).

To remove fine fuels such as light grass or weeds, you can use hand tools such as a fire broom or a steel fire rake to sweep them away. Special-use tools designed for cutting, digging, and scraping can be used to create a fire control line in light brush (**FIGURE 21-14**). Examples of tools used for this purpose are the **McLeod** (a combination hoe and rake tool) and the **Reinhart** (a scraping tool that looks like an oversized garden hoe).

FIGURE 21-14 Special-use hand tools can be used to create a fire control line in light brush.

Courtesy of Jeff Pricher.

When working in heavier brush, a **hazel hoe** (also known as an **adze hoe**) can be used to grub out brush to create a fire control line. The **Pulaski axe** combines an adze and an axe for brush removal. A **council rake** is a long-handled rake constructed with hardened triangular-shaped steel teeth that is used for digging, for rolling burning logs, for cutting grass and small brush, and for raking a fire control line down to soil with no subsurface fuel (mineral soil).

In some situations, saws are used to remove heavy brush and trees from the fire. The types of saws employed for this purpose range from handsaws to gasoline-powered chainsaws. Other powered equipment used to remove fuel from wildland fires includes tractors, plows, and bulldozers.

Setting a backfire is another technique that can be used to remove the fuel from a wildland fire. A **backfire** is a fire that is intentionally set by firefighters along the inner edge of a fire control line in an attempt to burn all the fuel in an area of vegetation. This creates a buffer area in which the fuel has already been burned. When the fire reaches this area, it is forced to cease or change direction. Backfires must be set at the right time and at the proper place to work correctly. The technique is potentially dangerous and should be conducted only by trained personnel when deemed necessary by the IC and after a coordinated plan has been established.

Firing out is another method of setting a fire to remove fuel. To **fire out** means setting a fire along the inner edge of a fire control line to consume the fuel between the fire control line and the fire's edge (**FIGURE 21-15**). This technique is used to extend a flank or other part of the fire to the fire control line or to clean up or strengthen the fire control line by

FIGURE 21-15 Firing out is a method of setting a fire to remove fuel from a wildland fire.

Courtesy of Jeff Pricher.

burning the fuel that did not completely burn at the fire perimeter. While conducting structure defense, firing out can be used when a safety zone is compromised and the crew needs to remove fuel to increase the safety zone. This technique is typically a smaller operation and involves burning less fuel than a backfire. To accomplish this technique, the firefighter uses a drip torch, fusee, or Very pistol (flare gun). As with setting a backfire, firing out should be conducted only by trained personnel when deemed necessary by the crew supervisor or IC.

SAFETY TIP

Once a fire has been set on the ground, it cannot be undone. When improperly set, backfires or the firing out technique can worsen a wildland fire, create a new fire, or even cause firefighter injuries or fatalities. These fires should be set only by firefighters who have had special training in the control of wildland fires.

Smothering the Fire

The third method are used to control and extinguish wildland fires is to smother the fire by removing oxygen. The smothering technique is most commonly used when firefighters are overhauling the last remnants of a wildland fire. During the overhaul process, water and mineral soil (dirt with no organic material that would serve as fuel for a still smoldering fire) that is dug up at the scene are often thrown on smoldering vegetation until it is cool to the touch to prevent flare-ups. This is referred to as dry mopping when performed without water and wet mopping when performed using water. Smothering is less useful during the more active phases of a fire, although some vegetative fires in grasses can be smothered using a special fire-beater sheet constructed of heavy rubber.

A **compressed air foam system (CAFS)** combines foam concentrate, water, and compressed air to produce a foam that will stick to vegetation and structures that are in the path of a wildland and ground cover fire. The foam helps extinguish the fire in two ways. First, it absorbs the fire's heat. It also breaks down the surface tension of brush and wood fibers, which allows the foam to penetrate deeper into the woody material, cooling the fuel, and preventing oxygen from combining with the fuel. Because the compressed air foam achieves better penetration of fuels and clings to the fuel, it increases the effectiveness of the water used, which means less water is needed. The better penetration also reduces the incidence of fires rekindling. CAFS can be delivered from a backpack fire extinguisher or from an appliance that is mounted on an apparatus.

In some locations, gel is used in place of foam to pretreat fuels in the path of a wildland or ground cover fire. Like foam, the gel sticks to vegetation and structures, absorbs the fire's heat, and prevents the fuel from burning. Unlike foam, gel cannot be pumped through a fire hose, so it must be introduced at the nozzle of a hose or backpack fire extinguisher. Depending on the type of gel, an application will remain effective for several hours. Make sure to follow the gel manufacturer's instructions if using this type of treatment.

To suppress a wildland fire, follow the steps in **SKILL DRILL 21-1**.

SKILL DRILL 21-1

Suppressing a Wildland Fire Firefighter I, NFPA 1010: 6.3.19

1. Don appropriate personal protective equipment (PPE). Identify safety and exposure risks. Protect exposures if necessary.

2. Construct a fire control line by removing fuel with hand tools.

SKILL DRILL 21-1 CONTINUED

Suppressing a Wildland Fire Firefighter I, NFPA 1010: 6.3.19

3. As an alternative to Step 2, extinguish the fire with a backpack pump extinguisher or a handline.

4. Overhaul the area completely to ensure complete extinguishment of the ground cover fire.

Voice of Experience

Several years ago, we responded to a wildland fire that had been sparked by a train—a common occurrence in our area. While the fire burned through tall grass, I was struck by how quickly the head of the fire moved, even over flat ground. The flames were moving so quickly and bringing with them such a surprising intensity that it made a direct attack nearly impossible.

Unfortunately, the apparatus used by our department was too heavy for the terrain, a fact we learned only after getting three brush units stuck in mud up to their axles. We lost situational awareness. Things had been dry for a period of time before this incident, and no one considered the ground condition. The flame fronts progressed toward the stuck apparatus. It was only the timely arrival of several tracked utility task vehicles (UTVs) with skid units and a bulldozer that saved our apparatus from being lost to the fire.

Size-up for wildland fires includes not just what is burning, but *where* is it burning. We must consider whether we can get resources to the location to mount a direct attack or if we will need to use some other method.

Matthew Marlow
Medway-Grapeville Fire District
Surprise, New York

Fire Apparatus Used for Wildland Fires

The fire apparatus used for wildland fire suppression include both structural fire apparatus and specialized wildland fire apparatus. Structural fire apparatus are commonly used because most small wildland fires are fought by departments whose primary mission is structural firefighting. For small wildland fires, structural fire pumpers carry enough water to mount an attack, and booster lines provide enough water for attacking. For larger wildland fires, preconnected 1¾-in. (44-mm) structural hose lines can be used.

Using structural fire apparatus for wildland fires has several drawbacks. In particular, most structural fire apparatus are not designed to pump water at the same time that the truck is moving forward. This means that the apparatus must stop and be shifted into pump gear before a stream of water can be pumped. It also makes the apparatus less efficient when a continuous stream of water is needed while the truck moves along the flank of a fire. In addition, most structural fire apparatus is designed for on-the-road use. As a consequence, trucks that lack all-wheel drive and underside protection may become stuck or damaged when used in off-road situations. Structural fire apparatus that meet the requirements of NFPA 1900, *Standard for Aircraft Rescue and Firefighting Vehicles, Automotive Fire Apparatus, Wildland Fire Apparatus, and Automotive Ambulances, 2024 Edition*, are not primarily designed for fighting wildland fires.

By contrast, specialized wildland fire apparatus are designed for fighting wildland fires. Because the terrain and conditions vary from one geographic location to another, many variations in wildland fire apparatus exist. The design of any such apparatus needs to be suited to the local wildland conditions. Wildland fire apparatus range from small pickup trucks or Jeep-type vehicles to large trucks. Most of this apparatus is designed with all-wheel drive, underside protection, and high road clearance, which enables the vehicles to travel off-road. Wildland fire apparatus normally pump between 10 and 550 gallons per minute (gpm; 38 and 2082 liters per minute [L/min]). They are equipped with a water tank that holds from 50 to 750 gal (189 to 2839 L) of water, and they are able to operate a water pump while moving. Wildland fire apparatus should meet the requirements of NFPA 1900.

Small wildland fire apparatus typically carry 200 to 300 gal (757 to 1136 L) of water. These smaller brush trucks can respond quickly and often can contain a fire before it grows into a larger blaze (**FIGURE 21-16**).

Larger wildland fire apparatus are often built on a midsized or large truck chassis (**FIGURE 21-17**). These trucks carry more water than small apparatus, which enables them to operate for a longer period of time without being refilled. Many of these vehicles have crew cabs that enable an entire crew of firefighters to travel to the fire scene in the truck.

A third type of fire apparatus used when fighting wildland fires is a tanker or water tender (**FIGURE 21-18**). These trucks are designed to transport water to the scene of a wildland fire. They are usually equipped with a pump that enables them to fill from a drafting site and to offload water when needed.

FIGURE 21-16 A smaller brush truck can respond quickly to wildland fires.

Courtesy of Jeff Pricher.

FIGURE 21-17 Larger wildland fire apparatus carry more water than small apparatus.

FIGURE 21-18 A water tanker or tender is used to transport water to wildland fires.

FIGURE 21-19 A wildland bulldozer is used to create fire control lines.

Courtesy of Jeff Pricher.

Some of these apparatus are suited for off-road use, whereas others must remain on a road.

Wildland fire apparatus also include tractors, plows, and bulldozers (**FIGURE 21-19**). These vehicles are used to create fire control lines.

Airplanes and helicopters are valuable resources for dropping water or fire retardant where the terrain makes it dangerous or impossible to fight a fire with ground-based firefighters. Aircraft can be used to attack the head of a fire, thereby slowing the growth and spread of the fire.

Safety in Wildland Firefighting

Safety is a relative term in firefighting since there are inherent unavoidable risks. The ideal way to manage risk is to apply established strategies and tactics and to implement safety standards for conducting the strategic and tactical operations. Over the last century, we have lost firefighters while fighting wildland fires in every state, with the exception of Massachusetts—over 1100 firefighters, in 49 states. The largest losses were in the state of California; losses in that state accounted for over 330 of the total. The fire service has identified many common situations that led to firefighter fatalities. It was also found that for every safety standard created, one or more firefighters died because that standard was violated.

This is why following established guidelines and safety standards when working directly on the fire line is so important. The *NWCG Incident Pocket Response Guild (IRPG)*, created by the National Wildfire Coordinating Group (NWCG), is an invaluable reference that every firefighter should carry with them. It describes safety zones and when to implement them. It lists the 10 Standard Fire Orders and 18 Watch Out Situations, as well as mnemonics to help firefighters remember safety procedures and planning guidelines. It also includes guidelines for how to mitigate risk when having to construct a fire line downhill—that is, from the top of a slope working toward the bottom and toward the fire—how to work around trees, and the procedure to refuse an unsafe order.

Whenever firefighters engage the fire at the head and the flame lengths are over 8 ft, they should be in a safety zone. In the wildland/urban interface environment, having good safety zones and maintaining situational awareness is especially crucial.

Safety Zones

Safety zones during wildland fire operations are important. For many years, a safety zone was defined as a place you could take refuge from the fire and not be injured without the use of PPE or a fire shelter. Later, the federal fire labs settled on a safety zone needing to be a minimum of four times the flame height. As this issue continued to be studied, later articles by fire behavior analysts suggested that this guideline needed to be increased to account for winds and steep slopes. It is accepted that an identified safety zone should be determined any time work is being conducted on an active fire line. If you are on the active fire line, the black interior areas of the fire are normally designated as the safety zone (when those areas are large enough to meet the criteria) when conducting direct attack on the perimeter of the fire. In addition, any time indirect attacks are conducted, firefighters should only engage the fire when working from a safety zone.

Current guidelines state that whenever firefighters engage the fire at the head of the fire and the flame lengths are higher than 8 ft (2.4 m), they should be working in a safety zone. As stated earlier, the size of a safety zone is four times the flame length to the nearest firefighter or the black areas. When high winds are present or the fire is on or near a slope, the size of the safety zone should probably be increased to increase the distance between the firefighters and the fire.

Crown Fires and Safety Zones

A **crown fire** is when the live and dead fuels in a tree (aerial fuels) off the ground are on fire. There are three types of crown fires (**FIGURE 21-20**):

- A **passive crown fire** is when an individual tree or group of trees catches fire and the runs all the way up the tree so the tree or trees look like a giant torch. This fire does not burn into the surrounding trees.

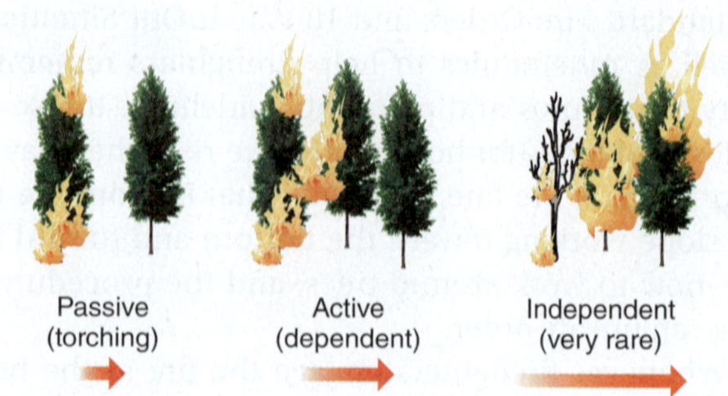

FIGURE 21-20 Active and independent crown fires are dangerous examples of extreme fire behavior.

Data from NWCG. 2007. "S-290, Intermediate Wildland Fire Behavior". https://www.nwcg.gov/publications/training-courses/s-290

- An **active crown fire** is when the ground fire burns beneath the trees and then the aerial fuels burn as the ground fire progresses. This type of crown fire is extremely dangerous because it spreads rapidly from the ground to the trees.
- An **independent crown fire** burns and moves rapidly from tree to tree, and the falling embers ignite a ground fire after the crown fire has passed. This type of crown fire is also dangerous because it spreads so quickly.

Crown fires often have over 80- to 100-ft (24- to 30-m) flame lengths with moderate to rapid rates of spread. In the forest, it is difficult to create adequate safety zones for this type of fire behavior. Many times firefighters have to drive a significant distance away from the fire to get to their safety zones. This takes time. Fortunately, active crown and independent crown fires are predictable. They occur only when crown spacing—the space between the tops of the trees—is less than 20 ft (6 m), and there is low relative humidity, high temperatures, and drought conditions (**FIGURE 21-21**). If these conditions are present, firefighters must disengage the firefighting operation and leave the area before the weather changes. Changes to consider for crown fire activity would include the inversion layer lifting, slope reversals, or peak burning conditions, normally considered between 1400 and 1700 hours. When the weather changes and cools at night, the fire becomes a ground fire and firefighters can resume firefighting activities. These tactics should continue until the weather changes and crown fires no longer occur.

FIGURE 21-21 Crown fires must be fought with indirect methods for firefighter safety. Even passive crown fires need to be evaluated for safety zones.

FIGURE 21-22 When responding to a wildland fire, it is important to keep informed on fire weather conditions and forecasts.

The 10 Standard Firefighting Orders

The **10 Standard Firefighting Orders** were developed in 1957 by the USFS after a task force investigated 16 tragic wildland fires that resulted in multiple firefighter fatalities. The 10 orders, which are meant to prevent further firefighter tragedies from occurring during wildland fires, are divided into three categories as follows (NWCG 2022, 122):

Fire behavior

1. Keep informed on fire weather conditions and forecasts (**FIGURE 21-22**).
2. Know what your fire is doing at all times.

3. Base all actions on the current and expected behavior of the fire.

Fireline safety

4. Identify escape routes and safety zones and make them known.
5. Post lookouts when there is possible danger.
6. Be alert. Keep calm. Think clearly. Act decisively.

Organizational control

7. Maintain prompt communications with your forces, your supervisor, and adjoining forces.
8. Give clear instructions, and ensure they are understood.
9. Maintain control of your forces at all times.

If 1 to 9 are considered, then . . .

10. Fight fire aggressively, having provided for safety first.

The 18 Watch Out Situations

The **18 Watch Out Situations** are an expansion of the 10 Standard Firefighting Orders. Outlined in the *IRPG*, the list addresses areas of specific concern in the wildland environment. The 18 watch out situations are as follows:

1. Fire not scouted and sized up
2. In country not seen in daylight
3. Safety zones and escape routes not identified
4. Unfamiliar with weather and local factors influencing fire behavior
5. Uninformed on strategy, tactics, and hazards
6. Instructions and assignments not clear
7. No communication link with crew members or supervisor
8. Constructing line without safe anchor point
9. Building fire line downhill with fire below
10. Attempting frontal assault on fire
11. Unburned fuel between you and the fire
12. Cannot see main fire and not in contact with anyone who can
13. On a hillside where rolling material can ignite fuel below
14. Weather becoming hotter and drier
15. Wind increases or changes direction
16. Getting frequent spot fires across line
17. Terrain and fuels make escape to safety zones difficult
18. Taking a nap near the fire line

The 18 Watch Out Situations can be used as a checklist to determine if an assignment is safe. If five or more of the situations exist, firefighters should reconsider engaging in the assignment.

Lookouts, Communications, Escape Routes, and Safety Zones (LCES)

Before attacking a wildland fire, firefighters should ensure that lookouts, communications, escape routes, and safety zones (**LCES**) are in place. The LCES mnemonic is derived from the core safety principles outlined in the 10 Standard Firefighting Orders and the 18 Watch Out Situations. Ensuring the components of LCES are in place before attacking a wildland fire reduces the risk associated with fighting these types of fires:

- *Lookouts* need to be stationed in a location from which they can see the firefighters working at the scene (**FIGURE 21-23**). If the lookout recognizes that the firefighters are becoming endangered by the fire or the hazards of the environment, they notify the firefighters so that they can reposition themselves to a safer area. Lookouts must be knowledgeable about the wildland environment and must be able to anticipate changes in the behavior of the fire and the weather.
- *Communications* on the wildland fire scene must be clear, concise, and prompt (**FIGURE 21-24**). All personnel should be on the same radio frequencies. Effective communications ensure that everyone is aware of approaching hazards.
- *Escape routes* provide firefighters with a path from a dangerous location to an area away from the fire. The effectiveness of an escape route depends on the behavior of the fire. Firefighters should make sure that more than one escape route is available to them at all times.

FIGURE 21-23 Lookouts need to be stationed in a location from which they can see the firefighters working at the scene.

Courtesy of Jeff Pricher.

FIGURE 21-24 Communications on the wildland fire scene must be clear, concise, and prompt to ensure that everyone is aware of approaching hazards.

Courtesy of Jeff Pricher.

- *Safety zones* are based on the behavior of the fire and the firefighter's location relative to the fire. A basic guideline is to choose an area that provides a separation of four times the flame height. This guideline is based on a three-person engine crew on flat ground with no wind. The size of the safety zone would need to be adjusted to take into account wind, location on a slope, and the number of resources.

Remember to plan for the components of LCES before firefighting in the wildland environment. Every firefighter must be informed of the escape routes and safety zones, and everyone should adhere to the operational plans.

TIP

LACES is an adaptation of the LCES mnemonic that is used in various parts of the country. The *A* stands for different terms depending on the area in which it is being used. It can stand for anchor points, attitude, or awareness.

SAFETY TIP

Certain topographical features, such as valleys, steep hillsides, and saddles, can cause sudden and rapid fire progression. Many firefighter fatalities and near misses have occurred in these types of terrain. Safety procedures, such as those covered by the LCES (Lookouts, Communications, Escape routes, Safety zones) mnemonic, must be rigidly enforced when operating in these areas.

Hazards of Wildland Firefighting

As previously mentioned, responses for wildland fires involve all of the hazards encountered by personnel responding to structural fires. In addition, they may involve some unique hazards specific to the wildland. For example, driving fire apparatus on poorly maintained or unimproved roads or trails, driving on steep terrain, and driving apparatus off-road greatly increase the chance of fire apparatus rollovers. Also, in wildland fires, often the roadway is obscured by smoke. Given these dangers, drivers must thoroughly understand the operating characteristics of their fire apparatus and operate the apparatus within the safe limits specific to its design.

When working in rough or steep terrain, wildland firefighters are at increased risk for falls. Rough ground often contains holes that are difficult to see in smoky conditions. Large trees have significant ground fuel around them. Close to stumps, root systems might be burning below the surface, weakening the ground so that firefighters fall into holes.

SAFETY TIP

It is important that drivers do not drive apparatus when there is no visibility.

The sharp tools involved in wildland firefighting can also cause injuries to firefighters. Carry these tools with the sharp edge facing the ground. If a tool has sharp edges on both ends, be sure to use a tool guard. Tool awareness is the key to injury prevention.

Other hazards of fighting wildland fires include burns, smoke inhalation, dehydration, and heat stroke. Because the PPE worn by wildland and ground cover firefighters provides less protection than the PPE worn by structural firefighters, firefighters must keep far enough from the heat of the fire to prevent burns.

Trees cause hazards to firefighters engaged in wildland firefighting. Always be alert for the hazards posed by falling trees. During a fire, the lower parts of a tree may burn away and weaken the support for the rest of the tree so that it falls. When a tree is burning in this manner, it is called a **snag** (**FIGURE 21-25**). Another area of concern is the tops of trees. In some instances, fire may burn inside a tree, from the base to the top, causing the top to break off. Branches broken at the tops of trees by lightning, snow, or wind sometimes get tangled in the branches above rather than falling to the ground. This type of branch is called a **widow maker**, and it can fall without a sound. One of the most serious and often overlooked hazards during a wildland fire is the use of chainsaws on trees. After vehicle accidents and medical-related occurrences, chainsaw use and hazard trees are the next leading cause of wildland firefighter fatalities and injuries (Wildland Fire Lessons

FIGURE 21-25 Snags are dead or live trees that are burning at the base and that can fall and kill firefighters. These large trees that are burned at the base create a significant hazard.

Courtesy of Bret Davidson.

SAFETY TIP

Serious and often overlooked hazards are hazard trees. It is important to look up in wooded areas to identify fire-weakened trees, loose branches, and your proximity to them. Notify others, and if necessary, have a qualified person cut down the tree.

SAFETY TIP

An ash-covered pit filled with hot embers can be left behind when stumps and large logs have finished burning. The ash often looks like stable ground and may ensnare unsuspecting firefighters. As you navigate through a stump-laden area, watch your step!

Learned Center, 2017). Only specially trained personnel should operate chainsaws. All federal hand crews have one or more qualified persons (called a **faller**). If you are not qualified for chainsaw use, flag or mark trees with known hazards that need to be cut down and the area around the trees so that others will not work around or underneath them.

Electrical transmission lines and other electrical wires may be present at the location of a wildland fire. Wires that drop on vegetation may ignite a wildland fire and pose an electrical hazard to firefighters. Many of these safety hazards can be difficult to see at night and in smoky conditions. Be alert for electrical hazards even in the middle of an area covered with vegetation with no buildings around.

Additionally, be alert for conduction caused by smoke and heat. Heat and smoke from a fire burning under a high-voltage transmission line can sometimes ionize (become electrically charged) above the fire, creating a conductor to the ground or across other conductors in the transmission lines. If possible, do not construct escape routes or anchor points underneath high-voltage transmission or power lines. Also, do not attack the fire when a heavy column of smoke is rising from the fire into the high-tension lines. Allow the fire to move past the high-tension lines, and then reengage the attack. Electricity can arc to the ground in heavy smoke.

Because many wildland fires occur during hot weather and because combating them involves strenuous physical activity, all firefighters must be alert to the possibility of dehydration and heat stroke. When fighting a wildland fire, make sure you drink enough water and take breaks when needed to give your body time to recover.

Personal Protective Clothing and Equipment

It is important for wildland firefighters to wear PPE appropriate for wildland firefighting (**FIGURE 21-26**). Specifically, they should be equipped with a protective one-piece jumpsuit or a coat, a shirt, and trousers that meet the requirements of NFPA 1977, *Standard on Protective Clothing and Equipment for Wildland Fire Fighting and Urban Interface Fire Fighting, 2022 Edition*. These garments should be constructed of a fire-resistant material, such as Nomex, to ensure that the clothing stops burning as soon as the heat and flame are removed. This reduces the chance of firefighters being burned. Wildland firefighters should also wear an approved helmet with a protective shroud, eye protection, gloves, and protective footwear. All of this equipment should meet the provisions of NFPA 1977.

FIGURE 21-26 PPE for a wildland firefighter.

Courtesy of Jeff Pricher.

Respiratory protection for wildland firefighting is usually limited to a filter mask. A **filter mask** filters out small particles of smoke; however, it does *not* protect against inhalation of heavy smoke or poisonous gases like self-contained breathing apparatus (SCBA). Check the manufacturer's instructions to determine the limits of each filter mask. Such masks should not be relied upon beyond the limits for which they are approved.

The PPE that is provided for wildland and ground cover firefighting varies from one location to another, depending on the types of wildland fires encountered and on the climate in a given location. Follow your local fire department's standards regarding PPE. The use of structural PPE for wildland fires can quickly result in rapid body overheating and dehydration. Given this risk, fighting wildland fires in structural PPE for long periods of time is dangerous and not recommended.

SAFETY TIP

Avoid the use of structural PPE for wildland firefighting. Wearing structural PPE for long periods of time will result in the retention of body heat and will greatly increase the chance of dehydration, heat exhaustion, and heat stroke.

FIGURE 21-27 A wildland fire shelter.

Courtesy of Anchor Industries, Inc.

Fire Shelters

One of the most important pieces of PPE for wildland and ground cover firefighters is the **fire shelter**, which is a thin, reflective-foil layer attached to a layer of fiberglass (**FIGURE 21-27**). They are designed to reflect approximately 95 percent of a fire's radiant heat for a short period of time. The reflective property allows a rapidly moving fire to pass over a firefighter who has deployed a fire shelter. This life-saving piece of equipment should be issued to all wildland and ground cover firefighters.

Fire shelters are carefully folded and carried in a protective pouch on the firefighter's belt. When firefighters are in danger of being overrun by a rapidly moving fire, they should try to get to a safe location. Only if they cannot escape should they use their fire shelters. To use a fire shelter, the firefighter finds the largest available clearing, scrapes away flammable litter, opens the shelter, lies face down on the shelter, and covers themselves with the shelter. When properly used by well-trained personnel, this equipment can save lives. As with all equipment, however, it is important to receive proper training to use a fire shelter safely. To use a fire shelter, follow the steps in **SKILL DRILL 21-2**. Note that the fire shelter pictured in Skill Drill 21-2 is a practice fire shelter, as indicated by its orange case. Fire shelters with orange cases are intended for training purposes only. Fire shelters intended for use in real fires come in a blue case.

SAFETY TIP

Following training with a fire shelter, take the time to properly repack the practice fire shelter so that it is ready to deploy realistically during the next training session. Follow the repacking instructions provided by the manufacturer.

SKILL DRILL 21-2

Using a Fire Shelter Firefighter I, NFPA 1010: 6.3.19

1. Pick the largest available clearing. Wear gloves and a hard hat, and cover your face and neck if possible. Scrape away flammable litter if you have time.

2. Pull the red ring to tear off the plastic bag covering the shelter.

Continues.

SKILL DRILL 21-2 CONTINUED

Using a Fire Shelter Firefighter I, NFPA 1010: 6.3.19

3. Grasp the handle labeled "Left Hand" in your left hand and the handle labeled "Right Hand" in your right hand.

4. Shake the shelter until it is unfolded. If it is windy, lie on the ground to unfold the shelter.

SKILL DRILL 21-2 CONTINUED

Using a Fire Shelter Firefighter I, NFPA 1010: 6.3.19

5. Slip your feet through the hold-down straps.

6. Slip your arms through the hold-down straps.

Continues.

SKILL DRILL 21-2 CONTINUED

Using a Fire Shelter Firefighter I, NFPA 1010: 6.3.19

7. Lie down in the shelter with your feet toward the oncoming fire. Push the sides out for more protection and keep your mouth near the ground.

Courtesy of Jeff Pricher.

SAFETY TIP

When working with fire shelters, remember the following key points:

- Inspect the shelter regularly. Do not open the shelter's clear plastic bag during inspections.
- Be sure you are on the ground in the shelter with your feet pointed toward the fire when the fire arrives.
- After the fire has cooled, stay in your shelter until you are instructed to exit it by your team leader or the fire scene manager; a second or third flame front may occur.
- Avoid falling branches by identifying a place to deploy where hazard trees are not present.

Strategic Considerations for the Wildland/Urban Interface

The **wildland/urban interface (WUI)** is the area where undeveloped land with vegetative fuels meets with human-made structures. A similar term is the **wildland/urban intermix**, which describes an area where vegetative fuels intermingle with human-made structures with no clear boundary between them.

The mixing of wildlands and developed areas occurs for a number of reasons. For example, people may want a house that is surrounded by the beauty of nature in the middle of the woods but that has modern conveniences. Additionally, overcrowding and the high expense of living in a big city have caused some people to find homes in remote areas located next to forests and uninhabited areas. The intermixing of wildlands and developed areas in this way has created massive fire problems in many parts of the United States. Wildland fires regularly ignite buildings and become structure fires. Structure fires, conversely, regularly ignite vegetative fuels and become wildland fires. In many cases, it is not possible to clearly separate one type of fire from the other. This phenomenon explains why most fire departments are involved in fighting some types of wildland fires.

The WUI presents a major challenge for firefighters. Because many people live in the WUI, fires in these areas present a significant life-safety hazard. Also, because so many structures are built in these areas, there is a huge potential for property loss whenever a fire occurs. Finally, many parts of the WUI do not have adequate municipal water systems.

The best offense against the huge problem of fire in the WUI is often vigorous prevention, so much of the effort geared toward reducing the loss from wildland fires is directed at prevention. Many communities work to involve homeowners by educating them about the risks their homes pose and the steps that they can take to reduce these risks. One important measure that has been implemented in many jurisdictions is prohibiting the use of flammable cedar shake roofs in houses in the WUI. Another important step is encouraging or requiring homeowners to maintain a defensible space around their structures. A **defensible space** is an area around a structure where combustible vegetation that can spread fire has been cleared, reduced, or replaced. It acts as a barrier between the structure and an advancing fire.

Many firefighters are responsible for both wildland and structural operations. Those departments that do not have a primary responsibility for wildland fires may still be called upon when resources are depleted or when a large fire directly impacts their jurisdiction. Structural firefighters need to be aware of structure triage guidelines during a wildland fire.

Structure triage guidelines describe how to categorize structures near a wildland fire based on the level of threat to them and the tactics that can be deployed to defend these threatened homes. There are three structure triage categories: not threatened, threatened defensible, and threatened non-defensible. To determine which category infrastructure or homes fit into, use the acronym S-FACTS, which stands for Survival, Fire environment, Access, Construction/clearance, Time constraints, and Stay or GO:

- *Survival* is the assessment of the structure based on whether there is a safety zone at the home or nearby. If the answer is no, you should immediately consider evacuating people. If time permits, use a prep-and-go tactic (discussed later).
- *Fire environment* affects the determination of whether residents have time to gather belongings before they must get to a safety zone or evacuate the area completely. You need to consider the current and expected fire behavior based on the fuel, weather, and topography.
- *Access* in and out before the fire threatens the location, as well as access out after the fire has passed, are important to consider. Many firefighters have been trapped by downed power lines or trees or damaged culverts and bridges after the fire has passed through an area. Blocked access routes may limit your availability to provide protection to more homes.
- *Construction* of the building you are protecting is important to identify. Shake shingle roofs or siding is more combustible than brick, block, or stucco. *Clearance* of combustible fuels is important for creating a safety zone. Also, decks and porches made of wood, the accumulation of debris, wood piles, and propane tanks near homes should be taken into consideration.
- *Time* is an important factor. If you have a day before the wildland fire gets to the location where the structures are located, a lot of work can be done. If you have limited time because the fire is approaching quickly, then strong consideration should be made to prep and leave the property. Spot fires and changes in the wind can make it difficult to accurately predict the ROS of an approaching fire. It is recommended that you provide twice the time needed to get to a safety zone or evacuate the area than you think you have, to ensure that you are in a safety zone when the fire reaches your location.
- When considering all of these factors, the last part is determine if you can *Stay or GO*. If you need to go, do so immediately.

SAFETY TIP

The information in this section is an introduction to these strategies and tactics. These operations are extremely dangerous for crews that do not normally work in the wildland environment.

WUI Tactics

There are three primary tactics firefighters use to defend homes in the WUI:

- *Check and go* is used to evacuate for life-safety reasons and there is no adequate safety zone at the home and there is no time to conduct any fuel reduction activities. In this case, immediately leave the home and get to a safety zone not at the home.
- *Prep and go* is used if there is no adequate safety zone at the home, but there is time to prep the home to potentially save it by using techniques to prevent or retard fire growth. Applying protective sprinklers, fire wrapping, foam, or gel have proven to be effective. Finally, clear as much fuel as possible from the outside of the structure.
- *Prep and defend* is used only if there is an adequate safety zone at the threatened home. If there is, you should implement a tactical PACE plan to use

techniques to prevent or retard fire growth, being prepared to go to the safety zone as soon as it is needed.

Tactical PACE When Prepping and Defending

The tactical PACE was developed to ensure that the leader's intent is clear under changing conditions during stressful events when miscommunication can easily occur (**FIGURE 21-28**). The intent is for the crew to discuss the four plans when staying and defending infrastructure. The tactical PACE plan is a good tool to provide firefighters with options and decision points when encountering the fire when it is the most dangerous.

The PACE plan should include the following:

- Primary plan: This is the most aggressive attack given expected fire behavior. An example is backing in the engine and pulling a hose line to the back of the structure and putting out the fire from the safety zone as the fire approaches.
- Alternative plan: If fire is more than was anticipated, this plan is a more defensive attack. An example is using the structure as a barrier or another natural barrier and fighting the fire from behind the barrier as the fire passes.
- Contingency plan: This plan is needed if the fire behavior becomes so intense that visibility is lost. If this happens, the crews should disengage and retreat to the predesignated **temporary refuge area (TRA)**, which is a place you can go to wait until fire behavior conditions improve so you can reassess the fire. A typical example of this is crews retreating to the cab of the apparatus or taking temporary refuge in the structure they are protecting.
- Emergency plan: This plan is enacted if the fire behavior becomes so intense that crews need to deploy their fire shelters. The location where the crew would meet and deploy the fire shelters should be identified as part of the plan.

WUI Strategies

When fire resources are assigned to protect homes during a wildland fire event, there are several WUI strategies that can be considered:

- *Bump and run* is using resources at the head of the fire. These resources can use any of the primary three tactics of check and go, prep and go, or prep and defend. These techniques require extreme caution because resources are located at the head of the fire where conditions can rapidly change.
- *Fire front following* is when resources with the primary strategy of defending homes move into an area after the head of the fire has passed.

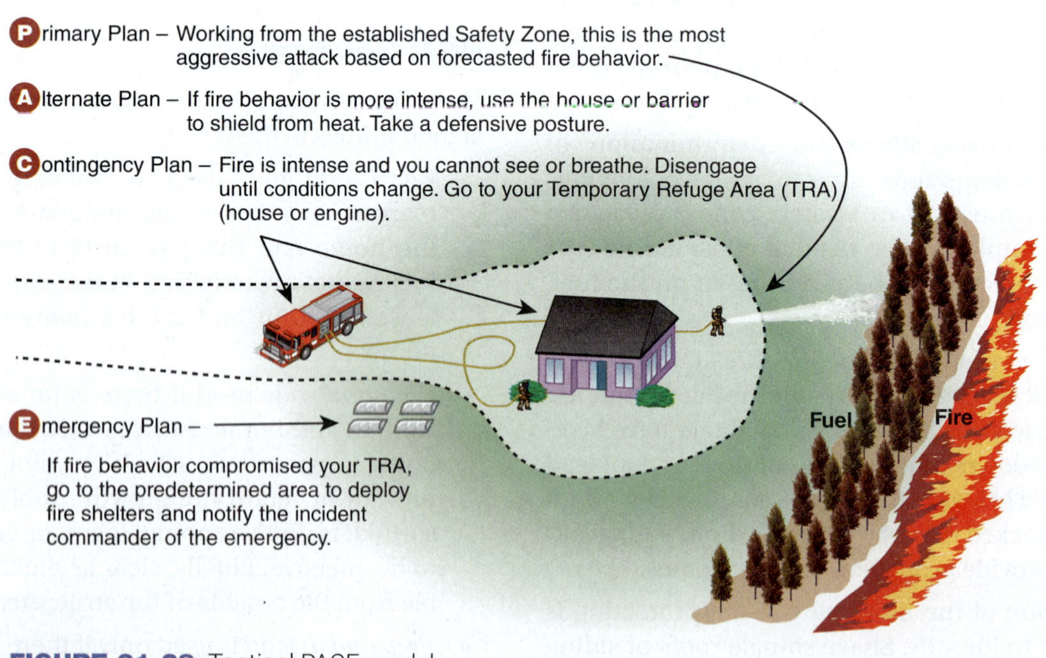

FIGURE 21-28 Tactical PACE model.

- *Tactical patrol* is crews stationed near homes after the head of the fire has passed, to conduct a constant patrol during the hours to days after the fire passes to ensure homes do not ignite from smoldering or creeping (very slow moving) fires. This is usually done in combination with other strategies.
- *Anchor and hold* is a strategy used when wildland fires come in contact with suburban areas with a good water supply. Engines set up in safety zones and use hydrants and hose streams to prevent a conflagration caused by home-to-home fire progression (**FIGURE 21-29**).

SAFETY TIP

Many areas that are prone to wildland fires have limited numbers of roads leading into and out of them. Be aware of the routes by which you can escape a rapidly moving fire or a fire that changes direction quickly. Do not let yourself get caught in a wildland fire with no escape route. As a rule of thumb, give yourself double the time you anticipate needing to get to your safety zone.

FIGURE 21-29 An example of the anchor-and-hold WUI strategy keeping homes downwind of the fire from burning.

CASE STUDY

You Are the Firefighter CONCLUSION

You have been dispatched to a wildland fire that is approaching your community. Normally, the state forestry service handles these calls, but they call local resources for assistance when they are out of resources, and that is what happened this time. Winds are high and the fire is moving rapidly. Your engine is assigned to a position on the edge of town, and your company has been assigned to evacuate people and protect as many homes as possible.

1. **How do you determine if you are in a safety zone?**

   Answer: Ask yourself, are you far enough away from the flames to not feel the effects of radiant or conventional heat? Do you have enough room between yourself and the flame length so you do not feel the effects of the radiant or convection heat?

2. **What does it mean to determine the fire behavior?**

   Answer: To determine the fire behavior, you have considered the fuel that is burning, the current weather conditions, and the topography in order to accurately predict the flame length and rate of spread of the fire.

3. **Based on the expected fire behavior, what strategy or tactics are effective and safe under these conditions?**

   Answer: Fall back to appropriate strategies based on the flame length and the tools and equipment that are available. For example, hand tools can only fight 4-ft (1-m) flames, so you would need to be defensive in the strategies you apply.

WRAP-UP

SUMMARY

KNOWLEDGE OBJECTIVES

- Discuss the history of wildland fire management.
  - Describe the role of fire labs in establishing the best practices in safely fighting wildfires. (p. 890)
- Identify the anatomy of a wildland fire and explain the effects of fuel, weather, and topography of the behavior of a wildland fire.
  - Describe the components of a wildland fire. (**NFPA 1010: 6.3.19**, pp. 890–891)
  - Define the components of the wildland fire triangle. (pp. 890–891)
  - Describe the effect of flame length on determining safety zones. (pp. 891–892)
  - Describe the importance of knowing the rate of spread in maintaining firefighter and bystander safety. (pp. 891–893)
  - Describe ground fuels, subsurface fuels, aerial fuels, and surface fuels. (**NFPA 1010: 6.3.19**, pp. 892–893)
  - Describe fuel group characteristics and how they impact the rate of spread. (**NFPA 1010: 6.3.19**, pp. 892–893)
  - Describe how weather conditions and topography influence the growth of wildland fires. (**NFPA 1010: 6.3.19**, pp. 893–896)
- Describe wildland terminology, parts of a wildland fire, and the strategy and tactics in suppressing wildland fires.
  - Describe the two strategies to extinguish wildfires. (**NFPA 1010: 6.3.19**, pp. 896–899)
  - Label the parts of a wildland fire. (**NFPA 1010: 6.3.19**, p. 896)
  - Describe how a direct attack is mounted on wildland fires. (**NFPA 1010: 6.3.19**, p. 897)
  - Describe how an indirect attack is mounted on wildland fires. (**NFPA 1010: 6.3.19**, pp. 897–898)
  - Describe how a parallel attack is mounted on wildland fires. (**NFPA 1010: 6.3.19**, p. 898)

  - Describe the methods and tools used to cool a fuel with water. (**NFPA 1010: 6.3.19**, pp. 899–900)
  - Describe the methods and tools used to remove a fuel from wildland fires. (**NFPA 1010: 6.3.19**, pp. 899–902)
  - Describe the methods and tools used to smother wildland fires. (**NFPA 1010: 6.3.19**, pp. 901–903)
- Describe the role of fire apparatus to suppress wildland fires.
  - Describe the characteristics of the fire apparatus used to suppress wildland fires. (**NFPA 1010: 6.3.19**, pp. 903–905)
- Describe wildland safety considerations.
  - Describe the role of safety zones in protecting the safety of firefighters and the impact of crown fires. (pp. 905–906)
  - Describe how the 10 Standard Firefighting Orders can be used to prevent future firefighter tragedies. (pp. 906–907)
  - Describe how the 18 Watch Out Situations can be used to determine if an assignment is safe. (p. 907)
  - Describe the role of each component of the LCES mnemonic. (pp. 907–908)
  - Describe the hazards associated with wildland firefighting. (**NFPA 1010: 6.3.19**, pp. 908–909)
  - Describe the personal protective clothing and equipment needed for wildland firefighting. (**NFPA 1010: 6.3.19**, pp. 909–910)
- Describe the triage categories, considerations, strategies, and tactics to defend structures during a wildland/urban interface fire.
  - Explain the problems created by the WUI. (pp. 914–915)
  - Describe the categories the acronym S-FACTS utilizes to assess a structure's risk in a WUI fire. (p. 915)
  - Explain the three primary tactics in the WUI strategy of fighting a WUI fire. (pp. 915–917)
  - Explain how to apply the strategic PACE model on a wildland fire and how to apply the tactical PACE model during structure defense. (p. 916)

SKILLS OBJECTIVES

- Demonstrate how to suppress wildland fires given hand tools and water delivery devices.
  - Suppress a wildland fire. (**NFPA 1010: 6.3.19**, pp. 902–903)
- Demonstrate how to deploy a fire shelter for firefighter survival.
  - Deploy a fire shelter. (**NFPA 1010: 6.3.19**, pp. 911–914)

KEY TERMS

active crown fire A crown fire that starts on the ground and then progresses to aerial fuels.

adze hoe See *hazel hoe.*

aerial fuels Fuels located more than 6 ft (2 m) off the ground, usually part of or attached to trees. Also called *canopy fuels.*

air stability The vertical movement of air in the atmosphere; unstable environments allow the free flow of air from the ground into the upper atmosphere, and stable air resists vertical air movement.

anchor, flank, tandem, and pincer attack A direct method of suppressing a wildland fire that involves two teams of firefighters establishing anchor points on each side of the fire and working toward the head of the fire until the fire gets "pinched" between them. Also called *pincer attack.*

anchor point A strategic and safe point from which to start constructing a fire control line. An anchor point is used to reduce the chance of firefighters being flanked by fire.

aspect The direction the side of the mountain is facing.

backfire A fire set along the inner edge of a fire control line to consume the fuel in the path of a wildland fire or change the direction of force of the fire's convection column. (NFPA 901)

black area An area that has already been burned and that is considered relatively safe for firefighters.

KEY TERMS CONTINUED

burning edge The active perimeter of the fire.

canopy fuels See *aerial fuels.*

chain A unit of measurement for length; 1 chain equals 66 ft (20 m).

chinook winds See *Foehn winds.*

compressed air foam system (CAFS) A foam system that combines air under pressure with foam solution to create foam. (NFPA 1900)

council rake A long-handled rake constructed with hardened triangular-shaped steel teeth that is used for raking a fire control line down to soil with no subsurface fuel, for digging, for rolling burning logs, and for cutting grass and small brush.

crown fire A fire in which live and dead fuels in a tree off the ground are burning.

defensible space An area, as defined by the authority having jurisdiction (typically a width of 30 ft [9 m] or more), between an improved property and a potential wildland fire where combustible materials and vegetation have been removed or modified to reduce the potential for fire on improved property spreading to wildland fuels or to provide a safe working area for firefighters protecting life and improved property from wildland fire. (NFPA 1140)

direct attack A method of wildland fire attack in which firefighters focus on containing and extinguishing the fire at its burning edge.

east winds See *Foehn winds.*

18 Watch Out Situations A list of potential situations published by the National Wildfire Coordinating Group (NWCG) that is used to assess whether or not a wildland firefighting assignment is safe to conduct.

faller A person qualified to use a chainsaw to cut down trees.

filter mask A face mask that filters out small particles of smoke but does not protect against inhalation of heavy smoke or poisonous gases.

fine fuel Fuel that ignites and burns easily, such as dried twigs, leaves, needles, grass, moss, and light brush.

finger A narrow point of fire whose extension is created by a shift in wind or a change in topography.

fire control line Comprehensive term for all constructed or natural barriers and treated fire edges used to control a fire. (NFPA 901)

fire line The crew of firefighters conducting a fire attack.

fire out A wildland firefighting technique in which a fire is set along the inner edge of a fire control line to consume the fuel between the fire control line and the fire's edge.

fire shelter An item of protective equipment configured as an aluminized tent utilized for protection, by means of reflecting radiant heat, in a fire entrapment situation. (NFPA 1550)

flame height The flame height measured from the ground to the tip of the flame.

flame length The flame height measured from the beginning of the flame to the tip of the flame.

flanking attack A direct method of suppressing a wildland or ground cover fire that involves placing a suppression crew on one flank of a fire.

flank of the fire The edge between the head and heel of the fire that runs parallel to the direction of the fire spread.

Foehn winds Winds caused by a high-pressure system of warm air replacing cold air and bringing with it unusually high winds in the opposite direction of the local winds. Also called *chinook winds, east winds,* and *Santa Ana winds.*

fuel moisture The amount of moisture present in a fuel, which affects how readily the fuel will ignite and burn.

green area An area of unburned fuels.

ground cover fire A fire that burns loose debris such as grass, leaves, and branches on the surface of the ground.

ground fuel All combustible materials such as grass, duff, loose surface litter, tree or shrub roots, rotting wood, leaves, peat, or sawdust that typically support combustion. (NFPA 1140)

hazel hoe A hand tool used to grub out heavy brush to create a fire control line. Also called *adze hoe.*

head of the fire The main or running edge of a fire; the part of the fire that spreads with the greatest speed.

heavy fuel Fuel of a large diameter, such as large brush, heavy timber, snags, stumps, branches, and dead timber on the ground, that ignites and is consumed more slowly by fire than light fuel.

heel of the fire The side opposite the head of the fire, which is often close to the area of origin. Also called *rear of the fire.*

independent crown fire A crown fire that moves rapidly from tree to tree burning aerial fuels and dropping falling embers to the ground, sometimes igniting a ground fire.

indirect attack A method of wildland fire attack in which the control line is located along natural fuel breaks, at favorable breaks in the topography, or at considerable distance from the fire, and the intervening fuel is burned out.

inversion layer A layer of warm air above a layer of cold air.

island An unburned area surrounded by fire.

LCES A mnemonic that stands for Lookouts, Communications, Escape routes, and Safety zones to remind firefighters to ensure that each of these components is in place before attacking a wildland fire to reduce the risks associated with fighting these types of fires.

McLeod A hand tool used for constructing fire control lines and overhauling wildland fires. One side of the head consists of a five-toothed to seven-toothed fire rake; the other side is a hoe.

parallel attack A method of attack in which the control line is located parallel to the fire edge, at a distance of about 5 to 50 ft (1.5 to 15 m) from the fire. The intervening fuel usually burns out as the fire control line moves alongside the fire but can also burn out with the main fire.

passive crown fire A crown fire that starts at the bottom of a tree or group of trees and then runs all the way up the tree or trees so they look like a giant torch, but does not spread to the surrounding trees.

pincer attack See *anchor, flank, tandem, and pincer attack.*

pocket A deep indentation of unburned fuel along the fire's perimeter, often found between a finger and the head of the fire.

Pulaski axe A hand tool that combines an adze and an axe for brush removal.

rate of spread (ROS) The speed, usually measured in chains per hour, at which the front of the fire moves.

rear of the fire See *heel of the fire.*

Reinhart A hand tool that looks like an oversized garden hoe used for constructing fire control lines and overhauling wildland fires.

relative humidity The ratio between the amount of water vapor in the gas at the time of measurement and the amount of water vapor that could be in the gas when condensation begins, at a given temperature. (NFPA 79)

safety zone An area of sufficient size and in a suitable location that is expected to protect fire personnel from known hazards without using fire shelters as the fire moves around them.

Santa Ana winds See *Foehn winds.*

slash Debris resulting from natural events such as wind, fire, snow, or ice breakage, or from human activities such as building or road construction, logging, pruning, thinning, or brush cutting.

slope The steepness of a mountain.

slope reversal The moment when flames burning downhill reach the bottom and start burning uphill.

snag A dead or live tree whose lower part is on fire, weakening the tree and presenting a significant falling hazard.

spot fire A new fire that starts outside areas of the main fire, usually caused by flying embers and sparks.

structure triage guidelines Guidelines that describe how to categorize structures near a wildland fire based on the level of threat to them and the tactics that can be deployed to defend these threatened homes.

subsurface fuels Partially decomposed matter that lies beneath the ground, such as roots, moss, and decomposed stumps.

surface fuels Fuels that are close to the surface of the ground, such as grass, leaves, twigs, needles, small trees, logging slash, and low brush. Also called ground fuels.

temporary refuge area (TRA) A predesignated area where firefighters can immediately take refuge or temporary shelter and short-term relief from the fire without using a fire shelter until fire conditions improve.

10 Standard Firefighting Orders A set of systematically organized rules developed by the U.S. Forest Service (USFS) task force to reduce danger to firefighting personnel.

timber litter Fuel that falls out of trees, such as pine needles, leaves, twigs, and small branches.

topography The land surface configuration. (NFPA 1140)

widow maker A broken top of trees or limbs of trees that have been damaged by wind or snow and are still hanging in the tree that can fall without making a sound.

KEY TERMS CONTINUED

wildland Land in an uncultivated, more or less natural state and covered by timber, woodland, brush, and/or grass. (NFPA 901)

wildland fire An unplanned fire burning in vegetative fuels. (NFPA 1140)

wildland fire behavior The flame length and rate of spread of a wildland fire given the fuel, weather, and topography.

wildland/urban interface (WUI) The line, area, or zone where structures and other human development meet or intermingle with undeveloped wildland or vegetative fuels. (NFPA 5000)

wildland/urban intermix An area where improved property and wildland fuels meet with no clearly defined boundary. (NFPA 5000)

wind The horizontal movement of air over the earth surface.

REVIEW QUESTIONS

1. List the federal agencies that have responsibility to suppress wildland fires.
2. Define wildland fire behavior.
3. Describe the difference in a strategic PACE and a tactical PACE for WUI structure defense.
4. List a direct attack method and an indirect method of attack.
5. Describe the features of specialized wildfire apparatus.
6. Describe the minimum distance for a safety zone from the head of the fire.
7. List the categories S-FACTS assesses to determine a structure's chance of survival.

DISCUSSION QUESTIONS

1. You arrive to a vegetation fire that is burning up a hill. What observations would determine your strategy and initial tactics to extinguish the fire?
2. Your engine is requested to assess protecting structures ahead of an oncoming wildfire. What considerations would determine the tactics used to protect the structures?
3. Your fire is burning in timber litter under trees and is going to move into 1-ft (30-cm) grass. Given the same weather conditions and topography, what changes in ROS and flame length do you predict will occur?

APPLYING THE CONCEPTS

Your engine has been dispatched to a car fire on a two-lane highway. Smoke is visible in the distance. Dispatch reported the car is on the side of the road and the driver is safely out of the vehicle. Flames are coming out of the engine block and flying embers have ignited the dried grass and twigs on the side of the road. When you arrive at the scene, the vehicle is fully involved, and the ground cover fire has spread from the dried vegetation along the edge of the road into the brush. Flames are about 5 ft (1.5 m) high, and the fire area is about 100 ft by 100 ft (30 m by 30 m). About 150 yards (137 m) away you see a grove of pine trees. Several residential structures are visible just beyond the trees, down a moderate slope.

1. Should your unit focus their attention on the vehicle or the brush fire? Why?
2. Wildland and ground cover fires behave differently than structure fires. What three factors influence wildland fire behavior—also known as the wildland fire triangle?
3. If you were the incident commander, what would you assess in the size-up to determine a course of action?

Presently the wind is calm, the fuel is generally live with areas of dead brush, and the fire is spreading at a slow rate. There is no immediate danger to the grove of trees or houses; however, a low-pressure system is forecast to arrive in 2 to 3 hours, bringing with it high winds. Your engine is carrying 1000 gallons (3785 liters) of water, but it is not designed for off-road operations and there is no additional water at the scene. Considering these factors, the IC radios for additional units to assist.

The IC decides on a direct (anchor, flank, and pincer) attack and breaks your unit into two teams. You are assigned to the team working the left flank.

4. What is the first thing your team will do? How will they proceed?
5. If there are not enough personnel for an anchor, flank, and pincer attack and a flanking attack is used, how does the IC determine which flank of the fire to fight?

The teams are advancing along the flanks, making progress pinching the head of the fire when the darkening sky announces the early arrival of the low-pressure system. Suddenly, gusts of wind push the head of the fire toward the tree line. Because the IC assessed the situation proactively, the wildland fire unit has just arrived with personnel, equipment, and expertise.

6. If the IC had not called for additional resources during the size-up and waited until now to do so, what challenges could arise?

REFERENCES

National Fire Protection Association (NFPA). 2020. NFPA 79, *Electrical Standard for Industrial Machinery.* 2021 Edition. Quincy, MA: NFPA.

National Fire Protection Association (NFPA). 2020. NFPA 901, *Standard Classifications for Fire and Emergency Services Incident Reporting.* 2021 Edition. Quincy, MA: NFPA.

National Fire Protection Association (NFPA). 2021. *NFPA 1140, Standard for Wildland Fire Protection.* 2022 Edition. Quincy, MA: NFPA.

National Fire Protection Association (NFPA). 2023. NFPA 1550. *Standard for Emergency Responder Health and Safety.* 2024 Edition. Quincy, MA: NFPA.

National Fire Protection Association (NFPA). 2023. NFPA 1900, *Standard for Aircraft Rescue and Firefighting Vehicles, Automotive Fire Apparatus, Wildland Fire Apparatus, and Automotive Ambulances.* 2024 Edition. Quincy, MA: NFPA.

National Fire Protection Association. 2023. NFPA 1977, *Standard on Protective Clothing and Equipment for Wildland Fire Fighting.* Quincy, MA: NFPA.

National Fire Protection Association (NFPA). 2023. NFPA 5000, *Building Construction and Safety Code®.* 2024 Edition. Quincy, MA: NFPA.

National Wildfire Coordinating Group (NWCG). 2007. Module 3: Fire Shelter in *S-130, Firefighter Training.* Accessed June 20, 2024. https://training.nwcg.gov/classes/s130/508%20Files/071231_s130_m3_508.pdf

National Wildfire Coordinating Group (NWCG). 2022. *NWCG Incident Response Pocket Guide (IRPG).* Accessed August 25, 2023. https://www.nwcg.gov/sites/default/files/publications/pms461.pdf.

U.S. Department of Agriculture Forest Service (USFS) and Aviation Management. 2019. "Implementation Guide for Aerial Application of Fire Retardant." Accessed December 21, 2023. https://www.fs.usda.gov/sites/default/files/2019-06/2019_afr_imp_guide.pdf.

Wildland Fire Lessons Learned Center. 2017. "2017 Incident Review Summary," Accessed June 20, 2024. https://lessons.wildfire.gov/annual-incident-review-summaries.

SECTION

2

Firefighter II

CHAPTER

22

Firefighter II

Establishing and Transferring Command

Note: To be NIMS compliant, in this chapter we will refer to the incident management system as the incident command system.

KNOWLEDGE OBJECTIVES

After studying this chapter, you will be able to:

- Outline the roles and responsibilities of a Firefighter II.
- Explain the function and organization of the Incident Command System (ICS) inside the National Incident Management System (NIMS).
- Understand how the sides of a building are identified at an incident.
- Explain how the ICS is implemented.
- Describe the components of a proper scene size-up and an incident action plan.
- Describe the importance of accurately documenting information in an incident report.
- Explain the crew resource management (CRM) concept.

SKILLS OBJECTIVES

After studying this chapter, you will be able to perform the following skills:

- Operate within the Incident Command System (ICS).
- Establish and transfer command of an incident.
- Complete and proofread an incident report.

ADDITIONAL NFPA STANDARDS

- **NFPA 901**, *Standard Classifications for Fire and Emergency Services Incident Reporting, 2021 Edition*
- **NFPA 1006**, *Standard for Technical Rescue Personnel Professional Qualifications, 2021 Edition*
- **NFPA 1026**, *Standard for Incident Management Personnel Professional Qualifications, 2024 Edition*
- **NFPA 1410**, *Standard on Training for Emergency Scene Operations, 2020 Edition*
- **NFPA 1550**, *Standard for Emergency Responder Health and Safety, 2024 Edition*
- **NFPA 1660**, *Standard for Emergency, Continuity, and Crisis Management: Preparedness, Response, and Recovery, 2024 Edition*
- **NFPA 2500**, *Standard for Operations and Training for Technical Search and Rescue Incidents and Life Safety Rope and Equipment for Emergency Services, 2022 Edition*
- **NFPA 3000**, *Standard for an Active Shooter/Hostile Event Response (ASHER) Program, 2024 Edition*
- **NFPA 5000**, *Building Construction and Safety Code, 2021 Edition*

CASE STUDY

You Are the Firefighter

Your engine company has been dispatched to the smell of smoke in a rural area near the edge of your jurisdiction. As you get closer to the location, you see smoke drifting across the road. When you turn the corner, you see flames in the second-story window of an old farmhouse. While you advance an attack line from the hose bed, your company officer does a 360-degree walk-around of the building. He gives a complete size-up over the radio, requests additional resources (including a tanker from the neighboring department), and establishes command. The driver/operator charges the attack line, and you hear a neighbor ask your officer if all of the occupants are outside.

1. What are the three incident priorities you must address at all emergency scenes?
2. How is the command structure likely to expand at this incident as additional resources arrive on the scene?
3. What are the benefits of operating under the Incident Command System during mutual aid incidents?

Introduction

Fire department operations involve the management of resources. Routine station duties, fire prevention activities, training, and emergency incident responses all require the safe, effective, and efficient use of fire department personnel, apparatus, and equipment. Most fire departments operate within a paramilitary-style command structure with various ranks of authority and responsibility. Ultimately, the fire chief is responsible for all of the operations of the department, but many of those responsibilities are delegated through the ranks so that the overall mission of the department can be accomplished. Routine tasks and duties are performed by the lower ranks; decision-making and strategic planning are carried out by officers of higher ranks. For example, the Firefighter I performs basic firefighter duties, such as using tools and hose properly; performing forcible entry, ventilation, and search and rescue operations; and attacking fires using basic firefighting methods, under direct supervision. The Firefighter II performs advanced skills, such as evaluating an emergency scene and establishing and transferring command to a more experienced firefighter, under general supervision. At each emergency incident, a command structure must be established so that everyone is working safely and effectively toward a common goal.

Roles and Responsibilities of the Firefighter II

The first step in understanding the organization of the fire service is to learn your roles and responsibilities. This is discussed in Chapter 1, *The Fire Service*. As a Firefighter II, you will receive a higher level of training that will allow you to more fully assist in mitigating emergency situations. For example, the Firefighter II coordinates and directs other firefighters and assumes a greater level of responsibility for incident management. The Firefighter I must work under direct supervision, but the Firefighter II works under general supervision to do the following:

- Perform an analysis of a scene.
- Determine the level of the Incident Command System (ICS) needed to safely and efficiently mitigate the emergency.
- Arrange and coordinate the ICS until command is transferred.
- Prepare reports.
- Communicate the need for assistance.
- Coordinate an interior attack line team.
- Extinguish an ignitable liquid fire.
- Control a flammable gas cylinder fire.
- Operate a thermal imaging camera (TIC).
- Locate the origin area of a fire.
- Protect evidence of fire cause and origin.
- Assess and disentangle victims from motor vehicle collisions.
- Assist special rescue team operations.
- Perform a fire safety survey.
- Present fire safety information.
- Prepare descriptions and drawings of areas in the community to be used by firefighters during an incident.

- Maintain fire power equipment.
- Perform annual service tests on fire hose.

Many Firefighter II duties involve interacting with the public, working with special rescue teams, and performing basic scene assessment and decision-making functions in the initial stages of an emergency incident.

Understanding the Incident Command System

The **chain of command** is a rank-based hierarchical structure that creates an orderly line of authority. Establishing a chain of command results in the upward and downward flow of information and assignments needed to effectively manage both routine operations and the most critical emergency incidents.

Whereas most routine, daily operations can be managed through the chain of command, standard operating procedures (SOPs), and other administrative tools, larger, more dynamic operations should be managed using a more formal system. An **incident management system (IMS)** is a formal system with defined roles and responsibilities for handling unexpected or emergency incidents.

When only one agency is involved in an incident, a single **incident commander (IC)** is responsible for all aspects of managing the incident. All personnel, apparatus, and resources work under the authority and direction of the IC.

During very complex incidents when multiple agencies are involved, a unified command may be established. A **unified command** creates a single set of incident goals and objectives and fosters mutual communication and cooperation among agencies. When a unified command is established, each responding agency assigns an IC to the unified command and these ICs work together to decide how the incident will be handled (**FIGURE 22-1**).

All fire department emergency operations and training exercises should be conducted within the framework of an ICS. An ICS provides a standard approach, structure, terminology, and operational procedure to organize and manage any operation, from a training session to an emergency scene. The same principles apply whether the situation involves a single engine company or hundreds of emergency responders from dozens of different agencies. These same principles also can be applied to many non-emergency events, such as large-scale public events. Planning, delegation, supervision, and communication are key components of an effective ICS.

FIGURE 22-1 A unified command involves many agencies directly in the decision-making process for a large incident.

Courtesy of Dean Moses.

History of the ICS

Prior to the 1970s, each individual fire department had its own method for commanding and managing incidents. Often, the organizational structure established to direct operations at an incident scene depended on the style of the chief on duty. Not surprisingly, this individualized approach did not work well when units from different districts or departments with mutual aid agreements responded to a major incident. In structural firefighting, the basic units are companies. Local procedures that were adequate for routine incidents were often ineffective and confusing when applied to large-scale incidents, incidents that had rapidly changing conditions, and incidents that required units that did not normally work together to collaborate.

This fragmented approach to managing emergency incidents is no longer considered acceptable. Over the past 30 years, formal incident management systems have been developed and refined. Today, the same basic approaches, organizational structures, and terminology are used by thousands of fire departments and emergency response agencies across the United States.

The move to develop a standard system began in the early 1970s after several large-scale wildland fires in southern California proved disastrous for both the fire service and residents of the region, in part due to a lack of organization and communication among responding agencies. To avoid a repeat of these disasters, a number of fire-related agencies at the local, state, and federal levels decided that better organization was necessary to effectively combat these costly fires. These agencies established an organization known as

FIGURE 22-2 ICS was first developed to coordinate efforts during large-scale wildland fires.

FIRESCOPE (**FI**re **RES**ources of **C**alifornia **O**rganized for **P**otential **E**mergencies) to develop solutions for all issues associated with large, complex emergency incident, including the following:

- Command and control procedures
- Resource management
- Terminology
- Communications

FIRESCOPE developed the first standard ICS. Originally, the ICS was intended only for large, multijurisdictional or multiagency wildland incidents involving more than 25 resources or operating units (**FIGURE 22-2**). It proved so successful, however, that it was applied to structural firefighting as well, and eventually became an accepted system for managing all emergency incidents across the United States.

Around the same time, the **fireground command (FGC)** system was developed in Phoenix, Arizona and adopted by many fire departments. The basic concepts of FGC were similar to those that formed the foundation of the ICS created by FIRESCOPE, with some differences in terminology and organizational structure. The FGC system was initially designed for day-to-day fire department incidents involving fewer than 25 fire suppression companies, but it could be expanded to meet the needs of much larger incidents.

During the 1980s, the ICS developed by FIRESCOPE was adopted by all federal and most state wildland firefighting agencies. The National Fire Academy (NFA) also used the FIRESCOPE ICS as the model ICS for its courses. Additional federal agencies, including the Federal Emergency Management Agency (FEMA) and the Federal Bureau of Investigation (FBI), adopted the same model for use during major disasters or terrorist events. This led to all federal agencies learning and using the same system.

During the 1980s, several federal regulations and consensus standards, including NFPA 1500, *Standard on Fire Department Occupational Safety, Health, and Wellness Program,* were adopted. These standards mandated the use of an ICS at emergency incidents. In 1990, NFPA 1561, *Standard on a Fire Department Incident Management System,* was issued. It identified the key components of an effective system and described the importance of using such a system at all emergency incidents. At that time, fire departments could choose to use either the FIRESCOPE ICS or the FGC system to meet the requirements of NFPA 1561.

NIMS Model

In the years that followed, fire service users of different ICs across the United States formed the National Fire Service Incident Management System Consortium to develop model procedure guides for implementing effective incident management systems at various types of incidents. In 2004, by order of President George W. Bush, the U.S. Department of Homeland Security (DHS) developed the **National Incident Management System (NIMS)** to provide a consistent, nationwide framework for incident management based partly on the previous work done by the fire service in developing the ICS. NIMS, which is administered by FEMA, is a standard approach to incident management that can be used by many different agencies to manage a wide range of emergency and nonemergency situations. The **Incident Command System (ICS)** is the part of NIMS that formally defines roles and responsibilities for managing emergency incidents of all sizes. The ICS blended the best aspects of FIRESCOPE's ICS and the FGC system. It defined common terminology and organizational systems and procedures so that everyone at an incident understands the commands that are given. This enables federal, state, and local governments; private sector and nongovernmental organizations; and all other organizations who assume a role in emergency management to work together effectively and efficiently.

The ICS can be used at any type or size of emergency incident and by any type or size of department or agency. Reflecting this change, the title of NFPA 1561 was changed to *Standard on Emergency Services Incident Management System and Command Safety.* NFPA 1026, *Standard for Incident Management Personnel Professional Qualifications,* was developed to define the various job performance requirements for each of the positions classified within the new ICS. In 2024, NFPA

1500 and NFPA 1561 were consolidated along with NFPA 1521, *Standard for Fire Department Safety Officer Professional Qualifications,* to create a new standard NFPA 1550, *Standard for Emergency Responder Health and Safety.*

Like the FIRESCOPE ICS, the NIMS model can be used regardless of the cause, size, complexity, or type of an incident, including acts of terrorism and natural disasters. Along with the basic concepts of flexibility and standardization, the NIMS principles are now taught in every incident management course. To encourage all agencies to adopt the same system, federal grant funding is often reserved for NIMS-compliant communities. As a result, you will likely receive NIMS-specific ICS training at your department in order to meet compliance requirements. FEMA provides three introductory courses that teach users how to work within the NIMS framework at any type of emergency incident (FEMA 2018). These courses are free of charge, and each course can be completed in approximately 3 hours, either online or in a classroom setting:

- IS-100: Introduction to the Incident Command System (ICS) provides an overview of the ICS, including its history, principles, and organizational structure. It also explains how ICS fits into the larger NIMS model of emergency preparedness, response, and recovery. There are no prerequisites for IS-100 Emergency Management Institute (EMI 2022a).
- IS-200: Basic Incident Command System for Initial Response provides training for personnel to function within the ICS at small incidents involving one or two resources, as well as at those incidents that start small but grow in complexity and size of operation. Completion of IS-100 is required before taking the IS-200 course (EMI 2022b).
- IS-700: An Introduction to the National Incident Management System (NIMS) is not required at the Firefighter II level, but it is recommended for a better understanding of how NIMS enhances community preparedness, communications and information sharing, resource management, and incident command during emergency incidents (EMI 2022c).

These courses are updated periodically and the updates are designated by a lowercase letter after the course number, such as IS-200.c. Be sure to take the latest available version of each course to ensure current information. You need to take these courses only once, but it is a good idea to periodically review the information. You can learn more about NIMS and the ICS courses on FEMA's website at www.fema.gov/emergency-managers/nims.

Organization of the ICS

The ICS identifies a full range of duties, responsibilities, and functions that are performed at emergency incidents, and it defines the relationships among those components. One of the key principles of the ICS is its adaptability. While some components are used at almost every incident, others are not, and the ICS organization can be easily expanded to manage the largest and most complex situations. Positions are staffed as they are needed. The only position that must be filled at every incident is command, which is the responsibility of an IC or unified command. Command decides which additional components are needed for the specific incident and assigns personnel as needed to perform those tasks.

To ensure consistency at emergency incidents, fire departments develop SOPs and then train and practice using the ICS. This approach increases safety and efficiency. Emergency scenes tend to be chaotic, so organizing operations at an incident often poses a serious challenge, particularly if the agencies involved use different terms to describe certain concepts and resources. One of the strengths of the ICS is its use of standard terminology. Specific terms apply to various parts of an incident organization.

To help clarify roles within the ICS, standard position titles are used to describe the personnel in charge of a component (**TABLE 22-1**).

Some departments may use slightly different terminology. If your department does this, you need to know both the terminology used by your department and the standard ICS terminology. For more information about ICS terminology, see NFPA 1026.

Command

The component that exists at every incident is **command**, which establishes the incident objectives and creates the plan for achieving them. At most incidents, the IC does not formally appoint personnel to each position below command. Instead, the IC informally fulfills the basic functions of each position. At more complex incidents, the IC needs to keep this span of control to between three and seven people or resources. If needed, the IC formally delegates the positions below command to qualified individuals.

As previously mentioned, command is the only position in the ICS that must be filled for every incident. There must always be someone to direct all emergency scene personnel, equipment, and operations. Command is established when the first unit arrives on the scene and is maintained until the last unit leaves the

TABLE 22-1 ICS Organization Levels

Level	Function	Leadership Title
Command	Responsible for command and control of the incident	Incident commander (IC) or unified command
Command staff	Responsible for incident safety, communications with involved agencies, and dissemination of information to the general public	Incident safety officer (ISO), liaison officer, and public information officer (PIO)
Section	Responsible for operations, planning, logistics, or finance/administration	Section chief
Branch	An optional subdivision of the operations section responsible for a specific function	Branch director
Division	A subdivision of a branch or section consisting of a company or crew working in a defined geographic area	Division supervisor
Group	A subdivision of a branch or section consisting of a company or crew working on the same task or function, but not necessarily in the same location	Group supervisor
Task force	Two or more single resources of different types who work together to accomplish a specific task	Task force leader
Strike team	Two or more single resources of the same type who work together to accomplish a specific task	Strike team leader
Crew	A single resource consisting of two or more firefighters working to accomplish a specific task without apparatus	Crew leader
Unit	Two or more resources who have responsibility for a specific activity during an incident	Unit leader

scene. Each of the other positions in the ICS is filled at the discretion of the IC. Initially, the IC is directly responsible for the following tasks:

- Establishing command
- Establishing the **incident action plan (IAP)**, which is an oral or written plan containing general objectives describing the strategy and tactics for achieving the objective
- Expanding the ICS organization as needed
- Coordinating with outside agencies
- Managing resources and coordinating resource activities
- Providing for scene safety
- Releasing information about the incident

If the incident becomes more complex, the IC may delegate some of these activities to other ICS positions, but the IC is still responsible for ensuring that all necessary activities are completed.

When an incident occurs within a single jurisdiction and there is no jurisdictional or functional agency overlap, a single IC is responsible for incident management. This does not mean that no other agencies respond and have a role in supporting the management of the incident. It simply means that these other agencies report to and work under the direction and authority of the IC.

When multiple agencies with overlapping jurisdictions or legal responsibilities respond to the same incident, a unified command is often established. As described earlier, under a unified command, each responding agency assigns an IC to the unified command, and the various ICs work as a team to create a single set of incident goals and objectives. A unified command provides several advantages. In this approach, representatives from each agency cooperate to share command authority. They work together and are directly involved in the decision-making process. Operating under a unified command helps to ensure cooperation, avoid confusion, and guarantee agreement on strategies and tactics needed to mitigate the emergency. Information about overlapping jurisdictions and unified command is discussed in greater detail later.

Over time, as various emergencies are stabilized, the strategic objectives become increasingly focused on a single jurisdiction or discipline. In the later stages of an incident that is managed by a unified command, command may be returned to a single IC from the agency most responsible for handling the rest of the events at the scene. For example, a mass-casualty incident resulting from an explosion would likely use a unified command, sharing command between fire, law enforcement, and EMS. Once all victims have been treated and transported from the scene, command would shift from a unified command to a single IC from the law enforcement agency while the crime scene is processed.

It is also acceptable, if all agencies and jurisdictions agree, to designate a single IC instead of establishing a unified command in multiagency and multijurisdictional incidents. If this decision is made, the IC should be carefully chosen.

Incident Command Post

The **incident command post (ICP)** is the headquarters location for the incident. Command functions are centered in the ICP, so command and all direct support staff should be located there. The location of the ICP should be clearly marked and announced to all personnel as soon as it is established. This is especially critical for incidents involving large structures or large geographic areas.

The ICP should be located in a protected location near the incident scene, but it should not be in the immediate vicinity of emergency operations. Possible locations for the ICP can be determined during the preincident survey of large structures. (A **preincident survey** is an examination of a property to gather information to develop a preincident plan. A **preincident plan** is data about locations in your community that describe the layout and hazards at the location and identifies response essentials.) For incidents involving large geographic areas, the ICP may be remote from some parts of the operational area. For major incidents, the ICP may be located in a special mobile command center vehicle or in a building. This choice of location enables the command staff to function without needless distractions or interruptions.

Command Staff

Individuals who are assigned to the **command staff** report directly to the IC. The incident safety officer, liaison officer, and public information officer are always part of the command staff (**FIGURE 22-3**). In addition, aides, assistants, and advisers may be assigned to work directly for members of the command staff. Often, these positions are filled by firefighters (sometimes officers) who serve as direct assistants to members of the command, and they cannot be delegated to work in other major sections of the ICS.

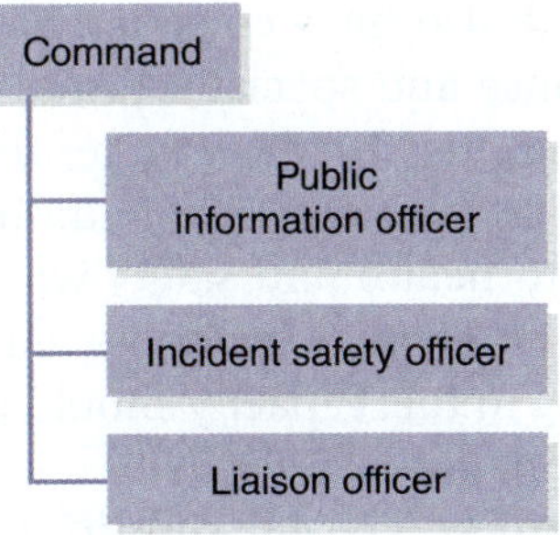

FIGURE 22-3 The command staff report directly to command.

Incident Safety Officer. The **incident safety officer (ISO)** is responsible for ensuring that safety issues are managed effectively at the incident scene. They serve as the eyes and ears of command, identifying and evaluating hazardous conditions, preventing and correcting unsafe practices, and ensuring that safety procedures are followed.

The ISO should be appointed early during an incident, especially if personnel are exposed to hazardous conditions or high-risk operations. If an incident becomes more complex and the number of resources at the scene increases, additional qualified personnel can be assigned as assistant ISOs.

Although the ISO is an adviser to command, they have the authority to stop or suspend operations when unsafe situations occur without checking with command first. This authority is clearly stated in national standards, including NFPA 1550 and NFPA 1026.

At the scene of a structure fire, the ISO should be an individual who is knowledgeable in fire behavior, building construction and collapse potential, firefighting strategy and tactics, hazardous materials, rescue practices, and departmental safety rules and regulations.

TIP

Several state and federal regulations require the assignment of an ISO at incidents involving hazardous materials and at certain technical rescue incidents.

The ISO should also have considerable experience in incident response and specialized training in occupational safety and health. Many larger fire departments have full-time ISOs who perform administrative functions relating to health and safety when they are not responding to emergency incidents. Firefighters interested in serving in this capacity should pursue training and certification under NFPA 1550.

Liaison Officer. The **liaison officer** is command's point of contact for representatives from assisting and cooperating agencies. An **assisting agency** helps with operations and provides personnel or other tactical resources to command. Fire departments, law enforcement agencies, EMS providers, and heavy equipment contractors are examples of assisting agencies. A **cooperating agency** such as the American Red Cross or the Salvation Army provide supplies and support but does not participate in the direct operations component of incident management.

During an incident, command may not have time to meet directly with everyone who comes to the ICP. The liaison officer functions as the representative of command. They coordinate with representatives from assisting and cooperating agencies. They also obtain and provide information and direct people to the proper location or authority. The liaison area should be adjacent to, but not inside, the ICP.

Public Information Officer. The **public information officer (PIO)** is responsible for gathering incident information and releasing it to the news media and other agencies and through social media (**FIGURE 22-4**). Especially in the event of a major incident, the public will want to know what has happened and what is being done to mitigate the incident. The PIO serves as the contact person for media requests, allowing command to concentrate on the incident. Like the area for the liaison officer, a media headquarters should be established near, but not in, the ICP. The information presented to the media by the PIO must be approved by command before it is released.

FIGURE 22-4 The PIO is responsible for gathering and releasing incident information to the media and other appropriate agencies.

Courtesy of Captain David Jackson, Saginaw Township Fire Department.

General Staff

When the incident is too large or too complex for just one person to manage effectively, the IC appoints personnel to oversee parts of the operation. Everything that occurs at an emergency incident can be divided among the following major functional components—each one is referred to as a **section**—within the ICS structure:

- Operations
- Planning
- Logistics
- Finance/administration

TIP

Occasionally, an incident might require a fifth section, Intelligence. The IC might decide this section is needed if there is suspected criminal or terrorist activity.

Resources operating within a section are led by a **section chief**. The section chiefs form the ICS **general staff** and report directly to command (**FIGURE 22-5**). Appointing section chiefs allows the IC to focus on the overall strategy, while the section chiefs focus on the specific tactics required to get the jobs done. Remember, only command must be established at an incident. The IC decides which, if any, of the four sections need to be activated, when to activate them, and who should be identified as the chief of each section. As each section chief is appointed, additional personnel and resources may then be assigned to work under the section chief's direction.

At a large incident, the four sections may operate from different locations, but the section chiefs will always remain in direct contact with command.

Operations Section

The **operations section** is responsible for taking direct action to control the incident. Personnel functioning

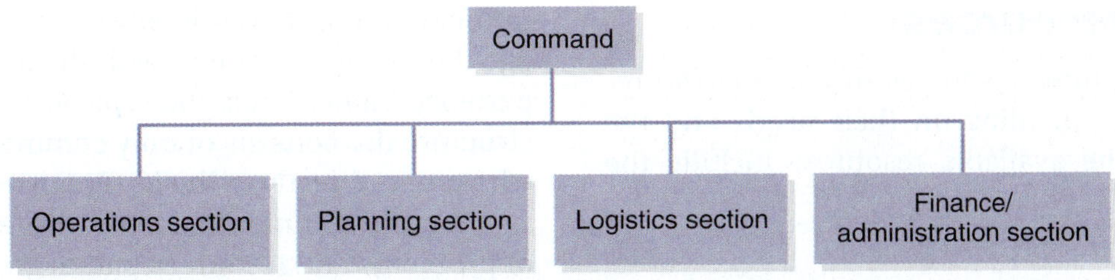

FIGURE 22-5 The five major functional components of the ICS. The four section chiefs report directly to command.

within the operations section perform tactical operations such as fighting the fire, rescuing trapped individuals, treating injured victims, and doing whatever else is necessary to alleviate the emergency situation. The **operations section chief** directly supervises the functions of the operations section.

Operations are conducted in accordance with an IAP that outlines the strategic objectives and how emergency operations will be conducted to achieve those objectives. At most incidents, the IAP is relatively simple and can be expressed in a few words or phrases. At large-scale incidents, however, the IAP can be a lengthy document that is regularly updated and used for daily briefings of the command staff.

Planning Section

The IC activates the **planning section** and appoints a **planning section chief** when information needs to be obtained, managed, and analyzed. The planning section collects, evaluates, and disseminates information relevant to the incident. To do this, the planning section uses preincident plans, building construction drawings, maps, aerial photographs, diagrams, reference materials, and status boards. The planning section is also responsible for developing, disseminating, and updating the IAP. Individuals assigned to planning examine the current situation, review available information, predict the probable course of events, and prepare recommendations for strategies and tactics. The planning section also keeps track of resources at large-scale incidents and provides command with regular situation and resource status reports. Basically, the planning section develops what needs to be done and by whom and identifies which resources are needed.

Logistics Section

The **logistics section** is responsible for providing supplies, services, facilities, and materials during an incident. The **logistics section chief** serves as the supply officer for the incident. Among the responsibilities of this section are the following:

- Keeping apparatus fueled
- Providing food, refreshments, and rehabilitation facilities for responders
- Obtaining the foam concentrate needed to fight a large flammable liquid fire
- Arranging for specialized equipment to remove a large pile of debris

In many fire departments, these logistic functions are routinely performed by permanent support services personnel and a logistics section is not needed. These support services personnel work in the background to ensure that the operations section has the resources needed to get the job done. However, if service and support requirements are very complex or extensive, the support services personnel need a manager. For example, an incident might require many resources or might continue for a long time. In that case, command needs to appoint a logistics section chief.

Finance/Administration Section

The **finance/administration section** is the fourth major ICS component managed directly by command. This section is responsible for the accounting and financial aspects of an incident, as well as any legal issues that may arise in its aftermath. This function is not staffed at most incidents because cost and accounting issues are typically addressed after the incident. Nevertheless, a **finance/administration section chief** may be assigned at large-scale and long-term incidents that require immediate fiscal management, particularly when outside resources must be procured quickly. A finance/administration section may also be established during a natural disaster or during a hazardous materials incident where reimbursement may come from the shipper, carrier, chemical manufacturer, or insurance company.

Assigning Resources

A section chief requires resources to conduct the required activities. Depending on their needs and the specific section, the available resources include the following:

- Single resources
- Groups and divisions
- Branches
- Task forces and strike teams
- Units

Single Resources

A **single resource** is an individual, a company, or a crew with an identified supervisor that can perform specific tasks at an incident or planned event (**FIGURE 22-6**). Recall that a **company** is a group of people who are assigned to a specific apparatus and who have specific responsibilities at an emergency scene under the command of the **company officer**. Within the ICS, a **crew** is a specific term that identifies a team of two or more firefighters working without apparatus. For example, members of an engine or ladder company who are assigned to operate inside a building are a crew. Additional personnel at the scene of an incident who are assembled to perform a specific task may also be called a crew. A crew has an assigned **leader** or company officer. Examples of a single resource are an engine and its crew, a ladder company, and a paramedic assigned to a base camp for medical care or rehabilitation on a structure fire.

Small-scale incidents can often be handled successfully by a single resource. The organizational structure implemented at an emergency incident starts small, with the arrival of the first unit or units. When a limited number of resources is involved, the IC can often manage the incident alone or with the assistance of an experienced aide. Thus, the typical ICS for a one-alarm structure fire consists of only command and resources who report directly to the IC (**FIGURE 22-7**).

At more complex incidents, the increasing number of problems and resources places greater demands on command and can quickly exceed command's **span of control**—the maximum number of individuals or resources that one person can supervise or manage effectively. According to FEMA, one person can effectively supervise only three to seven people or resources. In order to remain effective, the IC needs to activate the ICS sections. A larger incident requires a more complex command structure to ensure that no details are overlooked and that personnel safety is not compromised.

Groups and Divisions

In the early stages of an incident, individual companies are often assigned to work in different areas or perform different tasks. As the incident grows and more companies are assigned to it, command can establish groups and divisions. A **group** refers to companies and/or crews working on the same task or function, although not necessarily in the same location. A **division** refers to companies and/or crews working in the same geographic area. In fact, the most frequently used ICS components in structural firefighting, outside of command and reporting resources, are groups and divisions.

Groups and divisions place several single resources under one leader designated the **group supervisor** or **division supervisor**. The assigned supervisor can directly observe and coordinate the actions of several crews (**FIGURE 22-8**). The primary reason for establishing groups and divisions is to maintain an effective span of control. At smaller incidents where sections

FIGURE 22-6 A single resource is an individual, a vehicle and its assigned personnel, or a crew or team of individuals with an identified supervisor who is assigned to perform specific tasks at an incident.

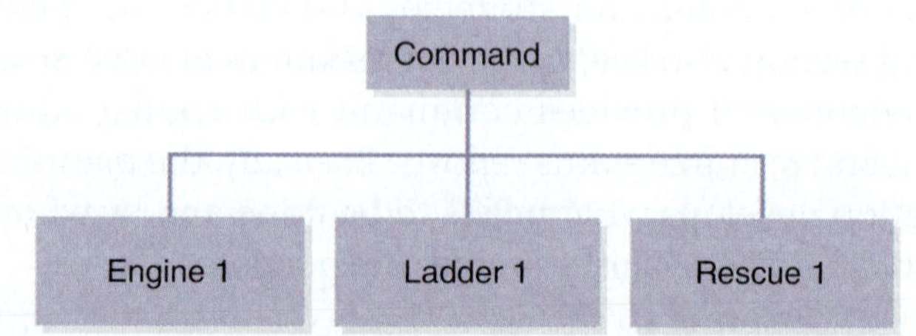

FIGURE 22-7 At a small-scale incident, the typical ICS command structure may consist of command and resources who report directly to command.

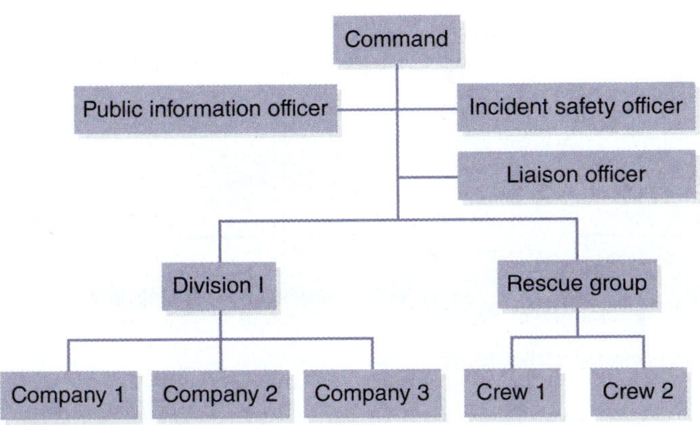

FIGURE 22-8 Divisions and groups are organized to manage the span of control.

have not been activated, group or division supervisors may report directly to the IC. At a larger incident where groups and divisions have been established as part of the operations section, the group and division supervisors report to the operations section chief.

Divisions are particularly useful when several resources are working near one another. This area could be a floor inside a building, the rear of the fire building, or a section of a wildland/ground cover fire. Divisions are most often employed during routine fire department emergency operations. By assigning all of the resources in one area to one supervisor, command is better able to coordinate their activities and ensure that all resources work together to achieve the objective noted in the IAP. For example, the division supervisor would coordinate the actions of a crew that is advancing a hose line into the area, a crew that is conducting search and rescue operations in the same area, and a crew that is performing horizontal ventilation.

Groups are useful when several resources are performing similar functions, such as ventilation, search and rescue, or water supply. Groups are responsible for performing an assignment, wherever it may be required, and often work across division lines. For example, the officer assigned to supervise several crews performing ventilation operations would use the radio designation "Ventilation Group."

Group supervisors and division supervisors have the same rank within the ICS. They are usually chief officers, but company officers also may be assigned to these positions. These supervisors are required to coordinate their actions and activities with one another. For example, a group supervisor must coordinate with the division supervisor when the group enters the division's geographic area, particularly if the group's assignment will affect the division's personnel, operations, or safety. The division supervisor, in turn, must be aware of everything that is happening within the division area and communicate with the group supervisor when the group's function occurs within the division area.

Task Forces and Strike Teams

It may be necessary at an incident to combine single resources under a supervisor using a common communications system. One way of doing this is to create a task force. A **task force** is two or more different types of single resources who work together under a **task force leader** to accomplish a specific task. For example, a task force may be composed of two engines and one truck company, two engines and two brush units, or one rescue company and four ambulances. Task forces may be assembled for a specific incident need, but they are often part of a fire department's standard dispatch philosophy. For example, a fire department in a large, urban city might dispatch a task force consisting of two engine companies, two truck companies, a rescue company, and a chief officer to all reported structure fires in a high-rise district. Some departments create task forces consisting of one engine and one brush unit for responses during wildland fire season. During that season, the brush unit responds with the engine company wherever it goes.

Another way of combining single resources under a supervisor is to create a strike team. A **strike team** is two or more single resources of the same type who work together under a **strike team leader**. For example, a strike team could be five engines (engine strike team), five trucks (truck strike team), or five ambulances (EMS strike team). Two or more ambulances and a supervisor are often combined to form an EMS strike team, to respond to multiple-casualty incidents or disasters. For example, rather than requesting 15 ambulances and establishing an organizational structure to supervise 15 single resources, command can request three EMS strike teams and can coordinate with three strike team leaders. Strike teams are also commonly used to combat wildland fires, incidents to which dozens or hundreds of companies may respond. During wildland fire season, many departments establish strike teams consisting of two or more engine companies that are dispatched and work together on major wildland fires (**FIGURE 22-9**). The assigned companies rendezvous at a designated location and then respond to the scene together. Each engine has a company officer and firefighters, but only one of the company officers is designated as the strike team leader.

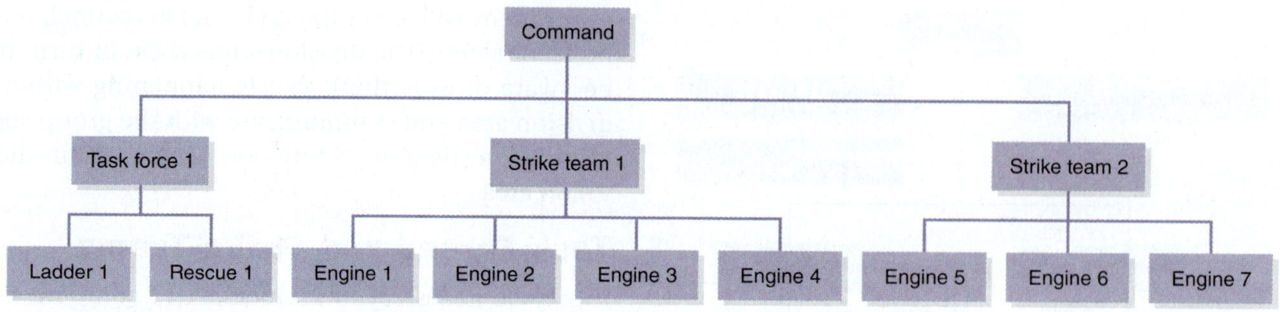

FIGURE 22-9 Example of the organization of a task force and engine strike teams.

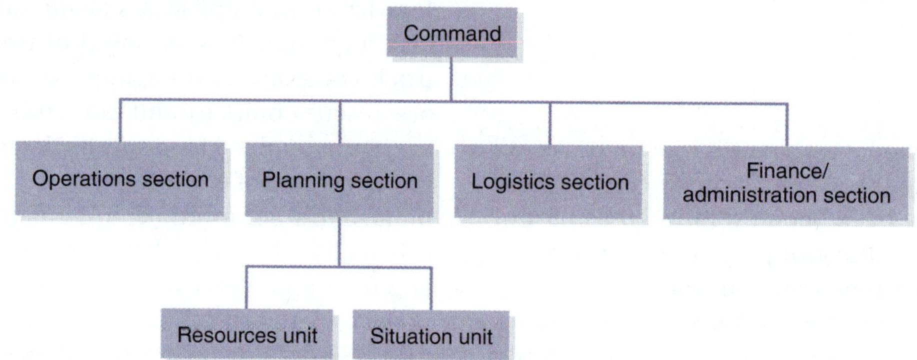

FIGURE 22-10 If activated, the resources unit and the situation unit report to the planning section chief.

Units

Within the ICS, a **unit** is a specific term that identifies two or more resources who have responsibility for a specific activity during an incident. They function only in certain areas of the incident under the supervision of a **unit leader**. The unit leader reports to a section chief at larger scale incidents where the planning, logistics, and finance/administration sections are activated (**FIGURE 22-10**).

Branches

At a major incident, several different activities may occur in separate geographic areas or involve distinct functions. Even after the establishment of divisions and groups, the span of control might still be a problem. For example, a structural collapse during a major fire may result in several trapped victims and many victims who need medical treatment and transportation. In this situation, the operations section chief would have multiple responsibilities that could exceed their span of control. To address this, one or more branches can be created. A **branch** is a higher level of combined resources than divisions and groups. Branches are created only in the operations and logistics sections. In this example, activating a fire suppression branch, a rescue branch, and an EMS branch in the operations section would address this problem (**FIGURE 22-11**). A **branch director** is responsible for each branch and reports to the operations or logistics section chief. Within each branch, several divisions or groups would report to the branch director.

Putting It All Together

All firefighters must understand the overall structure of ICS as well as the basic roles and responsibilities of each position within the ICS organization. As an emergency incident develops, a firefighter could start in logistics, move to operations, and eventually assume a command position. Understanding how ICS works enables firefighters to see how the different positions work together, allowing them to focus on their specific roles without being overwhelmed by the entire incident (**FIGURE 22-12**).

To fulfill your role as a member of the incident command team, follow the steps in **SKILL DRILL 22-1**.

Command
Public information officer
Incident safety officer
Liaison officer
Operations section
Planning section
Logistics section
Finance/ administration section
Suppression branch
Rescue branch
Medical branch
Division 1
Division 2
Group 1
Group 2
Triage strike force
Ambulance strike force

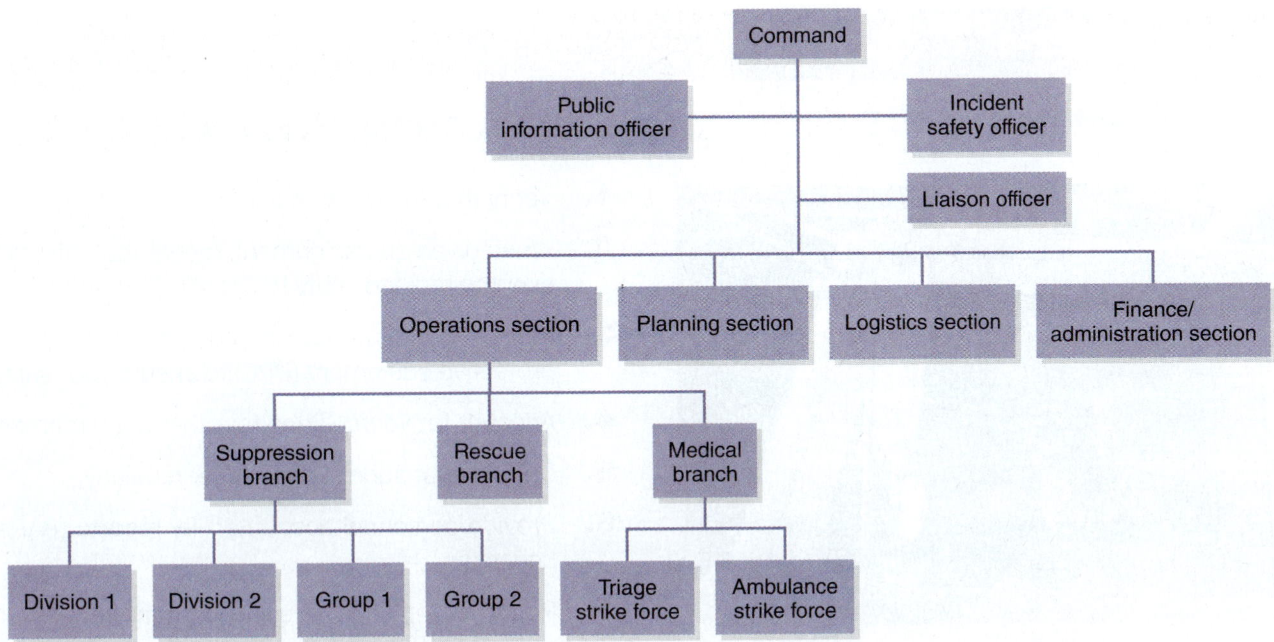

FIGURE 22-11 Creating branches within the operations section is another way to manage the span of control during a large incident.

Command
Public information officer
Incident safety officer
Liaison officer
Operations section
Planning section
Logistics section
Finance/ administration section
Suppression branch
Rescue branch
Medical branch
Documentation unit
Resources unit
Service branch
Support branch
Procurement unit
Cost unit
Division 1
Division 2
Group 1
Group 2
Ambulance strike force
Triage strike force
Crew 1
Crew 2
Supply unit
Facilities unit
Communication unit
Medical unit

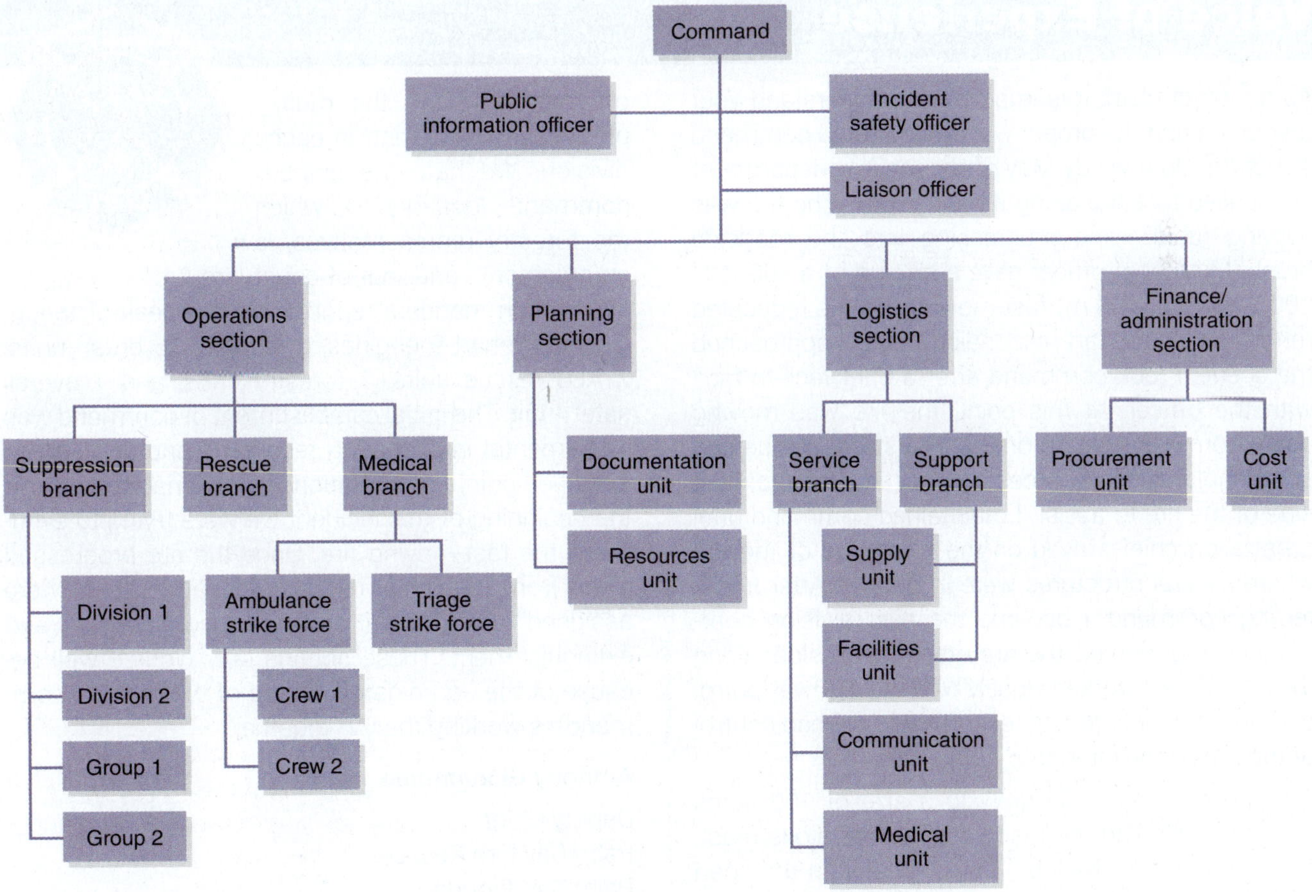

FIGURE 22-12 Example of an organization chart for the ICS at a complex incident.

Courtesy of Toby Ballard.

SKILL DRILL 22-1

Operating within the ICS Firefighter II, NFPA 1010: 7.1.1 and 7.1.2

1. Verify that the ICS is in use.
2. When given an assignment, repeat that information over the radio to verify it.
3. Assess the scene, hazards, equipment, and personal protective equipment (PPE) to ensure your safety.
4. Account for yourself and for other team members.
5. Update your supervising officer regularly.
6. Provide personnel accountability reports (PARs) as necessary.
7. Report completion of each assignment to your supervising officer.

Voice of Experience

As a district chief, it is important to understand your resources and the proper way to develop a command structure. On a windy May afternoon, my department was called to a fire along the interstate. The fire was moving rapidly and progressing into the roadside brush. My first-in officer gave a size-up of a 100 ft × 100 ft (30 m × 30 m) fast-moving fire. He requested multiple units to start responding. When I approached the scene, I took command after a brief face-to-face with the officer. At this point, the fire was moving away from our position on the interstate. I requested another alarm of resources to proceed from the west side of the fire to assist. I maintained command until a battalion chief arrived on the west side of the fire where several structures were in danger. After transferring command, I became the east division commander and worked the fire on the east side, along the interstate. I worked closely with the highway patrol and out-of-county resources to ensure we had control of the fire on the interstate.

The fire eventually grew to a size of 100 acres (40 ha). Assigning the resources into divisions (east, south, and west divisions) helped maintain the span of control among the multiple resources working in each division. We had one unified command location to which the forestry representatives, law enforcement, and out-of-county resource commanders reported. At the peak of the incident, we had 9 engines, 6 tenders, 16 brush units, 2 ALS rescue units, 3 forestry units, and 1 overall safety unit. The initial establishment of command was instrumental in allocating resources appropriately to identified points of operations in designated areas. At the beginning of this incident we were trying to extinguish the fast-moving fire. Once the fire progressed away from our initial interstate location, units were assigned for structural protection, confinement, and extinguishment. These actions only worked well because of the coordinated efforts of the division commanders working the IAP together.

Anthony Gianantonio

Deputy Chief
Palm Bay Fire Rescue
Palm Bay, Florida

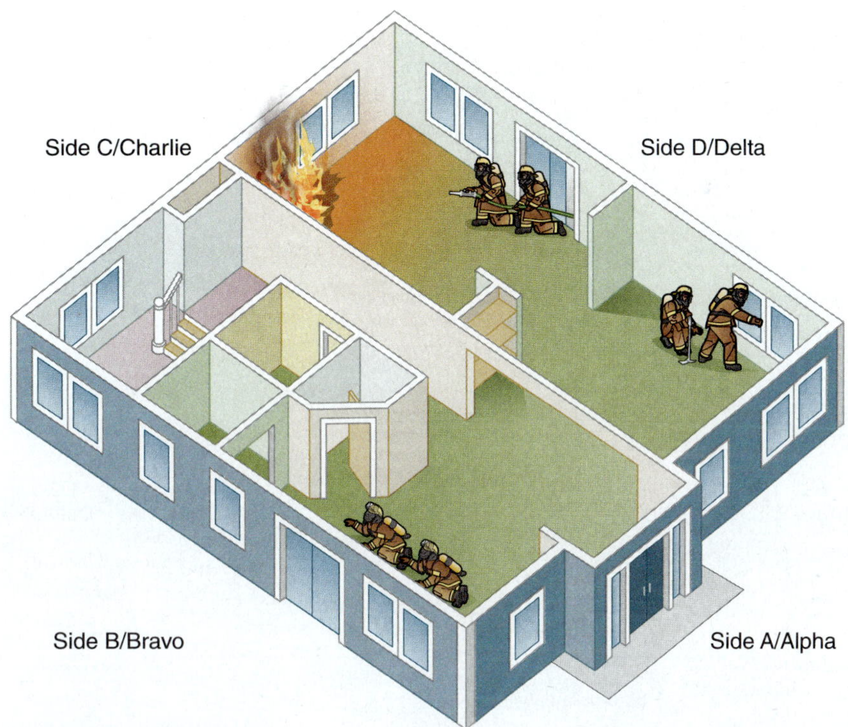

FIGURE 22-13 Standard method of quickly identifying the sides of a building at a fire scene.

Location Designators

The ICS uses a standard system to identify the different parts of a building at a fire scene. Every firefighter must be familiar with this terminology.

The exterior sides of a building are generally known as Sides A, B, C, and D. Letter designators make it easier to have a miscommunication, however, because many letters sound alike, especially in a loud environment such as an emergency scene. For example, B and D may be misheard. For this reason, some departments instead use the military alphabet—Alpha, Bravo, Charlie, and Delta—when naming the sides of a building.

The front—the side that shows the street number, that is, the address side—of the building is **Side A/Side Alpha**, with **Side B/Side Bravo**, **Side C/Side Charlie**, and **Side D/Side Delta** following in a clockwise direction around the building (**FIGURE 22-13**). If the address side is not designated as Side A/Alpha for some reason, it is crucial that that the initial arriving IC relay this information to all incoming units. If this happens, the side designated as Side A/Alpha is known as the **tactical Side A/Alpha**. Companies working on Side A/Alpha are designated "Division A/Alpha," and the radio designation for their supervisor is "Division A/Alpha." Similar terminology is used for companies working the other three sides of the building.

Any object, such as a building, structure, or vehicle, in areas adjacent to a burning building is called an **exposure**. Exposures are identified using the same letter as the adjacent side of the building. A firefighter facing Side A/Alpha can see the adjacent building on the left (Exposure B1/Bravo 1) and the building on the right (Exposure D1/Delta 1). If the burning building is part of a row of buildings, the buildings to the left are called Exposures B1/Bravo 1, B2/Bravo 2, and so on (**FIGURE 22-14**).

Firefighters working inside a building are divided into divisions based on the floor on which they are working. For example, firefighters working on the fifth floor would be in Division 5, and the radio designation for the officer assigned to that area would be Division 5. Crews performing different tasks on the fifth floor would all be part of Division 5.

Implementing the ICS

To an outsider, the ICS might appear to be a large, complicated organizational model that involves a complex set of SOPs. By contrast, to individual firefighters working within this system, ICS is simple and uncomplicated.

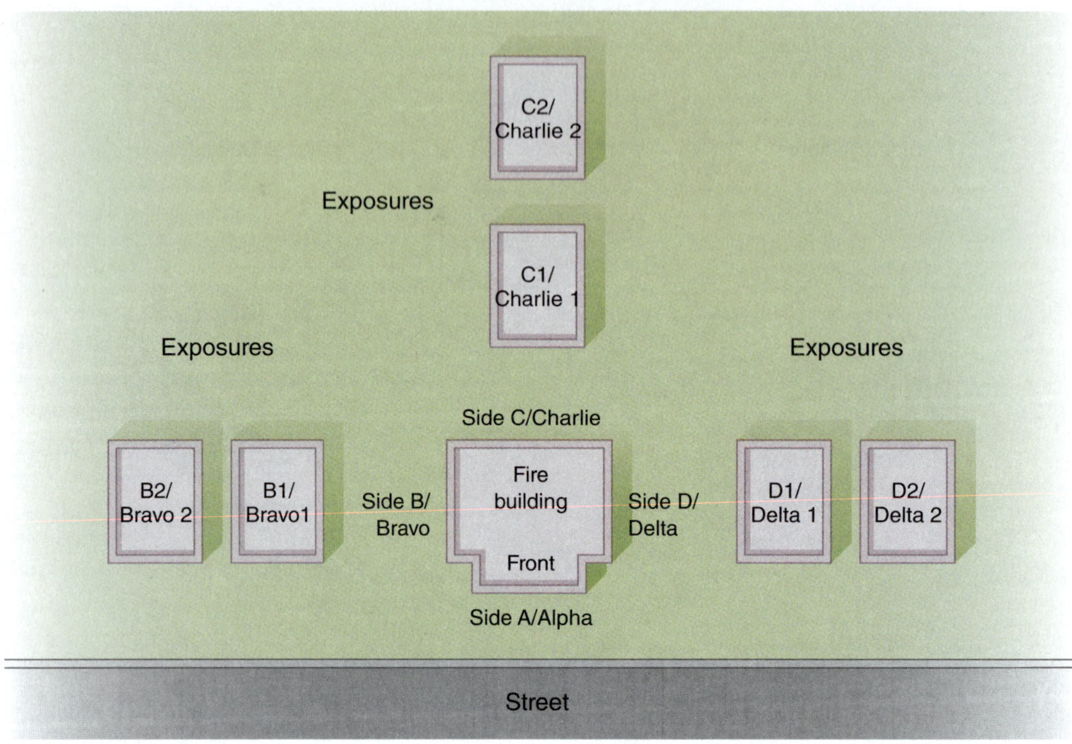

FIGURE 22-14 Standard method of quickly identifying the exposures at a fire scene.

Certain components of the ICS are used at every incident and training exercise. The following three basic principles always apply:

- Command is established at every incident and is maintained from the time that the first unit arrives until the last unit leaves. The identity of command may change, but there is always one individual in the case of single command or a group of individuals from different agencies or jurisdictions in the case of unified command who are in charge of the incident and responsible for everything that happens. SOPs may dictate who assumes command at different types of incidents.
- At any given time, each firefighter reports to only one supervisor. A firefighter's supervisor is usually a company officer. A company officer directly supervises a small crew of firefighters, such as an engine company or ladder company. If a company officer is not available, the position of company officer can be filled by an acting officer, who is a firefighter temporarily designated as an officer.
- Company officers report to command. If only one company is present at a scene, that company officer is command. This may change if additional companies and officers arrive and another officer assumes that role. At a small incident, the company officer may report directly to command. At a large incident, several layers of supervision may separate a company officer and command.

Establishing Command

The first firefighters to arrive at an emergency incident are the foundation of the ICS structure. Rarely is a high-ranking chief sent to a fire to evaluate the situation, design the organization structure, and order the resources that will be needed. Instead, the ICS builds its organization from the bottom up, around the units that take initial action.

If a chief officer is the first to arrive on scene, that officer automatically establishes command and begins executing command responsibilities. Otherwise, the company officer in charge of the first-arriving unit is responsible for establishing command, taking initial action, and exercising all of the authority that accompanies that position. This officer is the initial IC and is responsible for managing the incident until they are relieved by a senior officer. If there is no officer with the first-arriving unit, the acting officer must function as the IC until an officer arrives and assumes command. The position of command must have an unbroken line of succession from the moment the first unit arrives on the scene until the incident is terminated.

The initial IC assesses the situation, determines incident priorities, and provides direction for their own crew as well as any other units that arrive. Any units that arrive after the first unit know that they will be taking their orders from the already established command until a higher-level officer assumes command.

The IC must decide whether to take action and directly supervise the initial attack crew or to concentrate on managing the incident as a stationary IC. This decision depends on many factors, including the nature of the incident, the resources on scene or expected to arrive quickly, and the ability of the crew to work safely without direct supervision.

If the incident is large and complicated, the best option for command is to establish an ICP and focus on sizing up the situation, directing incoming units, and requesting additional resources. Command's own crew can be assigned to work with an acting officer or to join forces with another company officer and crew. If the situation is less critical, command might be able to function as both a company officer and command simultaneously, at least temporarily.

Most fire departments have written procedures that specify who will function as command in certain situations. If multiple units respond from the same station and arrive together, there should be an established protocol for which officer establishes command. If a company officer arrives only seconds ahead of a chief officer, the chief officer should establish command from the outset.

The individual who initially establishes command must formally announce that fact over the radio. This announcement eliminates any possible confusion over who is in charge. The initial announcement of command confirms that command has been established at an incident and identifies the IC. If no one announces that they are the IC, all responders should realize that command has not been established. The announcement of command also reinforces command's personal commitment to the position through a conscious personal act and a standard organizational act.

In addition to announcing who is taking command, the initial announcement should include a report describing facts from the initial size-up and the initial actions being taken, the **command designation**—that is, the name of the incident—and at larger incidents, the location of the ICP should also be announced. Individual fire department procedures vary in the way they name incidents. The first officer to assume command should establish a command designation that clearly identifies the location of the incident, such as "Engine 10 will be Seventh Avenue Command." This announcement reduces confusion on the radio and establishes a continuous identity for command, regardless of who holds that position during the incident. There can be only one "Seventh Avenue Command" at a time, so there is no confusion about who is in charge of the incident. When anyone needs to talk to command, a call to "Seventh Avenue Command" should be answered by the individual who is in command of the incident. Identifying the name of command also prevents confusion if multiple emergency incidents are operating on the same radio frequency.

The following is an example of the radio report that the first-arriving unit at a single-family dwelling fire might deliver:

> Engine 4 is on the scene at 309 Central Avenue, with smoke showing from Side A of a single-family dwelling. Engine 4 is laying a supply line and will make an offensive attack through the front door. Engine 4 is establishing Central Avenue command.

In smaller volunteer departments, the first-arriving firefighter may arrive on the scene in their personal vehicle. In this situation, the firefighter should provide the same information as described, with assignments for incoming units as they arrive. Be sure to follow your local SOPs for establishing the ICS and giving size-up information.

The initial report should include the information in **SKILL DRILL 22-2**.

Transferring Command

Transfer of command occurs when one person relinquishes the supervisory responsibility and authority of an incident to another person, usually a superior officer. For example, the first-arriving company officer would transfer command to the first-arriving chief officer, who would later transfer command to a higher-ranking officer during a major incident. The officer relinquishing command provides an assessment of the situation to the incoming officer. The transfer is transmitted over the radio to all units, indicating who has command and the effective time of transfer.

Some departments require the transfer of command when a higher-ranking officer arrives at an incident. Other departments give the higher-ranking officer the option of assuming command or leaving the existing personnel in the command assignment. When a higher-ranking officer arrives at the scene of an emergency incident and takes charge, that officer assumes the moral and legal responsibility of managing the overall operation. To eliminate communication errors

SKILL DRILL 22-2

Establishing Command Firefighter II, NFPA 1010: 7.1.1 and 7.1.2

Make the initial report over the radio to establish command and include the following information:

- The unit or individual who is establishing command
- The initial size-up report
- The initial actions being taken
- Command designation, and at larger incidents, the location of the ICP

and confusion, ideally command would be transferred no more than once. If command has been transferred a few times already and a higher-ranking officer arrives on scene, that officer may choose to not assume command and instead remain in a support role for the current IC. This helps eliminate the confusion caused by too many command transfers.

In addition to transferring command to a higher-ranking officer, command may be transferred at different points during an emergency incident for several other important reasons. A first-arriving company officer can usually direct the initial operations of two or three additional companies, but as situations become more complex, the problems of maintaining a span of control increase rapidly. A company officer's primary responsibility is to supervise one crew and ensure that those crew members operate safely. When three or more companies are operating at an incident, it is better to have a chief officer assume command, allowing company officers to focus on supervising their own companies.

As more companies are assigned to an incident, the command structure must expand. That is, it must grow to maintain an effective span of control. Additional chief officers may be assigned to the incident, and a command may be transferred to a higher-ranking officer. A transfer of command may also be required if the situation is beyond the training and experience of the current command.

FIGURE 22-15 The incoming command needs to be briefed before assuming this role.

Established procedures must be followed when command is transferred. One of the most important requirements of command transfer is the accurate and complete exchange of incident information (**FIGURE 22-15**). The officer who is relinquishing command needs to give incoming command a current situation status report that includes the following information:

- Tactical priorities
- Action plans

- Hazardous or potentially hazardous conditions
- Accomplishments
- Assessment of effectiveness of operations
- Current status of resources:
  - Assigned or working
  - Available for assignment
  - En route
  - Additional resource requirements

If transfer of command occurs early during an incident, the transfer of information may be brief. For example, the first-arriving company officer might have been in command for only a few minutes and have little information to report when command is transferred to the first-arriving chief officer. The chief may have heard all of the exchanges over the radio and know what the current situation is. Conversely, if the company officer has been command for several minutes, there may be a significant amount of information to report, such as the current assignments of all first-alarm companies. Whether the information is minimal or substantial, the transfer must be accurate and complete.

Each fire department establishes specific procedures for transferring command. In some cases, the transfer of command may be done via radio. At complex incidents, however, the transfer of command should be done face-to-face to avoid unnecessary radio traffic. This also gives both the incoming and outgoing ICs the opportunity to see facial expressions and body language when describing the incident and the situations of which they are about to take command. When done in person, the incoming IC should report to the ICP for a situational briefing and the formal transfer of command. Once command has been transferred, the identity of the new IC should be announced to all on-scene personnel.

As discussed previously, command is maintained for the entire duration of an incident. In the later stages of an incident, after the situation is under control, command might be transferred to a lower-level officer. A downward transfer of command requires the same type of briefing and exchange of information as an upward transfer of command. The officer in charge of the last company remaining on the scene would be command. When that company leaves the scene, the command function is terminated.

The steps involved in transferring command during an emergency incident are described in **SKILL DRILL 22-3**.

SKILL DRILL 22-3

Transferring Command Firefighter II, NFPA 1010: 7.1.2

1. Following departmental procedures, transfer command in a face-to-face meeting if possible; otherwise, transfer command over the radio.
2. Communicate the following to the incoming IC:
   - Tactical priorities
   - IAP
   - Hazardous conditions
   - Potentially hazardous conditions
   - Accomplishments
   - Assessment of the effectiveness of operations
   - Status of resources
   - Need for additional resources
3. Formally announce the transfer of command over the radio.

Communicating in the ICS

One of the benefits of working within the ICS is the ability to effectively communicate critical information to all personnel at the emergency scene. The IC can use the chain of command to provide initial and updated situational information, make assignments, and alert personnel of safety concerns. In turn, personnel can notify the IC of changing fire conditions, provide task progress reports, or request additional resources through that same chain of command. For example, firefighters may be assigned to perform search and rescue operations on the third floor of an apartment building. The IC gives the assignment to the company officer or to the Firefighter II leading a crew, either face-to-face or over the radio. The company officer or the crew leader repeats the assignment information to confirm that they have received and understand the assignment. The company officer or the crew leader keeps the IC informed about crew location and safety conditions, and lets the IC know if fire or building conditions change while working on the assignment. The IC may also ask the company officer or the crew leader for periodic personnel accountability report (PAR) updates, especially if the environment is hazardous.

A common way to communicate by radio is the "Hey you, it's me" method. This method helps ensure that radio traffic reaches the appropriate person and is properly understood. In the following example of the "Hey you, it's me" method, E1 stands for the crew leader of Engine 1:

IC: Engine 1 (*Hey you*), Elm Street Command (*it's me*).

E1: Go ahead Elm Street Command (*you*), this is Engine 1 (*me*).

IC: Engine 1, I want you to exit the building and give me a PAR upon exiting.

E1: Engine 1 copies. We will exit the building and report a PAR upon exiting.

In this example, the IC was clear about which unit they were trying to reach. That crew leader then acknowledged that they heard the communication. Once the IC gave their orders, the unit crew leader repeated the orders to confirm that they accurately understood the command. Using this method helps reduce miscommunications.

You should always follow your department's SOPs for communicating over the radio and all safety rules when working on an assignment. Depending on your role in the ICS, keep the IC or the appropriate supervisor informed of progress made and request additional tools, equipment, and personnel, as needed. Once the assignment has been completed, or, if you are unable to complete the assignment, notify the appropriate supervisor and wait for further instructions. The IC or supervisor may send additional resources, reassign your crew to a different section, assign your crew to rehabilitation, or tell them to go to the location where resources are placed while waiting for an assignment.

Size-Up

A formal part of establishing the ICS is **size-up**, which is the rapid evaluation and analysis of an incident to determine which resources need to be deployed and which actions can be undertaken safely. It is always the first step in making plans to bring the situation under control.

A complete size-up consists of two parts. The **initial size-up** is usually conducted by the first-arriving company officer, who serves as the IC until a higher-ranking officer arrives on the scene and assumes command. The initial size-up is also known as a **windshield size-up** because it is generally done by the engine officer before they get out of the engine. The IC uses the size-up information to develop an initial IAP and to set the stage for the actions that follow, including identifying the appropriate actions for first unit on scene and determining whether additional resources are needed. Size-up is one of the most important steps that needs to be taken at an emergency scene. It is often the key factor in determining whether an emergency operation is a success or a failure.

Once an initial size-up report has been given, the IC needs to complete the second part of a complete size-up—a **360-degree walk-around** (or simply **360**) of the building. The IC should carry a TIC and a tool while carrying this out. The 360 should be completed as soon as possible even if it requires another engine company to do it and relay the information to the initial IC.

As more complete and detailed information becomes available, the IC evaluates and updates both the initial size-up and the initial IAP. At a major incident, the size-up process often continues through several stages. During the incident, command must consider the effectiveness of the initial plan, the impact that responders are having on the problem, and any changing circumstances at the incident.

Although officers usually perform the size-up, all firefighters must understand how to gather and process information, how to formulate an operational plan, and how to use this information to change plans during the operation. If no officer is present on the first-arriving unit, a firefighter must establish command and conduct the preliminary size-up until an officer arrives. Individual firefighters are often asked to obtain information and to report their observations to command for ongoing size-up. They should routinely make observations during incidents to maintain their personal awareness of the situation and to develop their personal competence. Effective size-up requires a combination of training, experience, and good judgment.

Managing Information during Size-Up

The size-up process requires a systematic approach to managing information. Emergency incidents are often complicated and chaotic, and they can change rapidly. The IC must look at a complicated situation, identify the key factors that apply to the specific incident, and develop an IAP based on facts, observations, realistic expectations, and assumptions based on their knowledge and experience. As the operation unfolds, the IC must continually evaluate, revise, update, supplement, and reprocess the size-up information to ensure that the IAP is still valid and to identify when the plan needs to be changed. Size-up relies on two basic categories of information: facts and probabilities. The IAP is refined as more factual information is obtained and as probable information is either confirmed or revised.

Facts. Facts are pieces of accurate information that can be obtained from a variety of sources, including prior knowledge, reliable sources of information, and immediate, onsite observations. Facts are found before, during, and after an incident. Proper use of all the facts can determine the success or failure of incident outcomes.

Some of the most important preincident facts are learned during inspections and preincident surveys. Both require firefighters to visit sites with a critical eye for things that would be important to know if an incident were to occur at that location. These facts are documented in the preincident plan. Preincident plans contain many facts about the structure, including details about a building's construction, layout, contents, special hazards, and fire protection systems. Maps, manuals, and other references could provide additional information. Chapter 25, *Fire and Life Safety Initiatives*, contains information about the contents and development of preincident plans.

Facts can also be obtained using electronic devices, including mobile computer data terminals, laptop computers, tablets, and smartphones. The IC can call up previously entered facts or access the Internet. Individuals with specific training, such as a building engineer or a utility representative, can add even more specific information. Finally, a seasoned officer will have built up a knowledge bank of valuable information based on experience, training, and direct observations.

Another important source of preincident facts is the initial dispatch information, which contains facts such as the location and nature of the situation. This information could be very general ("a reported building fire in the vicinity of Central Avenue and Main Street") or very specific ("a smoke alarm activation in room 3102 of the First National Bank Building at 173 South Main Street").

During an incident, some of the most critical facts for successful operations come from the initial size-up and the follow-up 360 reports given by the initial IC. The facts in these two reports provide responders with the information needed to effectively and safely do their jobs on scene. The initial size-up report should paint an accurate picture of the incident to all incoming units. The information in these reports varies, but for structure fires, they almost always include the building size and occupancy type, fire conditions and location, initial actions of the first due company, water supply, whether more resources are needed, and who is in command. The 360 follow-up report should add information the IC finds when conducting the 360, including the following:

- Whether there are possible victims
- Location or extent of the fire
- Location of entrances or egresses from the building
- Locations of utility shut-offs
- Whether there is a basement
- Whether there are any fire suppression systems
- Location of threatened exposures
- Any other potential hazards

Once this information is gathered, the 360 follow-up report should be given over the radio by the officer. If the 360 is not completed due to hazards, obstructions, or the size of the building, this information should also be relayed to incoming units. It is possible that the information gathered on the 360 might indicate a need to alter tactics or the number of resources needed on scene. If changes need to be made, that should also be communicated in the report.

An experienced fire officer will have a basic knowledge of the community. The officer may remember the types of occupancies, construction characteristics of typical buildings, and specific information about particular buildings from previous incidents or preincident planning visits. Basic facts about a building's size, layout, construction, and occupancy can often be observed upon arrival, if they are not known in advance. In particular, the officer must consider the size, height, and construction of the building during the size-up. The IAP for a single-story, wood-frame dwelling on an acre (4047 square meters) of land will be quite different from that for a steel-frame, high-rise tower on a crowded urban street. Knowledge of building construction is vital when performing an initial size-up. Review Chapter 6, *Building Construction*, for construction details that will help you perform an accurate size-up.

Smoke reading is a vital part of scene size-up because this process helps the IC predict where the fire might go. It is described further in Chapter 5, *Fire Behavior*. The IC should observe where smoke is visible, how much is apparent, what color it is, and how it moves. In addition, the IC should note the weather conditions, especially if it is windy. This information should be one of the facts collected.

After the initial size-up and 360 and after command writes the initial IAP, command still needs to continue to gather as much factual information as possible about the incident. Because command often is located at a fixed ICP outside the building, company officers need to report their observations from their different locations. For example, a company operating a hose line inside the building reports on interior conditions, and the ventilation crew on the roof, who would have a very different perspective, reports on conditions there. Regular progress reports from companies working in different areas will provide updated information about the situation. Command may request that an officer or a firefighter gather specific information that will help them develop or modify the IAP. This information and progress reports enable command to judge whether an operational plan is effective or needs to be changed.

After the fire has been extinguished, more facts will be gathered by fire investigators. To find the cause of the fire, investigators will use the facts gathered prior to, during, and after the incident. All of these facts will be used to present the most accurate understanding possible of all the stages of the fire.

Probabilities. Probabilities are events and outcomes that can be reasonably assumed, predicted, or anticipated based on facts, observations, common sense, and previous experiences. Firefighters frequently use probabilities to anticipate or predict what is likely to happen in various situations. The IAP is also based on probabilities, predicting where the fire is likely to spread and anticipating potential problems. It is often necessary to rely on probabilities because firefighters rarely have all the facts necessary to create an IAP based on facts alone.

An IC must be able to quickly identify the probabilities that apply to a given situation. For example, a fire in an apartment building in the middle of the night will probably involve occupants who need to be rescued. In such circumstances, the IC would assign additional crews to search for potential victims, even if no factual information indicates that any occupants need to be rescued. Similarly, a fire burning on the top floor of a structure in a row of attached buildings has a high probability of spreading to the adjoining buildings through the cockloft. (In this case, the IAP should include opening the roof above the fire and sending crews into the exposed occupancies to check for fire extension in the cockloft.)

Effective size-up requires a good knowledge of fire behavior to evaluate the probabilities for the spread of a fire. The concepts of convection, conduction, and radiation enable an IC to predict how a fire will extend in a particular situation. By observing the combination of smoke and fire conditions in a particular type of building, the IC can identify a range of possibilities for fire extension within the structure or to other exposed buildings. Using these probabilities, the IC can predict what is likely to happen and include tactics to control the probable situation effectively.

The IC must also evaluate the potential for collapse of a burning structure. The building's construction, the location and intensity of the fire, and the length of time the structure has been burning are all factors that must be considered when making this determination. Houses built after 1990 using lightweight construction techniques, for example, tend to collapse in a much shorter period of time than houses built in earlier eras. When the possibility of collapse exists, the IC should not permit firefighters to enter the building, or if firefighters are already in the building, the IC should immediately order all firefighters to exit the building.

Resources

Resources include all of the means that are available to fight a fire or conduct emergency operations at any other type of emergency incident. A fire department's basic resources are its personnel and apparatus. Resource requirements depend on the size and type of incident. Resource availability depends on the capacity of a fire department to deliver firefighters, fire apparatus, equipment, water, and other items that can be used at the scene of an incident. Firefighting resources are usually defined as the number of engine companies, ladder companies, special units, and command officers required to control a particular fire.

An IC should be able to request the required number of companies and know that each unit will arrive with the appropriate equipment and the necessary firefighters to perform a standard set of functions at the emergency scene. The IC should also know how many and which types of companies are available to respond, how they are staffed and equipped, and how long they will take to arrive. This information might have to be updated at the incident. This is especially important in areas served by volunteer firefighters because the number of firefighters available to respond may vary at different times.

Water supply is a critical resource. It is rarely a problem if the fire occurs in a district that has fire hydrants and a strong, reliable municipal water system with adequately sized water mains. In an area without fire hydrants, however, water supply may be severely limited. Even when a static water supply is available, it takes time to establish a water supply from such a source. If the water supply is limited to mobile water supply apparatus, the amount of water that can be delivered is limited. In this situation, the IC would need to call for an additional mobile water supply apparatus to be dispatched to the scene.

Resources include more than firefighters, fire apparatus, and a water supply. A fire in a storage facility where flammable liquids are stored will require large quantities of foam and the equipment to apply it effectively. A hazardous materials incident might require special monitoring equipment, chemical-protective clothing, and bulk supplies to neutralize or absorb a spilled substance. A building collapse might require heavy equipment to move debris and a structural engineer to determine where firefighters can work safely. The fire department must have these supplies available or be able to obtain them quickly when they are needed. At large-scale incidents, needed resources include food and fluids for firefighters, fuel for the apparatus, and other supplies. Law enforcement agencies and organizations such as the Red Cross and the Salvation Army often provide these support resources.

During size-up, the IC determines which resources are needed to control the specific situation and whether those resources will be available when needed. An IAP will be effective only if the necessary resources can be assembled quickly. Ideally, a fire department will be able to dispatch enough firefighters and apparatus to control any situation within its jurisdiction. In reality, however, many fire departments do not have an adequate number of apparatus or personnel on duty to meet the emergency response needs of their communities. Most departments have established mutual aid agreements to assist surrounding jurisdictions if a situation requires more resources than the individual community can provide. In some areas, hundreds of firefighters and apparatus can be assembled to respond to a large-scale incident. In more remote areas, the resources available to fight a fire can be very limited.

If resources are insufficient or delayed, a fire can become too large to be controlled by available resources. If a delay occurs, the IC must anticipate how much the fire will grow and where it will spread. If the needed resources cannot be obtained, the IC must develop a realistic plan using the available resources to gain control of the situation. For example, if only one or two apparatus and a few firefighters respond to a structure fire, the IC will need to determine whether they will offensively attack the fire, defensively flow available water on the fire, or protect the exposures while the original building burns.

Resources must be organized to support efficient emergency operations. Firefighters must be properly equipped and organized in companies. Equipment and procedures must be standardized. The ICS must be used to manage all of the resources that could be used at a large-scale incident, and the communications system must enable the IC to coordinate operations effectively.

Incident Action Plan

Command is responsible for developing the strategic incident objectives on which the IAP will be based. Remember that IAPs describe strategic objectives and the tactics for achieving those objectives at an incident. The **strategy** is the "big picture" plan of what has to be done. "Stop the fire from extending into Exposure D" is a strategic objective. **Tactics** are the general steps that are taken to achieve the strategic objectives, such as "Make a trench cut in the roof and get ahead of the fire with hose lines on the top floor." A **task** is the specific assignment that will achieve the tactical objectives. "Ladder 1 will go to the roof to make the trench cut; Engines 3 and 5 will take the hose lines to the third floor; and Ladder 4 will pull ceilings to provide access into the cockloft" are all tasks. Command develops the IAP and orders and releases incident resources.

Based on information gathered during size-up, command develops an IAP based on the current conditions, priorities, and resources to outline the steps needed for beginning operations to control the situation. The plan is revised and expanded as necessary. Ultimately, the responsibility of the IC to direct incoming resources in a manner that will best mitigate the emergency.

Incident Priorities

Every incident has three strategic priorities—life safety, incident stabilization, and property conservation, in that order—and the IAP should be based on these priorities (**FIGURE 22-16**).

The first priority is life safety. This includes keeping firefighters and other emergency responders safe, as well as rescuing victims.

The second priority is to stabilize the incident. The incident must be stabilized before the fire can be fully extinguished or any hazardous material can be cleaned up. Incident stabilization may include protecting exposures from impinging fire and containing hazardous fuel spills.

The lowest priority is conserving property. Once life safety is ensured and the incident is not expanding or becoming more threatening, actions can be taken to prevent further property damage. These actions include extinguishing the fire, covering furnishings with salvage tarps, and covering ventilation holes in the roof to prevent further water damage.

This system of priorities clearly establishes that the highest priority in any emergency situation is saving lives. All initial emergency scene operations must focus on this priority. If there are no immediate life safety issues, the other two incident priorities can be addressed.

These priorities are not separate and exclusive, of course. Often, more than one incident priority can be addressed simultaneously, and certain activities help achieve more than one objective. For example, attacking a fire may simultaneously reduce the risk to trapped occupants, limit fire and smoke spread, and prevent further fire damage to the structure. If a direct attack on the fire brings it under control very quickly, the incident priorities of life safety, incident stabilization, and property conservation can all be addressed simultaneously.

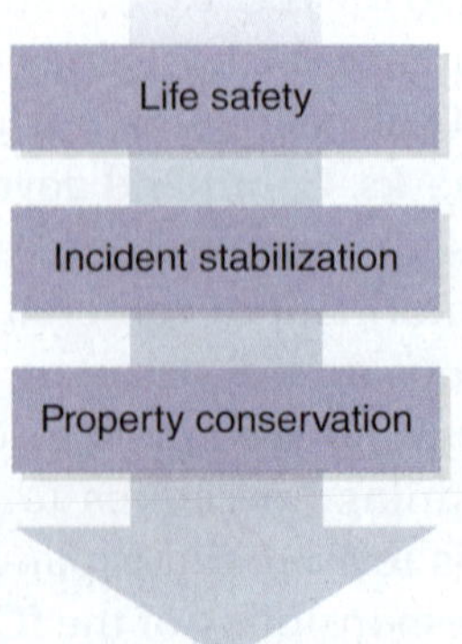

FIGURE 22-16 Strategic priorities at an incident.

Courtesy of Toby Ballard.

RECEO–VS

To help fire departments and command achieve the three incident priorities, Chief Lloyd Layman (Parkersburg, WV) created the acronym RECEO in the 1950s. RECEO stands for **R**escue, **E**xposure protection, **C**onfinement of the fire, **E**xtinguishment, and **O**verhaul. These strategies provide a general guideline for fireground commanders to systematically address the three incident priorities. Through the years, RECEO was modified by different departments to better meet incident strategies. A common version used in recent years is **RECEO–VS**, which adds Ventilation and Salvage to the original acronym under the property conservation umbrella (**TABLE 22-2**).

Although this acronym provides a suggested order of strategic objectives, the IC needs to determine if and how each of these objectives fits into the IAP. These objectives serve as a guide to help the IC make difficult decisions, particularly if not enough resources are available to address every objective. If a decision must be made between saving lives and saving property, saving lives comes first. After rescue is completed and if the fire is still spreading, the IC should then place exposure protection ahead of salvage and overhaul. A similar set of priorities can be established for any emergency situation. Saving lives is always more important than protecting property.

Rescue. Life safety is the first incident priority at a fire or any other emergency incident. The need for rescue depends on many circumstances. For example, the number of people in danger is likely to vary based on the type of occupancy and the time of day. A commercial building that is crowded during the workday might have few, if any, occupants at night. At night, rescue is more likely to be needed in a residential occupancy, such as a house, an apartment building, or a hotel.

The degree of risk to the lives of the building's occupants must also be evaluated. A fire that involves one apartment on the 10th floor of a high-rise apartment building could threaten the lives of both the residents on that floor and the residents who live directly above the fire. In contrast, residents below the fire are probably not in significant danger, and residents several floors above the fire might be safer staying in their apartments until the fire is extinguished instead of walking down smoke-filled stairways.

If the size-up suggests that people may be trapped in a burning building, the IC must then determine whether it is safe to send firefighters into the building. The IC should conduct a type of risk–benefit analysis called survivability profiling to make this decision. A **risk–benefit analysis** is an assessment that weighs

TABLE 22-2 RECEO–VS

Incident Priorities	RECEO–VS Acronym	Meaning
Life safety	**R**escue	The removal of victims from a life-threatening situation
Incident stabilization	**E**xposure protection	The protection of surrounding structures from a fire
	Confinement	The containment of a fire to one area or the prevention of further areas from becoming involved in the fire
Property conservation	**E**xtinguishment	The complete extinguishment of the fire
	Overhaul	The process of ensuring that the fire is extinguished completely
	Ventilation	The process of removing smoke, heat, and the products of combustion from a structure
	Salvage	The removal or protection of property that could be damaged during firefighting or overhaul operations

the risks to be taken versus the benefits of those risks. **Survivability profiling** is a risk–benefit analysis that weighs the viability and survivability of potential victims based on the current conditions in the structure. Victim survivability declines as room temperatures approach 212°F (100°C). Death occurs at temperatures of 350°F (175°C) in approximately 3 minutes. If an occupant breathes superheated air and gases, burns to the respiratory system may result in death within minutes.

In today's fires, interior temperatures can rise higher than 500°F (160°C) in just 3 to 4 minutes. Flashover produces temperatures over 1000°F (540°C) and often occurs before the first firefighters arrive on scene. Remember that firefighters, even with full PPE, cannot survive for more than a few seconds in a flashover. If an environment is not tenable for a firefighter in full PPE, it is very unlikely that trapped victims are alive. It is not acceptable to risk the lives of firefighters to try to save the lives of victims who have already died or will not survive.

Recent research indicates that due to modern construction methods and materials, victim survivability decreases dramatically with each passing minute after ignition. Survivability profiling helps the IC to decide whether to commit firefighters to interior search and rescue and fire suppression operations to save live victims. Firefighters should engage in interior operations only if there is a reasonable probability that trapped civilians might be alive and if interior conditions do not place the firefighters at an unacceptable risk (FirefighterNation 2011).

SAFETY TIP

It is not acceptable to risk the lives of firefighters to try to save the lives of victims who are already dead.

Often, the best way to protect lives is to extinguish the fire quickly. For this reason, efforts to control the fire are usually initiated at the same time as rescue operations. For example, hose lines may be used to protect exit paths and keep the fire away from victims during search and rescue operations. Chapter 17, *Search and Rescue*, discusses specific tactics and techniques for these operations.

Exposure Protection. The second incident priority is to stabilize the incident. Usually, this means mitigating the incident using an outside-to-inside approach and protecting exposures from impinging fire. For structural fire incidents, this is accomplished by controlling the fire so it does not extend beyond the room, area, or building of origin. Many historic conflagrations developed because the original fire spread quickly from the building of origin, ultimately involving dozens—or hundreds—of buildings. To prevent this, the IC must start by making sure the fire is not expanding. In some cases, the IC must look ahead of the fire and identify a place to stop its spread. For example, if flames are extending quickly through the cocklofts in a row of attached buildings, the IC might place hose lines ahead of the fire to stop its progress.

The IC must sometimes weigh potential losses when deciding where to attack a fire. If a fire in a vacant building threatens to spread to an adjacent occupied building, the IC will usually assign companies to protect the exposure before attacking the main body of fire. If, however, a fire in an occupied building might spread to a vacant building, the IC's decision might be to attack the fire first and worry about controlling the spread later.

Confinement. After ensuring that the fire is not actively extending into any exposed areas, the third priority is to continue to stabilize the incident and focus on confining the fire to a specific area. For example, if the fire is burning in only one room, the objective should be to ensure that it does not spread beyond that room. But if more than one room is involved, the objective might be to contain the fire to the single involved apartment or floor level. Sometimes the objective might be to confine the fire to one building, particularly if multiple buildings are attached or in close proximity to the building of origin. To accomplish this, the IC will define a safety perimeter so that the fire does not expand beyond that area. Firefighters on the perimeter must be alert to any indications that the fire is spreading to those limits.

Thermal imaging devices can detect hidden sources of heat, and they are valuable tools for finding fires in void spaces. The principles and use of thermal imaging devices are covered more fully in Chapter 17, *Search and Rescue.*

Extinguishment. The third incident priority is property conservation. To achieve this, various actions including extinguishing the fire, ventilation, and salvage and overhaul operations are undertaken.

Depending on the size of the fire and the risk involved, the IC will mount either an offensive or a defensive operation to extinguish a fire. If the fire is not too large, the attacking firefighters can mount an offensive operation and apply enough extinguishing agent to extinguish the fire quickly and efficiently. Extinguishing the fire using an offensive operation not only preserves property—the third incident priority—it often also satisfies the other two priorities—life safety and incident stabilization—at the same time.

If the fire is too large or too dangerous to extinguish with an offensive operation, and if the IC has determined that there is no potential for saving lives or property, the IC will implement a defensive operation. Sometimes a defensive strategy is effective in extinguishing the fire. Other times, a defensive strategy simply keeps the fire from spreading to exposed properties (**FIGURE 22-17**). For more information about fire suppression operations, see Chapter 14, *Fire Suppression,* and Chapter 23, *Advanced Fire Suppression.*

FIGURE 22-17 A defensive strategy extinguishes the fire and can keep the fire from spreading to exposures.

In some situations, all firefighters may be withdrawn from the area and the fire allowed to burn itself out. These situations generally involve potentially explosive or hazardous materials that represent an extreme danger to firefighters. Remember that preserving property is the last priority and should not be a goal if it puts the lives of firefighters at risk.

Each extinguishment strategy poses some risk to firefighters, so it is important that the IC consider the risks and benefits of each strategy. Sending firefighters into an unoccupied building poses a risk to firefighters with the only potential benefit being to save property. Sending firefighters into an occupied building to save lives has great benefit but only if there is a chance to save the occupants and the operation does not pose an unacceptable risk to the firefighters. Each IC must evaluate the risk versus the benefit of each action to be taken at an emergency scene.

Ventilation. Ventilation contributes to the third incident priority—property conservation. Ventilation can be a stand-alone strategy to control the spread of heat and smoke in the structure, or it may be used in combination with other tactics to support the safe removal of occupants or assist with an interior, offensive fire attack. Ventilation operations are covered in depth in Chapter 16, *Ventilation.*

Salvage and Overhaul. Salvage operations are conducted to protect property by preventing avoidable property damage and losses (**FIGURE 22-18**). Salvage operations are often aimed at reducing smoke and water damage to the structure and contents once the fire is under control.

The overhaul process is conducted after a fire is under control, with the goal being to completely extinguish

FIGURE 22-18 Salvage operations contribute to the third incident priority, property conservation, by preventing avoidable property losses.

© Jones & Bartlett Learning. Photographed by Glen E. Ellman.

FIGURE 22-19 Overhaul is conducted after a fire is under control; its goal is to completely extinguish any remaining pockets of fire.

© Glen E. Ellman

any remaining fire (**FIGURE 22-19**). The IC is responsible for ensuring that the fire is completely extinguished before terminating operations. These operations are covered in depth in Chapter 18, *Salvage and Overhaul.*

S.L.I.C.E.–R.S.

RECEO–VS provides guidance for the ongoing operations and command priorities at the scene of a structure fire. The mnemonic **S.L.I.C.E.–R.S.** provides guidance for initial engine company operations upon arrival at a structure fire scene. S.L.I.C.E.–R.S. was developed by the International Society of Fire Service Instructors (ISFSI) and incorporates the Underwriters Laboratories/National Institute of Standards and Technology (UL/NIST) recommendations for initial engine company operations upon arrival at a structure fire scene. S.L.I.C.E.–R.S. provides the first-arriving officer with a short list of objectives prior to the arrival of additional resources:

- **S**ize-up
- **L**ocate the fire
- **I**dentify and control the flow path
- **C**ool from a safe location
- **E**xtinguish the fire
- **R**escue
- **S**alvage

Note that the last two actions—rescue and salvage—are actions of opportunity. An **action of opportunity** is an action that is taken when needed at any point during the incident. The first five objectives—S.L.I.C.E.—are to be done in sequence. After conducting the size-up, the officer attempts to locate the fire and control the flow path of air to the fire to minimize fire and smoke spread. The transitional attack of cooling the fire from the exterior with a short burst of water limits the growth of the fire and immediately cools the interior. This stabilizes the incident and makes interior conditions more tenable for any victims who might still be alive. Although it is not always possible to cool the fire from an exterior window or door, it should be attempted if it can be accomplished quickly before making entry into the structure. A few seconds of water from a straight stream can dramatically reduce the intensity of the fire. Finally, if conditions justify an interior fire attack, the hose should be advanced into the structure to fully extinguish the fire and begin overhaul operations.

Each of these objectives is directly linked to the three incident priorities (life safety, incident stabilization, and property conservation) and should be attempted if conditions allow. During the first few minutes on the fireground, firefighters should look for opportunities to assist trapped occupants or salvage undamaged property. As with other fireground, mnemonics, S.L.I.C.E.–R.S. is only as valuable as the firefighters who thoughtfully implement it.

Characteristics of the ICS

Effective management of incidents requires an organizational structure to provide both a hierarchy of authority and responsibility, and formal channels for communications. Under the ICS, the specific responsibilities and authority of everyone in the organization are clearly stated, and all relationships are well defined. Important characteristics of an ICS include the following:

- Recognized jurisdictional authority and responsibility
- Applicable to all risk and hazard situations
- Applicable to both day-to-day operations and major incidents
- Unity of command
- Span of control
- Modular organization
- Common terminology
- Integrated communications
- Incident action plans
- Designated incident facilities
- Resource management

Jurisdictional Authority

An IC must be identified at every emergency incident, even if a single engine company or EMS crew responds to and mitigates the emergency. In an effective ICS, everyone working at the incident knows who the IC is and that they are in charge of the incident and responsible for all of the operations required to mitigate the emergency situation. Usually, the role of IC is filled by the ranking fire department, law enforcement, or EMS officer on the scene, depending on the type of emergency. Local protocols determine which agency has jurisdictional authority for various types of incidents. For example, the ranking law enforcement officer may serve as the IC during a localized civil disturbance, while a fire department company officer may fill that role during a routine structural fire incident.

As discussed earlier, the command function is structured as either a single IC or a unified command among multiple agencies. When an incident occurs within a single jurisdiction, and when there is no jurisdictional or functional agency overlap, a single IC should be identified and designated as having overall incident management responsibility by the appropriate jurisdictional authority.

Identifying the appropriate IC becomes more complicated, however, when several jurisdictions are involved or when multiple agencies within a single jurisdiction have authority for various aspects of the incident. As incidents become increasingly complex, it is unlikely that any one individual has the knowledge and expertise required to effectively develop, implement, and evaluate strategic objectives for a major incident. Such a situation involves both military and civilian agencies as well as multiple levels of government. Chaos would ensue if each affected jurisdiction claimed to be in charge of the incident and attempted to issue orders to the other jurisdictions. When responsibilities of responders overlap, a unified command may be used instead of a single IC. The function of command, whether conducted by an individual IC or through a team of individuals functioning as the unified command, is a management and leadership position. This role is responsible for setting strategic objectives, maintaining a comprehensive understanding of the impact of an incident, and identifying the strategies required to manage that impact effectively.

Unified command provides a framework that allows agencies with different legal, geographic, and functional responsibilities to coordinate, plan, and interact effectively. Unified command, through a consensus-based approach, addresses the challenges that were encountered when multiple organizations responded to the same incident and established separate, but concurrent, incident management structures. This individual approach at incidents that require multiple agencies to respond—which is still routine in many jurisdictions—leads to inefficiency and duplication of effort, resulting in ineffective incident management, on-scene conflict, and substantial safety issues. Unified command is crafted specially to address these issues.

The introduction of the NIMS ICS sparked tremendous discussion and debate related to the broader emphasis given to the concept of unified command. A multiagency or multijurisdictional response to an incident is a routine occurrence. Unified command is a critical evolution of the ICS system that recognizes this important reality of incident management. To be effective in implementing unified command, agencies and individuals need to worry less about the structural change and more about the changes required in organizational culture and interagency relationships.

The concept of unified command represents a clear departure from the traditional view of a single IC. Unfortunately, it is frequently misunderstood and difficult for some organizations to implement. When unified command is established, you should be less concerned with who is in charge and more concerned with what is required to safely and effectively manage the incident. For unified command to be effective, the various agencies at the incident must collaborate to establish a single set of integrated incident objectives and designate priorities to achieve those objectives. This approach produces a single IAP that clearly defines the various agency and jurisdictional responsibilities for incident management.

The unified command model of managing an incident using a single organizational structure limits duplication and frees resources for overall incident management duties. In addition, unified command ensures that the incident will be managed from a single ICP. Finally, unified command allows for a single planning section, which helps prevent multiple, redundant requests for similar items and the inefficient use of limited supplies and resources. If a unified command structure is to be effectively implemented and used, all responding agencies must understand and support the purpose of functioning under this type of ICS.

All-Risk, All-Hazard System

The ICS has evolved into an all-risk, all-hazard system. Basically, it can be applied to manage resources at all types of incidents, including fires, floods, tornadoes, plane crashes, earthquakes, hazardous materials incidents, mass-casualty incidents, and any other type of emergency situation (**FIGURE 22-20**). Systems similar the ICS have also been used to manage many non-emergency events, such as large-scale public events, that have similar requirements for command, control, and communication. The flexibility of the ICS enables the management structure to expand as needed, using whatever components are necessary. The operations of multiple agencies and organizations can be integrated smoothly in the management of the incident.

FIGURE 22-20 ICS can be used to manage different types of emergency incidents involving several agencies.

© Mark C. Ide

Everyday Applicability

An ICS can and should be used for everyday operations, as well as for major incidents. Command should be established at every incident, whether it is a trash bin fire, a motor vehicle collision, or a building fire (**FIGURE 22-21**). Regular use of the system ensures familiarity with standard procedures and terminology and increases the users' confidence in the system. Frequent use of ICS for routine situations makes it easier to apply to larger incidents.

Unity of Command

Unity of command is the concept that each person reports to only one direct supervisor, each supervisor answers to only one boss, and so on. All orders and assignments come directly from that supervisor, and all reports are made to the same supervisor. This approach eliminates the confusion that results when a person receives orders from multiple supervisors. Unity of command reduces delays in solving problems and reduces the potential for life and property losses. By ensuring that each person has only one supervisor, unity of command increases overall accountability, prevents freelancing, improves the flow of communication up and down the chain of command, assists with the coordination of operational issues, and enhances the safety of the entire situation.

Span of Control

As described earlier, span of control is the maximum number of personnel or resources that one person can supervise or manage effectively. The recommended number is three to seven people or resources. This

FIGURE 22-21 Command should be established at every incident, no matter how the size.

© Rick McClure/AP Photo

applies to all levels of the ICS—from the strategic level, to the operational/tactical level, to the task level—that is, individual companies or crews. Because of the dynamic nature and rapid pace of emergency incidents, an individual who has command or supervisory responsibilities in an ICS normally should not directly supervise more than five people. The actual span of control should depend on the complexity of the incident and the nature of the work being performed. For example, at a complex incident involving hazardous materials, the span of control might be only three; during less intense operations, the span of control could be as high as seven.

Modular Organization

The ICS has often been characterized as an organizational toolbox, where only the tools needed for the specific incident are used. These tools consist of position titles, job descriptions, and an organizational structure that defines the relationships between positions. Some positions and functions are used frequently. Other positions and functions are needed only for complex or unusual situations. Any position can be activated simply by assigning someone to it. For example, a small structure fire can usually be managed by an IC directly supervising four or five company officers, each of whom supervises their respective company firefighters. At a larger fire, when more companies respond, the ICS structure would expand to include groups or divisions, each under the direction of a supervisor. At the same time, company officers would be assigned to perform specific ICS functions, such as ISO or planning section chief. At even more complex incidents, the additional levels and positions within the ICS structure would be filled in the same manner.

An ICS is not a rank-oriented system. The best qualified person should be assigned at the appropriate level for each situation, even if that means a lower-ranking individual is temporarily assigned to a higher-level position in the command structure. For complex incidents, however, personnel in command and section chief positions should have experience and special training.

Common Terminology

The ICS promotes the use of common terminology both within an organization and among all agencies involved in emergency incidents. This means that each word has a single definition, and no two words used in managing an emergency incident have the same definition. Everyone uses the same terms to communicate the same concepts, so everyone has the same understanding of what is meant. Each job comes with one set of defined responsibilities, and everyone knows who is responsible for each duty. Common terminology is particularly important for radio communications.

Even though the ICS defines common terminology, there may be some variations across the country. For example, in many fire departments, the term *tanker* refers to a mobile water supply apparatus. But the NIMS and many western U.S. states that experience wildland fire incidents use the term *tanker* to mean an aircraft that drops an extinguishing agent on a fire. Therefore, they usually refer to the mobile water supply apparatus as a *tender*. Without common terminology and understanding the variations in your department or area, a request for a tanker could result in the operations section chief getting an aircraft instead of the mobile water supply apparatus they actually need. Another example is the use of by some fire departments of the "ten codes" commonly used by law enforcement (for example, 10-4 means acknowledgment or OK). Be aware that the ten codes are not always used consistently in all jurisdictions and many agencies do not use ten codes at all. If you are working on an incident with multiple departments or agencies it is best to use plain text. It is important to understand and use the terminology used by departments in your area.

TIP

FEMA's *Resource Typing Library Tool* lists the proper terminology for the common resources used at large-scale emergencies. This tool is available through the FEMA website at https://rtlt.preptoolkit.fema.gov/Public.

Integrated Communications

Under the NIMS model, all ICS participants must be able to communicate within the command structure. Using an integrated communications system ensures that everyone operating at an emergency can communicate with both supervisors and subordinates. The ICS at an incident must support communication up and down the chain of command at every level. This ensures that a message will move efficiently through the system, from command down to the lowest level and from the lowest level up to command. Within a fire department, the primary means of communicating at the incident scene is by radio (**FIGURE 22-22**).

FIGURE 22-22 Integrated communications is essential for successful operations during an emergency.

TIP

If a mayday call is transmitted on the radio, all other radio traffic should stop immediately. In some agencies, the person who transmits the mayday provides a LUNAR report: their **L**ocation in the building/at the incident, the **U**nit to which they are assigned, their **N**ame, the amount of **A**ir they have in their cylinder or the **A**ssignment to which they were last assigned, and the **R**esources needed to get them out of the mayday situation. In other agencies, the person reporting the mayday reports who (they are), what (the situation is), and where (they are located) instead of providing a LUNAR report. The steps for calling a mayday are explained more fully in Chapter 19, *Firefighter Survival*.

Incident Action Plans

An ICS ensures that everyone involved in the incident is following one overall IAP. Different components of the organization may perform different functions, but all of their efforts contribute to meeting the same goals and objectives. At smaller incidents, the IC develops the IAP and verbally communicates the incident priorities, objectives, strategies, and tactics to all of the operating units. At larger, complex incidents, a formal, written IAP is written and updated as needed by the IC, or unified command is used. If unified command is established, key representatives from all participating agencies meet regularly to develop and update this plan. In both large and small incidents, all personnel involved in the incident understand what their specific roles are and how they fit into the overall plan.

TIP

Incoming units and personnel need to be patient and give the IC enough time to perform a proper and thorough size-up. This step is essential in creating an effective IAP. If personnel are impatient and do not wait for the IAP, then freelancing could occur, disrupting operations, leading to confusion, and endangering the lives of both firefighters and civilians.

Designated Incident Facilities

The ICS designates the locations where specific functions are performed. For example, command is always based at the ICP. The **staging area**—the location where resources are placed while waiting for tactical assignment—the **incident base**—the location where logistics are coordinated—and the **helispot**—the location where helicopters land and take off—are additional designated areas where specific functions take place. The facilities required for each incident are established according to the specific IAP or a predefined ICS plan.

Resource Management

The ICS provides a standard method of **resource management** for assigning and keeping track of the resources involved in the incident. Engine companies, ladder companies, and other units are dispatched to the incident and assigned specific functions. The ICS structure allows command and section chiefs to keep track of the various company assignments through the chain of command and span of control.

At large-scale incidents, units are often dispatched to a staging area rather than going directly to the incident location. The staging area should be close to the incident scene. A number of units can be held in reserve at this location, ready to be assigned if needed.

Personnel are the most vital resource the fire service has. Under the ICS, the IC uses a personnel accountability system to track each member at the scene. Chapter 19, *Firefighter Survival*, discusses personnel accountability systems in detail.

Incident Reports

Every time a fire department unit responds to an incident, several key pieces of information must be documented in an **incident report**. This report includes where and when the incident occurred, who was involved, and what happened. Incident reports for fires should include the following details:

- Origin of the fire
- Extent of damage
- Any injuries or fatalities
- Actions taken by fire department personnel

The property owner and/or occupant is a primary source of information for the incident report. Any bystanders or eyewitnesses should also be questioned about what they observed. At a minimum, the name, birthdate, address, and telephone number of any involved persons; the address of the incident; and any relevant information about the specific incident should be recorded. The model number and serial number of any equipment owned by the occupant or owner that was damaged by or involved in the fire should be recorded as well. This information should be shared with any authorities, such as the fire marshal or law enforcement personnel, who are investigating the origin and cause of a fire or any other type of emergency incident investigation.

Incident reports can be completed on paper, but many fire departments now use computers to complete them and store the information in a database. Several commercial incident reporting software packages are available, and many large fire departments use custom-designed software.

National Fire Incident Reporting System (NFIRS)

Many fire departments in the United States create their incident reports using the **National Fire Incident Reporting System (NFIRS)**. NFIRS is a national reporting system used to compile and analyze incident reports from fire departments to help identify statistical trends, faulty equipment and products, civilian and firefighter injuries and fatalities, dollar loss filtered by type of property, and many other pieces of valuable information at local, state, and national levels. This information is used to develop programs designed to reduce the loss of life and property by fire. It also gives a clear picture of the fire problem in the United States.

NFIRS was developed by the U.S. Fire Administration (USFA) in partnership with the National Fire Information Council (NFIC). It is available to all states and territories to help them develop a more usable database of fire statistics. NFIRS consists of the following 11 modules designed to collect information about different types of incidents:

- Basic fire
- Structure fire
- Civilian fire casualty
- Fire service casualty
- EMS
- Hazardous materials
- Wildland fire
- Apparatus or resources
- Personnel
- Arson

The modules expand data collection beyond fires to encompass the full range of fire department activities.

NFIRS uses a code system to categorize data such as incident type, actions taken, and property use in the incident report. For example, after a fire incident, the type of fire must be noted. A building fire is code 111, a cooking fire is code 113, and a grass fire is code 143. The NFIRS Complete Reference Guide lists these data codes in an appendix. The Coding Questions Guide also provides coding information for NFIRS incident reports. Both of these documents are available at www.usfa.fema.gov/nfirs. If your department has its own incident reports and coding procedures, you should follow department SOPs.

NFIRS incident reports can be completed using the following:

- Free software provided by the USFA—training resources, system tools, and frequent technical updates are provided on the USFA website to assist users
- Commercially developed software that is NFIRS compliant
- Paper forms provided by NFIRS

Refer to your department's SOPs for specific incident reporting procedures.

Consequences of Incomplete and Inaccurate Incident Reports

From a legal standpoint, records and reports are vital parts of the incident. Information must be complete, clear, and concise, because these records can become admissible evidence in a court of law. When an incident report is filled out, care should be taken to ensure the data are entered accurately and completely. In addition, incident reports should be proofread to check to make sure there are no errors. Careless data entry of statistical fields such as type of property use, type of material first ignited, or floor of origin may undermine the purpose of collecting these data. The USFA relies on accurate data to identify national and regional trends in fire cause and loss.

Inaccurate coding of data such as fire origin and cause can create legal problems for the property owner and for your department. Fire reports are public records under the federal Freedom of Information Act, so they may be viewed by attorneys, insurance companies, the news media, or the public. If a fatality or loss occurs, incomplete or inaccurate reports can be used to prove that the fire department was negligent, and the department, the fire chief, and others may be held accountable. Careful, accurate documentation of all aspects of the incident is also a vital component of NIMS and ICS compliance, and it must be performed with the same level of discipline and professionalism as the emergency response.

Crew Resource Management

The fire service has long operated under the paramilitary chain of command concept; senior-level officers give orders to lower ranks who carry out those orders without question. Although many fire departments now use a more participative management model for routine, daily operations, the strict adherence to orders is still the norm for emergency operations.

Recent investigations of firefighter line-of-duty deaths (LODDs) have identified common factors that contribute to injuries and deaths at emergency scenes. Some of these factors are related to a fireground model of following orders without providing critical information to the IC. The International Association of Fire Chiefs (IAFC) adapted the **crew resource management (CRM)** concept from the aviation industry. CRM was developed to help reduce aircraft accidents caused by human error. The investigations into firefighter LODDs found that many of the same types of errors identified in the aircraft incidents were also found to be a factor in fire department LODDs. To limit these human factors, CRM seeks to enhance emergency scene communication, maintain situational awareness, strengthen decision-making, and improve teamwork. While clear lines of authority are maintained on emergency scenes, CRM encourages communication and input from all crew members to assess the situation and to develop the safest and most effective method to complete the assignment. Ultimately, the decision-making authority rests with the officers, but under CRM, crew members share their observations and their suggestions to accomplish the assigned objective.

Implementing CRM requires a change of SOPs, department-wide training for all personnel, and adopting a new culture that promotes a teamwork approach to safety. Safety is everyone's responsibility, and CRM provides the organizational framework to encourage communication and input from all personnel while maintaining clear lines of authority for emergency scene operations. More information about CRM can be found through the IAFC website (www.iafc.org).

CASE STUDY

You Are the Firefighter CONCLUSION

Your engine company has been dispatched to the smell of smoke in a rural area near the edge of your jurisdiction. As you get closer to the location, you see smoke drifting across the road. When you turn the corner, you see flames in the second-story window of an old farmhouse. While you advance an attack line from the hose bed, your company officer does a 360-degree walk-around of the building. He gives a complete size-up over the radio, requests additional resources (including a tanker from the neighboring department), and establishes command. The driver/operator charges the attack line, and you hear a neighbor ask your officer if all of the occupants are outside.

1. **What are the three incident priorities you must address at all emergency scenes?**

   Answer: The three incident priorities are life safety, incident stabilization, and property conservation. While there will be other considerations at an emergency scene, these three things must always be addressed, in that order.

2. **What is the Incident Command System, and how can it accommodate additional resources arriving on scene?**

   Answer: The ICS is a national system for creating safe and efficient operations at incident scenes no matter what the size. The ICS uses an organizational chart that has the ability to expand or contract based on the needs of the incident. The command component must always be activated, but other positions can be activated as the need arises.

3. What are the primary benefits of operating under the Incident Command System during mutual aid incidents?

Answer: The ICS is a standardized way of operating on many different incidents, and training is available to all agencies. As a result, agencies from different locations or with different specialties share a common understanding of the positions needed to operate efficiently and safely. In addition, the ICS promotes creating a unified command for large, complex incidents that require many responding agencies. Unified command allows the different agencies with different areas of expertise to participate in command decisions.

WRAP-UP

SUMMARY

KNOWLEDGE OBJECTIVES

- Outline the roles and responsibilities of a Firefighter II. (**NFPA 1010: 7.1.1**, pp. 928–929)
- Explain the function and organization of the Incident Command System (ICS) inside the National Incident Management System (NIMS).
  - Explain the chain of command. (p. 929)
  - Describe the difference between an IC and unified command. (p. 929)
  - Understand the history of the ICS. (pp. 929–930)
  - Describe the National Incident Management System (NIMS). (pp. 930–931)
  - Define the ICS. (**NFPA 1010: 7.1.1**, p. 931)
  - Explain how the ICP is established and used. (**NFPA 1010: 7.1.2**, pp. 931–932)
  - Describe the command component of the ICS. (**NFPA 1010: 7.1.2**, pp. 931–934)
  - Explain the organization of the ICS and how that makes it adaptable. (**NFPA 1010: 7.1.1**, pp. 931–941)
  - Describe the chain of command among the various components of the ICS. (**NFPA 1010: 7.1.2**, p. 933)
  - Describe the command staff and their responsibilities. (**NFPA 1010: 7.1.2**, pp. 933–934)
  - Describe the four sections in the ICS and their responsibilities. (**NFPA 1010: 7.1.2**, pp. 934–935)
  - Describe single resources. (**NFPA 1010: 7.1.2**, p. 936)
  - Explain the difference between groups and divisions. (**NFPA 1010: 7.1.2**, pp. 936–937)
  - Describe branches. (**NFPA 1010: 7.1.2**, p. 938)
  - Explain the difference between task forces and strike teams. (**NFPA 1010: 7.1.2**, p. 937)
  - Define units. (**NFPA 1010: 7.1.2**, p. 938)
- Describe how the sides of a building are identified at an incident (**NFPA 1010: 7.1.2**, p. 941)
- Explain how the ICS is implemented.
  - Describe the three principles of the ICS that always apply at an incident. (**NFPA 1010: 7.1.2**, pp. 941–942)
  - Explain how to establish command of the ICS until command of the incident is transferred. (**NFPA 1010: 7.1.1**, pp. 942–943)
  - Explain how to transfer command of a scene within the ICS. (**NFPA 1010: 7.1.1**, pp. 943–945)
  - Explain the importance of communicating progress on an assigned task to the person above you in the chain of command. (**NFPA 1010: 7.2.2**, p. 946)
- Describe the components of a proper scene size-up and an incident action plan.
  - Define a complete size-up. (**NFPA 1010: 7.1.2**, pp. 946–947)
  - Describe the process of performing an initial size-up. (**NFPA 1010: 7.1.2**, pp. 946–947)
  - Describe the process of performing a 360-degree walk-around. (**NFPA 1010: 7.1.2**, pp. 946–947)
  - Describe the two basic categories of information used in the size-up process. (**NFPA 1010: 7.1.2**, pp. 947–948)
  - Explain how the size-up process determines the resources required at the emergency incident. (**NFPA 1010: 7.1.2**, pp. 947–949)

- Explain how the size-up process can be used to determine whether additional resources are needed. (**NFPA 1010: 7.2.2**, pp. 947–949)
- Explain the need for requesting additional resources to complete a task. (**NFPA 1010: 7.2.2**, pp. 947–949)
- Describe IAPs, strategies, tactics, and tasks. (**NFPA 1010: 7.1.2**, pp. 949–950)
- List the three incident priorities from which an incident action plan (IAP) is based. (**NFPA 1010: 7.1.2**, p. 950)
- Describe the acronym RECEO–VS and how it provides a general guideline for incident commanders to systematically address the incident priorities. (**NFPA 1010: 7.1.2**, pp. 950–952)
- Describe the acronym S.L.I.C.E.–R.S. and how it provides initial engine company operations a short list of objectives prior to the arrival of additional resources. (**NFPA 1010: 7.1.2**, p. 953)
- Describe the characteristics of the Incident Command System (ICS). (pp. 954–957)
- Describe the importance of accurately documenting information in an incident report.
  - Explain the importance of an incident report. (**NFPA 1010: 7.2.1**, pp. 957–959)
  - Describe how to collect the necessary information for a thorough incident report. (**NFPA 1010: 7.2.1**, pp. 957–959)
  - Describe the resources that list the codes used in incident reports. (**NFPA 1010: 7.2.1**, pp. 957–958)
  - Explain how to submit incident reports to the NFIRS. (**NFPA 1010: 7.2.1**, p. 958)
  - Explain the consequences of an incomplete or inaccurate incident report. (**NFPA 1010: 7.2.1**, pp. 958–959)
- Describe the goals of crew resource management (CRM) in the fire service. (p. 959)

SKILLS OBJECTIVES

- Operate within the Incident Command System (ICS). (**NFPA 1010: 7.1.2**, p. 940)
- Establish command of an incident. (**NFPA 1010: 7.1.2**, p. 944)
- Transfer command of an incident to another firefighter. (**NFPA 1010: 7.1.2**, p. 945)
- Complete and proofread an incident report. (**NFPA 10101: 7.2.1**, pp. 957–959)

KEY TERMS

360 See *360-degree walk-around.*

360-degree walk-around An evaluation and analysis of an incident conducted by the incident commander (IC) or someone designated by the IC after the initial size-up is completed. Also called *360.*

action of opportunity An action taken when needed at any point during an incident.

assisting agency An agency or organization providing personnel, services, or other resources to the agency that has direct responsibility for incident management. See also *cooperating agency.* (NFPA 1026)

assisting organization See *assisting agency.*

branch A supervisory level established in either the operations or logistics function to provide a span of control. (NFPA 1550)

branch director A person in a supervisory level position in either the operations or logistics function to provide a span of control. (NFPA 1550)

chain of command A rank-based, hierarchical structure that creates an orderly line of authority.

command The act of directing and/or controlling resources by virtue of explicit legal, agency, or delegated authority. (NFPA 1026)

command designation The name of the incident.

command staff The public information officer, incident safety officer, and liaison officer, all of whom report directly to the incident commander (IC) and are responsible for functions in the incident management system that are not a part of the function of the line organization. (NFPA 1550)

company The basic firefighting organizational unit staffed by various grades of firefighters under the supervision of an officer and assigned to one or more specific pieces of apparatus. (NFPA 1410)

company officer The individual responsible for command of a company, a designation not specific to any

KEY TERMS CONTINUED

particular fire department rank (can be a firefighter, lieutenant, captain, or chief officer, if responsible for command of a single company). (NFPA 1026)

cooperating agency An agency supplying assistance other than direct suppression, rescue, support, or service functions to the incident management efforts (Red Cross, law enforcement agency, telephone company, etc.). (NFPA 1026)

crew Within the ICS, a specific term that identifies a team of two or more firefighters. (NFPA 1550)

crew resource management (CRM) A program focused on improved situational awareness, sound critical decision-making, effective communication, proper task allocation, and successful teamwork and leadership. (NFPA 1550)

division That organizational level having responsibility for operations within a defined geographic location. (NFPA 1026)

division supervisor A person in a supervisory level position responsible for a specific geographic area of operations at an incident. (NFPA 1550)

exposure An object such as a building, structure, or vehicle in an area adjacent to a burning building.

finance/administration section Section responsible for all costs and financial actions of the incident or planned event, including the time unit, procurement unit, compensation/claims unit, and the cost unit. (NFPA 1026)

finance/administration section chief The general staff position responsible for directing the finance/administrative function. It is generally assigned on large-scale or long-duration incidents that require immediate fiscal management.

fireground command (FGC) An incident management system (IMS) developed in the 1970s for day-to-day fire department incidents (generally handled with fewer than 25 units or companies).

FIRESCOPE Fire Resources of California Organized for Potential Emergencies; an organization of agencies established in the early 1970s to develop a standardized system for managing fire resources at large-scale incidents such as wildland fires.

general staff A group of incident management personnel organized according to function and reporting to the incident commander (IC), normally consisting of the operations section chief, planning section chief, logistics section chief, and finance/administration section chief. (NFPA 1026)

group Established to divide the incident management structure into functional assignments of operation. (NFPA 1026)

group supervisor A person in a supervisory level position responsible for a functional area of operation. (NFPA 1550)

helispot A heliport where no refueling, maintenance, repair, or storage of helicopters is permitted. (NFPA 5000)

incident action plan (IAP) A verbal or written plan containing incident objectives reflecting the overall strategy and specific control actions, where appropriate, for managing an incident or planned event. (NFPA 1026)

incident base The location at an incident where logistics are coordinated.

incident commander (IC) The individual responsible for all incident activities, including the development of strategies and tactics and the ordering and release of resources. (NFPA 1026)

incident command post (ICP) The field location at which the primary tactical-level, on-scene incident command functions are performed. (NFPA 1026)

Incident Command System (ICS) The combination of facilities, equipment, personnel, procedures, and communications operating within a common organizational structure that has responsibility for the management of assigned resources to effectively accomplish stated objectives pertaining to an incident or training exercise. (NFPA 2500)

incident management system (IMS) A system that defines the roles and responsibilities to be assumed by responders and the standard operating procedures to be used in the management and direction of emergency incidents and other functions. (NFPA 1410)

incident report A document prepared by fire department personnel on a particular incident. (NFPA 901)

incident safety officer (ISO) A member of the command staff responsible for monitoring and assessing safety hazards and unsafe situations and for developing measures for ensuring personnel safety. (NFPA 1550)

initial size-up Usually conducted by the first-arriving company officer, a rapid evaluation and analysis of an incident upon arrival at the incident and before the 360-degree walk-around is conducted. Also called *windshield size-up* because it is generally performed by the engine officer before they get out of the engine.

leader The individual responsible for command of a task force, strike team, or functional unit. (NFPA 1026)

liaison officer A member of the command staff responsible for coordinating with representatives from cooperating and assisting agencies. (NFPA 1550)

logistics section Section responsible for providing facilities, services, and materials for the incident or planned event, including the communications unit, medical unit, and food unit within the service branch and the supply unit, facilities unit, and ground support unit within the support branch. (NFPA 1026)

logistics section chief The general staff position responsible for directing the logistics function. It is generally assigned on complex, resource-intensive, or long-duration incidents.

National Fire Incident Reporting System (NFIRS) A national reporting system used to compile and analyze incident reports from fire departments to help identify statistical trends, faulty equipment and products, civilian and firefighter injuries and fatalities, dollar loss filtered by type of property, and many other pieces of valuable information at local, state, and national levels.

National Incident Management System (NIMS) A system mandated by Homeland Security Presidential Directive 5 (HSPD-5) that provides a systematic, proactive approach guiding government agencies at all levels, the private sector, and nongovernmental organizations to work seamlessly to prepare for, prevent, respond to, recover from, and mitigate the effects of incidents, regardless of cause, size, location, or complexity, so as to reduce the loss of life or property and harm to the environment. (NFPA 1026)

operations section Section responsible for all tactical operations at the incident or planned event. (NFPA 1026)

operations section chief The general staff position responsible for managing all operations activities. It is usually assigned when complex incidents involve more than 20 single resources or when command staff cannot be involved in all details of the tactical operation.

planning section Section responsible for the collection, evaluation, dissemination, and use of information related to the incident situation, resource status, and incident forecast. (NFPA 1026)

planning section chief The general staff position responsible for planning functions and for tracking and logging resources. It is assigned when command staff members need assistance in managing information.

preincident plan A document developed by gathering general and detailed data that is used by responding personnel in effectively managing emergencies for the protection of occupants, responding personnel, property, and the environment. (NFPA 1660)

preincident survey An examination of a property to gather information to develop a preincident plan.

public information officer (PIO) A member of the command staff responsible for interfacing with the public and media or with other agencies with incident-related information requirements. (NFPA 1026)

RECEO–VS An acronym developed for use by the incident commander (IC) for accomplishing tactical priorities on the fire ground; stands for Rescue, Exposure protection, Confinement, Extinguishment, Overhaul, Ventilation, and Salvage.

resource management Under the NIMS, includes mutual aid agreements; the use of special federal, state, local, and tribal teams; and resource mobilization protocols. (NFPA 1026)

risk–benefit analysis A decision made by a responder based on a hazard identification and situation assessment that weighs the risks likely to be taken against the benefits to be gained for taking those risks. (NFPA 1006)

section In the Incident Command System (ICS), the four components of incident management that report directly to command.

section chief In the Incident Command System (ICS), the person in a supervisory level position responsible for commanding a section.

Side A Normally the front or main entrance/access to the building and usually the side bearing the building address. For buildings with an unusual side A, side A will be identified by the incident commander (IC). Also called *Side Alpha.* (NFPA 3000)

Side Alpha See *Side A.*

Side B The adjacent side of the building or structure clockwise from Side A. Also called *Side Bravo.* (NFPA 3000)

Side Bravo See *Side B.*

Side C The adjacent side of the building or structure clockwise from Side B, generally the back of the building or structure. Also called *Side Charlie.* (NFPA 3000)

Side Charlie See *Side C.*

KEY TERMS CONTINUED

Side D The adjacent side of the building or structure clockwise from Side C. Also called *Side Delta.* (NFPA 3000)

Side Delta See *Side D.*

single resource An individual, a piece of equipment and its personnel, or a crew or team of individuals with an identified supervisor that can be used on an incident or planned event. (NFPA 1026)

size-up The process of gathering and analyzing information to help fire officers make decisions regarding the deployment of resources and the implementation of tactics. (NFPA 1410)

S.L.I.C.E.–R.S. An acronym intended to be used by the first-arriving company officer to accomplish important strategic goals on the fireground.

span of control The maximum number of personnel or activities that can be effectively controlled by one individual (usually three to seven).

staging area Location established where resources can be placed while they await a tactical assignment. (NFPA 1026)

strategy The general plan or direction selected to accomplish incident objectives. (NFPA 1550)

strike team Specified combinations of the same kind and type of resources, with common communications and a leader. (NFPA 1026)

strike team leader A person in a supervisory level position responsible for commanding a strike team.

survivability profiling A risk–benefit analysis that weighs the viability and survivability of potential fire victims based on the current conditions in the structure.

tactical Side A/Alpha At an incident, Side A/Alpha of a building when it is not the front or address side.

tactics Deploying and directing resources on an incident to accomplish the objectives designated by strategy. (NFPA 1026)

task A specific job behavior or activity. (NFPA 1010)

task force A group of resources with common communications and a leader that can be pre-established and sent to an incident or planned event or formed at an incident or planned event. (NFPA 1026)

task force leader A person in a supervisory level position responsible for commanding a task force.

transfer of command The formal procedure for transferring the duties of an incident commander (IC) incident scene. (NFPA 1026)

unified command A team effort that allows all agencies with jurisdictional responsibility for an incident or planned event, either geographic or functional, to manage the incident or planned event by establishing a common set of incident objectives and strategies. (NFPA 1026)

unit Within the ICS, a specific term that identifies two or more resources who have responsibility for a specific activity at an incident.

unit leader A person in a supervisory level position responsible for commanding a task force.

unity of command The concept by which each person within an organization reports to one, and only one, designated person. (NFPA 1026)

windshield size-up See *initial size-up.*

REVIEW QUESTIONS

1. Why was the first, standardized ICS developed?
2. What three courses do you need to take if your department is NIMS compliant?
3. Which ICS position must be filled at every incident?
4. How many people can one person safely and effectively supervise according to the ICS rule for span of control?
5. What are the components of the ICS?
6. How are the sides of a building identified at an incident?
7. What is transfer of command?
8. Who usually conducts the initial size-up?
9. What are the three strategic priorities at an emergency incident and what order are they prioritized?
10. What makes a unified command effective?
11. What is the NFIRS?
12. What is a main difference between the ICS and the CRM concept?

DISCUSSION QUESTIONS

1. Why is a proper transfer of command critical to incident operations?
2. How is it possible for the ICS to work on both simple and very complex incidents?
3. Describe the sources of information before, during, and after an incident.

APPLYING THE CONCEPTS

Your crew is dispatched to a fire in a large, detached garage of a residence in a densely populated neighborhood. The dispatch center reported that a neighbor who was walking their dog saw the fire and placed the call. The source of the fire is unknown. The neighbor said there is a lot of smoke, and flames are visible inside. No one is believed to be inside the structure. The neighbor does not know specifically what is in the garage but said the owner uses it to store old cars. You are the engine officer of the first engine to arrive at the scene.

1. Based on the information you have been given, what must you do when you arrive on the scene?
2. What are your primary goals when conducting an initial size-up?
3. In this situation, what potential hazards and options should you consider?

The fire appears to be contained inside the garage. However, because there is a large tree overhanging the garage, this is a densely populated neighborhood, and there are 15-mph winds, you call for more resources. You are in the midst of doing the 360-degree walkaround when the homeowner frantically runs up to you shouting, “My babies are inside! You have to save them!” You ask him to clarify if his children are inside. “No, not my kids . . . my cars! My Porsche 356, 928, and 911. I’m restoring them. They’re irreplaceable and you have to get them out. NOW!”

4. Do you immediately respond to the homeowner’s plea and begin a rescue operation?
5. Good decisions come from good information. What questions can you ask the homeowner to gather crucial information to complete the 360?

The homeowner says there are chemicals he is using to restore the cars, like paint solvents, paint strippers, and rust inhibitors, as well as gas cans and a propane tank in the garage. You immediately convey this information—and all the information you gathered in the 360—to the dispatch center and request additional hazardous material resources. Just then, the battalion chief arrives.

6. What information do you relay to the battalion chief to ensure a smooth command transfer?
7. Why is the accurate and detailed pass down of information so important?

REFERENCES

Emergency Management Institute (EMI). 2022a. “IS-100.C: Introduction to the Incident Command System, ICS 100.” Federal Emergency Management Agency (FEMA). Last modified August 8, 2022. Accessed March 24, 2023. https://training.fema.gov/is/crslist.aspx?lang=en.

Emergency Management Institute (EMI). 2022b. “IS-200.C: Basic Incident Command System for Initial Response, ICS-200.” Federal Emergency Management Agency (FEMA). Last modified August 8, 2022. Accessed March 24, 2023. https://training.fema.gov/is/crslist.aspx?lang=en.

Emergency Management Institute (EMI). 2022c. “IS-700.B: An Introduction to the National Incident Management System.” Federal Emergency Management Agency (FEMA). Last modified August 8, 2022. Accessed March 24, 2023. https://training.fema.gov/is/crslist.aspx?lang=en.

Federal Emergency Management Agency (FEMA). n.d.a. “ICS Management: Span of Control.” Accessed June 8, 2018. https://training.fema.gov/is/crslist.aspx?lang=en.

Federal Emergency Management Agency (FEMA). n.d.b. “Resource Typing Library Tool.” Accessed February 11, 2023. https://rtlt.preptoolkit.fema.gov/Public.

FirefighterNation. 2011. “Survivability Profiling Takes Size-Up to a New Level.” October 31, 2011. Accessed April 23, 2023. https://www.firefighternation.com/firerescue/survivability-profiling-takes-size-up-to-a-new-level/#:~:text=Survivability%20profiling%20is%20defined%20as,smoke%20conditions%20inside%20burning%20structures.

International Association of Fire Chiefs (IAFC). n.d. “Crew Resource Management.” Accessed March 4, 2023. https://www.iafc.org/topics-and-tools/resources/resource/crew-resource-management.

REFERENCES CONTINUED

International Society of Fire Service Instructors (ISFSI). n.d. "Principles of Modern Fire Attack: 8-hour Course." Accessed March 2, 2023. http://www.isfsi.org/p/cl/et/cid=1000.

National Fire Protection Association (NFPA). 2019. *NFPA 1410, Standard on Training for Emergency Scene Operations.* 2020 Edition. Quincy, MA: NFPA.

National Fire Protection Association (NFPA). 2020. *NFPA 901, Standard Classifications for Fire and Emergency Services Incident Reporting.* 2021 Edition. Quincy, MA: NFPA.

National Fire Protection Association (NFPA). 2020. *NFPA 1006, Standard for Technical Rescue Personnel Professional Qualifications.* 2021 Edition. Quincy, MA: NFPA.

National Fire Protection Association (NFPA). 2020. *NFPA 3000, Standard for an Active Shooter/Hostile Event Response (ASHER) Program.* 2021 Edition. Quincy, MA: NFPA.

National Fire Protection Association (NFPA). 2021. *NFPA 2500, Standard for Operations and Training for Technical Search and Rescue Incidents and Life Safety Rope and Equipment for Emergency Services.* 2022 Edition. Quincy, MA: NFPA.

National Fire Protection Association (NFPA). 2022. *NFPA 5000, Building Construction and Safety Code.* 2021 Edition. Quincy, MA: NFPA.

National Fire Protection Association (NFPA). 2023. *NFPA 1026, Standard for Incident Management Personnel Professional Qualifications.* 2024 Edition. Quincy, MA: NFPA.

National Fire Protection Association (NFPA). 2023. *NFPA 1550, Standard for Emergency Responder Health and Safety.* 2024 Edition. Quincy, MA: NFPA.

National Fire Protection Association (NFPA). 2023. *NFPA 1660, Standard for Emergency, Continuity, and Crisis Management: Preparedness, Response, and Recovery.* 2024 Edition. Quincy, MA: NFPA.

National Fire Protection Association (NFPA). 2023. *NFPA 3000, Standard for an Active Shooter/Hostile Event Response (ASHER) Program.* 2024 Edition. Quincy, MA: NFPA.

National Fire Protection Association (NFPA). 2023. *NFPA 1010: Standard for Professional Qualifications for Firefighters, 2024 Edition.* Quincy, MA: NFPA.

Shaw, Steven. 2015. "FHExpo: Survivability Profiling—Size-up On Steroids." Firehouse, July 15, 2015. Accessed March 4, 2023. https://www.firehouse.com/safety-health/news/12094782/survivability-profiling-starts-with-sizeup.

U.S. Fire Administration (USFA). 2015. *National Fire Incident Reporting System Complete Reference Guide.* January 2015. Accessed March 2, 2023. https://www.usfa.fema.gov/downloads/pdf/nfirs/NFIRS_Complete_Reference_Guide_2015.pdf.

U.S. Fire Administration (USFA). 2016. *NFIRS 5.0 Coding Questions Manual.* April 2016. Accessed March 2, 2023. https://www.usfa.fema.gov/downloads/pdf/nfirs/NFIRS_Coding_Questions_2016.pdf.

U.S. Fire Administration (USFA). 2023. "National Fire Incident Reporting System." Last reviewed January 23, 2023. Accessed March 2, 2023. https://www.usfa.fema.gov/nfirs/.

CHAPTER

23

Firefighter II

Advanced Fire Suppression

KNOWLEDGE OBJECTIVES

After studying this chapter, you will be able to:

- Describe the basic fireground decision processes that are made in order to coordinate an interior attack in a structure fire.
- Describe how tactics are implemented during fire suppression operations.
- Describe the fundamentals of flammable gases and the methods to suppress flammable gas fires.
- Describe the methods by which foam is generated and prevents or controls a hazard.
- Describe how an annual hose test is performed.

SKILLS OBJECTIVES

After studying this chapter, you will be able to perform the following skills:

- Utilize a thermal imager to assess fire conditions in a structure.
- Utilize the appropriate fire suppression attack technique based on the information obtain during size-up.
- Prepare and apply foam.
- Describe how an annual hose test is performed.

ADDITIONAL NFPA STANDARDS

- **NFPA 10**, *Standard for Portable Fire Extinguishers, 2022 Edition*
- **NFPA 11**, *Standard for Low-, Medium-, and High-Expansion Foam, 2021 Edition*
- **NFPA 18**, *Standard on Wetting Agents, 2021 Edition*
- **NFPA 460**, *Standard for Aircraft Rescue and Firefighting Services at Airports, 2024 Edition*
- **NFPA 1145**, *Guide for the Use of Class A Foams in Fire Fighting, 2022 Edition*
- **NFPA 1550**, *Standard for Emergency Responder Health and Safety, 2024 Edition*
- **NFPA 1700**, *Guide for Structural Fire Fighting, 2021 Edition*
- **NFPA 1710**, *Standard for the Organization and Deployment of Fire Suppression Operations, Emergency Medical Operations, and Special Operations to the Public by Career Fire Departments, 2020 Edition*
- **NFPA 1720**, *Standard for the Organization and Deployment of Fire Suppression Operations, Emergency Medical Operations, and Special Operations to the Public by Volunteer Fire Departments, 2020 Edition*
- **NFPA 1801**, *Standard on Thermal Imagers for the Fire Service, 2021 Edition*
- **NFPA 1900**, *Standard for Aircraft Rescue and Firefighting Vehicles, Automotive Fire Apparatus, Wildland Fire Apparatus, and Automotive Ambulances, 2024 Edition*
- **NFPA 1962**, *Standard for the Care, Use, Inspection, Service Testing, and Replacement of Fire Hose, Couplings, Nozzles, and Fire Hose Appliances*

CASE SUDY

You Are the Firefighter

You are the crew leader on Engine 29, which was dispatched at 1150 as first-due to a house fire. The second-due company, Engine 1, has an officer who will take command upon their arrival, but they are 5 to 7 minutes out. It is a weekday in early fall, and the weather report you reviewed at lineup that morning told you that the temperature is 60 F (15 degrees C) degrees, with winds blowing 13 to 15 miles per hour (mph; 21–24 kilometers per hour [km/h]) with no other impending weather concerns. You start filling in your crew on the information from the computer-aided dispatch (CAD), which says that the homeowner states that everyone is out of the house and that the fire started in the kitchen. On the map you note the location of a hydrant, and based on Engine 1's direction of travel and their response time, you tell your driver/operator to perform a forward lay from the hydrant and advise Engine 1 of the hydrant location. You ask Engine 1 to complete the water supply. Upon your arrival, the homeowner confirms everyone is out of the house. As you begin your size-up, you make note of several facts: the front door is open and a significant amount of dark, pressurized, turbulent smoke is pouring out; the house is an approximately 80- × 60-foot (ft; 24- × 18-meter [m]), two-story, colonial-style home and is set back at least 75 ft (23 m); and the wind is blowing from the back of the house toward the street. You provide your initial size-up report over the radio, "Engine 29 is on the scene. Two-story, large, single-family, colonial-style dwelling, Type V construction, smoke showing from the Alpha side on the first floor. Homeowner reports everyone out of the structure. I'll be out taking a lap around the structure."

1. Where is the flow path and what are your considerations?
2. What is your first tactical action as the initial incident commander?
3. To which side of the house should you direct the first crew with a handline, and what type of attack should they conduct? What about the second crew?
4. Because a command officer is not expected on scene for 5 to 7 minutes, what are the next command-level actions you should consider taking?

Introduction

The Firefighter II may need to act as either the initial incident command (IC) or as a crew leader of a crew working on scene within the framework of the Incident Command System (ICS). Both scenarios mean they may get involved in tactical decision making at a wide variety of incidents. Both scenarios means they may get involved in tactical decision making at a wide variety of incidents. Therefore, it is crucial that they have a broader base of knowledge and understanding of this process. Understanding all the elements that come into play when making decisions that can have grave consequences results in saving civilian lives and ensuring the safety of fellow firefighters.

The IC or the acting IC must conduct a risk–benefit analysis and reassess every time the situation changes. There will be times when no clear-cut, absolute answer can be found. However, by understanding the foundational elements of decision making, you have the best chance to make effective decisions in the dynamic, high-paced environment of the emergency scene. In any scenario, your tactical, decision-making skills will be required to ensure achieving the goals of the mission and safely mitigating the event.

Tactical Decision Making and Size-Up

The start of the decision-making and size-up process begins with a decision on strategy. A **strategy** is a description of the tactics to use to achieve the broad goals of the incident. **Tactics** are the specific actions taken to achieve those goals. Strategy drives tactics, not the other way around. It is critical, particularly for the first unit on scene, to determine the strategy for the operation. Making tactical decisions absent a strategy and a plan can lead to bad outcomes.

Developing an incident action plan (IAP) can be misinterpreted as a process where leadership sits down to create a multi-page document. Establishing an effective IAP upon arrival at a dynamic event starts with preplanning, clear procedural guidelines, and on-scene decision making. The decision maker needs to bring these elements together with a plan. That plan can be as short and concise as brief radio transmissions and instructions to crew members.

Once that plan is decided upon, the tactical assignments support the strategy chosen. The acronyms RECEO-VS—rescue, exposure protection, confinement of the fire, extinguishment, overhaul, ventilate, and salvage—and S.L.I.C.E.-R.S.—size-up, locate the fire, identify and control the flow path, cool from a safe location, extinguish the fire, rescue, and salvage—are often used to help crew leaders and ICs achieve the three incident priorities of life safety, incident stabilization, and property conservation. These acronyms are discussed in greater detail in Chapter 22, *Establishing and Transferring Command.*

It is understood that if there is no officer on the first-arriving apparatus, the first-arriving crew leader will be the initial IC responsible for creating the IAP, and that person needs to be ready to make tactical decisions in preparation for, and while conducting, firefighting operations. For example, before initiating an interior attack, the crew leader should conduct an effective and complete size-up (including a 360), request additional resources if needed, and potentially initiate operations. They must then be prepared to brief the arriving officer (company or chief) and transition command to them, then carry on with their duties as a crew leader or in whatever capacity dictated by the IC.

Tactically, when advancing a hose line into a building, the crew leader should be the first one in, leading the firefighter operating the nozzle and the crew to the seat of the fire. If the crew is met at the point of entry with a large volume of fire, the crew leader should be next to the nozzle firefighter. This puts the crew leader in the best position to ensure crew member safety by allowing them to monitor the progress of the fire attack and ensure the attack is effective by providing direction to the firefighter operating the nozzle. Transitioning between tactical decision making and task-level action is the hallmark of an effective crew leader.

The crew leader/IC must evaluate many factors to decide whether to use an offensive or defensive operation at a particular fire (see Chapter 14, *Fire Suppression*). If the risk factors are too great, a defensive operation is the only acceptable option. If the decision is made to conduct an offensive operation, then the decision must be made as to which type of attack—direct, indirect, combination, or transitional—to initiate after considering both the safety issues and the potential effectiveness of the operation. Factors to be evaluated before making a decision for an offensive operation include the following:

- Are there viable occupants who can be rescued?
- What are the risks versus the potential benefits?
- Is it safe to send firefighters into the building? Do hostile fire conditions prohibit entry?
- Are there indicators of potential collapse?
- Is the building made of lightweight construction?
- Are there enough firefighters on the scene to mount an interior attack so that lines and appliances can be deployed effectively and the two-in/two-out rule can be observed?
- Is an adequate water supply available?
- Are hazardous materials present?

As a Firefighter II/crew leader, you may be responsible for determining the type of fire attack that will be used or for helping translate the tactical priority into task-level action. You should understand the factors that affect that tactical decision-making process, whether as a crew member or as a crew leader. These decisions are critical to success on the fireground and member safety, and involve multiple, complex challenges.

Risk–Benefit Analysis for Life Safety

Once the crew leader/IC has enough information, they must rapidly consider the risks to the firefighters and weigh them against the potential benefits. As they balance the safety of firefighters against other fireground priorities, they must always keep in mind the three strategic priorities of every incident—life safety, incident stabilization, and property conservation, in that order. Although no active firefighting activity is ever risk-free, firefighters are willing to risk much to save much. However, firefighters should not assume too much risk for limited potential benefits. A risk–benefit analysis weighs the risks versus the benefits of taking those risks. A comprehensive size-up identifies critical life safety issues that impact the development of the IAP. For example, if there are victims in the burning building, the IC must assess the risks associated with aggressive search and rescue operations by asking several questions:

- Given the current conditions, are the victims still alive?
- Are fire conditions so severe that any interior operation is unsafe?
- If victims are likely still alive, how will the rescue attempt be made? Can firefighters rescue the victims without taking an unacceptable risk?
- Where will initial handlines be placed to best protect firefighters and make rescue possible?

For example, conditions from the front of a burning structure may suggest full involvement on multiple floors. But a complete size-up including a 360-degree walk-around could indicate that entry from the back of the structure may be possible, even if only for a brief period. With a credible report of victims inside, entry into the immediately dangerous to life or health (IDLH) scene may be warranted. This is why effective decision making is critical.

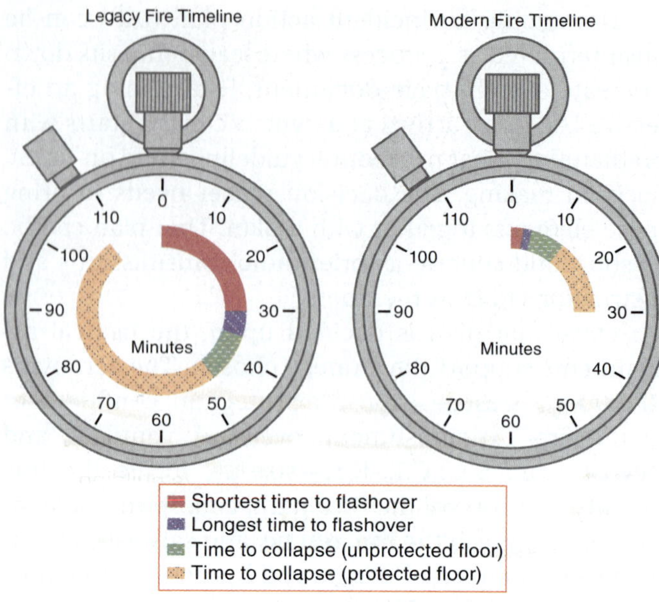

FIGURE 23-1 There is a significant difference in the time to flashover and the time to failure between modern and legacy construction. This is another issue that needs to be considered during the decision-making process.

Structural Collapse

Firefighters who are assigned to interior fire attack, search and rescue, or vertical ventilation operations should always be aware of building conditions. Structural collapse poses a significant risk to these firefighters. Modern homes use lightweight, engineered structural materials to support large spans. These modern construction techniques generally lend themselves to shorter times to collapse when compared to legacy construction (**FIGURE 23-1**). This is especially true if the point of origin is in the basement or attic where there is a higher probability of having unprotected structural components.

Localized collapse issues, such as a section of drywall or suspended ceiling, can be just as significant as complete structural failure. Firefighters conducting roof operations need to be aware of signs of imminent roof collapse such as fire showing around roof vents, melting snow, and evaporating or steaming water. Firefighters conducting an interior attack need to watch for signs of potential floor collapse such as sagging or soft floors or heavy fire or dark, turbulent smoke below the floor. These signs indicate that the structural members are being compromised by fire below them, and it is not safe to continue roof or interior operations.

General building collapse is also a concern during an exterior attack. Many firefighters have been injured or killed while working outside of burning buildings. Firefighters should immediately retreat and stay out of the **collapse zone**—the area around a structure where debris could fall if the building collapses—if they notice any of the following signs:

- Visible indicators of sagging of floors or roof supports
- Indicators that the roof assembly is separating from the walls, such as the appearance of fire or smoke near the roof edges
- Cracked, crumbling, or bulging walls (**FIGURE 23-2**)
- Smoke or water coming through masonry walls
- Structural failure of any portion of the building, even if it is some distance from the firefighters
- Heavy fire conditions below any interior operation
- Sudden increase in the intensity of the fire from the roof opening
- Noises associated with structural failure such as popping or cracking
- Any of these signs plus high heat indicators on a thermal imaging camera

Other factors that may contribute to building collapse include heavy snow on the roof, master streams flowing a high volume of water over a long period of time, and any building that is under construction or being renovated.

FIGURE 23-2 This crack indicates serious building instability.

FIGURE 23-3 Turbulent smoke flow.

SAFETY TIP

Signs of collapse may not always be present, so firefighters need to rely on their knowledge of building construction and fire dynamics to keep themselves safe.

Fire Conditions

Although every fire incident presents a certain level of risk to responding firefighters, some fire scenes are more dangerous than others. Interior fire and rescue operations are generally more dangerous than exterior operations. Fires above grade, such as a fire in an attic, and fires in basements make fireground operations more difficult and significantly increase the risks to firefighters. Smoke and fire conditions vary greatly, depending on the size and shape of the compartment, fuel type and fuel load, ventilation openings, and many other factors (discussed in Chapter 5, *Fire Behavior*). Firefighters must constantly monitor the current fire and smoke conditions and be aware of any changes. A gradual increase in heat, smoke, or visible flame indicates that current suppression operations are not effective. A rapid increase in any of these conditions, such as thick, turbulent smoke, may indicate a potential hostile fire event and a corresponding rapid increase in interior temperatures (**FIGURE 23-3**). A rollover or a rapidly lowering smoke layer may indicate that a flashover is imminent. Notify your crew immediately if you notice any rapidly deteriorating smoke, flame, or heat conditions. If this happens, particularly if you do not have a handline, all personnel should evacuate the building immediately until it is determined to be safe to resume interior operations.

Thermal Imager

A **thermal imager (TI)**, more commonly referred to as a **thermal imaging camera (TIC)**, is a camera that converts infrared radiation into visible light. NFPA 1801, *Standard on Thermal Imagers for the Fire Service, 2021 Edition*, describes the requirements for TICs used in the fire service. The TIC is one of the most critical tools used on the incident scene with respect to decision making. Tactical thermal imaging adds an additional perspective, particularly in a zero-visibility IDLH environment. The information gleaned with a TIC can assist with the following:

- More accurate size-up: A building that is not pushing smoke can be deceiving, but the TIC can still read a heat signature from the exterior, identifying the potential fire location.
- Identifying flow paths: After entering the IDLH, a sweep of the ceiling and floor with a TIC can identify heat currents, particularly in zero-visibility environments, to help determine temperature and the relative speed and direction of flow paths.
- Identifying the most efficient line and stream placement for fire attack: Particularly in zero-visibility environments, a TIC assists the crew leader/IC in instructing the nozzle firefighter where to direct the stream quickly and efficiently to achieve a quicker knockdown. This will also limit wasted water, especially when operating solely on tank water.
- Creating quicker search patterns and allowing for faster victim removal: In zero-visibility environments, the capability of a TIC to delineate a space and display heat signatures of a body allows a search team to move directly to the victim. Compared to conducting

a blind small- or wide-area search using traditional search methods, using a TIC vastly decreases the time required to find and extricate a victim. After finding a victim, the firefighters must exit the structure as quickly as possible. A TIC allows firefighters to find a means of egress out of the IDLH more rapidly, increasing the victim's chance of survival.

- Identifying the location of missing or lost persons in wilderness or other outdoor environments: Although the TIC is designed for interior suppression environments, it can also be used in wilderness or other outdoor environments for search and rescue.
- More precise overhaul operations ensuring minimal property loss: The absolute goal for effective overhaul is to identify and suppress all hidden and remaining fire. In a post-fire environment, a TIC's ability to differentiate heat levels is critical. The naked eye, especially in an unlit, burned out compartment, cannot clearly see where heat, that can smolder and rekindle hours later, may be present.
- Identifying the location of hazardous materials, and other incident types: The capability of a TIC to distinguish relative temperature differences means a firefighter can scan a barrel or container and "see" the level of liquid inside the container or look at a utility line (electrical or gas) and check to see if it is hot or cool, depending on the situation being investigated. At motor vehicle crashes, a firefighter can use a TIC to scan the perimeter of a tree line to look for additional victims who might have been ejected on a dark night. The TIC can also be used to look at the vehicle for overhaul.

Like every tool, the TIC has both strengths and weaknesses that must be considered. When used correctly, it is an invaluable aid for making effective decisions. Although firefighters should strive not to become overly reliant on this device, a TIC combined with a fundamental knowledge of the basics of fire behavior, fire dynamics, and building construction will provide firefighters with additional information so that they can attack a fire more efficiently, locate and rescue victims much more quickly, and conduct a more thorough overhaul. Knowledge of how the TIC works will ensure a firefighter's ability to interpret the images on the screen and translate that into action on the incident scene.

Operating Procedures and Limitations

Although the TIC can greatly increase situational awareness on the incident scene, there are important characteristics that must be understood to take full advantage of this tool. One of those is the TIC's field of view. The **field of view (FOV)** is the area observable at any given moment and is measured vertically and horizontally in degrees. A critical consideration for the FOV of the TIC is that it can be diminished by as much as 90 percent compared to the human eye. Consider the difference in viewing a panorama from a mountain top with your eyes versus a pair of binoculars. The TIC's limited FOV can lead to "tunnel vision" when operating in the IDLH environment. Therefore, in addition to using your other senses constantly (eyesight without the TIC, hearing, feeling/touch), it is vital to constantly scan all six sides—the four sides and the top and the bottom—of the compartment in which the firefighter is operating. Additionally, the TIC can be turned sideways to allow the firefighter to see both the floor and the ceiling in a hallway or upon making entry to a burning structure (**FIGURE 23-4**). Conversely, by holding the TIC in the standard orientation, the horizontal field of

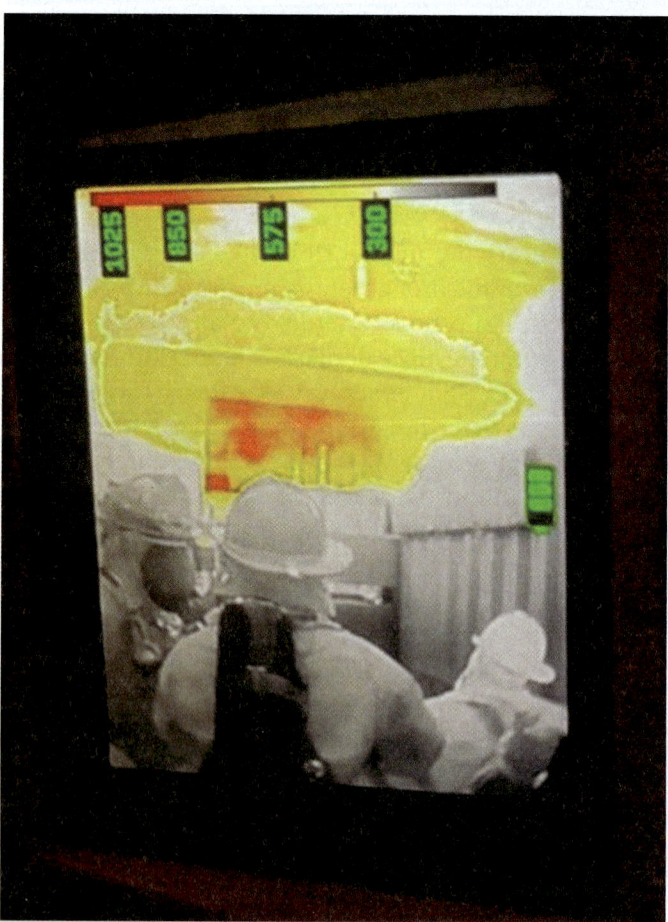

FIGURE 23-4 The field of view through the TIC screen is limited compared to the field of view through human eyes. By scanning constantly and turning the TIC sideways it can help overcome "tunnel vision."

Courtesy of Andrew Starnes.

view can be maximized to see the width of a structure or a large room.

Modes and Color Palettes

To use a TIC effectively, you need to be able to interpret the images and data shown . The images on the TIC screen are not actually images. Instead, they show the heat signatures of objects. Typically, darker colors indicate relatively cooler areas and lighter colors indicate relatively hotter areas. TICs also operate in one of two modes: high sensitivity mode or low sensitivity mode. This can greatly affect the imagery on the screen and, as one example, can create a situation where a person next to a fire may not be visible. Understanding the operating characteristics of the brand and model of TIC being used is critical to ensure its effectiveness in decision making.

Emissivity

One of the most important technical aspects of a TIC and how it is used is understanding emissivity. **Emissivity** is a description of the amount of energy radiated from a material's surface. In other words, it is measuring the reflections, not the object itself. A material with low emissivity can be considered to be a perfect mirror because it reflects all the radiated energy. Examples of materials with low emissivity that reflect heat and provide false readings are tin metal exteriors or roofing, stainless steel, and, in particular, glass. The TIC cannot see through glass, but it will be able to differentiate the heat signature convecting from a window exterior (**FIGURE 23-5**). In general, a surface that is somewhat rough in texture has a higher emissivity, and a surface that is shiny or smooth in color and texture has a low emissivity. This is why it is important for a firefighter to not rely solely on a TIC when making decisions on the fireground.

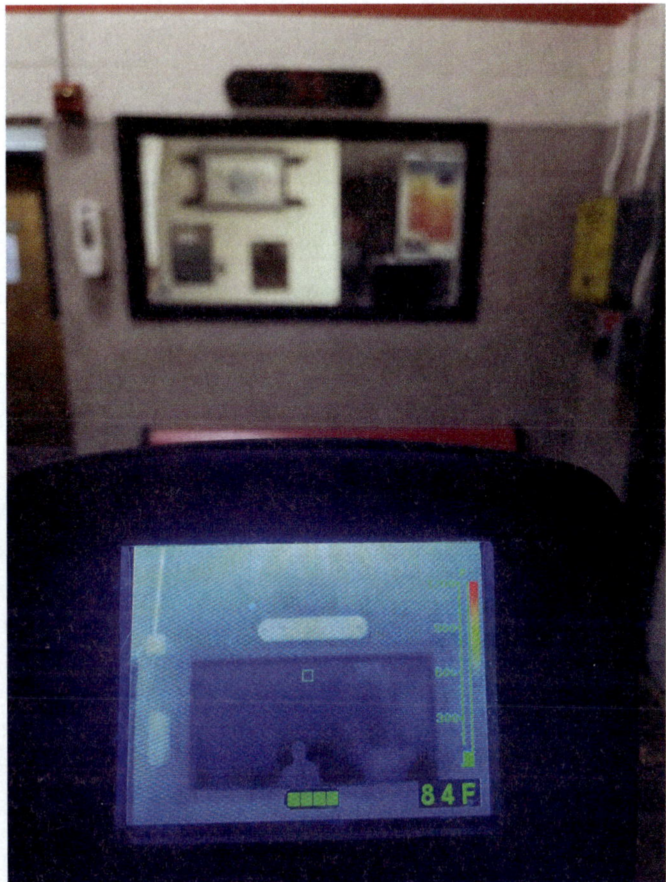

FIGURE 23-5 The TIC cannot see through glass because of its low emissivity, but can differentiate its convecting heat signature.

Courtesy of Andrew Starnes.

Using the TIC to Aid in Decision Making

With knowledge of the operating characteristics and limitations of the TIC, a firefighter—particularly a crew leader or the IC who has the responsibility at the incident scene to make tactical decisions—can take full advantage of its capabilities. The most obvious aid is to fireground decision making. Conducting a lap of the structure and ensuring a full 360-degree walk-around while looking both with the naked eye and the TIC allows you to understand where the strongest heat signature is presenting itself. This information, coupled with reading smoke, can help you determine the most effective point of attack for the first hose line and best entry for search crews to use. In the IDLH environment, particularly in zero-visibility, the TIC is invaluable in helping to identify flow paths, heat signatures from bodies, and loss of structural integrity in the floors or ceiling areas. All of this information will foster better decision making at a much more rapid pace than without a TIC. During emergencies involving utilities, a TIC can identify overheated circuit breakers or motors in heating, ventilation, and air conditioning (HVAC) systems. For hazardous materials calls, a TIC can be used to help determine surface temperatures of vessels, help differentiate product levels in containers, and identify locations of leaks. At vehicle crashes, a TIC can be used to scan the area on a dark night for additional victims who may have been ejected. A TIC can also help search and rescue efforts in wildland environments. In all cases, the TIC helps ensure situational awareness for firefighters and enhances effective decision making, resulting in lives saved and safe operations for crews. To use a TIC, follow the steps in **SKILL DRILL 23-1**.

SKILL DRILL 23-1

Using a TIC at a Structure Fire Firefighter II, NFPA 1010: 7.3.3

1. Don full personal protective equipment (PPE), including self-contained breathing apparatus (SCBA). Report to the IC, and then check into the personnel accountability system. Based on initial conditions, update the IC or call for additional resources if needed.
2. Turn on the TIC and allow it to go through its warm-up cycle. Check the status of the battery and ensure it is sufficiently charged for the incident need.
3. Conduct a full 360-degree walk-around of the exterior of the building with the TIC in low sensitivity mode. As you walk, point the TIC at the structure to identify any significant heat signatures and note their location, along with the location of smoke or fire visible to the eye.
4. If you are part of a team that will be entering the structure for fire attack or search and rescue, attach the SCBA face piece and start the air flow on your SCBA prior to entering the structure. Use the TIC to identify the spot temperature at the point of entry and any heat signatures nearby. Enter the structure if it is safe to do so.
5. Enter the IDLH area and move the TIC in an orderly fashion from left to right and up and down to scan all six sides (the four sides and the top and the bottom) of the structure. Note spot temperatures, and look for heat signatures, flow paths, and potential victims. Alternate this scan with putting the camera down and scanning with your eyes, as well as listening and feeling conditions in the structure, to avoid tunnel vision. Consider turning the TIC sideways to obtain a longer vertical view depending on the incident needs.
6. Use the TIC to direct you and your crew to the seat of the fire or the location of victims. Then use the TIC to find a safe egress route out of the structure.
7. When you are finished using the TIC, shut it off. When you return to the station, decontaminate the device and recharge it so that it is ready for its next use.

Water Supply Plan

Before beginning interior operations, a water supply should be planned for and established as quickly as possible. If the water supply is interrupted while in the IDLH environment, firefighters may suddenly find themselves in extreme danger. Initial interior operations may begin with water from the tank on the engine, but that supply is usually limited to 500 to 1000 gallons (gal; 1892 to 3785 liters [L]) depending on the apparatus. In particular, interior fire suppression in large buildings, high-rise buildings, and buildings with heavy fire or smoke require an effective plan that ensures a quick transition to an unlimited water supply. It is incumbent on the initial crew leader/IC to ensure that the plan is communicated to all units on the scene and to also prepare for a secondary water supply should the event scale up or require larger fire flows.

Staffing

In addition to having enough equipment and water resources, NFPA 1710, *Standard for the Organization and Deployment of Fire Suppression Operations, Emergency Medical Operations, and Special Operations to the Public by Career Fire Departments*, and NFPA 1720, *Standard for the Organization and Deployment of Fire Suppression Operations, Emergency Medical Operations, and Special Operations to the Public by Volunteer Fire Departments*, specify the minimum number of firefighters needed for safe and effective interior firefighting operations. Firefighters should not commit to interior operations without having enough trained firefighters on the scene to perform all of the essential fireground operations. Refer to the "11 Rules of Engagement for Firefighter Survival" (found in Chapter 19, *Firefighter Survival*), and your department's specific standard operating procedures (SOPs) before beginning any interior fireground operations.

Ventilation

Remember that ventilation of any kind impacts flow paths within the structure significantly, and that prior to final extinguishment, ventilation may cause fire growth. Ventilation must be coordinated with fire suppression efforts to ensure that it supports the IAP. Uncoordinated or improper ventilation can cause fire conditions to worsen and create a more hostile environment for both occupants and firefighters.

As extinguishment is being achieved, a decision to relieve interior smoke conditions vertically must be balanced against the potential to ventilate horizontally. To properly conduct the risk–benefit analysis of vertical ventilation, you need to be sure that the structural components over which the ventilation team will be working are not compromised. If the structural components of the roof were directly impinged upon, particularly in a modern structure, the tactic of vertical ventilation will be ineffective and possibly dangerous to the crew. The IC must determine the best course of action based on current conditions and a stringent risk–benefit analysis.

Like vertical ventilation, horizontal ventilation needs to be coordinated with the attack crew. The structure should not be vented horizontally until the attack crew has achieved knockdown and extinguishment. This will make the interior more tenable for both civilians and firefighters post-extinguishment. Ventilation prior to full extinguishment risks a high potential for fire growth and endangers crews and civilians in the interior.

SAFETY TIP

Because of modern lightweight construction and highly combustible contents, many structure fires are likely to be ventilation limited when you arrive on the scene. Remember that even the act of opening the front door can introduce additional oxygen to the fire, contributing to a rapid fire growth.

Implementation of Tactics

Once the size-up process is complete, the crew leader/IC develops the IAP with an initial determination of the strategy and communicates it to the crews on scene and those en route to the fireground. Typically, tactical objectives will be translated into task-level actions by the crews as dictated by operational guidelines and best practices.

Hose Line and Stream Selection

Once an offensive or defensive strategy is chosen, a hose line with the appropriate gallonage, length, and nozzle to implement the tactic must be selected and deployed. The volume and location of the fire will determine the volume of water required and the maneuverability needed. In general, smaller attack lines—1½ to 2 inches (in.; 38 to 51 millimeters [mm])—are lighter and more maneuverable, so they are often selected for an interior, offensive attack, particularly if the crew needs to move the line throughout a structure. Larger 2½-in. (64-mm) attack lines provide a greater flow of water for larger volume fires, but due to their weight, may not be as maneuverable in the interior. If the operation is defensive, master stream devices able to flow 350 to 1500 gallons per minute (gpm; 1325 to 5678 liters per minute [L/min]) or more may require 2½- to 3-in. (50- to 76-mm) lines. These larger lines inhibit maneuverability and require greater forethought when locating them. Efficiency and effectiveness is largely determined by the crew's level of training, which must be a point of consideration for the crew leader.

Once the line is selected and deployed, determining which type of stream—solid, straight, or fog—will be the most effective depends on the volume of fire, the size of the fire compartment, the reach required, and the nozzle selected. In each case, the crew leader must be prepared to direct the firefighter on the nozzle to ensure effective suppression efforts and avoid creating unintended hazards with poor selection and employment.

Interior Attack

Interior fire attack can be conducted on many different scales. An interior attack is often used to fight a fire that is burning in only one room. This kind of fire may be controlled quickly by one smaller handline. Larger fires that require more water can be attacked by two or more small handlines working together or by one or more larger handlines. Fires that involve multiple rooms, large spaces, or concealed spaces are more complicated and require more extensive coordination. The basic techniques for attacking larger fires are similar to the techniques used when extinguishing smaller fires. In all cases, the leader of a crew conducting an interior attack must stay mobile enough to direct the crew, ensure accountability, control hose movement, fix kinks as needed, and control the door to control airflow reaching the fire. The crew leader should also always maintain **crew integrity**—that is, they must ensure that the crew stays together and works as a team, and that no single firefighter goes off on their own. If a firefighter is hurt or needs assistance, the other member or members of the crew can provide assistance.

To coordinate an interior attack, follow the steps in **SKILL DRILL 23-2**.

SKILL DRILL 23-2

Coordinating an Interior Attack Firefighter II, NFPA 1010: 7.3.2

1. Don full PPE, including SCBA. Report to the IC, and then check into the personnel accountability system. Perform size-up and update the IC or call for additional resources if needed. Ensure an adequate water supply and appropriate backup resources are available. Select and deploy the appropriate hose line for the point of attack and the type of attack designated in the IAP. Communicate the point of attack, the type of attack, and the means of egress to the team, and then proceed to work as a team.

2. Maintain crew integrity at all times, and monitor air supply of all members. Notify command of changing fire or smoke conditions.

3. Coordinate fire attack, search and rescue, and ventilation operations in your assigned area.

4. Once the fire has been knocked down, coordinate overhaul and ensure complete extinguishment of the fire. Exit the hazard area, account for all members of the team, and report to the IC.

FIGURE 23-6 When a risk–benefit decision dictates exterior, defensive operations, the goal is to minimize further fire spread and loss of property while ensuring firefighter safety.

Courtesy of the Indianapolis Fire Department.

Exterior Attack

If fire conditions are such that there is little or no chance of occupant survival, or the building has sustained substantial damage, the IC will likely determine that a defensive operation is the only option to stabilize the incident (**FIGURE 23-6**). An exterior attack is typically used in a defensive operation. (This type of attack is discussed further in Chapter 16, *Fire Suppression*.) Initial fire streams should protect exposures to keep the fire from spreading. Master stream appliances may be used to apply thousands of gallons of water through doors, windows, or holes in the roof in the attack. During these operations, particular attention must be paid to establishing collapse zones in order to ensure firefighter safety.

Exterior operations are most effective when the fire stream (typically a master stream for defensive operations) is on the same level as the fire. For example, if fire is in the attic, an aerial device positioned to stream water horizontally through a gable vent of the attic is more efficient than an aerial device extended 60 ft (18 m) above the vent and aiming down at a 90-degree angle, particularly if the roof is still intact. Exterior fire attacks may take several hours to fully control and extinguish the fire, especially if part of the building collapses. Fires can continue to burn under piles of concrete, brick, or steel for hours or even days. If the IC determines that an exterior fire attack has substantially improved interior fire conditions, and that structural integrity has not been compromised, they may change the strategy to an offensive, interior operation using handlines for overhaul and final extinguishment. Before this occurs, the IC must conduct a risk–benefit analysis to ensure that there is a significant benefit to sending members into the interior. If the decision is made, all exterior fire streams must be shut down, safety precautions must be strictly observed, and interior crews must constantly look for signs of standing pools of water that could overload the structure, electrical or fuel gas hazards, structural damage, or impending building collapse.

Flammable Gases

Broadly speaking, flammable gases burn in the presence of an oxidant when provided with a source of ignition. Flammable gases that burn under the right circumstances include the following:

- Liquefied petroleum gas (LPG), also known as propane
- Methane, also called natural gas
- Hydrogen
- Butane
- Ethylene
- Acetylene
- Ammonia

While firefighters may encounter all of these gases in a wide variety of circumstances, the most common are natural gas and propane. In their natural state, both natural gas and propane are odorless and colorless. Because these gases are flammable and leaks can result in an explosion, the chemical mercaptan, which smells like sulfur or rotten eggs, is added so that after the gases leave production facilities, their presence can be identified by the sulfur smell (rotten egg odor). Without mercaptan, they cannot be identified unless a gas meter is present. Propane and natural gas are highly explosive and fill void spaces if not controlled quickly.

Natural Gas

Natural gas is a fossil energy source that formed deep beneath the earth's surface. It is a mixture of light hydrocarbons and other compounds including carbon dioxide (CO_2), helium, hydrogen sulfide, and nitrogen. Although the exact composition of natural gas is not constant, the primary component—typically, at least 90 percent—of natural gas is methane. Methane is highly flammable. It burns easily and almost completely and emits little air pollution. In addition, because it is lighter than air, natural gas rises if it escapes, so it dissipates from the site of a leak rather than collecting in that spot. In a structure, that could result in the gas at the ceiling or in upper floors. Natural gas is typically brought into the residential and commercial settings via a network of pipelines that start out with pressures as high as 1500 pounds per square inch (psi; 10,300 kilopascals [kPa]) in mains as large as 42 in. (1.1 m) in diameter, ultimately arriving inside a structure at pressures as low as 0.25 psi (1700 kPa) in lines that are only 1/2 in. (13 mm) in diameter. When arriving at the scene of a fire fed by natural gas, a key tactical objective is to identify the shut-off valve and shut down the fuel supply (**FIGURE 23-7**). In a residence, this is at the meter. The meter should be outside the building, but in some older buildings, it might be in the basement. If the fire is in the street because a line ruptured (for example, by road work), the shut-off valve will be somewhere along the main and usually firefighters will need to wait for the arrival of gas company personnel to shut off the flow.

It is a myth that shutting off the valve before the fire is extinguished will "suck" the fire into the pipeline. In all gas-fed fire situations, directing water at the gas-fed flame and extinguishing the gas-fed fire before the flow is shut off will create an even more hazardous situation. Because the escaping gas is no longer burning, it will collect in the atmosphere until an ignition source reignites the gas, creating an explosion with potentially devastating consequences. If you cannot access the shut-off valve, it is better to cool exposures, limit fire spread as much as is possible, and await the arrival of gas company personnel to shut off valves to the affected area.

FIGURE 23-7 The flow of gas must be shut off at a fire fed by natural gas before the flames are extinguished.

SAFETY TIP

As a general rule, do *not* extinguish fires fed by natural gas or propane until the flow is shut off. Burning gases create known combustion, but a large build-up of gas could find an ignition source and combust rapidly resulting in an explosion.

Flammable Gas in Pressurized Containers

Many flammable gases, such as propane, are stored in different types and sizes of containers. Each gas and container type presents a unique challenge when it is threatened by fire. Firefighters must quickly and carefully size up the scene to determine how, or if, they should approach the flammable gas container. Because any flammable gas stored in a container must be pressurized, they are inherently more dangerous.

The most common type of gas stored in pressurized containers is propane, which is used for heating and cooking throughout the world (**FIGURE 23-8**). There are millions of propane cylinders and tanks of various sizes throughout the United States and many of these tanks are found in unexpected locations, such as car trunks and dumpsters.

The boiling point—the temperature at which a substance changes from a liquid to a gas—of propane is –44°F (–42.2°C). When propane is placed into a storage cylinder under pressure (between 100 to 200 psi [690 to 14,500 kPa]), its boiling point is raised and it becomes a liquid. Storing propane as a liquid is very efficient because it has an expansion ratio of 270:1. This means that 1 cubic foot (ft^3; 0.02 cubic meter [m^3]) of liquid propane is converted to 270 ft^3 (7.64 m^3) of gaseous propane when it is released into the atmosphere. Put simply, a large quantity of propane fuel can be stored in a comparatively small container.

Inside a propane container, the space above the level of the liquid propane is filled with propane gas.

A.

B.

C.

FIGURE 23-8 **A.** Propane tank outside a residence. **B.** Valve for a typical 10 pound (453 gram) cylinder. **C.** Propane cylinder for a portable camping stove.

A: © VDB Photos/Shutterstock. B: © knowlesgallery/iStock/Getty Images Plus/Getty Images. C: © Sadie Mantell/Shutterstock

As the contents of the cylinder are used, the liquid level drops and the space occupied by vapor increases. A connection to a hose, tubing, or piping allows the propane gas to flow from the vapor space in the cylinder to its destination. In portable tanks, this connection is the most likely place a leak will occur.

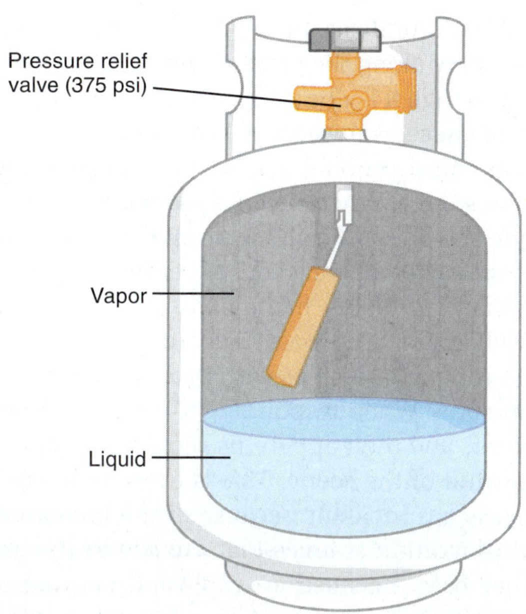

FIGURE 23-9 The pressure relief valve in a propane cylinder opens automatically when the pressure reaches 375 psi (2585 kPa).

© Jones & Bartlett Learning

As mentioned earlier, propane gas containers come in a variety of sizes and shapes. They have capacities ranging from a few ounces to thousands of gallons. The cylinder itself is usually made of steel or aluminum, although some newer ones are made from composite fiber. A discharge valve keeps the gas from escaping into the atmosphere and controls the flow of gas into the system where it is used. This valve should be easily visible and accessible and should be clearly marked to indicate the direction in which it should be turned or moved to reach the closed position. In the event of a fire, closing the valve should stop the flow of the propane and extinguish the fire.

Propane cylinders are equipped with a pressure relief valve within the vapor space that opens automatically when the pressure reaches 375 psi (2586 kPa) to allow excess pressure to escape, thereby preventing an explosion if the tank becomes overheated (or overfilled) (**FIGURE 23-9**). Propane cylinders must be stored in an upright position so that the pressure relief valve remains within the vapor space. If the cylinder is placed on its side, the pressure relief valve could fall below the liquid level, and if a fire occurs, heating the tank and causing an increase in pressure, the pressure relief valve would release liquid propane. The liquid propane would vaporize—change to gas—when exposed to the air, and would expand by the 270:1 ratio, creating a huge cloud of potentially explosive gas.

Unlike natural gas, propane gas is heavier than air, so it will flow along the ground and accumulate in low areas. Although it is nontoxic, propane can displace oxygen and cause asphyxiation. Wind and HVAC systems can also cause propane gas to migrate unexpectedly from the source. When using a gas meter to check for propane gas, be sure to check storm drains, basements, and other low-lying areas for concentrations of the gas.

Propane is highly flammable. Firefighters should be familiar with the basic hazards and characteristics of propane as well as with procedures for fighting propane fires. When responding to a reported gas leak, firefighters and their apparatus should be staged uphill and upwind of the scene. This is even more critical at a propane gas incident because of the increased likelihood of ignition sources that are low to the ground. In either case, because an explosion can happen at any time, firefighters should wear full PPE and SCBA at these incidents. Depending on the type and size of the leak, an evacuation of the area may be necessary.

Fighting fires involving propane or other flammable gas cylinders requires careful analysis and logical procedures. If the gas itself is burning because of a pipe or regulator failure, the best way to extinguish the fire is to close the main discharge valve at the cylinder. If the fire is extinguished and the fuel continues to leak, there is a high probability that it will reignite explosively because the escaping propane hugs the ground or sinks to lower levels of the structure and can finds an ignition source, such as a water heater. Do not attempt to extinguish the flames unless the source of the fuel has been shut off or all of the fuel has been consumed. If the fire is heating the storage tank, use hose streams to cool the cylinder, being careful not to extinguish the fire.

BLEVE stands for a boiling liquid/expanding vapor explosion, which is an explosion that occurs when pressurized liquid fuel, such as propane or butane, stored inside a closed vessel is exposed to a source of high heat (this phenomenon is described in detail in Chapter 4, *Fire Behavior*). As the heat causes the liquid fuel to convert to its gaseous form, the vapor pressure inside the tank increases until gas escapes from the pressure relief valve. If the fire is not extinguished quickly, more and more gas escapes, the level of the liquid fuel inside the tank drops, and the surface of the tank begins to weaken due to extreme temperatures. If the internal pressure exceeds the strength of the container, the container can catastrophically rupture, and any propane still in its liquid form will vaporize and ignite in an expanding fireball (**FIGURE 23-10**).

FIGURE 23-10 Propane cylinder failure due to fire.

The best way to prevent a BLEVE when a pressurized container is exposed to fire is to direct streams water onto the tank from a safe distance. The water should be directed at the part of the tank being heated while avoiding putting water directly on the flame itself. Also stream water on the upper part of the tank to cool the gas vapors but avoid the pressure relief valve. Rapidly releasing gas through the relief valve causes extreme cooling and adding water can create an ice plug that impedes the function of the relief valve and lead to rapid overpressurization of the cylinder. The firefighters operating these streams should work from shielded positions or use remote-controlled or unmanned master stream appliances. Horizontal tanks are designed to fail at the ends if a catastrophic failure occurs, so firefighters should operate from the sides of the tank.

Keep in mind that, if the pressure relief valve is open, the flammable gas container is under stress. Exercise extreme caution in this scenario. As the gas pressure is relieved, it can make a roaring sound. If the sound rises in frequency, an explosion could be imminent and evacuation should be ordered. Continue applying water from a distance until the relief valve resets itself.

Unless a remote discharge valve is available, the flow of gas can be stopped only if it is safe to approach the cylinder. Firefighters should inspect the integrity of the cylinder from a distance before they make any attempt to approach and close the discharge valve. If the container is damaged or the discharge valve is missing, the fuel should be allowed to burn off, while hose streams continue to cool the tank from a safe distance.

SAFETY TIP

If you need to approach a horizontal LPG tank, always approach it from the sides, not the ends.

FIGURE 23-11 A master stream appliance applying water to cool down and protect a flammable gas container that is exposed to fire.

To approach a flammable gas fire, two teams of firefighters, each with a 1¾-in. (44-mm) hose lines with an adjustable nozzle, working together, should start by using a straight stream to cool the tank from a distance. The team leader should be located between the two nozzle operators. On the command of the leader, the crews move forward, remaining together at all times. Then they should carefully approach the tank while adjusting the nozzle to a wide fog pattern. Upon reaching the discharge valve, the team leader in the center closes the discharge valve, stopping the flow of gas. Any remaining fire may then be extinguished by normal means. After the fire is extinguished, continue the flow of water as a protective curtain and to reduce sources of ignition. If the fire is extinguished prematurely, the teams should advance to close the discharge valve as quickly as possible.

SAFETY TIP

Always approach and retreat from these types of fires while facing the objective with water flowing in case of reignition.

Unmanned master stream appliances should be used to protect large flammable gas containers that are exposed to heavy fire. As with smaller cylinders, the water should be directed at the part of the tank being heated, but avoid putting water directly on the flame itself (**FIGURE 23-11**). Also stream water on the upper part of the tank to cool the gas vapors, but do not stream water onto the pressure relief valve.

SAFETY TIP

If the propane container is located next to a fully involved building or a fire that is too large to control, strongly consider evacuating the area.

Continue to cool liquid-fuel containers until either the discharge valve is closed or all fuel escapes from the container. Keep unmanned master streams in place until the incident is declared to be under control by the IC.

To suppress a flammable gas cylinder fire, follow the steps in **SKILL DRILL 23-3**.

Liquid-Fuel Fires

Class B fires involving combustible and flammable liquids are among the most dangerous and challenging types of fires to extinguish. Because these liquids produce huge amounts of heat energy, they can be dangerous to approach, are difficult to extinguish, and may accelerate fire growth in Class A fuels. Firefighters can encounter liquid-fuel fires in almost any type of occupancy. Many fires involving a vehicle, including planes, trains, ships, and trucks, as well as cars, may involve a

SKILL DRILL 23-3

Suppressing a Flammable Gas Cylinder Fire Firefighter II, NFPA 1010: 7.3.4

1. Don full PPE and SCBA. Using a straight stream from as far away as possible, direct water at the part of the tank being heated and at the upper part of the tank to cool the tank until the pressure relief valve resets. Avoid putting water directly on the flame itself or on the relief valve.

2. Two teams of firefighters, using a minimum of two 1¾-in. (44-mm) hose lines, advance toward the side of the tank using a straight stream. Do not approach the tank from either end. The team leader is located between the two nozzle operators. The team leader coordinates the advance toward the cylinder.

3. As the firefighters approach the side of the tank, gradually adjust the nozzles to a wide fog pattern, and overlap the fog streams as you reach the tank.

4. When the cylinder is reached, the two teams operating the handlines isolate the discharge valve from the fire with their fog streams while the leader closes the discharge valve, eliminating the fuel source and extinguishing the fire.

SKILL DRILL 23-3 CONTINUED

Suppressing a Flammable Gas Cylinder Fire Firefighter II, NFPA 1010: 7.3.4

5. Both teams continue to apply water to the cylinder to cool the metal, with the goal of preventing possible tank failure and a subsequent BLEVE.

6. As cooling continues, both teams slowly back away from the cylinder while adjusting the nozzles to a straight stream as they retreat.

combustible or flammable liquid. In a liquid-fuel fire, the liquid itself does not burn (explained further in Chapter 4, *Fire Behavior*). Instead, what burns is the vapors that evaporate from the surface of the liquid and mix with air. Different liquids have different flash and fire points. For example, gasoline produces flammable vapors at temperatures as low as −45°F (–42.7°C), so as a result, gasoline is extremely volatile at ambient temperatures and pressures.

The skills used in suppressing small flammable liquid-fuel fires are presented in Chapter 9, *Portable Fire Extinguishers*. Special tactics must be used when extinguishing larger liquid-fuel fires, and Class B extinguishing agents, such as foam or dry chemicals, may be needed.

Liquid-fuel fires are classified as either two-dimensional or three-dimensional. A **two-dimensional liquid-fuel fire** is a fire that is burning only on the top of a spill, pool, or open container of liquid. A **three-dimensional liquid-fuel fire** is a fire in which burning liquid fuel is dripping, spraying, or flowing over the edges of a container. A two-dimensional liquid-fuel fire can usually be controlled by applying the appropriate Class B foam to the burning surface. When the foam solution is discharged, the foam flows across the surface of the liquid and creates a seal that stops the fuel from vaporizing. The foam also separates the oxygen in the atmosphere from the fuel, extinguishing the fire. Class B foam contains water, and as the water drains from the foam blanket, the water will cool the liquid fuel and further reduce the possibility of reignition. A three-dimensional liquid-fuel fire is difficult to extinguish with a foam stream because the foam usually cannot establish an effective seal between the fuel and the oxygen. Dry chemical or gaseous extinguishing agents may be more effective than foam in controlling these kinds of fires. Dry chemical and gaseous agents can also be used to extinguish two-dimensional fires, but they do not provide protection against reignition. In some cases, a fire can be extinguished with a dry chemical, and then the surface covered with foam to prevent reignition. Chapter 24, *Systems of Fire Detection, Suppression, and Smoke Control*, discusses most of these special hazard-suppression agents in more detail.

It is important to determine which liquid fuel is involved so that you can select the appropriate extinguishing agent. You also need to know whether the vapors are lighter or heavier than air. Vapor density is the weight of a gas compared to air, and gases with a vapor density less than 1 are lighter than air and rise to the top of a confined space or rise into the atmosphere, while gases with a vapor density greater than 1 are heavier than air and sink to the ground (discussed further in Chapter 5, *Fire Behavior*). Vapors that linger and pool near the ground can ignite if exposed to hot surfaces or open flames even after the liquid-fuel fire has been extinguished.

SAFETY TIP

Firefighters should avoid standing in pools of flammable liquids or contaminated runoff, because their PPE will absorb the flammable product and become contaminated. In cases of serious contamination, the PPE itself can become flammable.

Firefighting Foam

Because water is so effective at cooling the surface of the burning fuels, it is the extinguishing agent of choice for many fires. Some fires, however, require the use of special extinguishing agents, such as incidents involving combustible and flammable liquids. Successful fire control and extinguishment at these incidents require not only the proper application of foam on the fuel surface, but also an understanding of the physical characteristics of foam.

Firefighting foam is produced by mixing **foam concentrate**—a concentrated liquid foam agent—with water to create a **foam solution**. When the foam solution is discharged, the foam solution is mixed with air to create the foam.

Different types of foam concentrate or even different brands of the same type of foam concentrate must never be mixed unless they are known to be compatible. Some concentrates react with other types or brands of foams and congeal in the storage containers or in the foam delivery system. This congealed foam can plug a foam delivery system and render it useless.

Although foam is a water-based extinguishing agent, its primary extinguishing mechanism is not cooling the surface of burning fuels. Most firefighting foams prevent ignition or extinguish fires by creating a **foam blanket** that separates the fuel vapors from the heat source and excludes oxygen in the air from the burning reaction. Some types of foam create foam blankets that are stable for a longer period of time than others. In addition, different types of foams have different levels of **burnback resistance**, which is the ability of the foam blanket to maintain its integrity and effectiveness when impinged by fire.

Once a foam blanket has been applied, it must not be disturbed. The fuel under the foam blanket is still capable of producing ignitable vapors. If the foam blanket is disturbed by wind, by someone walking through the liquid, or by hose streams breaking up the foam blanket, ignitable vapors may be released and could be reignited easily.

Firefighting foams are rated for either Class A or Class B fires or both (see Chapter 9, *Portable Fire Extinguishers*). Within each classification there are many types of foam. Each class and type of foam extinguishes fire in a unique way, depending on its intended application. Firefighters must become familiar with the specific types of foam used by their fire department and understand how to select and apply them based on the specific hazard faced. They should read and follow the manufacturer's instructions before using foam.

Standard PPE provides appropriate protection when working with foam agents, but it should be thoroughly washed and dried after contact with foam. Follow the instructions from both the PPE and foam manufacturers. All equipment should be thoroughly cleaned with clear water after use with foam. In addition, firefighters should wash their skin thoroughly after coming into contact with foam and completely clean all equipment with clear water after foam use.

SAFETY TIP

Some foam concentrates have been associated with health hazards, ranging from mild skin irritation to association with respiratory issues and cancer. Always adhere to the handling instructions provided by the manufacturers, as well as local SOPs.

Class A Foam

Class A foam is used to fight Class A fires, that is, fires involving ordinary, solid, combustible materials, such as wood, paper, and textiles, as well as organic materials such as grass, hay, and straw. Class A foam is particularly useful for protecting buildings in rural areas during forest and brush fires when the water supply is limited. It can also be used on a brush fire after extinguishment to allow for additional water penetration and reduce the likelihood of the fire rekindling.

Class A foam contains a solution of water and Class A foam concentrate. Because Class A foam is a water-based fire suppression agent, it is effective for cooling Class A combustibles below their ignition temperatures. It also increases the effectiveness of water as an extinguishing agent because it acts as a **surfactant**—a substance that reduces the surface tension of water, thereby increasing the concentrate's ability to penetrate, mix with, and spread. A concentrate that reduces the surface tension of water is commonly referred to as a **wetting agent** or **wet water**. This reduction in surface tension allows the water to penetrate dense materials and tightly stacked ordinary combustibles, such as large hay bales or deep-seated fires in pallets, instead of running off the surface. It allows more heat to be absorbed by the water. And because foam tends to stick to both horizontal and vertical surfaces, it also keeps water in contact with unburned fuel to prevent initial ignition or reignition. In addition to cooling the fuel, Class A foam applied directly to a burning fuel surface insulates the fuel from radiant heat, excludes oxygen, and reduces vapor production.

Class A foams are not designed to resist hydrocarbons and will break down, thereby eliminating their effectiveness. Additionally, they cannot be used on liquid-fuel fires because they do not adhere to the liquid-fuel surface. If Class A foam is used on a liquid-fuel fire, the foam blanket can be easily disrupted, and may break down quickly. If this happens, there is a risk of reignition.

Class A foam concentrates are usually formulated to be mixed with water to form solutions at concentrations of 0.1 percent to 1 percent—that is, at a ratio of 1 gallon or liter of concentrate to 999 gallons or liters of water to 1 gallon or liter of concentrate to 99 gallons or liters of water. This range makes it possible to produce a wetter foam that has good penetrating properties or a drier foam that is more effective when applied as a foam blanket.

Class B Foam

Class B foam is used to fight Class B fires involving flammable and combustible liquids. It can also be used to fight Class A fires. When applied to burning liquid fuel, the blanket of foam bubbles floats on the surface of the fuel. This foam blanket excludes oxygen from the surface of the burning fuel and separates the fuel from the fire (**FIGURE 23-12**). Additionally, because the foam bubbles contain some water, the surface of the fuel is cooled, further reducing vapor production of some liquid fuels.

When using foam to extinguish a liquid-fuel fire, it is critical to apply enough foam to cover the liquid surface fully. If the foam does not completely cover the surface, the fire will continue to burn. Also, all firefighting foam degrades over time, so you need to apply enough foam to maintain the foam blanket. The foam also needs to be applied relatively quickly. If it is applied too slowly, the column of hot gases, flames, and smoke rising above the fire will keep the foam from covering the surface. If this happens, the heat of the fire may destroy any foam that has already been applied.

FIGURE 23-12 A foam blanket excludes air and separates the flames from the fuel surface.

Class B foam can also be applied to a liquid-fuel spill to prevent a fire. A foam blanket floating on the surface of liquid fuel inhibits the production of vapors that might otherwise be ignited. The rate of foam application is not as critical in this case because there is no fire working to destroy the foam while it is being applied.

The compatibility of foam agents with other extinguishing agents needs to be considered, as well. For example, some combinations of dry chemical extinguishing agents and foam agents can cause an adverse reaction. Firefighters need to become familiar with all of the types of extinguishing agents used by their department and read all the information available from the manufacturers of the agents to ensure that they understand which agents are compatible.

Like Class A foam concentrates, Class B foam concentrates are mixed with water to form the foam solution. Most Class B foam concentrates are designed to be used in either 1 percent, 3 percent, or 6 percent solutions. For example, a 3 percent foam solution is created by mixing 3 gallons or liters of foam concentrate with 97 gallons or liters of water to produce 100 gallons or liters of foam solution. The same foam concentrate might be designed to be used as a 3 percent solution to extinguish a fire involving ordinary hydrocarbons but as a 6 percent solution for a fire involving polar solvents. Firefighters need to determine the type of fuel involved in the incident so that they can prepare the foam using the correct ratio of concentrate to water.

Types of Class B Foam

Different types of Class B foam are formulated to be effective on different types of liquid fuels. Firefighters must ensure that they use the proper foam for the specific hazard because some liquid fuels are incompatible with some foam formulations and will destroy the foam before the foam can control the fire. For example, an alcohol-resistant foam must be used if the incident involves a polar solvent. A **polar solvent** is a water-soluble flammable liquid (such as alcohols, acetone, esters, and ketones) that readily mixes with water. An ordinary foam would break down quickly if it came in contact with this type of product.

The major categories of Class B foam concentrate are as follows:

- Protein foam
- Fluoroprotein foam
- Film-forming fluoroprotein (FFFP)
- Aqueous film-forming foam (AFFF)
- Alcohol-resistant foam

Protein foam is made from animal by-products. They are effective on petroleum-based fires such as gasoline or diesel fuel fires. They are applied in 3 percent or 6 percent solutions, and they spread slowly over the surface of burning fuels.

Fluoroprotein foam is made from the same base materials as protein foam but include additional fluorochemical surfactant additives. These additives allow the foam to produce a fast-spreading film. This type of foam is used for hydrocarbon vapor suppression and extinguishment of deep-seated fires.

Film-forming fluoroprotein foam (FFFP) is a fluoroprotein foam with film-forming fluorinated surfactants, which makes this type of foam capable of forming a water solution film on the surface of most flammable liquid fuels. Like fluoroprotein foams, FFFP foams are fast-spreading, and they act as a barriers to exclude air and prevent vaporization.

Aqueous film-forming foam (AFFF) is a synthetic-based foam that is particularly suitable for fires involving gasoline and light hydrocarbon fuel spills, and it can also be used on actively burning pools of flammable liquids. It contains film-forming fluorinated surfactants, which makes it capable of forming a film on the surface of most flammable liquid fuels. The film forms a seal across a surface quickly, and the foam has excellent vapor suppression capabilities.

Alcohol-resistant foam has properties similar to AFFF, but it is formulated so that alcohols and other polar solvents do not dissolve it. Regular protein foams cannot be used on these types of products.

Dangers of Fluorinated Foams

Some components in Class B foam concentrates that have been widely used in the past are being phased out because of environmental and health concerns. More concerning for firefighters is that some foam concentrates contain perfluoroalkyl and polyfluoroalkyl substances (PFAS), which are carcinogens. Certain types of PFAS can accumulate and stay in the human body for long periods of time, and long-term exposure to PFAS in high concentrations causes a buildup in the body. This buildup may have negative health effects, including an increased risk of thyroid disease and testicular, kidney, and bladder cancers. Because of environmental and health concerns, newer foam concentrates have been developed that are equally effective but lack the undesirable properties.

Class B foams are sophisticated firefighting agents that require ongoing education and training to use them effectively and safely. With different foam types formulated for specific fuel types, it is critical for firefighters to receive continual instruction to ensure proper foam selection and application. As foam formulas evolve to address health and environmental concerns about fluorinated surfactants, departments must provide training on new guidance and alternative foams to smoothly transition usage. Trusted resources like the National Fire Protection Association's Fire Protection Research Foundation offer invaluable knowledge as technology progresses. The cancer risks associated with PFAS exposure highlight the need for continuing education on containment, decontamination, disposal, and other safe handling protocols. As an increasingly complex and critical firefighting tool, Class B foams demand up-to-date mastery through formal training and education programs to maximize firefighter safety while maintaining efficacy.

In May 2022, the Fire Protection Research Foundation (FPRF) completed a project focused on developing a strategic roadmap for fire departments transitioning away from legacy AFFF foams. The multiphase project involved extensive research and analysis on regulations, foam performance, health impacts, remediation, and best practices. A major component was a virtual workshop attended by over 500 individuals globally. The roadmap emphasizes the need for continued guidance and resources to assist fire departments in this complex transition. Key focus areas include assessing replacement foam capabilities, new education/training approaches, foam containment/disposal protocols, minimizing firefighter exposures, and developing model procedures and strategies for new agents. The project synthesized current knowledge into recommended best practices while identifying remaining knowledge gaps requiring further research. Fire departments can utilize the roadmap findings, available on the FPRF website, as an ongoing reference covering foam replacement challenges and potential solutions. As research continues, the roadmap provides a framework for safely phasing out PFAS-containing foams with suitable alternatives to protect firefighters, communities, and the environment.

Foam Apparatus

Some fire departments operate apparatus that is specifically designed to produce and apply foam. The most common examples are found at airports where these foam apparatus are used for aircraft rescue and firefighting (**FIGURE 23-13**). Additionally, specialized foam units can be found in departments where bulk fuel storage or other large industrial facilities exist. These large vehicles carry the foam concentrate and water onboard and are designed to quickly apply large quantities of foam to a liquid-fuel fire. Remote-control monitors can be used to apply foam while the vehicles are in motion. If these types of apparatus are close to your department, you should train with them and know the best way to use their capabilities before an emergency occurs.

FIGURE 23-13 Special foam apparatus are used for aircraft rescue and firefighting.

Courtesy of Tim Olk.

Creating Foam

To create and spray foam, fire departments need equipment to combine the foam concentrate and water, and either a system that injects compressed air into the mixture or special air-aspirating nozzles to apply the foam. Most engine companies carry the necessary equipment to place at least one foam attack line into operation. Some departments have apparatus with onboard tanks of foam concentrate and a built-in system for combining the foam concentrate and water. Other departments have foam apparatus available for situations when large quantities of foam are needed.

Backup Resources

When attempting to use foam to extinguish a liquid-fuel fire, it is critical that enough foam concentrate is available to complete the job. If the flow of foam is interrupted while additional foam supplies are obtained, the fire can destroy the foam that has already been applied. Foam manufacturers provide specific formulas that enable firefighters to calculate how much foam is required to extinguish fires of different sizes. Most fire departments have contingency plans to deliver large quantities of foam to the scene of a major incident.

This foam could be located on designated vehicles or kept in storage at fire stations. Backup sources often include airport crash vehicles or petroleum facilities that have their own foam apparatus and often keep large quantities of foam in storage. Manufacturers of foam products also have emergency programs to deliver large quantities of foam to the scene of exceptionally large-scale incidents.

Compressed Air Foam Systems

Compressed air foam (CAF) is produced by injecting compressed air into a stream of water that has been mixed with 0.1 percent to 1.0 percent foam. This results in a highly compacted foam with small bubbles. CAF has excellent surface-adherence properties, so it can be a good choice for preventive foam to be applied to exposure buildings. CAF is also an excellent foam for overhaul and for deep-seated Class A fires.

CAF is produced by a CAF system (CAFS), which is a combination of a centrifugal fire pump, a foam metering device to inject the foam concentrate on the discharge side of the pump, and an air compressor to inject compressed air bubbles into the foam solution. This solution is scrubbed by its movement through the attack line, which means it is mixed aggressively. As it exits the nozzle—typically a smooth bore for best results—the compressed air expands, producing a finished foam product. Hose lines containing CAF are lighter than hose lines that are completely filled with water because the CAF consists of a mixture of water and a significant amount of air. The finished foam product is usually discharged at a rate between 40 to 125 gpm (151 to 473 L/min) as measured at the nozzle tip before the foam expands.

Foams of different consistencies can be produced by adjusting the ratio of air to water. When CAF is used for interior structural firefighting, a wetter foam is often used. Wetter foams, produced by decreasing the ratio of air to water, have better surface penetration properties and drain more quickly in the presence of heat. Because CAF adheres to most surfaces, it is used to protect exposures during structural fires and to coat the sides and roofs of buildings during a wildland fire as a precautionary measure, and when prepared as a drier foam, it will remain in place for a longer period of time than many other foams. Drier foams, produced by increasing the ratio of air to water, produce more durable bubbles and have longer drain times.

Like every tool on the fireground, there are pros and cons to CAFS. Each department must decide what best suits their needs and what will work best in their community.

Foam Proportioning Equipment

A **foam proportioner** mixes the foam concentrate into the fire stream in the proper percentage. The two types of proportioners—eductors and injectors—are available in a wide range of sizes and capacities.

Foam Eductors. A foam **eductor** draws foam concentrate from a container or storage tank into a moving stream of water or into a hose line (**FIGURE 23-14**). Eductors operate on the Venturi effect, informally referred to as the suction effect, much like a garden hose–end sprayer applies fertilizer or weed killer from a container attached to the end at the nozzle. As the water stream passes through a narrowing in the eductor, the increase in water velocity reduces the water pressure as it flows into the eductor. Foam concentrate is introduced into the eductor using a metering device that sets the percentage of foam concentrate educted into the stream (**FIGURE 23-15**). Eductors are usually designed to work at a predetermined pressure and flow rate.

FIGURE 23-14 The most common type of eductor used in the fire service is the portable in-line eductor.

FIGURE 23-15 The metering device controls the amount of foam concentrate educted into the water.

Two types of eductors are used by the fire service: bypass eductors and in-line eductors. Bypass eductors are permanently mounted devices on an engine that can be used for water application or foam application, depending on what is required at the incident scene. The most common type of eductor is the portable in-line eductor, which is an eductor that is placed in the attack line at a coupling. The portable in-line eductor is sized to work with a 1½-in. (38-mm) or 1¾-in. (44-mm) attack line. It uses water pressure to draw foam concentrate from a portable container into the stream. As water flows through the eductor, foam concentrate is drawn into the stream through a tube called the pickup tube to create the proper foam solution. The pickup tube and the container of foam concentrate should be at similar elevations so that the Venturi effect will occur. The foam solution is then discharged through a foam nozzle to produce the finished foam. A **foam nozzle** is a nozzle specifically designed to help with aeration of the foam solution as it exits the hose line and complete the process of becoming a finished foam product. Be sure to follow the manufacturer's instructions or refer to your department's SOPs for proper foam eductor operation.

To place an in-line eductor foam line in service, follow the steps in **SKILL DRILL 23-4**.

Foam Injectors. A **foam injector** adds foam concentrate under pressure to the water stream. Most injector-based proportioning systems can work across a range of flow rates and pressures. A meter measures the flow rate and pressure of the water and adjusts the injector to add the proper amount of foam concentrate. This type of system is often installed on special foam apparatus.

SKILL DRILL 23-4

Operating an In-Line Foam Eductor Firefighter II, NFPA 1010: 7.3.1

1. Don all PPE. Make sure all necessary equipment is available, including an in-line foam eductor and the correct nozzle. Ensure that enough foam concentrate is available to suppress the fire. Deploy an attack line, and replace the nozzle with the foam nozzle if necessary.

2. Place the foam concentrate container next to the eductor, read the container label to check the percentage at which the foam concentrate should be used, and set the metering device on the eductor accordingly.

Continues.

SKILL DRILL 23-4 CONTINUED

Operating an In-Line Foam Eductor Firefighter II, NFPA 1010: 7.3.1

3. Place the in-line eductor in the hose line according to the manufacturer's instructions and your department's SOPs.

4. Place the pickup tube from the eductor into the foam concentrate, keeping both items at similar elevations. Charge the hose line with water per your department SOPs or as directed by the manufacturer.

5. Flow water through the hose line until foam starts to come out of the nozzle. The hose line is now ready to be advanced to the fuel.

Batch Mixing

Foam concentrate can be poured directly into an apparatus booster tank to produce a foam solution through a technique called **batch mix** (**FIGURE 23-16**). For example, if the booster tank has a capacity of 500 gal (1893 L), and you are using 3 percent foam concentrate, 15 gal (56 L) of the 3 percent foam concentrate should be added. If you are adding 6 percent foam concentrate to the same size tank, 30 gal (113 L) of foam concentrate should be added. It might be necessary to drain sufficient water from the tank first to make room for the foam concentrate. After the concentrate is added, circulate the solution through the pump to mix it before it is discharged. Note that if a Class A foam solution is batch mixed, it retains its effectiveness for only 24 hours, so it must be used within that time frame.

FIGURE 23-16 Batch mixing is not typically the preferred method to mix foam concentrate and water.

Premixed Foam Solution

Foam fire extinguishers are filled with a premixed foam solution and pressurized with compressed air or nitrogen. Many fire departments use these extinguishers for small flammable liquid spills at the scene of a motor vehicle accident. In addition, some apparatus are equipped with large tanks holding 50 or 100 gal (189 or 379 L) of premixed foam solution, and they operate in the same manner.

Foam Application

Foam can be applied to a fire or a liquid-fuel spill using portable extinguishers, handlines, master stream appliances, and a variety of automatic fire suppression systems for special applications. Most fire departments apply foam through handlines or master stream appliances mounted on fire apparatus.

Foam Expansion Ratios

Foam has a wide range of expansion rates, depending on the aeration—the amount of air mixed into the stream and the size of the bubbles produced—applied to the foam solution. The expansion rate is determined by the volume of foam concentrate compared to the volume of the finished foam as it is applied to the fire or hazard. NFPA 11, *Standard for Low-, Medium-, and High-Expansion Foam*, identifies three categories of foam based on their expansion rates.

Low-expansion foam has an expansion ratio of less than 20:1. This means that 1 ft^3 (1 m^3) of foam concentrate will produce about 20 ft^3 (or 29 m^3) of finished foam. Low-expansion foam has little air entrained into the foam solution, which results in smaller bubble size. This type of foam is often produced using standard adjustable fog-stream nozzles, where the air is entrained by the flowing stream and mixed into the foam solution. Although a fog-stream nozzle may be used to apply AFFF or Class A foam, many manufacturers recommend using specific air-aspirating nozzles to produce a higher-quality foam blanket. Most fire departments carry low-expansion foam for use on two-dimensional liquid-fuel spills such as gasoline or diesel fuel.

Medium-expansion foam has an expansion ratio between 20:1 and 200:1. It is produced using an air-aspirating nozzle on a handline or a foam-generating nozzles in an automatic fire suppression system. These nozzles introduce more air into the stream and produce a consistent bubble structure. Although some departments use medium-expansion foam for manual firefighting operations, it is typically used in fire suppression systems designed to protect three-dimensional hazards.

High-expansion foam has an expansion ratio between 200:1 and 1000:1. It contains a much higher proportion of air, which produces large bubbles. High-expansion foam is used in fire suppression systems designed to protect high-value equipment in an enclosed space. to provide a thick layer of foam several feet high. These fixed fire protection systems use a high-expansion foam generator to introduce large quantities of air into the discharge stream, which provides a thick layer of foam that completely fills entire rooms or a large, enclosed space with foam. The foam covers the liquid-fuel hazard and provides insulating protection to high-value equipment. These systems are most likely to be found in aircraft hangars or other large industrial areas.

Nozzles for Applying Foam

Nozzles are an important part of all foam operations. The proper nozzle is needed in order for firefighters to be able to produce a good quality foam blanket. The primary difference between the types of nozzles used to discharge foam is the manner in which air is introduced into the stream of foam solution to produce the desired expansion ratio of foam and consistency of air bubbles. Nozzle types used with foam include the following:

- Medium- and high-expansion foam-generating nozzles
- Master stream foam nozzles
- Air-aspirating foam nozzles
- Smooth-bore nozzles
- Fog-stream nozzles

The advantages and disadvantages of each type of nozzle are dependent upon the application technique, the foam proportioning equipment, the foam solution, and the type of fire being suppressed.

Foam Application Techniques Using Handlines

Class A foam is most effective if it is applied directly to the surface of the burning solid fuels or to fuels you want to protect from radiant heat. Many departments use foam nozzles for Class A foam application. Some departments that use Class A foam in very low concentrations—0.25 percent to 0.5 percent concentration—use a fog-stream nozzle. This provides the heat absorption capabilities of a fog stream with the benefits of foam in one application.

Firefighters must use caution when using foam to suppress liquid-fuel fires. When applying foam to a liquid-fuel fire using a handline, the correct application techniques must be used to produce the desired quality of foam and to successfully blanket the surface of a burning liquid-fuel fire or a liquid spill.

Applying a Class B foam stream directly onto a pool of burning liquid fuel may cause the fuel to splash and spread the fire. Plunging a stream into an existing foam blanket allows vapors to escape, which can result in fire spread, reignition, or a flare-up. To avoid these problems, three methods are used to apply foam blankets on Class B fires: the roll-in method, the bounce-off method, and the rain-down method. Whichever method is used, the foam must be applied carefully to avoid disrupting the existing foam blanket or splashing fuel outside of the container or pool. These methods can also be used to apply Class A foam when it is not possible or safe to apply the foam directly to burning objects.

Roll-On Method. The **roll-on method**, also called the **roll-in method** or **sweep method**, is used to apply foam to a pool of flammable product on open ground. To perform this technique, direct the foam stream onto the ground in front of the burning product and move the stream back and forth in a slow, steady, horizontal motion to push the foam forward gently until the area is covered. The energy of the stream pushes the foam blanket across the surface of the fuel—or rolls it onto the surface of the fuel. It is important to push the foam slowly and gently so that the foam blanket is not disturbed. Avoid aiming the nozzle down toward the surface of the spill to prevent splashing fuel. The firefighter applying the stream might need to move to different positions to be sure that the entire surface of the fuel is covered by the foam blanket. To perform the roll-in method of applying foam, follow the steps in **SKILL DRILL 23-5**.

Bounce-Off Method. The **bounce-off method**, also called the **bank-shot method** or **bank-down method**, is used at fires where the firefighter can deflect the foam stream off an object and let the foam flow down onto the burning surface. This method could be used to apply foam to an open-top storage tank or a rolled-over transport vehicle, for example. To do this technique, the firefighter sweeps the foam stream back and forth against the object and the foam flows down and spreads back across the surface. As with all foam application methods, it is important to let the foam blanket flow gently onto the surface of the flammable liquid to form a blanket. To perform the bounce-off method of applying foam, follow the steps in **SKILL DRILL 23-6**.

Rain-Down Method. The **rain-down method** is used when there is no object available to deflect the foam onto the burning surface or for applying foam to the edge of a large area of burning liquid fuel from a safe distance. To perform this technique, the firefighter lofts the foam stream into the air above the fire and lets it fall down gently onto the burning surface. Because the stream breaks apart above the fuel surface as it falls, the foam does not disrupt an existing foam blanket or cause the fuel to splash.

As you apply the foam, carefully observe how the foam blanket is building up, and direct the stream so that the entire surface is covered.

The rain-down method is a good choice for initial application if fire conditions make it difficult to approach the fuel surface. Once a portion of the burning liquid is covered with foam, the roll-in or bounce-off method may be used if it is safe to approach the fire.

To perform the rain-down method of applying foam, follow the steps in **SKILL DRILL 23-7**. In this example, foam is being applied as a protective coating.

SKILL DRILL 23-5

Performing the Roll-In Method of Applying Foam Firefighter II, NFPA 1010: 7.3.1

1. Open the nozzle and test to ensure that foam is being produced. Move within a safe range of the fuel product or tank, and then open the nozzle. Direct the stream of foam onto the ground just in front of the pool of product.

2. Move the stream back and forth in a slow, steady, horizontal motion. Allow the foam to roll across the top of the pool of the fuel product or tank until it is completely covered.

SKILL DRILL 23-6

Performing the Bounce-Off Method of Applying Foam Firefighter II, NFPA 1010: 7.3.1

1. Open the nozzle and test to ensure that foam is being produced. Move within a safe range of the fuel product or tank, and then open the nozzle. Direct the stream of foam onto a solid structure such as a wall or metal tank so that the foam is directed off the object and onto the pool of product or the tank. Allow the foam to flow across the top of the pool of product or the tank until it is completely covered. The foam may need to be bounced off several areas of the solid object to completely cover the pool or the tank.

SKILL DRILL 23-7

Performing the Rain-Down Method of Applying Foam Firefighter II, NFPA 1010: 7.3.1

1. Open the nozzle and test to ensure that foam is being produced.

2. Move within a safe range of the fuel product or tank, and then open the nozzle.

3. Direct the stream of foam into the air so that the foam gently falls onto the surface of the fuel product or tank.

4. Allow the foam to flow across the surface of the fuel product or tank until it is completely covered.

Hose Testing and Records

As discussed in Chapter 11, *Fire Hose, Fire Appliances and Tools, and Nozzles*, each length of hose should be tested at least annually, according to the procedures listed in NFPA 1962, *Standard for the Care, Use, Inspection, Service Testing, and Replacement of Fire Hose, Couplings, Nozzles, and Fire Hose Appliances.* The Firefighter II hose-testing procedures are complicated and require special equipment. This equipment must be operated according to the manufacturer's instructions.

To perform an annual service test on fire hose, follow the steps in **SKILL DRILL 23-8**.

A hose record is a written history of each individual length of fire hose. Each length of hose should be identified with a unique number stenciled or painted on it. A hose record should contain information such as the following:

- Hose size, type, and manufacturer
- Date when the hose was manufactured, purchased, and tested
- Any repairs that have been made to the hose

Some fire departments keep hose records on cards or paper files; others keep the records in a database in a computer system.

SKILL DRILL 23-8

Performing an Annual Service Test on a Fire Hose Firefighter II, NFPA 1001: 7.5.5

1. Don turnout gear. Connect up to 300 ft (91 m) of hose to a hose test gate valve on the discharge valve of a fire department pumper or hose tester.
2. Attach a nozzle to the end of each hose. Slowly fill each hose with water at 50 psi (345 kPa), and remove kinks and twists in the hose.
3. Open the nozzles to purge air from the hose, discharging the water away from the test area. Close the nozzles once the air is purged. Measure and record the length of each section of hose.
4. Mark the position of each hose coupling on the hose. This will help determine if slippage occurs during the test (Step 7). Check each coupling for leaks. If leaks are found behind the coupling, remove the hose from service. If the leak is in front of the coupling, tighten the leaking coupling. If the leak continues, replace gaskets if necessary after shutting down the hose line.
5. Close each hose test gate valve.
6. Ensure that all firefighters are clear of the test area. Increase the pressure on the hose to the pressure required by NFPA 1962, and maintain that pressure for 5 minutes. Monitor the hose and couplings for leaks as the pressure increases during the test. Close the gate valves and open the nozzles to bleed off the pressure. Uncouple and drain the hose.
7. Inspect the marks placed on the hose jacket near the couplings to determine whether slippage occurred.
8. Tag hose that failed.
9. Mark hose that passed. Record the results in the departmental logs.

CASE STUDY

You Are the Firefighter CONCLUSION

You are the crew leader on Engine 29, which was dispatched at 1150 as first-due to a house fire. The second-due company, Engine 1, has an officer who will take command upon their arrival, but they are 5 to 7 minutes out. It is a weekday in early fall, and the weather report you reviewed at lineup that morning told you that the temperature is 60 degrees F (15 degess C), with winds blowing 13 to 15 mph (21–24 km/h) with no other impending weather concerns. You start filling in your crew on the information from the CAD which says that the homeowner states that everyone is out of the house and that the fire started in the kitchen. On the map you note the location of a hydrant, and based on Engine 1's direction of travel and their response time, you tell your driver/operator to perform a forward lay from the hydrant and advise Engine 1 of the hydrant location. You ask Engine 1 to complete the water supply. Upon your arrival, the homeowner confirms everyone is out of the house. As you begin your size-up, you make note of several facts: the front door is open and a significant amount of dark, pressurized, turbulent smoke is pouring out; the house is an approximately 80- × 60-ft (24- × 18-m), two-story, colonial-style home and is set back at least 75 ft (23 m); and the wind is blowing from the back of the house towards the street. You provide your initial size-up report over the radio, "Engine 29 is on the scene. Two-story, large, single-family, colonial-style dwelling, Type V construction, smoke showing from the Alpha side on the first floor. Homeowner reports everyone out of the structure. I'll be out taking a lap around the structure."

1. Where is the flow path and what are your considerations?

   Answer: **With wind speeds of 13 to 15 mph (21–24 km/h), the crew leader must take the wind direction into account when selecting a strategy and type of attack. Winds blowing from the Charlie side toward the Alpha side mean that conducting an interior attack from the Alpha side puts the members in a potentially bad position. Knowing that the fire started in the kitchen and seeing the open front door, it is logical to assume the front door is the exhaust ventilation point, and that a door or window on Charlie side has failed or been left open and is the inlet ventilation point. This should be interpreted to mean that the flow path is unidirectional from the Charlie side to the Alpha side on the first floor.**

2. What is your first tactical action as the initial incident commander?

   Answer: **Because it was reported that everyone is out of the house, you can focus on attacking the fire. Quickly conduct a 360-degree size-up to confirm that your crew will not be operating over the fire—that is, that the basement is not involved—and to identify the point of attack.**

3. **Which side of the house should you direct the first crew with a handline to and what type of attack should they conduct? What about the second crew?**

   Answer: **The first crew should advance a hose line to the Charlie side and conduct a transitional attack by directing stream into the house toward the seat of the fire and the ceiling to start the cooling process. The second line should then be placed depending on the success or needs of the initial line.**

4. **Because an officer is not expected on scene for 5 to 7 minutes, what are the next command-level actions you should consider taking?**

   Answer: **Given the size and potential volume of the fire and structure, request additional resources quickly. Use the personnel accountability system and ICS procedures to ensure accountability of all units and personnel on the incident scene, and track unit assignments and locations at all times and be prepared to transfer command upon arrival of the next IC.**

WRAP-UP

SUMMARY

KNOWLEDGE OBJECTIVES

- Describe the basic fireground decision processes that are made in order to coordinate an interior attack in a structure fire.
  - List the factors that the incident commander evaluates when determining whether to perform a defensive operation or an offensive operation. (**NFPA 1010: 7.3.2**, pp. 968–970)
  - List the indicators of a potential structural collapse. (**NFPA 1010: 7.3.2**, pp. 970–971)
  - Describe the signs that indicate hazards may be increasing during suppression operations. (**NFPA 1010: 7.3.2**, p. 971)
  - Describe how a thermal imaging camera can be used in the size-up process. (**NFPA 1010: 7.3.3**, pp. 971–974)
  - Describe the importance of securing a water supply and staffing before beginning suppression operations. (p. 974)
  - Describe how ventilation operations can impact fire growth in the structure. (**NFPA 1010: 7.3.2**, pp. 974–975)
  - Explain how ventilation is coordinated with fire suppression operations. (**NFPA 1010: 7.3.2**, pp. 974–975)
- Describe how tactics are implemented during fire suppression operations.
  - Describe the considerations in selecting the proper hoseline and nozzle to achieve the goals of suppression operations. (**NFPA 1010: 7.3.2**, p. 975)
  - Describe why and how an interior attack is performed. (**NFPA 1010: 7.3.2**, p. 975)
  - Describe the importance of crew integrity. (**NFPA 1010: 7.3.2**, p. 975)
  - Describe why and how an exterior attack is performed. (**NFPA 1010: 7.3.2**, p. 977)
- Describe the fundamentals of flammable gases and the methods to suppress flammable gas fires.
  - Describe the characteristics of natural gas fires (**NFPA 1010: 7.3.4**, pp. 977–978)
  - Describe the characteristics of flammable gas cylinder fires. (**NFPA 1010: 7.3.4**, pp. 978–981)
  - Describe the hazards presented by flammable gas cylinder fires. (**NFPA 1010: 7.3.4**, pp. 978–981)
  - Describe a boiling liquid/expanding vapor explosion (BLEVE). (**NFPA 1010: 7.3.4**, p. 980)
  - Describe how to suppress a flammable gas cylinder fire. (**NFPA 1010: 7.3.4**, pp. 982–983)
  - Describe the characteristics of combustible or flammable liquid fires. (**NFPA 1010: 7.3.4**, pp. 981, 983–984)
  - Describe the hazards presented by combustible or flammable liquid fires. (**NFPA 1010: 7.3.4**, pp. 981, 983–984)
- Describe the methods by which foam is generated and prevents or controls a hazard.
  - Describe the characteristics of Class A foam. (**NFPA 1010: 7.3.1**, pp. 984–985)
  - Describe the characteristics of Class B foam. (**NFPA 1010: 7.3.1**, pp. 985–986)
  - List the major categories of Class B foam concentrate. (**NFPA 1010: 7.3.1**, p. 986)
  - Describe the characteristics of protein foam. (**NFPA 1010: 7.3.1**, p. 986)
  - Describe the characteristics of fluoroprotein foam. (**NFPA 1010: 7.3.1**, p. 986)
  - Describe the characteristics of aqueous film-forming foam. (**NFPA 1010: 7.3.1**, p. 986)
  - Describe the characteristics of alcohol-resistant foam concentrate. (**NFPA 1010: 7.3.1**, p. 986)
  - Describe the characteristics of compressed air foam. (p. 988)
  - Describe how foam proportioner equipment works with foam concentrate to produce foam. (**NFPA 1010: 7.3.1**, pp. 987–991)
  - Describe how foam is applied to fires. (**NFPA 1010: 7.3.1**, pp. 991–994)
- Describe how an annual hose test is performed. (**NFPA 1010: 7.5.5**, p. 995)

SKILLS OBJECTIVES

- Utilize a thermal imager to assess fire conditions in a structure.
  - Use a thermal imager at a structure fire. (**NFPA 1010: 7.3.3**, p. 974)
- Utilize the appropriate fire suppression attack technique based on the information obtain during size-up.
  - Coordinate an interior attack. (**NFPA 1010: 7.3.2**, p. 976)
- Prepare and apply foam.
  - Suppress a flammable gas cylinder fire. (**NFPA 1010: 7.3.4**, pp. 982–983)
  - Operate an in-line foam eductor. (**NFPA 1010: 7.3.1**, pp. 989–990)
  - Perform the roll-in method of applying foam. (**NFPA 1010: 7.3.1**, p. 993)
  - Perform the bounce-off method of applying foam. (**NFPA 1010: 7.3.1**, p. 993)
  - Perform the rain-down method of applying foam. (**NFPA 1010: 7.3.1**, p. 994)
- Describe how an annual hose test is performed. (**NFPA 1010: 7.5.5**, p. 995)

KEY TERMS

alcohol-resistant foam A concentrate used for fighting fires on water-soluble materials and other fuels destructive to regular, AFFF, or FFFP foams, as well as for fires involving hydrocarbons. (NFPA 11)

aqueous film-forming foam (AFFF) A solution based on fluorinated surfactants plus foam stabilizers to produce a fluid aqueous film for suppressing liquid-fuel vapors. (NFPA 10)

bank-down method See *bounce-off method.*

bank-shot method See *bounce-off method.*

batch mix The manual addition of foam concentrate to a water storage container or tank to make foam solution. (NFPA 1145)

bounce-off method A foam application method that applies the stream onto a nearby object, such as a wall, instead of directly onto the surface of the fire. Also called *bank-shot method* or *bank-down method.*

burnback resistance The ability of the foam blanket to maintain its integrity and effectiveness when impinged by fire.

compressed air foam (CAF) A homogenous foam produced by the combination of water, foam concentrate, and air or nitrogen under pressure (NFPA 11).

crew integrity The concept of keeping a crew together and working as a team so that no single firefighter goes off on their own.

collapse zone The area around a structure where debris could fall if the building collapses.

eductor A device placed in a hose line or a discharge pipe that incorporates a Venturi and proportions foam concentrate or other firefighting agents into the water stream. (NFPA 1900)

emissivity A description of the amount of energy radiated from a material's surface.

field of view (FOV) The area observable at any given moment and is measured vertically and horizontally in degrees.

film-forming fluoroprotein foam (FFFP) A protein-foam solution that uses fluorinated surfactants to produce a fluid aqueous film for suppressing liquid-fuel vapors. (NFPA 10)

fluoroprotein foam A protein-based foam concentrate, with added fluorochemical surfactants, that forms a foam showing a measurable degree of compatibility with dry chemical extinguishing agents and an increase in tolerance to contamination by fuel. (NFPA 460)

foam blanket A covering of foam over a surface to insulate, prevent ignition, or extinguish the fire. (NFPA 1145)

foam concentrate Foam firefighting agent as received from the manufacturer that must be diluted with water to make foam solution (NFPA 1900).

foam injector A device installed on a fire pump that meters out foam by pumping or injecting it into the fire stream.

foam nozzle A nozzle specifically designed to help with aeration of the foam solution as it exits the hose line and complete the process of becoming a finished foam product.

foam proportioner A device or method to add foam concentrate to water to make foam solution. (NFPA 1900)

foam solution A homogeneous mixture of foam concentrate and water in the proper proportions. (NFPA 1900)

polar solvent A water-soluble, flammable liquid that readily mixes with water.

protein foam A protein-based foam concentrate that is stabilized with metal salts to make a fire-resistant foam blanket. (NFPA 460)

rain-down method A foam application method that directs the stream into the air above the fire and allows it to gently fall on the surface.

roll-in method See *roll-on method.*

roll-on method A foam application method that involves sweeping the stream just in front of the target. Also called *roll-in method* or *sweep method.*

strategy A description of the tactics to use to achieve the broad goals of the incident.

surfactant A compound that lowers the surface tension (or interfacial tension) between two liquids, between a gas and a liquid, or between a liquid and a solid, and which can act as detergents, wetting agents, emulsifiers, foaming agents, and dispersants. (NFPA 1700)

sweep method See *roll-in method.*

tactics The specific actions taken to achieve those goals.

thermal imaging camera (TIC) See also *thermal imager (TI).*

thermal imager (TI) Special electronic equipment that creates a picture based on heat produced by a person or object. (NFPA 1801)

three-dimensional liquid-fuel fire A fire in which liquid fuel is dripping, spraying, or flowing over the edges of a container.

two-dimensional liquid-fuel fire A fire that is burning only on the top surface of a spill, pool, or open container of liquid.

wet water See *wetting agent.*

wetting agent A concentrate that, when added to water, reduces the water's surface tension and increases its ability to penetrate, mix with and spread. (NFPA 18)

REVIEW QUESTIONS

1. What is the difference between a strategy and tactics?
2. What are the signs of potential structural collapse?
3. What is crew integrity?
4. Under what circumstances should you let a flammable gas fire continue to burn rather than extinguish it?
5. Is methane lighter or heavier than air?
6. What is the expansion ratio of propane?
7. Which area of a propane tank should be cooled if fire is impinging on it?
8. Three-dimensional liquid-fuel fires should be extinguished using what agents?
9. What is the primary extinguishing mechanism of foam?
10. What is the most common type of eductor used in the fire service?
11. What is the expansion ratio of medium-expansion foams?
12. What can happen if you apply Class B foam directly onto a pool of burning liquid fuel?

DISCUSSION QUESTIONS

1. Why is effective tactical decision making on the incident scene so critical for a firefighter who needs to act as the crew leader or IC?
2. What makes any individual an effective tactical decision maker? Is seniority the biggest predictor of success?
3. Describe a situation (in any environment) where you observed someone making effective decisions in emergency conditions. Did those decisions lead to a good outcome? What knowledge, skills, or attributes did the decision maker demonstrate that contributed to the outcome?
4. Describe the key differences in dealing with a natural gas versus a propane gas emergency.
5. What types of foam resources—type and brand of foam concentrate, type of appliances, and types of apparatus—are used in your department? Are they all located in fire stations in your department or does your department have agreements with other agencies?

APPLYING THE CONCEPTS

You are the crew leader of a four- person engine company dispatched to a reported exterior fire involving a grill. Information provided indicates the call was made by a 13-year-old who lives at the residence. Your company is the only resource dispatched. You see a large volume of thick gray smoke showing from side Charlie as you pull up to the two-story, split-foyer structure. A group of teenagers are in the front yard and come running up to you before you can step out of the engine. They begin yelling that the grill blew up when they were trying to make hamburgers. You ask if anyone is hurt and if anyone is in the house. The boy replies, "Oh, yeah! My sister's inside in her room!"

1. What are your immediate priorities for action? What do you do first?

You get out of the engine with full PPE and SCBA. While conducting your 360, you see an active fire originating from the grill on the ground-floor patio underneath the deck on side Charlie. The sliding door to the house is open, fire is impinging on the deck, and exterior wall appears to have cedar shakes. You can see it is a propane grill, but the tank is intact. You request a working fire dispatch and communications assigns you a channel.

2. What are your actions at this point?
3. You are the Incident Commander for this event. what are your priorities until additional resources arrive? Upon arrival of those resources, what assignments would you give them?
4. What size hose line should you select, where should it be placed, and should the attack team be directed via the exterior or interior to get to side Charlie?

You direct the crew to apply the stream along the siding and deck. You are able to visualize that the fire is burning from the supply line to the grill.

5. What characteristics of propane do firefighters need to be aware of and concerned about?
6. What safety precautions must you and your crew take before approaching the fire?

REFERENCES

Back, Gerard G., Noah J. Lieb, Edward Hawthorne, and Casey Grant. 2022. "Firefighting Foams: Fire Service Roadmap" [Report]. Quincy, MA: National Fire Protection Association/Fire Protection Research Foundation.

National Fire Protection Association (NFPA). 2019. *NFPA 1710, Standard for the Organization and Deployment of Fire Suppression Operations, Emergency Medical Operations, and Special Operations to the Public by Career Fire Departments.* 2020 Edition. Quincy, MA: NFPA.

National Fire Protection Association (NFPA). 2019. *NFPA 1720, Standard for the Organization and Deployment of Fire Suppression Operations, Emergency Medical Operations, and Special Operations to the Public by Volunteer Fire Departments.* 2020 Edition. Quincy, MA: NFPA.

National Fire Protection Association (NFPA). 2020. *NFPA 11, Standard for Low-, Medium-, and High-Expansion Foam.* 2021 Edition. Quincy, MA: NFPA.

National Fire Protection Association (NFPA). 2020. NFPA 18, *Standard on Wetting Agents.* 2021 Edition. Quincy, MA: NFPA.

National Fire Protection Association (NFPA). 2020. NFPA 1700, *Guide for Structural Fire Fighting.* 2021 Edition. Quincy, MA: NFPA.

National Fire Protection Association (NFPA). 2020. *NFPA 1801, Standard on Thermal Imagers for the Fire Service.* 2021 Edition. Quincy, MA: NFPA.

National Fire Protection Association (NFPA). 2021. NFPA 10, *Standard for Portable Fire* Extinguishers. 2022 Edition. Quincy, MA: NFPA.

National Fire Protection Association (NFPA). 2021. *NFPA 1145, Guide for the Use of Class A Foams in Fire Fighting.* 2022 Edition. Quincy, MA: NFPA.

National Fire Protection Association (NFPA). 2023. NFPA 460, *Standard for Aircraft Rescue and Firefighting Services at Airports.* 2024 Edition. Quincy, MA: NFPA.

National Fire Protection Association (NFPA). 2023. NFPA 1900, *Standard for Aircraft Rescue and Firefighting Vehicles, Automotive Fire Apparatus, Wildland Fire Apparatus, and Automotive Ambulances.* 2024 Edition. Quincy, MA: NFPA.

Science Direct. n.d. "Natural Gas." Accessed June 5, 2023. https://www.sciencedirect.com/topics/engineering/natural-gas.

U.S. Fire Administration (USFA). 2020. "The Hidden Dangers in Firefighting Foam." Coffee Break Bulletin, February 11, 2020. Accessed June 5, 2023. https://www.usfa.fema.gov/blog/cb-021120.html.

Washington Township OEM. 2014. "Propane Tank Explosion. Video, March 1, 2014." Accessed June 5, 2023. https://youtu.be/Lr15rPHEmeQ.

CHAPTER

24

Firefighter II

Systems of Fire Detection, Suppression, and Smoke Control

KNOWLEDGE OBJECTIVES

After studying this chapter, you will be able to:

- Describe fire protection systems and the basic components and functions of fire alarm systems and define the various types of unwanted alarms.
- Describe residential smoke and carbon monoxide alarms.
- Explain the purpose of the emergency control function interface.
- Describe the types of fire department notification systems.
- Describe the types of fire suppression systems, including the main components and operation of each system.
- Describe different types of special hazard systems and identify hazards associated with these systems that firefighters may encounter.
- Describe the operation and application of smoke control systems.
- Explain the importance of inspecting, testing, and maintaining fire protection systems and the process of documenting these processes.

SKILLS OBJECTIVES

There are no skills objectives for Firefighter II candidates. NFPA 1010 contains no Firefighter II Job Performance Requirements for this chapter.

ADDITIONAL NFPA STANDARDS

- **NFPA 1**: *Fire Code, 2021 Edition*
- **NFPA 2**: *Hydrogen Technologies Code, 2023 Edition*
- **NFPA 12**: *Standard on Carbon Dioxide Extinguishing Systems*
- **NFPA 13**: *Standard for the Installation of Sprinkler Systems*
- **NFPA 14**: *Standard for the Installation of Standpipe and Hose Systems*
- **NFPA 25**: *Standard for the Inspection, Testing, and Maintenance of Water-Based Fire Protection Systems, 2023 Edition*
- **NFPA 72**: *National Fire Alarm and Signaling Code*
- **NFPA 92**: *Standard for Smoke Control Systems*
- **NFPA 853**: *Standard for the Installation of Stationary Fuel Cell Power Systems, 2020 Edition*
- **NFPA 1140**: *Standard for Wildland Fire Protection, 2022 Edition*
- **NFPA 2001**: *Standard on Clean Agent Fire Extinguishing Systems, 2022 Edition*

CASE STUDY

You Are the Firefighter

Your crew is updating the department's preincident plan for a large industrial facility in your city. The building's safety officer is showing you the state-of-the-art fire alarm system and suppression systems. The fire alarm system includes both manual and automatic initiating devices, as well as a comprehensive emergency notification system with the main fire alarm control unit located in the security office. The building is fully sprinklered, and a clean agent suppression system protects the large computer room. Class III standpipes are located strategically throughout the building. Although the facility uses a lot of hazardous processes in its operations, you feel that the fire protection systems provide a high level of protection for both the occupants and responding firefighters if an emergency occurs.

1. What information should be included in the preincident plan?
2. What are the advantages of a Class III standpipe system? Are there any disadvantages?
3. What safety measures should responding firefighters take if the clean agent system activates?
4. What can your department do to ensure the alarm and suppression systems are maintained functional?

Introduction

Modern fire prevention and building codes require most new structures to have some sort of fire protection system installed. A **fire protection system** is a system that detects a fire condition, alerts occupants, sometimes alerts the fire department, and sometimes controls or extinguishes a fire. Most of North America follows the applicable National Fire Protection Association (NFPA) codes and standards for these systems; however, there are some state and local jurisdictions that amend the requirements to make them more stringent.

Understanding how these systems operate is important both to keep firefighters and building occupants safe and to provide effective customer service to the public. From a safety standpoint, firefighters need to understand the operational capabilities and limitations of each type of fire protection system. Also, a building with a fire protection system will have different working conditions during a fire than a building with no fire protection system. Firefighters need to know how to fight a fire in both types of buildings. They also need to know how to operate fire protection systems to protect life and property. Most people have no idea how the fire protection systems in their building work. Firefighters who understand how fire protection systems operate can explain them to owners and occupants. They can educate the occupants that a fire alarm signal should be always treated as a true emergency until someone from the fire department arrives on the scene and indicates otherwise.

Fire Protection Systems

Fire protection systems include the following:

- The **fire alarm system**, which activates a fire alarm signal when a fire condition is detected to alert occupants, and sometimes the fire department as well, that there is a fire emergency and often identifies the location of the fire for responders. A **fire alarm signal** is an audible tone and/or flashing lights that indicate that the alarm was activated.
- A **fire detection system**, which detects a fire and then activates a fire alarm system, a fire suppression system, or both.
- A **fire suppression system**, which discharges extinguishing agents in the presence of a fire.
- A **smoke control system**, which contains smoke in an area to prevent it from moving to other areas.

In most commercial occupancies, the fire detection and fire alarm systems are combined in a single system. A fire detection system recognizes when a fire occurs. This causes the fire detection system to activate the fire alarm system, which activates the fire alarm signal to alert building occupants. Some fire detection systems also activate fire suppression systems. NFPA 72, *National Fire Alarm and Signaling Code*, describes minimum standards for fire alarm and detection systems.

Occasionally, a fire alarm system is activated when there is no emergency event. The term *false alarm* is frequently used by the public to describe all fire alarm activations that are not associated with a true emergency, even though, in some of these situations, the alarm system operated properly. The NFPA uses the term **unwanted alarm** to describe the activation of a fire alarm that is not the result of a potentially hazardous condition.

Fire Alarm Systems

Fire alarm and detection systems range from a simple smoke alarm for a small single-family private home, to complex fire detection, smoke control, and fire suppression systems for high-rise buildings. Many fire alarm and detection systems in large buildings also control other fire protection and building systems to help protect occupants and control the spread of fire and smoke. Although these systems can be complex, they generally include the same basic components as the simpler versions. A fire alarm system has three basic components:

- **Alarm initiating device**, which is a device that when operated either automatically or manually causes the alarm notification appliances to operate.
- **Alarm notification appliance**, which is the device that generates the fire alarm signal.
- **Fire alarm control unit (FACU)**, sometimes called the **fire alarm control panel (FACP)**, which serves as the "brain" of the system and links the alarm initiating device to the alarm notification appliance as well as perform other essential functions.

Fire Alarm Control Unit

The FACU manages and monitors the operation of the fire alarm system (**FIGURE 24-1**). In addition to linking the alarm initiating devices and the alarm notification appliances, the FACU also manages the primary and backup power supplies for the system. It includes an **annunciator**, which is a display that contains one or more indicator lamps, alphanumeric displays, or other means to provide status information about a circuit, condition, or location that activated the alarm so that responding fire personnel know what actions to take. It may perform additional functions as well, such as notifying the public safety communications center in the jurisdiction or acting as a monitoring center, where trained personnel are notified that an alarm has been activated so that they can initiate the appropriate action. The FACU might also be connected to other fire protection and building systems, such as elevators, locks on stairwell doors, and fans.

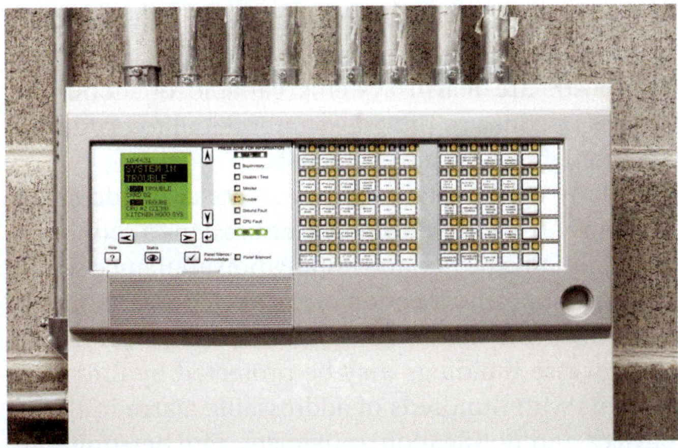

FIGURE 24-1 Modern FACUs indicate the zone or specify the exact location where an alarm was initiated, as well as the type of initiating device that activated the fire alarm signal.

The exact functions of the FACU depends on the age of the system and the manufacturer. For example, the FACU in an older system might indicate only that a fire alarm signal has been activated.

An older system that simply sounds an audible alarm when the system is activated is a **noncoded system**. These systems provide no information about the type or location of the initiating device that caused the alarm. Responding personnel need to search the entire building to locate the source of the alarm.

A **coded system** is an older fire alarm system that provides limited information through audible or visible coded signals. The signals are sounds or flashing lights in coded patterns that indicate the floor or zone where the fire alarm signal originated. For example, the alarm initiating device on the fourth floor of an apartment building may be coded as four bells followed by two bells. To interpret the coded signals, occupants and responding personnel need to refer to a posted chart. Noncoded and coded systems are no longer installed, but these systems are still in use in some older buildings.

Modern fire alarm systems divide a building or complex into zones. A **zone** is a defined area in the building or complex. A **zoned system** identifies the type or location of initiating devices by identifying the zone where each device is located. The zones are indicated by lights in the building or labeled on the FACU. Early zoned systems had a limited number of zones, so alarm initiating devices were categorized by floor, area, or type. For example, one type of alarm initiating device might be Zone 1, a second type of alarm initiating device would be Zone 2, and a third type of alarm initiating device might be Zone 3. As technology improved, fire alarm systems capable of identifying dozens of zones to provide greater detail to occupants and firefighters became available.

Many modern fire alarm systems are addressable systems. An **addressable system** assigns a number to each initiating device and notification appliance so that the exact type and location of the device and appliance are shown on the FACU. Large industrial complexes or high-rise buildings may be protected by fire alarm systems with hundreds of addressable alarm initiating devices and notification appliances. Fire personnel responding to alarms in these buildings can identify the exact location and specific conditions before they even enter the building.

In addition to the information displayed on the FACU or communicated by audible or visible devices, many buildings have one or more **remote annunciators** near the building entrances that display the information shown on the annunciator at the FACU (**FIGURE 24-2**). This allows firefighters to determine the type and location of the activated alarm device as they enter the building. Annunciators also allow firefighters to access some of the functions of the FACU, such as the ability to silence the alarm or reset the system. Responding firefighters use remote annunciators to gain important information about the device that initiated the alarm without having to locate the FACU.

Depending on the building's size and floor plan, a fire alarm may sound throughout the building or only in the areas where a fire is detected. In high-rise buildings, fire alarm systems often alert only the occupants on the same floor as the activated alarm as well as occupants on the floors immediately above and below that floor. If emergency responders think it is necessary, they can manually activate alarms on other floors from the FACU. Some fire alarm systems include a public address feature, which enables fire department personnel to provide specific instructions or information to occupants in different areas.

Fire alarm systems are usually powered by a dedicated 120-volt circuit on the building's electrical system. In addition to the primary power source, NFPA 72 requires a secondary power source for all fire alarm systems (**FIGURE 24-3**). Most fire alarm systems use batteries for backup power. Some buildings have an automatic generator to power essential equipment in the event of a power failure, and in these buildings, that generator provides the secondary power source for the fire alarm system.

In a large building, the FACU may be programmed to perform additional functions. For example, in the event of a fire, it may automatically shut down or change the operation of air-handling systems, recall elevators to the ground floor, and unlock stairwell doors so that anyone in an exit stairwell can enter an occupied floor.

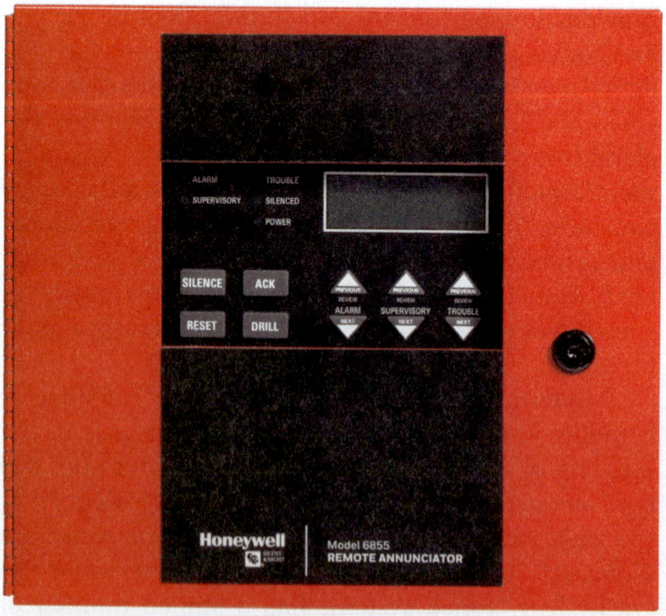

FIGURE 24-2 A remote annunciator allows firefighters to quickly determine the type and location of the activated alarm device.

Courtesy of Honeywell/Fire-Lite Alarms.

FIGURE 24-3 Fire, building, and life safety codes require that alarm systems have a backup power supply, which becomes activated automatically when the normal electrical power is interrupted. Most systems use batteries as the backup power supply.

Courtesy of A. Maurice Jones, Jr.

Another function of the FACU is to monitor the status of the alarm system. When the system is in **normal ready status**, it means that it is functioning properly and there are no faults (problems) in the system and no problems with any other fire protection or building system integrated with the FACU. A fault within the system, such as loss of primary or secondary power, low or no battery power, a broken wire, a device that has become inoperable or damaged, or a device that was removed, is called a **trouble condition** and is indicated by a trouble signal. A **trouble signal** indicates that a component of the system is not operating properly and requires attention from the building owner. The FACU also monitors changes in the normal ready status of other fire protection systems and building systems that are integrated with the FACU. These types of changes, called a **supervisory condition**, are indicated by a **supervisory signal**. Examples of supervisory conditions are a dry pipe sprinkler system that has air pressure that is too low or too high, a control valve that should be open but is closed, or if the temperature of the water in a fire suppression system storage tank drops below 40°F (4.44°C).

Trouble and supervisory signals do not activate the building's fire alarm, nor do they trigger a response from the fire department. They make a sound and turn on a light at the at the annunciator unit on the FACU and at the remote annunciator unit if there is one. They may also transmit a notification to an on- or off-site monitoring location where the signal can be received and acted upon. If the alarm is not acknowledged and the issue that caused the alarm is not corrected within a specified amount of time, the alarm will sound again.

FACUs should always be locked to prevent unauthorized access to the system. If an alarm activates, it should not be silenced or reset until the source of the alarm has been identified and checked by firefighters to ensure that the situation is under control. In some systems, the alarm initiating device or devices must be manually reset before the entire system can be reset. Other systems automatically reset after the problem has been resolved. Whether a system needs to be manually reset by a firefighter should be included in the preincident plan.

Alarm Initiating Devices

Alarm initiating devices are operated either manually by a person, automatically when a fire condition is detected, or automatically by a fire suppression system.

Manual Initiating Devices

Manual initiating devices allow building occupants to activate the fire alarm system by pulling a lever or turning a switch. The most common manual initiating device is the **manual fire alarm box** or **manual pull station**, which has a handle that is pulled to send an electrical signal that activates the alarm.

Pull stations can be either single-action or double-action devices. A **single-action pull station** activates the alarm as soon as the lever, toggle, or handle is pulled (**FIGURE 24-4A**). A **double-action pull station** does not activate the alarm until two steps are performed. The person must move a flap, lift a cover, or break a piece of glass to reach the lever, toggle, or handle that they then pull to activate the alarm (**FIGURE 24-4B**). Double-action pull stations are designed to reduce the number of unwanted alarms caused by someone accidentally or maliciously activating the alarm.

Both single-action and double-action pull stations may be covered by a plastic cover to deter malicious alarms and to protect the pull station from physical damage (**FIGURE 24-5**). Lifting the cover triggers a loud supervisory alarm at that specific location, but it does not activate the fire alarm system. Snapping the cover back into place resets this alarm. These covers are often used in areas where malicious alarms occur frequently, such as schools and college dormitories.

Once activated, a manual pull station should stay in the "activated" position until it is manually reset. This enables responding firefighters to determine which pull station initiated the alarm. Although older pull stations with breakable glass rods or panes are still in use, most new pull stations use a resettable locking mechanism to indicate that the pull station has been activated. Resetting the pull station requires a special key, screwdriver, or wrench. Pull stations must be reset before the building alarm system can be reset at the FACU.

Automatic Initiating Devices

Automatic initiating devices detect fire conditions, such as smoke, heat, or flowing water, and initiate the alarm when one of these conditions is detected, even if the building is unoccupied. Because automatic initiating devices need to detect a condition before activating the alarm, they are called detectors. Some automatic initiating devices initiate an alarm when they detect flow or discharge in automatic fire suppression systems.

A.

B.

FIGURE 24-4 Manual pull stations can be either single-action or double action. **A.** A single-action pull station. **B.** A double-action pull station.

A. © rob casey/Alamy Stock Photo; **B.** Courtesy of Honeywell/Fire-Lite Alarms.

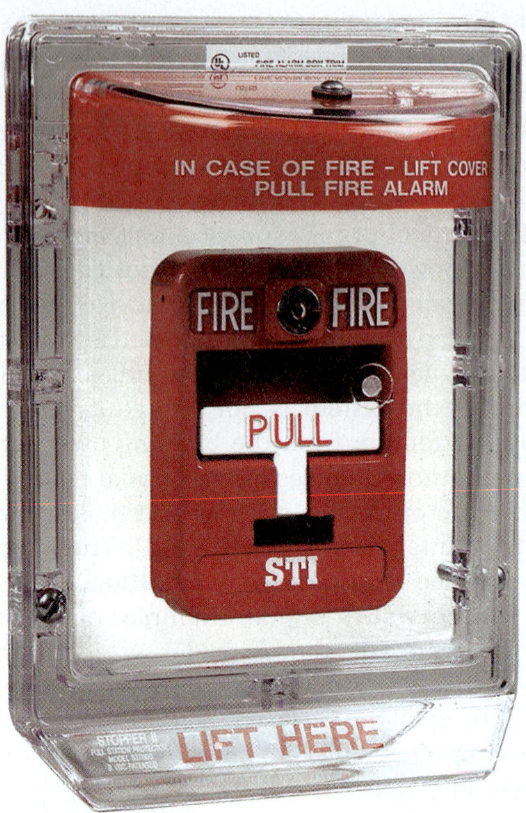

FIGURE 24-5 Clear plastic covers, designed to prevent malicious false alarms, need to be lifted before the alarm can be activated. Lifting the cover triggers a supervisory alarm at the location of the pull station but does not initiate the fire alarm.

Courtesy of STI-USA.

Smoke Detectors. A **smoke detector**, which activates the alarm when it detects smoke, is the most common type of automatic initiating device. It may be a stand-alone device, or it may be part of a comprehensive fire alarm system. Smoke detectors are commonly found in fire alarm systems in schools, hospitals, businesses, and commercial occupancies. Smoke detectors connected to a fire alarm system are powered by a low-voltage circuit, and they send a signal to the FACU when they are activated. After the smoke condition clears, the alarm system can be reset at the FACU. If the detector is damaged by smoke or heat, it must be replaced.

The most common type of smoke detector is the **spot-type detector**, which is an individual unit that can be spaced throughout an occupancy so that each detector covers a specific area (**FIGURE 24-6**). They have an internal sensing chamber that contains the detection device. The units may be in individual rooms

FIGURE 24-6 Spot-type smoke detector installed in a commercial building.

Courtesy of A. Maurice Jones, Jr.

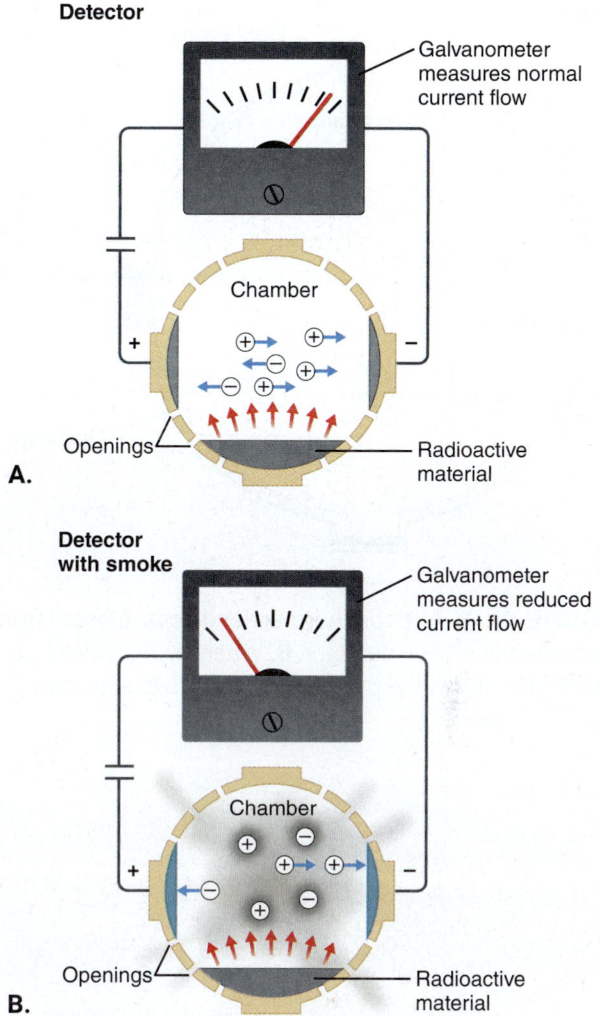

FIGURE 24-7 **A.** Under normal conditions, a small electric current is created inside the chamber in the ionization detector. **B.** When tiny smoke particles or gas enter the chamber—often before smoke is visible—the electric current is interrupted, which activates the alarm.

© Jones & Bartlett Learning

or spaced at intervals along the ceiling in larger areas. Spot-type detectors are usually installed in residential and commercial settings where conditions are not likely to produce unintentional or nuisance alarms.

Smoke detectors come in a variety of designs and styles designed for different applications. Each type of smoke detector is tested by an independent testing laboratory for use in specific applications based on the size, height, contents, and occupancy type of the protected area. The most commonly installed types are ionization and photoelectric spot-type detectors.

An **ionization smoke detector** works on the principle that burning materials release many different products of combustion, including electrically charged microscopic particles. An ionization detector senses the presence of these invisible charged particles (ions). This type of detector contains a minuscule amount of radioactive material inside its inner sensing chamber, which changes air molecules in the chamber into electrically charged molecules called ions. The positive ions move to a plate on one side of the chamber and the negative ions move to a plate on the other side. This radioactive material releases charged particles into the chamber and creates a small electric current flowing between the two plates, which is monitored by a galvanometer, an instrument that measures electrical current (**FIGURE 24-7A**). When even tiny smoke particles or invisible smoke in the form of a gas enters the chamber, they neutralize the charged particles, which reduces or interrupts the electric current. When the galvanometer in the detector senses this interruption, the alarm is activated (**FIGURE 24-7B**).

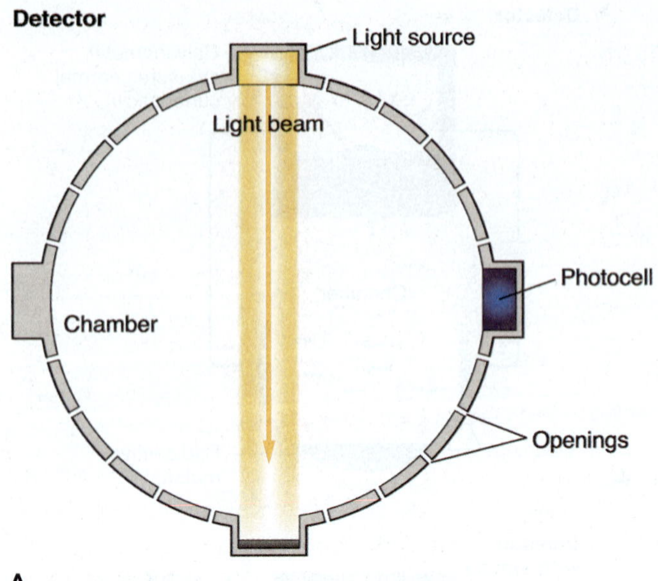

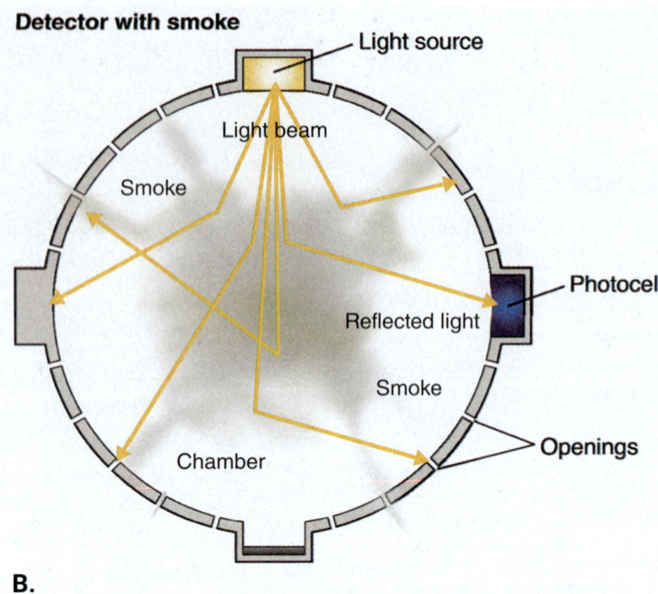

FIGURE 24-8 **A.** Under normal conditions, a beam from an LED light is emitted in a straight line across the inside of the photoelectric smoke detector. **B.** When visible smoke particles enter the detector, the beam is scattered, and when the reflected light hits the photocell, the alarm is activated.

FIGURE 24-9 Photoelectric smoke detectors react more quickly than ionization smoke detectors to a slow-burning or smoldering fire, such as a cigarette in furniture, which usually produces a large quantity of visible smoke.

Courtesy of Honeywell/Fire-Lite Alarms.

FIGURE 24-10 Beam detectors are used in large open spaces.

Courtesy of Honeywell/Fire-Lite Alarms.

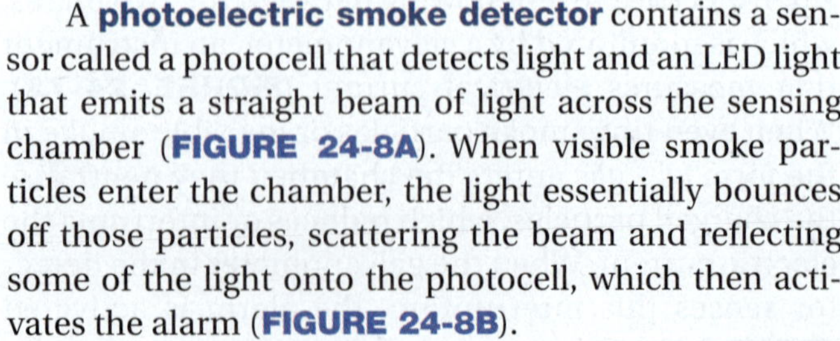

A **photoelectric smoke detector** contains a sensor called a photocell that detects light and an LED light that emits a straight beam of light across the sensing chamber (**FIGURE 24-8A**). When visible smoke particles enter the chamber, the light essentially bounces off those particles, scattering the beam and reflecting some of the light onto the photocell, which then activates the alarm (**FIGURE 24-8B**).

Most ionization and photoelectric smoke detectors look alike (**FIGURE 24-9**). Refer to the label on each detector to determine its type and applications. In addition, an ionization detector must have a label or engraving stating that it contains a small amount of radioactive material.

A **projected beam-type detector**, usually just called a **beam detector**, is a photoelectric smoke detector used to protect large open areas such as churches, atriums, auditoriums, airport terminals, and indoor sports arenas. In these types of facilities, smoke may not rise high enough to reach spot-type smoke detectors on the ceiling, so one or more beam detectors provide protection for the entire area (**FIGURE 24-10**).

A beam detector has two components: a sending unit, which projects a narrow beam of light across the open area, and a receiving unit, which measures the intensity of the light when the beam strikes the receiver. When smoke interrupts the light beam, the receiver detects the drop in light intensity and activates the fire alarm system. Most beam detectors are set to respond to a specific **obscuration rate**, or percentage of light blocked. If the light is completely blocked, such as when a solid object is moved across the beam, the trouble or supervisory signal will activate, but the fire alarm will not be activated.

Ionization smoke detectors react more quickly than photoelectric smoke detectors to fast-burning fires, such as a fire in a trash can because usually those types of fires produce little visible smoke at first. By comparison, photoelectric smoke detectors are more responsive to slow-burning or smoldering fires, such as a fire caused by a cigarette caught in a couch, which usually produces a large quantity of visible smoke. Also, ionization detectors are more susceptible than photoelectric smoke detectors to unwanted alarms from common activities, such as light smoke from cooking or steam from a shower, so they are not suitable for use near kitchens or bathrooms.

When properly installed and maintained, both types of smoke detectors are effective for detecting fires and alerting occupants early enough for them to escape safely. **TABLE 24-1** summarizes the differences between ionization and photoelectric smoke detectors.

Both ionization and photoelectric smoke detectors are acceptable life safety devices. Combination ionization/photoelectric smoke detectors are also available. These alarms quickly react to both fast-burning and smoldering fires. They are, however, prone to the same unwanted alarms as regular ionization smoke detectors.

Heat Detectors. A **heat detector** detects an abnormally high temperature; a sudden, unexpected temperature rise; or both. If these conditions occur, the detector initiates the alarm. Although heat detectors are very reliable and less prone to unwanted alarms than smoke detectors, they do not react quickly to incipient fires. This means that they can be used to provide protection for property, but they should not be relied on to provide life safety protection.

Heat detectors are generally used in areas where smoke alarms or detectors cannot be used. For example, smoke detectors should not be used in dusty

TABLE 24-1 Comparison of Ionization and Photoelectric Smoke Detectors

	Ionization Smoke Detector	Photoelectric Smoke Detector
How it works	A small amount of radioactive material ionizes air inside the detector, allowing an electric current to flow between two plates. Smoke entering the detector reduces or interrupts the electric current, which causes the FACU to initiate the alarm.	An LED light inside the detector creates a straight beam of light. Smoke entering the detector scatters the light and causing some of the light to be reflected into a photocell. When this happens, the FACU initiates the alarm.
Effectiveness	Has a somewhat faster response to small smoke particles from flaming fires than photoelectric alarms. Can also detect invisible smoke in the form of fire gases.	Has a considerably faster response to larger smoke particles from smoldering fires than ionization alarms.
Susceptibility to false alarms	May cause an unwanted alarm to be initiated if placed too close to a kitchen where smoke from a cooking appliance can interrupt the electric current or too close to a bathroom or any area in which there are changes in humidity and atmospheric pressure.	Less likely to cause unwanted alarms to be initiated when installed near an area where food is cooked or in areas subject to changes in humidity and atmospheric pressure.

environments because the detector cannot differentiate between smoke and dust particles. Smoke detectors should also not be used in areas that experience extreme cold or heat because they may fail to operate or produce nuisance alarms. Single-station heat alarms are sometimes installed in unoccupied areas of buildings that do not have fire alarm systems, such as attics, storage rooms, boiler rooms, and manufacturing areas. (A **single-station alarm** is a detector that includes a sensor, an alarm notification appliance, and a device that activates the alarm when the sensor detects a problem condition.)

Newer models have an indicator light that shows which device was activated. You may come across heat detectors that were installed 30 or more years ago and are still in service. Unfortunately, these older units lack an indicator light that shows which device was activated, so tracking down the cause of an alarm initiated by one of these devices may be difficult.

Heat detectors are designed to operate at a fixed temperature or to react to a rapid increase in temperature. The temperature or rate of temperature increase that causes a heat detector to activate is called the **temperature rating**. A **fixed-temperature detector** is a heat detector that activates when the temperature rating is reached (**FIGURE 24-11**). A typical temperature rating for a heat detector installed in a light hazard occupancy, such as an office building, is 135°F (57°C). Fixed-temperature detectors typically include a metal alloy that melts when the temperature rating is reached. This melting alloy releases a lever-and-spring mechanism, which closes or opens a circuit to initiate the alarm. Most fixed-temperature heat detectors must be replaced after they have been activated, even if the activation was accidental.

A **rate-of-rise detector** is a heat detector that activates when the temperature of the surrounding air increases by more than a set amount in a given period of time. A typical temperature rating might be "greater than 12°F (6.7°C) in 1 minute." If the temperature increase occurs more slowly than this rate, the rate-of-rise detector will not activate. Only a temperature increase greater than the rate specified by the temperature rating will activate the detector and initiate the fire alarm. Rate-of-rise detectors should not be located in areas that normally experience rapid changes in temperature, such as near garage doors in heated parking areas in a cold climate.

Some rate-of-rise heat detectors have a bimetallic strip made of two metals that bend at different rates

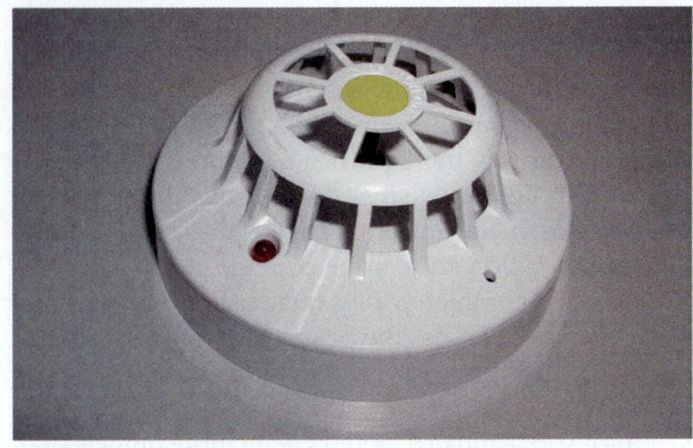

FIGURE 24-11 A fixed-temperature detector initiates an alarm when the temperature rating is reached.

Courtesy of Firetronics Pte Ltd.

when exposed to heat. A rapid increase in temperature causes the strip to bend unevenly, which closes a contact to trigger the alarm. Another type of rate-of-rise detector has an air chamber and diaphragm mechanism. As air in the chamber heats up, the pressure increases. Gradual increases in pressure are released through a small hole, but a rapid increase in pressure presses on the diaphragm and activates the alarm. Most rate-of-rise detectors are self-restoring, so they do not need to be replaced after an activation unless they were directly exposed to a fire.

Rate-of-rise detectors generally respond more rapidly to most fires than do fixed-temperature detectors. However, a slow-burning fire, such as a smoldering couch, might not activate a rate-of-rise detector until the fire is well established. Combination rate-of-rise and fixed-temperature detectors balance the faster response of the rate-of-rise detector with the reliability of the fixed-temperature heat detector.

Either fixed-temperature or rate-of-rise detectors can be configured as spot-type or line-type detectors. A **line-type detector** is wire or a sealed tube strung along the ceiling of large open areas. An increase in temperature anywhere along the line will activate the detector. Line-type detectors are often found in churches, warehouses, and industrial or manufacturing applications. They are also used in cable trays, which hold cable suspended from a ceiling or under a floor.

One type of wire line-type detector has two wires separated by an insulating material. When heat melts the insulation, the wires short circuit and activate the alarm. The damaged section of insulation must be replaced after the detector is activated. Another type of wire line-type detector measures changes in the electrical resistance of a single wire as it heats up. If the resistance increases to greater than a set amount, the detector is activated. This device is self-restoring and does not need to be replaced after activation unless it is directly exposed to a fire.

A tube line-type detector is a sealed metal tube filled with air or a nonflammable gas. When the tube is heated, the internal pressure increases and activates the alarm. Like the single-wire line-type detector, this device is self-restoring and does not need to be replaced after activation unless it is directly exposed to a fire.

Flame Detectors. A **flame detector** detects radiant energy in the form of infrared or ultraviolet light waves produced by a flame (**FIGURE 24-12**). These devices can quickly recognize even a very small fire. Flame detectors are typically found in places such as aircraft hangars, fuel loading racks, or specialized industrial settings where early detection and rapid reaction to a fire are critical. Flame detectors are also used in explosion suppression systems where they detect and suppress an explosion as it is occurring.

Flame detectors are both complicated and expensive. Normal infrared or ultraviolet light sources, such as the sun or a welding operation, can initiate an unwanted alarm. Some flame detectors use combined infrared and ultraviolet technologies to reduce unwanted alarms. This type of detector is not activated unless both infrared and ultraviolet light are detected.

Gas Detectors. A **gas detector** detects the presence of a specific gas that is created by combustion or that is used in a facility. These detectors are often found in commercial or industrial buildings. Depending on the system, a gas detector may be programmed to activate the building's fire alarm system or a separate alarm. Some gas detectors require regular calibration to operate properly. The most common type of gas detector you will encounter is a carbon monoxide detector.

Air Sampling Detectors. An **air sampling-type detector** continuously captures air samples and measures the concentrations of specific gases or products of combustion. This type of detector draws air samples through tubing into a sampling unit where they are analyzed using specialized technologies (**FIGURE 24-13**). A **duct detector** is an air sampling-type detector installed in the air ducts of buildings. It samples the air moving through the air distribution system to detect smoke. If a duct detector detects smoke, it initiates an alarm or a supervisory signal and shuts down the air handling system.

More complex air sampling-type detection systems are sometimes installed in special hazard areas. The detectors draw air samples from rooms, enclosed spaces, or equipment cabinets and pass the samples through gas analyzers. The analyzers identify smoke particles, products of combustion, and concentrations of other gases associated with a dangerous condition. These detectors identify problems much faster than conventional smoke or gas detectors because they actively draw in the air samples rather than passively waiting for smoke or gas to enter. For this reason, air sampling-type detectors are most

FIGURE 24-12 Flame detectors are specialized devices that detect the radiant energy produced by a flame.

FIGURE 24-13 Air sampling detector.

often used in areas that hold valuable contents or sensitive equipment.

Carbon Monoxide Detectors. A **carbon monoxide detector** is a gas detector that detects the presence of carbon monoxide and initiates either an alarm or a supervisory signal when the concentration of carbon monoxide becomes high enough to pose a health risk to the occupants of the building. The alarm is triggered by a high concentration of carbon monoxide that accumulates in a short period of time or by a lower concentration of carbon monoxide that occurs over a longer period of time. Carbon monoxide detectors are sometimes in the same spot-type detector unit as a smoke or heat detector (**FIGURE 24-14**). These combination detectors have one audible or visible alarm for fire and a different audible or visible alarm for elevated levels of carbon monoxide.

Carbon monoxide detector activations need to be investigated by trained personnel using gas detection devices. Investigators need to determine the level of the gas and the cause of the alarm activation. Any time you respond to an alarm initiated by a carbon monoxide detector, you need to check all building occupants to determine whether they are experiencing the following symptoms of carbon monoxide poisoning:

- Headache
- Nausea
- Disorientation
- Unconsciousness
- Flulike symptoms experienced by multiple people in the same location

Alarm Notification Appliances

Audible alarm notification appliances such as bells, horns, and electronic speakers produce a sound when the fire alarm is activated. Newer systems also incorporate visual alarm notification appliances.

Older systems used a variety of sounds as notification appliances, but this inconsistency often led to confusion about whether the sound was actually a fire alarm. NFPA 72 now requires a distinctive, standardized evacuation signal consisting of three tones followed by a short pause called the **temporal-3 pattern** or **T-3 signal**. Even smoke alarms designed for residential occupancies are required to use this sound pattern. As a result, people are now more likely to recognize a fire alarm immediately.

Some public buildings also play a recorded evacuation announcement in conjunction with the temporal-3 pattern. This recorded message is played through the fire alarm speakers and provides instructions on how to safely evacuate the site (**FIGURE 24-15**). In facilities such as airport terminals, this announcement is recorded in multiple languages. Such a system may also include a public address feature that allows fire department or building security personnel to provide specific instructions, information about the emergency, or notice when the alarm condition is terminated.

FIGURE 24-14 Multi-criteria detectors use four sensing technologies (photoelectric, infrared, heat, and CO) to accurately detect fire and CO alarm conditions.

Courtesy of Honeywell/Fire-Lite Alarms.

FIGURE 24-15 A speaker alarm notification appliance.

Courtesy of Honeywell/Fire-Lite Alarms.

FIGURE 24-16 This alarm notification appliance has both a loud horn and a high-intensity strobe light.

© Jones & Bartlett Learning

FIGURE 24-17 Automatic sprinkler systems use an electric flow switch to activate the building's fire alarm system.

© Jones & Bartlett Learning

Many new fire alarm systems incorporate visual notification devices, such as high-intensity strobe lights, as well as audio devices (**FIGURE 24-16**). Visual devices alert hearing-impaired occupants to a fire alarm and are useful in noisy environments where an audible alarm might not be heard.

Alarm Initiation by Fire Suppression Systems

Fire suppression systems initiate an alarm when they are activated. For example, some buildings have an **automatic sprinkler system**, which is a system of pipes filled with water under pressure that discharges when the system is activated. Automatic sprinkler systems are usually connected to the fire alarm system and initiate the alarm if water starts flowing through the system (**FIGURE 24-17**). If there is no fire but water is being discharged, this ensures that someone is made aware of this situation so that the water can be shut off and the reason for the discharge investigated. Any other types of fire suppression systems in a building should also be connected to the building's fire alarm system.

Unwanted Alarms

As the number of fire detection and alarm systems increases, so does the number of unwanted alarms. Unwanted alarms can actually cause a dangerous and deadly situation. If a fire alarm sounds frequently, occupants may become complacent and may neglect to take appropriate action in a real emergency. Such was the case in January 2000, when three students died in their dormitory rooms at Seton Hall University. The students ignored the alarm, which signaled a real fire emergency, because they had become used to hearing unwanted alarms on a regular basis. Fire department personnel who understand how fire detection and alarm systems operate can advise building owners how to reduce the chances of further unwanted alarms when these situations occur.

Four distinct types of unwanted alarms are possible: malicious alarms, nuisance alarms, unintentional alarms, and unknown alarms.

- A **malicious alarm** occurs when an individual deliberately activates a fire alarm when there is no fire. Manual fire alarm boxes are popular targets for pranksters. Causing a malicious alarm is an illegal act.
- A **nuisance alarm** is an alarm that is caused by environmental conditions, improper installation or maintenance, or the malfunction of one or more alarm systems. For example, a nuisance alarm could occur if construction or remodeling is occurring in a structure and construction dust enters the smoke detector. It can also occur if a person is smoking a cigarette under a smoke detector.

The smoke detector functions properly—it detects smoke and activates the alarm system—but there is no real emergency.

- An **unintentional alarm** is an alarm activated by a person acting without malice but accidentally operating an alarm initiating device. For example, a technician filling a fire suppression system might initiate an unintentional alarm that results in the activation of a waterflow device.
- An **unknown alarm** is an unwanted activation of an alarm device where the cause cannot be identified.

Regardless of the underlying cause, all nuisance alarms have the same results: They waste fire department resources, and they may delay legitimate emergency responses. Frequent nuisance alarms at the same site can desensitize building occupants to the alarm system so that they might not respond appropriately to a real emergency. Therefore, all nuisance alarms should be investigated to determine the cause of the alarms and to correct the problem.

Many nuisance alarms could be avoided by relocating a detector or using a different type of detector in a particular location. For example, if an alarm system is activated whenever it rains, water could be leaking into the wiring or a system component. Detectors subjected to a buildup of dust, dirt, or other debris may become more sensitive and initiate nuisance alarms.

In environments where detectors are potentially subject to nuisance alarm activations, such as at an industrial facility where solvent or paint is sprayed or large quantities of dust are generated, the fire alarm system can require verification of the detected fire condition. A **cross-zoned system** requires two or more initiating devices to activate before the alarm sounds. If one initiating device activates, a trouble signal sounds. If a second initiating device activates, the fire alarm sounds, and an alarm signal is sent to the monitoring station. These devices are also sometimes used to activate fire suppression systems.

Systems with an **alarm verification feature** delay sounding the alarm for 30 to 60 seconds after an initiating device activates. During this time, the system shows a trouble or "pre-alarm" condition at the FACU. After the preset interval, the system rechecks the initiating device. If the condition has cleared, the system resets to normal. If the detector is still sensing smoke or if a second detector activates, the fire alarm sounds.

Both cross-zoned systems and systems with an alarm verification feature are designed to prevent nuisance alarms after brief exposures to light smoke or dust. For more information, see the NFPA's *Fire Service Guide to Reducing Unwanted Fire Alarms* (available for free at catalog.nfpa.org/Fire-Service-Guide-to-Reducing-Unwanted-Fire-Alarms-PDF-P9419.aspx).

Residential Fire and Carbon Monoxide Alarms

Current building and fire codes require the installation of fire alarms in all new and existing residential dwelling units. The most common type of residential fire alarm is a single-station smoke alarm that detects smoke (**FIGURE 24-18**). Millions of single-station smoke alarms have been installed in private dwellings and apartments.

Building codes and NFPA 72 require new residences and residences undergoing an alteration to have at least one smoke alarm on each level of the dwelling, including basements and habitable attics; one smoke alarm in the corridor or hallway outside each sleeping area, and one smoke alarm in each bedroom. Older residences built before these requirements were established may not have enough smoke alarms to provide early detection and notification. Firefighters should recommend that occupants of dwellings where these requirements are not met install additional smoke alarms to improve their protection. Some jurisdictions require that when a home is sold or ownership is transferred, the seller must install an adequate number of smoke alarms. Read and understand your department's

FIGURE 24-18 A single-station smoke alarm.

Courtesy of Kidde Residential and Commercial Division.

policies before making recommendations to homeowners and occupants in your community about fire protection equipment.

Residential smoke alarms have either ionization smoke detectors, photoelectric smoke detectors, or both built into them. When properly installed and maintained, both ionization and photoelectric smoke alarms are effective for detecting fires and alerting occupants early enough for them to escape safely. Like ionization detectors in large fire alarm systems, ionization smoke alarms and combination ionization/photoelectric alarms are not suitable for use near kitchens or bathrooms because the smoke from cooking or steam from a bathroom could trigger nuisance alarms. The NFPA Educational Messages Advisory Committee recommends using both ionization and photoelectric fire alarms or combination ionization/photoelectric alarms in residential settings and placing them in the most appropriate setting.

Smoke alarms in existing structures can be either battery-powered, connected to the structure's 120-volt electrical system, or both. Smoke alarms connected to the 120-volt power source are called hard-wired because they are permanently connected with wires to the power source. Current building codes require smoke alarms in all newly constructed dwellings to be hard-wired and powered by the building's electrical system. Each smoke alarm must also contain a backup battery in case electrical service is disrupted. The smoke alarms in new construction must also be interconnected, either wirelessly or hardwired, so all alarms will sound if one alarm detects smoke. The single-station smoke alarms in many older residences are not hard-wired or connected to the other smoke alarms in the residence.

TIP

Hard-wired smoke detectors become inoperable if someone turns the power off at the circuit breaker, so the circuit should remain on at all times.

Smoke alarm batteries should be replaced twice per year even if they do not indicate a low battery unless they have a nonreplaceable long-life lithium battery that lasts up to 10 years. Some communities recommend replacing smoke alarm batteries when Daylight Saving Time begins and ends, regardless of the battery status. Because older smoke alarms may become overly sensitive and generate frequent nuisance alarms or develop decreased sensitivity and fail to activate in the event of a fire, NFPA 72 recommends that all smoke alarms be replaced 10 years after the date of manufacture. (The date of manufacture is stamped on each smoke alarm.)

Similar to smoke alarms, building codes and NFPA 72 require carbon monoxide alarms to be installed in new residences where either the dwelling unit contains a fuel-fired appliance or the dwelling unit has an attached garage with an opening to the dwelling unit. Residential carbon monoxide alarms sound an audible or visual alarm when the concentration of carbon monoxide becomes high enough to pose a health risk to the occupants of the building. Like carbon monoxide detectors in large fire alarm systems, the alarm is triggered by a high concentration of carbon monoxide that accumulates in a short period of time or by a lower concentration of carbon monoxide that occurs over a longer period of time. Some units are equipped with digital displays that indicate the level of carbon monoxide detected. Smart combination smoke and carbon monoxide alarms can connect to a home's wireless network and smart devices to provide warning from a fire or from elevated levels of carbon monoxide even when occupants are not home (**FIGURE 24-19**).

Like any appliance, smoke alarms can malfunction. Most problems with smoke alarms are caused by a lack of power, dirt in the sensing chamber, or defective equipment. Unfortunately, in the United States, many battery-powered smoke alarms contain a dead battery. Some people do not change batteries in smoke and carbon monoxide alarms or replace the devices when recommended, resulting in inoperable alarms.

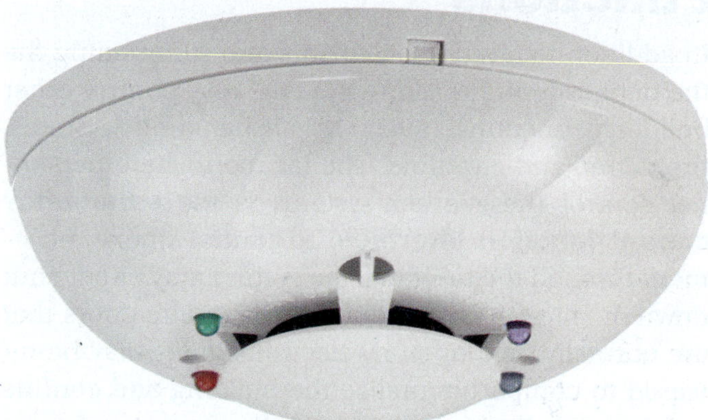

FIGURE 24-19 Combination smoke and carbon monoxide alarm with smart technology to provide notifications even when occupants are not home.

Courtesy of Honeywell/Fire-Lite Alarms.

People also sometimes remove smoke alarm batteries to use for another purpose or to prevent nuisance alarm activations.

Most smoke alarms containing batteries will alert occupants that the battery is low by emitting a chirp every few seconds. This indicates that the battery needs to be replaced. Many hard-wired smoke alarms with backup batteries chirp to indicate that the backup battery is low, even if the 120-volt power source is operational. This ensures that the smoke alarm always has an operational backup system.

Smoke alarms may generate nuisance alarms or may not activate properly if dust or an insect becomes lodged in the sensing chamber. To prevent this, alarms should be periodically cleaned by removing the cover and gently vacuuming the chamber. In some instances, a puff of air from a can of compressed air might remove the dust or insect; however, it could instead push the contaminant deeper into the detection chamber. Follow the manufacturer's instructions for proper cleaning and maintenance.

Some home smoke alarms are part of home security systems. Like commercial systems, these home fire alarm/security systems have an alarm control unit and require a passcode to set or reset the system. They may also be monitored by contract monitoring companies that receive the signals and initiate the appropriate action or response.

Understanding the basic functions and troubleshooting of simple smoke alarms enables firefighters to respond to citizens who call the fire department when they encounter problems with their smoke alarms. Most important, it enables you to keep smoke alarms operational.

Other Fire Alarm Functions

In addition to alerting occupants and summoning the fire department, fire alarm systems may control other building functions, such as air-handling systems, fire doors, and elevators. The location where personnel control these other systems is the **emergency control function interface**. To control smoke movement through the building, the system may either shut down or start up air-handling systems. Fire doors that are normally held open by electromagnets may be released to compartmentalize the building and confine the fire to a specific area. Doors allowing reentry from exit stairwells into occupied areas may be unlocked. Elevators may be summoned, or recalled, to a predetermined floor, usually the main lobby, so they can be used by fire crews.

Responding fire personnel must understand which building functions are controlled by the fire alarm system, for both safety and fire suppression reasons. This information should be noted in the preincident plan and should be available in printed form or on a graphic display at the FACU location.

Fire Department Notification

The fire department should always be notified when a fire alarm system is activated. In some cases, a person must make a telephone call to the fire department's emergency communications center, usually by calling 911. In other cases, the fire alarm system is connected directly to the fire department's emergency communications center or to a remote location where someone on duty calls the fire department. Fire alarm systems are classified into the following three main categories:

- Protected premises (local) fire alarm system
- Public emergency alarm reporting system
- Supervising station alarm system

These classifications describe how the fire department is notified of an alarm (**TABLE 24-2**).

Fire Suppression Systems

Fire suppression systems include automatic sprinkler systems, standpipe systems, and special hazard suppression systems. Most newly constructed commercial buildings incorporate at least one type of fire suppression system, and increasing numbers of single-family dwellings are being built with automatic sprinkler systems as well. Understanding how these systems work is important because they can affect fire behavior. In addition, firefighters should know how to interface with the fire suppression system and how to shut down a system to prevent unnecessary damage.

Automatic Fire Sprinkler Systems

The most common type of fire suppression system is the automatic sprinkler system. Automatic sprinklers are reliable and effective, with a history of more than 100 years of successfully controlling fires. When properly designed, installed, inspected, tested, and maintained, these systems are extremely effective in controlling or extinguishing fires. NFPA 13, *Standard*

TABLE 24-2 Fire Department Notification Systems

Type of System	Description
Protected premises (local) fire alarm system	This fire alarm system sounds an alarm in the building where it was activated. These systems are found in older buildings. No signal is sent out of the building—someone onsite must call the fire department to respond. Buildings with this type of system must have notices posted requesting occupants to call the fire department and report the alarm after they exit (**FIGURE 24-20**).
Public emergency alarm reporting system	In jurisdictions with fire alarm boxes, this fire alarm system sounds an alarm in the building and transmits a signal to a master fire alarm box located outside the building. The master fire alarm box then transmits the alarm to the fire department communications center. If the jurisdiction eliminates fire alarm boxes, any public emergency alarm reporting systems in the jurisdiction need to be replaced.
Supervising station alarm systems	This type of fire alarm system communicates between the protected property and the fire department or a monitoring location staffed 24 hours per day. The person on duty at the fire department or monitoring location interprets the alarm signals and takes the appropriate action to handle the alarm condition. There are three types of supervising station alarm systems—remote, proprietary, and central station.
Remote supervising station alarm system	This type of supervising station alarm system sounds an alarm in the building and transmits a signal via a phone line or radio signal to the fire department or to an off-site monitoring location. If the signal is sent to an off-site location, someone on duty is responsible for calling the fire department.
Proprietary supervising alarm system	This type of supervising station alarm system sounds an alarm in the building and transmits a signal to a monitoring location owned and operated by the facility's owner. The monitoring location is often located on the same premises as the fire alarm system but sometimes is located at a remote location belonging to the same owner (**FIGURE 24-21**). Depending on the nature of the alarm and predetermined arrangements with the local fire department, facility personnel either respond and investigate or immediately report the alarm to the fire department, usually by phone or a direct line. Proprietary systems are often installed at facilities where multiple buildings belong to the same owner, such as a university or an industrial complex.
Central station service alarm system	This type of supervising station alarm system sounds an alarm in the building and transmits a signal to an off-site alarm monitoring facility (the central station) operated by a third-party. This facility may be located in the same city as the facility or in a different part of the country. Alarm monitoring facilities contract with multiple building owners to monitor alarm systems (**FIGURE 24-22**). Depending on the nature of the alarm and predetermined arrangements with the local fire department, facility personnel either respond and investigate or immediately report the alarm to the appropriate fire department. These facilities also often conduct testing and maintenance and keep records for the protected property. Usually, fire alarm systems are connected to the central station through leased or standard telephone lines. The use of either cell phone or radios, however, is becoming more common, and in remote areas without telephone lines, may be the only available transmission method.

Courtesy of Chris DeSantis.

FIGURE 24-20 Buildings with a protected premises (local) fire alarm system should post notices requesting occupants to call the fire department and report the alarm after they exit.

for the Installation of Sprinkler Systems, describes minimum standards for automatic sprinkler systems.

Unfortunately, few members of the general public have an accurate understanding of how automatic sprinklers work. In movies and on television, when one sprinkler is activated, the entire system begins to discharge water. This inaccurate portrayal of how automatic sprinkler systems operate has made people hesitant to install automatic sprinklers because they fear the system will cause unnecessary water damage. The reality is quite different. In most automatic sprinkler systems, the sprinkler heads open one at a time as they are heated to their operating temperature. In most cases, only one or two sprinkler heads are needed to control the fire, producing minimal water damage.

FIGURE 24-21 In a proprietary system, fire alarms from several buildings are connected to a single monitoring site owned and operated by the buildings' owner.

The basic operating principles of an automatic sprinkler system are simple. A system of water pipes is installed throughout a building to deliver water to every area. Depending on the design and occupancy of the building, these pipes may be placed above or below the ceiling. Sprinkler heads are located along this system of pipes so that each sprinkler head covers a particular square footage of floor area. The heat of a fire in that area activates the sprinkler head, which then discharges water onto the fire. When a sprinkler head is activated, water starts flowing through the pipes, and the flowing water activates the building's fire alarm system. Automatic sprinkler systems are so effective that, in many instances, the sprinklers have often controlled or extinguished the fire by the time the firefighters arrive.

The overall design of automatic sprinkler systems can be complex, especially in large buildings. This

TIP

Here are some popular myths about sprinklers:

- All the sprinklers in the building will release water and flood the entire building when the sprinkler system is activated.
- Activating a fire alarm manually will set off every sprinkler in the building.
- If a smoke detector is activated, all of the sprinklers in the building will release water.
- Water discharged from a sprinkler system will do more damage than a fire.

These are all wrong!

FIGURE 24-22 A central station monitors alarm systems at many locations.

Courtesy of www.acimonitoring.com, Doug Beaulieu.

complexity is somewhat deceiving, however, because even the largest systems include just a few major components: automatic sprinkler heads, piping and fittings, waterflow alarms, various types of valves, and a water supply, which may or may not include a fire pump (**FIGURE 24-23**). Most sprinkler systems are connected directly to a public water supply system.

Automatic Sprinkler Heads

The **automatic sprinkler head**, which is commonly referred to simply as a **sprinkler head**, is the working end of a sprinkler system. In most systems, the sprinkler heads detect the heat of a fire and activate the sprinkler system to apply water to the fire. Sprinkler heads are composed of the following elements (**FIGURE 24-24**):

- The frame, which supports the rest of the components
- An opening called the orifice through which the water flows; standard sprinkler heads have a ½-inch (in.; 12.7-millimeter [mm]) orifice, but other sizes are available for special applications.
- A heat-sensitive release mechanism, which holds a cap in place over the orifice until heat activates it and the mechanism is released. When this happens, the water pushes the cap out of the way as it discharges onto the fire (**FIGURE 24-25**).
- A deflector, which directs the water in a spray pattern

Sprinkler heads manufactured before the mid-1950s were designed to direct part of the water stream

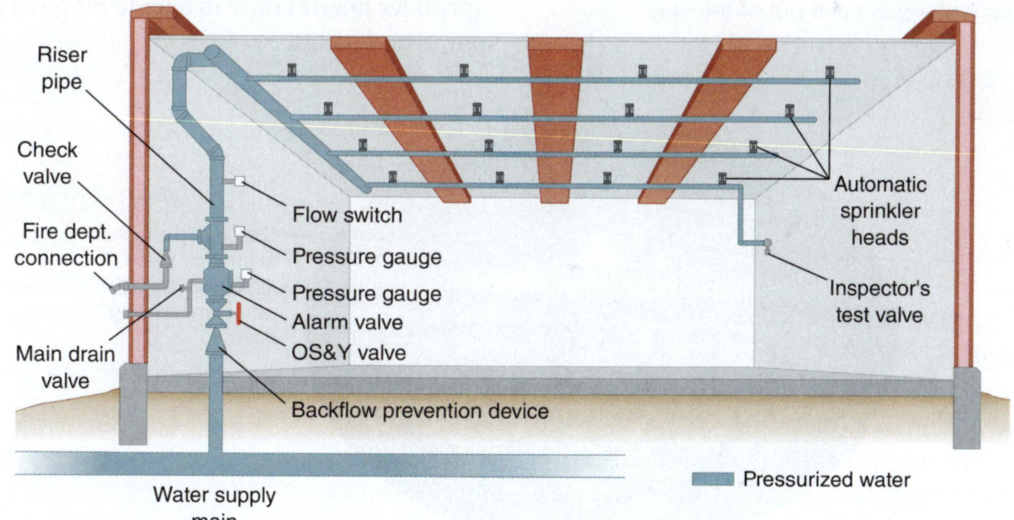

FIGURE 24-23 The basic components of an automatic sprinkler system include sprinkler heads, piping, control valves, and a water supply.

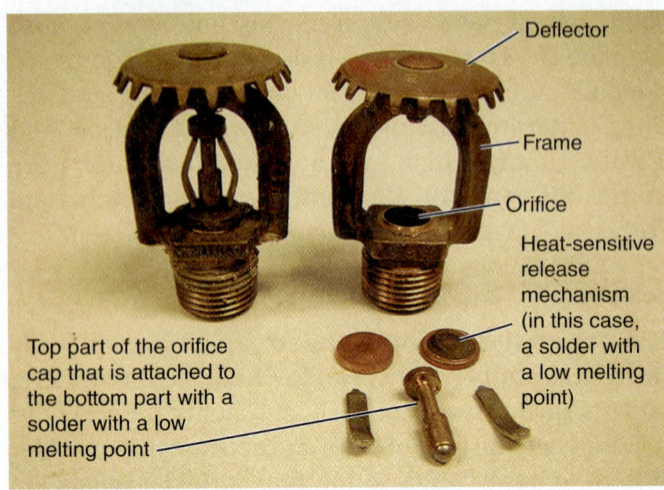

FIGURE 24-24 Automatic fire sprinkler heads consist of a frame, a heat-sensitive release mechanism, an orifice, an orifice cap, and a deflector.

FIGURE 24-25 When the temperature that activates the release mechanism is reached, the mechanism releases the cap, and the flowing water pushes it out of the way.

up toward the ceiling. At the time, it was believed that this action helped cool the area and extinguish the fire. Sprinkler heads with this design are called old-style sprinklers. Although they are no longer manufactured, many of them remain in service today. Automatic sprinklers manufactured after the mid-1950s deflect the entire water stream down toward the fire. These types of heads are referred to as new-style heads or standard spray heads. New-style heads can replace old-style heads, but the reverse is not true. Due to different coverage patterns, old-style heads should not be used to replace any new-style heads.

Sprinkler heads come in several styles. They are categorized according to the type of release mechanism used, the intended mounting position, and the temperature that activates the release mechanism. Sprinkler heads can also be designed for special applications, such as covering large areas, discharging the water in extra-large droplets, or discharging the water as a fine mist. Some sprinkler heads have protective coatings to help prevent corrosion. Fire protection engineers and design professionals consider these characteristics when they design a system and select the appropriate heads. It is important to ensure that the proper heads are installed and that any replacement heads are of the same type.

Release Mechanisms. There are three types of release mechanisms for automatic sprinkler heads. A **fusible-link sprinkler head** has a solder with a low melting point linking two pieces of metal as the release mechanism (**FIGURE 24-26**). When the melting temperature is reached, the solder melts, which breaks the link and releases the cap. Figure 24-25 shows a fusible-link sprinkler head broken apart. Fusible-link sprinkler heads come in a wide range of styles and temperature ratings.

FIGURE 24-26 In fusible-link sprinkler heads, two pieces of metal are linked together by an alloy, such as solder.

A **frangible-bulb sprinkler head** has a glass bulb that holds the cap in place (**FIGURE 24-27**). The bulb is filled with glycerin or alcohol and contains a small air bubble. As the bulb is heated, the liquid expands, and the pressure increases until the glass breaks, which releases the cap. The temperature at which the glass breaks is determined by the volume and composition of the liquid, the thickness of the glass, and the size of the air bubble.

A **chemical-pellet sprinkler head** has a plunger mechanism and a small chemical pellet that holds the cap in place. When the temperature reaches the melting point of the chemical pellet, the pellet melts and the resulting liquid compresses the plunger, which releases the cap.

An **early-suppression fast-response sprinkler head (ESFR)** is much larger than other types of sprinkler heads and is used in large warehouses and distribution facilities where early fire suppression is important (**FIGURE 24-28**). This type of sprinkler head often has large orifices to discharge large volumes of water onto a fire. They also have an efficient heat collector that identifies the preset temperature quickly, which means they activate more quickly than other types of sprinkler heads, speeding up the response and ensuring the rapid release of water.

A **deluge head** does not have a cap, so it also does not have a release mechanism. The orifice is always open and the water release is controlled by valves in the pipes (**FIGURE 24-29**). Deluge heads are used only in deluge sprinkler systems, which are discussed later in this chapter.

FIGURE 24-28 The ESFR sprinkler head is larger than a standard sprinkler head and discharges about four to five times the amount of water as a standard head.

FIGURE 24-27 Frangible-bulb sprinkler heads activate when the liquid expands and the resulting pressure in the bulb breaks the glass. The color of the glass indicates the temperature required to break the glass.

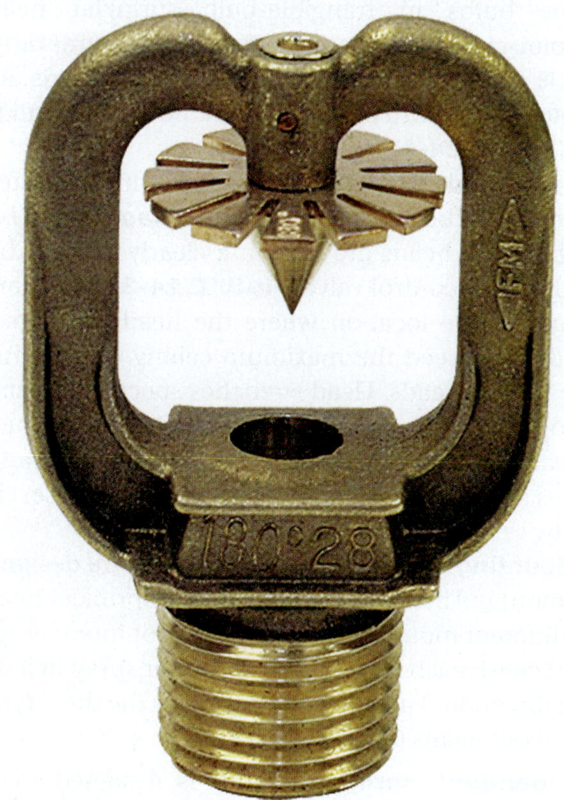

FIGURE 24-29 A deluge sprinkler head has no release mechanism so the orifice is always open.

Temperature Ratings. Sprinkler heads are rated according to their **release temperature**, which is the temperature at which the release mechanism activates. A typical temperature rating for sprinkler heads in a light-hazard occupancy, such as an office building, is 135°F to 175°F (57°C to 76°C). The temperature rating of a sprinkler head must match the anticipated ambient air temperatures. For example, sprinkler heads used in areas with warmer ambient air temperatures need to have higher ratings. If the rating is too low for the ambient air temperature, accidental activations may occur. Conversely, if the rating is too high, the system may react too slowly, and the fire may grow too quickly for the sprinklers to control it. The temperature rating is stamped on the body of the sprinkler head.

Sprinkler heads also have a **maximum ceiling temperature**. This temperature is the maximum temperature to which the sprinkler head can be exposed. It is usually about 25°F (14°C) degrees below the lower end of the temperature rating. If a sprinkler head is exposed to temperatures higher than the maximum ceiling temperature, they might not perform correctly. This applies to sprinkler heads installed in a sprinkler system and to sprinkler heads in storage.

The bulbs in frangible-bulb sprinkler heads are color-coded to identify the temperature rating (**TABLE 24-3**). Some fusible-link sprinkler heads also are color-coded with paint on the frame of the sprinkler head (**TABLE 24-4**).

Spare heads that match those used in the system should always be available onsite in a **head box**. Usually, the spare heads are kept in a clearly marked box near the main control valve (**FIGURE 24-30**). The temperature of the location where the heads are stored should not exceed the maximum ceiling temperature of the type of heads. Head wrenches specifically made for replacing all of the different heads should also be in the head box. Having spare heads and wrenches handy enables sprinkler systems to be returned to full service quickly.

Mounting Position. Sprinkler heads are designed to be mounted in one of three positions. Sprinkler heads with different mounting positions are not interchangeable because each type directs the water spray in a different direction. The following describes the three types of sprinkler heads (**FIGURE 24-31**):

- A **pendent sprinkler head** is designed to be mounted on the underside of the sprinkler piping, hanging down toward the floor. (*Pendent* means hanging down.) These heads are commonly marked *SSP*, which stands for *standard spray pendent*. The deflector on this type of sprinkler head changes the

TABLE 24-3 Temperature Rating for Frangible-Bulb Sprinkler Heads

Maximum Ceiling Temperature	Temperature Rating	Color Codes
100°F (38°C)	135 (57°C)	Orange
120°F (49°C)	155°F (68°C)	Red
150°F (66°C)	175°F (79°C)	Yellow
150°F (66°C)	200°F (93°C)	Green
225°F (107°C)	250 to 300°F (121 to 149°C)	Blue
300°F (149°C)	325 to 375°F (163 to 191°C)	Purple
375°F (191°C)	400 to 475°F (204 to 246°C)	Black
475°F (246°C)	500 to 575°F (260 to 302°C)	Black
625°F (329°C)	650°F (343°C)	Black

Modified from National Fire Protection Association (NFPA). 2021. *NFPA 13, Standard for the Installation of Sprinkler Systems*. 2022 Edition. Table 7.2.4.1(a), pp. 13–50. Quincy, MA: NFPA.

TABLE 24-4 Temperature Rating for Fusible-Link Sprinkler Heads

Maximum Ceiling Temperature	Temperature Rating	Color Code
100°F (38°C)	135 to 170°F (57 to 77°C)	Uncolored or black
150°F (66°C)	175 to 225°F (79 to 107°C)	White
225°F (107°C)	250 to 300°F (121 to 149°C)	Blue
300°F (149°C)	325 to 375°F (163 to 191°C)	Red
375°F (191°C)	400 to 475°F (204 to 246°C)	Green
475°F (246°C)	500 to 575°F (260 to 302°C)	Orange
625°F (329°C)	650°F (343°C)	Orange

Modified from National Fire Protection Association (NFPA). 2021. *NFPA 13, Standard for the Installation of Sprinkler Systems*. 2022 Edition. Table 7.2.4.1(b), pp. 13–50. Quincy, MA: NFPA.

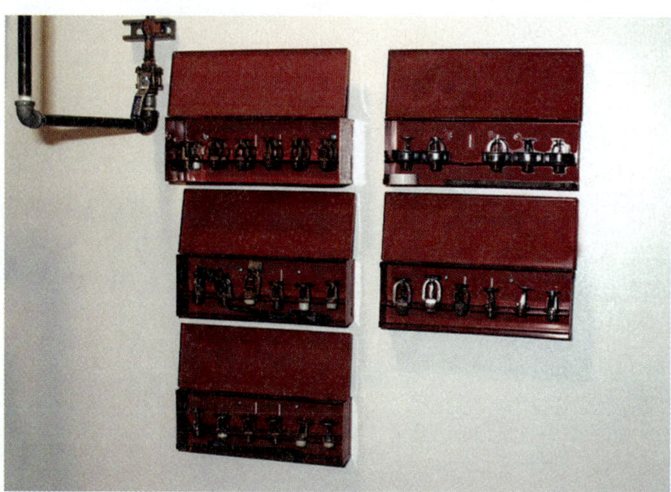

FIGURE 24-30 Spare sprinkler heads should be kept in a head box near the main sprinkler system valve.

stream from spraying straight down to water that sprays in an umbrella pattern. Pendent sprinkler heads are the most common type of sprinkler head and are used in homes and many businesses where ceilings are installed.

- An **upright sprinkler head** is designed to be mounted on the top side of the sprinkler piping. These heads are usually marked *SSU* for *standard spray upright*. The deflector on upright sprinkler heads deflect the stream 180 degrees so that the water is directed down towards the floor. Upright sprinkler heads are typically used in areas where there are no ceilings installed.
- A **sidewall sprinkler head** is designed for horizontal mounting, so that the stream is projected out and down from the top of a wall. The deflector on this type of sprinkler head has a shield on the top so that the water spray is directed in an arch pattern out from the wall and down toward the floor. Sidewall sprinkler heads are typically installed in areas where there are no ceilings but for esthetic reasons, exposed piping with upright sprinkler heads is not desirable.

Sprinkler Piping

Sprinkler piping is the network of pipes in a sprinkler system that delivers water to the sprinkler heads, and it includes the following:

- **Branch lines**: pipes that supply sprinkler heads
- **Cross mains**: pipes that supply branch lines
- **Risers**: vertical supply pipes
- **Feed mains**: pipes that supply cross mains from the water source, either directly or through risers

A.

B.

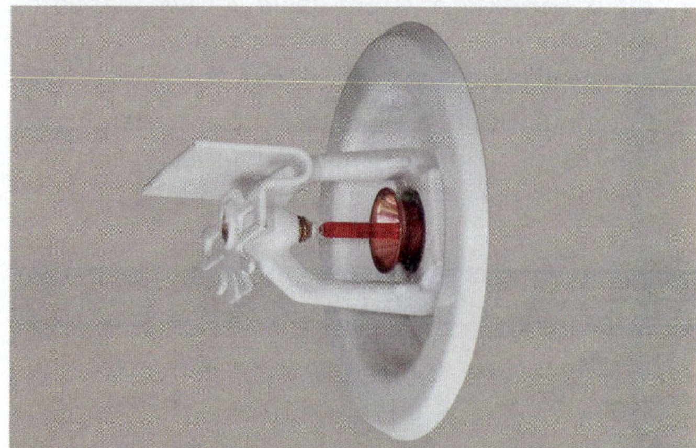

C.

FIGURE 24-31 Sprinkler head mounting positions. **A.** Pendent. **B.** Upright. **C.** Sidewall.

FIGURE 24-32 Most new sprinkler systems use steel pipes.

© Eduard Darchinyan/Shutterstock

Although sprinkler pipes are usually made of steel, other materials can be used as well. For example, plastic pipe is often used in residential sprinkler systems. To join pipes together, a **fitting** is used so that the piping can change direction or transition between different sizes (**FIGURE 24-32**). The diameter of the pipes is determined by the fire protection engineer when the system is designed. Near the main control valve, pipes have a larger diameter. As the pipes approach the sprinkler heads, the diameter usually decreases.

Valves

A sprinkler system includes several types of valves that control the flow of water, including the following:

- **Main water supply control valve**, a manually controlled valve that allows water to enter the system from the water source
- **Main sprinkler system valve**, a valve that automatically opens and signals an alarm when a sprinkler head is activated
- Additional valves throughout the system used for testing and servicing and other specialized functions

SAFETY TIP

All valves that are normally open in a fire protection system should be in the full open position at all times.

Main Water Supply Control Valves. Every sprinkler system needs at least one main water supply control valve that allows water to enter the system. This main water supply control valve must be an **indicating valve**, meaning that the position of the valve indicates whether the valve is open or closed. Three of the most common indicating valves used as main water supply valves are the following:

- Outside screw and yoke (OS&Y) valve
- Butterfly valve
- Post indicator valve (PIV)

FIGURE 24-33 The screw or stem on an OS&Y valve indicates whether the valve is open or closed. The screw on this OS&Y valve is out, indicating that this valve is open.

Courtesy of Chris DeSantis.

An **outside screw and yoke (OS&Y) valve**—also called an **outside stem and yoke valve**—has a threaded stem that moves in and out as the valve is opened or closed. If the stem is out, the valve is open; if the stem is in, the valve is closed (**FIGURE 24-33**). OS&Y valves are often found in a mechanical room in the building, where the water supply for the sprinkler system enters the building. In warmer climates, they may be found outside.

A **butterfly valve** has an internal disk the same diameter as the pipe that either blocks the pipe and prevents water from flowing into the system or is rotated

FIGURE 24-34 The yellow butterfly valve indicator is parallel to the pipe, indicating that the valve is open.

Courtesy of Chris DeSantis.

A.

B.

FIGURE 24-35 **A.** A ground PIV is mounted in the ground to open or close an underground valve. **B.** A wall PIV is mounted on the outside wall of a building.

A. © Jones & Bartlett Learning; **B.** © Jones & Bartlett Learning. Courtesy of MIEMSS.

to allow the water to flow past it. The valve can be rotated to any position to control the water flow. When it is rotated a full 90 degrees, the valve is completely open for maximum water flow. An external indicator on the stem shows if the valve is open or closed. If the indicator is perpendicular to the pipe, the valve is closed. If the indicator is parallel (in line) with the pipe, the valve is open (**FIGURE 24-34**).

A **post indicator valve (PIV)** is, as its name implies, a post with an indicator that reads either "OPEN" or "SHUT" (**FIGURE 24-35**). A PIV is usually located in an open area outside the building and controls an underground valve, but the post can also be mounted on a wall. Opening or closing a PIV requires a wrench, which is usually attached to the side of the valve.

It is critically important that automatic sprinkler systems always remain charged with water and ready to operate, which means the main water supply control valve must always be locked or supervised in the open position. This ensures that the water supply to the sprinkler system is never shut off unless the proper people are notified that the system is out of service. If this valve is closed and the fire alarm system is activated, water from the water supply cannot get past the closed water supply control valve. According to the NFPA, 57 percent of sprinkler system failures were caused because the main water supply control valve was shut off (Ahrens 2021, 6–7).

As an alternative to locking the main water supply control valve open, the valve may be electronically supervised by a supervisory switch (**FIGURE 24-36**). If someone opens or closes the valve, the supervisory switch sends a supervisory signal to the FACU, indicating a change in the valve position. If the change has not been authorized, the cause of the signal can be investigated and the problem corrected.

FIGURE 24-36 A supervisory switch activates an alarm if someone attempts to close a valve that should remain open.

Main Sprinkler System Valves. Main sprinkler system valves are usually installed on the main riser, above the water supply control valve. The type of main sprinkler system valve used depends on the type of sprinkler system installed. Options include an alarm valve, a dry pipe valve, a preaction valve, or a deluge valve.

The primary functions of an **alarm valve** are to signal an alarm when a sprinkler head is activated and to prevent nuisance alarms caused by pressure variations and surges in the water supply to the system. This valve has a clapper mechanism that remains in the closed position until a sprinkler head opens. The closed clapper prevents water from flowing out of the system and back into the public water mains when water pressure drops. When a sprinkler head is activated, the clapper opens fully and allows water to flow freely through the system.

The **dry pipe valve**, **preaction valve**, and **deluge valve** are specialized main sprinkler valves used in dry pipe, preaction, and deluge sprinkler systems, which are described later in this chapter. These valves function both as an alarm valve and as a dam, holding back the water until the sprinkler system is activated. When the system is activated, the valves open fully so that water can enter the sprinkler piping.

Additional Valves. Sprinkler systems are equipped with a variety of other valves. Several smaller valves, usually located near the main water supply control valve, include drain valves, test valves, and connections to alarm devices. All of these valves should be properly labeled.

In larger facilities, the sprinkler system may be divided into zones, with a zone valve controlling the flow of water to each zone. A **zone valve** allows the water supply to specific areas to be shut down without turning the entire system off. This design makes maintenance easy and can prove extremely valuable when a fire occurs. After a fire is extinguished, water flow to the affected area can be shut off so that the activated heads can be replaced. Fire protection in the rest of the building is unaffected by this shutdown.

FIGURE 24-37 In tall buildings, a fire pump may be needed to maintain appropriate pressure in the sprinkler system.

Courtesy of Chris DeSantis.

Sprinkler System Water Supplies

The water used in an automatic sprinkler system may come from a municipal water system, from onsite storage tanks, or from alternative water sources such as storage ponds or rivers. Whatever its source, the water supply must be able to satisfy the demands of the sprinkler system as well as meet the needs of the fire department in the event of a fire. The preferred water source for a sprinkler system is a municipal water supply if one is available. If the municipal supply cannot meet the water pressure and volume requirements of the sprinkler system, alternative supplies must be established.

Fire pumps are used when the water comes from a static source such as a storage pond. Because most municipal water supply systems do not provide enough pressure to control a fire on the upper floors of a high-rise building, fire pumps may also be used to boost the pressure in the sprinkler systems in those tall buildings (**FIGURE 24-37**). The fire pumps turn on automatically when the sprinkler system activates or when the pressure drops below a preset level. Depending on the height of the building, a series of fire pumps may be needed to provide adequate pressure to the upper floors.

A large industrial complex could have more than one water source, such as a municipal system and a backup storage tank (**FIGURE 24-38**). Multiple fire pumps can provide water to the fire protection systems in different areas through underground pipes. Private hydrants may also be connected to the same underground system.

Each sprinkler system also has a **fire department connection (FDC)**. The FDC usually has two or more 2½-in. (66-mm) female couplings or one large-diameter hose coupling mounted on an outside wall or placed near the building (**FIGURE 24-39**). It connects directly into the sprinkler system after the main control valve or system valve. The FDC allows the department's engine to pump water into the sprinkler system. The FDC may be used as either a supplement or the main source of water to the sprinkler system if the regular supply is interrupted or if a fire pump fails.

Each fire department should establish a standard operating procedure (SOP) for first-arriving companies that specifies how to connect to the FDC and when to charge the system.

FIGURE 24-38 An elevated storage tank ensures that sufficient water and adequate pressure will be available to fight a fire.

FIGURE 24-39 The FDC provides a point of connection to the sprinkler or standpipe system to deliver additional water and boost the pressure in the system.

FIGURE 24-40 A water-motor gong.

In large facilities, a single FDC may be used to deliver water to all fire protection systems in the complex. The water from this connection flows into the private underground water mains instead of into each system. Water pumped into this type of FDC should come from a source that does not service the complex, such as a public hydrant on a different grid.

Waterflow Alarms

All sprinkler systems are required to be equipped with a method for sounding an alarm when water begins flowing in the pipes. This type of warning is important both in an actual fire and in an accidental activation. Without these alarms, the occupants or the fire department might not be aware of the sprinkler activation. If a building is unoccupied, the sprinkler system could continue to discharge water long after a fire is extinguished, leading to extensive water damage.

The alarm for many older systems is sounded by a **water-motor gong**, which is an audible alarm notification device that makes noise as long as water is flowing (**FIGURE 24-40**). When the sprinkler system is activated and the main alarm valve opens, some water is fed through a pipe to a gong located on the outside of the building. This gong alerts people outside the building that there is water flowing. The water-motor gong is a mechanical alarm that functions even if there is no electricity.

Accidental soundings of water-motor gongs are rare. If a water-motor gong is sounding, water is almost certainly flowing from the sprinkler system somewhere in the building. Fire companies that arrive and hear the distinctive sound of a water-motor gong know that there is a fire or that something else is causing the sprinkler system to flow water.

Newer sprinkler systems are connected to the building's fire alarm system by either an electric **waterflow alarm device**—a device that activates when a specific water flow is detected in the pipes—or an electric pressure switch that activates when full water pressure is detected. These initiating devices cause the FACU to alert the building's occupants of their activation. In a supervised system, the fire department is also notified. Unlike water-motor gongs, waterflow alarm devices and pressure switches can be accidentally triggered by water pressure surges in the system. To reduce the risk of accidental activations, these devices usually have an alarm verification feature where the waterflow condition must persist for a period of time, usually 30 to 60 seconds (adjustable), after the device activates, or they do not sound an alarm.

Types of Automatic Sprinkler Systems

Automatic sprinkler systems are classified into five categories:

- Wet pipe
- Dry pipe
- Preaction
- Deluge
- Water mist

Although many buildings may use the same type of system to protect the entire facility, it is not uncommon to see two or three types of systems combined in one building. Some facilities use a wet pipe sprinkler system to protect most of the structure but implement a dry pipe sprinkler or preaction system in a specific area. In some cases, a dry pipe sprinkler or preaction system will branch off from the wet pipe sprinkler system.

Wet Pipe Sprinkler Systems. A **wet pipe sprinkler system** is the most common and least expensive type of automatic sprinkler system. As its name implies, the piping in a wet pipe system is always filled with water. When a sprinkler head activates, water is immediately discharged onto the fire. There are potential drawbacks to a wet pipe system, however. If a wet pipe system sprinkler head or pipe is damaged by a forklift or other equipment, water immediately flows from the system. Also, because water is in the piping at all times, wet pipe systems cannot be used in locations where pipes could freeze, such as on loading docks and in unheated warehouses.

A **dry pendent sprinkler head** extends from a wet sprinkler pipe to small, unheated areas, such as walk-in freezers (**FIGURE 24-41**). The head has an elongated

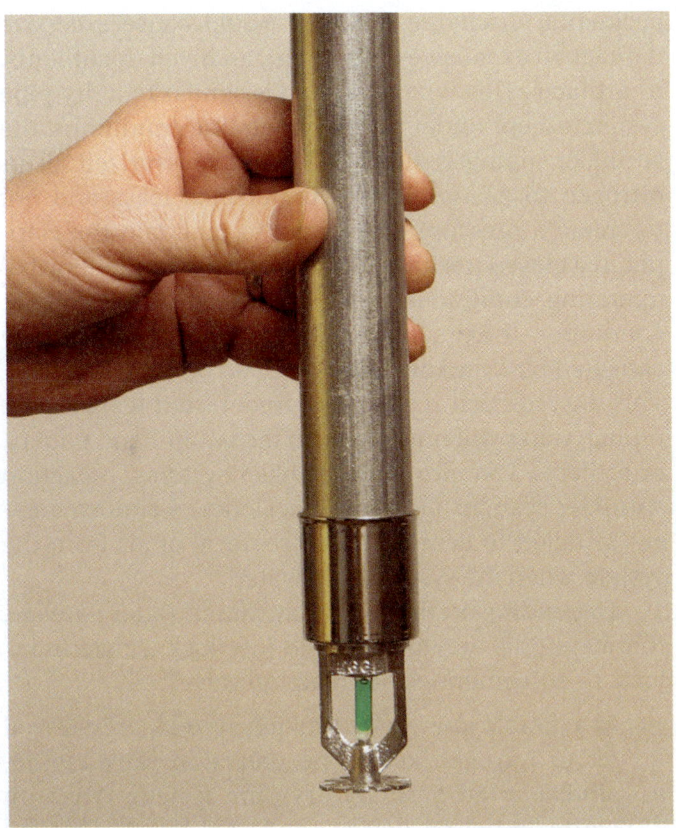

FIGURE 24-41 A dry-pendent sprinkler head can be used to protect a freezer box.

© Jones & Bartlett Learning

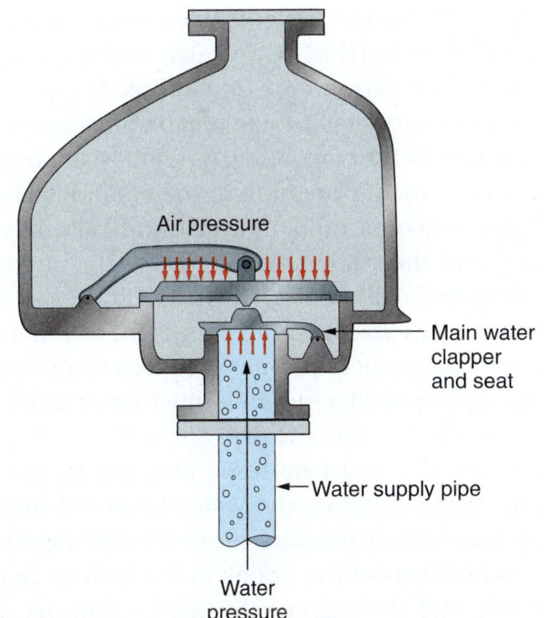

FIGURE 24-42 Water pressure on one side of the dry pipe valve is held down by air pressure on the other side of the valve.

© Jones & Bartlett Learning

neck, usually 6 to 18 in. (15 to 65 centimeters [cm]) long, that extends up and connects to the wet pipe sprinkler piping in a heated space above the small, unheated area. The bottom part of the dry-pendent head resembles a standard sprinkler head and is mounted inside the unheated area. The vertical neck section is filled with air and capped at each end. The top cap prevents water from entering the lower section, where it would freeze. The bottom cap acts just like the cap on a standard sprinkler head. When the head is activated and the lower cap drops out, a device inside the neck releases the upper cap, and water from the wet pipes flows down the neck of the dry sprinkler head and through the orifice. The entire dry-pendent head assembly must be replaced after it has been activated.

Large, unheated areas, such as a loading dock, can be protected with a wet pipe system that has an antifreeze loop. An **antifreeze loop** is a small section of the wet pipe sprinkler system filled with a glycol or glycerin solution instead of water. A check valve separates the antifreeze loop from the rest of the sprinkler system. A **check valve** prevents the flow of liquid in one direction. In an antifreeze loop, the closed check valve prevents the antifreeze from flowing out of the antifreeze loop and into the rest of the system. When a sprinkler head in the unheated area is activated, the antifreeze is discharged, the check valve opens, the water flows into the antifreeze loop and is discharged. (Large unheated areas can also be protected by dry pipe sprinkler systems.)

Dry Pipe Sprinkler Systems. A **dry pipe sprinkler system** operates much like a wet pipe sprinkler system, except that the pipes are filled with pressurized air instead of water. Dry pipe systems are used in facilities that may experience below-freezing temperatures, such as unheated warehouses, garages, or attics.

In a dry pipe sprinkler system, a dry pipe valve keeps water from entering the pipes until the air pressure is released when a sprinkler head is activated. A small compressor is used to maintain the air pressure in the system, and the clapper is held in the closed position by the air pressure inside of the pipe (**FIGURE 24-42**). When a sprinkler head opens, the air escapes. As the air pressure in the pipe drops, the water pressure on the other side of the clapper forces it open, and water begins to flow into the pipes. When the water reaches the open sprinkler head, it is discharged onto the fire.

Dry pipe sprinkler systems do not eliminate the risk of water damage from accidental activation. If a sprinkler head breaks, the air pressure will drop and water will flow, just as in a wet pipe sprinkler system.

Dry pipe sprinkler systems have a low air pressure switch connected to the FACU to alert building management personnel if the air pressure drops. The activation of this supervisory signal could mean one of two things: The compressor is not working, or there is an air leak in the system. If the air pressure in the system is too low, the clapper will open, and the system will fill with water. At that point, the system would essentially function as a wet pipe sprinkler system, which could freeze in low temperatures. In this scenario, the system would have to be drained and reset to prevent the pipes from freezing.

Dry pipe sprinkler systems must be drained after every activation so that the dry pipe valve can be reset. The clapper also must be reset, and the air pressure must be restored before the water is turned back on. During the warmer months, moisture develops in the pipes and collects at the drains in the system. Starting in the fall and throughout the winter months, these drains must be purged of water so that it doesn't freeze and crack the pipe, which will activate the system. Some dry pipe systems use pressurized nitrogen instead of air to reduce the possibility of corrosion from condensed moisture in the piping system.

Because the pipes in dry pipe sprinkler systems are filled with air, there is a delay between the activation of a sprinkler head and the flow of water from the head. The pressurized air that fills the system must escape through the open head before the water can flow. Large systems, however, can take several minutes for the air to be purged so water can flow from the heads. To compensate for this problem, two additional devices are used: accelerators and exhausters.

An **accelerator** speeds up the removal of air from a dry pipe system and is installed on the system side at the dry pipe valve. The rapid drop in air pressure caused by an open sprinkler head activates the accelerator, which assists in evacuating the air from the system to allow the water to start flowing more quickly. This quickly eliminates the pressure differential, opening the dry pipe valve and allowing the water pressure to force the remaining air out of the piping.

An **exhauster** is installed on the system side of the dry pipe valve, often at a remote location in the building. Like an accelerator, the exhauster monitors the air pressure in the piping. If it detects a drop in pressure, it opens a large-diameter portal, allowing the air in the pipes to escape. The exhauster closes when it detects water, diverting the flow to the open sprinkler heads. Large systems may have multiple exhausters located in different sections of the piping.

Preaction Sprinkler Systems. A **preaction sprinkler system** is installed in locations where the accidental discharge of water would cause excessive damage to expensive equipment, archived documents, or artifacts. This type of system is similar to a dry pipe system except under normal operating conditions, the sprinkler piping is supervised with pressurized air or nitrogen to ensure against undetected leaks and not to create a pressure differential to maintain the clapper in a closed position. This means that under normal operating conditions—that is, when there is no fire condition—the preaction valve cannot be accidentally activated by a broken sprinkler head or damaged pipe. This ensures that if there are undetected leaks in the piping, water will not drip onto the equipment, papers, or artifacts. Like dry pipe sprinkler systems, preaction sprinkler systems also have accelerators and exhausters installed to accelerate the removal of air from the system when the system is activated.

Depending on the risk of unwanted water damage, these systems may use double interlock, single interlock, or non-interlock or preaction valves:

- In a double interlock preaction system, pressurized air or nitrogen keeps the preaction valve closed, similar to an ordinary dry pipe system. When an alarm initiating device sends an alarm signal to the FACU, the FACU unlatches the preaction valve, but the air pressure is maintained and keeps the valve shut until it is released by a sprinkler head activation.
- In a single interlock preaction sprinkler system, an alarm initiating device sends an alarm signal to the FACU, which then releases the preaction valve so that water can enter the piping system and be discharged through a sprinkler head that has opened due to high temperatures.
- In a non-interlock preaction system, water flows into the system if the air pressure drops or an alarm initiating device send an alarm signal to the FACU.

Deluge Sprinkler Systems. A **deluge sprinkler system** is a dry pipe system that allows water to flow from all of the sprinkler heads at once when the alarm system is activated (**FIGURE 24-43**). Deluge systems are designed to flow large volumes of water and are used in special applications such as aircraft hangars or industrial processes where rapid fire suppression is critical. In some cases, foam concentrate is added to the water so that the system will discharge a foam blanket over the hazard. Deluge systems are also used for special hazard applications, such as liquid propane gas loading stations. In these situations, a heavy deluge of water is needed to protect exposures from a large fire that ignites rapidly.

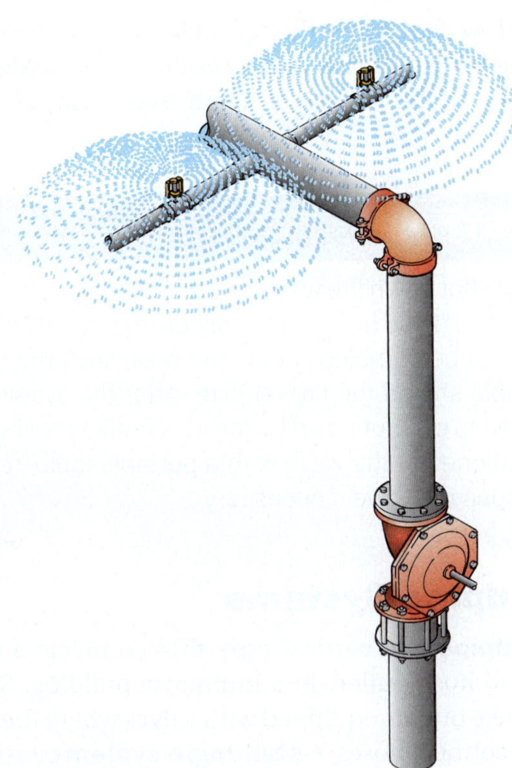

FIGURE 24-43 Water flows from all of the heads in a deluge system as soon as the system is activated.

In a deluge system, all of the sprinkler heads are always open. When the alarm is activated, the FACU opens the deluge valve in the area where the fire condition was detected, allowing water to flow into the system and through all of the sprinkler heads at once.

Some deluge systems use a separate dry pipe system called a **pilot line** instead of the FACU to open the deluge valve. When a sprinkler head or fixed temperature release device on the pilot line activates, the air pressure in the pilot line drops, and the water pressure opens the deluge valve to the deluge system.

TIP

Foam suppression systems can be permanently installed and use the same sprinkler valves and components in combination with special equipment to produce a foam solution that protects special hazards such as flammable liquid storage tanks, aircraft hangers, and chemical plants.

Water Mist Systems. A **water mist system** discharges a fine spray of water particles from specialized sprinkler heads and spray nozzles. The particles extinguish by cooling, displacing the oxygen, or blocking the radiant heat. Water mist systems can be wet pipe, dry pipe, preaction, or deluge.

Water mist systems operate at high pressures and are classified into three categories based on the maximum working pressure of the system. A low-pressure system exposes the system distribution piping to pressures less than 175 pounds per square inch (psi; 1207 kilopascals [kPa]). The intermediate-pressure system exposes the system distribution piping to pressures between 175 psi (1207 kPa) and 500 psi (3447 kPa), and the high-pressure system exposes the system distribution piping to pressures in excess of 500 psi (3447 kPa) and, in many installations, actually requires pressures to be over 1000 psi (6895 kPa).

The use of water mist systems has become more prevalent because of the following apparent advantages:

- A water mist system discharges much less water than a conventional sprinkler system.
- Water is not hazardous or toxic.
- A water mist system is self-contained and does not need to rely on a water main.
- Unlike an ordinary sprinkler system, a water mist system is listed and approved for complete fire extinguishment, not just for control over a fire.

There are disadvantages as well, including the following:

- A water mist system has a finite amount of water, so reserve water storage tanks are needed.
- The need for testing and approval of each specific application; for example, if a system is installed to protect a generator, the system must be specifically rated for that purpose.
- Water mist systems require exceptional water quality without impurities to avoid the nozzles clogging from particulate matter in the water.

Water mist systems protect many different high-value facilities and equipment, including computer rooms, telecommunication equipment areas, laboratories, archives, museums, and historic buildings. In addition, these systems are widely used to protect marine and offshore facilities and equipment, including ships, offshore platforms, turbine rooms, and personnel accommodation areas.

Residential Sprinkler Systems

Residential sprinkler systems are becoming more common. Some jurisdictions require residential sprinkler systems in every newly constructed home. The design of and theory underlying residential sprinkler systems

FIGURE 24-44 Special residential sprinkler heads are used in residential systems. They open more quickly than commercial heads and discharge less water.

are similar to those of commercial systems, but with some significant variations. The primary objective of a **residential sprinkler system** is to activate quickly and reduce the heat, flames, and smoke so occupants can evacuate safely. Quick response residential sprinkler heads are used to control fires at the incipient stage (**FIGURE 24-44**).

Residential systems typically use smaller piping and sprinkler heads with smaller orifices that discharge less water. To control costs, plastic pipe may be used instead of metal pipe. These systems typically protect high-risk areas such as bedrooms, kitchens, living rooms, and corridors. Small areas such as closets and bathrooms may not be covered.

Residential systems are usually wet pipe sprinkler systems. They are typically connected to the domestic water supply from the public water main. Residential sprinkler systems do not have a fire department connection, and they are not required to be connected to a fire alarm system, unless required by the local jurisdiction. Fire codes usually require the system to include a waterflow alarm.

The responding fire companies in the jurisdiction may not be familiar with every home that has a sprinkler system, but firefighters should know both how these systems work and how to shut down a system that has been activated. Usually, the main shut-off valve for the sprinklers can be found near the location where water enters the house. If the house has a basement, the shut-off valve may be located near other utility controls. If a separate shut-off valve for the sprinkler system cannot be found, the main water shut-off for the house can be used to deactivate the sprinkler system. Installing working smoke alarms and a residential sprinkler system reduces the chance of death from a home fire by 82 percent.

SAFETY TIP

Do not shut down the water to a sprinkler system until ordered to do so by the incident commander (IC). The fire must be completely out, and hose lines must be available should the fire reignite. After the system is ordered to be shut down by the IC, a firefighter should be stationed at the valve with a portable radio, ready to reopen the valve if necessary.

Standpipe Systems

A **standpipe** is a vertical pipe that connects a water supply to hose outlets in a multilevel building. Standpipes have outlets equipped with valves where firefighters can connect hoses. A **standpipe system** consists of a water supply and a network of horizontal and vertical piping built into a structure to carry the water to the vertical standpipes. The water may be supplied from the municipal water system, fire department connections, or both. Standpipe systems are required in all new high-rise buildings, and they are found in many other structures as well. Standpipes are also installed to carry water to large bridges and to supply water to limited-access highways that are not equipped with fire hydrants (**FIGURE 24-45**).

Standpipes are found in buildings both with and without sprinkler systems. In many newer buildings, sprinklers and standpipes are combined into a single system. In older buildings, the sprinkler and standpipe systems may be separate systems. NFPA 14, *Standard for the Installation of Standpipe and Hose Systems*, describes minimum standards for standpipe systems.

Standpipe Classes

Standpipes are categorized into three classes. **Class I standpipe** is designed for use by fire department personnel only. Each outlet has a 2½-in. (65-mm) male coupling and a valve to open the water supply after the attack line is connected (**FIGURE 24-46**). Sometimes the connection is located inside a cabinet, which may or may not be locked. Responding fire personnel carry the hose into the building with them. A Class I standpipe system must be able to supply an adequate volume of water with sufficient pressure to operate fire department attack lines.

FIGURE 24-45 Standpipe outlets allow fire hose to be connected inside a building.

FIGURE 24-46 A Class I standpipe provides water for fire department hose lines.

Class II standpipe is for use by specially trained building occupants or firefighters to attack fires at the incipient stage only. The outlets are usually equipped with a length of 1½-in. (38-mm) single-jacket hose connected to the system (**FIGURE 24-47**). Class II standpipe outlets are frequently connected to the water piping system in the building rather than to an outside water supply. The intent of Class II systems is to enable trained occupants to attack a fire at the incipient stage before the fire department arrives. However, keep in mind that most building occupants are not trained to attack fires safely. If a fire cannot be controlled with a portable fire extinguisher, it is usually safer for the occupants to simply evacuate the building and call the

FIGURE 24-47 Class II standpipes are intended to be used by building occupants to attack fires at the incipient stage.

fire department. Class II standpipes may be useful at facilities such as refineries and military bases, where workers are trained as an in-house fire brigade.

Responding fire personnel should only use Class II standpipes for fighting a fire if the fire is in its incipient stage. The water flow through these systems may not be adequate to control a developed fire. In addition, the hose and nozzles may be of inferior quality and lack maintenance and testing. Instead of using a hose and nozzle that may not be reliable or adequate, firefighters should always use department-issued equipment. Some jurisdictions require existing Class II standpipes to be replaced with Class III systems.

Class III standpipe has the features of both Class I and Class II standpipes in a single system. It has 2½-in. (65-mm) outlets for fire department use like a Class I system, as well as smaller outlets with attached hose for occupant use like a Class II system. For the same reason, responding personnel should not use a Class II standpipe system, in a Class III system, they should use only the 2½-in. (65-mm) outlets, even if they use a smaller diameter hose.

TIP

The occupant hose may have been removed—either intentionally or by vandalism—in many facilities, so the system functions as a Class I system.

Water Flow in Standpipe Systems

Standpipes are designed to deliver the required water flow to each floor. The requirements depend on the code requirements in effect when the building was constructed.

Flow-restriction devices (**FIGURE 24-48A**) or pressure-reducing valves (**FIGURE 24-48B**) are often installed at the outlets to limit the water flow or pressure to a preset amount. A vertical column of water, such as the water in a standpipe riser, exerts a backpressure opposed to the direction of the water flow. In a tall building, this backpressure can be dangerously high on the lower floors due to the force exerted by gravity. If a hose line is connected to an outlet without a flow restrictor or a pressure-reducing valve, the water pressure could rupture the hose, and the excessive nozzle pressure could make the line difficult or dangerous to handle. This could also happen if an installed flow restrictor or pressure-reducing valve malfunctions. Similarly, a sprinkler system fed by a standpipe system might not be able to handle the water pressure from the standpipe, so a flow restrictor or a pressure-reducing valve would need to be installed to avoid overpressurizing the sprinkler system. If a flow-restrictor or pressure-reducing valve is improperly calibrated, the water flow and pressure may not be adequate for firefighting or they may not perform properly

Like all parts of a fire suppression system, flow restrictors and pressure-reducing valves must be

A.

B.

FIGURE 24-48 A flow restrictor or a pressure-reducing valve on a standpipe outlet may be necessary on lower floors to avoid problems caused by backpressure and overpressurization of sprinkler systems fed from the standpipe. **A.** A flow restrictor. **B.** A pressure-reducing valve.

Courtesy of Dixon Valve and Coupling.

properly installed and maintained. An improperly adjusted pressure-reducing valve could severely restrict the pressure to a hose line. Likewise, an improperly adjusted flow-restriction device could limit the flow of water to fight a fire to an unacceptable level.

The flow and pressure capabilities of a standpipe system should be determined during preincident planning. Many older standpipe systems deliver water at a pressure of only 65 psi (448 kPa) or less at the top of the building. Combination nozzles are designed to operate at 75 to 100 psi (517 to 689 kPa). Many fire departments use low-pressure combination nozzles for fighting fires in high-rise buildings or require the use of only smoothbore nozzles when operating from a standpipe system.

Engine companies that respond to buildings equipped with standpipes should carry a kit that includes the appropriate hose and nozzle, a spanner wrench, and any required adapters. This kit should also include tools to adjust the settings of pressure-reducing valves or to remove restrictors that are obstructing flows. To determine exactly what should be included in this kit, the preincident plans for high-rise buildings should include an evaluation of the building's standpipe system, including the anticipated flow rates and pressures. Fire personnel can then make the correct decisions about the appropriate nozzles and tactics to use during an emergency response to those buildings.

High-rise buildings often incorporate complex systems of risers, storage tanks, and fire pumps to deliver the needed flows to upper floors. The details of these systems should be obtained during preincident planning surveys. Department procedures dictate how responding units will supply the standpipes with water as well as how crews should use the standpipes inside the building.

Types of Standpipe Systems and Water Supplies

Standpipe systems and sprinkler systems are supplied with water in essentially the same way. A **wet standpipe system** has water in the piping system at all times, whereas a **dry standpipe system** contains air or nitrogen until water is needed for fire suppression.

Many wet standpipe systems in modern buildings are connected to a public water supply and are equipped with an electric or a diesel fire pump to provide additional pressure. Some also have a water storage tank that serves as a backup supply. In these systems, the FDC on the outside of the building can be used to increase the flow rate, boost the water pressure, or provide additional water from an alternative source. An **automatic wet standpipe system** provides adequate volume and pressure at the hose valve at all times. The fire department needs only to connect a hose to the hose valve and open the valve to provide water for fire suppression. A **manual wet standpipe system**, however, needs the pump on the engine to be connected to the FDC to provide the required volume and pressure to the system.

Dry standpipe systems are found in many older buildings. Like dry pipe sprinkler systems, they are also used if freezing weather is a problem, such as in open parking structures, bridges, and tunnels. An **automatic dry standpipe system** is connected to a water supply, and water is kept out of the piping system by a dry pipe valve, similar to the valve in dry pipe sprinkler systems. In such systems, water flows through the system only after the alarm system opens the dry pipe valve. A **manual dry standpipe system** does not have a permanent connection to a water supply, so as with a manual wet standpipe system, the FDC must be connected to the pump on the engine to pump water into the system. If a fire occurs in a building with manual dry standpipes, connecting the hose lines to the FDC and charging the system with water are high priorities.

Special Hazard Suppression Systems

In certain situations, specialized suppression systems are needed. These kinds of systems are often used in areas where water would not be an acceptable extinguishing agent. For example, water is not the extinguishing agent of choice for areas containing sensitive electronic equipment or contents such as computers, valuable books, or documents. Water is also incompatible with materials such as flammable liquids, water-reactive chemicals, or liquefied gases at cryogenic temperatures because violent reactions could occur, endangering lives and property. Chapter 9, *Portable Fire Extinguishers*, describes specialized extinguishing agents in greater detail. Types of special hazard extinguishing agents include the following:

- Wet chemical agents
- Dry chemical agents
- Gaseous agents, including halogenated agents, clean agents such as halocarbons, and carbon dioxide

Dry Chemical and Wet Chemical Extinguishing Systems

Dry chemical and wet chemical extinguishing systems are the most common specialized agent systems. A **dry chemical extinguishing system** uses the same types of finely powdered agents as a portable dry chemical fire extinguisher. The agent is kept in self-pressurized tanks or in tanks with an external supply of pressurized inert gas that provides pressure and fluidizes the dry chemical agent when the system is activated. The chemical agent flows through the piping system and is discharged by nozzles directed at the target area. Some gas stations have dry chemical systems that protect the areas where gas is dispensed (**FIGURE 24-49**). These systems are also installed inside buildings to protect areas where flammable liquids are stored or used.

Similar to portable wet chemical fire extinguishers, a **wet chemical extinguishing system** is used in commercial kitchens and food product manufacturing facilities to protect the cooking areas and exhaust systems (**FIGURE 24-50**). This system uses a liquid Class K extinguishing agent, which is much more effective on cooking fats than dry chemicals. Wet chemicals are also easier to clean up after a discharge, so the kitchen can resume operations more quickly after the system has discharged. Older commercial kitchens may still have a dry chemical extinguishing system.

Although the ductwork in commercial kitchens should be cleaned regularly, it is not unusual for a kitchen fire to extend into the exhaust system. Therefore, many suppression systems in commercial kitchens discharge the extinguishing agent into the ductwork above the exhaust hood as well as onto the cooking surface. This helps prevent a fire from igniting any grease build-up inside the ductwork and spreading throughout the exhaust system.

FIGURE 24-49 Dry chemical extinguishing systems are installed at many self-service gasoline filling stations.

In both dry chemical and wet chemical extinguishing agent systems, a fusible-link detection device or another automatic initiating device are placed above the target hazard to activate the system (**FIGURE 24-51**). A manual pull station is also provided so that workers can activate the system if they discover

FIGURE 24-50 Wet chemical extinguishing systems are used in most commercial kitchens.

FIGURE 24-51 Fusible-link detection devices can be used to activate a specialized extinguishing system.

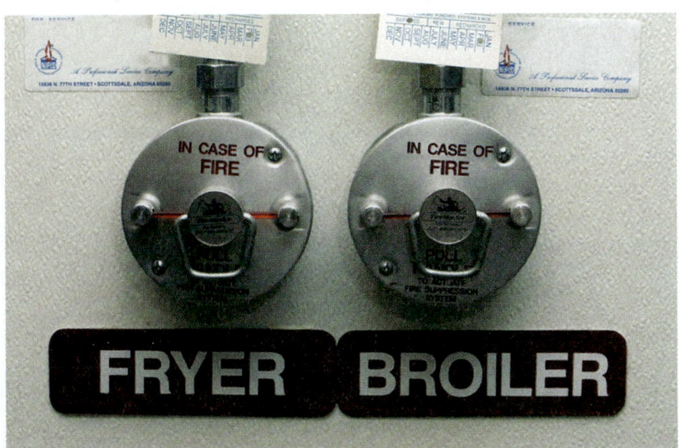

FIGURE 24-52 Most special extinguishing systems can be manually activated.

a fire (**FIGURE 24-52**). Open nozzles are located over the target areas to discharge the agent onto a fire. When the system is activated, the extinguishing agent flows out of all the nozzles, similar to a deluge system.

Most dry and wet chemical suppression systems are connected to the building's fire alarm system. Many are designed to shut down electricity, fuel gases, burners, exhaust fans, or other equipment.

Gaseous Suppression Systems

A **gaseous suppression system** is often installed in areas where computers or sensitive electronic equipment is used or where valuable documents are stored. Gaseous agents include halogenated agents, halocarbon agents, and inert gases such as carbon dioxide. These gases are stored under pressure as liquids, and they are discharged as gases through an engineered piping and nozzle system. Gaseous agents suppress fires by three different suppression mechanisms: They reduce the oxygen level in a room or building by displacing ambient air, they cool the fire, or they chemically inhibit the growth of fire.

Gaseous suppression systems are a type of **clean agent extinguishing system**, which extinguish fires using clean agents that do not leave a residue. Because of this, gaseous agents can be very effective in certain applications, such as protecting areas that contain computers, telecommunications equipment, and other sensitive electronic equipment, because they can extinguish a fire without causing significant damage to the room or contents. NFPA 2001, *Standard on Clean Agent Fire Extinguishing Systems*, describes minimum standards for clean agent fire extinguishing systems.

In some clean agent systems, inert gases are released into a room or building and displace enough ambient air to lower the oxygen content below the level required for combustion. In other systems, halogenated extinguishing agents extinguish fires by interrupting the chemical chain reaction of the fire or by cooling the fuel. Some halogenated extinguishing agents use both mechanisms. Clean agents are stored in pressurized containers as liquids. When the system is activated, the liquid vaporizes and is discharged as a gas. Most clean agents are used in total flooding applications where an entire room or enclosed area is flooded with a specific concentration of the agent. Some systems apply the agent directly to a hazardous process or piece of equipment.

Fire alarm initiating devices installed in the hazard area activate the system, although a manual discharge button is also provided with most installations. Discharge is usually delayed 30 to 60 seconds after the detector is activated to allow workers to evacuate the area. During this delay (the pre-alarm period), a manual override switch can be used to stop the discharge. In some systems, the manual override switch must be pressed until the fire alarm system is reset; releasing the override switch too soon causes the system to discharge.

If the system does activate, the clean agent extinguishing system completely discharges within a few seconds, depending on the design specifications. Firefighters entering the area must use self-contained breathing apparatus (SCBA) until the area has been properly ventilated. Although the gaseous agents used in clean agent extinguishing systems are not considered immediately dangerous to life and health (IDLH), some gaseous agents release toxic by-products when heated to extreme temperatures, and others reduce the ambient oxygen level to dangerously low levels. A strong Incident Command System (ICS) is essential, and firefighters entering the area must use strict accountability procedures to ensure the safety of all personnel.

Clean agent extinguishing systems should be connected to the building's fire alarm system. If the system has a preprogrammed delay, the pre-alarm should activate the building's fire alarm system. The area where the clean agent system is installed needs to be indicated as a special hazard suppression zone on the FACU or annunciator. This notification alerts firefighters that they are responding to a situation where a clean agent has discharged.

Halogenated Extinguishing Agent Systems. Halogenated agent extinguishing systems are one type of clean agent extinguishing system. Halogenated extinguishing agents interrupt the chemical reaction between fuel and oxygen. This process is further explained in Chapter 9, *Portable Fire Extinguishers*. Halogenated extinguishing agents are made from halocarbons that fall into two categories: Halons and halocarbons that do not deplete the ozone layer. Halons were widely used until the 1990s to protect areas containing computers, telecommunications equipment, and other sensitive electronic equipment. However, because Halons are harmful to the ozone layer, they can no longer be manufactured or imported.

However, Halons are the extinguishing agent in many existing systems still in use today. Halons can be reclaimed—removed from existing systems that are being taken out of service and processed to remove contaminants—and reclaimed Halon can be used to recharge a system that has discharged; eventually, this supply will be depleted. Once a Halon system is discharged and there is no reclaimed Halon available, the Halon system must be replaced with a new, different type of system. Old systems that are no longer used must be decommissioned by a certified company that reclaims the existing Halon agent according to strict environmental policies.

Carbon Dioxide Extinguishing Systems. Carbon dioxide (CO_2) extinguishes a fire by displacing the oxygen around the fire with CO_2 gas, effectively smothering the fire. Chapter 9, *Portable Fire Extinguishers*, discusses CO_2 fire extinguishers in further detail. A **carbon dioxide (CO_2) extinguishing system**, which is similar in design to a system that uses inert gases, is a fire suppression system that protects an area in a building by flooding the area with CO_2. Minimum concentrations of 34 percent CO_2 in the area are required to reduce oxygen concentration below 14 percent, so large quantities of the agent are required for total flooding applications. Higher CO_2 concentrations may be required depending on the hazard protected (**FIGURE 24-53**). NFPA 12, *Standard on Carbon Dioxide Extinguishing Systems*, describes minimum standards for carbon dioxide extinguishing systems.

FIGURE 24-53 Carbon dioxide extinguishes a fire by displacing the oxygen in the room and smothering the fire.
Courtesy of Chemetron Fire Systems.

Carbon dioxide extinguishing systems should be connected to the building's fire alarm system and identified on the FACU or annunciator. Firefighters must use extreme caution when entering a building in which a CO_2 system has discharged. Oxygen levels will be dangerously low; therefore, full personal protective equipment (PPE) and SCBA are required. Because CO_2 is heavier than air, it tends to settle at the floor or sink to rooms below the fire area. Firefighters must perform search and rescue operations in the discharge area and on the floors below to look for occupants who may have become incapacitated by the low-oxygen environment. Preincident planning can help responding personnel plan for emergency response procedures if the CO_2 system has discharged. As with other gaseous agents, a strong ICS and personnel accountability system are essential for firefighter safety.

Smoke Control Systems

Fires in buildings often generate large amounts of smoke and superheated toxic gases that can quickly disable, injure, and kill occupants. Smoke also limits visibility, which can delay or prevent occupants from evacuating. Studies indicate that high temperatures and smoke levels below 6 feet (ft; 1.8 meters [m]) above the floor significantly reduce the likelihood of occupants escaping from a fire. An example of this is the disastrous MGM Grand Casino and Hotel fire in 1980 that claimed the lives of 85 people. Although the fire started in the first-floor casino and didn't spread to the upper floors, most of the deaths occurred on the upper floors due to smoke inhalation. More than half of the victims on the upper floors were overcome by smoke that filled the stairways, corridors, and elevator shafts.

Smoke and other by-products of combustion can move throughout a building due to convection, the stack effect, wind, and the flow path. Many buildings now incorporate smoke control systems into their overall fire protection system. Smoke control systems control, alter, and limit the spread of smoke and gases and, in some cases, remove the smoke from the building. Coupled with comprehensive fire alarm and fire suppression systems, smoke control systems provide a safer environment for both occupants and firefighters. These systems may also reduce property damage in areas remote from the fire area. NFPA 92, *Standard for*

Smoke Control Systems, 2021 Edition, provides performance objectives and design fundamentals for passive and active methods of smoke control.

There are two types of smoke control systems: smoke containment systems and smoke management systems. **Smoke containment systems** are mechanical systems that keep smoke and other gases in the area or floor of fire origin. **Smoke management systems** are natural or mechanical systems that move or exhaust smoke that escapes from the fire area and threatens egress routes, elevator shafts, stairwells, and refuge areas. Smoke containment and smoke management are achieved through building design that passively controls smoke and active smoke control systems.

A **passive smoke control system** uses building features such as fire walls, automatic fire doors, a **draft curtain** (a non-combustible barrier that permanently hangs from the ceiling or is deployed when smoke or fire is detected to block smoke), and other smoke barriers to limit the spread of smoke (**FIGURE 24-54**). These building components are designed by fire protection engineers based on the building's size, design, occupancy type, and fire hazards. Passive smoke management compartmentalizes larger spaces to confine smoke to the room or area of origin. Passive smoke control does not require human intervention because it is built into the design of the building.

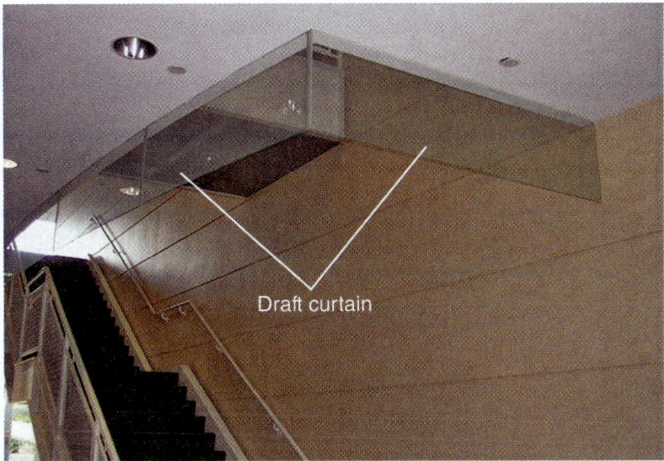

FIGURE 24-54 Passive smoke control uses construction barriers, such as a draft curtain, to contain and minimize the amount of smoke leakage from a fire area.

Some elements in a passive smoke control system interact with the FACU, such as automatic fire doors. When the fire alarm system is activated, automatic fire doors unlock, and that is controlled by the FACU.

An **active smoke control system** uses mechanical systems to prevent the migration of smoke from the fire area to uninvolved floors or areas of the building and to move smoke away from occupants or from egress routes. One type of active smoke control creates pressure differences between the fire area and areas around the fire area (sometimes called a pressure sandwich). When smoke is detected, the air flow through the heating, ventilation, and air-conditioning (HVAC) system is increased in uninvolved areas of the building, in egress routes, refuge areas, elevators shafts, and other vertical shafts. This means those areas are under higher air pressure than the area where the fire is, effectively creating a barrier for the smoke in the fire area. For example, if there is smoke in a corridor and the smoke control system increased the air pressure in a stairwell, when you open the door to the stairwell to exit the corridor, air rushes from the pressurized stairwell into the corridor, and not the other way around. Another type of active smoke control uses fans to blow smoke out of the fire area to the outdoors. This type of system is used in large open areas, such as an atrium, an indoor arena, and an airport terminal.

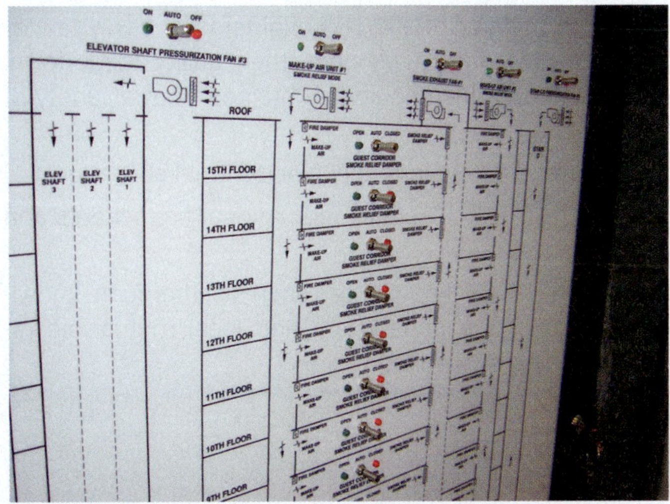

FIGURE 24-55 A smoke control panel has status indication and control functions for each major smoke control system.

Active smoke control systems usually interface with the FACU and are activated when the fire alarm system is activated. Some sophisticated smoke control systems allow firefighters or building personnel to manually control specific functions of the smoke control system from a smoke control panel (**FIGURE 24-55**). Smoke control systems must be inspected and maintained like any other fire protection system to ensure that they will function as designed during a fire emergency.

Voice of Experience

Of the many skills I have learned while training to be and working as a firefighter, one of the most valuable is maintaining calm while under extreme duress. If a firefighter who is traumatized panics, it will have a detrimental effect on their performance, will endanger the lives of team members, and can endanger everyone on the incident. Firefighters must be able to function as a team, which is one reason why the camaraderie is so strong in the fire service. You must be willing to trust your fellow firefighters with your life. You must train extensively to develop a sense of calm so that you can function effectively when everyone and everything around you appears to be in a chaotic state. Firefighters routinely operate in environments that endanger life and property. The old adage that "firefighters run in when everyone else is running away" still holds true.

In addition, the best firefighters are always seeking knowledge to enhance their ability to perform. This knowledge should be obtained from a variety of sources:

- College-based instruction and formal education
- Local, regional, state, and national fire schools and academies
- Institutional knowledge, such as thorough knowledge of the department and local mutual aid resources available
- Experiential learning by hands-on practical application of techniques
- Community knowledge: pertinent information about community demographics, water mains, water sources, utilities, building design and construction, potential community hazards, traffic routes, evacuation routes, and location of fire protection systems

The importance of each firefighter's contribution to the overall mission of public safety and the department's role must be extolled. Firefighters are called when people do not know what to do; it may be a seemingly insignificant operation such as water in the basement, or it may involve a devastating fire with multiple fatalities. The only way a firefighter can effectively respond to this task is through remaining calm and constantly learning.

Charles Garrity
Berkshire Community College
Pittsfield, Massachusetts

Inspection, Testing, Maintenance, and Documentation

Proper function of fire alarm, fire suppression, and smoke control systems is critical to ensure the safety of building occupants and responding firefighters during an emergency. Both building and fire codes require these systems to be inspected and tested regularly and maintained so they remain functional. When systems are not maintained, there may be an impairment, which is a condition in which a fire detection or suppression system or portion thereof is out of order and this results in the fire system not functioning during a fire event. Property owners or their designated representatives are responsible for ensuring that regular inspections, testing, and maintenance are conducted. They are also responsible for maintaining the initial records of system installation as well as all records of system inspections, tests, and maintenance.

Every fire protection system and smoke control system is required to undergo an installation acceptance test to ensure the system operates as required by code. In most instances this initial test requires operating each system component at full capacity to determine whether they function as required. Many municipalities require these initial tests to be witnessed by the fire code official. Once the system is accepted, continued inspections, testing, and maintenance are required. **Inspection** of fire suppression, detection, and smoke control systems is a visual examination of the system to verify that it appears to be in operating condition and is free of physical damage. **Testing** is a procedure used to determine the operational status of a component or system by conducting periodic physical checks, such as waterflow tests and fire pump tests. **Maintenance** is the work performed to repair, replace, and service systems to ensure that the equipment operates properly. The frequency of these events varies greatly based on the individual components of each system.

If a system is found to be out of service, the fire department shall be notified immediately. Sometimes, depending on the severity of the impairment, the building will be evacuated until the system has been restored to service.

Proper documentation and record keeping for every fire protection system is important for not only ensuring the systems are maintained in working order, but for understanding the way each system functions and operates, as well as how it communicates and connects with the other fire protection and building systems in a building or complex. These records should include all the procedures or activities performed to inspect, test, or maintain the system, their results, as well as the name of the qualified person who performed the activity and that person's organization, the required frequency of the activity, and the date the activity was conducted. The records must be kept for 1 year past the date of the next test. They also must be made available to the fire inspector upon request. The original system installation drawings, acceptance test records, and manufacturer data sheets must be kept for the life of the system.

CASE STUDY

You Are the Firefighter CONCLUSION

Your crew is updating the department's preincident plan for a large industrial facility in your city. The building's safety officer is showing you the state-of-the-art fire alarm system and suppression systems. The fire alarm system includes both manual and automatic initiating devices, as well as a comprehensive emergency notification system with the main fire alarm control unit located in the security office. The building is fully sprinklered, and a clean agent suppression system protects the large computer room. Class III standpipes are located strategically throughout the building. Although the facility uses a lot of hazardous processes in its operations, you feel that the fire protection systems provide a high level of protection for both the occupants and responding firefighters if an emergency occurs.

1. **What information should be included in the preincident plan?**

   Answer: Preincident plans should include all information necessary for fire department personnel to perform fire suppression, mitigation, and life safety responses at a facility. This information should include the type of occupancy, number of persons in the building or structure, locations of egress points, hydrants, fire department connections, types of fire detection and suppression systems, and how to properly operate these systems.

2. **What are the advantages of a Class III standpipe system? Are there any disadvantages?**

   Answer: Class III standpipes can be utilized by both trained building occupants as well as fire department personnel. They provide larger hose connections to deliver higher flow rates than Class II systems.

3. **What safety measures should responding firefighters take if the clean agent system activates?**

   Answer: Firefighters entering the area must use self-contained breathing apparatus (SCBA) until the area has been properly ventilated and must use strict accountability procedures to ensure safety of all personnel.

4. **What can your department do to ensure the alarm and suppression systems are maintained functional?**

   Answer: Firefighters can review the buildings records of inspections, testing, and maintenance of alarm and suppression systems to ensure they are maintained functional.

WRAP-UP

SUMMARY

KNOWLEDGE OBJECTIVES

After studying this chapter, you will be able to:

- Describe the components of fire protection systems and the role of each component. (**NFPA 1010: 7.5.3**, pp. 1002–1003)
- Describe the basic functions of fire alarm systems and how firefighters interact with the basic components of fire alarm systems.
  - Describe the basic components and functions of a fire alarm system. (**NFPA 1010: 7.5.3**, p. 1003)
  - Describe the function of the FACU and the annunciator. (**NFPA 1010: 7.5.3**, pp. 1003–1005)
  - Describe the function of and the differences between noncoded, coded, zoned, and addressable systems. (**NFPA 1010: 7.5.3**, pp. 1003–1005)
  - Describe the purpose of trouble and supervisory signals and a remote annunciator. (**NFPA 1010: 7.5.3**, pp.1003–1005)
  - Describe the difference between a smoke alarm and a smoke detector. (**NFPA 1010: 7.5.3**, pp. 1005–1009)
  - Explain the difference between manual and automatic alarm initiating devices. (**NFPA 1010: 7.5.3**, pp. 1005–1012)
  - Describe the types of manual pull stations. (**NFPA 1010: 7.5.3**, p. 1005)
  - Describe the types of detectors used in automatic alarm initiating devices and indicate where each type is most suitable. (**NFPA 1010: 7.5.3**, pp. 1005–1012)
  - Describe the basic types of alarm notification appliances and the temporal-3 pattern. (**NFPA 1010: 7.5.3**, pp. 1012–1013)
  - Explain how an alarm is initiated by a fire suppression system. (**NFPA 1010: 7.5.3**, p. 1013)
  - Describe the three types of unwanted alarms. (pp. 1013–1014)
- Describe how residential smoke and carbon monoxide alarms operate. (**NFPA 1010: 7.5.3**, pp. 1014–1016)
- Explain the purpose of the emergency control function interface. (**NFPA 1010: 7.5.3**, p. 1016)
- Describe the types of fire department notification systems. (**NFPA 1010: 7.5.3**, pp. 1016–1017)
- Describe the types of fire suppression systems, including the main components and operations of each system.
  - Describe the parts of an automatic sprinkler head. (**NFPA 1010: 7.5.3**, pp. 1019–1023)
  - Describe three common types of release mechanisms used in automatic sprinkler heads. (**NFPA 1010: 7.5.3**, pp. 1019–1023)
  - Explain the difference between release temperatures and maximum ceiling temperatures. (**NFPA 1010: 7.5.3**, p. 1022)
  - Identify the three mounting positions and the type of sprinkler head used for each. (**NFPA 1010: 7.5.3**, pp. 1022–1023)
  - Identify the four types of pipes used in sprinkler piping. (**NFPA 1010: 7.5.3**, pp. 1023–1024)
  - Describe three types of indicating valves used as main water supply control valves. (**NFPA 1010: 7.5.3**, pp. 1024–1026)
  - Identify the four types of main sprinkler system valves and list other valves used in sprinkler systems. (**NFPA 1010: 7.5.3**, p. 1026)
  - Describe water supply for sprinkler systems and the FDC. (**NFPA 1010: 7.5.3**, pp. 1026–1028)
  - Describe the water-motor gong and the waterflow alarm device. (**NFPA 1010: 7.5.3**, p. 1028)
  - Describe the operation and application of the following types of automatic sprinkler systems (**NFPA 1010: 7.5.3**, pp. 1028–1031):
    - Wet pipe system
    - Dry pipe system
    - Preaction system
    - Deluge system
    - Water mist systems
  - Describe the differences between commercial and residential sprinkler systems. (**NFPA 1010: 7.5.3**, pp. 1032–1033)
  - Describe standpipe systems and how they are used. (**NFPA 1010: 7.5.3**, pp. 1032–1035)

   - Identify the three classes of standpipes and explain the differences among them. (**NFPA 1010: 7.5.3**, pp. 1032–1035)
   - Explain the purpose of flow restrictors and pressure-reducing valves in a standpipe. (**NFPA 1010: 7.5.3**, pp. 1032–1035)
   - Explain the difference between wet and dry standpipe systems and the difference between automatic and manual standpipe systems. (**NFPA 1010: 7.5.3**, p. 1035)
- Describe different types of special hazard systems and identify the hazards associated with these systems that firefighters may encounter.
   - Describe the differences between dry chemical and wet chemical extinguishing systems and explain when each type is used. (**NFPA 1010: 7.5.3**, pp. 1035–1037)
   - Describe gaseous suppression systems and explain why they are clean agent extinguishing systems. (**NFPA 1010: 7.5.3**, pp. 1037–1038)
   - Describe halogenated extinguishing systems and how they differ from Halon extinguishing systems. (**NFPA 1010: 7.5.3**, pp. 1037–1038)
   - Explain the restrictions on Halon extinguishing systems. (**NFPA 1010: 7.5.3**, pp. 1037–1038)
   - Describe carbon dioxide extinguishing systems and explain when they are used. (**NFPA 1010: 7.5.3**, p. 1038)
- Describe the operation and application of smoke control systems.
   - Explain the difference between smoke containment systems and smoke management systems. (**NFPA 1010: 7.5.3**, pp. 1038–1039)
   - Describe passive and active smoke control systems. (**NFPA 1010: 7.5.3**, pp. 1038–1039)
- Explain the importance of inspecting, testing, and maintaining fire protection systems and the process of documenting these processes.
   - Describe the differences between inspections, testing, and maintenance of systems (pp. 1040–1041)
   - Identify the types of records that need to be maintained for systems (**NFPA 1010: 7.5.3**, pp. 1040–1041)

SKILLS OBJECTIVES

There are no skills objectives for Firefighter II candidates. NFPA 1010 contains no Firefighter II Job Performance Requirements for this chapter.

KEY TERMS

accelerator A device that allows the water to flow more quickly in a dry pipe valve by evacuating the air from the system.

active smoke control system A mechanical smoke control system that is activated when the fire alarm system is activated and that prevents the migration of smoke from the fire area to uninvolved floors or areas of the building by creating pressure differences between smoke control zones or floors or by exhausting smoke from large open areas to move smoke away from occupants or from egress routes.

addressable system A fire alarm system in which each initiating device and notification appliance is assigned a number so that the exact type and location of the devices and appliances are shown on the fire alarm control unit (FACU).

air sampling-type detector A detector that consists of a piping or tubing distribution network that runs from the detector to the area(s) to be protected. An aspiration fan in the detector housing draws air from the protected area back to the detector through air sampling ports, piping, or tubing. At the detector, the air is analyzed for fire products. (NFPA 72)

alarm initiating device A device that, when operated either automatically or manually, causes the alarm notification device to operate.

alarm notification appliance A device that generates a fire alarm signal.

alarm valve A valve that signals an alarm when a sprinkler head is activated and prevents nuisance alarms caused by pressure variations and surges in the water supply to the system.

KEY TERMS CONTINUED

alarm verification feature A feature of automatic fire detection and alarm systems to reduce nuisance alarms wherein smoke detectors report alarm conditions for a minimum period of time or confirm alarm conditions within a given time period after being reset, in order to be accepted as a valid alarm initiation signal. (NFPA 72)

annunciator A unit containing one or more indicator lamps, alphanumeric displays, or other equivalent means in which each indication provides status information about a circuit, condition, or location. Also called *remote annunciator.* (NFPA 72)

antifreeze loop A small section of a wet pipe sprinkler system that is filled with a glycol or glycerin solution instead of water and that has a check valve separating the antifreeze loop from the rest of the sprinkler system.

automatic dry standpipe system A standpipe system permanently attached to a water supply capable of supplying the system demand at all times, containing air or nitrogen under pressure, the release of which (as from opening a hose valve) opens a dry pipe valve to allow water to flow into the piping system and out of the opened hose valve. (NFPA 14)

automatic sprinkler head The part of the sprinkler system through which water is discharged and that consists of a frame, the orifice, a heat-sensitive release mechanism that holds a cap in place over the orifice, and a deflector that directs the water in a spray pattern. Also called *sprinkler head.*

automatic sprinkler system A sprinkler system of pipes with water under pressure that allows water to be discharged immediately when a sprinkler head operates. (NFPA 853)

automatic wet standpipe system A standpipe system containing water at all times that is attached to a water supply capable of supplying the system demand at all times and that requires no action other than opening a hose valve to provide water at hose connections. (NFPA 14)

beam detector See *projected-beam type detector.*

branch lines In a sprinkler system, the pipes that supply the sprinkler heads.

butterfly valve A sprinkler control valve consisting of a disk that either blocks the water and prevents it from flowing into the system or is rotated 90 degrees to allow the water to flow past it and that has an external indicator on the stem that is parallel or in-line with the piping to signify the valve is open or perpendicular to the piping to indicate the valve is closed.

carbon dioxide (CO_2) extinguishing system A fire suppression system designed to protect either a single room or series of rooms by flooding the area with carbon dioxide.

carbon monoxide detector A device having a sensor that responds to carbon monoxide gas that is connected to an alarm control unit. (NFPA 72)

central station service alarm system A system or group of systems in which the operations of circuits and devices are transmitted automatically to, recorded in, maintained by, and supervised from a listed central station that has competent and experienced servers and operators who, upon receipt of a signal, take such action as required by this Code. Such service is to be controlled and operated by a person, firm, or corporation whose business is the furnishing, maintaining, or monitoring of supervised alarm systems. (NFPA 72)

check valve A valve that prevents the flow of liquid in one direction.

chemical-pellet sprinkler head A sprinkler head activated with the cap held in place by a chemical pellet that liquefies at a preset temperature, and when the pellet melts, the liquid compresses a plunger which releases the cap.

Class I standpipe A standpipe system designed for use by fire department personnel only, with outlets and valves to control the flow of water and a 2½-in. (65-mm) male coupling for fire hose the personnel carry into the building with them.

Class II standpipe A standpipe system with outlets that have a length of 1½-in. (38-mm) single-jacket hose and a nozzle already connected for use by trained building occupants or firefighters and that are used to attack fire at the incipient stage only.

Class III standpipe A combination standpipe system that has features of both Class I and Class II standpipes.

clean agent extinguishing system A fire suppression system that expels an electrically nonconducting, volatile, or gaseous clean extinguishing agent that does not leave a residue upon evaporation.

coded system A fire alarm system that has audible or visible notification devices that indicate the location of the initiation device.

cross mains In a sprinkler system, the pipes that supply the branch lines.

cross-zoned system A fire alarm system that requires the activation of two or more initiating devices before the alarm sounds.

deluge head A sprinkler head that has no cap or release mechanism so that the orifice is always open and the water release is controlled by valves in the pipes.

deluge sprinkler system A sprinkler system employing open sprinklers or nozzles that are attached to a piping system that is connected to a water supply through a valve that is opened by the operation of a detection system installed in the same areas as the sprinklers or the nozzles. When this valve opens, water flows into the piping system and discharges from all sprinklers or nozzles attached thereto. (NFPA 13)

deluge valve A type of system actuation valve that is opened by the operation of a detection system installed in the same areas as the spray nozzles or by remote manual operation supplying water to all spray nozzles. (NFPA 13)

double-action pull station A manual alarm initiating device that requires the user to take two steps, such as moving a flap, lifting a cover, or breaking a piece of glass, and then activating the alarm.

draft curtain A fixed or automatically deployable barrier that protrudes downward from the ceiling to channel, contain, or prevent the migration of smoke.

dry chemical extinguishing system An automatic fire suppression system that discharges a dry chemical agent.

dry pendent sprinkler head A sprinkler head with an elongated neck that extends from a wet sprinkler pipe in a heated space to a small, unheated area, such as a walk-in freezer, and that is capped at both ends to prevent water from entering the unheated area.

dry pipe sprinkler system A sprinkler system employing automatic sprinklers that are attached to a piping system containing air or nitrogen under pressure, the release of which (as from the opening of a sprinkler) permits the water pressure to open a valve known as a dry pipe valve, and the water then flows into the piping system and out the opened sprinklers. (NFPA 13)

dry pipe valve The valve assembly in a dry pipe sprinkler system that prevents water from entering the system until the air pressure is released and acts as an alarm valve.

dry standpipe system A standpipe system designed to have piping contain water only when the system is being utilized. (NFPA 25)

duct detector A type of smoke detector that samples the air through the air distribution system for a building, and when it detects smoke, it initiates an alarm or a supervisory signal and shuts down the air handling system.

early-suppression fast-response sprinkler head (ESFR) A large sprinkler head designed to activate quickly and suppress a fire in its early stages.

emergency control function interface The location in a building where personnel control building functions such as air-handling systems, fire doors, and elevators.

exhauster A device that accelerates the removal of the air from a dry pipe or preaction sprinkler system by opening a large diameter portal until it detects water flow.

feed mains In a sprinkler system, the pipes that supply the cross mains from the water source, either directly or through risers.

fire alarm control panel (FACP) See *fire alarm control unit (FACU)*.

fire alarm control unit (FACU) A component of the fire alarm system, provided with primary and secondary power sources, which receives signals from initiating devices or other fire alarm control units and processes these signals to determine part of all of the required fire alarm system output function(s). Also called *fire alarm control panel (FACP)*. (NFPA 72)

fire alarm signal A signal that results from the manual or automatic detection of a fire alarm condition. (NFPA 72)

fire alarm system A system or portion of a combination system that consists of components and circuits arranged to monitor and annunciate the status of fire alarm or supervisory signal-initiating devices and to initiate the appropriate response to those signal. (NFPA 72)

fire department connection (FDC) A connection through which the fire department can pump supplemental water into the sprinkler system, standpipe, or other system furnishing water for fire extinguishment to supplement existing water supplies. (NFPA 13)

fire detection system A system that senses the presence of fire, smoke, or heat and activates a fire suppression system and/or an automatic alarm system. (NFPA 2)

KEY TERMS CONTINUED

fire protection system Any fire alarm device or system or fire-extinguishing device or system, or combination thereof, that is designed and installed for detecting, controlling, or extinguishing a fire or otherwise alerting occupants, or the fire department, or both, that a fire has occurred. (NFPA 1)

fire suppression system A system that discharges extinguishing agents in the presence of a fire.

fitting In a sprinkler system, a connector that joins the pipes together.

fixed-temperature detector A device that responds when its operating element become heated to a predetermined level. (NFPA 72)

flame detector A radiant energy–sensing fire detector that detects the radiant energy emitted by a flame. (NFPA 72)

frangible-bulb sprinkler head A sprinkler head with the cap held in place by a glass bulb filled with glycerin or alcohol and a small air bubble; when the liquid is heated, it vaporizes and the air pressure increases until the glass bulb breaks, releasing the cap.

fusible-link sprinkler head A sprinkler head with the cap held in place by two pieces of metal held together by solder with a low melting point, and when the solder melts, the link breaks, releasing the cap.

gas detector A device that detects the presence of a specified gas concentration. Gas detectors can be either spot-type or line-type detectors. (NFPA 72)

gaseous suppression system A system often installed in areas where computers or sensitive electronic equipment is used or where valuable documents are stored that stores gaseous agents including halogenated agents and inert gases such as carbon dioxide under pressure as liquids.

head box A box at a protected property that contains spare sprinkler heads and the head wrenches needed to install all of the types of heads.

heat detector A fire detector that detects either an abnormally high temperature, an abnormal rate of temperature rise, or both. (NFPA 72)

indicating valve A valve that has components that show if the valve is open or closed. (NFPA 1)

inspection A visual examination of a system or portion thereof to verify that it appears to be in operating condition and is free of physical damage. (NFPA 25)

ionization smoke detector A smoke detector that detects small smoke particles when an electric current created by a small amount of radioactive material in the detector is disrupted, activating the alarm.

line-type detector A device in which detection is continuous along a path. Typical examples are rate-of-rise pneumatic tubing detectors, projected beam smoke detectors, and heat-sensitive cable. (NFPA 72)

main sprinkler system valve A valve that automatically opens and signals an alarm when a sprinkler head is activated.

maintenance Work, including, but not limited to, repair, replacement, and service, performed to ensure that equipment operates properly. (NFPA 72)

main water supply control valve A manually controlled valve that allows water to enter the sprinkler system from the water source.

malicious alarm An unwanted activation of an alarm initiating device caused by a person acting with malice. (NFPA 72)

manual dry standpipe system A standpipe system with no permanently attached water supply that relies exclusively on the fire department connection to supply the system demand. (NFPA 14)

manual fire alarm box A manually operated device used to initiate a fire alarm signal. Also called *manual pull station.*

manual pull station See *manual fire alarm box.*

manual wet standpipe system A standpipe system containing water at all times that relies exclusively on the fire department connection (FDC) to supply the system demand. (NFPA 14)

maximum ceiling temperature The maximum temperature to which a sprinkler head can be exposed, while either installed in a system or in storage.

noncoded system An alarm system that provides no information indicating the type or location of the initiating device that was activated.

normal ready status The state of a system when it is functioning properly and there are no faults (problems) in the system and no problems with any other fire protection or building system integrated with the FACU.

nuisance alarm An unwanted activation of a signaling system or an alarm initiating device in response to a stimulus or condition that is not the result of a potentially hazardous condition. (NFPA 72)

obscuration rate A measure of the percentage of light that is blocked between a sender and a receiver unit.

outside screw and yoke (OS&Y) valve A sprinkler control valve with a threaded stem that extends out when the valve is open and retracts in when the valve is closed. Also called *outside stem and yoke valve.*

outside stem and yoke (OS&Y) valve See *outside screw and yoke valve.*

passive smoke control system A smoke control system that uses building features such as fire-rated walls, automatic fire doors, draft curtains, and other smoke barriers to limit the spread of smoke and compartmentalize larger spaces to confine the smoke to the room or area of origin.

pendent sprinkler head A sprinkler head designed to be mounted on the underside of sprinkler piping hanging down toward the floor so that the water stream is directed in a downward direction, and is usually marked *SSP*, which stands for *standard spray pendent.*

photoelectric smoke detector A smoke detector detects visible smoke particles when a light beam in the chamber is disrupted by the smoke particles and the light is reflected onto a photocell, activating the alarm.

pilot line A separate dry pipe system in a deluge sprinkler system in which the air pressure drops when a sprinkler head activates or a fixed temperature device releases so that water can flow through the pilot line and through a deluge valve.

post indicator valve (PIV) A sprinkler control valve that has an indicator on a post mounted in the ground or on a wall that reads either "OPEN" or "SHUT" depending on its position.

preaction sprinkler system A sprinkler system employing automatic sprinklers that are attached to a piping system that contains air that might or might not be under pressure, with a supplemental detection system installed in the same areas as the sprinklers. (NFPA 13)

preaction valve The valve assembly in a preaction sprinkler that prevents water from entering the system until a manual emergency release is activated, a fire sprinkler head activates, a fire detector activates, or a combination of sprinkler head and initiating device activation takes place.

projected beam-type detector A type of photoelectric light obscuration smoke detector wherein the beam spans the protected area. Also called *beam detector.* (NFPA 72)

proprietary supervising alarm system An installation of an alarm system that serves contiguous and noncontiguous properties, under one ownership, from a proprietary supervising station located at the protected premises, or at one of multiple noncontiguous protected premises, at which trained, competent personnel are in constant attendance. (NFPA 72)

protected premises (local) fire alarm system A fire alarm system located at the protected premises. (NFPA 72)

public emergency alarm reporting system A system of alarm initiating devices, transmitting and receiving equipment, and communication infrastructure (other than a public telephone network) used to communicate with the communications center to provide any combination of manual or auxiliary alarm service. (NFPA 72)

rate-of-rise detector A device that responds when the temperature rises at a rate exceeding a predetermined value. (NFPA 72)

release temperature The temperature at which the release mechanism in an automatic sprinkler head activates.

remote annunciators Displays near a building entrance that display the information shown on the annunciator at the FACU.

remote supervising station alarm system A protected premises fire alarm system (exclusive of any connected to a public emergency reporting system) in which alarm, supervisory, or trouble signals are transmitted automatically to, recorded in, and supervised from a remote supervising station that has competent and experienced servers and operators who, upon receipt of a signal, take such action as required by this Code. (NFPA 72)

residential sprinkler system A sprinkler system that protects dwelling units and that have quick response sprinkler heads so that they can activate quickly and reduce the heat, flames, and smoke, so occupants can evacuate quickly.

risers In a sprinkler system, vertical supply pipes.

sidewall sprinkler head A sprinkler that is designed for horizontal mounting so that the water spray is discharged horizontally into a room, but with a deflector that has a shield on the top so that the water spray is directed downward.

single-action pull station A manual alarm initiating device that requires the user to take a single step, such as pulling a lever, toggle, or handle, to activate the alarm.

KEY TERMS CONTINUED

single-station alarm A detector comprising an assembly that incorporates a sensor, control components, and an alarm notification appliance in one unit operated from a power source either located in the unit or obtained at the point of installation. (NFPA 72)

smoke containment systems Mechanical systems that keep smoke and other gases in the area or floor of origin.

smoke control system An engineered system that includes all methods that can be used singly or in combination to modify smoke movement. (NFPA 92)

smoke detector A device that detects visible or invisible particles of combustion. (NFPA 72)

smoke management systems Natural or mechanical systems that move or exhaust smoke that escapes from the fire area and threatens egress routes, elevator shafts, stairwells, and refuge areas.

spot-type detector A device in which the detecting element is concentrated at a particular location. (NFPA 72)

sprinkler head See *automatic sprinkler head.*

sprinkler piping The network of piping in a sprinkler system that delivers water to the sprinkler heads.

standpipe A pipe and attached hose valves and hose (if provided) used for conveying water to various parts of a building for fire-fighting purposes. (NFPA 1140)

standpipe system An arrangement of piping, valves, hose connections, and associated equipment installed in a building or structure, with the hose connections located in such a manner that water can be discharged in streams or spray patterns through attached hose and nozzles, for the purpose of extinguishing a fire, thereby protecting a building or structure and its contents in addition to protecting the occupants. (NFPA 25)

supervising station alarm system An alarm system that communicates between the protected property and the fire department or an alarm monitoring facility that is staffed 24 hours per day by personnel who receive, interpret, and take the appropriate action to handle the alarm condition.

supervisory condition An abnormal condition in connection with the supervision of other systems, processes, or equipment. (NFPA 72)

supervisory signal A signal that results from the detection of a supervisory condition. (NFPA 72)

T-3 signal See *temporal-3 pattern.*

temperature rating The temperature or rate of temperature increase that causes a heat detector to activate.

temporal-3 pattern A standard fire alarm audible signal for alerting occupants of a building consisting of three tones followed by a short pause. Also called *T-3 signal.*

testing A procedure used to determine the operational status of a component or system by conducting periodic physical checks, such as waterflow tests, fire pump tests, alarm tests, and trip tests of dry pipe, deluge, or preaction valves. (NFPA 25)

trouble condition An abnormal condition in a system that results from a fault. (NFPA 72)

trouble signal A signal that results from a trouble condition. (NFPA 72)

unintentional alarm An unwanted activation of an alarm initiating device caused by a person acting without malice. (NFPA 72)

unknown alarm An unwanted activation of an alarm initiating device or system output function where the cause has not been identified. (NFPA 72)

unwanted alarm Any alarm that occurs that is not the result of a potentially hazardous condition. (NFPA 72)

upright sprinkler head A sprinkler head designed to be installed on the top side of the sprinkler piping with a deflector that deflects the stream 180 degrees so that the water is directed down to the floor and is usually marked *SSU,* which stands for *standard spray upright.*

waterflow alarm device An attachment to the sprinkler system that detects a predetermined water flow and is connected to a fire alarm system to initiate an alarm condition. (NFPA 13)

water mist system A distribution system connected to a water supply or water and atomizing media supplies that is equipped with one or more nozzles capable of delivering water mist intended to control, suppress, or extinguish fires and that has been demonstrated to meet the performance requirements of its listing and [the applicable] standard. (NFPA 25)

water-motor gong An audible alarm notification device in a sprinkler system that makes noise as long as water is flowing at a certain pressure through the sprinkler system.

wet chemical extinguishing system A system often installed in commercial kitchens that discharges a liquid Class K extinguishing agent.

wet pipe sprinkler system A sprinkler system employing automatic sprinklers attached to a piping system containing water and connected to a water supply so that water discharges immediately from sprinklers opened by heat from a fire. (NFPA 13).

wet standpipe system A standpipe system having piping containing water at all times. (NFPA 25)

zone A defined area within a protected premise. A zone can define an area from which a signal can be received, an area in which a signal can be sent, or an area in which a form of control can be executed. (NFPA 72)

zone valve A valve that controls the flow of water into one zone in a sprinkler system that is divided into zones.

zoned system A fire alarm system that identifies the type or location of the initiating devices by identifying the zone where each device is located.

REVIEW QUESTIONS

1. What is the difference between a single-action pull station and a double-action pull station?
2. Which type of smoke alarm is recommended by NFPA to detect both smoldering fires and flaming fires?
3. Which alarm initiating device does not provide reliable life safety protection because it does not react quickly enough to incipient fires?
4. In residences, what locations are required to be equipped with at least one smoke alarm?
5. What are the two ways a residential carbon monoxide alarm is triggered?
6. What is the emergency control function interface?
7. What are the three types of supervising station alarm systems?
8. Which type of valve has a stem that indicates the valve is open if the stem is out?
9. Which type of sprinkler head is designed to be mounted on the underside of the supply piping?
10. Which type of sprinkler system is typically used in unheated warehouses that may experience below-freezing temperatures?
11. Which sprinkler head does not contain a cap or release mechanism?
12. What is the primary objective of residential sprinkler systems?
13. What classification of standpipe is designed for use by fire department personnel only?
14. What is the difference between a manual and an automatic standpipe system?
15. What is the difference between passive and active smoke control systems?
16. What is the procedure used to determine the operational status of a component or system by conducting periodic physical checks?

DISCUSSION QUESTIONS

1. What are some ways to prevent nuisance alarms in structures with a fire alarm system?
2. List the three mounting positions for sprinkler heads and explain why they are not interchangeable.
3. What are some of the problems firefighters can experience if flow restrictors or pressure-reducing valves are not properly installed in standpipes?
4. Why is record keeping of fire alarm system inspections, testing, and maintenance important to firefighters?

APPLYING THE CONCEPTS

At 1900, your fire company is en route to a nearby automotive parts manufacturing facility for report of a fire alarm. Dispatch reported that the call was initiated by the facility's fire alarm system and indicated "plating room heat detector." While on the way, dispatch provides an update that they received an additional alarm indicating a "water flow." Your engine pulls up to the front of the facility, and as you are getting out, a car pulls up. It's the security guard asking why you are here. After hearing the reason, the security says, "I've been here all evening and I haven't seen any sign of fire. That system is probably acting up again."

1. Based on the information from dispatch and the security guard, is this situation more likely a false alarm or a real emergency? Why?
2. Modern fire alarm systems like the one at this facility provide information about the alarm location and type of alarm. How do noncoded and coded alarm systems challenge firefighters? Where might they be used?

Your crew continues to do a size-up of the outside of the building. The security guard says everyone has left for the day, and an empty parking lot confirms this. No smoke is visible, and no exposures are nearby. Your department did preplanning on this facility, so you know the building layout, hazards, and location of the fire department connection (FDC). You also know it has a Class I automatic wet standpipe system and a dry chemical extinguishing system.

3. How does knowing about the standpipe system influence water supply planning and hose line selection?
4. If it was a manual wet standpipe system, how would this affect water supply planning?
5. What does the presence of a dry chemical extinguishing system tell you?

The IC confirms there are no hazardous materials in the plating room where the alarm was triggered. The IC instructs your crew to proceed into the building with 2½-in. (64-mm) supply lines and 1¾-in. (44-mm) attack lines. You locate the standpipe connector cabinet and connect the supply line. Two crew members remain to open the supply valve while you and your team connect the attack lines and advance to the plating room. As you get closer you can see light smoke, but no flames. It looks like the activated sprinkler system suppressed the fire.

6. What are your priorities in this situation?
7. How does the design of a complex, sophisticated automatic sprinkler system in a facility like this differ from a simple automatic system in a small building?

REFERENCES

Ahrens, Marty. 2021. *US Experience with Sprinklers.* National Fire Protection Association (NFPA). October 2021. Accessed April 4, 2023. https://www.nfpa.org.

National Fire Protection Association (NFPA). 2012. *Fire Service Guide to Reducing Unwanted Alarms.* Accessed December 7, 2022. https://catalog.nfpa.org.

National Fire Protection Association (NFPA). 2018. *NFPA 14: Standard for the Installation of Standpipe and Hose Systems.* 2019 Edition. Quincy, MA: NFPA.

National Fire Protection Association (NFPA). 2019. *NFPA 853: Standard for the Installation of Stationary Fuel Cell Power Systems.* 2020 Edition. Quincy, MA: NFPA.

National Fire Protection Association (NFPA). 2020. *NFPA 1: Fire Code.* 2021 Edition. Quincy, MA: NFPA.

National Fire Protection Association (NFPA). 2020. *NFPA 92: Standard for Smoke Control Systems.* 2021 Edition. Quincy, MA: NFPA.

National Fire Protection Association (NFPA). 2021. *NFPA 13: Standard for the Installation of Sprinkler Systems.* 2022 Edition. Quincy, MA: NFPA.

National Fire Protection Association (NFPA). 2021. *NFPA 15: Standard for Water Spray Fixed Systems for Fire Protection.* 2022 Edition. Quincy, MA: NFPA.

National Fire Protection Association (NFPA). 2021. *NFPA 72: National Fire Alarm and Signaling Code.* 2022 Edition. Quincy, MA: NFPA.

National Fire Protection Association (NFPA). 2021. *NFPA 1140: Standard for Wildland Fire Protection.* 2022 Edition. Quincy, MA: NFPA.

National Fire Protection Association (NFPA). 2021. *NFPA 2001, Standard on Clean Agent Fire Extinguishing Systems, 2022 Edition.* Quincy, MA: NFPA.

National Fire Protection Association (NFPA). 2022. *NFPA 2: Hydrogen Technologies Code.* 2023 Edition. Quincy, MA: NFPA.

National Fire Protection Association (NFPA). 2022. *NFPA 25: Standard for the Inspection, Testing, and Maintenance of Water-Based Fire Protection Systems.* 2023 Edition. Quincy, MA: NFPA.

CHAPTER

25

Firefighter II

Fire and Life Safety Initiatives

KNOWLEDGE OBJECTIVES

After studying this chapter, you will be able to:

- Define community risk reduction and the role of the fire service in maintaining overall community safety.
- Educate the public on hazard prevention and mitigation.
- Identify fire and life safety hazards that exist in residential properties and explain ways to mitigate them.
- Conduct preincident surveys and create preincident plans for non-residential locations such as commercial buildings, transportation centers, and other public locations.

SKILLS OBJECTIVES

After studying this chapter, you will be able to perform the following skills:

- Prepare presentations for educating the public about community risk reduction and fire and life safety, and conduct fire station tours.
- Conduct a fire safety survey at a residence.
- Conduct preincident surveys of commercial properties and other public locations, identifying tactical priorities and completing a preincident plan for that location.

ADDITIONAL NFPA STANDARDS

- **NFPA 1**, *Fire Code, 2021 Edition*
- **NFPA 72**, *National Fire Alarm and Signaling Code, 2022 Edition*
- **NFPA 170**, *Standard for Fire Safety and Emergency Symbols, 2021 Edition*
- **NFPA 470**, *Hazardous Materials/Weapons of Mass Destruction (WMD) Standard for Responders, 2022 Edition*
- **NFPA 801**, *Standard for Fire Protection for Facilities Handling Radioactive Materials, 2020 Edition*
- **NFPA 1030**, *Standard for Professional Qualifications for Fire Prevention Program Positions, 2024 Edition*
- **NFPA 1300**, *Standard on Community Risk Assessment and Community Risk Reduction Plan Development, 2020 Edition*
- **NFPA 1452**, *Guide for Training Fire Service Personnel to Conduct Community Risk Reduction for Residential Occupancies, 2020 Edition*
- **NFPA 1660**, *Standard for Emergency, Continuity, and Crisis Management: Preparedness, Response, and Recovery, 2024 Edition*

CASE STUDY

You Are the Firefighter

You entered the fire service to help your community, and you love the excitement of responding to the different types of calls for assistance. You like working with your crew, and even the daily station duties are enjoyable. A group of grade school children will be visiting your station this afternoon. You and your crew are cleaning the apparatus and equipment and discussing the visit before they arrive. As you discuss what you will talk to the children about, your captain says, "We will all go home having saved lives today." After receiving a few questioning glances from the crew, your captain continues, "Fire safety education is the easiest way to make sure people understand the dangers of fire and what to do to survive a fire in their home."

1. What did your captain mean, and what is your role in fire prevention?
2. What are the key points you will discuss with the group of students?
3. Other than conducting fire station tours, how can you assist your department's fire safety education program?
4. How does fire prevention affect firefighter safety?

Introduction

Firefighters across the country will tell you that they joined the fire service to save lives and property. Training, staffing, equipment selection, and many of the other topics you have learned about all align with that mission. However, it is important to view your role in the fire service as an all-inclusive, comprehensive firefighter whose mission is to save lives and property. Fire prevention and public education outreach provide an opportunity for fire service personnel to meet citizens where they are in their fire and life safety knowledge and help them live and work in safer environments. Today's fire departments go further by educating the public and working to reduce the number of all types of emergencies by identifying the potential risks, offering suggestions for reducing those risks, and creating plans and procedures to mitigate those risks. Fires can be prevented by conducting inspections to enforce fire codes and taking actions to mitigate risks before an incident occurs. Public safety education outreach allows you to extend that mission to the time before risks turn into emergencies. Fire prevention, education, and mitigation efforts to avoid all types of emergency incidents are an effective means of avoiding the loss of lives and property. For example, educating the public about fire risks and other hazards such as fall prevention, downed power lines, gas leaks, cardiopulmonary resuscitation (CPR), child car seat safety, bike helmets, and many other topics help limit the loss of life, injuries, and property damage.

Fire and Life Safety Education

One of the goals of community risk reduction (CRR) is to provide the public with fire and life safety education so that they know how to prevent loss of life, injuries, and property damage from occurring and to teach them how to react in an appropriate manner if an emergency does occur. CRR educators work to create a change in behavior that results in greater safety for their community. NFPA 1030, *Standard for Professional Qualifications for Fire Prevention Program Positions*, provides performance objectives for fire and life safety educators at two levels plus a management level.

Educational messages are more effective when they are tied to real-life experiences that have happened or could happen. Default fire safety messaging needs to be evaluated and reframed to meet the needs of the community, and more specifically, the audience, being targeted. The age and comprehension level of the audience need to be considered in every educational opportunity. Having three to five messages identified for each audience and shared with all personnel in the fire department will help with consistent messaging to the community.

When educating both children and adults, the best method is usually to use storytelling and analogies. This does not mean a firefighter should tell stories about tough calls they have been on. It means using visual images that are familiar to your audience to help them retain the new information you are giving them by tying it to that visual image.

FIGURE 25-1 Using a visual image that people are familiar with to tie new information helps with retention.

Courtesy of Becki Rowan-White.

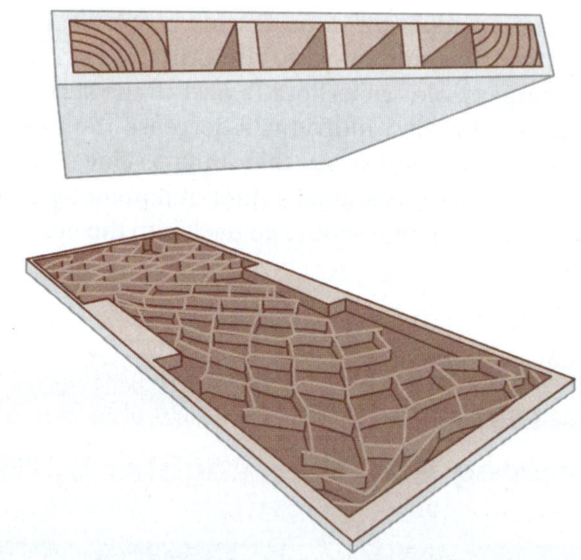

FIGURE 25-2 Hollow core doors have chambers or separators, so feeling the door with the back of a hand to check to see if it feels hot is not useful.

Courtesy of Becki Rowan-White.

For example, to explain the need for and placement of smoke alarms, sleeping with bedroom doors closed, and having established meeting places in case of fire, you could explain that smoke and fire are like water in a bathtub. When you open the faucet, the water runs down in a column, hits the bottom of the tub and spreads to fill the bathtub. Smoke and fire does the same thing, except upside down. Smoke goes up in a column (like the water goes down), reaches the ceiling and spreads out (like the water filling the tub), and then banks down (like the water getting deeper in the tub) (**FIGURE 25-1**). Explaining smoke and fire behavior in this visual manner can aid in explaining why smoke alarms are placed on the ceiling or high on walls, why you should sleep with bedroom doors closed, and why we say to crawl low under smoke.

Understanding and Updating Fire Safety Messages

The fire service has been advancing their response to emergencies based on research and new equipment, yet the same educational messages have continued to be delivered without an analysis of how effective those messages are. It's time to rethink these hand-me-down messages that have been constant for decades in the fire service. A good example of this is the Stop, Drop, and Roll message, which tends to be an old stand-by for fire departments, yet the only time this message should be used is when clothes are on fire. Oftentimes, students are taught to do the action, but they are never taught they only do this when their clothes catch fire. This should not be a default message for all students, but rather only be used for campfire safety or other similar times when there is a serious potential for clothes catching on fire. Looking deeper, we should be educating students how to avoid being in situations when their clothes have a potential to catch on fire.

Another example of an outdated message is feeling the door with the back of your hand. Residential doors are often hollow-core doors with spacers, which inhibit the potential that the inside of the door would have radiant heat from the outside (**FIGURE 25-2**). The other factor is that smoke and heat rise, causing the heat that they are feeling for to be very high on the door, rather than low to the ground where they have been instructed to feel. The best thing to do is to crack open the door and look for smoke. If there is smoke, close the door, turn on the light, and go out or wait near the window, depending on what level the room is on and whether egress is possible without aid.

A better safety message to focus on is escape planning, referred to as **Exit Drills in the Home (EDITH)**. Many fire deaths could be prevented if everyone learned a few simple rules about what to do in the event of a fire in the home.

The first part of EDITH is identifying two escape routes from each bedroom in case there is smoke, heat, or flames present in the primary escape route. The secondary escape route may be through a window or through a different door. During a fire, occupants should stay low and crawl to the primary exit. If that exit is impassable, they should crawl to the secondary exit. By staying low, individuals decrease the amount of heat, smoke, and gases they inhale. Once outside the house, residents should gather at a predetermined meeting spot. No one should go back into the house for any reason. At this point, they should make sure that the fire department has been called or make the call themselves.

Recommend that the occupants practice EDITH regularly. Some drills should be held in the daytime, and some should take place at night. To teach EDITH, follow the steps in **SKILL DRILL 25-1**.

SKILL DRILL 25-1

Teaching EDITH Firefighter II, NFPA 1010: 7.5.2

1. Encourage the audience to have a plan for safety and exit.
   - Describe the importance of having a sufficient number of working smoke alarms throughout the house and making a quick exit when a fire occurs.
   - Explain why residents should sleep with bedroom doors closed.
   - To keep an audience of children engaged, have them draw a map of each level of their house, showing all windows and all doors.
   - Tell your listeners—occupants of a home during a residential fire safety survey or an audience in a public setting—to plan two escape routes from each room and to make sure all doors and windows are operable and accessible.
   - Instruct the audience to designate a specific location outside the residence as a meeting place.

2. Tell the audience to practice the plan with all family members and caretakers, including those with disabilities. Instruct students to roll out of bed and crawl to the exit when they hear a smoke alarm or smell smoke.

SKILL DRILL 25-1 CONTINUED

Teaching EDITH Firefighter II, NFPA 1010: 7.5.2

3. Instruct listeners to follow the pre-established escape route that they practiced and proceed to the designated meeting place.

4. Instruct the listeners to ensure that the fire department has been called only after exiting the residence and emphasize the importance of never reentering a burning building.

Steps 1, 2, and 4: © Jones & Bartlett Learning. Photographed by Glen E. Ellman; **Step 3:** Courtesy of U.S. Fire Administration.

Differentiating between Fire Safety and Fire Prevention

Fire prevention has struggled throughout the years when it comes to justifying the return on investment during budget evaluations. It is extremely difficult to quantify fires and injuries that did not happen because they were successfully prevented. This means that prevention budgets are often reduced in favor of initiatives that have more tangible outcomes.

As the fire service branches into many different safety areas, such as emergency medical services (EMS), handling hazardous materials spills, mass-casualty and active shooter incidents, and mental health crisis intervention, the messages have changed from just fire prevention and into life safety and injury prevention. What was traditionally referred to as fire prevention falls into three categories: fire prevention, fire safety and survival, and miscellaneous, with most of it falling into the safety and survival category.

Fire prevention messages should influence actions that will prevent fires and injuries from happening. Examples include the following:

- Cooking fire safety—when you cook, stay and look
- Wear bike helmets
- Install and use car seats
- Keep matches and lighters away from children

Most of the messages disseminated by the fire service actually fall into the safety and survival category. Examples include the following:

- Smoke alarm placement, maintenance, batteries, etc.
- Fire extinguisher placement, use, and maintenance
- Exit Drills in the Home (EDITH) and Two Ways Out
- 3-foot zone: Keep children 3 feet from fireplaces, stoves, fire pits, grills, etc.
- Stop, drop, and roll

If you truly educate the public about fire prevention, you will see a corresponding reduction in the number of fires in your community. The life and safety messages keep people safe if a fire occurs, but these messages do not directly impact the number of fires in your community. If the goal of your fire prevention bureau is to reduce the number of fires in your community, you need to make sure your messaging is about fire prevention. Alternately, you can expand your goals to include the safety and survival of your residents, which will expand the return on investment.

Some messages do not fall into either the fire prevention or the safety and survival category. These messages are generally informational in nature, as opposed to trying to convince people to change their behavior, for example:

- Descriptions of fire behavior
- Explanation of the fire triangle
- Public relations in the form of fire gear, fire equipment, and fire apparatus

Resources for Educational Messaging and Community Data

There are several sources for data about your community. After every run, department officers spend time completing reports that ultimately go to the National Fire Incident Reporting System (NFIRS). The data from these reports help the U.S. Fire Administration (USFA) and the National Fire Protection Association (NFPA) develop fire safety messaging at the national level. Find out more about the NFIRS at www.usfa.fema.gov/nfirs. Whether your department sends its data to the NFIRS, completing quality reports is an essential part of having accurate data on incidents in your community.

Another way to learn more about your community demographics is to access census data from the U.S. Census Bureau for your area. That data can provide information about your first-due area, your entire city, a larger region, or national statistics. Start by visiting their website at data.census.gov.

If you want to see which educational strategies have worked in other areas, Vision 20/20 Community Risk Reduction is a great place to start. They compile data about demographics, fire and EMS incidents, and model community risk reduction programs for other departments to access. This data can be found at www.strategicfire.org/discovery.

Underwriters Laboratories (UL) Fire Safety Research Institute (FSRI) conducts research and provides education about fire-related issues, from how fires burn and firefighting tactics to firefighter health and safety. They have a Public Safety Education Advisory Group (PSEAG) that works on developing and packaging fire and life safety messaging based on the science of how fires burn and how the tactics firefighters use impact life safety for both the firefighters and the occupants. The Close Before You Doze fire safety messaging—an educational campaign to inform the public about the importance of closing the bedroom door when sleeping—was a direct result of a ventilation study conducted by the FSRI. The results of this study changed how fire departments conduct primary and secondary searches during fire suppression efforts. This collaboration among scientific research, operational tactics, and educational messaging is an exciting breakthrough in fire and life safety education. For more resources to use to help your department educate the public, visit closeyourdoor.org/first-responders/toolbox.

FIGURE 25-3 Fire Is Everyone's Fight logo.

Courtesy of the U.S. Fire Administration and FEMA. The use of this logo does not constitute endorsement by the U.S. Fire Administration or FEMA.

Any fire that is prevented creates a safer situation for firefighters. The Federal Emergency Management Agency (FEMA) encourages all fire departments to join the Fire Is Everyone's Fight initiative. Fire Is Everyone's Fight is a national initiative that unites the fire service and life safety organizations to reduce home fire injuries, deaths, and property loss by changing how people think about fire and fire prevention (**FIGURE 25-3**). USFA fire safety and prevention outreach materials are available on the USFA website at usfa.fema.gov/prevention/outreach; you can even create your own using their templates.

The NFPA has an educational messages advisory committee (EMAC) that provides recommendations to NFPA public education staff and the fire service for updating and revising fire safety education messaging. The EMAC Desk Reference is a free document available through NFPA that can be used to identify messages for your community or for use on social media. You can access this at www.nfpa.org/Public-Education/Teaching-tools/Educational-messaging.

Many other national and worldwide organizations have a shared mission of reducing injuries and deaths outside of fire safety, including the following:

- Safe Kids International, which works to reduce traffic injuries, drownings, falls, burns, poisonings, and more
- Centers for Disease Control and Prevention (CDC), which conducts research and provides health information against health threats
- Red Cross, which not only provides disaster assistance, but also health and safety education and materials

To teach fire safety to the public, follow the steps in **SKILL DRILL 25-2**.

SKILL DRILL 25-2

Teaching Fire Safety to the Public Firefighter II, NFPA 1010: 7.5.2

1. Plan the presentation.
   - Identify the message (or messages) you want to convey.
   - Identify the audience.
   - Prepare answers for anticipated questions.
   - Decide the method you will use to convey the message(s), such as stories and analogies; visual media (posters, slide shows, or videos); or educational materials from the USFA, NFPA, or other fire safety groups, and prepare any elements that you will need.
   - Prepare a written form of your presentation in a document or in presentation software.
2. Practice your presentation so that you are able to speak comfortably and with authority.
3. After giving the presentation, note how well the audience responded, and incorporate any feedback you received for future presentations.

Community Risk Reduction

Fire prevention is specifically focused on preventing fires. Fire prevention is further described in Chapter 1, *The Fire Service*. **Community risk reduction (CRR)**, on the other hand, is a comprehensive approach to reduce the overall incidence and impact of all types of emergencies within a community by identifying the potential risks and creating plans and procedures to mitigate those risks. Understanding the risks inherent in your community because of both controllable and uncontrollable factors, such as geographic features, local weather patterns, local industry, and community layout, is essential in order for the needs of your community to be met in emergencies.

CRR used to be viewed as a program exclusive to the fire prevention program in the fire department. It is now viewed as an overarching goal above all of the public safety departments in a community—police, fire, EMS, and emergency management—as well as the public works and community planning departments (**FIGURE 25-4**).

Other areas of city, county, and state government as well as businesses and organizations in a community have a vested interest in CRR. Establishing a dialogue among all the various agencies, departments, and organizations is a good first step in creating a CRR plan. Organizations that collaborate can promote CRR initiatives that complement each other rather than compete with each other. Examples include the following:

- Set up a safety camp: Meet with schools, recreation facilities, and the parks and recreation division in your community and set up a 1-day or a week-long camp focused on teaching kids about safety. Topics can include fire safety, babysitting, household poisons, electrical safety, bike safety, water safety, stranger danger, and many other topics depending on the age of the children in attendance. A side benefit of including firefighters and police officers as

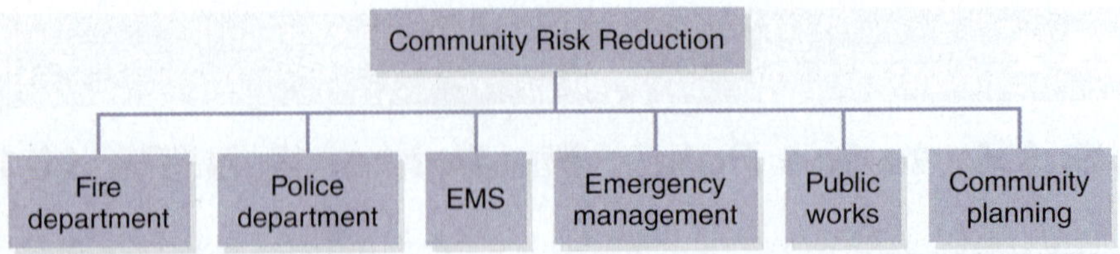

FIGURE 25-4 Community risk reduction is a community-wide concern.

Courtesy of Becki Rowan-White.

counselors is that it can help establish relationships between those agencies and youth in the community, resulting in long-range recruiting opportunities.

- Develop a home safety survey: Meet with members of law enforcement to get their opinions on safety issues that homeowners and renters experience, and then add their suggestions to the list of fire safety issues in your area.
- Establish partnership with other organizations: If you establish partnerships with organizations such as the Red Cross or Safe Kids, you can capitalize on the resources they provide to the public for home safety. You can also establish partnerships with local hardware stores that sell smoke and carbon monoxide alarms to provide information, education, and discounts to the public on the alarms. You should also establish partnerships with other agencies that provide care. For example, is you establish a partnership with a mental health agency, they can provide support when someone is experiencing a mental health crisis.

Programs like these are the heart of CRR. They both meet the mission of the fire department by saving lives and property and strengthen relationships with other community organizations for the benefit of risk reduction and well-being for the whole community.

TIP

Most fires are caused by situations that could have been avoided by taking preventive actions. A key element in reducing the number of injuries and deaths is proactive fire safety education and enforcement.

A lack of CRR efforts can result in death and serious injury for civilians. These same conditions and actions are also dangerous and fatal for firefighters. Preventing these deaths and injuries is why CRR is a primary concern of the fire service. The Everyone Goes Home Program and the 16 Firefighter Life Safety Initiatives reinforce this commitment and describe the steps that need to be taken to change the current culture of the fire service to help make it a safer work environment.

Initiative 14 states, "Public education must receive more resources and be championed as a critical fire and life safety program." Initiative 15 states, "Advocacy must be strengthened for the enforcement of codes and the installation of home fire sprinklers." In other words, every fire that is prevented reduces the risks to firefighters and the public.

Depending on the layout of the fire department, Firefighter I and Firefighter II personnel may have limited responsibilities for formal CRR activities. However, citizens view fire service personnel as subject matter experts on fire and life safety, so all fire service personnel need to understand the importance of their role in preventing fires and injuries in their communities. Whether inside or outside of the fire service, CRR initiatives are focused around the **5 Es of community risk reduction (CRR)**: education, enforcement, engineering, economic incentives, and emergency response, also called the **5 Es of prevention** (**FIGURE 25-5**). These are also discussed in Chapter 1, *The Fire Service.*

Education

The first, and usually least expensive, strategy is education. Generally, people are unaware of safety issues that cause fires and serious injuries. Education can be divided into three levels: public relations, public information, and public education (**FIGURE 25-6**).

The first level is public relations. Everything firefighters do is public relations, either good or bad. If firefighters show up at an event, they don't even have to talk to anyone to demonstrate good public relations. They provide education to everyone present by the way they dress, act, and represent themselves and their department. Likewise, responding to calls, driving in department vehicles, actions or inactions on scene, and behavior while wearing department clothing, whether on- or off-duty, all provide opportunities for education. It is very important to always conduct yourself in a professional manner, dress professionally, and act courteously, and to remember you are providing a service to your community. Individuals and departments have faced serious, career-ending situations because members of the public formed poor opinions about an event without the individuals or organizations having the opportunity to provide context.

FIGURE 25-5 Vision 20/20 is a collaborative project focused on CRR across the country.
Courtesy of Vision 20/20 Project.

The second level is public information. At this level, you provide information. Citizens will still assess you from a public relations perspective, but you have the opportunity to share a message with your words and actions. Informational flyers, social media posts, information on websites or printed on department vehicles, and answers given when questions are asked by community members or the media are all examples of public information.

The top level is public education. This happens when the public information prompts changes in behavior or affirms behavior on behalf of the recipient. The information is educational and useful, prompting a change of the recipients' normal routine. **TABLE 25-1** lists examples of activities and how information becomes public relations, public information, and public education.

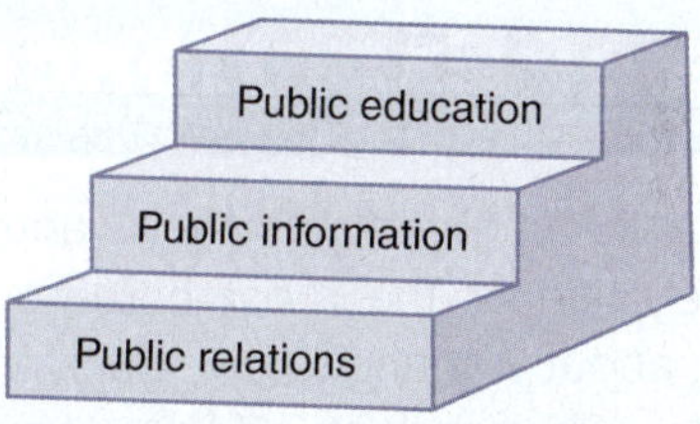

FIGURE 25-6 All of our actions are public relations. When we step up our efforts, we reach the higher level of public information, and good information is educational.
Courtesy of Becki Rowan-White.

TABLE 25-1 Comparing Public Relations, Public Information, and Public Education

Activity	Public Relations	Public Information	Public Education
Posting messages on social media	Shows department in a positive light; for example, posting a photo of a firefighter interacting with children at an elementary school	Provides a safety message; for example, adding a description of the message the firefighters brought to the students, such as talking to the students about how important it is to sleep with the bedroom door closed	Expands the safety message to include why the action should be taken or changed; for example, explaining that sleeping with the door closed gives people time to react to the sound of their smoke alarms
Distributing stickers, helmets, and other items that include the department name or logo and a safety message	Keeps the department name visible to the public	Keeps the safety message visible to the community	Message that ties into the usefulness of the item will reinforce the knowledge
Sharing safety messages as a phrase or slogan	Including the department name or logo reinforces the connection to the department	Stating the message as a phrase or slogan helps people remember it	Messages that provide context, either through an image or via a written description, of the benefits of the behavior change reinforce the message and how it relates to the receiver

Courtesy of Becki Rowan-White.

When you are speaking to the public, pay attention to the language that you use. Make sure that it will be understood by your audience and avoid using firefighting terms and jargon. When speaking to young children and older adults, speak at a slower pace.

Enforcement

The enforcement strategy is the "teeth" of the 5 Es—change your behavior or potentially deal with the consequences. Law enforcement has several successful examples of this. Speed limit signs are educational messages for drivers; traffic tickets are the enforcement. Enforcement is usually tied to federal, state, or local laws and ordinances. In the fire service, enforcement is accomplished through the fire code. Most departments take an educational approach to fire code enforcement—they meet with business owners or occupants to explain the codes and their importance, and they provide an opportunity for the owner or occupant to comply. Failure to comply with the fire code results in a fee or fine.

Enactment of Fire Codes

In the United States, fire and building codes are created by the NFPA and the International Code Council (ICC). Fire codes are discussed further in Chapter 1, *The Fire Service*. The following definitions are useful to clarify terms related to codes:

- A **code** is a set of legally adopted rules that specify the minimum requirements and standard practices in an industry or profession.

Voice of Experience

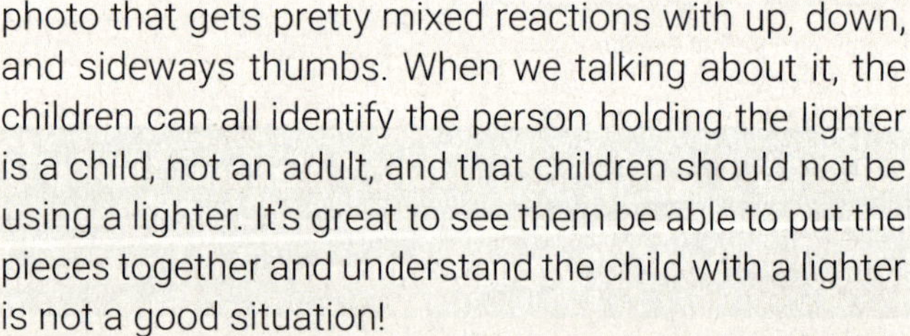

When you ask firefighters why they wanted to become a firefighter, one of the most common answers is something about wanting to give back and serve their community. Although fast, effective fire suppression is definitely a service to the community, wouldn't it be an even greater service if we could help people to prevent that fire from ever happening in the first place? I have met plenty of firefighters who have lamented that those of us who work in prevention are taking away their fun by reducing the number of fires or who will moan and groan when asked to do a public education event. However, I am pretty certain, if they really think about it, all of those firefighters would truly prefer to talk to a preschool class about not touching matches and lighters than respond to a fatal fire caused by a child playing with a lighter.

One of my favorite activities to do with young children, around first grade, is to ask them if they think fires are "good" or "bad." The majority will usually answer that fires are bad, because they think that is the answer I want to hear. I then show them a series of photos of different fires and ask them to give me a "thumbs-up" if they think it's a "good" fire, a "thumbs-down" if they think it's a "bad" fire, and a "sideways-thumb" if they aren't sure whether it's good or bad. By participating in a non-verbal way and all at the same time, it's a great way to get all the children involved, and quickly and easily assess their understanding. Most of the photos are very obvious for the children: car fires, house fires, wildland fires are all "bad", whereas birthday candles, fireworks, and fireplaces are all "good" fires. Even with the "good" fires, we talk about how they could turn into a "bad" fire. The last photo I show is of a child holding a lighter. This is the one photo that gets pretty mixed reactions with up, down, and sideways thumbs. When we talking about it, the children can all identify the person holding the lighter is a child, not an adult, and that children should not be using a lighter. It's great to see them be able to put the pieces together and understand the child with a lighter is not a good situation!

I think one of the challenges of fire prevention is that it is harder to quantify the impact. It is easy to see the impact of good firefighters at a fire scene. But how do you quantify the people who never had a fire because they learned from a firefighter that they should stay in the kitchen when cooking, or clean out their clothes dryer vent, or how to safely dispose of recreational fire ashes? We can save far more lives and property through prevention than are saved through emergency response. If we truly mean what we say about joining the fire service because we want to give back and serve the community, fire prevention needs to be a priority.

Bethany Brunsell

Assistant Fire Chief
Golden Valley Fire Department
Golden Valley, Minnesota

- A **fire code** specifies practices and procedures to prevent fires from occurring, requires structures to be constructed in a way that will help prevent fires from spreading by both suppressing fire and preventing it from spreading, and protects lives in the event of a fire by specifying how occupants will be evacuated.
- A **building code** specifies how structures are designed, constructed, and remodeled so that buildings are safe for the people who live and work in them. Building codes reference fire codes and require that buildings meet the established fire code for the municipality.

These codes are reviewed and adopted by state, local, or provincial governments and are enforced by the courts. Generally, the codes in a community include a building code, an electrical code, a plumbing code, a mechanical code, and a fire code. These codes and regulations must be reviewed and coordinated to avoid conflicting requirements and to ensure a safe community. Some states establish both minimum and maximum building and fire codes—called **mini-maxi codes**—that must be followed exactly by local jurisdictions, with no local amendments. Other states adopt codes that establish the minimum level of fire protection, but allow local jurisdictions to adopt more stringent codes if they think it is necessary.

Building codes establish safety requirements that apply to the construction of new buildings and almost always include fire protection requirements intended to minimize the development and spread of fire and smoke. Fire department personnel are usually involved in the enforcement of both fire and building codes. Every firefighter needs to understand the effects of the relevant codes and their involvement in establishing public safety requirements. Chapter 6, *Building Construction*, covers issues that are closely related to building codes. Generally, fire codes apply to all buildings, new or old, although older buildings may not comply with a particular code if the code was not retroactively applied. In most cases, only a change of ownership or occupancy type requires compliance with newer codes. In rare circumstances, a building may be required to comply with newer codes due to a major loss-of-life fire in a similar occupancy.

Some code requirements apply to all buildings, such as exit requirements, occupancy load limits, and electrical requirements. Some occupancy types are required to have fire alarm systems, sprinkler systems, or both. In addition, fire codes include regulations for dispensing flammable liquids, for storing them in approved storage containers, and for prohibiting their use or storage in some buildings. Regulations that prohibit parking in a fire lane or obstructing a fire hydrant are usually found in the fire code, as well. All of these requirements are intended to promote fire safety and protect both the community and firefighters. Because code adoptions and amendments vary from jurisdiction to jurisdiction, even within the same region, firefighters should be aware of local requirements.

Inspection and Code Enforcement

After a jurisdiction adopts a fire code, it must be able to enforce it. Regular inspections ensure compliance. Inspectors have the authority to note any violations and require their correction. If violations are not corrected, a series of increasing penalties can be imposed. If an individual or organization fails to correct the violation within a specified time, they can be fined. In some cases, a jail sentence can be imposed for serious violations.

Fire codes usually specify the types of occupancies that are subject to inspection and the frequency of the inspections. For example, some fire codes require all schools in a jurisdiction to be inspected twice each year. Sometimes the frequency of inspections is discretionary. The agency responsible for performing inspections and enforcing codes is usually named in the fire code; this varies in different communities. Sometimes the responsibility belongs to specific individuals or positions within the fire department, such as fire inspectors or fire marshals. Other fire departments train all firefighters to conduct inspections and take code enforcement actions. In some communities, building inspectors or special code enforcement personnel are given this responsibility. Generally, newly assigned firefighters are not directly involved in code enforcement activities, but they should understand how code enforcement is handled in the community.

Fire codes authorize local jurisdictions to conduct fire inspections in all buildings during the construction phase and prior to issuing an occupancy permit. Most jurisdictions inspect commercial occupancies annually. Although fire and building codes apply to both commercial buildings and private residences, most jurisdictions do not inspect single- and two-family residences after they are built and occupied unless there is a reason to believe the residence is unsafe. The occupant can request a voluntary inspection to identify hazardous conditions, however. Many fire departments offer this public service as part of a comprehensive fire prevention program. This type of voluntary inspection is called a **fire safety survey**. The legal requirements of the fire code are not applied when a fire safety survey is conducted, which means violations are not cited and fines are not imposed. Every firefighter should know how to conduct a home fire safety survey in a private dwelling, including knowing what to look for during the survey and how to communicate the survey findings. This information is presented in detail later in this chapter in the "Fire and Life Safety" section.

Engineering

The engineering strategy is usually outside the scope of the fire department. Firefighters and citizens benefit from engineering advancements in smoke and carbon monoxide alarms, air bags and seat belts in vehicles, equipment used to monitor facilities and extinguish fires, personal protective equipment (PPE), fire alarm systems, sprinkler systems, etc. Advancements in data analysis is an excellent example of engineering in action. Data analysis can identify trends proactively as well as reactively. Data analysis has become a more focused effort in evaluating the effectiveness of fire department operations, emergency medical operations, and CRR. This use of technology aids a department's response to its community.

Economic Incentives

The economic incentive strategy is generally tied to education, enforcement, or engineering. The incentive usually involves either discounts, or fees and fines, to motivate an individual or organization to start or stop an action. For example, an insurance company that gives a discount on the premium to businesses with sprinkler systems is a crossover between economic incentives and engineering. Some jurisdictions work with small businesses or homeowners by providing grants for retrofitting sprinkler systems into existing buildings, reducing the financial impact on those who want to increase their protection from fire. An example of a successful economic incentive is the Distressed Properties Initiative in the Surrey (British Columbia, Canada) Fire Service. This initiative guarantees that property owners secure their property so that no one can enter it, even if the property is abandoned. If the property owner does not secure their property, the city secures it and bills the owner. This program has reduced the fires related to distressed properties by two-thirds (Griffioen 2020, 1–2).

TIP

Learn more about using fire statistics on the U.S. Fire Administration website at usfa.fema.gov/statistics.

Emergency Response

Like the education strategy, public relations factors into the emergency response strategy. Most people believe that bad things, such as fires and major injuries, won't happen to them. When a neighbor or neighboring business has an emergency, it brings awareness to that neighborhood that if the emergency affected those around them, it could easily happen to them. Speaking to the media during an emergency provides an excellent platform for sharing how others can avoid having a similar thing happen to them. Always work in an action message, such as, "The homeowners are all safe because they had working smoke alarms, please make sure you check yours," or "This business had a fire sprinkler system, which reduced the fire and smoke damage. Hopefully they will be up and running soon." Canvasing the neighborhood or posting on social media after residential fires can provide more effective platforms for education then randomly selecting an area or posting on a schedule that is unrelated to current events.

Conducting Fire Station Tours to Promote Residential Home Safety

Another popular fire safety education activity is a fire station tour. They present a unique opportunity to help people learn how the fire department operates and how they can help prevent fires. Both children and adults enjoy the opportunity to tour local fire stations, and a fire station tour is an excellent opportunity to promote fire safety education. Remember, station tours present a public relations opportunity. If you are prepared, you can expand a station tour to relaying public information, and if you are well-informed you can use this as a public education opportunity. Most fire departments have a set format for conducting fire station tours. Your station officer is in charge of all activities and should teach you how to conduct tours in your station.

Before tours arrive, check to make sure any areas the visitors will walk through are safe. Identify and remove any hazards, especially tripping hazards. If hazards cannot be removed, cordon them off with caution tape and advise visitors to avoid them.

Explain to the visitors what to expect and what they should do if the station receives an alarm during the tour. Visitors should not wander around the station while firefighters respond to an emergency call. This is particularly important if the tour guide might have to leave in response to a call.

SAFETY TIP

Fire stations can be potentially dangerous for the general public. Before conducting a tour, ask yourself the following questions:

- Is the area safe?
- Am I showing the fire service in a positive light?
- Will there be enough firefighters to supervise the tour group?
- Are the planned activities age appropriate?

Always remember that when you conduct a tour, you are representing your department to the visitors. A professional appearance and presentation project a positive image of the department to the community. Try to leave every tour group with both a message and materials. First, reinforce a simple prevention or safety message that expands their awareness of fire safety and helps them to deal with potential emergency situations. Then, give them printed reference materials (**FIGURE 25-7**). These materials could range from coloring books for young children to a home fire prevention checklist for adults.

FIGURE 25-7 Always give visitors age-appropriate printed fire prevention or safety materials.
Courtesy of the Federal Emergency Management Agency.

Making the Tour Educational

Some tours start in the dayroom or a meeting room where the firefighter conducting the tour provides a general overview of the fire station and the department's expectations for visitors during the tour. This is usually followed by a tour of the fire apparatus. Alternatively, a tour can start with an inspection of the apparatus and end in a meeting room with a question-and-answer session and information about the department and fire prevention techniques. Showing a fire prevention video or a video about the operation of your department is a good way to start or finish a tour and stimulate questions from your audience.

The tour format will vary depending on the age and interests of the group, which could consist of small children, teenagers, adults, or senior citizens. The tour plan, including your presentation, should be designed to meet the needs of each group. The weather may also be an important factor if you are planning any outdoor demonstrations.

If the tour includes residential areas of the station, residential fire safety messages should be given at that time. In the kitchen, discuss cooking safety; in the bedrooms, discuss smoke alarms, EDITH, and sleeping with the doors closed. In the dayroom, discuss extension cord use and EDITH again—it is important that people understand they should identify the exits from each room. If the tour does not include residential areas, come up with some other way, such as using posters, to expand your reach of this important topic to more audiences.

A major draw during a station tour is the department's equipment and response capabilities. As you are sharing the apparatus, equipment, and PPE with the attendees, remember to work in educational messaging such as the following:

- Seat belt use
- Keeping tools clean, maintained, and secured
- Being aware of distractions when driving
- The importance of meeting places so emergency responders can find people when they arrive

Cater Your Educational Messages

Many fire safety messages are directed toward a young audience. However, teenagers and adults also need to hear these messages. Teenagers are ready for lessons that they can apply in everyday life. Many teens work as babysitters, so prevention messages for this audience can be geared toward cooking safety and evacuating children in the event of a fire. Teens are a high-risk group for motor vehicle accidents, so messages about seat belt use, using cell phones in hands-free mode, safe driving practices, the dangers of texting while driving and drinking and driving, and actions to take if they see or hear an emergency vehicle while driving are appropriate. Some teenagers might be interested in becoming firefighters or joining a firefighter explorer group. This age group can be a good source of recruits for both career and volunteer departments.

Adults are probably more interested in home fire safety, general emergency response, and available EMS. Remind smokers not to smoke in bed or to empty ashtrays into trash cans. Many seniors use oxygen concentrators or supplemental oxygen, so stress the fire hazards associated with the use of medical oxygen. Some seniors may forget to turn off burners, or they may leave cooking food unattended. They also may be worried about their ability to move quickly if a fire starts in their home. Tailor your prevention messages to the specific concerns of the group you are addressing.

TIP

Adults often ask about sources of funding for the fire department. If you give station tours, you should understand how your fire department is funded. Be prepared to answer questions on how taxpayer money is used to provide the myriad services the fire department provides.

The steps for conducting a fire station tour are listed in **SKILL DRILL 25-3**.

Residential Fire Safety Surveys

Each year, more civilians die from fire in their homes than in any other setting. According to the NFPA, residential fires make up one quarter (26%) of reported fires, yet account for 72% of reported civilian fire injuries and 75% of civilian fire deaths (Ahrens and Maheshwari 2021, 1). To combat this, many fire departments offer to conduct fire safety surveys in private dwellings. A home

SKILL DRILL 25-3

Conducting a Fire Station Tour Firefighter II, NFPA 1010: 7.5.2

1. Introduce yourself to the tour participants and explain the mission of the fire department. Show participants the layout of the fire station and the various types of personal protective equipment (PPE). Consider using an educational video presentation.

2. Tour the living areas of the fire station, if they are accessible. Remember to work in residential fire safety messaging.

SKILL DRILL 25-3 CONTINUED

Conducting a Fire Station Tour Firefighter II, NFPA 1010: 7.5.2

3. Explain the purpose of a variety of apparatus, tools, equipment, and PPE, keeping all explanations age appropriate, periodically checking to see if anyone has any questions.

4. Conclude the tour in a meeting room. Encourage questions from participants. Hand out appropriate fire safety materials.

Steps 1, 3, and 4: © Jones & Bartlett Learning. Photographed by Glen E. Ellman; **Step 2:** Courtesy of Keith Muratori.

fire safety survey helps identify fire and life-safety hazards and provides the occupants with recommendations for making their home safer.

In most jurisdictions, residential fire safety surveys cannot be conducted without the occupant's permission. Usually, the survey is done in response to a request by the occupant for an appointment. Fire departments may use public service announcements on the radio and television to increase public awareness of the need for a fire-safe home and the availability of a fire safety survey. In many communities, the fire department offers to check home smoke alarms and to conduct a home fire safety survey at the same time. Some fire departments install smoke alarms free of charge in a dwelling as a community service.

Conducting voluntary home fire safety surveys is an excellent way to focus on residential fire and life safety topics at an individual level with homeowners and occupants. During the survey, point out hazards, explain the reasons for making recommendations, and answer questions the occupant may have. Many fire departments provide educational materials that address

FIGURE 25-8 Always look like a professional when you interact with the public.

safety issues such as the installation and maintenance of smoke alarms, the role of residential fire sprinklers in preventing fire deaths, and the selection and use of portable fire extinguishers.

If individual residences are not surveyed, many fire departments will survey neighborhoods making note of addresses, access routes, and hydrant locations. Knowing which styles and types of private residences are present in your jurisdiction will help you prepare for emergency incidents.

Getting Started

To conduct a fire safety survey, you should present a neat, professional image (**FIGURE 25-8**). Bring plenty of handouts and items you can leave with the occupant that reinforce the concepts that you want to emphasize. When you arrive, identify yourself by name, title, and organization. Inform the occupant of the purpose of your visit and the format of the survey. Always remember that you are a guest in a private home and respect the privacy of the people who live there. You are there to help teach this person how to protect their family and property in the event of a fire.

During the survey, concentrate on the hazards that most often cause residential fires, such as cooking equipment, heating equipment, electrical wiring, electrical appliances, smoking materials, candles, and other open flames. Some hazards depend on the local climate. For example, wood-burning stoves and fireplaces are more common in colder climates. Some departments address additional safety issues during fire safety surveys such as swimming pool safety, slip-and-fall hazards, and the value of carbon monoxide alarms.

SAFETY TIP

According to the NFPA, cooking was the leading cause of home fires and home fire injuries, while smoking was the leading cause of home fire deaths (Hall 2023, 1).

Also, consider the occupants. In a residence with young children, flammable cleaners and matches and lighters should be stored high up where the children can't reach. A senior citizen or an occupant with a physical or mental disability might not be able to access an exit through a window if the door is blocked. It is a good idea to always ask the occupants if there are oxygen tanks or concentrators in the home.

During a survey, explain why different situations are considered potential fire hazards. Your goal should be to educate the occupant about common fire hazards. A person who understands why an overloaded electrical circuit is a fire hazard will be more likely to avoid overloading circuits in the future.

Conduct the survey in a systematic fashion for both the inside and the outside of the home. Some firefighters begin by inspecting the outside of the house, whereas others prefer to start with the interior. Always observe the SOPs and procedures of your department.

Identifying Outside Hazards

On the exterior of a residential occupancy, you can check for several hazards, including the following:

- Make sure that the house number or address is clearly visible from the street so emergency responders can identify the correct house quickly (**FIGURE 25-9**).
- Look for accumulated trash that could present a fire hazard or obstruct an exit.
- Note any combustible materials that are located close to the house. Remember to check the garage, especially an attached garage.
- Point out shrubs and vegetation near doors or windows that need to be trimmed or removed to allow clear access if a resident needs to exit through that opening during an emergency. Additionally, trees and shrubs should be cleared to allow large fire apparatus access to the residence.
- Recommend that fireplaces and chimneys be inspected annually by a qualified technician.

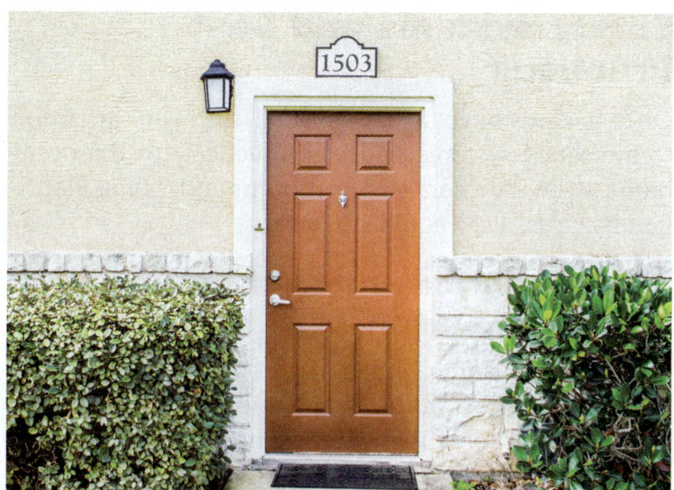

FIGURE 25-9 The home's address should be clearly visible from the street.

© LI Cook/Shutterstock

SAFETY TIP

While security bars on windows and double-cylinder door locks are great deterrents to burglars, they can prove deadly for occupants trying to escape from a house fire. Occupants should always be able to quickly release security bars and open locked doors from the inside.

Identifying Inside Hazards

During an interior inspection, systematically inspect each room for fire hazards. As you walk through the house, help the occupant identify alternate escape routes that can be used in case of fire. Look for primary and alternate exits from each room and discuss these points of egress with the occupant.

Bedrooms

Bedrooms typically have a lot of combustible materials such as bedding, curtains, clothing, books, and toys. If an ignition source comes into contact with these combustibles, a fire can ignite and quickly spread. Stress the importance of keeping ignition sources away from combustible materials. It is especially important to discuss the risk of children playing with lighters and matches. In addition, explain why it is important to sleep with bedroom doors closed to limit fire growth and spread, and to have an adequate number of smoke alarms positioned throughout the house.

Kitchens

From 2015 to 2019, almost half (49%) of reported home structure fires were caused by cooking (Ahrens and Maheshwari 2021, 5). These fires are often caused by leaving food cooking on the stove unattended, combustibles stored on or near the burners, and faulty electrical appliances. In the kitchen, look for signs of frayed wires on appliances, combustible towels or potholders stored too close to the stove, and cooking oils stored near the stove. Nothing that can be ignited, such as plastic food containers or paper packaging, should ever be placed on a cooking surface, even if the burners are turned off and the surface is cold. Many fires have resulted from someone inadvertently turning on the wrong burner and igniting something that was left on the stovetop.

Recommendations for kitchen safety include the following:

- Do not leave anything cooking unattended on a stove.
- Remind residents that if there is a grease fire, they should cover it to smother the fire and to never use water to try to put it out.
- Keep all combustible materials, cleaning supplies, cooking oils, and aerosols away from the stove.
- Do not place anything that could ignite on a cooking surface, even when it is turned off.
- Do not place towel racks near the stove.
- Do not overload electrical outlets or extension cords.
- Keep electrical cords properly maintained and replace damaged appliances.
- Keep the range exhaust hood clean and in good working order.

Living Rooms

The living room category includes formal living rooms, family rooms, entertainment rooms, and home offices. Common causes of living room fires include careless use of smoking materials, fires in fireplaces, overloaded electrical circuits, and portable heating appliances.

Look for indications of careless smoking, such as burn holes in upholstery. If the room contains a fireplace, a wood stove, or a portable heater, ensure that no combustible materials are stored nearby. A fireplace should have a screen to keep sparks and hot embers from escaping. If solid fuels are used, the chimney or flue pipe should be professionally inspected at least once each year. Stress the importance of placing ashes in metal containers with a tight-fitting lid and locating this container outside, away from all combustible materials.

Next, look for improperly installed home entertainment equipment, such as multiple devices plugged into a light-duty extension cord. Also look for overloaded electrical outlets and extension cords in home offices with computers, printers, scanners, and other devices. Stress the importance of using power strips that contain circuit breakers to prevent overloading of extension cords.

Garages, Basements, and Storage Areas

Explain the importance of good housekeeping and the need to clear junk out of garages, basements, and storage areas. These areas are often filled with large quantities of unneeded materials. Furnaces and water heaters that operate with an open flame are often located in a basement or garage. They can ignite flammable materials that are stored too close to them.

Storage of gasoline and other flammable liquids is a major concern because an open flame or pilot light can easily ignite flammable vapors. Gasoline and other flammable liquids should be stored only in approved containers and in outside storage areas or outbuildings. Propane tanks, such as those used in gas grills, should also be stored outside or in outbuildings. Small quantities of flammable and combustible liquids, such as paint, thinners, varnishes, and cleaning fluids, should be stored in closed metal containers away from heat sources. Oily or greasy rags should be stored in closed metal containers. Recommend that occupants maintain fully charged fire extinguishers in basements and garages.

Identifying Other Hazards

During a home fire safety survey, be on the lookout for potential non-fire safety hazards in the home. Poisoning, drowning, burns, falls, toppling furniture, and hoarding are all common causes of injuries in the home. Simple steps can prevent some of these emergencies including the following:

- Properly enclose pools.
- Properly secure furniture, televisions, and appliances to prevent tip-overs.
- Store medicines and chemicals where children cannot access them.
- Add handrails to stairs and install guards on balconies if there are none.
- Do not set the hot water temperature higher than 120°F (49°C).

Safety Systems and Best Practices

Once hazards have been identified, turn your attention to life safety systems and best practices for the occupants. Even if owners and occupants address all identified hazards, a fire could still occur.

Smoke Alarms and Carbon Monoxide Alarms

An important life-saving device is the smoke alarm. Smoke alarms fall in the engineering category of the 5 Es, but education about proper installation and maintenance is critical as well. Smoke alarms save lives, but only if they are working properly. Working smoke alarms alert residents to a fire and give them more time to escape from a burning building.

Examine the interior of the residence to see if smoke alarms are installed to meet the specifications in NFPA 72, *National Fire Alarm and Signaling Code.* There should be one smoke alarm on each level of the dwelling, including basements and habitable attics, one smoke alarm in the corridor or hallway outside each sleeping area, and one smoke alarm in each bedroom. Older residences, built before these requirements were established, may not have enough smoke alarms to provide early detection and notification. If the residence does not have the minimum recommended number and placement of smoke alarms, recommend that the occupants install additional smoke alarms to improve their protection. Interconnected smoke alarms offer the best protection. When smoke alarms are interconnected, when one alarm detects smoke and is activated, all of the connected smoke alarms activate as well. This notifies occupants in all areas of the home. Remind residents that it is best not to install smoke alarms close to kitchens, fireplaces, garages, windows, exterior doors, or heating or air-conditioning ducts, as placing smoke alarms in these locations may result in false alarms or delayed activation. Detailed information about smoke alarms and how they work is in Chapter 24, *Systems of Fire Detection, Suppression, and Smoke Control.*

Alarms should be tested once per month by pressing the test button. Stress the importance of keeping smoke alarms in working order. Some residential smoke alarms are operated by an alkaline or a lithium battery. Others are hard-wired to the electrical system, and some of these have a backup battery. Owners and occupants should identify the types of batteries in the smoke alarms in the residence.

Smoke alarms powered by alkaline batteries should have their batteries changed annually. Many communities participate in the Change Your Clock, Change Your Batteries campaign, which encourages residents to replace the alkaline batteries in their smoke alarms when Daylight Saving Time begins. This campaign used to recommend changing smoke alarm batteries twice a year—both when Daylight Saving Time begins and when it ends. This is not necessary! The current recommendation is to change alkaline batteries once every 12 months. Also, because many areas are considering doing away with the practice of switching to Daylight Savings Time and because many clocks update their time display automatically, it's time to consider a change in the traditional educational recommendations for changing smoke alarm batteries. For example, you could encourage your residents to change alkaline smoke alarms batteries when they change their calendars at the beginning of the year.

Many new smoke alarms contain an extended-life, 10-year battery. These batteries do not need to be changed. NFPA 72 requires that this type of smoke alarm be replaced if it fails the monthly test.

NFPA 72 requires that all smoke alarms be replaced 10 years after the manufacture date even if they pass the monthly test. This is because the sensor in the alarm has a 10-year life span. Even if a smoke alarm passes the monthly test, the alarm may not function correctly after 10 years.

In addition to checking monthly to see if smoke alarms are functional and replacing alkaline batteries annually, make sure owners and occupants understand the following maintenance requirements to ensure that smoke alarms remain operational:

- Circuit breakers that supply current to hard-wired smoke alarms must always be kept on.
- If hard-wired smoke alarms contain backup batteries that require regular replacement, add these batteries to the list when changing batteries in other smoke alarms annually.
- Vacuum the chamber inside the smoke alarms regularly to prevent the build-up of dust or insects, which can cause an alarm to sound when there is no fire.

Carbon monoxide (CO) alarms are required in most states and should be discussed alongside smoke alarm messaging. CO is a colorless, odorless, gas that is a biproduct of incomplete combustion. Common sources in the home include water heaters, natural gas stoves and dryers, fireplaces (both gas and traditional wood-fired), and vehicles and other small engines in the garage area. Because CO is colorless and odorless, there is no natural warning of CO poisoning until occupants experience symptoms. According to the Centers for Disease Control and Prevention (CDC), "the most common symptoms of CO poisoning are headache, dizziness, weakness, upset stomach, vomiting, chest pain, and confusion. CO symptoms are often describe as 'flu-like'" (CDC 2023).

CO alarms are available as stand-alone alarms or can be combined with a smoke alarm in a combination system. If a home you are inspecting contains this type of alarm, it is important to explain the function of this alarm to the occupants. Homeowners should always replace their alarms based on the manufacturer's recommendations, which is usually listed on the back of the alarms. It is important to note that most CO alarms need to be replaced more frequently than smoke alarms, usually around 7 years after the date of manufacture.

Teaching these simple facts about the installation and maintenance of smoke alarms helps families keep their homes safe in the event of a fire. To educate homeowners or groups on the installation and maintenance of smoke alarms, follow the steps listed in **SKILL DRILL 25-4**.

Exit Plans, Exit Drills, and Meeting Places

After ensuring a residence has working smoke alarms, the next step is to discuss with the occupants what to do when they go off. First, stress the importance of keeping bedroom doors closed during sleeping hours, as this step prevents smoke, heat, and flames from reaching bedrooms (Close before You Doze). Then stress the importance of alerting other occupants to the presence of a fire. Next, remind residents of the steps in EDITH. Help them identify two escape routes from each room so that if smoke, heat, or flames are present in the primary escape route, they have another way out. The secondary escape route may be through a window or through a different door. You should test each door and window to make sure that it is operable and accessible. Explain the benefits of closing doors as they exit the rooms and the residence itself. This action may help to smother the fire and limit damage.

All occupants, when alerted to the presence of a fire, should roll out of bed, stay low, and crawl toward the designated primary exit. If that exit is impassable, they should crawl to the secondary exit. Once outside the house, the residents should go to the predetermined meeting spot. At this point, they should make sure that

SKILL DRILL 25-4

Installing and Maintaining Smoke Alarms Firefighter II, NFPA 1010: 7.5.1, 7.5.2

1. Explain the importance of properly installing and maintaining smoke alarms as recommended by NFPA 72.

2. Stress the importance of testing smoke alarms once per month using the test button.

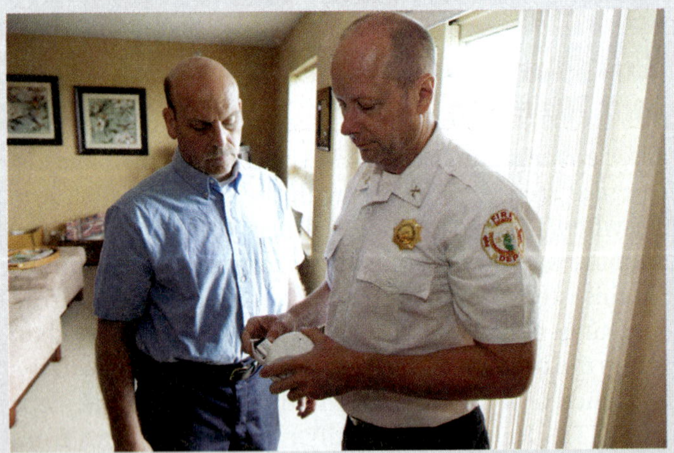

3. Explain that alkaline batteries in smoke alarms should be changed once per year and all smoke alarms should be replaced 10 years after the date of manufacture. Combination smoke/CO alarms should be replaced according to the manufacturer's recommendation.

4. Emphasize the need to clean smoke alarms regularly to prevent an alarm sounding when there is no fire.

the fire department has been called or make the call themselves. Remind residents that no one should go back into the house for any reason. Remind them to practice EDITH regularly. Some drills should be held in the daytime, and some should take place at night.

Portable Fire Extinguishers

A third important topic to discuss during a home safety survey is the selection, placement, and use of portable fire extinguishers. Many adults will have questions about which types of fire extinguishers they should have in their home, kitchen, garage, car, or boat (**FIGURE 25-10**). You should be familiar with the information you learned in Chapter 9, *Portable Fire Extinguishers,* so you can answer any questions that arise.

Provide recommendations for where to place fire extinguishers as follows:

- Kitchen: Cooking is one of the leading causes of home fires, so residents should have an extinguisher in the kitchen to slow or stop the spread of a kitchen fire. Make sure the extinguishers are placed where would be accessible during an emergency. They should not be stored under the sink or in the back of a cupboard.

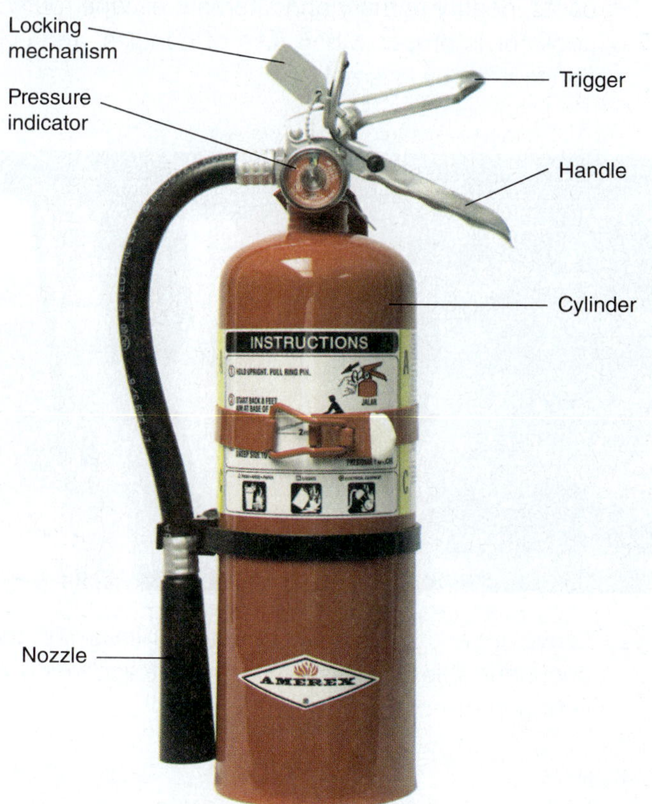

FIGURE 25-10 Encourage homeowners to purchase an approved ABC-rated fire extinguisher.

Courtesy of Amerex Corporation.

- Garage: There are a variety of potential hazards present in most garages, including small engines, fuel, tools that spark, and so on. Again, make sure the extinguisher is placed in a location that would be accessible during an emergency.
- Utility room: Having an extinguisher located near the dryer, heater, or other common appliances could help slow or stop the spread of a mechanical fire.

When discussing portable fire extinguishers with the public, emphasize that they are only recommended for adult use on small fires. Children should concentrate on knowing the escape routes and getting out immediately.

Residential Automatic Sprinkler Systems

A fourth residential home safety topic concerns the importance of residential fire sprinkler systems in preventing residential fire deaths. Statistics show that the combination of working smoke alarms and residential fire sprinklers can reduce the risk of fire death by approximately 82% (USFA 2021). With these statistics in mind, there is a national movement in the fire service to have fire sprinklers installed in all new residential construction. Make sure that any fire alarm or suppression equipment in the home is properly installed, maintained, and fully operational. Reference your department's policies before making recommendations on fire protection equipment. Most jurisdictions do not allow recommending a specific manufacturer's product. For more information about automatic sprinkler systems, see Chapter 24, *Systems of Fire Detection, Suppression, and Smoke Control.*

Closing a Fire Safety Survey

During the fire safety survey, listen carefully to any questions from the occupants. These questions will enable you to address any special concerns or correct inaccurate information. Bring brochures and a list of resources to leave with the occupants.

After you have completed a fire safety survey or any other type of inspection, you must file a report according to your department's policies. If you identified hazards that require further action or follow-up, discuss them with your company officer or a designated representative from the code enforcement office.

To conduct a fire safety survey, follow the steps in **SKILL DRILL 25-5**.

SKILL DRILL 25-5

Conducting a Fire Safety Survey Firefighter II, NFPA 1010: 7.5.1

1. Check for a visible house number. Look outside the house for accumulated trash, overgrown shrubs, and blocked exits. Explain the purpose of the fire safety survey to the occupant.

2. Check inside the house for properly working smoke alarms, fire extinguishers, and fire sprinkler systems. Stress the importance of proper cooking procedures and the safe storage of cooking oils and flammable objects. Explain the safe use of fireplaces, heating stoves, and portable heaters. Help building occupants identify primary and alternate escape routes. Look for improper wiring, use of candles, and evidence of careless smoking.

3. Complete the inspection form and review the results of the fire safety survey with the building occupants. Describe the steps that need to be taken to minimize the identified hazards. Give a copy of the inspection form to the occupants.

4. Leave fire and life safety brochures or links with the occupant. File or upload your report according to your department's policies.

Preincident Plans for Commercial Properties and Other Public Locations

A **preincident plan** is a set of data about each location in your community that describes the layout and hazards at the location and identifies response essentials, such as the type of construction, occupancy type, and potential issues with access and egress. Preincident planning is closely related to other CRR activities. It provides responding crews with vital information so that fireground operations are safer and more effective. The objective is to make valuable information available quickly during an emergency incident that otherwise would not be readily evident or easily determined. A good preincident plan helps the incident commander (IC) make better command decisions and allows firefighters to be made aware of hidden dangers and prepare for them.

TIP

With fire prevention, the goal is to identify hazards and minimize or correct them so that fires do not occur. With preincident planning, the person creating the plan assumes that a fire will occur and compiles information that responding firefighters would need in order to reduce the risk to everyone involved.

Preincident planning is typically done for commercial properties that are particularly large or that present unusual risks to firefighters or to the public. This type of property is called a **target hazard**. Examples of target hazards include the following:

- High-rise buildings
- Hospitals
- Nursing homes and assisted-living facilities
- Large apartment buildings
- Hotels and rooming houses
- Schools
- Public-assembly occupancies
- Lumberyards
- Manufacturing plants
- Shopping centers
- Warehouses
- Airports
- Rail yards
- Transportation centers

Most fire departments are not able to create a preincident plan for every individual property in their jurisdiction. A preincident plan should be prepared for every property that poses a high life safety hazard to its occupants or presents safety risks for responding firefighters. Preincident planning should also occur for properties that have the potential to create a large fire or conflagration (a large fire involving multiple structures).

With a preincident plan, you will know hydrant locations, entry and egress points, which hazards to anticipate, hours of occupation, and more. A preincident plan puts all of this information at your fingertips, either on paper, a tablet, or on a mobile data terminal (MDT) in the apparatus for use en route to the fire scene (**FIGURE 25-11**). Preincident planning is performed under the direction of a fire officer.

FIGURE 25-11 An example of a preincident plan on a mobile data terminal.

FIGURE 25-12 Preincident plan information is supplied to the IC at the emergency scene to help the IC make informed decisions when an emergency incident occurs at the location.

The information that goes into a preincident plan is gathered during a **preincident survey**. This survey is usually completed by one of the crews that would respond to an emergency incident at the location. A preincident survey can often uncover issues that indicate the need for a fire inspection. For this reason, many departments conduct preincident surveys simultaneously with fire inspections. If this happens, different members of the department should be assigned to conduct the inspection and to complete the preincident survey. This separation of duties allows each firefighter to concentrate on the objectives of one specific assignment. Each department has its own SOPs for conducting preincident surveys. Make sure you understand the policies in your department.

At the emergency scene, the IC uses the information gathered prior to the incident to direct the emergency operations much more effectively (**FIGURE 25-12**). The completed preincident plan can also be used in training activities to help firefighters become familiar with the properties within their jurisdiction. The information should be made available to all units that might respond to an incident at that location. Consider inviting neighboring fire departments to train with you and become familiar with the buildings in your jurisdiction.

The amount and the nature of the information provided for a property depends on the size and complexity of the property, the types of risks present, and the hazards or challenges likely to be encountered. A preincident plan usually includes the following:

- One or more diagrams that show details, including:
  - Building location
  - Building layout with the location and nature of any special hazards to the public or to firefighters highlighted
  - Height and overall dimensions of the building
  - Types of contents in different areas of the building
- Type of construction
- Occupancy type
- Access routes
- Entry points
- Floor plans, including stairway and elevator locations
- Lockbox locations
- Utility shut-off locations
- Exposures
- Information about built-in fire detection and suppression systems
- Fire department connections (FDCs)
- Hydrant locations or alternative water supplies

All of the information must be presented in a uniform and understandable format. Most departments have a form or template to guide the team conducting the preincident survey. For more information, see NFPA 1660, *Standard for Emergency, Continuity, and Crisis Management: Preparedness, Response, and Recovery.*

As with a home fire safety survey, the preincident survey of a building should be conducted with the knowledge and cooperation of the property owner or occupant. Making this initial contact enables the fire department to explain the purpose and the importance of the preincident survey, to clarify that the information is needed to prepare firefighters in the event that an emergency occurs at the location, and to schedule an acceptable time.

The team members who conduct the survey should dress and conduct themselves in a manner appropriate to the department's mission. A representative of the property should accompany the survey team to answer questions and provide access to different areas. Every effort should be made to obtain complete, accurate, and useful information.

Much like the home safety survey, the preincident survey is conducted in a systematic fashion, following a uniform format. Begin with the outside of the building, gathering all of the necessary information about the building's geographic location, external features, and access points. Next, survey the inside of the building to collect information about each interior area. A good, systematic approach starts at the roof and works down through the building, examining every level of the structure, including the basement. If the property is large and complicated, it may be necessary to make more than one visit to ensure that all of the required information is obtained and recorded accurately.

The firefighters conducting the survey should prepare sketches or drawings to show the building layout and the location of important features such as exits. In some cases, the building owner can provide the survey team with a copy of a plot plan or a floor plan. It takes practice and experience to learn how to sketch the required information while doing the survey and then to convert the information to a final drawing. It is essential that the sketches accurately show the scale and location of important features at the time of the inspection so the incident commander (IC) can make the most informed decision during an emergency response. Some departments use computer software to create and store these diagrams. That software can create graphics more quickly and accurately than is possible with hand drawings. Drawings and other graphics are usually prepared at the fire station, where the fire officer takes the time to organize the information properly and to generate reports and forms that can be used to develop the preincident plan. Completed drawings should use standard, easily understood symbols as shown in NFPA 170, *Standard for Fire Safety and Emergency Symbols* (**FIGURE 25-13**).

All information in preincident plans is confidential. The general public is not at liberty to see the information collected in your fire department's preincident plans. Keep all information that you learn about an occupancy during a preincident plan survey confidential.

The same set of basic information must be collected for each property that is surveyed. Most fire departments use standard forms to record the survey information. To conduct a preincident survey, follow the steps in **SKILL DRILL 25-6**. Additional information should be gathered for properties that are unusually large, are complicated, or might be the site of a particularly hazardous situation.

Preincident Planning for Response and Access

The first phase of an emergency incident is the response phase. Building layout and access information is particularly important during the response phase of an emergency incident. The preincident plan should provide this information. For instance, a preincident plan might identify the most efficient route for the apparatus to take to the building, as well as identifying an alternate route if the time of day and local traffic patterns might affect the primary route. An alternate route should also be identified if the primary route requires crossing railroad tracks, drawbridges, or other routes that could potentially be blocked.

Access to the Exterior of the Building

When conducting a preincident survey, firefighters should ensure that the building address is clearly visible to save time in locating the property during an emergency incident. If the building is part of a complex, the best route to each building, section, or apartment should be indicated on a map. The preincident plan should address all possible issues related to access to the exterior of the building. Points of access into the building must be noted, particularly if the building has multiple entrances or if access points are not easily seen from the street. In some cases, different entrances should be used, depending on the location of the emergency within the building. In other cases, the first-due unit should always respond to a designated entrance where the fire alarm annunciator panel is located or to meet a security guard who can act as a guide.

Diagrams should clearly indicate points at which gates, fences, or other barriers block access to parts of a building. The preincident survey should also note whether the topography makes one or more sides of a building inaccessible to fire apparatus or if an underground parking area will not support the weight of fire apparatus.

Ask yourself the following questions as you evaluate access to the exterior of the building during your survey:

- Do several roads lead to the building or just a few?
- Where are the hydrants located?
- Where are the FDCs for automatic sprinkler and standpipe systems located?
- Do any security barriers limit access to the site?
- Are there fire lanes to provide access to specific areas?
- Are any barricades, gates, or other obstructions so narrow or low that they would prevent passage of fire apparatus?
- Are there bridges or underground structures that will not support the weight of fire apparatus?
- Do any gates require keys or a code to gain entry?

Preincident Plan Symbols

Emergency Exit

Emergency Exit Use of Arrows

Manual Station—Pull Station/ Fire Alarm Box

Area of Refuge

Automated External Defibrillator (AED)

Fire Extinguisher

Fire Hose or Standpipe

Ordinary Combustibles

Flammable Liquids

Electrical Equipment

Combustible Metals

Fire Department Automatic Sprinkler Connection—Siamese

Fire Department Automatic Sprinkler Connection—Single

Fire Department Standpipe Connection

Fire Department Combined Automatic Sprinkler/Standpipe Connection

Fire Hydrant (All Types)

Automatic Sprinkler Control Valve

Electric Panel or Electric Shutoff

Gas Shutoff Valve

Fire-Fighting Hose or Standpipe Outlet

Fire Extinguisher

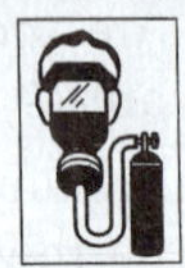

Self-Contained Breathing Apparatus (SCBA)

Opening in wall

Rated fire door in wall (less than 3 hours)

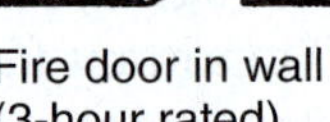

Fire door in wall (3-hour rated)

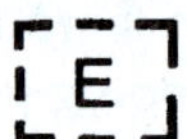

Elevator in combustible shaft

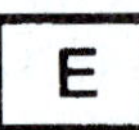

Elevator in noncombustible shaft

Open hoistway

Escalator

Stairs in combustible shaft

Stairs in fire-rated shaft

Stairs in open shaft

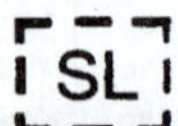

Skylight

FIGURE 25-13 Examples of preincident plan symbols listed in NFPA 170.

SKILL DRILL 25-6

Conducting a Preincident Survey Firefighter II, NFPA 1010: 7.5.3

1. Schedule the preincident survey in advance.
2. Make contact with a responsible person.
3. Present a neat and professional image.
4. Identify yourself by name, title, and department.
5. Ensure that a representative from the facility accompanies you during the survey.
6. Take notes and pictures as needed; start outside as if responding, then move inside.
7. Note the building location.
8. Note the size of the building (height, number of stories, length, width).
9. Identify the building construction.
10. Identify the building use and occupancy (occupancy type and time occupied).
11. Note any life hazards.
12. Note the access points to the interior of the building and the location of the lockbox containing the keys to the building, if applicable.
13. Note the location of the fire alarm annunciator panel.
14. Note the utility shut-off locations.
15. Assess the apparatus access locations to the building.
16. Note fire hydrant locations, FDCs, and alternative water supplies.
17. Note ventilation concerns.
18. Record information about built-in fire detection, suppression, and alarm systems.
19. Sketch floor plans and take photos.
20. Note the elevator and stairway locations.
21. Review exit plans and exit locations.
22. Identify any special hazards and hazardous materials.
23. Note the building exposures.
24. Anticipate the type of incident expected.
25. Identify any special resources needed.
26. Complete and file the department's preincident survey form.

- Will it be necessary to cut fences to gain access to the site?
- Does the site include natural barriers such as streams, lakes, or rivers that might limit access?
- Does the topography limit access to any parts of the building?
- Might the landscaping or the presence of snow prevent access to certain parts of the building?

Access to the Interior of the Building

As part of preincident planning for response and access, during the survey, ask yourself the following questions about access to the interior of the building:

- Is there a lockbox containing keys to the building? Where is the lockbox located? Is it clearly visible? Do the keys in the lockbox work?
- Are key codes needed to gain access to the building? Who has them? Are they in the lockbox?
- Does the building have security guards? Is a guard always on duty? Does the guard have access to all areas of the building?
- Is a key holder available to respond to the alarm within a reasonable amount of time? How can this person be reached?
- Where is the fire alarm annunciator panel located?
- Is the fire alarm annunciator panel properly programmed so you can quickly determine the exact location of an alarm?

Preincident Planning for Scene Size-Up

The second phase of an emergency incident is the size-up. The preincident survey must obtain information about the building to assist with the size-up. This information includes the building's construction type, height, area, use, and occupancy, as well as the presence of hazardous materials or other risk factors. The locations of other buildings or structures that might be jeopardized by a fire in the building should be recorded as well.

Knowing about the fire protection system and water supply in a building is also important for size-up. The preincident survey should identify areas that are protected by automatic sprinklers or other types of fire suppression systems, the locations of standpipes, and the locations of firewalls and other features designed to limit the spread of a fire. Likewise, the survey should note any areas where fire protection is lacking, such as an area without sprinklers in a building that otherwise has a sprinkler system.

Features inside a building that would allow a fire to spread but that are not readily visible from the outside should be noted as well. A common attic space over several occupancies, unprotected openings between floors, or buildings connected by overhead passages or conveyor systems are examples of these features.

Access and egress features, including stairways, exit doors, and elevators, should be diagrammed and labeled. These features are integral in order to identify the easiest access to the upper floors for rescue and fire control activities.

Construction

Building construction type is an important factor to identify in the preincident survey. Although they might have similar appearances, two buildings might be different construction types and, therefore, would react differently during a fire. Preincident planning enables you to take the time that is needed to determine the exact type of building construction—Type I, II, III, IV, or V—and identify any problem areas. (Building construction is discussed in detail in Chapter 6, *Building Construction.*) If an emergency occurs, the IC can check the preincident plan for information on construction, instead of hazarding a guess.

Building Use and Occupancy

The second major consideration during size-up—and hence during a preincident survey—is the building's use and occupancy. In each use and occupancy, different types of problems, concerns, and hazards may be present.

Knowing the building's use and occupancy can help firefighters determine the number of occupants and predict their ability to escape if a fire occurs. It is also a key factor in determining the probable contents of the building. These factors can have a major effect on the problems and hazards encountered by firefighters responding to an emergency incident at the location.

Many large building complexes contain multiple uses and occupancies under one roof. For example, a shopping mall usually includes both retail stores and restaurants. The mall might also be part of an even larger complex that includes offices, residential condominiums, a hotel, a movie theater, an underground parking garage, and a subway station. Each use and occupancy has very different characteristics and, therefore, different risk factors.

TIP

A building can have multiple uses during its lifetime or can incorporate more than one use within the same structure.

The use and occupancy of a building may change over time. An outdated factory may be transformed into a residential building, or an unused school may be converted into an office building. A warehouse that once stored concrete blocks may now be filled with combustible foam-plastic insulation or hazardous materials such as swimming pool chemicals. Given this propensity for change, preincident plans should be checked and updated on a regular basis.

Exposures

A preincident survey should identify any potential exposures for the property that is being evaluated. In addition to the size and construction type, you should evaluate the fire load of the property being surveyed, the distance to the exposures, and the ease of ignition of the exposures by radiation, convection, or conduction of heat.

Built-In Fire Protection Systems

The preincident survey should identify built-in fire detection and suppression systems on the property. These systems may include automatic sprinklers, standpipes, fire alarms, and fire detection systems, as well as systems designed to control or extinguish special hazard fires. Some buildings have automatic smoke control or exhaust systems. Most high-rise buildings have systems that control elevator function during an emergency. These systems are covered in detail in Chapter 24, *Systems of Fire Detection, Suppression, and Smoke Control.*

Information to note includes the following:

- Type and location of automatic sprinkler heads and relevant information in the maintenance records of the sprinkler system
- Location of FDCs and standpipes and whether the standpipes are in all or only certain stairwells or are located in corridors
- Type of fire alarm and fire detection systems, the location of the fire alarm control panel and the remote annunciator, and information about how and where the system is monitored
- Types and locations of special hazards; fire suppression systems, such as a wet- or dry-chemical fire suppression system; details about the method of operation; and so on, should be recorded.

Water Supply

During the preincident survey, use your department's approved method to calculate the water flow required to fight a fire involving the entire building. and identify the water supply source. It is important to understand the correlation between how much water is needed to fight a fire and the amount of water and pressure available in the municipal system. If the water is supplied from municipal hydrants, identify the location of the hydrants closest to the building and the number of feet (or meters) of hose needed to deliver water to the fire. For large buildings, locate enough hydrants to supply the volume of water required to control a fire. Determine whether the hydrants provide the required flow. In some cases, it may be necessary to use hydrants that are supplied by different water mains of sufficient size to achieve the needed water flow (**FIGURE 25-14**). If this is the case, identify the hydrants on the different water mains.

If the water needs to be obtained from a static source such as a lake or a dry hydrant, identify drafting sites in the preincident plan. You need to note whether the water supply is available year-round or only seasonally. It is also important to measure the distance from the water source to the building to determine whether large-diameter hose and additional engines will be needed for relay pumping. The preincident plan should also outline the operation that would be required to deliver water to the fire.

If mobile water supply apparatus will be used to deliver water, the preincident plan must identify the sites for filling the mobile water supply apparatus and for discharging their loads (**FIGURE 25-15**). The preincident plan should also identify how many mobile water supply apparatus will be needed based on the total distance they must travel, the quantity of water each vehicle can transport, and the time it takes to empty and refill the tank.

A large industrial or commercial complex may have its own water supply system that provides water for automatic sprinkler systems, standpipes, and private hydrants. These systems usually include storage tanks or reservoirs as well as fixed fire pumps to deliver the water under pressure.

The details and arrangement of the private water supply system must be determined during the preincident survey. The survey should indicate how much water is stored on the property and where the tanks are located. If a fixed fire pump is part of the sprinkler system, its location, capacity in gallons per minute (or liters per minute), and power source (electricity or a diesel engine) should be noted. It is also important to confirm that these private water systems and fixed fire pumps are being maintained in good operating condition.

In many cases, the same private water main provides water for both the sprinkler system and the

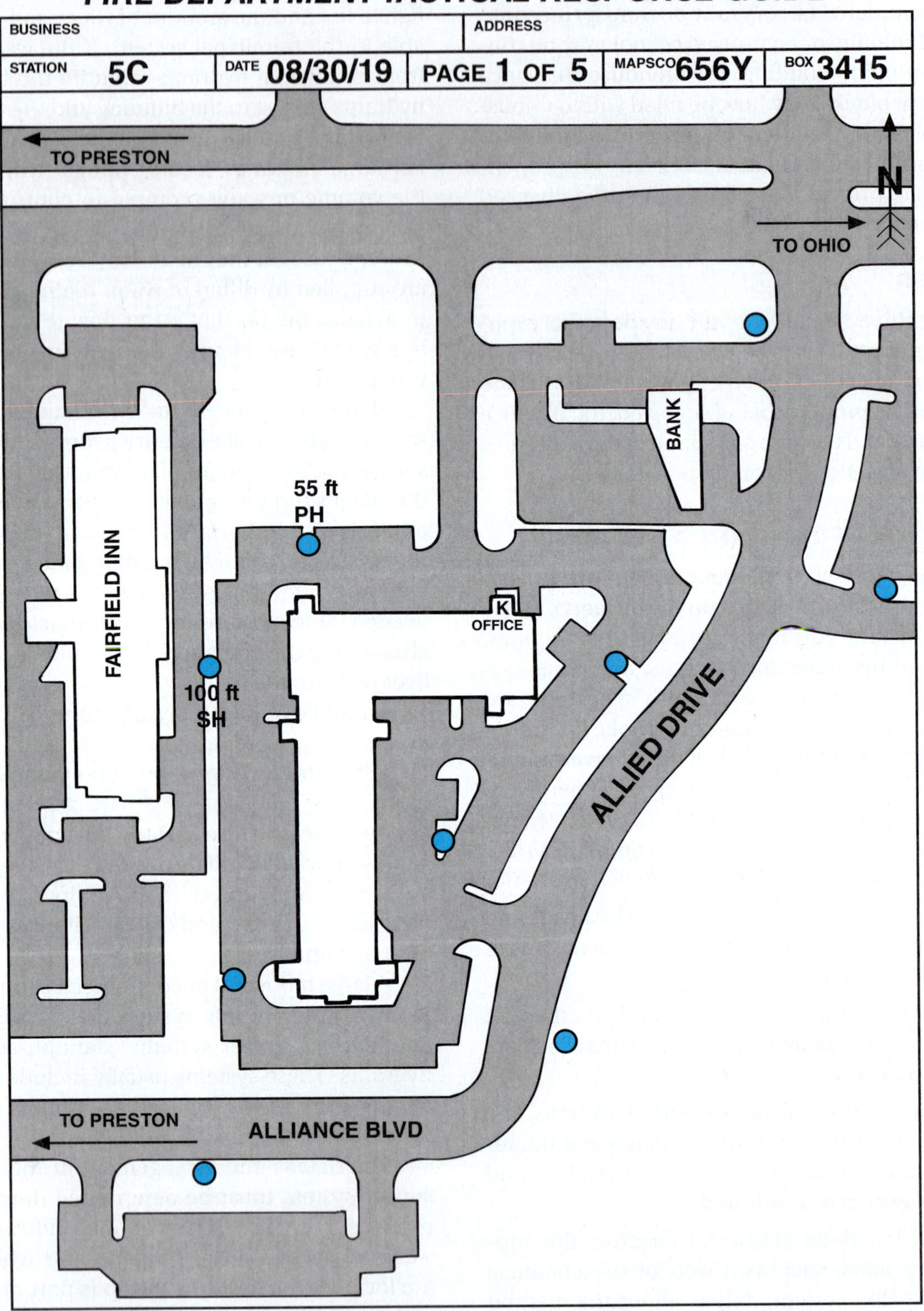

FIGURE 25-14 A diagram showing hydrant locations around a building. *PH* indicates the primary hydrant for connecting to the FDC. *55 ft* indicates that the primary hydrant is located 55 ft (17 m) from the FDC. *SH* indicates the secondary hydrant for connecting to the FDC. *100 ft* indicates that the secondary hydrant is located 100 ft (30 m) from the FDC.

FIGURE 25-15 The points where mobile water supply apparatus can be filled and where they can discharge their loads must be identified in the preincident plan.

© Jim Smalley

FIGURE 25-16 Considerations for the use of ladders should include identifying the best locations to place ground ladders and use aerial apparatus.

Courtesy of David Traiforos.

private hydrants. If the fire department uses the private hydrants, the system may not be able to simultaneously supply both automatic sprinkler systems and fire hose effectively. The preincident plan for these sites should note whether public, off-site hydrants should be used instead of the private hydrants.

Utilities

During an emergency incident, it may be necessary to turn off utilities such as electricity or natural gas as a safety measure. The preincident survey should note the locations where utilities can be shut off including electricity, natural gas, propane gas, fuel oil, other energy sources, and water. Contact information for the appropriate utility company should also be noted. In some cases, special knowledge or equipment may be required to disconnect utilities. This information must be noted in the preincident plan, along with the procedure for contacting the appropriate individual or organization.

Electricity may be supplied by overhead wires or through underground cables. Overhead wires can be deadly if a ground ladder or aerial ladder comes in contact with them. The preincident plan should show the locations of high-voltage electrical lines and equipment that could prove dangerous to firefighters. If electricity is supplied by underground cables, the location where the electricity is shut off may be located inside the basement of a building or in an underground vault. These locations should be noted during the survey.

If propane gas or fuel oil is stored on the property, the preincident plan should show the location and note the capacity of each tank. The presence of an emergency generator should be noted as well, along with the fuel source and a list of equipment powered by the generator.

Preincident Planning for Emergency Operations

The third phase of an emergency incident is the operations at the incident. The purpose of the preincident plan is to be as informed as possible in the event of an emergency incident at the location described in the plan. Different aspects of emergency operations can benefit from the information on the preplan:

- Forcible entry: When conducting the preincident survey, examine both exterior and interior access and try to anticipate where firefighters might encounter problems. Identify locations where forcible entry may be required and add this information to the site diagrams and building floor plans. Also note which tools would be needed to gain entry. This can save time during the actual emergency.
- Search and rescue: Firefighters who are conducting search and rescue operations need to know the locations in the building where occupants are likely to be as well as the locations of exits.
- Ladder placement: The preincident survey is an excellent opportunity to identify the best locations for placing ground ladders or using aerial apparatus (**FIGURE 25-16**). The length of ladder needed to reach a roof or entry point should be noted. Pay careful attention to any electrical wires and other obstructions that might not be visible at night or in a smoky atmosphere.

- Ventilation: It is important to consider ventilation during the preincident survey. What would be the best means to provide ventilation? How useful are the existing openings for ventilation? Are there windows and doors that would be suitable for horizontal ventilation? Where could fans be placed? What is the best way to reach the roof? Are there ventilators or skylights that could be easily removed or bulkhead doors that could be opened easily? Would multiple ceilings need to be punctured to allow smoke and heat to escape?
- Smoke control: Determine whether the heating, ventilation, and air-conditioning (HVAC) system can be used to remove smoke without circulating it throughout the building. Many buildings with sealed windows have controls that enable the fire department to set the HVAC system to deliver outside air to some areas and exhaust smoke from other areas. Instructions for controlling the HVAC system should be included in the preincident plan.

Preincident Planning for Locations with Hazardous Materials

One of the most important reasons for developing a preincident plan is to identify any special hazards at specific locations and to provide information that would be valuable during an emergency incident. This information includes chemicals or hazardous materials that are stored or used on the premises, structural conditions that could result in a building collapse, industrial processes that pose special hazards, high-voltage electrical equipment, and confined spaces. Preincident plans warn firefighters of potentially dangerous situations and could include detailed instructions about what to do in those circumstances—information that could ultimately save the lives of firefighters.

Firefighters who are conducting a preincident survey should always look for special hazards and obtain as much information as possible at that time. If the information is not readily available, it may be necessary to contact specialists for advice or conduct further research before completing the preincident plan.

When conducting a preincident survey, you should obtain a complete description of all hazardous materials stored, used, or produced at the property. This assessment includes an inventory of the types and quantities of hazardous materials on the premises. It should also include information about where they are located, how they are used, and how they are stored. The appropriate actions and precautions for firefighters in the event of a spill, leak, fire, or other emergency incident should also be listed.

Many jurisdictions require businesses that store or use hazardous materials to obtain a fire department permit. They may also require that hazardous materials specialists conduct special inspections. If the quantity of hazardous materials on hand exceeds a specified limit, federal and state regulations require a property owner to provide the local fire department with current inventories and **safety data sheets (SDSs)** (**FIGURE 25-17**). Firefighters conducting preincident surveys should learn where the SDSs are stored, ensure that the required information has been provided, and verify that it is up-to-date.

Firefighters should expect to encounter hazardous materials at certain types of occupancies, such as chemical companies, garden centers, swimming pool supply stores, hardware stores, and laboratories. Of course, hazardous materials can also be found in unexpected locations, so firefighters should always be alert for their presence. Some communities require placards on the outside of any building that contains hazardous materials.

In case of fire, some hazardous materials require use of special suppression techniques or extinguishing agents. This information should be noted in the preincident plan. The plan should also contain contact information about individuals or organizations that can provide advice if an incident occurs.

Preincident Planning for Locations Requiring Special Consideration

Preincident planning should extend beyond planning for fires and other types of emergency incidents that could occur in buildings. Specifically, a preincident plan should anticipate the types of incidents that could occur at locations such as the following:

- Airports
- Bridges
- Construction sites
- Electrical transmission lines (**FIGURE 25-18**)
- Gas or liquid fuel transmission pipelines
- Railroad lines (**FIGURE 25-19**)
- Ships and waterways (**FIGURE 25-20**)
- Subways (**FIGURE 25-21**)
- Tunnels (**FIGURE 25-22**)

For example, detailed preincident plans prepared for airports should note the various types of aircraft

SAMPLE
SAFETY DATA SHEET

Issuing Date January 5, 2015 **Revision Date** June 12, 2015 **Revision Number** 1

1. IDENTIFICATION OF THE SUBSTANCE/PREPARATION AND OF THE COMPANY/UNDERTAKING

Product identifier

Product Name	**XXXXX Regular-Bleach[1]**

Other means of identification

EPA Registration Number	5813-100

Recommended use of the chemical and restrictions on use

Recommended use	Household disinfecting, sanitizing, and laundry bleach
Uses advised against	No information available

Details of the supplier of the safety data sheet

Supplier Address
The XXXXX Company
1221 Broadway
Oakland, CA 94612

Phone: 1-510-XXX-XXXX

Emergency telephone number

Emergency Phone Numbers	For Medical Emergencies, call: 1-800-446-1014 For Transportation Emergencies, call Chemtrec: 1-800-424-9300

Page 1 / 10

FIGURE 25-17 Safety data sheets (SDSs) were formerly referred to as material safety data sheets (MSDSs).

Courtesy of the Occupational Safety & Health Administration.

FIGURE 25-18 Electrical transmission lines.

© Jones & Bartlett Learning

FIGURE 25-19 Railroad.

© Stan Tess/Alamy Stock Photo

FIGURE 25-20 Ships and waterways.

© GeoStock/Photodisc/Getty Images

FIGURE 25-21 Subway station.

© Arthur S. Aubry/Photodisc/Getty Images

FIGURE 25-22 Tunnel.

© gyn9038/iStock/Getty Images Plus/Getty Images

that use the airport. In addition, the airport facilities also require extensive preincident planning because of their size, the number of people who may be present, and security issues. Similar planning should be performed for bridges, tunnels, and any other locations where complicated situations could occur.

Occupancy Considerations in a Preincident Plan

Each type of occupancy has unique features that should be taken into account when preparing a preincident plan and when conducting a survey.

High-Rise Buildings

High-rise buildings present special problems during an emergency incident. It can be difficult to deliver adequate water pressure to the upper floors. Also, high-rise buildings have a large number of occupants. As a consequence, major fires in high-rise buildings often result in injuries, fatalities, and millions of dollars in property losses.

A preincident survey of a high-rise building should identify both the building construction and any special features that have been installed. For example, some high-rise buildings have firefighter air-replenishing systems (FARS) on different floors so that firefighters don't need to lug multiple bottles of air up the stairs. Standpipe systems might be hidden mid-floor because it is too far to the stairwell. You should note all of the systems that are present in the building and describe how they function.

The fire protection features of a high-rise building vary depending on the age of the building and the building and fire code requirements in the jurisdiction. Older high-rise buildings are generally constructed of noncombustible materials and designed to confine a fire within a limited area with fire walls, self-closing doors, etc. Newer buildings generally include fire alarm systems, automatic sprinkler and smoke detection systems, emergency generators, elevator control systems, smoke control systems, and building control stations where all of these systems are monitored and controlled. The details of each system must be documented in the preincident plan.

Assembly Occupancies

Assembly occupancies, such as theaters, nightclubs, stadiums, churches, hotel meeting rooms, and arenas, have frequently been the location of major loss of life. These structures are often very large and have complicated designs. They may be equipped with complex systems to manage emergency situations. Gaining access to the location of the fire or emergency situation may prove difficult, however, when all of the occupants are trying to evacuate at the same time.

SAFETY TIP

Modern high-rise buildings are often equipped with multiple fire protection systems—but these systems must be maintained properly to be effective. One Meridian Plaza (a modern high-rise office building in Philadelphia, Pennsylvania) had both a standpipe system and a partial sprinkler system on a few floors. When a fire broke out in this building in 1991, firefighters attached their hose lines to the standpipe system. Because the pressure-reducing valves on the standpipe outlets had been set incorrectly, the firefighters were unable to obtain sufficient pressure to control the fire. Three firefighters died, and the fire burned several floors of the building, which eventually had to be demolished. The fire was stopped only when it reached one of the floors that was equipped with a sprinkler system.

Healthcare Facilities

Healthcare facilities, such as hospitals, nursing homes, assisted-living facilities, surgery centers, and ambulatory healthcare centers, require special preincident planning. Hospitals are often very large and include many different areas, ranging from operating rooms to walk-in clinics. The most challenging problem during an emergency incident at a healthcare facility is protecting nonambulatory patients. **Nonambulatory patients** cannot move themselves to an area of safety during an emergency, owing to their physical condition, medical treatment, or other factors.

A **defend in place** philosophy is used when designing fire protection for these facilities. This approach presumes that patients will not be able to escape from a fire without assistance and that there may not be enough staff present to move all of the patients out of the structure. Instead, the occupants may be relocated to a safe place within the structure. Therefore, the facility itself is designed to protect the patients from fire. Most healthcare facilities utilize fire-resistant construction; have fire detection and automatic sprinkler systems; and are compartmentalized with fire-safe doors, smoke barriers, etc.

If it becomes necessary to move patients from a dangerous area, the preferred approach is often **horizontal evacuation**, which is moving patients from a dangerous area on a floor to a safe area on the same floor. It is much easier to wheel a bed to another area on the same floor level than to carry patients down stairs.

Detention and Correctional Facilities

By their very nature, detention and correctional facilities are designed to ensure that the occupants cannot leave the building, even under normal conditions. Additionally, security concerns can make it difficult for firefighters to gain rapid access to the building or for occupants and rescuers to exit the facility. The preincident plan must consider the technicalities of removing the inmates from a facility or relocating them within the facility.

Multifamily Residential Occupancies

Preincident plans are usually prepared for multifamily residential properties such as apartment complexes and condominiums. In addition to the standard preincident plan information, when developing a preincident plan for residential occupancies, it is helpful to identify sleeping areas and to include information about occupants with disabilities.

Staying Current and Updated

The first step in being a good responder to your community is understanding the community, the people, the layout, the buildings, the hazards, and the calls to which you are likely to respond. Look at emergency response as an outcome of education and prevention. How can you reduce the impact of civilian and firefighter injuries and deaths through preparation and CRR? Understand the messages you are delivering and stay up-to-date on any changes that are made to those messages. Delivering outdated—and in some cases false and dangerous—messages reflects poorly on you and your department, and could lead to causing the emergencies you are trying to reduce in the first place.

Set up a rotation for getting into the buildings in your response area. If your fire department has a prevention division that does the inspections, take some time to do a walkthrough as a company and identify the information laid out in the preincident plan section so you and your crew will be prepared. Remember buildings and occupancies change, so keep the routine and visit the buildings frequently.

Use your data to identify and predict trends in calls and call types to train on so your crew is ready to respond in an efficient manner. Although the specific details about buildings recorded in preincident plans are confidential, the trends you identify as a result of collecting the data can be shared. Pass the information to other organizations in the community and to the public at large to help reduce calls to those types of incidents and to save members in your community from having to suffer the effects of an emergency.

CASE STUDY

You Are the Firefighter CONCLUSION

You entered the fire service to help your community, and you love the excitement of responding to the different types of calls for assistance. You like working with your crew, and even the daily station duties are enjoyable. A group of grade school children will be visiting your station this afternoon. You and your crew are cleaning the apparatus and equipment and discussing the visit before they arrive. As you discuss what you will talk to the children about, your captain says, "We will all go home having saved lives today." After receiving a few questioning glances from the crew, your captain continues, "Fire safety education is the easiest way to make sure people understand the dangers of fire and what to do to survive a fire in their home."

1. **What did your captain mean, and what is your role in fire prevention?**

   Answer: If you give the public (children and adults) the information needed to keep them aware of fire dangers, they will take action to prevent issues that lead to fire in their homes.

2. **What are the key points you will discuss with the group of students?**

   Answer: Age-appropriate fire safety messages for young children include smoke alarms (having them, what they sound like, what to do when they go off), sleeping with your bedroom door closed, and having a meeting place and home escape plan.

3. **Other than conducting fire station tours, in what ways can you assist your department's fire safety education program?**

   Answer:

   - Keep the concept of community risk reduction at the forefront of your mind when you are on calls and when you are doing routine work in the community.
   - Be knowledgeable about fire safety and prevention messaging and sharing that information with the public on and off calls.
   - Be involved in the community and share your passion for the fire service with young people who might be the next generation in the fire service.

4. **How does fire prevention affect firefighter safety?**

   Answer: If more people are aware of common causes of fires and take measures to reduce those risks in their homes and the community, there will be fewer fires, which means less chance for firefighters to be hurt or injured. If more fires are discovered earlier due to smoke alarm usage, fires will not be as developed, which allows the structure to be less damaged prior to fire department arrival.

WRAP-UP

SUMMARY

KNOWLEDGE OBJECTIVES

- Educate the public on hazard prevention and mitigation.
  - Identify the best practices for educating the public on hazard prevention and mitigation. (**NFPA 1010: 7.5.1, 7.5.2**, pp. 1054–1059)
  - Identify opportunities to educate the public through outreach programs developed locally or nationally. (pp. 1054–1059)
  - Describe current fire safety messages. (**NFPA 1010: 7.5.2**, pp. 1055–1056)
  - Explain how to teach EDITH. (**NFPA 1010: 7.5.1, 7.5.2**, pp. 1055–1057)
  - Describe the difference between fire prevention and fire safety and survival messages. (**NFPA 1010: 7.5.2**, pp. 1057–1058)
- Define community risk reduction and the role of the fire service in maintaining overall community safety.
  - Describe the two Firefighter Life Safety Initiatives that relate specifically to public education and fire prevention. (pp. 1058–1059)
  - Define CRR and the piece the fire service plays in overall community safety. (pp. 1059–1064)
  - Describe the characteristics of a CRR program. (pp. 1059–1064)
  - Identify how the fire service partners with public and private entities to identify and mitigate hazards in your community. (pp. 1059–1064)
  - Describe the 5 Es of CRR. (p. 1060)
- Conduct fire station tours.
  - Describe how to prepare to give a tour of the fire station. (**NFPA 1010: 7.5.2**, pp. 1064–1066)
  - Explain how to make a fire station tour educational and appropriate for the age group on the tour. (**NFPA 1010: 7.5.2**, pp. 1065–1066)
  - Describe the steps in conducting a fire station tour. (**NFPA 1010: 7.5.2**, pp. 1064–1067)
- Identify fire and life safety hazards that exist in residential properties and explain ways to mitigate them.
  - Explain the importance of residential fire safety surveys and gathering knowledge about residential neighborhoods. (**NFPA 1010: 7.5.1**, pp. 1066–1068)

KNOWLEDGE OBJECTIVES CONTINUED

- Describe some of the considerations to keep in mind during a residential fire safety survey. (**NFPA 1010: 7.5.1**, pp. 1066–1068)
- Recognize hazards during a fire safety survey of a residence. (**NFPA 1010: 7.5.1**, pp. 1068–1070)
- Explain the current code requirements for the placement of smoke alarms. (**NFPA 1010: 7.5.1, 7.5.2**, pp. 1070–1071)
- Describe smoke alarm maintenance, including how to vacuum smoke alarm chambers, how often occupants should test smoke alarms, how often batteries should be changed in smoke alarms with alkaline batteries, and how often smoke alarms should be replaced. (**NFPA 1010: 7.5.1, 7.5.2**, pp. 1070–1072)
- Describe best practices for evacuating a residence during a fire. (**NFPA 1010: 7.5.1, 7.5.2**, pp. 1071–1073)
- Explain the importance of occupants in a residence practicing exit drills and establishing a meeting place outside. (**NFPA 1010: 7.5.1, 7.5.2**, pp. 1071–1073)
- Describe how to select and use portable fire extinguishers and where to place them in a residential home. (**NFPA 1010: 7.5.1, 7.5.2**, p. 1073)
- Explain the importance of residential automatic sprinkler systems. (**NFPA 1010: 7.5.1, 7.5.2**, p. 1073)
- Complete the department fire safety survey forms and file according to the department SOPs. (**NFPA 1010: 7.5.1**, p. 1073)

- Conduct preincident surveys and create preincident plans for non-residential locations such as commercial buildings, transportation centers, and other public locations.
  - List the typical target hazards that may be found in a community. (**NFPA 1010: 7.5.3**, pp. 1075–1077)
  - Describe why and for which types of properties a preincident survey is conducted. (**NFPA 1010: 7.5.3**, pp. 1075–1087)
  - List the general information that is gathered during a preincident survey. (**NFPA 1010: 7.5.3**, pp. 1075–1087)
  - Describe how to prepare to conduct a preincident survey. (**NFPA 1010: 7.5.3**, pp. 1075–1077)
  - Describe the information included in any sketches or drawings created during the preincident survey. (**NFPA 1010: 7.5.3**, pp. 1075–1078)
  - Describe the symbols commonly used in preincident plans. (**NFPA 1010: 7.5.3**, pp. 1077–1078)
  - Describe your department's requirements for conducting a preincident survey and preparing a preincident plan. (**NFPA 1010: 7.5.3**, pp. 1075–1079)
  - List the information that is gathered during a preincident survey for response and access to both the exterior and interior of a property. (**NFPA 1010: 7.5.3**, pp. 1077–1080)
  - Describe the information in the preincident plan that is used to assist the IC in making a rapid and correct size-up during an emergency incident. (**NFPA 1010: 7.5.3**, pp. 1080–1083)
  - Explain why the locations of utilities are noted on the preincident plan. (p. 1083)
  - Describe how preincident planning leads to efficient fireground operations during emergency response. (**NFPA 1010: 7.5.3**, pp. 1083–1084)
  - List the types of information gathered during a preincident survey at properties where hazardous materials are stored or used. (**NFPA 1010: 7.5.3**, p. 1084)
  - Describe and list the types of locations that require special consideration during preincident planning. (**NFPA 1010: 7.5.3**, pp. 1084–1087)
  - List the types of locations that require special considerations in preplanning. (**NFPA 1010: 7.5.3**, pp. 1084–1087)
  - Describe how to keep preincident plans up to date. (pp. 1087–1088)

SKILLS OBJECTIVES

- Prepare presentations for educating the public about community risk reduction and fire and life safety and conduct fire station tours.
  - Perform a public fire safety education presentation on Exit Drills In The Home (EDITH). (**NFPA 1010: 7.5.1, 7.5.2**, pp. 1056–1057)

- Teach fire safety to the public. (**NFPA 1010: 7.5.1, 7.5.2**, p. 1059)
- Give a public education tour of a fire station. (**NFPA 1010: 7.5.1, 7.5.2**, pp. 1066–1067)
- Install and maintain smoke alarms. (**NFPA 1010: 7.5.1, 7.5.2**, p. 1072)
- Conduct a fire safety survey at a residence. (**NFPA 1010: 7.5.1**, p. 1074)
- Conduct a preincident survey of a commercial properties or other public locations, identifying tactical priorities, and complete a preincident plan for that location. (**NFPA 1010: 7.5.3**, p. 1079)

KEY TERMS

5 Es of community risk reduction (CRR) Education, enforcement, engineering, economic incentives, and emergency response. Also called *5 Es of prevention.*

5 Es of prevention See *5 Es of community risk reduction (CRR).*

building code A regulation that specifies how structures are designed, constructed, and remodeled so that buildings are safe for people who live and work in them.

code A standard that is an extensive compilation of provisions covering broad subject matter or that is suitable for adoption into law independently of other codes and standards. (NFPA 1)

community risk reduction (CRR) A process to identify and prioritize local risks, followed by the integrated and strategic investment of resources to reduce their occurrence and impact. (NFPA 1452)

defend in place The operational response in which the action is to relocate the affected occupants to a safe place within the structure during an emergency.

Exit Drills in the Home (EDITH) A public fire and life safety education program that teaches occupants how to safely exit a house in the event of a fire or other emergency and meet other household members at a pre-established meeting place.

fire code A code that ensures a minimum level of fire safety in the home and workplace environments by preventing fires and protecting lives and property in the event of a fire.

fire prevention Measures taken toward avoiding the inception of fire. (NFPA 801)

fire safety survey The process of observing and recording conditions of an occupied structure for basic fire and life safety hazards. (NFPA 1010)

horizontal evacuation The process of moving occupants from a dangerous area on a floor to a safe area on the same floor.

mini-maxi codes Building and fire codes that legally specify both the minimum and maximum requirements.

nonambulatory patient A term describing individuals who cannot move themselves to an area of safety owing to their physical condition, medical treatment, or other factors.

preincident plan A document developed by gathering general and detailed data that is used by responding personnel in effectively managing emergencies for the protection of occupants, responding personnel, property, and the environment. (NFPA 1660)

preincident survey An examination of a property to gather information to develop a preincident plan.

safety data sheets (SDSs) Formatted information, provided by chemical manufacturers and distributors of hazardous products, about a chemical's composition, physical and chemical properties, health and safety hazards, emergency response, and waste disposal of the material. (NFPA 470)

target hazard A property that is particularly large or that presents unusual risks to firefighters or to the public.

REVIEW QUESTIONS

1. What type of agencies should fire departments collaborate with as they promote CRR initiatives?
2. Which of the 5 Es of CRR includes social media outreach?
3. What are the best practices for educating both children and adults about fire and life safety?
4. What should you always remember when you conduct a fire station tour?

REVIEW QUESTIONS CONTINUED

5. What are preincident plans and preincident surveys?
6. What is a target hazard?
7. What is the cause of approximately half of all residential fires?
8. Why is a defend in place strategy planned when designing fire protection for healthcare facilities?

DISCUSSION QUESTIONS

1. What are some ways you can use outreach to enhance your public image and/or increase your recruiting efforts?
2. How are home safety surveys and commercial preincident plans similar and how do they differ? Why are they important during fire and rescue response?
3. List information that should be included in the preincident plan for a commercial building.
4. How can you use call response data to enhance your on-shift or full department training sessions?

APPLYING THE CONCEPTS

Your station received a call for a fire alarm at a residence in an apartment complex. In fact, recently your station has received multiple calls from different units in this same complex. Each time the crew has determined the alarm was triggered by burning food or an issue with the cooking appliance. Despite educating residents after each call, the problem persists. When you arrive at the scene you determine that, again, burning food triggered the fire alarm.

1. Based on this information, how should your crew proceed?

On the way back to the station, you and your crew discuss this latest call, questioning what is at the root of these cooking-related calls. It is decided that the station's focus should shift from being reactive to proactive and that you and your crew will develop a plan for the apartment complex to identify risks, create procedures to mitigate those risks, and provide future-responding crews with vital information to be safer and more effective.

2. As you begin to develop a risk reduction plan, where can you look for historical information?
3. What are some easy and inexpensive ways to mitigate risks and reduce these incidents in the future?

A few weeks later, your station receives a call for a fire alarm at a different unit in the same apartment complex. Because your station has been creating a plan for this complex, you arrive armed with vital information and a proactive mindset. You recognize a few of the residents and the property manager from an education event your station held at the complex last week. They are happy to see you and help your crew locate the apartment where the alarm was triggered. Once again you determine the alarm was caused by burning food and you educate the resident on safe cooking practices.

4. Why is it important to continue reaching out to residents and offering education?
5. What additional steps should you take to map the building and preplan for future calls?
6. What are the benefits and drawbacks of a preincident plan?

REFERENCES

Ahrens, Marty. 2020. *Home Cooking Fires.* National Fire Protection Association, July 2020. Accessed January 16, 2023. https://www.nfpa.org.

Ahrens, Mary, and Radhika Maheshwari. 2021. *Home Structure Fires.* National Fire Protection Association, October 2021. Accessed January 16, 2023. https://www.nfpa.org.

Centers for Disease Control and Prevention (CDC). 2023. "Carbon Monoxide Poisoning: Frequently Asked Questions." Last reviewed March 27, 2023. Accessed March 31, 2023. https://www.cdc.gov/co/faqs.htm.

Griffioen, Mark. 2020. *Distressed Properties Initiative.* Presentation at "Model Performance in Community Risk Reduction" symposium hosted by Vision 20/20, February 17–20, 2020. Accessed January 11, 2023. https://strategicfire.org/wp-content/uploads/2020/02/4.-Griffioen-Program-Overview-Handout-1.31.20-Final.pdf.

Hall, Shelby. 2023. *Home Structure Fires.* National Fire Protection Association, April 2023. Accessed August 29, 2023. https://www.nfpa.org.

National Fallen Firefighters Foundation (NFFF). n.d. "16 Firefighter Life Safety Initiatives." Everyone Goes Home. Accessed January 9, 2023. https://www.everyonegoeshome.com/16-initiatives.

National Fire Protection Association (NFPA). n.d. "Home Fire Sprinklers." Accessed January 16, 2023. https://www.nfpa.org.

National Fire Protection Association (NFPA). n.d. Top Fire Causes. Accessed January 16, 2023. https://www.nfpa.org.

National Fire Protection Association (NFPA). 2019. *NFPA 801, Standard for Fire Protection for Facilities Handling Radioactive Materials.* 2020 Edition. Quincy, MA: NFPA.

National Fire Protection Association (NFPA). 2019. *NFPA 1300, Standard on Community Risk Assessment and Community Risk Reduction Plan Development.* 2020 Edition. Quincy, MA: NFPA.

National Fire Protection Association (NFPA). 2019. *NFPA 1452, Guide for Training Fire Service Personnel to Conduct Community Risk Reduction for Residential Occupancies.* 2020 Edition. Quincy, MA: NFPA.

National Fire Protection Association (NFPA). 2020. *NFPA 1, Fire Code.* 2021 Edition. Quincy, MA: NFPA.

National Fire Protection Association. 2020. *NFPA 99: Health Care Facilities Code.* 2021 Edition. Quincy, MA: NFPA.

National Fire Protection Association (NFPA). 2020. *NFPA 170, Standard for Fire Safety and Emergency Symbols.* 2021 Edition. Quincy, MA: NFPA.

National Fire Protection Association (NFPA). 2020. *NFPA Educational Messages.* 2020 Edition: *Desk Reference for the Fire Service and Fire and Life Safety Educators.* Quincy, MA: NFPA.

National Fire Protection Association (NFPA). 2021. *NFPA 72, National Fire Alarm and Signaling Code.* 2022 Edition. Quincy, MA: NFPA.

National Fire Protection Association (NFPA). 2021. *NFPA 470, Hazardous Materials/Weapons of Mass Destruction (WMD) Standard for Responders.* 2022 Edition. Quincy, MA: NFPA.

National Fire Protection Association (NFPA). 2023. *NFPA 1030, Standard for Professional Qualifications for Fire Prevention Program Positions.* 2024 Edition. Quincy, MA: NFPA.

National Fire Protection Association (NFPA). 2023. *NFPA 1660, Standard for Emergency, Continuity, and Crisis Management: Preparedness, Response, and Recovery.* 2024 Edition. Quincy, MA: NFPA.

National Fire Protection Association (NFPA). 2024. NFPA 1010: *Standard on Professional Qualifications for Firefighters.* 2024 Edition. Quincy, MA: NFPA.

Safe Kids Worldwide. Home page. Accessed April 1, 2023. https://www.safekids.org.

Underwriters Laboratories (UL) Fire Safety Research Institute (FSRI). 2023. "Toolbox." Close before You Doze. Accessed January 14, 2023. https://closeyourdoor.org/first-responders/toolbox.

U.S. Census Bureau. n.d. "Explore Census Data." Accessed January 14, 2023. data.census.gov.

U.S. Fire Administration (USFA). 2020. "Fire Is Everyone's Fight." Last reviewed April 15, 2020. Accessed January 15, 2023. https://www.usfa.fema.gov/ax/materials_fief.

U.S. Fire Administration (USFA). 2022. "Statistics." Last reviewed September 27, 2022. Accessed January 11, 2023. https://www.usfa.fema.gov/statistics.

U.S. Fire Administration (USFA). 2023. "Fire Is Everyone's Fight." Last reviewed January 6, 2023. Accessed January 15, 2023. https://www.usfa.fema.gov/prevention/outreach.

U.S. Fire Administration (USFA). 2023. "National Fire Incident Reporting System." Last reviewed January 11, 2023. Accessed January 11, 2023. https://www.usfa.fema.gov/nfirs/.

Vision 20/20. n.d. Home page. Accessed January 14, 2023. https://strategicfire.org.

CHAPTER

26

Firefighter II

Vehicle Rescue and Extrication

KNOWLEDGE OBJECTIVES

After studying this chapter, you will be able to:

- Describe a vehicle's anatomy.
- Describe the types and characteristics of alternative-fuel vehicles.
- List the tools and equipment utilized at a motor vehicle accident.
- Describe the best practices to follow when responding to, arriving at, and sizing up the scene.
- Describe basic site operations at a motor vehicle accident.
- Describe the components of an incident action plan for a motor vehicle accident.
- Describe how vehicle stabilization is performed.
- Describe how to safely access and extricate a victim from a motor vehicle accident.
- Describe the steps in terminating a motor vehicle incident.

SKILLS OBJECTIVES

After studying this chapter, you will be able to perform the following skills:

- Safely size-up and stabilize a motor vehicle accident.
- Access and extricate a victim from a motor vehicle accident.

ADDITIONAL NFPA STANDARDS

- **NFPA 556**, *Guide on Methods for Evaluating Fire Hazard to Occupants of Passenger Road Vehicles, 2024 Edition*
- **NFPA 1006**, *Standard for Technical Rescue Personnel Professional Qualifications, 2021 Edition*
- **NFPA 1951**, *Standard on Protective Ensembles for Technical Rescue Incidents, 2020 Edition*
- **NFPA 1960**, *Standard for Fire Hose Connections, Spray Nozzles, Manufacturer's Design of Fire Department Ground Ladders, Fire Hose, and Powered Rescue Tools, 2024 Edition*
- **NFPA 1970**, *Standard on Protective Ensembles for Structural and Proximity Firefighting, Work Apparel and Open-Circuit Self-Contained Breathing Apparatus (SCBA) for Emergency Services, and Personal Alert Safety Systems (PASS), 2024 Edition*
- **NFPA 2500**, *Standard for Operations and Training for Technical Search and Rescue Incidents and Life Safety Rope and Equipment for Emergency Services, 2022 Edition*

CASE STUDY

You Are the Firefighter

It has been storming all afternoon. Your station has just finished dinner when the alarm sounds for a motor vehicle accident involving a car that has slid off the road. As you pull out of the station, the rain gets heavier, the wind gets stronger, and it is now dark outside. When you arrive on the scene, you see a passenger car resting on the driver's side in the ditch. One passenger was ejected from the vehicle and is lying motionless by the car. Your officer directs you to check the vehicle for any occupants still inside while he performs a detailed, 360-degree size-up. As you approach the vehicle, you recognize features identifying it as an electric-drive vehicle. The driver is still in the driver's seat and a passenger appears to be unconscious in the back seat compartment. The officer establishes command, calls for additional resources, and assigns the ambulance crew to assess the ejected victim. He directs you to stabilize the vehicle and to set up the equipment for an extensive extrication.

1. What safety concerns should you have in this situation? How would you address those concerns?
2. Before making access to assess the victims, what steps should be taken?
3. What extrication equipment will you need?

Introduction

Motor vehicle accidents (MVAs) are a common occurrence in almost all jurisdictions. According to the National Highway Traffic Safety Administration (NHTSA), there was a 22 percent decrease in the number of motor vehicle crashes from 2019 to 2020, and a 16 percent increase from 2020 to 2021 (Stewart 2023, 7). The decrease from 2019 to 2020 was most likely due to the travel restrictions imposed due to the COVID-19 pandemic and the reduced number of vehicles on the road. As people returned to their normal activities, the number of crashes rose and will likely continue to trend upward.

As a firefighter, you will respond to many MVAs during your career. Most MVAs involve minimal damage with minor injuries. Some, however, will require careful planning and operations to safely extricate trapped and injured occupants from severely damaged vehicles, often under difficult circumstances. NFPA 1006, *Standard for Technical Rescue Personnel Professional Qualifications, 2021 Edition*, Chapter 8, Common Passenger Vehicle Rescue, identifies the minimum job performance requirements (JPRs) for fire service and other emergency response personnel who perform technical rescue operations. Although this chapter provides the knowledge and skills a firefighter needs to assist with vehicle extrication operations, it will not prepare you to work as part of a special extrication team. Firefighters who are members of special rescue teams or who may be expected to perform challenging vehicle extrication operations should complete a course in rescue techniques that meets the requirements of NFPA 1006 and the requirements of NFPA 2500, *Standard on Operations and Training for Technical Search and Rescue Incidents and Life Safety Rope and Equipment for Emergency Services, 2022 Edition.*

Vehicle Anatomy and Composition

Because head-on (frontal), side, and roll-over crashes account for the majority of crashes on America's roadways, the NHTSA performs these crash tests on every make and model of car sold in the United States. The NHTSA then rates motor vehicles based on how they perform during a crash test. To comply with established energy conservation and passenger safety laws, car manufacturers are constantly looking for ways to improve vehicle performance and fuel efficiency while at the same time maintaining or improving crash worthiness. This is often accomplished by using lighter, yet stronger materials, such as an **alloy**—materials composed of two or more metals—and by engineering integrated safety features into the vehicle's structural components. The use of alloys and nonmetals in today's motor vehicles on the road is a common industry standard. For example, carbon is used to produce plastic polymers such as **carbon fiber reinforced polymer (CFRP)**, which is an extremely strong, lightweight plastic that is reinforced with woven strands of carbon bound together with a polymer, such as an epoxy, and then formed into the desired shape. Carbon is also alloyed with iron to create carbon steel, which is found in the structural frame and components of some vehicles. New compounds and **advanced high-strength**

steel—alloyed steel with a minimum tensile strength of 65 kilopounds per square inch (ksi; 440 megapascals [MPa])—are being discovered and developed to replace conventional steels. **Tensile strength** is the resistance of a material to breaking under tension. Steel is classified by its tensile strength and its **yield strength**, the maximum stress that can be applied to steel before it permanently deforms. Other metals such as aluminum and magnesium can be alloyed and are commonly found in vehicle structural and body components.

When dealing with an MVA where extrication is required, you might encounter areas in the vehicle that are difficult to spread or cut due to the use of advanced high-strength steel. The use of high-pressure hydraulic tools is becoming more of a need rather than a luxury when it comes to vehicle rescue and extrication due to the higher forces required to overcome high-strength steels. Knowing where the weak or unreinforced areas of a vehicle are located will help you access trapped passengers. Adapting and fully understanding the capabilities and limitations of rescue tools in relation to modern vehicle construction is imperative for successful rescue operations.

Motor Vehicle Components

To reduce confusion and minimize the potential for mistakes at extrication scenes, it is important to use standardized terminology when referring to specific parts of vehicles. The left side of a vehicle is on your left as you sit in the vehicle. In the United States and Canada, the driver's seat is on the left side of the vehicle. The right side of a vehicle is where the passenger's seat is located. Always refer to left and right as they relate to the vehicle; do not refer to left and right from where you might be standing. To avoid confusion when referring to the vehicle sides, use "driver-side" and "passenger-side" respectively as a point of reference.

In general, vehicles consist of three main compartments: the engine compartment, the passenger compartment, and the trunk (cargo area). In most common passenger vehicles, the front of a vehicle is where the engine compartment is located. The hood covers the engine compartment. The structure that divides the engine compartment from the passenger compartment is called the **firewall** or **bulkhead**. Another structural component found in the front of passenger vehicles is the **upper rails**, which are two parallel beams, one on each side of a vehicle's front end, that support and attach the **strut tower**—the front suspension system—to the frame, hold the hood in place, and support the front fenders. The passenger compartment, which is sometimes called the occupant cage or the occupant compartment, includes the front and back seats. The trunk is usually located at the rear of the vehicle. The trunk may be exposed to the back seats, as in a station wagon, or enclosed, as in a sedan. In some vehicles, the trunk is located in the front (sometimes referred as the frunk) or they may have additional cargo space in the engine compartment.

Posts

Vehicles contain vertical posts, or pillars, that support the roof and form the upright columns of the passenger compartment. Each **post** or **pillar** is labeled alphabetically starting from the front and continuing toward the back of the vehicle as A, B, and C (**FIGURE 26-1**). The

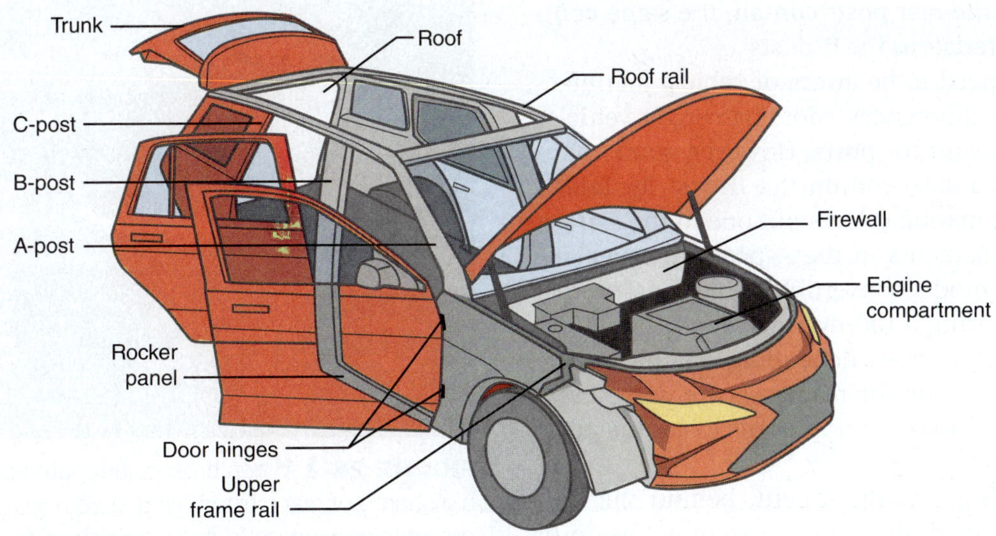

FIGURE 26-1 The anatomy of a vehicle.

A-post is located closest to the front of the vehicle; the two A-posts form the sides of the windshield and extend down alongside the firewall. Attached to the lower section of the A-post are the door hinges. For this reason, the A-post is sometimes called the **hinge pillar**.

In four-door vehicles, the **B-post** is located between the front and rear doors of the vehicle. In some vehicles, such as convertibles, these posts do not reach all the way to the roof of the vehicle. Because this area is frequently impacted in a collision, manufacturers reinforce the B-post from the **rocker panel**—the section of a vehicle's frame located below the doors, between the front and rear wheels—to the roof rail using alloyed metals. This may present a difficult challenge to rescuers when they attempt to cut or spread the B-post due to the high tensile and yield strength of these alloys. In most vehicles, the B-post also houses the seat belt mechanism that retracts and takes up slack from the seat belt at the lower end of the B-post close to the rocker panel. On the upper section of the B-post, most vehicles have an adjustable anchor for the seat belt that attaches to a slide track to adjust to the height of the occupant. The slide track is normally made of heavy-gauge or high-strength steel and covers several inches in the upper part of the B-post. This makes it difficult to cut the upper part of the B-post with low-pressure hydraulic tools or reciprocating saws. Rescuers should try to avoid cutting in this area if they are not using high-pressure hydraulic tools.

In four-door vehicles, the **C-post** is located behind the rear doors—specifically, behind the rear passenger windows. In two-door vehicles, the C-posts are the rear posts. Some vehicles, such as an SUV or a wagon-type vehicle, have four or more sets of posts that support the roof. In these larger vehicles, the posts located between the B-posts and the rear posts contain the same components and materials as the B-posts.

Firefighters need to be aware of vehicle manufacturer and design differences. Most passenger vehicles have the same layout for posts. However, some manufacturers place a steel rod on the B-post for added safety, and other manufacturers put one in the A-post. Steel rod reinforcements in the A-post are especially common in late model convertible vehicles to protect the occupants during a roll-over accident. Rear posts might be wide or narrow, depending on the vehicle make and model. When the rear posts are wide, a rescuer might need to make multiple cuts to cut the posts all the way through.

In the passenger compartment, behind the firewall, is the vehicle's dash. A transverse metal beam or bar called the **dash reinforcement bar** runs the entire length of the dash. Attached to the dash bar are one or two metal brackets located in the center console area. These metal brackets attach to the floor pan and are designed to stabilize the dash and reinforce the occupant compartment in the event of a crash.

Glass

Safety glass is a general term that describes glass designed not to break and, if broken, does not shatter into long, sharp shards. In automobiles, the most common types of safety glass are laminated safety glass and tempered safety glass. **Laminated safety glass (LSG)**, or simply **laminated glass**, consists of two layers of glass bonded together with a clear polymer film. This type of glass does not shatter when broken. Instead, when it is broken it cracks in a star shape, or, with stronger forces, a pattern resembling a spider web (**FIGURE 26-2**). **Tempered safety glass**, or simply **tempered glass**, is created by exposing ordinary glass to extreme heat and then rapidly cooling it. As a result, the outer surface of the glass goes into compression (pushing inward) and the center goes into tension (pushing outward).

FIGURE 26-2 When broken, laminated glass prevents big shards of glass from flying at vehicle occupants and prevents occupants from being ejected from the vehicle.

Courtesy of David Sweet.

This gives tempered glass its added strength. When tempered glass breaks, the opposing forces holding the glass together suddenly release, and the glass breaks into small, mostly dull-edged pieces.

Ordinary glass produces long, sharp shards when broken. Laminated and tempered glass are used in vehicles because if they break, they do not shatter into shards that can cut passengers. In automobiles, laminated glass is primarily found in the windshield so when it is broken, shards of glass don't fly in on the occupants. It also aids in preventing occupants from being ejected from the vehicle during a crash. The passenger and rear windows are usually made of tempered glass because it is less expensive than laminated glass to manufacture. Recently, however, in an effort to increase safety from side impact and roll-over crashes, some vehicle manufacturers have started using laminated glass for side and rear windows because of its ability to retain passengers inside the vehicle. U.S. federal law FMVSS 226, or the "Ejection Mitigation" law, requires that all vehicles manufactured after September 2018 use laminated glass for windows in the first three rows of seats, and in a portion of the cargo area behind the first or second rows in most passenger cars and trucks.

Some motor vehicles also have components made of dual-pane glass, similar to the glass found in windows in residential homes, to help with soundproofing and insulation from the elements. Polycarbonate and ballistic-type glass (armored glass) may be found in some motor vehicles.

When managing glass at emergency incidents involving motor vehicles, the firefighter must be able to identify the type of glass. Knowing the type of glass determines which type of tool is required to break the glass, if required, and how to safely remove it.

Frames

Motor vehicles usually have one of two types of frames: body-over-frame or unibody construction.

Body-over-frame construction consists of two large beams, referred to as frame rails, that run the length of the vehicle and are tied together by cross member beams. This type of frame construction is sometimes referred to as **ladder frame construction** because the frame resembles a ladder (**FIGURE 26-3**). The beams form the load-bearing frame of the vehicle. The engine and transmission are attached to this frame, and then the body of the vehicle is placed over the frame and attached to it. This type of frame is found primarily in trucks and larger SUVs; it is rarely present in smaller passenger cars.

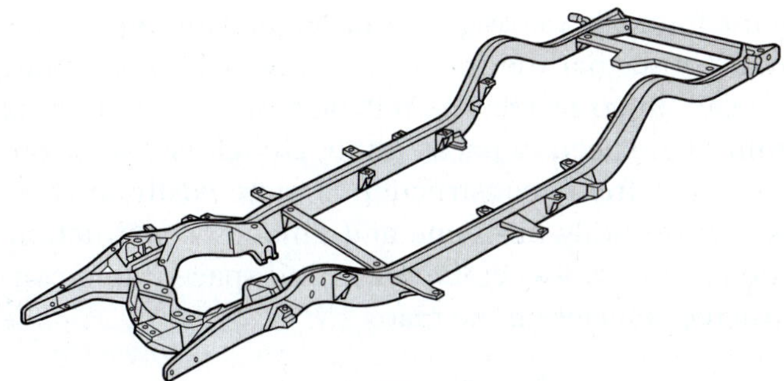

FIGURE 26-3 The two large beams tied together by cross member beams are indicative of a basic body-over-frame construction.

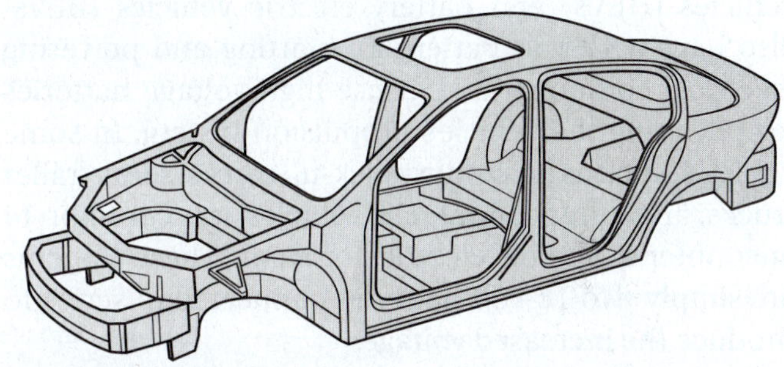

FIGURE 26-4 Unibody frame construction combines the vehicle body and the frame into a single component.

Unibody frame construction, which is used for most modern passenger cars, combines the vehicle body and the frame into a single component, thereby eliminating the rail beams used in body-over-frame construction. By folding multiple thicknesses of metal together, a column that is strong enough to serve as the frame for a lightweight vehicle is formed. A subframe section that consists of two short beams is attached to the passenger compartment at the front and the rear of the vehicle (**FIGURE 26-4**). This subframe houses the main components of the vehicle including the suspension system, steering, braking, transmission, and engine. Unibody construction allows auto manufacturers to produce lighter-weight vehicles. When extricating a person from a vehicle with a unibody frame, remember that these vehicles do not have frame rails like body-on-frame vehicles. The basic extrication techniques discussed in this chapter must be adapted for each type of vehicle.

A third, less common type of vehicle frame construction is **space frame construction**. A space frame is made up of multiple tubes welded into a rigid, but light, cage-like structure (a roll cage). When completed, the vehicle's outer panels are attached independently to the frame. It was originally developed for the race car industry so that the frame would be lighter, enabling the cars to go faster. The roll cage protects the driver from high-speed impacts. Although lighter and stronger, space frame construction is more costly in comparison to body-on-frame and unibody construction. Therefore, it is less common to find space frame constructed vehicles on the road.

Electrical Systems

Conventional vehicles (vehicles that run on gasoline or diesel fuel) have a 12-volt battery that is used for starting the car and powering electrical components in the vehicle. Electric-drive vehicles such as hybrid electric vehicles (HEVs) and battery electric vehicles (BEVs) also have a 12-volt battery for starting and powering electrical components, but use high-voltage batteries for powering the vehicle's propulsion system. In some vehicles, such as heavy duty pick-up trucks, semi-trailer trucks, and other commercial vehicles, it is common to encounter a 24-volt electrical system. These systems are simply two 12-volt batteries connected in series to produce the increased voltage.

Cabling is color-coded depending on voltage. Automotive wiring that contains less than 30 volts is considered low-voltage and is color-coded red or black. Voltage between 30 to 60 volts is considered medium-voltage and is usually color-coded yellow or blue. Anything greater than 60 volts DC is considered high-voltage and is color-coded orange, per the Society of Automotive Engineers (SAE) standards.

The standard 12-volt battery used in conventional vehicles is a lead acid battery. There are several designs of this type of battery, but the standard is the **wet cell battery**, sometimes called a **flooded cell battery**, which contains an active electrolyte solution consisting of sulfuric acid and water, a highly corrosive mixture. When these batteries are overcharged or damaged, they pose a hazard to firefighters and other responders on scene. When they are overcharged, highly explosive gases of hydrogen and oxygen can be generated. If a lead acid battery shows signs of overheating, such as emitting gas or the battery case bulging, immediately ventilate the area. When a wet cell battery spills its contents, as commonly occurs in a collision, the corrosive electrolyte solution causes a health hazard to responders and others near the spilled liquid. If a spill occurs, immediately contain the spill by applying an absorbent (soil, sand, or commercial absorbent) to reduce the chances of becoming exposed to the fluid.

An important step when preparing to initiate extrication operations is disabling the vehicle. After turning the motor off, the next step is often locating and disconnecting the low-voltage battery. Traditionally, the location of the 12-volt battery has been under the hood, somewhere in the engine compartment. In modern vehicles, especially in alternative-fuel vehicles, the low-voltage battery location is unpredictable. Besides the engine compartment, it can be located in the trunk, under the rear passenger seats, or in the fender wells. During an MVA incident, the driver might be able to tell responders the location of the battery in their vehicle. The next best option is to reference the vehicle manufacturer's **Emergency Response Guide** (not to be confused with the *Emergency Response Guidebook* [*ERG*] used for hazardous materials incidents), which is a reference book created by a vehicle's manufacturer that provides important information about the vehicle, such as information about hybrid/electric, fuel-cell, and alternative-fuel systems, to educate and assist emergency response personnel responding to incidents involving the specific vehicles. Vehicle ERGs also provide important information that can assist emergency response personnel dealing with specific types and models of vehicles. Information such as location and number of air bags, inflators, electronic control unit, 12-volt battery, high-voltage battery and components, and structural reinforced areas can all be found in the vehicle ERG. Some vehicle manufacturers make available a short version of the ERG called a **Quick Response Guide (QRG)** for some vehicles. Many of these reference materials are readily available online and can be downloaded. Organizations such as the National Fire Protection Association (NFPA) and the Energy Security Agency (ESA) both provide vehicle ERGs online free of charge. These documents can be found at the NFPA's website. Firefighters must take advantage of these invaluable resources to better prepare themselves and their respective agencies to respond to vehicle rescue and extrication incidents.

Supplemental Restraint Systems

A **supplemental restraint system (SRS)** can be found in most passenger vehicles on the road today. An SRS usually includes a series of air bags and seat belts. Air bags are found in many locations in modern vehicles. They originally were installed only in the driver's and front passenger's seat in the 1990s. By the early 21st century, they started appearing in other locations in vehicles, such as in the doors (side-impact air bags) and even within the seats of the vehicle. The locations of air bags are labeled with abbreviations or acronyms, such as SRS, SIR (supplemental inflatable restraint), HPS

FIGURE 26-5 A labeling system using acronyms is generally used to indicate that an air bag system is present. The label shown in this figure indicates there is a head protection system in the vehicle.

Courtesy of David Sweet.

FIGURE 26-6 Front air bags cushion the occupants during a crash by allowing gas to escape through holes in the back.

© saravuth/Shutterstock

(head protection system), SIPS (side-impact protection system), IC (inflatable curtain), and, perhaps the most helpful, "AIR BAG." These labels are usually located near the location where the air bag will deploy and may be embossed, raised, or sewn into the plastic, cloth, or leather material (**FIGURE 26-5**). Identifying the location of these components is critical to avoid the hazards they present.

The **inflator** is the system that fills air bags when a crash occurs. Inflators use either a stored compressed gas or a gas-generating system to inflate air bags. Beginning in 2006, all passenger vehicles and light-duty trucks sold in the United States are required to use advanced smart air bag systems with an **electronic control unit (ECU)**, which is generally located in the center of the vehicle. The ECU is the "brains" for the SRS and consists of a small processing unit that receives information such as crash severity, occupant's weight, proximity to the air bag, seat belt usage, and seat position to determine the deployment force of the air bag. The smart air bag system uses a dual-stage or multistage inflation algorithm to decide whether to inflate the air bag at full or reduced force depending on the factors and conditions analyzed by the ECU.

Several components make up an air bag system, including the air bag, igniter (a miniature explosive device), air bag control unit (contained within the ECU), propellant, inflator, and sensors. In a crash, the air bag control unit collects information and decides if one or more air bags should deploy. For example, in a frontal collision, an abrupt deceleration equivalent to the vehicle crashing into a stationary object at 14 miles per hour (mph); (23 kilometers per hour [km/h]) or greater. If this condition is detected, an electrical current is sent to the igniter which ignites the propellant in the air bag inflator. A chemical reaction produces gas—usually nitrogen gas—that instantly inflates the bag at a rate of approximately 200 to 250 mph (322 to 402 km/h). The inflation process takes approximately 30 to 40 milliseconds, faster than the blink of an eye. The occupant, still traveling forward at the speed the vehicle was moving before the collision, makes contact with the fully inflated air bag. As the occupant continues to move forward, the air bag "cushions" the impact by allowing air to escape through vent holes in the back of the bag (**FIGURE 26-6**). Side-impact air bags follow a similar algorithm, but their speed of inflation is faster—only 10 to 20 milliseconds. Side-impact air bags stay inflated longer than front-impact air bags to protect the occupants in case of secondary impacts or roll-over. If the ECU detects an imminent roll-over condition, it signals side-impact and head protection systems to deploy.

Driver-side, passenger-side, and side-impact air bags are different sizes and have different deployment distances within the vehicle. Typical deployment distances are 5 to 15 inches (in.; 13 to 38 centimeters [cm]) for side-impact air bags, 10 in. (25 cm) for driver-side air bags, and 20 to 25 in. (51 to 64 cm) for passenger-side air bags. When working inside a vehicle that has been involved in a traffic accident, it is important that you maintain a safe distance from the air bag deployment zones to avoid serious injury in case of inadvertent deployment. A good rule of thumb is the 10-10-20 rule (25-25-50 rule): Maintain a distance of 10 in. (25 cm) from side-impact air bags, 10 in. (25 cm) from the driver-side air bag, and 20 in. (50 cm) from passenger-side air bag. Note that these distances are only recommendations and do not guarantee your safety.

The seat belt pretensioning system, also part of the SRS, reduces the slack in the seat belt when an MVA is detected (**FIGURE 26-7**). The system is activated in conjunction with the vehicle air bags, but it also acts independently. The system operates at the belt buckle attachment or at the anchor attachment. Seat belt assemblies can be housed in any post, under the seats, or in the center console.

Convertibles have roll-over protection systems (ROPS). These systems consist of deployable roll bars that are concealed until activated by sensor detection. Roll bars can extend up to 20 in. (51 cm) in some models.

Firefighters need to be familiar with the different types of SRS devices found in modern vehicles.

FIGURE 26-7 The seat belt pretensioning system tightens, or takes up slack in the seat belt when a crash is detected.

Courtesy of Aaron Miranda.

Although these safety devices can provide an extra level of protection during an MVA, unexpected deployment during the extrication process can cause serious injuries to both occupants and rescuers. As part of the size-up procedure, firefighters should be made aware of any undeployed air bags or ROPS, especially those near trapped occupants or rescuers. Rescuers should remove trim around the vehicle's doors and posts to identify the presence of these SRS components. Before making any cuts, use a small prying tool to remove interior trim and look for hidden pistons, canisters, or electrical wiring.

Alternative-Fuel Vehicles

Most of the vehicles on the road today are conventional vehicles, but alternative-fuel vehicles are becoming more common. The different types of alternative-fuel vehicles are summarized in **TABLE 26-1**. They are also described further in Chapter 14, *Fire Suppression*.

Each of these unique types of alternative-fuel vehicles presents tactical and safety challenges for firefighters conducting firefighting and extrication operations. Electric-drive vehicles are especially hazardous due to their powerful batteries and high-voltage electrical systems. Most of electric-drive vehicles are designed so that the high-voltage system is isolated once the low-voltage electrical system is disconnected (see Chapter 14, *Fire Suppression*). Bright orange, high-voltage cables are routed from a high-voltage battery pack to the inverter, DC-DC converter, and electric motor. In electric-drive vehicles, the high-voltage traction battery supplies direct current (DC) power. However, most electric motors require alternating current (AC) power to run. The **inverter** takes DC from the high-voltage battery and converts it to AC to power the electric motor. The **DC-DC converter** converts DC from one voltage level to another. In electric-drive vehicles, DC from the high-voltage battery is routed through the DC-DC converter and reduced to lower DC voltages to power the headlights, interior lights, radio, fans, SRS, and other low-voltage systems within the vehicle. In addition, when the high-voltage battery in an electric-drive vehicle is charging using an outside source, it needs to convert the AC from the charging station to DC. An on-board charger or the DC-DC converter converts AC to DC so that the battery can receive a charge. Some electric-drive vehicles have a combination inverter/converter.

Most of the high-voltage components are routed along the underside of the vehicle and under the hood, but usually not in areas where cuts will be made during extrication operations. Avoid touching, cutting,

TABLE 26-1 Types of Alternative-Fuel Vehicles

Vehicle Type	Description
Battery electric vehicles (BEV)	These vehicles use an electric motor that is powered by a high-voltage battery that must be recharged from an external source such as an electrical outlet or a charging station.
Extended range electric vehicle (EREV)	These are electric vehicles that use an onboard fuel-powered engine that acts as a generator to recharge the high-voltage battery when it is near depletion, thus extending the driving range. The fuel-powered engine does not assist in the vehicle's propulsion. The EREV can also be plugged into an external source to recharge the high-voltage battery.
Hybrid electric vehicles (HEV)	These vehicles use both a battery-powered electric motor and a conventional liquid-fueled engine for propulsion. The battery powers the electric motor. The gasoline-powered engine provides auxiliary power and recharges the battery.
Plug-in hybrid electric vehicles (PHEV)	These vehicles function the same as HEVs with the added benefit of being able to be recharge the high-voltage battery when the vehicle is parked by plugging in to an external power source.
Blended liquid fuel–powered vehicles	These vehicles use a blend of liquid fuels, often called flex fuel (labeled as E85 at most gas stations). The most common combination is gasoline and another flammable liquid, such as methanol or ethanol. Diesel fuel may be blended with other combustibles to produce a blended diesel fuel, and gasoline may be blended with biofuels, derived from agriculturally produced products such as grains, grasses, or recycled animal oils.
Compressed gas–powered vehicles	These vehicles use one of three forms of compressed gas: compressed natural gas (CNG), liquefied natural gas (LNG), and liquefied petroleum gas (LPG; also called propane).
Fuel cell–powered vehicles	These vehicles use fuel cells that generate electricity through a chemical reaction between hydrogen gas and oxygen gas. These vehicles have a hydrogen gas storage tank.

or piercing these cables and components, as they can cause electrical shock.

To learn more about a specific electric-drive vehicle and to know the steps required to disable the electrical systems, review the manufacturer's manual and the ERGs offered by the NFPA and ESA so that you can respond safely to MVAs involving electric-drive vehicles and other alternative-fuel vehicles.

Tools and Equipment

This section provides a brief overview of the more common tools and equipment used during vehicle rescue and extrication incidents. For more information and a better understanding of how these tools work and their various applications in the fire service, see Chapter 7, *Firefighter Tools and Equipment*.

Personal Protective Equipment

An important component of safety is wearing the proper personal protective equipment (PPE) before engaging in vehicle rescue and extrication operations. Firefighters should follow local standard operating procedures (SOPs) to select the appropriate PPE for vehicle rescue incidents. In general, firefighters engaged in vehicle rescue operations must wear the same PPE components as they wear for structural firefighting, plus hearing protection and clothing to protect against bodily fluids. Respiratory protection may be necessary if personnel will be operating in an area that is deemed immediately dangerous to life and health (IDLH) or is suspected of containing airborne contaminants.

Structural firefighting protective clothing that is compliant with NFPA 1970, *Standard on Protective Ensembles for Structural and Proximity Firefighting, Work Apparel and Open-Circuit Self-Contained Breathing Apparatus (SCBA) for Emergency Services, and Personal Alert Safety Systems (PASS), 2024 Edition*, meets and in most cases supersedes the minimum PPE requirements for vehicle rescue and extrication. However, this type of PPE tends to be bulky, hot, and difficult to work in for extended periods of time. A better choice of PPE for rescue operations is clothing and equipment that is compliant with NFPA 1951, *Standard on Protective Ensembles for Technical Rescue Incidents, 2020 Edition*. Protective clothing that meets the requirements of NFPA 1951 is generally lighter weight than protective clothing suitable for structural firefighting.

The type of incident, the hazards at the scene, the duration of the incident, the type of equipment and tools being used, and the weather all affect the choice of appropriate PPE. For example, most rescues involving vehicles happen in areas where traffic poses a hazard to firefighters and other responders. Some jurisdictions mandate the use of high-visibility safety vests for responders operating in or near traffic. Unless contraindicated, it is always recommended that, even if not required by law, a high-visibility safety vest be worn when operating at vehicle rescue and extrication incidents unless you will be exposed to fire conditions. This will alert drivers and other vehicle operators of your presence, especially in low-light or dark conditions. An SCBA might also be required to be worn in some vehicle extrication incidents where the potential for a respiratory hazard exists. Regardless of the type of PPE worn, firefighters must follow the local standard operating procedures or guidelines (SOPs or SOGs) for PPE for vehicle rescue and extrication incidents.

Power Tools

Power tools are tools that are powered by an energy source other than manual labor, such as hydraulic, electric, or pneumatic power. There are a multitude of types of power tools used throughout the world today, all with different capabilities, advantages, and disadvantages depending on the manufacturer and the intended application. NFPA 1960, *Standard for Fire Hose Connections, Spray Nozzles, Manufacturer's Design of Fire Department Ground Ladders, Fire Hose, and Powered Rescue Tools, 2024 Edition,* outlines the minimum requirements for design, performance, testing, and certification of powered rescue tools. Firefighters need to be familiar with the different tools available for vehicle rescue and extrication and how to operate them. It is not uncommon for power tools to fail during an incident, and rescuers should be prepared to adapt and overcome obstacles.

Hydraulic Power Tools

Hydraulic tools use pressurized fluid to exert force and allow you to apply several tons of force on a small area. Due to the power and speed of operation, hydraulic rescue tools are superior to other tools when used properly and for the right application. This is why most fire departments that respond to vehicle rescues are usually equipped with a set of hydraulic rescue tools. The major disadvantage of hydraulic tools is that they are much heavier than electric and hand tools and usually have limited maneuverability in tight spaces. But other than that, hydraulic rescue tools are the preferred choice for most rescuers operating at a vehicle extrication incident. The main types of hydraulic rescue tools used at these incidents are as follows (**FIGURE 26-8**):

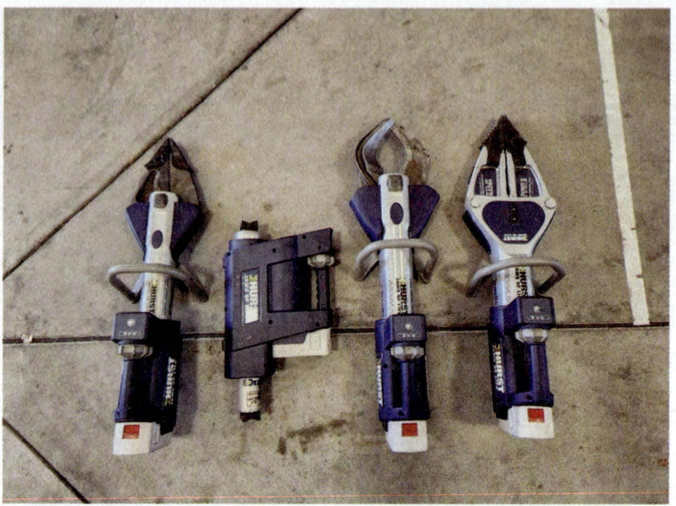

FIGURE 26-8 Battery-operated hydraulic rescue tools, including cutter, spreader, ram, and combination tool.

Courtesy of Aaron Miranda.

- Hydraulic spreader, used for prying, spreading, and crushing
- Hydraulic cutter, used to cut or shear metal
- Hydraulic ram, used for pushing
- Hydraulic combination tool, a combination of a hydraulic spreader and a hydraulic cutter

The power source for hydraulic rescue tools varies depending on the manufacturer. Conventional hydraulic rescue tools have a power unit that most commonly uses gasoline to operate. These power units come in different sizes and have the ability to operate multiple tools simultaneously. Most modern hydraulic units are high-pressure tools that operate at 10,500 pounds per square inch (psi; 72,395 kilopascals [kPa]) and use a mineral oil as their hydraulic fluid. Some older generation hydraulic power tools still in use are low-pressure and operate at 5000 psi (34,437 kPa) and use a synthetic hydraulic fluid. To avoid damaging the tools by adding the wrong type of hydraulic fluid, reference the manufacturer's owner's manual when performing maintenance on the equipment.

Most manufacturers of hydraulic rescue tools also offer battery-powered hydraulic rescue tools that have a self-contained pump operated by an electric motor. These tools offer the same capabilities as conventional hydraulic tools in a more portable package. Battery-power hydraulic rescue tools require minimal setup time and can be deployed and operated extremely quickly. They are also quieter than conventional hydraulic tools, which allows for better communication on scene. One disadvantage of battery-powered hydraulic tools is that because they have self-contained pumps, they tend to be bulkier and heavier than conventional hydraulic rescue tools. Also, they are limited by their battery capacity and in some cases will stop working without any warning. If your department uses battery-powered tools, it is imperative that additional replacement batteries are available to finish the task at hand.

Electric Power Tools

Electrically powered tools are powered by rechargeable batteries, portable or vehicle-mounted generators, or standard household current. Like battery-powered hydraulic tools, battery-powered electric tools are more portable than conventional electric tools and do not require a nearby power source, but they are limited by their battery capacity, so you should have backup batteries available. The most commonly used electric tool in vehicle rescue and extrication is the reciprocating saw. When paired with the right blade, reciprocating saws are an excellent alternative tool for cutting vehicle posts and other vehicle components. Circular saws and portable scene lights are two other electrically powered tools used in vehicle extrication and rescue.

Pneumatic Power Tools

Pneumatic tools operate using air under pressure as the power source. Pneumatic tools that rescuers might use include air impact wrenches, air shores, rescue air-lift bags, and air chisels (**FIGURE 26-9**). The source of the compressed air for pneumatic tools can be from SCBA cylinders, air compressors, or a vehicle-mounted system. Each pneumatic tool has a specific operating pressure, so it is important that the air supply source is able to deliver the required pressure for the tool to properly operate. Although pneumatic tools are less powerful, slower, and usually more labor-intensive compared to other power tools, if they are designed specifically for the task at hand and there is no equivalent hydraulic or electric option available, they can be the best choice in that particular set of circumstances.

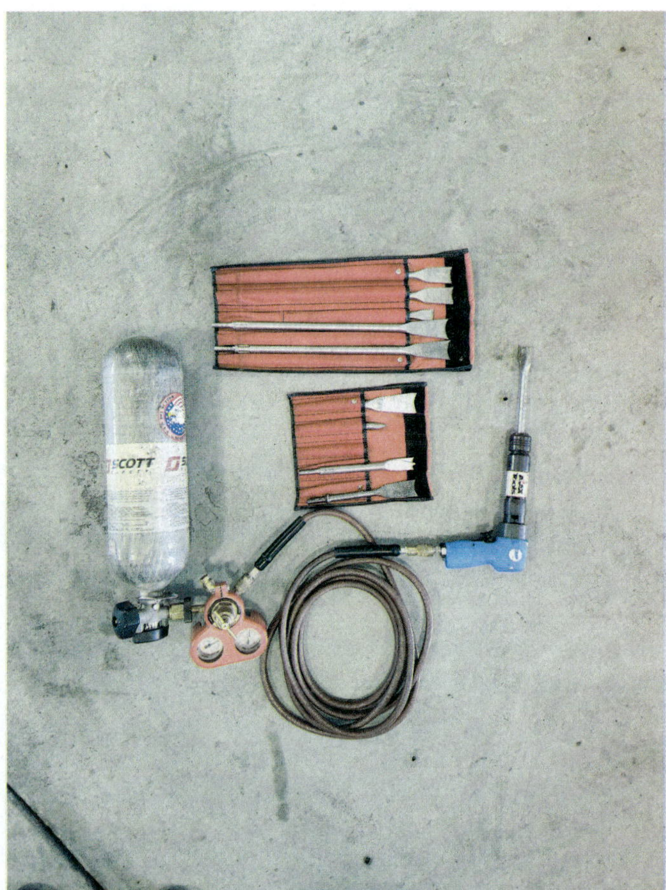

FIGURE 26-9 Pneumatic chisels, also referred to as air chisels, are used to cut sheet metal or hardened steel.

Courtesy of Aaron Miranda.

Hand Tools

A variety of hand tools are manufactured specifically for vehicle rescue, but conventional automotive and building construction tools as well some firefighting hand tools can be used to aid in extrication operations. Remember that hand tools extend or multiply the actions of your body to increase your effectiveness performing specific functions, and most of these tools operate using simple mechanical principles. Hand tools for vehicle rescue and extrication include striking, leverage, cutting, lifting, pushing, and pulling tools. The hand tools that are used the most often for vehicle rescue applications are the sledgehammer, rubber mallet, flat-head axe, hacksaw, bolt cutters, cable cutters, pry bar, Halligan tool, window or center punch, folding knife, and seat belt cutters, to name a few.

A versatile hand tool that can be used for lifting, pulling, and pushing is the First Responder Jack (FRJ) manufactured by Hi-Lift (**FIGURE 26-10**). The FRJ is a hand-operated tool that uses a mechanical ratcheting system. It is rated to be able to support about 4660 pounds (lb; 2118 kilograms [kg])—this rating is called the **working load limit**. The FRJ can be used as a jack or in place of a hydraulic spreader or a hydraulic ram.

Stabilization and Lifting Tools

Before you can access and remove victims from a vehicle, the vehicle must be stabilized. The goal of vehicle stabilization is to expand the vehicle's base so it has a

FIGURE 26-10 Most tasks that require the use of a hydraulic spreader or ram can also be accomplished with the FRJ.

Courtesy of Aaron Miranda.

wider footprint. The wider footprint means the vehicle is more stable and unwanted movement is prevented. In some circumstances, in addition to stabilizing the vehicle to prevent movement, it will also be necessary to lift the vehicle or remove objects from on top of a vehicle to gain access to and remove victims.

Cribbing

Cribbing is basically a stack of wood or composite materials that supports and stabilizes objects. Cribbing is sometimes arranged in a crisscross pattern to increase the stability of the stack. It is the most commonly used tool for stabilization tasks. Treated 4- by 4-inch (in.;10- by 10-centimeter [cm]) boards that are 18 to 24 in. (46 to 61 cm) in length are often used, but other dimensions are available. The most commonly type of wood used for cribbing is southern yellow pine or Douglas fir because of its ability to compress and lock in place under a load. When using this type of soft wood, the load-bearing capacity of each contact point is estimated at 6000 lb (2727 kg). Composite cribbing offers some advantages over wood because it is lighter and will last many times longer. It is also resistant to chemicals such as fuel, oil, and other substances. However, composite cribbing is more expensive to replace and may fail without warning, as opposed to wood cribbing, which tends to crack prior to failure.

Rescue Struts

A **strut** is an adjustable-length structural support, made from a metal or composite beam, that is used to stabilize and prop up something in danger of collapse. In vehicle stabilization, struts are used to transfer the load of the vehicle onto a wider base. Commercial struts, made of steel, aluminum, or a composite material, are usually telescoping devices that extend to various lengths to accommodate a multitude of stabilization scenarios. Struts have a wide range of load-bearing capacity depending on the manufacturer, ranging from 4000 to 88,000 lb (1814 to 39,916 kg).

Struts can also be constructed of wood, commonly 4- by 4-in. (10- by 10-cm) sections in various lengths ranging from 3 to 6 feet (ft; 0.9 to 1.8 meters[m]) or greater. A base and head attachment can be added to wood struts to be able to use accessories such as chain hooks, chains, and ratchet straps. In most cases, a ratchet strap or other tensioning device is used together with a strut to create the tension buttress system (**FIGURE 26-11**). Ratchet straps are usually a minimum of 2 in. (5 cm) wide and have a working load limit of 3330 lb (1510 kg). Other sizes and rated load capacities for ratchet straps are available to use for supporting a strut base. Firefighters should use the ratchet strap size and load rating recommended by the manufacturer of the strut system for the type of load being supported.

Rescue-Lift Air Bags

A **rescue-lift air bag** is an inflatable bladder made of rubber or synthetic material that is pneumatically filled with compressed air to lift an object or to spread one or more objects away from each other to assist in freeing a victim (**FIGURE 26-12**). Rescue-lift air bags are never used to shore or stabilize a vehicle by themselves; cribbing is always required in conjunction with rescue-lift air bags.

Rescue-lift air bags are often used to lift a vehicle or object off a victim. Firefighters should use extreme caution when applying this technique and should follow all safety precautions outlined in the manufacturer's instructions for the rescue-lift air bag. Cribbing must be used in conjunction with rescue-lift air bags

FIGURE 26-11 Struts and ratchet straps are used to create a tension buttress system that supports and stabilizes a vehicle.
Courtesy of Aaron Miranda.

FIGURE 26-12 A high-pressure, rescue-lift air bag and components.
Courtesy of David Sweet.

and whenever firefighters are lifting a load because instability can occur from weight shifts, or the rescue-lift air bags may fail under a load.

Several types of pneumatic rescue-lift air bags are available: low pressure, medium pressure, and high pressure. None of these devices should be used without first properly chocking the wheels opposite of the rescue-lift air bag and cribbing the vehicle or object as it is lifted. These safety precautions will prove invaluable in case the rescue-lift air bag suffers a catastrophic failure.

As they age, rescue-lift air bags become more prone to failure, so they should be tested regularly. Replace rescue-lift air bags per your department's SOPs or the manufacturer's recommendations.

Low-Pressure Rescue-Lift Air Bags. Low-pressure rescue-lift air bags are commonly used for recovery operations and are sometimes used for vehicle rescue operations. These devices come in many sizes and shapes; square air bags, however, offer greater stability. Because of their lightweight construction, low-pressure rescue-lift air bags can be less stable (at least until they are fully inflated) than high-pressure rescue-lift air bags, which have a lower lift height and stronger construction.

Medium-Pressure Rescue-Lift Air Bags. Medium-pressure rescue-lift air bags are designed so that they contain either two or three cells. For general vehicle rescue, these devices are not appropriate. Medium-pressure lift air bags are more suitable for aircraft, medium or heavy truck, or bus rescue, and for recovery work.

High-Pressure Rescue-Lift Air Bags. High-pressure rescue-lift air bags are the type used most frequently by firefighters. These devices have a very sturdy construction and are generally made of vulcanized rubber mats that are reinforced by steel or other material woven into a fiber mat and then covered with rubber. Traditional high-pressure bags tend to round out and decrease surface contact when inflated. Newer designs called flat-form air-lift bags retain their flat profile in the center as they are inflated. This eliminates rolling and shifting that can commonly occur with traditional bags. Multi-cell high-pressure air-lift bags utilize a unique lifting system that offers a distinct height advantage over traditional bags. They commonly consist of two cell bags joined together and preconnected, or two bags locked together with a threaded connector, creating a multi-cell system. When using any of the available high-pressure air bag systems, chocking and cribbing remain essential actions to provide a safe working environment.

FIREFIGHTER TIP

- Do not exceed the maximum amount of high-pressure rescue-lift air bags that can be stacked on top of one another per the manufacturer's recommendations.
- Do not use a rescue-lift air bag to pull a steering column.
- Do not use a rescue-lift air bag to stabilize a vehicle; cribbing must be the primary stabilizer.
- Never operate the rescue-lift air bag system without proper training and complete understanding of how the system works.
- When stacking rescue-lift air bags, the largest size should be on the bottom and the smallest on top.
- It may be necessary to place a sheet of protective material (plywood, grip mat, mud flap, etc.) on the ground under the bottom air bag or above the top air bag to protect them from sharp objects. Do not use boards or plywood between rescue-lift air bags.
- Refer to your department's SOPs and the manufacturer's instructions for effective and safe lifting techniques.
- Clean rescue-lift air bags by following the manufacturer's recommendations.
- Test rescue-lift air bags regularly.
- Never store a rescue-lift air bag near gasoline.

Responding to the Scene

When responding to the scene of a MVA, firefighters need to maintain a high level of awareness and consider the possibility of the incident being upgraded to a vehicle rescue. This will place the firefighter in the right mindset and state of readiness in case of a confirmed rescue. Listen carefully to the dispatch information for key words, such as "vehicle roll-over" or "multiple vehicles involved." These types of crashes have a higher potential for victim entrapment and severe injuries and may require additional resources. Consider other factors such as the time of day, the incident location, the speed of travel (highway or residential roads), and weather conditions to heighten your suspicions of the possibility of a rescue.

FIGURE 26-13 Scan the area to identify hazards at a scene of an MVA.

Courtesy of Aaron Miranda.

Arrival and Size-Up of the Scene

The first step in the extrication process is arrival and size-up. After arriving at the scene of an MVA, it is important to assess the hazards present and to determine the types of vehicles involved, the scope of the incident, the number and severity of injuries, and the need for additional resources. Scan the entire area for hazards. Did the vehicle hit an electrical source or rupture a gas line (**FIGURE 26-13**)? Hazards are everywhere, and a complete size-up might save everyone on the scene from further injury. After performing a size-up, the scene needs to be stabilized. This step consists of reducing, removing, or mitigating the hazards at the scene. The order in which these hazards are addressed will depend on the specific conditions at a scene and the amount of risk that each hazard poses. The scene needs to be stabilized to make it safe for firefighters and rescuers to operate prior to doing any type of work on a vehicle involved in an MVA.

The incident commander (IC) will usually perform a size-up of the scene by conducting a 360-degree walk-around of the scene. During this size-up, the IC evaluates the hazards present and determines the number of victims. Using the information obtained from the scene size-up, the IC can create an action plan and call for additional resources, if needed.

FIGURE 26-14 Many fire departments place an apparatus at an angle to the MVA.

Traffic Hazards

The hazards posed by traffic need to be handled quickly before they lead to additional MVAs or injuries. Deciding where to position emergency vehicles should take into account the safety of emergency workers, the victims, and the motorists traveling along the road. If possible, position emergency vehicles so that they ensure the safety of the responders and victims without disrupting traffic any more than necessary. Do not hesitate, however, to request that the road be closed, if necessary. Sometimes the most important action to take at the scene of an MVA is to slow, stop, or divert the flow of traffic before proceeding with additional actions. Remember—safety first!

Position large emergency vehicles so that they provide a barrier against motorists who fail to recognize or heed emergency warning lights. Many departments place apparatus at an angle to the MVA and pointing away from oncoming traffic (**FIGURE 26-14**). If the emergency apparatus is struck from behind while in this position, it should be pushed to the side of the accident. Avoid positioning emergency vehicles in a manner that confuses oncoming motorists. Make sure that the apparatus headlights do not shine in the direction of approaching drivers to avoid blinding them. Place traffic cones or flares to direct motorists away from the accident (**FIGURE 26-15**).

SAFETY TIP

Be sure to look for leaking fuels before using flares.

FIGURE 26-15 Traffic cones or flares can be placed to direct motorists away from the MVA.

Firefighters need to work together with law enforcement officials to control traffic in a manner that is safe for emergency workers, victims of the MVA, and motorists. If law enforcement officials are not on the scene when you arrive, verify that they are aware of the incident and that they have been dispatched.

SAFETY TIP

Extrications are dynamic events, especially on roadways, and firefighters must be constantly aware of what is going on around them. Maintain situational awareness at all times.

As mentioned previously, PPE must be worn at all MVAs. Unless you are exposed to or are likely to be exposed to fire conditions, you should wear a high-visibility safety vest that is compliant with the American National Standards Institute/International Safety Equipment Association (ANSI/ISEA) 107-2015 standard for Type P vests (**FIGURE 26-16**). These break-away vests are designed to provide high visibility to oncoming traffic. They are easily removed if they become entangled during the extrication process.

Before exiting fire apparatus at an emergency scene, be alert for vehicles that might cause injury to firefighters. Do not assume that motorists will heed the warning lights. Let law enforcement personnel coordinate traffic control.

Crowd Control

When you arrive on scene, bystanders may want to get involved and help. Sometimes it may be difficult to distinguish between those involved in the incident and

FIGURE 26-16 Break-away vests are designed to provide high visibility to oncoming traffic.

Courtesy of Aaron Miranda.

those who are just passing by. These well-intentioned Good Samaritans can be a dangerous liability and create problems for the victim, themselves, or rescue workers. These individuals may be trying to pull victims from the vehicle as you arrive on scene. They may yell at you that the vehicle is going to catch on fire or that the people in the vehicles are dying in an attempt to get you to move faster. This only adds to the pressure and stress already created by the incident. Request that law enforcement remove individuals not involved in the incident if they do not remove themselves voluntarily. It is imperative that crowd control be maintained during rescue operations to protect responders from individuals who might attempt to enter the emergency site. This will provide the necessary space so that emergency personnel can operate without worrying about bystanders interfering with the rescue efforts and/or being injured.

Fire Hazards

Motor vehicles use a variety of fuels and lubricants that might pose fire hazards. In addition, vehicles may carry combustible objects and ignitable liquids in the trunk or passenger area. Look for spilled fuel and other ignitable substances. A short in the electrical system or a damaged battery may cause a post-MVA fire by releasing sparks and igniting spilled fuel. If the vehicle is an electric-drive vehicle and the high-voltage battery is damaged, a thermal imager or a surface thermometer should be used to determine if the high-voltage battery is producing excessive heat above the manufacturer's recommended operating temperatures. Battery pack temperature above 200°F (93.3°C) or a rapid increase in temperature is a sign of thermal runaway—when the cells in the battery short circuit, overheat, and ignite. Sparks, smoke, and gurgling or popping sounds coming from the high-voltage battery are also signs of thermal runaway. These fires may trap the occupants of the vehicle and require rapid fire suppression.

If fuel or other ignitable liquids are spilled, or if the potential for a high-voltage battery fire exists at the scene of an MVA, a charged hose line should be advanced to the vehicle. This hose line should be at least 1½ in. (38 mm) in diameter and be staffed by a firefighter in full firefighting PPE and SCBA. MVAs that pose large fire hazards or actual fires may require additional fire suppression resources, which should be requested as soon as possible. This is especially true with lithium-ion battery fires, which require large amounts of water to suppress the fire and keep the temperature within the battery down to stop the propagation process. Small fuel spills can be mitigated by using an absorbent material to remove the fuel from the area around the damaged vehicle. Some departments advance a charged hose line on the scene of all MVAs. Refer to your department SOPs for the proper procedures in your jurisdiction. See Chapter 14, *Fire Suppression*, for more information on extinguishing vehicle fires and on thermal runaway.

Electrical Hazards

Downed and low-hanging power lines represent an electrical hazard. Look closely to determine whether the MVA has damaged any electrical power poles. Downed and low-hanging power lines may be difficult to see, and they may energize other objects such as fences, guardrails, and guy wires (cables designed to add stability to a structure). Energized objects create a deadly hazard for rescuers, the victims of the MVA, and bystanders.

Stabilizing electrical hazards posed by downed electrical wires is essential before firefighters attempt to approach the MVA site. If power lines are on or near the vehicle, do not approach the scene until electrical power has been disconnected by the utility provider. At times it may be necessary to instruct victims of an MVA to remain in their vehicles until the power can be turned off. Although it is frustrating to delay extrication and treatment because of downed power lines, this step is essential to avoid potentially life-threatening injuries to both rescuers and victims of the MVA.

Electric-drive vehicles also present electrical hazards. Their high-voltage batteries and electrical cables require special handling because they may pose an electrical or fire hazard. Do not touch any of these components; only specially trained personnel should handle the high-voltage system during an MVA involving an electric-drive vehicle.

Disconnecting the vehicle's low-voltage electrical system can mitigate electrical hazards posed by a damaged vehicle (described in Chapter 14, *Fire Suppression*). Follow your department's SOP for disconnecting the electrical system. Firefighters must weigh the need to disconnect the electrical system against the advantage of being able to operate electrically powered components of the vehicle.

SAFETY TIP

High-intensity discharge (HID) headlights are becoming more common on vehicles as standard equipment or as after-market modifications. HID headlights discharge several thousand volts of electricity. If the HID bulb is broken or removed, you may be exposed to an electrical hazard.

Other Hazards

Environmental conditions can lead to unique hazards at the scene of an MVA. MVAs that occur in rain, sleet, or snow, for example, present an added hazard for the rescuers and the victims. Weather-related issues must be considered for vehicle extrication operations in extreme heat, cold, rain, or snow. Both the victims and the rescuers must be protected from heat and direct sunlight on hot days. Provide shade for victims and rescuers in the vehicles and watch for signs of heat-related illness. In cold weather, cover victims with blankets. To prevent slips and falls, use caution while walking, working, and moving victims when the roads are wet or icy. Sprinkling sand or an oil absorbent material may help to give solid footing to rescuers. Be alert for oncoming vehicles that may slide into the operations area. Consider calling for additional resources so crews can be rotated to a climate-controlled environment for rehabilitation.

Many MVAs occur at night. It is important to provide adequate lighting so that rescuers can work quickly and safely. (Chapter 18, *Salvage and Overhaul*, describes the steps required to light an emergency scene.)

Assume that all vehicles are carrying hazardous materials until proven otherwise. Assess the scene for hazardous material placards, unusual odors, or leaking liquids. The Pipeline and Hazardous Materials Safety Administration (PHMSA), a part of the U.S. Department of Transportation (DOT), publishes the *Emergency Response Guidebook (ERG)*, which contains information about hazardous chemicals and recommendations for initial actions at an incident involving these chemicals. The *ERG* is distributed free of charge to emergency responders.

In a collision involving an electric-drive vehicle, if the high-voltage battery sustained direct physical damage, harmful or flammable gases may be released. It is recommended that personnel use a multi-gas meter when approaching the vehicle to detect the presence of toxic or combustible gases. If the high-voltage battery is damaged, or if personnel detect the same things that could indicate thermal runaway—increasing temperature, sparks, smoke, and gurgling or popping sounds coming from the high-voltage battery—the vehicle must be ventilated immediately by opening all doors, rolling down or breaking the windows, and opening the cargo compartment to prevent gas build-up.

Be especially alert for the presence of infectious bodily fluids. Be prepared for the presence of blood and exercise universal precautions. Specifically, do not let blood or other bodily fluids come in contact with your skin. Wear PPE, including gloves and eye protection, to protect yourself from the contaminated fluids and sharp objects that may be present at the site of an MVA. If you or your clothes become contaminated, report the contamination, document it, and then clean and wash the affected clothes and equipment.

Threats of violence may be present at the scene of some MVAs. In particular, intoxicated people or those who are upset with other motorists may pose a threat to you or to other people present at the scene. Be alert for weapons that are carried in vehicles.

Occasionally, animals become a hazard at the scene of an MVA. In particular, dogs and other family pets may be protective of their owners and threaten rescuers. In incidents involving animals, it may be necessary to secure the animals before proceeding with other activities. If the dog's owner is injured, remove the dog from the MVA site, and place it with an uninjured family member or other responsible person. If horses or other farm animals are involved in an MVA, they may need care, and you may need to call in specialized resources to assist with this type of incident. The preincident plan should identify who in your community is responsible for caring for large animals at rescue scenes.

MVAs often create a variety of sharp objects, such as mangled pieces of metal, plastic, and glass, that pose significant hazards for rescuers and victims. To reduce the chance of injury, rescuers should wear proper personal protective clothing. Sharp edges may be covered with blankets to reduce the possibility of injury to the victims or rescuers. Be especially careful when moving victims, making sure that they are protected from contact with any sharp objects.

To perform a scene size-up at an MVA, follow the steps in **SKILL DRILL 26-1**.

SKILL DRILL 26-1

Performing a Scene Size-Up at a Motor Vehicle Accident Firefighter II, NFPA 1010: 7.4.1

1. Position emergency vehicles to protect the MVA scene and the rescuers. Take any additional actions needed to prevent further MVAs.

2. Perform a quick initial assessment as you arrive on the scene, establish command, and give a brief initial radio report. Make sure all personnel wear proper PPE, including SCBA if conditions require. Establish fire suppression protection if fluids are released or if extrication may be required; a minimum of one 1½-in. (38-mm) hose line should be in place.

3. Perform a 360-degree walk-around of the vehicle(s) and entire scene to identify potential hazards including signs of thermal runaway in EVs. Approach vehicles from the side or at a 45-degree angle from the front, and look for hazards above and below the vehicle. Determine the stabilization equipment needed to prevent further movement of the vehicle involved in the incident. Identify the vehicle's fuel type or drive system (conventional, hybrid, all electric, or alternative-fuel vehicle). Scan for ejected and walking wounded victims. Identify additional vehicles involved in the accident.

SKILL DRILL 26-1 CONTINUED

Performing a Scene Size-Up at a Motor Vehicle Accident Firefighter II, NFPA 1010: 7.4.1

4. Determine the number of patients, the severity of their injuries, and the amount of entrapment. Determine interior hazards, such as supplemental restraint system (SRS) components and their locations. Give an updated report and call for additional resources if needed.

5. Establish a secure working area and an equipment staging area. Direct personnel to perform initial tasks such as immobilizing, disabling, and stabilizing the vehicle(s).

Site Operations

After the initial size-up and mitigating any immediate hazards, a detailed assessment of the scene needs to be completed and the vehicles need to be identified, immobilized, disabled, and stabilized before operations can begin.

Inner and Outer Surveys

Inner and outer surveys are detailed 360-degree inspections of the scene that are completed simultaneously by two or more personnel, depending on the size of the incident and number of vehicles involved. Personnel move in the same or opposite direction from each other, ensuring that every area encompassing the scene is investigated. These surveys provide additional information to personnel on scene about hazards, types of vehicles involved, number of victims, the level and type of entrapment, any possible ejections, and any additional resources needed.

The **inner survey**, also known as the **inner circle**, consists of a "tight," 360-degree inspection of the vehicles involved. Usually, the first-arriving company officer or the most experienced personnel perform this task. They approach the vehicle from the side or at a 45-degree angle from the front and move around the vehicle, inspecting the driver side, rear, passenger side, and front. The inspection should include the top of the vehicle and undercarriage on all sides. Personnel performing the inner survey should maintain a distance approximately 3 to 5 ft (0.9 to 1.5 m) from the vehicle and avoid touching it if there are any electrical hazards that have not been cleared yet. The goal of the inner survey is to positively identify the vehicle's drive system (conventional or alternative fuel), determine the locations of any victims and whether they are entrapped, identify entry and exit points, and determine vehicle stability for access to and removal of the victim(s). In addition, any IDLH hazards identified need to be identified immediately and all members ordered to withdraw until the hazard has been mitigated or made safe to continue.

For most incidents, the **outer survey**, or **outer circle**, extends from the perimeter of the inner survey outward to a distance sufficient to cover where the impact first occurred and the current resting point of the vehicle(s). Factors such as speed of travel and type of roadway (highway vs. residential roadway) could cause the survey to extend as far as 100 ft (30.5 m) or greater. The goal of the outer survey is to be able to identify IDLH hazards, victims who might have been ejected, any additional vehicles involved in the incident, **walking wounded** (victims who were involved in the accident in some way and who have injuries, but who are not immediately identifiable), and any other imminent or potential hazards that might have been missed during the initial size-up. Again, any IDLH hazards identified need to be identified immediately and all members ordered to withdraw until the hazard has been mitigated or made safe to continue. Also, it is recommended that personnel surveying the outer circle use a thermal imager as they conduct the search, especially at night or in low light conditions, to help find victims who might have been ejected and are now concealed by debris, brush, or darkness.

After completing the inner and outer surveys, a more accurate picture of the incident is obtained. Personnel then meet to discuss their findings and compile their information to determine the objectives for incident strategy.

Identify, Immobilize, Disable

Before entering vehicles to start medical care and disentanglement of victims, the vehicle must be made safe or deemed safe to enter. The steps for doing this are to identify the type of vehicles involved, immobilize the vehicles to prevent inadvertent movement, and disable the vehicle. These steps are described further in Chapter 14, *Fire Suppression*.

Identify

First, identify the vehicles to determine the type of drive system—are they conventional or alternative-fuel vehicles? If any of the vehicles are alternative-fuel vehicles, firefighters need to be aware of the many additional hazards they pose to firefighters and other personnel on scene, especially if the vehicles are electric-drive. Electric-drive vehicles have silent powered drive—that is, the vehicle is on, but no engine noise is produced to alert those nearby that the vehicle is on. Additional hazards of electric-drive vehicles include the possibility of electrical shock, exposure to toxic or flammable gases, battery fire, and re-ignition.

FIGURE 26-17 Do not assume that a vehicle is not an alternative-fuel vehicle because it does not look modern.

Courtesy of Aaron Miranda.

It is often difficult to differentiate conventional vehicles from alternative-fuel vehicles from a distance. Do not assume that because a vehicle does not look modern, it is not an electric-drive vehicle or some other type of alternative-fuel vehicle. Consumers can modify and retrofit their vehicles to convert them from conventional to alternative-fuel vehicles (**FIGURE 26-17**). For this reason, firefighters and rescuers should approach all vehicles involved in an MVA as if they are alternative-fuel vehicles until proven otherwise. This includes approaching the vehicle at a 45-degree angle whenever possible to avoid being hit or run over by inadvertent vehicle movement.

To positively identify alternative-fuel vehicles, look for formal and informal identifiers. Formal identifiers include badges or other labels and models that are available only as alternative-fuel vehicles such as all Tesla vehicles and the Toyota Prius. If you do not see badges or labels that identify the vehicle as an alternative-fuel vehicle, do not assume that it is a conventional vehicle. The badges and labels might not be visible, or they might have fallen off or gotten destroyed in the crash. Another formal identifier might be the license plate. Some states require that electric-drive vehicles have a license plate identifying them as such, and other states offer this option.

Informal identification methods include visual clues such as electrical cables, battery vents, charging ports, and the instrument panel. Medium- (blue) and high-voltage (orange) cables under the hood, near the high-voltage battery, and on the underside of the vehicle are indicative of an electric-drive vehicle (**FIGURE 26-18**). Medium- and high-voltage cables are usually not in areas considered points to cut during extrication. Also, because high-voltage battery packs produce a lot of heat and require adequate ventilation,

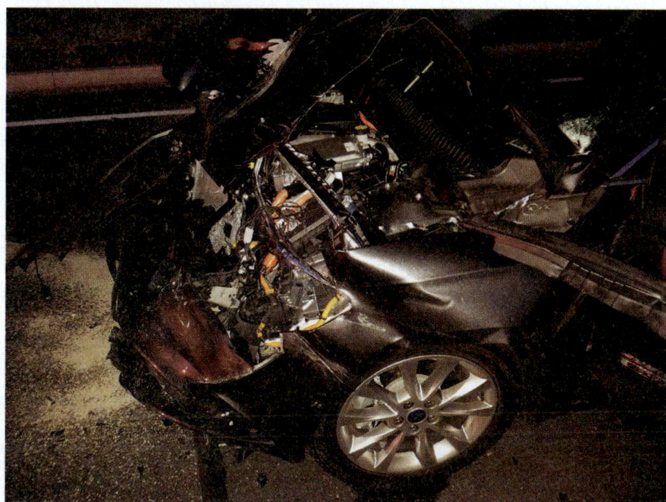

FIGURE 26-18 Orange high-voltage cables might be exposed after a car crash, identifying the vehicle as an electric-drive vehicle.

Courtesy of Aaron Miranda.

FIGURE 26-19 Charging ports on PHEVs and BEVs have a cover similar to that of a gas filling port and are usually located at the front of the vehicle or in the front or rear fenders.

Courtesy of Aaron Miranda.

battery vents near the battery that allow airflow around the case indicate an electric-drive vehicle. Finally, a charging port indicates a PHEV or a BEV. The charging port will have a cover similar to that of a gas filling port and is usually located in the front of the vehicle or in the front or rear fenders (**FIGURE 26-19**).

Immobilize

After identifying the type of vehicle, the next step is to immobilize it. As soon as it is deemed safe to touch the vehicle, deploy wheel chocks in front and behind one or more wheels. Wheel chocks should be strategically

FIGURE 26-20 Some vehicles have push-button shifters and electronically operated parking brakes. Make sure you engage the parking brake before disabling the low-voltage electric system.

Courtesy of Aaron Miranda.

placed to prevent forward and rearward movement but not interfere with stabilization efforts. The next step in the immobilization efforts is to place the vehicle in "park" and to engage the parking brake. If the driver is still inside the vehicle and is able to follow commands, ask them to perform these tasks. Some modern vehicles have some unique design features such as joystick shifters that return to the same position regardless of the gear selected or a push-button electronic shifter. Also, some vehicles have electronically operated parking brakes. Make sure the car is in Park and the parking brake is set before you disable the low-voltage electrical system, otherwise they will not engage (**FIGURE 26-20**).

Disable

The third step before beginning to extricate victims is to disable the vehicle's electrical system to prevent accidental starting, electrical shock, or fires. Alternate disabling methods involve removing fuses. Before disabling the vehicle, and if needed, roll down windows, move seats, unlock the doors, and open the trunk, as these systems will not work once the low-voltage power has been removed.

The first step in the disabling process is to shut off the vehicle ignition if accessible. If the driver is still inside the vehicle and able to follow commands, ask them to turn off the motor and hand you the key. If you need to perform these tasks, turn the key and remove it if it is a traditional key ignition, or push the start button and remove the proximity key from the vehicle if it can be located. In electric-drive vehicles, shutting off the ignition disengages the high-voltage systems, including the high-voltage battery and the DC-DC converter.

To prevent the high-voltage system from accidentally reenergizing, the low-voltage system needs to be disconnected. Remove a 6-in. (15-cm) section of the cable to the negative terminal (black cable) by making two cuts with a wire cutter. (Cut the cable to the negative terminal first to avoid producing sparks.) This procedure is described further in Chapter 14, *Fire Suppression*. For additional safety, do the same thing to the cable connected to the positive terminal (red cable). Make sure to do this to all 12-volt batteries in vehicles with multiple batteries. Once this is done, the vehicle's electrical system is disabled, including the SRS.

Firefighters should attempt to secure the key from the driver and any other occupant that also might have a key. Manufacturers recommend moving proximity keys at least 16 ft (4.8 m) away from the vehicle. If proximity keys cannot be located, follow normal shutdown procedures by pushing the button to turn it off and then disconnecting the low-voltage system. After disconnecting the low-voltage system, even if the car key is in the car, the vehicle will not start if you push the ignition button again.

If access to the ignition is not available, alternative methods for disabling passenger vehicles exist; however, these alternative methods vary by model. The best resource available to determine the correct method for disabling a vehicle is to refer to the manufacturer's *Emergency Response Guide*. For example, high-voltage batteries in electric-drive vehicles have manual service disconnects. These manual switches are generally there for automotive service technicians to be able to disable the high-voltage battery when performing maintenance on the vehicle. Recommendations for use by emergency responders and required equipment can vary by manufacturer. Consult the manufacturer's ERG before using the manual service disconnect to disable the vehicle's high-voltage system. In the absence of the manufacturer's ERG, the following methods work for most vehicles. First, disconnect the low-voltage system. Second, locate the high-voltage fuse(s) or the **relay** or relays—switch(es) that when closed complete a circuit and allow electrical current to flow—and pull them. This will prevent low-voltage electricity from reaching the high-voltage relays, keeping the circuit open and preventing the flow of electricity from the high-voltage battery.

Regardless of how the vehicle was shut down and disabled, remember that the high-voltage battery will still retain its charge. Always treat high-voltage components as if they are still energized.

SAFETY TIP

The DC-DC converter works like an alternator in a conventional vehicle—it powers the low-voltage components. In electric-drive vehicles, when the vehicle is on, the DC-DC converter supplies low-voltage power to keep the high-voltage relays closed. Simply disconnecting the low-voltage system does not open the high-voltage relays in many passenger vehicles. Whenever possible, the vehicle's ignition must first be turned off and then the low-voltage system can be disconnected.

To immobilize and disable the electrical system of common passenger and electric-drive vehicles, follow the steps in **SKILL DRILL 26-2**.

Incident Action Plan

After all the information obtained during the inner and outer surveys is gathered, the incident action plan (IAP) can be developed to guide the incident management process. The strategic parts of the IAP for vehicle extrication consist of the following three parts:

- **Primary access**, which is **plan A**, victim access, or the initial entry plan
- **Secondary access**, which is **plan B**, victim removal, or the extrication plan
- **Emergency escape route**, also called the **emergency escape plan**, which is the preplanned route that rescuers use to move themselves and the victim or victims to a safety zone or low-risk area

During primary access, rescuers use existing openings—that is, doors and windows—to create a pathway to the trapped and/or injured victim or victims so that a medical provider can access the victim or victims and initiate emergency trauma care and so that firefighters can provide other support functions from within the vehicle. During secondary access, rescuers

SKILL DRILL 26-2

Immobilize and Disable the Electrical System of a Vehicle Following a Motor Vehicle Accident Firefighter II, NFPA 1010: 7.4.1

1. Immobilize the vehicle by chocking the wheels. Chock both sides of one tire to prevent the vehicle from rolling by placing one wheel chock in front of a wheel and a second wheel chock in back of the wheel. If possible, chock more than one wheel. If you can access the controls, set the parking brake, and place the vehicle in park.

2. If possible, lower all automatic windows, move seats, unlock the doors, and open the trunk before shutting the motor off and disabling the electrical system. Shut the motor off by turning the key and removing it if it is a traditional key ignition or by pushing the start button and removing the proximity key (if possible). If you locate the proximity key, move it as far away from the vehicle as possible—at least 16 ft (4.8 m)—to prevent the possibility of an unintentional restart.

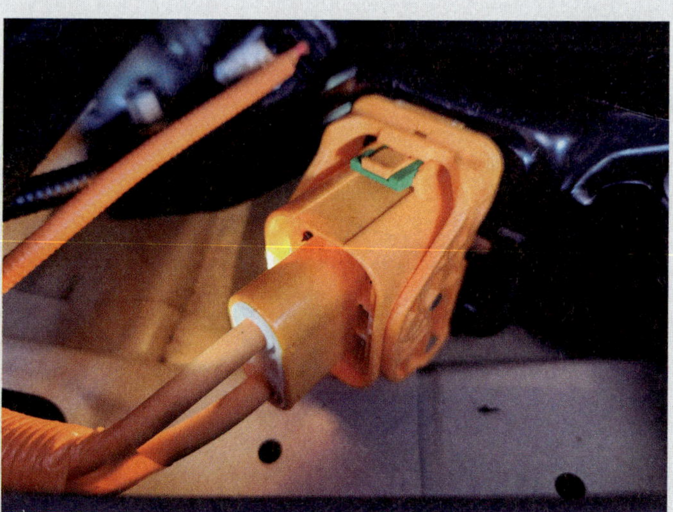

3. Determine the location of the 12-volt, low-voltage battery. Whenever possible, remove a section at least 6 in. (15 cm) long of the cable attached to the negative terminal (black cable) by making two cuts with wire cutters, and then repeat this with the cable attached to the positive terminal (red cable).

Locate the fuse box in the vehicle, identify the fuse(s) that control the low-voltage system, and remove them. Using this alternative method when the low-voltage battery cannot be located or accessed can prove difficult or time consuming, however. It is always recommended that the manufacturer's *Emergency Response Guide* be consulted to determine the best option for alternative shutdown methods.

In electric-drive vehicles, if the previous steps cannot be accomplished, some vehicle manufacturers recommend removing the high-voltage battery manual service disconnect (see the photo in Step 3). Consult the manufacturer's *Emergency Response Guide* before using the manual service disconnect to disable the vehicle's high-voltage system.

create openings that provide a pathway for removing or extricating the trapped or injured victims. The emergency escape route is mainly for the safety and well-being of rescuers when things do not go as planned, for example, if the scene becomes hazardous. It is also used when the victim's condition deteriorates based on information provided by emergency medical personnel on scene.

Vehicle Stabilization

Before disentanglement and extrication operations can begin, the vehicle needs to be made safe for rescuers to enter by stabilizing it to prevent any unwanted movement. Unstable vehicles pose serious challenges and risks to both rescuers and the victims involved in an MVA. The complexity of this depends on how the vehicles rest after a collision. Vehicles can present on all four wheels, on their roof or side, or even on top of or underneath other vehicles or objects. The goal of vehicle stabilization is to expand the vehicle's base and lower its center of gravity so it has a balanced footprint. Vehicles must be stabilized in the position they are found so that movement is prevented in all directions (**FIGURE 26-21**). There are a variety of stabilization devices and cribbing configurations available that will accomplish this goal.

Vehicle stabilization consists of two parts: primary and secondary stabilization. The goal of **primary stabilization** is to quickly stabilize the vehicle(s) sufficiently to make it safe for a medical provider to access the victim and initiate emergency trauma care as well as other support functions from within the vehicle. **Secondary stabilization**, sometimes called **supplemental stabilization**, supports and reinforces primary stabilization so that when the structural components of the vehicle are modified or destroyed during the disentanglement tasks, the vehicle does not collapse or crumple onto the rescuers and/or victims. Not all vehicles will require secondary stabilization. For example, if a vehicle is resting on its wheels, usually only primary stabilization is required. Primary and secondary stabilization are integral components of the primary and secondary access parts of the IAP.

Primary Stabilization

Primary stabilization involves using quick points of contact on the low side of the vehicle to widen its base. For vehicles that end up on their wheels, after they are immobilized with wheel chocks or other equipment to prevent them from rolling horizontally, they must also be stabilized vertically. This is usually accomplished with cribbing. If the vehicle is not stabilized vertically, it may still be able to move because of the rocking motion that the suspension system allows when rescuers get into the vehicle to assess and extricate victims. This type of instability can cause further injuries to the victims of the MVA (**FIGURE 26-22**).

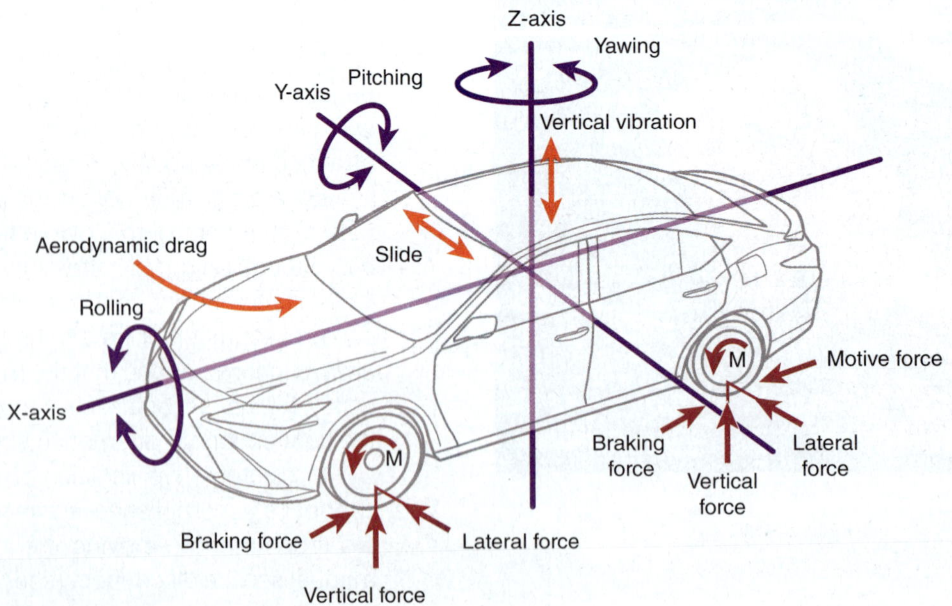

FIGURE 26-21 Vehicles need to be stabilized to prevent longitudinal, lateral, vertical, roll, pitch, and yaw movements.

FIGURE 26-22 The application of cribbing or other stabilization tools under solid areas of the vehicle prevents it from moving vertically.

FIGURE 26-23 Step chocks are cribbing shaped like stair steps used to vertically stabilize a vehicle.

FIGURE 26-24 A box crib can be constructed to stabilize a high clearance vehicle.

FIGURE 26-25 Wedges are used to snug loose cribbing when the void space is not big enough to accommodate a regular-sized crib.

The suspension system of most vehicles can be stabilized with step chocks. A **step chock** is prefabricated cribbing shaped like stair steps placed under both sides of a vehicle. On one side of the vehicle, place a step chock toward the front of the vehicle and a second step chock toward the rear of the vehicle under solid areas (**FIGURE 26-23**). Repeat this process on the other side of the vehicle.

If step chocks are not available or are not the right size and there is a need for a more stable platform than can be accomplished using a simple cribbing stack, rescuers can build a **box crib**—cribbing, usually wood, placed in layers usually at 90-degree angles to each preceding layer to create a box-like structure (**FIGURE 26-24**). Box cribs are used under high-clearance vehicles. Once step chocks or box cribbing are in place, the tires can be deflated to allow the vehicle to rest on top of them to create a stable vehicle.

A **wedge**, which is wedge-shaped piece of cribbing, is used to fill the void between the crib and the object as it is stabilized or raised (**FIGURE 26-25**). Wedges should be the same width as the cribbing, with the tapered end no less than ¼ in. (6 mm) thick. Ends that are less than ¼ in. (6 mm) will commonly fracture under a load.

For vehicles that are resting on their roofs or sides, a combination of wedges or step chocks on each end and on the sides of the vehicle can quickly accomplish primary stabilization.

To stabilize a vehicle resting on all four wheels following an MVA, follow the steps in **SKILL DRILL 26-3**.

SKILL DRILL 26-3

Stabilizing a Vehicle Resting on All Four Wheels Following a Motor Vehicle Accident Firefighter II, NFPA 1010: 7.4.1

1. Wearing PPE, including eye protection, assess the scene for hazards to rescuers and victims and complete the inner and outer surveys. Complete the steps to identify, immobilize, and disable the vehicle, including setting wheel chocks. Place cribbing and step chocks as needed under solid areas on both sides of the vehicle. Add wedges as needed to create a firm support for the vehicle.

2. After the cribbing and/or step chocks have been placed, consider deflating the tires for added stability.

Step 1: Courtesy of Aaron Miranda. **Step 2:** © Jones & Bartlett Learning. Photographed by Glen E. Ellman.

Secondary Stabilization

For a vehicle that has come to rest on its wheels on a flat, stable surface, secondary stabilization might not be necessary. After an MVA, some vehicles might come to rest on their roofs or sides. Vehicles that have overturned and vehicles resting on their side are very unstable and present a risk to both victims and rescuers. Due to the more rounded or oval shape of modern vehicles, the footprint of these vehicles is reduced when they are not resting on all four wheels. Placing the slightest amount of weight on them can cause them to move or roll over. Secondary stabilization provides additional stabilization that complements primary stabilization.

Secondary stabilization can be accomplished by building additional box cribbing to support the vehicle or by using a strut stabilization system to increase the footprint of the vehicle. If done correctly, secondary stabilization prevents unintentional vehicle movement and provides added safety during extrication. It is important to place secondary stabilization in a manner that does not impede or interfere with extrication efforts. This requires careful planning and a thorough understanding of the overall extrication plan.

Secondary stabilization can be accomplished by using struts to capture the high side of the vehicle and distributing the load onto a wider base, thereby lowering its center of gravity. Most strut stabilization systems use a ratchet strap or other tensioning device. The tips or head of the struts are secured to the high side of the vehicle. Then a ratchet strap is attached to the base plate on the strut and a low anchor point on the vehicle or to the base of another strut on the opposite side of the vehicle. After the strap is tightened, the strut and the strap create a **tension buttress system**, which adds tension to the object being stabilized (the vehicle), locking the object in place with a diagonal force that lowers the object's center of gravity by increasing the object's entire footprint. The number of struts required depends on the complexity of the rescue, the position the vehicle is in, and the surface conditions. Generally, a minimum of two struts is required (**FIGURE 26-26**).

FIGURE 26-26 Struts are used for structural support to stabilize and reinforce a vehicle.

Courtesy of Edward Monahan.

Vehicles that end up on top of other vehicles or end up with objects on top of them after an MVA also require secondary stabilization. Marrying, or joining them together, is a fast way to stabilize these vehicles by creating a more solid, unitary body. Marrying vehicles is best accomplished using industrial-grade ratchet straps. Then the vehicle on the bottom must also be stabilized with respect to the ground to prevent independent movement.

Gaining Access to the Victim (Plan A)

There are a few ways to gain access to a victim in a vehicle after an MVA. The three most common are to open a door, break a tempered glass window (usually all the windows except the windshield are tempered glass), or force a door open.

Open the Door

After stabilizing the vehicle, the simplest way to access a victim of an MVA is to open a door. Try all of the doors first, even if they appear to be badly damaged. The first rule of forcible entry—"Try before you pry"—applies to vehicles as well as to structures. It is an embarrassing

Voice of Experience

When approaching a vehicle that has been involved in an accident or fire, firefighters must ensure that they use precautions while working around an air bag, especially if the air bag has not activated. Several years ago my crew was dispatched to a vehicle fire. We arrived to find a single vehicle that had missed the turn at a T intersection and was sitting slightly over a 2-ft (0.6-m) embankment. The vehicle had little damage; however, the engine compartment was heavily involved in fire. A witness stated the occupant had left the vehicle on foot and was walking away.

After the hood was forced and the fire was extinguished, we decided to open the passenger compartment to release the smoke from the vehicle. Finding that both doors were locked, I noticed the passenger side window was down enough to reach in and unlock the door. While reaching in the window, the passenger-side air bag deployed, likely from exposure to the fire. Luckily, no injury occurred to me or my crew but at the moment it happened, I honestly thought I had lost my arm. An air bag can deploy in 25 to 50 milliseconds—nearly 200 mph. I do not know anyone who is that fast and would be able to get out of the way in time.

As you approach a vehicle where the air bag has not deployed, it is imperative that you disconnect the power supply, remembering that some vehicles have a capacitor that may take time to drain. I teach everyone in my crew to treat every air bag, regardless of whether it has deployed or not, like it is loaded and could go off at any time. Never place yourself between the patient and the air bag. Remember, you cannot help others if you become a victim.

Deputy Chief Toby Martin

Roanoke County Fire and Rescue

Roanoke County, Virginia

FIRE DEPT.

FIGURE 26-27 Three tools can be used to break tempered glass. One is the spring-loaded center punch.

waste of time and energy to open a jammed door with heavy rescue equipment, only to discover that another door can be opened easily and without any special equipment.

Attempt to unlock and open the least damaged door first. Make sure the locking mechanism is released. Then try to open the door using both the outside and inside door handles at the same time. This technique often forces a jammed locking mechanism to open the door, even when the door appears to be badly damaged.

Remove the Glass

If a victim's condition is serious enough to require immediate care and you cannot enter through a door, consider breaking a window. If you must break a window to unlock a door or gain access, cover the victim if possible, and then try to break the window that is farthest away from the victim. If the victim's condition warrants your immediate entry, however, do not hesitate to break the closest window.

The side and rear windows of most cars are made of tempered glass, which breaks easily into small, mostly dull-edged pieces when hit with a sharp, pointed object such as a spring-loaded center punch, a multipurpose tool, or the point of a pick-head axe (**FIGURE 26-27**). Small pieces of tempered glass do not usually pose a danger to victims trapped in cars. Tell emergency medical services (EMS) personnel if a victim is covered with broken glass so that they can notify the hospital emergency department. To break tempered glass using a spring-loaded center punch or a striking tool, follow the steps in **SKILL DRILL 26-4**.

SAFETY TIP

If you use something other than a spring-loaded center punch to break the window, always aim for a low corner, away from the victim.

Some newer vehicles have side and rear windows made of special laminated glass, called enhanced protective glass (EPG), which cannot be easily broken with a sharp object. This glass provides more protection during an MVA. Windshields and EPG cannot be broken with a spring-loaded center punch. When laminated glass is struck by a sharp stone or by a spring-loaded center punch, a small chip or crack appears, but the structure of the glass remains intact. Because of this special construction, it is necessary to remove the windshield in one large piece. If you encounter laminated glass when trying to break a window, or if you need to cut the windshield to gain access to the victim or to completely remove the vehicle's roof, use a serrated tool such as glass handsaw, a reciprocating saw, or another sharp cutting tool such as battery-powered glass shears to cut along the bottom of the window as close as possible to the windowsill. If using a striking tool such as an axe, use extreme caution, and strike the lowest corner, away from the victim. Avoid forceful strikes to prevent punching through and potentially hitting the victim or personnel working inside the vehicle. After breaking the window, use your gloved hands to pull the remaining glass out of the window frame so that it does not fall onto the victims or injure the rescuers. Next, try to unlock and open the door.

SAFETY TIP

Sawing through laminated glass produces large amounts of glass particles. It is highly recommended that personnel wear respiratory protection such as an N-95 or higher particle-filtering mask, in addition to eye protection, whenever managing glass at an MVA.

Breaking the rear window will sometimes provide an opening large enough to enable a rescuer to gain access to the victim if there is no other rapid means

SKILL DRILL 26-4

Breaking Tempered Glass Firefighter II, NFPA 1010: 7.4.1

1. Don PPE, including eye protection. Assess for hazards to rescuers and victims and complete the inner and outer surveys. Complete the steps to identify, immobilize, and disable the vehicle. Ensure stability of the vehicle by using appropriate step chocks and cribbing. Select a tool for breaking tempered glass. Communicate the action plan to the victim.

2. Ensure that the victim and other firefighters are as protected as possible.

3. Warn personnel and victims that you will be breaking the glass by verbally announcing "Breaking glass!" Place the tool in the lower corner of the window and sharply apply pressure until the spring is activated. If using a striking tool, strike the lowest corner, away from the victim.

4. Using a hand tool or other object or your gloved hand, remove loose glass around the window opening.

for doing so. The simple techniques of opening a door and breaking the rear window will enable firefighters to gain access to most MVA victims, even those in an upside-down vehicle.

Force the Door

If firefighters cannot gain access by the previously mentioned methods, they must use heavier extrication tools to gain access to the victim. The most commonly used technique for gaining access to a vehicle following an MVA is door displacement, which means forcing the door open using tools (**FIGURE 26-28**). The opening and displacement of vehicle doors may be difficult and somewhat unpredictable. As mentioned earlier, doors on newer vehicles are sometimes hard to force, and several attempts may be necessary.

When it is necessary to force a door to gain access to the victim, choose a door that will not endanger the safety of the victim. For example, do not force a door open if the victim is leaning against it.

To force a door, firefighters typically use a prying tool to bend the sheet metal at the edge of the door near the locking mechanism. Once this metal is exposed, they can then insert a hydraulic spreading tool into the space between the door and the post, and then pry the hinges or locking mechanism apart. After a door is removed, place it away from the vehicle where it will not be a safety hazard to rescuers.

Powered hydraulic tools are the most efficient and widely used tools for opening jammed doors. Jammed doors can be opened by releasing the door either from the latch side or the hinge side of the door. The decision regarding which method to use will depend on the structure of the vehicle and the type of damage the door has sustained.

FIGURE 26-28 A hydraulic spreader is often used to force a door open.

The first step in opening a vehicle door is to create a **purchase point** where the spreader of the hydraulic tool can be inserted. A purchase point can be created by using a pry bar or other hand tool to bend the metal away from the hinges or locking mechanism. A hydraulic spreader tool can also be used in several ways to pinch or spread the metal surface of the door to expose the hinges and locking mechanism. Firefighters should train with a special rescue team to learn different methods to create a purchase point.

To expose the door hinges using a hydraulic spreader, perform a **wheel well crush**, also referred to as a **fender crush**. To do this, locate an area between the strut tower and the firewall. Hold the spreader so that the arms open vertically, and then open the spreader arms and position the top tip on top of the wheel well. Close the arms, making sure that the bottom arm clears the tire and the strut tower and position it on the bottom section of the wheel well. Continue closing the arms to crush and crease the wheel well panel, pulling it away from the door and thereby exposing the hinges. This will create a purchase point to be able to remove the door by attacking the hinges (**FIGURE 26-29**).

When separating a door from the hinge side after creating a purchase point near the top hinge, place the spreader tips above the top hinge and open the tips to separate the door from the hinge. Use the spreader tips or cutting tool to separate the bottom hinge.

To expose the latching mechanism using hydraulic spreaders, perform a **vertical spread maneuver**. Hold the spreader so that the arms open vertically. Position one tip of the spreader arm so that it rests on top of the windowsill close to the door handle. Open the spreader arms until the top tip makes contact with the bottom of the roof rail. Using the roof rail as support, activate the hydraulic spreader to displace the door outward. This will create a purchase point and expose the door's latching mechanism as well as the hinges (**FIGURE 26-30**).

Once rescuers have gained access to the hinges or latch, they can place the spreader in a position so that it is not in the pathway that the door will take when the hinges or latch break. Rescuers should not stand in a position that might put them in danger. To ensure that the door does not swing open violently, rescuers can use a long pry bar or rope to limit the movement of the door when it opens.

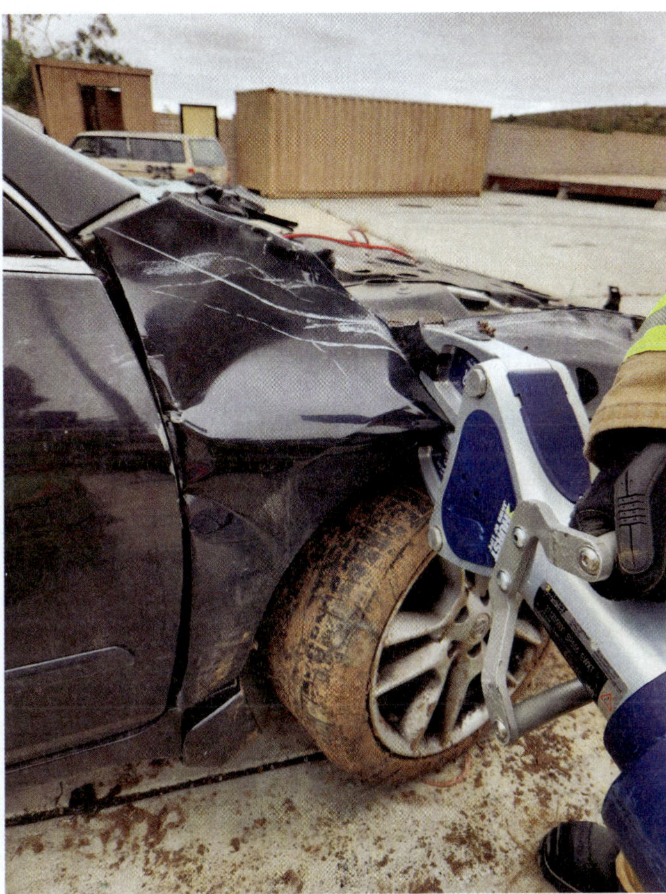

FIGURE 26-29 The wheel well crush technique exposes the hinges from the front door.

Courtesy of Aaron Miranda.

FIGURE 26-30 The vertical spread maneuver displaces the door outward and exposes the door latch.

Courtesy of Aaron Miranda.

To finish forcing the door, insert the closed tips of the spreader between the inner skin of the door and the doorjamb just above the latch or above the hinges. Activate the spreader to spread the tips until the latch or the hinge separates. If you separated the door from the latch, then once the latch has separated, start to separate the hinges of the door. Some hydraulic tools are capable of cutting door hinges. Check the manufacturer's recommendations for the proper operation of the hydraulic tool. If you separated the door starting from the hinge side, once the hinges have been separated, hyperextend the door to expose the latch and spread or cut the doorjamb.

To force a door following an MVA, follow the steps in **SKILL DRILL 26-5**.

Provide Initial Medical Care

As soon as you have secured access to the victim, begin to provide emergency medical care, per your department's EMS protocols. A caregiver should remain with

TIP

As vehicle manufacturers strive to make new vehicles stronger and lighter, they have begun using increasing quantities of high-strength steel. These stronger steels, in turn, require more robust rescue tools. Tool manufacturers continue to increase the capacity of rescue tools to cut and distort the metals in newer vehicles. By studying individual vehicle manufacturers' specifications, you can determine the easiest places to cut or distort specific vehicle parts.

TIP

The company officer or another Firefighter II should coordinate extrication and rescue operations to increase efficiency and help prevent the actions of one team from interfering with another.

SKILL DRILL 26-5

Forcing and Removing a Vehicle Door Firefighter II, NFPA 1010: 7.4.1

1. Don PPE, including eye protection. Assess scene for hazards and complete the inner and outer surveys. Complete the steps to identify, immobilize, and disable the vehicle. Ensure stability of the vehicle by using appropriate chocks and cribbing. Retrieve and set up the required tools. Check the equipment for readiness.

2. Communicate the action plan to the victim and minimize hazards to both the rescuers and the victim. Create a purchase point by performing a wheel well crush to expose the hinges or a vertical spread to expose the latch. Engage hand tools or a power tool to force the door, using good body mechanics.

3. If the latch was exposed, insert the closed tips of the hydraulic spreader between the doorjamb just above the latch. Once the latch has separated from the door frame, start to separate or cut the hinges of the door starting from the top hinge.

4. If removing the door starting from the hinge side, separate or cut the hinges from the door starting from the top and working down to the bottom hinge. Separate or cut the doorjamb or latch and remove the door completely.

the victim and provide both emotional comfort and physical care until extrication is complete. The IC coordinates victim care provided by the medical team with the operations conducted by the extrication team to ensure the safety of everyone on the scene. Although it might be necessary to delay one part of these processes for a short period, all personnel should work toward the goal of getting the victim stabilized and removed from the vehicle as quickly and safely as possible.

To gain access and provide medical care to a victim in a vehicle following an MVA, follow the steps in **SKILL DRILL 26-6**.

Disentangling and Extricating the Victim (Plan B)

The second part of the process is to extricate the victim. Before you can do that, you need to remove those parts of the vehicle that are trapping the victim—in other words, you need to disentangle the victim. The goal of disentangling the victim is to remove the sheet metal and plastic from around the victim, not to "cut the victim out of the vehicle." Before beginning disentanglement, study the situation. What is trapping the victim in the vehicle? Perform only those disentanglement procedures that are necessary to remove the victim safely from the vehicle. The order in which these procedures are carried out is dictated by the conditions at the scene.

As you work to disentangle the victim from the vehicle, be sure to protect the victim by covering them with a blanket or by using a backboard. Be sure the victim understands what is being done. Provide emotional support because the sounds made by extrication procedures may frighten the victims.

The five most commonly performed disentangling procedures are as follows:

- Displace the seat
- Remove the glass
- Relocate the steering wheel
- Displace the dashboard
- Displace the roof

SKILL DRILL 26-6

Providing Medical Care to a Victim in a Vehicle Firefighter II, NFPA 1010: 7.4.1

1. Enter the vehicle through the best access point. Begin initial medical care per your department's EMS protocols and communicate the victim's condition and level of entrapment to the outside rescuers.

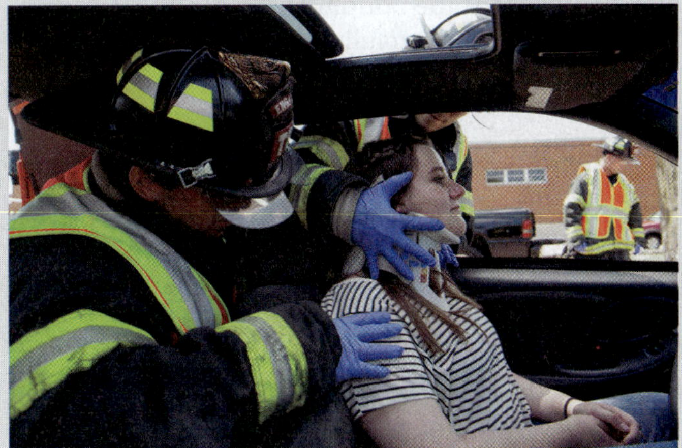

2. Begin packaging the victim for extrication.

To learn additional methods of disentangling a victim, firefighters should take an approved extrication course.

Displace the Seat

In frontal and rear-end MVAs, the vehicle may become compressed (**FIGURE 26-31**). As the front of the vehicle collapses, the space between the steering wheel and the seat becomes smaller. In some cases, the driver may be trapped between the steering wheel and the front seat. Displacing the seat can relieve pressure on the driver and give rescuers more space for removal.

If it is necessary to displace a seat, start with the simplest steps. Many times you can gain some room by moving the seat backward on its track. This maneuver works well with short drivers who have their seat forward. To move a seat back, first make sure that the victim is supported. With manually operated seats, release the seat-adjusting lever and carefully slide the seat back as far as it will go. Attempt to use this simple method first.

For cars that have electrically adjustable seats, use this feature to move the seat back to give the victim more room or lower the seat to help disentangle the victim. Make sure that the victim is adequately supported before attempting these maneuvers.

As a last resort, use a manual hydraulic spreader or a powered hydraulic tool to move the seat back. Place one tip of the hydraulic tool on the bottom of the seat but avoid pushing on the seat channel that is attached to the floor of the vehicle. Place the other tip of the spreader at the bottom of the A-post doorjamb. Support the victim carefully. If the seat is a manually operated seat, engage the seat adjustment lever. Open the spreader in a careful and controlled fashion.

FIGURE 26-31 In frontal MVAs, the vehicle may become compressed.

In some cases, it may be helpful to remove the back of the seat. To do this, cut the upholstery away from the bottom of the seat back where it joins the main part of the seat. You can then use a reciprocating saw or a hydraulic cutter to cut the supports for the seat back. Be certain that the victim is supported and protected throughout this procedure.

Remove the Glass

If not already done during the victim access phase (primary access), a second technique that is often part of the disentanglement process is the removal of glass. Also, if the roof of a vehicle must be removed, all of the glass must first be removed from the vehicle. Removing the rear window, the side glass, or the windshield may improve communication between rescue personnel inside the vehicle and personnel outside the vehicle. Sometimes rescuers can simply roll a window down to gain this advantage. Open windows also provide a good route for passing medical care supplies to the inside caregiver. Removing a windshield provides extra space when administering emergency medical care to an injured victim.

The procedure for breaking tempered glass windows was described earlier in this chapter. Removing windshields requires a different procedure. The windshields of most passenger vehicles are glued in place with a strong, plastic-type glue. The windshield must be removed with a special glass saw, reciprocating saw, or by carefully using an axe. Windshield removal can create a significant amount of glass dust, which can be an inhalation hazard to victims and rescuers. Consult your department's SOPs for taking protective measures against this hazard.

To remove a windshield, first make sure that the victim is protected from flying glass. One rescuer makes a purchase point at the top center of the windshield and cuts along the top of the windshield to the nearest A-post. The rescuer continues cutting down the side of the windshield close to the A-post (**FIGURE 26-32**). Finally, the rescuer cuts along the bottom of the windshield to the center, and then stabilizes the half of the windshield that has been cut free.

A second rescuer then starts on the opposite side of the windshield where the initial cuts were made and cuts the second half of the windshield, following the same sequence of cuts used by the first rescuer. When the second rescuer has completed cutting the windshield, the entire windshield is lifted out of its frame

FIGURE 26-32 Removing the windshield by cutting close to the A-post.

and placed under the vehicle or in another safe place where it will not present a safety hazard.

To remove a windshield from a vehicle, follow the steps in **SKILL DRILL 26-7**.

Relocate the Steering Wheel

During an MVA, any compression of the front part of the vehicle may push the steering wheel back into the victim's abdomen or chest. Relocating the steering wheel can help to disentangle a victim from the vehicle when other techniques such as displacing the dash are not practical or possible.

To relocate the steering wheel, use hand tools rather than power tools. Hand tools such as an FRJ and chain package or a come along allow the rescuer to have complete control over the movement and to feel the forces

SKILL DRILL 26-7

Removing the Windshield from a Vehicle Firefighter II, NFPA 1010: 7.4.1

1. Wearing PPE, including a mask and eye protection, select a tool for removing laminated glass. Ensure that the victim and other firefighters are as protected as possible. Using a pointed tool, create a purchase point by making a hole on the center top of the windshield. Using a serrated tool, insert the blade and begin a steady and continuous cut as close as possible to the window frame toward the A-post. Continue the cut vertically alongside the A-post and toward the bottom of the windshield. Repeat this cut on the opposite side of the windshield.

2. With the assistance of a second rescuer, support the windshield from the top to prevent it from falling in toward the victim as the final cut is made. Complete a horizontal cut along the bottom and push the windshield outward, toward the hood. Remove the windshield away from the vehicle and place it in a designated area or debris pile.

FIGURE 26-33 The FRJ can be used in conjunction with a chain package to relocate the steering wheel.

Courtesy of Edward Monahan.

FIGURE 26-34 One method of removing a steering wheel is to cut the column of the steering wheel.

being applied (**FIGURE 26-33**). If you use a hydraulic tool along with a rated chain package, you will not have control over the amount of force being applied to the steering wheel assembly when relocating it.

If additional space is required after relocating the steering wheel assembly, a section of the steering wheel can be cut. Hydraulic cutters and reciprocating saws are the tools of choice, but hand tools such as a hacksaw or bolt cutters can also cut steering wheel spokes and the steering wheel rim. If you think you need to cut a section of the steering wheel, first make sure that the steering wheel is not still impinging on the victim. The energy created by the cutting tools can be transferred to the victim before the steering wheel section being cut releases, and this can potentially cause further trauma to the victim.

One method of removing a steering wheel is to cut the spokes as close to the center hub of the steering wheel as possible. A second method is to cut the rim of the steering wheel. The steering wheel rim can be removed completely, or one section can be cut and removed. Note that cutting the steering wheel rim leaves sharp edges that present a safety hazard. These sharp edges must be covered to prevent injury both to the victim and to rescuers. A third method is to cut the column (**FIGURE 26-34**).

Some vehicles use a solid steel rod for the steering wheel column. When cutting the steering wheel column using hydraulic cutters, the steel rod tends to fracture by the force and cutting action, projecting the steering wheel assembly toward the victim. Support the steering wheel assembly with webbing or other material to prevent it from striking the victim.

SAFETY TIP

In most vehicles, the steering wheel contains an air bag. There is also an air bag in the dash for the front passenger. Some vehicles contain side-mounted or curtain air bags that provide lateral protection for occupants, and under-dash air bags for knee protection are also common in newer vehicles. If the air bags did not deploy during the MVA, they present a hazard for both the occupant of the vehicle and for rescue personnel because they could deploy if wires are cut or if they become activated during the rescue operation. If the air bag did not deploy, disconnect the battery and allow the air bag capacitor to discharge. The time required to discharge the capacitor varies for different models of air bags. If the air bag did deploy, it may still present a hazard because some air bags deploy in two stages.

If any air bags did not deploy, keep the following in mind:

- Do not place an object between the victim and an undeployed air bag.
- Do not attempt to cut the steering column.
- Never position yourself in front of an undeployed air bag.

Displace the Dashboard

During frontal MVAs, the dashboard is often pushed down or toward the seat. When a victim is trapped by the dash, it is necessary to displace this component (**FIGURE 26-35**). Two techniques are commonly used

FIGURE 26-35 When a victim is trapped by the dash, it is necessary to displace the dashboard.
Courtesy of Aaron Miranda.

to displace the dash: the dash roll and the dash lift. With the **dash roll**, a hydraulic ram or spreading tool rolls the dash and firewall forward and away from the victim. With the **dash lift**, the dash and the firewall are lifted up and away from the victim. The objective with both procedures is to displace the dashboard up and away from the victim. (These techniques are also sometimes used to displace the steering column up and away from a victim.)

Both types of dash displacement require a cutting tool such as a hacksaw, a reciprocating saw, an air chisel, or a hydraulic cutter. A cutting tool is needed to make one or more cuts on the A-post and other structural members of the vehicle. The dash roll requires fewer cuts than the dash lift and can often be accomplished more quickly; however, it may not provide as much room to remove the victim.

The first step of dash displacement is to remove the door on the side being displaced. If you need more room to access the dash, both doors and the roof can be removed. Whenever cutting or spreading near or next to the victim, if possible, place a protective hard cover between the victim and the part being cut or spread to shield the victim from any flying debris and to prevent accidentally harming them.

When performing a dash roll or a dash lift, you will need to make relief cuts to weaken the structural integrity of the vehicle body and to release tension. On the side where the victim is located, make a deep horizontal cut perpendicular to the A-post to avoid cutting the fuel line or high-voltage lines in case they are run within the rocker panel. If performing a dash lift, you will need to make an additional cut on the A-post parallel to the first and remove a section out of the A-post so that the tips of the hydraulic spreader can fit in the gap. In addition, a cut on the upper A-post between the firewall and the roofline and a cut on the upper frame rail between the strut tower and the firewall must be made.

Once the dash has been displaced away from the victim, cribbing and wedges are utilized to fill the gap created during the displacement to hold the dash in the proper position and to prevent it from collapsing back down on the victim. Only then can the spreading tool be removed.

Performing dash displacement requires careful monitoring of the dashboard's movement to be certain that it is moving away from the victim and is not causing any additional harm to the victim. Throughout this process, make sure that the victim is protected and that someone keeps the victim informed about what is happening. Communication is important during this procedure.

To perform a dash roll to displace the dashboard of a vehicle following an MVA, follow the steps in **SKILL DRILL 26-8**. To perform a dash lift to displace the dashboard of a vehicle, follow the steps in **SKILL DRILL 26-9**. There are variations to each of these techniques; be sure to train with your department's vehicle extrication personnel to learn the best practices in your department.

SAFETY TIP

Always expose and inspect the area to be cut, spread, or pushed. The phrases *peel and peek* and *expose and cut* mean remove or pull back the plastic molding or materials covering roof posts, roof rails, lower door posts, and any other vehicle component before you cut, spread, or displace them. After exposing the area, look for air bag inflator cylinders, seat belt pretensioning systems, and high-voltage wires before performing any maneuver. When operating tools close to the center console, consider the location of the air bag ECU and avoid crushing or cutting in this area without first exposing it. The vehicle's low-voltage (12-volt) power should be disabled to deactivate the SRS before performing disentanglement maneuvers.

TIP

Cribbing should be placed under the vehicle whenever downward pressure is expected. This will maximize the displacement of the dash.

SKILL DRILL 26-8

Performing a Dash Roll Firefighter II, NFPA 1010: 7.4.1

1. Wearing PPE, including eye protection, communicate with the victim and ensure that both the victim and rescuers are protected from hazards. Remove the front door on the side performing the dash roll. Lift and cut the vehicle's hood at the hinge attachment closest to the dash on the side to be displaced. If you see a gas-filled piston (the hinge for the hood), detach it from the base or cut the steel rod (do not cut the cylinder). If necessary, do the same on the opposite side to remove the hood completely. Remove the front wheel well fender panel on the side where you will be performing the dash roll to expose the upper frame rail. Cut through the upper frame rail between the strut tower and the firewall. Make a horizontal relief cut at the bottom of the A-post close to where the firewall meets the rocker panel. If the roof has not been removed, make a relief cut on the upper A-post as close as possible to the roofline.

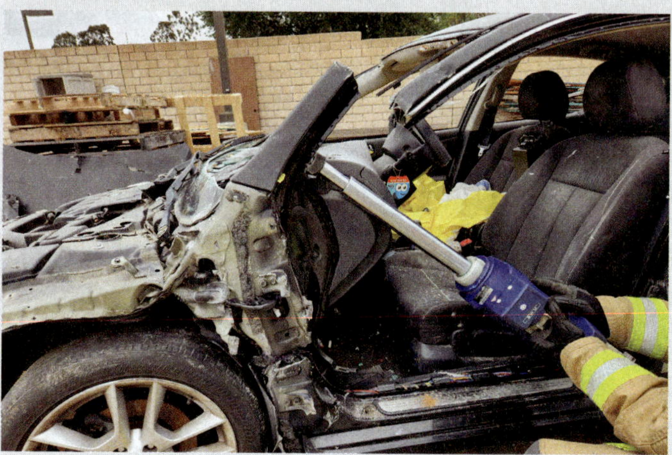

2. Place the base of a high-lift jack or hydraulic ram at the base of the B-post where the rocker panel and the B-post meet and place the tip of the jack or ram toward the top of the A-post at the bend in the A-post, and then extend the jack or ram to roll the dash and firewall up and away from the victim. Place cribbing under the rocker panel as the dash is being rolled.

3. To prevent the dash from returning to its original position, place the hydraulic spreader arms between the gap created by the A-post cut and open the arms of the hydraulic spreader to hold the dash in position. Alternatively, fill the gap with wedges. Remove the hydraulic ram or high-lift jack.

SKILL DRILL 26-9

Performing a Dash Lift Firefighter II, NFPA 1010: 7.4.1

1. Wearing PPE, including eye protection, communicate with the victim and ensure that both the victim and rescuers are protected from hazards. Remove the front door on the side where you will be performing the dash lift.

2. Lift and cut the vehicle's hood at the hinge attachment closest to the dash on the side to be lifted. If you see a gas-filled piston (the hinge for the hood), detach it from the base or cut the steel rod (do not cut the cylinder). If necessary, do the same on the opposite side to remove the hood completely.

3. Remove the front wheel well fender panel on the side where you will be performing the dash lift to expose the upper rail. Cut through the upper rail between the strut tower and the firewall. If the roof has not been removed, cut out a 2- to 4-in. (5- to 10-cm) section on the upper A-post between the roofline and the dash, again on the side where you will be performing the dash lift.

Continues.

SKILL DRILL 26-9 CONTINUED

Performing a Dash Lift Firefighter II, NFPA 1010: 7.4.1

4. On the lower part of the A-post, make a deep relief cut between the top and bottom hinges. Make sure to cut through the firewall. If necessary, make multiple deep cuts toward the wheel well and firewall to release tension and to create a gap.

5. Place the tips of a hydraulic spreading tool vertically in the A-post relief cut to lift the dash and firewall up and away from the victim. Place cribbing under the rocker panel as the dash is being lifted.

6. Maintain contact with the hydraulic spreading tool to prevent the tool from shifting while removing the victim.

Courtesy of Aaron Miranda.

Displace the Roof

Displacing the roof of a vehicle has several advantages (**FIGURE 26-36**):

- It enables equipment to be easily passed to the emergency medical providers.
- It increases the amount of space for performing medical care.
- It increases the visibility and space for performing disentanglement.
- It improves the fresh air supply.
- The increased space helps to reduce the feeling of panic caused by the confined space of the vehicle.
- It provides a large exit route for the victim.

FIGURE 26-36 Removing the roof provides several advantages, including creating a large exit route for the victim.

TIP

When removing the roof of a vehicle, remember to cut all of the seat belts so that they will not interfere with the operation.

A partial roof displacement can be accomplished by cutting the A-posts and folding the roof back toward the rear of the vehicle. This technique is beneficial when it is necessary to displace a portion of the roof that is impinging on the victim or to gain quick access to provide medical care to the victim. Partial roof displacement provides only limited space; therefore, it is often preferable to remove the entire roof.

To displace a roof, you can use hand tools such as hacksaws, air chisels, and manual hydraulic cutters and power tools such as reciprocating saws and powered hydraulic cutters. Remember that if the A-posts are made of high-strength steel, they may be difficult to cut with older cutting tools. In addition, most newer vehicles include additional reinforcement beams for the seat belts. Older hydraulic cutters may not be able to sever these beams, so cuts must be made above the seat belts.

The first step in displacing the roof is to ensure the safety of both the rescuers and the victim inside of the vehicle. In particular, as rescuers cut the posts that support the roof, they must support the roof to keep it from falling on the victim. Also, the glass must be removed from all windows and the windshield to prevent it from falling on the victim.

Prior to cutting any materials, remove interior trim around the posts to identify the presence and location of any hidden pistons, canisters, electrical wiring, or SRS devices, such as air bags and seat belt pretensioners, and address these hazards accordingly. Cut the vehicle posts farthest away from the victim at a level that will ensure the least amount of post remains in place after the roof is removed. When cutting the wider rear posts, cut them at the narrowest point of the post. As each post is cut, a rescuer needs to support that post. Cut the post closest to the victim last because the roof provides some protection for the victim during this process.

Working together, several rescuers should remove the roof and place it away from the vehicle, where it does not pose a safety hazard. They should then cover the sharp ends of the cut posts with a protective material such as duct tape or a commercial device.

To remove the roof of a vehicle following an MVA, follow the steps in **SKILL DRILL 26-10**.

Removing and Transporting the Victim

After a victim has been disentangled, they can be removed from the vehicle. During this phase, the victim needs to be stabilized and packaged in preparation for removal. Follow your department's EMS protocols for victim stabilization and packaging. The definitive treatment of trauma victims occurs at a hospital; therefore, only those steps needed to stabilize and prevent further injury to the victim should be performed.

Develop a plan for the victim's removal and make sure that a clear exit pathway is available. The actual victim removal should be directed by a designated person—usually the rescuer controlling cervical-spine stabilization—and conducted with clear commands. Ensure that an adequate number of rescuers are available to assist with victim removal and confirm that everyone involved in the process understands the commands that will be given.

Try to make the removal process as seamless as possible. Position the ambulance cot close by so that the victim can be placed on the cot without any delay. The victim should be moved to an ambulance as soon as possible.

Transport the victim to an appropriate medical facility as soon as the initial stabilization has been completed. If your department uses a helicopter service to transport victims, you also need to be familiar with the procedures used to load victims into helicopters.

TIP

Although the IC or incident safety officer (ISO) is responsible for rescuer safety, all personnel on the scene should be aware of hazards that may further injure victims or rescuers.

SKILL DRILL 26-10

Removing the Roof of a Vehicle Firefighter II, NFPA 1010: 7.4.1

1. Wearing PPE, including eye protection, retrieve and set up the required tools. Check the equipment for readiness. Assess the vehicle for stabilization and hazards. Prior to cutting any materials, remove interior trim around the posts to identify the presence and location of any hidden pistons, canisters, electrical wiring, or SRS devices, such as air bags and seat belt pretensioners, and address these hazards accordingly. Communicate with the victim and minimize hazards to both the rescuers and the victim. Remove any remaining glass.

2. Engage the appropriate tools to remove the roof, while using good body mechanics.

3. Control the vehicle roof and protect against sharp edges at all times.

4. Remove the roof to a safe location. Return the tools to the staging area upon completion of the tasks.

Terminating an Incident

Terminating an incident includes two parts: removing the damaged vehicle and equipment from the scene, and ensuring the scene is left in a safe condition. After you have secured the scene and packed your equipment, it is important to return to the station and fully inventory, clean, service, and maintain all of the equipment (per the manufacturer's instructions) to prepare it for the next call. Some items will need repair, but most will need simple maintenance before they are placed back on the fire apparatus and considered in service.

Just because the victim has been removed from the wreckage does not mean the scene can be abandoned. Law enforcement will secure the scene by conducting their on-scene investigation. A potential crime scene, such as a homicide, should be managed by law enforcement personnel, who will preserve and secure any evidence and close off the scene or roadway. Once a scene has been released by law enforcement, the removal of vehicles needs to be coordinated with a tow agency.

In some cases, a vehicle may need to be turned right-side up. This may cause sparks or a fire. In addition, the high-voltage batteries in electric-drive vehicles may cause a delayed fire. Be aware of smoke, sparks, irritating fumes, and popping or gurgling sounds from the battery. These are warning signs of a damaged high-voltage battery that is undergoing a thermal event and is at risk for fire. This risk can last for hours or even days after the incident. In addition, there is a risk of toxic off-gassing. The IC should instruct a firefighter to be proactive and stand by in PPE with a charged hose line.

Any fluid hazards that have spilled from the vehicle, such as gasoline, motor oil, transmission fluid, or radiator fluid that was either captured with an absorbent or not, will be removed by the towing agency. Towing agencies must be licensed to transport and dispose of any hazardous substances; fire rescue agencies are typically not licensed to do so. Ask the tow agency representative if you may assist the towing agency by placing any vehicle parts, such as doors, roofs, and fenders, back in a heavily damaged vehicle. The tow agency may have procedures or a policy of their own for stowing and transporting loose objects. Being considerate of their preferences will help to maintain a good working relationship with the tow agency for future endeavors such as acquiring vehicles for use in an extrication class. It is advisable to carry a contact list of various private and public organizations or businesses that can offer a particular resource to be utilized on the incident, such as public works/utilities, the DOT, or a heavy equipment company.

TIP

Electric-drive vehicles should be towed using a flatbed truck. Towing them with the drive wheels on the ground could result in an electrical fire. Law enforcement officers may take command of the scene to conduct an MVA or fatality investigation after the victims have been extricated. In the event that law enforcement investigators take possession of the vehicle, brief them on the hazards associated with electric-drive vehicles.

Organizations such as the Energy Security Agency (ESA) offer on-scene hybrid/electric vehicle support free of charge to first and second responders. The ESA can perform risk analysis services and provide guidance to make the system safe for release and recommend isolation procedures if the vehicle is at high-risk for thermal runaway. For more information, go to their website at energysecurityagency.com.

CASE STUDY

You Are the Firefighter CONCLUSION

It has been storming all afternoon. Your station has just finished dinner when the alarm sounds for a motor vehicle accident involving a car that has slid off the road. As you pull out of the station, the rain gets heavier, the wind gets stronger, and it is now dark outside. When you arrive on the scene, you see a passenger car resting on the driver's side in the ditch. One passenger was ejected from the vehicle and is lying motionless by the car. Your officer directs you to check the vehicle for any occupants still inside while he performs a detailed, 360-degree size-up. As you approach the vehicle, you recognize features identifying it as an electric-drive vehicle. The driver is still in the driver's seat and a passenger appears to be unconscious in the back seat compartment. The officer establishes command, calls for additional resources, and assigns the ambulance crew to assess the ejected victim. He directs you to stabilize the vehicle and to set up the equipment for an extensive extrication.

1. **What safety concerns should you have in this situation? How would you address those concerns?**

   Answer: In addition to weather, consider the time of day, location, and road conditions and proceed cautiously. Position emergency vehicles so that they provide a barrier from other motorists and illuminate the scene to improve visibility. A vehicle resting on its side or roof is much more unstable than one resting on all four wheels. Consider available stabilization equipment and request additional resources if necessary. Electric-drive vehicles pose additional safety concerns such as silent drive and high-voltage electrical power as well as the potential for battery fires. Make sure to immobilize the vehicle and disable the electrical system following manufacturer's recommendations. Deploy a minimum of a 1½-in (38-mm) hose line in case a fire ignites.

2. **Before making access to assess the victims, what steps should be taken first?**

   Answer: A complete inner and outer scene survey (360-degree) must be completed. The vehicle needs to be immobilized, disabled, and stabilized before making access.

3. **What extrication equipment will you need?**

   Answer: Due to the complexity of the incident, technical-level vehicle extrication is required. Depending on available rescue equipment already on scene and level of personnel training, more resources might be requested to respond to the scene to assist in extricating the victims. Cribbing, rescue struts, a full set of hydraulic rescue tools, reciprocating saws, and First Responder Jack are just some of the tools and equipment needed to extricate the victims in this scenario.

WRAP-UP

SUMMARY

KNOWLEDGE OBJECTIVES

- Describe a vehicle's anatomy.
  - Describe the major structural components of a motor vehicle. (**NFPA 1010: 7.4.1**, pp. 1096–1102)
  - Identify the two types of motor vehicle frames. (**NFPA 1010: 7.4.1**, pp. 1099–1100)
  - Describe electrical systems and supplemental restraint systems. (**NFPA 1010: 7.4.1**, pp. 1100–1103)
- Describe the types and characteristics of alternative-fuel vehicles. (pp. 1102–1103)
- List the tools and equipment utilized at a motor vehicle accident.
  - Describe the specialized personal protective equipment utilized at a motor vehicle accident. (**NFPA 1010: 4.7.1**, pp. 1103–1108)
  - List the types of power tools utilized at a motor vehicle accident. (**NFPA 1010: 4.7.1**, pp. 1104–1105)
  - Describe the purpose of cribbing. (**NFPA 1010: 7.4.1**, p. 1106)
  - Describe the purpose of rescue struts. (**NFPA 1010: 7.4.1**, p. 1106)
  - Describe the types and purpose of rescue-lift air bags. (**NFPA 1010: 7.4.1**, pp. 1106–1107)
- Describe the best practices to follow when responding to, arriving at, and sizing up the scene.
  - List the hazards to look for when arriving on the scene of a vehicle extrication situation. (**NFPA 1010: 7.4.1**, pp. 1108–1113)
- Describe basic site operations at a motor vehicle accident.
  - Describe how to perform inner and outer surveys at a motor vehicle accident. (pp. 1113–1114)
  - Describe how to identify, immobilize, and disable a vehicle. (**NFPA 1010: 4.7.1**, pp. 1114–1116)
- Describe the components of an incident action plan for a motor vehicle accident. (pp. 1116–1118)
- Describe how vehicle stabilization is performed.
  - List the hazards to look for when stabilizing the scene of a vehicle extrication situation. (**NFPA 1010: 7.4.1**, pp. 1118–1121)
- Describe how to safely access and extricate a victim from a motor vehicle accident.

- Describe how to gain access to a victim of a motor vehicle accident. (**NFPA 1010: 7.4.1**, pp. 1121–1135)
- Describe how to disentangle a victim of a motor vehicle accident. (**NFPA 1010: 7.4.1**, pp. 1127–1136)
- Describe how to remove and transport victims of a motor vehicle accident. (**NFPA 1010: 7.4.1**, p. 1135)
- Describe the steps in terminating a motor vehicle incident. (p. 1137)

SKILLS OBJECTIVES

- Safely size-up and stabilize a motor vehicle accident.
  - Perform scene size-up at a motor vehicle accident. (**NFPA 1010: 7.4.1**, pp. 1112–1113)
  - Immobilize and disable the electrical system of a vehicle following a motor vehicle accident. (**NFPA 1010: 7.4.1**, p. 1117)
  - Stabilize a vehicle following a motor vehicle accident. (**NFPA 1010: 7.4.1**, p. 1120)
- Access and extricate a victim from a motor vehicle accident.
  - Break tempered and laminated glass. (**NFPA 1010: 7.4.1**, p. 1123)
  - Force open and remove a vehicle door. (**NFPA 1010: 7.4.1**, p. 1126)
  - Provide medical care to a victim. (**NFPA 1010: 7.4.1**, p. 1127)
  - Remove a vehicle's windshield. (**NFPA 1010: 7.4.1**, p. 1129)
  - Perform a dash roll. (**NFPA 1010: 7.4.1**, p. 1133)
  - Perform a dash lift. (**NFPA 1010: 7.4.1**, p. 1134)
  - Remove the roof of a vehicle. (**NFPA 1010: 7.4.1**, p. 1136)

KEY TERMS

A-post One of the two posts closest to the front of a motor vehicle that form the sides of the windshield and extend down alongside the firewall. Also called *hinge pillar*.

advanced high-strength steel An alloyed steel with a minimum tensile strength of 65 ksi (440 MPa).

alloy A material composed of two or more metals or chemical elements.

B-post One of two posts located between the front and rear doors of a motor vehicle.

body-over-frame construction A type of vehicle frame resembling a ladder, which is made up of two parallel rails called frame rails joined by a series of cross member beams. Also called *ladder frame construction*.

box crib Cribbing, usually wood, placed in layers at 90-degree angles to each other to create a box-like structure.

bulkhead See *firewall*.

C-post One of two posts located behind the rear doors of a motor vehicle.

carbon fiber reinforced polymer (CFRP) An extremely strong, lightweight plastic reinforced with woven strands of carbon bound together with a polymer, such as an epoxy, and then formed into the desired shape.

converter A device that converts alternating current (AC) to direct current (DC).

cribbing Short lengths of timber or composite materials, usually 4 by 4 in. (101.60 by 101.60 mm) and 18 to 24 in. (457 to 609 mm) long that are used in various configurations to stabilize loads in place or while a load is moving. (NFPA 1006)

dash lift A technique for displacing the dash in which the dash and firewall are lifted up and away from the victim after the hood is removed.

dash reinforcement bar A metal beam or bar that runs the entire length of the dash in a motor vehicle to reinforce the occupant compartment in the event of a crash.

dash roll A technique for displacing the dashboard in a motor vehicle in which a hydraulic ram or spreading tool rolls the dash and firewall up and away from the victim.

DC-DC converter A device that converts direct current (DC) from one voltage level to another.

electronic control unit (ECU) The small processing unit generally located in the center of a motor vehicle that controls the supplemental restraint system.

emergency escape plan See *emergency escape route*.

emergency escape route A preplanned and understood route that rescuers use to move to a safety zone or other low-risk area. Also called *emergency escape plan*. (NFPA 1006)

KEY TERMS CONTINUED

Emergency Response Guide A reference book created by a vehicle's manufacturer that provides important information about the vehicle, such as information about hybrid/electric, fuel-cell, and alternative-fuel systems, to educate and assist emergency response personnel responding to incidents involving the specific vehicle.

fender crush See *wheel well crush.*

firewall The structural component in a vehicle that separates the engine compartment from the passenger compartment. Also called *bulkhead.*

flooded cell battery See *wet cell battery.*

hinge pillar See *A-post.*

inflator In a motor vehicle, the system that inflates air bags using either stored compressed gas or a gas-generating system.

inner circle See *inner survey.*

inner survey A tight, 360-degree survey of a vehicle involved in a motor vehicle crash conducted from approximately 3 to 5 ft (0.9 to 1.5 m) away from the vehicle to identify the vehicle's drive system (conventional or alternative fuel), determine the locations of any victims and whether they are entrapped, identify entry and exit points, determine vehicle stability for access to and removal of the victim(s), and identify hazards that are immediately dangerous to life and health (IDLH). Also called *inner circle.*

inverter A device that converts direct current (DC) to alternating current (AC).

ladder frame construction See *body-over-frame construction.*

laminated glass See *laminated safety glass.*

laminated safety glass (LSG) Two layers of glass bonded together with a clear polymer film that prevents the glass from shattering when broken. Also called *laminated glass.*

outer circle See *outer survey.*

outer survey A survey that extends from the perimeter of the inner survey outward to a distance sufficient to cover where the impact first occurred and the resting point of the vehicle or vehicles to identify hazards that are immediately dangerous to life and health (IDLH), victims who might have been ejected, any additional vehicles involved in the incident, walking wounded, and any other imminent or potential hazards that might have been missed during the initial size-up. Also called *outer circle.*

pillar See *post.*

plan A See *primary access.*

plan B See *secondary access.*

post One of the vertical support members of a vehicle that holds up the roof and forms the upright columns of the passenger compartment in a vehicle. Also called *pillar.*

primary access The existing openings of doors and/or windows that provide a pathway or entry to the trapped and/or injured victims(s). Also called *plan A.* (NFPA 1670)

primary stabilization The technique of stabilizing a vehicle using cribbing, step chocks, box cribs, and wedges to make it safe for a medical provider to access the victim and initiate emergency trauma care as well as other support functions from within the vehicle.

purchase point A small opening made to enable better tool access in forcible entry.

Quick Response Guide (QRG) A short version of a vehicle's Emergency Response Guide (ERG).

relay A switch in cabling or wiring that closes to complete the circuit and allows current to flow when it is activated.

rescue-lift air bag An inflatable bladder made of rubber or synthetic material that is pneumatically filled with compressed air to lift an object or spread one or more objects away from each other to assist in freeing a victim.

rocker panel The section of a motor vehicle's frame located below the doors, between the front and rear wheels.

secondary access Openings created by rescuers that provide a pathway to remove or extricate the trapped or injured victims. (NFPA 1670)

secondary stabilization Stabilization techniques that support and reinforce primary stabilization techniques so that when the structural components of the vehicle are modified or destroyed during the disentanglement tasks, the vehicle does not collapse or crumple onto the rescuers and/or victims. Also called *supplemental stabilization.*

space frame construction The vehicle frame construction made up of multiple tubes welded into a rigid, light cage-like structure (a roll cage).

step chock A prefabricated cribbing assembly shaped like stair steps.

strut An adjustable-length structural support, made from a metal or composite beam, that is used to stabilize and prop up something in danger of collapse.

strut tower The front suspension system in a vehicle.

supplemental restraint system (SRS) A system in a motor vehicle that uses supplemental restraint devices such as air bags and seat belt pretensioning systems to enhance safety in conjunction with properly applied seat belts.

supplemental stabilization See *secondary stabilization.*

tempered glass See *tempered safety glass.*

tempered safety glass A type of safety glass that is heat-treated so that, under stress or fire, it breaks into small pieces that are not as dangerous as the shards of glass that are created when untreated glass breaks. Also called *tempered glass.*

tensile strength The resistance of a material to breaking under tension.

tension buttress system A system consisting of a strut and a ratchet strap that adds tension to the object being stabilized, locking the object in place with a diagonal force that lowers the object's center of gravity by increasing the object's entire footprint.

unibody frame construction The frame construction most commonly used in passenger cars that combines the vehicle body and the frame into a single component, thereby eliminating the rail beams used in body-over-frame vehicles.

upper rails Two parallel beams, one on each side of a vehicle's front end, that support and attach the strut tower system to the frame, hold the hood in place, and support the front fenders.

vertical spread maneuver A technique for exposing the latch or hinges of a vehicle door using a spreading tool, typically a hydraulic spreader, by separating the top of the door from the roof rail.

walking wounded Victims who were involved in an accident in some way and who have injuries, but who are not immediately identifiable.

wedge Material used to tighten or adjust cribbing and shoring systems. (NFPA 1006)

wet cell battery A lead acid battery that contains an active electrolyte solution of sulfuric acid and water. Also called *flooded cell battery.*

wheel well crush A technique using a hydraulic spreader to expose the hinges of a vehicle door by pulling the wheel well panel away from the door. Also called *fender crush.*

working load limit The rating that identifies the maximum load a jack can safely support.

yield strength The maximum stress that can be applied to steel before it permanently deforms.

REVIEW QUESTIONS

1. What type of motor vehicle frame construction combines the vehicle body and the frame into a single component?
2. What is the difference between a battery electric vehicle (BEV), a hybrid electric vehicle (HEV), and a plug-in hybrid electric vehicle (PHEV)?
3. What is cribbing most commonly used for?
4. What is a rescue-lift air bag?
5. In addition to dispatch information, what other factors must you consider, heightening your suspicions of the possibility of a rescue?
6. How should large emergency vehicles be positioned at an MVA so that they provide a barrier against motorists who fail to recognize or heed emergency warning lights?
7. What is the name of the "tight" 360-degree inspection of the vehicles involved in an MVA performed by the first-arriving officer or most experienced rescuer who arrives on scene?
8. What is the difference between primary and secondary access of victims trapped in a vehicle at an MVA?
9. What is cribbing?
10. What equipment is used in conjunction with ratchet straps to stabilize a vehicle that is on its side or roof?
11. What are the three most common ways to gain access to a victim trapped in a car?
12. What is the difference between the dash roll and the dash lift?
13. Who usually directs the removal of a victim from a vehicle after they are disentangled?
14. What are the two parts of terminating an incident?

DISCUSSION QUESTIONS

1. Why is the fact that high-strength steels are being used in vehicle construction significant to firefighters?
2. What are the advantages of having laminated and tempered glass windows in motor vehicles? Are there any disadvantages?

APPLYING THE CONCEPTS

It's 1730 and you and your crew are en route to a multiple-vehicle accident at a busy intersection. Dispatch reports four vehicles are involved and at least one is overturned. At least some of the drivers exited their vehicles safely, but other victims might be trapped. Traffic is at a crawl near the accident and police have just arrived to direct traffic. As your engine pulls up to the scene you see three vehicles with heavy body damage resting upright, and one vehicle lying on its roof and hood in the oncoming lane of traffic. The overturned vehicle appears to be an electric vehicle with white smoke coming from it. The other vehicles appear to be traditional gas-powered vehicles.

1. Motor vehicle accidents (MVAs) can be extremely dynamic and dangerous incidents. As you are en route to the scene, how can you place yourself in the right mindset to approach the incident in a state of readiness?
2. As the first-arriving fire company, what are your crew's initial priorities?

Your engine parks at an angle to the accident scene to block it from traffic. You and your crew don structural fire PPE and put on high visibility traffic vests. The company officer commands you and your partner to do an outer survey while the officer and a veteran firefighter do an inner survey.

3. What is an inner survey? How is it done? What information does it gather?
4. What is an outer survey? How large is the survey area? What is the objective?
5. When doing an outer survey, what type of tool helps firefighters find victims?
6. What do the battery's white gas cloud, popping sounds, and high temperature mean?
7. Should your crew immediately begin extrication operations of the trapped victim? Why or why not?

REFERENCES

Energy Security Agency (ESA). n.d. "Emergency Response Guides." Accessed October 16, 2023. https://energysecurityagency.com/erg/.

Energy Security Agency (ESA). n.d. "Risk Analysis and Guidance for First Responders." Accessed September 28, 2023. https://energysecurityagency.com/hybrid-electric-vehicle-risk-analysis-for-fire-service/.

National Fire Protection Association (NFPA). 2016. 1670, *Standard on Operations and Training for Technical Search and Rescue Incidents*, 2017 Edition. Quincy, MA: NFPA.

National Fire Protection Association (NFPA). 2020. *NFPA 1006, Standard for Technical Rescue Personnel Professional Qualifications.* 2021 Edition. Quincy, MA: NFPA.

National Fire Protection Association (NFPA). 2021. *NFPA 2500, Standard for Operations and Training for Technical Search and Rescue Incidents and Life Safety Rope and Equipment for Emergency Services.* 2022 Edition. Quincy, MA: NFPA.

Stewart, Timothy. 2023. Overview of Motor Vehicle Traffic Crashes in 2021 (Report No. DOT HS 813 435). National Highway Traffic Safety Administration (NHTSA), April 2023. Accessed September 17, 2023. https://crashstats.nhtsa.dot.gov/Api/Public/ViewPublication/813435.

Sweet, David. 2022. *Vehicle Rescue and Extrication: Principles and Practice, Revised Second Edition.* Burlington, MA: Jones & Bartlett Learning.

U.S. Department of Transportation, Transport Canada, and Secretariat of Communications and Transport of Mexico. *2020 Emergency Response Guidebook.* Accessed October 16, 2023. https://www.phmsa.dot.gov/sites/phmsa:dot.gov/files/2021-01/ERG2020-WEB.pdf.

CHAPTER

27

Firefighter II

Assisting Special Rescue Teams

KNOWLEDGE OBJECTIVES

After studying this chapter, you will be able to:

- Describe the role of the firefighter in a technical rescue operation.
- Explain how to safely assist during technical rescue operations.
- Describe the role of the firefighter in operating, cleaning, and maintaining tools and equipment.

SKILLS OBJECTIVES

After studying this chapter, you will be able to perform the following skills:

- Assist during technical rescue operations.
- Maintain generators.

ADDITIONAL NFPA STANDARDS

- **NFPA 70**, *National Electric Code, 2024 Edition*
- **NFPA 470**, *Hazardous Materials/Weapons of Mass Destruction (WMD) Standard for Responders, 2022 Edition*
- **NFPA 1006**, *Standard for Technical Rescue Personnel Professional Qualifications, 2021 Edition*
- **NFPA 1550**, *Standard for Emergency Responder Health and Safety, 2024 Edition*
- **NFPA 1900**, *Standard for Aircraft Rescue and Firefighting Vehicles, Automotive Fire Apparatus, Wildland Fire Apparatus, and Automotive Ambulances, 2024 Edition*
- **NFPA 1951**, *Standard on Protective Ensembles for Technical Rescue Incidents, 2020 Edition*
- **NFPA 1970**, *Standard on Protective Ensembles for Structural and Proximity Firefighting, Work Apparel and Open-Circuit Self-Contained Breathing Apparatus (SCBA) for Emergency Services, and Personal Alert Safety Systems (PASS), 2024 Edition*
- **NFPA 1989**, *Standard on Breathing Air Quality for Emergency Services Respiratory Protection, 2019 Edition*
- **NFPA 2500**, *Standard for Operations and Training for Technical Search and Rescue Incidents and Life Safety Rope and Equipment for Emergency Services, 2022 Edition*

CASE STUDY

You Are the Firefighter

Your engine company is dispatched to assist a rescue company with the rescue of a 51-year-old electrical worker who collapsed in an underground electrical vault at the bottom of a utility hole. While en route, you receive additional information that the victim's crew can no longer communicate with him. Upon your arrival, witnesses state that two additional crew members entered the vault to retrieve the victim. Their voices can no longer be heard, and they are not responding to calls on their radio.

1. What hazards might be present at this incident?
2. What types of rescue operations may be necessary at this incident?
3. What might be your role at this incident?

Introduction

Over the years, the number of fires and the related human and property losses from fires in the United States has decreased. This decrease is a result of improved fire prevention measures, fire safety education, building codes, and fire safety design. At the same time, the mission of fire departments has expanded. As fire departments take on additional responsibilities, such as emergency medical services (EMS), hazardous materials response, and technical rescue response, they become all-hazards agencies. These departments must be prepared to deal with any type of emergency.

Most technical rescue incidents require many firefighters. Not all firefighters need specialized training, but all firefighters need to know how to assist in a technical rescue incident. This chapter will not make you an expert in the skills needed to handle all rescue situations. It will, however, teach you how to play an essential part in the rescue effort by assisting specially trained personnel and working under their direction.

Training in technical rescue areas is conducted at three levels: awareness, operations, and technician:

- Awareness level: This training serves as an introduction to the topic, with a focus on recognizing hazards, securing the scene, and calling for appropriate assistance. There is no use of rescue skills at the awareness level.
- Operations level: This level trains firefighters to work in an incident's "warm zone" (the area directly around the hazard area). This training will allow you to directly assist those conducting the rescue operation and to use certain skills and procedures to conduct the rescue.
- Technician level: This level trains you to be directly involved in the rescue operation. Training at this level includes learning to use specific techniques and specialized equipment, caring for victims during the rescue, and managing the incident and all personnel at the scene.

Most of the training you receive as a beginning firefighter focuses on preparing you at the awareness level so that you can identify hazards and secure the scene to prevent further injuries.

Types of Rescues Encountered by Firefighters

A **technical rescue incident (TRI)** is a rescue incident that requires firefighters who have been trained and equipped with specialized tools and techniques and work together as a team within the Incident Command System (ICS) (**FIGURE 27-1**). NFPA 1006, *Standard for Technical Rescue Personnel Professional Qualifications*, and NFPA 2500, *Standard for Operations and Training for Technical Search and Rescue Incidents and Life Safety Rope and Equipment for Emergency Services*, break technical rescue incidents into the following categories:

1. Vehicle rescue
2. Machinery rescue
3. Animal technical rescue
4. Rope rescue

FIGURE 27-1 Most fire departments respond to a variety of special rescue situations that require additional training for safe operations.

5. Tower rescue
6. Structural collapse
7. Confined-space rescue
8. Trench rescue
9. Cave rescue
10. Mine and tunnel rescue
11. Wilderness rescue
12. Helicopter rescue
13. Surface water rescue
14. Swiftwater rescue
15. Dive rescue
16. Ice rescue
17. Surf rescue
18. Watercraft rescue
19. Floodwater rescue

Fire departments typically only prepare for the types of technical rescues they are likely to encounter. Departments may rely on other agencies to provide the trained personnel, specialized equipment, and team coordination necessary to perform some types of technical rescues (**FIGURE 27-2**).

Responders trained to the awareness level must understand the hazards and requirements of these special types of rescues. The first emergency unit to arrive at a rescue incident is often a fire department engine or truck company. The initial actions taken by that company may determine the safety of both the victims and the rescuers, and they will impact how efficiently the rescue is completed.

FIGURE 27-2 Some personnel are trained in specific rescue skills, such as rope rescue.

SAFETY TIP

A firefighter should never attempt to perform technical or dangerous rescue skills without the proper training, equipment, and team working within a fire department command structure.

Steps of Technical Rescue

Although technical rescue situations take many different forms, all rescuers take the following basic steps to perform special rescues in a safe, effective, and efficient manner:

1. Preparation
2. Response
3. Arrival and size-up
4. Stabilization
5. Access
6. Disentanglement
7. Removal
8. Transport
9. Security of the scene and preparation for the next call
10. Post-incident analysis

Preparing for the Response

You can prepare for responses to technical rescue incidents by training with the specialized rescue teams with whom you will work. This experience will allow you to learn about their techniques and equipment. It will also help you understand what you can do to assist them and what you should *not* do. It is also helpful to train with other fire departments in your area. This training will better enable you to respond to a mutual aid call. In addition, you will be able to learn about the types of rescue equipment to which other departments have access, as well as the training levels of their personnel.

Most technical rescue scenes will be busy and noisy, and communication will be difficult. Using specific terminology makes communicating with other rescuers easier and more effective. To properly follow commands, firefighters must understand the exact language being used. For example, if you are assisting with a rope rescue by pulling on a haul line, the specific command to pull is "haul," and the specific command to cease pulling is "stop." Firefighters on the line are listening for these commands. They should not respond to or use other words or phrases such as "go," "now," "hold up," "ok," or "whoa." Training with the technical rescue team will help prepare firefighters for the terminology, tools, and techniques they will encounter on these specialized calls.

Prior to a technical rescue call, firefighters should consider questions such as the following:

- Does the department have the personnel and equipment to handle a TRI from start to finish?
- Does the department meet National Fire Protection Association (NFPA) and Occupational Safety and Health Administration (OSHA) standards for technical rescue calls?
- Which equipment and personnel will the department send on a technical rescue call?
- Do members of the department know the hazards located within the department's response area, and have they visited these areas with local representatives?

Responding to the Incident

A department should have a specific dispatch protocol for a TRI. If your agency has a **technical rescue team**, it will usually respond with a rescue unit supported by other equipment and personnel, such as a medic unit, an engine company, a truck company, and a chief. The makeup of the response team is based on the ability of the agency or department to provide resources for the response. Even if a technical rescue team is available, it may not be trained or equipped to handle all types of incidents. Your department should have a plan for obtaining additional resources and trained personnel to assist in handling incidents when necessary.

Arrival and Size-Up

The first company officer to arrive on the scene should assume command. A rapid and accurate size-up is needed to avoid placing would-be rescuers in danger and to identify additional resources that may be needed. An essential part of any rescue is identifying all hazards. It is also important to determine whether a TRI is a rescue event or a body recovery operation. After these determinations have been made, the incident commander (IC) can decide whether to call for additional resources and order other actions to stabilize the incident.

Do *not* rush into the incident scene until you have adequately assessed the situation. You may unintentionally make the situation worse. For example, a firefighter approaching a trench collapse may cause further failure. A firefighter entering a swiftly flowing river might be quickly knocked down and carried downstream. A firefighter climbing down into a well to evaluate an unconscious victim may be overcome if there is not enough oxygen. Stop and think about the dangers that may be present. Do not make yourself part of the problem.

Be alert for dangers such as electrical hazards and oxygen-deficient or poisonous atmospheres. If the incident hazards exceed your level of training and preparation, do not attempt to rescue victims or mitigate the incident. Report the size-up information over the radio and request the appropriate resources. Then, establish a safe perimeter, or boundary, around the incident to prevent other people from becoming victims. When the special rescue team arrives, give updated information to the officer, and wait for instructions to assist within your level of training.

SAFETY TIP

At many fire and emergency scenes, fire line tape is set up to keep bystanders back. Emergency responders operating at such a scene routinely ignore this barrier by simply ducking under or climbing over the tape. This complacency toward fire line tape as a warning device creates a potential hazard for responders at scenes where the tape is placed to warn both bystanders and responders. All fire line tape should be considered a hazard warning to firefighters unless and until it is proven to be for bystanders only.

Stabilizing the Incident

Once the size-up is complete, the necessary resources are on the way, and the scene is safe to enter, begin to stabilize the incident. Establish an outer perimeter to keep the public and media out of the staging area. When the special rescue team arrives, they will establish a smaller perimeter directly around the rescue site. A rescue area is an area that surrounds the incident site (e.g., a collapsed structure, a collapsed trench, or a hazardous spill area). The size of the rescue area depends on the hazards. The following three control zones should be established:

- Hot zone: This area is for entry teams and special rescue teams only. The hot zone immediately surrounds the dangers of the site (e.g., hazardous materials releases) and is established to protect personnel outside the zone. Technician-level personnel are typically the only ones to operate in this area.
- Warm zone: This area is for properly trained and equipped personnel only. The warm zone is where personnel and equipment decontamination and hot zone support take place. This is where operations-level personnel assist.
- Cold zone: This area is for staging vehicles and equipment. The cold zone contains the command post. The public and the media should be kept clear of the cold zone. Awareness-level personnel stay in this zone.

To determine the size and scope of each of these zones, firefighters should identify and evaluate the hazards at the scene. Observe the geographic area and weather and wind conditions. Note the routes of access and exit, and consider evacuation problems and transport distances. Firefighters should enter only those zones for which they are adequately trained.

The most common method of establishing the control zones for an emergency incident site is to use police or fire line tape along with traffic cones and other markers. Once the fire department has established the three zones, responders should ensure that access to the zones is controlled. Scene control activities are sometimes assigned to law enforcement personnel.

Lockout and tagout systems should be used to ensure a safe environment (**FIGURE 27-3**). These systems are intended to ensure that equipment is de-energized and is not turned back on, in order to prevent injury to victims and rescue personnel.

In some emergency situations, all power must be turned off and locked out. Switches and valves must be tagged with labels to protect personnel and emergency response workers from accidental start-up. Locks and tags are used for everyone's protection against electrical dangers. For your safety and that of others, never remove or ignore a lock or tag. Always be alert for electrical hazards. They are not always easy to recognize.

FIGURE 27-3 Lockout and tagout systems are intended to ensure that equipment has been de-energized and cannot be turned back on accidentally.

As part of stabilizing the scene, the atmosphere should be monitored to identify any environments that are immediately dangerous to life and health (IDLH). The next steps involve evaluating the incident and planning how to rescue victims safely. For example, in trench rescue, these measures might include setting up ventilation fans for air flow, setting up lights for visibility, and protecting a trench from further collapse.

Gaining Access to the Victim

After the scene is stabilized, the special rescue team needs to consider how to gain access to the victim. Victims may be buried under dirt, suspended by ropes, or trapped by debris. Gaining access may not be as simple as removing what is blocking the way. Hazards such as loose soil, an explosive or poisonous atmosphere, or dangerous animals may require more specialized tools and equipment.

Special rescue personnel will need to communicate with victims during the rescue to make sure they are not injured further by the rescue operation. Even if they are not injured, victims need to be reassured that the team is working as quickly as possible to free them.

Emergency medical care should be started as soon as there is safe access to the victim. EMS personnel must be integrated into ongoing operations during rescue incidents. Their main functions are to treat victims and to stand by in case a rescue team member needs medical assistance. As soon as a rescue area or scene is secured and stabilized, EMS personnel must be allowed access to the victims for medical assessment and stabilization. If it is not possible to adequately stabilize the scene, the special rescue team may try to move the victims to a safe location for treatment. Some fire departments have cross-trained paramedics to the technical rescue level so that they can enter hazardous areas and provide direct assistance to the victim. Throughout the course of the rescue operation, EMS personnel must continually monitor and ensure the stability of the victims. That means they must be allowed access to the victims.

Disentangling the Victim

Once precautions have been taken and the reason for any entrapment has been identified, the victim needs to be freed as safely and efficiently as possible. For example, in a trench incident, this step would include digging either with a shovel or by hand to free the victim. In a vehicle or machinery entrapment, disentanglement may entail cutting or disassembling the machinery or vehicle to free the victim (see Chapter 26, *Vehicle Rescue and Extrication*). A special rescue team member should remain with the victim to direct the rescuers who are performing the disentanglement.

SAFETY TIP

Technical rescue incidents may last longer than structure fires. Pay attention to your physical state and that of your team members. If you are tired, ask the IC to send your crew to the rehabilitation area.

Removing the Victim

After the victim has been disentangled, efforts will shift to removing the victim (**FIGURE 27-4**). This may simply amount to having someone assist the victim up a ladder. In other cases, special rescue teams may use Stokes baskets, stretchers, or other immobilization devices to remove an injured or unresponsive victim from a trench, a confined space, or an elevated point. Once the victim is out of the hazard area, the Firefighter II may assist EMS personnel with preparing victims for transport. **Packaging** is the process of preparing the victim for movement as a unit. Refer to your local EMS protocols for specific victim packaging and treatment before moving victims from special rescue situations.

FIGURE 27-4 Special rescuers removing a victim from a confined space.

The primary objective for victim rescue, transfer, and removal is to complete the process as safely and efficiently as possible. The rescuer must use good body mechanics and victim packaging, removal, and transportation skills.

Transporting the Victim

After the victim has been removed from the hazard area, EMS will transport them to an appropriate medical facility. The type of transport used will depend on the severity of the victim's injuries and the distance to the medical facility. For example, if a victim is critically injured or if the rescue is taking place far away from the hospital, air transportation may be used rather than a ground ambulance.

In rough-terrain rescues, four-wheel drive, high-clearance vehicles may be required to transport victims on stretchers to a waiting ambulance. Snowmobiles with attached sleds can be used to transport victims down a snowy mountain. Helicopters are increasingly used for quick evacuation from remote areas, but weather conditions sometimes limit their use.

Transporting victims who have been injured in a hazardous materials incident may raise additional challenges. It is extremely important that adequate decontamination occurs prior to transport. Placing a poorly decontaminated victim inside an ambulance or helicopter and then closing the doors can put both the victim and the rescue personnel in danger. Any toxic gases or vapors given off by the victim or the victim's clothing can contaminate the inside of the transport vehicle, causing injury to the EMS personnel. In such an incident, any bags containing contaminated clothing or other personal effects must be properly sealed if they are going with the victim. Receiving hospitals should be notified that victims who have been involved in a hazardous materials exposure incident will be arriving. This will alert the hospital's triage teams so that they can take the necessary precautions when victims begin to arrive.

Because special rescue incidents are often extended-duration events, it is important to pay attention to the rehabilitation needs of rescue personnel. It is often necessary to rotate personnel. This reduces the chance of injury or sickness for the rescuers and provides a safer rescue for the victim. Rehabilitation is especially important when temperatures are very high or very low and when rescuers are subjected to wet conditions. It is better to plan for rehabilitation and not need it than to not have rehabilitation available when it is necessary. For more information on rehabilitation, see Chapter 20, *Firefighter Rehabilitation*.

Security of the Scene and Preparation for the Next Call

Once the rescue is complete, the scene must be stabilized by the rescue crew to ensure that no one else becomes injured. In a trench incident, this includes filling the trench with dirt and roping off the surrounding area.

In a hazardous materials incident, the clean-up of equipment and personnel takes place after the hazardous materials incident has been completely controlled and all victims have been treated and transported. Trained disposal crews should clean the site. All equipment, protective gear, and clothing, as well as the rescue personnel themselves, must be decontaminated.

In a vehicle rescue incident, the vehicle is removed. Usually, it is transported by a tow truck to a storage lot. The street or roadway is cleaned of all vehicle debris and any spilled fluids, including gasoline.

In an industrial setting, the supervisor of the facility is responsible for securing the scene. However, the technical rescue team must follow up with the supervisor or the safety coordinator to ensure that further problems are prevented.

After you have secured the scene and packed up your equipment, return to the station and fully inventory, clean, service, and maintain all of the equipment to prepare it for the next call. Although some items may need repair, most will require only simple maintenance before being placed back on the apparatus and considered in service.

Back at the station, as you prepare for the next call, complete all reports and document the rescue incident. Adequate reporting and accurate recordkeeping ensure the continuity of quality care. They also guarantee proper transfer of responsibility and ensure the fire department fulfills local, state, and federal reporting requirements. In addition, the reports can be used to evaluate response times, equipment usage, and other areas of administrative interest.

Post-Incident Analysis

As with any type of call, the best way to prepare for the next rescue call is to review the last one. Identify the strengths and weaknesses in your response. What did you do well? What could have been done better? Would any other equipment have made the rescue safer or easier? If a death or serious injury occurred during the call, a critical incident stress management session may take place to assist firefighters. Reviewing a TRI with everyone involved will allow each person to learn from the call and make the next call more successful.

General Rescue Scene Procedures

At any incident scene—whether a fire, EMS, or technical rescue call—your safety, the safety of your company, and the public's safety are most important. At a TRI, many issues need to be considered during the response. Many firefighters have been injured or killed during special rescue incidents because they did not consider their own safety. While the impulse may be to immediately approach the victim or the accident area, it is critically important to slow down and evaluate the situation.

In confined-space rescue incidents, for example, potential hazards may include deep or isolated spaces, multiple hazards (such as water or chemicals), failure of essential equipment or services, and environmental conditions (such as snow or rain). In all rescue incidents, firefighters should consider the potential general hazards and risks posed by utilities, hazardous materials, confined spaces, and environmental conditions, as well as IDLH hazards.

Approaching the Scene

Often, the information received on the emergency call will be incomplete. You will have to evaluate the incident with the information available. Once on the scene, you can better identify life-threatening hazards and take the appropriate corrective measures to mitigate those dangers. Careful size-up and coordination of rescue efforts are essential to avoid further injuries to the victims and to ensure the safety of the firefighters.

At a technical rescue scene, it is everyone's responsibility to identify unsafe conditions and ensure they are made right. For instance, even if they are not apparatus driver/operators, firefighters must also consider if they and their apparatus are properly positioned. At a motor vehicle collision in a roadway, for example, apparatus should be positioned to protect the work area. Likewise, a firefighter may not be directly responsible for operating equipment or securing part of the scene, but they should recognize when the use of a tool or technique or the failure of a piece of life safety equipment risks injury to a firefighter, victim, or bystander.

Mitigating Utility Hazards

Look for any downed electrical wires near the scene (**FIGURE 27-5**). Is any of the equipment or machinery present energized so that it might present a hazard to the victim or the rescuers? Is a natural gas or propane fuel service connected to the structure? Firefighters of every level should be prepared to identify, report, and avoid hazards as necessary. Before utilities are disconnected by the utility provider, firefighters must assume that gaseous fuels and energized electrical systems and equipment are present.

FIGURE 27-5 Downed electrical wires present a hazard at many types of rescue scenes.

Courtesy of Lara Shane/FEMA Photo News.

Rescue operations involving energized electrical lines or other utility hazards require the assistance of trained personnel. Many firefighters have been injured or killed because they did not recognize the hazards at these types of incidents. Do not rush to assist victims who have come into contact with energized electrical equipment or have been overcome by gaseous fuels. Watch for falling utility poles. A damaged pole may bring other poles down with it. Do not touch any wires, power lines, or other electrical sources until they have been de-energized by a power company representative. This includes any conductive materials that may be in contact with electrical lines or equipment. For example, metal fences that become energized will be energized for their entire unbroken length, as will water or wet materials in contact with electricity.

Both natural gas and liquefied petroleum gas are nontoxic. However, they are classified as asphyxiants because they displace breathing air. In addition, both gases are flammable. If a call involves leaking gas, follow your standard operating procedures (SOPs) to report and mitigate it. If a victim has been overcome by leaking gas, wear full personal protective equipment (PPE) and positive-pressure self-contained breathing apparatus (SCBA) when making the rescue. Remove the victim from the hazardous atmosphere before beginning treatment.

Providing Scene Security

During a technical rescue, bystanders, family members, and even other rescuers may enter an unsafe scene and become additional victims. The IC should coordinate with law enforcement to help secure the scene and control access. The fire department should use a strict personnel accountability system to control access to the rescue scene.

You will often be asked to establish barriers to secure a rescue incident scene. When emergency incidents occur on streets or highways, vehicle warning lights, traffic cones, flares, and fusees are often used to warn oncoming traffic. When crowd control is an issue, plastic barrier tape marked with the words "Caution" or "Fire Line" is often used to keep out nonessential personnel. The steps for establishing a barrier are listed in **SKILL DRILL 27-1**.

Using Protective Equipment

Firefighting gear is designed to protect the body from the high temperatures of fire. It does, however, restrict movement. Technical rescues require the ability to move around freely, so standard firefighting gear does

SKILL DRILL 27-1

Establishing a Barrier Firefighter II, NFPA 1010: 7.4.2

1. Respond safely to the emergency scene. Place the emergency vehicle in a safe position that protects the scene. Don the appropriate PPE.

2. Perform a size-up to assess for hazards. Secure the scene.

3. Call for needed assistance. Use appropriate devices to establish a barrier, following the orders of the IC.

not always work well in a TRI. For this reason, most specialized teams also carry items such as harnesses; smaller, lighter helmets; and jumpsuits that are easier to move in than structural firefighting protective gear (**FIGURE 27-6**).

For personnel operating in a water rescue hazard zone, the minimum PPE includes a Coast Guard–approved **personal flotation device (PFD)**, thermal protection, a helmet appropriate for water rescue, a cutting device, a whistle that works in water, and contamination protection (if necessary).

Other equipment items that can be easily carried by firefighters include binoculars, chalk or spray paint for marking searched areas, a compass for wilderness rescues, first-aid kits, a whistle, a hand-held global positioning system (GPS), and Cyalume-type light sticks. The IC and the technical rescue team will help determine which PPE you will need to wear while assisting in the rescue operation. For more information, see NFPA 1951, *Standard on Protective Ensembles for Technical Rescue Incidents.*

Using the Incident Command System

The first-arriving officer immediately establishes command and starts using the ICS. Establishing the ICS early is critically important because many TRIs will eventually become complex and require many assisting units. Without the ICS in place, it will be difficult, if not impossible, to ensure effective scene operations and the safety of the rescuers. For more information on the ICS, see Chapter 22, *Establishing and Transferring Command.*

FIGURE 27-6 Technical rescue technicians wear different types of protective equipment that allow mobility while still providing an adequate level of protection for the incident.

Ensuring Accountability

Accountability should be practiced at all emergencies, no matter how small. A personnel accountability system helps protect rescuers by tracking all personnel on the scene, including their identities, assignments, and locations. This system ensures that only rescuers who have been given specific assignments are operating within the hot zone. By using a personnel accountability system and working within the ICS, an IC can track the resources at the scene, make appropriate assignments, and ensure that every person at the scene operates safely. Review the section on personnel accountability in Chapter 19, *Firefighter Survival*.

Making Victim Contact

Staying in constant communication with the victim is important. If possible, assign someone to talk to the victim while other firefighters carry out the rescue. The victim could be sick or injured and is probably frightened. If the firefighter is calm, their reassuring demeanor will calm the victim. To help keep a victim calm, do the following:

- Make and keep eye contact with the victim.
- Tell the truth. Lying destroys trust and confidence. You may not always tell the victim everything, but if the victim asks a specific question, answer it truthfully.
- Communicate at a level the victim can understand.
- Be aware of your own body language.
- Always speak slowly, clearly, and distinctly.
- Use the victim's name.
- If a victim is hard of hearing, speak clearly and directly to the victim so that they can read your lips.
- Allow time for the victim to answer or respond to your questions.
- Try to make the victim comfortable and relaxed whenever possible.

Many victims at TRIs require medical care, but this care should be given only if it can be done safely. Do not become another victim during a rescue attempt. Many firefighters have been killed while entering a TRI to help a victim who was already deceased.

Assisting Rescue Crews

Every TRI is different. If you have the role of assisting a technical rescue team, the most important thing you can do to prepare—other than using the ICS—is training with the team. By training with your department's or region's technical rescue teams, you will better understand how the team members operate and how you can assist them. The more knowledge you have, the more you will be able to do.

At any TRI, follow the orders of the company officer who receives direction from the IC. Many tasks will involve moving equipment and objects from one place to another. Other tasks will involve protecting the team and victims. For instance, you may need to pull on a haul line, treat medical victims to your level of training, assist EMS personnel with preparing a victim for transport, and assist a victim up or down a ladder, among other things. Do not take these tasks lightly. They might not seem important, but they are essential to the effort. Without scene security, traffic control, protective hose lines, and people to move and maintain equipment, the rescue cannot happen.

You may also be asked to retrieve specialized tools and equipment at any type of special rescue incident. You must become familiar with the different tools used by the specialized rescue teams in your department. The steps for identifying and retrieving rescue tools are listed in **SKILL DRILL 27-2**.

As you read the information that follows about assisting at specific types of rescue scenes, keep in mind the three guidelines for safely approaching the TRI scene:

- Approach the scene cautiously.
- Position the apparatus properly.
- Assist the specialized team members as needed.

SKILL DRILL 27-2

Identifying and Retrieving Rescue Tools Firefighter II, NFPA 1010: 7.4.2

1. Respond safely to the emergency scene.
2. Place the emergency vehicle in a safe position that protects the scene.
3. Don the appropriate PPE.
4. Perform a size-up to assess for hazards.
5. Secure the scene.
6. Call for needed assistance.
7. Receive the request from the special rescue team to retrieve rescue tools.
8. Identify and retrieve the rescue tools.

Categories of Specialized Technical Rescue

NFPA 1006 and NFPA 2500 list many different categories of specialized technical rescue. Many problems and hazards, as well as the tools and techniques to deal with them, are shared between categories. For example, firefighters will encounter physical hazards at both vehicle and machinery rescue scenarios. Both types of scenarios require ensuring that vehicles and equipment are properly stabilized and that energized equipment is safely de-energized and shut down. Likewise, both vehicle and machinery rescues require similar approaches to disentangle the victim, disassemble vehicle or machinery parts, or dissect (cut) metal.

In other technical rescue incidents, tools and techniques from several different categories may be required. For example, an incident where people are trapped in an elevator may generally fall under the category of machinery rescue but may require tools and techniques from vehicle and rope rescue to complete the mission. While the categories found in NFPA 1006 and NFPA 2500 are useful, be aware that many technical rescue incidents will fall into more than one category.

Vehicle and Machinery Rescue

A wide variety of rescue situations involve vehicles and machinery. The basic actions commonly taken at the scene of motor vehicle accidents are covered in Chapter 26, *Vehicle Rescue and Extrication.* This section focuses on other types of rescues involving vehicles and machinery, such as motorcycles, trucks, buses, trains, mass-transit vehicles, aircraft, farm machinery, agricultural implements, construction machinery, and industrial machinery.

Safe Approach

Hazards present at vehicle and machinery rescues can include flammable liquids, chemical hazards, unstable or moving vehicles or machines, and unstable surfaces, as well as physical, electrical, and other types of hazards.

Traffic control is crucial for most vehicle incidents that occur in or around a roadway. **Traffic incident management (TIM)** is the way police, fire, rescue, EMS, and highway agencies can work together to identify and respond to roadway incidents and control traffic in an area of emergency operations. Chapter 26, *Vehicle Rescue and Extrication*, discusses traffic control in greater detail. Just as different departments handle technical rescues differently, agencies handle TIM in various ways, but they still adhere to fundamental principles. These principles include establishing an area for firefighters to work at the incident, making the work area and the people in it visible, protecting the work area, diverting traffic around the work area, and safely breaking down the work area and restoring traffic flow when the incident is complete.

How You Can Assist

At a vehicle or machinery rescue call, you may need to assist in extricating and treating the victim. Protecting the victim with boards, blankets, or other materials can allow the technical rescue team to operate tools on the vehicle or machine without causing further injury to the victim. Many hydraulic tools are heavy, so you may need to help support the tools while technicians operate the equipment.

You may also be called upon to do the following:

- Assist with TIM.
- Establish the perimeter of the rescue incident and control site security.
- Obtain information from witnesses.
- Retrieve rescue tools and equipment from the fire apparatus or rescue trucks.
- Console family members of victims.
- Keep bystanders out of the way.
- Assist in moving items in the way of the rescue team.
- Troubleshoot equipment.
- Set up power tools.
- Stand by with a hose line.
- Extinguish a fire.

Tools Used

Many tools are required for a successful vehicle or machinery rescue. Simple tools such as a spring-loaded punch, for example, can safely break out a vehicle window so you can access a victim. Tools and items that may be required for a vehicle or machinery rescue include the following:

- PPE
- Hydraulic tools (e.g., spreaders, cutters, rams)
- Electric tools (e.g., reciprocating and whizzer saws)
- Cutting torch
- Air chisels
- Cribbing
- Saber saw
- Windshield cutter
- Spring-loaded punch
- Chains
- Rescue-lift air bags
- SCBA air cylinders
- Basic hand tools (e.g., screwdriver and pliers)
- Flat-head axe and Halligan tool
- Come along winch
- Portable generators, pumps, and compressors
- Seat belt cutters
- Hand lights and scene lighting
- Hose lines (protection)
- EMS equipment

Rope and Tower Rescue

Rope rescue skills are the most versatile and widely used technical rescue skills. A rope rescue may be performed to remove an ill or injured victim or to remove a victim from a position of peril. In Chapter 8, *Ropes and Knots,* you learned how to tie some basic knots and to use ropes to lift and lower tools. This section provides an introduction to the more complex parts of rope rescue. However, an approved rope rescue course is required to become an expert in rope rescue skills. These training courses will build on the skills you have already learned. They cover the technical skills needed to raise and lower people using specialized equipment, such as anchors, harnesses, victim litters or baskets, mechanical advantage systems, friction and clutch devices, and, of course, various types of rope and webbing (**FIGURE 27-7**).

FIGURE 27-7 Rope rescues require intense technical training.

Rope Rescue Incidents

Most rope rescue incidents involve people who are trapped in locations that cannot be easily reached, such as a mountainside, an electrical tower, or the outside of a building (**FIGURE 27-8**). Rescuers often must raise or lower themselves to reach the victim. Once rescuers reach the victim, they must stabilize them and determine how to get them to safety. Sometimes the victim must be lowered or raised to a safe location. Extreme cases could involve more complicated operations, such as moving the victim horizontally or transporting them in a basket lowered by a helicopter.

The type and number of ropes used in a rope rescue depend on the situation. There is almost always a primary rope that bears the weight of the rescuer (or rescuers) as they try to reach the victim. The rescuers often have a second line attached to them, known as a belay line, which serves as a backup if the main line fails. In some cases, additional lines might be needed to raise or lower the victim.

FIGURE 27-8 Ropes are invaluable when a victim is trapped in an inaccessible area, such as outside a high building.

Rope rescue incidents can be divided into low-angle and high-angle operations. **Low-angle operations** are situations where the slope of the ground over which the firefighters are working is less than 45 degrees. In these low-angle operations, firefighters depend on the ground for their primary support, and the rope system assists with support and movement (**FIGURE 27-9**). An example of a low-angle system is a rope stretched from the top of an embankment and used for support by rescuers who are carrying a victim up an incline. In this scenario, the rope is the secondary means of support and movement for the rescuers. The primary means of support is the rescuers' contact with the ground.

Low-angle operations are used when the scene only requires ropes to assist in pulling or hauling up a victim or rescuer. This assistance may be necessary on unstable surfaces such as ice, snow, dirt, mud, or rocky embankments. In such an incident, a rope will be tied to the rescuer's harness. The rescuer will then climb the embankment on their own, using the rope only to make sure they do not fall. Low-angle operations may also include life lines deployed during ice or water rescues.

Ropes can also be used to assist in carrying a Stokes basket to transport a victim. Using ropes in this situation aids the rescuers and frees them from having to carry all of the weight over rough terrain. Rescuers at the top of the embankment can help to pull up or lower the basket using a rope system.

High-angle operations are situations where the slope of the ground is greater than 45 degrees and rescuers or victims depend on **life safety rope** for support rather than on a fixed surface such as the ground. These types of rescue may involve victims trapped on

FIGURE 27-9 Low-angle rescue operations.

towers, such as communication, broadcast, and electrical line towers. High-angle rescue techniques are used to raise or lower a victim when other means of raising or lowering are not available. You will need to complete a rope rescue training course before you are ready to perform high-angle rescues. That training will allow you to quickly and safely perform the complex tasks required in these incidents.

Safe Approach

If you respond to an incident that may require a rope rescue operation, consider your safety and the safety of those around you. Extensive setup may be necessary, and a significant amount of equipment may need to be assembled before starting the rescue. Protect your safety by remaining away from the area under the victim and away from any loose materials that may fall or slide. Likewise, when you are above the victim, stay back from the edge unless you are securely tied off. Use extreme caution so you do not cause any loose debris to fall on the victim. Work to control the scene by moving the bystanders and friends of the victim to an area where they will not be injured.

How You Can Assist

Rope rescues can be labor-intensive operations, and successful rope rescues require a cohesive team effort. By remembering the fundamental rope skills you have learned, you may be able to assist with many phases of the rescue operation. For instance, a technical rescue team member might ask you to tie knots and prepare anchors. Do not be offended if the team members check your work. It is standard for firefighters to check a system two or three times to ensure its safety before use.

Avoid stepping on ropes. Any damage or friction to the outer layers of a rope can decrease its tensile strength and may cause a catastrophic failure. Keep everyone clear of areas where falling objects could cause injuries. Keep in mind the importance of following the ICS. Work to complete your assigned tasks, which may include the following:

- Recognizing the need for specialized low or high-angle teams and equipment
- Identifying hazards
- Initiating contact and establishing communication with victims
- Initiating control and management of the scene
- Establishing or coordinating with the incident command structure

Tools Used

Rope rescues rely almost completely on the equipment used. Some of the tools and equipment that you will need to identify and use are listed here:

- PPE (helmet, rescue gloves)
- Personnel harnesses
- Stokes basket
- Harness
- Rescue rope
- Webbing
- Prusik cord
- Edge protection materials
- Basic hardware such as D-rings and carabiners
- Ascending and descending hardware (racks, pulleys)

In addition to the rope itself, several hardware components might be used. Firefighters, for example, often use a **carabiner** (**FIGURE 27-10**). This device is used to connect one rope to another rope or to other hardware such as an anchor plate, swivel, or pulley. Several types of carabiners are available. You should know how to operate the type used by your department. Only a few are recognized for use in the fire service and in rescue operations. Refer to NFPA 1983, *Standard on Life Safety Rope and Equipment for Emergency Services.*

A **harness** is a piece of rescue or safety equipment made of webbing and worn by a person. It is used to secure the person to a rope or to a solid object. Two types of harnesses—Class II and Class III—can be used by rescuers, depending on the circumstances:

- Class II harness (seat harness) fastens around the rescuer's waist and legs and has a design load

FIGURE 27-10 Carabiners.

FIGURE 27-11 Class II harness (seat harness).

of 600 pounds (272 kilograms). It is used to support a firefighter, particularly in rescue situations (**FIGURE 27-11**).

- Class III harness (full-body harness) fastens around the rescuer's waist and thighs as well as their shoulders. It is the most secure type of harness and is often used to support a firefighter who is being raised or lowered on a life safety rope (**FIGURE 27-12**).

Harnesses need to be cleaned and inspected regularly, just as life safety ropes do. Follow the manufacturer's instructions for cleaning and inspecting harnesses.

Confined-Space, Structural Collapse, and Trench Rescue

A **confined space** is an enclosed area that is not designed for people to occupy for extended periods of time. Confined spaces have limited openings for entrance and exit and may have limited ventilation to provide air circulation and exchange. They can occur in natural, farm, commercial, and industrial settings. For example, grain silos, industrial pits, tanks, and below-ground structures, such as underground electrical vaults, are all confined spaces. Likewise, cisterns, well casings, sewers, and septic tanks are confined spaces found in residential settings.

Confined-space incidents are among the most dangerous scenes you may encounter. Confined spaces may be oxygen deficient and contain poisonous gases or combustible atmospheres. There is also the potential for slips, trips, falls, and entrapment, as well as deep water and collapse hazards. The National Institute for Occupational Safety and Health (NIOSH) reports that initial rescuers account for about 60 percent of all confined-space deaths (NIOSH 86-110, 1986).

FIGURE 27-12 Class III harness (full-body harness).

It is often difficult to rescue an unconscious or injured person from a confined space because of the poor ventilation and limited entry or exit area. For this reason, ropes are often used to remove an injured or unconscious victim (**FIGURE 27-13**).

Structural collapse is the sudden and unplanned fall of all or part of a building (**FIGURE 27-14**). Such collapses may occur because of fires, vehicle accidents, explosions, rain, wind, snowstorms, earthquakes, tornadoes, and activities related to construction, renovation, or demolition.

All building components, working together, provide a stable building. This concept is discussed further in Chapter 6, *Building Construction*. Consider the type

FIGURE 27-13 Ropes are often used to remove an injured or unconscious victim from a confined space.

© Jones & Bartlett Learning. Photographed by Glen E. Ellman.

FIGURE 27-14 Structural collapse is a sudden and unplanned fall of all or part of a building.

Courtesy of Dave Gatley/FEMA.

of building construction when determining the potential for collapse. When any part of a building becomes weakened, the dynamics of the building change. For this reason, firefighters should always be alert for signs

FIGURE 27-15 Trench rescue.

Courtesy of Captain David Jackson, Saginaw Township Fire Department.

of a possible building collapse. A partial building collapse is extremely hazardous to rescuers because of the potential for secondary collapse during search and rescue operations.

Rescues in collapsed trenches are often complicated and involve many different skills, such as shoring, air quality monitoring, confined-space operations, and rope rescue. Trench and excavation rescues become necessary when earth has been removed for placement of a utility line or for other construction and the sides of the excavation collapse, trapping a worker (**FIGURE 27-15**). **Entrapment** can occur when children play around a pile of sand or earth that collapses. Many entrapments occur because required safety precautions were not taken.

Whenever a collapse has occurred, the collapsed structure or material becomes unstable and prone to further collapse. Earth and sand are heavy, and a person who is partly entrapped in these materials cannot simply be pulled out. Instead, the victim must be carefully dug out. This step can be taken only after **shoring**, which involves the use of temporary supports, has stabilized the sides of the excavation.

Vibration or additional weight on top of displaced earth increases the risk of a **secondary collapse**—that is, a collapse that occurs after the initial collapse.

A secondary collapse can be caused by equipment vibration, people standing at the edge of the trench, or water eroding the soil. Do not attempt to rescue someone who is trapped by dirt or sand. You may quickly become another victim! Safe removal of trapped persons requires a special rescue team trained and equipped to erect shoring to protect the rescuers and the entrapped person from secondary collapse.

Cave, tunnel, and mine rescue operations are often lengthy, complicated incidents that require trained rescue teams and specialized equipment. Never attempt to enter a cave or tunnel for rescue operations without having the proper training and equipment. All confined spaces, either purpose-built or the result of a collapse or structural failure, should be considered to be IDLH until proven otherwise.

SAFETY TIP

The two greatest hazards in a confined space are the lack of oxygen and the presence of poisonous gases or asphyxiants. If rescuers fail to ensure that a confined space contains a safe atmosphere, both the rescuers and the original victim are at greater risk of injury and death. Treat all confined-space incidents with extreme caution. Do not enter a confined space without proper training. Instead, secure the area, gather as much information as possible, and wait for confined-space technical rescuers to arrive on the scene.

Safe Approach

As you approach a confined-space rescue scene, look for bystanders who might have information about the emergency. Do not assume that a victim has simply suffered a medical problem. Instead, presume an IDLH atmosphere is present at any confined-space call and that breathing protection is required for anyone entering the space. IDLH atmospheres may also be corrosive, explosive, or dangerous in other ways.

At any collapse incident, consider the need to shut off utilities and the possibility of further collapse. Even a well-trained engineer cannot always determine the stability of a structure simply by looking at it.

Walking close to the edge of a trench or excavation collapse can trigger a secondary collapse. Stay away from the edge of the collapse. Keep all workers and bystanders at a distance from the site. Vibration from equipment and machinery can cause additional collapses, so shut off all heavy equipment. Likewise, vibrations caused by nearby traffic can cause collapse, so it may be necessary to stop or divert traffic away from the scene.

Soil that has been removed from the excavation and placed in a pile is called the **spoil pile**. This material is unstable and may collapse if placed too close to the excavation. Avoid disturbing the spoil pile.

Make verbal contact with the trapped person, if possible, but do not place yourself in danger while doing so. At a trench rescue scene, if you need to approach the trench, do so from the narrow end where the soil will be more stable. It is best, however, not to approach the trench unless absolutely necessary. Stay out of a trench unless you have been properly trained in trench rescue techniques.

Provide reassurance by letting the trapped person know that a trained rescue team is on the way. By moving people away from a collapsed structure or the edges of the excavation, shutting down machinery, and establishing contact with the victim, you start the rescue process.

How You Can Assist

Your main role in confined-space rescues is to secure the scene and prevent other people from entering the confined space until additional rescue resources arrive. If the incident has occurred at an industrial site, determine if an OSHA-required confined-space entry permit was prepared. If so, locate the entry supervisor and attendant listed on the permit and have them stand by at the space. If both persons cannot be accounted for, they may have become additional victims inside the confined space.

As additional highly trained personnel arrive, your company may provide help by giving the rescuers a situation report using the ICS. Items of importance that should be included in a situation report are a description of any rescue attempts that have been made, the exposures, any hazards present, the extinguishment of fires, the facts and probabilities of the scene, the situation and resources of the fire company, the identities of any hazardous materials present, and an evaluation of the progress made so far.

Collapse and confined-space rescues can be complex and take a long time to complete. You may be asked to assist by bringing rescue equipment to the scene, staging tools and supplies, maintaining a charged hose line, or providing crowd control. As the rescue team starts to work, your company may be assigned to carry out additional tasks, ranging from unloading lumber for shoring to assisting with cutting timbers at a safe distance away from the entrapment. Depending on weather conditions or the length of the incident, a rehabilitation group may need to be set up.

Implementing the ICS quickly will help make this type of rescue go smoothly. If someone in your company has special skills, such as carpentry, they should make this fact known because this skill is valuable at a trench rescue. Measuring and cutting timber and shores for a trench operation are important aspects of the rescue.

Like confined spaces, trenches may have an IDLH atmosphere inside because of the presence of hazardous gases such as carbon monoxide, methane, or sewer gases. Flammable vapors from gas-powered equipment or gas cans may also be present in the trench. Setting up ventilation fans can improve victim and rescuer survivability in these scenarios.

It may be necessary to pump water out of a collapsed area. If the collapse occurred because of a ruptured pipe or during a rainstorm, for example, rising water in the trench may endanger the trapped victim and cause additional soil to collapse. Removing this water may help stabilize the situation.

Tools Used

One of the key tools used in any confined-space rescue is a **supplied-air respirator** system (**FIGURE 27-16**). These devices are like SCBA except that instead of carrying the air supply in a cylinder on your back, you are connected by a hose line to an air supply located outside the confined space. This kind of system provides the rescuer with a continuous supply of air that is not limited by the capacity of a back-mounted air cylinder.

Many other tools and pieces of equipment can be used during a confined-space rescue operation, including the following:

- Non-structural firefighting PPE, such as a hard-hat style helmet, work gloves, a rescue jumpsuit, harness, work boots, knee pads, elbow pads, and eye protection
- Air lines
- Air carts
- Breathing air compressors
- Extra SCBA air cylinders
- Tripod and winch for raising and lowering rescuers and victims
- Pry bar or manhole cover remover
- Rescue rope
- Personnel harnesses
- Gas monitors
- Ventilation fans
- Explosion-proof lights
- Radios
- Hard-wired communications equipment
- Lockout and tagout kits
- Personnel accountability boards
- Rope rescue equipment

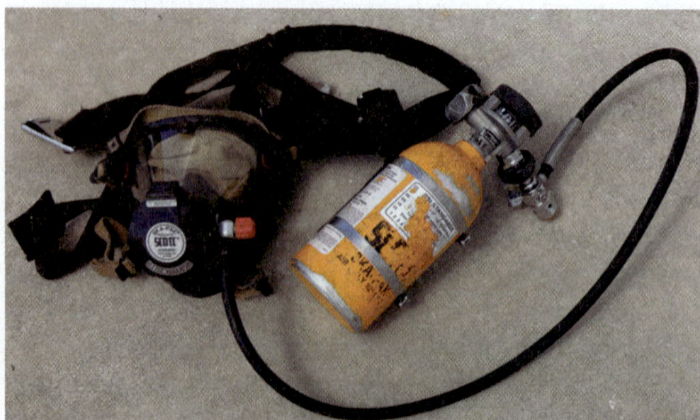

FIGURE 27-16 Supplied-air respirator system.

Tools and equipment used in collapse rescue include the following:

- Hydraulic, pneumatic, and wood shores
- Lumber and plywood for shoring
- Cribbing
- Power cutting tools and saws
- Carpentry hand tools
- Shovels
- Buckets for moving soil
- Rescue rope, harnesses, webbing, and associated hardware
- Utility rope
- Water pumps
- Ladders
- Shovels
- Extrication equipment such as Stokes baskets or backboards
- Harness sets for Stokes baskets
- Medical equipment

Wilderness and Helicopter Search and Rescue

Wilderness and helicopter search and rescue (SAR) missions consist of two parts: search and rescue. Search is defined as looking for a lost or overdue person. Rescue, in the SAR context, is defined as removing a victim from a hostile environment.

SAR missions may be required for a variety of reasons. Small children may wander off and be unable to find their way back to a known place. People who are living with cognitive impairment may fail to remember where they are going and become lost. People who are hiking, hunting, or taking part in other wilderness activities may become lost because they do not have the proper training or equipment for that activity or because the weather changes unexpectedly. Others may need to be rescued because they become sick or injured.

SAR scenarios may cover such large areas that it is necessary to use helicopters to assist with either the search, the rescue, or both. Helicopters can travel quickly over large distances and can land in a variety of locations. They are also extremely versatile equipment platforms that can provide specialized lighting, cameras, and rescue equipment.

Safe Approach

As a firefighter, you may respond to a possible SAR mission as part of your fire company. In some communities, SAR is the responsibility of law enforcement personnel. In others, it becomes the responsibility of volunteer search and rescue groups.

The term "wilderness" may be used to describe a variety of environments, such as forests, mountains, deserts, parks, and animal refuges. Depending on the terrain and environmental factors, the wilderness can be as little as a few minutes into the backcountry or a few feet off the roadway. For example, rescuers may be able to quickly reach an injured hiker in the woods, but it still may take a long time and a great deal of resources to remove them due to factors such as the terrain, their injuries, and the weather. Terrain hazards may include cliffs, caves, wells, mines, avalanches, and rockslides.

When you participate in SAR missions, prepare for the weather conditions by bringing suitable clothing. Do not exceed your physical limitations, and do not get into situations in the wilderness that are beyond your ability to handle. Call for a special wilderness rescue team, based on the needs of the situation and your local protocols.

How You Can Assist

The fire service teaches its members to work in a buddy system. Wilderness rescues are another situation in which you should never go alone. By working in teams of at least two and having a radio, teams can be carefully deployed and assigned to a search area. The more knowledge you have about the search area, the more vital your role is during the search. Knowing your response area and the places where a child or lost person might hide can be very helpful.

A well-coordinated team can be an effective SAR force. On the other hand, an unorganized group of people will produce increasing chaos. Do not enter the search area before the search team arrives. If the team will be using dogs, your scent will distract them.

Tools Used

Each specialized rescue team will bring its own equipment to the scene. Many of the same tools, equipment, and personal protective clothing used for other specialized rescue situations are also used in SAR emergencies:

- Personal clothing appropriate for the environment
- Water
- Food
- Lighting and hand lights
- Thermal imaging device
- Communications equipment, including radios
- Medical equipment
- Maps, compass, and GPS
- Shelter
- Rescue rope, harnesses, webbing, and associated hardware
- Extrication equipment such as a Stokes basket and harness
- Flare gun and flares, whistles, or other signaling devices

Water Rescue

Most fire departments are called to perform some type of water rescue (**FIGURE 27-17**). Types of water rescue identified in NFPA 1006, *Standard for Technical Rescue Personnel Professional Qualifications*, include surface water rescue, swiftwater rescue, dive rescue, ice rescue, surf rescue, floodwater rescue, and watercraft rescue. Each of these is a specialized category of technical rescue on its own; however, there is a great deal of overlap in both the hazards they present and the fundamental tools and techniques used to manage them.

Ropes can be used in a variety of ways during water rescue operations. For example, a rescuer on the shore may throw a rope to a person in the water and pull the victim to safety. Ropes may also be stretched across a stream or river to facilitate a rescue in a swiftwater or floodwater scenario (**FIGURE 27-18**). A boat might be tethered to the rope, and rescuers on shore might maneuver the boat using a series of ropes and pulleys.

Voice of Experience

Several years ago, I was involved in a technical rescue in one of the mountainous regions of our response area. A hiker had fallen an unknown distance while scrambling up a hillside. On our arrival, it became clear the rescue would be technically complex and involved. The victim appeared to have an unstable pelvic fracture and possible spinal injuries. Just getting to the victim involved a significant hike—climbing rocky ledges, negotiating tangled brush, and scrambling up loose hillsides.

While the victim was being packaged, we planned the rescue. We decided to carry the victim approximately 300 yards (274 meters) to where a high-angle rope system could be set up. The victim would be lowered down a sandstone slope and then transitioned to a sloping highline to clear a cliff and thick underbrush. The bottom of the highline would bring the victim right to the ambulance.

Three teams of rescuers, hundreds of feet of rope, and many pounds of equipment were required to accomplish the rescue. It took nearly 2 hours before the victim was loaded into the ambulance. In the end, the victim remembered only one detail from the rescue. She remembered and appreciated the rescuer who took the time to massage her feet when they became cold. I continue to use this story as an example of the importance of the "little things" during a rescue.

Shane Baird
Durango Fire Rescue
Durango, Colorado

FIGURE 27-17 Ice rescue is a type of water rescue.

© Steven Matto/Shutterstock

FIGURE 27-18 Ropes ensure rescuer safety during water rescues.

Courtesy of Captain David Jackson, Saginaw Township Fire Department.

Fire departments should be prepared for all types of water rescue incidents that may occur in their communities. Fixed bodies of water in your community are easy to identify. Identifying the areas in your community where flooding is likely to occur requires preplanning. Other common water rescue scenarios include cars rolling into lakes or streams, cars going off bridges, people falling into bodies of water, and swimmers getting into trouble.

Safe Approach

When responding to water rescue incidents, your safety and the safety of other responders are the primary concern. Firefighters should use PPE designed for water rescue, not structural firefighting. Any time you are within 10 feet (3 meters) of the water, you should be wearing a Coast Guard–approved PFD. Shoes that provide solid traction are preferred to fire boots in this situation.

Do not exceed your level of training. If you cannot swim, you should not operate around or near water. Similarly, training as a stillwater lifeguard does not prepare someone to enter flowing water with a strong current, such as is found in a river, stream, or ocean.

Many people who are in trouble in the water are close to shore. Trained swiftwater rescuers may use a throw rope, a pike pole, or a ladder to reach a person. Do not use a boat unless you are trained in its use. Ensure that bystanders do not try to rescue the victim and place themselves in a situation where they may need to be rescued as well.

Ice rescues are common in colder climates. Many departments in colder regions have developed specialized equipment for these scenarios. Throwing a rope or flotation device may be helpful, initially. If the body of water is small enough, two firefighters can slide a rope across the ice to the victim by walking around the shore. It may be necessary to tie ropes together if one is not long enough. In some cases, ladders can be extended to the victim or used to distribute the weight of the rescuer on ice-covered water. Many other devices can also be used to keep a victim above the ice and pull them to safety. Rescuers in these situations need special ice rescue suits to prevent hypothermia and provide flotation. If your department is involved in ice rescues, you need to receive training in these specialized procedures.

TIP

All firefighters should be trained in the use of water rescue throwlines. This rope is designed to be kept in a special throw bag that can be thrown to a victim struggling in water or on ice. The rescuer secures one end of the rope before throwing the bag. Once the victim has grasped the bag, the rescuer can pull the victim to shore.

How You Can Assist

As with any other technical rescue incidents, your main role is to secure the scene and prevent other people from entering the area until specialized rescuers arrive. At water rescues, you may also be called upon to assist by keeping victims in sight, retrieving equipment, and assisting with tending ropes and equipment. You can also assist special rescue teams by relaying communications.

Tools Used

Firefighters should become familiar with the following tools used in water rescue operations:

- PFD
- Self-contained underwater breathing apparatus (SCUBA)
- Helmet
- Waterproof whistle
- Water rescue throwline
- Boat
- Rescue rope, webbing, and associated hardware
- Lighting and hand lights
- Gloves
- Goggles or other eye protection
- Suitable footwear for the environmental and rescue conditions

Animal Rescue

Many kinds of rescue situations can involve animals. Examples include the collapse of a dairy barn, a motor vehicle collision involving a chicken truck, a horse stuck in deep mud, or a deer that has broken through thin ice. Such incidents require special knowledge of how to handle animals as well as the technical rescue knowledge and skills to handle the crash, collapse, or entrapment.

Safe Approach

At a rescue scene involving one or more animals, the firefighter should consider the components of a standard size-up as well as the scope of the rescue scene, the number and locations of the animals involved, and any special resources that will be needed. Special consideration must be given to securing a perimeter to protect both the animals and people that may injure or be injured by them.

How You Can Assist

Like many other types of technical rescue, animal rescues typically involve a great deal of gear and equipment. You may be called upon to retrieve or move gear and equipment, as well as to operate tools and equipment, including hauling on lines if ropes are involved in the rescue. You may also be called on to assist in rounding up and monitoring animals in enclosures as well as calming them and moving them to safe locations.

Tools Used

Tools and items that may be required for animal rescues are typically the same as for the primary rescue type involved. That is, extrication tools will be involved in an animal rescue that requires extrication, and ropes and related gear will be required in an animal rescue that involves hauling. Confined-space equipment will be required if animals are found in confined spaces.

Tools and Equipment

Power tools and equipment are used at an incident scene for lighting, ventilation, salvage, and overhaul. Powerful emergency scene lighting equipment requires a separate source of electricity, often supplied by a generator. As a Firefighter II, you may be responsible for cleaning, maintaining, and testing this equipment to ensure that it works properly at an emergency scene. This section discusses lighting equipment, generators, and power tools, including the use, cleaning, and maintenance of this equipment. All tools and equipment must be properly maintained so they are ready for use when needed. This is discussed further in Chapter 7, *Firefighter Tools and Equipment*. Use power tools and equipment only after you have received the necessary training. Read and follow the instructions supplied by the manufacturer.

Lighting Equipment and Electrical Generators

Powerful lighting equipment requires a separate source of electricity, generally a 120-volt alternating current (AC) delivered through power cords. The electricity for lighting equipment can be supplied by a **generator**, an **inverter**, or a building's electrical system, if the power has not been interrupted.

Power inverters convert 12-volt direct current (DC) from a vehicle's electrical system to 120-volt AC power. An inverter can provide a limited amount of AC current and is typically used to power one or two lights mounted directly on the apparatus. Some vehicles have additional outlets for operating portable lights or small electrically powered tools. Most power inverters, however, do not produce enough power to operate high-intensity lighting equipment, large power tools, or ventilation fans. Connecting devices that draw too much current can seriously damage an inverter.

Electrical generators are powered by gasoline or diesel engines. They can be either portable or permanently mounted on fire apparatus. Portable generators are small enough to be removed from the apparatus and carried to an incident scene. These devices come in various sizes and can produce as much as 6000 watts (6 kilowatts) of power. A portable generator has a small gasoline-or diesel-powered motor with its own fuel tank.

Apparatus-mounted generators have much larger power capacities, sometimes exceeding 20 kilowatts. The smaller units, which are similar to portable generators, can be permanently mounted in a compartment on the apparatus. Larger generators are usually mounted directly on the vehicle. They often have diesel engines and draw fuel directly from the vehicle's fuel tank. Permanently mounted generators can also be powered by the apparatus engine through a power takeoff and a hydraulic pump.

Another potential power source for lighting is a building's normal power supply. This is often not an option if a serious fire has occurred in the building because the power will be disconnected for safety reasons. However, if the fire is minor, the power might not have to be interrupted and can be used for lighting. Alternatively, power can sometimes be obtained from a nearby building. When drawing power from a building, be aware that you might not know the capacity of the circuit from which you are obtaining electricity. Do not overload a household electrical circuit.

SAFETY TIP

Never run a portable generator in a confined space or an area without adequate ventilation unless you and everyone else in the area are wearing SCBA. The exhaust from the generator contains carbon monoxide, which can kill you.

Cleaning and Maintenance of Electrical Equipment

It is important to properly maintain lighting equipment and generators to keep them in good operating condition. To clean lighting equipment, follow the manufacturer's instructions. In general, the directions specify how to clean the floodlights and search lights with a mild soap or detergent. Always use a soft cloth to clean and polish the lens. Do not use strong solvents. They may damage plastic or rubber components.

When cleaning and maintaining a generator, follow the manufacturer's instructions. Avoid overfilling the fuel tank on portable generators. When a fuel tank is too full, the fuel can overflow onto the hot engine and ignite. An overfull tank can also leak fuel as the fuel expands, especially if the tank was filled in a cool environment and then brought into a warmer place. Spilled fuel can also damage paint under certain circumstances.

Clean the external parts of the generator with a mild soap or detergent as recommended by the manufacturer. Remove any dirt, vegetation, or charred materials. Do not use strong solvents. They can damage rubber and plastic parts of the generator and can remove paint.

If a cleaning solvent is recommended by the manufacturer, follow the directions for its use on the product label. Read the safety data sheet before working with any solvent. Check whether the product is flammable. If it is, be sure to follow the safety recommendations for its use. Only use flammable solvents in a well-ventilated area that is not close to any flame or heat source.

Test and run generators on a weekly or monthly basis to confirm that they will start, run smoothly, and produce power. Run gasoline-powered generators for 15 to 30 minutes to reduce any deposit build-up that could damage the spark plugs and make the generator hard to start.

SAFETY TIP

Solvents can irritate or burn the skin, eyes, and lungs. Some solvents are known to cause cancer and should be avoided. Solvents can be absorbed through the lungs as well as through the skin. For those that can be absorbed through skin, check whether chemical-resistant gloves are required during their use. Also check whether eye protection or respiratory protection is required.

At the same time, inspect and test other electrical equipment on the apparatus. Take junction boxes out and check them for cracked covers or broken outlets. Plug them in to make sure the ground-fault circuit interrupter (GFCI) works properly. Check power cords for tears in the protective covering, mechanical damage, fraying, heat damage, or burns. Inspect plugs for loose or bent prongs. Inspect and test the operation of all power tools and equipment on the apparatus. Check lighting fixtures to confirm that they are in good condition and that the bulbs are working.

After shutting down the generator, refill the fuel tank. Follow the manufacturer's recommendations for generator engine maintenance, including regular oil changes. To conduct a weekly/monthly generator test, follow the steps in **SKILL DRILL 27-3**.

Power Tools and Equipment

Some rescue incidents require heavy machinery to move dirt or large sections of collapsed buildings, but special rescue teams often carry smaller tools and equipment for more precise rescue operations. Along with hand tools, power tools are used to move, bend, and distort metal to gain access to and to disentangle a

SKILL DRILL 27-3

Conducting a Weekly/Monthly Generator Test Firefighter II, NFPA 1010: 7.5.4

1. Remove the generator from the apparatus compartment, or open all doors as needed for ventilation. Install the grounding rod, if needed.

Continues.

SKILL DRILL 27-3 CONTINUED

Conducting a Weekly/Monthly Generator Test Firefighter II, NFPA 1010: 7.5.4

2. Check the oil and fuel levels, and start the generator. Connect the power cord or junction box to the generator, connect a load such as a fan or lights, and make sure the generator attains the proper speed, as specified by the manufacturer. Check the voltage and amperage gauges to confirm the generator is operating efficiently.

3. Run the generator under load for 15 to 30 minutes. Turn off the load, and listen as the generator slows down to idle speed. Allow the generator to idle for about 2 minutes before turning it off. Disconnect all power cords and junction boxes; clean all power cords, plugs, adaptors, GFCIs, and tools. Replace them in proper storage areas. Allow the generator to cool for 5 minutes. Refill the generator with fuel and oil, as needed, and return the generator to its compartment. Fill out the appropriate paperwork.

victim from a vehicle or machinery. Hydraulic rams and hydraulic spreaders can apply a large amount of force to displace the vehicle or machinery so that the victim can be removed. Powered hydraulic rams and spreaders use a hydraulic pump powered by an electric motor or a gasoline engine.

A variety of power tools for cutting or severing are also available. Some are air powered (pneumatic), some are electrically powered, some are fuel powered, and others are hydraulically powered. Air chisels are powered by pressurized air from portable SCBA air cylinders or from a compressed air source mounted on a vehicle.

Several types of power saws are used in special rescue operations. Some electric saws require power cords, and others operate on battery packs. Different types of power saws produce different types of motion. For example, reciprocating saws move back and forth quickly, and rotary saws move in a circular motion. Whenever you use power and hydraulic tools, cover the victim for protection.

Cleaning and Maintenance of Power Tools and Equipment

Test power tools and equipment frequently, and have these items serviced regularly by a qualified shop. Keep records of all inspections and maintenance performed on power tools and equipment. Preventive maintenance will help to ensure that tools and equipment operate properly when needed. If a power tool or piece of equipment does not work properly, report this fact to your officer.

Fill each tool with the proper fuel, remembering that some tools operate on gasoline and others use various gasoline and oil mixtures. Many small engines operate on gasoline, for example, but others are designed to burn diesel fuel. Fuels have a limited storage life, especially fuels containing ethanol. If the fuel is not used within a certain period of time, it may be necessary to drain and refill the equipment with fresh fuel. Check the manufacturer's recommendations before fueling a piece of equipment.

After returning from an incident, clean, inspect, and record maintenance data for all your overhaul tools. To ensure that power equipment is left in a "ready state" for immediate use at the next incident, use the following checklist:

- All debris should be removed, and the tools should be clean and dry.
- All fuel tanks should be filled completely with fresh fuel.
- Any dull or damaged blades and chains should be replaced.
- Belts should be inspected to ensure that they are tight and undamaged.
- All guards should be securely in place.
- All hydraulic hose should be cleaned and inspected.
- All power cords should be inspected for damage.
- All hose fittings should be cleaned, inspected, and tested to ensure a tight fit.
- Tools should be started to ensure that they operate properly.
- Tanks on water vacuums should be emptied, washed, cleaned, and dried.
- The hose and nozzles on wet vacuums should be cleaned and dried.

It is essential to read the manufacturer-provided manuals and follow all instructions on the care and inspection of power tools and equipment. Keep all manufacturers' manuals in a safe and easily accessible location. Refer to these documents when cleaning and inspecting the tools and equipment.

Learn the proper procedure for reporting a problem with a power tool and taking it out of service. Remember—your safety depends on the quality of your tools and equipment. Every tool and piece of equipment must be ready for use before you respond to an emergency incident.

CASE STUDY

You Are the Firefighter CONCLUSION

Your engine company is dispatched to assist a rescue company with the rescue of a 51-year-old electrical worker who collapsed in an underground electrical vault at the bottom of a utility hole. While en route, you receive additional information that the victim's crew can no longer communicate with him. Upon your arrival, witnesses state that two additional crew members entered the vault to retrieve the victim. Their voices can no longer be heard, and they are not responding to calls on their radio.

1. What hazards might be present at this incident?

Answer: This scenario describes a confined-space technical rescue incident. Potential hazards at this incident scene include electricity as well as a possible asphyxiant, toxic, or flammable atmosphere. Additional hazards could include water and other physical hazards, such as broken pipes, glass, and equipment.

2. **What types of rescue operations may be necessary at this incident?**

   Answer: Confined space will be the primary technical rescue mode at this incident scene. However, this incident will likely also involve rope rescue tools and techniques. The potential for structural collapse must also be considered and planned for.

3. **What might be your role at this incident?**

   Answer: Working within the ICS, firefighters at this technical rescue incident are likely to be assigned to establish a perimeter to keep bystanders out. They may also assist in staging tools, supplies, and equipment needed by the technical rescue crew. If rope rescue tools and techniques are required, firefighters may also be asked to assist in setting up and operating equipment under the direction of the technical rescue specialists.

WRAP-UP

SUMMARY

KNOWLEDGE OBJECTIVES

- Describe the role of the firefighter in a technical rescue operation.
  - Define the types of special rescues encountered by firefighters. (**NFPA 1010: 7.4.2**, pp. 1144–1145)
  - Describe the steps of a special rescue operation. (**NFPA 1010: 7.4.2**, pp. 1145–1149)
- Explain how to safely assist during technical rescue operations.
  - Describe the general procedures at a special rescue scene, including what to do in the case of utility hazards. (**NFPA 1010: 7.4.2**, pp. 1149–1152)
  - Describe how to assist special rescue crews. (**NFPA 1010: 7.4.2**, pp. 1152–1153)
  - Describe how to safely approach and assist at a vehicle or machinery rescue incident. (**NFPA 1010: 7.4.2**, pp. 1153–1154)
  - List the types of incidents that might require a rope rescue. (pp. 1154–1157)
  - Describe how to safely approach and assist at a rope rescue incident. (**NFPA 1010: 7.4.2**, p. 1156)
  - Describe the hardware components used during a rope rescue. (**NFPA 1010: 7.4.2**, pp. 1156–1157)
  - Describe how to safely approach and assist at a confined-space rescue incident. (**NFPA 1010: 7.4.2**, pp. 1157–1160)
  - Describe how to safely approach and assist at a structural collapse rescue incident. (**NFPA 1010: 7.4.2**, p. 1159)
  - Describe how to safely approach and assist at a trench and excavation rescue incident. (**NFPA 1010: 7.4.2**, pp. 1157–1160)
  - Describe how to safely approach and assist at a wilderness and helicopter search and rescue incident. (**NFPA 1010: 7.4.2**, pp. 1160–1161)
  - Describe how to safely approach and assist at a water or ice rescue incident. (**NFPA 1010: 7.4.2**, pp. 1161–1163)
  - Describe how to safely approach and assist an animal rescue incident. (**NFPA 1010: 7.4.2**, p. 1163)
- Describe the role of the firefighter in operating, cleaning, and maintaining tools and equipment.
  - Describe the types of generators used to power lighting equipment. (p. 1164)
  - Describe how generators operate. (p. 1164)
  - Describe how to clean and maintain lighting equipment. (**NFPA 1010: 7.5.4**, pp. 1164–1165)
  - Describe how to maintain generators. (**NFPA 1010: 7.5.4**, pp. 1164–1165)
  - Describe how to maintain power equipment and power tools. (**NFPA 1010: 7.5.4**, pp. 1165–1167)

SKILLS OBJECTIVES

- Assist during technical rescue operations.
  - Establish a barrier. (NFPA 1010: 7.4.2, p. 1151)
  - Identify and retrieve rescue tools. (NFPA 1010: 7.4.2, p. 1153)
- Maintain generators.
  - Conduct a weekly/monthly generator test. (NFPA 1010: 7.5.4, pp. 1165–1166)

KEY TERMS

carabiner An auxiliary equipment system item; a load-bearing connector with a self-closing gate used to join other components of life safety rope. (NFPA 1983)

confined space An area large enough and so configured that a member can bodily enter and perform assigned work, but which has limited or restricted means for entry and exit. A confined space is not designed for continuous human occupancy. (NFPA 1550)

entrapment A condition in which a victim is trapped by debris, soil, or other material and is unable to extricate themself.

generator An electromechanical device for the production of electricity. (NFPA 1901)

harness An equipment item; an arrangement of materials secured about the body used to support a person. (NFPA 1970)

high-angle operations Rope rescue operations where the angle of the slope is greater than 45 degrees. In this scenario, rescuers depend on life safety rope rather than a fixed support surface such as the ground.

inverter Equipment that is used to change the voltage level or waveform, or both, of electrical energy. Commonly, an inverter (also known as a power conditioning unit [PCU] or power conversion system [PCS]) is a device that changes DC input to AC output. Inverters may also function as battery chargers that use alternating current from another source and convert it into direct current for charging batteries. (NFPA 70)

life safety rope Rope dedicated solely for the purpose of supporting people during rescue, firefighting, other emergency operations, or during training evolutions. (NFPA 1983)

lockout and tagout systems Methods of ensuring that electricity and other utilities have been shut down and switches are "locked" so that they cannot be switched on. These systems are used to prevent the flow of power or gases into the area where rescue is being conducted.

low-angle operations Rope rescue operations on a mildly sloped surface (less than 45 degrees) or flat land. In this scenario, firefighters depend on the ground for their primary support, and the rope system is a secondary means of support.

packaging The process of securing a victim in a transfer device, with regard to existing and potential injuries or illness, so as to prevent further harm during movement. (NFPA 1006)

personal flotation device (PFD) A device manufactured in accordance with U.S. Coast Guard specifications that provides supplemental flotation for persons in the water. (NFPA 1006)

secondary collapse A subsequent collapse in a building or excavation. (NFPA 1006)

shoring A structure such as a metal hydraulic, pneumatic/mechanical, or timber system that supports the sides of an excavation and is designed to prevent cave-ins.

spoil pile A pile of excavated soil next to the excavation or trench. (NFPA 1006)

supplied-air respirator An atmosphere-supplying respirator for which the source of breathing air is not designed to be carried by the user. This is also known as an air-line respirator. (NFPA 1989)

technical rescue incident (TRI) Complex rescue incident requiring specially trained personnel and special equipment to complete the mission. (NFPA 2500)

technical rescue team A group of rescuers specially trained in the various disciplines of technical rescue.

traffic incident management (TIM) A coordinated method of identifying and responding to roadway incidents, including establishing safe work areas and clearing the area when the incident is complete.

wilderness and helicopter search and rescue (SAR) The process of locating and removing a victim from the wilderness.

REVIEW QUESTIONS

1. What is an "all-hazards" agency?
2. What is a technical rescue incident (TRI)?
3. According to the basic steps of technical rescue, firefighters must properly position their vehicles and perform a size-up upon arrival. What is the next basic step to be performed?
4. Who is responsible for identifying unsafe conditions at a technical rescue scene?
5. Name three tasks that a firefighter may be asked to perform to assist a technical rescue team.
6. Name three categories of technical rescue.
7. A Firefighter II may be responsible for what aspects of tool and equipment maintenance, as it pertains to technical rescue?

DISCUSSION QUESTIONS

1. What level of training allows firefighters to work in the "warm zone" of a technical rescue incident?
2. What categories of technical rescue incidents may involve animals?
3. Why is it important to use specific commands and terminology at a TRI?

APPLYING THE CONCEPTS

At 1120, your four-person engine is en route to a call for a structural collapse at an industrial facility on the outskirts of your small, tight-knit community. Dispatch reports a truck collided into the A-B corner of the structure, causing the concrete wall to collapse into a work area. Three workers are reported to be trapped under heavy debris. A technical rescue team is on scene and your crew will assist at the incident. Your engine pulls up and you see a gaping hole in the corner of the building.

1. As an awareness-level firefighter, what are your primary responsibilities as you arrive on the scene of a technical response incident (TRI)?
2. What can you do **before** a TRI happens so that you're better prepared when one occurs?

The incident area is chaotic and noisy. EMS is giving medical care to the truck driver and injured workers. Police are redirecting traffic. The rescue team established the hot zone and are preparing their equipment, and the IC just completed a thorough size-up. Your company officer assigns you and your partner to set up boundaries of the cold zone with fire tape.

3. What is the cold zone?
4. With all that's going on and lives at risk, why is establishing the cold zone a priority?

As you and your partner finish setting the fire tape, your company officer requests you bring a ventilation fan to the team in the warm zone. You grab the fan and as you're walking to the structure you hear a "weird" sound coming from the building. "You should call it in," your partner says.

5. Is your partner right? Should you call it in or are they overreacting?

Fearing additional structural collapse, the IC calls for the rescue team to retreat. Just as team members are moving to safety, another section of the wall collapses and part of the roof gives way. One of the rescue team members is partially trapped under the rubble.

6. Do you rush over to help the rescue team free the trapped responder?

REFERENCES

National Fire Protection Association (NFPA). 2017. *NFPA 1983, Standard on Life Safety Rope and Equipment for Emergency Services.* 2017 Edition. Quincy, MA: NFPA.

National Fire Protection Association (NFPA). 2020. NFPA 1006, *Standard for Technical Rescue Personnel Professional Qualifications.* 2021 Edition. Quincy, MA: NFPA.

National Fire Protection Association. 2020. *NFPA 1951, Standard on Protective Ensembles for Technical Rescue Incidents.* 2020 Edition.

National Fire Protection Association (NFPA). 2020. *NFPA 1989, Standard on Breathing Air Quality for Emergency Services Respiratory Protection.* 2019 Edition.

National Fire Protection Association (NFPA). 2021. *NFPA 470, Hazardous Materials/Weapons of Mass Destruction Incidents (WMD) Standard for Responders.* 2022 Edition. Quincy, MA: NFPA.

National Fire Protection Association (NFPA). 2021. *NFPA 1006, Standard for Technical Rescue Personnel Professional Qualifications.* 2021 Edition. Quincy, MA: NFPA.

National Fire Protection Association (NFPA). 2021. *NFPA 2500, Standard for Operations and Training for Technical Search and Rescue Incidents and Life Safety Rope and Equipment for Emergency Services.* 2022 Edition. Quincy, MA: NFPA.

National Fire Protection Association (NFPA). 2023. *NFPA 1550, Standard for Emergency Responder Health and Safety.* 2024 Edition. Quincy, MA: NFPA.

National Fire Protection Association (NFPA). 2023. *NFPA 1970, Standard on Protective Ensembles for Structural and Proximity Firefighting, Work Apparel and Open-Circuit Self-Contained Breathing Apparatus (SCBA) for Emergency Services, and Personal Alert Safety Systems (PASS).* 2024 Edition. Quincy, MA: NFPA.

National Fire Protection Association (NFPA). 2022. *NFPA 70, National Electric Code.* 2023 Edition. Quincy, MA: NFPA.

National Fire Protection Association (NFPA). 2024. *NFPA 1900, Standard for Aircraft Rescue and Firefighting Vehicles, Automotive Fire Apparatus, Wildland Fire Apparatus, and Automotive Ambulances.* 2024 Edition. Quincy, MA: NFPA.

Occupational Safety and Health Administration (OSHA). 2023. "Confined Spaces—Hazards and Solutions." Occupational Safety and Health Administration. Accessed October 23, 2023. https://www.osha.gov/confined-spaces/hazards-solutions.

CHAPTER

28

Firefighter II

Understanding Fire Investigation

KNOWLEDGE OBJECTIVES

After studying this chapter, you will be able to:

- Describe why fires are investigated and the role of the Firefighter II during a fire investigation.
- Describe the role of the Firefighter II in a fire that has caused injuries or fatalities.
- List the types of evidence that may be found at a fire scene, how to preserve evidence, and how to maintain the chain of custody.
- Describe the potential participants involved in a formal investigation and the laws that impact the securing of a fire scene for an investigation.
- Describe how the point of origin and cause of a fire is determined.
- Describe the crime of arson.

SKILLS OBJECTIVES

After studying this chapter, you will be able to perform the following skill:

- Demonstrate how to preserve evidence on the fireground.

ADDITIONAL NFPA STANDARDS

- **NFPA 556**, *Guide on Methods for Evaluating Fire Hazard to Occupants of Passenger Road Vehicles, 2024 Edition*
- **NFPA 921**, *Guide for Fire and Explosion Investigations, 2021 Edition*
- **NFPA 1033**, *Standard for Professional Qualifications for Fire Investigator, 2022 Edition*

CASE STUDY

You Are the Firefighter

You are dispatched to a reported house fire. It is around noon on a weekday. Weather conditions are sunny and mild. While en route, you are told that there is a 14-year-old burn victim. When you arrive at the scene, after a 2-minute response, you see that the second floor of a single-family dwelling is over 50 percent involved and is venting from two windows on the Delta side. Your officer quickly coordinates EMS care for the 14-year-old who is in the front yard with their parents, who are clearly upset. The officer links back up with you as you advance the line up the stairwell and begin extinguishing the fire. You push down the hallway to the back bedroom and come across what looks like a melted plastic gas container. The fire is well developed, but you eventually get a knock on it with your crew.

1. **Given the scenario, what information do you think would be of value to a fire investigator?**
2. **After overhaul and salvage is finished, what tools and equipment could you use to assist the fire investigator in processing the fire's room of origin?**
3. **How does a fire origin and cause investigation differ from an arson investigation?**

Introduction

Firefighters usually reach a fire scene before a trained fire investigator arrives. As a result, firefighters are often able to observe fire patterns and identify potential evidence that the fire investigator can use to determine how and where the fire started. Firefighters assist fire investigators both by recalling and reporting what they observed and by identifying and preserving evidence. The actions and observations of firefighters are a vital part of determining the origin and cause of the fire, a process known as fire investigation.

Determining the causes of fires allows fire departments to take steps to prevent future fires. For example, a fire department might develop an education program to reduce accidental fires, such as those caused by food left unattended on a hot stove. A series of fires could point to a product defect or construction errors, such as a design flaw in a chimney flue or the improper installation of chimney flues in a housing development. Identifying fires that were intentionally set could lead to the arrest of the person responsible for the crimes.

Firefighters must understand the basic principles of fire investigation to assist with an official investigation. Determining the origin and cause of a fire is often difficult because important evidence may be consumed by the fire or destroyed during fire suppression or salvage and overhaul operations. Fire investigators rely on firefighters to observe, capture, and retain information. Firefighters must also preserve evidence until it can be documented and examined by the investigator. Evidence that is lost can never be replaced. The firefighter must prioritize fully extinguishing a fire and ensuring effective overhaul. However, they should also be aware that unnecessary post-incident destruction may ruin any chance to determine origin and cause. Simply preserving as much of the scene as possible can do more to assist the investigator than nearly all other actions.

The firefighter should make note of what they witnessed upon their arrival and during fireground operations. Key things to note and document after the incident include the fire conditions when they arrived at the scene, who they saw, who opened or closed doors, and what tasks they performed. Although the investigation report will be written by a trained fire investigator, the information provided by firefighters is vital.

TIP

Fire departments can use both local and national data about the cause of fires to achieve multiple goals. For example, fire cause data could support laws requiring the use of residential sprinklers or those prohibiting the improper storage of combustible landscape or construction materials close to multifamily buildings. Fire cause data that help to identify faulty appliances, equipment, or automobiles can be used to raise consumer awareness and prompt product recalls that prevent future injury or death.

Why Do We Investigate Fires?

Every fire should be investigated, even if the origin and cause seem obvious. Determining the fire's origin and cause may help prevent future fires. The National Fire Protection Association (NFPA) and the U.S. Fire Administration track fire statistics from local fire department incident reports, insurance investigations, and product safety investigation testing facilities. This information, which comes from thorough, competent fire investigations, can be used to develop fire prevention and education programs. An investigation may also identify who is financially responsible for property damage, injuries, or deaths. In some cases, an investigation may produce evidence that is used in criminal prosecution.

The objective of a fire investigation is to identify the origin and cause of the fire. This investigation should not be confused with the criminal investigation of **arson**. Although the definition and levels of arson vary by state, the crime of arson generally involves the malicious act of burning property with a criminal intent.

Fire investigators must be careful to avoid **expectation bias** when conducting the investigation. Expectation bias can lead a fire investigator to make assumptions about the cause of a fire and to ignore evidence that does not match what they expected to find. For example, initial reports of a fire that occurred during a lightning storm may lull a fire investigator into assuming that lightning was the cause. This could potentially cause them to miss key pieces of evidence that could indicate otherwise. The evidence and the facts alone must be used to determine the origin and cause of the fire. If investigators determine that a crime has been committed, a separate criminal investigation must be conducted by law enforcement personnel.

Who Conducts Fire Investigations?

In most jurisdictions, the chief of the fire department has the legal responsibility to determine the causes of fires. In small communities, the fire chief may personally examine the scene to determine the cause. In other small jurisdictions, one or two fire department members may be trained to conduct these investigations. In large fire departments, the fire chief usually delegates this responsibility to a special unit staffed with trained fire investigators.

Sometimes a special unit conducts fire investigations for several area fire departments under a mutual aid agreement that determines when those unit members should be called in to investigate. For investigations involving serious injuries, deaths, or high financial loss, an investigator from a state-level fire marshal's office may be called for assistance.

Although the personnel responsible for conducting fire investigations may vary by jurisdiction, they must always meet the job performance requirements (JPRs) established by NFPA 1033, *Standard for Professional Qualifications for Fire Investigator*. This ensures that each fire investigation is conducted in a safe, thorough, and professional manner by trained and qualified personnel.

In some jurisdictions, a fire investigator may be dispatched to all working structure fires. In others, they may be dispatched only when the damage exceeds a certain level or when the incident involves injuries or fatalities. Some departments may require the incident commander (IC) to conduct a preliminary investigation and decide whether a fire investigator is needed. If the cause of the fire is evident and accidental, the IC would be responsible for gathering the information and filing the necessary reports. If the cause cannot be determined or appears to be intentional, the IC would then call for a fire investigator.

Although firefighters are not responsible for conducting the actual fire investigation, they are often the source of valuable information for the investigator because fire suppression and overhaul operations usually begin before the fire investigator arrives. The firefighter may see or hear something that could be relevant to the investigation. If so, they must share their observations with the fire investigator after extinguishing the fire. Avoid moving, altering, or destroying evidence (**spoliation**) unless such action is necessary to meet the incident priorities of life safety, incident stabilization, and property conservation. If evidence must be altered in any way before the investigation begins, firefighters must report this to the investigator. Firefighters may be interviewed or asked to prepare a written report documenting their observations.

By understanding the fire investigation process and the critical role that investigations play in ensuring the safety of a community, the firefighter can more effectively assist the investigator. In a sense, the firefighter starts the investigatory process when firefighting operations begin.

The Firefighter's Role

A firefighter's primary focus is on saving lives and property. However, you will inevitably make observations and gather information as you perform your duties on the incident scene. The fire investigator will usually contact the firefighters who were at the incident, particularly those who first arrived, to record their observations. Investigators rely heavily on firefighters to ensure they have a thorough understanding of the incident scene. The observations and information you provide to the investigator will supplement what they find during their investigation.

Dispatch and Response

During dispatch and response, develop a mental image of the scene you expect to encounter. Note the time of day and the weather conditions. As the incident progresses, make a mental note of the building, fire and smoke conditions, bystanders, and any other details you observe. This information could help the fire investigator determine the origin and cause of the fire.

Time of Day

The time of day and the type of occupancy can indicate the number and type of people who might be at an incident. Offices, stores, and other business places are filled with people during the day but are mostly empty at night. Conversely, restaurants and nightclubs are likely to be crowded after dark. Residential occupancies are more likely to be occupied at night and on weekends and holidays.

Weather Conditions

Note whether the day is hot, cold, cloudy, or clear and whether conditions in the burning structure match the weather. On a cold day, windows should be closed; on a hot day, the furnace should be off.

Lightning, heavy snow, ice, flooding, fog, or other hazardous conditions may delay the fire department's arrival and make a firefighter's job more difficult. Wind direction and speed help determine the natural path of fire spread. Being aware of these conditions could help you to recognize if the fire behaved in an unusual way.

Arrival and Size-Up

Size-up operations can provide valuable information for fire investigators. Pay attention to the fire conditions, the building characteristics, the positions of doors and windows, and any vehicles and people arriving or leaving the scene.

Description of the Fire

The fire investigator will compare the dispatcher's description with the actual fire conditions. If the fire intensified dramatically in a short time, an accelerant may have been used. During fire suppression operations, note whether flames are visible or whether only smoke is apparent. Also observe the quantity, color, and source of the smoke. Does the fire appear to be burning in one place or in multiple locations?

People on the Scene

The appearance and behavior of people and vehicles at the scene of a fire can provide valuable clues. Note the attitude and dress of the owner and/or the occupants of the building, as well as any other people at the scene of a fire. Normally, the owner or occupants of a building will be distressed. They will also usually be wearing clothing appropriate to the time of day. For example, at 4 AM, you would expect to see residents in pajamas. As victims evacuate the building, make note if someone is fully dressed in the middle of the night. Someone else may stand out to you because their behavior and demeanor are quite different from those of the other victims. Notify the investigator if anyone's dress, behavior, or words do not appear to be appropriate for the situation.

Sometimes fire investigators will photograph the crowd at a fire scene, particularly if a series of similar fires has occurred recently. Later, they can compare the photographs and look for the same faces at different incidents (**FIGURE 28-1**). Fire investigators may also want to know about someone who eagerly volunteered multiple theories or too much information.

Unusual Items or Conditions

Pay attention to any unusual items found at the site or anything about the property itself that seems out of the ordinary. Make note of whether windows and doors

FIGURE 28-1 Most persons at the fire scene are intent on watching the firefighters at work.

were open or closed upon your arrival or if firefighters forced entry into the building. Gasoline cans, forcible entry tools, and a damaged hydrant or sprinkler connection should all be reported to the investigator.

Entry

Before you enter the burning structure, note any evidence of prior entry, such as shoeprints leading into or out of the structure or tracks from vehicle tires. Note whether the windows and doors are intact, whether they are locked or unlocked, and whether any unusual barriers limit access to the structure. Also note any evidence of forced entry by others.

Tools that were used to force open a window or door may leave marks. Likewise, cut or torn edges of wood, metal, or glass may indicate forced entry. Later, fire investigators might be able to determine whether glass was broken by heat or by mechanical means. Note any evidence of a burglary. In some cases, a fire may have been set to destroy evidence of another crime.

TIP

Photographs or video can be used to document conditions before the arrival of fire investigators. However, specific procedures must be followed for these to be admissible in court as evidence. Follow your department's standard operating procedures (SOPs) regarding fire scene photos or video.

Search and Rescue

As you enter the building to perform search and rescue or interior fire suppression activities, consider the location and extent of the fire. Be especially alert for signs of multiple fires in a building. Multiple fires increase the risk of injury and death to both the building's occupants and responding firefighters. They may also indicate that the fire was intentionally set.

The location and condition of the building's contents may provide clues for the fire investigator. Inform the fire investigator about things such as barriers in doors and windows, the absence of inventory in a warehouse, or insurance papers or deeds left out on a desk.

The team responsible for shutting off electrical power to the building should identify whether the circuit breakers were on or off when they arrived. The location of any people found in the building should be noted as well. When turning off electrical power, turn off main breakers, if possible, and do not disturb breakers that are "tripped."

Ventilation

The ventilation crew should note whether the windows and doors of the building were open or closed and whether they were locked or unlocked. They should also note smoke conditions and any unusual odors.

Smoke Conditions

Note the color, volume, and velocity of smoke upon arrival and report it to the investigator. This information may help the investigator determine the extent of fire development when fire crews first arrived on the scene.

Unusual Odors

Self-contained breathing apparatus protects firefighters from hazardous vapors and toxic odors. Nevertheless, sometimes an odor is so strong you can smell it as you arrive on the scene or before you put on respiratory protection. Other odors may linger after the fire has been extinguished. Cooking fires, overheated light ballasts (which regulate the amount of current flowing to the light fixture), and burning electrical equipment produce distinctive odors that are familiar to experienced firefighters. Other common odors familiar to firefighters include gasoline, kerosene, paint thinner, lacquers, turpentine, linseed oil, furniture polish, and natural gas.

Odors often linger in the soil under a building without a basement, particularly if an accelerant has been used and if the ground is wet. Concrete, brick, and plaster will all retain vapors after a fire has been extinguished. If you notice any of these types of odors, inform the fire investigator.

Effects of Ventilation

Fire department ventilation operations can dramatically influence the behavior of a fire and alter fire patterns. Certain ventilation situations can cause a fire to rapidly grow or change the flow path. Share information about ventilation operations with the fire investigator so that they can correctly interpret fire patterns.

Suppression Operations

During fire suppression operations, note the fire's location and behavior, including any charring and fire patterns you observe. Note any evidence indicating that devices were used to start the fire, and report any obstacles you encounter. This information might help the fire investigator determine the origin and cause of the blaze.

FIGURE 28-2 While fighting the fire, be aware of streamers or trailers.

Courtesy of Joseph Whittaker, Nassau County Fire Marshals.

Behavior of the Fire

During the fire attack, observe the behavior of the fire and note how it reacts when an extinguishing agent is applied. Look for unusual flame colors, sounds, or reactions. For example, when water is applied to fires caused by ignitable liquids, the liquid typically floats and continues to burn, causing the fire to spread. Fires that rekindle in the same area and those that flare up when water is applied could indicate the presence of an ignitable liquid. Fires in multiple locations and those fueled by accelerants are more dangerous for you and your fellow firefighters.

Incendiary Devices, Trailers, and Accelerants

While fighting the fire, be aware of streamers or **trailers** (combustible materials, either solid or liquid, positioned to spread the fire) (**FIGURE 28-2**). Note any combustible materials such as wood, paper, or rags in unusual locations. Note containers of ignitable liquids that are found in unexpected locations. For example, a can of paint thinner in the garage is common. If it is found on the kitchen counter, however, it should be reported to the investigator. Very intense heat or rapid fire spread might indicate that an accelerant was used.

FIGURE 28-3 Incendiary devices are items used to initiate an incendiary fire or explosion.

An **incendiary device** (a device used to start an incendiary fire or explosion) can include unusual items in unlikely places, such as a packet of matches tied to a bundle of combustible fibers or attached to a mechanical device. Other incendiary devices may use an ignitable liquid in a glass container along with some sort of cloth or rag that serves as a fuse (**FIGURE 28-3**). Often, incendiary devices fail to ignite, or they burn out without igniting other materials. Even those that ignite may not be completely consumed by the fire.

SAFETY TIP

If you encounter an incendiary device that has not ignited, notify your fire officer or fire investigator and exit the structure or area immediately! Do not make any attempt to move or disable the device.

Condition of Fire Protection Systems

If the building is equipped with a fire alarm or fire suppression system, firefighters should note whether the system operated properly or was disabled (**FIGURE 28-4**). Note if the outside screw and yoke (OS&Y) valves or post indicator valves (PIVs) were open or closed (see Chapter 10, *Water Supply Systems*). Blocked or covered sprinkler heads, missing or disabled smoke detectors or notification appliances, and physical damage to any fire protection system should be

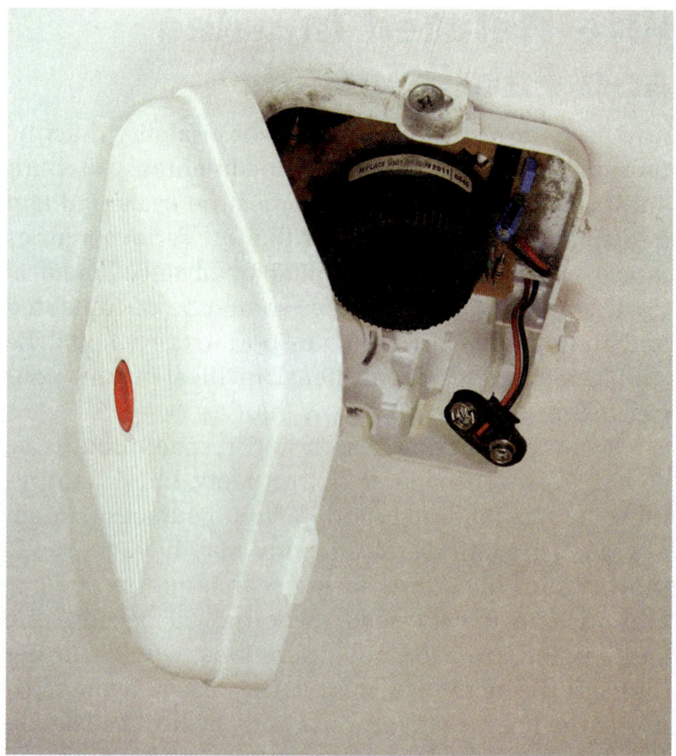

FIGURE 28-4 Fire-setters may sabotage fire protection systems to delay notification of a fire.

reported to the investigator. Note whether an intrusion alarm is sounding when you arrive on the scene.

Obstacles or Structural Modifications

Use extra caution if you find unusual locks, debris, or other obstacles blocking entrance doors, hallways, stairwells, or access to fire protection systems. These obstacles may be the result of renovation or demolition activities, or they may have been intentionally placed to get in the way of firefighters. In these situations, look for unexpected holes in floors, ceilings, and walls, which may have been cut to promote fire spread or hinder firefighters. Report any of these conditions to the fire investigator.

Contents

Make note of anything unusual about the contents of the building. The absence of personal items in a residence may indicate that they were removed before the fire was intentionally set. Empty boxes in a warehouse may belong there, or they may indicate that valuable contents were removed prior to the fire. In some case, the contents of a building are removed and replaced by less expensive or nonfunctioning items. In such cases, neighbors may report that they observed items being removed or replaced prior to the fire. Check whether appliances are plugged in at the time of the fire. Also note if any contents appear to have been protected from fire, smoke, or water damage.

Charring and Other Fire Patterns

Charring in unusual places, such as open floor space away from any likely accidental ignition source, could indicate that the fire was deliberately set. Char on the underside of a door or a low horizontal surface, such as a tabletop, could indicate that a pool of an ignitable liquid was placed below the item.

Salvage and Overhaul

Firefighters can often be most helpful to a fire investigation during the salvage and overhaul phase. At this point, activity at the scene has typically slowed down, and firefighters can take a more measured, intentional approach. With a slower pace and increased visibility, both firefighters and fire investigators can get a better look at the scene. Firefighters should continue to look for fire patterns and evidence while conducting salvage and overhaul—in a way that allows for evidence to be identified and preserved. If not done carefully, the overhaul process can quickly destroy valuable evidence.

If possible, the fire investigator will observe and document an area before overhaul begins. They can often quickly identify potential evidence and can help direct or guide the overhaul operations so that evidence is properly preserved. Evidence found during overhaul should be left where it is found, untouched and undisturbed, until the fire investigator examines it. Evidence that must be removed from the scene should be properly identified, documented, photographed, packaged, and placed in a secure location.

Fire suppression personnel and fire investigators must work as a team to ensure that the fire is completely extinguished and properly overhauled while searching for and preserving evidence. If overhaul operations must be done before the arrival of the fire investigator, the IC or company officer must thoroughly document the scene before, during, and after overhaul activities. Photographs or video of the scene can be especially helpful to the investigator, as long as standard local procedures are followed regarding how they are taken and transferred to ensure that the chain of custody is maintained.

In short, be careful not to destroy or damage evidence during salvage and overhaul. Avoid throwing materials into a pile. Use low-velocity hose streams to avoid breaking up and scattering debris needlessly. Thermal imaging devices can be used to find hot spots without tearing apart the interior structure. If it is clearly necessary to ensure a void space is clear of smoldering fire, use as small as an inspection hole as possible instead of tearing down an entire section of wall or ceiling. Avoid damaging electrical wiring in the room of origin. Faulty electrical wiring may have caused the fire, or the wiring may have been damaged by the fire. The investigator may need to remove entire circuits to use them for **arc mapping**, a process that helps the investigator determine if and where electrical circuits were energized at the time of the fire.

Watch for evidence that was shielded from the fire and is lying beneath burned debris. For example, a wall clock may have fallen during the fire and been covered with debris. If the clock is near the area of origin and stopped around the time the fire broke out, it could be an important piece of evidence.

TIP

While searching for hidden fires, avoid altering the area of origin as much as possible. Consider opening walls from the other side to avoid damaging fire patterns in the room of origin. This will leave valuable fire pattern evidence intact and aid the fire investigator in their examination.

Non-Structural Fires

Firefighters are frequently called to fires in automobiles, recreational vehicles, and boats. Many of the investigational techniques used for structural fires apply to these types of fires. Look for fire effects, fire patterns, and any physical evidence that should be reported to the fire investigator. Be careful to avoid destroying evidence or contaminating the fire scene until the investigation is complete. Depending on the amount of damage and whether the cause was accidental, the company officer may be tasked with completing the investigation and report. This makes the firefighter's assistance and understanding of the process even more critical.

Wildland fires should be investigated. However, the size of the scene and the evidence collected at these fires may be different from the evidence gathered at a building fire. Follow your department's SOPs for identifying and preserving evidence at a wildland fire.

Fire-Related Injuries and Fatalities

Any fire that results in an injury or a fatality must be thoroughly investigated and documented. All burn and smoke inhalation injuries must be evaluated and treated at the scene. Victims with more serious injuries must be transported to a hospital for advanced medical treatment. If possible, injuries should be documented and photographed prior to transport to the hospital. Be sure to follow your department's medical privacy policies before photographing any injuries.

While conducting search and rescue operations, always err on the side of victim safety, but do not remove bodies that are found in circumstances that are obviously not survivable. A body may have to be removed from the scene during search and rescue operations. Or, it may need to be moved because fire suppression or overhaul operations would further damage the body. No matter the reason, you must notify the fire investigator if a body is moved. The fire investigator must document the original location and position of any victims, especially in relation to the fire and the exits.

Clothing removed from any victim should be preserved as evidence. It may contain traces of ignitable liquids. Fire patterns on clothing may also be useful to the investigation. If the clothing is removed in the ambulance or at the hospital, personnel should be assigned to collect it and keep this evidence as intact as possible. Document anything lying under a victim's body after it is removed. This is often a protected area and may reveal important evidence.

Victims who do not escape the fire should not be disturbed, and no attempts should be made to try to identify them. Fire investigators will work with other investigators, possibly including police, coroner, or medical examiner, to examine the victims and the surrounding area for items of evidentiary value.

After the Fire is Out

After suppression operations are complete and the scene is considered stable, the investigatory process can develop further. The fire investigator will now be more involved, with the firefighters continuing to assist. For example, firefighters may be asked to assist in digging out the fire scene using shovels, rakes, and debris buckets to help identify the area of origin. "Digging out" is a term used to describe the process of carefully looking for evidence within the debris. Sometimes the entire fire scene must be closely examined to gather evidence and determine the cause of the fire. This may involve the use of scene lights and portable fans.

The fire investigator will take extensive photographs and video of the fire scene as it first appears. They will then begin to remove and inspect the debris, layer by layer, working from the top of the pile down to the bottom. The type of evidence uncovered and its location within the layers of debris can provide clues about how the fire ignited and progressed.

Removing and inspecting the layers of debris can help the fire investigator establish the sequence in which items burned. The fire investigator may also be able to determine whether an item burned from the top down or from the bottom up and how long it burned.

Did the fire start at a low point and burn up, or did burning items fall from above and ignite combustible materials below? Did the fire spread along the ceiling or along the floor? Are rags containing the residue of an ignitable liquid found under the furniture? Why are papers that should have been in a metal filing cabinet stacked on the floor and partially burned? All of these questions are ones the fire investigator may consider as part of their investigation.

Systematically digging out the debris can often uncover the exact point of origin and identify the cause of a fire. If circumstances or eyewitness accounts indicate that a fire may have been deliberately set, the fire investigator may request that firefighters help examine the entire area. The investigator will explain what to look for, how to search, and what to do with any potential evidence.

When you find possible evidence, stop and inform the fire investigator so that they can examine it in place. It is the fire investigator's job to document, photograph, and remove any potential evidence, whether or not it supports their initial hypothesis, or theory, about how the fire started. To determine the point of origin and cause of a fire, the fire investigator must evaluate all evidence gathered at the scene and from other sources.

Collecting and Processing Evidence

Evidence refers to all the information gathered and used by a fire investigator in determining the cause or point of origin of a fire. Evidence can be used in a legal process to establish a fact or prove a point. To be admissible in court, it must be gathered and processed under strict procedures.

Preservation of Evidence

Firefighters have a responsibility to preserve evidence that could indicate the cause or point of origin of a fire. They are not, however, responsible for deciding whether the evidence they find is relevant to the investigation. Too much evidence is better than too little, so no piece of potential evidence should be ignored.

Evidence is often found during the salvage and overhaul phases of a fire. Salvage and overhaul must be performed carefully. Often, these phases can be delayed until a fire investigator has examined the scene. Firefighters who find potential evidence should leave it in place, if possible. Do not move debris any more than is necessary, and never discard debris until the fire investigator gives their approval to do so. As mentioned, it is the fire investigator's job to decide whether the evidence is relevant, not the firefighter's.

If critical evidence could be damaged, destroyed, or altered in any way during fire suppression activities, attempt to cover it with a salvage cover, tarp, or some other type of protection. Use barrier tape to keep others from accidentally walking through evidence. These methods may prove to be ineffective, however, if no indication of their purpose is given.

Some evidence may need to watched over by personnel or moved to a secure location. Before moving an object, be sure that witnesses are present, that a location sketch is drawn, and that a photograph is taken. Immediately notify the IC or investigator of the object's original location and condition.

Fire investigators must use clean tools and special containers to store evidence and prevent any type of spoliation, including cross-contamination. Fire investigators must always wash their tools between taking samples to ensure that material from one piece of evidence is not unintentionally transferred to another piece. Fire investigators must also change their gloves each time they take a sample of evidence. Evidence should only be put into clean containers (**FIGURE 28-5**). Before entering the fire scene, fire investigators often wash their boots to keep from transporting any contamination into the fire scene.

Chain of Custody

Evidence relating to the cause of a fire may be presented in court for both criminal and civil cases. To be admissible in court, physical evidence must be handled according to certain set standards. Because the cause of the fire may not be known when evidence is first collected, all evidence should be handled using the same procedures.

Chain of custody (also known as chain of evidence or chain of possession) is a legal term that describes the process of maintaining continuous possession and control of evidence from the time it is discovered until

FIGURE 28-5 Fire investigators must place evidence in clean containers.

it is presented in court. Every step in the collection, movement, storage, and examination of the evidence must be properly documented.

For example, if a gasoline can is found at a fire scene, the investigator must record the person who found the can, along with where and when the can was found. Photographs should be taken to show the condition of the gasoline can and where it was found. In court, the fire investigator must be able to show that the gas can presented is the same can that was found at the fire site.

The person who takes initial possession of the evidence must keep it under their personal control until the item is turned over to another official. Each transfer of possession must be properly recorded.

If evidence is examined in a laboratory, the laboratory tests must be documented. If evidence is stored, documentation must indicate where and when it was placed in storage, whether the storage location was secure, and when the evidence was removed from that location. Often, evidence is maintained in a secured evidence locker to ensure that only authorized personnel have access to it.

Everyone who had possession of the evidence must be able to attest that it was not contaminated, damaged, or changed in any way. The documentation for chain of custody must establish that the evidence was never out of the control of the responsible agency and that no one could have tampered with it.

Firefighters are frequently the first link in the chain of custody. The firefighter's responsibility in protecting the integrity of this chain is relatively simple: Report everything to a supervisor, and disturb nothing needlessly. The person who finds the evidence should remain with it until the material is turned over to a company officer or to the fire investigator. Remember, you could be called as a witness to testify if you discover a piece of evidence.

The fire investigator's SOPs for collecting and processing evidence include the following steps:

- Take photographs of each piece of evidence as it is found and collected. If possible, photograph the item exactly as it was found, before it is moved or disturbed.
- Sketch, mark, and label the location of the evidence at the fire scene. Sketch the scene as near to scale as possible.
- Place evidence in appropriate containers to ensure it is secure and to prevent contamination. Unused paint cans with lids that automatically seal when closed are good containers for transporting evidence. Glass mason jars sealed with a sturdy sealing tape can be used to transport smaller quantities of materials. Plastic containers and plastic bags should not be used to hold evidence containing petroleum products, as these chemicals may deteriorate the plastic. Paper bags can be used to collect dry clothing or metal articles, matches, or papers. Soak up small quantities of liquids with either a sterile cellulose sponge or sterile cotton batting.
- Tag all evidence at the fire scene. The evidence container should be labeled with the date, time, location, and investigator's name.
- Record the time when the evidence was found, the location where it was found, and the name of the person who found it. Keep a record of each person who handled the evidence.
- Keep a constant watch on the evidence until it can be stored in a secure location. Evidence that must be moved temporarily should be put in a secure place that is only accessible to authorized personnel.
- Preserve the chain of custody in the handling of evidence. A broken chain of custody may result in the evidence being inadmissible in court.

One person should be responsible for collecting and taking custody of all evidence at a fire scene, no matter who discovers it. If someone else must seize the evidence, that person must photograph, mark, and contain the evidence properly and turn it over to the evidence collector as soon as possible. The evidence collector must also document all evidence according to department policies. A log of all photographs taken should be recorded at the scene as well.

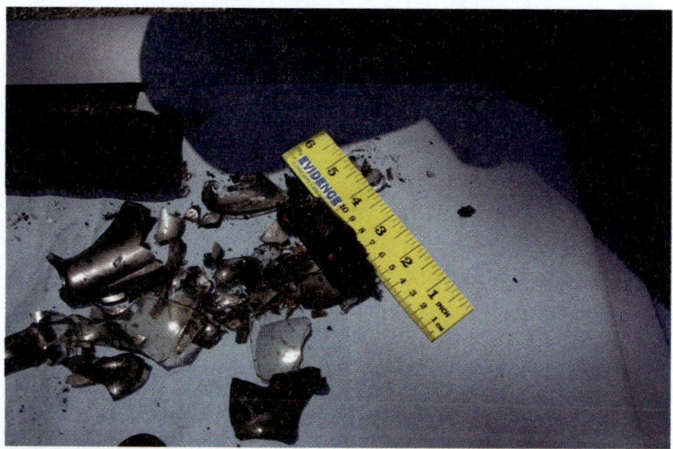

FIGURE 28-6 Physical evidence at a fire scene.

Courtesy of Mike Dalton.

Types of Evidence

Physical evidence (also called **real evidence**) consists of items that can be observed, photographed, measured, collected, examined in a laboratory, and produced in court to prove or demonstrate a point (**FIGURE 28-6**). Fire investigators can gather physical evidence at the fire scene, such as an empty gasoline can or a photograph of a fire pattern on a wall, to explain how the fire started or to document how it burned.

Demonstrative evidence is anything that can be used to validate a theory or to show how something could have occurred. To demonstrate how a fire could spread, a fire investigator might use photographs, diagrams, charts, or a computer model of the burned building.

Documentary evidence includes any type of written document, record, report, or other information that documents key facts. This may include bank records, alarm system activation reports, insurance policies, and credit card receipts. Each of these documents is used by the investigator to support their theory about how, when, and why the fire ignited.

Testimonial evidence is the fourth type of evidence gathered during the investigation (**FIGURE 28-7**). Testimonial evidence may be documented verbal or written statements previously given by firefighters or bystanders. More formal testimonial evidence includes affidavits, interrogatories, and court testimony that documents observations and statements made by lay and expert witnesses.

Evidence used in court can be classified as either direct or circumstantial. **Direct evidence** includes facts that can be observed or reported firsthand. Testimony from an eyewitness who saw a person ignite a fire and a video recording from a security camera showing the person starting the fire are examples of direct evidence.

FIGURE 28-7 Testimonial evidence includes documented statements given by firefighters or bystanders, affidavits, interrogatories, and court testimony made by lay and expert witnesses.

© Guy Cali/Corbis/Getty Images

Circumstantial evidence is based on inference and not on personal observation of a crime or other activity. For example, an investigator may infer that two or more unrelated fires that occurred at the same time and in the same vicinity could be incendiary fires. Or, suppose a fire victim is found in a chair in the living room, surrounded by ashtrays. If no other probable cause can be determined, the investigator may infer that careless use of smoking materials was the cause of the fire. The evidence in these cases is circumstantial because the fire investigator did not actually observe how the fires started. Fire investigators often must use circumstantial evidence along with other types of evidence to determine the cause of the fire.

Identifying Witnesses

Although firefighters may not interview witnesses, they can identify potential witnesses. People who were on the scene when firefighters arrived could have valuable information about the fire. If a firefighter learns something that might be related to the cause of the fire, they should pass this information on to a supervisor or to the fire investigator.

Interviews with witnesses should be conducted by the fire investigator or by law enforcement personnel. If the fire investigator is not on the scene or does not have the chance to interview a witness, the firefighter should get the witness's name, address, and telephone number. A witness who leaves the scene without providing this information could be difficult or impossible to locate later.

Firefighters have a primary responsibility to save lives and property. Until the fire is under control, they must focus on fighting the fire and not on investigating its cause. Even so, firefighters should pay attention to the scene and make mental notes about their observations. They must tell the fire investigator if they notice anything unusual. Information about the fire should be shared only with the fire investigator and only in private.

Do not make any statements of accusation, opinion, or probable cause to anyone other than the fire investigator. Comments that are overheard by the property owner, the occupant, a news reporter, or a bystander can complicate the work of the fire investigator who is trying to obtain complete and accurate information. A witness who is trying to be helpful might report an overheard comment as a personal observation. In this way, inaccurate information can generate a rumor, which then becomes a theory, which then turns into a reported "fact" as it passes from person to person.

Never make jokes at the scene. Careless and unauthorized remarks could embarrass the fire department. Statements to news reporters about the fire's cause should be made only by the department's public information officer (PIO) after the fire investigator and IC have agreed on their accuracy and validity. Refer all civilians to the IC or PIO for information about the fire (**FIGURE 28-8**).

FIGURE 28-8 Refer all civilians and news reporters to the PIO for information about the fire.

Documentation

The firefighter will not typically have a formal role in conducting or coordinating the investigation. However, they may be asked questions by the investigator and might be called to give testimony in court. In either case, the firefighter may have to document their findings. The firefighter should always be prepared to do so—particularly if the incident ended in injury, loss of life, significant property loss, or under unusual circumstances.

The Formal Investigation

After the suppression units have completed their work and have finished assisting the investigators, they may turn the scene over fully to the investigators. At this point, the investigators are responsible for controlling the scene. This step in the process ensures that evidence is secure from the first arrival of fire department units until the investigators release the property back to the owners. This phase allows for an even more methodical approach by the investigators to determine the point of origin and the sequence of events that caused the fire to start.

Investigatory Authority

If the fire investigation concludes that a crime may have been committed, a criminal investigation is conducted. Significant training, education, and experience are required to effectively investigate these types of events, whether criminal in nature or not. Statute authority determines who will be responsible for conducting the formal investigation.

Whether a fire investigator has **police powers** and can conduct a criminal investigation depends on state and local laws. In some jurisdictions, law enforcement personnel are trained as fire investigators, and they conduct the entire investigation. In other jurisdictions, fire department and law enforcement personnel work together throughout the investigation. Successful fire investigations often depend on good working relationships between different organizations. Firefighters need to understand how fire investigations are conducted in their community.

Investigation Assistance

Most fire departments and government agencies have limited time, personnel, and resources available for fire investigations. To handle larger investigations, a state fire marshal or similar authority may set up an investigations unit that focuses on major incidents and supports local investigators on larger fires. Many investigators working for the state fire marshal's office also serve as commissioned law enforcement officers so that they can enforce the laws and ordinances of the state. Because these investigators cannot always reach the fire scene quickly, the local fire department must be prepared to conduct a thorough preliminary investigation and to protect the scene and preserve evidence.

Federal resources are available for major investigations. The U.S. Bureau of Alcohol, Tobacco, Firearms and Explosives (ATF) has individual agents who can assist local jurisdictions with fire investigations. The ATF maintains response teams across the country that respond to large-scale incidents in the United States. Each of these teams includes about 15 agents. ATF teams use equipment that ranges from simple tools for digging out a fire scene to highly technical equipment for fully documenting large scenes or high-profile criminal investigations.

Private Investigation Entities

Insurance companies often investigate fires to determine the validity of a claim for damages or to identify factors that might help prevent future fires. The cost of an investigation is more than offset by the amount the company can save by identifying a fraudulent claim.

Some insurance companies employ their own fire investigators. Others retain the services of independent fire investigators. With any of these additional investigators, the firefighter's role remains the same with the potential, as determined by local policy, for additional questions or interviews to assist in a particular area of the examination.

Product manufacturers and testing laboratories also use their own investigators to conduct **destructive analysis investigations** that validate or refute claims of faulty equipment. This analysis results in the contamination, spoliation, or destruction of the evidence, so the entire process must occur after the initial fire investigation. This process should also be thoroughly documented by the analyst.

Outside fire investigators often have valuable experience and can provide critical technical support to determine the cause of a fire. If a private investigator finds evidence of a crime, they must provide that information to the municipal or other governmental fire investigator of record.

Each state has adopted some type of **arson reporting immunity law** that generally protects private investigators from claims filed by private citizens or entities for sharing information that may lead to criminal charges. These laws also provide limitations on which information must be shared and how that information must be shared.

Legal Authority to Enter, Secure, and Transfer Property

The Fourth Amendment protects private citizens and businesses from unlawful search and seizure without probable cause or a judge-issued search warrant. In most states, however, the **exigent circumstances rule** establishes that the need to provide emergency medical care, extinguish a fire, or take action in any other emergency situation to be of greater value to the community than an individual's Fourth Amendment rights.

Simply stated, exigent circumstances are situations in which there is an urgent need for action. The exigent circumstances rule grants firefighters, law enforcement officers, and EMS providers the right of entry into a private building without first obtaining consent or a warrant and without being held liable for unlawful entry. Although laws vary, most states allow the fire department to maintain control of the building after the fire has been extinguished to determine the origin and cause of the fire and, if necessary, to determine whether a crime has been committed.

The Federal Emergency Management Agency's incident command model (outlined in Chapter 22,

TIP

The exigent circumstances rule was established by *Michigan v. Tyler* (1978). In the case, a local Michigan fire department responded to a fire at a furniture store shortly after midnight on January 21, 1970. Although arson was suspected, the fire investigation was put on hold due to the smoke and steam at the fire scene. Approximately 5 hours later, the investigator returned and continued the investigation, finding evidence that was used during the trial. On February 16th, state police investigators went to the scene, performed an additional investigation, and obtained additional evidence.

The U.S. Supreme Court found the initial investigator's return to the scene to be a continuation of the fire investigation, but subsequent reentries by the state police were determined to be a violation of the Fourth Amendment. The court concluded that once a fire department leaves a property or delays an investigation without a legitimate cause, written consent or an administrative warrant must be obtained to reenter the property. Failing to do so may mean that any evidence obtained in such a search is inadmissible in court. Additionally, fire department officials could be charged with unlawful entry.

Establishing and Transferring Command) also grants the responding agency's IC "statutory responsibility" to mitigate the incident. This allows the IC to maintain legal authority over the property during fire suppression operations and to conduct a thorough investigation after the fire has been extinguished.

There is no open-ended authority to enter a fire scene and investigate. The investigation must take place within a "reasonable time" after the fire is extinguished. Usually, it must be conducted immediately after the fire has been extinguished. Any delay in beginning the investigation must be justified for the investigation to be considered part of the exigent circumstances surrounding the fire.

Briefly delaying the investigation due to bad weather or poor nighttime visibility is usually acceptable, as long as the fire department maintains control of the property until the investigation is complete. In the meantime, the fire department should limit salvage and overhaul operations to only those that are necessary to stabilize the fire scene.

The fire scene should be extensively photographed, starting with the area with the least amount of damage and proceeding toward the area of possible origin. Multiple photographs of the point of origin should be taken from various angles. Any physical evidence should be photographed exactly where it was found. If weather, traffic, or other factors could destroy evidence, take steps to preserve these items in the best way possible.

Controlling the incident scene during this time is critical. The building and premises must be properly secured and guarded until the fire investigator has finished gathering evidence and documenting the fire scene. To secure the property, close off the area using fire or police line tape (**FIGURE 28-9**). A member of the fire department or law enforcement agency should remain at the scene to keep nonessential personnel out of the area and prevent unauthorized persons from entering. Fires, explosions, or other emergency incidents that cause serious injuries, deaths, or substantial property damage may take days, weeks, or even months to fully investigate. These investigations require around-the-clock security to ensure that the scene is not compromised.

FIGURE 28-9 Fire or law enforcement personnel should remain on scene until the investigation is completed.

The fire department has the authority to deny access to a building for as long as necessary to conduct a thorough investigation and ensure public safety. After receiving permission from the fire investigator, a fire officer or a firefighter should accompany anyone who enters the premises until the scene is released. The fire department should keep a log of anyone who enters the property, the times of their entry and departure, and a description of any items taken from the scene.

After the investigator has documented the scene, collected evidence, and interviewed witnesses, the property is released to the owners. Before doing so, the investigator must make sure that the building is properly secured and that no threats to public safety exist. Fire departments can secure and protect the premises in several ways. Lock and secure doors, windows, and gates as much as possible. Use scene tape to cordon off dangerous areas or mark them with signs. Shut off all utilities and seal any openings in the roof to prevent additional water damage. Some fire departments have contracts with local companies that provide 24-hour services to board up and secure windows and doors to prevent unauthorized entry.

The fire department's authority ends at this time. Afterward, the fire investigator will need a search warrant or written consent to search the site. Any unguarded evidence left on the scene will likely be ruled inadmissible if the case goes to court.

Note that there are some limits to the exigent circumstances rule. If firefighters or law enforcement officers do not remain on the scene until the fire investigator arrives, the right of entry to conduct fire investigation is generally forfeited unless permission or a warrant is obtained. In most cases, firefighters may not maintain control of the property for an unreasonable length of time. In other words, the investigation must be conducted as soon as possible given the safety conditions, weather, daylight, and resources.

A search warrant is usually needed for lengthy, complicated investigations that require the scene be secured indefinitely. Additionally, firefighters must be aware of any state or local laws pertaining to right of access by the owners or occupants while the scene is under the control of the fire department.

Determining the Point of Origin and Fire Cause

The primary purpose of a fire investigation is to determine the sequence of events that brought together the ignition source and the initial fuel. The first step is to determine the point where ignition occurred. The investigator must then identify how and why ignition occurred. For this reason, fire investigations are often referred to as origin and cause investigations.

NFPA 921, *Guide for Fire and Explosion Investigations,* is used by many fire investigators as a guide for conducting fire investigations. NFPA 921 recommends using the **scientific method** to determine the origin and cause of a fire. The scientific method uses a seven-step systematic process to test one or more hypotheses, or proposed explanations, against the evidence collected during the investigation (**FIGURE 28-10**). The investigator must look at the situation objectively to be sure that the evidence is convincing and fully explains the situation. If there is more than one hypothesis for the observations, each hypothesis must be evaluated against the evidence. The hypothesis with the highest degree of certainty will then be identified as the cause of ignition.

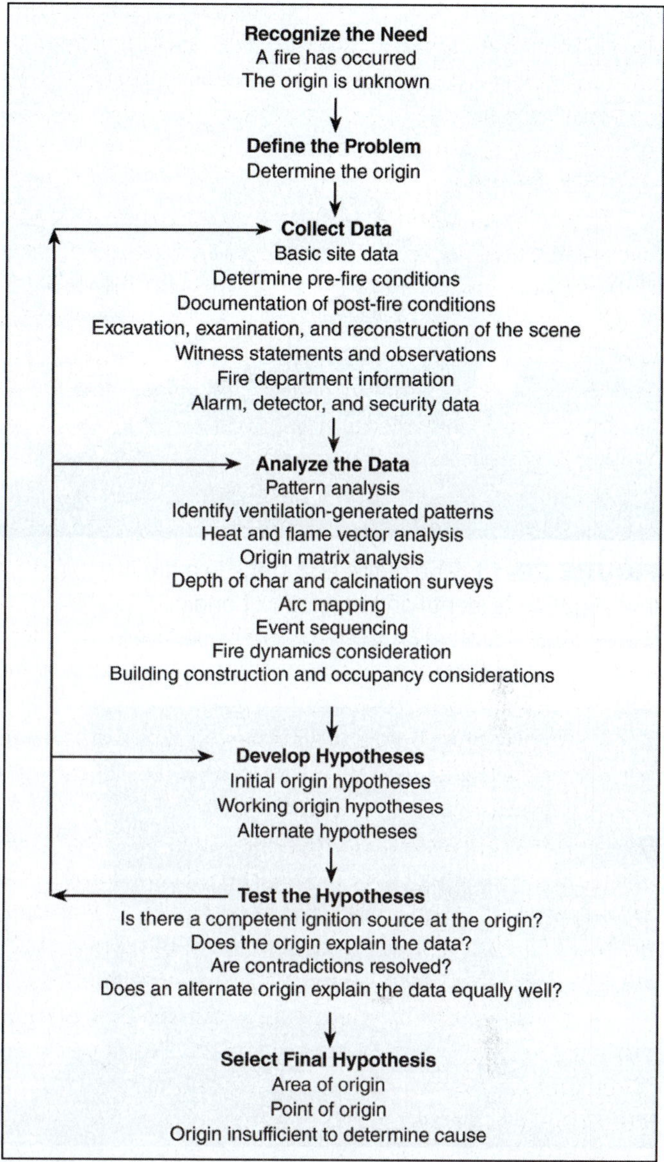

FIGURE 28-10 The scientific method's seven-step process should be used to determine the origin and cause of a fire.

Identifying the Point of Origin

One of the first steps in a fire investigation is identifying the **point of origin** (**FIGURE 28-11**). This step requires a careful, systematic analysis of the fire scene. The investigator will use notes, photographs, videos, sketches, diagrams, and other tools to permanently document the entire investigation. These tools help the investigator to remember the scene, specific evidence, and the process used to conduct the investigation. They may also be used by the investigator if they are called to testify in a civil or criminal court case.

When working to identify the point of origin, the fire investigator will usually begin by examining the building's exterior. They will look for signs that the fire began outside the building before looking inside. They will also make note of key details, such as the overall size, construction, layout, and occupancy of the building. The investigator will document the extent of damage that is visible from the exterior. They will also look for any openings that might have created drafts that influenced the fire spread. The investigator will examine the condition of outside utilities, such as the electrical power connection and gas meter, as well.

The search for evidence of the fire's point of origin continues inside the building, beginning with the area of least damage and moving systematically to the area of heaviest damage. The area of heaviest damage usually indicates where the fire was burning for the longest time or where it burned the hottest (**FIGURE 28-12**). Areas with less damage were probably not as heavily involved in the fire or were involved for shorter periods.

Fire Effects and Fire Patterns

Fire effects are the observable changes found on objects, surfaces, and construction materials from heat,

FIGURE 28-11 One of the first steps in a fire investigation is identifying the point of origin.

Courtesy of Captain David Jackson, Saginaw Township Fire Department.

FIGURE 28-12 The area of heaviest damage usually indicates where the fire burned for the longest time or burned the hottest.

Courtesy of Jamie Novak, Novak Investigations Inc. and the St. Paul Fire Department.

flame, smoke, and other products of combustion. Fire effects can be as simple as the charring of wood or as complex as the chemical change in a specific material. Each fire effect can be observed and evaluated separately. For example, the analysis of a single piece of melted plastic may help the investigator determine the approximate temperature of the room during the fire. If multiple objects or separate areas of the room show varying degrees of the same fire effects, **fire patterns** may be observed (**FIGURE 28-13**). Fire patterns show the relative degree of fire exposure time and temperature of the affected objects. For example, several wooden studs may show different degrees of charring.

FIGURE 28-13 If multiple objects or separate areas of the room show varying degrees of the same fire effects, fire patterns may be observed.

Courtesy of Robert Schaal.

FIGURE 28-14 Depth of char.

Courtesy of Charles B. Hughes/Unified Investigations & Sciences, Inc.

The studs with the heaviest charring were probably exposed to the fire for the longest time or exposed to the highest temperatures.

The **depth of char** is a fire effect that was once believed to indicate exactly how long a material was exposed to fire. Newer research indicates that this method is unreliable (NFPA 921). Charring may be deepest at the point of origin. However, the presence of an **ignitable liquid** and combustible materials, as well as the effects of ventilation, can also influence charring.

Still, when considered as a fire pattern, the relative depth of char on similar materials can indicate which area of a room or which side of a piece of furniture was exposed for the longest time or to the hottest fire (**FIGURE 28-14**). An area that was exposed to a lower intensity of fire or heat or exposed for shorter amounts of time will display shallow charring. Conversely, areas

FIGURE 28-15 Fire patterns often indicate the area or point of origin.

Courtesy of Charles B. Hughes/Unified Investigations & Sciences, Inc.

that were exposed to higher intensities of heat and fire or exposed for longer periods of time will display deeper charring and greater destruction of material.

Because heat rises, the flow of heated gases from a fire will often move up and out from the point of origin. This upward, outward flow can usually be recognized, even when the entire building was involved in the fire. The damage will often spread outward from the room or area where the damage is most severe. Often the point of origin is directly below the most heavily damaged area on the ceiling, where the heat of the fire was most intensely concentrated. A charred V, U, or hourglass pattern on a wall often indicates that fire spread up and out from something at the base. In this case, the point of origin may be found in a pile of charred debris at the base of the pattern (**FIGURE 28-15**).

Fire patterns can be helpful in identifying the point of origin, but they are not conclusive evidence. An experienced fire investigator knows that many factors can influence fire patterns, including ventilation, fire suppression efforts, and the burning materials themselves. A fire pattern on a wall could indicate that an easily ignited and intensely burning fuel source was present at that location. It might also indicate that something fell from a higher level and burned on the floor. Such patterns may be caused by direct flame contact, heat, smoke, soot, or a combination of these elements. The patterns produced by the fire may be obvious on any surface and may indicate movement away from the point of origin. Patterns may be present on wall and ceiling surfaces, appliances and furnishings, doors and windows, and even fire victims.

Once the fire investigator identifies the approximate point of origin, the search for clues to the specific cause can begin. This effort involves identifying the source of ignition, the fuels that were first ignited, and the sequence of events that brought the two together.

Determining Fire Cause

A fire investigation attempts to identify where, how, why, and sometimes when a fire ignited. Some fires have simple causes that are easily identified and understood; others result from a complex sequence of events that must be examined carefully. For example, if a fire started in a kitchen, potential causes could include the following:

- The stove
- Any of the electrical appliances
- Light fixtures
- Smoking materials
- Cleaning supplies
- Other ignition sources, such as candles

In some cases, the cause of a fire is never determined with a high degree of certainty.

Ignition Source

After the area of origin is determined, the investigator must search for a **competent ignition source**. To be considered "competent," the ignition source must have enough energy and be capable of transferring that energy to the first fuel long enough to heat the fuel to its ignition temperature.

In some cases, no obvious ignition source can be found, so the investigator must look for other evidence to establish the cause of the fire. In other cases, one or more possible ignition sources may be found in the area of origin. Each of these possible ignition sources must be evaluated to determine if it could have ignited the fuel.

First Fuel Ignited

To determine if a heat source is competent, the investigator must also identify the first fuel that ignited. Often, the fuel is obvious, such as in the case of a fire in a recycling bin full of paper. In other cases, many different types of fuel are close to the ignition source. The investigator must try to determine which fuels, if any, could have been ignited by the ignition source.

Oxidizing Agent

Most fires use the oxygen in the air in the combustion process, but some fire scenes may have special oxidizing agents that contribute to fire spread (see Chapter 5, *Fire Behavior*). Medical oxygen cylinders, pool sanitizers, and other chemical oxidizing agents at an incident scene should be reported to the investigator.

Ignition Sequence

The fire tetrahedron is used to explain the chemical reaction that allows a fire to ignite and continue to burn (described further in Chapter 5, *Fire Behavior*). The fuel and ignition source (heat) must come together, in the presence of oxygen (or an oxidizing agent), to cause ignition and sustained burning. The fire investigator must try to determine the exact sequence of events that brought those elements together. For example, if the investigator determines that an unattended candle ignited items on a table, they must try to explain how the combustible items came into contact with the candle's flame.

Levels of Certainty

The last step in determining the cause of a fire involves establishing the level of certainty of a hypothesized fire cause. In most cases, it is not possible to determine the cause with 100 percent certainty unless surveillance video or credible witnesses can confirm the facts. Usually, after all the evidence has been evaluated, one or more hypotheses are tested to determine which has the highest level of certainty. A hypothesis with less than a 50 percent certainty level is considered a **possible cause**. If more than one hypothesis is determined to be possible, the one with the highest level of certainty is listed as the official **probable cause**. A probable cause must have a level of certainty of 50 percent or more. If no cause can be determined or if no cause can be considered probable, the fire should be classified as undetermined.

The process of eliminating possible causes and documenting the reasons for rejecting them is just as important as properly documenting the hypothesis stating the cause of the fire.

Classifications of Fire Cause

After the point of origin has been identified and the cause of the fire has been determined, the investigator must assign the fire cause to one of four different cause classifications: accidental, natural, incendiary, or undetermined.

Accidental Fire Causes

An **accidental fire** occurs because of human action or inaction that does not involve malicious intent. People may do things that result in a fire without intending for the fire to start or get out of control. Forgetting about a pan of grease on the burner, refueling a hot lawn mower, or igniting a trash fire that suddenly spreads to a nearby building are all examples of the ways that accidental fires can start.

Natural Fire Causes

Natural events often cause unwanted fires—some of which can be quite disastrous. A **natural fire** is started without direct human intervention or action. They can be caused by earthquakes and tornadoes, which can rupture natural gas pipes and knock down high-voltage power lines. Lightning and high winds cause many natural fires.

Incendiary Fire Causes

An **incendiary fire** is ignited by a deliberate and intentional act. For a fire to be classified as incendiary, the person that committed the acts must have known or should have known that igniting the fire was wrong. Fires intentionally started by small children or by people lacking the ability to understand the consequences of their actions are usually classified as accidental fires. The determining factor for an incendiary fire is that the person intentionally committed an act that was known to be wrong.

Undetermined Fire Causes

Fires that lack a probable cause with more than 50 percent certainty must be classified as an **undetermined fire**. Even if the investigator feels strongly about a hypothesis, they must determine the origin, cause, and cause classification based solely on the evidence. Under no circumstances should a fire be classified as "suspicious." Fires classified as undetermined may later be reclassified as accidental, natural, or incendiary, if additional evidence is collected.

The Crime of Arson

The term "arson" has been misused and generalized for many years. Although incendiary fires and arson are often related, they are not the same. Fires classified as incendiary are those determined to have been intentionally started by someone who knew that it was wrong. Arson, however, is a specific crime that, in many states, has different definitions, degrees, and penalties.

Some arsonists are responsible for multiple fires in a specific location or over a certain time frame. Three repetitive fire-setting behaviors have been identified, with specific definitions for each one (Douglas, 2013). A **serial arsonist** is a person who sets three or more fires in succession, with an emotional "cooling-off" period between each fire. A **spree arsonist** sets three or more fires in separate locations with no lengthy cooling-off period between fires. A **mass arsonist** sets three or more fires at or near the same location during a limited time period.

All fire investigators attempt to identify the origin and cause of a fire. However, only those investigators with law enforcement authority can determine whether a crime has been committed, and if so, which crime. In some jurisdictions, the fire investigator is authorized to conduct both the fire investigation and the criminal investigation. In other jurisdictions, fire investigators determine origin and cause while law enforcement officers investigate criminal activity. Firefighters should avoid using the term "arson" on the fire scene, even in casual conversation. Your role as a firefighter is to identify and preserve potential evidence, share any relevant observations with the investigator, and assist with the investigation as directed by the IC or fire investigator. Determining the origin, cause, cause classification, and any criminal activity is the responsibility of other personnel. According to the U.S. Fire Administration, arson motives can be classified into six distinct categories:

- Vandalism
- Excitement
- Revenge
- Crime concealment
- Profit
- Extremism

Each of these motives can result in different criminal charges based on the facts of the individual case.

Youth Fire-Setters

Children tend to have a fascination with fire. Most children will quickly learn to respect fire, but some children become drawn to fire. If this fascination becomes too strong, the child may begin to play with fire or intentionally start fires for different reasons. Experts in the field recognize three distinct age categories of **youth fire-setters**.

Child fire-setters are children from 2 to 6 years of age. They are often curious about fire. They lack the mental capacity to understand the consequences of their actions, but they often realize that they should not be playing with matches or lighters. As a result, they often play with fire in hidden locations like closets, basements, and garages where many combustible fuels are present. Child fire-setters may be injured or killed by playing with fire. Fires started by young children are usually reported as accidental fires.

Juvenile fire-setters are 7 to 13 years old. Juvenile fire-setters often start fires in or around their own homes or schools.

Adolescent fire-setters are typically 14 to 16 years old. Because these fire-setters are older, the fires they ignite are often more severe. Adolescent fire-setters may start fires in schools, churches, outbuildings, vacant homes, and vacant lots.

There are preventive education and support programs to help treat the underlying causes of their fire-setting behavior.

CASE STUDY

You Are the Firefighter CONCLUSION

You are dispatched to a reported house fire. It is around noon on a weekday. Weather conditions are sunny and mild. While en route, you are told that there is a 14-year-old burn victim. When you arrive at the scene, after a 2-minute response, you see that the second floor of a single-family dwelling is over 50 percent involved and is venting from two windows on the Delta side. Your officer quickly coordinates EMS care for the 14-year-old who is in the front yard with their parents, who are clearly upset. The officer links back up with you as you advance the line up the stairwell and begin extinguishing the fire. You push down the hallway to the back bedroom and come across what looks like a melted plastic gas container. The fire is well developed, but you eventually get a knock on it with your crew.

1. Given the scenario, what information do you think would be of value to a fire investigator?

Answer: In this scenario, information that would be of value to a fire investigator includes the initial size-up observation noting a significant volume of fire following a quick on-scene time. Other relevant information is that there

was a school-aged child at home in the middle of the day and, based on their injuries, apparently near the fire. Also of note is that the fire appears to be in the bedroom area of a typical single-family dwelling and that there was a melted plastic gas container in an area of the house that is an atypical location for storage.

2. **After overhaul and salvage is finished, what tools and equipment could you use to assist the fire investigator in processing the fire's room of origin?**

   Answer: To assist the fire investigator in processing the fire's room of origin, you could use tools such as shovels, rakes, and debris buckets to help dig out areas of the fire scene. Other tools that may be useful include salvage covers, tarps, scene lights, and portable fans.

3. **How does a fire origin and cause investigation differ from an arson investigation?**

   Answer: An origin and cause investigation involves determining just those two elements (origin and cause). Depending on the circumstances, that type of investigation may be conducted by a fire department unit on scene. An arson investigation is a criminal investigation that must be conducted by trained fire investigators and, potentially, law enforcement personnel.

WRAP-UP

SUMMARY

KNOWLEDGE OBJECTIVES

- Describe why fires are investigated and the role of the Firefighter II during a fire investigation.
  - Explain the reasoning for conducting a fire investigation. (**NFPA 1010: 7.3.5**, p. 1175)
  - Describe the role of the Firefighter II in fire investigation. (**NFPA 1010: 7.3.5**, pp. 1175–1180)
  - Describe the evidential items and conditions that may be observed during fireground operations. (**NFPA 1010: 7.3.5**, pp. 1176–1180)
- Describe the role of the Firefighter II in a fire that has caused injury or fatalities. (p. 1180)
- List the types of evidence that may be found at a fire scene, how to preserve evidence, and how to maintain the chain of custody.
  - Describe techniques for preserving fire scene evidence. (**NFPA 1010: 7.3.5**, p. 1181)
  - Explain the chain of custody. (**NFPA 1010: 7.3.5**, pp. 1181–1182)
  - List the types of evidence that may be found at a fire scene. (**NFPA 1010: 7.3.5**, p. 1183)
  - Describe how potential witnesses are identified. (pp. 1183–1184)
- Describe the potential participants involved in a formal investigation and the laws that impact the securing of a fire scene for an investigation.
  - Describe the role and relationship of the Firefighter II to criminal investigators and insurance investigators. (**NFPA 1010: 7.3.5**, pp. 1184–1185)
  - Describe how to assist fire investigators with processing a fire scene. (**NFPA 1010: 7.3.5**, pp. 1184–1185)
  - Describe the exigent circumstances rule. (pp. 1185–1186)
  - Explain the importance of protecting a fire scene to aid in origin and cause determination. (**NFPA 1010: 7.3.5**, pp. 1185–1186)
  - Describe the steps needed to secure a property. (**NFPA 1010: 7.3.5**, p. 1186)
- Describe how the point of origin and cause of a fire is determined.
  - Describe how the point of origin of a fire is determined. (**NFPA 1010: 7.3.5**, pp. 1187–1190)
  - Describe how the cause of a fire is determined. (**NFPA 1010: 7.3.5**, pp. 1189–1190)
  - Describe the four classifications of fire cause. (p. 1190)
- Describe the crime of arson.
  - Describe three repetitive fire-setting behaviors. (pp. 1190–1191)
  - List the six motives of arson. (pp. 1190–1191)

SKILLS OBJECTIVES

- Demonstrate how to protect evidence at a fire scene.
  - Protect evidence. (**NFPA 1010: 7.3.5**, p. 1181)

KEY TERMS

accidental fire A fire for which the cause does not involve a human act with the intent to ignite or spread a fire. (NFPA 556)

adolescent fire-setters Fire-setters who are typically 14 to 16 years old. Adolescent fire-setters may start fires in schools, churches, outbuildings, vacant homes, and vacant lots.

arc mapping The systematic evaluation of the electrical circuit configuration, spatial relationship of the circuit components, and identification of electrical arc sites to assist in the identification of the area of origin and analysis of the fire's spread. (NFPA 921)

arson The crime of maliciously and intentionally, or recklessly, starting a fire or causing an explosion. (NFPA 921)

arson reporting immunity law A law that generally requires private investigators to provide information regarding possible criminal activity upon the written request of the authority having jurisdiction. These laws also provide immunity from liability for disclosing such information as provided for by the law.

chain of custody The trail of accountability that documents the possession of evidence in an investigation from the time it is discovered until it is presented in court.

child fire-setters Fire-setters who are typically 2 to 6 years old.

circumstantial evidence Evidence that is based on logical inference rather than personal observation of a crime or other activity.

competent ignition source An ignition source that has sufficient energy and is capable of transferring that energy to the fuel long enough to raise the fuel to its ignition temperature. (NFPA 921)

demonstrative evidence Any type of evidence that can be used to validate a theory or show how something could have occurred; examples include diagrams, photographs, maps, X-rays, visible tests, and demonstrations.

depth of char A fire effect that, when evaluated as a pattern on identical fuels, may be used to determine locations within a structure that were exposed to higher intensities of heat and fire or exposed to a heat source for longer periods of time.

destructive analysis investigation An investigation that uses the methodical deconstruction of evidence to determine specific component conditions, functionality, or failures as they relate to fire investigation.

direct evidence Facts that can be observed or testimony of witnesses who observe acts or detect something through their five senses or through surveillance equipment, such as a security camera.

documentary evidence Any type of written record or document that is relevant to the case.

exigent circumstances rule A legal condition that allows emergency service providers to enter, search, seize, and control private property and to investigate the cause of a fire without consent or warrant while lawfully performing emergency operations.

expectation bias Any preconceived determination or premature conclusions as to the cause of a fire without having examined or considered all relevant evidence.

fire effects The observable or measurable changes in or on a material as a result of a fire. (NFPA 921)

fire patterns The visible or measurable physical changes, or identifiable shapes, formed by a fire effect or group of fire effects. (NFPA 921)

ignitable liquid Any liquid or the liquid phase of any material that is capable of fueling a fire, including a flammable liquid, combustible liquid, or any other material that can be liquefied and burned. (NFPA 556)

incendiary device A device or mechanism used to start an incendiary fire or explosion.

incendiary fire A fire that is intentionally ignited in an area or under circumstances where and when there should not be a fire. (NFPA 921)

juvenile fire-setters Fire-setters who are typically 7 to 13 years.

mass arsonist A person who sets three or more fires at the same site or location during a limited period of time.

natural fire A fire that is started without direct human intervention or action, such as a fire resulting from lightning, an earthquake, or wind.

KEY TERMS CONTINUED

physical evidence A physical or tangible item that proves or disproves a particular fact or issue; also referred to as *real evidence.*

point of origin The exact physical location within the area of origin where a heat source and a fuel first interact, resulting in a fire or explosion. (NFPA 921)

police powers A legal authority based on the Tenth Amendment that grants state (and local) authorities the authority to establish and enforce laws protecting the greater good of the community.

possible cause A hypothesis determined to have less than 50 percent probability of being true.

probable cause A hypothesis determined to have greater than 50 percent probability of being true.

real evidence See *physical evidence.*

scientific method The systematic pursuit of knowledge involving the recognition and definition of a problem; the collection of data though observation and experimentation; analysis of the data; the formulation, evaluation, and testing of hypotheses; and, where possible, the selection of a final hypothesis. (NFPA 921)

serial arsonist A person who sets three or more fires with an emotional cooling-off period between fires.

spoliation Loss, destruction, or material alteration of an object or document that is evidence or potential evidence in a legal proceeding by one who has the responsibility to preserve it. (NFPA 921)

spree arsonist A person who sets three or more fires at separate locations with no emotional cooling-off period between fires.

testimonial evidence Documented verbal or written statements previously given by firefighters or bystanders, as well as affidavits, interrogatories, and court testimony that documents observations and statements made by lay and expert witnesses.

trailers Solid or liquid fuels used to intentionally spread or accelerate the spread of a fire from one area to another. (NFPA 921)

undetermined fire A classification of fire used when the cause cannot be proven to an acceptable level of certainty.

youth fire-setters Recognized classifications of minors who ignite fires for various reasons. Based on age youth fire-setters include child fire-setters, juvenile fire-setters, and adolescent fire-setters.

REVIEW QUESTIONS

1. Why must fire investigators be careful to avoid expectation bias?
2. In most jurisdictions, who has legal authority to investigate fire cause?
3. What are possible indicators that a fire protection system has been tampered with?
4. In which phase of a suppression operation can firefighters be most helpful in assisting the investigation process?
5. What terms describes the result of an investigator who wears the same boots from another scene without cleaning them?
6. What rule allows first responders to enter a structure without consent or a warrant during an emergency?
7. What are fire patterns?
8. Which fire cause classification describes fires ignited by deliberate and intentional acts?
9. What type of arsonist sets three or more fires in separate locations with no lengthy cooling-off period between fires?
10. What term describes fire-setters who are between the ages of 2 and 6 years of age?

DISCUSSION QUESTIONS

1. What types of observations should a firefighter be making during a fire incident?
2. Describe the exigent circumstances rule and why it is important.
3. Explain why maintaining constant control of an incident scene is critical to a fire investigation.

APPLYING THE CONCEPTS

At 1400, your engine company is en route to a report of a fire in an office building in the commercial area of your town. Dispatch reports a fire on the fourth floor of the five-story building and the call was made by one of the employees. Someone activated the fire alarm and workers are evacuating. It's a new building, and you and your department preplanned it last year. Your engine arrives on scene within 5 minutes. Many workers and bystanders are crowded on the sidewalk outside the building. Police just arrived and are controlling the crowd and redirecting traffic. Thick, dark smoke is visible outside the building.

1. As a firefighter, your primary job is to save lives and property. But another part of your job is to help fire investigators. While you're in the engine on the way to the scene and as you arrive, what types of things do you consider and observe that help do both?
2. In this situation, when you read the scene, what might the large volume of smoke tell you?

After the 360 is completed, and the water supply is established at the FDC, the IC assigns you and your crew to interior fire suppression. Your crew connects a 1¾-inch (44-millimeter). handline to the standpipe in the stairwell and proceeds onto the fourth floor. You hear the fire alarm going off and see heavy, black smoke and fire coming from multiple offices. You proceed toward the fire, cooling the environment as you go. As you get closer to the offices, you see that document storage boxes on top of the desks in each of the offices are on fire. You recall from preplanning that this building has a suppression system but notice the sprinklers haven't been activated.

3. What raises your concerns about this fire?
4. If you suspect this fire might have been purposely set, how do you respond?
5. What can happen to evidence if you ignore the unusual circumstances and just focus on putting out the fire?

The fire's been knocked down and a crew is inside conducting overhaul. The fire investigator has been interviewing your crew members and wants to talk to you.

6. Why are the observations of each of the first responding firefighters especially important to fire investigators?
7. What types of evidence do investigators gather, and what's an example of each?

REFERENCES

Douglas, John E., Ann W. Burgess, Alan G. Burgess, and Robert K. Ressler. 2013. *Crime Classification Manual, Third Edition.* Hoboken, NJ: John Wiley & Sons, Inc.

National Fire Protection Association (NFPA). 2020. *NFPA 921, Guide for Fire and Explosion Investigations.* 2021 Edition. Quincy, MA: NFPA.

National Fire Protection Association (NFPA). 2023. *NFPA 1033, Standard for Professional Qualifications for Fire Investigator.* 2022 Edition. Quincy, MA: NFPA.

National Fire Protection Association (NFPA). 2024. *NFPA 556, Guide on Methods for Evaluating Fire Hazard to Occupants of Passenger Road Vehicles.* 2024 Edition. Quincy, MA: NFPA.

U.S. Fire Administration (USFA). 2022. "Fire Investigation: The First Responder's Role." Last reviewed September 15, 2022. Accessed October 23, 2023. https://www.usfa.fema.gov/a-z/arson-fire-investigation/fire-investigations-first-responders.html.

SECTION

3

Hazardous Materials Awareness Level

NOTE: **NFPA 1010**, *Standard on Professional Qualifications for Firefighters*, 2024 Edition 6.1 in Chapter 6: Firefighter I (**NFPA 1001**) states that all firefighter candidates must meet the requirements defined in Chapter 7 and Sections 9.2 and 9.6 of **NFPA 470, 2022 Edition**. Content within these chapters meets the intent of **NFPA 470, 2022 Edition**, which includes chapter 5, "Professional Qualifications for Hazardous Materials/WMD Awareness Level Personnel (**NFPA 1072**)."

CHAPTER

29

Hazardous Materials Awareness Level

Hazardous Materials Regulations, Standards, and Laws

NOTE: Content within this chapter meets the intent of **NFPA 470, 2022 Edition**, which includes chapter 5, "Professional Qualifications for Hazardous Materials/WMD Awareness Level Personnel (**NFPA 1072**)."

KNOWLEDGE OBJECTIVES

After studying this chapter, you will be able to:

- Identify the difference between hazardous materials/weapons of mass destruction incidents and other emergencies.
- Identify the location of both the emergency response plan and/or standard operating procedures.
- Define the terms *hazardous materials* (or *dangerous goods*, in Canada) and *weapons of mass destruction*.
- Understand the difference(s) between the standards and federal regulations that govern hazardous materials response activities.
- Describe the different levels of hazardous materials training: awareness, operations, technician, specialist, and incident commander.
- Explain the need for a planned response to a hazardous materials incident.

SKILLS OBJECTIVES

There are no skills objectives for awareness-level personnel for this chapter.

CASE STUDY

You Are the Firefighter

At 1330 hours, your engine company responds to a report of a suspicious odor at a small plastics manufacturing company. When you arrive, you see a cargo delivery truck parked at the loading dock, with the motor still idling. From your vantage point, a few hundred feet away, you notice a liquid leaking from the back of the truck. On one side of the truck, you can see a black-and-white, diamond-shaped placard that reads "CORROSIVE," with the number 8 at the bottom. The facility security guard reports that the truck has been left unattended for at least an hour.

1. Is this an incident? How would you go about analyzing the scene to better understand the problem? What other pieces of information would you want?
2. You are trained to the awareness level. Would this level of training allow you to put on chemical protective equipment, enter the hazardous area, attempt to determine the nature of the leak, and fix the problem?
3. What reference sources would you want to access for more information on the placard?

Introduction

Firefighters, law enforcement personnel, and emergency medical services (EMS) routinely respond to a variety of incidents. These incidents may include structural fires, emergency medical calls, automobile accidents, confined-space rescues, water rescues, and acts of terrorism. All of these, and other situations, may involve hazardous substances that threaten lives, property, and/or the environment. When an incident clearly involves a hazardous material, or when you suspect the presence of a hazardous materials release, the nature of the incident changes, and so must the mentality of the responder (**FIGURE 29-1**).

Hazardous materials incidents are handled in a more deliberate fashion than structural firefighting, and your level of training will dictate the actions you can take to resolve the situation. Unknown substances could be present, or the setting in which the release occurred might be such that it takes time to get a full picture of the incident. For example, train derailments, which have a large incident footprint and typically require many response agencies to be involved, may also include dangerous chemical substances. These types of factors do not mean that hazardous materials incidents are more or less complicated than fighting fires, but they do mean that responders often take more time to get oriented to the potential hazards and to define a rational approach to solving the situation without becoming exposed to the released substance. If a rescue is required during a hazardous materials incident, or if the situation is imminently dangerous in some other way and requires quick action, events may move quickly.

FIGURE 29-1 The ability to recognize a potential hazardous materials/weapons of mass destruction incident is a critical first step to ensuring your safety. Notice the types of containers adjacent to the spill, the color of the liquid on the floor, and the presence or absence of any reaction between the liquid and the flooring material. Any or all of these observations may be key to forming initial impressions of the scene.

All personnel and responders must understand that actions taken at hazardous materials/weapons of mass destruction (WMD) incidents are largely dictated by the chemicals or hazards involved; environmental influences such as wind, rain, and temperature; and the way the chemicals behave during the release. In short, the nature and circumstances of the response as a whole dictate the decisions made by the personnel and responders at the incident.

TIP

The goal of the responder is to favorably change the outcome of the hazardous materials/WMD incident. This is accomplished through sound planning and by establishing safe and reasonable response objectives based on the level of training. Don't do it if you're not trained to do it!

Additionally, personnel operating at the scene of a hazardous materials/WMD incident must be conscious of the potential or actual law enforcement aspect of the incident. Especially where terrorist or other criminal acts are suspected, responders should be mindful of evidentiary issues associated with the incident. Being mindful of potential evidence at a hazardous materials/WMD event may assist later efforts to identify, capture, and prosecute the person(s) responsible for the act. To that end, all responders on the scene must be cognizant of the impact that their presence and actions will have on potential evidence. Although evidence preservation should not impede the efforts to eliminate the problem or slow life-saving operations, all responders should be diligent in remembering that their actions and observations may play a vital role in the successful prosecution of a criminal suspect.

When responding to hazardous materials/WMD incidents, make a conscious effort to change your perspective. Slow down, think about the problem and available resources, and then take well-considered actions to solve it.

Additionally, initial and ongoing actions may be guided by your authority having jurisdiction (AHJ) and local or emergency response plans and/or standard operating procedures (SOPs). An AHJ is the "organization, office, or individual responsible for enforcing the requirements of a code or standard, or for approving equipment, materials, an installation, or a procedure" (NFPA 470). Basically, the AHJ is the governing body that sets operational policy and procedures for the jurisdiction in which you operate. Every responder should have knowledge of and access to response plans and be trained on how to implement the actions specified in those plans in accordance with their level of training. For example, the AHJ might identify a set of tasks that responders would be expected to perform in their course of duty and match the training competencies to address those tasks.

From a broad perspective, the goals of this section are to help you do the following:

- Recognize the presence of a hazardous materials incident.
- Take initial actions, including establishing scene control zones.
- Implement the incident command system (ICS).
- Use basic reference sources, such as the *Emergency Response Guidebook* (*ERG*).
- Select personal protective clothing based on the anticipated task(s).
- Implement product control measures when needed (for operations and hazardous materials technicians).
- Perform appropriate decontamination.

Ultimately, this section will help you understand where you fit into a full-scale hazardous materials response.

The chapters in this section focus on awareness-level personnel. The rest of the chapters focus on operations-level responders, including those responders assigned mission-specific responsibilities. There is some overlap between awareness and operations, but the intent is to make a clear delineation between the two levels.

What Is a Hazardous Material?

The first points to understand, before diving into the standards and regulations that govern hazardous materials response, are the definitions of *hazardous material* and *weapon of mass destruction.*

The U.S. **Department of Transportation (DOT)**—the government agency that publicizes and enforces rules and regulations that relate to the transportation of many hazardous materials—defines a **hazardous material** as any substance or material that is capable of posing an unreasonable risk to human health, safety, or the environment. This definition includes hazardous wastes, marine pollutants, and elevated-temperature materials. It also includes illicit laboratories, environmental crimes, or industrial sabotage. Adding to that, NFPA 470 defines a hazardous material as "[m]atter (solid, liquid, or gas) or energy that when released is capable of creating harm to people, the environment, and property, including weapons of mass destruction (WMD)." For brevity, this section will refer to hazardous materials and WMD simultaneously. In United Nations model codes and regulations, hazardous materials are called *dangerous goods*.

Around the world, the term **weapon of mass destruction (WMD)** may have different definitions, abbreviations, and acronyms. In general terms, a WMD can be thought of as "any weapon or material that is designed to cause death or serious injury or damage to buildings, structures, or the environment, such as an explosive or incendiary bomb, rocket, or grenade . . . containing or delivering a toxic or dangerous chemical, biological agent, toxin, or vectors; or a weapon designed to release dangerous levels of radiation" (NFPA 470). The most common way of describing WMDs is with the acronym CBRN—that is, as chemical, biological, radiological, or nuclear weapons or materials.

To make things simple, a hazardous material can be almost anything, depending on the situation. Milk, for example, is not routinely regarded as a hazardous substance, but 5000 gallons of milk leaking into a creek does, in fact, pose an unreasonable risk to the environment (**FIGURE 29-2**). A release of ethylene glycol (antifreeze) from a motor vehicle accident may present a low hazard to responders, but a high hazard to certain animals. A large chlorine gas release also fits the definition, as would any substance used as a WMD or with criminal intent. Regarding the threat of terrorism, it would be unwise to believe that your jurisdiction is immune to deliberate criminal acts. Such an event could happen anywhere, at any time, with any type of substance.

FIGURE 29-2 A hazardous material can be found anywhere. This photo, for example, was taken at a food packaging plant that has a large ammonia system and numerous chemicals in a storage area. Not a typical backdrop for significant chemical use and storage!

Manufacturing processes sometimes generate hazardous wastes. A **hazardous waste** is what remains after a process or manufacturing activity has used a substance and the material is no longer pure. Hazardous waste can be just as dangerous as pure chemicals. It can also consist of a mixture of several chemicals, which may make it difficult to determine how the substance will react when it is released or if it encounters other chemicals. The illegal production of methamphetamine, for example, may produce a dangerous mixture of chemical wastes.

Regulations and Standards

Regulation government

To understand where you fit in when it comes to hazardous materials/WMD incidents, you must first recognize some of the regulatory drivers that apply to hazardous materials response, beginning with the difference between a regulation and a standard.

A **regulation** is a mandate that is issued and enforced by a governmental body. For example, the **Occupational Safety and Health Administration (OSHA)**, which is part of the U.S. Department of Labor, is the U.S. federal agency that regulates worker safety and, in some cases, responder safety. The **Environmental Protection Agency (EPA)** is the U.S. federal agency that ensures safe manufacturing, use, transportation, and disposal of hazardous substances.

Standards are "documents that contain mandatory . . . requirements . . . in a form . . . suitable for reference by another standard or code or for adoption into law" (NFPA 1). Standards are issued by nongovernmental entities and are generally consensus based. A standard may be voluntary, meaning that an agency such as a fire department is not required to adopt and follow the standard completely, or it could be mandatory, which would carry the weight of law. If an agency does adopt a voluntary standard, then it must meet all the requirements in the standard. For example, organizations such as the National Fire Protection Association (NFPA)—the association that develops and maintains nationally recognized, minimum, consensus

Standard non-gov

Voice of Experience

Over the last 30 years in which I have been involved in hazardous materials response, many changes have occurred. From the promulgation of the first regulations to the advent of the NFPA, it has been to the benefit of the first responder to understand how standards and regulations affect what we do.

One of the most interesting parts of the job is the exhilaration of a response, but *how* we respond is truly a result of what we have learned from the standards and regulations that we work with. These regulations and standards are outgrowths of previous laws that established the basis for hazardous materials/WMD response, such as the Resource Conservation and Recovery Act of 1976 (RCRA); the Comprehensive Environmental Response, Compensation, and Liability Act of 1980 (CERCLA); and the Superfund Amendments and Reauthorization Act of 1986 (SARA).

Several years ago, while I was responding to an incident, I found that knowing the rules helped to ensure we played by those rules. While working at an incident involving the spill of a nasty product, the state environmental and labor agencies began asking us questions about appropriate transportation, appropriate protection, clean-up, and disposal of the hazard. All of that information was listed on the safety data sheet in the section that described spill, waste, and disposal considerations for the material. This information was available because of the RCRA, CERCLA, and SARA regulations. We had followed all the rules and knew their importance to the responder. As responders, we must understand the *why* of what we do, and ensuring familiarity with knowledge of the standards and regulations is a good way to protect ourselves throughout a hazardous materials/WMD incident.

Glen Rudner

Hazardous Materials Compliance Officer,
Alabama Division
Norfolk Southern Railroad
Birmingham, Alabama

standards covering many areas of fire safety and hazardous materials—issue voluntary, consensus-based standards that any member of the public can comment on before committee members agree to adopt them. The technical committee responsible for periodically revising any NFPA standard meets regularly to revise, update, and possibly change the standard, reviewing and acting on public comments during the revision process. Once a standard is finalized, agencies may choose to adopt it.

OSHA regulations require minimum standards for federal and private-sector workers. However, OSHA cannot require states to apply these regulations to state and local government workers. Some states create their own state plan, which must at least meet the OSHA requirements and be approved by OSHA. Some state plans extend the protection to state and local government employees, and some do not. States without a state plan must, in addition to meeting OSHA regulations, meet the regulations for hazardous waste operations described in Title 40, "Protection of the Environment," of the Code of Federal Regulations, Part 311, "Worker Protection," usually referred to as *40 CFR 311*. The **Code of Federal Regulations (CFR)** is a collection of permanent rules published in the *Federal Register* that represent broad areas of interest governed by federal regulation; these rules are organized into 50 titles, which are updated annually. States with a state plan must include regulations for hazardous waste operations that meet the minimum regulations described in 40 CFR 311.

NFPA standards governing hazardous materials/WMD response come from the NFPA's Technical Committee on Hazardous Materials Response Personnel. This committee includes more than 30 members from private industry, the fire service, law enforcement, professional organizations, and governmental agencies. In 2021, this group consolidated three published standards, especially important to personnel who may be called upon to respond to hazardous materials/WMD incidents, into one new master standard: NFPA 470, *Hazardous Materials/Weapons of Mass Destruction (WMD) Standard for Responders*. Beginning in 2022, NFPA 470 integrated and replaced

NFPA 472, *Standard for Competence of Responders to Hazardous Materials/Weapons of Mass Destruction Incidents*; NFPA 1072, *Standard for Hazardous Materials/Weapons of Mass Destruction Emergency Response Personnel Professional Qualifications*; and NFPA 473, *Standard for Competencies for EMS Personnel Responding to Hazardous Materials/Weapons of Mass Destruction Incidents.*

Responders should understand the relationship between OSHA regulations and NFPA standards when it comes to hazardous materials/WMD response. OSHA regulations are the law that governs hazardous materials/WMD responders; NFPA standards are guidelines that agencies can choose to adopt. NFPA 470 is clear that responders shall receive any additional training to meet applicable DOT, EPA, OSHA, and other state, local, or provincial occupational health and safety regulatory requirements.

Levels of Training

Generally speaking, the former document NFPA 472 outlined training competencies for all hazardous materials responders, while the former NFPA 1072 identified their minimum professional qualifications, referred to as job performance requirements (JPRs). JPRs are written statements that describe specific job tasks, list the items necessary to complete those tasks, and define measurable or observable outcomes and evaluation areas for the tasks. JPRs are based on the requisite knowledge and requisite skills described in the standard.

The NFPA's latest revisions leave the intent and linkage between competency and JPRs unchanged. That is, they have simply been relocated to the new consolidated document, NFPA 470. In total, NFPA 470 contains 48 chapters with revised and updated content throughout, including the following, which are relevant to this text:

- Competencies for hazardous materials/WMD awareness-level personnel are found in NFPA 470, chapter 4.
- JPRs for hazardous materials/WMD awareness-level personnel (formerly found in NFPA 1072) are now found in NFPA 470, chapter 5.
- Competencies for hazardous materials/WMD operations-level responders are found in NFPA 470, chapter 6.
- JPRs (formerly found in NFPA 1072) for operations-level responders are now found in NFPA 470, chapter 7.
- Competencies for operations-level responders assigned mission-specific responsibilities are found in NFPA 470, chapter 8.
- JPRs for operations-level responders assigned mission-specific responsibilities (formerly NFPA 1072) are now found in NFPA 470, chapter 9.

There are many other chapters in NFPA 470 with content specific to hazardous materials technicians, and mission-specific competencies for technicians, incident commanders, safety officers, and other specialist employees. NFPA 475, *Recommended Practice for Organizing, Managing, and Sustaining a Hazardous Materials/Weapons of Mass Destruction Response Program*, remains a stand-alone document and is not contained in NFPA 470. The intent of NFPA 475 is to establish common criteria for the organization, management, programmatic elements, deployment of personnel, and resources for those entities responsible for the hazardous materials/WMD emergency preparedness function.

In the United States, NFPA standards as well as EPA and OSHA regulations are important to those personnel called upon to respond to hazardous materials/WMD incidents. The OSHA regulation containing the hazardous materials response competencies is commonly referred to as **HAZWOPER (HAZardous Waste OPerations and Emergency Response)**. The complete HAZWOPER regulation can be found in CFR, Title 29, 1910.120, and specifics for emergency response are located in subsection *q*. The training levels found in HAZWOPER are similar to the training levels found in the NFPA 470 standard and are identified as awareness, operations, technician, specialist, and incident commander. (*Specialist* is recognized only in the OSHA HAZWOPER regulation.) The content in NFPA 470 is scheduled to be updated every 5 years, the same as NFPA 472. This is a much more frequent revision cycle than that applied to the OSHA HAZWOPER regulation. This rapid revision cycle may be the source of some differences in the definitions of the levels of training presented here.

The following descriptions provide a broad overview, as found in NFPA 470 and the OSHA HAZWOPER regulation, of the different levels of hazardous materials/WMD responders and their training competencies. When reading NFPA 470, it is important to understand that the standard is organized to first spell out the *tasks* that a responder (awareness, operations, technician, incident commander) may be called upon to perform at the scene. The subsequent *training competencies* follow. To be clear, NFPA 470 and the HAZWOPER regulations are not "how to respond" documents. They provide no

Tactics achieve strategy

direction on how to plug leaking containers, use detection and monitoring devices, or decide which level of protection to wear in a certain situation. Instead, they are intended to provide guidance on the competencies associated with the various training levels.

The professional qualifications, as mentioned earlier, are intended to identify the JPRs that personnel should be able to perform to carry out their assigned job duties relative to their level of training. JPRs are to be accomplished in accordance with the requirements of the AHJ. Personnel at the awareness level must meet the JPR requirements defined in NFPA 470, chapter 5. Operations-level responders must also meet the JPR requirements defined in chapter 7 of NFPA 470. The mission-specific competencies for operations-level responders are found in NFPA 470, chapter 9.

Responders employ a **strategy**—that is, a "general course of action, direction, or plan to accomplish incident objectives" (NFPA 470). The two main operational strategies of awareness-level personnel are nonintervention mode or defensive mode. **Nonintervention mode** is a strategic decision whereby responders do not operate near the hazardous materials container. Instead, their efforts focus on public protection actions only, and they allow the container or product to take its natural course. The nonintervention route is taken when the adverse risk to emergency responders is greater than the benefit of taking action. **Defensive mode** is the mode of operation in which direct contact with the material or container is avoided and efforts are focused on controlling or limiting the effects of the release. Responders operating in defensive mode subject themselves to medium risk.

An **operations-level responder**, unlike awareness-level personnel, may choose to operate in an offensive mode. This occurs when responders have direct contact with the material or container and take aggressive action to control the release. **Offensive mode** carries a higher risk to the responders than they typically encounter. *Offensive mode is beyond the scope of awareness-level responders.*

Once a strategy and an operational mode are chosen, responders decide on the **tactics**—"specific actions or tasks taken to achieve strategies"—to use (NFPA 470).

To stay focused on the intent of this text—that is, to provide awareness-level personnel and operations-level responders with appropriate content consistent with the current edition of NFPA 470—we will discuss in detail only the competencies, tasks, and JPRs that awareness-level personnel and operations-level responders are expected to perform. A general overview of the training competencies for the other response levels—technician, specialist, and incident commander—is provided, but is not intended to be definitive or comprehensive.

Awareness-Level Personnel

Awareness-level personnel are those persons "who, during the course of their normal duties, could encounter an emergency involving hazardous materials/WMD and who are expected to recognize the presence of the hazardous materials/WMD, protect themselves, call for trained personnel, and secure the scene" (NFPA 470). Per NFPA 470, a person with awareness-level training is not considered to be a *responder*. Instead, these individuals are referred to as awareness-level *personnel*. Persons receiving this level of training are *not* typically called to the scene to respond; rather, awareness-level personnel, such as public works employees or fixed-facility security personnel, function in support roles.

Tasks that awareness-level personnel may be expected to perform on the scene include the following duties:

- Analyze the incident to detect the presence of hazardous materials/WMD.
- Use the *ERG* to identify the name; United Nations/North American Hazardous Materials Code (UN/NA) identification number; type of marking, label, or placard; and container shape of the hazardous material/WMD involved.
- Identify potential hazards from the current edition of the *ERG*, the safety data sheet (SDS), shipping papers, or other approved reference sources.
- Initiate and implement protective actions consistent with the AHJ emergency response plan, SOPs, SDS, and the current edition of the *ERG*.
- Initiate the notification process and communicate as necessary.

The items in this list are all described in Chapter 30, *Recognizing and Identifying the Hazards*, in this section. As with the awareness level, the operations-level JPRs are based on requisite knowledge and requisite skills.

OSHA's HAZWOPER view of the awareness level is somewhat different than that defined in NFPA 470, mostly because it designates awareness-level personnel as "responders." Per OSHA (1910.120[q] [6][i]), "first responders at the awareness level are individuals who are likely to witness or discover a hazardous substance release and who have been trained to initiate a response sequence by notifying the proper authorities of the release.

They would take no further action beyond notifying the authorities of the release." Based on the OSHA HAZWOPER regulation, first responders at the awareness level should have sufficient training or experience to objectively demonstrate competency in the following areas:

- An understanding of what hazardous substances are and the risks associated with them
- An understanding of the potential outcomes of an incident
- The ability to recognize the presence of hazardous substances
- The ability to identify the hazardous substances, if possible
- An understanding of the role of the first responder awareness individual in the response plan
- The ability to determine the need for additional resources and to notify the communication center

Operations Level

Per NFPA 470, operations-level responders are those persons who are tasked to respond "to hazardous materials/WMD incidents for the purpose of taking action "to protect nearby persons, the environment, or property from the effects of the release" (NFPA 470). These persons may also have competencies that are specific to their response mission, expected tasks, and equipment and training as determined by the AHJ, that is, the mission-specific competencies listed earlier in the chapter.

The phrase "tasked to respond" may raise a question about your role as a responder to incidents involving a hazardous material/WMD. To clarify this point, consider this scenario: If your local response system—professional or volunteer—is activated (e.g., if a 911 call is made), and your agency responds to the scene to provide on-scene emergency service at a hazardous materials/WMD incident, you are viewed as a responder. Personnel termed responders may include fire and rescue assets, law enforcement personnel, EMS workers (both BLS and ALS providers), experts or other employees from private industry, and other allied professionals.

NFPA 470 continues to expand the scope of an operations-level responder by separating the operations-level suite of competencies into two distinct categories: core competencies and mission-specific competencies. The core competencies are based on those vital tasks that operations-level personnel should perform on the scene of a hazardous materials/WMD incident. Some of those tasks are listed here:

- Analyze the scene of a hazardous materials/WMD incident to determine the scope of the incident.
- Survey the scene to identify containers and materials involved.
- Collect information from available reference sources.
- Predict the likely behavior of a hazardous material.
- Estimate the potential harm the substances might cause.
- Plan a response to the release, including selection of the correct level of personal protective clothing.
- Perform decontamination.
- Preserve evidence.
- Evaluate the status and effectiveness of the response.

This abbreviated list serves as only an illustration of the core competencies. Keep in mind that the core training competencies are designed to describe the skills, knowledge, and abilities to safely accomplish the tasks listed. Core competency training is required of all operations-level responders on the scene, no matter what their function. One of the goals of the NFPA 470 standard is to better match the expected tasks that may be required of the responder with the training that the responder should receive.

In addition to undertaking core competency training, an individual AHJ may find the need to do more training based on an identified or anticipated mission-specific need. To that end, those responders who are expected to perform additional missions, beyond the core competencies, must be trained to carry out those mission-specific responsibilities. These mission-specific competencies are *nonmandatory* and should be viewed as optional. NFPA 470 is designed to allow each AHJ to pick and choose the training program that makes the most sense for its jurisdiction. To that end, operations-level responders may end up performing a limited suite of technician-level skills, but do not have the broader knowledge and abilities of a hazardous materials technician.

NFPA 470 provides a mechanism to ensure that operations-level responders, including those with mission-specific training, do not go beyond their level of training and equipment. This is done by using technician-level personnel to provide direct guidance to the operations-level responder operating on a hazardous materials/WMD incident. Operations-level responders are expected to work under the direct control of a **hazardous materials technician** or allied professional who can continuously assess and/or observe the actions of the operations-level responder *and* provide immediate feedback. A hazardous materials technician or an allied professional may provide guidance through direct visual observation or through assessments communicated by the operations-level responder(s).

The goal of mission-specific competencies is to provide operations mission-specific responders with the knowledge and skills needed to perform the assigned responsibilities in a safe and effective manner. Examples of those specialty areas include those listed here (NFPA 470):

- Personal protective equipment (PPE)
- Mass decontamination
- Technical decontamination
- Evidence preservation and public safety sampling
- Product control
- Detection, monitoring, and sampling
- Victim rescue and recovery
- Illicit laboratory incidents
- Radiological hazard–specific tasks
- Disablement/disruption of improvised explosive devices (IEDs), improvised WMD dispersal devices, and operations at improvised explosives laboratories
- Diving in contaminated water environments
- Evidence collection

This text discusses all of these mission-specific competencies except for disablement/disruption of IEDs, improvised WMD dispersal devices, and operations at improvised explosives laboratories, and diving in contaminated water environments.

Per the OSHA HAZWOPER regulation, first responders at the operations level are individuals who respond to releases or potential releases of hazardous materials incidents as part of their normal duties for the purpose of protecting nearby persons, property, or the environment from the effects of the release (**FIGURE 29-3**). The OSHA HAZWOPER regulation mandates that the operations-level responders must be trained to respond in a defensive fashion, without trying to stop the release directly, while avoiding contact with the released substance. Their function is to contain the release from a safe distance, keep it from spreading, and prevent or reduce the potential for human exposures. First responders at the operations level are expected to have received at least 8 hours of training or to have sufficient experience to objectively demonstrate competency in the following areas, in addition to those listed for the awareness level:

- Knowledge of the basic hazard and risk assessment techniques
- Knowledge of how to select and use the proper PPE

FIGURE 29-3 Law enforcement and fire department operations-level responders at a joint task force training exercise.

- An understanding of basic hazardous materials terms
- Knowledge of how to perform basic control, containment, and/or confinement operations within the capabilities of the resources and PPE available with their unit
- Knowledge of how to implement basic decontamination procedures
- An understanding of the relevant SOPs and termination procedures

Technician/Specialist Level

Hazardous materials technicians are those persons who respond "to hazardous materials/WMD incidents using a risk-based response process to analyze a problem involving hazardous materials/WMD, plan a response to the problem, implement the planned response, evaluate progress of the planned response and adjust accordingly, and assist in terminating the incident" (NFPA 470).

These persons may have additional competencies that are specific to their response mission, expected tasks, and equipment and training as determined by the AHJ. A number of very detailed training competencies and JPRs for technician-level training are outlined in chapters 10 and 11 of NFPA 470. Technician-level personnel are integral to the NFPA 470 standard because they are, in many cases, intended to "supervise" the activities of on-scene operations-level responders.

Per the OSHA HAZWOPER regulation, individuals at the technician level respond to hazardous material releases or potential releases for the purpose of

FIGURE 29-4 Hazardous materials technician during a product control training exercise.

stopping the release (**FIGURE 29-4**). Conceptually, this is similar in intent to NFPA 470, in that a technician will likely assume a more proactive role than a responder at the operations level. Technicians typically function at a higher level in terms of their cognitive approach to the response. They should be proficient at implementing a comprehensive, risk-based approach to solving the problem. They will approach the point of release to plug, patch, or otherwise mitigate the problem.

Per the OSHA HAZWOPER regulation, hazardous materials technicians should have received at least 24 hours of training equal to the first responder operations level. In addition, technicians should have the competency, knowledge, and understanding necessary to fulfill the following duties:

- Implement the employer's emergency response plan.
- Classify, identify, and verify known and unknown materials by using field survey instruments and equipment.
- Function within an assigned role in the ICS.
- Select and use proper specialized chemical PPE.
- Understand hazard and risk assessment techniques.
- Perform advanced control, containment, and/or confinement operations within the capabilities of the resources and PPE available with the unit.
- Understand and implement decontamination procedures.
- Understand termination procedures.
- Understand basic chemical and toxicological terminology and behavior.

The **hazardous materials specialist** is identified only in the OSHA HAZWOPER standard. "Hazardous materials specialists are individuals who respond with and provide support to hazardous materials technicians" (OSHA 1970). They act as the incident-site liaison with federal, state, local, and other government authorities. This level of responder receives more specialized training than a hazardous materials technician. This individual's "duties parallel those of the hazardous materials technician, however, [the specialist's] duties require a more directed or specific knowledge of the various substances they may be called upon to contain" (OSHA 1970). Practically speaking, however, the two levels are not dramatically different.

Incident Commander

The **incident commander (IC)** is the person "responsible for all incident activities, including the development of strategies and tactics and the ordering and the release of resources" (NFPA 470). The IC must receive any additional training necessary to meet applicable governmental occupational health and safety regulations and the specific needs of the jurisdiction.

The OSHA HAZWOPER regulation requires specific IC hazardous materials–level training for those persons assuming command of a hazardous materials incident requiring action beyond the operations level. Individuals trained as ICs should have at least operations-level training as well as additional training specific to commanding a hazardous materials incident. ICs, who will assume control of the incident scene beyond the first responder awareness level, must receive at least 24 hours of training equal to the first responder operations level and have competency in the following areas:

- Know and implement the employer's ICS.
- Know how to implement the employer's emergency response plan.
- Know and understand the hazards and risks associated with chemical protective clothing.
- Know how to implement the local emergency response plan.
- Know how to initiate the state emergency response plan and the Federal Regional Response Team.
- Know and understand the importance of decontamination procedures.

For more complete information regarding hazardous materials/WMD ICs, refer to chapters 12 and 13 of NFPA 470.

In addition to the initial training requirements for all response levels listed earlier, OSHA regulations require annual refresher training of sufficient content and duration to ensure that responders maintain their competencies or that they demonstrate competency in those areas at least yearly. Consult your local agency for more specific information on refresher training, other hazardous materials laws, regulations, and regulatory agencies.

Other Governmental Agencies

In addition to OSHA and the EPA, several other governmental agencies are concerned with various aspects of hazardous materials/WMD response. The DOT, for example, promulgates and publishes laws and regulations that govern the transportation of goods by highway, rail, pipeline, air, and, in some cases, marine transport.

In addition to creating standardized training for hazardous materials response and hazardous waste site operations, the **Superfund Amendments and Reauthorization Act of 1986 (SARA)** created a method and standard practice for a local community to understand and be aware of the chemical hazards in that community. Under SARA Title III, the **Emergency Planning and Community Right-to-Know Act (EPCRA)** requires a business that handles certain types and amounts of chemicals to report the storage type, quantity, and storage methods to the fire department and the local emergency planning committee. This activity may be quite complex and should be undertaken only by those personnel who fully understand the requirements and the process.

The **local emergency planning committee (LEPC)** gathers and disseminates information about hazardous materials to the public. These voluntary organizations are made up of members of industry, transportation, media, fire and law enforcement agencies, and the public at large; they are established to meet the requirements of EPCRA and SARA. Essentially, LEPCs ensure that local resources are adequate to respond to a chemical event in the community. Responders should be familiar with their local LEPC and know how their department works with this committee.

In addition, each state has a **State Emergency Response Commission (SERC)**. The SERC acts as the liaison between local and state levels of authority. Its membership includes representatives from agencies such as the fire service, law enforcement services, and elected officials. The SERC is charged with the collection and dissemination of information relating to hazardous materials emergencies.

Preplanning

It is a mistake to assume that a response to a hazardous materials/WMD incident begins when the alarm sounds. In reality, the response begins with initial training, continuing education, and preplanning activities at target hazards and other potential problem areas throughout the jurisdiction or response district (**FIGURE 29-5**). A **target hazard** includes any occupancy type or facility that presents a high potential for loss of life or serious impact to the community resulting from a fire, explosion, or chemical release.

Preplanning activities enable agencies to develop logical and appropriate response procedures for anticipated incidents. Planning should focus on the real threats that exist in your community or adjacent communities where you could be assisting. Preplanning at major target hazards should include discussions and information sharing with the LEPC.

Once the threats have been identified, fire departments, law enforcement agencies, public health offices, and other governmental agencies should determine how they will respond and work together in case of a large-scale incident. In many cases, the move toward interoperability before an incident will make the actual response work run more smoothly. Remember this important point: People making good decisions solve problems effectively. When people are acquainted with each other before the event happens, they tend to work together better. Get to know your peers at whatever level you operate.

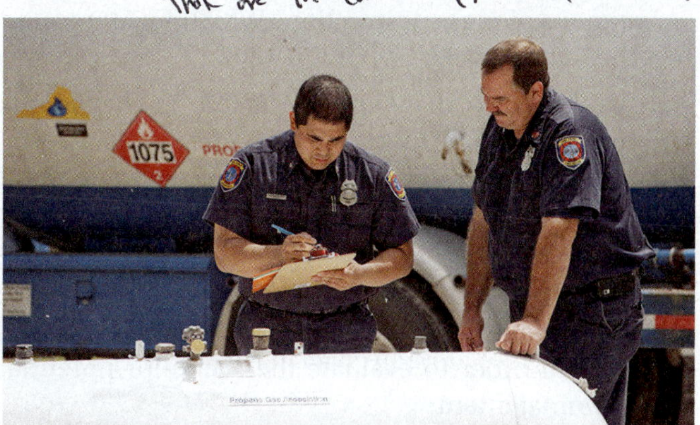

FIGURE 29-5 Conduct preincident planning activities at target hazards throughout the jurisdiction—they could pay off during an actual incident!

CASE STUDY

You Are the Firefighter CONCLUSION

At 1330 hours, your engine company responds to a report of a suspicious odor at a small plastics manufacturing company. When you arrive, you see a cargo delivery truck parked at the loading dock, with the motor still idling. From your vantage point, a few hundred feet away, you notice a liquid leaking from the back of the truck. On one side of the truck, you can see a black-and-white, diamond-shaped placard that reads "CORROSIVE," with the number 8 at the bottom. The facility security guard reports that the truck has been left unattended for at least an hour.

1. **Is this an incident? How would you go about analyzing the scene to better understand the problem? What other pieces of information would you want?**

   Answer: Based on the initial observations, it is reasonable to classify this as an incident. You could make further attempts to locate the driver and note any phone numbers or business names on the vehicle or trailer. Using the *ERG* to correlate the corrosive placard to an action guide will offer additional tactical guidance.

2. **You are trained to the awareness level. Would this level of training allow you to put on chemical protective equipment, enter the hazardous area, attempt to determine the nature of the leak, and fix the problem?**

   Answer: No, awareness-level personal are not considered to be responders per NFPA 470. Actions taken at this level are focused on remaining at a safe distance from the release, securing the scene, and notifying other, more highly trained responders per the policies and procedures of the AHJ.

3. **What reference sources would you want to access for more information on the placard?**

   Answer: The *ERG* is a good source of information in the early stages of an incident. The *ERG* can provide more information based on the placard or label, as well as providing guidance on identifying the type of container profile and offering some suggested initial actions.

WRAP-UP

SUMMARY

KNOWLEDGE OBJECTIVES

- Identify the difference between hazardous materials/weapons of mass destruction (WMD) incidents and other emergencies. (pp. 1201–1202)
  - Understand that actions taken at hazardous materials/WMD incidents are largely dictated by the chemicals or hazards involved; environmental influences such as wind, rain, and temperature; and the way the chemicals behave during the release.
  - Understand the policies and procedures of the AHJ in order to evaluate the scene for potential criminal intent.
- Define the terms *hazardous materials* (or *dangerous goods*, in Canada) and *weapons of mass destruction.* (**NFPA 470: 5.2.1**, pp. 1201–1202)
  - Understand the differences between a hazardous material, hazardous waste, and weapon of mass destruction.
- Identify the location of both the emergency response plan and/or standard operating procedures. (**NFPA 470: 5.1.3**, pp. 1202–1204)
- Understand the difference(s) between the federal regulations and standards and that govern hazardous material response activities. (pp. 1202–1204)
  - Understand the role of the AHJ, OSHA, EPA and the DOT as these agencies relate to hazardous materials response and planning.
  - Understand the roles of LEPC and SERC in hazardous materials response and planning.
- Describe the different levels of hazardous materials training: awareness, operations, technician, specialist, and incident commander. (**NFPA 470: 5.1.1**, **5.1.2**, **5.1.3**, pp. 1204–1209)
  - Identify the specific differences between the OSHA HAZWOPER law and NFPA 470 as they relate to the various levels of training.

- Explain the need for a planned response to a hazardous materials incident. (p. 1209)
  - Understand the role of the AHJ in outlining the goals and objectives of hazardous materials response.
  - Understand the various roles and responsibilities between responders and law enforcement at the scene of a hazardous materials incident.
  - Understand the importance of preplanning and identifying target hazards that pose unique or extraordinary hazards to responders and the public.
  - Understand that fire departments, law enforcement agencies, public health offices, and other governmental agencies should determine how they will respond and work together in case of a large-scale incident.

SKILLS OBJECTIVES

There are no skills objectives for awareness-level personnel for this chapter.

KEY TERMS

awareness-level personnel Personnel who, during the course of their normal duties, could encounter an emergency involving hazardous materials/weapons of mass destruction (WMD) and who are expected to recognize the presence of the hazardous materials/WMD, protect themselves, call for trained personnel, and secure the scene. (NFPA 470)

Code of Federal Regulations (CFR) A collection of permanent rules published in the *Federal Register* that represent broad areas of interest governed by federal regulation and that are organized into 50 titles, which are updated annually.

defensive mode The mode of operation in which direct contact with the material or container is avoided and responders' efforts focus on controlling or limiting the effects of the release.

Department of Transportation (DOT) The U.S. government agency that publicizes and enforces rules and regulations that relate to the transportation of many hazardous materials.

Emergency Planning and Community Right-to-Know Act (EPCRA) Legislation that requires a business that handles chemicals to report on those chemicals' type, quantity, and storage methods to the fire department and the local emergency planning committee.

Environmental Protection Agency (EPA) The U.S. federal agency that ensures safe manufacturing, use, transportation, and disposal of hazardous substances.

hazardous material Matter (solid, liquid, or gas) or energy that when released is capable of creating harm to people, the environment, and property, including weapons of mass destruction (WMD) as defined in 18 US Code, Section 2332a, as well as any other criminal use of hazardous materials, such as illicit labs, environmental crimes, or industrial sabotage. (NFPA 470)

hazardous materials specialist Defined in the OSHA HAZWOPER regulation only, a hazardous materials specialist responds with, and provides support to, hazardous materials technicians and acts as the incident-site liaison with federal, state, local, and other government authorities regarding site activities.

hazardous materials technician A person who responds to hazardous materials/weapons of mass destruction (WMD) incidents using a risk-based response process to analyze a problem involving hazardous materials/WMD, plan a response to the problem, implement the planned response, evaluate progress of the planned response and adjust accordingly, and assist in terminating the incident. (NFPA 470)

hazardous waste A substance that remains after a process or manufacturing plant has used some of the material and the substance is no longer pure.

HAZWOPER (HAZardous Waste OPerations and Emergency Response) The federal OSHA regulation that governs hazardous materials waste site and response training (specifics to emergency response can be found in 29 CFR 1910.120[q]).

incident commander (IC) The individual responsible for all incident activities, including the development of strategies and tactics and the ordering and the release of resources. (NFPA 470)

local emergency planning committee (LEPC) A committee comprising members of industry, transportation, the public at large, media, and fire and law enforcement agencies, which gathers and disseminates

KEY TERMS CONTINUED

information on hazardous materials stored in the community and ensures that there are adequate local resources to respond to a chemical event in the community.

nonintervention mode The operating mode in which responders do not operate near the hazardous materials container and focus their efforts on public protection actions only, allowing the container or product to take its natural course.

Occupational Safety and Health Administration (OSHA) Part of the Department of Labor, the U.S. federal agency that regulates worker safety and, in some cases, responder safety.

offensive mode The mode of operation in which responders have direct contact with the material or container and take aggressive action to control the release.

operations-level responder A person who responds to hazardous materials/weapons of mass destruction (WMD) incidents for the purpose of implementing or supporting actions to protect nearby persons, the environment, or property from the effects of the release. (NFPA 470)

regulation A mandate issued and enforced by a governmental body, such as the Occupational Safety and Health Administration (OSHA) or the Environmental Protection Agency (EPA).

State Emergency Response Commission (SERC) A group that acts as a liaison between the local and state levels by collecting and disseminating information relating to hazardous materials emergencies. SERC includes representatives from agencies such as the fire service, law enforcement services, and elected officials.

strategy The general course of action, direction, or plan to accomplish incident objectives. (NFPA 470)

Superfund Amendments and Reauthorization Act of 1986 (SARA) One of the first U.S. laws to affect how fire departments respond in a hazardous materials emergency.

tactics Specific actions or tasks taken to achieve strategies involving the deployment and directing of resources on an incident. (NFPA 470)

target hazard Any occupancy type or facility that presents a high potential for loss of life or serious impact to the community resulting from a fire, explosion, or chemical release.

weapon of mass destruction (WMD) Any weapon or material that is designed to cause death or serious injury or damage to buildings, structures, or the environment, such as an explosive or incendiary bomb, rocket, or grenade containing or delivering a toxic or dangerous chemical, biological agent, toxin, or vectors; or a weapon designed to release dangerous levels of radiation. (NFPA 470)

REVIEW QUESTIONS

1. Why should hazardous materials incidents be handled in a more deliberate fashion than structural firefighting?
2. What is a hazardous material according to the DOT?
3. What is the most common way of describing WMDs?
4. What is the difference between a regulation and a standard?
5. Do NFPA 470 and OSHA's HAZWOPER describe awareness-level personnel and operations-level responders in the same manner?
6. What are the roles of OSHA, EPA, LEPC, and SERC in hazardous materials response and planning?
7. How do preplanning activities enable agencies to develop logical and appropriate response procedures for anticipated incidents?

DISCUSSION QUESTIONS

1. What is meant by the phrase "tasked to respond" and how is that phrase impacted by the differences found in OSHA's definition of awareness-level training as opposed to the NFPA 470 perspective on awareness-level personnel?
2. Is it mandatory that an AHJ provide mission-specific training to all operations-level responders?
3. Why is preincident planning and identifying target hazards especially important in communities where target hazards include a potential for a hazardous waste release?

APPLYING THE CONCEPTS

Your crew is dispatched to a biotechnology research company for an odor investigation. Your initial information from the dispatch center reported an unusual odor in one of the laboratories. A security guard meets you at the street and relays the location of the spill of an unknown liquid. He reports that a scientist called the security desk to report finding a broken 1-gallon amber-colored glass container in laboratory #206. He tells you that the lab is evacuated and there are no injuries or exposures, and it is unknown how the container was broken or any other history of the event. You and your crew are trained to the hazardous materials awareness level.

1. Based on the preceding information, does your crew have the right level of training to enter the lab, identify the liquid, and clean up the spill?
2. What are your next steps?

Two scientists emerge from one of the laboratories in street clothes. They are familiar with the incident and report that the chemical is hydrochloric acid (HCl). They want the spill cleaned up as rapidly as possible because they have a time-sensitive solution in the lab. "We have all the equipment that you need to contain this spill here," they say. "Can't you just get in there and get it all cleaned up"?

3. You now have a chemical name for the spilled substance. Which of the reference sources would provide the most complete information in this scenario?
4. If the lab has all of the equipment required to contain the spill, should you attempt to do so? If yes, how would you do this? If no, what should you do next?

A lab technician approaches and has additional information. She states that the two scientists didn't know that the room where the spill occurred also contains sodium cyanide crystals. You consult your sources and learn that if the HCl and sodium cyanide combine, a highly toxic vapor that can be absorbed through the skin will be produced. It is unknown if the two substances have actually combined. You inform the impatient scientists that due to the potential health hazards, you need to bring in a regional response team of hazardous materials technicians. It will take approximately 1 hour for the team to arrive. In the meantime, the plan is to isolate the laboratory and deny entry to all employees.

5. How do you reconcile the two different sources of information from the scientists and the lab technician?
6. Would you consider making a quick entry in full structural firefighter's gear and SCBA to take a look and see if the two substances have combined? Why or why not?

REFERENCES

National Fire Protection Association (NFPA). 2020. *NFPA 1, Fire Code.* 2021 Edition. Quincy, MA: NFPA.

National Fire Protection Association (NFPA). 2021. *NFPA 470, Hazardous Materials/Weapons of Mass Destruction (WMD) Standard for Responders.* 2022 Edition. Quincy, MA: NFPA.

Occupational Safety and Health Administration (OSHA). 1970. "1910.120 - Hazardous Waste Operations and Emergency Response." Accessed September 15, 2021. https://www.osha.gov/laws-regs/regulations/standardnumber/1910/1910.120.

"Worker Protection." Code of Federal Regulations, title 40, part 311 (2020): 393–394.

CHAPTER

30

Hazardous Materials Awareness Level

Recognizing and Identifying the Hazards

NOTE: Content within this chapter meets the intent of **NFPA 470, 2022 Edition**, which includes chapter 5, "Professional Qualifications for Hazardous Materials/WMD Awareness Level Personnel (**NFPA 1072**)."

KNOWLEDGE OBJECTIVES

After studying this chapter, you will be able to:

- Describe how to approach a scene size-up when potentially hazardous materials are involved.
- Identify and describe the types of containers that are often used to contain hazardous materials.
- Describe the purpose and types of various transportation and facility markings for hazardous materials.
- Identify and describe the four routes of entry that harmful substances can take in the human body.

SKILLS OBJECTIVES

After studying this chapter, you will be able to:

- Use the *Emergency Response Guidebook* (*ERG*).

CASE STUDY

You Are the Firefighter

During an odor investigation in a light-industrial area of your district, you are directed by a citizen to an abandoned open-head 55-gallon steel drum just off the city street. The person points you to a local shopkeeper, who made the 911 call. The shopkeeper reports that the drum must have been illegally dropped off overnight, and that he smelled an unusual odor when he walked by it. You notice a green label on the side of the drum reading "nonhazardous waste." You and the other members of the engine company are trained to the hazardous materials awareness level (based on NFPA 470). Even though you are not tasked to respond to a hazardous materials incident, you now find yourself on the scene of a potential hazardous materials incident.

1. What initial actions would you take based on your level of training?
2. What does the label on the drum signify?
3. How would you go about obtaining more information on the potential contents and potential hazards of the contents of the drum? What notifications would you make?

Introduction

More than 4 billion tons of hazardous materials are shipped annually in the United States by rail, highway, sea, and air. Per the **Chemical Abstracts Service (CAS)**, which produces the largest databases on chemical information, including the CAS Registry (www.cas.org), more than 182 million organic and inorganic substances are registered for use in commerce in the United States, and several thousand new substances are introduced each year. The bulk of the new chemical substances are industrial chemicals, household cleaners, and lawn care products. Given the myriad substances available, responding to almost any situation could end up including a hazardous material or potentially hazardous material. Thus, even though awareness-level personnel are not considered to be responders per NFPA 470, it is possible to find yourself in the midst of a hazardous materials incident. To that end, it is critical to be familiar with the emergency response plans and other standard operating procedures (SOPs) developed by the authority having jurisdiction (AHJ).

Chemicals are used and/or stored in warehouses, hospitals, laboratories, industrial occupancies, residential garages, bowling alleys, home improvement centers, garden supply stores, restaurants, and scores of other facilities or businesses in your response area. So many different chemicals exist in so many different locations that you could encounter almost anything during any type of incident.

Identifying the kinds and quantities of hazardous materials used and stored by local facilities should be an integral part of any comprehensive community response plan. Additionally, the AHJ should develop emergency response plans and other methods to identify high-hazard occupancies and provide personnel with guidance on response procedures and other SOPs to handle hazardous materials incidents at those locations. Also, responding to a false alarm or some other type of incident can be a good opportunity to walk through a location for the purpose of preplanning a future response.

Basic Scene Size-Up

When you develop an emergency response, you should use a **risk-based response process**, which is a "systematic process based on the facts, science, and circumstances of the incident, by which responders analyze a problem involving hazardous materials/weapons of mass destruction (WMD) to assess hazards and consequences, develop an incident action plan (IAP), and evaluate the effectiveness of the plan" (NFPA 470). An **incident action plan (IAP)** is "an oral or written plan approved by the incident commander (IC) containing the incident objectives reflecting the overall strategy for managing the incident" (NFPA 470). This plan must be carried out as safely as possible, should not involve contacting the released substance in any way, and above all must be well thought out. As awareness-level personnel, the IAP should be in the nonintervention or defensive mode.

At any hazardous materials incident, the first action should always be to approach the scene from a safe location and direction. The traditional rules of staying uphill and upwind are a good place to start. It may

FIGURE 30-1 Notice the details at the scene, and think about what they mean.

make sense to use binoculars and view the scene from a safe distance, looking for labels, placards, container shapes, and other clues that could help awareness-level personnel understand the scene. Be sure to question anyone involved in the incident—a wealth of information may be available to you if you simply ask the right person. Take enough time to assess the scene and interpret other clues, such as dead animals near the release, discolored pavement, dead grass, visible vapors or puddles, and any other indicators that may help identify the presence of a hazardous material. A wet area near unidentified containers on an asphalt parking lot may not initially seem significant, but perhaps it is a hot day, and the "wet" area still looks wet an hour after the containers were noticed—well past the point that water would have evaporated. Perhaps the substance is a hydrocarbon, and the wet-looking pavement is actually a solvent that has permeated the asphalt (**FIGURE 30-1**).

TIP

First-responding firefighters, law enforcement personnel, or representatives of other allied agencies should place a high priority on analyzing the scene from a safe distance.

In some cases, it may be possible to detect the presence of a hazardous materials incident based on information relayed in the initial dispatch, from persons on the scene, or based on your own knowledge of the response area. Clues that are seen or heard may also provide valuable information from a distance, enabling you to take precautionary steps. Vapor clouds at the scene, for example, are a signal to move yourself and others away to a place of safety; the sound of an alarm from a toxic gas sensor in a chemical storage room or laboratory may also serve as a warning to retreat. Some highly vaporous and odorous chemicals—chlorine and ammonia, for example—may be detected by smell a long way from the actual point of release. These and other clues may alert awareness-level personnel to the presence of a hazardous atmosphere.

In other cases, you may have to put on your detective hat and search for clues that may indicate whether a hazardous substance is present. At all times, departmental SOPs and your level of training, along with your information-gathering efforts at the scene, should guide your initial and ongoing actions. Use your senses, but do so carefully to avoid becoming contaminated or exposed. The senses that are typically safe to employ on a regular basis are those of sight and sound. Generally, the farther you are from the incident when you notice a problem, the safer you will be. Using any of your senses that brings you close to the chemical should be done with caution or should be avoided. When it comes to hazardous materials incidents, "leading with your nose" is not a good tactic—but using binoculars from a distance is.

TIP

When you think about all of the places a hazardous materials incident could occur, do not limit your thinking. Explore your response district—you may be surprised by how many kinds of clues you find.

A thorough and thoughtful size-up will help clarify the problem you are facing and help you to decide whether taking an offensive action is within your scope of training. Once you have a basic idea of what happened and have determined whether danger may be present, you can begin to formulate a plan for addressing the incident. That plan begins with taking the basic actions known by the mnemonic SIN:

- Safety
- Isolate
- Notify

Safety

Scene size-up, though important in any incident, is especially important during hazardous materials incidents. The ability to read the scene is a critical skill, and

FIGURE 30-2 Personnel and responders make decisions based on the available clues.

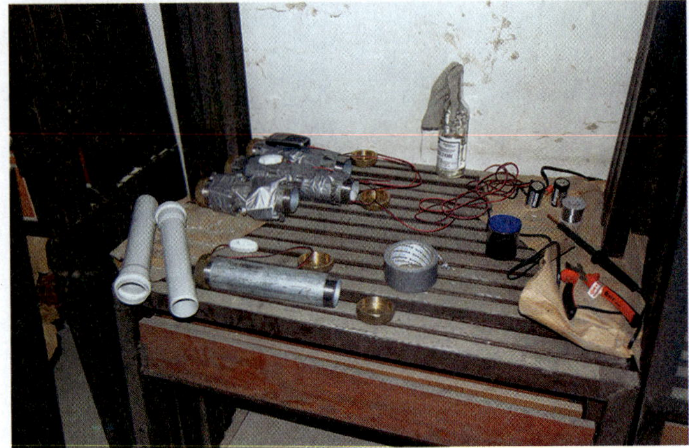

FIGURE 30-3 Pay attention. Situational awareness is key. With awareness-level training, what dangers do you see in the photo and what would you do if you encountered this situation?

personnel and responders must interpret the available clues and weave them together to make informed decisions and operate safely. It is not enough to simply scan the incident scene—you must train yourself to stop for a moment and pay attention to all of the clues that may be present (**FIGURE 30-2**).

SAFETY TIP

Pay attention to the scene and think before you act: It could save your life.

Looking at something is nothing more than pointing your eyes in the right direction; *seeing*, by contrast, entails taking in the visual clues and piecing them together to form a conclusion. This is the basis for situational awareness (**FIGURE 30-3**). The scenario described in the "You Are Firefighter" feature at the beginning of the chapter, for example, contains a few clues. An **open-head drum** (in which the lid is secured by a bolted clasp-type ring that circles the entire head of the drum) typically contains solid materials or types of substances that cannot be poured or pumped easily. Moreover, steel drums do not usually contain corrosives. The **label** also provides some preliminary information, although it may not be reflective of the substance(s) inside. Meanwhile, the shopkeeper mentioned detecting an odor, which the personnel should follow up and better understand. Each point does not tell the whole story, but collectively they provide a place from which to start your size-up.

If a chemical incident occurs at a fixed facility, attempt to locate key personnel early in the incident. Generally speaking, most fixed facilities that use and/or store a significant number of chemicals will have an environmental health and safety (EH&S) department. In many cases, EH&S departments employ certified industrial hygienists, chemists, chemical engineers, and/or certified safety professionals to ensure safe work practices are maintained at the site. These industry experts may be a valuable source of information for understanding the chemical inventory of the facility, the ventilation systems, high-hazard chemical storage areas, container profiles, and specialized areas such as clean rooms or refrigeration systems. Additionally, representatives of the EH&S department can connect personnel and responders with **safety data sheets (SDS)** ("formatted information, provided by chemical manufacturers and distributors of hazardous products, about chemical composition, physical and chemical properties, health and safety hazards, emergency response, and waste disposal of the material" [NFPA 470]), site security, maintenance workers, and other important employees at the facility.

It is important to find the right people—those who have the keys, unique site knowledge, and instructions on how to get around the facility. In many cases, facilities have written response plans with current contact information and/or on-duty representatives who can assist in the event of an emergency. Be prepared, however, to find outdated phone lists, personnel who are no longer in a particular position, inaccurate inventories, and other shortcomings that may hamper your ability to get timely, accurate, and actionable information. Also, you may be speaking with a person who is not familiar with the substance, the process, or the area where a release occurred, and who may not have all the answers you are seeking when trying to determine a course of action.

Although it is not possible to know each chemical by name, it is possible to identify many of the more common commodities through available identification systems, such as placards, labels, and other signage. Alternatively, other methods, such as detection devices, eyewitness accounts, visible indicators, and container profiles, may be used to identify the presence of a hazardous material and gain an understanding of the potential hazards. Knowing how to access the proper source of information will help awareness-level personnel interpret visual clues that may signal the possible presence of a hazardous materials incident. You must train yourself to take the time to look at the whole scene so that you can identify the available critical indicators and fit them together with what is known about the problem.

The following steps are suggested initial actions for awareness-level personnel. Your AHJ may also provide more specific and detailed procedures in a variety of response plans.

- Stay upwind, uphill, and out of the problem.
- Obtain a briefing from those involved in the incident prior to acting. These individuals may include bystanders, law enforcement personnel, emergency medical services (EMS) responders, facility representatives, and other responders.
- From a safe distance, attempt to make a positive identification of the released substance. If possible, obtain the correct SDS and shipping papers, and identify markings or labels on the container or transport vehicle. Also consult the ***Emergency Response Guidebook (ERG)*** or some other suitable reference source. The *ERG* is a reference book developed by the U.S. Department of Transportation (DOT), Transport Canada, and the Secretariat of Transport and Communications, Mexico. It is "written in plain language, to guide emergency [personnel and] responders in their initial actions at an incident scene" (NFPA 470). It is distributed free of charge to every public response vehicle in the United States. Consulting the *ERG* may allow you to identify potential hazards of the substance involved.

Isolate

After ensuring your safety, the next step is to isolate and deny entry to the scene. The first operational priority is to separate the people from the problem—life safety is always the first consideration. This is typically accomplished by establishing hazard control zones, areas at hazardous materials/WMD incidents within an established perimeter that are designated based on safety and the degree of hazard. **Hazard control zones** allow on-scene personnel to quickly identify the area of highest contamination and prevent accidental entry by untrained or unprotected public safety personnel or civilians.

If people are exposed to or threatened by the release, do not move past this step until you have addressed the life safety issues. This could include evacuating affected people from the environment; having potential victims shelter in place until a transient problem, such as a fast-moving vapor cloud or other situation, passes; decontaminating the scene; and rendering medical care.

Isolation and denial of entry to hazard control zones may be accomplished by having law enforcement personnel provide a physical presence, using scene control identifiers such as barrier tape that reads "Danger" or "Caution," or creating physical barriers with fences or other materials.

With hazardous materials incidents, it is important to establish clear and visible command. Establish the command post in an area where you are protected from both the incident and the weather, and where you have access to communications and technical reference materials. The command post is typically staffed by responders trained beyond the awareness level.

Next, determine your response objectives and begin to formulate a basic IAP. This plan must be carried out as safely as possible, should not involve contacting the released substance in any way, and—above all—must be well thought out. Remember, for awareness-level personnel, this plan should focus on a nonintervention or defensive mode.

Another isolation objective may be to identify and remotely secure potential ignition sources when flammable liquids and gases have been released. Common examples of ignition sources include open flames from pilot lights or other sources, arcs that occur when electrical switches are turned on or off, static electricity, and smoking materials.

Notify

Awareness-level personnel will likely need to notify and seek the assistance of other responders—for example, operations-level responders, hazardous materials technicians or other responders with additional training, law enforcement, and other technical experts. Large-scale incidents usually draw upon the resources of many agencies. Awareness-level personnel should know the basic notification procedures for reporting hazardous materials emergencies and requesting assistance from local and regional authorities. For example, you may need to notify regulatory agencies such as the local fish and game agency, the state-level Office of Emergency Services, or county-level agencies such as an air quality control board. The emergency response plan of the AHJ should include a current and comprehensive contact list of local, state, and federal resources available and help you identify the key players in your jurisdiction.

It is not possible in this text to identify all types of communications equipment and procedures necessary for making such notifications. Therefore, awareness-level personnel should be familiar with all communications equipment, radio frequencies, and protocols for using the communications equipment provided by the AHJ. SOPs will help you identify the various notifications that need to be made and should define the points of contact for local, state, and federal resources that might be called upon for assistance during hazardous materials emergencies.

FIGURE 30-4 Drums may be constructed of many different types of materials, including polyethylene, cardboard, or stainless steel.

FIGURE 30-5 An abandoned steel drum. Notice the configuration of the holes on top.

Basic Container Recognition

In basic terms, a **container** is "a receptacle, piping, or **pipeline** used for storing or transporting material of any kind" (NFPA 470). Often the container type, size, and material of construction provide important clues about the nature of the substance inside.

For example, a **drum** is a barrel-like storage vessel used to store a wide variety of substances, including food-grade materials, corrosives, flammable liquids, and grease. Drums may be constructed of low-carbon steel, polyethylene, cardboard, stainless steel, nickel, or other materials (**FIGURE 30-4**). Gasoline or waste solvents (from legitimate or illegitimate processes) may be stored in a 55-gallon steel drum with two capped openings (2 inches [51 mm] and ¾ inch [19 mm]) on the top (**FIGURE 30-5**). Sulfuric acid, at 97 percent concentration, could be found in a polyethylene drum that might be black, red, white, or blue. In most cases, there is no correlation between the color of the drum and the possible contents. The same sulfuric acid might also be found in a 1-gallon amber glass container. Hydrofluoric acid, by contrast, is incompatible with silica (glass) and would be stored in a plastic container. Steel or polyethylene drums, bags, high-pressure gas cylinders, railroad tank cars, plastic buckets, aboveground and underground storage tanks, cargo tanks, and pipelines are all representative examples of how hazardous materials are used, stored, and shipped.

Some very recognizable chemical containers, such as 55-gallon drums and compressed gas cylinders, can be found in almost every type of manufacturing facility. Stainless steel containers may hold particularly dangerous chemicals. Liquids that must be kept cold are often stored in Thermos-like Dewar containers, which are designed to maintain the appropriate temperature (**FIGURE 30-6**).

Personnel and responders should not rely solely on the type of container when making a determination

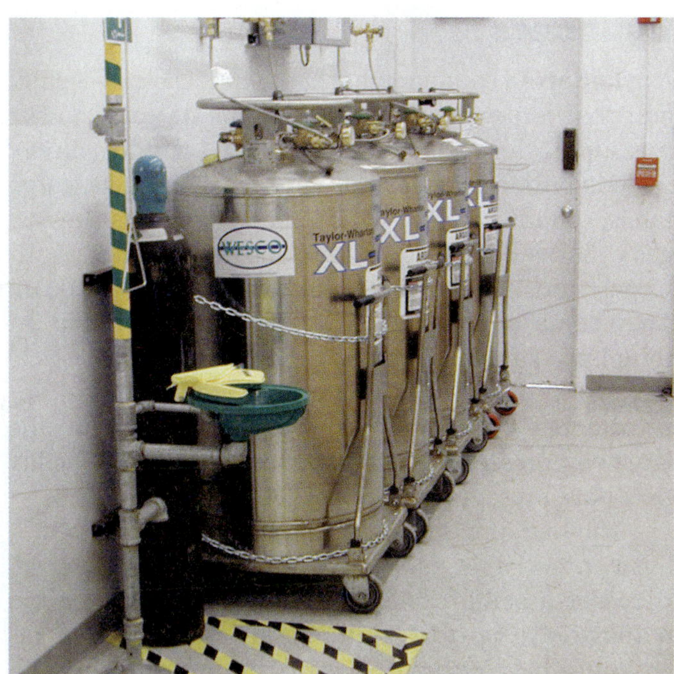

FIGURE 30-6 A series of Dewar containers stored adjacent to a compressed gas cylinder.

© Jones & Bartlett Learning. Courtesy of Rob Schnepp.

FIGURE 30-7 A bung wrench is used to operate the openings on the top of a closed-head drum.

© Jones & Bartlett Learning. Photographed by Glen E. Ellman.

about its contents because there are numerous examples of finding substances in the wrong type of container. The container might also lack any legitimate markings to alert personnel to the possible contents. In any case, it is important to look closely at a container and form an opinion about the material inside.

The following subsections describe a few types of containers with which awareness-level personnel should be familiar. For more information on other types of containers, rail cars, and road trailers, see Section 4, *Hazardous Materials Operations Level*, in this text.

Drums

Drums are easily recognizable, barrel-like containers. As mentioned earlier, drums are used to store a wide variety of substances, and may be constructed of a variety of materials. Generally, the nature of the chemical dictates the construction of the storage drum. Steel utility drums, for example, hold flammable liquids, cleaning fluids, oil, and other noncorrosive chemicals. Polyethylene drums are used for corrosives such as acids, bases, oxidizers, and other materials that cannot be stored in steel containers. Cardboard drums hold solid materials such as soap flakes, sodium hydroxide pellets, and food-grade materials. Stainless steel or other heavy-duty drums generally hold materials too aggressive (i.e., too reactive) for either plain steel or polyethylene.

A **closed-head drum** has a permanently attached lid with one or more small openings. This opening is called a **bung** (see FIGURE 30-5). Typically, these openings are threaded holes sealed by caps that can be removed only by using a special tool called a bung wrench (**FIGURE 30-7**). Closed-head drums usually have one 2-inch [51-mm] bung and one ¾-inch [19-mm] bung. The larger bung is used to pump product from the drum; the smaller bung functions as a vent.

As mentioned earlier, an open-head drum has a removable lid fastened to the drum with a ring (**FIGURE 30-8**). The ring is tightened with a clasp or a threaded nut-and-bolt assembly. These containers typically contain a product in solid form. This is an example of the type of drum described in the opening scenario.

Carboys

Some corrosives and other types of chemicals are transported and stored in a vessel called a carboy (**FIGURE 30-9**). A **carboy** is a glass, plastic, or steel container that holds 5 to 15 gallons of product. Glass carboys are often placed in protective wood, foam, fiberglass, or steel boxes to help prevent breakage. For

FIGURE 30-8 An open-head drum has a lid that is fastened with a ring; the ring is tightened with a clasp or a nut-and-bolt assembly.

Courtesy of Globalindustrial.com.

FIGURE 30-9 A carboy is used to transport and store corrosive chemicals.

Courtesy of EMD Chemicals, Inc.

example, nitric acid, sulfuric acid, and other strong acids are often transported and stored in thick glass carboys protected by a wooden or polystyrene crate to shield the glass container from damage during normal shipping.

Cylinders

A **cylinder** is a container that has a circular cross-section and that is designed to store liquids or gases under pressure higher than 40 psi (276 kPa). Cylinders do *not* include portable tanks, multiunit tank car tanks, cargo tanks, or tank cars (NFPA 1). Uninsulated compressed gas cylinders are used to store substances such as nitrogen, argon, helium, and oxygen (**FIGURE 30-10**). They come in a range of sizes and have variable internal pressures. An oxygen cylinder used for medical purposes, for example, has a pressure reading of approximately 2000 psi when full. By comparison, the large compressed gas cylinders found at a fixed facility may have pressure readings of 5000 psi or greater.

The high pressures exerted by these cylinders create a potential for danger. If the cylinder is punctured, the valve assembly fails, or the cylinder falls over and damages the valve, a rapid release of compressed gas or liquid may occur that turns the cylinder into an unpredictable missile. Also, if the cylinder is heated rapidly, it could explode with tremendous force, spewing product and metal fragments over long distances. Compressed gas cylinders have pressure-relief valves, but those valves may not be sufficient to relieve the pressure created during a fast-growing fire (**FIGURE 30-11**).

A propane cylinder is a compressed gas cylinder. Propane cylinders have lower pressures (200–300 psi)

FIGURE 30-10 Compressed gas cylinders in storage—what clues do you see that might indicate an unsafe situation?

© Jones & Bartlett Learning. Courtesy of Rob Schnepp.

FIGURE 30-11 Compressed gas cylinder failure due to heat generated by fire.

FIGURE 30-12 Small propane cylinder failure due to heat generated by fire.

than compressed gas cylinders and contain a liquefied gas. Liquefied gases such as propane are subject to the phenomenon known as **BLEVE (boiling liquid/expanding vapor explosion)**. BLEVEs occur when pressurized liquefied materials, such as propane or butane inside a closed vessel, are exposed to a source of high heat (**FIGURE 30-12**). Water, and other liquids capable of expanding, also has the potential to BLEVE. Chapter 31, *Properties and Effects*, in this text discusses BLEVE in more detail.

The low-pressure **Dewar container** is another commonly encountered cylinder type (**FIGURE 30-13**). As mentioned previously, a Dewar container is designed to hold a **cryogenic liquid**, or **cryogen**, which is a liquid that has a boiling point lower than –130°F (–90°C) at an absolute pressure of 14.7 psi (101.3 kPa).

FIGURE 30-13 A small Dewar container.

Typical cryogens include oxygen, helium, hydrogen, argon, and nitrogen. Under normal atmospheric conditions, each of these substances is a gas. A complex process turns them into liquids when they reach a specific low temperature, and these liquids can be stored and used for long periods of time. Nitrogen, for example, becomes a liquid at –320°F (–196°C) and must be kept at that temperature if it is to remain in a liquid state.

Cryogens pose a substantial threat if the Dewar container fails to maintain the low temperature of the cryogenic liquid. Cryogens have large expansion ratios—even larger than the expansion ratio of propane (270:1). Cryogenic helium, for example, has an expansion ratio of approximately 750:1. If one volume of liquid helium is warmed to room temperature and vaporized in a totally enclosed container, it can generate a pressure of more than 14,500 psi (99,974 kPa). To counter this possibility, cryogenic containers usually have two pressure-relief devices: a pressure-relief valve and a frangible (easily broken) metal disk.

SAFETY TIP

Cryogens are stored in a liquid state. Beware of skin exposures! Significant injuries, like those associated with thermal burns, can occur when skin meets one of these liquids.

Transportation and Facility Markings and Information Sheets

The presence of markings on buildings, packages, boxes, and containers often enables personnel to identify a released chemical. When used correctly, marking systems and information sheets indicate the presence of a hazardous material and provide clues about the substance. Marine pollutants and environmentally hazardous substances, for example, pose a risk to aquatic life and the marine ecosystem. Transportation markings denoting those substances may not be seen often, especially in areas without significant bodies of water. You may also find other transportation markings, such as those for elevated-temperature materials (e.g., asphalt and molten sulfur or other liquids transported at temperatures above 212°F [100°C] or 464°F [240°C]), consumer commodities intended for retail sale, and inhalation hazards, in your jurisdiction (**FIGURE 30-14**).

Voice of Experience

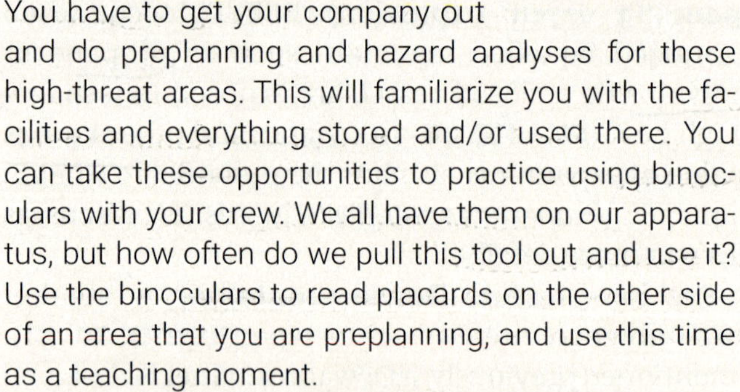

Your company is dispatched to a vehicle accident with injuries. While you are responding, dispatch advises they have received several calls stating that one of the vehicles is smoking and possibly on fire. Upon arrival on scene, you see a full-sized pickup truck with a camper shell, with a weird-colored smoke surrounding the back of the truck. There are two people beside the truck, on their knees and coughing violently.

Is this a typical motor vehicle collision (MVC)? Could that be dust coming from the air bags? Or could this be something a bit more dangerous? What do you do?

In today's world, emergency services personnel may encounter a hazardous materials incident anywhere and at any time. Many years ago, when the fire department responded to a structure fire, that is normally what they arrived to find. But today, every structure fire has the potential to be a hazardous materials incident. Think about the products that people keep under their sinks or in their garages. What happens to these chemicals when exposed to fire, heat conditions, or even water? How will they react?

These are just a few reasons that being able to recognize and identify hazardous materials is so important. All personnel and responders have to be hyperaware of their environment at all times. By knowing what to look for and being able to identify these problems, you can protect yourself, your crew, and the general public. You can also make prompt notifications to get the properly trained personnel responding to the scene.

By knowing and understanding your district, you will have an idea of the possible issues that you may face as soon as you hear the dispatch information. Recognizing that a call is in an area that contains hazardous materials doesn't happen by chance. You have to get your company out and do preplanning and hazard analyses for these high-threat areas. This will familiarize you with the facilities and everything stored and/or used there. You can take these opportunities to practice using binoculars with your crew. We all have them on our apparatus, but how often do we pull this tool out and use it? Use the binoculars to read placards on the other side of an area that you are preplanning, and use this time as a teaching moment.

The main goal is to make sure we all go home at the end of shift. By having situational awareness, preplanning your districts, and using all available tools in your toolbox, you can help make that happen.

Revisiting our previous scenario, what started as a dispatch to a normal, everyday MVC turned into a long, drawn-out hazardous materials response. The aforementioned pickup truck contained a mobile methamphetamine lab. Thankfully, sharp eyes and good judgment prevailed, the incident was contained, and there were no serious injuries.

Steven Marsh

Captain/Departmental Training Officer
Kurtz Industrial Fire Services, Tennessee
Operations
Chattanooga, Tennessee

A.

B.

C.

D.

FIGURE 30-14 Examples of markings indicating hazardous chemicals in a container. **A.** Environmental hazard. **B.** Inhalation hazard. **C.** Elevated temperature. **D.** Commodity.

A-D: Courtesy of the U.S. Department of Transportation; **C:** Reproduced from U.S. Department of Transportation 2020. DOT Chart 17: Hazardous Materials Markings, Labeling and Placarding Guide. https://www.phmsa.dot.gov/sites/phmsa.dot.gov/files/2021-11/USDOT%20Chart%2017.pdf.

Safety Data Sheets

A common source of information about a chemical is the SDS specific to that substance (**FIGURE 30-15**). Essentially, an SDS provides basic information about the chemical make-up of a substance, the potential hazards it presents, appropriate first aid in the event of an exposure, and other pertinent data for safe handling of the material. Although the SDS is not a definitive response tool, it may be a key resource for understanding the chemical and physical properties of a substance. The Occupational Safety and Health Administration (OSHA) requires that an SDS be available to workers who need to be around hazardous materials, whether they are used or stored.

An SDS typically includes the following details (the following list serves as an example and is not a complete list) (NFPA 470, 6.2.2(1), pages 470–425):

- Hazard identification
- Composition/information on ingredients
- First aid measures
- Firefighting measures
- Accident release measures
- Handling and storage
- Exposure controls/personal protection
- Toxicological information
- Physical and chemical properties

When responding to a hazardous materials incident at a fixed facility, personnel should ask the site representative for an SDS for the spilled material. All facilities that use or store chemicals are required by law to have an SDS on file for each chemical used or stored in the facility. Many sites, but especially those that stock many different chemicals, may keep this information archived in a computer database. An SDS can also be obtained from staffed national resource centers or on the transporting vehicle.

TIP

The Globally Harmonized System of Classification and Labelling of Chemicals (GHS) uses a standard methodology to define and classify hazards posed by chemical substances, along with a standard method—the SDS—to communicate the hazards. Information about the GHS can be found in the *ERG*.

The National Fire Protection Association 704 Marking System

The National Fire Protection Association (NFPA) has developed its own system for identifying hazardous materials. NFPA 704, *Standard System for the Identification of the Hazards of Materials for Emergency Response*, outlines a marking system characterized by a set of diamonds that are found on the outside of buildings, on doorways to chemical storage areas, and on fixed storage tanks. This marking system is designed for fixed-facility use. Personnel can use the NFPA diamonds to understand the broad hazards posed by chemicals stored in a building or part of a building.

The **NFPA 704 hazard identification system** uses a diamond-shaped symbol of any size, which is itself broken into four smaller diamonds, each representing a property or characteristic of a substance or group of substances (**FIGURE 30-16**). The blue, red, and yellow diamonds each contain a numerical rating in the range of 0–4, with 0 being the least hazardous and 4 being the most hazardous (**TABLE 30-1**).

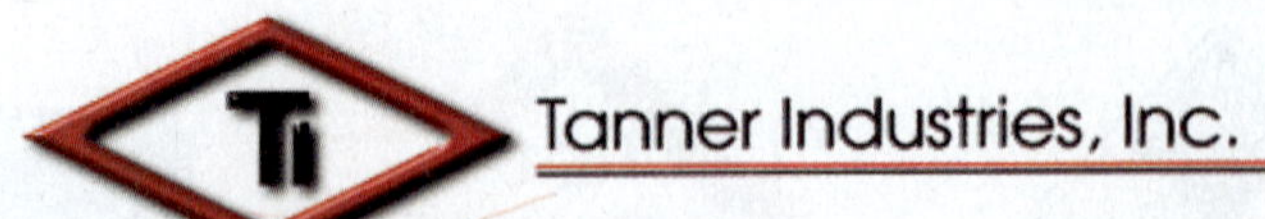

SAFETY DATA SHEET

Section 1. Identification

Product Name:	**Ammonia, Anhydrous**
Synonyms:	Ammonia
CAS REGISTRY NO:	7664-41-7
Supplier:	Tanner Industries, Inc. 735 Davisville Road, Third Floor Southampton, PA 18966
Website:	www.tannerind.com
Telephone (General):	215-322-1238
Corporate Emergency Telephone Number:	**800-643-6226**
Emergency Telephone Number:	**Chemtrec: 800-424-9300**
Recommended Use:	Various Industrial / Agricultural

Section 2. Hazard(s) Identification

Hazard: Acute Toxicity, Corrosive, Gases Under Pressure, Flammable Gas, Acute Aquatic Toxicity

Classification:
Acute Toxicity, Inhalation (Category 4)
Skin Corrosion / Irritation (Category 1B)
Serious Eye Damage / Irritation (Category 1)
Gases Under Pressure (Liquefied gas)
Flammable Gases (Category 2)
Acute Aquatic Toxicity (Category 1)

Note: (1 - Most Severe / 4 - Least Severe)

Pictogram:

Signal word: **Danger**

Hazard statements:
Harmful if inhaled.
Causes severe skin burns and serious eye damage.
Flammable gas.
Contains gas under pressure; may explode if heated.
Very toxic to aquatic life.

Precautionary statements:
Avoid breathing gas/vapors.
Use only outdoors or in well-ventilated area.
Wear protective gloves, protective clothing, eye protection, face protection.
Keep away from heat, sparks, open flames and other ignition sources. No smoking.

FIGURE 30-15 An example of an SDS for anhydrous ammonia (first page only).

Courtesy of Tanner Industries, Inc., Southhampton, PA.

A.

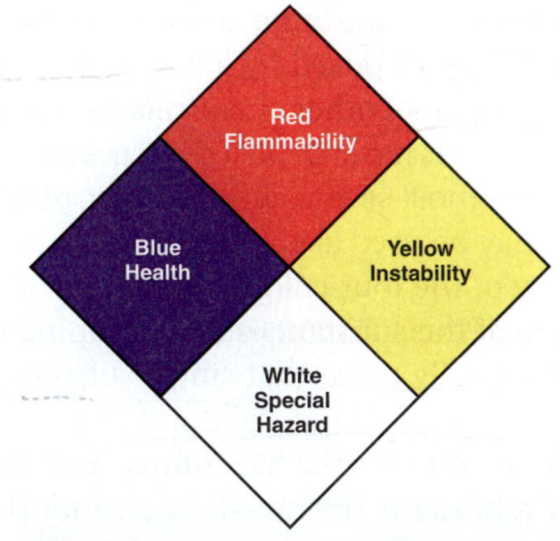

B.

FIGURE 30-16 **A.** Example of a placard using the NFPA 704 hazard identification system for fixed-facility use. **B.** Each color used in the diamond represents a particular property or characteristic.

TABLE 30-1 Hazard Levels in the NFPA Hazard Identification System

	Flammability Hazards (Red Diamond)		Instability Hazards (Yellow Diamond)
4	Materials that will rapidly or completely vaporize at atmospheric pressure and normal ambient temperature or that are readily dispersed in air and that will burn readily.	4	Materials that in themselves are readily capable of detonation or of explosive decomposition or reaction at normal temperatures and pressures.
3	Liquids and solids that can be ignited under almost all ambient temperature conditions.	3	Materials that in themselves are capable of detonation or explosive decomposition or reaction but require a strong initiating source, or that must be heated under confinement before initiation.
2	Materials that must be moderately heated or exposed to relatively high ambient temperatures before ignition can occur.	2	Materials that readily undergo violent chemical change at elevated temperatures and pressures.
1	Materials that must be preheated before ignition can occur.	1	Materials that in themselves are normally stable but can become unstable at elevated temperatures and pressures.
0	Materials that will not burn under typical fire conditions.	0	Materials that in themselves are normally stable, even under fire exposure conditions.
	Health Hazards (Blue Diamond)		**Special Hazard (White Diamond)**
4	Materials that under emergency conditions can be lethal.		SA indicates simple asphyxiant. W indicates water reactivity. OX indicates oxidizer.
3	Materials that under emergency conditions could cause serious or permanent injury.		
2	Materials that under emergency conditions could cause incapacitation or possible residual injury.		
1			Materials that under emergency conditions can cause serious irritation.
0			Materials that under emergency conditions would offer no hazard beyond that of ordinary combustible material.

See the NFPA 704 standard for more detailed information on the criteria that qualify materials as 0-1-2-3-4 in each of the four quadrants.
Data from National Fire Protection Association (NFPA.) 2020. *NFPA 704: Standard System for the Identification of the Hazards of Materials for Emergency Response.* 2022 Edition. Quincy, MA: NFPA.

The blue diamond (at the nine o'clock position) indicates the health hazard posed by a material alone or perhaps within a group of other chemicals. When an NFPA diamond represents hazards posed by several different substances, the most severe characteristic of any of the substances may be used as the basis for the hazard level within any of the four colored diamonds. For example, if any one of the substances in a grouping of chemicals could be fatally toxic, that single substance causes a 4 to appear in the blue diamond. All other substances could be much less hazardous, but the one causing the 4 represents the health hazard for the group. The same logic holds true for the red diamond at the 12 o'clock position, which indicates flammability, and the yellow diamond at the three o'clock position, which indicates reactivity.

The six o'clock position on the symbol represents special hazards and has a white background. The special hazards in use include W, OX, and SA. W indicates unusual reactivity with water and is a caution about the use of water in either firefighting or spill control response. OX indicates that the material is an oxidizer. SA indicates that the material is a simple **asphyxiant** gas (nitrogen, helium, neon, argon, krypton, or xenon). For complete information on the NFPA 704 system, consult www.NFPA.org/704.

To accurately compare the DOT marking system and NFPA 704 system, remember this important difference:

- The DOT hazardous materials marking system is used when materials are being transported from one location to another.
- The NFPA 704 hazard identification system is designed for fixed-facility use.

Hazardous Materials Information System

The **Hazardous Materials Information System (HMIS)** is a color-coded marking system that identifies the hazard level of chemicals in a container so that personnel can work safely around the chemicals. In Canada, a similar system, the Workplace Hazardous Materials Information System (WHMIS), is used. HMIS has helped employers comply with the Hazard Communication standard established by OSHA. The HMIS is similar to the NFPA 704 marking system and uses a numerical hazard rating with colored horizontal columns (**FIGURE 30-17**).

FIGURE 30-17 The HMIS helps employers comply with the Hazard Communication standard.

The HMIS is more than just a label; it is a method used by employers to give their personnel necessary information to work safely around chemicals and includes training materials to inform workers of chemical hazards in the workplace. The HMIS is not required by law, but rather is a voluntary system that employers choose to use to comply with OSHA's Hazard Communication standard. Although use of the HMIS is voluntary, informing personnel about hazardous materials is not. If an employer chooses not to use the HMIS, another system that complies with OSHA's requirements must be used. In addition to describing the chemical hazards posed by a substance, the HMIS provides guidance about the personal protective equipment (PPE) that employees need to use to protect themselves from workplace hazards. Letters and icons specify the different levels and combinations of protective equipment required.

Personnel and responders must understand the fundamental difference between the NFPA 704 marking system and HMIS. NFPA 704 is intended to be understood by all levels of hazardous materials personnel and responders; HMIS is intended for the employees of a facility. Although HMIS is not a response information tool, it can give clues about the presence and nature of the hazardous materials found in the facility.

Military Hazardous Materials/Weapons of Mass Destruction Markings

The U.S. military has developed its own marking system for hazardous materials. The military system serves primarily to identify detonation, fire, and special hazards.

In general, hazardous materials within the military marking system are divided into four categories based on the relative detonation and fire hazards they pose:

- Division 1 materials are considered mass detonation hazards and are identified by a number 1 printed inside an orange octagon (**FIGURE 30-18A**).

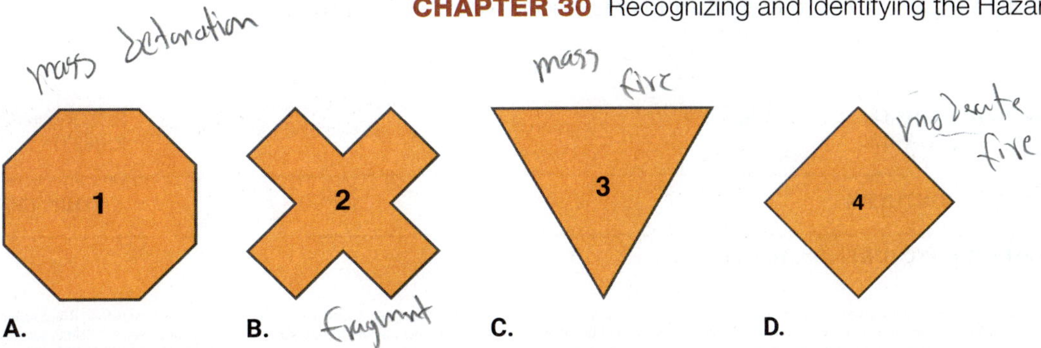

FIGURE 30-18 Military hazardous materials/WMD markings. **A.** Mass detonation hazards. **B.** Explosion-with-fragment hazards. **C.** Mass fire hazards. **D.** Moderate fire hazards.

- Division 2 materials have explosion-with-fragment hazards and are identified by a number 2 printed inside an orange X (**FIGURE 30-18B**).
- Division 3 materials are mass fire hazards and are identified by a number 3 printed inside an inverted orange triangle (**FIGURE 30-18C**).
- Division 4 materials are moderate fire hazards and are identified by a number 4 printed inside an orange diamond (**FIGURE 30-18D**).

Chemical hazards in the military system are depicted by colors. Toxic agents, such as sarin or mustard, are identified by the color red. Harassing agents, such as tear gas and smoke producers, are identified by yellow. White phosphorus is identified by white. Specific PPE requirements are identified using pictograms. Military shipments containing hazardous materials/WMD are not required, by exception, to be placarded with the DOT marking system.

Shipping Papers

Shipping papers are required by the DOT whenever materials are transported from one place to another (NFPA 498). They identify the names and addresses of the shipper, the receiver, and the material being shipped, and they specify the quantity and weight of each part of the shipment. Additionally, shipping papers allow the reader to match the chemical name found on the shipping papers with the mode of transportation. Shipping papers for road and highway transportation are called a **bill of lading**, or **freight bill**, and are located in the cab of the vehicle (**FIGURE 30-19**). Drivers transporting chemicals are required by law to have a set of shipping papers on their person or within easy reach inside the cab at all times.

A bill of lading and other types of shipping papers may provide additional information about a hazardous substance, such as its packaging group designation. The **packaging group designation** is another system used by shippers to identify special handling requirements or hazards. Some DOT hazard classes require shippers to assign packaging groups based on the material's flash point and toxicity. A packaging group designation may signal that the material poses a greater hazard than similar materials in a hazard class. There are three packaging group designations:

- Packaging group I: *High danger*
- Packaging group II: *Medium danger*
- Packaging group III: *Minor danger*

The shipping papers for railroad transportation are called a **waybill** (**FIGURE 30-20**). A list of the contents in every car on the train is called the **consist**, or **train list**. The conductor, engineer, or a designated member of the train crew will have a copy of the consist. If the incident happens in a railyard, then you may find a waybill as well.

On a marine vessel, shipping papers are called the **dangerous cargo manifest** (**FIGURE 30-21**). The manifest is generally kept in a tube-like container in the wheelhouse, in the custody of the captain or master.

For air transport, the **air bill** serves as the shipping papers (**FIGURE 30-22**). It is kept in the cockpit and is the pilot's responsibility.

Pipelines

Of all the various methods used to transport hazardous materials, the high-volume pipeline is the one that is most rarely involved in emergencies. A pipeline is "a length of pipe including pumps, valves, flanges, control devices, strainers, and/or similar equipment for conveying fluids" and gases over potentially long distances (NFPA 470). In many areas, large-diameter pipelines transport natural gas, gasoline, diesel fuel, and other products from delivery terminals to distribution facilities. Pipelines are often buried underground but may be aboveground in remote areas.

STRAIGHT BILL OF LADING
ORIGINAL - NOT NEGOTIABLE

BOL/Reference No.
RSI82715

CARRIER: NORFOLK SOUTHERN

Date: 12/23/2008

Shipper: RSI LOGISTICS, INC (OKEMOS, MI US)

The property described below, in apparent good order, except as noted (contents and condition of packages unknown), marked, consigned, and destined as indicated below, which said carrier (the word carrier being understood throughout this contract as meaning any person or corporation in possession of the property under the contract) agrees to carry to its usual place of delivery at said destination, if on its route, otherwise to deliver to another carrier on the route to said destination. It is mutually agreed, as to each carrier of all or any said property, that every service to be performed hereunder shall be subject to all the terms and conditions of the Uniform Domestic Straight Bill of Lading set forth (1) in Official, Southern, Western and Illinois Freight Classification in effect on the date hereof, if this is a rail or a rail-water shipment, or (2) in the applicable motor carrier classification or tariff if this is a motor carrier shipment

Shipper hereby certifies that he is familiar with all the terms and conditions or the said bill of lading, including those on the back thereof, set forth in the classification or tariff which governs the transportation of this shipment, and the said terms and conditions are hereby agreed to by the shipper and accepted for himself and his assigns.

Consignee Information: CONSIGNEE DEER PARK, TX Address: City: DEER PARK, TX US	
Route: NS-ESTL-BNSF	
Origin Switch Route:	
Destination Switch Route: HUSTN-PTRA	Rail Car No: GATX290861

For assistance in any transportation emergency involving chemicals, phone CHEMTREC, day or night, Toll Free 1-800-424-9300

DESCRIPTION		*WEIGHT
ONE TANK CAR	Contains: Methyl Esters STCC#2899415 BIODIESEL-15, Biodiesel Sales Order Contract No: RSI82715 Sales Order Contract No: AAT122308-4 Purchase Order Contract No: AAT122308-4	(Sub. To Correction) 204400 Lbs.

SEAL NUMBERS:

Gross

Tare

Net

Weighed By: ______________________

If charges are to be prepaid, write or stamp here, "To be Prepaid"
Prepaid

Subject to Section 7 of the conditions of applicable bill of lading, if this shipment is to be delivered to the consignee without recourse on the consignor, the consignor shall sign the following statement:: *The carrier shall not make delivery of this shipment without payment of freight and all other lawful charges.*

Not In Effect

* This is to certify that the above named materials are properly classified, described, packaged, marked, and labeled, and are in proper condition for transportation, according to the applicable regulations of the Department of Transportation.

FIGURE 30-19 A bill of lading, or freight bill.

Courtesy of RSI Logistics, Inc.

WAYBILL
NON-NEGOTIABLE

WAYBILL NO.

SHIP DATE:

SHIPPER NUMBER	SHIPPER REFERENCE NUMBER	CONSIGNEE REFERENCE NUMBER	P.O. NUMBER
SHIPPER		CONSIGNEE	
STREET ADDRESS		STREET ADDRESS	
STREET ADDRESS		STREET ADDRESS	
CITY, STATE AND ZIP CODE		CITY, STATE AND ZIP CODE	
CONTACT	PHONE NUMBER	CONTACT	PHONE NUMBER

3rd PARTY NUMBER	SHIPPING CO. LIABILITY IS LIMITED TO $50 PER SHIPMENT OR 50 CENTS PER POUND (U.S. DOLLARS), WHICHEVER IS HIGHER SUBJECT TO A MAXIMUM LIABILITY OF $25,000, UNLESS A HIGHER VALUE IS DECLARED AND APPLICABLE CHARGES (FOR DECLARING A VALUE ON WAYBILL) ARE PAID PRIOR TO SHIPPING (SUBJECT TO THE TERMS AND CONDITIONS ON REVERSE SIDE, AND THE SERVICE CONDITIONS FOUND IN THE SHIPPING CO. SERVICE CONDITIONS POLICY).
3rd PARTY NAME	
STREET ADDRESS	
CITY, STATE AND ZIP CODE	DECLARED VALUE
CONTACT NAME	$

(SUBJECT TO CORRECTION)

NO. OF PIECES	TYPE	HM	KIND OF PACKAGE, DESCRIPTION OF ARTICLES, SPECIAL MARKS & EXCEPTIONS	ACTUAL WEIGHT	LENGTH	WIDTH	HEIGHT

Special Instructions:

☐ SATURDAY DELIVERY ☐ SUNDAY DELIVERY ☐ APPOINTMENT DELIVERY ☐ INSIDE DELIVERY ☐ RESIDENTIAL DELIVERY

Services Requested:

☐ SAME DAY/ NEXT FLIGHT OUT ☐ NEXT DAY AM ☐ NEXT DAY PM ☐ 2ND DAY ☐ ECONOMY DEFERRED (3-5 DAYS)

FOR CHARTER AIR, TIME DEFINITE, OR GUARANTEED SERVICE, PLEASE CALL 1-800-000-XXXX FOR AVAILABILITY.

QUOTE NUMBER	DIM WEIGHT

SHIPPER CERTIFICATION: Shipper certifies by its signature, its agreement to all of the foregoing terms and conditions, and further certifies that the above named materials are properly classified, described, packaged, marked and labeled, and are in proper condition for transportation according to the applicable regulations of the DOT.

SHIPPER REPRESENTATIVE

SIGNATURE X ____________ Print Name X ____________ Date ________

PICKED UP BY:	RECEIVED BY:	RECEIVED BY CONSIGNEE IN GOOD ORDER UNLESS NOTED BELOW:
DRIVER SIGNATURE ________	CONSIGNEE SIGNATURE ________	# S/W SKIDS DEL'D INTACT ____
PLEASE PRINT ________	PLEASE PRINT ________	# SKIDS DEL'D: ____
COMPANY ________	COMPANY ________	☐ GOOD ORDER ☐ SHORT ☐ OVER ☐ DAMAGED
DATE ________	DATE ________	DESCRIBE EXCEPTIONS:
TIME ________	TIME ________	

All rules as contained in Shipping Co. Services Conditions Policy will apply. Terms and conditions stated on any Bill of Lading used to transfer goods for carriage, other than an Shipping Co. Waybill, will be null and void. Quotes are based on the information provided and are only an estimate. Final charges are based on actual shipment pieces, weight, dimensions, and services performed as a requirement for delivery. Any changes in actual shipment details will affect the final charges.

Shipping Co. is a certified participant in compliance with the Transportation Security Administration Regulations, Part 109, a Federal Security program.

1 - SHIPPER'S COPY

FIGURE 30-20 A waybill.

Courtesy of a private source.

CARGO MANIFEST

AIR — AIRCRAFT DATA: CARRIER | A/C NO. | A/C MODEL | DEST CODE | REF | DESTINATION | MISSION DATA: NO. | SU | DATE | ALW WT | ALW CU | MANIFEST ID: STA | FY | TY | NO. | PAGE NO.

SUR-FACE — POE | DATE SAILED | VOYAGE DOCUMENT NO. | POD | REF | VESSEL NAME | STATUS | SUST | TRUCK NO. | REMARKS | PAGE NO.

DOC ID	VEHICLE TRAILER OR CNTNR NUMBER	YR / MAKE / CNTNR NUMBER / COMMODTY DESCRIP	COM CODE	CAR-GO EXC	VOYAGE DOC NO. / AIRDM / TRNS PT	PORT OF DISCH	TYPE PACK	TRANSPORTATION CONTROL NUMBER	CONSIGNEE / SERV / ACTVTY ADDRESS	PRIORITY	NAME / AMMO LOT NO./NOMEN / DIMENSIONS / RDD / PROJ / STOW LOC / TRANS ACCT	IDENTIFICATION NO. OR REMARKS	PIECES	WEIGHT	CUBE

ITEMS HAVE BEEN LOADED: | ITEMS HAVE BEEN RECEIVED EXCEPT AS CIRCLED NOTED ON REVERSE SIDE | TOTALS | 0 | 0.00 | 0.00

DATE | SIGNATURE OF LOADING AGENT | DATE | SIGNATURE OF UNLOADING AGENT | DATE | SIGNATURE OF RECEIVING AGENT

DD FORM 1385, NOV 78 REPLACES EDITION OF 1 APR 66 WHICH MAY BE USED Adobe Professional 7.0

FIGURE 30-21 A dangerous cargo manifest.

Courtesy of the U.S. Department of Defense.

Pipeline incidents may present you with challenges and hazards not typically encountered at most hazardous materials incidents. Subject-matter experts from the company that owns the pipeline may be required to assist hazardous materials response teams from the local jurisdiction in such cases. These incidents, like rail incidents, could have far-reaching implications and present all types of incident response personnel with a challenging set of circumstances.

The **pipeline right-of-way** is an area, patch of land, or roadway that extends a certain number of feet on either side of the pipe itself. The company that owns the pipeline maintains this area. The company is also responsible for placing warning signs at regular intervals along the length of the pipeline. These pipeline warning signs must include a warning symbol, the pipeline owner's name, and an emergency contact phone number (**FIGURE 30-23**).

Information about the pipe's contents and owner is also often found at the vent pipes. These inverted J-shaped tubes provide pressure relief or natural venting during maintenance and repairs. **Vent pipes** are clearly marked and are located approximately 3 feet (1 m) above the ground.

Pipeline emergencies are complicated events that typically require specially trained responders who are trained well above the awareness level. If you suspect an incident involving a pipeline, contact the owner of the pipeline immediately. The company will dispatch a crew to assist with the incident.

U.S. DOT Marking System

In the United States, the DOT created a marking system consisting of placards and labels—that is, the **U.S. Department of Transportation (DOT) marking system**—that shippers must use to identify hazardous materials that are transported. (This system is also used in Canada by Transport Canada.) Placards are diamond-shaped indicators (10¾ inches [27 cm] on each side) that are placed on all four sides of highway transport vehicles, railroad tank cars, and other forms of transportation carrying hazardous materials (**FIGURE 30-24**). Labels are smaller versions (4-inch [10-cm] diamond-shaped indicators) of placards. Labels are placed on the four sides of individual boxes and smaller packages being transported (**FIGURE 30-25**).

Set your tabulator stops here

STAPLE DOCUMENTS ABOVE PERFORATION

Line-up here

Shipper's Name and Address

Shipper's Account Number

Not Negotiable

Air Waybill

Issued by

Copies 1, 2 and 3 of this Air Waybill are originals and have the same validity.

Consignee's Name and Address

Consignee's Account Number

It is agreed that the goods described herein are accepted in apparent good order and condition (except as noted) for carriage SUBJECT TO THE CONDITIONS OF CONTRACT ON THE REVERSE HEREOF. ALL GOODS MAY BE CARRIED BY ANY OTHER MEANS INCLUDING ROAD OR ANY OTHER CARRIER UNLESS SPECIFIC CONTRARY INSTRUCTIONS ARE GIVEN HEREON BY THE SHIPPER, AND SHIPPER AGREES THAT THE SHIPMENT MAY BE CARRIED VIA INTERMEDIATE STOPPING PLACES WHICH THE CARRIER DEEMS APPROPRIATE. THE SHIPPER'S ATTENTION IS DRAWN TO THE NOTICE CONCERNING CARRIER'S LIMITATION OF LIABILITY. Shipper may increase such limitation of liability by declaring a higher value for carriage and paying a supplemental charge if required.

Issuing Carrier's Agent Name and City

Accounting Information

Agent's IATA Code | Account No.

Airport of Departure (Addr. of First Carrier) and Requested Routing

Reference Number | Optional Shipping Information

To	By First Carrier (Routing and Destination)	to	by	to	by	Currency	CHGS Code	WT/VAL PPD	WT/VAL COLL	Other PPD	Other COLL	Declared Value for Carriage	Declared Value for Customs

Airport of Destination | Requested Flight/Date | Amount of Insurance

INSURANCE - If carrier offers insurance, and such insurance is requested in accordance with the conditions thereof, indicate amount to be insured in figures in box marked "Amount of Insurance".

Handling Information

SCI

These commodities, technology or software were exported from the United States in accordance with the Export Administration Regulations. Ultimate destination

Diversion contrary to U.S. law prohibited.

No. of Pieces RCP	Gross Weight	kg lb	Rate Class / Commodity Item No.	Chargeable Weight	Rate / Charge	Total	Nature and Quantity of Goods (incl. Dimensions or Volume)

Prepaid	Weight Charge	Collect
	Valuation Charge	
	Tax	
	Total Other Charges Due Agent	
	Total Other Charges Due Carrier	
Total Prepaid		Total Collect
Currency Conversion Rates		CC Charges in Dest. Currency
For Carriers Use only at Destination		Charges at Destination

Other Charges

Shipper certifies that the particulars on the face hereof are correct and that **insofar as any part of the consignment contains dangerous goods, such part is properly described by name and is in proper condition for carriage by air according to the applicable Dangerous Goods Regulations.**

Signature of Shipper or his Agent

Executed on (date) | at (place) | Signature of Issuing Carrier or its Agent

Total Collect Charges

APPERSON K0419 (10/03) WHSE. #05640

FIGURE 30-22 An air bill.

Courtesy of Apperson Print Resources Inc.

FIGURE 30-23 A pipeline warning sign provides information about the pipe's contents and the owner's name and contact information.

© Photodisc/Getty Images

FIGURE 30-24 A placard is a large diamond-shaped indicator that is placed on all sides of transport vehicles that carry hazardous materials.

© Mark Winfrey/Shutterstock

Placards, labels, and markings are intended to give you a general idea of the hazard inside a container or cargo tank. A **placard** identifies the broad hazard class based on the chemical family (flammable, poison, corrosive) to which the material inside belongs.

The nine chemical families (hazard classes), and their respective divisions recognized in the *ERG*, are outlined here (Transport Canada, U.S. Department of Transportation, and Secretariat of Communications and Transport of Mexico, 2020):

- DOT Class 1: Explosives
  - Division 1.1: Explosives that have a mass explosion hazard
  - Division 1.2: Explosives that have a projection hazard but not a mass explosion hazard
  - Division 1.3: Explosives that have a fire hazard and either a minor blast hazard or a minor projection hazard or both, but not a mass explosion hazard
  - Division 1.4: Explosives with no significant blast hazard
  - Division 1.5: Very insensitive explosives with a mass explosion hazard
  - Division 1.6: Extremely insensitive articles that do not have a mass explosion hazard
- DOT Class 2: Gases
  - Division 2.1: Flammable gases
  - Division 2.2: Nonflammable, nontoxic gases
  - Division 2.3: Toxic* gases
- DOT Class 3: Flammable liquids (and combustible liquids in the United States)
- DOT Class 4: Flammable solids, substances liable to spontaneous combustion, and substances that, on contact with water, emit flammable gases
  - Division 4.1: Flammable solids, self-reactive substances and solid desensitized explosives
  - Division 4.2: Substances liable to spontaneous combustion
  - Division 4.3: Substances that, when in contact with water, will emit flammable gases
- DOT Class 5: Oxidizing substances and organic peroxides
  - Division 5.1: Oxidizing substances
  - Division 5.2: Organic peroxides
- DOT Class 6: Toxic substances and infectious substances
  - Division 6.1: Toxic substances
  - Division 6.2: Infectious substances
- DOT Class 7: Radioactive materials
- DOT Class 8: Corrosive substances
- DOT Class 9: Miscellaneous hazardous materials/products, substances, or organisms

*The words "poison" and "poisonous" are synonymous with the word "toxic."

The *ERG* organizes chemicals into the nine basic hazard classes, or families defined by DOT. The members of each family exhibit similar properties. There is also a "Dangerous" placard, which indicates that the same load contains materials from more than one hazard class (**FIGURE 30-26**).

A label on a box inside a delivery truck, for example, relates only to the potential hazard or hazard class inside that package. In some cases, a four-digit United Nations

(UN) identification number may be required on some placards. This number identifies the specific material being shipped; a list of UN numbers is included in the *ERG*. You may see references to UN numbers that also include the letters NA and listed as UN/NA. NA stands for North America, but as a point of reference, UN and NA numbers are identical. These placards and labels can be viewed from a distance with binoculars, or shipping papers or other reference sources may provide the four-digit UN identification number or the name of the material, or you may see the colors on the placard. All of these methods can be used to determine the appropriate guide for managing a released material.

In most cases, the package or cargo tank must contain a certain amount of hazardous material before a placard is required. However, keep this point in mind: The absence of a placard does not mean no hazardous cargo is present. For example, the "1000-pound rule" applies to some explosives, flammable and nonflammable gases, flammable/combustible liquids, flammable solids, air-reactive solids, oxidizers and organic peroxides, poison solids, corrosives, and miscellaneous (class 9) materials. Placards are required for these materials only when the shipment weighs more than 1000 pounds. (More information on the 1000-pound rule can be found in CFR Title 49, Subtitle B, Chapter 1, Subchapter C, Part 172, Subpart F.)

Conversely, some chemicals are so hazardous that shipping any amount of them requires the use of labels or placards. These materials include some explosives, poisonous gases, water-reactive solids, and high-level radioactive substances. Personnel at the scene should seek additional specifics about any material in question by consulting the appropriate response agency or using

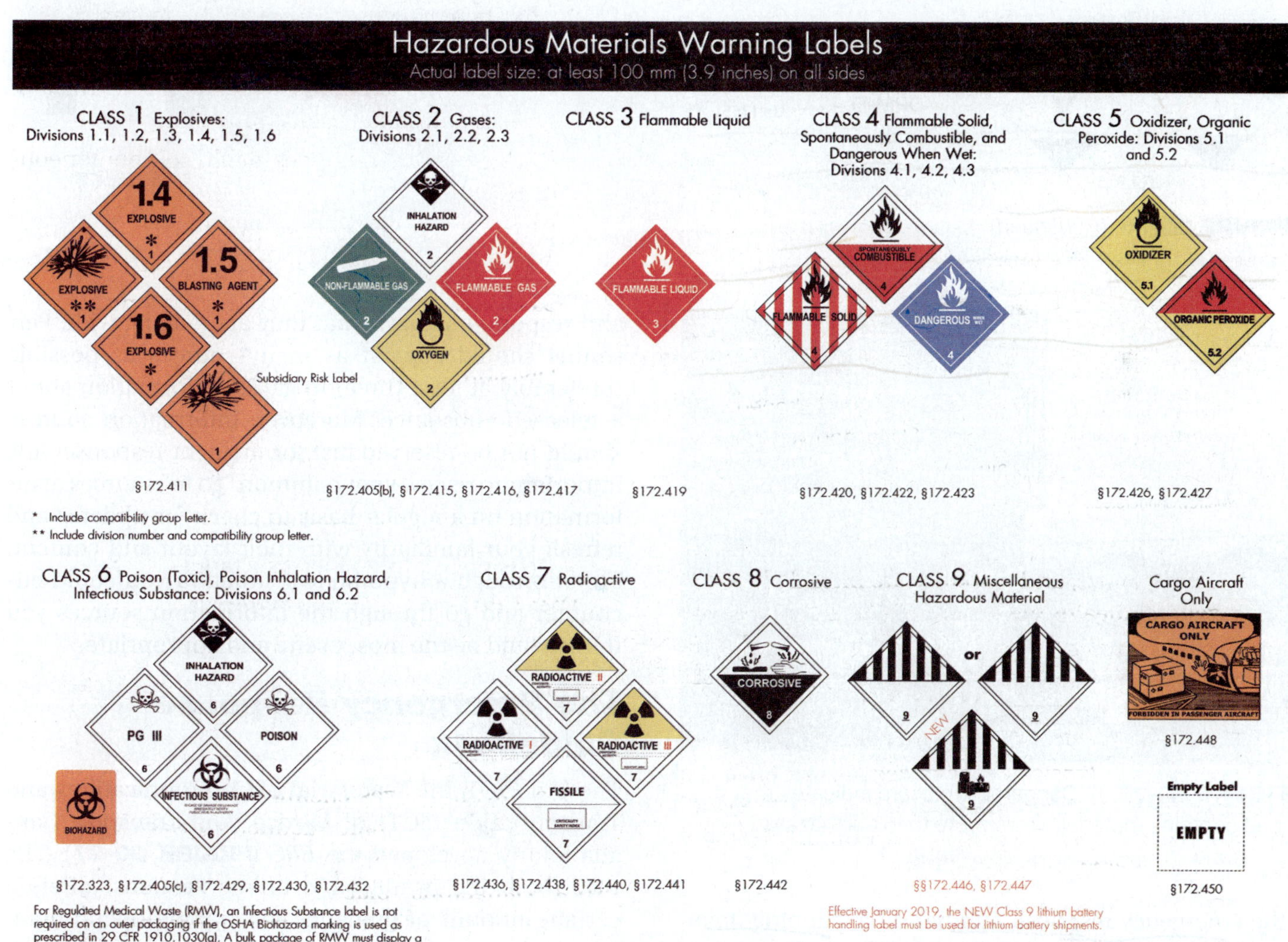

FIGURE 30-25 A label is a smaller version of the placard and is placed on boxes or smaller packages that contain hazardous materials.

Courtesy of the U.S. Department of Transportation.

HAZARDOUS MATERIALS MARKINGS

Package Orientation (Red or Black) — or — §172.312(a)

Keep Away from Heat — keep away from heat — §172.317

OVERPACK — §173.25(a)(4)

Fumigant Marking — DANGER — THIS UNIT IS UNDER FUMIGATION WITH ___ APPLIED ON — DO NOT ENTER — §172.302(g) and §173.9

INHALATION HAZARD — §172.313(a)

HOT — §172.325

§172.332(a)

Biological Substances, Category B — UN3373 — §173.199 (a)(5)

NEW Lithium battery handling marking, Transition December 31, 2018 — CAUTION! — For more information, call xxx.xxx.xxxx — §173.185

Marine Pollutant — §172.322

Limited Quantity — All other Modes — Air Only — Y — §172.315

ORM-D, Transition December 31, 2020 — CONSUMER COMMODITY — ORM-D — UN1755 — §172.316

Excepted Quantity — §173.4a(g)

Marking of IBCs — ... kg max — §178.703(b)(7)(i)

FIGURE 30-25 *(Continued)*

Courtesy of the U.S. Department of Transportation.

FIGURE 30-26 A "Dangerous" placard indicates that the load contains materials from more than one hazard.

the emergency response number on a shipping document, if applicable, to gather more information.

Thus, hazardous materials can be identified in several ways, so that if a spill or release occurs, personnel and responders know what they are dealing with. Personnel should consult as many sources as possible (preferably at least three) to gather information about a released substance. Moreover, information sources should not be reserved just for incident response: It is important to review your common "go to" sources of information on a regular basis to check for updates and refresh your familiarity with their layout and content. Think through a hypothetical situation you could encounter and go through the information sources you think would be the most useful and appropriate.

The *Emergency Response Guidebook*

The U.S. DOT, the Secretariat of Communications and Transportation (SCT) of Mexico, and Transport Canada jointly developed the *ERG* (**FIGURE 30-27**). The *ERG* (both the printed and online versions) offers a certain amount of guidance for anyone operating at a hazardous materials incident. This reference source merges the DOT labels and placards with the nine hazard classes and the UN identification system to provide

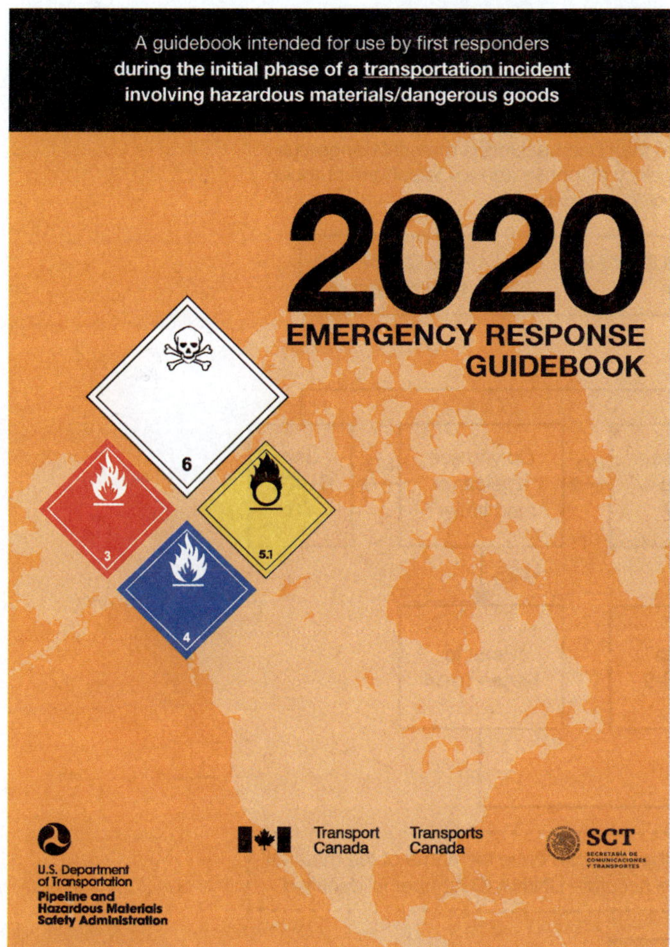

FIGURE 30-27 The *Emergency Response Guidebook (ERG)* is a reference that can be used as the basis for your initial actions at a hazardous materials incident.

Reproduced from U.S. Department of Transportation, Transport Canada, and Secretariat of Communications and Transport of Mexico. 2020. *2020 Emergency Response Guidebook*. Pipeline and Hazardous Materials Safety Administration, U.S. Department of Transportation. https://www.phmsa.dot.gov/sites/phmsa.dot.gov/files/2021-01/ERG2020-WEB.pdf

guidance on initial response actions regarding issues such as the following:

- Isolation and protective action distances
- Recommended PPE for general hazard classes
- Fire control measures
- Information on the GHS
- Hazard identification numbers on some types of intermodal containers
- Pipeline information
- Many other good sources of on-scene guidance

Information on specific substances appears in the yellow-, blue-, and orange-colored sections of the *ERG* (more on this later in the section). Note that the *ERG* does not list all chemicals that could potentially be shipped by land, sea, air, or rail. The guide, which is revised and issued every 4 years and provides information on more than 4000 chemicals, is intended to help you decide which preliminary actions to take.

It is important for all users of the ERG *to know and understand how to use the guide and to know which new information has been updated in each revision cycle.* This text does not go into the details of each revision cycle. Rather, it focuses on the fundamentals of the guide and how it is used by awareness-level personnel and other responders.

Using the *ERG*

At a hazardous materials/WMD incident, if it is possible to identify the UN identification number, the chemical name, marking, placard, label, or perhaps the container shape in the *ERG*, users of the *ERG* may be able to determine the initial emergency actions to be taken as well as make some initial choices about PPE (**FIGURE 30-28**). For example, based on the recommendations in the *ERG*, would a normal work uniform provide adequate protection? Is chemical-protective clothing and equipment indicated? Is SCBA needed? Is structural firefighting–protective clothing adequate, or is **high-temperature–protective clothing**—"protective clothing designed to protect the wearer for short-term high-temperature exposures" (NFPA 470)—indicated? Answering these types of questions also assists awareness-level personnel in determining whether the nonintervention or defensive mode of operation is indicated.

As mentioned earlier, the *ERG* is divided into four colored sections: yellow, blue, orange, and green.

- **Yellow section:** More than 4000 chemicals are found in this section, listed numerically by their four-digit UN number (ID No.). Use the yellow section when the UN number is known or can be identified (**FIGURE 30-29**). Each entry also includes the chemical name. Entry number 1005, for example, identifies "ammonia, anhydrous." This section is useful because it links the UN number to the chemical name of the substance and also directs the user toward the orange section, emergency action guide number (Guide No.). For example:

ID No.	Guide No.	Name of Material
1005	125	Ammonia, anhydrous

Some substances are highlighted in green in this section. Those highlighted substances are either a toxic inhalation hazard, a chemical warfare agent, or a water-reactive material and should be further referenced in the green section.

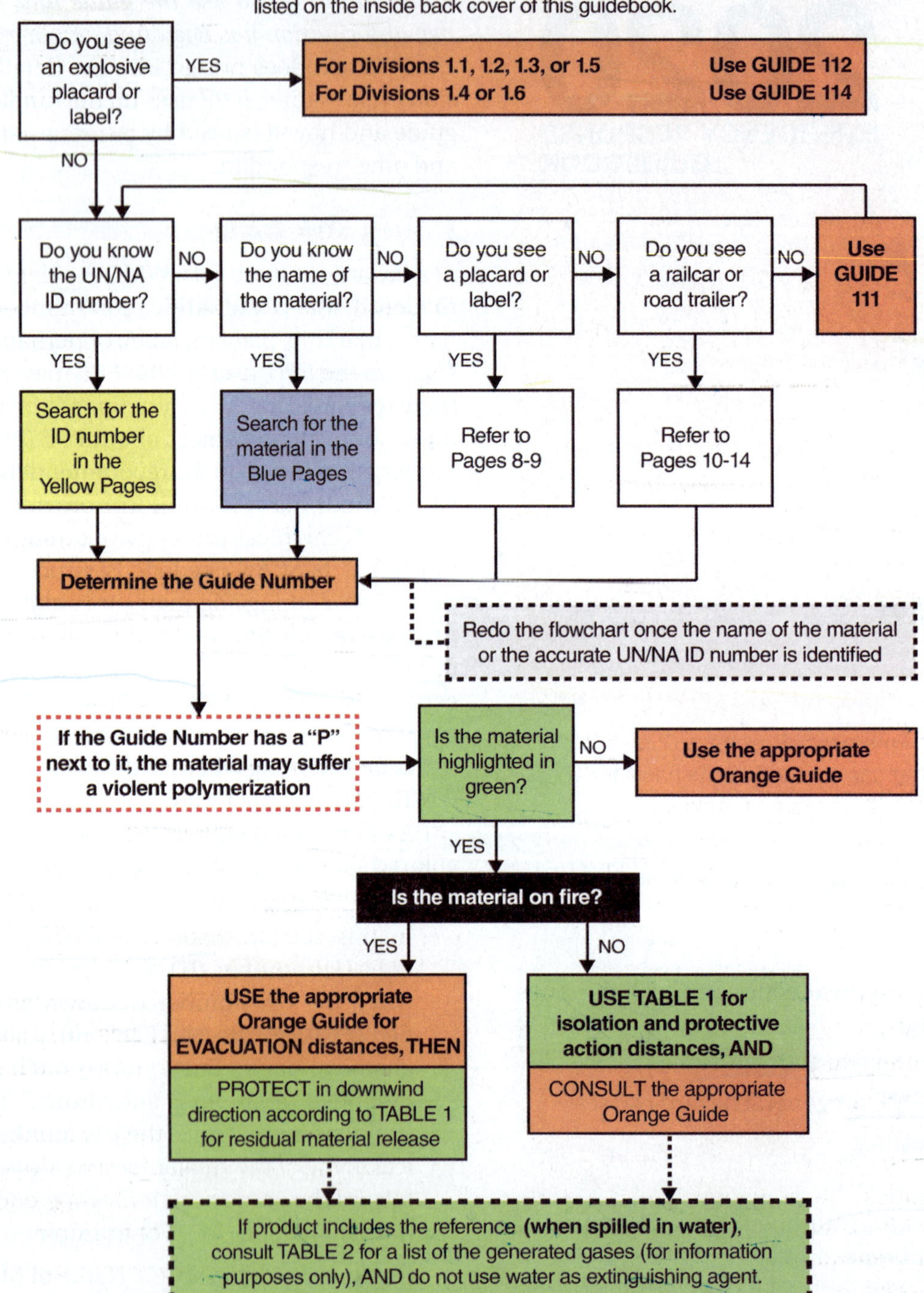

FIGURE 30-28 Flowchart illustrating how to use the *ERG*.

Reproduced from U.S. Department of Transportation, Transport Canada, and Secretariat of Communications and Transport of Mexico. 2020. *2020 Emergency Response Guidebook*. Pipeline and Hazardous Materials Safety Administration, U.S. Department of Transportation. https://www.phmsa.dot.gov/sites/phmsa.dot.gov/files/2021-01/ERG2020-WEB.pdf

ID No.	Guide No.	Name of Material
—	117	AC
—	154	Adamsite
—	112	Ammonium nitrate-fuel oil mixtures
—	158	Biological agents
—	112	Blasting agent, n.o.s.
—	153	Buzz
—	153	BZ
—	159	CA
—	125	CG
—	125	CK
—	153	CN
—	153	CS
—	154	CX
—	151	DA
—	153	DC
—	154	DM
—	125	DP
—	151	ED
—	112	Explosives, division 1.1, 1.2, 1.3 or 1.5
—	114	Explosives, division 1.4 or 1.6
—	153	GA
—	153	GB
—	153	GD
—	153	GF
—	153	H
—	153	HD
—	153	HL
—	153	HN-1
—	153	HN-2
—	153	HN-3
—	153	L (Lewisite)
—	153	Lewisite
—	152	MD
—	153	Mustard
—	153	Mustard Lewisite
—	152	PD
—	119	SA
—	153	Sarin
—	153	Soman
—	153	Tabun
—	153	Thickened GD
—	153	Toxins
—	153	VX
1001	116	Acetylene, dissolved
1002	122	Air, compressed
1003	122	Air, refrigerated liquid (cryogenic liquid)
1005	125	Ammonia, anhydrous
1005	125	Anhydrous ammonia
1006	120	Argon
1006	120	Argon, compressed
1008	125	Boron trifluoride
1008	125	Boron trifluoride, compressed
1009	126	Bromotrifluoromethane
1009	126	Refrigerant gas R-13B1
1010	116P	Butadienes, stabilized
1010	116P	Butadienes and hydrocarbon mixture, stabilized
1010	116P	Hydrocarbon and butadienes mixture, stabilized
1011	115	Butane
1012	115	Butylene

Page 29

FIGURE 30-29 Use the yellow section of the *ERG* when the UN number (ID No.) is known or can be identified.

Reproduced from U.S. Department of Transportation, Transport Canada, and Secretariat of Communications and Transport of Mexico. 2020. *2020 Emergency Response Guidebook*. Pipeline and Hazardous Materials Safety Administration, U.S. Department of Transportation. https://www.phmsa.dot.gov/sites/phmsa.dot.gov/files/2021-01/ERG2020-WEB.pdf

Name of Material	Guide No.	ID No.
Aluminum borohydride	135	2870
Aluminum borohydride in devices	135	2870
Aluminum bromide, anhydrous	137	1725
Aluminum bromide, solution	154	2580
Aluminum carbide	138	1394
Aluminum chloride, anhydrous	137	1726
Aluminum chloride, solution	154	2581
Aluminum dross	138	3170
Aluminum ferrosilicon powder	139	1395
Aluminum hydride	138	2463
Aluminum nitrate	140	1438
Aluminum phosphide	139	1397
Aluminum phosphide pesticide	157	3048
Aluminum powder, coated	170	1309
Aluminum powder, pyrophoric	135	1383
Aluminum powder, uncoated	138	1396
Aluminum remelting by-products	138	3170
Aluminum resinate	133	2715
Aluminum silicon powder, uncoated	138	1398
Aluminum smelting by-products	138	3170
Amines, flammable, corrosive, n.o.s.	132	2733
Amines, liquid, corrosive, flammable, n.o.s.	132	2734
Amines, liquid, corrosive, n.o.s.	153	2735
Amines, solid, corrosive, n.o.s.	154	3259
2-Amino-4-chlorophenol	151	2673
2-Amino-5-diethylaminopentane	153	2946
2-Amino-4,6-dinitrophenol, wetted with not less than 20% water	113	3317
2-(2-Aminoethoxy)ethanol	154	3055
N-Aminoethylpiperazine	153	2815
Aminophenols	152	2512
Aminopyridines	153	2671
Ammonia, anhydrous	125	1005
Ammonia, solution, with more than 10% but not more than 35% Ammonia	154	2672
Ammonia, solution, with more than 35% but not more than 50% Ammonia	125	2073
Ammonia solution, with more than 50% Ammonia	125	3318
Ammonium arsenate	151	1546
Ammonium bifluoride, solid	154	1727
Ammonium bifluoride, solution	154	2817
Ammonium dichromate	141	1439
Ammonium dinitro-o-cresolate, solid	141	1843
Ammonium dinitro-o-cresolate, solution	141	3424
Ammonium fluoride	154	2505
Ammonium fluorosilicate	151	2854
Ammonium hydrogendifluoride, solid	154	1727
Ammonium hydrogendifluoride, solution	154	2817
Ammonium hydrogen sulfate	154	2506
Ammonium hydrogen sulphate	154	2506
Ammonium hydroxide	154	2672
Ammonium hydroxide, with more than 10% but not more than 35% Ammonia	154	2672
Ammonium metavanadate	154	2859
Ammonium nitrate, liquid (hot concentrated solution)	140	2426

Page 97

FIGURE 30-30 The same chemicals listed in the yellow section of the *ERG* are found in the blue section, listed alphabetically by name.

Reproduced from U.S. Department of Transportation, Transport Canada, and Secretariat of Communications and Transport of Mexico. 2020. *2020 Emergency Response Guidebook*. Pipeline and Hazardous Materials Safety Administration, U.S. Department of Transportation. https://www.phmsa.dot.gov/sites/phmsa.dot.gov/files/2021-01/ERG2020-WEB.pdf

- **Blue section:** The same chemicals listed in the yellow section are also found in the blue section, but in the blue section, they are listed alphabetically by name (**FIGURE 30-30**). Use the blue section when you know the name of the substance but not the UN number (ID No.). As in the yellow section, each entry includes the emergency action guide number (Guide No.), found in the orange section, and the UN identification number (ID No.). This section links the UN number with the name of the substance and to the associated emergency action guide number (Guide No.), just in a different way than the yellow section.
- **Orange section:** The orange section describes the hazards associated with the chemicals listed the yellow and blue sections. In this section, the chemicals are organized by emergency action guide number (**FIGURE 30-31**). The general hazard class, fire/explosion hazards, health hazards, and basic emergency actions, based on the hazard class, are provided.
- **Green section:** The green section organizes the chemicals numerically by UN number. The chemicals included in this section are those highlighted in green in the yellow and blue sections. These chemicals include water-reactive materials that produce toxic gases (e.g., calcium phosphide and trichlorosilane); **toxic inhalation hazards (TIH)**, which are gases or volatile liquids that are extremely toxic to humans; chemical warfare agents (CWAs); and dangerous water-reactive materials (WRMs; e.g., anhydrous ammonia, sarin, and sodium cyanide). These gases or volatile liquids are extremely toxic to humans and pose a hazard to health during

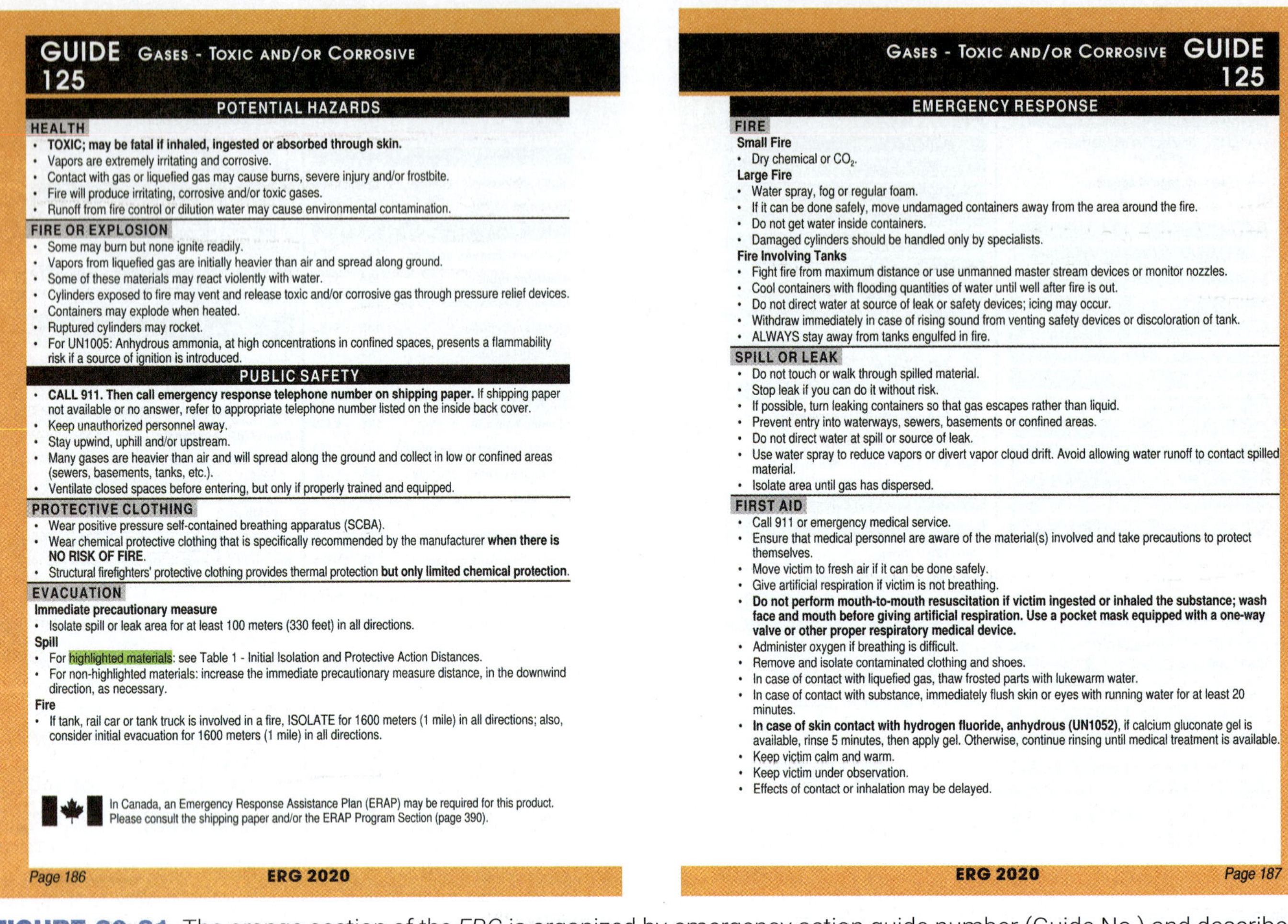

GUIDE 125 GASES - TOXIC AND/OR CORROSIVE

POTENTIAL HAZARDS

HEALTH
- **TOXIC; may be fatal if inhaled, ingested or absorbed through skin.**
- Vapors are extremely irritating and corrosive.
- Contact with gas or liquefied gas may cause burns, severe injury and/or frostbite.
- Fire will produce irritating, corrosive and/or toxic gases.
- Runoff from fire control or dilution water may cause environmental contamination.

FIRE OR EXPLOSION
- Some may burn but none ignite readily.
- Vapors from liquefied gas are initially heavier than air and spread along ground.
- Some of these materials may react violently with water.
- Cylinders exposed to fire may vent and release toxic and/or corrosive gas through pressure relief devices.
- Containers may explode when heated.
- Ruptured cylinders may rocket.
- For UN1005: Anhydrous ammonia, at high concentrations in confined spaces, presents a flammability risk if a source of ignition is introduced.

PUBLIC SAFETY
- **CALL 911. Then call emergency response telephone number on shipping paper.** If shipping paper not available or no answer, refer to appropriate telephone number listed on the inside back cover.
- Keep unauthorized personnel away.
- Stay upwind, uphill and/or upstream.
- Many gases are heavier than air and will spread along the ground and collect in low or confined areas (sewers, basements, tanks, etc.).
- Ventilate closed spaces before entering, but only if properly trained and equipped.

PROTECTIVE CLOTHING
- Wear positive pressure self-contained breathing apparatus (SCBA).
- Wear chemical protective clothing that is specifically recommended by the manufacturer **when there is NO RISK OF FIRE.**
- Structural firefighters' protective clothing provides thermal protection **but only limited chemical protection.**

EVACUATION

Immediate precautionary measure
- Isolate spill or leak area for at least 100 meters (330 feet) in all directions.

Spill
- For highlighted materials: see Table 1 - Initial Isolation and Protective Action Distances.
- For non-highlighted materials: increase the immediate precautionary measure distance, in the downwind direction, as necessary.

Fire
- If tank, rail car or tank truck is involved in a fire, ISOLATE for 1600 meters (1 mile) in all directions; also, consider initial evacuation for 1600 meters (1 mile) in all directions.

In Canada, an Emergency Response Assistance Plan (ERAP) may be required for this product. Please consult the shipping paper and/or the ERAP Program Section (page 390).

Page 186 **ERG 2020**

GASES - TOXIC AND/OR CORROSIVE GUIDE 125

EMERGENCY RESPONSE

FIRE

Small Fire
- Dry chemical or CO_2.

Large Fire
- Water spray, fog or regular foam.
- If it can be done safely, move undamaged containers away from the area around the fire.
- Do not get water inside containers.
- Damaged cylinders should be handled only by specialists.

Fire Involving Tanks
- Fight fire from maximum distance or use unmanned master stream devices or monitor nozzles.
- Cool containers with flooding quantities of water until well after fire is out.
- Do not direct water at source of leak or safety devices; icing may occur.
- Withdraw immediately in case of rising sound from venting safety devices or discoloration of tank.
- ALWAYS stay away from tanks engulfed in fire.

SPILL OR LEAK
- Do not touch or walk through spilled material.
- Stop leak if you can do it without risk.
- If possible, turn leaking containers so that gas escapes rather than liquid.
- Prevent entry into waterways, sewers, basements or confined areas.
- Do not direct water at spill or source of leak.
- Use water spray to reduce vapors or divert vapor cloud drift. Avoid allowing water runoff to contact spilled material.
- Isolate area until gas has dispersed.

FIRST AID
- Call 911 or emergency medical service.
- Ensure that medical personnel are aware of the material(s) involved and take precautions to protect themselves.
- Move victim to fresh air if it can be done safely.
- Give artificial respiration if victim is not breathing.
- **Do not perform mouth-to-mouth resuscitation if victim ingested or inhaled the substance; wash face and mouth before giving artificial respiration. Use a pocket mask equipped with a one-way valve or other proper respiratory medical device.**
- Administer oxygen if breathing is difficult.
- Remove and isolate contaminated clothing and shoes.
- In case of contact with liquefied gas, thaw frosted parts with lukewarm water.
- In case of contact with substance, immediately flush skin or eyes with running water for at least 20 minutes.
- **In case of skin contact with hydrogen fluoride, anhydrous (UN1052),** if calcium gluconate gel is available, rinse 5 minutes, then apply gel. Otherwise, continue rinsing until medical treatment is available.
- Keep victim calm and warm.
- Keep victim under observation.
- Effects of contact or inhalation may be delayed.

ERG 2020 Page 187

FIGURE 30-31 The orange section of the *ERG* is organized by emergency action guide number (Guide No.) and describes the hazards associated with the chemicals in the guide.

transport; in fact, any material listed in the green section is extremely hazardous.

The green section of the *ERG* provides the initial isolation distances for certain materials, including both small and large spills (**FIGURE 30-32**). Additionally, it provides guidance on how to gauge the size of the release. For example, a small spill is a leak from one small package; a small leak in a large container (up to 208 liters or 55 gallons); a small cylinder leak; or any small leak, even one in a large package. A large spill is a large leak or spill (greater than 208 liters or 55 gallons) from a larger container or package; a spill from a number of small packages; or anything from a 1-ton cylinder, tank truck, or railcar.

The green section also offers recommendations on isolation distances and the size and shape of protective action zones. This section is useful when it is necessary to protect people from TIH vapors resulting from a release. The initial isolation zones may be used to define an area surrounding an incident where persons may be exposed to potentially dangerous or life-threatening concentrations of the vapor both upwind and downwind from the release. The orange-bordered guides differ from the green section in that the orange-bordered guides are relevant to evacuation distances required to protect against fragmentation hazard from a large container if it should fail due to explosion. The rationale is that if a certain material becomes involved with fire, the TIH hazard may be less than the fire or explosion hazard.

To use the *ERG*, follow the steps in **SKILL DRILL 30-1**.

TABLE 1 - INITIAL ISOLATION AND PROTECTIVE ACTION DISTANCES

ID No.	Guide	NAME OF MATERIAL	SMALL SPILLS (From a small package or small leak from a large package) First ISOLATE in all Directions Meters (Feet)	SMALL SPILLS Then PROTECT persons Downwind during DAY Kilometers (Miles)	SMALL SPILLS Then PROTECT persons Downwind during NIGHT Kilometers (Miles)	LARGE SPILLS (From a large package or from many small packages) First ISOLATE in all Directions Meters (Feet)	LARGE SPILLS Then PROTECT persons Downwind during DAY Kilometers (Miles)	LARGE SPILLS Then PROTECT persons Downwind during NIGHT Kilometers (Miles)
—	153	Soman (when used as a weapon)	60 m (200 ft)	0.4 km (0.3 mi)	0.7 km (0.5 mi)	300 m (1000 ft)	1.8 km (1.1 mi)	2.7 km (1.7 mi)
—	153	Tabun (when used as a weapon)	30 m (100 ft)	0.2 km (0.1 mi)	0.2 km (0.1 mi)	100 m (300 ft)	0.5 km (0.4 mi)	0.6 km (0.4 mi)
—	153	Thickened GD (when used as a weapon)	60 m (200 ft)	0.4 km (0.3 mi)	0.7 km (0.5 mi)	300 m (1000 ft)	1.8 km (1.1 mi)	2.7 km (1.7 mi)
—	153	VX (when used as a weapon)	30 m (100 ft)	0.1 km (0.1 mi)	0.1 km (0.1 mi)	60 m (200 ft)	0.4 km (0.2 mi)	0.3 km (0.2 mi)
1005 1005	125 125	Ammonia, anhydrous Anhydrous ammonia	30 m (100 ft)	0.1 km (0.1 mi)	0.2 km (0.1 mi)	Refer to table 3		
1008 1008	125 125	Boron trifluoride Boron trifluoride, compressed	30 m (100 ft)	0.2 km (0.1 mi)	0.7 km (0.5 mi)	400 m (1250 ft)	2.3 km (1.4 mi)	5.1 km (3.2 mi)
1016 1016	119 119	Carbon monoxide Carbon monoxide, compressed	30 m (100 ft)	0.1 km (0.1 mi)	0.2 km (0.1 mi)	200 m (600 ft)	1.2 km (0.7 mi)	4.3 km (2.7 mi)
1017	124	Chlorine	60 m (200 ft)	0.3 km (0.2 mi)	1.4 km (0.9 mi)	Refer to table 3		
1026	119	Cyanogen	30 m (100 ft)	0.1 km (0.1 mi)	0.4 km (0.3 mi)	60 m (200 ft)	0.3 km (0.2 mi)	1.1 km (0.7 mi)
1040 1040	119P 119P	Ethylene oxide Ethylene oxide with Nitrogen	30 m (100 ft)	0.1 km (0.1 mi)	0.2 km (0.2 mi)	Refer to table 3		
1045 1045	124 124	Fluorine Fluorine, compressed	30 m (100 ft)	0.1 km (0.1 mi)	0.2 km (0.1 mi)	100 m (300 ft)	0.5 km (0.3 mi)	2.3 km (1.4 mi)
1048	125	Hydrogen bromide, anhydrous	30 m (100 ft)	0.1 km (0.1 mi)	0.2 km (0.2 mi)	150 m (500 ft)	1.0 km (0.6 mi)	3.4 km (2.1 mi)
1050	125	Hydrogen chloride, anhydrous	30 m (100 ft)	0.1 km (0.1 mi)	0.3 km (0.2 mi)	Refer to table 3		

Page 298

steps you take to preserve the health and safety of emergency responders and the public. People in this area should be evacuated and/or sheltered-in-place. Consult pages 289-291.

(6) Initiate protective actions beginning with those closest to the spill site and working away in a downwind direction. When a water-reactive TIH (PIH in the US) producing material is spilled into a river or stream, the source of the toxic gas may move with the current or stretch from the spill point downstream for a large distance.

In the figure below, the spill is located at the center of the small black circle. The larger circle represents the initial isolation zone around the spill. The square (the protective action zone) is the area in which you should take protective actions.

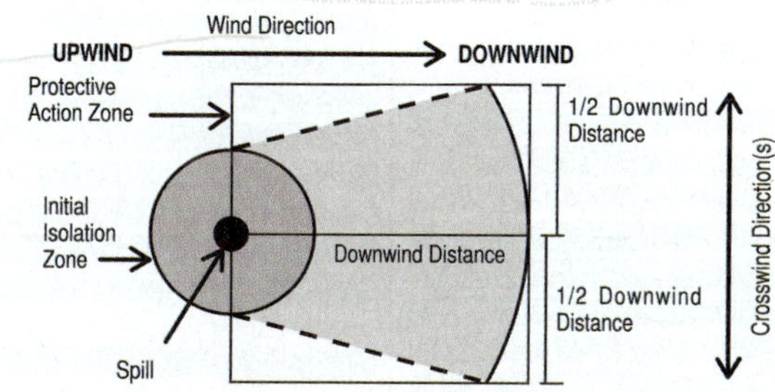

Note 1: For factors that may change the protective action distances, see "Introduction to Green Tables" (page 286).

Note 2: When a product in Table 1 has the mention (when spilled in water), you can refer to Table 2 for the list of gases produced when these materials are spilled in water. The TIH gases indicated in Table 2 are for information purposes only.

For more information on the material, safety precautions and mitigation procedures, call the emergency response telephone number listed on the shipping paper or the appropriate response agency as soon as possible.

Page 295

FIGURE 30-32 The green section of the *ERG* is organized numerically by UN number and provides the initial isolation distances, protective actions, and gauging the size of a release for certain materials.

Reproduced from U.S. Department of Transportation, Transport Canada, and Secretariat of Communications and Transport of Mexico. 2020. *2020 Emergency Response Guidebook*. Pipeline and Hazardous Materials Safety Administration, U.S. Department of Transportation. https://www.phmsa.dot.gov/sites/phmsa.dot.gov/files/2021-01/ERG2020-WEB.pdf

SKILL DRILL 30-1

Using the *Emergency Response Guidebook* NFPA 470: 5.2.1, 5.3.1

1. Identify the chemical name and/or the chemical ID number for the placard.

Continues.

SKILL DRILL 30-1 CONTINUED

Using the *Emergency Response Guidebook* NFPA 470: 5.2.1, 5.3.1

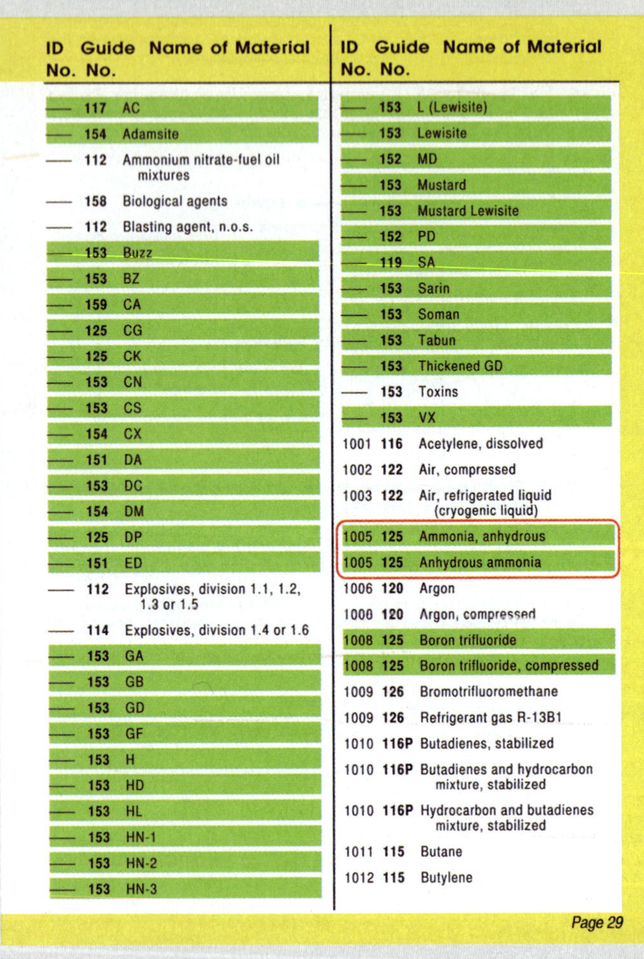

ID No.	Guide No.	Name of Material
—	117	AC
—	154	Adamsite
—	112	Ammonium nitrate-fuel oil mixtures
—	158	Biological agents
—	112	Blasting agent, n.o.s.
—	153	Buzz
—	153	BZ
—	159	CA
—	125	CG
—	125	CK
—	153	CN
—	153	CS
—	154	CX
—	151	DA
—	153	DC
—	154	DM
—	125	DP
—	151	ED
—	112	Explosives, division 1.1, 1.2, 1.3 or 1.5
—	114	Explosives, division 1.4 or 1.6
—	153	GA
—	153	GB
—	153	GD
—	153	GF
—	153	H
—	153	HD
—	153	HL
—	153	HN-1
—	153	HN-2
—	153	HN-3

ID No.	Guide No.	Name of Material
—	153	L (Lewisite)
—	153	Lewisite
—	152	MD
—	153	Mustard
—	153	Mustard Lewisite
—	152	PD
—	119	SA
—	153	Sarin
—	153	Soman
—	153	Tabun
—	153	Thickened GD
—	153	Toxins
—	153	VX
1001	116	Acetylene, dissolved
1002	122	Air, compressed
1003	122	Air, refrigerated liquid (cryogenic liquid)
1005	125	Ammonia, anhydrous
1005	125	Anhydrous ammonia
1006	120	Argon
1006	120	Argon, compressed
1008	125	Boron trifluoride
1008	125	Boron trifluoride, compressed
1009	126	Bromotrifluoromethane
1009	126	Refrigerant gas R-13B1
1010	116P	Butadienes, stabilized
1010	116P	Butadienes and hydrocarbon mixture, stabilized
1010	116P	Hydrocarbon and butadienes mixture, stabilized
1011	115	Butane
1012	115	Butylene

Page 29

Reproduced from U.S. Department of Transportation, Transport Canada, and Secretariat of Communications and Transport of Mexico. 2020. *2020 Emergency Response Guidebook*. Pipeline and Hazardous Materials Safety Administration, U.S. Department of Transportation. https://www.phmsa.dot.gov/sites/phmsa.dot.gov/files/2021-01/ERG2020-WEB.pdf

2. Look up the material name in the appropriate section of the *ERG*. Use the yellow section to obtain information based on the UN identification number. Use the alphabetized blue section to obtain information based on the chemical name. *Note any green highlights, which indicate the substance also has an entry and recommendations in the green section of the guide.*

SKILL DRILL 30-1 CONTINUED

Using the *Emergency Response Guidebook* NFPA 470: 5.2.1, 5.3.1

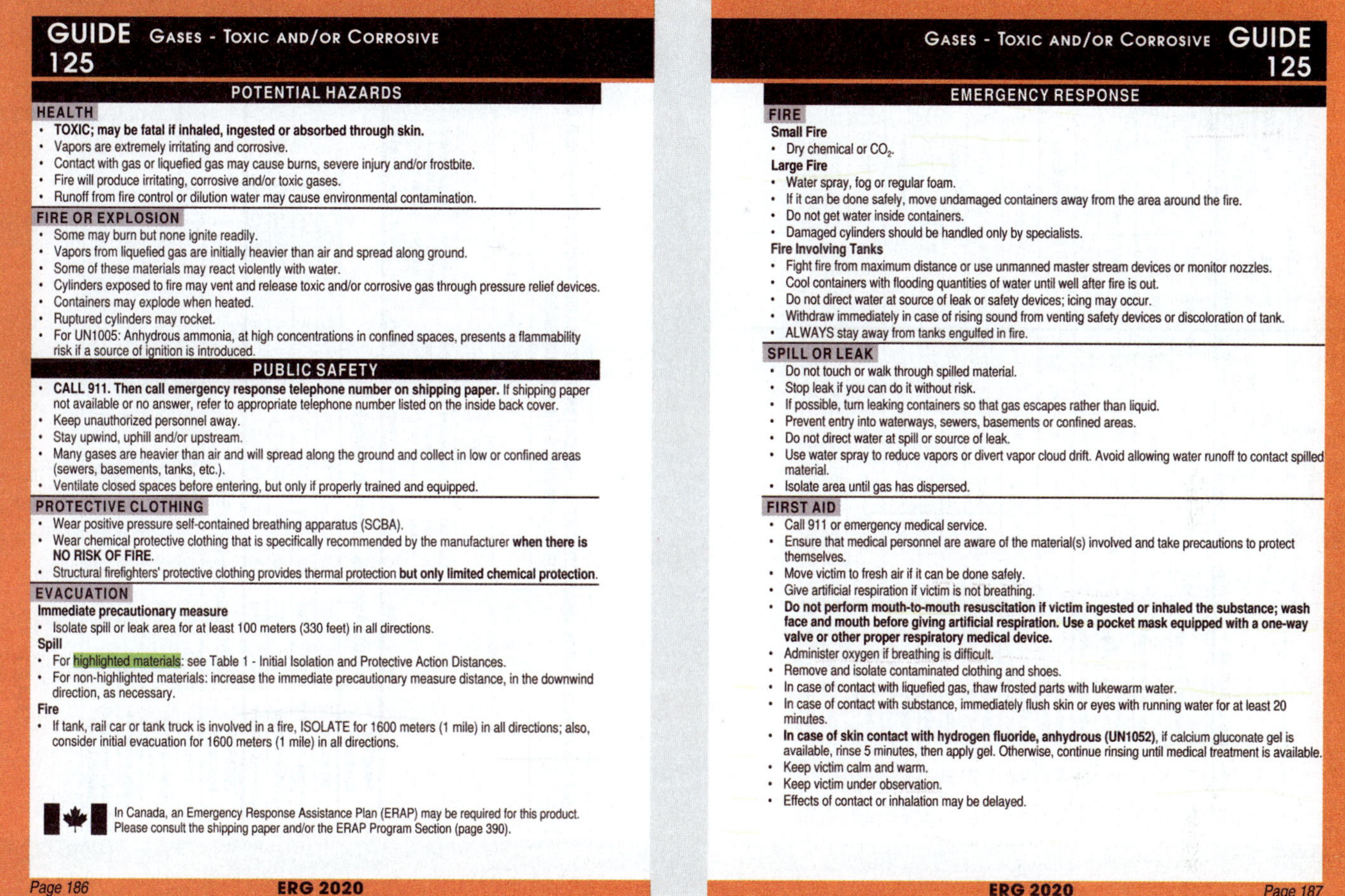

GUIDE 125 GASES - TOXIC AND/OR CORROSIVE

POTENTIAL HAZARDS

HEALTH
- **TOXIC; may be fatal if inhaled, ingested or absorbed through skin.**
- Vapors are extremely irritating and corrosive.
- Contact with gas or liquefied gas may cause burns, severe injury and/or frostbite.
- Fire will produce irritating, corrosive and/or toxic gases.
- Runoff from fire control or dilution water may cause environmental contamination.

FIRE OR EXPLOSION
- Some may burn but none ignite readily.
- Vapors from liquefied gas are initially heavier than air and spread along ground.
- Some of these materials may react violently with water.
- Cylinders exposed to fire may vent and release toxic and/or corrosive gas through pressure relief devices.
- Containers may explode when heated.
- Ruptured cylinders may rocket.
- For UN1005: Anhydrous ammonia, at high concentrations in confined spaces, presents a flammability risk if a source of ignition is introduced.

PUBLIC SAFETY
- **CALL 911. Then call emergency response telephone number on shipping paper.** If shipping paper not available or no answer, refer to appropriate telephone number listed on the inside back cover.
- Keep unauthorized personnel away.
- Stay upwind, uphill and/or upstream.
- Many gases are heavier than air and will spread along the ground and collect in low or confined areas (sewers, basements, tanks, etc.).
- Ventilate closed spaces before entering, but only if properly trained and equipped.

PROTECTIVE CLOTHING
- Wear positive pressure self-contained breathing apparatus (SCBA).
- Wear chemical protective clothing that is specifically recommended by the manufacturer **when there is NO RISK OF FIRE**.
- Structural firefighters' protective clothing provides thermal protection **but only limited chemical protection**.

EVACUATION

Immediate precautionary measure
- Isolate spill or leak area for at least 100 meters (330 feet) in all directions.

Spill
- For highlighted materials: see Table 1 - Initial Isolation and Protective Action Distances.
- For non-highlighted materials: increase the immediate precautionary measure distance, in the downwind direction, as necessary.

Fire
- If tank, rail car or tank truck is involved in a fire, ISOLATE for 1600 meters (1 mile) in all directions; also, consider initial evacuation for 1600 meters (1 mile) in all directions.

In Canada, an Emergency Response Assistance Plan (ERAP) may be required for this product. Please consult the shipping paper and/or the ERAP Program Section (page 390).

Page 186 ERG 2020

GASES - TOXIC AND/OR CORROSIVE GUIDE 125

EMERGENCY RESPONSE

FIRE

Small Fire
- Dry chemical or CO_2.

Large Fire
- Water spray, fog or regular foam.
- If it can be done safely, move undamaged containers away from the area around the fire.
- Do not get water inside containers.
- Damaged cylinders should be handled only by specialists.

Fire Involving Tanks
- Fight fire from maximum distance or use unmanned master stream devices or monitor nozzles.
- Cool containers with flooding quantities of water until well after fire is out.
- Do not direct water at source of leak or safety devices; icing may occur.
- Withdraw immediately in case of rising sound from venting safety devices or discoloration of tank.
- ALWAYS stay away from tanks engulfed in fire.

SPILL OR LEAK
- Do not touch or walk through spilled material.
- Stop leak if you can do it without risk.
- If possible, turn leaking containers so that gas escapes rather than liquid.
- Prevent entry into waterways, sewers, basements or confined areas.
- Do not direct water at spill or source of leak.
- Use water spray to reduce vapors or divert vapor cloud drift. Avoid allowing water runoff to contact spilled material.
- Isolate area until gas has dispersed.

FIRST AID
- Call 911 or emergency medical service.
- Ensure that medical personnel are aware of the material(s) involved and take precautions to protect themselves.
- Move victim to fresh air if it can be done safely.
- Give artificial respiration if victim is not breathing.
- **Do not perform mouth-to-mouth resuscitation if victim ingested or inhaled the substance; wash face and mouth before giving artificial respiration. Use a pocket mask equipped with a one-way valve or other proper respiratory medical device.**
- Administer oxygen if breathing is difficult.
- Remove and isolate contaminated clothing and shoes.
- In case of contact with liquefied gas, thaw frosted parts with lukewarm water.
- In case of contact with substance, immediately flush skin or eyes with running water for at least 20 minutes.
- **In case of skin contact with hydrogen fluoride, anhydrous (UN1052)**, if calcium gluconate gel is available, rinse 5 minutes, then apply gel. Otherwise, continue rinsing until medical treatment is available.
- Keep victim calm and warm.
- Keep victim under observation.
- Effects of contact or inhalation may be delayed.

ERG 2020 Page 187

3. Determine the correct emergency action guide to use for the chemical identified.

Continues.

SKILL DRILL 30-1 CONTINUED

Using the *Emergency Response Guidebook* NFPA 470: 5.2.1, 5.3.1

TABLE 1 - INITIAL ISOLATION AND PROTECTIVE ACTION DISTANCES

ID No.	Guide	NAME OF MATERIAL	SMALL SPILLS (From a small package or small leak from a large package) First ISOLATE in all Directions Meters (Feet)	SMALL SPILLS Then PROTECT persons Downwind during DAY Kilometers (Miles)	SMALL SPILLS Then PROTECT persons Downwind during NIGHT Kilometers (Miles)	LARGE SPILLS (From a large package or from many small packages) First ISOLATE in all Directions Meters (Feet)	LARGE SPILLS Then PROTECT persons Downwind during DAY Kilometers (Miles)	LARGE SPILLS Then PROTECT persons Downwind during NIGHT Kilometers (Miles)
—	153	Soman (when used as a weapon)	60 m (200 ft)	0.4 km (0.3 mi)	0.7 km (0.5 mi)	300 m (1000 ft)	1.8 km (1.1 mi)	2.7 km (1.7 mi)
—	153	Tabun (when used as a weapon)	30 m (100 ft)	0.2 km (0.1 mi)	0.2 km (0.1 mi)	100 m (300 ft)	0.5 km (0.4 mi)	0.6 km (0.4 mi)
—	153	Thickened GD (when used as a weapon)	60 m (200 ft)	0.4 km (0.3 mi)	0.7 km (0.5 mi)	300 m (1000 ft)	1.8 km (1.1 mi)	2.7 km (1.7 mi)
—	153	VX (when used as a weapon)	30 m (100 ft)	0.1 km (0.1 mi)	0.1 km (0.1 mi)	60 m (200 ft)	0.4 km (0.2 mi)	0.3 km (0.2 mi)
1005 1005	125 125	Ammonia, anhydrous Anhydrous ammonia	30 m (100 ft)	0.1 km (0.1 mi)	0.2 km (0.1 mi)	Refer to table 3		
1008 1008	125 125	Boron trifluoride Boron trifluoride, compressed	30 m (100 ft)	0.2 km (0.1 mi)	0.7 km (0.5 mi)	400 m (1250 ft)	2.3 km (1.4 mi)	5.1 km (3.2 mi)
1016 1016	119 119	Carbon monoxide Carbon monoxide, compressed	30 m (100 ft)	0.1 km (0.1 mi)	0.2 km (0.1 mi)	200 m (600 ft)	1.2 km (0.7 mi)	4.3 km (2.7 mi)
1017	124	Chlorine	60 m (200 ft)	0.3 km (0.2 mi)	1.4 km (0.9 mi)	Refer to table 3		
1026	119	Cyanogen	30 m (100 ft)	0.1 km (0.1 mi)	0.4 km (0.3 mi)	60 m (200 ft)	0.3 km (0.2 mi)	1.1 km (0.7 mi)
1040 1040	119P 119P	Ethylene oxide Ethylene oxide with Nitrogen	30 m (100 ft)	0.1 km (0.1 mi)	0.2 km (0.2 mi)	Refer to table 3		
1045 1045	124 124	Fluorine Fluorine, compressed	30 m (100 ft)	0.1 km (0.1 mi)	0.2 km (0.1 mi)	100 m (300 ft)	0.5 km (0.3 mi)	2.3 km (1.4 mi)
1048	125	Hydrogen bromide, anhydrous	30 m (100 ft)	0.1 km (0.1 mi)	0.2 km (0.2 mi)	150 m (500 ft)	1.0 km (0.6 mi)	3.4 km (2.1 mi)
1050	125	Hydrogen chloride, anhydrous	30 m (100 ft)	0.1 km (0.1 mi)	0.3 km (0.2 mi)	Refer to table 3		

Page 298

TABLE 3 - INITIAL ISOLATION AND PROTECTIVE ACTION DISTANCES FOR LARGE SPILLS FOR DIFFERENT QUANTITIES OF SIX COMMON TIH (PIH in the US) GASES

TRANSPORT CONTAINER	First ISOLATE in all Directions Meters (Feet)	Then PROTECT persons Downwind during DAY Low wind (< 6 mph = < 10 km/h) km (Miles)	DAY Moderate wind (6-12 mph = 10 - 20 km/h) km (Miles)	DAY High wind (> 12 mph = > 20 km/h) km (Miles)	NIGHT Low wind (< 6 mph = < 10 km/h) km (Miles)	NIGHT Moderate wind (6-12 mph = 10 - 20 km/h) km (Miles)	NIGHT High wind (> 12 mph = > 20 km/h) km (Miles)
UN1005 Ammonia, anhydrous: Large Spills							
Rail tank car	300 (1000)	1.9 (1.2)	1.5 (0.9)	1.1 (0.6)	4.5 (2.8)	2.5 (1.5)	1.4 (0.9)
Highway tank truck or trailer	150 (500)	0.9 (0.6)	0.5 (0.3)	0.4 (0.3)	2.0 (1.3)	0.8 (0.5)	0.6 (0.4)
Agricultural nurse tank	60 (200)	0.5 (0.3)	0.3 (0.2)	0.3 (0.2)	1.4 (0.9)	0.3 (0.2)	0.3 (0.2)
Multiple small cylinders	30 (100)	0.3 (0.2)	0.2 (0.1)	0.1 (0.1)	0.7 (0.5)	0.3 (0.2)	0.2 (0.1)
UN1017 Chlorine: Large Spills							
Rail tank car	1000 (3000)	10.1 (6.3)	6.8 (4.2)	5.3 (3.3)	11+ (7+)	9.2 (5.7)	6.9 (4.3)
Highway tank truck or trailer	600 (2000)	5.8 (3.6)	3.4 (2.1)	2.9 (1.8)	6.7 (4.3)	5.0 (3.1)	4.1 (2.5)
Multiple ton cylinders	300 (1000)	2.1 (1.3)	1.3 (0.8)	1.0 (0.6)	4.0 (2.5)	2.4 (1.5)	1.3 (0.8)
Multiple small cylinders or single ton cylinder	150 (500)	1.5 (0.9)	0.8 (0.5)	0.5 (0.3)	2.9 (1.8)	1.3 (0.8)	0.6 (0.4)

TABLE 3

"+" means distance can be larger in certain atmospheric conditions

Page 351

4. Identify the primary hazard (e.g., health hazards), potential fire and explosion hazards, recommended protective clothing, and evacuation recommendations. (For the substance used in this exercise, you should find a Table 1 recommendation of initial isolation and protective action distances as well as isolation distances if the substance is involved in fire. Also, take note of the firefighting recommendation, handling spills or leaks, and first aid measures.)

5. If necessary, identify the isolation distance and the protective actions required for the chemical substance in the green section.

Harmful Substances' Routes of Entry into the Human Body

Even though awareness-level personnel are not considered to be incident responders, it is important for them to understand the pathways by which chemical substances can enter the human body and potentially cause harm. For any chemical or other harmful substance to injure a person or animal, it must first get into or onto the body. Chemical substances can enter the human body in four ways (**FIGURE 30-33**):

- Inhalation: Through the lungs
- Absorption: By permeating the skin
- Ingestion: Via the gastrointestinal tract
- Injection: Through cuts or other breaches in the skin

The following sections discuss potential routes of entry into the human body of chemical or other harmful substances and methods used to protect against these agents.

Inhalation Exposure

An **inhalation exposure** occurs when harmful substances enter the body through the respiratory system. The lungs are a direct point of access to the bloodstream, so they can quickly transfer an airborne substance into the circulatory system and onward to the rest of the body. In addition, the lungs cannot be decontaminated, so any exposure will result in some type of harm that cannot be addressed in the same way as an exposure to skin.

The respiratory system is vulnerable to attack from a wide range of substances, from corrosive materials such as chlorine and ammonia, to solvent vapors such as gasoline and acetone, to superheated air from a fire, to any other material finding its way into the air. In addition to gases and vapors, small particles of dust, fiberglass insulation, asbestos, or soot from a fire can become lodged in sensitive lung tissue, causing substantial irritation.

Given these facts, it is imperative that personnel, if properly trained and equipped to do so, wear appropriate respiratory protection when operating in the presence of airborne contamination. Fortunately, most firefighters have ready access to an excellent form of respiratory protection—namely, the positive-pressure, open-circuit self-contained breathing apparatus (SCBA). This equipment is by far the single most important piece of PPE that firefighters have at their disposal. Sometimes firefighters may need to use other forms of respiratory protection depending on the specific respiratory hazard they are facing while performing a given task. For example, full-face and half-face air-purifying respirators (APRs) offer specific degrees of protection if the chemical hazard present is known and the appropriate filter canister is used (**FIGURE 30-34**).

APRs do not provide oxygen, however. Thus, if the oxygen content of the atmosphere in the work area is known or suspected to be abnormal, full- or half-face

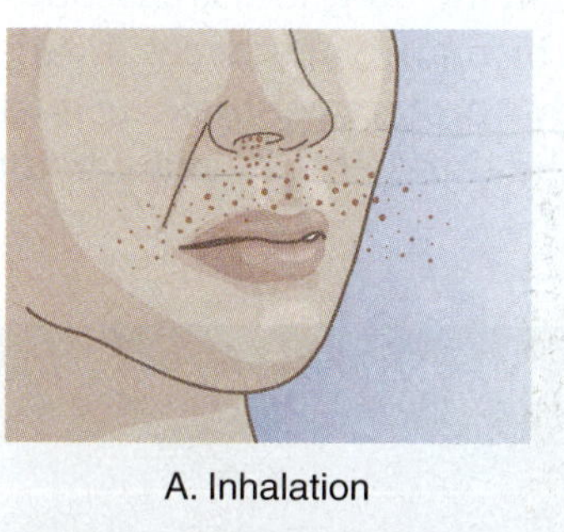

A. Inhalation

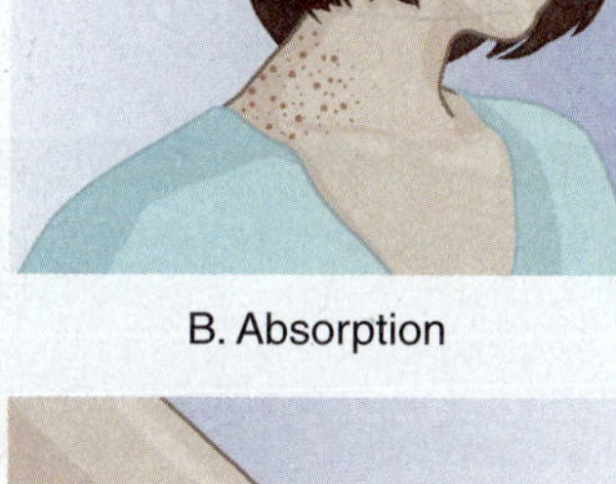

B. Absorption

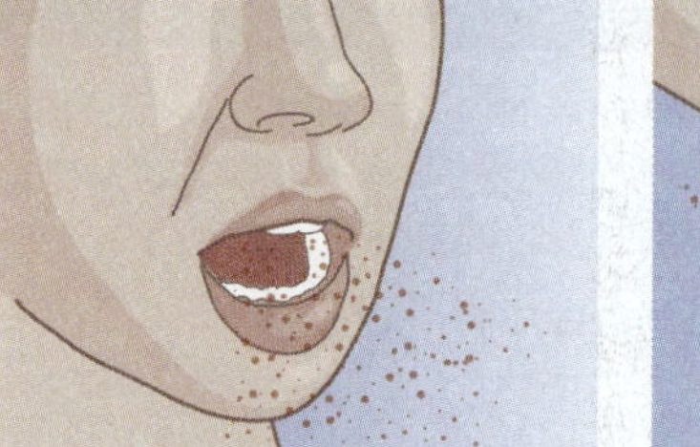

C. Ingestion

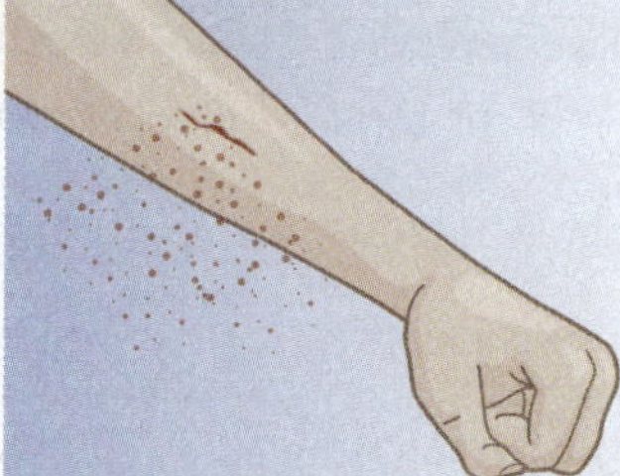

D. Injection

FIGURE 30-33 The four ways a chemical substance can enter the body.

FIGURE 30-34 Air-purifying respirators offer protection against known and characterized airborne chemical hazards.

respirators are not a viable option. Per OSHA, any work environment containing less than 19.5 percent oxygen is considered to be oxygen deficient and requires the use of SCBA or a supplied-air respirator.

Respirators are lighter than SCBA, are more comfortable to wear, and usually allow personnel to work for longer periods because they are not dependent on a limited source of breathing air. SCBAs certainly offer a higher level of respiratory protection, but in the right circumstances, APRs may represent a viable form of respiratory protection.

When considering protection against airborne contamination, it is important to understand the origin, concentration, and potential impact of the contamination relative to the oxygen levels in the area. In short, the appropriate type of respiratory protection is determined by looking at the overall situation, including the nature of the contaminant. In some cases, the anticipated particle size of the contamination may dictate the level of respiratory protection employed (**TABLE 30-2**).

Anthrax spores offer an excellent example to illustrate this point. Weaponized anthrax spores typically vary in size from 0.5 micron to 1 micron. Based on that size range, a typical full-face APR with a nuisance dust filter, or a surgical-type mask, would not offer sufficient protection against this hazard. Anyone operating in an area contaminated with anthrax should wear SCBA or, at a minimum, a full-face APR with P100 filtration (which filters out greater than 99 percent of 0.3-micron or larger particles).

Anthrax exposures also illustrate another important pair of terms and definitions that personnel should understand: **infectious** and **contagious**. Anthrax is a pathogenic microorganism capable of causing an illness (infectious). A person with an illness caused by an anthrax exposure, however, cannot pass the illness along to another person (contagious). In other words, anthrax is not contagious. Conversely, smallpox is both infectious and contagious, which is why this pathogen poses such a high risk in the event of an outbreak.

Consider this scenario: A container of gaseous helium is leaking inside a poorly ventilated storage room. Helium is nonflammable; the main threat it poses is the possibility of oxygen deficiency. In this case, SCBA is the appropriate respiratory protection based on the anticipated hazard. In many cases, when hazardous materials technicians respond to incidents, they use air-monitoring devices to characterize the work area prior to entry. When this step is taken, the choice of respiratory protection is based on more definitive information.

Particle size also determines where the inspired contamination will eventually end up. The larger particles that make up visible mists will be captured in the nose and upper airway, for example, whereas smaller particles may work their way deeper into the lung (**TABLE 30-3**).

Respiratory protection is one of the most important PPE components. In all cases where airborne contamination is encountered, take the time to understand the nature of the threat, evaluate the respiratory protection available, and decide whether it provides adequate protection. When protecting the lungs, "good enough" is not an option.

Absorption Exposure

The skin is the largest organ in the body and is susceptible to the damage inflicted by many substances. In addition to serving as the body's protective shield against heat, light, and infection, the skin helps regulate body

TABLE 30-2 Particle Sizes of Common Types of Respiratory Hazards

Type of Particle	Size
Fume	< 1 micron
Smoke	≤ 1 micron
Dust	≥ 1 micron
Fog	> 40 microns
Mist	< 40 microns

TABLE 30-3 Location of Respiratory Trapping by Particle Size

Particle Size	Respiratory Area Affected
< 7 microns	Nose
5–7 microns	Larynx
3–5 microns	Trachea and bronchi
2–3 microns	Bronchi
1–2.5 microns	Respiratory bronchioles
0.5–1 micron	Alveoli

temperature, stores water and fat, and serves as a sensory center for painful and pleasant stimulation. Without this important organ, human beings would not be able to survive.

When discussing chemical exposures, however, absorption is not limited only to the skin. **Absorption exposure** occurs when substances travel through body tissues until they reach the bloodstream. The eyes, nose, mouth, and, to a certain degree, the intestinal tract are also part of the equation. The eyes, for example, can absorb a large amount of any liquids and vapors that encounter these sensitive tissues. This absorption is particularly problematic because the eyes connect directly to the optic nerve, which allows the chemical to follow a direct route to the brain and the central nervous system.

Although the skin functions as a shield for the body, that shield can be pierced by many chemicals. An aggressive solvent such as methylene chloride (found in paint stripper), for example, is readily absorbed through the skin. A secondary hazard associated with this chemical occurs when the body attempts to metabolize the substance after it is absorbed. A by-product of that metabolism reaction is carbon monoxide, a cellular asphyxiant. Asphyxiants are substances that prevent the body from using oxygen at the cellular level, thereby causing suffocation. With solvents such as methylene chloride, the initial chemical is broken down to form another substance that is potentially a greater health hazard than the original chemical because it can cause asphyxiation. Methylene chloride is also suspected to be a **carcinogen**—a human cancer-causing agent.

Absorption hazards are not just limited to solvents. Hydrofluoric acid, for example, poses a significant threat to life when it is absorbed through the skin. This unique corrosive can bind with certain substances in the body (predominantly calcium). Secondary health effects occurring after exposure can include muscular pain and potentially lethal cardiac arrhythmias.

In the field, personnel must constantly evaluate the possibility of chemical contact with their skin and eyes. Structural firefighting turnout gear provides little or no protection against liquid chemicals. Consult the *ERG* or an SDS for response guidance when deciding whether turnout gear is appropriate for the hazard you are facing. In the event the turnout gear does not offer adequate protection, you may be operating beyond the scope of awareness-level training.

Ingestion Exposure

In addition to absorption through the skin, chemicals can be brought into the body through **ingestion exposure**—that is, through the gastrointestinal tract. The water, nutrients, and vitamins the body requires are predominantly absorbed in this manner. For example, at a structure fire, firefighters generally have an opportunity to rotate out of the building for rest and refreshment. When they do so, they may not always take the time to wash up prior to eating or drinking. This leads to a high probability of spreading contamination from the hands to the food and subsequently to the intestinal tract. If you do not think about every situation where you might become exposed, you may put yourself in harm's way.

Injection Exposure

Chemicals brought into the body through open cuts and abrasions qualify as an **injection exposure**. To protect yourself from this route of exposure, begin by realizing when you are in a compromised state. Any cuts or open wounds should be addressed before reporting for duty. If they are significant, you may be excluded from operating in contaminated environments. Open wounds act as a direct portal to the bloodstream and subsequently to muscles, organs, and other body systems. If a chemical substance encounters this open portal, the health effects could be immediate and pronounced. Remember—intact skin is a good protective shield. Do not go into battle if your shield is not up to the task.

Managing Exposure to Harmful Substances

The damage that a hazardous material/WMD will inflict on human beings, animals, or the environment is a function of the physical and chemical properties of the released substance, as well as the conditions under which it was released and the duration of the exposure. Among other factors, characteristics such as the concentration of the material, the temperature of the material at the time of its release, and the pressure under which the substance was released affect both the release parameters and the potential health effects on those exposed to the material. Additionally, the age, sex, genetics, and underlying medical conditions of the exposed person will have some bearing on their outcome. Chemical exposures are complicated events because of the number of variables that may be present.

When it comes to rendering medical care to persons exposed to harmful substances, guidance can be found in NFPA 470, chapter 46, *Competencies for Hazardous Materials/WMD Basic Life Support (BLS) Responders*, and chapter 47, *Competencies for*

Hazardous Materials/WMD Advanced Life Support (ALS) Responders. This standard outlines the basic set of hazardous materials response skills that all EMS responders, regardless of their scope of practice, need to work safely at a hazardous materials/WMD scene and deliver effective patient care. In most cases, that care will be performed in a safe area (cold zone) away from the hazard, after decontamination. There are very few circumstances where definitive advanced life support must be rendered in the hot zone. EMS responders, however, should understand the nature of the incident and look at the scene with a critical eye to pick up the clues that might assist with defining the nature of the exposure.

CASE STUDY

You Are the Firefighter CONCLUSION

During an odor investigation in a light-industrial area of your district, you are directed by a citizen to an abandoned open-head 55-gallon steel drum just off the city street. The person points you to a local shopkeeper, who made the 911 call. The shopkeeper reports that the drum must have been illegally dropped off overnight, and that he smelled an unusual odor when he walked by it. You notice a green label on the side of the drum reading "nonhazardous waste." You and the other members of the engine company are trained to the hazardous materials awareness level (based on NFPA 470). Even though you are not tasked to respond to a hazardous materials incident, you now find yourself on the scene of a potential hazardous materials incident.

1. **What initial actions would you take based on your level of training?**

   Answer: As awareness-level personnel, the on-scene engine company should maintain a safe distance from the drum, isolate the immediate area, and prevent civilians and other responders from entering the secured perimeter. Also, the company officer should request from the dispatch center that a higher level of hazardous materials responders be dispatched to the scene. These responders may be within the AHJ or requested from a neighboring agency.

2. **What does the label on the drum signify?**

   Answer: While the label may identify the contents as a nonhazardous material, it is not a 100 percent accurate indicator. It is not uncommon for illegally dumped hazardous materials to be in improperly stored or labeled containers. Hazardous materials technicians or specialists should be called in to sample and identify the contents.

3. **How would you go about obtaining more information on the potential contents and potential hazards of the contents of the drum? What notifications would you make?**

   Answer: There is no additional information to be gained from the label, but the responders should take note of the fact the material is in a steel drum. This may indicate that the material inside is not corrosive, provided there is no visible signs of degradation of the steel or obvious corrosion. Responders could also make an effort to question other residents or business owners to determine if there are any additional clues as to the origin of the drum. Responders could also investigate the surrounding streets to see if public or private security cameras captured any information on suspicious vehicles or persons in the area prior to the discovery of the drum by the shopkeeper.

WRAP-UP

SUMMARY

KNOWLEDGE OBJECTIVES

- Describe how to approach a scene size-up when potentially hazardous materials are involved. (**NFPA 470: 5.2.1, 5.3.1, 5.4.1**, pp. 1216–1220)
  - Understand the concept of using a risk-based response process to analyze a problem.
  - Understand the basic elements and actions of the acronym SIN [Safety—Isolate— Notify].
- Identify and describe the types of containers that are often used to contain hazardous materials. (**NFPA 470: 5.2.1**, pp. 1220–1224)
  - Link container type to basic types of chemicals that may be contained inside.
- Describe the purpose and types of various transportation and facility markings for hazardous material. (**NFPA 470: 5.2.1, 5.3.1**, pp. 1224–1229)
  - Determine initial tactics and strategies for safe approach and isolation of potential hazardous materials.
  - Determine initial health and safety concerns when transportation and facility markings are observed by responders.
  - Understand the benefits and limitations of the *Emergency Response Guidebook (ERG)*.
- Identify and describe the four routes of entry harmful substances take in the human body. (**NFPA 470: 5.2.1, 5.3.1**, pp. 1245–1247)
  - Identify basic types of personal protective equipment and safety precautions to eliminate or reduce an accidental exposure.
  - Understand basic actions and resources available to responders to better manage an acute chemical exposure.

SKILLS OBJECTIVES

- Use the *Emergency Response Guidebook* (*ERG*). (**NFPA 470: 5.2.1, 5.3.1**, pp. 1241–1244)

KEY TERMS

absorption exposure Exposure to a hazardous material in which substances travel through body tissues until they reach the bloodstream.

air bill The shipping papers on an airplane.

asphyxiant A material that prevents the body from using oxygen at the cellular level, causing the victim to suffocate.

bill of lading The shipping papers used for transport of chemicals over roads and highways; also called *freight bill.*

BLEVE (boiling liquid/expanding vapor explosion) An explosion that occurs when pressurized liquefied materials, such as propane or butane, inside a closed vessel are exposed to a source of high heat.

bung An opening on top of a closed-head drum that is typically sealed with a threaded cap.

carboy A glass, plastic, or steel storage container, ranging in volume from 5 to 15 gallons.

carcinogen A human cancer-causing agent.

Chemical Abstracts Service (CAS) A division of the American Chemical Society that provides hazardous materials personnel and responders with access to the CAS Registry, an enormous collection of chemical substance information.

closed-head drum A drum with a lid that is permanently attached to the drum; the lid has one or more small openings called bungs.

KEY TERMS CONTINUED

consist A list of the contents of every car on a train; also called *train list.*

contagious Capable of transmitting a disease.

container A receptacle, piping, or pipeline used for storing or transporting material of any kind. (NFPA 470)

cryogen See *cryogenic liquid.*

cryogenic liquid A fluid, such as liquid helium, liquid nitrogen, or liquid argon, that has a boiling point lower than –130°F (–90°C) at an absolute pressure of 14.7 psi (101.3 kPa); also called *cryogen.*

cylinder A container that has a circular cross-section and that is designed to store liquids or gases under pressure higher than 40 psi (276 kPa); this definition does not include portable tanks, multiunit tanks, car tanks, cargo tanks, or tank cars. (NFPA 1)

dangerous cargo manifest The shipping papers on a marine vessel.

Dewar container A cylinder container designed to preserve the temperature of cryogenic liquids.

drum A barrel-like storage vessel constructed of low-carbon steel, polyethylene, cardboard, stainless steel, nickel, or other materials, which is used to store a wide variety of substances, including food-grade materials, corrosives, flammable liquids, and grease.

Emergency Response Guidebook (ERG) The reference book, written in plain language, to guide emergency responders in their initial actions at the incident scene, specifically the *Emergency Response Guidebook* from the U.S. Department of Transportation, Transport Canada, and the Secretariat of Transport and Communications, Mexico, used to guide personnel and responders in their initial actions at the incident scene. (NFPA 470)

freight bill See *bill of lading.*

hazard control zones The areas at hazardous materials/WMD incidents within an established perimeter that are designated based on safety and the degree of hazard.

Hazardous Materials Information System (HMIS) A color-coded marking system that identifies the hazard level of chemicals in a container so that personnel can work safely around chemicals.

high-temperature–protective clothing Protective clothing designed to protect the wearer from short-term high-temperature exposures, such as protective clothing ensembles used by aircraft firefighters to fight fires in flammable liquids. (NFPA 470)

incident action plan (IAP) An oral or written plan approved by the incident commander (IC) containing incident objectives reflecting the overall strategy for managing an incident for a specific time frame and target location. (NFPA 470)

infectious Capable of causing an illness by entry of a pathogenic microorganism.

ingestion exposure Exposure to a hazardous material from swallowing the substance.

inhalation exposure Exposure to a hazardous material from breathing the substance into the lungs.

injection exposure Exposure to a hazardous material from the substance entering the body through cuts or other breaches in the skin.

label A smaller version of a placard (4 inches [10 cm] on each side) that is required to be placed on the four sides of individual boxes and smaller packages being transported.

NFPA 704 hazard identification system A hazardous materials marking system designed for fixed-facility use. It uses a diamond-shaped symbol of any size, which is itself broken into four smaller diamonds, each representing a particular property or characteristic of the material.

open-head drum A drum with a lid that is secured by a bolted clasp-type ring, which circles the entire head of the drum.

packaging group designation A label that describes a substance according to the degree of hazards it presents.

pipeline A length of pipe including pumps, valves, flanges, control devices, strainers, and/or similar equipment for conveying fluids. (NFPA 470)

pipeline right-of-way An area, patch, or roadway that extends a certain number of feet on either side of a pipeline and that may contain warning and informational signs about hazardous materials carried in the pipeline.

placard A diamond-shaped indicator (10¾ inches [27 cm] on each side) that is required to be placed on all four sides of highway transport vehicles, railroad tank cars, and other forms of transportation carrying hazardous materials to identify the substance being transported.

risk-based response process A systematic process based on the facts, science, and circumstances of the incident, by which responders analyze a problem involving hazardous materials/weapons of mass destruction (WMD) to assess the hazards and consequences, develop an incident action plan (IAP), and evaluate the effectiveness of the plan. (NFPA 470)

safety data sheet (SDS) Formatted information, provided by chemical manufacturers and distributors of hazardous products, about a chemical's composition, physical and chemical properties, health and safety hazards, emergency response, and waste disposal of the material. (NFPA 470)

shipping papers A shipping order, bill of lading, manifest, or other shipping document serving a similar purpose and containing the information required by regulations of the U.S. Department of Transportation (DOT). (NFPA 498)

toxic inhalation hazards (TIH) Gases or volatile liquids that are extremely toxic to humans.

train list See *consist.*

U.S. Department of Transportation (DOT) marking system A system of labels and placards that is used when materials are being transported from one location to another in the United States and Canada by Transport Canada.

vent pipes Inverted J-shaped tubes that allow for pressure relief or natural venting of a pipeline for maintenance and repairs.

waybill Shipping papers for railroad transport.

REVIEW QUESTIONS

1. What is one of the first actions awareness-level personnel should take at a hazardous materials incident?
2. Define drum, carboy, and cylinder.
3. Describe SDS, the NFPA 704 marking system, HMIS, the military hazardous materials/WMD marking system, pipeline warning signs, and the U.S. DOT marking system.
4. How might a responder use the U.S. DOT marking system?
5. What is the *ERG*?
6. What are the four ways a chemical substance can enter the human body?
7. At a hazardous materials incident, where is patient care usually provided?

DISCUSSION QUESTIONS

1. How can awareness-level personnel and operations-level responders use the various marking systems?
2. What are the three parts of the SIN mnemonic and how are they used?
3. Why is using respiratory protection so important?

APPLYING THE CONCEPTS

Your crew is dispatched to a biotechnology research company for an odor investigation. Your initial information from the dispatch center reported an unusual odor in one of the laboratories. Upon your arrival, a security guard meets you at the street and tells you that a spill of an unknown liquid occurred in laboratory #206. You and your crew are trained to the hazardous materials awareness level and are certified in basic life support (BLS).

1. Based on the preceding information, does your crew have the right level of training to enter the lab, identify the liquid, and clean up the spill?
2. If you are able to get a chemical name for the spilled substance, which of the reference sources would provide the most complete information in this scenario?

The security guard further reports that a scientist called the security desk to report finding a broken, 1-gallon (4-liter), amber-colored, glass container in laboratory #206. He tells you that the lab is evacuated and there are no injuries or exposures, and it is unknown how the container was broken or any other history of the event.

APPLYING THE CONCEPTS CONTINUED

3. How would you determine the initial operational priorities for this scenario?
4. Would propane, hydrofluoric acid, or sulfuric acid most likely be found in the amber glass container described in the scenario? Why?

As you are securing the entrance and exits from lab #206, another employee quickly approaches and tells you a coworker who had been in the lab has a small splash of liquid on the front of his pants, just below the knee, and is complaining of a burning sensation on both legs. The employee states that the coworker was escorted to a nearby locker room and directed to remove his pants and get into the shower to rinse off. That employee is now wrapped in a towel and sitting on a bench in the locker room stating he does not want to go to the hospital. The advanced life support (ALS) ambulance is 20 minutes away from arriving.

5. In your opinion, based on your speculation of the contents of the amber colored bottle, do you believe the decontamination action is appropriate?
6. Based on the long arrival time for the ambulance and the fact the employee is resisting medical transport, would you cancel the incoming ambulance?

REFERENCES

National Fire Protection Association (NFPA). 2017. *NFPA 498: Standard for Safe Havens and Interchange Lots for Vehicles Transporting Explosives.* 2018 Edition. Quincy, MA: NFPA.

National Fire Protection Association (NFPA). 2020. *NFPA 1: Fire Code.* 2021 Edition. Quincy, MA: NFPA.

National Fire Protection Association (NFPA). 2020. *NFPA 704: Standard System for the Identification of the Hazards of Materials for Emergency Response.* 2022 Edition. Quincy, MA: NFPA.

National Fire Protection Association (NFPA). 2021. *NFPA 470: Hazardous Materials/Weapons of Mass Destruction (WMD) Standard for Responders.* 2022 Edition. Quincy, MA: NFPA.

U.S. Department of Transportation, Transport Canada, and Secretariat of Communications and Transport of Mexico. 2020. *2020 Emergency Response Guidebook.* Pipeline and Hazardous Materials Safety Administration, U.S. Department of Transportation. https://www.phmsa.dot.gov/sites/phmsa.dot.gov/files/2021-01/ERG2020-WEB.pdf.

SECTION

4

Hazardous Materials Operations Level

NOTE: **NFPA 1010**, *Standard on Professional Qualifications for Firefighters*, 2024 Edition 6.1 in Chapter 6: Firefighter I (**NFPA 1001**) states that all firefighter candidates must meet the requirements defined in Chapter 7 and Sections 9.2 and 9.6 of **NFPA 470, 2022 Edition**. Content within these chapters meets the intent of **NFPA 470, 2022 Edition**, which includes chapter 7, "Professional Qualifications for Hazardous Materials/WMD Operations Level Personnel (**NFPA 1072**)."

CHAPTER

31

Hazardous Materials Operations Level

Properties and Effects

NOTE: Content within this chapter meets the intent of **NFPA 470, 2022 Edition**, which includes chapter 7, "Professional Qualifications for Hazardous Materials/WMD Operations Level Responders (**NFPA 1072**)."

KNOWLEDGE OBJECTIVES

After studying this chapter, you will be able to:

- Describe states of matter and their physical and chemical changes.
- Understand the ways containers can breach and release contents.
- Discuss the critical characteristics of flammable liquids.
- Understand how to determine the concentration of a released hazardous material.
- Describe ionizing and nonionizing radiation.
- Describe toxic products of combustion.
- Understand the concepts of hazard, exposure, and contamination.
- Describe how hazardous material exposure can lead to chronic and/or acute health effects.

SKILLS OBJECTIVES

There are no skills objectives for operations-level responders for this chapter.

CASE STUDY

You Are the Firefighter

Just after midnight, you receive a call for an explosion and a fire at a local landscaping company. Upon arrival at the site, you find a working fire in a storage shed located behind the main building. The wooden shed is fully involved, so you begin an indirect attack on the fire from the outside of the shed. As the fire is being knocked down, you receive an order over the radio to shut down the attack and move away from the building. About the same time, you experience an itchy feeling on the back of your neck. Other crew members are also complaining of itching and burning sensations around their wrists and necks. Some lower-floor residents of an adjacent multistory apartment complex are complaining of eye irritation and asking about a strange odor in the air.

1. Which types of chemicals might be found in this kind of occupancy?
2. Where could you obtain accurate technical information on the products stored in this building?
3. Which actions should be taken to address the complaints of burning and itching skin among firefighters as well as the complaints of the residents of the apartment complex?

Introduction

To safely mitigate hazardous materials/weapons of mass destruction (WMD) incidents, it is important to understand the **chemical and physical properties** of the substances involved. Chemical and physical properties are the characteristics of a substance that are measurable, such as vapor density, flammability, corrosivity, and water reactivity. However, you do not have to be a chemist to safely respond to hazardous materials incidents. In most cases, being an astute observer, referring to your incident response plan and/or standard operating procedures (SOPs), consulting the appropriate reference sources, and correctly interpreting and understanding the visual clues presented to you will provide enough information to take basic actions at the incident.

Pesticide bags are a good example of providing good information when you know what to look for. Pesticide bags must be labeled with specific information, and responders can learn a great deal from the label, including the following details (**FIGURE 31-1**):

- Pesticide name
- Active ingredients
- Hazard statement
- Total amount of product in the container
- Manufacturer's name and address
- Environmental Protection Agency (EPA) registration number, which provides proof that the product was registered with the EPA
- EPA establishment number, which shows where the product was manufactured

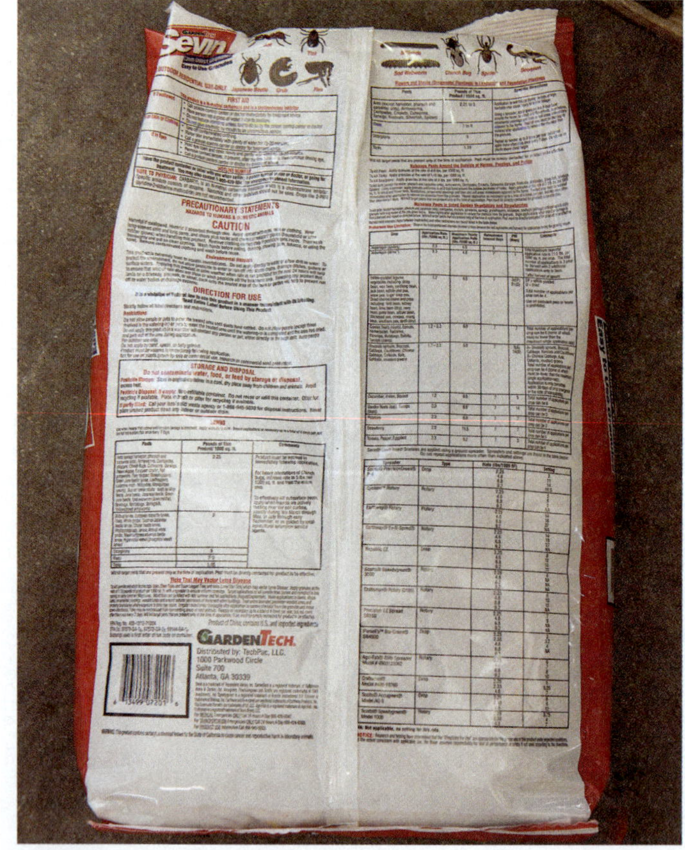

FIGURE 31-1 A pesticide label.

- Signal words to indicate the relative toxicity of the material, such as:
  - Danger—Poison: Highly toxic by all routes of entry
  - Danger: Severe eye damage or skin irritation

- Warning: Moderately toxic
- Caution: Minor toxicity and minor eye damage or skin irritation

- Practical first-aid treatment description
- Directions for use
- Agricultural use requirements
- Precautionary statements such as mixing directions or potential environmental hazards
- Storage and disposal information
- Classification statement on who may use the product

In addition, every pesticide label must carry the statement "Keep out of reach of children." In Canada, the pest control product (PCP) number can also be found on the pesticide label.

This chapter will guide you in learning basic key terms that will help you digest the information contained in reference sources such as safety data sheets (SDS) and other written and electronic sources of information. SDS may be obtained from site facility representatives, found online, or carried by those responsible for shipping hazardous materials. SDS are described in Chapter 30, *Recognizing and Identifying the Hazards*, in this section.

Responders may find a wealth of technical information in an SDS, such as the following:

- Identification, including supplier identifier and emergency telephone number
- Hazard identification
- Composition/information on ingredients
- First-aid measures
- Firefighting measures
- Accidental release measures
- Handling and storage requirements
- Exposure controls/personal protection
- Physical and chemical properties
- Stability and reactivity
- Toxicological information
- Ecological information (nonmandatory)
- Disposal considerations (nonmandatory)
- Transport information (nonmandatory)
- Regulatory information (nonmandatory)
- Other information

Additionally, by making a phone call to resources such as the Chemical Transportation Emergency Center (CHEMTREC) (or, in Canada, the Canadian Transport Emergency Centre [CANUTEC], and in Mexico, the Emergency Transportation System for the Chemical Industry, Mexico [SETIQ]), the responder may be able to obtain valuable information about a substance. The phone numbers for CHEMREC, CANUTEC, and SETIQ can be found in the current version of the *Emergency Response Guidebook* (*ERG*). These resources are available 24 hours per day and can provide responders with critical information for incidents involving hazardous materials and dangerous goods. By utilizing these sources of information, a responder can access live information from product specialists or access volumes of technical data from more than 6 million SDS.

Physical and Chemical Changes

An important first step in understanding the hazard(s) associated with any chemical involves identifying the state of matter, or physical state, of the substance. The **state of matter** defines the substance as a solid, liquid, or gas (**FIGURE 31-2**). The gas state is sometimes referred to as **vapor** (NFPA 1700), and the process of a solid or liquid changing to a gas is called **vaporization**.

If you know the state of matter and other physical properties of the chemical, you can begin to predict what the substance will do if it escapes, or has escaped, from its containment vessel. For example, it would be vital to know if a released gas is heavier or lighter than air and how that physical property relates to the environmental factors at the time of the incident, along with the other characteristics of the incident scene. Imagine how topography would affect the release of a heavy gas such as propane, perhaps in the setting of a trench rescue or other below-grade incident: The physical properties of propane could be a complicating factor when trying to perform a safe rescue in such a scenario.

Another critical part of comprehending the nature of the release is identifying the reason(s) why the containment vessel failed. Potential ways that containers

FIGURE 31-2 The state of matter identifies the hazard as a solid, liquid, or gas.

could breach include disintegration, runaway cracking, closures opening up, punctures, splits, or tears. In many cases, responders focus on the fact that a substance is being released rather than understanding why the product is escaping its container. It is one thing to notice that a container is leaking or generating a cloud from a puncture, crack, split, or tear; it is equally important, however, to figure out what type of stress caused the vessel to fail in the first place. Is the product release caused by a thermal influence—heat from inside or outside the container that may be the root cause of a catastrophic container failure? Maybe an errant forklift driver struck a drum, or a valve failure occurred. In these examples, physical damage allows the release of the container contents (**FIGURE 31-3**).

Generally speaking, three types of stress can cause a container to fail:

- Thermal: Heat created from fire or cold generated by environmental factors or substances such as cryogenics could cause a breach of the container.
- Chemical: The interaction of incompatible chemicals and/or the physical and chemical properties of a substance and how those substances interact inside or outside a container may lead to overpressure, disintegration, or other kinds of failures of any type of container.
- Mechanical: Falling debris, shrapnel, firearms, explosives, forklift puncture, and the like are all examples of how mechanical means can cause container failure.

FIGURE 31-3 Four examples of ways containers can release their contents. **A.** Rapid relief. **B.** Spill or leak. **C.** Violent rupture. **D.** Detonation.

These influences often result in predictable types of container failures, such as the ones listed in **FIGURE 31-4**.

Responders must also be aware of the possibility of exposure to hazardous materials. **Exposure** is "the process by which people, animals, the environment, property, and equipment are subjected to or come in contact with a hazardous material/WMD" (NFPA 470). The duration of exposure often affects the level of hazard. An exposure risk is considered short term if it lasts for seconds or minutes; medium term if it lasts for hours; or long term if its duration is days, weeks, or longer. Thus, incidents may last anywhere from seconds and minutes to several days and, in extreme cases, months or years. Tactics and strategies may vary greatly depending on the projected duration of the event. Responders

A.

B.

C.

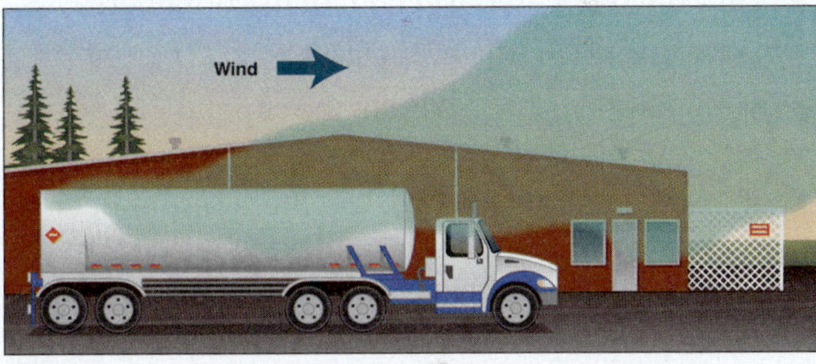

D.

E.

F.

G.

FIGURE 31-4 **A.** Cloud. **B.** Cone. **C.** Hemispheric release. **D.** Plume. **E.** Pool. **F.** Stream. **G.** Irregular dispersion.

must determine or estimate the duration of exposure and link that to the other information gathered.

After identifying as many facts and circumstances of the incident as possible, use those clues to form a hypothesis, establish **incident objectives** (incident-specific statements describing intended outcomes within a given operational period [NFPA 470]), and develop an **action plan** (a detailed proposal "to determine strategies and tactics within an incident action plan (IAP) . . . that will safely accomplish incident objectives, favorably influence outcomes, and increase the safety of responders and the public" [NFPA 470]). Consider the following questions as you do this:

- What is leaking?
- What are the chemical and physical properties of the material?
- What are the potential associated hazards and behavior, and how might those hazards affect people, animals, the environment, and property?
- How did the material get out of its container?
- How long might the event last?
- What happens if we handle the situation in nonintervention mode?
- If intervention is needed, are there responders on scene who are trained to handle the incident in defensive mode or offensive mode?
- Is the proper personal protective equipment (PPE) available?
- Are additional resources necessary?
- What is the likelihood of a positive outcome—that is, the chance of safely resolving the problem?

At any hazardous materials incident, it is important to link together all the bits of information to make an informed decision about how to handle the problem.

Chemicals can undergo a physical change—that is, transform from one state of matter to another—when they are subjected to environmental influences such as heat, cold, and pressure. For example, when water is frozen, it undergoes a **physical change** from a liquid to a solid. The actual chemical make-up of the water (H_2O) is still the same, but the state in which it exists is different.

Like water, propane is a nonrefrigerated liquified compressed gas, which is subject to physical change based on environmental influences such as heat and cold. If a propane cylinder is exposed to heat, the compressed liquefied propane inside changes state and becomes gaseous propane. In this case, the chemical make-up of propane has not changed; rather, the propane has physically changed state from a liquid to a gas.

When liquefied propane becomes gaseous propane, the pressure inside the vessel increases because the expansion ratio of propane is 270 to 1. The **expansion ratio** is a description of the volume increase that occurs when a compressed liquefied gas (e.g., propane, chlorine, and oxygen) changes to a gas. An expansion ratio of 270 to 1 means that when you compress 270 gallons of propane gas so that it becomes a liquid, it occupies only 1 gallon of space. Likewise, if heat causes 1 gallon of liquid propane to vaporize, the vapor will occupy 270 gallons. If this uncontrolled expansion takes place faster than the relief valve can vent the gas, a boiling liquid/expanding vapor explosion (BLEVE) could occur. It is possible, however, for any liquid in a sealed container to BLEVE if an external source of heat pushes the substance past its boiling point. Even water, if contained in a sealed container, can BLEVE.

The classic example of a BLEVE is a fire occurring below or adjacent to a propane tank. Once the liquid inside the vessel begins to boil, large volumes of vapor are generated within the vessel. Ultimately, the vessel fails catastrophically if it is unable to relieve the pressure through the safety valve. The ensuing explosion throws fragments of the vessel in all directions. When the pressurized liquefied material is flammable, a tremendous fireball is generated. Additionally, an overpressure blast wave is created because of the rapidly expanding vapor released by the vessel failure.

If fire is impinging on a vessel that contains pressurized liquefied materials, responders should carefully evaluate the risks of attempting to fight the fire (**FIGURE 31-5**). BLEVEs have claimed the lives of many firefighters throughout the years—and this history

FIGURE 31-5 Responders should take great care when handling incidents involving compressed liquefied gases.

should serve as a reminder of the dangers inherent in these types of incidents.

Chemical reactivity (also known as chemical change) describes the ability of a substance to undergo a transformation at the molecular level, which is usually accompanied by a release of some form of energy. Chemical reactivity can cause a chemical reaction—a **chemical change** or degradation—and if this occurs inside a container, it may cause the entire container to breach. It is important to understand that chemical change is different from physical change. Physical change is a change in state, whereas a chemical change results in an alteration of the material's chemical nature. Steel rusting and wood burning are examples of chemical changes.

The reactive nature of any substance can be influenced in many ways, such as by mixing it with another substance (e.g., adding an acid to a base) or by applying heat. A **polymer**—a large molecule made up of long repeating chains of smaller molecules—is a good example of a reactive substance. **Polymerization** is the process of reacting **monomers** (molecules that can be bonded to like molecules to form a larger polymer) together in a chain reaction to form polymers. Many plastics in use in the home and in industry are produced by the process of polymerization. Some substances, such as white phosphorus, are reactive with air; other substances, such as metallic sodium, are reactive with water. Responders should also be aware that some chemical reactions release heat.

To relate the concepts of physical and chemical change to a hazardous materials incident, think back to the "You are the Firefighter" feature at the beginning of the chapter. Assume for a moment that the owner of the landscaping company wadded up some rags soaked with linseed oil and left them in a corner of the shed. The rags spontaneously ignited due to the heat generated by the oil as it dried and started a fire that ultimately caused the failure of a small propane cylinder (the explosion that occurred prior to your arrival). The fire also melted the fusible plug on a chlorine cylinder (at approximately 160°F [71°C]), causing a release of chlorine gas; chlorine gas is 2.5 times heavier than air.

Looking closely at each event in this sequence, you can see the effect of physical and chemical changes. Linseed oil is an organic material that generates heat as it decomposes (chemical change). That heat, in the presence of oxygen in the surrounding air, ignited the rags, which in turn ignited other **combustibles** (materials such as wood, paper, and plastic that can ignite and burn [NFPA 1]). The surrounding heat caused the propane inside the tank to expand (physical change) until it overwhelmed the ability of the relief valve to handle the build-up of pressure. The cylinder ultimately exploded. These events are all consistent with the expected behavior based on the incident conditions. At the same time, the escaping chlorine gas mixed with the moisture in the air and formed an acidic mist (a chemical change). A slight breeze carried the smoke and mist toward the apartment complex. The firefighters began to feel itchy and burning sensations where the acidic mist contacted their moist skin. Because chlorine vapors are heavier than air, most of the residents complaining about eye irritation are located on the bottom floor of the complex.

Critical Characteristics of Flammable Liquids

When looking at the fire potential of a flammable liquid, several important aspects must be considered. Among them are flash point, ignition temperature, and flammable range. More important than memorizing the definitions of these terms, however, is having an understanding of the relationships between these and other physical characteristics of a flammable liquid. Keep in mind that all materials must be in a gaseous or vapor state prior to flaming combustion: Solids and liquids do not burn, but rather give off a gas or vapor that is ultimately ignited. Think of a log burning in a fireplace. If you look closely, the fire is not directly on the log but rather appears slightly above the surface. The reason for this phenomenon is that the wood must be heated to the point at which the molecular bonds of the wood break and a gas is created. The log, then, does not burn; instead, the gas produced by heating the log starts and sustains the process of combustion. Conceptually, liquid fuels such as gasoline, diesel fuel, and other hydrocarbon-based fuels behave the same way as the log in terms of producing flammable vapors when heated. One way or another, vapor production must occur before there can be fire. This vapor production must be factored in when estimating the probability of fire during a release.

Flash Point

The **flash point** is the minimum temperature at which a liquid or solid gives off sufficient vapors such that, when an ignition source (e.g., a flame, electrical equipment, lightning, or even static electricity) is present, the vapors will ignite, resulting in a flash fire. The flash fire involves only the vapor phase of the liquid (as in the example of the log) and will go out once the vapor fuel is consumed.

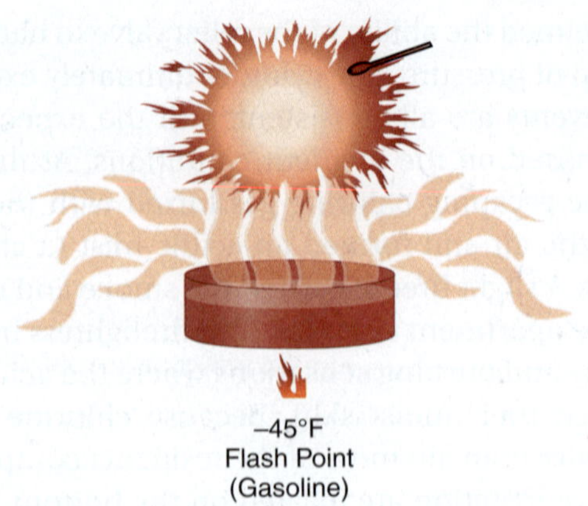

FIGURE 31-6 Responders should always be mindful of ignition sources at flammable/combustible liquid incidents.

To illustrate how such an event occurs, consider the following. The flash point of gasoline is –45°F (–43°C). When the temperature of gasoline reaches –45°F (–43°C) because of heat from an external source or from the surrounding environment, the gasoline gives off sufficient flammable vapors to ignite but not support combustion. In nearly all circumstances, when gasoline is spilled or otherwise released, the temperature of the external environment is well above the flash point of gasoline, creating the potential for ignition (**FIGURE 31-6**). Diesel fuel, by comparison, has a much higher flash point than gasoline—in the range of approximately 120°F (49°C) to 140°F (60°C), depending on the fuel grade. In either case, once the temperature of the liquid surpasses its flash point, the fuel will give off sufficient flammable vapors to support combustion.

The **fire point** is the lowest temperature at which sustained combustion of the vapor will occur. It is usually only slightly higher than the flash point for most materials (NFPA 1).

Ignition Temperature

The **ignition temperature** is another important temperature landmark for flammable/combustible liquids. From a technical perspective, you can think of the ignition temperature as the minimum temperature at which a fuel, when heated, will ignite in the presence of air and continue to burn.

A direct relationship exists between temperature and vapor production. Simply put, when the temperature increases, the vapor production of any flammable liquid increases, leading to a higher concentration of vapors. Therefore, even liquids with low flash points can be expected to produce a significant amount of flammable vapors at all but the lowest ambient temperatures.

From a responder's perspective, it is important to realize that when a liquid fuel is heated (with any type of heat) beyond its ignition temperature, it will ignite without an external ignition source. Think of a pan full of cooking oil on the stove. For illustrative purposes, assume that the ignition temperature of the oil is 300°F (149°C). What would happen if the burner was set on high and left unattended so that the oil was heated past 300°F (149°C)? Once the temperature of the oil exceeds its ignition temperature, it will ignite; there is no need for an external ignition source. In fact, this scenario is a common cause of stove fires.

Flammable Range

Flammable range is another important term to understand. Defined broadly, the flammable range is an expression of an air/fuel mixture, defined by upper and lower percentage limits, that reflects the amount of flammable vapor mixed with a given volume of air. Gasoline will serve as our example. The flammable range for gasoline vapors is 1.4 percent to 7.6 percent. These two percentages, called the **lower explosive limit (LEL)** and the **upper explosive limit (UEL)**, respectively, define the boundaries of the air/fuel mixture necessary for gasoline to burn properly. LEL is "the minimum concentration of combustible vapor or gas in" an air/fuel mixture "above which propagation of flame will occur on contact with an ignition source" (NFPA 115). UEL is the maximum amount of combustible vapor or gas that can be present in the air if the air/fuel mixture is to be flammable or explosive. If a given gasoline/air mixture falls between the LEL (1.4 percent) and the UEL (7.6 percent) and that mixture encounters an ignition source, a flash fire will occur.

The concept of automobile carburetion capitalizes on the notion of flammable range—the carburetor is the place where gasoline and air are mixed. When the mixture of gasoline vapors and air occurs in the right proportions, the car runs smoothly. When there is too much fuel and not enough air, the mixture is fuel rich. When there is too much air and not enough fuel, the carburetion is fuel lean. In either of the latter cases, optimal combustion is not achieved, and the motor does not run correctly.

Understanding flammable range as it relates to hazardous materials response relies on the same line of thinking. If gasoline is released from a rolled-over cargo tank, the vapors from the spilled liquid will mix with the surrounding air. If the mixture of vapors and air falls between the UEL and the LEL and those vapors reach an ignition source, a flash fire will occur. Ultimately,

TABLE 31-1 Flammable Ranges of Common Gases

Gas	Flammable Range
Hydrogen	4.0%–75%
Natural gas	5.0%–15%
Propane	2.5%–9.0%

the entire volume of gasoline will likely burn. As with carburetion, if too much or too little fuel is present, the mixture will not adequately support combustion.

Generally speaking, the wider the flammable range, the more dangerous the material is (**TABLE 31-1**). This relationship reflects the fact that the wider the flammable range, the more opportunity there is for an explosive mixture to find an ignition source.

Vapor Pressure

Vapor pressure is the pressure exerted by a liquid (NFPA 1). For our purposes, the definition of vapor pressure will pertain to liquids held inside any type of closed container. All liquified compressed gas cylinders have a small vapor space called the **headspace** above the liquid. When liquids are held in a closed 55-gallon drum or a 4-liter glass bottle, for example, some amount of pressure develops inside in the headspace above the liquid (**FIGURE 31-7**). All liquids—even water—will develop a certain amount of pressure in the airspace between the top of the liquid and the container. Vapor pressure is measured in pounds per square inch, absolute (psia). The psia of vapor pressure is determined by using the test method specified by ASTM D 323, *Standard Test Method for Vapor Pressure of Petroleum Products (Reid Method)*.

The key point to understanding vapor pressure is this: The vapors released from the surface of any liquid must be contained if they are to exert pressure. Essentially, the liquid inside the container will vaporize until the molecules given off by the liquid reach equilibrium with the liquid itself. Equilibrium is a balancing act between the liquid and the vapors—some molecules turn to vapor, whereas others leave the vapor phase and return to the liquid phase. The equilibrium occurs as the molecules constantly exchange between the liquid and vapor phase at a steady rate.

Carbonated beverages illustrate this concept. Inside the basic cola drink is the formula for the soda and

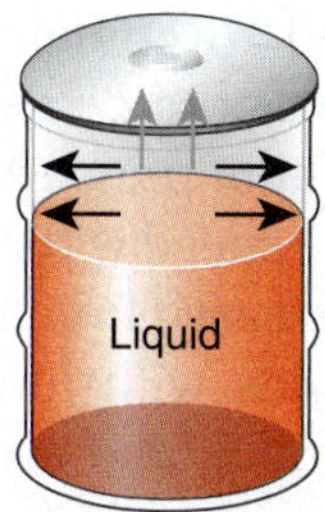

FIGURE 31-7 The vapor in the headspace is exerting pressure in all directions above the liquid inside the drum.

a certain amount of carbon dioxide (CO_2) molecules. The bubbles in the soda are composed of CO_2. When the soda sits in an unopened can or plastic bottle on the shelf, the balancing act of CO_2 is happening: CO_2 from the liquid becomes gaseous CO_2 above the liquid; at the same time, some of the CO_2 dissolves back into the liquid. The pressure inside the bottle can be verified by feeling the rigidity of the container—until you open it. Once the lid is removed, the CO_2 is no longer in a closed vessel, and it escapes into the atmosphere. With the pressure released, the sides of the bottle can be squeezed easily. Ultimately, the soda goes flat when all the CO_2 has escaped.

With the soda example in mind, consider the technical definition of vapor pressure: The vapor pressure of a liquid is the pressure exerted by its vapor until the liquid and the vapor are in equilibrium. Again, this process occurs inside a closed container, and temperature has a direct influence on the vapor pressure. For example, if heat impinges on a drum of acetone, the pressure above the liquid will increase and perhaps cause the drum to fail. If the temperature is dramatically reduced, the vapor pressure will drop. This is true for any liquid held inside a closed container. What happens if the container is opened or spilled onto the ground to form a puddle? The liquid still has a vapor pressure, but

TIP

Two specific points of temperature (68°F [20°C]) and pressure (14.7 psi) are commonly referred to as *normal temperature and pressure (NTP)*. These values represent standard conditions under which chemical and physical properties can be tested and documented. Another term, *standard temperature and pressure (STP)*, as defined by the International Union of Pure and Applied Chemistry (IUPAC), describes an ideal testing environment for a variety of chemical and physical processes: 14.7 psi at 32°F (0°C).

it is no longer confined to a container. Liquids with high vapor pressures will evaporate much more quickly than will liquids with low vapor pressures. Vapor pressure directly correlates to the speed with which a material will evaporate once it is released from its container.

Additionally, liquids with low flash points, such as gasoline, ethyl alcohol, and acetone, typically have higher vapor pressures and higher ignition temperatures. Consider the case of gasoline, whose flash point is –45°F (–43°C). At a vapor pressure of more than 275 mm Hg (compare this value to the vapor pressure of ethyl alcohol, which is 40 mm Hg at approximately 68°F [20°C]), gasoline has an ignition temperature of approximately 475°F (246°C). By contrast, flammable/combustible liquids with high flash points typically have lower ignition temperatures and lower vapor pressures. A grade of diesel—#2 grade diesel, for example—may have a flash point of 125°F (52°C) and a corresponding ignition temperature of approximately 500°F (260°C); this is clearly a much narrower range than that described in the gasoline example. The vapor pressure of #2 grade diesel fuel is very low—much lower than the vapor pressure of water. Several other examples of flash points of flammable/combustible liquids, illustrating the relationships among flash point, ignition temperature, and vapor pressure are listed in **TABLE 31-2**.

When consulting reference sources, be aware that the vapor pressure may be expressed in pounds per square inch, atmospheres, **torr**, or millimeters of mercury, and note that most sources give the vapor pressures of substances at a temperature of 68°F (20°C). Each expression of pressure is a valid point of reference in incident response. **Pounds per square inch (psi)** is a unit of pressure that expresses the pounds of force per square inch of area. An **atmosphere (atm)** is a unit of pressure equal to the average atmospheric pressure at sea level. The term **millimeters of mercury (mm Hg)** is commonly found in reference books; it is defined as the pressure exerted at the base of a column of mercury that is exactly 1 millimeter in height. The torr unit is named after the Italian physicist Evangelista Torricelli, who discovered the principle of the mercury barometer in 1644. In his honor, the torr was equated to 1 mm Hg. Another common expression of pressure is **bar**. The conversion from bar to psi is 1 bar = 14.7 psi.

Conversion factors allow for calculations to convert between one unit of measurement and another, but some of these factors are very complex. For the purposes of simplicity and incident response, the important point is to have some frame of reference to understand the concept of vapor pressure and to recognize how that concept will affect the release of chemicals into the environment. To understand the relationships among the various units of pressure, use the following comparison:

14.7 psi = 1 atm = 760 torr = 760 mm Hg = 1 bar

Another method of comparison is to take the values from the reference books and compare those values to the behavior of substances with which you may be familiar. The following example uses units of mm Hg to compare three common substances: water, motor oil, and isopropyl alcohol. The vapor pressure of water at room temperature is approximately 25 mm Hg. Standard 40-weight motor oil has a vapor pressure of less than 0.1 mm Hg at 68°F (20°C)—it is practically vaporless at room temperature. Isopropyl alcohol, by contrast, has a high vapor pressure, 30 mm Hg, at room temperature. Again, temperature has a direct correlation to vapor pressure, but all things being equal, these three substances give you a good starting point for making comparisons.

Boiling Point

Boiling point is the "temperature at which the vapor pressure of a liquid equals the surrounding

TABLE 31-2 Flash Point, Vapor Pressure, and Ignition Temperature

	Flash Point	Vapor Pressure	Ignition Temperature
Water	N/A	25 mm Hg at 68°F (20°C)	N/A
Gasoline	−45°F (−43°C)	275–400 mm Hg at 70°F (21°C)	475°F (246°C)
Acetone	−4°F (−20°C)	400 mm Hg at 104°F (40°C)	869°F (465°C)
#2 grade diesel	125°F (52°C)	< 2 mm Hg at 68°F (20°C)	500°F (260°C)

All readings within the table are closed cup results. These results were created with closed cup testing and are not always accurate within the atmosphere.

atmospheric pressure"; in other words, it is the temperature at which a liquid will continually give off vapors in sustained amounts and if held at that temperature long enough, will turn completely into a gas (NFPA 1). For example, at sea level, an atmospheric pressure of 14.7 psi is exerted on every surface of every object, including the surface of the water. The boiling point of water at sea level is 212°F (100°C). At this temperature, water molecules have enough kinetic energy—energy created by an object in motion—to overcome the downward force of the surrounding atmospheric pressure (**FIGURE 31-8**). At temperatures less than 212°F (100°C), there is insufficient heat to create enough **kinetic energy** to allow the water molecules to escape. At higher elevations above sea level, such as Denver, Colorado, which is at 5280 feet above sea level, the atmospheric pressure is lower—12 psi. Thus, water boils faster and at a lower temperature (about 202°F [94°C]) in Denver.

To demonstrate how heated liquids inside closed containers might create a deadly situation, we will look at a completely benign and very common example—popcorn. Popcorn pops when the trapped water inside the kernel of corn is heated beyond its boiling point. When the popcorn kernels are heated, the small amount of water inside each kernel eventually exceeds its boiling point and expands to 1700 times its original water volume. The kernel then "pops." A kernel of corn has no relief valve, so a rapid build-up of pressure will cause it to breach. Unpopped kernels are those with insufficient water inside the kernel to permit this expansion. If the kernel of corn was filled with typical motor oil, which has an approximate boiling point of more than 500°F (260°C), it would take much more heat energy to pop the kernel. Applying this concept to flammable liquids, it becomes apparent that substances with low boiling points are dangerous because they have the potential to produce large volumes of flammable vapor at relatively low temperatures.

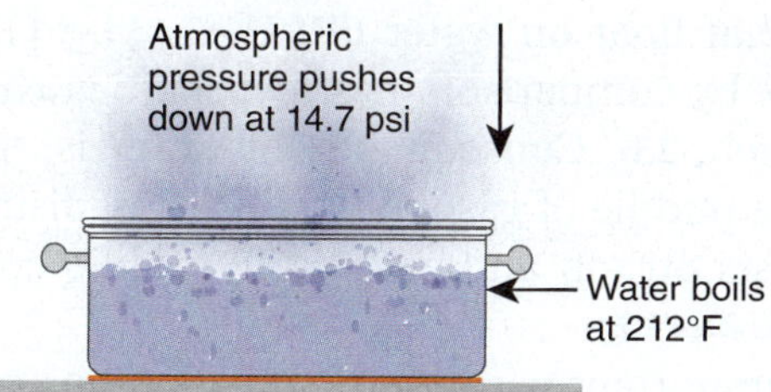

FIGURE 31-8 At sea level, atmospheric pressure pushes down at 14.7 psi. When water is heated to 212°F (100°C), enough energy is created to cause the water to boil, which in turn allows water molecules to escape in the form of steam.

TIP

The molecular weight of a flammable liquid dictates the behavior of the substance once it is released. A molecule of motor oil, for example, is much larger than a molecule of isopropyl alcohol. Some general conclusions about the relationships among vapor pressure, boiling point, flash point, ignition temperature, and heat output can be linked based on the molecular size of a chemical compound. Specifically, as the molecular weight of a substance increases:

↓ Vapor pressure decreases.
↑ Boiling point increases.
↑ Flash point increases.
↓ Ignition temperature decreases.
↑ Heat output increases.

Vapor Density

In addition to identifying the flammability of a vapor or gas, responders must determine whether the vapor or gas is heavier or lighter than air. In essence, you must know the vapor density of the substance in question. **Vapor density** is the weight of a vapor or gas compared to an equal volume of dry air measured under the same conditions (**FIGURE 31-9**).

Basically, vapor density is a question of comparison: What will the gas or vapor do when it is released in the air? Will it float upward into the ventilation system or upward into the air because it is less dense than air, or will it collect in low spots in the topography or somewhere inside a building because it is denser than

FIGURE 31-9 Cylinder A: Vapor density less than 1. Cylinder B: Vapor density greater than 1.

SAFETY TIP

In the absence of reliable reference sources in the field, you can use the following mnemonic to remember many common lighter-than-air gases: 4H MEDIC ANNA. This mnemonic translates as follows:

H: Hydrogen
H: Helium
H: Hydrogen cyanide
H: Hydrogen fluoride
M: Methane
E: Ethylene
D: Diborane
I: Illuminating gas (methane/ethane mixture)
C: Carbon monoxide
A: Ammonia
N: Neon
N: Nitrogen
A: Acetylene

If you encounter a leaking propane cylinder, for example, just refer to 4H MEDIC ANNA. Propane is not on that list, so by default, you can assume that propane is heavier than air. Likewise, chlorine and butane do not appear on the list: Both are heavier than air. Again, if the substance is not on this list, it is most likely heavier than air.

air? These are important response considerations for hazardous materials incidents and may influence the severity of the incident.

You can find a chemical's vapor density by consulting a good reference source such as an SDS. Air has a set vapor density value of 1.0. Vapor density is expressed as numbers relative to air density, such as 1.2, 0.59, or 4.0. Vapors with a density value greater than 1 are denser than air and will sink, whereas vapors with a density value less than 1 are less dense than air and will rise. The vapor density of propane gas, for example, is 1.55. This means propane gas is heavier than air; its vapor density value is greater than 1.0. (Chlorine gas is also denser than air.) Substances such as acetylene, natural gas, and hydrogen are lighter than air and have a vapor density value less than 1.0.

Specific Gravity

Specific gravity is to liquids what vapor density is to gases and vapors—namely, a comparison value. In this case, the comparison is between the weight of a liquid chemical and the weight of water. The specific gravity of water is assigned a value of 1.0. Materials that float on water have a specific gravity value less than 1.0 (**FIGURE 31-10**). Materials that sink in water and remain below the surface of the water have a specific gravity value greater than 1.0. Consult a comprehensive reference source to find the specific gravity of the chemical in question.

FIGURE 31-10 A substance floating on water has a specific gravity less than 1.0.

Most flammable liquids float on water. Gasoline, diesel fuel, motor oil, and benzene are all examples of liquids that float on water (**FIGURE 31-11**). Carbon disulfide, by comparison, has a specific gravity of approximately 2.6. Consequently, if water is gently applied to a puddle of gasoline and carbon disulfide, the water rests on top and covers the puddle completely (**FIGURE 31-12**).

It is important to carefully consider the application of water in all cases involving spilled or released chemicals because the results from this intervention could be something worse than the initial problem. Complications such as chemical reactivity or incompatibility, contaminated runoff, and unwanted increase in the surface area of the spill could occur.

FIGURE 31-11 Gasoline floating on water.

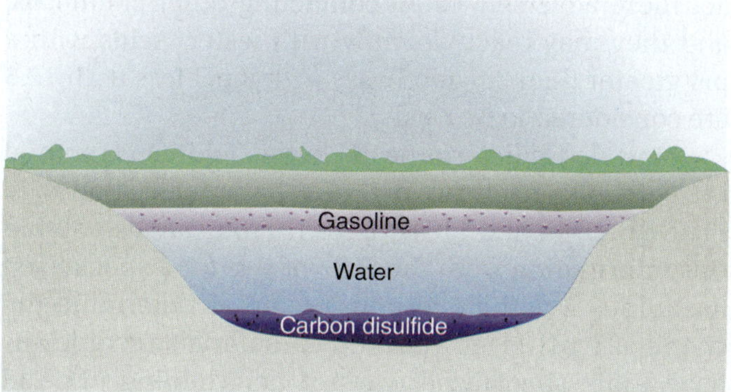

FIGURE 31-12 Gasoline floats on water, but carbon disulfide does not.

Water Solubility

When discussing the concept of specific gravity, it is also necessary to determine if a chemical will mix with water. **Water solubility** describes the ability of a substance to dissolve in water. Water is the predominant agent used to extinguish fire, but when you are dealing with chemical emergencies, it may not always be the best and safest choice to mitigate the situation. That caveat applies because water is a tremendously aggressive solvent and can react violently with certain chemicals. A solvent is "a substance (usually liquid) capable of dissolving or dispersing another substance" (NFPA 96).

Concentrated sulfuric acid, metallic sodium, and magnesium are just a few examples of substances that will adversely react with water. If water is applied to burning magnesium, for example, the heat of the fire will break apart the water molecule, creating an explosive reaction. Adding water to concentrated sulfuric acid would be like throwing water on a pan full of hot oil—popping and spattering will occur.

In other circumstances, water is a friendly ally when you are attempting to handle a chemical incident. Depending on the chemical involved, fog streams operated from handlines may knock down vapor clouds. Additionally, heavier-than-water flammable liquids can be extinguished by gently applying water to the surface of the liquid. In some cases, water may be an effective way to dilute a chemical, thereby rendering it less hazardous.

Corrosivity (pH)

Corrosivity is a property of a material that reflects its ability to cause damage on contact with skin, eyes, or other parts of the body. The technical definition, as found in the U.S. Code of Federal Regulations, includes the language, "irreversible damage to human skin at the site of contact" (49 CFR 173.136). Such materials are often also damaging to clothing, rescue equipment, and other physical objects in the environment.

Corrosives are a complex group of chemicals that should not be taken lightly. The tens of thousands of corrosive chemicals used in general industry, semiconductor manufacturing, and biotechnology can be further categorized into two classes: **acid** and **base**. The most common way to identify acids and bases is by their pH. In simple terms, think of **pH** as the "power of hydrogen." pH is a measurement of the presence of dissolved hydrogen ions (H^+) in a substance (technically defined as the potential of hydrogen).

The pH scale goes from 0 to 14. An acid is a substance with a pH less than 7. A base is a substance with a pH greater than 7. A substance with a pH of 7 is neutral (**FIGURE 31-13**). From a field perspective, pH can be viewed as a measurement of corrosive strength, which can loosely be translated into a certain degree of hazard. In simple terms, we can use pH to judge how aggressive a corrosive substance might be.

Common acids, such as sulfuric, hydrochloric, phosphoric, nitric, and acetic (vinegar) acids, have a predominant amount of hydrogen ions (H^+) in the solution and, therefore, will have pH values less than 7 (**FIGURE 31-14**). Chemicals that are bases, such as sodium hydroxide, potassium hydroxide, sodium carbonate, and ammonium hydroxide, have a predominant amount of hydroxide (OH^-) ions in the solution and will have pH values greater than 7 (**FIGURE 31-15**).

The middle of the pH scale (7) is where a chemical is considered to be neutral—that is, neither acidic nor basic. A pH of 7 is neutral because the concentration of

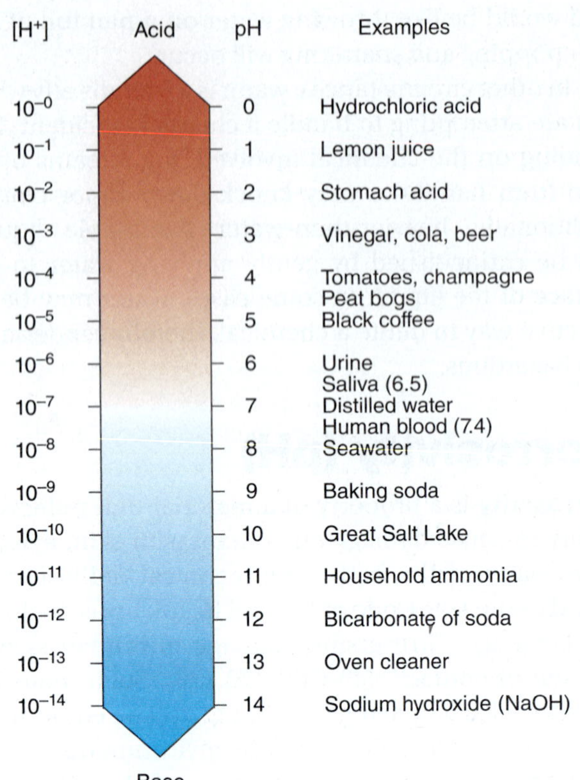

FIGURE 31-13 The pH scale.

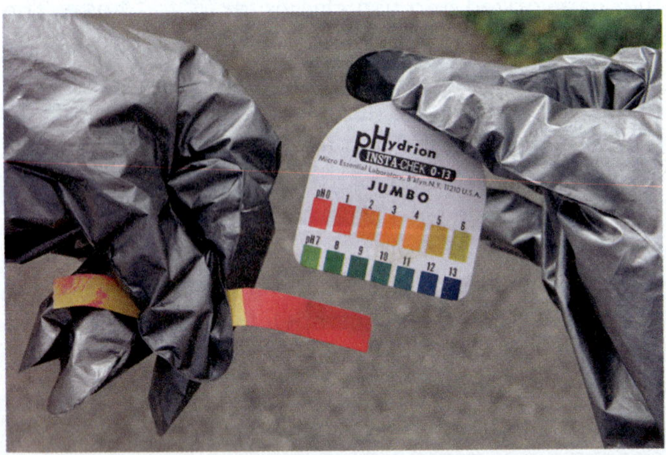

FIGURE 31-14 An acid.

hydrogen ions (H^+) in such a material is exactly equal to the concentration of hydroxide (OH^-) ions produced by dissociation of the water. At this value, a chemical will not harm human tissue.

Generally, pH values of 2.5 or less, and those of 12.5 or greater, are considered to be strong. In practical terms, this designation means that strong corrosives (i.e., acids and bases) will react more aggressively with metallic substances such as steel and iron; they will cause more damage to unprotected skin; they will react more adversely when contacting other chemicals; and they may react violently with water. Acids with a pH greater than 2.5 and bases with a pH less than 12.5 are considered to be weak.

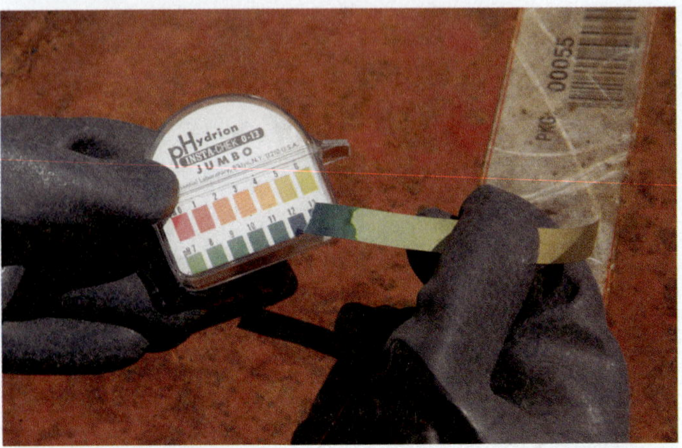

FIGURE 31-15 A base.

How do we determine pH in the field? One method is to use specialized pH test paper, as shown in FIGURE 31-14 and FIGURE 31-15. You can also obtain this information from the SDS, or call for a specialized hazardous materials response team to determine the chemical's pH level. Hazardous materials technicians have more specialized ways of determining pH and may be a useful resource when responders are handling corrosive incidents.

At the operations level, it is critical to understand that incidents involving corrosives may not be as straightforward as other types of chemical emergencies. The materials themselves are more complicated and, in many cases, the tactics employed to deal with them could be outside your scope of responsibilities.

Determining the Concentration of a Released Hazardous Material

It is possible that a hazardous materials/WMD incident might involve something other than a pure substance. In such a case, responders might encounter a released substance as a solution. A solution is a mixture of at least one chemical substance—the solute—dissolved in a solvent. Salt water is a good example of a **solution**. Salt is the **solute** and water is the **solvent**, and when mixed together, the two substances create a uniform solution.

FIGURE 31-16 A concentrated solution contains a large amount of solute for a given amount of solution.

FIGURE 31-17 A dilute solution contains a small amount of solute for a given amount of solution.

Concentration, from the perspective of a chemist, refers to the amount of solute in solvent or the ratio of solute in a given amount of total solution. From a practical standpoint, a **concentrated solution** of any kind contains a large amount of solute for a given amount of solution (**FIGURE 31-16**). A **dilute solution**, by contrast, contains a small amount of solute for a given solution (**FIGURE 31-17**). Going back to the salt water example, the presence of a lot of salt in the solution would make it concentrated, whereas just a little salt in the water would make it dilute. Either way the salt water is still a solution, but its concentration can vary.

Solutions are not always liquids, however. Thus, there are other situations where concentration comes into play—namely, when gases are released into the air.

Consider a natural gas release. If a gas heater in a tightly sealed house were to fail in some way, resulting in a sustained release of natural gas (the solute), a high concentration of gas could build up inside the house (with the air/gas mixture representing the total solution). The consequence of that accumulation could be an explosion if the gas were to reach the proper proportions and find an ignition source. If the windows or doors of the house were opened, however, the concentration of natural gas would decrease, perhaps becoming so diluted that it would pose no significant fire or health hazard. Typically, concentrations of gases are expressed as percentages, which is when flammable range comes into play. In this example, the flammable range of natural gas is 5 percent to 15 percent, so if the concentration of the natural gas vapors in the house is between these two values, combustion would occur if an ignition source is present.

In the preceding example, responders may be called upon to determine the airborne concentration of natural gas within the house. When this step is necessary, specialized detection and monitoring equipment may be required. Using detection and monitoring equipment properly requires some technical expertise, a lot of common sense, and a commitment to continual training. It is a mistake for responders to believe that they can simply turn on a machine, point it in some direction, and expect it to solve the problem. A reading from a gas detector, taken out of context, may cause an entire response to head off in the wrong direction, leading to an unsafe decision or a series of inefficient tactics. The responder must interpret the information the instrument is providing and make decisions based on the information. Always remember that using a detector or monitor entails more than just reading the screen or waiting for an alarm to sound.

When responding to incidents involving corrosives, it is also important to know the concentration of the released substance. Concentration, when discussing corrosives, is an expression of how much of the acid or the base (the solute) is dissolved in a solvent (usually water) to form the solution. Again, as with flammable range, this value is generally expressed as a percentage. Sulfuric acid at a concentration of 97 percent is concentrated, whereas the sulfuric acid found in car batteries, at a concentration of approximately 30 percent, is dilute.

Do not confuse concentration of the solution with the strength of the solution. The words **strong** and **weak** do not correspond to *concentrated* and *dilute*, respectively. The strength of a corrosive refers to the degree of ionization that occurs in a solution, which is determined by the solution's pH. A strong acid such as hydrochloric acid (HCl) is strong even if it is found in a diluted solution. Acetic acid (vinegar) is a weak acid, and it remains a weak acid even when it occurs in high concentrations. In general, strong corrosives will react more vigorously with incompatible materials such as organic substances and will be more aggressive when they contact metallic objects such as metal shelving, shovels, and other items.

Ionizing and Non-ionizing Radiation

Most firefighters have not been trained on the finer points of handling incidents involving radioactive materials and, therefore, have many misconceptions about radiation. **Radiation** is energy transmitted through space in the form of electromagnetic waves or energetic particles. Fundamentally, you should understand that you cannot escape radiation. It is all around you, every day. During your life, you receive radiation from the sun and the soil, when taking a plane ride, or by having an X-ray. Radiation has been around since the beginning of time. Our focus here, however, is not on background radiation—the kind of radiation you receive from the sun or the soil; rather, it is on the occupational exposures encountered in the field.

For the most part, the health hazards posed by radiation are a function of two factors:

- The amount of radiation absorbed by your body has a direct relationship to the degree of damage done.
- The exposure time and distance to the radiation will ultimately affect the extent of the injury.

To understand radiation, you need to understand a little bit about chemical elements. An **element** is a substance that cannot be chemically broken down into simpler substances. The periodic table of elements summarizes all the known elements that are found in all the chemical compounds on Earth (**FIGURE 31-18**). Those elements, in turn, are made up of atoms. An **atom** is the smallest unit of a chemical element. The **nucleus** is the center of an atom. The types of particles found in the nucleus are the **proton**, a particle with a positive (+) electrical charge, and the **neutron**, a particle with no electrical charge. Orbiting the nucleus are **electron** particles, which have a negative (–) electrical charge.

All atoms of any given element have the same number of protons—that number is what defines the element. For example, hydrogen has 1 proton in its nucleus, helium has 2, lithium has 3, and so on. The number of protons is called the **atomic number** of the element and corresponds to the number of that element in the periodic table; thus, hydrogen is element 1, helium is element 2, and lithium is element 3.

The stability of an atom is controlled by the balance of neutrons and protons within the nucleus. Most atoms have the same number of protons and neutrons, but atoms can sometimes have a different number of neutrons in the nucleus. An **ion** is an atom with more protons than electrons (a positive ion, also called a **cation**) or an atom with more electrons than protons (a

TIP

Isotopes are atoms with the same number of protons but different numbers of neutrons.

PERIODIC TABLE OF THE ELEMENTS

Key: Element — hydrogen; Atomic Number — 1; Symbol — H; *Atomic Mass — 1.01

Metals · Metalloids · Nonmetals

Period	1 IA	2 IIA	3 IIIB	4 IVB	5 VB	6 VIB	7 VIIB	8 VIII	9 VIII	10 VIII	11 IB	12 IIB	13 IIIA	14 IVA	15 VA	16 VIA	17 VIIA	18 VIIIA
1																		helium 2 He 4.00
2	lithium 3 Li 6.94	beryllium 4 Be 9.01											boron 5 B 10.81	carbon 6 C 12.01	nitrogen 7 N 14.01	oxygen 8 O 16.00	fluorine 9 F 19.00	neon 10 Ne 20.18
3	sodium 11 Na 22.99	magnesium 12 Mg 24.31											aluminum 13 Al 26.98	silicon 14 Si 28.09	phosphorus 15 P 30.97	sulfur 16 S 32.07	chlorine 17 Cl 35.45	argon 18 Ar 39.95
4	potassium 19 K 39.10	calcium 20 Ca 40.08	scandium 21 Sc 44.96	titanium 22 Ti 47.88	vanadium 23 V 50.94	chromium 24 Cr 52.00	manganese 25 Mn 54.94	iron 26 Fe 55.85	cobalt 27 Co 58.93	nickel 28 Ni 58.69	copper 29 Cu 63.55	zinc 30 Zn 65.39	gallium 31 Ga 69.72	germanium 32 Ge 72.61	arsenic 33 As 74.92	selenium 34 Se 78.96	bromine 35 Br 79.90	krypton 36 Kr 83.80
5	rubidium 37 Rb 85.47	strontium 38 Sr 87.62	yttrium 39 Y 88.91	zirconium 40 Zr 91.22	niobium 41 Nb 92.91	molybdenum 42 Mo 95.94	technetium 43 Tc (99)	ruthenium 44 Ru 101.07	rhodium 45 Rh 102.91	palladium 46 Pd 106.42	silver 47 Ag 107.87	cadmium 48 Cd 112.41	indium 49 In 114.82	tin 50 Sn 118.71	antimony 51 Sb 121.75	tellurium 52 Te 127.60	iodine 53 I 126.90	xenon 54 Xe 131.29
6	cesium 55 Cs 132.91	barium 56 Ba 137.33	lanthanum 57 La 138.91	hafnium 72 Hf 178.49	tantalum 73 Ta 180.95	tungsten 74 W 183.85	rhenium 75 Re 186.21	osmium 76 Os 190.2	iridium 77 Ir 192.22	platinum 78 Pt 195.08	gold 79 Au 196.97	mercury 80 Hg 200.59	thallium 81 Tl 204.38	lead 82 Pb 207.2	bismuth 83 Bi 208.98	polonium 84 Po (209)	astatine 85 At (210)	radon 86 Rn (222)
7	francium 87 Fr (223)	radium 88 Ra (226)	actinium 89 Ac (227)	rutherfordium 104 Rf (261)	dubnium 105 Db (262)	seaborgium 106 Sg (263)	bohrium 107 Bh (262)	hassium 108 Hs (265)	meitnerium 109 Mt (266)	ununnilium 110 Uun (269)	unununium 111 Uuu (272)	ununbium 112 Uub (277)	ununtrium 113 Uut (284.18)	flerovium 114 Fl (289.19)	ununpentium 115 Uup (288.19)	livermorium 116 Lv (293)	ununseptium 117 Uus (294)	ununoctium 118 Uuo (294)

Series														
Lanthanide Series	cerium 58 Ce 140.12	praseodymium 59 Pr 140.91	neodymium 60 Nd 144.24	promethium 61 Pm (147)	samarium 62 Sm 150.36	europium 63 Eu 151.97	gadolinium 64 Gd 157.25	terbium 65 Tb 158.93	dysprosium 66 Dy 162.50	holmium 67 Ho 164.93	erbium 68 Er 167.26	thulium 69 Tm 168.93	ytterbium 70 Yb 173.04	lutetium 71 Lu 174.97
Actinide Series	thorium 90 Th 232.04	protactinium 91 Pa (231)	uranium 92 U 238.03	neptunium 93 Np (237)	plutonium 94 Pu (244)	americium 95 Am (243)	curium 96 Cm (247)	berkelium 97 Bk (247)	californium 98 Cf (251)	einsteinium 99 Es (252)	fermium 100 Fm (257)	mendelevium 101 Md (258)	nobelium 102 No (259)	lawrencium 103 Lr (260)

*Note: For radioactive elements, the mass number of an important isotope is shown in parentheses; for thorium and uranium, the atomic mass of the naturally occurring radioisotopes is given.

FIGURE 31-18 The periodic table of elements.

negative ion, also called an **anion**). Different forms of the same atom are known as **isotopes** of that element.

For example, all atoms of the element carbon have 6 protons within their nucleus. If you were to take all of the carbon atoms in the world, you would find that nearly 98.9% of carbon atoms will also have 6 neutrons in their nucleus. We refer to these atoms as the isotope carbon-12 (C-12), where 12 is the total number of protons and neutrons within the nucleus. In contrast, nearly 1.1% of all carbon atoms have 7 neutrons in their nucleus, so this isotope is carbon-13 (C-13). Both of these isotopes of carbon are stable; the energy balance between the number of protons and the number of neutrons within the nucleus is all that is needed to hold that nucleus together for eternity. However, one out of every trillion atoms of natural carbon has 8 neutrons in its nucleus; this isotope is therefore called carbon-14 (C-14). Carbon-14 is not stable. The imbalance between the numbers of protons and neutrons creates an excess of energy that the nucleus will eventually eject to reach a stable configuration. This natural and spontaneous process by which unstable isotopes of an element decay to a different state and emit radiation in the form of particles or waves is called **radioactivity**.

There are two forms of radiation that an operations-level responder needs to understand: **ionizing radiation** and non-ionizing radiation. Ionizing radiation is a form of radiation that includes particles capable of producing ions. Ionizing radiation has enough energy to remove electrons from other atoms. The process of removing electrons from atoms or molecules is called **ionization**. Ionizing radiation's ability to remove electrons from atoms and break apart molecular bonds is what makes it potentially hazardous. When atoms are ionized, the chemical properties of those atoms are altered. This is how radiation can damage a cell; it ionizes the atoms and changes the resulting chemical behavior of the atoms and/or molecules in the cell. If a person receives a sufficiently high dose of radiation and many cells are damaged, there may be noticeable—observable—health effects.

Non-ionizing radiation is a form of radiation that comes from electromagnetic waves. The electronic waves can cause a disturbance of activity at the atomic level, but they do not have sufficient energy to break bonds and create ions. Typical non-ionizing waves include sound waves, radio waves, and microwaves. These forms of radiation do not have enough energy to remove an electron from an atom.

Small radiation detectors are available that can be worn on turnout gear. These detectors sound an alarm when dangerous levels of radiation are encountered and alert responders to leave the scene and call for more specialized assistance.

When it comes to gaining information from the labels found on radioactive material shipping packages, responders should seek to identify the following items:

- The type or category of the label
- The contents (isotopes) of the container
- The activity level of the contained substance (rate of disintegration or decay of that substance measured in becquerels)
- The transport index (maximum dose equivalent rate at 1 meter from the surface of the package)
- The criticality safety index (a number used to provide control over the accumulation of packages or freight containers containing material capable of undergoing **fission**, which is the splitting of an atom's nucleus into two or more pieces after capturing a low-energy neutron)

These basic pieces of information found on a radiation label will go a long way toward helping you understand the hazards posed by the material.

Toxic Products of Combustion

Toxic products of combustion are the hazardous chemical compounds released when a material decomposes under heat. Some of these compounds are carcinogens. Recall that the process of combustion is a **chemical reaction** and, like other reactions, will generate gases (many of them toxic) and particulates as by-products of the fuel's decomposition. Have you thought about what is in the smoke you may be breathing in or taken a moment to think about the toxic gases liberated during a residential structure fire? An easy way to think about this issue is to apply a well-worn phrase: Garbage in, garbage out. This phrase reflects the idea that whatever objects are involved in the fire (chairs, tables, and sofas, for example) decompose and release a host of chemical by-products.

Notable substances found in most fire smoke include soot (carcinogen), carbon monoxide, carbon dioxide, polycyclic aromatic hydrocarbons, benzene (carcinogen), water vapor, formaldehyde (carcinogen), cyanide compounds, chlorine compounds, and many oxides of nitrogen. Each of these substances is unique in its chemical make-up, and most are toxic to humans, even in small doses. For example, carbon monoxide affects the body's ability to transport oxygen. When it is present in the body in excessive amounts, the red blood cells cannot carry oxygen to the other cells of the body, so the person may die from tissue asphyxiation. Cyanide compounds also adversely affect oxygen uptake in the body and are often a cause or contributing cause of smoke-related illness and death. Formaldehyde is found in many plastics and resins; it is one of the many components of smoke that causes eye and lung irritation. The oxides of nitrogen, which include nitric oxide, nitrous oxide, and nitrogen dioxide, are deep lung irritants that may cause a serious medical condition called pulmonary edema (fluid build-up in the lungs).

Hazard, Exposure, and Contamination

When responders operate at hazardous materials incidents, it is vital to identify the potential hazards and risks to minimize the potential for exposure. This begins by keen observation of the scene to determine whether a hazardous substance has been released or has the potential to escape its container. Early in the incident, it is important to determine whether the pool of responders has the right training, equipment, and protective gear to positively influence the outcome of the incident. In some cases, a "no-go" situation may exist if the problem exceeds the responders' ability to solve it safely within the boundaries of their training. It is just as important to determine what cannot be done as it is to decide which actions to take. Additionally, responders must establish a safe perimeter around the problem, keeping it secure from accidental entrance and possible human exposures. In fact, the most important initial actions taken at a hazardous materials incident revolve around identifying the problem and taking steps to limit the spread of contamination and/or human exposures.

Hazard and Exposure

Responders must correlate the hazard and exposure in the situation and understand the difference between the two. **Hazard** is the capability of a material to cause harm; *exposure* is contact by people, animals, the environment, property, or equipment with a hazardous material/WMD (NFPA 470).

Contamination

Exposures cause **contamination**, which is the transfer of "a hazardous material or the hazardous component of a WMD from its source to people, animals, the environment, or equipment" (NFPA 470). Contaminated people and things can then expose additional people and things to the hazardous material. When a chemical has been released and physically comes in contact with people, the environment, and everything around it, either intentionally or unintentionally, the residue of that chemical is called a **contaminant**. In some cases, the process of removing contaminants is complex and requires a significant effort to complete. In other cases, the contaminants are easily eliminated. For now, simply understand that when a chemical escapes its container, it may potentially get into or onto something (exposed persons, animals, PPE, or tools and equipment), and you will need a system to safely and efficiently remove that chemical or otherwise address the hazard.

Secondary Contamination

Secondary contamination, also known as **cross-contamination**, occurs when a person or object transfers the contaminant or the source of contamination to another person (whether responder or civilian), object, or environment by direct contact. Responders who become contaminated and then subsequently handle tools and equipment, or who touch door handles or other responders, spread the contamination. Additionally, contaminated victims who are improperly handled by unprotected responders also run the risk of spreading the contamination (**FIGURE 31-19**).

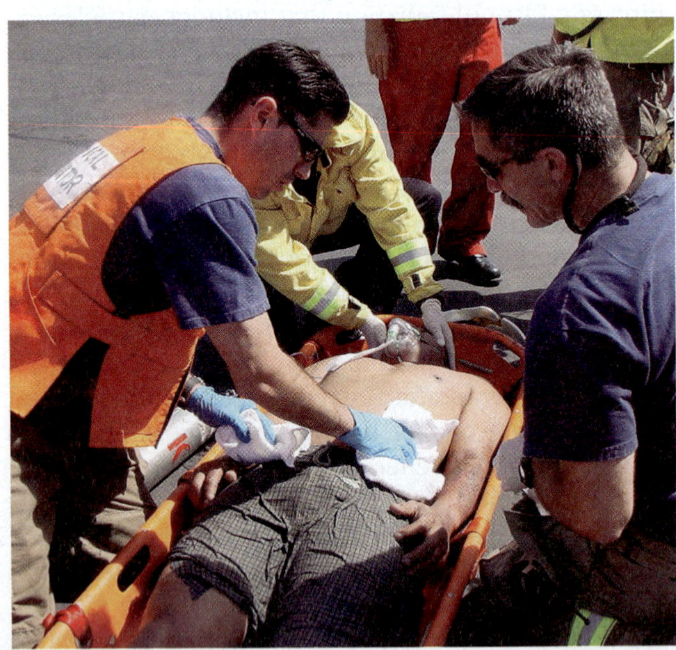

FIGURE 31-19 Responders should ensure contaminated victims are completely decontaminated prior to rendering care.

Secondary contamination may happen in several ways:

- A contaminated victim comes into physical contact with another person.
- A bystander or responder encounters a contaminated object from the hot zone (the area immediately around and adjacent to the incident).
- A decontaminated responder reenters the decontamination area and encounters a contaminated person or object.

The possibility of spreading contamination brings up a point that all responders should understand: The cleaner responders stay during the response, the less decontamination they will need later. Many responders have the misperception that PPE is worn to enable you to contact the product. In fact, quite the opposite is true: PPE, including specialized chemical-protective clothing and equipment, is worn to protect you in the event that you cannot avoid product contact. The less contamination encountered or spread, the easier the decontamination will be.

Chronic and Acute Health Effects

A **chronic health effect** (also known as a **chronic health hazard**) is an adverse health effect that occurs gradually over time after long-term exposure to a hazard. Chronic health effects may appear either after long-term or **chronic exposure**—long-term exposure to a hazardous material or materials occurring over the course of many months or years—or following multiple short-term exposures that occur over a shorter period. These exposures can involve all routes of entry into the body and may result in such health problems as cancer, permanent loss of lung function, or repetitive skin rashes. For example, inhaling asbestos fibers for years without wearing respiratory protection can result in a form of lung cancer called asbestosis.

Chemicals that pose a hazard to health after relatively short exposure periods are said to cause acute health effects, which are observable health problems, such as eye irritation, coughing, dizziness, and skin burns. With such a **subchronic exposure**, the effects occur either immediately after a single exposure or as long as several days or weeks after an exposure. Essentially, **acute exposure** is a "right now" exposure that produces some observable condition, such as eye irritation, coughing, dizziness, or skin burns.

Skin irritation and burning after a dermal exposure to sulfuric acid would be classified as **acute health effects**. Other chemicals, such as formaldehyde, are also capable of causing acute health effects. Among other things, formaldehyde (which is also classified as a human carcinogen) is a **sensitizer** that can cause an immune system response (i.e., an allergic reaction) when it is inhaled or absorbed through the skin. This reaction resembles other allergic reactions, in that the initial exposure produces mild health effects, but subsequent exposures cause much more severe reactions such as breathing difficulties and skin irritation. OSHA (1970) defines a sensitizer as a chemical that causes a substantial proportion of exposed people or animals to develop an allergic reaction in normal tissue after repeated exposure to the chemical.

It is also important to understand the difference between contagious and infectious exposures. An **infectious** substance is known to contain, or is reasonably suspected of containing, a microorganism or particles that can cause disease in humans or animals. Many biological agents (disease-causing bacteria, viruses, and other agents) we consider as potential terrorist agents are infectious. **Contagious** means that a bacteria or virus can be transmitted from person to person. Malaria serves as a good example of the difference between infectious and contagious. A person cannot "catch" malaria from another person (it is not contagious), but a single bite from an infected mosquito could potentially cause a fatal case of malaria (it is infectious).

A **toxin** is a poisonous substance produced from the metabolic processes of living plants, animals, or microorganisms. A **poison** is a substance capable of causing adverse health effects in humans and animals. **Toxicity** is a measure of the degree to which something is toxic or poisonous (NFPA 326). This term can also refer to the adverse effects a substance may have on a whole organism such as a human, a bacterium, or a plant; to a substructure such as a cell; or on a specific organ such as the liver, kidneys, or lungs. To understand the risks posed by a material's toxicity, responders must consider the physical and chemical properties of the substance causing the illness or injury, as well as the **dose–response relationship** that exists when a person is exposed to any substance. This important correlation refers to the response (signs, symptoms, illness, or injury) that a specific dose might provoke from the human body. The magnitude of the response depends on several factors, including

Voice of Experience

This chapter includes a significant number of terms that you will read about, and in your training, you will hear similar textbook definitions. One of my lessons, which unfortunately occurred to me later in my career, was a simple one-word question: Why? Our hazardous materials team had easy access to a chemist, who routinely responded to incidents, so I was able to gather advice from him, and, in an effort to save time, I never really asked the "why" question. Our team had moved into a very conservative methodology without doing much risk assessment. We had paralyzed ourselves with analysis, but we didn't make the leap from the information to the actual risk.

We were called to an incident where a tractor trailer was carrying a bunch of super sacks (see Chapter 32, *Understanding the Hazards*, in this text for description of super sacks). One of the sacks that contained catechol had a small hole, and the powder product was leaking out. I asked our chemist what his suggestion would be for protective clothing. We had looked at the SDS and had determined that the chemical didn't present much risk to our responders. The chemist said the plant workers wore dust masks with a coverall. He said that he would recommend that our folks wear an encapsulated chemical suit to deal with the product. I was a bit taken aback by this recommendation, which was way more protection that I thought we would use based on what their folks wore. So I asked the magic question: Why? Our chemist said, "I have been waiting for years for you to ask that question, and we need to sit down for a bit for me to provide you the answer you need."

I took his recommendation, and we handled the incident without any issues. We got back to the station and our chemist joined me in my office. The short answer to the "why" question on this incident was easy: When catechol gets wet, it turns anything it touches purple, and he didn't want our SCBA to turn purple. The longer answer was more complicated, and we had a long discussion about the street definitions for the various chemical properties and what they really should mean to me.

We determined that we needed to change to a risk-based response, and the use of chemical and physical properties played a major factor in that transformation. The one term that I realized was the key to safe hazardous materials response is vapor pressure. If you were going to give me only one aspect of a chemical, I would want to know its vapor pressure. The higher the vapor pressure, the more risk the chemical presents. The lower the vapor pressure, the less risk it presents. Obviously, you want to know more about a chemical, but once you get a good understanding of some basic terms and, more importantly, how materials can impact you on the street, you can safely respond.

We developed a new training program and spent a considerable amount of time going over real-life scenarios and the implications of various chemical properties. One such property is flash point. Most people think that flash point is related to ambient temperature, but it is actually the temperature of the liquid that is the key to establishing your risk.

Another term that is used in a cavalier way is *reactive*, as in "the material is reactive." There are actually various degrees of reactivity. Sodium hydroxide (commonly called lye) is noted as being violently water-reactive. When something is described as violently reactive, I am thinking about potential loss of limbs and horrific bodily damage. Homeowners buy lye (through its trade name Drano) every day and pour it in their stopped-up sinks, which are full of water. The violent reaction is a rapid increase in heat of approximately 20°F when the lye hits the water. That's not quite the violent category one would think. By comparison, oleum, which is concentrated sulfuric acid that is infused with sulfur trioxide, is a very aggressive acid. Its SDS also notes that it has a violent reaction with water. When water hits oleum, it creates an instant acid vapor cloud, which can travel for a considerable distance. The oleum immediately increases in temperature by 300°F. It bubbles and spatters, presenting some risk to anyone nearby. A straight water stream into a pool of oleum can cause an explosion-like reaction. That is what I consider to be violent, so you have to understand the various aspects of the risks.

The point of all this is to ask "why." Ask your instructor, your officers, and other responders. Do more than memorize the terms—learn about their potential impact on you when you are out on the street.

Chris Hawley
Fire Specialist (Retired)
Baltimore County Fire Department
Baltimore County, Maryland

the concentration of the hazardous substance and the duration of the exposure. The actual dose is the amount taken up by a person through one of the four routes of entry.

Cyanide, for example, is a harmful substance that can enter the body by all routes of entry. If cyanide gets into the body via the lungs, for example, it would be useful to understand at what airborne concentration (dose) the exposure might produce adverse health effects (response). The **lethal dose (LD)** of a material is a single dose that causes the death of a specified number of test animals in a group exposed by any route other than inhalation. The **lethal concentration (LC)** is the concentration of a material in air that, based on laboratory tests (inhalation route), is expected to kill a specified number of test animals in a group when administered over a specified period of time.

Several subcategories of LD and LC exist, each depicting a benchmark concentration of a particular exposure that will be harmful or fatal to a certain percentage of the test population. These values are commonly found in SDS, electronic databases, and reference books. In most cases, these values are derived from animal studies, so the results may not exactly correspond to the levels and concentrations that would be harmful to humans.

Many chemical substances on the market today have established LD or LC values. Typically, if the substance is a vapor or gas, it is expressed as an LC value. If the substance poses a dermal threat or is harmful when ingested, its toxicity is expressed as an LD value.

For any given substance, the LD is the lowest dosage per unit of body weight (typically stated in terms of milligrams per kilogram [mg/kg] of body weight) of a substance known to have resulted in fatality in an animal species. The median lethal dose (LD_{50}) of a toxic material is the dose required to kill half (50 percent) of the members of a tested population. For example, the LD_{50} of sodium cyanide (based on testing done on laboratory rats) is approximately 6.4 mg/kg. LD_{50} figures are frequently used as a general indicator of a substance's toxicity. The LD_{hi} or LD_{100} is the absolute dose of a toxic material required to kill all (100 percent) of the members of a tested population.

The LC_{lo} is the lowest LC of a material reported to cause death in an animal species, when administered via the inhalation route; it is the lowest LC for gases, dusts, vapors, and mists. The LC_{50} is the concentration of a material in air that is expected to kill 50 percent of an animal test group when administered via the inhalation route. Some literature reports the LC_{50} of carbon monoxide to be approximately 3700 parts per million (ppm) for a 1-hour exposure time frame. Carbon monoxide levels have been found to approach 10,000 ppm—well above the LC_{50}. The LC_{hi} or LC_{100} is the absolute concentration of a toxic material required to kill all (100 percent) of the members of a tested population when the substance is administered via the inhalation route.

The values expressed here figure prominently into OSHA's view when it comes to identifying the relative toxicity of a particular substance. The following information reflects OSHA's view of exposure levels considered to be toxic and highly toxic.

The label *toxic*, as defined by OSHA 29 CFR 1910.1200, is assigned to any chemical that falls into one of these three categories:

- A chemical that has an LD_{50} of more than 50 mg/kg but not more than 500 mg/kg of body weight when administered orally to albino rats weighing between 200 and 300 g each
- A chemical that has an LD_{50} of more than 200 mg/kg but not more than 1000 mg/kg of body weight when administered by continuous contact for 24 hours (or less if death occurs within 24 hours) with the bare skin of albino rabbits weighing between 2 and 3 kg each
- A chemical that has an LC_{50} in air of more than 200 ppm but not more than 2000 ppm by volume of gas or vapor, or more than 2 mg/L but not more than 20 mg/L of mist, fume, or dust, when administered by continuous inhalation for 1 hour (or less if death occurs within 1 hour) to albino rats weighing between 200 and 300 g each

OSHA 29 CFR 1910.1200 describes *highly toxic* materials as follows:

- A chemical that has an LD_{50} of 50 mg/kg or less of body weight when administered orally to albino rats weighing between 200 and 300 g each
- A chemical that has an LD_{50} of 200 mg/kg or less of body weight when administered by continuous contact for 24 hours (or less if death occurs within 24 hours) with the bare skin of albino rabbits weighing between 2 and 3 kg each
- A chemical that has an LC_{50} in air of 200 ppm by volume or less of gas or vapor, or 2 mg/L or less of mist, fume, or dust, when administered by continuous inhalation for 1 hour (or less if death occurs within 1 hour) to albino rats weighing between 200 and 300 g each

CASE STUDY

You Are the Firefighter CONCLUSION

Just after midnight, you receive a call for an explosion and a fire at a local landscaping company. Upon arrival at the site, you find a working fire in a storage shed located behind the main building. The wooden shed is fully involved, so you begin an indirect attack on the fire from the outside of the shed. As the fire is being knocked down, you receive an order over the radio to shut down the attack and move away from the building. About the same time, you experience an itchy feeling on the back of your neck. Other crew members are also complaining of itching and burning sensations around their wrists and necks. Some lower-floor residents of an adjacent multistory apartment complex are complaining of eye irritation and asking about a strange odor in the air.

1. **Which types of chemicals might be found in this kind of occupancy?**

   Answer: Many different types of chemicals may be found in this type of occupancy, including pesticides, solvents, fuels, fertilizers, and oils.

2. **Where could you obtain accurate technical information on the products stored in this building?**

   Answer: Ideally, there would be SDS for chemical inventory and/or an NFPA 704 diamond on the building. However, it is very common for small business like this to have no formal chemical inventory. If the building owner or other responsible person is available, they may be able to provide some information on the substances present.

3. **Which actions should be taken to address the complaints of burning and itching skin among firefighters as well as the complaints of the residents of the apartment complex?**

   Answer: First, immediately leave the immediate area and switch to a defensive attack to reduce the exposure. Next, firefighters should remove bunker gear, and all affected skin areas of firefighters and civilians should be rinsed with a basic water-based rinse, capturing the run-off if possible. If advanced life support medical services are available in the jurisdiction, dispatch that resource to the scene for further assessment of the affected people. Additionally, it may be necessary to dispatch a hazardous materials team to monitor the perimeter for airborne contaminants and to make positive ID of the substances once the fire is under control.

WRAP-UP

SUMMARY

KNOWLEDGE OBJECTIVES

- Describe states of matter and their physical and chemical changes. (**NFPA 470: 7.2.1**, pp. 1257–1261)
  - Understand the ways containers can breach and release contents. (**NFPA 470: 7.2.1**, pp. 1257–1259)
  - Understand the causes and effects of a BLEVE.
- Discuss the critical characteristics of flammable liquids. (pp. 1261–1265)
  - Define flash point, ignition temperature, flammable range, vapor pressure, and boiling point.
- Understand the definition of vapor density. (pp. 1265–1266)
  - Understand how the vapor density of a substance affects the nature of a release.
  - Identify heavier than air gases by using the mnemonic 4H MEDIC ANNA.
- Understand the definition of specific gravity. (pp. 1266–1267)
  - Understand how the specific gravity of a substance affects the nature of a release.

- Understand how the use of water to mitigate a release may be impacted the specific gravity of a substance.
- Understand the definition of water solubility. (p. 1267)
  - Understand how the water solubility of a substance affects the nature of a release.
  - Understand the benefits and limitations of using water to mitigate a release.
- Understand the definition of corrosivity. (pp. 1267–1268)
  - Understand the pH scale and how to identify an acid and a base.
  - Understand the impact of corrosivity on incident mitigation efforts when dealing with corrosives.
- Understand how to determine the concentration of a released hazardous materials. (pp. 1268–1269)
  - Understand the definition of a solution.
  - Understand the difference between concentrated and dilute solutions.
- Describe ionizing and non-ionizing radiation. (**NFPA 470: 7.2.1**, pp. 1269–1271)
  - Define the term *isotope*.
  - Identify a basic set of steps to understand information found on radioactive transportation labels.
- Describe toxic products of combustion. (**NFPA 470: 7.2.1**, pp. 1271–1272)
  - Understand the basic health effects posed by carbon monoxide exposure.
  - Understand the basic health effects posed by exposure to cyanide compounds.
- Understand the concepts of hazard, exposure, and contamination. (**NFPA 470: 7.3.1**, pp. 1272–1273)
  - Understand the need to take no action at an incident that exceeds the responder's level of training, personal protective equipment or other health and safety aspects of the release.
  - Understand the need for basic protective actions at a hazardous materials release.
- Describe how hazardous material exposure can lead to chronic and/or acute health effects. (pp. 1273–1275)
  - Understand the definition of contamination.
  - Understand the definition of secondary contamination.

SKILLS OBJECTIVES

There are no skills objectives for operations-level responders for this chapter.

KEY TERMS

acid A material with a pH value less than 7.

action plan A detailed proposal to determine strategies and tactics within an incident action plan that will safely accomplish incident objectives, favorably influence outcomes, and increase the safety of responders and the public. (NFPA 470)

acute exposure An exposure that produces instantly observable conditions such as eye irritation, coughing, dizziness, and skin burns.

acute health effects Observable health problems, such as eye irritation, coughing, dizziness, and skin burns, caused by relatively short exposures to a hazardous substance.

anion A negatively charged ion.

atmosphere (atm) A unit of pressure equal to the average atmospheric pressure at sea level.

atom The smallest unit of a chemical element.

atomic number The number of an element in the periodic table of elements that corresponds to the number of protons in the atom.

bar A unit of pressure equal to 14.7 psi.

base A material with a pH value greater than 7.

boiling point The temperature at which the vapor pressure of a liquid equals the surrounding atmospheric pressure. (NFPA 1)

cation A positively charged ion.

chemical and physical properties Measurable characteristics of a substance, such as its vapor density, flammability, corrosivity, and water reactivity.

chemical change See *chemical reactivity.*

chemical reaction Any chemical change or chemical degradation.

KEY TERMS CONTINUED

chemical reactivity The ability of a chemical to undergo an alteration in its chemical make-up, usually accompanied by a release of some form of energy; also called *chemical change.*

chronic exposure Long-term exposure to a hazardous material or materials, occurring over the course of many months or years.

chronic health effect See *chronic health hazard.*

chronic health hazard An adverse health effect that occurs gradually over time after long-term exposure to a substance; also called *chronic health effect.*

combustibles Materials that, in the form in which they are used and under the conditions anticipated, will ignite and burn. (NFPA 1)

concentrated solution A solution with a large amount of solute.

concentration The amount of solute in a given amount of solvent or the ratio of solute in a given amount of total solution.

contagious The property of being able to be transmitted from person to person.

contaminant The residue of a released substance that can be transferred to people, animals, the environment, or equipment.

contamination The transfer of a hazardous material or the hazardous component of a weapon of mass destruction (WMD) from its source to people, animals, the environment, or equipment. (NFPA 470)

corrosivity The property of a material reflecting its ability to cause damage on contact with skin, eyes, or other parts of the body.

cross-contamination See *secondary contamination.*

dilute solution A solution that contains a small amount of solute.

dose–response relationship The correlation between the amount of a toxic substance taken in by a person (the dose) and the signs, symptoms, illness, or injury that the dose might provoke from the human body (the response) when a person is exposed to any substance.

electron In an atom, a negatively charged particle that orbits the nucleus.

element A substance that cannot be chemically broken down into simpler substances.

expansion ratio A description of the volume increase that occurs when a liquid changes to a gas.

exposure Contact with a hazardous material/weapon of mass destruction (WMD) by people, animals, the environment, property, or equipment. (NFPA 470)

fire point The lowest temperature at which a liquid will ignite and achieve sustained burning when exposed to a test flame in accordance with ASTM D 92, *Standard Test Method for Flash and Fire Points by Cleveland Open Cup Tester.* (NFPA 1)

fission The splitting of an atom's nucleus into two or more pieces after capturing a low-energy neutron.

flammable range An expression of an air/fuel mixture, defined by upper and lower percentage limits, that reflects the amount of flammable vapor mixed with a given volume of air.

flash point The minimum temperature at which a liquid or solid gives off sufficient vapors such that, when an ignition source (e.g., a flame, electrical equipment, lightning, or even static electricity) is present, the vapors will ignite, resulting in a flash fire.

hazard A condition or material capable of causing harm to life, health, property, or the environment. (NFPA 470)

headspace The space above any liquid placed in a closed vessel that contains vapor.

ignition temperature The minimum temperature at which a fuel, when heated, will ignite in the presence of air and continue to burn.

incident objectives Incident-specific statements describing intended outcomes within a given operational period. (NFPA 470)

infectious The property of being able to cause disease in humans or animals.

ion An atom with more protons than electrons (a positive ion or cation) or more electrons than protons (a negative ion or anion).

ionization The process of removing electrons from atoms or molecules.

ionizing radiation A form of radiation that includes particles capable of producing ions.

isotopes Different forms of the same atom.

kinetic energy Energy generated by an object in motion.

lethal concentration (LC) The concentration of a material in air that, based on laboratory tests of the inhalation route to the body, is expected to kill a specified number of test animals in a group when administered over a specified period of time.

lethal dose (LD) A single dose that causes the death of a specified number of test animals in a group exposed by any route other than inhalation.

lower explosive limit (LEL) The minimum concentration of combustible vapor or gas in a mixture of the vapor or gas and gaseous oxidant above which propagation of flame will occur on contact with an ignition source. (NFPA 115)

millimeters of mercury (mm Hg) A unit of pressure equal to the pressure exerted at the base of a column of mercury that is exactly 1 millimeter in height; 760 mm Hg is equal to 14.7 psi.

monomers Molecules that can be bonded to like molecules to form a larger polymer.

neutron In the nucleus of an atom, a particle that does not have an electrical charge.

non-ionizing radiation A form of radiation that comes from electromagnetic waves; it can cause a disturbance of activity at the atomic level, but does not have sufficient energy to break bonds and create ions.

nucleus The center of an atom.

pH A measurement of the amount of dissolved hydrogen ions (H^+) in a solution.

physical change A transformation in which a material changes its state of matter, for instance, from a liquid to a solid.

poison A substance capable of causing adverse health effects in humans and animals.

polymer A larger molecule made up of long repeating chains of molecules; often used to describe different types of plastics.

polymerization The process of reacting monomers together in a chain reaction to form polymers.

pounds per square inch (psi) A unit of pressure that expresses the pounds of force per square inch of area.

proton In the nucleus of an atom, a positively charged particle that orbits the nucleus.

radiation The energy transmitted through space in the form of electromagnetic waves or energetic particles.

radioactivity The natural and spontaneous process by which unstable isotopes of an element decay to a different state and emit radiation in the form of particles or waves.

secondary contamination The transfer of a contaminant or the source of contamination out of the hot zone to people, animals, the environment, or equipment; also called *cross-contamination.*

sensitizer A chemical that causes a large percentage of people or animals to develop an allergic reaction after repeated exposure to it.

solute A substance dissolved in a solvent.

solution A mixture of at least one solute dissolved in a solvent.

solvent A substance (usually liquid) capable of dissolving or dispersing another substance. (NFPA 96)

specific gravity The weight of a liquid compared to water.

state of matter The physical state of a material—solid, liquid, or gas.

strong A description of acids with a pH of 2.5 or less or bases with a pH of 12.5 or greater.

subchronic exposure Exposure that causes effects to occur either immediately after a single exposure or as long as several days or weeks after an exposure.

torr Named after Evangelista Torricelli, a unit of pressure equal to 1 mm Hg; 760 torr is equal to 14.7 psi.

toxicity The degree to which a substance is harmful to humans. (NFPA 326)

toxic products of combustion Hazardous chemical compounds that are released when a material decomposes under heat.

toxin A substance created by a plant or an animal that may be harmful to humans depending on the exposure.

upper explosive limit (UEL) The maximum amount of combustible vapor or gas that can be present in the air if the air/fuel mixture is to be flammable or explosive.

vapor The gas state of a substance, particularly those that are normally liquids or solids at ordinary temperatures. (NFPA 1700)

vapor density The weight of a vapor or gas compared to an equal volume of dry air.

vaporization The process of a solid or liquid changing to a gas.

KEY TERMS CONTINUED

vapor pressure The pressure, measured in pounds per square inch, absolute (psia), exerted by a liquid, as determined by ASTM D 323, *Standard Test Method for Vapor Pressure of Petroleum Products (Reid Method).* (NFPA 1)

water solubility The ability of a substance to dissolve in water.

weak A description of acids with a pH greater than 2.5 or bases with a pH less than 12.5.

REVIEW QUESTIONS

1. What are the three states of matter?
2. What is the difference between physical and chemical change?
3. Why does water boil faster and at a lower temperature at high elevations?
4. What is vapor density?
5. Define specific gravity.
6. What is water solubility?
7. How do you determine the corrosivity of a substance?
8. What is the difference between a concentrated and dilute solution?
9. What is the difference between ionizing and non-ionizing radiation?
10. What is the definition of toxic byproducts of combustion?
11. What is the difference between hazard and exposure?
12. Describe the difference between acute and chronic health effects.

DISCUSSION QUESTIONS

1. Why is flash point an important characteristic of a flammable liquid?
2. Why is it important to know the vapor density of a released liquid?
3. Why is it important to know the specific gravity of a released liquid?
4. What is secondary contamination and why is important to avoid?

APPLYING THE CONCEPTS

Your crew is called to respond to a possible leaking container in the outdoor hazardous waste accumulation area at a nearby biotechnology research facility. It is winter, and the outside air temperature is 41°F (5°C). Upon arrival, you are met by a member of the site emergency response team (ERT). The person identifies herself as the incident commander (IC) of the ERT and tells you that a white polyethylene drum was knocked over and appears to be leaking. All members of your crew are trained to the operations level.

1. Based on the preceding information, does your crew have the right level of training to enter the facility, identify the liquid, and clean up the spill?
2. Based on your knowledge of containers, what type of material is likely in the container?

The IC from the ERT says that the drum contains 190-proof ethyl alcohol that is used in one of the company's processes. The SDS given to you by the IC contains the information listed below:

Chemical name: ETHYL ALCOHOL 190 proof

Appearance: Clear, colorless liquid

Flash point: 65.3°F (18.5°C)

Flammable range: LOWER: 3.3% UPPER: 19%

Solubility: Soluble in water

Specific gravity: 0.8

pH: Neutral

Vapor density (air = 1): 1.59

Health effects: Hazardous in case of skin contact (irritant) or eye contact (irritant). Noncorrosive to skin. Noncorrosive to eyes. Noncorrosive to lungs.

Fire hazard: Highly flammable in the presence of open flames and sparks or heat. Slightly flammable to flammable in the presence of oxidizing materials.

3. Based on the information provided by the SDS, at or above what temperature would you be concerned about the risk of fire if an uncontrolled ignition source was nearby?
4. If vapors are being produced, which specification would you consult to find out the percentage of air/fuel mixture required for the vapors to ignite?

As you approach the outdoor hazardous waste accumulation area, you notice an alcohol-like odor in the air. You direct everyone to stop and move back to a safe distance. The crew begins to isolate the area and stretch yellow caution tape around the perimeter. It is clear there is a large puddle of product on the ground.

5. Based on the SDS, would you expect the vapors generated from the release to pose a risk of flash fire?
6. Would it be useful to do a pH check on this substance?

REFERENCES

Code of Federal Regulations (CFR). "Characteristics of Corrosivity." CFR, title 40, part 261.22. Accessed September 23, 2021. https://www.ecfr.gov/current/title-40/chapter-I/subchapter-I/part-261/subpart-C/section-261.22.

Code of Federal Regulations (CFR). "Class 8 - Definitions." CFR, title 49, part 173.136. Accessed September 23, 2021. https://www.ecfr.gov/current/title-49/subtitle-B/chapter-I/subchapter-C/part-173/subpart-D/section-173.136.

National Fire Protection Association (NFPA). 2019. *NFPA 115: Standard for Laser Fire Protection.* 2020 Edition. Quincy, MA: NFPA.

National Fire Protection Association (NFPA). 2019. *NFPA 326: Standard for the Safeguarding of Tanks and Containers for Entry, Cleaning, or Repair.* 2020 Edition. Quincy, MA: NFPA.

National Fire Protection Association (NFPA). 2020. *NFPA 1: Fire Code.* 2021 Edition. Quincy, MA: NFPA.

National Fire Protection Association (NFPA). 2020. *NFPA 96: Standard for Ventilation Control and Fire Protection of Commercial Cooking Operations.* 2021 Edition. Quincy, MA: NFPA.

National Fire Protection Association (NFPA). 2020. *NFPA 1700: Guide for Structural Firefighting.* 2021 Edition. Quincy, MA: NFPA.

National Fire Protection Association (NFPA). 2021. *NFPA 470: Hazardous Materials/Weapons of Mass Destruction (WMD) Standard for Responders.* 2022 Edition. Quincy, MA: NFPA.

Occupational Safety and Health Administration (OSHA). "1910.1200 - Hazard Communication." United States Department of Labor. Accessed September 27, 2021. https://www.osha.gov/laws-regs/regulations/standardnumber/1910/1910.1200.

CHAPTER

32

Hazardous Materials Operations Level

Understanding the Hazards

NOTE: Content within this chapter meets the intent of **NFPA 470, 2022 Edition**, which includes chapter 7, "Professional Qualifications for Hazardous Materials/WMD Operations Level Responders (**NFPA 1072**)."

KNOWLEDGE OBJECTIVES

After studying this chapter, you will be able to:

- Identify and describe common types of hazardous materials containers.
- Describe the ways in which hazardous materials are transported.
- Identify resources that can provide technical chemical information.
- Identify different types of potential terrorist incidents and response options.

SKILLS OBJECTIVES

There are no skills objectives for operations-level responders for this chapter.

CASE STUDY

You Are the Firefighter

Your engine company is dispatched to the scene of a leaking drum at a local lumberyard. It appears to have been hit by a forklift, causing a small tear toward the bottom, and a substance is slowly leaking and an usual odor is reported. The dirt around the drum looks wet. The foreman of the lumberyard tells you that the drum holds waste oil, and the forklift driver on duty denies hitting the drum. From a safe distance, you can tell that the drum is made of steel and has a 2-inch (51-mm) bung and a ¾-inch (19-mm) bung on the top. You can see a small red label on the side of the drum that says "flammable," with the number 3 at the bottom.

1. What type of material is most likely stored in the drum (e.g., corrosive, solvent, explosive)? Does the description of the drum provide any other useful information? If so, what?
2. What does the label on the drum signify, and does the report of an unusual odor make sense given the information you have?
3. How would you go about obtaining more information on the potential contents of the drum?

Introduction

Even the most common chemical substances and the containers they are held in can become dangerous when released or involved in an accident. For example, we tend to think of gasoline tankers and railcars as valuable components of transportation and commerce, but if they are involved in a serious accident, or are used to create harm in the hands of determined terrorists, these common modes of transport can become deadly weapons. As another example, demolition companies use explosives to bring down unneeded structures quickly and safely, but terrorists could use the same explosives to bring down an occupied building, deliberately killing many people. When it comes to hazardous materials incidents, it is the intent that distinguishes between an accident and a terrorist event, not the chemical or the container involved. This chapter therefore offers a combined overview of a wide variety of containers and their physical characteristics, the common substances found in those containers, and a broad overview of several classes of terrorism agents.

Operations-Level Container Recognition

Lots of information can be gained about an incident by knowing some basic things about containers. Some commonly encountered containers include drums, carboys, and compressed gas cylinders. This section describes several types of containers and some general characteristics of the container, such as the materials used in its construction, its shape, and its typical internal pressures. This discussion is not meant as a definitive reference for any of the container types described. Rather, it is intended to help the responder form some initial impression of the hazards and potential harm posed by the chemical and physical properties of the materials inside.

Bulk Fixed-Facility Pressure Containers

The operations-level responder may encounter very large containers at fixed facilities. **Bulk packaging**, or large-volume containers, is defined by its internal capacity based on the following measures (excluding a vessel or a barge) (NFPA 470):

- A maximum liquid capacity greater than 119 gallons (450 liters)
- A maximum net mass greater than 882 pounds (400 kg) as a receptacle for a solid
- A water capacity greater than 1001 pounds (454 kg) as a receptacle for a gas

In contrast, **nonbulk packaging** is defined as packaging that has "a liquid capacity of 119 gallons (450 liters) or less, a solids capacity of 882 pounds (400 kg) or less, or a compressed gas water capacity of 1001 pounds (454 kg) or less" (NFPA 470). Essentially, nonbulk packaging includes all types of containers other than bulk containers.

TABLE 32-1 Common Bulk Storage Vessels, Locations, and Contents

Tank Shape	Common Locations	Hazardous Materials Commonly Stored
Underground tanks	Residential, commercial	Fuel oil, combustible liquids
Covered floating roof tanks	Bulk terminal and storage	Highly volatile flammable liquids
Cone roof tanks	Bulk terminal and storage	Combustible liquids
Open floating roof tanks	Bulk terminal and storage	Flammable and combustible liquids
Dome roof tanks	Bulk terminal and storage	Combustible liquids
Pressure horizontal tanks	Industrial storage and terminal	Flammable gases, chlorine, ammonia
Pressure spherical tanks	Industrial storage and terminal	Liquid propane gas, liquid nitrogen gas
Cryogenic liquid storage tanks	Industrial and hospital storage	Oxygen, liquid nitrogen gas

In general, bulk storage containers are found in occupancies that rely on and need to store large quantities of a substance. Most manufacturing facilities have at least one type of bulk storage container (**TABLE 32-1**). Often these bulk storage containers are surrounded by a secondary containment system to help control an accidental release. **Secondary containment** is an engineered method to control spilled or released product if the main containment vessel fails, consisting of a device or structure, such as dikes, curbing, and double-walled tanks. A 5000-gallon (18,927-liter) vertical storage tank, for example, may be surrounded by a series of short walls commonly referred to as a **dike** or **berm** that form a catch basin around the vessel.

Secondary containment basins typically can hold the entire volume of the tank, along with a percentage of the water flowed from hose lines or sprinkler systems in the event of fire and, in the case of outdoor storage, a certain amount of rainfall. Many storage vessels, including 55-gallon (208-liter) drums, may have secondary containment systems. If facilities that use bulk storage containers in your response area have secondary containment systems, it will be easier to handle leaks when they occur, but it is important to understand their capabilities and capacities.

Bulk storage tanks are usually made of aluminum, steel, or plastic and can be either pressure or nonpressure containers. Nonpressure tanks have a working pressure between 2.65 and 4 psi and are usually made of steel or aluminum. They commonly hold flammable or combustible materials, such as gasoline, ethanol, oil, or diesel fuel, but may also hold solids. A **pressure tank** has an internal pressure of more than 15 **psig** (pounds per square inch gauge, which is a measurement of pressure inside the container shown on an attached pressure gauge in relation to atmospheric pressure). When preplanning for fixed facilities in your response area, pay attention to the types of tanks you may encounter during an incident (**FIGURE 32-1**).

Liquified Compressed Gas Cylinders

Another typical type of bulk storage vessel is the liquified compressed gas cylinder. This common type of cylinder can contain a variety of volumes, but in all cases is a nonrefrigerated, pressurized tank. The typical, high-volume, liquified compressed gas cylinder has rounded ends and large vents or pressure-relief stacks. The most common aboveground storage vessel of this type typically contains liquified propane or liquified ammonia (both are **anhydrous**, which means there is no water in the substances). Such containers can hold a few hundred gallons to several thousand gallons of product. In most cases, 10 to 15 percent of the total container capacity is vapor in the headspace above the liquid (**FIGURE 32-2**). Smaller versions of pressure containers may be found in the form of portable propane cylinders and vehicle-mounted cylinders (**FIGURE 32-3**).

A.

B.

C.

D.

FIGURE 32-1 Examples of large-volume storage tanks. **A.** Spherical fixed-facility pressure containers. **B.** Bulk fixed-facility tanks. **C.** Bulk fixed-facility cryogenic container. **D.** A ton chlorine container after a leak.

FIGURE 32-2 In most cases, 10 to 15 percent of the total container capacity is vapor.

Ton Containers

A **ton container** is bulk packaging that commonly holds liquified compressed gases such as chlorine and sulfur dioxide, although you may sometimes encounter refrigerant fluorocarbon gases such as trichlorofluoromethane (Freon-11) and dichlorodifluoromethane (Freon-12) stored in ton containers. Regardless of the material inside, these vessels hold approximately 2000 pounds (907 kg) of product (hence the name, *ton container*) and are approximately 8 feet (2.4 m) in length and 3 feet (1 m) in diameter. When the weight of the cylinder is added, a full ton chlorine cylinder can weigh approximately 3500 pounds (1588 kg). All ton containers, other than those containing phosphine, have a pressure relief valve called a **fusible plug**. Figure 32-1D shows a ton container after a leak. The Y

A.

B.

FIGURE 32-3 **A.** Portable propane cylinder (nonrefrigerated liquified compressed gas) in a storage yard. **B.** Vehicle-mounted pressure container (propane) in use on a forklift.

A.

B.

FIGURE 32-4 **A.** Portable intermodal tank. **B.** Markings on a portable intermodal tank.

cylinder seen in Figure 32-1E is used for high-purity gases such as silane and phosphine.

Intermodal Tanks

An **intermodal tank (IM or IMO)** is a bulk packaging container that is used as both a shipping and a storage vessel. IM tanks hold between 5000 and 6300 gallons (18,927 to 23,848 liters) of product and can be either pressure or nonpressure containers. They can also be used to ship and store cryogenic liquids, such as liquid nitrogen and liquid argon. In most cases, an IM tank is shipped to a facility, where it is stored and used, and then returned to the shipper for refilling. IM tanks can be shipped by air, sea, or land (**FIGURE 32-4**).

Typically, an IM tank is surrounded by a box-like steel framework that facilitates efficient stacking and shipping. Several types of IM tanks are available:

- IM-101 portable tanks (IMO type 1 internationally) have a 6300-gallon (23,848-liter) capacity, with internal working pressures between 25.4 pounds per square inch (psi) and 100 psi. These containers typically carry mild corrosives, food-grade products, and flammable liquids (**FIGURE 32-5**).
- IM-102 portable tanks (IMO type 2 internationally) have a 6300-gallon (23,848-liter) capacity, with internal working pressures between 14.7 psi and 25.4 psi. They primarily hold nonhazardous materials, but may also contain flammable liquids and corrosives (**FIGURE 32-6**).
- Pressure IM tanks (IMO type 5 internationally or U.S Department of Transportation [DOT] Spec 51) are high-pressure vessels with internal pressures in the range of 100–600 psi. These intermodal containers commonly hold liquefied compressed gases such as propane and butane (**FIGURE 32-7**).
- Cryogenic IM tanks (IMO type 7 internationally) are low-pressure containers in transport but can be pressurized to 600 psi. This kind of container commonly holds cryogenic materials that have temperatures less than –150°F (–101°C), such as liquefied oxygen, nitrogen, or helium (**FIGURE 32-8**).
- **Multiple element gas containers (MEGCs)** consist of several high-pressure tubes attached to a frame. The tubes are individually specified and have

FIGURE 32-5 IM-101 portable tank (IMO type 1 internationally).

Courtesy of UBH International Ltd.

FIGURE 32-6 IM-102 portable tank (IMO type 2 internationally).

Courtesy of UBH International Ltd.

FIGURE 32-7 Pressure IM tank (IMO type 5 internationally).

Courtesy of UBH International Ltd.

FIGURE 32-8 Cryogenic IM tank (IMO type 7 internationally).

A.

B.

FIGURE 32-9 Multiple element gas containers (MEGCs) consist of several high-pressure tubes attached to a frame.

Courtesy of Bill Hand.

working pressures that range as high as 5000 psi. The products commonly carried in MEGCs include hydrogen and oxygen (**FIGURE 32-9**).

Intermediate Bulk Containers

An **intermediate bulk container (IBC)** is a bulk storage container so named because the volumes stored inside it fall between what is typically found in drums or bags and what is in cargo tanks. Their capacities are greater than 119 gallons (450 liters) but less than 793 gallons (3002 liters). IBCs are either flexible intermediate bulk containers (FIBCs) or rigid intermediate bulk containers (RIBCs). You may hear RIBCs referred to as "totes" and FIBCs called "sacks" or "super sacks." Super sacks are much larger than a normal bag and may hold solid materials weighing anywhere from 500 pounds (227 kg) to several thousand pounds (**FIGURE 32-10**). The fabric used for their construction ranges from woven cloth to polyethylene, and the IBCs may be palletized for transport. Solid materials transported and stored in super sacks may include oxidizers, corrosives, and flammable solids. Totes can be found as a high-density polyethylene tank inside a rigid stainless steel frame or perhaps a square metal tank around 4 feet (1.2 m) wide and 6 feet (1.8 m) tall. Both types have bottom-mounted discharge valves and are commonly found at fixed facilities. Totes may hold a wide variety of substances, such as solvents, corrosives, or select oxidizers.

Shipping and storing totes can create hazards. For example, these containers often are stacked atop one another and moved with a forklift. A mishap with the loading or moving process can compromise the tote (**FIGURE 32-11**). Because totes have no secondary containment system, any leak has the potential to create a large puddle. Additionally, the steel webbing around the tote makes it difficult to access and patch leaks.

Another type of unique container used for chemical storage is the flexible bladder. These vessels are typically portable and used for short-term storage and dispensing of low-hazard materials (**FIGURE 32-12**).

Transporting Hazardous Materials

Transportation of hazardous materials occurs most often on the roadway. Indeed, even when another primary mode of transport is used, roadway vehicles often transport the shipments from the rail station, airport, or dock to the point where it will be used. Responders must become familiar with all types of chemical transport vehicles they might encounter during a transportation emergency.

Roadway Transportation

Per the Code of Federal Regulations, 49 CFR 171.8(2), or local jurisdictional regulations, a **cargo tank** is bulk packaging that "is permanently attached to or forms a part of a motor vehicle or is not permanently attached to any motor vehicle and which, by reason of its size, construction, or attachment to a motor vehicle, is loaded or unloaded without being removed from the motor vehicle."

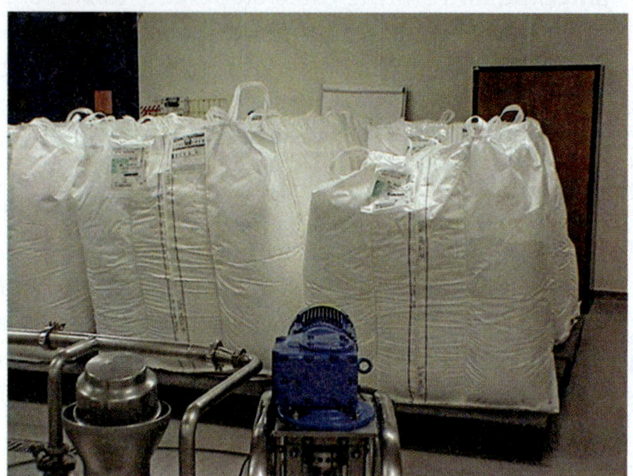

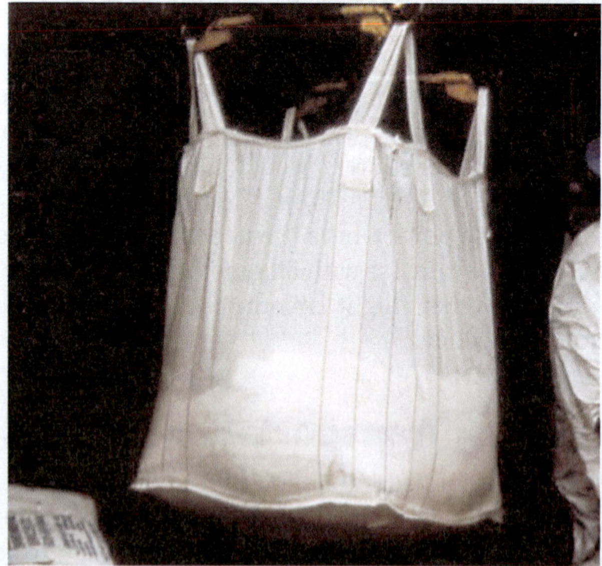

FIGURE 32-10 Examples of flexible intermediate bulk containers (FIBCs), also known as super sacks.

FIGURE 32-11 Rigid intermediate bulk containers (RIBCs), also known as totes, awaiting use at a fixed facility.

FIGURE 32-12 Flexible bladder.

One of the most common and reliable transportation vessels is the **nonpressure liquid cargo tank**, commonly referred to as an **MC-306/DOT 406 cargo tank**. The MC designation indicates "motor carrier," and the DOT designation indicates that a cargo tank meets a certain set of specifications enforced by the U.S. DOT (**FIGURE 32-13**). These cargo tanks frequently carry liquid food-grade products, gasoline, or other flammable and combustible liquids. The oval-shaped tank is pulled by a diesel (or liquefied natural gas [LNG] or compressed natural gas [CNG]) tractor and can carry 6000 to 10,000 gallons (22,712 to 37,854 liters) of product. The **nonpressure tank** is loaded and offloaded

FIGURE 32-13 The nonpressure liquid cargo tank (MC-306/DOT 406) typically hauls flammable and combustible liquids.

Courtesy of Polar Tank Trailer L.L.C.

FIGURE 32-14 The nonpressure liquid cargo tank (MC-306/DOT 406) has a remote emergency shut-off valve as a safety feature.

Courtesy of Glen Rudner.

through valves at the bottom of the tank. These cargo tanks have several safety features, including full roll-over protection and remote emergency shut-off valves (**FIGURE 32-14**).

A vehicle that is like a low-pressure liquid cargo tank is the **low-pressure chemical cargo tank** or the **MC-307/DOT 407 chemical hauler**. It has a round or horseshoe-shaped tank and can hold 6000 to 7000 gallons (22,712 to 26,498 liters) of liquid (**FIGURE 32-15**). The MC-307/DOT 407 chemical hauler, which is a tractor-drawn tank, is used to transport flammable liquids, mild corrosives, and poisons. This type of cargo tank may be insulated (horseshoe) or uninsulated

FIGURE 32-15 A low-pressure liquid cargo tank, the MC-307/DOT 407 chemical hauler carries flammable liquids, mild corrosives, and poisons.

Courtesy of Polar Tank Trailer L.L.C.

FIGURE 32-16 The MC-312/DOT 412 corrosives tanker is commonly used to carry corrosives such as concentrated sulfuric acid, phosphoric acid, and sodium hydroxide.

Courtesy of National Tank Truck Carriers Association.

(round) and may have a higher internal working pressure than the nonpressure liquid cargo tank—in some cases, up to 35 psi. Cargo tanks that transport corrosives may have a rubber lining to prevent corrosion of the tank structure.

A chemical cargo tank, commonly referred to as an **MC-312/DOT 412** corrosive tank, is often used to carry corrosives such as concentrated sulfuric acid, phosphoric acid, and sodium hydroxide (**FIGURE 32-16**). This cargo tank has a smaller diameter than the nonpressure liquid cargo tank or low-pressure **chemical cargo tank** and is often identifiable by the presence of several heavy-duty reinforcing rings around the tank. The rings provide structural stability during transportation and in the event of a roll-over. The inside of this type of chemical cargo tank operates at approximately 15 to 25 psi and holds approximately 5000 gallons (18,927 liters). These cargo tanks have substantial roll-over protection to reduce the potential for damage to the top-mounted valves.

A **pressure cargo tank**, commonly referred to as the MC-331, carries materials such as ammonia, propane, Freon, and butane (**FIGURE 32-17**). The liquid volume that **MC-331** tanks carry varies, from a 1000-gallon (3785-liter) delivery truck to a full-size 11,000-gallon (41,640-liter) cargo tank. The cargo tank has rounded ends, typical of a pressure vessel, and is commonly constructed of steel or stainless steel with a single tank compartment. The cargo tank operates at approximately 300 psi, with typical internal working pressures being near 250 psi. These cargo tanks are equipped with spring-loaded relief valves that traditionally operate at 110 percent of the designated maximum working pressure. A significant explosion hazard arises if fire impinges on a cargo tank, however. The nature of most materials carried in these tanks means a threat of explosion exists because of the relief valve's inability to keep up with the rapidly building internal pressure. Responders must use great care when dealing with this type of transportation emergency.

FIGURE 32-17 The MC-331 cargo tank carries materials such as ammonia, propane, Freon, and butane.

A **cryogenic liquid cargo tank**, commonly referred to as the **MC-338**, operates much like the cryogenic intermodal container described earlier and carries many of the same substances (**FIGURE 32-18**). This low-pressure tank relies on tank insulation to maintain the low temperatures required for the cryogens it carries. A box-like structure containing the tank control valves is typically attached to the rear of the tank. Special training is required to operate valves on this and any other tank. An untrained individual who attempts to operate the valves may disrupt the normal operation of the tank, thereby compromising its ability to keep the liquefied gas cold and creating a potential explosion hazard. Cryogenic cargo tanks have a relief valve near the valve control box. From time to time, small puffs of white vapor will be vented from this valve. Responders should understand that this is a normal occurrence—the valve is working to maintain the proper internal pressure. In most cases, this vapor is not indicative of an emergency.

FIGURE 32-18 The MC-338 cryogenic tank maintains the low temperatures required for the cryogens it carries.

A compressed gas **tube trailer** carries compressed gases such as hydrogen, oxygen, helium, and methane (**FIGURE 32-19**). Essentially, they are high-volume transportation vehicles that are made up of several individual cylinders banded together and affixed to a trailer. The DOT, however, does not view tube trailers as cargo tanks. The individual cylinders on the tube trailer are compressed gas cylinders. These large-volume cylinders operate at working pressures of 3000 to 5000 psi. One trailer may carry several different gases in individual tubes. Typically, a valve control box is found toward the rear of the trailer, and each individual cylinder has its own relief valve. These trailers can frequently be seen at construction sites or at facilities that use large quantities of compressed gases.

FIGURE 32-19 A compressed gas tube trailer.

A **dry bulk cargo trailer** is commonly seen on the road. They carry dry bulk goods such as powders, pellets, fertilizers, or grain (**FIGURE 32-20**). These tanks

FIGURE 32-20 A dry bulk cargo tank carries dry goods such as powders, pellets, fertilizers, or grain.

Courtesy of Polar Tank Trailer L.L.C.

are not pressurized, but may use pressure to offload the product. Dry bulk cargo tanks are generally V-shaped with rounded sides that funnel the contents to the bottom-mounted valves.

Some trailers carry more than one type of substance. Responders should always be aware of mixed-cargo trailers, as they may present multiple hazards. To understand the markings found on cargo tanks, the responder should have a basic understanding of not only the container profiles, but also written information found on the tank and tank specification plates (**FIGURE 32-21**).

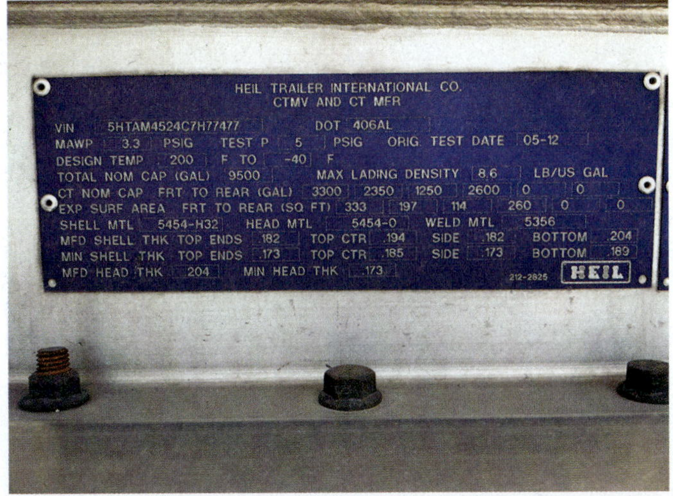

A.

B.

FIGURE 32-21 Cargo tank trucks. **A.** Specification plate. **B.** Inspection markings.

© Jones & Bartlett Learning

Railroad Transportation

Railroads move millions of carloads of chemicals and other freight each year in the United States, with relatively few accidents. Even so, responders should recognize that when rail incidents do occur, they can create unique and significant hazards. Railcars include passenger cars, freight cars, and tank cars, some of which can carry more than 30,000 gallons (113,562 liters) of product. Hazardous materials incidents involving railroad transportation have the potential to be large-scale emergencies.

Operations-level responders should recognize the five basic types of rail tank cars:

- Nonpressure tank cars
- Pressure tank cars
- Cryogenic liquid tank cars
- Gondolas
- Boxcars

Each has a distinctive profile that can be recognized from a distance. Additionally, rail tank cars are usually labeled on both sides with, among other things, the owner of the car, the car's capacity, and the specification. With dedicated haulers, the chemical name is often clearly visible on both sides of the rail tank car.

A **nonpressure tank car** (also referred to as a **general-service rail tank car**) typically carries general industrial chemicals and consumer products such as corn syrup, flammable and combustible liquids, and mild corrosives. Many nonpressure tank cars have visible valves and piping without protective housing on top of the car and internal vapor pressures less than 25 psi (**FIGURE 32-22**). Older nonpressure tank cars, as well as many seen on today's rail system, may have a protective housing that covers the top-mounted valves and bottom outlet valves (**FIGURE 32-23**). Nonpressure tank cars may hold volumes ranging from 4000 to 40,000 gallons (15,142 to 151,416 liters).

FIGURE 32-22 Nonpressure tank cars have visible valves and piping.

Courtesy of Chris Hawley.

FIGURE 32-24 Pressure tank car.

Courtesy of Chris Hawley.

FIGURE 32-23 An example of an older-style, out-of-service, nonpressure tank car.

FIGURE 32-25 Cryogenic liquid tank car.

A **pressure tank car** transports materials such as propane, ammonia, ethylene oxide, and chlorine. These cars have internal working pressures ranging from 100 to 600 psi and are equipped with top-mounted fittings for loading and unloading. The fittings are protected by a sturdy, easily identified protective housing that sits atop the rail tank car (**FIGURE 32-24**). Unfortunately, the high volumes carried in these cars can generate long-duration, high-pressure releases that can prove difficult to control. For example, a liquid or vapor release from the valve arrangements on a chlorine tank car requires a special kit and specific training to stop the leak. Additionally, pressure rail tank cars may be insulated or uninsulated, depending on the material being transported, and chlorine and ammonia railcars have federal security alarms in protective housing to prevent chemicals from being stolen and terrorist activities from occurring.

A **cryogenic liquid tank car** is the most common of the **special-use railcars** (i.e., **boxcar**, flat car, cryogenic tank car, or corrosive tank car) that emergency responders may encounter (**FIGURE 32-25**). These tank cars highlight an important concept for the emergency responder: The hazard will be unique to the railcar and its contents, and the responder will need to pay attention to the details and features of each type. Other examples of special-use railcars include the railway **gondola**, which is used to carry items such as lumber, scrap metal, coal, and pipes, and boxcars, which carry consumer goods, industrial supplies, and a multitude of boxes and palletized goods (**FIGURE 32-26**). Freight containers can be used to transport goods on trucks, ships, and railcars and may pose a challenge to a responder due to the wide variety of goods and materials that may be transported inside them (**FIGURE 32-27**).

FIGURE 32-26 One example of a special-use railcar is the boxcar, typically used for carrying freight.

FIGURE 32-27 Examples of freight containers in the background.

The train crew will have information about all the cars on the train. This information is typically found with the conductor or with the engineer in the lead locomotive. The train crew must retain control of the paperwork at all times and will have knowledge about the availability of a specialized response team from the railroad, access to a 24/7 call center for the railroad, and information about the location and contents of the railcars; thus, they will be able to help responders understand the scope and impact of the derailment. Incidents involving hazardous materials carried on railcars can have significant effects on communities adjacent to the incident and may pose a significant risk to life, property, or the environment (**FIGURE 32-28**).

Collecting Hazard and Response Information

The responsible person for any chemical substances during transportation should maintain the source(s) of information about hazardous cargo. When encountering a chemical release on any mode of transportation, take time to find the responsible person and track down these valuable pieces of information. Additionally, the person who finds this information for you, or has it in their possession, is likely to be a great resource to connect you with subject-matter experts from the company or may have some level of knowledge about the material or the mode of transportation that may prove valuable.

CHEMTREC

The **Chemical Transportation Emergency Center (CHEMTREC)**, which is operated by the American Chemistry Council, is a clearinghouse of technical chemical information. Since 1971, this emergency call center has served as an invaluable information resource on a 24-hour basis for first responders from all disciplines who are called upon to respond to chemical incidents (NFPA 470). CHEMTREC can provide responders with technical chemical information via telephone, fax, or another electronic medium. It also offers a phone conferencing service that will put a responder in touch with thousands of shippers, subject-matter experts, and chemical manufacturers.

When calling CHEMTREC, be sure to have the following basic information ready:

- Name of the chemical(s) involved in the incident (if known)
- Name of the caller and a callback telephone number
- Location of the actual incident or problem
- Shipper or manufacturer of the chemical (if known)
- Container type
- Railcar or vehicle markings or numbers
- Shipping carrier's name
- Recipient of material
- Local conditions and exact description of the situation

When speaking with CHEMTREC personnel, be very clear about the name of the substance and spell it out if necessary; if using a third party, such as a dispatcher, it is vital that you confirm all spellings to avoid misunderstandings. One number or letter out of place could throw off all subsequent research. When in doubt, be sure to obtain clarification. Be sure to have access

RAILROAD TANK CARS

Railroad Tank Car Nomenclature. The railroad industry uses tank car test pressure as the criterion for differentiating between pressure and nonpressure tank cars. Non-pressure tank cars have a test pressure of 100 psig or less, while pressure tank cars have a test pressure greater than 100 psig.

When describing a tank car, the "B-end" is used as the initial reference point. The B-end is where the hand brake wheel is located; numbers 1 through 4 indicate the wheels.

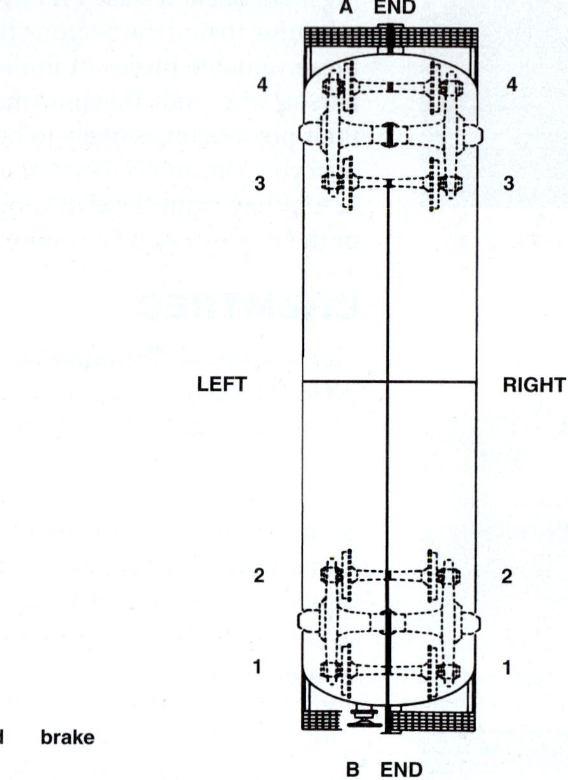

Railroad Tank Car Markings. The markings on railroad tanks cars can be used to gain knowledge about the tank itself and its contents. This information would be useful in evaluating the condition of the container. These markings include the following:

- *Commodity Stencil.* Tank cars transporting anhydrous ammonia, ammonia solutions with more than 50% ammonia, a Division 2.1 material (flammable gas), or a Division 2.3 material (poison gas) must have the name of the commodity marked on both sides of the tank in 4-inch (102-mm) minimum letters.
- *Reporting Marks and Number.* Railroad cars are marked with a set of initials and a number (e.g., GATX 12345) stenciled on both sides (left end as you face the tank) and both ends of the car. Some shippers and car owners also stencil these markings on top of tank cars to assist in identification in an accident or derailment scenario. New tank cars are also stenciled on top for tank car verification during loading operations, as plant personnel can easily verify the car initial and number without climbing down from the rack.
  These markings can be used to obtain information about the contents of the car from the railroad, the shipper, or CHEMTREC. The last letter in the reporting marks has special meaning:
  - "X" indicates a railcar is not owned by a railroad (for a rail car, the lack of an "X" indicates railroad ownership).
  - "Z" indicates a trailer.
  - "U" indicates a container.

FIGURE 32-28 Derailments can have significant effects on communities adjacent to the incident and may pose a significant risk to life, property, or the environment. Having access to the information found stenciled on rail cars may be valuable to the responders.

• *Capacity Stencil*. The capacity stencil shows the volume of a tank car in gallons (and sometimes liters), as well as in pounds (and sometimes kilograms). These markings are found on the ends of the car under the reporting marks. For certain tank cars (e.g., DOT-105, DOT-109, DOT-112, DOT-114, and DOT-111A100W4), the water capacity/water weight of the tank car is stenciled near the center of the car.

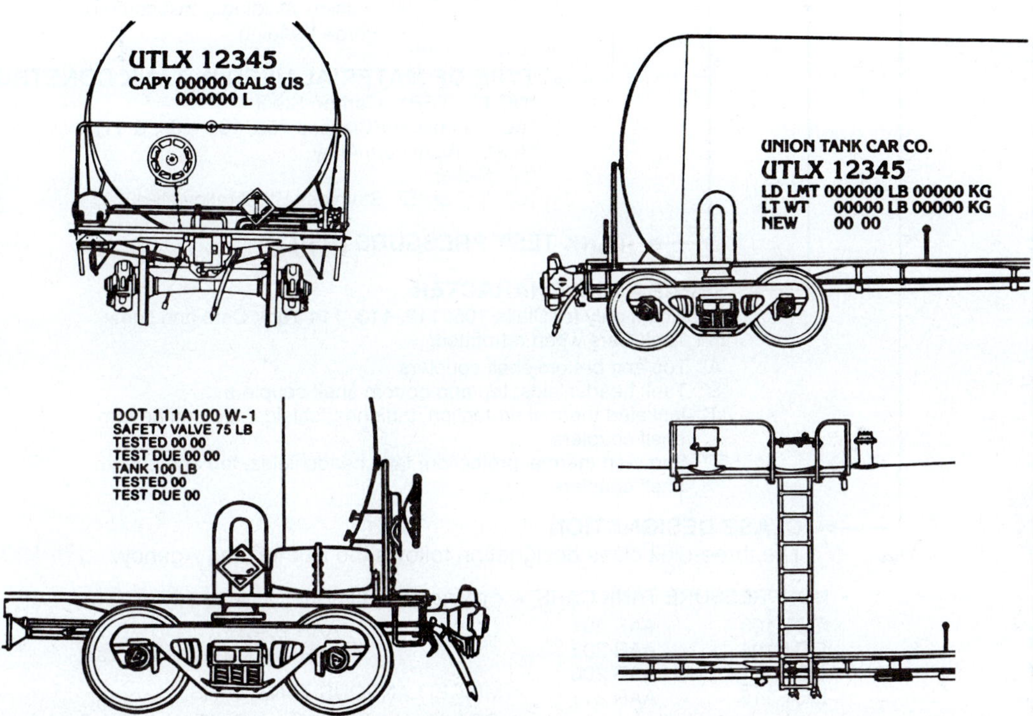

• *Specification Marking.* The specification marking indicates the standards to which a tank car was built. These markings will appear on both sides of the tank car (right end as you face the tank). The specification marking is also stamped into the heads of the tank, where it is not readily visible. The markings provide the following information:
  • Approving authority (e.g., DOT – Department of Transportation, AAR – Association of American Railroads, ICC – Interstate Commerce Commission [authority changed to DOT in 1966]), CTC – Canadian Transport Commission, TC – Transport Canada)
  • Class number: three numbers that follow the approving authority designation
  • Separator/delimiter character (significant in certain tank cars)
  • Tank test pressure
  • Type of material used in construction: Most tank cars are carbon steel and no designation appears. When other construction materials are used (e.g., aluminum, nickel, alloy steel), designations are used.
  • Type of weld used
  • Fittings/material/lining

• *Specification Plate.* Tank cars ordered after 2003 will have a plate on the A-end right bolster and the B-end left bolster (i.e., the structural cross-member that cradles the tank) that provides information about the tank car's characteristics. Although not designed as an emergency response tool, the plate will provide the following information:
  • Car builder's name
  • Builder's serial number
  • Certificate of construction/exemption
  • Tank specification tank shell material/head material
  • Insulation materials
  • Insulation thickness
  • Underframe/stub sill type
  • Date built

FIGURE 32-28 Continued

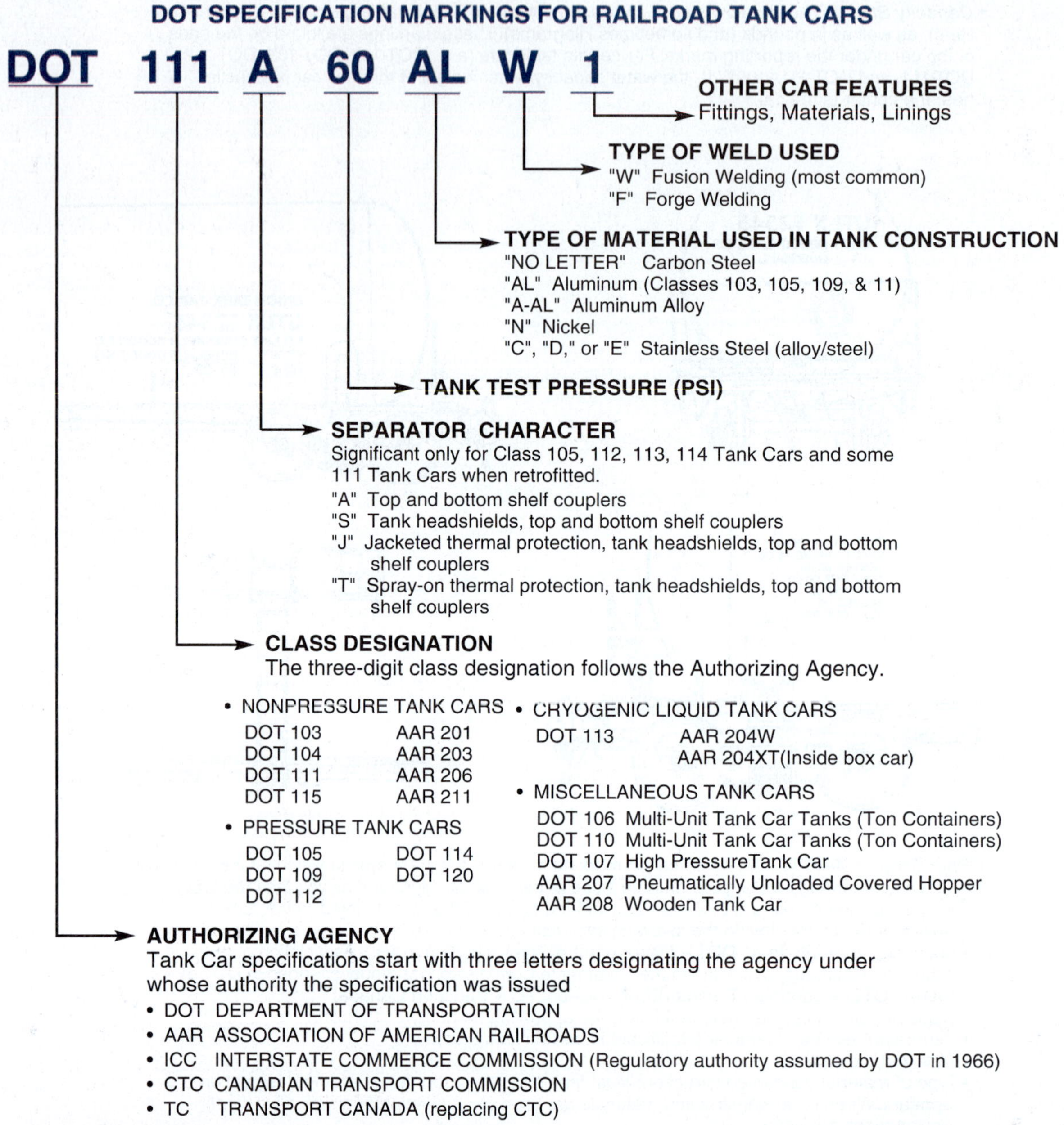

DOT SPECIFICATION MARKINGS FOR RAILROAD TANK CARS

DOT 111 A 60 AL W 1

OTHER CAR FEATURES
Fittings, Materials, Linings

TYPE OF WELD USED
"W" Fusion Welding (most common)
"F" Forge Welding

TYPE OF MATERIAL USED IN TANK CONSTRUCTION
"NO LETTER" Carbon Steel
"AL" Aluminum (Classes 103, 105, 109, & 11)
"A-AL" Aluminum Alloy
"N" Nickel
"C", "D," or "E" Stainless Steel (alloy/steel)

TANK TEST PRESSURE (PSI)

SEPARATOR CHARACTER
Significant only for Class 105, 112, 113, 114 Tank Cars and some 111 Tank Cars when retrofitted.
"A" Top and bottom shelf couplers
"S" Tank headshields, top and bottom shelf couplers
"J" Jacketed thermal protection, tank headshields, top and bottom shelf couplers
"T" Spray-on thermal protection, tank headshields, top and bottom shelf couplers

CLASS DESIGNATION
The three-digit class designation follows the Authorizing Agency.

- NONPRESSURE TANK CARS
  DOT 103 AAR 201
  DOT 104 AAR 203
  DOT 111 AAR 206
  DOT 115 AAR 211
- PRESSURE TANK CARS
  DOT 105 DOT 114
  DOT 109 DOT 120
  DOT 112
- CRYOGENIC LIQUID TANK CARS
  DOT 113 AAR 204W
  AAR 204XT(Inside box car)
- MISCELLANEOUS TANK CARS
  DOT 106 Multi-Unit Tank Car Tanks (Ton Containers)
  DOT 110 Multi-Unit Tank Car Tanks (Ton Containers)
  DOT 107 High PressureTank Car
  AAR 207 Pneumatically Unloaded Covered Hopper
  AAR 208 Wooden Tank Car

AUTHORIZING AGENCY
Tank Car specifications start with three letters designating the agency under whose authority the specification was issued
- DOT DEPARTMENT OF TRANSPORTATION
- AAR ASSOCIATION OF AMERICAN RAILROADS
- ICC INTERSTATE COMMERCE COMMISSION (Regulatory authority assumed by DOT in 1966)
- CTC CANADIAN TRANSPORT COMMISSION
- TC TRANSPORT CANADA (replacing CTC)

FIGURE 32-28 Continued

to a fax machine or electronic medium so that you can receive the information about the chemical sent from CHEMTREC.

The Canadian equivalent of CHEMTREC is the **Canadian Transport Emergency Centre (CANUTEC)**, which is located in Ottawa. This organization serves Canadian responders (in French and English) in much the same way that CHEMTREC serves responders in the United States. CANUTEC may be called 24 hours per day for emergency situations (NFPA 470).

The Mexican equivalent of CHEMTREC and CANUTEC is the **Emergency Transportation System for the Chemical Industry, Mexico (SETIQ)**. SETIQ also may be called 24 hours per day (NFPA 470).

Phone numbers for all three of these agencies can be found in the *Emergency Response Guidebook* (*ERG*).

National Response Center

The **National Response Center (NRC)** is a federal agency maintained and staffed by the U.S. Coast Guard. The NRC should always be notified if a hazard discharges into the environment. There are established, complex reporting requirements for different chemicals based on the reportable quantity (RQ) for each chemical. The shipper or the owner of the chemical has the ultimate legal responsibility to make this call, but by doing so themselves, response agencies will ensure their reporting bases are covered.

Potential Terrorist Incidents

The threat of terrorism has changed the way public safety agencies operate. In small towns and major metropolitan areas alike, the possibility exists that on any day, any agency could find itself in the eye of a storm involving intentionally released chemical substances, biological agents (disease-causing bacteria or other viruses that attack the human body), or an attack on buildings or people using explosives. No area of any country is immune to such attacks. Today, responders must recognize that they are a part of a much larger response mechanism in the United States; they must also understand that any incident involving terrorism will require the assistance and cooperation of countless local, state, and federal resources. Responders

Voice of Experience

The department that I work for provides regional response to several counties for response to hazardous material incidents. This includes covering multiple interstate highways that are critical links for infrastructure in our state.

One summer afternoon, our team was requested to assist with a motor vehicle incident on Interstate 77 near the town of Chelyan, West Virginia. The operator of a tandem tractor trailer suffered a medical emergency, which led to a crash in which the operator was ejected from the cab of his vehicle. The vehicle continued on for some distance until the rear trailer turned over on its side, dislodging and spilling the contents onto the highway.

Prior to this, our hazardous materials team members had always been warned about the potential hazards with vehicles that delivered packages from different shippers. The concern was that the shippers often send all types of different items without having any markings on the packages. The vehicle involved in this incident was, in fact, a freight truck carrying all types of packages, so our responders went in with additional concerns and cautions.

Upon arrival of our units, we surveyed the scene and performed an initial size-up. The trailers were located on a bridge 50 to 70 feet (15 to 21 m) over a small creek and local road. One trailer was upright; the other was on its side with the doors open. There was one ambulance parked in the middle of a spilled material, but no responders were near the vehicles themselves.

The incident commander (IC) advised us that the patient had already been transported, but that the transport was done by another emergency medical services (EMS) unit due to the initial unit being contaminated with the unknown material.

As with any hazardous material identification, our first request was for the bill of lading from the cab. The IC had advised that he already looked in the cab and found nothing, and with the driver not being available, they were unsure what chemical was involved.

For many years, our members have always been instructed to do a good scene survey including a 360-degree walkaround of the area, as long as they do not put themselves into any hazards. As one of our members was performing this survey, he noticed a large amount of paper in the creek below the incident.

A team member was sent down to collect this material. He returned with a wadded-up ball of paper that, when stretched out, was 20 to 30 feet (6 to 9 m) of continuous sheet feed–style paper. This was the bill of lading.

It took a small amount of time, but by comparing shipping package numbers and cross-referencing them with the bill of lading, we were able to identify that the primary hazards in this incident were xylene and ethanol. Once these chemicals were identified, we were able to provide the proper containment and decontamination for the personnel and units involved.

As a side note, further investigation into the contents of the trailer revealed the cargo included live vaccines. Again, this was proof showing that there could be anything in a freight shipment.

Robert "Les" Smith

RESA/Charleston (West Virginia) Fire Department
Charleston, West Virginia

within every jurisdiction should think about the locations of potential targets for terrorists; the general and specific hazards posed by chemical, biological, and **radiological agents** (materials that emit radiation); possible indicators of illicit laboratories; and basic operational guidelines for dealing with explosive events and identifying the possible indicators of secondary devices. It is vital that responders continually maintain good situational awareness, but potential terrorist events require an even higher level of attention. Changing conditions or the possibility of secondary devices or additional attacks should always be considered.

Terrorists can turn the most ordinary objects into powerful weapons. For example, we tend to think of gasoline tankers and commercial airliners as valuable elements of transportation and commerce, but in the hands of determined terrorists, these devices can turn into deadly weapons. Designating an incident as an act of terrorism reflects the intent of the attacker more than the use of a certain device. For example, demolition companies use explosives to bring down unneeded structures quickly and safely; terrorists could use the same explosives to bring down an occupied building, deliberately killing many people. Again, it is the intent that separates an accident from a terrorist event.

Although bombings are the most frequently encountered terrorist acts, responders also must be aware of the threats posed by other potential weapons. Shooting into a crowd or using a vehicle to attack pedestrians in a crowded area could cause devastating carnage. The release of a **biological agent** into a subway system could cause numerous people to become ill or die, simultaneously creating a public panic response that could overload the local fire department, EMS, law enforcement, and hospitals. Given the breadth of these threats, planning should consider the full range of possibilities.

Potential targets for terrorist activities include both natural landmarks and human-made structures. These sites can be classified into three broad categories: infrastructure targets, symbolic targets, and civilian targets (**FIGURE 32-29**). The following sections will help

A.

B.

C.

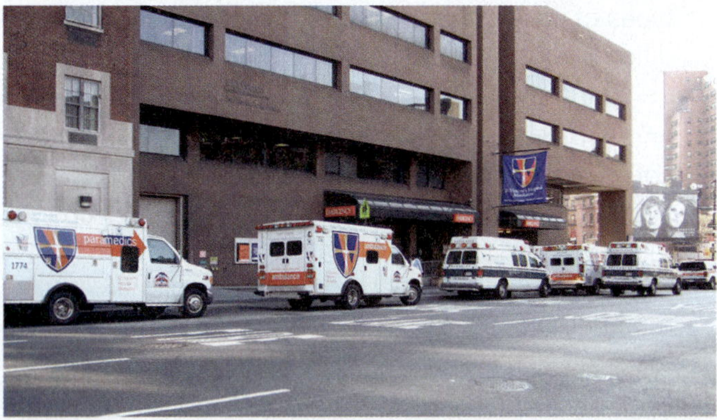

D.

FIGURE 32-29 Subways **(A)**, airports **(B)**, bridges **(C)**, and hospitals **(D)** are all vulnerable to attack by terrorists who seek to interrupt a country's infrastructure.

you build a foundation to become a more informed responder when it comes to dealing with potential terrorist incidents.

TIP

Preincident planning at infrastructure, symbolic, and civilian targets should consider the possibility of a terrorist attack.

Responding to Terrorist Incidents

The roles of public safety responders in handling terrorist events involve many of the same functions that responders perform on a day-to-day basis. Regardless of the intent behind the incident, responders are required to make risk-based decisions about tactics and strategy, personal protective equipment (PPE), victim rescue and/or evacuation, decontamination, and information obtained during detection and monitoring activities—the entire gamut of actions required to solve the problem.

What is different during an incident with criminal intent is the landscape in which that incident is handled and the interagency cooperation that must occur. An incident with criminal intent, causing widespread destruction and/or numerous casualties, requires the involvement of many local, state, and federal law enforcement agencies; emergency management agencies; allied health agencies; and perhaps the military. It is critical that all of these agencies train together and work together in a coordinated and cooperative manner. A mass-casualty incident (MCI) involving a weapon of mass destruction (WMD) could quickly overwhelm the capabilities of your agency, neighboring agencies, and the local healthcare system, in addition to quickly becoming national and international news. Given this reality, it is important to understand which assets and capabilities of those regional agencies may be available to assist with a large-scale incident.

Additionally, preserving potential evidence is an important factor for responders to consider. This concern may be difficult to address during an active incident but should be a consideration for all responders.

In many cases, high-impact events may also require some form of after-action care for the responders in terms of mental health. This assistance will be up to the authority having jurisdiction (AHJ) to implement, but all responders should at least be aware of the need for these types of services after the incident.

In most cases, the first-responder emergency units will not be dispatched for a known WMD or terrorist incident. Rather, the initial dispatch report might cite an explosion, a possible hazardous materials incident, a single person with difficulty breathing, or multiple victims with similar symptoms. Emergency responders will rarely know that a terrorist incident has occurred until personnel on the scene begin to piece together information gained from their own observations and from interviews with witnesses.

If appropriate precautions are not taken, the initial responders may find themselves in the middle of a dangerous situation before they realize it. For this reason, initial responders should take note of any factors that suggest the possibility of a terrorist incident and immediately implement appropriate procedures. The possibility of a terrorist incident should be considered when responding to any location that has been identified as a potential terrorist target. It could be difficult to determine the true nature of the situation until a scene size-up is conducted.

Initial Actions

Responders should approach a known or potential terrorist incident just as they would a hazardous materials incident. If possible, apparatus and personnel should approach the scene from a position that is uphill and upwind. Emergency responders should don PPE, including self-contained breathing apparatus (SCBA).

The first units to arrive should establish an outer perimeter to control access to and from the scene. They should deny access to all persons except emergency responders, and they should prevent potentially contaminated individuals from leaving the area before they have been decontaminated. The perimeter must surround the affected area, with the goal being to keep people who were not initially involved from becoming additional victims. Later-arriving units should be staged an appropriate distance away from the incident.

This operation sounds quite orderly while you are reading this text, but rest assured a real-world scene would be chaotic. There will be widespread panic, and you will probably be overwhelmed. It will take some time and many more responders before you get a handle on the scene. Expect this kind of tumultuous environment if the incident is significant.

Incident command should be established in a safe location, which could be as far as 3000 feet (nearly 1 km) away from the actual incident scene. The incident command post must be set up outside the area of possible contamination and beyond the distance where a secondary device may be planted. The initial task should be to determine the nature of the situation, the types of hazards that could be encountered, and the magnitude of the problems that must be faced.

An initial reconnaissance (recon) team should be sent out to quickly examine the involved area and to determine how many people are involved. Proper use of PPE, including SCBA, is essential for the recon team, and the initial survey must be conducted very cautiously, albeit as rapidly as possible. The possibility that chemical, biological, or radiological agents are involved cannot be ruled out until qualified personnel with appropriate instruments and detection devices have surveyed the area. Responders should begin their reconnaissance mission from a safe distance, working inward toward the scene, while taking care not to touch any liquids or solids or to walk through pools of liquid. Emergency responders who become contaminated must not leave the area until they have been decontaminated.

A process of elimination may be required to determine the nature of the situation. Occupants and witnesses should be asked if they observed any unusual packages or detected any strange odors, mists, or sprays. The presence of many casualties with no outward signs of trauma could indicate a possible **chemical agent** exposure. In such a case, victims' symptoms might include trouble breathing, skin irritation, or seizures. A visible vapor cloud would be another strong indicator of a chemical release. Such a release could have occurred as the result of either an accident or a terrorist attack. The presence of dead or dying animals, insects, or plant life might also point to a chemical agent release as the culprit.

When approaching the scene of an explosion, responders should consider the possibility of a terrorist bombing incident and remain vigilant for secondary explosive devices. They should note any suspicious packages and notify the incident commander (IC) immediately if any such items are discovered. Responders who have not been specially trained should never approach a suspicious object. Instead, **explosive ordnance disposal (EOD) personnel**—personnel who are trained to detect, identify, evaluate, render safe, recover, and dispose of unexploded explosive devices—should examine any suspicious articles and disable them.

The guidelines presented in this section are general recommendations, of course. You must also rely on your agency's standard operating procedures, your training, and your experience to take the appropriate actions. A terrorist event may never happen in your jurisdiction—but if it does, it will require every bit of skill you have to operate safely and effectively.

Interagency Coordination

If a terrorist incident is suspected, the IC, if not already a member of a law enforcement agency, should consult immediately with local law enforcement officials. In many cases, a unified command—including law enforcement (local, state, and federal), fire department, and EMS representatives—should be established. If there are casualties, or if a mass-casualty situation is evident, the IC should notify area hospitals and activate the MCI medical plan (if one exists). Typically, local hospitals will begin to conduct an open-bed count and communicate with other hospitals about the availability of specialized medical services. Depending on the nature of the event, specialists like tactical law enforcement (i.e., special weapons and tactics [SWAT] teams) and bomb technicians may be involved in the mitigation of the event (**FIGURE 32-30**).

State emergency management officials should be notified as soon as possible, which will help ensure a quick response by both state and federal resources to a major incident. Regional emergency operation centers (REOCs) and state-level emergency operation centers (EOCs) may be established, depending on the severity of the incident. Large-scale search-and-rescue incidents could require the response of urban search and rescue (USAR) task forces activated through the Federal Emergency Management Agency (FEMA). Medical response teams, such as disaster medical assistance teams (DMATs), may be needed for incidents involving large numbers of people. The Centers for Disease

FIGURE 32-30 Tactical law enforcement teams may be required to render the area safe for fire and EMS responders to mitigate additional problems and provide patient care.

FIGURE 32-31 An EOC is set up in a predetermined location for large-scale incidents.

Control and Prevention (CDC) may be requested to release items from the Strategic National Stockpile (SNS) when large caches of life-saving pharmaceuticals such as antidotes and medical supplies are needed.

An EOC can help coordinate the actions of all involved agencies in a large-scale incident, particularly if terrorism is involved. This center is usually set up in a predetermined remote location and is staffed by experienced command and staff personnel (**FIGURE 32-31**). The IC, who remains at the scene of the incident, should provide detailed situation reports to the EOC and request additional resources as needed.

Responders must remember that a terrorist incident is also a crime scene. To avoid destroying important evidence that could lead to a conviction of those responsible for perpetrating the attack, responders should not disturb the scene any more than is necessary. Where possible, law enforcement personnel should be consulted prior to overhaul and before the removal of any material from the scene. Responders should also realize that one or more terrorists could be among the injured. Be alert for threatening behavior, and make note of anyone who seems determined to leave the scene.

Chemical Agents

Chemical agents are toxic substances that are intended to cause harm to people or animals. Indicators of possible criminal or terrorist activity involving chemical agents may vary depending on the complexity of the operation, and there is no single indicator that may tip you off to the presence of such illicit activities. Sometimes overt indicators may be visible, such as chemical-type gloves, chemical suits, respirators, and marked or unmarked containers made of various materials in a variety of shapes and sizes. For example, glass containers are prevalent at such locations but may or may not appear to be obvious hazards or threatening in any way.

The chemicals may give off unexplained odors that are out of character for the surroundings. Residual chemicals (liquid, powder, or gas form) may also be found in the area. Chemistry books or other reference materials may be seen, as well as materials that are used to manufacture chemical weapons (e.g., scales, thermometers, or torches). The chemicals may or may not leave some type of easily identifiable signature such as an odor, liquid or solid residue, or dead insects or foliage. The main point is that you must always be on the lookout for items that may appear out of context with the setting—always pay attention to your surroundings. The routine medical or garage fire may quickly bring you up close and personal with a very dangerous situation.

Persons working around the illicit materials may themselves become exposed and exhibit symptoms of chemical exposure—for example, irritation to the eyes, nose, and throat; difficulty breathing; tightness in the chest; nausea and vomiting; dizziness; headache; blurred vision; blisters or rashes; disorientation; or even convulsions.

Chemical weapons can be disseminated in several ways. For example, intentionally releasing chlorine gas inside a building or a crowded gathering place could cause many injuries and deaths. To ensure broader distribution of a chemical agent, however, the agent might be added to an explosive device or mechanically dispersed. Crop-dusting aircraft, truck-mounted spraying units, or machine- or hand-operated pump tanks could all be used to disperse an agent.

The extent of dissemination of a toxic gas or suspended liquid particles depends on wind direction, wind speed, air temperature, and humidity at the time of the material's release. Because these factors can change quickly, it is difficult to predict the exact direction that might be taken by a chemical release cloud. Hazardous materials teams use computer models to predict the pathway of a toxic cloud. They also have the training and equipment to safely handle these situations.

Nerve Agents

A **nerve agent** is a toxic chemical agent that attacks the central nervous system. These weapons were first developed in Germany before World War II. Nerve agents resemble some pesticides (organophosphates) but are much more toxic—in some cases, 100 to 1000 times more toxic than similar pesticides. Exposure to these substances can result in injury or death within minutes.

FIGURE 32-32 In their normal states, nerve agents are liquids.

© AbleStock

In their normal states, most nerve agents are liquids (**FIGURE 32-32**). To be an effective weapon, the liquid must either be dispersed in aerosol form or be broken down into fine droplets so that it can be inhaled or absorbed through the skin. In their pure state, liquid nerve agents are colorless. In an impure state as a result of the manufacturing process, liquid nerve agents may be yellowish to brown. Tabun, an extremely toxic substance, is reported as having a faint fruity odor.

Pouring a liquid nerve agent such as **V-agent (VX)**—an oily liquid that can persist for several weeks—onto the floor of a crowded building may not affect large numbers of people because at the ambient temperature it is not highly vaporous. A more effective method to disperse this agent would be to aerosolize it in some fashion and distribute it across the widest populated area possible. In such a scenario, the effectiveness of the agent would depend on how long it stayed in the air and how widely it became dispersed throughout the building. VX is persistent because it has a low vapor pressure and does not evaporate quickly. In contrast, **sarin**, another chemical warfare agent, is considered nonpersistent because it evaporates at about the same rate as water. Common nerve agents, their method of contamination, and specific characteristics are listed in **TABLE 32-2**.

TABLE 32-2 Common Nerve Agents

Nerve Agent	Route of Exposure	Characteristics
Tabun (GA)	Skin contact Inhalation	Nonpersistent
Soman (GD)	Skin contact Inhalation	Nonpersistent
Sarin (GB)	Skin contact Inhalation	Nonpersistent
V-agent (VX)	Skin contact	Persistent

© Jones & Bartlett Learning

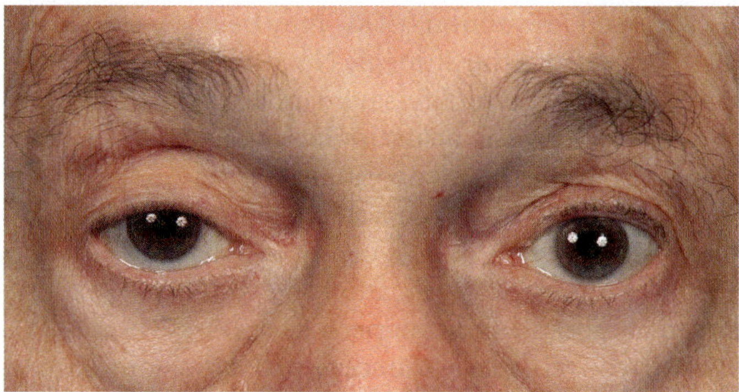
A.

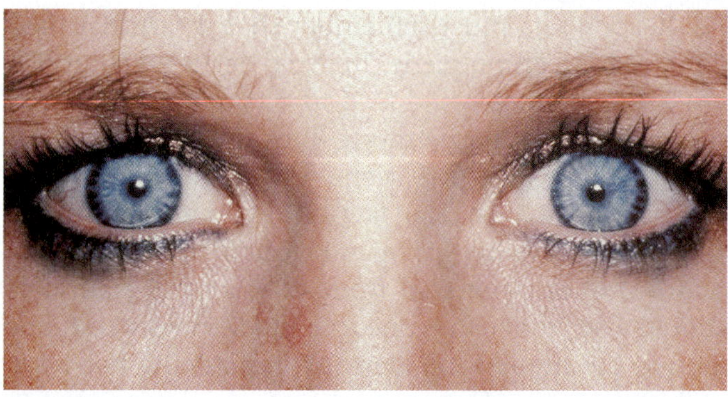
B.

FIGURE 32-33 Pupil responses. **A.** Normal. **B.** Pinpoint.

© Jones & Bartlett Learning

When a person is exposed to a nerve agent, symptoms of that exposure will become evident within minutes. The symptoms may include pinpoint pupils, runny nose, drooling, difficulty breathing, tearing, twitching, diarrhea, convulsions or seizures, and loss of consciousness (**FIGURE 32-33**). The same symptoms are seen in individuals who have been exposed to pesticides.

Several mnemonics can help you remember the symptoms of a nerve agent exposure. The mnemonic used most often in the emergency response community is SLUDGEM:

- **S:** Salivation (drooling)
- **L:** Lacrimation (tearing)
- **U:** Urination
- **D:** Defecation
- **G:** Gastric upset (upset stomach, vomiting)
- **E:** Emesis (vomiting)
- **M:** Miosis (pinpoint pupils)

Keep in mind that such mnemonics represent a very basic and limited way to identify a nerve agent exposure. From a medical standpoint, an exposure of this type involves both the sympathetic and parasympathetic nervous systems—each of which manifests its own unique set of signs and symptoms. Advanced life support providers (paramedics, for example) should have in-depth knowledge when it comes to recognizing the signs and symptoms of a nerve agent exposure. Additionally, many Internet and print resources are available for further study on the subject.

In terms of medical treatment for a nerve agent exposure, the most common field-level treatment is DuoDote, an auto-injector that contains 2.1 mg of atropine and 600 mg of 2-PAM (2-pyridine aldoxime methyl chloride, also known as pralidoxime), delivered as a single dose through one needle (**FIGURE 32-34**). Each drug in the auto-injector has a specific target and acts independently to reverse the effects of a nerve agent exposure. Atropine, for example, reverses **muscarinic effects** such as a runny nose, salivation, sweating, bronchoconstriction, bronchial secretions, nausea, vomiting, and diarrhea; essentially, atropine is intended to deal with the SLUDGEM effects of a nerve agent exposure. Atropine dosing is guided by the patient's clinical presentation and should be given until secretions are dry or drying and ventilation becomes less labored. The other drug, 2-PAM, decreases muscle twitching, improves muscle strength, and allows the patient to breathe better.

The antidote medications can be quickly injected into a person who has been contaminated by a nerve agent. Peak atropine levels are reached approximately 5 minutes after administration; peak levels of 2-PAM are achieved in 15 to 20 minutes. For the emergency responder, this delay means that you should not expect

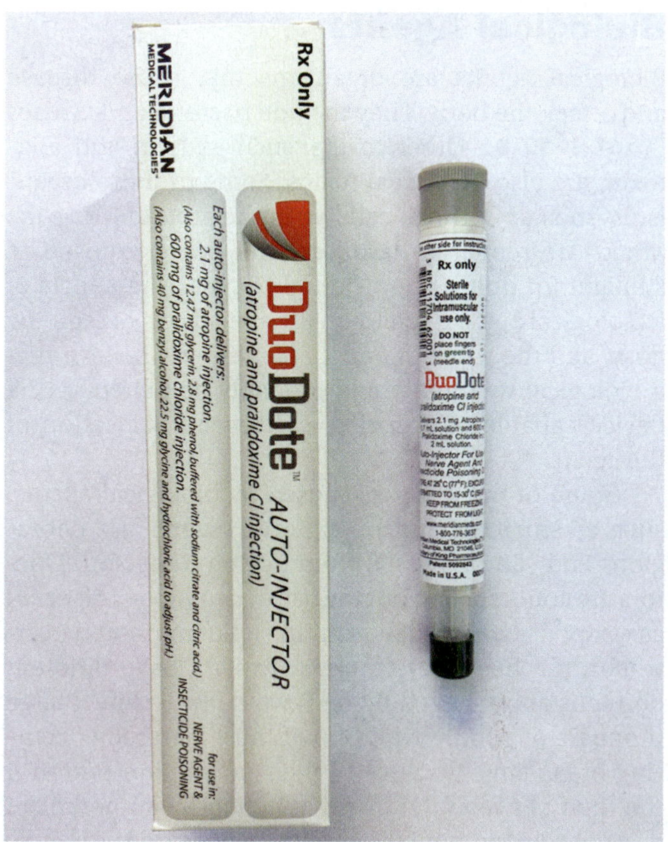

FIGURE 32-34 A DuoDote auto-injector.

to administer this antidote to a person exhibiting serious signs and symptoms of a nerve agent exposure and have the patient recover right away or even at all. Auto-injectors are not an immediate cure-all. In fact, they may be ineffective if the victim has suffered a significant exposure to a nerve agent or pesticide.

DuoDote kits, which are intended for patients between 18 and 55 years of age, have been provided to many fire departments' hazardous materials teams. In addition, many law enforcement agencies and EMS units around the United States carry these antidotes on their response vehicles.

When it comes to treating victims of nerve agent exposure, emergency responders should understand that all nerve agents "age" once they are absorbed by the body. Thus, after a certain period of time (which varies depending on the agent), the administration of the DuoDote kit may be largely ineffective. Soman, for example, has an aging time of approximately 2 minutes. Sarin's aging time is approximately 3 to 4 hours. The other nerve agents have longer aging times. Follow your local protocols before administering any antidote.

Biological Agents

Biological agents are organisms that cause disease and attack the body. They include bacteria and viruses (**TABLE 32-3**). Other toxins, such as ricin and aflatoxin, are also biological toxins. Some of these organisms, such as anthrax, can live in the ground for years; others are rendered harmless after being exposed to sunlight for only a short period of time. The effects of a biological agent depend on the specific organism, the dose, and the route of entry. Most experts believe that a biological weapon would probably be paired with a device capable of ensuring widespread dispersion of the agent.

Some of the diseases caused by biological agents, such as smallpox and pneumonic plague, are contagious and can be passed from person to person. Doctors are concerned about the use of contagious diseases as weapons because the resulting epidemic could overwhelm the healthcare system. Experts have different opinions about how difficult it would be to infect large numbers of people with one of these naturally occurring organisms. Because of their incubation period—the time between the initial infection by an organism and the development of symptoms by a victim—people would not begin to show signs of being infected until 2 to 17 days after exposure to these organisms.

People working in a lab may eventually exhibit symptoms consistent with the biological weapons with which they are working. Biological agents are associated with a delayed onset of symptoms—usually days to weeks after the initial exposure. In fact, the biggest difference between a chemical incident and a biological incident is typically the speed of onset of the health effects from the involved agents. Because most biological agents are odorless and colorless, there are usually no outward indicators that the agents have spread. Again, there may or may not be overt indicators of criminal activities at the scene—be aware and attentive!

Indicators of incidents that may potentially involve biological agents might include chemicals or production equipment such as Petri dishes, vented hoods, Bunsen burners, pipettes, microscopes, and incubators. Reference manuals, such as microbiology or biology textbooks, or handwritten research or directions—maybe in a foreign language—may be present. Containers used to transport biological agents may include metal cylindrical cans or red plastic boxes or bags, probably without specific biological hazard labels. PPE, including respirators, chemical or biological suits, and latex gloves, may be present at the scene, as well as excessive amounts of antibiotics to protect the people working with the agents (**FIGURE 32-35**). Other potential indicators may include abandoned spray devices and unscheduled or unusual sprays being disseminated (especially if outdoors at night).

Anthrax

Anthrax is an infectious disease caused by the bacterium *Bacillus anthracis.* These bacteria are typically found around farm animals such as cows and sheep. For use as a weapon, the bacteria must be cultured to develop anthrax spores. The spores, once weaponized into an ultra-fine powder, can then be dispersed in a variety of ways (**FIGURE 32-36**). Approximately 8000 to 10,000 spores are typically required to cause an anthrax infection. Spores infecting the skin cause cutaneous anthrax, ingested spores cause gastrointestinal anthrax, and inhaled spores cause inhalational anthrax. Anthrax has an **incubation period** of 2 to 6 days. The disease can be successfully treated with a variety of antibiotics if it is diagnosed early enough.

The threat posed by anthrax-related terrorism is quite real. In 2001, four letters containing anthrax were mailed to locations in New York City; Boca Raton, Florida; and Washington, D.C. Five people died after being exposed to the contents of these letters, including two

TABLE 32-3 Bacteria Versus Viruses

	Description	Dispersion	Examples
Bacteria	Single-cell microscopic organisms with a nucleus and cell wall	May form a spore	Anthrax Plague Tularemia
Viruses	Submicroscopic agents; DNA or RNA coated with protein	Require a host to live and reproduce	Smallpox Viral hemorrhagic fever (VHF)

FIGURE 32-35 Indicators of biological agents might potentially include production or containment equipment, such as Petri dishes, vented hoods, Bunsen burners, pipettes, microscopes, and incubators.

© Jones & Bartlett Learning. Courtesy of Rob Schnepp.

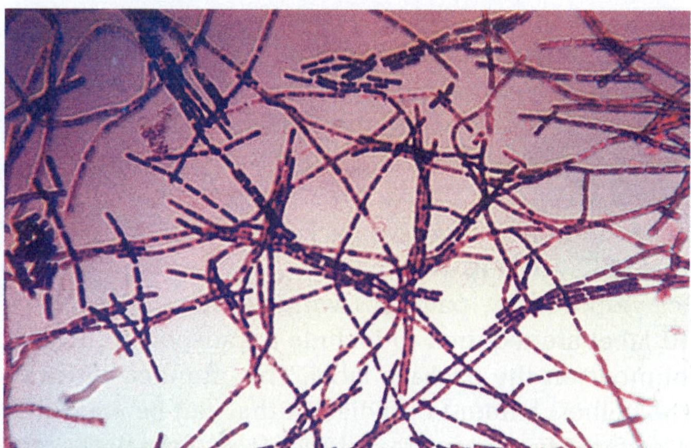

FIGURE 32-36 Anthrax spores can be dispersed in a variety of ways.

Courtesy of the Centers for Disease Control and Prevention.

postal workers who were exposed as the letters passed through postal sorting centers. Several major government buildings had to be shut down for months to be decontaminated. These incidents followed shortly after the terrorist attacks of September 11, 2001, and they caused tremendous public concern. In the wake of these incidents, emergency personnel had to respond to thousands of incidents involving suspicious packages and citizens who believed that they might have been exposed to anthrax.

Today, presumptive field tests are available that hazardous materials teams can use to determine whether the threat of anthrax or other biological agents is legitimate.

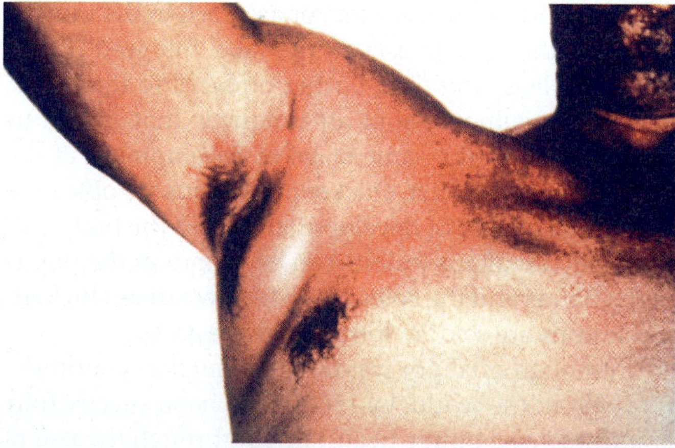

FIGURE 32-37 A bubo—one of the symptoms of the plague—consists of a swollen, painful lymph node.

Courtesy of the Centers for Disease Control and Prevention.

The gold standard to positively identify anthrax (and other biological agents), however, is not a field test. Anthrax must be cultured in a lab, by qualified microbiologists, to be positively identified. Your agency should have established procedures for handling suspicious powders and getting samples to a qualified laboratory. The local Federal Bureau of Investigation (FBI) office can provide guidance on the different types of labs available through the Laboratory Response Network and how to work with law enforcement to get a sample to the appropriate lab.

Plague

Plague is caused by *Yersinia pestis,* a bacterium that is commonly found in rodents. These bacteria are most often transmitted to humans by fleas that feed on infected animals and then bite humans.

The two main forms of plague are bubonic and pneumonic. Individuals who are bitten by fleas generally develop bubonic plague, which attacks the lymph nodes (**FIGURE 32-37**). Pneumonic plague can be contracted by inhaling the bacterium.

Yersinia pestis can survive for weeks in water, moist soil, or grains. These bacteria might also be cultured for distribution as a weapon in aerosol form. Inhalation of the aerosol form would put the target population at risk for pneumonic plague.

The incubation period for plague ranges from 2 to 6 days. This disease can be treated with antibiotics.

Smallpox

Smallpox is a highly infectious and often fatal disease caused by *Variola,* a virus; it kills approximately 30 percent of all people who become infected. Smallpox first presents with small red spots or as a rash in the mouth. The rash then progresses to the face, followed by the arms and legs, and then farther outward to the hands and feet. Smallpox lesions are unique in that all lesions appear to be in the same stage of development at the same time. In contrast, chickenpox lesions are observed in different stages of development across the body. Additionally, smallpox lesions can be found on the palms of the hands and the soles of the feet, whereas chickenpox lesions are seldom found on these areas.

Although smallpox was once routinely encountered throughout the world, by 1980 it had been successfully eradicated as a public health threat through the use of an extremely effective vaccine. Officially, two countries (the United States and Russia) have maintained cultures of the disease for research purposes. It is possible, however, that international terrorist groups may have acquired the virus.

The smallpox virus could potentially be dispersed over a wide area in an aerosol form; however, widespread broadcasting of this agent may not be necessary to cause a devastating outbreak. Infecting a small number of people could lead to a rapid spread of the disease throughout a targeted population, given smallpox's highly contagious nature: The disease is easily spread by direct contact, droplet, and airborne transmission. Patients are considered highly infectious and should be quarantined until the last scab has fallen off (**FIGURE 32-38**). The incubation period for smallpox is between 4 and 17 days (the average is 12 days).

Currently, there are millions of people who have never been vaccinated for smallpox. Millions more have reduced immunity because decades have passed since their last immunization.

Radiological Agents

Indicators of radiological agents may include production or containment equipment, such as lead or stainless steel containers (with or without labels), and explosives that may be used to disperse the radioactive source, along with containers (e.g., pipes), caps, fuses, gunpowder, timers, wire, and detonators. PPE present may include radiological-protective suits and respirators. Radiation monitoring equipment such as Geiger counters or radiation pagers may be present, as may similar radiation detection devices used by responders to alert them to the presence of a potential threat (**FIGURE 32-39**). As with other locations where nefarious activities may be present, responders must put the potential threat in context. There may or may not be a valid explanation for the presence of something suspicious.

A major difference between an illicit location where radioactive substances are present and a legitimate operation may be the way the substances are packaged and stored. Legitimate sources are well marked, tracked, and regulated. When radiological agents are shipped, the package type is dictated by the degree of radiation activity inside the package—that is, the labeling is driven by the *amount of radiation that can be measured outside the package*. Three varieties of labels are found on radioactive packages: White I, Yellow II, and Yellow III (**FIGURE 32-40**).

Additionally, packages with a Yellow II or Yellow III label are required to include a transport index (TI) number on the package label. This number indicates the highest amount of radiation that can be measured 1 meter (3 feet) away from the surface of the package.

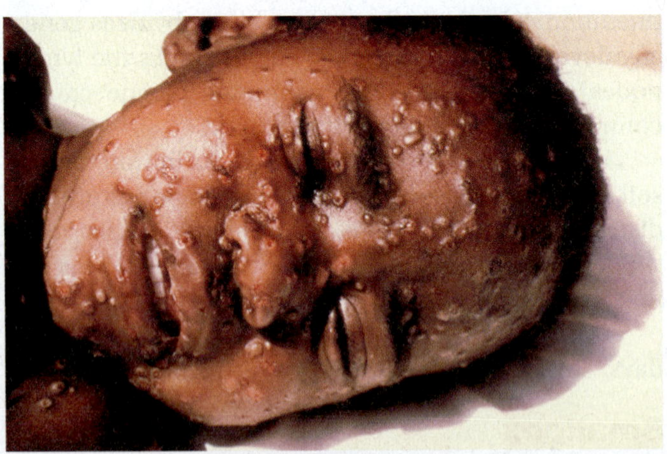

FIGURE 32-38 Smallpox is a highly contagious disease with a mortality rate of approximately 30 percent.

Courtesy of the Centers for Disease Control and Prevention.

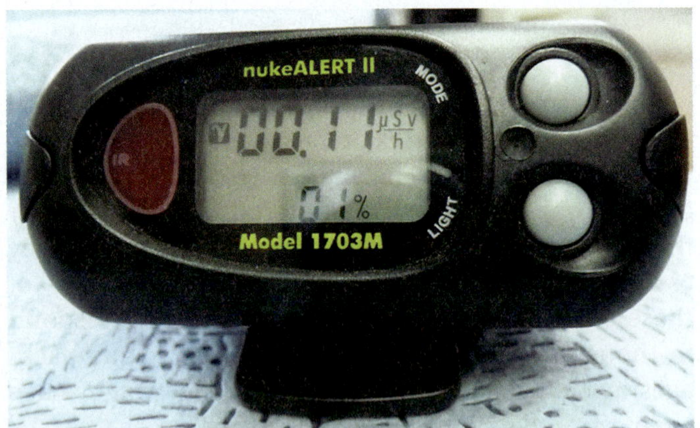

FIGURE 32-39 The personal radiation detector pictured here is capable of providing exposure rate readings as well as accumulated radiation dose.

FIGURE 32-40 **A.** White I label. **B.** Yellow II label. **C.** Yellow III label.

Courtesy of the U.S. Department of Transportation.

Responders must be able to recognize situations where radioactive materials might be encountered. Industries that routinely use radioactive materials include food testing labs, hospitals, medical research centers, biotechnology facilities, construction sites, and medical laboratories. For the most part, visual indicators (signs or placards) are used at sites to indicate the presence of radioactive substances, although this is not always the case.

The key is to be able to suspect, recognize, and understand when and where you may encounter radioactive sources. If you suspect a radiation incident at a fixed facility, you should initially consult with the radiation safety officer of the facility. This person is responsible for the use, handling, and storage procedures for all radioactive materials at the site. The radiation safety officer likely will be a tremendous resource to you and will know exactly what is being used at the facility. If the incident is not at a fixed site, the presence of radiation may never be apparent. Radioactive isotopes cannot be detected by sight, smell, taste, or any of the other senses. Therefore, if you have any suspicion that the incident involves radiation, it will be necessary to call a hazardous materials team or some other resource with radiation detection capabilities.

Significant incidents involving radiation are rare, largely due to the comprehensiveness of the regulations for using, storing, and transporting significant radioactive sources. This is not to say that these incidents will never happen; however, the regulations have helped considerably in keeping the number of incidents low. Most of the incidents you may encounter will involve low-level radioactive sources and can be handled safely. These low-level sources are typically found in Type A packaging, a packaging method that is unique to radioactive substances and is used for materials such as radiopharmaceuticals and other low-level emitters.

Radiological Packaging

The types of containers and packages most commonly used to store radioactive materials are divided into five major categories: excepted-range radioactive packaging; industrial radioactive packaging; and Type A (**FIGURE 32-41**), Type B (**FIGURE 32-42**), and Type C packaging (**FIGURE 32-43**).

Excepted packaging meets only general design requirements for any hazardous material package. Low-level radioactive substances are commonly shipped in excepted packages, which may be constructed out of heavy cardboard. Excepted packaging is authorized for limited quantities of radioactive material that would pose a very low hazard if released in an accident. Materials identified as excepted packaging include consumer goods such as smoke detectors. They are required to have the letters "UN" and the appropriate four-digit UN identification number marked on the outside of the package.

Industrial packaging is used in certain shipments usually categorized as radioactive waste. Contaminated medical equipment may serve as an example because it likely (but not always) contains a non-life-endangering amount of radioactivity. DOT regulations require that these packages allow no identifiable release of the material to the environment during normal transportation and handling. Three categories of industrial packages are distinguished: IP-1, IP-2, and IP-3, with IP-3 being the strongest. The package category will be marked on the exterior of the package.

Type A packaging is designed to protect the internal radiological contents during normal transportation and in the event of a minor accident. Such packaging is characterized by having an inner containment vessel made of glass, plastic, or metal and external packaging materials made of polyethylene, rubber, or vermiculite. Examples of materials typically shipped in Type A packages include nuclear medicines (radiopharmaceuticals),

A.

B.

FIGURE 32-41 Type A packaging.

FIGURE 32-42 Type B packaging.

FIGURE 32-43 Type C packaging.

radioactive waste, and radioactive sources used in industrial applications. Type A packaging and its radioactive contents must meet standard testing requirements designed to ensure that the package retains its containment integrity and shielding under normal transport conditions and may be transported by air. Type A packages must withstand moderate degrees of heat, cold, reduced air pressure, vibration, impact, water spray, drop, penetration, and stacking tests. The consequences of a release of the material in one of these packages would not be significant because the quantity of material in this package is so small.

Type B packaging is far more durable than Type A packaging and is designed to prevent a release in the case of extreme accidents during transportation. More dangerous radioactive sources might be found in Type B packaging. Some of the tests that Type B containers must undergo include heavy fire, pressure from submersion, and falls onto spikes and unyielding surfaces. Type B packages include small drums and heavily shielded casks weighing more than 100 metric tons. This type of containment vessel contains materials such as spent nuclear fuel, high-level radioactive waste, and high concentrations of other radioactive material such as cesium and cobalt. Type B packages are designed to protect their contents from greater exposure; the amount of protection is based on the potential severity of the hazard. These package designs must withstand all Type A tests and a series of tests that simulate severe or "worst-case" accident conditions. Accident conditions are simulated by performance testing and engineering analysis. Life-endangering amounts of radioactive material are required to be transported in Type B packages.

Type C packaging is intended for the transport by air of radioactive material, with no restriction on the total amount of radioactivity contained within the package. Its contents would include materials such as spent nuclear fuel. While this type of packaging does exist, there are no Type C packages currently approved for use within the United States.

Illicit Laboratories

Many indicators of possible criminal or terrorist activity involving illicit laboratories may be evident to responders. Many of the same materials used to manufacture homemade explosives are used to make illicit drugs. For example, terrorist paraphernalia may include terrorist training manuals, ideological propaganda, and documents indicating affiliation with known terrorist groups. Locations with certain characteristics are also commonly sites of illicit (clandestine) laboratories—for example, basements with unusual or multiple vents, buildings with heavy security, buildings with obscured windows, and buildings with odd or unusual odors. Personnel working in illegal laboratory settings may exhibit a certain degree of unusual or suspicious behavior; for instance, they may be nervous and have a high level of anxiety. In addition, they may be protective of the laboratory and not want to allow anyone to access the area for any reason, or they may rush people out of the area as soon as possible.

Equipment that may be present in illicit laboratory areas includes surveillance materials (e.g., photographs, maps, blueprints, or time logs of the target hazard locations), non-weapon supplies (e.g., identification badges, uniforms, and decals that would be used to allow the terrorist to access target hazards), and weapon-related supplies (e.g., timers, switches, fuses, containers, wires, projectiles, and gunpowder or fuel). Security weapons such as guns, knives, and booby trap systems may also be present.

Drug laboratories are by far the most common type of clandestine laboratory encountered by responders. These laboratories are typically very primitive and can be found in hotel rooms, in cars, and in even smaller settings. Materials used to manufacture the drugs often consist of everyday items (e.g., jars, bottles, glass cookware, coolers, and tubing) that have been modified to produce the illicit drugs (**FIGURE 32-44**).

Specific chemicals and materials found at the scene may include large quantities of cold tablets (ephedrine or pseudoephedrine), hydrochloric or sulfuric acid, paint thinner, drain cleaners, iodine crystals, table salt,

FIGURE 32-44 Items typically found at clandestine drug laboratories often include everyday items.

aluminum foil, and batteries. The strong smell of urine or unusual chemical smells such as ether, ammonia, or acetone are common indicators of clandestine drug manufacturing. Illicit drug laboratories should be considered significant hazardous materials scenes because the inexperienced chemists who run them take many shortcuts and disregard typical safety protocols to increase production.

The examples listed here are intended to give you an idea of some of the things that might tip you off to the presence of suspicious or illicit activity. It is not an exhaustive list, however, and you may encounter an illicit lab without any of these indicators. Remember, the people carrying out such activities are not interested in being obvious in their behaviors or actions.

Explosives

Indicators of possible criminal or terrorist activity involving explosives typically include materials that fit into four major categories: protective equipment, production and containment materials, explosive materials, and support materials. Protective equipment may include rubber gloves, goggles and face shields, and perhaps even fire extinguishers. Production and containment equipment may include funnels, spoons, threaded pipes, caps, fuses, timers, wires, detonators, and concealment containers such as briefcases, backpacks, or other innocuous-looking packages (**FIGURE 32-45**). Explosive materials may include gunpowder, gasoline, fertilizer, solvents, oxidizers, and similar materials (**FIGURE 32-46**).

FIGURE 32-45 Pipe bombs come in many shapes and sizes.

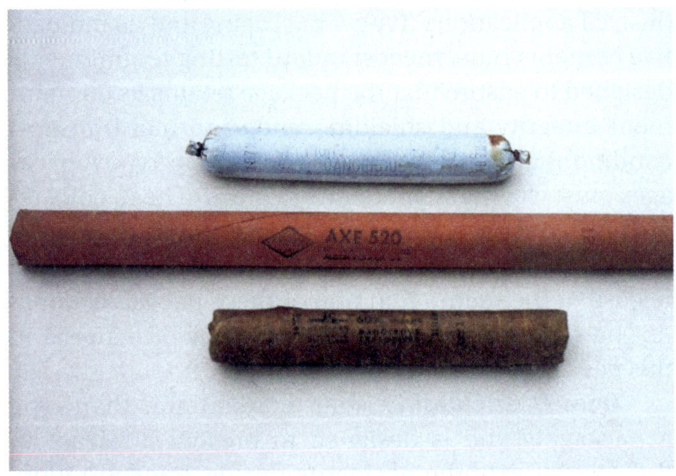

FIGURE 32-46 Every year, thousands of pounds of explosives are stolen from their rightful owners.

Support materials may include explosive reference manuals, Internet-based reference materials, and military information.

If you encounter an active bomb-making operation, it is important to note what you see, get yourself out of danger, secure an appropriate amount of real estate, and call for your local EOD team (**FIGURE 32-47**). This is not a situation that untrained responders should investigate (**FIGURE 32-48**).

The Dirty Bomb

In recent years, the **radiation dispersal device (RDD)**, also known as a dirty bomb, has emerged as a source of serious concern in terms of terrorism. Although not

FIGURE 32-47 EOD personnel are trained and have the necessary tools and equipment to evaluate and disable or disrupt a device or suspected device.

FIGURE 32-48 The first responders who arrive at the scene of an explosion should establish a command post in a safe location and begin the process of setting up a unified command with law enforcement.

considered a WMD, an RDD has been described by the National Fire Protection Association (NFPA) as "a device designed to spread radioactive material through a detonation of conventional explosives or other [non-nuclear] means" (NFPA 470).

Packing radioactive material around a conventional explosive device could contaminate a wide area, with the size of the affected area ultimately depending on the amount of radioactive material and the power of the explosive device. Only a few radioactive sources can be used effectively in an RDD. It is also possible for a criminal to construct a nonexplosive RDD, disseminating radioactive material via pressure sprayers or air-handling systems in buildings. To limit this threat, radioactive materials, even in small amounts, are kept secure and protected. Such materials are widely used in industry and health care, and a criminal could potentially construct a **dirty bomb** with just a small quantity of stolen radioactive material.

Other types of radiological devices that may be used as weapons include improvised nuclear devices and radiation exposure devices:

- Improvised nuclear device (IND): A weapon fabricated from fissile material (highly enriched uranium or plutonium) capable of producing a nuclear explosion. A generally accepted successful yield in the 10- to 20-kiloton range—the equivalent to 10,000–20,000 tons of dynamite—in addition to high levels of radiation, would make this a very lethal device.
- Radiation exposure device (RED): Radioactive material in a sealed container located where persons nearby would receive a direct exposure. Although not considered a device that would cause widespread death or sickness, an RED could constantly impact a steady stream of persons over a given period of time. If the container failed, then contamination might occur because the material could come into contact with victims.

Secondary Devices

A **secondary device** is an explosive or incendiary device designed to harm those responders summoned to the scene for some other reason. Terrorists who want to injure responding personnel with a secondary device or attack will typically make the initial attack very dramatic to draw responders into proximity with the scene. The secondary attack usually takes place as the responders begin to treat victims of the initial attack.

Indicators of potential secondary devices may include "trip devices" such as timers, wires, or switches. Common concealment containers, such as briefcases, backpacks, boxes, or other common packages, may also be present; uncommon concealment containers may include pressure vessels (propane tanks) or industrial chemical containers (chlorine storage containers). Terrorists may watch the site where the primary device is placed, as part of preparing to manually activate the secondary devices.

Responders can use the EVADE mnemonic to help think critically about the presence of secondary devices:

- **E:** Evaluate the scene for likely areas where secondary devices can be placed.
- **V:** Visually scan operating areas for a secondary device before providing patient care.
- **A:** Avoid touching or moving anything that can conceal an explosive device.
- **D:** Designate and enforce scene control zones.
- **E:** Evacuate victims, other responders, and nonessential personnel as quickly and safely as possible.

CASE STUDY

You Are the Firefighter CONCLUSION

Your engine company is dispatched to the scene of a leaking drum at a local lumberyard. It appears to have been hit by a forklift, causing a small tear toward the bottom, and a substance is slowly leaking and an usual odor is reported. The dirt around the drum looks wet. The foreman of the lumberyard tells you that the drum holds waste oil, and the forklift driver on duty denies hitting the drum. From a safe distance, you can tell that the drum is made of steel and has a 2-inch (51-mm) bung and a ¾-inch (19-mm) bung on the top. You can see a small red label on the side of the drum that says "flammable," with the number 3 at the bottom.

1. **What type of material is most likely stored in the drum (e.g., corrosive, solvent, explosive)? Does the description of the drum provide any other useful information? If so, what?**

   Answer: Based on the material of construction—steel—a good first assumption to make is that the drum should or does contain a solvent-based material. The fact that is has two bungs as described indicates the material inside has a low enough viscosity to be pumped if necessary. You should always examine the exterior of the drum for additional information and markings to confirm your assumptions.

2. **What does the label on the drum signify, and does the report of an unusual odor make sense given the information you have?**

   Answer: The label signifies a flammable liquid, which would have a flash point at or below 199°F (93°C). Most oils have much higher flash points, and do not give off vapors at ambient air temperatures to create a significant odor. There appear to be some things that don't add up in the narrative. A steel drum configured as described, labeled flammable, with a perceptible odor, would likely not contain waste oil.

3. **How would you go about obtaining more information on the potential contents of the drum?**

   Answer: There may be enough inconsistency in the story that further questioning of the responsible person will not yield additional actionable information. In this case, based on the level of training of the on-scene crew, it may be reasonable to isolate and deny entry and call for hazardous materials technicians or specialists to further evaluate the drum and contents and mitigate the incident.

WRAP-UP

SUMMARY

KNOWLEDGE OBJECTIVES

- Identify and describe common types of hazardous materials containers. (**NFPA 470: 7.2.1**, pp. 1284–1290)
  - Define bulk and nonbulk packaging parameters.
  - Identify bulk fixed-facility pressure containers, liquified compressed gas cylinders, ton containers, intermodal tanks, and intermediate bulk containers.
- Describe the ways in which hazardous materials are transported. (**NFPA 470: 7.2.1**, pp. 1290–1294)
  - Identify MC-306/DOT 406, MC-307/DOT 407, MC-312/DOT 412, MC-331, and MC-338 cargo tanks, and compressed gas tube trailers.
  - Identify the five basic types of rail tank cars: nonpressure tank cars, pressure tank cars, cryogenic liquid tank cars, gondolas, and boxcars.
- Identify resources that can provide technical chemical information. (**NFPA 470: 7.2.1**, pp. 1295–1298)
  - Understand the process to contact and utilize the resources available through CHEMTREC, CANUTEC, or SETIQ.
  - Understand the role of the National Response Center (NRC).
- Identify different types of potential terrorist incidents and response options. (**NFPA 470: 7.2.1, 7.3.1**, pp. 1299–1314)
  - Understand the need and general process for interagency coordination at a terrorist event.
  - Identify the various agents that may be used in a terrorist attack (chemical agents, biological agents, radiological agents, illicit laboratories, explosives, and secondary devices).

SKILLS OBJECTIVES

There are no skills objectives for operations-level responders for this chapter.

KEY TERMS

anhydrous Without water.

anthrax An infectious disease spread by the bacterium *Bacillus anthracis*, which is typically found around farm animals, such as cows and sheep.

berm See *dike*.

biological agent An organism that causes acute disease or long-term damage to the human body.

boxcar A railway car that carries consumer goods, industrial supplies, and a multitude of boxes and palletized goods.

bulk packaging Any packaging, including transport vehicles, having a liquid capacity of more than 119 gallons (450 liters), a solids capacity of more than 882 pounds (400 kg), or a compressed gas water capacity of more than 1001 pounds (454 kg). (NFPA 470)

Canadian Transport Emergency Centre (CANUTEC) Operated by Transport Canada, an organization that provides emergency response information and assistance on a 24-hour basis for responders to hazardous materials/weapons of mass destruction incidents. (NFPA 470)

cargo tank Bulk packaging that is permanently attached to or forms a part of a motor vehicle or is not permanently attached to any motor vehicle and which, by reason of its size, construction, or attachment to a motor vehicle, is loaded or unloaded without being removed from the motor vehicle.

chemical agent A toxic substance that is intended to cause harm to people or animals.

chemical cargo tank A tank, characterized by several heavy-duty reinforcing rings around the tank, that holds approximately 6000 gallons (22,712 liters) of product and often carries aggressive (highly reactive) acids such as concentrated sulfuric and nitric acid; also called *MC-312/DOT 412 corrosive tank*.

KEY TERMS CONTINUED

Chemical Transportation Emergency Center (CHEMTREC) A public service of the American Chemistry Council, which provides emergency response information and assistance on a 24-hour basis for responders to hazardous materials/weapons of mass destruction incidents. (NFPA 470)

cryogenic liquid cargo tank A low-pressure tank designed to maintain the low temperature required by the cryogens it carries and that has a box-like structure containing the tank control valves typically attached to the rear of the tanker; also called *MC-338.*

cryogenic liquid tank car A low-pressure railway tank car that maintains the low temperature required by the cryogens it carries.

dike A series of short walls that form a catch basin around a container; also called *berm.*

dirty bomb See *radiation dispersal device (RDD).*

dry bulk cargo trailer A trailer designed to carry dry bulk goods such as powders, pellets, fertilizers, or grain; it is generally V-shaped with rounded sides that funnel toward the bottom.

Emergency Transportation System for the Chemical Industry, Mexico (SETIQ) An organization in Mexico that provides emergency response information and assistance on a 24-hour basis for responders to emergencies involving hazardous materials/weapons of mass destruction. (NFPA 470)

excepted packaging Packaging, which may be constructed out of heavy cardboard, that meets only general design requirements for any hazardous material used to transport materials such as low-level radioactive substances.

explosive ordnance disposal (EOD) personnel Personnel trained to detect, identify, evaluate, render safe, recover, and dispose of unexploded explosive devices.

fusible plug A pressure relief valve on a ton container.

general-service rail tank car See *nonpressure tank car.*

gondola A railway car that carries items such as lumber, scrap metal, coal, and pipes.

incubation period The time between the initial infection by an organism and the development of symptoms in a victim.

industrial packaging Packaging used to transport materials that present a limited hazard to the public or the environment. It is classified into three categories, based on the strength of the packaging—IP-1, IP-2, or IP-3, with IP-3 being the strongest.

intermediate bulk container (IBC) A bulk storage container with a volume greater than 119 gallons (450 liters) but less than 793 gallons (3002 liters); these volumes are between the typical volume of drums or bags and the volume of cargo tanks.

intermodal tank (IM or IMO) A bulk storage container that holds between 5000 and 6000 gallons (18,927 and 22,712 liters) of product, can be either pressure or nonpressure, serves as both a shipping and storage vessel, and can be shipped by all modes of transportation—air, sea, or land.

low-pressure chemical cargo tank A rounded (uninsulated) or horseshoe-shaped (insulated) tank with an internal working pressure up to 35 psi that is capable of holding 6000 to 7000 gallons of flammable liquid, mild corrosives, and poisons; also called *MC-307/DOT 407 chemical hauler.*

MC-306/DOT 406 cargo tank See *nonpressure liquid cargo tank.*

MC-307/DOT 407 chemical hauler See *low-pressure chemical cargo tank.*

MC-312/DOT 412 corrosive tank See *chemical cargo tank.*

MC-331 See *pressure cargo tank.*

MC-338 See *cryogenic liquid cargo tank.*

multiple element gas containers (MEGCs) Several individually specified, high-pressure tubes, with working pressures that range as high as 5000 psi, that are attached to a frame and that commonly carry substances such as hydrogen and oxygen.

muscarinic effects Effects such as a runny nose, salivation, sweating, bronchoconstriction, bronchial secretions, nausea, vomiting, and diarrhea.

National Response Center (NRC) An agency maintained and staffed by the U.S. Coast Guard that should always be notified if a hazard discharges into the environment.

nerve agent A toxic chemical agent that attacks the central nervous system.

nonbulk packaging Packaging that has a liquid capacity of 119 gallons (450 liters) or less, a solids capacity of 882 pounds (400 kg) or less, or a compressed gas water capacity of 1001 pounds (454 kg) or less. (NFPA 470)

nonpressure liquid cargo tank A nonpressure tank that typically carries between 6000 and 10,000 gallons of a product such as gasoline or other flammable and combustible liquids; also called *MC-306/DOT 406 cargo tank.*

nonpressure tank A bulk storage tank that has a working pressure between 2.65 and 4 psi and that is usually made of steel or aluminum.

nonpressure tank car A railcar that is equipped with a nonpressure tank that typically carries general industrial chemicals and consumer products such as corn syrup, flammable and combustible liquids, and mild corrosives; also called *general service rail tank car.*

plague An infectious disease caused by the bacterium *Yersinia pestis,* which is commonly found on rodents and is usually transmitted to humans by fleas.

pressure cargo tank A pressure tank with rounded ends used to transport materials such as ammonia, propane, Freon, and butane, which carries volumes from 1000 gallons (3785 liters) to 11,000 gallons (41,640 liters) cargo tank, and is commonly constructed of steel or stainless steel with a single tank compartment; also called *MC-331.*

pressure tank A bulk storage tank that has an internal pressure greater than 15 psig.

pressure tank car A railcar used to transport materials such as propane, ammonia, ethylene oxide, and chlorine; it has an internal working pressure ranging from 100 to 600 psi and is equipped with top-mounted fittings for loading and unloading that are protected by a protective housing that sits atop the tank car.

psig Pounds per square inch gauge; a measurement of pressure inside a container shown on an attached pressure gauge, in relation to atmospheric pressure.

radiation dispersal device (RDD) A device designed to spread radioactive material through a detonation of conventional explosives or other [non-nuclear] means; also called *dirty bomb.* (NFPA 470)

radiological agent A material that emits radiation, such as X-rays and radioactive isotopes, in excess of normal background radiation levels.

sarin A chemical warfare agent.

secondary containment Any device or structure that prevents environmental contamination when the primary container or its appurtenances fail.

secondary device An explosive or incendiary device designed to harm emergency responders who have responded to an initial event.

smallpox A highly infectious disease caused by the *Variola* virus, which kills approximately 30 percent of people who become infected.

soman A nerve agent that is both a contact and a vapor hazard; it has the odor of camphor.

special-use railcars Boxcars, flat cars, cryogenic tank cars, or corrosive tank cars.

tabun A nerve agent that disables the chemical connections between nerves and target organs.

ton container Bulk packaging that commonly holds compressed liquefied gases.

tube trailer A high-volume transportation vehicle made up of several individual compressed gas cylinders that operate at working pressures of 3000 to 5000 psi and are banded together and affixed to a trailer.

Type A packaging Packaging designed to protect its internal radiological contents during normal transportation and in the event of a minor accident.

Type B packaging Packaging more durable than Type A packaging, which is designed to prevent a release of a radiological hazard in the case of extreme accidents during transportation.

Type C packaging Packaging used when radioactive substances must be transported by air, with no restriction placed on the total amount of radioactivity contained within the package.

V-agent (VX) An oily liquid, contact hazard, nerve agent that can persist for several weeks.

REVIEW QUESTIONS

1. Define bulk packaging and nonbulk packaging.
2. What is the definition of cargo tank?
3. What are the five basic types of rail tank cars?
4. List at the nine basic pieces of information you should have available when calling CHEMTREC, CANUTEC, OR SETIQ.
5. Which agencies should be involved in responding to a hazardous materials incident that involves criminal intent?

DISCUSSION QUESTIONS

1. What is a key difference between a MC-306/DOT 406 and a MC-307/DOT 407 cargo tank, and why is it important to responders?
2. What is a key benefit to calling CHEMTREC, CANUTEC or SETIQ when a chemical name is known to responders?
3. Describe how the tactics and strategies of handling an incident with criminal intent might differ from an incident without criminal intent.

APPLYING THE CONCEPTS

It is a rainy afternoon when your truck company is dispatched to a motor vehicle accident on a local freeway. Your estimated time to arrive on scene is 7 minutes. En route, your dispatch center informs you that this is a vehicle versus tanker truck with possible injuries.

1. What considerations do you have regarding the nature of the vehicles involved?
2. Does the presence of rain influence your thinking about what you may find on scene?

Upon arrival, you find that the tanker truck is a large tractor trailer rig pulling a single cargo trailer with rounded ends. The tank is painted white and appears to be a pressure vessel. There is minor damage to the rear of the cargo trailer and moderate front-end damage to a four-door passenger vehicle located behind the trailer. All occupants of the passenger vehicle self-extricated and are standing on the side of the road with what appear to be minor injuries. The driver of the truck is also out of the vehicle, speaking with an officer of the highway patrol. You have a crew of yourself (company officer) plus two firefighters, and one of the firefighters is a paramedic. All crew members are trained to the operations level.

3. Which type of cargo trailer is this likely to be?
4. Which substances would most likely be carried in the cargo tank described in the scenario?

After further investigation, you find no significant damage to the trailer and no apparent leaks. The passenger vehicle is leaking radiator fluid and brake fluid, but otherwise, there are no other leaking fluids. All patients have refused medical treatment or transport and signed the appropriate medical forms. You take this opportunity to do a walk-around of the cargo trailer to review the features of an MC-331 tank with your crew.

5. Based on the level of training of your crew and the policy and procedures of the AHJ, is cleaning up the fluids from the passenger vehicle within your scope of practice?
6. List three features that are unique to this type of cargo trailer.

REFERENCES

Asociacion National de la Industria Quimica (ANIQ). "Sistema de Emergencias en Transporte para la Industria Quimica (SETIQ)." Accessed October 5, 2021. https://aniq.org.mx/webpublico/Notas/Nota.asp?id=79.

Canada Transport Emergency Centre (CANUTEC). Accessed October 5, 2021. https://tc.canada.ca/en/dangerous-goods/canutec.

Chemical Transportation Emergency Center (CHEMTREC). "Emergency Responders." Accessed October 5, 2021. https://www.chemtrec.com/chemtrec-services/emergency-responders.

National Fire Protection Association (NFPA). 2021. *NFPA 470: Hazardous Materials/Weapons of Mass Destruction (WMD) Standard for Responders.* 2022 Edition. Quincy, MA: NFPA.

CHAPTER

33

Hazardous Materials Operations Level

Estimating Potential Harm and Planning a Response

NOTE: Content within this chapter meets the intent of **NFPA 470, 2022 Edition**, which includes chapter 7, "Professional Qualifications for Hazardous Materials/WMD Operations Level Responders (**NFPA 1072**)."

KNOWLEDGE OBJECTIVES

After studying this chapter, you will be able to:

- Explain how to estimate the potential harm or severity of an incident.
- Describe how to plan an initial response.
- Explain how exposures might be affected by various types of hazardous materials incidents.
- Describe how to report the size and scope of the incident.
- Describe how to select personal protective equipment for an incident.
- Explain the role of respiratory protection.
- Describe the basic types of decontamination.

SKILLS OBJECTIVES

After studying this chapter, you will be able to:

- Perform emergency decontamination.

CASE STUDY

You Are the Firefighter

Your paramedic ambulance and local engine company have been called to a semiconductor fabrication facility for a report of shortness of breath. Both units arrive at the reception area and are led to an employee break room. As you approach the seated patient, you notice an odor of ammonia. The laboratory manager is standing next to the patient—a laboratory technician—who is doubled over in the chair. The laboratory manager tells you that the technician was splashed in the face with approximately 100 mL of ammonium hydroxide while pouring chemicals into an instrument. He states that he helped the laboratory technician to an eyewash station immediately after the incident and then escorted him to the break room. The laboratory manager tells you he has seen this sort of injury before. In his opinion, the laboratory technician doesn't need to go to the hospital; he just needs some oxygen. The paramedic begins an assessment and finds that the laboratory technician has reddened skin over his entire face and is complaining of shortness of breath. You are the officer on the engine company, and your four-person crew is trained to the operations level.

1. What are the initial response objectives for the engine crew and ambulance once the ammonium hydroxide exposure is discovered?
2. What are the initial objectives for treatment and transport of the patient?
3. Are the ambulance and engine crew at risk of being exposed or contaminated to potentially toxic levels of ammonium hydroxide from the patient?

Introduction

It is important to have a set of basic priorities to guide your decision making at the scene of a hazardous materials/weapons of mass destruction (WMD) incident. To that end, the first response objective should be to ensure your own safety while operating at the scene. You will be of no good to anyone if you become part of the problem! At a minimum, you must arrive at the scene in a safe manner and make sure that you and your crew do not become a liability during the incident.

After ensuring your own safety, your next objective should be to address the potential life safety of those persons affected or potentially affected by the incident. In the chapter-opening scenario, your crew responded for one type of emergency—shortness of breath—and ended up facing a completely different kind of problem—a medical issue related to a chemical exposure. This new set of parameters will require you to quickly shift gears. You now have an exposed victim and a potentially exposed laboratory manager. Are additional response personnel or equipment required to handle the situation based on the location and/or form of the release?

You can smell ammonia:

- Does that mean you and other personnel are in danger, too?
- Has the patient been adequately decontaminated to the point you can safely render care?
- Who performed the decontamination, and where?
- Should more decontamination be done?
- Should you back out of the area without treating the victim and call for a hazardous materials/WMD response team?
- Can the exposed person wait for a team of specially trained hazardous materials responders to perform decontamination?
- Do you have enough information and a decision-making process to weigh the risks and benefits of safely rendering patient care to a chemically exposed person?

Such real-world challenges present a set of complicated questions and require responders to carefully consider the correct course of action and identify effective strategies and tactics for incident mitigation. This chapter will help you to think through identifying response options and to estimate and plan for the challenges you may encounter at a hazardous materials/WMD incident.

Estimating the Potential Harm or Severity of the Incident

Hazardous materials/WMD incident response objectives should be based on the need to protect and reduce the threat to life, property, critical systems, and the environment. In some cases, the people and the problem are one and the same, as in the chapter-opening scenario. Once the threat to life has been handled, the incident becomes a matter of reducing the impact to the property that may be affected and minimizing environmental complications.

TIP

It is important to separate the people from the problem as soon and as safely as possible.

To have some frame of reference for the degree of harm a substance may inflict, responders should have a basic understanding of some commonly used health and safety terms and definitions. Two main organizations establish and publish the toxicological data typically used by hazardous materials/WMD responders to understand the health effects of an identified pool of chemical substances: the **American Conference of Governmental Industrial Hygienists (ACGIH)**, a scientific organization that researches and publishes information in the field of industrial hygiene, and the Occupational Safety and Health Administration (OSHA).

For more than 80 years, the ACGIH has been a respected and trusted source for occupational health and industrial hygiene guidelines and information. Its best-known committee, the Threshold Limit Values for Chemical Substances (TLV-CS) Committee, was established in the 1940s. This committee first introduced the concept of **threshold limit value (TLV)** in 1956. The TLV is the point at which a hazardous material/WMD begins to affect a person. Today, TLVs have been published for more than 600 chemical substances. Any values indicated with a TLV are established by the ACGIH.

OSHA was created in 1971 with three goals: to improve worker safety, to conduct research, and to publish toxicological data. OSHA's **permissible exposure limit (PEL)**, for example, is conceptually the same as ACGIH's TLV. The PEL is the established standard limit of exposure to a hazardous material. This standard was established based on the maximum time-weighted concentration at which 95 percent of exposed, healthy adults suffer no adverse effects over a 40-hour workweek. You might see both terms in a reference source. Nevertheless, there is an important distinction between the ACGIH and OSHA standards—ACGIH sets guidelines; OSHA standards are the law.

It is common to see toxicological values expressed in units of parts per million (ppm), parts per billion (ppb), and, in some cases, parts per trillion (ppt). One part per million (ppm) means there is one particle of a given substance for 999,999 other particles, and one part per billion (ppb) is equivalent to one particle in 999,999,999 other particles. Typically, the amount of airborne contamination encountered with releases of gases such as arsine, chlorine, and ammonia will be expressed in this manner. Arsine, for example, has an OSHA-established PEL of 0.05 ppm. This is a very small amount compared to the OSHA PEL for chlorine, which is 1 ppm. Comparatively speaking, arsine is a far more toxic substance than chlorine.

Another way to express contamination levels for substances other than gases, such as fibers and dusts, is through units of milligrams per cubic meter (mg/m^3). For simplicity and better understanding of this text, keep in mind that you may see toxicological data expressed in several different ways. Regardless of the units used, the *lower* the value, the *more toxic* the product is.

The **threshold limit value/short-term exposure limit (TLV/STEL)** is the maximum concentration of a hazardous material that a person can be exposed to in 15-minute intervals, up to four times per day, without experiencing irritation or chronic or irreversible tissue damage. A minimum 1-hour rest period should separate any exposure to this concentration of the material. The lower the TLV/STEL concentration, the more toxic the substance is. The **threshold limit value/time-weighted average (TLV/TWA)** is the maximum airborne concentration of material to which a person can be exposed for 8 hours per day, 40 hours per week, with no ill effects. As with the TLV/STEL, the lower the TLV/TWA, the more toxic the substance is. The **threshold limit value/ceiling (TLV/C)** is the concentration of a hazardous material that should not be exceeded because exposure, even for an instant, can cause ill effects. Again, the lower the TLV/C, the more toxic the substance is.

The **threshold limit value/skin (TLV/skin)** indicates that direct or airborne contact with a material could potentially result in significant exposure from absorption through the skin, mucous membranes, and eyes. This designation is intended to suggest that appropriate measures should be taken to minimize skin absorption so that the TLV/skin is not exceeded.

The **recommended exposure level (REL)** is a comparable value established by the **National Institute for Occupational Safety and Health (NIOSH)**. NIOSH is part of the U.S. Department of Health and Human Services and is charged with providing information, training, research, and education in the field of occupational safety and health so as to ensure that individuals have a safe and healthy work environment. As mentioned earlier, the PEL is the standard limit of exposure to a hazardous material established and enforced by OSHA, while TLV/TWA is the standard established by ACGIH. These three terms—PEL, REL, and TLV/TWA—measure the maximum, time-weighted concentration of material to which 95 percent of healthy adults can be exposed without suffering any adverse effects over a 40-hour workweek.

The designation **immediately dangerous to life and health (IDLH)** means that an atmospheric concentration of a toxic, corrosive, or asphyxiant substance poses an immediate threat to life or could cause irreversible or delayed adverse health effects. Three types of IDLH atmospheres are distinguished: toxic, flammable, and oxygen deficient. Individuals exposed to atmospheric concentrations below the IDLH value (in theory) could escape from the atmosphere without experiencing irreversible damage to their health, even if their respiratory protection fails. Individuals who have a risk of exposure to atmospheric concentrations equal to or higher than the IDLH value must use positive-pressure self-contained breathing apparatus (SCBA) or equivalent protection.

Identifying and measuring the levels of airborne contamination require specific detection and monitoring instruments, along with training to interpret the results. With the appropriate equipment, responders will be able to measure concentrations of specific chemicals. Once the exposure values are understood, they can be applied at the scene of a hazardous materials/WMD emergency. For example, exposure guidelines can be used to identify three basic atmospheres or environments that might be encountered at a hazardous materials/WMD emergency:

- Green environments could be considered low hazard and low risk. These environments would not require any personal protective equipment (PPE) beyond a normal work uniform.
- Yellow environments are transitional, whereby the hazard is increasing, and some level of PPE is required (at least splash protection and some level of respiratory protection).
- Red environments pose a high hazard and high risk; therefore, you must wear the highest level of skin and respiratory protection. An exposure to unprotected skin and/or lungs could be fatal. These are above-IDLH environments where toxic chemicals are present. You can certainly work around lethal concentrations of released chemicals safely if you use the correct type and level of PPE.

Note that these are rough guidelines and serve only to illustrate an actual or potential level of risk.

In many cases, responders have no control over the hazard—the genie may be out of the bottle before your arrival! You do have control over your risk, however, as you determine how you will interact with the problem.

- What is the risk of taking no action?
- Do you have the proper type and level of protection that allows for safe entry into a contaminated atmosphere?

All exposure guidelines share a common goal: to ensure the safety and health of people working with or responding to incidents involving chemical substances.

Plan an Initial Response

Planning a response boils down to understanding the nature of an incident and determining a course of action that will favorably change the outcome. On the surface, this seems like a straightforward, uncomplicated task. In truth, the decision to act can be a weighty one, fraught with many pitfalls and dangers. When planning an initial hazardous materials/WMD incident response, it is important to be mindful of the safety of the responding personnel. The responders are there to isolate, contain, and/or remedy the problem—not to become part of it. Proper incident planning will keep responders safe and provide a means to control the incident effectively, preventing further harm to persons or property.

Initial Call for Help

The information obtained from the initial call for help is used to determine the safest, most effective, and fastest route to the hazardous materials/WMD scene. Choose a route that approaches the scene from an upwind and upgrade direction so that natural wind currents blow the hazardous material vapors away from arriving responders (**FIGURE 33-1**). A route that places

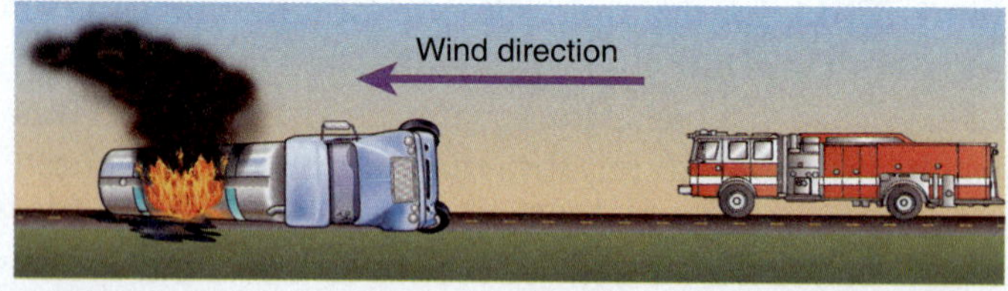

FIGURE 33-1 Approach a hazardous materials incident cautiously; choose a route that approaches from an upwind direction.

FIGURE 33-2 The response to a spill of a solid hazardous material will differ from the response to a liquid-release or vapor-release incident.

the responders uphill as well as upwind of the site is also desirable so that a liquid or vapor hazardous material flows away from responders.

Responders need to know as much as possible about the material involved. Is the material a solid, a liquid, or a gas? Is it contained in a drum, a barrel, or a pressurized tank? Is the spill still in progress (dynamic) or has it ceased (static)? The response to a spill of a solid hazardous material will differ from the response to a liquid-release incident or a vapor-release incident (**FIGURE 33-2**). A solid may be easily contained, whereas a gas release can be widespread and constantly moving, depending on the gas's characteristics and the weather conditions.

The characteristics of the affected area near the location of the spill or leak are also important factors in planning the response to an incident. If an area is heavily populated, evacuation procedures may be established very early in the incident. If the area is sparsely populated and rural, isolating the area from anyone trying to enter the location may be the top priority. A high-traffic area such as a major highway would necessitate immediate rerouting of traffic, especially during rush hours.

Resources for Determining the Size of the Incident

It is vital to understand the incident as a whole. To do so, responders must factor in results obtained from detection and monitoring devices, reference sources, bystander information, the current environmental conditions surrounding the incident, and other information to get a clear picture of what is going on and what is likely to happen next. Consider situational factors such as land use; weather conditions, overhead and underground wires and pipelines; locations of storm or sewer drains; potential ignition sources; adjacent land use; and, if inside a building, the locations of floor drains, ventilation ducts, and air returns. (Note: This list is intended to illustrate conditions that should be noted when surveying a scene, but does not represent all factors to consider.)

Response Objectives

Response objectives should be measurable, flexible, and time sensitive. They should also be based on the chosen strategy and carried out with clear and achievable tactics. Some examples might include the following:

- A three-person team will construct a dirt berm around the drain at the south end of the leaking tanker to protect the adjacent waterway. This task will need to be completed in the next 30 minutes.
- A two-person team, wearing a Level B ensemble, will immediately enter the steel door on the east side of the building and shut down the ventilation system.

In some cases, several incident objectives may be developed to solve a problem. To be effective and meaningful, however, those objectives need to be tied to the reason you chose to act in the first place. Again, if you do not understand why you are taking action, reevaluate the situation so that you can better understand the problem. In all cases, responders must constantly evaluate the effectiveness of strategies and tactics employed to accomplish the objectives in the incident action plan.

Typically, response objectives fall into one of three main categories: offensive, defensive, and nonintervention. (These categories were introduced in Chapter 29, *Hazardous Materials Regulations, Standards, and Laws*, in this text.) With offensive actions, responders take action to mitigate the issue. Offensive operations typically take place in the area closest to the release, where PPE is required to operate. Defensive actions take place some distance away from the point of release, in areas free of contamination. Examples of defensive actions include diking and damming in an area well ahead of the spill or release, or stopping the flow of a substance remotely from a valve or shut-off. Nonintervention occurs when the hazard is too great, or when it makes more sense and is safer to allow the incident to self-stabilize. Allowing a compressed gas cylinder to vent off the pressure until it is empty is an example of nonintervention.

TIP

The mode of operations should be identified in the verbal or written incident action plan (IAP). The main components of an IAP include specific objectives, clear work assignments for the responders, the resources needed to handle the incident, an organizational chart representing the key players on the scene, communications channels or methods, and a medical plan to follow if responders or civilians become exposed or injured during the response.

With any response, it is also vital to identify the facts and circumstances that indicate the chosen course of action is effective and that incident conditions are improving.

Protective Actions at the Operations Level

Evaluating the threat to life is the number one response priority at a hazardous materials incident. If there is no life threat—either immediate or anticipated—the severity of the incident is diminished. This is not to say that property or the environment is unimportant; rather, it is simply an acknowledgment that you, as a responder, must consider the threat to life before anything else.

Life safety actions include ensuring your own safety and searching for, and possibly rescuing, persons who were immediately exposed to the substance or those who may now be in harm's way. For example, an incident commander (IC) might be faced with a decision of whether to evacuate people ahead of an advancing vapor cloud or when an explosive device is discovered or some other potentially hazardous condition exists. This is not an easy decision, because efforts to relocate people bring up many complications.

Evacuation

Sometimes a decision must be made to evacuate or rescue people in danger. **Evacuation** is the removal or relocation of those individuals who may be affected by an approaching release of a hazardous material. In those instances, responders may need to consult printed and electronic reference sources for guidance on evacuation distances and other safety information.

A good way to practice before a response is needed is to imagine a chemical and a credible location or condition in which the chemical might be released and use the *Emergency Response Guidebook* (*ERG*) as a reference source. Choose a United Nations (UN) identification number for the chemical, and check whether it is highlighted and found in the green section of the *ERG* (as described in Chapter 30, *Recognizing and Identifying the Hazards*, in this text). Next, determine the recommended emergency actions and PPE that might be required to handle the incident. Also, it is useful to imagine the release occurring in several different areas of your jurisdiction.

Take a few minutes to think about initial isolation distances or other protective actions that could be taken. To that end, responders should understand that the "initial isolation" distances (the distance at which all persons should be considered for evacuation in all directions) and the "protective action" distances (the downwind distance over which some form of protective actions might be required) are based on the nature of the material, the environmental conditions of the release, and the size of the release. The *ERG* also offers suggested stand-off distances for improvised explosive devices (IEDs) and potential boiling liquid/expanding vapor explosion (BLEVE) situations (**FIGURE 33-3**).

If people need to be evacuated, a risk-based method of decision making should be employed. The IC should weigh the severity of the threat against the potential effort, impact, and resource requirements of moving people from one location to another (**FIGURE 33-4**). Evacuating a predominantly ill and elderly population from a care facility, for example, may take a long time, have negative effects on the evacuees, and require many resources. Given these concerns, the IC must decide if the hazard merits such a move or if other methods might be used to protect this type of population.

The IC must consider many factors before making decisions concerning evacuation. Some of those factors include the nature and duration of the release; the nature of the evacuees, such as their age, underlying health status, and mobility; the ability to support the evacuees with basic services such as food, water, and shelter; and transportation challenges. When considering evacuating a significantly large group of people, it may be helpful to consider the host of "shuns" associated with the task. The "shuns" is a wordplay on the ending of all the following terms:

- Contamination (the trigger for the evacuation)
- Communication
- Transportation
- Nutrition
- Sanitation
- Habitation
- Compassion

WARNING: The data given are approximate and should only be used with extreme caution. These times can vary from situation to situation. LPG tanks have been known to BLEVE within minutes. Therefore, never risk life based on these times.

BLEVE (USE WITH CAUTION)

Capacity		Diameter		Length		Propane Mass		Minimum time to failure for severe torch	Approximate time to empty for engulfing fire	Fireball radius		Emergency response distance		Minimum evacuation distance		Preferred evacuation distance		Cooling water flow rate	
Litres	(Gallons)	Meters	(Feet)	Meters	(Feet)	Kilograms	(Pounds)	Minutes	Minutes	Meters	(Feet)	Meters	(Feet)	Meters	(Feet)	Meters	(Feet)	Litres/min	USgal/min
100	(26.4)	0.3	(1)	1.5	(4.9)	40	(88)	4	8	10	(33)	90	(295)	154	(505)	307	(1007)	97	26
400	(106)	0.61	(2)	1.5	(4.9)	160	(353)	4	12	16	(53)	90	(295)	244	(801)	488	(1601)	195	51
2000	(528)	0.96	(3.2)	3	(9.8)	800	(1764)	5	18	28	(92)	111	(364)	417	(1368)	834	(2736)	435	115
4000	(1057)	1	(3.3)	4.9	(16.1)	1600	(3527)	5	20	35	(115)	140	(459)	525	(1722)	1050	(3445)	615	163
8000	(2113)	1.25	(4.1)	6.5	(21.3)	3200	(7055)	6	22	44	(144)	176	(577)	661	(2169)	1323	(4341)	870	230
22000	(5812)	2.1	(6.9)	6.7	(22)	8800	(19400)	7	28	62	(203)	247	(810)	926	(3038)	1852	(6076)	1443	381
42000	(11095)	2.1	(6.9)	11.8	(38.7)	16800	(37037)	7	32	77	(253)	306	(1004)	1149	(3770)	2200	(7218)	1994	527
82000	(21662)	2.75	(9)	13.7	(45)	32800	(72310)	8	40	96	(315)	383	(1257)	1435	(4708)	2200	(7218)	2786	736
140000	(36984)	3.3	(10.8)	17.2	(56.4)	56000	(123457)	9	45	114	(374)	457	(1499)	1715	(5627)	2200	(7218)	3640	962

A.

Improvised Explosive Device (IED)
SAFE STAND-OFF DISTANCE

	Threat Description	Explosives Capacity[1]		Mandatory Evacuation Distance[2]		Shelter-in-Place Zone		Preferred Evacuation Distance[3]	
High Explosives (TNT Equivalent)	 Pipe Bomb	5 lbs	2.3 kg	70 ft	21 m	71 - 1,199 ft	22 - 365 m	+1,200 ft	366 m
	Suicide Bomber	20 lbs	9 kg	110 ft	34 m	111 - 1,699 ft	35 - 518 m	+1,700 ft	519 m
	Briefcase/Suitcase	50 lbs	23 kg	150 ft	46 m	151 - 1,849 ft	47 - 563 m	+1,850 ft	564 m
	Car	500 lbs	227 kg	320 ft	98 m	321 - 1,899 ft	99 - 579 m	+1,900 ft	580 m
	SUV/Van	1,000 lbs	454 kg	400 ft	122 m	401 - 2,399 ft	123 - 731 m	+2,400 ft	732 m
	Small Delivery Truck	4,000 lbs	1,814 kg	640 ft	195 m	641 - 3,799 ft	196 - 1,158 m	+3,800 ft	1,159 m
	Container/Water Truck	10,000 lbs	4,536 kg	860 ft	263 m	861 - 5,099 ft	264 - 1,554 m	+5,100 ft	1,555 m
	Semi-Trailer	60,000 lbs	27,216 kg	1,570 ft	475 m	1,571 - 9,299 ft	476 - 2,834 m	+9,300 ft	2,835 m

[1] Based on the maximum amount of material that could reasonably fit into a container or vehicle. Variations possible.

[2] Governed by the ability of an unreinforced building to withstand severe damage or collapse.

[3] Governed by the greater of fragment throw distance or glass breakage/falling glass hazard distance. These distances can be reduced for personnel wearing ballistic protection. Note that the pipe bomb, suicide bomb, and briefcase/suitcase bomb are assumed to have a fragmentation characteristic that requires greater stand-off distances than an equal amount of explosives in a vehicle.

B.

FIGURE 33-3 The *Emergency Response Guidebook (ERG)* offers suggested stand-off distances for **(A)** potential boiling liquid/expanding vapor explosion (BLEVE) situations and **(B and C)** improvised explosive devices (IEDs).

Reproduced from U.S. Department of Transportation, Transport Canada, and Secretariat of Communications and Transport of Mexico. 2020. *2020 Emergency Response Guidebook*. Pipeline and Hazardous Materials Safety Administration, U.S. Department of Transportation. https://www.phmsa.dot.gov/sites/phmsa.dot.gov/files/2021-01/ERG2020-WEB.pdf

Improvised Explosive Device (IED)
SAFE STAND-OFF DISTANCE

	Threat Description	LPG Mass / Volume[1]		Fireball Diameter[2]		Safe Distance[3, 4]	
LPG - Butane or Propane	Small LPG Tank	20 lbs / 5 gal	9 kg / 19 L	40 ft	12 m	160 ft	48 m
	Large LPG Tank	100 lbs / 25 gal	45 kg / 95 L	69 ft	21 m	276 ft	84 m
	Commercial/Residential LPG Tank	2,000 lbs / 500 gal	907 kg / 1,893 L	184 ft	56 m	736 ft	224 m
	Small LPG Truck	8,000 lbs / 2,000 gal	3,630 kg / 7,570 L	292 ft	89 m	1,168 ft	356 m
	Semitanker LPG	40,000 lbs / 10,000 gal	18,144 kg / 37,850 L	499 ft	152 m	1,996 ft	608 m

[1] Based on the maximum amount of LPG that could reasonably fit into a container or vehicle. Variations possible.

[2] Assuming efficient mixing of the flammable gas with ambient air.

[3] Determined by U.S. firefighting practices wherein safe distances are approximately 4 times the flame height.

[4] This table is for a loaded LPG tank with explosives on the exterior. Note that an LPG tank filled with high explosives would require a significantly greater stand-off distance than if it were filled with LPG.

C.

FIGURE 33-3 Continued

You may identify more "shuns" that are specific to your authority having jurisdiction (AHJ)—be creative and complete when thinking about evacuation.

Once the evacuation order is given, firefighters and law enforcement personnel may be called upon to assist in the physical relocation of residents to a safe area. Their work may include such actions as traveling to homes and informing residents that they must relocate to a temporary shelter.

Before an evacuation order is given, a safe area with suitable facilities should be established. In many cases, large venues such as schools, fairgrounds, and sports arenas are used as evacuation shelters. The evacuation area or shelter should be located close enough to the exposure for the evacuation to be practical, but far enough away from the incident to be safe. Depending on the time of day and the season, it may take a considerable amount of time to evacuate even a small residential area. Temporary evacuation areas may be needed to shelter residents until evacuation sites or structures indicated in your community's emergency response plan are open and accessible. The security of the evacuation facility is an important consideration and should be accomplished with the assistance of law enforcement.

TIP

Evacuation has significant challenges even when the operation is properly planned. Plan carefully and consider all options before proceeding with an evacuation. It is a significant undertaking!

FIGURE 33-4 The IC must weigh the severity of the threat against the time, personnel, and other logistical challenges required to carry out an evacuation.

Access should be monitored and controlled to ensure the maximum amount of security and shielding from the media, onlookers, or persons interested in taking advantage of the evacuees in any way.

Transportation to temporary evacuation areas must be arranged for all populations when an evacuation is ordered. As part of this effort, the needs of elderly persons, physically challenged individuals, and persons with special needs should be considered. Accommodations for pets and other animals should also be considered. It would be a mistake to assume all evacuees are able or motivated to move on their own.

Initial evacuation distances may be derived from the *ERG* (**FIGURE 33-5**). Note that the *ERG* does not give complete, detailed information for all conditions that responders might potentially encounter, such as weather extremes, road conditions, or actual conditions encountered at the scene.

GUIDE 115 GASES - FLAMMABLE (INCLUDING REFRIGERATED LIQUIDS)

POTENTIAL HAZARDS

FIRE OR EXPLOSION
- EXTREMELY FLAMMABLE.
- Will be easily ignited by heat, sparks or flames.
- Will form explosive mixtures with air.
- Vapors from liquefied gas are initially heavier than air and spread along ground.

CAUTION: Hydrogen (UN1049), Deuterium (UN1957), Hydrogen, refrigerated liquid (UN1966), Methane (UN1971) and Hydrogen and Methane mixture, compressed (UN2034) are lighter than air and will rise. Hydrogen and Deuterium fires are difficult to detect since they burn with an invisible flame. Use an alternate method of detection (thermal camera, broom handle, etc.)
- Vapors may travel to source of ignition and flash back.
- Cylinders exposed to fire may vent and release flammable gas through pressure relief devices.
- Containers may explode when heated.
- Ruptured cylinders may rocket.

HEALTH
- Vapors may cause dizziness or asphyxiation without warning.
- Some may be irritating if inhaled at high concentrations.
- Contact with gas or liquefied gas may cause burns, severe injury and/or frostbite.
- Fire may produce irritating and/or toxic gases.

PUBLIC SAFETY
- CALL 911. Then call emergency response telephone number on shipping paper. If shipping paper not available or no answer, refer to appropriate telephone number listed on the inside back cover.
- Keep unauthorized personnel away.
- Stay upwind, uphill and/or upstream.
- Many gases are heavier than air and will spread along the ground and collect in low or confined areas (sewers, basements, tanks, etc.).

PROTECTIVE CLOTHING
- Wear positive pressure self-contained breathing apparatus (SCBA).
- Structural firefighters' protective clothing provides thermal protection but only limited chemical protection.
- Always wear thermal protective clothing when handling refrigerated/cryogenic liquids.

EVACUATION

Immediate precautionary measure
- Isolate spill or leak area for at least 100 meters (330 feet) in all directions.

Large Spill
- Consider initial downwind evacuation for at least 800 meters (1/2 mile).

Fire
- If tank, rail car or tank truck is involved in a fire, ISOLATE for 1600 meters (1 mile) in all directions; also, consider initial evacuation for 1600 meters (1 mile) in all directions.
- In fires involving Liquefied Petroleum Gases (LPG) (UN1075), Butane (UN1011), Butylene (UN1012), Isobutylene (UN1055), Propylene (UN1077), Isobutane (UN1969), and Propane (UN1978), also refer to BLEVE – SAFETY PRECAUTIONS (Page 366).

In Canada, an Emergency Response Assistance Plan (ERAP) may be required for this product. Please consult the shipping paper and/or the ERAP Program Section (page 390).

Page 166 ERG 2020

FIGURE 33-5 Sample page from the orange section of the *Emergency Response Guidebook (ERG)*.

Reproduced from U.S. Department of Transportation, Transport Canada, and Secretariat of Communications and Transport of Mexico. 2020. *2020 Emergency Response Guidebook*. Pipeline and Hazardous Materials Safety Administration, U.S. Department of Transportation. https://www.phmsa.dot.gov/sites/phmsa.dot.gov/files/2021-01/ERG2020-WEB.pdf

In severe weather, evacuation can be challenging at best. Flooded areas, heavy rains, and high winds present dangers to the evacuees as well as to responders. Even in perfect circumstances, evacuation of residents will be a process that is measured in hours, not minutes. Preplanning for evacuations is key to successful and safe evacuations.

A good way for the IC to determine the area to be evacuated, including how far to extend actual evacuation distances, is to use detection and monitoring devices to identify areas free of airborne contamination.

SAFETY TIP

When carrying out an evacuation, make provisions for protecting yourself in case the wind shifts, or you unexpectedly encounter the hazardous material.

Shelter-in-Place

Shelter-in-place is a method of safeguarding people in a hazardous area by keeping them in an enclosed atmosphere, usually inside structures. Local emergency plans should identify facilities where vulnerable populations might be found, such as schools, churches, hospitals, nursing homes, and apartment complexes. When residents are sheltered-in-place, they remain indoors with windows and doors closed. All ventilation systems are turned off to prevent outside air from being drawn into the structure.

The toxicity of the hazardous material and the amount of time available to avoid the oncoming threat are major factors in the decision of whether to evacuate or to use a sheltering-in-place strategy. The expected duration of the incident is also a factor in determining whether sheltering-in-place is a viable option. The longer the release is expected to continue, the more time the material has to enter or permeate into protected areas. Short-term events or transient vapor clouds might dictate a sheltering-in-place approach.

Another safety-related action to be taken at a hazardous materials/WMD incident is the establishment of an **area of safe refuge**. This location serves as an area where people are held temporarily until they can be safely decontaminated, treated, or removed.

Isolate and Deny Entry

Isolation of the hazard area by setting up a perimeter around the contaminated atmosphere is one of the first actions responders must take at a hazardous materials/WMD incident. Responders and civilians alike must be kept at a safe distance from the release site. This is a vital first step in beginning to establish safe work zones and identify the areas of high hazard.

TIP

Initial isolation zones and protective actions are just that—initial. You must constantly evaluate the conditions and adjust your tactics and strategy accordingly.

Imagine arriving first on scene as an engine company (all members are trained to the operations level) to find an MC-306/DOT 406 cargo trailer with a large active leak as the result of a motor vehicle accident. It is the middle of the day in a commercial area of town; there are several blocks of closely spaced occupied buildings.

- What material is likely to be in the container, and how much?

- Would you evacuate the buildings, shelter-in-place, or attempt to stop the leak?
- What resources could you use to determine the evacuation distances?
- Would you consider the buildings to be exposures that require action to protect?

Perhaps your operational priority is to isolate and deny entry to the area and begin evacuating the adjacent buildings. Many considerations could arise in this hypothetical scenario, but if isolation of the problem is chosen, it may be accomplished in several ways. Law enforcement officers are often posted a safe distance from the release to create a secure perimeter. Other public safety personnel, such as firefighters, may serve the same function, although it is important not to waste the skills of a cadre of trained hazardous materials/WMD responders by assigning them to guard doors, other points of ingress or egress from a building, or other contaminated areas.

In many cases, responders will stretch a length of barrier tape across roadways, doors, or other access points. Care must be taken, however, not to rely solely on this method of scene control. Quite often, public safety responders or the public will not respect the boundaries of areas marked with barrier tape. If barrier tape is used, it should still be backed up by a human presence (**FIGURE 33-6**). Also, keep in mind that the precise type of isolation efforts undertaken will be driven by the nature of the released chemical and the environmental conditions.

Once the hazard is isolated, access is denied to all but the small group of responders who are trained and equipped to enter the contaminated atmosphere. Isolating a contaminated atmosphere is always conjoined in some way with **denial of entry** (the policy that restricts access to all but essential personnel) to the site. Practically speaking, one action should not exist without the other. Typically, site access control is established to control the movement of personnel into and out of a contaminated area.

FIGURE 33-6 Law enforcement can be a great asset to help ensure your restricted area stays restricted.

Search and Rescue

The search for and rescue of people can be more complicated in the setting of a hazardous materials incident. In a structure fire, it is understood that smoke and flame pose serious threats to life, and firefighters are properly protected to conduct reasonable operations to perform search and rescue. In a hazardous materials incident, all response personnel (fire, law enforcement, and emergency medical services [EMS]) must first recognize and identify the released substance. This may take some time and is not as straightforward as with a typical structure fire. As in other situations at a hazardous materials incident, the IC should use a risk-based decision process to understand the hazards as they relate to the possibilities of victim survivability and responder safety. Only then should personnel consider undertaking search and rescue.

Ultimately, the IC must determine whether entry into a contaminated environment to perform search and rescue is a worthy endeavor. If the released substance is highly toxic, unprotected victims may not have survived the initial exposure. If it is determined that a search can be safely performed, rescue teams wearing proper PPE may then enter the hot zone to look for or retrieve victims. The victims should be removed to the warm zone, where they can be decontaminated and turned over to EMS providers for transport to a medical facility. In nearly all cases, definitive medical care should be provided to victims who have been decontaminated and removed to the cold zone. NFPA 470, Chapter 46, *Competencies for Hazardous Materials/WMD Basic Life Support (BLS) Responders,* and Chapter 47, *Competencies for Hazardous Materials/WMD Advanced Life Support (ALS) Responders,* offer additional guidance on the medical management of chemical exposures.

Consider Exposures

When considering the potential consequences of a hazardous materials/WMD incident, the on-scene crews must consider how exposures might be affected. In firefighting, the term *exposures* typically applies to those physical areas adjacent to the fire that might become involved if the fire is left unchecked. For our purposes here, exposures include property, structures, or environments that are subject to influence, damage, or injury due to contact with a hazardous material/WMD. The number of exposures is determined by the location of the incident, the physical and chemical properties of the released substance, and the amount of progress that has been made in protecting those exposures by isolating the release site or by taking protective actions such as evacuation or sheltering-in-place. Incidents in urban areas are likely to have a greater potential for exposures; consequently, more resources will likely be needed to protect those exposures from the hazardous materials/WMD.

Reporting the Size and Scope of the Incident

Reporting the estimated physical size of the area affected by a hazardous materials/WMD incident is accomplished by using information available at the scene. If a vehicle is transporting a known amount of material, for example, an estimate of the size of the release might be made by subtracting the amount remaining in the container from the maximum capacity of the container. This can be computed by looking at the shipping papers to see whether deliveries have been made. To "see" into containers such as railroad tank cars, steel drums, or cargo tanks and estimate their remaining contents, responders may use thermal imagers (TIs) (**FIGURE 33-7**).

For example, you might use a TI to investigate a steel drum discovered in a vacant lot. A quick look at the scene might reveal some wet-looking soil around the base of the drum. By using the TI, you might be able to determine the percentage of liquid remaining and make an educated guess about how much could have leaked. Of course, these estimations may be quite rough, especially when it is unknown how much a given vessel may have contained prior to a release; however, it does allow for some "worst-case scenario" estimations.

FIGURE 33-7 Thermal imagers allow a responder to estimate how much material remains in a container. In this photo, the whitish color at the bottom of the drum is the amount of liquid remaining.

Depending on its size, the extent of the release may be expressed in units as small as square feet or as large as square miles. There are no hard-and-fast rules here: Be as accurate and as clear as possible when communicating with other responders or assisting agencies or when contacting call centers such as CHEMTREC, the Canadian Transport Emergency Centre (CANUTEC), or the Emergency Transportation System for the Chemical Industry, Mexico (SETIQ) for assistance. Remember that the safety of responders is paramount to maintaining an effective response to any hazardous materials/WMD incident.

Personal Protective Equipment

The determination of the PPE needed is based on the hazardous material involved, the specific hazards present, and the physical state of the material, along with a consideration of the tasks to be performed by the operations-level responder. In the realm of hazardous materials/WMD response, the selection and use of chemical-protective clothing may have the greatest direct impact on responder health and safety. Without the proper PPE, responders place themselves at risk of suffering harmful exposures. Corrosives such as concentrated sulfuric acid and hydrochloric acid, as well as caustic substances such as sodium hydroxide, for example, can damage the skin. The human body is also susceptible to the adverse effects of solvents such as methylene chloride and toluene, which may penetrate the skin and cause systemic health effects. Some poisons—the nerve agent VX, for example—are highly toxic and can be fatal in small doses.

Skin protection, however, is just one factor to be considered when discussing PPE. Anyone planning to work in a contaminated atmosphere must also place a high priority on respiratory protection. Respiratory protection is so important that it can be viewed as the defining element of PPE. The head-to-toe ensemble is not complete until the hazards have been identified and the respiratory protection has been properly matched to both the hazard and the garment. Keep in mind that PPE is not intended to function as an impenetrable suit of armor: It has limitations!

Several types and levels of PPE may be selected for use at a hazardous materials/WMD incident. These levels are spelled out in detail in OSHA's Hazardous Waste Operations and Emergency Response (HAZWOPER) regulations; in NFPA 1990, *Standard for Protective Ensembles for Hazardous Materials and CBRN Operations*; and in the emergency response plans for the AHJ.

To avoid redundancy on the topic of PPE for operations-level responders and the mission-specific competency for PPE, the specifics of each type and level of PPE can be found in Chapter 3, *Personal Protective Equipment,* and in Chapter 35, *Hazardous Materials Responder Personal Protective Equipment,* in this text. This chapter does, however, discuss respiratory protection in detail.

In addition to selecting PPE for offensive actions, responders must be aware of the procedures for cleaning, disinfecting, and inspecting PPE. These procedures may vary from manufacturer to manufacturer, so it is important for all responders to understand what is required to maintain the PPE used in their jurisdiction.

Some types of PPE—primarily reusable garments—are required to be tested at regular intervals and after each use. Individual manufacturers will have well-defined procedures for the maintenance, testing, and inspection of their equipment. Prior to purchasing any PPE, the AHJ should understand what is required in terms of its maintenance and upkeep, including the cleaning and disinfection of the PPE.

Most types of chemical-protective garments should be stored in a cool, dry place, free of significant temperature swings and/or high levels of humidity. If repairs are required, consult the manufacturer prior to performing any work. There is a risk that the garment will not perform as expected if it has been modified or repaired incorrectly.

Excessive-Heat Disorders Related to PPE

PPE, while an essential part of operating at a hazardous materials incident, does pose a safety concern in terms of creating increased body temperature, even if the ambient temperature is not extreme. To that end, hazardous materials responders operating in protective clothing should be aware of the signs and symptoms of dehydration, heat exhaustion, and heat stroke. If the body is unable to adequately disperse heat because an ensemble of PPE covers it, serious short- and long-term medical issues could result.

Most heat-related illnesses are typically preceded by **dehydration**, which is a medical condition that occurs when the body loses more water than it takes in. It is important to stay hydrated so that you can function at your maximum capacity. As a frame of reference, athletes should consume approximately 500 mL of fluid (water) prior to an event and 200–300 mL at regular intervals. Responders are occupational athletes—so keep up on your fluids!

Heat exhaustion is a mild form of shock that arises when the circulatory system begins to fail because the body is unable to dissipate excessive heat and becomes overheated. With heat exhaustion, the body's core temperature rises, followed by weakness and sweating. A person suffering from heat exhaustion may become dizzy and have episodes of blurred vision. Other signs and symptoms of heat exhaustion include acute fatigue, headache, and muscle cramps; however, the signs and symptoms may vary among individuals. Any individual experiencing heat exhaustion should be removed at once from the heated environment, rehydrated with electrolyte solutions (perhaps by intravenous methods), and kept cool. If not properly treated, heat exhaustion may progress to a potentially fatal condition.

Heat stroke is a severe and potentially fatal condition resulting from the failure of the body's temperature-regulating capacity. It is caused by exposure to the sun or working in high temperatures. Reduction or cessation of sweating is an early symptom. The body temperature can rise to 105°F (40.6°C) or higher and be accompanied by a rapid pulse; hot, red-looking skin; headache; confusion; unconsciousness; and possibly seizures. Heat stroke is a true medical emergency that requires immediate transport to a medical facility.

When treating heat-related illnesses, it is important to avoid pouring cold water on the victim, or otherwise placing the victim in an unusually cold environment. The extreme swing in temperature may have adverse effects on the individual's recovery. Using tepid water for drinking and cooling the skin is the safest approach to take. Keep in mind that rehydration by mouth is much slower than rehydration by intravenous line.

To combat heat stress while their personnel are wearing PPE, many response agencies employ some form of cooling technology under the garment. These technologies include, but are not limited to, air-, ice-, and water-cooled vests, along with phase-change cooling technology. Many studies have been conducted on each form of cooling technology. Each is designed to accomplish the same goal: to reduce the effects of heat stress on the human body.

Cold-Temperature Exposures

Responders at hazardous materials incidents may be exposed to two types of cold temperatures: those caused by the released materials and those caused by the operating environment, including ambient air temperatures or conditions such as rain, snow, or other adverse cold-weather conditions. Hazardous materials such as liquefied gases and cryogenic liquids may expose responders to the same low-temperature hazards as those created by cold-weather environments. Exposure to severe cold for even a short period of time may cause severe injury to body surfaces, especially to the ears, nose, hands, and feet.

Two environmental factors influence the extent of cold injuries: temperature and wind speed. Because still air is a poor heat conductor, responders working in low temperatures with little wind can endure these conditions for longer periods (if their clothing remains dry). However, when low temperatures are combined with significant winds, wind chill occurs. As an example, if the temperature with no wind is 20°F (–7°C), it will feel like –5°F (–21°C) when the wind speed is 15 miles per hour.

Regardless of the temperature, anyone will perspire while wearing chemical-protective clothing. Wet clothing extracts heat from the body as much as 240 times faster than dry clothing does. It may also lead to hypothermia, a condition in which the core body temperature falls below 95°F (35°C). Hypothermia is a true medical emergency.

Responders must be aware of the dangers of frostbite, hypothermia, and impaired ability to work when temperatures are low or when they are working in wet clothing. All personnel should wear layered clothing and be able to warm themselves in heated shelters or vehicles.

The layer of clothing next to the skin, especially the socks, should be kept dry. The combination of cold and wet softens the skin, causing numbness, tingling, and, in some cases, peeling skin.

Responders should carefully schedule their work and rest periods and monitor their physical working conditions. For example, warm or cool shelters should be available where responders may don and doff their protective clothing.

Respiratory Protection

To discuss respiratory protection, it is important to acknowledge the performance of protective ensembles and garments (including respiratory protection) specific to WMD, which is described in NFPA 1990. This standard covers the performance of the garments and factors affecting the performance requirements for respiratory protection gear. This consideration is

Voice of Experience

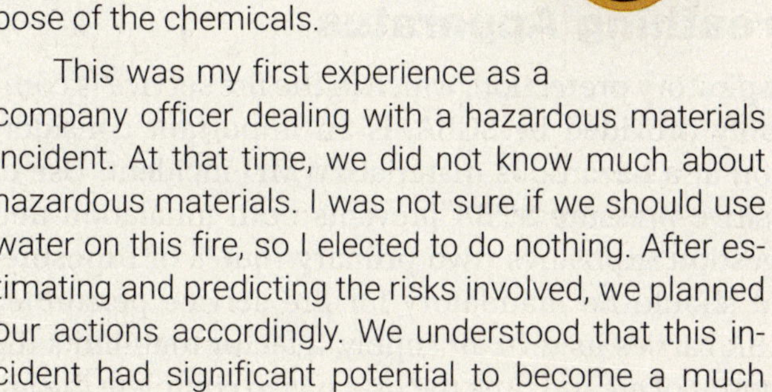

My engine company was dispatched to a small fire in a garbage dump. As we arrived on the scene, we could see a small fire burning about 200 yards (183 m) away. We dismounted and walked closer to determine whether to pull a line or try to hit the fire with the master stream. My engine carried 1000 gallons of water, so we felt confident that we could handle a refuse fire.

We were approximately 20 feet (6 m) away when bright red smoke began billowing from the burning pile of junk. I had never seen anything like it before. We immediately pulled back, and I ordered the crew to don SCBA. You do not have to be a chemist to understand that junk fire smoke is not typically bright red.

Common sense prevailed, and we exercised caution by moving the engine back toward the entrance road to stay upwind of the red smoke. I called dispatch and reported that we had a possible hazardous materials fire. I was told to stand by for a hazardous materials officer. When the hazardous materials officer arrived, it was decided that we should let the fire burn out. We later learned that someone had piled old tires around abandoned drums of unknown liquid and added a few gallons of gasoline. It was a cheap (and fast) way to dispose of the chemicals.

This was my first experience as a company officer dealing with a hazardous materials incident. At that time, we did not know much about hazardous materials. I was not sure if we should use water on this fire, so I elected to do nothing. After estimating and predicting the risks involved, we planned our actions accordingly. We understood that this incident had significant potential to become a much larger event.

This call convinced me that I needed to know more about hazardous materials. Shortly after the incident, I registered for a class on hazardous materials chemistry. I was later appointed to be the coordinator to start the hazardous materials team for our county.

Rick Emery
Lake County Hazardous Materials Team (Retired)
Vernon Hills, Illinois

critical when it comes to WMD events because some of the chemicals may cause the components of an air-purifying respirator (APR) or SCBA to fail, thereby exposing the responder to a highly toxic environment. Essentially, NFPA 1990 acknowledges that the entire ensemble is only as good as the individual components. Based on that criterion, NFPA requires that all components be certified to perform in a CBRN (chemical, biological, radiological, and nuclear) environment. NFPA 1990 typifies a thought process about PPE that should extend beyond the standard—namely, chemical-protective clothing should be thought of as a system.

The CBRN performance requirements for SCBA and APRs were born out of the terrorist attacks that occurred on September 11, 2001. In response to the growing threat of terrorism, and in addition to meeting the requirements of NFPA 1981, *Standard on Open-Circuit Self-Contained Breathing Apparatus (SCBA) for Emergency Services*, and NFPA 1986, *Standard on Respiratory Protection Equipment for Tactical and Technical Operations*, any SCBA or APR with a NIOSH CBRN certification must have passed a battery of tests that measured their performance against sarin and sulfur mustard. From a practical standpoint, these tests are intended to ensure that the components of the respiratory protection will stand up to the aggressive nature of these chemicals. To reiterate, NFPA 1990 is intended to serve as an integrated performance guideline, factoring in both the garment and the respiratory protection used.

FIGURE 33-8 SCBA carries its own air supply, which limits the amount of available work time to complete the job.

Positive-Pressure Self-Contained Breathing Apparatus

Respiratory protection, which in the fire service is commonly provided by SCBA, is an important consideration at a hazardous materials/WMD incident. Use of positive-pressure SCBA prevents both inhalation and ingestion exposures (two primary routes of exposure) and should be mandatory for fire service personnel. SCBA carries its own air supply, a factor that limits the amount of air and time the user has to complete the job (**FIGURE 33-8**).

The extra weight and reduced visibility are other factors to consider when choosing to wear SCBA. As with any piece of PPE, there are as many positive benefits as there are negative points to consider when determining whether SCBA is appropriate. Any responder called upon to wear an SCBA should be fully trained by the AHJ prior to operating in a contaminated environment. All responders should follow the manufacturers' recommendations for using, cleaning, filling, and servicing the units they use.

Supplied-Air Respirators

A **supplied-air respirator (SAR)**, also referred to as a **positive-pressure air-line respirator (with escape units)**, use an external air source such as a compressor or a compressed air cylinder. A hose connects the user to the air source and provides air to the face piece. SARs are useful during extended operations such as decontamination, clean-up, and remedial work.

SARs are equipped with a small "escape cylinder" of compressed air. Escape cylinders typically provide the user with approximately 5 minutes of breathing air. SARs may be less bulky and weigh less than SCBA, but the length of the air hose may limit movement, and there is potential for physical damage or perhaps chemical damage to the hose if it were to contact a released product (**FIGURE 33-9**).

Closed-Circuit SCBA

Some hazardous materials/WMD response teams use a form of respiratory protection referred to as a **closed-circuit self-contained breathing apparatus**. Commonly referred to as a "rebreather," this type of unit can

FIGURE 33-9 A supplied-air respirator (SAR) is less bulky than self-contained breathing apparatus (SCBA) but is limited by the length and structural integrity of the air hose.

be used when long work periods are required. The basic operating principle for this equipment is different than that for the SCBA, in that exhaled air is scrubbed free of carbon dioxide, supplemented with a small amount of oxygen, and "rebreathed" by the wearer. No exhaled air is released to the outside environment, making this type of unit a closed-circuit system. The earliest rebreathers were developed in the mid-1800s and were used primarily by mine workers.

Air-Purifying Respirators

An **air-purifying respirator (APR)** is a filtering device or particulate respirator that removes particulates, vapors, and contaminants from the air before it is inhaled (NFPA 1984). They should be worn only in atmospheres where the type and quantity of the contaminants are known and where sufficient oxygen for breathing is available. APRs should not be used in an IDLH atmosphere. These devices may be appropriate for operations involving volatile solids and for remedial clean-up and recovery operations where the type and concentration of contaminants are verifiable (**FIGURE 33-10**).

FIGURE 33-10 Air-purifying respirators (APRs) can be used only where there is sufficient oxygen in the atmosphere and the work environment is known and characterized.

APR devices range from full–face piece, dual-cartridge masks, to half-mask, face piece–mounted cartridges with no eye protection. APRs do not have a separate source of air, but rather filter and purify ambient air before it is inhaled. The models used in environments containing hazardous gases or vapors are commonly equipped with an absorbent material that soaks up or reacts with the gas. Consequently, cartridge selection is based on the expected contaminants. Particle-removing respirators use a mechanical filter to separate the contaminants from the air. Both types of devices require that the ambient atmosphere contain a minimum of 19.5 percent oxygen.

APRs are easy to wear, but they do have some drawbacks. Because the filtering cartridges are specific to the expected contaminants, these devices will be ineffective if the contaminant changes suddenly, possibly endangering the lives of responders. The air must also be continually monitored for both the known substance and the ambient oxygen level throughout the incident. For these reasons, APRs should not be employed at hazardous materials/WMD incidents until qualified personnel have tested the ambient atmosphere and determined that the devices can be used safely.

Powered Air-Purifying Respirators

A **powered air-purifying respirator (PAPR)** is similar in function to the standard APR described earlier but includes a small fan to help circulate air into the mask (**FIGURE 33-11**). The battery-powered fan unit is worn around the waist. The fan draws outside air through the filters and into the mask via a low-pressure hose (NFPA 1984). PAPRs are not considered to be true positive-pressure units like SCBA because it is possible for the wearer to "out-breathe" the flow of supplied air, thereby creating a negative-pressure situation inside the mask, and possibly allowing contaminants to enter the face mask due to a poor seal with the face. The main advantages of the PAPR are that it diminishes the wearer's work of breathing, helps reduce fogging in the mask, and provides a constant flow of cool air across the face.

Physical Capability Requirements

Hazardous materials/WMD response operations subject responders to a great deal of physiological and psychological stress. During the incident, personnel may be exposed to both chemical and physical hazards. They may face life-threatening emergencies, such as fire and explosions, or they may develop heat or cold stress while wearing protective clothing or working under extreme temperatures. For these reasons, every emergency response organization should have a comprehensive health and safety management program.

FIGURE 33-11 Powered air-purifying respirators (PAPRs) in use during a training exercise.

The components of a health and safety management system for hazardous materials responders are outlined in the OSHA HAZWOPER regulation. Briefly, a health and safety program should include the following broad elements:

- Medical surveillance, including pre-employment screening and periodic medical examinations
- Medical monitoring on scene, with treatment plans for acute on-scene illness and injury
- Thorough recordkeeping of all elements of the program
- A mechanism to periodically review the entire process

The medical surveillance piece of the overall program is the cornerstone of an effective health and safety management system for responders. The two primary objectives of a medical surveillance program are to determine whether an individual can perform their assigned duties, including use of personal protective clothing and equipment, and to detect any changes in body system functions caused by physical or chemical exposures.

As part of the medical surveillance program, responders should be examined by a physician once per year or biennially based on the physician's recommendations. During this examination, the physician may, among other things, perform a routine exam based on the expected tasks the employee may perform, including wearing PPE; conduct a health questionnaire; take X-rays; and perform a respiratory function test to measure lung capacity and function. The physician may also evaluate an individual's fitness for wearing SCBA and other respiratory protection devices. The specific requirements for any responder who is assigned to any duty where any form of respiratory protection will be used can be found in 29 CFR 1910.134, OSHA's respiratory protection standard. Additional guidance may be found in the NIOSH/OSHA/U.S. Coast Guard (USCG)/U.S. Environmental Protection Agency (EPA) document entitled *Occupational Safety and Health Guidance Manual for Hazardous Waste Site Activities.*

Medical monitoring and support differ in several ways from a medical surveillance program. Medical monitoring is the on-scene evaluation of response personnel who may experience adverse effects because of exposure to heat, cold, stress, or hazardous materials.

Such monitoring can quickly identify problems so they can be treated in a timely fashion, thereby preventing severe adverse effects, and maintaining the optimal health and safety of on-scene personnel. Medical monitoring criteria are covered in depth in NFPA 470, chapter 46, *Competencies for Hazardous Materials/WMD Basic Life Support (BLS) Responders*, and chapter 47, *Competencies for Hazardous Materials/WMD Advanced Life Support (ALS) Responders*.

Decontamination

Even though there should be no intentional contact with the hazardous material involved in a hazardous materials/WMD incident, a procedure or a plan must be established to decontaminate anyone who becomes contaminated (**FIGURE 33-12**). Remember that contamination is the transfer of "a hazardous material, or the hazardous component of a WMD, from its source to people, animals, the environment, or equipment, which can act as a carrier" (NFPA 470). (See Chapter 31, *Properties and Effects*, in this text for more information.)

Decontamination, then, is "the physical and/or chemical process of reducing and preventing the spread and effects of contaminants to people, animals, the environment, or equipment involved at hazardous materials/WMD incidents" (NFPA 470). The overall goal of any decontamination is to make personnel, equipment, and supplies safe by reducing or, in some cases, eliminating, the offending substances. Proper decontamination is essential at every hazardous materials/WMD incident to ensure the safety of personnel and property.

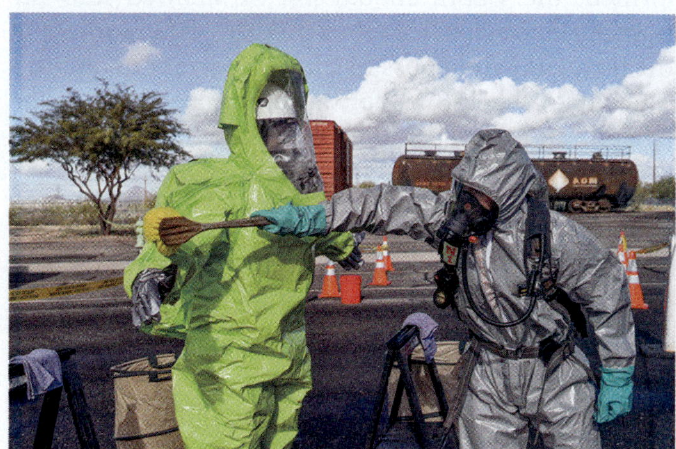

FIGURE 33-12 There must be a plan in place for decontamination at every hazardous materials/WMD incident.

TIP

Dry decontamination is accomplished without water and includes removing outer clothing and/or wrapping victim in any available fabric such as blankets, canvas tarps, etc.

These definitions of contamination and decontamination, when placed into context, should frame the concept and need for decontamination for any responder to a hazardous materials incident. Although many factors and influences surround the task of decontamination, the relationship of contamination to decontamination is clear—if contamination of PPE is suspected or confirmed, it must be removed efficiently in a *systematic* fashion. In short, decontamination is a process-driven activity, determined by the nature of the contaminant.

There are various types of decontamination, ranging from emergency decontamination to mass decontamination.

To begin, it is important to understand the different types of decontamination processes and when to employ each. **Emergency/field-expedient decontamination** is "the process of immediately reducing contamination of people in potentially life-threatening situations with or without the formal establishment of a decontamination corridor" (NFPA 470). The goal is to reduce the effect of an exposure and get a victim clean enough to receive medical care from first responders and, if needed, be admitted to a receiving hospital. Victims who receive emergency decontamination often must remove some or all of their clothing to reduce the harm posed by the contaminant.

SAFETY TIP

All attempts should be made to understand the physical and chemical properties of a contaminant prior to beginning decontamination. Some hazardous materials are water reactive and may require a special decontamination process.

Mass decontamination is emergency decontamination on large numbers of people with the same goal as emergency decontamination—to remove the contaminants as quickly as possible to reduce the health effects of a chemical exposure (NFPA 470). Basically, the difference between mass decontamination and emergency/field-expedient decontamination is the number of people who are affected.

FIGURE 33-13 An example of a gross decontamination setup. This apparatus might be placed at the entrance of the technical decontamination line.

Technical decontamination is "the planned and systematic process of reducing contamination to a level that is as low as reasonably achievable" (NFPA 470). Technical decontamination may involve several stations or steps using water or a special cleaning solution. It takes place in a defined area from which noncontaminated persons are restricted from entering.

Gross decontamination is decontamination that takes place as soon as possible within a controlled area (NFPA 470). It consists of a prewash, most often accomplished by mechanical removal of the contaminant from tools, equipment, PPE, or vehicles, or initial rinsing from hand-held hose lines, emergency showers, or other nearby sources of water, before technical decontamination takes place (**FIGURE 33-13**).

Emergency/ Field-Expedient Decontamination

Emergency/field-expedient decontamination needs to occur immediately in potentially life-threatening situations. The major difference between formal decontamination and emergency/field-expedient decontamination is that emergency/field-expedient decontamination is intended to quickly separate as much of the contaminant as possible from the individual to minimize exposure and injury; indeed, that is its sole purpose. Thus, emergency decontamination is the process of quickly reducing or removing the bulk of contaminants from a person or animal as rapidly as possible. This procedure is undertaken in potentially life-threatening situations without the establishment of a more formal and detailed decontamination process, although such a process may follow later. Emergency decontamination usually involves removing contaminated clothing and dousing the victim with large quantities of water.

TIP

By removing the appropriate clothing of a contaminated victim, you have significantly reduced the exposure.

If an emergency/field-expedient decontamination area has not been designated, responders should isolate the exposed victims in a contained area and establish an appropriate location. If possible, try to prevent the run-off from getting into drains, streams, or ponds; instead, divert the stream of water into an area where it can be treated or disposed of later. Emergency/field-expedient decontamination can easily be accomplished from a fire engine. Simply pull a handline, maneuver it into a circle, throw a tarp over the middle of the circle, and pull a booster line or other small handline to accomplish the decontamination. This technique is a quick way to handle a conscious victim of a chemical exposure or a responder who may have contacted a chemical and requires PPE to be decontaminated.

If adequate decontamination has not been performed, the victim should not be allowed into the transport ambulance or the emergency department at the hospital. When possible, it is important to obtain a safety data sheet (SDS) for the substance to which the person was exposed and to make sure the information goes to the hospital with the patient.

Always ensure your own safety first before attempting decontamination of others. Avoid touching contaminated victims and/or entering contaminated environments without the proper level of protection. You will be no help in resolving the incident if you become a victim, too.

Refer to the scenario at the beginning of the chapter. Would the victim be a candidate for emergency decontamination? If so, how could that process be accomplished at a fixed facility? Most fixed facilities that use chemicals will have emergency safety showers and eyewash stations available to the employees. These sites may be the most easily accessible locations in which responders can perform emergency decontamination. The water is clean and readily available; also, employees of such a facility have typically been trained to use the safety showers (**FIGURE 33-14**).

An effective emergency decontamination operation depends on an understanding of the contaminant and its chemical and physical properties. To perform emergency decontamination, follow the steps in **SKILL DRILL 33-1**.

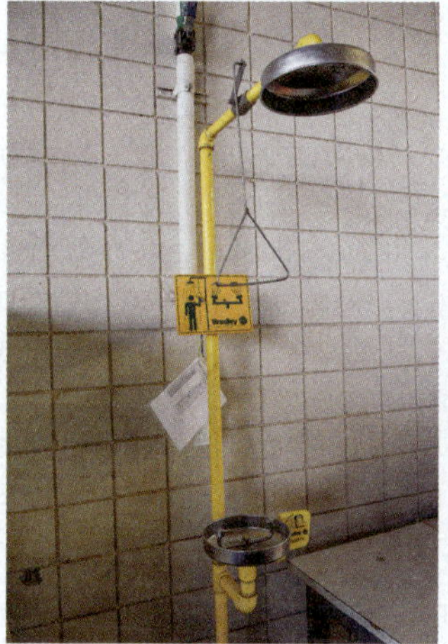

A.

B.

FIGURE 33-14 **A.** Indoor safety showers are in rooms that have no drains. Be prepared for a large clean-up process after activating a safety shower. **B.** Some safety showers also have eyewash stations.

SKILL DRILL 33-1

Performing Emergency/Field-Expedient Decontamination NFPA 470: 7.3.1, 7.4.1, 7.5.1

1. Confirm that you (the responder) have the appropriate PPE to protect against the contaminant. Stay clear of the product, and avoid physical contact with it if possible. Try to contain run-off by directing the victim out of the hazard zone and into a suitable location for decontamination. It is not imperative to capture the run-off when you are in the process of performing emergency decontamination on an exposed person—removing the contaminant from the person is more important than capturing the run-off. Instruct the victim in removing contaminated clothing. If the victim is capable of removing their own clothing, this is the preferred method. It ensures the assisting responder will stay as clean as possible.

2. Rinse the victim with copious amounts of water. Avoid using water that is too warm or too cold; room-temperature water is best. Provide or obtain medical treatment for the victim and arrange for the victim's transport.

CASE STUDY

You Are the Firefighter CONCLUSION

Your paramedic ambulance and local engine company have been called to a semiconductor fabrication facility for a report of shortness of breath. Both units arrive at the reception area and are led to an employee break room. As you approach the seated patient, you notice an odor of ammonia. The laboratory manager is standing next to the patient—a laboratory technician—who is doubled over in the chair. The laboratory manager tells you that the technician was splashed in the face with approximately 100 mL of ammonium hydroxide while pouring chemicals into an instrument. He states that he helped the laboratory technician to an eyewash station immediately after the incident and then escorted him to the break room. The laboratory manager tells you he has seen this sort of injury before. In his opinion, the laboratory technician doesn't need to go to the hospital; he just needs some oxygen. The paramedic begins an assessment and finds that the laboratory technician has reddened skin over his entire face and is complaining of shortness of breath. You are the officer on the engine company, and your four-person crew is trained to the operations level.

1. **What are the initial response objectives for the engine crew and ambulance once the ammonium hydroxide exposure is discovered?**

   Answer: Patient care is top priority in this situation. Emergency decontamination was performed prior to arrival but the presence of the ammonia odor may indicate further decontamination might be needed. A responder should return to the eyewash station to confirm whether the patient's clothing is contaminated. Also, the engine company crew should locate the laboratory where the spill occurred, isolate and deny entry, and size up the scene for the next steps once patient care is complete. As with any response, adherence to local policies and procedures is a must.

2. **What are the initial objectives for treatment and transport of the patient?**

   Answer: Confirm that the patient is fully decontaminated, provide basic life support (or advanced life support if required), and transport to the hospital. Send an SDS to the hospital if possible. At minimum, provide the EMS personnel with the exact chemical name, and in this case, the concentration of the ammonium hydroxide.

3. **Are the ambulance and engine crew at risk of being exposed or contaminated to potentially toxic levels of ammonium hydroxide from the patient?**

   Answer: Ammonium hydroxide has a relatively low odor threshold (starting at approximately 50 ppm), providing good warning properties of its presence. If the odor is not irritating to the eyes or mucous membranes of the patient or any responder, the airborne levels will most likely not be in the toxic range.

WRAP-UP

SUMMARY

KNOWLEDGE OBJECTIVES

- Explain how to estimate the potential harm or severity of an incident. (**NFPA 470: 7.2.1**, pp. 1320–1322)
  - Describe the categories that measure the level of risk an agent poses to the human body.
- Describe how to plan an initial response. (**NFPA 470: 7.2.1**, pp. 1322–1324)
  - Describe the tasks that are performed during the initial response.

- Explain how exposures might be affected by various types of hazardous materials incidents. (**NFPA 470: 7.4.1**, pp. 1324–1329)
  - List the five protective actions at the operations level.
- Describe how to report the size and scope of the incident. (**NFPA 470: 7.3.1**, p. 1329)
- Describe how to select personal protective equipment for an incident. (**NFPA 470: 7.3.1, 7.4.1**, pp. 1329–1331)
  - Describe the risks of excessive heat exposure and cold-temperature exposure when wearing personal protective equipment.
- Identify and describe the types of personal protective equipment needed for hazardous materials incidents. (**NFPA 470: 7.3.1**, pp. 1329–1331)
- Explain the role of respiratory protection. (**NFPA 470: 7.3.1, 7.4.1**, pp. 1331–1335)
  - Describe the five types of respiratory protection that may be utilized at a hazardous materials incident.
- Describe the basic types of decontamination. (**NFPA 470: 7.3.1, 7.4.1, 7.5.1**, pp. 1335–1336)

SKILLS OBJECTIVES

- Perform emergency decontamination. (**NFPA 470: 7.3.1, 7.4.1, 7.5.1**, pp. 1336–1337)

KEY TERMS

air-purifying respirator (APR) A respirator that removes specific air contaminants by passing ambient air through one or more air-purification components. (NFPA 1984)

American Conference of Governmental Industrial Hygienists (ACGIH) A scientific organization that researches and publishes information in the field of industrial hygiene.

area of safe refuge An area where people are held temporarily until they can be safely decontaminated, treated, or removed.

closed-circuit self-contained breathing apparatus Self-contained breathing apparatus that recycles the user's exhaled air by removing carbon dioxide and generating fresh oxygen.

decontamination The physical and/or chemical process of reducing and preventing the spread and effects of contaminants to people, animals, the environment, or equipment involved at hazardous materials/weapons of mass destruction (WMD) incidents. (NFPA 470)

dehydration A medical condition that occurs when the body loses more water than it takes in.

denial of entry A policy under which, once the perimeter around a release site has been identified and marked, responders limit access to all but essential personnel.

emergency/field-expedient decontamination The process of immediately reducing contamination of people in potentially life-threatening situations with or without the formal establishment of a decontamination corridor. (NFPA 470)

evacuation The removal or relocation of those individuals who may be affected by an approaching release of a hazardous material.

gross decontamination The phase of the decontamination process that takes place as soon as possible and before technical decontamination within a decontamination corridor, during which significant reduction of the amount of surface contamination is removed with a prewash most often accomplished by mechanical removal of the contaminant or initial rinsing from handheld hose lines, emergency showers, or other nearby sources of water. (NFPA 470)

heat exhaustion A mild form of shock that arises when the circulatory system begins to fail because the body is unable to dissipate excessive heat and becomes overheated.

heat stroke A severe and potentially fatal condition resulting from the failure of the body's temperature-regulating capacity caused by exposure to the sun or high temperatures.

immediately dangerous to life and health (IDLH) The atmospheric concentration of any toxic, corrosive, or asphyxiant substance such that it poses an immediate threat to life or could cause irreversible or delayed adverse health effects.

KEY TERMS CONTINUED

isolation of the hazard area The process of identifying a perimeter around a contaminated atmosphere so that responders and civilians are kept at a safe distance from the release site.

mass decontamination The physical process of reducing or removing surface contaminants from large numbers of victims in potentially life-threatening situations in the fastest time possible. (NFPA 470)

National Institute for Occupational Safety and Health (NIOSH) Part of the U.S. Department of Health and Human Services; the department charged with providing information, training, research, and education in the field of occupational safety and health so as to ensure that individuals have a safe and healthy work environment.

permissible exposure limit (PEL) The established standard limit of exposure to a hazardous material based on the maximum time-weighted concentration at which 95 percent of exposed, healthy adults suffer no adverse effects over a 40-hour workweek.

positive-pressure air-line respirator (with escape units) See *supplied-air respirator (SAR).*

powered air-purifying respirator (PAPR) A respirator that uses a powered blower to force the ambient air through one or more air-purifying components to the respiratory inlet covering. (NFPA 1984)

recommended exposure level (REL) The maximum time-weighted concentration of material to which 95 percent of healthy adults can be exposed without suffering any adverse effects over a 40-hour workweek.

shelter-in-place A method of safeguarding people in a hazardous area by keeping them in an enclosed atmosphere, usually inside structures.

supplied-air respirator (SAR) A respirator with a hose that connects the user to a remote air source such as a compressor or storage cylinder; also called *positive-pressure air-line respirator (with escape units).*

technical decontamination The planned and systematic process of reducing contamination from responder PPE to a level that is as low as reasonably achievable. (NFPA 470)

threshold limit value (TLV) The point at which a hazardous material or weapon of mass destruction begins to affect a person.

threshold limit value/ceiling (TLV/C) The concentration of hazardous material that should not be exceeded because exposure, even for an instant, can cause ill effects to a person.

threshold limit value/short-term exposure limit (TLV/STEL) The maximum concentration of hazardous material that a person can be exposed to in 15-minute intervals, up to four times a day, with a minimum of 1 hour between exposures; the lower the TLV/STEL value, the more toxic the substance is.

threshold limit value/skin (TLV/skin) The concentration at which direct or airborne contact with a material could result in possible and significant exposure from absorption through the skin, mucous membranes, and eyes.

threshold limit value/time-weighted average (TLV/TWA) The maximum airborne concentration of a material to which a person can be exposed for 8 hours a day, 40 hours a week, and not suffer any ill effects.

REVIEW QUESTIONS

1. Describe the hazardous materials/WMD incident response objectives when responding to a hazardous materials incident.
2. What is the difference between PEL and TLV?
3. Why should responders choose a route to approach a hazardous materials incident that is upwind and upgrade from the incident site?
4. What must happen before firefighters can perform search and rescue?
5. What units are used when the extent of a hazardous materials release is assessed?
6. What should hazardous materials responders be aware of in order to avoid serious short- and long-term medical issues resulting from increased body temperature when wearing PPE?
7. Identify five types of respiratory protection.
8. Identify four types of decontamination.
9. What is the intent of emergency/field-expedient decontamination?

DISCUSSION QUESTIONS

1. What are some of the situational factors that should be considered during the size-up of a hazardous materials incident?
2. Is evacuation always the best strategy? Why or why not?
3. Under what circumstances might and incident commander choose to employ defensive operations at a hazardous materials release?

APPLYING THE CONCEPTS

It is 0900 hours when your three-person engine crew is dispatched to a potential chemical spill at a nearby manufacturing facility. The entire crew is certified in basic life support (BLS) and trained to the operations level.

1. Is there enough information to make any general assessments based on the initial dispatch?
2. What initial actions could you take?

Upon your arrival, the IC of the site emergency response team reports that a worker intentionally mixed sodium cyanide and sulfuric acid in a 5-gallon container inside a closed 20- × 20-foot (6- × 6-m) storage room to create cyanide gas. The IC gives you a handwritten note from the worker, which indicates that the chemical release was an apparent suicide attempt. There is a security camera monitoring the room, and the IC tells you the worker is on the floor next to the bucket, not moving. The entire building has been evacuated.

3. Based on this information, what is your operational priority?
4. As an operations-level responder with no additional mission-specific competency training, is it within your level of training to safely enter the storage room and assess the worker?

The IC states that the room is well ventilated and urges you to attempt an immediate rescue. He hands you two SDSs—one for sodium cyanide and one for 97 percent sulfuric acid. As you read the documents, the IC is getting more agitated, demanding you go in and try to save the worker. He states repeatedly that you have SCBA and are not doing your job if you do not go in and starts videoing you with his smartphone.

5. Based on his statement, would you attempt to make entry with SCBA and full structural firefighter PPE?
6. If the IC continues to be aggressive, what actions can you take?

REFERENCES

National Fire Protection Association (NFPA). 2016. *NFPA 1986, Standard on Respiratory Protection Equipment for Tactical and Technical Operations.* 2017 Edition. Quincy, MA: NFPA.

National Fire Protection Association (NFPA). 2018. *NFPA 1981, Standard on Open-Circuit Self-Contained Breathing Apparatus (SCBA) for Emergency Services.* 2019 Edition. Quincy, MA: NFPA.

National Fire Protection Association (NFPA). 2021. *NFPA 470, Hazardous Materials/Weapons of Mass Destruction (WMD) Standard for Responders.* 2022 Edition. Quincy, MA: NFPA.

National Fire Protection Association (NFPA). 2021. *NFPA 1984, Standard on Respirators for Wildland Fire-Fighting Operations and Wildland Urban Interface Operations.* 2022 Edition. Quincy, MA: NFPA.

National Fire Protection Association (NFPA). 2021. *NFPA 1990, Standard for Protective Ensembles for Hazardous Materials and CBRN Operations.* 2022 Edition. Quincy, MA: NFPA.

National Institute for Occupational Safety and Health (NIOSH), Occupational Safety and Health Administration (OSHA), U.S. Coast Guard (USCG), and U.S. Environmental Protection Agency (EPA). "Occupational Safety and Health Guidance Manual for Hazardous Waste Site Activities." Centers for Disease Control and Prevention. October 1985. Accessed December 6, 2021. https://www.cdc.gov/niosh/docs/85-115/default.html.

CHAPTER

34

Hazardous Materials Operations Level

Implementing the Planned Response

NOTE: Content within this chapter meets the intent of **NFPA 470, 2022 Edition**, which includes chapter 7, "Professional Qualifications for Hazardous Materials/WMD Operations Level Responders (**NFPA 1072**)."

KNOWLEDGE OBJECTIVES

After studying this chapter, you will be able to:

- Size up an incident.
- Identify and describe the safety procedures at a hazardous materials incident.
- Explain the role of the operations-level responder in implementing a planned response.
- Identify and describe the components of the incident command system.

SKILLS OBJECTIVES

There are no skills objectives for operations-level personnel for this chapter.

CASE STUDY

You Are the Firefighter

Local law enforcement is requesting a response for a suspicious package leaking a liquid. The package is located in a small storeroom on the ground floor of a four-story office building. While you are en route, the dispatcher announces this update: "Engine 40, law enforcement on scene is reporting several people complaining of eye irritation and nausea. There are also reports of an unusual odor on the second and third floors. A full building evacuation is in progress."

1. Based on the initial reports, what would your initial actions be upon arrival?
2. Working with the on-scene law enforcement is important. What would you do to facilitate a coordinated approach to managing and jointly commanding this incident?
3. Does your agency have the capabilities to communicate via radio with other public safety agencies (various law enforcement agencies, for example)?

Introduction

Scene control is important at all incidents. At a hazardous materials incident, however, scene control is paramount because it influences both scene security and personnel accountability. Typically, the starting point for implementing any response is sizing up the situation and taking control of the affected area. **Size-up** is the process of rapidly evaluating the critical visual indicators of an incident (what is happening now), analyzing that information based on your training and experience (what might happen next), and arriving at a conclusion that will serve as the basis for forming and implementing an action plan (what you are going to do about the situation).

An **action plan** is a detailed proposal describing the strategies and tactics within the incident action plan (IAP) that will safely accomplish the incident objectives, favorably influence outcomes, and increase the safety of both responders and the public. Sometimes the action plan includes an aggressive offensive posture—that is, attack the problem. In other cases, a defensive or nonintervention posture is appropriate—that is, isolate the scene, protect exposures, and allow the incident to stabilize on its own. The posture chosen should be driven by prudent decisions based on an accurate size-up and your level of training. Thus, the initial actions taken at the scene set the tone for the response and are critical to the overall success of the effort.

Reconnaissance at an incident is a visual scan and exploration of a physical area to gain actionable information to guide incident decision making. Size-up is a form of incident reconnaissance, intended to help responders form an initial impression of the incident and formulate an action plan, and is always a work in progress. In other words, as the incident progresses, everyone on the scene should be constantly sizing up their own piece of the problem. All responders should maintain good situational awareness and understand the actions they are taking. Remember that responder safety and life safety of the public are the most important priorities.

Incidents are seldom black-and-white problems; more typically, they involve many shades of gray that require you to *think*. It is always essential that you understand your job and know how to use the tools at your disposal both effectively and appropriately.

TIP

Think before you act or give a command to others to act. Do not put yourself or others at undue risk or become part of the problem. Your safety is the number one priority!

TIP

Situational awareness refers to the degree of accuracy to which a person's perception of the current operating environment mirrors reality. In essence, situational awareness is the act of processing all the information available to you and understanding how it fits together in the big scheme of things.

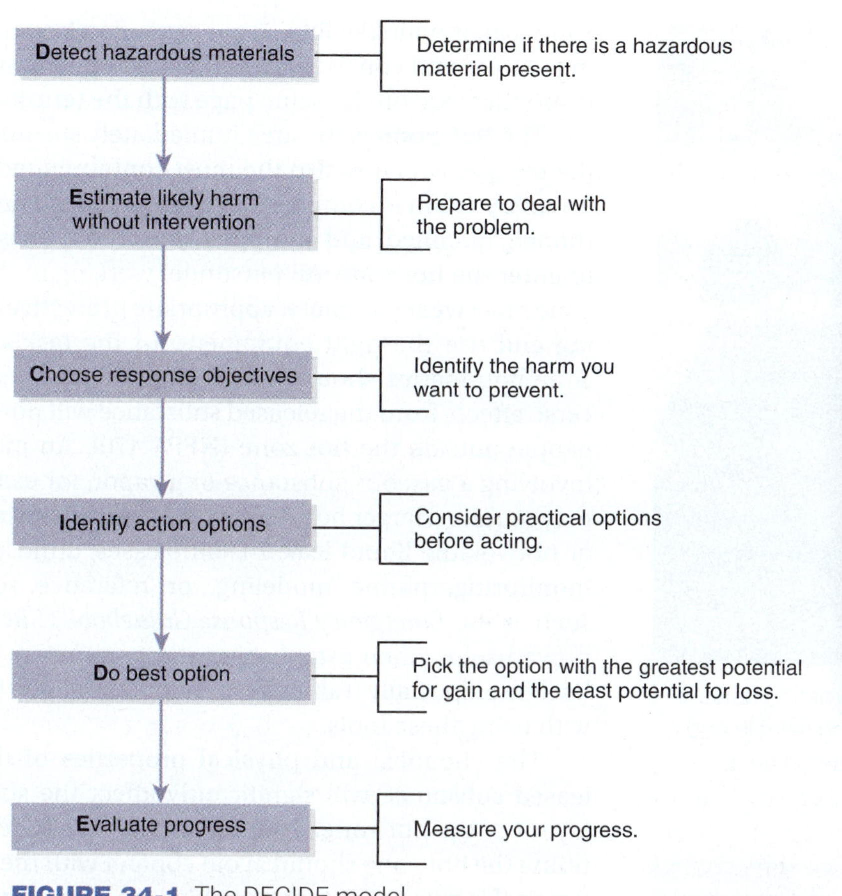

FIGURE 34-1 The DECIDE model.

Data from the DECIDE model by Ludwig Benner.

TIP

Do not let pressure from incoming units push you to make hasty decisions or assignments. Take time to think and formulate a well-considered plan of attack.

It is important to follow a decision-making algorithm when facing a hazardous materials incident (**FIGURE 34-1**). Use this algorithm as a loose guide for developing an action plan and to focus your thinking.

Scene Control Procedures

Managing a hazardous materials incident by establishing **control zones** and limiting access to the incident site helps reduce the number of civilians and public service personnel who may be exposed to the released substance. **Hazard control zones**, or simply control zones, are areas designated at a hazardous materials/ weapons of mass destruction (WMD) incident based on the chemical and physical properties of the released material, the environmental factors at the time of the release, and the general layout of the scene (NFPA 470). Of course, isolating a city block in a busy downtown area of a large city presents far different challenges than isolating the area around a rolled-over cargo tank on an interstate highway. Each situation is different, requiring you to be flexible and thoughtful about how you secure the area. Securing access to the incident helps ensure that responders arriving after the first-due units do not accidentally enter a contaminated area.

If the incident takes place inside a structure, the best place to control access is at the normal points of ingress and egress—doors and windows. Once they are secured so that no unauthorized personnel can enter the building, appropriately trained response crews can begin to isolate other areas as appropriate.

The same concept applies to outdoor incidents. The goal is to secure logical access points around the hazard. Begin by controlling intersections, on/off-ramps, service roads, or other access points (including pedestrian routes) to the scene. Law enforcement personnel should block off streets, close intersections and sidewalks, or redirect traffic as needed (**FIGURE 34-2**).

FIGURE 34-2 Law enforcement officers might assist at the hazardous materials incident by diverting vehicle and foot traffic at a safe distance outside the hazard area.

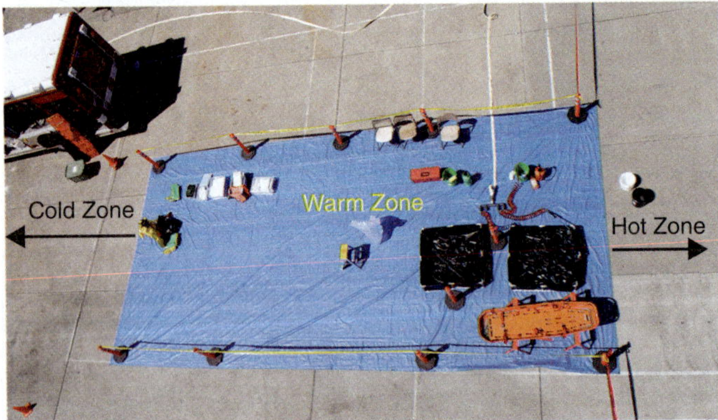

FIGURE 34-3 Control zones designated with an established decontamination area.

Typically, control zones at hazardous materials incidents are labeled as hot, warm, or cold (**FIGURE 34-3**). You may also hear other terms used, such as **exclusionary zone** (hot zone), **contamination reduction zone** (warm zone), or **outer perimeter** (cold zone). Both sets of terms are correct and common, so make sure you understand the terminology used in your jurisdiction. Be prepared to discover that different jurisdictions use terminology and setup procedures unlike the ones used by your agency. If you understand the concepts behind the actions and remember that safety is the main focus, the act of setting up and naming your zones can remain flexible. Avoid confusion by performing regular and consistent training and making sure all responders are on the same page with the terminology.

The **hot zone** is the area immediately surrounding the release, which is also the most contaminated area. An **entry team** is composed of personnel who are fully trained, qualified, and equipped, and who are assigned to enter the hot zone. All personnel working in the hot zone must wear complete, appropriate protective clothing and use the right equipment for the task(s). Hot zone boundaries should be set large enough that adverse effects from the released substance will not affect people outside the hot zone (NFPA 470). An incident involving a gaseous substance or a vapor, for example, may require a larger hot zone than one involving a solid or nonvolatile liquid leak. In some cases, atmospheric monitoring, plume modeling, or reference sources such as the *Emergency Response Guidebook* (*ERG*) may prove useful when establishing the parameters of the hot zone. Specially trained responders should be tasked with using these tools.

The chemical and physical properties of the released substance will significantly affect the size and layout of the hot zone. Additionally, all responders entering the hot zone should avoid contact with the product to the greatest extent possible—an important goal that should be clearly understood by the personnel entering this area. Adhering to this policy makes the job of decontamination easier and reduces the risk of secondary contamination.

Personnel accountability is important, so access into the hot zone must be limited to only those persons necessary to control the incident. All personnel, tools, and equipment must be decontaminated when they leave the hot zone. This practice ensures that contamination is not inadvertently spread to "clean" areas of the scene.

The **warm zone** is where personnel and equipment transition into and out of the hot zone. It contains control points for access to the hot zone as well as the **decontamination corridor** (NFPA 470). The decontamination corridor, typically used for responder decontamination, is a controlled area within the warm zone where decontamination is normally performed (NFPA 470). Only the minimal amount of personnel and equipment necessary to perform decontamination, or to support those operating in the hot zone, should be permitted in the warm zone.

Generally, personnel working in the warm zone can use personal protective equipment (PPE) whose protection is rated one level lower than that of the PPE used in the hot zone. For example, if personnel in the hot zone are wearing Level A ensembles, personnel in the warm zone should don at least Level B protection. This recommendation is not a hard-and-fast rule, however. You must understand the hazard well enough to make that determination on a case-by-case basis.

TIP

The primary functions of warm zone activities are decontamination and providing ready support to personnel operating in the hot zone.

Beyond the warm zone is the cold zone, which is "the hazard control zone of hazardous materials/WMD incidents that contains the incident command post and such other support functions as are deemed necessary to control the incident" (NFPA 470). The **cold zone** is a safe area where personnel do not need to wear any special protective clothing for safe operation. Personnel staging; the command post; emergency medical services (EMS) providers; and the area for medical monitoring, support, and treatment after decontamination are all located in the cold zone. Typically, support personnel for the incident operate in this area.

It is not uncommon to set large control zones at the onset of an incident, only to discover later that their boundaries may have been established too liberally. At the same time, control zones should not be defined too narrowly (**FIGURE 34-4**). As the incident commander (IC) gets more information about the specifics of the chemical or material involved, the control zones may be changed. Ideally, these zones will be established in the right place, geographically speaking, the first time. This may be accomplished by understanding the state of matter or the nature of the material, by undertaking visual reconnaissance, or perhaps based on the results of air monitoring. In all cases, you should be prepared to expand or contract control zones as necessary. Wind shifts are a common reason why control zones are modified during the incident. If there is a prevailing wind pattern or a predicted shift based on the time of day in your area, factor that consideration into your decision making when it comes to control zones.

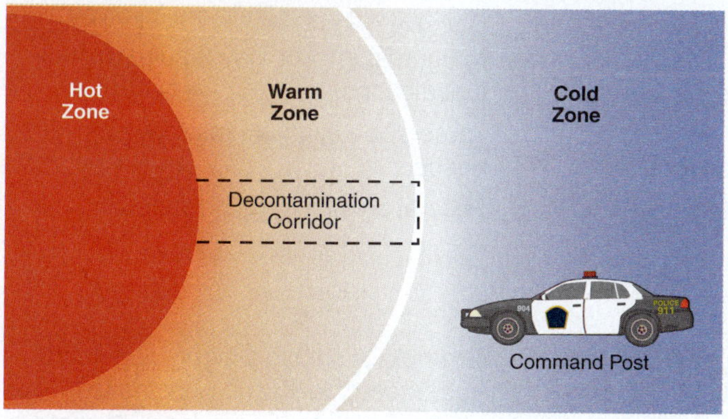

TIP

It is generally quicker, easier, and safer to reduce the size of the control zones than to enlarge them.

Response Safety Procedures

Isolation of the release area is one of the first measures that operations-level responders should implement (**FIGURE 34-5**). For example, access to the scene can be controlled by stretching banner tape across roads to divert traffic away from a spill or to prevent civilians or other responders from entering or reentering a contaminated building or approaching a rolled-over cargo tank. Law enforcement or site security personnel can be posted in safe areas to prevent bystanders from wandering into a contaminated area. At some incidents, scene control may be one of your biggest response challenges and achieving it could require a coordinated effort among fire, law enforcement, EMS, and other agencies. Be generous with your scene control zones but take care not to mark off more real estate than you can control!

Responders should consult the *ERG* for guidance on protective actions or as a starting point for gathering

Hot Zone
Warm Zone
Cold Zone
Decontamination Corridor
Command Post

FIGURE 34-4 Control zones spread outward from the hot zone.

FIGURE 34-5 Isolating the area of the release is a vital step in gaining control of the incident.

information about a released substance. Other initial actions may include evacuating people from the immediate area, implementing a shelter-in-place strategy, rendering emergency medical care in a safe location away from the release, and surveying the scene to detect indicators of a hazardous materials/WMD release.

These basic protective actions should be taken immediately, while maintaining your own safety. For the first several minutes, not even the first responders themselves should enter the area unless the substance can be accurately identified and the responders are adequately staffed, properly protected, and appropriately trained to take action. During this period of time, responders can begin to gather necessary scene information and take initial steps to identify the materials involved (size-up). Ideally, size-up can be completed from a safe distance, away from potential contamination. Any actions taken should follow the predetermined local response plan or your agency's response guidelines. Additionally, each authority having jurisdiction (AHJ) should have clearly defined policies and procedures for reporting the status of the response through the normal chain of command. This is a vital factor for ensuring that everyone on the incident is informed of the current conditions and any changes (e.g., the incident is escalating or getting worse, the incident is under control) that might necessitate modifying the IAP. In concert with passing incident status reports up the chain of command, the AHJ should identify the methods for immediate notification of the IC and other response personnel about critical incident conditions at the incident. Also, if any responders notice an unsafe situation, this information should be communicated up the chain of command as soon as possible.

As a responder, you should be familiar with all emergency response plans for hazardous materials/WMD incidents that may occur within your jurisdiction. If none exists, as a responder you should default to the scope of practice established by your training when you are called into service, regardless of the nature of the incident.

It is common for a jurisdiction to predetermine response levels and response configurations to anticipated hazards. To that end, agencies use a wide variety of methods to identify and denote the severity of certain types of incidents. Establishing these thresholds takes the guesswork out of determining the type and number of resources required to handle a problem, as well as the levels of training required for the responding personnel. **TABLE 34-1** illustrates how an agency might choose to denote the different levels of response. For obvious reasons, the role of the operations-level responder will differ depending on the severity of the incident.

A quick and easy format you can use for communicating your status is the CAN report. CAN—an acronym for "conditions, actions, and needs"—is an easy way to remember the process for giving a quick, concise briefing or update. *Conditions* refer to your current status or perhaps the status of the incident (such as progress reports; signs or other indicators that the status of the incident is improving, deteriorating, or staying the same;

TABLE 34-1 Hazardous Materials Incident Levels

Level I: Lowest Level of Threat	Level II: Medium Level of Threat	Level III: Highest Level of Threat
A small amount of a low-toxicity/low-hazard substance is involved. The incident can usually be handled by a single resource such as an engine company. Appropriate level of protection includes turnout gear and self-contained breathing apparatus (SCBA).	An organized hazardous materials team is needed. Additional chemical-protective clothing will be required. Civilian evacuations may be required. Decontamination may need to be performed.	Highly toxic chemicals are involved. Mitigation efforts may require multiple jurisdictions. Large-scale evacuations may be needed. Federal agencies will be called in.
Example: A small gasoline spill from a motor vehicle accident.	Example: A tanker carrying sulfuric acid has overturned in a tunnel and is leaking onto the freeway.	Example: A ship in a highly populated harbor catches fire and begins to release chlorine vapors from its cargo area.

or circumstances where it would be prudent to withdraw from offensive activities). *Actions* are what you are doing, what is being done by the team, or what is occurring with the incident (largely based on the policies and procedures of the AHJ). *Needs* include any additional resources or actions that are needed to accomplish a task. Usually, needs include requests for more people or more time.

The CAN report allows you (no matter what job function you have) to distill a message down to its most basic parts. CAN reports are useful to use in the following circumstances:

- You are briefing others about a task or the status of the incident.
- The IC is requesting information from individuals or teams engaged in the incident.
- The IC, group leader, or any other leadership function needs to broadcast information to all responders within a group or on the scene. Safety messages or general information can be disseminated via a CAN report.

The following is an example of a CAN report from an entry team to the entry team leader:

- Conditions: "Entry team 1 to entry team group leader. Be advised we have made access to the control room at the rear of the building and have identified the leaking valve."
- Actions: "We have confirmed that this is the valve identified in the safety briefing and are starting to work on stopping the leak."
- Needs: "It looks like we need a bigger pipe wrench, too, but we don't have one that size in the toolbox. We will come back to the edge of the hot zone in 5 minutes and pick up a different one."

A CAN report can contain any relevant content, but it is a good format to help keep communications crisp, clear, and concise.

Another important initial safety action might consist of identifying and securing any possible ignition sources—especially when the incident involves the release of flammable materials. To ensure that they do not create an unintentional ignition source, responders should use only radios and other electrical devices that have been certified as intrinsically safe (**FIGURE 34-6**). With intrinsically safe electrical devices, the thermal and electrical energy within the device always remains low enough that a flammable atmosphere could not be ignited. Typically, a portable radio that is certified to be intrinsically safe must be specifically wired and used in conjunction with an intrinsically safe battery.

FIGURE 34-6 All intrinsically safe radios and batteries are marked by the factory with a specific label denoting them as such.

Safety Briefings

Typically, before significant actions are taken at a hazardous materials incident, the IC should ensure that a written site safety and control plan is completed, and a verbal safety briefing is performed. The purpose of the safety briefing is to inform (at a minimum) all responders of the health hazards that are known or anticipated, the incident objectives, emergency medical procedures, radio frequencies and emergency signals, a description of the site, and the PPE to be worn. Each AHJ should develop templates for site safety plans and verbal safety briefings, which should also include the protocols and/or procedures for using approved communications tools and equipment provided by the AHJ and the policies and procedures for contacting and cooperating with outside agencies. In an incident where different entities are operating (e.g., law enforcement, fire, and EMS), perhaps at the scene of an incident with criminal intent, all roles and responsibilities between the specific agencies should be clearly identified. The use of a standardized form means that the safety plan can be completed in a timely manner; standardized formats also keep safety briefings on track and provide a logical format that everyone can follow.

There are many ways to conduct a safety briefing. Chiefly, a good safety briefing should be carried out in the manner that its name implies—it should be brief. It should be complete enough to provide the responders on the scene with relevant information, yet not overly detailed or filled with useless or "nice to know" information (**FIGURE 34-7**).

TIP

A written site safety and control plan may be suspended if a rescue becomes necessary. A plan is still required, but it may take the form of a short verbal plan and a rapid safety briefing.

FIGURE 34-7 All personnel must be briefed before approaching the hazard area or entering the hot zone.

At small incidents where the incident management structure remains simple, the IC may be responsible for both putting together a site safety plan and conducting the safety briefing. At larger incidents, additional officers may be needed. An **incident safety officer (ISO)** (also referred to as a **safety officer**) is "a member of the command staff responsible for monitoring and assessing safety hazards or unsafe situations and for developing measures for ensuring personnel safety" (NFPA 1026). A **hazardous materials safety officer**, sometimes referred to as an **assistant safety officer (ASO)**, is the person who works within an incident management system (IMS) in the **hazardous materials branch** or group to ensure that recognized safe practices are followed at hazardous materials/WMD incidents (NFPA 470). If an ISO or a hazardous materials safety officer is appointed, that person is responsible for participating in the preparation and implementation of the site safety plan. The roles of these officers are discussed in more detail later in this chapter.

The depth and scope of the safety briefing should have a direct correlation with the severity and complexity of the incident. For significant incidents, where the released substances are highly toxic or where other significant health hazards are present, the site safety plan and briefing may be comprehensive and documented in writing.

The IC may establish predetermined trigger points for evaluating the current status of the planned response, which may in turn lead to withdrawal of responders from the hot zone. In case of a lack of progress toward meeting the incident objectives, the IC may later decide to abandon the current plan of action and withdraw to a safe distance, set a defensive perimeter, and wait for additional resources to arrive—or choose nonintervention and allow the hazardous materials incident to run its course. Typically, these decisions are made when the offensive actions are not effective in mitigating the problem, the selected PPE is found to be incorrect or ineffective, the tools and equipment required to solve the problem are ineffective or unavailable, or the problem is simply too complex to handle with the available resources. A reasonable and prudent IC knows when the responders are "outgunned" and does not make a foolhardy decision to risk the health and safety of the personnel. In the event that a withdrawal is necessary, the IC should include evacuation signals in the pre-entry briefing.

It is also important, within the context of the safety briefing, to discuss incident communications, including procedures for interacting with the media or other outside entities that provide public information. It is common for communication to be disrupted or otherwise hampered during an incident. To avoid undue communications complications, the IC may designate command and tactical channels for the incident. Entry teams may be instructed to communicate on a dedicated radio channel to ensure they are not cut off or excluded from making a critical radio transmission by other radio chatter. In this case, the backup team should also be monitoring the same radio channel for situational awareness. They may learn about the progress the entry team is making on the mitigation effort or be quickly notified if the entry team gets in trouble and needs rescue assistance. Radio transmissions are an excellent way to communicate the status of the mitigation efforts and to gauge the success or failure of those efforts (**FIGURE 34-8**).

If things are not going as planned, the entry team may contact the entry team leader, who may have a face-to-face conversation with the safety officer or hazardous materials safety officer (ASO). That conversation, carried out through the normal chain of command, may result in a change of strategy or tactics, modification of the incident objectives, or a complete abandonment of the effort. In any case, by observing the proper chain of command, all parties with management and safety interests at the scene will be included

FIGURE 34-8 Radio communications should be clear, concise, specific, and accurate. Think about what you will say before you push the button to talk!

in the appropriate discussions. This approach also helps to ensure that coordinated communications with the public take place.

In some jurisdictions, the safety officer conducts most of the safety briefing; in others, the IC or hazardous materials group supervisor conducts the safety briefing in conjunction with the safety officer. Regardless of how it is accomplished, a site safety and control plan should be completed (and approved by the IC), and a safety briefing should be conducted before responders attempt to enter the incident site. Make sure that you are familiar with and follow the standard operating procedures in your jurisdiction.

In today's world, the threat of terrorism or other incidents with criminal intent drives the need for complete and meaningful safety briefings. Briefings at these types of incidents may also include procedures for operating at a crime scene or evidence collection procedures. It is imperative to understand the needs of law enforcement, especially as those needs relate to beginning an investigation or collecting evidence for a possible prosecution. Operating at a crime scene can present a complicated set of circumstances, but maintaining good working relationships with all agencies on the scene (ideally, with these relationships having been established before an event occurs) will create better operational efficiency during the response. Refer to the standard operating procedures and policies established by your agency to fully understand your role during a hazardous materials incident that involves criminal intent.

A written site safety and control plan should also include information about excessive heat disorders or cold stress that responders may encounter while engaging in their work. Working in PPE in ideal conditions is challenging enough on its own. When temperature extremes are anticipated, their effects on the responders should be acknowledged and understood by everyone operating at the scene.

The Buddy System and Backup Personnel

It is *not* an accepted practice at a hazardous materials incident to allow only one responder to don a chemical-protective ensemble and enter a contaminated environment alone. The risk of something going wrong is too great, and operating alone should *never* be allowed. Instead, it is an accepted practice to implement the **buddy system**—a system in which two responders work as a team for safety purposes—for those personnel entering contaminated areas. The simplest expression of the buddy system is for no fewer than two responders to enter a contaminated area. There should be no deviation from this practice under any circumstance. The use of the buddy system is so important, in fact, that it is required by the Occupational Safety and Health Administration's (OSHA) Hazardous Waste Operations and Emergency Response (HAZWOPER) regulation. It is the responsibility of the IC not only to limit the number of responders working in the contaminated area, but also to prevent responders from working alone. More than two responders might enter a contaminated area, but never fewer than two. To work alone is to take an undue risk!

On a hazardous materials incident, a backup team should also be established. A **backup team** is a team consisting of two or more responders equipped with approved PPE or chemical-protective clothing (CPC), who are assigned by the IC to perform emergency removal of a stricken entry team member from the hot zone. The backup team typically wears the same level of protection as the initial entry team and provides another layer of safety for those entering the contaminated areas. Like the use of the buddy system, the use of a backup team is required by the OSHA HAZWOPER regulation and may be called upon if the entry team declares a mayday. A **mayday** is a "situation at a hazardous materials/WMD incident where an entry team member is unable to safely exit the hot zone, or an event that cannot be resolved by the entry team member within 30 seconds" (NFPA 470). The concept is much like the practice of establishing a rapid intervention team at a structure fire. Backup personnel must be in place, dressed in appropriate PPE (up to the point that respiratory protection can be quickly donned), and ready to spring into action whenever personnel are operating in the hot zone.

SAFETY TIP

A backup team should never be more than a minute or so from being fully dressed and ready for action.

The Incident Command System

In the 1970s, a series of devastating wildland fires occurred in Southern California, requiring the services of numerous local and state resources for the responses. That rash of fires highlighted the need for a better way to organize and manage large numbers of agencies and resources called to a major incident. Many of the agencies involved in the responses to those fires agreed to form a working group, subsequently named FIRESCOPE (Fire Resources of Southern California Organized for Potential Emergencies). The FIRESCOPE group began its work by identifying several problem areas common to almost all major or complex incidents: ineffective communications, span of control challenges, a lack of a common command structure, the inability to track personnel working on the scene, and the inability to effectively coordinate on-scene resources.

Span of control refers to how many responders or activities can be effectively managed by one supervisor. Typically, an acceptable ratio for emergency operations is one supervisor to five responders. This span of control is flexible and depends on issues such as the criticality of the mission or task being performed, the skill level of the supervisor, and the skill levels of the responders.

The FIRESCOPE consortium developed a standardized, yet flexible component of an IMS called the **incident command system (ICS)**, which enables "effective and efficient on-scene incident management by integrating organizational functions, tactical operations, incident planning, incident logistics, and administrative tasks within a common organizational structure" (NFPA 470). Key benefits of using the ICS include the following:

- Common terminology
- Consistent organizational structure
- Consistent position titles
- Common incident facilities

The ICS, which was originally designed for managing wildland fires, subsequently evolved into an all-risk management structure suitable for managing resources at all types of fires, natural disasters, technical rescues, and any other type of incident of virtually any size. When properly used, it provides a strong organizational framework through which to achieve the operational goals and objectives. The OSHA HAZWOPER regulation requires that the ICS be implemented in response to all hazardous materials incidents. Now more than ever, it is critical that all responders—regardless of the patch on their uniform or the nature of their job, or even their geographic location—understand the need for this cohesive incident management tool.

TIP

The OSHA HAZWOPER regulations mandate the use of the ICS at hazardous materials incidents.

The ICS can be expanded to handle an incident of any size and complexity. Hazardous materials incidents can be extremely complex, with local, state, and federal responders and agencies all becoming involved in an incident with criminal intent or of long duration. The basic ICS consists of five functions: command, operations, planning, logistics, and finance/administration (**FIGURE 34-9**).

Command

Command is established when the first unit arrives on the scene and is maintained until the last unit leaves the scene. The concept of incident command is the first main principle of incident response and the cornerstone of the ICS. Without leadership and a process for organizing and directing personnel and resources, the effort to solve a problem may end up being as chaotic as the problem itself. A certain chain of command must exist from the top to the bottom of the incident organizational structure. In this structure, everyone either

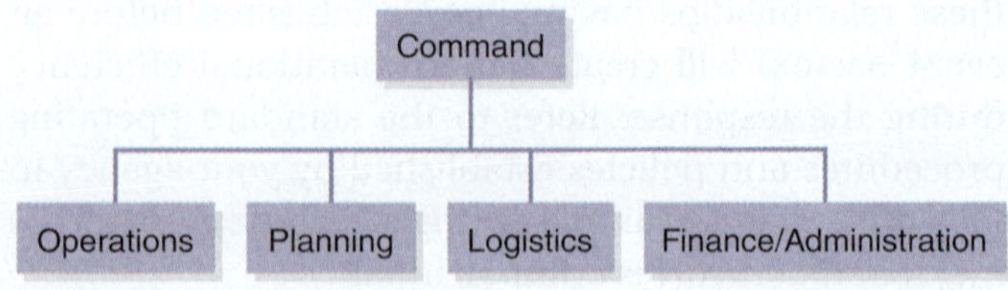

FIGURE 34-9 General staff functions of the incident command system.

reports to or is responsible for someone else. Command is directly responsible for at least the following tasks:

- Determining strategy
- Selecting incident tactics
- Creating the action plan
- Developing the ICS organization
- Managing resources
- Coordinating resource activities
- Providing for scene safety
- Releasing information about the incident
- Coordinating with outside agencies

Unified Command

When multiple agencies with overlapping jurisdictions or legal responsibilities are involved in the same incident, a team effort called a **unified command** "allows all agencies with jurisdictional responsibility for an incident or planned event, either geographic or functional, to manage the incident or planned event by establishing a common set of incident objectives and strategies" (NFPA 1026). A unified command provides several advantages. Using this approach, representatives from various agencies cooperate to share command authority (**FIGURE 34-10**). Oil spills on navigable waterways; wildfires that threaten or occur in a combination of state, federal, and local lands; hazardous materials incidents on public roadways or at fixed facilities; and perhaps active shooter incidents represent situations requiring the establishment of a unified command.

FIGURE 34-10 A unified command involves many agencies directly involved in the decision-making process for a large incident.

Incident Command Post

Regardless of whether there is a single IC or the incident is run under a unified command, the command function is always physically located in an **incident command post (ICP)**. The ICP is the field location where the IC is located and where primary tactical-level, on-scene incident command functions, such as coordination, control, and communications, are performed (NFPA 1026). If additional resources of any kind are required to respond to the incident, the ICP is where those requests originate. The emergency response plan of the AHJ should outline those potential outside agencies that may be called upon to respond to certain circumstances. The plan's command and all direct support staff should be located at the ICP. Ideally, the ICP should be sited in an area where it is not threatened by the incident and that has the necessary infrastructure (e.g., communications, technology support, bathrooms, meeting space) to support sustained operations if required.

During a hazardous materials incident, the ICP should be established uphill and upwind of the incident, keeping in mind the potential for predicted changes in wind direction based on the time of day.

SAFETY TIP

It is important to consider the prevailing winds of the area when establishing the location of the incident command post.

Command Staff

The IC is the person in charge of the entire incident and should be qualified to be in this position. The OSHA HAZWOPER regulation (1910.120[q][6][v]) states that an on-scene IC, who will assume control of the incident scene beyond the first-responder awareness level, should receive at least 24 hours of training equal to the first-responder operations level and, in addition, have competency in the following areas:

- Know and be able to implement the jurisdiction's ICS
- Know how to implement the jurisdiction's emergency response plan

- Know and understand the hazards and risks associated with responders working in CPC
- Know how to implement the local emergency response plan
- Know about the state emergency response plan and the Federal Regional Response Team
- Know and understand the importance of decontamination procedures

The **command staff** is responsible for functions in the IMS and consists of the ISO, the liaison officer, and the **public information officer (PIO)** (NFPA 1561). OSHA requires the IC and safety officer positions to be filled during a hazardous materials response. The command staff report directly to the IC and are critical to the effective management of a hazardous materials/WMD incident (**FIGURE 34-11**).

Incident Safety Officer. The OSHA HAZWOPER regulation (1910.120[q][3][vii]) defines the role of the ISO at a hazardous materials incident as follows:

> The individual in charge of the ICS shall designate a safety officer, who is knowledgeable in the operations being implemented at the emergency response site, with specific responsibility to identify and evaluate hazards and to provide direction with respect to the safety of operations for the emergency at hand.

The OSHA HAZWOPER regulation, in 1910.120(q)(3)(viii), goes on to describe the authority of the ISO at a hazardous materials incident:

> When activities are judged by the safety officer to be an IDLH [immediately dangerous to life and health] and/or to involve an imminent danger condition, the safety officer shall have the authority to alter, suspend, or terminate those activities. The safety official shall immediately inform the individual in charge of the ICS of any actions needed to be taken to correct these hazards at the emergency scene.

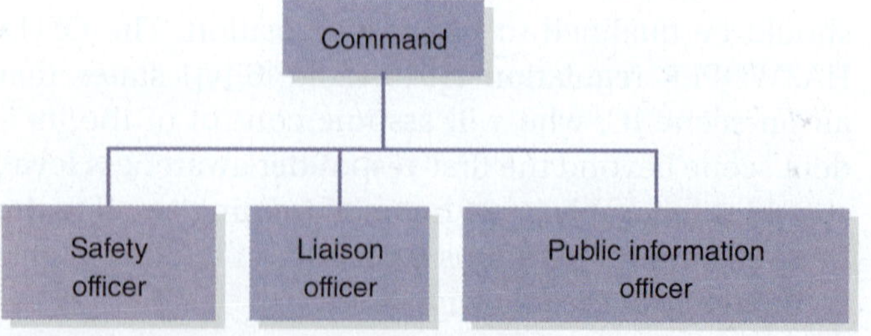

FIGURE 34-11 The command staff members report directly to the incident commander.

A hazardous materials safety officer is responsible for only the safety of the hazardous materials team. When a hazardous materials branch or hazardous materials group has been established, or when the incident requires a dedicated hazardous materials response, the safety officer may appoint a hazardous materials safety officer to the hazardous materials branch or group. This safety officer reports directly to the ISO, who in turn reports directly to the IC.

TIP

There is only one ISO for the entire incident, but that position may have many assistants.

Liaison Officer. The **liaison officer** is the "member of the command staff responsible for coordinating with representatives from cooperating and assisting agencies" (NFPA 1561). Essentially, the liaison officer acts as the point of contact for cooperating and assisting agencies on the scene. The basic distinction between a cooperating agency and an assisting agency relates to the financial stake each has in the management or outcome of the incident. Typically, if an agency has a financial interest/responsibility for the incident, it is considered an assisting agency; all others are considered cooperating agencies. There is much gray area here, however, and you must follow your jurisdiction's direction when determining the distinction between an assisting agency and a cooperating agency.

On a hazardous materials incident, the liaison officer deals with agency representatives (A reps) from federal agencies, state and local resources, and any other outside agency with an interest in the management or outcome of the incident. Ideally, these representatives should have the authority to make decisions on behalf of the agency they represent. It is extraordinarily cumbersome to deal with agency representatives who must constantly go back to key decision makers within their own organization to get approvals. This function is critical when the interests of many jurisdictions are at stake. Liaison officers can find themselves quite busy dealing with questions and providing information to agency representatives during an incident.

Public Information Officer. The public information officer (PIO) is the "member of the command staff responsible for interfacing with the public and media or with other agencies with incident-related information requirements" (NFPA 1026). The PIO functions as a point of contact for the media or any other entity seeking information about the incident (**FIGURE 34-12**). As with all the other command staff positions, only one person should fill this role. Although many people may work with or for the PIO, they should serve as assistants. This chain of command is necessary to streamline communications and reduce redundancies in the management system.

FIGURE 34-12 The public information officer functions as a point of contact for the media.

The value of the PIO should not be underestimated. Keeping the media (and others) well informed is an important piece of successfully managing almost any type of incident. A wise IC will acknowledge that fact and incorporate good media relations into the incident objectives.

General Staff Functions

When the incident is too large or too complex for just one person—the IC—to manage effectively, the IC may assign other individuals to oversee parts of the incident. Everything that occurs at an incident can be divided among the four major functional components within ICS:

- Operations
- Planning
- Logistics
- Finance/administration

Operations

The **operations section**, which is typically led by an **operations section chief** on larger incidents, carries

Voice of Experience

One spring morning at approximately 0900 hours, we responded, along with multiple other agencies, to a neighboring department for an overturned tanker truck that failed to negotiate a turn in the road. At the time of our response, it was unknown what type of chemical we were dealing with.

After I checked in with the incident commander, he assigned me as ISO. Once we were in place, all operational individuals were briefed on the suspected chemical and its properties, along with the response plan. Our first responsibility was to verify the identity of the chemical involved. Next, we established the hot, warm, and cold zones. A decontamination corridor was also established, with constant air monitoring.

An entry team, with backup teams in place, identified the damage to the truck tanker and the approximate amount of the chemical spilled. They also conducted diking and absorption activities in containing the spilled chemical.

Through the post briefing of the entry team, it was determined that the safest and most efficient means to mitigate the situation would be to transfer the chemical to another tanker truck. The incident commander, along with the operations section chief, developed an incident action plan, and the incident was mitigated exactly as planned. Although this incident lasted approximately 15 hours and involved many agencies, including private clean-up contractors, we could ensure life safety by establishing control zones, delegate tasks through use of the incident command system, and take protective actions such as conducting a safety briefing and utilizing backup teams. Remember to utilize all your resources when dealing with a potential hazardous materials release.

Jeffry J. Harran

Lake Havasu City Fire Department

Lake Havasu City, Arizona

out the objectives developed by the IC and is responsible for all tactical operations at the incident (NFPA 1026). The operations section chief directs and manages the resources assigned to their section. These resources could include fire units of any type, law enforcement resources, emergency medical units, airborne resources such as helicopters and fixed-wing aircraft, and hazardous materials response resources. This position is usually assigned when complex incidents involve more than 20 single resources or when the IC cannot be involved in all details of the tactical operation. When incident operations require many responders, the operations section can be further divided into groups and divisions.

Groups and Divisions. Organizational units such as groups and divisions are established to aggregate single resources or crews under one supervisor. The primary reason for establishing groups and divisions is to maintain an effective span of control.

The term "group" is specific as it applies to the ICS: A **group** is assembled to relieve span of control issues and is considered to consist of functional assignments that may not be tied to any one geographic location. A **division** usually refers to companies and crews that are working in the same geographic location (NFPA 1561). Groups and divisions place several single resources under the control of one supervisor, effectively reducing the IC's span of control. **FIGURE 34-13** provides an overview of how divisions and groups are integrated into the overall command structure. The names and functions can be tailored to the specific needs of the incident. The important thing to ensure is that divisions and groups understand their functions and where they fit into the ICS organizational chart.

Hazardous Materials Branches. If necessary, during a large-scale incident, special responders may be placed under the operations section when several different activities occur in separate geographic locations or involve distinct functions. These resources are identified by their specific discipline (e.g., fire, law enforcement, EMS, hazardous materials) and are designated as a **branch**. At a major incident, the IC can establish branches to place a higher-level supervisor (a hazardous materials branch director) in charge of several divisions and/or groups. The hazardous materials branch, when established within an overall IAP, handles "the mitigation and control of the hazardous materials/ WMD portion of an incident" (NFPA 470).

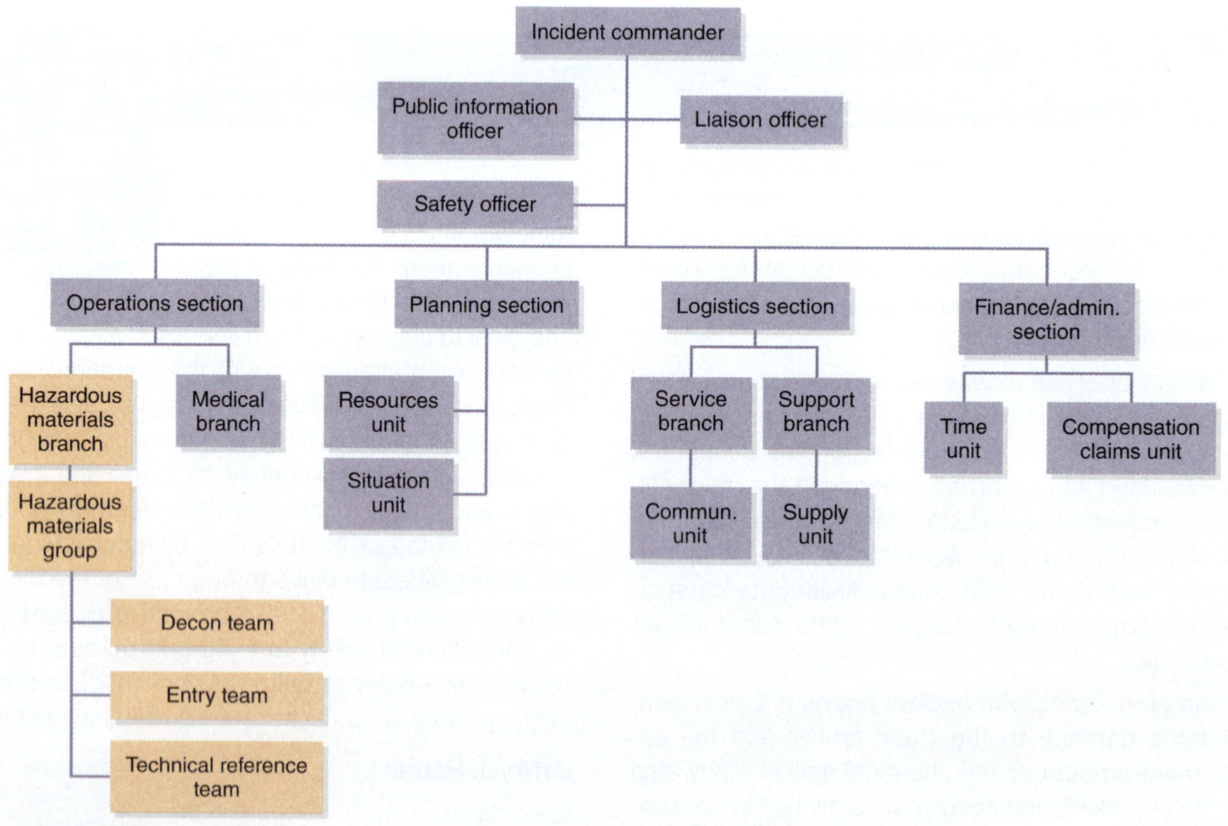

FIGURE 34-13 An overview of the hazardous materials branch.

Modified from FEMA.

The hazardous materials branch director reports to the operations section chief or directly to the IC if the operations section chief position is not assigned. Responsibilities of this branch director include obtaining a briefing from the IC, staffing the hazardous materials branch functions as required, ensuring that scene control zones are established, ensuring that a site safety plan is developed, maintaining accountability, ensuring that proper PPE is worn, and ensuring that the operational objectives are being met. Within the hazardous materials branch, it is common practice to also establish a hazardous materials group, led by a hazardous materials group supervisor.

When a hazardous materials group supervisor is appointed under the hazardous materials branch director, that position directs and manages other job functions such as decontamination group supervisor or team leader, entry group supervisor or team leader, technical reference group, or other functions specific to the hazardous materials portions of the response. The management structure and the ICS positions that are ultimately assigned are based on the needs associated with the specific incident.

Planning

The **planning section** is "responsible for the collection, evaluation, dissemination, and use of information related to the incident" (NFPA 1026). The planning section, which is led by a **planning section chief**, acts as the central point for collecting information on the situation status (sit-stat) of the event, tracking and logging on-scene resources and disseminating the written IAP. On large-scale incidents, the planning section chief facilitates the incident briefings and planning meetings.

Logistics

The **logistics section** can be viewed as the support side of an incident management structure. This section within the ICS is responsible for providing facilities, services, and materials necessary for the incident response (NFPA 1026). Logistics is headed by a **logistics section chief**, who is responsible for providing the bulk of the support functions for an incident. This position is generally assigned on long-duration or resource-intensive, complex incidents.

Logistical support includes food, sleeping facilities, transportation needs, sanitation facilities, showers, and all other requests for resources needed to manage the incident. If someone on the incident needs a truckload of sand, for example, the logistics section is the group that makes it happen. When resources are needed, the requesting person follows the appropriate chain of command to route the request to logistics. It is then up to logistics personnel to determine whether they will be able to fill the request. Typically, during a planning meeting at a large incident, the operations section chief, planning section chief, ISO, IC, and a small core of other key management positions meet to discuss the logistical needs of the incident based on the operational objectives. The logistics section then determines whether those needs can be met and, if so, under what time frame.

Finance

The **finance/administration section** tracks the costs related to the incident, handles procurement issues, records the time that responders are on the incident for billing purposes, and keeps a running tally of costs for the incident (NFPA 1026). Today, cost is certainly a factor in every response, even when it comes to operational objectives. Large-scale incidents can cost millions of dollars to handle, and it is common to see incident objectives tied to financial constraints. Ideally, an incident should be handled in the most cost-effective way possible, to minimize the financial impact on the jurisdiction in which the incident occurs.

CASE STUDY

You Are the Firefighter CONCLUSION

Local law enforcement is requesting a response for a suspicious package leaking a liquid. The package is located in a small storeroom on the ground floor of a four-story office building. While you are en route, the dispatcher announces this update: "Engine 40, law enforcement on scene is reporting several people complaining of eye irritation and nausea. There are also reports of an unusual odor on the second and third floors. A full building evacuation is in progress."

1. **Based on the initial reports, what would your initial actions be upon arrival?**

   Answer: Ensuring your own safety and that of other responders is a common first operational priority at a hazardous materials incident, closely followed by addressing and injuries or potential threats to human life. In this case, working with law enforcement and representatives of the building, the evacuation should be completed followed by further isolating the affected floors plus at least one floor above and below. As civilians are evacuated and ill civilians are treated, you can collect additional intelligence on the release by matching signs and symptoms to potential chemicals and determining the location of the release based on the locations of those most significantly exposed.

2. **Working with the on-scene law enforcement is important. What would you do to facilitate a coordinated approach to managing and jointly commanding this incident?**

   Answer: Unified command is the best method for joining all the key decision makers from each discipline.

3. **Does your agency have the capabilities to communicate via radio with other public safety agencies (various law enforcement agencies, for example)?**

   Answer: This question is best answered with local knowledge and communications policies, procedures, and capabilities of the AHJ. Not all jurisdictions have the ability to communicate outside their own sphere of influence.

WRAP-UP

SUMMARY

KNOWLEDGE OBJECTIVES

- Size up an incident. (**NFPA 470: 7.2.1, 7.4.1**, pp. 1344–1345)
  - Understand the operational needs for establishing control zones at a hazardous materials incident.
  - Understand how to use the CAN communications format.
- Identify and describe the safety procedures at a hazardous materials incident. (**NFPA 470: 7.4.1, 7.6.1**, pp. 1345–1351)
  - Understand the concept of situational awareness.
  - Describe how to use the *ERG* to implement protective actions.
- Explain the role of the operations-level responder in implementing a planned response. (**NFPA 470: 7.1.1, 7.1.2, 7.1.3, 7.1.4, 7.4.1**, pp. 1347–1351)
  - Describe how the operations-level responder can use AHJ established threat levels to standardize response actions back on the situation encountered.
  - Understand the need and value of conducting safety briefings prior to taking action at a hazardous materials incident.
- Identify and describe the components of the incident command system. (**NFPA 470: 7.4.1**, pp. 1352–1357)
  - Understand the need and requirements for using the backup team and buddy system at a hazardous materials incident.

SKILLS OBJECTIVES

There are no skills objectives for operations-level personnel for this chapter.

KEY TERMS

action plan A detailed proposal describing the strategies and tactics within an incident action plan that will safely accomplish the incident objectives, favorably influence outcomes, and increase the safety of responders and the public.

assistant safety officer (ASO) See *hazardous materials safety officer*.

backup team Required by OSHA's HAZWOPER regulation, a team composed of two or more responders equipped with approved personal protective equipment or chemical-protective clothing and assigned by the incident commander to provide emergency removal of a stricken entry team member from the hot zone.

branch A main supervisory level implemented at large scale incidents; divisions and groups report up through their respective branches.

buddy system A system in which two responders work as a team for safety purposes.

cold zone The control zone of hazardous materials /weapons of mass destruction (WMD) incidents that contains the incident command post and such other support functions as are deemed necessary to control the incident. Also called *outer perimeter*. (NFPA 470)

command staff The public information officer (PIO), incident safety officer (ISO), and liaison officer who report directly to the incident commander (IC) and who are responsible for functions in the incident management system that are not a part of the function of the line organization. (NFPA 1561)

contamination reduction zone See *warm zone*.

control zones See *hazard control zones*.

decontamination corridor A controlled area within the warm zone where decontamination is normally performed. (NFPA 470)

division A supervisory level established to divide personnel at an incident into geographic areas of operations.

entry team A team of fully qualified and equipped responders who are assigned to enter the designated hot zone.

exclusionary zone See *hot zone*.

finance/administration section The section in an incident command system responsible for all costs and financial actions of the incident or planned event, including the time unit, procurement unit, compensation /claims unit, and cost unit. (NFPA 1026)

group Personnel assembled to relieve span of control issues by taking over functional assignments that may not be tied to any one geographic location.

hazard control zones The designated areas established at a hazardous materials/weapons of mass destruction (WMD) incident based on the chemical and physical properties of the released material, the environmental factors at the time of the release, and the general layout of the scene. (NFPA 470)

hazardous materials branch The function within an overall incident management system (IMS) that deals with the mitigation and control of the hazardous materials/weapons of mass destruction (WMD) portion of an incident. (NFPA 470)

hazardous materials safety officer The person who works within an incident management system (IMS) in the hazardous materials branch or group to ensure that recognized hazardous materials/weapons of mass destruction (WMD) safe practices are followed at hazardous materials/WMD incidents. Also called *assistant safety officer (ASO)*. (NFPA 470)

hot zone The control zone immediately surrounding hazardous materials/weapons of mass destruction (WMD) incidents, which extends far enough to prevent adverse effects of hazards on personnel outside the zone, and where only personnel who are trained, equipped, and authorized to do assigned work are permitted to enter. Also called *exclusionary zone*. (NFPA 470)

incident command post (ICP) The field location at which the primary tactical-level, on-scene incident command functions are performed. (NFPA 1026)

incident command system (ICS) A component of an incident management system (IMS) designed to enable effective and efficient on-scene incident management by integrating organizational functions, tactical operations, incident planning, incident logistics, and administrative tasks within a common organizational structure. (NFPA 470)

incident safety officer (ISO) The member of the command staff responsible for monitoring and assessing safety hazards or unsafe situations and for developing measures for ensuring personnel safety. Also called *safety officer*. (NFPA 1026)

liaison officer The member of the command staff responsible for coordinating with representatives from cooperating and assisting agencies. (NFPA 1561)

KEY TERMS CONTINUED

logistics section The section in an incident command system responsible for providing facilities, services, and materials for the incident or planned event, including the communications unit, medical unit, and food unit within the service branch and the supply unit, facilities unit, and ground support unit within the support branch. (NFPA 1026)

logistics section chief The general staff position responsible for directing the logistics function; a role generally assigned on complex, resource-intensive, or long-duration incidents.

mayday Any situation at a hazardous materials/weapons of mass destruction (WMD) incident where an entry team member is unable to safely exit the hot zone, or an event that cannot be resolved by the entry team member within 30 seconds. (NFPA 470)

operations section The section in an incident command system responsible for all tactical operations at an incident or planned event, including up to 5 branches; 25 divisions or groups; and 125 single resources, task forces, or strike teams. (NFPA 1026)

operations section chief The general staff position responsible for managing all operations activities; a role usually assigned when complex incidents involve more than 20 single resources or when command staff cannot be involved in all details of the tactical operation.

outer perimeter See *cold zone.*

planning section The section in an incident command system responsible for the collection, evaluation, dissemination, and use of information related to the incident situation, resource status, and incident forecast. (NFPA 1026)

planning section chief The general staff position responsible for planning functions and for tracking and logging resources assigned when command staff members need assistance in managing information.

public information officer (PIO) The member of the command staff responsible for interfacing with the public and media or with other agencies with incident-related information requirements. (NFPA 1026)

reconnaissance A visual scan and exploration of a physical area to gain actionable information to guide incident decision making.

safety officer See *incident safety officer.*

size-up The process of rapidly evaluating the critical visual indicators of an incident (what is happening now), analyzing that information based on your training and experience (what might happen next), and arriving at a conclusion that will serve as the basis to form and implement an action plan (what you are going to do about the situation).

span of control The maximum number of personnel or activities that can be effectively controlled by one individual (usually three to seven).

unified command A team effort that allows all agencies with jurisdictional responsibility for an incident or planned event, either geographic or functional, to manage the incident or planned event by establishing a common set of incident objectives and strategies. (NFPA 1026)

warm zone The control zone at hazardous materials/weapons of mass destruction (WMD) incidents where personnel and equipment decontamination and hot zone support take place. Also called *contamination reduction zone.* (NFPA 470)

REVIEW QUESTIONS

1. What is an action plan?
2. What are the three control zones?
3. What is the buddy system?
4. What are the four key benefits of using an ICS?
5. When should unified command be implemented?
6. How does OSHA define the role of the incident safety officer (ISO)?

DISCUSSION QUESTIONS

1. Why is isolation of the release area at a hazardous materials incident one of the first measures that responders should implement?
2. Why is the establishment of a backup team a critical safety aspect of a hazardous materials response?
3. Why is the CAN report a useful communications tool?

APPLYING THE CONCEPTS

It is 0400 hours when your engine company is dispatched to a 100-year-old manufacturing plant for a report of a slowly leaking 2000-gallon (7570-liter) aboveground storage tank. Your four-person crew is trained to the operations level.

1. Is there enough information to make any general assessments of the severity of the problem based on the initial dispatch?
2. What initial actions should you take?

Upon arrival, the on-scene security guard directs you to an old-looking aboveground storage tank approximately 100 yards (91 meters) from the security building. He states the tank has been there for years, and management knows that it leaks but has done nothing about it. He has been told it might contain polychlorinated biphenyls (PCBs), but he doesn't know if this is true. You direct the driver to approach the tank slowly. From a distance you see what looks to be wet asphalt adjacent to the storage tank. You stop the apparatus and consult the *ERG* for PCBs. According to the *ERG*, PCBs are extremely toxic, and a hot zone of at least 330 feet (100 meters) should be established. After doing some additional research, you learn that PCBs are a type of toxic chemical that was banned in the United States in the late 1970s but are still found in some older devices and products.

3. How would you go about verifying the report of the security guard?
4. Based on what you learned from your research of PCB, do you have any suspicions about the validity of the report from the security guard?

After consulting an online reference source, you learn that this chemical is very harmful to the environment, and it can also cause skin irritation, liver damage, nausea, dizziness, and eye irritation. It is a thick, oily liquid with a low vapor pressure at ambient atmospheric temperature.

5. Based on what you know or suspect at this point, is it within your scope to take offensive action to approach the tank with the goal of stopping a leak?
6. What actions can an operations-level responder take?

REFERENCES

National Fire Protection Association (NFPA). 2019. *NFPA 1561, Standard on Emergency Services Incident Management System and Command Safety.* 2020 Edition. Quincy, MA: NFPA.

National Fire Protection Association (NFPA). 2021. *NFPA 470, Hazardous Materials/Weapons of Mass Destruction (WMD) Standard for Responders.* 2022 Edition. Quincy, MA: NFPA.

National Fire Protection Association (NFPA). 2023. *NFPA 1026, Standard for Incident Management Personnel Professional Qualifications.* 2024 Edition. Quincy, MA: NFPA.

Occupational Safety and Health Administration (OSHA). 1970. "1910.120 - Hazardous Waste Operations and Emergency Response." Accessed September 15, 2021. https://www.osha.gov/laws-regs/regulations/standardnumber/1910/1910.120

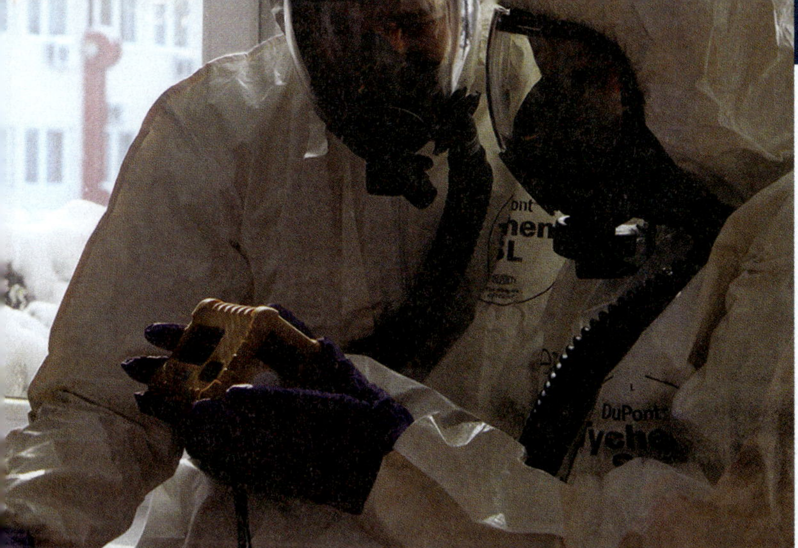

SECTION

5

Hazardous Materials Operations Level: Mission Specific

NOTE: **NFPA 1010**, *Standard on Professional Qualifications for Firefighters*, 2024 Edition 6.1 in Chapter 6: Firefighter I (**NFPA 1001**) states that all firefighter candidates must meet the requirements defined in Chapter 7 and Sections 9.2 and 9.6 of **NFPA 470, 2022 Edition**. Content within these chapters meets the intent of **NFPA 470, 2022 Edition**, which includes chapter 9, "Professional Qualifications for Hazardous Materials/WMD Operations Level Responders Assigned Mission-Specific Responsbilities (**NFPA 1072**)."

CHAPTER

35

Hazardous Materials Operations Level: Mission Specific

Hazardous Materials Responder Personal Protective Equipment

NOTE: Content within this chapter meets the intent of **NFPA 470, 2022 Edition**, which includes chapter 9, "Professional Qualifications for Hazardous Materials/WMD Operations Level Responders Assigned Mission-Specific Responsibilities (**NFPA 1072**)."

KNOWLEDGE OBJECTIVES

After studying this chapter, you will be able to:

- Discuss the similarities and differences in how single-use and reusable personal protective equipment (PPE) are used.
- Explain how to select, use, and maintain PPE.
- Explain how PPE needs are determined.
- Identify and describe specific PPE for hazardous materials response.
- Describe OSHA/EPA levels of chemical-protective clothing.
- Explain the safety considerations when wearing PPE.
- Explain the inclusion of PPE in reporting and documenting the incident.

SKILLS OBJECTIVES

After studying this chapter, you will be able to:

- Don and doff hazardous materials ensembles.

CASE STUDY

You Are the Firefighter

Your engine company arrives on the scene of a vehicle accident involving a small passenger vehicle and an MC 331 cargo tank truck carrying 20 tons of anhydrous ammonia. The truck rolled over on its side and slid down the highway for approximately 100 feet (30 m). There are no injuries to the three people in the passenger vehicle, but the driver of the truck is pinned inside the cab. You notice the smell of ammonia in the air but see no visible signs of a product release. The regional hazardous materials team, fully staffed with technician-level responders, is also on scene. Your company officer confers with the hazardous materials team and then directs you and another firefighter to don your self-contained breathing apparatus (SCBA) and full turnout gear and evaluate the driver for injuries.

1. **Would full structural firefighter turnout gear and SCBA offer adequate protection in this situation?**
2. **Based on your level of training—operations level—would you be qualified to perform this task?**
3. **Describe the steps you would need to take, including obtaining information about ammonia, to complete your assignment.**

Introduction

This chapter addresses the mission-specific competencies for personal protective equipment (PPE) found in NFPA 470, chapter 8, section 8.2. The details of PPE job performance requirements (JPRs) are found in NFPA 470, chapter 9, section 9.2.1. The operations-level responder assigned to perform mission-specific responsibilities at hazardous materials/weapons of mass destruction (WMD) incidents must be trained to meet all responsibilities at the awareness level and the operations level. Additionally, such a responder must operate under the guidance of a hazardous materials technician, an allied professional, or standard operating procedures (SOPs).

Emergency responders should be familiar with the policies and procedures of the authority having jurisdiction (AHJ) to ensure a consistent risk-based approach to selecting the proper PPE for an expected task or set of tasks. Additionally, all responders charged with responding to hazardous materials/WMD incidents should be proficient with their local procedures for technical decontamination as well as the manufacturers' guidelines for maintenance, testing, inspection, storage, and documentation procedures for the PPE provided by the AHJ.

While this edition of the text was being written, many NFPA documents were undergoing a consolidation process. For example, NFPA 472, 473, and 1072 are now *consolidated* into NFPA 470. By this point, the new layout and format of the NFPA documents should be somewhat familiar to the reader.

Another revision effort that directly affects the hazardous materials/WMD responder relates to the area of PPE. The 2022 edition of NFPA 1990, like NFPA 470, *consolidates* NFPA 1991, *Standard on Vapor-Protective Ensembles for Hazardous Materials Emergencies and CBRN Terrorism Incidents*; NFPA 1992, *Standard on Liquid Splash–Protective Ensembles and Clothing for Hazardous Materials Emergencies*; and NFPA 1994, *Standard for Protective Ensembles for First Responders to Hazardous Materials Emergencies and CBRN Terrorism Incidents.* NFPA 1999, *Standard on Protective Clothing and Ensembles for Emergency Medical Operations,* will not be covered in this chapter, as it is more appropriate to emergency medical response.

SAFETY TIP

All responders who may be called upon to wear any type of PPE should read and understand the performance-based requirements for PPE as well as the manufacturer's specifications and procedures for the maintenance, testing, inspection, cleaning, and storage of PPE provided by the AHJ.

NFPA 1990 is intended to be the main reference standard that specifies the minimum design, performance, testing, documentation, and certification requirements for PPE used by responders during hazardous materials/WMD incidents and CBRN (chemical, biological, radioactive, nuclear) terrorism incidents. The standard covers three main types of PPE:

- Vapor-protective ensembles and ensemble elements
- Liquid splash–protective ensembles and ensemble elements
- Hazardous materials and CBRN-protective ensembles and ensemble elements

NFPA 1990 provides users, manufacturers, test laboratories, and certification authorities with one source of information for product performance requirements for hazardous materials/WMD and CBRN ensembles and ensemble elements. The technical committees responsible for these standards spent much time and effort to consolidate these standards in hopes of making the information easier to access and understand. This chapter does not go into detail on the specifics of NFPA 1990 and NFPA 1891; however, it does offer an overview of each and a correlation back to the familiar Occupational Safety and Health Administration (OSHA)/Environmental Protection Agency (EPA) PPE designations (Levels A, B, C, and D) found in OSHA Title 29, Code of Federal Regulations (CFR) 1910.120. NFPA 1990 and NFPA 1891 are intended to be used in harmony to create a uniform set of definitions to provide consistency between the standards.

TIP

Every AHJ must examine its most common hazards and hazardous materials/WMD incident response mission requirements closely and select appropriate performance standards for the anticipated use cases. All PPE should be employed in accordance with federal OSHA standards, including those in 29 CFR 1910, Subpart H—"Hazardous Materials" (including 29 CFR 1910.120—"Hazardous Waste Operations and Emergency Response"), and 29 CFR 1910, Subpart I—"Personal Protective Equipment" (including 29 CFR 1901.134—"Respiratory Protection"), which include requirements for safety and health plans, medical evaluation, and training.

Single-Use Versus Reusable PPE

Much of the chemical-protective equipment on the market today is intended for a single use (i.e., it is disposable) and is usually discarded along with the other hazardous wastes generated by the incident. Consequently, single-use PPE should be decontaminated to the point that it is safe for the responder to remove but not so extensively that the garment is completely free of contamination. Single-use PPE is generally less expensive than reusable gear and needs to be restocked and/or replenished after the incident.

Reusable garments, in contrast, are required to be maintained, inspected, and tested at regular intervals and after each use. For example, vapor-protective ensembles, "multiple elements of compliant protective clothing and equipment that when worn together provide protection from some, but not all, risks of vapor, liquid splash, and particulate environments during hazardous materials/WMD incident operations" (NFPA 470), are required to be pressure tested, usually upon receipt from the manufacturer, after each use, and annually (**FIGURE 35-1**). Individual manufacturers will have well-defined procedures for this activity. Prior to purchasing any type of PPE, the AHJ should understand the maintenance and upkeep requirements for that equipment.

It is important to understand that PPE is not intended to be purchased and completely left alone until it is needed. A level of care and attention must be devoted to the barrier between you and the released substance. Also, keep in mind that both single-use and reusable garments have a shelf life. This time frame should be noted and adhered to. Do not use PPE that

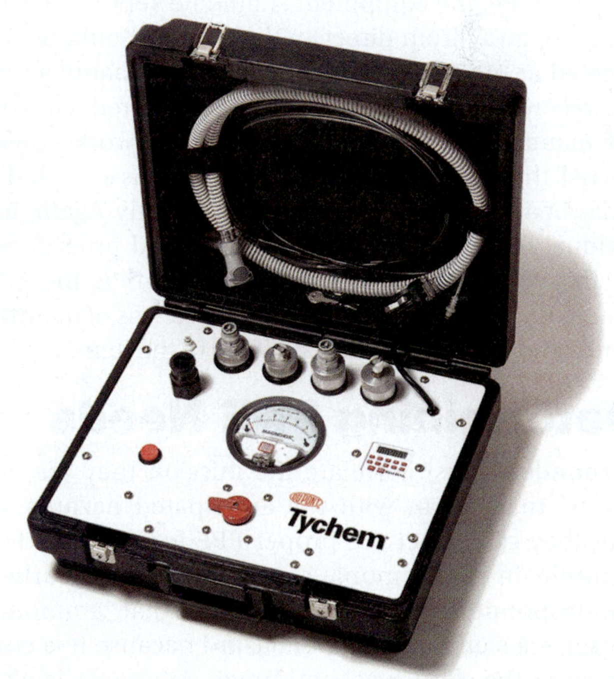

FIGURE 35-1 Vapor-protective suit testing kit.

Courtesy of the DuPont Company.

is beyond its shelf life—you may not be able to depend on it! Before any PPE is used, it should be subjected to a thorough visual inspection to ensure that piece of equipment is absolutely response ready.

Maintaining PPE

Chemical-protective garments are tested in accordance with the manufacturer's recommendation or following the method outlined in OSHA's Hazardous Waste Operations and Emergency Response (HAZWOPER) standard (29 CFR 1910.120, Appendix A). In addition, NFPA 1891 provides detailed guidance on selection, care, and maintenance of such equipment.

A pressure test, for example, is accomplished by using a specially designed kit to pump a certain amount of air into the suit and then leaving the suit pressurized for a specified period of time. At the end of that period, if the garment has lost more than a certain percentage (usually 20 percent) of pressure, it is assumed the suit has a leak. Often, leaks can be located by gently spraying or brushing the inflated suit with a solution of soapy water. Small bubbles will begin to form at the leak, alerting you to its presence and location. Any garment with a leak should be removed from service until the defect is identified and repaired in accordance with the manufacturer's specifications.

Chemical-protective equipment should be stored in a cool, dry place that is not subject to significant temperature extremes and/or high levels of humidity. Furthermore, the equipment should be kept in a clean location, away from direct sunlight, and should be inspected at regular intervals based on the manufacturer's recommendations. If repairs are required, consult the manufacturer prior to beginning any work—there is a risk that the garment will not perform as expected if it has been modified or repaired incorrectly. Again, individual manufacturers have well-defined procedures for this activity; prior to purchasing any PPE, the AHJ should understand what is required in terms of maintenance and upkeep, and the shelf life of the item.

Determining PPE Needs

Responders must correlate the mission they are expected to perform with the anticipated hazards so that they can select the proper PPE for the task. For example, in the ammonia scenario described earlier, the responders should understand that ammonia presents a significant health hazard because it is corrosive to the skin, eyes, and lungs. Ammonia is also flammable at concentrations of approximately 15 to 25 percent (by volume) in a mixture with air. Exposure to a concentration of approximately 300 parts per million (ppm) is considered to be immediately dangerous to life and health (IDLH). If the possibility of exposure to an ammonia concentration exceeding 300 ppm exists, an NFPA 1981–compliant self-contained breathing apparatus (SCBA) is required. SCBA units that comply with the current version of NFPA 1981, *Standard on Open-Circuit Self-Contained Breathing Apparatus (SCBA) for Emergency Services*, are positive-pressure, CBRN-certified units that maintain a pressure inside the face piece, in relation to the pressure outside the face piece, such that the pressure is positive during both inhalation and exhalation. This is a very important feature when operating in airborne contamination.

The OSHA HAZWOPER standard states that all employees engaged in emergency response who are exposed to hazardous substances *shall* wear a positive-pressure SCBA. Furthermore, the incident commander (IC) is *required* to ensure the use of SCBA. It is not just a good idea—it is the law. Additionally, all responders should follow manufacturers' recommendations for using, maintaining, testing, inspecting, cleaning, and filling the SCBA unit. Be sure to document all of these activities so there is a record of what has been done to the unit.

Ideally, your AHJ will determine the PPE available for use in situations like the ammonia release. It will also identify the procedures and requirements for selecting and using PPE on the incident scene as part of the action plan.

Although SCBA will protect responders from suffering an inhalation exposure, its use is only one piece of the PPE equation. Another question must be answered in this scenario: Will structural firefighting gear provide sufficient skin protection? Knowing that ammonia presents a flammability hazard is important—and the firefighting gear would address that potential—but this choice of equipment still does not address the hazard of skin irritation. In this example, the release occurred outdoors: Would you make a different choice, perhaps deciding that some level of chemical-protective clothing is indicated, if the release occurred indoors, inside a poorly ventilated room? Responders should always consider the impact of the operating environment as part of the hazard evaluation. To that end, gathering information on the scene is pivotal to determining what PPE to choose, including the type and level of respiratory protection, chemical-protective clothing, boots, gloves, and so on.

The responders tasked with undertaking a medical reconnaissance mission in the ammonia scenario should balance the hazards and the risk of the mission with the potential gains. This scenario is based on an outdoor release, with unknown variables of wind speed, direction, and ambient air temperature, which may positively or negatively influence the decision to approach the cab of the truck. Unfortunately, it is impossible to decide on the right course of action based on a few sentences in this text—you must make that decision on the street when the emergency occurs. Not all structural firefighters' gear is created equal, and not all types are intended to function in an environment that may contain a hazardous material. Well-worn, 10-year-old structural firefighters' gear will certainly not provide the same level of protection as a new set of gear that meets the current performance standards found in NFPA 1970, *Standard on Protective Ensembles for Structural and Proximity Firefighting, Work Apparel and Open-Circuit Self-Contained Breathing Apparatus (SCBA) for Emergency Services, and Personal Alert Safety Systems (PASS)*.

In summary, it is incumbent on every responder to understand the hazards that may be present on a scene and to appreciate how those hazards may affect the PPE requirements and the mission the responders are tasked with carrying out.

Specific PPE for Hazardous Materials Response

Different levels of PPE may be required at different hazardous materials/WMD incidents depending on the identified mitigation strategy and associated tactics found in the incident action plan (IAP). This section reviews the protective qualities of various ensembles, from those offering the least protection to those providing the greatest protection.

At the lowest end of the spectrum are street clothing and normal work uniforms, which offer the least amount of protection in a hazardous materials emergency. Normal clothing (or flame-resistant coveralls) may prevent a noncaustic powder from coming into direct contact with the skin, for example, but it does not offer significant protection against many other hazardous materials. Such clothing is often used in industrial applications, such as oil refineries, rail yards, or city public works facilities, as a general work uniform (**FIGURE 35-2**). Police officers and emergency medical services (EMS) providers typically wear this level of "protection." Most often, distance from the hazard is the best level of protection with this PPE.

FIGURE 35-2 A Nomex jumpsuit.

Courtesy of the DuPont Company.

SAFETY TIP

The AHJ must properly outfit all responders expected to respond to a hazardous materials incident. The current OSHA HAZWOPER regulations [29 CFR 1910.120 (q)(3)(iii)] (and many local jurisdictional regulations) require the IC to ensure that the personal protective clothing worn at a hazardous materials emergency is appropriate for the hazards encountered.

The next higher level of protection is provided by structural firefighting protective clothing (sometimes referred to as bunker or turnout gear) (**FIGURE 35-3**). Such an ensemble includes a helmet, coat, pants, boots, gloves, a fire-resistant hood, and a personal alert safety system (PASS) device. SCBA is typically worn with structural firefighting protective clothing. Standard firefighting turnout gear is not technically considered "chemical protection," because the fabric may break down when exposed to chemicals and may not provide complete protection from the harmful gases, vapors, liquids, and dusts that could be encountered during hazardous materials incidents.

Returning to the ammonia scenario, since it is outdoors, it may be safe and reasonable to carry out the patient assessment mission wearing this level of protection—again, based on a full risk assessment. Keep in mind that structural firefighting gear is primarily intended to

FIGURE 35-3 Standard structural firefighting protective clothing.

FIGURE 35-4 High temperature–protective equipment protects the wearer from high temperatures during a short-term exposure.

protect the wearer from thermal hazards (predominantly encountered during firefighting) and mechanical hazards such as broken glass or other sharp objects.

Responders wearing **high temperature–protective clothing** may best address unusually high thermal hazards, such as those posed by aircraft fires. This type of PPE shields the wearer during short-term exposures to high temperatures (NFPA 470) (**FIGURE 35-4**). Sometimes referred to as a *proximity suit*, high temperature–protective clothing allows the properly trained firefighter to work in extreme fire conditions. It provides protection against high temperatures only; in other words, it is not designed to protect the firefighter from hazardous materials.

Chemical-Protective Clothing and Equipment

Chemical-protective clothing (CPC) is unique in that it is designed to shield or isolate the wearer from the hazards encountered during hazardous materials/WMD operations. Such equipment is not intended to provide high levels of protection from prolonged exposure to thermal hazards (heat and cold) or to protect the wearer from injuries that may result from torn fabric, chemical damage, or other mechanical damage (tears and abrasion) to the suit. Furthermore, not all CPC is the same, and each type, brand, and style may offer varying degrees of protection and chemical resistance.

To help you safely estimate the chemical resistance of a particular garment, manufacturers supply compatibility charts with their protective equipment (**FIGURE 35-5**). These charts are designed to assist you in choosing the right CPC for the incident at hand, based on specific performance testing and validation. Performance testing is not covered in detail in this text. As an overview, suffice it to say a standard battery of test chemicals is used to determine the ability of a suit material to resist chemical penetration (for seams, closures, and other aspects of the CPC) and permeation, along with physical resistance to abrasions, tears, and other damage. Compatibility charts are available to aid responders in matching the anticipated chemical hazard to the resistance characteristics of the garment. Note that the

FIGURE 35-5 An example of a compatibility chart.

test battery of chemicals does not reflect every possible chemical; instead, specific chemicals are chosen to represent a range of chemical exposure possibilities.

Chemical-resistant material is fabric specifically designed to inhibit or resist the passage of chemicals into and through the material by the process of penetration, permeation, or degradation. Clothing made of chemical-resistant material is designated as CPC. Storage conditions, temperature, and resistance to cuts, tears, and abrasions are all factors that affect the chemical resistance of materials. Other factors include flexibility, shelf life, and sizing criteria. The bottom line is that chemical-resistant materials are specifically designed to inhibit or resist the passage of chemicals into and through the material.

Penetration is "the movement of a material through a suit's closures, such as zippers, buttonholes, seams, flaps, or other design features of chemical-protective clothing and through punctures, cuts, and tears" (NFPA 470). To reduce the threat of a penetration-related suit failure, responders should carefully evaluate their CPC prior to entering a contaminated atmosphere. Checking the garment fully—that is, performing a visual inspection of all its components—before donning the CPC is paramount. Just because the suit or gloves came out of a sealed package does not mean they are perfect! Also, a lack of attention to detail could result in zippers not being fully closed and tight. Poorly fitting seams around the ankles and wrists could allow chemicals to defeat the integrity of the garment.

Use of the buddy system is beneficial in this setting because it creates the opportunity to have a trained set of eyes examine parts of the suit that the wearer cannot see. Prior to entering a contaminated atmosphere, or periodically while working, each member of the entry team should quickly scan the ensemble of the other member(s) to determine whether any other responder has suffered damage, discoloration, or other insult that may jeopardize the wearer's health and safety.

Permeation is "a chemical action involving the movement of chemicals, on a molecular level through intact material" (NFPA 470). It differs from penetration in that permeation occurs through the material itself rather than through openings in the material. Permeation may be impossible to identify visually, but it is important to note the initial status of the garment and to determine whether any changes have occurred (or are occurring) during the course of an incident.

Degradation is another point to consider when it comes to the failure of CPC. **Degradation** is "a chemical action involving the molecular breakdown of a protective clothing material or equipment due to contact with a chemical" (NFPA 470). It may be evidenced by visible signs such as charring, shrinking, swelling, color changes, or dissolving. When chemicals are highly aggressive, or when the suit fabric is a poor match for the suspect substance, fabric degradation is possible. If the suit dissolves or rapidly fails in some other way, the possibility of the wearer suffering an injury is high.

Types of Chemical-Protective Clothing

CPC can be constructed as a single- or multi-piece garment. A single-piece garment may or may not completely enclose the wearer and is often found as a coverall-type garment. A multi-piece garment typically includes a jacket, pants, an attached or detachable hood, and perhaps attached fabric to cover the feet.

Single- or multi-piece garments can be designated as liquid splash–protective ensembles—"compliant protective clothing and equipment products that when worn together provide protection from some, but not all, risks of hazardous materials/WMD emergency incident operations involving liquids" (NFPA 470). Vapor-protective ensembles are built as fully encapsulated one-piece garments with attached gloves and suit fabric that covers the feet and equipment; when worn together, these garments provide protection from some, but not all, risks of vapor, liquid splash, and particulate environments during hazardous materials/WMD incident operations. Commonly used suit fabrics include butyl rubber, Tyvek, Saranex, polyvinyl chloride (PVC), and Viton. Chemical-resistant boots should be worn to offer protection from abrasion and mechanical hazards.

Chemical-protective equipment suited for law enforcement and other missions is becoming more popular and finding its way into traditional hazardous materials response. These protective ensembles typically offer protection against liquid and particulate forms of CBRN agents and are much cooler and more comfortable to wear for extended periods of time. Your AHJ will determine which type of CPC may be available and should provide all performance data for those ensembles.

SAFETY TIP

No single chemical-protective garment (vapor or splash) on the market will protect you from every possible chemical or other hazardous substance.

A **vapor-protective ensemble** offers full body protection from highly toxic environments and requires the wearer to use an air-supplied respiratory device such as an SCBA (**FIGURE 35-6**). Optional requirements may apply to NFPA 1990–compliant garments, such as limited protection against flash fires or liquified gas protection (protection against propane and butane, for example). In a vapor-protective ensemble, the wearer is completely zipped inside the protective "envelope," leaving no skin (or the lungs) accessible to the outside. If the ammonia scenario described at the beginning of the chapter were occurring in a different location—such as inside a poorly ventilated storage area within an ice-making facility—vapor-protective clothing might be required. Ammonia aggressively attacks the skin, eyes, and mucous membranes such as in the eyes and mouth and can cause severe and irreparable damage to the lungs. Hydrogen cyanide would be another example of a chemical substance that would require this level of protection. Hydrogen cyanide can be fatal if inhaled or absorbed through the skin, so the use of a fully encapsulating suit is required to adequately protect the wearer.

FIGURE 35-6 Vapor-protective ensembles retain body heat. Wearing this level of protection increases the risk of heat-related issues.

A **liquid splash–protective ensemble** is designed to protect the wearer from chemical splashes (**FIGURE 35-7**). Equipment that meets this standard has been tested for penetration against a battery of chemicals. *The tests do not include any gases, because this level of protection is not considered to be vapor protection.*

Responders may choose to wear liquid splash–protective clothing based on the anticipated hazard posed by a particular substance. Such clothing does not provide total body protection from gases or vapors, and it should not be used for incidents involving liquids that emit vapors known to affect or be absorbed through the skin. This level of protection may consist of several pieces of clothing and equipment designed to protect the skin and eyes from chemical splashes. Some agencies, depending on the situation, choose to have their personnel wear liquid splash protection over or under structural firefighting clothing.

Operations-level responders wear liquid splash–protective clothing for a variety of reasons, including entering the hot zone, performing decontamination, or constructing isolation barriers such as dikes, diversions, retention areas, or dams.

FIGURE 35-7 Liquid splash–protective clothing is worn whenever there is the danger of chemical splashes.

FIGURE 35-8 A vapor-protective ensemble (Level A) envelops the wearer in a totally encapsulating suit.

OSHA/EPA Levels of Protection

This section outlines the OSHA/EPA guidelines for Levels A, B, and C chemical-protective clothing. Specifics on PPE selection, levels of protection, and maintenance can be found in CFR 1910.120 and in Appendix A of that standard. Some information contained in NFPA 1990 and OSHA 1910.120 has similarities, but is not the same. The procedures for the **donning**—the process of putting on an ensemble of PPE—and **doffing**—the process of taking off an ensemble of PPE—of each level of chemical-protective equipment are described in skill drills in the following sections.

Decontamination is also linked to PPE use and must be understood by all responders operating at the scene. Refer to your AHJ when it comes to all aspects of PPE use and decontamination processes and procedures.

Level A

A **Level A ensemble** consists of a fully encapsulating vapor-protective ensemble that completely envelops both the wearer and the selected respiratory protection (**FIGURE 35-8**). Level A should be used when the hazardous material identified requires the highest level of protection for the skin, eyes, and lungs. Such an ensemble is effective against vapors, gases, mists, and dusts and is typically indicated when the operating environment exceeds IDLH values for skin absorption.

PPE is an area where the NFPA and OSHA/EPA intersection may be confusing: Level A protection must meet the performance requirements outlined in NFPA 1990 and be worn with SCBA that has been certified as meeting NFPA 1981 or NFPA 1986, *Standard on Respiratory Protection Equipment for Tactical and Technical Operations.* Assessment of the overall performance of the vapor-protective garment consists of testing liquid tightness, inward leakage, and abrasion resistance, among other tests. Additionally, NFPA specifies testing for flash fire protection.

Responders should approach potential or actual thermal environments with extreme caution. In this setting, a Level A ensemble more than addresses the skin and lung threat; thus, the temperature extreme is the key factor that must be addressed during a risk assessment. In all cases, a potentially flammable atmosphere should be considered an extremely dangerous situation. In such circumstances, Level A suits, even with the flash fire component of the suit in place, provide very limited protection.

SAFETY TIP

Keep in mind that flash fire testing is not intended to make the garment flameproof.

Essentially, each part of the Level A ensemble must pass a particular set of testing challenges prior to receiving certification based on the NFPA testing standards. Users of any garment that meets the performance requirements set forth by the NFPA standard can rest assured that the garment will withstand reasonable insults from most mechanical-type hazards encountered on the scene. Of course, achieving a third-party independent certification based on the NFPA standards does not mean the suit is invincible and cannot fail; it simply means that the equipment will hold up under "normal" conditions. It is up to the user to be aware of the hazards and avoid situations that may cause the garment to fail. Also, direct contact between the suit fabric and a cryogenic material, such as liquid nitrogen or liquid helium, may result in immediate suit failure.

It can be quite difficult to have a good field of vision when wearing a vapor-protective ensemble, which increases the possibility that the wearer may unknowingly bump into sharp objects or rub against materials that might puncture or abrade the suit's vapor protection. Therefore, some forethought about the operating environment should occur well before entering the contaminated atmosphere. As always, a risk-versus-benefit thought process should prevail. The "best" level of protection is the one that is the most appropriate for the hazard and the mission. Remember, thorough detection and monitoring actions will help you determine the nature of the operating environment.

To don a vapor-protective ensemble (Level A), follow the steps in **SKILL DRILL 35-1**. To doff a vapor-protective ensemble (Level A), follow the steps in **SKILL DRILL 35-2**.

SKILL DRILL 35-1

Donning a Vapor-Protective (Level A) Ensemble NFPA 470: 9.2.1

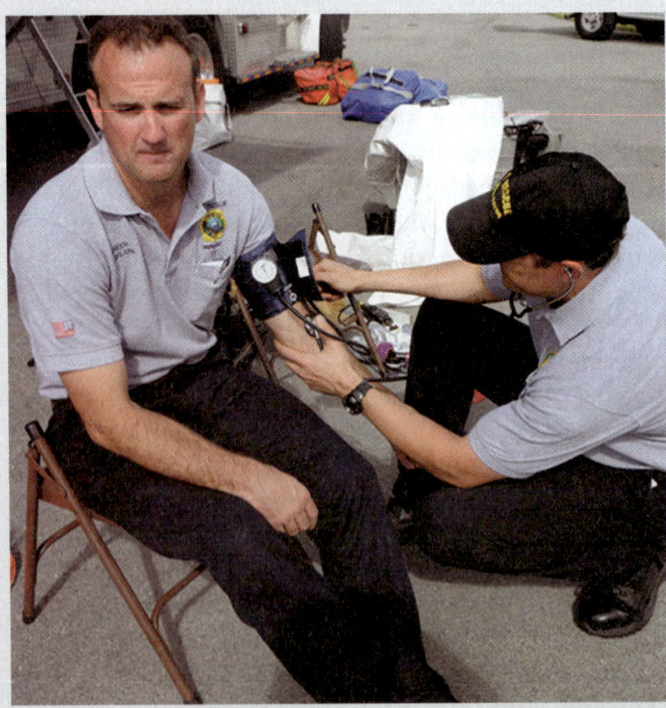

1. Conduct a pre-entry briefing, medical monitoring, and equipment inspection. Ensure that the selected level of protection is appropriate for the mission.

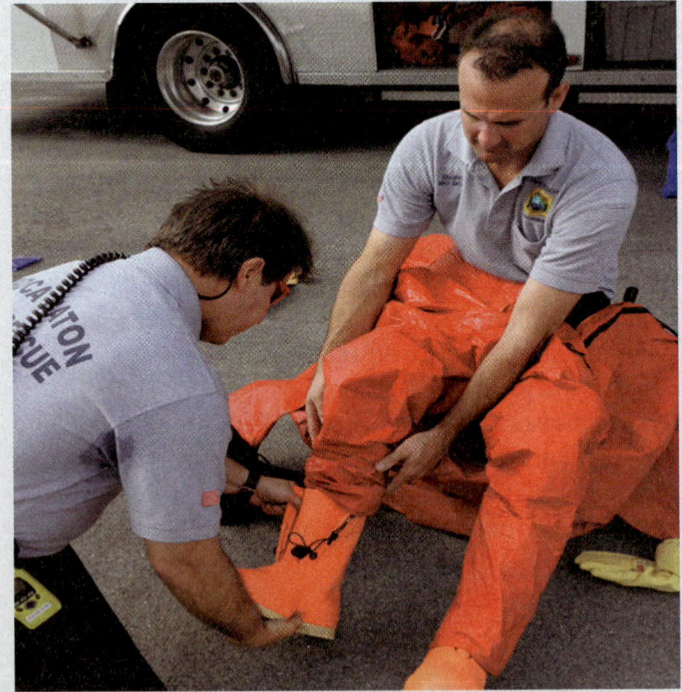

2. While seated, pull on the suit to waist level; pull on the chemical boots over the top of the chemical suit. Fold the suit boot covers over the tops of the boots.

SKILL DRILL 35-1 CONTINUED

Donning a Vapor-Protective (Level A) Ensemble NFPA 470: 9.2.1

3. Stand up and don the SCBA frame and SCBA face piece, but do not connect the regulator to the face piece.

4. Place the helmet on your head.

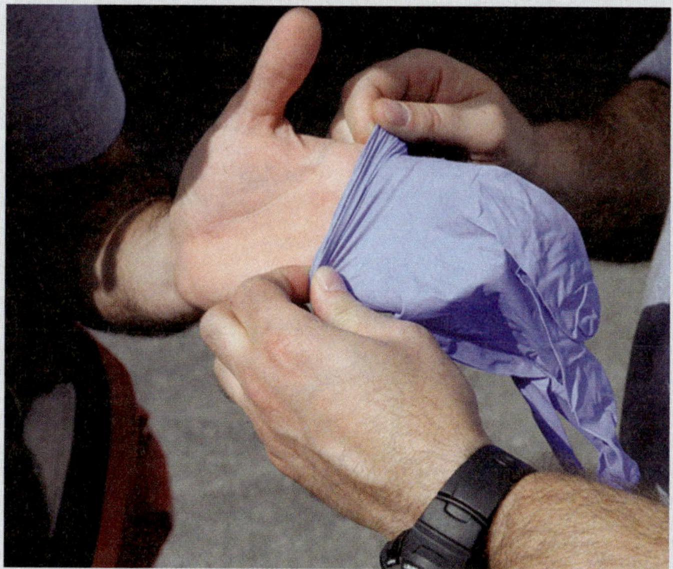

5. Don the inner gloves.

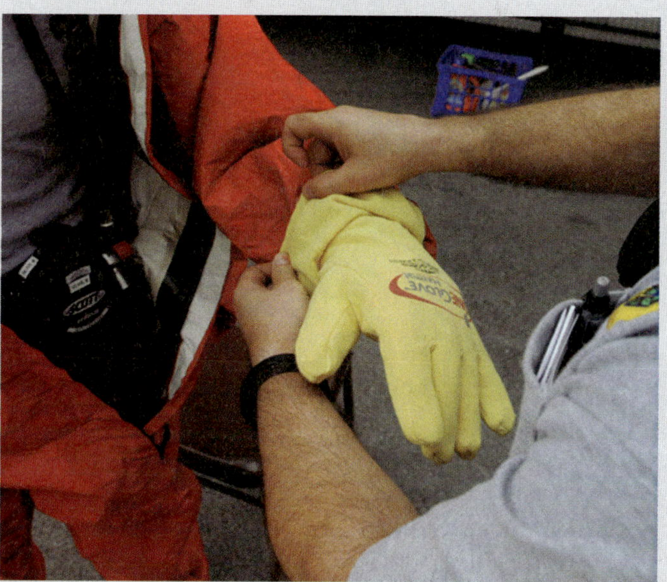

6. With assistance, complete donning the suit by placing both arms in the suit, pulling the expanded back piece over the SCBA, placing the chemical suit over your head, and donning the outer chemical gloves (if required).

Continues.

SKILL DRILL 35-1 CONTINUED

Donning a Vapor-Protective (Level A) Ensemble NFPA 470: 9.2.1

7. Instruct the assistant to connect the regulator to the SCBA face piece and ensure air flow.

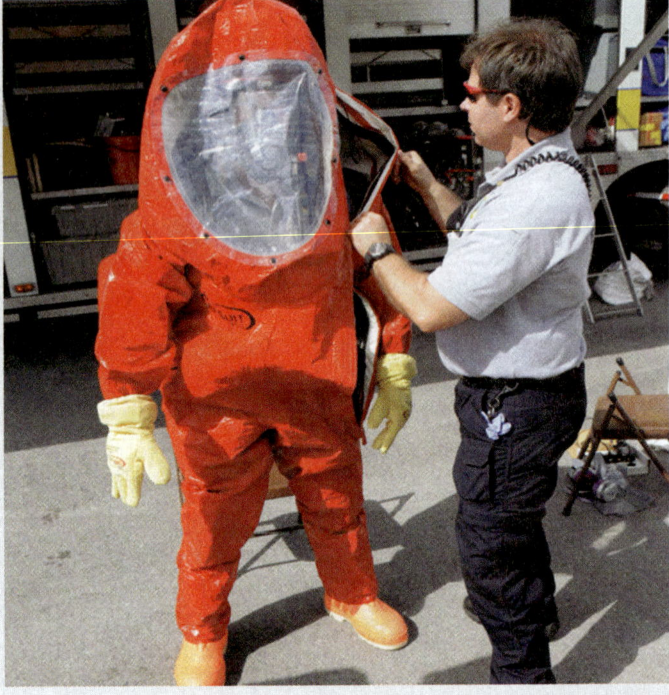

8. Instruct the assistant to close the chemical suit by closing the zipper and sealing the splash flap.

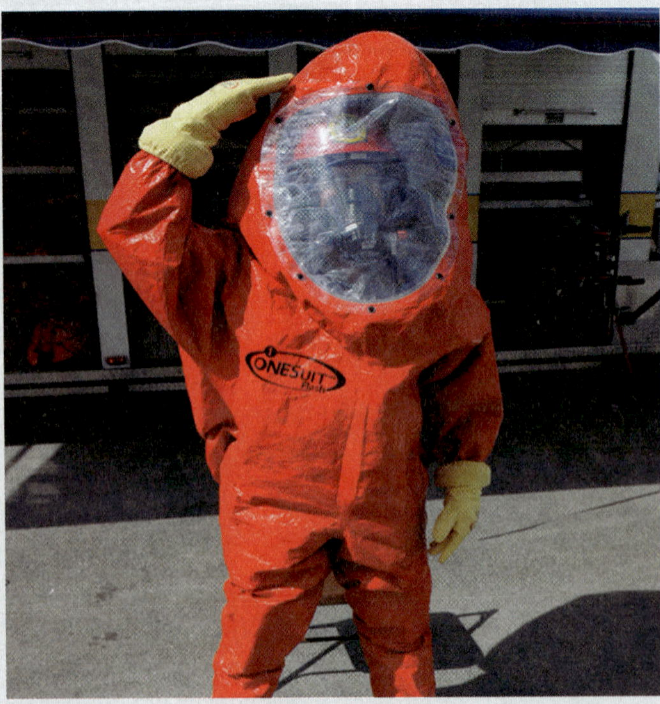

9. Review hand signals and indicate that you are okay.

SKILL DRILL 35-2

Doffing a Vapor-Protective (Level A) Ensemble NFPA 470: 9.2.1

1. After completing the established decontamination procedure for the mission, proceed to the clean area for suit doffing. Pull your hands out of the outer gloves and your arms from the sleeves, and cross your arms in front inside the suit.

2. Instruct the assistant to open the chemical splash flap and suit zipper.

3. Instruct the assistant to begin at the head and roll the suit down and away until the suit is below waist level.

4. Instruct the assistant to complete rolling the suit from the waist to the ankles; step out of the attached chemical boots and suit.

Continues.

SKILL DRILL 35-2 CONTINUED

Doffing a Vapor-Protective (Level A) Ensemble NFPA 470: 9.2.1

5. Doff the SCBA frame. The face piece should be kept in place while the SCBA frame is doffed.

6. Take a deep breath and doff the SCBA face piece; carefully peel off the inner gloves, and walk away from the clean area. Go to the rehabilitation area for medical monitoring, rehydration, and personal decontamination shower. After use, ensure that the garment is inspected, maintained, and stored (or disposed of) in accordance with the manufacturer's guidelines and the AHJ's policies and procedures.

Level B

A **Level B ensemble**—that is, a liquid splash–protective ensemble—consists of a single- or multipiece chemical-protective garment, boots, gloves, and SCBA (**FIGURE 35-9**). The SCBA components, tested to NFPA 1981 and NFPA 1986 criteria, should be considered as well when wearing liquid splash protection. This type of protective ensemble should be used when the type and atmospheric concentration of identified substances require a high level of respiratory protection but a moderate level of skin protection. The kinds of gloves and boots chosen will depend on the physical and chemical properties of the identified chemical. As with Level A ensembles, specific performance-based testing standards are used to certify this level of CPC and the associated equipment.

Level B protection is the workhorse of hazardous materials response—it is a very common level of protection and is often chosen for its versatility. Personnel initially processing a clandestine drug laboratory, performing preliminary missions for reconnaissance, or engaging in detection and monitoring duties commonly wear such an ensemble. However, the typical Level B ensemble provides little or no flash fire protection.

FIGURE 35-9 A liquid splash–protective ensemble (Level B) provides a high level of respiratory protection but less skin protection.

Thus, it should be viewed in the same manner as the Level A ensemble when it comes to potentially flammable environments and other considerations of use such as protection from mechanical hazards (e.g., cuts, tears, abrasions).

Garments and ensembles that are worn for Level B protection should comply with the performance requirements found in the pertinent sections of NFPA 1990.

SAFETY TIP

According to the OSHA HAZWOPER regulation, Level B is the minimum level of protection to be worn when operating in an unknown environment.

You may also encounter single-piece garments that are worn as Level B protection. These suits, referred to in the field as encapsulating Level B garments, are not constructed to be "vapor tight" like Level A garments. Encapsulating Level B garments do not have vapor-tight zippers, seams, or one-way relief valves around the hood. Although the encapsulating Level B suit may look a lot like a Level A ensemble, it is not constructed similarly and will not offer the same level of protection.

To don and doff a Level B encapsulated CPC ensemble, follow the same steps found in Skill Drill 35-1 and Skill Drill 35-2. Remember, the difference between the Level A ensemble and the Level B encapsulating ensemble is not the procedure—it is the construction and performance of the garment.

To don a liquid splash–protective (Level B) nonencapsulated CPC ensemble, follow the steps in **SKILL DRILL 35-3**. To doff a liquid splash–protective (Level B) nonencapsulated CPC ensemble, follow the steps in **SKILL DRILL 35-4**.

Level C

A **Level C ensemble**—also a liquid splash–protective ensemble—is appropriate when the type of airborne contamination is known, its concentration is measured, and the criteria for using an air-purifying respirator (APR) are met. Typically, Level C ensembles are worn with a National Institute of Occupational Safety and Health (NIOSH)–approved CBRN APR or a powered air-purifying respirator (PAPR). As with Levels A and B, specific performance-based testing standards are used to certify this type of CPC. The complete liquid splash–protective ensemble consists of standard work clothing, CPC, chemical-resistant gloves, and a form of respiratory protection other than an SCBA or supplied-air respirator (SAR) system (**FIGURE 35-10**).

Level C equipment is appropriate when potentially harmful dermal or respiratory exposure is unlikely. In many cases, Level C ensembles are worn in low-hazard situations such as clean-up activities lasting hours or days, *once an area is fully characterized and the hazards are found to be low enough to allow this level of protection*, or after responders mitigate the problem to the extent that they can dress down to this lower level to complete the mission. Many law enforcement agencies have provided their officers with Level C ensembles to be carried in the trunk of patrol cars. Based on the mission of perimeter scene control, this may be a prudent level of protection.

The garment selected for the Level C ensemble must meet the appropriate performance requirements listed in NFPA 1990. Respiratory protection may be provided by a half-face (with eye protection) or full-face mask.

SKILL DRILL 35-3

Donning a Liquid Splash–Protective (Level B) Nonencapsulated Ensemble NFPA 470: 9.2.1

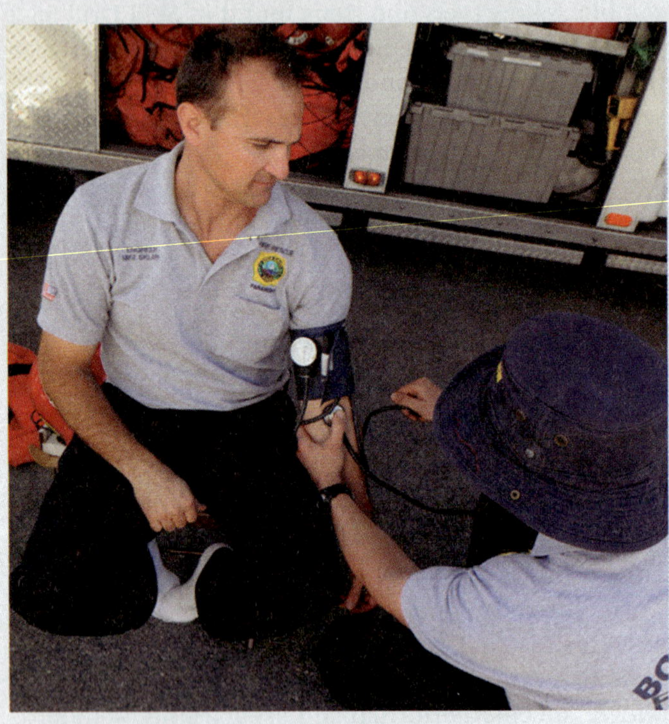

1. Conduct a pre-entry briefing, medical monitoring, and equipment inspection. Ensure that the selected level of protection is appropriate for the mission.

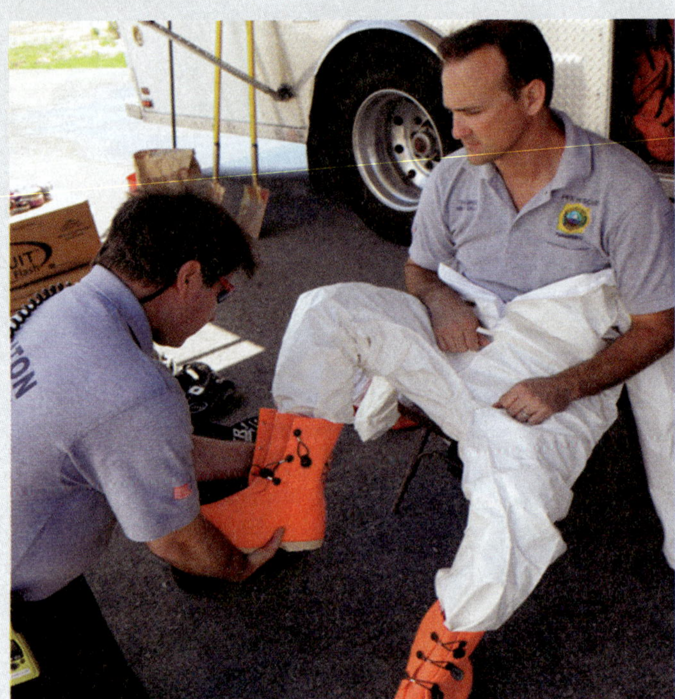

2. Sit down, and pull on the suit to waist level; pull on the chemical boots over the top of the chemical suit.

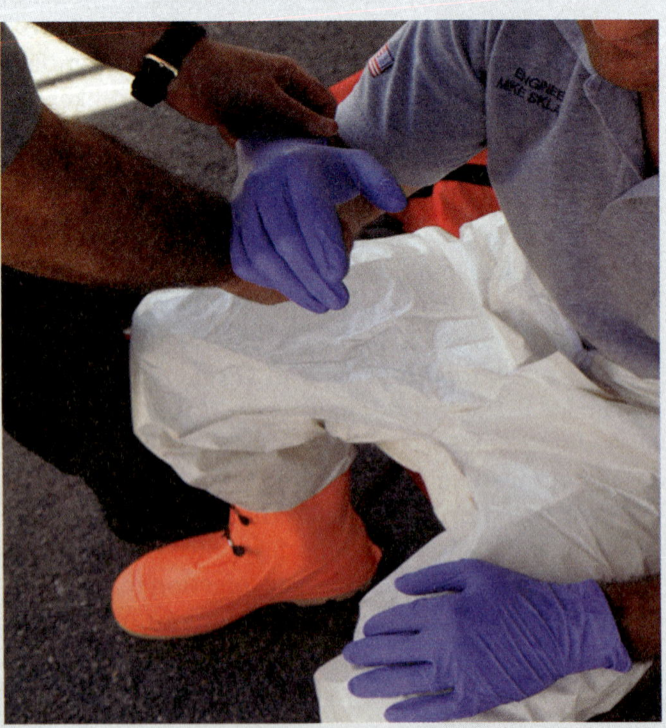

3. Don the inner gloves.

SKILL DRILL 35-3 CONTINUED

Donning a Liquid Splash–Protective (Level B) Nonencapsulated Ensemble NFPA 470: 9.2.1

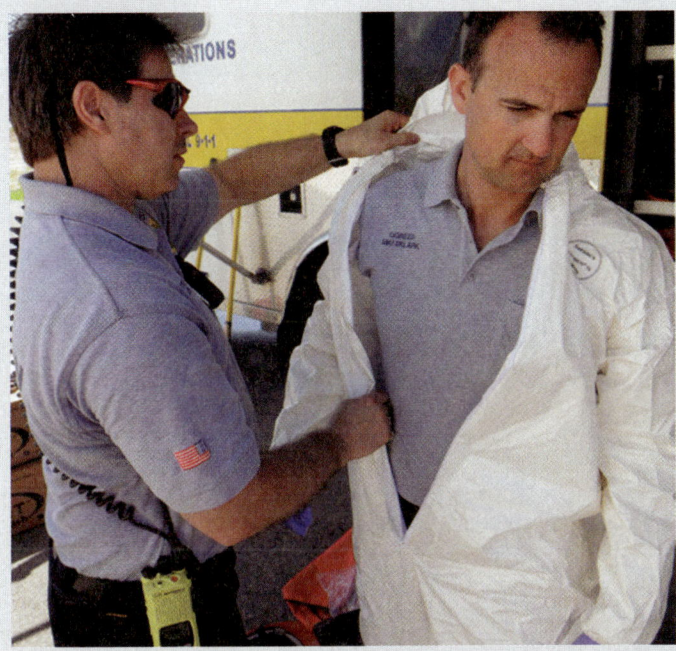

4. With assistance, complete donning the suit by placing both arms in the suit and pulling the suit over your shoulders. Instruct the assistant to close the chemical suit by closing the zipper and sealing the splash flap.

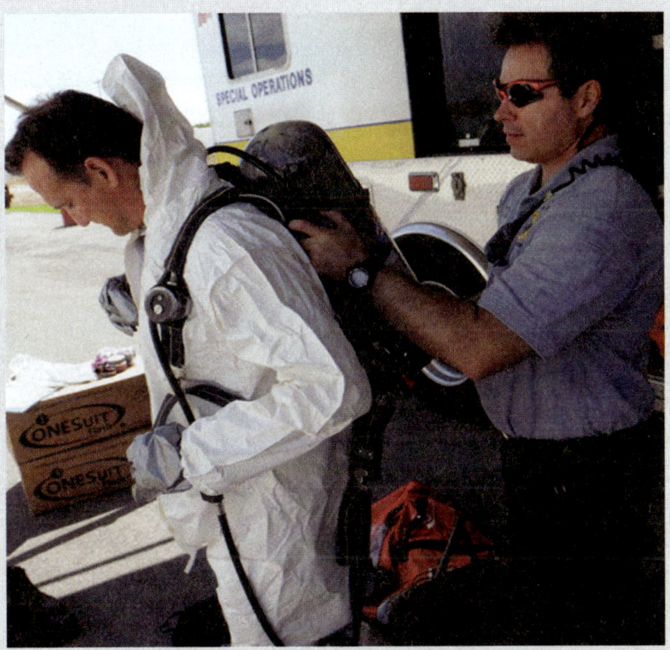

5. Don the SCBA frame and SCBA face piece, but do not connect the regulator to the face piece.

6. With assistance, pull the hood over your head and the SCBA face piece. Place the helmet on your head. Put on the outer gloves (over or under the sleeves, depending on the AHJ requirements for the incident). Instruct the assistant to connect the regulator to the SCBA face piece, and ensure you have air flow. Review hand signals and indicate that you are okay.

SKILL DRILL 35-4

Doffing a Liquid Splash–Protective (Level B) Nonencapsulated Ensemble NFPA 470: 9.2.1

1. After completing the wash/rinse cycle, proceed to the clean area for PPE doffing. The SCBA frame is removed first. The unit may remain attached to the regulator while the assistant helps the responder out of the PPE, or the air supply may be detached from the regulator, leaving the face piece in place to provide face and eye protection while the rest of the doffing process is completed.

2. Instruct the assistant to open the chemical splash flap and suit zipper.

SKILL DRILL 35-4 CONTINUED

Doffing a Liquid Splash–Protective (Level B) Nonencapsulated Ensemble NFPA 470: 9.2.1

3. Remove your hands from the outer gloves and your arms from the sleeves of the suit. Cross your arms in front inside the suit. Instruct the assistant to begin at the head and roll the suit down and away until the suit is below waist level.

4. Sit down and instruct the assistant to complete rolling down the suit to the ankles; step out of the attached chemical boots and suit.

Continues.

SKILL DRILL 35-4 CONTINUED

Doffing a Liquid Splash–Protective (Level B) Nonencapsulated Ensemble NFPA 470: 9.2.1

5. Doff the SCBA face piece and helmet

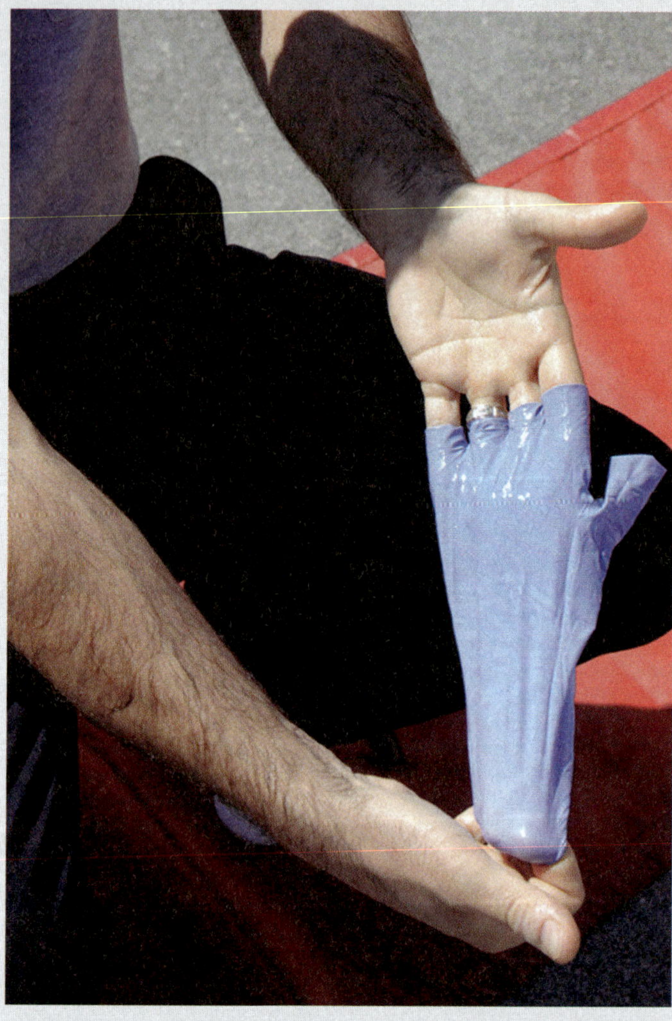

6. Carefully peel off the inner gloves and go to the rehabilitation area for medical monitoring, rehydration, and personal decontamination shower. After use, ensure that the garment is inspected, maintained and stored (or disposed of) in accordance with the manufacturer's guidelines and the AHJ's policies and procedures.

FIGURE 35-10 A Level C, liquid splash–protective ensemble includes chemical-protective clothing, gloves, and boots, as well as NIOSH–approved respiratory protection (air-purifying respirator or powered air-purifying respirator).

To don a liquid splash–protective ensemble (Level C), follow the steps in **SKILL DRILL 35-5**. To doff a liquid splash–protective ensemble (Level C), follow the steps in **SKILL DRILL 35-6**.

Level D

A **Level D ensemble** offers the lowest level of protection. It typically consists of coveralls, work shoes, hard hat, gloves, and standard work clothing. This type of equipment should be used only when the atmosphere contains no known hazard and when work functions preclude splashes, immersion, or the potential for unexpected inhalation of or contact with hazardous levels of chemicals. Level D protection should be used when the situation involves nuisance contamination (such as dust) only. It should not be worn on any site where respiratory or skin hazards exist.

To don a Level D ensemble, follow the steps in **SKILL DRILL 35-7**. When doffing a Level D ensemble, the procedure is simply a reversal of the donning process. Because no chemical contact is expected with Level D, it is a nonhazardous process.

Correlating NFPA 1990 and OSHA/EPA PPE Ensembles

The correlation shown in **TABLE 35-1** is reproduced from Annex material found in NFPA 1990 and intended to summarize the linkage between the various NFPA 1990 ensemble descriptions based on performance standards and the OSHA/EPA PPE designations found in Title 29, CFR 1910.120, "Hazardous Waste Operations and Emergency Response." *Even though NFPA 1991, 1992, and 1994 are incorporated into NFPA 1990, you will see specific chapter references to those specific performance standards in NFPA 1990.*

For clarity in this chapter, think of the correlation between NFPA 1990 and the OSHA/EPA designations like this:

NFPA 1991/NFPA 1994 = OSHA/EPA Level A = vapor-protective ensembles

NFPA 1992 and certain classes of NFPA 1994 = OSHA/EPA Level B/Level C = liquid splash–protective ensembles

Safety

There are many hazards associated with wearing PPE. These hazards are best addressed by understanding the NFPA performance requirements for chemical-protective ensembles and the safety considerations that arise when wearing PPE.

SAFETY TIP

As exemplified by the CBRN abbreviation, many hazards can be encountered during the course of a hazardous materials incident. Given this possibility, multiple layers or multiple types of protection may have to be used in some situations. You should also understand the working environment and match the right garment to the anticipated hazards.

Responder Safety

Working in PPE is a hazardous proposition on two different levels. First, simply by wearing PPE, the responder acknowledges that some degree of danger exists: If there were no hazard, there would be no need for the PPE! Second, wearing the PPE puts inherent mental

SKILL DRILL 35-5

Donning a Liquid Splash–Protective (Level C) Ensemble
NFPA 470: 9.2.1

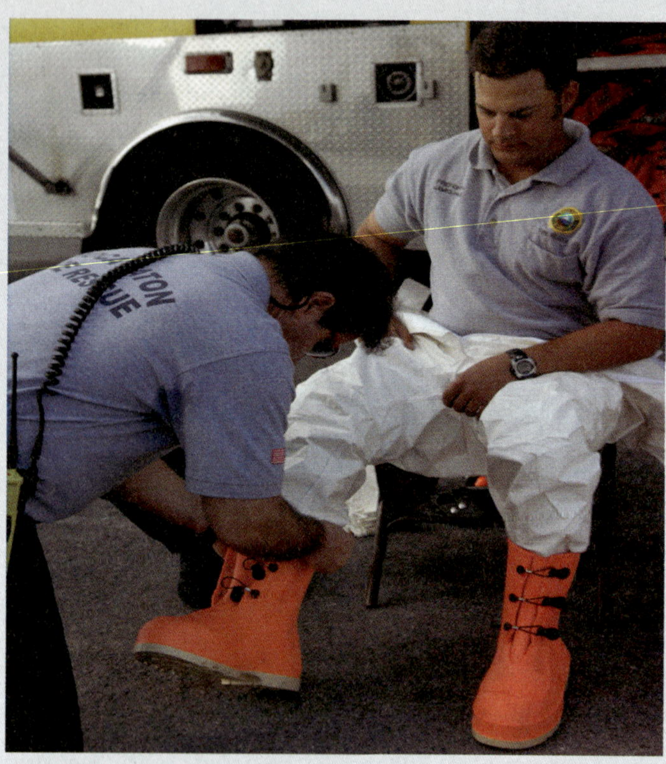

1. Conduct a pre-entry briefing, medical monitoring, and equipment inspection. Ensure that the selected level of protection is appropriate for the mission. While seated, pull on the suit to waist level; pull on the chemical boots over the top of the chemical suit.

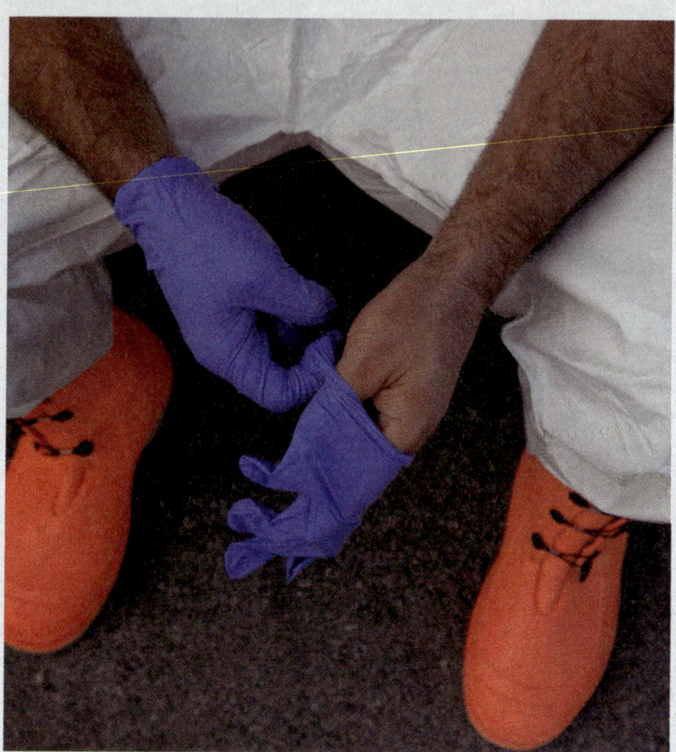

2. Don the inner gloves.

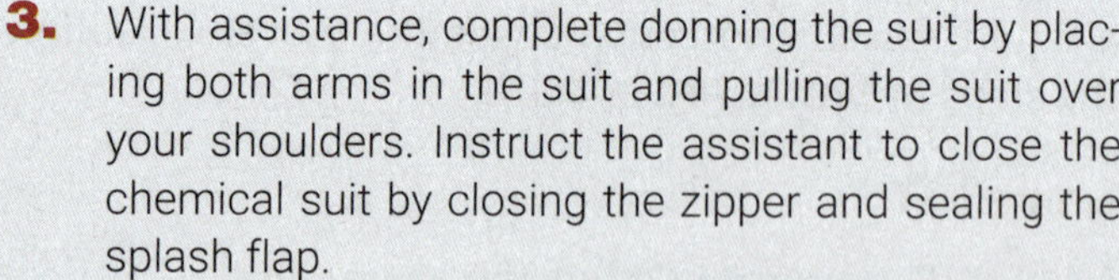

3. With assistance, complete donning the suit by placing both arms in the suit and pulling the suit over your shoulders. Instruct the assistant to close the chemical suit by closing the zipper and sealing the splash flap.

SKILL DRILL 35-5 CONTINUED

Donning a Liquid Splash–Protective (Level C) Ensemble
NFPA 470: 9.2.1

4. Don the APR/PAPR face piece. With assistance, pull the hood over your head and the APR/PAPR face piece. Place the helmet on your head. Pull on the outer gloves. Review hand signals and indicate that you are okay.

and physical stresses on the responder, separate and apart from the conditions imposed by the operating environment. The next sections are devoted to raising your awareness of the issues that may arise from the very protective clothing and equipment intended to keep you safe.

The next section, which covers in-suit cooling measures, addresses one of those risks by examining the various cooling technologies that may be used to reduce the effects of overheating inside a chemical-protective garment.

Responders should also be aware that their field of vision will be compromised by the face piece of an SCBA or APR and even by the encapsulating garment. This factor may result in the responder slipping in a puddle of spilled chemicals or tripping on something. Moreover, the face piece often fogs up at some point, further limiting the responder's vision. This creates many problems, such as the inability to read labels, see other responders, see the screens on detection and monitoring devices, or quickly find an escape route in the event of an unforeseen problem in the hot zone. Wearing bulky PPE, such as an encapsulating suit, may inhibit the mobility of the wearer to the point that bending over becomes difficult or reaching for valves above head level is taxing. Furthermore, when gloves become contaminated with chemicals (especially solvents), they become slippery, making it difficult to effectively grip tools, handrails, or ladder rungs. All in all, the environment inside the PPE can be just as challenging as the conditions outside the suit.

To mitigate some of the potential safety considerations that arise when wearing PPE, responders can employ a variety of safety procedures and training. To begin, conducting a pre-entry medical evaluation is important to catch the medical indicators that may signal a responder should not wear PPE.

Further guidance on pre-entry medical evaluation can also be found in NFPA 470, chapters 46 and 47, in the EMS basic life support and advanced life support sections. Keep in mind that the medical monitoring station may serve many purposes at the scene of a hazardous materials event. The primary role of the medical monitoring station is to evaluate the medical status of

SKILL DRILL 35-6

Doffing a Liquid Splash–Protective (Level C) Ensemble NFPA 470: 9.2.1

1. After completing decontamination, proceed to the clean area. As with level B, the assistant opens the chemical splash flap and suit zipper. Remove your hands from the gloves and your arms from the sleeves. Remove the helmet. Instruct the assistant to begin at the head and roll the suit down below waist level. Instruct the assistant to complete rolling down the suit and to take the outer boots and suit away. The assistant helps remove the inner gloves. Remove the APR/PAPR.

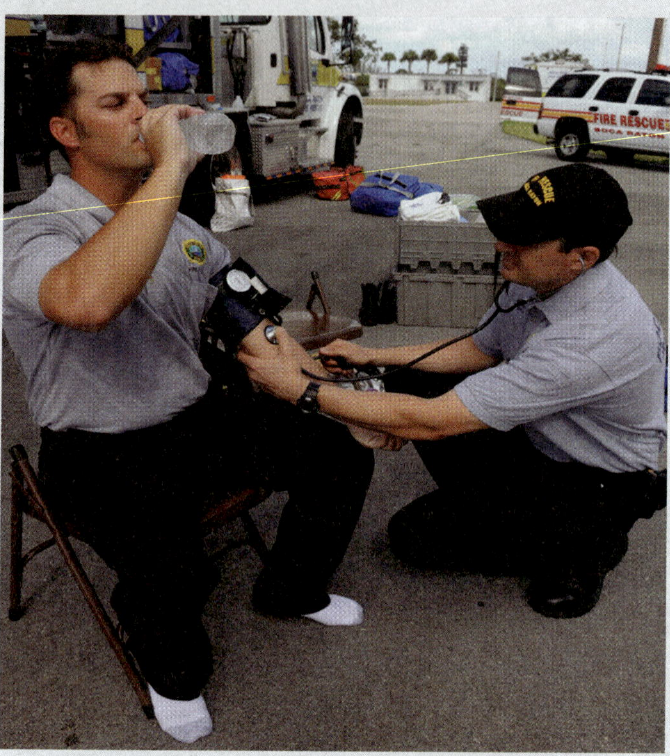

2. Go to the rehabilitation area for medical monitoring, rehydration, and personal decontamination shower. After use, ensure that the garment is inspected, maintained, and stored (or disposed of) in accordance with the manufacturer's guidelines and the AHJ's policies and procedures.

the entry team, the backup team, and those personnel assigned to decontamination duties. At larger incidents, a medical group or team may be required to obtain basic physiological information from each responder and plan to provide care in the event a responder becomes a patient.

The use of the buddy system is another way that responders can mitigate some of the risks that may be encountered at the scene of a hazardous materials/WMD incident. Recall that the OSHA HAZWOPER regulation requires the use of the buddy system.

SAFETY TIP

Remember to take rehabilitation breaks throughout the hazardous materials incident. Wearing any type of PPE requires a great deal of physical energy and mental concentration. Responders should also acknowledge the psychological stress that wearing PPE may present. Claustrophobia is a common problem when wearing chemical-protective equipment, especially encapsulated suits.

SKILL DRILL 35-7

Donning and Doffing a Level D Ensemble NFPA 470: 9.2.1

1. Conduct a pre-entry briefing, medical monitoring, and equipment inspection. Ensure that the selected level of protection is appropriate for the mission. Don the Level D suit. Don boots. Don safety glasses or chemical goggles. Don a hard hat. Don gloves, a face shield, and any other required equipment.

2. Doff all PPE. No chemical contact is expected with Level D, so no particular precautions are needed when doffing this ensemble.

In-Suit Cooling Technologies

Hazardous materials responders operating in protective clothing should be aware of the signs and symptoms of heat exhaustion, heat stress, heat stroke, dehydration, and illness caused by extreme cold. The most common maladies striking anyone wearing PPE are heat related (refer to Chapter 33, *Estimating Potential Harm and Planning a Response*, for more specifics). If the body is unable to disperse heat because it is covered by an ensemble of PPE, serious short- and long-term medical issues could occur.

TABLE 35-1 NFPA 1990 Ensemble Descriptions and Corresponding OSHA/EPA PPE Designations

NFPA 1990 Ensemble Description*	Equivalent OSHA/EPA PPE Designations
NFPA 1991 worn with NFPA 1981– or NFPA 1986–certified SCBA	OSHA/EPA Level A
NFPA 1994 Class 1 worn with NFPA 1981– or NFPA 1986–certified SCBA	OSHA/EPA Level A
NFPA 1994 Class 2/2R worn with NFPA 1981– or NFPA 1986–certified SCBA	OSHA/EPA Level B
NFPA 1992 worn with NFPA 1981– or NFPA 1986–certified SCBA	OSHA/EPA Level B
NFPA 1994 Class 2/2R worn with NIOSH CBRN APR or PAPR	OSHA/EPA Level C
NFPA 1992 worn with NIOSH CBRN APR or PAPR	OSHA/EPA Level C
NFPA 1994 Class 3/3R worn with NFPA 1981– or NFPA 1986–certified SCBA	OSHA/EPA Level B
NFPA 1994 worn with NIOSH CBRN APR or PAPR	OSHA/EPA Level C
NFPA 1994 Class 4/4R worn with NFPA 1981– or NFPA 1986–certified SCBA	OSHA/EPA Level B
NFPA 1994 Class 4/4R worn with NIOSH CBRN APR or PAPR	OSHA/EPA Level C
NFPA 1994 Class 5 worn with NFPA 1981– or NFPA 1986–certified SCBA	OSHA/EPA Level B

*See NFPA 1990 for specifics on Classes 1 through 5.
Data from National Fire Protection Association (NFPA). 2022. *NFPA 1990, Standard for Protective Ensembles for Hazardous Materials and CBRN Operations*, 2023 Edition. Quincy, MA: NFPA; and Occupational Safety and Health Administration (OSHA) 1970. "1910.120 - Hazardous Waste Operations and Emergency Response." Accessed September 15, 2021. https://www.osha.gov/laws-regs/regulations/standardnumber/1910/1910.120

Most heat-related illnesses are typically preceded by dehydration. It is important for responders to stay hydrated so that they can function at their maximum capacity. As a frame of reference, athletes should consume approximately 500 mL (16 oz) of fluid (water) prior to an event and 200–300 mL (6–10 oz) of fluid at regular intervals during the event. Responders can be considered occupational athletes—so keep up on your fluids!

In an effort to combat heat stress while wearing PPE, many response agencies employ some form of cooling technology under the garment. These technologies include, but are not limited to, air-, ice-, and water-cooled vests, along with phase-change cooling technology. Many studies have been conducted on the various forms of cooling technology. Each of these approaches is designed to accomplish the same goal—to reduce the impact of heat stress on the human body. As mentioned earlier, the same suit that seals you up against the hazards also seals in the heat, defeating the body's natural cooling mechanisms.

Air-cooled vests work similarly to the fluid-chilled systems described later in this section. As the cooler air passes by the skin, heat is drawn away—by convection—from the body and released into the atmosphere. Forced-air systems are designed to function as the first level of cooling the body would naturally employ. Typically, these systems are lightweight and provide long-term cooling benefits, but mobility is limited because the umbilical is attached to an external, fixed compressor.

SAFETY TIP

Approximately 90 percent of all body heat is generated by the organs and muscles located in your torso.

Ice-cooled or gel-packed vests are commonly used due to their low cost, ready portability, and unlimited "recharging" ability (i.e., refreezing the packs). These vest-like garments are designed to be worn around the torso. The principle underlying this approach is that the ice-chilled vest absorbs the heat generated by the body. On the downside, this technology is bulkier and heavier than the aforementioned systems, and it may cause discomfort to the wearer due to the nature of the ice-cold vest near the skin. Additionally, the cold temperature near the skin may actually fool the body into

Voice of Experience

In my years at the St. Joseph County Airport Public Safety Division, being cross-trained as a certified law enforcement police officer, a certified firefighter and aircraft rescue firefighter, emergency medical technician, hazardous materials technician, and hazardous materials instructor, all of the basics were covered for responding to any type of police, fire, medical, airport, or hazardous materials incident.

While on duty, I was called to assist the city fire department on an incident involving radioactive materials. Upon arriving at the emergency incident scene, the department's incident commander informed me that there was a spill of radioactive materials at a construction site. The spill involved the breaking of a portable density gauge metering instrument that was being used to measure the density of core samples of asphalt. Even though the fire department had radiological monitoring instruments on its rig, not one of the hazardous materials response team members or any other firefighters at this scene knew how to read or use the instruments.

To assist those responders, I quickly checked the batteries of their radiological monitors and survey instruments. I performed the standard operational checks of their radiological survey meters, charged their personal dosimeters, and set the radiological monitoring probes in the correct position for monitoring and surveying. I assessed the hazardous materials team's Level A protective gear, as well as the backup team's PPE, and attached their personal dosimeters to their turnout PPE. I then finished donning my own PPE, put on an SCBA, attached a personal dosimeter to my turnout collar, and accompanied the team to the incident area so that I could read and monitor the radiological readings on the instruments for them.

It was determined that this hazardous materials incident involved a moisture density gauge that used two different radioactive sources to produce two different types of radiation. One of the radioactive sources, cesium-136, emits gamma-ray photon radiation to determine density; the other radioactive source, americium-241 (combined with nonradioactive beryllium), emits neutron radiation to determine moisture content.

It was also determined that there were no injured victims at the scene. Proper exclusion zones were set up to consider both contamination and exposure control. A decontamination corridor was established, and all decontamination protocols were followed.

Although early responders generally do not have extensive training or equipment to perform radioactive contamination surveys, they can significantly reduce the spread of contamination by applying relatively simple measures of contamination control. It is useful to have a checklist of information to be transmitted when requesting assistance or reporting a radiological incident. Knowing the principles of protection is vital: recognizing the hazards, time, distance, and shielding. First responders should be very familiar with their radiological monitoring instruments, including how to perform the basic operational check, how to accurately read the survey meters, and what the instruments' limitations are.

Judith Hoffmann (RET), CECM, EMT
Officer
Hoffmann Consulting
South Bend, Indiana

thinking it is cold instead of hot, thereby encouraging retention of even more heat.

Fluid-chilled systems operate by pumping ice-chilled liquids (water is often used, so these systems are referred to as "water-cooled") from a reservoir, through a series of tubes held within a vest-like garment, and back to the reservoir (**FIGURE 35-11**). Mobility may be limited with some varieties of this system, because the pump may be located away from the garment. Some systems incorporate a battery-operated unit worn on the hip, but the additional weight may increase the body's workload and generate more heat, thereby defeating the purpose of the cooling vest.

Phase-change cooling technology operates in a similar fashion to the ice- or gel-packed vests (**FIGURE 35-12**). The main difference between the two approaches is that the temperature of the material in the phase-change packs is chilled to approximately 60°F (15.5°C), and the fabric of the vest is designed to wick perspiration away from the body. The packs

FIGURE 35-11 A fluid-chilled or water-cooled system.

FIGURE 35-12 Phase-change cooling technology.

typically "recharge" more quickly than those of an ice- or gel-packed vest. Even though the temperature of the phase-change pack is higher than the temperature of an ice- or gel-packed vest, it is sufficient to absorb the heat generated by the body.

Communication and Hand Signals

Wearing PPE not only presents physical challenges but also, in many cases, compromises communications. Communications are often problematic on emergency scenes, so take whatever steps you can to minimize these problems before anyone enters a contaminated atmosphere. Prior to entry, all radio communications should be sorted out and tested. This includes all approved, available, and provided AHJ communications equipment. The AHJ must ensure that all responders are competent in the operation of its preferred communications equipment.

To back up radio communication, all responders on the scene should have a method to communicate by universally accepted hand signals. These hand signals could be used to rapidly share messages about problems with an air supply, a suit problem, or any other problem that might occur in the hot zone.

Your safety briefing and the IAP should always include a review of hand signals. Everyone should know the basic communication "fall back" hand signal to ensure that in the event of an in-suit emergency (e.g., loss of breathing air, SCBA cylinder failure, or acute illness and injury), anyone having visual contact with the responders will quickly understand that a problem is occurring. **FIGURE 35-13** illustrates common hand signals used by responders. Your AHJ may have different hand signals; it is the responsibility of all responders to know how to use them correctly.

Reporting and Documenting the Incident

As with any other type of incident, documenting the activities carried out during a hazardous materials/WMD incident is an important part of the response. Many responders may pass through the scene, and it could be quite difficult to sort everything out when it comes time to reconstruct the events for an accurate and legally defensible incident report. Good documentation after the incident is directly correlated with how well organized the response was.

Along with the formal written accounts of the event, some agencies require that personnel fill out exposure records, which include information such as the name of the substances involved in the incident and the level of protection used. This information, coupled with a comprehensive medical surveillance program, provides a method to chronicle the exposure history of the responders over a period of time. Consult your AHJ for the exact details and procedures for reporting and documenting the incident.

FIGURE 35-13 **A.** A hand on top of the head making a tapping motion indicates "I'm okay." **B.** Hands across the throat indicate "air problems." **C.** Hands over the head in a waving motion or both hands tapping the head is a signal for "trouble." This could indicate a suit problem or that something is amiss with the task or situation on the part of the responder or responders.

CASE STUDY

You Are the Firefighter CONCLUSION

Your engine company arrives on the scene of a vehicle accident involving a small passenger vehicle and an MC 331 cargo tank truck carrying 20 tons of anhydrous ammonia. The truck rolled over on its side and slid down the highway for approximately 100 feet (30 m). There are no injuries to the three people in the passenger vehicle, but the driver of the truck is pinned inside the cab. You notice the smell of ammonia in the air but see no visible signs of a product release. The regional hazardous materials team, fully staffed with technician-level responders, is also on scene. Your company officer confers with the hazardous materials team and then directs you and another firefighter to don your self-contained breathing apparatus (SCBA) and full turnout gear and evaluate the driver for injuries.

1. **Would full structural firefighter turnout gear and SCBA offer adequate protection in this situation?**

   Answer: While SCBA is a good level of respiratory protection, firefighter PPE may not be adequate protection against skin exposure. If the driver is visible from a distance and conscious, it could be a good indicator of the airborne level of ammonia in the general area. If driver is experiencing no signs and symptoms of exposure, approaching may be indicated. Technician-level responders on scene can perform a quick assessment of the airborne environment around the cab of the truck to help assess the risk.

2. **Based on your level of training—operations level—would you be qualified to perform this task?**

   Answer: Yes. The entry team is not being asked to do product control, and there are technicians on scene to guide the actions of the operations-level personnel.

3. **Describe the steps you would need to take, including obtaining information about ammonia, to complete your assignment.**

   Answer: In addition to the steps listed in the answer the first question, additional steps may include using the *Emergency Response Guidebook* or other electronic database to research ammonia.

WRAP-UP

SUMMARY

KNOWLEDGE OBJECTIVES

- Discuss the similarities and differences in how single-use and reusable personal protective equipment (PPE) are used. (**NFPA 470: 9.2.1**, pp. 1367–1368)
  - Understand that reusable PPE must be maintained, inspected, and tested at regular intervals.
  - Understand the need to be aware of the maintenance and upkeep of any PPE purchased by the AHJ.
- Explain how to select, use, and maintain PPE. (**NFPA 470: 9.2.1**, p. 1368)
  - Understand the role that NFPA 1990 plays in PPE maintenance.
  - Understand the general storage requirement for PPE.
- Explain how PPE needs are determined. (**NFPA 470: 9.2.1**, pp. 1368–1369)
  - Describe the correlation between the expected mission to be performed and the selection of proper PPE for that mission.
  - Understand the guidance in the OSHA HAZWOPER standard when it comes to wearing SCBA at a hazardous materials incident.
- Identify and describe specific PPE for hazardous materials response. (**NFPA 470: 9.2.1**, pp. 1369–1384)
  - Understand the benefits and limitations of using chemical compatibility charts.
  - Understand how permeation, penetration, and degradation factor into CPC failure.
- Describe OSHA/EPA levels of chemical-protective clothing. (pp. 1370–1372)
  - Describe the general use case parameters in which vaper-protective and splash-protective garments could be used.
  - Understand the role of NFPA 1981 and NFPA 1896 as it relates to vapor-protective and splash-protective PPE.
- Explain the safety considerations when wearing PPE. (**NFPA 470: 9.2.1**, pp. 1385–1392)
  - Understand the role of NFPA 470, chapters 46 and 47, as it relates to medical evaluation prior to donning PPE.
  - Understand the benefits and limitations of cooling technologies for PPE.
- Explain the inclusion of PPE in reporting and documenting the incident. (**NFPA 470: 9.2.1**, pp. 1392–1393)
  - Understand the need to include all responders and the type of PPE worn during a response.
  - Understand the need and benefit of personal exposure records over the course of a career.

SKILLS OBJECTIVES

- Don and doff hazardous materials ensembles. (**NFPA 470: 9.2.1**, pp. 1374–1389)
  - Don a vapor-protective (Level A) ensemble.
  - Doff a vapor-protective (Level A) ensemble.
  - Don a liquid splash–protective (Level B) nonencapsulating ensemble.
  - Doff a liquid splash–protective (Level B) nonencapsulating clothing ensemble.
  - Don a liquid splash–protective ensemble (Level C).
  - Doff a liquid splash–protective ensemble (Level C).
  - Don a Level D protective ensemble.

KEY TERMS

chemical-resistant material Fabric specifically designed to inhibit or resist the passage of chemicals into and through the material by the process of penetration, permeation, or degradation.

degradation A chemical action involving the molecular breakdown of a protective clothing material or equipment due to contact with a chemical. (NFPA 470)

doffing The process of taking off an ensemble of personal protective equipment.

donning The process of putting on an ensemble of personal protective equipment.

high temperature–protective clothing Protective clothing designed to protect the wearer for short-term high temperature exposures. (NFPA 470)

Level A ensemble Personal protective equipment consisting of a totally encapsulating suit that includes a self-contained breathing apparatus and that provides the highest level of protection against vapors, gases, mists, and even dusts.

Level B ensemble Personal protective equipment consisting of a single- or multi-piece chemical-protective garment, boots, gloves, and a self-contained breathing apparatus, which is used when the type and atmospheric concentration of substances require a high level of respiratory protection but a moderate level of skin protection.

Level C ensemble Personal protective equipment consisting of a NIOSH-approved CBRN (chemical, biological, radioactive, nuclear) air-purifying respirator or a powered air-purifying respirator, which is used when the type of airborne substance is known, the concentration is measured, the criteria for using an air-purifying respirator are met, and skin and eye exposure is unlikely.

Level D ensemble Personal protective equipment typically consisting of coveralls, work shoes, hard hat, gloves, and standard work clothing, which is used when the atmosphere contains no known hazard, and work functions preclude splashes, immersion, or the potential for unexpected inhalation of or contact with hazardous levels of chemicals.

liquid splash–protective ensemble Compliant protective clothing and equipment products that when worn together provide protection from some, but not all, risks of hazardous materials/weapons of mass destruction (WMD) incident operations involving liquids. (NFPA 470)

penetration The movement of a material through a suit's closures, such as zippers, buttonholes, seams, flaps, or other design features of chemical-protective clothing, and through punctures, cuts, and tears. (NFPA 470)

permeation The chemical action involving the movement of chemicals, on a molecular level through intact material. (NFPA 470)

vapor-protective ensemble Multiple elements of compliant protective clothing and equipment that when worn together provide protection from some, but not all, risks of vapor, liquid splash, and particulate environments during hazardous materials/weapons of mass destruction (WMD) incident operations. (NFPA 470)

REVIEW QUESTIONS

1. What is the difference between single-use and reusable PPE?
2. Where should representatives of the AHJ go to find definitive information on use, storage, and maintenance of PPE?
3. How do responders select the proper PPE for the task when responding to hazardous materials incidents?
4. Define permeation.
5. Describe the key components of a Level B (liquid splash–protective) ensemble.
6. Is Level D considered to be a chemical-protective ensemble?
7. Should donning and doffing PPE be performed in a specific order?
8. In NFPA 470, which chapters provide responders with guidance on pre-entry medical evaluation?
9. What is the most common type of in-suit cooling technology?
10. At a minimum, what three pieces of information should be included in exposure records to provide a chronicle of the exposure history of the responders over a period of time?

DISCUSSION QUESTIONS

1. Under what conditions could Level C protection be used?
2. What is the key difference between Level B protective ensembles and Level C protective ensembles?
3. What are some of the safety considerations that responders should be aware of when wearing PPE?
4. What are some factors that should be assessed when determining the appropriate PPE to use at a hazardous materials incident?

APPLYING THE CONCEPTS

The chief of your fire company asked you to give a brief presentation to the town council about the new Level A suits the company is planning to purchase. You decide to use the example of an ammonia release at a local ice-making facility to underscore the reasons why your fire company needs this level of protection.

1. Which laws and/or standards would be appropriate to reference the various aspects of your new Level A suits?
2. Where might you find legislative information to justify the need for these suits?

In addition to the garments, there is a plan to purchase a forced-air cooling system to be worn inside the vapor-protective garment. You are asking the town council for four cooling units.

3. Provide a description of forced-air cooling technology.
4. How would you justify the request for four cooling units?

You will also share with the council that these are reusable garments and likely to become an ongoing commitment of time and funds for maintenance and upkeep.

5. What references would you use to outline a plan to address those needs?
6. Although maintenance and upkeep are necessary to maintain the PPE and associated equipment in a response ready mode, eventually the garments will need to be replaced. How would you address this during your presentation?

REFERENCES

National Fire Protection Association (NFPA). 2016. *NFPA 1986, Standard on Respiratory Protection Equipment for Tactical and Technical Operations.* 2017 Edition. Quincy, MA: NFPA.

National Fire Protection Association (NFPA). 2017. *NFPA 1971, Standard on Protective Ensembles for Structural Fire Fighting and Proximity Fire Fighting.* 2018 Edition. Quincy, MA: NFPA.

National Fire Protection Association (NFPA). 2018. *NFPA 1981, Standard on Open-Circuit Self-Contained Breathing Apparatus (SCBA) for Emergency Services.* 2019 Edition. Quincy, MA: NFPA.

National Fire Protection Association (NFPA). 2021. *NFPA 470, Hazardous Materials/Weapons of Mass Destruction (WMD) Standard for Responders.* 2022 Edition. Quincy, MA: NFPA.

National Fire Protection Association (NFPA). 2021. *NFPA 1891, Standard on the Selection, Care, and Maintenance of Hazardous Materials, CBRN, and Emergency Medical Operations Clothing and Equipment.* 2022 Edition. Quincy, MA: NFPA.

National Fire Protection Association (NFPA). 2021. *NFPA 1990, Standard for Protective Ensembles for Hazardous Materials and CBRN Operations.* 2022 Edition. Quincy, MA: NFPA.

Occupational Safety and Health Administration (OSHA). 1970. "1910.120 - Hazardous Waste Operations and Emergency Response." Accessed September 15, 2021. https://www.osha.gov/laws-regs/regulations/standardnumber/1910/1910.120.

CHAPTER

36

Hazardous Materials Operations Level: Mission Specific

Product Control

NOTE: Content within this chapter meets the intent of **NFPA 470, 2022 Edition**, which includes chapter 9, "Professional Qualifications for Hazardous Materials/WMD Operations Level Responders Assigned Mission-Specific Responsibilities (**NFPA 1072**)."

KNOWLEDGE OBJECTIVES

After studying this chapter, you will be able to:

- Describe the differences between containment and confinement as product control strategic options.
- Describe the various control options available to an operations-level responder.
- Describe the recovery phase of a hazardous materials incident.

SKILLS OBJECTIVES

After studying this chapter, you will be able to:

- Control a hazardous materials release.

CASE STUDY

You Are the Firefighter

Your engine company has been dispatched to assist a rescue company on the scene of a rolled-over diesel truck pulling a flatbed trailer. The trailer was carrying several lengths of large steel pipe, which are now lying on the road. The fuel tanks on the diesel truck are leaking. Upon your arrival, your officer receives an assignment to protect a curbside drain located downslope from the leaking tanker. He assigns the task to you and two other firefighters. There is a slight breeze blowing away from you, back toward the leaking tanker. The product is confirmed to be diesel fuel. The concrete roadway ahead of the spill is completely dry. You have access to three plastic shovels and several bags of loose absorbent.

1. Would full structural firefighter turnout gear and self-contained breathing apparatus offer adequate protection in this situation?
2. Based on your level of training (i.e., operations core competencies and mission-specific competencies), would you be qualified to perform this task?
3. Describe the steps you would need to take, including obtaining information about diesel fuel, to complete the assignment.

Introduction

This chapter addresses the mission-specific competencies for product control found in NFPA 470, chapter 8, section 8.6. The details of product control job performance requirements (JPRs) are found in NFPA 470, chapter 9, section 9.6.1. The operations-level responder assigned to perform mission-specific responsibilities of product control at hazardous materials/WMD incidents must be trained to meet all responsibilities at the awareness level and the operations level, and all requirements in NFPA 470, chapter 8, section 8.2, *Personal Protective Equipment (PPE)*. Additionally, such a responder must operate under the guidance of a hazardous materials technician, an allied professional, or standard operating procedures (SOPs).

TIP

Some of the mission-specific competencies described in this section are taken from the competencies required of hazardous materials technicians. That does not mean, however, that operations-level responders, with a mission-specific competency in product control or any other mission-specific competency, are replacements for hazardous materials technicians. Operations-level responders should operate under the guidance of a hazardous materials technician, an allied professional, or SOPs.

Control Options

It is not uncommon for a hazardous materials incident to require some form of product control. NFPA 470 defines **control** as "the procedures, techniques, and methods used in the mitigation of hazardous materials/weapons of mass destruction (WMD) incidents, including containment, extinguishment, and confinement." Scenarios like the one described in the chapter-opening vignette are relatively common, as are incidents involving leaking drums, spills into waterways, and other types of releases involving flammable liquids or gases. Many of these incidents require emergency responders to intervene (control the release) by shutting off valves or applying loose absorbents or various kinds of foam to mitigate the situation.

In most cases, the best course of action is to confine the problem to the smallest area possible. **Confinement** is "those procedures taken to keep a material, once released, in a defined or local area" (NFPA 470). This goal can usually be accomplished by damming or diking a material, or by suppressing a vapor with an appropriate type of foam. **Containment**, by contrast, refers to "actions that stop the hazardous material from leaking or escaping its container" (NFPA 470). Examples of containment include patching or plugging a breached container or righting an overturned container to stop a slow leak. Sometimes it is necessary to first stop the leak (contain) and then confine whatever has been released. In other cases, it may be impossible to stop the leak, so all that can be done is to confine the hazardous material as much as possible. In situations where ignition of a material has occurred, extinguishment may be the best course of action for incident containment.

There is no clear-cut sequence or single best way to handle every problem involving product control. You must size up the situation as you would any other problem and employ the best strategy and tactics available to safely handle the incident. When considering a control option, certain factors must be evaluated, such as the maximum quantity of material that can be released and the likely duration of the incident without intervention.

In these situations, as with any other type of release, it is vital for responders to understand the nature of the release; to have access to sources of information to understand the chemical and physical properties of the released substance; to wear the appropriate level of PPE for the situation; and to have all the tools and equipment—including air monitoring and detection equipment—available to accomplish the task at hand. Additionally, as with any other response to a hazardous materials incident, an incident action plan (IAP) with incident objectives should be developed and followed based on the policies and procedures of the authority having jurisdiction (AHJ). This includes safety measures based on the nature of the incident. In many cases, product control measures bring responders near the released product. To work safely in a contaminated (or potentially contaminated) atmosphere, responders must stay informed about their working environment. Readings from monitoring and detection devices should always be interpreted in the context of the specific event. Additionally, all emergency responders should be familiar with the policies and procedures of the local jurisdiction to ensure that they employ a consistent approach to the selected control option.

The most challenging aspect of mitigating a hazardous materials emergency is arriving at a solution that can be employed quickly and safely, while minimizing the potential negative effects on people, property, and the environment. If that sounds like a tall order, it is. Handling a hazardous materials incident is a bit like a chess game: One move sets up and (either positively or negatively) influences the next move or series of moves. It is vital to use a risk-based response process when choosing to employ any control option. Remember—a solid response objective should be well thought out and realistic, considering both the positive and negative effects of the actions taken. Think creatively and know which resources are available in your jurisdiction—and be prepared to think outside the box (**FIGURE 36-1**).

FIGURE 36-1 Bales of hay procured from a local feed store. In this incident, the bales were used to absorb some of the downstream flow of a gasoline spill (floating on water) from a nonpressure (MC-306/DOT-406) cargo trailer.

Sometimes, no action is the safest course of action. Unfortunate as it may seem, occasionally it may be prudent to create a safe perimeter and let the problem stabilize on its own. Perhaps the situation is unusually extreme, or the responders cannot be properly protected, or the product will evaporate quickly, cool down, or solidify. Choosing not to intervene may be challenging for a culture of aggressive firefighters, but there are times when taking a hands-off approach may be the correct course of action. As an example, responders may need to let a fire in a gasoline tanker burn itself out if not enough foam is available to fight the fire effectively.

Similarly, when incompatible chemicals are mixed, it may be wise to let the reaction run its course before intervening. Imagine a spill of a highly volatile flammable liquid like isopropyl alcohol on a hot summer day. It would likely evaporate before you could make an entry and attempt to clean it up. Think about the chemical and physical properties of the released material and how those properties may be used to your advantage regarding confinement or containment (**FIGURE 36-2**).

As emergency responders approach a hazardous materials incident, they should be aware of natural control points—areas in the terrain or places in a structure

FIGURE 36-2 Some extremely cold materials like cryogenic nitrogen may be left to evaporate naturally without taking an offensive action to clean them up.

where materials might be contained or confined. Natural barriers that might be used include doors to a room, doors to a building, designated areas for secondary containment, and curbed areas of roadways. A number of control measures may be available if responders are thinking creatively and are aware of their surroundings.

Absorption and Adsorption

Absorption is the process whereby a spongy material (e.g., soil or loose absorbents such as vermiculite, clay, or peat moss) or specially designed spill pads are used to soak up a liquid hazardous material (**FIGURE 36-3**). The contaminated mixture of absorbent material and chemicals is then collected, and the materials are disposed of together.

Absorption minimizes the spread of liquid spills, but it is effective only on flat, relatively nonporous surfaces. Although absorbent materials such as soil and sawdust are inexpensive and readily available, they become hazardous materials themselves once they come

FIGURE 36-3 Spill pads are often used to soak up a liquid hazardous material.

in contact with the spilled liquid; therefore, they must be disposed of properly. The process of absorption does not change the chemical properties of the involved substance, but rather is just a method to collect the substance for subsequent containment.

The technique of absorption may prove challenging for personnel because it requires them to be near the spilled material, operating in an offensive mode. It also involves the addition of material to a spilled product, which adds volume to the spill. Sometimes the absorbent material may react with certain hazardous substances, so it is important to determine whether the specific absorbent substance is compatible with the hazardous material. Hydrofluoric acid, for example, is not compatible with the silica-based absorbents commonly found in some types of spill booms.

Some absorbent materials repel water while still absorbing a spilled liquid that does not mix with water—an effective property under the right circumstances, such as when the goal is to mitigate an oil spill on a body of water (**FIGURE 36-4**). In these situations, spill booms are deployed as floating barriers used to obstruct passage. These booms are typically constructed of an outer mesh covering that is filled with a material such as polypropylene or silica. Many different sizes and types of spill booms are on the market today. Some will float on water, whereas others are designated for use only on dry land (**FIGURE 36-5**). It is up to your AHJ to decide which types of absorbents are appropriate for your jurisdiction.

Another example where the technique of absorption may prove challenging for personnel is an incident involving nitric acid at a concentration higher than 72 percent. In this situation, the nitric acid becomes

FIGURE 36-4 Gasoline floating on water in a creek.

A.

B.

FIGURE 36-5 **A.** A spill boom deployed by boats to contain a fuel spill in a harbor. **B.** A spill boom can also be used to confine a liquid on the ground.

such an aggressive oxidizer that using an organic-based absorbent may result in the acid igniting the absorbent material. To reiterate the point made earlier, not all absorbent materials are created equal, and using the wrong one might end up unduly complicating your situation.

The opposite of absorption is adsorption. In **adsorption**, the contaminant adheres to the surface of an added material—such as silica or activated carbon—rather than combining with it (as in absorption). In some cases, the process of adsorption can generate heat—a key point you should consider when use of adsorbent materials is proposed. The concept of adsorption is analogous to the mechanism underlying Velcro: Adsorbents are "sticky" and grab onto whatever substance they are designed to be used for. A commonly encountered adsorbent material is activated carbon, which is found in the filter cartridges used in air-purifying respirators.

To use absorption/adsorption to manage a hazardous materials incident, follow the steps in **SKILL DRILL 36-1**.

Damming

Damming is a containment technique that is used when liquid is flowing in a natural channel or depression and its progress can be stopped by blocking the channel with a temporary dam. Three types of dams are typically used in such circumstances: a complete dam, an overflow dam, or an underflow dam.

A **complete dam** is a dam placed across a small stream or ditch to completely stop the flow of materials through the channel. This type of dam is used only in areas where the stream or ditch is basically dry and the amount of material that needs to be controlled is relatively small.

For hazardous materials spills in streams or ditches with a continuous flow of water, an overflow or underflow dam must be constructed. An **overflow dam** is used to contain materials heavier than water (i.e., materials whose specific gravity is greater than 1.0). It is

SKILL DRILL 36-1

Using Absorption/Adsorption to Manage a Hazardous Materials Incident NFPA 470: 9.6.1

1. Decide whether absorption or adsorption is best suited for use with the spilled product. Assess the location of the spill and stay clear of any spilled product. Use detection and monitoring devices to determine whether airborne contamination is present. Consult reference sources to obtain information on the physical and chemical properties of the spilled material.

2. Apply the appropriate material to control the spilled product.

SKILL DRILL 36-1 CONTINUED

Using Absorption/Adsorption to Manage a Hazardous Materials Incident NFPA 470: 9.6.1

3. Maintain control of the absorbent/adsorbent materials and take appropriate steps for their disposal.

constructed by building a dam base up to a level that holds back the flow of water. A polyvinyl chloride (PVC) pipe or hard suction hose is installed at a slight downward angle to allow the water to flow over the released liquid, thereby trapping the heavier material at a low level at the base of the dam (**FIGURE 36-6**). It is important to ensure that the piping has sufficient capacity to allow the water to drain without overtopping the dam.

An **underflow dam** is used to contain materials lighter than water (its specific gravity is less than 1.0) and is basically constructed in the opposite manner from the overflow dam. The piping on the underflow dam is installed near the bottom of the dam at a slight upward angle so the water flows under the dam, thereby allowing the materials floating on the water to accumulate at the top of the dam area (**FIGURE 36-7**).

If you build an underflow dam to handle materials that are lighter than water, it is critical to ensure that you have sufficient piping to allow enough water to flow past the dam without flowing over the top of the dam. If there is insufficient flow capacity, the water behind the dam will continue to rise, with the lighter-than-water hazardous material floating on top, and will eventually flow over the top of the dam, rendering the containment procedure useless.

To construct an overflow dam, follow the steps in **SKILL DRILL 36-2**. To construct an underflow dam, follow the steps in **SKILL DRILL 36-3**.

Diking

Diking is the placement of a material such as sand, dirt, loose absorbent, or concrete so as to form a barrier that will keep a hazardous material in liquid form from entering an unwanted area or to hold the material in a specific location. Before contemplating construction of a dike, you must confirm that the material used to create the dike will not react adversely with the spilled material.

It may be necessary to construct a series of dikes if there is a concern about the amount of product being released. A wise incident commander (IC) will consider backing up the original structure to ensure that the product will be controlled effectively.

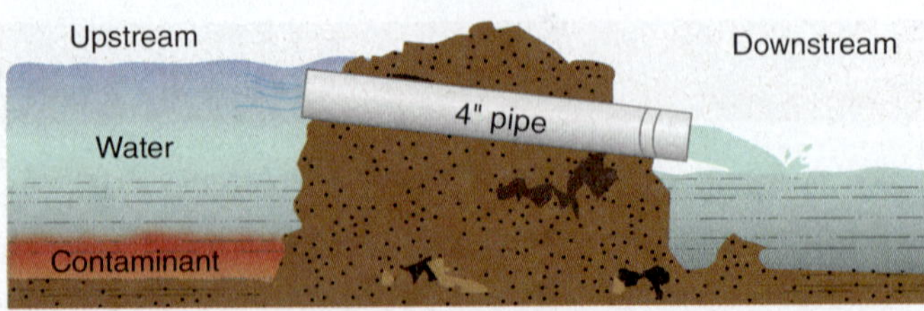

A.

B.

C.

FIGURE 36-6 **A.** An overflow dam is used to contain materials that are heavier than water. **B.** The piping at or near the surface of the retained water flow. **C.** A wider view of the overflow dam showing where the water can flow through the dam above the heavier-than-water liquid held back at the base of the dam.

A: © Jones & Bartlett Learning; **B:** © Jones & Bartlett Learning. Courtesy of Rob Schnepp; **C:** © Jones & Bartlett Learning. Photographed by Glen E. Ellman.

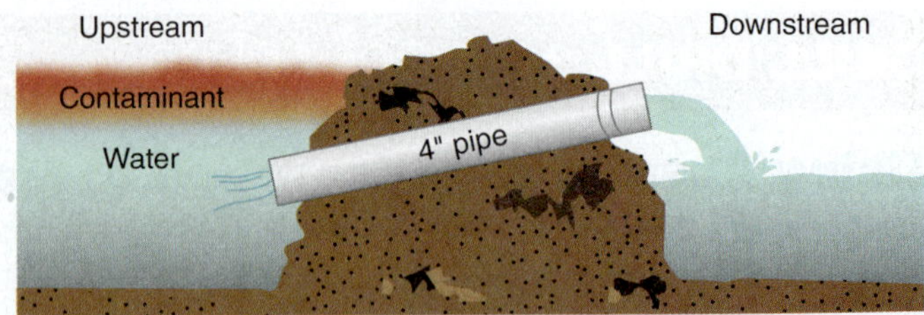

A.

B.

C.

FIGURE 36-7 **A.** An underflow dam is used to contain materials that are lighter than water. **B.** The piping below the surface of the water, allowing the lighter-than-water liquid to float on the surface and be retained by the dam. **C.** Water flowing through the dam.

A: © Jones & Bartlett Learning; **B:** © Jones & Bartlett Learning. Photographed by Glen E. Ellman; **C:** © Jones & Bartlett Learning. Photographed by Glen E. Ellman.

To construct a dike, follow the steps in **SKILL DRILL 36-4**.

Dilution

Dilution is the addition of water or another substance to weaken the strength or concentration of a hazardous material (typically a corrosive). Dilution can be used only when the identity and properties of the hazardous material are known with certainty. One concern with dilution is that water applied to dilute a hazardous material may simply increase the total volume; if the volume increases too much, it may overwhelm the containment measures implemented by responders. For example, if 1 gallon (3.8 L) of water was added to 3 gallons (11.3 L) of spilled hydrochloric acid (pH = 3), the result of this

FIGURE 36-8 When available, consider using remote shut-off valves to mitigate the problem.

carry flammable and combustible liquids and Class B poisons. These single-shell aluminum tanks may hold as much as 9200 gallons (34,000 L) of product at atmospheric pressure. The tanks have an oval/elliptical cross-section and overturn protection that serves to protect the top-mounted fittings in the event of a roll-over. In the event of fire, a fusible link is designed to melt at a predetermined temperature (typically around 250°F [121°C]); its collapse causes the closure of the internal product discharge valve. A baffle system within the cargo tank compartment reduces the surge of the liquid contents. A remote shut-off valve is located at the front of the cargo tank on the driver's side or at the rear of the cargo tank on the passenger side (**FIGURE 36-9**).

Low-pressure (MC-307/DOT-407) cargo tanks carry chemicals such as flammable and combustible liquids as well as mild corrosives and poisons. These vehicles may carry as much as 7000 gallons (26,400 L) of product and are outfitted with many safety features like those found on nonpressure cargo tanks (**FIGURE 36-10**).

FIGURE 36-9 The remote shut-off valve is typically found near the front of the cab, adjacent to the driver's door, or at the rear of a nonpressure (MC-306/DOT-406) cargo tank.

FIGURE 36-10 The remote shut-off valve is typically found near the front of the cab, adjacent to the driver's door, or at the rear of a low-pressure (MC-307/DOT-407) cargo tank.

High-pressure (MC-331) cargo tanks carry compressed liquefied gases such as anhydrous ammonia, propane, butane, and liquefied petroleum gas (LPG). The tanks, which are not insulated, have rounded ends, baffles, and a circular cross-section. High-pressure cargo tanks have a carrying capacity between 2500 and 11,500 gallons (9500 to 43,500 L). These types of cargo tanks have remote shut-off valves located at both ends of the tank, internal shut-off valves, a rotary gauge depicting product pressure, and top-mounted vents (**FIGURE 36-11**).

A.

B.

FIGURE 36-11 **A.** Intermodal tanks have emergency shut-offs that close the bottom outlet valve. Typically, the shut-off valves are located on one side of the container near the discharge end. **B.** The high-pressure MC-331 cargo tank has remote shut-off valves at both ends of the tank, internal shut-off valves, a rotary gauge depicting product pressure, and two top-mounted vents.

SAFETY TIP

Participating in product control operations may require a great deal of physical energy and mental concentration. Remember to take rehabilitation breaks throughout the incident to minimize the effects of heat stress and fatigue. Refer to NFPA 470, chapters 46 and 47, *Competencies for Hazardous Materials/WMD Basic Life Support (BLS) Responders*, and *Competencies for Hazardous Materials/WMD Advanced Life Support (ALS) Responders*, respectively, for more information on heat stress.

Vapor Dispersion and Suppression

When a hazardous material produces a vapor that collects in an area or increases in concentration, vapor dispersion or suppression measures may be required.

Vapor dispersion is the process of lowering the concentration of vapors by spreading the vapors out. Vapors can be dispersed with hose streams set on fog patterns, large displacement fans (while being mindful of the risk of the fan itself becoming a source of ignition), or other types of mechanical ventilation found in fixed hazardous materials handling systems. Before dispersing vapors, consider the consequences of this action. If the vapors are highly flammable, an attempt to disperse them may ignite the vapors. The dispersed vapors can also contaminate areas outside the hot zone. Vapor dispersion with fans or a fog stream should be attempted only after the hazardous material is safely identified and in weather conditions that will promote the diversion of vapors away from populated areas. Those conditions must be constantly monitored during the entire operation.

Vapor suppression is the process of controlling fumes or vapors that are given off by certain materials, particularly flammable liquids, to prevent their ignition. It is accomplished by covering the hazardous material with foam or some other material, or by reducing the temperature of the material. For example, gasoline gives off vapors that present a danger because they are flammable. To control these vapors, a blanket of firefighting or vapor-suppressing foam is layered on the surface of the liquid.

Not all vapor-suppressing foams are appropriate for all types of applications. Much like the concept of choosing the right PPE to protect you from a hazardous substance, responders must choose the type of foam that is designed to work in the specific situation. For example, Class A foam is typically used to fight fires involving ordinary combustible materials such as wood and paper. In most situations involving hazardous materials or WMD agents, Class A foam is *not* the most appropriate type of foam to use. Class B foam, by contrast, is used to fight fires and suppress vapors involving flammable and combustible liquids such as gasoline and diesel fuel. This type of foam seals the surface of a spilled liquid and prevents the vapors from escaping, thereby reducing the danger of vapor ignition. If the vapors are burning, a gently applied foam blanket will also spread out over the surface of the liquid and work to smother the fire.

Responders should be aware of the nature of the materials that are released and the way in which foam concentrates behave when they are applied. Take the time to become familiar with the various types of foam concentrates that may be used by your agency.

Reducing the temperature of some hazardous materials will also suppress vapor formation. Unfortunately, there is no easy way to accomplish this goal except in cases of small spills.

Foam concentrates should be applied in accordance with the policies and procedures of the AHJ, and after consulting with a hazardous materials technician or another subject-matter expert in the field.

To use vapor dispersion to manage a hazardous materials release, follow the steps in **SKILL DRILL 36-8**.

To perform vapor suppression using foam, follow the steps in **SKILL DRILL 36-9**.

Always remember that as foam is applied, more volume is added to the spill. All of the following foams suppress the ignition of flammable vapors and may be

SKILL DRILL 36-8

Using Vapor Dispersion to Manage a Hazardous Materials Incident
NFPA 470: 9.6.1

1. Determine the viability of a dispersion operation. Use the appropriate monitoring instrument to determine the boundaries of a safe work area. Ensure that ignition sources in the area have been removed or controlled.

2. Apply water from a distance to disperse vapors. Monitor the environment until the vapors have been adequately dispersed.

SKILL DRILL 36-9

Using Vapor Suppression to Manage a Hazardous Materials Incident
NFPA 470: 9.6.1

1. Determine the viability of a vapor suppression operation. Use the appropriate instrument to determine a safe work area. Ensure that ignition sources in the area have been removed or controlled. Apply foam from a safe distance to suppress vapors. Monitor the environment until the vapors have been adequately suppressed.

used to extinguish fires in flammable or combustible liquids:

- **Aqueous film-forming foam (AFFF)** is "a concentrate based on fluorinated surfactants plus foam stabilizers to produce a fluid aqueous film for suppressing hydrocarbon fuel vapors and usually diluted with water to a 1 percent, 3 percent, or 6 percent solution" (NFPA 11). This type of foam is designed to form a blanket over spilled flammable liquids to suppress vapors or on actively burning pools of flammable liquids. Foam blankets are designed to prevent a fire from reigniting once extinguished. Most AFFF foam concentrates are biodegradable. AFFF is usually applied by way of in-line foam eductors, foam sprinkler systems, and portable or fixed proportioning systems.
- **Alcohol-resistant foam concentrate** is a type of concentrate used for fighting fires on water-soluble materials and for fires involving hydrocarbons. These types of concentrates have properties like those of AFFF, except that they are formulated so that alcohols such as methyl alcohol and isopropyl alcohol, as well as other polar solvents, will not dissolve the foam (NFPA 11). (Regular protein foams cannot be used on these types of products.)
- **Fluoroprotein foam** contains protein products mixed with synthetic fluorinated surfactants. It is often referred to as **film-forming fluoroprotein foam (FFFP)**, because the surfactants produce a fluid film for suppressing hydrocarbon fuel vapors. These foams can be used on fires or spills involving gasoline, oil, or similar products. Fluoroprotein foams rapidly spread over the fuel, ensuring fire knockdown and vapor suppression. Fluoroprotein foams, like many of the synthetic foam concentrates and other types of special-purpose foams, are resistant to polar solvents such as alcohols, ketones, and ethers.
- **Protein foam** concentrates are made from hydrolyzed proteins (animal by-products), along with stabilizers and preservatives, and make a fire-resistant foam blanket. These types of foams are very stable and possess good expansion properties. They are quite durable and resistant to reignition when used on Class B fires or spills involving nonpolar substances such as gasoline, toluene, oil, and kerosene.
- **High-expansion foam** is used when large volumes of foam are required for spills or fires in warehouses, tank farms, and hazardous waste facilities.

It is not uncommon to have a yield of more than 1000 gallons (3785 L) of finished foam from every gallon of foam concentrate used. The expansion is accomplished by pumping large volumes of air through a small screen coated with a foam solution. Because of the large amount of air in the foam, high-expansion foam is referred to as dry foam. By excluding oxygen from the fire environment, this agent smothers the fire, leaving very little water residue.

There are several ways to apply foam. Foam concentrates (other than high-expansion foam) should be gently applied or bounced off another adjacent object so that they flow down across the liquid and do not directly upset the burning surface. Foam can also be applied in a rain-down method by directing the stream into the air over the material and letting the foam gently fall onto the surface of the liquid, as rain would. To review how to apply foam, see Chapter 23, *Advanced Fire Suppression*.

Recovery

The **recovery phase** of a hazardous materials incident occurs when the imminent danger to people, property, and the environment has passed or is controlled, and clean-up begins. During the recovery phase, a variety of local, state, and federal agencies may become involved in cleaning up the site, determining the responsible party, and implementing cost-recovery methods. The recovery phase in large-scale incidents can go on for days, weeks, or even months and may require large amounts of resources and equipment (**FIGURE 36-12**). As part of recovery, reporting and documentation should be completed in accordance with the AHJ's policies and procedures for incidents requiring product control.

In some situations, there is a clear transition between the emergency phase and the clean-up phase. The decision to declare a change from the emergency phase to the recovery phase is typically made by the IC. For example, a distinct hand-off between on-scene public safety responders and private-sector commercial clean-up companies would fit this description. It is not uncommon for public-sector responders to hand off the clean-up operations but remain on scene to ensure that the operation is carried out properly. While public safety personnel and responders remain on the scene, they should maintain their vigilance for safety to ensure that no new hazards are created, and no new injuries are incurred during the recovery phase.

FIGURE 36-12 The recovery phase involves clean-up, determination of the responsible party, and implementation of cost recovery.

Courtesy of Captain David Jackson, Saginaw Township Fire Department.

At other times, the initial responders perform both the emergency response and the clean-up. For example, responders often handle in their entirety small gasoline spills resulting from auto accidents or spills at gas stations.

TIP

The ultimate goal of the recovery phase is to return the property or site of the incident to its preincident condition and to return the facility or mode of transportation to the responsible party.

The transition from emergency response to clean-up should be carried out in a manner consistent with your agency's SOPs and the guidelines determined by your AHJ. The recovery phase also includes completion of the records necessary for documenting the incident. Additionally, responders should be aware of cost-recovery policies and procedures in their AHJ. In some instances, the spiller or party responsible for the release may be financially responsible for the costs incurred to mitigate the problem.

TIP

After performing any product control procedure, follow your local procedures for technical decontamination.

Voice of Experience

There are times when all of the aspects of what you have learned just come together. This happened to me on one very cold February night.

I had been an officer for a couple of years at the time of this incident, and I was feeling comfortable in my knowledge base. We were called to a report of a multiple-vehicle crash, including one semi-trailer truck, at a particular mile marker on Interstate 25. The semi was reported to be leaking diesel fuel. We assembled our plan en route, including scene safety, security, care for patients, patching the hole in the tank, and throwing down "kitty litter" and pads to clean up what had spilled.

We did not realize that this particular mile marker included a bridge over a lake. The semi was on its side and the low-pressure MC-307/DOT-407 was leaking diesel fuel from the top of the tanker trailer. The fuel had not begun dripping off the side of the bridge, but we knew we had less than 10 minutes before it did.

While two firefighters checked the scene and assessed the patients, another firefighter grabbed our supply of kitty litter and pads to create a small diversion dam and a small sponge area to give us a little time. We knew it would not hold, and we needed more material to make it last. Help was quickly called—for the department's complete stock of kitty litter, public works personnel with lots of sand (which I was told would be at least 45 minutes in arriving), and the regional hazardous materials team.

Using the kitty litter we had, we built a long diversion barrier to allow the diesel fuel to slowly flow down the length of the bridge toward its end, away from its edge. We also used the natural traffic ruts in the interstate to help us with this task. Where the flow reached the end of the bridge, we had several firefighters working with picks and axes to break up the ground. These firefighters created a dam to retain the flow in place with the dirt they were able to break up.

Once the public works department arrived, we used the equipment and truckloads of dirt they brought to reinforce both our diversion barrier and the dam. This completely contained the spill until the clean-up and offloading crews arrived, along with representatives of the regional hazardous materials team. Upon their arrival, we turned the scene over to them and went back in service (another crew remained on scene to help with offloading the tanker).

On the way back, it occurred to us that the information we learned in hazardous materials classes really does work, and that multiple techniques and scenarios can easily apply to just one incident (recognition, diking, absorption, diversion, containment, damming, recovery phase, and retention). By taking the appropriate steps, remaining flexible, and using all the knowledge we had gained, we prevented this incident from becoming a serious environmental nightmare.

Philip Oakes
Laramie County Fire District #4
Cheyenne, Wyoming

CASE STUDY

You Are the Firefighter CONCLUSION

Your engine company has been dispatched to assist a rescue company on the scene of a rolled-over diesel truck pulling a flatbed trailer. The trailer was carrying several lengths of large steel pipe, which are now lying on the road. The fuel tanks on the diesel truck are leaking. Upon your arrival, your officer receives an assignment to protect a curbside drain located downslope from the leaking tanker. He assigns the task to you and two other firefighters. There is a slight breeze blowing away from you, back toward the leaking tanker. The product is confirmed to be diesel fuel. The concrete roadway ahead of the spill is completely dry. You have access to three plastic shovels and several bags of loose absorbent.

1. **Would full structural firefighter turnout gear and self-contained breathing apparatus offer adequate protection in this situation?**

   Answer: Yes, structural firefighter's PPE is adequate protection, and the flash point of diesel is much higher than the ambient air temperature, so there is little to no vapor production or threat of respiratory exposure.

2. **Based on your level of training (i.e., operations core competencies and mission-specific competencies), would you be qualified to perform this task?**

   Answer: Yes, it is within your scope to carry out the identified task.

3. **Describe the steps you would need to take, including obtaining information about diesel fuel, to complete the assignment.**

   Answer: After consulting the *ERG* or other suitable reference source and wearing full firefighter's protective gear, apply selected material well ahead of the diesel fuel to absorb the spill, or divert the flow of the material, or construct a collection basin, or take some other appropriate confinement action.

WRAP-UP

SUMMARY

KNOWLEDGE OBJECTIVES

- Describe the differences between containment and confinement as product control strategic options. (pp. 1398–1399)
- Describe the various control options available to an operations-level responder. (**NFPA 470: 9.6.1**, pp. 1398–1413)
  - Absorption and adsorption
  - Damming
  - Diking
  - Dilution
  - Diversion
  - Retention
  - Remote valve shut-off
  - Vapor dispersion and suppression
- Describe the recovery phase of a hazardous materials incident. (p. 1413)

SKILLS OBJECTIVES

- Control a hazardous materials release. (**NFPA 470: 9.6.1**, pp. 1398–1413)
  - Use absorption/adsorption to manage a hazardous materials incident.
  - Construct an overflow dam.
  - Construct an underflow dam.
  - Construct a dike.
  - Use dilution to manage a hazardous materials incident.
  - Construct a diversion.
  - Use retention to manage a hazardous materials incident.
  - Use vapor dispersion to manage a hazardous materials incident.
  - Use vapor suppression to manage a hazardous materials incident.

KEY TERMS

absorption The process of applying a material that soaks up and holds a hazardous material in a sponge-like manner, so as to collect the hazardous material for subsequent disposal.

adsorption The process of applying a material, such as silica or activated carbon, whose surface a hazardous material adheres to, so as to collect the hazardous material for subsequent disposal.

alcohol-resistant foam concentrate A concentrate used for fighting fires on water-soluble materials and other fuels destructive to regular foams, aqueous film-forming foams (AFFF), synthetic fluorine-free foams (SFFF), or film-forming fluoroprotein foams (FFFP), as well as for fires involving hydrocarbons. (NFPA 11)

aqueous film-forming foam (AFFF) A concentrate based on fluorinated surfactants plus foam stabilizers to produce a fluid aqueous film for suppressing hydrocarbon fuel vapors and usually diluted with water to a 1 percent, 3 percent, or 6 percent solution. (NFPA 11)

complete dam A dam placed across a small stream or ditch to completely stop the flow of materials through the channel.

confinement Those procedures taken to keep a material, once released, in a defined or local area. (NFPA 470)

containment The actions taken to stop a release of a material and keep it in its container or to reduce the amount being released. (NFPA 470)

control The procedures, techniques, and methods used in the mitigation of hazardous materials/weapons of mass destruction (WMD) incidents, including containment, extinguishment, and confinement. (NFPA 470)

damming The product-control process used when liquid is flowing in a natural channel or depression, and its progress can be stopped by constructing a barrier to block the flow.

diking The placement of materials to form a barrier that will keep a hazardous material in liquid form from entering an area or that will hold the material in an area.

dilution The process of adding a substance—usually water—to weaken the concentration of another substance.

diversion The process of redirecting spilled or leaking material to an area where it will have less impact.

film-forming fluoroprotein foam (FFFP) See *fluoroprotein foam.*

fluoroprotein foam A protein-based foam concentrate based on fluorinated surfactants plus foam stabilizers to which fluorochemical surfactants have been added to produce a fluid film for suppressing hydrocarbon fuel vapors. Also called *film-forming fluoroprotein foam (FFFP).*

high-expansion foam A foam created by pumping large volumes of air through a small screen coated with a foam solution to produce foams with expansion ratios ranging from 200:1 to approximately 1000:1.

overflow dam A dam that includes piping installed at a slight downward angle, which allows the water to flow over the released liquid, and with enough capacity that the water drains without overtopping the dam, thereby trapping the heavier-than-water material held behind at the base of the dam.

protein foam A protein-based foam concentrate made from hydrolyzed proteins (animal by-products) and stabilized with metal salts to make a fire-resistant foam blanket.

recovery phase The stage of a hazardous materials incident after the imminent danger has passed, when clean-up and the return to normalcy have begun.

remote valve shut-off A valve that provides a way to remotely shut down a system or isolate a leaking fitting or valve.

retention The process of creating a defined area to contain a hazardous materials release.

underflow dam A dam that includes piping installed at a slight upward angle, which allows the water to flow below the released liquid, and with enough capacity that the water drains without causing the lighter-than-water material to overtop the dam, thereby trapping the lighter-than-water material behind the top of the dam.

vapor dispersion The process of lowering the concentration of vapors by spreading them out, typically with a water fog from a hose line.

vapor suppression The process of controlling vapors given off by hazardous materials by covering the product with foam or other material or by reducing the temperature of the material, thereby preventing vapor ignition.

REVIEW QUESTIONS

1. What is the definition of product control?
2. Define the terms *confinement* and *containment*.
3. Define the terms *absorption* and *adsorption*.
4. Under what circumstances might you build an underflow dam?
5. Which control option redirects the flow of a liquid away from an endangered area to an area where it will have less impact?
6. Which protective action should be considered at transportation emergencies or incidents at fixed facilities, such as a large-scale ammonia release or a tank carrying Class B poisons?
7. Define the terms *vapor dispersion* and *vapor suppression*.
8. Under what circumstances would Class B foam be used for product control?
9. When does the recovery phase of a hazardous materials incident occur?

DISCUSSION QUESTIONS

1. What are some factors to consider when deciding which product control option to implement in a given situation?
2. What are some of the things to consider when deciding whether to use a complete dam, an overflow dam, or an underflow dam?
3. During the recovery phase of large-scale hazardous materials incidents that require multiple agencies to work together, what are some issues that need to be considered and addressed?

APPLYING THE CONCEPTS

Your engine company has arrived on the scene of a MC-306/DOT-406 flammable liquid cargo tanker parked on the side of the interstate. Upon arrival, you notice the beginnings of a stream of released liquid flowing from the underside of the cargo tank, away from your location.

1. Based on the description of the cargo trailer, what types of materials might you expect to find upon arrival?
2. Would you expect this kind of cargo trailer to carry its load under high pressure or low pressure?

The driver approaches you and confirms the product to be gasoline and states that he has no idea why the leak is occurring. The air temperature is approximately 90°F (32°C), and the winds are calm. You are considering the options available to contain the leak.

3. Where would be the most likely places to find the remote shut-off valves on this type of cargo tanker?
4. What would you expect to be the maximum volume of the cargo tank to be, assuming it is full?

There are approximately 500 gallons (1900 L) of liquid retained in a catch basin. What are the next steps to take?

5. Which type of foam might be applied effectively to this type of product (gasoline) to suppress the vapors?
6. Imagine that the leak worsens, to the point that the flowing liquid has now reached an adjacent 15-foot-wide (4.6-m-wide) concrete drainage canal with water flowing in it. If your crew needed to build a dam to capture the flowing gasoline, which type of dams would be appropriate given the physical properties of gasoline?

REFERENCES

National Fire Protection Association (NFPA). 2020. *NFPA 11, Standard for Low-, Medium-, and High-Expansion Foam.* 2021 Edition. Quincy, MA: NFPA.

National Fire Protection Association (NFPA). 2021. *NFPA 470, Hazardous Materials/Weapons of Mass Destruction (WMD) Standard for Responders.* 2022 Edition. Quincy, MA: NFPA.

Appendix A

NFPA 1010 and NFPA 470 Correlation Grids for Firefighter I and II

NFPA 1010: Standard for Professional Qualifications for Firefighters 2024 Edition		
Job Performance Requirement	Chapter(s)	Learning Objective(s)
Firefighter I		
6.1: General		
6.1.1	1, 2	▪ Explain the mission of the fire service. (pp. 4–5) ▪ Explain the difference between career, volunteer, and combination fire departments. (pp. 11–13, 16–20, 20–21) ▪ Describe the fire department's standard operating procedures (SOPs) and rules and regulations as they apply to the Firefighter I. (pp. 22–23) ▪ Explain the concept of governance and describe how policies and standard operating procedures affect it. (p. 14) ▪ Describe the organization of the fire service. (pp. 16–20) ▪ List the different types of fire department companies and describe their functions. (pp. 16–17) ▪ Describe the common roles of firefighters within the organization of fire department. (pp. 17–18) ▪ Describe the specialized response roles within the fire department. (pp. 18–19) ▪ Explain the basic structure of the chain of command within the fire department. (p. 19) ▪ Describe a situation in which you will interact with other organizations within your community. (pp. 19–20) ▪ Define the four basic management principles used to maintain organization within the fire department. (p. 18) ▪ Describe how to organize a fire department in terms of staffing, function, and geography. (pp. 16–20) ▪ Outline the roles and responsibilities of a Firefighter I. (pp. 22–23) ▪ Describe the 16 firefighter life safety initiatives and the role of NFPA 1500 in promoting health and safety. (p. 37) ▪ Describe some of the organizations that set the regulations, standards, and procedures intended to ensure a safe working environment for the fire service. (pp. 36–38) ▪ Describe the connection between physical fitness and firefighter safety. (pp. 38–39) ▪ Describe the components of a well-rounded physical fitness program. (pp. 38–39) ▪ Explain the practices firefighters should take to promote optimal physical and mental health and well-being. (pp. 38–45) ▪ Explain the role of a critical incident stress debriefing in preserving the mental well-being of firefighters. (pp. 42–43) ▪ List signs and symptoms of behavioral and emotional distress. (pp. 43–45) ▪ Describe the purpose of an employee assistance program. (p. 45)

Job Performance Requirement	Chapter(s)	Learning Objective(s)
6.1.2	1, 3, 8	▪ Describe how to locate information in departmental documents and standard operating procedures. (p. 14) ▪ Locate information in departmental documents and standard operating procedures. (p. 14) ▪ Describe how to access specific information in fire service code and standards documents. (pp. 8–11) ▪ Describe the procedure for donning personal protective clothing. (pp. 72–74) ▪ Describe the procedure for doffing personal protective clothing. (pp. 72, 75–76) ▪ List the complete sequence of donning PPE. (pp. 72–74) ▪ Describe how to inspect the condition of PPE. (pp. 78–79) ▪ Describe how to properly maintain PPE. (pp. 78–80) ▪ Describe why thoroughly cleaning PPE immediately after it has been exposed to smoke or fire conditions is an important step in reducing your chance of developing cancer. (pp. 79–80) ▪ Don approved personal protective clothing. (pp. 73–74) ▪ Doff approved personal protective clothing. (pp. 75–76) ▪ Tie a safety knot. (p. 287) ▪ Tie a figure eight knot. (p. 288) ▪ Tie a half hitch. (pp. 289–290) ▪ Tie a clove hitch in the open. (p. 291) ▪ Tie a clove hitch around an object. (pp. 292–293) ▪ Tie a figure eight on a bight. (p. 294) ▪ Tie a figure eight follow-through. (p. 295) ▪ Tie a bowline. (pp. 296–297) ▪ Tie a sheet bend. (pp. 298–299) ▪ Tie a figure eight bend. (pp. 300–301) ▪ Tie a water knot. (pp. 301–302) ▪ Hoist an axe. (pp. 303–304) ▪ Hoist a pike pole. (p. 305) ▪ Hoist a ladder. (pp. 306–307) ▪ Hoist a charged hose line. (pp. 308–309) ▪ Hoist an uncharged hose line. (pp. 310–311) ▪ Hoist an exhaust fan or power tool. (pp. 311–312)
6.2: Communications		
6.2.1	4	▪ Explain the methods of receiving emergency and non-emergency fire department communications. (pp. 139–140) ▪ Determine if a communication is emergent or non-emergent. (p. 144) ▪ Explain the procedures for transmitting the emergency information to a dispatch center. (p. 145) ▪ Describe the procedures for handling emergency calls. (pp. 139–140) ▪ Explain the importance of following department SOPs for receiving and processing communications. (p. 146) ▪ Identify the information to be obtained when taking a report of an emergency to enable necessary assistance to be dispatched. (p. 146) ▪ Identify other modes of fire service communication. (pp. 147–152) ▪ Identify radio departmental procedures and codes for using fire department radios. (pp. 152–154) ▪ Send and receive messages over the fire department radio. (p. 154)

Job Performance Requirement	Chapter(s)	Learning Objective(s)
6.2.2	4	■ Describe the basic principles of effective radio communication. (pp. 149–152) ■ Identify radio departmental procedures and codes for using fire department radios. (pp. 152–154) ■ Describe when to use plain language and how ten-codes are implemented in fire service communications. (pp. 152–153) ■ Outline the information provided in size-up and progress reports. (pp. 153–155) ■ Recognize routine traffic, emergency traffic, and emergency evacuation signals. (pp. 155–156) ■ Receive a phone call, and obtain, route, and document information according to department procedures. (p. 147) ■ Send and receive messages over the fire department radio. (p. 154)
6.2.3	4, 19	■ Recognize routine traffic, emergency traffic, and emergency evacuation signals. (pp. 155–156) ■ Describe the information that should be included in a mayday call for emergency assistance. (pp. 845–848) ■ Send and receive messages over the fire department radio. (p. 154) ■ Determine if a radio communication is routine or emergency traffic. (p. 155) ■ Initiate a mayday call for emergency assistance. (pp. 847–848)
6.3: Fireground Operations		
6.3.1	3, 19	■ Explain how a personal alert safety system (PASS) helps to ensure firefighter safety. (pp. 97–98) ■ List the complete sequence of donning PPE. (pp. 72–74) ■ List the respiratory hazards posed by smoke and fire. (pp. 80–87) ■ List the conditions that require respiratory protection or self-contained breathing apparatus (SCBA). (pp. 80–90) ■ Describe the limitations of SCBA. (pp. 90–92) ■ Describe the physical and psychological limitations of an SCBA user. (p. 92) ■ List and describe the major components of SCBA. (pp. 92–97) ■ Don an SCBA from an apparatus seat mount. (pp. 92–97) ■ Don an SCBA from an apparatus compartment mount. (p. 102) ■ Don an SCBA from a storage case using the over-the-head method. (pp. 103–104) ■ Don an SCBA from a storage case using the coat method. (pp. 103–104) ■ Don a face piece. (pp. 107–108) ■ Doff an SCBA. (pp. 109–111) ■ Replace an SCBA air cylinder. (pp. 118–119) ■ Replace an SCBA air cylinder on another firefighter. (pp. 120–121) ■ Describe the devices on an SCBA that can assist the user in air management. (pp. 92–97) ■ Explain the breathing techniques used to conserve air supply. (p. 97) ■ Describe air management procedures. (pp. 844–845) ■ Use the backhanded swim technique to escape through a wall. (pp. 851–852) ■ Use the forward swim technique to escape through a wall. (p. 853)

Job Performance Requirement	Chapter(s)	Learning Objective(s)
6.3.2	2	▪ Describe how to safely mount an apparatus. (p. 48) ▪ Describe how to safely ride a fire apparatus to an emergency scene. (pp. 48–49) ▪ Describe how to safely dismount an apparatus. (p. 49) ▪ Describe the hazards and safety measures associated with riding apparatus. (pp. 48–49) ▪ List the NFPA standards that require firefighters to wear safety belts while riding in a fire apparatus. (p. 48) ▪ List the prohibited practices when riding in a fire apparatus to an emergency scene. (pp. 48–49) ▪ Mount an apparatus safely. (p. 48)
6.3.3	2	▪ Describe how to manage traffic safely at an emergency scene. (pp. 49–50) ▪ Describe the considerations for hazard and scene control. (p. 56) ▪ List the common hazards at an emergency incident. (p. 56) ▪ Describe the measures firefighters follow to ensure electrical safety at an emergency incident. (pp. 56–59) ▪ Dismount from an apparatus safely. (p. 49)
6.3.4	15	▪ Describe the situations and circumstances that require forcible entry into a structure. (pp. 596–597) ▪ Describe the basic components of a door. (p. 597) ▪ Describe the basic classifications of doors by opening type and how the opening type affects forcible entry operations. (pp. 597–607) ▪ Explain the differences between a solid-core and a hollow-core door. (pp. 598–600) ▪ Describe the major components of a lockset. (pp. 607–608) ▪ Describe the five major types of locksets and how the lockset type affects forcible entry operations. (pp. 609–618) ▪ Describe the basic configurations of window construction. (pp. 618–620) ▪ Describe the common types of window frames and glass. (pp. 622–624) ▪ Describe how the window type and glass affects forcible entry operations. (pp. 620–624) ▪ Describe the types of wall construction. (pp. 624–626) ▪ Explain the differences between load-bearing and nonbearing walls. (p. 624) ▪ Describe the materials used in exterior and interior walls. (pp. 624–626) ▪ Describe how security bars and gates affect forcible entry operations. (pp. 626–628) ▪ List the types of tools used in forcible entry. (pp. 629–630) ▪ List the special-use and lock tools used in forcible entry. (pp. 631–635) ▪ Describe the types of forcible entry techniques utilized to gain entry via a door. (pp. 635–677) ▪ Describe the types of forcible entry techniques utilized to gain entry via window. (pp. 677–678) ▪ Describe how to force entry through security gates and windows. (pp. 677–678) ▪ Force an inward-swinging door set in a wood frame with a tubular deadbolt using the sledgehammer method. (pp. 636–637)

Job Performance Requirement	Chapter(s)	Learning Objective(s)
		▪ Force an inward-swinging door set in a wood frame using the stutter-step method. (pp. 638–639) ▪ Use the GSF technique and a Halligan fork on an outward-swinging door. (p. 641) ▪ Use the GSF technique and a Halligan adze on an outward-swinging door. (p. 642) ▪ Use the GSF technique and a Halligan fork on an inward-swinging door. (pp. 643–644) ▪ Use the GSF technique with a Halligan adze on an inward-swinging door. (p. 645) ▪ Use the GSF technique on an inward-swinging door with a Hydra-Ram or hydraulic spreaders with door-opening tips. (pp. 646–647) ▪ Force a bored lockset with an S&D or Rex tool. (pp. 648–650) ▪ Force a bored lockset with an A tool or Halligan adze tool. (pp. 650–652) ▪ Force a mortise lock with a K tool, an S&D tool, or a Rex tool. (pp. 653–655) ▪ Force a mortise lock with locking pliers or angled slip-joint pliers. (pp. 655–656) ▪ Force a rim lockset with an S&D tool, a Rex tool, or a K tool. (pp. 657–659) ▪ Use a shove knife on a bored or mortise lockset in an outward-swinging door. (pp. 660–661) ▪ Use a J tool on panic hardware on an outward-swinging door. (p. 662) ▪ Cut a latch with a rotary saw. (pp. 664–665) ▪ Cut a padlock with a rotary saw. (p. 667) ▪ Cut a hockey puck lock with a rotary saw. (pp. 668–669) ▪ Make a triangle cut in an overhead door with a rotary saw. (pp. 670–671) ▪ Make an overhead door rectangle box cut with a rotary saw. (pp. 672–673) ▪ Make a rectangle hinged cut in an overhead door with a rotary saw. (p. 674) ▪ Cut a rolling steel door with a rotary saw using the slice and pull method. (pp. 675–676) ▪ Force a wood frame single- and double-hung windows. (p. 678) ▪ Force a vinyl-frame single- and double-hung windows. (p. 679) ▪ Breach a stud wall. (pp. 680–681)
6.3.5	19	▪ List the common hazard indicators that should alert firefighters to a potentially life-threatening situation. (p. 839) ▪ Explain how to maintain team integrity during emergency operations. (pp. 840–841) ▪ Define personnel accountability system. (pp. 841–842) ▪ Describe the types of personnel accountability systems and how they function. (pp. 841–842) ▪ Explain how a personnel accountability report is taken. (p. 842) ▪ Describe how to find a safe location while awaiting rescue. (p. 844) ▪ Describe air management procedures. (pp. 844–845) ▪ Describe how to initiate emergency communications procedures. (pp. 845–848) ▪ Describe common self-rescue techniques. (pp. 848–858) ▪ Perform a self-rescue using a hose line. (p. 849)

Job Performance Requirement	Chapter(s)	Learning Objective(s)
6.3.6	12	■ List and describe the parts of a ladder. (pp. 432–433) ■ Categorize the different types of ladders. (pp. 433–439) ■ Specify the hazards associated with ladders. (pp. 446–474) ■ Itemize the measures firefighters should take to ensure safety when working with and on ladders. (p. 481) ■ Cite the factors and guidelines used to select the appropriate ladder from the fire apparatus. (pp. 446–448) ■ Determine appropriate ladder placement for common fireground tasks. (pp. 460–461) ■ Carry a ladder using the one-firefighter shoulder carry. (pp. 451–452) ■ Carry a ladder using the two-firefighter shoulder carry. (pp. 452–453) ■ Carry a ladder using the three-firefighter shoulder carry. (pp. 452–453) ■ Carry a ladder using the two-firefighter suitcase carry. (pp. 454–455) ■ Carry a ladder using the three-firefighter suitcase carry. (pp. 454–455) ■ Carry a ladder using the three-firefighter flat carry. (pp. 456–457) ■ Carry a ladder using the four-firefighter flat carry. (pp. 456–457) ■ Carry a ladder using the three-firefighter flat-shoulder carry. (pp. 458–459) ■ Carry a ladder using the four-firefighter flat-shoulder carry. (pp. 458–459) ■ Tie the halyard. (p. 463) ■ Raise a ladder using the one-firefighter flat raise. (pp. 464–465) ■ Raise a ladder using the two-firefighter flat raise. (pp. 467–468) ■ Raise a ladder using the two-firefighter beam raise. (pp. 469–470) ■ Raise a ladder using the three-firefighter flat raise. (pp. 471–472) ■ Raise a ladder using the four-firefighter flat raise. (pp. 473–474) ■ Climb a ladder while carrying a tool. (p. 476) ■ Use a leg lock to work from a ladder. (pp. 477–478) ■ Dismount a ladder. (pp. 479–480)
6.3.7	14	■ Describe the characteristics of vehicle fires. (p. 572) ■ Describe the types of motor vehicles. (pp. 573–577) ■ Describe the different types of alternative fuels that power motor vehicles. (pp. 573–577) ■ Describe the tactics used to suppress vehicle fires. (p. 572) ■ Describe the tactics used to suppress fires in the passenger area of a vehicle. (p. 580) ■ Describe the tactics used to suppress fires in the engine compartment of a vehicle. (pp. 580–581) ■ Describe the tactics used to suppress fires in the trunk of a vehicle. (p. 581) ■ Describe how to overhaul a vehicle fire. (pp. 581–584) ■ Extinguish a vehicle fire. (pp. 581–583) ■ Immobilize a vehicle. (p. 587)

Job Performance Requirement	Chapter(s)	Learning Objective(s)
6.3.8	14	▪ Describe the characteristics of a master stream appliance. (pp. 556–559) ▪ Describe the characteristics of a portable monitor. (pp. 557–558) ▪ Describe the characteristics of a deck gun. (p. 558) ▪ Describe the characteristics of elevated master stream appliances. (p. 558) ▪ Describe the tactics used to protect exposures. (pp. 559–560) ▪ Describe the characteristics of fires in confined spaces. (pp. 560–561) ▪ Describe the characteristics of fires in stacked or piled materials. (p. 568) ▪ Describe the tactics used to suppress fires in stacked or piled materials. (p. 568) ▪ Describe the characteristics of fires in lumberyards. (p. 568) ▪ Describe the tactics used to suppress fires in lumberyards. (p. 568) ▪ Describe the characteristics of fires in trash containers. (pp. 569–570) ▪ Describe the tactics used to suppress fires in trash containers. (pp. 569–570) ▪ Describe the importance of preserving evidence. (p. 572) ▪ Perform the one-firefighter method for operating a large handline. (pp. 555–556) ▪ Perform the two-firefighter method for operating a large handline. (p. 557) ▪ Operate a deck gun. (p. 559) ▪ Deploy and operate a portable monitor. (p. 558)
6.3.9	17, 19	▪ Describe the mission of search operations. (p. 744) ▪ Describe the mission of rescue operations. (p. 744) ▪ Describe the information that impacts a search and rescue size-up. (p. 746) ▪ Explain how search operations are coordinated. (pp. 742–749) ▪ Describe the role of a fire officer during search operations. (pp. 742–749) ▪ Describe the priorities of search operations. (pp. 748–749) ▪ Explain how firefighters maintain safety through risk management. (pp. 749–750) ▪ List the tools and equipment used in search and rescue operations. (pp. 750–751) ▪ Describe the methods firefighters use to determine whether an area is tenable. (p. 751) ▪ Describe the objectives of a primary search. (pp. 751–753) ▪ Describe the search patterns commonly used in primary search operations. (pp. 751–754) ▪ Explain how thermal imaging devices are used during search operations. (pp. 754–755) ▪ Describe a primary search using the standard search method. (pp. 755–756) ▪ Describe a primary search using the oriented search method. (pp. 756–758) ▪ Describe a primary search using the oriented-vent-enter-isolate-search (O-VEIS) method. (p. 760) ▪ Describe a primary search using the team search method. (pp. 760–762) ▪ Describe the objectives of a secondary search. (pp. 762–763) ▪ List the major types of rescue methods. (pp. 764–765) ▪ Describe the concept of sheltering-in-place. (p. 788) ▪ Describe how to assist a victim to an exit. (pp. 765–772) ▪ List the common types of simple victim carries performed during rescue operations. (pp. 765–772) ▪ Describe why a ground ladder rescue may be required. (pp. 772–786) ▪ Describe the advantages of using aerial ladders and platforms during rescue operations. (pp. 786–787)

Job Performance Requirement	Chapter(s)	Learning Objective(s)
		▪ Conduct a primary search using the standard search method. (p. 757) ▪ Conduct a primary search using the oriented search method. (pp. 758–759) ▪ Conduct a primary search using the oriented-vent-enter-isolate-search (O-VEIS) method. (p. 760) ▪ Conduct a primary search using the team search method. (pp. 760–762) ▪ Conduct a secondary search. (pp. 763–764) ▪ Perform a one-person walking assist. (p. 766) ▪ Perform a two-person walking assist. (p. 767) ▪ Perform a two-person extremity carry. (p. 768) ▪ Perform a two-person seat carry. (p. 769) ▪ Perform a two-person chair carry. (p. 770) ▪ Perform a cradle-in-arms carry. (p. 771) ▪ Perform a clothes drag. (p. 772) ▪ Perform a blanket drag. (p. 773) ▪ Perform a standing drag. (pp. 774–775) ▪ Perform a webbing sling drag. (p. 776) ▪ Perform a firefighter drag. (p. 777) ▪ Perform a one-person emergency drag from a vehicle. (p. 778) ▪ Perform rescues by ladder to conscious and unconscious victims. (pp. 779–787) ▪ Rescue a conscious victim from a window. (pp. 779–781) ▪ Rescue an unconscious victim from a window. (pp. 782–783) ▪ Rescue an unconscious child or small adult from a window. (pp. 784–785) ▪ Rescue a large adult from a window. (pp. 786–787) ▪ Describe common techniques for rescuing a downed firefighter. (pp. 848–858) ▪ Describe how a rapid intervention pack or emergency breathing safety system can provide an emergency air supply to a downed or trapped firefighter. (pp. 858–864) ▪ Use the backhanded swim technique to escape through a wall. (pp. 851–852) ▪ Use the forward swim technique to escape through a wall. (p. 853) ▪ Rescue a downed firefighter using the firefighter's SCBA straps. (p. 855) ▪ Rescue a downed firefighter using a drag rescue device. (p. 856) ▪ Rescue a downed firefighter as a two-person team. (pp. 857–858) ▪ Supply air to a downed firefighter using the low-pressure hose from a rapid intervention pack. (pp. 860–861) ▪ Supply air to a downed firefighter using the emergency breathing safety system. (pp. 862–863)
6.3.10	2, 3, 5, 11, 13, 14	▪ List the major causes of death and injury in firefighters. (pp. 34–35) ▪ List the respiratory hazards posed by smoke and fire. (pp. 80–87) ▪ List the three states of matter. (pp. 164–165) ▪ Describe the characteristics of couplings. (pp. 388–396) ▪ List the common types of couplings. (pp. 388–396) ▪ Describe the characteristics of attack hose. (pp. 397–398) ▪ Discuss the differences between smooth-bore nozzles and fog-stream nozzles. (pp. 401–405) ▪ Describe the characteristics of wyes. (p. 413) ▪ Describe the characteristics of hose jackets. (p. 416)

Job Performance Requirement	Chapter(s)	Learning Objective(s)
		▪ List the common types of hose damage and how to prevent them. (pp. 418–419) ▪ Describe the common techniques used to deploy attack hose. (pp. 514–516) ▪ Describe the general procedures that are followed during attack line evolutions. (pp. 508–527) ▪ Describe the procedures to follow when advancing attack hose. (pp. 514–524) ▪ Describe how to extend an attack line. (pp. 524–527) ▪ Describe how to connect and advance an attack line from a standpipe outlet. (p. 528) ▪ Describe how to replace a damaged section of attack hose. (pp. 528–531) ▪ Describe the objectives of a defensive operation. (p. 541) ▪ Describe the operations performed during a defensive operation. (p. 541) ▪ Describe the objectives of an offensive operation. (pp. 541–542) ▪ Describe the operations performed during an offensive operation. (pp. 541–542) ▪ Describe the objectives of a direct attack. (p. 542) ▪ Describe the objectives of an indirect attack. (p. 542) ▪ Describe the objectives of a combination attack. (p. 546) ▪ Describe the objectives of a transitional attack. (pp. 548–550) ▪ Describe the operations performed during a transitional attack. (pp. 548–553) ▪ Describe the techniques used to operate large handlines. (pp. 554–556) ▪ Describe the characteristics of concealed-space fires. (p. 562) ▪ Describe the tactics used to suppress concealed-space fires. (p. 562) ▪ Describe the characteristics of basement fires. (pp. 562–564) ▪ Describe the tactics used to suppress basement fires. (pp. 562–564) ▪ Describe the characteristics of fires above ground level. (p. 564) ▪ Describe the tactics used to suppress fires above ground level. (p. 564) ▪ Describe the characteristics of attic fires. (pp. 564–566) ▪ Describe the tactics used to suppress attic fires. (pp. 564–566) ▪ Describe the characteristics of fires in large buildings. (pp. 566–567) ▪ Describe the tactics used to suppress fires in large buildings. (pp. 566–567) ▪ Describe the characteristics of fires in buildings under construction, renovation, or demolition. (p. 567) ▪ Describe the tactics used to suppress fires in buildings under construction, renovation, or demolition. (p. 567) ▪ Describe the characteristics of chimney fires. (pp. 570–571) ▪ Describe the tactics used to suppress fires in chimneys. (pp. 570–571) ▪ Perform the one-firefighter foot-tilt method of coupling a fire hose. (p. 392) ▪ Perform the two-firefighter method of coupling a fire hose. (p. 393) ▪ Perform the one-firefighter knee-press method of uncoupling a fire hose. (p. 394) ▪ Perform the two-firefighter stiff-arm method of uncoupling a fire hose. (p. 395) ▪ Uncouple a hose with a spanner wrench. (pp. 395–396) ▪ Operate a smooth-bore nozzle. (pp. 407–408) ▪ Operate a fog-stream nozzle. (pp. 409–410) ▪ Advance an accordion load. (pp. 509–510) ▪ Perform a working hose drag. (pp. 506–507) ▪ Perform a shoulder carry. (pp. 507–508)

Job Performance Requirement	Chapter(s)	Learning Objective(s)
		■ Deploy and advance a minuteman hose load. (pp. 517–518) ■ Deploy and advance a preconnected flat hose load. (pp. 518–519) ■ Deploy and advance a triple-layer hose load. (pp. 519–520) ■ Deploy and advance wyed lines. (p. 521) ■ Advance an uncharged attack line up the stairway. (p. 523) ■ Advance a charged attack line down a stairway. (p. 524) ■ Advance an uncharged attack line up a ladder. (p. 525) ■ Operate an attack line from a ladder. (p. 526) ■ Connect and advance an attack line from a standpipe outlet. (p. 530) ■ Replace a damaged section of hose. (p. 531) ■ Perform a direct attack. (p. 544) ■ Perform an indirect attack. (p. 545) ■ Perform a combination attack. (pp. 546–547) ■ Perform a transitional attack. (pp. 551–552) ■ Locate and suppress concealed-space fires. (p. 563) ■ Extinguish an outside trash fire or other outside Class A fire. (pp. 569–570)
6.3.11	5, 16	■ Explain the concept of the fire triangle. (p. 167) ■ Explain the concept of the fire tetrahedron. (p. 167) ■ Describe the chemistry of combustion. (p. 168) ■ Describe the by-products of combustion. (pp. 168–169) ■ Define flow path and describe how it influences the growth of a building fire. (pp. 169–172) ■ Describe the importance of the following characteristics in solid-fuel fires: composition of fuel, amount of fuel, and configuration of fuel. (pp. 175–176) ■ Describe the four stages of fire development: incipient stage, growth stage, fully developed stage, and decay stage. (pp. 176–182) ■ Describe the conditions that lead to a backdraft. (p. 180) ■ Describe the importance of including ventilation considerations in a size-up. (p. 694) ■ Describe how the location, size, and stage of fire affect ventilation operations. (pp. 694–695) ■ Describe how the characteristics of different construction types affect ventilation operations. (pp. 695–698) ■ List the three Ws in timing ventilation with suppression. (p. 698) ■ Describe steps that can be taken to minimize the risk of backdrafts. (p. 698) ■ Describe steps that can be taken to minimize the risk of flashovers. (p. 699) ■ List the two basic types of ventilation. (pp. 699–700) ■ Explain how horizontal ventilation removes contaminated atmosphere from a structure. (pp. 700–701) ■ List the two methods of horizontal ventilation. (pp. 700–701) ■ Explain how natural ventilation removes contaminated atmosphere from a structure. (pp. 701–705) ■ Describe the techniques used to provide natural ventilation to a structure. (pp. 701–705) ■ Explain how mechanical ventilation removes contaminated atmosphere from a structure. (pp. 705–708) ■ Describe the techniques used to provide mechanical ventilation to a structure. (pp. 705–708)

Job Performance Requirement	Chapter(s)	Learning Objective(s)
		▪ Describe how negative-pressure ventilation removes contaminated atmosphere from a structure. (pp. 705–706) ▪ Describe the techniques used to provide negative-pressure ventilation to a structure. (pp. 705–706) ▪ Describe how positive-pressure ventilation removes contaminated atmosphere from a structure. (pp. 706–708) ▪ Describe the techniques used to provide positive-pressure ventilation to a structure. (pp. 706–708) ▪ Describe how hydraulic ventilation removes contaminated atmosphere from a structure. (pp. 708–710) ▪ Describe the techniques used to provide hydraulic ventilation to a structure. (pp. 708–710) ▪ Break glass with a hand tool. (p. 702) ▪ Break glass with a ladder. (pp. 703–704) ▪ Deliver negative-pressure ventilation. (p. 706) ▪ Deliver positive-pressure ventilation. (p. 709) ▪ Perform hydraulic ventilation. (p. 710)
6.3.12	5, 12, 16	▪ Explain how fires are spread by conduction, convection, and thermal radiation. (pp. 169–172) ▪ Describe the conditions that cause thermal layering. (p. 178) ▪ Describe the conditions that lead to rollover. (p. 178) ▪ Describe the conditions that lead to flashover. (pp. 178–179) ▪ Describe how vertical ventilation removes contaminated atmosphere from a structure. (pp. 711–718) ▪ Describe how to ensure firefighter safety during vertical ventilation operations. (pp. 711–713) ▪ Identify the warning signs of roof collapse. (p. 713) ▪ Describe the components and characteristics of roof assemblies. (pp. 713–716) ▪ List the differences in solid-beam construction and lightweight construction in roofs. (pp. 713–715) ▪ Explain how roof construction affects fire resistance. (p. 715) ▪ List the basic types of roof design. (pp. 715–716) ▪ Describe the characteristics of flat roofs. (pp. 715–716) ▪ Describe the characteristics of pitched roofs. (pp. 715–716) ▪ Describe the characteristics of curved roofs. (p. 716) ▪ Describe the techniques of vertical ventilation. (pp. 716–718) ▪ List the tools utilized in vertical ventilation. (pp. 718–720) ▪ List the types of roof cuts utilized in vertical ventilation operations. (pp. 720–728) ▪ Describe the characteristics of a rectangular or square cut. (pp. 720–721) ▪ Describe the characteristics of a seven, nine, eight (7, 9, 8) rectangular cut. (pp. 721–722) ▪ Describe the characteristics of a louver cut. (pp. 722, 724–725) ▪ Describe the characteristics of a triangular cut. (pp. 722, 725–726) ▪ Describe the characteristics of a peak cut. (pp. 722, 726–727) ▪ Describe the characteristics of a trench cut. (p. 728) ▪ Describe the special considerations in ventilating basements. (p. 729) ▪ Describe the special considerations in ventilating concrete roofs. (p. 729)

Job Performance Requirement	Chapter(s)	Learning Objective(s)
		■ Describe the special considerations in ventilating metal roofs. (pp. 729–730) ■ Describe the special considerations in ventilating high-rise buildings. (pp. 730–731) ■ Describe the special considerations in ventilating windowless buildings. (p. 731) ■ Describe the special considerations in ventilating large buildings. (pp. 731–732) ■ Place a roof ladder. (pp. 479–480) ■ Operate a power saw. (pp. 718–720) ■ Make a rectangular cut to deliver vertical ventilation. (pp. 721–722) ■ Make a seven, nine, eight (7, 9, 8) rectangular cut to deliver vertical ventilation. (pp. 723–724) ■ Make a louver cut to deliver vertical ventilation. (pp. 724–725) ■ Make a triangular cut to deliver vertical ventilation. (pp. 725–726) ■ Make a peak cut to deliver vertical ventilation. (pp. 726–727) ■ Make a trench cut to deliver vertical ventilation. (p. 728)
6.3.13	18	■ Describe the purpose of overhaul operations. (p. 820) ■ List the concerns that must be addressed to ensure the health of firefighters who are performing overhaul. (p. 820) ■ List the indicators of possible structural collapse. (p. 825) ■ Explain how to preserve structural integrity during overhaul. (p. 825) ■ Describe how to preserve evidence during overhaul operations. (pp. 825–826) ■ Explain how firefighters determine overhaul locations. (pp. 826–829) ■ List the tools that are used for overhaul operations. (pp. 829–830) ■ Describe the general techniques used in overhaul operations. (pp. 829–832) ■ Open a ceiling to check for fire using a pike pole. (pp. 830–831) ■ Open an interior wall to check for fire. (pp. 831–832)
6.3.14	18	■ Explain the purpose of salvage operations. (pp. 799–800) ■ List the tasks involved in a salvage operation. (pp. 799–800) ■ Describe how forcible entry operations affect salvage operations. (p. 800) ■ Describe the safety precautions that need to be considered when performing salvage. (pp. 800–801) ■ List the tools used to perform salvage operations. (p. 801) ■ Describe the salvage techniques commonly used to prevent water damage. (p. 801) ■ Describe the general procedures for preventing excess water damage from fire sprinklers. (pp. 801–811) ■ List the equipment used to shut down fire sprinklers. (pp. 801–803) ■ Describe the identifying characteristics of a main control valve of a fire sprinkler system. (pp. 803–804) ■ Describe the general procedures and equipment used to remove excess water from a structure. (pp. 808–811) ■ Describe the general procedures and equipment used to limit smoke and heat damage. (pp. 811–819) ■ Describe how to maintain salvage covers. (pp. 812–819) ■ Explain when fire investigators should become involved in salvage operations. (pp. 825–826) ■ Use a sprinkler wedge to shut down a sprinkler head. (p. 803) ■ Use a sprinkler stop to shut down a sprinkler head. (p. 804)

Job Performance Requirement	Chapter(s)	Learning Objective(s)
		▪ Close and reopen a main OS&Y valve. (pp. 806–807) ▪ Close and open a main butterfly valve. (p. 807) ▪ Close and open a main post indicator valve. (p. 808) ▪ Construct a water chute. (pp. 809–810) ▪ Construct a water catch-all. (pp. 810–811) ▪ Fold a salvage cover for one-firefighter deployment. (pp. 812–813) ▪ Fold a salvage cover for two-firefighter deployment. (p. 814) ▪ Fold and roll a salvage cover. (pp. 815–816) ▪ Perform a one-firefighter salvage cover roll. (pp. 816–817) ▪ Perform a salvage cover shoulder toss. (p. 818) ▪ Perform a salvage cover balloon toss. (p. 819)
6.3.15	10, 11, 13	▪ Describe the characteristics of dry barrel hydrants. (p. 363) ▪ Describe the characteristics of wet barrel hydrants. (pp. 361–362) ▪ Describe the equipment and procedures that are used to access static sources of water. (pp. 374–376) ▪ Describe the advantages of a portable tank system. (pp. 376–378) ▪ Describe the characteristics of a mobile water supply apparatus. (pp. 378–380) ▪ List the two types of fire hose. (pp. 386–387) ▪ Explain how fire hose is constructed. (pp. 387–388) ▪ Describe the characteristics of single-jacket hose. (p. 387) ▪ Describe the characteristics of multiple-jacket hose. (p. 387) ▪ Describe the characteristics of rubber-covered hose. (p. 387) ▪ Describe how supply hose is utilized in the field. (pp. 396–399) ▪ Describe the two types of suction hose. pp. 396–399) ▪ List the common hose appliances used in conjunction with fire hose. (pp. 412–417) ▪ Describe the characteristics of adapters and reducers. (pp. 412–413) ▪ Describe the characteristics of water thieves. (p. 413) ▪ Describe the characteristics of Siamese connector. (pp. 413–414) ▪ Describe the types of valves used to control water in pipes or hose lines. (p. 414) ▪ Describe the characteristics of master stream appliances. (pp. 414–416) ▪ Describe the characteristics of hose bridges. (p. 417) ▪ Describe the procedures used to connect supply lines to a fire hydrant. (pp. 488–490) ▪ Describe the common types of supply line evolutions. (pp. 488–508) ▪ Describe the common techniques used to load supply hose. (pp. 497–502) ▪ Describe the common techniques to offload a supply hose. (pp. 506–508) ▪ Operate a dry barrel fire hydrant. (pp. 365–366) ▪ Shut down a dry barrel fire hydrant. (p. 367) ▪ Operate a wet barrel fire hydrant. (pp. 368–369) ▪ Shut down a wet barrel fire hydrant. (p. 369) ▪ Assist the pump driver/operator with drafting. (pp. 376–377) ▪ Set up a portable tank. (p. 379)

Job Performance Requirement	Chapter(s)	Learning Objective(s)
		▪ Attach a soft sleeve hose to a fire hydrant. (pp. 489–490) ▪ Perform a forward hose lay. (pp. 492–494) ▪ Perform a reverse hose lay. (pp. 495–496) ▪ Perform a split hose lay. (p. 497) ▪ Attach a fire hose to a four-way hydrant valve. (pp. 498–499) ▪ Connect a hose line to supply a fire department connection. (p. 529)
6.3.16	9	▪ State the primary purposes of fire extinguishers. (pp. 320–321) ▪ Define the classes of fire and select the type of fire extinguisher used for each. (pp. 321–324) ▪ Explain the classification and rating system for fire extinguishers. (pp. 322–324) ▪ Explain the labeling system for fire extinguishers. (pp. 322–324) ▪ Describe the three risk classifications for area hazards. (pp. 324–326) ▪ Describe the types of agents and operating systems used in fire extinguishers. (pp. 328–337) ▪ Select the proper class of fire extinguisher. (pp. 337–339) ▪ Describe the basic steps of fire extinguisher operation. (pp. 340–342) ▪ Explain the basic steps of inspecting, maintaining, recharging, and hydrostatic testing of fire extinguishers. (pp. 347–348) ▪ Transport the fire extinguisher to the location of the fire. (pp. 339–340) ▪ Extinguish a Class A fire with a stored-pressure water-type fire extinguisher. (pp. 341–342) ▪ Extinguish a Class A fire with a multipurpose dry-chemical fire extinguisher. (p. 342) ▪ Extinguish a Class B flammable liquid fire with a dry-chemical fire extinguisher. (p. 343) ▪ Extinguish a Class B flammable liquid fire with a stored-pressure foam fire extinguisher. (p. 344) ▪ Operate a carbon dioxide fire extinguisher. (p. 345) ▪ Operate a halogenated agent–type fire extinguisher. (p. 346) ▪ Operate a dry-powder fire extinguisher. (p. 346) ▪ Operate a wet-chemical fire extinguisher. (p. 347)
6.3.17	18	▪ Describe the types of lights used to illuminate exterior and interior scenes. (pp. 796–797) ▪ Describe the equipment used to illuminate an emergency scene. (pp. 797–799) ▪ Describe the safety precautions to take when working with lighting equipment. (p. 799) ▪ Describe how to operate lighting equipment to light exterior and interior scenes. (p. 799) ▪ Illuminate an emergency scene. (pp. 797–799)
6.3.18	2, 14	▪ Describe when gas service should be shut off. (pp. 559–560) ▪ Describe when the electrical service should be shut off. (pp. 560–561) ▪ Describe when water service should be shut off. (p. 561) ▪ Describe the hazards posed by electrical fires. (pp. 567–568) ▪ Shut off gas utilities. (p. 560) ▪ Shut off electric utilities. (p. 561)

Job Performance Requirement	Chapter(s)	Learning Objective(s)
6.3.19	21	▪ Describe the components of a wildland fire. (pp. 890–891) ▪ Describe ground fuels, subsurface fuels, aerial fuels, and surface fuels. (pp. 892–893) ▪ Describe fuel group characteristics and how they impact the rate of spread. (pp. 891–893) ▪ Describe how weather conditions and topography influence the growth of wildland fires. (pp. 893–896) ▪ Describe the two strategies to extinguish wildfires. (pp. 896–899) ▪ Label the parts of a wildland fire. (pp. 896–903) ▪ Describe how a direct attack is mounted on wildland fires. (p. 897) ▪ Describe how an indirect attack is mounted on wildland fires. (pp. 897–898) ▪ Describe how a parallel attack is mounted on wildland fires. (p. 898) ▪ Describe the methods and tools used to cool a fuel with water. (pp. 896–898) ▪ Describe the methods and tools used to remove a fuel from wildland fires. (pp. 899–901) ▪ Describe the methods and tools used to smother wildland fires. (pp. 901–903) ▪ Describe the characteristics of the fire apparatus used to suppress wildland fires. (pp. 903–905) ▪ Describe the hazards associated with wildland firefighting. (pp. 908–909) ▪ Describe the personal protective clothing and equipment needed for wildland firefighting. (pp. 909–910) ▪ Suppress a wildland fire. (pp. 902–903) ▪ Deploy a fire shelter. (pp. 911–914)
6.3.20	8	▪ Describe the four primary types of fire service rope. (pp. 272–273) ▪ List the two types of life safety rope and their minimum breaking strength. (p. 273) ▪ Describe the characteristics of escape rope and fire escape rope. (p. 274) ▪ Describe the characteristics of water rescue throwlines. (p. 274) ▪ Describe the characteristics of utility ropes. (p. 274) ▪ Describe the characteristics of webbing. (pp. 274–276) ▪ List the disadvantages of natural fiber ropes. (p. 276) ▪ List the advantages of synthetic fiber ropes. (p. 276) ▪ List the disadvantages of synthetic fiber ropes. (p. 276) ▪ List common types of synthetic fibers that are used in fire service rope. (pp. 277–278) ▪ Describe how twisted ropes are constructed. (p. 277) ▪ Describe how braided ropes are constructed. (p. 277) ▪ Describe how kernmantle ropes are constructed. (p. 278) ▪ Explain the differences between dynamic kernmantle rope and static kernmantle rope. (p. 278) ▪ List the four components of the rope maintenance formula. (p. 278) ▪ Describe how to preserve rope strength and integrity. (pp. 278–284) ▪ List the terminology used to describe the parts of a rope when tying knots. (pp. 284–286) ▪ List the terminology used to describe the bends in rope that are formed when a knot is tied. (pp. 284–286)

Job Performance Requirement	Chapter(s)	Learning Objective(s)
		▪ List the 10 common types of knots that are used in the fire service. (pp. 285–286) ▪ Describe the characteristics of a safety knot. (pp. 286–287) ▪ Describe the characteristics of a figure eight knot. (p. 288) ▪ Describe the characteristics of a half hitch. (pp. 289–290) ▪ Describe the characteristics of a clove hitch. (pp. 290–293) ▪ Describe the characteristics of a figure eight on a bight. (pp. 293–294) ▪ Describe the characteristics of a figure eight follow-through. (pp. 293–295) ▪ Describe the characteristics of a bowline. (pp. 295–297) ▪ Describe the characteristics of a sheet bend. (pp. 298–299) ▪ Describe the characteristics of a figure eight bend. (pp. 300–301) ▪ Describe the characteristics of a water knot. (pp. 300–302) ▪ Describe the methods used to hoist tools or equipment. (pp. 302–312) ▪ Describe the general process of hoisting tools or equipment. (pp. 302–312) ▪ Tie a safety knot. (p. 287) ▪ Tie a figure eight knot. (p. 288) ▪ Tie a half hitch. (pp. 289–290) ▪ Tie a clove hitch in the open. (p. 291) ▪ Tie a clove hitch around an object. (pp. 292–293) ▪ Tie a figure eight on a bight. (p. 294) ▪ Tie a figure eight follow-through. (p. 295) ▪ Tie a bowline. (pp. 296–297) ▪ Tie a sheet bend. (pp. 298–299) ▪ Tie a figure eight bend. (pp. 300–301) ▪ Tie a water knot. (pp. 301–302) ▪ Hoist an axe. (pp. 303–304) ▪ Hoist a pike pole. (p. 305) ▪ Hoist a ladder. (pp. 306–307) ▪ Hoist a charged hose line. (pp. 308–309) ▪ Hoist an uncharged hose line. (pp. 310–311) ▪ Hoist an exhaust fan or power tool. (pp. 311–312)
6.3.21	18	▪ Describe the common methods of air monitoring at the fire scene. (pp. 821–824) ▪ Use a multi-gas air monitoring device. (pp. 821–822)
6.4: Rescue Operations		▪ Not applicable for Firefighter I
6.5: Preparedness and Maintenance		
6.5.1	3, 7, 8, 12, 16, 18	▪ Describe how to properly maintain PPE. (pp. 78–80) ▪ Describe why thoroughly cleaning PPE immediately after it has been exposed to smoke or fire conditions is an important step in reducing your chance of developing cancer. (pp. 79–80) ▪ Describe the importance of SCBA inspections and SCBA operational testing. (pp. 112–117) ▪ Explain how to inspect an SCBA to ensure that it is operation ready. (pp. 115–117)

Job Performance Requirement	Chapter(s)	Learning Objective(s)
		■ Describe the importance of properly maintaining tools and equipment. (pp. 262–264) ■ Describe the supplies needed to clean and inspect hand tools. (pp. 262–263) ■ Explain the importance of replacing tools in their assigned locations. (p. 262) ■ Identify procedures, including reporting requirements, for removing a damaged tool from service. (p. 262) ■ Describe why it is important for you to know where tools are stored. (p. 262) ■ Describe how to clean rope. (pp. 279–280) ■ Describe how to inspect rope. (pp. 280–282) ■ Describe how to store rope properly. (pp. 282–284) ■ Describe how to keep an accurate rope record. (pp. 280–282) ■ Inspect ladders. (pp. 441–443) ■ Maintain ladders. (pp. 441–443) ■ Clean ladders. (p. 442) ■ Describe when, where, and who performs service testing on ladders. (p. 444) ■ Explain how to ensure that ventilation equipment is in a state of readiness. (pp. 732–735) ■ Describe how to maintain salvage covers. (pp. 812–819) ■ Perform a visible inspection of an SCBA. (pp. 113–114) ■ Perform an operational inspection of an SCBA. (pp. 115–117) ■ Refill an SCBA air cylinder from a compressor or a cascade system. (p. 123) ■ Clean an SCBA. (pp. 124–125) ■ Clean and inspect hand tools. (p. 263) ■ Care for fire department ropes. (p. 280) ■ Clean fire department ropes. (p. 281) ■ Inspect fire department ropes. (p. 283) ■ Place a life safety rope in a rope bag. (p. 284) ■ Clean, inspect, and maintain a ladder. (pp. 445–446) ■ Perform a readiness check on a power saw. (pp. 732–734) ■ Maintain a power saw. (pp. 734–735)
6.5.2	11	■ Describe how to clean and maintain hose. (pp. 419–420) ■ Describe the importance of a hose inspection. (pp. 420, 422) ■ Clean and dry hose. (p. 421) ■ Inspect and mark a defective hose. (p. 422)
Chapter 7: Firefighter II		
7.1: General		
7.1.1	22	■ Outline the roles and responsibilities of a Firefighter II. (pp. 928–929) ■ Define the ICS. (p. 931) ■ Explain the organization of the ICS and how that makes it adaptable. (pp. 931–940) ■ Explain how to establish command of the ICS until command of the incident is transferred. (pp. 942–943) ■ Explain how to transfer command of a scene within the ICS. (pp. 943–945)

Job Performance Requirement	Chapter(s)	Learning Objective(s)
7.1.2	22	▪ Describe single resources. (p. 936) ▪ Explain the difference between groups and divisions. (pp. 936–937) ▪ Describe branches. (p. 938) ▪ Explain the difference between task forces and strike teams. (p. 937) ▪ Define units. (p. 938) ▪ Describe the command component of the ICS. (pp. 931–940) ▪ Explain how the ICP is established and used. (p. 933) ▪ Describe the command staff and their responsibilities. (pp. 933–934) ▪ Describe the four sections in the ICS and their responsibilities. (pp. 934–935) ▪ Describe the chain of command among the various components of the ICS. (p. 929) ▪ Understand how the sides of a building are identified at an incident. (p. 941) ▪ Describe the three principles of the ICS that always apply at an incident. (pp. 941–942) ▪ Define a complete size-up. (pp. 946–947) ▪ Describe the process of performing an initial size-up. (pp. 946–947) ▪ Describe the process of performing a 360-degree walk-around. (pp. 946–947) ▪ Describe the two basic categories of information used in the size-up process. (pp. 947–948) ▪ Explain how the size-up process determines the resources required at the emergency incident. (pp. 947–949) ▪ Describe IAPs, strategies, tactics, and tasks. (pp. 949–950) ▪ List the three incident priorities from which an incident action plan (IAP) is based. (p. 950) ▪ Describe the acronym RECEO–VS and how it provides a general guideline for incident commanders to systematically address the incident priorities. (pp. 950–952) ▪ Describe the acronym S.L.I.C.E.–R.S. and how it provides initial engine company operations a short list of objectives prior to the arrival of additional resources. (p. 953)
7.2: Communications		
7.2.1	22	▪ Explain the importance of an incident report. (pp. 957–959) ▪ Describe how to collect the necessary information for a thorough incident report. (pp. 957–959) ▪ Describe the resources that list the codes used in incident reports. (pp. 957–958) ▪ Explain how to submit incident reports to the NFIRS. (p. 958) ▪ Explain the consequences of an incomplete or inaccurate incident report. (pp. 958–959) ▪ Operate within the Incident Command System (ICS). (pp. 929–931) ▪ Establish command of an incident. (pp. 931–934) ▪ Transfer command of an incident to another firefighter. (pp. 943–945) ▪ Complete and proofread an incident report. (pp. 957–959)
7.2.2	22	▪ Explain the importance of communicating progress on an assigned task to the person above you in the chain of command. (p. 946) ▪ Explain how the size-up process can be used to determine whether additional resources are needed. (pp. 947–949) ▪ Explain the need for requesting additional resources to complete a task. (pp. 947–949)

Job Performance Requirement	Chapter(s)	Learning Objective(s)
7.3: Fireground Operations		
7.3.1	23	▪ Describe the characteristics of combustible or flammable liquid fires. (pp. 981–984) ▪ Describe the hazards presented by combustible or flammable liquid fires. (pp. 981–984) ▪ Describe the characteristics of Class A foam. (pp. 984–985) ▪ Describe the characteristics of Class B foam. (pp. 985–986) ▪ List the major categories of Class B foam concentrate. (p. 986) ▪ Describe the characteristics of protein foam. (p. 986) ▪ Describe the characteristics of fluoroprotein foam. (p. 986) ▪ Describe the characteristics of aqueous film-forming foam. (p. 986) ▪ Describe the characteristics of alcohol-resistant foam concentrate. (p. 986) ▪ Describe the characteristics of compressed air foam. (p. 988) ▪ Describe how foam proportioner equipment works with foam concentrate to produce foam. (pp. 987–991) ▪ Describe how foam is applied to fires. (pp. 991–994) ▪ Operate an in-line foam eductor. (pp. 989–990) ▪ Prepare the appropriate type of foam for application to a flammable liquid fire. (p. 993) ▪ Perform the roll-in method of applying foam. (p. 993) ▪ Perform the bounce-off method of applying foam. (p. 993) ▪ Perform the rain-down method of applying foam. (p. 994)
7.3.2	23	▪ List the factors that the incident commander evaluates when determining whether to perform a defensive operation or an offensive operation. (pp. 968–970) ▪ Describe how a thermal imager camera can be used in the size-up process. (pp. 971–974) ▪ Coordinate an interior attack. (p. 976) ▪ Describe how ventilation operations can impact fire growth in the structure. (pp. 974–975) ▪ Explain how ventilation is coordinated with fire suppression operations. (pp. 974–975) ▪ Coordinate an interior attack. (p. 976)
7.3.3	23	▪ Describe how a thermal imager camera can be used in the size-up process. (pp. 971–974) ▪ Utilize a thermal imager camera. (p. 974)
7.3.4	23	▪ Describe the characteristics of natural gas fires. (pp. 977–978) ▪ Describe the characteristics of flammable gas cylinder fires. (pp. 978–981) ▪ Describe the hazards presented by flammable gas cylinder fires. (pp. 978–981) ▪ Describe a boiling liquid/expanding vapor explosion (BLEVE). (p. 980) ▪ Describe how to suppress a flammable gas cylinder fire. (pp. 982–983) ▪ Suppress a flammable gas cylinder fire. (pp. 982–983)

Job Performance Requirement	Chapter(s)	Learning Objective(s)
7.3.5	28	▪ Explain the reasoning for conducting a fire investigation. (p. 1175) ▪ Describe the role of the Firefighter II in fire investigation. (pp. 1175–1180) ▪ Describe the evidential items and conditions that may be observed during fireground operations. (pp. 1176–1180) ▪ Describe the techniques for preserving fire scene evidence. (p. 1181) ▪ Explain the chain of custody. (pp. 1181–1182) ▪ List the types of evidence that may be found at a fire scene. (p. 1183) ▪ Describe the role and relationship of the Firefighter II to criminal investigators and insurance investigators. (pp. 1184–1185) ▪ Describe how to assist fire investigators with processing a fire scene. (pp. 1184–1185) ▪ Explain the importance of protecting a fire scene to aid in origin and cause determination. (pp. 1185–1186) ▪ Describe the steps needed to secure a property. (p. 1186) ▪ Describe how the point of origin of a fire is determined. (pp. 1187–1190) ▪ Describe how the cause of a fire is determined. (pp. 1189–1190) ▪ Protect evidence. (p. 1181)
7.4: Rescue Operations		
7.4.1	26	▪ Describe the major structural components of a motor vehicle. (pp. 1096–1102) ▪ Identify the two types of motor vehicle frames. (pp. 1099–1100) ▪ Describe electrical systems and supplemental restraint systems. (pp. 1100–1103) ▪ Describe the specialized personal protective equipment utilized at a motor vehicle accident. (pp. 1103–1108) ▪ List the types of power tools utilized at a motor vehicle accident. (pp. 1104–1105) ▪ Describe the purpose of cribbing. (p. 1106) ▪ Describe the purpose of rescue struts. (p. 1106) ▪ Describe the purpose of rescue-lift air bags. (p. 1106) ▪ List the hazards to look for when arriving on the scene of a vehicle extrication situation. (pp. 1108–1113) ▪ List the hazards to look for when stabilizing the scene of a vehicle extrication situation. (pp. 1118–1121) ▪ Describe how to gain access to a victim of a motor vehicle accident. (pp. 1126–1135) ▪ Describe how to disentangle a victim of a motor vehicle accident. (pp. 1127–1136) ▪ Describe how to remove and transport victims of a motor vehicle accident. (p. 1135) ▪ Perform scene size-up at a motor vehicle accident. (pp. 1112–1113) ▪ Immobilize and disable the electrical system of a vehicle following a motor vehicle accident. (p. 1117) ▪ Stabilize a vehicle following a motor vehicle accident. (p. 1120) ▪ Break tempered and laminated glass. (p. 1123) ▪ Force open and remove a vehicle door. (p. 1126) ▪ Provide medical care to a victim. (p. 1127) ▪ Remove a vehicle's windshield. (p. 1129) ▪ Perform a dash roll. (p. 1133) ▪ Perform a dash lift. (p. 1134) ▪ Remove the roof of a vehicle. (p. 1136)

Job Performance Requirement	Chapter(s)	Learning Objective(s)
7.4.2	27	▪ Define the types of special rescues encountered by firefighters. (pp. 1144–1145) ▪ Describe the steps of a special rescue. (pp. 1145–1149) ▪ Describe the general procedures at a special rescue scene, including what to do in the case of utility hazards. (pp. 1149–1152) ▪ Describe how to safely approach and assist at a vehicle or machinery rescue incident. (pp. 1153–1154) ▪ Describe how to safely approach and assist at a rope rescue incident. (p. 1156) ▪ Describe the hardware components used during a rope rescue. (pp. 1156–1157) ▪ Describe how to safely approach and assist at a confined-space rescue incident. (pp. 1157–1160) ▪ Describe how to safely approach and assist at a trench and excavation rescue incident. (pp. 1157–1160) ▪ Describe how to safely approach and assist at a structural collapse rescue incident. (p. 1159) ▪ Describe how to safely approach and assist at a wilderness and helicopter search and rescue incident. (pp. 1160–1161) ▪ Describe how to safely approach and assist at a water or ice rescue incident. (pp. 1161–1163) ▪ Describe how to safely approach and assist an animal rescue incident. (p. 1163) ▪ Establish a barrier. (p. 1151) ▪ Identify and retrieve rescue tools. (p. 1153)
7.5: Fire and Life Safety Initiatives, Preparedness, and Maintenance		
7.5.1	25	▪ Identify the best practices for educating the public on hazard prevention and mitigation. (pp. 1052–1056) ▪ Explain the importance of residential fire safety surveys and gathering knowledge about residential neighborhoods. (pp. 1066–1070) ▪ Describe some of the considerations to keep in mind during a residential fire safety survey. (p. 1066) ▪ Recognize hazards during a fire safety survey of a residence. (pp. 1068–1070) ▪ Explain the current code requirements for the placement of smoke alarms. (pp. 1070–1073) ▪ Describe smoke alarm maintenance, including how to vacuum smoke alarm chambers, how often occupants should test smoke alarms, how often batteries should be changed in smoke alarms with alkaline batteries, and how often smoke alarms should be replaced. (pp. 1070–1071) ▪ Describe the best practices for evacuating a residence during a fire. (pp. 1070–1073) ▪ Explain the importance of occupants in a residence practicing exit drills and establishing a meeting place outside. (pp. 1071–1073) ▪ Explain how to teach EDITH. (pp. 1055–1057) ▪ Describe how to select and use portable fire extinguishers and where to place them in a residential home. (p. 1073) ▪ Explain the importance of residential automatic sprinkler systems. (p. 1073) ▪ Complete the department fire safety survey forms and file according to the department SOPs. (p. 1073) ▪ Perform a public fire safety education presentation on Exit Drills In The Home (EDITH). (pp. 1056–1057) ▪ Teach fire safety to the public. (p. 1059) ▪ Give a public education tour of a fire station. (pp. 1066–1067) ▪ Install and maintain smoke alarms. (p. 1072) ▪ Conduct a fire safety survey at a residence. (p. 1074)

Job Performance Requirement	Chapter(s)	Learning Objective(s)
7.5.2	25	▪ Identify the best practices for educating the public on hazard prevention and mitigation. (pp. 1052–1056) ▪ Describe current fire safety messages. (pp. 1055–1056) ▪ Describe the difference between fire prevention and fire safety and survival messages. (pp. 1057–1058) ▪ Describe how to prepare to give a tour of the fire station. (pp. 1064–1066) ▪ Explain how to make a fire station tour educational and appropriate for the age group on the tour. (pp. 1064–1066) ▪ Describe the steps in conducting a fire station tour. (pp. 1064–1066) ▪ Explain the current code requirements for the placement of smoke alarms. (pp. 1070–1073) ▪ Describe smoke alarm maintenance including how to vacuum smoke alarm chambers, how often occupants should test smoke alarms, how often batteries should be changed in smoke alarms with alkaline batteries, and how often smoke alarms should be replaced. (pp. 1070–1071) ▪ Describe best practices for evacuating a residence during a fire. (pp. 1070–1073) ▪ Explain the importance of occupants in a residence practicing exit drills and establishing a meeting place outside. (pp. 1071–1073) ▪ Explain how to teach EDITH. (pp. 1055–1057) ▪ Describe how to select and use portable fire extinguishers and where to place them in a residential home. (p. 1073) ▪ Explain the importance of residential automatic sprinkler systems. (p. 1073) ▪ Perform a public fire safety education presentation on Exit Drills In The Home (EDITH). (pp. 1056–1057) ▪ Teach fire safety to the public. (p. 1059) ▪ Give a public education tour of a fire station. (pp. 1066–1067) ▪ Install and maintain smoke alarms. (p. 1072)
7.5.3	24, 25	▪ Describe the components of fire protection systems and the role of each component. (pp. 1002–1003) ▪ Describe the basic components and functions of a fire alarm system. (p. 1003) ▪ Describe the function of the FACU and the annunciator. (pp. 1003–1005) ▪ Describe the function of and the differences between noncoded, coded, zoned, and addressable systems. (pp. 1003–1005) ▪ Describe the purpose of trouble and supervisory signals and a remote annunciator. (pp. 1003–1005) ▪ Describe the difference between a smoke alarm and a smoke detector. (pp. 1005–1009) ▪ Explain the difference between manual and automatic alarm initiating devices. (pp. 1005–1012) ▪ Describe the types of manual pull stations. (p. 1005) ▪ Describe the types of detectors used in automatic alarm initiating devices and indicate where each type is most suitable. (pp. 1005–1012) ▪ Describe the basic types of alarm notification appliances and the temporal-3 pattern. (pp. 1012–1013) ▪ Explain how an alarm is initiated by a fire suppression system. (p. 1013) ▪ Describe how residential smoke and carbon monoxide alarms operate. (pp. 1014–1016) ▪ Explain the purpose of the emergency control function interface. (p. 1016) ▪ Describe the types of fire department notification systems. (pp. 1016–1017) ▪ Describe the parts of an automatic sprinkler head. (pp. 1019–1023)

Job Performance Requirement	Chapter(s)	Learning Objective(s)
		■ Describe three common types of release mechanisms used in automatic sprinkler heads. (pp. 1019–1023) ■ Explain the difference between release temperatures and maximum ceiling temperatures. (p. 1022) ■ Identify the three mounting positions and the type of sprinkler head used for each. (pp. 1022–1023) ■ Identify the four types of pipes used in sprinkler piping. (pp. 1023–1024) ■ Describe three types of indicating valves used as main water supply control valves. (pp. 1024–1026) ■ Identify the four types of main sprinkler system valves and list other valves used in sprinkler systems. (p. 1026) ■ Describe water supply for sprinkler systems and the FDC. (pp. 1026–1028) ■ Describe the water-motor gong and the waterflow alarm device. (p. 1028) ■ Describe the operation and application of the following types of automatic sprinkler systems: wet pipe system, dry pipe system, preaction system, deluge system, water mist systems. (pp. 1028–1031) ■ Describe the differences between commercial and residential sprinkler systems. (pp. 1032–1035) ■ Describe standpipe systems and how they are used. (pp. 1032–1035) ■ Identify the three classes of standpipes and explain the differences among them. (pp. 1032–1035) ■ Explain the purpose of flow restrictors and pressure-reducing valves in a standpipe. (pp. 1032–1035) ■ Explain the difference between wet and dry standpipe systems and the difference between automatic and manual standpipe systems. (p. 1035) ■ Describe the differences between dry chemical and wet chemical extinguishing systems and explain when each type is used. (pp. 1035–1037) ■ Describe gaseous suppression systems and explain why they are clean agent extinguishing systems. (pp. 1037–1038) ■ Describe halogenated extinguishing systems and how they differ from Halon extinguishing systems. (pp. 1037–1038) ■ Explain the restrictions on Halon extinguishing systems. (pp. 1037–1038) ■ Describe carbon dioxide extinguishing systems and explain when they are used. (p. 1038) ■ Explain the difference between smoke containment systems and smoke management systems. (pp. 1038–1039) ■ Describe passive and active smoke control systems. (pp. 1038–1039) ■ Identify the types of records that need to be maintained for systems. (pp. 1040–1041) ■ List the typical target hazards that may be found in a community. (pp. 1075–1077) ■ Describe why and for which types of properties a preincident survey is conducted. (pp. 1075–1087) ■ List the general information that is gathered during a preincident survey. (pp. 1075–1085) ■ Describe how to prepare to conduct a preincident survey. (pp. 1075–1085) ■ Describe the information included in any sketches or drawings created during the preincident survey. (pp. 1075–1077) ■ Describe the symbols commonly used in preincident plans. (pp. 1077–1078) ■ Describe your department's requirements for conducting a preincident survey and preparing a preincident plan. (pp. 1075–1079) ■ List the information that is gathered during a preincident survey for response and access to both the exterior and interior of a property. (pp. 1077–1080)

Job Performance Requirement	Chapter(s)	Learning Objective(s)
		▪ Describe the information in the preincident plan that is used to assist the IC in making a rapid and correct size-up during an emergency incident. (pp. 1080–1083) ▪ Describe how preincident planning leads to efficient fireground operations during emergency response. (pp. 1083–1084) ▪ List the types of information gathered during a preincident survey at properties where hazardous materials are stored or used. (p. 1084) ▪ Describe and list the types of locations that require special consideration during preincident planning. (pp. 1084–1087) ▪ List the types of locations that require special considerations in preplanning. (pp. 1084–1087) ▪ Conduct a preincident survey of a commercial properties or other public locations, identifying tactical priorities, and complete a preincident plan for that location. (pp. 1087–1088)
7.5.4	27	▪ Describe how to clean and maintain lighting equipment. (pp. 1164–1165) ▪ Describe how to maintain generators. (pp. 1165–1166) ▪ Describe how to maintain power equipment and power tools. (p. 1167) ▪ Conduct a weekly/monthly generator test. (pp. 1165–1166)
7.5.5	23	▪ Describe how an annual hose test is performed. (p. 995) ▪ Perform an annual hose test. (p. 995)

NFPA 470: Hazardous Materials/Weapons of Mass Destruction (WMD) Standard for Responders, 2022 Edition		
Job Performance Requirement	**Chapter(s)**	**Learning Objective(s)**
5.1: General		
5.1.1	29	▪ Describe the different levels of hazardous materials training: awareness, operations, technician, specialist, and incident commander. (pp. 1204–1209)
5.1.2	29	▪ Describe the different levels of hazardous materials training: awareness, operations, technician, specialist, and incident commander. (pp. 1204–1209)
5.1.3: General Skill Requirements	29, 30	▪ Identify the location of both the emergency response plan and/or standard operating procedures. (pp. 1202–1204) ▪ Describe the different levels of hazardous materials training: awareness, operations, technician, specialist, and incident commander. (pp. 1204–1209)
5.1.4: General Skills Requirements. (Reserved)		

Job Performance Requirement	Chapter(s)	Learning Objective(s)
5.2: Recognition and Identification		
5.2.1	29, 30	▪ Define the terms *hazardous materials* (or *dangerous goods*, in Canada) and *weapons of mass destruction*. (p. 1204) ▪ Describe how to approach a scene size-up when potentially hazardous materials are involved. (pp. 1216–1220) ▪ Identify and describe the types of containers that are often used to contain hazardous materials. (pp. 1220–1223) ▪ Describe the purpose and types of various transportation and facility markings for hazardous material. (pp. 1224–1225) ▪ Identify and describe the four routes of entry harmful substances take in the human body. (pp. 1245–1247) ▪ Use the *Emergency Response Guidebook* (*ERG*). (pp. 1241–1244)
5.2.2		
5.3: Initiate Protective Actions		
5.3.1	30	▪ Describe how to approach a scene size-up when potentially hazardous materials are involved. (pp. 1216–1220) ▪ Describe the purpose and types of various transportation and facility markings for hazardous material. (pp. 1224–1225) ▪ Identify and describe the four routes of entry harmful substances take in the human body. (pp. 1245–1247) ▪ Use the *Emergency Response Guidebook* (*ERG*). (pp. 1241–1244)
5.4: Notification		
5.4.1	30	▪ Describe how to approach a scene size-up when potentially hazardous materials are involved. (pp. 1216–1220)
Chapter 7: Professional Qualifications for Hazardous Materials/WMD Operations Level Responders (NFPA 1072)		
7.1: General		
7.1.1	34	▪ Explain the role of the operations-level responder in implementing a planned response. (pp. 1347–1351)
7.1.2	34	▪ Explain the role of the operations-level responder in implementing a planned response. (pp. 1347–1351)
7.1.3	34	▪ Explain the role of the operations-level responder in implementing a planned response. (pp. 1347–1351)
7.1.4 General Knowledge Requirements.	34	▪ Explain the role of the operations-level responder in implementing a planned response. (pp. 1347–1351)

Job Performance Requirement	Chapter(s)	Learning Objective(s)
7.2: Identify Potential Hazards		
7.2.1	31, 32, 33, 34	▪ Describe states of matter and their physical and chemical changes. (pp. 1257–1261) ▪ Understand the ways containers can breach and release contents. (pp. 1257–1261) ▪ Describe ionizing and non-ionizing radiation. (pp. 1269–1271) ▪ Describe toxic products of combustion. (pp. 1271–1272) ▪ Identify and describe common types of hazardous materials containers. (pp. 1284–1290) ▪ Describe the ways in which hazardous materials are transported. (pp. 1290–1294) ▪ Identify resources that can provide technical chemical information. (pp. 1295–1298) ▪ Identify different types of potential terrorist incidents and response options. (pp. 1299–1314) ▪ Explain how to estimate the potential harm or severity of an incident. (pp. 1320–1322) ▪ Explain how exposures might be affected by various types of hazardous materials incidents. (pp. 1322–1324) ▪ Describe how to report the size and scope of the incident. (p. 1329) ▪ Size up an incident. (pp. 1344–1345)
7.3: Identify Tactics		
7.3.1	31, 32, 33	▪ Understand the concepts of hazard, exposure, and contamination. (pp. 1272–1273) ▪ Describe how to select personal protective equipment for an incident. (pp. 1329–1331) ▪ Explain the role of respiratory protection. (pp. 1331–1335) ▪ Identify different types of potential terrorist incidents and response options. (pp. 1299–1314) ▪ Describe the basic types of decontamination. (pp. 1335–1336)
7.4: Action Plan Implementation		
7.4.1	34	▪ Size up an incident. (pp. 1344–1345) ▪ Identify and describe the safety procedures at a hazardous materials incident. (pp. 1345–1351) ▪ Explain the role of the operations-level responder in implementing a planned response. (pp. 1347–1351) ▪ Identify and describe the components of the incident command system. (pp. 1352–1357)
7.5: Emergency Decontamination		
7.5.1	33	▪ Describe the basic types of decontamination. (pp. 1335–1336) ▪ Perform emergency decontamination. (pp. 1336–1337)

Job Performance Requirement	Chapter(s)	Learning Objective(s)
7.6: Progress Evaluation and Reporting		
7.6.1	34	▪ Identify and describe the safety procedures at a hazardous materials incident. (pp. 1345–1351)
Chapter 9: Professional Qualifications for Hazardous Materials/WMD Operations Level Responders Assigned Mission-Specific Responsibilities (NFPA 1072)		
9.2: Personal Protective Equipment		
9.2.1	35	▪ Discuss the similarities and differences in how single-use and reusable personal protective equipment (PPE) are used. (pp. 1367–1368) ▪ Explain how to select, use, and maintain PPE. (p. 1368) ▪ Explain how PPE needs are determined. (pp. 1368–1369) ▪ Identify and describe specific PPE for hazardous materials response. (pp. 1369–1384) ▪ Explain the safety considerations when wearing PPE. (pp. 1385–1392) ▪ Explain the inclusion of PPE in reporting and documenting the incident. (pp. 1392–1393) ▪ Doff and don hazardous materials ensembles. (pp. 1374–1389)
9.6: Product Control		
9.6.1	36	▪ Describe the various control options available to an operations-level responder. (pp. 1398–1413) ▪ Control a hazardous materials release. (pp. 1398–1413)

Glossary

5 Es of community risk reduction (CRR) Education, enforcement, engineering, economic incentives, and emergency response. Also called *5 Es of prevention.*

5 Es of prevention See *5 Es of community risk reduction (CRR).*

10 Standard Firefighting Orders A set of systematically organized rules developed by the U.S. Forest Service (USFS) task force to reduce danger to firefighting personnel.

18 Watch Out Situations A list of potential situations published by the National Wildfire Coordinating Group (NWCG) that is used to assess whether or not a wildland firefighting assignment is safe to conduct.

360-degree walk-around An evaluation and analysis of an incident conducted by the incident commander (IC) or someone designated by the IC after the initial size-up is completed. Also called *360.*

360 See *360-degree walk-around.*

A

absorption The process of applying a material that soaks up and holds a hazardous material in a sponge-like manner, so as to collect the hazardous material for subsequent disposal.

absorption exposure Exposure to a hazardous material in which substances travel through body tissues until they reach the bloodstream.

accelerator A device that allows the water to flow more quickly in a dry pipe valve by evacuating the air from the system.

accidental fire A fire for which the cause does not involve a human act with the intent to ignite or spread a fire. (NFPA 556)

accordion hose load A method of loading hose on a vehicle that results in a hose appearance that resembles accordion sections. This is achieved by standing the hose on its edge and laying it side to side in the hose bed.

acid A material with a pH value less than 7.

action of opportunity An action taken when needed at any point during an incident.

action plan A detailed proposal to determine strategies and tactics within an incident action plan that will safely accomplish incident objectives, favorably influence outcomes, and increase the safety of responders and the public. (NFPA 470)

active crown fire A crown fire that starts on the ground and then progresses to aerial fuels.

active smoke control system A mechanical smoke control system that is activated when the fire alarm system is activated and that prevents the migration of smoke from the fire area to uninvolved floors or areas of the building by creating pressure differences between smoke control zones or floors or by exhausting smoke from large open areas to move smoke away from occupants or from egress routes.

activity logging system A device that keeps a detailed record of every incident and activity that occurs.

acute exposure An exposure that produces instantly observable conditions such as eye irritation, coughing, dizziness, and skin burns.

acute health effects Observable health problems, such as eye irritation, coughing, dizziness, and skin burns, caused by relatively short exposures to a hazardous substance.

adapter Any device that allows fire hose couplings to be safely interconnected with couplings of different sizes, threads, or mating surfaces, or that allows fire hose couplings to be safely connected to other appliances. (NFPA 1960)

addressable system A fire alarm system in which each initiating device and notification appliance is assigned a number so that the exact type and location of the devices and appliances are shown on the fire alarm control unit (FACU).

adjustable fog-stream nozzle See *fog-stream nozzle.*

adjustable-gallonage fog nozzle A nozzle that allows the operator to select a desired flow from several settings. Also called *variable-flow fog nozzle.*

adjustable wrench An open-ended wrench whose opening can be adjusted to accommodate bolts of different sizes.

adolescent fire-setters Fire-setters who are typically 14 to 16 years old. Adolescent fire-setters may start fires in schools, churches, outbuildings, vacant homes, and vacant lots.

adsorption The process of applying a material, such as silica or activated carbon, whose surface a hazardous material adheres to, so as to collect the hazardous material for subsequent disposal.

Advanced Emergency Medical Technician (AEMT) Emergency medical services (EMS) personnel who can do everything an EMT can do, and who have advanced training in specific areas of advanced life support, including IV therapy, interpretation of cardiac rhythms, and advanced airway management.

advanced high-strength steel An alloyed steel with a minimum tensile strength of 65 ksi (440 MPa).

adze The curved or straight wedge part of a Halligan tool.

adze hoe See *hazel hoe.*

aerial See *aerial fire apparatus.*

aerial apparatus See *aerial fire apparatus.*

aerial fire apparatus A vehicle equipped with an aerial ladder, elevating platform, or water tower that is designed and equipped to support firefighters and rescue operations by positioning personnel, handling materials, providing continuous egress, or discharging water at positions elevated from ground. (NFPA 1900)

aerial fuels Fuels located more than 6 ft (2 m) off the ground, usually part of or attached to trees. Also called *canopy fuels.*

aerial ladder A self-supporting, turntable-mounted, power-operated ladder of two or more sections permanently attached to a self-propelled automotive fire apparatus and designed to provide a continuous egress route from an elevated position to the ground. (NFPA 1900)

aerosol An intimate mixture of a liquid or a solid in a gas; the liquid or solid, called the dispersed phase, is uniformly distributed in a finely divided state throughout the gas, which is the continuous phase or dispersing medium. (MED) (NFPA 99)

A-frame hoist See *ladder A-frame.*

A-frame ladder A ladder with two ladder sections connected with a joint so that it forms an A-shaped structure when it is set up to climb on and that closes to fold flat for storage.

air-aspirating nozzle A nozzle that draws air into the water stream, creating an aerated or foamy spray that increases the surface area of water droplets, allowing for better heat absorption and faster cooling of a fire, and when used with firefighting foam solutions, aerate the foam mixture.

air bill The shipping papers on an airplane.

air-cylinder pressure gauge The device on an SCBA that measures and displays pressure readings to indicate the quantity of breathing air available.

air cylinder See *breathing air cylinder.*

air-line respirator See *supplied-air respirator.*

airport firefighter The Firefighter II who has demonstrated the skills and knowledge necessary to function as an integral member of an aircraft rescue and firefighting (ARFF) team. (NFPA 1010)

air-purifying respirator (APR) A respirator that removes specific air contaminants by passing ambient air through one or more air-purification components. (NFPA 1984)

air sampling-type detector A detector that consists of a piping or tubing distribution network that runs from the detector to the area(s) to be protected. An aspiration fan in the detector housing draws air from the protected area back to the detector through air sampling ports, piping, or tubing. At the detector, the air is analyzed for fire products. (NFPA 72)

air stability The vertical movement of air in the atmosphere; unstable environments allow the free flow of air from the ground into the upper atmosphere, and stable air resists vertical air movement.

alarm initiating device A device that, when operated either automatically or manually, causes the alarm notification device to operate.

alarm notification appliance A device that generates a fire alarm signal.

alarm valve A valve that signals an alarm when a sprinkler head is activated and prevents nuisance alarms caused by pressure variations and surges in the water supply to the system.

alarm verification feature A feature of automatic fire detection and alarm systems to reduce nuisance alarms wherein smoke detectors report alarm conditions for a minimum period of time or confirm alarm conditions within a given time period after being reset, in order to be accepted as a valid alarm initiation signal. (NFPA 72)

alcohol-resistant foam A concentrate used for fighting fires on water-soluble materials and other fuels destructive to regular, AFFF, or FFFP foams, as well as for fires involving hydrocarbons. (NFPA 11)

alcohol-resistant foam concentrate A concentrate used for fighting fires on water-soluble materials and other fuels destructive to regular foams, aqueous film-forming foams (AFFF), synthetic fluorine-free foams (SFFF), or film-forming fluoroprotein foams (FFFP), as well as for fires involving hydrocarbons. (NFPA 11)

alloy A material composed of two or more metals or chemical elements.

alternative-fuel vehicles Vehicles that use anything other than a petroleum-based motor fuel (gasoline or diesel fuel) to propel a motorized vehicle.

ambulance A vehicle used for out-of-hospital medical care and patient transport that provides a driver's compartment; a patient compartment to accommodate an emergency medical services provider (EMSP) and at least one patient located on the primary cot positioned so that the primary patient can be given emergency care during transit; equipment and supplies at the scene as well as during transport; safety, comfort, and avoidance of aggravation of the patient's injury or illness; two-way radio communication; and audible and visual warning devices. (NFPA 1900)

American Conference of Governmental Industrial Hygienists (ACGIH) A scientific organization that researches and publishes information in the field of industrial hygiene.

ammonium phosphate See *monoammonium phosphate.*

anchor, flank, tandem, and pincer attack A direct method of suppressing a wildland fire that involves two teams of firefighters establishing anchor points on each side of the fire and working toward the head of the fire until the fire gets "pinched" between them. Also called *pincer attack.*

anchor point A strategic and safe point from which to start constructing a fire control line. An anchor point is used to reduce the chance of firefighters being flanked by fire.

anhydrous Without water.

anion A negatively charged ion.

annealed The process of forming standard window glass.

annunciator A unit containing one or more indicator lamps, alphanumeric displays, or other equivalent means in which each indication provides status information about a circuit, condition, or location. Also called *remote annunciator.* (NFPA 72)

anthrax An infectious disease spread by the bacterium *Bacillus anthracis,* which is typically found around farm animals, such as cows and sheep.

antifreeze loop A small section of a wet pipe sprinkler system that is filled with a glycol or glycerin solution instead of water and that has a check valve separating the antifreeze loop from the rest of the sprinkler system.

A-post One of the two posts closest to the front of a motor vehicle that form the sides of the windshield and extend down alongside the firewall. Also called *hinge pillar.*

apparatus See *fire apparatus.*

aqueous film-forming foam (AFFF) A solution based on fluorinated surfactants plus foam stabilizers to produce a fluid aqueous film for suppressing liquid-fuel vapors. (NFPA 10)

arc mapping The systematic evaluation of the electrical circuit configuration, spatial relationship of the circuit components, and identification of electrical arc sites to assist in the identification of the area of origin and analysis of the fire's spread. (NFPA 921)

area of origin A structure, part of a structure, or general geographic location within a fire scene, in which the "point of origin" of a fire or explosion is reasonably believed to be located. Also called *fire compartment, fire seat,* or *seat of the fire.* (NFPA 901)

area of safe refuge An area where people are held temporarily until they can be safely decontaminated, treated, or removed.

arson The crime of maliciously and intentionally, or recklessly, starting a fire or causing an explosion. (NFPA 921)

arson reporting immunity law A law that generally requires private investigators to provide information regarding possible criminal activity upon the written request of the authority having jurisdiction. These laws also provide immunity from liability for disclosing such information as provided for by the law.

aspect The direction the side of the mountain is facing.

asphyxiant A material that prevents the body from using oxygen at the cellular level, causing the victim to suffocate.

assembly Various construction materials joined together to form a component of a structure.

assist Rescue technique in which the victim is responsive and able to walk but needs support.

assistant chief A midlevel chief who often has a functional area of responsibility, such as training, or who is responsible for a group of battalions or districts and who answers directly to the fire chief. Also called *deputy chief* or *division chief.*

assistant safety officer (ASO) See *hazardous materials safety officer.*

assisting agency An agency or organization providing personnel, services, or other resources to the agency that has direct responsibility for incident management. See also *cooperating agency.* (NFPA 1026)

assisting organization See *assisting agency.*

atmosphere (atm) A unit of pressure equal to the average atmospheric pressure at sea level.

atmosphere-supplying respirator (ASR) A respirator that supplies the respirator user with breathing air from a source independent of the ambient atmosphere and includes self-contained breathing apparatus (SCBA) and supplied-air respirators (SARs). (NFPA 1970)

atom The smallest unit of a chemical element.

atomic number The number of an element in the periodic table of elements that corresponds to the number of protons in the atom.

A tool A special-use tool used for pulling bored locksets.

attack engine An engine used to pump water through attack lines at the fireground.

attack hose Hose designed to be used by trained firefighters and fire brigade members to combat fires beyond the incipient stage. Also called *attack line.* (NFPA 1962)

attack line See *attack hose.*

attack line evolutions The delivery of water from an attack engine to a handline, which discharges the water onto the fire. Also called *attack line operations.*

attack line operations See *attack line evolutions.*

attic ladder See *folding ladder.*

authority having jurisdiction (AHJ) An organization, office, or individual responsible for enforcing the requirements of a code or standard or for approving equipment, materials, an installation, or a procedure. (NFPA 1)

auto-exposure See *vertical fire extension.*

autoignition Initiation of combustion by heat but without a spark or flame. (NFPA 921)

automatic-adjusting fog nozzle A nozzle that can deliver a wide range of water stream flows. As the pressure at the nozzle increases or decreases, an internal spring-loaded piston moves in or out to adjust the size of the opening.

automatic dry standpipe system A standpipe system permanently attached to a water supply capable of supplying the system demand at all times, containing air or nitrogen under pressure, the release of which (as from opening a hose valve) opens a dry pipe valve to allow water to flow into the piping system and out of the opened hose valve. (NFPA 14)

automatic location identification (ALI) The automatic display at the PSAP of the caller's telephone number, the address/location of the telephone, and supplementary emergency services information about the location from which a call originates. (NFPA 1225)

automatic number identification (ANI) A series of alphanumeric characters that informs the recipient of the source of an event or power supply module. (NFPA 1225)

automatic sprinkler head The part of the sprinkler system through which water is discharged and that consists of a frame, the orifice, a heat-sensitive release mechanism that holds a cap in place over the orifice, and a deflector that directs the water in a spray pattern. Also called *sprinkler head.*

automatic sprinkler system A sprinkler system of pipes with water under pressure that allows water to be discharged immediately when a sprinkler head operates. (NFPA 853)

automatic wet standpipe system A standpipe system containing water at all times that is attached to a water supply capable of supplying the system demand at all times and that requires no action other than opening a hose valve to provide water at hose connections. (NFPA 14)

awareness-level personnel Personnel who, during the course of their normal duties, could encounter an emergency involving hazardous materials/weapons of mass destruction (WMD) and who are expected to recognize the presence of the hazardous materials/WMD, protect themselves, call for trained personnel, and secure the scene. (NFPA 470)

awning windows Windows that have one large or multiple medium-size panels that do not overlap when they are closed. The window is operated by a hand crank from the corner of the window. The hinge is on the top.

axes Cutting tools that have a wide cutting blade that can be used to chop into a wall, roof, or door.

axial loads Loads that bear directly through the center of the structure.

B

backdraft A deflagration (explosion) resulting from the sudden introduction of air into a confined space containing oxygen-deficient products of incomplete combustion. (NFPA 1403)

backfire A fire set along the inner edge of a fire control line to consume the fuel in the path of a wildland fire or change the direction of force of the fire's convection column. (NFPA 901)

backpack fire extinguisher A portable fire extinguisher usually consisting of a 5-gal (19-L) water tank that is worn on the user's back and features a hand-powered piston pump for discharging the water and that is primarily used to fight brush and grass fires.

backup team Required by OSHA's HAZWOPER regulation, a team composed of two or more responders equipped with approved personal protective equipment or chemical-protective clothing and assigned by the incident commander to provide emergency removal of a stricken entry team member from the hot zone.

bale A shut-off valve on a nozzle that is a handle that you pull to control the flow of water from the nozzle.

balloon-frame construction An older type of wood frame construction in which the wall studs extend vertically from the basement of a structure to the roof without any fire stops.

ball valves Valves used on nozzles, gated wyes, and engine discharge gates. They consist of a ball with a hole in the middle of the ball.

band saw An electrically powered saw that has a toothed metal blade stretched over two pulleys in a loop.

Bangor ladder A ladder equipped with tormentor poles, or staypoles, that stabilize the ladder during raising and lowering operations.

bank-down method See *bounce-off method.*

bank-shot method See *bounce-off method.*

bar A unit of pressure equal to 14.7 psi.

barrel The upright steel casing that is the main part of a fire hydrant.

base A material with a pH value greater than 7.

base distance The distance from the base of the building to the butt of the ladder when the ladder is raised.

base section See *bed section.*

base ladder See *butt.*

base station A stationary radio transceiver with an AC or DC power supply. (NFPA 1225)

batch mix The manual addition of foam concentrate to a water storage container or tank to make foam solution. (NFPA 1145)

battalion chief Usually the first level of chief, the person in charge of running calls and supervising multiple stations or districts within a city.

battering ram A tool made of hardened steel with handles on the sides used to force doors and to breach walls. Larger versions may be used by as many as four people; smaller versions are made for one or two people.

battery electric vehicle (BEV) A vehicle propelled solely by an electric motor that is powered by a high-voltage battery.

beam The main structural side of a ground ladder. (NFPA 1960)

beam detector See *projected-beam type detector.*

beam raise A method of raising a ladder by laying the ladder on one beam and them raising it by lifting the tip of the ladder while resting the butt of the ladder on the single beam.

bed section The lowest or widest section of an extension ladder. Also called *base section.* (NFPA 1960)

bend A knot that joins two ropes or webbing pieces together. (NFPA 2500)

berm See *dike.*

bib The lower part of the protective hood that is part of the structural firefighting ensemble.

bight The open loop in a rope or piece of webbing formed when it is doubled back on itself. (NFPA 1006)

bill of lading The shipping papers used for transport of chemicals over roads and highways; also called *freight bill.*

biological agent An organism that causes acute disease or long-term damage to the human body.

black area An area that has already been burned and that is considered relatively safe for firefighters.

black fire A hot, high-volume, high-velocity, turbulent, ultra-dense black smoke that can produce charring and cause heat damage to steel and concrete and indicates an impending flashover or autoignition.

blended liquid fuel–powered vehicle A vehicle powered by a blend of liquid fuels.

BLEVE (boiling liquid/expanding vapor explosion) An explosion that occurs when pressurized liquefied materials, such as propane or butane, inside a closed vessel are exposed to a source of high heat.

blitz attack See *transitional attack.*

block creel construction Rope constructed without knots or splices in the yarns, ply yarns, strands or braids, or rope. (NFPA 2500)

blood sugar See *glucose.*

body-over-frame construction A type of vehicle frame resembling a ladder, which is made up of two parallel rails called frame rails joined by a series of cross member beams. Also called *ladder frame construction.*

boiling liquid/expanding vapor explosion (BLEVE) An explosion that occurs when pressurized liquefied materials (e.g., propane or butane) in a closed container are exposed to a source of high heat, releasing the fuel which instantly vaporizes and ignites.

boiling point The temperature at which the vapor pressure of a liquid equals the surrounding atmospheric pressure. (NFPA 1)

bolt cutter A cutting tool used to cut through thick metal objects such as bolts, locks, and wire fences.

bonnet The top of a hydrant.

booster hose A non-collapsible hose used under positive pressure having an elastomeric or thermoplastic tube, a braided or spiraled braided reinforcement, and an outer protective cover. Also called *booster line.* (NFPA 1962)

booster line See *booster hose.*

bored locksets The most common fixed locks in use today. The locks and handles are placed into a predrilled hole in the door. The outside of the doorknob will usually have a key-in-knob lock; the inside will usually have a keyway, a button, or another type of locking/unlocking mechanism.

bounce-off method A foam application method that applies the stream onto a nearby object, such as a wall, instead of directly onto the surface of the fire. Also called *bank-shot method* or *bank-down method.*

bowstring truss A truss that is curved on the top and straight on the bottom.

boxcar A railway car that carries consumer goods, industrial supplies, and a multitude of boxes and palletized goods.

box crib Cribbing, usually wood, placed in layers at 90-degree angles to each other to create a box-like structure.

box-end wrench A hand tool used to tighten or loosen bolts. The end is enclosed, as opposed to an open-end wrench. Each wrench is a specific size, and most have ratchets for easier use.

B-post One of two posts located between the front and rear doors of a motor vehicle.

braided reinforcement A hose reinforcement consisting of one or more layers of interlaced spiraled strands of yarn or wire, with a layer of rubber between each braid. (NFPA 1962)

braided rope Rope constructed by intertwining strands in the same way that hair is braided.

branch A supervisory level established in either the operations or logistics function to provide a span of control. (NFPA 1550)

branch director A person in a supervisory level position in either the operations or logistics function to provide a span of control. (NFPA 1550)

branch line See *secondary feeder.*

branch lines (sprinkler system) In a sprinkler system, the pipes that supply the sprinkler heads.

breakaway fire nozzle A nozzle with a tip that can be separated from the shut-off valve.

breathing air cylinder The pressure vessel or vessels that are an integral part of the SCBA and that contain the breathing gas supply; can be configured as a single cylinder or other pressure vessel or as multiple cylinders or pressure vessels. (NFPA 1970)

Bresnan distributor nozzle A nozzle that can be placed in confined spaces such as cellars or basements. The nozzle spins, spreading water over a large area.

British thermal unit (BTU) The amount of heat energy required to raise 1 pound of water at sea level by 1 degree Fahrenheit.

brush company See *wildland company.*

buddy breather See *emergency breathing safety system (EBSS).*

buddy system A system in which two responders work as a team for safety purposes.

building code A regulation that specifies how structures are designed, constructed, and remodeled so that buildings are safe for people who live and work in them.

bulk packaging Any packaging, including transport vehicles, having a liquid capacity of more than 119 gallons (450 liters), a solids capacity of more than 882 pounds (400 kg), or a compressed gas water capacity of more than 1001 pounds (454 kg). (NFPA 470)

bulkhead See *firewall.*

bung An opening on top of a closed-head drum that is typically sealed with a threaded cap.

bunker coat See *structural firefighting protective coat.*

bunker gear See *structural firefighting protective clothing.*

bunker pants See *structural firefighting protective trousers.*

bunny tool See *rabbet tool.*

burnback resistance The ability of the foam blanket to maintain its integrity and effectiveness when impinged by fire.

burning edge The active perimeter of the fire.

butterfly valve (sprinkler system) A sprinkler control valve consisting of a disk that either blocks the water and prevents it from flowing into the system or is rotated 90 degrees to allow the water to flow past it and that has an external indicator on the stem that is parallel or in-line with the piping to signify the valve is open or perpendicular to the piping to indicate the valve is closed.

butterfly valves (fire pump) Valves that are found on the large pump intake connections where the suction hose connects to the suction side of the fire pump.

butt The end of the beam that is placed on the ground, or other lower support surface, when ground ladders are in the raised position. Also called *heel* or *base.* (NFPA 1960)

butt plate An alternative to a simple butt spur; a swiveling plate with both a spur and a cleat or pad that is attached to the butt of the ladder.

butt spurs That component of ground ladder support that remains in contact with the lower support surface to reduce slippage. (NFPA 1960)

C

call box A system of telephones connected by phone lines, radio equipment, or cellular technology to a communications center or fire department.

call classification and prioritization The process of assigning a response category based on the nature of the reported problem.

call arrival The process of receiving a call for service and obtaining the information necessary to initiate a response.

calorie The amount of heat energy required to raise 1 gram of water (at sea level) by 1 degree Celsius.

Canadian Transport Emergency Centre (CANUTEC) Operated by Transport Canada, an organization that provides emergency response information and assistance on a 24-hour basis for responders to hazardous materials/weapons of mass destruction incidents. (NFPA 470)

candidate A person who applies to become a firefighter.

canopy fuels See *aerial fuels.*

cap gauge An outlet cap with a built-in pressure gauge.

captain The second rank of promotion in the fire service, between the lieutenant and the battalion chief. Captains are responsible for managing a fire company and for coordinating the activities of that company among the other shifts.

carabiner An auxiliary equipment system item; a load-bearing connector with a self-closing gate used to join other components of life safety rope. (NFPA 1983)

carbon dioxide (CO_2) A nontoxic gas produced when sufficient oxygen is available for complete combustion that can displace oxygen in the atmosphere. Also, a colorless, odorless, electrically nonconductive inert gas that is a suitable medium for extinguishing Class B and Class C fires. (NFPA 10)

carbon dioxide (CO_2) extinguishing system A fire suppression system designed to protect either a single room or series of rooms by flooding the area with carbon dioxide.

carbon dioxide (CO_2) fire extinguisher A fire extinguisher that uses carbon dioxide gas as the extinguishing agent. It is rated for use on Class B and C fires.

carbon fiber reinforced polymer (CFRP) An extremely strong, lightweight plastic reinforced with woven strands of carbon bound together with a polymer, such as an epoxy, and then formed into the desired shape.

carbon monoxide (CO) A toxic gas produced through incomplete combustion.

carbon monoxide detector A device having a sensor that responds to carbon monoxide gas that is connected to an alarm control unit. (NFPA 72)

carboy A glass, plastic, or steel storage container, ranging in volume from 5 to 15 gallons.

carcinogen A human cancer-causing agent.

cargo tank Bulk packaging that is permanently attached to or forms a part of a motor vehicle or is not permanently attached to any motor vehicle and which, by reason of its size, construction, or attachment to a motor vehicle, is loaded or unloaded without being removed from the motor vehicle.

carpenter's handsaw A saw designed for cutting wood.

carry Technique that physically moves a victim who is conscious and responsive but incapable of standing or walking.

carryall A piece of heavy canvas with handles, which can be used to tote debris, ash, embers, and burning materials out of a structure.

cartridge/cylinder-operated fire extinguisher A fire extinguisher in which the expellant gas is in a separate container from the agent storage container. (NFPA 10)

cascade system A method of piping air tanks together to allow air to be supplied to the SCBA fill station using a progressive selection of tanks, each with a higher pressure level. (NFPA 1900)

case-hardened steel Steel created in a process that hardens the outer portion of a steel component while the inner core remains soft.

casement windows Windows in a steel or wood frame that open away from the building via a crank mechanism. These windows have a side hinge.

cation A positively charged ion.

ceiling hook A tool with a long wooden or fiberglass pole that has a metal point with a spur at right angles at one end. It can be used to probe ceilings and pull down plaster lath material.

ceiling jet A strong, turbulent convection current that rises to the ceiling and travels along it.

cellar nozzle A nozzle used to fight fires in cellars or basements and other inaccessible places that spreads water in a wide pattern.

central station service alarm system A system or group of systems in which the operations of circuits and devices are transmitted automatically to, recorded in, maintained by, and supervised from a listed central station that has competent and experienced servers and operators who, upon receipt of a signal, take such action as required by this Code. Such service is to be controlled and operated by a person, firm, or corporation whose business is the furnishing, maintaining, or monitoring of supervised alarm systems. (NFPA 72)

chafing block A sturdy rubber, plastic, or wooden block placed under a fire hose where it lays on the ground or rests against hard surfaces to raise the hose off the ground and provide a smooth contact surface and protect the hose from abrasion, friction, and wear against rough surfaces like asphalt, concrete, or debris.

chain A unit of measurement for length; 1 chain equals 66 ft (20 m).

chain of command A rank-based, hierarchical structure that creates an orderly line of authority.

chain of custody The trail of accountability that documents the possession of evidence in an investigation from the time it is discovered until it is presented in court.

chainsaw A power saw that uses the rotating movement of a chain equipped with sharpened cutting edges. It is typically used to cut through wood.

charge To fill with water under pressure.

charged hose line A hose line filled with water and under pressure from the pump.

check valve A valve that prevents the flow of liquid in one direction.

Chemical Abstracts Service (CAS) A division of the American Chemical Society that provides hazardous materials personnel and responders with access to the CAS Registry, an enormous collection of chemical substance information.

chemical agent A toxic substance that is intended to cause harm to people or animals.

chemical and physical properties Measurable characteristics of a substance, such as its vapor density, flammability, corrosivity, and water reactivity.

chemical cargo tank A tank, characterized by several heavy-duty reinforcing rings around the tank, that holds approximately 6000 gallons (22,712 liters) of product and often carries aggressive (highly reactive) acids such as concentrated sulfuric and nitric acid; also called *MC-312/DOT 412 corrosive tank*.

chemical change See *chemical reactivity*.

chemical energy Potential energy in molecular bonds and the kinetic energy created by a chemical reaction.

chemical-pellet sprinkler head A sprinkler head activated with the cap held in place by a chemical pellet that liquefies at a preset temperature, and when the pellet melts, the liquid compresses a plunger which releases the cap.

chemical reaction Any chemical change or chemical degradation.

chemical reactivity The ability of a chemical to undergo an alteration in its chemical make-up, usually accompanied by a release of some form of energy; also called *chemical change*.

chemical-resistant material Fabric specifically designed to inhibit or resist the passage of chemicals into and through the material by the process of penetration, permeation, or degradation.

Chemical Transportation Emergency Center (CHEMTREC) A public service of the American Chemistry Council, which provides emergency response information and assistance on a 24-hour basis for responders to hazardous materials/weapons of mass destruction incidents. (NFPA 470)

chief's bugle See *chief's trumpet.*

chief's trumpet An obsolete amplification device that was a precursor to a bullhorn and that enabled a chief officer to give orders to firefighters during an emergency. Also called *chief's bugle.*

child fire-setters Fire-setters who are typically 2 to 6 years old.

chimney nozzle A nozzle that has a 45-degree elbow attached to a long discharge pipe that sprays water upward.

chinook winds See *Foehn winds.*

chisel A metal tool with one sharpened end that is used to break apart material in conjunction with a hammer, mallet, or sledgehammer.

chronic exposure Long-term exposure to a hazardous material or materials, occurring over the course of many months or years.

chronic health effect See *chronic health hazard.*

chronic health hazard An adverse health effect that occurs gradually over time after long-term exposure to a substance; also called *chronic health effect.*

circumstantial evidence Evidence that is based on logical inference rather than personal observation of a crime or other activity.

clapper mechanism See *clapper valve.*

clapper valve A mechanical device installed within a piping system that allows water to flow in only one direction. Also called *clapper mechanism.*

Class A fire A fire in ordinary combustible materials, such as wood, cloth, paper, rubber, and many plastics. (NFPA 1)

Class A foam concentrate A concentrate that when combined with water reduces the surface tension of the water and creates a foam. Also called *wet water.*

Class A foam fire extinguisher A fire extinguisher that contains a solution of water and Class A foam concentrate.

Class B fire A fire in flammable liquids, combustible liquids, petroleum greases, tars, oils, oil-based paints, solvents, lacquers, alcohols, and flammable gases. (NFPA 1)

Class C fire A fire that involves energized electrical equipment. (NFPA 1)

Class D fire A fire in combustible metals, such as magnesium, titanium, zirconium, sodium, lithium, and potassium. (NFPA 1)

Class I standpipe A standpipe system designed for use by fire department personnel only, with outlets and valves to control the flow of water and a 2½-in. (65-mm) male coupling for fire hose the personnel carry into the building with them.

Class II standpipe A standpipe system with outlets that have a length of 1½-in. (38-mm) single-jacket hose and a nozzle already connected for use by trained building occupants or firefighters and that are used to attack fire at the incipient stage only.

Class III standpipe A combination standpipe system that has features of both Class I and Class II standpipes.

Class K fire A fire in a cooking appliance that involves combustible cooking media (vegetable or animal oils and fats). (NFPA 1)

claw bar A tool with a pointed claw-hook on one end and a forked- or flat-chisel pry on the other end. It is often used for forcible entry.

clean agent Electrically nonconducting, volatile, or gaseous fire extinguishant that does not leave a residue upon evaporation. (NFPA 10)

clean agent extinguishing system A fire suppression system that expels an electrically nonconducting, volatile, or gaseous clean extinguishing agent that does not leave a residue upon evaporation.

clean-agent fire extinguisher A fire extinguisher that uses a halogenated extinguishing agent. Also called *halogenated-agent fire extinguisher.*

clean air The outside air that replaces a contaminated atmosphere.

Clemens hook A multipurpose tool that can be used for several forcible entry and ventilation applications because of its unique head design.

climbing angle The angle that provides the best combination of strength, stability, and vertical reach for accessing the roof or for venting operations.

closed-circuit self-contained breathing apparatus (closed-circuit SCBA) A recirculation-type SCBA in which the exhaled gas is rebreathed by the wearer after the carbon dioxide has been removed from the exhalation gas and the oxygen content within the system has been restored from sources such as compressed breathing air, chemical oxygen, liquid oxygen, or compressed gaseous oxygen. Also called *rebreather.* (NFPA 1970)

closed-head drum A drum with a lid that is permanently attached to the drum; the lid has one or more small openings called bungs.

closet hook A type of pike pole intended for use in tight spaces, commonly 2 to 4 ft (0.6 to 1.2 m) in length.

code A standard that is an extensive compilation of provisions covering broad subject matter or that is suitable for adoption into law independently of other codes and standards. (NFPA 1)

Code of Federal Regulations (CFR) A collection of permanent rules published in the Federal Register by the executive departments and agencies of the U.S. federal government. Its 50 titles represent broad areas of interest that are governed by federal regulation. Each volume of the CFR is updated annually.

coded system A fire alarm system that has audible or visible notification devices that indicate the location of the initiation device.

cold smoke Smoke in a room where a fire sprinkler system was activated so that the smoke is cooled and banked down to the floor.

cold zone The control zone of hazardous materials/weapons of mass destruction (WMD) incidents that contains the incident command post and such other support functions as are deemed necessary to control the incident. Also called *outer perimeter.* (NFPA 470)

collapse zone The area that is exposed to trauma, debris, and/or thrust should a building or part of a building collapse. (NFPA 1550)

collapsible fire hose Fire hose typically made from synthetic materials that make the hose flexible and foldable.

combination attack An indirect attack followed by a direct attack.

combination hose load A hose loading method used when one long hose line is needed; the end of the last length of a hose in one bed of a split hose bed is coupled to the beginning of the first length of hose in the opposite bed.

combination ladder A ground ladder that is capable of being used both as a stepladder and as a single or extension ladder. Also called *multipurpose ladder.* (NFPA 1960)

combination wrench A hand tool with an open-end wrench on one end and a box-end wrench on the other.

combustibility The property describing whether a material will burn and how quickly it will burn.

combustibles Materials that, in the form in which they are used and under the conditions anticipated, will ignite and burn. (NFPA 1)

combustion A chemical process of oxidation that occurs at a rate fast enough to produce heat and usually light in the form of either a glow or a flame. (NFPA 1)

come along A hand-operated tool used for dragging or lifting heavy objects that uses pulleys and cables or chains to multiply a pulling or lifting force.

command The act of directing and/or controlling resources by virtue of explicit legal, agency, or delegated authority. (NFPA 1026)

command designation The name of the incident.

command staff The public information officer (PIO), incident safety officer (ISO), and liaison officer who report directly to the incident commander (IC) and who are responsible for functions in the incident management system that are not a part of the function of the line organization. (NFPA 1561)

communications center See *public safety communications center.*

community risk reduction (CRR) A process to identify and prioritize local risks, followed by the integrated and strategic investment of resources to reduce their occurrence and impact. (NFPA 1452)

company The basic firefighting organizational unit staffed by various grades of firefighters under the supervision of an officer and assigned to one or more specific pieces of apparatus. (NFPA 1410)

company officer The individual responsible for command of a company, a designation not specific to any particular fire department rank (can be a firefighter, lieutenant, captain, or chief officer, if responsible for command of a single company). (NFPA 1026)

competent ignition source An ignition source that has sufficient energy and is capable of transferring that energy to the fuel long enough to raise the fuel to its ignition temperature. (NFPA 921)

complete dam A dam placed across a small stream or ditch to completely stop the flow of materials through the channel.

compressed air foam (CAF) A homogenous foam produced by the combination of water, foam concentrate, and air or nitrogen under pressure (NFPA 11).

compressed air foam system (CAFS) A foam system that combines air under pressure with foam solution to create foam. (NFPA 1900)

compressed gas–powered vehicle A vehicle powered by one of three forms of compressed gas (compressed natural gas (CNG), liquefied natural gas (LNG), and liquefied petroleum gas (LPG or propane).

compressive force The force that pushes a material together.

compressor A device used for increasing the pressure and density of a gas. (NFPA 853)

computer-aided dispatch (CAD) A combination of hardware and software that provides data entry, makes resource recommendations, and notifies and tracks those resources before, during, and after fire service alarms, preserving records of those alarms and status changes for later analysis. (NFPA 1225)

concentrated load A load focused in one specific area of a building.

concentrated solution A solution with a large amount of solute.

concentration The amount of solute in a given amount of solvent or the ratio of solute in a given amount of total solution.

concrete A mixture of cement, aggregates such as sand and gravel, and water.

conduction Heat transfer to another body or within a body by direct contact. (NFPA 921)

confined space An area large enough and so configured that a member can bodily enter and perform assigned work, but which has limited or restricted means for entry and exit. A confined space is not designed for continuous human occupancy. (NFPA 1550)

confinement Those procedures taken to keep a material, once released, in a defined or local area. (NFPA 470)

consist A list of the contents of every car on a train; also called *train list.*

contagious Capable of transmitting a disease.

container A receptacle, piping, or pipeline used for storing or transporting material of any kind. (NFPA 470)

containment The actions taken to stop a release of a material and keep it in its container or to reduce the amount being released. (NFPA 470)

contaminant The residue of a released substance that can be transferred to people, animals, the environment, or equipment.

contaminated atmosphere The products of combustion that must be removed from a building.

contamination The transfer of a hazardous material or the hazardous component of a weapon of mass destruction (WMD) from its source to people, animals, the environment, or equipment. (NFPA 470)

contamination reduction zone See *warm zone.*

contemporary construction Buildings constructed since about 1970 that incorporate lightweight construction techniques and engineered wood components. These buildings exhibit less resistance to fire than older buildings.

control The procedures, techniques, and methods used in the mitigation of hazardous materials/weapons of mass destruction (WMD) incidents, including containment, extinguishment, and confinement. (NFPA 470)

control zones See *hazard control zones.*

convection Heat transfer by circulation within a medium such as a gas or a liquid. (NFPA 921)

convection column See *plume.*

conventional vehicles Vehicles that use internal combustion engines for power.

converter A device that converts alternating current (AC) to direct current (DC).

cooperating agency An agency supplying assistance other than direct suppression, rescue, support, or service functions to the incident management efforts (Red Cross, law enforcement agency, telephone company, etc.). (NFPA 1026)

coping saw A saw designed to cut curves in wood.

corrosivity The property of a material reflecting its ability to cause damage on contact with skin, eyes, or other parts of the body.

council rake A long-handled rake constructed with hardened triangular-shaped steel teeth that is used for raking a fire control line down to soil with no subsurface fuel, for digging, for rolling burning logs, and for cutting grass and small brush.

coupling One set or pair of connection devices attached to a fire hose that allow the hose to be interconnected to additional lengths of hose or adapters and other firefighting appliances. (NFPA 1960)

C-post One of two posts located behind the rear doors of a motor vehicle.

crew A collective term used casually to refer to a group of firefighters in a department with similar duties or responsibilities. Within the ICS, a specific term that identifies a team of two or more firefighters. (NFPA 1550) See also *company* and *unit.*

crew integrity The concept of keeping a crew together and working as a team so that no single firefighter goes off on their own.

crew resource management (CRM) A program focused on improved situational awareness, sound critical decision-making, effective communication, proper task allocation, and successful teamwork and leadership. (NFPA 1550)

cribbing Short lengths of timber or composite materials, usually 4 by 4 in. (101.6 by 101.6 mm) and 18 to 24 in. (457 to 609 mm) long that are used in various configurations to stabilize loads in place or while a load is moving. (NFPA 1006)

critical incident stress debriefing (CISD) A post-incident meeting designed to assist rescue personnel in dealing with psychological trauma as the result of an emergency. (NFPA 1006)

critical incident stress management (CISM) A program designed to reduce acute and chronic effects of stress related to job functions. (NFPA 450)

cross-band repeater A repeater that boosts a weak signal and then retransmits it over a different radio band.

cross-contamination See *secondary contamination.*

cross mains In a sprinkler system, the pipes that supply the branch lines.

cross-zoned system A fire alarm system that requires the activation of two or more initiating devices before the alarm sounds.

crowbar A straight bar made of steel or iron with a forked chisel on the working end that is suitable for performing forcible entry.

crown fire A fire in which live and dead fuels in a tree off the ground are burning.

cryogen See *cryogenic liquid.*

cryogenic liquid A fluid, such as liquid helium, liquid nitrogen, or liquid argon, that has a boiling point lower than –130°F (–90°C) at an absolute pressure of 14.7 psi (101.3 kPa); also called *cryogen.*

cryogenic liquid cargo tank A low-pressure tank designed to maintain the low temperature required by the cryogens it carries and that has a box-like structure containing the tank control valves typically attached to the rear of the tanker; also called *MC-338.*

cryogenic liquid tank car A low-pressure railway tank car that maintains the low temperature required by the cryogens it carries.

curtain wall Nonbearing walls that separate the inside and outside of the building but are not part of the support structure for the building.

curved roof A roof with a curved shape.

cutting torch A torch that produces a high-temperature flame capable of heating metal to its melting point, thereby cutting through an object. Because of the high temperatures (5700°F [3148°C]) that these torches produce, the operator must be specially trained before using this tool.

cylinder A container that has a circular cross-section and that is designed to store liquids or gases under pressure higher than 40 psi (276 kPa); this definition does not include portable tanks, multiunit tanks, car tanks, cargo tanks, or tank cars. (NFPA 1)

cylinder (fire extinguisher) The body of the fire extinguisher where the extinguishing agent is stored. Also called *container.*

D

damming The product-control process used when liquid is flowing in a natural channel or depression, and its progress can be stopped by constructing a barrier to block the flow.

dangerous cargo manifest The shipping papers on a marine vessel.

dash lift A technique for displacing the dash in which the dash and firewall are lifted up and away from the victim after the hood is removed.

dash reinforcement bar A metal beam or bar that runs the entire length of the dash in a motor vehicle to reinforce the occupant compartment in the event of a crash.

dash roll A technique for displacing the dashboard in a motor vehicle in which a hydraulic ram or spreading tool rolls the dash and firewall up and away from the victim.

DC-DC converter A device that converts direct current (DC) from one voltage level to another.

dead-end water main A water main that is supplied from only one direction.

dead latch A manually operated latch bolt that is squared off that does not allow a door to be closed when slammed. (NFPA 80)

dead load Dead loads consist of the weight of all materials of construction incorporated into the building including but not limited to walls, floors, roofs, ceilings, stairways, built-in partitions, finishes, cladding and other similarly incorporated architectural and structural items, and fixed service equipment including the weight of cranes. (NFPA 5000)

decay stage The stage of fire development within a structure characterized by either a decrease in the fuel load or available oxygen to support combustion, resulting in lower temperatures and lower pressure in the fire area. (NFPA 1410)

deck gun A device that is permanently mounted on and operated from a vehicle and equipped with a piping system that delivers water to the gun.

decontamination The physical and/or chemical process of reducing and preventing the spread and effects of contaminants to people, animals, the environment, or equipment involved at hazardous materials/weapons of mass destruction (WMD) incidents. (NFPA 470)

decontamination corridor A controlled area within the warm zone where decontamination is normally performed. (NFPA 470)

deep-seated fire A fire burning below the surface.

defend in place The operational response in which the action is to relocate the affected occupants to a safe place within the structure during an emergency.

defensible space An area, as defined by the authority having jurisdiction (typically a width of 30 ft [9 m] or more), between an improved property and a potential wildland fire where combustible materials and vegetation have been removed or modified to reduce the potential for fire on improved property spreading to wildland fuels or to provide a safe working area for firefighters protecting life and improved property from wildland fire. (NFPA 1140)

defensive attack See *defensive operation.*

defensive mode The mode of operation in which direct contact with the material or container is avoided and responders' efforts focus on controlling or limiting the effects of the release.

defensive operation Actions that are intended to control a fire by limiting its spread to a defined area, avoiding the commitment of personnel and equipment to dangerous areas. Also called *defensive attack* or *defensive strategy.* (NFPA 1550)

defensive strategy See *defensive operation.*

degradation A chemical action involving the molecular breakdown of a protective clothing material or equipment due to contact with a chemical. (NFPA 470)

dehydration A state in which the body's fluid losses are greater than fluid intake. If left untreated, dehydration may lead to shock and even death.

deluge head A sprinkler head that has no cap or release mechanism so that the orifice is always open and the water release is controlled by valves in the pipes.

deluge sprinkler system A sprinkler system employing open sprinklers or nozzles that are attached to a piping system that is connected to a water supply through a valve that is opened by the operation of a detection system installed in the same areas as the sprinklers or the nozzles. When this valve opens, water flows into the piping system and discharges from all sprinklers or nozzles attached thereto. (NFPA 13)

deluge valve A type of system actuation valve that is opened by the operation of a detection system installed in the same areas as the spray nozzles or by remote manual operation supplying water to all spray nozzles. (NFPA 13)

demonstrative evidence Any type of evidence that can be used to validate a theory or show how something could have occurred; examples include diagrams, photographs, maps, X-rays, visible tests, and demonstrations.

denial of entry A policy under which, once the perimeter around a release site has been identified and marked, responders limit access to all but essential personnel.

Department of Transportation (DOT) The U.S. government agency that publicizes and enforces rules and regulations that relate to the transportation of many hazardous materials.

deploy hose To remove hose from the hose bed or other storage location. Also called *pull hose.*

depth of char A fire effect that, when evaluated as a pattern on identical fuels, may be used to determine locations within a structure that were exposed to higher intensities of heat and fire or exposed to a heat source for longer periods of time.

deputy chief See *assistant chief.*

destructive analysis investigation An investigation that uses the methodical deconstruction of evidence to determine specific component conditions, functionality, or failures as they relate to fire investigation.

Dewar container A cylinder container designed to preserve the temperature of cryogenic liquids.

diagonal cutter See *wire cutter.*

digital radio A radio that transmits information via radio waves using digital data or analog (voice) signals that have been converted to a digital signal and compressed.

dike A series of short walls that form a catch basin around a container; also called *berm.*

diking The placement of materials to form a barrier that will keep a hazardous material in liquid form from entering an area or that will hold the material in an area.

dilute solution A solution that contains a small amount of solute.

dilution The process of adding a substance—usually water—to weaken the concentration of another substance.

dimensional lumber Lumber cut to nominal sizes, then dried and cut down to new standard sizes that are smaller than the nominal sizes. See also *solid lumber.*

direct attack Firefighting operations involving the application of extinguishing agents directly onto the burning fuel. (NFPA 1145)

direct attack (wildland fire) A method of wildland fire attack in which firefighters focus on containing and extinguishing the fire at its burning edge.

direct evidence Facts that can be observed or testimony of witnesses who observe acts or detect something through their five senses or through surveillance equipment, such as a security camera.

direct-line phone A phone that connects two predetermined points and does not require the user on either end to dial to cause the phone at the other end to ring. Also called *ring-down phone.*

dirty bomb See *radiation dispersal device (RDD).*

discharge cap See *hydrant cap.*

dispatch To send resources to an address or incident location for a specific purpose. (NFPA 450)

dispatcher See *telecommunicator.*

distributor See *distributor pipe.*

distributor pipe The smallest-diameter underground water main pipes in a water distribution system that deliver water to local users within a neighborhood. Also called *distributor.*

diversion The process of redirecting spilled or leaking material to an area where it will have less impact.

division That organizational level having responsibility for operations within a defined geographic location. (NFPA 1026)

division chief See *assistant chief.*

division supervisor A person in a supervisory level position responsible for a specific geographic area of operations at an incident. (NFPA 1550)

documentary evidence Any type of written record or document that is relevant to the case.

doff The process of properly removing a member's PPE and respiratory protection to limit additional contamination and exposure. (NFPA 1700)

doffing The process of taking off an ensemble of personal protective equipment.

dog See *pawl.*

don The process of properly dressing in full PPE, ensuring all exposed skin and airway are protected. (NFPA 1700)

donning The process of putting on an ensemble of personal protective equipment.

door frame The upright or vertical parts of a door system onto which a door is secured.

door system The door, door frame, and lockset.

dose–response relationship The correlation between the amount of a toxic substance taken in by a person (the dose) and the signs, symptoms, illness, or injury that the dose might provoke from the human body (the response) when a person is exposed to any substance.

double-action pull station A manual alarm initiating device that requires the user to take two steps, such as moving a flap, lifting a cover, or breaking a piece of glass, and then activating the alarm.

double-female adapter A hose adapter that is used to join two male hose couplings.

double-hung windows Windows that have two movable panels, or sashes, that can move up and down.

double-jacket hose A hose constructed with two layers of woven fibers.

double-male adapter A hose adapter that is used to join two female hose couplings.

double-pane glass A window design that traps air or inert gas in the space between two pieces of glass to help insulate a house.

draft The use of suction to move a liquid (such as water) from a vessel or source that is below the intake of a pump. (NFPA 1910)

draft curtain A fixed or automatically deployable barrier that protrudes downward from the ceiling to channel, contain, or prevent the migration of smoke.

drafting hydrant See *dry hydrant.*

drag rescue device (DRD) A fabric handle integrated just below the collar at the back of the protective coat that a rescuer can grab to drag an incapacitated firefighter to safety.

dress To tighten and remove twists, kinks, and slack from the rope after tying a knot.

driver/operator A person qualified to operate a fire apparatus. Also called *engineer* or *technician.* (NFPA 1910)

drum A barrel-like storage vessel constructed of low-carbon steel, polyethylene, cardboard, stainless steel, nickel, or other materials, which is used to store a wide variety of substances, including food-grade materials, corrosives, flammable liquids, and grease.

dry barrel hydrant A type of hydrant with the main control valve below the frost line between the footpiece and the barrel. Also called *frost-proof hydrant.* (NFPA 24)

dry bulk cargo trailer A trailer designed to carry dry bulk goods such as powders, pellets, fertilizers, or grain; it is generally V-shaped with rounded sides that funnel toward the bottom.

dry chemical A powder composed of very small particles, usually sodium bicarbonate, potassium bicarbonate, or ammonium phosphate based with added particulate material supplemented by special treatment to provide resistance to packing, resistance to moisture absorption (caking), and the proper flow capabilities. (NFPA 10)

dry chemical extinguishing system An automatic fire suppression system that discharges a dry chemical agent.

dry-chemical fire extinguisher A fire extinguisher that uses a dry-chemical extinguishing agent and is usually rated for use on Class B and C fires, and sometimes on Class A fires.

dry hydrant An arrangement of pipe permanently connected to a water source other than a piped, pressurized water supply system that provides a ready means of water supply for firefighting purposes and that utilizes the drafting (suction) capability of a fire department pump. Also called *drafting hydrant.* (NFPA 1142)

dry pendent sprinkler head A sprinkler head with an elongated neck that extends from a wet sprinkler pipe in a heated space to a small, unheated area, such as a walk-in freezer, and that is capped at both ends to prevent water from entering the unheated area.

dry pipe sprinkler system A sprinkler system employing automatic sprinklers that are attached to a piping system containing air or nitrogen under pressure, the release of which (as from the opening of a sprinkler) permits the water pressure to open a valve known as a dry pipe valve, and the water then flows into the piping system and out the opened sprinklers. (NFPA 13)

dry pipe valve The valve assembly in a dry pipe sprinkler system that prevents water from entering the system until the air pressure is released and acts as an alarm valve.

dry powder Solid materials in powder or granular form designed to extinguish Class D combustible metal fires by crusting, smothering, or heat-transferring means. (NFPA 10)

dry-powder fire extinguisher A fire extinguisher that uses solid materials in powder or granular form to extinguish Class D combustible metal fires by crusting, smothering, or heat-transferring means.

dry standpipe system A standpipe system designed to have piping contain water only when the system is being utilized. (NFPA 25)

drywall hook A specialized version of a pike pole that can remove drywall more effectively because of its hook design.

dual hose lines Two parallel hose lines laid at the same time.

dual-path pressure reducer A feature that automatically provides a backup method for air to be supplied to the regulator of an SCBA if the primary passage malfunctions.

duckbill padlock breaker A tool with a point that can be inserted into the shackles of a padlock. As the point is driven farther into the lock, it gets larger and forces the shackle out of the lock body.

duct detector A type of smoke detector that samples the air through the air distribution system for a building, and when it detects smoke, it initiates an alarm or a supervisory signal and shuts down the air handling system.

dump valve A large opening from the water tank of a mobile water supply apparatus for unloading purposes. (NFPA 1900)

dynamic rope A rope typically used for climbing that is designed to be elastic and stretch when loaded.

E

early-suppression fast-response sprinkler head (ESFR) A large sprinkler head designed to activate quickly and suppress a fire in its early stages.

east winds See *Foehn winds.*

eccentric loads Loads that are off center.

eductor A device placed in a hose line or a discharge pipe that incorporates a Venturi and proportions foam concentrate or other firefighting agents into the water stream. (NFPA 1900)

electrical energy Energy is produced by an electrical charge.

electric-drive vehicle A vehicle that is powered solely by a high-voltage battery or by a combination of a high-voltage battery and a liquid fuel.

electrolytes Certain salts and other chemicals that are dissolved in body fluids and cells. Proper levels of electrolytes need to be maintained for good health and strength.

electron In an atom, a negatively charged particle that orbits the nucleus.

electronic control unit (ECU) The small processing unit generally located in the center of a motor vehicle that controls the supplemental restraint system.

element A substance that cannot be chemically broken down into a simpler substance.

elevated master stream appliance A nozzle mounted on the end of an aerial device that is capable of delivering large amounts of water into a fire or exposed building from an elevated position.

elevated platform An apparatus that includes a passenger-carrying platform (bucket) attached to the tip of a boom or ladder.

elevated water storage tower An aboveground water storage tank that is designed to maintain pressure on a water distribution system.

elevation pressure The amount of pressure created by gravity. Also called *head pressure.*

emergency breathing safety system (EBSS) A device on an SCBA that allows users to share their available air supply in an emergency situation. Also called *buddy breather.* (NFPA 1970)

emergency control function interface The location in a building where personnel control building functions such as air-handling systems, fire doors, and elevators.

emergency escape plan See *emergency escape route.*

emergency escape route A preplanned and understood route that rescuers use to move to a safety zone or other low-risk area. Also called *emergency escape plan.* (NFPA 1006)

emergency/field-expedient decontamination The process of immediately reducing contamination of people in potentially life-threatening situations with or without the formal establishment of a decontamination corridor. (NFPA 470)

emergency incident rehabilitation A function on the emergency scene that cares for the well-being of the firefighters. It includes relief from climatic conditions, rest, cooling or warming, rehydration, calorie replacement, medical monitoring, member accountability, and release.

emergency medical dispatcher (EMD) Personnel specifically trained and certified in interviewing techniques, pre-arrival instructions, and call prioritization. (NFPA 450)

Emergency Medical Responder (EMR) Emergency medical services (EMS) personnel who have basic training for providing initial medical assistance, have training in bleeding control and CPR, and often perform in an assistant role within the ambulance.

emergency medical services (EMS) company A company that may include medical units and first-response vehicles and that responds to and assists in the transport of medical and trauma victims to medical facilities for further treatment. Also called *emergency medical services (EMS) squad* or *squad.*

emergency medical services (EMS) personnel Personnel responsible for administering care to people who are sick and injured.

emergency medical services (EMS) squad See *emergency medical services (EMS) company.*

Emergency Medical Technician (EMT) Emergency medical services (EMS) personnel who can do everything an emergency medical responder (EMR) can do, and who have training in basic emergency care skills, including oxygen therapy, bleeding control, CPR, automated external defibrillation, use of basic airway devices, and assisting patients with certain medications.

emergency message See *emergency traffic.*

Emergency Planning and Community Right-to-Know Act (EPCRA) Legislation that requires a business that handles chemicals to report on those chemicals' type, quantity, and storage methods to the fire department and the local emergency planning committee.

Emergency Response Guide A reference book created by a vehicle's manufacturer that provides important information about the vehicle, such as information about hybrid/electric, fuel-cell, and alternative-fuel systems, to educate and assist emergency response personnel responding to incidents involving the specific vehicle.

***Emergency Response Guidebook* (ERG)** The reference book, written in plain language, to guide emergency responders in their initial actions at the incident scene, specifically the *Emergency Response Guidebook* from the U.S. Department of Transportation, Transport Canada, and the Secretariat of Transport and Communications, Mexico, used to guide personnel and responders in their initial actions at the incident scene. (NFPA 470)

emergency scene The area encompassed by the incident and the surrounding area needed by the emergency forces to stage apparatus and mitigate the incident. Also called *on-scene.* See also *fireground* and *fire scene.* (NFPA 901)

emergency traffic An urgent message that takes priority over all other communications except a mayday call. Also called *emergency message.*

Emergency Transportation System for the Chemical Industry, Mexico (SETIQ) An organization in Mexico that provides emergency response information and assistance on a 24-hour basis for responders to emergencies involving hazardous materials/weapons of mass destruction. (NFPA 470)

emergency vehicle technician (EVT) The individual who repairs and performs service on emergency vehicles.

emissivity A description of the amount of energy radiated from a material's surface.

employee assistance programs (EAPs) An employer-sponsored service designed for personal or family problems, including mental health, substance abuse, various addictions, marital problems, parenting problems, emotional problems, or financial or legal concerns. (NFPA 450)

end-of-service-time-indicator (EOSTI) A warning device on an SCBA that alerts the user that the reserved air supply is being utilized. (NFPA 1970)

endothermic A chemical reaction that absorbs heat.

energy The ability to do work.

engine See *pumper.*

engine company A piece of fire apparatus along with firefighters that have the primary responsibility to deliver a fire stream or streams to extinguish the fire in coordination with ventilation (truck company) and rescue operations. (NFPA 1700)

engineer See *driver/operator.*

engineered wood Manufactured building material made of smaller pieces of wood held together with glue or adhesive; examples include plywood, fiberboard, oriented strand board (OSB), and particle board. See also *manufactured board, manmade wood,* or *composite wood.*

entrain To encircle, draw along, and transport.

entrapment A condition in which a victim is trapped by debris, soil, or other material and is unable to extricate themself.

entry team A team of fully qualified and equipped responders who are assigned to enter the designated hot zone.

Environmental Protection Agency (EPA) The U.S. federal agency that ensures safe manufacturing, use, transportation, and disposal of hazardous substances.

escape rope Rope dedicated solely for the purpose of supporting people during emergency self-escape (self-rescue); not intended for use in a hazardous environment involving fire or fire products; not classified as a life safety rope. (NFPA 2500)

evacuation The removal or relocation of those individuals who may be affected by an approaching release of a hazardous material.

evacuation signal A distinctive signal intended to be recognized by the occupants as requiring evacuation of the building.

evolution A set of prescribed actions that result in an effective fireground activity. (NFPA 1410)

examination opening A small opening in a roof used to determine how large an area is involved, whether a fire is spreading, and in which direction it is moving, and to evaluate conditions under the roof and verify the proper location for a ventilation opening.

excepted packaging Packaging, which may be constructed out of heavy cardboard, that meets only general design requirements for any hazardous material used to transport materials such as low-level radioactive substances.

exclusionary zone See *hot zone.*

exhauster A device that accelerates the removal of the air from a dry pipe or preaction sprinkler system by opening a large diameter portal until it detects water flow.

exigent circumstances rule A legal condition that allows emergency service providers to enter, search, seize, and control private property and to investigate the cause of a fire without consent or warrant while lawfully performing emergency operations.

Exit Drills in the Home (EDITH) A public fire and life safety education program that teaches occupants how to safely exit a house in the event of a fire or other emergency and meet other household members at a pre-established meeting place.

exothermic A chemical reaction that produces heat.

expansion ratio A description of the volume increase that occurs when a liquid changes to a gas.

expectation bias Any preconceived determination or premature conclusions as to the cause of a fire without having examined or considered all relevant evidence.

explosion A violent and pressurized release of energy.

explosive limits See *flammable range.*

explosive ordnance disposal (EOD) personnel Personnel trained to detect, identify, evaluate, render safe, recover, and dispose of unexploded explosive devices.

exposure Contact with a hazardous material/weapon of mass destruction (WMD) by people, animals, the environment, property, or equipment. (NFPA 470)

exposure (fire incident) An object such as a building, structure, or vehicle in an area adjacent to a burning building.

extension ladder A non-self-supporting ground ladder that consists of two or more sections traveling in guides, brackets, or the equivalent arranged so as to allow length adjustment. (NFPA 1960)

exterior attack A fire attack conducted from outside a structure.

exterior wall A wall—often made of wood, brick, metal, or masonry—that makes up the outer perimeter of a building. Exterior walls are often load-bearing.

extinguishing agent A material used to stop the combustion process. Extinguishing agents may include liquids, gases, dry-chemical compounds, and dry-powder compounds.

extra hazard area An occupancy where the total amount of Class A combustibles and Class B flammables is greater than expected in occupancies classed as ordinary (moderate) hazards and the combustibility and heat release rate of the materials are high. Also called *high hazard area.*

F

face piece The part of the SCBA consisting of the face mask and exhalation valve that delivers breathing air to the firefighter and protects the face from high temperatures and smoke.

faller A person qualified to use a chainsaw to cut down trees.

Federal Communications Commission (FCC) The federal regulatory authority that oversees radio communications in the United States.

feed mains In a sprinkler system, the pipes that supply the cross mains from the water source, either directly or through risers.

fender crush See *wheel well crush.*

field of view (FOV) The area observable at any given moment and is measured vertically and horizontally in degrees.

film-forming fluoroprotein (FFFP) foam A protein-foam solution that uses fluorinated surfactants to produce a fluid aqueous film for suppressing liquid fuel vapors. (NFPA 10)

filter mask A face mask that filters out small particles of smoke but does not protect against inhalation of heavy smoke or poisonous gases.

finance/administration section The section in an incident command system responsible for all costs and financial actions of the incident or planned event, including the time unit, procurement unit, compensation/claims unit, and cost unit. (NFPA 1026)

finance/administration section chief The general staff position responsible for directing the finance/administrative function. It is generally assigned on large-scale or long-duration incidents that require immediate fiscal management.

fine fuel Fuel that ignites and burns easily, such as dried twigs, leaves, needles, grass, moss, and light brush.

finger (wildland fire) A narrow point of fire whose extension is created by a shift in wind or a change in topography.

fingers Individual, adjustable vanes or protrusions on the face of a fog-stream nozzle that are responsible for shaping the water stream into a specific pattern.

fire The visible result of combustion.

fire alarm box A device connected via an underground cable to a municipal fire alarm system.

fire alarm control panel (FACP) See *fire alarm control unit (FACU).*

fire alarm control unit (FACU) A component of the fire alarm system, provided with primary and secondary power sources, which receives signals from initiating devices or other fire alarm control units and processes these signals to determine part of all of the required fire alarm system output function(s). Also called *fire alarm control panel (FACP).* (NFPA 72)

fire alarm signal A signal that results from the manual or automatic detection of a fire alarm condition. (NFPA 72)

fire alarm system A system or portion of a combination system that consists of components and circuits arranged to monitor and annunciate the status of fire alarm or supervisory signal-initiating devices and to initiate the appropriate response to those signal. (NFPA 72)

fire and life safety educator (FLSE) An individual who has demonstrated the ability to coordinate, create, administer, prepare, deliver, and evaluate educational programs and information.

fire apparatus A vehicle designed to be used under emergency conditions to transport personnel and equipment or to support the suppression of fires and mitigation of other hazardous situations. Also called *apparatus.* (NFPA 1010)

fire attack The process of cooling the fire gases and suppressing a fire.

fireball A burst of flames that rapidly ignites available flammable vapors but is not under pressure.

fire barrier wall A wall, other than a fire wall, having a fire resistance rating. (NFPA 5000)

firebreak A swath where the fuel (trees and brush) are removed.

fire chief The highest-ranking officer in charge of a fire department. (NFPA 1550)

fire code A code that ensures a minimum level of fire safety in the home and workplace environments by preventing fires and protecting lives and property in the event of a fire.

fire compartment The area of origin when the fire is in a structure.

fire control line Comprehensive term for all constructed or natural barriers and treated fire edges used to control a fire. (NFPA 901)

fire department An organization providing rescue, fire suppression, and related activities, including any public, governmental, private, industrial, or military organization engaging in this type of activity. (NFPA 1010)

fire department connection (FDC) A connection through which the fire department can pump supplemental water into the sprinkler system, standpipe, or other system furnishing water for fire extinguishment to supplement existing water supplies. (NFPA 13)

fire department ground ladder Any portable ladder specifically designed for fire department use in rescue, firefighting operations, or training. Also called *ground ladder.* (NFPA 1960)

fire detection system A system that senses the presence of fire, smoke, or heat and activates a fire suppression system and/or an automatic alarm system. (NFPA 2)

fire door assembly Any combination of a fire door, a frame, hardware, and other accessories that together provide a specific degree of fire protection to the opening. (NFPA 80)

fire effects The observable or measurable changes in or on a material as a result of a fire. (NFPA 921)

fire escape rope Rope dedicated solely for the purpose of supporting people during emergency self-escape (self-rescue) from an immediately hazardous environment involving fire or fire products; not classified as a life safety rope. (NFPA 2500)

fire extension The spreading of the fire to adjacent areas from the seat of the fire.

firefighter A member of a fire department who is assigned to do routine cleaning and maintenance, place hose line to extinguish fires, and assist with a public fire prevention program.

firefighter assist and search team (FAST) See *rapid intervention crew/company (RIC).*

firefighter breathing air replenishment system (FBAR) A system that allows firefighters to refill their SCBA cylinders on any floor in a high-rise building.

firefighter down CPR (FD-CPR) A technique for doing CPR that allows firefighters to begin providing immediate life-saving care to a firefighter in cardiac arrest while their SCBA and gear are removed in the most efficient way.

Firefighter I A person, at the first level of progression as defined in Chapter 6 of NFPA 1010, who has demonstrated the knowledge and skills to function as an integral member of a firefighting team under direct supervision in hazardous conditions. (NFPA 1010)

Firefighter II A person, at the second level of progression as defined in Chapter 7 of NFPA 1010, who has demonstrated the skills and depth of knowledge to function under general supervision. (NFPA 1010)

fireground Another name for *emergency scene* when the incident is a fire. Also called *fire scene.*

fireground command (FGC) An incident management system (IMS) developed in the 1970s for day-to-day fire department incidents (generally handled with fewer than 25 units or companies).

fire hook A tool used to pull down burning structures.

fire hose A flexible conduit used to convey water or other extinguishing agents (NFPA 1960).

fire hose appliance A piece of hardware (excluding nozzles) generally intended for connection to fire hose to control or convey water. Also called *hose appliance.* (NFPA 1962)

fire hose tool A device that assists firefighters with handling, manipulating, connecting, and using a fire hose.

fire house See *fire station.*

fire hydraulics The physical science of how water flows through a pipe or hose.

fire inspector An individual who conducts fire code inspections and applies codes and standards. (NFPA 1030)

fire investigator An individual who has demonstrated the skills and knowledge necessary to conduct, coordinate, and complete an investigation. (NFPA 1030)

fire line The crew of firefighters conducting a fire attack.

fire load The total energy content of combustible materials in a building, space, or area including furnishing and contents and combustible building elements expressed in MJ. (NFPA 557)

fire mark A plaque displayed on a building with the name or logo of a fire insurance company informing firefighters that the building was insured by that insurance company, which means that insurance company would pay the firefighters for extinguishing the fire.

fire marshal A person designated to provide delivery, management, and/or administration of fire protection– and life safety–related codes and standards, investigations, education, and/or prevention services for local, county, state, provincial, federal, tribal, or private sector jurisdictions as adopted or determined by that entity. (NFPA 1030)

fire out A wildland firefighting technique in which a fire is set along the inner edge of a fire control line to consume the fuel between the fire control line and the fire's edge.

fire patterns The visible or measurable physical changes, or identifiable shapes, formed by a fire effect or group of fire effects. (NFPA 921)

fireplug Historically speaking, a plug installed to control water accessed from wooden pipes, but today is slang for fire hydrant.

fire point The lowest temperature at which a liquid will ignite and achieve sustained burning when exposed to a test flame in accordance with ASTM 92, Standard Test Method for Flash and Fire Points by Cleveland Open Cup Tester. Also called *flame point.* (NFPA 1)

fire police officer An individual officially deployed who provides scene security, directs traffic, and conducts other duties as determined by the authority having jurisdiction (AHJ). (NFPA 1091)

fire prevention Measures taken toward avoiding the inception of fire. (NFPA 801)

fire protection engineer A member of the fire department or an employee of an architectural firm who is responsible for reviewing building plans and working with building owners to ensure that the design of and systems for fire detection and suppression meet applicable codes and function as needed.

fire protection system Any fire alarm device or system or fire-extinguishing device or system, or combination thereof, that is designed and installed for detecting, controlling, or extinguishing a fire or otherwise alerting occupants, or the fire department, or both, that a fire has occurred. (NFPA 1)

fire resistance The measure of the ability of a material, product, or assembly to withstand fire or give protection from it. (NFPA 251)

fire resistance ratings Ratings that are assigned to materials and building components based on the length of time that the material or component can maintain its integrity when subjected to fire.

fire safety survey The process of observing and recording conditions of an occupied structure for basic fire and life safety hazards. (NFPA 1010)

fire scene Another name for *emergency scene* when the incident is a fire. Also called *fireground.*

FIRESCOPE Fire Resources of California Organized for Potential Emergencies; an organization of agencies established in the early 1970s to develop a standardized system for managing fire resources at large-scale incidents such as wildland fires.

fire seat See *area of origin.*

fire separation A horizontal or vertical fire resistance–rated assembly of materials that have protected openings and are designed to restrict the spread of fire. (NFPA 45)

fire shelter An item of protective equipment configured as an aluminized tent utilized for protection, by means of reflecting radiant heat, in a fire entrapment situation. (NFPA 1550)

fire station A building that houses fire apparatus and equipment for a geographic area within a fire department. Also called *fire house.*

fire stream Stream of water or extinguishing agents.

fire suppression The activities involved in controlling and extinguishing fires. (NFPA 1500)

fire suppression system A system that discharges extinguishing agents in the presence of a fire.

fire tetrahedron A geometric shape used to depict the four components—fuel, oxygen, heat, and chemical chain reactions—required for a fire to occur.

fire triangle The three components—fuel, oxygen, and heat—required for combustion.

fire wall A wall separating buildings or subdividing a building to prevent the spread of fire and having a fire-resistance rating and structural stability. (NFPA 5000)

firewall The structural component in a vehicle that separates the engine compartment from the passenger compartment. Also called *bulkhead.*

fire warden An individual charged with enforcing fire regulations in colonial America.

fire watch The assignment of a person or persons to an area for the express purpose of notifying the fire department, the building occupants, or both of an emergency; preventing a fire from occurring; extinguishing small fires; protecting the public from fire and life safety dangers. (NFPA 1)

fire window A window assembly rated in accordance with NFPA 257 and installed in accordance with NFPA 80. (NFPA 5000)

fission The splitting of an atom's nucleus into two or more pieces after capturing a low-energy neutron.

fitting In a sprinkler system, a connector that joins the pipes together.

fixed-gallonage fog nozzle A nozzle that delivers a set number of gallons per minute (liters per minute) as per the nozzle's design, no matter what pressure is applied to the nozzle.

fixed-temperature detector A device that responds when its operating element become heated to a predetermined level. (NFPA 72)

flake out See *lay.*

flame detector A radiant energy–sensing fire detector that detects the radiant energy emitted by a flame. (NFPA 72)

flame height The flame height measured from the ground to the tip of the flame.

flame impingement Flames in direct contact with the surface of a material transferring radiant heat.

flame inhibitor A chemical extinguishing agent that reacts with the fuel to chemically disrupt the combustion process.

flame length The flame height measured from the beginning of the flame to the tip of the flame.

flameover See *rollover.*

flame point See *fire point.*

flammable range The range in concentration between the lower and upper flammable limits. Also called *explosive limits.* (NFPA 67)

flange On an I-beam ladder, the top and bottom of a beam.

flank of the fire The edge between the head and heel of the fire that runs parallel to the direction of the fire spread.

flanking attack A direct method of suppressing a wildland or ground cover fire that involves placing a suppression crew on one flank of a fire.

flashover A transition phase in the development of a compartment fire in which surfaces exposed to thermal radiation reach ignition temperature more or less simultaneously, and fire spreads rapidly throughout the space, resulting in full room involvement or total involvement of the compartment or enclosed space. (NFPA 921)

flash point The minimum temperature at which a liquid or a solid emits vapor sufficient to form an ignitable mixture with air near the surface of the liquid or the solid. (NFPA 115)

flat bar A specialized type of prying tool made of flat steel with prying ends suitable for performing forcible entry.

flat-head axe A tool that has a head with an axe on one side and a flat head on the opposite side.

flat hose load A hose loading method in which the hose is laid flat and stacked on top of the previous section.

flat raise A method of raising a ladder by positioning the ladder on the ground with the butt near the building and then lifting the tip to lean it against the building. Also called *rung raise.*

flat roof A horizontal roof; often found on commercial or industrial occupancies.

flooded cell battery See *wet cell battery.*

floodlight A light that illuminates a broad area.

floor runner A piece of canvas or plastic material, usually 3 to 4 ft (91 to 122 cm) wide in various lengths, that is used to protect flooring from dropped debris and dirt from shoes and boots.

flow meter A device that measures water flow in gallons (or liters) per minute from an orifice.

flow path The movement of heat and smoke from the higher pressure within the fire area toward the lower pressure areas accessible via doors, window openings, and roof structures. (NFPA 1410)

flow rate The quantity of water flowing, usually measured in gallons (or liters) per minute.

fluoroprotein foam A protein-based foam concentrate, with added fluorochemical surfactants, that forms a foam showing a measurable degree of compatibility with dry chemical extinguishing agents and an increase in tolerance to contamination by fuel. Also called *film-forming fluoroprotein foam (FFFP).* (NFPA 460)

fly section The upper section of an extension ladder or any section of an aerial telescoping device beyond the base section. (NFPA 1960 and 1900)

foam blanket A covering of foam over a surface to insulate, prevent ignition, or extinguish the fire. (NFPA 1145)

foam concentrate Foam firefighting agent as received from the manufacturer that must be diluted with water to make foam solution (NFPA 1900).

foam injector A device installed on a fire pump that meters out foam by pumping or injecting it into the fire stream.

foam nozzle A nozzle specifically designed to help with aeration of the foam solution as it exits the hose line and complete the process of becoming a finished foam product.

foam proportioner A device or method to add foam concentrate to water to make foam solution. (NFPA 1900)

foam solution A homogeneous mixture of foam concentrate and water in the proper proportions. (NFPA 1900)

Foehn winds Winds caused by a high-pressure system of warm air replacing cold air and bringing with it unusually high winds in the opposite direction of the local winds. Also called *chinook winds, east winds,* and *Santa Ana winds.*

fog stream A stream of water that is flowed in the form of small water droplets. (NFPA 1700)

fog-stream nozzle A nozzle that is placed at the end of a fire hose and can be adjusted to produce a straight stream or to separate the water into droplets to produce a variety of fog streams. Also called *spray nozzle* or *adjustable-fog-stream nozzle.*

fog ventilation See *hydraulic ventilation.*

folding ladder A single-section ladder with rungs that can be folded or moved to allow the beams to be brought into a position touching or nearly touching each other. Also called *attic ladder.* (NFPA 1960)

foot See *heel (v.).*

forces Fundamental physical presences (e.g., gravity).

forcible entry Techniques used by fire personnel to gain entry into buildings, vehicles, aircraft, or other areas of confinement when normal means of entry are locked or blocked. (NFPA 440)

forestry fire hose A hose designed to meet specialized requirements for fighting wildland fires. (NFPA 1960)

forward hose lay A method of laying a supply line where the line starts at the water source and ends at the attack engine. Also called *straight hose lay.*

four-way hydrant valve A specialized type of valve that can be placed on a hydrant and that allows another engine to increase the supply pressure without interrupting flow.

frangible-bulb sprinkler head A sprinkler head with the cap held in place by a glass bulb filled with glycerin or alcohol and a small air bubble; when the liquid is heated, it vaporizes and the air pressure increases until the glass bulb breaks, releasing the cap.

freelancing The dangerous practice of acting independently of command instructions.

freight bill See *bill of lading.*

Fresno ladder A narrow, two-section extension ladder that has no halyard. Because of its limited length, it can be extended manually.

friction loss The reduction in pressure resulting from the water being in contact with the side of the hose. This contact requires force to overcome the drag that the wall of the hose creates.

friction point A place where a hose line encounters resistance, such as a corner, a doorway, and stairs.

frostbite A localized condition that occurs when the layers of the skin and deeper tissue freeze. (NFPA 704)

frost-proof hydrant See *dry barrel hydrant.*

fuel A material that will maintain combustion under specified environmental conditions. (NFPA 53)

fuel cell–powered vehicle A vehicle powered by fuel cells.

fuel cells Cells that generate electricity through a chemical reaction between hydrogen gas and oxygen gas to produce water and in the process produce electricity to propel a vehicle.

fuel-limited fire A fire in which the heat release rate and fire growth are controlled by the characteristics of the fuel because there is adequate oxygen available for combustion. (NFPA 1410)

fuel moisture The amount of moisture present in a fuel, which affects how readily the fuel will ignite and burn.

fugitive gas Gaseous fuel that escapes from the system.

fully developed stage The stage of fire development where heat release rate has reached its peak within a compartment. (NFPA 1410)

fully encapsulated suit A protective suit that completely covers the firefighter, including the breathing apparatus, and does not let any vapor or fluids enter the suit. It is commonly used in hazardous materials emergencies.

fusible-link sprinkler head A sprinkler head with the cap held in place by two pieces of metal held together by solder with a low melting point, and when the solder melts, the link breaks, releasing the cap.

fusible plug A pressure relief valve on a ton container.

G

gas The physical state of a substance that has no shape or volume of its own and will expand to take the shape and volume of the container or enclosure it occupies. (NFPA 921)

gas detector A device that detects the presence of a specified gas concentration. Gas detectors can be either spot-type or line-type detectors. (NFPA 72)

gaseous suppression system A system often installed in areas where computers or sensitive electronic equipment is used or where valuable documents are stored that stores gaseous agents including halogenated agents and inert gases such as carbon dioxide under pressure as liquids.

gate valve A valve found on hydrants and sprinkler systems with a rotating a spindle that causes the gate to move slowly across the opening.

gated wye A valved device that splits a single hose into two separate hose lines, allowing each hose to be turned on and off independently.

general-service rail tank car See *nonpressure tank car.*

general staff A group of incident management personnel organized according to function and reporting to the incident commander (IC), normally consisting of the operations section chief, planning section chief, logistics section chief, and finance/administration section chief. (NFPA 1026)

general use life safety rope A life safety rope with a diameter that is at least 7/8 in. (11 mm) but not larger than 5/8 in. (16 mm), with a minimum breaking strength of 8992 lbf (40 kN).

generator An electromechanical device for the production of electricity. (NFPA 1901)

geographic information systems (GIS) A system of computer software, hardware, data, and personnel to describe information tied to a spatial location. (NFPA 450)

glass blocks Thick pieces of glass that are similar to bricks or tiles.

global positioning systems (GPS) A satellite-based radio navigation system composed of three segments: space, control, and user. (NFPA 1900)

glucose The source of energy for the body. One of the basic sugars, it is the body's primary fuel, along with oxygen. Also called *blood sugar.*

gondola A railway car that carries items such as lumber, scrap metal, coal, and pipes.

governance The framework and procedures for managing and operating an organization.

gravity-feed system A water distribution system that depends on gravity to provide the required pressure. The system storage is usually located at a higher elevation than the end users of the water.

green area An area of unburned fuels.

gripping pliers A hand tool with a pincer-like working end that can be used to bend wire or hold smaller objects.

gross decontamination The phase of the decontamination process that takes place as soon as possible and before technical decontamination within a decontamination corridor, during which significant reduction of the amount of surface contamination is removed with a prewash most often accomplished by mechanical removal of the contaminant or initial rinsing from handheld hose lines, emergency showers, or other nearby sources of water. (NFPA 470)

ground cover fire A fire that burns loose debris such as grass, leaves, and branches on the surface of the ground.

ground fault interrupter (GFI) A device that senses when there is a problem with an electrical ground and interrupts the current, shutting down both the power source and the equipment it is feeding.

ground fuel All combustible materials such as grass, duff, loose surface litter, tree or shrub roots, rotting wood, leaves, peat, or sawdust that typically support combustion. (NFPA 1140)

ground ladder See *fire department ground ladder.*

group Established to divide the incident management structure into functional assignments of operation. (NFPA 1026)

group supervisor A person in a supervisory level position responsible for a functional area of operation. (NFPA 1550)

growth stage The stage of fire development where the heat release rate from an incipient fire has increased to the point where heat transferred from the fire and the combustion products are pyrolyzing adjacent fuel sources and the fire begins to spread across the ceiling of the fire compartment (rollover). (NFPA 1410)

guide A strip of metal or wood that serves to guide a fly section during extension.

gusset plate Connecting plate made of a thin sheet of steel used to connect the components of the truss.

gypsum A naturally occurring material consisting of calcium sulfate and water molecules.

gypsum board The generic name for a family of sheet products consisting of a noncombustible core primarily of gypsum with paper surfacing. (NFPA 5000)

H

hacksaw A cutting tool designed for use on metal. Different blades can be used for cutting different types of metal.

Halligan See also *Halligan tool.*

Halligan bar See also *Halligan tool.*

Halligan tool A prying tool that incorporates a sharp tapered pick, a blade (either an adze or wedge), and a fork; it is specifically designed for use in the fire service. See also *Halligan* and *Halligan bar.*

halocarbon See *halogenated hydrocarbon.*

halogen A family of elements (chemicals listed in the periodic table), including fluorine (F), bromine (Br), iodine (I), and chlorine (Cl), that are chemically related.

halogenated-agent fire extinguisher A fire extinguisher that uses a halogenated extinguishing agent. Also called *clean-agent fire extinguisher.*

halogenated extinguishing agent A liquefied gas extinguishing agent that extinguishes fire by chemically interrupting the combustion reaction between fuel and oxygen. Halogenated agents leave no residue. (NFPA 402)

halogenated hydrocarbon A hydrocarbon in which at least one hydrogen atom of the hydrocarbon is replaced by a halogen. Also called *halocarbon.*

halyard Rope used on extension ladders for the purpose of raising a fly section(s). (NFPA 1960)

hammer A striking tool.

handle The grip used for holding and carrying a portable fire extinguisher.

hand light A small, portable light carried by firefighters to improve visibility at emergency scenes; it is often powered by rechargeable batteries.

handline A hose and nozzle that can be held and directed by hand. (NFPA 11)

handline nozzle A nozzle with a rated discharge of less than 350 gpm (1325 L/min).

handsaw A manually powered saw designed to cut different types of materials. Examples include hacksaws, carpenter's handsaws, keyhole saws, and coping saws.

hard suction hose A short section of supply hose that is used to draft water from a static source such as a river, lake, or portable drafting basin to the suction side of the fire pump on a fire department engine or into a portable pump.

hardware The parts of a door or window that enable it to be locked or opened.

harness An equipment item; an arrangement of materials secured about the body used to support a person. (NFPA 1970)

hazard A condition or material capable of causing harm to life, health, property, or the environment. (NFPA 470)

hazard control zones The designated areas established at a hazardous materials/weapons of mass destruction (WMD) incident based on the chemical and physical properties of the released material, the environmental factors at the time of the release, and the general layout of the scene. (NFPA 470)

hazardous material Matter (solid, liquid, or gas) or energy that when released is capable of creating harm to people, the environment, and property, including weapons of mass destruction (WMD) as defined in 18 US Code, Section 2332a, as well as any other criminal use of hazardous materials, such as illicit labs, environmental crimes, or industrial sabotage. (NFPA 470)

hazardous materials branch The function within an overall incident management system (IMS) that deals with the mitigation and control of the hazardous materials/weapons of mass destruction (WMD) portion of an incident. (NFPA 470)

hazardous materials company A company that responds to and controls scenes where hazardous materials have spilled or leaked and whose members wear special suits and are trained to deal with most chemicals.

Hazardous Materials Information System (HMIS) A color-coded marking system that identifies the hazard level of chemicals in a container so that personnel can work safely around chemicals.

hazardous materials safety officer The person who works within an incident management system (IMS) in the hazardous materials branch or group to ensure that recognized hazardous materials/weapons of mass destruction (WMD) safe practices are followed at hazardous materials/WMD incidents. Also called *assistant safety officer (ASO).* (NFPA 470)

hazardous materials specialist Defined in the OSHA HAZWOPER regulation only, a hazardous materials specialist responds with, and provides support to, hazardous materials technicians and acts as the incident-site liaison with federal, state, local, and other government authorities regarding site activities.

hazardous materials technician A person who responds to hazardous materials/weapons of mass destruction (WMD) incidents using a risk-based response process to analyze a problem involving hazardous materials/WMD, plan a response to the problem, implement the planned response, evaluate progress of the planned response and adjust accordingly, and assist in terminating the incident. (NFPA 470)

hazardous waste A substance that remains after a process or manufacturing plant has used some of the material and the substance is no longer pure.

hazel hoe A hand tool used to grub out heavy brush to create a fire control line. Also called *adze hoe.*

HAZWOPER (HAZardous Waste OPerations and Emergency Response) The federal OSHA regulation that governs hazardous materials waste site and response training (specifics to emergency response can be found in 29 CFR 1910.120[q]).

head box A box at a protected property that contains spare sprinkler heads and the head wrenches needed to install all of the types of heads.

head of the fire The main or running edge of a fire; the part of the fire that spreads with the greatest speed.

head pressure See *elevation pressure.*

headspace The space above any liquid placed in a closed vessel that contains vapor.

heads-up display (HUD) Visual display of information and system conditions status that is visible to the wearer. (NFPA 1970)

heat cramp A painful muscle spasm, usually in muscles in the leg or abdomen, that occurs suddenly during or after physical exertion.

heat detector A fire detector that detects either an abnormally high temperature, an abnormal rate of temperature rise, or both. (NFPA 72)

heat energy The potential energy of a combustible material and the kinetic energy released when heat is applied to the material. Also called *thermal energy.*

heat exhaustion A mild form of shock that arises when the circulatory system begins to fail because the body is unable to dissipate excessive heat and becomes overheated.

heat flux The measure of the rate of heat transfer to a surface, typically expressed in kilowatts per meter squared (kW/m^2) or BTU/ft^2. (NFPA 268)

heat-induced tear (HIT) A tear in a nonpressurized container that occurs when the container is exposed to direct or indirect flame impingement for a prolonged period, causing the structural integrity of the container to fail.

heat rash Itchy rash on skin that is wet from sweating; seen after prolonged sweating.

heat release rate (HRR) The rate at which heat energy is generated by burning. (NFPA 921)

heat sensor label A label that changes color at a preset temperature to indicate a specific heat exposure. (NFPA 1960)

heat stratification See *thermal layering.*

heat stress The condition that occurs when the body cannot get rid of excess heat, possibly leading to heat rash, heat cramps, heat exhaustion, and heat stroke.

heat stroke A severe, sometimes fatal condition that occurs when the body's temperature-regulating capacity fails and that is indicated by a reduction or cessation of sweating, body temperature of 105°F (40.5°C) or higher, rapid pulse, hot and dry skin, headache, confusion, unconsciousness, and convulsions.

heat transfer The movement of heat energy from a hotter medium to a cooler medium by conduction, convection, or radiation.

heavy fuel Fuel of a large diameter, such as large brush, heavy timber, snags, stumps, branches, and dead timber on the ground, that ignites and is consumed more slowly by fire than light fuel.

heel (n.) See *butt.*

heel (v.) To use your weight to keep the base of a ladder from slipping by placing the toes or instep of one foot on one of the beams and the heel of that foot on the ground, placing the toes of one foot or both feet on the bottom rung and the heel(s) on the ground, or kneeling in front of the ladder. Also called *foot.*

heel of the fire The side opposite the head of the fire, which is often close to the area of origin. Also called *rear of the fire.*

helispot A heliport where no refueling, maintenance, repair, or storage of helicopters is permitted. (NFPA 5000)

Higbee indicator An indicator on both the male and female threaded couplings that indicates where the threads start. These indicators should be aligned before firefighters start to thread the couplings together.

high-angle operations Rope rescue operations where the angle of the slope is greater than 45 degrees. In this scenario, rescuers depend on life safety rope rather than a fixed support surface such as the ground.

high-expansion foam A foam created by pumping large volumes of air through a small screen coated with a foam solution to produce foams with expansion ratios ranging from 200:1 to approximately 1000:1.

high hazard area See *extra hazard area.*

highline system A system of using rope or cable suspended between two points for movement of persons or equipment over an area that is a barrier to the rescue operation, including systems capable of movement between points of equal or unequal height. (NFPA 1006)

high-temperature–protective clothing Protective clothing designed to protect the wearer from short-term high-temperature exposures, such as protective clothing ensembles used by aircraft firefighters to fight fires in flammable liquids. (NFPA 470)

hinge pillar See *A-post.*

hitch A knot that attaches to or wraps around an object so that when the object is removed, the knot will fall apart. (NFPA 2500)

hockey puck padlock A type of padlock with hidden shackles that cannot be forced open through conventional methods.

hollow-core door A door with a sheet metal covering or wood or fiberglass veneer over a light framework. These doors are sometimes left completely hollow, but are usually filled with cardboard baffles, gypsum, or foam.

hook ladder See *roof ladder.*

horizontal evacuation The process of moving occupants from a dangerous area on a floor to a safe area on the same floor.

horizontal ventilation The opening or removal of windows or doors on any floor of a fire building to create flow paths for fire conditions. (NFPA 1410)

horn The tapered discharge nozzle of a carbon dioxide fire extinguisher.

horseshoe hose load A hose loading method in which the hose is laid on its edge around the perimeter of the hose bed so that it resembles a horseshoe.

hose appliance See *fire hose appliance.*

hose bed The main storage area on an apparatus for carrying hose.

hose bridge A device that protects a hose when it is necessary for a vehicle to drive over a hose. Also called *hose ramp.*

hose clamp A device used to compress a fire hose to stop water flow.

hose hoist See *hose roller.*

hose jacket A device used to stop a leak in a fire hose or to join hose lines that have damaged couplings.

hose liner The inside portion of a hose that is in contact with the flowing water; also called *hose inner jacket.*

hose ramp See *hose bridge.*

hose record A written history of each individual length of fire hose.

hose roller A device that is placed on the edge of a roof and is used to protect hose as it is hoisted up and over the roof edge. Also called *hose hoist.*

hose size An expression of the internal diameter of the hose. (NFPA 1962)

hose tool See *fire hose tool.*

hot zone The control zone immediately surrounding hazardous materials/weapons of mass destruction (WMD) incidents, which extends far enough to prevent adverse effects of hazards on personnel outside the zone, and where only personnel who are trained, equipped, and authorized to do assigned work are permitted to enter. Also called *exclusionary zone.* (NFPA 470)

Hux bar A multipurpose tool that can be used for several forcible entry and ventilation applications because of its unique design. It also may be used as a hydrant wrench.

hybrid building A building that does not fit entirely into any of the five construction types because it incorporates building materials of more than one type.

hybrid electric vehicle (HEV) A vehicles that use both a battery-powered electric motor and a liquid-fueled engine to propel a vehicle.

hydrant cap The cover on a fire hydrant outlet that is in place when the hydrant is not in use. Also called *discharge cap.*

hydrant wrench A hand tool that is used to operate the valves on a hydrant; it also may be used as a spanner wrench. Some models are plain wrenches, whereas others have a ratchet feature.

Hydra-Ram A one-piece integrated hydraulic forcible entry tool.

hydraulic cutters See *hydraulic shears.*

hydraulic shears A lightweight, hand-operated tool that can produce up to 10,000 lb (4500 kg) of cutting force. See also *hydraulic cutters.*

hydraulic spreader A lightweight, hand-operated tool that can produce up to 10,000 lb (4500 kg) of prying and spreading force.

hydraulic tool A power tool that uses pressurized fluid to exert force.

hydraulic ventilation Ventilation that relies on the movement of air caused by a fog stream that is placed 2 to 4 ft (0.6 to 1.2 m) in front of an open window. Also referred to as *fog ventilation.*

hydroflurocarbon (HFC) A common type of halocarbon used in fire suppression.

hydrogen cyanide (HCN) An extremely toxic gas produced by the incomplete combustion of many common plastic-based materials.

hydrophobic The quality of repelling or being unable to mix with water.

hydrostatic test A test performed by filling pressure-containing components completely with water or other incompressible fluid while expelling all contained air, closing or capping all open ports of the pressure-containing components, and then raising and maintaining the contained pressure to pressurize the pressure-containing components to a prescribed value through an externally supplied pressure-generating device. (NFPA 1900)

hydrostatic testing Pressure testing of a fire extinguisher to verify its strength against unwanted rupture. (NFPA 10)

hypothermia A condition in which the internal body temperature falls below 95°F (35°C), usually a result of prolonged exposure to cold or freezing temperatures.

I

I-beam ladder A ladder beam constructed of one continuous piece of I-shaped metal or fiberglass to which the rungs are attached.

I-beams Beams that look like an uppercase letter *I.*

ignitable liquid Any liquid or the liquid phase of any material that is capable of fueling a fire, including a flammable liquid, combustible liquid, or any other material that can be liquefied and burned. (NFPA 556)

ignition temperature Minimum temperature a substance should attain in order to ignite under specific test conditions. (NFPA 402)

immediately dangerous to life and health (IDLH) Any condition that would pose an immediate or delayed threat to life, cause irreversible adverse health effects, or interfere with an individual's ability to escape unaided from a hazardous environment. (NFPA 1700)

impact load A load imposed quickly on a structure.

incendiary device A device or mechanism used to start an incendiary fire or explosion.

incendiary fire A fire that is intentionally ignited in an area or under circumstances where and when there should not be a fire. (NFPA 921)

incident action plan (IAP) An oral or written plan approved by the incident commander (IC) containing incident objectives reflecting the overall strategy for managing an incident for a specific time frame and target location. (NFPA 470)

incident base The location at an incident where logistics are coordinated.

incident command post (ICP) The field location at which the primary tactical-level, on-scene incident command functions are performed. (NFPA 1026)

Incident Command System (ICS) The combination of facilities, equipment, personnel, procedures, and communications operating within a common organizational structure that has responsibility for the management of assigned resources to effectively accomplish stated objectives pertaining to an incident or training exercise. (NFPA 2500)

incident commander (IC) The individual responsible for all incident activities, including the development of strategies and tactics and the ordering and release of resources. (NFPA 1026)

incident management system (IMS) A system that defines the roles and responsibilities to be assumed by responders and the standard operating procedures to be used in the management and direction of emergency incidents and other functions. (NFPA 1410)

incident objectives Incident-specific statements describing intended outcomes within a given operational period. (NFPA 470)

incident report A document prepared by fire department personnel on a particular incident. (NFPA 901)

incident safety officer (ISO) The member of the command staff responsible for monitoring and assessing safety hazards or unsafe situations and for developing measures for ensuring personnel safety. Also called *safety officer*. (NFPA 1026)

incipient stage The early stage of fire development where the fire's progression is limited to a fuel source and the thermal hazard is localized to the area of the burning material. (NFPA 1410)

incomplete combustion A combustion process during which the fuel is not completely consumed, usually due to a limited supply of oxygen.

incubation period The time between the initial infection by an organism and the development of symptoms in a victim.

independent crown fire A crown fire that moves rapidly from tree to tree burning aerial fuels and dropping falling embers to the ground, sometimes igniting a ground fire.

indicating valve A valve that has components that show if the valve is open or closed. (NFPA 1)

indirect attack Firefighting operations involving the application of extinguishing agents to reduce the build-up of heat released from a fire without applying the agent directly onto the burning fuel. (NFPA 1145)

indirect attack (wildland fire) A method of wildland fire attack in which the control line is located along natural fuel breaks, at favorable breaks in the topography, or at considerable distance from the fire, and the intervening fuel is burned out.

industrial packaging Packaging used to transport materials that present a limited hazard to the public or the environment. It is classified into three categories, based on the strength of the packaging—IP-1, IP-2, or IP-3, with IP-3 being the strongest.

infectious The property of being able to cause disease in humans or animals.

inflator In a motor vehicle, the system that inflates air bags using either stored compressed gas or a gas-generating system.

ingestion exposure Exposure to a hazardous material from swallowing the substance.

inhalation exposure Exposure to a hazardous material from breathing the substance into the lungs.

initial attack apparatus Fire apparatus with a fire pump of at least 250 gpm (946 L/min) capacity, water tank, and hose body, whose primary purpose is to initiate a fire suppression attack on structural, vehicular, or vegetation fires and to support associated fire department operations. Also called *quick attack apparatus*. (NFPA 1900)

initial size-up Usually conducted by the first-arriving company officer, a rapid evaluation and analysis of an incident upon arrival at the incident and before the 360-degree walk-around is conducted. Also called *windshield size-up* because it is generally given by the engine officer before they get out of the engine.

injection exposure Exposure to a hazardous material from the substance entering the body through cuts or other breaches in the skin.

inner circle See *inner survey*.

inner survey A tight, 360-degree survey of a vehicle involved in a motor vehicle crash conducted from approximately 3 to 5 ft (0.9 to 1.5 m) away from the vehicle to identify the vehicle's drive system (conventional or alternative fuel), determine the locations of any victims and whether they are entrapped, identify entry and exit points, determine vehicle stability for access to and removal of the victim(s), and identify hazards that are immediately dangerous to life and health (IDLH). Also called *inner circle*.

inspection A visual examination of a system or portion thereof to verify that it appears to be in operating condition and is free of physical damage. (NFPA 25)

intentional ventilation Ventilation that is planned and done on purpose.

interior attack A fire attack conducted inside a structure.

interior finish The exposed surfaces of walls, ceilings, and floors within buildings. (NFPA 5000)

interior wall A wall inside a building that divides a large space into smaller areas.

intermediate bulk container (IBC) A bulk storage container with a volume greater than 119 gallons (450 liters) but less than 793 gallons (3002 liters); these volumes are between the typical volume of drums or bags and the volume of cargo tanks.

intermodal tank (IM or IMO) A bulk storage container that holds between 5000 and 6000 gallons (18,927 and 22,712 liters) of product, can be either pressure or nonpressure, serves as both a shipping and storage vessel, and can be shipped by all modes of transportation—air, sea, or land.

International Code Council (ICC) An international association that creates standards and codes for buildings.

International Fire Service Accreditation Congress (IFSAC) A national organization that accredits or recognizes emergency service certification systems.

interoperability The ability to communicate across different radio bands and between different agencies.

intumescent coating Paint-like coatings of mineral materials or cement-like materials to insulate steel from heat in case of fire.

inversion layer A layer of warm air above a layer of cold air.

inverter Equipment that is used to change the voltage level or waveform, or both, of electrical energy. Commonly, an inverter (also known as a power conditioning unit [PCU] or power conversion system [PCS]) is a device that changes DC input to AC output. Inverters may also function as battery chargers that use alternating current from another source and convert it into direct current for charging batteries. (NFPA 70)

ion An atom with more protons than electrons (a positive ion or cation) or more electrons than protons (a negative ion or anion).

ionization smoke detector A smoke detector that detects small smoke particles when an electric current created by a small amount of radioactive material in the detector is disrupted, activating the alarm.

ionization The process of removing electrons from atoms or molecules.

ionizing radiation A form of radiation that includes particles capable of producing ions.

irons A combination of tools, usually consisting of a Halligan tool and a flat-head axe, that are commonly used for forcible entry.

island An unburned area surrounded by fire.

isolation of the hazard area The process of identifying a perimeter around a contaminated atmosphere so that responders and civilians are kept at a safe distance from the release site.

isotopes Different forms of the same atom.

J

J tool A tool that is designed to fit between double doors that are equipped with push bars or panic bars.

jalousie windows Windows made of small slats of tempered glass, which overlap each other when the window is closed. Often found in trailers and mobile homes, jalousie windows are held together by a metal frame and operated by a small hand wheel or crank found in the corner of the window.

joists Beams that hold up a floor.

joule A measure of heat energy equal to 0.4 calorie.

junction box A device that attaches to an electrical cord to provide additional outlets.

juvenile fire-setters Fire-setters who are typically 7 to 13 years old.

K

K tool A tool that is used to remove mortise and rim lockset cylinders from structural doors so the locking mechanism can be unlocked.

Kelly tool A steel bar with two main features: a large pick and a large chisel or fork.

kerf cut A cut that is the width and depth of the saw blade. It is used to inspect cockloft spaces from the roof.

kern In a kernmantle rope, the center or core of the rope.

kernmantle rope Rope made of two parts, the kern and the mantle.

keyhole saw A saw designed to cut keyhole circles in wood and drywall.

key tools Three simple tools—a ¼-in. (6-mm) slotted screwdriver, needle-nose pliers, and a bent-end key tool—often carried together as a set; used to manipulate most locks after the cylinder is pulled or removed.

kinetic energy Energy generated by an object in motion.

knock down To attack the fire to the point where it is almost extinguished and it is unlikely to extend further.

knot A fastening made by tying rope or webbing in a prescribed way. (NFPA 2500)

L

label A smaller version of a placard (4 inches [10 cm] on each side) that is required to be placed on the four sides of individual boxes and smaller packages being transported.

ladder A-frame An A-shaped structure formed by using rope to attach two straight ladders at the tip in an A-shape and that can be used as a makeshift lift when raising a trapped person. Also called *A-frame hoist*.

ladder belt A compliant equipment item that is intended for use as a positioning device for a person on a ladder. (NFPA 2500)

ladder company See *truck company*.

ladder frame construction See *body-over-frame construction*.

ladder lock See *pawl*.

ladder pipe A monitor that attaches to the rungs of a vehicle-mounted aerial ladder. (NFPA 1960)

ladder truck See *aerial fire apparatus*.

laid rope See *twisted rope*.

laminar smoke flow The smooth or streamlined movement of smoke.

laminated glass Also known as safety glass. The lamination process places a thin layer of plastic between two layers of glass so that the glass does not shatter and fall apart when broken. See also *laminated safety glass*.

laminated safety glass (LSG) Two layers of glass bonded together with a clear polymer film that prevents the glass from shattering when broken. Also called *laminated glass*.

laminated wood Pieces of wood that are glued together.

large-diameter hose (LDH) A hose 3.5 in. (89 mm) or larger that is designed to move large volumes of water to supply master stream appliances, portable hydrants, manifolds, standpipe and sprinkler systems, and fire department pumpers from hydrants and in relay. (NFPA 1410)

latch A component of a lockset; a moveable bolt that protrudes from the edge of the door and slides into the strike plate when the door is closed.

lath Thin strips of wood used to make the supporting structure for roof tiles.

lay To arrange hose on the ground. Also called *flake out*.

LCES A mnemonic that stands for Lookouts, Communications, Escape routes, and Safety zones to remind firefighters to ensure that each of these components is in place before attacking a wildland fire to reduce the risks associated with fighting these types of fires.

leader The individual responsible for command of a task force, strike team, or functional unit. (NFPA 1026)

legacy construction An older type of construction that used dimensional lumber and was built before about 1970.

lethal concentration (LC) The concentration of a material in air that, based on laboratory tests of the inhalation route to the body, is expected to kill a specified number of test animals in a group when administered over a specified period of time.

lethal dose (LD) A single dose that causes the death of a specified number of test animals in a group exposed by any route other than inhalation.

Level A ensemble Personal protective equipment consisting of a totally encapsulating suit that includes a self-contained breathing apparatus and that provides the highest level of protection against vapors, gases, mists, and even dusts.

Level B ensemble Personal protective equipment consisting of a single- or multi-piece chemical-protective garment, boots, gloves, and a self-contained breathing apparatus, which is used when the type and atmospheric concentration of substances require a high level of respiratory protection but a moderate level of skin protection.

Level C ensemble Personal protective equipment consisting of a NIOSH-approved CBRN (chemical, biological, radioactive, nuclear) air-purifying respirator or a powered air-purifying respirator, which is used when the type of airborne substance is known, the concentration is measured, the criteria for using an air-purifying respirator are met, and skin and eye exposure is unlikely.

Level D ensemble Personal protective equipment typically consisting of coveralls, work shoes, hard hat, gloves, and standard work clothing, which is used when the atmosphere contains no known hazard, and work functions preclude splashes, immersion, or the potential for unexpected inhalation of or contact with hazardous levels of chemicals.

liaison officer A member of the command staff responsible for coordinating with representatives from cooperating and assisting agencies. (NFPA 1550)

lieutenant The first level of officer and the person who is usually responsible for a single fire company on a shift.

life safety rope Rope dedicated solely for the purpose of supporting people during rescue, firefighting, other emergency operations, or during training evolutions. (NFPA 2500)

light-emitting diodes (LEDs) Electronic semiconductors that emit a single-color light when activated. LEDs are used for operational displays in SCBA.

light energy Energy produced by electromagnetic waves packaged in discrete bundles called photons.

light hazard area An occupancy where the quantity, combustibility, and heat release of the materials is low, and the majority of materials are arranged so that a fire is not likely to spread. Also called *low hazard area*.

line-of-sight system See *simplex communication system*.

line-type detector A device in which detection is continuous along a path. Typical examples are rate-of-rise pneumatic tubing detectors, projected beam smoke detectors, and heat-sensitive cable. (NFPA 72)

liquid Matter that has a specific volume but does not have a specific size or shape.

liquid splash–protective ensemble Compliant protective clothing and equipment products that when worn together provide protection from some, but not all, risks of hazardous materials/weapons of mass destruction (WMD) incident operations involving liquids. (NFPA 470)

live load The load produced by the use and occupancy of the building or other structure, which does not include construction or environmental loads such as wind load, snow load, rain load, earthquake load, flood load, or dead load. Live loads on a roof are those produced (1) during maintenance by workers, equipment, and materials and (2) during the life of the structure by movable objects such as planters and by people. (NFPA 5000)

load-bearing wall A wall that supports any vertical load in addition to its own weight or any lateral load.

loaded-stream fire extinguisher A stored-pressure water-type fire extinguisher that uses an alkali metal salt as a freezing-point depressant.

loads The forces to which a structure is subjected due to superimposed weight or pressures

local emergency planning committee (LEPC) A committee comprising members of industry, transportation, the public at large, media, and fire and law enforcement agencies, which gathers and disseminates information on hazardous materials stored in the community and ensures that there are adequate local resources to respond to a chemical event in the community.

lock body The structural housing of a padlock that holds the main locking mechanisms and secures the shackle.

lock box A small safe containing the keys to a building's exterior doors and possibly interior doors.

locking mechanism A device that locks a fire extinguisher's trigger to prevent its accidental discharge.

lockout and tagout systems Methods of ensuring that electricity and other utilities have been shut down and switches are "locked" so that they cannot be switched on. These systems are used to prevent the flow of power or gases into the area where rescue is being conducted.

lockset A standard doorknob lock, deadbolt lock, sliding latch, or padlock.

logistics section The section in an incident command system responsible for providing facilities, services, and materials for the incident or planned event, including the communications unit, medical unit, and food unit within the service branch and the supply unit, facilities unit, and ground support unit within the support branch. (NFPA 1026)

logistics section chief The general staff position responsible for directing the logistics function; a role generally assigned on complex, resource-intensive, or long-duration incidents.

loop A piece of rope formed into a circle by crossing the rope.

loop knot A knot that forms a secure loop in the end of a rope.

louver cut A cut that is made using power saws and axes to cut along and between roof supports so that the sections created can be tilted into the opening.

low-angle operations Rope rescue operations on a mildly sloped surface (less than 45 degrees) or flat land. In this scenario, firefighters depend on the ground for their primary support, and the rope system is a secondary means of support.

lower explosive limit (LEL) The minimum concentration of a combustible vapor or combustible gas in a mixture of the vapor or gas and gaseous oxidant, above which propagation of flame will occur on contact with an ignition source. (NFPA 115)

lower flammable limit (LFL) See *lower explosive limit.*

low hazard area See *light hazard area.*

low-pressure chemical cargo tank A rounded (uninsulated) or horseshoe-shaped (insulated) tank with an internal working pressure up to 35 psi that is capable of holding 6000 to 7000 gallons of flammable liquid, mild corrosives, and poisons; also called *MC-307/DOT 407 chemical hauler.*

low-volume nozzle A nozzle that flows 40 gallons per minute (151 L/min) or less.

lug Protrusion or indentation on a hose coupling that aids in securing and tightening the connection between two couplings.

M

main sprinkler system valve A valve that automatically opens and signals an alarm when a sprinkler head is activated.

maintenance Work, including, but not limited to, repair, replacement, and service, performed to ensure that equipment operates properly. (NFPA 72)

main water supply control valve A manually controlled valve that allows water to enter the sprinkler system from the water source.

malicious alarm An unwanted activation of an alarm initiating device caused by a person acting with malice. (NFPA 72)

mallet A short-handled hammer.

mantle In a kernmantle rope, the braided covering that protects the kern from dirt and abrasion. Also called *sheath.*

manual dry standpipe system A standpipe system with no permanently attached water supply that relies exclusively on the fire department connection to supply the system demand. (NFPA 14)

manual fire alarm box A manually operated device used to initiate a fire alarm signal. Also called *manual pull station.*

manual pull station See *manual fire alarm box.*

manual wet standpipe system A standpipe system containing water at all times that relies exclusively on the fire department connection (FDC) to supply the system demand. (NFPA 14)

masonry Built-up unit of construction or combination of materials such as clay, shale, concrete, glass, gypsum, tile, or stone set in mortar. (NFPA 5000)

mass arsonist A person who sets three or more fires at the same site or location during a limited period of time.

mass decontamination The physical process of reducing or removing surface contaminants from large numbers of victims in potentially life-threatening situations in the fastest time possible. (NFPA 470)

mass lumber Large-scale engineered wood products designed to replace structural materials such as concrete and steel in smaller structures such as single-family dwellings. See also *mass timber.*

mass timber See *mass lumber.*

master stream appliance A device that discharges high-volume water streams, usually between 350 gpm (1325 L/min) and 1500 gpm (5678 L/min), though much larger capacities are available. Also called *master stream device.*

master stream device See *master stream appliance.*

master stream nozzle A nozzle with a rated discharge of 350 gpm (1325 L/min) or greater. (NFPA 1960)

matter Anything that has mass and volume.

maul A specialized striking tool, weighing 6 lb (3 kg) or more, with an axe on one side of the head and a sledgehammer on the other side.

maximum ceiling temperature The maximum temperature to which a sprinkler head can be exposed, while either installed in a system or in storage.

mayday Any situation at a hazardous materials/weapons of mass destruction (WMD) incident where an entry team member is unable to safely exit the hot zone, or an event that cannot be resolved by the entry team member within 30 seconds. (NFPA 470)

MC-306/DOT 406 cargo tank See *nonpressure liquid cargo tank.*

MC-307/DOT 407 chemical hauler See *low-pressure chemical cargo tank.*

MC-312/DOT 412 corrosive tank See *chemical cargo tank.*

MC-331 See *pressure cargo tank.*

MC-338 See *cryogenic liquid cargo tank.*

McLeod A hand tool used for constructing fire control lines and overhauling wildland fires. One side of the head consists of a five-toothed to seven-toothed fire rake; the other side is a hoe.

mechanical energy The potential energy stored because of the position of an object or the kinetic energy of an object in motion.

mechanical ventilation A process of removing heat, smoke, and gases from a fire area by using exhaust fans, blowers, air-conditioning systems, or smoke ejectors. (NFPA 440)

medium-diameter hose (MDH) A hose 2½-in. (64-mm) or 3-in. (76-mm) in diameter most often used as attack hose, but can be used as supply hose.

mill construction See *Type IV construction.*

millimeters of mercury (mm Hg) A unit of pressure equal to the pressure exerted at the base of a column of mercury that is exactly 1 millimeter in height; 760 mm Hg is equal to 14.7 psi.

mini-maxi codes Building and fire codes that legally specify both the minimum and maximum requirements.

minimum breaking strength (MBS) The result of subtracting three standard deviations from the mean result of the lot being tested using the formula in 28.2.5.2. (NFPA 2500)

minuteman hose load A hose loading method that allows a single fire fighter to deploy and advance the necessary amount of hose from the shoulder and avoids having to maneuver the hose lines around obstacles and helps to prevent sharp kinks.

mobile data terminals (MDTs) Technology that allows firefighters to receive data while in the fire apparatus or at the station.

mobile radio A two-way radio that is permanently mounted in a fire apparatus.

mobile repeater A repeater used at an incident scene to boost signals at the scene by capturing the weak signals from portable radios and rebroadcasting them.

mobile water supply apparatus A vehicle designed primarily for transporting (pickup, transporting, and delivering) water to fire emergency scenes to be applied by other vehicles or pumping equipment. Also called *tanker, tender,* or *water tender.* (NFPA 1900)

moderate hazard area See *ordinary hazard area.*

molecular bond The connection between atoms in a molecule.

molecule Atoms chemically bonded together.

monoammonium phosphate A finely ground substance that looks like yellow talcum powder that is used as an extinguishing agent in dry-chemical fire extinguishers that are rated for Class A, B, and C fires. Also called *ammonium phosphate.*

monomers Molecules that can be bonded to like molecules to form a larger polymer.

mortar An adhesive produced by mixing sand, lime, water, and cement.

mortise lockset Rectangular-shaped lockset that fits into a rectangular cavity within a door. Often found in commercial buildings or older homes.

multiple element gas containers (MEGCs) Several individually specified, high-pressure tubes, with working pressures that range as high as 5000 psi, that are attached to a frame and that commonly carry substances such as hydrogen and oxygen.

multiple-jacket A construction consisting of a combination of two separately woven reinforcements (double jacket) or two or more reinforcements interwoven. (NFPA 1962)

multiplex channel Simultaneous transmission of multiple data streams, most often voice signals, in either or both directions over the same frequency.

multipurpose dry-chemical fire extinguisher A fire extinguisher that uses a monoammonium phosphate–based extinguishing agent that is effective on fires involving ordinary combustibles, such as wood or paper, and fires involving flammable liquids and that is rated to fight Class A, B, and C fires.

multipurpose hook A long pole with a wooden or fiberglass handle and a metal hook on one end used for pulling.

multipurpose ladder See *combination ladder.*

multi-tool A compact, pocket-size, multiple-function tool that combines several different tools, such as a knife, scissors, wire cutters, pliers, and screwdriver.

municipal fire alarm system A network of fire alarm boxes and emergency telephones on street corners or in public places

municipal water system A system having water pipes servicing fire hydrants and designed to furnish, over and above domestic consumption, a minimum of 250 gpm (946 L/min) at 20 psi (138 kPa) residual pressure for a 2-hour duration. (NFPA 1140)

muscarinic effects Effects such as a runny nose, salivation, sweating, bronchoconstriction, bronchial secretions, nausea, vomiting, and diarrhea.

N

National Board on Fire Service Professional Qualifications A national organization that accredits or recognizes emergency service certification systems. Also called *Pro Board.*

National Fire Incident Reporting System (NFIRS) A national reporting system used to compile and analyze incident reports from fire departments to help identify statistical trends, faulty equipment and products, civilian and firefighter injuries and fatalities, dollar loss filtered by type of property, and many other pieces of valuable information at local, state, and national levels.

National Fire Protection Association (NFPA) A nonprofit association that develops and maintains nationally recognized minimum consensus standards and fire codes for fire safety and handling hazardous materials.

National Incident Management System (NIMS) A system mandated by Homeland Security Presidential Directive 5 (HSPD-5) that provides a systematic, proactive approach guiding government agencies at all levels, the private sector, and nongovernmental organizations to work seamlessly to prepare for, prevent, respond to, recover from, and mitigate the effects of incidents, regardless of cause, size, location, or complexity, so as to reduce the loss of life or property and harm to the environment. (NFPA 1026)

National Institute for Occupational Safety and Health (NIOSH) Part of the U.S. Department of Health and Human Services; the department charged with providing information, training, research, and education in the field of occupational safety and health so as to ensure that individuals have a safe and healthy work environment.

National Response Center (NRC) An agency maintained and staffed by the U.S. Coast Guard that should always be notified if a hazard discharges into the environment.

natural fire A fire that is started without direct human intervention or action, such as a fire resulting from lightning, an earthquake, or wind.

natural ventilation The flow of air or gases created by the difference in the pressures or gas densities between the outside and inside of a vent, room, or space. (NFPA 853)

negative pressure The condition that exists when the air pressure in one area is lower than the atmospheric pressure in another area.

negative-pressure ventilation Ventilation that relies on electric fans to pull or draw the smoke, heat, and air by creating negative pressure in a structure or an area.

nerve agent A toxic chemical agent that attacks the central nervous system.

neutral plane The interface at a vent, such as a doorway or a window opening, between the hot gas flowing out of a fire compartment and the cool air flowing into the compartment where the pressure difference between the interior and exterior is equal.

neutron In the nucleus of an atom, a particle that does not have an electrical charge.

New York hook See *roofman's hook.*

Next Generation 911 (NG911) A system designed to replicate traditional Enhanced 911 (E911) features and functions and provide additional capabilities to provide access to emergency services from all connected communications sources and provide multimedia data capabilities for public safety answering points (PSAPs) and other emergency service organizations. (NFPA 1225)

NFPA 704 hazard identification system A hazardous materials marking system designed for fixed-facility use. It uses a diamond-shaped symbol of any size, which is itself broken into four smaller diamonds, each representing a particular property or characteristic of the material.

nominal lumber Lumber that is true to its stated dimensions.

nonambulatory patient A term describing individuals who cannot move themselves to an area of safety owing to their physical condition, medical treatment, or other factors.

nonbearing wall Any wall that is not a bearing wall. (NFPA 5000)

nonbulk packaging Packaging that has a liquid capacity of 119 gallons (450 liters) or less, a solids capacity of 882 pounds (400 kg) or less, or a compressed gas water capacity of 1001 pounds (454 kg) or less. (NFPA 470)

noncoded system An alarm system that provides no information indicating the type or location of the initiating device that was activated.

non-collapsible fire hose Fire hose typically made from durable materials such as PVC or synthetic rubber so that it maintains its shape and structure. Also called *rigid fire hose.*

nonintervention mode The operating mode in which responders do not operate near the hazardous materials container and focus their efforts on public protection actions only, allowing the container or product to take its natural course.

non-ionizing radiation A form of radiation that comes from electromagnetic waves; it can cause a disturbance of activity at the atomic level, but does not have sufficient energy to break bonds and create ions.

non-load-bearing wall Walls that only support their own weight and do not support any of the dead or live load of the building. See also *partition wall* or *nonbearing wall.*

nonpressure liquid cargo tank A nonpressure tank that typically carries between 6000 and 10,000 gallons of a product such as gasoline or other flammable and combustible liquids; also called *MC-306/DOT 406 cargo tank.*

nonpressure tank A bulk storage tank that has a working pressure between 2.65 and 4 psi and that is usually made of steel or aluminum.

nonpressure tank car A railcar that is equipped with a nonpressure tank that typically carries general industrial chemicals and consumer products such as corn syrup, flammable and combustible liquids, and mild corrosives; also called *general service rail tank car.*

normal operating pressure The observed static pressure in a water distribution system during a period of normal demand.

normal ready status The state of a system when it is functioning properly and there are no faults (problems) in the system and no problems with any other fire protection or building system integrated with the FACU.

nose cups Inserts inside the face piece of an SCBA that fit over the user's mouth and nose.

nozzle A constricting appliance attached to the end of a fire hose or monitor to increase the water velocity and form a stream. (NFPA 1960)

nuclear energy Potential energy stored in the nucleus of an atom or the kinetic energy released by splitting the nucleus of an atom into two smaller nuclei (fission) or by combining two small nuclei into one large nucleus (fusion).

nucleus The center of an atom.

nuisance alarm An unwanted activation of a signaling system or an alarm initiating device in response to a stimulus or condition that is not the result of a potentially hazardous condition. (NFPA 72)

O

obscuration rate A measure of the percentage of light that is blocked between a sender and a receiver unit.

occupancy The purpose for which a building or other structure, or part thereof, is used or intended to be used. (NFPA 5000)

Occupational Safety and Health Administration (OSHA) The U.S. federal agency that regulates worker safety and, in some cases, responder safety. It is part of the U.S. Department of Labor.

offensive attack See *offensive operation.*

offensive mode The mode of operation in which responders have direct contact with the material or container and take aggressive action to control the release.

offensive operation Actions generally performed in the interior of involved structures that involve a direct attack on a fire to directly control and extinguish the fire. Also called *offensive attack* or *offensive strategy.* (NFPA 1550)

offensive strategy See *offensive operation.*

off-gas To emit harmful chemicals in the form of a gas.

on-scene See *emergency scene.*

open-circuit self-contained breathing apparatus (open-circuit SCBA) An SCBA in which the exhaled air is released into the atmosphere and is not reused.

open-end wrench A hand tool that is used to tighten or loosen bolts. The end is open, as opposed to a box-end wrench. Each wrench is a specific size.

open-head drum A drum with a lid that is secured by a bolted clasp-type ring, which circles the entire head of the drum.

operating spindle The handle, doorknob, or keyway of a door that turns the latch to open it.

operating stem The steel rod extending from the top of a dry barrel hydrant to the hydrant valve that firefighters can turn using the stem nut at the top to open and close the main valve.

operations-level responder A person who responds to hazardous materials/weapons of mass destruction (WMD) incidents for the purpose of implementing or supporting actions to protect nearby persons, the environment, or property from the effects of the release. (NFPA 470)

operations section The section in an incident command system responsible for all tactical operations at an incident or planned event, including up to 5 branches; 25 divisions or groups; and 125 single resources, task forces, or strike teams. (NFPA 1026)

operations section chief The general staff position responsible for managing all operations activities; a role usually assigned when complex incidents involve more than 20 single resources or when command staff cannot be involved in all details of the tactical operation.

ordinary dry-chemical fire extinguisher A dry-chemical fire extinguisher rated for only Class B and Class C fires.

ordinary hazard area An area that contains more Class A and Class B materials than a light hazard area and the combustibility and heat release rate of the materials is moderate. Also called *moderate hazard area.*

oriented search method A primary search method in which a search team consisting of an officer or team leader and one to three searchers conduct a systematic search where the searchers search the rooms using a left-handed or right-handed search pattern along the outside wall of rooms and the officer or team leader stays at the door to direct the search.

oriented-vent-enter-isolate-search method (O-VEIS) A primary search method in which a search team consisting of an officer or team leader and one searcher approach a window, determine that the room is tenable, open the window so that the searcher can quickly enter the room, close the door, and search the room.

orifice In fire hydraulics, an opening whose diameter is used in calculations.

outer circle See *outer survey.*

outer perimeter See *cold zone.*

outer survey A survey that extends from the perimeter of the inner survey outward to a distance sufficient to cover where the impact first occurred and the resting point of the vehicle or vehicles to identify hazards that are immediately dangerous to life and health (IDLH), victims who might have been ejected, any additional vehicles involved in the incident, walking wounded, and any other imminent or potential hazards that might have been missed during the initial size-up. Also called *outer circle.*

outlet An opening on a fire hydrant through which water is discharged.

outside screw and yoke (OS&Y) valve A sprinkler control valve with a threaded stem that extends out when the valve is open and retracts in when the valve is closed. Also called *outside stem and yoke (OS&Y) valve.*

outside stem and yoke (OS&Y) valve See *outside screw and yoke (OS&Y) valve.*

overflow dam A dam that includes piping installed at a slight downward angle, which allows the water to flow over the released liquid, and with enough capacity that the water drains without overtopping the dam, thereby trapping the heavier-than-water material held behind at the base of the dam.

overhaul A firefighting term involving the process of final extinguishment after the main body of the fire has been knocked down. All traces of fire must be extinguished at this time. (NFPA 1700)

oxidation Reaction with oxygen either in the form of the element or in the form of one of its compounds. (NFPA 53)

P

packaging group designation A label that describes a substance according to the degree of hazards it presents.

packaging The process of securing a victim in a transfer device, with regard to existing and potential injuries or illness, so as to prevent further harm during movement. (NFPA 1006)

padlocks The most common types of locks on the market today, portable locks built to provide regular-duty or heavy-duty service. Several types of locking mechanisms are available, including keyways, combination wheels, and combination dials.

parallel attack A method of attack in which the control line is located parallel to the fire edge, at a distance of about 5 to 50 ft (1.5 to 15 m) from the fire. The intervening fuel usually burns out as the fire control line moves alongside the fire but can also burn out with the main fire.

parallel chord truss A truss in which the top and bottom chords are parallel.

Paramedic Emergency medical services (EMS) personnel who can do everything an advanced emergency medical technician (AEMT) can do, and who have extensive training in advanced life support, including administering drugs, cardiac monitoring, inserting advanced airways, manual defibrillation, and other advanced assessment and treatment skills.

parapet The part of a wall entirely above the roofline. (NFPA 5000)

partition wall A non-load-bearing interior wall that spans horizontally or vertically from support to support. The supports may be the basic building frame, subsidiary structural members, or other portions of the partition system. See also *non-load-bearing wall.*

party wall A wall constructed on the line between two properties.

PASS Acronym for the steps involved in operating a portable fire extinguisher: Pull pin, Aim nozzle, Squeeze trigger, Sweep across burning fuel.

passive crown fire A crown fire that starts at the bottom of a tree or group of trees and then runs all the way up the tree or trees so they look like a giant torch, but does not spread to the surrounding trees.

passive smoke control system A smoke control system that uses building features such as fire-rated walls, automatic fire doors, draft curtains, and other smoke barriers to limit the spread of smoke and compartmentalize larger spaces to confine the smoke to the room or area of origin.

pawl A device attached to a fly section(s) to engage ladder rungs near the beams of the section below for the purpose of anchoring the fly section(s). Also called *ladder lock, dog,* or *rung lock.* (NFPA 1960)

peak cut A ventilation opening that runs along the top of a pitched roof.

pendent sprinkler head A sprinkler head designed to be mounted on the underside of sprinkler piping hanging down toward the floor so that the water stream is directed in a downward direction, and is usually marked *SSP,* which stands for *standard spray pendent.*

penetration The movement of a material through a suit's closures, such as zippers, buttonholes, seams, flaps, or other design features of chemical-protective clothing, and through punctures, cuts, and tears. (NFPA 470)

permeation The chemical action involving the movement of chemicals, on a molecular level through intact material. (NFPA 470)

permissible exposure limit (PEL) The established standard limit of exposure to a hazardous material based on the maximum time-weighted concentration at which 95 percent of exposed, healthy adults suffer no adverse effects over a 40-hour workweek.

personal alert safety system (PASS) A device that continually monitors for lack of movement of the wearer and automatically activates an alarm signal, indicating the wearer is in need of assistance; can also be manually activated to trigger the alarm signal. (NFPA 1970)

personal flotation device (PFD) A device manufactured in accordance with U.S. Coast Guard specifications that provides supplemental flotation for persons in the water. (NFPA 1006)

personal protective equipment (PPE) The full complement of garments fire fighters are required to wear while on an emergency scene, including turnout coat, protective trousers, firefighting boots, firefighting gloves, a protective hood, self-contained breathing apparatus (SCBA), a personal alert safety system (PASS) device, and a helmet with eye protection. (NFPA 1010)

personnel accountability report (PAR) Periodic reports verifying the status of responders assigned to an incident or planned event. (NFPA 1026)

personnel accountability system A system that readily identifies both the location and function of all members operating at an incident scene. (NFPA 1550)

pH A measurement of the amount of dissolved hydrogen ions (H^+) in a solution.

phosgene A gas formed from incomplete combustion of many common household products.

photoelectric smoke detector A smoke detector that detects visible smoke particles when a light beam in the chamber is disrupted by the smoke particles and the light is reflected onto a photocell, activating the alarm.

physical change A transformation in which a material changes its state of matter, for instance, from a liquid to a solid.

physical evidence A physical or tangible item that proves or disproves a particular fact or issue; also referred to as *real evidence.*

pick-head axe A tool that has a head with an axe on one side and a pointed end ("pick") on the opposite side.

piercing nozzle A nozzle that can be driven through sheet metal or other material to deliver a water stream to that area.

pike pole A pole with a sharp point ("pike") on one end coupled with a hook. It is used to make openings in ceilings and walls. Pike poles are manufactured in different lengths for use in rooms of different heights.

pillar See *post.*

pilot line A separate dry pipe system in a deluge sprinkler system in which the air pressure drops when a sprinkler head activates or a fixed temperature device releases so that water can flow through the pilot line and through a deluge valve.

pincer attack See *anchor, flank, tandem, and pincer attack.*

pin lug A lug that looks like a small cylinder that extends outward from a hose coupling.

pinning To make a connection using bolts or rivets between building materials.

pipeline A length of pipe including pumps, valves, flanges, control devices, strainers, and/or similar equipment for conveying fluids. (NFPA 470)

pipeline right-of-way An area, patch, or roadway that extends a certain number of feet on either side of a pipeline and that may contain warning and informational signs about hazardous materials carried in the pipeline.

pipe wrench A wrench having one fixed grip and one movable grip that can be adjusted to fit securely around pipes and other tubular objects.

pitched chord truss A type of truss typically used to support a sloping roof.

pitched roof A roof with sloping or inclined surfaces.

Pitot gauge A type of gauge that is used to measure the velocity pressure of water that is being discharged from an opening. It is used to determine the flow of water from a hydrant or nozzle.

placard A diamond-shaped indicator (10¾ inches [27 cm] on each side) that is required to be placed on all four sides of highway transport vehicles, railroad tank cars, and other forms of transportation carrying hazardous materials to identify the substance being transported.

plague An infectious disease caused by the bacterium *Yersinia pestis,* which is commonly found on rodents and is usually transmitted to humans by fleas.

plan A See *primary access.*

plan B See *secondary access.*

planning section The section in an incident command system responsible for the collection, evaluation, dissemination, and use of information related to the incident situation, resource status, and incident forecast. (NFPA 1026)

planning section chief The general staff position responsible for planning functions and for tracking and logging resources assigned when command staff members need assistance in managing information.

plaster hook A long pole with a pointed head and two retractable cutting blades on the side.

plate glass A thicker type of annealed glass that is commonly used for large windows in commercial buildings.

platform-frame construction Construction technique for building the frame of the structure one floor at a time. Each floor has a top and bottom plate that acts as a firestop.

plug-in hybrid electric vehicle (PHEV) A vehicle that uses both a battery-powered electric motor and a liquid-fueled engine to propel the vehicle and whose battery can be recharged by connecting it to an external power source.

plume The column of hot gases, flames, and smoke rising above a fire. Also called *convection column, thermal updraft,* or *thermal column.* (NFPA 921)

pocket A deep indentation of unburned fuel along the fire's perimeter, often found between a finger and the head of the fire.

point of origin The exact physical location within the area of origin where a heat source and a fuel first interact, resulting in a fire or explosion. (NFPA 921)

poison A substance capable of causing adverse health effects in humans and animals.

polar solvent A water-soluble, flammable liquid that readily mixes with water.

police powers A legal authority based on the Tenth Amendment that grants state (and local) authorities the authority to establish and enforce laws protecting the greater good of the community.

policies Formal statements that outline expectations for performance and procedures in different circumstances, but usually require personnel to make judgments to determine the best course of action within the stated study.

polymer A larger molecule made up of long repeating chains of molecules; often used to describe different types of plastics.

polymerization The process of reacting monomers together in a chain reaction to form polymers.

portable monitor A monitor that can be lifted from a vehicle-mounted bracket and moved to an operating position on the ground by not more than two people. (NFPA 1960)

portable radio A battery-operated, hand-held transceiver. (NFPA 1225)

portable tanks Folding or collapsible tanks that are used at the fire scene to hold water for drafting.

positive pressure The condition that exists when the air pressure in one area is higher than the atmospheric pressure in another area.

positive-pressure air-line respirator (with escape units) See *supplied-air respirator (SAR).*

positive-pressure attack Ventilation that controls the flow of products of combustion before the fire is controlled by introducing clean air into a structure, which pushes the contaminated atmosphere out.

positive-pressure fan A large, powerful fan that is used to create positive-pressure ventilation by forcing clean air into a structure.

positive-pressure ventilation Ventilation that relies on fans to push or force clean air into a structure after a structure fire has been controlled.

possible cause A hypothesis determined to have less than 50 percent probability of being true.

post One of the vertical support members of a vehicle that holds up the roof and forms the upright columns of the passenger compartment in a vehicle. Also called *pillar.*

post indicator valve (PIV) A sprinkler control valve that has an indicator on a post mounted in the ground or on a wall that reads either "OPEN" or "SHUT" depending on its position.

post-traumatic stress disorder (PTSD) A behavioral disorder that develops after a person has experienced a critical incident; characterized by reexperiencing the event and overresponding to stimuli that recalls the event; symptoms include depression, startle reactions, flashback phenomena, and dissociative episodes (e.g., amnesia of the event).

potential energy Energy stored by an object as a result of its position or condition.

pounds per square inch (psi) A unit of pressure that expresses the pounds of force per square inch of area.

power saw A saw that usually is powered by an electric motor or a gasoline engine. The three primary types of mechanical saws are chainsaws, rotary saws, and reciprocating saws.

powered air-purifying respirator (PAPR) An air-purifying respirator that uses a powered blower to force the ambient air through one or more air-purifying components to the respiratory inlet covering. (NFPA 1984)

preaction sprinkler system A sprinkler system employing automatic sprinklers that are attached to a piping system that contains air that might or might not be under pressure, with a supplemental detection system installed in the same areas as the sprinklers. (NFPA 13)

preaction valve The valve assembly in a preaction sprinkler that prevents water from entering the system until a manual emergency release is activated, a fire sprinkler head activates, a fire detector activates, or a combination of sprinkler head and initiating device activation takes place.

preconnect See *preconnected attack line.*

preconnected attack line Attack hose that travels on the engine already equipped with a nozzle and connected to a pump discharge outlet so it is ready for immediate use. Also called *preconnect.*

preconnected flat hose load A hose loading method in which the hose is laid flat and stacked on top of the previous section, similar to the flat hose load but with one or two loops sticking out of the load to make it easier to deploy.

pre-habilitation The concept of ensuring that a firefighter sustains physical readiness at all times through adequate hydration, nutrition, rest, and awareness of the state of their body and the surrounding environment.

preincident plan A document developed by gathering general and detailed data that is used by responding personnel in effectively managing emergencies for the protection of occupants, participants, responding personnel, property, and the environment. (NFPA 1660)

preincident survey An examination of a property to gather information to develop a preincident plan.

pressure cargo tank A pressure tank with rounded ends used to transport materials such as ammonia, propane, Freon, and butane, which carries volumes from 1000 gallons (3785 liters) to 11,000 gallons (41,640 liters) cargo tank, and is commonly constructed of steel or stainless steel with a single tank compartment; also called *MC-331.*

pressure indicator A gauge on a pressurized portable fire extinguisher that indicates the internal pressure of the expellant.

pressure tank A bulk storage tank that has an internal pressure greater than 15 psig.

pressure tank car A railcar used to transport materials such as propane, ammonia, ethylene oxide, and chlorine; it has an internal working pressure ranging from 100 to 600 psi and is equipped with top-mounted fittings for loading and unloading that are protected by a protective housing that sits atop the tank car.

primary access The existing openings of doors and/or windows that provide a pathway or entry to the trapped and/or injured victims(s). Also called *plan A.* (NFPA 1670)

primary feeder The largest-diameter water main pipe in a water distribution system that carries the greatest amounts of water. Also called *trunk line.*

primary search A quick search of the structures likely to contain survivors. (NFPA 2500)

primary stabilization The technique of stabilizing a vehicle using cribbing, step chocks, box cribs, and wedges to make it safe for a medical provider to access the victim and initiate emergency trauma care as well as other support functions from within the vehicle.

private water system A privately owned water system that operates separately from the municipal water system.

Pro Board See *National Board on Fire Service Professional Qualifications.*

probable cause A hypothesis determined to have greater than 50 percent probability of being true.

projected beam-type detector A type of photoelectric light obscuration smoke detector wherein the beam spans the protected area. Also called *beam detector.* (NFPA 72)

projected windows Windows that project outward on a top hinge or that pivot at the middle of the window. They are usually found in older warehouses or factories.

proprietary supervising alarm system An installation of an alarm system that serves contiguous and noncontiguous properties, under one ownership, from a proprietary supervising station located at the protected premises, or at one of multiple noncontiguous protected premises, at which trained, competent personnel are in constant attendance. (NFPA 72)

protected premises (local) fire alarm system A fire alarm system located at the protected premises. (NFPA 72)

protection plates Reinforcing material placed on a ladder at chafing and contact points to prevent damage from friction and contact with other surfaces.

protein foam A protein-based foam concentrate that is stabilized with metal salts to make a fire-resistant foam blanket. (NFPA 460)

proton In the nucleus of an atom, a positively charged particle that orbits the nucleus.

pry bar A specialized prying tool made of a hardened steel rod with a tapered end that can be inserted into a small area.

psig Pounds per square inch gauge; a measurement of pressure inside a container shown on an attached pressure gauge, in relation to atmospheric pressure.

public emergency alarm reporting system A system of alarm initiating devices, transmitting and receiving equipment, and communication infrastructure (other than a public telephone network) used to communicate with the communications center to provide any combination of manual or auxiliary alarm service. (NFPA 72)

public information officer (PIO) A member of the command staff responsible for interfacing with the public and media or with other agencies with incident-related information requirements. (NFPA 1026)

public safety answering point (PSAP) A facility equipped and staffed to receive emergency and non-emergency calls requesting public safety services via telephone and other communication devices. (NFPA 1225)

public safety communications center A building or portion of a building that is specifically configured for the primary purpose of providing emergency communications services or public safety answering point (PSAP) services to one or more public safety agencies under the authority or authorities having jurisdiction. Also called *communications center.* (NFPA 1225)

Pulaski axe A hand tool that combines an adze and an axe for brush removal.

pull hose See *deploy hose.*

pump tank fire extinguisher A nonpressurized, manually operated water-type fire extinguisher that is rated for use on Class A fires. Discharge pressure is provided by a hand-operated, double-acting piston pump.

pumper Fire apparatus with a permanently mounted fire pump of at least 750 gpm (1300 L/min) capacity water tank, and a hose body whose primary purpose is to control structural and associated fires. (NFPA 1900)

purchase point A small opening made to enable better tool access in forcible entry.

pyrolysis A process in which material is decomposed, or broken down, into simpler molecular compounds by the effects of heat alone; pyrolysis often precedes combustion. (NFPA 921)

Q

quick attack apparatus See *initial attack apparatus.*

Quick Response Guide (QRG) A short version of a vehicle's Emergency Response Guide (ERG).

quint See *quint apparatus.*

quint apparatus Fire apparatus with a permanently mounted fire pump, a water tank, a hose storage area, an aerial device or elevating platform with a permanently mounted waterway, and a complement of ground ladders. Also called *quint.* (NFPA 1710)

R

rabbet tool A hydraulic spreading tool designed to pry open doors that swing inward. See also *bunny tool.*

radiation The energy transmitted through space in the form of electromagnetic waves or energetic particles.

radiation dispersal device (RDD) A device designed to spread radioactive material through a detonation of conventional explosives or other [non-nuclear] means; also called *dirty bomb.* (NFPA 470)

radioactivity The natural and spontaneous process by which unstable isotopes of an element decay to a different state and emit radiation in the form of particles or waves.

radiological agent A material that emits radiation, such as X-rays and radioactive isotopes, in excess of normal background radiation levels.

rafters Joists that are mounted in an inclined position to support a roof.

rail The top or bottom piece of a trussed beam assembly used in the construction of a trussed ladder. Also, the top and bottom surfaces of an I-beam ladder. Each beam has two rails.

rain-down method A foam application method that directs the stream into the air above the fire and allows it to gently fall on the surface.

rapid intervention crew/company (RIC) A minimum of two fully equipped personnel on site, in a ready state, for immediate rescue of disoriented, injured, lost, or trapped rescue personnel. Also called *rapid intervention team (RIT), firefighter assist and search team (FAST),* and *rescue assist team (RAT).* (NFPA 1550)

rapid intervention crew/company universal air connection (RIC UAC) A system that allows emergency replenishment of breathing air to the SCBA of disabled or entrapped fire or emergency services personnel. (NFPA 1970)

rapid intervention pack (RIC pack) A portable air supply that provides an emergency source of breathing air for a single firefighter who has run out of air or whose air supply is insufficient to safely exit from an IDLH atmosphere.

rapid intervention team (RIT) See *rapid intervention crew/company (RIC).*

rate-of-rise detector A device that responds when the temperature rises at a rate exceeding a predetermined value. (NFPA 72)

rate of spread (ROS) The speed, usually measured in chains per hour, at which the front of the fire moves.

real evidence See *physical evidence.*

rear of the fire See *heel of the fire.*

rebar Solid reinforcing bars.

rebreather See *closed-circuit self-contained breathing apparatus (closed-circuit SCBA)*

RECEO–VS An acronym developed for use by the incident commander (IC) for accomplishing tactical priorities on the fireground; stands for Rescue, Exposure protection, Confinement, Extinguishment, Overhaul, Ventilation, and Salvage.

recessed lug A lug that is circular indentation and requires a specially designed spanner wrench called a booster hose wrench to engage.

reciprocating saw A saw that is powered by an electric motor or a battery motor and whose blade moves back and forth.

recirculation The phenomenon that occurs when a smoke ejector is not completely sealed so that exhausted air is drawn back into the building.

recommended exposure level (REL) The maximum time-weighted concentration of material to which 95 percent of healthy adults can be exposed without suffering any adverse effects over a 40-hour workweek.

reconnaissance A visual scan and exploration of a physical area to gain actionable information to guide incident decision making.

recovery phase The stage of a hazardous materials incident after the imminent danger has passed, when clean-up and the return to normalcy have begun.

recruit A candidate whose application to become a firefighter is accepted.

rectangular cut The most common vertical ventilation opening that is usually 4-ft by 4-ft (1.2-m by 1.2-m) in size. Also called *square cut.*

reducer A fitting used to connect a small hose line or pipe to a larger hose line or pipe. (NFPA 1142)

regulation A mandate issued and enforced by governmental bodies such as the U.S. Occupational Safety and Health Administration (OSHA) and the U.S. Environmental Protection Agency (EPA).

regulator purge/bypass valve A device or devices designed to bypass a regulator.

rehab See *rehabilitate.*

rehabilitate To restore someone or something to a condition of health or to a state of useful and constructive activity. Also referred to as *rehab.*

reinforced concrete Concrete embedded with steel for additional strength.

Reinhart A hand tool that looks like an oversized garden hoe used for constructing fire control lines and overhauling wildland fires.

rekindle A return to flaming combustion after apparent but incomplete extinguishment. (NFPA 921)

relative humidity The ratio between the amount of water vapor in the gas at the time of measurement and the amount of water vapor that could be in the gas when condensation begins, at a given temperature. (NFPA 79)

relay A switch in cabling or wiring that closes to complete the circuit and allows current to flow when it is activated.

release temperature The temperature at which the release mechanism in an automatic sprinkler head activates.

remote annunciators Displays near a building entrance that display the information shown on the annunciator at the FACU.

remote-controlled hydrant valve A valve that is attached to a fire hydrant to allow the operator to turn the hydrant on without flowing water into the hose line.

remote pressure gauge The device on an SCBA that measures and displays pressure readings to indicate the quantity of breathing air available and that is located on the shoulder strap or in another location where it can be seen by the user while the SCBA is in use.

remote supervising station alarm system A protected premises fire alarm system (exclusive of any connected to a public emergency reporting system) in which alarm, supervisory, or trouble signals are transmitted automatically to, recorded in, and supervised from a remote supervising station that has competent and experienced servers and operators who, upon receipt of a signal, take such action as required by this Code. (NFPA 72)

remote valve shut-off A valve that provides a way to remotely shut down a system or isolate a leaking fitting or valve.

repeater A combination of a radio receiver and transmitter that receives radio signals, usually over multiple channels, and retransmits—repeats—them to a wider geographic location.

repeater channel A radio channel that transmits to a repeater.

rescue Those operations directed at locating and removing endangered persons and treating the injured at an emergency incident. (NFPA 1951)

rescue angle The angle that provides a more gradual and controllable slope than the climbing angle for both firefighters to safely climb out of a window and onto the ladder or to remove a victim.

rescue apparatus Apparatus that carry an extensive array of regular and specialized tools and equipment that are used to rescue victims.

rescue assist team (RAT) See *rapid intervention crew/company (RIC).*

rescue company A piece of fire apparatus along with firefighters that are generally utilized for search and rescue at fire incidents. (NFPA 1700)

rescue knife A spring-assisted folding knife that can be used with one hand; often includes a seat belt cutter and a window breaker.

rescue-lift air bag An inflatable bladder made of rubber or synthetic material that is pneumatically filled with compressed air to lift an object or spread one or more objects away from each other to assist in freeing a victim.

rescue technician See *technical rescuer.*

resetting the fire See *transitional attack.*

residential sprinkler system A sprinkler system that protects dwelling units and that have quick response sprinkler heads so that they can activate quickly and reduce the heat, flames, and smoke, so occupants can evacuate quickly.

residual pressure The pressure that exists in the distribution system, measured at the residual hydrant at the time the flow readings are taken at the flow hydrants. (NFPA 24)

resource management Under the NIMS, includes mutual aid agreements; the use of special federal, state, local, and tribal teams; and resource mobilization protocols. (NFPA 1026)

response Immediate and ongoing activities, tasks, programs, and systems to manage the effects of an incident that threaten life, property, operations, or the environment. (NFPA 1600).

retention The process of creating a defined area to contain a hazardous materials release.

reverse hose lay A method of laying a supply line where the supply line starts at the attack engine and ends at the water source.

rex tool A 24-in. (61-cm) steel shaft with a large lock-pulling wedge welded onto one end; used to pull mortise and rim cylinders and key-in-knob locks.

rigging The process of building a system to move or stabilize a load. (NFPA 1006)

rigid fire hose See *non-collapsible fire hose.*

rim lockset A lock mounted to the interior surface of a door.

ring-down phone See *direct-line phone.*

risers In a sprinkler system, vertical supply pipes.

risk-based response process A systematic process based on the facts, science, and circumstances of the incident, by which responders analyze a problem involving hazardous materials/weapons of mass destruction (WMD) to assess the hazards and consequences, develop an incident action plan (IAP), and evaluate the effectiveness of the plan. (NFPA 470)

risk–benefit analysis A decision made by a responder based on a hazard identification and situation assessment that weighs the risks likely to be taken against the benefits to be gained for taking those risks. (NFPA 1006)

rocker lug A rectangular-shaped, bevel-edged lug that extends from a coupling. Also called *rocker pin.*

rocker panel The section of a motor vehicle's frame located below the doors, between the front and rear wheels.

rocker pin See *rocker lug.*

roll-in method See *roll-on method.*

roll-on method A foam application method that involves sweeping the stream just in front of the target. Also called *roll-in method* or *sweep method.*

rollover The condition in which unburned fuel (pyrolysate) from the originating fire has accumulated in the ceiling layer to a sufficient concentration (i.e., at or above the lower flammable limit) that it ignites and burns. This can occur without ignition of, or prior to the ignition of, other fuels separate from the origin. Also called *flameover.* (NFPA 921)

roof covering The membrane, which may also be the roof assembly, that resists fire and provides weather protection to the building against water infiltration, wind, and impact. (NFPA 5000)

roof decking The rigid portion of roof between the roof supports and the roof covering.

roof hooks The spring-loaded, retractable, curved metal pieces that allow the tip of a roof ladder to be secured to the peak of a pitched roof.

roof ladder A single ladder equipped with hooks at the top end and/or the bottom of the ladder. Also called *hook ladder.* (NFPA 1960)

roofman's hook A long pole with a solid metal hook used for pulling. See also *New York hook.*

rope A compact but flexible, torsionally balanced, continuous structure of fibers produced from strands that are twisted, plaited, or braided together, and that serve primarily to support a load or transmit a force from the point of origin to the point of application. (See also 3.3.153.2, Life Safety Rope.) (NFPA 1006)

rope bag A bag used to protect and store rope so that the rope can be easily and rapidly deployed without kinking.

rope record A record for each piece of rope that includes a history of when the rope was placed in service, when it was inspected, when and how it was used, and which types of loads were placed on it.

rope search method See *team search method.*

rotary control valve A shut-off valve on a nozzle that is a dial that you rotate to control the flow of water from the nozzle.

rotary saw A saw that is powered by an electric motor or a gasoline engine and that uses a large rotating blade to cut through material. The blades can be changed depending on the material being cut.

round turn A piece of rope looped to form a complete circle with the two ends parallel.

rubber-covered hose Hose whose outside covering is made of rubber, which is said to be more resistant to damage. Also referred to as *rubber-jacket hose.*

run cards Information prepared in advance and stored that describes a predetermined response to an emergency.

run The act of a unit travelling to an incident.

running end The part of a rope that is not used to form a knot.

rung The ladder crosspieces, on which a person steps while ascending or descending. (NFPA 1960)

rung lock See *pawl.*

rung raise See *flat raise.*

S

S&D tool A double-headed lock-puller tool; large head pulls rim cylinders, mortise cylinders, and key-in-knob/lever locksets; small head pulls tubular deadbolts.

safe haven See *safe location.*

safe location A location remote or separated from the effects of a fire so that such effects no longer pose a threat. Also called *safe haven.* (NFPA 101)

safety data sheet (SDS) Formatted information, provided by chemical manufacturers and distributors of hazardous products, about a chemical's composition, physical and chemical properties, health and safety hazards, emergency response, and waste disposal of the material. (NFPA 470)

safety knot A knot used to back up another knot that has a tendency to become loose when not continuously loaded by or that can slip when loaded by securing the leftover working end of the rope.

safety officer See *incident safety officer.*

safety zone An area of sufficient size and in a suitable location that is expected to protect fire personnel from known hazards without using fire shelters as the fire moves around them.

salvage The process of protecting the contents within a building during and following the fire incident. (NFPA 1700)

salvage cover A large square or rectangular sheet made of heavy canvas or plastic material that is spread over furniture and other items to protect them from water runoff and falling debris.

San Francisco hook A multipurpose tool that can be used for several forcible entry and ventilation applications because of its unique design, which includes a built-in gas shut-off and directional slot.

Santa Ana winds See *Foehn winds.*

saponification The process of converting the fatty acids in cooking oils or fats to soap or foam; the action caused by a Class K fire extinguisher.

sarin A chemical warfare agent.

sash frames Part of the window system that moves to open or close the window by sliding or swinging out; frames can be made of wood, metal, or vinyl. See also *sashes.*

sash locks Components that prevent the window from being opened from the outside.

sashes See *sash frames.*

SCBA harness The backpack or frame for mounting the working parts of the SCBA and the straps and fasteners used to attach the SCBA to the firefighter.

SCBA regulator The part of the SCBA that reduces the high pressure in the cylinder to a usable lower pressure and controls the flow of air to the user.

scientific method The systematic pursuit of knowledge involving the recognition and definition of a problem; the collection of data though observation and experimentation; analysis of the data; the formulation, evaluation, and testing of hypotheses; and, where possible, the selection of a final hypothesis. (NFPA 921)

scraper bar A small tool, usually 9 to 10 in. (23 to 25 cm) in length designed for scraping paint and pulling wood trim and molding.

screwdriver A tool used for turning screws.

search Land-based efforts to find victims or recover bodies. (NFPA 1951)

search and rescue The process of searching a building for a victim and extricating the victim from the building.

search line See *search rope.*

search rope A rope with an anchor point to keep firefighters in contact with each other and with the egress route from a burning building during an interior attack or a search and rescue operation. Also called *search line.*

seat belt cutter A specialized cutting device that cuts through seat belts.

seat of the fire See *area of origin.*

secondary access Openings created by rescuers that provide a pathway to remove or extricate the trapped or injured victims. (NFPA 1670)

secondary collapse A subsequent collapse in a building or excavation. (NFPA 1006)

secondary containment Any device or structure that prevents environmental contamination when the primary container or its appurtenances fail.

secondary contamination The transfer of a contaminant or the source of contamination out of the hot zone to people, animals, the environment, or equipment; also called *cross-contamination.*

secondary device An explosive or incendiary device designed to harm emergency responders who have responded to an initial event.

secondary feeder The smaller-diameter water main pipe in a water distribution system that connects a primary feeder to a distributor. Also called *branch line.*

secondary loss Property damage that occurs due to smoke, water, or other measures taken to extinguish the fire or mitigate the emergency incident.

secondary search A detailed, systematic search of an area. (NFPA 2500)

secondary stabilization Stabilization techniques that support and reinforce primary stabilization techniques so that when the structural components of the vehicle are modified or destroyed during the disentanglement tasks, the vehicle does not collapse or crumple onto the rescuers and/or victims. Also called *supplemental stabilization.*

section In the Incident Command System (ICS), the four components of incident management that report directly to command.

section chief In the Incident Command System (ICS), the person in a supervisory level position responsible for commanding a section.

self-contained breathing apparatus (SCBA) An atmosphere-supplying respirator that supplies a respirable air atmosphere to the user from a breathing air source that is independent of the ambient environment and designed to be carried by the user. (NFPA 1970)

self-expelling agent An agent that has sufficient vapor pressure at normal operating temperatures to expel itself from a fire extinguisher.

sensitizer A chemical that causes a large percentage of people or animals to develop an allergic reaction after repeated exposure to it.

serial arsonist A person who sets three or more fires with an emotional cooling-off period between fires.

set screw A screw that is used to secure an object by pressure or friction, usually within or against another object.

seven, nine, eight (7, 9, 8) rectangular cut A ventilation opening that is usually about 8 ft by 4 ft (1.2 m by 2.4 m) in size; it is primarily used for large commercial buildings with flat roofs.

shackle The U-shaped part of a padlock that runs through a hasp and then is secured back into the lock body.

shear force A force that occurs when materials slide past each other.

sheath See *mantle.*

shelter-in-place A method of safeguarding people in a hazardous area by keeping them in an enclosed atmosphere, usually inside structures.

shipping papers A shipping order, bill of lading, manifest, or other shipping document serving a similar purpose and containing the information required by regulations of the U.S. Department of Transportation (DOT). (NFPA 498)

shock load An instantaneous load that places a rope under extreme tension, such as when a falling load is suddenly stopped as the rope becomes taut.

shoring A structure such as a metal hydraulic, pneumatic/mechanical, or timber system that supports the sides of an excavation and is designed to prevent cave-ins.

shove knife A forcible entry tool used to trip the latch of outward-swinging doors.

shut-off valve A valve whose primary function is to operate in either a fully shut off or a fully open condition. (NFPA 1960)

Siamese connector A hose appliance that allows two hose to be connected together and flow into a single hose.

Side A Normally the front or main entrance/access to the building and usually the side bearing the building address. For buildings with an unusual side A, side A will be identified by the incident commander (IC). Also called *Side Alpha.* (NFPA 3000)

Side Alpha See *Side A.*

Side B The adjacent side of the building or structure clockwise from Side A. Also called *Side Bravo.* (NFPA 3000)

Side Bravo See *Side B.*

Side C The adjacent side of the building or structure clockwise from Side B, generally the back of the building or structure. Also called *Side Charlie.* (NFPA 3000)

Side Charlie See *Side C.*

Side D The adjacent side of the building or structure clockwise from Side C. Also called *Side Delta.* (NFPA 3000)

Side Delta See *Side D.*

sidewall sprinkler head A sprinkler that is designed for horizontal mounting so that the water spray is discharged horizontally into a room, but with a deflector that has a shield on the top so that the water spray is directed downward.

simplex communication system A radio system that allows communication to flow in only one direction at a time. Also called *line-of-sight system* or *talk-around channel.*

single-action pull station A manual alarm initiating device that requires the user to take a single step, such as pulling a lever, toggle, or handle, to activate the alarm.

single-doughnut hose roll A hose roll that has both female couplings on the outside of the roll and the male coupling is protected by the hose rolled on top of it and that is used when the hose will be put into use directly from its rolled state.

single-jacket hose A construction consisting of one woven jacket. (NFPA 1962)

single ladder See *straight ladder.*

single-pane glass windows Windows that have only one pane of annealed glass.

single resource An individual, a piece of equipment and its personnel, or a crew or team of individuals with an identified supervisor that can be used on an incident or planned event. (NFPA 1026)

single-station alarm A detector comprising an assembly that incorporates a sensor, control components, and an alarm notification appliance in one unit operated from a power source either located in the unit or obtained at the point of installation. (NFPA 72)

size-up The process of gathering and analyzing information to help fire officers make decisions regarding the deployment of resources and the implementation of tactics. (NFPA 1410)

slam latch An angled latch bolt that is spring-loaded. After release by physical action, it returns to its operating position and automatically engages the strike plate when the door is slammed or is returned to the closed position. (NFPA 80)

slash Debris resulting from natural events such as wind, fire, snow, or ice breakage, or from human activities such as building or road construction, logging, pruning, thinning, or brush cutting.

sledgehammer A hammer that can be one of a variety of weights and sizes.

S.L.I.C.E.-R.S. An acronym intended to be used by the first-arriving company officer to accomplish important strategic goals on the fireground.

slider windows Windows that slide open horizontally.

slope The steepness of a mountain.

slope reversal The moment when flames burning downhill reach the bottom and start burning uphill.

small-diameter hose (SDH) A hose 1½-in. (38-mm) or 1¾-in. (44-mm) hose that is most often used as the primary attack hose for most fires.

smallpox A highly infectious disease caused by the *Variola* virus, which kills approximately 30 percent of people who become infected.

smoke The airborne solid and liquid particulates and gases evolved when a material undergoes pyrolysis or combustion, together with the quantity of air that is entrained or otherwise mixed into the mass. (NFPA 1404)

smoke containment systems Mechanical systems that keep smoke and other gases in the area or floor of origin.

smoke control system An engineered system that includes all methods that can be used singly or in combination to modify smoke movement. (NFPA 92)

smoke detector A device that detects visible or invisible particles of combustion. (NFPA 72)

smoke ejector A mechanical device, similar to a large fan, that can be used to force heat, smoke, and gases from a post-fire environment and draw in fresh air. (NFPA 440)

smoke explosion A violent release of energy that occurs when smoke travels away from its source to a void area or other area separate from the fire compartment and comes in contact with a source of ignition without any change to the ventilation profile.

smoke management systems Natural or mechanical systems that move or exhaust smoke that escapes from the fire area and threatens egress routes, elevator shafts, stairwells, and refuge areas.

smoke particles The unburned, partially burned, and completely burned substances found in smoke.

smooth-bore nozzle A nozzle that produces a solid stream, or a solid column of water.

snag A dead or live tree whose lower part is on fire, weakening the tree and presenting a significant falling hazard.

socket wrench A wrench that fits over a nut or bolt and uses the ratchet action of an attached handle to tighten or loosen the nut or bolt.

soft sleeve hose A short section of large-diameter supply hose that is used to provide water from the large steamer outlet (the large-diameter port) on a fire hydrant or other pressurized water source to the suction side of the fire pump. Also called *soft suction hose.*

soft suction hose See *soft sleeve hose.*

softening the target See *transitional attack.*

solar panels A general term for thermal collectors or photovoltaic (PV) modules. (NFPA 780)

solar photovoltaic (PV) system A power system designed to convert solar energy into electrical energy; may also be referred to as a PV system or a solar power system.

solid Matter that has a specific size and shape; one of the three states of matter.

solid beam ladder A ladder beam constructed of a solid rectangular piece of material (typically wood), to which the ladder rungs are attached.

solid-core door A door design that consists of solid wood panels or wood pieces inside the door. This construction creates a stronger door that may be fire rated.

solid lumber See *dimensional lumber.*

solid stream A solid column of water.

solute A substance dissolved in a solvent.

solution A mixture of at least one solute dissolved in a solvent.

solvent A substance (usually liquid) capable of dissolving or dispersing another substance. (NFPA 96)

soman A nerve agent that is both a contact and a vapor hazard; it has the odor of camphor.

soot Black particles of carbon produced in a flame. (NFPA 1700)

sound To strike a roof or floor system for stability with a tool.

space frame construction The vehicle frame construction made up of multiple tubes welded into a rigid, light cage-like structure (a roll cage).

spalling Chipping or pitting of concrete or masonry surfaces. (NFPA 921)

span of control The maximum number of personnel or activities that can be effectively controlled by one individual (usually three to seven).

spanner wrench A type of tool used to couple or uncouple hose by turning the rocker lugs or pin lugs on the connections.

special-use railcars Boxcars, flat cars, cryogenic tank cars, or corrosive tank cars.

specific gravity The weight of a liquid compared to water.

speed socket wrench A long, curved handle with a socket wrench that fits over a nut or bolt at the end. To tighten or loosen the nut or bolt, the handle is rotated.

split hose bed A hose bed arranged to enable the engine to lay out either a single supply line or two supply lines simultaneously.

split hose lay A scenario in which the attack engine forward lays a supply line from an intersection to the fire, and the supply engine reverse lays a supply line from the hose left by the attack engine to the water source.

spoil pile A pile of excavated soil next to the excavation or trench. (NFPA 1006)

spoliation Loss, destruction, or material alteration of an object or document that is evidence or potential evidence in a legal proceeding by one who has the responsibility to preserve it. (NFPA 921)

spot fire A new fire that starts outside areas of the main fire, usually caused by flying embers and sparks.

spotlight A light designed to project a narrow, concentrated beam of light.

spot-type detector A device in which the detecting element is concentrated at a particular location. (NFPA 72)

spray nozzle See *fog-stream nozzle.*

spree arsonist A person who sets three or more fires at separate locations with no emotional cooling-off period between fires.

spring-loaded center punch A spring-loaded punch used to break automobile glass.

sprinkler head See *automatic sprinkler head.*

sprinkler piping The network of piping in a sprinkler system that delivers water to the sprinkler heads.

sprinkler stop A mechanical device inserted between the deflector and the orifice of a sprinkler head to stop the flow of water.

sprinkler wedge A piece of wedge-shaped wood placed between the deflector and the orifice of a sprinkler head to stop the flow of water.

squad See *emergency medical services (EMS) company.*

square cut See *rectangular cut.*

stack effect The vertical air flow within buildings caused by the temperature-created density differences between the building interior and exterior or between two interior spaces. (NFPA 92)

staging area Location established where resources can be placed while they await a tactical assignment. (NFPA 1026)

stand-alone communications center A public safety communications center that serves and dispatches a single emergency response agency.

standard Documents, the main text of which contain only mandatory provisions using the word "shall" to indicate requirements and that is in a form generally suitable for mandatory reference by another standard or code or for adoption into law. Nonmandatory provisions are not to be considered a part of the requirements of a standard and shall be located in an appendix or annex, footnote, informational note, or other means as permitted in the NFPA Manuals of Style. (NFPA 1)

standard operating guidelines (SOGs) Written organizational directives that establish or prescribe specific operational or administrative methods to be followed routinely, which can be varied due to operational need in the performance of designated operations or actions. (NFPA 1550)

standard operating procedures (SOPs) Written organizational directives that establish or prescribe specific operational or administrative methods to be followed routinely for the performance of designated operations or actions. (NFPA 1550)

standard search method A primary search method in which a search team consisting of at least two firefighters conducts a systematic search using a left-handed or right-handed search pattern along the outside wall of rooms.

standing part The part of a rope between the working end and the running end.

standpipe A pipe and attached hose valves and hose (if provided) used for conveying water to various parts of a building for fire-fighting purposes. (NFPA 1140)

standpipe system An arrangement of piping, valves, hose connections, and associated equipment installed in a building or structure, with the hose connections located in such a manner that water can be discharged in streams or spray patterns through attached hose and nozzles, for the purpose of extinguishing a fire, thereby protecting a building or structure and its contents in addition to protecting the occupants. (NFPA 25)

State Emergency Response Commission (SERC) A group that acts as a liaison between the local and state levels by collecting and disseminating information relating to hazardous materials emergencies. SERC includes representatives from agencies such as the fire service, law enforcement services, and elected officials.

state of matter The physical state of a material (solid, liquid, or gas).

static pressure The pressure that exists at a given point under normal distribution system conditions measured at the residual hydrant with no hydrants flowing. (NFPA 24)

static rope A rope that stretches very little under load.

static water source A water source that is not under pressure, such as a pond, a lake, a stream, or a swimming pool.

staypole A pole attached to each beam of the base section of extension ladders, which assist in raising the ladder and help provide stability of the raised ladder. Also called *tormentor.* (NFPA 1960)

steamer port The large-diameter port on a fire hydrant.

steel An alloy or a combination of two or more metals of iron and carbon.

stem nut The large nut at the top of the operating stem in a dry barrel hydrant that is turned to open the hydrant valve.

step chock A prefabricated cribbing assembly shaped like stair steps.

stop A piece of material that prevents the fly sections of a ladder from becoming overextended, leading to collapse of the ladder. Also called *stop block.*

stop block See *stop.*

storage hose roll See *straight hose roll.*

stored-pressure fire extinguisher A fire extinguisher in which both the extinguishing agent and expellant gas are kept in a single container and that includes a pressure indicator or gauge. (NFPA 10)

stored-pressure water-type fire extinguisher A fire extinguisher in which water or a water-based extinguishing agent is stored under pressure.

storefront windows Large windows that do not move, slide, or open found in many commercial occupancies; typically made of tempered glass or plate glass.

Storz hose coupling A hose coupling that has the property of being both the male and the female coupling. It is connected by engaging the lugs and turning the coupling a one-third turn.

straight hose lay See *forward hose lay.*

straight hose roll A hose roll that has the male coupling at the center of the roll and the female coupling on the outside of the roll and that is used for general handling and transportation of hose as well as to prepare hose to be stored on a rack. Also called *storage hose roll.*

straight ladder A single-section, fixed-length ground ladder. Also called *single ladder* or *wall ladder.*

straight stream A water stream that flows from a solid-bore nozzle or a stream that flows from a combination nozzle with the stream setting placed in the narrowest stream setting that is available. (NFPA 1700)

strategy The general course of action, direction, or plan to accomplish incident objectives. (NFPA 470)

strike plate A component of a lockset; a metal plate with a hole in it that is mounted on the door frame.

strike team Specified combinations of the same kind and type of resources, with common communications and a leader. (NFPA 1026)

strike team leader A person in a supervisory level position responsible for commanding a strike team.

strip cut See *trench cut.*

strong A description of acids with a pH of 2.5 or less or bases with a pH of 12.5 or greater.

structural firefighting The activities of rescue, fire suppression, and property conservation in buildings or other structures, vehicles, railcars, marine vessels, aircraft, or like properties. (NFPA 1010)

structural firefighting protective clothing All of the clothing elements of the structural firefighting protective ensemble.

structural firefighting protective coat The element of the protective ensemble that provides protection to the upper torso and arms, excluding the hands and head. Also called *bunker coat* or *turnout coat.* (NFPA 1970)

structural firefighting protective ensemble Multiple elements of compliant protective clothing and equipment that when worn together provide protection from some risks, but not all risks, of emergency incident operations. (NFPA 1970)

structural firefighting protective footwear The element of the protective ensemble that provides protection to the foot, ankle, and lower leg. (NFPA 1970)

structural firefighting protective glove The element of the protective ensemble that provides protection to the hand and wrist. (NFPA 1970)

structural firefighting protective helmet The element of the protective ensemble that provides protection to the head. (NFPA 1970)

structural firefighting protective hood The interface element of the protective ensemble that provides limited protection to the coat/helmet/SCBA facepiece interface area. (NFPA 1970)

structural firefighting protective trousers The element of the protective ensemble that provides protection to the lower torso and legs, excluding the ankles and feet. Also called *bunker pants* or *turnout pants.* (NFPA 1970)

structure triage guidelines Guidelines that describe how to categorize structures near a wildland fire based on the level of threat to them and the tactics that can be deployed to defend these threatened homes.

strut An adjustable-length structural support, made from a metal or composite beam, that is used to stabilize and prop up something in danger of collapse.

strut tower The front suspension system in a vehicle.

stud The vertical member of the frame of a structure.

stud space The hollow space between two vertical members in a stud wall.

stud walls Walls constructed of vertical framing members called studs that run from floor to ceiling on each floor or from foundation to attic in older balloon-frame construction; can be load-bearing or non-load-bearing.

subchronic exposure Exposure that causes effects to occur either immediately after a single exposure or as long as several days or weeks after an exposure.

subsurface fuels Partially decomposed matter that lies beneath the ground, such as roots, moss, and decomposed stumps.

suction hose A hose that is designed to prevent collapse under vacuum conditions so that it can be used for drafting water from below the pump (lakes, river, wells, etc.) (NFPA 1960)

Superfund Amendments and Reauthorization Act of 1986 (SARA) One of the first U.S. laws to affect how fire departments respond in a hazardous materials emergency.

supervising station alarm system An alarm system that communicates between the protected property and the fire department or an alarm monitoring facility that is staffed 24 hours per day by personnel who receive, interpret, and take the appropriate action to handle the alarm condition.

supervisory condition An abnormal condition in connection with the supervision of other systems, processes, or equipment. (NFPA 72)

supervisory signal A signal that results from the detection of a supervisory condition. (NFPA 72)

supplemental restraint system (SRS) A system in a motor vehicle that uses supplemental restraint devices such as air bags and seat belt pretensioning systems to enhance safety in conjunction with properly applied seat belts.

supplemental stabilization See *secondary stabilization.*

supplied-air respirator (SAR) An atmosphere-supplying respirator for which the source of breathing air is not designed to be carried by the user. Also called *air-line respirator.* (NFPA 1852)

supply engine An engine used to pump water to an attack engine using supply lines that may be connected to a pressurized water source such as a fire hydrant or to a static (unpressurized) water source.

supply hose Hose designed for the purpose of moving water between a pressurized water source and a pump that is supplying attack lines. Also called *supply line.* (NFPA 1960)

supply line See *supply hose.*

supply line evolutions The process of laying supply hose to deliver water from a water supply source to an attack engine. Also called *supply line operations.*

supply line operations See *supply line evolutions.*

support person A fire department member who is not a firefighter but who assists members of a fire department performing duties in environments that are not hazardous.

surface fuels Fuels that are close to the surface of the ground, such as grass, leaves, twigs, needles, small trees, logging slash, and low brush. Also called ground fuels.

surfactant A compound that lowers the surface tension (or interfacial tension) between two liquids, between a gas and a liquid, or between a liquid and a solid, and which can act as detergents, wetting agents, emulsifiers, foaming agents, and dispersants. (NFPA 1700)

survivability profiling A risk–benefit analysis that weighs the viability and survivability of potential fire victims based on the current conditions in the structure.

sweep To use your arm or a tool to swipe across areas to feel for victims or objects.

sweep method See *roll-in method.*

swivel A ring around female couplings that you turn to secure the female coupling around the male coupling without twisting the hose.

swivel gasket An O-shaped piece of rubber inside the swivel section of a female hose coupling that forms a seal that stops water from leaking when a male hose coupling is tightened against it.

system A collective term for the assemblies and materials that together make up a building component.

T

T-3 signal See *temporal-3 pattern.*

tabun A nerve agent that disables the chemical connections between nerves and target organs.

tactical Side A/Alpha At an incident, Side A/Alpha of a building when it is not the front or address side.

tactics Deploying and directing resources on an incident to accomplish the objectives designated by strategy. (NFPA 1026)

tag line A rope that personnel on the ground can use to guide an object that is being hoisted or lowered.

talk-around channel See *simplex communication system.*

tamper seal A retaining device that breaks when the locking mechanism is released.

tanker See *mobile water supply apparatus.*

target hazard Any occupancy type or facility that presents a high potential for loss of life or serious impact to the community resulting from a fire, explosion, or chemical release.

task A specific job behavior or activity. (NFPA 1010)

task force A group of resources with common communications and a leader that can be pre-established and sent to an incident or planned event or formed at an incident or planned event. (NFPA 1026)

task force leader A person in a supervisory level position responsible for commanding a task force.

team integrity The concept of keeping a crew together and working as a team so that no single firefighter goes off on their own. Also called *crew integrity.*

team search method A primary search method in which a search team consisting of an officer or team leader who acts as the anchor person and two or more searchers conduct a systematic search where the anchor person ties the rope to a stationary object and the searchers search the rooms while holding on to the rope so that they can quickly exit by following the rope out if needed.

technical decontamination The planned and systematic process of reducing contamination from responder PPE to a level that is as low as reasonably achievable. (NFPA 470)

technical rescue incident (TRI) Complex rescue incident requiring specially trained personnel and special equipment to complete the mission. (NFPA 2500)

technical rescue team A group of rescuers specially trained in the various disciplines of technical rescue.

technical rescuer A person who is trained to perform or direct a technical rescue. Also called *rescue technician.* (NFPA 1006)

technical use life safety rope A life safety rope with a diameter that is at least 3/8 in. (9.5 mm) but not larger than 1/2 in. (12.5 mm), with a minimum breaking strength of 4496 lbf (20 kN) and that is used by highly trained rescue teams who deploy to technical environments such as mountainous or wilderness terrain.

technician See *driver/operator.*

teeth Serrations or grooves around the fog-stream nozzle's orifice that help to break up the water stream and create a fog pattern.

telecommunicator An individual whose primary responsibility is to receive, process, or disseminate information of a public safety nature via telecommunication devices. Also called *dispatcher.* (NFPA 1225)

telephone interrogation The phase in a 911 call during which the telecommunicator asks questions to obtain vital information such as the location of the emergency.

temperature The measurement of the movement of molecules used to describe how hot or cold something is.

temperature rating The temperature or rate of temperature increase that causes a heat detector to activate.

tempered glass A type of safety glass that is heat treated so that it will break into small pieces that are not as dangerous. See also *tempered safety glass.*

tempered safety glass A type of safety glass that is heat-treated so that, under stress or fire, it breaks into small pieces that are not as dangerous as the shards of glass that are created when untreated glass breaks. Also called *tempered glass.*

temporal-3 pattern A standard fire alarm audible signal for alerting occupants of a building consisting of three tones followed by a short pause. Also called *T-3 signal.*

temporary refuge area (TRA) A predesignated area where firefighters can immediately take refuge or temporary shelter and short-term relief from the fire without using a fire shelter until fire conditions improve.

ten-codes A system of predetermined coded messages, such as "What is your 10-20?" used by responders over the radio.

tender See *mobile water supply apparatus.*

tendons Steel cables under tension.

tensile force The force that pulls a material apart.

tensile strength The resistance of a material to breaking under tension.

tension buttress system A system consisting of a strut and a ratchet strap that adds tension to the object being stabilized, locking the object in place with a diagonal force that lowers the object's center of gravity by increasing the object's entire footprint.

testimonial evidence Documented verbal or written statements previously given by firefighters or bystanders, as well as affidavits, interrogatories, and court testimony that documents observations and statements made by lay and expert witnesses.

testing A procedure used to determine the operational status of a component or system by conducting periodic physical checks, such as waterflow tests, fire pump tests, alarm tests, and trip tests of dry pipe, deluge, or preaction valves. (NFPA 25)

tether line A smaller line, typically thinner and more lightweight than the main search rope, attached to the main search rope during large area searches to serve as a secondary line to aid in navigation and orientation within the search area. Also called *tag line.*

thermal column See *plume.*

thermal conductivity The ability of a material to conduct heat.

thermal degradation Damage caused by heat transfer.

thermal energy See *heat energy.*

thermal imager device An electronic device that detects differences in temperature based on infrared energy and then generates images based on those data. It is commonly used in smoke-filled environments to locate victims as well as to search for hidden fire during size-up and overhaul.

thermal imager (TI) Special electronic equipment that creates a picture based on heat produced by a person or object. (NFPA 1801)

thermal imaging camera (TIC) See also *thermal imager (TI).*

thermal layering The phenomenon of gases forming into layers according to their temperatures. Also called *heat stratification.*

thermal radiation The means by which heat is transferred to other objects.

thermal runaway In a high-voltage battery, the condition when the temperature in the cell continues to rise, producing gas, and eventually igniting.

thermal updraft See *plume.*

thermoplastic material Plastic material capable of being repeatedly softened by heating and hardened by cooling and, that in the softened state, can be repeatedly shaped by molding or forming. (NFPA 5000)

thermoset material Plastic material that, after having been cured by heat or other means, is substantially infusible and cannot be softened and formed. (NFPA 5000)

threaded hose coupling A type of coupling that requires a male fitting and a female fitting to be screwed together.

three-dimensional liquid-fuel fire A fire in which liquid fuel is dripping, spraying, or flowing over the edges of a container.

threshold limit value (TLV) The point at which a hazardous material or weapon of mass destruction begins to affect a person.

threshold limit value/ceiling (TLV/C) The concentration of hazardous material that should not be exceeded because exposure, even for an instant, can cause ill effects to a person.

threshold limit value/short-term exposure limit (TLV/STEL) The maximum concentration of hazardous material that a person can be exposed to in 15-minute intervals, up to four times a day, with a minimum of 1 hour between exposures; the lower the TLV/STEL value, the more toxic the substance is.

threshold limit value/skin (TLV/skin) The concentration at which direct or airborne contact with a material could result in possible and significant exposure from absorption through the skin, mucous membranes, and eyes.

threshold limit value/time-weighted average (TLV/TWA) The maximum airborne concentration of a material to which a person can be exposed for 8 hours a day, 40 hours a week, and not suffer any ill effects.

throw bag A water rescue system that includes 50 ft to 75 ft (15.24 m to 22.86 m) of water rescue rope, an appropriately sized bag, and a closed-cell foam float. (NFPA 1006)

throwline A floating rope that is intended to be thrown to a person during water rescues or as a tether for rescuers entering the water. (NFPA 2500)

tie rod Typically found on wood ladders, a metal rod that runs from one beam of the ladder to the other to keep the beams from separating.

tiller truck A tractor-trailer apparatus that requires a second operator positioned at the back of the trailer to steer the back wheels. Also called a *tractor-drawn aerial.*

timber litter Fuel that falls out of trees, such as pine needles, leaves, twigs, and small branches.

time marks Status updates provided to the communications center every 10 to 20 minutes. Such an update should include the type of operation, the progress of the incident, the anticipated actions, and the need for additional resources.

tip The very top of the ladder.

ton container Bulk packaging that commonly holds compressed liquefied gases.

topography The land surface configuration. (NFPA 1140)

tormentor See *staypole.*

tormentor ladder See *Bangor ladder.*

torr Named after Evangelista Torricelli, a unit of pressure equal to 1 mm Hg; 760 torr is equal to 14.7 psi.

torsional loads Loads that create twisting.

tower ladder company See *truck company.*

tower ladder truck See *aerial fire apparatus.*

toxic inhalation hazards (TIH) Gases or volatile liquids that are extremely toxic to humans.

toxic products of combustion Hazardous chemical compounds that are released when a material decomposes under heat.

toxicity The degree to which a substance is harmful to humans. (NFPA 326)

toxin A substance created by a plant or an animal that may be harmful to humans depending on the exposure.

tractor-drawn aerial See *tiller truck.*

traffic incident management (TIM) A coordinated method of identifying and responding to roadway incidents, including establishing safe work areas and clearing the area when the incident is complete.

trailer (rope) A label that travels the entire length of a life safety rope under the outer sheath that identifies the rope as a life safety rope.

trailers Solid or liquid fuels used to intentionally spread or accelerate the spread of a fire from one area to another. (NFPA 921)

training officer The person designated by the fire chief with authority for overall management and control of the organization's training program. (NFPA 1401)

train list See *consist.*

transfer of command The formal procedure for transferring the duties of an incident commander (IC) incident scene. (NFPA 1026)

transitional attack An offensive fire attack initiated by an exterior, indirect handline operation into the fire compartment to initiate cooling while transitioning into interior direct fire attack in coordination with ventilation operations. Also called *blitz attack, softening the target,* and *resetting the fire.*

trench cut A roof cut that is made from one load-bearing wall to another load-bearing wall and that is intended to prevent horizontal fire spread in a building. Also called *strip cut.*

triangular cut A triangle-shaped ventilation cut in the roof decking that is made using a saw or an axe.

trigger The button or lever used to discharge the agent from a portable fire extinguisher.

triple-layer hose load A hose loading method in which the hose is folded back onto itself to reduce the overall length to one-third before loading the hose into the hose bed. This load method reduces deployment distances.

triple-pane glass A window design that traps air or inert gas in the two spaces between three pieces of glass to help insulate a house.

trouble condition An abnormal condition in a system that results from a fault. (NFPA 72)

trouble signal A signal that results from a trouble condition. (NFPA 72)

truck See *aerial fire apparatus.*

truck company A company of firefighters who are equipped with one or more pieces of aerial fire apparatus. Also called *ladder company* or *tower ladder company.* (NFPA 1700)

trunk line See *primary feeder.*

trunked radio system A repeater system that uses a computerized shared bank of frequencies to make the most efficient use of radio resources.

truss A collection of lightweight structural components joined in a triangular configuration that can be used to support either floors or roofs.

truss block A piece of wood or metal that ties the two rails of a trussed beam ladder together and serves as the attachment point for the rungs.

trussed beam ladder A ladder beam constructed of top and bottom rails joined by truss blocks that tie the rails together and support the rungs.

TTY/TDD systems User devices that allow speech- and/or hearing-impaired citizens to communicate over a telephone system. TTY stands for teletype, and TDD stands for telecommunications device for the deaf; the displayed text is the equivalent of a verbal conversation between two hearing persons.

tube trailer A high-volume transportation vehicle made up of several individual compressed gas cylinders that operate at working pressures of 3000 to 5000 psi and are banded together and affixed to a trailer.

tubular deadbolt Surface- or interior-mounted lock on or in a door with a bolt that provides additional security, a type of bored lockset.

turbulent smoke flow The agitated, boiling, and angry movement of smoke caused by rapid molecular expansion of the gases within the smoke and the restrictions of the box containing the smoke.

turnout coat See *structural firefighting protective coat.*

turnout gear See *structural firefighting protective clothing.*

turnout pants See *structural firefighting protective trousers.*

twisted rope Rope constructed of fibers twisted into strands, which are then twisted together. Also called *laid rope.*

two-dimensional liquid-fuel fire A fire that is burning only on the top surface of a spill, pool, or open container of liquid.

two-in/two-out rule A safety procedure that requires a minimum of two personnel to enter a hazardous area and a minimum of two backup personnel to remain outside the hazardous area during the initial stages of an incident.

two-way radios Portable communication devices used by firefighters. Every firefighting team should carry at least one radio to communicate distress, progress, changes in fire conditions, and other pertinent information.

Type A packaging Packaging designed to protect its internal radiological contents during normal transportation and in the event of a minor accident.

Type B packaging Packaging more durable than Type A packaging, which is designed to prevent a release of a radiological hazard in the case of extreme accidents during transportation.

Type C packaging Packaging used when radioactive substances must be transported by air, with no restriction placed on the total amount of radioactivity contained within the package.

Type I construction (fire-resistive) The type of construction in which the fire walls, structural elements, walls, arches, floors, and roofs are of approved noncombustible or limited-combustible materials that have a specified fire resistance.

Type II construction (noncombustible) The type of construction in which the fire walls, structural elements, walls, arches, floors, and roofs are of approved non-combustible or limited-combustible materials without fire resistance.

Type III construction (ordinary) The type of construction in which exterior walls and structural elements that are portions of exterior walls are of approved non-combustible or limited-combustible materials and in which fire walls, interior structural elements, walls, arches, floors, and roofs are entirely or partially of wood of smaller dimensions than required for Type IV construction or are of approved non-combustible, limited-combustible, or other approved combustible materials. (NFPA 14)

Type IV construction (heavy timber) The type of construction in which fire walls, exterior walls, and interior bearing walls and structural elements that are portions of such walls are of approved non-combustible or limited-combustible materials. Other interior structural elements, arches, floors, and roofs are constructed of solid or laminated wood or cross-laminated timber without concealed spaces within allowable dimensions of the building code. (NFPA 14)

Type V construction (wood frame) The type of construction in which structural elements, walls, arches, floors, and roofs are entirely or partially of wood or other approved material. (NFPA 14)

U

ultrahigh-frequency (UHF) band Radio frequencies between 300 and 3000 MHz.

uncharged hose line A hose line that is not filled with water and not under pressure from a pump.

underflow dam A dam that includes piping installed at a slight upward angle, which allows the water to flow below the released liquid, and with enough capacity that the water drains without causing the lighter-than-water material to overtop the dam, thereby trapping the lighter-than-water material behind the top of the dam.

undetermined fire A classification of fire used when the cause cannot be proven to an acceptable level of certainty.

unibody frame construction The frame construction most commonly used in passenger cars that combines the vehicle body and the frame into a single component, thereby eliminating the rail beams used in body-over-frame vehicles.

unified command A team effort that allows all agencies with jurisdictional responsibility for an incident or planned event, either geographical or functional, to manage the incident or planned event by establishing a common set of incident objectives and strategies. (NFPA 1026)

unintentional alarm An unwanted activation of an alarm initiating device caused by a person acting without malice. (NFPA 72)

unintentional ventilation Ventilation that occurs when a window or door fails because of the fire or when a window or door is left open by mistake.

unit A collective term used casually to refer to a group of firefighters in a department with similar duties or responsibilities. Within the ICS, a specific term that identifies two or more resources who have responsibility for a specific activity at an incident. See also *company* and *crew.*

unit leader A person in a supervisory level position responsible for commanding a task force.

unit selection The process of determining exactly which unit or units should be dispatched after a call to 911 is received by a communications center, based on the location and classification of the incident.

unity of command The concept by which each person within an organization reports to one, and only one, designated person. (NFPA 1026)

unknown alarm An unwanted activation of an alarm initiating device or system output function where the cause has not been identified. (NFPA 72)

unwanted alarm Any alarm that occurs that is not the result of a potentially hazardous condition. (NFPA 72)

upper explosive limit (UEL) The maximum amount of combustible vapor or gas that can be present in the air if the air/fuel mixture is to be flammable or explosive.

upper flammable limit (LFL) See *upper explosive limit.*

upper rails Two parallel beams, one on each side of a vehicle's front end, that support and attach the strut tower system to the frame, hold the hood in place, and support the front fenders.

upright sprinkler head A sprinkler head designed to be installed on the top side of the sprinkler piping with a deflector that deflects the stream 180 degrees so that the water is directed down to the floor and is usually marked *SSU,* which stands for *standard spray upright.*

U.S. Department of Transportation (DOT) marking system A system of labels and placards that is used when materials are being transported from one location to another in the United States and Canada by Transport Canada.

utility rope Rope used for securing objects, hoisting equipment, or securing a scene to prevent bystanders from being injured; utility rope must never be used in life safety operations.

V

V-agent (VX) An oily liquid, contact hazard, nerve agent that can persist for several weeks.

vapor The gas state of a substance, particularly those that are normally liquids or solids at ordinary temperatures. (NFPA 1700)

vapor density The weight of a vapor or gas compared to an equal volume of dry air.

vapor dispersion The process of lowering the concentration of vapors by spreading them out, typically with a water fog from a hose line.

vapor pressure The pressure, measured in pounds per square inch, absolute (psia), exerted by a liquid, as determined by ASTM D 323, *Standard Test Method for Vapor Pressure of Petroleum Products (Reid Method).* (NFPA 1)

vapor-protective ensemble Multiple elements of compliant protective clothing and equipment that when worn together provide protection from some, but not all, risks of vapor, liquid splash, and particulate environments during hazardous materials/weapons of mass destruction (WMD) incident operations. (NFPA 470)

vapor suppression The process of controlling vapors given off by hazardous materials by covering the product with foam or other material or by reducing the temperature of the material, thereby preventing vapor ignition.

vaporization The process of a solid or liquid changing to a gas.

variable-flow fog nozzle See *adjustable-gallonage fog nozzle.*

ventilation The controlled and coordinated removal of heat and smoke from a structure, replacing the escaping gases with fresh air. (NFPA 1410)

ventilation-limited fire A fire in which the heat release rate and fire growth are regulated by the available oxygen within the space. (NFPA 1410)

vent pipes Inverted J-shaped tubes that allow for pressure relief or natural venting of a pipeline for maintenance and repairs.

ventilation saws Cutting tools designed for roof ventilation that have a different cutting chain than chainsaws used to cut wood; also may have a depth gauge on its bar.

vertical fire extension Fire spread from a lower floor to a higher floor through an exterior window. Also called *auto-exposure.*

vertical reach The height a ladder can reach.

vertical spread maneuver A technique for exposing the latch or hinges of a vehicle door using a spreading tool, typically a hydraulic spreader, by separating the top of the door from the roof rail.

vertical ventilation The vertical venting of structures involving the opening of bulkhead doors, skylights, scuttles, and roof cutting operations to release smoke and heat from inside the fire building. (NFPA 1410)

very high-frequency (VHF) band Radio frequencies between 30 and 300 MHz; the VHF spectrum is further divided into high and low bands.

Voice over Internet Protocol (VoIP) Technology that converts a person's voice into a digital signal that can be sent via the Internet to another device.

voice recording system A system that records communications over the phones and the radio.

voting receiver A device in a repeater system that receives signals from radio transmissions and then sends the strongest signal to the repeater.

W

walk-in A person who comes to a fire station seeking assistance rather than calling 911. Also called *door banger.*

walking wounded Victims who were involved in an accident in some way and who have injuries, but who are not immediately identifiable.

wall ladder See *straight ladder.*

warm zone The control zone at hazardous materials/weapons of mass destruction (WMD) incidents where personnel and equipment decontamination and hot zone support take place. Also called *contamination reduction zone.* (NFPA 470)

water catch-all A salvage cover that has been folded to form a container to hold water until it can be removed.

water chute A salvage cover that has been folded to direct water flow out of a building or away from sensitive items or areas.

water curtain nozzle A nozzle used to deliver a flat screen of water that forms a protective sheet of water to protect exposures from fire.

water flow The volume of water moving through a pipe, hose, or nozzle over a period of time, usually expressed in gallons (liters) per minute (gpm or L/min).

waterflow alarm device An attachment to the sprinkler system that detects a predetermined water flow and is connected to a fire alarm system to initiate an alarm condition. (NFPA 13)

water hammer The surge of pressure that occurs when a high-velocity flow of water is abruptly shut off. The pressure exerted by the flowing water against the closed system can be seven or more times that of the static pressure. (NFPA 1962)

water main A generic term for any underground water pipe.

water mist fire extinguisher A fire extinguisher containing distilled or de-ionized water and employing a nozzle that discharges the agent in a fine spray. (NFPA 10)

water mist system A distribution system connected to a water supply or water and atomizing media supplies that is equipped with one or more nozzles capable of delivering water mist intended to control, suppress, or extinguish fires and that has been demonstrated to meet the performance requirements of its listing and [the applicable] standard. (NFPA 25)

water-motor gong An audible alarm notification device in a sprinkler system that makes noise as long as water is flowing at a certain pressure through the sprinkler system.

water pressure The application of force by one object against another. When water is forced through the distribution system, it creates water pressure.

water shuttle operations A method of transporting water from a source to a fire scene using a number of mobile water supply apparatus.

water solubility The ability of a substance to dissolve in water.

water supply A source of water for firefighting activities. (NFPA 1140)

water tender See *mobile water supply apparatus.*

water thief A device that has a 2½-in. (64-mm) inlet and a 2½-in. (64-mm) outlet in addition to two 1½-in. (38-mm) outlets. It is used to supply more than one hose from one source.

water tower Aerial devices that deliver streams of water and contain high-capacity pre-piped waterways.

water vacuum See *wet vacuum.*

waybill Shipping papers for railroad transport.

weak A description of acids with a pH greater than 2.5 or bases with a pH less than 12.5.

weapon of mass destruction (WMD) Any weapon or material that is designed to cause death or serious injury or damage to buildings, structures, or the environment, such as an explosive or incendiary bomb, rocket, or grenade containing or delivering a toxic or dangerous chemical, biological agent, toxin, or vectors; or a weapon designed to release dangerous levels of radiation. (NFPA 470)

wear course The layer of the roof that is sound enough to walk on.

web On an I-beam ladder, the part the connects the top and bottom flanges on a beam.

web members A system of diagonal and sometimes vertical members between two parallel chords.

webbing Woven material of flat or tubular weave in the form of a long strip. (NFPA 2500)

wedge Material used to tighten or adjust cribbing and shoring systems. (NFPA 1006)

wet barrel hydrant A type of hydrant that is intended for use where there is no danger of freezing weather and where each outlet is provided with a valve and an outlet. (NFPA 24)

wet cell battery A lead acid battery that contains an active electrolyte solution of sulfuric acid and water. Also called *flooded cell battery.*

wet-chemical extinguishing agent Normally, an aqueous solution of organic or inorganic salts or a combination thereof that forms an extinguishing agent. (NFPA 10)

wet-chemical extinguishing system A system often installed in commercial kitchens that discharges a liquid Class K extinguishing agent.

wet-chemical fire extinguisher A fire extinguisher containing a wet-chemical extinguishing agent for use on Class K fires.

wet pipe sprinkler system A sprinkler system employing automatic sprinklers attached to a piping system containing water and connected to a water supply so that water discharges immediately from sprinklers opened by heat from a fire. (NFPA 13).

wet standpipe system A standpipe system having piping containing water at all times. (NFPA 25)

wet vacuum A device similar to a wet/dry shop vacuum cleaner that can pick up liquids. It is used to remove water from buildings. See also *water vacuum.*

wet water See *Class A foam concentrate* or *wetting agent.*

wetting agent A concentrate that, when added to water, reduces the water's surface tension and increases its ability to penetrate, mix with and spread. (NFPA 18)

wetting-agent fire extinguisher A fire extinguisher that expels water combined with a concentrate to reduce the surface tension and increase its ability to penetrate and spread.

wheeled fire extinguisher A portable fire extinguisher equipped with a carriage and wheels intended to be transported to the fire by one person. (NFPA 10)

wheel well crush A technique using a hydraulic spreader to expose the hinges of a vehicle door by pulling the wheel well panel away from the door. Also called *fender crush.*

widow maker A broken top of trees or limbs of trees that have been damaged by wind or snow and are still hanging in the tree that can fall without making a sound.

wilderness and helicopter search and rescue (SAR) The process of locating and removing a victim from the wilderness.

wildland Land in an uncultivated, more or less natural state and covered by timber, woodland, brush, and/or grass. (NFPA 901)

wildland apparatus A four-wheel drive vehicle used to transport firefighters closer to wildfires over rough, uneven terrain and that carries a tank of water and a pump that enables them to pump water while the truck is moving and special firefighting equipment such as portable pumps, rakes, shovels, and chainsaws.

wildland company A company who fight vegetation fires where larger pumpers cannot gain access and who are equipped with four-wheel drive vehicles and special fire-fighting equipment. Also called *brush company*.

wildland fire An unplanned fire burning in vegetative fuels. (NFPA 1140)

wildland fire behavior The flame length and rate of spread of a wildland fire given the fuel, weather, and topography.

wildland/urban interface (WUI) The line, area, or zone where structures and other human development meet or intermingle with undeveloped wildland or vegetative fuels. (NFPA 5000)

wildland/urban intermix An area where improved property and wildland fuels meet with no clearly defined boundary. (NFPA 5000)

window frames Components that attach the window system to the building.

wind The horizontal movement of air over the earth surface.

windshield size-up See *initial size-up*.

wire cutter A hand tool used to cut wire and small diameter cable. See also *diagonal cutter*.

wired glass A glazing material with embedded wire mesh.

wood truss An assembly of small pieces of wood or wood and metal.

wooden beam Load-bearing member assembled from individual wood components.

working end The part of the rope used for forming a knot.

working load limit The rating that identifies the maximum load a jack can safely support.

wrench A hand tool that comes in several sizes and is used to tighten or loosen bolts.

wye A device used to split a single hose into two or more separate lines.

Y

yield strength The maximum stress that can be applied to steel before it permanently deforms.

youth fire-setters Recognized classifications of minors who ignite fires for various reasons. Based on age, youth fire-setters include child fire-setters, juvenile fire-setters, and adolescent fire-setters.

Z

zone A defined area within a protected premise. A zone can define an area from which a signal can be received, an area in which a signal can be sent, or an area in which a form of control can be executed. (NFPA 72)

zone valve A valve that controls the flow of water into one zone in a sprinkler system that is divided into zones.

zoned system A fire alarm system that identifies the type or location of the initiating devices by identifying the zone where each device is located.

Index

B

C

D

E

F

G

M

Q

R

S

W